Contents

For a more detailed Table of Contents, see page xv

Corrosion in Specific Environments 1

Introduction to Corrosion in Specific Environments 5

Corrosion in Fresh Water Environments
Corrosion in Potable Water Distribution and Building Systems 8
Corrosion in Service Water Distribution Systems 12
Rouging of Stainless Steel in High-Purity Water 15
Corrosion in Wastewater Systems 23

Corrosion in Marine Environments
Corrosion in Seawater 27
Corrosion in Marine Atmospheres 42
Corrosion of Metallic Coatings 61
Performance of Organic Coatings 69
Marine Cathodic Protection 73

Corrosion in Underground Environments
External Corrosion Direct Assessment Integrated with Integrity Management 79
Close-Interval Survey Techniques 84
Corrosion of Storage Tanks 89
Well Casing External Corrosion and Cathodic Protection 97
Stray Currents in Underground Corrosion 107
Corrosion Rate Probes for Soil Environments 115
Cathodic Protection of Pipe-Type Power Transmission Cables 122

Corrosion in Military Environments
Corrosion in the Military 126
Military Specifications and Standards 136
Corrosion Control for Military Facilities 141
Ground Vehicle Corrosion 148
Armament Corrosion 151
High-Temperature Corrosion in Military Systems 156
Finishing Systems for Naval Aircraft 171
Military Coatings 180
U.S. Navy Aircraft Corrosion 184
Military Aircraft Corrosion Fatigue 195
Corrosion of Electronic Equipment in Military Environments 205
Microbiologically Influenced Corrosion in Military Environments 211
Service Life and Aging of Military Equipment 220

Corrosion in Specialized Environments
Corrosion in Supercritical Water—Waste Destruction Environments 229
Corrosion in Supercritical Water—Ultrasupercritical Environments for Power Production 236
Corrosion in Cold Climates 246
Corrosion in Emissions Control Equipment 251

Corrosion in Recreational Environments 257
Corrosion in Workboats and Recreational Boats 265
Corrosion of Metal Artifacts and Works of Art in Museum and Collection Environments 279
Corrosion of Metal Artifacts Displayed in Outdoor Environments 289
Corrosion of Metal Artifacts in Buried Environments 306
Chemical Cleaning and Cleaning-Related Corrosion of Process Equipment 323

Corrosion in Specific Industries 331

Introduction to Corrosion in Specific Industries 337

Corrosion in the Nuclear Power Industry
Introduction to Corrosion in the Nuclear Power Industry 339
Corrosion in Boiling Water Reactors 341
Corrosion in Pressurized Water Reactors 362
Effect of Irradiation on Stress-Corrosion Cracking and Corrosion in Light Water Reactors 386
Corrosion of Zirconium Alloy Components in Light Water Reactors 415
Corrosion of Containment Materials for Radioactive-Waste Isolation 421

Corrosion in Fossil and Alternative Fuel Industries
Introduction to Corrosion in Fossil and Alternative Fuel Industries 438
High-Temperature Corrosion in Gasifiers 441
Corrosion in the Condensate-Feedwater System 447
Corrosion of Flue Gas Desulfurization Systems 461
Corrosion of Steam- and Water-Side of Boilers 466
Corrosion of Steam Turbines 469
Fireside Corrosion in Coal- and Oil-Fired Boilers 477
High-Temperature Corrosion in Waste-to-Energy Boilers 482
Corrosion of Industrial Gas Turbines 486
Components Susceptible to Dew-Point Corrosion 491
Corrosion of Generators 497
Corrosion and Erosion of Ash-Handling Systems 499
Corrosion in Portable Energy Sources 501
Corrosion in Fuel Cells 504

Corrosion in the Land Transportation Industries
Automotive Body Corrosion 515
Automotive Exhaust System Corrosion 519
Engine Coolants and Coolant System Corrosion 531
Automotive Proving Ground Corrosion Testing 538
Corrosion of Aluminum Components in the Automotive Industry 545
Electric Rail Corrosion and Corrosion Control 548
Corrosion in Bridges and Highways 559

Corrosion in the Air Transportation Industry
Corrosion in Commercial Aviation 598

Corrosion in the Microelectronics Industry
Corrosion in Microelectronics 613
Corrosion in Semiconductor Wafer Fabrication 623
Corrosion in the Assembly of Semiconductor Integrated Circuits 629
Corrosion in Passive Electrical Components 634
Corrosion and Related Phenomena in Portable
    Electronic Assemblies 643

Corrosion in the Chemical Processing Industry
Effects of Process and Environmental Variables 652
Corrosion under Insulation 654
Corrosion by Sulfuric Acid 659
Corrosion by Nitric Acid 668
Corrosion by Organic Acids 674
Corrosion by Hydrogen Chloride and Hydrochloric Acid 682
Corrosion by Hydrogen Fluoride and Hydrofluoric Acid 690
Corrosion by Chlorine 704
Corrosion by Alkalis 710
Corrosion by Ammonia 727
Corrosion by Phosphoric Acid 736
Corrosion by Mixed Acids and Salts 742
Corrosion by Organic Solvents 750
Corrosion in High-Temperature Environments 754

Corrosion in the Pulp and Paper Industry
Corrosion in the Pulp and Paper Industry 762

Corrosion in the Food and Beverage Industries
Corrosion in the Food and Beverage Industries 803

Corrosion in the Pharmaceutical and Medical Technology Industries
Material Issues in the Pharmaceutical Industry 810
Corrosion in the Pharmaceutical Industry 813
Corrosion Effects on the Biocompatibility of Metallic
    Materials and Implants 820
Mechanically Assisted Corrosion of Metallic Biomaterials 826
Corrosion Performance of Stainless Steels, Cobalt, and
    Titanium Alloys in Biomedical Applications 837
Corrosion Fatigue and Stress-Corrosion Cracking in
    Metallic Biomaterials 853
Corrosion and Tarnish of Dental Alloys 891

Corrosion in the Petroleum and Petrochemical Industry
Corrosion in Petroleum Production Operations 922
Corrosion in Petroleum Refining and Petrochemical
    Operations 967
External Corrosion of Oil and Natural Gas Pipelines 1015
Natural Gas Internal Pipeline Corrosion 1026
Inspection, Data Collection, and Management 1037

Corrosion in the Building Industries
Corrosion of Structures 1054

Corrosion in the Mining and Metal Processing Industries
Corrosion of Metal Processing Equipment 1067
Corrosion in the Mining and Mineral Industry 1076

Gallery of Corrosion Damage 1083
Selected Color Images 1085

Reference Information 1095
Corrosion Rate Conversion 1097
Metric Conversion Guide 1098
Abbreviations and Symbols 1101
Index 1105

ASM Handbook®

Volume 13C
Corrosion: Environments and Industries

Prepared under the direction of the
ASM International Handbook Committee

Stephen D. Cramer and Bernard S. Covino, Jr., Volume Editors

Charles Moosbrugger, Project Editor
Madrid Tramble, Senior Production Coordinator
Diane Grubbs, Editorial Assistant
Pattie Pace, Production Coordinator
Diane Wilkoff, Production Coordinator
Kathryn Muldoon, Production Assistant
Scott D. Henry, Senior Product Manager
Bonnie R. Sanders, Manager of Production

Editorial Assistance
Joseph R. Davis
Elizabeth Marquard
Heather Lampman
Marc Schaefer
Beverly Musgrove
Cindy Karcher
Kathy Dragolich

Materials Park, Ohio 44073-0002
www.asminternational.org

Copyright © 2006
by
ASM International®
All rights reserved

No part of this book may be reproduced, stored in a retrieval system, or transmitted, in any form or by any means, electronic, mechanical, photocopying, recording, or otherwise, without the written permission of the copyright owner.

First printing, September 2006

This book is a collective effort involving hundreds of technical specialists. It brings together a wealth of information from worldwide sources to help scientists, engineers, and technicians solve current and long-range problems.

Great care is taken in the compilation and production of this Volume, but it should be made clear that NO WARRANTIES, EXPRESS OR IMPLIED, INCLUDING, WITHOUT LIMITATION, WARRANTIES OF MERCHANTABILITY OR FITNESS FOR A PARTICULAR PURPOSE, ARE GIVEN IN CONNECTION WITH THIS PUBLICATION. Although this information is believed to be accurate by ASM, ASM cannot guarantee that favorable results will be obtained from the use of this publication alone. This publication is intended for use by persons having technical skill, at their sole discretion and risk. Since the conditions of product or material use are outside of ASM's control, ASM assumes no liability or obligation in connection with any use of this information. No claim of any kind, whether as to products or information in this publication, and whether or not based on negligence, shall be greater in amount than the purchase price of this product or publication in respect of which damages are claimed. THE REMEDY HEREBY PROVIDED SHALL BE THE EXCLUSIVE AND SOLE REMEDY OF BUYER, AND IN NO EVENT SHALL EITHER PARTY BE LIABLE FOR SPECIAL, INDIRECT OR CONSEQUENTIAL DAMAGES WHETHER OR NOT CAUSED BY OR RESULTING FROM THE NEGLIGENCE OF SUCH PARTY. As with any material, evaluation of the material under end-use conditions prior to specification is essential. Therefore, specific testing under actual conditions is recommended.

Nothing contained in this book shall be construed as a grant of any right of manufacture, sale, use, or reproduction, in connection with any method, process, apparatus, product, composition, or system, whether or not covered by letters patent, copyright, or trademark, and nothing contained in this book shall be construed as a defense against any alleged infringement of letters patent, copyright, or trademark, or as a defense against liability for such infringement.

Comments, criticisms, and suggestions are invited, and should be forwarded to ASM International.

Library of Congress Cataloging-in-Publication Data

ASM International

ASM Handbook
Includes bibliographical references and indexes
Contents: v.1. Properties and selection—irons, steels, and high-performance alloys—v.2. Properties and selection—nonferrous alloys and special-purpose materials—[etc.]—v.21. Composites

1. Metals—Handbooks, manuals, etc. 2. Metal-work—Handbooks, manuals, etc. I. ASM International. Handbook Committee.
II. Metals Handbook.
TA459.M43 1990 620.1´6 90-115
SAN: 204-7586

ISBN-13: 978-0-87170-709-3
ISBN-10: 0-87170-709-8

ASM International®
Materials Park, OH 44073-0002
www.asminternational.org

Printed in the United States of America

Multiple copy reprints of individual articles are available from Technical Department, ASM International.

Foreword

This work, *Corrosion: Environments and Industries*, is application driven. The best practices in segments of industry with respect to materials selection, protection of materials, and monitoring of corrosion are presented. The challenges of local environments encountered within these industries, as well as large-scale environmental challenges, are documented. The choice of solutions to these challenges can be found.

Just as the environment affects materials, so also corrosion and its by-products affect the immediate environment. Nowhere is the immediate effect of more concern than in biomedical implants. We are pleased with the new information shared by experts in this field.

As we recognize the energy costs of producing new materials of construction, the creation of engineered systems that will resist corrosion takes on added importance. The importance and costs of maintenance have been discussed for many of the industrial segments—aviation, automotive, oil and gas pipeline, chemical, and pulp and paper industries, as well as the military. The consequences of material degradation are addressed as the service temperatures of materials are pushed higher for greater efficiency in energy conversion. As engineered systems are made more complex and the controlling electronics are made smaller, the tolerance for any corrosion is lessened.

ASM International is deeply indebted to the Editors, Stephen D. Cramer and Bernard S. Covino, Jr., who envisioned the revision of the landmark 1987 *Metals Handbook,* 9th edition, Volume 13. The energy they sustained throughout this project and the care they gave to every article has been huge. The resulting three Volumes contain 281 articles, nearly 3000 pages, 3000 figures, and 1500 tables—certainly impressive statistics. Our Society is as impressed and equally grateful for the way in which they recruited and encouraged a community of corrosion experts from around the world and from many professional organizations to volunteer their time and ability.

We are grateful to the 200 authors and reviewers who shared their knowledge of corrosion and materials for the good of this Volume. They are listed on the next several pages. And again, thanks to the contributors to the preceding two Volumes and the original 9th edition *Corrosion* Volume.

Thanks also go to the members of the ASM Handbook Committee for their involvement in this project and their commitment to keep the information of the *ASM Handbook* series current and relevant to the needs of our members and the technical community. Finally, thanks to the ASM editorial and production staff for the overall result.

Reza Abbaschian
President
ASM International

Stanley C. Theobald
Managing Director
ASM International

Policy on Units of Measure

By a resolution of its Board of Trustees, ASM International has adopted the practice of publishing data in both metric and customary U.S. units of measure. In preparing this Handbook, the editors have attempted to present data in metric units based primarily on Système International d'Unités (SI), with secondary mention of the corresponding values in customary U.S. units. The decision to use SI as the primary system of units was based on the aforementioned resolution of the Board of Trustees and the widespread use of metric units throughout the world.

For the most part, numerical engineering data in the text and in tables are presented in SI-based units with the customary U.S. equivalents in parentheses (text) or adjoining columns (tables). For example, pressure, stress, and strength are shown both in SI units, which are pascals (Pa) with a suitable prefix, and in customary U.S. units, which are pounds per square inch (psi). To save space, large values of psi have been converted to kips per square inch (ksi), where 1 ksi = 1000 psi. The metric tonne ($kg \times 10^3$) has sometimes been shown in megagrams (Mg). Some strictly scientific data are presented in SI units only.

To clarify some illustrations, only one set of units is presented on artwork. References in the accompanying text to data in the illustrations are presented in both SI-based and customary U.S. units. On graphs and charts, grids corresponding to SI-based units usually appear along the left and bottom edges. Where appropriate, corresponding customary U.S. units appear along the top and right edges.

Data pertaining to a specification published by a specification-writing group may be given in only the units used in that specification or in dual units, depending on the nature of the data. For example, the typical yield strength of steel sheet made to a specification written in customary U.S. units would be presented in dual units, but the sheet thickness specified in that specification might be presented only in inches.

Data obtained according to standardized test methods for which the standard recommends a particular system of units are presented in the units of that system. Wherever feasible, equivalent units are also presented. Some statistical data may also be presented in only the original units used in the analysis.

Conversions and rounding have been done in accordance with IEEE/ASTM SI-10, with attention given to the number of significant digits in the original data. For example, an annealing temperature of 1570 °F contains three significant digits. In this case, the equivalent temperature would be given as 855 °C; the exact conversion to 854.44 °C would not be appropriate. For an invariant physical phenomenon that occurs at a precise temperature (such as the melting of pure silver), it would be appropriate to report the temperature as 961.93 °C or 1763.5 °F. In some instances (especially in tables and data compilations), temperature values in °C and °F are alternatives rather than conversions.

The policy of units of measure in this Handbook contains several exceptions to strict conformance to IEEE/ASTM SI-10; in each instance, the exception has been made in an effort to improve the clarity of the Handbook. The most notable exception is the use of g/cm^3 rather than kg/m^3 as the unit of measure for density (mass per unit volume).

SI practice requires that only one virgule (diagonal) appear in units formed by combination of several basic units. Therefore, all of the units preceding the virgule are in the numerator and all units following the virgule are in the denominator of the expression; no parentheses are required to prevent ambiguity.

Preface

Corrosion, while silent and often subtle, is probably the most significant cause of physical deterioration and degradation in man-made structures. The 2004 global direct cost of corrosion, representing costs experienced by owners and operators of manufactured equipment and systems, was estimated to be $990 billion United States dollars (USD) annually, or 2.0% of the $50 trillion (USD) world gross domestic product (GDP) (Ref 1). The 2004 global indirect cost of corrosion, representing costs assumed by the end user and the overall economy, was estimated to be $940 billion (USD) annually (Ref 1). On this basis, the total cost of corrosion to the global economy in 2004 was estimated to be approximately $1.9 trillion (USD) annually, or 3.8% of the world GDP. The largest contribution to this cost comes from the United States at 31%. The next largest contributions were Japan, 6%; Russia, 6%; and Germany, 5%.

ASM Handbook Volume 13C, *Corrosion: Environments and Industries* is the third and final volume of the three-volume update, revision, and expansion of *Metals Handbook,* 9th edition, Volume 13, *Corrosion,* published in 1987. The first volume—Volume 13A, *Corrosion: Fundamentals, Testing, and Protection*—was published in 2003. The second volume—Volume 13B, *Corrosion: Materials*—was published in 2005. These three volumes together present the current state of corrosion knowledge, the efforts to mitigate corrosion's effects on society's structures and economies, and a perspective on future trends in corrosion prevention and mitigation. Metals remain the primary focus of the Handbook. However, nonmetallic materials occupy a more prominent position, reflecting their wide and effective use to solve problems of corrosion and their frequent use with metals in complex engineering systems. Wet (or aqueous) corrosion remains the primary environmental focus, but dry (or gaseous) corrosion is also addressed, reflecting the increased use of elevated- or high-temperature operations in engineering systems, particularly energy-related systems, where corrosion and oxidation are important considerations.

Volume 13C recognizes, as did Volumes 13A and 13B, the diverse range of materials, environments, and industries affected by corrosion, the global reach of corrosion practice, and the levels of technical activity and cooperation required to produce cost-effective, safe, and environmentally-sound solutions to materials problems. As we worked on this project, we marveled at the spread of corrosion technology into the many and diverse areas of engineering, industry, and human activity. It attests to the effectiveness of the pioneers of corrosion research and education, and of the organizations they helped to create, in communicating the principles and experience of corrosion to an ever-widening audience. Over 50% of the articles in Volume 13C are new. Looking back over the three volumes, 45% of the articles are new to the revised Handbook, reflecting changes occurring in the field of corrosion over the intervening 20 years. Authors from 14 countries contributed articles to the three Handbook volumes.

Volume 13C is organized into two major Sections addressing the performance of materials in specific classes of environments and their performance in the environments created by specific industries. These Sections recognize that materials respond to the laws of chemistry and physics and that, within the constraints of design and operating conditions, corrosion can be minimized to provide economic, environmental, and safety benefits.

The first Section is "Corrosion in Specific Environments," addressing distinct classes of environments where knowledge of the general attributes of the environment provides a "generic" framework for understanding and solving corrosion problems. By the nature of this approach, solutions to problems of corrosion performance and corrosion protection are viewed as spanning industries. The specific environments addressed in Volume 13C are fresh water, marine (both atmospheric and aqueous), underground, and military, with an eclectic mix of other environments included under specialized environments.

The second Section is "Corrosion in Specific Industries," addressing corrosion performance and corrosion protection in distinct environments created by specific industries. The specific industries addressed in Volume 13C are nuclear power, fossil energy and alternative fuels, petroleum and petrochemical, land transportation, commercial aviation, microelectronics, chemical processing, pulp and paper, food and beverage, pharmaceutical and medical technology, building, and mining and mineral processing. Corrosion issues in the energy sector receive considerable attention in this Section. In addition, there is substantial overlap between this Section and topics addressed in military environments in the first Section.

Supporting material is provided at the back of the Handbook. A "Corrosion Rate Conversion" includes conversions in both nomograph and tabular form. The "Metric Conversion Guide" gives conversion factors for common units and includes SI prefixes. "Abbreviations and Symbols" provides a key to common acronyms, abbreviations, and symbols used in the Handbook.

Many individuals contributed to Volume 13C. In particular, we wish to recognize the efforts of the following individuals who provided leadership in organizing subsections of the Handbook (listed in alphabetical order):

Chairperson	Subsection title
Alain A. Adjorlolo	Corrosion in Commercial Aviation
Vinod S. Agarwala	Corrosion in Military Environments
Hira Ahluwalia	Corrosion in the Chemical Processing Industry
Denise A. Aylor	Corrosion in Marine Environments
Bernard S. Covino, Jr.	Corrosion in Specialized Environments
Stephen D. Cramer	Corrosion in Fresh Water Environments
	Corrosion in Specialized Environments
	Corrosion in the Pharmaceutical and Medical Technology Industries
Harry Dykstra	Corrosion in the Pulp and Paper Industry
Dawn Eden	Corrosion in the Petroleum and Petrochemical Industry
Barry Gordon	Corrosion in the Nuclear Power Industry
Donald L. Jordan	Corrosion in the Land Transportation Industries
Russell Kane	Corrosion in the Petroleum and Petrochemical Industry
Brajendra Mishra	Corrosion in the Mining and Metal Processing Industries
Bert Moniz	Corrosion in the Food and Beverage Industry
Seshu Pabbisetty	Corrosion in the Microelectronics Industries
Kevin T. Parker	Corrosion in Underground Environments
Larry Paul	Corrosion in the Fossil and Alternative Fuel Industries
Robert L. Ruedisueli	Corrosion in Marine Environments
John E. Slater	Corrosion in the Building Industry

These knowledgeable and dedicated individuals generously devoted considerable time to the preparation of the Handbook. They were joined in this effort by more than 200 authors who contributed their expertise and creativity in a collaboration to write and revise the articles in the Handbook, and by the many reviewers of their articles. These volunteers built on the contributions of earlier Handbook authors and reviewers who provided the solid foundation on which the present Handbook rests.

For articles revised from the 1987 edition, the contribution of the previous author is acknowledged at the end of the article. This location in no way diminishes their contribution or our gratitude. Authors responsible for the current revision are named after the title. The variation in the amount of revision is broad. The many completely new articles presented no challenge for attribution, but assigning fair credit for revised articles was more problematic. The choice of presenting authors' names without comment or with the qualifier "Revised by" is solely the responsibility of the ASM staff.

We thank ASM International and the ASM staff for their skilled support and valued expertise in the production of this Handbook. In particular, we thank Charles Moosbrugger, Gayle Anton, Diane Grubbs, and Scott Henry for their encouragement, tactful diplomacy, and many helpful discussions. We are most grateful to the National Energy Technology Laboratory (formerly the Albany Research Center), U.S. Department of Energy, for the support and flexibility in our assignments that enabled us to participate in this project. We especially thank our supervisors, Jeffrey A. Hawk and Cynthia A. Powell, for their gracious and generous encouragement throughout the project.

Stephen D. Cramer, FNACE
Bernard S. Covino, Jr., FNACE
National Energy Technology Laboratory
U.S. Department of Energy

REFERENCE

1. R. Bhaskaran, N. Palaniswamy, N.S. Rengaswamy, and M. Jayachandran, Global Cost of Corrosion—A Historical Review, *Corrosion: Materials,* Vol 13B, *ASM Handbook,* ASM International, Materials Park, OH, 2005, p 621–628

Officers and Trustees of ASM International (2005–2006)

Reza Abbaschian
President and Trustee
University of California Riverside

Lawrence C. Wagner
Vice President and Trustee
Texas Instruments

Bhakta B. Rath
Immediate Past President and Trustee
U.S. Naval Research Laboratory

Paul L. Huber
Treasurer and Trustee
Seco/Warwick Corporation

Stanley C. Theobald
Secretary and Managing Director
ASM International

Trustees

Sue S. Baik-Kromalic
Honda of America

Christopher C. Berndt
James Cook University

Dianne Chong
The Boeing Company

Roger J. Fabian
Bodycote Thermal Processing

William E. Frazier
Naval Air Systems Command

Pradeep Goyal
Pradeep Metals Ltd.

Richard L. Kennedy
Allvac

Frederick J. Lisy
Orbital Research Incorporated

Frederick Edward Schmidt, Jr.
Engineering Systems Inc.

Members of the ASM Handbook Committee (2005–2006)

Jeffrey A. Hawk
(Chair 2005–; Member 1997–)
General Electric Company

Larry D. Hanke
(Vice Chair 2005–; Member 1994–)
Material Evaluation and
Engineering Inc.

Viola L. Acoff (2005–)
University of Alabama

David E. Alman (2002–)
U.S. Department of Energy

Tim Cheek (2004–)
International Truck & Engine
Corporation

Lichun Leigh Chen (2002–)
Engineered Materials Solutions

Craig Clauser (2005–)
Craig Clauser Engineering Consulting Inc.

William Frazier (2005–)
Naval Air Systems Command

Lee Gearhart (2005–)
Moog Inc.

Michael A. Hollis (2003–)
Delphi Corporation

Kent L. Johnson (1999–)
Engineering Systems Inc.

Ann Kelly (2004–)
Los Alamos National Laboratory

Alan T. Male (2003–)
University of Kentucky

William L. Mankins (1989–)
Metallurgical Services Inc.

Dana J. Medlin (2005–)
South Dakota School of Mines and
Technology

Joseph W. Newkirk (2005–)
Metallurgical Engineering

Toby Padfield (2004–)
ZF Sachs Automotive of America

Frederick Edward Schmidt, Jr. (2005–)
Engineering Systems Inc.

Karl P. Staudhammer (1997–)
Los Alamos National Laboratory

Kenneth B. Tator (1991–)
KTA-Tator Inc.

George F. Vander Voort (1997–)
Buehler Ltd.

Previous Chairs of the ASM Handbook Committee

R.J. Austin
(1992–1994) (Member 1984–1985)

L.B. Case
(1931–1933) (Member 1927–1933)

T.D. Cooper
(1984–1986) (Member 1981–1986)

C.V. Darragh
(1999–2002) (Member 1989–2005)

E.O. Dixon
(1952–1954) (Member 1947–1955)

R.L. Dowdell
(1938–1939) (Member 1935–1939)

Henry E. Fairman
(2002–2004) (Member 1993–2005)

M.M. Gauthier
(1997–1998) (Member 1990–2000)

J.P. Gill
(1937) (Member 1934–1937)

J.D. Graham
(1966–1968) (Member 1961–1970)

J.F. Harper
(1923–1926) (Member 1923–1926)

C.H. Herty, Jr.
(1934–1936) (Member 1930–1936)

D.D. Huffman
(1986–1990) (Member 1982–2005)

J.B. Johnson
(1948–1951) (Member 1944–1951)

L.J. Korb
(1983) (Member 1978–1983)

R.W.E. Leiter
(1962–1963) (Member 1955–1958,
1960–1964)

G.V. Luerssen
(1943–1947) (Member 1942–1947)

G.N. Maniar
(1979–1980) (Member 1974–1980)

W.L. Mankins
(1994–1997) (Member 1989–)

J.L. McCall
(1982) (Member 1977–1982)

W.J. Merten
(1927–1930) (Member 1923–1933)

D.L. Olson
(1990–1992) (Member 1982–1988,
1989–1992)

N.E. Promisel
(1955–1961) (Member 1954–1963)

G.J. Shubat
(1973–1975) (Member 1966–1975)

W.A. Stadtler
(1969–1972) (Member 1962–1972)

R. Ward
(1976–1978) (Member 1972–1978)

M.G.H. Wells
(1981) (Member 1976–1981)

D.J. Wright
(1964–1965) (Member 1959–1967)

Authors and Contributors

Alain A. Adjorlolo
The Boeing Company

Vinod S. Agarwala
Naval Air Systems Command, U.S. Navy

Hira S. Ahluwalia
Material Selection Resources, Inc.

Peter L. Andresen
General Electric Global Research

Zhijun Bai
Syracuse University

Wate Bakker
Electric Power Research Institute

Donald E. Bardsley
Sulzer Process Pumps Inc.

John A. Beavers
CC Technologies

Graham Bell
M.J. Schiff & Associates

J.E. Benfer
NAVAIR Materials Engineering Competency

David Bennett
Corrosion Probe Inc.

Henry L. Bernstein
Gas Turbine Materials Association

James Brandt
Galvotec Corrosion Services

S.K. Brubaker
E.I. Du Pont de Nemours & Company, Inc.

Sophie J. Bullard
National Energy Technology Laboratory

Kirk J. Bundy
Tulane University

Jeremy Busby
University of Michigan

Sridhar Canumalla
Nokia Enterprise Systems

Clifton M. Carey
American Dental Association Foundation

Bryant "Web" Chandler
Greenman Pedersen, Inc.

Norm Clayton
Naval Surface Warfare Center, Carderock Division

M. Colavita
Italian Air Force

Everett E. Collier
Consultant

Pierre Combrade
Framatome ANP

Greg Courval
Alcan International Limited

Bernard S. Covino, Jr.
National Energy Technology Laboratory

William Cox
Corrosion Management Ltd.

Stephen D. Cramer
National Energy Technology Laboratory

J.R. Crum
Special Metals Corporation

Chester M. Dacres
DACCO SCI, Inc.

Phillip Daniel
Babcock & Wilcox Company

Michael Davies
Cariad Consultants

Stephen C. Dexter
University of Delaware

James R. Divine
ChemMet, Ltd., PC

Joe Douthett
AK Research

Harry Dykstra
Acuren

Dawn C. Eden
Honeywell Process Solutions

Teresa Elliott
City of Portland, Oregon

Paul Eyre
DuPont

F. Peter Ford
General Electric Global Research (retired)

Aleksei V. Gershun
Prestone Products

Jeremy L. Gilbert
Syracuse University

William J. Gilbert
Branch Environmental Corp.

Barry M. Gordon
Structural Integrity Associates, Inc.

R.D. Granata
Florida Atlantic University

Stuart L. Greenberger
Bureau Water Works, City of Portland, Oregon

Richard B. Griffin
Texas A&M University

I. Carl Handsy
U.S. Army Tank-Automotive & Armaments Command

Gary Hanvy
Texas Instruments

William H. Hartt
Florida Atlantic University

Robert H. Heidersbach
Dr. Rust, Inc.

Drew Hevle
El Paso Corporation

Gordon R. Holcomb
National Energy Technology Laboratory

W. Brian Holtsbaum
CC Technologies Canada, Ltd.

Ronald M. Horn
General Electric Nuclear Energy

Jack W. Horvath
HydroChem Industrial Services, Inc.

Wally Huijbregts
Huijbregts Corrosion Consultancy

Herbert S. Jennings
DuPont

David Johnson
Galvotec Corrosion Services

Otakar Jonas
Jonas, Inc.

D.L. Jordan
Ford Motor Company

Russell D. Kane
Honeywell Process Solutions

Ernest W. Klechka, Jr.
CC Technologies

Kevin J. Kovaleski
Naval Air Warfare Center, Aircraft Division

Angel Kowalski
CC Technologies

Lorrie A. Krebs
Anderson Materials Evaluation, Inc.

Ashok Kumar
U.S. Army Engineer Research and Development Center (ERDC) Construction Engineering Research Laboratory (CERL)

Steven C. Kung
The Babcock & Wilcox Company

Kenneth S. Kutska
Wheaton (IL) Playground District

George Y. Lai
Consultant

Jim Langley
USA Cycling Certified Mechanic

Michael LaPlante
Colt Defense LLC

R.M. Latanision
Exponent

Jason S. Lee
Naval Research Laboratory, Stennis Space Center

René Leferink
KEMA

Clément Lemaignan
CEA France

Lianfang Li
W.R. Grace, Inc.

E.L. Liening
The Dow Chemical Company

Brenda J. Little
Naval Research Laboratory, Stennis Space Center

Joyce M. Mancini
Jonas, Inc.

W.L. Mathay
Nickel Institute

Ronald L. McAlpin
Gas Turbine Materials Associates

R. Daniel McCright
Lawrence Livermore National Laboratory

Sam McFarland
Lloyd's Register

Spiro Megremis
American Dental Association

Joseph T. Menke
U.S. Army TACOM

Michael Meyer
The Solae Company

William Miller
Sulzer Process Pumps Inc.

B. Mishra
Colorado School of Mines

D.B. Mitton
University of North Dakota

Bert Moniz
DuPont

William G. Moore
National Electric Coil

Max D. Moskal
Mechanical and Materials Engineering

Ned Niccolls
Chevron Texaco

Randy Nixon
Corrosion Probe Inc.

J.J. Pak
Hanyang University, Korea

Rigo Perez
Boeing Company

Lyle D. Perrigo
U.S. Arctic Research Commission (retired)

Frank Pianca
Ontario Ministry of Transportation

Joseph Pikas
MATCOR Inc.

Jerry Podany
Paul Getty Museum

David F. Pulley
Naval Air Warfare Center, Aircraft Division

Jianhai Qiu
Nanyang Technological University

Richard I. Ray
Naval Research Laboratory, Stennis Space Center

Raúl B. Rebak
Lawrence Livermore National Laboratory

Craig Reid
Acuren

John Repp
Corrpro/Ocean City Research Corp.

P.R. Roberge
Royal Military College of Canada

Ralph (Bud) W. Ross, Jr.
Consultant

Alberto A. Sagüés
University of South Florida

K.K. Sankaran
Boeing Company

Adrian Santini
Con Edison of New York

Daniel P. Schmidt
The Pennsylvania State University

Peter M. Scott
Framatome ANP

L.A. Scribner
Becht Engineering

K. Anthony Selby
Water Technology Consultants, Inc.

Lyndsie S. Selwyn
Canadian Conservation Institute

Barbara A. Shaw
The Pennsylvania State University

Wilford W. Shaw
The Pennsylvania State University

David A. Shifler
Naval Surface Warfare Center, Carderock Division

Stan Silvus
Southwest Research Institute

Douglas Singbeil
Paprican

Prabhakar Singh
Pacific Northwest National Laboratory

James Skogsberg
Chevron Texaco

John E. Slater
Invetech Inc.

John S. Smart III
John S. Smart Consulting Engineers

Herbert Smith
Boeing Company

Narasi Sridhar
Southwest Research Institute

Sridhar Srinivasan
Honeywell Process Solutions

L.D. Stephenson
U.S. Army Engineer Research and Development Center (ERDC) Construction Engineering Research Laboratory (CERL)

Kenneth R. St. John
The University of Mississippi Medical Center

John Stringer
Electric Power Research Institute

Mats Ström
Volvo Car Corporation

Khuzema Sulemanji
Texas Instruments

Windsor Sung
Massachusetts Water Resources Authority

Barry C. Syrett
Electric Power Research Institute

A.C. Tan
Micron Semiconductor Asia

J.L. Tardiff
Ford Motor Company

Ramgopal Thodla
General Electric Company

Mercy Thomas
Texas Instruments

Chris Thompson
Paprican

Neil G. Thompson
CC Technologies

Jack Tinnea
Tinnea & Associates, LLC

Arthur H. Tuthill
Tuthill Associates

John Tverberg
Metals and Materials Consulting Engineers

Jose L. Villalobos
V&A Consulting Engineers

Puligandla Viswanadham
Nokia Research Center

Nicholas Warchol
U.S. Army ARDEC

Gary S. Was
University of Michigan

Angela Wensley
Angela Wensley Engineering

Paul K. Whitcraft
Rolled Alloys

Peter M. Woyciesjes
Prestone Products

Zhenguo G. Yang
Pacific Northwest National Laboratory

Te-Lin Yau
Yau Consultancy

Lyle D. Zardiackas
University of Mississippi Medical Center

Shi Hua Zhang
DuPont

Reviewers

Ralph Adler
U.S. Army

Hira Ahluwalia
Material Selection Resources, Inc.

Todd Allen
University of Wisconsin

Anton Banweg
Nalco Company

Sean Barnes
DuPont

Gregory A. Bates
Solae Company

Franceso Bellucci
University of Naples "Federico II"

Ron Bianchetti
East Bay Municipal Utility District

Timothy Bieri
BP

Francine Bovard
Alcoa

Robert L. Bratton
Nuclear Materials Disposition and Engineering

Mike Bresney
AGT

Stanley A. Brown
FDA

Stephen K. Brubaker
DuPont

Kirk J. Bundy
Tulane University

Juan Bustillos
Dow Chemical

Gary M. Carinci
TMR Stainless

Tom Chase
Chase Art Services

Tim Cheek
International Truck & Engine Corp.

Lichun Leigh Chen
Engineered Materials Solutions

Jason A. Cline
Spectral Sciences, Inc.

Desmond C. Cook
Old Dominion University

Thomas Cordea
International Truck and Engine Corporation

Robert A. Cottis
UMIST

Irv Cotton
Arthur Freedman Associates, Inc.

Bruce Craig
MetCorr

Larry Craigie
American Composites Manufacturers Association

Jim Crum
Special Metals Corporation

Phil L. Daniel
Babcock & Wilcox

Craig V. Darragh
The Timken Company

Chris Dash
Conoco Phillips Alaska, Inc.

Michael Davies
CARIAD Consultants

Guy D. Davis
DACCO SCI, Inc.

Sheldon Dean
Dean Corrosion Technology

John Devaney
Hi-Rel Laboratories, Inc.

John B. Dion
BAE Systems

John Disegi
Synthes (USA)

Roger Dolan
Dolan Environmental Services, Inc.

Gary Doll
The Timken Company

David E. Dombrowski
Los Alamos National Laboratory

R. Barry Dooley
Electric Power Research Institute

Timothy Eckert
Electric Power Research Institute

Dave Eden
InterCorr International

Peter Elliott
Corrosion and Materials Consultancy, Inc.

Henry "Ed" Fairman
Cincinnati Metallurgical Consultants

Robert Filipek
AGH University of Science and Ceramics

Brian J. Fitzgerald
ExxonMobil Chemical Company

John Fitzgerald
ExxonMobil Chemical Company

Gerald S. Frankel
The Ohio State University

Peter Furrer
Novelis Technology AG

Brian Gleeson
Iowa State University

John J. Goetz
Thielsch Engineering

Martha Goodway
Smithsonian Center for Materials Research and Education

Gary Griffith
Mechanical Dynamics & Analysis, LLC

Carol Grissom
SCMRE

John Grubb
Allegheny Ludlum Technical Center

Charlie Hall
Mears Group

Nadim James Hallab
Rush University Medical Center

Larry D. Hanke
Materials Evaluation and Engineering, Inc.

Jeffrey A. Hawk
General Electric Company

M. Gwyn Hocking
Imperial College London

Paul Hoffman
CIV NAVAIR

Mike Holly
General Motors Corp.

Glenn T. Hong
General Atomics, San Diego, CA

Merv Howells
Honeywell Airframe Systems

Fred H. Hua
Bechtel SAIC Co., LLC

Dennis Huffman
The Timken Company

Kumar Jata
CIV USAF AFRL/MILL

Carol Jeffcoate
Honeywell Airframe Systems

David Jensen
Eli Lilly and Company

Anders Jenssen
Studsvik Nuclear AB, Sweden

Paul Jett
Smithsonian Institute

Randy C. John
Shell Global Solutions (US) Inc.

Kent Johnson
Engineering Systems Inc.

Joanne Jones-Meehan
Naval Research Laboratory

Donald L. Jordan
North American Engineering

Robert Kain
LaQue Center for Corrosion

Don Kelley
Dow Chemical

Srinivasan Kesavan
FMC Corporation

Naeem A. Khan
Saudi Arabian Oil Company

Jonathan K. Klopman
Marine Surveyor NAMS-CMS

Ernest Klechka
CC Technologies

David Kolman
U.S. Department of Energy
Los Alamos National Laboratory

Lou Koszewski
U.S. Tank Protectors Inc.

David Kroon
Corrpro Companies

Roger A. LaBoube
University of Missouri-Rolla

Gregg D. Larson
Exelon Nuclear

Kevin Lawson
Petrofac Facilities Management Ltd.

Thomas W. Lee
Jabil Circuit, Inc.

William LeVan
Cast Iron Soil Pipe Institute

E.L. Liening
Dow Chemical

Scott Lillard
U.S. Department of Energy
Los Alamos National Laboratory

Huimin Liu
Ford Motor Company

Gary A. Loretitsch
Puckorius & Associates, Inc.

Stephen Lowell
Defense Standardization Program Office

Digby MacDonald
Pennsylvania State University

William L. Mankins
Metallurgical Services Inc.

Florian B. Mansfeld
University of Southern California

William N. Matulewicz
Wincon Technologies, Inc.

Craig Matzdorf
U.S. Navy

Graham McCartney
University of Nottingham

Bruce McMordie
Sermatech

Gerald H. Meier
University of Pittsburgh

Joseph T. Menke
U.S. Army TACOM

Ronald E. Mizia
Idaho National Engineering &
Environmental Laboratory

Raymond W. Monroe
Steel Founders' Society of America

Jean Montemarano
Naval Surface Warfare Center, Carderock Division

Robert E. Moore
Washington Group International

Sandra Morgan
International Truck and Engine Corporation

Bill Mullis
Aberdeen Test Center

M.P. Sukumaran Nair
FACT, Ltd.

Larry Nelson
GE Global Research Center

Karthik H. Obla
National Ready-Mixed Concrete Association

David Olson
Colorado School of Mines

Michael R. Ostermiller
Corrosion Engineering

Toby V. Padfield
ZF Sachs Automotive of America

Larry Paul
ThyssenKrupp VDM USA Inc.

Steven J. Pawel
U.S. Department of Energy
Oak Ridge National Laboratory

Fred Pettit
University of Pittsburgh

G. Louis Powell
Y-12 National Security Complex

Raúl Rebak
Lawrence Livermore National Laboratory

Michael Renner
Bayer Technology Services GmbH

Chris Robbins
Health & Safety Executive

Elwin L. Rooy
Elwin L. Rooy and Associates

Marvin J. Rudolph
DuPont

Brian Saldnaha
DuPont

Sreerangapatam Sampath
Army Research Laboratory

Philip J. Samulewicz
Ambiant Air Quality Services, Inc.

B.J. Sanders
BJS and Associates

Jeff Sarver
The Babcock & Wilcox Company

Frederick Edward Schmidt, Jr.
Engineering Systems Inc.

Michael Schock
U.S. Environmental Protection Agency

Robert J. Shaffer
DaimlerChrysler Corporation

C. Ramadeva Shastry
International Steel Group, Inc.

Robert W. Shaw
U.S. Army Research Office

Theresa Simpson
Bethlehem Steel Corp.

Robert Smallwood
Det Norske Veritas

Gaylord D. Smith
Special Metals Corporation

Vernon L. Snoeyink
University of Illinois

Donald Snyder
Atotech R & D Worldwide

Gerard Sorell
G. Sorell Consulting Services

Andy Spisak
EME Associates

David L. Sponseller
OMNI Metals Laboratory, Inc.

Roger Staehle
Staehle Consulting

Karl Staudhammer
Los Alamos National Laboratory

Jan Stein
Electric Power Research Institute

Martin L. Stephens
DaimlerChrysler Corp.

John Stringer
Electric Power Research Institute (retired)

Henry Tachick
Dairyland Electrical Industries, Inc.

Ken Tator
KTA Tator Inc.

Oscar Tavares
Lafarge North America Inc.

Michael Tavary
Dow Chemical

George J. Theus
Engineering Systems Inc.

Dominique Thierry
Technopôle de Brest Rivoalon

A.C. Tiburcio
US Steel Research

Whitt L. Trimble
Fluor Corporation

Arthur Tuthill
Tuthill Associates, Inc.

Elma van der Lingen
MINTEK

Krishna Venugopalan
DePuy, Inc.

Mike Wayman
University of Alberta

Alan Whitehead
GE – Power Systems (retired)

James Whitfield
U.S. Navy (CIV NAVAIR)

Gary S. Whittaker
Whittaker Materials Engineering Associates, LLC

Roger Wildt
RW Consulting Group

David Willis
BlueScope Steel Research

Tim Woods
U.S. Navy (CIV NAVAIR)

Ernest Yeboah
Orange County Sanitary District

Shi Hua Zhang
DuPont

Contents

Corrosion in Specific Environments 1

Introduction to Corrosion in Specific Environments
   Stephen D. Cramer 5
      Corrosion in Freshwater Environments 5
      Corrosion in Marine Environments 5
      Corrosion in Underground Environments 5
      Corrosion in Military Environments 6
      Corrosion in Specialized Environments 7

Corrosion in Fresh Water Environments

Corrosion in Potable Water Distribution and Building Systems
   Windsor Sung 8
      Theoretical Considerations 8
      Mitigation against Corrosion 10
      Additional Considerations 11

Corrosion in Service Water Distribution Systems
   K. Anthony Selby 12
      Typical System Designs 12
      Typical Water Qualities 13
      Corrosion Mechanisms in Service Water Systems 13
      Corrosion Challenges in Service Water Systems 13
      Corrosion Control in Service Water Systems 14
      Deposit Control 14

Rouging of Stainless Steel in High-Purity Water
   John C. Tverberg 15
      Pharmaceutical Waters 15
      Chlorides 16
      Passive Layer 17
      Surface Finish 18
      Rouge Classification 18
      Castings 20
      Cleaning and Repassivation 21

Corrosion in Wastewater Systems
   Jose L. Villalobos, Graham Bell 23
      Predesign Surveys and Testing 23
      Material Considerations 24

Corrosion in Marine Environments

Corrosion in Seawater
   Stephen C. Dexter 27
      Consistency and the Major Ions 27
      Variability of the Minor Ions 30
      Effect of Pollutants 37
      Influence of Biological Organisms 38
      Effect of Flow Velocity 40

Corrosion in Marine Atmospheres
   Richard B. Griffin 42
      Important Variables 42
      Modeling of Atmospheric Corrosion—ISO CORRAG
        Program 51
      Corrosion Products 57
      Atmospheric Corrosion Test Sites 57

Corrosion of Metallic Coatings
   Barbara A. Shaw, Wilford W. Shaw, Daniel P. Schmidt 61
      Thermal Sprayed Coatings 61
      Hot Dip Coatings 65
      Electroplated Coatings 66
      Methods of Protection 66

Performance of Organic Coatings
   R.D. Granata 69
      Surface Preparation 69
      Topside Coating Systems 70
      Immersion Coatings 72

Marine Cathodic Protection
   Robert H. Heidersbach, James Brandt, David Johnson,
      John S. Smart III 73
      Cathodic Protection Criteria 73
      Anode Materials 73
      Comparison of Impressed-Current and Sacrificial Anode
        Systems 74
      Cathodic Protection of Marine Pipelines 74
      Cathodic Protection of Offshore Structures 75
      Cathodic Protection of Ship Hulls 77

Corrosion in Underground Environments

External Corrosion Direct Assessment Integrated with Integrity
   Management
   Joseph Pikas 79
      Four Step ECDA Process 79
      Step 1: Preassessment (Assessment of Risk
        and Threats) 79
      Step 2: Indirect Examinations 81
      Step 3: Direct Examination 81
      Step 4: Post Assessment 82

Close-Interval Survey Techniques
   Drew Hevle, Angel Kowalski 84
      CIS Equipment 84
      Preparation 85
      Procedures 86
      Dynamic Stray Current 87
      Offshore Procedures 87
      Data Validation 87
      Data Interpretation 88

Corrosion of Storage Tanks
  Ernest W. Klechka, Jr. 89
    Soil Corrosivity 89
    Cathodic Protection 90
    Data Needed for Corrosion Protection Design 90
    Soil-Side Corrosion Control 92
    Aboveground Storage Tanks 93
    Underground Storage Tanks 95
    Monitoring ASTs and USTs 95

Well Casing External Corrosion and Cathodic Protection
  W. Brian Holtsbaum 97
    Well Casing Corrosion 97
    Detection of Corrosion 97
    Cathodic Protection of Well Casings 99

Stray Currents in Underground Corrosion
  W. Brian Holtsbaum 107
    Principles of Stray Current 107
    Consequences of Stray Current 109
    Interference Tests 109
    Mitigation 111

Corrosion Rate Probes for Soil Environments
  Bernard S. Covino, Jr., Sophie J. Bullard 115
    Nonelectrochemical Techniques—Principles of
      Operation 115
    Electrochemical Techniques—Principles of Operation 116
    Nonelectrochemical Techniques—Examples of Uses
      in Soils 117
    Electrochemical Techniques—Examples of Uses
      in Soils 117
    Potential Sources of Interference with Corrosion
      Measurements 119

Cathodic Protection of Pipe-Type Power Transmission Cables
  Adrian Santini 122
    Resistor Rectifiers 122
    Polarization Cells 122
    Isolator-Surge Protector 123
    Field Rectifiers 123
    Stray Currents 123

Corrosion in Military Environments

Corrosion in the Military
  Vinod S. Agarwala 126
    Introduction 126
    Military Problems 127
    Corrosion Control and Management 132
    Long-Term Strategy to Reduce Cost of Corrosion 134

Military Specifications and Standards
  Norm Clayton 136
    Types of Documents and Designations 136
    Format of Specifications 138
    Sources of Documents 139
    Notable Specifications, Standards, and Handbooks 139
    Department of Defense Corrosion Policy 140

Corrosion Control for Military Facilities
  Ashok Kumar, L.D. Stephenson, Robert H. Heidersbach 141
    The Environment 141
    Case Studies 141
    Emerging Corrosion-Control Technologies 144

Ground Vehicle Corrosion
  I. Carl Handsy, John Repp 148
    Background 148
    Requirements for Corrosion Control 148
    Procurement Document 148
    Testing Systems to Meet the Army's Needs 149
    Supplemental Corrosion Protection 149
    Improved Maintenance Procedures 150
    Considerations for Corrosion in Design 150

Armament Corrosion
  Nicholas Warchol 151
    Overview of Design, In-Process, Storage, and In-Field
      Problems 151
    Design Considerations 151
    In-Process Considerations 152
    Storage Considerations 154
    In-Field Considerations 154

High-Temperature Corrosion in Military Systems
  David A. Shifler 156
    High-Temperature Corrosion and Degradation
      Processes 156
    Boilers 156
    Diesel Engines 161
    Gas Turbine Engines 162
    Incinerators 164

Finishing Systems for Naval Aircraft
  Kevin J. Kovaleski, David F. Pulley 171
    Standard Finishing Systems 171
    Compliant Coatings Issues and Future Trends 173

Military Coatings
  Joseph T. Menke 180
    Electroplating 180
    Conversion Coating 181
    Supplemental Oils 181
    Paint Coatings 182
    Other Finishes 183

U.S. Navy Aircraft Corrosion
  John E. Benfer 184
    Environment 184
    Aircraft Alloys 184
    Aircraft Inspection 185
    Prevention and Corrosion Control 186
    Examples of Aircraft Corrosion Damage 189

Military Aircraft Corrosion Fatigue
  K.K. Sankaran, R. Perez, H. Smith 195
    Aircraft Corrosion Fatigue Assessment 195
    Causes and Types of Aircraft Corrosion 196
    Impact of Corrosion on Fatigue 197
    Corrosion Metrics 198
    Investigations and Modeling of Corrosion/Fatigue
      Interactions 199
    Methodologies for Predicting the Effects of Corrosion
      on Fatigue 201
    Recent Development and Future Needs 203

Corrosion of Electronic Equipment in Military Environments
  Joseph T. Menke 205
    An Historical Review of Corrosion Problems 205
    Examples of Corrosion Problems 206

Microbiologically Influenced Corrosion in Military Environments
   Jason S. Lee, Richard I. Ray, Brenda J. Little 211
      General Information about Microorganisms 211
      Atmospheric Corrosion 211
      Metals Exposed to Hydrocarbon Fuels 213
      Immersion 213
      Burial Environments 217
Service Life and Aging of Military Equipment
   M. Colavita 220
      Reliability and Safety of Equipment 220
      Aging Mechanisms 220
      Management 223
      Prevention and Control 225
      Prediction Techniques 227

Corrosion in Specialized Environments

Corrosion in Supercritical Water—Waste Destruction
Environments
   R.M. Latanision, D.B. Mitton 229
      The Unique Properties of Supercritical Water 229
      The Economics and Benefits of SCWO 229
      Facility Design Options 230
      Materials Challenges 230
      Mitigation of System Degradation 233
      The Future 233
Corrosion in Supercritical Water—Ultrasupercritical
Environments for Power Production
   Gordon R. Holcomb 236
      Water Properties 236
      Steam Cycle 236
      Steam Cycle Chemistry 237
      Materials Requirements 237
      Boiler Alloys 238
      Turbine Alloys 240
      Corrosion in Supercritical Water 240
      Efficiency 242
      Benefits 243
      Worldwide Materials Research 244
Corrosion in Cold Climates
   Lyle D. Perrigo, James R. Divine 246
      Cold Climates 246
      Corrosion Control Techniques and Costs 246
      Design 247
      Transportation and Storage 248
      Construction 248
      Operations and Maintenance 249
Corrosion in Emissions Control Equipment
   William J. Gilbert 251
      Flue Gas Desulfurization 251
      Waste Incineration 252
      Bulk Solids Processing 253
      Chemical and Pharmaceutical Industries 254
Corrosion in Recreational Environments
   Lorrie A. Krebs, Michael LaPlante, Jim Langley,
     Kenneth S. Kutska 257
      Corrosion in Boats 257
      Corrosion of Firearms 257
      Bicycles and Corrosion 259
      Public Playground Equipment 260
      Free Rock Climbing 262
Corrosion in Workboats and Recreational Boats
   Everett E. Collier 265
      Hulls, Fittings, and Fastenings 265
      Metal Deck Gear 266
      Equipment 267
      Propulsion Systems 268
      Electrical and Electronic Systems 271
      Plumbing Systems 273
      Masts, Spars, and Rigging 274
Corrosion of Metal Artifacts and Works of Art in Museum and
Collection Environments
   Jerry Podany 279
      Common Corrosion Processes 279
      Pollutants 280
      The Museum as a Source of Corrosion 280
      Plastic and Wood 281
      Sulfur 281
      Corrosion from Carbonyl Compounds 282
      Past Treatments 285
      Preservation 286
Corrosion of Metal Artifacts Displayed in Outdoor Environments
   L.S. Selwyn, P.R. Roberge 289
      Environmental Factors Causing Damage 289
      Corrosion of Common Metals Used Outdoors 293
      Preservation Strategies 298
      Conservation Strategies for Specific Metals 299
      New Commissions 301
Corrosion of Metal Artifacts in Buried Environments
   Lyndsie S. Selwyn 306
      The Burial Environment 306
      Corrosion of Metals during Burial 309
      Corrosion after Excavation 314
      Specific Corrosion Problems after Excavation 314
      Archaeological Conservation 315
      Conservation Strategies 315
Chemical Cleaning and Cleaning-Related Corrosion of Process Equipment
   Bert Moniz, Jack W. Horvath 323
      Chemical Cleaning Methods 323
      Chemical Cleaning Solutions 324
      Chemical Cleaning Procedures 326
      On-Line Chemical Cleaning 328
      Mechanical Cleaning 329
      On-Line Mechanical Cleaning 330

Corrosion in Specific Industries 331

Introduction to Corrosion in Specific Industries
   Hira S. Ahluwalia 337
      Corrosion in the Nuclear Power Industry 337
      Corrosion in Fossil and Alternative Fuel Industries 337
      Corrosion in the Land Transportation Industries 337
      Corrosion in Commercial Aviation 337
      Corrosion in the Microelectronics Industry 338
      Corrosion in the Chemical Process Industry 338
      Corrosion in the Pulp and Paper Industry 338
      Corrosion in the Food and Beverage Industry 338

Corrosion in the Pharmaceutical and Medical Technology
   Industries 338
Corrosion in the Petroleum and Petrochemical Industry 338
Corrosion in the Building Industries 338
Corrosion in the Mining and Metal Processing
   Industries 338

Corrosion in the Nuclear Power Industry

Introduction to Corrosion in the Nuclear Power Industry
   Barry M. Gordon 339
Corrosion in Boiling Water Reactors
   F. Peter Ford, Barry M. Gordon, Ronald M. Horn 341
      Background to Problem 341
      EAC Analysis 343
      Prediction of EAC in BWRs 350
      Mitigation of EAC in BWRs 354
      Summary of Current Situation and Commentary
         on the Future 356
Corrosion in Pressurized Water Reactors
   Peter M. Scott, Pierre Combrade 362
      PWR Materials and Water Chemistry Characteristics 362
      General Corrosion, Crud Release, and Fouling 363
      PWR Primary Side Stress-Corrosion Cracking 367
      Irradiation-Assisted Corrosion Cracking of Austenitic
         Stainless Steels 375
      Steam Generator Secondary Side Corrosion 375
      Intergranular Attack and IGSCC at the Outside Surface 377
      Corrosion Fatigue 380
      External Bolting Corrosion 381
Effect of Irradiation on Stress-Corrosion Cracking and Corrosion
in Light Water Reactors
   Gary S. Was, Jeremy Busby, Peter L. Andresen 386
      Irradiation Effects on SCC
         Gary S. Was, Peter L. Andresen 387
            Service Experience 388
            Water Chemistry 391
            Radiation-Induced Segregation 393
            Microstructure, Radiation Hardening, and
               Deformation 398
            Radiation Creep and Stress Relaxation 403
      Mitigation Strategies
         Peter L. Andresen, Gary S. Was 404
            Water Chemistry Mitigation—BWRs 404
            Water Chemistry Mitigation—PWR Primary 405
            LWR Operating Guidelines 405
            Design Issues and Stress Mitigation 405
            New Alloys 405
      Irradiation Effects on Corrosion of Zirconium Alloys
         Jeremy Busby 406
Corrosion of Zirconium Alloy Components in Light Water Reactors
   Clément Lemaignan 415
      Zirconium Alloys 415
      Corrosion of Zirconium Alloys in Water without
         Irradiation 416
      Heat Flow Conditions 417
      Impact of Metallurgical Parameters on Oxidation
         Resistance 417
      Hydrogen Pickup and Hydriding 417

      Oxide Morphology and Integrity 418
      Effect of Irradiation on Microstructure and Corrosion 418
      LWR Coolant Chemistry 419
      Corrosion of Fuel Rods in Reactors 419
Corrosion of Containment Materials for Radioactive-Waste Isolation
   Raúl B. Rebak, R. Daniel McCright 421
      Time Considerations 421
      Environmental and Materials Considerations 422
      Reducing Environments 422
      Oxidizing Environments 425

Corrosion in Fossil and Alternative Fuel Industries

Introduction to Corrosion in Fossil and Alternative Fuel Industries
   John Stringer 438
      Fuels 438
      Energy Conversion 438
      Efficiency 439
High-Temperature Corrosion in Gasifiers
   Wate Bakker 441
      Corrosion Mechanism and Laboratory Studies 441
      Long-Term Performance of Materials in Service 444
Corrosion in the Condensate-Feedwater System
   Barry C. Syrett, Otakar Jonas, Joyce M. Mancini 447
      Corrosion of Condensers
         Barry C. Syrett, Otakar Jonas 447
            Types of Condensers 447
            Cooling Water Chemistry 448
            Corrosion Mechanisms 448
            Biofouling Control 452
            Other Problems 452
            Corrosion Prevention 452
      Corrosion of Deaerators
         Otakar Jonas, Joyce M. Mancini 452
            Deaerator Designs 452
            Corrosion Problems and Solutions 453
      Corrosion of Feedwater Heaters
         Otakar Jonas, Joyce M. Mancini 456
            Types of Feedwater Heaters 456
            Materials 456
            Water and Steam Chemistry 457
            Corrosion Problems 457
Corrosion of Flue Gas Desulfurization Systems
   W.L. Mathay 461
      Flue Gas Desulfurization (FGD) Technology 461
      FGD Corrosion Problem Areas 461
      Materials Selection for FGD Components 462
      Future Air Pollution Control Considerations 463
Corrosion of Steam- and Water-Side of Boilers
   Phillip Daniel 466
      Chemistry-Boiler Interactions 466
      Corrosion Control and Prevention 466
      Common Fluid-Side Corrosion Problems 466
Corrosion of Steam Turbines
   Otakar Jonas 469
      Steam Turbine Developments 469
      Major Corrosion Problems in Steam Turbines 469
      Turbine Materials 471
      Environment 472

Design	473
Solutions to Corrosion Problems	474
Monitoring	474
Further Study	474

Fireside Corrosion in Coal- and Oil-Fired Boilers
Steven C. Kung 477
    Waterwall Corrosion 477
    Fuel Ash Corrosion 478
    Prevention of Fireside Corrosion 479

High-Temperature Corrosion in Waste-to-Energy Boilers
George Y. Lai 482
    Corrosion Modes 482
    Corrosion Protection and Alloy Performance 483

Corrosion of Industrial Gas Turbines
Henry L. Bernstein, Ronald L. McAlpin 486
    Corrosion in the Compressor Section 486
    Corrosion in the Combustor and Turbine Sections 487

Components Susceptible to Dew-Point Corrosion
William Cox, Wally Huijbregts, René Leferink 491
    Dew Point 491
    Components Susceptible to Attack 491
    Mitigation of Dew-Point Corrosion 494
    Guidance for Specific Sections of the Plant 496

Corrosion of Generators
William G. Moore 497
    Retaining-Ring Corrosion 497
    Crevice-Corrosion Cracking in Water-Cooled Generators 497

Corrosion and Erosion of Ash-Handling Systems 499
    Fly Ash Systems 499
    Wet Bottom Ash Systems 499
    Mitigating the Problems 500
    The Future 500

Corrosion in Portable Energy Sources
Chester M. Dacres 501
    Battery Types 501
    Corrosion of Batteries 501
    Corrosion of Fuel Cells 502

Corrosion in Fuel Cells
Prabhakar Singh, Zhenguo Yang 504
    Fuel Cell Types 504
    Corrosion Processes in Fuel Cell Systems 506
    Materials and Technology Status 511

Corrosion in the Land Transportation Industries

Automotive Body Corrosion
D.L. Jordan, J.L. Tardiff 515
    Forms of Corrosion Observed on Automobile Bodies 515
    Corrosion-Resistant Sheet Metals 516
    Paint Systems 517

Automotive Exhaust System Corrosion
Joseph Douthett 519
    High-Temperature Corrosion 520
    Cold End Exhaust Corrosion 522

Engine Coolants and Coolant System Corrosion
Aleksei V. Gershun, Peter M. Woyciesjes 531
    Antifreeze History 531
    Cooling System Functions 531
    Corrosion 533
    Engine Coolant Base Components and Inhibitors 535
    Engine Coolant Testing 536
    Automotive, Light-Duty versus Heavy-Duty Antifreeze/Coolant 536

Automotive Proving Ground Corrosion Testing
Mats Ström 538
    When To Use Complete Vehicle Testing 538
    When Complete Vehicle Testing is Less Than Adequate 538
    Real-World Conditions the Tests are Aimed to Represent 538
    Elements of a Complete Vehicle Corrosion Test 539
    Evaluation of Test Results 539

Corrosion of Aluminum Components in the Automotive Industry
Greg Courval 545
    Stress-Induced Corrosion 545
    Cosmetic Corrosion 545
    Crevice Corrosion 545
    Galvanic Corrosion 546

Electric Rail Corrosion and Corrosion Control
Stuart Greenberger, Teresa Elliott 548
    Stray-Current Effects 548
    Electric Rail System Design for Corrosion Control 548
    Electric Rail Construction Impacts on Underground Utilities 551
    Utility Construction and Funding 553
    Monitoring and Maintenance for Stray-Current Control 556

Corrosion in Bridges and Highways
Jack Tinnea, Lianfang Li, William H. Hartt, Alberto A. Sagüés, Frank Pianca, Bryant 'Web' Chandler 559
    A Historical Perspective and Current Control Strategies
    Jack Tinnea 559
        History 559
        Current Corrosion-Control Strategies 560
        Terminology 560
    Concrete: Implications for Corrosion
    Jack Tinnea 560
        Cement Chemistry 561
        Additives to Concrete 563
        Aggressive Ions 564
    pH and Corrosion Inhibition
    Lianfang Li 565
        pH of Concrete Pore Water 565
        Chloride Threshold 565
        Applications 566
    Modes of Reinforcement Corrosion
    Jack Tinnea 566
        General Corrosion 566
        Localized Corrosion 567
    Prestressing Steel
    William H. Hartt 569
        Types of Prestressed Concrete Construction 569
        Categories of Prestressing Steel 569
        Performance of Prestressing Steel in Concrete Highway Structures 569
    Posttensioned Grouted Tendons
    Alberto A. Sagüés 570

Corrosion Inspection
   Jack Tinnea 571
      Corrosion Condition Surveys 571
      Assessment of Concrete Quality and Cover 572
      Visual Inspection and Delamination Survey 573
      Reinforcement Potentials 573
      Concrete Resistivity 573
      Chloride and Carbonation Profiles 574
      Corrosion Rate Testing and Other Advanced
         Techniques 575
      Inspection of Steel Elements 575
Corrosion-Resistant Reinforcement
   Jack Tinnea 575
      Approaches to Corrosion Resistance 576
      Epoxy-Coated Reinforcement 576
      Stainless Steels and Microcomposite Alloys 578
      Galvanized Reinforcement 580
Performance of Weathering Steel Bridges in North America
   Frank Pianca 580
      Weathering Steel as a Material 580
      Rate of Corrosion of Weathering Steel 580
      Performance of Weathering Steel 581
      Recommendations and Considerations on the
         Use of Weathering Steel 581
Coatings
   Bryant 'Web' Chandler 582
      Barrier Coatings for Steel 582
      Concrete Sealers 583
Electrochemical Techniques: Cathodic Protection,
Chloride Extraction, and Realkalization
   Jack Tinnea 583
      Cathodic Protection 584
      Electrochemical Chloride Extraction (ECE)
         and Realkalization 590

Corrosion in the Air Transportation Industry
Corrosion in Commercial Aviation
   Alain Adjorlolo 598
      Corrosion Basics 598
      Commonly Observed Forms of Airplane Corrosion 599
      Factors Influencing Airplane Corrosion 600
      Service-Related Factors 605
      Assessing Fleet Corrosion History 606
      Airworthiness, Corrosion, and Maintenance 607
      New Fleet Design: Establishing Rule-Based Corrosion
         Management Tools 610
      New Airplane Maintenance 611

Corrosion in the Microelectronics Industry
Corrosion in Microelectronics
   Jianhai Qiu 613
      Characteristics of Corrosion in Microelectronics 613
      Common Sources of Corrosion 614
      Mechanisms of Corrosion in Microelectronics 616
      Corrosion Control and Prevention 620
      Corrosion Tests 620
Corrosion in Semiconductor Wafer Fabrication
   Mercy Thomas, Gary Hanvy, Khuzema Sulemanji 623
      Corrosion During Fabrication 623
      Corrosion Due to Environmental Effects 626
Corrosion in the Assembly of Semiconductor Integrated Circuits
   A.C. Tan 629
      Factors Causing Corrosion 629
      Chip Corrosion 630
      Oxidation of Tin and Tin Lead Alloys (Solders) 630
      Mechanism of Tarnished Leads (Terminations) 630
      Controlling Tarnished Leads at the Assembly 633
Corrosion in Passive Electrical Components
   Stan Silvus 634
      Halide-Induced Corrosion 634
      Organic-Acid-Induced Corrosion 636
      Electrochemical Metal Migration (Dendrite Growth) 638
      Silver Tarnish 640
      Fretting 641
      Metal Whiskers 641
Corrosion and Related Phenomena in Portable Electronic Assemblies
   Puligandla Viswanadham, Sridhar Canumalla 643
      Forms of Corrosion Not Unique to Electronics 643
      Forms of Corrosion Unique to Electronics 645
      Corrosion of Some Metals Commonly Found in
         Electronic Packaging 646
      Examples from Electronic Assemblies 647
      Future Trends 650

Corrosion in the Chemical Processing Industry
Effects of Process and Environmental Variables
   Bernard S. Covino, Jr. 652
      Plant Environment 652
      Cooling Water 652
      Steam 652
      Startup, Shutdown, and Downtime Conditions 653
      Seasonal Temperature Changes 653
      Variable Process Flow Rates 653
      Impurities 653
Corrosion under Insulation
   Hira S. Ahluwalia 654
      Corrosion of Steel under Insulation 654
      Corrosion of Stainless Steel under Insulation 655
      Prevention of CUI 656
      Inspection for CUI 658
Corrosion by Sulfuric Acid
   S.K. Brubaker 659
      Carbon Steel 659
      Cast Irons 660
      Austenitic Stainless Steels 660
      Higher Austenitic Stainless Steels 662
      Higher Chromium Fe-Ni-Mo Alloys 662
      High Cr-Fe-Ni Alloy 662
      Nickel-Base Alloys 663
      Other Metals and Alloys 664
      Nonmetals 665
Corrosion by Nitric Acid
   Hira S. Ahluwalia, Paul Eyre, Michael Davies, Te-Lin Yau 668
      Carbon and Alloy Steels 668
      Stainless Steels 668
      Other Austenitic Alloys 670
      Aluminum Alloys 670

Titanium	671
Zirconium Alloys	671
Niobium and Tantalum	672
Nonmetallic Materials	672

Corrosion by Organic Acids
L.A. Scribner 674
    Corrosion Characteristics 674
    Formic Acid 675
    Acetic Acid 676
    Propionic Acid 678
    Other Organic Acids 679

Corrosion by Hydrogen Chloride and Hydrochloric Acid
J.R. Crum 682
    Effect of Impurities 682
    Corrosion of Metals in HCl 682
    Nonmetallic Materials 686
    Hydrogen Chloride Gas 687

Corrosion by Hydrogen Fluoride and Hydrofluoric Acid
Herbert S. Jennings 690
    Aqueous Hydrofluoric Acid 690
    Anhydrous Hydrogen Fluoride 698

Corrosion by Chlorine
E.L. Liening 704
    Handling Commercial Chlorine 704
    Dry Chlorine 704
    Refrigerated Liquid Chlorine 706
    High-Temperature Mixed Gases 706
    Moist Chlorine 706
    Chlorine-Water 708

Corrosion by Alkalis
Michael Davies 710
    Caustic Soda—Sodium Hydroxide 710
    Corrosion in Contaminated Caustic and Mixtures 721
    Soda Ash 723
    Potassium Hydroxide 723

Corrosion by Ammonia
Michael Davies 727
    Aluminum Alloys 727
    Iron and Steel 727
    Stainless Steels 730
    Alloys for Use at Elevated Temperatures 730
    Nickel and Nickel Alloys 731
    Copper and Its Alloys 732
    Titanium and Titanium Alloys 733
    Zirconium and Its Alloys 733
    Niobium and Tantalum 733
    Other Metals and Alloys 733
    Nonmetallic Materials 733

Corrosion by Phosphoric Acid
Ralph (Bud) W. Ross, Jr. 736
    Corrosion of Metal Alloys in H_3PO_4 736
    Resistance of Nonmetallic Materials 739

Corrosion by Mixed Acids and Salts
Narasi Sridhar 742
    Nonoxidizing Mixtures 743
    Oxidizing Acid Mixtures 747

Corrosion by Organic Solvents
Hira S. Ahluwalia, Ramgopal Thodla 750
    Classification of Organic Solvents 750
    Corrosion in Aprotic (Water Insoluble) Solvent
        Systems 750
    Corrosion in Protic (Water Soluble) Solvent Systems 751
    Importance of Conductivity 752
    Corrosion Testing 752

Corrosion in High-Temperature Environments
George Y. Lai 754
    Oxidation 754
    Carburization 756
    Metal Dusting 757
    Nitridation 758
    High-Temperature Corrosion by Halogen and Halides 759
    Sulfidation 759

Corrosion in the Pulp and Paper Industry

Corrosion in the Pulp and Paper Industry
Harry Dykstra 762
    Areas of Major Corrosion Impact 762
    Environmental Issues 763
    Corrosion of Digesters
        Angela Wensley 763
            Batch Digesters 763
            Continuous Digesters 765
            Ancillary Equipment 767
    Corrosion Control in High-Yield Mechanical Pulping
        Chris Thompson 767
            Materials of Construction and Corrosion
                Problems 768
    Corrosion in the Sulfite Process
        Max D. Moskal 768
            The Environment 769
            Construction Materials 769
            Sulfur Dioxide Production 769
            Digesters 770
            Washing and Screening 770
            Chloride Control 770
    Corrosion Control in Neutral Sulfite Semichemical Pulping
        Chris Thompson 770
            Materials of Construction 770
    Corrosion Control in Bleach Plants
        Donald E. Bardsley, William Miller 771
            Stages of Chlorine-Based Bleaching 771
            Nonchlorine Bleaching Stages 772
            Process Water Reuse for ECF and Nonchlorine
                Bleaching Stages 772
            Selection of Materials for Bleaching Equipment   773
            Oxygen Bleaching 774
            Pumps, Valves, and the Growing Use of Duplex
                Stainless Steels 774
    Paper Machine Corrosion
        Angela Wensley 775
            Paper Machine Components 775
            White Water 776
            Corrosion Mechanisms 777
    Suction Roll Corrosion
        Max D. Moskal 779
            Corrosion 779

    Operating Stresses 779
    Manufacturing Quality 780
    Material Selection 780
    In-Service Inspection 780
  Corrosion Control in Chemical Recovery
    David Bennett, Craig Reid 780
    Black Liquor 780
    Chemical Recovery Tanks 782
    Additional Considerations for Tanks in
      Black Liquor, Green Liquor, and White
      Liquor Service 783
    Lime Kiln and Lime Kiln Chain 783
  Corrosion Control in Tall Oil Plants
    Max D. Moskal, Arthur H. Tuthill 784
  Corrosion in Recovery Boilers
    Douglas Singbeil 785
    Recovery Boiler Corrosion Problems 786
  Corrosion Control in Air Quality Control
    Craig Reid 793
    Materials of Construction 793
  Wastewater Treatment Corrosion in Pulp and Paper Mills
    Randy Nixon 794
    Wastewater System Components and Materials
      of Construction 794
    Parameters Affecting Wastewater Corrosivity 794
    Corrosion Mechanisms 795

Corrosion in the Food and Beverage Industries
Corrosion in the Food and Beverage Industries
  Shi Hua Zhang, Bert Moniz, Michael Meyer 803
    Corrosion Considerations 803
    Regulations in the United States 803
    Corrosivity of Foodstuffs 804
    Contamination of Food Products by Corrosion 805
    Selection of Stainless Steels as Materials of
      Construction 805
    Avoiding Corrosion Problems in Stainless Steels 807
    Stainless Steel Corrosion Case Studies 807
    Other Materials of Construction 807
    Corrosion in Cleaning and Sanitizing Processes 808

Corrosion in the Pharmaceutical and Medical Technology Industries
Material Issues in the Pharmaceutical Industry
  Paul K. Whitcraft 810
    Materials 810
    Passivation 811
    Electropolishing 811
    Rouging 812
Corrosion in the Pharmaceutical Industry 813
    Materials of Construction 813
    Corrosion Failures 815
Corrosion Effects on the Biocompatibility of Metallic Materials
  and Implants
  Kenneth R. St. John 820
    Origins of the Biocompatibility of Metals and Metal
      Alloys 820
    Failure of Metals to Exhibit Expected Compatibility 821
    Metal Ion Leaching and Systemic Effects 822

    Possible Cancer-Causing Effects of Metallic
      Biomaterials 823
Mechanically Assisted Corrosion of Metallic Biomaterials
  Jeremy L. Gilbert 826
    Iron-, Cobalt-, and Titanium-Base Biomedical
      Alloys 826
    Surface Characteristics and Electrochemical
      Behavior of Metallic Biomaterials 826
    The Clinical Context for Mechanically Assisted
      Corrosion 827
    Testing of Mechanically Assisted Corrosion 832
Corrosion Performance of Stainless Steels, Cobalt, and Titanium
  Alloys in Biomedical Applications
  Zhijun Bai, Jeremy L. Gilbert 837
    Chemical Composition and Microstructure of
      Iron-, Cobalt-, and Titanium-Base Alloys 837
    Surface Oxide Morphology and Chemistry 839
    Physiological Environment 840
    Interfacial Interactions between Blood and
      Biomaterials 841
    Coagulation and Thrombogenesis 841
    Inflammatory Response to Biomaterials 841
    General Discussion of Corrosion Behavior of
      Three Groups of Metallic Biomaterials 841
    Corrosion Behavior of Stainless Steel, Cobalt-Base
      Alloy, and Titanium Alloys 844
    Biological Consequences of *in vivo* Corrosion and
      Biocompatibility 847
Corrosion Fatigue and Stress-Corrosion Cracking in Metallic
  Biomaterials
  Kirk J. Bundy, Lyle D. Zardiackas 853
    Background 853
    Metallic Biomaterials 855
    Issues Related to Simulation of the *in vivo* Environment,
      Service Conditions, and Data Interpretation 861
    Fundamentals of Fatigue and Corrosion Fatigue 863
    Corrosion Fatigue Testing Methodology 867
    Findings from Corrosion Fatigue Laboratory Testing 868
    Findings from *in vivo* Testing and Retrieval Studies
      Related to Fatigue and Corrosion Fatigue 871
    Fundamentals of Stress-Corrosion Cracking 873
    Stress-Corrosion Cracking Testing Methodology 878
    Findings from SCC Laboratory Testing 880
    Findings from *in vivo* Testing and Retrieval Studies
      Related to SCC 882
    New Materials and Processing Techniques for CF
      and SCC Prevention 883
Corrosion and Tarnish of Dental Alloys
  Spiro Megremis, Clifton M. Carey 891
    Dental Alloy Compositions and Properties 891
    Tarnish and Corrosion Resistance 892
    Interstitial versus Oral Fluid Environments and
      Artificial Solutions 894
    Effect of Saliva Composition on Alloy Tarnish
      and Corrosion 896
    Oral Corrosion Pathways and Electrochemical
      Properties 897
    Oral Corrosion Processes 899

Nature of the Intraoral Surface 902
Classification and Characterization of Dental Alloys 904

Corrosion in the Petroleum and Petrochemical Industry

Corrosion in Petroleum Production Operations
Russell D. Kane 922
Causes of Corrosion 922
Oxygen 923
Hydrogen Sulfide, Polysulfides, and Sulfur 923
Carbon Dioxide 924
Strong Acids 925
Concentrated Brines 926
Stray-Current Corrosion 926
Underdeposite (Crevice) Corrosion 926
Galvanic Corrosion 927
Biological Effects 927
Mechanical and Mechanical/Corrosive
Effects 927
Corrosion Control Methods 928
Materials Selection 928
Coatings 932
Cathodic Protection 933
Types of Cathodic Protection Systems 933
Inhibitors 937
Nonmetallic Materials 941
Environmental Control 942
Problems Encountered and Protective Measures 944
Drilling Fluid Corrosion 944
Oil Production 946
Corrosion in Secondary Recovery Operations 953
Carbon Dioxide Injection 955
Corrosion of Oil and Gas Offshore
Production Platforms 956
Corrosion of Gathering Systems, Tanks,
and Pipelines 958
Storage of Tubular Goods 962
Corrosion in Petroleum Refining and Petrochemical Operations
Russell D. Kane 967
Materials Selection 967
Corrosion 974
Environmentally Assisted Cracking (SCC, HEC, and
Other Mechanisms) 987
Velocity-Accelerated Corrosion and Erosion-Corrosion 999
Corrosion Control 1002
Appendix: Industry Standards
James Skogsberg, Ned Niccolls,
Russell D. Kane 1005
External Corrosion of Oil and Natural Gas Pipelines
John A. Beavers, Neil G. Thompson 1015
Differential Cell Corrosion 1016
Microbiologically Influenced Corrosion 1017
Stray Current Corrosion 1017
Stress-Corrosion Cracking 1018
Prevention and Mitigation of Corrosion and SCC 1020
Detection of Corrosion and SCC 1023
Assessment and Repair of Corrosion and SCC 1023
Natural Gas Internal Pipeline Corrosion
Sridhar Srinivasan, Dawn C. Eden 1026
Background to Internal Corrosion Prediction 1026
Real-Time Corrosion Measurement and Monitoring 1031
Inspection, Data Collection, and Management
Sam McFarland 1037
Inspection 1037
Noninvasive Inspection 1040
Data Collection and Management 1045
Appendix: Review of Inspection Techniques 1047
Visual Inspection 1047
Ultrasonic Inspection 1047
Radiographic Inspection 1049
Other Commonly Used Inspection Techniques 1050

Corrosion in the Building Industries

Corrosion of Structures
John E. Slater 1054
Metal/Environment Interactions 1054
General Considerations in the Corrosion of Structures 1054
Protection Methods for Atmospheric Corrosion 1058
Protection Methods for Cementitious Systems 1060
Case Histories 1062

Corrosion in the Mining and Metal Processing Industries

Corrosion of Metal Processing Equipment
B. Mishra 1067
Corrosion of Heat Treating Furnace Equipment 1067
Corrosion of Plating, Anodizing, and Pickling
Equipment 1071
Corrosion in the Mining and Mineral Industry
B. Mishra, J.J. Pak 1076
Mine Shafts 1077
Wire Rope 1077
Rock Bolts 1078
Pump and Piping Systems 1078
Tanks 1079
Reactor Vessels 1079
Cyclic Loading Machinery 1079
Corrosion of Pressure Leaching Equipment 1079

Gallery of Corrosion Damage 1083

Selected Color Images 1085
Fundamentals of Corrosion 1085
Evaluation of Corrosion Protection Methods 1085
Forms of Corrosion in Industries and Environments 1085

Reference Information 1095

Corrosion Rate Conversion 1097
Metric Conversion Guide 1098
Abbreviations and Symbols 1101
Index 1105

Corrosion in Specific Environments

Introduction to Corrosion in Specific Environments 5
    Corrosion in Freshwater Environments 5
    Corrosion in Marine Environments 5
    Corrosion in Underground Environments 5
    Corrosion in Military Environments 6
    Corrosion in Specialized Environments 7

Corrosion in Fresh Water Environments
Corrosion in Potable Water Distribution and Building Systems 8
    Theoretical Considerations 8
    Mitigation against Corrosion 10
    Additional Considerations 11
Corrosion in Service Water Distribution Systems 12
    Typical System Designs 12
    Typical Water Qualities 13
    Corrosion Mechanisms in Service Water Systems 13
    Corrosion Challenges in Service Water Systems 13
    Corrosion Control in Service Water Systems 14
    Deposit Control 14
Rouging of Stainless Steel in High-Purity Water 15
    Pharmaceutical Waters 15
    Chlorides 16
    Passive Layer 17
    Surface Finish 18
    Rouge Classification 18
    Castings 20
    Cleaning and Repassivation 21
Corrosion in Wastewater Systems 23
    Predesign Surveys and Testing 23
    Material Considerations 24

Corrosion in Marine Environments
Corrosion in Seawater 27
    Consistency and the Major Ions 27
    Variability of the Minor Ions 30
    Effect of Pollutants 37
    Influence of Biological Organisms 38
    Effect of Flow Velocity 40
Corrosion in Marine Atmospheres 42
    Important Variables 42
    Modeling of Atmospheric Corrosion—ISO
        CORRAG Program 51
    Corrosion Products 57
    Atmospheric Corrosion Test Sites 57
Corrosion of Metallic Coatings 61
    Thermal Sprayed Coatings 61
    Hot Dip Coatings 65
    Electroplated Coatings 66

    Methods of Protection 66
Performance of Organic Coatings 69
    Surface Preparation 69
    Topside Coating Systems 70
    Immersion Coatings 72
Marine Cathodic Protection 73
    Cathodic Protection Criteria 73
    Anode Materials 73
    Comparison of Impressed-Current and Sacrificial Anode
        Systems 74
    Cathodic Protection of Marine Pipelines 74
    Cathodic Protection of Offshore Structures 75
    Cathodic Protection of Ship Hulls 77

Corrosion in Underground Environments
External Corrosion Direct Assessment Integrated
    with Integrity Management 79
    Four Step ECDA Process 79
    Step 1: Preassessment
        (Assessment of Risk and Threats) 79
    Step 2: Indirect Examinations 81
    Step 3: Direct Examination 81
    Step 4: Post Assessment 82
Close-Interval Survey Techniques 84
    CIS Equipment 84
    Preparation 85
    Procedures 86
    Dynamic Stray Current 87
    Offshore Procedures 87
    Data Validation 87
    Data Interpretation 88
Corrosion of Storage Tanks 89
    Soil Corrosivity 89
    Cathodic Protection 90
    Data Needed for Corrosion Protection Design 90
    Soil-Side Corrosion Control 92
    Aboveground Storage Tanks 93
    Underground Storage Tanks 95
    Monitoring ASTs and USTs 95
Well Casing External Corrosion and Cathodic Protection 97
    Well Casing Corrosion 97
    Detection of Corrosion 97
    Cathodic Protection of Well Casings 99
Stray Currents in Underground Corrosion 107
    Principles of Stray Current 107
    Consequences of Stray Current 109
    Interference Tests 109
    Mitigation 111

Corrosion Rate Probes for Soil Environments 115
    Nonelectrochemical Techniques—Principles of
        Operation 115
    Electrochemical Techniques—Principles of Operation 116
    Nonelectrochemical Techniques—Examples of Uses
        in Soils 117
    Electrochemical Techniques—Examples of Uses
        in Soils 117
    Potential Sources of Interference with Corrosion
        Measurements 119
Cathodic Protection of Pipe-Type Power Transmission Cables 122
    Resistor Rectifiers 122
    Polarization Cells 122
    Isolator-Surge Protector 123
    Field Rectifiers 123
    Stray Currents 123

Corrosion in Military Environments
Corrosion in the Military 126
    Introduction 126
    Military Problems 127
    Corrosion Control and Management 132
    Long-Term Strategy to Reduce Cost of Corrosion 134
Military Specifications and Standards 136
    Types of Documents and Designations 136
    Format of Specifications 138
    Sources of Documents 139
    Notable Specifications, Standards, and Handbooks 139
    Department of Defense Corrosion Policy 140
Corrosion Control for Military Facilities 141
    The Environment 141
    Case Studies 141
    Emerging Corrosion-Control Technologies 144
Ground Vehicle Corrosion 148
    Background 148
    Requirements for Corrosion Control 148
    Procurement Document 148
    Testing Systems to Meet the Army's Needs 149
    Supplemental Corrosion Protection 149
    Improved Maintenance Procedures 150
    Considerations for Corrosion in Design 150
Armament Corrosion 151
    Overview of Design, In-Process, Storage, and
        In-Field Problems 151
    Design Considerations 151
    In-Process Considerations 152
    Storage Considerations 154
    In-Field Considerations 154
High-Temperature Corrosion in Military Systems 156
    High-Temperature Corrosion and Degradation
        Processes 156
    Boilers 156
    Diesel Engines 161
    Gas Turbine Engines 162
    Incinerators 164
Finishing Systems for Naval Aircraft 171
    Standard Finishing Systems 171
    Compliant Coatings Issues and Future Trends 173
Military Coatings 180
    Electroplating 180
    Conversion Coating 181

    Supplemental Oils 181
    Paint Coatings 182
    Other Finishes 183
U.S. Navy Aircraft Corrosion 184
    Environment 184
    Aircraft Alloys 184
    Aircraft Inspection 185
    Prevention and Corrosion Control 186
    Examples of Aircraft Corrosion Damage 189
Military Aircraft Corrosion Fatigue 195
    Aircraft Corrosion Fatigue Assessment 195
    Causes and Types of Aircraft Corrosion 196
    Impact of Corrosion on Fatigue 197
    Corrosion Metrics 198
    Investigations and Modeling of Corrosion/Fatigue
        Interactions 199
    Methodologies for Predicting the Effects of
        Corrosion on Fatigue 201
    Recent Development and Future Needs 203
Corrosion of Electronic Equipment in Military Environments 205
    An Historical Review of Corrosion Problems 205
    Examples of Corrosion Problems 206
Microbiologically Influenced Corrosion in Military Environments ... 211
    General Information about Microorganisms 211
    Atmospheric Corrosion 211
    Metals Exposed to Hydrocarbon Fuels 213
    Immersion 213
    Burial Environments 217
Service Life and Aging of Military Equipment 220
    Reliability and Safety of Equipment 220
    Aging Mechanisms 220
    Management 223
    Prevention and Control 225
    Prediction Techniques 227

Corrosion in Specialized Environments
Corrosion in Supercritical Water—Waste Destruction
    Environments 229
    The Unique Properties of Supercritical Water 229
    The Economics and Benefits of SCWO 229
    Facility Design Options 230
    Materials Challenges 230
    Mitigation of System Degradation 233
    The Future 233
Corrosion in Supercritical Water—Ultrasupercritical Environments
    for Power Production 236
    Water Properties 236
    Steam Cycle 236
    Steam Cycle Chemistry 237
    Materials Requirements 237
    Boiler Alloys 238
    Turbine Alloys 240
    Corrosion in Supercritical Water 240
    Efficiency 242
    Benefits 243
    Worldwide Materials Research 244
Corrosion in Cold Climates 246
    Cold Climates 246
    Corrosion Control Techniques and Costs 246
    Design 247
    Transportation and Storage 248

Construction 248	Sulfur 281
Operations and Maintenance 249	Corrosion from Carbonyl Compounds 282
Corrosion in Emissions Control Equipment 251	Past Treatments 285
Flue Gas Desulfurization 251	Preservation 286
Waste Incineration 252	Corrosion of Metal Artifacts Displayed in Outdoor
Bulk Solids Processing 253	Environments 289
Chemical and Pharmaceutical Industries 254	Environmental Factors Causing Damage 289
Corrosion in Recreational Environments 257	Corrosion of Common Metals Used Outdoors 293
Corrosion in Boats 257	Preservation Strategies 298
Corrosion of Firearms 257	Conservation Strategies for Specific Metals 299
Bicycles and Corrosion 259	New Commissions 301
Public Playground Equipment 260	Corrosion of Metal Artifacts in Buried
Free Rock Climbing 262	Environments 306
Corrosion in Workboats and Recreational Boats 265	The Burial Environment 306
Hulls, Fittings, and Fastenings 265	Corrosion of Metals during Burial 309
Metal Deck Gear 266	Corrosion after Excavation 314
Equipment 267	Specific Corrosion Problems after Excavation 314
Propulsion Systems 268	Archaeological Conservation 315
Electrical and Electronic Systems 271	Conservation Strategies 315
Plumbing Systems 273	Chemical Cleaning and Cleaning-Related Corrosion
Masts, Spars, and Rigging 274	of Process Equipment 323
Corrosion of Metal Artifacts and Works of Art in	Chemical Cleaning Methods 323
Museum and Collection Environments 279	Chemical Cleaning Solutions 324
Common Corrosion Processes 279	Chemical Cleaning Procedures 326
Pollutants 280	On-Line Chemical Cleaning 328
The Museum as a Source of Corrosion 280	Mechanical Cleaning 329
Plastic and Wood 281	On-Line Mechanical Cleaning 330

Introduction to Corrosion in Specific Environments

Stephen D. Cramer, National Energy Technology Laboratory, U.S. Department of Energy

ENVIRONMENT is an explicit element of all corrosion processes. It is the sum of all those factors external to the corroding metal or alloy (and associated corrosion films) that affect the corrosion process. It includes the fluids that render charge transfer reactions possible. It makes possible the delivery of reactants to corrosion sites and the removal of products of corrosion reactions. It provides the medium through which transport of ionic species between anodic and cathodic sites occurs. It connects the atomic- and molecular-level processes of corrosion with the macroscopic processes of chemical processing, construction, energy production, electronics, food processing, manufacturing, medical technology, mining, and transportation. In doing so, it engages chemical, biochemical, and mechanical processes in affecting the corrosion process.

Complex technological processes often involve many and varied environments that affect corrosion performance, corrosion protection, and corrosion control. While these environments may share similarities with others when organized by unit operation or process, they are typically treated on the basis of the industry, specifically its needs and conditions. However, there are environments where the knowledge required to solve corrosion problems spans industries, and corrosion practices translate from one industry to another with regard to these environments. These are the subject of this section and include: freshwater environments, marine environments, and underground environments. Military environments are included here as well, as military weapons systems and technology must be capable of operating at the extremes of the physical world. Specialized environments are also included, representing less-well-known environments with more limited applications, but with important impacts on human activities.

Corrosion in Freshwater Environments

This environment is characterized by waters that generally come from precipitation, are not salty, contain minimal quantities of dissolved solids, especially sodium chloride, and is usually defined as containing less than 1000 mg dissolved solids per liter (mg/L) (equivalent to 1000 ppm). Water containing 500 mg/L (ppm) or more of dissolved solids is generally undesirable for drinking and many industrial uses. *Potable water and building water* systems are characterized by waters containing low levels of dissolved solids, some chemicals added for public health reasons, and low residence times in the distribution system. Ductile iron (usually lined inside) is favored for water mains, while copper is the choice for bringing water from the main to the customer. However, use of plastic in lieu of copper is increasing. *Ultrahigh purity water* systems are needed in laboratories and high-technology manufacturing processes. *Service water* systems are auxiliary water systems typically using "raw" or untreated water for cooling in fossil-fuel and nuclear power plants. The primary corrosion challenges are related to the chemistry of the "raw" water, stagnant conditions, flow variations, and temperature variations. *Wastewater* can contain substantial levels of dissolved solids (including chlorides and sulfates), suspended solids, and biomaterials. Biochemical oxygen demand (BOD), chemical oxygen demand (COD), pH, Langelier index (related to scale formation and corrosion), and sulfide generation are important considerations in selecting materials for service in wastewater and atmospheric service in wastewater treatment plants.

Corrosion in Marine Environments

Such conditions occur in seawater and atmospheric environments associated with the world's oceans. Seawaters are salty, containing substantial quantities of dissolved solids, especially the chloride and sulfate salts of sodium, magnesium, calcium, and potassium. In more than 97% of the seawater, the concentration of dissolved solids is between 33,000 and 37,000 mg/L (ppm). Microorganisms and dissolved gases are other important constituents of seawater. Brackish waters, found at the margins of seawater and freshwater, have dissolved solids concentrations between 1000 and 35,000 mg/L (ppm). *Seawater* corrosion exposures include full-strength open ocean water, coastal seawater, brackish and estuarine waters, and bottom sediments. Seawater is a biologically active medium and biofouling contributes to the complexity of corrosion processes in seawater, particularly to the occurrence of localized corrosion processes. Corrosion in *marine atmospheres* is distinguished by the presence of airborne contaminants, particularly chlorides, by the availability of moisture in fogs, dew, and precipitation, and by the distance from the sea. Nickel alloys and stainless steels have good corrosion resistance in marine atmospheres.

Sacrificial *metallic coatings* applied by thermal spray, hot dipping, or electroplating can add up to 20 years service life to steel structures in marine atmospheres. Aluminum and zinc coatings are the primary coatings in use and thickness, composition, and microstructure of the coating are the key variables affecting service life of the coated structure. *Organic coatings* are the principal means of corrosion control for ship hulls and topsides and for the splash zones on offshore structure and can be used with sacrificial metallic coatings to extend service life. Corrosion protection of marine pipelines is usually achieved through the use of protective coatings and supplemented by using *cathodic protection*. Sacrificial anodes are often chosen for offshore platforms because they are simple, rugged, and become effective immediately on platform launch. The primary corrosion protection for ship hulls is provided by coatings, augmented by cathodic protection to protect areas of coatings holidays and damage.

Corrosion in Underground Environments

Detection of corrosion in underground environments relies on a variety of electrochemical inspection and monitoring techniques for determining the condition of structures in or on the ground and, in many cases, unavailable for visual inspection. *External corrosion direct assessment* (ECDA) is a structured process intended for use to assess and manage the impact of external corrosion on the integrity of underground pipelines. It integrates field measurements with the physical characteristics, environmental factors, and operating history of pipelines. It includes nonintrusive, aboveground examinations with pipeline physical examinations (direct assessment) at sites identified by assessment of the indirect examinations. *Close interval survey* techniques involve a series of structure-to-electrolyte potential measurements on a buried or submerged structure, most often a pipeline. Close interval surveys are used to assess the performance and operation of cathodic protection systems. Additional benefits include identifying areas of inadequate CP or excessive polarization, locating medium-to-large defects in coatings on existing or newly constructed pipelines, locating areas of stray-current pickup and discharge, identifying possible shorted casings and defective electrical isolation devices, locating possible high-pH stress-corrosion cracking risk areas, and locating areas at risk of external corrosion.

Storage tanks are designed to store products economically and in an environmentally safe way. Internal corrosion can be controlled by a combination of protective coatings and cathodic protection. Soil-side external corrosion can be mitigated by cathodic protection with and without the use of protective coatings. Aboveground steel storage tanks are designed to last for 20 to 30 years.

Well casing corrosion above depths of 60 m (200 ft) is typically due to oxygen reduction enhanced by chlorides and sulfates, while below this depth corrosion is caused by carbon-dioxide-rich formation water. Cathodic protection of well casings has proven to be an effective means of minimizing corrosion on the casing provided the proper amount of current is applied and maintained. Isolating the casing from surface facilities eliminates the macro corrosion cell between the casing and these structures and provides a means for controlling and measuring the protection current to the casing.

Direct Current (dc) stray currents accelerate the corrosion of structures where a positive current leaves the structure to enter the earth or an electrolyte. Stray currents also corrupt potential measurements that are being taken to establish a cathodic protection criterion. Stray currents (direct and alternating) can be controlled by alteration of the source, addition or adjustment of cathodic protection, and use of reverse current switches and of mitigation bonds. *Corrosion rate probes* for soil environments can employ both electrochemical and nonelectrochemical techniques for corrosion measurement. Corrosion rate probes allow continuous measurement of external corrosion rates as a function of time, as well as a way to monitor changes in soil corrosivity with time. *Pipe-type power transmission cables* provide power in cities at voltage levels that can be used by both industrial and residential customers. Cathodic protection is necessary to prevent corrosion damage that would allow the loss of pressurized dielectric fluid from the pipe. The dc potential of the pipe must be more negative (approximately -0.5 V) than the earth around it. All acceptable CP systems for doing this must be capable of safely conducting high alternating current (ac) fault currents to ground.

Corrosion in Military Environments

Corrosion is a major concern of the U.S. Department of Defense (DoD) and is estimated to cost at least $20 billion (U.S.) annually. This concern is reflected in the DoD Corrosion Policy initiated in 2003 (Ref 1) providing guidance to procurement personnel, maintenance units, and service personnel who must see that limited resources are efficiently used and that the readiness of the military is not compromised by materials failures. Military assets include more than 350,000 ground and tactical vehicles, 15,000 aircraft, 1000 strategic missiles, 300 ships, and facilities worth roughly $435 billion (U.S.). Since the military does not choose where its next battle must be fought, military assets must perform reliably and effectively at the extremes of the physical world. Embedded in these extremes is damage due to corrosion, wear, and the synergistic effects of corrosion and wear. The primary means for ensuring that raw materials, commodities, materials of construction, and manufactured equipment and systems meet the needs of the military services is the DoD system of *military specifications and standards*. These are supplemented by the standards and practices of NACE International, ASM International, and ASTM International, and other professional organizations.

Corrosion problems associated with *military facilities* and installations are similar to those encountered at civilian facilities. They represent some of the most costly and pervasive maintenance and repair problems in the services. *High-temperature corrosion and oxidation* of metals and alloys occurs in many military applications, including power plants (coal, oil, natural gas, and nuclear), land-based gas turbine and diesel engines, gas turbine engines for aircraft, marine gas turbine engines for shipboard use, waste incineration, high-temperature fuel cells, and missile components. Predicting corrosion performance in these applications is difficult because of the variety of materials used, the operational demands placed on the materials, and because the materials often degrade by more than one mechanism. *Armament corrosion* protection relies heavily on coatings systems and regular maintenance to prevent damage, not only in service, but during the extended periods of storage common to these systems. The Army has one of the largest *ground vehicle fleets* in the world, having an average vehicle age of more than 17 years (well past the manufacture's corrosion warranty for commercial vehicles), and relies on accelerated corrosion testing to identify improved materials, coatings systems, corrosion inhibitors, design, and maintenance practice to ensure continued satisfactory long-term vehicle performance.

Military coatings systems typically have multifunctional performance requirements, ranging from corrosion protection, oxidation resistance, wear resistance, camouflage, spectral reflectance, adhesion, weatherability, and mar resistance. The chemical agent resistant coating (CARC) systems, in use since 1983, provide a wide selection of coatings for corrosion protection in military environments. Traditional coatings for *military aircraft* include inorganic pretreatments, epoxy primers, and polyurethane topcoats. Chromate continues to be eliminated from aluminum aircraft pretreatments and from sealers for anodized aluminum; low- and no-VOC (volatile organic coating) polymer binder systems are increasingly being used on military aircraft.

Navy aircraft experience severe corrosion conditions in operational service. Life-cycle costs are high due to increased maintenance and decreased component/system reliability as a result of cumulative corrosion damage. Cleaning, washing, inspection, surface treatments, and coatings form the core of navy aircraft maintenance. Historical data show that *fatigue and corrosion* cause 55 and 25% of aircraft failures, respectively, while corrosion contributes to only a small fraction of the fatigue failures. However, corrosion damage can result in an initial flaw that dramatically reduces the predicted fatigue life of critical components and renders noncritical components as critical.

The corrosion of *electronic equipment* is most greatly affected by: temperature, moisture, biological growth, rain, salt spray, dusts, shock, and vibration. The Navy found that 40 to 50% of corrosion-damaged printed circuit boards could be returned to service by simply cleaning corrosion products from the contacts. More than 90% of failed removable electronic assemblies could be returned to weapons service when repaired by trained technicians following recommended practices.

Microorganisms, including both bacteria and fungi, can accelerate corrosion reactions and change reaction mechanisms in atmospheric, hydrocarbon/water, immersed, and burial environments encountered in military operations. The main consequence of biofilm formation on cathodically protected carbon steel is to increase the current density necessary to polarize the steel to the protected potential. Shrinking military budgets and escalating weapons systems costs have led to increasing efforts to extend the *service life* of aging military equipment (characterized by decreasing structural strength and increasing maintenance costs). Effective life-cycle cost modeling combines traditional operational and support cost elements with an expert analysis system and demonstrates built-in flexibility and ease of updating with new information.

Corrosion in Specialized Environments

Specialized environments address an eclectic mix of environments that are important to human activities. Water is supercritical above its vapor-liquid critical point, 374 °C (706 °F) and 22 MPa (3.191 ksi). Supercritical water has unique solvating, transport, and compressibility properties compared to liquid water and steam. These properties are finding growing commercial applications. Two of these addressed here are waste destruction using *supercritical water* and the use of *ultrasupercritical water* (temperatures above 565 °C, or 1050 °F) for power production.

Corrosion in *cold climates* challenges the conditions typically addressed in atmospheric corrosion. Solar heating can produce water at ambient temperatures below the freezing point; melt water can concentrate corrosive ions; extreme climatic conditions make maintenance and repairs difficult. These factors place unexpected corrosion demands on structures expected to serve with little or no maintenance for very long times.

Emission-control equipment operates in an environment that brings together acidic gases, particulates, water vapor (and condensed water), and large volumes of exhaust gas. Design and materials selection are crucial factors in mitigating corrosion damage under such conditions.

Corrosion of *recreational equipment* has little visibility to the consuming public through efforts by the wide and diverse recreational equipment community to produce products that are safe and meet public expectations for performance and service. Corrosion is discussed as it relates to four consumer products: recreational boats, firearms, playground equipment, and bicycles. *Workboats, traditional, and recreational boats* function in one of the more corrosive environments commonly encountered. Materials selection, galvanic corrosion, cathodic protection, and design are critical elements in reducing the impact of corrosion on boat service life, performance, and safety.

Cultural resources and artifacts are aesthetically and historically important objects that people choose to preserve for the present and future generations. The significant challenges to doing so are discussed in three papers addressing *museum environments, outdoor environments,* and the recovery of artifacts from *buried environments*.

Process equipment may need to be chemically cleaned to operate properly, efficiently, and according to specifications. Such cleaning should be addressed during the equipment design to minimize the effects of corrosion on materials damage and service life.

REFERENCE

1. Under Secretary of Defense (Acquisition, Technology, and Logistics), Memorandum for Secretaries of the Military Departments, *Corrosion Prevention and Control,* U.S. Department of Defense, Washington D.C., Nov 12, 2003

Corrosion in Potable Water Distribution and Building Systems

Windsor Sung, Massachusetts Water Resources Authority

WATER IS ESSENTIAL TO LIFE. It is no exaggeration to say that civilizations rose and fell with their ability to procure and convey water of good quality to their citizens. Construction of the needed conveyances tends to be resource extensive, so the chosen construction materials need to last. Early favored materials were wood, stone, and brick. Now materials selected include metals, cement, and plastics. As metals are strong, durable, and ductile, they have been the material of choice in modern times. Iron has been most extensively used of the metals that are used for water distribution. Others include aluminum, copper, and lead. The iron is in the form of pig iron, cast iron, steel, ductile iron, or galvanized iron. Modern water utilities favor the use of ductile iron (usually lined internally) for water mains. The metal of choice for bringing water from the main to the consumer tends to be copper. The use of plastic in lieu of copper has grown on the consumer end, albeit slowly.

Table 1 summarizes the water quality effects from interaction of water with different materials that are commonly used in water distribution pipes (Ref 1). This article focuses on the internal corrosion of iron and copper in potable water as these are still the prevalent materials. External corrosion of pipes is a serious problem that is addressed in the articles "Stray Currents in Underground Corrosion" and "Corrosion Rate Probes for Soil Environments" in this Volume.

While the deterioration of water quality from a public health perspective is of primary concern, there are other issues to consider. Corrosion of iron pipes tends to form deposits that over time may decrease the capacity of transmitting water, increase the pressure drop, and thus increase pumping costs. The deposits may provide safe harbor for micro-organisms, including pathogenic ones from disinfectants in the water. Sudden changes in flow velocity and/or pressure (as when fire hydrants are opened) may dislodge deposits, causing customer complaints over discolored water. Some chemicals added for mitigating water corrosion may cause scaling in heat exchangers, and some of these chemicals can be toxic to human or ecological health. These are some of the reasons why one needs to understand and mitigate corrosion in potable water and building systems.

Theoretical Considerations

Electrochemical reactions are almost always the cause of corrosion of metals in contact with water. Metals in their elemental state are unstable in the presence of water and dissolved oxygen. For corrosion to occur, all the components of an electrochemical cell are needed: an anode, a cathode, and an electrolyte to complete the circuit. The electrochemical reactions are sometimes called reduction-oxidation (redox) reactions. Thermodynamics and equilibrium concepts are useful for understanding the phenomena.

At the anode, the metal loses electrons and is oxidized. The cathode is where the electron is gained and the electron gainer (or acceptor) can be dissolved oxygen, which is then reduced. It is not clear what factors influence the distribution of anodic and cathodic areas on the surfaces of pipes. The Nernst equation is commonly used to describe redox reactions theoretically, but its practical use is limited. See the articles

Table 1 Corrosion and water-quality problems caused by materials in contact with drinking water

Material	Corrosion type	Tap water quality deterioration
Cement-based		
Asbestos cement(a)	Uniform corrosion	Calcium dissolution. Increased pH values (up to 12). For asbestos-cement pipes, in unstable waters, pH increases and asbestos fibers can be found in the water. Surface roughening, strength reduction, and pipe failure. Aluminum release
Concrete		
Cement mortar(b)		
Iron		
Cast	Uniform corrosion	Rust tubercles leading to blockage of pipe
Ductile	Graphitization	Iron and suspended particles release
	Pitting under unprotective scale	
Steel	Pitting	Rust tubercles (blockage of pipe). Iron and suspended particles release
Galvanized steel	General pitting corrosion	Excessive zinc, lead, cadmium, iron release, and blockage of pipe
Copper	Uniform corrosion	Copper release
	Localized attack	Pipe perforation and subsequent leakage from pipes
	Cold-water pitting (type I)	
	Hot-water pitting (type II)	
	Other types of localized attack	
	Microbiologically influenced corrosion	Leakage from pipes and sporadic blue deposit release. Rupture of pipes and fittings and consequently leakage
	Corrosion fatigue	
	Erosion corrosion	Leakage from pipes
Brass	Erosion and impingement attack	Penetration failures of piping
	Dezincification	Blockage of pipes and fittings
	Stress corrosion	Lead and zinc release
Lead		
Lead pipe	Uniform corrosion	Lead release
Lead-tin solder	Uniform corrosion	Lead and cadmium release
Plastic	Possible degrading by sunlight and microorganisms	Taste and odor

(a) No internal lining such as tar. (b) Used as internal lining of iron and steel materials. Source: Ref 1

"Electrode Potentials" and "Potential versus pH (Pourbaix) Diagrams" in Volume 13A, *ASM Handbook*, 2003. One of the useful associated parameters from the Nernst equation is the standard electrode potential for classification purposes. The galvanic series is a compilation of standard electrode potentials. Older literature lists the galvanic series as anodic (oxidation) reactions, for example, $Zn(s) = Zn^{2+} + 2e^-$. Metals with high positive potentials (in this case the standard potential is +0.76 V) are more readily oxidized in this convention. The preferred convention today is to write cathodic (reduction) reactions, $Zn^{2+} + 2e^- = Zn(s)$. The standard potential referenced to the standard hydrogen potential (SHE) in this case is −0.76 V, and metals with large negative potentials are more readily oxidized. This is an unfortunate point of confusion, but cannot be avoided since conventions do change with time. A partial galvanic series of metals relevant to water distribution is shown in Table 2. More easily corroded metals are on top, and the bottom ones are mostly inert and are sometimes referred to as noble metals. This series shows how galvanic protection works, which is the use of a more readily corroded metal such as zinc as a sacrificial anode in order to protect material such as iron pipe. It also explains galvanic corrosion, which occurs when two dissimilar metals are in contact with each other and electrons flow preferentially from one to the other.

The reduction of oxygen in water produces hydrogen ions, which impacts pH. In addition, ferrous iron is unstable with respect to ferric iron in the presence of oxygen. Therefore, the corrosion of iron is accompanied by pH changes and formation of iron oxide and oxyhydroxides. These reactions are most conveniently summarized by E_H-pH diagrams, also known as Pourbaix diagrams. The redox potential of a half reaction (written as a reduction reaction) in reference to the SHE is the definition for E_H. Just as pH is defined as $-\log\{H^+\}$ where $\{H^+\}$ is the activity of the aqueous proton, $p\varepsilon$ is defined as $-\log\{e^-\}$ where $\{e^-\}$ is the activity of the free aqueous electrons and is related to E_H. Figure 1 (Ref 2) shows such a diagram for ferric hydroxide with ionic activities of dissolved species set to be 1 μmol. As such, it is a graphical summary of thermodynamic information. Exact locations of the lines depend on the free energy of formation of the species and phases in question. The stability field of water is outlined by the two diagonal lines (identified by regions O_2 H_2O, H_2 H_2O), although in practice it would be rare for potable water to have a pH of less than 6 and greater than 10. The diagram shows that the stable phase for iron under mildly oxidizing environments and typical pH values is ferric hydroxide.

Figure 2 shows a 1.5 m (60 in.) cast iron water main that has been in contact with water. Mushroomlike ferric tubercles have formed on its interior surface. Detailed analyses of similar deposits have shown the presence of ferrous solids such as ferrous carbonate (siderite), mixed ferrous-ferric solids (green rust, magnetite), various ferric oxyhydroxides (goethite, lepidocrocite), and amorphous ferric hydroxide. In principle, each distinct solid would have different energies of formation and occupy different regions in the E-pH diagram. Increasing levels of chloride and sulfate enhance corrosion by increasing conductance of the electrolyte. Their presence appears to be important for the formation of green rust. Micro-organisms also play an important role in enhancing corrosion of iron water mains.

Copper plumbing tube has been standard for water distribution in buildings in the United States for the last half century. It has a relatively high standard reduction potential and the corrosion mechanisms are very different from that of iron. Pitting corrosion is an increasing problem with copper pipes leading to pinhole leaks. NACE International organized a symposium on the topic (Ref 3) and concluded that the process is not well understood. There is some evidence pointing to poor workmanship, in particular the overuse of flux as related to the development

Table 2 Partial galvanic series of metals relevant to potable water systems

Anode	Reaction	Standard reduction potential (SHE) at 25 °C (77 °F), V
Aluminum	$Al^{3+} + 3e^- = Al(s)$	−1.68
Zinc	$Zn^{2+} + 2e^- = Zn(s)$	−0.76
Iron	$Fe^{2+} + 2e^- = Fe(s)$	−0.44
Nickel	$Ni^{2+} + 2e^- = Ni(s)$	−0.25
Tin	$Sn^{2+} + 2e^- = Sn(s)$	−0.14
Lead	$Pb^{2+} + 2e^- = Pb(s)$	−0.13
Copper	$Cu^{2+} + 2e^- = Cu(s)$	+0.34
Gold	$Au^{2+} + 2e^- = Au(s)$	+1.50

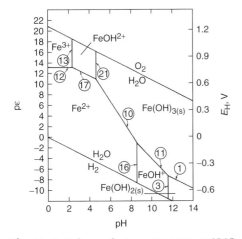

Fig. 1 E-pH diagram for iron-water system at 25 °C (77 °F). It is a graphic representation of thermodynamic stability. The potential E_H is expressed in reference to the standard hydrogen potential. $p\varepsilon$ is defined as $-\log ae^-$ where ae^- is the activity of the free aqueous electrons and is related to E_H.

Fig. 2 A 1.5 m (60 in.) cast iron pipe with tuberculation. Courtesy of Terry Bickford

of pits. There is also some indication that chlorination and chloramination may also impact copper corrosion. Chloramination involves the use of ammonia to bind free chlorine, and ammonia is known to complex copper effectively. Copper oxide/hydroxide solubility exhibits the classic U-shaped curve with respect to pH. Increasing pH beyond 9 promotes the formation of copper hydroxide complexes and increases copper solubility.

The traditional way that water chemists deal with corrosion control of potable water systems is to calculate indices such as the Langelier saturation index (LSI). See the articles "Modeling Corrosion Processes" and "Corrosion Inhibitors in the Water Treatment Industry" in Volume 13A, *ASM Handbook*, 2003. The LSI measures the degree of saturation of the water with respect to calcium carbonate scale. Corrosive water has a negative LSI, which means it will tend to dissolve calcium carbonate scale. Noncorrosive water has a positive LSI, which means it has a tendency to form calcium carbonate scale. In this sense, noncorrosive water may not be a good thing, since calcium carbonate scale is less soluble with rising temperatures. Water with a positive LSI will have a tendency to form scales in hot water heaters. However, scaling can be a problem in cold water, too. The LSI calculation is not complicated and references such as Ref 2 can be consulted. Inputs include the amount of calcium, pH, and alkalinity, as well as the total dissolved solids (TDS). However, the LSI is not very useful for determining other corrosion issues.

The LSI applies solubility reactions, which are described by solubility product, written as K_{sp}. K_{sp} of calcium carbonate is written as the product $\{Ca^{2+}\} \times \{CO_3^{2-}\}$ and has the numerical value of 5×10^{-9} (M^2) at 25 °C (77 °F) (Ref 4). $\{Ca^{2+}\}$ is the activity or ideal concentration of calcium ions, and $\{CO_3^{2-}\}$ is the carbonate ion activity. Equilibrium constants are functions of temperature. The activity is related to actual concentration modified by an activity coefficient (function of ionic strength, which is related to total dissolved solids). If the ion activity product (IAP) of the calcium multiplied by the carbonate ion (related to alkalinity and pH) exceeds the solubility product, the solution is supersaturated and there is a tendency for scale to form (i.e., the solution has a positive LSI). The solution has a negative LSI if the IAP is less than the solubility product or there is a tendency for preformed calcium carbonate scale to dissolve. Ideally, the LSI should be close to 0. Figure 3 gives an alternate way of showing basically same idea as a LSI calculation. It depicts the amount of calcium that would be in equilibrium with the given pH at two alkalinity values (5 and 30 mg/L as calcium carbonate). It is calculated for a temperature of 25 °C (77 °F) and does not include activity corrections (the TDS term). Decreases in temperature and increases in TDS will increase the equilibrium calcium value so this figure actually shows conservative results. When alkalinity is relatively low at 5 mg/L, calcium concentration has to be in excess of 400 mg/L to form scale at pH of 8. This water has a negative LSI and would be termed corrosive. The amount of calcium necessary to be scale forming at a pH of 9 and an alkalinity of 30 mg/L is only about 8 mg/L.

Another index found in corrosion literature is the Larson Index (LaI). The Larson index compares the ratio of the sum of sulfate and chloride to the bicarbonate ion (all expressed as equivalents per liter). Empirical observations show a Larson Index of less than 0.7 is desirable for corrosion control.

Mitigation against Corrosion

There are two major ways of mitigation against corrosion in potable water systems. The first is to line the pipe surface physically such that water and dissolved oxygen (or other electron acceptors) cannot reach the metal surface. Common lining materials include concrete, asbestos-cement, and epoxies. Concrete is alkaline and the pH of stagnant water in contact with concrete can exceed 10. The alkaline water can react with copper plumbing through hydrolysis reactions, and consumers may complain of copper staining such as green and blue water. Improperly installed or cured epoxy liners can deteriorate and leach organic compounds into water, causing taste and odor complaints.

The second method is to add chemical inhibitors to alter water chemistry to mitigate against corrosion. See the article "Corrosion Inhibitors in the Water Treatment Industry" in Volume 13A, *ASM Handbook*, 2003. In many cases the added chemical causes a scale to form on the pipe surface, so it behaves like a liner. Changing water quality to achieve a positive LSI promotes the formation of a calcium carbonate (calcite) scale to protect against corrosion. This is usually achieved by adding sodium or potassium hydroxide, calcium hydroxide, and or sodium carbonate to increase pH and alkalinity. Incidentally, increased pH and alkalinity is one of the options for lowering lead levels in potable water, and the mechanism may involve the formation of a basic lead carbonate scale. Another commonly used chemical for corrosion control is zinc orthophosphate. The common recommendation is to dose enough chemical to achieve a zinc concentration of about 2 mg/L at a pH between 7.5 and 7.8. The K_{sp} for $Zn_3(PO_4)_2$ is about 10^{-32} (M^5) (Ref 4) and is equal to the product $\{Zn^{2+}\}^3 \times \{PO_4^{3-}\}^2$. The amount of total phosphorus is related to the amount of zinc added (if the raw water phosphate concentration is negligible). The fraction of total phosphorus that is phosphate is a function of pH and the acidity constants of phosphoric acid. Figure 4 is a plot of the theoretical amount of zinc that is in equilibrium with zinc phosphate scale (with the given solubility product), with the zinc and phosphate coming from the chemical addition alone. The agreement with the general rule of thumb to achieve zinc in the mg/L range at pH less than 8 shows that solubility calculations were used to generate the recommended doses. In some cases, far lower zinc levels are needed, and there are indications that zinc hydroxycarbonate scales are formed on cements (which may eventually convert to zinc silicates).

The use of phosphorus can include phosphoric acid as well as polyphosphates (long chain phosphate molecules), and polyphosphate/orthophosphate blends. Polyphosphates work well for sequestration of iron and manganese, but have been shown to increase lead and copper releases in soft water. Polyphosphates can also be detrimental to cement and cement linings in soft waters. When phosphates are used for corrosion control, sometimes $FePO_4$ scale may form. In other cases, the phosphate ion is adsorbed onto preexisting iron oxide scales. In fact, a whole different class of chemicals similar to phosphate is available for cooling system use, such as chromates, nitrites, and molybdates. They have been described as passivation chemicals, and the most likely reaction mechanism is their preferential adsorption onto cathodic sites, thus blocking the contact of electron acceptors such as dissolved oxygen onto the electrochemical cell. These chemicals are toxic and not used for potable water conditioning. Their discharge can sometimes pose a challenge for wastewater operations; for example, molybdenum tends to accumulate on biosolids and can exceed regulatory limits for beneficial reuse. Organic compounds such as amines can be used

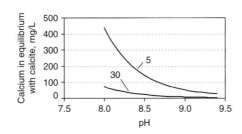

Fig. 3 Amount of calcium in equilibrium with calcite at alkalinity values of 5 and 30 as a function of pH. This is a graphic interpretation of the Langelier saturation index.

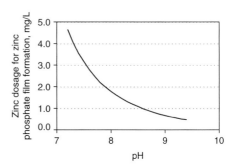

Fig. 4 Calculated zinc dosage for the formation of zinc phosphate scale as a function of pH. The zinc levels shown exceed those recommended by some potable water standards.

for conditioning pipe surfaces for nonpotable use. It is assumed that the organic compound is adsorbed onto the pipe surface in thin layer coatings. Aromatic triazoles are effective corrosion inhibitors for copper and its alloys (Ref 5).

Silicates are also used for corrosion control. They can form a coating on pipes, but more likely they act like polyphosphates as sequestering agents. The exact mechanism is not fully understood, but it has been suggested that this works by adsorption again onto preexisting particulates. The surface coverage causes charge reversal and the suspension becomes stable and colloidal (less than micron size). The colloidal nature of the suspension "masks" the physical appearance of turbidity and color. Long-term use of silicates can convert carbonate and oxide scales to metal silicates.

Additional Considerations

Lead continues to be of concern even though potable water intake is not a major contributor of body burden lead. Progress has been made to replace lead pipes and goosenecks, as well as banning lead solder. However, there still remain a significant number of lead pipes that are not easily located or removed. It is now recognized that some plumbing fixtures may also contribute lead even though they may be advertised as lead-free. It has been shown that brasses used frequently in faucets, valves, and fittings leach lead into high-purity water (Ref 6). Producing a water quality to minimize lead remains a challenging goal for every utility, and consumers are well advised to minimize lead sources within their home. There is now considerable interest in limiting the use of lead in brasses, and interest in clarifying the use of term "lead-free" in advertising plumbing products.

This discussion has relied heavily on the use of equilibrium chemistry. Corrosion concerns are driven by the kinetics of reaction and often limited by mass transfer. For example, the U.S. Environmental Protection Agency lead rule specifies collection of first-draw samples after the water has been stagnant in the pipes for 6 to 8 h. Sometimes this is an insufficient time for a system to reach equilibrium after perturbations in water chemistry.

There has been much progress made in corrosion measurements including the use of coupons, corrosion meters, rotating annular reactors with coupon inserts to study the relation between micro-organisms and corrosion rate, and polarization scans. Research involving the use of corrosion cells suitable for on-line monitoring of corrosion and metal release processes using the corrosion potential stagnation/flow (CPSF) method appear to be promising (Ref 7).

Reference 8 is suggested for a more detailed and in-depth treatment of internal corrosion and control in water systems.

REFERENCES

1. *Internal Corrosion of Water Distribution Systems,* 2nd ed., American Water Works Association Research Foundation, 1996
2. V.L. Snoeyink and D. Jenkins, *Water Chemistry,* John Wiley & Sons, 1980
3. "Plumbing Tube," A4056-XX/01, Copper Symposium 2001, The Copper Development Association, 2001; accessible at www.copper.org/environment/NACE02122
4. W. Stumm and J.J. Morgan, *Aquatic Chemistry,* 3rd ed., Wiley Interscience, 1996
5. P.A. Schweitzer, Ed., *Corrosion and Corrosion Control Handbook,* 2nd ed., Marcel Dekker, 1987
6. J.I. Paige and B.S. Covino, Jr., Leachability of Lead from Selected Copper-Base Alloys, *Corrosion,* Vol 48 (No. 12), 1992, p 1040–1046
7. G. Kirmeyer, et al., *Post-Optimization Lead and Copper Control Monitoring Strategies,* AWWARF, 2004
8. M.R. Schock, Internal Corrosion and Deposition Control, Chapter 17, *Water Quality and Treatment: A Handbook of Community Water Supplies,* 5th ed., R.D. Letterman, Tech. Ed., AWWA and McGraw-Hill, 1999

Corrosion in Service Water Distribution Systems

K. Anthony Selby, Water Technology Consultants™, Inc.

SERVICE WATER SYSTEMS are auxiliary cooling systems in fossil-fueled and nuclear power plants. They are separate from the steam surface condenser cooling system that condenses the main process steam for reuse in the cycle. Service water systems cool a wide range of plant components, some common to most power plants regardless of fuel type, including turbine lubricating oil coolers, generator hydrogen coolers, and pump lubricating oil coolers. Corrosion in service water systems in electric utility plants is a significant problem. This is especially true in nuclear plants because some service water systems are safety related and are required for the safe shutdown of the plant. Those systems required for a safe shutdown are labeled "safety-related," "essential," or "emergency cooling." The majority of the service water systems in nuclear plants do not fall into this category and are labeled "nonsafety-related" or "nonessential," but their unavailability may challenge continued operation of the plant.

Corrosion mechanisms in service water systems are not unique but may be exacerbated by design features and operating modes. The cost of corrosion-related failures in power plants is significant. Replacement or repair costs can be substantial because of the inaccessibility of piping, much of which may be underground. In addition, repairs to service water systems may necessitate plant shutdowns, the loss of production capacity, and the cost of purchasing replacement power.

In a nuclear power plant, safety-related and non-safety-related systems may use the same source of cooling water or may have separate sources. By design the safety-related service water system is a redundant and secure water source representing the ultimate heat sink for the plant.

In the United States, the Nuclear Regulatory Commission (NRC) has closely scrutinized safety-related service water systems in nuclear plants. In 1989, in response to a number of problems associated with these systems across the nuclear industry in the previous decade, the NRC issued "Generic Letter (GL) 89-13" (Ref 1). The overwhelming majority of the problems were the result of biological macrofouling, human error, or corrosion. In order to address these areas, GL 89-13 recommended that nuclear plants establish ongoing programs that would evaluate the reliability and operability of their safety-related service water systems, against the design requirements of those systems, which were specified in the laws that governed the NRC licenses for those plants. Today, it is recognized that GL 89-13 has been successful in focusing the proper level of attention on these systems. Additionally, in view of the increasing economic importance of operating these systems past their original design lifetimes, the nuclear industry has recognized that corrosion and corrosion control are increasingly playing more significant roles in service water system reliability.

Typical System Designs

Most service water systems use raw water for cooling purposes. Raw water is defined as untreated water such as that provided by a lake or river. It can also refer to the same water used in a cooling tower. Two typical designs are shown in Fig. 1 and 2. There are numerous variations on these designs.

In Fig. 1, the raw water from a lake or river passes through the service water system and is then used as makeup to the cooling tower. The cooling tower recirculating water removes heat from the main surface condenser.

In Fig. 2, there is no cooling tower. Both the service water system and circulating water system (surface condenser cooling water) are

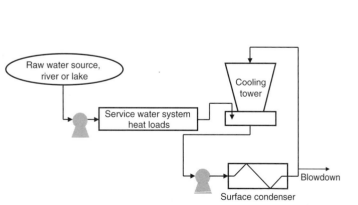

Fig. 1 Typical arrangement where raw water is used in service water system and as makeup to the cooling tower. Because the condenser loop is an open-recirculating system, make-up water is needed to replace evaporation and blowdown.

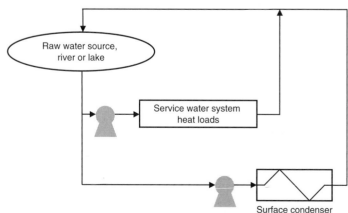

Fig. 2 In a once-through service water and circulating water system without a cooling tower large quantities of water are circulated through the systems and back to the source.

"once-through" from the lake or river and back again.

Some power plants have a service water system consisting of a closed loop of treated water that is cooled by a raw water heat exchanger (Fig. 3). This is typical of plants that utilize seawater or brackish water for cooling.

Materials of construction in service water systems vary from plant to plant. In many cases piping is constructed of unlined carbon steel. In some cases, cement lined carbon steel, lined (epoxy coated) carbon steel, stainless steels, copper alloys, or titanium are used. Stainless steels are typically 300 series austenitic grades but other alloys such as 6% molybdenum austenitic stainless steel are sometimes used.

Heat exchanger tubing is usually constructed of copper alloys (copper, copper-nickel, brasses), stainless steel, or titanium. Two of the most common copper alloys are 90-10 copper-nickel and admiralty brass. Stainless steels are typically 300 series austenitic grades but other alloys are sometimes used. These alloys can include 6% molybdenum austenitic stainless steel and high chromium-molybdenum ferritic stainless steel. Carbon steel is not used for power plant service water heat exchanger tubing.

Typical Water Qualities

The waters used in service water systems can have wide variations in constituents and impurities. Some typical values are shown in Table 1.

Corrosion Mechanisms in Service Water Systems

Corrosion mechanisms applicable to service water systems include general corrosion, concentration cell corrosion which includes crevice, tuberculation, and under-deposit corrosion (UDC), microbiologically influenced corrosion (MIC), galvanic corrosion, stress corrosion cracking (SCC), and dealloying. General corrosion rates vary greatly because some waters are much more aggressive than others. Localized forms of corrosion, pitting, concentration cell corrosion and MIC are of particular concern because of the impact on metal integrity.

General corrosion, concentration cell, and MIC are the corrosion mechanisms of greatest concern with regard to carbon steel components. Copper alloy components can suffer from pitting, dealloying, and SCC. Stainless steels in the 300 series can suffer from pitting and SCC and are also susceptible to MIC failure.

Corrosion Challenges in Service Water Systems

The primary corrosion challenges in service water systems are related to the basic characteristics of oxygen saturated raw water and the system design. Necessary design results in climatic and temperature induced flow variations, and redundant equipment requires cross-connecting piping that undergoes stagnant or, even worse, intermittent flow conditions.

An example is a turbine lubricating oil cooler. Typically, the oil flows through the shell side of one 100% capacity heat exchanger at a constant rate. The service water flows through the heat exchanger tubes and that flow is throttled to maintain the oil temperature within a specified range. For large base-loaded fossil and nuclear plants, fully redundant heat exchangers are provided in order to maintain operational flexibility. These redundant piping systems remain stagnant when not in service, assuming that their isolation valves are leak tight. Under warm weather conditions, the service water flow control valves for the in-service heat exchanger may operate completely open. In many cases the control valves have one or more bypass lines installed for summer conditions. Under cold weather conditions, the control valves throttle flow to maintain oil temperatures above recommended minimums. This means that flow velocities are high during warm weather and low during cold weather. Operation at low cooling water velocities and intermittent and trickle flow conditions allows for the accumulation of sediment and encourages the formation of other deposits in the piping and heat exchangers. The accumulation of sediments and deposits contributes to UDC and tuberculation through a mechanism of oxygen concentration differential cell corrosion.

Another challenge is the presence of microorganisms in the raw water that may contribute to MIC, which can cause failures in service water piping and heat exchangers. Types of bacteria related to MIC include sulfate reducing bacteria (SRB), acid-producing bacteria (APB), and iron-depositing bacteria. See the article "Microbiologically Influenced Corrosion" in Volume 13A and "Microbiologically Influenced Corrosion in Military Environments" in this Volume. Macrobiological growth can also cause problems in service water systems. In fresh water systems,

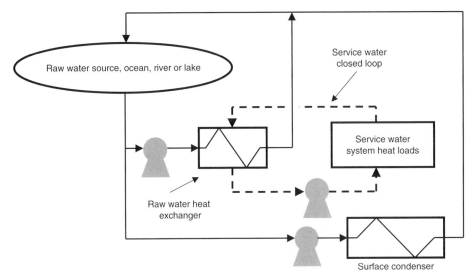

Fig. 3 A closed loop service water system typical for plants using sea water as the prime coolant. As the service loop is a closed-recirculating system, little make-up water is needed.

Table 1 Typical water quality in service water systems

Constituent	Fresh surface water	Fresh ground water	Brackish water	Saline water
Total dissolved solids, mg/L	80–1500	100–1500	1500–3000	>3000
Conductivity, μS/cm	120–2000	150–2000	2000–4000	>4000
Total hardness CaCO$_3$, mg/L(a)	5–300	5–300	50–1000	50–2000
Calcium hardness CaCO$_3$, mg/L(b)	3–200	3–200	30–800	9–30–800
Magnesium hardness CaCO$_3$, mg/L(b)	2–100	2–100	20–200	20–1200
Total alkalinity CaCO$_3$, mg/L(a)	2–350	2–350	20–500	20–800
Chloride Cl, mg/L	5–1000	5–1000	1000–10000	>10000
Sulfate SO$_4$, mg/L	5–200	5–200	50–500	50–1000
Silica SiO$_2$, mg/L	0.3–20	0.3–100	1–50	1–50
Total iron Fe, mg/L	0–5	0–5	0–5	0–5
Total manganese Mn, mg/L	0–2	0–2	0–2	0–2
Total phosphorus PO$_4$, mg/L	0–5	0–1	0–5	0–5
Ammonia NH$_3$, mg/L	0–5	0–2	0–5	0–5
Total suspended solids, mg/L	0–300	0–10	0–300	0–300

(a) These constituents are typically expressed as CaCO$_3$ (calcium carbonate) to facilitate calculations. (b) These constituents are either expressed as the ion (Ca and Mg) or as CaCO$_3$. The use of CaCO$_3$ facilitates calculations.

Asiatic clams (*Corbicula fluminea*), zebra mussels (*Dreissena polymorpha*), and bryozoans are the predominant fouling species. In seawater systems, foulants consist of barnacles, oysters, hydroids, bryozoans, mussels, and others.

Corrosion Control in Service Water Systems

Techniques for controlling corrosion in service water systems consist of the addition of corrosion inhibitors to control general corrosion and the addition of biocides to control microbiological growth and prevent MIC.

Some power plants utilize corrosion inhibitors. The decision to use corrosion inhibitors is based on the severity of a corrosion problem, corrosion history, an evaluation of potential benefits, and plant economics. Once-through systems are more costly to treat than recirculating systems. The most common corrosion inhibitors for carbon steel are phosphates and polyphosphates. Often these are used in conjunction with zinc salts. Copper alloy corrosion inhibitors include filming azoles such as tolyltriazole or benzotriazole. See the article, "Corrosion Inhibitors in the Water Treatment Industry" in Volume 13A for details on treatment of water in cooling systems. Biocides for control of microbiological growth include both oxidizing and nonoxidizing biocides. Oxidizing biocides are usually chlorine or bromine compounds. These act by destroying the organic material. Nonoxidizing biocides are chemicals that "kill" organisms via a metabolic mechanism. The choice of a biocide and the application mechanism depends on design characteristics, past history, economics, and environmental regulation requirements.

Deposit Control

The accumulation of deposits in piping and heat exchangers is a function of raw water and flow characteristics. In some cases, organic polymers are applied to disperse suspended solids particles and keep them moving through the system. In most cases these "silt dispersants" are low molecular weight water-soluble polymers such as polyacrylates.

REFERENCE

1. "Service Water System Problems Affecting Safety-Related Equipment," Generic Letter 89-13, United States Nuclear Regulatory Commission, July 1989

SELECTED REFERENCES

- S. Borenstein, *Microbiologically Influenced Corrosion Handbook,* Industrial Press Inc., 1994
- W. Dickinson and R. Peck, Manganese-Dependent Corrosion in the Electric Utility Industry, *Proceedings of Corrosion 2002,* April 2002 (Denver, CO), Paper 02444
- C. Felder and D. Cubicotti, "Microbiologically Influenced Corrosion of Carbon and Stainless Steel Weld and Base Metal—4-Year Field Test Results," presented at the EPRI Service Water Systems Reliability Improvement Seminar, July 1993
- H. Herro and R. Port, *The Nalco Guide to Cooling Water System Failure Analysis,* McGraw-Hill, 1993
- G. Kobrin, Ed., *A Practical Manual on Microbiologically Influenced Corrosion,* NACE International, Houston, TX, 1993
- R. Lutey and A. Stein, "A Review and Comparison of MIC Indices (Models)," presented at the 62nd International Water Conference, 2001
- K. Selby, Ed., A Review of Chemical Treatment Programs for Control of Fouling and Corrosion in Service Water Systems, *Proceedings of the 6th EPRI Service Water Systems Reliability Improvement Seminar,* July 1993 (Philadelphia, PA)
- K. Selby, G. Larson, M. Enrietta, and W. Rund, Pitting Corrosion of Helical-Wound Solder-Finned 90-10 Copper-Nickel Hydrogen Cooler Heat Exchanger Tubing, *Proceedings of the 14th EPRI Service Water System Reliability Improvement Seminar,* June 2002 (San Diego, CA)

Rouging of Stainless Steel in High-Purity Water

John C. Tverberg, Metals and Materials Consulting Engineers

MATERIALS OF CONSTRUCTION for equipment and piping in pharmaceutical processing plants must be resistant to corrosion from the high-purity water, the buffer solutions used in preparation of the products, and the cleaning solutions used to maintain the purity of the product. Most of the bioprocessing equipment is made of type 316L stainless steel. This alloy is selected because of its good corrosion resistance and ease of fabrication. When the buffer solutions and the final product become too corrosive for type 316L, a 6% M alloy, such as AL-6XN, 25-6Mo, or 1925hMo, is used. In very severe applications, alloys C-22 or C-276 may be used. The composition and UNS numbers of these alloys are given in Table 1.

Type 316L stainless steel is adequate for most high-purity water applications. However, under some conditions, an orange, red, magenta, blue, or black oxide coating forms on the surface. This condition is called rouging, so named because of the red variety that resembles cosmetic rouge. The orange, red, or magenta condition occurs in pure water, and the blue and black varieties appear in pure steam environments. Rouging does not appear to be a problem with the higher-nickel alloys, such as alloys C-22 or C-276. The classification of rouge and the mechanism of formation were first described in a paper given at the Validation Council's Institute for International Research seminar in 1999 (Ref 1).

Two phenomena are responsible for rouge formation: the very high purity of the water and any contaminants that may be in it. What makes high-pure water so aggressive? In its pure form, water is nearly a universal solvent. The ionization constant for water at room temperature, $K_w = [H_3O^+][OH^-] = 1.0 \times 10^{-14}$, is very slight. However, this value increases by nearly 100 times as the temperature approaches boiling. Each of the ion groups competes for association with other ions and will react with nearly everything to satisfy this driving force. As a result, pure water is extremely reactive.

Pharmaceutical Waters

Water used in the preparation of pharmaceuticals must undergo stringent purification. The system must be validated to assure that the purity requirements are met. The three quality standards for water are the United States Pharmacopoeia (USP) (Ref 2), water for injection (WFI), and high-purity water. Corresponding clean steam is made from these waters. Table 2 presents the requirements for USP 24 pharmaceutical water (Ref 3).

The primary water used in pharmaceutical production is WFI. To qualify for WFI, the purified water (water from reverse osmosis, deionization, or softening) must pass through an evaporation stage and meet the requirements in Table 2. High-purity water is made from WFI using a second distillation.

There are various techniques for producing this water. In general, the more complete the treatment, the fewer troublesome impurities enter the WFI stream. Table 3 presents the steps for preparing WFI.

Elimination of steps in the water treatment process or poor quality control can result in deposition and corrosion problems downstream. For example, overcharging the brine softeners can result in chloride spikes that can cause pitting or crevice corrosion of the stainless steel. Elimination of the softening operation may cause calcium deposits to form on the reverse osmosis membranes. Failure to remove the chlorine and chloramines in the water can harm the reverse osmosis membranes and, under some conditions, allow chloramines to enter the WFI. Evaporation will not remove chloramine. Failure to include reverse osmosis and/or electrolytic deionization or mixed-bed deionizers in the treatment will allow volatile salts to be carried over in the preparation of the WFI. Neither a still nor a vapor compression unit will remove all of these salts. Some of the compounds that are carried over in steam are:

- Iron magnesium hydroxide silicate
- Sodium iron silicate
- Calcium silicate
- Calcium aluminum silicate hydroxide
- Magnesium octahydride silicate
- Magnesium silicate hydrate
- Sodium metasilicate
- Sodium aluminum silicate
- Sodium chlorohexaaluminum silicate
- Potassium aluminum silicate
- Potassium trisodium aluminum silicate
- Magnesium silicate hydrate
- Magnesium octahydride silicate

Table 1 Composition of alloys used in bioprocessing equipment

Alloy	UNS No.	C, max	Mn, max	P, max	S, max	Si, max	Ni	Cr	Mo	N	Fe	Co, max	V, max	Cu	W
316L	S31603	0.035	2.00	0.040	0.005–0.017(a)	0.75	10.00–15.00	16.00–18.00	2.00–3.00	...	bal	...	...	...	...
AL-6XN(b)	N08367	0.030	2.00	0.040	0.030	1.00	23.50–25.50	20.00–22.00	6.00–7.00	0.18–0.25	bal	...	...	0.75	...
25-6Mo(c)	N08926	0.020	2.00	0.03	0.01	0.5	24.00–26.00	19.00–21.00	6.0–7.0	0.15–0.25	bal	...	...	0.5–1.5	...
1925hMo(d)	N08925	0.020	1.00	0.045	0.030	0.50	24.0–26.0	19.0–21.0	6.0–7.0	0.1–0.2	bal	...	...	0.8–1.5	...
C-22(e)	N10276	0.015	0.50	0.02	0.02	0.08	bal	20.0–22.5	12.5–14.5	...	2.0–6.0	2.5	0.35	...	2.5–3.5
C-276	N06022	0.010	1.0	0.04	0.03	0.06	bal	14.5–16.5	15.0–17.0	...	4.0–7.0	2.5	0.35	...	3.0–4.5

(a) According to ASME BPE-2002. (b) AL-6XN is a trademark of Allegheny Ludlum Company. (c) 25-6Mo is a trademark of Special Metals Company. (d) 1925hMo is a trademark of VDM. (e) C-22 is a trademark of Haynes International.

Silica is especially troublesome because it precipitates on stainless steel surfaces and can trap other contaminants that are in the water. Silica deposits can form concentration cells and set the stage for chloride crevice corrosion. Acid cleaning will not remove it; it must be removed mechanically. Silica is usually present as the silicate, either as an ion or when agglomerated as colloidal particles. Reverse osmosis will remove most silica, but there is a possibility of fouling the membranes. Passing the water through a series of mixed-bed deionizers effectively removes most of the silica.

Chlorides

Chloride is the primary contaminant in water that attacks stainless steel. Chloride destroys the passive layer by dissolving chromium and allowing reaction of water with iron, forming the red oxide Fe_2O_3, named hematite. The chemical reaction appears to take place in two phases. The first involves dissolution of chromium by the chloride ion; the second is oxidation of iron after the passive layer is dissolved:

$$Cr^0 + 3Cl^- \rightarrow CrCl_3 + 3e^- \quad (Eq\ 1)$$

$$2Fe^0 + 3H_2O \rightarrow Fe_2O_3 + 3H_2\uparrow \quad (Eq\ 2)$$

According to Ref 4, "this gelatinous precipitate adheres loosely to the iron" component of the stainless steel and "influences further corrosion in two ways": it retards corrosion "because it reduces the mobility of ions migrating to anodes and cathodes" that exist within the alloy formed by the "minute alloying elements that exist within the corrosion cell"; and it "accelerates corrosion by blocking off certain areas of the iron from access to the oxygen." Therefore, "oxide is removed most energetically from those areas where rust (oxide) has accumulated and the supply of oxygen is the most limited." When this occurs, pitting usually results. This may be an explanation for the red gelatinous oxide often found in WFI systems.

Chloride can come from a number or sources: chloramines from the disinfection of the water supply, from brine used to recharge the sodium zeolite softeners, or from additives such as sodium or potassium chloride and hydrochloric acid used in the preparation of the pharmaceuticals. Figure 1 illustrates a pit in the casing of a hot WFI centrifugal pump caused by chloramine, identified using x-ray photoelectron spectroscopy.

Chloramines are formed by the reaction of hypochlorous acid and ammonia. There are three species of chloramine: monochloramine (NH_2Cl) formed above pH 7 and which predominates at pH 8.3; dichloramine ($NHCl_2$) formed at pH 4.5; and nitrogen trichloride (NCl_3), sometimes called trichloramine, formed below pH 4.5. Between pH 4.5 and 7, both dichloramine and monochloramine exist. They are very volatile; therefore, distillation will not remove them.

Three chloride-induced corrosion mechanisms affect stainless steel in the pharmaceutical systems: pitting corrosion, crevice corrosion, and chloride stress-corrosion cracking. Pitting and crevice corrosion are fairly common, and stress-corrosion cracking occasionally occurs in stills, heat exchangers, and hot water systems. Three charts (Ref 5) are helpful in determining the effect of chlorides on stainless steel. Figure 2 illustrates the pitting relationship between chlorides, pH, and molybdenum content. If the pH and chloride content are above the molybdenum content curves, the alloy will pit. If they are below the molybdenum content curves, the alloy is safe to use. Figure 3 is the crevice corrosion relationship between the alloy, expressed as the pitting resistance equivalent number (PREN), and the critical crevice temperature. The PREN includes nitrogen content and is related to the composition of the alloy. The PREN is equal to $\%Cr + 3.3\%Mo + 16\%N$. Table 4 lists the PREN for a number of common alloys based on the minimum allowed alloy content. This relationship (Fig. 3) gives an indication of the temperature for onset of crevice

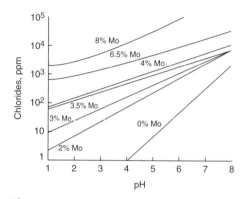

Fig. 2 Pitting corrosion as a function of chloride content, pH, and molybdenum content of austenitic iron-chromium-nickel alloys. Temperature range 65 to 80 °C (150 to 180 °F). Pitting is not a problem below the line but may be severe above the line. Source: Ref 5

Table 2 USP 24 pharmaceutical-grade water

Organics	<0.5 ppm TOC(a)
Conductivity	<1.3 μs/cm at 25 °C (75 °F)
Endotoxin	
Purified water	No specification
Water for injection	<0.25 EU/mL(b)
Bacteria	
Purified water	<100 cfu/mL(c)
Water for injection	<10 cfu/mL

(a) TOC, total organic carbon. (b) EU, endotoxin units. (c) cfu, colony-forming units

Table 3 Water treatment process for water for injection (WFI)

Feedwater from municipality or private water source

Step	Process details
Primary filtration	Multimedia filter
Hardness reduction	Zeolite softening
Chlorine removal	Sulfite injection + activated carbon filtration + ultraviolet light
Primary purification	Two-Pass reverse osmosis + electrolytic deionization or mixed-bed deionization
WFI production	Evaporation still or vapor compression

Fig. 1 Chloride pit on the casing of a centrifugal pump used for hot water for injection. The pit was caused by chloramines.

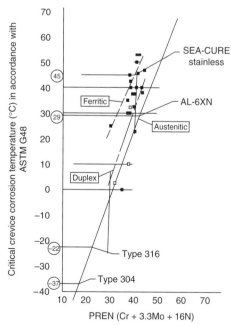

Fig. 3 Critical crevice corrosion temperature as a function of the pitting resistance equivalent number (PREN) and alloy type. Crevice corrosion will not occur below the temperature indicated but will above. Tests made in 6% ferric chloride. Source: Ref 5

corrosion, known as the critical crevice temperature. If the temperature and PREN are above the alloy type line (austenitic, duplex, or ferritic), the alloy will crevice-corrode. If it is below, the alloy is safe to use. The greater the interval between the critical crevice temperature, and the operating temperature, the greater the potential for crevice corrosion to occur. Figure 4 illustrates the probability of stress-corrosion cracking of a nickel alloy in a chloride environment. Three conditions must be met for this to take place: nickel content in the range of 6 to 25%; a residual tensile stress that exceeds the yield strength; and the proper environmental conditions of pH, chloride content, and threshold temperature. Each alloy has a threshold temperature above which it will crack but below which it will not. Table 5 lists the approximate threshold temperatures for some of the more common alloys.

Table 4 Pitting resistance equivalent number (PREN) for various alloys

Metallurgical category	Alloy	PREN, min
Austenitic	304, 304L	18.0
	304N, 304LN	19.6
	316, 316L	22.6
	316N, 316LN	24.2
	317, 317L	27.9
	317LMN	31.8
	AL-6XN	42.7
	625	46.4
	C-276	73.9
Duplex	2205	30.5
Ferritic	SEA-CURE stainless(a)	49.5
	430	16.0
	439	17.0
	444	23.3

(a) SEA-CURE is a registered trademark of Crucible Materials Corporation, UNS S44660.

Passive Layer

Stainless steels and chromium-containing nickel alloys derive their corrosion resistance from a chromium-rich passive layer. This passive layer is extremely thin, on the order of 10 to 100 atoms thick (Ref 6). It is composed mainly of chromium oxide, which prevents further diffusion of oxygen into the base metal. However, chromium is also stainless steel's Achilles heel, and the chloride ion is the problem. Chloride combines with the chromium in the passive layer to form soluble chromium chloride. As the chromium dissolves, active iron is exposed on the surface, which reacts with the environment to form rust. Alloying elements such as molybdenum minimize this reaction.

Two conditions must be met for a chromium-containing alloy to be passive (Ref 7). First, the chromium content on the surface must be greater than the iron content. Second, both the chromium and iron must be present as oxides. To meet the first condition, iron must be removed from the surface. To meet the second condition, an oxidation-reduction reaction must occur: the metals must be oxidized, and the passivating solution must be reduced. Chemical passivation is required; air passivation does not yield a stable passive layer. Today (2006), two acid combinations are used for passivation: 20% nitric acid at 20 to 50 °C (70 to 120 °F) for 30 to 60 min; and 10% citric acid + 5% ethylenediamine tetraacetic acid (EDTA) at 75 to 80 °C (170 to 180 °F) for 5 to 16 h. Technically, citric acid is not a passivating acid. It preferentially dissolves the iron, and the EDTA, a chelating agent, keeps iron ions in solution. The low-pH hot water oxidizes the chromium and remaining iron to form the passive layer, the composition of which approximates that of chromite spinel ($FeO \cdot Cr_2O_3$). It appears that citric acid passivation treatments yield higher chromium/iron ratios in the passive layer than does nitric acid. The mechanism for nitric acid passivation is described in Ref 6 to 8.

The quality of the passive layer is normally expressed as a chromium/iron ratio. Several methods are used to measure the composition, including Auger electron spectroscopy (AES) and x-ray photoelectron spectroscopy (XPS), also known as electron spectroscopy for chemical analysis. Both methods employ sputtering with ionized argon to remove layers of atoms. This allows progressive analyses of succeeding layers of the metal and oxide film so that a composition profile, or depth profile, can be made. Both methods report the composition in atomic percent. Table 6 compares these two methods with energy-dispersive spectroscopy, sometimes called microprobe analysis. Figure 5 presents an AES depth profile of a passivated type 316L surface. In this scan the chromium/iron ratio is 1.5, indicating a passive surface. Figure 6 illustrates the depth profile of type 316L with an outstandingly high chromium/iron ratio of 7.7, as determined by XPS. By comparison, Fig. 7 shows a depth profile for a type 316L tube with an extremely poor chromium/iron ratio of 0.13. This material will rust in a humid environment.

Several terms need to be defined to understand the convention in interpreting depth profile results:

- Maximum chromium/iron ratio generally occurs at the depth where oxygen reaches its maximum concentration, usually 0.3 to 1.5 nm, or 1 to 5 atoms, from the surface
- Depth of passive layer is defined as the depth where the oxygen content is half that of the difference between the maximum and base metal content.
- Carbon content is the sum of all carbonaceous materials on the surface, including carbon dioxide, residual isopropanol from cleaning, carbonates from the water or reaction with carbon dioxide, surfactants, and any other organic compound. Most industrial materials are in the 50 to 60 at.% C range.
- Oxygen content includes that from air, occluded carbon dioxide, organic compounds, and the passive layer. The most meaningful numbers occur away from the metal surface.

Table 5 Approximate threshold temperatures for chloride stress-corrosion cracking in chloride-enriched water

Alloy	UNS No.	Threshold temperature °C	°F
304	S30400	20	70
304L	S30403	20	70
316	S31600	50	125
316L	S31603	50	125
AL-6XN	N08367	230	450
C-276	N10276	>400	>750
C-22	N06022	>400	>750

Table 6 Comparison of analytical techniques

Characteristic	Auger electron spectroscopy	X-ray photoelectron spectroscopy	Energy-dispersive spectroscopy
Probe beam	Electrons	X-rays	Electrons
Detection beam	Auger electrons	Photoelectrons	X-rays
Element range	3–92	2–92	5–92
Detection depth, μm	0.003	0.003	1
Detection limits	1×10^{-3}	1×10^{-4}	1×10^{-5}
Accuracy, %	30	30	10
Identify organics	No	Some	No
Identify chemical state	Some	Yes	No

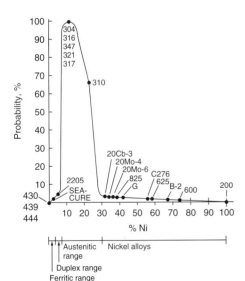

Fig. 4 Probability of chloride stress-corrosion cracking occurring as a function of the nickel content of the alloy. Cracking will not occur below the stress-corrosion cracking threshold temperature but will above (Table 5). Source: Ref 5

Not all stainless steel strip is created equal when it comes to the passive layer. Two compositions may be essentially identical, but one may have a chromium/iron ratio of 1.5 and the other a chromium/iron ratio of only 0.2. This difference arises from the use, or nonuse, of nitric acid in descaling the strip. If the strip is descaled using a nitric-hydrofluoric acid bath, the chromium/iron ratio will be high. If the strip is grit blasted or descaled in a sulfuric-hydrofluoric acid bath, the ratio will be low. It appears that starting with a low chromium/iron ratio condemns the finished part to a continued lifetime with a low chromium/iron ratio.

Surface Finish

Pharmaceutical equipment is normally specified with either mechanically polished or electropolished surfaces. Mechanically polished surfaces are either 0.635 or 0.508 μm (25 or 20 μin.) R_a (average roughness), and electropolished surfaces are either 0.381 or 0.254 μm (15 or 10 μin.) R_a. Figure 8 illustrates a mechanically polished surface with a 0.508 μm (20 μin.) R_a finish. Polishing debris is seen in the grit lines, and metal laps fold over some of the grit lines. This debris leads to rouging of the system. Figure 9 is the same surface after a hot nitric acid passivation treatment. Much of the debris is gone, and many of the laps have been dissolved. Figure 10 is a much finer polish, 0.203 μm (8 μin.) R_a, that shows finer grit lines with occluded polishing debris. An electropolished surface is depicted in Fig. 11. This surface is very smooth; in fact, it is possible to see the grains. The white specks in this micrograph are pits where manganese sulfide inclusions were dissolved during electropolishing. Because this material was purchased to the American Society of Mechanical Engineers biopharmaceutical equipment requirements, the sulfur content was approximately 0.012%.

All surface finishes should receive a nitric acid passivation, even electropolished surfaces. This treatment has two benefits: it raises the chromium/iron ratio and removes surface contaminants. Table 7 presents data obtained from different surface finishes and with various passivation techniques. Although citric acid passivation tends to yield higher chromium/iron ratios, it does not dissolve as much polishing debris from the grit lines.

Rouge Classification

Rouge is iron oxide. The different colors result from different valences and degrees of hydration. If rouge absolutely is not allowed, the alloy of

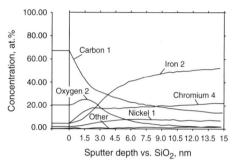

Fig. 5 Auger electron spectroscopy depth profile of a type 316L stainless steel surface. The exposed metal surface is on the left, and the composition with depth from the surface changes as one moves to the right. The base metal composition is reached at approximately 12.5 nm, or 35 atoms, from the surface. In this example, the chromium/iron ratio is 1.5.

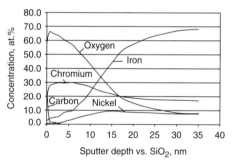

Fig. 6 X-ray photoelectron spectroscopy depth profile of a type 316L stainless steel surface. The base metal composition is reached at approximately 35 nm, or 100 atoms, from the surface. In this example, the chromium/iron ratio is 7.7, an outstanding value.

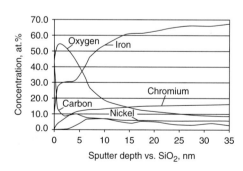

Fig. 7 X-ray photoelectron spectroscopy depth profile chart of a type 316L stainless steel surface with an extremely poor chromium/iron ratio of only 0.13. This material will show rust in only a few hours in a humid environment.

Fig. 8 Mechanically polished surface with a 0.508 μm (20 μin.) R_a finish. The dark deposits in the grit lines are residual polishing debris. This debris typically leads to class 1 rouging of the stainless steel surfaces. SEM; original magnification 500×

Fig. 9 Same surface as in Fig. 8 after a hot nitric acid passivation treatment. Many of the laps on the grit lines have been dissolved and much of the polishing debris removed.

Fig. 10 Mechanically polished surface with a profilometer reading of 0.203 μm (8 μin.) R_a. Note the polishing debris in the grit lines. SEM; original magnification 500×

Fig. 11 Electropolished surface with a profilometer reading of 0.102 μm (4 μin.) R_a. The white spots are pits from which manganese sulfide inclusions were leached out. The grains are clearly visible. SEM; original magnification 450×

construction should be changed to one that contains little or no iron. Not all rouge is the same, and its formation differs accordingly. There are three general classes of rouge.

Class 1 Rouge. These oxides originate elsewhere in the system and are deposited on stainless steel surfaces. They generally are held onto the surface by electrostatic attraction. The chromium/iron ratio under the deposited oxides is unaltered from that of the original passivated stainless steel. Usually, they can be removed by wiping or ultrasonic cleaning. The chemical form is Fe_2O_3 (hematite) or one of the hydroxides. Sources include erosion and/or cavitation from pump components or spray balls, residual debris from mechanical as-polished surfaces, corroding iron components in the system, and deposition from iron held in solution in the water. Metal particles from polishing debris are oxidized in the WFI and deposited on the surface. The particles and/or oxides have a composition that matches that of the stainless steel from which they originated, for example, polishing debris or particles from the pump impeller.

The lowest valence state for class 1 rouge is the yellow-orange oxide FeO(OH), hydrated ferrous oxide. According to Ref 9, this form corresponds to the mineral limonite, which is yellowish-brown to orange in color. Its formation can be chemically stated as:

$$2Fe^0 + 4H_2O \rightarrow 2FeO(OH) + 3H_2 \uparrow \quad (Eq\ 3)$$

According to Ref 10, "rusting occurs only if both oxygen and water are present. Iron will not rust in dry air or oxygen-free water." Therefore, oxygen must be present in the WFI, or a corrosion cell must be present to liberate oxygen at the anode.

The following reaction is the oxidation of the hydrated ferrous oxide to ferric oxide (Fe_2O) to produce the red or magenta color:

$$2FeO(OH) \rightarrow Fe_2O_3 + H_2O \quad (Eq\ 4)$$

Alternatively, the reaction could be:

$$2Fe^0 + 3H_2O \rightarrow Fe_2O_3 + 3H_2 \uparrow \quad (Eq\ 5)$$

These reactions allow oxidation to a higher valence state without a decrease in pH by formation of gaseous molecular hydrogen. According to Ref 9, the red oxide hematite can exist in several forms: an "earthy" form, usually amorphous, and a crystalline form. Its crystalline form is rhombohedral. It can exist as crystals, specular hematite, compact columnar, or fibrous. When chromium is present, it can be octahedronal and is a form of chromite spinel ($FeO \cdot Cr_2O_3$). When nickel is present, it is called trevorite ($NiO \cdot Fe_2O_3$), also octahedral. Figure 12 illustrates the general form of the orange amorphous deposit. When examined at higher magnification (Fig. 13), this amorphous-appearing deposit has a rhombohedral crystal form. Figure 14 illustrates the growth of crystals in the polishing striations and on the surface. These are the areas that had deposits of polishing debris. Figure 15 illustrates twinned rhombohedral crystals attached to the surface but not appearing to be growing on the surface. The twinned structure may be growing by taking iron from the water.

Class 2 Rouge. This type of rouge forms in situ on the stainless steel surface. By its formation, the chromium/iron ratio is altered, usually decreased by dissolution of the chromium and formation of ferrous/ferric hydroxide/oxide, usually Fe_2O_3. There are at least two subclasses of class 2 rouge.

Class 2A. This forms in the presence of chlorides. This type of rouge can be removed only by chemical or mechanical means. The chromium/iron ratio under the rouge approaches that of the base metal composition, and tubercles or crystal growths are observed on the surface. The metal surface under the tubercle, when removed, is bright silver, representing an active corrosion site. Figure 16 illustrates rouge that grows in the presence of chloride. The corrosion mechanism was described earlier. Figure 17 illustrates acicular crystals growing from the surface, possibly from chloride micropits. Rhombohedral crystals can then grow from these seeds, using the iron from the metal or from that in the water. Figure 18 illustrates the fibrous structure that develops from the acicular crystals.

Class 2B. This rouge forms on unpassivated or improperly passivated surfaces where the passive layer is inadequate to prevent the diffusion of oxygen to the metal below the passive layer. Both the thickness of the passive layer and the chromium/iron ratio have an effect. A classic example is air passivation of a mechanically abraded surface. Such a passive layer is on the order of 1 nm (3 atoms) thick and is easily

Table 7 Comparison of type 316L chromium/iron ratios using various polishing and passivation techniques

Polishing method	Chromium/iron ratio
Electropolish, no passivation	0.8
Electropolish, nitric passivation	1.8
Electropolish, citric passivation	2.5
Heavy mechanical polish, no passivation	0.3
Heavy mechanical polish, nitric passivation	0.8
Heavy mechanical polish, citric passivation	1.1
Light mechanical polish, no passivation	0.33
Light mechanical polish, nitric passivation	1.0
Light mechanical polish, citric passivation	1.4

Fig. 12 Class 1 rouge that is transported from other locations and deposited on the stainless steel surface. It is amorphous in structure, and the color varies from orange to red. The primary mineral form is hematite, Fe_2O_3. SEM; original magnification 450×

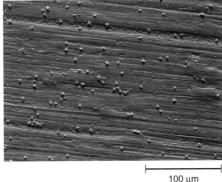

Fig. 14 Rhombohedral crystals forming on the surface of a mechanically polished tube with a high chromium/iron ratio. This appears to be class 1 originating from polishing debris. Note the high density of small crystals in the polishing striations. SEM; original magnification 450×

Fig. 13 Highly magnified view of the amorphous class 1 rouge crystals. They appear to be rhombohedral in form. SEM; original magnification 20,000×

Fig. 15 These twinned rhombohedral crystals do not appear to be growing on the surface. They are associated with residual polishing debris. Class 1 rouge. SEM; original magnification 7500×

destroyed. By comparison, a properly electropolished and chemically passivated surface may be over 30 nm thick. The formation of class 2B rouge is strongly influenced by the surface preparation of the stainless steel. Tuthill and Avery (Ref 11) subjected different metal surfaces to high-purity water. They found that mechanically polished surfaces without passivation had the most iron dissolved in the water, and properly passivated electropolished surfaces had none. Table 8 is a summary of their findings. Figure 19 illustrates crystals attached to a stainless steel surface with a lower chromium/iron ratio.

Class 3 Rouge. This rouge forms in the presence of high-temperature steam, usually above 120 °C (250 °F). The color is black. The chemical composition is totally different from that of class 1 or class 2 rouge: Fe_3O_4 rather than Fe_2O_3. The mineralogical classification is magnetite rather than hematite, and the morphology is decidedly crystalline on both the substrate and the crystals projecting above the substrate.

The black color is due to the formation of iron sesquioxide (Fe_3O_4), or magnetite. According to Ref 9, this oxide is commonly, if not always, formed under conditions of high temperature. It is of the form $FeFe_2O_4$ or $FeO \cdot Fe_2O_3$, in which the ferrous iron occasionally may be replaced with nickel, forming trevorite ($NiO \cdot Fe_2O_3$). Reference 4 states that "this oxide is formed at very high temperatures by the action of air, steam, or carbon dioxide upon iron." The reaction involves two oxidation steps:

$$Fe^0 + H_2O \rightarrow FeO + H_2 \uparrow \quad (Eq\ 6)$$

$$2FeO + H_2O \rightarrow Fe_2O_3 + H_2 \uparrow \quad (Eq\ 7)$$

$$FeO + Fe_2O_3 \rightarrow Fe_3O_4 \quad (Eq\ 8)$$

The two oxides coexist in the same octahedral unit cell because of the limited diffusion of oxygen.

There are three subclassifications of this rouge form, depending on the nature of the stainless steel surface.

Class 3A. This form of rouge occurs on properly passivated and electropolished surfaces. Its appearance is glossy black, very adherent (cannot be rubbed off), very stable, and can be removed only by chemical dissolution or mechanical abrasion. It forms at higher steam temperatures, usually above 150 °C (300 °F). It completely covers the surface with continuous crystals, as Fig. 20 illustrates. The crystals probably contain chromium, because both magnetite and chromite are octahedrons and both form the spinel class of minerals.

Class 3B. This form of rouge occurs on unpassivated or mechanically polished surfaces and at lower temperatures, usually in the range of 105 to 120 °C (220 to 250 °F). Its appearance is powdery black, sometimes with sparkles. It can be rubbed off but can be completely removed only by chemical dissolution or mechanical abrasion. It is magnetite or chromite spinel.

Class 3C. This form of rouge forms on surfaces subjected to high-temperature or hot electrolytic solutions. It is a precursor to either class 3A or 3B rouge. The color is iridescent gold, red, or blue. The color is related to the oxide thickness (Ref 12). Figure 21 illustrates the depth of oxygen as a function of color. The control is the standard silver electropolished film, the gold color is the fully oxidized chromite spinel, and the blue is the onset of magnetite formation.

Castings

Castings have a highly variable surface, depending on the amount of mechanical polishing. Cast type 316L stainless steel is designated

Fig. 16 Class 2A rouge that forms in the presence of chlorides. When the tubercles are broken off, a bright silver spot is under them, indicating an active chloride corrosion cell. SEM; original magnification 450×

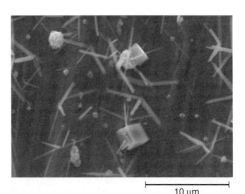

Fig. 17 Acicular crystals growing from the surface of stainless steel exposed to chloramines at steam temperatures. The acicular crystals appear to be growing from the surface of the stainless steel, perhaps from a chloride micropit. They appear to be the start of class 2A rouge. The large rhombohedrons may be from polishing debris or may have grown from dissolved iron in the water. SEM; original magnification 5000×

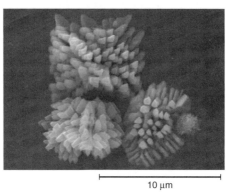

Fig. 18 Fibrous rhombohedral crystals growing on the surface. Probably class 2A rouge that originated from chloride micropits on the stainless steel surface. SEM; original magnification 7250×

Table 8 Effect of type 316L stainless steel surface finish on iron release
Deionized water in 24 h

Specimen surface finish	Iron, ng/cm²
360 grit (0.254 µm, or 10 µin., R_a) MP only(a)	1190
180 grit (0.635 µm, or 25 µin., R_a) MP only	1090
180 grit (0.635 µm, or 25 µin., R_a) MP + full electroplish	990
2B strip finish (~0.254 µm, or ~10 µin., R_a) + HNO₃ passivation, 49 °C (120 °F)(b)	7
2B strip + HNO₃ + HF passivation, 49 °C (120 °F)	0
180 grit MP + full electropolish + HNO₃ passivation, 49 °C (120 °F)	0

(a) MP, mechanical polish. (b) 2B, cold rolled with polished rolls, No. 2 finish, bright. Source: Ref 11

Fig. 19 These rhombohedrons appear to be growing from the surface. The stainless steel had a lower chromium/iron ratio. This appears to be class 2B rouge. SEM; original magnification 20,000×

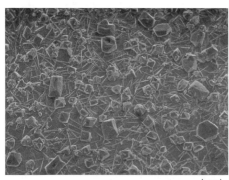

Fig. 20 Class 3A rouge. This is a black, glossy rouge that forms on the surface of electropolished stainless steel in high-temperature steam. The crystals completely cover the surface. The crystal form is octahedral, and the mineral is magnetite. SEM; original magnification 450×

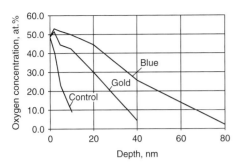

Fig. 21 Oxygen content as a function of depth for various iridescent color rouge films. The control is the standard silver electropolished surface finish. The darker the color, the thicker the oxide film.

as CF-3M. The cast version has lower manganese and nickel and higher chromium and silicon than the wrought version. This promotes a higher delta-ferrite content in the casting. In general, the silicon content is at its maximum, 1.5%, to promote greater fluidity of the molten metal. This high silicon content further promotes the formation of delta ferrite. Delta ferrite is the high-temperature form of the iron-rich compound that solidifies first from the melt, forming the basic dendritic structure. Normally, a skin of ferrite that is low in chromium and high in silicon and iron is formed on the casting. If the delta-ferrite content of the casting is less than 8%, as determined from the DeLong-Schaefler (Ref 13) or Welding Research Council (Ref 14) diagrams, the ferrite can be dissolved by a long-term solution anneal. If the heat treatment is inadequate or if the ferrite content is greater than 8%, ferrite remains in the structure. Ferrite will corrode preferentially, resulting in pits on the surface. If machining or grinding, as in the case of valve components, removes the high-ferrite skin, the substrate surface composition becomes the nominal composition of the alloy. A high chromium/iron ratio results (Fig. 22) if this surface is electropolished and passivated.

Investment casting is used to produce complex shapes such as pump impellers. Because the surface is so smooth, little or no grinding is performed; the surfaces are simply buffed. The ferrite-rich skin is not removed, and the resulting chromium/iron ratio is very low, 0.06 (Fig. 23). Such a surface will be subject to corrosion and cavitation/erosion and will be a source of class 1 rouge. Figure 24 illustrates the surface of an impeller vane that has been subjected to cavitation and/or erosion. The delta-ferrite ridges (the white phase) stand in relief because they represent the harder material, and the interdendritic austenite (the dark phase) has eroded away.

Cleaning and Repassivation

Badly rouged surfaces should be acid cleaned. Class 1 rouge needs to removed with a mild citric acid + EDTA solution. Class 2 rouge should be taken to bare metal with a primary cleaning using oxalic acid, formic acid, or ammonium bifluoride, followed with a passivation treatment using citric acid + EDTA. Class 3 rouge is different. Once it forms, it is best to leave it alone. An alternative procedure is to clean with citric acid to remove any occluded iron oxides.

When should one repassivate? The only way to know the condition of the passive layer is to perform a depth profile analysis. This is a destructive test, requiring approximately 1 cm^2 (0.16 in.^{2}) material from the system. This material can come from a spool piece, a blind flange, or any component, except a casting, that has been exposed to the environment and has been part of

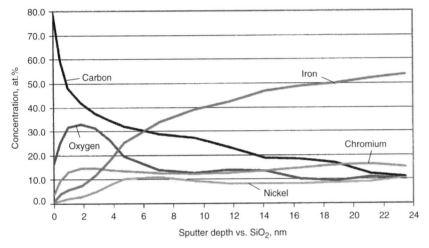

Fig. 22 Depth profile for a CF-3M (type 316L) casting that has been machined, electropolished, and nitric acid passivated. The chromium/iron ratio on the surface is 4.1.

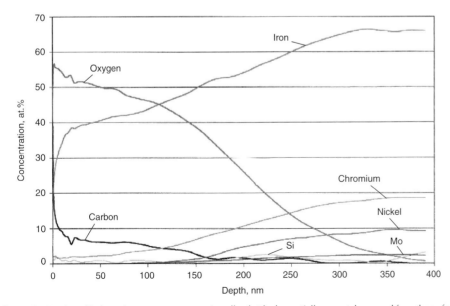

Fig. 23 Depth profile for an investment cast pump impeller that had essentially no metal removed from the surface. The chromium/iron ratio is low (0.06), and the metal does not meet nominal composition until a depth of 350 nm is reached. This surface will be subject to corrosion and cavitation/erosion. It can be a source of class 1 rouge in the system.

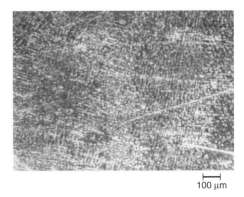

Fig. 24 Investment cast impeller surface roughened by cavitation/erosion in water for injection service. The white ridges are harder delta ferrite, and the dark areas are the softer interdendritic austenite. Original magnification 100×

the system since the original passivation. In general, if the surface has not been altered or compromised since the original passivation, it is not necessary to repassivate. This includes class 1 rouge. The surface can be derouged using citric acid, but it does not need repassivation. The difference in these two treatments is time. Derouging is rather fast, whereas passivation is a long-term process. Class 3A rouge should never be removed, because the only way to remove it is to use very harsh reagents that will destroy the surface finish and passive layer. In addition, it will form again in several months, probably as unstable rouge.

The surface should be cleaned and repassivated under these conditions:

- New components have been added to the system.
- Welding has taken place anywhere in the system.
- Class 2A rouge is present.
- Class 2B rouge is present.
- Class 3B rouge is present under some conditions.

Class 3B should be addressed if the surface was not passivated or if the original passivation was not adequate. The only way to determine if the passivation was inadequate is to perform a depth profile analysis.

Summary

To prevent or at least minimize rouge formation in WFI, high-purity water, or clean steam systems, the following procedures should be followed:

- Use and maintain the best water treatment process available to produce the best-quality water possible. Maintain a water database that charts the critical elements. The incoming water database, up to the deionizer, should include, as a minimum, calcium, iron, magnesium, manganese, silicon (silica), copper, zinc, aluminum, chloride, dissolved solids, pH, temperature, conductivity, and the Langelier saturation index. These values can be reported in parts per million (ppm or mg/L). The deionized, reverse osmosis, WFI, high-purity water, and clean steam condensate should be analyzed for calcium, potassium, iron, nickel, chromium, chloride, silicon, magnesium, manganese, aluminum, copper, zinc, conductivity, pH, and temperature. These values should be reported in parts per trillion (ppt) and updated monthly.
- Use electropolished components wherever possible. Specify the components to be passivated at the factory, using nitric acid.
- Use care during welding to prevent heat discoloration on the heat-affected zone. Discoloration, at the worst, should be a light straw color. Ideally, the heat-affected zone should be silver.
- Thoroughly clean and passivate the system prior to start-up. This will passivate the welds in the system.
- Use proper design and installation practices for all piping systems. This includes positive drain angles, not placing reducers in the inlet to centrifugal pumps, proper steam traps, elimination of dead legs, and so on.
- Choose the alloy of construction based on the nature of the product to be contained in the lines, vessels, and other components, not based on the price. Use Fig. 2 to 4 to aid in making the proper selection.
- Make certain that no iron, brass, copper-nickel, or other materials that corrode easily are in the system.
- Use submicron filters where critical operations take place.

REFERENCES

1. J.C. Tverberg and J.A. Ledden, Rouging of Stainless Steel in WFI and High Purity Water Systems, *Conference Proceedings, Preparing for Changing Paradigms in High Purity Water*, Oct 27–29, 1999, The Validation Council, A Division of The Institute for International Research, New York
2. *United States Pharmacopoeia*, The United States Pharmacopeial Convention, Inc., Rockville, MD
3. *National Formulary*, Section 24, The United States Pharmacopoeial Convention, Inc., Rockville, MD
4. W.F. Ehret, *Smith's College Chemistry*, 6th ed., D. Appleton-Century Company, New York, 1947, p 637
5. J.C. Tverberg, Stainless Steel in the Brewery, *Tech. Q.*, Vol 38 (No. 2), Master Brewers Association of the Americas, Wauwatosa, WI, 2001, p 67–83
6. J.C. Tverberg and S.J. Kerber, Effect of Nitric Acid Passivation on the Surface Composition of A 270 Type 316L Mechanically Polished Tubing, *Proceedings of the International Pharmaceutical Exposition and Conference, Interphex*, March 17–19, 1998, Reed Exhibition Companies, p 55–65
7. J.C. Tverberg and S.J. Kerber, Effect of Nitric Acid Passivation on the Surface Composition of a Mechanically Polished Type 316L Sanitary Tube, *Eur. J. Parent. Sci. (London)*, Vol 3 (No. 4), 1998, p 117
8. J.C. Tverberg, Conditioning of Stainless Steel for Better Performance, *Stainl. Steel World*, Vol 11 (No. 3), April 1999, p 36–41
9. E.S. Dana, *A Textbook of Mineralogy*, 14th ed., John Wiley & Sons, New York, Oct 1951, p 483, 491, 505
10. J.E. Brady, *General Chemistry, Principles and Structure*, John Wiley and Sons, New York, 1990, p 703
11. A. Tuthill and R. Avery, *ASME BPE*, American Society of Mechanical Engineers, June 2004
12. J.C. Tverberg and S.J. Kerber, Color Tinted Electropolished Surfaces: What Do They Mean?, *Proceedings of the International Pharmaceutical Exposition and Conference*, April 15–17, 1997, Reed Exposition Companies, p 255–262
13. Delta Ferrite Content (DeLong-Schaefler Diagram), Fig. NC-2433.1-1, *ASME Boiler and Pressure Vessel Code*, Section III, Division 1 NC, NC2000, 1988 ed., p 25
14. Weld Metal Delta Ferrite Content (Welding Research Council Diagram), Fig. NC-2433.1-1, *ASME Boiler and Pressure Vessel Code*, Section III, Division 1 NC, NC2000, 1995 ed., p 28

Corrosion in Wastewater Systems

Jose L. Villalobos, V&A Consulting Engineers
Graham Bell, M.J. Schiff & Associates

METALLIC AND CONCRETE STRUCTURES and equipment at wastewater treatment plants are exposed to corrosive environments that can lead to corrosion, degradation, and ultimately failure (Ref 1). Options are available for eliminating or minimizing corrosive impacts arising from these exposures, including proper materials selection, coatings, and electrochemical methods, such as cathodic protection. It is also possible, in some cases, to modify the environment to make it less corrosive.

Predesign Surveys and Testing

The environments in wastewater treatment plants can be classified as buried, fluid, and atmospheric exposures.

Soil Testing. Prior to the detailed design of a wastewater treatment plant or construction of new capital improvement projects, a corrosion survey should be performed at the project site. This survey should include soil resistivity testing, chemical analysis of soil and/or water samples, and inspection of any facilities that may already exist at the proposed treatment plant location. An investigation should be conducted to determine if existing cathodically protected utilities are located at the proposed project location. In situ soil resistivity measurements should be made in a grid pattern at distances that will provide representative data. The testing intervals should not exceed 300 m (1000 ft) in either direction and to the excavation depth. Soil samples should be collected at representative locations and at locations of low soil resistivity for chemical analysis and laboratory resistivity testing. Half-cell potential surveys and current requirement testing should be performed at locations where cathodic protection will be installed on existing facilities. Data regarding the interpretation of soil properties for the development of corrosion control strategies are available in several recent publications (Ref 2–4).

Testing of the wastewater to be transported is critical for materials selection and corrosion control for submerged exposures and vessel interiors. Many of the same parameters determined for soils should be determined for water or other fluids to be transported or contained. These include pH, chloride ion content, and resistivity. Other parameters to be investigated include dissolved oxygen, the biochemical oxygen demand, the chemical oxygen demand, levels of suspended solids, and the Langelier saturation index.

Dissolved oxygen (DO) is the oxygen dissolved in water, wastewater, or other liquid, usually expressed in milligrams per liter (mg/L) or parts per million (ppm).

Biochemical oxygen demand (BOD) is the quantity of oxygen used in the biochemical oxidation of organic matter in a specified time (5 days, unless otherwise indicated) and at a specified temperature.

Chemical oxygen demand (COD) is a measure of the oxygen-consuming capacity of both organic and inorganic matter present in water or wastewater. It is expressed as the amount of oxygen consumed by this matter in a specified test (Ref 5).

The Langelier index gives an indication of the tendency of a fluid to deposit minerals on the interior surfaces of pipelines or to corrode the interior surfaces of the pipeline. A negative value for this index, less than -0.5, is indicative of a corrosive environment, whereas a positive index, greater than 0.5, indicates a tendency for calcium carbonate scale to form.

Atmospheric Testing. Wastewater treatment plants are situated in marine, industrial, and rural atmospheric environments. In marine atmospheric environments, the combination of wet/dry cycling in the presence of chlorides found in airborne salts creates the potential for significant atmospheric corrosion activity on all exposed metal surfaces. In addition to marine exposures, plant processes and sanitary sewers are likely to generate airborne substances, such as hydrogen sulfide, that will also be deleterious to metallic and concrete structures. The proximity of wastewater treatment plants to major highways and industries may expose materials at the plant to high levels of sulfur dioxide that can cause corrosion of certain metals and deterioration of coating systems. Testing should be conducted for relative humidity values, atmospheric salt content, and hydrogen sulfide (Ref 6).

Hydrogen Sulfide Testing. Levels of suspended solids, DO, BOD, and COD affect the ability of wastewater streams to produce hydrogen sulfide. Hydrogen sulfide, when oxidized, will degrade concrete and corrode exposed steel. Hydrogen sulfide gas is generated after the available oxygen in the wastewater is depleted by aerobic (oxygen-consuming) bacteria. Once the oxygen is depleted, anaerobic (nonoxygen-consuming) bacteria, also present in the wastewater, become active. Under these conditions, the anaerobic bacteria reduce sulfate to hydrogen sulfide. Most of the production of hydrogen sulfide occurs in the slime layer that coats the submerged pipe walls of sewer pipelines. The anaerobic bacteria also metabolize sulfur from proteins and other elements found in the wastewater. Most wastewater carries large quantities of sulfur compounds (Ref 7).

Once hydrogen sulfide is formed, it remains concentrated in the wastewater until the saturation level is reached. At the saturation level, the hydrogen sulfide is released from the wastewater and enters the airspace inside the sewer above the waterline. Another group of aerobic bacteria, which lives on the walls of the sewer above the flowline, reacts with hydrogen sulfide and oxygen to form sulfuric acid. Sulfuric acid is corrosive to steel, concrete, and copper. The following terms are important for understanding and analyzing the conditions for generation of hydrogen sulfide.

Dissolved Sulfide. The sum of all sulfur species, that is, sulfide, hydrosulfide, and hydrogen sulfide, measured in a wastewater sample collected in the field, using a LaMotte Pomeroy test kit, is referred to as the dissolved sulfide.

Total Sulfides. The sum of soluble and insoluble sulfur species is referred to as total sulfides. Total sulfides are measured in the field from a wastewater sample collected from the sewer system using another version of the LaMotte Pomeroy test kit. The measurement of total sulfides in the field indicates the magnitude of the sulfur species present in a wastewater.

Sulfide Generation. The primary parameters that influence the production of sulfides within the piping system that transports the wastewater to the treatment facility are:

- *Temperature of the wastewater:* As the temperature increases, the rate of biological reactions increases, and the concentration of sulfides will also increase.
- *Dissolved oxygen concentration:* As DO increases in the wastewater, the concentration of sulfides will be reduced. A DO concentration of approximately 1 mg/L is generally adequate to keep the concentration of total sulfides to less than 1 mg/L.
- *Wetted surface area:* As the depth of wastewater increases in a sewer pipe, the wetted surface area of the pipe is also increased. This wetted surface area is the location of the sulfide-producing microorganisms. As the wetted surface area increases, so does the number of microorganisms within the pipe, which in turn increases the total sulfides that are generated.
- *Sewer slopes:* A less sloped sewer pipeline will reduce the flow velocity, which increases sedimentation and diminishes the rate at which oxygen becomes dissolved in the wastewater. This promotes anaerobic conditions and provides microorganisms with ample time and nutrient to feed and generate sulfide. Hydraulic conditions that cause splashing or hydraulic jumps promote the release of hydrogen sulfide.

Material Considerations

Materials selection is based on many engineering and economic parameters in addition to corrosion resistance. The corrosion performance of these materials in the various environments follows.

Concrete pipes, basins, and tanks are common in facilities.

Soil Exposure. Concrete is typically a durable material in underground service. Unlike metals, it is rarely affected by electrolytic corrosion. Attack on concrete is likely when soils are acidic. Barrier coatings may be required to protect concrete surfaces in low-pH environments. In areas of high water-soluble sulfate ion concentrations, modification of the water-cement ratio and use of type V cements (sulfate resistant), admixtures, and barrier coatings should be considered for all buried concrete structures.

Fluid/Atmospheric Exposure. Concrete structures can be attacked by biologically formed sulfuric acid. Sulfuric acid formation is greatest where hydrogen sulfide is readily available and where aerobic conditions and moisture are present to oxidize the sulfides to sulfuric acid. These conditions exist where biological slimes are present on the pipe walls and where aerobic conditions and moisture exist above the wastewater flow.

Wastewater contains a variety of sulfates in solution. Sulfate-reducing bacteria present in wastewater convert these sulfates to sulfides and related compounds. Dissolved sulfide concentrations of 0.4 ppm in the waste streams are considered detrimental. Some of these sulfides are released from solution into the airspace over the water surface. Aerobic bacteria, commonly found in the crown of sewers or in the top of structures such as wet wells, can then convert the hydrogen sulfide to sulfuric acid. Concrete is severely attacked by sulfuric acid. Rapid destruction of the concrete and subsequent exposure of the aggregate and reinforcing steel can occur. A typical example of hydrogen-sulfide-induced corrosion that has progressed to the point of exposing the reinforcing steel in a gravity sewer pipe is shown in Fig. 1.

The use of bare concrete is not recommended at locations where high hydrogen sulfide concentrations may occur. These areas include channels, manholes, wet wells, and surfaces of structures that are not exposed to the atmosphere, such as interior pipe surfaces and covered digesters. Protection of the concrete with a polyvinyl chloride sheet lining ensures the durability of structures, both on the submerged surfaces and within the airspace above the fluid surface. This is particularly vital when submergence varies, continuously exposing the surface to wet and dry conditions. Reinforced concrete structures are subject to cracking and spalling due to corrosion of reinforcing steel. Because the volume of corrosion product is much greater than that of the reinforcing steel itself, great pressure is exerted on the concrete, which may cause it to crack and eventually spall. Sewage pumping stations are especially susceptible to this type of corrosion activity.

Fig. 1 Exposed reinforcing steel in gravity sewer pipe

Exposure of concrete surfaces to chlorides and hydrogen sulfide should be minimized through the use of coatings.

Steel and Ductile Iron. For all exposures, steel and ductile iron should be electrically isolated from dissimilar metals to prevent the formation of galvanic corrosion cells.

Soil Exposure. Low resistivities and high chloride concentrations in the soil may lead to corrosion of buried steel and ductile iron pipelines or structures. Cathodic protection should be considered for all buried steel and ductile iron pipelines or structures. Where cathodic protection is not provided, corrosion-monitoring equipment should be incorporated into the design to allow the operating staff to monitor the condition of the pipelines or structures. Non-welded joints must be bonded for electrical continuity.

Steel and ductile iron structures should be well coated, and where the potential for corrosion is high, based on resistivity and soil chemical analysis, they should be cathodically protected.

Fluid Exposure. Steel and ductile iron are subject to corrosion when exposed to fluids at a wastewater facility. Corrosion of steel and ductile iron is most severe in the splash zone, where atmospheric oxygen hastens the corrosion process. Submerged steel and ductile iron should be coated with a material suitable for use in immersed exposures. Figure 2 illustrates a sample of a submerged steel structural member that has lost the protection provided by the coating system. Due to the corrosive nature of the wastewater treatment process streams, cathodic protection can be an effective means of corrosion control of steel and ductile iron surfaces. Cathodic protection should be used in combination with suitable coating systems.

Atmospheric Exposure. Corrosion of steel and ductile iron structures exposed to the atmosphere at wastewater facilities is likely due to atmospheric chloride and hydrogen sulfide levels. Hydrogen sulfide attacks metallic surfaces rapidly. Uncoated steel and ductile iron will not endure in outdoor ambient conditions

Fig. 2 Delaminating coating on steel rake arm

where hydrogen sulfide is present and are usually coated in less harsh atmospheres as well. Steel and ductile iron structures of particular concern, such as floating digester covers, must be coated.

Aluminum is a part of structures and indoor and outdoor equipment.

Soil Exposure. Aluminum is not recommended for underground applications. Soil becomes more aggressive to aluminum as moisture and dissolved salt levels increase. Contact with elevated levels of chlorides and sulfates or low-pH environments will be detrimental to buried aluminum. Dry, sandy, and well-aerated soils are less likely to be corrosive to aluminum.

Fluid Exposure. Aluminum is attacked by both acids and bases. When exposed to fluids, aluminum is most stable in the pH range from 4 to 8.5. Contact with solutions with pH greater than 8.5 or less than 4 will cause corrosion of the aluminum.

Severe pitting of aluminum can occur where iron or copper ions are in a solution in contact with aluminum. A galvanic couple is established, and aluminum is attacked at localized areas. Chloride ions can also lead to pitting of aluminum. Pitting is most likely to occur in crevices, welds, and other stagnant areas. The introduction of chemicals, such as ferric chloride, into the process stream may cause severe degradation of aluminum, and failure could occur rapidly. Aluminum is not recommended for use in fluid exposures, due to the wide range of pH values found in wastewater facilities.

Atmospheric Exposure. Aluminum has excellent resistance to atmospheric sulfides and other pollutants due to a tightly adherent oxide film. Destruction of the oxide film will expose a very reactive surface. If the protective oxide film is disturbed, the presence of salts, including chlorides, can cause rapid pitting of aluminum. Nonanodized aluminum can suffer severe corrosion in marine environments. Electrical coupling to iron, stainless steel, or copper will accelerate the deterioration of aluminum.

Aluminum is stable only in a band of pH values. Aluminum handrails to be installed in concrete (pH 12 to 13) should be placed in plastic shields that are cast into the concrete. In addition, a sealant should be placed between the plastic shield and the aluminum. Under these conditions, aluminum will provide acceptable performance. Aluminum is not recommended in areas where contact with chlorine, sodium hydroxide, and other strong acids and bases may occur.

Anodized aluminum has shown excellent performance in limited applications at wastewater facilities. Anodized aluminum is recommended for use in areas such as electrical switchgear enclosures.

Aluminum is an electrically active metal (potential of -1.66 V versus standard hydrogen electrode); therefore, its use in water is limited. If coupled to any of the common engineering alloys (steel, iron, stainless steel, or copper and its alloys), aluminum will become the anode and will galvanically corrode. Aluminum must be electrically isolated from dissimilar materials.

Copper and brass are often used for potable water lines and fittings and in most electrical equipment.

Soil Exposure. Copper and brass typically perform quite well in underground applications where the pH is neutral to alkaline and the concentration of aggressive ions, such as chloride and sulfate, is low. Electrical isolation of copper from most materials commonly used in wastewater plant construction is required in buried service.

Copper is subject to changes in corrosion resistance with changes in temperature. Electrolytic corrosion can occur on hot and cold water lines buried in a common trench. To prevent corrosion, the two lines should be electrically isolated from each other. Electrical isolation can be accomplished with the use of insulating-type couplings. Copper or brass piping should not be used without a tape wrap coating and cathodic protection in environments where high chloride concentrations, high sulfate concentrations, and low pH values are present. Copper or brass should be electrically isolated from other structures if used in an aggressive environment. Connections of copper service lines to plastic mains should be accomplished using brass tapping saddles.

Fluid Exposure. Copper and brass may be subject to corrosion in wastewater environments. The presence of sulfides and ammonia compounds in the wastewater can lead to dissolution of cuprous compounds. If copper is coupled to a less noble metal, such as steel or aluminum, galvanic corrosion of the less noble metal may result. Because copper is a fairly soft material, it is also subject to erosion-corrosion, that is, accelerated by high fluid velocities, high temperatures, and abrasive particulate matter.

Copper and brass should not be used in any process flow stream that will allow exposure of the metal to solutions carrying residual chlorine (2 ppm or more). This is especially critical in reclaimed water systems, because chlorine can cause severe corrosion of copper and brass. Any process stream containing more than 0.1 ppm of hydrogen sulfide should be prevented from contacting copper or brass. Brasses containing over 15% Zn may suffer dezincification. Dezincification is especially prevalent in stagnant, acidic solutions.

Atmospheric Exposure. Copper and brass typically have excellent atmospheric corrosion resistance; however, they are actively attacked by hydrogen sulfide. Due to the probability of the presence of hydrogen sulfide in the atmosphere at a wastewater treatment plant, copper and brass should not be used without protective coatings. Because it may be impossible to eliminate the use of copper alloys, particularly in electrical systems such as switches and fuses, extreme care should be taken to prevent rapid deterioration of these items.

One means of reducing corrosion of copper and brass is to use conduits and sealed junction boxes (typically, National Electrical Manufacturer's Association type 4X). These boxes should be constructed of either stainless steel or a noncorroding, nonmetallic material (e.g., glass-reinforced plastic, polycarbonate, or polyester). Conduit entrances should be sealed tightly to the boxes, and conduits themselves should be internally sealed to prevent transmission of corrosive gases between boxes. Vapor corrosion inhibitors may also be required in certain applications. Additional protection may be provided by the use of spray- or dip-type corrosion-inhibiting coatings.

Stainless steel is available for piping, enclosures, and apparatuses for harsh environments.

Soil Exposure. Stainless steel is most often used in situations where protection from contamination of the material carried in the pipe is the prime concern. Generally, stainless steel should not be used in a buried exposure in wastewater treatment plants. Stainless steel pipe derives its corrosion resistance from a tough oxide coating on the surface of the pipe. Direct burial of the pipe, even in reasonably aerobic soils, presents the threat of differential oxygen corrosion cells. Therefore, if stainless steel is to be directly buried, it should be installed with a carefully placed uniform backfill.

Fluid Exposure. Stainless steel is typically resistant to corrosion in flowing waters. The austenitic grades (300 series) of stainless steel typically have the best performance. Type 316L (UNS S31603) alloy has provided the best overall performance in most wastewater treatment plant exposures. Pitting of stainless steel may occur in stagnant or slow-moving water. Oxidizing metal salts, such as ferric chloride, may also attack stainless steel. Type 316L alloy is more resistant to chlorine, hypochlorous acid, and hypochlorite ions than other alloys that may be used in the process streams. Type 316L stainless alloy is also resistant to hydrogen sulfide and other organic materials likely to be found in wastewater environments.

Cathodic protection of stainless steel is an option for preventing pitting. Pitting has been encountered in many aqueous environments. By electrically coupling stainless steel to a large immersed structure made of steel, zinc, or other metal that is more anodic to stainless steel, pitting will be reduced or eliminated. Stainless steel will increase the magnitude of corrosion deterioration of the structure to which it is bonded. Electrical contact should be eliminated where the anode/cathode ratio is not favorable, that is, small anode to large cathode.

Atmospheric Exposure. Stainless steel has been used successfully in both outdoor and indoor applications. Of the various types of stainless steel, the austenitic grades (typically, 302, 304, and 316) generally have the best corrosion resistance. Type 316 alloy—although more expensive than the others—is the most resistant to pitting. Overall, stainless steel has demonstrated excellent corrosion resistance in severe atmospheric environments.

Fig. 3 Corroded roof vent

Coatings used for wastewater facilities must be resistant to moisture, atmospheric sulfides, sunlight, and atmospheric chlorides. Several generic coating types are available (Ref 8, 9). Generic coating types include alkyds, coal tar epoxies, zinc-rich coatings, polyester resins, polyurethanes, and urethanes. All coatings must meet local, state, and federal air quality regulations. The manufacturer's recommendations for surface preparation should be strictly followed to ensure a quality coating system.

Soil Exposure. Coatings for service in soil must be abrasion resistant, must be suitable for long-term exposure to moisture, and must have excellent dielectric properties. Coal tar epoxy and tape wrap are good coatings for ductile iron, steel, and other metallic materials. These coatings may be applied in the field; however, they are typically shop-applied. Cathodic protection may also be used in conjunction with coatings on steel and ductile iron pipe. Cathodic protection may provide corrosion protection to the pipeline at areas of coating damage and coating imperfections. The use of a coating will reduce the current requirements of the cathodic protection system.

Fluid Exposure. Coating of structures is an effective means of protecting the structure from corrosive environments typically found in wastewater facilities. Coating of concrete structures in contact with the process flow can prevent corrosion caused by sulfuric acid in the splash zone. Coating of metal structures can reduce or eliminate corrosion due to galvanic coupling and exposure to corrosive chemicals commonly found in wastewater facilities.

For severe chlorine service, as in a chlorination apparatus, elastomeric urethanes or high-solids epoxy coatings on steel provide good service. These coatings are available in a number of colors and are resistant to chlorine solutions. Urethane coatings form a glossy, tough, attractive surface with excellent abrasion resistance and high resistance to water and effluent chemicals.

High-performance polymorphic resin coatings are also available. Polymorphic resin coatings are based on a polyester resin and cure extremely quickly. They are resistant to chemicals and are suitable for application on concrete.

Atmospheric Exposure. Coatings for atmospheric service should be specified to provide corrosion control and easy maintenance. Coatings for interior piping, equipment, and metalwork should be high-solids epoxies. Coatings for exterior service should be aliphatic polyurethanes to provide long-term corrosion protection and color retention. As illustrated in Fig. 3, corrosion due to atmospheric exposures at roof vents is a common occurrence on water and recycled water storage tanks. Each of these coatings systems has excellent resistance to chemical and process spillage.

REFERENCES

1. G. Tchobanoglous, *Wastewater Engineering Treatment/Disposal/Reuse,* 2nd ed., Metcalf & Eddy, 1979
2. A.W. Peabody, *Peabody's Control of Pipeline Corrosion,* 2nd ed., NACE International, 2001
3. W. von Baeckmann, W. Schwenk, and W. Prinz, *Handbook of Cathodic Corrosion—Protection Theory and Practice of Electrochemical Protection Processes,* 3rd ed., Gulf Publishing, 1997
4. "Control of External Corrosion on Underground or Submerged Metallic Piping Systems," RP01-69, NACE International, 2002
5. *Standard Methods for the Examination of Water and Wastewater,* 20th ed., American Public Health Association, New York, 1999
6. W.H. Ailor, *Atmospheric Corrosion,* John Wiley & Sons, 1982
7. *Sulfide in Wastewater Collection and Treatment Systems,* Manual 69, American Society of Civil Engineers, 1989
8. C.G. Munger, *Corrosion Prevention by Protective Coatings,* NACE International, 1997
9. L.M. Smith, *Generic Coating Types—An Introduction to Industrial Maintenance Coating Materials,* 95-08, SSPC: The Society for Protective Coatings, 1996

SELECTED REFERENCES

- J. Morgan, *Cathodic Protection,* 2nd ed., NACE International, 1987
- R. Nixon and R. Drisko, *The Fundamentals of Cleaning and Coating Concrete,* SSPC: The Society for Protective Coatings, 2001
- D. Peckner and I.M. Bernstein, *Handbook of Stainless Steels,* McGraw-Hill, 1977

Corrosion in Seawater

Stephen C. Dexter, University of Delaware

ALTHOUGH SEAWATER is generally considered to be a corrosive environment, it is not widely understood just how corrosive it is in comparison to natural freshwaters. Figure 1 shows the corrosion rate of iron in aqueous sodium chloride (NaCl) solutions of various concentrations. The maximum corrosion rate occurs near 3.5% NaCl—the approximate salt content of average full-strength seawater (Ref 1).

The general marine environment includes a great diversity of subenvironments, such as full-strength open ocean water, coastal seawater, brackish and estuarine waters, bottom sediments, and marine atmospheres. Exposure of structural materials to these environments can be continuous or intermittent, depending on the application. Structures in shallow coastal or estuarine waters are often exposed simultaneously to five zones of corrosion. Beginning with the marine atmosphere, the structure then passes down through the splash, tidal, continuously submerged (or subtidal), and subsoil (or mud) zones. The relative corrosion rates often experienced on a steel structure passing through all of these zones are illustrated in Fig. 2.

The major chemical constituents of seawater are consistent worldwide. The minor constituents, however, vary from site to site and with season, storms, and tidal cycles. These minor constituents include dissolved trace elements and dissolved gases. In addition, seawater contains dissolved organic materials and living microscopic organisms. Frequently, the minor chemical constituents of seawater, together with the organic materials and living organisms, are the rate-controlling factor in the corrosion of structural metals and alloys.

Because of its variability, seawater is not easily simulated in the laboratory for corrosion-testing purposes. Stored seawater is notorious for exhibiting behavior as a corrosive medium that is different from that of the water mass from which it was taken. This is due in part to the fact that the minor constituents, including the living organisms and their dissolved organic nutrients, are in delicate balance in the natural environment. This balance begins to change as soon as a seawater sample is isolated from the parent water mass, and these changes often have a large effect on the types of corrosion experienced and the corrosion rate.

Variations in the chemistry of open ocean seawater tend to take place slowly (over time periods of 3 to 6 months) and over horizontal and vertical distances that are large in comparison to the dimensions of most marine structures. Such gradual changes may produce an equally gradual change in the corrosion rate of structural materials with season and location. However, they are unlikely to produce sharp changes in either corrosion mechanism or rate. Such gradual changes are relatively easy to measure and monitor.

On the other hand, changes that take place over periods of hours to days and over distances of centimeters to meters can occur as the result of point inputs of various chemical pollutants or the attachment of micro- and macroscopic marine plants and animals to the surface of a structure. The chemical changes produced by the attachment of biological fouling organisms take place directly at the metal/water interface where the corrosion occurs, not in the bulk water. This means that the chemical environment in which the corrosion reactions occur in the presence of a micro- or macrofouling film may bear little resemblance to that of the bulk water.

It is these types of effects, which can be produced quickly and can lead to sharp chemical gradients over short distances, that often result in the onset of localized corrosion. Crevice Corrosion, beginning under the base of an isolated barnacle on stainless steel, is an example of this type of influence. Whether the fouling film is composed of microscopic bacteria or large sedentary fouling organisms is often less important than whether the film provides complete or spotty coverage of the metal surface. Almost invariably, a spotty film, or one that forms in discrete colonies of organisms with bare metal in between, will be more likely to induce structurally significant corrosion than a film that produces a continuous and homogeneous layer.

The general properties of ocean water and their effects on corrosion are discussed in the next section. The major and minor features, including the effects of variability, pollutants, and fouling organisms, are covered.

Consistency and the Major Ions

The concentrations of the major constituents of full-strength seawater are shown in Table 1. Major constituents are considered to be those that have concentrations greater than approximately 0.001 g/kg of seawater and are not greatly affected by biological processes. The behavior of these major ions and molecules is said to be conservative, because their concentrations bear a relatively constant ratio to

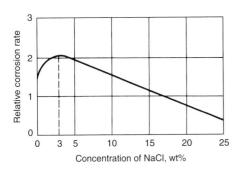

Fig. 1 Effect of NaCl concentration on the corrosion rate of iron in aerated room-temperature solutions. Data are complied from several investigations. Source: Ref 1

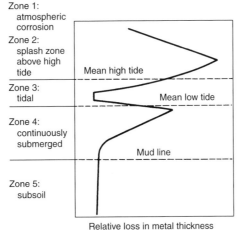

Fig. 2 Zones of corrosion for steel piling in seawater, and relative loss of metal thickness in each zone. Source: Ref 2

each other over a wide range of dilutions. Although most of the known elements can be found dissolved in seawater, the ions and molecules listed in Table 1 account for over 99% of the total dissolved solids. Moreover, the conservative nature of these major ions means that all of their concentrations can be calculated if the concentration of any one of them, or the total salt content (salinity) of the water, is measured.

Salinity and Chlorinity. The most commonly measured property of seawater is its salinity. Salinity, S, in parts per thousand (ppt or ‰), historically has been defined as the total weight in grams of inorganic salts in 1 kg of seawater when all bromides and iodides are replaced by an equivalent quantity of chlorides and all carbonates are replaced by an equivalent quantity of oxides. Salinity is usually determined by measuring either the chlorinity or the electrical conductivity of the seawater. Chlorinity, Cl, is defined as the mass in grams of silver required to precipitate the halogens in 0.3285234 kg of seawater. This is nearly equal to the mass of chloride in the seawater sample. Chlorinity is related to salinity by:

$$S = 1.80655 \, Cl \quad \text{(Eq 1)}$$

where S and Cl are measured in parts per thousand.

If pure water is the only substance added to or removed from seawater, the concentration of any ion, χ, from Table 1, at a salinity other than 35‰ can be calculated from the following relationship:

$$[\chi] \text{ at salinity } S = [\chi] \text{ at 35‰ salinity} \times S/35‰ \quad \text{(Eq 2)}$$

where the brackets denote concentration, and S is again given in parts per thousand. For example, using Eq 2 and Table 1, the concentration of sodium ion in seawater of 20‰ salinity is:

$$[Na^+] \text{ at 20‰} = (0.468 \text{ mol/kg of seawater}) \times 20‰/35‰$$
$$= 0.2667 \text{ mol } Na^+/\text{kg of seawater}$$

Table 1 Concentrations of the most abundant ions and molecules in seawater of 35‰ salinity

Density of seawater: 1.023 g/cm^3 at 25 °C (75 °F)

Ion or molecule	Concentration	
	m mol/kg of seawater	g/kg of seawater
Na^+	468.5	10.77
K^+	10.21	0.399
Mg^{2+}	53.08	1.290
Ca^{2+}	10.28	0.412
Sr^{2+}	0.09	0.008
Cl^-	545.9	19.354
Br^-	0.84	0.067
F^-	0.07	0.0013
HCO_3^-	2.30	0.140
SO_4^{2-}	28.23	2.712
$B(OH)_3$	0.416	0.0257

Source: Ref 3

This relationship may be less accurate at very low salinities (<10‰ for Ca^{2+} and HCO_3^- and <5‰ for the other major ions). The relationship may also lose accuracy in grossly polluted seawater. In this case, the concentration of the ion of interest must be known in both seawater and in the solution being added.

The only processes that affect the concentrations of the major ions over seasonal and shorter time periods are evaporation, precipitation, and river discharge. Because water is the only substance added to or removed from seawater by the first two of these processes, they affect the absolute concentrations of each ion but have no affect on the concentration ratios.

Effect of River Discharge. In contrast, river discharge does add constituents other than pure water. Therefore, the conservative behavior of the major ions may not hold in the low-salinity regions near river outlets. To a first approximation, river water is a 0.4 millimolar (m mol/L) solution of calcium bicarbonate as shown in Table 2 (Ref 4). At a salinity of 10‰, the calcium concentration calculated using Eq 2 and Table 1, but ignoring the river input, will be 10% too low. The error for bicarbonate ion will be even larger. Errors in concentrations of the other major ions, calculated by ignoring the river input, however, will be much less because of their relatively high concentrations in seawater compared to those in river water.

Salinity Variations. The total salt content of open ocean seawater varies from 32 to 36‰. It can rise above that range in the tropics or in enclosed waterways, such as the Red Sea, where evaporation exceeds freshwater input. It will be lower than that range in estuaries and bays, where there is appreciable dilution from river input. Salinity variations in the surface waters of the Pacific Ocean are shown in Fig. 3 (Ref 5). The data for these figures, as well as the surface water data for other variables to follow, were taken from the *Russian Atlas of the Pacific Ocean* (Ref 6). Many additional sources of information on the chemical properties of seawater are available. See, for example, the recent book by Pilson and the references therein (Ref 7).

Table 2 Concentrations of the most abundant ions and molecules in average river water

Ion or molecule	Concentration	
	m moles/L of river water	mg/L of river water
Na^+	0.274	6.3
K^+	0.059	2.3
Mg^{2+}	0.171	4.1
Ca^{2+}	0.375	15.0
Cl^-	0.220	7.8
HCO_3^-	0.958	58.4
SO_4^{2-}	0.117	11.2
NO_3^-	0.016	1.0
Fe^{2+}	0.012	0.67
$Si(OH)_4$	0.218	20.9

Source: Ref 4

Salinity variations with depth at given locations in the Atlantic and Pacific Oceans are shown in Fig. 4. The locations from which these and other data on variations with depth were taken are illustrated in Fig. 5. It should be noted that the open ocean salinity variations with horizontal location and depth are quite small.

Effect of Salinity on Corrosion. The main effects of salinity on corrosion result from its influence on the conductivity of the water and from the influence of chloride ions on the breakdown of passive films. Specific conductance varies with temperature and chlorinity, as indicated in Table 3 (Ref 8). The high conductivity of seawater means that the resistance of the electrolyte plays a minor role in determining the rate of corrosion reactions and that surface area relations play a major role.

Two examples will serve to illustrate this point. First, galvanic corrosion in freshwater systems tends to be localized near the two-metal junction by the high resistivity of the electrolyte. In seawater, however, anodes and cathodes that are tens of meters apart can operate; therefore, the galvanic corrosion is much more spread out and is less intense at the junction. In the second example, a large area of cathodic metal, such as stainless steel, will produce more severe galvanic attack on an anodic metal in seawater than in freshwater, because high conductivity allows the entire area of stainless steel to participate in the reaction. Similarly, pitting corrosion tends to be more intense in seawater, because large areas of boldly exposed cathode surface are available to support the relatively small anodic areas at which pitting takes place.

The second effect of salinity on corrosion in seawater is related to the role of chloride ions in the breakdown of passivity on active-passive metals such as stainless steels and aluminum alloys. The higher the salinity of the water, the more rapidly chloride ions succeed in penetrating the passive film and initiating pitting and crevice corrosion at localized sites on the metal surface.

The open ocean salinity changes shown in Fig. 3 and 4 have very little effect on the processes of galvanic, pitting, and crevice corrosion. The much larger salinity changes found in coastal and brackish waters can have a substantial effect on both the susceptibility to and the intensity of localized corrosion. These coastal salinity changes have undoubtedly contributed to the variability in reported pitting and crevice corrosion rates with season and location or with time at a given location. For alloys that corrode uniformly, variations in corrosion rate due to salinity changes are small compared to those caused by changes in oxygen concentration and temperature.

Temperature. When all other factors are held constant, an increase in temperature increases the corrosivity of seawater. If the dissolved oxygen concentration is held constant, the corrosion rate of low-carbon steel in seawater will approximately double for each 30 °C (55 °F) increase in temperature.

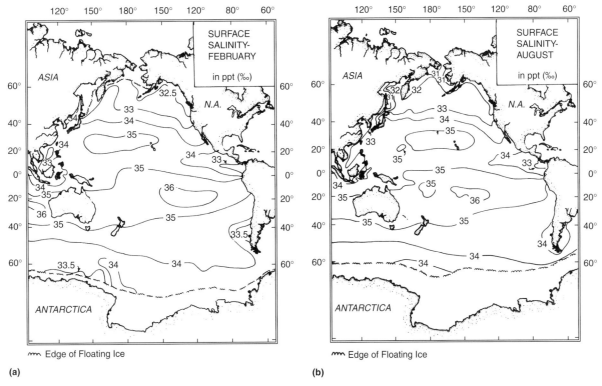

Fig. 3 Pacific Ocean surface salinity (‰) for (a) February and (b) August. Source: Ref 5

Open ocean temperature variations are shown for surface waters in Fig. 6 and, as a function of depth, in Fig. 7. From these data alone, one would expect corrosion rates in tropical surface waters to be approximately twice those in the polar regions or in deep water. Corrosion rates are usually higher in warm surface waters than in cold deep waters, as illustrated in Fig. 8 and 9 (Ref 9, 10), but the picture is not nearly as simple as temperature alone would indicate. For surface waters, the saturation level of dissolved oxygen increases as the temperature decreases, and the effects of dissolved oxygen on the corrosion rate are often stronger than those of temperature. Short-term local temperature fluctuations and the effects of biofouling and scaling films must also be considered.

The historical database for seawater immersion corrosion of carbon steels and low-alloy steels taken at a variety of sites has recently been analyzed, and a model for general corrosion has been proposed (Ref 11). Results from the modeling effort have revealed (Ref 11, 12) that for fully aerated seawater, the general trend is for increasing corrosion rates with average water temperature, as shown in Fig. 10. There is some indication that the temperature at the time of initial immersion may be more important than subsequent seasonal temperature fluctuations (Ref 11, 12).

Data for the temperature and salinity variations with depth at the four coastal locations in Fig. 5 are shown in Fig. 11 to 14. The temperature and salinity profiles for Cook Inlet (station 1) and for the Oregon coast (station 3) given in Fig. 11(a,b) and 12(a,b), respectively, show only small differences with depth and season. No reliable data are available for the

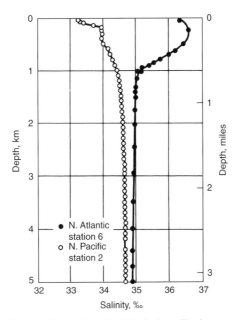

Fig. 4 Comparison of salinity-depth profiles for open ocean sites 2 and 6 (see Fig. 5 for site locations). Source: Ref 5

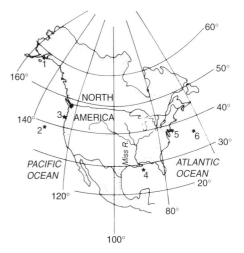

Station number	Location	Chart coordinates
1	Cook Inlet	58° 57′ N 152° 59′ W
2	Northeast Pacific	38° 21′ N 133° 38′ W
3	West coast, U.S.	44° 39′ N 124° 39′ W
4	Gulf of Mexico	28° 12′ N 88° 26′ W
5	East coast, U.S.	39° 20′ N 73° 40′ W
6	Northwest Atlantic	36° 44′ N 64° 28′ W

Fig. 5 Station positions for Fig. 4, 7, 11–14, 17, and 29. Source: Ref 5

winter months at the Cook Inlet station. The differences shown here are inconsequential with regard to corrosion.

Larger differences are shown for the Gulf of Mexico (station 4) and the New Jersey coast (station 5) in Fig. 13 and 14. The temperature differences with depth and season shown for these two locations are large enough to influence both corrosion rate and calcareous deposition. The salinity changes shown will have little influence. The seasonal differences in the surface waters of the Gulf of Mexico disappear at greater depths, while those off the New Jersey coast persist all the way to the bottom.

Variability of the Minor Ions

This section considers how the variability of the minor constituents of seawater affects corrosion. Variabilities in seawater properties with horizontal location and depth are presented for the dissolved gases, oxygen and carbon dioxide, and for various pH values. Each of these properties has a range over which it typically varies in the marine environment. The effect on corrosion of each property as it changes within this range is considered.

Dissolved Oxygen. Many of the minor constituents that are important to corrosion processes are dissolved gases such as carbon dioxide and oxygen. Their concentrations are not conservative (that is, constant), because they are influenced by air-sea exchange as well as by biochemical processes. The concentration of dissolved oxygen in surface waters is usually within a few percent of the equilibrium saturation value with atmospheric oxygen at a given temperature. The solubility of oxygen in seawater varies inversely with both temperature and salinity, but the effect of temperature is greater. If the absolute temperature T (°K) and salinity S (‰) are known, the solubility of oxygen can be calculated from the relationship:

$$\ln [O_2] = A_1 + A_2\left(\frac{100}{T}\right) + A_3 \ln\left(\frac{T}{100}\right) + A_4\left(\frac{T}{100}\right) + S\left[B_1 + B_2\left(\frac{T}{100}\right) + B_3\left(\frac{T}{100}\right)^2\right] \quad \text{(Eq 3)}$$

Table 3 Specific conductance of seawater as a function of temperature and chlorinity
Conductivity: S/m

Chlorinity, ‰	Temperature, °C (°F)					
	0 (32)	5 (40)	10 (50)	15 (60)	20 (70)	25 (75)
1	0.1839	0.2134	0.2439	0.2763	0.3091	0.3431
2	0.3556	0.4125	0.4714	0.5338	0.5971	0.6628
3	0.5187	0.6016	0.6872	0.7778	0.8702	0.9658
4	0.6758	0.7845	0.8958	1.0133	1.1337	1.2583
5	0.8327	0.9653	1.1019	1.2459	1.3939	1.5471
6	0.9878	1.1444	1.3063	1.4758	1.6512	1.8324
7	1.1404	1.3203	1.5069	1.7015	1.9035	2.1121
8	1.2905	1.4934	1.7042	1.9235	2.1514	2.3868
9	1.4388	1.6641	1.8986	2.1423	2.3957	2.6573
10	1.5852	1.8329	2.0906	2.3584	2.6367	2.9242
11	1.7304	2.0000	2.2804	2.5722	2.8749	3.1879
12	1.8741	2.1655	2.4684	2.7841	3.1109	3.4489
13	2.0167	2.3297	2.6548	2.9940	3.3447	3.7075
14	2.1585	2.4929	2.8397	3.2024	3.5765	3.9638
15	2.2993	2.6548	3.0231	3.4090	3.8065	4.2180
16	2.4393	2.8156	3.2050	3.6138	4.0345	4.4701
17	2.5783	2.9753	3.3855	3.8168	4.2606	4.7201
18	2.7162	3.1336	3.5644	4.0176	4.4844	4.9677
19	2.8530	3.2903	3.7415	4.2158	4.7058	5.2127
20	2.9885	3.4454	3.9167	4.4114	4.9248	5.4551
21	3.1227	3.5989	4.0900	4.6044	5.1414	5.6949
22	3.2556	3.7508	4.2614	4.7948	5.3556	5.9321

Source: Adapted from Ref 8

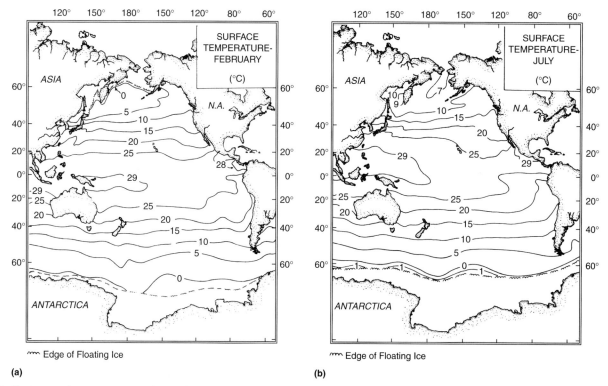

Fig. 6 Pacific Ocean surface temperature (°C) for (a) February and (b) July. Source: Ref 5

where oxygen concentration is given in milliliters per liter (mL/L), and salinity S is in parts per thousand (‰). The constants A_1 through B_3 are given in Table 4 (Ref 7, 13). Table 5 lists the equilibrium oxygen saturation levels in milliliters per liter as a function of temperature and salinity calculated from Eq 3 and Table 4.

Generally, the surface waters of the ocean are in equilibrium with the oxygen in the atmosphere at a specific temperature. Two sets of conditions, however, can lead to these waters becoming substantially supersaturated with oxygen. The first of these conditions is oxygen production due to photosynthesis by microscopic marine plants.

During high growth periods, intense photosynthesis can produce concentrations as high as 200% saturation for periods of up to a few weeks. Such oxygen levels are most often found in nearshore regions as a transient phenomenon.

The second condition that may cause oxygen supersaturation is the entrainment of air bubbles due to wave action. This factor usually will not cause greater than approximately 10% supersaturation, because vigorous wave action also promotes re-equilibration with the atmosphere.

Oxygen Variability. The distribution of dissolved oxygen in the surface waters of the Pacific Ocean is shown in Fig. 15(a) for the months of January through March and in Fig. 15(b) for July through September. Comparison with Fig. 6 reveals that the highest concentrations of oxygen coincide with the lowest temperatures; this agrees with the oxygen solubility data given in Table 5.

Surface waters are either saturated or supersaturated with oxygen at atmospheric conditions. In contrast, deep waters tend to be isolated from the atmosphere above them. Waters currently in the deep ocean are thought to have formed thousands of years ago in cold, polar regions. At the time of formation, these waters were cold, dense, and rich in oxygen. Due to their high density, they sank down and spread out in horizontal layers throughout the deep portions of the world ocean. Thus, the oxygen concentrations of deep ocean waters today have no direct relation to oxygen in the atmosphere above them. Deep waters are usually high in dissolved oxygen according to the location of their origin. Waters at intermediate depths are often undersaturated due to consumption of oxygen during the biochemical oxidation of organic matter. Figure 16 shows horizontal maps of dissolved oxygen concentration in the Pacific Ocean at depths of 500 and 1000 m (1640 and 3280 ft). The level of oxygen is generally lower at these depths, especially in the northeastern Pacific.

This decrease in oxygen concentration with depth is shown more clearly in Fig. 17. The oxygen profiles for the open Atlantic and Pacific stations both go through a minimum at intermediate depths and increase again at great depths. In the Atlantic Ocean, the surface oxygen concentrations are usually lower, and the oxygen minimum is not as intense as in the Pacific. In addition, the oxygen concentrations in the deep Atlantic are higher than those in the deep Pacific, and they can be even higher than those in the Atlantic surface waters for reasons given previously.

Figure 18 shows the depth of the dissolved oxygen minimum in the Pacific (solid contours) and the concentration of oxygen at that depth (dotted contours). The depth of the oxygen minimum ranges from 400 m (1310 ft) in the equatorial eastern Pacific to over 2400 m (7875 ft) in the central south Pacific. The concentration of oxygen at the depth of the minimum ranges from 0.01 to 0.40 mg·atm/L (1 mg·atm/L = 12.2 mL/L = 16 ppm at 25 °C, or 75 °F). Oxygen profiles with depth for the four coastal stations were shown in Fig. 11 to 14.

Effect of Oxygen on Corrosion. The corrosion rate of active metals (for example, iron and steel) in aerated electrolytes such as seawater at constant temperature is a direct linear function of the dissolved oxygen concentration, as shown in Fig. 19. When oxygen and temperature vary

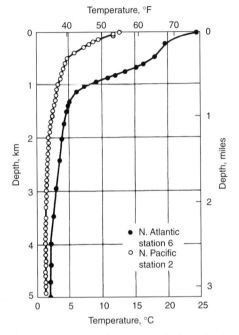

Fig. 7 Comparison of temperature-depth profiles for open ocean sites 2 and 6 (see Fig. 5 for site locations). Source: Ref 5

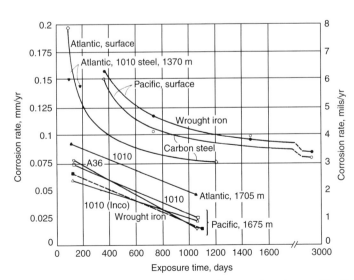

Fig. 8 Corrosion rates of carbon steels and wrought iron in the Atlantic and Pacific Oceans at various depths. Source: Ref 9, 10

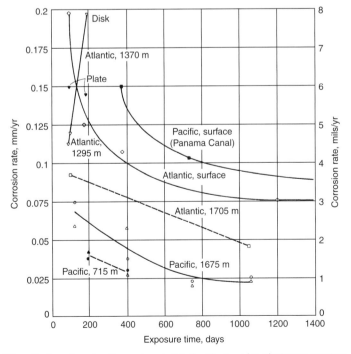

Fig. 9 Corrosion rates of low-carbon steels in the Atlantic and Pacific Oceans at various depths. Source: Ref 9, 10

together, as they do in the marine environment, the oxygen effect tends to predominate. This trend is illustrated by data for the corrosion rate of steel at various depths in the Pacific Ocean in Fig. 20. The corrosion rate decreases with dissolved oxygen down to the oxygen minimum, then increases again with oxygen at greater depths, despite a continuing decrease in temperature. The corrosion rates of nickel and nickel-copper alloys are somewhat less affected by oxygen concentration, as shown in Fig. 21. The effect of oxygen on copper alloys depends on the flow velocity. Figure 22 shows that there is very little effect of oxygen on copper alloys exposed in quiet open ocean water. At a flow velocity of 1.8 m/s (6 ft/s), however, increasing oxygen has a marked accelerating effect (Fig. 23) on the corrosion rate.

In contrast, the effect of oxygen on the corrosion rates of active-passive metals, such as aluminum alloys, stainless steels, and other corrosion-resistant alloys, can be quite variable (Fig. 24). For such alloy systems, high oxygen concentrations tend to promote healing of the passive film and thus retard initiation of pitting corrosion. On the other hand, high oxygen favors a vigorous cathodic reaction and tends to increase the rate of pit and crevice propagation after initiation.

For all alloy systems, the conditions most conducive to corrosion are those in which differences in dissolved oxygen are allowed to develop between two regions of the wetted metal surface. This can lead to an oxygen concentration cell, with potential differences as large as 0.5 V. The portion of metal surface on which the oxygen concentration is lowest becomes the anode and is subject to localized corrosion. Differences in dissolved oxygen concentration of this type are unlikely to occur over short distances within the water itself. Instead, they are usually caused by localized deposits or structural design factors that create oxygen-shielded regions on the metal surface.

These effects also can lead to pitting corrosion of active metals such as carbon and low-alloy steels. The average uniform corrosion rates of carbon and low-alloy steels in a wide variety of marine environments are found to range from 0.050 to 0.125 mm/yr (2 to 5 mils/yr), slowly decreasing with time of exposure. Data from the Panama Canal zone showed that, although the average penetration rate was 0.068 mm/yr (2.7 mils/yr), the penetration by pitting was some five to eight times higher (Fig. 25).

Differences in dissolved oxygen from point to point along the metal surface caused by spotty biofouling films can contribute to the pitting rate.

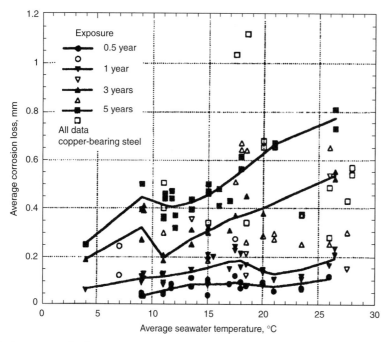

Fig. 10 Average corrosion loss versus average seawater temperature for copper-bearing steels, showing data points and trend lines. Solid points are considered by author of Ref 11 to have a strong correlation to aerated open sea conditions. Open points have a weaker correlation. Source: Ref 11. (Copyright NACE International 2002, used with permission)

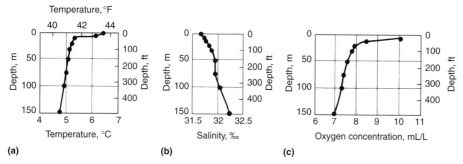

Fig. 11 Variation of (a) temperature, (b) salinity, and (c) dissolved oxygen concentration with depth at Cook Inlet (station 1, Fig. 5) for May 1968. Source: Ref 5

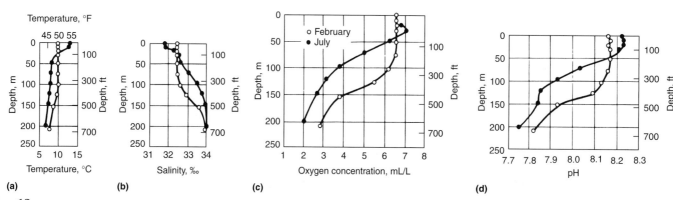

Fig. 12 Variation of (a) temperature, (b) salinity, (c) dissolved oxygen, and (d) pH with depth and season off the Oregon coast (station 3, Fig. 5). Source: Ref 5

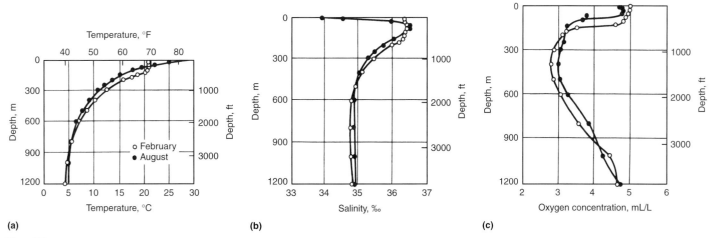

Fig. 13 Variation of (a) temperature, (b) salinity, and (c) oxygen concentration with depth and season in the Gulf of Mexico (station 4, Fig. 5). Source: Ref 5

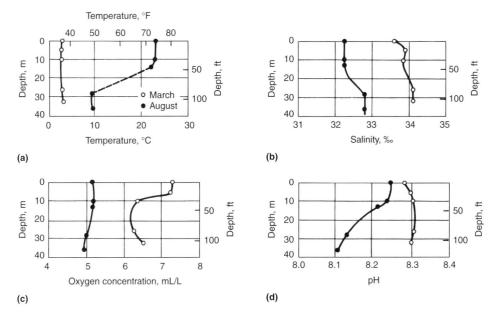

Fig. 14 Variation of (a) temperature, (b) salinity, (c) dissolved oxygen, and (d) pH with depth and season off the New Jersey and Delaware coasts (station 5, Fig. 5). Source: Ref 5

Table 4 Constants for use with Eq 3

These values can be used with Eq 3 to calculate oxygen concentration relative to air at 1 atm total pressure and 100% relative humidity

Constant	Value
A_1	−173.4292
A_2	249.6339
A_3	143.3483
A_4	−21.8492
B_1	−0.033096
B_2	0.014259
B_3	−0.0017000

Source: Ref 13

Table 5 Solubility of oxygen in seawater as a function of temperature and salinity

Solubility values were calculated using Eq 3

Temperature		Oxygen solubility (mL/L) at indicated salinity (‰)					
°C	°F	0	8	16	24	31	36
0	32	10.22	9.70	9.19	8.70	8.27	7.99
5	41	8.93	8.49	8.05	7.64	7.28	7.04
10	50	7.89	7.52	7.14	6.79	6.48	6.28
15	60	7.05	6.72	6.40	6.10	5.83	5.65
20	70	6.35	6.07	5.79	5.52	5.29	5.14
25	75	5.77	5.52	5.27	5.04	4.84	4.70
30	85	5.28	5.06	4.84	4.63	4.45	4.33

Source: Ref 5

In contrast to the effects of a spotty film, complete coverage of the surface by hard-shelled, sedentary fouling organisms can lead to a marked decrease in the overall corrosion rate by acting as a diffusion barrier against dissolved oxygen reaching the metal surface. In the case of aluminum and stainless alloys, point-to-point differences in oxygen concentration can lead to both pitting and crevice corrosion.

Dissolved Carbon Dioxide and pH. The concentration of carbon dioxide is less affected by air-sea interchange than the concentration of dissolved oxygen, because the carbon dioxide system in seawater is buffered by the presence of bicarbonate and carbonate ions. Carbon dioxide is a weak acid and undergoes two ionizations in aqueous solutions (Ref 7):

$$CO_2 + H_2O = H^+ + HCO_3^- \quad \text{First ionization}$$
(Eq 4)

$$HCO_3^- = H^+ + CO_3^{2-} \quad \text{Second ionization}$$
(Eq 5)

Surface seawater usually has a pH value greater than 8 because of the combined effects of air-sea exchange and photosynthesis. At this pH, 93% of the total inorganic carbon is present as HCO_3^-, 6% as CO_3^{2-}, and 1% as CO_2. Bicarbonate ion accounts for at least 85% of the total inorganic carbon under all naturally-occurring conditions. However, the relative concentrations of CO_2 and CO_3^{2-} vary greatly depending on pH. The CO_3^{2-} concentration is relatively high in surface waters, and surface waters are nearly always supersaturated with respect to the calcium carbonate phases calcite and aragonite. This supersaturation favors deposition of calcareous scales on metal surfaces undergoing cathodic protection, as is discussed later.

Relationship Among CO_2, Oxygen, and pH. The concentrations of carbon dioxide and oxygen are closely coupled and related to the pH of seawater through the processes of photosynthesis and biochemical oxidation, as represented in the following general reaction (Ref 7):

$$CH_2O + O_2 \underset{\text{Biochemical oxidation (respiration)}}{\overset{\text{Photosynthesis}}{\longleftrightarrow}} CO_2 + H_2O \quad \text{(Eq 6)}$$

where CH_2O represents a typical carbohydrate molecule. During decomposition of organic material in seawater, Eq 6 proceeds from left to right, dissolved oxygen is consumed, and CO_2 is produced. Production of CO_2, in turn, makes the water more acidic (that is, lower pH) and decreases the saturation state with respect to carbonates.

Variability of pH. The pH of the Pacific surface waters ranges from 8.1 to 8.3, and its general distribution for the months of January to March is shown in Fig. 26. Distributions of pH at depths

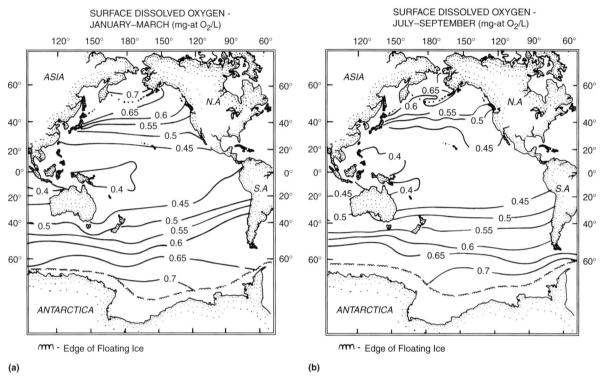

Fig. 15 Pacific Ocean surface dissolved oxygen (mg·atm/L) for (a) January through March and (b) July through September. (1 mg·atm/L = 12.2 mL/L). Source: Ref 5

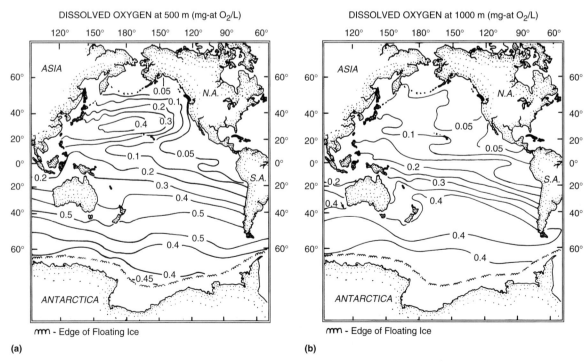

Fig. 16 Pacific Ocean dissolved oxygen (mg·atm/L) at depths of (a) 500 m (1640 ft) and (b) 1000 m (3280 ft). (1 mg·atm/L = 12.2 mL/L). Source: Ref 5

of 500 and 1000 m (1640 and 3280 ft) are shown in Fig. 27 and 28, respectively. In comparing these to the dissolved oxygen distributions at the same depths (Fig. 15 and 16), it should be noted that the trends for the two variables are similar. For example, at a depth of 500 m (1640 ft), the region of maximum dissolved oxygen, centered on 180° longitude between 20 and 40° north latitude in Fig. 16(a), is reproduced closely for pH in Fig. 27.

Profiles of pH with depth for the two open ocean locations are shown in Fig. 29. A comparison of the corresponding pH and oxygen profiles from Fig. 17 and 29 reveals the closely coupled nature of their relationship through the carbon dioxide system. The oxygen and pH minima are reached at the same depth for a given location, as was predicted. The deep North Pacific water is from 0.15 to 0.40 pH units more acidic than that in the North Atlantic, primarily because of the increased oxidation of organic matter in the North Pacific.

Profiles of pH for the coastal waters off Oregon and New Jersey were shown in Fig. 12 and 14, respectively. The close correlation between the shapes of the oxygen and pH profiles in both winter and summer for the Oregon data in Fig. 12 is particularly striking. Upon close examination, the oxygen and pH profiles in Fig. 14 do not appear to be closely related in the manner seen earlier. In March, the water column is well mixed down to the bottom, and the changes with depth of all four variables are small. In August, however, the dissolved oxygen profile is nearly independent of depth, while the pH and temperature profiles show substantial changes. Based on salinity and temperature, the oxygen saturation levels during August are approximately 5.2 mL/L in the surface waters and 6.5 mL/L in the deep water. The oxygen profile for August shows that the surface waters are nearly saturated, while in the deep waters, biological activity has used up enough oxygen and produced enough CO_2 to decrease the pH—but not enough to produce a strong oxygen minimum. This indicates the danger inherent in assuming that a pH minimum will always correspond to a similar minimum in oxygen. The two profiles may not correspond closely in shape when the biological demand for oxygen is not sufficiently intense to produce a strong oxygen minimum or when there is a strong temperature gradient.

Effects of pH on Corrosion and Calcareous Deposition. The pH of open ocean seawater ranges from approximately 7.5 to 8.3. Changes within this range have no direct effect on the corrosion of most structural metals and alloys. The one exception to this general statement is the effect of pH on aluminum alloys. A decrease in pH from the surface water value of 8.2 to a deep water value of 7.5 to 7.7 causes a marked acceleration in the initiation of both pitting and crevice corrosion (Ref 14). This effect accounts for the reported increase in corrosion of aluminum alloys in the deep ocean (Ref 15).

Although variations in seawater pH have little direct effect on corrosion of alloys other than aluminum, they do have an indirect effect through their influence on calcareous deposition.

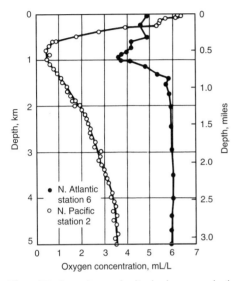

Fig. 17 Comparison of dissolved oxygen-depth profiles for open ocean stations 2 and 6 (see Fig. 5). Source: Ref 5

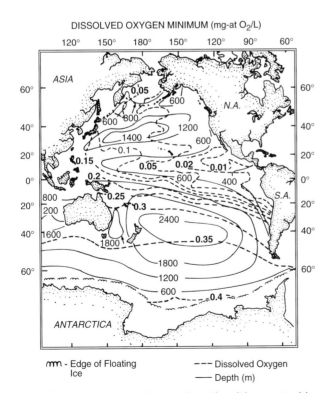

Fig. 18 Depth (meters) of the dissolved oxygen minimum in the Pacific (solid contours) and the value of the minimum in mg·atm/L (dashed contours). (1 mg·atm/L = 12.2 mL/L). Source: Ref 5

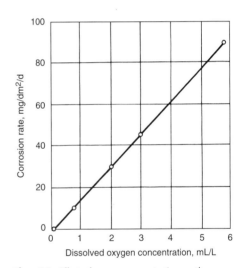

Fig. 19 Effect of oxygen concentration on the corrosion of low-carbon steel in slowly moving water containing 165 ppm $CaCl_2$. The 48 h test was conducted at 25 °C (75 °F). Source: Ref 1

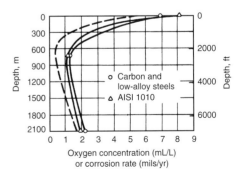

Fig. 20 Corrosion of steels versus depth after 1 year of exposure compared to the shape of the dissolved oxygen profile (dashed line). Source: Ref 2

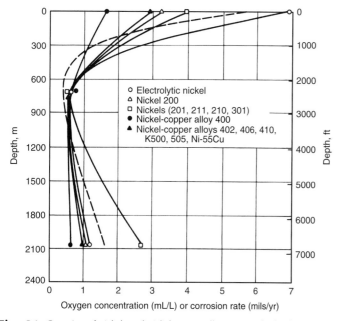

Fig. 21 Corrosion of nickels and nickel-copper alloys versus depth after 1 year of exposure compared to the shape of the dissolved oxygen profile (dashed line). Source: Ref 2

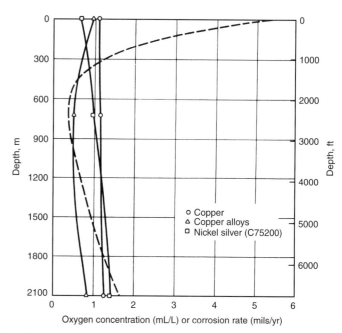

Fig. 22 Corrosion of copper alloys versus depth after 1 year of exposure compared to the shape of the dissolved oxygen profile (dashed line). Source: Ref 2

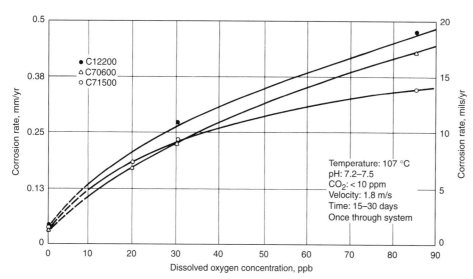

Fig. 23 Effect of dissolved oxygen in seawater on the corrosion rate of three Copper Development Association copper alloys. Source: Ref 9

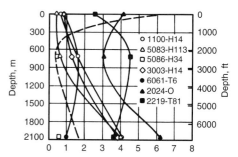

Fig. 24 Corrosion rates of aluminum alloys versus depth after 1 year of exposure compared to the shape of the dissolved oxygen profile (dashed line). Source: Ref 2

The surface waters of most of the world's oceans are 200 to 500% supersaturated with respect to the calcium carbonate species calcite and aragonite (Ref 7). Thus, precipitation of carbonate-type scales will occur readily on any solid surface where there is an elevated pH in the water at the interface.

Scale precipitation is most likely to occur in the elevated-pH regime adjacent to cathodically protected surfaces, where OH⁻ ions are produced by the cathodic reactions involving reduction of dissolved oxygen (Ref 16) and breakdown of water. The predominant calcareous specie precipitated in warm surface waters is aragonite and, at interface pH values above 9.3 (as experienced during cathodic protection) (Ref 16), brucite ($Mg(OH)_2$).

For many years, the marine cathodic protection industry has relied on the buildup of calcareous scales to make cathodic protection more economical. The scale deposit on cathodically protected steel is normally composed of an initial magnesium-rich inner layer, followed by a thicker outer layer of aragonite (Ref 17). The higher the pH at the water/metal interface, the more brucite is favored and the lower the calcium-magnesium ratio of the deposit will be (Ref 16, 17). A lower calcium-magnesium ratio, in turn, makes the scale less dense and less protective. Thus, a high level of cathodic protection applied in the early stages of immersion, as is sometimes done to accelerate scale buildup, can be counterproductive in terms of scale quality if it is maintained continuously at the same high level. It has been found that the most protective deposits are formed by the so-called rapid polarization approach, in which a high initial current is applied to encourage rapid surface coverage by the magnesium-rich phase, brucite, followed by a lower current to form the more protective aragonite (Ref 18, 19). This has led to the development of dual anodes composed of a thin outer layer of magnesium over an inner core of aluminum (Ref 19, 20).

In deep waters, where the temperature and pH are both lower than at the surface, calcareous deposits do not form spontaneously under ambient conditions, and it has often been difficult to form deposits even under cathodic protection conditions. This is partly because the deep waters—below 300 m (985 ft) in the Atlantic and 200 m (655 ft) in the North Pacific—are undersaturated in carbonates as a result of low pH and high pressure (Ref 7). At the low temperatures of the deep water, calcite is the predominant calcium carbonate phase. At first, this would seem to be beneficial, because calcite forms a dense, protective film. However, calcite formation is strongly inhibited by the free magnesium ions that are abundant in seawater (Ref 16). Recent tests on deep water deployments found that while brucite and aragonite deposits could be formed in deep water, they were less protective than those formed in surface waters, and the rapid polarization technique was not effective (Ref 21, 22).

Effect of Pollutants

The ratios of the major ions are not affected by pollution of the water as long as the salinity remains above 5 to 10‰. The relations between the major, conservative ions will hold constant, except perhaps in a confined waterway with poor tidal flushing in which a pollutant containing a large concentration of one of the major ions is introduced in quantities approaching that of the waterway itself.

In contrast, the concentrations of the minor constituents of seawater may be radically changed by pollution. This is an important fact, because it is usually the minor ions and dissolved gases that determine the corrosion rate. Concentrations of heavy metals; nutrients such as nitrates and phosphates; dissolved organics; and dissolved gases such as oxygen, carbon dioxide, and hydrogen sulfide are particularly sensitive to pollution.

Effects Related to Dissolved Oxygen. Pollutants containing organic material usually increase the utilization of dissolved oxygen in the water. As the organics become oxidized, oxygen concentrations fall, carbon dioxide concentrations rise, and the water becomes more acidic. If the pH does not fall below 4, these conditions often result in a decrease in the corrosivity of the water toward carbon and low-alloy steels. During the first half of this century, for example, the upper Delaware estuary in the Chester-Philadelphia, PA, area was sufficiently polluted that the yearly mean dissolved oxygen in the Delaware River was nearly zero. Consequently, the corrosion rates of industrial steel structures in that waterway were very low during that period. As political pressure directed toward cleaning up the river mounted during the 1950s and 1960s, the yearly mean dissolved oxygen began to recover. By the mid-1970s, the oxygen concentrations had increased enough that "lace-paper" conditions were being noted on sheet steel and H-pilings in the area. In contrast, the corrosion rate for steel structures in low-oxygen (or even anoxic) waters and sediments can increase if certain types of bacteria are active. For active-passive metals such as aluminum and stainless steels, which undergo localized corrosion at pits and crevices, a decrease in the bulk

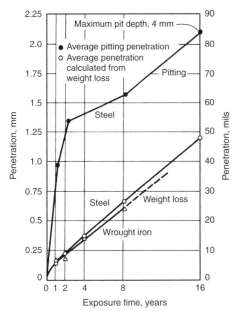

Fig. 25 Corrosion of carbon steel and wrought iron continuously immersed in seawater. Average penetration rate was 0.068 mm/yr (2.7 mils/yr) for steel; that of wrought iron was 0.061 mm/yr (2.4 mils/yr). Source: Ref 9

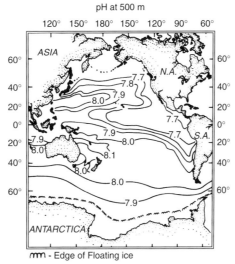

Fig. 27 Pacific Ocean pH at a depth of 500 m (1640 ft). Source: Ref 5

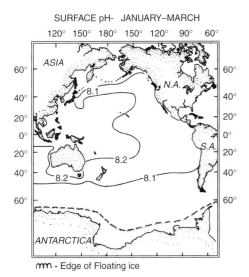

Fig. 26 Pacific Ocean surface pH for the period January to March. Source: Ref 5

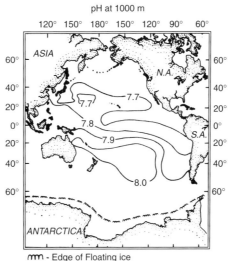

Fig. 28 Pacific Ocean pH at a depth of 1000 m (3280 ft). Source: Ref 5

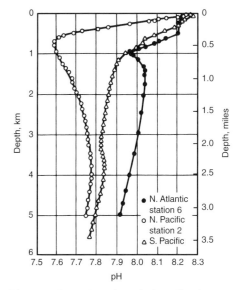

Fig. 29 Comparison of pH-depth profiles for open ocean sites 2 and 6 (see Fig. 5). Note that the data for the South Pacific are highest at the surface but are intermediate at depths greater than 500 m (1640 ft). Source: Ref 5

water oxygen concentration can produce either an increase or decrease in the corrosion rate. Localized oxygen and other chemical concentration cells along the metal surface can be more important for these alloys than the bulk water values.

Sulfides. Hydrogen sulfide and various sulfates are frequent components of organic pollutants. Sulfates themselves are not particularly detrimental except that they can be reduced to sulfides by the action of sulfate-reducing bacteria. The effects of these bacteria are considered later. Hydrogen sulfide may reach levels of 50 ppm or higher in severely polluted estuarine or harbor waters. Bottom muds in harbors and salt marshes rich in decomposing organic matter may also have high sulfide concentrations.

Penetration by pitting corrosion of low-carbon steel panels in the polluted seawater of the San Diego harbor was several times higher than the uniform penetration rates usually experienced (Ref 9) (Table 6). Similarly, the corrosion rate of several copper alloys used in condenser service was 3 to 10 times higher in polluted than in clean seawater (Ref 9) (Table 7), and as little as 4 ppm of hydrogen sulfide seriously increased the corrosion rate of copper alloys, as shown in Table 8 (Ref 9).

Sulfide films are well known to form on copper alloys in polluted waters. These films can be very harmful. Under most conditions, the sulfide film is itself cathodic to the bare copper alloy surface. This makes the film effective in accelerating pitting corrosion at any break in the film, and the effects on corrosion are known to persist long after the polluted water has been removed. For this reason, it is important to remove sulfide films from copper alloys, even when the source of pollution has been eliminated (Ref 23). The sulfide film will continue to accelerate pitting corrosion as long as it remains on the metal surface, even in clean water. It is usually recommended that the first exposure of copper alloys be in clean, rather than sulfide-polluted, seawater whenever possible. Experience has shown that if sulfide films are allowed to form before other corrosion product films on the copper alloys of a marine condenser or piping system, they can be very difficult to remove. Moreover, even after cleaning, traces of the sulfide film are likely to plague that system throughout its service life (Ref 23).

Heavy Metals. Nominally unpolluted seawater contains nearly every known element, most of them in very small concentrations. For example, the copper concentration in clean seawater is approximately 0.2 ppb (Ref 24). This does not normally cause corrosion problems for any of the common marine structural alloys (Ref 23). At elevated copper concentrations, however, aluminum alloys can suffer accelerated corrosion. The copper concentration in the water can be elevated by copper-containing pollutants, by leaching from copper-base antifouling paints, or by corrosion of copper alloys (Ref 23).

Acceleration of aluminum corrosion by copper corrosion products has often been observed in seawater piping systems having copper alloy pumps, even when the aluminum piping is not in direct electrical contact with the copper alloy (Ref 23). In freshwater, copper concentrations as low as 0.05 ppm have been found to accelerate aluminum corrosion. In seawater, the threshold concentration below which copper contamination has no effect seems to be approximately 0.03 ppm, as shown in Fig. 30 (Ref 24). Copper accumulates on the aluminum surface by electrochemical deposition and provides an efficient cathode; this depolarizes the aluminum and can lead to the initiation of pitting corrosion (Ref 23, 24).

A similar effect has sometimes been observed for iron and steel corrosion products generated upstream of aluminum components in desalination plants. However, the effect of iron contamination is not as strong or as consistent as that of copper.

Influence of Biological Organisms

Seawater is a biologically active medium that contains a large number of microscopic and macroscopic organisms. Many of these organisms are commonly observed in association with solid surfaces in seawater, where they form biofouling films. Because the influence of both micro- and macrofouling organisms on corrosion has been dealt with in detail in the article "Microbiologically Influenced Corrosion" in *ASM Handbook*, Volume 13A, 2003, only a brief description is given here. Immersion of any solid surface in seawater initiates a continuous and dynamic process, beginning with adsorption of nonliving, dissolved organic material and continuing through the formation of bacterial and algal slime films and the settlement and growth of various macroscopic plants and animals (Ref 25–27). This process, by which the surfaces of all structural materials immersed in seawater become colonized, adds to the variability of the marine environment in which corrosion occurs. The rate of biofilm formation is a function of nutrient concentrations, velocity of water flow, and temperature (Ref 28).

Bacterial Films. The process of colonization begins immediately upon immersion, with the adsorption of a nonliving organic conditioning film. This conditioning film is nearly complete within the first 2 h of immersion, at which time the initially colonizing bacteria begin to attach in substantial numbers. The microbial, or primary, slime film develops over the first two weeks of immersion in most natural seawaters, providing highly variable degrees of coverage of the metal surface (Ref 27, 29). Biofilms are typically composed of pillar- and mushroom-shaped cell

Table 6 Pitting of low-carbon steel submerged in the San Diego harbor (polluted seawater)
Penetration rate averaged 0.056 mm/yr (2.2 mils/yr) for this exposure

Exposure time, days	Number of panels	Penetration			
		Average of five deepest pits per panel		Deepest pit per panel	
		mm	mils	mm	mils
155	6	0.33–0.61	13–24	0.46–0.75	18–30
361	12	0.5–1.34	20–53	0.74–1.5	29–60
552	6	0.81–1.04	32–41	0.66–1.3	26–50

Source: Ref 9

Table 7 Corrosion of copper alloy condenser tubes in polluted and clean seawater
Velocity: 2.3 m/s (7.5 ft/s). Test duration: 64 days

Alloy	CDA/UNS designation	Corrosion rate			
		Clean seawater		Polluted seawater(a)	
		mm/yr	mils/yr	mm/yr	mils/yr
90Cu-10Ni	C70600	0.075	3	0.86	34
70Cu-30Ni	C71500	0.13	5	0.66	26
2% Al brass	C68700	0.075	3	0.56	22
6% Al brass	C60800	0.13	5	0.53	21
Arsenical admiralty brass	C44300	0.33	13	0.89	35
Phosphorus deoxidized copper	C12200	0.36	14	2.7	105

CDA, Copper Development Association. (a) Contained 3 ppm hydrogen sulfide. Source: Ref 9

Table 8 Effect of hydrogen sulfide in seawater on corrosion of copper condenser tube alloys
64 day test in seawater flowing at 2.3 m/s (7.5 ft/s). Test temperature: 27 °C (80 °F)

Alloy	Corrosion rate			
	Clean seawater		Seawater plus 4 ppm H_2S	
	mm/yr	mils/yr	mm/yr	mils/yr
Phosphorus deoxidized copper	0.36	14	0.38	15
Admiralty brass	0.33	13	0.89	35
70Cu-30Ni	0.13	5	0.66	26

Source: Ref 9

clusters separated by water channels that allow nutrients in and waste products out (Ref 30–32).

Microorganisms in the biofilm change the chemistry at the metal/liquid interface in a number of ways that have an important bearing on corrosion. As the biofilm grows, the bacteria in the film produce a number of by-products. Among these are organic acids (Ref 33), hydrogen sulfide (Ref 34, 35), and protein-rich polymeric materials commonly called slime. Formation of biofilms can also change the pH at the metal surface (Ref 36–38).

Moreover, the development of a microbial biofilm, an example of which is shown in Fig. 31, results in a heterogeneous distribution of microorganisms both parallel and perpendicular to the metal surface (Ref 39). This creates a heterogeneous distribution of the chemistry from point to point along the metal surface. The result is not only a different chemistry at the metal surface from that of the bulk water but also a highly variable chemistry along the surface on a scale of tens to hundreds of micrometers (Ref 40, 41). Such chemical concentration cells increase the likelihood that the corrosive attack will be localized rather than uniformly distributed.

Two chemical species, oxygen and hydrogen, that are often implicated (or even rate-controlling) in corrosion are also important in the metabolism of the bacteria. A given bacterial slime film can be either a source or a sink for oxygen or hydrogen. Coupled together with the heterogeneous nature of the chemistry and distribution of biofilms, this means that they are capable of inducing oxygen and other chemical concentration cells (Ref 40, 41).

Under anaerobic (no oxygen) conditions, such as those found in marshy coastal areas and many sea bottom sediments, in which all the dissolved oxygen in the mud is used in the decay of organic matter, the corrosion rate of steel is expected to be very low. Under these conditions, however, sulfate-reducing bacteria use hydrogen produced at the metal surface in reducing sulfates from decaying organic material to sulfides, including H_2S. The sulfides combine with iron from the steel to produce an iron sulfide (FeS) film, which is itself corrosive. The bacteria thus transform a benign environment into an aggressive one in which steel corrodes quite rapidly in the form of pitting (Ref 42–45).

Even under open ocean conditions at air saturation, the presence of a heterogeneous bacterial slime film can result in anaerobic conditions at selected sites along the metal surface (Ref 40, 41, 46). This creates anaerobic microniches where sulfate-reducing bacteria can flourish, encouraging localized attack.

In all of these cases, the biofilm is able to make substantial changes to the chemistry of the electrolyte and its distribution at the water/metal interface. In doing this, microbes facilitate electrochemical reactions not predicted by thermodynamic analysis of the bulk water chemistry (Ref 47). Another example of this lies in the ability of biofilms formed from river water, estuarine, and coastal seawater environments to introduce manganese reduction (in addition to that of dissolved oxygen) as a cathodic reaction supporting corrosion (Ref 48–50). Manganese redox cycling, in which microbes in marine biofilms reoxidize manganese species reduced at

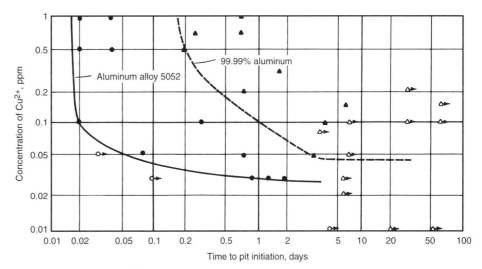

Fig. 30 Effect of adding Cu^{2+} ion to seawater on the time to pit initiation for aluminum alloy 5052 and 99.99% Al. Solid points represent conditions under which pitting started; open points indicate conditions under which no pitting occurred. Source: Ref 24

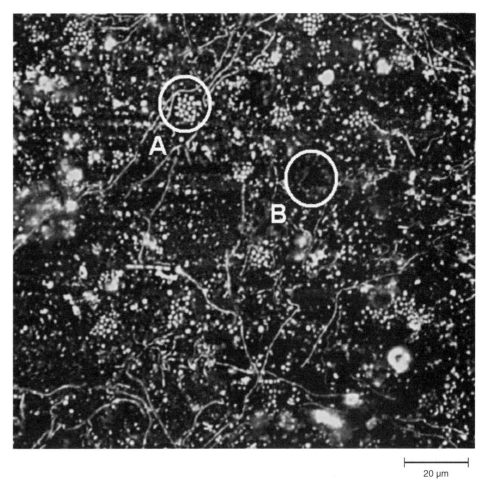

Fig. 31 Laser confocal microscope image of the variability in distribution and types of microorganisms in a 2 week old biofilm grown on a stainless steel substratum in Lower Delaware Bay coastal seawater. The chemistry at the metal surface within a microcolony, as shown at location "A," will be quite different from that in either the bulk seawater or at location "B." Source: Ref 39

the cathode of a corrosion cell, is thought to be responsible for the ennoblement of passive alloys, acceleration of crevice corrosion of stainless steels (Ref 51, 52), and acceleration of galvanic corrosion for copper, steel, and aluminum alloys coupled to stainless steel cathodes (Ref 53). The heterogeneity of the biological community within microbial biofilms also can produce anaerobic areas rich in biogenic sulfides only a few tenths of a millimeter away from other areas having nearly air-saturated concentrations of oxygen or partially deaerated areas rich in biologically produced manganese compounds (Ref 41). The potential difference between such areas can be as high as 500 mV (Ref 39). In comparison, differential aeration cells are quite weak. Even for an oxygen concentration differential of 10^4 between the aerated and deaerated areas, the potential difference is only approximately 60 mV. Thus, the type of attack as well as the corrosion rate may depend more on the details of the electrolyte chemistry at the interface than on the ambient bulk seawater chemistry. Additional details about many aspects of biological corrosion can be found in the article "Microbiologically Influenced Corrosion" in *ASM Handbook,* Volume 13A, 2003.

Macrofouling Films. Within the first 2 or 3 days of immersion, the solid surface, already having acquired both conditioning and bacterial films, begins to be colonized by the larvae of macrofouling organisms. A heavy encrustation of these organisms can have a number of undesirable effects on marine structures. Both weight and hydrodynamic drag on the structure will be increased by the fouling layer. Interference with the functioning of moving parts may also occur.

In terms of corrosion, the effects of the macrofouling layer are similar to those of microfouling. If the macrofoulers form a continuous layer, they may decrease the availability of dissolved oxygen at the metal/water interface and can reduce the corrosion rate. If the layer is discontinuous, they may induce oxygen or chemical concentration cells, leading to various types of localized corrosion. Fouling films may also break down protective paint coatings by a combination of chemical and mechanical action.

Effect of Flow Velocity

An increase in velocity of flow is generally regarded as causing an increase in average corrosion rates (Ref 54). The historical database of information on velocity effects on marine corrosion has recently been reviewed (Ref 55). Average seawater velocities below 0.25m/s (0.82 ft/s) (nominally laminar flow) increased the instantaneous corrosion rate of steel proportional to the square root of velocity. At higher flow velocities in the turbulent flow range, the instantaneous corrosion rate increased as the square of the velocity. The gradual buildup with time of both corrosion product scales and biofilms on the metal surface provides a shielding effect against the influence of velocity. Thus, the effects of velocity are most important in the early stages of immersion. Removal of such films by erosion or abrasive action from ice or suspended sediments is expected to restore the effect of flow velocity. It was also concluded that average wave action had an effect roughly equivalent to a flow velocity of 0.1 to 0.15 m/s (0.3 to 0.5 ft/s) as long as the wave action did not remove corrosion product or biological films (Ref 55).

REFERENCES

1. H.H. Uhlig and R.W. Revie, *Corrosion and Corrosion Control,* 3rd ed., Wiley-Interscience, 1985, p 108
2. F.L. LaQue, *Marine Corrosion, Causes and Prevention,* Wiley-Interscience, 1975
3. J.P. Riley and G. Skirrow, Ed., *Chemical Oceanography,* Vol 2, 2nd ed., Academic Press, 1975
4. D.A. Livingstone, Chemical Composition of Rivers and Lakes, *Data of Geochemistry,* U.S. Geological Survey, Prof. Paper 440, Chapter G, M. Fleischer, Ed., 1963
5. S.C. Dexter and C.H. Culberson, Global Variability of Natural Sea Water, *Mater. Perform.,* Vol 19 (No. 19), 1980, p 16–28
6. C.G. Gorshkov, *Atlas of the Oceans—Pacific Ocean,* Ministry of Defence of the USSR, Military Sea Transport (in Russian; See also the *Atlas of the Mediterranean Sea*)
7. M.E.Q. Pilson, *An Introduction to the Chemistry of the Sea,* Prentice Hall, 1998
8. B.D. Thomas, T.G. Thompson, and C.L. Utterback, *J. du conseil,* Vol 9, 1934, p 28–35
9. W.K. Boyd and F.W. Fink, "Corrosion of Metals in Marine Environments," MCIC Report 78-37, Metals and Ceramics Information Center, Battelle Columbus Laboratories, 1978
10. J.A. Beavers, G.H. Koch, and W.E. Berry, "Corrosion of Metals in Marine Environments," MCIC Report 86-50, Metals and Ceramics Information Center, Battelle Columbus Laboratories, 1986
11. R.E. Melchers, Effect of Temperature on the Marine Immersion Corrosion of Carbon Steels, *Corrosion,* Vol 58 (No. 9), 2002, p 768–782
12. R.E. Melchers, Modeling of Marine Immersion Corrosion for Mild and Low Alloy Steels—Parts 1 and 2, *Corrosion,* Vol 59 (No. 4), 2003, p 319–344
13. D.R. Kester, Dissolved Gases Other Than CO_2, *Chemical Oceanography,* Vol 1, 2nd ed., J.P. Riley and G. Skirrow, Ed., Academic Press, 1973, p 498
14. S.C. Dexter, Effect of Variations in Seawater upon the Corrosion of Aluminum, *Corros. J.,* Vol 36 (No. 8), 1980, p 423–432
15. H.T. Rowland and S.C. Dexter, Effects of the Seawater Carbon Dioxide System on the Corrosion of Aluminum, *Corros. J.,* Vol 36 (No. 9), 1980, p 458–467
16. S.C. Dexter and S.-H. Lin, Calculation of Seawater pH at Polarized Metal Surfaces in the Presence of Surface Films, *Corrosion,* Vol 48 (No. 1), 1992, p 50
17. K.D. Mantel, W.H. Hartt, and T.Y. Chen, *Corrosion,* Vol 48, 1992, p 489–500
18. W.H. Hartt, S. Chen, and D.W. Townley, *Corrosion,* Vol 54, 1998, p 317–322
19. S. Rossi, P.L. Bonora, R. Pasinetti, L. Benedetti, M. Draghetti, and E. Sacco, *Corrosion,* Vol 54, 1998, p 1018–1025
20. W.H. Hartt and S. Chen, *Corrosion,* Vol 56, 2000, p 3–11
21. S. Chen and W.H. Hartt, *Corrosion,* Vol 58, 2002, p 38–48
22. S. Chen, W. Hartt, and S. Wolfson, *Corrosion,* Vol 59, 2003, p 721–732
23. F.L. LaQue *Marine Corrosion, Causes and Prevention,* Wiley-Interscience, 1975, p 122–123
24. S.C. Dexter, *J. Ocean Sci. Eng.,* Vol 6 (No. 1), 1981, p 109–148
25. W.G. Characklis and K.C. Marshall, Ed., *Biofilms,* John Wiley and Sons, 1990, p 779
26. L.V. Evans, Ed., *Biofilms: Recent Advances in their Study and Control,* Harwood Academic Publishers, 2000, p 466
27. J.D. Costlow and R.C. Tipper, Ed., *Marine Biodeterioration: Proceedings of the Symposium,* Naval Institute Press, 1984
28. A. Ohashi, T. Koyama, K. Syutsubo, and H. Harada, *Wat. Sci. Tech.,* Vol 39 (No. 7), 1999, p 261–268
29. K.C. Marshall, *Interfaces in Microbial Ecology,* Harvard University Press, 1976
30. Z.P. Lewandowski, P. Stoodley, and S. Altobelli, *Wat. Sci. Tech.,* Vol 3, 1995, p 153–162
31. D. de Beer and P. Stoodley, *Wat. Sci. Tech.,* Vol 32, 1995, p 11–18
32. D. de Beer, P. Stoodley, and Z. Lewandowski, *Wat. Res.,* Vol 30, 1996, p 2761–2765
33. D.H. Pope, *A Study of Microbiologically Influenced Corrosion in Nuclear Power Plants and a Practical Guide for Countermeasures,* Electric Power Institute, 1986
34. C.A.H. Von Wolzogen Kuhr and L.S. Vad der Vlugt, Water, *Den Haag,* Vol 18, 1934, p 147–165
35. D.T. Ruppel, S.C. Dexter, and G.W. Luther, *Corrosion,* Vol 57, 2001, p 863–873
36. F.L. LaQue, *Marine Corrosion,* John Wiley and Sons, 1975, p 332
37. R.G.J. Edyvean and L.A. Terry, *Int. Biodeterior. Bull.,* Vol 19, 1983, p 1–11
38. S.C. Dexter and P. Chandrasekaran, Direct Measurement of pH within Marine Biofilms on Passive Metals, *Biofouling,* Vol 15 (No. 4), 2000, p 313–325
39. K. Xu, "Effect of Biofilm Heterogeneity on Corrosion Behavior of Passive Alloys in

Seawater," Ph.D. dissertation, University of Delaware, 2000, p 101, 169–175
40. K. Xu, S.C. Dexter, and G.W. Luther III, Voltammetric Microelectrodes for Biocorrosion Studies, *Corrosion,* Vol 54 (No. 10), 1998, p 814
41. S.C. Dexter, K. Xu, and G.W. Luther III, Mn Cycling in Marine Biofilms: Effect on Rate of Localized Corrosion, *Biofouling,* Vol 19 (Supplement), 2003, p 139–149
42. P.F. Sanders and W.A. Hamilton, Biological and Corrosion Activities of SRB in Industrial Process Plant, *Biologically Induced Corrosion,* S.C. Dexter, Ed., NACE International, 1986, p 47–68
43. I.B. Beech, S.A. Campbell, and F.C. Walsh, Marine Microbial Corrosion, *A Practical Manual on Microbiologically Influenced Corrosion,* Vol 2, J. Stoecker, Ed., NACE International, 2001, p 11.3–11.14
44. T. Gehrke and W. Sand, "Interactions between Microorganisms and Physicochemical Factors Cause MIC of Steel Pilings in Harbours (ALWC)," Paper 03557, Corrosion 2003, NACE International, 2003
45. R.A. King, J.D.A. Miller, and J.F.D. Stott, Subsea Pipelines: Internal and External Biological Corrosion, *Biologically Induced Corrosion,* S.C. Dexter, Ed., NACE International, 1986, p 268–274
46. J.W. Costerton and G.G. Geesy, The Microbial Ecology of Surface Colonization and of Consequent Corrosion, *Biologically Induced Corrosion,* S.C. Dexter, Ed., NACE International, 1985, p 223
47. M. McNeil, B. Little, and J. Jones, *Corrosion,* Vol 47 (No. 9), 1991, p 674–677
48. D.T. Ruppel, S.C. Dexter, and G.W. Luther, *Corrosion,* Vol 57, 2001, p 863–873
49. P. Linhardt, Failure of Chromium-Nickel Steel in a Hydroelectric Power Plant by Manganese-Oxidizing Bacteria, *Microbially Influenced Corrosion of Materials,* E. Heitz et al., Ed., Springer-Verlag, 1996
50. B.H. Olesen, R. Avci, and Z. Lewandowski, *Corros. Sci.,* Vol 42, 2000, p 211–227
51. S.C. Dexter, Effect of Biofilms on Crevice Corrosion, *Proc. COR/96 Topical Research Symposium on Crevice Corrosion,* NACE International, 1996, p 367–383
52. H.-J. Zhang and S.C. Dexter, Effect of Biofilms on Crevice Corrosion of Stainless Steels in Coastal Seawater, *Corrosion,* Vol 51 (No. 1), 1995, p 56–66
53. S.C. Dexter and J.P. LaFontaine, Effect of Natural Marine Biofilms on Galvanic Corrosion, *Corrosion,* Vol 54 (No. 11), 1998, p 851
54. F.L. LaQue, Behavior of Metals and Alloys in Sea Water, *The Corrosion Handbook,* H.H. Uhlig, Ed., John Wiley, 1948, p 391
55. R.E. Melchers, *Corrosion,* Vol 60, 2004, p 471–478

SELECTED REFERENCES

- S.A. Campbell, N. Campbell, and F.C. Walsh, Ed., *Developments in Marine Corrosion, Proc. Ninth International Congress on Marine Corrosion and Fouling,* The Royal Society of Chemistry, Cambridge, U.K., 1998 (See also the series of proceedings volumes from the International Congress on Marine Corrosion and Fouling.)
- W.G. Characklis and K.C. Marshall, Ed., *Biofilms,* John Wiley and Sons, 1990
- J.D. Costlow and R.C. Tipper, Ed., *Marine Biodeterioration: Proceedings of the Symposium,* Naval Institute Press, 1984
- G.R. Edwards, W. Hanzalek, S. Liu, D.L. Olson, and C. Smith, Ed., *International Workshop on Corrosion Control for Marine Structures and Pipelines,* American Bureau of Shipping, 2000
- L.V. Evans, Ed., *Biofilms: Recent Advances in Their Study and Control,* Harwood Academic Publishers, 2000
- D.A. Jones, *Principles and Prevention of Corrosion,* 2nd ed., Prentice Hall, 1996
- M.E.Q. Pilson, *An Introduction to the Chemistry of the Sea,* Prentice Hall, 1998
- M. Schumacher, Ed., *Seawater Corrosion Handbook,* Noyes Data Corporation, 1979
- H.H. Uhlig and R.W. Revie, *Corrosion and Corrosion Control,* 3rd ed., Wiley-Interscience, 1985

Corrosion in Marine Atmospheres

Richard B. Griffin, Texas A&M University

THE ANNUAL COST OF CORROSION has been estimated to be 3.1% of the gross national product for the United States. According to a recent study, the 1998 cost for the United States was $276 billion. See the article "Direct Costs of Corrosion in the United States" in *Corrosion: Fundamentals, Testing, and Protection,* Volume 13A of *ASM Handbook,* 2003. A substantial part of the total is due to atmospheric corrosion (Ref 1). Buildings, automobiles, bridges, storage tanks, ships, and other items that must be repaired, coated, or replaced represent some of the costs attributed to atmospheric corrosion in the economy.

Truly, worldwide interest exists in this topic. Several factors contribute to marine-atmospheric corrosion, with the local environment being the single most important factor. The most aggressive condition is a warm tropical coastal location with prevailing onshore winds that carry both Cl^- and SO_4^{2-} to the site. Moving inland (decreasing Cl^-) and decreasing SO_4^{2-} concentrations will decrease the extent of marine-atmospheric corrosion.

Both nickel alloys and stainless steels have very good-to-excellent resistance to marine-atmospheric corrosion. Changing the composition of plain carbon steel to that of weathering steel will increase the resistance to marine-atmospheric corrosion. Coating plain carbon steels can improve resistance to marine-atmospheric corrosion.

Modeling has become a very important method of assessing a local environment without having to develop long-term corrosion data. However, the results must be carefully used, and long-term data should be developed. Results from the models provide a very useful means of making comparisons. For steels, in particular, the results are acceptable.

The International Standards Organization (ISO) has made a significant contribution to the study of atmospheric corrosion. Standards have been developed and applied to sites located around the globe. The standards are in the process of being updated, and the reader should check for the latest revisions. In addition, ASTM International is active in establishing standards for atmospheric corrosion.

Typically, atmospheric corrosion is divided into the categories listed in Table 1, which includes the corrosion rates for low-carbon steel at a variety of locations with different atmospheric conditions (Ref 2). The International Standards Organization has established a set of corrosion standards that enable the corrosivity of a location to be described in terms of the time of wetness, sulfur dioxide, and chloride levels (Ref 3, 4). The marine or marine-industrial environments are generally considered to be the most aggressive. Important variables associated with atmospheric corrosion in marine atmospheres are chloride and sulfur dioxide content, location, alloy content, and exposure time. These are examined in this article, and the ISO CORRAG program is discussed. In addition, corrosion comparisons of metal alloys are included.

Important Variables

A number of factors, such as moisture, temperature, winds, airborne contaminants, alloy content, location, and biological organisms, contribute to atmospheric corrosion.

Moisture. For corrosion to occur by an electrochemical process, an electrolyte must be present. An electrolyte is a solution that will allow a current to pass through it by the diffusion of anions (negatively charged ions) and cations (positively charged ions). Water that contains ions is a very good electrolyte. Therefore, the amount and availability of moisture present is an important factor in atmospheric corrosion. For ferrous materials beyond a certain critical relative humidity (RH), there will be an acceleration of the atmospheric corrosion rate. The critical RH is 60% for iron in an atmosphere free of sulfur dioxide (Fig. 1a). For magnesium under similar conditions, 90% RH is critical (Fig. 1b) (Ref 5). The critical RH is not a constant value; it depends on the hygroscopicity (tendency to absorb moisture) of the corrosion products and the contaminants. The type of moisture is significant. For example, rain may help wash contaminants from surfaces, while dew and fog allow surfaces to become wet without the washing action of the rain.

One of the measures of moisture is time of wetness (TOW). As Fig. 2 shows, corrosion rate increases as TOW increases (Ref 6). In addition, Fig. 2 illustrates the importance of a contaminant; when sulfur dioxide (SO_2) levels increase, a corresponding increase in the overall corrosion rate occurs. The ISO 9223 quantifies the TOW (τ); details are discussed later in this article (Ref 3). However, the severity of the marine environment is related to the salt content of the atmosphere that contacts the material surface, which is usually more corrosive than rainfall without Cl^-.

Table 1 Types of atmospheres and corrosion rates of low-carbon steel
Test duration: 2 years

Atmosphere	Location	Corrosion rate mm/yr	mils/yr
Marine	Point Reyes, CA	0.5	19.71
Severe	25 m (80 ft) lot, Kure Beach, NC	0.53	21.00
Industrial	Brazos River, TX	0.093	3.67
Mild	250 m (800 ft) lot, Kure Beach, NC	0.146	5.73
Rural	Esquimalt, BC, Canada	0.013	0.53
Industrial	East Chicago	0.084	3.32
Marine	Bayonne, NJ	0.077	3.05
Urban	Pittsburgh, PA	0.03	1.20
Suburban (semi-industrial)	Middletown, OH	0.029	1.13
Rural	State College, PA	0.023	0.90
Marine	Esquimalt, BC, Canada	0.013	0.53
Desert	Phoenix, AZ	0.0046	0.18

Source: Ref 2

For acid rain conditions, there appears to be no significant increase in corrosion rate. A study conducted in Sweden from October to November 1976 for carbon steels showed an increase in corrosion rates with increasing SO_2; however, the incidences were relatively infrequent. The study also showed that the corrosion rates measured for a longer time do not seem to be influenced by the incidences of acid rain (Ref 6). Similar results were obtained in a British study of the atmospheric corrosion of zinc (Ref 7).

Airborne Contaminants. After TOW, the second most important factor in atmospheric corrosion is the contaminants found in the air. These can be natural or manmade, such as airborne moisture carrying salt from the sea, or SO_2 put into the atmosphere by coal-burning utility plants. The importance of the atmospheric SO_2 level on the corrosion rate of zinc has been seen (Fig. 2). As the parts per million of SO_2 increase, the weight loss of zinc increases. Other important contaminants are chlorides (Cl^-), carbon dioxide (CO_2), nitrogen oxides (NO_x), and hard dust particles (for example, sand or minerals).

Chlorides. There is a direct relationship between atmospheric salt content and measured corrosion rates. The amount of sea salts measured off the coast of Nigeria illustrates this relationship between salinity and corrosion rate (Ref 8). This is shown in Fig. 3, in which salinity of 10 gm/m^2/d results in a corrosion rate of less than 0.1 g/dm^2/mo, while a salinity of 1000 gm/m^2/d results in a corrosion rate of almost 10 g/dm^2/mo. At the LaQue Center for Corrosion Technology (Wrightsville Beach, NC) test site at Kure Beach, NC, a similar effect has been observed for carbon steel. The corrosion rate at the site 25 m (80 ft) from the mean tide line (this site is now called the oceanfront lot) was 1.19 mm/yr (47 mils/yr), while at the 250 m (800 ft) location (this site is now called the near-ocean lot), the corrosion rate for the same material was 0.04 mm/yr (1.6 mils/yr).

Chlorides are contained within droplets formed from seawater that have been entrained in the air. The droplets will evaporate and leave a residue of salt on the surface. From a corrosion standpoint, the droplets bring both water and chloride to a surface. The distance that droplets are carried inland will depend on the size of the droplets and the air currents.

The average atmospheric chloride levels collected in rainwater for the United States are shown in Fig. 4 (Ref 6). The highest levels occur along the coast of the Atlantic Ocean, Pacific Ocean, and the Gulf of Mexico. The maximum corrosion rate is related to the maximum chloride in the atmosphere. This will be related to the distance inland, the height above sea level, and the prevailing winds. The chlorides of calcium and magnesium are hygroscopic and have a tendency to form liquid films on metal surfaces, which increases TOW.

Sulfur Dioxide. The presence of SO_2 in the atmosphere lowers the critical RH while increasing the thickness of the electrolyte film and increasing the aggressiveness of the environment. For carbon steel, the effect of SO_2 levels from three Norwegian test sites is shown in Fig. 5. These data and Fig. 2 illustrate that as SO_2 concentrations are increased, the corrosion rate increases. A summary of Scandinavian data for carbon steel and zinc showed the following relationships between corrosion rate and SO_2 concentrations:

$$r_{St} = 5.28[SO_2] + 176.6 \quad \text{(Eq 1a)}$$

$$r_{Zn} = 0.22[SO_2] + 6 \quad \text{(Eq 1b)}$$

where r is the atmospheric corrosion rate in g/m^2/yr, and [SO_2] represents the concentration of SO_2 in μg/m^3 (Ref 9). Similar types of relationships have been shown for other alloy systems and locations, as is described later in this article.

Sulfur dioxide is very acidic and will dissolve in water and form sulfuric acid in the presence of oxygen:

$$SO_2 + H_2O \rightarrow H_2SO_3 \quad \text{(Eq 2a)}$$

$$2H_2SO_3 + O_2 \rightarrow 2H_2SO_4 \quad \text{(Eq 2b)}$$

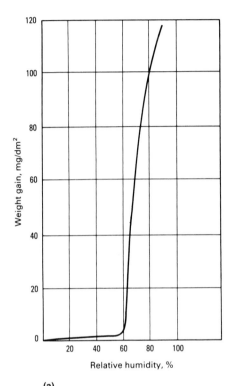

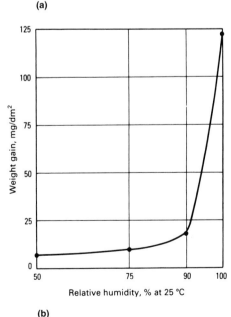

Fig. 1 Corrosion rates of iron and magnesium as a function of relative humidity. (a) For iron, the critical relative humidity is 60%. (b) For magnesium, corrosion rate increases significantly at a critical relative humidity of approximately 90%. Source: Ref 5

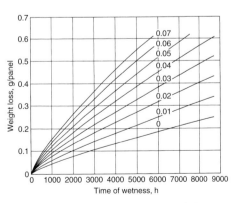

Fig. 2 The increase in corrosion rate of zinc as a function of time of wetness and SO_2 concentration. Numbers on lines are ppm SO_2. Source: Ref 6

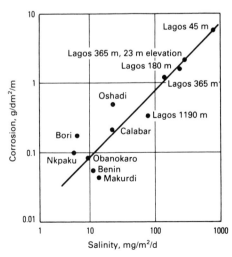

Fig. 3 Atmospheric corrosion of mild steel as a function of salinity at various sites in Nigeria. Source: Ref 8

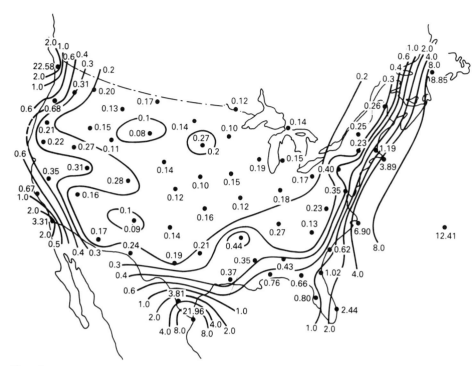

Fig. 4 Average chloride concentration (mg/L) in rainwater in the United States. Source: Ref 6

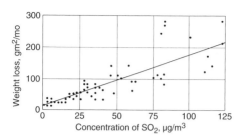

Fig. 5 Effect of SO_2 concentration on the corrosion rate of carbon steel at three Norwegian sites. Source: Ref 9

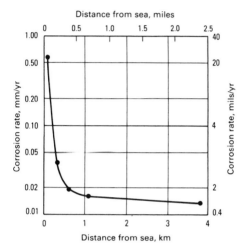

Fig. 7 Corrosion rate of carbon steel as a function of distance from the sea at Aracaju, Brazil. Source: Ref 12

In marine environments, SO_2 often appears as a result of the burning of sulfur-containing fuels that are not properly controlled.

Carbon Dioxide. The opinion of the majority of investigators is that carbon dioxide (CO_2) has an effect on the corrosion of metals. Carbon dioxide in the presence of water forms carbonic acid (Eq 3 and 4). A pH <7 may be obtained with atmospheric CO_2 in equilibrium with pure water (Ref 10):

$$CO_2 + H_2O \rightarrow HCO_3^- \qquad (Eq\ 3)$$

$$HCO_3^- \rightarrow H^+ + CO_3^{2-} \qquad (Eq\ 4)$$

Carbonates are found in corrosion products on a number of metals. The presence of CO_2 is important for zinc to be able to form a protective carbonate layer. Carbon dioxide does not have nearly the same level of significance in atmospheric corrosion as SO_2 and Cl^-.

Location. The site where materials are located is a very important variable in atmospheric corrosion. The distance from the sea and the height above sea level are both significant.

Distance from the Sea. Figure 3 shows the effect of moving inland along the coast of Nigeria from the 45, 365, and 1190 m (50, 400, and 1300 yd) sites at Lagos. From studies done on Barbados, the effect of distance is confirmed by the map of the island shown in Fig. 6 (Ref 11). This represents one of the worst conditions: tropical beach, on-shore winds, and facing a large, uninterrupted stretch of ocean. Similarly, at a site in Aracaju, Brazil, low-carbon steel samples were tested at five sites from approx-

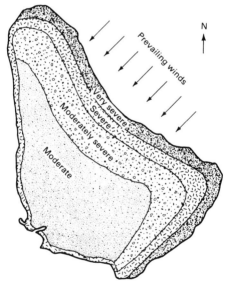

Fig. 6 Estimates of marine-atmosphere corrosivity at various locations on the island of Barbados in the West Indies. Based on CLIMAT data. Source: Ref 11

imately 0.1 to almost 4 km (0.06 to 2.5 miles) from the sea. There was a rapid falloff in the corrosion rate as the testing sites were moved inland (Fig. 7). By approximately 1.5 km (0.9 miles) inland, the corrosion rate had reached a value that showed it was basically independent of the marine atmosphere (Ref 12). The formation of aerosol droplets as a function of wind and surf zones for the distance ~400 m to 600 m (440 to 660 yd) inland is characterized by an exponential expression and describes nicely the variation in corrosion rate (Ref 13).

The height above sea level of specimens is also important. In Fig. 8(a) the corrosion rate of carbon steel specimens in the 25 m (80 ft) oceanfront lot at Kure Beach, NC, varied from approximately 360 μm/yr (14 mils/yr) at a height of 5 m (16.5 ft) to 600 μm/yr (24 mils/yr) at a height of approximately 8 m (26 ft). There is considerably less corrosion for the carbon steel at the Kure Beach, NC, 250 m (800 ft) near-ocean test site (Fig. 8b), where the corrosion rate for carbon steel varies from approximately 50 μm/yr (2 mils/yr) to a maximum of approximately 230 μm/yr (9 mils/yr). Here, the average chloride content is 100 mg/m²/d, while at the ocean-front lot it is approximately 400 mg/m²/d (Ref 14).

In the splash zone (see Fig. 2 in the article "Corrosion in Seawater" in this Volume), the highest corrosion rate is slightly above mean high tide. This zone not only has high chloride content but also is alternately wet and dry. As the height above the sea increases, the corrosion rate decreases, because the specimen is not as wet as often.

Orientation. Another corrosive factor is the orientation of a material with respect to the earth's surface. Results for a 1 year exposure of iron specimens placed vertically and inclined at

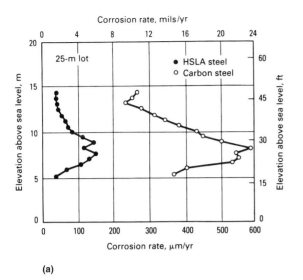
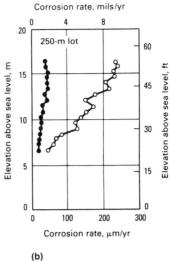

Fig. 8 Effect of elevation above sea level for carbon and high-strength, low-alloy (HSLA) steels at Kure Beach, NC. (a) 25 m (80 ft) lot. (b) 250 m (800 ft) lot. Source: Ref 14

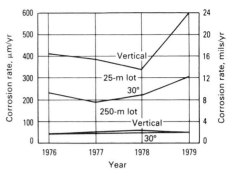

Fig. 9 Effect of specimen orientation on corrosion rates of iron specimens exposed vertically and at an angle of 30° to the horizontal. Results of one-year test at Kure Beach, NC. Source: Ref 14

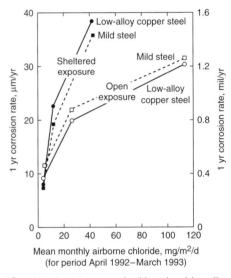

Fig. 10 Corrosion rates of mild steel and low-alloy copper steel versus site mean level of airborne chloride. Source: Ref 15

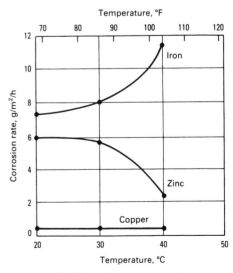

Fig. 11 The effect of temperature on the corrosion rates of iron, zinc, and copper. Source: Ref 5

an angle of 30° with respect to the ground are shown in Fig. 9 (Ref 14). The spread in the data is much greater for the Kure Beach 25 m (80 ft) test lot than for the 250 m (800 ft) test lot. In both cases, the vertical specimens showed a higher corrosion rate. This was attributed to the formation of a nonuniform, less protective oxide in the vertical position than in the 30° position. It is also possible that the 30° samples have the chloride deposits cleaned from their surfaces more easily than the vertical specimens. Ratios of the corrosion rate in the vertical position to that in the 30° position are given in Table 2 for five sites (Ref 14). In the vertical position, the corrosion rate is greater on the side facing the sea than on the side facing land. At the 25 m (80 ft) lot at Kure Beach, steel pipe specimens corroded at the rate of 850 μm/yr (33.5 mils/yr) facing the ocean, as compared to 50 μm/yr (2 mils/yr) facing away from the ocean over a 4.5 year period.

The corrosion rate was measured on the skyward and groundward side of specimens that are parallel to the earth's surface. Tests conducted at Kure Beach showed that the skyward side corroded at a greater rate after 3 months. However, after 6 months of testing, the rates were identical. Similarly, for an AZ31B magnesium alloy in a 30 day test, the skyward-facing specimens lost more material than the groundward-facing ones.

For corrosion tests performed on mild steel and low-alloy copper steel in Australia, the 1 year corrosion rates for sheltered and open exposures are shown in Fig. 10. The chloride content has a significant effect. For example, at 40 mg/m²/d, the corrosion rate for the sheltered locations was approximately 38 μm/yr (1.5 mil/yr), while for the open exposure it was approximately 20 μm/yr (0.8 mil/yr). It is interesting to note that the alloys swapped positions when comparing open and sheltered sites (Ref 15).

Table 2 Comparison of atmospheric corrosion rates for specimens held vertically and inclined at 30° to the horizontal

Location	Corrosion rate ratio, vertical/30°
Kearny, NJ	1.25
Vandergrift, PA	1.26
South Bend, PA	1.20
25 m (80 ft) lot, Kure Beach, NC	1.41
250 m (800 ft) lot, Kure Beach, NC	1.25

Source: Ref 14

Temperature affects the RH, TOW, and the kinetics of the corrosion process. For atmospheric corrosion, the presence of moisture as determined by TOW is probably the most important role of temperature. Figure 11 shows the effect of temperature on iron, zinc, and copper (Ref 5). Increasing temperature over the range of 20 to 40 °C (70 to 100 °F) while holding the chloride content (16 mg/m³) and RH (80%) constant results in three distinct patterns: corrosion rate increases for iron, decreases for zinc, and remains constant for copper.

The temperature of interest may not be the average daily temperature. It may be more important to know the dewpoint temperature or the surface temperature. From an atmospheric corrosion standpoint, minimizing the TOW reduces the corrosion rate.

Sunlight influences the degree of wetness and affects the performance of coatings and plastics. Sunlight may also stimulate photosensitive corrosion reactions on such metals as copper and steel. Ultraviolet (UV) light and photo-oxidation can cause embrittlement and surface cracks in polymers. This can be avoided by the addition of UV absorbers (for example, carbon black).

Wind. The direction and velocity of the wind affect the rate of accumulation of particles on

46 / Corrosion in Specific Environments

metal surfaces. Also, wind disperses the airborne contaminants and pollutants. Figure 6 shows that the corrosion rate zones widen from an ocean beach facing the prevailing wind. The effect caused by the chloride ions being carried inland is illustrated in Fig. 7, which shows an increased corrosion rate at 1 km (0.6 mile) inland. Stronger prevailing winds can carry the airborne contaminants even further inland. A marine site may be even more aggressive due to the prevailing winds bringing industrial pollutants, particularly SO_2, to the marine site.

Time. For many materials, there is a decrease in the corrosion rate as time increases. This decrease is associated with the formation of protective corrosion layers. Figure 12 provides an example of this for low-carbon steel at eight sites in South Africa (Ref 16). An initial increase in the atmospheric corrosion rates occurs, followed by a slowing down of the corrosion rates as corrosion products form on the alloy surface. This is true only for sites C through G. For site A and B, the corrosion rate is sufficiently high to prevent the formation of a protective layer; therefore, a constant very high corrosion rate was maintained. The effect of marine-atmospheric corrosion on tensile strength is shown in Fig. 13 for low-carbon steel and three aluminum alloys. The initial rate of loss in ultimate tensile strength is highest, but as time continues, the rate of loss decreases except for the low-carbon steel at Kure Beach and Point Judith (Ref 17).

Starting Date. The initial variation in the corrosion rate may depend on when the tests were started. Figure 14 compares the measured weight losses for iron and zinc in tests started at two different dates (Ref 6). Over a 60 day test, the variation in corrosion rate for zinc is much larger than that for iron. Similarly, for iron specimens at the Kure Beach oceanfront lot (25 m, or 80 ft), there are variations of hundreds

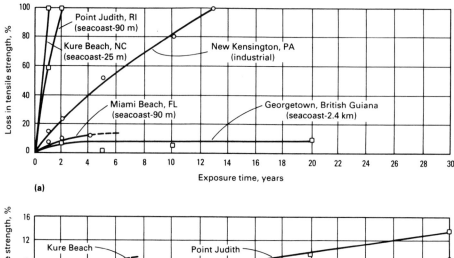

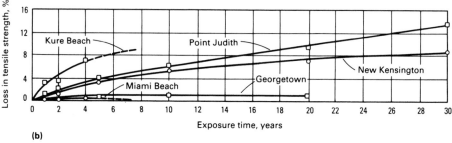

Fig. 13 Loss in tensile strength as a function of time for (a) 1.6 mm (1/16 in.) low-carbon steel and (b) aluminum alloys of the same thickness at five test sites. Data in (b) are averages for aluminum alloys 1100, 3003, and 3004. Source: Ref 17

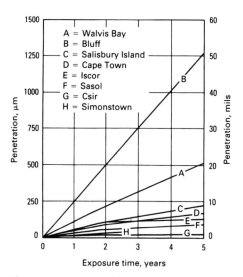

Fig. 12 Change in corrosion rate as a function of time for eight South African sites. Source: Ref 16

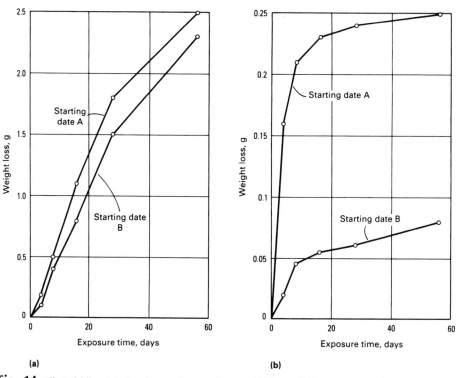

Fig. 14 Effect of different starting dates on the corrosion rate of (a) iron and (b) zinc. Source: Ref 6

of micrometers per year in corrosion rates as measured on samples exposed vertically for 1 and 2 years each. This is shown in Fig. 15 for iron calibration specimens tested from 1949 through 1979 (Ref 14).

Site Variability. Large variations in atmospheric corrosion rate occur within a particular type of region. An example would be the corrosion behavior of steel and zinc in different tropical environments. Figure 16 shows the average penetration for steel in a 1 year test at tropical sites (Ref 18). For zinc under similar conditions, the average penetration varied from 31 to 11 µm (1.2 to 0.4 mils). As Fig. 13 shows, there is a wide variation in the loss of tensile strength between the four seacoast locations.

Temperature, tropical-marine sites, and inland sites are compared in Fig. 17 for zinc and copper. The zinc corrodes more rapidly at the tropical-marine site; however, the reverse is true at the inland site, where the corrosion rate at State College, PA, is higher than at Miraflores, Panama. Overall, the long-term (15 to 20 years) rates for copper are similar at both marine and inland sites (Ref 18).

A similar comparison for carbon and low-alloy steels (Fig. 18) illustrates that the tropical environment has a higher overall corrosion rate. Figure 18(a) compares the stabilized corrosion rate for carbon steel of 20 µm/yr (0.8 mil/yr) at Cristobal, Panama, to 16 µm/yr (0.63 mil/yr) at Kure Beach, NC, near-ocean lot (250 m, or 800 ft). Low-alloy steel exhibited a similar increased rate of corrosion, as shown in Fig. 18(b). The stabilized corrosion rate is the slope of the average penetration corrosion loss-time curve. The values are given next to the curves in Fig. 18 and 19. The same pattern is exhibited for carbon steel at inland sites (Fig. 18c) (Ref 18).

A comparison of 1, 2, and 4 year corrosion rates for aluminum, copper, zinc, and iron is given in Table 3 as part of the ISO CORRAG program. The five sites used were Kure Beach (KB, 250 m, or 800 ft, lot), Newark-Kearny (NK), Point Reyes (PR), Panama Canal Zone (PCZ), and Los Angeles (LA-USC). Plate- and helix-shaped specimens were used. Generally, the helix exhibited larger corrosion loss than the plate. Table 4 provides the environmental data for the five sites (Ref 19).

Alloy Content. The particular alloy composition can make a significant difference in the marine-atmospheric corrosion rate of a material. For steels, a comparison of carbon steels, low-alloy steels, and steels with 5% alloying elements is shown in Fig. 19 for marine and inland exposure (Ref 18). In each case, the long-term corrosion rate is greater for the marine environment. Additionally, Fig. 19 shows the accelerated corrosion that occurs in the first 1 to 3 years and the constant rate associated with long-term atmospheric corrosion. Very similar results (Fig. 12) have been reported for a study done in South Africa at eight sites that are classified as rural to severe marine.

The results of 15.5 year studies of low-alloy steels conducted at the Kure Beach, NC, near-ocean lot (250 m, or 800 ft) are shown in Fig. 20, in which the mass loss per unit area is plotted as a function of the total alloy content. If alloy additions of approximately 2 wt% are considered, then the mass loss per area is reduced from greater than 40 mg/dm^2 to less than 8 mg/dm^2 (Ref 8).

The significance of chromium as an alloying element is shown in Fig. 21 for atmospheric

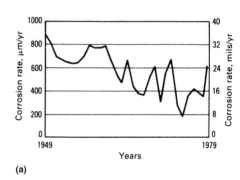

Fig. 15 Corrosion of iron calibration specimens tested for (a) 1 year and (b) 2 years at the 25 m (80 ft) lot at Kure Beach, NC. Source: Ref 14

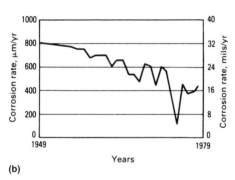

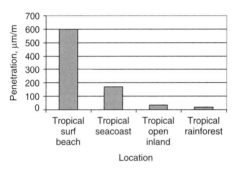

Fig. 16 Variation in corrosion rate after 1 year exposure of steel at four different tropical sites. Source: Ref 18

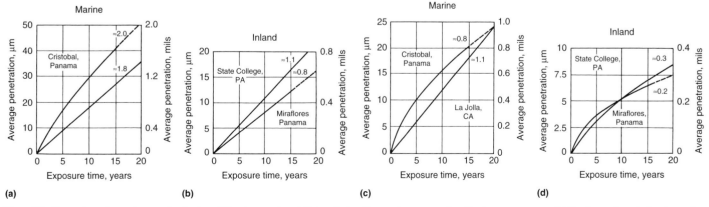

Fig. 17 Comparison of corrosion rates for zinc (a and b) and copper (c and d) at tropical and temperate exposure sites. Numbers on curves are stabilized corrosion rates in micrometers per year. Source: Ref 18

48 / Corrosion in Specific Environments

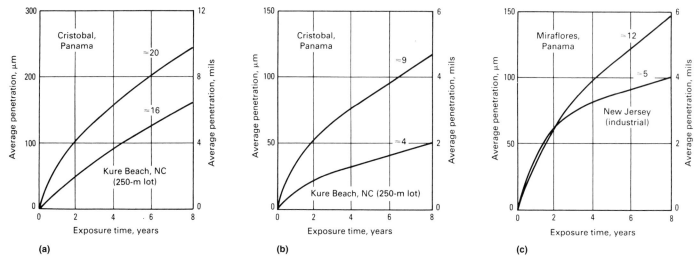

Fig. 18 Comparison of corrosion rates of steels at temperate and tropical exposure sites. Numbers on curves are stabilized corrosion rates in micrometers per year. (a) Carbon steel, marine exposure. (b) Low-alloy steel, marine exposure. (c) Carbon steel, inland exposure. Source: Ref 18

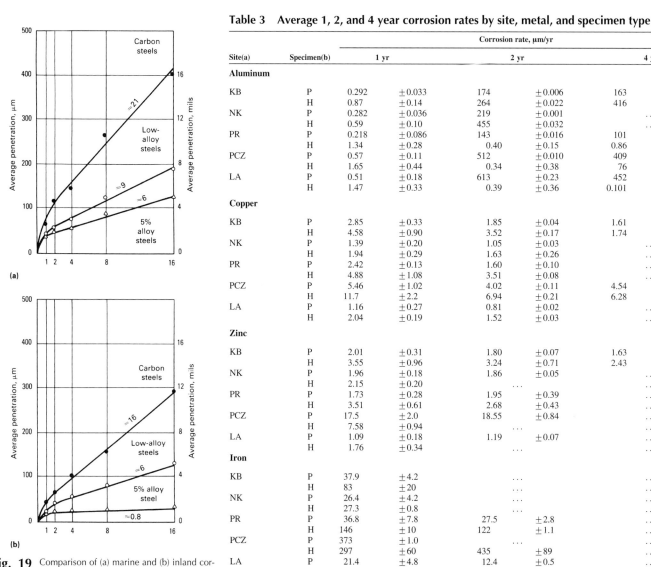

Fig. 19 Comparison of (a) marine and (b) inland corrosion rates for carbon steel, low-alloy steels, and 5% alloy steels at the Naval Research Laboratory test site in Panama. Numbers on curves are stabilized corrosion rates in micrometers per year. Source: Ref 18

Table 3 Average 1, 2, and 4 year corrosion rates by site, metal, and specimen type

Site(a)	Specimen(b)	\multicolumn{6}{c}{Corrosion rate, μm/yr}
		1 yr		2 yr		4 yr	
Aluminum							
KB	P	0.292	±0.033	174	±0.006	163	±0.003
	H	0.87	±0.14	264	±0.022	416	±0.017
NK	P	0.282	±0.036	219	±0.001	...	
	H	0.59	±0.10	455	±0.032	...	
PR	P	0.218	±0.086	143	±0.016	101	±0.014
	H	1.34	±0.28	0.40	±0.15	0.86	±1.04
PCZ	P	0.57	±0.11	512	±0.010	409	±0.017
	H	1.65	±0.44	0.34	±0.38	76	±0.24
LA	P	0.51	±0.18	613	±0.23	452	±0.25
	H	1.47	±0.33	0.39	±0.36	0.101	±0.028
Copper							
KB	P	2.85	±0.33	1.85	±0.04	1.61	±0.04
	H	4.58	±0.90	3.52	±0.17	1.74	±0.01
NK	P	1.39	±0.20	1.05	±0.03	...	
	H	1.94	±0.29	1.63	±0.26	...	
PR	P	2.42	±0.13	1.60	±0.10	...	
	H	4.88	±1.08	3.51	±0.08	...	
PCZ	P	5.46	±1.02	4.02	±0.11	4.54	±0.05
	H	11.7	±2.2	6.94	±0.21	6.28	±0.08
LA	P	1.16	±0.27	0.81	±0.02	...	
	H	2.04	±0.19	1.52	±0.03	...	
Zinc							
KB	P	2.01	±0.31	1.80	±0.07	1.63	±0.06
	H	3.55	±0.96	3.24	±0.71	2.43	±0.14
NK	P	1.96	±0.18	1.86	±0.05	...	
	H	2.15	±0.20	...		...	
PR	P	1.73	±0.28	1.95	±0.39	...	
	H	3.51	±0.61	2.68	±0.43	...	
PCZ	P	17.5	±2.0	18.55	±0.84	...	
	H	7.58	±0.94	...		...	
LA	P	1.09	±0.18	1.19	±0.07	...	
	H	1.76	±0.34	...		...	
Iron							
KB	P	37.9	±4.2	...		...	
	H	83	±20	...		...	
NK	P	26.4	±4.2	...		...	
	H	27.3	±0.8	...		...	
PR	P	36.8	±7.8	27.5	±2.8	...	
	H	146	±10	122	±1.1	...	
PCZ	P	373	±1.0	...		...	
	H	297	±60	435	±89	...	
LA	P	21.4	±4.8	12.4	±0.5	...	
	H	19.2	±0.3	...		...	

(a) KB, Kure Beach (250 m, or 800 ft, lot); NK, Newark-Kearny; PR, Point Reyes; PCZ, Panama Canal Zone; LA, Los Angeles. (b) P, plate-shaped specimens; H, helix-shaped specimens. Source: Ref 19

Table 4 Environmental data for the ISO CORRAG exposures

	Kure Beach (250 m, or 800 ft, lot)(a)				Newark-Kearny(b)				Point Reyes(c)				Panama Canal Zone(d)				Los Angeles(e)			
Exposure code	Temp., °C	TOW, %	SO_2, mg/m^2	NaCl, mg/m^2	Temp., °C	TOW, %	SO_2, mg/m^2	NaCl, mg/m^2	Temp., °C	TOW, %	SO_2, mg/m^2	NaCl, mg/m^2	Temp., °C	TOW, %	SO_2, mg/m^2	NaCl, mg/m^2	Temp., °C	TOW, %	SO_2, mg/m^2	NaCl, mg/m^2
11	19.01	50.0	0	129	12.35	22.7	27.3	NA	14.26	45.2	NA	NA	27.32	82.6	0	517	16.71	45.6	11.6	NA
12	17.78	48.8	0	117	12.99	23.6	26.2	NA	14.26	44.2	NA	NA	26.83	81.6	0	532	16.76	43.2	11.6	NA
13	17.55	45.2	0	149	11.45	18.4	29.7	NA	14.07	46.3	NA	NA	26.35	83.7	NA	605	16.46	41.8	8.1	NA
14	17.40	47.4	0	193	13.06	22.1	26.7	NA	13.70	49.7	NA	NA	26.18	87.5	NA	724	16.44	44.0	28.1	NA
15	17.74	49.6	0	242	13.11	26.0	27.6	NA	13.52	53.2	NA	NA	26.42	91.9	NA	764	16.71	41.4	6.3	NA
16	18.03	50.6	0	266	12.99	28.2	26.8	NA	NA	NA	NA	NA	26.75	91.7	NA	723	17.42	38.1	6.3	NA
21	18.00	46.4	0	162	13.64	24.2	26.4	NA	14.10	45.7	NA	NA	26.77	83.2	NA	554	16.47	43.6	10.4	NA
41	18.17	47.5	0	166	NA	NA	NA	NA	NA	NA	NA	NA	26.74	86.1	NA	629	16.82	41.2	8.3	NA
1X1	...	...	...	...	...	...	...	...	...	...	...	...	27.24	88.4	0	93	...	...	...	...

NA, not available. (a) SO_2 data from sulfation plates; NaCl from chloride candle; temperature and time of wetness (TOW) from weather station. (b) SO_2 from the hourly max. concentration; temp. and TOW from Newark International Airport weather station. (c) Temp. and TOW from San Francisco International Airport weather station. (d) SO_2 from sulfation plates; NaCl from chloride candle; temp. and TOW from local measurement. 1X1 is 6 mo data for steel. (e) SO_2 from average hourly max. concentration; temp. and TOW from Los Angeles International Airport weather station. Source: Ref 19

Table 5 Chemical analyses for stainless steels exposed at Kure Beach beginning May 14, 1941

	Composition, wt%							
Alloy	C	Ni	Cr	Si	Mn	S	P	Other
301	0.11	8.14	17.74	0.47	1.40	0.014	0.015	...
302	0.10	10.05	18.61	0.41	0.39	0.003(a)	0.020	...
304	0.07	8.92	18.39	0.38	0.41	0.013	0.010	...
308	0.07	10.74	20.38	0.38	0.63	0.012	0.020	...
309	0.09	13.60	23.62	0.38	1.15	0.017	0.020	...
310	0.07	19.77	24.12	0.39	1.46	0.008	0.018	...
316	0.08	13.16	17.82	0.39	1.52	0.016	0.017	2.81Mo
317	0.05	14.13	18.55	0.39	1.70	0.018	0.027	3.5Mo
321	0.05	9.66	18.65	0.53	0.54	0.015	0.015	0.48Ti
347	0.07	11.23	18.64	0.24	0.56	...	...	0.78Nb
430	0.05	0.32	17.10	0.31	0.30	0.018	0.018	...

(a) Considering it unlikely that 1940s commercial melting practice could produce a stainless steel heat with this low a sulfur content, the authors suspect a typographical error in the original report from which these data were obtained. Source: Ref 21

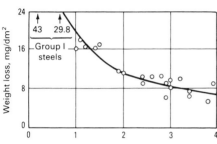

Fig. 20 Corrosion data for 25 low-alloy steels tested over a 15.5 year period at the Kure Beach, NC, 250 m (800 ft) lot. Source: Ref 8

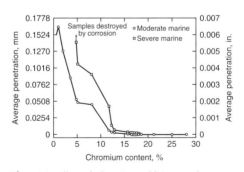

Fig. 21 Effect of chromium addition on the atmospheric corrosion of steels. Source: Ref 20

Table 6 Ranking of austenitic stainless steels according to 15 year pit depths

Stainless grade	Average pit depth		Average R_a at 60 years(a)		60 year mass change, g	Kure Beach, NC, average corrosion rate(b)	
	µm	mils	µm	mils		µm	mils
347	86	3.4	0.8	0.030	−0.06	<0.03	<0.001
321	66	2.6	0.5	0.020	−0.06	<0.03	<0.001
308	41	1.6	0.6	0.025	No data	<0.03	<0.001
301	41	1.6	0.5	0.021	−0.03	<0.03	<0.001
302	31	1.2	No specimen		No specimen	<0.03	<0.001
304	28	1.1	0.8	0.032	−0.07	<0.03	<0.001
309	28	1.1	0.8	0.030	−0.02	<0.03	<0.001
317	28	1.1	0.3	0.012	−0.03	<0.03	<0.001
316	25	1.0	0.3	0.010	−0.01	<0.03	<0.001
310	10	0.4	0.3	0.012	−0.03	<0.03	<0.001

(a) R_a, surface roughness. (b) Results of a 15 year test at the Kure Beach, NC, 250 m (800 ft) lot. Source: Ref 21

corrosion conditions classified as moderate and severe marine (Ref 20). Above 12 or 12.5 wt% Cr, the atmospheric corrosion becomes negligible; lower chromium levels result in a rapid increase in the corrosion rate. Table 5 provides compositions for 11 type 300- and 400-series stainless steels tested at the near-ocean lot (250 m, or 800 ft), Kure Beach, NC, over a 15 year period (Ref 21). Table 6 lists the 15 year pit depths, the surface roughness (R_a) at 60 years, the average corrosion rate, and the 60 year mass change for the 10 austenitic (300-series) alloys listed in Table 5. The average pit depth varied, for a 15 year study, from 86 µm/yr (3.4 mils/yr) for type 347 (UNS S34700) to 10 µm/yr (0.4 mil/yr) for type 310 (UNS S31000). (It is unlikely that the actual pitting rates would be linear with respect to time, but these data are as reported.) The mass loss during the 60 years of exposure is low, with a maximum of 0.07 g and a minimum of 0.01 g. The average corrosion rates

Table 7 Composition of test panels

Steel type	Composition, wt%								
	C	Mn	P	S	Si	Cu	Ni	Cr	V
Carbon	0.046	0.38	0.012	0.022	0.016	0.014	0.012	0.025	<0.01
Copper-bearing	0.042	0.35	0.002	0.012	0.004	0.26	0.014	0.014	<0.01
ASTM A242 (COR-TEN A)	0.11	0.31	0.092	0.020	0.42	0.30	0.31	0.82	<0.01
ASTM A588 grade A (COR-TEN B)(a)	0.13	1.03	0.006	0.019	0.25	0.33	0.015	0.56	0.038

(a) Pre-1978 composition. Source: Ref 22

Table 8 Estimated 50 year corrosion penetrations

Based on regression analysis of 16 year data, except as noted

Steel type	Orientation	Urban industrial		Rural		Moderate-marine	
		μm	mils	μm	mils	μm	mils
ASTM A242	30° S	45	1.8	84	3.3	200	7.9
	30° N	57	2.2	165	6.5	288	11.3
	90° S	51	2	102	4	217	8.5
	90° N	73	2.9	178	7	367	14.4
	Average	*57*	*2.2*	*132*	*5.2*	*268*	*10.5*
ASTM A588	30° S	69	2.7	170	6.7	342	13.5
	30° N	104	4.1	264	10.4	421	16.6
	90° S	96	3.8	200	7.9	369	14.5
	90° N	136	5.4	302	11.9	513	20.5
	Average	*101*	*4*	*234*	*9.2*	*411*	*16.2*
Copper-bearing	30° S	96	3.8	290	11.4	641	25.2
	30° N	138	5.4	324	12.8	794	31.3
	90° S	124	4.9	300	11.8	767	30.2
	90° N	164	6.5	420	16.5	1094	43.1
	Average	*130*	*5.2*	*334*	*13.1*	*824*	*32.4*
Carbon	30° S	120	4.7	306	12	1586(a)	62.4(a)
	30° N	151	5.9	338	13.3	2066(a)	81.3(a)
	90° S	122	4.8	322	12.7	1348(a)	53.1(a)
	90° N	155	6.1	413	16.2	5092(a)	200.5(a)
	Average	*137*	*5.4*	*345*	*13.6*	*2523(a)*	*99.3(a)*

(a) Based on 8 year data. Source: Ref 22

were equal to or less than 0.03 μm/yr (0.001 mil/yr). From the data in Table 6, the alloys with the best resistance are type 309 (UNS S30900), 317 (UNS S31700), 316 (UNS S31600), and 310 (Ref 21). For austenitic stainless steel alloys, it is important to avoid sensitization, resulting in intergranular attack, and the buildup of chloride ions on 304L and 316L under a load, which potentially may lead to stress-corrosion cracking.

Another category of steels that is of interest for marine-atmospheric corrosion applications are weathering steels. Four weathering steels and their compositions are listed in Table 7 (Ref 22). The predicted 50 year corrosion penetration results are given in Table 8 for three environments. Each material has four orientations. The highest average predicted penetrations (2523 μm, or 99.3 mils) are for the carbon steel, which has the lowest copper alloy content, in the moderate-marine environment. In contrast, the predicted average value for ASTM A242, under the same conditions, is approximately one-tenth of the carbon steel value (268 μm, or 10.5 mils) (Ref 22). Using the ISO corrosivity recommendations, the moderate-marine site would be a C_4 or C_5 site (Ref 3). Figure 22 shows a comparison of the aforementioned weathering steels at marine, rural, mountaintop, and rooftop sites. Of those four sites, the marine environment exhibits the highest corrosion (Ref 23). Additional sources of information are available in the ASTM standards listed in Table 9.

Another important group of materials for resisting atmospheric corrosion is coated materials. In Table 10, the corrosion losses for

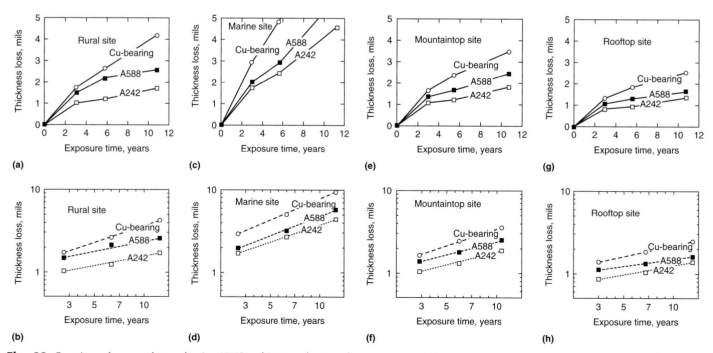

Fig. 22 Corrosion performance of copper-bearing, A588B, and A242 weathering steels. Locations: (a) and (b), rural; (c) and (d), marine; (e) and (f), mountaintop; and (g) and (h), rooftop. Linear plots: (a), (c), (e), and (g). Logarithmic plots: (b), (d), (f), and (h). Source: Ref 23

Table 9 ASTM standards related to atmospheric corrosion

Standard number	Title of standard
G 101	Standard Guide for Estimating the Atmospheric Corrosion Resistance of Low-Alloy Steels
G 50	Standard Practice for Conducting Atmospheric Corrosion Tests on Metals
B 826	Standard Test Method for Monitoring Atmospheric Corrosion Tests by Electrical Resistance Probes
G 92	Standard Practice for Characterization of Atmospheric Test Sites
B 808	Standard Test Method for Monitoring of Atmospheric Corrosion Chambers by Quartz Crystal Microbalances
G 33	Standard Practice for Recording Data from Atmospheric Corrosion Tests of Metallic-Coated Steel Specimens
B 810	Standard Test Method for Calibration of Atmospheric Test Chambers by Change in Mass of Copper Coupons
G 91	Standard Practice for Monitoring Atmospheric SO_2 Using the Sulfation Plate Technique
D 6675	Standard Practice for Salt-Accelerated Outdoor Cosmetic Corrosion Testing of Organic Coatings on Automotive Sheet Steel
B 537	Standard Practice for Rating of Electroplated Panels Subjected to Atmospheric Exposure
G 84	Standard Practice for Measurement of Time-of-Wetness on Surfaces Exposed to Wetting Conditions as in Atmospheric Corrosion Testing
G 1	Standard Practice for Preparing, Cleaning, and Evaluating Corrosion Test Specimens
G 140	Standard Test Method for Determining Atmospheric Chloride Deposition Rate by Wet Candle Method
G 113	Standard Terminology Relating to Natural and Artificial Weathering Tests of Nonmetallic Materials
D 4097	Standard Specification for Contact-Molded Glass-Fiber-Reinforced Thermoset Resin Corrosion-Resistant Tanks
D 1435	Standard Practice for Outdoor Weathering of Plastics
D 1014	Standard Practice for Conducting Exterior Exposure Tests of Paints and Coatings on Metal Substrates

(a) Date suffix is not given; current revision is recommended.

Table 10 Corrosion losses for galvanized and 55% Al-Zn alloy-coated specimens with less than 5% rust after 21 years

Site	Description	Material	Corrosion loss, μm (mil) Average	Replicates	Standard deviation
Point Reyes, CA	Mild marine, on hills overlooking Pacific Ocean	G60	11.7 (0.46)	1	...
		G90	13.2 (0.52)	3	0.5
		AZ55	4.3 (0.17)	3	0.5
State College, PA	Rural with acid rain	AZ55	6.1 (0.24)	3	0.5
Newark-Kearny, NJ	Industrial	AZ55	9.9 (0.39)	3	0.5
Kure Beach, NC	Severe marine, 25 m (80 ft) from Atlantic Ocean	AZ55	19.8 (0.78)	1	...

Source: Ref 24

Table 11 Metallic-coated steel sheet test materials

Material designation	Coating metal	Coating mass (triple-spot total for both sides), g/m²	Nominal coating thickness (per side) μm	mil	Sheet thickness mm	in.
G60 galvanized (ASTM A653)	Zinc	180	13	0.5	1.2	0.05
G90 galvanized (ASTM A653)	Zinc	275	19	0.7	1.0	0.04
AZ55 55% Al-Zn coated (ASTM A792)	55% Al-Zn alloy	165	22	0.9	0.5	0.02
T2 100 aluminum coated (ASTM A463)	Aluminum	300	48	1.9	0.8	0.03

Source: Ref 24

Table 12 Nominal composition of evaluated nickel alloys

Alloy	UNS No.	Composition, wt% Ni	Cu	Cr	Mo	Fe	Al	Ti
200	N02200	99.5	...	...	...	...	...	...
400	N04400	66.5	31.5	...	...	...	...	...
600	N06600	76.0	...	15.5	...	8.0	...	...
625(a)	N06625	61.0	...	21.5	9.0	2.5	0.2	0.2
800	N08800	32.5	0.4	21.0	...	46.0	0.4	0.4
825	N08825	42.5	2.2	21.5	3.0	30.0	0.1	0.9

(a) Also 3.6 wt% Nb

galvanized and 55% Al-Zn-coated steel specimens exposed for 21 years are listed (Ref 24). Only specimens with less than 5% rust were used for the corrosion-loss calculations. The maximum loss was 19.8 μm (0.8 mil) or less than 1 μm/yr (0.04 mil/yr) over a 21 year period at the worst site, Kure Beach, and approximately 20% of that value for the AZ55 coating at Point Reyes. The details on the coating specimens are in Table 11.

Nickel alloys are considered to have excellent atmospheric corrosion resistance. Six nickel alloys were tested at five sites: Kure Beach, NC; Kearny, NJ; Point Reyes, CA; State College, PA; and Panama Canal Zone, Panama. The compositions of the alloys are given in Table 12; they range from Ni 200 (UNS N02200) with 99.5% Ni to alloy 800 (UNS N08800) with 32.5% Ni and 67.5% alloying elements (Ref 25). The results after 20 years of corrosion are shown in Table 13, which refers to Table 10 for the test sites. All six of the alloys corroded less than 0.001 mm/yr (0.04 mil/yr). The deepest pit was found on the N02200 alloy, and it was only 0.046 mm (1.8 mil) deep (an average of the four deepest pits). With increasing alloy content and avoiding the seacoast environment, the pitting was <0.01 mm (0.4 mil) in depth. The mechanical properties, after 20 years of exposure, are tabulated in Table 14 (Ref 25).

Table 15 provides an excellent tabulation of general corrosion, pitting, and loss of tensile strength for a wide variety of metals and alloys, including aluminum, copper, carbon steels, coated steels, and stainless steels (Ref 18).

Exposure Time. One of the difficulties with atmospheric-corrosion testing is the length of time required for the tests. For steels, while a reasonable estimate of long-term corrosion performance may be made from short-term data, these estimates must be used very cautiously, because the short-term results may not be representative. It is best to have long-term data available. Fortunately, there are considerable long-term data available for a number of alloy steels (Ref 26). Table 15 provides data for a tropical seacoast site, Cristobal, Panama, and includes up to 16 years of data for many different alloys, including steels.

A power-law relationship is often used for describing long-term corrosion data (Ref 27):

$$M = aT^n \qquad \text{(Eq 5)}$$

where M is the mass loss per unit area, T is the exposure time, n is the mass loss exponent (Table 17) or slope, and a is the mass loss during the first year.

Table 16 provides data from test sites in Europe, Central America, and the United States where exposure times ranged from 12 to 30 years for steel, zinc, and copper (Ref 28). The table includes experimental data and corrosion rates predicted using ISO 9224 and the power-law representation. The data in Table 16 show better agreement between predicted and measured rates for steel and zinc than for copper. The range of n-values used in the calculations for different ISO corrosivities is shown in Table 17 (Ref 28).

Modeling of Atmospheric Corrosion—ISO CORRAG Program

The ISO organization has developed an atmospheric-corrosion classification scheme.

Table 13 American Society for Testing and Materials 1976 atmospheric test program of 20 year exposure results for corrosion rates and pit depths

Location	Alloy	Average mass loss, mg/dm²	Average corrosion rate mdd(a)	mm/yr	mil/yr	Average of four deepest pits mm	mil
Kure Beach, NC	N02200	455.1	0.06	<0.001	<0.039	0.0463	1.824
	N04400	444.6	0.06	<0.001	<0.039	0.0081	0.319
	N06600	23.95	0.0035	<0.001	<0.039	0.0232	0.914
	N06625	2.6	0.00033	<0.001	<0.039	0.0076	0.299
	N08800	26.4	0.0036	<0.001	<0.039	0.0211	0.831
	N08825	20.36	0.003	<0.001	<0.039	0.0169	0.666
Kearny, NJ (industrial)	N02200	698.2	0.1	<0.001	<0.039	0.0230	0.906
	N04400	652.5	0.09	<0.001	<0.039	0.0178	0.701
	N06600	0.1	0.00002	<0.001	<0.039	0.0091	0.359
	N06625	0	0	<0.001	<0.039	0.0005	0.020
	N08800	0.31	0.00005	<0.001	<0.039	0.0120	0.473
	N08825	0	0	<0.001	<0.039	0	0
Point Reyes, CA (west coast marine)	N02200	87.07	0.01	<0.001	<0.039	0.0184	0.725
	N04400	118.7	0.017	<0.001	<0.039	0.0124	0.489
	N06600	5.27	0.0007	<0.001	<0.039	0.0111	0.437
	N06625	0.97	0.00013	<0.001	<0.039	0	0
	N08800	5.87	0.0007	<0.001	<0.039	0.0098	0.386
	N08825	5.3	0.0006	<0.001	<0.039	0.0046	0.181
State College, PA (rural)	N02200	178.8	0.02	<0.001	<0.039	0.00523	0.206
	N04400	211.9	0.03	<0.001	<0.039	0.0066	0.260
	N06600	0.1	0.0004	<0.001	<0.039	0.0048	0.189
	N06625	1.63	0.0002	<0.001	<0.039	0.0034	0.134
	N08800	0.007	0	<0.001	<0.039	0.0001	0.004
	N08825	2.5	0.0004	<0.001	<0.039	0.0099	0.390
Panama Canal Zone (tropical)	N02200	248.4	0.03	<0.001	<0.039	0.043	1.69
	N04400	234.9	0.03	<0.001	<0.039	0.0115	0.453
	N06600	7.4	0.001	<0.001	<0.039	0.032	1.26
	N06625	0.6	0.00008	<0.001	<0.039	0.0036	0.142
	N08800	10.13	0.001	<0.001	<0.039	0.0096	0.378
	N08825	4.2	0.0006	<0.001	<0.039	0.0013	0.051

(a) mg/dm/day. Source: Ref 25

Table 14 American Society for Testing and Materials 1976 atmospheric test program of 20 year exposure results for mechanical properties

Alloy	Test sites	Ultimate tensile strength, MPa (ksi) average 20 years	% loss	Elongation, % in 50.8 mm (2.0 in.) average 20 years	% loss
N02200	Initial	478.8 (69.4)	...	36.0	...
	Kure Beach, NC (east coast marine)	477.2 (69.2)	0.3	35.1	2.5
	Kearny, NJ (industrial)	477.3 (69.2)	0.3	36.8	0
	Point Reyes, CA (west coast marine)	475.9 (69.0)	0.6	35.7	0.8
	State College, PA (rural)	478.8 (69.4)	0	36.2	0
	Panama Canal Zone (tropical)	476.2 (69.0)	0.5	35.6	1.1
N04400	Initial	536.6 (77.8)	...	40.0	...
	Kure Beach, NC (east coast marine)	536.6 (77.8)	0	39.3	1.8
	Kearny, NJ (industrial)	534.7 (77.5)	1.9	38.7	3.3
	Point Reyes, CA (west coast marine)	536.1 (77.7)	0.09	38.6	3.5
	State College, PA (rural)	536.2 (77.7)	0.07	38.8	3.0
	Panama Canal Zone (tropical)	538.7 (78.1)	0	39.1	2.3
N06600	Initial	664.9 (96.4)	...	40.0	...
	Kure Beach, NC (east coast marine)	666.6 (96.7)	0	36.9	7.8
	Kearny, NJ (industrial)	659.3 (95.6)	0.8	39.4	1.5
	Point Reyes, CA (west coast marine)	662.5 (96.1)	0.4	39.1	2.3
	State College, PA (rural)	660.4 (95.8)	0.7	38.6	3.5
	Panama Canal Zone (tropical)	675.5 (97.9)	0	37.4	6.5
N06625	Initial	858.0 (124.4)	...	54.0	...
	Kure Beach, NC (east coast marine)	861.7 (124.9)	0	53.0	1.9
	Kearny, NJ (industrial)	858.3 (124.4)	0	51.0	5.6
	Point Reyes, CA (west coast marine)	862.2 (125.0)	0	52.5	2.8
	State College, PA (rural)	864.3 (125.3)	0	52.8	2.2
	Panama Canal Zone (tropical)	859.4 (124.6)	0	53.3	1.3
N08800	Initial	594.4 (86.2)	...	39.0	...
	Kure Beach, NC (east coast marine)	593.4 (86.0)	0.2	38.3	1.8
	Kearny, NJ (industrial)	592.2 (72.8)	0.4	32.7	3.3
	Point Reyes, CA (west coast marine)	596.4 (86.5)	0	32.9	2.8
	State College, PA (rural)	597.7 (86.7)	0	32.9	2.8
	Panama Canal Zone (tropical)	594.0 (86.1)	0.1	38.1	2.3
N08825	Initial	782.7 (113.5)	...	32.0	...
	Kure Beach, NC (east coast marine)	785.7 (113.9)	0	30.0	6.3
	Kearny, NJ (industrial)	779.5 (113.0)	0.4	30.4	5.0
	Point Reyes, CA (west coast marine)	788.6 (114.3)	0	31.2	2.5
	State College, PA (rural)	789.8 (114.5)	0	30.7	4.0
	Panama Canal Zone (tropical)	789.5 (114.5)	0	31.1	2.8

Source: Ref 25

The outline for this is shown in Fig. 23 for ISO categories 9223 to 9226 (Ref 29). The ISO scheme considers TOW, SO_2, and Cl^- content. Table 18 defines the ISO parameters, where "P" represents Cl^-, "S" represents SO_2, and τ represents the TOW (Ref 30). The ISO classifications for "P" and "S" are given in Table 19, and TOW is shown in Table 20 (Ref 30). The description of the five ISO 9223 categories C_1 through C_5 is given in Table 21 for carbon steel, zinc, copper, and aluminum (Ref 30). As an example, for corrosion rates of steel <1.3 μm/yr (0.05 mil/yr), the corrosivity is categorized as very low, while for corrosion rates between 80 and 200 μm/yr (3.2 and 7.9 mils/yr), the corrosivity is given as very high. Table 22 provides a list of sites from around the world that have corrosion rates listed, and for 27 of the sites, their corrosivity category as determined from ISO 9223 is included (Ref 31).

Considerable effort has been made through ISO to determine 1 year corrosion rates for steel, zinc, copper, and aluminum. Equations modeling this behavior are given as follows. Equations 6 through 9 are based on data taken from 1 year of exposure and may be used only for classification purposes (Ref 30).

The dose-response functions are as follows for carbon steel (C_{St}, $N = 119$, $R^2 = 0.87$), zinc (C_{Zn}, $N = 116$, $R^2 = 0.78$), copper (C_{Cu}, $N = 114$, $R^2 = 0.81$), and aluminum (C_{Al}, $N = 108$, $R^2 = 0.61$):

$$C_{St} = 0.085 SO_2^{0.56} \, TOW^{0.53} \exp\{f_{St}\}$$
$$+ 0.24 Cl^{0.47} \, TOW^{0.25} \exp\{0.049T\}$$
$$f_{St}(T) = 0.098(T - 10) \text{ when } T \leq 10\,°C$$
$$f_{St}(T) = -0.087(T - 10) \text{ when } T > 10\,°C$$

(Eq 6)

$$C_{Zn} = 0.0053 SO_2^{0.43} \, TOW^{0.53} \exp\{f_{Zn}\}$$
$$+ 0.00071 Cl^{0.68} \, TOW^{0.30} \exp\{0.11T\}$$
$$f_{Zn}(T) = 0 \text{ when } T \leq 10\,°C$$
$$f_{Zn}(T) = -0.032(T - 10) \text{ when } T > 10\,°C$$

(Eq 7)

$$C_{Cu} = 0.00013 SO_2^{0.55} \, TOW^{0.84} \exp\{f_{Cu}\}$$
$$+ 0.0024 Cl^{0.31} \, TOW^{0.57} \exp\{0.030T\}$$
$$f_{Cu}(T) = 0.047(T - 10) \text{ when } T \leq 10\,°C$$
$$f_{Cu}(T) = -0.029(T - 10) \text{ when } T > 10\,°C$$

(Eq 8)

$$C_{Al} = 0.00068 SO_2^{0.87} \, TOW^{0.38} \exp\{f_{Al}\}$$
$$+ 0.00098 Cl^{0.49} \, TOW^{0.38} \exp\{0.057T\}$$
$$f_{Al}(T) = 0 \text{ when } T \leq 10\,°C$$
$$f_{Al}(T) = -0.031(T - 10) \text{ when } T > 10\,°C$$

(Eq 9)

where C_M is corrosion attack in a micrometer of metal (M) after 1 year (μm/yr) of exposure,

Table 15 Corrosion data for noncoupled metal panels exposed at the U.S. Naval Research Laboratory tropical seacoast site at Cristobal, Panama

Metal or alloy	Surface(a)	General corrosion Average penetration(b), μm (mils)					Final corrosion rate(c), μm (mils)	Pitting Average deepest 20 pits(d), μm (mils)		Deepest pit, μm (mils)	Loss in tensile strength(e), %	
		1 year	2 years	4 years	8 years	16 years		8 years	16 years		8 years	16 years
Magnesium alloys												
AZ31X	...	28 (1.1)	48 (1.9)	91 (3.6)	201 (7.9)	381 (15)	23 (0.9)	178 (7)	559 (22)	864 (34)	25	47
AZ61X	...	12 (0.47)	33 (1.3)	...	157 (6.2)	304 (12)	19 (0.75)	177 (6.9)	466 (18.3)	533 (21)	28	32
Aluminum alloys												
1100	...	<0.3 (0.01)	1 (0.04)	<0.3 (0.01)	0.5 (0.02)	2.8 (0.11)	<0.3 (0.01)	<125 (4.9)	<125 (4.9)	<125 (4.9)	<1	<1
6061-T6	...	0.8 (0.3)	1.5 (0.06)	2.0 (0.08)	0.8 (0.03)	2.8 (0.11)	0.3 (0.01)	<125 (4.9)	<125 (4.9)	<125 (4.9)	1	<1
2024-T6	...	0.8 (0.3)	1.0 (0.04)	0.5 (0.02)	0.5 (0.02)	3.3 (0.13)	<0.3 (0.01)	125 (4.9)	125 (4.9)	125 (4.9)	1	1
Zinc (99.5%)	...	5.8 (0.23)	9.1 (0.36)	17 (0.67)	28 (1.1)	41 (1.6)	1.8 (0.07)	<125 (4.9)	<125 (4.9)	381 (15)	3	3
Iron												
Low-copper ingot	Pickled	101 (4)	207 (8.1)	794 (31.2)	...	...	...	...	...	...	...	...
ASTM K	Pickled	52 (2)	79 (3.1)	128 (5)	210 (8.3)	...	19 (0.75)	762 (30)	...	...	...	...
Aston wrought	Pickled	70 (2.8)	99 (3.9)	177 (7)	281 (11.0)	475 (18.7)	24 (0.94)	737 (29)	1346 (53)	1549 (61)	...	...
Aston wrought	Mill scale	69 (2.7)	138 (5.4)	168 (6.6)	282 (11.1)	403 (15.9)	...	1041 (41)	1245 (49)	1549 (61)	...	...
Carbon steel												
0.24% C	Pickled	64 (2.5)	122 (4.8)	144 (5.7)	259 (10.2)	402 (15.8)	21 (0.83)	863 (33.9)	1295 (51)	3124 (123)	...	...
0.24% C	Mill scale	66 (2.6)	114 (4.5)	141 (5.6)	278 (10.9)	401 (15.78)	...	940 (37)	1321 (52)	3124 (123)	...	...
0.24% C	Machined	50 (2)	78 (3)	126 (5)	173 (6.8)	270 (10.6)	12 (0.47)	355 (13.9)	457 (17.9)	991 (39)	...	...
Copper-bearing	Pickled	55 (2.2)	78 (3)	116 (4.6)	222 (8.7)	345 (13.6)	19 (0.75)	787 (30.9)	762 (30)	1676 (66)	...	...
Low-alloy steel												
Cu, Ni	Pickled	44 (1.7)	60 (2.4)	79 (3.1)	127 (5)	198 (7.8)	10 (0.4)	301 (11.9)	356 (14)	432 (17)	...	...
Cu, Cr, Si	Pickled	43 (1.65)	57 (2.2)	79 (3.1)	130 (5.1)	204 (8)	10 (0.4)	305 (12)	457 (17.9)	889 (35)	...	...
Cu, Ni, Mn, Mo	Pickled	44 (1.7)	61 (2.4)	76 (3)	124 (4.9)	188 (7.4)	9.7 (0.38)	305 (12)	406 (16)	914 (36)	...	...
Cr, Ni, Mn	Pickled	42 (1.6)	57 (2.2)	71 (2.8)	115 (4.5)	160 (6.3)	7.9 (0.31)	305 (12)	330 (13)	737 (29)	...	...
Nickel steel (2% Ni)	Pickled	39 (1.5)	51 (2)	66 (2.6)	95 (3.7)	146 (5.7)	6.6 (0.26)	279 (10.9)	330 (13)	483 (19)	...	...
Nickel steel (5% Ni)	Pickled	34 (1.3)	47 (1.85)	58 (2.3)	90 (3.5)	136 (5.4)	6.4 (0.25)	305 (12)	305 (12)	381 (15)	...	...
Chromium steel (3% Cr)	Pickled	50 (2)	63 (2.5)	77 (3.03)	116 (4.6)	169 (6.7)	7.7 (0.3)	457 (17.9)	609 (24)	1600 (63)	...	...
Chromium steel (5% Cr)	Pickled	41 (1.6)	47 (1.85)	55 (2.2)	90 (3.5)	113 (4.4)	5.1 (0.2)	279 (10.9)	330 (13)	483 (19)	...	...
Cast steel (0.27% C)	Machined	44 (1.7)	63 (2.5)	90 (3.5)	140 (5.5)	217 (8.5)	11 (0.43)	356 (14)	432 (17)	914 (36)	...	...
Cast iron-gray (3.2% C)	Machined	39 (1.5)	56 (2.2)	88 (3.46)	133 (5.2)	196 (7.7)	8.1 (0.32)	356 (14)	457 (17.9)	940 (37)	...	...
Cast iron												
Austenitic (18% Ni)	Machined	25 (1)	34 (1.3)	44 (1.7)	113 (4.4)	233 (9.2)	15 (0.6)	558 (21.9)	1041 (41)	1499 (59)	...	...
Stainless steels												
Type 410	...	1.0 (0.04)	1.0 (0.04)	1.5 (0.06)	1.0 (0.04)	4.6 (0.18)	0.3 (0.01)	<125 (4.9)	<125 (4.9)	<125 (4.9)	<1	<1
Type 430	...	0.5 (0.02)	1.0 (0.04)	1.0 (0.04)	1.0 (0.04)	2.0 (0.08)	<0.3 (0.01)	<125 (4.9)	<125 (4.9)	<125 (4.9)	<1	<1
Type 301	...	0.3 (0.01)	<0.3 (0.01)	<0.3 (0.01)	0.3 (0.01)	0.5 (0.02)	0.3 (0.01)	<125 (4.9)	<125 (4.9)	<125 (4.9)	<1	<1
Type 321	...	<0.3 (0.01)	<0.3 (0.01)	<0.3 (0.01)	0.3 (0.01)	0.5 (0.02)	<0.3 (0.01)	<125 (4.9)	<125 (4.9)	<125 (4.9)	<1	<1
Type 316	...	<0.3 (0.01)	<0.3 (0.01)	<0.3 (0.01)	<0.3 (0.01)	<0.3 (0.01)	<0.3 (0.01)	<125 (4.9)	<125 (4.9)	<125 (4.9)	<1	<1
α-β Brass												
Muntz metal (1/4% As)	...	1.8 (0.07)	2.3 (0.091)	3.6 (0.14)	5.8 (0.23)	11 (0.43)	0.8 (0.03)	<125 (4.9)	<125 (4.9)	<125 (4.9)	4	8
Naval	...	1.5 (0.06)	2.0 (0.08)	3.3 (0.13)	5.3 (0.21)	9.9 (0.38)	0.5 (0.02)	<125 (4.9)	<125 (4.9)	<125 (4.9)	3	7
Manganese bronze	...	4.6 (0.18)	4.8 (0.19)	7.6 (0.3)	8.4 (0.33)	15 (0.6)	0.8 (0.03)	<125 (4.9)	<125 (4.9)	<125 (4.9)	6	8
α brass												
Cu-30Zn	...	1.3 (0.05)	1.8 (0.07)	2.8 (0.11)	4.6 (0.18)	8.4 (0.33)	0.5 (0.02)	<125 (4.9)	<125 (4.9)	<125 (4.9)	5	4
Cu-20Zn	...	2.0 (0.08)	2.8 (0.11)	4.1 (0.16)	5.8 (0.23)	9.4 (0.37)	0.5 (0.02)	<125 (4.9)	<125 (4.9)	<125 (4.9)	2	3
Cu-10Zn	...	3.0 (0.12)	3.6 (0.14)	5.6 (0.22)	7.8 (0.31)	12 (0.47)	0.5 (0.02)	<125 (4.9)	<125 (4.9)	<125 (4.9)	2	3
Bronze												
Aluminum (5%)	...	2.0 (0.08)	2.8 (0.11)	3.8 (0.15)	5.8 (0.23)	9.9 (0.38)	0.5 (0.02)	<125 (4.9)	<125 (4.9)	<125 (4.9)	1	2
Phosphor	...	5.1 (0.2)	7.4 (0.29)	10 (0.4)	15 (0.6)	24 (0.95)	1.0 (0.04)	<125 (4.9)	<125 (4.9)	<125 (4.9)	6	3
Silicon	...	7.9 (0.31)	10 (0.4)	17 (0.67)	28 (1.1)	48 (1.9)	2.3 (0.09)	<125 (4.9)	<125 (4.9)	<125 (4.9)	2	3
Cast bronze												
Tin (8%)	Machined	4.6 (0.18)	8.9 (0.35)	11 (0.43)	14 (0.55)	21 (0.83)	1.0 (0.04)	<125 (4.9)	<125 (4.9)	<152 (6)	...	...
Ni-Sn (6% Ni)	Machined	3.3 (0.13)	4.6 (0.18)	7.4 (0.29)	11 (0.43)	16 (0.63)	0.5 (0.02)	125 (4.9)	125 (4.9)	125 (4.9)	...	...
Copper (99.9%)	...	4.3 (0.17)	5.8 (0.23)	9.7 (0.38)	14 (0.55)	20 (0.78)	0.8 (0.03)	<125 (4.9)	<125 (4.9)	<125 (4.9)	4	5
Copper/nickel (70/30)	...	0.8 (0.03)	1.5 (0.06)	3.0 (0.1)	5.8 (0.23)	10 (0.4)	0.5 (0.02)	<125 (4.9)	<125 (4.9)	<125 (4.9)	<1	1
Monel 400	...	1.0 (0.04)	1.0 (0.04)	1.8 (0.07)	3.0 (0.1)	5.6 (0.22)	0.3 (0.01)	<125 (4.9)	<125 (4.9)	<125 (4.9)	<1	2
Nickel (99%)	...	0.2 (0.008)	0.5 (0.02)	0.8 (0.03)	1.5 (0.05)	5.0 (0.2)	<0.3 (0.01)	<125 (4.9)	<125 (4.9)	<125 (4.9)	<1	<1
Lead (99%)	...	1.5 (0.06)	3.4 (0.13)	6.3 (0.25)	11 (0.43)	20 (0.8)	1.3 (0.05)	<125 (4.9)	<125 (4.9)	<125 (4.9)	<1	<1
Coated steels												
Galvanized	...	6.6 (0.26)	...	15 (0.6)	24 (0.95)	...	...	<125 (4.9)	...	...	...	...
Zn sprayed	...	1.5 (0.06)	13 (0.51)	14 (0.55)	17 (0.67)	...	...	127 (5)	...	...	...	...
Pb coated	...	2.0 (0.08)	5.1 (0.2)	...	9.1 (0.36)	...	...	<125 (4.9)	...	...	...	...
Al sprayed	...	<0.3 (0.01)	<0.3 (0.01)	<0.3 (0.01)	<0.3 (0.01)	<0.3 (0.01)	<0.3 (0.01)	<125 (4.9)	<125 (4.9)	<125 (4.9)	...	...

(a) All specimens were degreased before exposure; any treatment prior to degreasing is listed. (b) Average penetration over a 4.23 dm^2 (65.6 in.^{2}) exposed area; calculations based on weight loss and density. (c) Rate after time-corrosion relation had stabilized; slope of the linear portion of the curve, usually after two to eight years. (d) Averages obtained by measuring the five deepest measurable (>125 μm, or 5 mils) penetrations on each surface of duplicate panels. (e) Percent loss in ultimate tensile strength for 1.59 mm (1/16 in.) thick metal. Source: Ref 18

Table 16 Comparisons between the real long-term (more than 10 years) atmospheric corrosion data and the values estimated according to ISO 9224 criteria and by applying the power-law (Eq 5) and n-range in Table 17

		Experimental data		ISO 9224 criteria			Power law		
				Range			Range		
Test site	Time of exposure, years	μm	mil	μm	mil	Satisfactory prediction	μm	mil	Satisfactory prediction
Low-carbon steel corrosion									
El Escorial	13	32.7	1.29	5.3–54.5	0.2–2.15	Yes	17–48	0.67–1.9	Yes
Madrid	16	90.0	3.55	59–156	2.3–6.15	Yes	103–313	4.06–12.3	No
Zaragoza	13	119.0	4.69	54.5–138	2.15–5.44	Yes	71–199	2.8–7.84	Yes
Praha Letnany	16	205	8.07	59–156	2.3–6.15	No	78–237	3.1–9.34	Yes
Hurbanovo	15	131	5.16	57.5–150	2.27–5.9	Yes	106–313	4.18–12.3	Yes
Bilbao	16	534	21.0	156–420	6.15–16.5	No	162–491	6.38–19.3	No
Usti	15	540	21.3	400–1450	15.8–57.1	Yes	437–987	17.2–38.9	Yes
East Chicago	30	244	9.61	240–700	9.46–27.6	Yes	180–703	7.09–27.7	Yes
Bayonne	18.1	278	11.0	169–462	6.66–18.2	Yes	176–562	6.93–22.1	Yes
Kearny	12	154	6.07	340–1180	13.4–46.49	No	360–758	14.2–29.9	No
Alicante, 100 m (35 ft)	16	173	6.82	59–156	2.3–6.15	No	64–194	2.5–7.64	Yes
Barcelona	13	315	12.4	138–360	5.44–14.2	Yes	110–306	4.33–12.1	No
Kure Beach, 25 m (80 ft)	12	776	30.6	340–1180	13.4–46.49	Yes	626–1320	24.7–52.0	Yes
Kure Beach, 250 m (800 ft)	30	502	19.8	65–180	2.6–7.09	No	78–303	3.1–11.9	No
Point Reyes	12	1028	40.5	340–1180	13.4–46.49	Yes	426–899	16.8–35.4	No
Miraflores	16	279	11.0	59–156	2.3–6.15	No	78–237	3.1–9.34	No
Cristobal	16	393	15.5	156–420	6.15–16.5	Yes	156–474	6.15–18.7	Yes
Zinc corrosion									
El Escorial	13	4.5	0.18	6.5–26	0.26–1.0	No	7.3–12.2	0.29–0.48	No
Madrid	16	16.5	0.65	8.0–32	0.32–1.3	Yes	13.2–23.0	0.52–0.91	Yes
Zaragoza	13	15.2	0.60	6.5–26	0.26–1.0	Yes	9.2–15.3	0.36–0.60	Yes
Praha Letnany	18	49.0	1.93	9.0–36	0.35–1.4	No	19.2–34.2	0.76–1.35	No
Hurbanovo	15	24.0	0.95	7.5–30	0.30–1.2	Yes	9.6–16.5	0.38–0.65	No
State College	20	22.6	0.89	10–40	0.39–1.6	Yes	15.4–28.0	0.61–1.1	Yes
Bilbao	16	70.6	2.78	64–160	2.5–6.3	Yes	36.6–63.7	1.44–2.51	No
Usti	18	88.0	3.47	72–180	2.8–7.1	Yes	34.8–62.0	1.37–2.44	No
New York	20	114.8	4.52	80–200	3.2–7.9	Yes	41.5–75.6	1.63–2.98	No
Sandy Hook	20	37.0	1.46	40–80	1.6–3.2	No	32.6–44	1.28–1.7	Yes
Alicante, 100 m (35 ft)	16	14.7	0.58	8.0–32	0.32–1.3	Yes	9.2–16.0	0.36–0.63	Yes
Miraflores	16	13.7	0.54	8.0–32	0.32–1.3	Yes	12.9–22.4	0.51–0.88	Yes
Cristobal	16	41.7	1.64	64–160	2.5–6.3	No	40.4–83.5	1.59–3.29	Yes
Copper corrosion									
El Escorial	13	2.6	0.10	1.3–18	0.05–0.71	Yes	3.7–10.5	0.15–0.41	No
Madrid	16	5.4	0.21	1.3–18	0.05–0.71	Yes	4.1–11.5	0.16–0.45	Yes
Zaragoza	13	5.9	0.23	21–48	0.83–1.9	No	5.7–17.2	0.22–0.68	Yes
Praha Letnany	18	18.2	0.72	1.8–23	0.07–0.90	Yes	5.5–17.5	0.22–0.69	No
Bilbao	16	24.4	0.96	48–80	1.9–3.2	No	22.6–39.4	0.89–1.55	Yes
Usti	18	30.4	1.20	54–90	2.1–3.5	No	26.6–47.5	1.05–1.87	Yes
Alicante, 30 m (10 ft)	16	11.9	0.47	48–80	1.9–3.2	No	21.6–37.6	0.85–1.48	No
Alicante, 100 m (35 ft)	16	6.1	0.24	21–48	0.83–1.9	No	7.0–21.1	0.28–0.83	No
Barcelona	13	10.6	0.42	39–65	1.5–2.6	No	13.4–22.3	0.53–0.88	No
Miraflores	16	6.9	0.27	21–48	0.83–1.9	No	8.0–24.3	0.32–0.96	No
Cristobal	16	20.1	0.79	48–80	1.9–3.2	No	21.6–37.7	0.85–1.49	No

Source: Ref 28

Table 17 Predictions for long-term atmospheric corrosion of low-carbon steel, zinc, and copper

The range of the exponent n in Eq 5 for each type of atmosphere and ISO corrosivity category (Ref 3) are shown.

	Rural-urban atmospheres (without marine component)		Industrial atmospheres (without marine component)		Marine atmospheres	
Material	ISO corrosivity category	Range of n in Eq 5	ISO corrosivity category	Range of n in Eq 5	ISO corrosivity category	Range of n in Eq 5
Low-carbon steel	C_1–C_3	0.3–0.7	C_4–C_5	0.3–0.7	C_1–C_5	0.6–0.9
Zinc	C_1–C_3	0.8–1.0	C_4–C_5	0.9–1.0	C_1–C_5	0.7–0.9
Copper	C_1–C_4	0.5–0.9	C_5	0.6–0.8	C_1–C_5	0.4–0.6

$f_M(T)$ is the function for the particular metal, N is the number of tests, and R^2 is the statistical term estimating goodness of fit; $R^2 = 1$ is a perfect fit.

Table 23 provides the description of the symbols and the intervals used in the previous equations. The equations include the TOW, SO_2, and the Cl^- deposition rate. The results are reasonable, and a comparison between the predicted and the observed values is shown in Fig. 24 (Ref 30).

Additionally, linear regression relationships are used to describe atmospheric corrosion. Equation 10 is an example (Ref 29):

$$\ln(r_{corr}) = b_0 + b_1[SO_2] + b_2\ln[Cl] + b_3\ln[TOW] \quad \text{(Eq 10)}$$

where r_{corr} is the corrosion loss per year (μm/yr), $[SO_2]$ is the yearly average of concentration of SO_2 (μg/m^3), [Cl] is the yearly average of deposition rate of chloride (μg/m$^2\cdot$d), TOW is the percentage of hours per year when the RH is greater than 80% at a temperature greater than 0 °C (%), and b_0, b_1, b_2, and b_3 are coefficients given in Table 24.

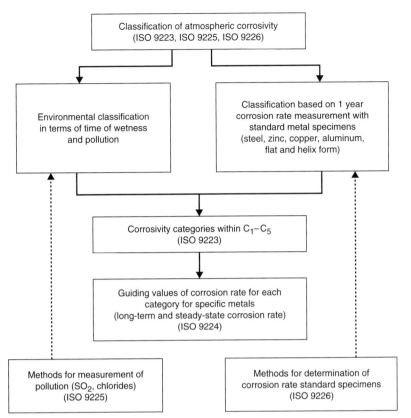

Fig. 23 Scheme for classification of atmospheric-corrosivity approach in ISO 9223 to 9226. Source: Ref 29

Equation 10 determines the 1 year corrosion loss for flat specimens. The results summarize all of the CORRAG data. For steel and copper, the R^2 values are 0.63 and 0.58, respectively. Zinc and aluminum have R^2 values less than 0.5. Given the environmental data for a particular location, Eq 10 may be used to predict the amount of corrosion in 1 year. In comparison, Eq 6 to 9 are an improved version of Eq 10.

Additional results from a study conducted in Vietnam are shown in Table 25 (Ref 32). The authors compared predicted corrosion rates with those measured for three models. The first model (a) used the average environmental data for the test years July 1995 through July 1998. Percent differences varied from −13 to 13.5%. For the second model (b), the same environmental data were used as in (a), except that the chloride concentration was ignored. For this case, the percent differences varied from +1 to −15%. The results suggested that for these tests and locations, chloride content was not a significant factor. For the third model (c), the authors used the average environmental data from January 1996 through December 1999, and for this case, the percent difference between measured and calculated varied from 0 to −19%.

For the Vietnamese study, the initial equation used to describe the 1 year corrosion is similar to

Table 18 Estimated corrosivity category for unalloyed carbon steel for time of wetness (TOW) category τ_3, τ_4, and τ_5 and different SO_2 (P_0 to P_3) and Cl (S_0 to S_3) categories

TOW chloride classification	τ_3			τ_4			τ_5		
	S_0–S_1	S_2	S_3	S_0–S_1	S_2	S_3	S_0–S_1	S_2	S_3
P_0, P_1	C_2–C_3	C_3–C_4	C_4	C_3	C_4	C_5	C_3–C_4	C_5	C_5
P_2	C_3–C_4	C_3–C_4	C_4–C_5	C_4	C_4	C_5	C_4–C_5	C_5	C_5
P_3	C_4	C_4–C_5	C_4	C_5	C_5	C_5	C_5	C_5	C_5

Definition of corrosivity categories C_1 to C_5 is given in Table 21. See Table 20 for wetness classifications τ_1–τ_5. See Table 19 for SO_2 classifications S_0–S_3

Table 19 Classification of sulfur compounds based on sulfur dioxide (SO_2) and airborne salinity contamination (Cl) according to ISO 9225

Class	Deposition rate, mg/m²·d	Concentration, mg/m³
Sulfur dioxide(a)		
P_0	<10	<12
P_1	10–35	12–40
P_2	36–80	41–90
P_3	80–200	91–250
Chloride(b)		
S_0	<3	...
S_1	3–60	...
S_2	61–300	...
S_3	>300	...

(a) Sulfation plate measurement. (b) Chloride candle measurement

Table 20 Classification of time of wetness from ISO 9223

Wetness class	Time of wetness	
	h/yr	%
τ_1	≤10	<0.1
τ_2	10–250	0.1–3
τ_3	250–2500	3–30
τ_4	2500–5500	30–60
τ_5	>5500	>60

Table 21 ISO 9223 corrosivity categories for carbon steel, zinc, copper, and aluminum based on corrosion rates

Corrosivity	Category	Carbon steel		Zinc		Copper		Aluminum	
		μm/yr	mil/yr	μm/yr	mil/yr	μm/yr	mil/yr	g/m²·yr	mil/yr
Very low	C_1	≤1.3	≤0.05	≤0.1	≤0.004	≤0.1	≤0.004	Negligible	
Low	C_2	1.3–25	0.05–1.0	0.1–0.7	0.004–0.028	0.1–0.6	0.004–0.02	≤0.6	...
Medium	C_3	25–50	1.0–2.0	0.7–2.1	0.028–0.08	0.6–1.3	0.02–0.05	0.6–2	...
High	C_4	50–80	2.0–3.2	2.1–4.2	0.08–0.17	1.3–2.8	0.05–0.11	2–5	...
Very high	C_5	80–200	3.2–7.9	4.2–8.4	0.17–0.33	2.8–5.6	0.11–0.22	5–10	...

Source: Ref 30

Table 22 Test sorting of 1 year corrosion loss of flat specimens (μm/yr) based on values submitted by member countries

Unalloyed steel			Zinc			Copper			Aluminum		
Test site	A	B	Test site	A	B	Test site	A	B	Test site	A	B
Panama CZ	373.0	C₅(a)	Panama CZ	17.50	C₅(a)	Panama CZ	5.46	C₅	Auby	1.70	C₃
Auby	106.0	C₅	Auby	5.60	C₅	Biarritz	3.69	...	Ostende (B)	1.50	...
Ostende (B)	99.3	...	Ostende (B)	5.10	...	Kopisty	3.30	...	Jubay-Antarctic	1.31	...
Biarritz	87.2	...	Salin de Gir.	4.60	...	Salin de Gir.	3.20	...	St. Denis	1.20	...
Okinawa	75.2	C₄	Biarritz	4.30	...	Ostende (B)	3.10	...	Biarritz	1.20	...
Salin de Gir.	73.0	...	Borregaard	3.80	C₄	Kure Beach	2.85	...	Ponteau Mart.	1.00	...
Ponteau Mart.	72.4	...	Kopisty	3.50	...	Kvarnvik	2.80	...	Paris	0.90	...
Kopisty	70.7	...	Okinawa	3.40	...	Ponteau Mart	2.70	C₄	Murmansk	0.80	...
Borregaard	61.7	...	Tananger	3.00	...	Res Triang Park	2.43	...	Salin de Gir.	0.70	...
Kvarnvik	61.6	...	Paris	3.00	...	Point Reyes	2.42	...	Kopisty	0.70	...
Tananger	59.6	...	Praha	2.80	...	Camet	2.23	...	St. Remy	0.70	...
Rye	58.5	...	Ponteau Mart.	2.60	...	Okinawa	2.10	...	Tananger	0.60	...
Praha	47.4	C₃	Rye	2.54	...	Jubay-Antarctic	2.04	...	Praha	0.60	...
St. Remy	44.1	...	Vladivostok	2.30	...	Batumi	2.00	...	Kvarnvik	0.60	...
Baracaldo	43.9	...	Birkenes	2.30	...	Kasperske Hory	2.00	...	Borregaard	0.60	...
Choshi	43.3	...	Bergen	2.10	...	Tananger	1.90	...	Panama CZ	0.57	C₂
Paris	41.7	...	Kure Beach	2.01	C₃	Auby	1.90	...	Los Angeles	0.56	...
Point Reyes	40.1	...	Newark	1.96	...	Rye	1.86	...	Tokyo	0.54	...
Tokyo	39.5	...	Kasperske Hory	1.90	...	St. Remy	1.80	...	Kasperske Hory	0.50	...
Fleet Hall	39.0	C₃	Jubay-Antarctic	1.87	C₃	Kattesand	1.70	C₄	Rye	0.42	C₂
Stratford	38.7	...	Kvarnvik	1.80	...	Murmansk	1.70	...	Boucherville	0.40	...
Kure Beach	37.9	...	Point Reyes	1.73	...	Vladivostok	1.40	...	Kattesand	0.40	...
Crowthrone	37.4	...	Stratford	1.67	...	Paris	1.40	...	Fleet Hall	0.36	...
St. Denis	37.2	...	Iugazu	1.62	...	Picherande	1.40	...	Choshi	0.33	...
Camet	36.8	...	Bergisch Glad.	1.60	...	Borregaard	1.40	...	Picherande	0.30	...
Jubay-Antarctic	36.6	...	Batumi	1.60	...	Newark	1.39	...	Vladivostok	0.30	...
Bergisch Glad.	36.2	...	Kattesand	1.50	...	Judgeford	1.36	...	Helsinki	0.30	...
Kattesand	35.2	...	St. Remy	1.50	...	Choshi	1.35	...	Bergisch Glad.	0.30	...
Helsinki	33.3	...	St. Denis	1.50	...	Praha	1.30	...	Stratford	0.29	...
Murmunsk	30.8	...	Tokyo	1.50	...	Birkenes	1.30	...	Kure Beach	0.29	...
Batumi	28.7	...	Boucherville	1.40	...	Baracaldo	1.20	C₃	Newark	0.28	...
Bergen	27.9	...	Choshi	1.40	...	St. Denis	1.20	...	Okinawa	0.26	...
Madrid	27.7	...	Fleet Hall	1.34	...	Los Angeles	1.16	...	Point Reyes	0.22	...
Lagoas	26.9	...	Helsinki	1.30	...	Stratford	1.13	...	Stockholm Vana	0.20	...
Newark	26.4	...	Oslo	1.30	...	Crowthorne	1.10	...	Oslo	0.20	...
Kasperske Hory	26.0	...	Camet	1.26	...	El Pardo	1.10	...	Lagoas	0.20	...
Vladivostok	25.9	...	Baracaldo	1.20	...	Boucherville	1.10	...	Baracaldo	0.20	...
Otaniemi	25.6	C₃	Crowthorne	1.10	C₃	Bergen	1.00	C₃	Camet	0.19	C₂
Oslo	25.2	...	Murmansk	1.10	...	Lagoas	1.00	...	Crowthorne	0.12	...
Stockholm Vana	24.4	C₂	Los Angeles	1.09	...	Fleet Hall	0.93	...	Res Triang Park	0.11	...
Boucherville	23.2	...	Buenos Aires	1.01	...	Iugazu	0.80	...	Birkenes	0.10	C₁
Res Triang Park	23.1	...	Lagoas	1.00	...	Otaniemi	0.80	...	Ahtari	0.10	...
Los Angeles	21.4	...	Otaniemi	0.90	...	Svanvik	0.80	...	Batumi	0.10	...
Svanvik	20.2	...	Picherande	0.90	...	Helsinki	0.70	...	Otaniemi	0.10	...
Birkenes	19.7	...	Res Triang Park	0.84	...	Ahtari	0.70	...	Bergen	0.10	...
Judgeford	19.3	...	Svanvik	0.80	...	Tokyo	0.66	...	Svanvik	0.10	...
Buenos Aires	16.2	...	Ahtari	0.70	C₂	Buenos Aires	0.64	...	Madrid	0.07	...
Picherande	16.1	...	Judgeford	0.66	...	Bergisch Glad.	0.60	C₂	Oymyakon	0.07	...
El Pardo	15.5	...	Stockholm Vana	0.60	...	Stockholm Vana	0.60	...	Judgeford	0.06	...
Ahtari	12.8	...	Madrid	0.60	...	Oslo	0.60	...	Iugazu	0.05	...
Iugazu	5.8	...	El Pardo	0.50	...	Madrid	0.50	...	Buenos Aires	0.05	...
San Juan	4.6	...	Oymyakon	0.40	...	San Juan	0.18	...	El Pardo	0.05	...
Oymyakon	0.8	C₁	San Juan	0.18	...	Oymyakon	0.09	C₁	San Juan	0.03	...

Note: A, corrosion loss (μm/yr); B, corrosivity categories (ISO 9223). (a) Corrosion rate exceeding the upper limit in C₅. Source: Ref 29

Table 23 Parameters used in dose-response functions, including symbol, description, interval measured in the program, and unit

All parameters are expressed as annual averages.

Symbol	Description	Interval	Unit
T	Temperature	−17.1–28.7	°C
TOW	Time of wetness	206–8760	h/yr
SO_2	SO_2 deposition	0.7–150.4	mg/m² · d
Cl^-	Cl^- deposition	0.4–699.6	mg/m² · d

Source: Ref 30

Table 24 Coefficients for the ISO CORRAG program linear regression analysis (Eq 10)

Metal	b_0	b_1	b_2	b_3	R^2
Fe	3.647	0.011	0.137	0.833	0.63
Zn	0.388	0.010	0.126	0.552	0.49
Cu	0.354	0.005	0.148	0.702	0.58
Al	−1.972	0.014	0.233	0.225	0.39

Source: Ref 29

Eq 10, except that a relative humidity term (RH) is included. Because the Cl^- content did not have much of an effect, the final equation follows and does not include the Cl^- term:

$$r_{\text{corr}} = -8.78T + 5.25RH + 0.0081TOW - 10.228 \quad \text{(Eq 11)}$$

$$R^2 = 0.94$$

where r_{corr} is the corrosion after 1 year of exposure [g/(mm² · yr)], T is the annual average

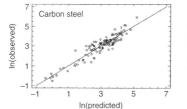 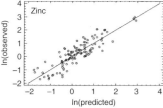 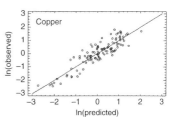 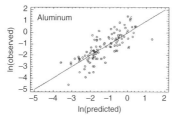

Fig. 24 Observed versus predicted values (logarithmic) for carbon steel (Eq 6), zinc (Eq 7), copper (Eq 8), and aluminum (Eq 9). Source: Ref 30

Table 25 A comparison between corrosion rates by experiment and calculations measured in Vietnam

	Corrosion rate, g/mm^2 · yr						
Test site	Observed	Calculated(a)	Error, %	Calculated(b)	Error, %	Calculated(c)	Error, %
Hanoi	240.36	243.291	1	243.044	1	230.668	−4
Doson	290.7	297.839	2	286.096	−2	281.994	−3
Danang	264.64	245.774	−8	244.588	−8	231.484	−14
Nhatrang	254.23	224.649	−13	217.219	−15	213.607	−19
HCM City	191.92	177.996	−8	177.526	−8	170.696	−12
Vungtau	191.23	206.284	7	201.056	5	192.049	0
Sontay	289.74	255.15	13.50	...	...	...	...

(a) Using the average environmental data July 1995 to July 1998 (during the exposure time). (b) Same as (a) but without chloride concentration. (c) Using the annual average environmental data Jan 1996 to Dec 1999. Source: Ref 32

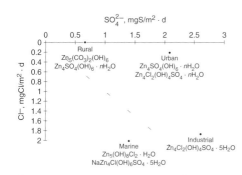

Fig. 25 Compositions of corrosion products on zinc as a function of sulfate on chloride. Source: Ref 33

Table 26 Results of measurements of contaminants in the rust by energy-dispersive spectroscopy (EDS) and by chemical analysis of water extracts after leaching in distilled water

	Thickness of the rust		Composition (by EDS), wt%		Soluble salts, mg/m^2	
Test site	μm	mils	Sulfur	Chlorine	Sulfates	Chlorides
Escorial	59.5	2.34	0.66	0.06	562	723
Madrid	93.0	3.66	0.74	0.41	974	1117
Bilbao	124.5	4.91	0.85	0.07	1410	756
Barcelona	111.0	4.37	0.78	1.03	1203	2237

Source: Ref 34

temperature (°C), RH is the annual average relative humidity (%), and TOW is the time of wetness (h/yr).

Corrosion Products

A study in Sweden examined the corrosion products formed on zinc panels that were rain sheltered. The locations had varying amounts of SO_4^{2-} and Cl^- in the atmosphere. The compositions of the different compounds formed on the zinc are shown in Fig. 25 (Ref 33). The upper left portion of the figure shows the compounds formed under low SO_4^{2-} and Cl^- concentrations, while the lower right portion of the figure identifies the compounds formed under the highest concentrations of SO_4^{2-} and Cl^-.

Examination of rust layers formed on steel in four locations in Spain is shown in Table 26 (Ref 34). The study examined the rust layer using energy-dispersive spectroscopy and analyzing soluble salts extracted from the rust layers. The thickness of the layers increased in going from Escorial, Spain (rural), to Barcelona, Spain (marine). Similarly, the chlorides showed a percentage increase in going from the rural to the marine environment.

Atmospheric Corrosion Test Sites

Table 27 provides a list of atmospheric corrosion sites throughout the world. Where available, the 2 year corrosion rates for low-carbon steel and zinc are given. Some of the sites have a marine corrosion index and/or an atmospheric corrosion index number after them. The higher the index numbers, the more aggressive the environment (Ref 2).

REFERENCES

1. C.L. Leygraf and T.E. Graedel, *Atmospheric Corrosion*, John Wiley & Sons, 2000, p 3
2. W.H. Ailor, Ed., *Atmospheric Corrosion*, John Wiley & Sons, 1982
3. "Corrosion of Metals and Alloys—Corrosivity of Atmospheres—Classification," ISO 9223: 1992, International Organization for Standardization
4. "Corrosion of Metals and Alloys—Corrosivity of Atmospheres—Guiding Values for the Corrosivity Categories," ISO 9224: 1992, International Organization for Standardization
5. P.W. Brown and L.W. Masters, Factors Affecting the Corrosion of Metals in the Atmosphere, *Atmospheric Corrosion*, W.H. Ailor, Ed., John Wiley & Sons, 1982, p 31
6. H. Guttman, Atmospheric and Weathering Factors in Corrosion Testing, *Atmospheric Corrosion*, W.H. Ailor, Ed., John Wiley & Sons, 1982, p 51
7. F.F. Ross and T.R. Shaw, Control of Atmospheric Corrosion Pollutants in Great Britain, *Atmospheric Corrosion*, W.H. Ailor, Ed., John Wiley & Sons, 1982, p 19
8. M. Schumacher, Ed., *Seawater Corrosion Handbook*, Noyes Data Corporation, 1979
9. L. Atteraas and S. Haagenrud, Atmospheric Corrosion in Norway, *Atmospheric Corrosion*, W.H. Ailor, Ed., John Wiley & Sons, 1982, p 873
10. *Betz Handbook of Industrial Water Conditioning*, 9th ed., Betz Laboratories, Inc., 1991
11. D.P. Doyle and T.E. Wright, Rapid Methods for Determining Atmospheric Corrosivity and Corrosion Resistance, *Atmospheric Corrosion*, W.H. Ailor, Ed., John Wiley & Sons, 1982, p 227

Table 27 Some marine-atmospheric corrosion test sites around the world
Corrosion rates of steel and zinc are also listed for some sites.

| Test site | Type of atmosphere | Distance from sea | | Corrosivity index(a) | | Corrosion rate from 2 year test | | | |
| | | | | | | Steel | | Zinc | |
		km	miles	MCI	ACI	mm/yr	mils/yr	mm/yr	mils/yr
United States									
Cape Canaveral, FL									
0.8 km (½ mile) from ocean	Marine	0.8	0.5	...	...	0.086	3.39	0.0011	0.045
55 m (60 yd), 9 m (30 ft) elevation	Marine	0.055	0.035	...	...	0.165	6.48	0.004	0.158
55 m (60 yd), ground level	Marine	0.055	0.035	...	...	0.44	17.37	0.0041	0.163
55 m (60 yd), 18 m (60 ft) elevation	Marine	0.055	0.035	...	...	0.131	5.17	0.0044	0.173
Point Reyes, CA	Marine	0.400	0.25	11	0.183	0.50	19.71	0.0015	0.060
Brazos River, TX	Industrial marine	...	...	...	...	0.093	3.67	0.0018	0.072
Daytona Beach, FL	Marine	...	...	...	...	0.295	11.63	0.0022	0.079
Kure Beach, NC									
250 m (800 ft)	Marine	0.244	0.15	...	...	0.145	5.73	0.0022	0.079
25 m (80 ft)	Marine	0.0244	0.015	11.4	...	0.53	21.00	0.0064	0.250
Miami, FL	Marine	4	2.5	5.9	0.04	...	...	...	...
Ormond Beach, FL	Marine	...	...	...	...	...	...	...	...
Battelle, Sequin, WA	Marine	0.030	0.018	6.9	0.07	...	...	...	...
Hickham AFB, HI	Marine	0.150	0.09	8.7	1.4	...	...	...	...
Panama									
Fort Amidor	Marine	...	...	...	...	0.014	0.57	0.0011	0.045
Miraflores	Marine	...	...	...	...	0.043	1.69	0.0026	0.104
Limon Bay	Marine	...	...	...	...	0.062	2.45	0.0026	0.104
Galeta Point	Marine	...	...	...	...	0.69	27.14	0.015	0.607
Canada									
Esquimalt, Vancouver Island, BC	Rural marine	...	...	...	...	0.013	0.53	0.0005	0.019
Cape Beale, NC	Marine	0.025	0.015	12.4	0.20	...	...	...	...
Chebucto Head, NS	Marine	0.100	0.06	13.0	1.2	...	...	...	...
Estevan Point, BC	Marine	0.400	0.25	8.4	0.02	...	...	...	...
Daniels Harbor, NF	Marine	0.150	0.09	17.5	0.11	...	...	...	...
Sable Island, NS	Marine	...	...	13.9	0.99	...	...	...	...
St. Vincents, NF	Marine	0.150	0.09	14.7	0.18	...	...	...	...
Deadmans Bay, NF	Marine	0.030	0.018	11.9	0.12	...	...	...	...
England									
Dungeness	Industrial marine	...	...	...	...	0.49	19.22	0.0036	0.143
Pilsey Island	Industrial marine	...	...	...	...	0.103	4.04	0.0057	0.223
Cornwall	Industrial marine	0.400	0.25	12.9	0.83	...	...	...	...
Ghana									
Tema	...	0.030	0.018	77.5	3.4	...	...	...	...
Benin									
Cotonou	...	0.150	0.09	17.6	0.67	...	...	...	...
Togo									
Lome	...	0.100	0.06	23.6	0.27	...	...	...	...
South Africa									
Durban, Salisbury Island	Marine	0.010	0.006	64.0	5.7	0.056	2.20	0.015	0.607
Dyeban Bluff	Severe marine	...	...	...	...	0.26	10.22	0.0032	0.126
Cape Town docks	Mild marine	...	...	...	...	0.047	1.84	0.0032	0.126
Walvis Bay military camp	Severe marine	...	...	...	...	0.11	4.33	0.063	2.483
Simonstown	Marine	...	...	...	...	0.016	0.63	0.0032	0.126
Nigeria									
Lagos									
45 m (50 yd)	Severe marine	0.046	0.03	...	...	...	...	...	...
365 m (400 yd)	Marine	0.366	0.23	...	...	...	...	...	...
1190 m (1300 yd)	Mild marine	1.189	0.74	...	...	...	...	...	...
Bahrain									
Sadad	Marine	0.800	0.5	11.2	0.30	...	...	...	...
Iran									
Shapour	Marine	0.010	0.006	5.2	0.15	...	...	...	...
Pakistan									
Karachi	Marine	0.060	0.037	33.8	4.1	...	...	...	...
Yemen									
Rasketenib	Marine	0.100	0.06	14.3	1.1	...	...	...	...

(continued)

(a) MCI, marine corrosivity index; determined by the weight loss of an aluminum wire/mild steel bolt couple. ACI, atmospheric corrosivity index; determined by the weight loss of an aluminum open-helical coil specimen or an aluminum wire/plastic bolt specimen. Source: Ref 2

Table 27 (continued)

| Test site | Type of atmosphere | Distance from sea | | Corrosivity index(a) | | Corrosion rate from 2 year test | | | |
| | | km | miles | MCI | ACI | Steel | | Zinc | |
						mm/yr	mils/yr	mm/yr	mils/yr
Japan									
Hitachi	Marine	1.0	0.62	5.2	0.41	...	...	...	...
Okinawa	Marine	0.500	0.31	26.2	1.8	...	...	...	...
Zushi	Marine	0.016	0.01	2.6	1.4	...	...	...	...
Australia									
Sydney (beach)	Marine	0.010	0.006	6.4	2.0	...	...	...	...
Sydney (D.S.L.)	...	3	1.8	7.1	1.3	...	...	...	...
New Zealand									
Phia	Marine	0.2	0.12	15.8	2.4	...	...	...	...
Greece									
Rafina	Marine	0.2	0.12	13.6	1.0	...	...	...	...
Rhodes	Marine	0.2	0.12	14.3	1.5	...	...	...	...
Netherlands									
Schagen	Marine	2.4	1.5	17.0	2.0	...	...	...	...
Spain									
Almeria	...	0.035	0.022	22.4	1.6	...	...	...	...
Cartagena	...	0.050	0.031	5.2	1.9	...	...	...	...
La Coruña	...	0.160	0.1	26.2	1.4	...	...	...	...
Barbados									
Holetown	Marine	0.075	0.047	59	0.42	...	...	...	...
Dominican Republic									
El Macao	Marine	0.100	0.06	30	0.27	...	...	...	...
Colombia									
Barranquilla	Marine	0.010	0.006	12.6	3.3	...	...	...	...
Cartagena	Marine	0.010	0.006	16.3	3.1	...	...	...	...
Galera Zamba	Marine	0.190	0.12	48.0	8.8	...	...	...	...
Santa Marta	Marine	0.060	0.037	1.9	0.06	...	...	...	...
Guatemala									
Pacific Beach	Marine	...	...	17.2	2.7	...	...	...	...
Uruguay									
Punta Del Este	Marine	0.040	0.025	48.3	0.79	...	...	...	...
Venezuela									
Carmaine Chico	...	0.800	0.5	39.0	0.53	...	...	...	...

(a) MCI, marine corrosivity index; determined by the weight loss of an aluminum wire/mild steel bolt couple. ACI, atmospheric corrosivity index; determined by the weight loss of an aluminum open-helical coil specimen or an aluminum wire/plastic bolt specimen. Source: Ref 2

12. A.C. Dutra and R. de O. Vianna, Atmospheric Corrosion Testing in Brazil, *Atmospheric Corrosion,* W.H. Ailor, Ed., John Wiley & Sons, 1982, p 755
13. S. Feliu, M. Morocillo and B. Chico, Effect of Distance from Sea on Atmospheric Corrosion Rate, *Corrosion,* Vol 55 (No. 9), 1999, p 883
14. S.W. Dean and E.C. Rhea, Ed., *Atmospheric Corrosion of Metals,* STP 767, American Society for Testing and Materials, 1982
15. G.A. King and D.J. O'Brien, The Influence of Marine Environments on Metals and Fabricated Coated Metal Products, Freely Exposed and Partially Sheltered, *Atmospheric Corrosion,* STP 1239, W.W. Kirk and H.H. Lawson, Ed., ASTM, 1995, p 167
16. B.G. Callaghan, Atmospheric Corrosion Testing in Southern Africa, *Atmospheric Corrosion,* W.H. Ailor, Ed., John Wiley & Sons, 1982, p 893
17. G. Sowinski and D.O. Sprowls, Weathering of Aluminum Alloys, *Atmospheric Corrosion,* W.H. Ailor, Ed., John Wiley & Sons, 1982, p 297
18. C.R. Southwell and J.D. Bultman, Atmospheric Corrosion Testing in the Tropics, *Atmospheric Corrosion,* W.H. Ailor, Ed., John Wiley & Sons, 1982, p 943
19. S.W. Dean, Analyses of Four Years of Exposure Data from the USA Contribution of ISO CORRAG Program, *Atmospheric Corrosion,* STP 1239, W.W. Kirk and H.H. Lawson, Ed., ASTM, 1995, p 56
20. M.J. Johnson and P.J. Pavlik, Atmospheric Corrosion of Stainless Steel, *Atmospheric Corrosion,* W.H. Ailor, Ed., John Wiley & Sons, 1982, p 461
21. R.M. Kain, B.S. Phull and S.J. Pikul, 1940 'til Now—Long-Term Marine Atmospheric Corrosion Resistance of Stainless Steel and Other Nickel Containing Alloys, *Outdoor Atmospheric Corrosion,* STP 1421, H.E. Townsend, Ed., ASTM International, 2002, p 343
22. S.K. Coburn, M.E. Komp, and S.C. Lore, Atmospheric Corrosion Rates of Weathering Steels at Test Sites in the Eastern United States—Effect of Environment and Test-Panel Orientation, *Atmospheric Corrosion,* STP 1239, W.W. Kirk and H.H. Lawson, Ed., ASTM, 1995, p 101
23. H.E. Townsend, Effects of Silicon and Nickel Contents on the Atmospheric Corrosion Resistance of ASTM A588 Weathering Steel, *Atmospheric Corrosion,*

STP 1239, W.W. Kirk and H.H. Lawson, Ed., ASTM, 1995, p 85
24. H.E. Townsend and H.H. Lawson, Twenty-One Year Results for Metallic-Coated Steel Sheet in the ASTM 1976 Atmospheric Corrosion Tests, *Outdoor Atmospheric Corrosion,* STP 1421, H.E. Townsend, Ed., ASTM International, 2002, p 284
25. E.L. Hibner, Evaluation of Nickel-Alloy Panels from the 20-Year ASTM G01.04 Atmospheric Test Program Completed in 1996, *Outdoor Atmospheric Corrosion,* STP 1421, H.E. Townsend, Ed., ASTM International, 2002, p 277
26. A.A. Bragard and E. Bonnarens, Prediction of Atmospheric Corrosion of Structural Steels for Short-Term Experimental Data, *Atmospheric Corrosion of Metals,* STP 767, S.W. Dean and E.C. Rhea, Ed., ASTM, 1982, p 339
27. S.W. Dean and D.B. Reiser, Analysis of Long-Term Atmospheric Corrosion Results from ISO CORRAG Program, *Outdoor Atmospheric Corrosion,* STP 1421, H.E. Townsend, Ed., ASTM International, 2002, p 3
28. M.J. Morcillo, J.S. Simancas, and S. Feliu, Long-Term Atmospheric Corrosion in Spain: Results after 13 to 16 Years of Exposure and Comparison with Worldwide Data, *Atmospheric Corrosion,* STP 1239, W.W. Kirk and H.H. Lawson, Ed., ASTM, 1995, p 195
29. D. Knotkova, V. Kucera, S.W. Dean and P. Boschek, Classification of the Corrosivity of the Atmosphere—Standardized Classification System and Approach for Adjustment, *Outdoor Atmospheric Corrosion,* STP 1421, H.E. Townsend, Ed., ASTM International, 2002, p 109
30. J. Tidblad, V. Kucera, A.A. Mikhailov, and D. Knotkova, Improvement of the ISO Classification System Based on Dose-Response Functions Describing the Corrosivity of Outdoor Atmospheres, *Outdoor Atmospheric Corrosion,* STP 1421, H.E. Townsend, Ed., ASTM International, 2002, p 73
31. D. Knotkova, P. Boschek, and K. Kreislova, Results of ISO CORRAG Program: Processing of One-Year Data in Respect to Corrosivity Classification, *Atmospheric Corrosion,* STP 1239, W.W. Kirk and H.H. Lawson, Ed., ASTM, 1995, p 38
32. L.T.H. Lien and P.T. San, The Effect of Environmental Factors on Carbon Steel Atmospheric Corrosion; The Prediction of Corrosion, *Outdoor Atmospheric Corrosion,* STP 1421, H.E. Townsend, Ed., ASTM International, 2002, p 103
33. I. Odnevall and C. Leygraf, Reaction Sequences in Atmospheric Corrosion of Zinc, *Atmospheric Corrosion,* STP 1239, W.W. Kirk and H.H. Lawson, Ed., ASTM, 1995, p 215
34. J. Simancas, K.L. Scrivener, and M. Morcillo, A Study of Rust Morphology, Contamination of Porosity by Backscattered Electron Imaging, *Atmospheric Corrosion,* STP 1239, W.W. Kirk and H.H. Lawson, Ed., ASTM, 1995, p 137

Corrosion of Metallic Coatings

Barbara A. Shaw, Wilford W. Shaw, and Daniel P. Schmidt, Department of Engineering Science and Mechanics, The Pennsylvania State University

A SACRIFICIAL COATING applied to a steel substrate can add 20 years or more of life to the substrate, depending on its thickness and composition. Different techniques to apply sacrificial coatings offer various characteristics that contribute to corrosion resistance. Several of these techniques and the corrosion attributes of the respective coatings are discussed in this article.

Thermal Sprayed Coatings

A wide variety of materials such as aluminum, zinc, and their alloys can be applied via thermal spraying and have proved to significantly extend substrate lifetimes cost effectively. Their effectiveness in combating corrosion has been evaluated in several long-term studies (Ref 1–5), which have revealed that depending on composition and thickness, these coatings are capable of providing complete protection of steel substrates for 50 years. In 1951, a Cambridge University study under the direction of T.P. Hoar was initiated on over 1500 steel panels thermal sprayed with the following materials (Ref 1):

Mixed powders and alloy powder coatings (0.08 mm, or 3.2 mils), wt%

90% Zn + 10% Al
80% Zn + 20% Al
70% Zn + 30% Al
60% Zn + 40% Al
50% Zn + 50% Al
40% Zn + 60% Al
30% Zn + 70% Al
20% Zn + 80% Al
10% Zn + 90% Al
22% Zn + 78% Sn
90% Zn + 10% Mg
80% Zn + 20% Mg
70% Zn + 30% Mg
90% Al + 10% Mg
80% Al + 20% Mg
70% Al + 30% Mg
60% Zn + 20% Al + 20% Mg
60% Al + 20% Al + 20% Mg

Single element powder coatings

0.08 mm (3.2 mils) Al
0.15 mm (6 mils) Al
0.08 mm (3.2 mils) Zn

Single element powder coatings (continued)

0.15 mm (6 mils) Zn
0.08 mm (3.2 mils) Mn
0.13 mm (5.2 mils) Mn
0.20 mm (8 mils) Mn

Dual-layer powder coatings

0.08 mm (3.2 mils) Zn + 0.08 mm (3.2 mils) Al
0.08 mm (3.2 mils) Al + 0.08 mm (3.2 mils) Zn

A report was issued after 34 years of marine exposure in the 250 m (800 ft) lot in Kure Beach, NC, revealing that the aluminum powder thermal coatings and the layered zinc/aluminum and aluminum/zinc powder thermal spray coatings showed no base metal corrosion or rust staining of the thermal spray coatings after 34 years. A comparison of the performance of the pure metal coatings evaluated in this study after 34 years of exposure is presented in Fig. 1. When observed in August 2001, almost 50 years after initial exposure at the 250 m (800 ft) marine atmospheric site, the 0.08 mm (3 mils) and 0.15 mm (6 mils) thermal spray aluminum coatings were fully intact with only a small amount of rust staining noted on the 0.08 mm (3 mils) panel at a cut edge and no rust or rust staining noted on the thicker coating as shown in Fig. 2.

Another, well-known study of thermal spray coatings was initiated slightly later in the 1950s by the American Welding Society (AWS). The AWS study included aluminum and zinc wire-flame-spray coated steel specimens with coating thicknesses of 0.08, 0.15, 0.23, 0.30, and 0.40 mm (3, 6, 9, 12, and 15 mils). Field exposures were conducted at a variety of atmospheric exposure sites and two seawater immersion sites. The study was scheduled to last 12 years, but because the coatings were doing so well, the exposure period was extended to 19 years. Results of this 19 year study are presented in a 1974 report (Ref 2). After 19 years of marine atmospheric exposure, the flame sprayed aluminum coated steel panels showed no rusting of the steel substrates. Over 4000 specimens were included in this study, which found that 0.08 to 0.15 mm (3 to 6 mils)-thick thermal sprayed aluminum coatings (either sealed or unsealed) provided complete protection to steel substrates

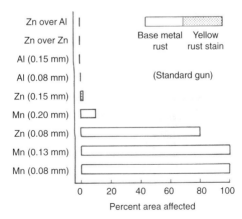

Fig. 1 Percent of area corroded on single-element powder thermal spray coatings after 34 years of marine atmospheric exposure in the 250 m (800 ft) lot at Kure Beach, NC. Source: Ref 1

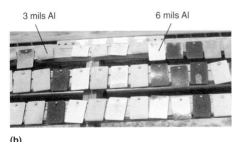

Fig. 2 Original T.P. Hoar study panels after over 48 years of exposure at the 250 m (800 ft) marine atmospheric exposure site in Kure Beach, NC. (a) View of test rack. (b) Closer view of thermal spray coatings and other panels.

in seawater, severe marine atmospheric, and industrial atmospheres. Some blistering and rust staining of the aluminum coating on unsealed panels were noted. Thermal sprayed zinc coatings were also capable of providing 19 years of protection to steel substrates, but a minimum of 0.30 mm (12 mils) was required for seawater exposures and 0.23 mm (9 mils) of unsealed zinc or 0.08 to 0.15 mm (3 to 6 mils) of sealed zinc for marine and industrial atmospheres.

A Navy study was undertaken in the mid 1980s of flame and arc-sprayed aluminum, zinc, prealloyed zinc-aluminum, and duplex zinc/aluminum coatings in marine atmospheric, splash and spray, and immersion environments. More than 600 coated panels were exposed. The thermal spray coatings were applied to both steel and 5086 aluminum substrates. The thermal spray coatings were tested in both sealed and painted conditions and as-deposited without any sealer or paint systems conditions. In addition, both commercial and military suppliers prepared specimens. The study revealed that when deposited under controlled conditions (using either military or commercial facilities), both flame and arc sprayed aluminum coatings (0.18 to 0.25 mm, or 7 to 10 mils, thick sealed and painted) provided excellent protection of steel substrates exposed to marine environments for the duration of the study (68 months). Figure 3 compares the condition of the flame and the arc sprayed aluminum-coated steel to painted steel after 42 months of severe marine atmospheric exposure (25 m, or 80 ft, lot at Kure Beach, NC) Visually, the painted steel, which had been scribed, was heavily corroded with blisters covering the majority of the exposed surface. On the other hand, the flame and arc-sprayed aluminum, which were also scribed, were free of base metal corrosion or degradation of the sprayed metal coatings. The appearance of the flame and arc-sprayed panels was the same at the end of the exposure period (after 68 months) (Ref 6).

The U.S. Army Corps of Engineers conducted a long-term study of coated pilings exposed in the waters of Buzzard's Bay Maine (the bottoms of the coated pilings were driven into the bottom of the bay, exposing the column to mud, immersion, tidal, splash and spray, and atmospheric conditions) (Ref 7). The 21 pilings contained coatings including zinc-rich primers with various topcoats, as-deposited flame sprayed zinc, as-deposited flame sprayed aluminum, and flame sprayed zinc and aluminum with sealers and topcoats. Bare steel was included as a control. After 18 years, the flame sprayed aluminum coatings with a vinyl wash primer and sealer performed the best of all the coatings evaluated in this study.

The ability of a sacrificial coating (thermal sprayed or otherwise deposited) to provide protection to the substrate at defects in the coating is of significant importance. Often, exposure testing includes damaged or scribed panels, and assessment of these damaged areas is included in the inspection protocol. One method to quantify the corrosion damage in the scribed area of top coated/scribed specimens is the "Navy Scribe and Bold Surface Inspection Practice" (Ref 8). This technique, described in Table 1, involves dividing the scribe into segments (as shown in the accompanying figure in Table 1), and at each segment the minimum and maximum lateral creepage of corrosion are measured (in mm) and then added together. This number is referred to as the *segment value creep* and is used to find the corresponding "rating" in Table 1. The next rating is made by measuring the minimum and maximum lateral creepage of corrosion over the entire scribe (in mm) and adding these numbers together. This number is referred to as the *maximum and minimum creep value* and is used to find the corresponding *rating number* in Table 1. A third rating is then made using ASTM D 1654 (Ref 9) by estimating the percent of the panel surface that exhibited corrosion blistering (a ½ in. band around the edge of the panel and corrosion associated with the scribe were excluded from the rating). The three ratings can then be averaged together or used separately to assess a coating systems ability to provide protection. The way in which the damage is introduced to test panels can have a significant impact on the type of results obtained. No matter what approach is used, a quantifiable, easily reproducible method in which the underlying substrate is exposed is needed.

Electrochemical techniques are also being used to assess the long-term durability and mechanisms of protection of sacrificial thermal spray coatings. These studies have identified the deleterious effects of porosity, oxide layers within the deposit, embedded grit blasting media, and thin areas in the coating. Electrochemical methods used to investigate thermal spray coatings typically include: corrosion potential versus time measurements, anodic and cathodic polarization, polarization resistance measurements, and most importantly, electrochemical impedance spectroscopy (EIS). Electrochemical impedance spectroscopy, especially the maximum impedance at lowest frequency, is useful for investigating degradation of sealed and or sealed and painted thermal spray coatings. In-situ electrochemical measurements are also now being conducted from time to time on field exposed panels such as the EIS interrogation of defect sites on the marine atmospheric exposures shown in Fig. 4.

Description of Thermal Spray Processes

Sprayed metal coatings can be defined as processes that deposit fine metallic materials onto a prepared substrate. Sprayed metal processes have these major advantages: a wide variety of materials can be used to make coatings, a coating can be applied without significant heating of the substrate, and the coating can be replaced without changing the properties or dimensions of the part. Sprayed metallic coatings are limited, however, only to substrates in direct view and size limitations of the equipment prohibit coating small deep cavities or crevices where the spray gun or nozzle will not fit.

Although all spray deposition processes can be classified as thermal spray technologies, there are two fundamental types: those that deposit the coating in the molten state, as illustrated in Fig. 5, and those that deposit the coating in a warm state. Conventional thermal spray processes, which deposit the materials in a molten state, include: high-velocity oxyfuel (HVOF), plasma, flame, laser cladding, and electric arc

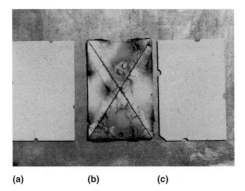

Fig. 3 Comparison of scribed, sealed, and painted thermal spray coatings on steel substrates to a scribed painted steel panel after 42 months of severe marine atmospheric exposure. (a) Flame-sprayed aluminum on steel, sealed/painted. (b) Painted steel panel (one coat MIL P24441 F150 primer followed by one coat of formula 150 and one coat formula 151). (c) Arc-sprayed aluminum on steel, sealed/painted

Table 1 Evaluation of thermal spray defects by using the "Navy Scribe and Bold Surface Inspection Practice"

See text for details. The accompanying figure shows the location of segments along a diagonal scribe. Source: Ref 8

1. Segment value creep, mm	2. Maximum and minimum creep, mm	Rating No.
0.0	0.0	10
0–1	0.0–0.5	9
1–2	0.5–1.0	8
2–4	1–2	7
4–8	2–4	6
8–12	4–6	5
12–16	6–8	4
16–24	8–12	3
24–32	12–16	2
32–40	16–20	1
40+	20+	0

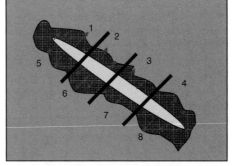

deposition. Conversely, high-velocity particle consolidation (HVPC) only warms the metal particles and uses supersonic velocities to accelerate particles onto the substrate surface. In the HVPC process, compressed gas at pressures between 1 and 3 MPa (145 and 435 psi) expand when passed through a de Laval nozzle yielding exit speeds up to 1200 m/s (4000 ft/s). Metallic powder is introduced into the gas flow accelerating particles to velocities between 18 and 1000 m/s (60 and 3300 ft/s) depending on particle size and the material. A gas heater is used to increase gas temperature. Exposing particles to heated gas increases ductility and improves deposition efficiency. Figure 6 shows a schematic of the HVPC equipment and a photograph of the robotic nozzle (Ref 10). Upon impact, the solid particles deform and bond with the substrate. Repeated impact causes particles to bond with particles already deposited, resulting in a uniform coating with very little porosity (values <3% are typical). The fundamental difference between HVPC and other thermal spray processes is that HVPC applies material particles in the solid state, whereas other conventional thermal spray methods such as wire flame spray and wire arc spray deposit particles in the molten state (Ref 10). The HVPC process has some advantages over conventional thermal spray processes. The lower deposition temperature is advantageous for some materials such as aluminum substrates and gives the ability to produce coating mixtures that otherwise may not be feasible (Ref 10). Oxide contamination within the coating, as pictured in Fig. 7, is greatly reduced. Extremely thick coatings, millimeters thick, are possible to produce. In addition, residual stresses are reduced and the powders can be recycled because they are not significantly heated (Ref 10). Limitations include: the process can use only powders, high deposition rates limit the substrate material that can be coated, high gas pressure and flow rates are required, the objects to be coated must be placed 25 mm (1 in.) from the exit plane of the de Laval nozzle, and one of the constituents must be ductile (Ref 10). As in other sprayed metal coating techniques, only areas in line-of-sight can be coated.

Hardness, Density, and Porosity. In comparison with paint, thermal spray coatings have a high hardness. This hardness is often less than that of the initial feedstock and is dependent on the coating material, equipment used, and choice of deposition parameters. Detonation spraying, HVOF, plasma spray, arc spray and flame spraying rank from the highest particle velocity process (highest density and hardness) to the lowest particle velocity (lowest density and hardness). Other factors, such as thermal spray process and application parameters, also have a significant influence on porosity.

The porosities of flame sprayed coatings may exceed 15% as a result of oxidation that occurs because of the oxidizing potential of the fuel-gas mixture in the flame spraying (Ref 11). An optical micrograph showing these oxide layers for a flame sprayed aluminum coating is presented in Fig. 7. This porosity can have a significant influence on the corrosion resistance of the coating. If left unsealed, surface connected porosity provides a path for aggressive species into the coating as illustrated in Fig. 8, which shows an electron probe dot-map revealing chloride along the oxide layers within unsealed aluminum spray (Ref 12).

Thickness. The sprayed metal coatings are normally applied to a thickness of 75 to 180 μm (3 to 7 mils) to provide adequate corrosion protection. The thickness of the coating is selected to limit interconnected porosity (too thin a coating) and to minimize thermal expansion mismatch (too thick a coating) with the substrate, which could result in bond-line separation (Ref 13). For marine applications, thermal sprayed aluminum coatings 180 to 250 μm (7 to 10 mils) thick are used in order to limit through porosity (Ref 14).

Adhesion of typical thermal spray coatings is similar to that of organic coatings with adhesion values ranging from 5.44 to 13.6 MPa (0.8 to 2 ksi) (Ref 15) when measured in accordance with ASTM D 451. However, specialized wear-resistant coatings with high tensile adhesive strengths in excess of 34 MPa (5 ksi) as measured by ASTM C 633 can be produced by the thermal spray processes with the higher particle velocities (Ref 15).

Imperfections. The arc spray and flame spray processes accelerate molten metal material

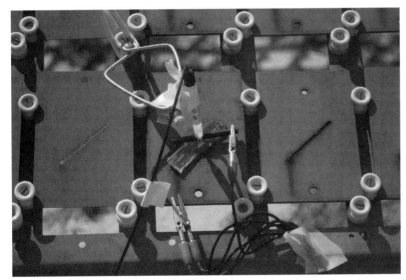

Fig. 4 In-situ electrochemical measurement on scribed area of a panel exposed to the marine atmosphere

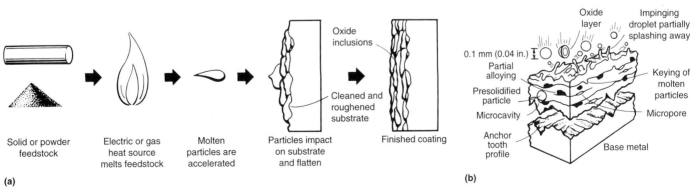

Fig. 5 Schematics showing (a) coating deposition in thermal spray processes and (b) the morphology of thermal spray coatings

64 / Corrosion in Specific Environments

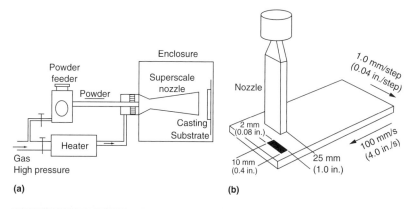

Fig. 6 Schematic illustration of (a) the high-velocity particle consolidation (HVPC) coating deposition process and (b) the nozzle placement with regard to substrate surface in HVPC. The nozzle and gas heat can be mounted on a robot for optimal deposition. (c) Close-up photograph of HVPC nozzle and gas heater mounted on a 6-axis robot. Source: Ref 10

from powder, wire, or rod in a gas stream and project the molten metal onto a suitably prepared substrate. Upon impact with the substrate, the molten droplets rapidly solidify to form a thin "splat" (Ref 16) as shown in Fig. 5. Thin overlapping and interlocking particles forming the coating characterize this splat. The coating is built up by successive impingement and interbonding among splats as illustrated in Fig. 5 and 7. Imperfections originate from "unmelted" particles that are not totally molten, large particles that are bigger than the median size, low velocities, oxidation of particles, and fragmented splats (Ref 16). Note that voids and oxides form an interconnected network within the coating, allowing the environment to eventually work its way through the coating, if the coating is not sealed and/or painted. In addition to the thickness and composition of these coatings, the microstructure can be extremely varied. Materials deposited may be in thermodynamically metastable states, and the grains within the splats may be sub-micron-size or even amorphous (Ref 17). This ultrafine-grained microstructure leads to an anisotropy of the coating in the direction perpendicular to the spray direction (Ref 16). Despite their being fine grained, the thermal sprayed microstructures have an abundance of imperfections comprising a variety of particle sizes, volume densities, morphologies, and sometimes orientations (Ref 16). Micrographs of a few common imperfections in thermal spray coatings are presented in Fig. 9.

Sealing and Topcoat. A standard practice is to seal the thermal spray coating with low-viscosity sealers because the metal coating is inherently porous. Once the thermal spray

Secondary electron image

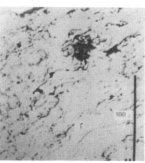
Backscattered electron image

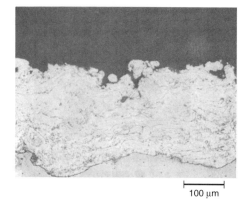

Fig. 7 Micrograph through a flame-sprayed aluminum coating showing oxide layers within the coating (thin dark lines)

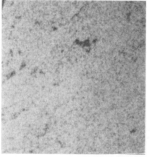
Aluminum Kα x-ray scan

Oxygen Kα x-ray scan

Chlorine Kα x-ray scan

Fig. 8 Electron microprobe x-ray scans of flame-sprayed aluminum coating cross sections after full immersion in filtered seawater for 15 months

sacrificial coatings are applied, the sealer is applied. The sealer is used to penetrate the pores of the spray metal coating and to resist migration of atmospheric corrosives through the sacrificial coating to the substrate material. The sealer is normally sprayed or brushed on, and it penetrates and fills the pores. Sealers should also be used in acidic or alkali environments (Ref 13). Vinyl and thinned epoxies are typical sealers. For high-temperature applications, a silicone alkyd sealer is used (Ref 18). One typical sealer is a two-component epoxy polyamide sealant designed to increase protection of metallic thermal spray coatings up to 150 °C (300 °F). It exhibits resistance to corrosion in industrial and marine atmospheres. This sealer meets governmental specifications designated in MIL-P-53030A (Ref 19). The sealer is both lead- and chromate-free and contains no more than 340g/L of volatile organic compounds. After the two-part epoxy is thoroughly mixed and diluted with 7 parts deionized water, it is applied by an airless sprayer to a recommended wet film thickness of 0.2 mm (8.0 mils), yielding a 0.05 mm (2.0 mils) dry thickness (Ref 19). Maximum performance by the sealer occurs when coating surfaces are clean, dry, and free of foreign matter. The sealer is dry to touch after 45 min and complete air cure takes 7 days (Ref 19). Finally, finish coating layers (water or solvent reducible primer/topcoat combinations) are applied. The primer is applied on top of the sealer to aid in adhesion of the paint to the remaining coating system. Available primers include water or solvent reducible, air-drying, and corrosion-inhibiting primers. A typical coating system, illustrating the various layers in the system, is presented in Fig. 10.

Hot Dip Coatings

Galvanizing has been used extensively for protection against marine environments and approximately 40 million tons of steel are hot-dip galvanized each year (Ref 20). The advantages associated with applying zinc by thermal spray versus galvanizing makes the former method attractive for certain applications and the latter attractive for others (Ref 21). Hot dip galvanizing produces a fully dense coating that is metallurgically bonded to the substrate. In galvanizing, the size of the part, heat distortion, ease of application, and the thickness and uniformity of the coating are factors that must be considered. The thickness of galvanized coatings can vary from less than 25 to 200 μm (<1 to 8 mils) and should be selected depending on the environment to be experienced and the desired lifetime. A range of zinc coating processes, including hotdipping, and their respective coating lifetimes are presented in Fig. 11 (Ref 20). Thick galvanizing and thermal spray were the

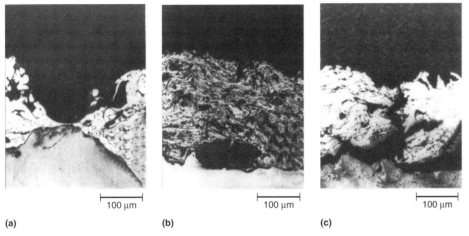

Fig. 9 Typical imperfections in flame/arc spray coatings. (a) Thin area in coating. (b) Imbedded blasting grit. (c) Void extending to substrate

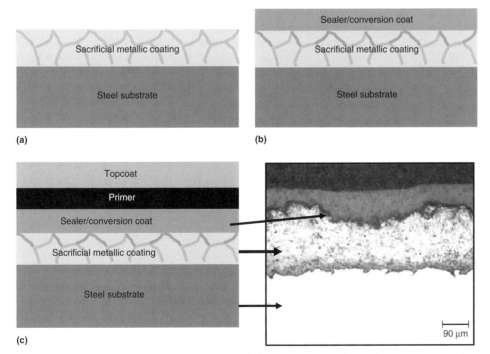

Fig. 10 Schematics of typical coating systems. These include (a) the "as-deposited" pure aluminum or zinc sacrificial metallic coating, without the addition of any organics, applied to SAE 1018 steel; (b) the pure sacrificial coating applied to SAF 1018 steel plus a sealer or conversion coat; and (c) the pure aluminum or zinc sacrificial coating applied to SAE 1018 steel plus a water or solvent-based sealer, primer, and a topcoat. 200×

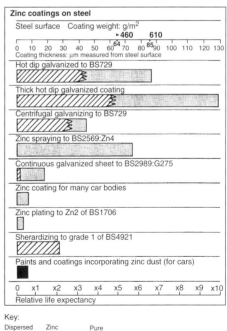

Fig. 11 Range of thickness for typical zinc coatings and their respective relative life expectancies. Source: Ref 20

only protection methods recommended by the British Standards Institution for providing long-term corrosion protection in a polluted marine atmosphere (Ref 21). The American Society for Testing and Materials (ASTM) exposed galvanized sheet specimens in two marine environments—Sandy Hook, NJ, and Key West, FL—in 1926 and reported that panels with a coating weight of 760 g/m² (2.5 oz/ft²) of zinc first showed rust after 13.1 and 19.8 years of exposure, respectively (Ref 22). An extensive study of the atmospheric corrosion of galvanized steel at the 250 m (800 ft) lot at Kure Beach, NC, resulted in predicted weight losses after 10 years of 103 g/m² and 55 g/m² for skyward and groundward marine exposures, respectively (Ref 23). Most investigators agree that the life of a zinc coating is roughly proportional to its thickness in any particular environment and is independent of the method of application.

Hot dip aluminum coatings, or aluminized coatings, are also used for the corrosion protection of steel in marine environments. An extensive comparative study was conducted on the atmospheric corrosion behavior of aluminized and galvanized steels (Ref 24, 25). Table 2 shows predicted 10 year weight losses of both of these coatings based on exposures conducted in the 250 m (800 ft) lot at Kure Beach, NC.

ASTM conducted a further comparison of the atmospheric-corrosion behavior of aluminized and galvanized panels. After 20 years of marine atmospheric exposure (250 m, or 800 ft, lot, Kure Beach, NC), many of the galvanized steel panels were showing rust, but consistently good results were reported for the aluminized coating, which showed only minor pinholes of rust (Ref 26). Since 1972, a commercially produced aluminum-zinc (55Al-1.5Si-43.5Zn) hot dip coating has also been available for the corrosion protection of steel. After 13 years, good long-term corrosion resistance has been reported for 55Al-1.5Si-43.5Zn hot dip coatings in industrial and marine atmospheres. Corrosion-time curves for 55Al-1.5Si-43.5Zn hot dip coatings in these atmospheric environments are presented in Fig. 12. The advantages of hot dip aluminum coatings are discussed in (Ref 27). The corrosion behavior of aluminum coatings obtained from aluminizing baths of various compositions was studied in laboratory tests. Aluminized coatings containing manganese were suggested as possible candidates for corrosion protection for coastal structures and deep-sea oil rigs. More information on hot dip galvanized and aluminized coatings are available in the articles "Continuous Hot Dip Coatings" and "Batch Process Hot Dip Galvanizing" in Volume 13A.

Electroplated Coatings

Electroplated zinc or cadmium is the standard coating used to provide corrosion protection to steel fasteners in the marine environment. The cadmium coating is used because of its hardness, close dimensional tolerance, and barrier to hydrogen permeation into or out of steels (Ref 28). The disadvantages of cadmium plating are its short life (for example, 4 months) in the marine atmospheric environment and concerns about occupational health due to the toxicity in the plating process. Zinc plating also has a short service life. A comparison of zinc and cadmium coatings in industrial and marine sites (Fig. 13) illustrates the importance of zinc in marine environments. Alternatives for zinc plating include ion vapor deposited aluminum and paints containing zinc or aluminum pigment in a ceramic binder. These coatings, including zinc with a potassium silicate binder and aluminum with a phosphate-chromate binder, exhibit excellent corrosion protection for fasteners (minimum 1 year marine protection) (Ref 29). They are normally applied by conventional hand spraying.

Methods for electroplating aluminum are still in development, although plating using an organic aprotic solvent is a promising process (Ref 30). In laboratory polarization and galvanic tests, ion-deposited aluminum coatings performed well, indicating their potential for use on aircraft fasteners (Ref 31). Corrosion tests of zinc and aluminum coatings used for aircraft fastener applications showed variable results (Ref 32). Aluminum and zinc pigmented coatings performed better than electroplated zinc, ion vapor deposited aluminum, and electroplated cadmium on steel fasteners in laboratory seawater immersion tests (Ref 28). However, hydrogen permeability through the coating, as well as the corrosion performance of the coating, must be considered for a given fastener application (Ref 31).

Methods of Protection

Aluminum Coatings. Corrosion protection of aluminum-coated steel arises from the excellent corrosion resistance of the bilayer protective film on the aluminum surface, the barrier properties of the aluminum layer, and the cathodic protection of the exposed steel with the aluminum coating acting as a sacrificial anode

Table 2 Predicted 10 year corrosion rates for galvanized and aluminized steel panels

Tested 250 m (800 ft) from the ocean at Kure Beach, NC

Coating	Predicted weight loss, g/m²	
	Skyward exposure	Groundward exposure
Galvanized	103.3	55.2
Type 1 aluminized (Al-Si)	17.8	20.1
Type 2 aluminized (pure aluminum)	11.6	17.9

Source: Ref 26

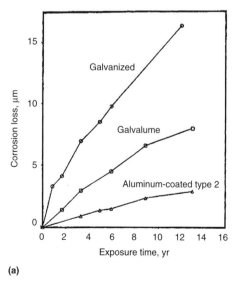
(a)

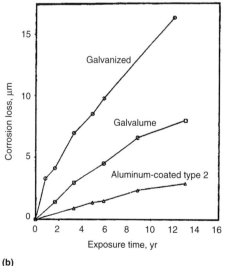
(b)

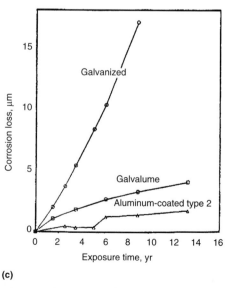
(c)

Fig. 12 Corrosion-time plots for hot dip zinc, zinc-aluminum (55Al-1.5Si-43.5Zn), and aluminum-coated steel in (a) marine atmosphere (Kure Beach, NC: 250 m, or 800 ft, lot), (b) severe marine atmosphere (Kure Beach, NC: 25 m, or 80 ft, lot), and (c) industrial atmosphere (Bethlehem, PA)

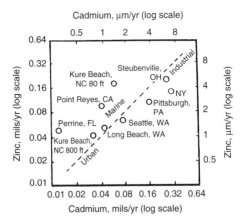

Fig. 13 A comparison of corrosion rates of zinc and cadmium in marine, urban, and industrial atmospheres. Source: Ref 20

(aluminum being more active than steel in the galvanic series). The sprayed aluminum coatings do not provide as much protection at defects as zinc coatings, but the protection that is provided will last longer. Aluminum owes its excellent corrosion resistance to the barrier oxide film that is strongly bonded to its surface. If damaged, this film often has the ability to immediately repassivate itself. The oxide film is composed of two layers: next to the metal surface is a thin amorphous oxide barrier layer and covering the barrier layer is a thicker, more permeable outer layer of hydrated oxides. The protective film in aluminum is stable in chloride-free environments at neutral pH values of 4 to 9. At higher or lower pH values and in the presence of chlorides, the protective passive film is subject to localized breakdown such as pitting. Once outside the passive window, aluminum corrodes in aqueous solutions because its oxides are soluble in many acids and bases.

Zinc Coatings. Corrosion protection in zinc-coated steel arises from the barrier action of the zinc layer, the secondary barrier action of the zinc corrosion products, and the cathodic protection of exposed steel with the coating acting as a sacrificial anode (Ref 33). First, the barrier action of the zinc is made possible because zinc is ten to a hundred times more corrosion resistant than steel in atmospheric environments (Ref 34). This initial film is backed up by the secondary barrier action of corrosion products. The corrosion products that form in the beginning are loosely attached to the surface. Gradually, they become more adherent and dense. This layer is a precipitate (zinc hydroxide). The third way that zinc metallic coatings provide corrosion protection is through cathodic protection. At voids in the coating, such as scratches and at edges, the zinc behaves as a sacrificial anode, thus providing galvanic protection (Ref 35). The zinc surface will preferentially corrode at a slow rate, thus protecting the steel. This is a result of the zinc's being a more active metal in both the electromotive force (EMF) and galvanic series compared with the iron or steel. With this three-way defense, zinc metallic coatings sacrificially protect structural steel in corrosive environments and extend the lifetime of equipment.

Three main features of a zinc coating that are pertinent to its effectiveness are thickness, composition, and microstructure. Coating thickness is a key factor in determining coated product performance. In general, thicker coatings provide greater corrosion protection, whereas thinner coatings tend to giver better formability (Ref 36). In one study (Ref 37), it was shown that plating thickness was the prime factor in assessing the corrosion performance of zinc coatings. The corrosion resistance of zinc-plated steel was found to be directly proportional to zinc thickness (Ref 37). It was also found (Ref 11) that the corrosion loss of hot-dip zinc coatings was considered to be linear, thus the lifetime of a zinc coating is proportional to its thickness.

In addition to thickness, another important feature of zinc coatings is their composition. In order to provide the most protection, zinc coatings must be as dense and as continuous and smooth as possible. Zinc coatings last longer if the corrosion of the surface is uniform. With a rough surface, localized corrosion is more evident and, therefore, a smooth coating is desired. The more compact and continuous the coating layers, the smaller the active surface area within the pores will be and thus the smaller the observed corrosion rate (Ref 38). Defects in the coating can always be present due to the substrate porosity, and these will be of different types and extents depending on the kind of deposition treatment employed. One study (Ref 39) showed that the presence of flaws (porosity and microcracks) was more important than the crystallographic orientations for the corrosion resistance of zinc coatings. These flaws hindered the barrier protection of zinc that the coating usually provides. In another study, it was found that zinc corrosion centers initially formed on the areas where the film was the weakest (thin, more porous) (Ref 37). This is all evidence that the most desirable composition for zinc coatings is a homogeneous, smooth, and dense coating.

A third feature that also plays a role in providing protection is the coating's microstructure. The study described in Ref 39 discovered that zinc crystal planes with different orientations corrode at different rates. This is associated with variances in zinc single crystal corrosion rates because of differences in planar packing density. The activation energy for dissolution was suggested to increase as the packing density increases (Ref 39). The researchers also showed a correlation between texture and corrosion behavior, suggesting that the 001 basal plane is the most resistant to corrosion (Ref 39). Another study (Ref 40) states that the characteristic texture of unalloyed zinc is basal planes parallel to the surface, which provides for bright, smooth crystals. This study also states that although coating corrosion resistance is particularly dependent on the zinc film chemical composition, crystallographic orientation also influences coating corrosion resistance. It further showed that the preferred crystallographic orientation (texture) depends mainly on external factors such as cooling rate gradient and surface conditions of the steel substrate (Ref 40). Impedance spectra provided information that the lower surface energy of bright crystals (due to their smooth surface and smaller amount of surface-segregated elements) caused the degree of corrosive attack to be significantly less (Ref 40). This all demonstrates the important roles that zinc thickness, composition, and microstructure have in providing corrosion protection of steel.

REFERENCES

1. R.M. Kain and E.A. Baker, "Marine Atmospheric Corrosion Museum Report on the Performance of Thermal Spray Coatings on Steel," STP 947, G.A. DiBari and W.B. Harding, Ed., American Society for Testing and Materials, 1987, p 211–234
2. "Corrosion Tests of Flame-Sprayed Coated Steel 19-Year Report," American Welding Society, Miami, FL, 1974
3. B.A. Shaw and D.M. Aylor, Barrier Coatings for the Protection of Steel and Aluminum Alloys in the Marine Atmosphere, in *Degradation of Metals in the Atmosphere*, T.S. Lee and S.W. Dean, Ed., American Society for Testing and Materials, 1988, p 206–219
4. B.A. Shaw, A.M. Leimkuhler, and P.J. Moran, Corrosion Performance of Aluminum and Zinc-Aluminum Thermal Spray Coating in Marine Environments, *Testing of Metallic and Inorganic Coatings,* STP 947, G.A. DiBari and W.B. Harding, Ed., American Society for Testing and Materials, 1987, p 246–264
5. B. Shaw and P. Moran, Characterization of the Corrosion Behavior of Zinc-Aluminum Thermal Spray Coatings, *Mater. Perform.*, Vol 24 (No. 11), Nov 1985, p 22–31
6. B.A. Shaw and A.G.S. Morton, Thermal Spray Coatings—Marine Performance and Mechanisms, *Thermal Spray Technology, National Thermal Spray Conference* (Cincinnati, OH), 1988, p 385–407
7. A. Beitelman, V.L. Van Blaricum, and A. Kumar, Performance of Coatings in Seawater: A Field Study, *Corrosion 93: The NACE Annual Conference and Corrosion Show,* National Association of Corrosion Engineers, 1993
8. S. Pikul, Navy Scribe and Bold Surface Inspection Practice, private communication with D. Schmidt, University Park, PA
9. "Standard Test Method for Evaluation of Painted or Coated Specimens Subjected to Corrosive Environments," ASTM D 1654, ASTM International
10. M.F. Amateau and T.J. Eden, High-Velocity Particle Consolidation Technology, *iMast Quarterly,* Vol 25 (No. 7), 2000, p 17–25
11. R.C. Tucker, Thermal Spray Coatings, *Surface Engineering,* Vol 5, *ASM Handbook,* ASM International, 1994, p 497–509

12. B.A. Shaw and P.J. Moran, Characterization of the Corrosion Behavior of Zinc-Aluminum Thermal Spray Coatings, *Corrosion 85* (Boston, MA), NACE International, 1985
13. *Metallized Coatings for Corrosion Control of Naval Ship Structures and Components,* National Academy Press, 1983
14. "Metal Sprayed Coatings for Corrosion Protection Aboard Naval Ships," MIL-STD-2138, U.S. Navy: Naval Sea Systems Command, Washington, DC, 1992
15. U.S.A.C.O.E. Army, Ed., *Thermal Spraying: New Construction and Maintenance,* U.S. Government, 1999
16. H. Herman, S. Sampath, and R. McCune, Thermal Spray: Current Status and Future Trends, *MRS Bull.,* Vol 25 (No. 7), 2000, p 17–25
17. D.E. Crawmer, Coating Structures, Properties, and Materials, *Handbook of Thermal Spray Technology,* J.R. Davis, Ed., ASM International/Thermal Spray Society, 2004, p 47–53
18. W. Cochran, Thermally Sprayed Aluminum Coatings on Steel, *Met. Prog.,* 1982, p 37–40
19. MIL-P-53030A, U.S. Army, U.S. Army Research Laboratory, 1992
20. F.C. Porter, *Corrosion Resistance of Zinc and Zinc Alloys,* Marcel Dekker, 1994
21. J.C. Bailey, U.K. Experience in Protecting Large Structures by Metal Spraying, *Eighth International Thermal Spray Conference,* American Welding Society, 1976
22. R.M. Burns and W.W. Bradley, *Protective Coatings for Metals,* 3rd ed., Reinhold, 1967
23. R. Legault and V. Pearson, Ed., Kinetics of the Atmospheric Corrosion of Galvanized Steel, in Atmospheric Factors Affecting the Corrosion of Engineering Metals, STP 646, S.K. Coburn, Ed., American Society for Testing and Materials, 1978, p 83–96
24. R. Legault and V. Pearson, *Corrosion,* Vol 34, 1978, p 349
25. R. Legault and V. Pearson, *Inland Steel Research Laboratories Report: The Atmospheric Corrosion of Galvanized and Aluminized Steel,* Inland Steel Research Laboratories, East Chicago, IN, 1976
26. D.E. Tonini, *Corrosion Test Results for Metallic Coated Steel Panels Exposed in 1960,* American Society for Testing and Materials, 1982, p 163–185
27. S. Marut'yan, I.A. Boyka, V. Bobrova, and I. Legkova, Influence of Manganese on the Corrosion Resistance of Hot-Aluminized Steel, *Prot. Met.,* Vol 18 (No. 2), 1982, p 181–182
28. B. Allen and R. Heidersbach, The Effectiveness of Cadmium Coatings as Hydrogen Barriers and Corrosion Resistant Coatings, *Corrosion/83,* National Association of Corrosion Engineers, 1983
29. D. Aylor, Anticorrosion Barriers: Chemistry and Applications, *Philadelphia Symposium,* American Chemical Society, Philadelphia, PA, 1984
30. J. Mazia, In Search of the Golden Fleece-Aluminum in Focus, *Met. Finish.,* Vol 80 (No. 3), 1982, p 75–80
31. M. El-Sherbiny and F. Salem, Surface Protection by Ion Plated Coatings, *Anti-Corros.,* Nov 1981, p 15–18
32. V. McLoughlin, The Replacement of Cadmium for the Coating of Fasteners in Aerospace Applications, *Trans. IMF,* Vol 57, Part 3, 1974, p 102–104
33. G. Parry, B.D. Jeffs, and H.N. McMurray, Corrosion Resistance of Zn-Al Alloy Coated Steels Investigated Using Electrochemical Impedance Spectroscopy, *Ironmaking Steelmaking,* Vol 25 (No. 3), 1998, p 210–215
34. D. Wetzel, Batch Hot-Dip Galvanized Coatings, *Surface Engineering,* Vol 5, *ASM Handbook,* ASM International, 1994, p 360–371
35. H.E. Townsend, Continuous Hot-Dip Coatings, *Surface Engineering,* Vol 5, *ASM Handbook,* ASM International, 1994, p 339–348
36. *Cathodic Protection,* www.corrosion-doctors.org/CP/Introduction.htm, accessed Jan 2006
37. S. Rajendran, S. Bharathi, C. Krishna, and T. Vasudevan, Corrosion Evaluation of Cyanide and Non-Cyanide Zinc Coatings Using Electrochemical Polarization, *Plat. Surf. Finish.,* Vol 84 (No. 3), 1997, p 59–62
38. L. Fedrizzi et al., Corrosion Protection of Sintered Metal Parts by Zinc Coatings, *Organic and Inorganic Coatings for Corrosion Prevention,* 1997, p 144–159
39. L. Diaz-Ballote and R. Ramanauskas, Improving The Corrosion Resistance of Hot-Dip Galvanized Zinc Coatings by Alloying, *Corros. Rev.,* Vol 17 (No. 5), 1999, p 411–422
40. P.R. Sere et al., Relationship between Texture and Corrosion Resistance in Hot-Dip Galvanized Steel Sheets, *Surf. Coat. Technol.,* Vol 122 (No. 2), 1999, p 143

SELECTED REFERENCES

- *Corrosion Tests of Flame-Sprayed Coated Steel 19-Year Report,* American Welding Society, Miami, FL, 1974
- R.M. Kain and E.A. Baker, "Marine Atmospheric Corrosion Museum Report on the Performance of Thermal Spray Coatings on Steel," G.A. DiBari and W.B. Harding, Ed., STP 947 American Society for Testing and Materials, 1987, p 211–234
- Metal Sprayed Coatings for Corrosion Protection Aboard Naval Ships, MIL-STD-2138, U.S. Navy: Naval Sea Systems Command, Washington, DC, 1992
- B.A. Shaw and D.M. Aylor, Barrier Coatings for the Protection of Steel and Alumunum Alloys in the Marine Atmosphere, *Degradation of Metals in the Atmosphere,* T.S. Lee and S.W. Dean, Ed., American Society for Testing and Materials, 1988, p 206–219
- H.E. Townsend, Continuous Hot-Dip Coatings, *Surface Engineering,* Vol 5, *ASM Handbook,* ASM International, 1994, p 339–348

Performance of Organic Coatings

Revised by R.D. Granata, Florida Atlantic University

ORGANIC COATINGS are the principal means of corrosion control for the hulls and topsides of ships and for the splash zones on permanent offshore structures. Most stationary offshore oil industry platforms are not painted below the waterline, and most marine pipelines are factory coated with special proprietary coatings (see the articles "Corrosion in Petroleum Production Operations," and "External Corrosion of Oil and Natural Gas Pipelines" in this Volume).

Figure 1 shows the marine environments that are destructive to shipboard coatings. Similar environments are found on offshore oil production platforms (Fig. 2), lighthouses, docks, and other marine structures.

Before the 1960s, most marine coatings were fairly simple and could be applied by laborers such as seamen or maintenance personnel. Although the advent of high-performance marine coatings in the 1960s changed this, the performance of marine coatings has improved to such an extent that topside coating lifetimes of 20 years have been experienced on some offshore oil production platforms. During the 1980s, environmental issues drove the development of compliant coatings based on new chemistries and technologies. The coatings that resulted required rapid assessment of performance expectations and field validations. Many high-performance coatings are now available for marine service. Some highly specific and detailed information related to preservation of naval vessels is contained in a Department of Navy, Commander Military Sealift Command document COMSC Instruction 4750.2C (Ref 1).

Surface Preparation

Proper surface preparation is the most important consideration in determining the performance of organic coating systems. Surface cleanliness and proper surface profile are both important. Surface preparation frequently accounts for two-thirds of total painting costs for offshore structures.

The Society for Protective Coatings (formerly the Steel Structures Painting Council, SSPC), NACE International (formerly the National Association of Corrosion Engineers), and standards groups in Sweden, Germany, the United Kingdom, and Japan have all issued standards for surface preparation. Examples of these are listed in Table 1. Wet abrasive blast cleaning and waterjetting (waterblasting) are not yet included in the standards, but are now being extensively used. Wet blasting is useful for dust control and for avoiding electrical sparking in class I (explosive) areas. Generally, a small amount of nitrite inhibitor is added to the water to prevent rerusting before priming.

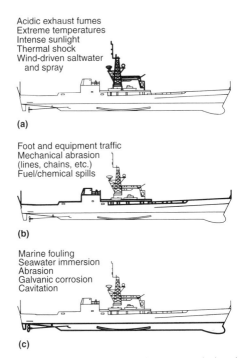

Fig. 1 Environments that are destructive to shipboard coatings. (a) Antennas and superstructures. (b) Deck areas. (c) Underwater hull

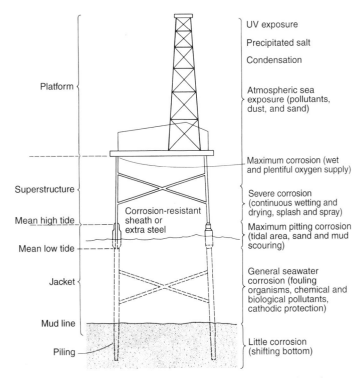

Fig. 2 Zones of severity of environment for a typical offshore drilling structure. UV, ultraviolet

Waterjetting can be used around rotating equipment, such as pumps, turbines, and generators; underwater; and in class I areas. No grit is used; therefore, few solids result that could harm equipment. Waterjetting with 34.5 to 138 MPa (5 to 20 ksi) of pressure will remove all but the most adherent paint and oxide scale and, once the surface is blown dry, provides an excellent surface for painting. Waterjetting with detergent in the water at lower pressure is a good alternative to solvent washing for preparing oily and greasy surfaces for painting.

Grit blasting is usually used for surface preparation for marine coatings. The severe corrosion exposure conditions in offshore and coastal locations require the best possible surface preparation.

Inorganic zinc primers, which are frequently used in marine applications, require white metal grit blasting to remove all surface contamination because inorganic zinc has both a chemical bond and a mechanical bond to the surface. Epoxy primers can be applied over commercial grade surfaces for land-based exposures, but require near-white metal surfaces to maintain performance offshore.

Table 2 lists the characteristics of several types of grit. Grit that is used offshore is not recoverable; this limits the economical choices. Some types are too expensive unless they can be recycled. Silica sand is generally not used because of the possibility of silicosis, its friable nature, and its rounded shape, which is not conducive to high productivity. Safety issues related to surface preparation should be given due consideration including: abrasive dust toxicity, for example, silicosis; surface coating and substrate toxicity, for example, lead, chromate, organotin coatings, and pigments; and, explosion hazard, for example, plastic and organic media dust.

Topside Coating Systems

Organic coatings are usually composed of three components: binders (resins), pigments, and solvents. Not all paints, however, have all three components. For example, solventless paints have been developed in response to environmental restrictions on the use of volatile solvents. Solventless paints can be applied at thicknesses to 13 mm ($1/2$ in.); such thick films would not be possible in a paint containing volatile solvents, because the thickness of the film would prevent solvent evaporation.

Paints can be classified by the type of binder or resin into categories:

- Air-drying oils (for example, linseed oil, alkyds)
- Lacquers (vinyls, chlorinated rubbers)
- Chemically cured coatings (epoxies, phenolics, and urethanes)
- Inorganic coatings (silicates)

The article "Organic Coatings and Linings" in Volume 13A of the *ASM Handbook* contains detailed information on the formulation of all of these types of organic coatings.

Primers

The primer is by far the most important coat in the protection of steel substrates. The primary function of subsequent coats is to protect the primer and to give color and a pleasing appearance.

Inhibitive primers contain substances that resist the effects of contaminants on the steel, such as rust and salt, and that resist disbondment under corrosive conditions and cathodic protection. Formerly, lead pigments and chromates were used as inhibitors, but this is no longer the case, because of tightened environmental regulations that prohibit or severely restrict their use. The inhibitors currently in use include a number of proprietary compounds, some of which enable coating service lifetimes approaching that associated with restricted lead and chromate pigments.

Zinc-Rich Primers. The introduction of coatings containing a high percentage of metallic zinc is an important development in protective coatings in the last 50 years. Zinc dust is loaded into both organic and inorganic binders to form primers that are extremely effective in the prevention of corrosion, underfilm creepage, and coating system failure.

Inorganic zinc-rich primers are based on various silicate binders. There are several

Table 1 Uses and applicable standards for various surface preparation techniques

Technique	Applicable standards	Uses
Solvent cleaning	SSPC-SP1	Used to remove oil, grease, dirt, soil, drawing compounds, and various other contaminants. Does not remove rust or mill scale. No visual standards are available.
Hand tool cleaning	SSPC-SP2	Used to remove loose rust, mill scale, and any other loose contaminants. Standard does not require the removal of intact rust or mill scale. Visual standards: SSPC-VIS 3—BSt3, CSt3, and DSt3(a)
Power tool cleaning	SSPC-SP3	Same as hand tool cleaning. Visual standards: SSPC-VIS 3—BSt3, CSt3, and DSt3(b)
White-metal blast cleaning	SSPC-SP5; NACE 1	Used when a totally cleaned surface is required; blast-cleaned surface must have a uniform, gray-white metallic color and must be free of all oil, grease, dirt, mill scale, rust, corrosion products, oxides, old paint, stains, streaks, or any other foreign matter. Visual standards: SSPC-VIS 1—ASa3, BSa3, CSa3, and DSa3(a); NACE 1
Commercial blast cleaning	SSPC-SP6; NACE 3	Used to remove all contaminants from surface, except the standard allows slight streaks or discolorations caused by rust stain, mill scale oxides, or slight, tight residues of rust or old paint or coatings. If the surface is pitted, slight residues of rust or old paint may remain in the bottoms of pits. The slight discolorations allowed must be limited to one-third of every square inch. Visual standards: SSPC-VIS 1—BSa2, CSa2, DSa2(a); NACE 3; SSPC-VIS 5/NACE VIS 9(d)
Brush-off blast cleaning	SSPC-SP7; NACE 4	Used to remove completely all oil, grease, dirt, rust scale, loose mill scale, and loose paint or coatings. Tight mill scale and tightly adherent rust and paint or coatings may remain as long as the entire surface has been exposed to the abrasive blasting. Visual standards: SSPC-VIS 1—BSa1, CSa1, DSa1(a); NACE 4
Pickling	SSPC-SP8	Used for complete removal of all mill scale, rust, and rust scale by chemical reaction, electrolysis, or both. No available visual standards
Near-white blast cleaning	SSPC-SP10; NACE 2	Used to remove all oil, grease, dirt, mill scale, rust, corrosion products, oxides, paint, or any other foreign matter. Very light shadows, very slight streaks, and discolorations caused by rust stain, mill scale oxides, or slight, tight paint or coating residues are permitted to remain but only in 5% of every square inch. Visual standards: SSPC-VIS 1—ASa2-1/2, BSa2-1/2, CSa2-1/2, and DSa2-1/2(a); NACE 2; SSPC-VIS 5/NACE VIS 9(d)
Power tool cleaning to bare metal	SSPC-SP11	Used to completely remove all contaminants from a surface using power tools. Equivalent to SSPC-SP5, "White-Metal Blast Cleaning." Requires a minimum 1 mil anchor pattern. Visual standard: SSPC-VIS 3(b)
Waterjetting	SSPC-SP12; NACE 5	Cleaning using high- and ultrahigh-pressure waterjetting prior to recoating. Visual standards: SSPC-SP4/NACE 7. Photos depict degrees of cleanliness and flash rusting(c)
Mechanical, chemical, or thermal preparation of concrete	SSPC-SP13; NACE 6	Prepares cementitious surfaces for bonded protective coating or lining systems
Industrial blast cleaning	SSPC-SP14; NACE 8	Used to clean painted or unpainted steel surfaces free of oil, grease, dust, and dirt. Traces of tight mill scale, rust, and coating residue are permitted. Visual standard: SSPC-VIS 1(a)
Commercial grade power tool cleaning	SSPC-SP 15	Used for power tool cleaning of steel surfaces while retaining or producing a minimum 25 μm (1 mil) surface profile. This standard requires higher degree of cleanliness than SSPC-SP 3, but allows stains of rust, paint, and mill scale not allowed by SSPC-SP 11.

(a) SSPC-VIS 1, "Guide and Reference Photographs for Steel Surfaces Prepared by Dry Abrasive Blast Cleaning" is the most commonly used standard for evaluating the cleanliness of a prepared surface. The use of these standards requires a determination of the extent of rust on the uncleaned steel; this is graded from A to D. (b) SSPC-VIS 3, "Visual Standard for Power- and Hand-Tool Cleaned Steel." (c) SSPC-VIS 4/NACE VIS 7, "Guide and Reference Photographs for Steel Surfaces Prepared by Waterjetting." (d) SSPC-VIS 5/NACE VIS 9, "Guide and Reference Photographs for Steel Surfaces Prepared by Wet Abrasive Blast Cleaning"

types of self-curing and postcuring primers. The binder serves as a strong, adherent matrix for the zinc metal. The zinc dust must be present in sufficient amounts to provide metal-to-metal contact between both the zinc particles and the steel surface. The zinc dust provides protection to the steel substrate in the same manner as in galvanizing. If a break develops in the coating, the zinc acts as a sacrificial anode and corrodes preferentially; this provides protection of the iron for long periods. Laboratory tests and field experience indicate that inorganic zinc-rich primers can at least double the life of a coating system and can often increase it tenfold. To be effective, however, inorganic zinc-rich primers must be applied to a clean surface.

Organic zinc-rich primers are alternatives to the inorganic zinc-rich coatings when conditions are not appropriate for inorganic zinc-rich coatings. Organic zinc-rich primers can be formulated with epoxy, urethane, vinyl, and chlorinated rubber binders. The most common binder used for marine applications is polyamide epoxy. Zinc-rich epoxies provide a lower degree of conductivity and cathodic protection than inorganic zinc, but impart several other desirable characteristics:

- Organic zinc-rich primers frequently may be applied over old paint, which makes them a good choice for maintenance painting.
- The good adhesion of the epoxy binder makes surface preparation requirements less stringent than those for inorganic zinc-rich coatings. Near-white metal surfaces are adequate for offshore applications, and commercial grade grit blasting can be used in less severe environments.
- The epoxy binder provides some protection to the zinc, and this allows moderate exposure of the primer to the marine environment without corrosion of the zinc and formation of zinc corrosion products. Zinc corrosion products can cause intercoat adhesion problems and paint blistering.

Topcoats

Topcoats for steel serve mainly to protect the primer and to add color and appearance. To serve this function, they must be:

- Barrier coatings impervious to moisture, salt, chemicals, solvents, and ion passage
- Strong and resistant to mechanical damage
- Of adequate color and gloss retention

Some of the most common topcoats in use are discussed below, and detailed information on each of these types is available in the article "Organic Coatings and Linings" in Volume 13A of the *ASM Handbook*.

Alkyds are the most common and versatile coatings in existence, but they are seldom used in severe marine applications, because of their poor performance over steel. This poor performance is due to the oil base of the alkyd. As corrosion proceeds on steel, hydroxyl ions (OH^-) are generated at cathode sites. Hydroxyl ions saponify (break down) the oils in the coating, and this results in coating failure.

Alkyds are the product of the reaction of a polybasic acid, a polyhydric alcohol, and a monobasic acid or oil. The number of possible combinations is large; therefore, a wide range of performance is available. Alkyds are used in marine service in relatively mild applications, such as interior coatings for cabins, quarters, engine rooms, kitchens, heads, and some superstructure applications.

Vinyls also have a broad range of desirable properties. Most vinyl resins are the product of the polymerization of polyvinyl chloride (PVC) and polyvinyl acetate (PVA). Vinyls are solvent-base coatings that form a tight homogeneous film over the substrate. They are easy to apply by brush, roller, and spray. Intercoat adhesion is excellent because of the solvent-base nature of the coating. Vinyls do not oxidize or age, and they are inert to acid, alkali, water, cement, and alcohol. They do soften slightly when covered with some crude oils. Vinyl coatings dry quickly and can be recoated in a short time (often in minutes), depending on the solvent used. Vinyl coatings are also flexible and can accommodate the motion of the steel beneath them, such as when a ship or platform is launched.

Vinyls were extensively used on ships and offshore platforms for many years, and they are still in use in many areas. However, they have given way to epoxies in most marine applications because vinyl coatings are relatively thin and are not very strong. Film thickness is typically only 50 µm (2 mils) per coat, and the coatings cannot withstand mechanical abuse. In addition, vinyls are not very effective for covering rough, previously corroded surfaces.

Chlorinated rubber coatings are based on natural rubber that has been reacted with chlorine to give a hard, high-quality resin that is soluble in various solvents. Chlorinated rubbers have been used for many years as industrial-type paints because of their low moisture permeability, strength, resistance to ultraviolet (UV) degradation, and ease of application. Chlorinated rubbers have found application on ships and containers, railroads, and as traffic paint for road stripes. For many years, chlorinated rubber coatings were used to paint ships because of their ease of application and repairability, tolerance of poor

Table 2 Properties of abrasives

Abrasive	Moh's hardness	Shape	Bulk density kg/m^3	Bulk density lb/ft^3	Color	Free silica, wt%	Degree of dusting	Reuse
Naturally occurring abrasives								
Sand								
Silica	5	Rounded	1600	100	White	90+	High	Poor
Mineral	5–7	Rounded	2000	125	Variable	5	Medium	Good
Flint	6.7–7	Angular	1280	80	Light gray	90+	Medium	Good
Garnet	7–8	Angular	2320	145	Pink	nil	Medium	Good
Zircon	7.5	Cubic	2965	185	White	nil	Low	Good
Novaculite	4	Angular	1600	100	White	90+	Low	Good
By-product abrasives								
Slag								
Boiler	7	Angular	1360	85	Black	nil	High	Poor
Copper	8	Angular	1760	110	Black	nil	Low	Good
Nickel	8	Angular	1360	85	Green	nil	High	Poor
Walnut shells	3	Cubic	720	45	Brown	nil	Low	Poor
Peach pits	3	Cubic	720	45	Brown	nil	Low	Poor
Corn cobs	4.5	Angular	480	30	Tan	nil	Low	Good
Manufactured abrasives								
Silicon carbide	9	Angular	1680	105	Black	nil	Low	Good
Aluminum oxide	8	Blocky	1920	120	Brown	nil	Low	Good
Glass beads	5.5	Spherical	1600	100	Clear	67	Low	Good
Metallic abrasives								
Steel shot(a)	40–50(b)	Round	...	...	...	...	...	Excellent
Steel grit (made by crushing steel shot)(a)	40–60(b)	Angular	...	...	...	...	...	Excellent

(a) Steel shot produces a peened surface, while steel grit produces an angular, etched type of surface texture. (b) Rockwell C hardness. Source: Ref 2

surface preparation, fast drying characteristics, and relatively good wear and abrasion resistance. They are still used on ships and have been the standard paint system for containers. Modern fleet owners, however, have phased out chlorinated rubbers in favor of higher-quality coating systems, such as epoxy, for reasons to be discussed below.

Epoxies. The combination of excellent adhesion (some can be applied underwater), good impact and abrasion resistance, high film builds (up to 6.4 mm, or $1/4$ in., on a wet, vertical surface), relatively low cost, and excellent chemical and solvent resistance has made epoxy coatings the workhorses of modern marine coatings. These properties result in service lifetimes of 7 to 12 years on ships, offshore platforms, and coastal applications when epoxy topcoats are applied over inhibited epoxy or inorganic zinc-rich primers. Because epoxies are chemically cured, a wide range of properties can be achieved by varying the molecular weight of the resin, the type of curing agent, and the type of pigments or fillers used.

Immersion Coatings

Immersion coatings for submerged marine service have far greater requirements than other organic coatings. They must resist moisture absorption, moisture transfer, and electroendosmosis (electrochemically induced diffusion of moisture through the coating). They also must be strong and have good adhesion.

Most ship hulls and many marine structures use cathodic protection to supplement the protection afforded by organic coatings (see the article "Marine Cathodic Protection" in this Volume). This is desirable because it is virtually impossible to apply and maintain a defect-free organic coating system on a large structure. When cathodic protection is used, the immersion coating must be able to resist the additional conditions imposed upon it by the cathodic potential (blistering) and resulting alkalinity (cathodic delamination).

Property Requirements

Barrier Properties. To be effective in seawater immersion service, an organic coating must have a low moisture vapor transfer rate as well as low moisture absorption. Moisture absorption is the molecular moisture absorbed into and held within the molecular structure of the coating. This property is not important to the effectiveness of the coating unless the moisture absorption lowers the dielectric characteristics of the coating and increases the passage of electrical current. Moisture vapor transfer, on the other hand, is important, particularly when the coating is exposed to an external current (as in cathodic protection). Generally, the lower the moisture vapor transfer rate of a coating, the more effective the coating.

Where electroendosmosis may be encountered, adhesion is also very important. Most organic coatings are negatively charged, and under cathodic protection, the cathode has an excess of electrons, which makes it negatively charged. This being the case, coatings with a high moisture vapor transfer rate or questionable adhesion would be more subject to damage and blistering by cathodic potentials.

Mechanical Properties. Coatings used on marine structures must be strong. Most damage to marine coating systems is mechanical, not a breakdown of the coating from exposure to seawater. Immersion coatings must have good impact and abrasion resistance and must be able to flex well enough to maintain contact with the steel substrate when it is bent. Rubbing by mooring ropes, chains, and crane wire ropes, as well as impact from cargo handling, work parties, and berthing operations, are major causes of damage.

Types of Immersion Coatings

Many of the common paint formulations can be used for immersion service, but the most common coatings in use have been coal tar epoxies and straight epoxies. Restrictions are being imposed on coal tar materials leading to decreasing usage.

Coal tar epoxies were introduced in 1955 and became the most common coatings in use on fixed marine structures (Ref 3). These thermosetting materials are available with a variety of setting temperatures and chemical curing systems. Coal tar epoxies require near-white surface preparation and are very adherent and abrasion resistant. They tend to be brittle and should not be used on flexible structures.

Straight epoxies (no coal tar) have been commercially available longer than coal tar epoxies (Ref 4). Epoxies are usually applied in thinner coats and are more expensive than coal tar epoxies. Epoxies have become the material of choice for immersion service because of their superior performance (Ref 2, 4). They have replaced chlorinated rubbers for most ship hull applications, and they are available in a variety of polyamide- or amine-based formulations (Ref 3). Detailed information on these and other coating materials is available in the article "Organic Coatings and Linings" in Volume 13A of the *ASM Handbook*.

Antifouling Topcoats. Most shipboard applications require antifouling (AF) topcoats. The formulations for these coatings are changing because of environmental legislation. Historically, copper-containing AF coatings became standard, but had only a 12 to 18 month service life. These coatings were nearly replaced by organotin compounds, for example, TBT (tributyltin), before the environmental impact was understood. Copper-containing AF coatings are returning to higher-frequency use as conventional, ablative, and self-polishing types, the latter two offering the prospects of longer service lifetimes (Ref 5).

ACKNOWLEDGMENT

This article was adapted from J.S. Smart III and R. Heidersbach, Organic Coatings, Vol 13, *ASM Handbook* (formerly 9th ed., *Metals Handbook*), ASM International, 1987, p 912–918.

REFERENCES

1. "Preservation Instructions for MSC Ships," COMSC 4750.2c, Department of Navy (www.msc.navy.mil/instructions/doc/47502c.doc, accessed Dec 2005)
2. H.S. Preiser, Jacketing and Coating, in *Handbook of Corrosion Protection for Steel Structures in Marine Environments*, American Iron and Steel Institute, 1981
3. S. Rodgers and R. Drisko, Painting Navy Ships, in *Steel Structures Painting Manual*, Vol 1, 2nd ed., *Good Painting Practice*, Steel Structures Painting Council, 1982
4. J. Smart, Marine Coatings, in *Marine Corrosion*, AIChE, Today Series, American Institute of Chemical Engineers, 1985
5. G. Swain, Redefining Antifouling Coatings, *J. Prot. Coat. Linings*, Vol 16 (No. 9), Sept 1999, p 26ff

Marine Cathodic Protection

Robert H. Heidersbach, Dr. Rust, Inc.
James Brandt and David Johnson, Galvotec Corrosion Services
John S. Smart III, John S. Smart Consulting Engineers

CATHODIC PROTECTION (CP) is an electrochemical means of corrosion control that is widely used in the marine environment. A detailed explanation of the principles of CP appears in the article "Cathodic Protection" in *ASM Handbook,* Volume 13A, 2003.

Cathodic protection can be defined as a technique of reducing or eliminating the corrosion of a metal by making it the cathode of an electrochemical cell and passing sufficient current through it to reduce its corrosion rate. All CP systems require:

- Voltage
- Current
- Anode
- Cathode
- Return circuit
- Electrolyte

Two types of CP systems are: impressed-current (active) systems (ICCP) and sacrificial anode (passive) systems (SACP). Both are common in marine applications. In recent years, hybrid systems—combinations of impressed-current and sacrificial anodes—have been used for very large marine structures.

Cathodic Protection Criteria

A number of criteria are used to determine whether or not the CP of a structure is adequate. These criteria, covered in NACE RP0176 (Ref 1), include potential measurements, visual inspection, and test coupons.

Potential Measurements. Reference 2 specifies a negative (cathodic) voltage of at least −0.80 V between the platform structure and a silver-silver chloride (Ag/AgCl) reference electrode contacting the water. Normally, voltage is measured with the protective current applied. The −0.80 V standard includes the voltage drop across the steel/water interface, but does not include the voltage drop in the water.

Application of the protective current should produce a minimum negative (cathodic) voltage shift of 300 mV. The voltage shift is measured between the platform surface and a reference electrode contacting the water; it includes the voltage drop across the steel/water interface but not the voltage drop in the water.

Visual inspection should indicate no progression of corrosion beyond limits acceptable for platform life (Ref 1).

Corrosion test coupons must indicate a corrosion type and rate that is within acceptable limits for the intended platform life (Ref 1).

A number of other criteria are used, but in practice, −0.80 V versus Ag/AgCl is the most common. Other reference electrodes that can be used for marine applications are listed in Table 1.

Anode Materials

The choice of anode material depends on whether active (ICCP) or passive (SACP) systems are under consideration. Sacrificial anodes must be naturally anodic to steel and must corrode reliably (avoid passivation) in the environment of interest. However, above all, sacrificial anodes should be inexpensive and durable. Impressed-current anodes rely on external voltage sources; therefore, they do not need to be naturally anodic to steel. They usually are cathodic to steel if not forced to assume anodic potentials by the impressed current. Additional information on materials for sacrificial and impressed-current anodes is available in the article "Cathodic Protection" in *ASM Handbook,* Volume 13A.

Sacrificial Anodes. Commercial sacrificial anodes are magnesium, aluminum, or zinc or their alloys. Table 2 lists the energy capabilities of sacrificial anode alloys. Magnesium anodes have not been popular for offshore applications since the 1980s because of improvements in aluminum and zinc anodes. Several operators have experimented with composite sacrificial anode systems for offshore platforms. These designs use aluminum or zinc anodes for long-term performance and have magnesium anodes that are intended to provide an initially high current density and polarize the platform quickly to the desired protection potential. Results from the limited applications of this composite design are mixed, and this concept remains controversial.

Aluminum anodes (aluminum-zinc alloys) are the preferred sacrificial anodes for offshore platform cathodic protection. This is because aluminum anodes have reliable long-term performance when compared to magnesium, which may be consumed before the platform has served its useful life. Aluminum also has better current/weight characteristics than zinc. Weight can be a major consideration for large offshore platforms.

The major disadvantage of aluminum for some applications—for example, the protection of painted ship hulls—is that aluminum is too corrosion resistant in many environments. Aluminum alloys will not corrode reliably onshore or in freshwaters. In marine environments, the chloride content of seawater depassivates some aluminum alloys and allows them to perform reliably as anode materials. Unfortunately, it is necessary to add mercury, antimony, indium, tin, or similar metals to the aluminum alloy to ensure that this depassivation occurs. Heavy-metal pollution concerns have led to bans on the use of mercury alloys in most locations.

Aluminum-zinc-indium anodes have become the most popular choice for pipeline bracelets in seawater or seamud. The greater current capacity allows for much lighter weight anodes to be handled and installed. Additionally, although the

Table 1 Reference electrodes used for cathodic protection systems on offshore structures

Type of electrode	Protection potential of steel, V
Ag/AgCl	−0.80 (or more negative)
Cu/CuSO$_4$	−0.85 (or more negative)
Zinc	+0.25 (or less positive)

Table 2 Energy characteristics of materials used for sacrificial anodes

	Energy capability		Consumption rate	
Material	A · h/kg	A · h/lb	kg/A · yr	lb/A · yr
Al-Zn-In	2535–2601	1150–1180	3.5–3.4	7.6–7.4
Al-Zn-Hg	2750–2840	1250–1290	3.2–3.1	7.0–6.8
Al-Zn-Sn	925–2600	420–1180	3.4–9.4	7.4–20.8
Zinc	815	370	10.8	23.7
Magnesium	1100	500	7.9	17.5

alloy operates at a lower efficiency, it should be the only alloy used at temperatures above 60 °C (140 °F).

Zinc anodes are used on ship hulls because, unlike aluminum, zinc will continue to perform when ships enter the brackish water or freshwater of harbors. Tankers with combination ballast/product tanks use zinc anodes because of their lower tendency to cause sparks if they fall from their supports and strike steel.

Zinc bracelet anodes are also used on pipelines. Again, they would be the preferred choice in brackish or freshwater and bottom mud. Because large-diameter marine pipelines must be buoyancy compensated, the increased weight of zinc can be an advantage. Caution: zinc passivates above 60 °C (140 °F).

Sacrificial anode manufacturing tolerances are established for dimensions, weight, and condition.

Cracks or casting shrink tears in the body of an aluminum anode do not affect performance. Even with good foundry practices a certain amount of cracking will occur. Cracks are permitted within the body of the anode wholly supported by the core. These cannot exceed 5 mm (~0.2 in.) wide or be full circumferential. Individual owners typically specify their crack criteria, but in lieu of that, the recommendation is that no more than 10 cracks per anode is acceptable. Cracks less than 0.5 mm (~0.02 in.) wide are not counted. If cracks occur in a section of the anode that is not supported with a core, it is cause for rejection as this section of the anode may fall away if the cracking propagates. Additionally, longitudinal cracks are not permitted except in the final "topping-up" metal.

Weight. Anode net weight for large offshore anodes (over 50 kg or 110 lb) should be +3% for any one anode. The total order weight should not be more than +2% of the nominal order weight and not less than the nominal order weight.

Identification. Each anode should be stamped with a unique heat casting number to identify it with its laboratory test performance and other compliance data. Additionally, clients may wish to specify a casting sequence number and foundry identification stamp.

Dimensions. Each anode should be within +3% of the specified length or +25 mm (1 in.), whichever is less. The width should be within +5% of the nominal mean width and the depth should be within +10% of the nominal mean depth. The anode straightness should not deviate more than 2% of the nominal length from the longitudinal axis of the anode. To confirm compliance, typically 10% of the anodes should be inspected for dimensional compliance.

Anode cores on large offshore-type standoff anodes should be made from schedule 80 steel pipe with the diameter determined by the weight of the anodes cast on it. Typically 50 mm (2 in.) pipe cores are used on anodes from 100 to 200 kg (220 to 440 lb). Anodes from 200 to 300 kg (440 to 660 lb) should have 75 mm (3 in.) pipe cores and anodes over 300 kg should have anode cores

of 100 mm (4 in.) pipe. Flush-mounted anodes typically contain cores made from flatbar.

Impressed-Current Anodes. Most materials used as impressed-current anodes are insoluble and corrosion resistant, with very low rates of consumption (Table 3). Exchange current density—the ability of a material to sustain high current densities with lower power consumption—is an important consideration for some applications.

High-silicon cast iron is the most corrosion-resistant, nonprecious alloy in commercial use as an impressed-current anode. Anodes made with this alloy are very strong, durable, and abrasion resistant. The major disadvantage in offshore applications is the high weight/current characteristics of high-silicon cast iron. Marine applications for high-silicon cast iron anodes include docks and similar coastal structures.

Precious metals have the advantage of high exchange current densities (Ref 1). The practical result is that a small precious metal surface is equivalent to thousands of times the surface area of other anode materials. Therefore, a small surface of platinum or palladium may be more economical than a much larger anode of less expensive material. Early precious metal anodes were alloys that, after a short period of use, became enriched on the surface with the precious metal component. Anodes of this type are still used in harsh environments, such as Cook Inlet, Alaska, and the North Sea, where anode sleds weighing several tons are necessary to withstand high ocean currents and storm conditions. For most other applications, smaller anodes, consisting of precious metals clad to stronger substrates (for example, platinum bonded to a niobium substrate), have gained acceptance. The extreme weight advantages of these anodes over other systems make them especially desirable for deep-water structures.

Mixed-metal oxide anodes consist of a titanium substrate or core, manufactured to ASTM grade 1 or 2, whose surface is activated with an electrocatalyst of a proprietary mix of precious metal oxides, typically iridium, tantalum, or ruthenium. These anodes have gained acceptance for cathodic protection applications. Mixed-metal oxide anodes have many of the same advantages as precious metal anodes (light weight, high current capacity, and they can be used in all electrolytes).

Polymer Anodes. Several suppliers are marketing polymer anodes. These contain embedded graphite conductors and are used in applications requiring low current densities, such as coastal concrete structures that use CP to reduce the corrosion of reinforcing steel. Most offshore applications require anodes with high current capacities so polymer anodes are not used.

Comparison of Impressed-Current and Sacrificial Anode Systems

Sacrificial anode cathodic protection systems are simpler than ICCP systems and require little or no maintenance except for periodic anode replacement. The capital cost of small systems is minimal, and they are often used for applications such as pipelines and boats. The capital cost of large systems is proportional to submerged surface area.

The capital costs of ICCP may be lower than those of SACP systems in applications such as large offshore platforms. Impressed-current systems are normally used where the low conductivity of the electrolyte (freshwater, concrete) makes sacrificial anodes impractical.

Cathodic Protection of Marine Pipelines

Corrosion control of marine pipelines is usually achieved through the use of protective coatings and supplemental CP. A variety of organic protective coatings can be used. They are usually applied in a factory so that the only field-applied coatings are at joints in pipeline sections. Larger marine pipelines have an outer "weight coating" of concrete. The CP system supplements these coatings and is intended to provide corrosion control at holidays (defects) in the protective coating.

Design Considerations. The average cathodic protection current density required to protect a marine pipeline will depend on the type of coating applied, the method used to coat field joints, the amount of damage inflicted on the coating during shipment and installation, whether or not burial is specified, and the location of the pipeline. Large-diameter pipelines can be protected by installing an ICCP system at one or both ends of the pipeline. Such a system would include a suitably sized transformer/rectifier unit and inert anodes, such as graphite or

Table 3 Typical current densities and consumption rates of materials used for impressed-current anodes

Material	Typical anode current density		Consumption rate	
	A/m^2	A/ft^2	$kg/A \cdot yr$	$lb/A \cdot yr$
Mixed-metal oxide	600	55.8	0.004	0.0088
Pb-6Sb-1Ag	160–215	15–20	0.045–0.09	0.1–0.2
Platinum (plated on substrate)	540–1080	50–100	6×10^{-6}	1.3×10^{-5}
Platinum (wire or clad)	1080–5400	100–500	10^{-5}	2.2×10^{-5}
Graphite	10.8–43	1–4	0.23–0.45	0.5–1.0
Fe-14Si-4Cr	10.8–43	1–4	0.23–0.45	0.5–1.0

high-silicon cast iron. Caution should be exercised to prevent overvoltage in seawater or marine mud.

Most marine pipelines are protected by the installation of bracelet-type zinc or aluminum alloy sacrificial anodes (Fig. 1). Electrical contact between the anode and the pipeline is made through an insulated copper cable bonded to the pipeline. Zinc anodes may either be high purity or alloyed. Aluminum anodes are usually fabricated from a proprietary Al-Zn-In alloy.

Design Procedures. Typically, bracelet anodes are spaced at a maximum of 150 m (500 ft) on small-diameter pipelines (<355 mm, or 14 in.) and 300 m (1000 ft) on larger pipelines. The current required for a segment of pipeline is calculated by using the current density required for the given environment, the surface area of the pipe segment, and the fraction of steel assumed to be bare. Anodes are then sized to fit the conditions; that is, the anode must have adequate weight to satisfy the relationship:

$$\frac{W}{C} > IL \quad \text{(Eq 1)}$$

where W is the anode weight (kg), C is the alloy consumption rate in kilogram/amp year (kg/A yr), I is the anode current output (A), and L is the desired design life in years. The anode nominal current output must exceed the required current. Anode current output I is determined from Ohm's law:

$$I = \frac{E}{R} \quad \text{(Eq 2)}$$

where E is the net driving voltage (V), and R is the anode-to-electrolyte resistance (Ω). Anode-to-electrolyte resistance R can be calculated using an empirical relationship, such as McCoy's equation:

$$R = \frac{0.315\rho}{\sqrt{A}} \quad \text{(Eq 3)}$$

where ρ is the electrolyte resistivity ($\Omega \cdot cm$), and A is the anode area (cm^2).

Data sheets from the manufacturer can also be consulted for information on the electrochemical properties of specific proprietary alloys. Anode geometry can be optimized by using a successive iteration technique, but most anode manufacturers offer a range of standard sizes, which can be optimized for specific applications. If the pipeline is to have a concrete weight coating for stabilization, the thickness of the anode should match the concrete thickness to facilitate installation. Anodes to be installed on nonweight-coated pipelines should have tapered ends so that the anodes do not hang up on the rollers of the lay barge.

Isolation. If the pipeline and the offshore production facilities are operated by separate parties, the pipeline is usually electrically isolated from the production facilities through the use of insulated flanges or monolithic isolation joints. This practice prevents loss of current from the pipeline cathodic protection system to the platform and facilitates recordkeeping. Pipelines are always electrically isolated from shore-based facilities, usually at the valve pit on the beach.

Cathodic Protection of Offshore Structures

Offshore oil production platforms are unusual because most platforms are not painted below the waterline. The cathodic protection system causes a pH shift in the water, which becomes more alkaline (higher pH). Most minerals are less soluble in alkalis than in near-neutral environments (neutral water has a pH of 7; the pH of seawater averages approximately 7.8). The higher pH near the cathode causes minerals to precipitate onto the steel surface and form a protective scale or calcareous deposit. Depending on such factors as water depth, temperature, and velocity, this protective scale may be calcium carbonate, magnesium hydroxide, or a mixture of these and other minerals.

The technology of offshore cathodic protection is rapidly changing. Many of these changes are required because offshore structures are now being built in deeper, colder water where mineral deposits are less likely to form. The formation of a mineral deposit in such conditions may require current densities as high as approximately 750 to 1000 mA/m^2 (70 to 93 mA/ft^2).

Another problem associated with deep-water platforms is that current density requirements change with depth. In recent years, several deep-water platforms have been found to be underprotected. These North American platforms (Gulf of Mexico and Santa Barbara Channel) were located in deep water, but the inadequate protection was at intermediate depths. Gases are more soluble at depth, and carbonate scales (calcareous deposits) are harder to deposit in deep, cold waters. This difficulty is offset by the fact that many deep waters have little dissolved oxygen and should therefore be less corrosive than shallow waters. It should be noted however, that Cook Inlet, the North Sea, and other stormy waters may be oxygen saturated (and presumably carbon dioxide saturated) all the way to the bottom.

Many operators prefer to use sacrificial anodes on offshore platforms because the SACP systems are simple and rugged. In addition, they become effective as soon as the platform is launched and do not depend on external electric power supplies. Surveys of the reliability of ICCP systems have led to the conclusion that they do not perform as well as sacrificial anode systems (Ref 3). Reasons for this lack of performance may be the poor or fragile design of some early impressed-current systems and the lack of ongoing maintenance.

Unfortunately, the weight of sacrificial anodes can be a serious consideration for deep-water platforms. Impressed-current systems are gaining acceptance. Some operators have introduced hybrid designs. In these designs, the primary cathodic protection system uses impressed current, and a sacrificial anode system is used to protect the platform after launching and before the electrical system on the platform becomes operational. The Murchison Platform has the most widely publicized hybrid cathodic protection system (Ref 4).

In the past, the inefficiencies associated with CP design were not serious. Water depths were shallow, and CaP systems were overdesigned to ensure satisfactory performance. This was justified based on economics. A typical CP system is only 1 to 2% of the total capital cost of a new platform, but a retrofit may cost as much as the platform itself. Early platforms in Cook Inlet and the North Sea were underdesigned, and the costs of retrofits led to the efforts that produced NACE RP0176 (Ref 1). Reference 5 and 6 detail some problems experienced with deep-water platforms.

Design Procedures. A typical cathodic protection design procedure for an offshore platform might consist of:

1. Selection of a proper maintenance current density; this will depend on the geographic location (see Table 4)
2. Calculation of surface areas of steel in mud and in seawater
3. Calculation of the total amount of anode material required to guarantee a desired life
4. Selection of an anode geometry. Initial current density for the Gulf of Mexico (calculated for a single anode from Dwight's equation; see the example below) should exceed 110 mA/m^2 (10 mA/ft^2), assuming a native potential of 0.28 V between bare steel

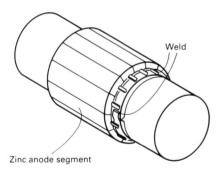

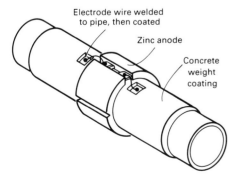

Fig. 1 Typical pipeline bracelet anodes

and aluminum anodes. See Tables 4 and 5 for resistivity of seawater.
5. Judicious distribution of anodes on the steel, assuming a throwing power of 7.6 m (25 ft) in line of sight and placing anodes within 3 m of all nodes

The criterion for complete cathodic protection is steel structure potential more negative than -0.80 V versus the Ag/AgCl reference electrode at any point on the structure.

Example 1: Sacrificial Anode Calculation. This is a typical method for calculation of galvanic anode current output and anodes required using initial, maintenance, and final current densities. This method has been commonly employed in the past for CP design, so practitioners tend to be familiar with it.

For a platform in the Gulf of Mexico, the number of anodes required for protection must satisfy three different calculations. There must be enough anodes to polarize the structure initially (initial current density I_{prot} from Table 4), to produce the appropriate current over the design life of the structure (mean current requirement, $\bar{I}$), and to produce enough current to maintain protection at the end of the 20 year design life (final current requirement, I_{prot20}). Assume that structure surface area needing protection is 3111 m^2 (33,484 ft^2) in water (A_w), 4458 m^2 (47,984 ft^2) in mud (A_m); ρ is 20 $\Omega \cdot$ cm (Table 4).

From a modification of Dwight's equation, the resistance of a cylindrically shaped anode to the electrolyte in which it is placed is equal to the product of the specific resistivity of the electrolyte and certain factors relating to the shape of the anode:

$$R = \frac{\rho K}{L}\left[\ln\frac{4L}{r} - 1\right] \quad \text{(Eq 4)}$$

where R is anode-to-electrolyte resistance (Ω), ρ is electrolyte resistivity ($\Omega \cdot$ cm) (see Table 4), K is 0.159 if L and r are cm, or 0.0627 if L and r are in., L is length of anode cm (in.), r is radius of anode cm (in.). For clarity, this example will be done in metric units. In this case, an anode with a square cross section (sides 21.6 cm) is being used, L is 274 cm, and mass per anode W is 330 kg.

For other than cylindrical shapes, $r = C/2\pi$, where C is cross-section perimeter. Thus, C is 86.4 cm and r is 13.7 cm. Based on this information, an anode made of an Al-Zn-In alloy is selected. This provides 0.28 V driving voltage between an aluminum-zinc anode of -1.08 V (Ag/AgCl in seawater reference) and polarized steel.

Substituting into Eq 4, the resistance is:

$$R = \frac{(20)(0.159)}{274}\left[\ln\frac{4 \times 274}{13.7} - 1\right] = 0.0392\ \Omega$$

To determine the current output from an anode, use Ohm's law:

$$I = \frac{E}{R}$$

$$I = \frac{0.28}{0.0392} = 7.14\ \text{A/anode} \quad \text{(Eq 5)}$$

The number of anodes (N) required for initial protection is:

$$N = \frac{I_{prot} \cdot A_w + I_{protmud} \cdot A_m}{I}$$
$$= \frac{(0.11\ \text{A/m}^2)(3111\ \text{m}^2) + (0.0215\ \text{A/m}^2)(4458\ \text{m}^2)}{7.14\ \text{A}}$$
$$= 61 \quad \text{(Eq 6)}$$

I_{prot} and $I_{protmud}$ are the initial current densities from Table 4.

In order to meet the mean current ($\bar{I}$) requirements, the weight loss of anodes over the 20 year design life is calculated based on the consumption rate (CR) given in Table 2.

$$\bar{I} = \bar{I}_w \cdot A_w + \bar{I}_m \cdot A_m$$
$$= (55\ \text{mA/m}^2)(3111\ \text{m}^2)$$
$$+ (10.8\ \text{mA/m}^2)(4458\ \text{m}^2) = 219\ \text{A} \quad \text{(Eq 7)}$$

Using CR 3.5 kg/A $\cdot$ yr:

$$W = (3.5\ \text{kg/A} \cdot \text{yr})(219\ \text{A})(20\ \text{yr}) = 15{,}330\ \text{kg}$$

The number of anodes (without considering efficiency and utilization) is:

$$N = \frac{15{,}330\ \text{kg}}{330\ \text{kg/anode}} = 46$$

Finally, to verify that sufficient protection is maintained at the end of the 20 year design life, a calculation is made similar to the initial current calculation, except with the reduced anode dimension that represents the anode at the end of its life. This anode design initially had an effective radius (r_i) of 13.7 cm (5.41 in.). There is an inert core with a radius (r_c) of 5.7 cm (2.25 in.). The effective (end-of-life) dimension (r_e) is:

$$r_e = r_i - (r_i - r_c) \times 0.9$$
$$= 13.7 - (13.7 - 5.7) \times 0.9$$
$$= 6.5\ \text{cm} \quad \text{(Eq 8)}$$

where 0.9 is a utilization factor for a standoff anode. Assuming there is no change in anode length, the final current output per anode is:

$$I = \frac{E}{R_f}$$

$$R_f = \frac{(20)(0.159)}{(274)}\left[\ln\frac{4 \times 274}{6.5} - 1\right]$$
$$= 0.0479\ \Omega$$

$$I = \frac{0.28\ \text{V}}{0.0479\ \Omega} = 5.85\ \text{A} \quad \text{(Eq 9)}$$

Table 4 Design criteria for cathodic protection systems

Production area	Environmental factors (a)				Typical design current density(d), mA/m^2 (mA/ft^2)		
	Water resistivity (b)(c), $\Omega \cdot$ cm	Temperature(c), °C	Turbulence factor (wave action)	Lateral water flow	Initial(e)	Mean(f)	Final(g)
Gulf of Mexico	20	22	Moderate	Moderate	110 (10)	55 (5)	75 (7)
U.S. West Coast	24	15	Moderate	Moderate	150 (14)	90 (8)	100 (9)
Cook Inlet	50	2	Low	High	430 (40)	380 (35)	380 (35)
Northern North Sea(d)	26–33	0–12	High	Moderate	180 (70)	120 (11)	120 (11)
Southern North Sea(d)	26–33	0–12	High	Moderate	150 (14)	100 (9)	100 (9)
Arabian Gulf	15	30	Moderate	Low	130 (12)	90 (8)	90 (8)
Australia	23–30	12–18	High	Moderate	130 (12)	90 (8)	90 (8)
Brazil	23–30	15–20	Moderate	High	180 (17)	90 (8)	90 (8)
West Africa	20–30	5–21	Low	Low	130 (12)	90 (8)	90 (8)
Indonesia	19	24	Moderate	Moderate	110 (10)	75 (7)	75 (7)
South China Sea	18	30	Low	Low	100 (9)	35 (3)	35 (3)
Sea mud	100	Same as water	N/A	N/A	21.5 (2)	10.8 (1)	10.8 (1)

(a) Typical values and ratings based on average conditions, remote from river discharge. (b) Water resistivities are a function of both chlorinity and temperature. See Table 5. (c) In ordinary seawater, a current density less than the design value suffices to hold the structure at protective potential once polarization has been accomplished and calcareous coatings are built up by the design current density. *Caution:* Depolarization can result from storm action. (d) Conditions in the North Sea can vary greatly from the northern to the southern area, from winter to summer, and during storm periods. (e) Initial current densities are calculated using Ohm's Law and a resistance equation, with the original dimensions of the anode. See example of this calculation, which uses an assumed cathode potential of -0.80 V (Ag/AgCl [seawater]). (f) Mean current densities are used to calculate the total weight of anodes required to maintain the protective current to the structure over the design life. See example. (g) Final current densities are calculated in a manner similar to the initial current density, except that the depleted anode dimensions are used. See example.

Table 5 Resistivity

Chlorinity(a), ppt	Resistivity $\Omega \cdot$ cm					
	at 0 °C (32 °F)	at 5 °C (41 °F)	at 10 °C (50 °F)	at 15 °C (59 °F)	at 20 °C (68 °F)	at 25 °C (77 °F)
19	35.1	30.4	26.7	23.7	21.3	19.2
20	33.5	29.0	25.5	22.7	20.3	18.3

(a) ppt, part per trillion.

The anodes required are:

$$N = \frac{I_{prot20} \cdot A_w + I_{protmud20} \cdot A_m}{I}$$

$$= \frac{(0.075 \text{ A/m}^2)(3111 \text{ m}^2) + (0.01076 \text{ A/m}^2)(4458 \text{ m}^2)}{5.85}$$

$$= 48 \quad \text{(Eq 10)}$$

where the final current densities are from Table 4. For this application, 61 anodes are needed for initial protection, 46 are required for mean current density, and 48 are required at end-of-life. The proper number to use is 61.

Dwight's equation is valid for long cylindrical anodes when $4L/r \geq 16$. For anodes when $4L/r < 16$ or for anodes that do not approximate cylindrical shapes, equations such as Crennell's/McCoy's (Eq 3) or other versions of Dwight's may better predict the actual current output of the anodes. Theoretically, for a deep-sea submerged cylindrical anode, a more accurate equation would be as shown in Eq 11; however, Eq 4 is more widely used in CP practice.

$$R = \rho \frac{K}{L}\left[\ln\left(\frac{2L}{r}\right) - 1\right] \quad \text{(Eq 11)}$$

Note:

- For practical designs and to ensure adequate current to protect the structure during the life of the anode, the length (L) and radius (r) should be selected to show the condition of the anode when it is nearly consumed. For an elongated anode, the change in length may be ignored.
- If the structure potential rises above the minimum protection potential of -0.80 volt (Ag/AgCl/seawater), E becomes less than 0.25 V. This decreases the anode current output and increases anode life.
- The anode net weight must be sufficient to provide the calculated current for the design life of the system, in accordance with the actual consumption rate of the anode material selected (Table 2).

Anode Distribution. A final consideration concerns the positioning of anodes about the structure. They are placed within 3 m (10 ft) of nodes (welded or cast locations with the highest structural loading), but elsewhere are assumed to protect steel in line of sight within a 7.6 m (25 ft) radius. Thus, areas shadowed by other structural elements may not be fully protected by any particular anode.

Computer-aided cathodic protection designs for offshore structures have been used by several organizations since the 1980s. These computer-aided designs are of two types. Computers have been used to make calculations such as wetted surface area versus anode consumption that are commonly needed for CP design. The computer is a time saver that allows a greater number of alternatives to be considered, but does not change the actual methodology of design.

An alternative approach is to use numerical techniques, such as finite element, finite difference, or boundary integral, to model the potential-current distribution field in the region of an offshore structure. Computers can be programmed to generate complex analyses of various alternative designs (Ref 7). In the past, these computerized designs had limited acceptance because of the expense associated with the inputting designs and the time delay caused by communications difficulties among the operator, the cathodic protection designer, and the computer expert. The increased memory capabilities of personal computers now allow these numerical programs to be run on them. Design engineers are now able to compare a number of design alternatives quickly and inexpensively. The same modeling techniques—finite element, finite difference, and boundary integral—that are used for structural design can be used for cathodic protection design (Ref 7–12). Figure 2 is a typical computer-generated plot of potential gradients around a node on an offshore platform. Comparisons of plots generated using different anode locations allow engineers to quickly determine the optimal locations for any portion of the platform. Color coding on the monitor display allows the engineer to identify quickly locations where additional CP current is needed.

Cathodic Protection of Ship Hulls

Ships normally have protective coatings as their primary means of corrosion control. Cathodic protection systems are then sized so that an adequate electric current will be delivered to polarize the structure to the desired level. This is done for new structures by estimating the percentage of bare steel that results from holidays in the protective coatings. Once the estimated amount of bare steel is determined, anodes are sized to provide adequate current densities for the design life of the system.

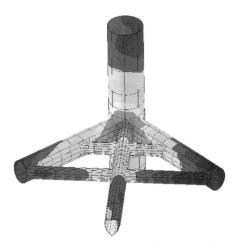

Fig. 2 Computer-generated plot of potential gradient distribution at the node of an offshore platform support structure

Ships are returned to drydocks; so the size (and weight) of anodes can be reduced from what would be necessary for permanent anodes on offshore pipelines or platforms since the anodes can be replaced. Anodes are concentrated near the bow and stern, where coating damage is most likely to occur. The stern is also the location where galvanic couples (for example, propeller to hull) are possible. Relatively small anodes are placed on ships in these locations. Small anodes are desirable to minimize the drag effects caused by turbulence due to anode protrusions.

Anode Materials. Aluminum anodes are available for ship hulls, but they can passivate and become inactive on ships that enter rivers or brackish estuaries. For this reason, zinc anodes are almost universally used in commercial service.

Impressed-current cathodic protection systems are used on very large ships. The galvanic couple between the propeller, the shaft, and the hull of the ship can cause significant corrosion problems. Modeling of the current requirements for cathodic protection near tanker propellers was one of the first applications of the computer in cathodic protection design (Ref 11, 12).

Impressed-current cathodic protection systems can produce overprotection in some cases. Organic coatings can disbond because of the formation of hydrogen gas bubbles underneath coatings. Coating disbondment can produce increased surface areas that require more cathodic protection and is controlled by placing dielectric shields between the impressed-current anode and the hull. Larger shields are sometimes fabricated from glass-reinforced epoxy, which is molded directly on the hull of the ship. At one time, coating disbondment was a major concern, but modern coatings are resistant to disbonding. Hydrogen embrittlement of steel due to cathodic protection is sometimes a concern. This has been a problem on case-hardened shafts, bolts, and other high-strength attachments. Most structural steels have relatively low strength as well as minimum susceptibility to hydrogen embrittlement.

REFERENCES

1. "Recommended Practice: Corrosion Control on Steel, Fixed Offshore Platforms Associated with Petroleum Production," RP0176, NACE International
2. B. Allen and R. Heidersbach, "The Effectiveness of Cadmium Coatings as Hydrogen Barriers and Corrosion Resistant Coatings," paper 230, Corrosion/83, National Association of Corrosion Engineers, April 1983
3. D. Boening, "Offshore Cathodic Protection Experience and Economic Reassessment," paper 2702, Offshore Technology Conference (Houston, TX), May 1976
4. E. Levings, J. Finnegan, W. McKie, and R. Strommen, "The Murchison Platform Cathodic Protection System," paper 4565,

Offshore Technology Conference (Houston TX), May 1983
5. J. Smart, Corrosion Failure of Offshore Steel Platforms, *Mater. Perform.,* May 1980, p 41
6. K. Fischer, P. Mehdizadeh, P. Solheim, and A. Hansen, "Hot Risers in the North Sea: A Parametric Study of CP and Corrosion Characteristics of Hot Steel in Cold Seawater," paper 4566, Offshore Technology Conference (Houston, TX), May 1983
7. R. Heidersbach, J. Fu, and R. Erbar, *Computers in Corrosion Control,* National Association of Corrosion Engineers, 1986
8. R. Strommen, "Computer Modeling of Offshore Cathodic Protection Systems Utilized in CP Monitoring," paper 4367, Offshore Technology Conference (Houston, TX), May 1982
9. R.A. Adey and J.M. Baynham, "Design and Optimisation of Cathodic Protection Systems Using Computer Simulation," paper 00723, Corrosion 2000, NACE International
10. M. Haroun, "Cathodic Protection Modeling of Nodes in Offshore Structures," M.S. thesis, Oklahoma State University, 1986
11. J. Fu, *Corrosion,* Vol 38 (No. 5), May 1982, p 295
12. J. Fu and S. Chow, *Mater. Perform.,* Vol 21 (No. 10), Oct 1982, p 9

External Corrosion Direct Assessment Integrated with Integrity Management

Joseph Pikas, MATCOR Inc.

EXTERNAL CORROSION DIRECT ASSESSMENT (ECDA) is a structured process intended for use by pipeline operators to assess and manage the impact of external corrosion on the integrity of underground pipelines. The process integrates field measurements with the physical characteristics, environmental factors, and operating history of a pipeline. ECDA includes primarily nonintrusive or above ground examinations, which are tailored to the pipeline or segment to be evaluated. In addition, physical examinations (direct assessments) of the pipeline at sites identified as potential areas of concern by analysis of the indirect examinations are included. A detailed description of the ECDA process is given in the NACE International standard (Ref 1).

ECDA represents a way to control integrity-related threats through a structured process while maintaining reasonable operating and maintenance costs. A pipeline vulnerable to multiple threats from external corrosion or third party intrusions (mechanical damage) may require direct assessment. One major factor in the control of costs is the proper selection and matching of tools and processes to the predetermined threats. ECDA is a continuous improvement process through the use of successive applications. An operator using established criteria can track the reliability of results when the ECDA process is applied. The effectiveness of this process is measured by being able to address with confidence the anomaly locations of concern, and by having a safe operating pipeline system. In the United States, pipeline systems are governed by the Office of Pipeline Safety (OPS) in accordance with integrity management rules.

Four Step ECDA Process

ECDA is a four step process that integrates data and information from pipeline, construction, and cathodic protection records, physical pipe examinations along the pipeline, and operating history. Through successive application of the four step process, a pipeline operator can identify and address where corrosion activity has occurred, is occurring, or may occur. Results from ECDA are used to prioritize future integrity-related actions. The four steps are:

1. *Preassessment:* Collects historic and current data to determine if ECDA is feasible, defines ECDA regions, and selects indirect inspection tools.
2. *Indirect examinations:* Conducts above ground inspection(s) to identify and define coating faults, anomalies, and corrosion activity.
3. *Direct examinations:* Evaluates indirect inspection data to select sites and then conducts excavations for pipe examinations.
4. *Postassessment:* Analyzes data collected from the previous three steps to assess the effectiveness of the ECDA process and determine re-assessment intervals. ECDA has a proactive advantage over alternative integrity assessment methodologies, such as pressure testing and in-line inspection, by identifying areas where defects could become an integrity concern in the future.

Pipeline Integrity and Data. The ECDA process requires that integrity and operating history data are integrated with data from indirect and direct examinations. This is a strength of the ECDA process. Analyzed together, the data can provide a required integrity confidence level or lead to recommendations for further action, such as additional field testing, in-line inspection runs (smart pigging), hydrostatic testing, reconditioning, or pipe replacement.

Although the focus of ECDA is to locate external corrosion, it is recognized that other conditions that adversely impact pipe integrity may be associated with coating faults. Such conditions should be addressed using appropriate remediation criteria covered in company operating procedures, or other industry documents such as ASME B31.8S (Ref 2) and API 1160 (Ref 3). External corrosion is only one of 22 threats (Ref 4) that can impact structural integrity. Therefore, ECDA only addresses part of an operator's overall integrity management program. Overall integrity should be established by considering all credible threats that can impact the pipeline or segment being evaluated.

Direct bell hole examination (direct visual examination of the pipe or coating) is used to prioritize corrosion defects for remediation and post-assessment determination of pipeline integrity and the safe operating pressure for a specified time interval. As digging is expensive, ECDA can improve location selection by integrating results from preliminary integrity analyses, statistical data analyses, and inferential measurements.

ECDA should be used for below-ground sections of transmission and distribution piping systems constructed from ferrous materials. It may be used in conjunction with or in place of other integrity assessment tools including in-line inspection, pressure testing, or other proven technology. The ECDA process applies to coated as well as bare pipe; however, all inspection methods do not apply to both coated and bare pipe and may require different interpretation based upon the particular application.

Step 1: Preassessment (Assessment of Risk and Threats)

The preassessment action step consists of the collection, analyses, and review of pipeline data to determine if the ECDA process can be applied over the pipeline or a segment of the pipeline.

ECDA specific data analysis includes many of the same data elements that are typically considered in the overall pipeline risk (threat) assessment. Depending on the operator's integrity management plan, operating and maintenance procedures, and ECDA data, analyses could also be performed in concert with a general risk/threat assessment effort.

Data elements are grouped into five categories as shown below. These data have been selected to provide for a comprehensive analysis and guidance for establishing ECDA feasibility and regions. A data table should contains these essential data elements, but additional elements may also be utilized. (Ref 1 provides guidance for the use of each element and its influence on the ECDA process in its first table). The five categories to be included in the analysis process are:

- *Pipe related:* such as wall thickness, diameter, specified minimum yield strength (SMYS), year of manufacture
- *Construction related:* such as year of installation, installation design and practices, depth of cover
- *Soils and other environmental factors:* such as soil type, drainage, topography
- *Cathodic protection (CP):* including CP type, test points, maintenance, coating
- *Pipeline operations:* including operating temperature, monitoring, excavation results

The preassessment determines if conditions exist where particular ECDA methods cannot be used or where ECDA is totally precluded. Analyses can also be used to estimate the likelihood and extent of existing corrosion and prioritize indirect assessment steps. Where ECDA feasibility cannot be established, analyses results can be used as guidance for selecting alternative integrity assessment methods.

One of the major objectives derived from collecting the data elements is to assist in determining if ECDA tools are applicable for determining pipeline integrity at locations where external corrosion has been determined to be a credible threat. Such a risk (or threat) assessment is required as an element of the integrity management plan. Risk/threat assessment is an accepted industry practice; it is a systematic method for integrating and using a variety of data elements that together provide an improved understanding of the nature and locations of risks along a pipeline or segment. Such models can also be used to prioritize the severity of risk. Risk is typically described as the product of the probability of failure (POF) and a measure of the consequence of failure (COF).

Data collected for ECDA applicability analyses in the preassessment step can be analyzed using similar models without necessarily including the COF term used for risk assessment. This may include one general model or several specific models such as for soils or corrosion. In any case, the methods used for such analyses must be consistently and systematically conducted to permit an accurate assessment of ECDA feasibility, selection of ECDA regions, and determining the appropriate methods to be used in each region. These analysis results can also be the basis for prioritization based on the estimated corrosion severity and extent at each location. Depending on the integrity management plan and risk assessment methods used by individual pipeline operators, ECDA specific data analyses can be accomplished as a subset of the overall pipeline risk assessment.

Tools are selected based on their ability to detect corrosion activity and/or coating holidays under normal pipeline conditions. Reference 1 provides an ECDA selection matrix to determine the tool choices similar to Table 1.

The data is used to define ECDA regions along the pipeline. An ECDA region is a section (or sections) of a pipeline determined with reference to the ECDA tool selection matrix (Table 1). A section is suitable for the sucessful application of the same two above ground indirect examination methods. Other ECDA regions must be defined where data analyses and Table 1 indicate that different pairs of ECDA methods are needed. ECDA regions only apply to selection of indirect examination methods.

Figure 1 provides an example of the definitions for ECDA regions. The pipeline operator must consider whether more than two indirect inspection tools are needed to detect corrosion activity reliably. The pipeline operator should consider all conditions that could significantly affect external corrosion where defining criteria

Table 1 ECDA tool selection matrix

Condition	Applicability(a)				
	Close interval survey (CIS)	Direct current voltage gradient (DCVG)	Alternating current voltage gradient (ACVG)	Soil contact with pipe locator (Pearson)	Pipeline current mapper (PCM)
Coating holidays	2	1, 2	1, 2	2	2
Anodic zones on bare pipe	2	NA	NA	NA	NA
Near river or water crossing	2	NA	NA	NA	2
Under frozen ground	NA	NA	NA	NA	2
Stray currents	2	1, 2	1, 2	2	2
Shielded corrosion activity(b)	NA	NA	NA	NA	NA
Adjacent metallic structures	2	1, 2	1, 2	NA	2
Near parallel pipelines	2	1, 2	1, 2	NA	2
Under high voltage AC transmission lines	2	1, 2	1, 2	2	NA
Shorted casing	2	1, 2	1, 2	2	2
Under paved roads(c)	2	1, 2	1, 2	2	2
Uncased crossing	2	1, 2	1, 2	2	2
Cased crossing	NA	NA	NA	NA	2
At deep burial locations(d)	2	1, 2	1, 2	2	2
Wetlands	2	1, 2	1, 2	2	2
Rocky terrain, rock ledges, rock backfill	NA	NA	NA	NA	2

(a) 1, applicable for small isolated coating holidays, typically <6.5 cm^2 (1 in.^{2}), and conditions that do not cause fluctuations in CP potentials under normal operating conditions. 2, applicable for large isolated or continuous coating holidays, or conditions that cause fluctuations in CP potentials under normal operating conditions. NA, not applicable without additional considerations. (b) An orifice through the soil is required. (c) Drilled holes required for paved areas. (d) All instruments lose sensitivity. Source: Ref 1

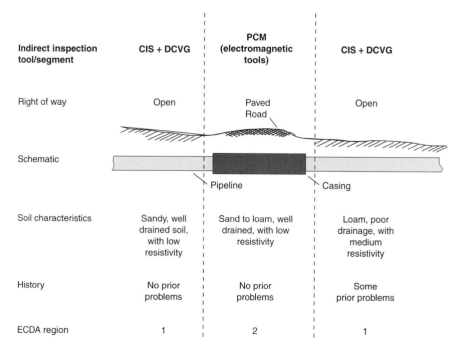

Fig. 1 Characteristics of external corrosion direct assessment (ECDA) regions. Inspect tool/segments: close interval survey, CIS; direct current voltage gradient, DCVG; pipeline current mapper, PCM

for ECDA regions. A single ECDA region does not need to be continuous. It can be broken along the pipeline, for example across rivers, paved roads, or parking lots.

The Fig. 1 example is a pipeline segment that indicated reasonably consistent conditions except that part of the segment has been paved so two regions are required with different measurement techniques.

The definitions of ECDA regions may be modified based on results from the indirect examination step and the direct examination step. The definitions made at this point are preliminary and are expected to be fine-tuned later in the total ECDA process. Once the ECDA regions have been defined, the operator should select a minimum of two indirect examination methods for each defined ECDA region. Depending on the specific conditions in an ECDA region as indicated by data analysis or sections of special concern, additional indirect examination methods may be required to achieve an adequate inspection reliability and/or resolve unexplained differences detected when comparing results from the two required indirect examinations. If the conditions determined by data analysis in a particular ECDA region are such that only one indirect examination method is applicable, ECDA methods should not be allowed and alternative in-line inspection (ILI), hydrostatic testing, or other proven methods must be used.

Demonstration of Feasibility. Under the OPS of the U.S. Department of Transportation, Gas Pipeline Integrity Rules, a pipeline operator must assess, analyze, and prove the conditions that allow ECDA. The pipeline integrity rule requires pipeline operators to develop an integrity management program for gas transmission pipelines located where a leak or rupture could do the most harm, called high consequence areas (HCAs). The rule requires pipeline operators to perform ongoing assessments of pipeline integrity, to improve data collection, integration, and analysis, to repair and remediate the pipeline as necessary, and to implement preventive and mitigation actions.

Conditions that require operator assessment are customer interruption or pipelines operating below a prescribed value of percent specified minimum yield strength (SMYS). Application of integrity data management systems and risk/threat assessment methodologies represent various approaches available to operators in order to fulfill the requirements for the direct assessment (DA) process. The Gas Pipeline Rule allows the use of DA methodologies for assessment of third party damage (this is mechanical damage that has occured with coating damage) as long as it is used in its entirety, can demonstrate effectiveness, and is documented with the following auditable steps:

- Methods and procedures used to establish ECDA feasibility
- Methods for ECDA region establishment and region definitions
- Data analyses and integration methods including any models used
- Results of general risk (threat) analyses establishing the areas where external corrosion is a credible threat in HCAs

Step 2: Indirect Examinations

The second ECDA process step includes above-ground, indirect examinations in each ECDA region established in the preassessment step and analysis of the data. Depending on the indirect examination results and their consistency, these data analyses may indicate the need for additional indirect or preliminary direct examinations. An indirect examination process includes a feedback loop to facilitate continuous improvement of the preassessment step.

Prior to conducting indirect examinations, the extent of each ECDA region must be identified and clearly marked. Measures must be used to assure a continuous indirect examination is achieved over the pipeline or segment being evaluated. This could include some examination overlap into adjacent ECDA regions.

Indirect examinations should be conducted using closely spaced intervals and no greater than a 90-day time interval with a detailed assessment of the area. Each indirect examination must be conducted and analyzed in accordance with industry and NACE accepted practices.

Two indirect examinations should be conducted over the entire length of each ECDA region or HCA in the segment being evaluated. Results from these inspections are then evaluated to establish that these results are complementary. When analyzing indirect examination results, the operator must be aware that some spatial error will likely be present when comparisons are made. Errors can cause difficulties when determining the similarity between two indirect examination results. This can occur due to location differences at the ground surface when conducting the indirect examinations or changes in burial depth. Such error may be reduced by using an increased number of above-ground reference points such as fixed pipeline features and additional above-ground markers. Other techniques including commercially available software based graphical overlay methods may also be used to resolve spatial errors.

Depending on the particular indirect examination method used, the operator should also attempt to determine if the areas that may contain corrosion are active or inactive.

All indirect examination actions should be thoroughly documented. This may include the following:

- Indirect location identification versus the actual condition as indicated by any preliminary direct examinations
- Analyses of indirect process effectiveness and any difficulties encountered
- Feedback to the preassessment stage for improvements, modifications
- ECDA region locations, boundaries for future reference
- Anomaly severity classification versus that determined by direct examinations as in Table 2

The anomaly classifications are:

- *Severe:* indications that the pipeline operator considers as having the highest likelihood of corrosion activity
- *Moderate:* indications that the pipeline operator considers as having possible or questionable corrosion activity
- *Minor:* indications that the pipeline operator considers inactive or as having the lowest likelihood of corrosion activity

Step 3: Direct Examination

Direct examination requires excavations to expose the pipe surface, metal-loss measurements, estimated corrosion growth rates, and measurement corrosion morphology at coating faults identified during indirect examination. The goal of these excavations is to collect enough information to characterize the external corrosion anomalies that may be present on the pipeline or segment being evaluated.

Excavations for direct examinations should be made at one or more locations from each ECDA region in which coating faults have been found. Where an extended length pipeline is included in a particular ECDA region, additional excavations must be considered. Excavations should be conducted based on the initial estimated severity and prioritization established during indirect examination with most severely corroded areas and active areas examined first. Additional severity prioritization should be also conducted during direct examination, which will provide validation data for the preassessment and indirect examination steps.

The ECDA process is primarily focused on external corrosion anomalies. However, it is recognized that other threats often associated with coating fault locations, such as mechanical damage and stress-corrosion cracking (SCC), may also be discovered during the direct examination step. The operator's pipeline risk (threat) assessment process will provide guidance as to the potential existence of anomalies other than external corrosion. The operator must address such anomalies based on criteria contained in the appropriate industry standards.

Data that should be acquired prior to and during each excavation and before any coating removal are:

- Pipe-to-soil potential
- Soil resistivity
- Soil sample collection
- Water sample collection
- Coating condition assessment

- Microbiological influenced corrosion (MIC) analysis
- Corrosion product analysis
- Photographic documentation

The following additional data should be acquired during each excavation after coating removal:

- Coating thickness
- Coating adhesion
- Pipe temperature under coating
- Coating condition (blisters, disbondment, etc.)
- Under-film liquid pH
- Corrosion sample analysis
- Coating backside contamination analysis

During pipe examination at locations determined by the severe indications as in Table 2, corrosion depth measurements and severity estimates should be made using ASME B31G (Ref 5) or remaining strength of corroded pipe calculations, RSTRENG software (Ref 6), are made to determine the integrity of the pipeline. Sufficient corrosion depth measurements should be made at each excavation to provide adequate data to make a statistically valid maximum depth estimate that exists in each ECDA region.

The pipeline operator must establish criteria for prioritizing the need for direct examination of each indication found during the indirect examination step. Prioritization is the process of estimating the need for direct examination of each indication based on the likelihood of current corrosion activity plus the extent and severity of previous corrosion. Different criteria may be required in different regions, as a function of the pipeline condition, age, history of cathodic protection, and location history.

The minimum number of prioritization levels is given in Table 3:

- *Immediate action required:* This priority category should include indicators that the pipeline operator considers to have ongoing corrosion activity and that, when coupled with prior corrosion, pose an immediate threat.
- *Scheduled action required:* This priority category should include indicators that the pipeline operator considers may have ongoing corrosion activity but that, when coupled with prior corrosion, do not pose an immediate threat.
- *Suitable for monitoring:* This priority category should include indicators that the pipeline operator considers inactive or as having the lowest likelihood of ongoing or prior corrosion activity.

Step 4: Post Assessment

Where the operator has measured corrosion rate data that is applicable to the ECDA region, such actual rates may be used for inspection interval determinations in the post assessment step.

As a minimum, soil resistivity may be used; however, more precise corrosion rate estimates can be obtained from other soil measurements in addition to resistivity.

If conditions exist that prevent a statistically valid sample from being obtained from a single ECDA region, data from excavations in other similar ECDA regions, as defined in the preassessment step, may be used. Alternatively, additional excavations may be performed to obtain the necessary data.

Post assessment sets re-inspection intervals, provides a validation check on the overall ECDA process, and provides performance measures for integrity management programs.

The first step in post assessment is to determine the remaining pipeline life from calculations of the possible corrosion defects at coating fault locations that were not subjected to direct examination. Operators will have mathematical models for life calculations. An estimate can be made of the remaining pipeline life using:

$$RL = \frac{0.85}{P}\left(\left(\frac{P_f}{SF_{DR}} - MAOP\right)\frac{t}{CR}\right)$$

where RL is remaining life (yr); P is applied pressure (psi); P_f is burst pressure calculated by RSTRENG (psi); SF_{DR} is design requirement safety factor; $MAOP$ is maximum allowable operating pressure (psi); 0.85 is a calibration factor; t is the wall thickness (in units compatable with CR); and CR is the corrosion growth rate per year.

This method of calculating expected remaining life is reasonably conservative for pipeline external corrosion. Where data is inadequate or deficient, the half life must be used as a default on the basis for repair and reassessment intervals.

If the half life is acceptable, one additional excavation is to be performed for validation purposes. This excavation is to be performed at the coating fault location that was estimated to contain the next most severe defect not previously subjected to direct examination. Corrosion severity at this location should be determined and compared with the maximum severity predicted during the direct examinations.

- If the actual corrosion defect severity is less than half of the maximum predicted severity, validation is completed.
- If the actual corrosion severity is between the maximum predicted severity and one half of the maximum predicted severity, double the predicted maximum severity and recalculate the half life.
- If the actual corrosion severity is greater than the maximum predicted severity, the ECDA process may not be appropriate and the operator must return to the preassessment stage.

ECDA validation may also be performed using historical data from prior excavations on the same pipeline. Prior excavation locations must be assessed to determine that they are equivalent to the ECDA region being considered and such a comparison is valid. If validity is established,

Table 2 Example of severity classification

	Inspection classification		
Inspection tool	Severe indications	Moderate indications	Minor indications
CIS	Two or more of the following must exist: • OFF potential less than −850 mV • ON potential less than −850 mV • Reduced potentials shifted in a positive direction	Two or more of the following must exist: • OFF potential less than −850 mV • ON potentials greater than −850 mV • Reduced ON potentials shifted in a positive direction	Any of the following can exist: • OFF potential at or near −850 mV • ON potential above −850 mV • Single spikes • Saw tooth patterns in both ON and OFF • Step potential
PCM	Greater than 20% change in 100 ft	Between 10 to 20% change in 100 ft	Less than 10% change in 100 ft
DCVG	36 to 100% IR anodic/anodic	16 to 35% IR cathodic/anodic	1 to 15% IR cathodic/neutral cathodic/cathodic
ACVG	Greater than 90 dB	50 to 89 dB	Less than 50 dB

Table 3 Example of indirect inspection categorization indication table

Priority level	Indicators
Immediate action required	• Severe indications in close proximity regardless of prior corrosion • Individual severe indications or groups of moderate indications in regions of moderate prior corrosion • Moderate indications in regions of severe prior corrosion
Scheduled action required	• All remaining severe indications • All remaining moderate indications in regions of moderate prior corrosion • Groups of minor indications in regions of severe prior corrosion
Suitable for monitoring	• All remaining indications

then maximum corrosion depths may be estimated from the prior data.

The last step in the post assessment stage is to set reinspection intervals. Repair intervals must be based on the expected half-life calculation or an equivalent method. The half-life (default data deficient) or equivalent (proof) can be used to prorate the repair interval.

ACKNOWLEDGMENT

Pipeline Research Committee International (PRCI), National Association of Corrosion Engineers International (NACE), Interstate National Gas Association of America (INGAA), American Gas Association (AGA), Gas Technology Institute (GTI), New England Gas Association, Battelle Institute, American Petroleum Institute (API), Office of Pipeline Safety (OPS) and many others contributed their efforts putting together the External Corrosion Direct Assessment Methodology.

REFERENCES

1. "Pipeline External Corrosion Direct Assessment Methodology," Standard Recommended Practice RP0502-2002, NACE International, 2002
2. "Managing System Integrity of Gas Pipelines," ANSI/ASME B31.8S, ASME, 2004
3. "Managing System Integrity for Hazardous Liquid Pipelines," Standard 1160, API, Washington, DC, 2001
4. PRCI's "22 Pipeline Threats, Categorized by Type" by Dr. John Kiefner
5. B31G, Manual for Determining Remaining Strength of Corroded Pipelines: Supplement to B31 Code-Pressure Piping, ASME, 1991, reaffirmed 2004
6. Rstreng, software available from Technical Toolboxes, accessed January 2005 at www.ttoolboxes.com

Close-Interval Survey Techniques

Drew Hevle, El Paso Corporation
Angel Kowalski, CC Technologies, Inc.

A CLOSE-INTERVAL SURVEY (CIS) is a series of structure-to-electrolyte direct current (dc) potential measurements performed at regular intervals for assessing the level of cathodic protection (CP) on pipelines and other buried or submerged metallic structures (Fig. 1). Within the industry, the terms close-interval survey (CIS) and close-interval potential survey (CIPS) are used interchangeably.

Types of CIS include:

- *On survey,* data collection with the CP systems in operation
- *Interrupted or on/off survey,* a survey with the CP current sources synchronously interrupted
- *Asynchronously interrupted survey,* a close-interval survey with the CP current sources interrupted asynchronously, using the waveform analyzer technique
- *Depolarized survey,* a close-interval survey with the CP current sources turned off for some time to allow the structure to depolarize
- *Native-state survey,* data collection prior to application of CP
- *Hybrid surveys,* close-interval surveys incorporating additional measurements such as lateral potentials, side-drain gradient measurements (intensive measurement surveys), or gradient measurements along the pipeline

The term CIS (or CIPS) does not refer to surveys such as cell-to-cell techniques used to evaluate the direction of current (hot-spot surveys, side-drain surveys) or the effectiveness of the coating (traditional direct current voltage gradient, DCVG).

Typical CIS graphs are shown for a fast-cycle interrupted survey combined with a depolarized survey to evaluate a minimum of 100 mV of cathodic polarization (Fig. 2) and a slow-cycle interrupted survey (Fig. 3). Close-interval survey is used to assess the performance and operation of a CP system in accordance with established industry criteria for CP such as those in NACE International Standard RP0169. The −850 mV criteria is indicated on Fig. 2 and 3.

Close-interval survey is one of the most versatile tools in the CP toolbox and, with new integrity assessment procedures, has become an integral part of pipeline integrity program.

See the article "External Corrosion Direct Assessment Integrated with Integrity Management" in this Volume.

Close-interval survey data interpretation provides additional benefits, including:

- Identifying areas of inadequate CP or excessive polarization
- Locating medium-to-large defects in coatings on existing pipelines
- Locating areas of stray-current pickup and discharge
- Identifying possible shorted casings
- Locating defective electrical isolation devices
- Detecting unintentional contact with other metallic structures
- Testing current demand and current distribution along a pipeline
- Locating possible high-pH SCC risk areas or as a component of a stress-corrosion cracking direct assessment (SCCDA)
- Locating and prioritizing areas of risk of external corrosion as part of an integrity-management program or as a component of an external corrosion direct assessment (ECDA)

Certain conditions can make the data from a CIS difficult or impossible to properly interpret or make the survey impractical to perform. Examples include areas of high contact resistance, such as pipe located under concrete or asphalt pavement, frozen ground, or very dry conditions. Contact resistance in paved areas may be reduced by drilling through the paving to permit electrode contact with the soil. Contact resistance in dry areas can be reduced by moistening the soil with water until the contact is adequate. Because this is often difficult or impractical, CIS should be scheduled, when possible, to avoid unfavorable seasons. Other impediments to close-interval surveys are adjacent buried or submerged metallic structures, surface conditions limiting access to the electrolyte, backfill with significant rock content or rock ledges, gravel, and dry vegetation. Telluric (natural currents near the surface of the earth) or other dynamic stray currents and high levels of induced alternating current (ac) can introduce errors into the potential measurements. Pipelines buried very deep, such as horizontal directional drill crossings under rivers and highways, do not allow a high-resolution survey. Pipelines with locations at which coatings cause electrical shielding do not allow valid potential measurements. Pipelines without electrical continuity, such as with some forms of mechanically coupled pipe, do not allow for close-interval surveys.

CIS Equipment

The equipment required to perform a CIS comprises:

- Voltmeter
- Reference electrode(s)
- Electrical connection to the pipeline

Figure 1 is a sketch of the equipment used in a CIS. Depending on the type of survey, additional equipment may be required, such as:

- Pipe locator
- Distance measuring device
- Data loggers
- Current interrupters
- Global positioning system (GPS) surveying equipment

Voltmeters. To determine a pipe-to-electrolyte potential value, a voltmeter must measure the potential drop across an external circuit resistance that varies depending on the type of environment. To compensate for these variable conditions, voltmeters must be built with high impedance or input resistance. Typical minimum input resistance is >20 MΩ. A higher external resistance requires a higher input resistance to maintain accuracy during measurements. Voltmeters with low internal resistance can cause significant errors when measuring pipe-to-soil potentials. Additional relevant characteristics of the voltmeter are range, resolution, and accuracy.

Reference Electrodes. The choice of reference electrodes is determined by the environment in which the electrode is placed. For onshore surveys the most commonly used is the copper/copper-sulfate electrode (CSE). Other electrodes such as the silver/silver-chloride and zinc/zinc-sulfate are used in environments such as seawater. The reference electrode calibration must be checked before and during the CIS. A common calibration method consists of measuring the potential difference between the working reference electrodes and a master electrode; differences less than 5 mV are typically acceptable. A reference electrode that fails a calibration check must be replaced or repaired before future use.

Electrical Connection. To perform a CIS, the voltmeter must be connected to the reference electrode and to the structure being inspected. The use of wire reels (relatively heavy gage wire on reels) or disposable wire (lighter gage wire) is necessary to maintain electrical contact with the structure and the voltmeter. This connection is normally made at test stations. A low-resistance current path is needed to minimize voltage drops in the metallic circuit. It is also important that the wire is properly insulated to avoid direct contact between the metallic conductor and the electrolyte.

Pipeline Location. Visual identification of the pipeline by aboveground appurtenances, casing vents, or pipeline markers may or may not be sufficient for accurate location of the pipeline. If visual identification does not provide accurate location of the pipeline, radio-frequency pipeline locators or other devices may be required. In congested pipeline rights-of-way, areas of deep cover, small diameter or poorly coated pipe, or areas of high ac background potentials, conductive location techniques, or other more accurate location techniques and equipment may be required.

Distance Measuring Device. When performing a CIS, it is important to register the location where the pipe-to-electrolyte potential was measured. This can be accomplished by determining the relative distance to an aboveground reference point (such as a test station, valve station, or road crossing) by chaining or using another distance measuring device. The use of the GPS equipment also allows recording of testing locations.

Current Interrupters. When an interrupted CIS is performed, the use of current interrupters is required. If more than one dc source is interrupted, the interrupters must be able to synchronize their interruption cycle so that they remain in the same state simultaneously.

Data Loggers. It is a common practice to use a field computer (data logger) with a built-in voltmeter to capture the pipe-to-soil potentials during a CIS. A data logger can store a large amount of data, including structure-to-electrolyte potentials, state of CP current (on/off), distance or chainage, geographical location, field comments, waveforms, and additional measurements. When a data logger is used on an interrupted CIS, it must be capable of recording the on and instant-off potentials at the specified interruption cycle. Data acquisition software and/or hardware for interrupted CIS must be adjusted to avoid recording transient potentials produced by a "spike effect."

Data loggers placed at a fixed test station to record structure-to-electrolyte potentials, named stationary loggers, are installed to detect the presence of dynamic stray current and verify the correct operation of current interrupters.

Preparation

Presurvey activities are essential for obtaining reliable results on a CIS.

Review and Analyze System Information. This pipeline system or other underground structure information includes:

- Structure-related data (geographical location, depth of burial, electrolyte resistivity, coating type, alignment drawings, locations of water crossings, types of terrain, locations of high-voltage AC power transmission lines and dc rail systems)
- Cathodic protection operating system data, test stations, electrical bonds
- Foreign CP systems
- Foreign underground structures in proximity to pipeline
- AC mitigation systems
- Electrical isolation devices

Equipment Calibration. The voltmeter and reference electrodes must be calibrated before starting the survey and records of the calibration should be maintained for data validation purposes.

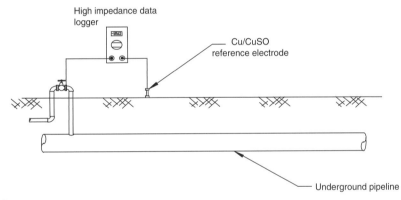

Fig. 1 Arrangement of close-interval survey components

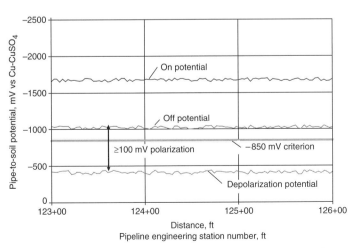

Fig. 2 Typical graph of fast-cycle interrupted survey combined with depolarized survey. Distances are measured from a specified starting point, 123+00 is 12,300 ft.

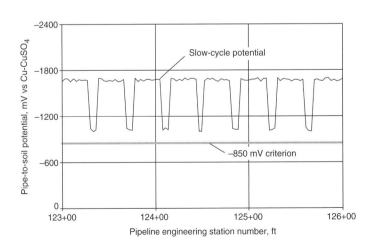

Fig. 3 Typical graph of slow-cycle interrupted close-interval survey. Distances are measured from a specified starting point, 123+00 is 12,300 ft.

Direct Current Source Influence Test. If an interrupted CIS is performed, a dc source influence test must be performed at the different segments of the structure to be surveyed. This should be performed by interrupting all known dc sources individually and determining their coverage by measuring the potential difference between the on and the off portions of the interrupting cycle. The operating conditions of the dc sources should be measured and recorded. After the dc source influence test is completed, a dc source interruption plan can be elaborated for each section of the structure and the location of the stationary data loggers can be determined.

Locating Underground Structures. In order to perform the survey accurately, the precise location of the underground structure must be determined. Normally, the survey operator with the data logger either follows directly behind the pipe locator, or a separate locating crew marks the structure route at intervals of 15 to 30 m (50 to 100 ft).

Procedures

After completing the preparation activities the surveyor should define:

- Start and end points
- Survey direction
- Survey potential reading interval
- Aboveground reference features
- Precise location of the underground structure

If an interrupted survey will be performed, the surveyor should also define:

- dc interruption cycle
- dc interruption plan

The following activities should be performed during the execution of a CIS.

Install a Stationary Data Logger. For every section of the structure a stationary data logger may be located at a test station near the midpoint of the segment to be tested that day. This data logger is connected to the structure (test station) and to a reference electrode. The data logger will record structure-to-electrolyte potentials for the duration of the CIS. The data recorded should contain the date of the survey, the pipe-to-soil potential, and the time. Stationary data loggers are normally configured to capture a reading every second. These data allow the determination of possible dynamic stray currents and any depolarization that may affect the interpretation of the CIS results.

Structure-to-Electrolyte Potential Measurements. The surveyor will measure and record potentials at the defined interval and also record all relevant information about survey conditions. The distance and potential are the critical data; if time is also attached to each potential reading, these data can be compared with the data from the stationary data logger. Aboveground features may be recorded and referenced to start point of the survey to aid in relocation of indications.

The location of the electrical connection to the structure should be documented. External conditions such as changes in soil conditions, terrain, land use, and so forth should also be recorded. This information is valuable when analyzing the survey results. When performing an interrupted survey it is a good practice to periodically measure the waveform of the interrupted cycle to confirm the synchronization of the current interrupters.

End of Survey. When the surveyor arrives at the end point of the survey, additional measurements should be obtained, such as metal voltage (*IR*) drop, and, if performing an interrupted survey, the waveform of the interrupted cycle. The initial and final waveform data serve as a validation of the recorded potentials.

Data Processing. After completing the field survey, the data are downloaded from the data logger or transferred from the written form to a computer with graphing software capabilities. The results of typical close-interval surveys are shown in Fig. 2 and 3.

Ohmic voltage drop (*IR* drop) correction of CIS may be achieved using a number of different methods. The most common method is the instant-off-potential method using synchronized-current interrupters installed at CP current sources. It is beneficial to measure both the uncorrected potential (the on potential) and the *IR* drop corrected potential in order to determine the level of *IR* drop correction. The magnitude of *IR* drop in CIS of pipelines with the current applied can aid in detecting locations of protective coating faults. Correction methods that apply *IR* drop to every reading provide the most accurate CIS potential data. Other methods include: CP coupons, stepwise current reduction, and waveform analysis.

IR drop may not be a significant concern when electrolyte, current densities, depth of burial, and coating condition are consistent, and the magnitude of the *IR* drop is known or considered to be negligible. *IR* drop correction information can also be applied to other surveys, such as future test point surveys.

External influences that can be current sources include:

- Foreign pipelines or other cathodically protected facilities that are electrically bonded to the pipeline being surveyed. These must be interrupted to measure potentials to the desired accuracy.
- Direct current transit and mine railway systems, and high-voltage dc power transmission can cause stray or induced currents through the electrolyte near the pipeline being surveyed. Additionally, long-line currents and telluric currents can cause currents along the pipeline. These currents cannot be interrupted, but can be measured by methods listed in the section "Dynamic Stray Current" in this article.

Current Interruption Cycle Time. Fast-cycle interruption is a term that usually refers to an interruption cycle in which the off cycle is less than 1 s. The advantage of fast-cycle interruption is that an instant-off reading can be obtained at each reading location without slowing data collection, providing more information. A disadvantage of fast-cycle interruption is that this procedure requires a fast voltmeter, precisely synchronized interruption, and data acquisition software that can correctly differentiate between the on and instant-off potentials and transitions and can record accurate potentials. Because of the difficulty in synchronizing interrupters operating on a very short cycle, slow-cycle current interruption has been historically more common. Advances in electronics and GPS time controlled equipment have made extremely accurate timekeeping more practical.

It is often not practical to use slower cycles to obtain an instant-off potential at each reading location because of the time required to obtain both an on and an instant-off-potential measurement. A disadvantage of the slow-cycle interruption is that depolarization may occur during the interruption of the CP. Slow-cycle current interruption may or may not differentiate between on and instant-off potentials during data collection. When the data are differentiated, separate and continuous plots of on and instant-off potentials are usually provided. Slow-cycle surveys that do not differentiate the potentials during the survey may differentiate the on and instant-off potentials by visual interpretation of a graph of the potentials (saw-tooth graph, Fig. 3).

There are various methods to confirm the operation of the current interrupters. Oscilloscopic waveforms may be used to show that the interrupters that influence a location are synchronized (Fig. 4). Waveforms do not indicate influencing current that is not interrupted. Lateral potentials and measurement of metallic *IR* drop may be used to obtain an indication of influencing CP current that is not being interrupted or foreign currents that are causing *IR* drops in the off cycle.

Metallic *IR* drop is the component of potential drop that occurs in the metallic path of the measurement circuit, primarily in the pipeline, under normal conditions. The magnitude of metallic *IR* drop represents the net current in the pipeline between the two test points that may be different from the current at specific points within the segment. Metallic *IR* drop should be measured when possible and recorded at the end of each survey run. Metal *IR* drop can be measured by direct metal-to-metal potential measurement or by taking the difference between the near-ground (NG) and far-ground (FG) potentials. Metallic *IR* drop should be measured and recorded when applicable for both the on and off cycles. Electrical connections should be made at every available contact point in order to minimize voltage drops in the metallic path. Metallic *IR* drop for depolarized or native-state surveys should be measured to determine whether all influencing CP has been deactivated.

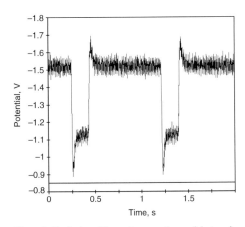

Fig. 4 Typical oscilloscopic wave form of fast-cycle interruption

Near-ground potentials should be measured and recorded at the start of every survey run, and NG and FG potentials should be measured and recorded at every contact point during the survey run and at the end of the survey run. The amount of current can be calculated when the resistance of the pipeline section is known. If the *IR* drop correction is effective, then theoretically there are no metallic *IR* drops in the off cycle (in the absence of long-line currents and of stray dynamic and static currents).

Lateral potentials or side-drain potentials should be measured and recorded at the start of each survey run. Lateral potentials are "on" and instant-off (when applicable) pipe-to-electrolyte potentials taken to either side of the pipeline. Side-drain potentials are the potential difference between two reference electrodes in the "on" and "off" cycles (when applicable): one located over the pipeline and the other at a distance to each side of the pipeline. Lateral potentials or side-drain potentials also may be measured and recorded at areas indicating possible problems. If the lateral or side-drain potentials indicate significant current to the pipe in the "off" cycle, an attempt should be made, when practical, to locate, determine the source of, and interrupt the influencing current. If significant errors are observed, the survey may be discontinued until the source of the error can be determined, and previously collected data should be evaluated for acceptable *IR* drop error. Errors that cannot be corrected should be noted in the CIS data. Lateral potentials or side-drain potentials for depolarized or native-state surveys should be measured to determine whether all influencing CP has been deactivated.

Dynamic Stray Current

Stray current is current through paths other than the intended circuit. Dynamic stray current refers to any stray current that is changing over time. Dynamic stray currents can come from many sources, including electric transit systems and telluric currents.

Recording the structure-to-electrolyte potentials over a time period, typically 24 h, can identify dynamic stray current. If deviations in the structure-to-electrolyte potentials are significant, stray-current correction of the survey results is warranted. Long-term data recordings of the structure-to-electrolyte potential at numerous locations are required to ascertain the influence of telluric currents on structure-to-electrolyte potential measurements. Telluric current effects on the structure-to-electrolyte potential are most significant at changes in direction of the pipeline or at electrical discontinuities, such as dielectric isolation devices.

One method of dynamic stray-current compensation is to correct the CIS potentials with the variation caused by dynamic stray current as recorded by stationary data logger(s). For the compensation to be effective, the structure-to-electrolyte potentials recorded in the CIS must be precisely synchronized with the stationary chart recorders, such as by use of the same time standard (such as universal coordinated time as provided by GPS).

The number and location of the static recorders required to effectively compensate for stray-current effects on the section of pipeline to be surveyed will vary. In areas of telluric current activity, stationary data loggers are typically connected to the pipeline at intervals not exceeding 5 km (3 mi). In areas of dynamic stray currents from dc traction systems, data loggers are typically connected to the pipeline at intervals not exceeding 2 km (1.25 mi).

Offshore Procedures

Close-interval survey can be performed on submerged pipelines, in marshy areas, and offshore using special equipment. Typically, the pipe location techniques and half-cell positioning are not as accurate as those for buried pipelines without considerable expense. Using visual or dead reckoning and dragging a reference electrode is the most inexpensive method of pipe location and electrode placement. Other methods include the use of divers, remotely operated vehicles (ROV), magnetometers, or electronic positioning that tracks the as-built coordinates of the pipelines. In marsh areas, other vehicles such as air boats and swamp buggies can be used.

Data Validation

The validation of CIS data requires the review and analysis of data gathered before, during, and after the survey. Because the main purpose of performing a CIS is usually to establish the CP level of an underground metallic structure and compare it with criteria established by industry standards, the accuracy of the pipe-to-soil potentials is extremely important. Before the results can be properly interpreted, the data obtained during the survey must be validated. Many factors can cause invalid CIS data, including:

- Missing data
- Improper stationing or distance measurement
- Excessive scatter or high contact resistance
- Inaccurate reference electrodes or voltmeter
- Broken wires/high-resistance connections
- Improper *IR* drop/interruption
- High induced ac
- Improper line location

Missing data may result in a misinterpretation of survey results. Typical information required in survey records includes:

- Company and location name
- Line identification and size
- Starting milepost/station number
- Location and operating condition of dc sources influencing the survey area
- Technician identification
- Equipment identification, e.g., voltmeter, data logger, and reference electrode serial number and description (for traceability to calibration records)
- Type of connections
- ac pipe-to-soil potential
- Near-ground pipe-to-soil potential (on and instant-off, if an interrupted survey)
- Left and right lateral pipe-to-soil potentials (on-off)
- Type of reference electrode used
- Survey direction
- Survey increment
- Waveform (if a fast-cycle interrupted run)
- Date and time of survey
- Description of survey conditions
- Ending milepost or station number
- Reason for ending run
- Far-ground potential reading with current applied
- Far-ground potential reading with current interrupted (if an interrupted survey)
- Near-ground reading with current applied, if ending at a connection to the structure
- Near-ground reading with current interrupted (if an interrupted survey), if ending at a connection to the structure
- Calculated or measured metal *IR* with current applied, if ending at a connection to the structure
- Calculated or measured metal *IR* with current interrupted (if an interrupted survey), if ending at a connection to the structure
- On and instant-off (if an interrupted survey) near-ground casing-to-soil potentials at casing vents
- On and instant-off (if an interrupted survey) near-ground potentials at each metallic foreign line crossing
- Bond current and polarity at each bond location
- Potentials on each side of insulating flanges
- Structure or pipe depth of burial

- Existence of buried foreign metallic structures in the vicinity of the surveyed line

It is also important to include the aboveground features encountered along the pipeline right of way during the survey, such as pipeline appurtenances, line markers, and physical features such as hills, creeks, ditches, fences, and street and highway names; the stationing at starting and ending connection points and at key physical features should be compared to the engineering stations when provided. Key physical features should be entered as comments into the data stream and engineering station may be reset to match stationing.

Regardless of the type of CIS, when the integrity of a structure is surveyed, the length of time at which the CP system has been operating at the conditions in which the CIS was performed is very important and should be included on the pipe-to-soil potential profiles.

As can be seen from the above considerations, a detailed execution plan must be developed well before an operator starts collecting the pipe-to-soil potentials.

Data Interpretation

After performing a CIS, the results of the validation will indicate if the survey data were acquired properly, if additional considerations may be required, or if some sections may require reinspection. It is not uncommon to resurvey portions due to errors such as scatter, problems with *IR* drop correction (such as an uninterrupted dc source) or dynamic stray current. Close-interval validated data may be compared with industry standards to determine if adequate CP levels exist.

ACKNOWLEDGMENT

This article is based on a draft standard developed by NACE International task group TG 279.

SELECTED REFERENCES

- F.J. Ansuini and J.R. Dimond, Factors Affecting the Accuracy of Reference Electrodes, *Mater. Perform.*, Vol 33 (No. 11), 1994, p 14
- T.J. Barlo, "Field Testing the Criteria for Cathodic Protection of Buried Pipelines," PR-208-163, Pipeline Research Council International, 1994
- "Control of External Corrosion on Underground or Submerged Metallic Piping Systems," Standard RP0169, NACE International, 2002
- R.A. Gummow, "Cathodic Protection Considerations for Pipelines with AC Mitigation Facilities," PR-262-9809, Pipeline Research Council International, 1999
- D.H. Kroon, "Wave Form Analyzer/Pulse Generator Technology Improves Close Interval Potential Surveys," paper No. 404, CORROSION/90, NACE, 1990
- D.H. Kroon, M. Mayo, and W. Parker, "Modification of the WaveForm Analyzer/Pulse Generator System for Close Interval Potential Survey," GRI-92/0332, Gas Technology Institute, Aug 1992
- D.H. Kroon and K.W. Nicholas, "Computerized Potential Logging—Results on Transmission Pipelines," paper No. 40, CORROSION/82, National Association of Corrosion Engineers, 1982
- R.J. Lopez, E. Ondak, and S.J. Pawel, Chemical and Environmental Influences on Copper/Copper Sulfate Reference Electrode Half-Cell Potential, *Mater. Perform.*, Vol 37 (No. 5), 1998, p 24
- "Measurement Techniques Related to Criteria for Cathodic Protection on Underground or Submerged Metallic Piping Systems," Standard TM0497, NACE International, 2002
- J.P. Nicholson, "Stray and Telluric Current Correction of Pipeline Close Interval Potential Data," Proc. Eurocorr 2003, Sept 28–Oct 2, 2003, European Federation of Corrosion, London, 2003
- R.L. Pawson, Close Interval Potential Surveys—Planning, Execution, Results, *Mater. Perform.*, Vol 37 (No. 2), 1998, p 16–21
- R.L. Pawson and R.E. McWilliams, "Bare Pipelines, the 100 mV Criterion and C.I.S. A Field Solution to Practical Problems," paper No. 587, CORROSION/2001, NACE International, 2001
- N.G. Thompson and K.M. Lawson, "Improved Pipe-to-Soil Potential Survey Methods," PR-186-807, Pipeline Research Council International, 1991
- N.G. Thompson and K.M. Lawson, "Causes and Effects of the Spiking Phenomenon," PR-186-006, Pipeline Research Council International, 1992
- N.G. Thompson and K.M. Lawson, "Most Accurate Method for Measuring an Off-Potential," PR-186-9203, Pipeline Research Council International, 1994
- N.G. Thompson and K.M. Lawson, "External Corrosion Control Monitoring Practices," PR-286-9601, Pipeline Research Council International, 1997
- N.G. Thompson and K.M. Lawson, "Impact of Short-Term Depolarization of Pipelines," PR-186-9611, Pipeline Research Council International, 1999

Corrosion of Storage Tanks

Ernest W. Klechka, Jr., CC Technologies Inc.

STEEL STORAGE TANKS are the primary means for storing large volumes of liquids and gaseous products. The stored fluid could be water, but it could be a volatile, corrosive, and flammable fluid requiring special precautions for storage as well. It is extremely important to maintain the integrity of on-grade and buried carbon steel storage tanks for economic and environmental reasons (Ref 1–3). For water storage tanks, internal corrosion can result in changes in color (turbidity) and taste that would be of importance for potable water. For boiler feed water storage tanks, corrosion products in the water can result in damage to the boiler. There may be specific requirements for the stored products including temperature, pressure, and control of contamination. Metal loss from internal and external corrosion can reduce the service life of the tank. External corrosion can occur because of contact with the soil and moisture in the soil. The main causes of internal corrosion are contact with corrosive storage products and water collected on the bottom of the tank (introduced with the other product or condensed from the air). Interior surface of some storage tanks are coated or lined to prevent contamination of the stored product and to extend the useful life of the storage tank. Typically, potable water storage tanks are internally coated to prevent contamination of the water. Crude oil storage tanks and fuel storage tanks are typically coated over the entire floor and a meter up the wall. Some chemicals cannot be suitably stored in carbon steel tanks. See articles dealing with specific chemicals in this Volume for selection of materials of construction.

An important consideration is the impact of storage tanks on the environment. Regulations have been formulated to address the possibility of leaks and spills, emissions for the tanks, seepage from tanks into the ground, and safety. These regulations define stringent standards that manufacturers and users must follow.

Another consideration is the fact that byproducts of internal corrosion can lead to reduced quality for the product stored in the tank. For example, the presence of iron oxides and water can be damaging to fuels, causing plugged filters and freezing in cold environments.

Storage tanks can be broadly divided on the basis of their installation into aboveground storage tanks (ASTs) and underground storage tanks (USTs). Aboveground tanks are common means for storing liquid hydrocarbon products such as crude oil, aviation fuels, diesel fuel, gasoline, and other refined products. Underground tanks are often used for dispensing and storage of home heating oil, gasoline, and diesel fuel.

Soil-side external corrosion of both ASTs and USTs can be mitigated by the use of cathodic protection (CP) with or without the use of protective coatings. Proper design, installation, and maintenance of CP systems maintain the integrity and increase the useful life of ASTs and USTs. High-quality dielectric protective coatings compatible with CP can be applied to properly prepared surfaces on the exterior of USTs and to the exterior tank bottom of ASTs (Ref 4). Aboveground steel tanks are typically designed for 20 to 30 years of useful life. Without CP or coatings, tank bottoms may have to be replaced after a few years (Ref 5).

Soil Corrosivity

Corrosion is generally worst where the tank is in contact with the soil. Resistivity of the backfill or sand under the AST or bedding and padding surrounding the UST is an important factor in determining how aggressive that environment is. Table 1 shows the relative degree of corrosion that can be expected based on soil resistivity.

Soil Characteristics. The pH and the presence of chloride and sulfate ions in the soil can have a significant effect on the corrosion rate of metals in soils. Chloride and sulfate contamination can be naturally occurring or the result of site contamination. Sulfate ions can be a source of food for microbes and can result in microbiologically influenced corrosion (MIC). Above 10,000 ppm, sulfate ions in the soil can have a severe effect on the degree of corrosion.

Chlorine ions are depassivating agents and cause pitting corrosion. ASTM D 512 (Ref 7) is a method to measure chloride ion concentration. As shown in Table 2, concentrations less than 500 ppm of chloride ions usually do not contribute significantly to corrosion; those above 5000 ppm can contribute to severe corrosion.

Sulfide ion in the soil can indicate the presence of anaerobic bacteria which can greatly accelerate the rate of corrosion. Corrosion rates in excess of 4 mm/yr (0.16 in./yr) have been measured in the presence of anaerobic bacteria. The test procedure shall satisfy the requirements of American Public Health Standard Method 4500, which is equivalent to U.S. EPA 376.2.

Acidity. As the soil pH decreases below 5.5 (acidic), the corrosion rate of steel increases very rapidly and can easily exceed 2 mm/yr (0.08 in./yr). When the soil pH is greater than 7.5 (alkaline), corrosion of carbon steel will be mitigated to very low levels. Cathodic protection will normally increase the pH near the cathode and conversely decrease the pH near the anode. Table 2 gives the general degree of corrosivity for chlorides, sulfates, and pH conditions.

Table 1 Approximate indication of soil resistivity versus degree of corrosivity

Soil resistivity, $\Omega \cdot cm$	Corrosivity
0–500	Very corrosive
500–1000	Corrosive
1000–2000	Moderately corrosive
2000–10,000	Mildly corrosive
>10,000	Negligible (very low levels of corrosion)

Source: Ref 6

Table 2 Effects of chlorides, sulfates, and pH on corrosion of buried steel

Concentration, ppm	Degree of corrosivity
Chloride ions	
>5000	Severe
1500–5000	Very corrosive
500–1500	Moderate
<500	Threshold
Sulfate ions	
>10,000	Severe
1500–10,000	Very corrosive
150–1500	Moderate
150	Negligible
pH	
<5.5 (acidic)	Severe
5.5–6.5	Moderate
6.5–7.5	Neutral
>7.5 (alkaline)	None

Source: Ref 6

Cathodic Protection

Cathodic protection is a proven method of controlling corrosion of buried or submerged metallic structures. Design, installation, and maintenance of CP for the exterior bottoms of carbon steel ASTs and the external surfaces of USTs can mitigate corrosion.

Cathodic protection can be applied to new or existing tanks, but cannot protect carbon steel surfaces that are not in contact with an electrolyte. During testing, the tank should be partially filled in order to ensure contact with the soil.

If coatings are used in conjunction with CP the coatings must be compatible with the CP, resist cathodic disbondment, resist abrasion, and be flexible enough to withstand filling and emptying of the storage tank. External tank bottom coatings for ASTs can be damaged during welding and weld repairs. Aboveground storage tanks with external tank bottom coatings should be repaired using nonweld repair techniques.

Galvanic (sacrificial) and impressed current methods, like a corrosion cell itself, requires four components: anode, cathode, an electric path, and an electrolyte. Galvanic CP uses an active metal (anode), such as zinc or magnesium, in electrical contact with a more noble metal (cathode), such as a carbon steel structure, in an electrolyte such as soil. The active metal corrodes, generating an electric current that protects the more noble metal. Impressed current cathodic protection (ICCP) uses an external power source to provide the direct current (dc) that flows from the anode and the cathode through the electrolyte (soil) and returns through an external circuit. System characteristics are compared in Table 3.

Cathodic protection systems should be operated continuously to maintain polarization. Access for testing and monitoring of the CP system must be considered during design, and these activities should be part of the operating procedures for the AST or UST.

The size, type, and location of anodes and reference electrodes are determined during design. Potential interference with external liners (for product release control) and buried piping should be considered. Electrical interference with other CP systems should be resolved. Grounding system for storage tanks can result in mixed potentials because of the copper used for grounding and can affect the initial potential of the carbon steel AST or UST. As a result of mixed potentials, the amount of current needed for CP may be increased to protect the grounding system (Ref 8).

Cathodic protection criteria are based on consensus industry standards. NACE International RP0169 (Ref 9) addresses underground or submerged metallic piping systems, RP0193 (Ref 2) addresses external corrosion of ASTs, while RP0285 (Ref 3) considers USTs. Corrosion control can be achieved at various levels of cathodic polarization depending on the environmental conditions. Based on RP0169, RP0285, and RP0193—all of which have the same criteria for CP—piping, AST tank bottoms, and USTs meet the criteria when the structure-to-soil potential meets one of these three criteria:

- A negative (cathodic) potential of at least 850 mV with the CP current applied. This potential shall be measured with respect to a saturated copper/copper sulfate reference electrode (CSE) contacting the electrolyte. Consideration must be given to voltage drops other than those across the structure-to-electrolyte boundary for valid interpretation of this voltage measurement. Consideration is understood to mean the application of sound engineering practice in determining the significance of voltage drops by methods such as measuring or calculating the voltage drop, reviewing the historical performance of the CP system, evaluating the physical and electrical characteristics of the tank bottom and its environment, and determining whether or not there is physical evidence of corrosion.
- A negative (cathodic) *IR*-free polarized potential of at least 850 mV relative to a CSE.
- A minimum of 100 mV of cathodic polarization between the carbon steel surface of the tank bottom and a stable reference electrode contacting the electrolyte. The formation or decay of polarization may be measured to satisfy this criterion. Although the 100 mV criterion can be a valuable approach to confirming CP, it is not popular in the CP industry. This is possibly because the test for the 100 mV polarization shift is more costly, requiring extra time for the development or decay of cathodic polarization than the tests for other criteria, or possibly because applications of the 100 mV polarization shift are not widely understood. See Ref 10 to 12 for more on the 100 mV criteria. See the article "Cathodic Protection" in Volume 13A for CP criteria in general.

The 100 mV polarization shift criterion may not be valid, especially where dissimilar metal couples such as copper ground grids and steel tanks are involved. The 100 mV polarization shift criterion also may be inappropriate for sulfate-reducing bacteria (SRB) containing soils, interference current, or telluric currents. More CP current may be needed to overcome the reduced pH caused by bacterial activity. Interference and telluric currents make interpretation of structure-to-soil potential difficult.

Some ASTs are built with a double steel bottom and the space between the two bottoms is filled with high-resistivity dry sand. In this environment, achieving a −850 mV CSE polarized potential is usually not practical. For ASTs with double bottoms, the 100 mV criteria is recommended.

For effective CP, the tank bottom of an AST must be in contact with the electrolyte (soil). During testing to the CP criteria, the tank should be partly filled (approximately 1/3 full) to ensure adequate contact between the tank bottom and the soil. Some additional time may be needed for testing to allow for polarization of the tank bottom. Underground storage tanks are usually in good electrical contact with the soil, provided the soil has been properly compacted.

Alternative Reference Electrodes. Occasionally, reference electrodes other than CSE are used to measure the structure-to-soil potential of a tank. These electrodes include standard calomel electrodes (SCEs), silver/silver chloride electrodes (Ag/AgCl), and zinc electrodes. Copper/copper sulfate reference electrodes are used for measuring potentials in soils. Standard calomel electrodes contain mercury and are often used in the laboratory, but seldom in the field. Chloride contamination of a CSE will result in an error in the measurement. Therefore, silver/silver chloride electrodes (Ag/AgCl) are used in seawater and brackish water. Zinc electrodes are used for their long life and can be used inside tanks, under ASTs, and next to or under USTs. Inside-of-tanks zinc reference electrodes can be used as a stationary reference electrode to test the structure-to-electrolyte potentials. Zinc reference electrodes are often buried with stationary CSE or Ag/AgCl reference electrodes under tanks or next to USTs to verify the accuracy of the stationary reference electrode. Conversion values are given in Table 4.

Data Needed for Corrosion Protection Design

Prior to designing a CP system (alone or in conjunction with a protective coating system), information should be gathered and evaluated on:

- Construction data for the tank, piping, and grounding systems: site plans and layout, detailed construction drawings, date of

Table 3 Cathodic protection system characteristics

	Galvanic (sacrificial)	Impressed current
External power required	None	Required
Driving voltage	Fixed, limited	Adjustable
Available current	Limited, based on anode size	Adjustable
Satisfied current requirement	Small	High
Suitable environment	Lower resistivity environments	Higher resistivity environments
Stray-current consideration	Usually no interference	Must consider interference with other structures

Source: Ref 6

construction, material specifications and manufacturer, joint design and construction (e.g. welded, riveted), containment membranes (impervious linings), double-wall or secondary bottoms, coating specifications
- Other existing or proposed CP systems
- Availability of electrical power (for ICCP)
- Backfill information: soil resistivity and type of tank backfill material
- History of the tank foundation
- Unusual environmental conditions, including soil contamination and weather extremes, local atmospheric condition
- Operating and maintenance history of the tank including leak history (internal or external corrosion)
- Water table and site drainage
- Type and levels of liquid contained in the tank
- Nearby structures
- Operating temperature (See article "Corrosion under Insulation" in this Volume if relevant)
- Electrical grounding systems

Predesign Site Assessment. For existing tanks, determine the extent of existing corrosion. Corrosion data and history may indicate that a new tank is needed or that major repairs are required (Ref 13, 14). Corrosion products "plugging" leaks may loosen and leak when a new CP system is applied. Field procedures for determining the extent of existing corrosion may include:

- Visual inspection
- Measurement of tank plate thickness (ultrasonic testing, coupon testing, physical measurement)
- Estimated general corrosion rates through electrochemical procedures
- Magnitude and direction of galvanic or stray current transferred to the tank through piping or other interconnections
- Soil characteristics: resistivity, pH, chloride ions concentration, sulfide ion concentration, moisture content
- Degree of corrosion deterioration based on comparison with data from similar facilities subject to similar conditions
- Data pertaining to existing corrosion conditions should be obtained in sufficient quality to permit reasonable engineering judgments. Statistical procedures should be used in the analysis, if appropriate.

Electrical isolation of structures must be compatible with electrical grounding requirements of applicable codes and safety requirements. If the tank bottom is to be cathodically protected, the use of alternatives to copper for electrical grounding materials, such as galvanized steel and galvanic anodes, should be considered. Electrical isolation of the tank from piping and other interconnecting structures may be necessary for effective CP or safety considerations.

Tank Electrical Characteristics. When examining an existing tank, several measurements should be made to determine the electrical characteristics of the tank:

- Tank-to-earth resistance tests
- Tank-to-grounding system resistance and potential tests
- Tank-to-electrolyte potential tests
- Electrical continuity tests for mechanical joints in interconnecting piping systems
- Electrical leakage tests for isolating fittings installed in interconnecting piping and between the tanks and safety grounding conductors

Soil-Resistivity Measurement. Soil resistivity is important to the design of either CP system. Soil-resistivity values measured at multiple locations are needed to determine the type of CP (galvanic or impressed current) required and the configuration for the anode system. Resistivity can be determined using the four-pin method described in ASTM G 57 (Ref 15) with pin spacing (a) corresponding to the depth of interest for burying the anodes (Fig. 1). As a general guideline, resistivity data should be obtained at a minimum of two locations per tank. For sand used as bedding around USTs and under ASTs, resistivity can be measured using a soil box and the same meter as in the four-pin method.

Soil resistivity is calculated using the resistance measured on the four-pin resistance meter and:

$$\rho = 2\pi a R$$

where ρ is soil resistivity ($\Omega \cdot cm$), a is the distance (in cm) between probes (and depth of interest), and R is the soil resistance (Ω) measured by the instrument.

If deep anode groundbeds are considered (Fig. 2), soil resistivity should be analyzed using procedures described by Barnes to determine conditions on a layer-by-layer basis (Ref 6). For a Barnes layer analysis, several measurements are made at increasing pin spacing centered on a fixed point between the two center pins. The average resistivity and the resistivity for each layer can be calculated.

On-site resistivity data can be supplemented with geological information from other sources including water-well drillers, oil and gas production companies, the U.S. Geological Survey Office, and other regulatory agencies.

Testing for Current. Cathodic protection current requirements can be established using test anode arrays simulating the type of groundbed planned. Test currents are applied using suitable dc sources. Test groundbeds can include driven rods, anode systems for adjacent CP systems, or other temporary structures that are electrically separated from the tank being tested. Small-diameter anode test wells may be appropriate and should be considered if extensive use of deep anode groundbeds are being considered.

The applied current is measured. On the tank, the initial potential and the potential shift is measured. Based on these measurements,

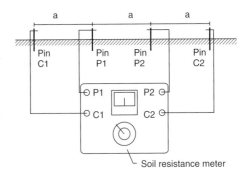

Fig. 1 Soil-resistivity testing by the four-electrode test method. Current is applied to the outside electrodes (pins C1 and C2), while potential is measured on the inside pins P1 and P2. The pins are placed in a line and equally spaced (a) to simplify resistivity calculation (in text). This resistivity is the average resistivity of a hemisphere of radius (a). Source: Ref 15

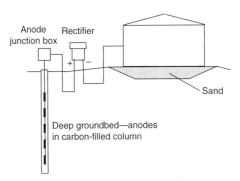

Fig. 2 Typical deep anode groundbed CP system. Groundbed can be 20 m (65 ft) to several hundred meters deep. Source: Ref 2

Table 4 Conversion of voltage measurements at 25 °C (77 °F)

Reference electrode used to measure potential	Electrode potential(a), V vs SHE	Measured potential, equivalent to −0.85 V vs CSE	Value added to measured potential to correct it to a potential vs CSE
Copper/copper sulfate (CSE)	+0.300	−0.850 V	0.000 V
Saturated calomel (SCE)	+0.241	−0.791 V	−0.059 V
Silver/silver chloride (Ag/AgCl) in seawater	+0.250	−0.800 V	−0.050 V
Zinc (Zn) in seawater	−0.800	+0.250 V	−1.100 V

(a) Standard hydrogen electrode (SHE), also called normal hydrogen electrode (NHE). See the article "Reference Electrodes" in Volume 13B for temperature coefficients.

the current needed to protect the tank can be calculated.

If test data are not available, the current required for an UST or AST can be calculated. Typically, for tanks with sand backfill the bare surface area of the tank can be protected using 10 to 20 mA/m^2 of CP current. See the article "Cathodic Protection" in Volume 13A for details on calculations.

Stray Currents. Underground structures are subject to dynamic and static stray currents. Dynamic stray currents vary in magnitude and often in direction and can be caused by welding shops, electrically powered rail transit, and improperly grounded or faulted electrical equipment. Dynamic stray currents can be manmade or natural in origin. Static stray currents can be caused by other CP systems. See the article "Stray Currents in Underground Corrosion" in this Volume. Stray currents caused by a pipeline CP system in close proximity to a tank farm are shown in Fig. 3. The location where the stray current leaves the tank and returns to the pipeline is where stray current corrosion will occur.

Stray currents can be reduced by bonding to the foreign CP systems, adding galvanic anode discharge points, adding or changing the location of current drains, moving anode groundbeds, or as a last resort by the addition of more CP. Bonding to the foreign structure allows the current to return to the foreign structure through the bond and not by discharging from the tank. Adding galvanic anodes at the discharge point gives the current a low resistance path for the stray current to discharge; the anode corrodes because of the discharge. Changing the current drain locations and changing groundbed locations can improve current distribution and reduce stray currents. Adding more CP can overcome the effects of stray currents, but may result in increased stray currents in other locations.

The presence of stray currents may result in CP current requirements that are greater than those required under natural conditions. Dynamic stray currents can be detected by monitoring fluctuations in the CP potential over a period of time, usually 24 h. Static stray currents caused by other CP systems can normally be detected by interrupting individual CP systems and monitoring the tank for CP potential fluctuations.

New CP designs should minimize electrical interference on structures not included in the protection system and any existing CP systems. Predesign test results can be analyzed to determine the possible need for stray current control provisions.

Intended Use of Storage Tank. Obviously, the materials stored in existing tanks and the intended use of new tanks will determine the design of internal corrosion-control measures. The coatings, lining, and inhibitors used to control corrosion of the internal surfaces of both ASTs and USTs must be compatible with the products to be stored. Frequently, CP is also applied inside surfaces of storage tanks.

The interior of water storage tanks, with their large surface areas, are frequently protected with compatible coatings and ICCP. Consumption of sacrificial anodes could result in contamination of the water. The use of inhibitors with water systems is discussed in "Corrosion Inhibitors in the Water Treatment Industry" in Volume 13A.

Crude oil storage tanks are frequently internally coated with two-part epoxy coatings over the entire floor and for the first meter (yard) up the wall to prevent corrosion because of water accumulation. Water accumulates as a result of water carried in the crude oil and because of condensation of the water vapor in the air pulled into the tank when the tank is emptied. Supplemental galvanic CP is frequently used in crude tanks.

The use of inhibitors in petroleum facilities is discussed in the articles "Corrosion Inhibitors for Oil and Gas Production" and "Corrosion Inhibitors for Crude Oil Refineries" in Volume 13A.

Soil-Side Corrosion Control

Generally a combination of CP and protective coatings is used. Sacrificial CP anodes are consumed as they generate electric current. The consumption is calculated as:

$$W = E \cdot CR \cdot I \cdot L$$

where W is the weight (mass) loss, E is efficiency, CR is the consumption rate (kg/A·yr), I is total current (A), and L is life or time (yr). Current efficiency for magnesium anodes is typically 50%, for aluminum anodes approximately 95%, and for zinc anodes between 90 and 95%. The CR for magnesium is 7.9 kg/A·yr, for aluminum 3.1 kg/A·yr, and for zinc 11.8 kg/A·yr.

Depending on the CP design, for crude tanks sacrificial anodes may need to be replaced as frequently as every 10 years, but they would be the choice where power is unavailable.

Impressed current cathodic protection is supplied with electric power, usually through a transformer and rectifier. The amount of current can be adjusted to account for changes in the coating condition and the surface area to be protected. As the external coatings deteriorate with time, the amount of current supplied can be increased.

All USTs are good candidates for CP systems. Many USTs are fabricated and coated with pre-designed sacrificial CP systems. For ASTs and USTs, if necessary, CP can be applied inside secondary containment.

Protective coatings are often the first line of defense against corrosion. For carbon steel USTs, a coal tar epoxy or other barrier-type coating material is often used to isolate the metal from the electrolyte.

The exterior surfaces of an AST tank bottom is normally not coated, leaving large surface areas to be protected by CP. Plates are normally arranged on the tank foundation and welded in place. As a result, relatively large CP current is needed and ICCP is typically used.

However, when the tank bottom exterior is coated it typically is coated with a coal tar epoxy or a two-component epoxy coating material. The external surface of AST bottoms can be coated by applying external coating to the bottom plate before welding, leaving the weld lanes bare. Up to 80% of the external tank bottom can be coated this way. Alternatively, the AST external bottom can be coated by lifting the tank and abrasive blasting and coating the bottom. Aboveground tanks with external bottom coatings should be repaired using nonwelding techniques such as glass reinforced overlays or adhesive bonding and coating of repair patches.

NACE Standard RP0169 (Ref 9) gives requirements and desired characteristics for coating in conjunction with CP:

- Effective electrical isolator. Because soil-side corrosion is an electrochemical process, reducing current flow by isolating the steel from the environment electrolyte reduces the uniform corrosion rate. The coating system should maintain constant electrical resistance over long periods of time to minimize changes in the CP current required.
- Effective moisture barrier. Water transfer through the coating can cause blistering and will contribute to coating failure and corrosion.
- Easily applied with a minimum of defects (holidays). Multiple coatings decrease number of through defects.
- Resists the development of holidays over time. After the coating is buried, minimal coating degradation and minimal damage from soil stresses and soil contamination occur.
- Good adhesion is needed to prevent water ingress and migration under the coating (undercutting).
- Toughness—the ability to withstand handling, storage, and installation with minimal damage—is required.
- The ability to withstand cathodic disbondment is required.
- The coating should be easy to repair.

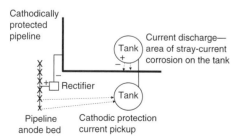

Fig. 3 Stray-current corrosion caused by electrically common tanks picking up current from pipeline protection anode bed. Corrosion is most severe where current density leaving tank is highest (arrows). Source: Ref 2

- The coating should be environmentally friendly, nontoxic, inert, and easily disposed of.

Coatings are important factors in CP engineering. If the structure is coated, it is necessary to protect only the exposed metal at holidays, greatly reducing the size and cost of the CP system.

Aboveground Storage Tanks

The effectiveness of the coatings and CP systems, internal and external, affect the life of the AST. When calculating the remaining life of an AST, the internal and external corrosion rates are considered (Ref 16).

External Aboveground Coatings for ASTs. External coatings must be able to withstand atmospheric corrosion, the temperature of the stored material, and flexing of the tank. External surfaces of ASTs are frequently primed with inorganic zinc, organic zinc, or a high-performance primer coating. Epoxy and polyurethane protective top coats are often used to provide atmospheric corrosion control for ASTs.

Frequently, aboveground storage tanks are painted white to reflect solar radiation. By minimizing solar heat gain, the amount of vapors released from a tank containing volatile hydrocarbon can be reduced.

Foundations. Aboveground storage tanks on ring-wall foundations with plastic liners to detect and contain leaks can be protected by anodes and reference electrodes installed in sand between the liner and the tank bottom (Ref 17). The following factors should be considered when evaluating the method of CP. Replacing galvanic anodes under the tank can be very expensive. The flexibility and adjustability of ICCP is desirable. Rectifiers for ICCP typically require bimonthly inspection.

For both systems, reference electrodes placed at the center and at several other locations beneath the tank bottom enable tank bottom potential monitoring. For small tanks, diameter less than 20 m (66 ft), tank-to-soil potentials can be measured at four locations around the tank periphery. For larger tanks, slotted casings can be installed underneath the tanks to measure the potential profile under the tank.

Foundation characteristics such as material of construction, thickness of ring walls, and water drainage are important in the assessment of the extent of existing corrosion.

For existing tanks, current requirement tests can be conducted to determine the amount of CP current needed to protect the tank. For new tank bottoms, an estimation of the amount of CP current needed to protect the tank can be calculated by using approximately 10 mA/m^2 for bare steel, or by measuring the current required on similar tanks in a similar environment. This initial value for polarizing the bare metal is considerably higher than the current density required to maintain protection.

Regulations. Containment of petrochemicals, petroleum, petroleum products, hazardous chemicals, hazardous waste, and similar regulated substances in ASTs is a concern for soil, air, surface water and groundwater contamination. Loss of integrity to the tank and releases of the regulated substances may be due to corrosion, structural defect, inadequate maintenance and repair, or improper installation or operation. In order to minimize impact to the environment, secondary containment is required. Secondary containment must be impermeable and can be concrete for smaller tanks or containment liners for larger tanks.

Industry standards for the design, operation, and maintenance of ASTs are given in Table 5. It is always good practice to use the current standard. Some regulations are part of the Federal Code, such as liquid pipeline breakout tanks, which are covered by 49 CFR Part 195.

Potable water tanks should comply with ANSI/NSF 61. American Water Works Association standards for storage tanks include: D100, Welded Steel Tanks for Water storage; D102, Coating Steel Water-Storage Tanks; D104, Automatically Controlled Impressed Current Cathodic Protection for the Interior of Steel Water Tanks.

Anode Systems. The three most common galvanic anodes used to protect tank bottoms are standard magnesium anodes, high potential magnesium anodes, and high-purity zinc anodes. The selection and use of these anodes should be based on the current requirements, soil conditions, temperature, and cost of materials.

High-purity zinc anodes should meet the requirement of ASTM B 418 type II anodes (Ref 24). The purity of the zinc greatly affects the performance of the galvanic anode. Zinc anodes should not be used if the soil temperature around the anode might exceed 49 °C (120 °F). Higher temperatures can cause passivation of the zinc anode. The presence of salts such as carbonates, bicarbonates, or nitrates in the electrolyte may also reduce the performance of zinc anode materials by causing passivation.

Galvanic anode performance may be enhanced in most soils by special backfill materials. Mixtures of gypsum, bentonite, and sodium sulfate are the most common packaged anode backfill materials.

Monitoring to verify CP must also verify that current is reaching the entire tank bottom. For tanks less than 20 m (66 ft) in diameter, measurements around the exterior of the tank may be sufficient to determine the level of the CP. For large-diameter tanks, permanent reference electrodes placed under the tank can help to determine the CP potential distribution on the exterior tank bottom. Slotted nonconductive tubes can also be placed under the tank to measure CP potential profiles under a tank.

Soil Contact. For the CP system to function properly, the tank bottom must be in contact with the soil. On large-diameter ASTs, the bottom acts as a diaphragm compressing the soil under the tank bottom. When the tank is emptied the bottom may bulge upward, losing contact with the soil. Once soil contact is lost, the CP system cannot protect the steel that is not in contact with the soil.

Vertically drilled anodes are distributed around the tank to give uniform current distribution (Fig. 4). The negative lead from the rectifier is connected to the tank. The positive lead from the rectifier is connected to the anodes through a junction box. The junction box allows for monitoring current measurement to the anodes. Reference electrodes should also be installed as part of the monitoring system. This system is effective for small-diameter tanks, less than 20 m (66 ft), as long as the anodes are installed above a secondary containment membrane.

Angle-drilled anodes are a variation of the vertically drilled anode system. In order to get more current under the tank, the anodes are placed under the tank in holes drilled at an angle

Table 5 Industry standards for storage tanks

Standard Number	Title	Ref
Aboveground storage tanks		
API standard 650	Welded Steel Tanks for Oil Storage	18
API RP651	Cathodic Protection of Aboveground Storage Tanks	19
API RP652	Lining of Aboveground Petroleum Storage Tank Bottoms	20
API RP653	Tank Inspection, Repair, Alteration, and Reconstruction	16
NACE RP0169	Control of External Corrosion on Underground or Submerged Metallic Piping Systems	9
NACE RP0163	External Cathodic Protection of On-grade Carbon Steel Metallic Storage Tanks	2
Underground storage tanks		
API Spec 12F	Specification for Shop Welded Tanks for Storage of Production Liquids	21
API RP1650	Set of Six Recommended Practices on Underground Petroleum Storage Tank Management Includes 1604, 1615, 1621, 1628, 1631, 1632	22
API RP1604	Recommended Practice for Abandonment or Removal of Used Underground Service Station Tanks	...
API RP1615	Installation of Underground Petroleum Storage Systems	23
API RP1621	Recommended Practice for Bulk Liquid Stock Control at Retail Outlets	...
API RP1628	Underground Spill Cleanup Manual	...
API RP1631	Interior Lining of Existing Steel Underground Storage Tanks	...
API RP1632	Cathodic Protection of Underground Petroleum Storage Tanks and Piping Systems	...
NACE RP0285	Corrosion Control of Underground Storage Tank Systems by Cathodic Protection	3

between 30° and 45° (Fig. 5). This system is effective on tanks typically less than 55 m (180 ft). Monitoring systems should also be installed when installing this type of system.

Deep-anode groundbeds are a common CP construction method used to distribute current over large areas uniformly. Deep-anode groundbeds can be drilled vertically from 20 to several 100 m deep. Typically graphite, cast iron, or mixed-metal-oxide anodes are used in deep-anode groundbeds. The negative return to the rectifier is connected to the tank (Fig. 2).

Deep-anode systems provide the greatest current distribution and are used on large storage tanks up to 100 m (330 ft) where current can access the tank bottom. These systems do not work well with secondary containment barriers because the current frequently must pass through the barrier, which is a high resistance path to get to the tank bottom.

Horizontally Installed Anode Groundbed. In order to distribute current under a tank, anodes can be horizontally installed under a tank. If the tank has a secondary containment lining, this method can be used. Anodes installed in a perforated nonconductive tube can be removed and replaced if necessary (Fig. 6).

These systems work well for all size tanks provided the anodes can be installed above any secondary containment barrier. If the system is installed outside of a secondary containment barrier, the current will have to pass through a high resistance path and will have a great deal of difficulty providing CP.

Double-Bottom Cathodic Protection Layout. Nonconductive containment liners also prevent the flow of CP current outside the barrier from reaching the external tank bottom.

Repairs to corroded tank bottoms can be accomplished by installing a double bottom. A nonconductive liner is placed over the existing tank bottom. An anode system is placed on top of the liner on the original tank bottom along with monitoring reference electrodes. High-resistivity sand is used to fill the space between the original floor and the new floor installed in the tank (Fig. 7).

In order to distribute the current evenly, the anodes are typically installed in a grid pattern as shown in Fig. 8. Either high-purity zinc or magnesium ribbons or rods are used as sacrificial anodes. For zinc anodes, the temperature of the space between the double tank bottoms should not exceed 49 °C (120 °F) to prevent passivation.

Impressed current anode design is more common for double-bottom applications. Typically metal-oxide-coated titanium ribbons or wires or other linear anodes are used for the anode material (Fig. 9). Leads from the rectifier are connected to each anode through a junction box where current to each anode can be measured.

Reference electrodes are installed in order to evaluate the distribution of current on the tank bottom (Fig. 9). They measure the polarized potential distribution under the tank.

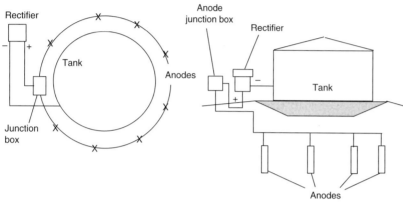

Fig. 4 Vertically drilled anode CP system, plan and elevation views. Source: Ref 2

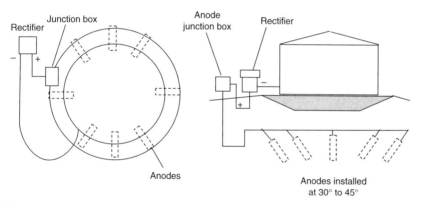

Fig. 5 Plan and elevation view of angle drilled anode CP system. Source: Ref 2

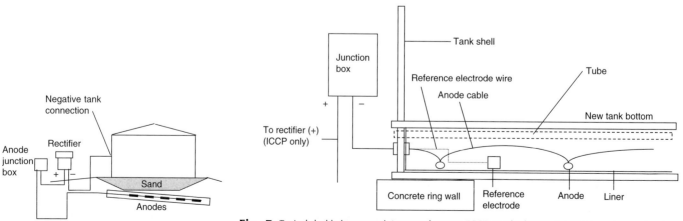

Fig. 6 Horizontally drilled anode system. Source: Ref 2

Fig. 7 Typical double bottom tank impressed current (ICCP) anodes layout. Sacrificial anodes can be placed in a similar fashion. Tube is a perforated nonmetallic tube for monitoring CP system.

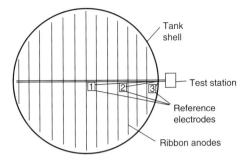

Fig. 8 Plan view of sacrificial anode grid pattern. Source: Ref 2

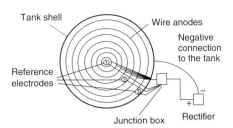

Fig. 9 Impressed current grid using wire anodes. Plan view. Source: Ref 2

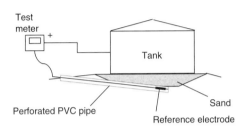

Fig. 10 Perforated nonmetallic pipe for monitoring the CP potentials under the tank bottom. Source: Ref 2

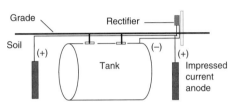

Fig. 11 Impressed current anode system for a UST

A perforated nonconductive tube can be directionally drilled and placed under the tank (Fig. 10). A permanent reference electrode can be installed in the tube, or a portable reference electrode can be used to measure the potential profile under the tank.

If necessary, water can be introduced into the perforated tube with the reference electrode in order to maintain good soil contact during testing.

Underground Storage Tanks

An underground storage tank is defined by the U.S. Environmental Protection Agency (EPA) as a tank and any underground piping connected to the tank that has at least 10% of its combined volume underground. Under the EPA UST program, a tank owner must notify a designated state or local agency of any tank storing petroleum or hazardous substances. Most regulated USTs store fuel for vehicles and are located at gas stations (Ref 23, 25).

Underground storage tanks have posed an environmental risk to groundwater because of the potential for leaks, overfill, or faulty equipment. In 1984, the EPA responded to the increasing risk by enacting a comprehensive regulatory program. These regulations, part of the Resource Conservation and Recovery Act, required owners or operators of underground storage tanks to be responsible for preventing, detecting, and cleaning up releases. The EPA has allowed states to implement and enforce UST regulations that either meet or exceed the EPA regulations. The industry standards listed in Table 5 for underground storage have served in part as the basis for the standards enacted by federal and state regulators.

Organic coatings may be applied to both the interior and exterior of underground steel tanks. Interior coatings prevent internal corrosion and extend the life of the tank. In the case of shop-assembled tanks, coating and linings are generally applied at the factory.

A high-quality dielectric coating should be applied to a properly prepared surface of the exterior areas of the UST including anode connections, attachments, and lifting lugs. Crevice or corner areas that restrict coating coverage should be seal welded prior to coating. Any type of coating used on a steel tank must have high dielectric properties. The dielectric coating isolates the tank electrically from the environment while reducing demands on the CP system. Other properties necessary in a dielectric coating are resistance to environmental fluids and the product being stored, impact/abrasion resistance, adhesion, and resistance to cathodic disbondment.

Cathodic protection systems for new UST systems may be: factory-fabricated galvanic, field-installed galvanic, or field-installed ICCP. The recommended practices with respect to field-installed systems are similar to those for existing UST systems; vertical sacrificial or impressed anodes can be installed. Consideration must be given to voltage drops other than those across the structure-to-electrolyte boundary, the presence of dissimilar metals, and the influence of other structures that may interfere with valid interpretation of structure-to-soil voltage measurements.

Typically, ICCP anodes are place around the UST as shown in Fig. 11. Typically, factory-installed sacrificial anodes are located at each end of the tank. Additional field sacrificial anodes can be installed as shown in Fig. 11 to provide better current distribution. Galvanic anode can be installed in a similar manner.

Monitoring ASTs and USTs

Once an ICCP system is operational, the rectifier should be monitored bimonthly, measuring the rectifier output voltage and current and a structure-to-soil potential at a representative test point. The rectifier voltage and current are useful for tracking increases in circuit resistance and forecasting the anode service life. The structure-to-soil potential helps to identify changes in the level of CP caused by changes in soil resistivity, degraded coating, passivated anodes, or depletion of anodes. Unexplained changes in the structure-to-soil potential can be used as a trigger for additional CP surveys to identify the cause of the change in structure-to-soil potential and to develop remedial measures (Ref 4, 25).

Potential measurements should be made at least annually. These measurements should be made around the perimeter of the tank and at all permanent reference electrodes. If slotted casings are installed under the tank bottom of an AST, a potential profile should also be measured annually. These measurements should be made around the perimeter of the UST and all permanent reference electrodes.

Potential measurements should not be taken through concrete or asphalt. Typically, soil contact may be established through at-grade openings, by drilling a small hole in the concrete or asphalt, or by contacting a seam of soil between concrete and asphalt.

REFERENCES

1. P.N. Cheremisinoff, Ed., *Storage Tanks*, Gulf Publishing, 1996, p 9–30
2. "External Cathodic Protection of On-grade Carbon Steel Metallic Storage Tanks," RP0193-2001, NACE International, 2001
3. "Corrosion Control of Underground Storage Tank Systems by Cathodic Protection," RP0285-2002, NACE International, 2002
4. R.A. Castillo, Cathodic Protection in Refineries, Chemical Plants, and Similar Complex Facilities, *Mater. Perform.*, May 2002, p 20–23
5. C.G. Munger, *Corrosion Prevention by Protective Coatings*, National Association of Corrosion Engineers, 1984, p 183–186
6. R.L. Bianchetti, Ed., *Peabody's Control of Pipeline Corrosion,* 2nd ed., NACE International, 2001
7. "Test Methods for Chloride Ion in Water," D 512, *Annual Book of ASTM Standards*, Vol 11.01, ASTM International
8. E.L. Kirkpatrick, Conflict Between Copper Grounding and CP in Oil & Gas Production Facilities, *Mater. Perform.*, Aug 2002, p 22–25
9. "Control of External Corrosion on Underground or Submerged Metallic Piping Systems," RP0169-2002, NACE International, 2002
10. W.B. Holtsbaum, Application and Misapplication of the 100-mV Criterion for

Cathodic Protection, *Mater. Perform.,* Jan 2003, p 30–32
11. M.A. Al-Arfaj, The 100-mV Depolarization Criterion for Zinc Ribbon Anodes on Externally Coated Tank Bottoms, *Mater. Perform.,* Jan 2002, p 22–26
12. L. Koszewski, "Application Of The 100 mV Polarization Criteria for Aboveground Storage Tank Bottoms," paper No. 591, CORROSION/2001, NACE International, 2001
13. L. Koszewski, Improved Cathodic Protection Testing Techniques for Aboveground Storage Tank Bottoms, *Mater. Perform.,* Jan 2003, p 24–26
14. L. Koszewski, Retrofitting Asphalt Storage Tanks with an Improved Cathodic Protection System, *Mater. Perform.,* July 1999, p 20–24
15. "Test Method for Field Measurement of Soil Resistivity Using the Wenner Four-Electrode Method," G 57, *Annual Book of ASTM Standards,* Vol 3.02, ASTM International
16. "Tank Inspection, Repair, Alteration, and Reconstruction," RP653, 3rd ed., American Petroleum Institute, Dec 2001
17. W.W.R. Nixon, Corrosion Control of Tank Bottoms within Spill Containment Systems, *Mater. Perform.,* March 2004, p 22–25
18. "Welded Steel Tanks for Oil Storage," Standard 650, 10th ed., American Petroleum Institute, Nov 1998
19. "Cathodic Protection of Aboveground Storage Tanks," RP651, 2nd ed., American Petroleum Institute, Dec 1997
20. "Lining of Aboveground Petroleum Storage Tank Bottoms," RP652, 2nd ed., American Petroleum Institute, Dec 1997
21. "Specification for Shop Welded Tanks for Storage of Production Liquids," Spec 12F, American Petroleum Institute, May 2000
22. "Set of Six API Recommended Practices on Underground Petroleum Storage Tank Management," RP1650, American Petroleum Institute, 1989
23. "Installation of Underground Petroleum Storage Systems," RP1615, 5th ed., March 1996/Reaffirmed, American Petroleum Institute, Nov 2001
24. "Specification for Cast and Wrought Galvanic Zinc Anodes," B 418, *Annual Book of ASTM Standards,* Vol 2.04, ASTM International
25. "Improved Inspections and Enforcement Would Better Ensure the Safety of Underground Storage Tanks," U.S. GAO Environment Protection, May 2001

Well Casing External Corrosion and Cathodic Protection

W. Brian Holtsbaum, CC Technologies Canada Ltd.

THE PORTION OF THE WELL OF CONCERN is that portion of the casing in contact with the formation either directly or through a cement barrier. It must be noted that where multiple casing strings are used, only that portion of each casing string in contact with the formation applies to this discussion.

Well Casing Corrosion

The corrosion mechanism will vary depending on the depth and the conditions at various parts of the casing. Gordon et al. (Ref 1) reported corrosion on well casings above a depth of 60 m (200 ft) that was due to oxygen enhanced by chlorides and sulfates in the soil while below that depth corrosion was caused by carbon-dioxide-rich formation water. These conclusions were based on scale analyses, sidewall core analyses, and soil analyses. In addition to these mechanisms, galvanic corrosion (especially if the casing is connected to surface facilities), anaerobic bacteria supported by drilling mud, and stray-current electrolysis are other possible causes of corrosion (Ref 2). Cementing the casing in place helps reduce the corrosion rate but does not eliminate it (Ref 3).

The procedure for predicting the probability and/or rate of corrosion is given in NACE RP0186 (Ref 4) and can be summarized:

1. Study the corrosion history of the well or other wells in the area (Ref 5).
2. Study the downhole environment, including the resistivity logs, different strata, drilling mud, and cement zones.
3. Inspect any casing that has been pulled (Ref 1).
4. Review the results of pressure tests.
5. Review the results of downhole wall thickness tests (Ref 1).
6. Review the results of casing potential profiles (CPP).
7. Review the oil/gas/water well maintenance records.

In a given area, after the first leak has occurred, the subsequent accumulated number of casing leaks often follows a straight-line relationship with time when presented on a semilog plot, that is, the log of the leaks versus time (Ref 5–7). This in effect means that the leak rate is increasing tenfold over equal periods of time. Repairs to the casing will alter this relationship as a repair often replaces several potential leaks; however, the leak rate will not be reduced to a tolerable level until cathodic protection is applied.

As part of many drilling programs it is common practice to pump cement into the annular space between the well borehole and the casing, usually to a point above the producing formation (sometimes from surface to producing formation depth and other times only portions of the casing strings are cemented) to achieve a seal. The cement in newer wells is often brought to the surface. However, in older wells, the cement was only sufficient to achieve a seal from the oil- and/or gas-bearing formation and therefore was brought from the bottom to a specific point along the casing. It should be noted that sections of casing pressed into the formation before cement injection will not necessarily have a cover of cement, or at the most, a very thin layer that is inadequate for corrosion control. Furthermore sections of casing not in the cement will continue to be exposed to the remains of the drilling mud.

The formation of corrosion cells can be:

- Local or pitting
- Between the cement and noncement sections of casing
- Between differential temperature zones
- Between brine formations and relatively inert rock
- Between the well casing and the surface facilities if there is a metallic connection

In addition, corrosive gases from a formation, such as carbon dioxide (CO_2) and hydrogen sulfide (H_2S) in an aqueous environment, can cause more aggressive attack.

Direct-current (dc) stray-current interference is another possible source of external corrosion. These may come from other cathodic protection systems, surface welding, or dc operated equipment. Alternating-current (ac) stray current in high current densities can also be a source of corrosion (Ref 8). Stray current accelerates corrosion on the casing if it discharges into the formation when returning to its source.

Detection of Corrosion

The two principal methods for detecting well casing corrosion include metal-loss (corrosion-monitoring) tools and casing current measurement. Both are described in this section.

Metal-Loss Tools

Casing monitoring tools for corrosion consist of three basic types: mechanical tools, electromagnetic tools, and ultrasonic tools (Ref 9).

The mechanical caliper tool is the oldest method where many "fingers" are spaced around a tool mandrel. When the tool is pulled past an anomaly, these fingers either extend into a defect or are pushed in by scale, a dent, or a buckle in the casing.

Electromagnetic tools consist of:

- High-resolution magnetic flux leakage and eddy-current devices
- An "electromagnetic thickness, caliper, and properties measurement" device

The source of magnetic flux comes from the electromagnet (or permanent magnet) in the tool. As the tool moves along the casing, the magnetic flux through the casing wall is constant until it is distorted by a change in the pipe wall thickness. The flux leakage induces current in sensing coils that is related to the penetration of the defect in the casing wall. A uniform thinning of the casing wall may be detected only as a defect at the beginning and end as there may be little change in flux leakage in between. Strictly, a magnetic flux tool cannot discriminate between a defect in the inside or the outside of the casing.

However, by adding a high-frequency eddy current that can be generated in the same tool, which induces a circulating current through the inner skin of the casing wall, discrimination between internal and external defects can be achieved. Sensing coils on the tool then detect the high-frequency field. A metal flaw or loss in the inside of the casing impedes the formation of circulating currents, and the change in this current is a measure of the surface quality and approximate vertical height of the defect. By comparing the defects from the electromagnetic to those obtained from the eddy-current signals in the tool, the external defects can be defined by a process of elimination.

The ultrasonic tool has transducers around the tool that act as both transmitters and receivers of an acoustic signal. The reflected signal is then analyzed for casing thickness, internal diameter, casing wall roughness, and defects. In addition, a cement evaluation can be included.

Tool Limitations. Since each tool has limitations, it may be necessary to run more than one tool depending on the type of flaw expected. In spite of the limitations, these tools can provide a reasonably accurate assessment of the casing metal loss; unfortunately, they can only detect corrosion damage after it has occurred.

Casing Current Measurement

According to Faraday's law (Eq 1), the metal loss due to corrosion is proportional to the dc current and the length of time that it leaves the metal and enters the electrolyte:

$$W = \frac{MtI}{nF} \quad \text{(Eq 1)}$$

where W is weight loss in grams (g); M is the atomic weight in grams (g); t is the time in seconds (s); I is the current in amperes (A); n is the number of electrons transferred per atom of metal consumed in the corrosion reaction; and F is Faraday's constant (96,500 coulombs per gram equivalent weight).

For steel, this equates to a metal loss of 9.1 kg/A-yr (20.1 lb/A-yr). If the current can be measured then the metal loss, as a measure of weight, for a given period of time can be calculated.

An axial current at any point in the casing can be calculated from Ohm's law (Eq 2) by measuring a voltage (microvolt, μV) drop across a known length of casing resistance:

$$I_p = \frac{V_2}{R_p} \quad \text{(Eq 2)}$$

where I_p is the axial current in casing (μA); V_2 is the axial voltage drop between two contact points along the casing pipe (μV); and R_p is the casing pipe wall resistance between the two contact points (Ω).

This voltage measurement is commonly called a casing potential profile (CPP), but the intent is to assess the axial and radial current profile in the casing. By determining an axial current value and direction between consecutive points in the casing, a radial current pickup or discharge can then be predicted in accordance with Kirchoff's current law, which states "the sum of the current at any junction must equal zero." Figure 1 illustrates three possible current measurement scenarios (A, B, and C); in all cases, the junction in Kirchoff's current law is at the center of each scenario.

The anodic or corroding sections of a casing are at the sections of current discharge, while the current pickup areas are cathodic and are not corroding. Scenario A of Fig. 1 shows the axial current increasing from 1.5 to 2.0 A; therefore, there must have been a 0.5 A pickup in between, indicating that this section is cathodic. The current of 2.0 A coming up the casing in scenario B is greater than the 1.5 A that continues up the casing; thus, 0.5 A must have discharged from the section in between the two points, causing this to be anodic or corroding. The current of 1.0 A that is coming downhole at the top of scenario C is in the reverse direction from the 1.5 A coming uphole; therefore, the current coming into the casing section from both ends must discharge from the pipe section somewhere in between. This section is therefore anodic and would be corroding.

By measuring the axial current at regular intervals along the casing, a complete current map along the casing can be obtained as shown in Fig. 2. Such a test is called a casing potential profile (CPP), and the plot in Fig. 2 is called an axial current profile. Both the amount and the direction of current have to be determined to predict a current pickup or discharge. An increasing slope coming uphole (equal to a negative change in depth per change in current going downhole) in Fig. 2 indicates a current pickup (cathodic section), while the reverse slope indicates a current discharge (anodic section). The amount of metal loss can be predicted for a given period of time on the assumption that the relative current will remain the same.

Determination of the amount of current pickup and discharge along the casing in Fig. 2 results in the radial current profile shown in Fig. 3. Referring to Fig. 2, the direction of net current flow at about "85% of depth" is in the downhole direction as it crosses the zero (0) current axis, while the current below that depth is coming uphole. This causes a current discharge centering

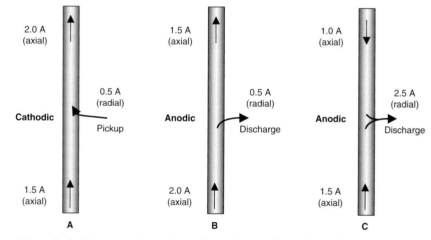

Fig. 1 Example of radial current pickup or discharge from axial current. Refer to the text for a discussion of scenarios A, B, and C.

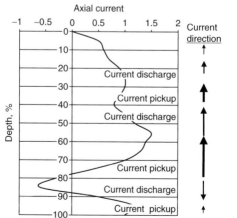

Fig. 2 Example of axial current profile in casing without cathodic protection

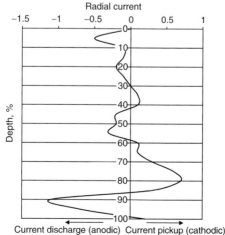

Fig. 3 Calculation of radial current profile from Fig. 2

at about 90% of depth as shown in Fig. 3. This is the same as scenario C in Fig. 1. In a similar fashion, the current at a depth between 0 and 25% and also 40 and 55% of depth is less than the current below, although in the same direction, which is the same as scenario B in Fig. 1. This also indicates a current discharge (anodic) area. The remainder of the casing in this example is picking up current and is cathodic, which fits the condition illustrated by scenario A in Fig. 1.

Limitations and Advantages of Casing Current Measurements. Casing current measurements, are only sensitive enough to measure long-line currents and do not detect local corrosion cells that exist between the spacing of the two contacts.

The advantage of this test is that macro-corrosion can be predicted before it occurs. However, the assumption that the current magnitude and location will stay the same can create a large error. The existence of local corrosion pits will be missed, and these can represent a large amount of the corrosion taking place (Ref 9, 10).

Cathodic Protection of Well Casings

At one time there was a concern that cathodic protection current applied at the surface would not reach the bottom of deeper well casings. Blount and Bolmer (Ref 11) conducted polarization tests with a reference electrode located at the top and the bottom of well casings and concluded that cathodic protection is feasible to a depth of at least 1000 m (3300 ft). Subsequent tests have shown that it is feasible to depths up to at least 3960 m (13,000 ft).

Two methods of determining the amount of cathodic protection current required are described in this section: a casing polarization (E log I) test and a CPP test. The first test attempts to predict when the casing becomes a polarized electrode, while the second test confirms if an adequate amount of cathodic protection current is being discharged from the anode bed(s) to ensure current is being picked up along the length of the casing being tested.

E log I Test (Tafel Potential)

The E log I test is a measurement of the polarized casing-to-soil (electrolyte) potential (E) compared to the logarithm of different increments of applied current (I). The casing-to-soil potential is measured with respect to a remote reference electrode, often a copper/copper-sulfate reference electrode (CSE). "Remote" in this case is a point where the electrical voltage gradient is zero. Polarization is considered to take place at the intersection of the two straight lines as shown at point "A" in Fig. 4. At the intersection of the upper straight line (point "B"), the curve becomes a hydrogen overvoltage curve and obeys the Tafel equation.

In the early years, the point where the two straight lines intersected (Fig. 4, point A) was taken as the current required for the protection of the well casing. This not only gave widely varying results depending on the relative slope of the two lines, but also provided current requirements that were found to be too low to protect the casings. The laboratory and field research of Blount and Bolmer (Ref 11) confirmed that the intersection of the upper portion of the Tafel slope with the curve was the point of corrosion control and proved to yield more consistent results (Fig. 4, point B). This point is normally used to establish a cathodic protection criterion for the casing.

A schematic of a typical E log I test is shown in Fig. 5. The test is conducted by impressing an increment of current for period of time and then measuring the "instant off" potential when the applied current is briefly interrupted. This process is repeated at increasing increments of current to a point beyond where the Tafel break in a plot between the instant off potential (E) and the logarithm of the current (log I) occurs (point B in Fig. 4). There has been extensive experimentation both in the laboratory and the field (Ref 11–13) comparing the current increments and the length of time at each increment to allow polarization to occur. The conclusion was that the best results occur when the increments of current and the time intervals between current increases are constant. A sufficient time interval must be established that ensures polarization will be complete before proceeding to the next current value. Although the current increment and time needs to be established for each E log I test, current increments of 0.5 A and time intervals of 10 min is often a practical combination. The time interval has been reduced to 5 min under certain circumstances where the well polarizes more quickly. It must be noted that too short of time intervals can yield an inaccurate higher current requirement as polarization may not be complete at given current values before the test current is increased incrementally.

Equipment (Fig. 6) can be set to automatically interrupt the current and record casing-to-electrolyte potentials continuously during the current interruption. In this way, the existence of a "spike" can be seen and the appropriate instant off casing-to-soil potential selected for each current interval. Furthermore, the current output

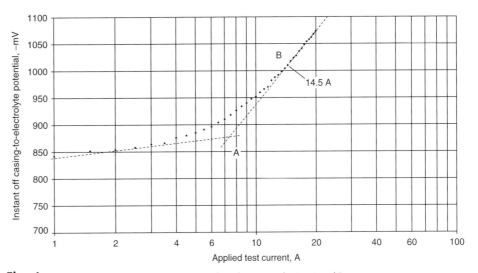

Fig. 4 An example of an E log I plot. Refer to text for a discussion of points A and B.

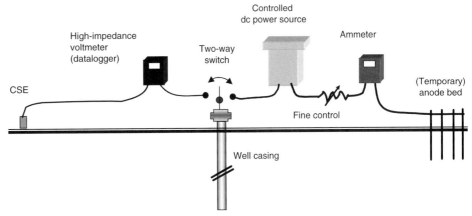

Fig. 5 Basic E log I test. CSE, copper/copper-sulfate reference electrode; dc, direct current.

can be controlled using silicon-controlled rectifiers (SCRs) to ensure that it remains constant during the test interval and that the desired fine incremental output control can be achieved.

Often a premature ending of the test occurs because the $E \log I$ profile was interpreted incorrectly as having straightened out. This variance is likely due to reactions that are taking place at different times or at different points as the test proceeds. To protect against stopping the test prematurely, a linear plot of E versus I should be made as the test proceeds to ensure that the test has left a straight-line relationship indicating that polarization is occurring. Often there is an early straight-line segment or the profile starts to leave the linear straight-line relationship only to return to the same slope. These early deviations are false indications as shown by the data from Fig. 4 plotted on a linear profile in Fig. 7. The Tafel break of interest in the $E \log I$ plot is beyond that determined by the linear plot (12.5 A) and becomes the criterion for protection for that well casing, as shown by point B in Fig. 4.

Subsequent $E \log I$ analysis by this method has compared favorably to the current requirement determined by CPP test results provided that the break (Fig. 4, point B) was selected after the straight-line relationship has ended on a linear plot. When this method is not used, an erroneous analysis of the $E \log I$ test can be expected (Ref 14).

Advantages and Limitations. An advantage of the $E \log I$ test is that it can be performed while the well is still in production. However, the casing still should be electrically isolated from all other structures for this test, or at least one must be able to measure the portion of the test current returning from the casing by perhaps using a clamp-on ammeter that can either fit around the wellhead or individually around all of the lines, instrument tubing, and conduit that connects to the well.

One disadvantage of the $E \log I$ test is the concern as to whether the test "sees" the lower part of the casing.

Casing Potential Profile

The CPP test for cathodic protection is similar to that described previously for predicting corrosion from casing current measurements except that now a current pickup is desired at all locations similar to that illustrated in Fig. 1 (scenario A). The casing has to be electrically isolated from all surface structures and the service rig during this test, otherwise the current returning at the wellhead must be measured.

The original CPP tool had two contacts that were 3 m (10 ft) to 7.6 m (25 ft) apart. The tool was stopped at regular intervals for microvolt (μV) measurements. Davies and Sasaki (Ref 13) describe a newer CPP tool (the CPET corrosion-protection evaluation tool) that has four rows of knife contacts that are 0.6 m (2 ft) apart between rows (Fig. 8). Measurements taken between the different rows of contacts include the pipe resistance, a voltage drop (μV_2) between the inner 0.6 m (2 ft) contacts, and another voltage drop (μV_6) across the outer contacts 1.8 m (6 ft) apart.

Pipe (Casing) Resistance Determination. Using a conventional four-pin resistance test (the same test is often used in conjunction with a resistivity measurement), the instrument

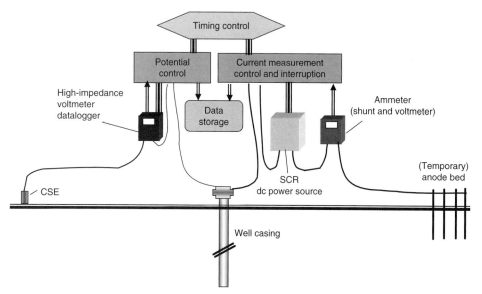

Fig. 6 Automatically controlled $E \log I$ test. CSE, copper/copper-sulfate reference electrode; dc, direct current; SCR, silicon-controlled rectifier

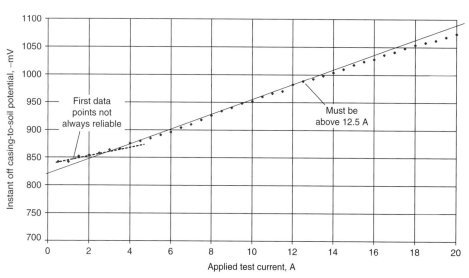

Fig. 7 Linear plot showing where the curve leaves a linear relationship. Tafel point on $E \log I$ must be at a higher current than the point that deviated from a linear straight line in this figure. Data generated from Fig. 4

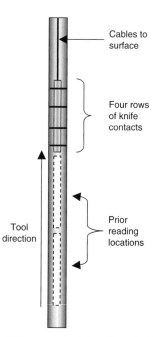

Fig. 8 CPET casing potential profile tool. CPET, corrosion protection evaluation tool

impresses a known current (I_{test}) between the outer contacts and measures the resulting voltage (V_2) across the inner 0.6 m (2 ft) contacts. This then allows the casing resistance between these 0.6 m (2 ft) contacts (R_2) to be calculated by using Ohm's law ($R_2 = V_2/I_{test}$).

Casing Axial Current Determination. Once the pipe resistance for the test point has been determined the axial current can then be calculated by $I_2 = V_2/R_2$. Identical measurements and calculations are made across all other sets of contacts and the results averaged. Normally, the results across the 0.6 m (2 ft) and 1.8 m (6 ft) rows are reported ($I_2 = V_2/R_2$ and $I_6 = V_6/R_6$). The radial current is then calculated between consecutive current measurements noting current direction.

It must be understood that the current in the casing when measured at any given point is the accumulation of all of the current pickup less any discharge on the casing below that point. Also the cathodic protection current direction has to be toward the top of the casing in order to return to the dc power source. Therefore, only when cathodic protection has been successfully applied does a plot of the axial casing current continually increase from the bottom to the top of the casing, thus indicating a continuous current pickup.

Figure 9 illustrates two cathodic protection trials with current applied. From the plots it can be seen that trial 1 did not eliminate all of the anodic areas. Thus, the applied current was increased until the anodic areas were eliminated as indicated by the axial current increasing continuously from the casing bottom to top, trial 2.

Trial 1 in Fig. 9 shows an axial current pickup at all but two sections. One current discharge is at approximately 55% of depth and the other at approximately 85% of depth; both of which are identified by "downward" slopes on the profile. The axial current at 85% of depth is in the downhole direction as it crosses the zero (0) current axis, while the current below is coming uphole. While the axial current at 55% of depth is in the same downward direction, the current above is less than that below, which also indicates a current discharge or an anodic section.

Since this was unsatisfactory, the current was increased for trial 2 (it must be noted that during an actual test, time must be given to ensure a steady state has been achieved after ampere adjustments before another log is run to obtain reliable results). Here, continuous axial current pickup occurred from bottom to top as shown by the positive slope in the accumulated current profile. The total current value established by this test now becomes the criterion for cathodic protection. It should be noted that errors can occur in this measurement due to poor contacts. However, this is the best technology available at the present time to determine the amount of cathodic protection current required to protect a well casing, or a portion of a well casing.

A partial CPET plot is shown in Fig. 10 that illustrates the axial current, radial current, and the casing thickness. The casing thickness is an estimate based on Faraday's law (Eq 1) and the assumption that the radial current discharge has remained the same over time. As a result, the casing thickness estimate may not be a true measure of the wall thickness remaining. Experience has shown that there is often quite a discrepancy between corrosion-prediction losses by this method when compared to actual metal-loss measurements.

Factors Influencing the use of CPP Tests. Even though CPP is probably the best means now available to establish the current required for a well casing, it is not often used. The main reasons are associated with the cost of running the tool, both direct and indirect costs. Some of the reasons include:

- In order to run the tool the well has to be taken out of service. This in itself limits the number of potential candidates unless there is a very urgent need to take a well out of service.
- Depending on the fluid in the well bores, many wells will have to be "killed" before the tool can be run.
- In order for the tool to make good contact with the casing, any scale or product buildup on the inside of the casing will have to be cleaned off before the tool is run.
- There are not many CPET tools available worldwide, and the older CPP tool is not available. Coordinating the work is therefore vital to ensure the well and the tool are available at the same time.
- A cathodic protection system: anodes, rectifier (or some other suitable dc power source), cabling, and so forth, must be constructed and operating in advance of the downhole log if the test is to verify a current requirement target.
- If the testing is to determine cathodic protection current requirements, then weeks or even months between runs might be necessary in order to allow a steady state to be achieved between output adjustments.
- Since completion practices for wells in the same producing area can vary significantly, multiple tests on multiple wells may be necessary to arrive at current return criteria that meet all of the well completion variations.

Mathematical Modeling of Total Current Requirement for Well Casing Cathodic Protection

Several mathematical models (Ref 15–19) have been developed to estimate the total current required to protect a well casing by cathodic protection that can be summarized:

- Current density
- An attenuation equation
- A modified attenuation equation
- A computerized equivalent circuit using formation resistivity, nonlinear polarization characteristics, and well casing information

The current density model applies an empirical current density to the surface area of other well casings of similar characteristics to the source of the empirical data to estimate the total current requirement of each well casing. The

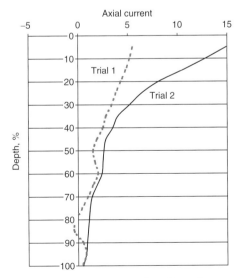

Fig. 9 Sample casing potential profile axial current profile

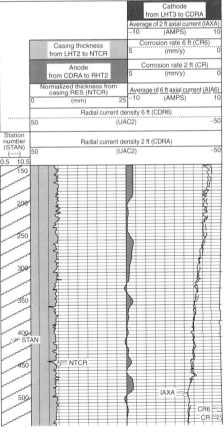

Fig. 10 CPET axial and radial current plot with a conventional rectifier and casing thickness. Total current is 15.3 A.

variations in well depth and completion such as the amount of the casing that was cemented between it and the formation and the quality of the cement can make this approach quite inaccurate. Verification by field tests on typical well casings in a given geographical area is advised.

Attenuation calculations modified from those used on pipelines were applied initially to casings to estimate a potential at a given depth based on the potential change at the surface. The relationship developed by Schremp and Newton (Ref 16) is given by Eq 3 to calculate the potential change at any given depth in the casing with the applied current source being interrupted:

$$e_x = e_o \exp\left[\frac{-1.648.7(r_1)(x_1)(I_1)\exp(-x_1/L_1)}{e_o}\right]$$

(Eq 3)

where e_o is the potential change at wellhead when applied current is momentarily interrupted (mV); e_x is the potential change at depth x_1 from the wellhead (mV); r_1 is the unit resistance of the innermost casing (Ω/m or Ω/ft); x_1 is the distance from wellhead (m or ft); I_1 is the current in the innermost casing (A); and L_1 is the length of innermost casing (m or ft).

A more sophisticated mathematical model was developed by Dabkowski (Ref 17), and a spreadsheet version was developed by Smith et al. (Ref 18).

Casing-to-Anode Separation

The spacing of the anode to the casing can also change the current required for a particular casing as illustrated by data from Blount and Bolmer (Ref 11) plotted in Fig. 11. The current requirement to protect the casing increases if the anode is brought too close to the casing. There is an optimum distance beyond which a further increase in distance is of no benefit. Hamberg et al. (Ref 7) also demonstrated a similar result in offshore well casings.

A comparison of the distribution of current in two similar casings that were 2600 m (8530 ft) (well casing "A") and 2475 m (8120 ft) (well casing "B") deep in the same area but with different casing-to-anode separations is shown in Fig. 12. The excess current being impressed onto the casing near the surface helps provide an understanding of this change in current requirement due to the casing-to-anode separation (Ref 14).

Blount and Bolmer (Ref 11) found that the anode bed should be at least 30 m (100 ft) from casings that were on the order of 1220 m (4000 ft) deep. This distance should be increased for deeper wells for optimum performance. The anode bed in either the E log I or the CPP test should therefore be located at a distance from the well casing similar to where the permanent anode bed will be installed.

Coated Casings

Coatings are available that are durable enough to withstand many of the rigors of a casing installation. Although significant coating damage is expected, Orton et al. (Ref 20) reported that the current requirement of a coated casing with bare couplings and no effort to repair coating damage can be reduced to less than 10% of that of a similar bare casing. A further benefit is that a reduction in the current requirement will also reduce the interference effects on nearby structures and casings as discussed below.

Cathodic Protection Systems

The cathodic protection system for a well casing requires the same consideration as that for a pipeline. There are two types of cathodic protection systems used for well casings and pipelines: sacrificial anode systems and impressed-current systems (see the article "cathodic protection" in Volume 13A for additional information).

Sacrificial Anode Systems. In the early years, a sacrificial anode system was often used for wells where a low current requirement was predicted. Sacrificial anode systems are still appropriate for more shallow wells with a low current requirement.

An impressed-current cathodic protection system is the most common type for well casings due to the amount of current typically required for protection. A separate installation (Fig. 13) is common at each well. If two or more wellheads are in close proximity, interference can result (Ref 21–23).

Power Sources. Where ac power is available, it is likely that a standard or pulse-type rectifier will be used as a dc power source. Otherwise, thermoelectric generators, solar, wind-powered generators, and engine-driven generators are all possible candidates for the dc power source.

Thermoelectric generators (TEG) have a limited power availability; therefore, the anode bed resistance should be kept low to obtain the required current. The available power from a TEG usually peaks at around 0.6 to 1.2 Ω and reduces as the circuit resistance increases. The manufacturer's technical information must be consulted. A clean regulated fuel source such as natural gas or propane is required.

Both solar- and wind-generated power need batteries as a backup power source to provide cathodic protection current when there is either no sun or wind, respectively. The use of solar is

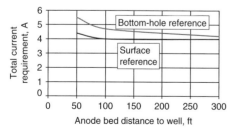

Fig. 11 An example of the change in current requirements with anode to well spacing. Source: Ref 11

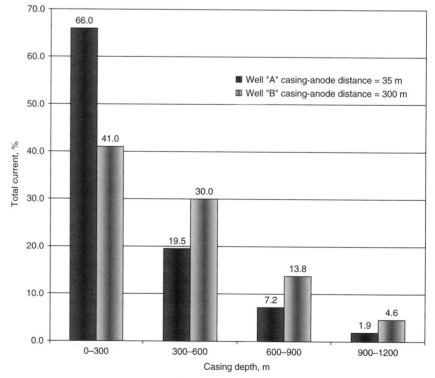

Fig. 12 An actual example of current distribution in similar casings but with different casing-to-anode distances. Source: Ref 18

less popular in the northern regions where there is a lack of sunlight in the winter, and wind power is not appropriate unless the area is historically windy.

Engine-generator systems are best used with an ac generator feeding a rectifier for dc output and control. Maintenance on dc generators has proved to be high in the past, resulting in many outages during the year. Although to a lesser degree than the dc generator, the ac generator also requires maintenance and regular inspections.

Pulse rectifiers provide a high-voltage dc pulse of short duration. The frequency of the pulse may be from 1000 to 5000 Hz, but the duty cycle is normally set in the range of 10 to 15%. Bich and Bauman (Ref 24) reported that total current requirements can be reduced to 50% or less using a pulse rectifier instead of a conventional rectifier, and more current will reach the lower portions of the casing. The improved performance is attributed to the waveform. However, Dabkowski (Ref 25) showed mathematically that the pulse from the rectifiers would attenuate to 0 at 500 to 1000 m (1640 to 3280 ft) from the casing top, suggesting that any improved performance is not due to the pulse. It has been the author's experience that cathodic protection with pulse rectifiers can be achieved down to 80% of the comparable current to a conventional rectifier, but not the significant reduction suggested by Bick and Bauman. Further work needs to be completed to validate any of these claims. The digital instrumentation measuring the pulse rectifier output is another factor in this comparison, as errors can be realized depending on the sampling rate. A major disadvantage of pulse rectifiers is noise interference, especially on communication equipment that may be servicing the well. This can be reduced by locating the pulse rectifier away from the electrical/communication building, not paralleling electrical cables, and using deep anodes.

Regardless of the power source, one negative cable must be connected to the well casing while a second negative cable is often run to the isolated surface facilities to assist in interference mitigation.

The anode bed design and location is largely dictated by the soil layer resistivity and the location of surface facilities and pipelines. If uniform low-resistivity soil conditions exist at a surface location that is sufficiently remote from the casing and other structures, a shallow anode type of anode bed can be used. Where high-resistivity conditions exist at the surface but more suitable strata exist underneath, a deep anode bed would be preferred. The latter anode bed will also tend to reduce interference with surface facilities, as the major portion of the anode gradient exists below pipeline and foundation depth. It must be noted that the same spacing between the casing and anode must be maintained whichever type of anode bed is used, as going deeper does not change the distance between the structures.

The anode bed should be located at an equal or greater distance than the temporary anode bed to the casing that was used during the current requirement test. However, a minimum spacing of 30 m (100 ft) from the well for shallow wells but preferably greater than 50 m (165 ft) should be maintained. The separation between anodes and structures not receiving current will vary depending on the voltage gradients in the soil but should be 100 m (300 ft) or more. Otherwise, provision for interference control discussed below must be considered.

Direct-Current Stray-Current Interference

Stray current can be defined as current in an unintended path. Many sources of current use the earth as part of their electrical circuit. Conductors in the earth such as well casings and pipelines provide opportune parallel paths for current intended for another purpose.

The area of stray-current pickup is similar to cathodic protection and not of concern. However, the manner by which that current returns to its original source is of concern. Should that current *leave* the casing to enter the soil, the casing in that location is anodic and accelerated corrosion occurs.

Stray-Current Pickup. A stray current may be picked up at the surface, in which case the current must discharge into the soil downhole to return to its source. Alternately, a current discharge near the surface to either facilities near the wellhead or to the surface casing may occur, in which case there will have to be a current pickup downhole. Both cases (Fig. 14) are a cause for concern as there is a current discharge occurring at some point along the casing.

Since an electronegative shift in casing-to-soil potentials occurs with the application of cathodic protection, a stray-current pickup at a lower depth with a discharge near the surface can be detected by an electropositive shift in casing-to-electrolyte potentials, with the reference electrode located near the wellhead, when the foreign dc power source(s) is energized. Conversely, a current pickup at the surface will be detected by an electronegative shift in potentials when the foreign current source comes on. The area of current discharge will then be at a point lower on the casing, and its location would have to be defined by a CPP log, or similar.

Stray-current pickup on pipeline systems away from the well casing can result in a stray-current discharge from the well casing if the two structures are continuous. In these cases, a current pickup is normally close to the anode bed while the discharge is near the wellhead. However, it is conceivable that the current pickup and discharge points can develop at other points, especially if varying coating qualities or vastly differing resistivities exist along the pipeline or casing.

Stray-Current Sources. The stray current may come from a relatively steady-state source such as another cathodic protection system (Ref 21–23) or a high-voltage dc power line ground, or it may come from a dynamic source such as a transit system, welding machines, dc mine equipment or, finally, from telluric current that is a natural source of stray current (Ref 26).

Interference Control. Interference can be controlled by:

- Providing a metallic return path for the stray current
- Moving the offending anode bed or ground
- Adjusting the current distribution in the foreign system
- Installing and/or adjusting a cathodic protection system on the well casing to counter the stray-current effects
- Using common cathodic protection systems (Ref 21)
- Balancing wellhead potentials

A well casing cathodic protection system can also cause interference on surface facilities or

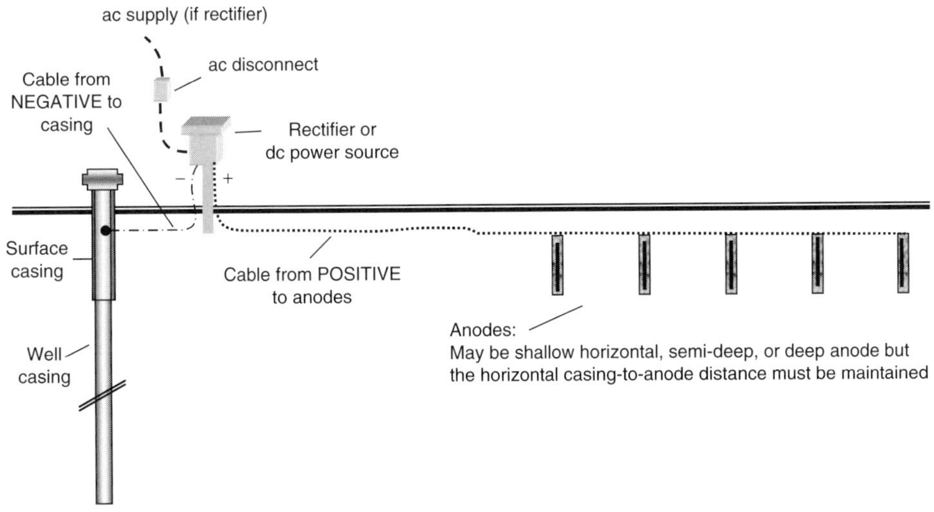

Fig. 13 Typical cathodic protection installation

pipelines. In this case, a second negative circuit is often provided in the rectifier to both control interference and assist with the protection of the surface facilities or pipelines.

Orton et al. (Ref 20) reported that a coated casing reduced the cathodic protection current requirement to 10% of a bare well casing. This in turn will reduce the tendency for mutual interference of nearby casings.

Isolation of Well Casings

The purpose of isolating a well casing from surface facilities is twofold: (a) it eliminates a macrocorrosion cell between the casing and the surface facilities, and (b) it allows the cathodic protection current distribution to be controlled between the well casing and the surface facilities.

In addition, an isolating feature allows the current impressed on the well casing to be directly measured in the connecting cable. If not isolated, a means of measuring the current return from the casing itself must be established, such as a clamp-on ammeter around the wellhead at the surface, to confirm that the "current" criterion is being met.

From a cathodic protection standpoint, the preferred location for this isolation is at the wellhead. However, some operators locate it a distance away in the event of a fire at the well so that the isolating material does not melt and complicate firefighting procedures. All tubing conduits and pipe supports must also be isolated if they are bypassing the isolating feature.

If the product from the well contains a large amount of brine, there is a risk of "internal" interference. This occurs where current picked up on the opposite side of the isolation uses the brine as a path around the isolation. In such a case, corrosion is seen only on one side of the isolating feature (Fig. 15A).

A "long-path" isolation, which consists of an isolating feature and an internally coated or lined section of pipe (Fig. 15B), can be used to reduce the internal interference. If this is not effective in controlling internal interference, the isolating feature should be omitted.

Commissioning and Monitoring

Inspection. A cathodic protection system must operate continuously to be effective. Regular inspection of the dc power supply to ensure that the required current is being provided in all circuits is necessary throughout the year. A more detailed inspection should be conducted annually. A description of the cathodic protection system operation and the records is given in NACE RP0186 (Ref 4).

Inspections of the dc power source should only be made by persons who are trained and qualified to work on electrical equipment. The use of strict safety practices including lockout/tagout procedures is especially necessary when working

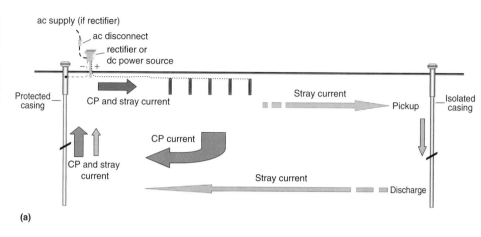

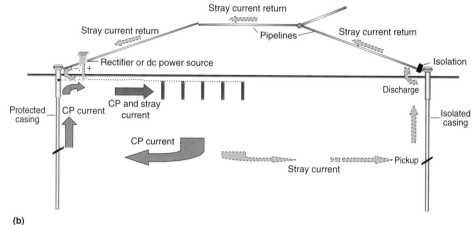

Fig. 14 Direct-current stray-current interference. (a) Stray-current pickup near top with discharge downhole. (b) Stray-current pickup downhole with discharge near top. CP, cathodic protection

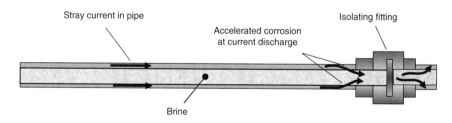

(a) Internal stray current interference through brine path across isolating fitting

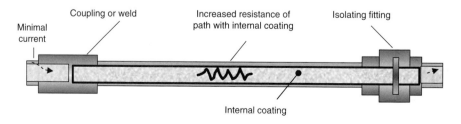

(b) "Long-path" increases resistance across isolating fitting to reduce stray current

Fig. 15 (a) Internal interference across an isolating feature and (b) reduced by a long-path isolating feature

on the rectifiers. The routine readings should include these measurements:

- dc power source current output
- dc power source voltage output
- dc power source adjustment setting (tap setting if applicable)
- dc current in secondary circuits
- dc interference control devices
- Power meter or fuel supply where applicable

The annual inspection should include:

- Completion inspection of the dc power source
  (a) Calibration of the dc power source current output
  (b) Calibration of the dc power source voltage output
  (c) Direct-current power source adjustment setting (tap setting if applicable)
  (d) Calibration of the dc in secondary circuits
- Measurement of the well-to-electrolyte potential
- Measurement of the surface facility structure-to-electrolyte potentials
- Testing the effectiveness of wellhead isolation, if applicable
- Measurement of the current returning from the casing at the wellhead with a clamp-on ammeter, if there is no isolation
- Confirmation that dc interference control devices are providing the necessary control
- Specialty tests applicable to the specific cathodic protection installation

Corrosion-control records are of paramount importance in an effective corrosion-control program. They will be used to establish a need for enhancements of the corrosion-control program and to ensure that the existing corrosion-control equipment is operating. The records should include but not be limited to:

Historical:

- Well completion data including casing sizes and lengths, cementing information and well total depth
- Corrosion leaks identifying well, depth, internal or external, date of failure compared to date of drilling and/or workover (s)
- Inspections of casing failures and corrosion products
- Electrical well logs (wall thickness, CPP identifying corrosion, and resistivity)
- Coating type and thickness, if applicable
- Drawing of well casing strings and lease equipment and piping
- System map of the field
- Location and type of electrical isolation

Cathodic Protection:

- Current requirement tests (CPP log(s), E log I test(s), and soil resistivity in layers near the surface)
- Design and drawings of cathodic protection installation detailing:
  (a) Well location
  (b) Piping and lease facilities
  (c) dc power source type, rating and location
  (d) Description of energy supply for dc power source
  (e) Cable type(s) and location
  (f) Cable to wellhead and piping connections
  (g) Anode beds type and location
  (h) Anode material type, spacing and depth
  (i) Backfill type and amount
  (j) Junction box and test station details

Interference Control:

- Records of all tests pertaining to interference on the well from other systems and on other systems from the well cathodic protection system
- List of owners and contacts involved in the interference control program
- Description of the method of interference mitigation, including control devices and target values of current and potential
- If bonds or directional devices are used, the location, type, resistance value, current, and current direction

All records must show the date, the name of the inspector or tester and, if different, the names of those who make recommendations. Any changes in current output must be correlated with other measurements taken.

Cathodic Protection Summary

For new wells, the use of an abrasion-resistant underground coating on those portions of the casing exposed to the strata should be considered as part of a corrosion-control program, as this will greatly reduce the amount of cathodic protection current required for protection. If coating is used, though, a cathodic protection system must be planned and implemented immediately, as a coating alone will concentrate corrosion at the coating holidays.

Prior to applying cathodic protection, a review of the existing well historical data should be made to assess the possibility of corrosion that will cause premature and costly failure repairs. Electrical logging tools, which are reasonably accurate, are available to assess the metal loss that has occurred and to predict the possibility of future corrosion.

Provided the proper amount of current is applied and maintained, cathodic protection of well casings has proved to be an effective means of minimizing corrosion on the casing. The cathodic protection current can be determined by various means; however, two of the more reliable results have to date been with CPP type of testing and polarization tests (E log I). The former test is difficult to perform in that the well has to be taken out of service, which usually results in few candidate wells in an older field. Also it may be necessary to perform multiple tests, with time provided between tests to allow for steady-state conditions to be achieved, which adds to the cost of the test. The E log I test must be correctly analyzed to identify the Tafel point on the profile; otherwise, a current less than that necessary may be defined as the criterion. Another option is to use a mathematical model; however, the validity of this option should be confirmed by tests at the start of the cathodic protection program.

Another factor in designing well casing cathodic protection systems is to remember that the amount of cathodic protection current required is also dependent on the spacing between the casing and the anodes, up to a certain distance, and that distance must be defined for each well. If the anodes are placed within that distance, the current requirement increases. Once a cathodic protection current requirement is established for a temporary anode bed, the same distance or greater should be used in the final cathodic protection design.

Isolation of the casing from other facilities is another important cathodic protection system design consideration. Isolating the well casing from surface facilities is preferred to eliminate the macrocorrosion cell between the casing and these structures without cathodic protection and to provide a means for controlling and measuring the cathodic protection current to the casing. However, if the product inside the isolation contains a large quantity of brine, either a "long-path" isolating fitting should be used to minimize internal interference, or in some cases the isolator may have to be removed entirely.

Generally, cathodic protection systems using conventional rectifiers are designed and installed for the protection of the casings, although pulse rectifiers have also been used. Particular attention has to be placed on the size and the location of the anode bed in order to achieve the required current output for the desired life of the anode bed.

Stray current must also be considered during the cathodic protection system design. Stray-current interference from other dc power sources will accelerate corrosion on the casing if it encourages a current discharge into the formation. A common source is from other cathodic protection systems in the same oil/gas field, but can also come from other sources not related to the oil/gas field. Several methods have been outlined to either avoid or minimize these interference effects. Any stray-current control device must be continuously inspected and maintained.

Detailed records must be kept on the history of the well, electrical logs, casing repairs, and on the operation of the corrosion-control equipment. These records must be able to stand up to future legal scrutiny.

REFERENCES

1. B.A. Gordon, W.D. Grimes, and R.S. Treseder, Casing Corrosion in the South Belridge Field, *Mater. Perform.*, March 1984, p 9
2. W.R. Lambert and G.G. Campbell, Cathodic Protection of Casings in the Gas Storage Wells, *Appalachian Underground Short Course*, Fourth Annual proceedings, West Virginia University, p 502

3. C. Brelsford, C.A. Kuiper, and C. Rounding, "Well Casing Cathodic Protection Evaluation Program in the Spraberry (Trend Area) Field," paper 03201, Corrosion 2003, NACE International
4. "Application of Cathodic Protection for Well Casings," RP0186, NACE International
5. W.F. Gast, A 20-Year Review of the Use of Cathodic Protection for Well Casings, *Mater. Perform.*, Jan 1986, p 23
6. W.C. Koger, Casing Corrosion in the Hugoton Gas Field, *Corrosion*, Oct 1956
7. A. Hamberg, M.D. Orton, and S.N. Smith, "Offshore Well Casing Cathodic Protection," paper 64, Corrosion/87, National Association of Corrosion Engineers. Reprinted from *Mater. Perform.*, March 1988, p 26
8. R.G. Wakelin, R.A. Gummow, and S.M. Seagall, "AC Corrosion—Case Histories, Test Procedures and Mitigation," paper 565, Corrosion/98, NACE International
9. B. Dennis, "Casing Corrosion Evaluations using Wireline Techniques," Schlumberger of Canada, Calgary, Alberta, Canada
10. B. Husock, Methods for Determining Current Requirements for Cathodic Protection of Well Casings—Review, *Mater. Perform.*, Jan 1984, p 39
11. F.E. Blount and P.W. Bolmer, Feasibility Studies on Cathodic Protection of Deep Well External Casing Surfaces, *Mater. Protect.*, Aug 1962, p 10
12. E.W. Haycock, Current Requirement for Cathodic Protection of Oil Well Casing, *Corrosion*, Nov 1957, p 767t
13. D.H. Davies and K. Sasaki, Advances in Well Casing Cathodic Protection Evaluation, *Mater. Perform.*, Aug 1989, p 17
14. W.B. Holtsbaum, "External Protection of Well Casings Using Cathodic Protection," Canadian Region Western Conference, National Association of Corrosion Engineers, Feb 20, 1989
15. J.K. Ballou and F.W. Schremp, Cathodic Protection of Oil Well Casings at Kettleman Hills, California, *Corrosion*, Vol 13 (No. 8), 1957, p 507
16. F.W. Schremp and L.E. Newton, paper 63, Corrosion/79, National Association of Corrosion Engineers
17. J. Dabkowski, "Assessing the Cathodic Protection Levels of Well Casings," American Gas Association, Jan 1983
18. S.N. Smith, A. Hamberg, and M.D. Orton, "Modified Well Casing Cathodic Protection Attenuation Calculation," paper 65, Corrosion 87, National Association of Corrosion Engineers
19. M.A. Riordan and R.P. Sterk, Well Casing as an Electrochemical Network in Cathodic Protection Design, *Mater. Protect.*, July 1963, p 58
20. M.D. Orton, A. Hamberg, and S.N. Smith, "Cathodic Protection of Coated Well Casing," paper 66, Corrosion/87, National Association of Corrosion Engineers
21. W.F. Gast, Well Casing Interference and Potential Equalization Investigation, *Mater. Protect.*, May 1974, p 31
22. G.R. Robertson, Effects of Mutual Interference Oil Well Casing Cathodic Protection Systems, *Mater. Protect.*, March 1967, p 36
23. R.F. Weeter and R.J. Chandler, Mutual Interference between Well Casings with Cathodic Protection, *Mater. Perform.*, Jan 1974, p 26
24. N.N. Bich and J. Bauman, Pulsed Current Cathodic Protection of Well Casings, *Mater. Perform.*, April 1995, p 17
25. J. Dabkowski, Pulsed Rectifier Limitations for Well Casing Cathodic Protection, *Mater. Perform.*, Oct 1995, p 25
26. D. Warnke and W.B. Holtsbaum, "Impact of Thin Film Coatings on Cathodic Protection," paper IPC 02-27325, ASME International Pipeline Conference, 2002

SELECTED REFERENCES

- "Application of Cathodic Protection for Well Casings," RP0186, NACE International
- W. von Baeckmann, W. Schwenk, and W. Prinz, Ed., *Cathodic Corrosion Protection*, Gulf Publishing, 1997, p 415–426

Stray Currents in Underground Corrosion

W. Brian Holtsbaum, CC Technologies Canada Ltd.

STRAY CURRENT can be defined as a current in an unintended path. In this case, it applies to stray electrical currents in structures that are underground or immersed in an electrolyte. Stray current can be from man-made sources or from natural sources (telluric). A dc stray current discharge will accelerate corrosion on a structure where a positive current leaves the structure to enter the earth or an electrolyte. Stray alternating current at high densities will also cause corrosion, although at a much lower rate than for dc stray current.

In addition to the consequences of accelerated corrosion, stray current corrupts the potential measurements that are being taken to establish a cathodic protection (CP) criterion.

Early stray current sources came from electric street railways. The first documented occurrence (Ref 1) in 1894 was due to a direct current (dc) powered railway installed in Richmond, VA, in 1888. In Boston, MA, during 1892, a negative cable laid between the tracks was bonded frequently to a parallel water main as the first documented attempt to stop stray-current corrosion. Unfortunately, this did not work so operators tried the opposite polarity with more disastrous effects because both the cable and pipe then corroded. Litigation against the transit operators followed in 1900, and the judgment found in favor of the claimant, with the courts limiting stray-current leakage.

The American Committee on Electrolysis was started in 1913 and published a comprehensive report in 1921, after which it became inactive. Local electrolysis committees were formed as early as 1913 to 1917 (Ref 2) to address the issue. Drainage of streetcar stray currents was practiced in Belgium beginning in 1932 (Ref 3). This practice was also recognized in Germany based on a report in 1939. With the increase in industrialization, other sources of stray current also became an issue, and today it has become a problem that is automatically reviewed in underground or immersed structures. As problems are resolved, the interest in electrolysis committees decreases, but several committees still exist to resolve problems from many different stray-current sources.

Principles of Stray Current

The basic electrical laws apply to stray current including, but not limited to:

- A closed electrical circuit must exist, especially as it applies to a parallel electrical circuit.
- Ohm's law relating to voltage, current, and resistance
- Direct current can go in only one direction in a conductor.
- Kirchoff's current law, where the sum of the current at a junction is equal to zero
- Faraday's law of metal weight loss related to current and time

Stray current applies to a parallel electrical circuit where the structure is a parallel path within another electrical circuit as depicted in Fig. 1. The point of current pickup in Fig. 1 is indicated by "A," while the point of discharge is at "B" and may be either a metallic or electrolytic path to a structure.

When a current approaches or leaves a buried or immersed structure, a voltage gradient is established around the structure that is dependent on the resistivity of the electrolyte and the amount of current (Ref 3). Equipotential lines perpendicular to the direction of current can be measured around the structure as illustrated in Fig. 2 and 3. If the gradient is negative with respect to remote earth, a cathodic gradient exists. An anodic gradient exists when the gradient is positive with respect to remote earth.

When a structure passes through a cathodic gradient, a current discharge, often called cathodic interference, can occur (Fig. 2). This current discharge must return to its source, but in this case, the controlling factor is the cathodic gradient.

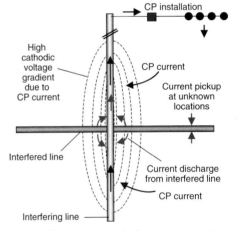

Fig. 2 Illustration of cathodic stray-current interference. CP, cathodic protection

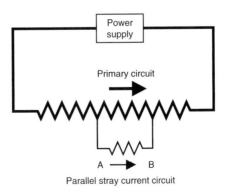

Fig. 1 Schematic of a parallel interference path

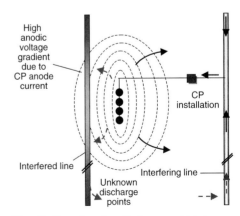

Fig. 3 Illustration of anodic stray-current interference. CP, cathodic protection

When a structure passes through an anodic gradient, a current pickup is encouraged and is often called anodic interference (Fig. 3). Again, this current must return to its source, and the manner by which this occurs is of the most concern. Both cathodic and anodic interference can occur at the same time.

If the stray current is dc, the latter case is of utmost significance, and it is important to detect even small current values. If the stray current is alternating current (ac), then a larger current density becomes critical.

Any current source that may use the earth as a path, either intentionally or inadvertently, can be a source of stray current. These can include, but are not limited to:

- Cathodic protection (CP) (Ref 3–5)
- High voltage dc (HVDC) power lines (Ref 6)
- Transit systems (Ref 1, 7–9)
- Direct current operated mining equipment
- Electric railways
- Welding, both onshore and offshore (Ref 10)
- Electroplating or battery-charging equipment with ground faults
- Natural (telluric) current (Ref 5, 11)
- High voltage ac (HVAC) power lines (Ref 12–14)

The first two sources are steady state interference sources. The remaining sources are more of a dynamic interference as they change in magnitude and often in direction. The CP examples in Fig. 2 and 3 are of a steady state type of interference, while the transit system in Fig. 4 is an example of a dynamic stray current.

In addition to these man-made stray currents, a naturally occurring stray current (telluric) influence such structures as pipelines. Telluric current is a naturally occurring current that results from geomagnetic fluctuations in the earth (Ref 11). The earth's magnetic field is generally from north to south but does vary throughout the world as shown in Fig. 5.

This magnetic field projects into outer space where it is affected by the "solar wind" consisting of solar plasma (high-energy protons, electrons, and other subatomic particles). The sun produces a stream of solar plasma of varying intensity with bursts of short wave radiation emitted with solar flares that varies in magnitude on cycles throughout the year and over a period of years.

The telluric current associated with the geomagnetic fluctuations tends to flow in the earth's crust. Should a pipeline be installed in the area of this telluric activity, a current can either be induced onto the pipe or may enter it by conduction. The potential change may be due only to changes in the earth's potential gradient, which does not reflect a current pickup or discharge. A major problem with telluric current is the inability to collect meaningful data when assessing the status of cathodic protection on a structure.

Induced ac voltages on parallel conductors have been recognized for many years but were originally a more common problem between communication lines and power lines. As utility corridors have become more common, power lines now occupy parallel right-of-ways with pipelines. This and the improved coatings on pipelines have resulted in induced ac voltages that are becoming an ever-increasing problem.

There has been a significant amount of study on the subject since the 1970s. The American Gas Association and the Electric Power Research Institute cosponsored a study (Ref 15) to develop a method of predicting voltages on pipelines. The Canadian Electrical Association (Ref 16) commissioned a study of problems with pipelines occupying joint-use corridors with ac transmission lines. Canadian Standards Association later issued a guide (Ref 14) for power line and pipeline owners that covers the safety of the personnel working on pipelines in the area of HVAC power lines. NACE International issued a similar recommended practice (Ref 13) dealing with this problem.

There are three mechanisms, capacitive, conductive (also called resistive), and inductive, by which voltages can be transferred to pipelines paralleling an electrical power transmission line.

Capacitive effects have to be considered on aboveground pipelines, especially those with no contact to the ground such as when pipelines are under construction and on skids. The problem becomes more severe as welding increases the length of the aboveground section of pipe. Precautionary measures are covered in Ref 15 and 16.

A conductive, or resistive, coupling takes place through the soil when a pipeline is in proximity to a power transmission line ground. If large fault currents or lightning strikes on the transmission line create a large ground fault current, it can enter the pipe. It will not only cause large potential gradients for the short duration of the fault, but it may cause damage to the pipe and/or coating if it enters at a high current density. Research (Ref 14) indicated that not only can molten pits occur, but cracks can develop around the molten area (assuming that penetration has not taken place).

An inductive coupling is caused by the changing magnetic field from the ac flow in the power transmission line and different distances to each phase conductor. An ac voltage will be induced on a pipeline in the vicinity of the power line,

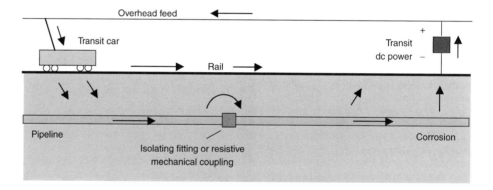

Fig. 4 Example of dynamic interference from a transit system

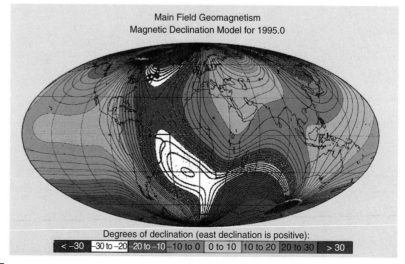

Fig. 5 World isomagnetic chart. Source: Ref 5

which passes through this changing magnetic field. These voltages will be essentially permanent on the pipeline but will vary somewhat with the actual load on the power line.

The prediction of induced ac voltages is complex and is covered in the literature (Ref 13). The voltage will peak at discontinuities between the pipeline and the power line and attenuate exponentially between the discontinuities to the point that portions of the pipeline may have little or no induced voltage. Discontinuities between the power line and the pipeline may be:

- A change in the transmission line to pipeline separation
- The end or beginning of the parallel exposure
- A change in the number of line conductors or pipelines in the common right-of-way

A parallel length between the discontinuities is necessary for an induced voltage on the pipeline. The better insulated the pipe is from the ground, which occurs with more effective pipeline coatings, the higher the induced voltage will be as low resistance grounding would reduce this voltage.

Discontinuities due to changes in the pipeline characteristics are sometimes more difficult to determine because the information is not readily apparent from drawings or file information. Changes in the characteristics can arise from any of the following reasons:

- A change in the coating conductivity
- An extreme change in soil resistivity (somewhat dependent on coating)
- A change in the pipe size or thickness
- An interruption in the pipe continuity (isolating features)

The most influential factors are the phase currents and their relative magnitudes, the length of the parallel section between the pipeline and power line, and the distance between the conductors and the pipeline in the relationship shown in Eq 1:

$$V_{ac} = f[I_{ac}, L, \frac{1}{D}] \quad \text{(Eq 1)}$$

where V_{ac} is the induced ac voltage, I_{ac} is the phase current, L is the length of parallel section, and D is the distance between conductor(s) and pipeline (note this may not apply in proximity of power line as voltage increases initially before decreasing in the perpendicular direction from the power line).

Consequences of Stray Current

A dc discharge from a metal into an electrolyte, such as illustrated in Fig. 2, causes corrosion at the following rates for these metals:

Iron	9.2 kg/A-yr (20 lb/A-yr)
Copper	10.4 kg/A-yr (23 lb/A-yr)
Lead	34.5 kg/A-yr (75 lb/A-yr)

This in turn has to be related to the surface area of discharge and the wall thickness of the structure.

For example, a 150 mm (6 in.) schedule 40 pipe weighs 8.62 kg/m (18.98 lb/ft) and has a 7.11 mm (0.280 in.) wall thickness. This suggests that in just over one year, a 1 A current discharging from a meter of this pipe would completely consume it. If a current of only 1 mA was discharged from a 10 mm (0.394 in.) diameter coating holiday on this pipe, it would likely leak in less than 7 weeks. Immediate action is therefore required whenever a stray-current effect is noted.

Although no corrosion occurs at the point of current pickup, this current must return to its source and therefore the area(s) where this current leaves the structure (Fig. 3) must be considered as the consequences will be similar to those described previously.

There also appears to be a relationship with ac and corrosion; however, it is not as well defined as that for dc (Ref 12). At a current density less than 20 A_{ac}/m^2, it appears that CP is able to control corrosion. Between 20A_{ac}/m^2 and 100 A_{ac}/m^2 (1.86 A_{ac}/ft^2 and 9.29 A_{ac}/ft^2) corrosion is unpredictable, but at greater than 100 A_{ac}/m^2 (9.29 A_{ac}/ft^2), corrosion can be expected.

Interference from steady state current sources will result in a current discharge and pickup at consistent locations. Dynamic interference sources that change both in magnitude and direction will continually change the locations of current pickup and discharge. Expected locations of current discharge are at areas where the structure is in a high cathodic gradient such as at a pipeline crossing, across poor continuity joints, across an isolating flange, or between isolated reinforcing steel.

An exception to stray current causing corrosion is when a structure is receiving adequate CP and has formed hydroxyl ions. The oxidation reaction may involve the oxidation of these hydroxyl ions to oxygen and water without involving the metal atoms.

Interference Tests

Current Mapping. Where it is possible to trace the stray current by a current mapping process, the location of current pickup and discharge can be determined readily. Current was measured at points in a pipeline in Fig. 6. By Kirchoff's current law, the sum of the current at a junction must equal zero. At junction A, the current increases from 0.5 A to 1.0 A; therefore, a current pickup of 0.5 A took place in between the pipeline measurements. The current reduced after junction B; therefore, a current discharge occurred in between. The current is in opposite directions at junction D; therefore, the entire amount had to discharge at that location. There was no change in pipeline current at junctions C and E; therefore, there was no current pickup or discharge at those locations.

These current values measured in the pipeline can be plotted on a graph as shown in Fig. 7, where the current going in the opposite direction is given a negative value. The bars are the actual current pickup and discharge in this figure. The current discharge and pickup sections can readily be noted by the pipeline current profile. A positive slope indicates a current pickup and a negative slope indicates a current discharge.

A second type of current mapping can be used where the current in the soil is determined by measuring the voltage gradient caused by the current in the soil (Ref 17). This is illustrated in Fig. 8. The data are then processed by a computer program to show the variations in stray current at any given location compared with another.

Structure-to-Electrolyte Potentials. If the source of the stray current can be interrupted, a shift in the structure-to-electrolyte potential will occur at the point of exposure. The shift will be in an electropositive direction if there is a current pickup and in an electropositive direction if there is a current discharge. Both shifts are of concern.

Combination of Current and Potentials. Where possible, a combination of both line current and potentials can be compared at different locations. Such procedures are often used for a dynamic stray-current situation, as in the transit system shown in Fig. 9, and are often called beta curves (Ref 18). In the past, the use of x-y plotters was popular, but data loggers are now more commonly used to gather data under these conditions.

The current in the pipeline at different locations can be determined by a current span as shown in Fig. 9, or by a clamp-on ammeter. The current at a given point in time can be plotted as shown in Fig. 10. Knowing the direction of current, the sections of current pickup and discharge can be determined. The current over a period of time at one location, however, will vary, in which case the current at one location can be plotted against the current at the next location over a given time period. Knowing the current direction, the slope of the line will indicate a current pickup (<45°), no

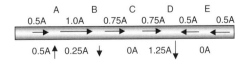

Fig. 6 Current mapping of a pipeline

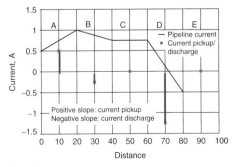

Fig. 7 Plot of pipeline current from Fig. 6

pickup/discharge (45°), or current discharge (>45°), as shown in the top of Fig. 11.

The next technique is an exposure survey in which the pipe-to-electrolyte potential is measured simultaneously with the current measured above. In this case, a current pickup is expected to correspond to an electronegative shift in potentials, or a current discharge corresponds to an electropositive shift in potentials. A plot of current against potential at each location will indicate a current pickup by a positive slope and a current discharge by a negative slope and the point of maximum exposure (Fig. 11). A near vertical slope suggests that there was neither a prevalent current pickup nor discharge.

A mutual survey is conducted by the measurement of the voltage between the interfering structure and the interfered structure and compared with the pipe-to-electrolyte potential measured at the same time. Correlation is indicated by a straight line on a plot of these measurements. This measurement does not reveal all points of current discharge but averages the condition between the measurement locations.

Telluric Current. For best results in potential measurement, it is desirable to wait for a quiet period of solar activity and complete the survey at that time. Before measuring pipe-to-electrolyte potentials, a geomagnetic forecast that is available throughout the world should be checked.

If a survey must be conducted in periods of high solar activity, special testing and compensation of measurements are necessary. One fundamental type of telluric survey involves the continuous measurement of structure-to-electrolyte potentials with a data logger at stationary locations within the test sections and the recording of structure-to-electrolyte potentials with a portable data logger that is time stamped with the stationary data loggers.

The data in Fig. 12 were obtained by a data logger and show the effects of telluric current at three different locations on a fusion bonded epoxy coated pipeline. One is approximately 2 km (1.24 miles) from the first, while the second is approximately 57 km (35 miles) away. The significance in this profile is that the potential variations are similar over time along a large section of well-coated pipe. If the potential at one location can be established, then by extrapolation, the potential at other locations can also be determined.

Another study (Ref 18) showed similar voltage fluctuations on two different gas distribution systems that were owned by different companies 700 km (435 miles) apart. This information was detected by remote monitoring potential equipment that has only recently been used. Heretofore, the belief was that telluric current primarily affected pipelines longer than 50 km (30 miles) in length. This new information shows that pipelines less than 5 km (3 miles) in length can also be affected. In addition, the area of influence due to this activity can be very large, and similar effects can be seen on completely separate systems.

Once the true potential at the stationary locations has been established, the true potential at the portable location can be approximated using Eq 2 and 3:

$$eb = \left[\frac{ea(c-b)}{c}\right] + \left[\frac{ec(b-a)}{c}\right] \quad \text{(Eq 2)}$$

where a is the first stationary potential location, b is the portable potential location, c is the second stationary potential location, ea is the error in potential at stationary data logger "a" at time "x," eb is the error in potential at portable data logger "b" at time "x," and ec is the error in potential at stationary data logger "c" at time "x."

Equation 2 then applies the correction to the measured potential at the portable data logger's location:

$$E_{p\ true} = E_{p\ measured} - eb \quad \text{(Eq 3)}$$

where $E_{p\ true}$ is the true potential at the portable data logger location, $E_{p\ measured}$ is the measured potential at the portable data logger location, and eb is the error in potential at portable data logger location.

If the potential closely follows that of the nearby stationary data logger position, then correcting only to the closest stationary data logger location would be accurate. Equation 4 is then used to compensate the measured potential in this case:

$$E_p - E_s - \Delta(E_{sa} - E_{pa}) \quad \text{(Eq 4)}$$

where E_p is the true potential at the portable data logger location, E_s is the true potential at the

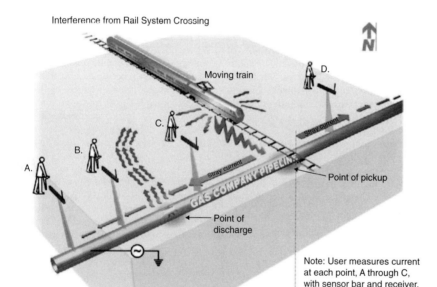

Fig. 8 Current mapping of stray current at a buried pipeline. Source: Ref 7

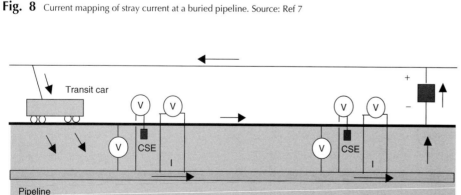

Fig. 9 Potential and current tests on a dc transit system

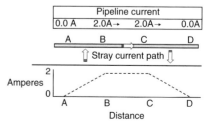

Fig. 10 Relationship of line current and direction

stationary location, E_{sa} is the stationary potential at time "a" during the data logging, and E_{pa} is the portable potential at time "a" during the data logging.

Hazardous ac Voltages. Alternating current voltages can occur on pipelines that parallel HVAC power lines by either a capacitive, inductive, or conductive coupling. The inductive coupling is relatively steady state and is the ac voltage normally measured.

A measurement of the ac voltage to ground can be made in a similar manner as a dc structure-to-electrolyte potential except that an ac voltmeter must be used, and the type of reference electrode is not critical. A major difference is that the ac voltage will vary over short periods of time depending on the power line load as illustrated in Fig. 13. A single measurement of the ac voltage on the pipeline may not reflect the most hazardous condition.

Figure 13 also demonstrates the effect of a grounding, but the resistance to ground was still not low enough to bring the voltages down to below 15 V_{ac} (Ref 13, 14) that is considered safe on the pipeline.

An explosion resulting from ac interference has been documented (Ref 19). The safety of the operating personnel and the public is the basis for attempting to predict and mitigate hazardous situations, which may arise when ac voltages are transferred to a pipeline.

Visual Inspection. The first indication of dc stray current may be by a visual inspection of corrosion. Stray current can be suspected by a lack of corrosion product and by its physical location. Corrosion at a pipeline crossing, near an isolating feature, or on one side of a mechanical joint are suspect.

Internal dc stray current interference may occur on one side of an isolating feature where there is a low resistivity product inside (Ref 20).

Mitigation

Eliminate or Minimize Source of Stray Current. The most effective means of controlling stray current is to relocate or reduce the exposure. The approach is unique for each exposure. Examples include:

- Anodic interference can be removed by relocating the anode bed or structure or by redistributing the current with additional anode beds.
- Readjustment of the current at the source if possible
- Cathodic interference may be reduced by recoating the structure with the high cathodic gradient to reduce the current at that point and, therefore, the gradient.
- Transit tracks can be isolated from the earth through isolation material between the rails and the ties and in switch connections (Ref 8).
- Bonds in transit tracks to reduce the resistance of the intended current return path (Ref 7, 8)
- Repair of equipment that is faulting to ground

Control Bonds. A metallic bond that is of low-enough resistance can provide a safe path for stray current to return to its source provided the potential of the "interfering" structure is more electronegative at the point of connection (Fig. 14). If the potential is more electropositive, the additional current transferred to the "interfered" structure adds to the amount that may be discharging through the electrolyte. That is, the interference condition will be compounded.

Advantages of a mitigation bond include:

- The bond will transfer the stray current safely if the resistance is correct.
- The bond is relatively inexpensive and can usually be installed quickly.
- The bond can be monitored relatively easily.

Figure 15 shows the use of mitigation bonds on a transit system.

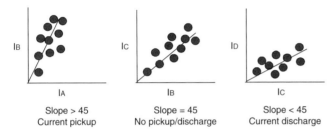

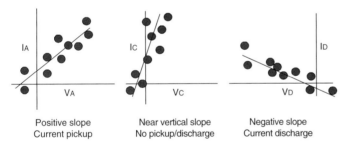

Fig. 11 Relationship of varying line current and pipe-to-electrolyte potentials over time

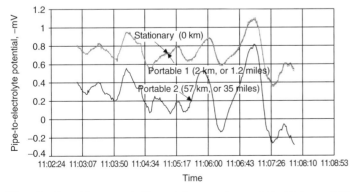

Fig. 12 Potentials measured with two stationary and one portable data logger at different locations along a fusion bonded epoxy-coated pipeline. Note that the Portable 1 profile is virtually identical to the Stationary profile 2 km (1.24 miles) away and similar in shape to Portable 2 profile 57 km (35 miles) away. Source: Ref 4

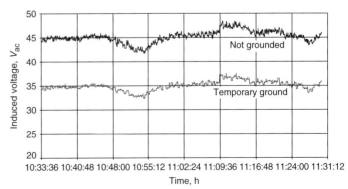

Fig. 13 Sample induced ac voltage on a pipe (top profile) and the effect of a temporary ground (lower profile)

Disadvantages of a mitigation bond include:

- Bonds can be destroyed by current surges.
- The bond is a critical bond and must be inspected bimonthly if on a regulated pipeline and should be monitored this frequently otherwise.
- The cathodic protection systems of the two structures become dependent on one another.
- Excessive potentials can occur before the stray current effect is controlled.
- Under dynamic stray current conditions, the potentials can reverse (see "Reverse Current Switches" subsequently).

Reverse current switches are used in a bond to allow current to go in the desired direction but to prevent a reverse current that could cause accelerated corrosion on the structure (Ref 21, 22). Ideally, a reverse current switch would have zero impedance in the direction of desired current and infinite impedance in the opposite direction and must have the capacity to control the desired amount of current.

A reverse current switch could consist of an electromagnetic relay, a diode, or a hybrid system. A potential-controlled rectifier installed in the bond can be used as a forced-current drainage bond to serve the same purpose.

An electromagnetic switch will close when sensors detect that the structure becomes more electropositive than a set point. The stray current passes through the relay contacts. The relay should open when the current passes through the zero point, which will reduce the arc burns on the contacts. The disadvantages of the relay include:

- They normally require a power source.
- They have a limited number of open/close cycles.
- They may have a slow response time relative to the changing stray-current frequency.

A diode is a solid state device that has low forward impedance with high reverse impedance. Germanium and copper oxide diodes have a low forward voltage drop and the expense of poor reverse voltage breakdown. Silicon diodes have high forward and high reverse voltage drop characteristics. A high amount of heat can be generated through a diode that must be dissipated, usually through a heat sink. Stray currents often have an ac component that will be rectified by the diode. The diode must be derated from a dc output at maximum current output, depending on the waveform.

Unfortunately, the waveform is not detected by an ammeter, thus an oscilloscope must be used to determine the phase angle. Finally, the possibility of induced ac voltages must also be considered in rating these devices. Diodes have a forward voltage drop that has to be exceeded before they conduct. This may be too great for low-voltage applications.

A hybrid system may consist of two different types of diodes (germanium and silicon) in parallel often with a resistor in series with the low forward impedance diode (germanium). This combination ensures a faster conduction, while the resistor ensures that the silicon diode will conduct more current. Another combination is a silicon diode in parallel with an electromagnetic relay. The relay can be much smaller as the diode conducts the higher current.

Another hybrid system consists of a diode in parallel with a tapless automatic potentially controlled rectifier. The rectifier can be set to conduct before the diode up to a given current after which the diode continues to carry the balance. These are custom designed to a particular application.

Cathodic Protection. It is possible that by enhancing the CP on the interfered structure, the stray current could be minimized. A sacrificial anode(s) can be installed at current discharge areas such that the discharge will go from the anode to the electrolyte. Care must be used in the location of sacrificial anodes to ensure an adequate current will discharge from them and because they could be a source of ac pickup. In one documented case, this in turn caused an explosion when the ac surge arced across an isolating flange (Ref 13).

Dividing the system into several electrical sections by installing isolation and protecting each section by independent CP systems has been used (Ref 14). This approach has to be taken very carefully because an interference problem can be established at any one of the new isolating features.

Mitigating Induced ac Voltage. Dangerous capacitive voltages can be diminished by ensuring that all aboveground pipe sections are adequately grounded. These include electrical grounds to aboveground pipelines, bonds around open sections of pipe that are attached with a bolted clamp, and temporary gradient mats attached to the pipe for personnel to work on.

Conductive or resistive couplings can be reduced by ensuring that the pipeline and electrical grounds are separated as far as possible. Canadian Standards Association (Ref 14) recommends a spacing of 10 m (33 ft). The use of temporary ground mats and bonds installed around "open" pipe sections will protect personnel.

Induced voltages can be predicted with mathematical models, and a permanent mitigation scheme can be devised that normally includes grids around above pipeline appurtenances and grounds at voltage "peaks." It

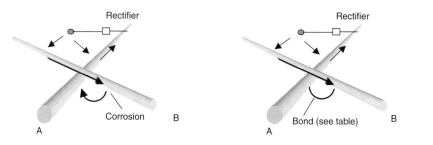

Case	Pipe-to-electrolyte potentials (−mV CSE) A	B	Bond current direction (+ to −)	Remarks
1	−800	−840	A to B	Bond will not control interference as direction in bond wrong. Must go from B to A
2	−890	−840	B to A	Bond may control interference if enough current can be drained.

Fig. 14 Mitigation bond showing when and when not practical

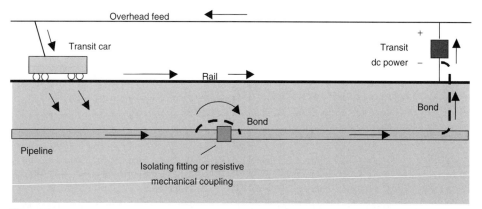

Fig. 15 Mitigation bonds for a transit system (reverse current switches often put in bond back to power supply)

Stray Currents in Underground Corrosion / 113

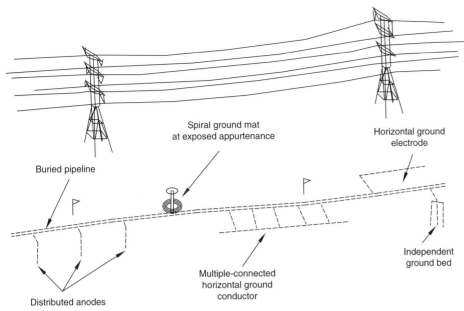

Fig. 16 Alternating current voltage protection during operations. Source: Ref 16, 22

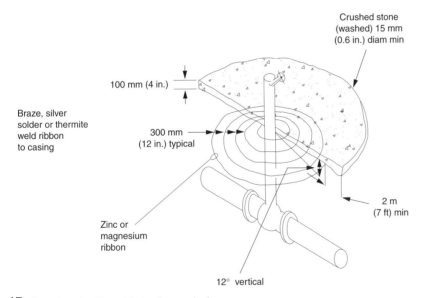

Fig. 17 Ground mat (gradient grid) at underground valve

should be noted that the reduction of a voltage peak at one location often causes an increase in voltage elsewhere.

Alternating current voltages on the pipeline in excess of 15 V_{ac} have been accepted by industry (Ref 13, 14) as being hazardous. Alternating current voltages can be mitigated by strategically placed electrical grounds to the pipeline, usually at the discontinuities. Voltages in excess of 50 V_{ac} have been measured on pipelines and that required a current drain to ground in excess of 50 A_{ac} to reduce the pipeline voltage to less than 15 V_{ac}. Direct current decouplers may be necessary in series with the ground cable to allow the passage of ac but to isolate dc and thus reduce an adverse effect of adding bare metal to the cathodic protection system.

Personnel working with CP must be trained to measure ac voltages safely because they are continuously taking electrical measurements on pipelines but expecting low voltages. A safe practice is to take the ac voltage measurement first.

Cathodic protection test stations need not have a gradient grid, but their terminals should be protected to prevent accidental contact from the public, and pipeline personnel need to be trained accordingly.

Pipeline personnel working in these areas need to be trained properly in the hazards of ac voltage on pipelines and the necessary safety procedures to include in maintenance, excavations, and repairs (Fig. 16). This is particularly important on the well-coated pipelines that are more common today. In addition, personnel contacting pipeline appurtenances within 20 km (12 miles) of an HVAC parallel section should be protected by the installation of a gradient grid around it (Fig. 17).

Where a dc isolation is required but an ac connection is necessary, a dc decoupler can be inserted in the bond. These can be in the form of an electrolytic polarization cell or an electronic decoupler. These allow dc to be maintained for CP while reducing ac voltage to a safe level.

REFERENCES

1. J.J. Meany Jr., A History of Stray Traction Current Corrosion in the United States, *Mater. Perform.*, 1974, p 20
2. R.M. Lawall, "A Cooperative Approach to Electrolysis Problems," National Association of Corrosion Engineers Annual Meeting, April 1948
3. W. von Baeckmann, W. Schwenk, and W. Prinz, *Handbook of Cathodic Corrosion Protection,* 3rd ed., Gulf Publishing Company, Houston Texas, 1997, p 18
4. M.E. Parker and E.G. Peattie, *Pipe Line Corrosion and Cathodic Protection,* 3rd ed., Gulf Publishing Company, Houston, Texas, 1984, p 22–25
5. D.H. Warnke and W.B. Holtsbaum, Impact of Thin Film Coatings on Cathodic Protection, *Proc. International Pipeline Conference 2002,* Paper IPC02-27325
6. A.L. Verhiel, HVDC Interference on a Major Canadian Pipeline Counteracted, *Mater. Perform.,* March 1972, p 37
7. F.A. Perry and M.I.E. Aust, "A Review of Stray Current Effects on a Gas Transmission Main in the Boston, Massachusetts Area," Paper presented at Corrosion/94, NACE International
8. W. Sidoriak, "D.C. Transit Stray Current Leakage Paths—Prevention and/or Correction," Paper presented at Corrosion/94, NACE International
9. R.E. Schaffer, Control of Stray-Current Effects from DC Powered Transit Systems, *Section Proceedings 85-DT-81,* American Gas Association, 1985, p 118
10. J.N. Britton, Stray Current Corrosion during Marine Welding Operations, *Mater. Perform.,* Feb 1991, p 30
11. Government of Canada, Geological Surveys, www.geo-orbit.org/sizepgs/magmapsp.html, accessed Feb 2006
12. R.A. Gummow, R.G. Wakelin, and S.M. Segall, "AC Corrosion—A New Challenge to Pipeline Integrity," Paper 566, presented at Corrosion/98, NACE International
13. "Mitigation of Alternating Current and Lightning Effects on Metallic Structures and Corrosion Control Systems," NACE RP-01-77
14. "Principals and Practices of Electrical Coordination between Pipelines and Electric

Supply Lines," Canadian Standards Association C22.3 No. 6-M1987
15. J. Dabkowski, A. Taflove, et al. (IIT Research Institute), *Mutual Design Considerations for Overhead AC Transmission Lines and Gas Transmission Pipelines,* American Gas Association and Electric Power Research Institute, Sept 1978
16. *Study of Problems Associated with Pipelines Occupying Joint-Use Corridors with AC Transmission Lines,* Canadian Electric Association, by BC Hydro and Power Authority, 1979
17. A. Kacicnik, D.H. Warnke, and G. Parker, Stray Current Mapping Enhances Direct Assessment (DA) of an Urban Pipeline, *Northern Area Western Conference* (Victoria, B.C., Canada), NACE International, Feb 2004
18. S. Croall, Telluric and HVDC Voltage Fluctuations on Distribution Pipelines, *Northern Area Western Conference* (Alberta, Canada), NACE International, Feb 2001
19. D.L. Caudill and K.C. Garrity, Alternating Current Interference—Related Explosions of Underground Industrial Gas Piping, *Mater. Perform.,* Aug 1998, p 17
20. R.B. Bender, Internal Corrosion of Large Diameter Water Pipes by Cathodic Protection, *Mater. Perform.,* Sept 1984, p 35
21. J.I. Munroe, "Optimization of Reverse Current Switches," Paper 142, presented at Corrosion/80, NACE International
22. J. Dabkowski and A. Taflove, Mutal Design Considerations for Overhead AC Transmission Lines and Gas Transmission Pipelines, *Prediction and Mitigation Procedures,* Vol 2, Electric Power Research Institue, EL-904, research project 742-1, PRC/AGA contract PR132-80, p 4-3, 4-7

SELECTED REFERENCES

- V. Ashworth and C.J.L. Booker, Ed., *Cathodic Protection Theory and Practice,* for Institution of Corrosion Science and Technology, Birmingham, U.K., by Ellis Horwood Limited, Chichester, U.K., 1986, p 180, 327–343
- W. von Baeckmann, W. Schwenk, and W. Prinz, Ed., *Cathodic Corrosion Protection,* Gulf Publishing Company, Houston, Texas, p 7, 18–23, 100–102
- M.E. Parker and E.G. Peattie, *Pipeline Corrosion and Cathodic Protection,* 3rd ed., Gulf Publishing Company, Houston, Texas, 1984, p 34, 100–124
- A.W. Peabody, *Control of Pipeline Corrosion,* 2nd ed., R.L. Bianchetti, Ed., NACE International, p 40, 211–236
- H.H. Uhlig, *The Corrosion Handbook,* John Wiley & Sons Inc., 1948, p 606–610

Corrosion Rate Probes for Soil Environments

Bernard S. Covino, Jr. and Sophie J. Bullard, National Energy Technology Laboratory

DESIGN ENGINEERS WORKING WITH BURIED OR PARTIALLY BURIED STRUCTURES must allow for the corrosion of their structures. Typical metal structures that are exposed to soil corrosion include bridge pilings, pipelines, buried storage tanks, and storage tank bottoms. Corrosion of these structures contributes significantly to the direct and indirect costs of corrosion. Direct costs include not only the damage to or the potential loss of the structure but also the costs of corrosion prevention methods such as coatings and cathodic protection. Indirect costs result when these measures fail to protect the structure and lead to a loss of or downgrading of services, such as the closure or derating of bridges or pipelines.

There is much known about the corrosivity of soil and a good general review is available in the article "Simulated Service Testing in Soil" in *Corrosion: Fundamentals, Testing, and Protection*, Vol 13A, of the *ASM Handbook* (Ref 1). Factors such as soil composition, structure and texture, soil electrical resistivity, and soil pH have been well characterized and correlated to the corrosion of metals. The American Water Works Association (AWWA) standard C-105 for Soil Corrosivity (Ref 2) assigns numerical values to different levels of resistivity, pH, redox potential, sulfides, and moisture in order to predict the level of corrosion in soils. At the present time, however, most of the information on the corrosion of metals in soils is acquired through gravimetric measurements of buried test specimens. Such tests typically yield accurate information on the corrosion rate of the specimen in one particular soil but only for a fixed period of time.

There are, however, times when it is important to know what has happened between the time mass loss specimens are buried and when they are retrieved. In other words, how does the corrosion of metals in soils change with time or how is it influenced by other factors? One way to do this is to use sensors or probes coupled with an appropriate technique for measurement of the corrosion behavior of the specimen in soil. Another use for soil corrosion rate sensors is to measure the relative changes in soil corrosivity with time. This may be used to determine, for example, when water levels or soil compositions change.

This article explores the use of several techniques for measuring the corrosion behavior of buried metals and the types of probes that were used. The discussion is divided between electrochemical and nonelectrochemical techniques for measuring the corrosion rates of buried probes. Principles of operation for all of the corrosion measuring techniques are covered first, followed by examples of their use from literature reports. Descriptions of the principles of operation of the techniques are brief because all have been well explained in the literature (Ref 3) and in the article "Methods for Determining Aqueous Corrosion Reaction Rates" in *Corrosion: Fundamentals, Testing, and Protection*, Vol 13A, of the *ASM Handbook* (Ref 4).

Nonelectrochemical Techniques—Principles of Operation

Electrical Resistance. The electrical resistance (ER) technique is the main nonelectrochemical technique used for measuring corrosion rate. It uses the electrical resistance of a thin piece of a metal test specimen as the sensor to monitor the loss of metal due to corrosion. For some applications, the sensor must be relatively thin in order to have sensitivity that is adequate to measure small changes in corrosion rate. This can lead to short sensor lifetimes. The principle of operation is that as the piece of metal becomes thinner, the resistance of the metal increases. While knowledge of the specific resistivity of the metal may be needed to calculate the change in thickness of the metal as the ER changes, the use of a half-bridge configuration can make that unnecessary. The half-bridge measurement configuration is used to provide temperature compensation, and it also removes resistivity from the calculation giving the change in area directly. Knowledge of the atomic mass and density of the metal then allows for the calculation of a mass change and, ultimately, a corrosion rate. The ER technique functions the same regardless of the corrosion mechanism and will give a constant readout of the gross change in cross-sectional area as a function of time. Note that the ER technique does not function well in pitting environments because corrosion pits could be interpreted as thinning of the sensor cross-sectional area and thus as a uniform corrosion rate.

Figure 1 shows a common configuration for a multiuse ER probe. The fixed-length ER probe has a wire, tube, or strip loop element of fixed length. While not configured for deep soil use, this probe can be adapted or reconfigured for those uses. Figure 2 shows a surface strip element ER probe that is more appropriate for

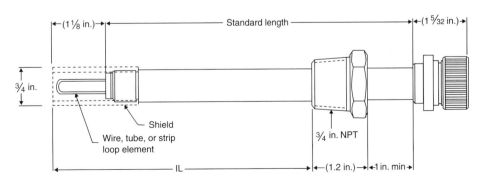

Fig. 1 Fixed-length electrical response probe schematic. Dimensions given in inches. IL, insertion length. NPT, American National Standard Taper Pipe Thread. Source: Metal Samples, Munford, AL

monitoring of underground pipelines and other structures. As shown, it includes a grounding wire for connection to a structure that is under cathodic protection (CP).

Electrochemical Techniques— Principles of Operation

In the context of the discussion in this article, electrochemical techniques are defined as those techniques that are able to measure the rate of one or more of the electrochemical reactions that are part of the corrosion process. While most electrochemical techniques are developed and first used in aqueous solutions in laboratories, many are applied in the field and used in soils and concrete. Many of the following techniques have been discussed elsewhere (Ref 4).

The linear polarization resistance (LPR) technique is based on the polarization of a test specimen in both the anodic and cathodic directions within ±20 mV of the open circuit corrosion potential (OCP). The LPR technique makes use of a simplification of the Butler-Volmer equation by Stern and Geary that results in the equation:

$$R_p = \frac{\Delta E}{\Delta i} = \frac{\beta_a \beta_c}{2.303(i_{corr})(\beta_a + \beta_c)} = \frac{B}{(i_{corr})} \quad \text{(Eq 1)}$$

where R_p is a resistance obtained from the LPR and electrochemical impedance spectroscopy

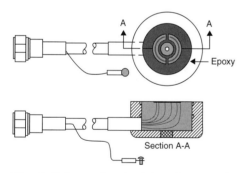

Fig. 2 Surface strip electrical resistance probe schematic. Source: Metal Samples, Munford, AL

(EIS) techniques; ΔE and Δi are the changes in potential and current density that are caused by applying either a potential or current, respectively; B is the Stern-Geary constant; β_a and β_c are the anodic and cathodic Tafel constants, respectively; and i_{corr} is the corrosion current density from which a corrosion rate may be calculated.

A typical three-electrode LPR probe is shown in Fig. 3. As configured here, the three electrodes are cylindrical and of equal surface area. This type of probe can be used equally well with most of the other electrochemical measurement techniques discussed subsequently (except for hydrogen permeation and potential probe).

Electrochemical Noise (EN). The EN technique involves the measurement of spontaneous changes in current and potential due to natural variations in the corrosion current and the corrosion potential. The corrosion rate is estimated from the resistance noise, R_n. Instability in the corrosion processes due to localized corrosion can also be identified by the technique. For this form of localized corrosion, the risk of pitting on the metal surface is derived from the EN and the harmonic distortion analysis (HDA) data. This value is termed the pitting factor (PF). The PF, which is calculated from the ratio of the standard deviation of the EN corrosion current divided by the average corrosion current from the LPR technique, refers to the risk of localized attack (pitting) on the metal surface and is always examined together with corrosion rate. The PF has a value between 0 and 1. Values of PF < 0.1 indicate a low probability of pitting. For PF = 0.1 to 1, the system will be in a pitting regime rather than a regime of general corrosion.

Harmonic distortion analysis (Ref 5) is an extension of the LPR technique that uses a low-amplitude, low-frequency sine wave to polarize the electrodes. Its use allows for the measurement of an HDA corrosion rate and, more importantly, the measurement of the Tafel constants, β_a and β_c, and the calculation of B, the Stern-Geary constant.

Electrochemical impedance spectroscopy involves the analysis of the impedance of a corroding metal as a function of frequency. Analysis of the data focuses on determining a value of the charge transfer resistance that is analogous to the polarization resistance (Ref 3). This technique also uses Eq 1 to calculate corrosion rate.

Galvanic current probes rely on the difference in corrosion potential between two dissimilar metals that are coupled together through a zero-resistance ammeter (ZRA) as shown in Fig. 4. Typically, the more noble metal becomes the cathode and the more active metal becomes the anode. If it is important to match one part of the probe couple to the structure of interest, it would be best to make the structure's metal the anode. While it may be difficult to find a couple that would simulate the corrosion of the structure and give quantitative corrosion rates for the structure, a more likely use would be to use the galvanic current probe as a qualitative indicator of soil corrosivity.

A recently developed multielectrode array corrosion probe (Ref 7, 8) measures a coupling current across a resistor placed between any two sensors in the array. The sensors are of identical composition, which would normally lead to no coupling current unless the sensors are in different environments or are experiencing different types or levels of corrosion. While individual corrosion currents can be measured for each pair of sensors in the array, it is the cumulative corrosion rate for the entire probe that would be used to monitor corrosion in soils.

The hydrogen permeation technique is used to detect the amount of hydrogen diffusing through a metal membrane. This hydrogen is typically produced by the corrosion reaction or by metal charging operations such as cathodic protection. While it is not always used to measure corrosion rate, it can be used to study soil corrosion as one investigator (Ref 9) did. He used it to attempt to measure the presence of microbiologically influenced corrosion.

Potential probes are sometimes used to assess the qualitative corrosion state of structures but cannot be used to measure corrosion rates.

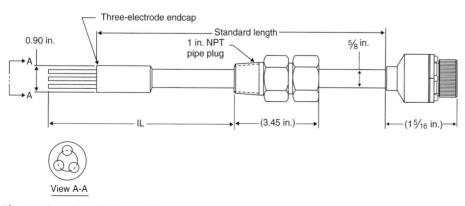

Fig. 3 Three-electrode linear polarization resistance probe schematic. Dimensions given in inches. IL, insertion length. NPT, American National Standard Taper Pipe Thread. Source: Metal Samples, Munford, AL

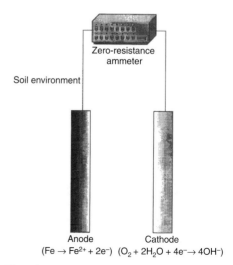

Fig. 4 Galvanic probe. Source: Ref 6

Nonelectrochemical Techniques—Examples of Uses in Soils

The ER technique was coupled to a thin film resistance probe to measure the corrosion rate of an underground pipeline (Ref 10). The corrosion probe consisted of a 1 μm thick carbon steel layer deposited onto a glass substrate and is shown as configured in Fig. 5. A titanium dioxide (TiO_2) interlayer was used to improve the adhesion of the steel to the glass. The final probe configuration consisted of two ER probes and one mass loss coupon. One of the drawbacks of this probe is that it could have a short lifetime due to the thin (1 μm) sensor element. Laboratory research did show, however, that the probe was responsive to corrosion rates ranging from 0.013 to 220 mm/yr (0.5 to 8660 mils/yr).

The same thin film ER probe was also used in a field test that was located near a liquefied natural gas (LNG) pipeline that had been coated with polyethylene (PE) and was cathodically protected. The corrosion rate measured by a probe that was not connected to the pipe or the CP system was 0.088 mm/yr (3.5 mils/yr), a value that was consistent with that reported for steel in soils. A probe attached to the pipeline, and thus the CP system, showed an order of magnitude lower corrosion rate, 0.008 mm/yr (0.3 mil/yr), as would be expected.

Another type of ER probe was used to determine the effectiveness of a coated underground pipeline CP system (Ref 11, 12). A recommendation for the area of the ER sensor was that it should be about the size of possible coating defects in the pipeline being monitored. For this study, that area was 25 cm^2 (3.9 $in.^2$). The thickness of the sensor should be appropriate to last for the design life of the pipeline. This requires some prior knowledge of the soil corrosivity, and for this study led to the selection of 0.64 mm (25 mils) for the sensor thickness for this probe. At one of the test sites the corrosion rate measured for the unprotected probe was 71 μm/yr (2.8 mils/yr) and 5 μm/yr (0.2 mil/yr) when the probe was connected to the CP system. Recommendations from this study (Ref 12) were that a minimum of 12 months monitoring was necessary in order to observe corrosion rate trends.

A specially constructed ER probe, made using a high molecular weight polymer with continuous micropores to absorb water, was used to measure soil moisture content (Ref 13). Soil moisture can cause decreases in soil resistivity, resulting in higher corrosion rates. This probe was found to work well in loam soils with 10 to 60% moisture.

A series of 20 ER probes were used at eight different stations of a PE-coated ductile iron pipeline (Ref 2). These probes were used in clean sand fill and in native soil. Probes were used both under and above the PE coating in the ungrounded (i.e., not connected to the pipe), grounded, and grounded and cathodically protected configurations. The ungrounded ER probes were observed to correctly predict the low corrosion rates under the PE coating and the lower corrosion rates in the clean fill as opposed to native soil. Problems occurred, however, when grounded probes had anomalously high corrosion rates, probably due to galvanic effects from the different compositions of the steel probes and the ductile iron pipe. A suggestion by the authors (Ref 2) was to use probes with the same composition as the equipment being monitored and to use ER probe corrosion rates in a comparative rather than absolute mode.

Electrochemical Techniques—Examples of Uses in Soils

Several investigators have designed and used multisensor probes (Ref 9, 14–18) (Fig. 6). Many were modifications or improvements (Ref 9, 14–16) on the original Novaprobe (Ref 17). Common features of these types of probes is that some include sensors for measuring hydrogen absorption, resistivity, and open circuit potential (OCP). They also typically include three sensors for using techniques such as LPR, EIS, Tafel, galvanostatic, and pulse. The Novaprobe (Ref 17) used only soil resistivity, soil redox potential, and pipe to soil potential to characterize corrosion susceptibility, not corrosion rate.

In addition to LPR and EIS corrosion rate measurements, one of the multisensor probes (Ref 14, 15) was used to detect sulfate-reducing environments that identified sulfate-reducing bacteria (SRB) activity. Large electrode capacitances (Ref 9) measured with EIS were interpreted to indicate the formation of porous iron sulfide scales during SRB corrosion. This has the possibility of being used as a SRB indicator. Hydrogen permeation measurements were used to evaluate the risk of hydrogen-assisted stress-corrosion cracking.

Another version of the multisensor probe (Ref 16) was used with EIS as the main corrosion measurement technique. With that probe/technique combination, the investigators were able to rank the soil corrosivity at four different test sites with fair agreement between the electrochemically determined corrosion rates and those of mass loss coupons. The electrochemical corrosion rates varied from 1.3×10^{-2} to 3.3×10^{-2} $g/m^2 \cdot h$ compared with 3.4×10^{-2} to 5.5×10^{-2} $g/m^2 \cdot h$ for mass loss coupons.

Fig. 5 Thin film electrical resistance probe. Source: Ref 10

Fig. 6 A modified Novaprobe showing the soil hydrogen permeation electrode (SHPE), four stainless steel (SMO) rings for measuring soil resistivity and redox potential, and two sets of three electrodes for conventional electrochemical measurements. Source: Ref 9

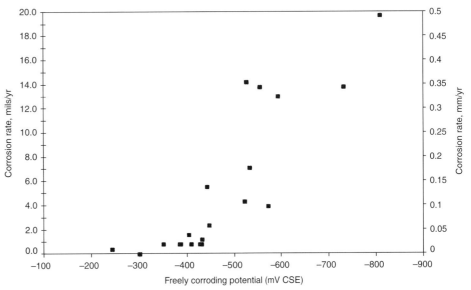

Fig. 7 Corrosion rates of unpolarized coupons (measured using electrochemical impedance spectroscopy) vs. freely corroding potential throughout the test section. Source: Ref 18

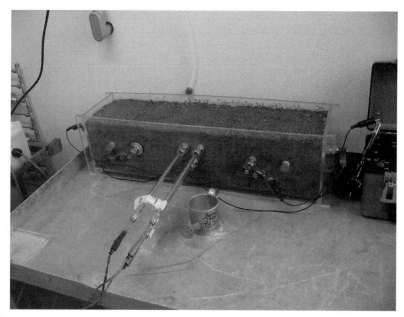

Fig. 8 Soil test cell with three corrosion electrodes and four resistivity electrodes

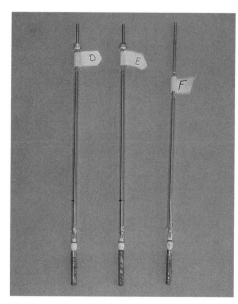

Fig. 9 Soil corrosion electrodes used in the soil test cell in Fig. 8.

A modified Novaprobe was coupled with EIS to measure soil corrosion rates (Ref 18). Investigators preferred the EIS technique over other electrochemical techniques because of the following reasons: (a) the EIS technique is nondestructive because the 10 mV signal is not large enough to drive either the anodic or the cathodic reactions; (b) the EIS equipment can change the polarization state of the working electrode from freely corroding to anodic or cathodic while measuring the corrosion rate; and (c) the measurements are instantaneous (<20 min/measurement). The EIS-determined corrosion rates gave a good correlation to OCP with corrosion rates the highest at more negative potentials (Fig. 7). Corrosion rates more than 0.025 mm/yr (1 mil/yr) occurred only where soil resistivity was less than 10,000 $\Omega \cdot cm$.

The EIS technique was also coupled with a probe made from sensors cut from an epoxy-coated steel pipeline (Ref 19). The necessity of having an appropriate equivalent circuit in order to interpret the EIS measurements was stressed in this investigation.

Laboratory studies of soil corrosion were conducted using the LPR/EN/HDA techniques to measure the simulated external corrosion of gas transmission pipelines (Ref 20). The test cell (Fig. 8) was a typical soil box as specified in ASTM G 57, "Field Measurement of Soil Resistivity Using the Wenner Four-Electrode Method." Three cylindrical electrodes of X42 gas transmission pipeline steel, Fig. 9, were inserted into the soil box as the corrosion rate monitoring electrodes. Four stainless steel electrodes were inserted at and near the ends of the test cell in order to measure soil resistivity, which could then be correlated with corrosion rate. Figure 10 shows the effect of added water and soil resistivity on the corrosion of X42 gas transmission pipe cylindrical electrodes as a function of time in soil + 1 wt% NaCl. In this environment, after the addition of only 300 mL (2 wt%) additional water, the resistivity dropped from 35,000 to 10,000 $\Omega \cdot cm$ and the corrosion rates began increasing. At approximately 2400 mL (~16 wt%) of added water, the resistivity dropped to 1000 $\Omega \cdot cm$ and the corrosion rates became constant, possibly due to the lack of availability of oxygen in the water-saturated soil. Data in Table 1 show a good agreement between the electrochemical corrosion rates and gravimetric corrosion rates from mass loss samples exposed at the same time.

A recently completed field study of soil corrosion coupled a three electrode probe, Fig. 11, with the LPR/EN/HDA techniques (Ref 21).

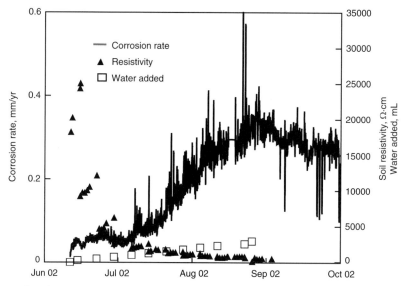

Fig. 10 Effect of added water on the soil resistivity and corrosion rate of X42 steel transmission pipeline electrodes in soil + 1 wt% NaCl

Table 1 Corrosion rates in soil compared using two different techniques

Environment	Electrochemical corrosion rates		Gravimetric corrosion rates	
	mm/yr	mils/yr	mm/yr	mils/yr
Soil	0.005	0.197	0.004	0.157
Soil + 1 wt% NaCl	0.349	13.74	0.364	14.33

This probe was made by encasing steel rod in epoxy. Three of these probes were buried at different locations: (a) between the CP anode and the protected pipeline (probe 1), (b) very near to the protected pipeline (probe 2), and (c) far from the protected pipeline (probe 3). Note that none of the soil probes were connected to the pipe or CP system. Figure 12 shows that two different tests were conducted using these probes: the first was during dry conditions (August 2003) and the second was during wet conditions (January 2004). Corrosion rates at the beginning of each test, where there was no applied CP and where pipeline potentials were the least negative, ranged from 0.02 to 0.03 mm/yr (0.79 to 1.2 mils/yr) for probes 1, 2, and 3. Figure 13 shows that the PF were on the order of 0.01 for probes 1, 2, and 3, indicating that pitting corrosion was not likely.

Galvanic corrosion rate probes (Ref 6) are claimed to be the least complicated next to mass loss coupons. Using carbon steel (CS)-stainless steel (SS) and CS-Cu galvanic pairs in the probes, the investigators showed a good correlation between galvanic current and mass loss of separate specimens (Fig. 14a, b).

Potential Sources of Interference with Corrosion Measurements

One of the concerns with measuring corrosion using any electrochemical or nonelectrochemical technique is whether there is some outside influence affecting this measurement. Stray currents from power sources and lines and electrical fields generated by CP systems are two sources of concern.

The presence of an impressed alternating current (ac) voltage on a thin film ER probe caused an increased corrosion rate despite the presence of CP (Ref 10). The ac signal varied from 1 to 4 V_{rms} and increased the corrosion rate from 0 to 0.008 mm/yr (0 to 0.31 mil/yr). The impressed ac voltage did not interfere in the measurement of the corrosion rate but rather altered the corrosion rate. The thinness of the probe allowed for the measurement of such a low corrosion rate.

The main reason for conducting the LPR/EN/HDA study (Ref 20, 21) described previously was to determine whether the electrical fields generated by CP of a pipeline caused interferences in the corrosion rate measurements. At the test site, CP was applied to a 50 mm (2 in.) diameter fusion bonded epoxy coated steel pipe using a high-silicon cast iron anode to change the pipeline potential. Electrochemical corrosion rate measurements were then made using the LPR/EN/HDA techniques at the three locations described above. Corrosion rates are shown in Fig. 12 as the pipeline potential was increased to approximately 2 V versus the copper/copper sulfate electrode. Comparing the corrosion rates before the application of CP to those at different levels of CP shows that there were no events in

Fig. 11 Three-electrode soil corrosivity probe

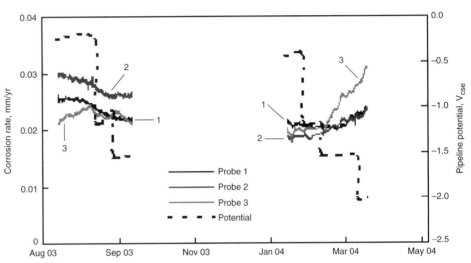

Fig. 12 Linear polarization resistance corrosion rates of the three soil probes (not connected to pipeline) and the pipeline potentials of the two experiments as a function of time

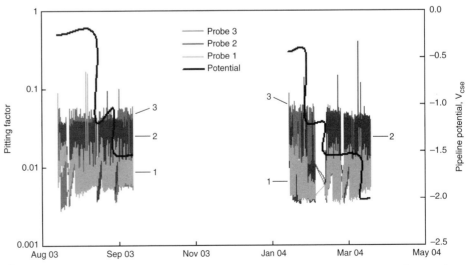

Fig. 13 Pitting factors of the three soil probes and the pipeline potentials of the two experiments as a function of time

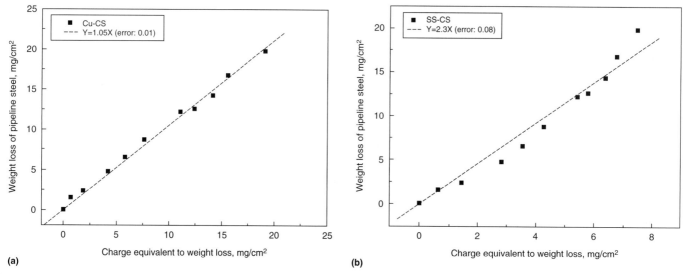

Fig. 14 Relationship between the weight loss as detected from the corrosion probe and measured weight loss for the specimens. (a) Carbon steel (CS)-Cu probe. (b) Carbon steel (CS)-stainless steel (SS) probe. Source: Ref 6

the corrosion rate-time data that could be correlated to the application of CP current, the increase in CP current, or any other cause. The same is true of the PF data in Fig. 13. The conclusion is electrochemical corrosion rate probes will have no interference problems when monitoring the corrosivity of the soil near a cathodically protected pipeline when using the LPR, EN, or HDA techniques. Note that if the probes were electrically connected to the CP system, the LPR, EN, and HDA measurements would have been affected.

Both the ER technique (Ref 12) and a multielectrode array galvanic current technique (Ref 7, 8) were used while coupled to a CP system. Corrosion rates decreased due to the CP, suggesting that neither technique was negatively affected by the presence of CP.

Summary

- A number of techniques have been identified as being suitable for monitoring corrosion in soils: the LPR, EN, HDA, ER, EIS, and galvanic current techniques.
- No interference was detected when using the LPR/EN/HDA techniques on soil corrosion probes located near cathodically protected structures.
- ER and galvanic current techniques were able to be used to measure corrosion rate of cathodically protected structures.
- Impressed ac voltages altered the corrosion rate measurements using the ER technique.

REFERENCES

1. M.K.A. Flitton and E. Escalante, Simulated Service Testing in Soil, *Corrosion: Fundamentals, Testing, and Protection,* Vol 13A, *ASM Handbook,* ASM International, 2003, p 497–500
2. M.J. Schiff and B. McCollom, "Impressed Current Cathodic Protection of Polyethylene-Encased Ductile Iron Pipe," *Mater. Perform.,* Vol 32 (No. 8), 1993, p 23–27
3. D.A. Eden and A. Etheridge, "Corrosion Monitoring as a Means of Effecting Control of CO_2 Corrosion," Paper 01057, presented at Corrosion 2001 (Houston, TX), NACE International, 2001
4. J.R. Scully and R.G. Kelly, Methods for Determining Aqueous Corrosion Reaction Rates, *Corrosion: Fundamentals, Testing, and Protection,* Vol 13A, *ASM Handbook,* ASM International, 2003, p 68–86
5. J. Devay and L. Meszaros, Study of the Rate of Corrosion of Metals by a Faradaic Distortion Method, Part I, *Acta Chim., Acad. Sci. Hung.,* Vol 100 (No. 1–4), 1979, p 183–202
6. Y.-S. Choi, M.-K. Chung, J.-G. Kim, "A Galvanic Sensor for Monitoring the Corrosion Damage of Buried Pipelines, Part 1: Laboratory Tests to Determine the Correlation of Probe Current to Actual Corrosion Damage," Paper 03438, presented at Corrosion 2003 (Houston, TX), NACE International, 2003
7. X. Sun, "Online Monitoring of Corrosion under Cathodic Protection Conditions Utilizing Couples Multielectrode Sensors," Paper 04094, presented at Corrosion 2004 (Houston, TX), NACE International, 2004
8. X. Sun, "Real-Time Monitoring of Corrosion in Soil Utilizing Coupled Multielectrode Array Sensors," Paper 05381, presented at Corrosion 2005 (Houston, TX), NACE International, 2005
9. L.V. Nielsen, B. Baumgarten, N.K. Bruun, L.R. Hilbert, C. Juhl, and E. Maahn, Determination of Soil Corrosivity Using a New Electrochemical Multisensor Probe, *Eurocorr' 98: Solutions to Corrosion Problems,* Oct 1998, p 230–235
10. Y. Kim, D. Won, H. Song, S. Lee, Y. Kho, "Utilization of Thin Film Electric Resistance Probe for Underground Pipeline Corrosion Rate Measurement, *Proceedings of the 14th International Corrosion Congress,* Corrosion Institute of South Africa, Kelvin, South Africa, Oct 1999, p 73
11. N.A. Khan, "Use of ER Soil Corrosion Probes to Determine the Effectiveness of Cathodic Protection," Paper 02104, presented at Corrosion 2002 (Houston, TX), NACE International, 2002
12. N.A. Khan, "Using Electrical Resistance Soil Corrosion Probes to Determine Cathodic Protection Effectiveness in High-Resistivity Soils," *Mater. Perform.,* Vol 43 (No. 6), 2004, p 20–25
13. C. Minte, H. Yui, and S. Chungteh, Study on Electric Resistance Type of Soil Moisture Content Sensor, *J. of Agriculture and Forestry,* Vol 51 (No. 1), 2002, p 15–27
14. L.V. Nielsen and N.K. Bruun, Screening of Soil Corrosivity by Field Testing: Results and Design of an Electrochemical Soil Probe, *Eurocorr' 96: Physical and Chemical Methods of Corrosion Testing,* European Federation of Corrosion, 1996, p 21
15. L.V. Nielsen, "Microbial Corrosion and Cracking in Steel: Assessment of Soil Corrosivity Using an Electrochemical Soil Corrosion Probe," Report NEI-DK-3281, Danmarks Tekniske University, 1998
16. M.C. Li, Z. Han, and C.N. Cao, "A New Probe for the Investigation of Soil Corrosivity," *Corrosion,* Vol 57 (No. 10), 2001, p 913–917
17. M.J. Wilmott, T.R. Jack, J. Guerligs, R.L. Sutherby, O. Diakow, and B. Dupuis, *Oil Gas J.,* April 3, 1995, p 54–58

18. S. Gabrys and G. Van Boven, "Use of Coupons and Probes to Monitor Cathodic Protection and Soil Corrosivity," *Corrosion Experiences and Solutions,* NACE International, 1998, p 43–59
19. G. Hammon and G. Lewis, "Electrochemical Impedance Spectroscopy Studies of Coated Steel Specimens in Soils," *Corros. Prev. Control,* March 2004, p 3–10
20. S.J. Bullard, B.S. Covino, Jr., J.H. Russell, G.R. Holcomb, S.D. Cramer, and M. Ziomek-Moroz, "Electrochemical Noise Sensors for Detection of Localized and General Corrosion of Natural Gas Transmission Pipelines," DOE/ARC-TR-03-0002, U.S. Dept. of Energy, Dec 2002
21. S.J. Bullard, B.S. Covino, Jr., S.D. Cramer, G.R. Holcomb, M. Ziomek-Moroz, M.L. Locke, M. Warthen, R.D. Kane, D.A. Eden, and D.C. Eden, "Electrochemical Noise Monitoring of Corrosion in Soil Near a Pipeline under Cathodic Protection," Paper 04766, presented at Corrosion 2004 (Houston, TX), NACE International, 2004

SELECTED REFERENCES

- S.A. Bradford, *CASTI Handbook of Corrosion Control in Soils,* co-published by CASTI Publishing, Inc. and ASM International, 2000
- J.H. Fitzgerald III, "Probes for Evaluating CP Effectiveness on Underside of Hot Asphalt Storage Tanks, *Mater. Perform.,* Vol 37 (No. 12), 1998, p 21–23
- Y. Miyata and S. Asakura, Corrosion Monitoring of Metals in Soils by Electrochemical and Related Method, Part 1: Monitoring of Actual Field Buried Metal Structures and Electrochemical Simulation with Corrosion Probes and Pilot Pieces, *Zairyo-to-Kankyo (Corros. Eng.),* Vol 46 (No. 9), 1997, p 541–551
- M. Yaffe and V. Chaker, Corrosion Rate Sensors for Soil, Water, and Concrete, *Innovative Ideas for Controlling the Decaying Infrastructure,* V. Chaker, Ed. (Houston, TX), NACE International, 1995

Cathodic Protection of Pipe-Type Power Transmission Cables

Adrian Santini, Con Edison of New York

PIPE-TYPE CABLES have reliably served the electric needs of large cities for many years. They are the method by which large amounts of electric power are brought into a city environment where high-voltage transmission towers cannot be constructed. These high-voltage cables are buried beneath city streets and supply power to a network of substations. In turn, the substations reduce the voltage to levels that can be used by industrial and residential customers throughout the city.

A pipe-type cable is made up of a steel pipe that contains three insulated conductors and pressurized dielectric fluid, which is used to cool the cables and maintain the integrity of their insulation.

The reliability of these cables is contingent upon all three components working together. Power cannot travel along the conductors unless the insulation is effective. The insulation cannot be effective without the pressurized fluid. The fluid cannot be pressurized unless the structural integrity of the pipe is maintained. It is for these reasons that cathodic protection (CP) is used to protect the pipe against corrosion. Failing to do so may result in dielectric fluid leaks, possible interruption of service to customers, as well as financial loss to the power transmission cable operator.

To cathodically protect a pipe-type cable, the impressed current method of cathodic protection is used. As is the case with any other buried or submerged metallic structure, it is necessary to keep its direct current (dc) potential more negative (approximately -0.5 V) than the surrounding earth. This can be accomplished by various methods, but for pipe-type cable all systems must have one thing in common. They must be capable of safely conducting high alternating current (ac) fault currents to ground. Such faults occur when the conductor comes in electrical contact with the inner surface of the pipe as a result of cable insulation failure. In such rare cases, the resultant ac fault current must be given a path to ground to protect other equipment and personnel. To do so, the pipe must be electrically continuous for its entire length and be connected to ground at each substation.

Resistor Rectifiers

Direct current isolation and ac conduction seem to be opposing requirements, but they can both be achieved in several ways. The oldest and perhaps the most widely used method is the resistor rectifier (Fig. 1).

In this method, a resistor bar is inserted in the cable connection between the pipe and the station ground mat. A rectifier (R) is connected across this bar, impressing a dc voltage on it. The resistance of the bar is very low (typically 0.004 Ω) to minimize pipe-to-ground voltage during ac faults. Therefore, the rectifier must circulate at least 125 amperes dc through R in order to obtain the required -0.5 V, pipe-to-ground. Since the station ground is connected in parallel with the resistor bar, a portion of the rectifier current will follow the ground path, be discharged from the station ground mat, travel through the earth, onto the pipe surface, and along the pipe back to the rectifier. This is the current that is actually required to protect the pipe. Unfortunately, this current will also corrode the substation ground mat and for this reason it must be minimized.

The magnitude of this current depends on the sum of the ground mat-to-ground resistance plus the resistance of the earth path plus the pipe-to-ground resistance. Since the ground mat-to-ground resistance cannot be increased and the earth resistance cannot be realistically changed, the only way to maximize the resistance of this path is to ensure that the resistance of the pipe coating is as high as practical. This is why pipe-type cable specifications often call for coating resistance values as high as 186,000 $\Omega \cdot m^2$ (2,000,000 Ω-ft^2). At such high coating resistance, the current requirement for cathodic protection is very low. For example, at 186,000 $\Omega \cdot m^2$ (2,000,000 Ω-ft^2), the pipe-to-ground resistance of a 25.4 cm (10 in.) pipe, whose length is 9.66 km (6 miles), is approximately 24 Ω. The current required (I_R) to shift the potential of such a structure would therefore be:

$$I_R = \frac{0.5 \text{ V}}{24 \text{ }\Omega} = 0.02 \text{ A}$$

This of course is an idealized value since it does not take into account the other resistances in the earth path. However, since these other resistances are small compared with the coating resistance, the current requirement would still be very small with all resistances considered.

Note that in this example, in order to have 0.02 A traveling in the ground path, 125 A must travel though the 0.004 Ω resistor to impress the same -0.5 V on it. One sees then that, in the resistor rectifier method of cathodic protection, most of the rectifier current must circulate through the resistor in order to produce the small amount of current that is actually required for CP.

Polarization Cells

If the resistor was eliminated, the rectifier could be much smaller and the number of rectifiers could be reduced. However, since the pipe must remain grounded, some other device must take the place of the resistor. This device would have to produce the same pipe-to-ground voltage drop for CP using a minimal amount of dc current, while allowing the cable to remain effectively grounded for ac fault current safety. One such device is an electrochemical device, the polarization cell (PC) (Fig. 2).

A typical PC consists of two electrodes, each composed of 14 nickel plates, immersed in a 30% solution of potassium hydroxide. Like the resistor bar, the PC is installed between the pipe and the station ground. Under normal conditions, with the rectifier turned on, dc current flows through the cell and causes a polarization film of hydrogen to be formed on the negative electrode. As this film builds, it increases resistance to further dc flow and eventually blocks all but a small leakage (milliamperes) of dc current. At this minimum level of dc current flow, the blocking voltage for each cell is approximately 1.0 V. It should be noted that the nominal blocking voltage of the polarization cell is 1.7 V; however, this value is reduced by the varying amounts of ac current that flows from the pipe, through the cell, to ground. This ac current is induced on the pipe by the unbalance in the loads on each of the three phase conductors in the pipe-type cable. This continuous ac flow acts to either break down the polarization film or prevent it from fully forming; for this reason, the blocking voltage across a polarization cell is much closer to 1.0 V. This means that, as the rectifier output increases, there is a corresponding increase in the voltage across the cell until 1.0 V is reached. Above this value, any additional dc from the rectifier will simply flow through the cell, corrode the positive plates, and *not* increase the pipe-to-ground voltage. This is why the current flow through the cell must be monitored to ensure that it is at its minimum.

Sometimes, it is impossible to keep the voltage across the cell below 1.0 V. In such cases, more than one PC may have to be installed in series in order to achieve the required blocking voltage that will minimize dc flow through each cell.

In addition, in order to maintain the ac fault current rating of the PC, the level and specific gravity of the potassium hydroxide solution must be kept at their specified values. The plates must be periodically inspected, as well, to ensure that they have not deteriorated thereby decreasing their surface area.

Isolator-Surge Protector

The high maintenance considerations led the industry to develop other devices to serve the same function as the PC; one of these is the isolator-surge protector (ISP) (Fig. 3). The ISP is a solid-state device, and the basic circuitry is shown in Fig. 4.

During normal operations, the capacitor blocks up to 12 V dc. If the voltage goes higher than 12 V, the gate circuit will turn on one of the two thyristors depending on polarity, dc current will begin to flow, and cathodic isolation will be interrupted until the voltage again drops below 12 V. For unbalance currents of up to 90 A ac (depending on the steady-state current rating selected), the current flows through the inductor and capacitor. If the ac increases above these levels, such as would occur during fault conditions, the voltage across the capacitor would exceed 12 V at which point the thyristors will turn on alternately every one-half cycle allowing the fault current to go to ground until the fault clears. When the ISP is subjected to lightning surges, the voltage developed across the inductor rises to a value that places the surge protector into conduction, thereby diverting most of the current to ground through the surge protector.

Although the blocking voltage of the ISP is 12 V dc, this amount is also reduced by the ac unbalance current that flows from the pipe, through the ISP, to ground. However, the reduction is rarely enough to render the ISP ineffective for dc isolation.

A more important factor to consider, when designing an ISP application, is the maximum ac unbalance current that is likely to flow through the ISP. This is because, if this current exceeds the ISP rating, the thyristors will turn on, shorting out the ISP and, in effect, eliminate cathodic protection on the pipe-type cable.

Field Rectifiers

In the previous discussion on resistor rectifiers, it was shown that if the pipe-type cable were grounded through a 0.004 Ω resistor bar, a large rectifier (most of whose output would circulate through the resistor) would be needed for adequate cathodic protection. If, however, polarization cells or ISPs were substituted for the resistor, this circulating current would be eliminated and the rectifier(s) would be sized only to supply the small amount of current required for cathodic protection. These rectifiers and their associated anode-groundbeds would be located in the "field" along the route of the pipe-type cable, allowing for better current distribution on the pipe surface.

These field rectifiers (Fig. 5) could also have their output current increased or decreased as required. This is not the case with resistor-rectifiers where, in order to increase the cathodic protection current, a proportional increase in the current flowing through the resistor bar would be required. In this example, this would mean that to increase the protective current from 0.02 to 0.04 A would require an additional 125 A flow through the resistor bar. For this reason, resistor-rectifiers were designed to provide minimum CP levels at each substation end of the pipe-type cable. These protection levels can only decrease as one moves farther away from the substation, even if one assumes that the coating is in excellent condition. This, of course, is never the case, and as the pipe coating expectedly deteriorates over time the cathodic protection levels decrease even further.

There is one other advantage to replacing resistor rectifiers and that is preservation of the substation ground mat. Since the resistor rectifiers use the substation ground mat as a "sacrificial" anode for discharging the required CP current, although the current is designed to be very small and the substation ground mat has a large surface area, current discharge will cause the mat to be damaged over time.

Stray Currents

Having discussed the various ways that CP can be applied to pipe-type cables, stray dc currents

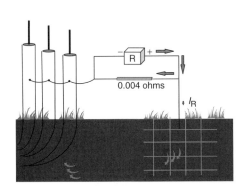

Fig. 1 Resistor rectifier. The rectifier (R) circulates dc current through two parallel paths, the 0.004 Ω resistor bar and the ground path. I_R is the current required to cathodically protect the pipes. The three individual pipes above ground entering the substation combine underground into a single pipe containing the three conductors.

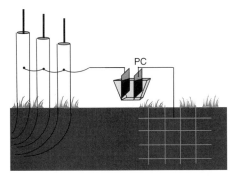

Fig. 2 Polarization cell. The polarization cell (PC) is an electrochemical device that blocks dc and passes ac current. It replaces the resistor bar and makes it possible to reduce the number and size of rectifiers. The three individual pipes above ground entering the substation combine underground into a single pipe containing the three conductors.

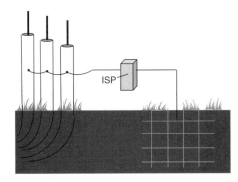

Fig. 3 Isolator surge protector (ISP). The ISP is a solid-state device that blocks dc and passes ac current. It also replaces the resistor bar and requires little maintenance. The three individual pipes above ground entering the substation combine underground into a single pipe containing the three conductors.

will, almost without exception, interfere with their protection. This is because pipe-type cables operate within large cities in close proximity to dc subways, commuter railroads, and trolley lines. A "cause and effect" summary of this problem is shown in Fig. 6.

Pipe-type feeders, like other buried metallic structures, pick up stray currents because they are exposed to voltage gradients in the earth caused by I_R (voltage) drops in the running rail of dc railroads. These I_R drops cannot be eliminated because the longitudinal resistance of the running rails can never be 0. Similarly, the voltage gradients in the soil due to these I_R drops can never be eliminated because the electrical resistance of the wooden or concrete ties, which support the running rails, cannot be infinite. Although all nearby buried metallic structures pick up these stray currents, pipe-type cables have a particular affinity for them. Their low longitudinal resistance, in effect, makes them good parallel conductors to the rails for returning these stray currents to the transit substations.

For any given voltage gradient in the earth, the amount of strays picked up depends on the pipe-to-ground resistance of the feeders. This resistance in turn depends on:

- *Method of grounding* the pipe-type feeder in the substation (resistor, PC, or ISP). Station ground mats can themselves pick up stray currents. If the resistor bar is still in place, these strays flow through the bar onto the pipe and cause the pipe to corrode at the point of discharge. Replacing the resistor with PC or ISPs virtually eliminates stray current pickup through station ground mats.

- *Accidental contacts* with other buried metallic structures. Periodic electrical surveys of these feeders are conducted to find the locations of any accidental metallic contacts between the feeder pipe and other buried metallic structures. These areas are excavated and the contacts cleared.

- *The coating quality* on the feeder pipe. Even if all stray pickup is eliminated through station ground mats, and by the elimination of accidental contacts, some strays will be picked up through the coating because its resistance cannot be infinite. In the case of two parallel forced-cooled feeders installed in one trench, two 25.4 cm (10 in.) pipes and two 12.7 cm (5 in.) pipes acting as one conductor for the purpose of calculating stray currents, a relatively high coating resistance of 200,000 $\Omega \cdot ft^2$ would result in a pipe to

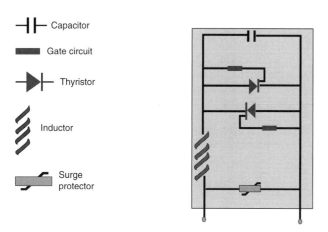

Fig. 4 Isolator surge protector components. The various components work together to keep the pipe-type cable effectively grounded during ac faults.

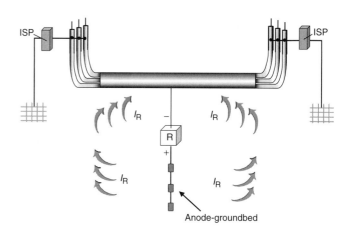

Fig. 5 Field rectifier and groundbed. Rectifiers are removed from their substation locations and are replaced by fewer, smaller rectifiers in the "field." The current required by the pipe (I_R) is impressed by the rectifier via anode-groundbeds. One rectifier as shown could theoretically protect the pipe, but in order to lower the current impressed by each rectifier, smaller rectifiers at multiple locations are preferred. The single underground pipe-type transmission cable with a mild steel outer wall branches into three separate pipes at the substation termination. These are generally stainless steel.

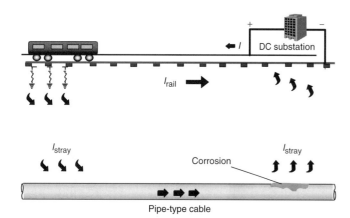

Fig. 6 Stray current interference. Most of the current that powers dc trains returns to the substation via the rails. A small portion of this current "strays" from the rails and is discharged into the ground. The pipe-type cable provides a low resistance return path. Localized corrosion can occur where the stray current is discharged by the pipe-type cable as it flows back to the train's dc substation.

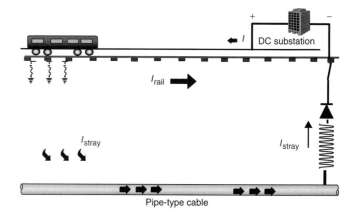

Fig. 7 Stray current drain bonds. Drain bonds prevent stray current picked up by the pipe-type cables from being discharged back into the ground. The drain bonds provide a metallic return path to the substation. The diode in the circuit prevents reverse currents.

ground resistance of approximately 5 Ω per mile. Since the stray current pickup areas can extend for several miles and since old dc rail systems often impress large voltage gradients in the soil, it is possible to pick up considerable amounts of stray current through the coating. In this example, pipes are 25.4 and 12.7 cm (10 and 5 in.), coating resistance is 186,000 $\Omega \cdot m^2$, and the resulting pipe to ground resistance is 3.1 Ω/km.

The strays that are picked up travel along the pipe until, due to the change in the polarity of the earth gradients, they are discharged back into the earth resulting in corrosion failures. Stray current discharge areas are generally more localized than pickup areas and occur in the vicinity of dc power stations or other grounded structures. These include bridges, tunnels, overpasses, and support structures for elevated lines. In each case, testing must be done over the route of the feeder to determine the extent of the exposure areas and appropriate action taken.

These remedial actions include the installation of:

- *Impressed current rectifier* systems to overcome the adverse effects of strays. Although these rectifiers can be effective in resolving these problems, they have one major disadvantage: they must be designed to correct the highest level of interference that occurs over a 24 h period. For transit systems these peak levels occur for only a small portion of the day, so the rest of the time the excess rectifier output is being wasted and more importantly can itself cause interference on other structure. Potential-controlled rectifiers have not proven effective in responding to the variations in potentials associated with dynamic stray currents.
- *Stray current drain bonds* to return stray currents to their source (Fig. 7). These bonds are the most effective way of mitigating the adverse effects of stray currents because, unlike rectifiers, they operate only when they need to. Their operation is completely dependent on the level of stray current activity so that during rush-hour periods they may operate at full capacity, but are essentially inactive during non-rush hours and evening or night hours. Their effectiveness also depends on their proper design and location. It is important that the affected transmission cable operator works closely with the local transit authority personnel, both directly and through local coordinating committees to choose the appropriate location for stray current drain bonds. In addition, the bonds must be designed with appropriate resistors so that the interference can be mitigated with minimal current flow and with diodes to ensure that the current flow is unidirectional, from the pipe to the transit authority substation.

More information on this general subject is contained in the article "Cathodic Protection" in *ASM Handbook* Volume 13A, 2003.

Corrosion in the Military

Vinod S. Agarwala, Naval Air Systems Command, U.S. Navy

CORROSION IN THE UNITED STATES MILITARY is a matter of serious concern; it is estimated to cost at least $20 billion per year, and the cost is rising. The U.S. Department of Defense (DoD) has been combating this for years and has now taken a proactive role to minimize these costs. Worldwide, the total annual cost of corrosion is estimated at over $1.9 trillion for 2004 or roughly 3.8% of the world gross domestic product (GDP). This is a heavy tax burden on any country's economy, whether small or large. Currently, the U.S. DoD is facing the cause and effects of corrosion in its budgetary plans where maintenance and repairs of weapon systems, support systems, and military infrastructures have taken a high priority over new acquisitions (Ref 1). In addition, national priorities under present worldwide conflicts have imposed new requirements whereby the aging fleet of weapons systems has to continue to perform in service even longer than their intended service lives.

The articles that follow in this Section of the Handbook provide an overview of the problems, concerns, and solutions the defense agencies and their contractors are dealing with on a routine basis. Attempts have been made to include the following relevant subject matters: military specifications and standards; corrosion of military facilities; ground vehicle corrosion; armament corrosion; design, in-process, and field corrosion problems; high-temperature corrosion/oxidation; military aircraft; engines and turbine blades in naval environments; protective coatings in military applications; corrosion fatigue in military aviation; corrosion of electronic and electrical systems; microbiologically influenced corrosion; and service life and aging of military equipment. These topics are intended to provide a brief account of the types and magnitude of corrosion problems the military attempts to mitigate. They also seek a reasonable way to estimate the corrosion damage and costs to provide a basis to develop policy for new acquisitions and new design and engineering specifications. In particular, this article provides a brief overview of some of these aspects and major issues and actions the U.S. military takes in corrosion control and mitigation.

Introduction

The effects of corrosion across all three services in the United States (Army, Navy, and Air Force) are immense (Ref 1, 2). The U.S. DoD owns a vast array of physical assets, ranging from ground vehicles, aircraft, ships, ammunition, and other support systems to infrastructures such as wharves, buildings, utilities, and many other stationary structures. Hence, there is a major budgetary concern in routine maintenance and operations (Ref 1–8). Also, as military systems are pressed into increasingly longer periods of service in various theaters of operation, limitations on the performance of the equipment and materials selected a few years ago are becoming increasingly evident. They mostly manifest themselves in terms of greater maintenance, repairs, spare parts, and other types of rework that hinder new acquisitions (Ref 2, 7–9).

Historically the effects of corrosion were ignored and considered inconsequential because, whenever problems occurred, maintenance was done by replacement of parts and without any care for how many hours it took to do it. In times of military conflict, the maintenance and repair practices adopted more often than necessary were to make sure that fleet readiness requirements were complied with, and the DoD appropriated the funds necessary for such practices (Ref 2). In the worst-case scenarios, adequate budget was provided to retire nonoperative, unsafe, and/or unreliable equipment without considerations of their life expectancy, and new systems were acquired in their place (Ref 7, 8).

Immense resources are usually needed to find the problems and then fix them to a level of reliability, safety, and predictability during military operations. These mandated actions contribute significantly to the total ownership cost of military systems, thus, leaving very few dollars for innovation and advanced material development. When these recurring costs became overwhelming, the U.S. General Accounting Office (GAO) mandated that the DoD take action to reduce corrosion costs (Ref 1). The DoD is now (2006) establishing the strategic policy that implements best-known corrosion engineering principles and practices in basic systems design and material and processes selection, based on a fundamental acceptance of the fact that defense materials operate in one of the most corrosion-susceptible environments (Ref 9).

DoD Directives on Corrosion. Under the direction of the U.S. Congress, the GAO examined the impact and the extent of corrosion problems within the DoD. The GAO investigations considered the $20 billion annual corrosion cost unacceptable and recommended that the DoD take action and adopt proper corrosion prevention and control practices in all weapons systems, support systems, and military infrastructures. The Congress also enacted law [Title 10 U.S.C. 2228] with a directive to DoD management to focus its effort on reducing life-cycle costs of their weapon systems, facilities, and infrastructures. The Congress also directed the GAO to monitor the DoD's progress on the development and implementation of overall corrosion prevention and control strategy. Thus, the Office of Corrosion Policy and Oversight was created by the Under Secretary of Defense for Acquisition, Technology, and Logistics, whose objectives are to institute policies that reduce costs and review "who is doing what" to reduce corrosion. However, the directives mentioned little or nothing about how to go about actually reducing corrosion in military systems. Recently, the Office of Corrosion Policy and Oversight issued two publications: "Corrosion Prevention and Control Planning Guidebook," Spiral No. 1 and No. 2. These publications provide program managers and design engineers the guidance on how to actually select materials in the design process that will institute corrosion prevention and control and enhance weapons system service life. It would be useful for the DoD to mandate corrosion control designs and engineering practices to enhance operational capabilities, sustainability, readiness, and safety in the design phase of weapons systems.

Almost all materials used in weapon systems are predictably susceptible to corrosion, stress-corrosion cracking (SCC), and corrosion fatigue (CF) (Ref 10, 11). They include predominantly aluminum, magnesium, and titanium alloys, steels, and graphite-reinforced composites. Stress-corrosion cracking and CF are corrosion-related premature failure phenomena that occur in high-strength structural components under internal and/or external stresses. Cracking due to CF occurs only under cyclic or fluctuating operating loads or stresses, while failure from static or slowly rising loads result in SCC and hydrogen embrittlement (HE) or environmentally induced cracking. The latter usually occurs with certain high-strength alloy systems such as steels and titanium alloys sensitive to hydrogen.

In addition, most defense weapon systems have numerous structural joints such as butt, overlaps, fastener, weld, and dissimilar metal joints, which make corrosion prevention and control an essential strategy to prevent crevice, pitting, and galvanic corrosion. The technologies that isolate joints from electrolytic conduction, coating systems that serve as corrosion-resistant barriers or sacrificial coatings, corrosion preventive compounds (CPCs) as temporary environmental masks, and alloys that are less susceptible to corrosion, SCC, CF and HE are required (Ref 11). In order to minimize maintenance costs, condition-based maintenance (CBM) using devices that can detect and monitor corrosion "structural health" as find-and-fix tools are implemented when it makes economic sense (Ref 12).

This article reviews corrosion problems in the DoD and discusses management and maintenance aspects of the present (2006) and past practices that address cost and readiness. It also describes future plans to institute corrosion prevention and control strategies under new DoD directives in engineering design, material selection, and fabrication processes for new acquisitions.

Military Problems

Corrosion is very indiscriminate and affects all military assets including 350,000 ground and tactical vehicles, 15,000 aircraft, 1,000 strategic missiles, 300 ships, and approximately $435 billion worth of facilities (Ref 2, 13). The cost of corrosion in the military is enormous (Ref 13). Although it is difficult to capture the indirect costs such as equipment downtime and reduced readiness and deployment capacity due to corrosion damage, the direct annual cost is between $10 and $20 billion. Corrosion initiated by land, air, or sea operational environments may have different magnitudes of problems, but in due course they become equally devastating in cause and effect with an immense drain on the economy. Although corrosion of warfighting machines is well acknowledged and even documented, what is not acknowledged are the ravages of corrosion on military-base facilities, shore and inland constructions, infrastructures such as gas, oil, steam and water pipelines, piers, docks and docking platforms at ports, hangers and runways, electrical and power lines and equipment, reinforced concrete structures, storage tanks, housing units, and so forth.

The Electrochemical Concept

Corrosion problems in the military environment are not different than those experienced in the civilian sector. The nature and perhaps the magnitude of corrosion, except when military equipment is deployed in combat during conflict, are almost the same. However, in order to appreciate where and why corrosion occurs, some understanding of corrosion fundamentals is necessary. The thermodynamics or free-energy states of materials used in the construction of equipment for both military and nonmilitary applications are always in the negative domain. Figure 1 illustrates "the corrosion cycle" in which all metallic materials exist, eventually returning to their lowest energy states of oxides/ores from which they were originally produced or extracted. All forms of corrosion are undesirable; however, those that are localized or preferential in nature, such as pitting, crevice, intergranular, and galvanic corrosion are the most insidious destroyers of the physical and mechanical properties of a material (Ref 14). Such forms of corrosion may lead to stress cracking, fatigue, fretting, cavitational and hydrogen-assisted damage, or embrittlement cracking, if not managed properly.

Pitting and crevice corrosion are the most frequently observed forms of corrosion in almost all weapon systems. In dynamic structural components they cause the most damage. Pitting is an autocatalytic process and often leads to serious cause-and-effect situations where fatigue is involved; it is particularly damaging to critical aircraft structures such as landing and arresting gears, hinges, and load-bearing struts. Figure 2 shows the reactions and mechanisms of how a material corrodes and forms a pit. The pit grows in a corrosive environment of acid and chloride by an autocatalytic action (shown in reactions 4 and 5), whereby new material corrodes to form chloride, which then hydrolyzes to form HCl and the cycle keeps repeating. This makes the pit grow deeper. In addition to metal dissolution, as shown in reactions 1 and 5, there is another (cathodic) reaction which occurs that involves consumption of electrons produced. This cathodic reaction produces hydrogen, first in the atomic state and then in the molecular state. In the atomic state, H is adsorbed and then absorbed at the apex of stress under mechanical load and lowers the ductility of the region by a process called embrittlement, creating a plastic zone. When the modulus of this plastic zone is exceeded by mechanical working (such as fatigue), a crack initiates and grows rapidly. In high-strength structural components, such as landing gear, this phenomenon may lead to catastrophic failure. The montage in Fig. 2 illustrates where oxides or corrosion products collect due to dissolution of metal in the pit and where hydrogen evolution and absorption occur causing HE and cracking.

Pit-Initiated Cracking. The cracking model, as shown in Fig. 2, illustrates the mechanism of pit formation and growth through corrosion action of chloride and acid by an autocatalytic process. Figure 3 shows the schematic of crack initiation and growth from the pit under cyclic mechanical loading. Under tensile loading (fatigue half cycle) the pit, as shown in Fig. 3(a), leads to the generation of a new surface in its bottom by slip deformation and produces a step D, shown in Fig. 3(b). This step D is very reactive and dissolves very fast, as shown in Fig. 3(c), and flattens out or creates a trough which under compression (second half cycle) pushes back the step to create a crack as illustrated in Fig. 3(d). This process repeats with every cycle and the conjoint actions of corrosion and fatigue lead to a much greater increase in crack growth rates than one would expect under pure fatigue mode. The illustration of such a pit-initiated service failure is shown Fig. 4. The collapse of the main landing gear of an F/A-18 aircraft strut (a load-bearing cylindrical support) of high-strength 300M alloy steel (UNS K44220) under its own weight was due to severe pitting and crevice corrosion found all around the barrel, and high-pulse load fatigue was a serious warning of the shortcomings in the inspection protocol and techniques (Ref 10).

Galvanic Corrosion. In addition to CF and SCC, galvanic corrosion of dissimilar metal joints such as in electrical and electronic boxes, and in structural assemblies caused by exposure

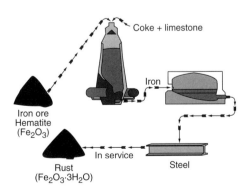

Fig. 1 The corrosion cycle illustrating the need for energy to convert oxides/ores to metallic form. If not protected in-service, the metal reverts back to oxides due to corrosion.

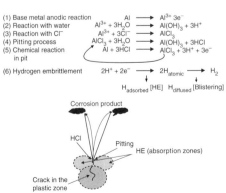

Fig. 2 A cracking model illustrating the mechanism of pit initiation and growth and hydrogen embrittlement (HE)

to wet environment, is the most widespread form of corrosion in almost all weapons systems. Although this problem can be easily solved by separation of the joints with a nonconductive sealant, gasket, or coating system, it continues to cause serious concerns in maintenance and repair. It is almost impossible to avoid dissimilar joints in which one metal is electrochemically more active than the other. The best practice is to choose materials that are as close to each other as possible according to the galvanic series. The objective is to avoid large potential drops, which are the driving force for corrosion to occur between the two metals (see Table 1).

Of course, materials selection for a component is often dictated by the engineering requirements of the physical and mechanical properties and rarely if ever by their electrochemical compatibility (Ref 15–17). However, one can always separate the anodes from the cathodes by using electrical insulators such as sealants and barrier coatings. For underwater systems, such as ships and submarines, in addition to coatings, cathodic protection is commonly used in the Navy. When designing a fastener system, unfavorable area ratios must always be avoided when using dissimilar metal combinations. Active alloys must not be used as small-area components in large surface area cathode materials, such as aluminum bolts, fasteners, and hinges in steel or titanium. Small area anodes would corrode much faster and preferentially, thus compromising the joint.

Aging Systems

Corrosion of military equipment and facilities is an ongoing problem that exacerbates with time and becomes more significant in yearly costs to protect the assets, affecting new procurement and maintenance. The DoD has a large aging fleet of aircraft, ships, land and amphibious combat vehicles, ammunitions, and submarines that require tremendous repair, modification, and upgrade costs to extend their life to perform into the 21st century. For both the Air Force and Navy, aircraft are high-cost items with almost 30% of the fleet more than 25 years old (Ref 8, 10, 11). Since corrosion is a time-dependent phenomenon, the goal of corrosion-control schemes is to slow down the rate of corrosion. The slower the corrosion kinetics the longer the service life. On airborne systems, limitations on the choice of materials and design engineering impose a design life of approximately 20 years. This of course changes and mostly depends on theater of operation, loads and stresses, the combat environment, and the time spent at sea. For naval aircraft, the aircraft carriers are their platforms and base of operation for approximately 90% of their life.

The average age of major battle force ships (carriers, destroyers, cruisers, amphibious ships, and submarines) was 14.5 years in 2000 compared with 13.6 years in 1980 (Ref 8). Today (2006) most ships continue to perform in service at a cost of approximately $1500/per steaming hour, a cost that has not changed much during the last 20 years, except during the early 1990s where reduction in the size of the fleet allowed the service to retire many older ships. Ship systems corrosion is controlled with generous use of coatings and periodic maintenance. The cost of maintenance of ship systems was nearly $15 million per ship in 1984. With the retirement of older ships, the cost has been reduced to $9 million per ship.

Land-based vehicles for the Army, which number in tens of thousands, although not as old as the fleet, suffer from extensive corrosion because of lack of concern for corrosion or its effect on vehicle performance (Ref 8). Consequentially, there have been poor choices made regarding selection of materials, engineering designs, combination of materials in joints, and use of inadequate coating systems to protect them from corrosion effects. In particular, most vehicles, such as light armored vehicles, Bradley and Abraham fighting vehicles systems, and high-mobility multipurpose wheeled vehicles (HMMWV) suffered seriously from galvanic and crevice type corrosion. As a result, their life expectancy was reduced from 10 to 3 years. It may have been called and "aging problem," but it is more an engineering design and corrosion-control problem. Construction materials in these vehicles are usually carbon steels and, if not protected, will corrode irrespective of the implied life or age. Aging of military facilities is another high-cost item and is discussed in the section "Facilities Problems" in this article.

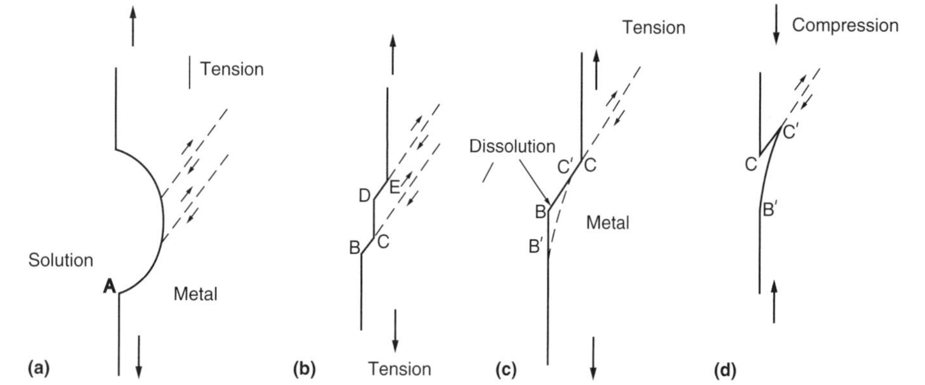

Fig. 3 Corrosion (active path dissolution) action and crack initiation under cyclic mechanical loading. Refer to text for a discussion of stages (a) through (d).

Fig. 4 Pit-initiated in-service failure of a landing gear due to dynamic stresses. The collapse of the high-strength 300M steel main landing gear load barrel was due to severe all-around pitting.

Table 1 Galvanic Series of most commonly used construction materials (metals and alloys) in seawater

Most noble (cathodic)
Platinum
Gold
Graphite
18-8 stainless steel (300 series)
Titanium
Stainless steel (400 series)
Copper-nickel alloys
Copper
Brass
Alloy steel (Cr, Ni, Mo, etc.)
Carbon steel
Aluminum alloys
Zinc
Magnesium
Most active (anodic)

Navy Problems

The U.S. Navy operates its oceangoing, airborne, and shorebound assets in one of the most corrosive environments known on earth, seawater, combined with marine species, airborne pollutants such as sulfur dioxide (SO_2), nitrogen dioxide (NO_2), and carbonaceous matter as exhaust produced from burning of fuel by ships and aircraft. Salt fog and salt spray seriously degrade the readiness of ships, aircraft, landing craft, shore/harbor facilities, and even land vehicles in oceanic transit. According to a 1993 estimate, the total direct cost of corrosion for all naval systems is $2 billion per year (Ref 2, 8). Corrosion prevention and control at all levels is one of the primary directives of the Naval Sea Systems Command (NAVSEA) and the Naval Air Systems Command (NAVAIR). The Office of Naval Research (ONR) directs programs toward development of new technology to reduce total ownership costs.

For ship systems, the largest and highest priority is in the protection of ship hulls (interior and exterior), decks (topside and well-deck), tanks (ballast and wastewater), and voids (cavities in hull walls) that are primarily steel structures (Ref 18). For NAVSEA, the primary defense against corrosion is the meticulous use of protective coatings everywhere; on underwater hull structures, cathodic protection schemes are also used. The types of coating systems used on a Navy ship vary with location and are shown in Fig. 5. Coatings are the biggest maintenance driver for ship systems as they cover nearly 7.1 million square feet of surface that include flight, freeboard, topside, and well-deck areas (Fig. 6) (Ref 19) and costs the U.S. Navy approximately $975 million a year (Ref 2). These coatings traditionally last less than 10 years, at which point the ship goes to dry dock to remove and apply new coatings. The maintenance-related dry-dock labor costs due to corrosion are about $4 million per year per ship in the fleet (Ref 4).

Most maintenance actions are done in tanks and voids that hold seawater as ballast, fuel, storage, potable water, sewage waste, sludge, and lubricating oils. The life of the epoxy coatings used in these areas is usually short, less than 3 years. The new advanced high-solid epoxy coating has increased the life to almost 10 years. Figure 7 demonstrates the performance of this new high-solid epoxy coating system where it has been used on a shipboard tank of the U.S.S. Ogden (Ref 18, 19).

The impact of corrosion on the life-cycle cost of military aircraft systems is very significant, nearly $3 billion per year (Ref 2, 19–25). For naval aircraft alone it is estimated as nearly $1 billion per year (Ref 2, 19–20). Since NAVAIR has varieties of aircraft in its inventory such as surveillance, strike, rotary wing, trainers, tankers, and transport, the corrosion problems also vary greatly and depend upon their mission and area of deployment. It must be noted that corrosion problems of Navy aircraft are exacerbated by several environmental and stress factors and sometimes work conjointly to cause catastrophic damage. Naval aircraft operate in the most severe corrosive environments. The service conditions include:

- Being stationed at sea for long periods of time on aircraft carrier flight decks
- Facing the challenging environments that contain chloride, sulfate, SO_2, and other marine seawater species (Fig. 8)
- Being subjected to high electromagnetic fields or electromagnetic interference (EMI) environment
- Being exposed to solar radiation effects
- Being serviced in a very limited space to perform adequate maintenance for corrosion protection

The impact of time (aging) on sustainment and readiness of aircraft is the next greatest challenge facing naval aviation. Here the issues are that almost 50% of the naval aircraft are more than 20 years old and require increasing maintenance to keep them in service. The biggest concern today is that these aircraft have been slated to continue in service for another 20 years (Ref 20). This has imposed a tremendous pressure on operational safety, readiness, and the maintenance budget. The added expense will take away the thrust from innovation that comes from research and development programs.

If one tries to identify the single most important issue on aging aircraft, it is structural

Fig. 6 Examples of coatings on surface ships. Topside (right side), freeboard (lower left), and well-deck (upper left) areas require very frequent stripping and repainting.

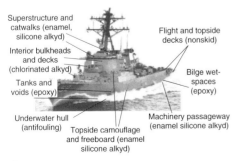

Fig. 5 Shipboard coatings are a major maintenance driver for corrosion control. Navy ships use a variety of organic coatings for interior and exterior applications.

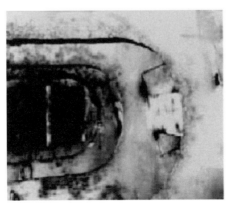
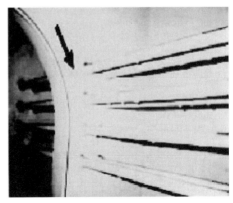

Fig. 7 Corrosion protection of shipboard tanks on the U.S.S. Ogden (LPD 5). Old coating technology after 3 years (left) and new high-solid coating after 6 years (right).

integrity where effects of corrosive environment make a serious impact on fatigue life of critical load-bearing structures and components of aircraft (Ref 13). Thus, the Navy is seriously looking into development and design of sensors that can be used to provide diagnostic and prognostic aircraft health management tools as discussed in Ref 15 and 19 and the article "In-Service Techniques—Damage Detection and Monitoring" in Volume 13A of the ASM Handbook (Ref 21). Under joint services, the structural prognostics and health management (SPHM) corrosion and strain monitoring program is expected to provide a predictive modeling tool that offers more accurate assessment of their expended service life and facilitates condition-based maintenance (Ref 22). Such schemes are in place now for the new Joint Strike Fighter.

Among 14 aircraft systems evaluated by NAVAIR, the major cost drivers by aircraft fall in the following order: surveillance (structures and avionics), strike (structures, subsystems, and landing gear tailhook), and rotary wing (dynamic components, pumps, landing gear, hydraulics). Major depot-level repairs were related to corrosion and aging (50%), obsolescence (14%), and design and item change (25%) (Ref 23). The order in which most repairs are done are: avionics, dynamic components, electrical/power systems, structures, subsystems, and engines (Ref 23). The Navy has three levels of maintenance for their aircraft: (1) organizational maintenance is performed on individual equipment at squadrons and include inspection, servicing, small parts replacements, and some assemblies; (2) intermediate maintenance is conducted on parts removed from the aircraft and includes calibration, repair, and replacement of damaged components at operational sites or wings; and (3) depot-level maintenance is conducted as an overhaul or major refurbishment and rebuilding of parts, subassemblies, and end items, including manufacture of parts, modifications, testing, and recycle. This is done at one of the three depots within the continental United States.

For naval aircraft, coatings are the first line of defense against corrosive aircraft carrier environments. That is why the Navy aircraft coating system has to be highly resistant to the operational environmental. The coating generally comprises pretreatments, such as a chemical conversion coating that serves as corrosion inhibitor and prepares the surface for bonding with paints. This pretreated surface is sprayed with epoxy polyimides containing corrosion-inhibitor additives and then with a topcoat of polyurethane containing filler materials for protection against ultraviolet (UV) and other radiations. Since coatings do wear or break down, repainting is one of the major repairs done at the organizational level, that is, at squadrons. Squadron repairs are limited to touch-up painting (Ref 19). Squadrons have also been supplied with CPCs for more generous use as a spray treatment on top of coatings to prolong service life.

The use of sealant on all joints and around fasteners is mandatory for all Navy aircraft. Sealants are adhesives and corrosion-inhibiting compounds formulated in the form of a paste, rope, or tape and are typically applied at lap and stringer-to-skin joints and around fastener holes. Their primary function is to eliminate crevice corrosion and isolate dissimilar metal joints to avoid galvanic corrosion (Fig. 9). In particular, steel and titanium fasteners on the exterior of the aircraft must always be installed with sealant. All internal structural crevices, cavities, and corners are suspect locations for moisture collection and hence must be protected by using a flexible sealant and then a coating. In most applications, polysulfide or room-temperature vulcanizing (RTV) elastomeric sealant, polythioether, and fluorinated resins such as Skyflex (W.L. Gore & Associates, Inc.) are commonly used. Polysulfide sealants, which have short life before they harden with time and under UV exposure, are problematic in providing crevice corrosion resistance and replacement repairs. The other two varieties stay more flexible. In applications where electrical joints or antenna mounts are involved, gasket-type seals are best suited. The commercially available gaskets like HiTak (Logis-Tech, Inc.) and Av-Dec (Aviation Devices and Electronic Components, LLC) have found much use in such applications. Often, preshaped gaskets of sealant materials are more economical than handheld sealant itself.

Air Force Problems

The major corrosion issues with the Air Force is related to aging of aircraft (Ref 24, 25). Although the Air Force did not ignore the corrosion problem for the past few decades, it had followed a "find and fix it" mentality. It did have programs in place to detect, quantify, mitigate, and address generalized, structural, and cosmetic corrosion for many years in their aircraft

Fig. 8 Naval aircraft carrier flight deck environment containing salt fog, hydrocarbons, and engine exhaust gases such as SO_2, NO_x, and so forth

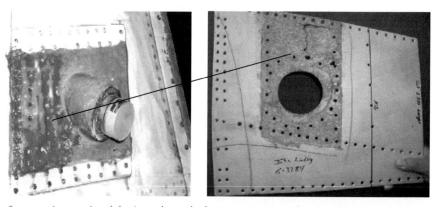

Fig. 9 Severe crevice corrosion of aluminum plate under the antenna mounts resulting from the absence of sealant in the joints

systems. Now (2006), the Air Force has a well-developed program called Aircraft Structural Integrity Program (ASIP) to monitor and control fatigue cracking.

Transport and tanker aircraft such as C-130, C-141, and KC-135 have experienced and continue to experience severe internal/hidden corrosion damage and problems of SCC in their primary structures. Legacy aircraft, such as the Navy P-3C, suffer with similar problems (Ref 26). The Air Force aircraft problems largely emanate from very little use of sealant in the joints and around fasteners and from accepting the impact of corrosion. This perspective originates primarily from the belief that land-based aircraft, as compared to those of the Navy, are not subjected to corrosion. Of course, even land-based environments are corrosive (humidity, industrial pollutants, acid rain, etc.), and corrosion damage will accumulate with time. Most Air Force aircraft corrode from inside out as they are not properly inspected for hidden corrosion (Fig. 10) (Ref 12, 24). This penalizes Air Force aircraft significantly in cost, readiness, and survivability (Ref 1, 2). An estimate shows that the Air Force spends nearly $900 million per year in direct corrosion maintenance costs (Ref 1, 12).

The present Air Force maintenance strategy includes comprehensive inspection for corrosion, fatigue, and stress cracking, repair of all corrosion damage, periodic washing and rinsing of aircraft, application of sealant in most joints, and use of preshaped gaskets in most electrical boxes, antenna mounts, and connectors to avoid environmental intrusions (Fig. 9). The Air Force has authorized the use of film tape type sealant to apply on spars, ribs, and decking structures.

To address the problems of SCC and CF in primary structures, airframes, and skins, alloy substitution to replace corrosion-prone materials such as aluminum alloy 7075-T6 has been established by the Air Force. A user-friendly software was developed to make use of MIL-HDBK-5 data (now MMPDS) for drop-in replacements of certain alloys. For example, alloys 7055-T76 and 7055-T74 are excellent replacement candidates with much better exfoliation resistance (ASTM Standard G 34, rating EB), and almost double the threshold for SCC (Fig. 11) (Ref 12, 24).

Army Problems

Corrosion problems in the Army are widespread (Ref 1, 2). Global conflict necessitates transport of warfighting machines and support equipment via ocean routes that expose them to the moisture and salinity of the marine environment (Ref 27–30). Ocean transits take up to 14 days on average, but equipment is often stored at the port of embarkation (port staging area) for several weeks or months waiting for a ship. Once shipped, it is off-loaded at the port of debarkation, awaiting cleanup and shipment to its assigned unit. This often takes months, and during this time equipment is subjected to continuous wetting and drying cycles of the salt spray from the shipboard environment. This creates damage to inner cavities of components that mostly go undetected and unremoved (Ref 30). In spite of the fact that all military equipment is protected and pretreated with corrosion prevention technologies before shipping, the removal of corrosion-causing elements deposited during oceanic transit becomes necessary (see Fig. 12). Although washing and rinsing is part of the scheduled maintenance policy, the availability of clear (fresh) water at destinations becomes a serious problem. The Army has a portable wash facility called "Bird Bath" that they install at deployment sites for rinsing and washing of all tactical ground vehicles and helicopters. Although rinse facilities are expensive to install and cost about $2.2 million each, they do have tremendous potential for controlling corrosion damage and more importantly on readiness (Ref 30).

The Army has nearly 340,000 units of tactical ground vehicles and ground-support equipment, and 2,770 helicopters and fixed-wing aircraft that require constant corrosion protection. It spends nearly $6.5 billion per year in corrosion repairs. Their major corrosion cost items are helicopters, HMMWVs, and howitzers (Ref 2, 27, 28). The Army helicopter corrosion repairs are very extensive and are performed on skin, structural frames, engine, transmission beams, rotor blades, and controls. This alone costs nearly $4 billion per year. It is primarily due to lack of or very little corrosion protection on most surfaces. The use of sealant material in laps, joints, and around fasteners, and the presence of inhibitors in rinse wash was not practiced much in the past. In 2000, the Army reported that 40% of their helicopter fleet was not combat-ready.

Corrosion is a major concern for all tactical ground vehicles, including Bradley fighting vehicle systems, Abraham tank systems, and HMMWVs. Among them, the HMMWVs were the most corrosion-prone vehicles (Ref 1, 2, 27). These are essentially light trucks designed to operate in an all-terrain environment. Poor corrosion prevention designs and inadequate material selection and corrosion repair requirements put them out of service in less than 12 months. The most commonly identified shortcomings in these vehicles are: the use of nongalvanized 1010 carbon steel with no protective coating, thousands of unprotected rivets forming galvanic couples, inadequate paint system, and no protection against chemical agents that could rapidly destroy coating systems. These vehicles cost the Army nearly $2 billion dollars per year in corrosion repairs and maintenance costs. The Marine Corps also uses these Army-procured vehicles and has had to face even worse consequences as they are operated in an even more corrosive environment.

The other significant corrosion cost contributors in the Army are the towed howitzer firing platforms. These 2300 firing cannon platforms suffered severe corrosion in unprotected dissimilar joints and areas on the platforms where water could collect. Their designs were so poor that the water-drain holes were not located at the lowest gravity point. The replacement cost for these platforms is $18,000 each with new designs (Ref 2).

Facilities Problems

There is a lack of awareness of corrosion problems associated with facilities at various military bases. The DoD has approximately 200 air bases, 40 naval ports and air stations, and numerous (in several hundred) Army and Marine

Fig. 10 Corrosion of the aircraft structure from the inside. A hidden corrosion phenomenon called "pillowing" occurs where corrosion products grow under the aluminum skin and puffs up the top surface of the thin metal sheet in the area around fasteners.

Fig. 11 Exfoliation corrosion resistance of alloy 7055-T765 (ASTM rating EB) is much superior to that for 7075-T65 (ASTM rating ED)

Fig. 12 Army shipment (equipment) via sea is exposed to the elements of marine environment during transportation

Corps facilities, which are located throughout the world (Ref 2). These extensive facilities consist of land and port infrastructures such as hangers, airfields, docks, waterfront piers, fuel storage tanks, barracks, housing, heat exchangers, electric and gas supplies, drinking water and sewer systems, bridges, and miles of railroads, pipelines, and concrete runways. These facilities experience corrosion issues that are similar to those in the civilian sector. One of the major high-cost issues is degradation of concrete, which costs in billions of dollars annually (Ref 1, 2, 31). Cracking of concrete airfields for all of the military poses severe safety hazards, impairs readiness, and increases maintenance costs.

One cause of this deterioration is the corrosive action of water with concrete (Ref 1, 31) or, more specifically, with alkali in the concrete that causes concrete to crack by expansion as shown in Fig. 13. This action is called alkali-silica reactivity (ASR), and it decomposes concrete through the thickness (bulk of pavement). The Navy has found ways to mitigate concrete corrosion problems by counteracting the effect of ASR through advanced formulation and improvement of cement. A Navy study proposes using 30% fly ash as a partial replacement of portland cement. It would allow immediate reduction in the material cost of concrete by half and double the concrete life (Ref 22). This alone is estimated to save the Navy $9.5 million per year (Ref 32). The deterioration of Channel Islands Air National Guard airfield is an example of concrete degradation due to moisture and salt (Fig. 13). It was built in 1978 and then repaired in the 1990s at a cost of $14 million as a result of ASR. This airfield showed extensive surface (apron) damage and cracking and concrete decomposition through the pavement thickness.

Corrosion of steel reinforcement is the most common form of deterioration in marine concrete structures (Ref 32, 33). The cost to repair damages of a single pier is often in the millions. An example of rebar corrosion and degradation of concrete of piers on a bridge in Hawaii is shown in Fig. 14. The recommended practice is to use epoxy-coated rebar in pier construction, which delays onset of rebar corrosion. With the help of ASTM Standard A 934 (epoxy-coated rebar), Navy Facilities Command has developed a user guidebook that provides specifications on epoxy-coated rebar technologies for all rebar-cement construction. In new constructions and where economically possible, the use of stainless steel rebar and application of cathodic protection have been also recommended.

Corrosion Control and Management

Although a number of corrosion control initiatives have been mentioned, the following are reiterated and specifically described to show the significance of the maintenance actions. Most of the military services have either implemented these initiatives or have plans to implement them in the near future.

Material Substitution. The most corrosion-prone aluminum alloys are 7075-T6 and 2024-T3. They are used in aircraft structures because of their high strength. Many newer, corrosion-resistant alloys, such as 7055-T7751, 7150-T77511, 2224-T3511, 2324-T39, and 2525-T3, developed in the past three decades, can now replace them (Ref 10–13). These newer alloys have high threshold to SCC, CF, and intergranular and pitting corrosion (Ref 12). Figure 15 shows examples of such alloys already in use on commercial jetliners such as the Boeing 777. In military applications, it is crucial that material substitutes in legacy transport, fuel carriers, and surveillance aircraft exactly match the mechanical properties of the materials they replace. Otherwise, it is possible to increase stress levels on abutting structures. Material substitutions have been recommended on the Air Force C-130 and C-5 and the Air National Guard C-141.

For the Navy's P-3C surveillance aircraft, several alloys with superior corrosion resistance were investigated as possible drop-in material substitutes for parts that needed replacement due to extensive corrosion repair history. Among them, alloys 7150-T77511, 7249-T76511, and 7055-T7XXX emerged as prime replacement candidates. These alloys were fully characterized by the Sustainment Life Assessment Program (SLAP) by full-scale testing before any such substitution was recommended (Ref 26). With these substitutions, the aircraft are now expected to provide 30 more years of service life.

Retrogression Re-Aging (RRA). In some cases, where substitution is cost prohibitive, a newly developed heat treatment process called retrogression and re-aging (RRA) has been studied and perhaps would be recommended. This process was developed to keep the original strength of T-6 temper to within 90% of the yield strength, but most importantly increase the resistance of the alloy to exfoliation corrosion and SCC. The RRA treatment consists of a retrogression phase (heat to 195 °C, or 385 °F, for 40 min and oil quench) and then a re-aging phase (heat to 120 °C, or 250 °F, for 24 h and then air cool). The Air Force is currently investigating

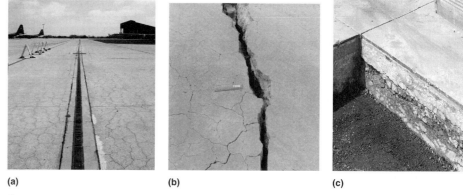

Fig. 13 Concrete decomposition at Channel Islands Air National Guard airfield from alkali-silica reactivity. (a) The airfield, which was built in 1978. (b) Surface (apron) damage and cracking. (c) Concrete decomposition through pavement thickness. The cost of repair was $14 million.

Fig. 14 Severe concrete degradation caused by rebar corrosion of piers of a NAVSTA bridge in Pearl Harbor, HI

this process using a special thermal blanket that can be placed on the actual aircraft skin of 7075-T6 alloy sheet metal and heated to approximately 200 °C (390 °F) for a short time and then rapidly cooled with a cold jacket and then re-aged by heating to 120 °C (250 °F) (Ref 34). The problem with such in situ application is that thermal conduction or heat sink by the structural frames attached to the skin is not quantifiable; hence, this treatment cannot be universally adopted over all the aircraft. The RRA parameters have to be determined for each location to get the optimum results.

Rinsing and Washing. The most effective procedure and closest to the immediate corrosion reduction initiative is washing and rinsing, in which a majority of aircraft and other weapon systems are washed and rinsed with waters containing corrosion inhibitors. The design of the facility depends on the types of aircraft, tanks, and ground vehicles. For helicopters the facility is like a "birdbath," and for fixed wings and ground equipment it is the overhead spray type installed in the hanger or in the field. The Army, Air Force, and Navy all have such facilities at their respective maintenance facilities. Most wash and rinse facilities include a freshwater wash-injection system with corrosion inhibitors as additives. Generally, rinse facilities use a closed-loop system where water is recycled after a filtration and replenished with the additives for the next wash and rinse. The additives generally comprise a surfactant, cleanser, a pH modifier, and multifunctional inhibitors.

Some of these closed-loop facilities are deployable so that they can be installed on-site in the regions of military conflict (Ref 30) where corrosion-causing species can be removed from the aircraft before they have time to initiate corrosion. This practice is also being used for tanks and ground vehicles, but not as frequently as for aircraft. In deployment conditions of extremely corrosive environments, washing and rinsing are most frequently recommended for all outdoor hardware systems.

Finding Corrosion and Corrosion-Assisted Damage. Recognizing the impact of corrosive operating environment on the integrity of military weapons systems, and the national economy, original equipment manufacturers and the DoD agencies have taken a significant interest in developing nondestructive inspection (NDI) and nondestructive evaluation (NDE) technologies. In the last three decades numerous innovative systems have been developed that play an important role in providing detection and monitoring of early signs of corrosive environments, corrosion, and corrosion-assisted damage. As the cost of corrosion damage continues to escalate in terms of maintenance man-hours, and lack of readiness, the demand on early detection devices to avoid major repairs and downtime is now increasing substantially. Condition-based maintenance has taken the place of scheduled or periodic maintenance. The following technologies are currently being used or introduced to address corrosion:

- Optically aided inspection—visual, borescope, fluorescence
- Thermal imaging
- Digital radiography
- Magneto-optic eddy current
- Guided wave ultrasonics
- Microwave scanning
- Electrochemical techniques—galvanic, resistance, and impedance type
- Eddy current—multifrequency, mobile automated ultrasonic scanner (MAUS)
- Neutron radiography
- Magnetic resonance spectroscopy
- Acoustic radio and thermal scanning
- Laser pulse-echo method
- Stress/strain sensing

The details of some of these techniques have been discussed in the *ASM Handbook* (see, for example Ref 14 and 21). Among these, the most recommended and preferred technologies are those that can be embedded or are the leave-in-place type and that can monitor the onset of corrosion and locate small cracks or pits as they occur (Ref 21). These early-warning devices detect corrosion damage before it becomes significant, minimize corrosion damage, allow small repairs that can be performed at organization or intermediate levels, and can extend depot-level maintenance by another few years (Ref 14). One of the devices called wireless intelligent corrosion sensor or ICS does just that. Through its thin-film galvanic sensor, it monitors the corrosive environment in hidden areas and active corrosion in joints and splices or under coatings. It is an autonomous stand-alone miniature device that measures, collects, stores, and then downloads the data at command through a handheld transponder or data-gathering device (Ref 25). This thin-film device can be attached in aircraft structural cavities or sandwiched in joints to detect the intrusion of corrosive environment.

Corrosion Removal and Repair. Once corrosion has been identified, removal of corrosion damage and the subsequent ability to make small repairs becomes an important initiative in the corrosion prevention and control program. This process requires the use of a kit that contains small repair tools, brushes, chemical tubes, and spray cans that can be applied at organization levels without having to wait until depot-level maintenance is scheduled. The kit is lightweight and contains only essential items for small repairs, such as a bristle disk for removal of corrosion products, touch-up pens, brushes and pads for surface pretreatment, conversion coating and primer, and topcoat. The kit allows in-field application to prevent spread of corrosion and extends the life of the structure. Programs are in place to extend this concept to repair materials other than aluminum. Whenever large surface areas are corroded, paint stripping becomes absolutely necessary. Under those conditions, immediately after paint stripping and removal of corrosion product, temporary corrosion protection schemes are needed before repainting and involve the use of CPCs.

Corrosion preventive compounds are dispersant liquids of solvent and sometimes water containing petroleum distillates, surfactants, and

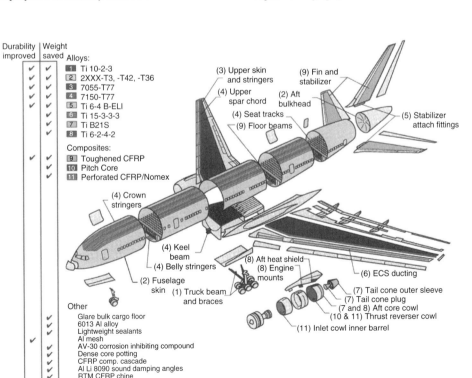

Fig. 15 Material substitution on Boeing 777 aircraft. Aluminum alloys 7075-T6 and 2024-T3 have been replaced by 7055-T7 and 2324-T3 and other more corrosion- and SCC-resistant aluminum alloys.

corrosion inhibitors. There are several types of CPCs on the market, and all of them are used primarily for temporary corrosion protection in interior structures and cavities. They must be reapplied frequently or in some timely manner to be effective. Some CPCs are also applicable on the exterior surfaces. Appropriateness of a CPC product is based on their physical and chemical attributes, such as viscosity, dryness, lubrication, volatile organic content, smell, environmental compliance, surface wetness, hydrophobicity, corrosion protection period, and inhibition efficiency. The application of CPCs is widely accepted in hard-to-reach areas of all types of equipment, aircraft missiles, doors, hinges, bulkheads of ships, electrical boxes, galleys, wheel wells, joints, voids, finishes, and so forth. Since most CPCs wash away with water, reapplication is always needed and the preferred periods are 12 to 15 months. It is essential that CPCs used in moving components contain some lubricants and do not carry tacky or waxy substances that cause seizing and result in loss of mobility. In the selection of CPCs, one must practice caution and seek to match CPC property with the part design and its functional and material requirements. The following is a partial list of some CPCs that are commercially available: Dinitrol compounds (several grades AV-15, 30, 40, etc.); Cor-Ban 35 and Cor-Ban 22; CorrosionX; ACF-50; LPS 2, LPS-3, Procyon; SuperCorrB; Amalguard (a Navy developed product).

Protection and Preservation by Humidity Control. The presence of moisture is essential to corrosion. It has been well established that control of relative humidity (RH) controls the magnitude of corrosion (Ref 35–37). Numerous studies have shown that in components such as electrical and electronic boxes, and electrical wires and connectors, the effects of high humidity are significant. For example, the resistance of nylon insulation on an electrical wire can drop from 10^{14} to 10^7 Ω if the RH is changed from 10% RH to 90% RH. All services have evaluated the benefits of controlling RH below 35% for their weapons systems and found it to be very effective in corrosion protection and preservation. Munitions and equipment stored in high humidity can experience severe corrosion and perhaps loss of functionality. Services have used temporary shelters such as makeshift hangers and clamshells and employed desiccant wheels to remove humidity from the interior of aircraft and missile systems and wheeled vehicles. Many foreign governments use humidity control as a maintenance technology for their weapon systems. The Europeans (The Netherlands, Belgium, Sweden, Italy, France, Germany, and United Kingdom) have been using this method to avoid corrosion since 1995 (Ref 35). Maintenance of low humidity, <35% RH, in weapon systems storage areas is essential to preservation and protection from corrosion. Low-humidity preservation also helps in readiness and quick deployment (Ref 37).

Corrosion Education and Training. Proper education and on-site training of technical personnel in the state-of-the-art corrosion science and engineering is necessary if a cost-effective corrosion control and maintenance program is to be implemented at DoD facilities. There are numerous sources for such training within the DoD such as the Tri-Service conferences, Army, Navy, and Air Force workshops, corrosion manuals and websites. In addition, schools such as Massachusetts Institute of Technology, University of Virginia, The Ohio State University, Case Western Reserve University, University of Southern California, University of Florida, University of Texas, Texas A&M and NACE International—the Corrosion Society—offer many corrosion courses such as corrosion basics, coating inspection, and cathodic protection. The DoD has also established a website that is available to approved members of the defense acquisition force. This website (www.dodcorrosionexchange.org, accessed Dec 2005) has an extensive database of military corrosion control products, specifications, guidebooks, and technology papers related to corrosion prevention and control.

Long-Term Strategy to Reduce Cost of Corrosion

The U.S. Congress has directed the DoD and its agencies to seek to mitigate the effects of corrosion and improve the department's coordination of corrosion prevention and control practices among the services. The long-term strategy includes:

- Consider life-cycle costs when new systems are procured
- Replace older equipment with new at a faster rate—accelerate modernization
- Reduce burden of maintaining underutilized infrastructure—close marginal facilities
- Establish a Corrosion Information Exchange Network, an Internet web-based tool for sharing best practices among the services, including construction of an *expert system*
- Revise/develop policy and regulations on design, acquisition, and maintenance of military equipment and make it specifically address corrosion prevention and mitigation
- Develop standardized methodologies for collecting and analyzing corrosion cost, readiness, and safety data. Increase use of industry standards for corrosion testing and evaluation of commercial products and processes
- Increase interaction with professional societies and private sector corrosion-focused organizations. Encourage NACE International to develop information-sharing network
- Continue to develop and test materials, processes, and treatments that reduce manpower, downtime, and costs associated with corrosion
- Develop a new or augmented strategy through survey of current corrosion control practices with metrics of cost reduction for each of the services. The purpose is to develop criteria for testing and use of best materials and processes from the collected consolidated database
- Mandate a requirement for corrosion education and training for all design engineers, maintenance personnel, and (even) program managers responsible for military hardware acquisition

An effort has been made to create a consortium of government, academia, and industry, which will address corrosion education as part of the existing DoD education forum and will be incorporated in its website. The intent of this consortium is to develop corrosion science and engineering education tools for the user community at large that contains lucid and comprehensive understanding of corrosion issues related to their sector interest including the impact of corrosion on the national productivity and the economy. A corrosion steering committee of experts from the DoD, academia, and industry has been established, and West Virginia University is spearheading the effort. NACE International has been selected as the major source for training and providing a database of technologies that could be available for general use.

Summary

Corrosion in the military has been traditionally treated as strictly a maintenance problem and has been argued as a "necessary evil." Unfortunately this has gone on for too long and is costing the U.S. military nearly $20 billion per year and jeopardizing the readiness of warfighters. The corrosion problems in all services are of a very generic nature and simply emanate from bad judgments, improper use of materials and processes, inadequate design engineering practices to meet corrosion science and engineering principles, and an overall lack of recognition of the fact that the corrosive environment has a severe impact on the performance of weapons systems.

The overwhelming consensus in the U.S. DoD is that there should be a new policy requiring prime contractors to improve material selection practices by paying more attention to corrosion resistance and engineering design concepts that prevent and/or control corrosion. Justification of additional acquisition cost to accommodate corrosion prevention that reflects on reduction of total ownership cost with some predictable metrics for savings should be acceptable. It is generally accepted that corrosion awareness training is vital for acquisition program managers of various military weapon systems so that they can implement corrosion prevention and control planning as an explicit part of performance-based acquisition and performance-based logistics.

REFERENCES

1. "Defense Management Opportunities to Reduce Corrosion Costs and Increase Readiness," U.S. General Accounting Office, GAO-30-753, July 2003
2. G.H. Koch, M.P.H. Brongers, N.G. Thompson, Y.P. Virmani, and J.H. Payer, "Corrosion Costs and Preventive Strategies in the United States," Report No. FHWA-RD-01-156, Federal Highway Administration, 2002
3. H. Mendlin, B.F. Gilp, L.S. Elliot, and M. Chamberlain, "Corrosion in DOD Systems: Data Collection and Analysis," MIAC Report, Metals Information Analysis Center, Aug 1995
4. "Economic Effects of Metallic Corrosion in the United States," NBS Special Publication, 511-1, SD Stock No. SN-003-003-01926-7, 1978, and Appendix B, NBS Special Publication, 511-2, SD Stock No. SN-003-003-01926-5, 1978
5. "Economic Effects of Metallic Corrosion in the United States—Update," Battelle Report, April 1995
6. R. Hays and R.L. Stith, Success Stories in Marine Corps, *Beating Corrosion*, Special Issue, *AMPTIAC Q.*, Vol 7 (No. 4), 2003, p 54–74
7. G. Cooke et al., "A Study to Determine the Annual Direct Cost of Corrosion Maintenance for Weapons Systems in US Air Force," Final Report, Feb 1998
8. "Paying for Military Readiness and Upkeep: Trends in O&M Spending," Jan 1997; "The Effects of Aging on the Costs of Operating and Maintaining Military Equipment," Aug 2001; Congressional Budget Office, Congress of the United States
9. "Best Practices: Setting Requirements Differently Could Reduce Weapons Systems Total Ownership Costs," GAO-03-57, U.S. General Accounting Office, Feb 2003
10. V.S. Agarwala, "Control of Corrosion and Service Life," Corrosion/2004, Proc. NACE International Conference, Preprint No. 4257, 2004
11. V.S. Agarwala, Corrosion and Aging Aircraft, *Can. Aeronaut. Space J.*, Vol 42 (No. 2), June 1996
12. D.T. Peeler, "Comprehensive Damage Management: Merging Environmental Exposure, NDI Assessment and Structural Analysis to Manage Structural Damage," Proc. Sixth International Aerospace Corrosion Control Symposium, Oct 9–11, 2002 (Amsterdam, The Netherlands), IIR-Greenline Communications Ltd., London, United Kingdom
13. G.H. Koch, "Cost of Corrosion in Military Equipment," Corrosion/2004, Proc. NACE International Conference, Preprint No. 4252, 2004
14. Forms of Corrosion (section), *Corrosion: Fundamentals, Testing, and Protection*, Vol 13A, *ASM Handbook*, S.D. Cramer and B.S. Covino, Jr., Ed., ASM International, 2003, p 187–248
15. D.H. Rose and H.A. Matzkanin, Improved Access to Corrosion Research Will Reduce Total Ownership Costs, *AMPTIAC Q.*, Vol 7 (No. 4), 2003, p 17–22
16. S. Chawla and R. Gupta, *Materials Selection for Corrosion Control*, ASM International, 1993, p 476
17. R.J. Bucci, C.J. Warren, and E.A. Starke, Jr., The Need for New Materials in Aging Aircraft Structures, *J. Aircraft*, Vol 37, 2000, p 122–129
18. "Preservation and Maintenance of U.S. Navy Ships," An Update for CINPACFLT N438 Staff, Naval Sea Systems Command, SEA 03M, Feb 1999
19. A. Kazanoff, S. Spadafora, and A. Perez, Success Stories: Navy, *Beating Corrosion*, Special Issue, *AMPTIAC Q.*, Vol 7 (No. 4), 2003, p 66–68
20. "U.S. Navy Aircraft Corrosion Prevention and Control Program," Report No. 97-181, Audit Report, Inspector General, 1997
21. V.S. Agarwala, In-Service Techniques—Damage Detection and Monitoring, *Corrosion: Fundamentals, Testing, and Protection*, Vol 13A, *ASM Handbook*, S.D. Cramer and B.S. Covino, Jr., Ed., ASM International, 2003, p 501–508
22. A.K. Kuruvilla, "Life Prediction and Performance Assurance of Structural Materials in Corrosive Environments," AMPTIAC Report, Contract No. SPO070097-D-4001, IIT Research Institute, Aug 1999
23. L. Stoll, "Aging Aircraft Cost Growth: Current Joint Service Research Results," Navair Fellows Lecture, Jan 2004, Naval Air Systems Command; "Analysis of Cost Growth for Depot Level Repairables of Selected Aircraft," Seventh DoD/FAA/NASA Conference on Aging Aircraft (New Orleans, LA), Sept 2003
24. "Aging of U.S. Air Force Aircraft," Final Report, National Materials Advisory Board, National Academy Press, 1997
25. V.S. Agarwala, "Aircraft Corrosion and Aging: Problems and Concerns," Keynote Paper No. 1, Proc. 15th International Corrosion Congress (Granada, Spain), Sept 2002
26. N. Phan, "P-3 Service Life Assessment Program—A Holistic Approach to Inventory Sustainment for Legacy Aircraft," Proc. 2003 Tri-Service Corrosion Conference (Las Vegas, NV), Nov 17–21, 2003, Naval Air Systems Command
27. Audit Report on Army "High-Mobility Multipurpose Wheeled Vehicle," Office of the Inspector General, U.S. Department of Defense, 1993
28. "Army Trucks Information and Delivery Delays and Corrosion Problems," GAO/NSIAD-99/2/6, Government Accounting Office, Jan 1999
29. J. Repp, "Corrosion Control of Army Vehicles and Equipment—Use of Existing Technologies for Corrosion Control," 2003 Army Corrosion Summit (Clearwater Beach, FL), http://www.armycorrosion.com, Sept 2003
30. H. Mills, Success Stories: Army, *Beating Corrosion*, Special Issue,' *AMPTIAC Q.*, Vol 7 (No. 4), 2003, p 62–65
31. A. Kumar, L.D. Stephenson, and G. Gerdes, "Corrosion Related Costs for Military Facilities," Corrosion/2004, Proc. NACE International Conference, Preprint No. 4269, 2004
32. M.D.A. Thomas, "Review of the Effects of Fly Ash and Slag on Alkali-Aggregate Reactions in Concrete," BRE Report 314, British Research Establishment, London, United Kingdom, 1996
33. L.J. Malvar, G.D. Cline, D.F. Burke, R. Rollings, T.W. Sherman, and J. Greene, Alkali-Silica Reaction Mitigation: State-of-the-Art and Recommendation, *ACI Mater. J.*, Vol 99 (No. 5), Sept–Oct 2002, p 480–489
34. D. Raizzene, P. Sjoblom, R. Rondeau, J. Snide, and D. Peeler, "Retrogression and Reaging of New and Old Aircraft Parts," Progress Report, United States Air Force, 2002
35. A. Timko and O.R. Thompson, Success Stories: Controlled Humidity Protection, *AMPTIAC Q.*, Vol 7 (No. 4), 2003, p 74–79
36. "Dry Air Technology for Defense Applications," Munters Incentive Group, Cambridge, United Kingdom, 1995
37. Naval Audit Service Report, "Dehumidification of In-Service Aircraft," Audit Report 025-95, 1995; U.S. Navy, "Preservation and Dehumidification," NAVAIR Tech. Manual 15-01-500

Military Specifications and Standards

Norm Clayton, Naval Surface Warfare Center, Carderock Division

FOR ANY TYPE OF INDUSTRY, it is critical to have specifications and standards that set the requirements for products, ranging from raw materials, commodities, and materials of construction, to pieces of equipment and large-scale systems. Specifications and standards also define construction and fabrication processes and testing procedures. Although international voluntary consensus standards organizations, such as the International Organization for Standardization (ISO) and ASTM International (ASTM), fulfill the need for standards for a vast array of products, many industries have their own sets of specifications and standards based on the unique performance requirements for that industry. These are issued through their respective professional societies or by individual companies. The military organizations in each nation are no exception, as military equipment experiences many unique combinations of environmental exposure and war-fighting requirements that do not exist in other industries. This article provides a perspective on United States (U.S.) Department of Defense (DOD) specifications, standards, handbooks, and related documents and their role in corrosion prevention and control activities in the U.S. military.

Corrosion-control activities in the DOD deal with *weapons systems* (ships, aircraft, tanks, artillery, ground vehicles, guns, missiles, ammunition, etc.) and the *facilities* infrastructure needed to support them (buildings, piers, airfields, fuel and water tanks, piping, etc.). Both the weapons systems and facilities communities rely heavily on specifications and standards prepared and issued by the various service entities within the DOD, predominantly the Army, Navy (including the Marine Corps), and Air Force, either separately or jointly. The great majority of these documents are designated as *military* specifications, standards, and handbooks, with the designation prefix "MIL," although there are some designated as Department of Defense documents with the prefix "DOD."

Keeping specifications up-to-date can be costly. For this reason and others, there has been a large effort in the DOD over the past ten years to evaluate the suitability of commercial specifications and standards (such as ASTM) and adopt them where suitable, canceling the associated military specification in the process. This effort has taken place under the broad initiatives of "Acquisition Reform" and the use of "commercial-off-the-shelf" (COTS) products. For ASTM specifications adopted by DOD, the phrase "This standard has been approved for use by agencies of the Department of Defense" appears under the title of the specification. This effort has been particularly successful in specifications for metallic materials; however, in other areas, such as paints and coatings, most military specifications have been retained by the various services. (For additional information on U.S. policy regarding federal agency use of voluntary consensus standards, refer to Public Law 104–113, "National Technology Transfer and Advancement Act of 1995," and the Office of Management and Budget (OMB) Circular A-119, "Federal Participation in the Development and Use of Voluntary Consensus Standards and in Conformity Assessment Activities" dated Feb 10, 1998.)

In addition to MIL, DOD, and ASTM specifications mentioned previously, the U.S. military corrosion-control communities make use of a wide variety of other standards to specify materials, coatings, sealants, inhibitors, abrasive blast media, and other chemical products, test methods, and manufacturing and production quality assurance (QA) requirements. These specifications may either be directly cited or be cited for specific requirements within a MIL specification. Notable examples include:

- Federal Specifications and Standards, and Commercial Item Descriptions issued by the U.S. General Services Administration (GSA) Federal Supply Service (FSS): specifications for a wide variety of products. One example, FED-STD-595B (Ref 1) is widely used within MIL specifications for paints and coatings to specify standardized color and gloss requirements
- SSPC—The Society for Protective Coatings: industrial painting surface preparation and painting requirements, specifications for abrasive blasting media, and certification programs for individuals and companies involved in industrial painting
- NACE International: coatings surface preparation standards issued jointly with SSPC, recommended practices for fluid tank design and cathodic protection systems, and certification programs for individuals and companies involved in industrial painting
- ISO: industrial painting surface preparation QA requirements
- SAE International: SAE Aerospace Material Specifications (AMS) and SAE J Series Ground Vehicle Standards

The remainder of this article discusses specifications, standards, and related documents created and issued by the DOD.

Types of Documents and Designations

U.S. military organizations use a variety of documents to provide *requirements* and *guidance* for corrosion-control products and processes in both the design and construction of new weapon systems and facilities and the maintenance of these systems and facilities. The distinction between *requirements* and *guidance* is important, especially in a contracting application. Military specifications and standards provide requirements, whereas military handbooks can only provide guidance. The main types of documents are briefly described in this section.

Specifications. MIL-STD-961 (Ref 2) provides the format and content requirements for DOD specifications. The Foreword section of MIL-STD-961 summarizes the purpose of DOD specifications as:

> The overall purpose of a specification is to provide a basis for obtaining a product or service that will satisfy a particular need at an economical cost and to invite maximum reasonable competition.... A good specification should do four things: (1) identify minimum requirements, (2) list reproducible test methods to be used in testing for compliance with specifications, (3) allow for a competitive bid, and (4) provide for an equitable award at the lowest possible cost.

Military specifications may include those that are widely used across several programs or applications and those that are unique to a single program or system that would have little or no potential for use with other programs or systems. Most corrosion-control products, such as paints, coatings, sacrificial anodes, inhibitors, cleaners, sealants, and so forth would fall into the first category and represent the widest and most publicly known use of DOD specifications. Specifications that provide broader corrosion-prevention requirements for the design, testing, construction, or maintenance of a specific weapon system or facility fall into the second category and may be issued by a single branch of the DOD. There are two main categories of specifications: detail specifications, designated as "MIL-DTL," and performance specifications, designated as "MIL-PRF." These categories indicate how the requirements in the specification are stated. Older specifications may be designated as "MIL-X," where the "X" is a single letter representing the first word of the document title. All specification identifiers are being replaced with the DTL and PRF designations as they are revised. MIL-STD-961 (Ref 2) describes the types of specifications succinctly (note that a general specification must still be designated as a performance or detail specification):

Detail specification: A specification that specifies design requirements, such as materials to be used, how a requirement is to be achieved, or how an item is to be fabricated or constructed. A specification that contains both performance and detail requirements is still considered a detail specification. Both defense specifications and program-unique specifications may be designated as a detail specification.

Performance specification: A specification that states requirements in terms of the required results with criteria for verifying compliance, but without stating the methods for achieving the required results. A performance specification defines the functional requirements for the item, the environment in which it must operate, and interface and interchangeability characteristics. Both defense specifications and program-unique specifications may be designated as a performance specification.

General specification: A specification prepared in the six-section format, which covers requirements and test procedures that are common to a group of parts, materials, or equipments and is used with specification sheets. Specification sheets are documents "...that specify requirements and verifications unique to a single style, type, class, grade, or model that falls within a family of products described under a general specification."

Military specifications dealing with corrosion-control products and processes are too numerous to list here, and such a list would rapidly become obsolete. Electronic searching of the on-line databases and sources of specifications described later in this article is the best way to find a specification of interest.

Qualified Products List (QPL). A detail or performance specification may or may not require that products be submitted to a qualifying activity in order for them to be listed on a QPL. When a QPL is not required, frequently there is an alternative requirement to pass first article testing or inspection. When a specification requires that products be listed on a QPL, it will be described in one of the first paragraphs in Section 3 "Requirements" in the specification. (See the section, "Format of Specifications" below.) The QPL for a product specification will have the same number as the parent specification, with a suffix to indicate the revision number of the QPL. For example, the tenth edition of the QPL for notional specification MIL-PRF-123 would be designated as QPL-123-10. A QPL listing will contain:

- Government designation: the type, class, grade, and so forth of the product as defined by the parent specification
- Manufacturer's designation: the commercial product name used by the manufacturer to identify the product
- Test or qualification reference: a citation for the test report or other document used to place the product on the QPL
- Manufacturer's name and address: self-evident

Standards. MIL-STD-962 (Ref 3) provides the format and content requirement for military standards. Military standards provide requirements that are to be used when the standard is specifically cited in a contract, purchase order, or other specification. Just because an active military standard exists does not indicate that it automatically applies to any military weapon system or facility; it must be specifically invoked. Table 1 lists active military standards pertaining to corrosion and coating as of the date of this article (2005). The "source" column indicates the lead service activity that is responsible for maintenance of the standard. In some cases, the scope of a standard may be limited to a single branch of the military, such as MIL-STD-1303 and the Navy. In others, the standard may be used and collaboratively updated by all branches of the military, such as MIL-STD-810.

For corrosion-prevention applications, Table 1 shows that there are three main categories of military standards:

- Standards that cover specific processes for the application of a type of coating or other surface treatment. MIL-STD-865, MIL-STD-1501, and MIL-STD-2138 are examples.
- Standards that provide a set of requirements for coatings or other corrosion-control methods to be used on certain types of military systems or facilities. MIL-STD-171, MIL-STD-1303, and MIL-STD-7179 are examples. These standards will themselves contain requirements that cite a variety of specific surface preparation and coating treatments according to commercial industry or other military specifications.
- Test methods that are to be used to evaluate or qualify military systems, equipment, or facilities, or to qualify specific types of products. MIL-STD-810, MIL-STD-2195, and MIL-STD-3010 are examples. Note that when a military standard for test methods is cited in a specification, specific test methods within the standard and their acceptance criteria generally must also be specified.

There are other miscellaneous standards in Table 1 that do not fall into the general categories described in the preceding list, such as

Table 1 Active military standards pertaining to corrosion and coating

Preparing organization	Number	Date(a)	Title
Army	MIL-STD-171E	June 23, 1989	Finishing of Metal and Wood Surfaces
Air Force	MIL-STD-810F	May 5, 2003	Environmental Engineering Considerations and Laboratory Tests
Air Force	MIL-STD-865C	Nov 1, 1988	Selective (Brush Plating), Electrodeposition
Air Force	MIL-STD-868A	March 23, 1979	Nickel Plating, Low Embrittlement, Electrodeposition
Air Force	MIL-STD-869C	Oct 20, 1988	Flame Spraying
Air Force	MIL-STD-889B	May 17, 1993	Dissimilar Metals
Navy	MIL-STD-1303C	April 11, 1994	Painting of Naval Ordnance Equipment
Air Force	MIL-STD-1501C	April 2, 1990	Chromium Plating, Low Embrittlement, Electrodeposition
Air Force	MIL-STD-1503B	Nov 13, 1989	Preparation of Aluminum Alloys for Surface Treatments and Inorganic Coating
Air Force	MIL-STD-1504B	June 8, 1989	Abrasive Blasting
Air Force	MIL-STD-1530B	Feb 20, 2004	Aircraft Structural Integrity Program (ASIP)
Navy	MIL-STD-1687A	Sept 23, 1994	Thermal Spray Processes for Naval Ship Machinery Application
Navy	MIL-STD-2073	May 10, 2002	Standard Practice for Military Packaging
Navy	MIL-STD-2138A	Aug 29, 1994	Metal Sprayed Coatings for Corrosion Protection Aboard Naval Ships (Metric)
Navy	DOD-STD-2187	Aug 20, 1987	Chemical Cleaning of Salt Water Piping Systems (Metric)
Navy	MIL-STD-2195A	Dec 17, 1993	Detection and Measurement of Dealloying Corrosion on Aluminum Bronze and Nickel-Aluminum Bronze Components, Inspection Procedure for
Army	MIL-STD-3003A	July 7, 2003	Vehicles, Wheeled: Preparation for Shipment and Storage of
Navy	MIL-STD-3010	Dec 30, 2002	Test Procedures for Packaging Materials
Navy	MIL-STD-7179	Sept 30, 1997	Finishes, Coatings, and Sealants, for the Protection of Aerospace Weapons Systems

(a) Base date of primary document; validation notices may be dated later

MIL-STD-889, MIL-STD-1530, and MIL-STD-3003. MIL-STD-889 is notable among them and is described in more detail later in this article.

Handbooks. MIL-STD-967 (Ref 4) provides the format and content requirement for military handbooks. Table 2 lists military handbooks dealing with corrosion control or coatings that are still active as of the date of this article (2005). Unlike military specifications and standards, military handbooks have evolved to become documents that are to be used for guidance purposes only, especially in the context of a DOD contract specification for equipment. If this is not clearly stated in the text of the handbook, then a standardized notice has been added to the cover page of the handbook, such as the one shown below for MIL-HDBK-1568:

NOTE: MIL-STD-1568B(USAF) has been re-designated as a handbook and is to be used for guidance purposes only. This document is no longer to be cited as a requirement. For administrative expediency, the only physical change from MIL-STD-1568B(USAF) is this cover page. If cited as a requirement, contractors may disregard the requirements of this document and interpret its contents only as guidance.

Despite being designated as guidance documents only, military handbooks dealing with corrosion prevention, and coating and related processes still may have a variety of uses, such as being used as basic reference documents and readily available training resources for military personnel responsible for maintenance of systems and facilities. However, these documents are not updated and kept up-to-date with current policies and technology as frequently as the more actively used military specifications and standards.

Military Drawings. While military specifications and standards as described previously are used by the U.S. military services to specify most corrosion-control products and processes, there is another significant class of documents broadly characterized as military drawings. These are detailed drawing sheets for piece parts of mechanical or electrical hardware and other items, which include dimensions, materials and finishes, and other technical manufacturing characteristics, and provide an accompanying part number system. These drawings may often be used to support a parent procurement specification. Examples of the various types of military drawings include:

- AN: Air Force-Navy Aeronautical Standard Drawing
- DESC: Defense Electronics Supply Center Drawings
- MS: Military Standard Drawing, or Military Specification Sheet

Service-Specific and Other Documents. In addition to the aforementioned documents, each branch of the DOD has an assortment of corrosion-control and coating documents that variously provide policy requirements or recommended guidance specific to the issuing branch of the DOD. These include documents termed Technical Manuals (TM), Technical Orders (T.O.), Technical Publications, Army Regulations (AR), Technical Repair Standards (TRS), Standard Maintenance Items, Project Peculiar Documents (PPD), and Preservation Process Instructions (PPI), to name a few. A great number of these documents are intended to convey corrosion-prevention and repair requirements to equipment or facility operating and maintenance personnel, whether they are uniformed service personnel or contracted companies. However, many are also used in new procurements of systems or facilities.

The U.S. Army Corps of Engineers and the Naval Facilities Engineering Command also maintain a set of technical manuals and guide specifications (GS) pertaining to corrosion prevention and control in the construction and maintenance of facilities. A number of these are listed in Table 3. For additional information on these specifications, refer to MIL-STD-3007 (Ref 5).

Format of Specifications

Like commercial industry standards such as ASTM, military specifications are required to follow a standard format. Military specifications use a six-section format using well-defined requirements in MIL-STD-961 (Ref 2). The six numbered sections and their content are briefly described in this section.

1. Scope: provides a concise description or abstract of what the specification covers, as well as any categories and subcategories of use or types of material, using terms such as Type, Class, Grade, or other distinction. A relatively new requirement for military specifications is that if the specification describes more than one part or item, and if U.S. Federal Supply System National Stock Numbers (NSNs) are to be

Table 2 Active military handbooks pertaining to corrosion and coating

Preparing organization	Number	Date(a)	Title
Army	MIL-HDBK-113C	April 24, 1989	Guide for the Selection of Lubricants, Functional Fluids, Preservatives and Specialty Products for Use in Ground Equipment Systems
Army	MIL-HDBK-205A	July 15, 1985	Phosphate and Black Oxide Coating of Ferrous Metals
Navy	MIL-HDBK-267A	March 3, 1987	Guide for Selection of Lubricants and Hydraulic Fluids for Use in Shipboard Equipment
Air Force	MIL-HDBK-310	June 23, 1997	Global Climatic Data for Developing Military Products
Army	MIL-HDBK-341	Feb 27, 1998	Coating, Aluminum and Silicon Diffusion, Process for
Army	MIL-HDBK-506	April 10, 1998	Process for Coating, Pack Cementation, Chrome Aluminide
Army	MIL-HDBK-509	Dec 7, 1998	Cleaning and Treatment of Aluminum Parts Prior to Painting
Army	MIL-HDBK-729	Nov 21, 1983	Corrosion and Corrosion Prevention Metals
Army	MIL-HDBK-735	Jan 15, 1993	Material Deterioration Prevention and Control Guide for Army Materiel, Part One, Metals
Navy	MIL-HDBK-1004/10	Jan 31, 1990	Electrical Engineering Cathodic Protection
Navy	MIL-HDBK-1015/1	May 31, 1989	Electroplating Facilities
Navy	MIL-HDBK-1110/1	Jan 17, 1995	Handbook for Paints and Protective Coatings for Facilities
Army	MIL-HDBK-1250(b)	Aug 18, 1995	Handbook for Corrosion Prevention and Deterioration Control in Electronic Components and Assemblies
Army	MIL-HDBK-1473A	Aug 29, 1997	Color and Marking of Army Materiel
Air Force	MIL-HDBK-1568	July 18, 1996	Material and Processes for Corrosion Prevention and Control in Aerospace Weapons Systems
Army	MIL-HDBK-1884	March 14, 1998	Coating, Plasma Spray Deposition
Army	MIL-HDBK-1886	March 27, 1998	Tungsten Carbide-Cobalt Coating, Detonation Process for
Army	MIL-HDBK-46164	Jan 2, 1996	Rustproofing for Military Vehicles And Trailers

(a) Base date of primary document; validation notices may be dated later. (b) MIL-HDBK-1250 is considered active, but not used for new design.

Table 3 Army Corps of Engineers and Naval Facilities Engineering Command documents

Number	Date	Title
Army Corps of Engineers Engineer Manuals (EM), Engineer Technical Letters (ETL), and Guide Specifications for Construction (CEGS)		
EM 1110-2-3400	April 1995	Painting: New Construction and Maintenance
ETL 1110-3-474	July 1995	Engineering and Design: Cathodic Protection
ETL 1110-9-10 (FR)	Jan 1991	Engineering and Design, Cathodic Protection System Using Ceramic Anodes
CEGS 16640	June 1997	Cathodic Protection System (Sacrificial Anode)
CEGS 16641	June 1997	Cathodic Protection System (Steel Water Tanks)
CEGS 16642	June 1997	Cathodic Protection System (Impressed Current)
Naval Facilities Maintenance and Operations Manuals (MO)		
MO-225	Aug 1990	Industrial Water Treatment
MO-307	Sept 1992	Corrosion Control

assigned to the items, then the scope section will provide the means for creating a Part Identification Number (PIN) based on the specification number.

2. Applicable Documents: lists all of the documents referenced in sections 3, 4, and 5 of the specification and the source that someone can use to obtain the documents.

3. Requirements: describes all of the requirements that the item(s) must meet in order to be acceptable under the specification. For each requirement, there should be an accompanying verification, or test cross-referenced and described in section 4. The type of specification—that is, detail, performance, or general—will determine the detail or performance nature of the requirements.

4. Verification: describes all of the tests or inspections that are needed to verify that the item(s) meets the requirements in section 3. The inspections or tests may be classed as first article, conformance, or qualification. This section may also provide sampling requirements and define the size of inspection lots.

5. Packaging: MIL-STD-961 provides a standard paragraph that must be used, stating that packaging shall be described in the contract or purchase order.

6. Notes: The notes in section 6 are only for general information or explanation. There is frequently an "intended use" paragraph in this section that can help a potential user of the specification determine which of the various types, classes, or grades of material covered by the specification is best suited to meet their needs. MIL-STD-961 now also requires the *Intended Use* paragraph to indicate what causes the product to be military-unique.

In addition to the aforementioned sections, some specifications may also have appendices that may or may not be mandatory parts of the specification. Frequently, appendices are used to describe some unique test or test apparatus required by section 4, when there is no suitable equivalent industry standard test method.

Sources of Documents

The definitive source for DOD and military specifications is through the Department of Defense Single Stock Point (DODSSP) for Military Specifications, Standards and Related Publications. The DODSSP is managed by the Document Automation and Production Service (DAPS), Building 4/D, 700 Robbins Ave., Philadelphia, PA 19111-5094. Documents in the DODSSP collection include:

- Military performance and detail specifications
- Military standards
- DOD-adopted non-government industry specifications and standards
- Federal specifications and standards and commercial item descriptions
- Data item descriptions (DID)
- Military handbooks
- Qualified products lists (QPL) and qualified manufacturer's lists (QML)
- U.S. Air Force and U.S. Navy aeronautical standards and design standards
- U.S. Air Force specifications bulletins

The Department of Defense Index of Specifications and Standards (DODISS) contains the complete list of documents in the DODSSP collection. This reference publication and all of the above documents are available to subscribers of an on-line database known as the "Acquisition Streamlining and Standardization Information System" (ASSIST). The ASSIST database is considered to be the official source of DOD specifications and standards. The documents are provided electronically in Adobe Portable Document Format (PDF). The ASSIST web page (2005) is http://assist.daps.dla.mil/online/start.

There is a complementary web site to ASSIST that has fewer features and only allows access to the electronic PDF format of the specifications. This site does not require an account or password and may be quicker. This web page is http://www.assistdocs.com.

In addition to being available through the ASSIST on-line database described previously, numeric and alphabetical lists of U.S. federal specifications, standards, and commercial item descriptions can be researched at the web site of the General Services Administration (GSA) Federal Supply Service (FSS): http://apps.fss.gsa.gov/pub/fedspecs.

Another popular source of DOD, Federal, and many other types of commercial and industry standards and specifications is the Information Handling Services (IHS), 15 Inverness Way East, Englewood, CO 80112. This is a commercial paid subscription service. For more information, see their web site at http://www.ihserc.com/index.html.

Notable Specifications, Standards, and Handbooks

While the number of military specifications, standards, and handbooks pertaining to corrosion control and coatings are slowly dwindling in favor of commercial standards, there are some that will continue to have significant use for years to come, as reference, guidance, or requirements documents. Several of those that find widespread use are described in this section.

MIL-HDBK-729: Corrosion and Corrosion Prevention; Metals. This is a 251-page general reference document that is freely available to DOD personnel, from equipment and facility maintainers, to designers and engineers, to program managers. Its primary value is as an educational tool tailored to corrosion in military equipment. It describes corrosion principles, the influences of different types of operating environments, and most of the common (and uncommon) forms of corrosion. It also discusses corrosion in each of the major alloy groups and corrosion-prevention techniques to match the different forms of corrosion discussed. Laboratory and field corrosion testing are also discussed. While this handbook has many figures and technical literature references, it is not intended to provide quantitative design data such as corrosion rates.

MIL-HDBK-310: Global Climatic Data for Developing Military Products. On the surface, this handbook does not purport itself to be a document concerned with corrosion control. However, it provides worldwide climate data that can affect corrosion and material performance, such as temperature, humidity, rainfall, and ozone concentrations. The data are presented for defined climatic regions such as "basic," "hot," "cold," and "coastal/ocean" and provides extremes and frequencies of occurrence. The data are intended for use in engineering analyses to set requirements, develop, and test military equipment. Therefore, it is used to support the use of MIL-STD-810, described below.

MIL-STD-810: Environmental Engineering Considerations and Laboratory Tests. This is an important, complex, and lengthy (549 pages) standard that is widely used in the acquisition of DOD equipment. (Equipment is often called "materiel" in military terms.) The standard was extensively revised in the "F" revision in 2000 and divided into two complementary parts. As described in the Foreword, the emphasis of these two parts is on "... tailoring a materiel item's environmental design and test limits to the conditions that the specific materiel will experience throughout its service life, and establishing laboratory test methods that replicate the effects of environments on materiel rather than trying to reproduce the environments themselves." Part One describes an Environmental Engineering Program (EEP) that focuses on the selection, design, and criteria for testing equipment under the specific environmental conditions, or stresses it is expected to be exposed to in service. A key element in Part One is providing guidance on the tailoring, or customizing, of the total program and the required testing. Part Two contains the laboratory test methods, which in turn provide for various degrees of tailoring depending on the specific test. The test methods cannot be specified simply by name or number; the equipment specification that invokes MIL-STD-810 must also provide the details of the test tailoring, such as frequency, cycle, or dwell times, temperatures, and so forth, and the associated acceptance criteria. Table 4 lists the test methods that are currently included in Part Two of MIL-STD-810.

MIL-STD-889: Dissimilar Metals. Many well-meaning writers of military equipment procurement specifications include a requirement that the design shall avoid dissimilar metal couples and galvanic corrosion or shall take steps to mitigate them. Unfortunately, in many cases that language is all that is used, leaving the definition of what exactly constitutes a dissimilar metal couple subject to the interpretation of the contractor or designer. Sometimes a minimum voltage potential difference between the metals is cited in the specification, above which a couple shall be considered to be dissimilar. However, experienced corrosion engineers know that it is not simply the potential difference on a galvanic series (typically only provided for seawater) that determines whether significant galvanic corrosion will be a design risk. Factors such as the anodic to cathodic area ratio (the "area effect" discussed elsewhere in this Volume), polarization and passivity, the nature of the electrolyte, and the temperature all should be considered. While not a substitute for a design review by a corrosion professional, MIL-STD-889 is sometimes cited in a contract specification to provide a better definition of dissimilar metals and to provide guidance for corrosion-protection methods that can be used when galvanic couples cannot be avoided. The dissimilar metals charts in *ASM Handbook,* Vol 13B, 2005, are based on this standard.

MIL-DTL-53072: Chemical Agent Resistant Coating (CARC) System; Application Procedures and Quality Control Inspection. A unique military requirement for the exterior coatings on ground equipment is that they be able to provide corrosion protection and camouflage, while resisting the absorption of harmful chemical warfare agents. Furthermore, these coatings should facilitate cleaning with caustic and other types of decontaminating solutions in the event that the equipment is exposed to these agents. The Army and the Marine Corps are the primary users of ground equipment, which includes tactical combat vehicles (tanks, artillery, armored personnel carriers, etc.), and logistics and engineering support equipment (trucks, trailers, cranes, bulldozers, etc.). However, both the Air Force and Navy also use this equipment, so the CARC coating system finds wide application. There are a number of pretreatments and coatings that comprise the total CARC system, which generally includes epoxy primers and polyurethane topcoats. MIL-DTL-53072 is the specification that covers the general requirements for application and inspection of the CARC system. From the scope section of the specification, it also:

> ... is intended for use as a guide in selection of the appropriate materials and procedures, and as a supplement to information available in ... referenced cleaning, pretreating, and coating specifications. The document also includes information on touchup/repair, health and safety guidelines, environmental restrictions, national stock numbers (NSN) for CARC and CARC-related materials, and application equipment and techniques.

Department of Defense Corrosion Policy

In 2003, the DOD initiated a new corrosion prevention and control policy aimed at "... implementing best practices and best value decisions for corrosion prevention and control in systems and infrastructure acquisition, sustainment, and utilization" (Ref 6). This policy requires that all major new U.S. military acquisition programs create a corrosion-control strategy and a "Corrosion Prevention and Control Plan" (CPCP). To accompany this policy, a planning guidebook has also been created by a team of corrosion-control professionals in DOD (Ref 7). This guidebook, and other information related to the DOD corrosion-prevention policy, can be found at the web site www.DoDcorrosionexchange.org.

REFERENCES

1. "Colors Used in Government Procurement," FED-STD-595B, Dec 15, 1989
2. "Department of Defense Standard Practice; Defense and Program-Unique Specifications Format and Content," MIL-STD-961E, Aug 1, 2003
3. "Department of Defense Standard Practice; Defense Standards Format and Content," MIL-STD-962D, Aug 1, 2003
4. "Department of Defense Standard Practice; Defense Handbooks Format and Content," MIL-STD-967, Aug 1, 2003
5. "Standard Practice for Unified Facilities Criteria and Unified Facilities Guide Specifications," MIL-STD-3007, Oct 1, 2004
6. "Corrosion Prevention and Control," Under-Secretary of Defense (Acquisition, Technology & Logistics) Memorandum for Secretaries of the Military Departments, Nov 12, 2003
7. "Corrosion Prevention and Control Planning Guidebook," Spiral 2, U.S. Dept. of Defense, PDUSD(AT&L), July 2004

Table 4 MIL-STD-810 environmental laboratory test methods

Method number(a)	Title
500.4	Low Pressure (Altitude)
501.4	High Temperature
502.4	Low Temperature
503.4	Temperature Shock
504	Contamination by Fluids
505.4	Solar Radiation (Sunshine)
506.4	Rain
507.4	Humidity
508.5	Fungus
509.4	Salt Fog
510.4	Sand and Dust
511.4	Explosive Atmosphere
512.4	Immersion
513.5	Acceleration
514.5	Vibration
515.5	Acoustic Noise
516.5	Shock
517	Pyroshock
518	Acidic Atmosphere
519.5	Gunfire Vibration
520.2	Temperature, Humidity, Vibration, and Altitude
521.2	Icing/Freezing Rain
522	Ballistic Shock
523.2	Vibro-Acoustic/Temperature

(a) Numbers as of MIL-STD-810F, Notice 2, Aug 30, 2002

Corrosion Control for Military Facilities

Ashok Kumar and L.D. Stephenson, U.S. Army Engineer Research and Development Center (ERDC)
Construction Engineering Research Laboratory (CERL)
Robert H. Heidersbach, Dr. Rust, Inc.

CORROSION DEGRADATION is the most costly and pervasive maintenance and repair problem in the U.S. Army. Studies have shown that the annual corrosion-related costs at U.S. Army installations are about 13 to 15% of the Maintenance of Real Property and Minor Construction costs. About half of these corrosion-related costs could be attributed to structural, electrical, and mechanical components in buildings and the other half to utilities. Anecdotal evidence suggests that this percentage has not increased with the aging of facilities since older facilities are being replaced with newer facilities incorporating improved corrosion-control technologies.

Public Law addresses "Military equipment and infrastructure prevention and mitigation of corrosion" and requires the Secretary of Defense to "develop and implement a long-term strategy to reduce corrosion and the effects of corrosion on the military equipment and infrastructure of the Department of Defense (DoD)" (Ref 1).

Components susceptible to corrosion include building structural components and utilities (that will be privatized at some installations) such as metal buildings, metal roofing, aircraft hangars, outdoor electrical sheet metal for air conditioners, electrical boxes, underground pipes (gas, water, steam, high-temperature hot water), pipes in buildings, boilers, chillers, condensate lines, water storage tanks, and so forth.

Title 10 of the Uniform Service Code, Section 2228 directs the DoD to actively pursue a department-wide approach to combat corrosion (Ref 1). In response to Congressional interest, the Secretary of Defense has designated the creation of an Office of Corrosion Policy and Oversight, and the Office of the Secretary of Defense (OSD) Corrosion Prevention and Control Program for weapons and facilities was initiated to demonstrate and implement emerging corrosion-control technology that can save up to 30% of the corrosion-related costs (Ref 2). Corrosion-control technologies being implemented under this program include:

- Coatings
- Cathodic protection (CP)
- Advanced (corrosion-resistant) materials selection and design
- Water treatment
- Remote corrosion assessment and management

The major benefit of the implementation of these corrosion-control technologies at Army installations is the extension of the service life of buildings and other structures.

Information presented here is from a variety of Army and Air Force installations throughout the United States and focuses on selected examples of corrosion and corrosion-prevention projects for infrastructure that have recently been completed or are currently underway. Many of the case studies are taken from the research program in which the authors have participated or have direct knowledge.

The Environment

Facilities and equipment operated by the U.S. Army are exposed to a wide variety of environmental conditions, including soils, waters, or atmospheres of varying corrosivity. The various types of resulting corrosion can create a costly maintenance and repair burden while adversely affecting Army operations. The major types of corrosion phenomena are (a) uniform corrosion, (b) pitting attack, (c) galvanic corrosion, (d) environmentally induced delayed failure (e.g., stress-corrosion cracking), (e) concentration-cell corrosion, (f) dealloying, (g) intergranular corrosion, and (h) various forms of erosion corrosion. It is not at all unusual for more than one form of corrosion to act on the same structure at the same time. For example, the steel components in a steam-heating system can be simultaneously subjected to conditions that cause uniform corrosion, pitting attack, galvanic corrosion, and the cavitation form of erosion corrosion.

It is important to understand that the characteristics of soils, waters, and atmospheres at Army installations can be expected to vary greatly and that this variation has important implications for corrosion-mitigation strategies. No two installations have identical environmental conditions, and corrosion-promoting conditions can even be expected to vary within the boundaries of a given installation. For example, soils at an inland installation may be even more naturally corrosive than the chloride-containing soils along an ocean. Many island and peninsular locations have atmospheric conditions that are severely corrosive on the windward sides but relatively mild on the leeward sides. There are also many inland locations where the atmosphere can be unusually aggressive due to the proximity of facilities such as paper mills or power plants burning high-sulfur coal. Also, the use of road salts to melt snow and ice in the higher latitudes is a contributing factor to corrosion of steel structures. For example, Fort Drum is an inland location in the northern United States, where road salts led to premature corrosion of steel doors, which have since been replaced by fiberglass-reinforced plastic (FRP).

Although protective coatings are an effective option for mitigating many varieties of atmospheric corrosion, they are basically useless in steam-heating systems. Similarly, methods for altering the internal environments in a steam-heating system to mitigate corrosion of the boiler, pipes, and heat exchangers should *not* be considered for corrosion control on the soil-side surfaces of the steel conduits and casings that house the steam and condensate lines. In the latter case, effective corrosion control can be achieved only by protective coatings used in conjunction with CP. However, neither of these corrosion-control techniques will work properly if wet and chemically aggressive insulation contacts the inside surfaces of the casings/conduits or the outside surfaces of the steam and condensate pipes.

References on corrosivity of soils, waters, and atmospheres at various military locations are available from websites maintained by the Army Corps of Engineers, Engineer Research and Development Center, Construction Engineering Research Laboratory (ERDC-CERL).

Case Studies

These case studies illustrate typical examples of the types of corrosion problems found on military installations. Most of the problems encountered are typical of those found on civilian installations, although some problems are unique to the military.

Fences. Secure fencing is more important at military installations than at many other locations. Sensitive equipment and ammunition is often stored near roads and other access points where military personnel and their dependents travel on a frequent basis. Airfields, firing ranges, and ammunition storage facilities are typical areas where fencing security is particularly important. Not only does the low-alloy steel chain-link fencing corrode, but the posts often show more corrosion of the structural steel, especially in severe environments, such as coastal atmospheres.

Replacement of these fences is expensive, and a typical military installation will have many miles of such fencing. Polyvinyl chloride (PVC)-clad galvanized steel chain-link fencing should be used in severely corrosive atmospheres. Posts, gates, and their accessories should have a coating system consisting of 27 mg of zinc per square centimeter (0.9 oz of zinc per square foot), a minimum of 76 µm (0.3 mil) of cross-linked polyurethane acrylic, and a minimum of 178 µm (7 mil) vinyl topcoat. For severely corrosive atmospheric conditions, the topcoat should be 381 µm (15 mils) thick. Chain-link fence systems may be fabricated from anodized aluminum alloys provided they have adequate strength (Ref 3, p 31).

Exfoliation of an aluminum guardrail (a form of fencing) on a public highway passing through a military installation can be caused by the rubbing of the aluminum against the support bolt. Fretting corrosion that starts at this location is often due to the daily expansion and contraction of the aluminum causing it to rub against the fixed bolt.

Structural steel is used for many purposes on military installations including buildings, bridges, electric utility poles, and support structures for various types of equipment. Figures 1 and 2 show an electric utility light pole within the confines of a wastewater treatment plant (WWTP) at an Army installation. The pole corroded near the base as well as along the vertical portions of the pole. Causes of the corrosion shown in these pictures include the humid environment and the attack of hydrogen sulfide from the WWTP. Poor surface preparation prior to application of protective coatings, shipping and handling damage prior to erection, and abrasion and submersion of the pole when water and sand or gravel accumulate near the base of the pole may have exacerbated the corrosion. Ultraviolet (UV) degradation of protective coatings is also a concern, even though modern protective coatings are more resistant to UV degradation. In many cases, it is possible to remove the deteriorated coating by sandblasting and, after proper surface preparation, recoat the surface with coal-tar epoxy coatings to protect against future corrosion (Ref 3, p 47).

It is also noted that poles, standards, and accessories (shafts, anchor belts, bracket arms, and other hardware) for such applications should be galvanized steel for relatively nonaggressive atmospheres. In more aggressive environments, such as within the confines of a WWTP, the same components should be aluminized steel, type 304 stainless steel, or type 201 stainless steel.

The highest structural loading on poles and guardrails occurs near the base of the structure where the maximum bending moment is located. Thus, corrosion on structural steel, which is commonly more pronounced near the base, occurs where the remaining metal will have the highest mechanical loading.

Figure 3 shows a guardrail that must be replaced because of corrosion damage. Figure 4 is a close-up of a guardrail that was attached to a concrete curb. The steel rail was inserted into a hole in the concrete and then soldered in place using a lead-base low-melting-point alloy. Removal of the lead alloy would create a major hazardous waste disposal problem. Current guidelines for dealing with lead in construction such as that shown in Fig. 3 allow the material to remain in place. Having identified the lead alloy prior to guardrail removal, the replacement was redesigned, so that the attachment was left in place and the new guardrail was bolted to the concrete curb. In early 2004, a positive chemical analysis to identify lead use in the attachment cost only a small fraction (<0.04%) of the cost for lead hazardous waste disposal of the lead. Current specifications require that guardrails and handrails be fabricated from a suitable anodized alloy such as 6061-T6 (Ref 3, p 37).

Metal Roofing. Many military installations have "temporary" buildings built during times when expansion of facilities takes priority over quality of construction. Many of these structures then remain in service for decades beyond their design life. Corrosion of sheet metal roofing is common on these buildings and the corrosion of

Fig. 1 Corroded low-carbon steel light pole due to corrosive atmosphere at a wastewater treatment plant on a military installation

Fig. 2 Close-up of corrosion at the base of the pole shown in Fig. 1

Fig. 3 Corrosion pattern on a mild steel pedestrian guardrail at a military installation exposed to a coastal atmosphere

Fig. 4 Close-up of the base of the vertical pole shown in Fig. 3. Note that the pole was soldered in place using a lead-base low-melting-point alloy.

roofs is a continuing concern. Coated galvanized steel is not appropriate for many coastal or severely corrosive atmospheres. The preferred material for roofing and siding is type 2 aluminized steel, with factory-applied oven-baked fluoropolymer enamel coatings applied to a minimum thickness of 25 μm (1 mil), although thicker coatings are preferred when the coatings are subjected to wind-blown sand environments, such as in the Middle East.

Military specifications for metal roofs and sidings have recently been updated by specifying ASTM D 5894, "Standard Practice for Cyclic Salt Fog/UV Exposure of Painted Metal, (Alternating Exposures in a Fog/Dry Cabinet and a UV Condensation Cabinet)" for 2016 h to evaluate candidate galvanized and zinc-aluminum alloys on steel, as well as the use of polymeric coatings (Ref 4). Polyvinylidiene fluoride (PVF2) coated and silicone-modified polyester (SMP) coated galvanized and zinc-aluminum alloys have been found to pass these tests and are acceptable for metal roofing. The coatings on both materials were 25 μm (1 mil) thick (Ref 5). Long-term appearance of PVF2-coated zinc-aluminum or galvanized steel substrates, as evaluated by the ASTM D 5894 test, is excellent. PVF2 is only slightly better than SMP. Galvalume (Zn-55 Al) coatings performed slightly better than galvanized coatings (Ref 5).

Doors and windows are locations where corrosion is frequently more severe than walls, roofs, and other components of a building. Figures 5 and 6 show corroded doorways near a wastewater treatment plant on a military installation. The replacement of a few doors is a fairly inexpensive project. However, since there are large numbers of doors on military installations any cost savings due to the use of more corrosion-resistant materials can result in major savings on maintenance costs. Fiberglass-reinforced plastic doors with type 304 stainless steel handles and kickplates are viable corrosion-resistant alternatives to steel doors, and they are commercially available in a variety of styles and colors, with special features such as being fireproof and resistant to certain industrial chemicals. If steel doors and frames are required (e.g., for security reasons), they should be aluminized with a factory-applied oven-baked fluoropolymer enamel coating. The benefits of implementing corrosion-resistant building components composed of materials such as FRP or aluminized steel and stainless steel hardware are extended service life, reduced maintenance cost, and enhanced appearance (Ref 3, p 31).

Electrical/Mechanical Systems. Heating, ventilation, and air conditioning (HVAC) systems are major sources of corrosion problems on military installations. Many, but not all, of these systems are located inside buildings. Channeling in a steam condensate return pipeline on a military installation (Fig. 7) is caused by acidic condensate. Condensate lines, like most liquid return lines, normally run only partially full, and corrosion occurs if inhibitors do not prevent pH shifts that cause the condensate to become acidic. Leakage of air (oxygen and carbon dioxide) into the return line causes this corrosion. (See the article "Corrosion in the Condensate-Feedwater System" in this Volume and "Corrosion Inhibitors in the Water Treatment Industry" in *ASM Handbook*, Vol 13A, 2003.) Military installations are limited in the choice of the chemicals they can use for treating steam/condensate systems, because their steam, unlike the steam in many industrial applications, is often used for cooking and other applications where toxic corrosion inhibitors cannot be used. Emerging corrosion-control programs are demonstrating nontoxic "green" corrosion inhibitors for use in steam systems at military installations. Industrial and commercial utility plants often use on-line monitoring and control systems to track water chemistry parameters and control chemical treatment and blowdown. Many of these systems, such as pH and conductivity controllers, are relatively easy to use and maintain. This technology reduces manpower for system monitoring and control of corrosion/scale, increases life cycle of heating and cooling systems, and reduces the use of environmentally sensitive chemicals (Ref 6).

Many military installations transport steam long distances. Steam leaked under the insulation may condense and wet the insulation, leading to additional corrosion. Corrosion under insulation is very hard to detect (Ref 7, 8). While corrosion can, and does, occur on the outside of steam lines, it is much more likely to occur on the outside of condensate return lines. Since condensate lines operate below the boiling point, water is much more likely to accumulate on the metal pipe surfaces underneath insulation on condensate lines than it is on steam lines. Concentration of salts caused by evaporation of moisture underneath the insulation can lead to stress-corrosion cracking on the outside of condensate lines. Similar problems occur on water lines. The corrosion shown on the pipe in Fig. 8 was caused by degradation of the elastomeric insulation used to protect the outside of the pipe. Neither the metal pipe nor substrate had been coated. Guidelines have been developed on how to specify and install insulation to prevent this

Fig. 6 Close-up of corrosion on doors shown in Fig. 5

Fig. 5 Corroded metallic doorway at a military installation water treatment plant due to chlorine atmosphere

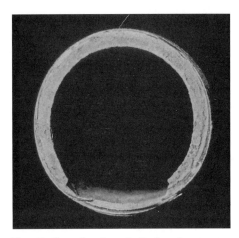

Fig. 7 Channeling ("grooving") of condensate return line due to carbon dioxide leaking into steam/condensate system

degradation. They include (a) avoiding coupling dissimilar metals together in the presence of conductive moisture and (b) properly sealing insulation against the intrusion of humid air.

Fuel Transport and Storage Systems. The most viable option for mitigating corrosion of underground fuel storage tanks exposed to aggressive soils is a combination of protective coatings and CP. The preferred system for underground fuel storage tanks (≥76,000 L, or 20,000 gal) is horizontal double-wall tanks with full 360° secondary containment, glass-fiber-reinforced polyester tanks and accessories manufactured according to ASTM D 4021-81 (Ref 9). Underground fuel storage tanks can be installed at service stations using horizontal double-wall steel tanks that are coated (a) internally with epoxy or urethane and (b) externally with high-performance coal-tar epoxy, or fiberglass/resin for maximum protection in very aggressive soils. Steel pipes should be similarly coated. Both the tanks and pipes and should be provided with CP (Ref 3, p 78–79).

Chemical Process Equipment. Most military installations have limited chemical processing responsibilities, and the numbers of chemically trained personnel, both military and civilian, are smaller than in many other industries. Nonetheless, chemical processing equipment, and the corrosion failures related to it, are present on many military installations.

Figure 9 shows a corroded valve control handle at a wastewater treatment plant. The economics of alternative materials and the proper selection of these materials are beyond the background of the engineering organizations on many military installations. Stainless steels and other corrosion-resistant alloys are an available economical alternative that can be implemented to minimize the costs and disruption associated with corrosion failures such as those shown in Fig. 9. Other recommendations for mitigation of corrosion at wastewater treatment plants include:

- The use of petrolatum tape coatings for exposed pipes/valves
- Restoration coatings for deteriorated concrete for large structures such as clarifiers
- Corrosion-resistant metals and polymers for electrical junction boxes
- Polymeric gaskets not susceptible to embrittlement
- Cathodic protection for immersed steel components, such as rake arms and metal weirs in clarifiers

Emerging Corrosion-Control Technologies

Protective Coatings and Linings. Many new coatings are under development in the civilian sector and are being applied on military installations. Figure 10 shows a potable water pipe contaminated by rusty water that services an Air Force Base hospital. It was repaired in situ by draining, drying, and abrasively blasting the interior followed by pumping an epoxy slug through the pipe, as shown in Fig. 11. The coating application to a thickness of 350 to 500 μm (14 to 20 mils) was successful in rehabilitating the piping system in question with a cost savings of more than 30% over the cost of replacement piping. The in situ pipe coating technology accommodates a variety of pipe lengths, diameters (>12.5 mm, or 0.5 in., diameter), bends, and materials. It eliminates corrosion and scaling, and lead and copper dissolution in potable water and energy piping systems, and restores full water pressure.

Figure 12 shows a deteriorated protective coating for a water deluge tank (part of a fire-suppression system at an Army airfield), where

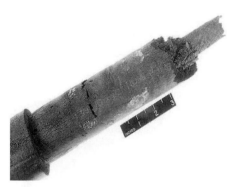

Fig. 8 Corrosion of hot water line after insulation was degraded by water intrusion

Fig. 9 Corroded control handle exposed to corrosive atmosphere at a wastewater treatment plant on a military installation

Fig. 10 Corroded water pipe contaminates water that services an Air Force Base hospital

Fig. 11 In situ epoxy coating applied to the inside of the pipe shown in Fig. 10 after abrasive blast cleaning

the original lead-base paint was overcoated to dry film thicknesses of about 190 μm (7.5 mils) with test coatings of two moisture-cured polyurethanes. Both overcoating systems performed well over a winter test period with no evidence of blistering, spalling, or peeling. This entire tank was then repainted with one of the moisture-cured polyurethane systems. The coating exhibited no evidence of cracking or peeling, no evidence of rust, and no evidence of corrosion for the past 4 years (Ref 10).

The benefit of implementing innovative overcoating technology, such as applications of moisture-cured polyurethane, is to extend the service life of steel structures, reduce maintenance cost, improve the performance of equipment, and eliminate the expense of removing lead-base paint. Overcoating can be significantly less expensive than other maintenance practices, particularly when the pre-existing coating contains lead or other hazardous materials. Deluge tanks are mission critical equipment at an Army airfield, and flights are grounded if these systems are not operable.

Cathodic protection is one of many conventional corrosion-control techniques used at DoD installations. Most CP systems are installed by civilian contractors. Since engineers at most DoD installations have limited experience in CP, it is important that agency-wide guidance is readily available and up-to-date.

Industry-standard storage tanks are widely used for storing liquids underground. They have factory-installed CP systems. Unfortunately, some tanks were provided (as recently as March 2003) with sacrificial zinc anodes for use on tanks in all types of soil and sacrificial magnesium anodes for use only in high-resistivity soil (see the article "Cathodic Protection" in *ASM Handbook*, Vol 13A, 2003). This application of zinc and magnesium anodes was contrary to conventional wisdom for the use of these anodes. Conventional practice allows magnesium anode usage in all soil conditions and the use of zinc anodes only in soils having very low resistivity (Ref 11). It is necessary that engineers at military installations be sufficiently familiar with these types of systems so that they can spot misuse of anodes or inappropriate CP practices. Obtaining instruction on these practices is a needed part of their training.

Impressed-current CP is used to prevent internal corrosion of the water side of potable water storage tanks by applying a negative potential to the structure from an external source. Remote monitoring units (RMUs) are being used to demonstrate the reliability of CP in controlling corrosion of elevated water towers. Cathodic protection systems for water storage tanks must be periodically tested in order to ensure proper performance (Ref 12). Remote monitoring units provide the ability to continuously monitor CP system performance data from remote locations using modem-equipped personal computers. Figure 13 shows a CP rectifier coupled to a RMU. Elevated water towers are frequently the largest structures at military installations that use impressed-current CP.

In the past, large and heavy iron-silicon and graphite anodes were required for CP systems, which made the anode vulnerable to debris and ice damage and prone to field installation problems, leading to numerous electrical shorts in the system. Ceramic-coated anodes, usually made by depositing mixed metal oxides onto titanium substrates, are an alternative to silicon-iron and graphite anodes. The ceramic anode makes CP available at one-half the life cycle cost of previous technologies and in a size reduction that permits installation in areas previously too small (Ref 13). One ampere of current supplied to the ceramic anode will stop corrosion on 46 m^2 (500 ft^2) of uncoated steel; that is, the consumption rate of conducting ceramic materials such as mixed metal oxides is 500 times less than the silicon-iron and graphite anodes. This has resulted in a smaller anode, weighing 50 times less, with the same life span and performance.

Advanced (Corrosion-Resistant) Materials Selection and Design. Industrial practice on the economical and safe use of existing materials changes with development of new materials and their application in different environments.

Coating degradation and corrosion can occur on steel vent pipes where a buried gas line crosses under a roadway. Conventional practice is to use steel casings to prevent natural gas distribution lines from excessive traffic loads. Viable alternatives include the use of plastic vent pipes for this application, which is being field tested as part of a technology demonstration at an Army installation. If the demonstration project is successful, plastic vent pipes can then be used nationwide at a significant cost savings.

Other advanced material applications include the use of stainless steels and aluminum where coated carbon steel had been used in the past, as well as widespread use of composite materials as substitutes for steel where structural loading requirements can be met.

Water Treatment. Many military installations transport steam over long distances and use the steam in applications where toxic corrosion inhibitors cannot be used. The conventional civilian practice of using neutralizing amines is precluded in many of these applications. The Army is evaluating the use of "green" corrosion inhibitors such as othoxalated soya amines for corrosion control on condensate systems. Successful demonstration of these inhibitors will reduce the "CO_2 channeling" corrosion shown in Fig. 7. Automated monitoring of steam and condensate chemistry is expected to reduce the incidence of this type of corrosion (Ref 6, p 48).

Inspection and Monitoring. Nondestructive analysis of equipment condition is a priority with many DoD installations. Figure 14 shows where a clay sewer line on a military installation was breached during the horizontal insertion of a plastic gas distribution line. Discovery of five breaches of the sewer lines led the installation to question whether damage to other buried utilities had gone undetected. A remote camera was utilized to inspect more than 6000 m (20,000 ft) of buried sewer lines in only four weeks.

The inspections revealed no additional breaches of the sewer line due to gas pipeline penetration. However, the inspection did result in the identification of a blockage in one of the sewer lines and allowed maintenance personnel

Fig. 12 Deluge tank (for fire-suppression system) with test patches overcoat applied to lower half of exterior surface. Later, the entire tank was overcoated with a moisture-cure polyurethane system

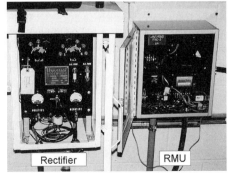

Fig. 13 Rectifier with remote monitoring system (RMU) for use as part of an elevated water tower cathodic protection system for water side corrosion protection.

Fig. 14 Clay sewer line breached as a new gas distribution line was laid, due to lack of knowledge of where the sewer lines were located. A remote camera inserted into the sewer lines was later utilized to ascertain that no additional breaches had occurred.

to excavate and repair only the affected section of that pipeline. This saved an estimated 8 man-years in labor cost and completed the inspection in one-third of the time it would have taken if maintenance personnel had to excavate the entire area affecting the 400 housing units.

Remote inspection technologies, such as the sewer inspection camera system provide rapid assessment of the status of pipelines and eliminate extensive digging to determine locations where leaks or blockages are believed to exist. These systems can also be used for preventive maintenance purposes, for example, to determine if the pipe is beginning to corrode, or areas where the pipe wall thickness is reduced, so that corrective action must be taken.

Leak Detection. Corrosion of pipes often results in leaks. Leaks under pipe insulation can often result in additional corrosion. Acoustic emission leak detection technology has been developed to accurately locate and estimate the size of leaks in metallic water and fuel pipes. The technology operates at 15,000 Hz and, using coincidence detection, is capable of detecting leaks as small as 0.4 L/h (0.1 gal/h). Three acoustic sensors, with a minimum of two sensors bracketing the leak, are required for leak location. The sensors may be up to 600 m (2000 ft) apart. Implementation of this technique takes only about 1 h to investigate 1.6 km (1 mile) of pipeline. Acoustic leak detection technology pinpoints leaks for a small fraction of the cost of excavation normally used to locate the leaks. Drawbacks of this technology are: (a) it requires cables to connect the instrument to the sensors, and (b) it does not work as well in plastic pipes because of high attenuation, or in steam lines because of inherent noise.

Below-Grade Moisture Mitigation. Below-ground concrete facilities, such as basements and elevator shafts may sustain structural damage due to chronic water seepage through the walls. Electro-osmotic pulse (EOP) technology offers an alternative to the trench-and-drain approach by mitigating water-related problems from the interior of affected areas without the cost of excavation. The EOP method uses pulses of electricity to reverse the flow of water seepage, actually causing moisture to flow out of the basement walls, away from the building. Prior to application of the EOP system, major sources of active water were addressed by locating and repairing cracks with epoxy injection and/or hydraulic cement. Once active sources of intrusion are repaired, the system is installed to prevent the water from reaching the repair joint. The EOP system is installed by inserting anodes (positive electrodes) into the concrete wall or floor on the inside of the structure and by placing cathodes (negative electrodes) in the soil directly outside the structure. The power supply typically consumes 30 W of power for every 9 m^2 (100 ft^2) treated. The EOP system is relatively easy to install compared to conventional waterproofing methods and costs about 40% less per linear foot of wall treated than traditional methods (Ref 6, p 48–50).

Conclusions

In summary, many corrosion problems associated with military installations are similar to those encountered in civilian installations. Military installations provide support for extensive training and preparedness for battle, and military culture revolves around execution of specific military missions. Corrosion of military facilities components and utilities, therefore, may be more likely to occur than in the civilian sector, since the military budgets provide more funds for maintenance of weapons and field equipment or training than for preventive maintenance for facilities. Also, installation maintenance personnel are frequently unaware of emerging technologies and best practices that could significantly reduce corrosion problems. Efforts associated with corrosion control must emphasize understanding of available corrosion-control techniques used in other industries and applying corrosion-monitoring techniques to critical military facilities.

REFERENCES

1. *Bob Stump National Defense Authorization Act for Fiscal Year 2003,* 10 U.S.C. 2228, Public Law 107-314, § 1067, Dec 2, 2002
2. Under Secretary of Defense for Acquisition, Technology & Logistics, Memorandum, *Corrosion Prevention and Control,* Nov 12, 2003, Appendix A
3. J.R. Myers, A. Kumar, and L.D. Stephenson, "Materials Selection for Army Installation Exposed to Severely Corrosive Environments," ERDC-CERL TR-02-5, Engineer Research and Development Center, Construction Engineering Research Laboratory, March 2003
4. "Standard Practice for Cyclic Salt Fog/UV Exposure of Painted Metal (Alternating Exposures in a Fog/Dry Cabinet and a UV Condensation Cabinet)," D 5894, *Annual Book of ASTM Standards,* ASTM International
5. T. Race, A. Kumar, and L.D. Stephenson, "Evaluation of Galvanized an Galvalume Paint Duplex Coatings System for Steel Building Panels," ERDC-CERL TR-0208, Engineer Research and Development Center, Construction Engineering Research Laboratory, Feb 2002
6. V. Hock et. al., Success Stories: Military Facilities Inter-Service Cooperation Reduces Infrastructure Corrosions, *AMPTIAC Q.,* Vol 7 (No. 4), 2003
7. P. Elliott, Designing to Minimize Corrosion, *Corrosion,* Vol 13A, *ASM Handbook,* ASM International, 2003, p 929–939
8. W. Pollock and C. Steely, *Corrosion under Wet Thermal Insulation,* NACE International, 1990
9. "Standard Specifications for Glass-Fiber-Reinforced Polyester Underground Petroleum Storage Tanks," D 4021-81, *Annual Book of ASTM Standards,* ASTM
10. T. Race, A. Kumar, R. Weber, and L.D. Stephenson, "Overcoating of Lead-Based Paint on Steel Structures," ERDC-CERL TR-03-5, Engineer Research and Development, Construction Engineering Research Laboratory, March 2003
11. "Cathodic Protection Anode Selection," Public Works Technical Bulletin 420-49-37, Dept. of the Army, U.S. Army Corps of Engineers, June 15, 2001
12. "Cathodic Protection for Steel Tanks," Unified Facility Guide Specifications UFGS 13111A, U.S. Army Corps of Engineers
13. "Cathodic Protection Using Ceramic Anodes," ETL 1110-9-10 (ER), Dept. of the Army, U.S. Army Corps of Engineers, Jan 5, 1991

SELECTED REFERENCES

- C. Allen, Design Systems to Prevent Corrosion under Thermal Insulation, *Mater. Perform.,* Vol 32 (No. 3), March 1993
- H.R. Amler and A.A.J. Bain, Corrosion of Metals in the Tropics, *J. Appl. Chem.,* Vol 5, 1955, p 47
- N.S. Berke, "The Effects of Calcium Nitrite and Mix Design on the Corrosion Resistance of Steel in Concrete: Part 2, Long-Term Results," paper No. 132, *Corrosion 87,* National Association of Corrosion Engineers
- R.T. Blake, *Water Treatment for HVAC and Potable Water Systems,* McGraw-Hill, 1980, p 44–45
- G.H. Brevoort, M.F. MeLampy, and K.R. Shields, Updated Protective Coating Costs, Products, and Service Life, *Mater. Perform.,* Vol 36 (No. 2), Feb 1997
- G. Byrnes, Preparing New Steel for Coating, *Mater. Perform.,* Vol 33 (No. 10), Oct 1994
- J. Carew, A. Al-Hashem, W.T. Riad, M. Othman, et al., Performance of Coating Systems in Industrial Atmospheres on the Arabian Gulf, *Mater. Perform.,* Vol 33 (No. 12), Dec 1994
- A. Cohen and J.R. Myers, Mitigating Copper Pitting Through Water Treatment, *J. Am. Water Works Assoc.,* Vol 79 (No. 2), Feb 1987, p 58–61
- A. Cohen and J.R. Myers, Pitting Corrosion of Copper in Cold Potable Water Systems, *Mater. Perform.,* Vol 34 (No. 10), Oct 1995, p 60–62
- A. Cohen and J.R. Myers, Overcoming Corrosion Concerns in Copper Tube Systems, *Mater. Perform.,* Vol 35 (No. 9), Oct 1996, p 53–55
- A. Cohen and J.R. Myers, Erosion-Corrosion of Copper Tube Systems by Domestic Waters, *Mater. Perform.,* Vol 37 (No. 11), Nov 1998, p 57–59
- I.J. Cotton, Oxygen Scavengers—The Chemistry of Sulfite under Hydrothermal

- Conditions, *Mater. Perform.*, Vol 26 (No. 3), March 1987
- B.A. Czaban, Foamed Insulation Around Carbon Steel Pipe Creates Site for Corrosion Failure, *Mater. Perform.*, Vol 32 (No. 5), May 1993
- S. Dean, Corrosion Testing of Metals under Natural Atmospheric Conditions, *Corrosion Testing and Evaluations: Silver Anniversary Volume*, STP 1000, R. Baboian and S.W. Dean, Ed., American Society for Testing Materials, 1990
- C.K. Dittmer, R.A. King, and J.D.A. Miller, Bacterial Corrosion of Iron Encapsulated in Polyethylene Films, *Br. Corros. J.*, Vol 10 (No. 1), 1975, p 47–51
- R.W. Dively, Corrosion of Culverts, *Mater. Perform.*, Vol 31 (No. 12), Dec 1992, p 47–50
- R.W. Drisko, Coatings for Tropical Exposures, *J. Prot. Coat. Linings*, Vol 16 (No. 3), March 1999, p 17–22
- J.R. Duncan and J.A. Balance, Marine Salts Contribution to Atmospheric Corrosion, *Degradation of Metals in the Atmosphere*, STP 965, S. Dean and T.S. Lee, Ed., American Society for Testing Metals, 1988, p 316–326
- "Final Report for Air Force Base Environmental Corrosion Severity Ranking System" F09603-95-D-0053, NCI Information Systems, 1998
- G.T. Halvorsen, Protecting Rebar in Concrete, *Mater. Prot.*, Vol 32 (No. 8), March 1993, p 31–33
- G. Illig, Protecting Concrete Tanks in Water and Wastewater Treatment Plants, *Water Eng. Manage.*, Vol 145 (No. 9), Sept 1998, p 50–57
- B.L. Jones and A.J. Sansum, Review of Protective Coating Systems for Pipe and Structures in Splash Zones in Hostile Environments, *Corros. Manage.*, No. 14, Oct/Nov 1996
- R.W. Lane, *Control of Scale and Corrosion in Building Water Systems*, McGraw-Hill, 1993, p 164–165, 203
- R. Leong, "Design Criteria for Construction in the U.S. Army Kwajalein Atoll," Memorandum CEPOD-ED-MP(415-10f), U.S. Army Construction Engineering Research Laboratory (CERL), June 24, 1991
- F.W. Lipfert, M. Bernarie, and M.L. Daum, "Derivation of Metallic Corrosion Functions for Use in Environmental Assessments," BNL 51896, Brookhaven National Laboratory, 1985
- "Materials of Construction: Gas Feeders," Capitol Controls Company, Inc., Colmar, PA, 1991
- Z.G. Matta, Protecting Steel in Concrete in the Persian Gulf, *Mater. Perform.*, Vol 33 (No. 6), June 1994
- G.S. McReynolds, Prevention of Microbiologically Influenced Corrosion in Fire Protection Systems, *Mater. Perform.*, Vol 37 (No. 7), July 1998
- "Metering Pumps," Capitol Controls Company, Inc., Colmar, PA, 1997
- F.S. Merritt and J.T. Rickets, Ed., *Building Design and Construction Handbook*, McGraw Hill, 1994
- J.R. Myers, Corrosion of Galvanized Steel, Potable Water Pipes, *Austral. Corros. Eng.*, Vol 17 (No. 5), May 1973, p 29–32
- J.R. Myers, *Cathodic Protection Design*, University of Wisconsin Press, 1996
- J.R. Myers, Inspector's Guide for Sacrificial-Anode-Type Cathodic Protection: Checklist, Part 2, *Underground Tank Technol. Update*, Vol 11 (No. 5), Sept/Oct 1997, p 9–12
- J.R. Myers, Inspector's Guide for Sacrificial-Anode-Type Cathodic Protection: Checklist, Part 1, *Underground Tank Technology Update*, Vol 11 (No. 4), July/August 1977, p 8–9
- J.R. Myers, H.B. Bomberger, and F.H. Froes, Corrosion Behavior and Use of Titanium and Its Alloys, *J. Met.*, Vol 36 (No. 10), Oct 1984, p 50–60
- J.R. Myers and A. Cohen, Conditions Contributing to Underground Copper Corrosion, *J. Am. Water Works Assoc.*, Vol 76 (No. 8), Aug 1984, p 68–71
- M.F. Obrecht and J.R. Myers, Potable Water Systems: Recognition of Cause Vital in Minimizing Corrosion, *Mater. Prot. Perform.*, Vol 11 (No. 4), April 1972, p 41–46
- P.H. Perkins, *Concrete Structures: Repair, Waterproofing, and Protection*, John Wiley & Sons, 1977, p 229–238
- J.S. Pettibone, Stainless Lampposts Should Last 100 Years, *Nickel*, Vol 9 (No. 2), Dec 1993
- F.C. Porter, Zinc Alloy Coatings on Steel, *Ind. Corros.*, Vol 18 (No. 4), June/July 1998, p 5–9
- D. Reichert, Corrosion of Carbon Steel under Set Insulation, *Mater. Perform.*, Vol 37 (No. 5), May 1998
- W.J. Rossiter, W.E. Roberts, and M.A. Streicher, Corrosion of Metallic Fasteners in Low-Sloped Roofs, *Mater. Perform.*, Vol 31 (No. 2), Feb 1992
- W. Harry Smith, *Corrosion Management in Water Supply Systems*, Van Nostrand-Reinhold, 1989, p 39–42
- H.E. Townsend, Twenty-Five-Year Corrosion Tests of 55% Al-Zn Alloy Coated Steel Sheet, *Mater. Perform.*, Vol 32 (No. 4), April 1993, p 68–71
- H.E. Townsend and A.R. Borzillo, Thirty-Year Atmospheric Corrosion Performance of 55% Aluminum-Zinc Alloy-Coated Sheet Steel, *Mater. Perform.*, Vol 35 (No. 4), April 1996, p 30–36
- R.S. Treseder, Ed., *Corrosion Engineer's Reference Book*, National Association of Corrosion Engineers, 1991, p 112
- A.H. Tuthill, Practical Guide to Using Marine Fasteners, *Mater. Perform.*, Vol 29 (No. 4), April 1990
- A.J. Walker, Coal-Tar Coatings, *Ind. Corros.*, Vol 11 (No. 5), Aug/Sept 1993
- *Welding and Brazing*, Vol 6, *Metals Handbook*, 9th ed., American Society for Metals, 1971, p 200–201

Ground Vehicle Corrosion

I. Carl Handsy, U.S. Army Tank-Automotive & Armaments Command
John Repp, Corrpro/Ocean City Research Corporation

THE U.S. ARMY has one of the largest tactical ground vehicle fleets in the world. These systems are continually being updated with the latest in weaponry, electronics, and fighting hardware. However, the basic structure of the vehicles remains largely unchanged. Most of this materiel was designed with automotive technologies for corrosion protection that were used in the 1970s and 1980s. These technologies cannot provide the level of corrosion protection necessary to maintain a vehicle for desired life of 15 to 25 years.

With a fleet of more than 120,000 vehicles for "High Mobility Multi-Wheeled Vehicles" (HMMWV or Humvees) alone, it is easy to see why deterioration due to corrosion is a major issue. As the average age of vehicles in the fleet is more than 17.9 years (Ref 1), which is 5 to 10 years longer than current commercial automotive standard warranties for corrosion, there is need for improved corrosion control to maintain a continually aging fleet.

An overall discussion of the Army's current position on corrosion control for wheeled tactical vehicles is presented here and includes:

- Army requirement for corrosion control
- Testing to meet the requirement
- Improving supplemental corrosion protection, the use of corrosion-inhibitive compounds, maintenance procedures, and design considering corrosion

Background

Wheeled tactical vehicles first saw widespread use after Word War I, following some initial limited use by the Marine Corps. At the time, the vehicles were manufactured using the same techniques and production lines as commercial automobiles. Today (2006), military vehicles are created with unique requirements, specifications, coatings, and equipment that are not common to commercial vehicles.

Army vehicles are developed and manufactured by contractors who specialize in making that specific item. Due to the unique requirements on these vehicles, hand assembly is needed along with automatic processes. The manufacturers do not always have large assembly plants like those of the U.S. automakers, and this sometimes limits the state-of-the-art technology that can be incorporated, such as hot-dip galvanizing, electrodeposition coatings, and other technologies that the automotive industry uses.

However, such technologies can often be found at subvendors, so leveraging their abilities allows manufacturers of wheeled tactical vehicles to improve the product without investment in costly infrastructure.

Requirements for Corrosion Control

The Army's requirements for corrosion control are based on protecting its materiel from deterioration due to operation under normal conditions. For ground vehicles, these requirements are often based on corrosion-control technologies developed by the commercial automotive industry. However, the tactical environment in which Army vehicles must operate is more severe, and so more robust technologies and more stringent requirements may be required. This is the case for vehicles deployed in Southwest Asia. The soil was an ancient sea bed and is full of salts and other minerals that are extremely hostile to coatings and metals. The weather extremes of high winds, abrasive sand, and temperatures ranging from daytime 53 °C (128 °F) to 15 °C (60 °F) at night play havoc on all equipment.

In addition to corrosion-control requirements for coatings, observability and chemical agent resistance are of paramount concern. To prevent detection by infrared (IR) and other scanners and to allow for decontamination after chemical agent exposure, the Army has developed a chemical agent resistant coating (CARC) system. This unique coating formulation reduces the IR signature of a vehicle, provides a dull flat finish, and can be cleaned using a highly basic decontamination solution. It is required on any Army tactical system, and it must be compatible with the corrosion-control methods.

In the past, corrosion control was not a primary concern as the tactical vehicle life expectancy was relatively short. However, for some current vehicles the life can be greater than 25 years. As such, better corrosion control is essential to producing an asset that can last for the specified life. The U.S. Army Tank-automotive and Armaments Command (TACOM) defines the corrosion prevention and control requirements in the procurement document.

Procurement Document

The requirements from a procurement document are summarized.

Corrosion Control Performance. The minimum service life in years of the vehicle, subsystem, or component is stated, and the operating conditions are given (high humidity, salt spray, gravel impingement, temperature range). The type and amount of maintenance to be given is stipulated.

A method of evaluating corrosion is given. The allowable level of corrosion is 0.1% of the surface (rust grade 8 per ASTM D 610, "Evaluating Degree of Rusting on Painted Steel Surfaces"). Further, a U.S. Army Corrosion Rating System is cited. There shall be no effect on form, fit, or function of any component due to corrosion.

Verification of Corrosion Control. The entire vehicle shall be evaluated for corrosion control by the accelerated corrosion test (ACT). The specified number of cycles that represents the vehicle service life is specified in the contract. For less than complete vehicles, the cyclic corrosion test per GM 9540P or equivalent such as the SAE J2334 shall be performed on the actual component for the number of cycles representing the service life (e.g., 160 cycles for the 20 year period of performance). All test panels and component parts shall be scribed per ASTM D 3359 prior to testing to validate performance of the paint or any other coatings. After completion of the test, the scribed area shall be scraped with a metal putty knife or equivalent to determine the extent of any coating undercutting/loss of adhesion of any coating and/or treatment. Alternative validation test methods must be approved by the government prior to fielding or manufacturing.

The pass/fail criteria for the ACT test and other tests is clearly defined. Any loss of form, fit, or function shall be considered a corrosion failure and requires the same type of corrective action during or after the ACT as any other failure occurring during or after the initial production test. Loss of coating adhesion or corrosion emanating from the scribe shall be limited to 3 mm maximum at any point at the scribe. There shall be no blistering of the coating film in excess of 5 blisters in any 24 square inch area. The maximum blister size is 1 mm. There shall be no more than 0.1% surface corrosion (ASTM D 610, rust grade 8) on any component part (exclusive of the scribe). In addition, there shall be no loss of original base metal thickness greater than 5% or 0.010 in., whichever is less. Expendable items (identified as exempted parts prior to the test) shall retain their function for their intended service life and are not subject to these criteria.

Notes of Guidance and Caution. The procurement document provides assistance to the vendor, such as:

- Corrosion control can be achieved by a combination of design features (as in TACOM Design Guidelines for Prevention of Corrosion in Combat and Tactical Vehicles, March 1988) or any automotive corrosion design guide such as SAE J447, material selection (e.g., composites, corrosion-resistant metal, galvanized steel), organic or inorganic coatings (e.g., zinc phosphate pretreatment, corrosion-resistant plating, E-coat, powder coating) and production techniques (e.g., coil coating, process controls, welding, inspection, and documentation).
- Corrosion protection for low-carbon sheet steel can be achieved by hot-dip galvanizing in accordance with ASTM A 123, or electrogalvanized 0.75 mil minimum thickness per ASTM B 633 (or minimum coating thickness of 0.75 mil on pregalvanized sheet 0.063 in. or less), with zinc phosphate pretreatment, epoxy prime preferably E-coat primer and CARC top coat. Alternate designs may be evaluated by comparison to a galvanized sample (as described previously) using ASTM D 522 Mandrel Bend Test and Accelerated Corrosion Test GM 9540P and gravelometer testing. Failure constitutes a defect such as extensive corrosion at scribe, chipping of coatings, loss of adhesion, or significant penetration of base material (per ASTM D 3359).
- Due to changes in climatic conditions and the development of newer materials and processes, all accelerated corrosion tests undergo a continuous adjustment to reflect these conditions. Therefore, modifications to the testing are to be expected over time. However, any changes need to be agreed upon with the government prior to testing.
- CARC coatings over steel is not expected to be sufficient corrosion protection to achieve 10 year service life. In marine environments such a system usually delivers only a 5 year performance.

The above requirements are capable of being met using already proven materials and processes for corrosion control (Ref 2). Using the processes and procedures already in use by commercial automotive manufacturers will help improve the corrosion resistance of military vehicles and make a design life of greater than 20 years achievable.

Testing Systems to Meet the Army's Needs

As required by the procurement contract, existing or new corrosion-control technologies used in a vehicle system need to be evaluated to determine their benefit. Accelerated corrosion test methods can demonstrate differences in performance of competing alternatives, identify areas requiring additional corrosion protection, and demonstrate the interaction between corrosion and operation of the vehicle.

Preproduction Testing. These initial tests are used to screen candidate materials to evaluate their inherent corrosion resistance. Most commonly, these are short-term aggressive tests performed in a laboratory corrosion chamber (see the article "Cabinet Testing" in Volume 13A). Traditionally, methods such as the ASTM B 117 salt spray (fog) test were used to compare relative performance, but they had very little if any relation to actual field use.

In the 1990s it was found that newer cyclic tests provide a better correlation to actual exposure environments. The GM9540P and SAE J2334 test methods are now commonly used to evaluate painted metals to determine relative corrosion resistance and select the best candidate system.

Cyclic corrosion tests are generically similar, although their exact makeup can vary. Corrosion specimens are exposed to a combination of corrosive electrolyte (salt-water solution), high temperature, high relative humidity (RH), and ambient conditions (nominally 70 °F, or 20 °C, <50% RH). These events are used to introduce corrosive species (e.g., chloride ions) to the samples, create conditions that accelerate corrosion (increase time of wetness, TOW), and "bake" the salts onto the specimens so they can be activated during TOW. Using combinations of these events over a period of time can accelerate levels of corrosion to represent years of exposure in a matter of weeks or months. Additionally, gravel impingement using a gravelometer is used in conjunction during a test to simulate events found in actual vehicle usage (Ref 3–5).

Prototype Testing. As major subsystems or complete vehicles are assembled into prototypes, more detailed evaluations can occur. These evaluations are used to determine if interactions exist between any of the components of these assemblies and if their normal operation is affected by corrosion. Prototype testing is performed by combining durability and corrosion inputs. For smaller subsystems, this can include periodic exercising of components during accelerated testing. For larger systems and vehicles, testing is performed using proving-ground-type accelerated corrosion tests (road tests).

A road test is a combination of driving mileage and corrosion inputs used to simulate the expected vehicle mission profile (Ref 6). A vehicle is run through road courses representative of various terrains (paved roads, gravel roads, cobblestone streets, cross-country trails) that the vehicle is designed to negotiate. Intermixed with these conditions are corrosion events to apply corrosive contaminants (electrolytes) to the vehicle and TOW. Operating this type of test exposes the vehicle to mechanical and corrosion stresses. This combination of tests can identify deficiencies in corrosion-control methods, which can then be remedied before large-scale production.

Analysis of Test Results. The nature of accelerated corrosion testing is such that a failure in the test increases the likelihood of observing the same failure in the field; however, *a lack of failure in the test does not mean a failure will not occur in the field*. This is the nature of accelerated testing, where the time for failures to occur is accelerated and not all failure mechanisms are accelerated at the same rate. This is why comparative testing is performed early in vehicle development, and road testing is used once all material choices have been made to identify any interactions between final assemblies.

Benefits. The results of accelerated corrosion tests are used as feedback to vehicle designers. These results can be used to improve the design of a vehicle, to identify other materials for certain systems, to improve maintenance requirements, or in cost-benefit analyses to identify trade-offs and value of adding additional corrosion protection. While it is often impractical to expect a tactical vehicle to last the desired 20+ years of service with no maintenance, accelerated tests can benchmark the relative life of specific systems and highlight maintenance activities that should be performed. This is used to develop the best possible system and to reduce life-cycle costs (LCC) to optimize service and performance.

Supplemental Corrosion Protection

Supplemental corrosion protection improves the corrosion resistance of a material. These methods can include:

- Galvanizing of steel
- Plating of metals
- Sacrificial coatings
- Organic coatings

Each of these can be used as part of a system to reduce corrosion. While individually each does increase service life, combinations of these are needed to reach the >20 year design life presently being requested of new wheeled tactical vehicles.

For example, a steel body using double-sided galvanized sheet steel and a CARC system will be protected more effectively than CARC or galvanizing alone. Using the above with a good pretreatment such as zinc phosphate, a high-performing primer such as E-Coat, followed by the top coat a 20+ year service life is economically achievable. The CARC system provides the first line of defense against contaminants. Without this coating, corrosion of the galvanized steel would begin immediately at voids. Conversely, if only the coating was used, once contaminants penetrated the CARC corrosion of steel substrate would begin immediately.

Corrosion-Inhibitor Compounds. For existing equipment, there may be components or locations (crevices, recesses, blind holes) that are vulnerable to corrosion attack. The entire vehicle may need extra protection during shipping or storage. Temporary inhibitive compounds may be used to reduce corrosive attack.

Corrosion-inhibitor compounds are most commonly liquid aerosols sprayed onto vehicles. Other forms include vapor-phase inhibitors, greases, and waxes. Most of these products are similar to other maintenance fluids used in motor pools and, as such, their use is implemented as maintenance procedures or in specialized service locations. However, similar to other lubricants and fluids, they need to be handled and applied with care. Some materials have been found to be detrimental to rubbers and plastics with prolonged exposure. Overspray can also be of concern, as this can attract dirt and contaminants and increase maintenance time by necessitating postapplication washing.

The U.S. Marine Corps have published guidance on the use of inhibitors with ground vehicles (Ref 7). These documents stress application of products to specific components and locations. This has helped alleviate some of the potential incompatibility issues. For example, certain inhibitors may reduce corrosion on one type of metal, but accelerate attack on others.

Improved Maintenance Procedures

Maintenance procedures can also be used to combat corrosion. More frequent lubrication, application of inhibitors, and repainting can reduce corrosion damage. Although these procedures do have benefits, excessive maintenance can be both time and readiness prohibitive. With steadily decreasing operating budgets and a need to have vehicles ready-to-go, continual maintenance is not practical. Often a compromise between maintenance and corrosion control needs to be developed and realistic maintenance goals established.

While maintenance can be used to reduce corrosion, it should not be relied upon as the major corrosion-control method. Emphasis should be placed on less labor-intensive methods.

Considerations for Corrosion in Design

Considerations for corrosion control during design of a vehicle goes beyond choosing proper base metals and coatings. It includes the geometry and manufacturing methods used to construct a vehicle. These methods are described in TACOM and Society of Automotive Engineers (SAE) guidance documents (Ref 8–10).

These documents stress using good construction practices and creating geometries that minimize water entrapment areas or promote drainage of those areas. Design of body panels and components should also minimize the use of sharp corners and edges, which reduce paint adhesion. Adhesives and seam sealants should be used along with continuous welds for joining to eliminate crevices and water seepage locations.

Conclusions

As new military vehicles are being produced and acquired, corrosion control is becoming a major component of the acquisition strategy. Requirements such as those discussed in this article are being used to improve the corrosion resistance of vehicles. Placing the focus on performance instead of materials allows manufacturers to select corrosion-control solutions that best work within their operations, yet provide the level of protection required. By looking to proven technologies already in use by commercial manufacturers, original equipment manufacturers can leverage this knowledge and improve their end product.

The Army has embraced accelerated corrosion test methods and evaluation techniques for tactical vehicles. These methods permit the demonstration of effective design choices. It provides the ability to evaluate new corrosion-control technologies as they become commercially viable for use on military vehicles.

REFERENCES

1. A.E. Holley, "Aging Systems 'Classic to Geriatric to Jurassic' When Will It Stop?" DoD Maintenance Symposium, Oct 29, 2002
2. J. Repp, "Corrosion Control of Army Vehicles and Equipment—Use of Existing Technologies for Corrosion Control," Army Corrosion Summit 2003, http://www.armycorrosion.com
3. C.H. Simpson, C.J. Ray, and B.S. Skerry, Accelerated Corrosion Testing of Industrial Maintenance Paints Using a Cyclic Corrosion Weathering Method, *J. Prot. Coat. Linings,* May 1991, p 28–36
4. Cleveland Society for Coatings Technology Technical Committee, Correlation of Accelerated Exposure Testing and Exterior Exposure Sites, *J. Coat. Technol.,* Oct 1994, p 49–67
5. B. Goldie, Cyclic Corrosion Testing: A Comparison of Current Methods, *Prot. Coat. Europe,* July 1996, p 23–24
6. J. Repp, Accelerated Corrosion Testing—Truth and Misconceptions, *Mater. Perform.,* Sept 2002, p 60–63
7. "Organizational Corrosion Prevention and Control Procedures for USMC Equipment," U.S. Marine Corps TM4795-12, Dec 1999
8. "Design Guidelines for Prevention of Corrosion in Combat and Tactical Vehicles," U.S. Army TACOM
9. "Prevention of Corrosion of Motor Vehicle Body and Chassis Components," Surface Vehicle Information Report, SAE International, 1994
10. "A Guide to Corrosion Protection for Passenger Care and Light Truck Underbody Structural Components," Auto/Steel Partnership, 1999

SELECTED REFERENCES

- R. Baboian, *Automotive Corrosion by Deicing Salts,* National Association of Corrosion Engineers, 1981
- R. Baboian, *Automotive Corrosion and Protection,* National Association of Corrosion Engineers, 1992
- F. Bouard, J. Tardiff, T. Jafolla, D. McCune, G. Courual, G. Smith, F. Lee, F. Lutze, and J. Repp, "Development of an Improved Cosmetic Corrosion Test by the Automotive and Aluminum Industries for Finished Aluminum Autobody Panels: Correlation of Laboratory and OEM Test Track Results," World Congress 2005, SAE International, April 2005
- I.C. Handsy and J. Repp, "Development and Use of Commercial Item Descriptions in Army Acquisition and Maintenance Activities," presented at *Corrosion 2002,* National Association of Corrosion Engineers, April 2002
- I.C. Handsy and J. Repp, "Corrosion Control of Army Vehicles and Equipment," presented at the Army Conference on Corrosion, Feb 2003
- J. Repp, "Comparison Testing of Environmentally Friendly CARC Coating Systems over Aluminum Substrates," presented at Joint Services Pollution Prevention Conference and Exhibition (San Antonio, TX), NDIA, Aug 1998
- J. Repp, "Update on the Development of SAE J2334 Accelerated Corrosion Test Protocol," presented at the Army Conference on Corrosion, March 2002
- J. Repp and T. Saliga, "Corrosion Testing of 42-Volt Electrical Components," presented at World Congress 2003, SAE International, March 2003

Armament Corrosion

Nicholas Warchol, U.S. Army ARDEC

ARMAMENT SYSTEMS comprise guns and ammunition ranging from the M-16 machine gun and ammunition (5.56 mm) to the 155 mm mortar rounds and M198 howitzers that fire the rounds. This includes weapon systems found on tanks and other mobile units, so the number of systems is large.

Armament systems, must meet specified requirements, including functionality, environmental, time, and cost requirements. Functional requirements are that the system performs its basic task, that includes the ability to fire a projectile, aim, and rotate. Another requirement is that there must be visual and spectral camouflage. Spectral camouflage refers to the infrared profile of the system and its ability to blend into the surrounding environment so the system is invisible to infrared-sighting equipment. This provides an extra level of tactical protection for the soldiers. The system must also be corrosion resistant and chemical-agent resistant. Chemical-agent resistance is the ability of the system to be decontaminated if it were to come in contact with chemical agents. These requirements are accomplished through the use of the chemical-agent resistant coating (CARC). It provides visual and spectral camouflage as well as corrosion and chemical-agent resistance. The CARC system consists of a primer and topcoat. The epoxy primer provides corrosion protection, while the urethane topcoat provides chemical-agent resistance and camouflage properties. Armament systems are exposed to some of the most severe environments on earth. Wars are not fought in a climate- and humidity-controlled environment. From arctic cold to desert heat the systems must be able to perform their function in all environments.

Overview of Design, In-Process, Storage, and In-Field Problems

Armaments corrosion problems must be looked at in four specific stages: design, in-process, storage, and in the field. To accurately understand the corrosion problems that are faced with today's (2006) armament systems, these aspects must be looked at individually and their effects analyzed over the useful life of the system. Design considerations include geometry, material selection, assembly, pretreatment, coatings, and working and storage environments. In-process corrosion concerns include: processing locations, in-process storage of parts, time between processing steps, and quality control of each processing step. How, where, and how long the systems will be stored before they are fielded must be considered. Finally, analysis of the in-field corrosion of the finish product should include: physical environments; repair of corrosion-protective coatings, shipment concerns, general corrosion-protection maintenance, and appropriate fixes and procedures that can be implemented by soldiers in-field to stop continued corrosion of armament equipment.

There are common corrosion problems associated with each stage in the life of an armament system. The three most common types of corrosion associated with design are uniform, galvanic, and crevice corrosion. The most common form of corrosion during processing is uniform corrosion of parts being exposed to corrosive environments before the corrosion protection is in place. The most common form of corrosion for equipment in storage is uniform corrosion. This is again from parts being exposed to corrosive environments or being stored for periods longer than the protection systems are designed. The three most common forms of corrosion found on in-field systems are crevice, galvanic, and uniform. All types of corrosion are evident in all the stages; the process by which the most common armament corrosion is addressed within the military to ensure functional equipment reaches the field is discussed with applicable examples.

Design Considerations

From a design standpoint, one must be aware of the eight types of corrosion and consciously design the system for corrosion resistance. The functional goals of the system must be established in a set of requirements determining what is to be accomplished by this part, how the system will work, how long the system will need to function at a time, and what are the physical requirements on the system. In many cases, with the designer's concern for the functional requirements of the system, corrosion is not a major consideration.

Material Selection. To adequately design a part to be corrosion resistant, the design engineer must first make good decisions in the materials selection process. When placing materials in a system, the design engineer must not only know the physical and mechanical properties of the materials, but also the susceptibility to corrosion of the material in specified environments of the system. For example, aluminum is often assumed to be a corrosion-resistant material, and for 1000-series aluminum this is generally correct. Different aluminum alloys have different corrosion susceptibilities. The design engineer must understand that if a material passivates when exposed to oxygen and it is placed in an environment that is absent of oxygen, then the corrosion resistance of the material is significantly reduced, if not completely destroyed.

Dissimilar Metals. Design engineers must also look at the interface of dissimilar metals within a system. Galvanic corrosion can destroy systems rapidly, especially in the case of a very large cathode in direct contact with a small anode. A galvanic series appropriate to the environment can be consulted, and all efforts should be made by the design engineer to use materials combinations that do not cause galvanic corrosion. See the compatibility chart based on MIL-F-14072D in the article "Corrosion in Microelectronics" (Table 6) in this Volume.

Design geometry can also play a large role in the susceptibility of a system to corrosion. Good practice is to eliminate crevices or seal crevices and joints. The design engineer must assume that water will get into parts or trap and pool on the surface of the system. The systems must be designed to drain water through holes, channels, or other devices. In pipes, the design engineer must prevent turbulent flow in joints in high-speed flow conditions. Bends, kinks, corners, and the internal features of the pipe all affect the flow and can increase erosion-corrosion within the system.

Coatings applied to the systems must also be researched and chosen depending on the specific requirements of the system. The previously mentioned CARC system is designed to be a 15 year coating that provides chemical-agent resistance, spectral and visual camouflage, as well as corrosion protection. It is a two-part system that is applied over a zinc phosphate coating. The epoxy primer, 25 to 50 μm (1 to 2 mils) thick, provides corrosion protection. The urethane topcoat is applied 50 to 75 μm (2 to 3 mils) thick and provides the camouflage and chemical-agent resistance.

Examples of Design-Related Problems. An example of design affecting the corrosion resistance of an engineered system is the M198 howitzer. There is an anodized 7079-T6 aluminum alloy ring gear that connects the upper carriage and the gun tube to the lower carriage and the trails. The ring gear allows the gun tube to rotate and is fastened to the upper and lower carriage with mounting bolts. Figure 1 shows the results of a poor design on the system. The upper carriage of the ring gear has become completely disconnected from the lower carriage and the gun tube has fallen to the ground. There are multiple problems with the design of this system. First, the material selected, 7079-T6 aluminum, is susceptible to stress-corrosion cracking (SCC) in the transverse direction. For SCC to occur, a susceptible material, a specific corrodent, and a sustained tensile load are needed. 7079-T6 aluminum has a transverse SCC threshold of 55 MPa (8 ksi). The 13 mm ($\frac{1}{2}$ in.) and 16 mm ($\frac{5}{8}$ in.) mounting bolts used to secure the ring gear to the upper and lower carriage, when proper torque is applied, produce 110 and 172 MPa (16 and 25 ksi) sustained tensile loads, respectively, at the countersink. This load is sufficient to produce SCC if a corrodent is present, and for aluminum alloys, 50% relative humidity is sufficient. In this case, all three criteria for SCC are present and the material experienced a large amount of SCC. Figure 2 shows SCC at the countersink of the ring gear.

A second design problem deals with the anodized coating of the ring gear. The anodized coating is applied to the aluminum to reduce the susceptibility of the material to corrosion. For the M198 howitzer, the seal used to finish the anodized coating was deleted on the drawing. Without the seal, the anodized coating does not protect the ring gear from pitting, and the ring gear surface experienced extreme pitting (Fig. 3). The pitting that occurred in the countersink provided initiation points for the SCC to propagate and accelerate the corrosion damage. Both the SCC and pitting could have been easily avoided. If 7075-T73 aluminum had been selected, SCC would have been avoided since 7075-T73 has a SCC threshold of 303 MPa (44 ksi). Pitting would have been prevented by simply requiring the seal to be placed on the anodized coating.

A third example of design affecting corrosion resistance is the copper rotating band found on 40 mm grenades. The copper bands are swaged onto the steel or aluminum grenade body. This creates a crevice beneath the rotating band as well as creates a dangerous galvanic couple between base metal and copper. It is also common to find galvanic corrosion of steel adjacent to the copper-rotating band as seen with the 105 mm cartridge (Fig. 4). Another problem is that machining lubricants can become trapped in the crevice between the body and the rotating band. This lubricant can then seep out of the crevice and react with the copper band causing discoloration (Fig. 5).

In-Process Considerations

In-Process Monitoring. If a part is not properly monitored during processing, there is no way to accurately determine the reliability of the resulting system. In-process corrosion will depend on the type of process and its sensitivity to changes in process variables. Where and for how long will the parts be stored between processing steps? Do the unfinished parts need to be transported for further processing? These are all questions that must be considered in the quality assurance (QA) program. Quality assurance uses

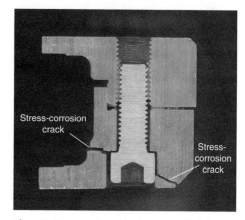

Fig. 2 Cut-away view of the ring gear and bolt showing stress-corrosion cracking. Source: Ref 1

Fig. 3 Severe pitting on the surface of the aluminum alloy ring gear. Source: Ref 2

Fig. 4 Galvanic corrosion at the interface of the copper rotating band and the steel base metal in a 105 mm cartridge. Source: Ref 3

Fig. 1 Results of ring gear failure in the M198 howitzer. Source: Ref 1

a system of quality assurance representatives (QARs), who are government employees who travel to vendors' plants to monitor the actions of contractors and subcontractors to ensure compliance.

Adherence to Specifications and Standards. To monitor processing, the QARs must know how the quality of parts being produced is monitored, requirements for the part, how finishes are applied, and the tests used to verify these requirements. Specifications and standards are cited in purchasing documents that contractors and subcontractors must follow. These specifications and standards also define the engineering requirements. The goal of a specification or a standard is to establish critical criteria to ensure proper function of a part or system.

There are many cases of contractors certifying that the specifications were met while not stating which tests were performed. In military contracting, a contractor or subcontractor must perform three steps after the parts have been produced to ensure acceptance by the military inspector. First the parts must be tested and data must be collected. Secondly, the data are presented in a certified test report (CTR). The CTR lists the tests run and the test procedure, displays the data collected, and provides proof that the work meets the requirements. Once this document has been created, a second document, the certificate of conformance (COC), can be issued. A COC states that the contractor completed all the necessary tests on the produced parts and has fulfilled the other contractual obligations such as documentation and shipping requirements.

Conflicting Technical Data. A major problem for in-process corrosion control is the existence of conflicting technical data. For example, there may be a requirement on a drawing that a certain test is to be run, but the document also cites another drawing that says that the test is not required. In this case, the QAR cannot check the contractor for the requirement on the primary drawing because there are conflicting data. These conflicting requirements can be corrected for future contracts, but a new solicitation or a change to the contract would be required to fix the current contract, so the parts may be shipped as is. In armament corrosion, the person who suffers is the soldier who receives parts that do not function as they are supposed to or do not last as long as they are needed.

Examples of In-Process-Related Problems. An example of an in-process corrosion problem is 155 mm M549A1 ammunition rounds that needed a complete repainting only 4 months after the initial painting. M549A1 is a steel projectile that is phosphated and then painted with enamel. The rounds were produced in California and then shipped to Iowa to be filled. The rounds that arrived in Iowa were rusted and required complete repainting. This was due to incomplete application of the phosphate pretreatment. The benefit of the phosphate coating is lost if complete and uniform coverage is not obtained. Without a properly applied pretreatment, the original paint coating was unable to protect the surface of the ammunition.

Another example of an in-process corrosion problem is the M119 howitzer firing platform. The firing platform was required to be 7075-T73 aluminum with chromate conversion pretreatment and painted using the CARC primer and topcoat system. After five years of service in Hawaii, the firing platforms failed due to exfoliation corrosion. It was determined that new platforms were required and again the T73 heat treatment was specified. The new platforms with this heat treatment failed after 2 years of service. Figure 6 shows the failed firing platform. Figure 7 is a close-up of the exfoliation corrosion on the firing platform. The T73 temper was designated because it is highly resistant to exfoliation. Then why did the parts exfoliate in only two years of service? They were not tested or documented during production to verify that the parts had in fact been treated to the T73 condition. To obtain a T73 temper, a part must first be placed in a T6 temper. If the parts are not adequately heated, they will not achieve the T73 state and will not be resistant to exfoliation corrosion. In this case, the parts and the temper recipe were never tested under ASTM B 209, "Standard Specification for Aluminum and Aluminum Alloy Sheet and Plate." Based on the premature failure of the supposed 7075-T73 parts, it is apparent the treatment was not sufficient. To verify the heat treatment of the platforms, tensile bars where cut from a failed platform and tested. The results indicated that the parts were in fact not in the T73 condition. If the parts had been properly monitored with the heat

Fig. 6 Failed M119 firing platform. Source: Ref 5

Fig. 7 Firing platform exfoliation corrosion. Source: Ref 5

Fig. 5 Grenade body showing discoloration of copper rotating bands resulting from exposure to trapped machining lubricant. Source: Ref 4

treatment plan recorded or the parts tested, the inadequate heat treatment would have been discovered and corrective action taken.

Storage Considerations

Storage Practices. The third stage that must be considered for armament corrosion is storage. In armament systems, parts can and do sit in storage for extended amounts of time. The goal of storage is to have systems on hand that can be deployed on short notice. In this case, the military must use processes to prevent degradation without affecting the readiness of the systems. In some cases equipment is stored in climate-controlled facilities. Other common practices include volatile corrosion inhibitor (VCI) packaging, rust-preventative oils, and hermetically sealed packaging. These systems are designed to preserve the integrity of the system and not reduce readiness. Despite best practices, the most common type of corrosion during storage is general corrosion, caused by failed or nonexistent corrosion protection.

Examples of Storage-Related Problems. Military storage is not a perfect system, and in many cases, the storage is longer than the protection scheme life, or the packaging is compromised. If the packaging is compromised and goes unnoticed, the protection is completely lost. Loss of protection can be as simple as a tear in the packaging, or wrapping the items in VCI packaging designed to protect the system for 2 years, but storing them for 5 years. There have also been examples of oils used to preserve equipment that are capable of unzipping heat-sealed packages. One must also consider how the parts or systems will be stored. If a system is stored outdoors, will personnel be available to inspect and perform maintenance on the storage system and will readiness be affected?

An example of how storage can affect the readiness of equipment is again the M198 howitzer. Howitzers were stored outside in the elements, with individuals monitoring the systems to ensure their readiness. The howitzers were placed in the "ready" position, meaning that the trails were lifted off the ground so the howitzers were ready to be towed (Fig. 8). The problem is the howitzer was not designed to be stored in this position. Drain holes in the lower carriage were placed in the back by the trails to allow water to drain from the system. However, in the "ready" position water does not drain from the lower carriage. Thus water accumulated inside the carriage and caused corrosion damage. Adding holes in the front of the lower carriage so water could drain from the system while in the "ready" position corrected the problem.

In-Field Considerations

The final stage in the life of an engineered system is in-field or in-service. This is the place recognized as the cause for degradation and failure and provides the true test to parts, systems, and corrosion protection.

Preventive Maintenance and Cleaning. In armament systems, maintenance and cleaning must be performed to realize the useful life of systems. Common maintenance activities include cleaning, oiling, paint touchups, and parts replacement. The design engineer generates the required maintenance procedures. The goal is to create a maintenance system that anticipates problems and provides adequate guidance on how to prevent or repair them. For each part in the system there are cleaning and replacement requirements that lay out what must be done and when they should be completed. These requirements can be long and comprehensive. With maintenance crews seeing multiple systems, the sheer volume of manuals to be studied and reviewed before maintenance is performed is a daunting task. In this case, most crews develop a system of general practices for cleaning and repairing parts. The other case is that the crews will be told to clean this system. The crew will then determine the best way to clean it. They could clean it by hand using solvents, then let it dry, and finally re-oil the equipment, or they could simply power-wash the equipment and then re-oil. The process of solvent cleaning and drying can take upwards of 4 h, while power-washing the parts and oiling with a water-displacing oil will take 5 min. It is easy to see which is done more often, and without guidance or properly reading the manuals the soldiers do not see the benefit associated with the other process. Figure 9 shows a soldier using a pressure washer to clean ammunition containers. The problem with this process is that the rubber seals on the storage containers are only watertight to 21 kPa (3 psi), and the soldier is washing the containers at a pressure of 690 kPa (100 psi).

The real problem with this situation is that systems are often designed to require a large

Fig. 8 M198 howitzer in "ready" position. Source: Ref 5

Fig. 9 Pressure washing of ammunition containers. Source: Ref 6

Table 1 M119 operator preventive maintenance and lubrication requirements

M119 subassemblies and components(a)	PMCS(b) B	D	A	W	M	Lubrication(c)(d) D	W	Q
(5) Wheel and tire assembly	X				X			
(3) Handbrake assembly	X							
(4) Firing platform	X	X		X	X		GAA/CLP	
(3) Gun barrel assembly	X					CLP	GAA/CLP	
(3) Recuperator recoil mechanism	X	X	X	X	X	CLP/OHT		
(4) Elevation gear assembly	X	X			X	GAA		GAA
(2) Traversing mechanism	X	X		X		GAA	GAA	WTR
(3) Breech mechanism	X	X		X		CLP	CLP	CLP
(1) Balancing gear assembly							GAA/CLP	
(2) Gun barrel support army assembly						CLP	CLP	
(5) Hand spike, jack strut and platform clamps						CLP		
(2) Buffer recoil mechanism and slide assembly	X	X	X	X		OHT		
						CLP	CLP	
(3) Saddle assembly	X			X			GAA	
(2) Trail assembly, gun carriage	X			X				
(1) Traveling stays						CLP		
(4) Trail end hydraulic brake assembly						GAA		
						BFS		
(2) Suspension						GAA		
(3) Cam assembly						GAA		
(2) Traveling lock clamp assembly						CLP		

(a) Numbers in parenthesis in the left-hand column represent the number of corroded parts per assembly. (b) Planned maintenance checks and services (PMCS) requirements: B, before operation; D, during operation; A, after operation; W, weekly; M, monthly. (c) Lubrication requirements: D, daily; W, weekly; Q, quarterly. (d) Lubrication subentries: CLP, cleaner, lubricant, and preservative; GAA, grease, automotive, and artillery; OHT, hydraulic fluid, petroleum base; BFS, brake fluid, silicon; WTR, wide temperature range. Source: Ref 7

amount of maintenance. There is monthly, weekly, and in some cases daily maintenance required to keep systems functioning. This process removes the design engineer from responsibility if the system fails, because it is not the designer's fault that the maintenance was not completed. If a part is designed to require little or no maintenance and the part fails, then the design is faulty. Table 1 shows the maintenance schedule for the M119 howitzer. It is apparent the M119 requires a large amount of maintenance to keep parts in working order. For example, the recuperator recoil mechanism has planned maintenance checks and services before, during, and after use. There are also weekly and monthly checks. Hydraulic fluid and cleaner, lubricant, and preservative (CLP) must be applied daily. The M119 howitzer has daily maintenance requirements for 9 of the 19 subassemblies in the system. For armament corrosion, this raises the question whether it is realistic to assume soldiers will be able to complete required maintenance for equipment in a war-fighting condition. If parts will not receive the maintenance, then they have simply been designed to fail.

Conclusions

Of the forms of corrosion, the ones that are experienced most in armament systems are general or uniform corrosion, galvanic corrosion, and crevice corrosion. These types of corrosion account for a large portion of the corrosion problems found in armament systems, but are not the only causes of corrosion. In this case, everyone involved with an armament system needs to be aware of the types of corrosion, their causes, and steps that can be taken to prevent degradation. If everyone involved in a system is consciously trying to avoid problems associated with these types of corrosion then the readiness of equipment will be drastically increased and the total cost to the government associated with these systems will be reduced.

REFERENCES

1. J. Menke, "Executive Summary of M198 Howitzer Ring Gear Failure," United States Army, 2003
2. J. Menke, "Failure Analysis of the Ring Gear Bearing in The M198 Howitzer," United States Army ARDEC, presented at NACE International Regional Conference, 2001
3. R.C. Ebel, "Cartridge 105 MM: APFSDS-T M735 Corrosion Study," Technical Report ARLCD-TR-80004, U.S. Army Armament Research, Development and Engineering Center, Large Caliber Weapon Systems Laboratory, June 1980
4. "Corrosion of Rotating Band M918 TP," Quality Deficiency Report 03-013, U.S. Army ARDEC, 2004
5. J. Menke, "M119 Firing Platform Failure Analysis #2," U.S. Army ARDEC, 2003
6. K.G. Karr and R.G. Terao, Ammunition Retrograde from SWA, *Ordnance*, May 1992, p 44–47
7. *Technical Maintenance Manual for Howitzer, Medium, Towed: 155-MM, M198*, Army TM 9-1025-211-20&P Department of the Army, 1991

SELECTED REFERENCES

- M.F. Ashby, *Materials Selection in Mechanical Design*, 2nd ed., Butterworth Heinemann, 2001
- J.W. Bray, Aluminum Mill and Engineered Wrought Products, *Properties and Selection: Nonferrous Alloys and Special-Purpose Materials*, Vol 2, *ASM Handbook*, ASM International, 1990, p 29–61
- R.B.C. Caulyess, Alloy and Temper Designation Systems for Aluminum and Aluminum Alloys, *Properties and Selection: Nonferrous Alloys and Special-Purpose Materials*, Vol 2, *ASM Handbook*, ASM International, 1990, p 15–28
- M.G. Fontana and R.W. Staehle, *Advances in Corrosion Science and Technology*, Vol 2, Plenum Press, 1972
- J. Gauspari, *I Know It When I See It*, AMACOM, 1985
- L. Gilbert, Briefing Notes "U.S. Army Corrosion Control" (Rock Island, IL), April 18, 1979
- H.P. Godard, W.B. Jepson, M.R. Bothwell, and R.L. Kane, *The Corrosion of Light Metals*, John Wiley & Sons, 1967
- G.A. Greathouse and C.J. Wessel, *Deterioration of Materials: Causes and Preventive Techniques*, Reinhold Publishing, 1954
- E.H. Hollingsworth and H.Y. Hunsicker, Corrosion of Aluminum and Aluminum Alloys, *Corrosion*, Vol 13, 9th ed., *Metals Handbook*, ASM International, 1987, p 583–609
- R.J. Landrum, *Fundamentals of Designing for Corrosion Control, A Corrosion Aid for the Designer*, NACE International, 1989
- G. Lorin, *Phosphating of Metals: Constitution, Physical Chemistry, and Technical Applications of Phosphating Solutions*, Finishing Publications Ltd., 1974
- J.L. Parham, "Silicone Resin Reversion in the Army's Night Sights," U.S. Army Missile Command Technical Report-RD-ST-93-2, April 1993
- V.R. Pludek, *Design and Corrosion Control*, The Macmillan Press Ltd., 1977
- Properties of Wrought Aluminum and Aluminum Alloys, *Properties and Selection: Nonferrous Alloys and Special-Purpose Materials*, Vol 2, *ASM Handbook*, ASM International, 1990, p 62–122
- J. Senske, J. Nardone, and K. Kundig, "Corrosion Survey for Large Caliber Projectiles," Special Report CPCR-2 Project 1L1612105AH84, U.S. Army Armament Research, Development and Engineering Center
- G. Shaw, Optimizing Paint Durability, Part I, *Prod. Finish.*, Nov 2004, p 42–46
- H.H. Uhlig and R.W. Revie, *Corrosion and Corrosion Control, An Introduction to Corrosion Science and Engineering*, 3rd ed., John Wiley & Sons, 1985

High-Temperature Corrosion in Military Systems

David A. Shifler, Naval Surface Warfare Center, Carderock Division

HIGH-TEMPERATURE CORROSION AND OXIDATION occur in various military applications. Aircraft, ships, vehicles, weapon systems, and land-based facilities require power that may be supplied by boilers, diesel engines, gas turbines, or any combination of the three power sources. High-temperature exposure of materials occurs in many applications such as power plants (coal, oil, natural gas, and nuclear), land-based gas turbine and diesel engines, gas turbine engines for aircraft, marine gas turbine engines for shipboard use, waste incineration, high-temperature fuel cells, and missile components. The service performances of boilers, diesels, and turbines can be affected by exposure to numerous environments and are affected by temperature, alloy or protective coating composition, time, and gas composition. Materials degradation can lead to problems that often bring about unscheduled outages resulting in loss of reliability, loss of readiness, decreased safety, and increased maintenance costs.

Predicting corrosion of metals and alloys or coated alloys is often difficult because of operational demands placed on a given power system, the range of the composition of corrosive gaseous or molten environments, and the variety of materials that may be used in a given power system. Moreover, corrosion prediction is further complicated because materials often degrade in a high-temperature environment of a given application by more than a single corrosion mechanism. High-temperature corrosion in boilers, diesel engines, gas turbines, and waste incinerators are discussed in this article.

High-Temperature Corrosion and Degradation Processes

There are a number of corrosion and degradation processes that may occur in boiler, diesel engines, gas turbine engines, and incinerators. The degree of degradation is dependent on the material being exposed and the specific environment and other conditions to which the material is exposed. High-temperature corrosion/degradation of materials may occur through a number of potential processes described in the article "High-Temperature Gaseous Corrosion Testing" in Volume 13A, of the *ASM Handbook* (Ref 1), either singly and/or in some combination with one another:

- Oxidation
- Carburization and metal dusting
- Sulfidation
- Hot corrosion
- Chloridation and other halogenization reactions
- Hydrogen damage, hydrogen embrittlement
- Molten salts
- Aging reactions such as sensitization
- Creep
- Erosion/corrosion
- Environmental cracking (stress–corrosion cracking and corrosion fatigue)

Boilers

Boilers may be used to supply heating and cooling, main power, or auxiliary power at a number of land-based military installations. Marine boilers may be used to heat water and to produce steam to generate main power, auxiliary power, or industrial services.

When the first boiler was installed in 1875, boilers were used to supply power for primary propulsion for U.S. naval ships. The need for higher power, lower weight, and smaller footprint designs for ships forced boiler advances that were later adopted for current stationary designs (Ref 2). The primary propulsion systems of most ships have progressed to diesels, gas turbines, and nuclear power systems, but boilers still play a vital role in select ship systems. Shipboard space is critical, and marine boilers must fit within a minimum engine room space and be accessible for operation, inspection, and maintenance. Marine boilers must be rugged to operate and absorb vibration and forces resulting from rolling and pitching in rough seas. Boiler design needs to be conservative so that continuous operation is ensured over a long period of time without extensive maintenance.

There are two types of boilers: fire tube and water tube boilers. In fire tube boilers, the hot gases are on the inside of the tubes and the water is on the outside, and the boiler is usually used without superheat. In water tube boilers, the water is on the inside of the tubes and the hot gases are on the outside. Only water tube boilers can be used in large installations. The conventional steam cycle used in larger water tube boilers is the Rankine cycle. The Rankine cycle consists of compression of liquid water by a boiler feed pump, heating to the saturation temperature in an economizer, evaporation in the furnace, expansion work in the steam turbine, and condensation of the exhaust steam in a condenser. Steam cycle efficiency can be improved by adding superheater tubes to heat the steam above saturation temperatures. Reheat cycles further improve boiler efficiency. For pressures below 22 MPa (3200 psia), a steam drum is required to provide a tank volume sufficient to separate water from steam. The separated water, together with boiler makeup feedwater from the economizer, is returned by unheated downcomer tubes. For natural circulation boiler design (boilers <20 MPa, or 2900 psia), the fluid temperature in the heated riser tubes remains constant. The driving force for natural circulation boilers is the difference in fluid density between the heated risers and the unheated downcomers with the steam drum at the top of the boiler unit (Ref 3).

There are several types of boilers used in land-based and sea-based military applications: integral-furnace naval boilers, auxiliary package boilers, and waste heat boilers (Ref 2). Boilers may be fired by using coal, special residual oil, or natural gas as the primary fuel. Discussion and diagrams of marine boilers can be found in several publications (Ref 2, 4). Steels, cast irons, stainless steels, and high-temperature alloys are used to construct various boiler components. Boiler construction materials conform to ASME International specifications (Ref 5, 6) or other pertinent specification bodies.

In all boiler tubes, adequate circulation must be provided to avoid critical heat flux or departure from nucleate boiling (DNB) when the rate of bubble nucleation on the boiler tube surfaces exceeds the rate by which the bubbles are swept away. The overall volume of steam becomes too great, and DNB forms a continuous film of steam on the metal surface. The DNB depends on many variables, including pressure, heat flux, and fluid mass velocity.

Design defects, fabrication defects, improper operation, and improper maintenance are some of the common causes for boiler failures. Elevated-temperature and corrosion failures are common failure modes for boilers. Additionally, mechanical failures due to phenomena such as fatigue or wear occur as well. Some of the most common failures modes for boilers used for steam generating include overheating, fatigue or corrosion fatigue, fireside or waterside/steam-side corrosion, stress-corrosion cracking, and defective or improper materials.

The Electric Power Research Institute (EPRI) described 22 mechanisms, shown in Table 1, that are primarily responsible for boiler tube failures experienced in electric utility power generation boilers (Ref 7). The major failure categories are: (a) stress rupture, (b) waterside corrosion, (c) fireside corrosion, (d) erosion, (e) fatigue, and (f) lack of quality control. The classifications are arbitrary and based on visual characteristics of attack morphology. Reference 8 describes these failure modes.

Stress Rupture Failures in Boiler Environments. Stress and temperature influence the useful life of the tube steel. The strength of the boiler tube within the creep range decreases rapidly when the tube metal temperature increases. Creep entails a time-dependent deformation involving grain sliding and atom movement. When sufficient strain has developed at the grain boundaries, voids and microcracks develop. With continued operation at high temperatures, these voids and microcracks will grow and coalesce to form larger and larger cracks until failure occurs. The creep rate will increase, and the projected time to rupture will decrease when the stress and/or the tube metal temperature increases. High-temperature creep generally results in a longitudinal, fish-mouthed, thick-lipped rupture that has progressed from overheating over a long period of time. High-temperature creep can develop from insufficient boiler coolant circulation, long-term elevated boiler gas temperatures, inadequate material properties, or as the result of long-term deposition. A thick, brittle magnetite layer near the failure indicates long-term overheating.

High-temperature creep failures can be controlled by restoring the boiler components to boiler design conditions or by upgrading the tube material (either with ferritic alloys containing more chromium or with austenitic stainless steels). Failure from overheating caused by internal flow restrictions or loss of heat-transfer capability can be eliminated by removal of internal scale, debris, or deposits through flushing or chemical cleaning.

Boiler tubes exposed to extremely high temperatures for a brief period—metal temperatures of 455 °C (850 °F) and often exceeding 730 °C (1350 °F)—also fail from overheating. The overheating may occur from a single event or a series of brief events. Short-term overheating, generally, is related to tube pluggage, insufficient coolant flow due to upset conditions, and/or overfiring or uneven firing patterns. Loss of coolant circulation may be caused by low drum water levels or by another failure located downstream in the same tube. Inadequate coolant circulation can result in DNB in horizontal or sloped tubes when steam bubbles forming on the hot tube surface interfere with the flow of water coolant to the surface. This restricts the flow of heat away from the tube. Normally, the short-term overheating failure will exhibit considerable tube deformation from bulging, metal elongation, and reduction of wall thickness. Normally, the rupture will be longitudinal, fish-mouthed, and thin-lipped. Often, the suddenness of the rupture bends the tube. Very rapid overheating may produce a thick-lipped failure. A metallurgical analysis can determine tube metal temperature at the moment of rupture. Heavy internal deposits often are absent from a short-term overheating failure.

Since short-term overheating failure is caused most often by boiler upsets, rectifying any abnormal conditions can eliminate these ruptures. If restricted coolant flow or plugged sections are responsible, the boiler should be inspected and cleaned. Boiler regions with high heat flux may be redesigned with ribbed or rifled tubing to alleviate film boiling. Boiler operation should be monitored to avoid rapid start-ups, excessive firing rates, low drum levels, and improper burner operation.

Waterside corrosion failures often occur due to contamination of boiler feedwater. Maintaining cleanliness of the boiler water is critical for long-term, low-maintenance operation of boilers. The mineral and organic substances present in natural water supplies vary greatly in their relative proportions, but principally comprise carbonates, sulfates and chlorides of lime, magnesia and sodium, iron and aluminum salts, silicates, mineral and organic acids, and the gases oxygen and carbon dioxide. Scale is formed from the carbonates and sulfates of lime and magnesia and from the oxides of iron, aluminum, and silicon, and it will result in:

- Reduction in the boiler efficiency because of the decreased rate of heat transfer
- Overheating and burning of tubes resulting in tube failure

Scales are dangerous long before they reach this thickness. A very thin scale can cause tube failure due to overheating. Scale has about 2% of the thermal conductivity of steel. A scale thickness of about 1 mm (0.04 in.) can be sufficient to reduce the heat-transfer rate to a dangerous point; when the water inside the tube cannot receive and carry away the heat fast enough from the tube alloy to keep it below its transformation temperature, the tubes "burn out."

Water and waterside chemistries are important in maintaining protection of internal tube surfaces, but they also can contribute to waterside corrosion problems such as caustic gouging, hydrogen damage, pitting, or dealloying. The change in pH can markedly affect the corrosion rate of steel by water. Upsets in water chemistry can affect the corrosion rate and the amount of deposition on the tube wall.

Caustic corrosion (also referred to as caustic gouging, caustic attack, or ductile gouging) results from deposition of feedwater corrosion products, in which sodium hydroxide can concentrate to high pH levels. The hydroxide solubilizes the protective magnetite layer and reacts directly with the iron (Ref 9, 10):

$$4NaOH + Fe_3O_4 = 2NaFeO_2 + Na_2FeO_2 + 2H_2O$$

$$Fe + 2NaOH = NaFeO_2 + H_2$$

Sodium hydroxide can concentrate under porous deposits through a mechanism called wick boiling. Deposition occurs at the highest heat input areas and accumulates at flow disruptions such as downstream of backing rings, around tube bends, and in horizontal or inclined tube regions. Caustic corrosion results in irregular wall thinning or gouging. Failures develop after critical wall loss. During DNB conditions, boiler water solids will develop at the metal surface, usually at the interface between the bubbles and the water. Corrosive solids will precipitate at the edges of the blanket with corresponding wall loss. The metal under the DNB blanket is largely intact. A visual examination often can identify caustic gouging (Ref 9). Ultrasonic examination can detect regions where caustic gouging may occur. Metallurgical examination may or may not reveal an overheated microstructure in the region affected

Table 1 Electric Power Research Institute classification of boiler tube failure modes

Failure mode	Causal factor
Stress rupture	Short-term overheating
	High-temperature creep
	Dissimilar metal welds
Erosion	Fly ash
	Falling slag
	Sootblower
	Coal particle
Waterside corrosion	Caustic corrosion
	Hydrogen damage
	Pitting (localized corrosion)
	Stress-corrosion cracking
Fatigue	Vibration
	Thermal
	Corrosion
Fireside corrosion	Low-temperature
	Waterwall
	Coal ash
	Oil ash
Lack of quality control	Maintenance cleaning damage
	Chemical excursion damage
	Materials damage
	Welding defects

Source: Ref 7

by caustic gouging. Analysis of the bulk deposit by energy-dispersive spectroscopy (EDS) analysis and/or x-ray mapping can distinguish regions of high sodium content caused by caustic gouging.

Caustic corrosion may be reduced by minimizing the entry of deposit-forming materials into the boiler and by performing periodic chemical cleanings to remove waterside deposits. Monitoring the water chemistry to reduce the amount of available free sodium hydroxide, preventing in-leakage of alkaline-producing salts into condensers, preventing DNB, using ribbed tubing in susceptible areas, and eliminating welds with backing rings or joint irregularities can reduce or negate caustic gouging.

Hydrogen damage also results from fouled heat-transfer surfaces. There is some disagreement as to whether hydrogen damage can occur only under acidic conditions or whether it can happen under alkaline and acidic conditions as well. Hydrogen damage may occur from the generation of atomic hydrogen during rapid corrosion of the waterside tube surface, although it may occur with little or no apparent wall thinning. The atomic hydrogen diffuses into the tube steel, where it reacts with tube carbides (Fe_3C) to form gaseous methane (CH_4) at the grain boundaries. Large methane gas molecules concentrate at the grain boundaries. When methane gas pressures exceed the cohesive strength of the grains, a network of discontinuous, intergranular microcracks is produced. Often, a decarburized tube microstructure observed by metallographic examination is associated with hydrogen damage. The cracks grow and link together to cause a thick-wall failure. Hydrogen damage has been incorrectly referred to in the literature and in practice as hydrogen embrittlement; actually, the affected ferrite grains have not lost their ductility. However, because of the microcrack network, a bend test will indicate brittlelike conditions.

Hydrogen damage and caustic gouging are experienced at similar boiler locations and, usually, under heavy waterside deposits. Hydrogen damage from low-pH conditions may be distinguished from high-pH conditions by considering the boiler water chemistry. A low-pH condition can be created when the boiler is operated outside of normal recommended water chemistry parameter limits. This is caused by contamination such as: (a) condenser in-leakage (e.g., seawater or recirculating cooling water systems incorporating cooling towers), (b) residual contamination from chemical cleaning, and (c) the inadvertent release of acidic chemicals into the feedwater system. A mechanism for concentrating acid-producing salts (DNB, deposits, waterline evaporation) must be present to provide the low-pH condition. Low pH will dissolve the magnetite scale and may attack the underlying base metal through gouging (Ref 9, 11):

$$M^+Cl^- + H_2O = MOH(s) + H^+Cl^-$$

Proper control of the water chemistry and removal of waterside deposits may eliminate hydrogen damage when concentrating boiler solids occur. History and details of caustic embrittlement and hydrogen damage and embrittlement are discussed elsewhere (Ref 11).

Passive film breakdown is followed by the formation of a concentration cell. At the anode, the metal oxidizes, while at the much larger cathode surface surrounding the pit, oxygen or hydrogen is reduced. Pit propagation is autocatalytic. The rapid metal dissolution within the pit tends to produce an excess of positive charges within the cavity, resulting in migration of chlorides, sulfates, thiosulfates, or other anions into the pit cavity to maintain electroneutrality. Hydrolysis of metal salts ($M^+Cl^- + H_2O = MOH + H^+Cl^-$) forms an insoluble metal hydroxide and a free acid, which decreases the pH. Low pH results in rapid corrosion (Ref 9, 11).

Boiler steels pit in the presence of moisture and oxygen. This generally occurs when the boiler is not operational and has not been completely dried or protected by a nitrogen blanket during shutdown. Pitting attack from dissolved oxygen in pools of condensation has been found in reheater tubing. Pitting in the economizer section materializes during start-ups and low-load operations because of high oxygen levels. Oxygen pitting can be corrected if hydrazine-treated water and a proper nitrogen blanket are used during shutdown.

The propagation of crevice corrosion is similar to the propagation of pitting. The crevice-corrosion initiation stage differs from pitting initiation because the crevice is an artificial pit. The anodic reaction will predominate within the crevice after oxygen or hydrogen reduction diminishes; the oxygen or hydrogen reduction reaction predominates outside the crevice. Waterside deposits, silt, sand, and shells can be found on the tube surfaces producing such crevices. Underdeposit attack is usually uniform on steel, cast iron, and most copper alloys.

Intergranular corrosion occurs at or adjacent to grain boundaries with little corresponding corrosion of the grains. Intergranular corrosion can be caused by impurities at the grain boundaries or enrichment or depletion of one of the alloying elements in the grain-boundary area. Austenitic stainless steels such as type 304 become sensitized or susceptible to intergranular corrosion when heated in the range of 510 to 790 °C (950 to 1450 °F). In this temperature range, $Cr_{23}C_6$ precipitates form and deplete chromium from the area along the grain boundaries (below the level required to maintain stainless properties). The chromium-depleted area becomes a region of relatively poor corrosion resistance. Controlling intergranular corrosion of austenitic stainless steels occurs through: (a) employing high-temperature solution heat treatment, (b) adding elements (stabilizers) such as titanium, niobium, or tantalum that are stronger carbide formers than chromium, and (c) lowering the carbon content below 0.03%. Other alloys such as aluminum, ferritic stainless steels, and bronzes are also affected by intergranular corrosion. A microscopic, metallographic examination and SEM/EDS can determine the degree of degradation. Several standard tests can evaluate the susceptibility of different alloys to intergranular corrosion.

In order to minimize corrosion associated with the watersides and steamsides of steam generators and other steam/water cycle components, a comprehensive chemistry program should be instituted. There are a wide variety of chemical treatment programs that are utilized for these systems depending on the materials of construction, operating temperatures, pressures, heat fluxes, contaminant levels, and purity criteria for components using the steam. Minimizing the transport of corrosion products to the steam generator is essential for corrosion control in the steam generator. Corrosion products from the feedwater deposit on steam generator tubes, inhibit cooling of tube surfaces, and provide an evaporative-type concentration mechanism of dissolved salts under boiler tube deposits. Therefore, the chemistry of water in contact with the tube surfaces is often much worse than in the bulk water. In general, the higher the pressure of the operating boiler, the more critical the control of water chemistry is for proper operation of the system.

Fireside corrosion failures are dependent on the type of fuel environment and the component metal temperature. The corrosiveness of the environment depends on the surface temperature and the condition of and/or the corrosive ingredients in the medium. External corrosion of steel piping can occur under wet fiberglass insulation at room temperature. Fiberglass contains soluble chlorides that can be leached out when the insulation becomes wet and can concentrate at the metal surface. This promotes corrosion.

Low-temperature or dew-point corrosion results from the condensation of sulfuric acid or other acidic flue gas vapors when the component temperature drops below the acid dew point or is operated below the acid dew point so that condensate will form a low-pH electrolyte on fly ash particles and produce acid smuts. Dew-point corrosion failures in boilers lead to stress rupture of the tube steel from loss of load-carrying material on the fireside surface. The external surface will be gouged or pitted. Sulfuric acid is formed when moisture reacts with sulfur trioxide. The acid dew point is related to the concentration of sulfur trioxide in the boiler flue gas (at 10 ppm SO_3, the dew point is approximately 140 °C, or 280 °F). Dew-point corrosion can be corrected by the injection of enough magnesium oxide to neutralize the acid. Minimizing excess oxygen will reduce the formation of SO_3. When SO_3 is maintained below 10 ppm, dew-point corrosion is often not a problem. Dew-point corrosion is thoroughly discussed in another source (Ref 12). The primary methods of control are: (a) keeping the metal at a temperature above the acid dew point, (b) specifying low-sulfur fuels, and (c) removing fireside deposition from metal surface immediately after boiler shutdown using high-pressure water sprays followed by neutralizing lime solutions.

Corrosive constituents in fuel at appropriate metal temperatures may promote fireside corrosion in boiler tube steel. For coal, lignite, oil, or refuse, the corrosive ingredients can form liquids (liquid ash corrosion) that solubilize the oxide film on tubing and react with the underlying metal to reduce the tube wall thickness. This can occur in the combustion zone (e.g., waterwall tubes) or convection pass (e.g., superheaters). Technologies to reduce NO_x emissions from the boiler also result in more reducing conditions in the furnace, which greatly increases the potential for liquid ash corrosion. This has been alleviated by installing air ports along the walls to create an air blanket and installing stainless steel and high-alloy weld overlays.

Waterwall fireside corrosion may develop when incomplete fuel combustion causes a reducing condition because of insufficient oxygen. Incomplete combustion causes the release of volatile sulfur and chloride compounds, which causes sulfidation and accelerated metal loss. Poor combustion conditions and steady or intermittent flame impingement on the furnace walls may favor an environment that forms sodium and potassium pyrosulfates ($Na_2S_2O_7$ or $K_2S_2O_7$), which have melting points below 427 °C (800 °F). The presence of chlorides lowers the melting temperature of the combined molten salts and increases the corrosion rate of steel. Pyrosulfate formation due to sufficient presence of SO_3 and alkalis can cause significant corrosion and metal wastage at temperatures from 400 to 595 °C (750 to 1100 °F) (Ref 13).

Metal attack occurs along the crown of the tube and may extend uniformly across several tubes. Corroded areas are characterized by abnormally thick iron oxide and iron sulfide scales. Visual, microscopic, and SEM/EDS analyses may identify the corrodent. Verification of waterwall fireside corrosion involves analyzing the fuel, the completeness of combustion, and the evenness of heat transfer. Carbon contents in the ash greater than 3% are indicative of reducing and corrosive combustion conditions. The ratio of carbon monoxide to carbon dioxide may be indicative of whether oxidizing or reducing conditions are prevailing, but local condition may differ considerably from the bulk environment.

The reducing conditions tend to lower the melting temperature of any deposited slag, which also increases the solvation of the normal oxide scales. A stable gaseous sulfur compound under reducing conditions is hydrogen sulfide (H_2S). Under the reducing environment, iron sulfide is the expected corrosion product of iron reacting with pyrosulfates. Sulfur prints will show the sulfide in the corrosion products around the metal surface penetrations.

Operating factors that may improve combustion conditions are: (a) better coal grinding, (b) amended fuel distribution, (c) increased and redistributed secondary air, and (d) supplemented air into the boiler. However, this may provide only a marginal improvement. A complete furnace modification may be required. Pad welding may be economical for low levels of fireside corrosion, while coatings may provide short-term protection.

Coal ash corrosion results when a molten ash of complex alkali-iron trisulfates forms on the external surfaces of reheater and superheater tubes in the temperature range of 540 to 705 °C (1000 to 1300 °F). Liquid trisulfates solubilize the protective iron oxide scale and expose the base metal to oxygen, which produces more oxide and subsequent metal loss, according to a mechanism proposed by Reid (Ref 13):

$$3K_2SO_4 + Fe_2O_3 + 3SO_3 = 2K_3Fe(SO_4)_3$$

$$9Fe + 2K_2Fe(SO_4)_3 = 3K_2SO_4 + 4Fe_2O_3 + 3FeS$$

$$4FeS + 7O_2 = 2Fe_2O_3 + 4SO_2$$

$$2SO_2 + O_2 = 2SO_3$$

$$3K_2SO_4 + Fe_2O_3 + 3SO_3 = 2K_3Fe(SO_4)_3$$

Alkali sulfates, originating from the alkalis in the fuel ash and the sulfur oxides in the furnace atmosphere are deposited on the metal oxide layer setting up a temperature gradient. The alkali sulfates attract fuel ash. The temperature of the fuel ash increases to a point where sulfur trioxide is released by thermal dissociation sulfur compounds within the ash. The sulfur trioxide then migrates toward the cooler metal surface. Reaction of sulfur compounds with the metal oxide forms alkali metal sulfates and dissolves the protective alloy film. Reaction of the alkali metal trisulfates can reform iron oxide and iron sulfide. The iron sulfide will react with available oxygen to SO_2 and SO_3. The net overall reaction is:

$$4Fe + 3O_2 = 2Fe_2O_3$$

which accounts for the metal loss. The greatest metal loss creates flat sites at the interface between the fly-ash-covered half and the uncovered half (2 and 10 o'clock positions to the gas flow). Corrosive coal typically contains a significant amount of sulfur and sodium and/or potassium compounds.

Visually, ferritic steel will exhibit shallow grooves (referred to as alligatoring or elephant hide). A sulfur print will reveal the presence of sulfides at the metal/scale interface. Coal ash corrosion reduces the effective wall thickness, thereby increasing tube stress. This combined with temperatures within the creep range may cause premature creep stress ruptures because of coal ash corrosion.

Fireside corrosion can also occur in oil-fired boilers. As in coal combustion, SO_2 and SO_3 form in relative amounts depending primarily on the temperature. In excess air, vanadium forms vanadium pentoxide (V_2O_5) and sodium forms sodium monoxide (Na_2O). Together, V_2O_5 and Na_2O form a range of compounds that melt to temperatures less than 540 °C (1000 °F). Oil ash corrosion is believed to be a catalytic oxidation of the material by reaction with V_2O_5. Sodium oxide also reacts with sodium trioxide to form sodium sulfate, which together with V_2O_5 also forms a range of low-melting-point liquids with a minimum temperature around 540 °C (1000 °F) (Ref 13, 14). The temperature range precludes waterwall damage by this corrosion mechanism. Superheaters and reheaters may be prone to oil ash corrosion.

Several corrective options to reduce failure caused by fireside corrosion are: (a) employing thicker tubes of the same material, (b) shielding tubes with clamp-on protectors, (c) coating with thermal sprayed, corrosion-resistant materials, (d) purchase of fuels with low impurity levels (limit levels of sulfur, chloride, sodium, vanadium, etc.), (e) purification of fuels (e.g., coal washing), (f) blending coals to reduce corrosive ash constituents, (g) replacing tubes with higher-grade alloys or coextruded tubing, (h) adjustment of operating conditions (e.g., increasing percent excess air, percent solids in recovery boilers, etc.), (i) redesigning affected sections to modify heat transfer, (j) adding fuel additives such as calcium sulfate ($CaSO_4$) or magnesium sulfate ($MgSO_4$) to bind SO_3 to form a less corrosive form, and (k) modification of lay-up practices.

Environmental cracking failures can occur with a wide combination of metals and alloys, stress fields and mode, and in various specific environments. Environmental cracking is defined as the spontaneous, brittle fracture of a susceptible material (usually quite ductile itself) under tensile stress in a specific environment over a period of time. Stress-corrosion cracking (SCC), hydrogen damage, and corrosion fatigue are some of the forms of environmental cracking that can lead to failure.

Stress-corrosion cracking results from the conjoint, synergistic interaction of tensile stress and a specific corrodent for the given metal. The tensile stress may be either applied (such as caused by internal pressure) or residual (such as induced during forming processes, assembly, or welding). Stress-corrosion cracking involves the concentration of stress and/or the concentration of the specific corrodent at the fracture site. Fractures due to SCC may be oriented either longitudinally or circumferentially, but the fractures are always perpendicular to the stresses. In boiler systems, carbon steel is specifically sensitive to concentrated sodium hydroxide, stainless steel is sensitive to concentrated sodium hydroxide, chlorides, nitrates, sulfates, and polythionic acids, and some copper alloys are sensitive to ammonia and nitrites. Stress-corrosion cracking failures produce thick-walled fractures that may be intergranular, transgranular, or both. Branching is often associated with SCC. Normally, gross attack of the metal by the corrodent is not observed in SCC failures. Failures caused by SCC are difficult to see with the naked eye. Metallographic and chemical analyses are performed to identify the constituents in the alloy and the bulk corrosion products. Analytical techniques such as EDS or Auger spectroscopy may be used to determine the source of the corrodent within the cracks. Magnetic particle or ultrasonic techniques can detect cracked regions in the boiler. A stress analysis can be conducted to determine if high applied or residual stresses are involved. A check of the background history of the failed component is particularly useful in assessing the contributing factors for this type of failure. Stress-corrosion cracking failures frequently have been experienced immediately after chemical cleanings and initial start-ups. Copper alloys, specifically brasses, are susceptible to SCC. Most steamside failures have occurred where high concentrations of ammonia and oxygen exist such as in the air-removal section of the condenser. Stress-corrosion cracking failures are found often at inlet or outlet ends where the tubes have been expanded into the tubesheet.

Stress-corrosion cracking may be reduced by either removing applied or residual stresses or avoiding concentrated corrodents. The reduction of corrodents by avoiding boiler upsets and in-leakage is generally the most effective means of diminishing or eliminating SCC. A change in tube metallurgy also may reduce SCC.

Boiler tube cyclic stresses are stresses periodically applied to boiler tubes that can reduce the expected life of the tube through the initiation and propagation of fatigue cracks. The environment within the boiler would suggest that corrodents would also interact with the tube to assist in this cracking process. The number of cycles required to produce cracking is dependent on the level of strain and the environment. Vibration and thermal fatigue failures describe the type of cyclic stresses involved in the initiation and propagation of these fractures.

Vibration fatigue cracks originate and propagate as a result of flow-induced vibration. This occurs where tubes are attached to drums, headers, walls, seals, or supports. Circumferential orientations are common to vibration fatigue cracks. Crack-initiation sites generally occur on the fireside (external) tube portions. Dye penetrant or magnetic particle techniques can detect crack sites. Metallographic examination can confirm fatigue cracking, which is often transgranular. Vibration restraints can be installed to eliminate gas-flow-induced vibration and vibration fatigue cracking.

Thermal fatigue cracks develop from excessive strains induced by rapid cycling and sudden fluid temperature changes in contact with the tube metal across the tube wall thickness. This can be caused by rapid boiler start-ups above proper operational limits. Water quenching by spraying from condensate in the soot-blowing medium and from the bottom ash hoppers also can induce thermal fatigue cracks. Sudden cooling of tube surfaces causes high tensile stresses because cooled surface metal tends to contract; however, this metal becomes restricted by the hotter metal below the surface. Procedures that can reduce these sudden cyclic temperature gradients will diminish or eliminate thermal fatigue cracking.

The combined effects of a corrosive environment and cyclic stresses of sufficient magnitude cause corrosion fatigue failures. Corrosion fatigue cracks may develop at stress concentration sites (stress risers) such as pits, notches, or other surface irregularities. Corrosion fatigue is commonly associated with rigid restraints or attachments. Corrosion fatigue cracking most frequently occurs in boilers that operate cyclically. The fracture surface of a corrosion fatigue crack will be thick-edged and perpendicular to the maximum tensile stress. Multiple parallel cracks are usually present at the metal surface near the failure. Microscopic examination will detect straight, unbranched cracks. The cracks often will be wedge-shaped and filled with oxide. The oxide serves to prevent the crack from closing and intensifies the stresses at the crack tip during tensile cycles, thereby assisting in the crack growth.

Corrosion fatigue cracking can be reduced or eliminated by: (a) controlling cyclic tensile stresses (reducing or avoiding cyclic boiler operation or extending start-up and shutdown times), (b) redesigning tube restraints and attachments where differential expansion could occur, (c) controlling water chemistry to reduce the formation of stress risers such as pits, and (d) removing residual stresses by heat treatment, if possible. Some further discussions on environmental cracking and corrosion fatigue can be found in other sources (Ref 15, 16).

Case History: Boiler Tube Failure (Ref 17). The roof tube section from a land-based boiler at a military base failed. Figure 1 shows that the tube failure consisted of a 25 mm (1 in.) longitudinal split along the hot side. This split coincided with local bulging of the tube at the hot side crown. The tube sample was cut along the longitudinal, membrane axis. A side view of the respective waterside deposits (Fig. 2) indicated that the hot deposit was 25 mm (1 in.) thick in some places along the sample, while the cold side deposits were 0.13 to 0.51 mm (0.005 to 0.020 in.) thick. The roof tube was low-alloy carbon steel, which conformed to ASME SA 192 or SA 210 chemical specifications. The oxidation limit and maximum allowable temperature for these carbon steels is 455 °C (850 °F).

A view of the cold side microstructure (180° and opposite from the failure) by SEM as shown in Fig. 3 attests to the lamellar nature of the pearlite colonies in the ferrite matrix. The waterside deposit in the vicinity of the failure exhibited several distinct layers of oxide (Fig. 4). There was also heavy oxidation and corrosion of the waterside surface. The deposit shown in this micrograph is 0.61 to 0.71 mm (0.024 to 0.028 in.) thick. Some of the layers are quite dense while others are very porous. One layer exhibits a dispersion of copper metal particles while still another layer revealed needlelike, fibrous oxides.

In Fig. 5, the tube microstructure near the failure (~13 mm, or 0.5 in., away) displayed complete spheroidization of the carbides from overheating. Voids developed at grain boundaries, particularly at three-grain junctions; these voids tend to be oriented perpendicular to the tube surface. The formation of these voids is due to long-term overheating and creep rupture. Some of the voids have grown and coalesced into larger voids and incipient microcracks; some of the largest voids have formed surface oxides. There is slight elongation of the grains as a result of tube swelling and expansion. The micrograph in Fig. 6 is a higher-magnification view that clearly shows that the carbides at former pearlite colonies have fully spheroidized. Some of these

Fig. 1 Roof tube "as received." Narrow 25 mm (1 in.) long failure is along the hot side crown. See also Fig. 2 through 6.

Fig. 2 Side view of tube sample reveals that hot side deposits are as much as 25 mm (1.0 in.) thick. Relatively little deposition was present on the cold side

carbides have dispersed into the grains. Other random carbides are present along the grain boundaries; voids have formed at some of these grain-boundary sites. Void formation and propagation occurred in a direction perpendicular to the hoop stress.

The waterside deposit was extremely heavy, >1100 mg/cm^2 (1000 g/ft^3) and unusual for a roof tube from a low-pressure boiler. The presence of high levels of copper in the tube deposit reportedly from the present, all-ferrous boiler system and the lack of any apparent contribution from the boiler water source tend to suggest that the deposits had formed long ago when copper alloys were used in the condenser or feedwater heater. The presence of the heavy deposits on the roof tube sample interfered with normal heat transfer. The horizontal roof tubes in the upper portion of the boiler are also prone to sluggish coolant flow. Both of these factors lead to increasing tube metal temperatures sufficient to cause creep rupture. The examination of the tube microstructure around the failure indicated a temperature of around 480 °C (900 °F). These deposits led to local overheating and creep rupture on the hot side crown of the roof tube. Waterside pitting accentuated the applied tube stresses and decreased the time to rupture. The boiler should be examined for other occurrences of heavy boiler deposits. Affected roof tube sections should either be removed or the boiler chemically or mechanically cleaned.

Diesel Engines

In the diesel engine, air is compressed adiabatically with a compression ratio typically between 14 to 1 and 20 to 1. This compression raises the temperature to the ignition temperature of the fuel mixture that is formed by injecting fuel into a chamber once the air is compressed. The ideal air-standard cycle is modeled as a reversible adiabatic compression followed by a constant pressure combustion process, then an adiabatic expansion as a power stroke and an isovolumetric exhaust.

A diesel engine designates a reciprocating engine where air is compressed within a cylinder to the extent that spontaneous ignition of fuel occurs, followed by burning a measured amount of oil-grade fuel. The compression ratio must be sufficiently high so that the air temperature at the end of compression will ignite the fuel when it is sprayed into the cylinder. Diesel engines are either two-stroke or four-stroke cycle, depending on the number of strokes required to complete a full cycle of operation. Compression ratios in the diesel engine range between 14 to 1 and 20 to 1. This high ratio causes increased compression pressures of 2760 to 4135 kPa (400 to 600 psi) and cylinder temperatures to reach 425 to 650 °C (800 to 1200 °F). At the proper time, the diesel fuel is injected into the cylinder by a fuel injection system, which usually consists of a pump, fuel line, and injector or nozzle. When the fuel oil enters the cylinder, it will ignite because of the high temperatures. Diesel engines are either water-cooled or air-cooled since a considerable amount of heat is generated in the cylinders and the temperature of the cylinder boundaries must be controlled to keep within safety limits. Exhaust temperatures of diesel engines are between 260 and 540 °C (500 and 1000 °F).

Today, diesel engines are used extensively in the military, serving as propulsion units for small boats, ships, and land vehicles. They are also used as prime movers in auxiliary machinery, such as emergency diesel generators, pumps, and compressors.

Medium-sized combatant ships and many auxiliary vessels are powered by large (~37,000 kW, or 50,000 brake horsepower, bhp) single-unit diesel engines or, for more economy and operational flexibility, by combinations of somewhat smaller engines. Diesel engines have relatively high efficiency at partial load, and much higher efficiency at very low partial load than steam turbines. They also have greater efficiency at high speeds than any of the other fossil-fueled plants. Thus they require the least weight of fuel for a given endurance. Other advantages include low initial cost and relatively low rpm, the latter resulting in small reduction gears. Additionally, diesel engines can be brought on-line from cold conditions rapidly. They are reliable and simple to operate and maintain, having a long history of active development for marine use.

In general, however, the use of diesels on intermediate-sized combatants and larger vessels requires that several smaller units be combined to drive a common shaft. This requirement results in severe space and arrangement problems. Among other disadvantages is the fact that periodic engine overhaul and progressive maintenance are required. These result in frequent down periods, which, because of the number of similar units, may increase the amount of necessary in-port maintenance time and decrease the amount of time the ship has full power available while at sea. Finally, the marine diesel has a high rate of lubricant oil consumption, which may approach 5% of the fuel consumption; thus large quantities of lubricant oil must be carried. Ship propulsion systems employ many different arrangements of engines, shafts, reduction gears, and propellers to suit the operating requirements of the ships they serve.

One of the operational characteristics of diesel engines is the ability to operate while burning a variety of fuels. Marine use of diesel fuels generally requires the use of NATO F-76 (Ref 18) diesel distillate, JP-5 turbine fuel, aviation (Ref 19), or JP-8 turbine fuel, aviation (Ref 20). When

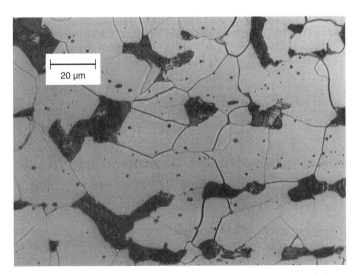

Fig. 3 Magnified view of cold side microstructure shows the lamellar carbides in the pearlite colonies. Spherical phases are small alloy constituents such as oxide and sulfide inclusions. Original magnification 1050×

Fig. 4 Hot side deposit in the vicinity of the failure displays several layers and waterside oxidation/corrosion. Some layers are dense, while others are very porous. One layer displayed needlelike, fibrous oxides. Another layer contains a number of particles of copper metal. Original magnification 210×

these fuels are unavailable and to minimize operational impact, various commercial fuel grades or contract Marine Gas Oil-type fuels may be purchased under a bunker contract (local purchase) by the Defense Energy Support Center (DESC). Fuels under the bunker contract meet the Navy Marine Gas Oil (MGO) Purchase Description (PD) (Ref 21). The Navy's MGO-PD fuel is subject to standard commercial practice quality control rather than the extensive handling and storage requirements specified for MIL-SPEC fuels. The effect of sulfur content in the fuel on engine wear and performance is minimized if the engine water jacket temperature is at least 60 °C (140 °F) (Ref 21). Saltwater-contaminated fuel has caused erosion in fuel nozzles and corrosion leading to burned pistons (Ref 22).

Contamination of combustion fuel in diesel engines can cause high-temperature corrosion. When burning ash-rich and sulfur-rich fuels in engines, the exhaust gases cause considerable corrosion of the alloys surfaces to which they come in contact. The degree of corrosion is dependent on the fuel composition, on the oxygen content in the exhaust gas, and the alloy temperature (Ref 23). Vanadium and sodium contamination can cause the formation of coke deposits on high-temperature components that reduce cooling engine capacity and lead to higher exhaust gas temperatures. The formation of molten sodium vanadates at temperatures as low as 535 °C (995 °F) may lead to the high-temperature corrosion of diesel engine components (Ref 23).

Cylinder wear is due mainly to friction, abrasion, and corrosion. Frictional wear occurs between the sliding surface of the cylinder liner and piston rings. The degree of wear will depend on the materials involved, the surface conditions, efficiency of the cylinder lubrication, piston speed, engine loading with corresponding pressures and temperatures, maintenances of the piston rings, combustion efficiency, and air or fuel contamination (Ref 24). If the cylinder is operated with excessive wear, the degradation may lead to gas blowpast. Gas blowpast may remove the lubricating oil film, the piston rings may distort and break, and piston slap may cause scuffing. Reduced compression causes incomplete combustion, which can lead to exhaust system fouling.

Corrosion occurs mainly in engines that burn heavy fuels that contain significant sulfur levels. The combustion of heavy sulfur-containing fuels forms acids. Sulfuric acid may condense and cause corrosion in the lower part of the liner if the jacket cooling water temperature is too low. Water vapor condenses on the liner and absorbs any sulfur species such as SO_3 or SO_2 to form the sulfuric acid. Increasing the jacket temperature above the dew-point temperature prevents this form of corrosion.

Abrasion may take place from the hard-particle products of mechanical wear, corrosion, and combustion. All fuels have a given ash content; heavy fuels tend to have a higher ash content.

Cylinder liners require adequate lubrication to reduce piston ring friction and wear. The oil film also acts as a gas seal between liners and rings and as a corrosion inhibitor. Special cylinder oil, having a high alkalinity, for engines burning heavy fuels, neutralize acids formed from the combustion of sulfur present in the fuel. Oils must maintain their viscosity at high temperatures, resist oxidation and carbonization, spread readily, yet be able to cling to working surfaces (Ref 24).

It is possible for fuel or the oil systems to be infected and contaminated by microorganisms. Their presence can generate organic acids that can lead to corrosion wear of metal surfaces and create sludge and slime that will plug oil filters. This contamination can be minimized if cleanliness and care are exercised.

Gas Turbine Engines

The gas turbine was developed generally for main propulsion and power for certain auxiliary systems of aircraft, ships, or certain other military platforms that require a higher power density than diesel engines can generate. In the case of gas turbines, fuel and air quality and the specific engine environment in which it operates influence the corrosion of turbine components significantly. The gas turbine operates on the principle of the Brayton cycle. A gas turbine has three major components: (1) compression of a gas (typically air), (2) addition of heat energy into the compressed gas either by directly firing or combusting the fuel in the compressed air or by transferring the heat through a heat exchanger into the compressed gas, followed by (3) expansion of the hot pressurized gases in a turbine to produce useful work (Ref 25). The compressor is a series of blades or airfoils, some rotating (rotors), some stationary (stators), that draw air in and compress it. The more rows of blades, the more the air pressure increases as it passes through the compressor stages. Typical pressures can be up to 40 times higher than atmospheric pressure.

The compressed air is pushed forward into a combustion chamber where fuel is injected with the air and it ignites. The flow and burn of the air/fuel mixture is controlled to ensure that the engine sustains a continuous flame. The expanding exhaust gases flow quickly toward the rear of the engine. For modern aircraft gas turbines, combustion temperatures above 2500 °C (4530 °F) are possible. This is well above the melting point of most alloys, so film cooling of the combustion chamber is necessary. The gases

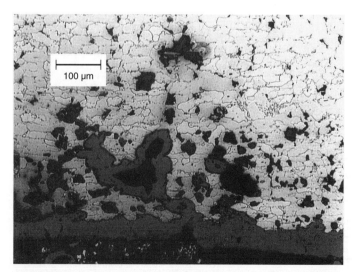

Fig. 5 Waterside surface, hot side, near the failure. Carbides in prior pearlite colonies have completely spheroidized from overheating. Creep voids have developed at grain boundaries; some of these voids have grown and coalesced. Original magnification 210×

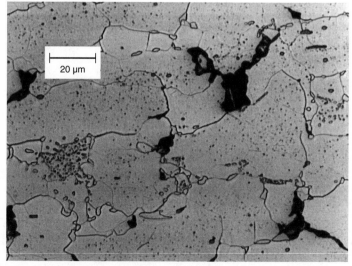

Fig. 6 Carbides are fully spheroidized from thermal degradation near failure. Voids (dark sites) have formed along the grain boundaries that are perpendicular to the direction of applied stress. Original magnification 1050×

that move out of the combustor exert force against the turbine blades (airfoils), which are connected to the compressor by shafts. The gases exiting the turbine create energy by spinning the turbine that powers the compressor to continue the process of pulling air into the engine.

Air from the compressor is used for cooling the turbine in order to maintain the metal temperatures within their design limits, while the gas flowing through the turbine may be as high as 1370 °C (2500 °F). Metal temperatures can be as high as 1150 °C (2100 °F) and require some type of protective coating. The requirement for cooling the turbine limits the ultimate thermal efficiency of the gas turbine, and technologies are being developed in the areas of materials including ceramics and enhanced cooling effectiveness in order to minimize the cooling air requirement. With more advanced materials and cooling technologies, increases in turbine inlet temperature are possible in order to increase the thermal efficiency of the cycle.

There are several types of gas turbine engines. The primary purpose of aircraft gas turbines is to generate high performance and power quickly at the expense of fuel economy. Afterburning, or reheat, increases exhaust velocity and engine thrust for short times to improve aircraft takeoff, climb, and maneuverability by injecting additional fuel in the jet pipe. Vectored thrust engines employ swiveling nozzles that can direct the gas stream from vertically downward for upward lift through an arc to horizontally rearward for conventional forward thrust. Turboshaft engines are used in helicopters, marine propulsion and auxiliary systems, and some land-based military assets. It uses a power turbine and a gearbox, though the power is transmitted directly to a variety of devices, such as the rotor system, ship or platform propulsion, or auxiliary systems.

Aeroengines use relatively pure fuels with low sulfur contents, and the air quality is generally good unless there are recurrent low altitude flights or operations over marine environments. Jet fuels JP-5 and JP-8 are substances used as aircraft fuels by the military. JP-5 and JP-8 is shorthand for jet propellants 5 and 8. JP-5 is the U.S. Navy's primary jet fuel, and JP-8 is one of the jet fuels used by the other services.

Both JP-5 and JP-8 are composed of a large number of chemicals, and both are colorless liquids that may change into gas vapor. They smell like kerosene, since kerosene is the primary component of both substances. They are made by refining either crude petroleum oil deposits found underground or shale oil found in rock.

High-temperature processes that may occur in gas turbines are dependent largely on environmental factors and the type of gas turbine engine. High-temperature processes in aircraft gas turbine engines are generally due to high-temperature oxidation or to high-temperature creep. Thermal barrier coatings (TBCs) are used to protect hot section alloys from catastrophic oxidation. There is strict compositional control of the nickel-base superalloys is used in the hot sections of aerogas turbines. These alloys are either directionally solidified or are single crystals to improve the creep strength and prolong blade life from creep damage. High-temperature corrosion at operating temperature does not occur often, but ingestion of sea salt in naval aircraft may result in pitting and molten salt corrosion if not water-washed often. Sand may cause abrasion of turbine components. The exposure to airborne dirt such as sand and ash may introduce silicates into the engine, commonly know as CMAS ($CaO-MgO-Al_2O_3-SiO_2$), which refers to the main chemical components of calcium, magnesium, aluminum, and silicon. Molten silicate deposits can penetrate TBCs in advanced gas turbine engines due to increasing operating temperatures. Molten CMAS partially penetrates the TBC during the hot phase where the critical temperature is 1240 °C (2265 °F). Upon cooling the CMAS solidifies, degrades the coatings, and promotes spallation of the infiltrated coating. Degradation appears to involve dissolution of the porous TBC columnar grains in the silicate melt and reprecipitation of dense globular zirconia crystals. The yttria stabilizer is influenced by the silicate melt leading to the formation of monoclinic zirconia that is more prone to erosion damage and spallation. Penetration of CMAS can be substantially reduced by adding alumina, which lowers the melting point of the quaternary mixture.

Shipboard marine gas turbines operate at lower metal temperatures (500 to 1000 °C, or 930 to 1830 °F) than observed in aircraft gas turbines. Shipboard gas turbine engines also use lower-grade fuels with 0.3 to 1.0 wt% S content. Unfiltered air intake for marine gas turbines may contain up to 2600 ppm Na_2SO_4, 19,000 ppm NaCl, and contain sand-derived deposits. Filtration may reduce the chloride and sulfate content from marine environmental air to ≤0.01 ppm, but that is sufficient to form Na_2SO_4 (as well as Na_2SO_4 directly created from sulfur in the fuel), which can lead to sulfidation and hot corrosion. Fuel quality may vary during the service life of the engine. The use of less refined, heavy fuels may contain up to 200 ppm V that, when combined with Na_2SO_4, accelerates hot corrosion reactions. Engine temperatures may also vary with operational loads by 400 to 500 °C (720 to 900 °F) on going from idle to full power (Ref 26). The engine may ingest solid particles (sand, ash, dust, and sea salt) and pyrolytic carbon from uncombusted or poor combustible conditions that contribute metal wastage by erosion or impact on turbine components. Carbon may also accentuate corrosion by reducing sodium sulfate to sulfides. Thermal cycling or transient conditions may further increase the corrosivity of salt deposits by cracking protective oxide layers. The complexity of the turbine operating environment and other high-temperature applications makes it impossible to develop simple tests.

The important factors that influence hot corrosion in marine gas turbines are composition of the gas, condensate composition and deposition rate, and temperature. In order to simulate gas turbine conditions in a low-velocity atmospheric pressure rig (LVBR), it is desirable to duplicate all the variables. However, it is not possible to duplicate them exactly due to the pressure differentials between the gas turbine (0.5 to 1.5 MPa, or 5 to 15 atm) and the burner rig at 101 kPa (1 atm) LVBR (Ref 27). The corrosion rate depends on the salt deposition rate, which is nonlinear, and the corrosion depends on the composition of the salt and the substrate alloy or coating (Ref 28).

Experiences in the 1960s and 1970s led to the discoveries of severe high-temperature corrosion in shipboard gas turbine engines that usually did not occur in aircraft turbine engines. Early observations noted severe corrosion attack on the first-stage blade and vane components of a shipboard marine gas turbine engine was sufficiently rapid to cause engine failure in several hundred hours (Ref 28). The ingestion of sea salt and the combustion of fuels containing some measure of sulfur by gas turbine engines operating in marine environments can lead to corrosion of hot section components, particularly turbine vanes (nozzles) and blades (buckets). This attack was documented in the early open literature as hot corrosion as discussed earlier (Ref 29–31).

Hot corrosion is a complex process involving both sulfidation and oxidation (Ref 32). Hot corrosion is a form of accelerated oxidation that affects alloys and coatings exposed to high-temperature gases contaminated with sulfur and alkali metal salts (Ref 33). These contaminants combine in the gas phase to form alkali metal sulfates; if the temperature of the alloy is below the dew point of the alkali sulfate vapors and above the sulfate melting points, molten sulfate deposits are formed (Ref 33). Molten sodium sulfate is the principal agent in causing hot corrosion (Ref 34, 35). Sulfide formation results from the interaction of the metallic substrate with a thin film of fused salt of sodium sulfate (Ref 36–38). Sulfur compounds come from two sources in a marine gas turbine engine: (a) sulfur from the combustion fuel that often ranges from 0.1 to 1 wt% (or more) depending on the grade of fuel, and (b) sulfate salts contained in the marine air that are ingested into the hot section of the turbine engine.

Chloride salts can act as a fluxing agent and dissolve protective oxide films or cause alloy oxides to fracture and spall. Air in a marine environment ingested into the combustion zone of any marine gas turbine engine will be laden with chlorides unless properly filtered. Sodium chloride has been viewed as an aggressive constituent in the hot corrosion of gas turbine components in the marine environment (Ref 39, 40).

Two general forms of sulfate hot corrosion exist. Type I, high-temperature hot corrosion (HTHC) occurs through multiple mechanisms. It is generally thought to transpire by basic fluxing and subsequent dissolution of the normally protective oxide scales by molten sulfate deposits that accumulate on the surfaces of high-temperature components such as hot section turbine blades and vanes. High-temperature hot corrosion usually occurs at metal temperatures ranging from 850 to 950 °C (1560 to 1740 °F). Type I hot corrosion involves general broad attack caused by internal sulfidation above 800 °C (1470 °F); alloy depletion is generally associated with the corrosion front. This basic fluxing attack may involve raising the Na_2O activity in the molten sulfate by formation of metal sulfides (Ref 41). Basic fluxing may also occur above 800 °C (1470 °F) when p_{SO_3} is low since the basicity of the molten sulfate deposits is controlled primarily by the partial pressure of sulfur trioxide, p_{SO_3}. As the reaction proceeds, the $SO_2 + O_2$ concentration determines if the sulfate-induced hot corrosion is sustained. Very small amounts of sulfur and sodium or potassium can produce sufficient Na_2SO_4 or K_2SO_4. In gas turbine environments, a sodium threshold level below 0.008 ppm by weight precluded type I hot corrosion (Ref 42).

Type II, low-temperature hot corrosion (LTHC) occurs in the temperature range of 650 to 750 °C (1200 to 1380 °F) where p_{SO_3} is relatively high or melts are deficient in the oxide ion concentration leading to acidic fluxing of metal oxides that results in pitting attack (Ref 43). Sulfides are found in the pitted area when nickel-base alloy are utilized (Ref 44). The LTHC may involve a gaseous reaction of SO_3 or SO_2 with CoO and NiO that results in pitting from the formation of low-melting mixtures of Na_2SO_4 and $NiSO_4$ or Na_2SO_4 and $CoSO_4$ in Ni-Cr, Co-Cr, Co-Cr-Al, and Ni-Cr-Al alloy systems (Ref 42, 45). The interaction of these oxidation products with salt deposits forms a complex mixture of salts with a lower melting temperature (Ref 46). When the salt mixture melts, the corrosion rate increases rapidly. If other reactants are added, melting temperatures of the resultant salts can be further lowered. High chromium content (>25 to 30 wt% Cr) is required, generally, for good corrosion resistance to hot corrosion. Nickel-base alloys with both chromium and aluminum show further improvement in hot corrosion resistance.

Acid-base oxide reactions with molten sulfate through the measurement of oxide solubilities as a function of Na_2O activity in fused Na_2SO_4 were examined by Rapp (Ref 47). Oxide solubility is dependent on Na_2O activity, which also serves to rank the acid-base character of individual oxides.

Other impurities such as vanadium ($\geq$0.4 ppm), phosphorus, lead, chlorides, and unburned carbon can be involved in lowering salt melting temperatures, altering the sulfate activity, or changing the solution chemistry and acidity/basicity that leads to accelerating hot corrosion. Vanadic hot corrosion appears to be potentially more complex because five compounds exist in the Na,V,O system (Ref 48). The high-temperature reaction of sulfate and vanadium with ceramic oxides involved a Na_2O-V_2O_5 system can be explained by Lewis acid-base chemistry (Ref 49). Basic zirconia (ZrO_2) stabilizing oxides such as yttria (Y_2O_3) do not react with Na_3VO_4 (or $3Na_2O$-V_2O_5), but do react with the V_2O_5 component of $NaVO_3$ (Na_2O-V_2O_5) and V_2O_5 itself to form YVO_4 (Ref 50). Acidic oxides such as Ta_2O_5 react with the Na_2O component of Na_2VO_4 and $NaVO_3$ to form sodium tantalates and yield α-$TaVO_5$ with V_2O_5. The vanadate that is most corrosive in the initiation of vanadic attack will depend on the acidity/basicity of the coating or alloy oxide. No reaction occurs when the acid-base properties of a stabilizing oxide is equal (Ref 51). The thermochemistry of vanadate and sulfate melts and reaction with different stabilizing oxides with SO_3-$NaVO_3$ was studied by Jones (Ref 52).

High chromium content (>25 to 30% Cr) is required for good resistance to hot corrosion. Nickel alloys with both chromium and aluminum show improved hot corrosion resistance. However, inspection of Ellingham phase stability diagrams for the systems M-Na-O-S, where M can be nickel, cobalt, iron, aluminum, or chromium, indicates that there are no combination of melt basicity and oxygen activity where these metals, absent of a protective oxide film, are stable in contact with fused sodium sulfate (Ref 53). The relative hot corrosion resistance of a number of alloys has been evaluated in incinerator environments (Ref 54–56).

Yttria-stabilized zirconia (YSZ) is attacked and destabilized by phosphorus impurities in fuel (Ref 57). Phosphoric anhydride, also known as anhydrous phosphoric acid (P_2O_5), reacts with basic Y_2O_3 to form the salt, YPO_4. Zirconia also synergistically reacts with sodium and P_2O_5 to form $NaZr_2(PO_4)_3$. YSZ thermal barrier coatings have been exposed to $PbSO_4$-Na_2SO_4 molten salts without observable destabilization or reaction with this ceramic (Ref 58). However, lead, as PbO, appears to cause TBC failures by reacting with chromium in the NiCrAlY bond coat to form $PbCrO_4$.

Case Study: Turbine Blade Failure. A case study of a marine gas turbine failure centered on the extent of corrosion that occurred under the platformg of a turbine blade. Figure 7 shows a failed turbine blade. Figure 8 show unfailed blades that displayed heavy deposition under the platform of the double-stemmed blade. The blade alloy was within material specifications and the corrosion-resistant coating was applied by thermal spray. Metallographic examination showed that the coating thickness under the platform and in the curved area of transition between the platform and the blade stem was either very thin (up to 40 μm thick) or nonexistent. The coating, if present, usually was porous or had entrained contamination under the platform due to lack of adequate spray deposition in these non-line-of-sight areas as shown in Fig. 9. MCrAlY coating thickness at other sites along the blade stem was 35 to 105 μm.

The corrosion that was observed under the platform, in all cases, was caused by type II, LTHC, which occurs in the temperature range of 650 to 730 °C (1200 to 1350 °F). In addition to the presence of hot corrosion, cracking was observed to initiate at several hot corrosion sites (Fig. 10). This was found to be advancing to various degrees through the stem of several blades. The pitting caused by the type II hot corrosion provided initiation sites for cracking to begin and reduced the overall undamaged cross section at the stem, thus increasing the applied stress to the corroded area. The cracking initially advanced via corrosion fatigue, but later, in some blade cases, by high-cycle fatigue (Fig. 11). Some cracking proceeded through both blade stem walls, causing some blades to break off during service. Many of the "unfailed" blades exhibited evidence of hot corrosion and varying degrees of corrosion fatigue/fatigue cracking.

Incinerators

Incinerators are used to manage wastes in an efficient manner rather than sending these waste materials to a landfill or accumulating the wastes aboard ship. Military (land-based) incinerators must comply with various regulations for operation and for management of residual wastes and the gaseous emissions. Compliance with local, national, and international (such as the International Treaty for the Prevention of Pollution for Ships/Marine Pollution Protocol, MARPOL) environmental regulations and statutes controlling overboard discharge of liquid wastes, sanitary wastes, and air emissions have become more stringent for ships deployed at sea.

Among the many issues involved, the reliability and design of incinerators will depend on the materials of construction. One of the primary operational factors is the corrosion resistance of candidate materials; the discussion of many of the other issues is beyond the scope of this article. Complex processes occur in incineration of solid wastes, which involve thermal and chemical reactions that occur at various times, temperatures, and locations. Pyrolysis via high reaction temperatures derived from plasma energy will also encompass complex reactions, dynamic gas flows, particulate and deposition effects, mechanical and chemical mixing, and residence times of the reaction species. The operational mode, the temperatures existing within the primary combustion chamber, secondary combustion, and other system components, composition of the waste stream, the chamber chemical environment, the waste stream abrasiveness, the aforementioned space constraints, safety, maintainability, and reliability will affect the choice of construction materials.

Incinerators also contain inherently complex high-temperature environments. The corrosiveness of the incinerator environment depends on the relative stream of wastes that are generated and the efficiency of removing potential wastes that have been proven to be severely corrosive, such as chlorine-containing plastics. High-temperature corrosion was observed in shipboard waste incinerators (Ref 55, 59).

Incinerator design generally considers two types of operating scenarios using two classes of materials. They are primarily refractory-lined incinerators and air-cooled or water-cooled alloy-lined incinerators.

A refractory-lined incinerator does not employ any active cooling and may operate at temperature up to 1500 °C (2730 °F). Most alloys will degrade in environments with prolonged service temperatures above 1000 °C (1830 °F). Refractories are mainly high-melting-point metallic oxides, but also include substances such as carbides, borides, nitrides, and graphite (Table 2). Maximum service temperatures will always be less than the melting points of pure ceramics because refractories usually contain minor constituents and actual incinerator environments are vastly different from simple oxygen atmospheres. Refractories are used to act as a thermal barrier between the high-temperature environment and an ambient temperature outside the chamber skin to maximize safety to personnel. Refractory suitability depends on its resistance to abrasion, maximum service temperature, corrosion resistance to liquids, gases and slag, erosion resistance, spalling resistance, resistance to thermal cycling, shock, and fatigue from high thermal gradients, oxidation or reduction reactions, its mechanical strength, and its inspection and maintenance requirements. Refractories are available in either shaped (bricks) or unshaped, monolithic forms such as castable plastic refractories, ramming mixes, ceramic fiber blankets, or gunning mixes.

There are often property trade-offs required to attain the desired optimum, serviceable refractory. Insulating refractories generally have lower density, strength, corrosion resistance, erosion resistance, and higher porosity and void fraction than stronger, less insulating refractories. Refractories are generally susceptible to thermal shock and spalling. Monolithic refractories are often more susceptible to thermal shock than shaped refractories. Dense refractories tend to have high thermal conductivities, low porosity, relatively high strength, and improved corrosion resistance, but tend to be susceptible to thermal shock. Insulating refractories have higher porosity, which improves thermal shock resistance, but provides numerous pathways for molten and gaseous materials to promote corrosion and spalling.

The corrosion resistance of refractories will be based on the thermodynamics of the potential corrosion reactions, the reaction rates and kinetics of these possible reactions, the composition and form of the refractory material, and the surface chemistry of the refractory with corrosive gaseous, liquid, or solid environments. As with other chemical reactions, the spontaneous direction of the possible corrosion reactions can be calculated, the determination of which will indicate which reactions are theoretically possible. The kinetics will be dependent on the specific environment (gaseous, liquid, or solid), surface chemistry, the material microstructure, material composition, refractory phase(s), and the temperature. All ceramic or refractories used in an oxygen-rich environment are oxides or develop a protective oxide on their surface (such as SiO_2 on SiC, or Si_3N_4). Though oxide ceramics are chemically inert and resist oxidation and reduction, they are not inert to certain molten salts. Ceramic corrosion between ceramic oxides and molten salt deposits are primarily due to oxide acid-base reactions (Ref 48).

To determine if a particular slag is acidic or basic, the lime-to-silica ratio in the slag is commonly used. As a general rule, the slag is basic if the CaO/SiO_2 ratio (or $CaO + MgO/SiO_2 + Al_2O_3$ ratio) is greater than one. If the ratio is less than one, the slag is considered acidic (Ref 60). Gases and liquids can penetrate deeply into high porous refractories that can exacerbate spalling and accelerated corrosion. Oxidizing gases such as NO_x, Cl_2, O_2, CO_2, H_2O, and SO_2, reducing gases (NH_3, H_2, C_xH_y, CO, and H_2S), and vapors of volatile elements and compounds may react with different refractories. Hydrodynamic mass transport by either convection or diffusion can markedly affect the corrosion rate of a refractory in an environment. Nonwetting refractories are relatively resistant to corrosion

Fig. 7 Failed shipboard turbine at the transition between blade and stem. See also Fig. 8 through 11.

Fig. 8 Two shipboard turbine blades. Pressure side shown facing out of the page. Arrows denote areas where heavy corrosion products are observed.

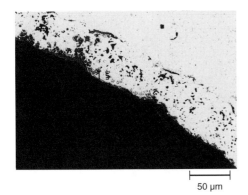

Fig. 9 Blade location along curvature at platform/stem transition. Poor coating quality at this site provides easy pathways for corrosion penetration. Original magnification 200×

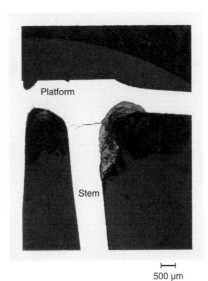

Fig. 10 Close-up of failed turbine blade. Cracking has initiated at platform/stem transition where corrosion has occurred. Original magnification 20×

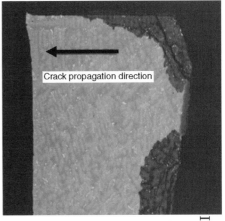

Fig. 11 Crack propagation direction of failed turbine blade. Corrosion propagates over 60% of the stem wall, thus increasing the applied stresses in the remaining stem wall. Fracture shows a sharp transgranular characteristic suggesting cleavage or high cycle fatigue. Original magnification 55×

with liquids or gases. High-purity refractories are more corrosion resistant because minor impurity components from secondary, phases and defect structures that can introduce significant corrosion are kept to a minimum; alkali impurities should be minimized as much as possible. High-purity refractories also improve corrosion resistance by avoiding low-melting eutectics. Solids, either as fine particulates or dust, can cause degradation of refractories through abrasion or deposition.

There are several limitations to the use of refractories in incinerators. First, a layer of a working refractory such high-purity alumina, silicon carbide, or alumina-chromia is required to be used since these refractories have the highest resistance to corrosion reactions with slag, liquids, or gases, are abrasion and spall resistant, have low porosity, and have a limited thermal shock resistance. Since these refractories also have a high thermal conductivity, an insulating refractory must be inserted between the chamber shell and the working refractory. To reduce the temperature from 1500 °C (2730 °F) on the inner wall surface to about 100 °C (212 °F) on the outer shell, the combined thickness of the two refractory layers could be 300 mm (12 in.) or more. This adds an unacceptable weight and space in applications where space is critical.

Second, refractories may crack from thermal shock and react with wastes to corrode or wear away, thus requiring repair. Eight or more hours may be required for the crucible chamber to cool before repairs can initiate. Repair would require the use of plastic refractories, mortar materials, or patching mixes for quick repairs or a more time-consuming reinstallation of refractory brick, which may limit the location of refractory-lined incinerators.

For additional information on high-temperature corrosion of refractory materials, see the article "Performance of Refractories in Severe Environments," in *Corrosion: Materials,* Vol 13B, of the *ASM Handbook.*

Alloy-lined incinerators reduce weight and size and improve reliability, operability, and maintainability while maintaining environmental compliance and maintaining safety. The gas temperature in the primary chamber may be up to 1200 °C (2190 °F) with possible hot spots above 1370 °C (2500 °F). An alloy-lined incinerator will be actively cooled by water. Thermal shock may occur during start-up or shutdown, particularly if the incinerator is not operated properly. Thermal shock may also be experienced if cooling water leaks occur. The chamber environment would contain N_2, CO_2, H_2O vapor, NO_x, and small quantities of alkalis and acid gases (HCl, SO_2). Particulates in the gas stream could cause erosion and abrasion damage. While conventional refractory-lined incinerator may be more than 300 mm (12 in.) thick, an alloy-lined incinerator wall may be 13 mm (0.5 in.) thick using a superalloy such as alloy 718 (UNS N07718). Table 3 shows possible alloys that may be used for alloy liners for incinerators.

Shipboard incinerators used for the disposal of blackwater (sewage) may utilize alloy 690 (Ni-30Cr-9Fe UNS N06690) operating at a peak temperature of 760 °C (1400 °F). The ability of the liner alloy to resist corrosion is dependent on its protective surface oxide film. Conditions that can damage the oxide and the substrate alloy may include: (a) alternating oxidizing and reducing environments at high temperatures, which may interfere with oxide formation and oxide maintenance resulting in porous films, (b) mechanical stresses such as residual stresses derived from fabrication and welding processes, cyclic thermal stresses that deform the alloy substrate because of rapid cooling rates, or crack or spall the oxide layer because of tensile or compressive stresses, (c) impingement/erosion that removes the oxide when fluid (gas or liquid) flow exceeds a threshold velocity, (d) reaction with corrodents, moisture condensation, (e) deposits of low-melting point mixtures, (f) flame impingement, (g) crevices or defects encouraging galvanic attack, (h) changing metal properties from prolonged exposure at high temperatures, (i) molten chlorides, and (j) molten alkali sulfates (Ref 60). Reaction of the slag with the chamber wall must also be considered. Corrosion in waste incinerators has also been discussed in detail previously (Ref 55, 56, 63–70).

Most metals and alloys in an oxidizing environment are protected to varying degrees by an oxide layer. Iron oxides alone are not protective above 550 °C (1020 °F) (Ref 71). Oxidation of carbon steels from air or steam forms an oxide scale along the metal surface that grows and thickens with time. The scales, depending on chemical composition, can be mildly insulating to very insulating. The scale growth rate increases with a rise in temperature. Chromium, aluminum, and/or silicon assist in forming scales that are more protective at higher temperatures (Ref 61). Depending on flow rate and the p_{O_2},

Table 2 Melting points of common refractories

Nomenclature	Formula	Melting point °C	Melting point °F
Corundum, alpha alumina	Al_2O_3	2054	3729
Baddeleyite, zirconia	ZrO_2	2700	4892
Periclase, magnesia	MgO	2852	5166
Chromic oxide, chromia	Cr_2O_3	2330	4226
Lime, calcia	CaO	2927	5301
Cristobalite, silica	SiO_2	1723	3133
Carborundum, silicon carbide	SiC	2700 (sublimes)	4892 (sublimes)
Silicon nitride	Si_3N_4	1800 (dissociates)	3272 (dissociates)
Boron nitride	BN	3000 (sublimes)	5432 (sublimes)

Source: Ref 60

Table 3 Candidate alloys for alloy-lined incinerators

UNS No.	C	Mn	P	S	Cr	Ni	Co	Mo	Si	Cu	Fe	Al	Ti	W	Nb + Ta	Others
K11789	0.17	0.40–0.65	0.30	0.30	1.00–1.50	...	...	0.45–0.65	0.50–0.80	...	bal	...	...	...	...	...
S31008	0.08	2.00	0.045	0.030	24.00–26.00	19.00–22.00	...	...	...	...	...	...	...	...	...	...
S31603	0.03	2.00	0.045	0.03	16.0–18.0	10.0–14.0	...	2.0–3.0	1.00	...	...	...	...	...	...	...
N06022	0.010	0.50	...	0.08	22	56	2.5	13	0.08	...	3	...	...	3	...	0.35 V
N06059	...	...	...	...	23 nom	59 nom	...	16 nom	...	...	1 nom	...	...	...	...	...
N06455	0.01	1.0	0.025	0.010	14.0–18.0	bal	...	14.0–17.0	0.08	...	3.0	...	0.70	...	...	...
N08366	0.03	2.00	0.030	0.003	20.0–22.0	23.5–25.5	...	6.0–7.0	0.75	...	...	...	...	...	...	...
N06600	0.10	1.00	...	0.015	14.00–17.00	72.0 min	...	...	0.50	0.50	6.00–10.00	...	...	...	...	...
N06601	0.10	1.0	...	0.015	21.0–25.0	58.0–63.0	...	...	0.50	1.0	bal	1.0–1.7	...	...	...	...
N06625	0.10	0.50	0.015	0.015	20.0–23.0	55.0 min	1.00	8.0–10.0	0.50	...	5.00	0.40	0.40	...	3.15–4.15	...
N06671	0.05 nom	...	...	...	48.0 nom	bal	...	...	...	...	...	...	0.35 nom	...	...	...
N06690	0.03 nom	...	...	...	30 nom	60.5 nom	...	...	...	...	9.5 nom	...	...	...	...	...
R30556	0.05–0.15	0.5–2.0	0.04	0.015	21–23	19–22.5	16–21	2.5–4.0	0.2–0.5	...	bal	0.10–0.50	...	2.0–3.5	0.30–1.25 Ta; 0.30 Nb	0.005–0.10 La; 0.001–0.10 Zr; 0.10–0.30 N; 0.04 B
N12160	0.05	0.5	...	...	28	37	30	1	2.75	...	3.5 max	...	0.45	1	1.0 max	...
N08120	0.05	0.7	...	...	25	37	3	2.5	0.6	...	33	0.1	0.1	2.5	0.7	0.004 B; 0.2 N
N07214	0.5 nom	0.5	...	...	16 nom	75 nom	...	...	0.2	...	3 nom	4.5 nom	...	...	...	0.1 Zr; 0.01 B; 0.01 Y nom

Source: Ref 61, 62

chromium oxide is stable up to 983 °C (1801 °F), above which the oxide volatilizes to gaseous CrO$_3$. Mechanical damage by spallation or cracking of the oxide film from cyclic oxidation of chromia and alumina oxides formed on nickel-base alloys is due to the mismatch of thermal expansion coefficients between the oxide and the base alloy (Ref 63). Small additions of lanthanum, yttrium, tantalum, ceria, zirconia, and/or niobia improve the scale adhesion and scale resilience to cyclic oxidation (Ref 71).

The incinerator alloy material reportedly may be protected from molten slag by an oxide/slag skull. However, in the event of possible contact with molten slag, it is important to know the corrosion resistance to slag of the candidate materials. Liquid metal corrosion may cause dissolution of a surface directly, by intergranular attack, or by leaching. Liquid metal attack may also initiate alloying, compound reduction, or interstitial or impurity reactions. Carbon and low-alloy steels are susceptible to various molten metals or alloys such as brass, aluminum, bronze, copper, zinc, lead-tin solders, indium, and lithium at temperatures from 260 to 815 °C (500 to 1500 °F). Plain carbon steels are not satisfactory for long-term usage with molten aluminum. Stainless steels are generally attacked by molten aluminum, zinc, antimony, bismuth, cadmium, and tin (Ref 72).

Nickel, nickel-chromium, and nickel-copper alloys generally have poor resistance to molten metals such as lead, mercury, and cadmium. In general, nickel-chromium alloys also are not suitable for use in molten aluminum (Ref 73). Liquid metal embrittlement (LME) is a special case of brittle fracture that occurs in the absence of an inert environment and at low temperatures (Ref 74). The selection of fabricating processes must be chosen carefully for nickel-base superalloys.

Case Study: Corrosion in a Waste Incinerator. A failed uncoated alloy 690 incinerator liner was removed from service. The incinerator liner had been in service for an estimated 264 h before being repaired with a patch of alloy 160 plate (13 mm, or 0.5 in., thick) tack welded over the alloy 690 incinerator liner hole. About 382 h of additional service were clocked on this incinerator before its removal. Figure 12 shows that ash deposits covered the entire inner alloy 690 incinerator liner surface and the degree of deterioration of the alloy 690 liner around the alloy 160 patch.

The uncoated liner developed a very large perforation on the liner directly under the burner head where the incinerator flame could impact the liner surface, but inspection of this liner also revealed other, smaller perforations at numerous locations and elevations. The plate had not experienced any appreciable, visible corrosion after its 265 h of shipboard incinerator service. The alloy 690 liner adjacent to the alloy 160 patch showed severe deterioration and thinning of the liner wall, which had reduced the liner thickness from 6.4 mm (0.25 in.) to less than 0.51 mm (0.020 in.). Figure 13 displays generalized, uniform corrosive penetration about 0.125 to 0.20 mm (0.005 to 0.008 in.) deep into the alloy 690 liner that was caused by sulfidation. The major, outer, unexposed portion of liner alloy at this site was undisturbed. The surface corrosion had initiated at the inner, exposed liner surface; corrodents penetrated and advanced along the grain boundaries and combined with intragranular species to form oxide/sulfide precipitates.

Sulfidation led to corrosion and thinning of the incinerator wall and the diffusion of sulfur into the liner weakened the overall alloy mechanical strength and led to loss of ductility of the local microstructure. This induced an increased susceptibility to fatigue or corrosion fatigue cracking from cyclic thermal stresses from incinerator operation as shown in Fig. 14. Sulfidation and hot corrosion caused a large perforation of the liner under the burner head. Remaining perforated and pitted sections of the uncoated liner displayed varying degrees of attack from sulfidation and

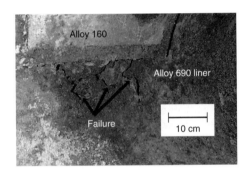

Fig. 12 Failed incinerator liner. Hole near alloy 160 patch shows that the original 6.4 mm (0.250 in.) wall thickness had been reduced to 1.3 mm (0.050 in.) or less in the general area. See also Fig. 13 through 15.

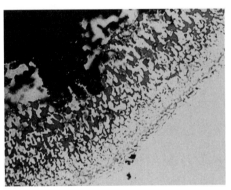

Fig. 13 Uncoated incinerator liner. Diffusion of sulfur species along the grain boundaries has led to sulfidation of the liner alloy. Sulfur species concentration decreases as it moves deeper into the alloy. Original magnification 250×

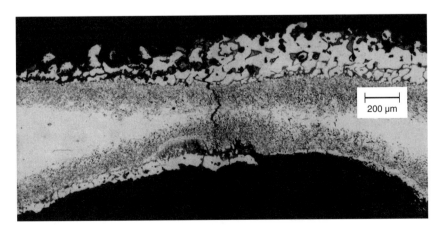

Fig. 14 Uncoated incinerator liner near failure. Corrosion penetration from both inner and outer surfaces accompanied by cracking of liner. Total wall thickness has been reduced to 350 μm (0.014 in.). Some grain reorientation and corrosion products are present. Original magnification 50×

Fig. 15 Uncoated incinerator liner. Diffusion of sulfur species extends about 250 to 300 μm (0.01 to 0.012 in.) into the liner alloy. Severe sulfidation and type II hot corrosion had penetrated about 175 μm (0.007 in.) into the liner alloy. Original magnification 250×

hot corrosion. Figure 15 shows that diffusion of sulfur penetrated about 0.25 to 0.30 mm (0.01 to 0.012 in.) into the liner alloy. Severe sulfidation and type II hot corrosion had penetrated about 0.175 mm (0.007 in.) into the liner alloy. Debris blocking the air cooling channel between the inner and outer liners raised the overall alloy liner temperature and acerbated the high-temperature corrosion. Sulfur species reacted with alloy 690 constituents at breaks in the surface oxide. Those sites affected by sulfidation were susceptible to cracking and accelerated wall loss and thinning. Alloy 160 appeared to offer some improvement in hot corrosion resistance over alloy 690.

REFERENCES

1. D.A. Shifler, High-Temperature Gaseous Corrosion Testing, *Corrosion: Fundamentals, Testing, and Protection,* Vol 13a, *ASM Handbook,* ASM International, 2003, p 658–663
2. S.C. Stultz and J.B. Kitto, Ed., Marine Applications, *Steam: Its Generation and Uses,* 40th ed., Babcock and Wilcox, chap 30, 1992, p 30-1 to 30-13
3. D.N. French, *Metallurgical Failures in Fossil-Fired Boilers,* 2nd ed., John Wiley & Sons, 1993, p 21–25
4. R.L. Harrington, *Marine Engineering,* The Society of Naval Architects and Marine Engineers (SNAME), 1971, p 78–129
5. ASME Boiler and Pressure Vessel Code, Section 1, Power Boilers, ASME International, 2004
6. ASME Boiler and Pressure Vessel Code, Section 2, Materials, ASME International, 2004
7. *Manual for Investigation and Correction of Boiler Tube Failures,* report CS-3942, RP1890-1, prepared by Southwest Research Institute for Electric Power Research Institute, 1985
8. D.A. Shifler, "The Role of Failure Analysis in Controlling Problems in Steam Boilers," Water Technologies '93, Association of Water Technologies, McLean, VA, Nov 1993
9. R.D. Port and H.M. Herro, *The Nalco Guide to Boiler Failure Analysis,* McGraw-Hill, 1991, chap. 4, 6, 14
10. R.D. Port, "Identification of Corrosion Damage in Boilers," paper 224, CORROSION/84, NACE International, 1984
11. D.A. Shifler and J.F. Wilkes, "Historical Perspectives of Intergranular Corrosion of Boiler Metal by Water," paper 44, Corrosion/93, NACE International, 1993
12. D.R. Holmes, Ed., *Dewpoint Corrosion,* Ellis-Horwood, London, 1985
13. W.T. Reid, *External Corrosion and Deposits: Boiler and Gas Turbines,* American Elsevier Publishing, 1971, p 121–143
14. D.N. French, *Metallurgical Failures in Fossil Fired Boilers,* 2nd ed., Wiley-Interscience, 1993, p 342–371
15. R.P. Gangloff and M.B. Ives, Ed., *Environmental-Induced Cracking of Metals: Proc. First International Conference on Enviromentally-Induced Cracking of Metals,* NACE International, 1990
16. O. Devereux, A.J. McEvily, and R.W. Staehle, Ed., *Corrosion Fatigue: Chemistry, Mechanics, and Microstructure,* NACE International, 1973
17. D.A. Shifler, "Investigation of a Roof Tube Failure from a Utility Boiler," CARDIVNSWC-TR-61-1999/13, Naval Surface Warfare Center, Carderock Div., June 1999
18. "Military Specification Fuel, Naval Distillate (NATO F-76)," MIL-F-16884K, Department of Defense, Nov 14, 2002
19. "Military Specification Turbine Fuel, Aviation (Grades JP-4, JP-5, and JP-8ST)," Department of Defense, MIL-T-5624T, Sept 18, 1998
20. "Military Specification Turbine Fuel, Aviation (Grade JP-8)," MIL-T-183133E, Department of Defense, April 1, 1999
21. Marine Gas Oil (MGO) Purchase Description, NAVSEA Serial No. 03M3/45, Naval Sea Systems Command, Aug 9, 1996
22. R.L. Harrington, *Marine Engineering,* The Society of Naval Architects and Marine Engineers (SNAME), 1971, p 246–279
23. S. Bludszuweit, H. Jungmichel, B. Buchholz, K. Prescher, and H.G. Bunger, "Mechanism of High Temperature Corrosion in Turbochargers of Modern Four-Stroke Marine Engines," Motor Ship Conference 2000, Amsterdam, March 29–30, 2000
24. A.J. Wharton, *Diesel Engines,* 3rd ed., Butterworth-Heinemann, Oxford, England, 1991, p 43–51
25. "The Modern Gas Turbine," Brochure TS22760, Rolls Royce PLC, Derby, England, 2000
26. S.R.J. Saunders, Correlation between Laboratory Corrosion Rig Testing and Service Experience, *Mater. Sci. Technol.,* Vol 2, 1986, p 282–289
27. D.J. Wortman, R.E. Fryxell, K.L. Luthra, and P. Bergman, Mechanisms of Low Temperature Hot Corrosion: Burner Rig Studies, *Proc. Fourth US/UK Navy Conference on Gas Turbine Materials in a Marine Environment,* Vol 1, U.S. Naval SEA Command, Annapolis, MD, 1979, p 379
28. K.L. Luthra, Simulation of Gas Turbine Environments in Small Burner Rigs, *High Temp. Technol.,* Vol 7, 1989, p 187–192
29. G.J. Danek, Jr., State-of-the-Art Survey on Hot Corrosion in Marine Gas Turbine Engines, *Naval Eng. J.,* Vol 77, Dec 1965, p 859
30. *Hot Corrosion Problems Associated with Gas Turbines,* from Symposium at 69th Annual Meeting ASTM (Atlantic City, NJ), 26 June-1 July 1966, ASTM Special Publication 421, Sept 1967
31. N.S. Bornstein and M.A. DeCrescente, "Sulfidation in Gas Marine Engines," paper 165, CORROSION/75, NACE International, 1975
32. A.U. Seybolt, P.A. Bergman, and M. Kaufman, "Hot Corrosion Mechanism Study, Phase II," Final Report under contract No. N600 (61533) 65595, MEL 2591, Assignment 87-122, prepared for Naval Ship Research and Development Center, Annapolis, MD, 30 June 1966 to 31 Oct 1967
33. S.R.J. Saunders, "Corrosion in the Presence of Melts and Solids," *Guidelines for Methods of Testing and Research in High Temperature Corrosion,* Working Party Report, H.J. Grabke and D.B. Meadowcroft, Ed., European Federation of Corrosion, No. 14, The Institute of Materials, London, 1995, p 85–103
34. E.L. Simons, G.V. Browning, and H.A. Liebhafsky, Sodium Sulfate in Gas Turbines, *Corrosion,* Vol 11 (No. 12), 1955, p 505t
35. R.K. Ahluwalia and K.H. Im, Aspects of Mass Transfer and Dissolution Kinetics of Hot Corrosion Induced by Sodium Sulphate, *J. Inst. Energy,* Vol 64 (No. 461), 1991, p 186
36. J.A. Goebel and F.S. Pettit, Na_2SO_4-Induced Accelerated Oxidation (Hot Corrosion) of Nickel, *Metall. Trans.,* Vol 1, (No. 7), 1970, p 1943
37. N.S. Bornstein and M.A. DeCrescente, The Relationship Between Compounds of Sodium and Sulfur and Sulfidation, *Trans. Metall. Soc. AIME,* Vol 245 (No. 9), 1969, p 1947–1954
38. N.S. Bornstein and M.A. DeCrescente, The Role of Sodium in the Accelerated Oxidation Phenomenon Termed Sulfidation, *Metall. Trans.,* Vol 2 (No. 10), 1971, p 2875
39. J.F. Condé and B.A. Wareham, *Proc.: Gas Turbine Materials in the Marine Environment Conference,* MCIC-75-27, 1974, p 73
40. D.A. Shores, D.W. McKee, and H.S. Spacil, *Proc.: Properties of High Temperature Alloys with Emphasis on Environmental Effects,* PV 77-1, Z.A. Foroulis and F.S. Pettit, Ed., The Electrochemical Society, 1976, p 649
41. J.A. Goebel, F.S. Pettit, and G.W. Goward, Mechanisms for the Hot Corrosion of Nickel-Base Alloys, *Metall. Trans.,* Vol 4, 1973, p 261–278
42. R.D. Kane, "High-Temperature Gaseous Corrosion, *Corrosion Fundamentals, Testing, and Protection,* Vol 13A, *ASM Handbook,* ASM International, 2003, p 228–235
43. K.L. Luthra and D.A. Shores, Mechanisms of Na_2SO_4 Induced Corrosion at 600°–900° C, *J. Electrochem. Soc.,* Vol 127, 1980, p 2202
44. K.T. Chiang, F.S. Pettit, and G.H. Meier, Low Temperature Corrosion, NACE-6, *High Temperature Corrosion,* R.A. Rapp, Ed., NACE International, 1983, p 519
45. H. Lewis and R.A. Smith, Corrosion of High Temperature Nickel-Base Alloys by

Sulfate-Chloride Mixtures, *First International Congress on Metallic Corrosion,* April 10–15, 1961, Butterworths, London, 1962, p 202

46. A.S. Khanna, *Introduction to High Temperature Oxidation and Corrosion,* ASM International, 2002, p 172–201
47. R.A. Rapp, Chemistry and Electrochemistry of Hot Corrosion of Metals, *Mater. Sci. Eng.,* Vol 87, 1987, p 319–327
48. R.L. Jones, Low-Quality Fuel Problems with Advanced Engine Materials, *High Temp. Technol.,* NRL Memorandum report 6252, Vol 6, Aug 9 1988, p 187
49. R.L. Jones, C.E. Williams, and S.R. Jones, Reaction of Vanadium Compounds with Ceramic Oxides, *J. Electrochem. Soc.,* Vol 133, 1986, p 227–230
50. R.L. Jones and R.F. Reidy, Development of Hot Corrosion Resistant Scandia-Stabilized Zirconia Thermal Barrier Coatings, *Proc. Elevated Temperature Coatings: Science and Technology I,* N.B. Dahotre, J.M. Hampikian, and J.J. Stiglich, Ed., The Minerals, Metals, and Materials Society, 1995, p 23
51. R.L. Jones, Oxide Acid-Base Reactions in Ceramic Corrosion, *High Temp. Sci.,* Vol 27, 1990, p 369–380
52. R.L. Jones, "Vanadate-Sulfate Melt Thermochemistry Relating to Hot Corrosion of Thermal Barrier Coatings," NRL/MR/6170-97-8103, Naval Research Laboratory, Oct 30, 1997
53. R.A. Rapp, Hot Corrosion of Materials: A Fluxing Mechanism?, *Corros. Sci.,* Vol 44, 2002, p 209–221
54. S. Mrowec, The Problem of Sulfur in High-Temperature Corrosion, *Oxid. Met.,* Vol 44, 1995, p 177
55. G. Wacker and W.L. Wheatfall, "Corrosion Problems Associated with Shipboard Waste Incineration Systems," *1976 Tri-Service Corrosion Conference* (Philadelphia, PA), Oct 26–28, 1976, p 185
56. P. Elliott, Materials Performance in High-Temperature Waste Combustion Systems, *Mater. Perform.,* Vol 32, 1993, p 82
57. S.C. Singhal and R.J. Bratton, Stability of a $ZrO_2(Y_2O_3)$ Thermal Barrier Coating in Turbine Fuel with Contaminants, *Trans. ASME: J. Eng. Power,* Vol 102 (No. 10), 1980, p 770–775
58. D.W. McGee, K.L. Luthra, P. Siemers, and J.E. Palko, *Proc. First Conference on Advanced Materials for Alternative Fuel Capable Directly Fired Heat Engines,* J.W. Fairbanks and J. Stringer, Ed., CONF-790749, National Technical Information Service, 1979, p 258
59. *Materials of Construction for Shipboard Waste Incinerators,* National Research Council, National Academy of Sciences, 1977, p 91–92
60. *Handbook of Refractory Practice,* 1st ed., Harbison-Walker/Indresco, 1992, p PR-12
61. Alloy Data, Carpenter Technology Corp., Carpenter Steel Division
62. *High Temperature Product Literature and Hastelloy Corrosion-Resistant Alloy Product Literature,* Haynes International, 1991–1996
63. P. Ganesan, G.D. Smith, and L.E. Shoemaker, "The Effects of Exclusions of Oxygen and Water Vapor Contents on Nickel-Containing Alloy Performance in a Waste Incineration Environment," Corrosion/91, paper 248, NACE International, 1991
64. H.H. Krause, High Temperature Corrosion Problems in Waste Incineration Systems, *J. Mater. Energy Syst.,* Vol 7, 1986, p 322
65. G.D. Smith, P. Ganesan, and L.E. Shoemaker, "The Effect of Environmental Excursions within the Waste Incinerator on Nickel-Containing Alloy Performance," Corrosion/90, paper 280, NACE International, 1990
66. C.T. Wall, J.T. Groves, and E.J. Roberts, How to Burn Salty Sludges, *Chem. Eng.,* Vol 7, April 14, 1975
67. P. Ganesan, G.D. Smith, and L.E. Shoemaker, "The Effects of Excursions of Oxygen and Water Vapor on Nickel Containing Alloy Performance in a Waste Incineration Environment," Corrosion/91, paper 248, NACE International, 1991
68. H.H. Krause, D.A. Vaugn, and P.D. Miller, Corrosion and Deposits from Combustion of Solid Waste, *J. Eng. Power (Trans. ASME),* Vol 95, 1973, p 45
69. P. Elliot, Practical Guide to High-Temperature Alloys, *Mater. Perform.,* Vol 28, 1989, p 57
70. F.H. Stott, G.C. Wood, and J. Stringer, Influence of Alloying Elements on the Development and Maintenance of Protective Scales, *Oxid. Met.,* Vol 44, 1995, p 113
71. M. Shütze, Mechanical Properties of Oxide Scales, *Oxid. Met.,* Vol 44, 1995, p 29
72. J.F. Grubb, T. DeBold, and J.D. Fritz, Corrosion of Wrought Stainless Steels, *Corrosion: Materials,* Vol 13B, *ASM Handbook,* ASM International, 2005, p 54–76
73. R.A. Smith and S.R.J. Saunders, *Corrosion,* 3rd ed., L.L. Sheir, R.A. Jarman, and G.T. Burstein, Ed., Butterworth-Heinemann, London, 1994, p 7:136
74. D.G. Kolman, Liquid Metal Induced Embrittlement, *Corrosion: Fundamentals, Testing, and Protection,* Vol 13A, *ASM Handbook,* ASM International, 2003, p 381–392

SELECTED REFERENCES

- R.A. Ainworth and G.A. Webster, *High Temperature Component Life Assessment,* Kluwer Academic Publishers, Dordrecht, Germany, 1994
- *Characterization of High Temperature Materials: Microstructural Characterization,* Vol 1, Maney Publishers, Leeds, U.K., 1989
- Committee on Coatings for High-Temperature Structural Materials, *Coatings for High-Temperature Structural Materials: Trends and Opportunities,* National Research Council, National Academies Press, 1996
- N.B. Dahotre, J.M. Hampikian, and J.J. Stiglich, Ed., *Elevated Temperature Coatings: Science and Technology I,* The Minerals, Metals, and Materials Society, 1995
- N.B. Dahotre and J.M. Hampikian, Ed., *Elevated Temperature Coatings: Science and Technology II,* The Minerals, Metals, and Materials Society, 1996
- N.B. Dahotre, J.M. Hampikian, and J.E. Morral, Ed., *Elevated Temperature Coatings: Science and Technology IV,* The Minerals, Metals, and Materials Society, 2001
- M.J. Donachie and S.J. Donachie, *Superalloys: A Technical Guide,* 2nd ed., ASM International, 2002
- G.R. Edwards, D.L. Olson, R. Visawantha, T. Stringer, and P. Bollinger, Ed., *Proc. Workshop on Materials and Practice to Improve Resistance to Fuel Derived Environmental Damage in Land and Sea Based Turbines,* Oct 22–23, 2002, Electric Power Research Institute, 2003
- P. Elliot, Practical Guide to High-Temperature Alloys, *Mater. Perform.,* Vol 28, 1989, p 57
- G.E. Fuchs, K.A. Dannemann, and T.C. Deragon, Ed., *Long Term Stability of High Temperature Materials,* The Minerals, Metals, and Materials Society, 1999
- S. Grainger and J. Blunt, Ed., *Engineering Coatings: Design and Application,* 2nd ed., Woodhead Publishing Ltd., Cambridge, U.K., 1999
- J.M. Hampikian and N.B. Dahotre, Ed., *Elevated Temperature Coatings: Science and Technology III,* The Minerals, Metals, and Materials Society, 1999
- P.Y. Hou, T. Maruyama, M.J. McNallan, T. Narita, E.J. Opila, and D.A. Shores, *High Temperature Corrosion and Materials Chemistry: Per Kofstad Memorial Symposium,* PV 99-38, The Electrochemical Society, 2000
- P.Y. Hou, M.J. McNallan, R. Oltra, E.J. Opila, and D.A. Shores, *High Temperature Corrosion and Materials Chemistry,* PV 98-9, The Electrochemical Society, 1998
- W.B. Johnson and R.A. Rapp, Ed., *Proc. Symposium on High Temperature Materials Chemistry V,* PV 90-18, The Electrochemical Society, 1990, p 1–98
- A.S. Khanna, *Introduction to High Temperature Oxidation and Corrosion,* ASM International, 2002
- G.Y. Lai, *High Temperature Corrosion of Engineering Alloys,* ASM International, 1990

- A.V. Levy, *Solid Particle Erosion and Erosion-Corrosion of Materials,* ASM International, 1995
- K.G. Nickel, Ed., *Corrosion of Advanced Ceramics: Measurement and Modeling,* Kluwer Academic Publishers, Dordrecht, Germany, 1994
- E.J. Opila, P.Y. Hou, T. Maruyama, B. Pieraggi, M.J. McNallan, E. Wuchina, and D.A. Shifler, *High Temperature Corrosion and Materials Chemistry IV,* PV 2003-12, The Electrochemical Society, 2003
- E.J. Opila, M.J. McNallan, D.A. Shores, and D.A. Shifler, *High Temperature Corrosion and Materials Chemistry III,* PV 2001-12, The Electrochemical Society, 2001
- L. Remy and J. Petit, Ed., *Temperature-Fatigue Interaction,* Vol 29, Elsevier International Series on Structural Integrity, Elsevier Science, 2002
- M. Schütze, W.J. Quaddakers, and J.R. Nicholls, Ed., *Lifetime Modelling of High Temperature Corrosion Processes: Proc. European Federation of Corrosion Workshop* (EFC 34), Maney Publishers, Leeds, U.K., 2001
- S.C. Singhal, Ed., *High Temperature Protective Coatings,* The Metallurgical Society of AIME, 1982
- P.F. Tortorelli, I.G. Wright, and P.Y. Hou, Ed., *John Stringer Symposium on High Temperature Corrosion,* ASM International, 2003
- Y. Tamarin, *Protective Coatings for Turbine Blades,* ASM International, 2002
- M.H. Van de Voorde and G.W. Meethan, *Materials for High Temperature Engineering Applications (Engineering Materials),* Springer Verlag, London, 2000

Finishing Systems for Naval Aircraft

Kevin J. Kovaleski and David F. Pulley, Naval Air Warfare Center, Aircraft Division

EXTERIOR COATINGS provide corrosion protection, camouflage, erosion resistance, markings, electrical grounding, electromagnetic shielding, as well as other specialized properties. Protective coatings serve as the primary defense against the corrosion of metallic alloys as well as the degradation of other materials (such as polymeric composites and plastics) (Ref 1). Traditional coatings for aircraft include inorganic pretreatments, epoxy primers, and polyurethane topcoats. Pretreatments provide some corrosion protection and prepare the surface for subsequent organic coatings. Primers normally contain high concentrations of corrosion inhibitors such as chromates and are designed to provide superior adhesion. Polyurethane topcoats enhance protection and durability while providing the desired optical effects, including camouflage.

More recently, alternative coatings have been developed, such as nonhexavalent chromate pretreatments and primers, flexible primers, advanced performance topcoats, self-priming topcoats, low volatile organic compounds (VOCs) coatings, temporary coatings, and multifunctional coatings. These new developments reflect trends in protective coatings technology, changes in aircraft operational requirements and capabilities, and concerns over environmental protection and worker safety. Environmental issues have created a drive toward coatings with ultralow/zero concentrations of VOCs and nontoxic corrosion inhibitors. In turn, these changes have led to concerns over long-term performance, especially protection against corrosion. Current protective coatings technology, future needs, and trends based on advancing technology, environmental concerns, and operational requirements are included in this article.

Coatings life is critical. An Air Force study (Ref 2) concluded that "The rate-controlling parameter for the corrosion of aircraft alloys, excluding the mechanical damage factor, is the degradation time of the protective coating system." Navy aircraft face a unique and harsh environment because they are deployed at coastal land bases or onboard aircraft carriers. The continuous proximity to saltwater and high humidity, combined with atmospheric impurities, results in one of the most corrosive natural environments. In addition, many operational and maintenance chemicals commonly used or found on aircraft, such as paint strippers, battery acid, deicing fluids, and cleaners, are corrosive. These effects are exaggerated with aging fleet aircraft that have flown many flights over long periods of time, adding fatigue as another factor. Considering the high cost of these aircraft (in addition to fewer numbers of new aircraft programs), materials protection is of the utmost importance.

While traditional coatings possess exceptional corrosion inhibition, adhesion, and durability characteristics, these coating systems have been identified as a major contributor to the generation of hazardous materials and hazardous waste for the Navy (Ref 3). Volatile organic compounds, such as carrier solvents plasticizers, and heavy metal compounds, such as corrosion inhibitors and colorants, are being severely regulated, and coating formulations are being drastically changed. These changes have led to concerns over long-term performance. The typical lifetime of these coatings for deployed aircraft is approximately 4 years, after which the coating system is removed. The aircraft surface is then cleaned, pretreated, and recoated.

Standard Finishing Systems

The following is a description of the specific finishing systems being used on Navy aircraft and future trends for these materials. More general information is provided in Ref 4 to 8.

Surface Pretreatment. The primary goal of surface preparation and pretreatment processes is to enhance the corrosion resistance and adhesion of subsequent organic coatings. Proper surface preparation is an important step in the protection of aluminum and is accomplished using materials such as alkaline cleaners, etchants, and deoxidizers. These materials remove organic contamination along with the existing surface oxide layer of the aluminum to prepare it for subsequent chemical pretreatments. MIL-S-5002, "Surface Treatments and Inorganic Coatings for Metal Surfaces of Weapon Systems," is the military specification for the surface preparation and pretreatment of virtually every Navy aircraft and weapon system. The two primary surface pretreatments for naval aircraft are chromate conversion coatings and anodic films. Chromate conversion coatings (CCCs) are excellent surface pretreatments for aluminum alloys. These materials chemically form a surface oxide film (typically 430 to 650 mg/m^2, or 40 to 60 mg/ft^2), which enhances the overall adhesion and corrosion protection of the finishing system applied over it. Typical CCC film performance requirements are covered by MIL-C-5541, "Chemical Conversion Coatings on Aluminum and Aluminum Alloys," and CCC material properties are described in MIL-DTL-81706, "Chemical Conversion Materials for Coating Aluminum and Aluminum Alloys." Anodize processes form a thicker oxide film (2200+ mg/m^2, or 200+ mg/ft^2) by electrochemical means, providing more protection against degradation in comparison to chemical conversion coatings. Anodize processes are performed in accordance with MIL-A-8625.

Primer. Epoxy resins are commonly used as binders in high-performance primers due to their exceptional adhesion and chemical resistance. The solvent-borne epoxy primer is manufactured and packaged as a two-component epoxy/polyamide system. One component contains an epoxy resin that is the product of a condensation reaction between epichlorohydrin and bisphenol-A. The second component contains a multifunctional polyamide resin. After mixing the two components, the reaction of epoxide and amide groups within the resins (R) ensues according to Fig. 1.

The product of this reaction is a highly crosslinked polymer that forms the matrix of the primer film. The chemical and mechanical properties of the epoxy matrix cause the primer to be adherent, chemically resistant, and durable.

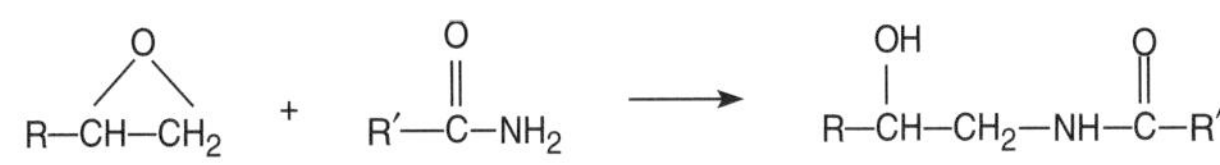

$$R-\underset{\underset{O}{\diagdown \diagup}}{CH-CH_2} \;+\; R'-\underset{\underset{NH_2}{|}}{\overset{\overset{O}{\|}}{C}} \;\longrightarrow\; R-\underset{\underset{OH}{|}}{CH}-CH_2-NH-\overset{\overset{O}{\|}}{C}-R'$$

Fig. 1 Chemical reaction for a typical two-part epoxy coating. R and R' are multifunctional polymeric units.

Hydroxyl groups on the solid epoxy are usually given credit for the excellent adhesion of these coatings. Detailed discussions of epoxy resin chemistry are found in the article "Organic Coatings and Linings" in *ASM Handbook*, Volume 13A, 2003, as well as in Ref 9 and 10.

The epoxide component of the primer contains various pigments, including titanium dioxide, strontium chromate, and extender pigments. Strontium chromate is the most critical of these pigments, because it is known as an exceptional corrosion inhibitor, especially for aluminum. Titanium dioxide in the primer enhances durability, chemical resistance, and opacity of the applied coating. The extender pigments can be silicas, silicates, carbonates, or sulfates. The extenders are normally inexpensive and provide a cost-effective component that fills the coating and reduces the gloss of the applied film. The surface irregularities that cause gloss reduction also act as anchors for a topcoat, thus enhancing inter-coat adhesion by improving the mechanical attachment. References 6 and 7 provide a comprehensive review of epoxy primer technology for aircraft applications.

On mixing the two components of the epoxy/polyamide primer, the curing reaction begins. After a dwell time of 30 min, the coating is suitable for spray application. The coating is applied to a dry-film thickness of 15 to 24 μm (0.6 to 0.9 mil). The coating is tack-free within 1 to 5 h and dry-hard within 6 to 8 h. For corrosion-prone areas, the primer may be applied up to double this thickness. If a topcoat is to be applied, it is usually accomplished within the tack-free to dry-hard time period to ensure proper adhesion. The primer attains sufficient dry-film properties within 7 to 14 days of application.

Adhesion of the primer is determined according to ASTM D 3359 by a tape test after 24 h immersion in distilled water (Ref 11). There should be no coating removal from the substrate. Adhesion is also characterized by a method that quantifies the force required to scrape the primer from the substrate, ASTM D 2197 (Ref 12). Typically, scrape-adhesion values of at least 3 kg (7 lb) are considered acceptable. Other adhesion tests employed in research and development laboratories are the tensile adhesion and Hesiometer knife-cutting tests (Ref 13). These sophisticated laboratory adhesion tests yield quantitative data but require more training and expertise to perform the tests and analyze the data, compared to the tape and scrape-adhesion tests.

Corrosion resistance is evaluated by applying the primer to chromate-conversion-coated aluminum substrates, such as 2024-T3 alloy. After curing for 7 to 14 days, the primer is scribed with an "X" so that the substrate is exposed. Resistance to 5% NaCl salt fog exposure (ASTM B 117) requires that no substrate corrosion or coating defects are produced after 2000 h exposure. Resistance to filiform corrosion is evaluated on primed and topcoated test panels. After exposure to hydrochloric acid for 1 h, the panels are exposed to high humidity for 1000 h. Generally, specimens should not exhibit any filiform growth from the scribe greater than 6.35 mm (0.25 in.), with the majority less than 3.175 mm (0.125 in.). Sulfur dioxide/salt fog exposure (ASTM G 85, annex 4), cyclic wet/dry exposure, and electrochemical impedance spectroscopy have also been used to evaluate the corrosion resistance of coating systems. The performance requirements for the standard solvent-borne epoxy primer are specified in specification MIL-PRF-23377, "Primer Coatings: Epoxy, High-Solids."

High-performance waterborne epoxy primer, which is currently used on the exterior surfaces of many military aircraft, was developed (Ref 14) and implemented in the Navy community in the late 1970s to mid-1980s. This primer is procured under specification MIL-PRF-85582, "Primer Coatings: Epoxy, Waterborne." The primer is supplied as a two-component epoxy/polyamide or epoxy/amine system. The resin systems are water-reducible, and film formation occurs via coalescence of resin particles and cross linking of the epoxy/polyamide or amine reactive groups. The pigments used are similar to those used in the solvent-borne primer system. Both strontium and barium chromates are used as the primary corrosion inhibitors in waterborne primers. Organic cosolvents and surface active agents are also used to enhance formulation and processing properties, such as water miscibility, dispersion stability, film formation, and surface finish.

Brittleness. The epoxy primer is brittle, especially at low temperatures (-51 °C, or -60 °F). This can result in extensive cracking of the paint system in highly flexed areas of the aircraft. Sealants, which are sometimes spray-applied between the primer and topcoat in aircraft finishing systems to increase overall coating flexibility, are soft, easily deformed, and difficult to apply and remove. An alternative is an organic coating that possesses the adhesion of a primer and the flexibility of a sealant, thus eliminating the logistical and application problems inherent in stocking and applying two materials instead of one. An elastomeric primer that provides these benefits has been characterized (Ref 15) and implemented on Navy aircraft. This technology conforms to specification TT-P-2760, "Primer Coating: Polyurethane, Elastomeric." This material is based on polyurethane resin technology, a pigment system containing strontium chromate for corrosion inhibition, and extender pigments for gloss control. Most of the requirements for this flexible primer are similar to those in the current epoxy primer specifications, with the exception of film flexibility. This requirement is significantly more stringent than the current epoxy primers: 80% versus 10% elongation, respectively. One of the major coating failure mechanisms on aircraft is cracking around fasteners, thus exposing bare metal. Application of this coating to numerous Navy and Air Force aircraft has resulted in fewer coating system failures due to cracking and chipping.

Topcoat. A high-performance topcoat, conforming to specification MIL-PRF-85285, "Coating: Polyurethane, Aircraft and Support Equipment," is applied to Navy aircraft in order to enhance protection against the operational environment. Aliphatic polyurethane coatings are ideal for this application due to their superior weather and chemical resistance, durability, and flexibility. These urethanes are two-component reactive materials. One component of the coating is a polyisocyanate resin or an isocyanate-terminated prepolymer based on hexamethylene diisocyanate. The second component contains hydroxylated polyester. On mixing, the isocyanate groups react with the hydroxyl groups of the polyester, as shown in Fig. 2.

The resulting polymer is flexible yet extremely durable and chemical resistant. Aliphatic isocyanates and polyesters are used in topcoats because they provide outstanding weather resistance compared to epoxies, whose aromatic groups degrade when exposed to ultraviolet light. References 16 to 19 provide more detailed discussions about polyurethane chemistry. When the two components are combined and the polyurethane reaction begins, the coating is ready for application (i.e., no induction time is required). This coating is normally spray-applied to a dry-film thickness of 50.8 ± 7.6 μm (2.0 ± 0.3 mils). The typical topcoat is set-to-touch and dry-hard (when cured at room temperature) within 2 and 8 h, respectively. Although the painted surface can be handled after 6 h without damage to the coating, full properties are normally not obtained until approximately 7 to 14 days.

The most critical performance requirements for topcoats are weather resistance, chemical resistance, and flexibility. Weather resistance is evaluated by laboratory exposure in an accelerated weathering chamber (Ref 20) for 500 h according to ASTM G 155. Exposure in this chamber consists of cycles of 102 min of high-intensity ultraviolet light (xenon arc) followed by 18 min of combined ultraviolet light and water spray. Although studies have shown that there is no precise correlation with outdoor exposure (Ref 21–23), the accelerated exposure does indicate if the coating is susceptible to ultraviolet and/or water degradation. Both accelerated and real-time weathering conditions cause only minimal changes in the color, gloss, and flexibility of high-performance aircraft topcoats.

$$R-N=C=O \; + \; R'-OH \; \longrightarrow \; R-NH-\overset{\overset{\displaystyle O}{\|}}{C}-O-R'$$

Fig. 2 Chemical reaction for a typical two-part polyurethane coating. R and R' are repeating polymer chains.

Chemical stability is evaluated by exposure to various operational fluids, such as lubricating oil, hydraulic fluid, and jet fuel, at elevated temperatures and/or extended durations. Aerospace topcoats are also subjected to heating at 121 °C (250 °F) for 1 h. Topcoats should show no defects other than slight discoloration after exposure to these conditions.

Flexibility requirements for polyurethane topcoats include impact and mandrel-bend tests. For high-gloss colors, 40% elongation of the coating after impact at room temperature and a 180° bend around a 2.54 cm (1.0 in.) cylindrical mandrel at −51 °C (−60 °F) are required without cracking of the film. Flexibility requirements for low-gloss colors are less stringent at low temperatures, because it is difficult to formulate flexible, low-gloss coatings due to the high pigment concentrations.

Advanced-Performance Topcoat. Aircraft coatings tend to degrade in their operational environment, as evidenced by discoloration, chalking, dirt accumulation, and corrosion. Continual corrosion repairs and cosmetic touchup yield a patchy appearance. The weight of excess paint reduces aircraft speed and range. Eventually, the aircraft must be sent to a depot for complete repainting and is therefore out of service for several months. Fluorinated polymers are known for their excellent environmental stability. A fluoropolymer resin can be combined into a conventional polyurethane topcoat to produce a coating with greatly improved weather resistance. In laboratory tests, such a coating showed superior resistance to fading and chalking after 3000 h of accelerated weathering. A durable topcoat would help to extend aircraft service life (time between rework cycles) and reduce maintenance costs. If the current 3 to 4 year life of these coatings could be increased to 8 years, half of the rework due to paint degradation would be eliminated, and maintenance costs would be greatly reduced. Boeing estimated a cost saving of $760,000 per depot cycle for a C-17 aircraft.

Self-priming topcoat (SPT) is a VOC-compliant, nonchromate, high-solids polyurethane coating that was designed to replace the current primer/topcoat paint system used on aircraft (Ref 24). This technology conforms to TT-P-2756, "Polyurethane Coating: Self-Priming Topcoat, Low Volatile Organic Compounds (VOC)." The SPT possesses the adhesion and corrosion-inhibition properties of a primer as well as the durability of a topcoat. The SPT effectively eliminates the need for a primer and thus reduces the application time and materials. In addition, the hazardous emissions and toxic wastes associated with current primers are eliminated. The specified thickness of this film is 50.8 to 66.0 μm (2.0 to 2.6 mils). It has been successfully applied to a variety of operational Navy aircraft, such as the F-14, F-18, AV-8, H-3, H-46, and P-3. These materials, however, are more sensitive to surface preparation and have raised issues regarding adhesion and corrosion protection.

Specialty coatings are used to address specific concerns.

Sealants. Although the current epoxy primers provide excellent adhesion and corrosion inhibition, they are brittle. This lack of ductility may result in cracking of the paint system on highly flexed areas of the aircraft. In order to improve the overall flexibility of the epoxy primer/polyurethane topcoat coating system, sealants are frequently incorporated into aircraft finishing systems. These sprayable materials are applied between the primer and topcoat up to 203 μm (8 mils) thickness and are primarily formulated from polysulfide, polyurethane, and polythioether binders. Their elastic nature minimizes cracking of the paint system. Critical requirements in these specifications are low-temperature flexibility (mandrel-bend tests), chemical resistance (fluid immersion at elevated temperatures), and corrosion resistance (5% NaCl salt spray tests). Although these sealants provide corrosion protection by the formation of a relatively impermeable barrier, some sealants also contain strontium chromate for corrosion inhibition. A detailed discussion of this technology can be found in Ref 25 and 26.

Rain-Erosion Coatings. Rain droplets and airborne debris (sand and dust) can impact aircraft leading edges and radomes during flight. The force of impact from these particles can erode the coating system and adversely affect the underlying substrate. The current primer/topcoat and SPT paint systems do not provide adequate protection against erosion, even when applied at two to three times its normal thickness.

The rain-erosion-resistant coating used on Navy aircraft is a two-component polyurethane material. One component consists of a pigmented, high-molecular-weight, polyether-type polyurethane. The other component contains a clear ketimine (blocked diamine) resin that acts as both a cross-linking agent and a chain extender. When combined, the two components form an elastomeric coating that can absorb and dissipate the energy of impacting rain droplets, thus preventing failure. Flexibility is characterized by a 0.635 cm (0.25 in.) mandrel bend at −51 °C (−60 °F) and tensile elongation of 450%. However, in order to exhibit this high elasticity, the polymer cross-link density is decreased, yielding reduced chemical resistance and weathering properties. In order to improve the finishing system durability, these materials are usually overcoated with the standard topcoat.

Although elastomeric coatings offer increased resistance to rain erosion, elastomeric tapes provide the optimal protection for Navy aircraft. These materials can be clear or pigmented polyurethane-based films supplied with or without an adhesive backing. Unlike coatings, these tapes are bonded to the surface. Early versions of these materials were clear aromatic-type polyurethanes. Although durable, these aromatic materials had poor weatherability. The newer aliphatic versions are extremely durable and have excellent weatherability.

High-temperature-resistant coatings are needed at various areas of Navy aircraft that are routinely subjected to elevated temperatures. The standard paint system was only designed to resist thermal exposures up to 176 °C (350 °F) for short durations. The only organic coatings able to resist high temperatures contain silicone binders. These silicone coatings contain aluminum flakes as pigment and are designed to withstand temperatures up to 650 °C (1200 °F). They can be applied by conventional air spray and are cured by heating to 204 °C (400 °F) for 1 h or upon reaching elevated temperatures during component operation. During the curing period, the binder system for this coating will oxidize, leaving a barrier layer of silicone oxide/aluminum to protect the underlying substrate from adverse conditions. Although this material provides adequate barrier protection, it is relatively soft and easily damaged in service.

Fuel Tank Coatings. Aviation fuels contain additives that may be corrosive. If left unprotected, fuel tanks would corrode and leak. In order to protect the tank interior, polyurethane fuel tank coatings are typically used. These highly cross linked chemically resistant coatings are two-component materials designed for application to nonferrous surfaces. The fluid resistance requirements for these materials are significantly more severe than those for the standard primer and topcoat. The higher degree of chemical resistance is necessary, because the fuel tank temperature can reach as high as 120 °C (250 °F) and the coating is not only subjected to various chemicals contained in aviation fuels but also to aircraft operational chemicals, saltwater, and dilute acidic solutions (Ref 27).

Compliant Coatings Issues and Future Trends

As the environmental consciousness of the world continues to increase, more efforts are being devoted to finding safe, compliant solutions to past, current, and future environmental problems.

Environmental Regulations and Hazardous Materials. One major factor affecting Naval aviation in recent years has been the Clean Air Act Amendment (CAAA) of 1990. This law significantly affects the type of materials and processes that will be approved for use in the future. In response to this situation, the Navy has expanded its efforts to reduce the amounts of hazardous materials generated from the cleaning, pretreating, plating, painting, and paint-removal processes used in both production and maintenance operations. The materials associated with these processes have been identified as major sources of hazardous waste by the Environmental Protection Agency (EPA) (Ref 28). Specifically, numerous research and development efforts have been established to address the environmental concerns with organic coatings. These efforts have two main thrusts: the development of low-VOC coatings and the development of nontoxic inhibited coatings. The aim of low-VOC coatings is to meet environmental regulations, especially California's Air Quality Management Districts rules and the CAAA Control Techniques Guideline (CTG) for the aerospace industry (one of 174 source categories). The development of nontoxic inhibited coatings is concerned with eliminating toxic heavy metal pigments such as the lead, chromate, and cadmium compounds used in protective primers and topcoats.

Low-VOC versions of military aircraft primers and topcoats have already been developed that comply with the CAAA Aerospace CTG, based on waterborne and high-solids formulations. Figure 3 summarizes the VOC content of coatings over time. Some solvents are legally designated by the EPA as neither VOCs nor hazardous air pollutants (HAPs), such as 1,1,1-trichloroethane and acetone. These exempt solvents, while solving a VOC problem, frequently create an ozone-depleting chemical or safety problem and are no longer perceived as a potential solution.

Nonchromated Pretreatments. One of the main environmental thrusts in the pretreatment area is the total elimination of hexavalent chromium. This toxic material has been used widely in the aforementioned processes because of its outstanding performance as a corrosion inhibitor for aluminum. This property is of particular importance to the Navy, due to the extensive use of aluminum in aircraft and weapon systems. Chromium (VI) is a known carcinogen, and regulatory agencies have recently enacted rules that limit or prohibit the use of this material, so alternatives are needed.

Nonchromated alkaline cleaners and nonchromated deoxidizers have been identified as acceptable alternatives to the current chromated processes. These materials have provided satisfactory performance in surface preparation and, in some cases, have been more cost-effective than their chromated predecessors.

Numerous nonchromated surface pretreatment materials have been investigated as replacements for the standard CCC. A summary of one such investigation is described in Ref 29. Three categories of alternatives have been studied in the Navy: inorganic nonchromated solutions, chromium (III)-base treatments, and sol-gel formulations. The first category includes solutions based on permanganate, cobalt amine, and ceric ion (among others) as the active corrosion-inhibiting agent. A Navy-developed chromium (III) treatment (Ref 30, 31) has shown corrosion-resistance and paint-adhesion properties comparable to CCCs in laboratory evaluations. Broader testing and in-service demonstrations of this technology are underway. Sol-gel formulations, although still in the initial phases of development, show promise. These materials are organic/inorganic polymers based on the hydrolysis and condensation of metal alkoxides. Figure 4 illustrates how, for example, silicon alkoxides (of general formula $Si(OR)_4$) can be reacted to form a barrier film strongly bonded to an oxidized aluminum surface. The chemistry of sol-gel films also allows a great deal of flexibility in modifying adhesive and other properties. While some of these alternatives have shown promise, a universal replacement for CCCs has not yet been achieved.

A self-contained pen has been approved by the Naval Air Systems Command and the Air Force for the repair and touch-up of CCCs. The pen significantly reduces waste related to conventional chromate wiping, because the formulation requires no rinsing and eliminates the need for brushes or rags. Nonchromate versions of the pen will be available for evaluation soon.

Chromic acid anodizing has been widely used in aerospace production and maintenance operations. Alternatives are sulfuric/boric acid anodize (SBAA) and thin-film sulfuric acid anodizing (TFSAA), as well as the standard type II sulfuric acid anodizing. General information on these processes can be obtained from Ref 32 to 35. Several Navy laboratory investigations are described in Ref 36 to 38. A full-scale successful demonstration of SBAA led to incorporation into specification MIL-A-8625 as type IC (sulfuric/boric acid anodize). Also, the laboratory studies have resulted in the definition of acceptable performance parameters for the TFSAA alternative that was incorporated into MIL-A-8625 as type IIB. The Navy is also investigating the performance of trivalent chromium pretreatment as a sealer for types IC, II, and IIB anodized aluminum.

Waterborne Technology. Water has long been used as a carrier for organic coatings. The polymers for these coatings are usually modified with hydrophilic groups and dispersed in water to form either solutions or emulsions. Most latex paints are based on thermoplastic resins that are suspended in water to form spherical particles. These particles, whether pigmented or neat, are usually covered with a thin layer of emulsifier to maintain a stable dispersion. When applied to a surface, these spheres coalesce into a continuous film as the water in the emulsion evaporates. This film formation mechanism tends to lead to longer drying times in humid environments. Effects of using water as the diluent include smoother surface finishes due to greater flowout, less overspray when using air application equipment (due to the higher density of water), and easier cleanup (usually accomplished with soap and water). Unfortunately, these coatings have some disadvantages. They are more sensitive to surface contamination, such as oils and greases. Also, these films tend to be porous, and their high affinity for water can lead to coating failure in high humidity.

Waterborne or water-reducible high-performance coatings are unique in the way that they employ resins that are usually not soluble in water. The resin exists in its own micellar phase (Fig. 5). Neutralized carboxylic groups and surfactants stabilize the particle. Excess amine and solvent distribute between the phases. Because the polymer exists in its own organic phase surrounded by water, the solvent distributes

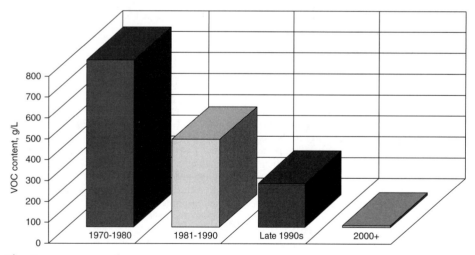

Fig. 3 Summary of volatile organic content (VOC) levels in standard military coatings over time

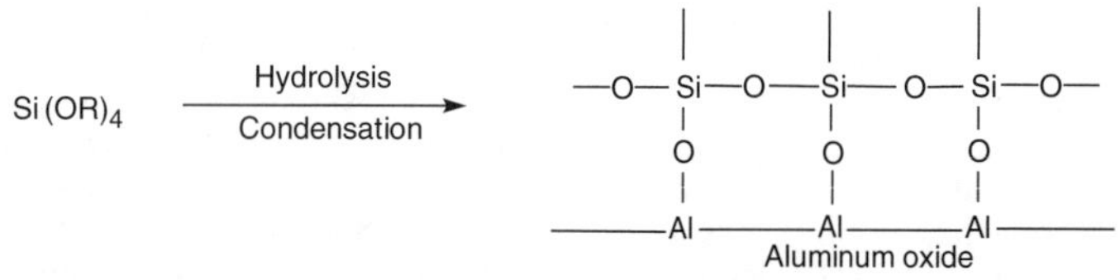

Fig. 4 Formation of a silicon oxide barrier film to bond to an oxidized aluminum surface

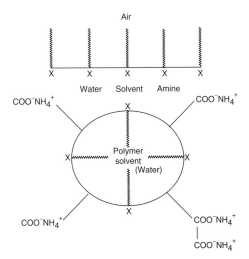

Fig. 5 Schematic diagram of a resin micelle (approximately 120 nm in diameter) in a waterborne coating

between the organic phase and the aqueous phase. This solvent, called the coalescing solvent or cosolvent, aids in film formation as the water evaporates by allowing binder and pigment particles to fuse in a continuous film (Ref 39). Because water is used as the primary liquid medium or as a diluent, formulations based on waterborne resins have much lower VOC levels than their solvent-borne counterparts. Recent advances in urethane and additive chemistries have shown that coatings approaching zero VOC are feasible (Ref 40).

One-component polyurethane dispersions have been in existence for some time. They generally consist of fully reacted polyurethane resins that are predominately thermoplastic. Because urethanes are not readily compatible with water, these systems are modified ionically and nonionically with hydrophilic groups to aid in the stability of the dispersions. After application, the films form by the coalescence of the long-chain urethanes. They tend to have lower cross-link densities and are not as chemically resistant as their solvent-borne counterparts.

Recently, researchers have been working with two-component waterborne polyurethane resins for high-performance coatings. One coating is based on an aliphatic polyol prepolymer and a polyisocyanate. The polyols are prereacted with diisocyanate resins and emulsifying agents to form a linear hydroxy-terminated prepolymer. The hydroxy-functional groups aid in the stabilization of the polyurethane dispersion. In addition, a water-dispersible polyisocyanate has been synthesized that has a preferential affinity for the polyol over the water competitor. The two components are mixed with an excess of isocyanate to form the final high-performance polyurethane product. Government laboratories and the commercial resin industry are investigating other variations on waterborne urethane chemistry. Coating manufacturers have begun to formulate finished products from this technology.

Zero-VOC Waterborne Topcoat. A zero-VOC topcoat was developed, based on a novel urethane chemistry that requires no cosolvent. Through manipulation of the polymer backbone chemistry and new rheological additives, a water-reducible polyurethane binder system contains virtually no organic solvents and emits no HAPs. The zero-VOC topcoat offers the potential for the Department of Defense (DoD) to go beyond environmental compliance in its painting operations. Successful implementation of this topcoat would result in the elimination of approximately 120 tons of VOCs per year, based on General Services Administration estimates of MIL-PRF-85285 topcoat usage throughout the DoD.

High-Solids Technology. Another method to attaining a reliable coating with lower VOCs is through the use of high-solids technology. Several paths to increase coating solids are possible. The first and most obvious reduction comes from simply lowering the solvent concentration. This approach shortens the pot life and significantly increases the resin viscosity when traditional raw materials are used. Also, the surface finish tends to be rougher from decreased flow characteristics. Another option is the use of reactive diluents. These materials are low-viscosity, low-molecular-weight compounds that act like a solvent for viscous resins. On curing, the reactive diluent becomes part of the polymer backbone and is not driven off as a VOC. Using low-molecular-weight resins can produce a high-solids coating with lower viscosity and better flowout. However, these materials tend to have short workable pot lives and lower flexibility when cured with traditional polyisocyanates. This lower flexibility is related to the increase in cross-link density resulting from the smaller backbone structures between functional groups. Using isocyanate-terminated prepolymers with a narrow molecular weight distribution as the isocyanate source produces low-VOC coatings with good performance and processing characteristics. These prepolymers yield coatings with lower viscosity, shorter drying time, and longer pot life.

Another approach uses blocked polymers. These yield a longer pot life but tend to be less mobile, with slower reaction rates. Their decreased reactivity leads to long drying times, which is not desirable. In addition, high-boiling solvents can be used to replace conventional solvents. By incorporating these materials, the applied films retain the solvent longer, providing a smoother surface finish. However, this solvent retention leads to longer drying times and can cause the coating to sag. This characteristic has also produced a new phenomenon where sharp edges are exposed as the coating dries. Finally, solvent retention can result in film porosity, thus decreasing the chemical resistance.

Reactive diluents were used in the development of a low-VOC epoxy topcoat. For more than 30 years, epoxy/polyamide systems have been used as the standard binder for coatings requiring superior chemical and corrosion resistance. These coatings use an epoxy resin cured with a standard high-viscosity polyamide curing agent. The epoxy resin is based on the glycidyl ether of bisphenol-A given in Fig. 6. Here, the value of "n" is between 2 and 3, and the epoxide equivalent weight (molecular weight divided by functionality) is approximately 500. In formulations containing such systems, 40 to 50 vol% solvent is necessary to attain a sprayable viscosity, significantly contributing to VOCs. Two reactive diluents were explored: one was monofunctional and the other was difunctional. The number of epoxide groups contained in the molecule defines the functionality. The monofunctional cresyl glycidyl ether (CGE) and the difunctional neopentyl glycol diglycidyl ether (NGDE) are pictured in Fig. 7.

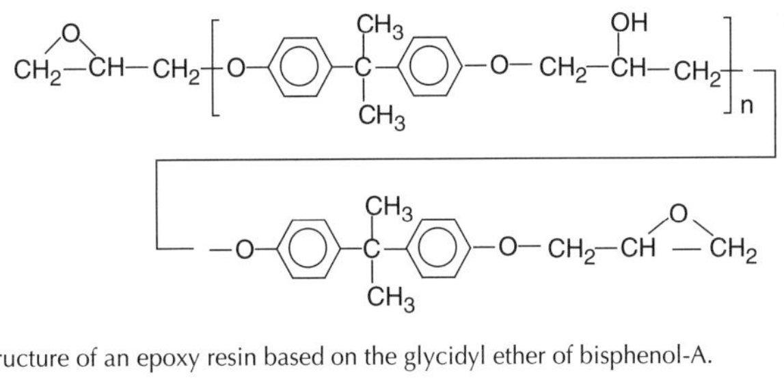

Fig. 6 Chemical structure of an epoxy resin based on the glycidyl ether of bisphenol-A.

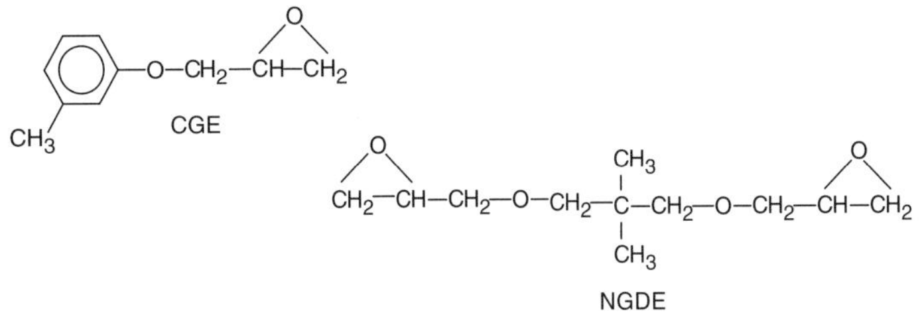

Fig. 7 Chemical structures of the reactive diluents cresyl glycidyl ether (CGE) and neopentyl glycol diglycidyl ether (NGDE)

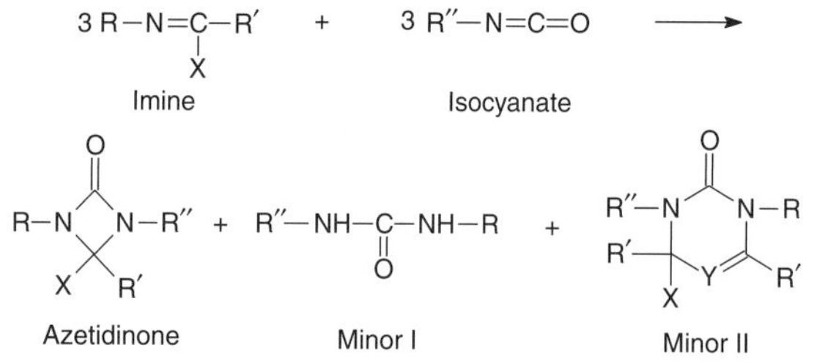

Fig. 8 Imine-isocyanate chemical reaction scheme. R, R', and R" are repeating polymer chains. If the imine is aldimine, X is H; if it is ketimine, X is methyl or greater. In minor II, Y = X − 2H; for example, if X is CH$_3$, then Y is CH.

Fig. 9 Application of nonchromated waterborne epoxy primer to a T-2 aircraft at a naval air depot

Coatings prepared with CGE were unacceptable due to poor surface properties and possible fluid permeation through missing chemical cross links. Two successful formulations were prepared using NGDE, exhibiting superior performance and reducing VOC content to 60% of the standard epoxy topcoat. The reactive diluent effort was summarized in a technical report (Ref 41).

Low-molecular-weight resins based on aldimine chemistry were employed to develop low-VOC coatings. Aldimines are reaction partners for isocyanates. They replace the polyols in the reaction scheme. Aldimines have viscosities that are so low that the standard polyester polyol is more than 50 times more viscous. This allows for significantly less solvent usage to reach a sprayable viscosity.

Aldimine use in binder systems resulted from studies by a resin manufacturer using ketimine-isocyanate reactions to produce a major product, azetidinone, and some minor products, according to the schematic in Fig. 8. The minor products reacted further with the isocyanates, resulting in coatings that varied in stability and color. Substituting aldimines for the ketimines resulted in the formation of the cross-linked azetidinone polymer only (no minor products). It demonstrated good color and stability, with 100% retention of solids.

Pigmented formulations were prepared at VOC levels of approximately 200 g/L. These formulations failed to produce workable coatings even after numerous attempts. Each individual approach has deficiencies that present a challenging problem to resin manufacturers.

A combination of these technologies may have the greatest potential for success.

Low-VOC Technology Status. Numerous military and commercial specifications have been written to cover materials based on these technologies. High-performance, VOC-compliant primers, topcoats, and self-priming topcoats are required to have a maximum VOC content of 340, 420, and 420 g/L, respectively. Waterborne and high-solids technologies have allowed for the development of protective coatings that contain significantly reduced levels of VOC. Some of these are currently under evaluation both in the laboratory and in the field. At present, it does not appear that VOC levels of less than 200 g/L can be attained with high-solids technology (without exempt solvents) using present application methods. Novel waterborne resins suggest that the development of zero-VOC coatings, which meet existing requirements, is plausible.

Nontoxic Inhibitive Primer. Until recently, chromates were virtually the sole source for active corrosion inhibition in aircraft coatings due to their outstanding performance for nearly all metals in a range of environments. Nontoxic inhibitors that have been investigated as alternatives include phosphates, borates, molybdates, nitrates, and silicates. The mechanisms by which these inhibitors work are not thoroughly understood. Proposed mechanisms for zinc phosphate include the adsorption of ammonium ions, complex formation on the exposed surface, passivation through a phosphating process, and anodic/cathodic polarization. Phosphates, borates, and silicates are generally regarded as anodic passivators that reduce the rate of corrosion by increasing anodic polarization. Molybdates also have been classified as anodic inhibitors and are particularly effective at limiting pitting corrosion. At high concentrations, the oxidizing action of molybdates is the main factor behind its corrosion-inhibiting properties. Molybdate ions migrate into anodic areas and accumulate there. Although these pigments individually provide some level of corrosion inhibition, a one-for-one substitution for chromates did not result in equivalent corrosion protection. However, synergistic effects from combinations of some inhibitors provide nearly equivalent performance (Ref 35, 42–44).

A nonchromated waterborne primer was qualified to MIL-P-85582, class N, following evaluation on several aircraft in the 1990s. There were no significant differences between the aircraft painted with the nonchromated primer and those painted at similar times with the standard primer. Application of the nonchromated primer is shown in Fig. 9.

In 1994, the Joint Group on Acquisition Pollution Prevention was chartered to coordinate joint service issues identified during the acquisition process. The group (now called Joint Group on Pollution Prevention) objectives are to reduce or eliminate hazardous materials, share technology, and avoid duplication of effort. Their product is a joint test protocol that establishes the critical requirements necessary to qualify alternative technologies. The Boeing Company organized a joint industry/government team to evaluate proprietary nonchromated epoxy primers for aircraft application. The Navy, Air Force, and National Defense Center for Environmental Excellence are participating.

Specifications MIL-PRF-23377 and MIL-PRF-85582 define the performance requirements for high-solids and waterborne primers, respectively. Paint manufacturers were asked to submit samples of nonchromated primers for evaluation under either of the specifications. Seven companies responded with 13 different products. The corrosion inhibitors were typically blends of several compounds described previously (phosphates, borates, molybdates). Primers were evaluated by the Navy for viscosity, pot life, drying time, surface finish, adhesion, flexibility, cure time, fluid resistance, corrosion resistance, and strippability. The nonchromated primers generally exhibited more corrosion in salt spray and filiform tests in comparison with control primers containing chromates. Only one candidate, a water-borne coating, met all of the qualification requirements.

Boeing conducted similar tests on these materials. They also included additional properties, such as sprayability, heat resistance, humidity resistance, thermal shock resistance, and compatibility with sealants, topcoats, and nondestructive inspection techniques. They confirmed the original results and recommended another waterborne primer as a second candidate for the service demonstration.

The demonstration included eight F/A-18 aircraft, two F-15 aircraft (wings only), an AV-8 aircraft, and several T-45 aircraft (touch-up applications). Either primer was applied to one side of each aircraft, and a standard chromated primer was applied to the other side. A MIL-PRF-85285 polyurethane topcoat was used over all of the primers.

The aircraft were based with operating squadrons at naval air stations in Beaufort, SC, Cecil Field and Tyndall Air Force Base, FL, and Lemoore, CA. Aircraft inspections over a 4 year period indicated no significant differences in corrosion resistance between the chromate and nonchromate primers. A joint test report for alternatives to chromate-containing primers for aircraft exterior mold-line skins was issued in February 1998. A later communication limited implementation to scuff-sand/overcoat operations only (Ref 45).

Touch-Up Paints. One approach used to reduce VOC and hazardous waste from painting operations has been the development of self-contained touch-up paint applicators. Maintenance personnel must treat corrosion by removing oxidation products and loose paint, then repairing the original finish. Aircraft paints, such as MIL-P-23377 or MIL-PRF-85582 epoxy primer and MIL-PRF-85285 polyurethane topcoat, are supplied in two-component kits. Each component is taken from a quart or gallon can, mixed in a specific volume ratio, and sprayed with air-atomized equipment. Workers tend to use more material than necessary to assure sufficient coverage. Excess paints must be processed as a hazardous waste. Improper mixing ratios often yield poor film properties. Spray application requires respiratory protection for all personnel in the area. Usage beyond damaged areas adds weight to the aircraft, thereby reducing its speed and range.

A unique kit designed to store, mix, and apply small quantities of two-component paints has been developed. Approximately $10\ cm^3$ ($0.6\ in.^3$) of the base and curing agent are contained in a clear plastic tube, separated by an impermeable barrier. When the barrier is displaced, the two components are easily mixed by shaking. A narrow brush on one end is used to dispense and apply the mixed material. A wider brush is also included. This kit, known as a Sempen, can be used to touch up areas of 0.1 to $0.2\ m^2$ (1 to $2\ ft^2$) and is shown schematically in Fig. 10.

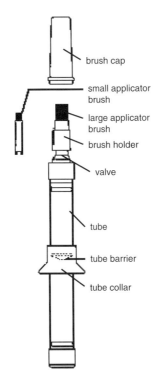

Fig. 10 Schematic diagram of the Sempen, a self-contained paint applicator pen

It has a number of advantages over spray application techniques using bulk materials:

- The small touch-up kit restricts maintenance personnel from mixing large quantities of paints that may be applied to excessive areas or disposed of as a hazardous waste.
- Brush application minimizes airborne concentrations of the toxic solvents, isocyanates, and the like in paints. Other personnel can work nearby without protective equipment.
- The individual components are premeasured to assure precise mixing and optimal properties.

A number of kits containing MIL-PRF-85582 epoxy primer or MIL-PRF-85285 polyurethane topcoat were supplied to operating squadrons for aircraft touchup. The materials were used during normal maintenance operations to repair painted surfaces that were cracked, chipped, or corroded. Personnel observed the ease of application, the appearance of the applied paint, and its durability over a 6 month period.

A one-component waterborne polyurethane topcoat has shown excellent performance. It is currently packaged in a 28 g (1 oz) jar with a brush applicator attached to the cap. A compatible one-component alkyd primer is also available in a similar container. A two-component high-solids polyurethane topcoat with unique properties has been developed. It is packaged in a 57 g (2 oz) jar with the curing agent in a plastic container underneath the cap. After mixing, it has an effective pot life of 7 days at ambient temperatures (30 days if refrigerated) but, when applied, will react with moisture in the air to dry within 2 h. The jar is designed for attachment to an aerosol sprayer. A number of two-component epoxy primers have also been packaged in the 57 g (2 oz) jars.

Maintenance personnel expend approximately 4 h per aircraft in the preparation, treatment, and cleanup associated with corrosion control. This time could be greatly reduced if they had access to all of the necessary materials in a single portable container. The container should include:

- Premoistened pads (such as Scotch-Brite, 3M Company) for surface preparation
- Wet and dry towelettes for cleaning
- Alodine Touch-N-Prep pens for chromate conversion coating
- Sempens and 28 g (1 oz) jars for brush application of paints
- Various 57 g (2 oz) jars for aerosol application of paints

Adhesive Films. Protective film technology has been used for specific applications in naval aviation for many years. One of the primary means of protecting aircraft radomes and leading edges from rain erosion is the use of elastomeric tapes. These tapes are flexible films bonded to painted surfaces with an attached adhesive. Recently, one manufacturer has developed an adhesive film aimed at replacing aircraft exterior topcoats. This material, also known as appliqué?, provides a durable weather-resistant finish and is intended for application over a standard corrosion-resistant primer. The material consists of a polymeric layer and an adhesive layer, with a plastic backing over the adhesive. After removing the backing, the film is bonded to the primer in adjoining or overlapping sections. During application, only simple measuring and cutting tools are needed. The work can be done in any enclosed area with minimal training. Nearby personnel can perform installation and maintenance work at the same time, because no safety or environmental hazards are present.

Aircraft demonstrations of appliqué? technology have shown two significant problems. The pressure-sensitive acrylic adhesive is thermoplastic and tends to lose adhesion at higher temperatures, above 80 °C (180 °F). In addition, mechanical removal of the aged appliqué? for corrosion repair is difficult. The adhesive often leaves a sticky residue that must be stripped with solvent.

Electrodeposition Coatings and Powder Coatings. Electrodeposition of paint (e-coat) works on principles similar to plating. Charged paint particles are electrically plated to an oppositely charged conductive substrate. The object to be coated is dipped into the electrodeposition tank, and the current is turned on. During the operation, the paint coats every conductive surface (regardless of shape) with a uniform film whose thickness can be controlled very accurately. The paint film then insulates the substrate, preventing any further deposition of the coating. More details can be found in Ref 46 to 48. Because waterborne coatings are used in the e-coat process, VOCs are kept to a minimum. Also, no flammability hazards exist, and no elaborate ventilation systems are required, as with conventional spray techniques. A significant hazard does exist, however, with the electric power required. The tank area must be enclosed and have fail-safe controls. The technique is also limited to small conductive parts in aircraft applications. Also, temperatures approximately 150 °C (300 °F) or higher are required to cure these coatings.

Corrosion testing of e-coated panels revealed that, although it had superior adhesion, the coating was easily undercut by corrosion products at defects in the film. This phenomenon was due to the lack of corrosion inhibitor in the e-coat. The incorporation of such inhibitors in the tank resulted in coating bath instabilities. This fact, along with the costs and workload needed to maintain the process line, resulted in the decision not to implement e-coat for naval aviation purposes.

Powder coatings can be applied by a variety of methods, including fluidized bed, dip coating, and electrostatic spray. With electrostatic spray, as the powder passes the high-voltage electrode at the tip of the spray gun, it picks up the electrostatic charge and is attracted to grounded work. There, the powder adheres and will remain until fused and cured in an oven. More

Fig. 11 Powder coating spray booth at a naval air depot

detailed descriptions of powder coatings and application techniques can be found in Ref 47 and 49.

As with e-coats, the film thickness of powder coatings is controlled by the insulating effect of the powder film as it builds up on the substrate. Overspray is considerable, but it can be collected and reused, so losses are low. Because the material is a solid, there are no VOCs. However, an explosion hazard may exist due to the fine dust overspray. Therefore, careful grounding of the booth and equipment is required. A powder coating spray booth with an electrostatic spray gun is pictured in Fig. 11. Typically, parts are cured in a convection oven at 190 °C (375 °F) for 15 min.

Because of concerns affecting substrate properties such as temper at elevated temperatures, investigations continue for powder coatings with lower cure temperatures. Materials that cure between 120 and 150 °C (250 and 300 °F) are under test to determine if any serious property trade-offs will occur on lowering the cure temperature. Cure temperature affects cross-link density. It will, therefore, be necessary to address such properties as fluid resistance, corrosion resistance (limited barrier properties due to fewer cross links), and adhesion. There is also an investigation into the use of powder coatings on components that are not temperature sensitive.

Paint Application Equipment. As part of the CAAA aerospace guide, conventional spray application equipment will no longer be authorized for applying paints. This equipment has a transfer efficiency of approximately 28%. The types of paint application equipment authorized will be similar to those specified by California's regulations that require minimum transfer efficiencies of 60 to 85% and maximum gun-tip air pressures of 0.007 MPa (10 psi). A number of alternative technologies have been proposed to meet this requirement. The only two spray application techniques authorized were electrostatic and high-volume, low-pressure (HVLP) spray guns. Both of these techniques have improved transfer efficiencies over conventional air spray. Roller, brush, dip, and other nonspray methods are also acceptable. Each of these techniques has its unique capabilities and limitations. Some methods can be used in combination (air-assisted airless with electrostatic) to yield even higher efficiencies. A study showed HVLP to be the most cost-effective, so this technology has been implemented at all naval air depots.

The method of cleaning spray equipment is also being regulated. The old solvent wash method, which generated large quantities of hazardous waste and was time-consuming, is being prohibited. An enclosed cleaning method, which captures the majority of the cleaning solvent, will be used. These enclosed cleaning operations take approximately one-fourth the time as the old method.

Nonchromated Sealants. The most effective inhibited sealants for fuel tanks were formulated from polysulfide polymers, soluble chromates, manganese dioxide curing agents, and other additives/fillers. Inhibited sealants are applied to prevent faying surface corrosion and dissimilar-metal corrosion and are also applied as corrosion-resistant coatings. Polythioether sealants have several advantages over polysulfide types, such as faster curing rates and a higher temperature range. Specification MIL-S-29574 has a type II category for a corrosion-inhibitive, fuel-resistant sealant.

A direct replacement for soluble chromate has not been found, so combinations of two or more compounds are necessary to make up new inhibitor packages for sealants. Inhibitors found to be effective replacements for chromates in paint primers and other corrosion-prevention materials were not directly transferable to sealants. Because of the interactions of the polymers, curing agents, adhesion promoters, fillers, and other ingredients, it was difficult and expensive to effectively formulate new inhibited nonchromated sealants with all the necessary properties. As a result, most of the reformulated inhibited sealants released by suppliers had limitations that needed to be resolved before even considering them for military use.

REFERENCES

1. R.C. Cochran et al., AGARD Report 785, 73rd AGARD Panel/Specialists Meeting on Aerodynamics and Aeroelasticity, North Atlantic Treaty Organization, Oct 1991
2. R.N. Miller, "Predictive Corrosion Modeling Phase I/Task II Summary Report," Air Force Wright Aeronautical Laboratories Report AFWAL-TR-87-4069, Wright-Patterson Air Force Base, OH, Aug 1987
3. NAVAIR RDT&E EQ Requirements Working Group Meeting (Arlington, VA), Naval Air Systems Command, 13–14 July 1999
4. S.J. Ketcham, *A Handbook of Protective Coatings for Military and Aerospace Equipment,* TPC Publication 10, National Association of Corrosion Engineers, 1983
5. "Finishes, Coatings, and Sealants for the Protection of Aerospace Weapons Systems," MIL-F-7179, Dec 1991
6. J.B. Lewin, *Aircraft Finishes, Treatise on Coatings, Vol 4, Formulations Part I,* R.R. Myers and J.S. Long, Ed., Marcel Dekker, 1975
7. A.K. Chattopadhyay and M.R. Zentner, *Aerospace and Aircraft Coatings,* Monograph Publication, Federation of Societies for Coatings Technology, May 1990
8. S. Wernick and R. Pinner, *The Surface Treatment and Finishing of Aluminum and Its Alloys,* 4th ed., Robert Draper, Ltd., 1972
9. C.R. Martens, Ed., *Technology of Paints, Varnishes, and Lacquers,* Reinhold Book Corp., 1968
10. J. Boxall and J.A. von Fraunhofer, *Concise Paint Technology,* Chemical Publishing, 1977
11. "Standard Test Methods for Measuring Adhesion by Tape Test," D 3359, ASTM, Sept 1987
12. "Standard Test Method for Adhesion of Organic Coatings by Scrape Adhesion," D 2197, ASTM, Oct 1986
13. W.K. Asbeck, *J. Paint Technol.,* Vol 43 (No. 556), 1971
14. R.A. Albers, U.S. Patent 4,352,898, Oct 1982
15. D.F. Pulley and S.J. Spadafora, "Elastomeric Primers and Sealants," Report NADC-83140-60, Naval Air Development Center, Warminster, PA, Nov 1983
16. J.H. Saunders and K.C. Frisch, *Polyurethanes: Chemistry and Technology,* John Wiley, 1962
17. G. Oertel, Ed., *Polyurethane Handbook,* Macmillan Publishing, 1985
18. T.A. Potter and J.L. Williams, *J. Coatings Technol.,* Vol 59 (No. 749), 1987, p 63–72
19. L.G.J. Van der Ven et al., The Curing of Polyurethane Coatings; Chemical and Physical Information Processed in a Mathematical Model, *Proceedings of the American Chemical Society Division of Polymeric Materials: Science and Engineering* (Toronto, Canada), 1988
20. "Standard Practice for Operating Light-Exposure Apparatus (Xenon-Arc Type) with and without Water for Exposure of Nonmetallic Materials," G 26, ASTM, Sept 1988
21. J.L. Scott, *J. Coatings Technol.,* Vol 49 (No. 633), 1977, p 27–36
22. J.A. Simms, *J. Coatings Technol.,* Vol 59 (No. 748), 1987, p 45–53

23. C.R. Hegedus and D.J. Hirst, Report NADC-88031-60, Naval Air Development Center, Warminster, PA, March 1987
24. C.R. Hegedus et al., "UNICOAT Program Summary," Naval Air Development Center, Warminster, PA, March 1990
25. F.R. Eirich, Ed., *Science and Technology of Rubber,* Academic Press, 1978
26. A. Damusis, Ed., *Sealants,* Reinhold, 1967
27. D.J. Hirst, "Evaluation of Corrosion Resistant Coatings for Environmental Control System Ducts," NADC-83109-60, Naval Air Development Center, Warminster, PA, Aug 1983
28. V. Boothe, presentation at U.S. Environmental Protection Agency Aerospace CAAA CTG Hearing (Durham, NC), May 1993
29. S.J. Spadafora, Report NAWCADWAR-92077-60, Naval Air Warfare Center Aircraft Division Warminster, Warminster, PA, 18 Aug 1992
30. F. Pearlstein and V.S. Agarwala (to the Navy), U.S. Patent 5,304,257, 19 April 1994
31. F. Pearlstein and V.S. Agarwala, *Plat. Surf. Finish.,* July 1994, p 50
32. T.C. Chang, Report MDC-K5784, McDonnell Douglas, 28 Jan 1991
33. Y. Moji, Report D6-55313TN, Boeing Aerospace Co., 6 Feb 1990
34. W.C. Cochran, *Electroplating Engineering Handbook,* L.J. Durney, Ed., Van Nostrand Reinhold Co., 1984
35. M. Howard, Report RHR-90-194, Rohr Industries, Dec 1990
36. S.J. Spadafora and F.R. Pepe, *Met. Finish.,* Vol 92 (No. 4), 1994
37. S.J. Spadafora and F.R. Pepe, A Comparison of Alternatives to Chromic Acid Anodizing, *Proceedings of 1994 Tri-Service Corrosion Conference* (Orlando Fl), Department of Defense, 21–23 June 1994
38. S. Cohen and S.J. Spadafora, Report NAWCADWAR-95023-43, Naval Air Warfare Center Aircraft Division, Warminster, PA, April 1995
39. C.R. Martens, *Waterborne Coatings: Emulsion and Water-Soluble Paints,* Van Nostrand Reinhold Co., 1981
40. D. McClurg, personal communication, 1994
41. K.J. Kovaleski, R.D. Granata, and S.J. Spadafora, Report NAWCADPAX-96-41-TR, Naval Air Warfare Center Aircraft Division, Patuxent River, MD, Dec 1995
42. T. Foster et al., *J. Coatings Technol.,* Vol 63 (No. 801), 1991
43. E.J. Carlson and J.F. Martin, "Environmentally Acceptable Corrosion Inhibitors: Correlation of Electrochemical Evaluation and Accelerated Test Methods," Federation of Societies for Coatings Technology 71st Annual Meeting and Paint Industries Show (Atlanta, GA), Oct 1993
44. W.J. Green and C.R. Hegedus (to the Navy), U.S. Patent 4,885,324, Dec 1989
45. Naval Air Systems Command (NAVAIR) letter 13150/Ser AIR-4.3.4/7.4629 of 18 April 2002
46. A. Brandau, Introduction to Coatings Technology, *Federation Series on Coatings Technology,* Federation of Societies for Coating Technology, 1990, p 26–27
47. S.B. Levinson, Application of Paints and Coatings, *Federation Series on Coatings Technology,* Federation of Societies for Coating Technology, 1988, p 36–39
48. B.N. McBane, Automotive Coatings, *Federation Series on Coatings Technology,* Federation of Societies for Coating Technology, 1987, p 14–20
49. J.H. Jilek, Powder Coatings, *Federation Series on Coatings Technology,* Federation of Societies for Coating Technology, 1991

Military Coatings

Joseph T. Menke, U.S. Army

MILITARY COATINGS are as divergent as electrodeposited chromium plating in gun barrels and the visual comouflage paints applied to tactical equipment such as tanks, aircraft, and ships. The chromium plating in a gun barrel provides high-temperature (approximately 1650 °C, or 3000 °F) oxidation resistance, wear, and corrosion resistance. This plating is a single coating with multifunctional performance requirements. The paint system currently in use is a chemical agent resistant coating (CARC). This paint system consists of an epoxy primer and urethane topcoat and is multifunctional. The paint system provides weatherability (ultraviolet protection), mar resistance, chemical agent resistance, corrosion resistance, and spectral reflectance in addition to the visual camouflage. Spectral reflectance is the ability of paint to hide a piece of equipment from infrared (IR) sighting devices. Green high spectral reflectance coatings are used on ground equipment to match the color of trees and grass. Low spectral reflectance coatings of light blue, gray, and white are used to match the spectral properties of water and sky.

The many coatings used on Department of Defense (DoD) systems are categorized in five major areas. These areas are obviously not all-inclusive but include:

- Electroplated coatings (applied with an electric current)
- Conversion coatings (applied without an electric current)
- Supplemental oils, waxes, lubricants
- Organic paint coatings
- Other finishes such as vacuum deposits (chemical and physical vapor deposition, ion vapor deposition, sputtering), mechanical plating, thermal spray coatings, and hot-dip coatings

Electroplating

Military coatings have many different performance applications. This article is limited to the corrosion performance of these coatings. Table 1 lists electroplated coatings. Typical direct current (dc) electroplated coatings used on steel surfaces include cadmium, zinc, chromium, nickel, copper, tin, lead, and aluminum. Anodizing and hardcoat are two processes used on aluminum surfaces that require dc electrical current. One critical characteristic of most electroplated deposits that can affect corrosion resistance is coating thickness. The minimum thickness is generally a requirement that is specified in the technical data. Another aspect that is crucial to corrosion behavior is whether the coating is anodic or cathodic to the base metal. Of the electroplated coatings, only zinc and cadmium are anodic (sacrificial) to steel. The other electroplated coatings must be thick enough to create a barrier to chemical reactants to provide any level of corrosion resistance to the substrate metal. Figure 1 shows what can happen when the plating is a cathode with respect to the substrate metal and is not thick enough to be a barrier coating. The small-anode/large-cathode area ratio results in serious pitting of the underlying steel. Tin, lead, and aluminum coatings are an exception. They corrode readily in the salt-spray test, as the environment is aggressive enough to keep these coatings from being cathodes. In normal outdoor environments, tin, lead, and aluminum react with oxygen in the air and form a protective oxide on their surface. This metal oxide becomes a cathode with respect to iron and steel, and pitting corrosion can result with these coatings.

Commercial zinc plating for mild environments is 5 μm (0.2 mils) thick and will rarely pass 24 h in the salt-spray test before red rust appears. This thickness class service condition is considered "for indoor use" per ASTM B 633 (Ref 2) and is unsuitable for outdoor military

Table 1 Test requirements and time to failure of electrodeposited coatings

Coating	Thickness μm	Thickness mils	Test minimum(a), h	Time to failure, h
Aluminum substrate				
Anodize	2.5–25	0.1–1	336	1500–7000
Hardcoat	51	2.0	...	100–500
Steel substrate				
Cadmium	5–13	0.2–0.5	...	200–600
Cadmium + chromate	5–13	0.2–0.5	96	2000–3000
Zinc	13	0.5	96	150–200
Zinc	25	1.0	192	200–300
Zinc + chromate	2.5–25	0.5–1.0	96	300–500
Chromium	5–50	0.2–2.0	...	30–100
Chromium	50–130	2.0–5.0	...	100–500
Nickel	13–25	0.5–1.0	...	50–100
Copper	13–25	0.5–1.0	...	10–20
Cu-Ni-Cr	10–51	0.4–2.0	...	50–80
Tin	7.6	0.3	24	30–50
Lead	13	0.5	48	...
Lead	25	1.0	96	...
Al-Ni flash(b)	10–25	0.4–1.0	...	1000–2000

(a) Test per ASTM B 117 (Ref 1). (b) Proprietary aluminum plating

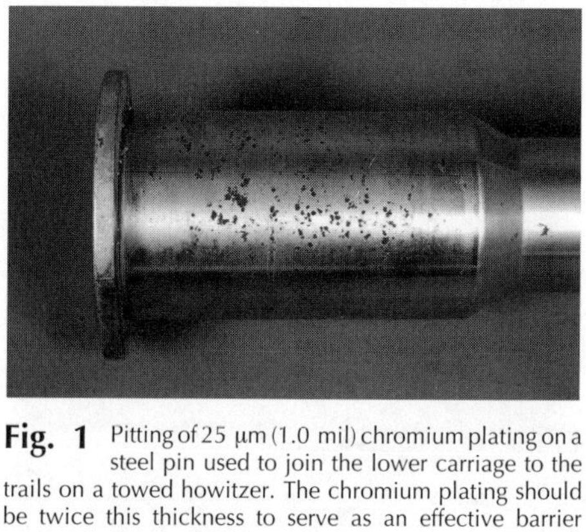

Fig. 1 Pitting of 25 μm (1.0 mil) chromium plating on a steel pin used to join the lower carriage to the trails on a towed howitzer. The chromium plating should be twice this thickness to serve as an effective barrier coating for this application

tropical environments. A minimum 12 μm (0.5 mils) zinc coating class is suggested for military hardware by this standard, and the service condition is classified as severe. Material specifications must be explicit as to service condition when citing the standard and the thickness verified.

Anodize and hardcoat finishes on aluminum convert the aluminum metal to a thicker aluminum oxide film (relative to the "natural" or air-formed film). As shown in Fig. 2, aluminum oxide is a cathode with respect to aluminum, so pitting corrosion occurs at defects in the coating. Not sealing the hardcoat improves the wear resistance, but significantly compromises the corrosion resistance.

Accelerated Corrosion Tests. Table 1 summarizes the minimum results specified in technical procurement documents for accelerated corrosion tests and compares these to actual times to failure where the information is available for the coating process. The accelerated corrosion test method is the 5% salt spray per ASTM B 117 (Ref 1). This is the test procedure by which the times to failure are determined. See the article "Cabinet Testing" in Volume 13A for much more on this test and similar tests.

Table 1 assists understanding the relative level of corrosion protection provided by various finishes under controlled and reproducible conditions. It is not a predictor of field performance. The minimum test requirement can be used as a process verification tool as it is what the process (properly applied) must meet to prove the process is under control. Process control test requirements are generally kept to a minimum so production acceptance times are as short as possible. The normal experience is that the time to failure is consistent under typical environmental conditions for systems exceeding the minimum requirement. The time to failure represents the actual performance of a coating.

A reference to automotive corrosion (Ref 3) has associated the 5% salt-spray test with expected performance in the field. Approximately 100 h exposure was determined to be equivalent to 1 year in service. This can be used only as a loose guide for comparison to estimate coating performance, because the variations in environmental factors, especially in the military environments, are so significant. Table 1 is consistent with experience that surfaces with 25 μm (1 mil) of chromium are generally rusty in a short period of time while surfaces with 76 μm (3 mils) of chromium are seldom found to be rusty. Table 1 also shows the beneficial effect of the chromate treatment on cadmium and zinc. The 96 h requirement to "white" corrosion product shows that the chromate treatment is protecting the plating that finally fails when the steel is exposed.

Conversion Coating

Conversion coatings used on military equipment include chromate conversion coatings on aluminum, cadmium, and zinc-base products such as galvanize or zinc die castings. Electroless nickel, zinc and manganese phosphate, and black oxide are considered conversion coatings on steel. See the articles "Phosphate Conversion Coating" and "Chromate and Chromate-Free Conversion Coatings" in Volume 13A.

Black sulfide coatings on stainless steel and naphthanate coatings on wood are also conversion-coating processes. The black oxide and phosphate coatings are generally used with a supplemental treatment of oil, wax, or some other organic coating.

Table 2 lists the various conversion coatings that are applied by immersion without current. The same salt-spray test is applied. Using 100 h test exposure as a measure of 1 year in the field, it can be seen that conversion coatings by themselves do not provide significant corrosion protection. The exceptions to this are the black coating on 300 series stainless steel, the chromate on aluminum, and the electroless nickel on steel. These coatings are more permanent and do not require supplemental oil for improved corrosion resistance. The electroless nickel is a shiny coating and would not be considered for exterior surfaces on weapons unless it could be made nonreflective. The class 1 (C.1), class 2 (C.2), and class 4 (C.4) black coatings in Table 2 refer to coatings produced to MIL-DTL-13924 (Ref 4). The classes in MIL-DTL-13924 describe the different blackening processes that are available for different substrate materials.

Supplemental Oils

The supplemental oils add significantly to the salt-spray protection of phosphate and black oxide coatings. These oil coatings are, however, washed off easily in a rain or evaporated in the hot sun and must be constantly reapplied to maintain the corrosion resistance in the field. Figure 3 shows a weapon component that came back from the desert and was found to contain

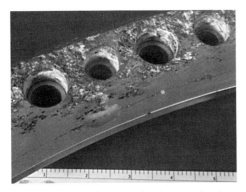

Fig. 2 A section of a ring gear bearing from a howitzer in Hawaii. The gear is hardcoated aluminum and shows severe pitting at defects in the coating and at a defect from a scratch (bottom center). The coating is cathodic with respect to the scratch that is anodic.

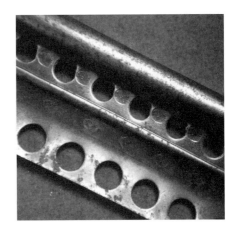

Fig. 3 Rusting on a phosphated part after service in a desert environment. The phosphate coating had been blasted off by blowing sand leaving a bare metal surface. Phosphate coatings are specified on 75 to 90% of the parts for small arms.

Table 2 Test requirements and time to failure of conversion coatings

Coating	Coating thickness			Test minimum(a), h	Time to failure, h
	g/m^2	μm	mils		
Aluminum substrate					
Chromate	>430	...	...	168	200–500
Electroless nickel	...	25.4	1.0	100	48–500
Steel substrate					
Black oxide, C.1	...	1.0–2.5	0.04–0.1	...	0.1–0.5
Zinc phosphate	>10.7	...	...	2	3–24
Zinc phosphate + oil	...	5.1–10.2	0.2–0.4	96	100–500
Manganese phosphate	>16.1	...	...	1.5	2–8
Manganese phosphate + oil	...	5.1–15.2	0.2–0.6	72	100–300
Cold phosphate	>10.7	...	...	2	3–10
Paint phosphate base	1.6–5.4	...	...	...	0.1–2.0
Electroless nickel	...	38	1.5	100	200–1000
300 type stainless steel substrate					
Black oxide, C.4	...	1.0–25	0.04–0.1	...	100–500
400 type stainless steel substrate					
Black oxide, C.2	...	1.0–25	0.04–0.1	96	3–8

(a) Test per ASTM B 117 (Ref 1)

almost no phosphate coating due to the abrasive action of blowing sand in the area. The phosphate coating is relatively soft, and the sand blasted the coating off the parts. Weapon parts became susceptible to rusting without the phosphate coating to hold the oil.

Supplemental oils provide a degree of corrosion protection that depends on the amount of rust preventive added to the formula and the film thickness of the oil. Levels of corrosion protection are shown in Table 3 and are normally determined on a sand or steel grit blasted, bare steel surface. The viscosity of these materials determines to some degree the thickness of the oil that will be retained on the surface. The MIL-PRF-32033 oil (Ref 5) is similar to the commercial product known as WD-40, and MIL-PRF-63460 (Ref 6) is the current field lubricant for small arms weapons. The film thickness of the oils in Table 3 is an approximation based on readings obtained with a wet film paint thickness tester. The MIL-PRF-3150 (Ref 7) and MIL-PRF-16173 (Ref 8) oils are primarily used for protecting metal parts inside various packaging schemes. They may be used over a phosphate coating or just over bare metal when used inside military packaging. Bare steel fasteners have a rust preventive oil specified on the drawings, but this protection may be lost prior to transportation to the field.

Paint Coatings

Paint coatings on military hardware should be considered painting systems that include surface preparation. The pretreatment for paint on aluminum may be the chemical film or anodize coating. The pretreatment for paint on steel is usually a light zinc phosphate coating as shown in Table 4. The CARC systems have been in use since 1985 and consist of a pretreatment, epoxy primer, and urethane topcoat as shown in Fig. 4. The test, tactical, and material requirements for the CARC system are also shown, and it should be obvious that this is a coating system. The pretreatment ensures good adhesion of the primer and assists in providing a significant level of corrosion protection when used in combination with the primer. For example, the primer on bare steel provides about 50 h of salt-spray corrosion protection at the specified thickness. When that same primer is applied over a light zinc phosphate pretreatment, up to 1500 h of salt-spray corrosion protection can result. Incomplete coverage of the phosphate coating and a primer thickness below the minimum are the two most common causes of premature corrosion on CARC painted items. When topcoat is applied over the primer, additional corrosion protection is available. See the article "Organic Coatings and Linings," in Volume 13A.

As mentioned, the CARC topcoat is not applied solely for corrosion protection. The topcoat provides visual camouflage, resistance to chemical agents and decontaminating solutions, and exhibits spectral reflectance to escape detection by IR sensors (film or sights). The topcoat must exceed a minimum thickness of 46 μm (1.8 mils) to provide the spectral reflectance properties.

Level of Required Protection. Ammunition items using alkyd enamel paint may have parts requiring only 24 or 48 h of accelerated corrosion testing. On the other hand, the paint specification requires the zinc phosphate and paint to go 150 h in the 5% salt-spray test. Thus, a three- to sixfold reduction in the required life of the paint is acceptable according to the design of many ammunition items. The paint thickness used on the item is less than recommended elsewhere to meet other functional requirements, so the corrosion resistance for the item is reduced. For example, the lacquer paint on 20 mm projectiles is applied at 13 to 20 μm (0.5 to 0.8 mils) and the part has a 24 h salt-spray requirement. It is difficult to get corrosion resistance with very thin coatings. That is why ammunition items are often kept in sealed containers until the time they are used for practice or combat. With the one-coat ammunition paint, coverage with the phosphate coating is even more important since no primer is used to provide additional resistance to corrosion.

The levels of corrosion protection achieved with the different types of painted coatings are shown in Table 4. A dry film lubricant coating is applied like paint. The only difference is that the dry film coating has a pigment in it that provides lubrication rather than color. Numerous areas on the M16 rifle have a dry film lubricant providing different functions. On the exterior of the hardcoated aluminum lower receiver extension tube, the dry lubricant provides resistance to crevice corrosion. On the interior of the aluminum tube, the dry film coating prevents galvanic corrosion by the precipitation hardened 17-7 (UNS S17700) stainless steel spring. The dry film coating also lubricates the spring as it cycles in the tube. On the interior of the hardcoated aluminum upper receiver, the dry film coating lubricates the forward and back motion of the bolt. It is also important to note that the bake-on dry film

Table 3 Test requirements and time to failure of supplemental oils
Steel substrate

Oil coating	Thickness		Test minimum(a), h	Time to failure, h	Ref
	μm	mils			
MIL-PRF-32033	51	2	...	3–8	5
MIL-PRF-3150	203	8	48	50–100	7
MIL-PRF-63460	127	5	100	125–200	6
MIL-PRF-16173 grade 3	127	5	168	200–500	8
MIL-PRF-16173 grade 1	254	10	336	350–800	8
Proprietary oil	381	15	...	2000–3000	...

(a) Test per ASTM B 117 (Ref 1)

Table 4 Test requirements and time to failure of various coatings
Steel substrate except where indicated

Coating	Thickness		Test minimum(a), h	Time to failure, h
	μm	mils		
CARC primer	20–30	0.8–1.2	...	48–96
Light zinc phosphate + CARC primer	20–30	0.8–1.2	336	350–1500
Light zinc phosphate + enamel paint	20–30	0.8–1.2	150	200–500
Light zinc phosphate + lacquer paint	23–28	0.9–1.1	120	150–250
Dry film lubrication, air cured	8–13	0.3–0.5	...	50–100
Heavy zinc phosphate + dry film lubrication, baked	12–25	0.5–1.0	100	500–1000
CARC topcoat	46–56	1.8–2.2	...	25–50
Light zinc phosphate + CARC primer + topcoat	76–127	3–5	336	2000–4000
Chromate + CARC primer(b)	15–20	0.6–0.8	2000	...

CARC, chemical agent resistant coating. (a) Test per ASTM B 117 (Ref 1). (b) Aluminum substrate

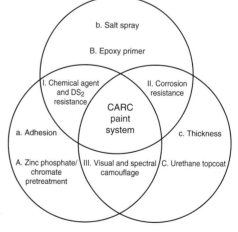

a. b. c. Test requirements
I. II. III. Tactical requirements
A. B. C. Material requirements

Fig. 4 A logical schematic of the relationship of materials (pretreatment, primer, and topcoat) used for chemical agent resistant coating (CARC) paint systems with the tactical and test requirements. DS_2 is decontamination solution.

lubricant is chemical agent and decontamination solution resistant, while the air-cure material is not resistant to chemical agents. It is apparent from the last entry in Table 4 that, when the pretreatment provides significant corrosion resistance by itself, the overall corrosion resistance of pretreatment and paint system shows significant improvement.

Other Finishes

The final category of finishes used on military equipment include deposits applied in a vacuum, mechanical plating, thermal spray, and hot-dip coatings. See the articles "Continuous Hot Dip Coatings," "Batch Process Hot Dip Galvanizing," "Thermal Spray Coatings," and "CVD and PVD Coatings" in Volume 13A.

Cadmium and ion vapor deposited (IVD) aluminum are two production coatings applied under vacuum conditions. Zinc, tin, aluminum, and lead-tin coatings are typically applied by the hot-dip (molten metal) process. Mechanically deposited coatings include cadmium, tin-cadmium, tin, and zinc. Thermal spray coatings include aluminum and zinc. The levels of corrosion protection achieved in this final category are shown in Table 5. One of the interesting aspects of thin aluminum coatings is its ability to protect steel in the salt-spray test but not in normal outdoor environments. In the salt-spray test (a very aggressive electrolyte), the aluminum remains an anode and protects the steel, as would be expected from looking at a galvanic series chart. However, a normal outdoor environment is not as aggressive and an oxide coating forms on the aluminum surface. This oxide coating changes the aluminum from an anode to a cathode with respect to steel, and rusting begins at defects or voids in the aluminum coating. Thus, the accelerated test information can be misleading when the test does not represent what is happening in the outdoor environment. Thermal spray coatings of aluminum 127 to 254 μm (5 to 10 mils) thickness are essentially barrier coatings and not sacrificial coatings. These coatings have been used on the decks of ships and have performed very well. Vacuum-deposited coatings are used to minimize the problems associated with hydrogen embrittlement that may be encountered during electroplating operation. Hydrogen-induced stress-corrosion cracking is a problem with electroplated coatings used on high-strength fasteners. Hydrogen generated during acid pickling and electroplating can enter the steel and cause embrittlement. If the hydrogen is not removed, it can cause premature failure in the field. Vacuum-deposited coatings do not generate hydrogen in the plating process and, as such, are the preferred process for critical coating applications on high-strength steel parts.

In summary, coatings in the five areas represent a selection of processes available for corrosion protection in military environments. Process tests ensure that the minimum requirements are met before the parts leave the manufacturer's plant. The coatings can provide the needed corrosion protection only when the parts are properly designed, the coating system is selected to meet the specified service requirements, and the processing has been controlled.

REFERENCES

1. "Standard Practice for Operating Salt Spray (Fog) Apparatus," B 117, ASTM International
2. "Standard Specification for Electrodeposited Coatings of Zinc on Iron and Steel," B 633, ASTM International
3. T.S. Doppke, Fasteners and Finishes, Part VI, *Prod. Finish.*, Sept 1998
4. *Coating, Oxide, Black, for Ferrous Metals*, MIL-DTL-13924, Document Automation and Production Services
5. *Lubricating Oil, General Purpose, Preservative (Water-Displacing, Low Temperature)*, MIL-PRF-32033, Document Automation and Production Services
6. *Lubricating Oil, Preservative, Medium*, MIL-PRF-3150, Document Automation and Production Services
7. *Lubricant, Cleaner and Preservative for Weapons and Weapons System*, MIL-PRF-63460, Document Automation and Production Services
8. *Corrosion Preventive Compound, Solvent Cutback, Cold Application*, MIL-PRF-16173, Document Automation and Production Services

SELECTED REFERENCES

- R. Baboian, *Automotive Corrosion by Deicing Salts*, National Association of Corrosion Engineers, 1981
- R. Baboian, *Corrosion Tests and Standards: Application and Interpretation*, American Society for Testing and Materials, 1995
- C.M. Cotell, J.A. Sprague, and F.A. Smidt, Jr., *Surface Engineering*, Vol 5, *ASM Handbook*, ASM International, 1994
- *Finishing '85*, Conf. Proc., Sept 16–19, 1985, Association for Finishing Processes of SME, 5-1 to 5-12
- G.A. Greathouse and C.J. Wessel, *Deterioration of Metals*, Reinhold Publishing, 1954
- H. Leidheiser, Jr., *Corrosion Control by Organic Coatings*, National Association of Corrosion Engineers, 1981
- F.A. Lowenheim, *Electroplating*, McGraw-Hill, 1978
- R. Winston Revie, Ed., *Uhlig's Corrosion Handbook*, 2nd ed., John Wiley & Sons, 2000

Table 5 Test requirements and time to failure of other finishes on steel

Coating	Thickness		Test minimum(a), h	Time to failure, h
	μm	mils		
Vacuum cadmium + chromate	5–13	0.2–0.5	96 (white)	2000–6000
Vacuum aluminum (class 1)	25	1.0	336 (red)	...
Vacuum aluminum (class 2)	13	0.5	192 (red)	...
Hot-dipped Pb-5Sn	8	0.3	48 (white)	...
Hot-dipped Pb-7Sn	8	0.3	96 (white)	...
Hot-dipped Pb-30Sn	8	0.3	72 (white)	...
Hot-dipped zinc	38	1.5	...	350–1000
Hot-dipped tin	18	0.7	24 (white)	...
IVD aluminum (bare)	8	0.3	168 (red)	...
	13	0.5	336 (red)	...
	25	1.0	504 (red)	...
IVD aluminum + chromate	8	0.3	336 (red)	...
	13	0.5	504 (red)	...
	25	1.0	672 (red)	...
Mechanical cadmium + chromate	8	0.3	96 (white)	500–1500
Mechanical zinc + chromate	13	0.5	96 (white)	200–500
Mechanical Sn-Cd + chromate	8	0.3	168 (white)	200–600
Thermal spray aluminum	127–254	5–10	...	...
Thermal spray zinc	127–254	5–10	...	...

(a) Test per ASTM B 117 (Ref 1). White is oxidation of protective coating; red is oxidation of base steel

U.S. Navy Aircraft Corrosion

John E. Benfer, Naval Air Systems Command, Research and Engineering Sciences Department

U.S. NAVY AIRCRAFT experience severe conditions in operational service that include: aircraft carrier based deployment and carrier landing (Fig. 1), maritime shore based deployment, and near sea level maritime missions. The mechanical loads experienced during flight have led to the development and selection of aerospace alloy systems with a primary emphasis on high strength, stiffness, and low specific gravity. Trade-offs associated with corrosion-resistance properties occurred during the initial concept and design phases of these aircraft during acquisition processes of the 1960s through early 1980s. As a result, currently fielded aircraft weapon systems experience high life-cycle costs due to increased maintenance and decreased component/system reliability from degradation associated with corrosion.

Aircraft structures begin to deteriorate from the time they leave the manufacturing facility, primarily due to mechanical and corrosion damage. Abrasion, wear, and fatigue mechanisms contribute to the mechanical damage phenomena and are affected by aircraft design and service loads. Corrosion damage in the form of pitting, exfoliation, and uniform metal loss phenomena via electrochemical oxidation is controlled by the aircraft design, the materials selected, and the operating environment. Interactions between mechanical and corrosion mechanisms can also occur resulting in corrosion fatigue, stress corrosion cracking, and fretting phenomena.

As an aircraft ages, the effects of corrosion become more severe due to cumulative damage. If corrosion and corrosion-damaged areas are not detected early and mitigated, a serious hazard to the structural integrity of the aircraft may result. Therefore, it is critical to identify and employ the latest technologies associated with cleaning, washing, inspection, treatment, and corrosion prevention into the maintenance cycles of all navy platforms. Failure to address corrosion control throughout the life cycle of the aircraft increases maintenance costs and reduces aircraft life, thus affecting the military's ability to achieve their operational commitment.

Environment

The operational environments of U.S. Navy aircraft vary widely, and therefore corrosion rates can also vary considerably. Geographic location affects temperature, humidity, and air quality of ground based squadrons. The severe conditions associated with aircraft carrier based squadrons (Fig. 2) create the most challenging corrosion-control issues within aviation. While flight conditions and landing and takeoff events are in themselves very severe operational environments, most navy aircraft spend a considerable amount of time on the ground. Therefore, the ground environment is an important consideration to overall corrosion-prevention strategies. It has been found through corrosion monitoring of military sites and in-flight environments that the above-deck aircraft carrier environment is three to four times more severe than any land based environment.

Atmospheric environments can be classified in three basic corrosion categories to represent a generic rating for the degree of corrosiveness. Climate, industry, and location form a basis of this generic rating to include the effects of temperature, humidity, pollution/impurities, pH, and salts. These categories can be reduced to many subcategories; however, most common descriptions, organized based on corrosion rate, are shown in Table 1.

Temperature directly affects the kinetics of corrosion rates. Higher temperatures also cause corrosive solutions to dry, which results in a higher concentration of impurities on aircraft surfaces. Humidity levels above 60% lead to formation of thin aqueous electrolyte films capable of creating local corrosion-reaction cells with increased corrosion rates. Cyclical occurrence of condensate contaminated with dissolved salts and atmospheric impurities followed by dry-off in high-temperature areas concentrates impurities and creates one of the most common mechanisms associated with general corrosion within the naval aviation maintenance community.

Aircraft Alloys

Corrosion-resistance properties at times are sacrificed to achieve the previously mentioned mechanical design requirements, formability, and cost, during development of aircraft weapon systems. The most widely used materials in airframe structures and components are aluminum, steel, titanium, and magnesium alloy systems.

Aluminum Alloys. With the exception of magnesium, aluminum by its position in the electrochemical series is thermodynamically one of the most reactive metals used in naval aircraft. In service, it is often resistant to corrosion because it is protected by oxide films that rapidly

Fig. 1 Navy aircraft making carrier landing

Fig. 2 U.S. Navy aircraft carrier deck illustrating operating environment of carrier based squadrons

Table 1 General corrosion rates associated with geographic location

	Type of atmosphere		
Rate of corrosion	Climate	Amount of industry	Location
High	Tropical	Industrial	Marine
Moderate	Temperate	Suburban	Inshore
Low	Arctic	Rural	Inland

form in atmospheric environments. This passive film, however, can result in localized pitting corrosion at defect areas within the oxides. Aerated halide solutions (e.g., chloride from salts) commonly result in pitting corrosion damage of aluminum alloys. This environment commonly occurs with maritime aircraft programs and aircraft deployed to aircraft carriers.

The corrosion resistance of aluminum is strongly influenced by its chemical composition as a consequence of impurities and alloying elements. Principal alloying elements include zinc, magnesium, silicon, copper, and manganese. Wrought heat treatable aluminum alloys are the most widely used in airframe construction with many different compositions and heat treatments to achieve final properties. The most important of these alloy systems are the 2xxx series (Al-Cu-Mg) and the 7xxx series (Al-Zn-Mg). These alloys are strengthened by precipitation-hardening treatments that involve heat treatment, rapid quenching, and natural aging at room temperature or artificial aging at elevated temperature.

2xxx series alloys have strong electrochemical effects due to the presence of copper. Variations of copper concentrations (heterogeneous segregation) in solid solution lead to variations in electrode potentials and thus local galvanic cells. Copper also has the tendency during electrochemical corrosion reactions to replate on aluminum surfaces and form small cathodic sites causing additional local galvanic cells. These alloys are susceptible to pitting and stress corrosion.

7xxx series alloys contain zinc and magnesium and include a great number of alloys that also contain copper (e.g., Al 7075). These alloys are strengthened by precipitation of solute-rich zones and are among the highest-strength materials available on the basis of strength-to-weight ratios. 7xxx series alloys are also more resistant to general corrosion than 2xxx series alloys. However, they are susceptible to stress-corrosion cracking and exfoliation corrosion. 7xxx series alloys with higher copper contents allow for higher aging temperatures without excessive loss of strength. The T7 heat treatment have been developed to improve resistance to exfoliation and stress-corrosion cracking.

Aluminum cladding is used on sheet products, such as fuselage skins, fuel tank divider webs, and other components. Cladding alloys are selected to be anodic to the core alloy and provide cathodic protection to the core. Cladding materials are metallurgically bonded via rolling processes to the core material. Corrosion progressing through the cladding will progress to the core interface and spread laterally and prevent perforation of the core material. Figure 3 illustrates the protective nature of aluminum cladding and the associated resistance to pitting corrosion of the core material.

Steel Alloys. Carbon and low-alloy steels used in airframe structures and components vary in susceptibility to corrosion and the modes of corrosive attack due to variations in composition and heat treatment. Carbon steels used in low-strength applications are prone to uniform corrosion in atmospheric conditions. Low-alloy steels (Ni-Cr-Mo) used in high-strength applications can be less susceptible to atmospheric corrosion as compared with carbon steel alloys. Low-alloy steels, however, are more prone to pitting, stress-corrosion cracking, and hydrogen embrittlement. Carbon and low-alloy steels require the application of inorganic coatings (e.g., ion vapor deposited aluminum or electroplated cadmium) and/or organic coatings (e.g., epoxy, polysulfide, or polyurethane) to resist corrosion in the environments associated with naval aviation.

Stainless steel alloys form a protective oxide film that provides excellent resistance to uniform corrosion and include austenitic, martensitic, and age-hardening groupings. The passive film formation on stainless steels leads these alloys to be very susceptible to localized pitting and crevice corrosion. Segregation of alloying elements due to improper heat treatment or thermal exposure can promote intergranular corrosion. Stress-corrosion cracking can occur in chloride environments, especially when aircraft are carrier based and exposed to cyclic wetting and drying of chloride-containing seawater.

Austenitic stainless steel (300 series) provides the greatest resistance to general corrosion due to the presence of chromium and nickel alloying elements. These alloys are not strengthened by heat treating and are used predominantly in low-strength applications. Martensitic stainless steel (400 series) are less resistant to general corrosion than the austenitic stainless steels due to compositional changes that provide increased strength through quench-and-temper heat treatments. Age-hardening stainless steels are hardenable through precipitation heat treatments and provide high strength combined with atmospheric and general corrosion resistance approaching the austenitic stainless steel alloys.

Titanium and titanium alloys provide low density, high strength-to-weight ratios, and high corrosion resistance. The excellent corrosion resistance of titanium is due to the formation of a protective titanium oxide film during atmospheric exposure. Titanium is cathodic to most alloys used in aviation with the exception of some passive stainless steel alloys. As a result, protective measures should be in place at areas in which titanium is coupled to dissimilar metal airframe structures.

Titanium alloys have excellent resistance to pitting, galvanic corrosion, crevice corrosion, erosion-corrosion, and corrosion fatigue in marine environments. However, many alloys are susceptible to stress-corrosion cracking in aqueous chloride environments. Liquids such as methanol, ethanol, and ethylene glycol have also been found to cause cracking. Titanium alloys are also highly susceptible to the reduction of fatigue life by fretting at interfaces between other metals.

Magnesium is the most corrosion-prone and reactive material used in the manufacturing of aircraft structural components. Unalloyed magnesium is not extensively used for structural purposes and, therefore, is alloyed with additions of aluminum, lithium, zinc, rhenium, thorium, and silver. Minor additions of cerium, manganese, and zirconium may also be used. These elements are selected to enhance response to strengthening treatments and improve corrosion resistance. Elements such as iron, nickel, cobalt, and copper have severely deleterious effects on the corrosion resistance of magnesium. They are typically controlled at low impurity tolerance limits to reduce this effect.

Aluminum and zinc are the most widely used alloying elements in magnesium because they have high solubility and give rise to some of the highest strengths. Magnesium alloy AZ91 (Mg-9Al-1Zn) is a commonly used cast alloy that exhibits good weldability and low porosity. All magnesium alloys are subject to general and pitting corrosion when exposed to natural environments and must be surface treated and protected. Breakdowns in corrosion-protection systems result in rapid corrosion of the magnesium alloys, especially where galvanic couples are present, that is, steel inserts, fasteners, and mounting areas. This is especially prevalent within maritime environments associated with naval aviation.

Aircraft Inspection

Aircraft must be inspected for signs of corrosion and to ascertain the condition of protective coatings used for corrosion prevention. Prior to flight, inspections involve visual checks of airframe surfaces and operational checks of engines and flight controls. Detailed maintenance inspections are usually performed at predetermined intervals based on flight hours or time. These intervals are set using prior history of corrosion-prone areas with highly susceptible

Fig. 3 Aluminum laminate. Al 7072 aluminum cladding providing corrosion protection of Al 7075-T6 core. Note lateral spread of corrosion at clad layer to prevent through wall failure of a P-3 fuel tank divider web. Courtesy of K. Himmelheber, Naval Aviation Depot—Jacksonville

and critical areas being inspected more frequently than areas of low susceptibility and criticality.

Visual inspections are carried out on external surfaces and internal structures visible through quick access doors and panels by direct line of sight or through the use of inspection mirrors. Aided visual inspections are carried out less frequently and incorporate more thorough techniques to include magnifiers and fiber-optic probes. Special inspections are carried out on known problem areas using additional nondestructive techniques to include x-ray, magnetic particle, eddy current, ultrasonic, and liquid penetrant inspections. These techniques provide more detailed inspection results and at times require either partial or complete disassembly of components to be evaluated.

Nondestructive inspections are designed and timed to detect all forms of structural deterioration at early stages to prevent in-service component failure. The first appearance of corrosion on unpainted surfaces may be in the form of small deposits or spots. On painted surfaces this appears as small blisters within the coating that vary in shape from round and circular to thin and filamentlike. Common areas include skin areas around fasteners, seams, lap joints, butt joints, and crevices that collect debris and have long times of wetness. Areas exposed to battery electrolytes, engine exhaust, and gunfire gases are also prone to corrosion. The form and location of corrosion on aircraft materials have characteristics as indicated in Tables 2 and 3. Examples of aircraft corrosion damage are shown in subsequent sections of this article.

Prevention and Corrosion Control

The primary consideration in the design and construction of naval aircraft is the balancing of safety, affordability, and environmental needs with operational and mission requirements. Aircraft are also required to perform reliably, have minimum maintenance requirements, and acceptable degradation rates to achieve maximum service life. Deterioration due to all anticipated forms of corrosion (uniform, pitting, galvanic, crevice, exfoliation, stress-corrosion cracking, corrosion fatigue, fretting, hydrogen embrittlement, weathering, and fungus growth) must be managed through the selection of suitable materials and processing methods and maintenance.

Materials selection and processing method choices to obtain the best value with regard to the control and mitigation of corrosion requires attention at the component, assembly, and system level. The design should also control the exposure to corrosive contaminants through the use of appropriate manufacturing processes and design geometries that prevent the collection and entrapment of contaminants. The proper selection of materials and use of design in conjunction with the application of protective coatings is the optimum approach to maximize corrosion performance. More can be done during design to improve corrosion performance than at any other time of product life cycle.

Cleaning. Aircraft cleaning is the first step in preventing corrosion. Cleaning requires a knowledge of the materials and methods needed to remove corrosive contaminants and fluids that tend to retain contaminants. Aircraft are cleaned on a regular basis (typically every 14 days while ashore and every 7 days while at sea) in order to:

- Prevent corrosion by removing salt deposits, other corrosive soils, and electrolytes
- Allow a thorough inspection to identify corrosion and corrosion damage

More frequent and thorough cleaning is necessary for a particular aircraft model that experiences:

- Excessive exhaust or gun blast soil that accumulates at impingement areas
- Damaged paint systems including peeling, cracking, flaking, or softening
- Leakage of oil, coolant, and hydraulic fluids
- Exposure to salt spray, saltwater, and other corrosive materials

Cleaning compounds and detergents used for washing aircraft work by dissolving soluble soils, emulsifying oily soils, and suspending solid soils. Moderately alkaline cleaners of pH < 10 are used and procured under MIL-PRF-85570 (Ref 1). This military specification contains a variety of detergent types that include:

- Type I—General-purpose aromatic solvent base
- Type II—General-purpose, nonsolvent base
- Type III—Abrasive spot cleaner
- Type IV—Rubberized spot cleaner
- Type V—Gel-type wheel well degreaser, low solvent base

Postcleaning procedures are required after aircraft washing to ensure that entrapped water is removed/drained and that lubricants are reapplied to minimize wash-induced corrosion effects. Postcleaning procedures are:

1. Remove covers from vents, tubes, air ducts, and so forth.
2. Remove tape from openings sealed prior to washing.
3. Clean all drain holes to allow proper drainage.
4. Inspect all water accumulation areas and drain/dry as required.
5. Lubricate in accordance with maintenance manuals to displace remaining moisture/water and allow proper operation of applicable system.

Lubrication has multiple purposes: preventing wear between moving parts, filling air spaces, displacing water, and providing a barrier against corrosive media. Conventional lubricants consist of a variety of grease-type materials that can be applied by lever or pressure-type grease guns, aerosol spray, squirt can, or applied by hand or brush. Solid-film lubricants are used where conventional lubricants are difficult to apply, or become contaminated with wear products or moisture. Typical applications are for flap tracks, hinges, turnbuckles, and cargo latches.

Corrosion-preventive compounds (CPCs) are mixtures of special additives in petroleum, solvent, lanolin, or wax bases and are used to temporarily protect metallic aircraft parts and components. They function by preventing corrosive materials from contacting and corroding bare metal surfaces. Many of these compounds also displace water, seawater, and other

Table 2 Appearance of corrosion product

Alloy system	Type and susceptibility of corrosion	Appearance of corrosion product
Aluminum	Pitting, intergranular, and exfoliation	White or gray powder
Titanium	Highly corrosion resistant. Extended contact with chlorinated solvents may result in degradation of structural properties.	No visible corrosion products
Magnesium	Highly susceptible to pitting	White powder, snowlike mounds, and white spots on surface
Low-alloy steels	Surface oxidation, pitting, and stress-corrosion cracking	Reddish-brown oxide (rust)
Corrosion-resistant steel (CRES)	Highly resistant to general corrosion. Susceptible to pitting, intergranular corrosion, and stress-corrosion cracking	Corrosion evident by rough surface with possible red, brown, or black stains
Nickel	Good corrosion resistance. Sometimes susceptible to pitting	Green powdery deposit
Copper	Surface and intergranular corrosion	Blue or blue-green powder deposit
Cadmium (sacrificial electroplated coating)	Good corrosion resistance. Will protect steel from attack	White-gray powdery product
Chromium (wear-resistant electroplated coatings)	Susceptible to pitting in chloride environments	No corrosion product. Steel corrosion may be visible at coating defects

Table 3 Typical locations of corrosion on aircraft

Location	Type of corrosion	Comments	Detection method(a)	Maintenance repairs(b)
Exterior fuselage				
Skin	Surface, pitting, and filiform	Paint failure around rivets and fasteners	V	RBR. Install patches if beyond negligible damage limit.
Doors	Surface, intergranular, and exfoliation	Paint failure induced by door use	V	RBR. Install patches if beyond negligible damage limit.
Access panels and attach area	Surface, intergranular, and exfoliation	Paint failure induced by maintenance, moisture intrusion	V	RBR. BAS
Nose section				
Nose radome latches	Galvanic (steel fasteners in aluminum structures) and intergranular	Paint failure induced by usage	V	RBR. BAS. Item replaced if damage is extensive
Nose wheel well	Galvanic (steel fasteners in aluminum structures) and intergranular	Paint failure from door actuation, moisture intrusion	V	RBR. Item replaced if damage is extensive
Windshield frames	Surface, intergranular, and exfoliation	Moisture intrusion from sealant failure	V	RBR. Install patches if beyond negligible damage limit. Item replaced if damage is extensive
Battery and vent area	Chemical attack	Battery failure	V	BAS.
Bulkheads	Surface and intergranular	Paint damaged by maintenance	V	RBR. Install patches if beyond negligible damage limit.
Heat-exchanger door	Intergranular and exfoliation	Moisture intrusion	V	RBR. BAS. Item replaced if damage is extensive
Control cables	Surface	...	V	Item replaced if damage is extensive
Forebody				
Longerons, frames, and stringers	Surface and intergranular	Moisture trapped in insulation	V	RBR. BAS. Item replaced if damage is extensive
Floor structures, lavatory floor, and substructure	Surface, chemical attack, and exfoliation	Paint failure from usage and spills	V	RBR. Install patches if beyond negligible damage limit. BAS. Item replaced if damage is extensive
Bomb bay	Galvanic (steel fasteners in aluminum structures) and surface	Paint failure by door operation and stores loading	V	RBR. BAS. Item replaced if damage is extensive
Midbody				
Floor structures	Surface, chemical attack, and exfoliation	Paint failure from usage and spills	V	RBR. Install patches if beyond negligible damage limit. BAS. Item replaced if damage is extensive
Fuselage to wing fitting	Stress-corrosion cracking	Paint failure due to wing flex and moisture entrapment	...	RBR. Item replaced if damage is extensive. Extensive repair
Aftbody				
Entry ladder and towage area	Surface and intergranular	Paint failure from usage	V	RBR. Item replaced if damage is extensive
Floor structures	Surface, chemical attack, and exfoliation	Paint failure from usage and spills	V	RBR. Install patches if beyond negligible damage limit. BAS. Item replaced if damage is extensive
Empennage—vertical and horizontal stabilizer				
Planks, spar caps, and webs	Intergranular and stress-corrosion cracking	Paint and sealant failure from airframe flex and vibration in flight	V, U	BAS. Item replacement if damage is extensive. Requires access holes cut and panel to be manufactured
Control surfaces and trim tabs	Surface and exfoliation between inner and outer skins	Moisture entrapment, paint and sealant failure from airframe flex and vibration in flight	V, U	BAS. Item replacement if damage is extensive. Cut out extensive damaged structure and insert a replacement section.
Latches	Galvanic (steel fasteners in aluminum structures) and intergranular	Paint failure induced by usage	V	RBR. BAS. Item replaced if damage is extensive
Rudder upper torque tube attach fitting and lower rib to beam fitting	Surface, exfoliation, and dissimilar metal galvanic (steel fasteners in aluminum structures)	Moisture entrapment, paint failure from flex, and vibration in flight	V, U	RBR. Item replaced if damage is extensive
Wing assembly				
Forward spar caps and web assembly	Intergranular, exfoliation, and stress-corrosion cracking	Inspection difficult, requires removal of leading edges	V or fuel leaks	RBR. BAS. Extensive repair. Cut out extensively damaged structure and insert a replacement section.
Aft spar caps and web assemblies	Intergranular, exfoliation, and stress-corrosion cracking	...	V or fuel leaks, confirmed U	RBR. BAS. Extensive repair. Cut out extensively damaged structure and insert a replacement section.
Upper and lower planks	Intergranular, exfoliation, and stress-corrosion cracking	...	V or fuel leaks, confirmed U	RBR. BAS. Extensive repair. Cut out extensively damaged structure and insert a replacement section.
Fillet panel and dome nut holes	Surface intergranular and galvanic (steel fasteners/nutplate in aluminum structures)	Paint failure induced by airframe flex and moisture intrusion		RBR. BAS.
Access door lands	Surface, intergranular, and stress-corrosion cracking	Paint failure induced by maintenance	V	RBR. BAS.
Mooring fitting attach and stores fitting holes	Surface, exfoliation, and galvanic (steel fasteners in aluminum structures)	Mooring fitting installed for extended periods of time, moisture entrapment	V	RBR. BAS. Cut out extensively damaged structure and insert a replacement section.
Flapwell panels	Surface and exfoliation between inner and outer skins	Moisture entrapment, paint and sealant failure from airframe flex and vibration in flight	V	BAS. Item replacement if damage is extensive. Cut out extensively damaged structure and insert a replacement section.

(continued)

(a) V, visual inspection; U ultrasonic nondestructive inspection. (b) RBR, remove corrosion product by blending and repaint; BAS, install backup angles and plates to restore strength

Table 3 (continued)

Location	Type of corrosion	Comments	Detection method(a)	Maintenance repairs(b)
Wing assembly (continued)				
Trailing edge	Surface and exfoliation between inner and outer skins	Moisture entrapment, paint and sealant failure from airframe flex and vibration in flight	V	BAS. Item replacement if damage is extensive. Cut out extensively damaged structure and insert a replacement section.
Nacelles				
Main wheel wells	Galvanic (steel fasteners in aluminum structures), intergranular	Paint failure from door actuation	V	RBR. Item replaced if damage is extensive
Nacelle attach angles	Intergranular	Paint failure from vibration	V	RBR. Item replaced if damage is extensive. Cut out extensively damaged structure and insert a replacement section.
Gearbox supports	Surface and intergranular	Moisture intrusion	V	RBR. Item replaced if damage is extensive
Lower longerons	Galvanic (aluminum extrusion with stainless steel webs), intergranular	Moisture intrusion	V	RBR. Item replaced if damage is extensive. Cut out extensively damaged structure and insert a replacement section.

(a) V, visual inspection; U ultrasonic nondestructive inspection. (b) RBR, remove corrosion product by blending and repaint; BAS, install backup angles and plates to restore strength

contaminants from the surfaces to be protected. Some CPCs also provide lubrication in addition to corrosion protection. CPCs range in appearance from thick black barrier coatings (MIL-PRF-16173, grade 1) (Ref 2) to light lubricating oils (MIL-PRF-32033) (Ref 3). CPCs can be categorized as:

- Water displacing
- Non-water-displacing
- Lubricating
- Hard film
- Soft film
- Thin film

Selection of these materials is based on specific functional requirements of the aircraft component. Thick heavy coatings of CPC materials provide the best corrosion protection and are longer lasting, but are difficult to remove. Hard films provide excellent barrier protection, but are prone to chipping or cracking under stress and vibration. Thin films provide lubricating properties, but require more frequent applications. Under certain circumstances, CPCs can be applied as a system with a thin water-displacing material applied as an initial layer and then followed with a thicker hard film for improved durability.

Corrosion removal involves cleaning all visible corrosion products from a damaged area. The type of process will vary based on the type of surface involved, the size of the area, and the degree of corrosion found. Corrosion should always be removed using the mildest technique available to ensure that additional damage is not incurred during treatment. Tools for the removal of corrosion include nonpowered abrasive mats, abrasive cloth, abrasive paper, metallic wool, wire brushes, pumice, and scrapers. Powered methods include pneumatic sanders, flap brushes, flap wheels, rotary files, and abrasive blasting. Other factors to consider are:

- Before attempting corrosion removal, surfaces must be cleaned and stripped of paint.
- Adjacent components and parts must be protected from damage that could be induced during the corrosion-removal process.
- Allowable repair limits must be identified. If the corrosion-removal process cannot be performed without exceeding repair limits, the component will require replacement.

Material compatibility must be considered for any corrosion-removal process to ensure that additional corrosion damage does not occur due to galvanic effects associated with cross-contamination. For mechanical removal, like materials should be used whenever possible. Aluminum wool (not steel wool) should be used to remove general corrosion products from aluminum surfaces, while brass wire brushes should be used for copper alloys. Abrasives for sanding and blending are typically manufactured from aluminum oxide or silicon carbide. Blast media are supplied as silicon dioxide (glass) beads or aluminum oxide. The more aggressive aluminum oxide blast media are typically used on ferrous alloys.

Critical structures require additional treatment following corrosion removal. Pitting damage on critical structures requires complete blending to prevent stress risers that may initiate fatigue or stress-corrosion cracks. Blending is typically performed mechanically by "dishing" out the process areas to a diameter 20 times the depth of corrosion pitting. Blending of pitting damage is not routinely performed on noncritical structures.

Surface Treatment. Chemical surface treatments are performed to modify or convert the surface of aluminum and magnesium aerospace alloy systems and produce a surface layer with greater corrosion resistance and improved adhesion properties, through the formation of protective oxides or chromate films. These treatments typically involve the use of chromic acid or chromium salts for magnesium and aluminum alloys. However, nonchromate-containing products are under development to ensure compliance with future environmental and safety regulatory requirements. When exposed to water/moisture, corrosion inhibitors can leach out, providing a self-healing property when the base alloy is exposed by scratching or abrasion. In addition to improved corrosion performance, chemical surface treatments provide improved surface adhesion of subsequently applied organic coating systems through a change in surface morphology and chemistry at the microscopic level that increases the surface area and number of bonding sites for subsequent coatings.

Thicker, harder, and more corrosion-resistant films can also be produced on aluminum and magnesium alloys through a process of electrolytic oxidation. These anodic (anodized) coatings are formed by oxidation in an electrolyte forming a thin, continuous oxide barrier film. Thickness is controlled and regulated by oxidation time and processing conditions. Anodic coatings contain microscopic pores that are sealed to enhance corrosion resistance, typically with a chromate solution. Anodic coatings are occasionally used to eliminate galling at faying surfaces, to retain lubricating fluids or greases, and to provide a surface for bonding adhesives and organic coating systems.

Anodic coatings can lead to the lowering of fatigue strength properties for metals. Such an effect can be reduced through peening processes that produce compressive stress at the metal surface or through the reduction of oxide thickness (thin-film anodizing). For certain fatigue-sensitive materials or components, chemical conversion coatings are generally used as an alternative to anodic coatings.

The performance requirements for control of conversion coatings and anodizing of magnesium and aluminum are contained within the specifications listed in Table 4.

Paint Systems. High-performance epoxy primers are used on navy aircraft due to their exceptional adhesion and chemical resistance in conjunction with their corrosion-resistance properties. The epoxide and hydroxyl groups of the epoxy resin react with polyfunctional amines or polyamides to form a highly corrosion-resistant thermoset polymer. These materials can be either solvent based or waterborne and utilize chromated corrosion-inhibiting pigments as their primary source of active corrosion inhibition. The use and disposal of carcinogenic chromate-containing materials is becoming severely restricted, and efforts to develop primers with nonchromate inhibitor compounds are being performed. The performance requirements of these aerospace primers are contained within the military specifications listed in Table 5.

High-performance polyurethane paints are used to topcoat external aircraft surfaces following epoxy priming. These topcoats enhance corrosion protection and provide desired optical properties. The fundamental urethane reaction is that of an isocyanate with an alcohol. The combination of different isocyanates with the large number of available polyols makes possible the formation of a wide variety of polymers with varying physical properties. Aliphatic polyurethane coatings are mainly used for naval aviation due to improved weatherability performance as compared to aromatic polyurethanes. The use of these topcoats is indicated in Table 5.

New topcoat materials have been developed that virtually eliminate volatile organic compounds (VOCs) in aircraft paints based on the class W waterborne formulation. These new materials have shown equivalent characteristics to the standard class H high solids formulation in laboratory and field testing. Advantages of these products are no/low VOCs, enhanced cleanability, and improved application characteristics.

Triglycidylisocyanurate (TGIC) polyester powder coating has recently been approved for use as an alternate material to MIL-PRF-85285 polyurethane topcoat. Powder coating provides the additional benefits of improved corrosion protection and toughness with ease of application, decreased processing times, and decreased costs associated with environmental, safety, and hazardous waste compliance. Powder coating materials require heat treatment for cure and, therefore, are not recommended for use with alloy systems in which metallurgical properties are negatively affected by powder coat bake temperatures of 190 to 200 °C (375 to 390 °F). Low-temperature cured powders of less than 120 °C (250 °F), however, are being investigated to expand usage across these alloy systems.

Sealants. The primary function of a sealant is to exclude moisture, contaminants, and fluids to protect the aircraft or substrate from an aggressive environment. The functional sealing requirements of aircraft are:

- Fuel tank sealing
- Environmental sealing
- Pressure sealing
- Sealing for severe corrosive environment
- Aerodynamic smoothing
- High-temperature and firewall sealing
- Sealing of windshields and canopies
- Sealing of honeycomb sandwich structure
- Electrical sealing and insulation

No single sealant will meet the major requirements of all applications. The selection of the most appropriate material should consider:

- Temperature and pressure
- Resistance to fluids
- Mechanical properties
- Adhesion
- Reparability

Mechanical exclusion of the environment is achieved by use of heavy films of low permeable material having excellent adhesion. This process, however, can have disadvantages with equipment and components that are difficult to properly clean in preparation for sealing. This can result in poor adhesion and the entrapment of residual salts within crevices that provide a permanent source of corrosion. The use of chemical inhibitors incorporated into a sealant can provide added protection. Moisture that is naturally permeable through the sealant material can dissolve some of these inhibitors to provide additional passivation of the metal surface.

Sealing compounds used within naval aviation include polysulfides, polyurethanes, polythioethers, and silicones. Polysulfides, polyurethanes, and polythioethers are two-component systems consisting of a base (containing the prepolymer) and the accelerator (containing the curing agent). When thoroughly mixed, the catalyst cures the prepolymer to a rubbery solid. Rate of cure and sealant properties depend on temperature and humidity along with the types of catalyst and prepolymer selected. Silicone sealing compounds are one-component systems that cure by reacting with moisture in the air. Some silicone sealing compounds produce acetic acid as a condensation reaction by-product during cure, which can lead to severe corrosion problems. These sealants should be avoided.

A new commercially available off-the-shelf (COTS) polyurethane conductive gasket has been identified that provides electrical bonding between aircraft substrate and the mounting base of aircraft antennas and static discharger mounting bases. This technology is used to seal and protect components against moisture and subsequent corrosion while at the same time providing a mechanism for bonding and grounding and eliminating airborne communication precipitation static (P-static) discrepancies caused by corrosion.

The performance requirements of sealant compounds commonly used within naval aviation are contained in Table 6.

Examples of Aircraft Corrosion Damage

Example 1: Localized Corrosion. In Fig. 4, filiform corrosion damage is observed on aluminum exterior surfaces of a P-3 aircraft. In Fig. 5, the filiform damage is more evident in

Table 4 Specified coating performance requirements

Class	Description
MIL-DTL-81706 Chemical Conversion Coating Materials for Aluminum (Ref 4)	
Class 1A	Maximum protection against corrosion
Class 3	Protection where low electrical resistance is required
Form I	Concentrated liquid
Form II	Powder
Form III	Premixed liquid
Form IV	Premixed thixotropic liquid
Form V	Premeasured powder, thixotropic
Form VI	Premixed liquid (ready for use in self-contained applicator)
Method A	Spray
Method B	Brush
Method C	Immersion
Method D	Applicator pen
AMS-M-3171 Pretreatment and Corrosion Prevention of Magnesium Alloys (Ref 5)	
Type I	Chrome pickle treatment
Type III	Dichromate treatment
Type IV	Galvanic anodizing treatment
Type VI	Chrome acid brush-on treatment
Type VII	Fluoride anodizing process with corrosion preventative treatment
Type VIII	Chromate treatment
MIL-A-8625 Anodic Coatings for Aluminum and Aluminum Alloys (Ref 6)	
Type I	Chromic acid anodizing
Type IB	Chromic acid anodizing, low voltage (22 ± 2 V)
Type IC	Nonchromic acid anodize, type I and IB alternative
Type II	Sulfuric acid anodize
Type IIB	Thin sulfuric acid anodize, type I & IB alternative
Type III	Hard anodic coating
Class 1	Non dyed
Class 2	Dyed

Table 5 Performance requirements for coatings

Class or type	Description
MIL-PRF-23377 Epoxy Primer (High-Solids) (Ref 7)	
Type I	Standard pigment
Type II	Low infrared reflective pigments
Class C	Strontium chromate based corrosion inhibitors
Class N	Nonchromate based corrosion inhibitors
MIL-PRF-85582 Epoxy Primer (Waterborne) (Ref 8)	
Type I	Standard pigment
Type II	Low infrared reflective pigments

Table 6 Aircraft sealing compounds

Specification and type	Form(a)	Curing temperature	Service temperature, °C (°F)	Peel strength(b), kg/cm (lb/in.)	Corrosion inhibitors	Fluid resistance(c)	Intended use
MIL-PRF-81733 Sealing and Coating Compound, Corrosion Inhibitive (Polysulfide) Type I (thin): brush or dip application Type II (thick): sealant gun or spatula Type III (sprayable): spray gun Type IV (spreadable): extended assembly	2	Room	−54 to 120 (−65 to 250)	2.7 (15)	Yes	Yes	Sealing faying surfaces and for wet installation of fasteners on permanent structure repair
AMS-S-8802 (supersedes MIL-S-8802), Sealing Compound, Temperature Resistant, Integral Fuel Tanks and Fuel Cell Cavities, High Adhesion (Polysulfide) Class A (thin): brush application Class B (thick): sealant gun or spatula Class C (spreadable): extended assembly times	2	Room	−54 to 120 (−65 to 250)	3.6 (20)	No	Yes	Used for fillet and brush sealing integral fuel tanks and fuel cell cavities. Not to be exposed to fuel or overcoated until tack-free
AMS-3276 (supersedes MIL-S-83430), Sealing Compound, Integral Fuel Tanks and General Purpose (Polysulfide) Class A (thin): brush application Class B (thick): sealant gun or spatula Class C (thick): extended assembly times Class D (thick): hole and void filling Class E (thick): for automatic riveting equipment	2	Room	−54 to 180 (−65 to 360)	3.6 (20)	No	Yes	High-temperature applications. Used for fuel tank sealing, cabin pressure sealing, aerodynamic smoothing, faying surface sealing, wet installation of fasteners, overcoating fasteners, sealing joints and seams, and nonstructural adhesive bonding. For fuel tank applications, treat bond surfaces with AMS 3100 adhesion promoter to enhance sealant adhesion.
PR-1773 (supersedes PR-1403G), Sealing Compound, Non-Chromate Corrosion Inhibitive Polysulfide Rubber Class B (thick): sealant gun or spatula	2	Room	−54 to 120 (−65 to 250)	...	Yes	Yes	Air Force preferred sealant for general-purpose, low-adhesion sealing of access door and form in place (FIP) gasket
AMS 3267 (supersedes MIL-S-8784), Sealing Compound, Low Adhesion Strength, Accelerator Required (Synthetic Rubber) Class A (thin): brush application Class B (thick): sealant gun or spatula	2	Room	−54 to 120 (−65 to 250)	0.36 (2 max)	No	Yes	Fillet and faying surface sealing of removable structure such as access doors, floor panels and plates, removable panels, and fuel tank inspection plates. Not for high-temperature areas or permanent structure
AMS 3374 (supersedes MIL-S-38249), Sealing Compound, One-Part Silicone, Aircraft Firewall (Synthetic Rubber) Type 1 (one-part silicone): cures on exposure to air Type 2 (two-part silicone): addition cured Type 3 (two-part silicone): condensation cured Type 4 (two-part polysulfide): high temperature resistant	1	Room	−54 to 204 (−65 to 400)	1.8 (10)	No	Yes	Sealing firewall structures exposed to very high temperatures against the passage of air and vapors. Cures on exposure to air
MIL-S-85420, Sealing Compound, Quick Repair, Low Temperature Curing, for Aircraft Structures (Polysulfide) Class A (thin): brush application Class B (thick): sealant gun or spatula	2	Low	−54 to 93 (−65 to 200)	1.8 (10)	No	Yes	Cold temperature (5 °C, or 40 °F) and quick repair sealing of aircraft structures. Use only with recommended adhesion promoter. Not to be used when temperature exceeds 25 °C (80 °F) or poor adhesion will result
AMS 3277 (supersedes MIL-S-29574), Sealing Compound, for Aircraft Structures, Fuel and High Temperature Resistant, Fast Curing at Ambient and Low Temperatures (Polythioether) Class A (thin): brush application Class B (thick): sealant gun or spatula Class C (thick): extended assembly times	2	Low and room	−62 to 150 (−80 to 300)	3.6 (20)	Yes	Yes	Multipurpose aircraft structure and integral fuel tank sealants with rapid ambient and low-temperature curing capabilities. Use only with recommended adhesion promoter.
MIL-A-46146, Adhesive—Sealants, Silicone, Room Temperature Vulcanizing (RTV), Non-corrosive (Synthetic Rubber) Type I: paste Type II: liquid Type III: high strength	1	...	−57 to 204 (−70 to 400)	...	No	...	Convenient one-component sealant for use with sensitive metals and equipment. Not to be used where resistance to fuels, oils, or hydraulic fluids is required.
AMS 3255, Skyflex Sealing Tape, Polytetrafluoroethylene, Expanded (ePTFE) Class 1: Continuous Ribbed, includes: GUA-1071-1 for <1 in. wide fay surfaces GUA-1001-2 for >1 in. wide fay surfaces GUA-1003-1 compensation tape	PG	...	−54 to 120 (−65 to 250)	(2 max)	No	Yes	Sealing of faying surfaces, access panels, floorboards, and windscreens. Not for fuel-soaked or high-temperature applications. Nonhazardous alternative to two-component sealants

(continued)

(a) 1, 1 component; 2, 2 component; PG, preformed gasket. (b) Minimum unless indicated. (c) Resists fuel, oil, and hydraulic fluid

Table 6 (continued)

Specification and type	Form(a)	Curing temperature	Service temperature, °C (°F)	Peel strength(b), kg/cm (lb/in.)	Corrosion inhibitors	Fluid resistance(c)	Intended use
AMS 3255 (continued)							
Class 1: Continuous Ribbed (continued)							
GSC-21-80767-00 for high moisture areas of floorboards and thicker fay surface gaps							
GUA-1401-1 for dry areas of floorboards							
Class 2: Continuous Non-Ribbed, includes:							
GUA-1057-1 for <1 in. wide fay surfaces, use as shim/barrier to resist minor chafing							
GUA-1059-1 for >1 in. wide fay surfaces, use as shim/barrier to resist minor chafing							
GUA-1301-1 for <1 in. wide fay surfaces with thick gaps							

(a) 1, 1 component; 2, 2 component; PG, preformed gasket. (b) Minimum unless indicated. (c) Resists fuel, oil, and hydraulic fluid

the macrograph. Corrective action included the removal of corrosion via glass bead blasting, cleaning, surface treatment, and reapplication of epoxy primer and polyurethane topcoat.

Example 2: Atmospheric Corrosion of Magnesium. A magnesium radar assembly (AZ-91C) was damaged from atmospheric exposure (Fig. 6). Corrosion was removed via glass bead blasting. MIL-C-81309 type III (Ref 8) corrosion-preventive compound was applied to unpainted surfaces to prevent recurrence.

Example 3: Pitting Corrosion. Aluminum alloy 7075-T6 seen in Fig. 7 experienced pitting corrosion damage. The damage to an EA-6B aircraft lower longeron (fuselage structural member) resulted from water ingress and entrapment through fastener holes. Repair required removal and replacement with a newly manufactured item. Corrective action included design modification of drain holes and improvements to coating system with the addition of polyurethane topcoat.

Example 4: Internal Exfoliation. Figure 8 shows corrosion of an aluminum housing associated with the Barostatic Release Unit (BRU) of a F-14 ejection seat. Corrosion damage was the result of moisture entrapment at O-ring areas and required replacement of component. Corrective action included the application of MIL-L-87177 corrosion-inhibiting lubricant.

Example 5: Exfoliation corrosion damage occurred on an aluminum P-3 integral fuel tank resulting from long-term exposure to moisture, salts, and fuel-system icing inhibitors (Fig. 9). Corrective action involved the mechanical removal of corrosion with powered abrasive sanders followed by surface treatment and application of AMS-C-27725 fuel tank coating.

Fig. 4 Filiform corrosion of an aluminum external surface adjacent to steel fasteners. Courtesy of J. Benfer, Naval Air Depot—Jacksonville

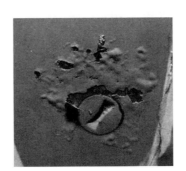

Fig. 5 Macrophotograph illustrating filiform corrosion of an aluminum aircraft surface adjacent to steel fastener. Courtesy of J. Benfer, Naval Air Depot—Jacksonville

Fig. 6 Surface corrosion of AZ-91C magnesium radar assembly resulting from atmospheric exposure. Courtesy of J. Benfer, Naval Air Depot—Jacksonville

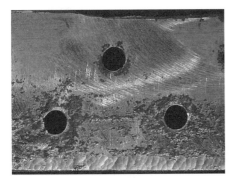

Fig. 7 Extensive pitting corrosion damage of an aluminum lower longeron from a EA-6B aircraft requiring removal and replacement with newly manufactured item. Courtesy of J. Benfer, Naval Air Depot—Jacksonville

Fig. 8 Internal exfoliation corrosion of a barostatic release unit associated with an F-14 ejection seat. Courtesy of J. Benfer, Naval Air Depot—Jacksonville

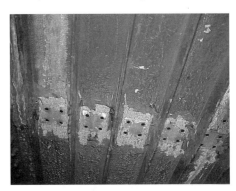

Fig. 9 Exfoliation corrosion of a integral fuel tank on a P-3 aircraft resulting from long-term exposure to moisture, salts, and fuel-system icing inhibitors. Courtesy of J. Benfer, Naval Air Depot—Jacksonville

Example 6: Fretting corrosion damage occurred on the interior of an aluminum electrical wiring conduit located within the integral fuel tank of an EA-6B aircraft (Fig. 10). Through-wall corrosion damage resulted in significant fuel leaks, requiring the rerouting of electrical wires. Corrective action included the removal of wire bundle ties and application of additional protective coatings to newly installed conduits.

Example 7: Saltwater Intrusion. H-60 helicopter floor frame (aluminum alloy) corrosion around a floorboard mounting hole due to saltwater intrusion is shown in Fig. 11. Preventive measures included improved protective coatings and sealants and routine application of corrosion-preventive compounds.

Example 8: Galvanic Corrosion. AV-8 aircraft avionics mounting bracket (nickel-plated aluminum alloy) suffered the corrosion damage seen in Fig. 12. The aluminum alloy underneath the nickel plate corroded as a result of exposure to salt-laden air. Galvanic corrosion of the underlying aluminum alloy caused the nickel plate to blister and flake. Corrective measures included thicker nickel plating and routine application of avionics-grade corrosion-preventive compounds.

Example 9: Galvanic corrosion occurred on F/A-18 dorsal scallops and wing substructures. Composite doors are fastened to the aluminum substructure with titanium or steel fasteners (Fig. 13). A more detailed view of mounting holes is seen in Fig. 14. Moisture migrated past the fastener head and set up a galvanic cell where corrosion began on the aluminum substructure around the fastener hole. To mitigate this corrosion, water-displacing corrosion-preventive compounds (CPCs) were applied to the shank of the fastener, and sealant was applied around its head during door installation. A design modification for new aircraft models eliminates this configuration.

Example 10: Exfoliation Corrosion. Figure 15 shows a T-45 access door corroded from moisture intrusion underneath electromagnetic interference (EMI) fingers (not shown). Direct metallic contact between access door and aircraft substructure for control of EMI through the use of conductive EMI fingers prevents the usage of many protective coating systems. Corrective action is the frequent application of avionics-grade corrosion-preventive compounds and a recommended design change to incorporate the replacement of EMI fingers with a conductive sealant.

Example 11: General atmospheric corrosion of EA-6B slat gearbox limit switches prevented proper electrical and mechanical operation and resulted in increased maintenance (Fig. 16). Steel corrosion and rusting is prevented through the application of MIL-L-87177 corrosion-inhibiting lubricant (Fig. 17).

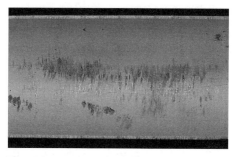

Fig. 10 Fretting corrosion damage on the internal walls of aluminum electrical wire conduit installed within the wing of an EA-6B aircraft. Courtesy of J. Benfer, Naval Air Depot—Jacksonville

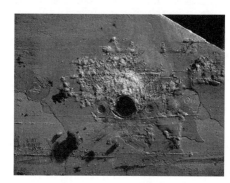

Fig. 11 Corrosion around an aluminum floorboard mounting hole due to saltwater intrusion. Courtesy of J. Whitfield, Naval Air Depot—Cherry Point

Fig. 12 Corrosion of a nickel-plated aluminum mounting bracket for AV-8 aircraft due to atmospheric exposure. Courtesy of J. Whitfield, Naval Air Depot—Cherry Point

Fig. 13 Galvanic corrosion of F/A-18 aircraft dorsal scallops resulting from composite doors attached to aluminum substructure with titanium and steel fasteners in the presence of moisture. Courtesy of S. Long, Naval Air Depot—North Island

Fig. 14 Galvanic corrosion of an F/A-18 aircraft wing substructure resulting from composite doors attached to aluminum substructure with titanium and steel fasteners in the presence of moisture. Courtesy of S. Long, Naval Air Depot—North Island

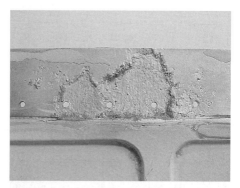

Fig. 15 Corrosion damage of a T-45 aircraft access panel. Corrosion initiates at and beneath electromagnetic interference (EMI) fingers (removed) needed to control the EMI. Courtesy of J. Benfer, Naval Air Depot—Jacksonville

Fig. 16 Slat gearbox limit switch for EA-6B aircraft following ASTM B 117 neutral salt-spray corrosion testing without MIL-L-87177 corrosion-inhibiting lubricant. Courtesy of J. Benfer, Naval Air Depot—Jacksonville

Fig. 17 EA-6B slat gearbox limit switch following ASTM B 117 neutral salt-spray corrosion testing with MIL-L-87177 corrosion-inhibiting lubricant. Courtesy of J. Benfer, Naval Air Depot—Jacksonville

Example 12: Exfoliation corrosion occurred on an aluminum T-45 cockpit kick plate angle. Corrosion initiated at the fastener hole area due to water intrusion within the cockpit (Fig. 18).

Example 13: High-temperature corrosion of a 300 series stainless steel union fitting located within the canopy ejection system of an EA-6B aircraft is shown in Fig. 19. High-temperature exposure during previous actuation of the canopy ejection system resulted in the oxidation of interior surface.

Example 14: General corrosion of a cadmium electroplated carbon steel pin associated with the locking pawl system of a F-14 ejection seat is shown in Fig. 20. Water intrusion and atmospheric exposure resulted in consumption of the protective cadmium electroplate. Corrective action was the use of improved corrosion-preventive compounds and redesign using corrosion-resistant stainless steel.

Example 15: Catastrophic failure of a F-14 nose landing gear cylinder seen in Fig. 21 was due to multiple fatigue crack initiation sites concentrated on the aft portion of the cylinder. The fatigue cracks initiated in a line of corrosion likely caused by contact of the thrust washer in the radius area. The contact removed the corrosion-protection system of paint and aluminum plating, permitting extensive corrosion around the cylinder circumference in the radius area.

Example 16: Hydrogen Embrittlement. F-14 arresting gear tail hook stinger catastrophically failed during an attempted arrested landing aboard a carrier. Failure occurred at the forward weld area and was the result of hydrogen embrittlement of the 300M high-strength steel. Figure 22 shows the component as received in the laboratory, and Fig. 23 is a macrograph of the fracture initiation area that was an area of high residual tensile strength.

Example 17: Damage Due to Improper Packaging. Newly manufactured EA-6B aluminum horizontal skin in Fig. 24 exhibited extensive corrosion as the result of water ingress from improper packaging and preservation during shipment from the manufacturer.

Conclusion

Aircraft corrosion is very expensive in terms of costs associated with inspection, maintenance and repair manpower, decreased aircraft availability, and loss of aircraft and human life. Aircraft operators and maintainers that operate in the harshest of conditions require the latest science and technology for corrosion control to include improved inspection, corrosion removal, and treatment processes. The application of good engineering practices with appropriate materials selection, design, and use of protective coating systems is necessary to ensure that naval aircraft weapon systems are both reliable and capable of performing their mission requirements.

REFERENCES

1. "Cleaning Compounds, Aircraft, Exterior," MIL-PRF-85570, U.S. Department of Defense, 2002

Fig. 18 Exfoliation corrosion of an aluminum T-45 cockpit kick plate angle resulting from water intrusion within cockpit areas. Courtesy of J. Benfer, Naval Air Depot—Jacksonville

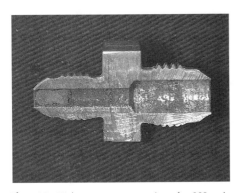

Fig. 19 High-temperature corrosion of a 300 series stainless steel union fitting within an EA-6B canopy ejection system. Courtesy of J. Benfer, Naval Air Depot—Jacksonville

Fig. 22 F-14 arresting hook gear as received following failure during carrier landing. Failure was the result of hydrogen embrittlement induced cracking at an area of high residual tensile strength. Courtesy of J. Yadon and K. Himmelheber, Naval Air Depot—Jacksonville

Fig. 20 General corrosion of a cadmium electroplated carbon steel pin associated with the locking pawl of a F-14 ejection seat. Courtesy of J. Benfer, Naval Air Depot—Jacksonville

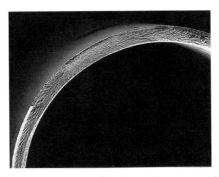

Fig. 23 Macrograph of the fracture initiation area of a catastrophically failed F-14 arresting gear tail hook. Courtesy of J. Yadon and K. Himmelheber, Naval Air Depot—Jacksonville

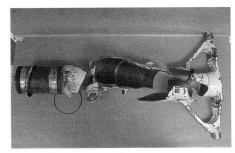

Fig. 21 Catastrophic failure of F-14 nose landing gear cylinder, caused by corrosion-induced fatigue cracking of high-strength steel. Courtesy of S. Binard, Naval Air Depot—Jacksonville

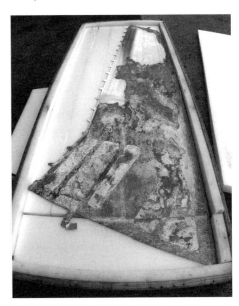

Fig. 24 EA-6B aluminum horizontal skin exhibiting extensive water-induced corrosion due to improper packaging and preservation during shipment from manufacturer. Courtesy of S. Bevan, Naval Air Depot—Jacksonville

2. "Corrosion Preventative Compound, Solvent Cutback, Cold-Application," MIL-PRF-16173, 1993
3. "Lubricating Oil, General Purpose, Preservative (Water-Displacing, Low Temperature)," MIL-PRF-32033, U.S. Department of Defense, 2001
4. "Chemical Conversion Materials for Coating Aluminum and Aluminum Alloys," MIL-DTL-81706, U.S. Department of Defense, 2004
5. "Magnesium Alloy, Process for Pretreatment and Corrosion Prevention On," SAE-AMS-M-3171, Society of Automotive Engineers, 1998
6. "Anodic Coatings for Aluminum and Aluminum Alloys," MIL-A-8625, U.S. Department of Defense, 2003
7. "Primer Coatings: Epoxy, High Solid," MIL-PRF-23377, U.S. Department of Defense, 2005
8. "Primer Coatings: Epoxy, Waterborne," MIL-PRF-85582, U.S. Department of Defense, 2002
9. "Coating: Polyurethane, Aircraft and Support Equipment," MIL-PRF-85285, U.S. Department of Defense, 2002

SELECTED REFERENCES

- *Aircraft Weapons Systems Cleaning and Corrosion Control,* NAVAIR 01-1A-509, Naval Air Systems Command, 2001
- A.S. Birks and R.E. Green, Jr., *Nondestructive Testing Handbook,* 2nd ed., American Society for Nondestructive Testing, 1991
- A.W. Brace, *Anodic Coating Defects, Their Causes and Cure,* Technicopy Books, England, 1992
- A.W. Brace, *The Technology of Anodizing Aluminum,* 3rd ed., Interall Srl, Modena, Italy, 2000
- W.D. Callister, Jr., *Materials Science and Engineering, An Introduction,* 4th ed., John Wiley & Sons, 1997
- *Corrosion Prevention and Control Planning,* Spiral No. 1, Department of Defence Guidebook, Director of Corrosion Policy and Oversight, 2003
- *Corrosion Prevention and Control Planning,* Spiral No. 2, Department of Defence Guidebook, Director of Corrosion Policy and Oversight, 2004
- H.V. Droffelaar and J.T.N. Atkinson, *Corrosion and Its Control,* NACE International, 1995
- L.J. Durney, *Electroplating Engineering Handbook,* 4th ed., Van Nostrand Reinhold, 1984
- *Finishes: Organic, Weapons System, Application and Control,* MIL-F-18264, U.S. Department of Defense
- M.G. Fontana, *Corrosion Engineering,* 3rd ed., McGraw-Hill, 1986
- *General Specifications for Finishes and Coating: Protection of Aerospace Weapon Systems, Structures and Parts,* MIL-STD-7179, U.S. Department of Defense
- C.H. Hare, *Paint Film Degradation Mechanisms and Control,* Society for Protective Coatings, 2001
- R.E. Hummel, *Understanding Materials Science,* Springer-Verlag, 1998
- D.A. Jones, *Principles and Prevention of Corrosion,* 2nd ed., Prentice-Hall, 1996
- R. Lambourne and T.A. Strivens, *Paint and Surface Coatings,* William Andrew Publishing, 1999
- *Materials and Processes for Corrosion Prevention and Control in Aerospace Weapons Systems,* MIL-HDBK-1568, U.S. Department of Defense
- C.G. Munger, *Corrosion Prevention by Protective Coatings,* NACE International, 1986
- *Nondestructive Inspection Methods,* NAVAIR 01-1A-16, Naval Air Systems Command, 2000
- P.A. Schweitzer, *Corrosion Resistance Tables—Metals, Nonmetals, Coatings, Mortars, Plastics, Elastomers, Linings, and Fabrics,* 3rd ed., Marcel Dekker, 1991
- D. Stoye and W. Freitag, *Paints, Coatings and Solvents,* 2nd ed., Wiley-VCH, 1993
- *Surface Treatments and Inorganic Coatings for Metal Surfaces of Weapons Systems,* MIL-S-5002, U.S. Department of Defense
- S. Wernick, R. Pinner, and P.G. Sheasby, *The Surface Treatment and Finishing of Aluminum and Its Alloys,* 5th ed., Vol 1 and 2, ASM International, 1987

Military Aircraft Corrosion Fatigue

K.K. Sankaran, R. Perez, and H. Smith, Boeing Company

CORROSION, FATIGUE, AND THEIR SYNERGISTIC INTERACTIONS are among the principal causes of damage to aircraft structures. Analysis of aircraft failure modes drawn from case histories from World War II showed that fatigue and corrosion accounted for 55 and 25% of the failures, respectively (Ref 1). A worldwide survey of 1885 aircraft accidents also showed that the proportion of fatigue failures directly attributed to the presence of corrosion was small (Ref 2). However, corrosion can still be a significant safety concern as it increases the stress in an airframe part by reducing the effective cross-sectional area of the structure carrying the load. Corrosion also causes stress concentrations, which can result in premature fatigue crack initiation. Therefore, the presence of corrosion reduces fatigue life and can result in a greater risk of failure of aircraft components. These problems are particularly relevant for "aging" aircraft, many of which, due to economic pressures, remain in service well beyond their initial design goals.

Corrosion is also a significant economic concern since the costs related to its detection, mitigation, repair, and maintenance are among the largest components of life-cycle costs for military systems. United States government studies conducted in 1996 and 2001 estimated the direct corrosion-related costs for military systems and infrastructure to be approximately $10 billion and $20 billion, respectively (Ref 3). These costs do not include the indirect costs resulting from the loss of opportunity to use the equipment while it is being repaired. The annual cost of corrosion for all aircraft systems in the United States in 1995 was estimated to be $13 billion. Of this total, the share for the military was nearly $3 billion to maintain its fleet of 15,000 aircraft, of which the Navy's share was about $1 billion for its fleet of about 4600 aircraft (Ref 4, 5). An Air Force study determined the annual cost of direct corrosion maintenance to be about $795 million in 1997. While the number of aircraft in the Air Force fleet decreased by 20% from 1990, the overall cost declined only 10% and aircraft-specific costs increased by 4% (Ref 6). Continuing assessments of the cost of corrosion to the Air Force show an increasing trend with the total costs increasing to about $1.1 billion in 2001 and $1.5 billion in 2004 (Ref 7).

Many of the military aircraft presently in service were built prior to the availability of improved corrosion-control systems and alloys with superior corrosion resistance. Thus, corrosion will remain a concern until these aircraft are retired from service. While considerable emphasis has been placed on the detection and mitigation procedures for corrosion in these aircraft, the management of its effects on their structural integrity has received only limited attention principally due to the difficulty of quantitatively predicting the future effects of corrosion.

The effect of corrosion can be addressed by its superimposition on the methods presently used to predict fatigue life, which consists of a crack initiation period (including crack nucleation and microcrack growth) and a crack growth period (covering the growth of a visible crack to failure) (Ref 8). The development and validation of corrosion metrics that can be used as inputs into analytical fatigue life prediction models is critical to the superimposition of corrosion effect. A fundamental understanding of the corrosion mechanisms and corrosion/fatigue interactions under conditions that simulate aircraft service environment is necessary to satisfy this need. In the context of applying these concepts and methods for predicting and managing aircraft corrosion fatigue, the key topics and their interrelationships that are discussed in this article are outlined in Fig. 1.

Aircraft Corrosion Fatigue Assessment

The safe and economical operation of aircraft requires an accurate quantitative assessment of the effects of corrosion on the structural integrity of components. This assessment must be conducted in the context of the different approaches used to:

- Manage aircraft structural integrity
- Schedule aircraft inspection intervals
- Perform repair and maintenance of aircraft in service

The safe-life, fail-safe, and damage-tolerance approaches to the fatigue design and life assessment of aircraft structures and the methods used to address the additional effects of corrosion are described in *Fatigue and Fracture*, Vol 19, *ASM Handbook* (Ref 9). Unlike the effects of fatigue alone, the future occurrence and effects of corrosion cannot be predicted deterministically. Accordingly, the U.S. Navy, which manages the structural integrity of its aircraft using a safe-life approach to meet the safety and readiness demands of carrier operations (Ref 10), bases its maintenance practice on a "find it and fix it" corrosion repair philosophy. However, low levels of corrosion causing insignificant impact on fatigue and residual strength may not require immediate repair. While delaying repairs that can wait until a later maintenance cycle results in cost savings and higher fleet readiness, unfortunately there are no validated guidelines currently available for determining if a corrosion repair can be postponed without compromising safety.

The U.S. Air Force manages the structural integrity of its aircraft using fail-safe and damage-tolerance approaches (Ref 11). The combination of corrosion damage and fatigue is a problem in structures, and this problem will grow as the aircrafts continue to age in service. Corrosive environments and the resulting damage increase the stress in structural members, provide fatigue crack initiation sites, and accelerate crack propagation. The result is a component with shorter fatigue life and reduced residual strength. A report issued by the National Materials Advisory Board of the National Research Council has provided a detailed assessment of the effects of corrosion on the safety limits and the inspection intervals for both fail-safe and safe-crack-growth-designed aircraft. It has recommended research and development programs to address these issues (Ref 11).

The subject of corrosion and its effects on fatigue and aircraft structural integrity have been discussed in comprehensive reviews (Ref 12, 13). The proceedings of a NATO Applied Vehicle Technology Panel workshop titled *Fatigue in the Presence of Corrosion*, contains many papers in the areas of in-service experience with corrosion fatigue, fatigue prediction methodologies in corrosive environments, and effects of corrosion/fatigue interactions on structural integrity (Ref 14).

To ensure structural integrity, the undamaged aircraft structure must withstand specified flight maneuvers and must be designed to sustain limited damage between scheduled inspection intervals (Ref 15). Loads anticipated during aircraft service and the resulting fatigue damage are well understood and used for predicting the life of components undergoing mechanical loading. Aircraft components are subjected to fatigue loading only during flight, but can experience corrosion throughout their calendar life. Since it is difficult to determine precisely the onset and progression of corrosion damage, its effects on the fatigue life of aircraft components cannot be accurately predicted. Accordingly, there have only been a very few attempts to incorporate the effects of corrosion in analytical models to predict the life of aircraft components in service (Ref 16). See the article "Predictive Modeling of Structure Service Life" in Volume 13A.

Prior Corrosion Damage as Equivalent Initial Flaw. Corrosion damage in aircraft structures results in an initial flaw size that can reduce the predicted life of a critical component and render noncritical components as critical (Ref 13). Therefore, approaches to incorporating the effects of corrosion have so far focused on developing and validating models for life prediction using data from fatigue tests conducted on relevant materials and corrosion environments to simulate in-service experience of aircraft structures.

Commonly used tests to obtain data for fatigue design and analysis include the stress-life, strain-life, and fatigue crack growth tests (Ref 17, 18). Data on the simultaneous influence of corrosion exposure and fatigue loading can be obtained by testing pristine specimens (specimens without prior corrosion exposure) in selected corrosive environments. While they can provide information on the mechanisms of corrosion fatigue crack nucleation and growth, it is difficult to accurately simulate the in-service interactions between corrosion and fatigue in these tests.

More useful data for modeling the effects of corrosion can be obtained from tests conducted in ambient environments using specimens already containing varying levels of corrosion damage caused by controlled exposure to corrosive environments prior to fatigue loading. The corrosion damage at the start of the test (prior corrosion) can be treated as an "equivalent flaw" whose effect on the fatigue life can be determined in these tests. Quantitative measures or metrics for corrosion can then be used as the "size" of the equivalent flaws in order to predict life in a manner similar to that used for predicting fatigue life in the absence of corrosion. The goal is to validate this approach to predicting the life of aircraft components containing measurable corrosion damage, which can be treated as "prior corrosion" and which can be detected during inspection. Achieving this goal requires the definition and validation of quantitative metrics for the various corrosion types experienced by aircraft, development of probabilistic models for the evolution of future corrosion and fatigue damage, followed by their use directly in fatigue life prediction models.

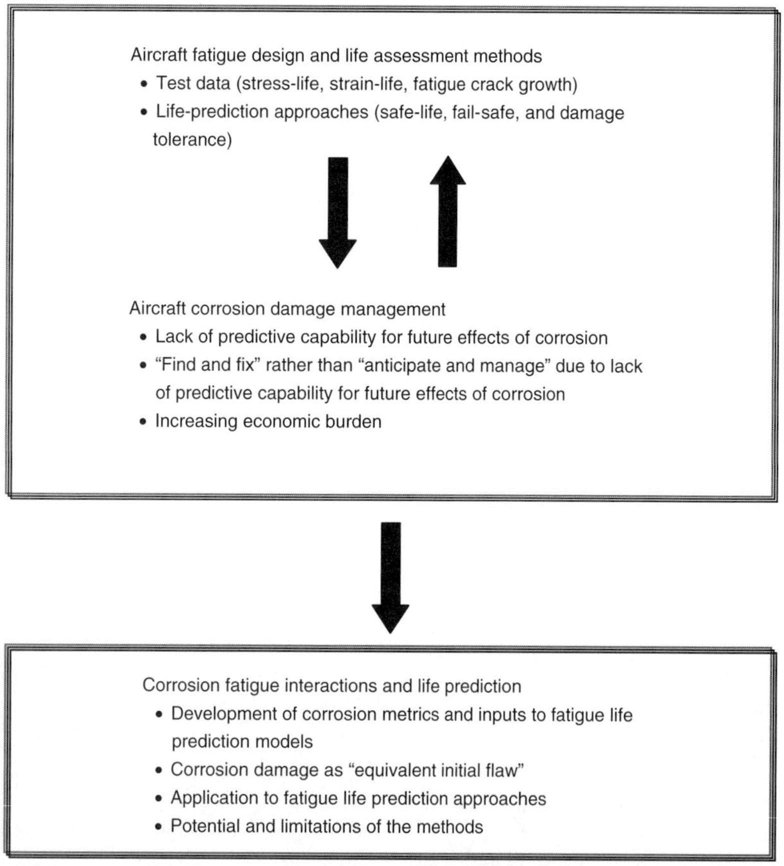

Causes and Types of Aircraft Corrosion

Military aircraft operate under harsh and widely varying conditions that differ significantly from commercial aircraft. Naval aircraft, such as the F/A-18 deployed in carriers, are subjected to the ocean environment as well as to the exhaust fumes in below-deck areas. Flexing of aircraft structure and maintenance activities can damage corrosion-protection systems, which can become scratched and can crack around fasteners. These processes allow the intrusion of water, salt, and other contaminants into the structural joints and initiate corrosion. The article "Corrosion in Commercial Aviation" in this Volume discusses the various types of corrosion observed in aircraft and provides numerous examples of corroded components.

The types of corrosive attack observed in aircraft structures include uniform, galvanic, pitting, filiform, fretting, intergranular, and exfoliation corrosion, and stress-corrosion cracking. These types of corrosion result in combinations of localized attack (pitting), bulk material loss (galvanic or exfoliation corrosion), and general surface roughening (uniform, filiform, or fretting corrosion). Stress-corrosion cracking can occur in susceptible alloys exposed to tensile stresses and corrosive environments. Crevice corrosion occurs when discrete areas on an alloy are physically isolated or occluded such as at joints in aircraft structures. Crevice corrosion can eventually cause surface grain separation in the surrounding part resulting in severe intergranular or exfoliation corrosion and a volumetric increase from the corrosion products (pillowing), which can increase the stress and crack growth rates (Ref 19).

Commonly used aircraft structural materials such as the high-strength aluminum alloys possess heterogeneous microstructures, which render them susceptible to corrosion. The problem is compounded when they are used in conjunction with dissimilar materials in the assembly of aircraft parts. For example, corrosion is common in aluminum alloy hinges used in military aircraft. Hinges contain copper alloy (beryllium copper) bushings at the lug holes and galvanic corrosion can occur at the bushing/aluminum-alloy interface (Fig. 2). Figure 3 shows an example of exfoliation corrosion around a fastener hole in a longeron (a major airframe structural member running from the front to the rear of an aircraft and oriented in the longitudinal axis of the fuselage). Longerons are typically fabricated

Fig. 1 Military aircraft corrosion fatigue status and assessment approach

from high-strength aluminum alloy extrusions. An example of crevice corrosion and pillowing is shown in Fig. 4, where a 2024-T4 aluminum alloy fuselage skin is spot welded to a 2024-T4 doubler. Pillowing due to the accumulation of the corrosion products in the joint can be seen.

Impact of Corrosion on Fatigue

To varying degrees, all the observed forms of corrosion damage can reduce the load-carrying capacity of the aircraft structure, initiate fatigue cracks, and increase fatigue crack growth rates. As discussed earlier, in the absence of a corrosive environment at present, fatigue cracks can nucleate in structures with prior corrosion damage. Fatigue cracks can also nucleate from simultaneously developing corrosion damage due to the concurrent influence of fatigue loading and corrosive environment. The results of a large number of investigations on the effects of prior corrosion or the presence of a corrosive environment on fatigue crack initiation and growth in high-strength aluminum alloys (commonly used aircraft structural materials) can be summarized in the context of their schematic stress-fatigue life (cycles to failure) behavior (Ref 12):

- Prior corrosion has a large effect on fatigue life and decreases the lives significantly at all stress levels. This is because, at any given stress level, the life of specimens with preexisting damage will always be lower than those without such damage.
- Prior corrosion is more detrimental than continuous corrosion at high stresses and relatively short lives. This is because at the high stress levels that result in low fatigue lives, sufficient corrosion damage may not develop in the available time in the environment prior to fatigue failure.
- The concurrent effect of a corrosive environment and fatigue loading is synergistic at lower stress levels and relatively longer fatigue lives because corrosion damage continues to accumulate with time. Thus in the presence of a corrosive environment, the fatigue life will continue to decrease more rapidly with increasing stress compared with the lives in the absence of a corrosive environment.

Since corrosion causes stress concentrations, prior corrosion will reduce the fatigue lives of materials. This is illustrated by the measured fatigue lives of 7075-T6 aluminum alloy specimens with and without prior corrosion damage (Fig. 5) (Ref 20). The corroded 7075-T6 specimens contained varying amounts of damage caused by exposing them to various durations in the prohesion salt-spray test (ASTM G 85, Annex 5). Prohesion used to describe this test was derived from "protection is adhesion." The measured lives are also compared with the values contained in the *Metallic Materials Properties Development and Standardization Handbook* (MMPDS, formerly MIL-Handbook-5) for the fatigue lives of bare 7075-T6 alloy without any prior corrosion damage (Ref 21). The MMPDS values are plotted for lives of specimens at values of stress-concentration factor, K_t, of 1 (smooth specimen with no surface stress concentration) and $K_t = 2$ (specimen containing a geometrical discontinuity that raises the local stress at the discontinuity to two times the nominal stress applied to the specimen). The R value of 0.1 is

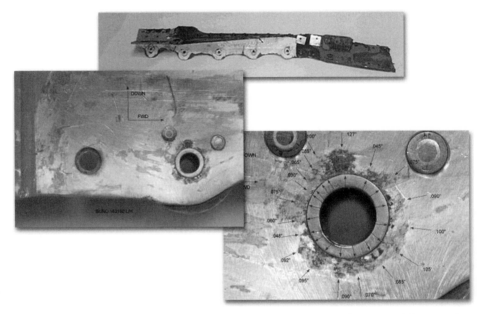

Fig. 3 Exfoliation corrosion around a fastener hole in a 7049-T73 aluminum alloy longeron. Radial arrows indicate measurements taken to assess damage.

Fig. 2 Corrosion of an aluminum alloy hinge (7050-T74) around a copper alloy bushing (UNS C17200, beryllium copper)

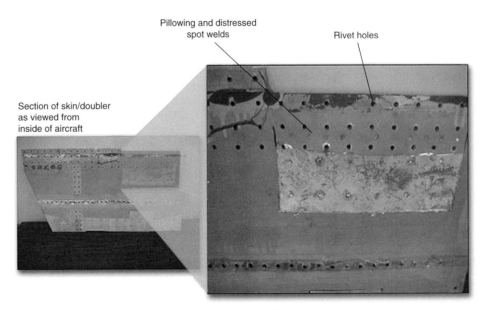

Fig. 4 Corrosion pillowing on aircraft skin at joint. Aluminum alloy 2024-T4 skin is joined to doubler by spot welding and with fasteners (2017-T3). Rivet holes for the lap joint are seen in the lower (inside) portion of the joint.

the ratio of the minimum stress to the maximum stress applied in each load cycle during the fatigue test.

A continuous decrease in fatigue life with increasing prior corrosion damage is observed. It is interesting to note that the measured fatigue lives for 7075-T6 containing various levels of prior corrosion damage are bounded by the design handbook values for noncorroded 7075-T6 at K_t values of 1 and 2. The observation that the magnitude of the reduction of the fatigue lives by prior corrosion is similar to that due to a geometric flaw or discontinuity suggests that equivalent stress-concentration factors attributable to corrosion damage could be used as a measure or metric to quantitatively describe the effects of corrosion.

In polycrystalline materials, preexisting damage will initiate a fatigue crack immediately upon the application of load (Ref 22). Therefore, in specimens with prior corrosion, a fatigue crack can be expected to initiate and grow from the first load cycle, whereas specimens without prior corrosion require time for corrosion damage to accumulate and to initiate and grow fatigue cracks. The effect of prior corrosion is thus more detrimental at higher fatigue stress levels. At lower stress levels corresponding to longer fatigue lives or larger number of cycles to failure, the effects of corrosion and fatigue can be expected to be synergistic.

Difference in Military and Commercial Aircraft. The nature of corrosion/fatigue interactions has broad implications on the life assessment of military aircraft, which typically fly fewer hours than commercial aircraft and accumulate fatigue damage at a slower rate but corrosion damage at a higher rate (Ref 12). Thus, the principal need is to quantify the structural performance of critical areas of aircraft with the assumed presence of prior corrosion in the representative environment that is present during service. Because corrosive environments such as salt water are typically absent during flight, many of the corrosion fatigue investigations have concentrated on the testing and modeling of the fatigue behavior of materials and components with prior corrosion but in-flight service in a benign environment. However, the need also exists for understanding and accounting for the effects of the corrosive environments during flight service, as such cases have been reported (Ref 23). Development and use of metrics to represent the different types of observed corrosion is necessary to satisfy these needs.

Corrosion Metrics

Corrosion detected during aircraft inspection is frequently described qualitatively as "light," "moderate," or "severe," which precludes a quantitative assessment of its effects on the structural integrity. The problem is being overcome by applying fracture mechanics principles and modeling damage from corrosion in a manner similar to that caused by fatigue cracking (Ref 24). Geometric parameters such as pit dimensions, surface roughness, loss of metal thickness, and volume increase due to pillowing can be used to quantitatively characterize types of corrosion. These metrics, in turn, have the potential to be integrated into corrosion fatigue models to assess structural integrity.

Pitting Corrosion Metrics. Pitting is a common form of corrosion experienced by structural materials such as the high-strength aluminum alloys used in the aerospace industry. This type of corrosion has received considerable attention because it initiates fatigue cracks. It is almost always present with other types of corrosion, and crack initiation can frequently be traced to a pitlike defect. Further, it is relatively easy to characterize quantitatively to obtain metrics for use in life prediction models.

In high-strength aluminum alloys, pit formation is associated with the various types of constituent particles in the microstructure (Ref 25). Depending on the relative electrochemical difference between the particles and the aluminum matrix, pits are nucleated either by the dissolution of the aluminum matrix or the constituent particles, which are typically equiaxed and 25 to 50 μm in size. Each pit is generally associated with a particle on the alloy surface and reaches a depth approximately equal to the particle size. However, particles are also often present in clusters, which nucleate pits that have been found to extend to a depth of 300 μm in 2024-T3 (Ref 26). The morphology of pitting lends itself to quantitative description by parameters such as density (number of pits per unit area), depth, and length. Both laboratory studies on aluminum alloy 2024-T3 (Ref 27, 28) and 7075-T6 (Ref 20), as well as analysis of failed aircraft parts (Ref 23) have demonstrated that pits are sites for fatigue crack initiation.

Documented cases of fatigue cracks initiating from corrosion pits in 7xxx series alloys have also shown that crack growth was promoted by the environment until failure of the components occurred. Analysis of the causes of failure of a Royal Australian Air Force aircraft showed that the trailing edge flap failed at its outboard hinge and twisted upward about the inboard hinge (Ref 23). During a fleet inspection, cracking was detected in the 7050-T74 (formerly 7050-T73652) hinge lug in two other Australian aircraft. Corrosion pitting was found in the majority of the lug hole bore surfaces, and fatigue cracks initiated from the pits in both aircraft. In one aircraft, the corrosion pits reached an average depth of 200 μm, and the largest pit was 290 μm deep. In the other aircraft, the average pit depth was 150 μm. The investigation concluded that the flap failure was caused by fatigue cracks initiating at corrosion pits.

Research concentrating on corrosion metrics for steels has received less attention than that for aluminum alloys in the aerospace industry. Cadmium plating typically protects steel alloys, and the breakdown of the plating results in the corrosion of steel. However, this mechanism of pitting is not the same as that observed in aluminum alloys, which possesses a passive layer and whose breakdown results in surface pitting. Nevertheless, the "equivalent flaw size" approach has been used to account for corrosion pitting in D6AC steel (UNS K24728 with nominal composition of Fe-0.46C-1.00Cr-1.00Mo-0.55Ni, a medium-carbon, ultrahigh-strength, quench-and-temper steel commonly used for highly stressed aircraft parts (Ref 29). This study was undertaken to incorporate pitting corrosion damage into fatigue life modeling and to avoid unnecessary maintenance of F-111 aircraft. Localized pitting damage was induced in the D6AC steel specimens by electrochemical means.

Surface Roughness Parameters. Production processes such as chemical milling and etching are used in the fabrication of military aircraft parts. These processes result in surface roughness similar to that caused by pitting corrosion. Thus, the standard surface roughness parameters measured from characterization methods such as profilometry (which are used in production processes) should also be applicable to corrosion damage characterization (Ref 30). Commonly used surface parameters include the roughness average, roughness height rating, root-mean-square roughness, and total indicator runout. Some of these parameters have been related to the observed dimensions of pits nucleating fatigue cracks in etched specimens of aluminum alloys (Ref 30).

Surface Corrosion and Thickness Loss. Thickness reduction caused by surface corrosion has been identified as a parameter for modeling corrosion damage. Increased stress caused by the reduced cross section has been incorporated in the stress-intensity factors used in fatigue crack growth predictions (Ref 31). While the decrease in cross section caused by the corrosion was reflected in the stress intensity, it did not completely account for the higher crack growth rates measured at lower stress intensity. Further work is needed to determine if other factors, such as embrittlement, affected the crack behavior.

Exfoliation Corrosion and Pillowing. Exfoliation is a severe form of intergranular corrosion in materials with directional grain

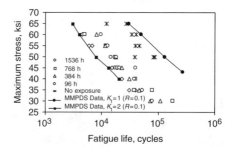

Fig. 5 Fatigue properties of unexposed and corroded, 0.080 in. thick 7075-T6 sheet. Data of various exposure times of prohesion corrosion spray test are compared to *MMPDS Handbook* data. Stress-concentration factor $K_t = 1$ is smooth specimen, $K_t = 2$ has a discontinuity that doubles stress. R is the ratio of minimum stress to maximum stress applied each cycle.

structure, typical of rolled or extruded high-strength aluminum alloys. It originates at exposed end grains, for example at fastener holes, and is characterized by flaking that causes leaflike bulging as the corrosion products accumulate (Ref 32). This type of corrosion leads to a loss of metal capable of carrying load. When exfoliation occurs in joints, the corrosion products cause a volumetric increase, or pillowing, as shown schematically in Fig. 6 and in Fig. 7. Extensive documentation of this phenomenon has shown that the skin volume increases by more than 300% (Ref 33). This in turn, causes high local stresses and rivet head fracture (environment-assisted cracking) where the rivet head meets the shank. The increase of corrosion product volume by exfoliation is typically 6.5 times greater than that predicted by thickness lost. Finite-element analysis has shown that pillowing causes significant stresses. Testing confirms that cracks initiate in the highly stressed areas predicted by the analysis (Ref 34).

Compared to pitting corrosion, the modeling of exfoliation corrosion has received much less attention, although exfoliation and pillowing are significant and escalating maintenance problems. This may be partly due to the difficulty of geometrically modeling the exfoliation corrosion morphologies as equivalent initial flaws for fatigue crack initiation, as well as the complex interactions of loading with the corrosion damage. Attempts have been made to identify a "corrosion process" zone with exfoliation corrosion and to relate the observed fatigue lives to the dimensions of this zone (Ref 35). An investigation of 7075-T6 specimens with exfoliation damage representative of damage occurring in service showed that fatigue lives decreased with increase of the depth of exfoliation (Ref 36).

The metrics used to describe corrosion damage are either highly qualitative (light, moderate, severe) or very detailed (individual pit dimensions) so as to be difficult to measure in the field. The challenge is to relate the metrics developed in the laboratory to damage parameters from inspection of in-service aircraft and to validate the relationships by correlation of life estimated from models using damage metrics with that observed in service. To enable such a correlation, it is necessary to understand the nucleation and growth of corrosion, nucleation and growth of fatigue cracks from corrosion sites, and the process of corrosion fatigue in detail.

Investigations and Modeling of Corrosion/Fatigue Interactions

It is difficult to accurately simulate in-service aircraft conditions in the laboratory. How the effects of corrosion/fatigue interactions during the relatively short military aircraft flight periods are related to those of corrosion alone during the relatively long ground times is not entirely known. Other factors such as type of accelerated corrosion test and environments, specimen geometries, loading conditions, corrosion-protection systems, and their degradation all need to be representative of service conditions.

Accelerated corrosion test methods, designed to approximate in a short time the corrosion-deterioration effects under normal long-term service conditions, have been developed for evaluating various types of corrosion such as pitting, crevice, and exfoliation. ASTM standards have been developed for performing such tests. In the aerospace industry, accelerated laboratory corrosion tests, principally spray tests, are used extensively to evaluate the corrosion susceptibility of alloys and coatings for screening of materials and quality control.

Several environments are currently being evaluated in spray corrosion tests with the intention of simulating service environments. The most commonly used of these is the neutral salt-spray environment consisting of a 5% NaCl solution (ASTM B 117). The SO_2/salt-spray test (ASTM G 85, Annex 4) is used to simulate the performance of alloys in environments such as those present in aircraft carriers. These environments are quite aggressive to many aluminum alloys. The recently developed prohesion test (ASTM G 85 Annex 5), a modified salt-spray test that combines a wet-dry cycle with a dilute environment of lower Cl^- ion concentration, is already recognized as providing a more realistic simulation of corrosion attack than the neutral salt-fog or the SO_2/salt-spray tests (Ref 37).

The validity of the accelerated tests for generating corrosion growth rate data for life prediction of aircraft structures remains to be verified. Nevertheless, these tests are useful in providing an understanding of the nature of corrosion/fatigue interactions in alloys and environments of interest, and considerable progress has been made in this regard. The premise here is that these tests can be useful if the corrosion damage obtained in these accelerated tests is similar to that found in service. Pitting and exfoliation corrosion, followed by crevice corrosion and pillowing, have received the most attention.

Pitting Corrosion and Effects on Fatigue. In studying the effects of corrosion on fatigue behavior, the influence of both preexisting corrosion as well as corrosive environments on fatigue crack initiation have been considered. For the latter case, corrosion/fatigue interactions have been modeled by the testing of pristine specimens in a corrosive environment during which fatigue and corrosion damage accumulate simultaneously. In these models, corrosion pits are considered as surface cracks whose growth rates are determined by the pitting kinetics (Ref 27, 38, 39). A fatigue crack nucleates from the corrosion pit either when the pit grows to a critical size at which the stress-intensity factor (K) reaches the threshold for fatigue cracking (Ref 38) or when the fatigue crack growth rate (da/dN) exceeds the pit growth rate (Ref 39). A recent study of alloy 2024-T3 showed that both these criteria are needed for adequately describing the transition from pitting to fatigue crack growth (Ref 27).

During the service life of an aircraft, the structure typically experiences corrosion between flights and fatigue loading during flight, thus pointing to the need for delineating the effects of prior corrosion on fatigue behavior. In this context, Harmsworth conducted the first quantitative study of the influence of preexisting corrosion on reducing the fatigue life of aluminum alloys (Ref 40). In this study, the amount of corrosion in alloy 2024-T4 was measured as a function of exposure time. Decreasing fatigue lives were observed with increasing corrosion exposure (pit depths).

Pitting-corrosion/fatigue interactions can be modeled by using pit dimensions as inputs into fatigue life prediction models. Quantitative measures such as pit dimensions and surface roughness have been used as metrics to describe corrosion, and their potential feasibility in combination with crack growth analysis tools in estimating fatigue life of components subjected to corrosion has been demonstrated (Ref 30). In this study, the measured fatigue properties of etched (pitted) 2124-T8 and 7050-T74 aluminum alloys compared reasonably well with those calculated analytically using equivalent initial flaw size assumptions. A series of criteria were also described for ranking various corrosion metrics to enable selection of the applicable parameter for structural analysis. Prior corrosion pitting was also found to reduce the fatigue

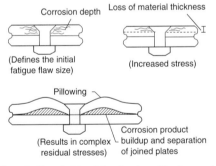

Fig. 6 Corroded lap joint showing pillowing mechanism

Fig. 7 Overall surface appearance of panels representing a fuselage skin joint. 7075-T6 alloy skin with 2017 fasteners was exposed 110 days in an ASTM copper-assisted salt solution spray test.

strength of aluminum alloys 2024-T3 and 2524-T3 at 10^5 cycles by approximately 40% (Ref 41). Crack growth analysis was used in this study to predict the fatigue lives (the number of cycles needed to grow the fatigue crack to its critical size) and to investigate the effects of the alloy compositions and pit morphology on the fatigue lives.

In a similar systematic investigation, the effects of preexisting localized surface pitting corrosion damage on the fatigue lives of alloy 7075-T6 were measured and compared with the predicted lives using measures of corrosion (metrics) obtained from a characterization of the corrosion damage (Ref 20). Pit depth histograms resulting from exposing bare 7075-T6 sheets for up to 1536 h exposure in the prohesion test (ASTM G 85 Annex 5) were determined, and the observations of damage evolution were found to be consistent with the nucleation/growth/decay nature of the pitting process. Fatigue crack growth models were used to predict the life (the number of cycles needed to grow the fatigue crack to its critical size) of 7075-T6 specimens containing various levels of pitting damage. The measured lives generally agreed with the predictions using the average pit size as the initial crack size. This result was explained as due to the pit size distributions offering a significant population of pits near the average size. A continuous decrease in fatigue life with increasing pit dimensions (increasing corrosion exposure) was observed. For the purposes of simulating corrosion/fatigue interactions, the prohesion spray test enables a more controlled and progressive evolution of corrosion damage, amenable to systematic characterization, compared with the commonly used but more aggressive neutral or acidified salt sprays.

While the salt spray and its modifications using standardized corrosion environments provide accelerated corrosion rates and are in common use to screen the relative corrosion susceptibility of materials, they may not be representative of the corrosive environments experienced by aircraft structure. To address this issue, the corrosion products found in crevices in aircraft joints have been analyzed and the chemistry of these products from one such joint was used to reconstitute a representative solution. Interestingly, the crevice environment for a 2024 joint in this case was found to be alkaline, and the alloy showed no tendency to pit when subjected to this environment in the laboratory. This finding suggests that the crevice environment in an aircraft at a given time may not be representative of the overall environmental history of the joint.

In all of these investigations, fractographic examination of broken corrosion fatigue specimens clearly show that the dominant fatigue cracks nucleate from corrosion pits, whose depths are typically in the 50 to 250 μm range.

Exfoliation, Crevice Corrosion, and Effects on Fatigue. The current "find it and fix it" corrosion repair philosophy requires that even the smallest amount of exfoliation damage should be removed, even if its effect on structural integrity is not known. Fatigue testing of specimens prepared from aluminum alloy 7178 wing skins removed from a commercial aircraft showed that specimens with 29% of exfoliation penetration through the skin lasted two lifetimes (twice the number of cycles that the structure was designed to operate without failure) without cracking, whereas specimens in which the exfoliation corrosion was repaired by grind-out had significantly lower fatigue lives (Ref 42). This study suggests that repair by grinding produces effects that may be more detrimental than exfoliation corrosion. Unfortunately, the level of acceptable corrosion itself is not known. It should be noted that these observations were from fatigue tests with compression-dominated loading spectra and may not hold true for tension-dominated loading spectra. In a different study, an evaluation of the effects of exfoliation of the 7075-T6 upper wing skin of the C141 aircraft showed that the fatigue lives decreased with increase of exfoliation depth (Ref 36).

In investigating the effects of corrosion of lap joints, varying amounts of pillowing must be obtained without the pillowing stresses causing failure of the fasteners. An example of pillowing resulting from the exposure of specimens representing a fuselage lap-splice joint is shown in Fig. 7. The specimens were fabricated from clad aluminum alloy 7075-T6 sheet (1.6 mm, or 0.063 in., nominal thickness) and 2017 aluminum alloy fasteners. Aluminum alloy products, particularly sheet and tube, are clad with a layer of a different aluminum alloy that is 80 to 100 mV more anodic than the core to provide cathodic protection to the core alloy. Alloys 7008, 7011, and 7002 are commonly used for cladding 7075. In clad alloys, corrosion progresses to the core/cladding interface and then laterally, thus eliminating perforation of thin products (Ref 43).

The clad 7075 skin lap-splice specimens with 2017 fastener specimens were exposed in a corrosion chamber using a modified ASTM copper-assisted salt solution (CASS) test. The result of metallographic sectioning of a specimen excised from the corroded panel is shown in Fig. 8. The amount of thinning varied from 0.051 to 0.127 mm (0.002 to 0.005 in.). This relatively low level of material thinning corresponds to an average of about 5% thickness loss. The use of the ASTM CASS test with a lap joint representative of fuselage skins has been shown to induce corrosion in the faying surfaces. While the cladding layer has been consumed, the 7075 core had not suffered corrosion damage in 110 days of continuous exposure (Fig. 8). Despite this low level of corrosion, the volume of corrosion products was sufficient to cause pillowing. Additional work is necessary in determining appropriate electrolytes that will provide varying levels of measurable corrosion in reasonable exposure times, from which metrics for exfoliation corrosion can be developed and the effects on fatigue behavior can be determined.

Effects of Corrosion Mitigation. Corrosion prevention compounds (CPCs) are being used more during aircraft maintenance for mitigation of future corrosion. While intended as temporary protection for regions of damaged coatings and exposed surfaces, presently CPCs are also being used in the manufacture of original equipment (Ref 44). Environmental regulations are also requiring the substitution of new nonchromated systems for the proven chromated systems. The effects of CPCs on fatigue behavior of aircraft joints are mixed and depend on the type of joints and loading conditions. Further research is required to reliably predict their effects. In a recent study, application of CPCs to 2024-T3 coupons with prior corrosion extended the remaining fatigue life by as much as a factor of up to 2.5 over specimens with similar corrosion but no CPCs (Ref 45). Similarly, prior corrosion had less effect on the fatigue performance of joints with CPCs in which the fasteners also remained tight.

Figure 9 presents the results of constant amplitude fatigue testing of chromate conversion coated 7075-T6 specimens further treated with primers and tested in four different conditions, treated with: a chromated primer, two different nonchromated primers, and no primer (tested as conversion coated) (Ref 46). For each of these

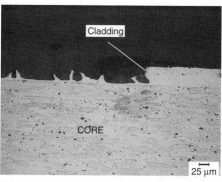

Fig. 8 Corroded lap joint of 7075-T6 skin with 2017 fasteners. (a) Specimen from panel exposed in the ASTM CASS spray test show pillowing and attack of cladding. (b) Micrograph cross section show loss of cladding, but no loss to the 7075-T6 core. Original magnification: 200×

four conditions, six specimens—three without any prior corrosion exposure and three with 1536 h exposure to the prohesion corrosion environment (ASTM G 85, Annex 5)—were fatigue tested at a maximum stress level of 55 ksi and R ratio of 0.1. The fatigue cycles to failure for specimens with and without the corrosion exposure are shown by the shaded and unshaded bars, respectively. Corrosion exposure resulted in a decrease in the fatigue lives of all specimens except those with one of the nonchromated primers. Development of new environmentally compliant systems is presently aimed primarily at achieving corrosion-protection performance approaching that of chromated systems. Their corrosion fatigue behavior under service conditions needs to be characterized to assess effects on structural integrity.

Methodologies for Predicting the Effects of Corrosion on Fatigue

The two most common fatigue life assessment methods used in the military aerospace industry are fatigue crack initiation and crack growth analysis.

Crack initiation analysis uses stress-life or strain-life curves to predict the number of flight hours for crack initiation, which is typically defined to occur when a fatigue crack of 0.25 mm (0.01 in.) length is present. The analysis of crack initiation is the basis for the safe-life design methodology (Ref 10), which defines the flight hours for crack initiation (the crack initiation life or the design life) for a component to be equal to a safety factor times the desired operating life (the number of desired flight hours without crack initiation) for that component. A safety factor of 4 is typically used if the airframe will be subjected to a full-scale fatigue test. Full-scale test articles are typically required to undergo two simulated operating lifetimes without crack indications. In other words, in this case, the airframe is designed to four lifetimes before crack initiation (presence of a 0.25 mm, or 0.01 in., length crack) and tested to two operating lifetimes. If a full-scale test will not be completed for a particular component, a larger safety factor can be required in the design criteria.

Major users of the safe-life design methodology include the U.S. Navy and the U.S. Army for designing dynamic aircraft components. The reasons for adopting this approach vary by application. In U.S. Navy aircraft carriers, for example, the ability to inspect aircraft for fatigue cracks and perform structural repairs is very limited when compared to land-based facilities. As a result, the U.S. Navy has adopted the safe-life approach to minimizing the occurrence of cracks larger than 0.01 in. in length.

An important parameter in the safe-life analysis is the stress-concentration factor, or K_t, which depends on the geometry of a discontinuity such as a hole or notch. Harmsworth documented the first incorporation of corrosion fatigue test data into the prediction of fatigue crack initiation by using test data to determine stress concentrations as a function of corrosion magnitude (Ref 40).

Recently, an attempt has been made to develop a corrosion fatigue life prediction procedure that accounts for variations in the magnitude of corrosion. The measured fatigue life data for specimens with varying levels of prior corrosion were used to perform an extensive probabilistic analysis to determine the corrosion-induced stress-concentration factor for aluminum alloy 7075-T6 (Ref 47). To supplement the measured test data, crack growth analysis was used to predict the number of cycles to initiate 0.01 in. long semielliptic cracks in the corroded specimens for which the measured pit sizes were used as the initial flaw dimensions. This analysis can provide a distribution of fatigue lives (equal to the number of cycles needed to initiate a 0.01 in. crack) for a given stress level corresponding to the distribution of the measured pit sizes for a given degree of corrosion damage. The analysis can then be repeated for various stress levels. This analysis was conducted for 7075-T6 specimens with an open-hole ($K_t = 3$) geometry without corrosion as well as with superimposed initial flaw sizes corresponding to the characteristic, 5th and 95th percentile values for the Weibull distribution of the pit dimensions for a given corrosion damage. The results are plotted (Fig. 10), which shows the predicted stress-life curves for 7075-T6 specimens with an open-hole geometry without corrosion ($K_t = 3$) curve and with three levels of initial flaw sizes (5%, char, 95% corresponding to the characteristic, 5th, and 95th percentiles values for the Weibull distribution of the pit dimensions for a given corrosion damage, in this case corresponding to the 96 h of corrosion exposure shown in Fig. 5.

The reduction in fatigue life is typically related to an increase in the stress-concentration factor. The concentration (or notch) factor associated with a given level of corrosion damage is then the ratio of the stresses at a particular life for the noncorroded specimen and the corroded specimen. Sets of stress values at a given fatigue life (cycle) can be read from plots such as those shown in Fig. 10. The ratios of the stress values corresponding to the noncorroded specimens to those for the three curves are used to obtain the required notch factors corresponding to the characteristic 5th and 95th percentiles values of the pit dimensions.

A metric required to characterize the level of corrosion on a particular specimen was chosen to be directly related to the size of the pits. The intent was to make the method valid for any corrosive environment and one that could then be used in the field to quantify the corrosion severity

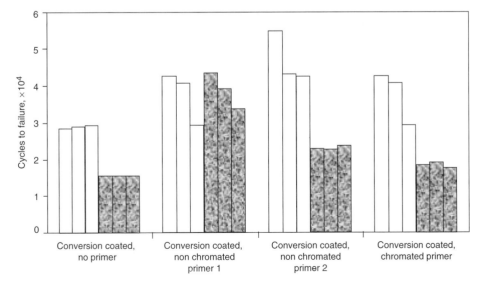

Fig. 9 Effect of various corrosion protection schemes on the fatigue behavior of 7075-T6 alloy. Six samples were fatigue tested from each protection scheme. Fatigue test maximum stress 380 MPa (55 ksi), $R = 0.1$. The white bars had no prior corrosion exposure. The three shaded bars in each set were exposed to 1536 h prohesion corrosion environment.

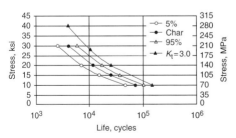

Fig. 10 Predicted fatigue lives for 7075-T6 alloy, open-hole geometry specimens with superimposed corrosion damage. Open-hole without corrosion damage, $K_t = 3.0$ curve. Char is the characteristic damage curve. 5% and 95% curves represent the 5th and 95th percentile of pit dimension in a Weibull distribution. Original units are ksi.

as well as to correlate with the corrosion notch factor. The pit norm metric is defined as:

$$\text{Pit norm} = \sqrt{l^2 + w^2 + d^2}$$

where l is pit length, w is pit width, and d is pit depth. The pit norm was computed for each of the individual pits measured on each corroded test specimen. A Weibull distribution was fit to each set of the pit norms, and the characteristic value was found. The corrosion notch factors computed from Fig. 10 were plotted against the characteristic pit norm for that data set. The results are shown in Fig. 11 from which the distribution of the notch factor for a given amount of corrosion can be determined. The pit norm is the metric used to characterize the corrosion, and the notch factor is the quantity used to assess the effect of that amount of corrosion on the life of the component.

To use the data in Fig. 11, the surface pitting corrosion is first characterized in terms of pit sizes by computing the pit size norm for a sufficient number of pits that adequately represents the extent of corrosion in a part. The best-fit Weibull distribution to the pit norm data defines the characteristic value. For this value, the effective notch factor and the range representing the 5th to 95th percentile of the expected distribution of the notch factor can then be determined. As can be observed in Fig. 11, the corrosion notch factor for aluminum alloy 7075-T6 varies from 1.0 (for limited corrosion) to 2.25 (for more severe corrosion). This ability to analyze corrosion with respect to some quantitative severity scale such as a "notch" factor has the potential to enable its incorporation in the assessment of structural integrity using safe-life methods.

A study of 2xxx series aluminum alloys also concluded that the effect of corrosion on a part can be modeled by a stress-concentration factor and that even light corrosion can reduce the crack initiation life by effectively increasing the local stress by 18% or more (Ref 48).

Crack growth analysis methods use fracture mechanics to predict the number of flight hours for the initial flaw to grow to failure. These techniques form the analytical basis of the damage tolerance analysis (DTA) design philosophy (Ref 9), developed by the U.S. Air Force in response to the crash of an F-111 aircraft in 1969. This approach centers on the assumption that a preexisting flaw is located at the critical area in a part. Because a flaw that can propagate under cyclic loading is present, there is no crack initiation period. The assumed initial flaw size is defined by the minimum inspectable flaw that can be detected. A typical initial flaw size used in industry is 1.25 mm (0.05 in.) in length. The predicted crack growth life (the flight hours needed to grow the crack to its critical length) is divided by a safety factor of two to define the inspection interval for the part.

Because the initial flaw used in DTA is typically much larger than the dimensions associated with corrosion damage, the effect of corrosion can be easily covered by DTA. The effect of environment can also be included during DTA by using the crack growth rate behavior for the environment and by representing the loss of thickness by a corresponding increase in stress.

Boeing has assessed the corrosion damage effects with respect to specific DTA. The corrosion model used for the analyses assumed corrosion damage limits for specific part details. Those limits were either the minimum thicknesses associated with repairs or an overall predefined maximum thickness loss. Both limits are defined by an operating stress equal to the design strength for the part. The outcome of this analysis was a revised inspection schedule accounting for corrosion damage. Details with unacceptable inspection intervals were identified for pre-engineered repairs. For the KC-135 aircraft, while the fuselage lap joints are not critical, some fuselage panels with spot welded joints, which are located in corrosion-prone areas, are being replaced due to corrosion damage. Replacement of those joints is again based on inspection results (visual clues confirmed by nondestructive inspection). One complex problem in this DTA is the uncertainty about future corrosion. A joint with light corrosion could be accepted without significant repairs needed. However, continuing corrosion could take the joint past an acceptable level of damage prior to the next programmed depot maintenance cycle. This corrosion state is a moving target that needs to be accounted for in the acceptance criteria for fuselage joints.

Although DTA can incorporate the effect of corrosion on a primary crack, its effects at multiple locations also need to be addressed. Widespread corrosion can lead to multisite damage. In the case of pitting corrosion, fatigue cracking from individual pits needs to be modeled in addition to the analysis of the dominant flaw in the DTA. This issue has been addressed in several ways such as by using pit growth models to predict when a pit is large enough to produce a crack that exceeds the fatigue crack growth threshold (Ref 39) or by modeling the deepest pit as the equivalent initial crack (Ref 24).

The fatigue damage and failure induced by pitting corrosion are composed of seven stages: pit nucleation, pit growth, transition from pit growth to short crack, short crack growth, transition from short crack to long crack, long crack growth, and fracture. In studying the effects of preexisting corrosion on fatigue behavior, pit nucleation and growth are not relevant in DTA, but the nature of transition from the pit to a short crack and short crack growth need to be clearly delineated. The transition from pit growth to short cracks covers the initiation of fatigue cracks from the pits. Short crack growth has been characterized by faster crack growth rates than observed for larger cracks (>0.25 mm, or >0.01 in.) at the same stress-intensity value, K (Ref 49). As short cracks continue to grow, they transition to long cracks that propagate in accordance with the standard da/dN versus ΔK curves for the particular alloy. Available life prediction methods are adequate once the cracks grow beyond the "short" crack stage.

In a study in which the transition from pit to a crack is assumed to occur when the crack growth rate exceeds the pit growth rate, the fatigue crack growth curve for long crack data was modified to account for the crack growth rate of the small cracks. This modification affected the lower end of the curve where the threshold stress-intensity factor changed from 2.4 to 1.2 MPa(m)$^{1/2}$ (Ref 38).

The U.S. Air Force is developing a holistic structural assessment method under an initiative to develop analytical tools to assess the effects of corrosion on aircraft structural integrity (Ref 50). Models for the effects of surface corrosion (pitting and general) and corrosion pillowing in joints as well as the growth of fatigue cracks from the corrosion are available in this method. In experiments conducted during the development of this method, the fatigue lives of 7075-T6 specimens were reduced significantly after natural corrosion exposure for 3 months with little further reduction upon exposure up to 12 months. This initial reduction of fatigue lives was attributed to the formation of an "influential" pit following, while further exposure had only a minor effect on fatigue life. Scanning electron

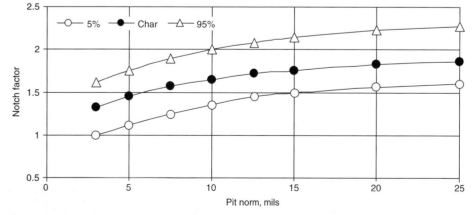

Fig. 11 Corrosion notch factor (corrosion severity) as a function of characteristic pit norm corrosion metric. Notch factors were computed for data of Fig. 10.

microscopy examination of the fracture surface showed that the pits at crack origin were typically 75 to 100 µm deep. Based on a holistic life assessment, it was suggested that the "influential" pit should be 100 to 150 µm deep with an aspect ratio of 3 to 1. Many of the other investigations discussed earlier also showed a large reduction of fatigue lives, however, at corrosion damage levels corresponding to much smaller measured pit depths.

Recent Development and Future Needs

In response to the significant cost and readiness impact resulting from corrosion, a new U.S. federal law was enacted in Dec 2002, which created a Department of Defense-wide corrosion prevention and mitigation program for both equipment and infrastructure. Several initiatives are underway, ranging from the formation of a "Corrosion Prevention and Control Integrated Product Team," and establishing a "Corrosion Information Exchange Network," to development and testing of materials, processes, and treatments to combat corrosion. Planning for corrosion prevention and control is now a requirement for most Department of Defense programs (Ref 7). This new corrosion policy can be expected to have an overall positive impact in mitigating the effects of corrosion/fatigue interactions.

Significant progress has been made in modeling the environments and corrosion growth experienced by various aircraft (Ref 11). A major difficulty in the development and validation of methods to assess the effects of corrosion is the lack of correlation between actual in-service corrosion damage and corrosion fatigue failures with those observed in the laboratory. Testing of structural components removed from aircraft under conditions that simulate the aircraft environment is invaluable in this regard. One difficulty the authors have experienced is that a component removed from the aircraft and made available for testing has already undergone a few maintenance cycles in which the corrosion has been typically removed and the component is cleaned or it is beyond repair and hence not useful for testing.

The nature of accelerated testing needs to be studied further. Whereas it is useful to study the corrosion effects on laboratory specimens, some meaningful link must be made back to the nature of the corrosion experienced in the field. It would be very helpful if time-scale references could be determined between life in the field and exposure in the laboratory. The ability to correlate damage observed in the field with standardized metrics and with the results of nondestructive inspection is also important. Techniques such as laser profilometry have been used recently to map and quantitatively characterize the surfaces of corroded specimens (Ref 51). The corrosion metrics determined from this technique were successfully used to predict the lives of 7075-T6 specimens with prior corrosion. Laser and white light profilometry and other similar techniques need to be developed and demonstrated for obtaining corrosion metrics from the field and for use in life prediction models.

ACKNOWLEDGMENT

The authors gratefully acknowledge support since 1996 under several contracts from the U.S. Air Force and U.S. Navy.

REFERENCES

1. S.J. Findlay and N.D. Harrison, *Mater. Today,* Vol 5 (No. 11), 2002, p 18–25
2. G.S. Campbell and R. Lahey, *Int. J. Fatigue,* Vol 6 (No. 1), 1984, p 25–30
3. "Opportunities to Reduce Corrosion Costs and Increase Readiness," report GAO-03-753, Report to Congressional Committees, U.S. General Accounting Office, July 2003, p 7
4. V.S. Agarwala and S. Ahmed, "Corrosion Detection and Monitoring—A Review," paper 00271, Corrosion 2000, Proceedings, National Association of Corrosion Engineers
5. V.S. Agarwala, *Naval Res. Rev.,* Vol 50 (No. 4), 1998, p 14–24
6. R. Kinzie and G. Cooke, Corrosion in USAF Aging Aircraft Fleets, *Fatigue in the Presence of Corrosion,* Proc., RTO-MP-18, NATO Research and Technology Organization Meeting, 1999, p 16-1 to 16-12
7. R.C. Kinzie and D.T. Peeler, Structural Damage Management: New Rules, New Tools, *Aging Aircraft 2005,* Proc., Eighth Joint NAS/FAA/DOD Conference on Aging Aircraft, Joint Council on Aging Aircraft, 2005
8. J. Schijve, Fatigue Crack Growth under Variable Amplitude Loading, *Fatigue and Fracture,* Vol 19, ASM Handbook, 1996, p 110–133
9. M.P. Kaplan and T.A. Wolff, Life Extension and Damage Tolerance of Aircraft, *Fatigue and Fracture,* Vol 19, ASM Handbook, 1996, p 557–565
10. M.E. Hoffman and P.C. Hoffman, *Naval Res. Rev.,* Vol 50 (No. 4), 1998, p 4–11
11. Final Report, Aging of U.S. Air Force Aircraft, Committee on Aging of U.S. Air Force Aircraft, National Materials Advisory Board Commission on Engineering and Technical Systems, National Research Council, National Materials Advisory Board Publication NMAB-488-2, National Academy Press, 1997
12. R.J.H. Wanhill, Aircraft Corrosion and Fatigue Damage Assessment, *Proc. U.S. Air Force ASIP Conf.,* 1995, p 983–1027
13. G.K. Cole, G. Clark, and P.K. Sharp, "The Implications of Corrosion with Respect to Aircraft Structural Integrity," Research report DSTO-RR-0102, AR-010-199, Aeronautical and Maritime Research Laboratory, Melbourne, Australia, March 1997
14. *Fatigue in the Presence of Corrosion,* Proc., NATO Research and Technology Organization Meeting (Corfu, Greece), 1998
15. U.G. Goranson, *Int. J. Fat.,* Vol 19 (Suppl. 1), 1997, p S3–S21
16. C.L. Brooks, S. Prost-Domasky, and K. Honeycutt, Corrosion is a Structural and Economic Problem: Transforming Metrics to a Life Prediction Method, RTO-MP-18, *Fatigue in the Presence of Corrosion,* Proceedings, NATO Research and Technology Organization Meeting, 1999, p 14-1 to 14-12
17. D.W. Cameron and D.W. Hoeppner, Fatigue Properties in Engineering, *Fatigue and Fracture,* Vol 19, ASM Handbook, 1996, p 15–26
18. S.S. Manson and G.R. Halford, *Fatigue and Durability of Structural Materials,* ASM International, 2006
19. J.P. Komorowski, N.C. Bellinger, and R.W. Gould, Local Stress Effects of Corrosion in Lap Splices, RTO-MP-18, *Fatigue in the Presence of Corrosion,* Proc., NATO Research and Technology Organization Meeting, 1999, p 5-1 to 5-8
20. K.K. Sankaran, R. Perez, and K.V. Jata, *Mater. Sci. Eng.,* Vol A297, 2001, p 223–229
21. "Metallic Materials Properties Development and Standardization," Report No. DOT/FAAAR-MMPDS-01, U.S. Department of Transportation, Federal Aviation Administration, 2003
22. K.J. Miller, *Mater. Sci. Technol.,* Vol 9, 1993, p 453–458
23. S. Barter, P.K. Sharp, and G. Clark, *Eng. Fail. An.,* Vol 1, 1994, p 255–266
24. A.F. Doerfler, R.J. Grandt, Jr., R.J. Bucci, and M. Kulak, A Fracture Mechanics Based Approach for Quantifying Corrosion Damage, Proc., *Tri-Services Conference on Corrosion,* 1994, p 433–444
25. G.S. Chen, M. Gao, and R.P. Wei, *Corrosion,* Vol 52 (No. 1), 1996, p 8–15
26. C.-M. Liao, G.S. Chen, and R.P. Wei, *Scr. Mater.,* Vol 35, 1996, p 1341–1346
27. G.S. Chen, K.-C. Wan, M. Gao, R.P. Wei, and S. Flournoy, *Mater. Sci. Eng.,* Vol A219, 1986, p 126–132
28. M. Khobaib, T. Matikas, and M.S. Donley, *J. Adv. Mater.,* Vol 35, 2003, p 3–8
29. T. Mills, P.K. Sharp, and C. Loader, "The Incorporation of Pitting Corrosion Damage into F-111 Fatigue Life Modeling," report DSTO-RR-0237, DSTO Aeronautical and Marine Research Laboratory, 2002
30. R. Perez, Corrosion/Fatigue Metrics, Proceedings, *19th Symposium of the International Committee on Aeronautical Fatigue* (Edinburgh, Scotland), Vol 1, 1997, p 215–229
31. P.S. Pao, S.J. Gill, and J.C.R. Feng, *Scr. Mater.,* Vol 43, 2000, p 391–402

32. J.R. Davis, *Corrosion of Aluminum and Aluminum Alloys,* ASM International, 1999, p 68
33. R.S. Piascik, R.G. Kelly, M.E. Inman, and S.A. Willard, Fuselage Lap Splice Corrosion, U.S. Air Force report WL-TR-96-4094, Proc., *1995 U.S. Air Force ASIP Conference,* 1996
34. N.C. Bellinger and J.P. Komorowski, The Effect of Corrosion on the Structural Integrity of Fuselage Lap Joints, U.S. Air Force Report WL-TR-96-4094, *1995 U.S. Air Force ASIP Conference,* 1996
35. K. Sharp, T. Mills, S. Russo, G. Clark, and Q. Liu, Effects of Exfoliation on the Fatigue Life of Two High-Strength Aluminum Alloys, Aging Aircraft 2000, Proc., Fourth Joint NAS/FAA/DOD Conference on Aging Aircraft, Joint Council on Aging Aircraft, 2000
36. N.C. Bellinger, A. Marincak, M. Harrison, and T. Reeb, Effect of Exfoliation Corrosion on the Structural Integrity of 7075-T6 Upper Wing Skins, Proc., 2003 U.S. Air Force ASIP Conference
37. S.P. Lyon, G.E. Thompson, and J.B. Johnson, Materials Evaluation Using Wet-Dry Mixed Salt Spray Tests, *New Methods for Corrosion Testing of Aluminum Alloys,* STP 1134, ASTM International, 1992, p 20
38. Y. Kondo, *Corrosion,* Vol 45, 1989, p 7–11
39. D.W. Hoeppner, Model for Prediction of Fatigue Lives based Upon a Pitting Corrosion Fatigue Process, *Fatigue Mechanisms,* STP 675, ASTM International, 1979, p 841–870
40. C.L. Harmsworth, "Effect of Corrosion on the Fatigue Behavior of 2024-T4 Aluminum Alloy," Technical Report 61-121, Aeronautical System Division, July 1961
41. G.H. Bray, R.J. Bucci, E.L. Colvin, and M. Kulak, Effects of Prior Corrosion on the S/N Fatigue Performance of Aluminum Alloys 2024-T3 and 2524-T3, *Effects of Environment on the Initiation of Crack Growth,* STP 1298, ASTM International, 1997, p 89–103
42. N.C. Bellinger, T. Foland, and D. Carmody, Structural Integrity Impacts of Aircraft Upper Exfoliation Corrosion and Repair Configurations, Aging Aircraft 2003, Proc., Seventh Joint NAS/FAA/DOD Conference on Aging Aircraft, Joint Council on Aging Aircraft, 2003
43. E.H. Hollingsworth and H.Y. Hunsicker, Corrosion Resistance of Aluminum and Aluminum Alloys, *Metals Handbook,* 9th ed., 1979, p 204–236
44. R.G. Kelly and R.C. Kinzie, Current State of Corrosion Prevention Compounds, Aging Aircraft 2003, Proc., Seventh Joint NAS/FAA/DOD Conference on Aging Aircraft, Joint Council on Aging Aircraft, 2003
45. R.C. Rice, Corrosion Fatigue Life Prediction of Bare and CPC-Protected 2024-T3 Sheet, Proc., 2003 U.S. Air Force ASIP Conference
46. J. Deffeyes, K. Hunter, R. Lederich, R. Perez, and K.K. Sankaran, "Effects of Corrosion Control Coatings on the Pitting Behavior of Aluminum Alloy 7075-T6," Corrosion 2001, National Association of Corrosion Engineers, 2001
47. H.G. Smith, R. Perez, K.K. Sankaran, M.E. Hoffman, and P.C. Hoffman, "The Effect of Corrosion on Fatigue Life—A Nondeterministic Approach," paper 2001-1376, AIAA
48. C. Paul and J. Gallagher, "Modeling the Effect of Prior Corrosion on Fatigue Life Using the Concept of Equivalent Stress Concentration," U.S. Air Force ASIP Conference, 2001
49. R.C. McClung, K.S. Chan, S.J. Hudak, Jr., and D.L. Davidson, Behavior of Small Fatigue Cracks, *Fatigue and Fracture,* Vol 19, *ASM Handbook,* 1996, p 153–158
50. E.J. Tuegel and T.B. Mills, Correlation of Holistic Structural Assessment Method with Corrosion-Fatigue Experiments, *Proc. Sixth Joint FAA/DoD/NASA Conference on Aging Aircraft* (San Francisco, CA), Sept 2002
51. M. Koul, *Corrosion,* Vol 59, 2003, p 563

SELECTED REFERENCES

- R.G. Ballinger, L.W. Hobbs, R.C. Lanza, and R.M. Latanision, "Environmental Degradation/Fatigue in Aircraft Structural Materials: Relationship between Environmental Duty/Component Life," Contract F49620-93-1-0291, Final Report, Air Force Office of Scientific Research, May 1997
- *Effects of the Environment on the Initiation of Crack Growth,* STP 1298, ASTM International
- Joint Council on Aging Aircraft, *Proceedings of Aging Aircraft Conferences* (2001, Orlando; 2003, New Orleans; and 2005, Las Vegas)
- A.K. Kuruvilla, "Corrosion Predictive Modeling for Aging Aircraft," Contract SPO700-97-D-4001, Critical Review and Technology Assessment, Defense Supply Center-Columbus, July 1999 (available from Advanced Materials and Processes Information Analysis Center, Rome, NY)
- A.K. Kuruvilla, "Life Prediction and Performance Assurance of Structural Materials in Corrosive Environments," Contract SPO700-97-D-4001, State of the Art Report, Defense Supply Center-Columbus, Aug 1999 (available from Advanced Materials and Processes Information Analysis Center, Rome, NY)
- Special Issue: Beating Corrosion—Sweeping Policy Change Overhauls DoD Acquisition and Sustainment, *AMPTIAC Quarterly,* Vol 7 (No. 4), 2003

Corrosion of Electronic Equipment in Military Environments

Joseph T. Menke, U.S. Army

CORROSION OF ELECTRONIC EQUIPMENT became a concern when electrical components left the 50% relative humidity and 20 °C (70 °F) controlled environment of the rooms that computers first occupied. Electronic equipment went from labs, banks, and offices to aircraft, tanks, ships, and missiles. These pieces of equipment experience the cold of winter, the heat of the desert, the salt spray of the ocean, the moisture-condensing conditions of landing aircraft, or the pressure spray washing conditions that maintenance operations may entail. The corrosivity of solder flux residues and etching solutions, and the cleanliness testing of printed circuit boards (PCBs) prior to conformal coating, became issues in the manufacturing process to lessen corrosion problems. Potting, hermetic sealing, conformal coating, and other operations were the design and manufacturing attempts for survival of the electronic components in adverse environments.

An Historical Review of Corrosion Problems

Some of the earliest information on the deterioration of electronic equipment is found in Greathouse and Wessel's treatise *Deterioration of Materials: Causes and Preventive Techniques,* published in 1954 (Ref 1). The chapter on electronic equipment identified the environmental concerns as temperature, moisture, biological growth, rain, salt spray, sand and dust, and shock and vibration. It further identified the deterioration of plastic insulating materials as a significant problem of the time. In 1958, military standard 441 was prepared by the Navy's Bureau of Ships, which essentially addressed the reliability of military electronic equipment (Ref 3).

Prior to the standard, military electronic equipment was "reliable" when procured, because all the testing was performed on new systems. The use of mean-time-to-failure data gathering for purposes of establishing replacement electronic parts was unique. Equipment was allowed to age in the environment, and maintenance consisted of changing parts or assemblies to keep the equipment operational. The basic corrosion concerns of the 1960s were the deterioration of the physical properties of the materials involved in electronic equipment. A report in 1967 compared the fungus resistance of vinyl polymer insulating cable coverings (Ref 4). A 1963 Bell Telephone Labs report described a procedure for reducing the embrittlement of gold-plated solder joint (Ref 5). A gold-tin intermetallic compound was formed during soldering, and the thickness of the gold plating had to be reduced to minimize the amount of gold penetrating/dissolving in the solder that produced the brittle joint.

In 1965, Rock Island Arsenal evaluated the corrosive effects of sleeving materials, tapes, varnishes, sealing compounds, and molded plastics on materials such as zinc, cadmium, steel, magnesium, and copper (Ref 6). In 1965, the cover of *Chemical and Engineering News* magazine (Ref 7) described corrosion as "a plague from space to razor blades." The article was an excellent review of corrosion; however, the reference to corrosion as a "plague" became commonplace in the electronics industry. "Red plague" was used to describe the red corrosion product observed on silver-coated copper wire (Ref 8). The silver plating was the cathode. Because small pores are generally present in the thin plating, galvanic corrosion of the copper occurred and was accelerated by the large cathode/small anode relationship. The resulting red corrosion product of cuprous oxide (Cu_2O) was formed in the presence of water and oxygen. "Purple plague" has been identified as an intermetallic compound that forms when a gold wire is bonded to an aluminum pad (Ref 9). Time and temperature can affect the solid solubility of gold and aluminum. When the compound $AuAl_2$ is formed, the purple color of the intermetallic is evident. An example of purple plague is shown in Fig. 1. "Blue plague" was used to describe the oxidation of tin coatings as a result of solder reflowing operations. "Green plague" described the green corrosion product of copper/copper chloride that forms on tin-plated copper wire and on bare copper surfaces when solder flux residues are not properly removed (Ref 10). "White plague" was used to describe general corrosion of solder material on PCBs. "Worm plague" was used to describe the selective leaching of lead from lead-tin solder joints that corroded in contact with antistatic materials impregnated in pink poly bags used for cushioning and packaging of PCBs. There are probably other "plagues" in the electronics industry, but it is important to note that the front cover of Ref 7 had a more lasting effect than the actual article on corrosion. In 1966, S.M. Arnold of Bell Laboratories reported the presence of tin whiskers (hairlike metallic growths) on tin-plated items of electrical equipment (Ref 2). It has not been determined to this day (2006) if surface oxides, intermetallics, or some other mechanism caused the residual stress necessary for whisker growth. Finally, in the October 1968 edition of *Metal Finishing,* various chemical etchants were evaluated for controlling the undercutting problems of nickel/gold coatings on copper for PCB manufacture (Ref 11). Control of the undercutting was essential to prevent trapping of the etchant residues on the PCB.

An interesting change began to occur in the 1970s. Actual components from electronic hardware began to be evaluated for corrosion resistance. In 1971, the Electronics Command (ECOM) investigated general corrosion caused by the effects of plastics decomposition products on beryllium-copper contacts (Ref 12). In 1974, ECOM investigated the effects of corrosion on waveguides (Ref 13). Galvanic corrosion, crevice corrosion, and stress-corrosion cracking were highlighted as specific corrosion problems associated with these components. In 1975, the Tank and Automotive Command highlighted the corrosion on components such as relays, starters, and motors (Ref 14). In 1979, a report was written on electrical contacts in submarine-based electronic equipment (Ref 15). This report was a first attempt at considering the effects of specific environments of a system (submarine) on component parts (electrical contacts.) It addressed the materials involved in the component parts and the specific types of corrosion that occurred. The 1970s also brought about different concepts of finishing contacts. The military was gold plating most of the high-reliability items, while industry was using tin plating (Ref 16, 17). This conflict often resulted in tin-plated contacts on PCBs being inserted into gold-plated beryllium-copper connectors. The resulting galvanic corrosion destroyed the tin-plated contacts and the electrical function of the board. Also, the classification of white tin corrosion products on solder joints as cosmetic contributed to the confusion of the situation.

In the 1980s, a more technical approach to electronics corrosion had evolved with the identification of the types of corrosion being experienced on electronics hardware. Electronics corrosion symposia were conducted at the University of Minnesota and the National Association of Corrosion Engineers annual corrosion conferences. The Army originally published MIL-STD-1250 as a corrosion prevention and deterioration control document for electronic components and assemblies in 1967. It was updated in 1992 and was changed to MIL-HDBK-1250 in 1995 (Ref 18). In 2001, it was designated "inactive for new design" as a result of acquisition reform. In 1983, the Navy published a document (Ref 19) entitled "Design Guidelines for Prevention and Control of Avionic Corrosion." It described the effects of the environment and the types of corrosion failure with Naval equipment. In 1988, the author was provided an "Air Force Lessons Learned" summary that detailed the major cause of electronic corrosion as a water intrusion problem (Ref 20). The Air Force papers cited water contamination in parts per million causing problems with hermetically sealed microcircuits. The Army was experiencing moisture in the ounces range in taillights and other black boxes. The Navy was also experiencing saltwater contamination in gallon quantities in the electronics onboard aircraft carriers. This situation typifies the electronic corrosion problems with military equipment.

Examples of Corrosion Problems

Importance of Design for Corrosion Control. Looking at individual pieces of equipment today (2006) can illustrate the electronic corrosion problems. The typical "electronic black box" is used to contain electrical equipment that provides various functions (Ref 21). The worst assumption that one can make is that the box is hermetically sealed. Webster's dictionary defines *hermetic* as fused, which implies soldered, brazed, or welded. It does not mean an O-ring seal, potted connector, or some similar organic material seal. Rubber seals, for example, often leak because of the physical properties of rubber. Unconfined rubber occupies a known volume, but under a load, the rubber tends to extrude outward. Knowing that rubber extrudes rather than compresses requires that the O-ring groove be large enough to allow for this extrusion. Otherwise, the rubber will extrude out of the groove during tightening, and a seal is not formed. Diurnal cycling can result in moisture condensation that produces a humidity cabinet environment inside the electronic box, vehicle taillight, or similar electronic devices. Two other adverse properties of rubber in service are ozone cracking and compression set. When these phenomena are experienced on supposedly sealed electronic devices, water can get in and cause serious corrosion problems.

Thus, the design should allow the box to breathe if it is not hermetically sealed. The main problem is that if the box is almost hermetically sealed, water can get in and, most likely, cannot get out. If the inside gets wet, a drain hole is one approach that can be used to allow the water to run out. The system can then breathe, and, if it gets wet, it can also dry out so that minimal corrosion may be encountered. A drain hole in the vehicle taillights would have resulted in a longer useful life for these components. Another approach may be used in lieu of or in conjunction with a drain hole. Volatile corrosion inhibitors (VCIs) will protect multimetal surfaces that are usually present in electronic devices. The VCIs can be used in semisealed units of various sizes for up to one year. Testing has shown that VCIs in a sealed plastic bag with PCBs can protect against 4000 h of humidity cabinet exposure, with no adverse effects on the PCB components. It should be noted that not every VCI is a multimetal formulation and that the older VCIs formulated for protecting bare steel are generally unsuited for electronic component protection.

The next aspect of an electronic box is the position of the PCBs. The PCBs should be positioned in the vertical plane, not the horizontal plane. In the horizontal plane, dust, dirt, salt, moisture, and even mold can accumulate on the surface of the PCB. When a drop of water bridges two traces on a board that are at different voltages, the high-voltage lead becomes the anode and starts to corrode. Figure 2 shows the result of an electrolyte bridging two circuits on an electronic board. The low-voltage lead becomes the cathode, and ions that went into solution can begin to plate out at the cathode. When the plating reaches the anode, a short circuit is created, and the board ceases to function properly. Often, military boards are given a conformal coating so that water drops are prevented from contacting the circuits. However, all organic coatings are permeable to water vapor. Pure water under a conformal coating does not cause an immediate problem, because pure water is nonconducting. If the board is contaminated with solder flux residues, circuit etching salts, or other conductive materials under the conformal coating, a failure can be expected, because a conductive path under the conformal coating is created. In addition, water can be trapped inside electrical connectors. Figure 3 shows the result of an electrolyte trapped inside an electrical connector. The Army has 690, 1380, 2760, and 5520 kPa (100, 200, 400, and 800 psi) steam cleaners in the field for use by soldiers to perform certain types of maintenance. These devices can blow water past most types of rubber or O-ring 35 kPa (5 psi) seals, trapping the water inside. Another way in which horizontally positioned boards become contaminated with electrolyte is through spillage and runoff of soda beverages.

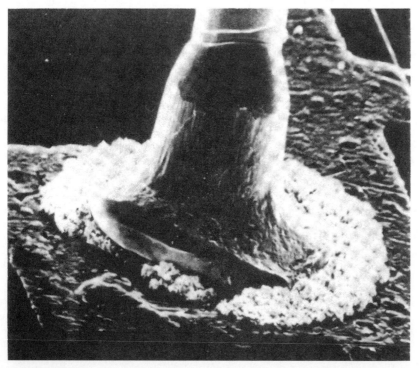

Fig. 1 Example of purple plague. The failure characteristics often associated with purple plague are increased electrical resistance and lower mechanical (bond) strength.

Coffee spillage was encountered on PCBs mounted in the flight control unit of a C-5 aircraft. Figure 4 shows the result of the coffee spill in the contact area. Naval electronic maintenance personnel have suggested that black boxes have a pointed top (shaped like a doghouse) to prevent liquid containers from being set on the devices.

The third aspect of the black box design is proper location of the connectors for insertion of the PCB. Locate the board connectors on the side or the top, not the bottom. In design, there is a natural tendency to locate the connectors in the bottom of the electrical box, especially if the access door to the unit is on the top. The original control units in the Bay Area Rapid Transit cars were an example of this design. Rundown of moisture and debris accumulates in the contact area in the bottom of the box. This results in electrolyte bridging and shorting between the individual connector tabs and the PCB contacts, causing corrosion and eventual failure of the system. With the connectors on the side or the top and the PCBs in a vertical position, corrosion due to moisture accumulation is minimized.

Proper positioning or location of the internal connectors is the next consideration for the feed-through connections. The design should locate the feed-through connectors on the side of the black box. Water accumulation in sump areas around a feed-through located in the bottom can completely corrode the highest-voltage lead, rendering the unit totally useless, as shown in Fig. 5. Moisture-proof (environmental) connectors can provide a false sense of security, based on the wicking nature of multistrand wire or coaxial leads. Any nick, slit, or hole that penetrates the sheathing and/or insulation allows moisture to enter the cable, wick along the strands, and enter the black box through the connector. Fasteners holding the feed-through in place are also potential paths for electrolyte to leak into the electrical box. Drip loops need to be included in the external lead wires coming to the feed-through to keep water out of the system.

Material Considerations. The basic material used for circuitry on PCBs and in general wiring is copper. Connectors may be made of phosphor bronze or beryllium-copper alloys. The coating applied to the connectors and circuits is generally gold plating, to provide solderability, conductivity, and corrosion resistance for maximum reliability. In other instances, tin plating, lead-tin plating (solder plating), or silver plating may be used on various components. It is essential to use similar metals on the connectors, PCB contacts, and circuits to prevent galvanic corrosion. Gold in contact with tin or solder plating will result in severe galvanic corrosion of the tin or solder plating. Figure 6 shows tin-plated contacts that were inserted into a gold-plated beryllium-copper connector in the horizontal position.

Another materials problem in electronic design that is unfamiliar to many electrical engineers is solid-solution phenomena or the formation of intermetallic compounds when certain materials are in contact with each other. Soldering of gold surfaces to tin-coated surfaces results in an intermetallic gold-tin compound that forms a brittle solder joint. The effects of solid-solution phenomena are best illustrated when gold and copper are in intimate contact with each other. This results when the copper circuitry or copper contacts are gold plated. Because the gold is typically 1 μm (40 μin.) or less in thickness, the gold is slowly diluted by the copper migrating (via solid solution) through the gold layer to the surface. As the copper content in the gold surface increases, a general blackening (copper oxidation) of the surface occurs. The typical cleaning technique prior to soldering for maintaining the board contacts is to use a pencil eraser (soft abrasive) to get rid of the black (copper) corrosion product. To prevent the solid solution and thus the subsequent corrosion from occurring in the first place, an antidiffusion coating of nickel or cobalt is required between the copper and gold. The same antidiffusion coating can be used between silver plating on copper wire or tin plating on copper wire to prevent corrosion of these coated wires. Figure 4 shows contact surfaces that were gold plated, with the rest of the circuits on the board left with the tin coating. While this can save money, the presence of the electrolyte and the horizontally positioned board allowed electrolyte to permeate the conformal coating, and galvanic corrosion attacked the tin at the gold-tin interface. In recent years, the amount of silver plating used on electrical equipment has been decreasing because of past problems with silver migration and the formation of silver whiskers. Silver can

Fig. 3 Effects of water being trapped inside a connector. The high-voltage pin (anode) is completely corroded off (right side), while the ground pins (cathodes) exhibit no corrosion. The pins in between are at lower voltages and show a lesser degree of corrosion.

Fig. 4 White corrosion products on tin-coated circuits and galvanic corrosion between the gold-tin contact/circuit interface resulting from a coffee spill

Fig. 5 Effects of water accumulation in the sump area of a black box. The arrow points to the high-voltage lead wire that has completely corroded, rendering the unit nonfunctional. Note that the connector is wrongly positioned horizontally inside the box.

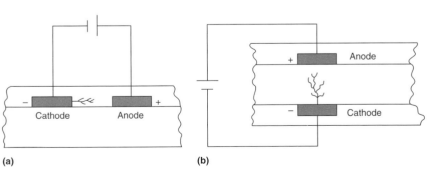

Fig. 2 Schematic showing an electrolyte bridging two circuits on an electronic board. (a) Transverse view of the board. (b) Top-down view of the circuits on the board. Source: Ref 22

plate out across circuits, similar to the shorting shown in Fig. 2, when the circuits are at different potentials. However, silver can also plate out across nonconductive surfaces by a dissolution/plating mechanism that also creates an electrical short between two conductors. In addition, most of the corrosion products of silver are conductive, while almost all other conductive metals have corrosion products that resist the flow of current. Figure 7 shows silver sulfide whiskers that formed on the surface of a copper pin that was plated with silver and gold (Ref 23). The silver migrates through the gold coating and reacts with sulfur-containing elements in the air, forming the silver sulfide whisker. This photo was taken of a connector procured for use in missile components. Substitution of a nickel undercoating for the silver coating eliminates this problem. In the past, alloying the tin coating with 1.5% Pb helped to eliminate the tendency for whiskers to occur. Due to the elimination of lead in electronic devices, reflowing the tin by heating (a form of stress relief) also seems to minimize the tendency for whisker formation. Because of the wide variety of metals that are used in the design of electronics, the properties and interactions of these materials must be better understood to effectively improve the reliability of electrical equipment.

Effects of Contaminants. Small quantities of not-so-obvious contaminants can also have disastrous effects on black boxes. These contaminants, probably the most elusive, are produced by materials that emit corrosive vapors. Do not use any material in electronic equipment that may emit corrosive vapors. Materials that emit corrosive vapors include chlorinated organic compounds such as polyvinyl chloride (PVC) insulation on electrical wiring. Polyvinyl chloride is the least expensive and most widely used commercial wiring insulation material available. This insulation decomposes with time at temperatures from ambient to 120 °C (250 °F) to form hydrogen chloride gas. Figure 8 shows how heating can accelerate the decomposition of the insulation material. The hydrogen chloride gas reacts with moisture to form hydrochloric acid, a very corrosive material even at low concentrations. A list of military specification insulation materials for wiring that contain PVC can be found in Ref 24.

Bubble pack cushioning materials seem to be used everywhere today (2006), and that includes electronic devices. These materials may contain

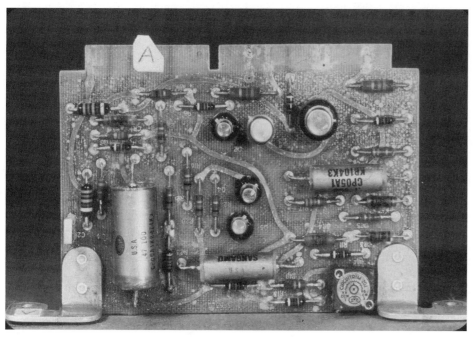

Fig. 6 Tin-plated contacts (a + A) that have been completely corroded as a result of being inserted into a gold-plated beryllium-copper connector. Electrolyte was retained as a result of the board being in the horizontal position.

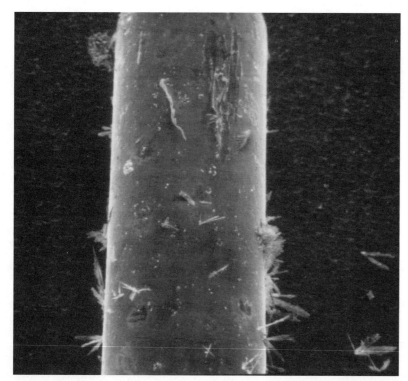

Fig. 7 Silver sulfide whiskers growing on a copper pin that was coated with silver plating followed by gold plating

Fig. 8 Two solder connections that were made without a heat sink. The solder area was hot enough to initiate the thermal decomposition of the polyvinyl chloride insulation. The continuing decomposition of the insulating material resulted in the release of hydrogen chloride gas to the inside of the electronic unit.

another chlorinated organic compound, polyvinylidene chloride. Polyvinylidene chloride has very low oxygen permeability and maintains the "cushy" properties of the bubble pack material. Dual-in-line packages (DIPs) with nickel- and gold-plated Kovar (Fe-29Ni-17Co) leads, wrapped in bubble pack and heat sealed in polyethylene craft paper, were found blistered and corroded after two years in storage. The corrosive environment caused rust on the Kovar substrate to bleed out onto the surfaces of the gold leads. Exposure of the DIPs with bubble pack wrapped around them and heat sealed in 0.13 mm (5 mils) polyethylene to a 100% humidity test for one month reproduced the corrosion problem. The testing included using pink polyethylene used for antistatic packaging purposes, and the results showed severe corrosion where the antistatic material touched the nickel- and gold-plated Kovar surface. Newer bubble pack materials use nylon in place of the polyvinylidene to eliminate the corrosive decomposition products of the older bubble pack materials.

Alkyd paints containing drying oils can liberate organic acids (mainly formic acid) in various quantities that will attack zinc, iron, cadmium, and other metals. It has been reported that these vapors may continue to be evolved in quantities sufficient to cause corrosion for 15 months. Catalytically cured paints (epoxies, urethanes, etc.) should be used, because corrosive vapors are not emitted during the curing of these materials. Certain room-temperature vulcanizing (RTV) silicone adhesive sealants react with moisture in the air and emit acetic acid vapors. These acetic-acid-emitting materials do not meet specification MIL-A-46106 (Ref 25) and should not be used for sealing or potting electronic units. Acetic acid vapors confined in a sealed electronic unit will corrode zinc, steel, cadmium, magnesium, copper, and aluminum.

To seal electronic units, the noncorrosive RTVs that evolve methanol on curing should be used. The noncorrosive RTVs meet military specification MIL-A-46146 (Ref 26). A listing of corrosive and noncorrosive RTVs is also available in Ref 24. Additional materials that may emit corrosive vapors are identified in Ref 27. Effects of air pollutants and other contaminants on the reliability of electronic items have also been the subject of papers authored by Frankenthal (Ref 28), Douthit (Ref 29), and others (Ref 30). The design of telephone exchange buildings (actually an oversized black box) incorporates various materials to avoid corrosive vapors. Cinder block is not used in the construction because it contains iron sulfide and can subsequently result in the formation of sulfurous acid vapors. White adhesives for floor tile and latex paints are not used because they contain morpholine and ammonia, respectively. The air is filtered to remove chlorides and makes only one pass through the building, because nitrous oxides are formed during switching operations. All of these considerations are very important in preventing the corrosion of the electrical switching equipment.

Summary. Figure 9 shows an electronic black box that has boards in the horizontal position, a feed-through connector on the bottom, PVC insulation on the internal wiring, no drain hole, and tin contacts in a gold-plated connector. The box was sealed with an O-ring seal that is ozone cracked at the corners, and painted externally with an alkyd paint. The unit value was in excess of $10,000 and required constant depot maintenance to keep it working.

In most cases with electronic equipment, the corrosion experienced is based on the individual properties of the material/finish exposed to some environment. The military environment includes any area of the world where troops may be deployed—cold climates, deserts (sand/dust), jungles (tropics), sea coastal, and marine environments. When the environmental concerns are added to the long-term life performance expectancies of military hardware (some systems today, 2006, are approaching 50 years), the corrosion resistance associated with the environmental exposure must be a real consideration during the initial design of the equipment. One of the few descriptions of various environments is found in ANSI/EIA-364-B (Ref 31). The environmental classes are:

Class number	Definition
1.0	Year-round filtered air conditioning with humidity control
1.1	Year-round air conditioning (nonfiltered) with humidity control
1.2	Air conditioning (non-year-round) with no humidity control
1.3	No air conditioning or humidity control with normal heating and ventilation
2.0	With normal ventilation but uncontrolled heating and humidity
2.1	Year-round exposure to heat, cold humidity, moisture, industrial pollutants, and fluids
3.0	Outdoor environment with moisture, salt spray, and weathering
4.0	Aircraft environment (uncontrolled)

The testing procedures associated with accepting parts intended for these environments are also described in this standard.

In the area of corrosion maintenance for electronic equipment, the Navy has prepared a document, NAVAIR 01-1A-509, for cleaning and corrosion control (Ref 32). The Navy found that 40 to 50% of the boards could be returned to service just by pulling out the board and reinserting it—in effect, cleaning the contacts of corrosion products. They also found that over 90% of the failed weapon removable assemblies could be returned to service when trained technicians followed the manual. This Navy document has been adopted for use by the Air Force and the Army in the maintenance of their electronic equipment. The Air Force has determined that using MIL-L-87177A (Ref 33) lubricant on the contacts can prevent deterioration. Testing of this lubricant by Battelle showed that this lubricant provides corrosion protection for the contacts, while some other contact lubricants were actually corrosive to the contact materials.

REFERENCES

1. G.A. Greathouse and C.J. Wessel, *Deterioration of Materials: Causes and Preventive Techniques,* Reinhold Publishing Corporation, 1954, p 650–701
2. S.M. Arnold, Repressing the Growth of Tin Whiskers, *Plating,* Jan 1966, p 96–99
3. "Reliability of Military Electronic Equipment," MIL-STD-441, U.S. Navy Bureau of Ships, June 20, 1958
4. S.H. Ross and F.J. Dougherty, Jr., "Fungus Resistance of Vinyl Polymer Insulating Cable Coverings," U.S. Army Frankford Arsenal Memorandum Report M68-12-1, Nov 1967
5. F.G. Foster, How to Avoid Embrittlement of Gold-Plated Solder Joints, *Prod. Eng.,* Aug 19, 1963
6. W.F. Garland, Effect of Decomposition Products from Electrical Insulation on Metal and Metal Finishes, *Proc. Sixth Electrical Insulation Conference, Materials and Application,* Sept 13–16, 1965 (New York, NY), IEEE Service Center
7. H. Leidheiser, Jr., Corrosion—Sometimes Good Is Really Mostly Bad, *Chem. Eng. News,* April 5, 1965, p 78–92
8. S. Peters, Review and Status of Red Plague Corrosion of Silver Plated Copper Wire, *Insul. Circuits,* May 1970, p 55–57
9. J. Agnew, Problem-Solving Production Techniques for Handling Semiconductor Surfaces, *Insul. Circuits,* Nov 1973, p 34
10. J.T. Menke, "Military Electronics—A History and Projection," Paper 328, Corrosion/89 (New Orleans, LA), National Association of Corrosion Engineers, 1989
11. E. Duffek and E. Armstrong, Printed Circuit Board Etch Characteristics, *Met. Finish.,* Oct 1968, p 63–69
12. S.J. Krumbein and A.J. Raffalovich, "Corrosion of Electronic Components by Fumes from Plastics," U.S. Army Electronics Command Technical Report ECOM-3468, Sept 1971

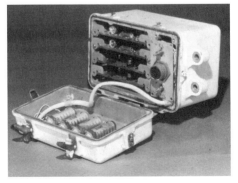

Fig. 9 Electronic black box that ceased to function in the tank in which it was mounted. The design of the box (see related text) violated the rules for building a reliable piece of equipment.

13. A.J. Raffalovich, Waveguide Corrosion, *Mater. Perform.*, Nov 1974, p 9–12
14. D.V. McClendon, G.F. Washington, and C. Jackson, Jr., Simultaneous Operational and Microbial Evaluation Parameters for Electrical Components, U.S. Army Tank and Automotive Command Technical Report 12021, April 1975
15. J.D. Guttenplan and L. Hashimoto, Corrosion Control for Electrical Contacts in Submarine Based Electronic Equipment, *Mater. Perform.*, Dec 1979, p 49–55
16. A.M. Olmedo, As Good as Gold, *Circuits Manuf.*, March 1980, p. 97–99
17. J.H. Whitley, The "Tin Commandments" of Contactor Materials Selection, *Circuits Manuf.*, June 1981, p 72–75
18. *Handbook for Corrosion Prevention and Deterioration Control in Electronic Components and Assemblies*, MIL-HDBK-1250A, U.S. Army Missile Command, Redstone Arsenal, AL, June 29, 1992
19. W.J. Willoughby, "Design Guidelines for Prevention and Control of Avionic Corrosion," Department of the Navy Document NAVMAT P 4855-2, June 1983
20. "Air Force Lessons Learned," Wright-Patterson Air Force Base, private communication, Feb 13, 1984
21. J.T. Menke, "The Anatomy of an Electronic Black Box," Paper 225, Corrosion/83 (Anaheim, CA), National Association of Corrosion Engineers, 1983
22. G.B. Fefferman and D.J. Lando, Testing Long-Term Corrosion of PC Materials, *Circuits Manuf.*, Nov 1978, p 10–14
23. S.B. Tulloch and R.G. Britton, Migration of Silver Through Gold Plating on Electrical Contacts, U.S. Army Missile Command Technical Report RL-72-19, Dec 1972
24. L.O. Gilbert, "Prevention of Material Deterioration: Corrosion Control Course," U.S. Army, Rock Island Arsenal, 1966
25. "Adhesive-Sealants, Silicone, RTV, One-Component," MIL-A-46106, Naval Air systems Command, Lakehurst, NJ, June 26, 1992
26. "Adhesive-Sealants, Silicone, RTV, Non-corrosive (For Use with Sensitive Metals and Equipment)," MIL-A-46146, Naval Air Systems Command, Lakehurst, NJ, Oct 28, 1992
27. P.D. Donovan, *Protection of Metals from Corrosion in Storage and Transit*, Halsted Press, 1986
28. R.P. Frankenthal, D.J. Siconolfi, and J.D. Sinclair, Accelerated Life Testing of Electronic Devices by Atmospheric Particles: Why and How, *J. Electrochem. Soc.*, Vol 140 (No. 11), Nov 1993, p 3129–3144
29. D.A. Douthit, "Avionics Systems, Reliability and Harsh Environments," American Helicopter Society 57th Annual Forum, May 9–11, 2001 (Washington, D.C.)
30. H.C. Shields and C.J. Weschler, Are Indoor Air Pollutants Threatening the Reliability of Your Electronic Equipment?, *Heat./Piping/Air Cond.*, May 1998, p 46–54
31. "Electrical Connector Test Procedures Including Environmental Classifications," ANSI/EIA-364-C, Electronic Industries Alliance, Engineering, Publications office, Washington, D.C., 1994
32. *Technical Manual, Aircraft Weapons Systems Cleaning and Corrosion Control*, NAVAIR 01-1A-509, Naval Air Systems Command, July 1, 1988
33. "Lubricants, Corrosion Preventive Compound, Water Displacing, Synthetic," MIL-L-87177, Ogden Air Logistics center. Ogden, UT, Feb 28, 1991

SELECTED REFERENCES

- W.H. Abbott, "Studies of Natural and Laboratory Environmental Reactions on Materials and Components," Battelle Columbus Division, Aug 1986
- C.R. Ailles and G. Neira, "Dormant Storage Effects on Electronic Devices," Report ARFSD-CR-89014, Science Applications International Corporation, 1989
- *Automotive Electronics Reliability Handbook*, Electronics Reliability Subcommittee, Society of Automotive Engineers, Inc., Feb 1987
- A.S. Brar and P.B. Narayan, *Materials and Processing Failures in the Electronics and Computer Industries: Analysis and Prevention*, ASM International, 1993
- J.P. Cook and G.E. Servais, Corrosion Can Cause Electronics Failure, *Automot. Eng.*, Aug 1984, p 36–41
- G.O. Davis, W.H. Abbott, and G.H. Koch, "Corrosion of Electrical Connectors," MCIC Report-86-C1, Metals and Ceramics Information Center, May 1986
- L.W. Ekman, Spacecraft Wire Harness Design Considerations, *Interconn. Technol.*, March 1994, p 26–30
- G. Fry, *Microcircuit Manufacturing Control Handbook: A Guide to Failure Analysis and Process Control*, 2nd ed., Integrated Circuit Engineering Corporation
- C. Harper, *Electronic Packaging and Interconnection Handbook*, 2nd ed., McGraw-Hill, 1997
- L.D. Kapner, Alternative Fluxes for Hi-Rel Assemblies, *Circuits Assem.*, Feb 1993, p 66–70
- R. Mroczkowski, Fretting, *Interconn. Technol.*, June 1993, p 32–34

Microbiologically Influenced Corrosion in Military Environments

Jason S. Lee, Richard I. Ray, and Brenda J. Little, Naval Research Laboratory, Stennis Space Center

MICROBIOLOGICALLY INFLUENCED CORROSION (MIC) designates corrosion due to the presence and activities of microorganisms. Microorganisms can accelerate rates of partial reactions in corrosion processes and/or shift the mechanism for corrosion (Ref 1). Most laboratory and field MIC studies have focused on bacterial involvement; however, other single-celled organisms, including fungi, can influence corrosion. This article focuses on MIC of military assets and is divided into atmospheric, hydrocarbon and water immersed, and buried environments. Individual mechanisms for MIC are discussed for specific examples. More general discussions of MIC are found in the articles "Microbiologically Influenced Corrision" and "Microbiologically Influenced Corrosion Testing" in *ASM Handbook*, Volume 13A, 2003.

General Information about Microorganisms

Liquid water is needed for all forms of life, and availability of water influences the distribution and growth of microorganisms. Water availability can be expressed as water activity (a_w) with values ranging from 0 to 1. Microbial growth has been documented over a range of water activities from 0.60 to 0.998. Microorganisms can grow in the temperature range in which water exists as a liquid, approximately 0 to 100 °C (32 to 212 °F). Microorganisms can grow over a range of 10 pH units or more (Ref 2). Many microorganisms can withstand hundredfold or greater variations in pressure. The highest pressure found in the ocean is slightly inhibitory to growth of many microorganisms. Heavy metal concentrations as low as 10^{-8} M can inhibit the growth of some microorganisms, while others may continue to grow at concentrations of a millionfold or greater. Microbial species show thousandfold differences in susceptibility to irradiation (Ref 3).

Microorganisms also require nutrients and electron acceptors. All organisms require carbon, nitrogen, phosphorus, and sulfur, in addition to trace elements. Microorganisms can use many organic and inorganic materials as sources of nutrients and energy (Ref 2). Organisms that require oxygen as the terminal electron acceptor in respiration are referred to as aerobes. Anaerobes grow in the absence of oxygen and can use a variety of terminal electron acceptors, including sulfate, nitrate, Fe^{+3}, Mn^{+4}, and others. Organisms that can use oxygen in addition to alternate electron acceptors are known as facultative anaerobes. Microbial nutrition and respiration are coupled and adapted to environmental conditions. Additionally, microorganisms living in consortia can produce growth conditions, nutrients, and electron acceptors not available in the bulk environment.

Atmospheric Corrosion

Because fungi are the most desiccant-resistant microorganisms and can remain active down to $a_w = 0.60$ (Ref 4), they are the microorganisms most frequently involved in atmospheric MIC. Most fungi are aerobes and are found only in aerobic habitats. Fungi are nonphotosynthetic organisms, having a vegetative structure known as a hyphae, the outgrowth of a single microscopic reproductive cell or spore. A mass of threadlike hyphae make up a mycelium. Mycelia are capable of almost indefinite growth in the presence of adequate moisture and nutrients so that fungi often reach macroscopic dimensions. Spores, the nonvegetative dormant stage, can survive long periods of unfavorable growth conditions (drought and starvation). When conditions for growth are favorable, spores germinate.

Biodeterioration due to fungi has been documented for the following nonmetallic military assets: cellulosics (paper, composition board, and wood); photographic film; polyvinyl chloride films; sonar diaphragm coatings; map coatings; paints; textiles (cotton and wool); vinyl jackets; leather shoes; feathers and down; natural and synthetic rubber; optical instruments; mechanical, electronic, and electric equipment (radar, radio, flight instruments, wire strain gages, and helicopter rotors); hammocks; tape; thermal insulation; and building materials. Fungi cause corrosion in atmospheric environments by: acid production, indirect dissolution of coatings, or direct degradation of coatings. Direct degradation is related to derivation of nutrients. Fungi assimilate organic material and produce organic acids including oxalic, lactic, formic, acetic, and citric (Ref 4).

Researchers isolated the following fungal genera from polyurethane-coated 2024 aluminum helicopter interiors (Ref 5): *Pestolotia, Trichoderma, Epicoccum, Phoma, Stemphylium, Hormoconis* (also known as *Cladosporium*), *Penicillium,* and *Aureobasidium* (Fig. 1). Several genera, including *Aureobasidium,* penetrated the polyurethane topcoat but not the chromate primer. The result was disbonding of the topcoat

Fig. 1 Interiors of H-53 helicopter showing fungal growth on polyurethane painted surfaces. (a) Overview. (b) Detail of cable penetration

(Fig. 2), with no corrosion of the base metal as long as the primer was intact. The biocidal properties of zinc chromate primer (Ref 6) were documented. None of the isolates in the study detailed in Ref 5 degraded the polyurethane coating directly as a sole source of nutrients; however, all grew on hydraulic fluid that accumulated on painted interiors during routine operations. Glossy finish polyurethane was colonized more rapidly than the same formulation with a flat finish. Aged paint fouled more rapidly than did new coatings. Laboratory tests demonstrated that in the presence of hydraulic fluid, all of the isolates caused localized corrosion of bare 2024 aluminum. It was demonstrated (Ref 5) that performance specification (MIL PRF 85570) for cleaning painted aircraft interiors is effective in removing fungal spores but does not kill fungal cells embedded in the paint. Fungal regrowth was observed within days of cleaning.

Numerous reports document fungal degradation of coatings and, in some cases, corrosion of the underlying metal (Ref 7–9) in atmospheric exposures. It was reported (Ref 10) that ship cargo holds coated with chlorinated rubber and carrying dry cereals and woods were severely corroded within months. Heavy pitting and reduced thickness of the steel plate were observed. Corrosion products were populated with viable fungi. It was demonstrated (Ref 10) that the fungi derived nutrients from degradation of protective coatings in addition to the cargo. Corrosion resulted from acidic metabolic by-products. Deterioration of the epoxy resin coating of ship holds filled with molasses, fatty oils, and other fluid cargoes was reported (Ref 11). Others (Ref 12) studied direct microbial degradation of coatings, such as Buna-N (a polymer of acrylonitrile and butadiene); polyurethane (a carbamate polymer); and a polysulfide. They found that both bacteria and fungi could degrade these coatings. Pitting of the underlying metal coincided with blisters and the presence of microorganisms. It was demonstrated (Ref 13) that malfunctioning of M483 155-mm howitzer shells stored in humid environments was directly related to fungal degradation of the lubricating grease used to facilitate the screw connection between the base and the body of the shell.

Indirect dissolution of protective greases by fungi was investigated (Ref 14). Protective greases are used to provide corrosion protection for seven-strand carbon steel cable used as wire rope and as highlines. Each cable is made of six strands wrapped around a central core. When cable is used as rope or highline, the cable is coated with thick maintenance grease, threaded onto wooden spools (Fig. 3a–c), and wrapped in brown paper and black plastic. The maintenance grease is applied to the cable to provide corrosion protection. Wire rope is stored on wooden spools for weeks to months before being used. In an investigation of localized corrosion on wire rope stored on wooden spools, fungal growth was observed on interiors of some wooden spools stored outdoors. Corrosion was most severe on wraps of wire in direct contact with the wooden spool flanges. *Aspergillus niger* and *Penicillium* sp. were isolated from wooden spool flanges (Fig. 4). Fungal isolates could not grow on the protective grease as the sole nutrient source. The isolates grew on wood and produced copious amounts of acids and CO_2. In all cases, localized corrosion was observed in areas where acidic condensate dissolved the maintenance grease and exposed bare areas of carbon steel.

Researchers (Ref 15) determined that 80% of lubricants used for protecting materials were contaminated with 38 biological agents (21 fungi and 17 bacteria) during storage and use, independent of climate or relative humidity. They identified the following species as those most frequently isolated from lubricating oils: *Aspergillus versicolor, Penicillium chrysogen, Penicillium verrucosum, Scopulariopsis brevicaulis, Bacillus subtilis,* and *Bacillus pumilis*. Microorganisms isolated from one particular lubricant could not always grow vigorously on others. Microbial growth in lubricants was accompanied by changes in color, turbidity, acid number, and viscosity. Acid number refers to the acid or base

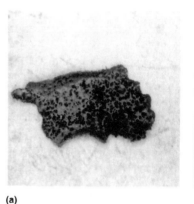
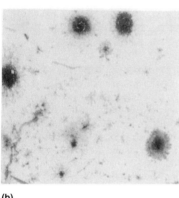

Fig. 2 A piece of disbonded polyurethane paint showing growth of fungi. (a) Top surface. (b) Underside showing that fungi had penetrated the coating

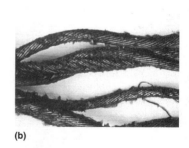

Fig. 3 Carbon steel wire rope. (a) Carbon steel highline being used to transfer equipment between ships at sea. (b) Seven-strand carbon steel wire rope with maintenance grease. (3) Typical wooden spool used to store wire rope

composition of lubricating oils and is also referred to as corrosion number. Biocides, including 4-caproyl, have been evaluated as additives to protect lubricants from mold formation. Biocides have limited lifetimes and limited effectiveness (Ref 16).

Metals Exposed to Hydrocarbon Fuels

One of the most persistent corrosion problems throughout the military is the result of microbial contamination and decomposition of hydrocarbon fuels during fuel transportation, storage, and use. Microbial interaction with hydrocarbon fuels is limited to water availability. Water is sparingly soluble in hydrocarbons. Therefore, microbial growth in hydrocarbons is concentrated at oil/water interfaces, emulsified water, and in separate water phases. The volume of water required for microbial growth in hydrocarbon fuels is extremely small. Because water is a product of the microbial mineralization of organic substrates, it is possible for microbial mineralization of fuel to generate a water phase for further proliferation. For example, *Hormoconis resinae,* the kerosene fungus, grew in 80 mg water per liter of kerosene, and after four weeks' incubation, the concentration of water increased more than tenfold (Ref 17).

The first step in microbial decomposition of hydrocarbons is an aerobic process that requires molecular oxygen. Researchers (Ref 18) compared degradation of hydrocarbons by bacteria and fungi. Bacteria showed decreasing abilities to degrade alkanes with increasing chain length. Filamentous fungi did not exhibit a preference for specific chain lengths. The first products of microbial oxidation of hydrocarbons are alcohols, aldehydes, and aliphatic acids.

An increase in the corrosivity of jet fuel (JP4) stored underground in unlined rock caverns in Sweden caused by sulfate-reducing bacteria (SRB) was documented (Ref 19). Several authors have documented the problem of MIC in aircraft fuel tanks. It was proposed (Ref 20, 21) that microorganisms influenced corrosion of aluminum fuel tanks by:

- Removing corrosion inhibitors, including phosphate and nitrate, from the medium
- Producing corrosive metabolites
- Establishing microcenters for galvanic activity, including oxygen concentration cells
- Removing electrons directly from the surface of the metals

Several investigators reported a decrease in bulk fuel pH due to metabolites produced during growth of fungi (Ref 21–25). One researcher (Ref 24) demonstrated a correlation between growth of *Cladosporium (Hormoconis)* and pH at fuel/water interfaces and measured pH values between 4.0 and 5.0 in the fuel. Fungal-influenced corrosion has been reported for carbon steel and aluminum alloys exposed to hydrocarbon fuels. Another investigator (Ref 22) demonstrated metal ion binding by fungal mycelia, resulting in metal ion concentration cells on aluminum surfaces. It was reported (Ref 24) that corrosivity increased with contact time due to accumulation of metabolites under microbial colonies attached to metal surfaces. Others (Ref 25) demonstrated that the metabolic products enhanced aqueous phase aggressiveness even after the life cycle of *Cladosporium (Hormoconis)* was completed.

Microbiologically influenced corrosion has been identified in engines, holding tanks, skegs, and oily waste tanks on surface ships due to microorganisms growing in water contaminated hydrocarbons. One study (Ref 26) determined that ship engine malfunction and corrosion were associated with MIC. The researchers identified both bacteria and fungi growing in engine lubricants and attributed the problem to a combination of mechanisms, including depletion of protective additives, acid production, and sulfide production. Progressive changes in the formulations of lubricating oils have introduced nitrogen, phosphorus, and sulfur which provide required nutrients for microbial growth. The current trend to produce environmentally benign engine oils means that the resulting formulations are more readily biodegraded. Slow-speed marine engines are at risk because they run for long periods of time at constant temperatures (37 to 55 °C, or 99 to 131 °F) conducive to microbial growth. Oil additives that encourage microbial growth include (Ref 26):

Metal soaps, e.g., barium sulphonates
Polyalkenyl succinimides
High-molecular-weight carboxylic acids
Metal dithiophosphates
Polyorganosiloxanes
Hindered phenols, e.g., 2.6 dietertiary butyl-4-methyl phenol
Aromatic amines-phenyl B naphthylamine
Alkyl phosphates
Alkyl-Aryl phosphates, e.g., tricresyl phosphate

Immersion

Immersion environments are those in which the surface is boldly exposed to an aqueous environment, in contrast to the previous examples in which water was the limiting factor for microbial growth. The most important factor controlling the distribution of microorganisms in immersion environments is the availability of nutrients. For example, organic nutrients and bacteria are most abundant in the upper layers of oceans, and both decrease with depth (Ref 27). Microbial biofilms develop on all surfaces in contact with aqueous environments (Ref 28). Chemical and electrochemical characteristics of the substratum influence biofilm formation rate and cell distribution during the first hours of exposure. Electrolyte concentration, pH, organic, and inorganic ions also affect microbial settlement. Biofilms produce an environment at the biofilm-surface interface that is radically different from that of the bulk medium in terms of pH, dissolved oxygen, and inorganic and organic species. In some cases, the presence of localized microbial colonies can cause differential aeration cells, metal concentrations cells, and under-deposit corrosion. In addition, reactions within biofilms can control corrosion rates and mechanisms. Reactions are usually localized and can include:

- Sulfide production
- Acid production
- Ammonia production
- Metal deposition
- Metal oxidation/reduction
- Gas production

Many of the problems of MIC of military assets exposed to aqueous environments are directly related to an operational mode that includes periods of stagnation. Heat exchangers, fire protection systems, holding tanks, and transfer lines are exposed to flow/no-flow cycles. The following statements are applicable for distilled/demineralized, fresh, estuarine, and marine waters. During stagnation, naturally occurring microorganisms form a biofilm, and aerobic microorganisms use the dissolved oxygen as the terminal electron acceptor. If the rate of respiration is faster than the rate of oxygen diffusion through the biofilm, the metal/biofilm interface becomes anaerobic, allowing anaerobic bacteria to produce corrosive metabolites, including acids and sulfides. The amount of sulfide that can be produced within a biofilm depends on the sulfate concentration and the numbers and activities of SRB. Seawater contains approximately 2 g/L of sulfate and a population of SRB whose numbers vary with nutrient concentration. Most sulfide films are not tenacious and are easily removed by turbulence. Introduction of flowing oxygenated water causes oxidation and/or disruption of surface deposits formed under stagnant conditions.

Copper alloys have a long history of successful application in seawater piping systems

Fig. 4 Fungi (white spots) growing on inside flange of wooden spool.

due to their corrosion resistance, antifouling properties, and mechanical properties. The corrosion resistance of copper in seawater is attributable to the formation of a protective film that is predominantly cuprous oxide, irrespective of alloy composition (Ref 29). Copper ions and electrons pass through the film. In seawater, copper ions dissolve and precipitate as $Cu_2(OH)_3Cl$ (Ref 30). Copper seawater piping systems are often exposed to polluted harbor water containing sulfides. In the presence of sulfides, copper ions migrate through the layer, react with sulfide, and produce a thick black scale. Failure of copper-nickel pipes in estuarine and seawaters can be associated with waterborne sulfides that stimulate pitting and stress corrosion cracking (Ref 31–34). 90Cu-10Ni suffered accelerated corrosion attack in seawater containing 0.01 ppm sulfide after a one day exposure (Ref 35). Galvanic relationships between normally compatible copper piping and fitting alloys become incompatible after exposure to sulfide-containing seawater (Ref 33).

Sulfides produced within biofilms have the same effect as waterborne sulfides on copper alloys. Alloying additions of nickel and iron into the highly defective p-type Cu_2O corrosion product film alters the structure (Ref 29) and results in a film that possesses low electronic and ionic conductivity. In an attempt to prevent sulfide-induced corrosion of copper-nickel piping, $FeSO_4$ treatments were evaluated. Corrosion of $FeSO_4$-treated pipes was compared with untreated pipes that had been cleaned according to military specifications (Ref 36). Batch $FeSO_4$ treatments did not result in a persistent increase in surface-bound iron. The authors found that the dissolved iron concentration in most harbor waters exceeded the amount of iron in the recommended batch $FeSO_4$ treatments. Ferrous sulfate (0.10 mg/L ferrous ion) treatments for 90Cu-10Ni and 70Cu-30Ni alloys were evaluated (Ref 37). Neither pretreatment before sulfide exposure nor intermittent treatment during sulfide exposure significantly reduced sulfide attack on either alloy. However, continuous treatment eliminated the attack on both alloys because the $FeSO_4$ removed sulfides from solution.

Nickel Alloys. Nickel 201, sometimes used for heat exchangers with distilled water, is vulnerable to microbiologically produced acids (Ref 38). The Ni-Cu alloys are used in seawater under conditions including high velocity (propeller shafts, propellers, pump impellers, pump shafts, and condensers), where resistance to cavitation and impingement is required. Under turbulent and erosive conditions, nickel-copper alloys are superior to predominantly copper alloys because the protective surface film remains intact. Nickel alloys are used extensively in highly aerated, high-velocity seawater applications. The formation of the protective film on nickel is aided by the presence of iron, aluminum, and silicon. However, under stagnant seawater conditions, nickel-copper alloys are susceptible to pitting and crevice corrosion attack where chlorides penetrate the passive film (Ref 39). Sulfides produced by SRB cause either a modification or breakdown of the oxide layer and dealloying (Ref 40). Another report (Ref 41) indicated that predominantly nickel alloys were susceptible to underdeposit corrosion and oxygen concentration cells. Other studies (Ref 42, 43) demonstrated pitting and denickelfication of nickel-copper tubes exposed in Arabian Gulf seawater with deposits of SRB.

Stainless Steels. The corrosion resistance of stainless steel is due to the formation of a thin passive chromium-iron oxide film. Crevice corrosion is the most problematic issue affecting the performance of stainless steels in seawater. Investigators (Ref 44) studied crevice corrosion of stainless steel beneath dead barnacles and proposed the following scenario:

- Decomposition of the barnacle by aerobic bacteria, including *Thiobacillus*, reduces pH within the barnacle shell.
- The acid penetrates the shell base and initiates a corrosion cell between the crevice area and the exposed SS substratum.
- Crevice corrosion initiates near the edge of the shell base and propagates inward.

Crevice corrosion is exacerbated in warm natural seawater where biofilms form rapidly. Pit propagation under barnacles is assisted by poor water circulation.

Several investigators (Ref 45–52) have documented the tendency for biofilms to cause a noble shift, or an ennoblement, in open-circuit potential of passive alloys exposed in marine environments. Alloys tested include, but are not limited to: UNS S30400, S30403, S31600, S31603, S31703, S31803, N08904, N08367, S44660, S20910, S44735, N10276, and R50250. The practical importance of ennoblement is increased probability of localized corrosion as E_{corr} approaches the pitting potential (E_{pit}) for stainless steels vulnerable to crevice corrosion, especially types 304 (UNS S30400) and 316 (UNS S31600). Investigators (Refs 53, 54) concluded that biofilms increased the propagation rate of crevice corrosion for UNS S31603, S31725 and N08904 by 1 to 3 orders of magnitude. They attributed ennoblement to an increase in kinetics of the cathodic reaction by the biofilms. Others (Ref 55, 56) demonstrated that ennoblement of electrochemical potential in the presence of a biofilm could be reconciled without reference to modified oxygen reduction mechanisms and without enhancement of cathodic processes. They concluded that the biofilm does not directly affect oxygen reduction near the equilibrium potential and far from the oxygen diffusion limiting current. They further concluded that H_2O_2 and manganese oxides did not play a direct role in the oxygen reduction process at potentials $>300 mV_{SCE}$. Instead, they demonstrated that anodic oxidation of organic material in biofilms produced currents corresponding to passive currents or higher. Oxidation of organic material affects the value of the corrosion potential and lowers the pH at the surface.

One of the most common forms of MIC attack in austenitic stainless steel is pitting at or adjacent to welds. The following observations were made for MIC in 304L (UNS S30403) and 316L (UNS S31603) weldments (Ref 57): both austenite and delta ferrite phases may be susceptible; and varying combinations of filler and base materials failed, including matching, higher-, and lower-alloyed filler combinations. Microsegregation of chromium and molybdenum with chemically depleted regions increases susceptibility to localized attack.

Candidate materials for a double hull vessel designed with permanent water ballast, 316L, Nitronic 50 (UNS S20910), and AL6XN (N08367) were evaluated for potential MIC in flowing and stagnant freshwater and seawater (Ref 58). No pitting was observed in AL6XN under any exposure condition after one year. Leaks were located at weld seams of Nitronic 50 and 316L stainless steels after 6 and 8 week exposures to stagnant and flowing seawater. A failed vertical weld in 316 stainless steel after an 8 week exposure to stagnant natural seawater is shown in Fig. 5(a). Figure 5(b) is the corresponding x-ray image indicating failure due to pitting. In all cases, large numbers of bacteria were associated with the corrosion products (Fig. 5c). Residual material in pits was typical of removal of iron. A scanning vibrating electrode technique was used to demonstrate that there were no persistent anodic sites in autogenous welds or heat affected zones of these materials exposed to sterile seawater. The spatial and causal relationship between bacteria and pitting in weldments in 300 series stainless steels is well documented (Ref 59, 60).

Ethylene glycol/water and propylene glycol/water mixtures were evaluated as permanent ballast waters for 316L double hull vessels (Ref 60). The compounds are attractive as ballast liquids because they will have minimal impact if the outer hull is breached and the glycols are released to the environment. Both have low volatility and are miscible with water. In terms of corrosion protection, propylene glycol-based mixtures were shown to protect against pitting of 316L in concentrations of 50% or higher when mixed with seawater (3.5% salinity). Slightly higher concentrations (55%) of ethylene glycol-based mixtures were required under the same conditions to prevent pitting. The compounds have little or no capacity to bind to particulates and will be mobile in soils or sediments. Glycol-based mixtures were also shown to be bacteriostatic at concentrations above 10%. Also, the low octanol/water partition coefficient and measured bioconcentration factors in a few organisms indicate low capacity for bioaccumulation.

Carbon Steel. Unexpectedly rapid localized corrosion of steel bulkheads and ship hull plating of tankers in marine harbor environments was documented (Ref 61). In each case, the localized attack was found beneath macrofouling layers. The biofilm at and around the corrosion sites was

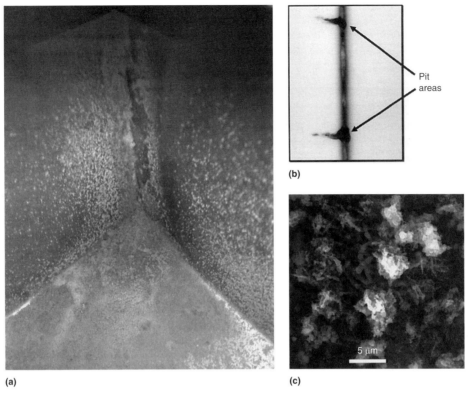

Fig. 5 Vertical weld in 316 stainless steel after exposure to stagnant natural seawater for 8 weeks. (a) Weldment. (b) X-ray of failed vertical weld. (c) Bacteria associated with corrosion products.

populated with a rich consortium of aerobic and anaerobic microorganisms, and the SRB population was elevated by several orders of magnitude above that in the biofilm remote to corrosion sites. Researchers (Ref 62) demonstrated a cycle of sulfur oxidation and reduction causing aggressive corrosion of steel pilings in a harbor. At low tide, the fouling layer was thoroughly aerated and thiobacilli produced oxidized sulfur species. High tide produced anaerobic conditions within the fouling layers and reduction of sulfur compounds.

Natural seawater has also been evaluated as a ballast fluid in unpainted 1020 carbon steel ballast tanks, and deoxygenation has been proposed as a method to reduce corrosion. However, it was determined (Ref 63) that corrosion of 1020 carbon steel coupons in natural seawater over a six month period was more aggressive under stagnant anaerobic conditions than stagnant aerobic conditions as measured by weight loss and instantaneous corrosion rate (polarization resistance) (Fig. 6). Under oxygenated conditions, a two-tiered oxide layer formed (Fig. 7a). The outer oxide layer was reddish-brown and contained numerous filamentous bacteria (Fig. 7b). The inner oxide was extremely adherent and resistant to acid cleaning. Under anaerobic conditions, a nontenacious sulfur-rich corrosion product with enmeshed bacteria formed on carbon steel surfaces (Fig. 7c, d). In anaerobic exposures, corrosion was more aggressive on horizontally oriented coupons compared with vertically oriented coupons. Bulk water chemistry and microbial populations were measured as a function of time. Both were dynamic despite the stagnant conditions.

Cathodically Protected Carbon Steel. In most cases, carbon steel used in seawater is cathodically protected or painted. It has been reported that cathodic protection retards microbial growth because of the alkaline pH generated at the surface. It was demonstrated (Ref 64) that a potential of -1.10 V SCE markedly decreased settlement of *Balanus cyprids* on painted steel surfaces, but it had no effect on settlement, sand-tube building, reproduction, or larval release of other species. Biofouling in seawater was retarded using pulsed cathodic polarization of steel (Ref 65).

Numerous investigators have demonstrated a relationship between marine fouling and calcareous deposits on cathodically protected surfaces (Ref 66–68); however, their interrelationships are not understood. Microbiological data for cathodically polarized surfaces are often confusing and impossible to compare because of differing experimental conditions (laboratory vs. field) and techniques used to evaluate constituents within the biofilm. Differences in organic content of seawater that produced differences in current density, electrochemistry, calcareous deposits, and biofilm formation have been reported (Ref 66).

The influence of a preexisting biofilm on the formation of calcareous deposits under cathodic protection in natural seawater was studied (Ref 67). It was shown that applied current densities up to 100 μA/cm^2 did not remove attached biofilms from stainless steel surfaces. Both natural marine and laboratory cultures changed the morphology of calcareous deposits formed under cathodic polarization at a current density of 100 μA/cm^2.

In one study (Ref 69), cathodic potentials to -1000 mV SCE caused a decrease in pH and an increase of SRB on carbon steel. At potentials more negative than -1000 mV SCE, the pH became more alkaline and SRB numbers decreased. A study of the influence of SRB in marine sediments using electrochemical impedance spectroscopy to monitor corrosion and lipid analysis as biological markers, complemented by chemical and microbiological analysis, showed that -880 mV SCE encouraged the growth of hydrogenase-positive bacteria in the sediment surrounding the metal and facilitated the growth of other SRB species (Ref 70).

Because the enumeration technique strongly influences the number of cells one is able to count, and because the number of cells cannot be equated to cellular activity, including sulfate reduction, some investigators have attempted to measure cellular activity directly on cathodically protected surfaces. One investigator (Ref 71) cathodically protected 50D mild steel (BS 4360) coupons exposed in the estuarine waters of Aberdeen Harbor using an imposed potential of -950 mV Cu:CuSO$_4$ and sacrificial anodes. Activities within biofilms were determined using a radiorespirometric method—a technique for studying microbial respiration using

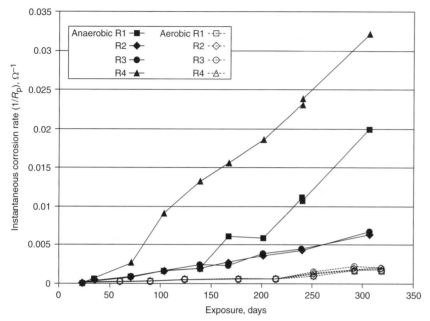

Fig. 6 Instantaneous corrosion rates using polarization resistance (R_p) for carbon steel exposed in stagnant aerobic or stagnant anaerobic natural seawater. Corrosion rates were higher for the anaerobic exposures (solid lines) than for aerobic exposure (broken lines). R1, R2, R3, and R4 refer to the vertical location of the electrode, with R1 being closest to the water surface and R4 located at the bottom of the experimental chamber.

Fig. 7 Carbon steel electrodes exposed to aerobic and anaerobic natural seawater for 290 days. (a) Aerobic. (b) Aerobic. Scanning electron micrograph of iron-oxide encrusted bacteria enmeshed in corrosion products. (c) Anaerobic. (d) Anaerobic. Scanning electron micrograph of sulfide-encrusted organisms enmeshed in corrosion products.

radiolabeled substrates or electron acceptors. Biofilms developed on all substrata—both unprotected and cathodically protected surfaces. The activities of aerobic and anaerobic bacteria, including SRB, were significantly greater on unprotected coupons. Furthermore, sulfide, a metabolic fingerprint of SRB activity, could be detected only in biofilms on unprotected coupons. These results show that a potential of −950 mV Cu:$CuSO_4$ does not prevent SRB from developing on cathodically protected surfaces. The lower activity of SRB within biofilms on cathodically protected coupons was not directly caused by any inhibitory effect of the cathodic potential. Instead, the greater activity of SRB on unprotected coupons was the result of production of an extensive corrosion film offering more favorable anaerobic conditions.

The NACE Standard (Ref 72), which is currently under revision, lists cathodic protection criteria for underground or submerged steel, cast iron, aluminum, and copper structures. Microbiologically influenced corrosion is cited as "one of several abnormal conditions which sometimes exist and where cathodic protection is ineffective or only partially effective." It is important to point out that in several studies (Ref 66–68), SRB were present on cathodically protected steels, but accelerated corrosion was not reported.

Thermodynamic data with iron in a pH 7 electrolyte saturated with hydrogen sulfide was studied (Ref 73). A potential of −1024 mV SCE was required to achieve cathodic protection. It was demonstrated (Ref 74) that −1024 mV was capable of providing cathodic protection in the presence of active SRB. The influence of cathodic protection on the growth of SRB and on corrosion of steel in marine sediments was investigated (Ref 75). The investigators concluded that a cathodic potential of −880 mV SCE did not appear to be sufficient for protection and that large amounts of cathodically produced hydrogen promoted the growth of SRB in the sediments surrounding the samples. Laboratory tests were conducted (Ref 76) in anaerobic, artificial sediments containing SRB. Results indicated that a polarization of −1024 mV SCE was adequate for corrosion protection. Cathodic protection current density was between 4.5 and 12 mA/ft^2. Another study (Ref 77) indicated that a cathodic potential of −1054 mV SCE lowered the corrosion rate of steel by 82.7%, even though protective potentials in the range −774 to −1134 mV SCE did not inhibit growth of SRB. It was concluded (Ref 78) that if anaerobic bacterial activity is suspected, a cathodic polarization shift of approximately 200 to 300 mV SCE is required for carbon steel protection. Cathodic protection was imposed on steel surfaces actively corroding in cultures of SRB (Ref 79), and it was concluded that cathodic protection in the presence of SRB decreased corrosion by a factor of 8 or 9.

It was shown (Ref 80) that cathodically protected stainless steel surfaces in artificial seawater can become colonized by aerobic, acid-producing bacteria. Formation of calcareous deposits and initial settlement of microorganisms resulted in decreased current density requirements to maintain a protection potential. Subsequent colonization and pH changes destabilized the calcareous deposits and dramatically increased the current density required to maintain the protected potential.

Hydrogen embrittlement of carbon steel is a form of corrosion involving the cathodic reaction. Atomic hydrogen generated in the cathodic reaction penetrates the steel resulting in the loss of ductility. A number of mechanisms have been postulated to account for the embrittlement effect, which, when combined with the presence of local regions of high stress, can result in severe cracking (Ref 81). Hydrogen embrittlement is enhanced by the high levels of hydrogen generated by SRB under certain conditions, as well as that generated by overly aggressive cathodic protection systems. If the levels of cathodic protection are increased too high to combat SRB corrosion, there is a danger that hydrogen embrittlement may be enhanced. The presence of H_2S, which can be produced by SRB, is known to retard formation of molecular hydrogen on the metal surface and to enhance adsorption of atomic hydrogen by the metal. Whenever algae provide conditions for SRB, they may also enhance hydrogen embrittlement.

In summary, bacteria can settle on cathodically protected surfaces. Cathodic potentials to -1074 mV SCE do not prevent biofilm formation. It has been suggested that actual cell numbers may be related to polarization potential, dissolved organic carbon, or to the enumeration technique. Numbers of SRB may be increased or decreased depending on exposure conditions. Carbon steel is considered protected when a potential of -924 mV SCE is achieved. In many cases, the potential is further reduced to -1024 mV SCE to protect the steel from corrosion caused from the activity of SRB. The decreased potential is not applied to prevent growth of SRB but is based on a theoretical level that will allow passivity of steel in a sulfide-rich environment produced by SRB. The main consequence of biofilm formation on protected surfaces appears to be an increase in the current density necessary to polarize the metal to the protected potential. The presence of large numbers of cells on cathodically protected surfaces does mean that in the event that cathodic protection is intermittent, discontinuous, or discontinued, the corrosion attack due to the microorganisms will be more aggressive.

Coated Carbon Steel. Although coatings alone do not prevent MIC, they delay the onset of MIC and other corrosion reactions. Many types of polymeric coatings can be subject to biodegradation. The attack is usually caused by acids or enzymes produced by bacteria or fungi. This often results in selective attack on one or more specific components of a coating system with consequent increase in porosity and water or other ion transport through the coating and the formation of blisters, breaches, and disbonded areas. Many of these effects have recently been reviewed (Ref 82). It was also demonstrated (Ref 83) that marine bacteria are attracted to corrosion products at coating defects. The microorganisms responsible for damage to coatings may or may not be involved in corrosion initiation under the damaged coating.

Titanium and Titanium Alloys. There are no case histories of MIC for titanium and its alloys. One investigator (Ref 84) reviewed mechanisms for MIC and titanium's corrosion behavior under a broad range of conditions. He concluded that at temperatures below 100 °C (212 °F) titanium is not vulnerable to iron/sulfur-oxidizing bacteria, SRB, acid-producing bacteria, differential aeration cells, chloride concentration cells, and hydrogen embrittlement. In laboratory studies, (Ref 85) corrosion of Grade 2 titanium (UNS 850400) was not observed in the presence of SRB or iron/sulfur-oxidizing bacteria at mesophilic (23 °C, or 73 °F) or thermophilic (70 °C, or 158 °F) temperatures. Using the model in Ref 86, one would predict that titanium would be immune to SRB-induced corrosion. There are no standard free energy reaction data for the formation of a titanium sulfide. If one assumes a hypothetical sulfide product to be titanium sulfide, the standard enthalpy of reaction is $+587$ kJ. While standard free energies of reaction are not identical to standard enthalpies of reaction, it is still unlikely that titanium will be converted to the sulfide under standard conditions of temperature and pressure.

Aluminum alloys were evaluated (Ref 87) for the impact of microorganisms on corrosion of aircraft in bilge and toilet areas. The researchers isolated numerous microbiological species and were able to cause corrosion of 7075, which is used in aircraft construction. However, the authors could not relate their experiments to actual aircraft.

Polymeric Composites. Microorganisms and their products can be responsible for changes in physical, chemical, and electrochemical properties of polymeric materials. Reference 88 demonstrated that under immersion conditions, epoxy and nylon coatings on steel were breached by mixed cultures of marine bacteria.

In laboratory experiments (Ref 89), it was demonstrated that epoxy resin and carbon fibers, either individually or in composite, were not degraded by sulfur/iron-oxidizing, hydrogen-producing, calcareous depositing, or SRB. Bacteria colonized resins, fibers, and composites but did not cause damage. Sulfate-reducing bacteria preferentially colonized vinyl ester composites at the fiber-resin interfaces, and hydrogen-producing bacteria appeared to disrupt the fiber-vinyl ester resin bonding with penetration of the vinyl ester resin.

It is standard practice to coat the surface of the filaments with a sizing chemical to provide a better bonding with the resin matrix and to prevent abrasion between individual fibers during shipping and handling. This treatment permits optimal stress transmission between filaments. Fiber sizing chemicals are starch-oil mixtures. The sizing materials are highly susceptible to biodegradation of strength resulting from abrasion between fibers and as a coupling agent to the matrix. The sizing materials are highly susceptible to biodegradation and can be expected to decompose in the presence of contaminating microorganisms (Ref 90).

Researchers (Ref 91, 92) investigated fungal degradation of polyimides used as insulators in electronic packaging. Growth of microorganisms on these polymers was found to result in loss of their dielectric properties. They also studied biodeterioration of fiber-reinforced composites, graphite sheets, and graphite fibers used in composite materials. They observed fungal penetration into composite resin and graphite sheets and concluded that fungi caused substantial damage to composites under conditions favorable to fungal growth. Investigators (Ref 93) demonstrated in the laboratory that mixed cultures of marine bacteria could penetrate three conductive caulks (PRC 1764, PI8500, and PI8505) used to secure antenna foundations to ship superstructures.

Burial Environments

Tapes and coatings for buried pipes and cables are susceptible to biodegradation and MIC (Ref 94). A recent study evaluated the potential for MIC of unexploded ordnances (UXO) buried in soil environments. Unexploded ordnances are military munitions that have been prepared for action but remain unexploded and constitute a potential hazard. The 1998 Defense Science Board estimated 1,400 individual sites contained UXO (Ref 95). The munitions corrode at varying site-specific corrosion rates. In most subsurface environments, microbial growth is limited by water. The likelihood that MIC will take place is directly related to water availability. Microorganisms concentrate at interfaces, including soil/surface interfaces. A survey determined that the microbial populations measured on the surface of UXO were sufficient to cause localized corrosion and that the most likely mechanism was microbial acid production. Microbially induced corrosion of carbon steel is independent of pH over pH values 4.5 to 9.5. In this range, the corrosion products maintain a pH of 9.5 next to the steel surface, regardless of the pH of the solution. At a pH of 4 or below, hydrogen evolution begins and corrosion increases rapidly. Fungi and acid-producing bacteria can reduce the pH locally to values below 4.0. The localized corrosion mechanism of the steel fragments was in many cases pitting, with pits inside pits, indicating multiple initiation sites. In other cases, tunneling was observed. Both types of localized corrosion are consistent with microbiological acid-induced corrosion.

ACKNOWLEDGMENT

Preparation of this chapter was funded under Office of Naval Research Program Element

0601153N, Work Unit No. 5052. NRL/BC/7303/04/0001.

REFERENCES

1. B. Little, P. Wagner, and F. Mansfeld, *Inter. Mater. Rev.*, Vol 36 (No. 6), 1991, p 253–272
2. L. Prescott, J. Harley, and D. Klein, *Microbiology*, 5th ed., McGraw-Hill, 2002
3. G. Geesey, "A Review of the Potential for Microbially Influenced Corrosion of High-Level Nuclear Waste Containers," Report prepared for Center for Nuclear Waste Regulatory Analyses, San Antonio, TX, 1993
4. J.E. Smith and D.R. Berry, *An Introduction to Biochemistry of Fungal Development*, Academic Press, 1974, p 85
5. D. Lavoie, B. Little, R. Ray, K. Hart, and P. Wagner, *Corrosion/97, Proceedings* (Houston, TX), NACE International, Paper 218, 1997
6. M. Stranger-Johannessen, The Antimicrobial Effect of Pigments in Corrosion Protective Paints, *Biodeterioration 7* (United Kingdom), The Bioremediation Society, 1987, p 372
7. W.J. Cook, J.A. Cameron, J.P. Bell, and S.J. Huang, *J. Polym. Sci.*, Polymer Letters Edition, Vol 19, 1981, p 159–165
8. D.S. Wales and B.F. Sager, *Proc. Autumn Meeting of the Biodeterioration Society* (Kew, U.K.) 1985, p 56
9. R.A. Zabel and F. Terracina, *Dev. Ind. Microbiol.*, Vol 21, 1980, p 179
10. M. Stranger-Johannessen, Fungal Corrosion of the Steel Interior of a Ship's Holds, *Biodeterioration VI, Proceedings* (Slough, U.K.), CAB International Mycological Institute, 1984, p 218–223
11. M. Stranger-Johannessen, *Microbial Problems in the Offshore Oil Industry*, John Wiley and Sons, 1986, p 57
12. R.J. Reynolds, M.G. Crum, and H.G. Hedrick, *Dev. Indust. Microbiol.*, Vol 8, 1967, p 260–266
13. B. Little and P. Wagner, Informal Report, Army Ammunition Plant, Stennis Space Center, MS, 1982
14. B. Little, R. Ray, K. Hart, and P. Wagner, *Mater. Perform.*, Vol 34 (No. 10), 1995, p 55–58
15. E.G. Toropova, A.A. Gerasimenko, A.A. Gureev, I.A. Timokhin, G.V. Matyusha, and A.A. Belousova, Khimiya I Tekhnologiya Topliv i Masel. Vol 11, 1988, p 22–24
16. M.S.W. Rodionova, L.V. Bereznikovskaya, Ye.I. Pannfilenck, A. Baygozhin, A.I. Latynina, and L.V. Sergryev, "Method for Protection of Lubricants from Accumulation of Biological Material," Technical Translation—Translated from 1965 Russian Document FTD-HT-23-1427-68, Patent 18948, U.S. Army Foreign Science and Technology Center, 1968, p 3
17. K. Bosecker, *Microbially Influenced Corrosion of Materials*, Springer Verlag, 1996, p 439
18. J.D. Walker, H.F. Austin, and R.R. Colwell, *J. Gen. Appl. Microbiol.*, Vol 21, 1975, p 27
19. R. Roffey, A. Edlund, and A. Norqvist, *Biodeterioration*, Vol 25, 1989, p 191–195
20. H.G. Hedrick, M.G. Crum, R.J. Reynolds, and S.C. Culver, *Electrochem. Technol.*, March–April 1967, p 75–77
21. H.A. Videla, P.S. Guiamet, S. DoValle, and E.H. Reinoso, *A Practical Manual on Microbiologically Influenced Corrosion*, NACE International, 1993, p 125
22. B.M. Rosales, Argentine-USA Workshop on Biodeterioration, *Proceedings* (La Plata, Argentina) CONICET-NSF, 1985, p 135
23. E.R. De Schiapparelli and B.R. de Meybaum, *Mater. Perform.*, Vol 19, 1980, p 47
24. M.F.L. De Mele, R.C. Salvarezza, and H.A. Videla, *Int. Biodeterior. Bull.*, Vol 15, 1979, p 39
25. B.R. de Meybaum and E.R. De Schiapparelli, *Mater. Perform.*, Vol 19, 1980, p 41
26. E.C. Hill, Microbial Corrosion in Ship Engines, *Proceedings, Conference on Microbial Corrosion* (London), Book 303, Metals Society, 1983, p 123–127
27. W.K.W. Li, J.F. Jellett, and P.M. Dickie, *Limnology and Oceanography*, Vol 40, 1995, p 1485–1495
28. W.G. Characklis and Kevin C. Marshall, Ed., *Biofilms*, John Wiley and Sons, 1990, p 796
29. R.F. North and M.J. Pryor, *Corros. Sci.*, Vol 10, 1970, p 297–311
30. A.M. Pollard, R.G. Thomas, and P.A. Williams, *Mineralogical Magazine*, Vol 53, p 557–563
31. J.P. Gudas and H.P. Hack, *Corrosion*, Vol 35 1979, p 57–73
32. K.D. Efrid and T.S. Lee, *Corrosion*, Vol 35 (No. 2), 1979, p 79–83
33. H.P. Hack, *J. Test. Eval.*, Vol 8 (No. 2), 1980, p 74–79
34. S. Sato and K. Nagata, *Light Met. Technical Reports*, Vol 19 (No. 3), 1978, p 1–12
35. C. Pearson, *Br. Corros. J.*, Vol 7, 1972, p 61–68
36. P.A. Wagner, B.J. Little, and L. Janus, An Investigation of Microbiologically Mediated Corrosion of Cu/Ni Piping Selectively Treated with Ferrous Sulfate, *Proceedings of Oceans '87 Conference* (Halifax, Canada), 1987, p 439–444
37. H.P. Hack and J.P. Gudas, Inhibition of Sulfide-Induced Corrosion of Copper-Nickel Alloys with Ferrous Sulfate, *Mater. Perform.*, Vol 18 (No. 3), 1979, p 25–28
38. B.J. Little, S.M. Wagner, S.M. Gerchakov, M. Walch, and R. Mitchell, *Corrosion*, Vol 42 (No. 9), 1986, p 533–536
39. W.Z. Friend, Nickel-Copper Alloys, *The Corrosion Handbook*, H.H. Uhlig, Ed., John Wiley and Sons, 1948, p 269
40. B. Little, P. Wagner, R. Ray, and M. McNeil, *J. Mar. Technol. Soc.*, Vol 24 (No. 3), 1990, p 10–17
41. M. Schumacher, Ed., *Seawater Corrosion Handbook*, Noyes Data Corp., Park Ridge, NJ, 1979, p 494
42. V.K. Guoda, I.M. Banat, W.T. Riad, and S. Mansour, *Corrosion*, Vol 49, 1993, p 63–73
43. V.K. Gouda, H.M. Shalaby, and I.M. Banat, *Corros. Sci.*, Vol 35 (No. 1–4), 1993, p 683–691
44. M. Eashwar, G. Subramanian, P. Chandrasekaran, and K. Balakrishnan, *Corrosion*, Vol 48 (No. 7), 1992, p 608–612
45. S.C. Dexter and G.Y. Gao, *Corrosion*, Vol 44 (No. 10), 1988, p 717–723
46. R. Johnson and E. Bardal, *Corrosion*, Vol 41 (No. 5), 1985, p 296
47. V. Scotto Dicentio and G. Marcenaro, *Corros. Sci.*, Vol 25 (No. 3), 1985, p 185–194
48. S. Motoda, Y. Suzuki, T. Shinohara, and S. Tsujikawa, *Corros. Sci.*, Vol 31, 1990, p 515–520
49. V. Scotto, "Electrochemical Studies on Biocorrosion of Stainless Steel in Seawater," EPRI Workshop on Microbial Induced Corrosion, G. Licina, Ed., Electric Power Research Institute, 1989, p B.1–B.36
50. A. Mollica, G. Ventura, E. Traverso, and V. Scotto, Cathodic Behavior of Nickel and Titanium in Natural Seawater, *7th Intl. Biodeterioration Symp.*, 1987 (Cambridge, U.K.)
51. S.C. Dexter and J.P. LaFontaine, *Corrosion*, Vol 54 (No. 11), 1998, p 851–861
52. S.C. Dexter and H.-J. Zhang, Effect of Biofilms on Corrosion Potential of Stainless Alloys in Estuarine Waters, *Proc. 11th Intl. Corrosion Congress*, April 1990 (Florence, Italy), p 4.333
53. H.-J. Zhang and S.C. Dexter, Effect of Marine Biofilms on Crevice Corrosion of Stainless Alloys, Paper 400, *CORROSION/92*, NACE International, 1992 (Houston, TX)
54. H.-J. Zhang and S.C. Dexter, Effect of Marine Biofilms on Initiation Time of Crevice Corrosion for Stainless Steels, Paper 285, *CORROSION/95*, NACE International 1995 (Houston, TX)
55. G. Salvago and L. Magagrin, *Corrosion*, Vol 57 (No. 8), 2001, p 680–692
56. G. Salvago and L. Magagrin, *Corrosion*, Vol 57 (No. 9), 2001, p 759–767
57. S.W. Borenstein, *Microbiologically Influenced Corrosion Handbook*, Industrial Press Inc., New York, NY, 1994, p 50–112
58. R.I. Ray, J. Jones-Meehan, and B.J. Little, A Laboratory Evaluation of Stainless Steel Exposed to Tap Water and Seawater, *Proceedings of Corrosion 2002 Research Topical Symposium on Microbiologically Influenced Corrosion*, NACE, 2002, p 133–144
59. G. Kobrin, *Mater. Perform.*, Vol 15 (No. 9), 1976, p 38–42

60. J.S. Lee, R.I. Ray, K.L. Lowe, J. Jones-Meehan, and B.J. Little, *Biofouling,* Vol 19, 2003, p 151–160
61. I.B. Beech, S.A. Campbell, and F.C. Walsh, Marine Microbial Corrosion, *A Practical Manual on Microbiologically Influenced Corrosion,* Vol 2, J. Stoecker, Ed., NACE International, Houston, TX, 2001, p 11.3–11.14
62. T. Gehrke and W. Sand, Interactions between Microorganisms and Physicochemical Factors Cause MIC of Steel Pilings in Harbours (ALWC), Paper 03557, *Proceedings Corrosion 2003* (Houston, TX), NACE, 2003
63. J.S. Lee, R.I. Ray, E.J. Lemieux, and B.J. Little, An Evaluation of Carbon Steel Corrosion under Stagnant Seawater Conditions, Paper 04595, *Proceedings Corrosion 2004* (Houston, TX), NACE, 2004
64. M. Perez, C.A. Gervasi, R. Armas, M.E. Stupak, and A.R. Disarli, *Biofouling,* Vol 8, 1994, p 27
65. E. Littauer and D.M. Jennings, The Prevention of Marine Fouling by Electrical Currents, *Proceedings 2nd International Congress Marine Corrosion and Fouling,* Technical Chamber of Greece, Athens, Greece, 1968
66. R.G.J. Edyvean, Interactions between Microfouling and the Calcareous Deposit Formed on Cathodically Protected Steel in Seawater, *6th International Congress on Marine Corrosion and Fouling,* Athens, Greece, 1984
67. S.C. Dexter and S.-H. Lin, *Mater. Perform.,* Vol 30 (No. 4), 1991, p 16
68. M.F.L. de Mele, Influence of Cathodic Protection on the Initial Stages of Bacterial Fouling, *NSF-CONICET Workshop, Biocorrosion and Biofouling, Metal/Microbe Interactions,* Mar del Plata, Argentina, 1992
69. G. Nekoksa and B. Gutherman, Determination of Cathodic Protection Criteria to Control Microbially Influenced Corrosion in Power Plants, *Proceedings Microbially Influenced Corrosion and Biodeterioration,* University of Tennessee (Knoxville, TN), 1991, p 6-1
70. J. Guezennec, *Biofouling,* Vol 3, 1991, p 339
71. S. Maxwell, *Mater. Perform.,* Vol 25 (No. 11), 1986, p 53
72. "Control of External Corrosion on Underground or Submerged Metallic Piping Systems," NACE Standard RP-01-69, NACE, Houston, TX, 1972
73. J. Horvath and M. Novak, *Corros. Sci.,* Vol 4, 1964, p 159
74. T.R. Jack, M.McD. Francis, and R.G. Worthingham, *Proceedings of the Int. Conf. on Biologically Induced Corrosion,* NACE, Houston, TX, 1986, p 339
75. J. Guezennec and M. Therene, *Proc. First European Federation of Corrosion Workshop on Microbiological Corrosion* (Sintra, Portugal), 1988, p 93
76. K.P. Fischer, *Mater. Perform.,* Vol 20 (No. 10), 1981, p 41
77. V.V. Pritula, G.A. Sapozhnikova, G.M.K. Mogilnitskii, M.I. Ageeva, and S.S. Kamaeva, Protection Potential of St-3 in Liquid Cultures of Soil Microorganisms, translated from *Zashch. Met.,* Vol 23 (No. 1), Plenum Press, 1987, p 133
78. T.J. Barlo and W.E. Berry, *Mater. Perform.,* Vol 23 (No. 9), 1984, p 9
79. I.B. Ulanovskii and A.V. Ledenev, Influence of Sulfate-Reducing Bacteria on Cathodic Protection of Stainless Steels, translated from *Zashch. Met.,* Vol 17 (No. 2), Plenum Press, 1981, p 202
80. B. Little, P. Wagner, and D. Duquette, *Corrosion,* Vol 44 (No. 5), 1988, p 270
81. L.A. Terry and R.G.J. Edyvean, *Botanica Marina,* Vol 24, 1981, p 177–183
82. B. Little, R. Ray, and P. Wagner, Biodegradation of Nonmetallic Materials, *A Practical Manual on Microbiologically Influenced Corrosion,* Vol 2, J.G. Stoecker, Ed., NACE International, Houston, TX, 2001, p 3.1
83. B. Little, R. Ray, P. Wagner, J. Jones-Mehan, C. Lee, and F. Mansfeld, *Biofouling,* Vol 13 (No. 4), 1999, p 301–321
84. R.W. Schultz, *Mater. Perform.,* Vol 30 (No. 1), 1991, p 58–61
85. B. Little, R.I. Ray, and P. Wagner, An Evaluation of Titanium Exposed to Thermophilic and Marine Biofilms, Paper 525, *Corrosion 93,* NACE, 1993
86. M.B. McNeil and A.L. Odom, Thermodynamic Prediction of Microbially Influenced Corrosion (MIC) by Sulfate-Reducing Bacteria (SRB), *Microbiologically Influenced Corrosion Testing,* J. Kearns and B. Little, Ed., ASTM STP 1232 (Philadelphia, PA), ASTM, 1994, p 173–179
87. A. Hagenauer, R. Hilpert, and T. Hack, *Weerkstoffe und Korrosion,* Vol 45, 1994, p 355–360
88. J. Jones-Meehan, M. Walch, B. Little, R. Ray, and F. Mansfeld, Effect of Mixed Sulfate-Reducing Bacterial Communities on Coatings, *Biofouling and Biocorrosion in Industrial Water Systems,* G. Geesey, Z. Lewandowski, and H-C Flemming, Ed., CRC Press, Inc., Boca Raton, FL, 1994, p 107
89. P. Wagner, B. Little, R. Ray, and W. Tucker, Microbiologically Influenced Degradation of Fiber Reinforced Polymeric Composites, Paper 255, *Corrosion 94,* NACE, 1994
90. B. Little, P. Wagner, R. Ray, and K. Hart, *Mater. Perform.,* Vol 35 (No. 2), 1996, p 79–82
91. J.-D. Gu, T.E. Ford, K.E.G. Thorp, and R. Mitchell, Microbial Degradation of Polymeric Materials, *Proceedings of the Tri-Service Conference on Corrosion* (Orlando, FL), 1994, p 291–302
92. K.E.G. Thorp, A. Crasto, J.-D. Gu, and R. Mitchell, Biodegradation of Composite Materials, *Proceedings of the Tri-Service Conference on Corrosion,* (Orlando, FL), 1994, p 303
93. J. Jones-Meehan, K.L. Vasanth, R.K. Conrad, M. Fernandez, B.J. Little, and R.I. Ray, Corrosion Resistance of Several Conductive Caulks and Sealants from Marine Field Tests and Laboratory Studies with Marine, Mixed Communities Containing Sulfate-Reducing Bacteria (SRB), *Microbiologically Influenced Corrosion Testing,* J. Kearns and B. Little, Ed., ASTM STP 1232 (Philadelphia, PA) ASTM, 1994, p 217–233
94. B. Little and P. Wagner, Chapter 14, *Peabody's Control of Pipeline Corrosion,* R. Banchetti, Ed., NACE International, p 273–284
95. "Report of the Defense Science Board Task Force on Unexploded Ordnance (UXO) Clearance, Active Range UXO Clearance, and Explosive Ordnance Disposal (EOD) Programs," Office of the Under Secretary of Defense for Acquisition and Technology, Washington, D.C., 1998

Service Life and Aging of Military Equipment

M. Colavita, Italian Air Force

MILITARY SYSTEMS AND EQUIPMENT are designed to have a certain service life and to perform one or more specific functions over a specified length of time or, in the case of one-shot devices, be capable of performing a function after storage or within a specific time interval after manufacture when operated and maintained according to some stated plan. Service life extension and aging of military systems are a continuously growing problem (Ref 1). The combination of shrinking post Cold War military budgets and escalating costs for development and acquisition of new military equipment have led to increasing efforts worldwide to extend the operation of the existing systems far beyond their original design lives (Ref 2, 3). This article is mainly devoted to aircraft, but most of the considerations and the criteria exposed can be extended to land and sea vehicles.

In a general sense, aging vehicles are characterized by the deterioration of structural strength properties and increasing maintenance costs (Ref 4, 5), whose common effect is to reduce the readiness of the vehicle (Ref 6). Some of these problems are time dependent, such as corrosion, which also depends heavily on the usage environment; others are usage dependent, such as in fatigue cracking, which is caused by the mechanical loads that are introduced into the structure, and also in electronic devices. Intermediate conditions are often met and the strong correlation existing between corrosion and fatigue must always be taken into account. A schematic representation of this correlation is given in Fig. 1. The accelerating effect on corrosion damage due to cracking is shown (Ref 7). Corrosion that starts acting after a given initial time (t_i) (e.g., due to coating breakdown) instead of following its natural rate trend (dashed line), strongly increases in the presence of a crack (curve AC); the synergic effect given by the combination of these two elements—corrosion and cracking—is also shown by the shift of the damage curve toward shorter time (from AC to AB) when the environment inside the crack moves from inert to aggressive.

To maintain structural integrity, steps must be taken toward the prevention, detection, repair, and prediction of the initiation and growth of structural damage (Ref 8). This matter is a frontier between safety and economy issues, and every evaluation must be carefully appraised in terms of risks and benefits. Otherwise the short-term cost savings will be offset by greater long-term costs, safety problems, or decreased system availability.

The primary focus is to maintain operational military capability by directing resources toward high payback technologies able to:

- Identify structural deterioration
- Prevent the threat to equipment safety or performance degradation and minimize structural deterioration
- Avoid the system becoming an excessive economic burden or adversely affecting force readiness
- Assist in replacement of components
- Assist in development of failure analysis and life-predictive tools for such problems as fatigue damage, corrosion, and stress-corrosion cracking

Taking into account that a uniform in-service life for all equipment components in an aged system is not achievable, it is therefore very important to record in-service operational and maintenance data for critical equipment in order to be able to assess the usage history and life consumption and define the point of retirement.

Reliability and Safety of Equipment

In agreement with the usual behavior of the expected failure for any population of mechanical or electronic devices, the reliability of equipment with time is defined by the curve in Fig. 2, widely known as the "bathtub curve" because of its shape (Ref 9). In particular, the reliability decreases as the rear field of the curve (wear-out) is achieved. This occurs because malfunctions caused by aging and/or wear increase. In this way, the spare part demand and the number of maintenance actions increase, which consequently result in higher operating costs and reduced availability.

A typical measure within a life-extension program would determine the optimal point of time when it would be more economical to replace an item with a new part or to introduce redesign or an upgrade instead of repairing the old item again and again.

It is important to include preventive maintenance actions for critical components and areas in the existing maintenance procedures to ensure that aging problems will be detected at an early stage (Ref 10) when the amount of effort required for repair is still acceptable and a safety critical situation is avoided.

Mainly, reliability of an aged component depends on several factors such as its usage history, repairs and life-extension concessions, obsolescence, availability of spare parts, and failure data already available.

Aging Mechanisms

Aging is a process where the structural and/or functional integrity of components will be continuously degraded by exposure to the environmental conditions under which it is operated. There are various mechanisms that alone, or in

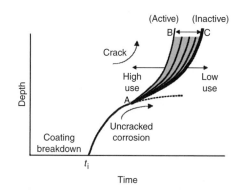

Fig. 1 Growth of corrosion and fatigue cracking with time

combination, are responsible for component aging:

- The exposure to normal or salty atmospheres, heat, water, oil, fuel, and so forth, which could lead to corrosion, overheating, melting, disbonding, or other material degradation processes, electrical interruptions, short circuits, and so forth
- The exposure to vibration and acoustic environments, which could lead to fatigue damages, wear, tear, and so forth
- The endurance to normal service conditions, which leads to leakage, wear, tear, and so forth

Additionally, improper installation of equipment, or incorrect maintenance activities, also can induce accidental damage or have a detrimental effect and assist the aging process, for example, scouring of wire bundles on the surrounding structure.

Factors that have a direct effect on the problems associated with military system aging are its usage, the service environment, and the inspection, test, and maintenance practices. In particular, the location of the equipment will determine corrosion problems (Ref 11). For example, military aircraft that are exposed to salt water environments, as on an aircraft carrier deck, will experience a much higher degree of corrosion than other military aircraft. The predisposition to corrosion also depends on factors such as corrosion resistance of the material, combination of materials, and surface protection and sealing.

Structural Parts

Corrosion usually plays the most important role in the degradation with time of structural components. Many types of corrosion can occur in military structures, but the general attack occurring on the surfaces in reciprocal contact of splice joints (faying surfaces) is probably the process most often observed on aged parts. It usually occurs when adhesive bond or sealant between the layers breaks down, allowing moisture ingress. Depending on the severity of the attack and the material corrosion resistance characteristics, it can then develop or be accompanied by pitting or exfoliation. The final effect of this degradation process is a reduction of the mechanical properties of the part, in particular of the fatigue life of the joint (Fig. 3, 4).

Another class of damage produced by corrosion of splice joints is "pillowing cracks," an environmentally assisted form of failure under sustained stress due to the deformation of the layers of the joint between the fasteners. It is caused by the increase in volume associated with aluminum alloy corrosion products. This volume increase was recently estimated at 6.45 (molecular volume ratio) times the volume of the parent material lost (Ref 12). This volume increase results in a significant increase in the stress in the vicinity of the rivet holes and is believed to act as a source of the multiside-damage (MSD) whose effects on structural integrity are well known (Ref 13).

In effect, pillowing produces a compressive stress in the rivet area on the outer surface or skin while a high tensile stress is present on the faying surface or inner skin, causing the growth of nonsurface breaking cracks with the suggested shape shown in Fig. 5. This circumstance strongly reduces the probability of detection before first linkup in a MSD situation occurs and therefore significantly increases the risk of an in-service failure. Attempts have been made to quantify this effect by means of mechanistic models (Ref 14), but the many parameters needed render the models unreliable so far, and a more empirical approach continues to be relied upon.

However, the effect of interaction between corrosion and fatigue (Ref 15), one of the main concerns in aging aluminum structures, suggests that some form of hydrogen-related failure is the corrosion mechanism causing environmentally induced cracking in the 2000 and 7000 series aircraft aluminum alloys (Ref 16). Actually, a difference in the effect of preexisting corrosion on fatigue crack initiation and short crack growth on these two alloys has been shown. In the case of 2024-T6, fractographic analysis (Fig. 6) and measured fatigue initiation time indicate that fatigue initiation is primarily a mechanical phenomenon. Corrosion acts as a stress-concentration factor, and geometrical consideration of the corrosion pit can be used to

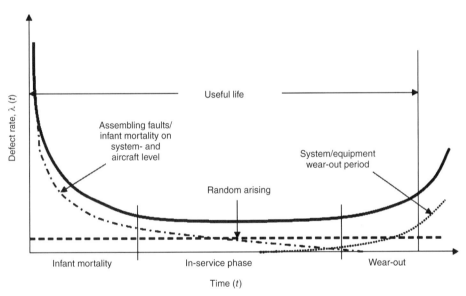

Fig. 2 Reliability of equipment as defined by the bathtub curve. The infant mortality phase can typically be 20% of useful life.

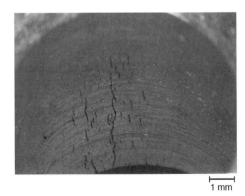

Fig. 3 Optical micrograph of a high stressed region in a B707 wing panel showing extensive corrosion fatigue cracking

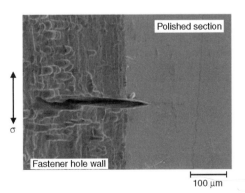

Fig. 4 Scanning electron micrograph of a section through an elongated damage in a B707 wing panel showing corrosion fatigue cracking

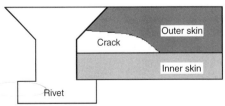

Fig. 5 Suggested shape of cracks under influence of pillowing

predict the subsequent fatigue behavior (Ref 17). In the case of alloy 7075-T6, the evidence suggests that other embrittling phenomena may affect the crack growth initiation (Ref 18).

When localized corrosion occurs, an aggressive environment is created inside the pit. The pH will be determined by the hydrolysis reaction $Al^{3+} + 3H_2O \rightarrow Al(OH)_3 + 3H^+$, and its usual value will be between 3 and 4. To explain the hydrogen embrittlement mechanism observed for the 7000 alloys, it has been suggested that second-phase particles act as sinks for dislocation transported hydrogen (Ref 19).

Engines

Aging of engine components may take many forms depending on component, engine type, and operating conditions (Ref 20). The damage may be external, affecting dimensions and surface finish as a result of erosion, wear, corrosion, or oxidation. In this case, the damage affects the aerodynamic performance and load-bearing capacity of gas path components. Additionally, surface cracks and notches induced by low-cycle fatigue (LCF), and fretting and foreign object damage (FOD) may eventually lead to high-cycle fatigue (HCF) failures. The damage may also be internal, affecting the microstructure of highly stressed and hot parts, as a result of metallurgical aging reactions, creep, and fatigue. In this case, the damage may reduce component strength and lead to component distortion. The accumulated damage may cause the initiation of flaws that lead to subsequent cracking and component failure (Ref 21).

Surface Damage. Erosion by ingested sand and other hard particulate matter can significantly alter compressor airfoil shape and surface finish. Such changes, in addition to reducing compressor efficiency, may lead to resonant excitation and HCF failures of airfoils. Pitting corrosion may develop in marine environments and both steel and titanium alloys are particularly susceptible to this form of damage. Corrosion pits provide sites for crack initiation and have been known to be responsible for HCF failures (Fig. 7).

Surface oxidation reduces the load-bearing capacity of turbine blades and vanes. In marine environments hot corrosion may rapidly destroy airfoils by surface melting, as evidenced by surface rippling (Fig. 8a) and metallographic section (Fig. 8b) for a MAR-M246 nickel-base superalloy turbine blade.

Turbine blades also often suffer tip oxidation, as protective coatings wear off rapidly due to tip rub. In the case of directionally solidified (DS) blades, tip oxidation may produce thermal fatigue cracking at the longitudinal grain boundaries that are embrittled as a result of boundary oxidation (Fig. 9).

Aging associated with fretting has been often observed in the dovetail base where fan and compressor blades come in contact with disks (Fig. 10).

Microstructural Damage. Internal microstructural damage is the result of metallurgical aging reactions and plastic strain accumulation. It is an insidious form of damage because, in contrast with surface damage, it is not readily detected by nondestructive techniques (NDT). Furthermore, its rate of accumulation is strongly influenced by service stresses and temperatures. Since there are uncertainties in the temporal variations of these parameters, the extent of damage accumulation cannot be easily predicted.

Time-dependant aging reactions occur primarily in hot parts, such as turbine blades and vanes; they are varied but invariably detrimental to mechanical properties. Coarsening of γ' precipitates, for instance, causes loss of creep strength in nickel-base superalloys (Fig. 11).

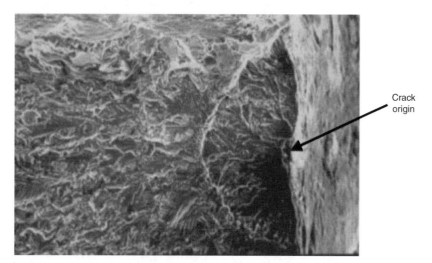

Fig. 6 Crack origin inside a rivet hole developed from an existing corrosion pit

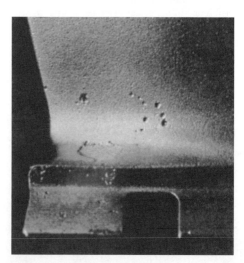

Fig. 7 Corrosion pits exceeding allowable limits in the root section of a compressor blade

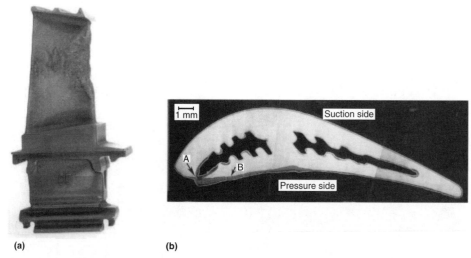

Fig. 8 Examples of engine component surface damage. (a) Evidence of hot corrosion damage on the pressure side of a MAR-M-246 turbine blade. (b) Metallographic section across the airfoil of the MAR-M-246 blade, showing evidence of hot corrosion damage penetrating the leading edge (B) right through to the internal cooling passage (A)

Subsystems

Aging consequences on subsystems are sometimes underestimated with respect to their frequency and relevance. For example, main landing gears should be carefully considered as one of the weakest points with respect to time-related performances on account of the severity of their service conditions and continuous exposure to harsh environmental conditions.

Catastrophic failures can also be insidiously produced when apparently nondangerous corrosion phenomena are ongoing. This happened in the case of the high-strength steel main landing gear (MLG) truck beam of a B707 tanker, which broke during taxiing (Fig. 12). The final failure was caused by stress-corrosion cracking (SCC) resulting from the high-static stress that the part usually undergoes. However, the original cause was the uniform corrosion phenomenon developed over 35 years of service on the bottom of its internal part, where moisture and humidity had collected (Ref 22). This particular damage was also due to a failure to review appropriate maintenance procedure. The aircraft, originally in civilian use, was maintained in the military on the same scheduled maintenance program, with a consequent large time delay between inspections.

In general, the introduction of materials such as the Aermet 100 Co-Ni low-carbon steel, as an alternative to the more traditional high-strength materials such as AISI 4340 and 300M, should be encouraged in order to increase the safety margin and minimize material sensitivity to hydrogen embrittlement, SCC, and corrosion fatigue (Ref 23).

Wiring is another critical subsystem to be carefully considered in terms of aging behavior. Indeed, greater attention is being paid to this subject since very dangerous failures have occurred (Ref 24). This specific phenomenon is due to a decay of the external insulation that starts cracking with the passage of time and leads to the corrosion of terminal ends (Ref 25). An example of cracked wiring is shown in Fig. 13. Sometimes corrosion occurs under the insulation, especially in the case of aluminum wiring (Ref 26).

In both cases the electrical resistance of the wiring is increased due to the presence of corrosion damage. Deterioration of the electrical system can cause a number of problems, some of which are extremely difficult to sort out. Corrosion damage can cause erroneous instrumentation readings, including faulty caution and warning indications. It can result in arcing (shorts), either between wires or between wires and the aircraft structure, resulting in fire and smoke.

Management

The management of an aging fleet is a complex task that can be achieved only with actions that include management of corrosion problems. In this sense, the introduction of more effective corrosion preventive measures is a cornerstone of any strategy. Some of the key measures are:

- Improved training in corrosion recognition and treatment
- Wider use of corrosion preventive compounds during regular maintenance
- Washing of aircraft with water containing inhibitors

Full-Scale Structural Testing

Full-scale structural testing is usually done in response to customer mandated requirements at the time of aircraft purchase. Subsequent fatigue improvement modifications are verified by component tests. However, in an aged fleet the discrepancy between tested and current configuration usually becomes too great, dictating the need to update the test database (Ref 27).

Furthermore, to take into account the effects of time on corrosion damage, it is often necessary to evaluate the integrity of the whole structure before extending operational life well beyond the original design life. It means that a renovated full-scale test is needed to verify the actual residual fatigue resistance.

While it is true that the cost of full-scale structural testing often appears prohibitive, the value of the data generated for use in management of service aircraft renders these tests very desirable. They can be summarized as:

- Experimental determination of residual strength of cracked and corroded structures

Fig. 9 Tip cracking of a directionally solidified MAR-M-246 turbine blade

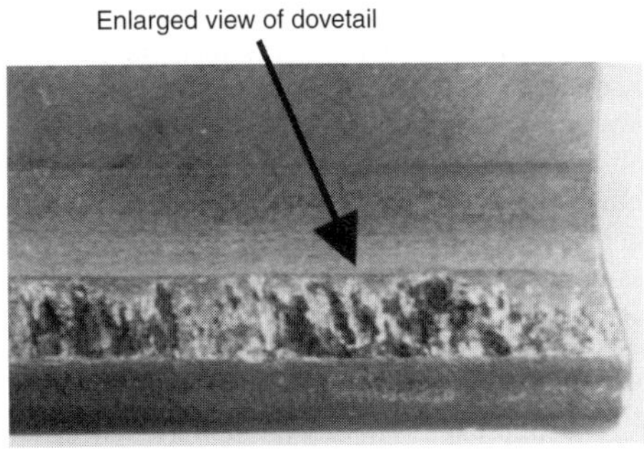

Fig. 10 Fretting fatigue damage along a titanium fan stage-2 blade dovetail experienced on F-400 engines

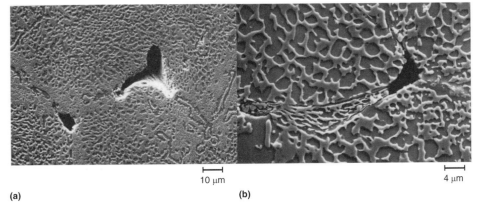

Fig. 11 Microstructures of nickel-base Alloy 713C turbine blades. (a) Original structure prior to service. (b) Coarsening of γ' precipitates and elimination of secondary γ' caused by 5000 h of service

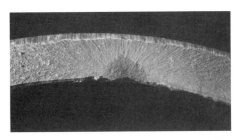

Fig. 12 Cross section of a stress-corrosion cracking failure occurred on high-strength steel AISI 4340 B707 truck beam, caused by a preexisting uniform corrosion phenomenon

 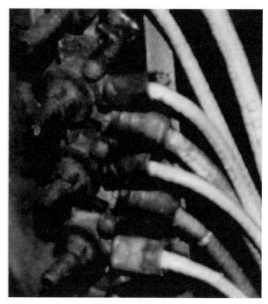

Fig. 13 Wiring insulation cracks (left) and corroded terminals on a C130 (right)

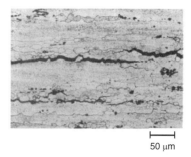 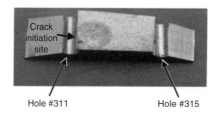

Fig. 14 Stress-corrosion cracking on the Macchi MB-326 7075-T6 stabilizer spar of RAAF aircraft used by the acrobatic team

- Experimental measurement of crack growth curves
- Strain survey verification of the assumed internal loads distribution
- Validation or replacement of the assumed crack location

When these tests are planned, special attention must be focused on sampling. Usually, to be conservative, the oldest aircraft of the fleet—the fleet leader—can be chosen as representative of the worst case. However, because of the primary role played by corrosion, the environment in which the aircraft has been operating throughout its life, the real load spectrum suffered, the maintenance history, and preservation strategies adopted are very important factors to be taken into account. In effect, the same aircraft, when operated by different nations in very different environments and perhaps with different missions will show a widespread range of residual mechanical deterioration properties. An example is the results of tests carried out on a Macchi MB-326 operated as a military trainer by the Italian Air Force (IAF) and in the acrobatic team by the Royal Australian Air Force (RAAF). The strong corrosiveness of the Australian environment resulting from the high salt content, and the elevated wind in conjunction with the high loads corresponding to the acrobatic use, led to SCC on the 7075-T6 stabilizer spar that did not occur in the Italian fleet (Ref 28). All the observed cracks on the RAAF aircraft initiated along the interior of rivet holes (Fig. 14).

Structure. Risk assessment is certainly one of the most powerful means to manage an aged fleet. It consists of a study of the elements that contribute to the reduction in airworthiness and results in an evaluation based on the acceptable risk. The analysis is founded on a previous knowledge of the degradation mechanisms involved in the aging process. With respect to this point, corrosion data collection has already proved to be an effective tool in risk assessment. Essentially, the complexity of the military systems has encouraged the introduction of appropriate corrosion data banks with two main objectives:

- Increasing the reliability of predictive analysis of failures
- Addressing the resources needed to solve the problems in agreement with a predefined scale of criticalities

The former objective is usually split into two branches, respectively, dedicated to understanding of the general trend and introduction of optimized inspection on specific items.

The introduction of corrosion data banks appears promising in both increasing the availability of the fleet and making it easier to achieve standard maintenance procedures among different nations that operate the same fleet.

Engine. For the purpose of life-cycle management, engine components can be classified as either durability-critical or safety-critical parts. The former parts are those for which aging deterioration only affects performance and fuel efficiency, resulting in a significant maintenance burden, but will not directly impair flight safety: they include cold and hot gas path components such as vanes and blades.

The latter parts are those for which fracture may result in loss of the military system and endanger human life because of noncontainment. They include most of the rotating compressor and turbine components such as wheels, disks, spacers, and shafts.

Durability-Critical Parts. These parts employ an "On Condition" maintenance philosophy, which means that a preventative process is used that allows deterioration of components by monitoring them for their continued compliance with a required standard. The parts are removed from service when damage limits exceed those dictated by design.

Safety-Critical Parts. These parts employ two different management approaches.

Safe Life Approach. All components are retired before a first crack is detectable. This methodology follows a "cycles to crack initiation" criterion, with a minimum safe life capability established statistically through extensive mechanical testing of test coupons and components under simulated service conditions. The statistic is based on the probability that 1 in 1000 components (-3σ) will have developed a detectable crack, typically 0.8 mm (0.03 in.) long, at retirement.

This approach is overly conservative, because components are retired with a significant amount of residual life, and its main benefit is that maintenance requirements are kept to a minimum. Under this approach the supply of parts that need to be replaced at the same time may be substantial and this can be a serious problem.

Damage Tolerance Approach. Fracture critical areas of all components are assumed to contain manufacturing or service-induced defects giving rise to growing cracks. In accordance with this approach, components are retired when a crack is detected.

This approach also assumes that components are capable of continued safe operation as the crack grows under service stresses and that the crack growth can be predicted from linear fracture mechanics or other acceptable methods. Of course, it is also required that cracks grow slowly enough to allow their detection through scheduled inspection.

The interval between inspections, or safe inspection interval (SII), is established from the time the crack grows from a size immediately below the detection limit (it will depend on the technique used to inspect the component) to a critical size. For an assumed geometry, the critical size is obtained by analysis from the fracture toughness of the material and the stress-intensity factor, using appropriate safety factors. This approach is schematically illustrated in Fig. 15, where SII is obtained through the application of either deterministic or probabilistic fracture mechanics (DFM or PFM).

In general terms, two different methods can be used to implement damage tolerance based life-cycle management for safety critical parts: ENSIP (Engine Structural Integrity Program, MIL-STD-1783), introduced in 1984 by the U.S. Air Force (USAF), and RCF (Retirement for Cause), as shown in Fig. 16. The RCF, based on periodic inspections until a crack is detected at which point the part is retired, allows life extension beyond LCF-based safe-life limits.

Practical Life-Enhancement Methods

Structural Parts. Various life-enhancement techniques are already available, some of them being commonly used on operational aircraft (material substitution, cold working, shot peening, and composite patches), and many others are still considered to require engineering supervision (e.g., friction stir welding, grid-lock, laser shock processing, active vibration suppression, laser-formed titanium, etc.).

Material substitution is probably the most common and effective countermeasure to the corrosion effects caused by aging. In effect, new alloys have been developed in the recent past to overcome low corrosion resistance, especially the low SCC resistance characteristic of many older metallic alloys. Typical is the case of aluminum alloy 7055 used to replace 7075, mainly in spar and lugs applications, or all the attempts made to modify the T6 thermal treatments on this alloy, responsible for its low SCC resistance, but preserving its high strength by retrogression and reaging (RRA) (Ref 29). Another example is the substitution of the current aluminum trailing edge on the C130 that has shown a tendency to corrode and debond from graphite/epoxy panels due to a combination of moisture from the surrounding environment, engine exhaust products, and sonic fatigue.

Fatigue cracks and damage originating at holes are also a strong life-limiting factor, due to the geometric stress-concentration effect that intensifies the magnitude of the applied load by factors of three or more. In this regard, the "hole cold expansion" (Fig. 17) is a practical method to repair corroded or fatigue cracked holes and at the same time to ensure continued airworthiness by inducing beneficial annular residual compressive stresses (Ref 30).

Engines. Life extension can be achieved by means of different strategies (Ref 31):

- Restoration of damaged components by repair or rework (e.g., electron beam weld repair of blades damaged by foreign objects (Ref 32), and weld tip repair of high pressure turbine blades (Ref 33), advanced braze repairs for nozzle guide vanes, and component rejuvenation through hot isostatic pressing)
- Improvement of the structural performance or damage tolerance of engine components, achieved by a range of different techniques, depending on the objective to be accomplished. For example, (a) retrofitting with new materials, when enhancement of component durability in terms of better oxidation behavior, creep strength, or thermal fatigue resistance is required, (b) addition or substitution of a protective coating to improve erosion (e.g., titanium nitride applied by physical vapor deposition) or hot corrosion (e.g., platinum aluminides) performances, and (c) ion implantation and chemical surface treatments in conjunction with shot peening and soft coatings and lubricants to reduce fretting fatigue damage by means of a surface modification treatment (Ref 34)
- Reuse of components under a damage tolerance-based life cycle management scheme, achieved using five steps: (1) determination of stress and temperature data, (2) identification of the fracture critical location in the component of interest, (3) determination of the stress-intensity factor, (4) generation of fracture mechanics data for safe inspection interval calculations, and (5) generation of probability of detection (POD) curves from nondestructive inspection (NDI) data

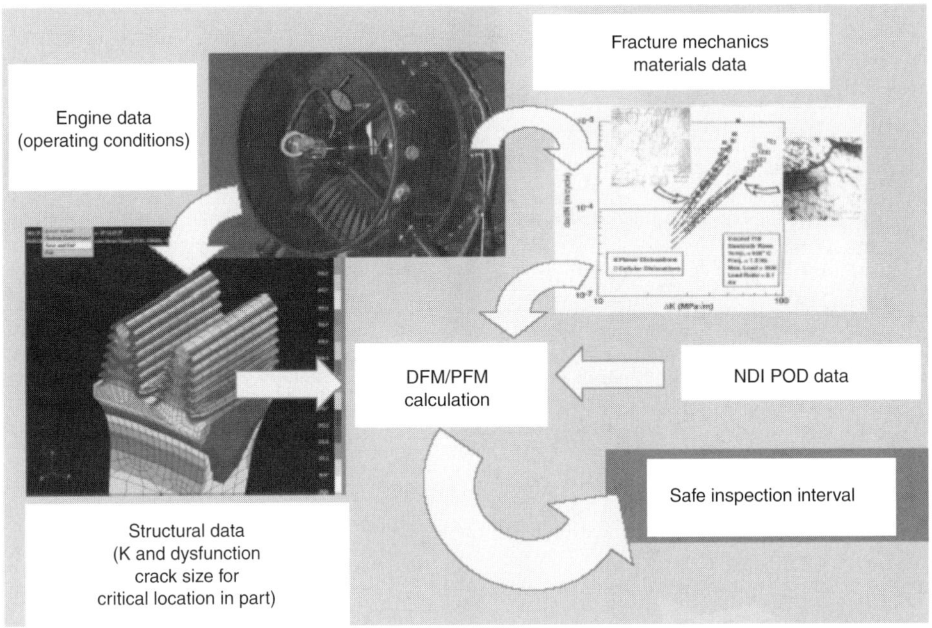

Fig. 15 Steps associated with safe inspection intervals (SII) calculation for damage tolerance based life management of critical engine parts

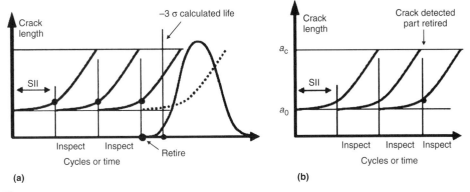

Fig. 16 Application of damage tolerance by ENSIP (a) and RFC (b) and relation to safe inspection interval (SII)

Prevention and Control

Control and prevention are both issues used to describe the procedures necessary to provide effective corrosion maintenance on military systems. In effect, they must always be considered as complementary because they provide a synergistic effect when each achieves its specific action (Ref 35).

Corrosion control usually includes:

- Corrosion detection
- Corrosion removal
- Renewing the protective systems

On the other hand, corrosion prevention is mainly devoted to:

- Material design
- Surface treatments, finishes, and coatings
- Corrosion inhibitors, compounds, and sealants
- Preservation techniques

The entire process, including all these phases, has been recently called corrosion surveillance (Ref 36) and indicates the increasing interest from systems operators in this matter. This interest is largely due to the growing number of aged fleets that no longer respond to a simple "find and fix" maintenance approach. In addition, the attention to corrosion prevention and control has strongly increased in the past decade because of the very important role played by environmental constraints (Ref 37). In particular, corrosion prevention of aging systems is strongly dependent on the application of new water-displacement compounds and advanced coating and new plating alternatives. Previously successful chromic anodize and cadmium and chromium plating are being replaced with newer, safer materials.

Nevertheless, taking into account the strong interconnections existing between corrosion and other structural failures that can affect highly engineered systems, the monitoring of a much more complex number of parameters not always directly linked to corrosion has recently been introduced to preserve the safety and readiness of military systems (Ref 38).

Usage Monitoring (UM) Systems

Usage monitoring systems are increasingly introduced on aging military systems to guarantee an optimized aircraft management or to monitor the events that can lead to damage (excessive G forces, hard landings, etc.). The general trend in UM systems development is unambiguously directed toward integration in the avionic structure, in order to use the information provided by the interface progress unit (IPU), and consequently to have all the data immediately available for calculation. The Eurofighter is a case in point (Ref 39), whose UM system (Fig. 18) is subdivided into:

- Engine UM
- Structural UM
- Secondary power system UM
- Aircraft system UM

Health Monitoring (HM) Systems

The development of microelectronics has shown new ways to monitor structures and systems by means of sensors. As a matter of fact, their reduced dimensions and increased computer performance (that now offers the possibility of fast signal processing) make them currently available at a reasonable price for the monitoring of weight-optimized structures (Ref 40). In effect, an integrated HM system allows the implementation of an effective on-condition inspections program, the main benefits being:

- Avoidance of expensive conventional inspections for life extension in difficult-to-access areas of the aircraft
- Increase of time between inspections
- Increase of aircraft availability
- Increase of useful life without direct repair measures

An aircraft health monitoring system consists of sensors, signal amplifier, and filter and signal processing that, used in conjunction, enables the detection of a signal eventually related to structural damage.

Three types of sensors are most commonly suitable: fiber optic, piezoelectric, and sensory foils. The first, interesting for its low cost, low weight, low power demand, and long lifetime, has already proven to be particularly effective in damage identification by means of acoustic emission and is easily embedded into fiber composite materials.

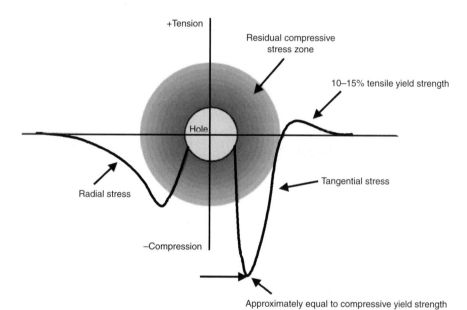

Fig. 17 Typical residual and circumferential stress distributions around a cold expanded hole

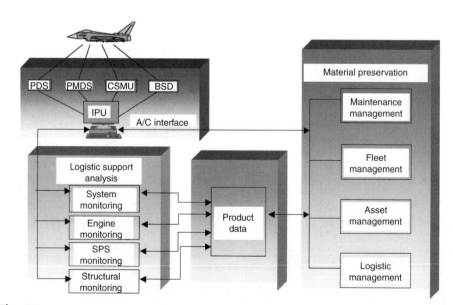

Fig. 18 Usage monitoring system data processing for the Eurofighter

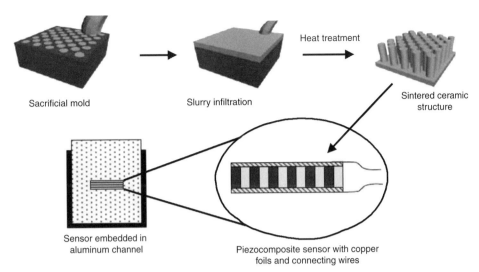

Fig. 19 Process for making the embedded piezoelectric ceramic-polymer composite sensors

Fig. 20 Thin-film bimetallic sensing element

Particularly interesting are the sensory foils, where piezoelectric sensors are embedded together with the wiring between two foils, bonded onto the structure, or embedded in a group structure as shown in Fig. 19 (Ref 41).

Wireless galvanic sensors (Ref 42–44) are a different class of sensors, specifically dedicated to the diagnostics of corrosion nucleation on aging aircraft (Fig. 20). In this case, an increase in the current circulating between the two electrodes embedded in the sensors is the early signal of an increase in corrosiveness of the local environment.

Prediction Techniques

Equivalent Precrack Size Approach. This is a risk-analysis approach based on the use of the probability of failure during a flight at different times throughout the life of the aircraft, where each structural location considered is assumed to have a single critical flaw whose size at a given time is grown from a probability distribution of equivalent initial flaw sizes (EIFS) or alternatively equivalent precrack sizes (EPS) until failure (Ref 45, 46). This approach is considered a very promising tool to predict fracture on aging structures where pitting corrosion plays an important role as a stress-concentration factor for future fatigue crack growth. By including all significant uncertainties, such as EPS on its own, possible load history, and material property variation, the risk of failure can be calculated (Ref 47).

The main uncertainties that govern the probability distribution at failure are usually due to the variability in:

- The EPS at the start of the fatigue life of the airframe
- Crack growth rates
- The fracture toughness of the material
- The maximum load in each flight
- The accuracy in calculating aircraft fatigue life consumed that relates to the amount of crack growth experienced based on the measured loading sequence

Life-Cycle Cost Modeling and Simulation. An effective cost-estimating methodology to forecast costs associated with maintaining an aging aircraft usually combines the traditional operational and support cost elements with an expert analysis system. The result of this combination provides the capability to evaluate multiple and simultaneous "what if?" scenarios in order to weigh the effects of operational changes such as the number of aircraft, personnel to aircraft ratio, or annual flying hours.

To deal with the unavoidable uncertainty of long-term forecast estimates, analysts have to provide a range of estimated costs; this is generally accomplished by the use of probability distributions taking advantage of Monte Carlo techniques. The main strength of life-cycle cost modeling and simulation is its built-in flexibility. The model can be updated with the very latest information as it becomes available over time.

Holistic Life-Prediction Methodology (HLPM). This is in effect a proactive structural integrity methodology that looks at the design as a whole, where concepts of failure processes and pathology pervade all the phases of the design itself (Ref 48). Here, probability of occurrence of specific corrosion, wear, fretting, and thermal degradation, acting singly or conjointly with fatigue, are acknowledged as part of the failure process (Ref 49). This is achieved by means of a constant evaluation, model, and test method development. Based on HLPM, the USAF has developed since the late 1990s the environmental and cyclic life interaction prediction software (ECLIPSE), which has been extensively exercised on lap-joint geometries and on early pitting-to-fracture components. The expected benefits from introduction of this program in conjunction with specific NDI activity on the aircraft maintenance life are currently under investigation.

The reader may wish to see "In-Service Techniques—Damage Detection and Monitoring" in *Corrosion*, Vol 13A, *ASM Handbook*.

REFERENCES

1. J.W. Lincoln, "Challenges for the Aircraft Structural Integrity Program," FAA/NASA Symposium on Advanced Structural Integrity Methods for Airframe Durability and Damage Tolerance, May 1994
2. T. Vogelfänger, "Common Understanding of Life Management Techniques for Ageing Air Vehicles," Proc. Specialists' Meeting on Life Management Techniques for Ageing Air Vehicles (Manchester, U.K.), NATO, Oct 2001
3. S.G. Sampath, "Aging Combat Aircraft Fleets—Long Term Implications," Introduction, AGARD Lecture Series 206, 1996
4. J.W. Lincoln and R.A. Melliere, Economic Life Determination for a Military Aircraft, *J. Aircraft*, Vol 36 (No. 5), 1999, p 737–742
5. R. Kinzie, "Cost of Corrosion Maintenance," Second Joint NASA/FAA/DoD Conference on Aging Aircraft, Proceedings (Williamsburg, VA), Aug 1998
6. http://www.gao.gov/atext/d03753.txt, May 2005
7. G. Clark, "Management of Corrosion in Aging Military Systems," Proc. Specialists' Meeting on Life Management Techniques for Ageing Air Vehicles (Manchester, U.K.), NATO, Oct 2001
8. Aging of U.S. Air Force Aircraft Final Report, Publication NMAB-488-2, Committee on Aging of U.S. Air Force Aircraft—National Materials Advisory Board, National Academy Press, 1997
9. http://www.weibull.com/hotwire/issue21/hottopics21.htm, May 2005
10. J.W. Lincoln, "Role of Nondestructive Inspections Airworthiness Assurance," RTO MP-10, RTO AVT Workshop on Airframe Inspection Reliability under Field/Depot Conditions (Brussels, Belgium), NATO, May 13–14, 1998
11. M. Colavita, "Occurrence of Corrosion in Airframes," RTO-EN-015 Lecture Series on Aging Aircraft Fleets: Structural and Other Subsystems (Sofia, Bulgaria), NATO, Nov 2000
12. J.P. Komorowski, N.C. Bellinger, R.W. Gould, D. Forsyth, and G. Eastaugh, Research in Corrosion of Ageing Transpost Aircraft Structures at SMPL, *Can. Aero. Space J.*, Vol 47 (No. 3), Sept 2001, p 289

13. R.J.H. Wanhill, T. Hattenberg, W. Van der Hoeven, and M.F.J. Koolloos, "Practical Investigation of Aircraft Pressure Cabin MSD and Corrosion," Proc. Fifth FAA/DoD/NASA Joint Conference on Aging Aircraft (Orlando, FL), Sept 10–13, 2001
14. S. Mahadevan and P. Shi, Corrosion Fatigue Reliability of Aging Aircraft Structures, *Progr. Struct. Eng. Mater.*, Vol 3 (No. 2), Aug 2001, p 188–197
15. D.W. Hoeppner, V. Chandrasekaran, and A.M.H. Taylor, Review of Pitting Corrosion Fatigue Models, ICAF '99 Proc. (Seattle, WA), July 12–16, 1999
16. G.M. Bond, I.M. Robertson, and K. Birnbaum, Effects of Hydrogen on Deformation and Fracture Processes in High-Purity Aluminum *Acta Metall.*, Vol 36, 1988, p 2193–2197
17. V. Chandrasekaran, A.M.H. Taylor, Y. Yoon, and D.W. Hoeppner, Quantification of Pit Parameters to Small Fatigue Cracks, ICAF '99 Proc. (Seattle, WA), July 12–16, 1999
18. G.H. Koch, On the Mechanisms of Interaction between Corrosion and Fatigue Cracking in Aircraft Aluminum Alloys, *Structural Integrity in Aging Aircraft*, AD-Vol 47, American Society of Mechanical Engineers, 1995, p 159–169
19. D.J. Duquette, "Mechanisms of Corrosion Fatigue of Aluminum Alloys," NTIS-AD-A09712, Rensselaer Polytechnic Institute, April 1981, p 15
20. J.-P. Immarigeon, W. Beres, P. Au, A. Fahr, W. Wallace, A.K. Koul, P.C. Patnaik, and R. Thamburaj, "Life Cycle Management Strategies for Aging Engines," RTO-MP-079 (II), Proc. Specialists' Meeting on Life Management Techniques for Ageing Air Vehicles (Manchester, U.K.), NATO, Oct 2001
21. M.M. Ratwani, A.K. Koul, J.-P. Immarigeon, and W. Wallace, "Aging Airframes and Engine, Future Aerospace Technologies in the Service of the Alliance," NATO-AGARD Conference Proc. 600, Dec 1997
22. M. Colavita and L. Aiello, "Aging Aircraft and Corrosion Management: A Review of the Italian Air Force Fleet," Proc. Fifth Int. Workshop in Aircraft Corrosion (Solomons Islands, MD), NACE, Sept 2002
23. V.S. Agrawala, Corrosion and Aging Aircraft, *Can. Aero. Space J.*, Vol 42 (No. 2), June 1996
24. C. Smith and R. Pappas, "Requirements for Risk Assessment Tools for Aircraft Electrical Interconnection Subsystems," RTO-MP-079 (II), Proc. Specialists' Meeting on Life Management Techniques for Ageing Air Vehicles (Manchester, U.K.), NATO, Oct 2001
25. A. Baig, P. Eng, and Y. Yan, "CC130 Aging Aircraft Initiatives," Proc. Fifth Joint FAA/DoD/NASA Conference on Aging Aircraft (Orlando, FL), Sept 10–13, 2001
26. http://www.termpapergenie.com/aging.html, May 2005
27. J.G. Bakuckas, Jr., C.A. Bigelow, and P.W. Tan, "Full-Scale Aircraft Structural Tests and Evaluation Research (FASTER) Facility," Proc. Third Joint FAA/DoD/NASA Conference on Aging Aircraft (Albuquerque, NM), Sept 1999
28. M. Colavita, E. Dati, and G. Trivisonno, "Aging Aircraft: In Service Experience on MB-326," RTO-AVT-018, Report of the Specialists' Meeting on Fatigue in the Presence of Corrosion (Corfú, Greece), NATO, Oct 1998
29. D. Raizenne, X. Wu, and C. Poon, "In-situ Retroregression and Re-aging of Al 7075-T6 Parts," Proc. Fifth Joint FAA/DoD/NASA Conference on Aging Aircraft (Ornaldo, FL), Sept 10–13, 2001
30. L. Reid, "Sustaining an Aging Aircraft Fleet with Practical Life Enhancements Methods," RTO-MP-079 (II), Proc. Specialists' Meeting on Life Management Techniques for Ageing Air Vehicles (Manchester, U.K.), NATO, Oct 2001
31. A.D. Boyd-Lee, D. Painter, and G.F. Harrison, "Life Extension Methodologies and Risk-Based Inspection in the Management of Fracture Critical Aeroengine Components," RTO-MP-079 (II), Proc. Specialists' Meeting on Life Management Techniques for Ageing Air Vehicles (Manchester, U.K.), NATO, Oct 2001
32. P. Azar, Li Ping, P.C. Patnaik, R. Thamburaj, and J.-P. Immarigeon, "Electron Beam Weld Repair and Qualification by Titanium Fan Blades for Military Gas Turbine Engines," RTO-MP-069(II), RTO-AVT Workshop on Cost Effective Applications of Titanium Alloys in Military Platforms (Loen, Norway), NATO, April 2001
33. J. Liburdi, "Enabling Technologies for Turbine Component Life Extension," RTO-MP-17, RTO-AVT Workshop on Qualification of Life Extension Schemes for Engine Components (Corfú, Greece), NATO, Oct 1998
34. A.K. Koul, L. Xue, W. Wallace, M. Bibby, S. Chakraverty, R.G. Andrews, and P.C. Patnaik, "An Investigation on Surface Treatments for Improving the Fretting Fatigue Resistance of Titanium Alloys," AGARD CP N. 589, Proc. Specialists' Meeting on Tribology for Aerospace Systems (Sesimbra, Portugal), NATO, Sept 1996
35. M. Colavita, "Prevention and Control in Corrosion," RTO-EN-015 Lecture Series on Aging Aircraft Fleets: Structural and Other Subsystems (Sofia, Bulgaria), NATO, Nov 2000
36. P.R. Roberge, M. Tullmin, L. Grenier, and C. Ringas, Corrosion Surveillance for Aircraft, *Mater. Perform.*, Dec 1996, p 50–54
37. S.J. Hartle and B.T.I. Stephens, "U.S. Environmental Trends and Issues Affecting Aerospace Manufacturing and Maintenance Technologies," NATO-AGARD Report 816 (Florence, Italy), Sept 1996
38. J.W. Lincoln, "Aging Systems and Sustainment Technology," RTO-MP-069(II), RTO-AVT Workshop on Cost Effective Applications of Titanium Alloys in Military Platforms (Loen, Norway), April 2001
39. S.R. Hunt and I.G. Hebden, Validation of the Eurofighter Typhoon Structural Health and Usage Monitoring System, *Smart Mater. Struct.* Vol 10, 2001, p 497–503
40. C. Boller and W.J. Staszewski, Aircraft Structural Health and Usage Monitoring, *Health Monitoring of Aerospace Structures*, John Wiley & Sons, Ltd., published online, Feb 10, 2004
41. P. Szary, R.K. Panda, A. Maher, and A. Safari, "Implementation of Advanced Fiber Optic and Piezoelectric Sensors—Fabrication and Laboratory Testing of Piezoelectric Ceramic-Polymer Composite Sensors for Weigh-in Motion Systems," FHWA 1999-029, Final Report, Federal Highway Administration, Dept. of Transportation, Feb 1999
42. V.S. Agarwala, "In-situ Corrosivity Monitoring of Military Hardware Environments," Paper 632, Corrosion 96, Conference Proc. (Orlando, FL), NACE, 1996
43. V.S. Agarwala, "Corrosivity Monitoring Wireless Intelligent Sensor," Proc. Fifth Intl. Workshop on Aircraft Corrosion (Solomons Islands, MD), NACE, Aug 2002
44. http://www.corrosion-doctors.org/Aircraft/NAWC-1.htm, May 2005
45. A.P. Berens, P.W. Hovey, and D.A. Skinn, "Risk Analysis for Aging Aircraft Fleets Vol 1—Analysis, Aircraft," Technical Report AFRLVA-WP-TR-91-3066 OH454 33-6533, Flight Dynamic Directorate, Wright Laboratory, Air Force Systems Command, 1991
46. S.D. Manning and J.N. Yang, "USAF Durability Design Handbook: Guidelines for the Analysis and Design of Durable Aircraft Structures," Technical Report AFFDL-TR-84-3027, Air Flight Laboratory, 1984
47. P. White, S. Barte, and L. Molent, "Probabilistic Fracigue Test Data," Sixth Joint FAA/DoD/NASA Aging Aircraft Conference, Sept 16–19, 2002
48. J.P. Komorowski, D. Forsyth, N.C. Bellinger, and W. Hoeppner, "Life and Damage Monitoring-Using NDI Data Interpretation for Corrosion Damage and Remaining Life Assessments," RTO-MP-079 (II), Proc. Specialists' Meeting on Life Management Techniques for Ageing Air Vehicles (Manchester, U.K.), NATO, Oct 2001
49. K. Honeycutt, C. Brooks, S. Prost-Domasky, and T. Mills, "Holistic Life Assessment Methods," Conference Proc. 16th Aerospace Structures and Materials Symposium (Montréal, Canada), April 2003

Corrosion in Supercritical Water—Waste Destruction Environments

R.M. Latanision, Exponent
D.B. Mitton, University of North Dakota

THE PRACTICALITY of supercritical water oxidation (SCWO) technology for destroying hazardous materials was explored in the early 1980s. Capitalizing on the work of earlier research, the possibility of extraordinarily high destruction efficiencies was recognized. Diverse types of aqueous waste that incorporate relatively dilute ($\approx$1 to 20%) quantities of organic components may successfully be treated by using this methodology. The two primary classes of waste that are being considered for destruction by SCWO are military and industrial wastes, including wastewater sludge. Military waste tends to present a severe challenge with respect to corrosion and solids handling, while the latter class is relatively innocuous (Ref 1).

Virtually any pumpable fluid may be treated by SCWO. These wastes (and the interested organizations) may include: human metabolic byproducts that could potentially be produced during long-term space flight (National Aeronautics and Space Administration, NASA); regulated wastes (U.S. Environmental Protection Agency, EPA); mixed low-level radioactive waste (U.S. Department of Energy, DoE); and explosives or nerve and blister agents destined for chemical demilitarization (U.S. Department of Defense, DoD). The DoD has a substantial quantity of both stockpiled waste (characterized and stored in a controlled environment) and nonstockpiled waste (all chemical agent materials outside the stockpile) that need to be destroyed. By exploiting the properties of water above its critical point, 374 °C (705 °F) and 22.4 MPa (221 atm) for pure water, this technology is capable of providing rapid and complete oxidation of the organic component, with high destruction efficiencies. The relatively low viscosity of supercritical water promotes mass transfer and mixing, while the low dielectric constant promotes dissolution of nonpolar organic materials. In conjunction with the high temperatures (which increase thermal reaction rates), these properties provide a medium in which mixing is fast, organic materials dissolve well and react quickly with oxygen, and salts precipitate (Ref 2). While the products of hydrocarbon oxidation are CO_2 and H_2O, heteroatoms are converted to inorganic compounds (usually acids, salts, or oxides in high-oxidation states). The formation of acids at elevated temperatures and pressures under very oxidizing conditions may result in severe corrosion. Thus, while SCWO is an effective process for the destruction of such waste, degradation of the materials of fabrication is a serious concern. The potential for identifying a universal construction material capable of maintaining system integrity when exposed to the highly aggressive conditions associated with the various sections in the SCWO process stream is extremely unlikely. Ultimately, the practicality of SCWO may be limited by the ability to control corrosion (Ref 3–5).

The Unique Properties of Supercritical Water

As presented in Fig. 1 (Ref 6), the properties of water undergo major changes in the vicinity of the critical point. In this regime, the viscosity (Fig. 1a) decreases with increasing temperature, and the density (Fig. 1b) is intermediate between that of liquid water (1.0 g/cm^3) and low-pressure water vapor (<0.001 g/cm^3). At SCWO conditions, water density is typically 0.1 g/cm^3, and consequently, the properties of supercritical water are significantly different from liquid water at ambient conditions. The dielectric constant (Fig. 1c) reflects the ability of water to support ionic reactions. At a pressure of 25 MPa (250 bar), the dielectric constant drops from approximately 80 at room temperature to 2 at 450 °C (840 °F), and the ionic dissociation constant (Fig. 1d) decreases from 10^{-14} at room temperature to 10^{-23} at supercritical conditions. These changes result in supercritical water acting essentially as a nonpolar dense gas with solvation properties approaching those of a low-polarity organic. Inorganic solubility is significantly decreased, and a concurrent increase in the solubility of organics and miscibility of many gases is seen (Ref 7). The solubility of NaCl, for example, decreases from approximately 37 wt% at 25 °C (75 °F) to only 120 ppm at 550 °C (1020 °F), while the solubility of organics and gases increases substantially above the critical point. With these solvation characteristics, when organic compounds and oxygen are dissolved in water above the critical point, kinetics are fast and the oxidation reaction proceeds rapidly to completion, with times required for complete reaction ranging from a few seconds to a few minutes for most wastes (Ref 2). At typical operating conditions (25 MPa, or 250 bar; 600 °C, or 1110 °F), destruction and removal efficiencies (DRE) can exceed 99.999%, with residence times on the order of a minute or less (Ref 8). For instance, the DRE recorded during kinetics studies of a number of chemical nerve agents, GB (sarin), VX, and mustard gas, have been shown to be on the order of 99.9999% or higher (Ref 9).

The Economics and Benefits of SCWO

The main technological advantages of SCWO include:

- System is entirely self-contained and can be shut down quickly
- High destruction efficiencies with low reactor residence times
- Relatively low operating temperature precludes the formation of nitrogen oxides
- Ability to capture the reaction products for analysis, storage, or further treatment
- Potential for a compact portable system that can be moved to the toxic waste site (Ref 2), thus circumventing public opposition to the transportation of hazardous waste

Supercritical water oxidation is a promising technology applicable to many dilute aqueous organic wastes (Ref 7, 10, 11) with total organic carbon content in the range of 1 to 20 wt%. As a result of the relatively low operating temperature, NO_x and SO_2 compounds are not produced (Ref 7). The latter may be particularly important during the destruction of explosives that produce nitrogen oxides during incineration (Ref 12). From the perspective of both economics and safety, the high pressure requirements of this technology are a potential expense. SCWO must still be able to demonstrate both economical competitiveness and system endurance. While this has been accomplished for benign wastes, this methodology is still comparatively expensive for aggressive feeds and currently does not appear to be competitive for private sector applications (Ref 1).

Facility Design Options

Current supercritical toxic waste systems typically incorporate either a downflow or tubular reactor configuration. Because it provides the lowest capital cost, a tubular reactor is preferred for relatively innocuous SCWO feeds. Conversely, a lined vessel is the preferred reactor configuration for aggressive feeds. The liner isolates the processing environment from the pressure boundary (Ref 1). Figure 2 presents a schematic diagram of a downflow vertical vessel waste treatment system based on SCWO technology (Ref 13). In this process, aqueous organic waste (1), which may be neutralized with a caustic solution (2) or have fuel (3) injected for startup, is initially pressurized from ambient to the pressure of the reaction vessel and pumped through a heat exchanger (4). This helps to bring the stream up to the desired temperature before reaching the reactor. At the head of the reactor (5), the stream is mixed with air or oxygen. In some cases, an oxidant such as hydrogen peroxide (H_2O_2), which catalytically decomposes to oxygen and water, may be used in preference to either air or oxygen; however, this is not economically advantageous. The reaction occurs in the top zone (6), where spontaneous oxidation of the organics liberates heat and raises the temperature to levels as high as 650 °C (1200 °F). Organic destruction occurs quickly, with typical reactor residence times of 1 min or less. As a result of their low solubility, salts precipitate and impinge on the lower zone (7), which is maintained at a temperature of approximately 200 °C (390 °F) by injecting water. The salts may be continuously taken off as a brine (8) or removed periodically as solids. The primary effluent (9) passes out of the top of the reactor into a separator (10), where the gas (11) and liquid (12) are quenched and separated; however, a portion of the liquid may be recycled (13).

The downflow and tubular reactor designs are by no means exclusive, and over the years, a number of alternative reactor arrangements have emerged. Design modifications typically attempt to limit the concerns associated with corrosion or plugging. As previously indicated, in its simplest form, a corrosion-resistant liner may be incorporated; however, more complex designs have included the dual-shell pressure-balanced vessel (Ref 14), a coaxial hydrothermal burner (Ref 15), and the transpiring wall reactor. In the transpiring wall reactor (Fig. 3), an attempt is made to reduce or prevent corrosion of the pressure-bearing wall and plugging of the reaction vessel through a reactor design modification (Ref 16). The latter incorporates an internal rinsing flow pattern (in the inward radial direction) superimposed on the main axial flow. The transpiring water forms a lower-temperature fluid boundary that prevents direct contact of the hot corrosive media with the pressure-bearing wall, thus reducing the potential for degradation. In addition, it is assumed that salts will not build up on the inner wall if the salt particulates cannot reach the wall or, alternatively, if they redissolve (as a result of the lower fluid temperature) prior to reaching the wall. The multiple inlets at the top of the reactor presented in Fig. 3 permit water and fuel (1), wastewater (2), and oxygen (3) to be added and mixed. The reactor has numerous side inlets (4) for the transpiring water (5). The annular channel between the inner wall of the vessel and the porous metal insert (6) can be divided into several chambers to permit independent control of the transpiring fluid flow rate.

In many industrial applications, increased throughput is frequently accomplished by means of scaleup; however, in the case of SCWO, it may be preferable to achieve this by running parallel systems (Ref 2).

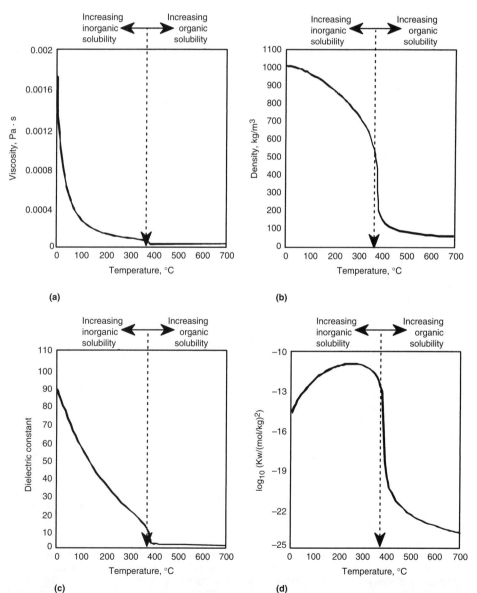

Fig. 1 The properties of water at 25 MPa (250 bar) in the vicinity of the critical point. Source: Adapted from Ref 6

Materials Challenges

For many waste streams, corrosion continues to be one of the central challenges to the full development of this technology. The system

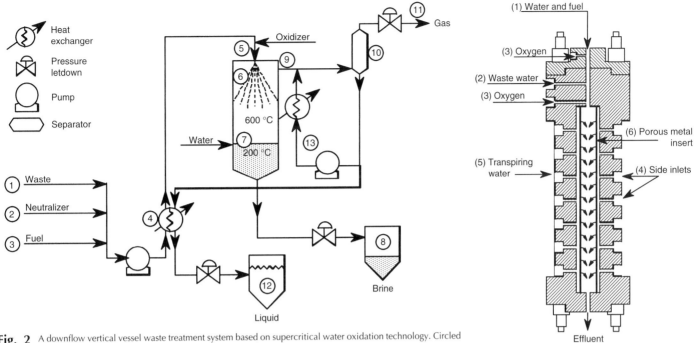

Fig. 2 A downflow vertical vessel waste treatment system based on supercritical water oxidation technology. Circled numbers are explained in text. Source: Adapted from Ref 13

Fig. 3 The main components of a transpiring wall reactor. The numbers are used in text. Source: Adapted from Ref 16

Table 1 Forms of corrosion for selected construction materials and environments

Material	Environment	Temperature °C	°F	Form of Corrosion	Reference
Iron-base alloys					
316	Deionized water	300–500	570–930	Uniform	17
	Deionized water	≈374	≈705	Some pitting	18
	Highly chlorinated organic	600	1110	Wastage, stress-corrosion cracking (SCC)	19
	pH higher than 12	500	930	SCC and other forms	17
Nickel-base alloys					
Alloys 625 and C276	Deionized water	300–500	570–930	Uniform and pitting	17
	Highly chlorinated organic	600	1110	Wastage	19
Alloys 625 and 718	Low-acid chloride and sulfate (with solid deposits)	600	1110	Pitting and crevice corrosion	20
Alloy 625	Methylene chloride, isopropyl alcohol, NaOH	580	1075	SCC and pitting	3
C276	Methylene chloride, isopropyl alcohol, NaOH	380	715	SCC and pitting	3
G30	CH_2Cl_2 plus H_2O_2	420	790	Acceptable corrosion behavior	21
	Highly chlorinated organic	600	1110	Moderate corrosion behavior	22
Titanium					
Grade 7	0–10 pH + Cl, NO_3, SO_4, PO_4(a)	500	930	Uniform	17
Grade 12	Low-acid chloride and sulfate (with solid deposits)	600	1110	Minor oxide spallation	20
Grade 2	Low-acid chloride and sulfate (with solid deposits)	600	1110	Pitting	20
Noble metals					
Pt, Pt-10Ir, Pt-30Rh	Low-acid chloride and sulfate (with solid deposits)	600	1110	No evident attack	20
Pt	Mustard simulant (HCl and H_2SO_4)	350	660	Serious attack	9
	Mustard simulant (HCl and H_2SO_4)	450 or 550	840 or 1020	Much more resistant to attack than at 350 (660)	9
Ceramics					
Alumina, SiC, Si_3N_4, zirconia	2–12 pH, 400 ppm Cl	350–550	660–1020	Uniform and other forms	17
Alumina, SiC, Si_3N_4, zirconia, sapphire	HCl plus H_2O_2	465	870	Generally low resistance to degradation	23

(a) Acids measurable as wt%

must accommodate an aggressive oxidizing environment over a wide temperature range at high pressure, with no loss of containment resulting from a corrosion-related breach of the pressure-bearing wall. A summary of selected materials exposed to various environments as well as the observed form of corrosion is presented in Table 1.

Figure 4 presents the potential-pH (Pourbaix) diagram for iron at a pressure of 50 MPa (500 bar) and 400 °C (750 °F). The lines representing the thermodynamic equilibrium for the various species are labeled. The two hatched regions indicate the range of potential and pH conditions within which SCWO systems, represented by the hatched rectangle, and supercritical thermal power plants (SCTPP), represented by the hatched oval, are likely to operate. In relation to both pH and oxidizing potential, it is clear that the conditions associated with SCWO are substantially more aggressive than those for SCTPP. Further, because the majority of SCWO systems are designed to operate with high partial pressure of oxygen, p_{O_2}, the upper region of the hatched rectangle best reflects SCWO conditions. On the other hand, SCTPP generally operate with low levels of oxygen, possibly even employing reducing agents such as hydrazine or ammonia, and it is estimated that the potential would be in the lower half of the oval representing these systems (Ref 24). Background on potential-pH diagrams can be found in the article "potential versus pH (Pourbaix) Diagrams" in *ASM Handbook*, Volume 13A, 2003.

Although they have been included as a base material in many research programs, stainless steels such as 316L are not likely to be used for fabrication of SCWO systems designed for aggressive waste streams. Generally, they exhibit unacceptable performance (Ref 5, 17, 18, 25), a susceptibility to stress-corrosion cracking (SCC) in acidic environments (Ref 19, 26), pitting and crevice corrosion in sludge (Ref 5), wastage on the order of 50 mm/yr (2000 mils/yr) (Ref 19), and SCC in highly chlorinated environments (Ref 19). Figure 5 presents a micrograph of a cracked section at the top of a 316L alloy U-bend sample exposed to a highly chlorinated oxidizing environment for approximately 66 h at 600 °C (1110 °F) (Ref 27). Nevertheless, binary iron-base alloys containing very high levels of chromium ($\approx$50%) may show some promise (Ref 28); however, additional testing is still required.

Conversely, as a result of a superior severe service history, high-nickel alloys have been selected for fabrication of many of the existing benchscale and pilot plant reactors. However, they are not without problems and, when exposed to aggressive conditions, may suffer significant degradation including, but not limited to, wastage on the order of 18.8 to 19.0 mm/yr (740 to 750 mils/yr), pitting, SCC, and dealloying (Ref 3, 9, 17–19, 22, 25, 29–36). Dealloying is seen in the cross section of a C276 alloy tube that was exposed to an aggressive chlorinated environment at a temperature estimated to be high but below the critical point (Fig. 6). While the macroscopic dimensions of the tube have not changed, the dealloyed layer has lost essentially all of the nickel component and is friable (Ref 31).

There is evidence for a correlation between chromium content and corrosion resistance for nickel alloys in SCWO systems (Ref 37), and certainly, the high-chromium alloys such as G30 ($\approx$30% Cr) generally exhibit reasonable corrosion resistance (Ref 5, 21, 38). In the case of binary nickel-chromium alloys exposed to oxidizing chlorinated supercritical conditions (500 °C, or 930 °F), research suggests that chromium levels above approximately 20% may actually be deleterious to corrosion behavior (Ref 39). This observation has recently been corroborated by additional research that also reveals the counterproductive nature of a high chromium content for binary nickel-chromium alloys with as much as 45% Cr (Ref 40). In this case, the authors observed a corrosion minimum in the region of 25% Cr. Ultimately, however, this trend may not be valid for multi-element alloys.

There is a body of literature indicating that the experimentally observed degradation can be more pronounced in the high-subcritical regime (Ref 30–33, 41) and that the aggressiveness of the solution may actually decrease above the critical point (Ref 42). This has even been noted in the case of SCC, which tends to be less probable at supercritical temperatures (Ref 20, 33).

A phenomenological approach has been adopted to explain the decrease in corrosion in the near-critical region (Ref 43). This treatment

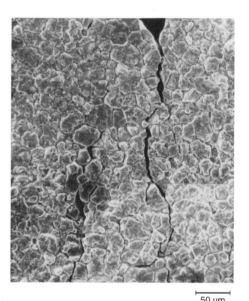

Fig. 5 A micrograph of a cracked section at the top of a 316L alloy U-bend sample exposed to a highly chlorinated oxidizing environment for approximately 66 h at 600 °C (1110 °F). Source: Ref 27

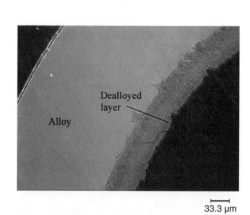

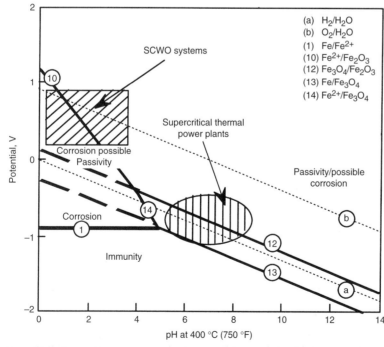

Fig. 4 Potential-pH diagram for iron in supercritical aqueous solution at 400 °C (750 °F) and 50 MPa (500 bar), showing the approximate regions in potential-pH space for the operation of supercritical water oxidation (SCWO) reactors and supercritical thermal power plants. Source: Adapted from Ref 24

Fig. 6 Dealloying of a C276 alloy tube after exposure to an aggressive chlorinated environment at a high subcritical temperature. Original magnification 300×

of the subject considers H^+ and O_2 as the only aggressive species, and Cl^--induced localized attack is not considered. By calculating the relative corrosion rate (R/R^0), based on changes in the hydrogen ion concentration and the density of water as a function of temperature over the temperature range associated with SCWO systems, an increase with increasing temperature is observed until a maximum is reached as the critical point is approached (Fig. 7). With a further increase in temperature, the relative corrosion rate decreases. This decrease is associated with a rapid reduction in the concentration of aggressive species as a function of an increasing supercritical temperature. The reduction in aggressive species reflects the combined effect of a decrease in the dissociation constant (resulting in a decrease in acid dissociation) and a decrease in the solution density.

It is essential, however, to consider other contributory factors not strictly related to changes in the solution. Experimental observations suggest that both time and system pressure influence the corrosion phenomena. For example, the upper temperature limit for severe corrosion depends on the pressure and thus the density of the solution, with higher densities favoring corrosion (Ref 41). Further, even at supercritical conditions, a substantial increase in corrosion may be observed after extended exposure times (for example, alloys 625 and 718 in Ref 20). Finally, while SCC appears to be more probable at subcritical temperatures (Ref 20, 33), SCC of alloy 625 has been reported for complex feeds at supercritical temperatures (580 °C, or 1075 °F) after extended exposure, on the order of 300 h (Ref 3).

There is an apparent dichotomy in the corrosion resistance data for titanium. Preliminary results indicated acceptable resistance to chlorinated, but not to nonchlorinated, acidic chemical agent simulant feeds (Ref 9). Outstanding performance of titanium has been reported at subcritical temperatures, while performance at supercritical temperatures is comparable to the nickel alloys (Ref 20); thus, titanium is viewed as a potential liner material (Ref 20, 44). Nevertheless, through-wall pitting of liners during destruction efficiency testing of a chlorinated waste has been reported (Ref 45). At elevated temperatures, potential problems with creep need to be considered for this material. In addition, the pyrophoric nature of titanium has caused some problems (Ref 1).

While incorporating noble metals or their alloys in the fabrication of SCWO systems would significantly increase cost, this may be one potential solution to severe corrosion problems for highly aggressive military waste streams. Although the corrosion resistance of platinum may be good at supercritical temperatures (Ref 20), in acidic chlorinated feeds it can suffer high rates of degradation at subcritical temperatures (Ref 9). Even in the low supercritical temperature range, corrosion rates can be higher than desirable (Ref 20). For systems incorporating a platinum liner, this would necessitate a potentially troublesome transition between platinum (supercritical temperature region) and a second material (subcritical temperature region). The potential for erosion or erosion-corrosion could be exacerbated in SCWO systems by entrained solids such as salts, oxides, and corrosion products (Ref 8). Erosion problems were recently encountered during testing of a platinum liner exposed to a VX hydrolysate feed, confirming the need to consider not only the potential for corrosion but also for flow-related phenomena in these systems (Ref 1).

The problems associated with the corrosion of various alloys has prompted research into ceramic materials; however, ceramics have generally exhibited less than satisfactory resistance (Ref 9, 46). Degradation was found to proceed predominantly either by homogeneous surface attack or by grain-boundary diffusion (Ref 23). In addition to the effect of system conditions such as temperature and feed stream composition, at least in the case of alumina, degradation also depends on the purity, porosity, and microstructure of the ceramic (Ref 46).

Mitigation of System Degradation

While corrosion continues to be a challenge, a number of control approaches have been adopted. Feed stream dilution could reduce the severity of corrosion, but the required dilution may be so large as to make this procedure economically unattractive (Ref 47).

Because it provides the lowest capital cost, a tubular reactor design is favored for relatively innocuous SCWO feeds. Alternatively, for more aggressive feeds, a lined vessel is the preferred reactor configuration. The liner isolates the processing environment from the pressure boundary and may consist of a removable or replaceable solid-wall corrosion-resistant barrier. The aspect ratio of the reactor must be suitable for practical mechanical incorporation of the liner (Ref 1). Currently, no universal liner material has been identified that will meet the requirements of all SCWO feeds; thus, liners must be selected to accommodate the character of individual feed streams. Titanium and platinum are the liner materials most frequently suggested for solid-wall liners; however, as previously indicated, they have limitations.

A transpiring wall can also be viewed as a type of liner, and this design has successfully been employed without corrosion of the transpiring wall platelets (fabricated from alloy 600) for several hundred hours of operation on chemical demilitarization feeds. Longer-term operational history is still desirable to confirm this trend (Ref 1).

Although feed neutralization (with NaOH, for example) has seen some success in SCWO systems for acidic feeds, it has been carried out without a full understanding of its effect on corrosion. Typically, feed neutralization involves stoichiometric quantities of neutralizer; however, there is evidence to suggest that the most favorable thermodynamics are obtainable by adjusting the feed stream chemistry (Ref 31). This would, however, necessitate the use of in situ monitoring. An additional disadvantage of neutralization is the production of salt species that can lead to plugging of the system. To some extent, for wastes that do not already contain salts, this can be circumvented by restricting neutralization to the cool-down section of the system (Ref 48). In this way, salts that would normally be formed in the supercritical regime do not deposit on the reactor walls. This modification has been adopted in Europe by Chematur Engineering (Ref 1).

Ultimately, to reduce degradation to an acceptable level in the various regions of SCWO systems handling aggressive influents, it may be necessary to adopt a synergistic approach incorporating feed chemistry control, reactor design modifications, and intelligent materials selection.

The Future

There is little doubt that this promising technology has both economic and environmental incentives; nevertheless, the corrosion of the materials of fabrication persists as a serious concern, and substantial research is needed to identify materials or methods to reduce degradation to an acceptable level. Some key areas for future research could include:

- Evaluate the effect of entrained solids on the endurance and life of constructional or liner materials
- Develop an in-depth understanding of the relationship between solid deposits (salts, for example) and the corrosion rate and mode

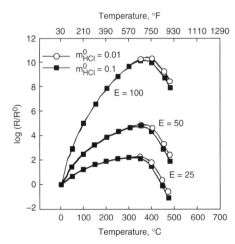

Fig. 7 The relative corrosion rate (R/R^0) at two different molal concentrations of HCl (m^0_{HCl}) for the temperature range associated with supercritical water oxidation systems at a pressure of 50 MPa (500 bar) and activation energies (E) of 25, 50, and 100 kJ/mol. Source: Adapted from Ref 43

- Assess innovative reactor designs that may provide potential solutions to materials degradation challenges
- Design and fabricate robust in situ potential, pH, and corrosion-monitoring electrodes
- Investigate the link between degradation and (i) the elemental alloy composition, (ii) various temperature zones, (iii) system pressure, and (iv) influent composition

With respect to material performance, feed chemistry control can provide a less challenging environment and Pourbaix diagrams can be used to guide adjustments for this purpose. There is however, a lack of data for these diagrams at higher operating temperatures. For the longer term, and in parallel with experimental efforts, development of appropriate instrumentation techniques and data collection for the generation of Pourbaix diagrams for materials of construction at higher temperatures is desirable.

ACKNOWLEDGMENT

The work reported in this article was performed while the authors were affiliated with The H.H. Uhlig Corrosion Laboratory at the Massachusetts Institute of Technology.

REFERENCES

1. "Supercritical Water Oxidation—Achievements and Challenges in Commercial Applications," a workshop sponsored by the Office of Naval Research (ONR), Arlington, VA, Aug 2001
2. R.W. Shaw, Supercritical Water Oxidation of Toxic Military Materials: Current Status, *Proc. of The International CW Demil Conference,* May 2000 (The Hague, Netherlands), OPCW and Dutch Ministry of Foreign Affairs, 2000
3. R.M. Latanision and R.W. Shaw, "Corrosion in Supercritical Water Oxidation Systems," Report MIT-EL 93-006, Massachusetts Institute of Technology, Sept 1993
4. D.B. Mitton, J.C. Orzalli, and R.M. Latanision, Corrosion Phenomena Associated with SCWO Systems, *Proc. of Third Int. Symp. on Supercritical Fluids,* Vol 3, Oct 1994 (Strasbourg, France), Institut National Polytechnique de Lorraine, p 43–48
5. A.J. Thomas III and E.F. Gloyna, "Corrosion Behavior of High-Grade Alloys in the Supercritical Water Oxidation of Sludges," Report CRWR 229, The University of Texas at Austin, Feb 1991, p 1–50
6. M. Hodes, P.A. Marrone, G.T. Hong, K.A. Smith, and J.W. Tester, *J. Supercrit. Fluids,* Vol 29, 2004, p 265–288
7. J.W. Tester, H.R. Holgate, F.J. Armellini, P.A. Webley, W.R. Killilea, G.T. Hong, and H.E. Barner, Supercritical Water Oxidation Technology: Process Development and Fundamental Research, *Proc. of ACS Symposium Series 518,* Oct 1991 (Atlanta, GA), American Chemical Society, 1993, p 35–76
8. J.W. Tester and J.A. Cline, *Corrosion,* Vol 55 (No. 11), 1999, p 1088–1100
9. K.W. Downey, R.H. Snow, D.A. Hazlebeck, and A.J. Roberts, Corrosion and Chemical Agent Destruction: Research on Supercritical Water Oxidation of Hazardous Military Wastes, *Proc. of ACS Symposium Series 608,* Nov 1994 (San Francisco, CA), American Chemical Society, 1995, p 313–326
10. M. Modell, *Standard Handbook of Hazardous Waste Treatment and Disposal,* McGraw-Hill, 1989, p 8.153–8.167
11. E.U. Franck, *High Temperature, High Pressure Electrochemistry in Aqueous Solutions,* National Association of Corrosion Engineers, 1976, p 109
12. C.A. LaJeunesse, B.E. Mills, and B.G. Brown, "Supercritical Water Oxidation of Ammonium Picrate," Report SAND95-8202·UC-706, Sandia National Laboratories, 1994
13. H.E. Barner, C.Y. Huang, T. Johnson, G. Jacobs, M.A. Martch, and W.R. Killilea, *J. Hazard. Mater.,* Vol 31, 1992, p 1–17
14. A.G. Fassbender, R.J. Robertus, and G.S. Deverman, The Dual Shell Pressure Balanced Vessel: A Reactor for Corrosive Applications, *Proc. of First Int. Workshop on Supercritical Water Oxidation,* Feb 1995 (Jacksonville, FL), Session VI: Innovative Reactor Design to Mitigate Corrosion Effects, U.S. Dept. of Energy, U.S. Dept. of Defense, U.S. Dept. of Commerce, and U.S. Army
15. H.L. La Roche, M. Weber, and Ch. Trepp, Rationale for the Filmcooled Coaxial Hydrothermal Burner (FCHB) for Supercritical Water Oxidation (SCWO), *Proc. of First Int. Workshop on Supercritical Water Oxidation,* Feb 1995 (Jacksonville, FL), Session VI: Innovative Reactor Design to Mitigate Corrosion Effects, U.S. Dept. of Energy, U.S. Dept. of Defense, U.S. Dept. of Commerce, and U.S. Army
16. M. Weber, B. Wellig, K. Lieball, and R. von Rohr, Operating Transpiring-Wall SCWO Reactors: Characteristics and Quantitative Aspects, Paper 370, *Proc. of Corrosion 01,* March 2001 (Houston, TX), National Association of Corrosion Engineers, 2001
17. S. Tebbal and R.D. Kane, Materials Selection in Hydrothermal Oxidation Processes, Paper 413, *Proc. of Corrosion 98,* March 1998 (San Diego, CA), National Association of Corrosion Engineers, 1998
18. N. Boukis, R. Landvatter, W. Habicht, G. Franz, S. Leistikow, R. Kraft, and O. Jacobi, First Experimental SCWO Corrosion Results of Ni-Base Alloys Fabricated as Pressure Tubes and Exposed to Oxygen Containing Diluted Hydrochloric Acid at $T \leq 450\ °C, P = 24$ MPa, *Proc. of First Int. Workshop on Supercritical Water Oxidation,* Feb 1995 (Jacksonville, FL), Session VIII: Materials Testing; Corrosion Experiments, U.S. Dept. of Energy, U.S. Dept. of Defense, U.S. Dept. of Commerce, and U.S. Army
19. D.B. Mitton, E.H. Han, S.H. Zhang, K.E. Hautanen, and R.M. Latanision, Degradation in Supercritical Water Oxidation Systems, *Proc. of ACS Symposium Series 670,* March 1996 (New Orleans, LA), American Chemical Society, 1997, p 242–254
20. G.T. Hong, D.W. Ordway, and V.A. Zilberstein, Materials Testing in Supercritical Water Oxidation Systems, *Proc. of First Int. Workshop on Supercritical Water Oxidation,* Feb 1995 (Jacksonville, FL), Session VIII: Materials Testing; Corrosion Experiments, U.S. Dept. of Energy, U.S. Dept. of Defense, U.S. Dept. of Commerce, and U.S. Army
21. S. Fodi, J. Konys, J. Haussell, H. Schmidt, and V. Casal, Corrosion of High Temperature Alloys in Supercritical Water Oxidation Systems, Paper 416, *Proc. of Corrosion 98,* March 1998 (San Diego, CA), National Association of Corrosion Engineers, 1998
22. D.B. Mitton, J.H. Yoon, J.A. Cline, H.S. Kim, N. Eliaz, and R.M. Latanision, *Ind. Eng. Chem. Res.,* Vol 39 (No. 12), 2000, p 4689–4696
23. N. Boukis, N. Claussen, K. Ebert, R. Janssen, and M. Schacht, *J. Eur. Ceram. Soc.,* Vol 17, 1997, p 71–76
24. D.D. Macdonald, Critical Issues in Understanding Corrosion and Electrochemical Phenomena in Super Critical Aqueous Media, Paper 484, *Proc. of Corrosion 04,* April 2004 (New Orleans, LA), National Association of Corrosion Engineers, 2004
25. D.B. Mitton, J.C. Orzalli, and R.M. Latanision, Corrosion in Supercritical Water Oxidation Systems, *Proc. of 12th ICPWS,* Sept 1994 (Orlando, FL), Begell House, 1995, p 638–643
26. Y. Watanabe, H. Abe, and Y. Daigo, Corrosion Resistance and Cracking Susceptibility of 316L Stainless Steel in Sulfuric Acid-Containing Supercritical Water, Paper 493, *Proc. of Corrosion 04,* April 2004 (New Orleans, LA), National Association of Corrosion Engineers, 2004
27. D.B. Mitton, S.H. Zhang, K.E. Hautanen, J.A. Cline, E.H. Han, and R.M. Latanision, Evaluating Stress Corrosion and Corrosion Aspects in Supercritical Water Oxidation Systems for the Destruction of Hazardous Waste, Paper 203, *Proc. of Corrosion 97,* March 1997 (New Orleans, LA), National Association of Corrosion Engineers, 1997
28. J. Konys, S. Fodi, A. Ruck, and J. Haussell, Comparison of Corrosion of Nickel and Chromium Based Alloys under Conditions of Supercritical Water Oxidation, Paper 253, *Proc. of Corrosion 99,* April 1999

(San Antonio, TX), National Association of Corrosion Engineers, 1999
29. B.C. Norby, "Supercritical Water Oxidation Benchscale Testing Metallurgical Analysis Report," Report EGG-WTD-10675, Idaho National Engineering Laboratory, Feb 1993
30. R.M. Latanision, *Corrosion*, Vol 51 (No. 4), 1995, p 270–283
31. D.B. Mitton, P.A. Marrone, and R.M. Latanision, *J. Electrochem. Soc.*, Vol 143 (No. 3), 1996, p L59–61
32. D.B. Mitton, J.C. Orzalli, and R.M. Latanision, Corrosion Studies in Supercritical Water Oxidation Systems, *Proc. of ACS Symposium Series 608*, Nov 1994 (San Francisco, CA), American Chemical Society, 1995, p 327–337
33. V.A. Zilberstein, J.A. Bettinger, D.W. Ordway, and G.T. Hong, Evaluation of Materials Performance in a Supercritical Wet Oxidation System, Paper 558, *Proc. of Corrosion 95*, March 1995 (Orlando, FL), National Association of Corrosion Engineers, 1995
34. T.T. Bramlette, B.E. Mills, K.R. Hencken, M.E. Brynildson, S.C. Johnston, J.M. Hruby, H.C. Feemster, B.C. Odegard, and M. Modell, "Destruction of DOE/DP Surrogate Wastes with Supercritical Water Oxidation Technology," Report SAND90-8229, Sandia National Laboratories, Nov 1990, p 1–35
35. S.F. Rice, R.R. Steeper, and C.A. LaJeunesse, "Destruction of Representative Navy Wastes Using Supercritical Water Oxidation," Report SAND94-8203·UC-402, Sandia National Laboratories, Oct 1993, p 1–35
36. P. Kritzer, N. Boukis, and E. Dinjus, *Corrosion*, Vol 54 (No. 10), 1998, p 824–834
37. K.M. Garcia and R. Mizia, "Corrosion Investigation of Multilayered Ceramics and Experimental Nickel Alloys in SCWO Process Environments," Report INEL-95/0017, Idaho National Engineering Laboratory, Feb 1995
38. D.B. Mitton, Y.S. Kim, J.H. Yoon, S. Take, and R.M. Latanision, Corrosion of SCWO Constructional Materials in Cl^- Containing Environments, Paper 257, *Proc. of Corrosion 99*, April 1999 (San Antonio, TX), National Association of Corrosion Engineers, 1999
39. N. Hara, S. Tanaka, S. Soma, and K. Sugimoto, Corrosion Resistance of Fe-Cr and Ni-Cr Alloys in Oxidizing Supercritical HCl Solution, Paper 358, *Proc. of Corrosion 02*, April 2002 (Denver, CO), National Association of Corrosion Engineers, 2002
40. C. Schroer, J. Konys, J. Novotny, and J. Hausselt, Corrosion in SCWO Plants Processing Chlorinated Substances: Aspects of the Corrosion Mechanisms and Kinetics for Binary Ni-Cr Alloys, Paper 496, *Proc. of Corrosion 04*, April 2004 (New Orleans, LA), National Association of Corrosion Engineers, 2004
41. P. Kritzer, N. Boukis, and E. Dinjus, Corrosion Phenomena of Alloy 625 in Aqueous Solutions Containing Sulfuric Acid and Oxygen under Subcritical and Supercritical Conditions, Paper 415, *Proc. of Corrosion 98*, March 1998 (San Diego, CA), National Association of Corrosion Engineers, 1998
42. S. Huang, K. Daehling, T.E. Carleson, M. Abdel-Latif, P. Taylor, C. Wai, and A. Propp, Electrochemical Measurements of Corrosion of Iron Alloys in Supercritical Water, *Proc. of ACS Symposium Series 406*, Nov–Dec 1988 (Washington, D.C.), American Chemical Society, 1989, p 287–300
43. L.B. Kriksunov and D.D. Macdonald, *J. Electrochem. Soc.*, Vol 142 (No. 12), 1995, p 4069–4073
44. D.A. Hazlebeck, K.W. Downey, D.D. Jensen, and M.H. Spritzer, Supercritical Water Oxidation of Chemical Agents, Propellants, and Other DOD Hazardous Wastes, *Proc. of 12th ICPWS*, Sept 1994 (Orlando, FL), Begell House, 1995, p 632–637
45. K.M. Garcia, "Supercritical Water Oxidation Data Acquisition Testing," Report INEL-96/0267, Idaho National Engineering Laboratory, Aug 1996, p 1–41
46. M. Schacht, N. Boukis, E. Dinjus, K. Ebert, R. Janssen, F. Meschke, and N. Claussen, *J. Eur. Ceram. Soc.*, Vol 18, 1998, p 2373–2376
47. C.M. Barnes, R.W. Marshall, Jr., R.E. Mizia, J.S. Herring, and E.S. Peterson, "Identification of Technical Constraints for Treatment of DOE Mixed Waste by Supercritical Water Oxidation," Report EGG-WTD-10768, Idaho National Engineering Laboratory, Oct 1993
48. P. Kritzer and E. Dinjus, *Chem. Eng. J.*, Vol 83, 2001, p 207–214

SELECTED REFERENCES

- Y. Daigo, Y. Watanabe, K. Sugahara, and T. Isobe, "Compatibility of Ni Base Alloys with Supercritical Water Applications: Aging Effects on Corrosion Resistance and Mechanical Properties," Paper 487, Proc. of Corrosion 04, April 2004 (New Orleans, LA), National Association of Corrosion Engineers, 2004
- H. Inoue, Y. Maeda, and T. Mizuno, "Corrosion-Potential Fluctuations and Polarization Resistance Measurements in High-Subcritical and Supercritical Water Oxidation Environment," Paper 491, Proc. of Corrosion 04, April 2004 (New Orleans, LA), National Association of Corrosion Engineers, 2004
- P. Kritzer, *J. Supercrit. Fluids*, Vol 29, 2004, p 1–29
- D.B. Mitton, J.H. Yoon, and R.M. Latanision, *Zairyo-to-Kankyo (Corros. Eng.)*, Vol 49 (No. 3), 2000, p 130–137
- S.C. Tucker and M.W. Maddox, *J. Phys. Chem. B*, Vol 102 (No. 14), 1998, p 2437–2453

Corrosion in Supercritical Water—Ultrasupercritical Environments for Power Production

Gordon R. Holcomb, National Energy Technology Laboratory

THE OPERATING TEMPERATURES AND PRESSURES of steam boilers and turbines have been increased to improve efficiencies in steam and power production. Improvements in materials properties, such as high-temperature strength, creep resistance, and oxidation resistance, have enabled this increase. From 1910 to 1960, there was an average increase in steam temperature of 10 °C per year, with a corresponding increase in plant thermal efficiency from less than 10 to 40% (Ref 1).

While there is no specific definition of ultra supercritical (USC), plants operating above 24 MPa (3.5 ksi) and 593 °C are currently deemed as such (Ref 2). As the state of the art advances, the conditions regarded as USC will probably increase. Plants operating below 24 MPa are subcritical, and those at or above 24 MPa are supercritical (SC) (Ref 2). The critical point of water is 22.1 MPa (3.2 ksi) not 24 MPa—the 24 MPa distinction is based more on the lack of an evaporation step in the steam cycle. Subcritical plants may, at some point in their steam cycle, have supercritical steam.

The first commercial boiler with a steam pressure above its critical value of 22.1 MPa (3.2 ksi) was the 125 MW Babcock & Wilcox universal pressure (UP) steam generator in 1957 at the Ohio Power Company's Philo 6 plant (Ref 3). A UP boiler can operate at both subcritical and supercritical conditions. It delivered steam at 34.1 MPa (5.0 ksi) and 621 °C, with two reheats of 566 and 538 °C. The pressure and temperatures of the primary, first reheat (if present), and second reheat (if present) are designated by a nomenclature such as 34.1 MPa/621 °C/566 °C/538 °C for the Philo 6 plant. In 1960, Eddystone 1 reached a world record in efficiency of 40%, operating at 34.5 MPa/649 °C/565 °C/565 °C (Ref 1, 2). The Eddystone 1 plant, and others of its generation, soon reduced operating temperatures and pressures primarily because of thermal fatigue issues within the boiler (Ref 1). Eddystone 1 continues to operate at 32.4 MPa/610 °C (Ref 2). It still uses the highest steam pressures and temperatures in the world. Since 1960, the overall trend of increasing temperatures and pressures has stopped and stabilized at approximately 538 °C and 24.1 MPa (3.5 ksi) for U.S. power plants (Ref 2).

Since the 1970s, advances in the high-temperature strength of ferritic steels have allowed increases of operating temperatures and pressures without the thermal fatigue issues of the austenitic steels that had to be used to obtain the required high-temperature strengths in the early 1960s. The term *ferritic steels,* as used here, refers to the equilibrium structure. In practice, a martensitic or partially martensitic structure is obtained from heat treating. Currently, the most efficient fossil power plants operate at 24 MPa/600 °C and are deemed state of the art. Operations at 28 MPa/630 °C are expected in the near term (Ref 2). Current research programs are aimed at increasing the operating conditions to as high as 760 °C and 37.9 MPa (5.5 ksi).

Water Properties

Several physical properties of water undergo large changes at or near its critical point (374 °C and 22.1 MPa, or 3.2 ksi) (Ref 4). Above 374 °C (705 °F), water vapor can no longer be compressed into a liquid. Density and dielectric constant decrease with increasing temperature and fall sharply at the critical temperature. The ionic product reaches a maximum at approximately 300 °C and also falls sharply at the critical temperature. Supercritical water is a low-density fluid that is a nonpolar solvent with high solubilities for organic compounds and gases and low solubilities for salts.

Steam Cycle

A simplified water path in a large supercritical utility boiler, with a single reheat, is illustrated in Fig. 1. Input water from the boiler feed pump is preheated in the economizer and then travels up the water walls and into the separator. Steam from the separator is fed into the primary and secondary superheaters. From there, the steam goes to the high-pressure turbine, then to the reheater, the intermediate-pressure turbine, the low-pressure turbine, and then to the condenser. From the condenser, water goes back to the boiler feed pump to then repeat the cycle. One possible physical layout and position of the components within a supercritical boiler is shown in Fig. 2 (Ref 5).

Supercritical boilers are typically once-through systems but with the capacity to separate water from steam in the separator (particularly during startup). Boilers that operate at constant pressure in either subcritical or supercritical temperatures are UP boilers. Benson boilers operate in both subcritical and supercritical conditions but with a sliding pressure capability.

The Rankine cycle, in terms of temperature and entropy, is shown in Fig. 3 for a subcritical system and in Fig. 4 for a SC system. In these figures, the critical temperature is at the maximum in the saturated-liquid and -vapor curves. Under the saturated lines is a wet mixture; to the left of the saturated-liquid line is compressed liquid, and to the right of the saturated-vapor line is superheated vapor. In Fig. 3, liquid water is compressed from points A to B. The temperature rise (A-B) is exaggerated in Fig. 3 and is typically on the order of 0.6 °C. The heating that takes place from points B to C occurs first in the economizer, then in the water walls (the constant temperature evaporation step), and then in the superheater. Expansion takes place within the turbines from points C to D. Lastly, unavailable heat is rejected to the atmosphere from points D to A. The SC cycle in Fig. 4 is similar (points B to E) but does not include the evaporation step, and so, energy is not needed to overcome the latent heat of evaporation. Also shown in Fig. 4 is a reheat step through points F to G. The unavailable heat is the area under the D-A step in Fig. 3 and the area under the H-A step in Fig. 4.

The thermal efficiency in converting heat into work in reversible Rankine cycles is the ratio of temperature-entropy area within the cycle to the combined areas of the cycle and unavailable heat. Supercritical systems allow the upper temperature to increase, which increases the cycle area and thus the thermal efficiency. Figure 4 shows how reheating also increases the thermal efficiency by raising the average maximum temperature.

The overall efficiency power generation station is the ratio of energy output to energy input and is a function of many factors: the thermal efficiency, the quality of coal, the condenser pressure, and the heating value (heat of reaction) used. Differing methods for calculating the heating value of fuel result in European plant efficiencies being cited as 3 to 5% higher than comparable plants in the United States.

An additional point is the distinction between a 1% increase in efficiency and a percentage improvement in efficiency. The first is absolute, and the second is relative. So, with a base efficiency of 37%, a 1% increase in efficiency corresponds to a 2.7% improvement (relative increase). For more information, see the subsequent section on efficiency.

Steam Cycle Chemistry

The control of water chemistry in the steam cycle of a power plant is important for corrosion control, deposition prevention, and higher cycle efficiency (Ref 6). There are three basic steps in controlling the water chemistry: removal of impurities (in makeup water, feedwater, and condensate), control of feedwater pH, and boiler water treatments (for once-through systems, these are feedwater treatments).

Impurity removal is by boiler blowdown, mechanical deaeration in the condenser and deaerator, and oxygen scavenging (usually with hydrazine) (Ref 6). For once-through systems, demineralization (condensate polishing) is also performed.

Feedwater pH is normally controlled by injection of ammonia or amines to maintain feedwater pH between 9.2 and 9.6, as measured at room temperature (8.8 to 9.2 when copper alloys are also present in the feedwater system) (Ref 6). These high pH values help protect carbon steel components at low temperatures. Ammonia and amines are weak bases, so they have little effect in raising the pH at high temperatures. In the absence of strong bases, such as phosphate or hydroxide, the pH at elevated temperatures is close to neutral.

In subcritical boilers, a variety of boiler water treatments (usually containing phosphates or hydroxides) is used that is not used in higher-pressure systems. For higher-pressure once-though systems, either an all-volatile treatment (AVT) or an oxygenated treatment (OT) is used (Ref 6). In the AVT, the pH is controlled as described previously, with ammonia or volatile amines. There are no phosphates or buffering salts, so there is little tolerance for contamination, and thus condensate polishers are required. The AVT was first used in supercritical boilers.

Oxygenated treatment (Ref 6) is relatively new in the United States but has been used elsewhere for many years. Ultrapure feedwater is required (<0.15 μS/cm cation conductivity). Oxygen, either in the form of O_2 gas or hydrogen peroxide, is added to the feedwater at a controlled rate. The oxygen promotes the formation of a dense and protective iron-oxide film on the boiler and feedwater piping. Condensate polishers are required in OT programs. Oxygenated treatment can be used alone, termed neutral water treatment, with oxygen controlled to ~200 ppb, or in combination with ammonia, termed combined water treatment, with oxygen controlled between 50 to 200 ppb and with ~200 ppb ammonia.

Materials Requirements

The components exposed to supercritical water in SC and USC power plants include high-pressure steam piping and headers, superheater and reheater tubing, and water wall tubing in the boiler, and high- and intermediate-pressure rotors, rotating blades, and bolts in the turbine section (Ref 7). A synopsis of the materials requirements for each of these is presented as follows.

Steam piping and headers require high-temperature creep strength. These are heavy-section components and are particularly subject to fatigue from thermal stresses. Ferritic steels have lower thermal expansion coefficients than austenitic steels and so are preferred with respect to thermal fatigue. Current ferritic steels are limited to 620 °C, while the theoretical limit is thought to be approximately 650 °C (Ref 7).

Superheater and reheater tubing requires high-temperature creep strength, thermal fatigue strength, weldability, fireside corrosion resistance, and steamside corrosion resistance. Ferritic steels are preferred due to their thermal fatigue resistance. However, high-temperature creep strength currently limits these alloys to 620 °C (650 °C theoretical limit). Fireside corrosion resistance further limits ferritic steels to approximately 593 °C which corresponds to a steam temperature of approximately 565 °C (Ref 7).

Waterwall tubing faces similar issues as superheater and reheater tubing but at lower temperatures, so that lower-alloyed materials are typically used.

High-pressure and intermediate-pressure rotors are large forgings that carry the rotating blades and transmit the mechanical energy. They are subject to centrifugal loads during operation and to thermal stresses during startups and

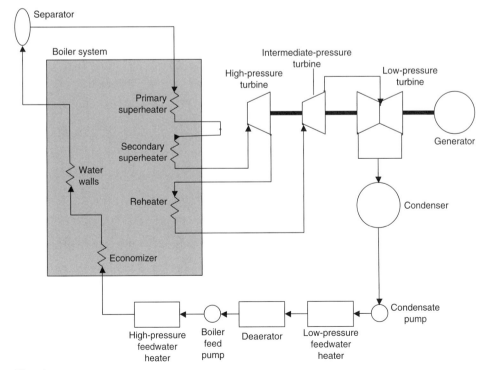

Fig. 1 Simplified water path in a large supercritical utility boiler, with a single reheat step

shutdowns. The key materials properties are creep strength, low-cycle fatigue strength, and fracture toughness (Ref 7).

Rotating blades require high-temperature strength, creep resistance, and a coefficient of thermal expansion similar to the rotor (Ref 7). In addition, the alloys must be able to be peened. In each row of blades, a circular cover is used to couple the blades together and to act as a seal. The tenon part of the blades protrudes from the cover and is peened into heads, which attach the blades to the cover (Ref 7).

Bolts require high-temperature strength, creep resistance, notch sensitivity resistance, and a coefficient of thermal expansion similar to the rotor. The bolts must remain tight between scheduled shutdowns (20,000 to 50,000 h) (Ref 7).

Boiler Alloys

The alloys used (or proposed to be used) in SC and USC boilers can be broadly classified as ferritic steels, austenitic steels, and nickel-base alloys. Ferritic steels are currently limited to temperatures below 620 °C, with a theoretical limit of approximately 650 °C; austenitic steels are thought to have applications from 620 to 675 °C, and nickel alloys are used above 675 °C (Ref 2). Table 1 lists candidate alloys for USC boilers and their preferred applications (Ref 8, 9).

Ferritic boiler steels developed over the past 40 years are based on simple 2Cr, 9Cr, and 12Cr steels. Figure 5 shows the evolution of these steels, grouped into four generations by 10^5 h

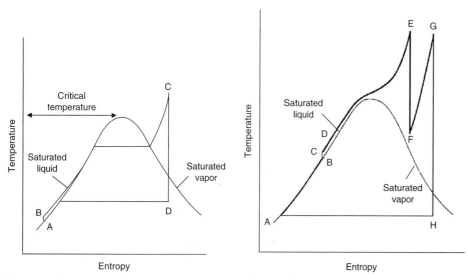

Fig. 2 Schematic of a supercritical sliding pressure Benson boiler, Tachibana-Wan No. 2, 25.0 MPa/600 °C/610 °C/610 °C. Source: Ref 5

Fig. 3 Rankine cycle for a subcritical boiler. Source: Ref 3

Fig. 4 Rankine cycle for a supercritical boiler. Source: Ref 3

Table 1 Candidate alloys for ultra supercritical power plant boilers

Trade designation	ASME code/code case	Preferred application(a)	Metal temperature of application(b), °C
Ferritic steels			
HCM2S	2199	WW	≤575
Tempaloy F-2W	...	WW	≤575
HCM12	...	WW	≤575
NF616 (ASME P92)	2179	P	≤620
HCN12A (ASME P22)	2180	P	≤620
E911	...	P	≤620
NF12	...	P	≤650
SAVE12	...	P	≤650
Austenitic steels			
SAVE25	...	T	620–675
NF709	...	T	620–675
HR3C	2113	T	620–675
Super304H	...	T	620–675
347HFG	2159	T	620–675
800HT	1987	T	620–675
HR120	2315	T	620–675
Nickel-base alloys			
Alloy 740	...	P, T	675–788
Alloy 230	2063	P, T	675–788
Alloy 625	1409	P, T	675–788
Alloy 617	1956	P, T	675–788
HR6W	...	P, T	675–788
45TM	2188	P, T	675–788

(a) WW, water walls; P, pipes and headers; T, superheater and reheater tubes. (b) Metal temperatures are based on creep limitations. Source: Ref 8, 9

creep-rupture strength at 600 °C (Ref 7, 10). The first generation was developed in the 1960s by additions of molybdenum, niobium, or vanadium and has a maximum metal use temperature of approximately 565 °C (Ref 7). Metal temperatures are typically at least 28 °C (50 °F) higher than steam temperatures for tubing. They are roughly the same for headers and piping. The alloy HCM9M has had extensive use in superheater and reheater tubes, replacing 18Cr-8Ni steels for reheater applications (Ref 10). The alloy HT91 has been used extensively in Europe for tubing, headers, and piping. Poor weldability has limited its use in the United States and Japan (Ref 10). The second generation was developed between 1970 and 1985 by optimization of carbon, niobium, and vanadium and has a maximum metal use temperature of approximately 593 °C (Ref 7). The alloy ASME T91 has been used extensively for heavy-section components such as steam pipes and headers (Ref 10). The third generation was developed between 1985 and 1995 by partially substituting tungsten for molybdenum and has a maximum metal use temperature of approximately 620 °C (Ref 7). The fourth generation is emerging by adding cobalt and increasing tungsten and is thought to have a maximum metal use temperature of approximately 650 °C (Ref 7).

The role of alloy additions is primarily to improve creep strength. Additions of tungsten, molybdenum, and cobalt are useful as solid-solution strengtheners. Some of the solid-solution strengthening from tungsten may be reduced by the formation of Laves phases (Ref 11). Additions of vanadium and niobium are useful as precipitation strengtheners and form fine carbide and nitride precipitates. Additional chromium provides solid-solution strength as well as corrosion resistance. The addition of nickel improves toughness at the expense of creep strength, which can be partially overcome by replacing nickel with copper (Ref 7). The overall creep strength with time is shown schematically in Fig. 6, which shows how the creep strength decreases with time from various strengthening mechanisms (Ref 11). The nominal compositions of ferritic boiler steels are given in Table 2.

Austenitic steels evolved as shown in Fig. 7, beginning from alloys 302 and 800H (Ref 7, 10). Austenitic steels have better oxidation and fireside corrosion resistance than ferritic steels and thus are primarily candidates for the finishing stages in superheater and reheater tubing (Ref 7).

Alloying additions of titanium and niobium were added to stabilize the steels for corrosion resistance. Later, titanium and niobium were reduced (understabilized) to promote creep strength. Additions of copper are for precipitation strengthening, and tungsten is for solid-solution hardening. Additions of nitrogen have

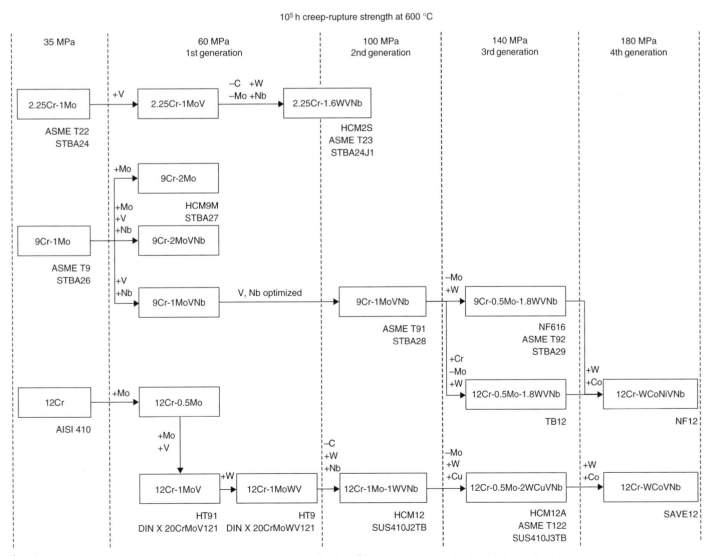

Fig. 5 The evolution of ferritic steels for boilers. Generations are categorized by the 10^5 h creep-rupture strength at 600 °C. Source: Ref 7, 10

helped to stabilize the austenitic structure (Ref 7). The nominal compositions of austenitic boiler steels are given in Table 3. Also in Table 3 are some candidate high-chromium/high-nickel alloys and superalloys for higher-temperature applications in boilers and turbine components. Alloy 740 has resulted from the THERMIE program (Table 4).

Turbine Alloys

The materials used in turbine applications are subject to the stresses of high-speed rotating equipment.

Ferritic Rotor Steels. The primary alloy used in the industry for rotors up to 545 °C is 1Cr-1Mo-0.25V. For higher temperatures, 12Cr steels are needed for increased creep strength and corrosion resistance. The first of the 12Cr alloys used for rotors was X21CrMoV121. The evolution of ferritic rotor steels from X21CrMoV121 for high- and intermediate-pressure steam turbines has loosely followed that of ferritic boiler steels. Alloy additions have been added and their amounts optimized, as shown in Fig. 8 (Ref 7). Nominal compositions of ferritic rotor steels are given in Table 5.

Turbine Blade Alloys. The ferritic steel 422 has been successfully used at temperatures up to 550 °C (Ref 7). Two possible ferritic alloys for use up to 630 °C are identified (Ref 14) and designated alloy C and alloy D. Alloy C is very similar to HR1200 and FN5 (Table 5). In the early 1990s, the Electric Power Research Institute identified four superalloys (M-252, Inconel 718, Refractory 26, and Nimonic 90) as good candidate materials for steam turbine blades (Ref 15). Although superalloys have been used extensively in gas turbines, they have not been evaluated with experience in steam turbines. An important consideration for the selection of these candidate superalloys was to have thermal expansion coefficients close to that of the 12Cr rotor steels. The four proposed superalloys are shown in Table 6, along with the ferritic alloys. Several candidate boiler alloys (Table 3) are also being considered for use in the turbine section (HR6W, alloy 230, alloy 617, and alloy 740) (Ref 16).

Bolting Materials. Bolting materials have requirements close to that of blade materials. Superalloys that have seen service in steam turbines as bolt materials are shown in Table 7. Of these, Refractory 26 and Nimonic 80A have seen the most service (Ref 7).

Corrosion in Supercritical Water

Steamside corrosion in supercritical power plants is important because of three factors (Ref 8):

- Section loss combined with high pressures can lead to component failures.
- Loss of heat transfer in water walls, superheaters, and reheaters due to the buildup of low-conductivity oxides. This leads to an increase in metal temperature that increases corrosion and creep.
- The buildup of thick oxides is much more prone to spallation, which increases the chance for tube blockages and steam turbine erosion.

Corrosion in steam has been studied intensively. Therefore, this review is limited to aspects that relate to supercritical conditions for the alloys of interest. Reference 17 lists a compilation of high-temperature steam oxidation publications since 1962 for advanced steam conditions; Ref 18 and 19 pertain to supercritical conditions.

McCoy and McNabb (Ref 18) examined several iron- and nickel-base alloys exposed for up to 22,000 h in supercritical steam (538 °C and 24.1 MPa, or 3495 psi) at the TVA Bull Run Steam Plant. They found that the corrosion rates of ferritic steels with chromium contents from 1.1 to 8.7% were all within a factor of 2 of each other. Higher chromium levels (17 to 25%) dropped the corrosion rate by an order of magnitude. This is consistent with findings at lower

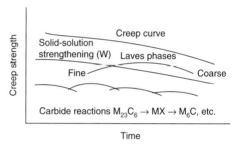

Fig. 6 Strengthening mechanisms and their effect on creep strength with time. Source: Ref 11

Table 2 Nominal compositions of candidate ferritic boiler steels

Name	Specification ASME	Specification JIS	C	Si	Mn	Cr	Mo	W	Co	V	Nb	B	N	Others
1.25Cr														
T11	T11	...	0.15	0.5	0.45	1.25	0.5	...	...	...	...	...	...	...
NFIH	...	...	0.12	...	...	1.25	1.0	...	...	0.20	0.07	...	...	...
2Cr														
T22	T22	STBA24	0.12	0.3	0.45	2.25	1.0	...	...	...	...	...	...	...
HCM2S	T23	STBA24J1	0.06	0.2	0.45	2.25	0.1	1.6	...	0.25	0.05	0.003	...	...
Tempaloy F-2W	...	...	...	...	...	2.0	0.6	1.0	...	0.25	0.05	...	...	...
9Cr														
T9	T9	STBA26	0.12	0.6	0.45	9.0	1.0	...	...	...	...	...	...	...
HCM9M	...	STBA27	0.07	0.3	0.45	9.0	2.0	...	...	...	...	...	...	...
T91	T91	STBA28	0.10	0.4	0.45	9.0	1.0	...	...	0.20	0.08	...	0.05	0.8Ni
E911	...	...	0.12	0.2	0.51	9.0	0.94	0.9	...	0.20	0.06	...	0.06	0.25Ni
NF616	T92	STBA29	0.07	0.06	0.45	9.0	0.5	1.8	...	0.20	0.05	0.004	0.06	...
12Cr														
HT91	(DIN X 20CrMoV121)		0.20	0.4	0.60	12.0	1.0	...	...	0.25	...	...	...	0.5Ni
HT9	(DIN X 20CrMoWV121)		0.20	0.4	0.60	12.0	1.0	0.5	...	0.25	...	...	...	0.5Ni
Tempaloy F12M	...	...	...	...	...	12.0	0.7	0.7	...	...	...	...	...	...
HCM12	...	SUS410J2TB	0.10	0.3	0.55	12.0	1.0	1.0	...	0.25	0.05	...	0.03	...
TB12	...	...	0.08	0.05	0.50	12.0	0.50	1.8	...	0.20	0.05	0.30	0.05	0.1Ni
HCM12A	T122	SUS410J3TB	0.11	0.1	0.60	12.0	0.4	2.0	...	0.20	0.05	0.003	0.06	1.0Cu
NF12	...	...	0.08	0.2	0.50	11.0	0.2	2.6	2.5	0.20	0.07	0.004	0.05	...
SAVE12	...	...	0.10	0.3	0.20	11.0	...	3.0	3.0	0.20	0.07	...	0.04	0.07Ta 0.04Nd

ASME, American Society of Mechanical Engineers; JIS, Japanese Industrial Standard. Source: Ref 7

pressures, where below 600 °C, the corrosion behavior is largely independent of chromium (2 to 12%) content (Ref 20, 21). The difference between 9% Cr and 12% Cr becomes significant at 650 °C (Ref 21, 22).

Paterson et al. (Ref 19) used in-plant measurements to compare the oxide scale thickness from superheater tubes and reheater tubes. The oxide scales were 45% thicker on the superheater tubes. Using the bulk steam pressures (P) experienced by the two types of tubing, it was calculated that the oxide growth rate was proportional to $P^{1/5}$. It was surmised that the pressure effect was due to differences in the partial pressure of oxygen at the different pressures. For example, using the STANJAN computer code (Ref 23), the oxygen partial pressure of steam at 593 °C derived from 200 ppb dissolved oxygen feedwater was calculated to be 1.91×10^{-5} and 2.75×10^{-5} at 17.2 and 24.8 MPa (2.5 and 3.5 ksi), respectively.

The phase stability diagram for iron in water, as functions of temperature and oxygen partial pressure, is shown in Fig. 9, based on calculations by Gaskell (Ref 24) and including the predicted conditions by Paterson et al. (Ref 19). This diagram helps to explain the scale morphologies found on low-alloy boiler steels. Typically, the scales consist of several layers that correspond to the decreasing oxygen activity found within the scale closer to the metal. Hematite (Fe_2O_3) may form as a thin outer layer. A duplex magnetite layer forms under the hematite. The outer magnetite layer is Fe_3O_4 and grows out from the original metal surface. The inner magnetite layer is $(Fe,Cr)_3O_4$ and grows inward from the original metal surface. The interface between the two magnetite layers corresponds to the original metal interface (Ref 25). Wustite (FeO) may form under the magnetite layer. Wustite formation appears to be a function of chromium content, temperature, and water vapor pressure (Ref 26). Wustite formation is associated with an increase in oxidation kinetics and may be responsible for the transition from parabolic to linear kinetics that occurs after many thousands of hours of service (Ref 8).

Hematite formation is also associated with increased spallation during shutdowns, due to its large thermal mismatch with magnetite. In addition to the excess oxygen conditions of Fig. 9, hematite can also be formed due to a lowering of the amount of iron arriving at the outer scale, which has been attributed to the voids that form at the base of the scale after many tens of thousands of hours of exposure (Ref 27).

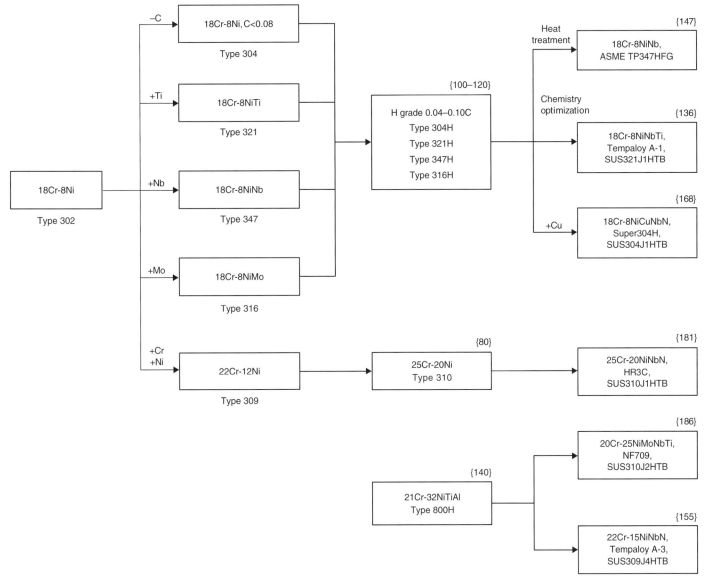

Fig. 7 The evolution of austenitic steels for boilers. {#} indicates 10^5 h creep-rupture strength at 600 °C, 2 Mpa. Source: Ref 7,10

Austenitic steels with 18Cr, such as 304 and 347, are similar but form a thin layer of Cr_2O_3 under the $(Fe,Cr)_3O_4$ layer (Ref 28). The Cr_2O_3 layer is quite protective and substantially reduces the oxidation kinetics. With 25% Cr or more, the morphology changes to just Cr_2O_3 on the surface of the alloy (Ref 28).

Spallation is a common occurrence for both ferritic and austenitic steels. As a rule, only the outer layer of the oxides detaches for ferritics, which can lead to small particle erosion of downstream components, including turbine blades. The thermal expansion mismatch between austenitic steels and their oxides is larger than for ferritic steels. So, the spallation of scales down to the bare metal is more common, and the larger particles can cause blockages of tubes, leading to short-term overheating failures of superheater and reheater tubing (Ref 27).

Efficiency

A discussion of the overall efficiency (or simply, efficiency) is helpful to understand the benefits of USC power generation. Efficiency is the ratio of energy output to energy input and is a function of many factors, for example, the thermal efficiency, the quality of coal, the condenser pressure, and the heating value (heat of reaction)

Table 3 Nominal compositions of candidate austenitic steels

Name or specification		Composition, wt%										
ASME	JIS	C	Si	Mn	Ni	Cr	Mo	W	Nb	Ti	B	Others
18Cr-8Ni												
304H	SUS304HTB	0.08	0.6	1.6	8.0	18.0	...	...	...	...	...	...
Super304H	SUS304J1HTB	0.10	0.2	0.8	9.0	18.0	...	...	0.40	...	...	3.0Cu 0.10N
321H	SUS321HTB	0.08	0.6	1.6	10.0	18.0	...	...	...	0.5	...	...
Tempaloy A-1	SUS321J1HTB	0.12	0.6	1.6	10.0	18.0	...	...	0.10	0.08	...	...
316H	SUS316HTB	0.08	0.6	1.6	12.0	16.0	2.5	...	...	...	...	...
347H	SUSTP347HTB	0.08	0.6	1.6	10.0	18.0	...	...	0.8	...	...	...
347HFG	...	0.08	0.6	1.6	10.0	18.0	...	...	0.8	...	...	...
15Cr-15Ni												
17-14CuMo	...	0.12	0.5	0.7	14.0	16.0	2.0	...	0.4	0.3	0.006	3.0Cu
Esshete 1250	...	0.12	0.5	6.0	10.0	15.0	1.0	0.2	...	0.06	...	1.0V
Tempaloy A-2	...	0.12	0.6	1.6	14.0	18.0	1.6	...	0.24	0.10	...	...
20–25Cr												
310	SUS310TB	0.08	0.6	1.6	20.0	25.0	...	...	...	...	...	...
310NbN (HR3C)	SUS310J1TB	0.06	0.4	1.2	20.0	25.0	...	...	0.45	...	...	0.2N
NF707(a)	...	0.08	0.5	1.0	35.0	21.0	1.5	...	0.2	0.1	...	...
Alloy 800H	NCF800HTB	0.08	0.5	1.2	32.0	21.0	...	...	...	0.5	...	0.4Al
Tempaloy A-3(a)	SUS309J4HTB	0.05	0.4	1.5	15.0	22.0	...	...	0.7	...	0.002	0.15N
NF709(a)	SUS310J2TB	0.15	0.5	1.0	25.0	20.0	1.5	...	0.2	0.1	...	...
High-Cr-high-Ni and Ni-base superalloys												
SAVE25(a)	...	0.10	0.1	1.0	18.0	23.0	...	1.5	0.45	...	...	3.0 Cu 0.2 N
CR30A(a)	...	0.06	0.3	0.2	50.0	30.0	2.0	...	...	0.2	...	0.03Zr
HR6W(a)	...	0.08	0.4	1.2	43.0	23.0	...	6.0	0.18	0.08	0.003	...
Alloy 230	...	0.10	0.4	0.5	57	22	2	14	...	...	Max 0.015	Max 3Fe Max 5Co 0.3Al 0.02La
Alloy 617	...	...	0.40	0.4	54.0	22.0	8.5	...	...	...	...	12.5Co 1.2Al
Alloy 740	...	...	...	...	bal	24	...	...	2	2	...	20Co Al

ASME, American Society of Mechanical Engineers; JIS, Japanese Industrial Standard. (a) Not ASME code approved. Source: Ref 7, 8

Table 4 Major international research and development efforts

Research effort	Time span	Targets	Notes
EPDC, Japan Electrical Power Development Company	1981–2000	30.0 MPa (4350 psi) 630 °C/630 °C	Materials development and component manufacture with 50 MW pilot plant operation
NIMS, Japan National Institute for Materials Science	1997–2007	650 °C	Ferritic steel development
EPRI, U.S.A. Electric Power Research Institute	1978–2003	...	Boiler and turbine thick-walled components; standardization and trial components in service. Validated NF616 (ASME P92) and HCM12A (ASME P122)
DoE Vision 21, U.S.A. Department of Energy	2002–2007	35.0 MPa (5075 psi) 760 °C	Materials development and qualification as part of the larger Vision 21 effort
COST 501, Europe Co-Operation in the Field of Science and Technology	1986–1997	530 °C/565 °C	Turbine and boiler materials development for all major components
COST 522, Europe	1998–2003	30.0 MPa (4350 psi) 620 °C/650 °C	Turbine and boiler materials development for all major components
THERMIE AD700, Europe	1998–2013	35.0 MPa (5075 psi) 700 °C/720 °C	Materials development and qualification, component design, and demonstration plant

Source: Ref 12, 13

used. There are two main ways of quantifying the heating value of the fuel: the gross or higher heating value (HHV), and the net or lower heating value (LHV). Efficiency is a function of the reciprocal of the heating value; that is, it is the energy-input part of the efficiency equation.

Water vapor is a combustion product for all hydrogen-containing fuels. The HHV is determined based on the final product water being condensed into water; the LHV is determined based on the final product water remaining a vapor (Ref 3). The HHV of a fuel can be directly determined using a bomb calorimeter, while the LHV of a fuel is usually calculated from the HHV using a formula (Ref 3) such as:

$$\text{LHV} = \text{HHV} - 23.96(\text{H}_2 \times 8.94) \quad (\text{Eq 1})$$

where LHV and HHV are in kilojoules per kilogram, and H_2 is the mass percent of hydrogen in the fuel. Because HHV is typically used in the United States and LHV in Europe, it is important to recognize the differences between them when comparing efficiencies. Plant efficiencies in Europe are usually cited as three to five percentage points higher than in the United States; the use of LHV instead of HHV accounts for approximately two of these percentage points (Ref 2).

Benefits

The driving force for increased operating temperatures and pressures has been increased efficiency in power generation. Recently, an

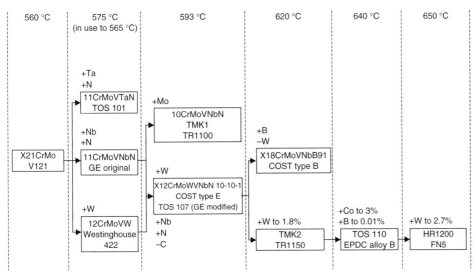

Fig. 8 The evolution of ferritic steels for rotors for use at higher temperatures. Source: Ref 7

Table 5 Nominal compositions of ferritic rotor steels

Alloy designation	Similar alloys	Composition, wt%											
		C	Si	Mn	Ni	Cr	Mo	V	Nb	Ta	N	W	Others
X21CrMoV121	...	0.23	...	0.55	0.55	11.7	1.0	0.30	...	...	...	...	...
11CrMoVTaN	TOS 101	0.17	...	0.60	0.35	10.6	1.0	0.22	...	0.07	0.05	...	...
11CrMoVNbN	...	0.16	...	0.62	0.38	11.1	1.0	0.22	0.57	...	0.05	...	...
GE original	...	0.19	0.30	0.50	0.50	10.5	1.0	0.20	0.085	...	0.06	...	...
12CrMoVW	Westinghouse 422	0.23	0.40	0.80	0.75	13.0	1.0	0.25	...	...	...	1.0	...
10CrMoVNbN	TMK1, TR1100	0.14	0.05	0.50	0.60	10.2	1.5	0.17	0.06	...	0.04	...	...
X12CrMoWVNbN 10-10-1	COST type E	0.12	0.01	0.4	0.75	10.5	1.0	0.19	0.05	...	0.06	1.0	...
TOS 107	GE modified	0.14	...	...	0.7	10.0	1.0	0.2	0.05	...	0.05	1.0	...
X18CrMoVNbB91	COST type B	0.18	0.06	0.07	0.12	9.0	1.5	0.25	0.05	...	0.02	...	0.10B
TMK2	TR1150	0.13	0.05	0.50	0.70	10.2	0.40	0.17	0.06	...	0.05	1.8	...
TR1200	...	0.12	0.05	0.50	0.8	11.2	0.3	0.20	0.08	...	0.06	1.8	...
TOS 110	EPDC alloy B	0.11	0.1	0.08	0.20	10.0	0.7	0.20	0.05	...	0.02	1.8	0.01B 3.0Co
HR1200	FN5	0.10	0.06	0.55	0.50	11.0	0.23	0.22	0.07	...	0.02	2.7	0.02B 2.7Co

Source: Ref 7

Table 6 Nominal compositions of proposed turbine blade alloys

Alloy designation	Similar alloys	Composition, wt%											
		Fe	Ni	Co	Cr	Al	Ti	Mo	W	C	Mn	Si	Others
M-252	...	...	bal	10.0	20.0	1.0	2.6	10.0	...	0.15	0.5	0.5	0.005B
Inconel 718	...	18.5	bal	...	18.6	0.4	0.9	3.1	...	0.04	0.2	0.3	5.0Nb
Refractory 26	...	16.0	bal	20.0	18.0	0.2	2.6	3.2	...	0.03	0.8	1.0	...
Nimonic 90	...	...	bal	16.5	19.5	1.45	2.45	...	...	0.07	0.3	0.3	0.003B 0.06Zr
Alloy C	HR1200, FN5	bal	0.5	2.7	11.0	...	...	0.23	2.7	0.10	0.55	0.06	0.07Nb 0.02B 0.22V
Alloy D	TOS 203	bal	0.6	1.0	10.5	...	...	0.10	2.5	0.11	0.5	0.05	0.10Nb 0.01B 0.2V 0.03N 0.2Re

Source: Ref 7

Table 7 Compositions of superalloy bolt materials used in steam turbines

Alloy designation or trade name	Composition, wt%							
	Fe	Ni	Co	Cr	Al	Ti	Mo	C
Incoloy 901	bal	40–45	Max 1.0	11–14	Max 0.35	2.0–3.0	5.0–7.0	Max 0.10
Refractory 26	bal	35–39	18–22	16–20	Max 0.25	2.5–3.5	2.5–3.5	Max 0.08
Inconel X750	5–9	bal	...	14–17	0.4–1.0	2.25–2.75	...	Max 0.08
PER 2B	Max 10.0	bal	13–20	19–23	1–2	1.6–3	...	Max 0.15
Nimonic 80A	Max 3.0	bal	Max 2.0	18–21	1.0–1.8	1.8–2.7	...	Max 0.10

Source: Ref 7

Table 8 Cost-effectiveness of methods to improve fossil fuel power plant efficiency

Rank	Method	Cost(a)
1	Reducing condenser back pressure	3.1
2	Increasing to eighth extraction point feedwater heater, raising feedwater temperature	3.8
3	Raising live steam and reheat temperatures	8.3
4	Raising live steam temperature	8.6
5	Using separate boiler feed pump turbine (BFPT) instead of main turbine-driven pump	9.6
6	Raising live steam pressure	25.1
7	Changing from single to double reheat	38.2
8	Using separate BFPT condenser	41

(a) Cost in terms of millions of U.S. dollars per net percent increase in low heating value efficiency. Low heating value is explained in text. Source: Ref 2

Table 9 Net plant efficiency improvement over a subcritical 16.5 MPa/538 °C/538 °C plant with an efficiency of 37% (high heating value)

Steam conditions	Power plant examples	Net percentage point increase in efficiency	Net plant efficiency, %
28.4 MPa/538 °C/538 °C	...	2.5	39.5
28.4 MPa/538 °C/566 °C	Schwarze Pump, 1998	2.9	39.9
28.4 MPa/566 °C/566 °C	...	3.6	40.6
28.4 MPa/566 °C/593 °C	Nanaoota 1, 1995; Noshiro 2, 1995; Haramachi 1, 1997; Millmerra, 2002	4.0	41.0
28.4 MPa/593 °C/593 °C	Matsuura 2, 1997; Misumi 1, 1998; Haramachi 2, 1998; Tchibana Bay, 2000; Bexback, 2002	4.5	41.5
31.0 MPa/593 °C/593 °C	Lubeck, 1995; Alvedore 1, 2000	4.9	41.9
31.0 MPa/593 °C/621 °C	Westfalen, 2004	5.2	42.2
31.0 MPa/593 °C/593 °C/593 °C	Nordjylland, 1998	6.5	43.5

Source: Ref 29

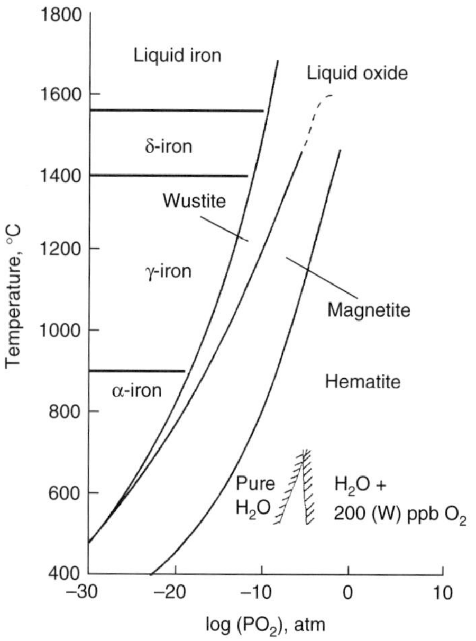

Table 10 Cost calculations comparing subcritical and supercritical plants at 80% capacity

Total cost is on a 20 yr, constant dollar basis.

Cost categories, $/MW/h	Subcritical	Supercritical, 566 °C	Supercritical, 593 °C
Capital	25.03	26.3	26.8
Operation and maintenance	6.2	6.3	6.3
Fuel	13.9	13.1	12.9
Total cost of electricity	45.1	45.7	46.0

Source: Ref 2

Fig. 9 The phase stability of iron as a function of temperature and oxygen partial pressure. Included are calculations of boiler conditions predicted from feedwaters of pure H_2O and H_2O with 200 ppb oxygen. Source: Ref 19, 24

additional recognized benefit has been decreased CO_2 emissions. Estimates of the cost-effectiveness of various ways to improve the efficiency of power plants are shown in Table 8. Increasing the steam temperature is one of the more cost-effective ways of increasing efficiency, while increasing the steam pressure is less effective. Improvements in efficiency for the newer generation of SC and USC power plants are shown in Table 9, as compared to a subcritical 16.5 MPa/538 °C/538 °C plant with an efficiency of 37% (HHV) (Ref 29).

For reduced CO_2 emissions, calculations (Ref 30) indicate that a subcritical 37% efficient 500 MW plant burning Pittsburgh No. 8 coal would produce approximately 850 tons of CO_2 per kilowatt/hour. Ultra supercritical plants at 43 and 48% efficiency would produce approximately 750 and 650 tons of CO_2 per kilowatt/hour, respectively.

Calculations (Ref 2) indicate that factoring in the cost of CO_2 emissions is necessary for making SC and USC power generation cost-effective, as shown in Table 10. In these calculations, the increased efficiencies are factored into the fuel savings, but the increased capital costs are more than the fuel savings. Factoring in the cost of CO_2 emissions, either by a carbon or CO_2 tax or by regulation, could shift the cost benefits toward the use of SC systems. This helps to explain why Europe and Japan (with CO_2 or energy taxes) have more ongoing construction of USC plants than does the United States.

Worldwide Materials Research

International materials research during the last 20 years has led to numerous new alloys for steam boilers and turbines and an increase in steam temperatures. Much of the efforts aimed at temperatures up to 650 °C is by improving on the ferritic steel 12Cr-1MoV. Major industrial research and development efforts in Japan, the United States, and Europe are briefly outlined in Table 4.

REFERENCES

1. B.B. Seth, U.S. Developments in Advanced Steam Turbine Materials, *Advanced Heat Resistance Steels for Power Generation*, Electric Power Research Institute, 1999, p 519–542

2. R. Viswanathan, A.F. Armor, and G. Booras, "Supercritical Steam Power Plants—An Overview," Proc. of Best Practices and Future Technologies. Oct 2003 (New Delhi, India), National Thermal Power Corporation's Center for Power Efficiency and Environmental Protection (CenPEEP) and the U.S. Agency for International Development (USAID), 2003
3. S.C. Stultz and J.B. Kitto, *Steam,* 40th ed., Babcock & Wilcox, 1992, p 2.13, 2.20, 9, 9.9
4. P. Kritzer, N. Boukis, and E. Dinjus, *J. Supercrit. Fluids,* Vol 15, 1999, p 205–227
5. H. Kimura, J. Matsuda, and K. Sakai, "Latest Experience of Coal Fired Supercritical Sliding Pressure Operation Boiler and Application for Overseas Utility," PGE2003BLR, http://www.bhk.co.jp/english/4tech/contents/pge2003paper_blr.pdf, Babcock-Hitachi K.K., 2003, accessed 20 Dec 2004
6. O. Jonas, "Effective Cycle Chemistry Control," Proc. of ESAA Power Station Chemistry Conference. May 2000 (Rockhampton, Queensland, Australia), Energy Supply Association of Australia, 2000
7. R. Viswanathan and W. Bakker, *J. Mater. Eng. Perform.,* Vol 10, 2001, p 81–101
8. J. Sarver, R. Viswanathan, and S. Mohamed, "Boiler Materials for Ultrasupercritical Coal Power Plants—Task 3, Steamside Oxidation of Materials—A Review of Literature," topical report, U.S. Department of Energy Grant DE-FG26-01NT41174 and Ohio Coal Development Office Grant Agreement D-0020, 2003
9. R. Viswanathan, R. Purgert, and U. Rao, Materials for Ultrasupercritical Coal Power Plants, *Materials for Advanced Power Engineering,* Vol 21, Part II, J. Lacomte-Becker, Ed., European Commission and University of Liége, 2002, p 1109–1130
10. F. Masuyama, New Development in Steels for Power Generation Boilers, *Advanced Heat Resistant Steels for Power Generation,* Book 798, R. Viswanathan and J.W. Nutting, Ed., IOM Communications, 1999, p 33–48
11. B. Scarlin, S.W. Amacker, and T.B. Gibbons, Materials for Advanced Steam Turbines and Boilers, *The Steam Turbine-Generator Today,* PWR-Vol 21, B.R. King, Ed., ASME, 1993, p 93–107
12. R. Blum and J. Hald, Benefit of Advanced Steam Power Plants, *Materials for Advanced Power Engineering,* Vol 21, Part II, J. Lacomte-Becker, Ed., European Commission and University of Liége, 2002, p 1007–1015
13. M. Staubli, K.-H. Mayer, T.U. Kern, R.W. Vanstone, R. Hanus, J. Stief, and K.-H. Schönfeld, COST 522—Power Generation into the 21st Century; Advanced Steam Power Plant, *Advances in Materials Technology for Fossil Power Plants,* R. Viswanathan, W.T. Bakker, and J.D. Parker Ed., University of Wales and Electric Power Research Institute, 2001, p 15–32
14. K. Muramatsu, Development of Ultra-Super Critical Plant in Japan, *Advanced Heat Resistant Steels for Power Generation,* Book 798, R. Viswanathan and J.W. Nutting, Ed., IOM Communications, 1999, p 543–559
15. Y. Yamada, A.M. Betran, and G.P. Wozney, "New Materials for Advanced Steam Turbines, Vol 1: Evaluation of Superalloys in Turbine Buckets," Report TR-100979, Electric Power Research Institute, 1992
16. G.R. Holcomb, B.S. Covino, Jr., S.J. Bullard, S.D. Cramer, M. Ziomek-Moroz, D.E. Alman, and T. Ochs, Ultrasupercritical Steamside Oxidation, Proceedings of the 29th International Technical Conference on Coal Utilization and Fuel Systems, April 2004 (Clearwater, FL), U.S. Department of Energy, 2004
17. I.G. Wright and B.A. Pint, "An Assessment of the High-Temperature Oxidation Behavior of Fe-Cr Steels in Water Vapor and Steam," Paper 02377, Proc. of CORROSION/2002, April 2002 (Denver, CO), NACE International, 2002
18. H.E. McCoy and B. McNabb, "Corrosion of Several Metals in Supercritical Steam at 538 °C," Report ORNL/TM-5781, Oak Ridge National Laboratory, 1977
19. S.R. Paterson, R.S. Moser, and T.W. Rettig, Oxidation of Boiler Tubing, Report TR-102101, *Interaction of Iron-Based Materials with Water and Steam,* R.B. Dooley and A. Bursik, Ed., Electric Power Research Institute, 1992, p 8-1–8-25
20. J.C. Griess, J.H. DeVan, and W.A. Maxwell, "Long-Term Corrosion of Cr-Ni Steels in Superheated Steam at 482 and 538 °C," Paper 81014, Proc. of CORROSION/81, April 1981 (Toronto, ON), National Association of Corrosion Engineers, 1981
21. F. Eberle and J.H. Kitterman, Behavior of Superheater Alloys, *High Temperature, High Pressure Steam,* American Society of Mechanical Engineers, 1968, p 67
22. P. Ennis, Y. Wouters, and W.J. Quaddakers, The Effect of Oxidation on the Service Life of 9–12% Cr Steels, *Advanced Heat Resistant Steels for Power Generation,* Book 798, R. Viswanathan and J.W. Nutting, Ed., IOM Communications, 1999, p 457–467
23. W.C. Reynolds, "Implementation in the Interactive Program, STANJAN Version 3," Department of Mechanical Engineering, Stanford University, Jan 1986
24. D.R. Gaskell, *Introduction to Metallurgical Thermodynamics,* McGraw-Hill, 1973, p 442
25. N.J. Cory and T.M. Herrington, *Oxid. Met.,* Vol 28, 1987, p 237–258
26. P.J. Grobner, C.C. Clark, P.V. Andreae, and W.R. Sylvester, "Steamside Oxidation and Exfoliation of Cr-Mo Superheater and Reheater Steels," Paper 80172, Proc. of CORROSION/80, March 1980 (Chicago, IL), National Association of Corrosion Engineers, 1980
27. B. Dooley, "The Importance of Oxide Growth and Exfoliation," Proc. of EPRI International Conference on Materials and Corrosion Experience for Fossil Power Plants, Nov 2003 (Isle of Palms, SC), Electric Power Research Institute, 2003
28. N. Otsuka and H. Fujikawa, *Corrosion,* Vol 47, 1991, p 240–248
29. R. Swanekamp, *Power,* Vol 146 (No. 4), 2002, p 32–40
30. G.S. Booras, R. Viswanathan, P. Weitzel, and A. Bennett, "Economic Analysis of Ultrasupercritical PC Plants," Paper 55.1, Proc. of Pittsburgh Coal Conference. Sept 2003 (Pittsburgh, PA), University of Pittsburgh, 2003

SELECTED REFERENCES

- *Parsons 2000 Advanced Materials for 21st Century Turbines and Power Plant, Proceedings of the Fifth International Charles Parsons Turbine Conference,* Institute of Material, 2000
- S.C. Stultz and J.B. Kitto, *Steam,* 40th ed., Babcock & Wilcox, 1992

Corrosion in Cold Climates

Lyle D. Perrigo, U.S. Arctic Research Commission (retired)
James R. Divine, ChemMet, Ltd., PC

THE FORMS OF CORROSION encountered in cold climates and the methods of control are the same as elsewhere in the world, although frequently the intensity or rate of corrosion may vary significantly. The effect and impact of corrosion often differ significantly. Although temperatures are generally lower, other variables, such as higher dissolved oxygen concentrations, abrasion, isolated internal and external environments, salt deposition, and maintenance difficulties, can create aggressive conditions and result in high corrosion rates. On the other hand, low-temperature quiescent environmental conditions and slow oxygen diffusion rates can lead to the exhaustion of oxygen adjacent to metallic objects and a consequent reduction of corrosion. Corrosion measurements taken at docks and piers in southeastern Alaska lend support to these views (Ref 1). These special aspects are the focus of this article, while more detailed information can be found elsewhere.

Cold Climates

Arctic and *Antarctic* are terms used to identify parts of the world above and below the Arctic and Antarctic Circles (66° 33' north and south latitudes), respectively. Those lines do not define or include all geographic areas where cold-climate effects are prevalent, nor do the terms *subarctic* and *subantarctic*, which some define as those parts of the world north and south of 60° latitude and running to the polar circles.

The geographic boundaries of the mission areas of the U.S. Army Cold Regions Research and Engineering Laboratory (CRREL) are used to define cold-climate areas in the Northern Hemisphere (Ref 2). The CRREL mission area lies north of line B in Fig. 1. Above line A, the approximate average temperature in the coldest month is −18 °C (0 °F) or lower, and above line B, the approximate average temperature of the coldest month is 0 °C (32 °F) or lower. Cold climate corrosion conditions can exist in the lower latitudes but are likely to be limited in scope.

Similarly, two lines define where cold weather conditions are important in the Southern Hemisphere. The northernmost is the summer isotherm, where the mean temperature of the warmest month is 10 °C (50 °F). The polar front, or southern convergence, is the line between two well-defined water masses: near-freezing Antarctic waters to the south and warmer waters to the north. This boundary wanders between latitudes 50° and 60° S. It is detectable by a sudden change in temperature and salinity. The southern convergence line is shown in Fig. 2 (Ref 3). Cold-climate corrosion problems are found in the lower latitudes of both hemispheres in the higher altitudes of mountainous regions in other parts of the world (Ref 4).

Another factor influencing corrosion, and particularly corrosion control, is permafrost. *Permafrost* is a term used to note the presence of frozen ground throughout the year. Most of the areas bordering the Arctic Ocean have continuous permafrost, which means that the entire area is underlaid with frozen ground. Permafrost can be more than approximately 300 m (1000 ft) thick (Ref 5). South of the artic continuous permafrost areas are discontinuous permafrost regions that contain patches of permafrost. The extent of permafrost in the Southern Hemisphere is not as well defined but does include all of Antarctica.

Corrosion Control Techniques and Costs

Using pipes and tanks as examples, corrosion environments exist both internally and externally to containment. Internal environments are those inside the systems and equipment and involve the fluids being handled in those systems. They

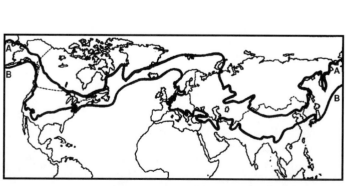

Fig. 1 Cold-climate areas in the Northern Hemisphere. For the coldest winter month above line A, the average temperature is −18 °C (0 °F) or lower, and above line B, the average temperature is 0 °C (32 °F) or lower. Source: Ref 2

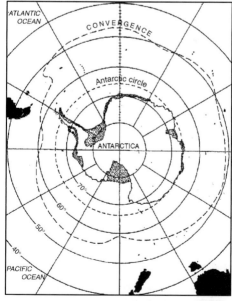

Fig. 2 Location of the southern convergence, or polar front, (dashed line) marks an abrupt change in water temperature and salinity. Source: Ref 3

may be affected by external temperature conditions but are generally no different from those in the temperate climes. Internal corrosion can be controlled primarily by selection of materials, design, process control, internal coatings, and inhibition. The external environments are subdivided into three categories: atmospheric, buried, and submerged. The methods used to control corrosion in external environments typically include suitable materials selection, anticorrosion design principles, cathodic protection (CP) techniques, and application of protective coatings.

The total annual cost of corrosion in the United States was recently estimated to be 3.1% of the gross national product, which, for 1998, was $276 billion (Ref 6). For additional information, see the article "Direct Cost of Corrosion in the United States" in *ASM Handbook,* Volume 13A, 2003, p 959–967. To establish a context for the impact of cold-climate corrosion, the cost in Alaska in 1997, using the value of 3.1%, would be slightly under a billion dollars per year. A detailed examination of corrosion control in the North would show added costs are associated with transportation, storage, weather constraints, ergonomics, and a number of other details. Thus, the percentage of the gross state product for Alaska, the Antarctica, the northern parts of Canada, Scandinavia, and Russia associated with corrosion is estimated to be in excess of 5%.

While the effect of cold-climate environmental impact on corrosion costs is not known exactly, designers, construction companies, operators, and maintenance people must recognize that these costs are significant. The need for corrosion control assistance by corrosion engineers with cold-climate experience is apparent.

Design

Engineers design systems, facilities, and equipment by using a combination of experience, available data and information, and calculations. Canadian and Scandinavian building specifications and standards recognize the need to apply different methods than those used in temperate zones. The author of design specifications for structures and equipment intended for cold climates needs to be cognizant of the unique requirements and not simply apply to national or industry standards that may not have consideration of cold climate within its scope. What follows in this section are examples of some of the major issues involved in designing systems with acceptable cold-climate corrosion performance. Topics include location, inappropriate measures, special factors, and ergonomics.

Location. Some of the highest marine corrosion rates in the world are found on offshore petroleum production platforms in Cook Inlet, AK, (Ref 7) where CP current densities of 376 to 430 mA/m^2 (35 to 40 mA/ft^2) are typically required to protect the structure. These rates are twice as high as those in the North Sea and six times greater than those in the Gulf of Mexico, despite the higher temperatures in the Gulf. The high rates are caused by high oxygen concentrations, abrasion from glacial silts, a tidal rise and fall of more than 9 m (30 ft), and the action of ice floes in the winter. Splash zone corrosion rates on docks are estimated to be as high as 762 to 889 μm/yr (30 to 35 mils/yr) in the Aleutian Islands (Ref 1).

The literature on uses of CP to control corrosion in cold-climate marine environments is substantial (Ref 1, 8). One problem is that itinerant specialists, sometimes unfamiliar with cold climates, determine the performance of CP systems installed in remote and rural locations. Because impressed current CP systems require more attention to ensure proper operation than do sacrificial anode systems, the latter are generally favored in cold climates. Background on CP systems is found in the article "Cathodic Protection" in *ASM Handbook,* Volume 13A, 2003. The environment and temperature are factors in the capacity and efficiency of all sacrificial anode alloys. High-resistivity soils affect the design of underground systems in the cold regions. Further, sacrificial zinc anodes become passive in high-resistive soils in 3 to 4 years, while magnesium anodes often provide needed protection for much longer periods. A compilation of papers, standards, and articles describing CP applications, guiding procedures, and monitoring techniques was published in the early 1980s (Ref 1, 8). Many of these papers relate to problems and conditions in cold-climate areas of Alaska, Canada, and Norway.

Antarctic studies (Ref 9, 10) show that ISO standard 9223 (Ref 11) does not properly define the time of wetness (TOW) as a means to determine corrosivity at that location. See the article "Atmospheric Corrosion" in *ASM Handbook,* Volume 13A, 2003, p 196–209 for a discussion of the relevance of TOW. The studies concluded that significant corrosion was evident on samples exposed at temperatures below 0 °C (32 °F), while the criterion for estimating TOW of ISO standard 9223 is only for temperatures greater than 0 °C (32 °F). Salt deposits, consisting of chlorides and sulfates, play a significant role in promoting corrosion under cold-climate conditions by being deliquescent and forming nearly invisible films of moisture. Except in the interiors of the cold regions, such as the dry valleys of central Antarctica, atmospheric corrosion rates in the cold regions can approximate those in more temperate zones. Ratcliffe (Ref 12) has noted extensive amounts of external corrosion on metal buildings, particularly under the insulation on those structures. In addition, acceleration of external corrosion due to abrasion from wind-driven glacial silt has been observed near the Antarctic coast. Figure 3 shows what occurred on the external surfaces of a Quonset hut. The costs for repairs to such structures are high because of the remoteness of the area and the limited periods during which work can be performed (Ref 12).

While atmospheric corrosion rates in inland areas of the cold-climate regions are low (Ref 13, 14), they are definitely not zero. Industrial coatings (epoxies, urethanes, and alkyls) are used to ensure adequate performance of materials and facilities (Ref 1). If effective performance is to be realized, they must be applied and cured at 10 to 32 °C (50 to 90 °F). Even summer temperatures may not reach the lower acceptable temperature. A temporary plastic-covered framework with high-velocity air heaters may be needed to provide the required temperature for surface preparation, application, and curing. This will increase time and costs. Extra emphasis must be placed on safety in this artificial environment.

Inappropriate Measures. Another factor to consider is that common techniques used to compensate for cold-climate effects may have adverse impacts on corrosion prevention. For example, utilidors are underground or aboveground structures that protect utilities from cold weather while providing general access for operations and maintenance. Underground utilidors may flood due to piping failures, overhead snow melts, or stream overflow during ice breakup and jamming in the spring. When thermal insulation is soaked, the pipes can suffer from corrosion under insulation (Ref 15, 16). The designer should specify immersion-service coatings for external piping surfaces used in underground utilidors to limit these effects, in accordance with NACE International recommended practices (Ref 16).

Marine atmospheres, such as those found in the Aleutian Islands, often induce high corrosion rates. Coatings may fail rapidly, and fasteners used in the installation of building siding must be properly selected and seated, or large blisters can be expected under the coating in 2 to 3 years (Ref 1). Special marine coatings must be used to avoid corrosion problems, especially in facilities that may not be accessible most of the year (Ref 15).

Industrial facilities and larger municipalities in cold regions often have most of their equipment in heated spaces, where maintenance can be done in a timely manner. In smaller facilities, however, equipment that requires some protection from the elements may be in unheated

Fig. 3 Corrosion on a metal building in Antarctica. Corrosion is also evident on the steel vise.
Source: Ref 12

enclosures. Because access is difficult, maintenance is often delayed until conditions are more favorable. A common result is that leaking fluids, although freezing quickly, will still soak insulation and ultimately cause corrosion problems.

Special Factors. Anticorrosion design or corrosion control design describes the use of layout, location, joining practices, orientation, avoidance of liquid holdup areas, and related topics to minimize corrosion (Ref 17). Although these techniques have been known for decades and are effective, these principles may not always be applied. User and operating organizations should require engineering design groups to address anticorrosion design factors, because of the potentially severe consequences of poor design in cold climates.

Two corrosion control design problems, often overlooked because they are not considered as corrosion issues, involve accessibility and safety. Corrosion control personnel must be able to reach operating systems and facilities and to use monitoring equipment to be effective in their work. Maintenance crews can do their jobs only if they have ready access to all items that may require repair or replacement during the operational lives of plants and facilities. Providing access in cold climates involves ways to avoid problems created by snow and ice, the effects of low temperatures, and the impact of winter darkness.

Gratings used around fuel storage tanks and the stairs to living quarters should be designed and constructed so that there are no crevices or holding areas that will retain deicing chemicals or salt-laden moist marine air (Ref 1). Chemicals in such crevices lead to accelerated corrosion, which in turn results in unsafe conditions. Dike systems surrounding tanks often trap melting snow and ice in the late spring. Water in the drains installed at low points in the diked area can remain frozen when other, more exposed ice and snow melt. The trapped water rises to the base of the tank, filling any crevice between the sand and the tank bottom. This is especially troublesome if oiled sand is used beneath storage tanks, because oiled sand impedes effective use of cathodic protection.

Ergonomics, the study of human factors or man-machine interface issues, addresses conditions that affect corrosion control work in cold weather and remote areas. Three factors to be considered in designing equipment and facilities are:

- Equipment used in monitoring work or to support such activities must be available when needed (Ref 1, 17). In cold regions, this may mean extensive warm-up periods or special arrangements to place the needed items in remote locations. The equipment must function reliably and in an effective manner, to avoid safety problems or costly delays. Electronic instruments must be operated within the temperature limits for which they are designed and calibrated. This means they may need to be housed in a warmed, insulated container or operated from a warm vehicle.
- It takes longer to inspect equipment, study failures, and conduct monitoring operations undertaken outside in cold weather. Designs should allow extra space for movement in bulky clothing, limit the number of monitoring stations located outside, and arrange equipment so that only a few movements are needed to make measurements or observations (Ref 18).
- Because corrosion coupon locations, sampling ports, and other monitoring equipment and facilities are located outside, the design must allow equipment to be operated with heavy mittens, bulky clothing, and limited visibility.

Transportation and Storage

A common cause of corrosion in cold climates is improper transportation and storage practices (Ref 19, 20). Most equipment and materials used in cold climates for construction, operations, and maintenance are transported by barges and ships from temperate-zone ports. Transportation routes through polar regions are matters of concern to corrosion engineers for three reasons:

- Several have notoriously bad weather and sea conditions.
- Many are open only for limited times each year.
- Special arrangements are needed to ensure that transit and storage do not adversely affect the corrosion performance of the cargo (Ref 1).

The spalling seen on the precast concrete panels in Fig. 4 occurred after three years. The panels were exposed to seawater from wave action during shipment. Poor packaging can be costly on the return route as well. An example is pumps that were shipped to Seattle after use in the *Exxon Valdez* oil spill cleanup in Prince William Sound in 1989. These items were placed in containers with wooden floors that allowed moisture to reach the pumps. With no vapor-phase inhibition, corrosion of the electrical equipment occurred in the containers. Extended storage near the docks after arrival in Seattle, with no inspection or remedial action, led to a total loss of the equipment (Ref 1).

Construction

Corrosion problems resulting from cold-region construction practices confront construction and maintenance personnel (Ref 17). Cognizant contractor personnel require special corrosion education courses, training, and certifications to eliminate such problems. Such training is required by Alyeska Pipeline Service Company, which requires inspectors to have NACE International certifications. Of great importance is how changes in design, procedures, or materials will impact corrosion and corrosion control of the completed facility (Ref 1). Detailed briefings at prejob conferences avoid later conflicts and produce excellent results.

Frequent checks of construction projects by a corrosion engineer and maintenance supervisors can be of benefit in two areas. First, observations may supplement the work of the project manager and inspectors. Second, the acquisition of general information will be useful after the system or facility begins operation. For the latter, observations made by a corrosion engineer during construction are particularly helpful in establishing a background to assess problems and conduct failure analyses. Maintenance personnel find that such observations create an awareness that makes future repairs easier and quicker. Some contractors discourage visitors on the basis that their questions distract personnel and cause delays. Nevertheless, the possible benefits in cold regions are significant. Giving corrosion engineers and maintenance supervisors access to make periodic observations should be included in the contract language.

Types of common on-site practices that may cause corrosion problems in cold regions are briefly discussed here. First, the use of readily available local rock and gravel, rather than adherence to the size and type specifications for gravel and rock, often occurs, which may affect the performance of buried, coated piping systems. Freeze-thaw cycles cause the ground to move. The irregular fill may increase the chances of mechanical coating damage, leading to local corrosion. Variations in the fill can create anodic and cathodic areas along the pipe that contribute to galvanic corrosion.

The short construction season and remote location create pressure to make substitutions when the needed and specified parts are not available or are damaged in transit. Substitutes often do not perform well and lead to early failures. Another problem caused by short construction seasons is a tendency to save time by relaxing the quality of surface preparation work

Fig. 4 Spalling on precast concrete panels occurred in three years after exposure to seawater during shipment. Reinforcing bars are exposed.

before coatings are applied. Less rigorous procedures lead to shorter service lives for the facilities (Ref 21).

Infrequent shipment to isolated sites, a common occurrence in cold regions, can also play a role in corrosion prevention work. Careful planning is necessary prior to construction to ensure that the correct materials are available when needed. Such an approach will minimize tendencies to substitute parts constructed from materials with poor corrosion performance. Proper storage of coating materials and other moisture- and temperature-sensitive materials at the site is required. The benefits of proper storage in a centralized facility are negated if these materials are placed in vehicles where temperatures may drop below critical temperatures. The use of coating materials after such handling will lead to poor coatings and inadequate corrosion protection. It is also necessary to ensure that work is done when the needed equipment is available. For example, cranes are not allowed to operate at very low temperatures.

Often, cold-climate building materials use precoated, insulated metal sheets for exterior (and often interior) walls. A lack of care in installing them can lead to early coating and panel failures.

Operations and Maintenance

Some corrosion problems arise from improper operations and maintenance practices. A lack of care in snowplowing to clear walkways and parking areas can jam snow and ice against the metal siding of buildings, opening seams and damaging coatings. When this occurs, moisture can enter and cause corrosion of the substrate, exposed by cracks and chips in the coating. The moisture behind the siding creates two problems. First, this moisture provides a medium for freeze-thaw damage that accentuates the mechanical damage. In addition, it may result in counterbuckling the external surface or damaging the wall beyond the insulation. Another effect is corrosion of the internal surface of the siding, because this surface is rarely coated, and the insulation keeps the surface wet. Maintenance personnel can take action to avoid such damage. Placing bollards near buildings where snowplows operate limits the approach of those vehicles.

Often, operating and maintenance personnel in cold regions find themselves with facilities and equipment that were designed without the use of anticorrosion design concepts (Ref 18). These systems have crevices and places where fluids stay in contact with the metals for appreciable periods of time. Sometimes, these problems can be avoided by simply creating ways to drain fluids from the holdup areas.

Minor changes to the exterior surface of an existing tank, such as installing a drip skirt, can limit the amount of corrosion. However, the metal should be joined to the tank with a continuous weld to prevent fluids and/or snow from reaching the tank bottom. If not, the interfaces between the bottom and the support members become sites where accelerated corrosion would occur. Channels and angles oriented with their webbing upward can collect snow and rain, leading to corrosion. Problems can be avoided or reduced by drilling a few holes through the webbing of these members, to serve as drains for any water that may temporarily collect.

ACKNOWLEDGMENTS

The authors gratefully acknowledge information provided by Guy Ratcliffe, Australian Antarctic Division, Kingston; and other input from Janet Hughes, National Gallery of Australia, Canberra; and George King, Sandringham, all of Australia, who provided extensive amounts of information about the effects of seasalts on atmospheric corrosion in Antarctica. Winston Revie, CANMET, Ottawa, Canada, chaired a symposium on cold-climate corrosion at the annual NACE International conference in 2002 that provided the authors with access to additional corrosion information.

REFERENCES

1. L.D. Perrigo, H.G. Byars, and J.R. Divine, *Cold Climate Corrosion,* NACE International, 1999
2. E.A. Wright, *1961–1986, CRREL's First 25 Years,* U.S. Army Cold Regions Research and Engineering Laboratory, Hanover, NH, 1986
3. A. Gurney, *Below the Convergence,* W.W. Norton & Company, New York, 1997
4. B. Stonehouse, *North Pole, South Pole: A Guide to the Ecology and Resources of the Arctic and Antarctic,* Prion, London, U.K., 1990
5. J.A. Heginbottom, J. Brown, E.S. Melnikov, and O.J. Ferrians, Jr., Circumarctic Map of Permafrost and Ground Ice Conditions, *Proceedings of the Sixth International Conference,* 5–9 July (South China University of Technology, Wushan, Guangzhous, China), 1993, p 1, 132–1,136
6. "Corrosion Cost and Preventive Strategies in the United States," final report, 99-01, NTIS PB2002106409, CC Technologies Laboratories, Inc., supported by Federal Highway Administration and NACE International, 2002
7. "Standard Recommended Practice: Corrosion Control of Steel, Fixed Offshore Platforms Associated with Petroleum," RPO 176-83, NACE International, 1983
8. "Cathodic Protection of Production Platforms in Cold Seawaters," NACE International, 1981
9. J. Hughes, G. King, and W. Ganther, "The Application of Corrosion Information for the Conservation of Historic Buildings and Artefacts (sic) in Antarctica," paper presented at the Northern Area Western Region Conference, 26–28 Feb 2001 (Anchorage, AK), NACE International
10. G. King and W. Ganther, "Studies in Antarctica Help to Better Define the Temperature Criterion for Atmospheric Corrosion," presented at Northern Area Western Region Conference, 26–28 Feb 2001 (Anchorage, AK), NACE International
11. "Corrosion of Metals and Alloys—Corrosivity of Atmospheres—Classification," ISO 9223, International Organization for Standardization
12. G. Ratcliffe, Australian Antarctic Division, Australian Government, personal communication, 15 Sept 2003
13. G.H. Biefer, "Survey of Atmospheric Corrosivity in the Canadian Arctic: Supplementary Results," Report MRP/PMRL 80-71(TR), Energy, Mines, and Resources Canada, Sept 1980
14. G.H. Biefer and J.G. Garrison, "Long-Term Corrosion Tests of Carbon and Weathering Steels in Ottawa and in the Arctic," Report MRP/PMRL 81-7(TR), Energy, Mines, and Resources Canada, Feb 1981
15. J.A. Beavers and N.G. Thompson, "Corrosion Beneath Disbonded Coatings," MP 36, NACE, 1997, p 13–19
16. "The Control of Corrosion Under Thermal Insulation and Fireproofing Materials—A Systems Approach," RPO 198-98, NACE International, 1998
17. T.T. McFadden and F.L. Bennett, *Construction in Cold Regions: A Guide for Planners, Engineers, Contractors, and Managers,* John Wiley and Sons, Inc., 1991
18. J. Abeysekera, "Preliminary Study Report—The Use of Protective Clothing and Devices in the Cold Environment," COLD-TECH 92-7, Lulea? University, Sweden, 1992
19. L.D. Perrigo III, Shipping and Storage Effects on Corrosion Performance, *Proceedings of the Fifth International Symposium on Cold Region Development,* H.K. Zubeck, C.R. Woolard, D.M. White, and T.S. Vinson, Ed., 4–10 May 1997 (Anchorage, AK), American Society of Civil Engineers in association with the International Association of Cold Regions Development Studies, 1997
20. T.R. Dart, *People and Logistics in Arctic Corrosion Control Activities, Proceedings of the Fifth International Symposium on Cold Region Development,* H.K. Zubeck, C.R. Woolard, D.M. White, and T.S. Vinson, Ed., 4–10 May 1997 (Anchorage, AK), American Society of Civil Engineers in association with the International Association of Cold Regions Development Studies, 1997

21. L.D. Perrigo and G.A. Jensen, Fundamentals of Corrosion Control Design, *North. Eng.,* Vol. 13 (No. 4), 1981, p 16–18, 28–34

SELECTED REFERENCES

- J.F. Henriksen, "Corrosion in SO_2 Polluted Cold Climate Atmospheres in Northern Norway and at the Border between Norway and Russia," Paper CldCli10, presented at the Northern Area Western Region Conference, 26–28 Feb 2001 (Anchorage, AK), NACE International
- J.D. Hughes, Ten Myths about the Preservation of Historic Sites in Antarctica and Some Implications for Mawson's Huts at Cape Denison, *Polar Rec.,* Vol 36 (No. 197), 2000, p 117–130
- J.D. Hughes, G.A. King, and D.J. O'Brien, Corrosivity in Antarctica—Revelations on the Nature of Corrosion in the World's Coldest, Driest, Highest, and Purest Continent, Paper 24, *Proceedings of the 13th International Corrosion Conference,* 25–29 Nov 1996, Australasian Corrosion Association, Melbourne, Australia
- A.W. Peabody, Corrosion Aspects of Arctic Pipelines, *Mater. Perform.,* Vol 18 (No. 5), 1979, p 30
- R.D. Seifert, "Concerns about Corrosion in Arctic and Subarctic Housing Technologies and Applications," Paper CldCli12, presented at the Northern Area Western Region Conference, 26–28 Feb 2001 (Anchorage, AK), NACE International

Corrosion in Emissions Control Equipment

William J. Gilbert, Branch Environmental Corp.

CORROSION PROBLEMS and materials selection for emissions control equipment can be difficult because of the varied corrosive compounds present and the severe environments encountered. Therefore, a number of the more common emissions control applications are discussed. More detailed information on the applications is available in the references cited at the end of this article.

Flue Gas Desulfurization

By far the most common cleaning application for flue gases is flue gas desulfurization (FGD). This section discusses the selection of materials of construction for FGD systems. More information on corrosion in FGD systems is available in the article "Corrosion of Flue Gas Desulfurization Systems" in this Volume.

These systems were introduced in the late 1960s to meet tightening restrictions on the release of sulfur emissions. The oil shortage of the mid-1970s and subsequent oil price increases led to the increased use of coal in new and renovated power plants. In virtually all cases, this meant the potential for increased sulfur emissions, so many more FGD systems were needed.

Flue gas desulfurization systems typically use wet scrubbing units with lime or limestone slurries for sulfur dioxide (SO_2) absorption. Initially, it was thought that the relatively mild pH and temperature conditions found within most of these systems would not present a significant corrosion problem. This was soon found not to be the case. Because the FGD system constituted up to 25% of the total capital and operating expenses of the power plant, it was imperative to determine the causes of the failure of the material.

Environment. The gases encountered by the FGD system are hot and contain SO_2 at significant levels, some sulfur trioxide (SO_3) as a result of the oxidation of SO_2 at high temperatures, and fly ash. Initially, these gases may be sent to a dry-dust collector, such as an electrostatic precipitator or fabric filter baghouse, for fly ash removal. The gases typically enter a wet scrubber (a venturi with separator) and are quenched as SO_2 is absorbed. The components that often have the severest problems, however, are the outlet duct and stack. Here, the condensates are more acidic, the gases are highly oxygenated, and the presence of chlorides and fluorides can cause serious corrosion problems. Nevertheless, throughout the entire system, corrosion can occur to various degrees and because of various factors.

Corrosion Factors. Four basic factors affect the severity and type of corrosion that occurs.

pH. The result of the reactions that take place within the scrubber is a slurry with a typical pH of 4 to 5. This is desirable, because it allows for good absorption of SO_2 and is acidic enough to reduce scale formation. Local pH values as low as 1 may exist from the concentration of chlorides entering the makeup liquid (with contributions from fluorides). Low-pH conditions with the presence of chlorides and fluorides limit the use of carbon steels, stainless steels, and a number of higher-nickel alloys (Fig. 1).

Gas Saturation. The dry flue gas is not severely corrosive. However, when the gas reaches its dewpoint, sulfuric (H_2SO_4) and sulfurous (H_2SO_3) acids can form. In addition, hydrochloric acid (HCl) is produced because of the presence of hydrogen chloride (formed at the elevated temperatures of combustion) plus the condensing water vapor. Again, carbon and stainless steels are affected.

Temperature. The problems caused by temperature excursions lessen the corrosion-resistant properties of synthetic coatings, fiberglass-reinforced plastics (FRPs), and thermoplastics, possibly to the point of complete destruction at high temperatures. This affects metals to a lesser extent but still can degrade the materials.

Erosion generally occurs as a result of fly ash carried by the gas impacting a surface in a relatively dry area of the system or the liquid slurry impinging on a wetted surface. In either case, areas susceptible to corrosion attack are produced.

General Materials Selection. An easily overlooked but critical aspect of materials selection is the ability of the manufacturer to construct the equipment with correct fabrication techniques. In particular, with regard to the use of high-nickel alloys, the welding recommendations of alloy producers should be precisely followed to maintain the corrosion resistance of the materials (Ref 2). This is, of course, true for any type of fabrication. The most careful materials selection process can be negated by poor fabrication practices.

Metals. Where pH is neutral or higher, austenitic stainless steels (types 304, 316, and 317, L grades preferred) perform well even at elevated temperatures. If pH is as low as 4 and chloride content is low (less than 100 ppm) but temperatures are above approximately 65 °C (150 °F), then Incoloy 825, Inconel 625, Hastelloy G3, and alloy 904L (UNS N08904) or their equivalents are usually acceptable. Table 1 lists compositions of alloys commonly used in FGD systems.

When chloride content is up to 0.1% and pH approaches 2, only Hastelloys C276, G, and G3, and Inconel 625 can be successfully used. The other alloys mentioned previously would be subjected to pitting and crevice corrosion. If a region is encountered with pH as low as 1 and chloride content above 0.1%, one of the only acceptable alloys is reported to be Hastelloy C276 or its equivalent. Generally, the higher the molybdenum content of an alloy, the more

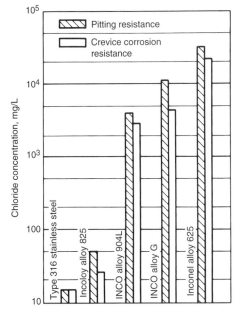

Fig. 1 Minimum levels of chloride that cause pitting and crevice corrosion in 30 days in SO_2-saturated chloride solutions at 80 °C (175 °F). Source: Ref 1

Table 1 Compositions of some alloys used in flue gas desulfurization systems

Alloy	UNS No.	C	Fe	Ni	Cr	Mo	Mn	Others
304L	S30403	0.03 max	bal	10.0	19.0	...	2.0 max	0.045 max P, 0.03 max S, and 1.00 max Si
316L	S31603	0.03 max	bal	12.0	17.0	2.5	2.0 max	1.00 max Si, 0.045 max P, and 0.03 max S
317L	S31703	0.03 max	bal	13.0	19.0	3.5	2.0 max	1.00 max Si, 0.045 max P, and 0.03 max S
625	N06625	0.10 max	5.0 max	bal	21.5	9.0	0.50 max	0.40 Al, 0.40 max Ti, 3.65 Nb, 0.015 max P, 0.015 max S, and 0.50 max Si
825	N08825	0.05 max	bal	42.0	21.5	3.0	1.0 max	0.8 Ti, 0.5 max Si, 0.2 max Al, 2.25 Cu, and 0.03 max S
G	N06007	0.05 max	19.5	bal	22.25	6.5	1.5	1.0 max Si, 2.125 Nb, 2.5 max Co, 2.0 Cu, 1.0 max W, and 0.04 max P
G3	N06030	0.15 max	19.5	bal	22.25	7.0	1.0 max	5.0 Co, 2.0 Cu, 0.04 max P, 1.0 max Si, 0.03 max S, 1.5 max W, and 0.50 max Nb + Ta
C276	N10276	0.02 max	5.5	bal	15.5	16.0	1.0 max	2.5 max Co, 0.03 max P, 0.03 max S, 0.08 max Si, and 0.35 max V
904L	N08904	0.02 max	bal	25.5	21.0	4.5	2.0 max	1.5 Cu, 1.0 max Si, 0.045 max P, and 0.035 max S

(a) Nominal composition unless otherwise specified

severe the corrosive environment it can withstand in the FGD system (Ref 3).

Nonmetals. Fiberglass-reinforced plastics can be used in almost any application in which temperatures do not exceed 120 °C (250 °F) (preferably 95 °C, or 205 °F), regardless of whether there are high chlorides or low pHs. The best choices would be premium grades of vinyl-ester and polyester resins. Polypropylene, chlorinated polyvinyl chloride (CPVC), and other thermoplastics can be used in such applications as mist elimination, where temperatures are lower. For example, polypropylene can withstand 80 °C (175 °F). Rubber linings can also be used where temperatures are suitable and mechanical damage can be avoided.

Waste Incineration

In a number of ways, the problems associated with materials for incinerator off-gas treatment equipment are similar to those of FGD systems. Depending on the wastes being burned, however, significantly higher gas temperatures as well as more varied and more highly corrosive compounds may be encountered. Materials selection can often be more demanding.

The importance of incineration for the treatment of domestic and industrial wastes has increased as sanitary landfills are less available and more expensive. At the same time, environmental safety regulations have limited the use of deep below-ground and sea-disposal sites for untreated wastes.

Incineration provides a viable, although not inexpensive, alternative that reduces the volume of solid waste by as much as 90%. The remaining material is generally inert and will not leach chemicals into the surrounding aquifer. The by-products of combustion contain gas and particulate contaminants that can be filtered or scrubbed. Incinerators are used to burn municipal solid waste, biomaterials such as hospital waste, low-level radioactive waste, industrial chemical waste, and sewage sludge. In general, the off-gases can be classified according to their corrosiveness in descending order as: industrial chemical, hospital, municipal solid, and sewage sludge.

Industrial Chemical. These gases are characterized by extremely high temperatures (1000 °C, or 1830 °F, is not uncommon) used to destroy the chemical. This can result in the formation of halogenated compounds from chlorine or fluorine in the raw material. Hydrogen bromide, sulfur oxides, phosphorus compounds, and oxides of nitrogen can also be produced, depending on the composition of the waste treated.

The typical treatment system uses a gas quench to saturate and cool the gases, a wet venturi scrubber (if particulates pose a problem), a packed tower absorber, exhaust fan, ducting, liquid piping, and liquid recirculation pumps. Figure 2 shows a standard system arrangement, and installed equipment is shown in Fig. 3.

Because of high temperatures, the presence of chlorides, and the fact that the gas becomes saturated with water vapor within the first-stage quench, very few materials have been successfully used for the quench construction. The major problem with many metals is not uniform attack but local pitting and crevice corrosion. In particular, chloride stress-corrosion cracking severely affects austenitic stainless steels.

The materials that perform well include Hastelloy C276, Inconel 625, and sometimes titanium for the high temperature. These materials have been used in other critical areas of the treatment system, such as fan wheels, dampers, liquid spray nozzles, and piping. Multiple-year service life has been reported with these alloys (Ref 4).

Refractory linings for the quench have been used with some success. Alumina or selected carbide bricks can be installed. Because the brick is an insulating material, the outer-wall temperature is low enough to allow the use of carbon steel, such as A285 grade C, for the structural outer components. Steel must be protected by using an epoxy or elastomer lining to prevent corrosion of the steel vessel. This can sometimes prove to be a more economical alternative to the use of high-nickel alloys, especially on larger-diameter systems. Because care must be taken to avoid thermal shock that can fracture the refractory material, refractory linings are usually employed on larger systems with slow start-up times rather than on the smaller hazardous waste incinerators, where batch operation is often the rule.

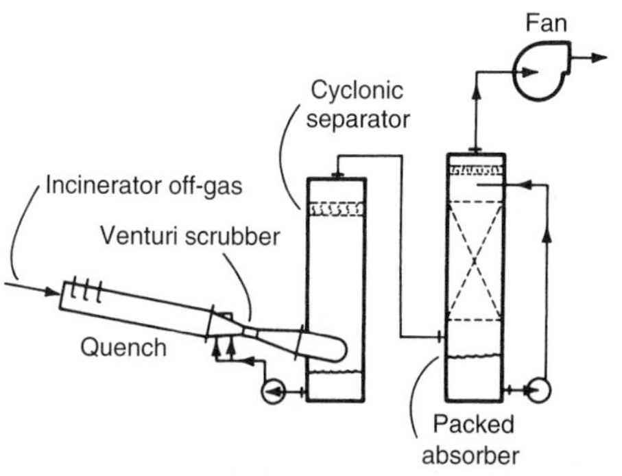

Fig. 2 Schematic of a general scrubber system arrangement

Fig. 3 A stainless steel venturi tube scrubber separator. Courtesy of Branch Environmental Corporation

Following the quench, temperatures are typically less than 95 °C (205 °F), but the price of satisfactory materials is increased. The materials used will often be those recommended for the chemical industry that produce the same chemicals as those expected in the gas. For example, phosphoric acid may readily be handled by using type 316 stainless steel in the scrubber following the quench.

Because the temperatures are relatively low, the concentrations of contaminants are low, and because cost is a design factor in pollution control equipment, it is not unusual to see FRP in the final equipment stage. That is, scrubber housings, pump bodies, and even fans can be made of FRP. A premium polyester or vinyl-ester resin can withstand even the most severe corrosive atmospheres at these milder temperatures. Even the presence of glass-attacking fluorides would not necessarily preclude the use of FRP. Synthetic veils are used to replace glass veils within the resin layers closest to the internally exposed surfaces.

Recirculating fluids, often alkaline because they are used to scrub acidic gases, can often be handled satisfactorily by type 316 stainless steel, fluoroplastic-lined pipe (for example, Teflon lined steel pipe), FRP, or CPVC. The choice is often based on plant preference: the metal pipe for mechanical strength, the metal pipe with plastic liner where more aggressive chemicals are present, or FRP/CPVC where cost is the most important consideration. In case the caustic feed system fails to operate, examination of the possible occurrence of a low-pH condition should also be considered. In many cases with high chlorides, this could result in pitting or damage to type 316 stainless steel.

Fiberglass-reinforced plastic ductwork is used to transport the gases in the milder-temperature areas of the system. Because polypropylene exhibits good resistance to most of the corrosives usually encountered, it is used for tower packing, mist eliminators, and spray nozzles. It is a particularly good choice for environments having the potential for severe fluoride attack.

Caution must be exercised when using plastics in the system following the quench. If the quench loses its liquid and there are no safeguards, a major part of the downstream equipment may be destroyed. Temperatures are typically monitored so that an emergency cooling liquid is injected into the quench to prevent disastrous temperature excursions if the normal liquid source is lost.

A more conservative design approach is to use a high-nickel alloy for the equipment directly downstream of the quench. In any case, this question must be addressed during the design phase of any incineration project.

Hospital Waste (Biowaste). The emission gas is in a lower temperature range (approximately 250 °C, or 480 °F) than for chemical incineration and can be handled by alternate technology such as a dry flue gas system. Where the gas is kept dry and above its dewpoint (the HCl dewpoint), carbon steel construction is very common. Where fabric filter dust collectors are included, they are usually made of steel, although some collectors are constructed of type 316 stainless steel to avoid rust or corrosion.

Where further cooling is required, the wet-dry interface will become very corrosive as the HCl is exposed to moisture. Following the particulate collection system, it would not be unusual to have a secondary quench stage that is made from Hastelloy C276 or equivalent types of materials. Once the temperature is reduced, the final scrubber will often be made of fiberglass, using vinyl-ester resins, polypropylene internals, and thermoplastic pipe.

A fan for such systems is usually an epoxy-coated steel with a stainless steel impeller. Where high levels of chloride are suspected the impeller may be constructed out of Hastelloy C.

Low-Level Radioactive Waste. In the nuclear power industry, waste materials such as rags and clothes are treated as low-level radioactive waste. While containing little or no radioactivity, they must be treated separately from general waste. To reduce the handling volume of this material, a small incinerator can be used that destroys the combustible portion of the material, leaving behind approximately 5% residual ash.

The gases present in a typical incinerator of this type are very similar to those of a hospital waste incinerator, although the chloride content is usually lower. The component materials would generally be the same.

Municipal Solid Waste. The by-products of municipal solid wastes can be similar to those found in chemical incineration. However, the levels of the contaminants—chlorides, for example—are always much lower. Today, the operating temperature for many of these systems is close to 1000 °C (1830 °F) in the incinerator itself. For conservation, there will almost always be a wasteheat recovery system. This is usually a wasteheat boiler to generate steam or hot water. This results in an exit gas temperature that is typically in the 260 °C (500 °F) range. The requirements for burning these wastes, which contain large portions of cellulose, result in lower off-gas temperatures than those for chemical incineration.

Nevertheless, corrosion problems are severe, and material choices are similar to that of industrial chemicals incineration. Reference 5 provides a ranking of metals based on corrosion tests in this service. Reference 6 shows the results of corrosion tests for a very wide range of alloys in six distinct incinerator system zones.

Sewage Sludge. Burning of sewage sludge presents the least corrosive discharge of the four types under discussion. This can be attributed to limited halogen compounds in the gas and lower temperatures (typically 315 to 650 °C, or 600 to 1200 °F) compared to industrial and hospital waste incinerators.

Types 304 and 316 stainless steels are suitable for construction in most areas of the system, including the quenching area, whether as a separate quench or part of the wet scrubber. Again, FRP, thermoplastics, and lined carbon steel can be used in the cooler regions. Typical materials are epoxy-coated carbon steel for the sump and lower portion of the scrubbers, with type 316 stainless steel used for portions where critical dimensions must be maintained, such as impingement trays.

The predominant contaminants in the environment are sulfur compounds, which may be present either as reduced sulfide compounds (H_2S plus mercaptans) or sulfur dioxide, depending on the operating temperatures within the sludge incinerator system. Usually, compounds are converted to sulfur dioxide prior to the scrubber. Chlorides can exist, but they normally originate from the water used for makeup. Their presence sometimes requires the use of high-nickel alloys for such components as fan wheels and pump impellers.

Erosion can be a significant problem in any of these systems. It can wear down critical moving mechanical components and equipment walls at points of liquid or gas impingement. It can contribute to corrosion. See the article "Gaseous Corrosion-Wear Interactions" in *ASM Handbook,* Volume 13A, 2003.

The erosion effects are not as severe as in FGD treatment equipment, but there are a number of areas of concern. The venturi throat and spray nozzles can suffer abrasion. Fan wheels and pump impellers, however, are usually the most critical areas in these systems. High-nickel alloys are used. Rubber lining can also be used, although generally not on fan wheels. Fiberglass-reinforced plastic can also be fabricated with silicon carbide impregnation for increased abrasion resistance.

Bulk Solids Processing

Bulk solids processing includes grain handling, foundries, coal handling, pneumatic conveying systems, and spray-drying systems, where there is the need to collect dust for air pollution control. In every case, fine particles of dust can become entrained in the exhaust air and must be removed prior to discharge of the air. The three most common types of dust collection equipment are fabric filters, electrostatic precipitators, and wet scrubbers.

Fabric Filters. The selection of materials of construction varies with the industry, but the dust handled is generally not severely corrosive. Carbon steel is the most common material in the vessel itself. The fabric used in the collector varies. The manufacturers of fabric filters have experience with selection for a specific application, so their expertise should be used in evaluating the initial and maintenance costs of alternate fabrics. Typically, most bulk-handling applications can be managed with the use of polypropylene for the filter bags. Currently, its cost is close to that of cotton filter bags, it does not rot when wet, and it offers relatively good corrosion resistance. The primary limitation is temperature.

Fans and stacks located downstream from a fabric filter are normally constructed of carbon steel. Because most of the dust has already been collected before it reaches this point, abrasion is not a major concern in the design of the downstream components.

Electrostatic precipitators are applied to a bulk solids application where high performance is required. This normally occurs in relatively dry services with inert particles. As such, the materials of construction are typically carbon steel.

Wet scrubbers are often used where the solids being handled are more reactive. For example, if there is a potential for an explosive mixture of the dust with air, the wet scrubber eliminates this problem. Wet scrubbers are also versatile and can simultaneously remove dust and gas.

The initial cost of scrubbers is partially determined by the need to recirculate a water-based solution. Care must be taken to ensure that this solution does not become corrosive or, if it does, to select the proper materials.

General nuisance dust collected by a scrubber does not cause a direct corrosion problem. Instead, the problems arise because of the need to minimize the wastewater from the scrubber. For example, if the inlet air is at ambient conditions, the scrubber will evaporate 122 L/h (31.8 gal/h) for every 10,000 m^3/h (5889 ft^3/min) of air flow. If the quantity of solids captured requires only 10% of the evaporation rate as a liquid bleed rate, then the dissolved solids in the water will be concentrated by a factor of 10. Thus, 200 ppm of chloride would suddenly become 2000 ppm of chloride. This is sufficient to cause corrosion concerns.

In most cases, fiberglass is used where abrasion is not particularly severe. In other cases, carbon steel is used, particularly for coal handling and other applications where abrasion is present.

Spray Dryers. Spray-drying applications typically require stainless steel. Either type 304 or 316 is used, depending on the particular compound being collected. The use of stainless steel arises from the need for product purity. Because the slurry is usually returned to the spray dryer, care must be taken to avoid potential corrosion problems.

Chemical and Pharmaceutical Industries

Chemical process and pharmaceutical industries experience a wide variety of corrosion problems. Many of the compounds used have severe effects on materials. For air pollution control equipment, the concentration of the compounds in the air (or liquid) is much lower than in the normal plant process flow. One of the differences of process operations compared to air pollution control is the operating pressures. Many chemical reactions are carried out under high pressures (several atmospheres) in reactors to improve economics or yield. However, in ventilation systems, the gas will be at atmospheric pressure, so more economical materials that do not require the mechanical strength to handle high pressure can be selected.

Many of the clean-up systems used in chemical and process plants are fabricated from fiberglass, because FRP has low initial cost and good corrosion resistance in a wide variety of services. The corrosion resistance of FRP is a function of both the resin content and the specific resin used in the laminate.

Chloro-Alkali Plants. The production of chlorine results in severe corrosion problems. Quantities of chlorine in the effluent gas are normally scrubbed using dilute caustic solutions. Specific types of resin must be employed for the FRP used in this very difficult application. Vinyl-ester resins are most commonly used to handle chlorine and chlorides. In addition to using a high-performance resin, the inner glass reinforcement is usually replaced with a synthetic veil to provide additional protection and to avoid any attack by hypochlorite or caustic on the inner liner. The fiberglass is used for the scrubbers handling chlorine removal, ductwork, fans, and stacks; even recycle pumps are manufactured of this material.

Where heat exchangers are used in the recycled solution, fiberglass is obviously not a practical material because of its low heat-transfer coefficient. For dilute hypochlorite solutions, such alloys as Hastelloy C276 or Inconel 625 have been used. Graphite can be used if the proper binder is selected. With the high heat-transfer coefficients of plate heat exchangers, constructions of Hastelloy C276 can be economical.

Polyvinyl chloride (PVC) or high-temperature PVC (CPVC) is another material that performs well in this service. These materials are sometimes selected for small units or for ductwork construction.

Nitric Acid Plants. In nitric acid (HNO_3) manufacture, stainless steel is the most common material of construction. Concentrated HNO_3 will affect many of the nonmetallic materials. Fiberglass-reinforced plastic is not as common or as easily accepted.

Type 304L stainless steel or CPVC can be used for many ventilation systems. Fiberglass-reinforced plastic could also be used when the acid is being neutralized. Relative costs are shown in Fig. 4. Because the cost of stainless steel is still relatively high compared to that of FRP, FRP should be considered where very dilute acid concentrations are involved. Type 304L stainless steel remains the best where high concentration and return to the process are involved.

In many cases, HNO_3 manufacture also produces oxides of nitrogen. Nitric oxide (NO) and nitrogen dioxide (NO_2) would be handled by the same materials of construction. Most commonly, NO and NO_2 are removed from the air by using either scrubber systems or selective catalytic reduction (SCR).

With scrubbers, the recycled solution becomes a dilute HNO_3 solution. A special wet-phase catalyst has been developed for use in this service. Type 304 stainless steel is used for portions of the vessel, although fiberglass can be used if the concentrations are very low.

In other HNO_3 facilities, SCR is used to eliminate the residual oxides from the air. The catalyst itself is normally a ceramic honeycomb structure with semiprecious metal oxides applied to the surface. These materials are selected by the manufacturers and would be compatible with stainless steel components, which are typically type 304L stainless steel, that are selected for ductwork, fans, and other auxiliaries.

Sulfuric Acid Plant. The most common problem area in emissions control equipment in a sulfuric acid plant is where acid mist is collected by fiber bed mist eliminators. Such units employ a glass mat held inside of a vessel operating at low velocities to remove submicron mists. Fiber bed unit shells for H_2SO_4 are either type 316 stainless steel or alloy 20Cb3. The relative economics suggest the use of type 316 stainless steel, although it may suffer a small amount of attack.

Alloy 20Cb3 can be used for most ranges of acid that would be encountered in air pollution control systems. Type 316 stainless steel can be used if the solution is weak enough. Also, if

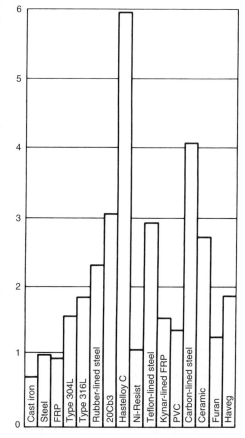

Fig. 4 Relative costs of scrubber materials. FRP, fiberglass-reinforced plastic; PVC, polyvinyl chloride

the temperatures are low enough, FRP should be considered. It is best to obtain coupons of the materials and to conduct some initial testing at dilute conditions before making a final decision. The industry practice is to use more and more FRP on these inorganic acid applications because of the low initial cost.

Sulfur Dioxide Service. Sulfur dioxide has requirements similar to sulfuric acid, even though the initial solution formed is usually a neutralized salt; that is, SO_2 is normally absorbed using an alkali solution such as lime or caustic. The solution formed is a sodium or calcium sulfite/sulfate mixture. It can be handled at low temperatures in fiberglass and at higher temperatures in type 316L stainless steel.

Because most of the SO_2 gas is removed in the air pollution control equipment, the downstream equipment can often be handled using liners rather than expensive alloys. In fans, an epoxy coating can be applied to the housing. For the wheel itself, a solid alloy is recommended because of the high speed involved. The combination of an epoxy liner and a stainless steel wheel can cost 25% less than a totally stainless steel fan.

Discharge stacks are often treated in the same manner. Where the stacks are large enough, a coating can be applied to a steel stack. Of course, the first choice may be a FRP stack if the temperature is low enough, because maintenance is eliminated. More information on materials of construction for the chemical processing industry is available in the articles on specific chemicals in this Volume.

The Fertilizer Industry. Several severe problems can occur in the manufacture of fertilizers. Trace quantities of fluorides in phosphates result in the formation of HF and tetrafluoride in the gas. Although these can be scrubbed out, the resulting solution is extremely corrosive to most metals and fiberglass.

The most common solution is to use FRP with synthetic veils rather than a glass lining. The FRP resin itself is not affected by the HF, but the internal glass could be. A small pinhole leading to the glass can result in catastrophic attack. This is avoided by substituting a synthetic veil for the glass.

Fiberglass is used throughout the industry as a standard. Polypropylene, PVC, and similar thermoplastics are also used. Nitrates and urea products are typically handled by using type 304 stainless steel. Concentrated HNO_3 can attack FRP-type materials.

Lime Kiln and Similar Kiln Operations. Lime kilns are found in several applications, including the pulp and paper industry. Lime and other kilns produce hot gas that contains dust, most often requiring a pollution control system. When possible, a dry collection system is used, because it allows the material to be collected in a form that can be returned directly to the kiln. Some products are too reactive for this technique, or the temperature of the kiln is too high, so wet scrubbers are then used.

Most of the scrubbers on kilns are made of carbon steel. The problem of corrosion resistance is usually minimal, because the solutions tend to be alkaline or at least neutral. The primary problem is usually abrasion resistance. The basic collectors are manufactured of carbon steel, and high-wear areas are often made of stainless steel. Some units use heat treated stainless steel, which is hardened for the wear-resistant areas. Another technique is to install a liner to protect areas of greater wear.

Pulp and Paper Industry. Most of the pollution control problems in the pulp and paper industry consist of either the organic sulfur compounds produced from digesting the pulp or the chlorine-related oxidizing agents produced from bleaching the pulp. The reduced sulfur compounds are generally handled in FRP construction. Temperature limitations are not normally a factor, because most of these applications are at temperatures of 80 °C (175 °F) or less. Chlorine or chlorine dioxide applications are handled by materials discussed previously in this article. More information on materials of construction for the pulp and paper industry is available in articles about corrosion in the pulp and paper industry in this Volume.

Groundwater Treatment—Emission Control. Groundwater sources can become contaminated through tank leakage or buried waste. For example, chlorinated hydrocarbons traditionally used to remove grease and oil from aircraft engines for maintenance have been found in nearby wells. Levels exceeding the Environmental Protection Agency standard of between 1 and 5 ppb for most chlorinated organic compounds must be addressed. To remove these, groundwater is usually passed through either activated carbon or an aeration column called an air stripper.

The volatile organic compound easily evaporates into the air in the air stripper. If it is present in a very small quantity, regulations allow it to be discharged directly into the atmosphere, where ultraviolet light will break down the compound within 24 h. When it is present in a significant quantity, a second-stage treatment system, such as a vapor phase carbon system or thermal/catalytic oxidation, is used for destruction. If the contaminated water were sent directly to a carbon system, the adsorption of the compound may be 1 to 2 wt%. From air, 25 wt% is collected, so the air stripper increases efficiency.

While air strippers are exposed to chlorinated compounds and other reactive chemical species, the concentration of these components is extremely low. Stainless steel, polypropylene, or fiberglass have been found to be successful at these low levels. For small liquid flows, stainless steel is a common material of construction. Fiberglass is the most common structural material for the housing for larger liquid flows. For either the air-side vapor phase carbon or the liquid-side carbon, the most common material of construction is steel with an epoxy or similar finish. The combination of corrosion-resistant lining and steel outer housing gives a good mechanical integrity at low initial cost.

Industrial wastewater treatment equipment, such as that treating the discharge from a scrubber, may require special materials, but the majority of groundwater treatment or industrial water treatment for emission control can be handled using epoxy-lined steel construction. Examples include settlers, flocculating tanks, retention tanks, and filtration equipment.

Metal fabrication plants may generate a combination of contaminants, because metal fabrication involves a very wide range of processes, products, and materials. It embraces any metal work that concerns itself with metal removal, forming, joining, and finishing, such as painting.

The treatment of bulk solids and the use of dust collectors were previously addressed.

Painting operations include spraying, dipping, and electrostatic finishing, among others. The coatings applied in these operations vary widely as to composition and physical properties. Where paint contains solvents, the volatile organic compounds released during the paint-drying process must be captured and treated. This usually requires the use of a thermal destruction system, such as a thermal, catalytic, or regenerative oxidizer (Fig. 5). For larger air flows, a regenerative thermal oxidizer is used. In this technology, the air is heated through a ceramic bed close to the operating temperature, at which point the gas is further heated with a direct flame to destroy the organic compounds. The exhaust gas then exits through a ceramic bed that captures its heat. The beds are alternated so that as the inlet bed cools, the gas switches direction and passes through the preheated former exhaust bed. Using this technology, a 90 to 95% thermal efficiency can be achieved.

Fig. 5 A catalytic oxidizer of stainless steel construction. Courtesy of Branch Environmental Corporation

The heat-transfer medium is virtually always a ceramic bed of either structured packing or random packing. The housing itself is typically carbon steel. It would be unusual to encounter any chlorides or sulfur compounds in painting facilities. The solvents are usually all organic, and the combustion products are typically carbon dioxide and water.

The face of the support structure for the ceramic is often made of heat-resistant stainless steel. As long as the housing is maintained at or above the temperature of the gas dewpoint, corrosion should not be a problem.

Wastewater—Odor Control. In handling municipal wastewater, the collection, separation, and biological treatment of solids can result in the formation of hydrogen sulfide (H_2S) or other organic sulfide compounds, including methyl mercaptan, methyl sulfide, and dimethyl disulfide. Especially during sludge dewatering, the released odors are usually captured in an air pollution control system consisting of one or more stages of counterflow packed tower absorbers. These absorption columns typically contain alkali solution and a bleach such as sodium hypochlorite to react with the compounds and effectively destroy them, with efficiencies over 99%.

While the air containing sulfide compounds will be somewhat corrosive, the scrubber solution containing sodium hypochlorite (bleach) is extremely aggressive and will attack most metal compounds. Fiberglass or other thermoplastics have found extensive use in wastewater plants for this reason (Fig. 6). The scrubber housing, recycle pump, and related components are typically made of fiberglass using vinyl-ester resins with synthetic veils. Where critical components must be made of metal and exposed to the solution, Hastelloy C has been used. Because there is always the chance that sulfides are present externally, items such as hardware are usually made of type 316 stainless steel.

Textile Mills. Finely divided lint is present in quantity in the air in the vicinity of various operations associated with handling yarn and cloth. Most modern mills have air tunnels to capture this lint and carry it to a central collection point. A filter constructed of galvanized steel with flat flexible stainless steel screens to support the raw filter media is used.

In finishing operations, especially in textile drying, the knitting oils may be driven out of the fabric during heating. These can result in a smokelike emission that is either captured with a cooler/fiber bed mist eliminator or destroyed through thermal oxidation. While there are no special chemicals of concern involved, the thermal oxidizer uses type 304 stainless steel to eliminate corrosion problems from the high temperature. The cooler/filter approach uses stainless steel for the first-stage scrubber and an epoxy coating for the remainder of the system to avoid corrosion from the condensed water.

During manufacture of synthetic fibers, the process can generate a number of chemicals that must be controlled, using the same techniques employed by the chemical process industry.

One of the largest concerns with textile processing is the wastewater treatment. The recovery and reuse requires pH adjustment, suspended solids removal, and removal of color. These steps typically do not involve extremely corrosive chemicals and can be handled using lined or coated steel construction.

The food industry uses large quantities of water. The canning industry uses 190 L (50 gal) of water per case, for instance. Every effort is made to clean the water, either for reuse or to meet water-release regulations.

Water can contain suspended or dissolved solids. The typical control steps of screening, flocculation, or filtering as described for wastewater treatment are used, and the same materials apply.

Odor control for air handling is also of concern. While the odor from coffee would not be considered offensive in small quantities, a coffee-roasting operation typically takes steps to provide mechanical separation of dust from the roasting operation. It may employ a catalytic or thermal oxidizer on the discharge to eliminate further odors. Because the emissions for most food operations are primarily hydrocarbons, the material of construction for the typical catalytic oxidizer is type 304 stainless steel.

ACKNOWLEDGMENT

This article is adapted from the article "Corrosion of Emission-Control Equipment" by W.J. Gilbert and R.J. Chironna in *Corrosion*, Volume 13, *ASM Handbook*, 1987, 1367–1370.

REFERENCES

1. J.R. Crum, E.L. Hibner, and R.W. Ross, Jr., "Corrosion Resistance of High-Nickel Alloys in Simulated SO_2-Scrubber Environments," Huntington Alloys, Inc.
2. F.G. Hodge, High Performance Alloys ... Make Wet Scrubbers Work, *Chem. Eng. Prog.*, Vol 74 (No. 10), 1978, p 84–88
3. R.W. Kirchner, Materials of Construction for Flue-Gas-Desulfurization Systems, *Chem. Eng.*, 19 Sept 1983, p 81–86
4. D.C. Agarwal and F.G. Hodge, "Material Selection Processes and Case Histories Associated with the Hazardous Industrial and Municipal Waste Treatment Industries," Cabot Corporation
5. R.W. Kirchner, Corrosion of Pollution Control Equipment, *Chem. Eng. Prog.*, Vol 71 (No. 3), 1975, p 58–63
6. H.D. Rice, Jr., and R.A. Burford, "Corrosion of Gas-Scrubbing Equipment in Municipal Refuse Incinerators," paper presented at the International Corrosion Forum, National Association of Corrosion Engineers, 19–23 March 1973

SELECTED REFERENCES

- G.L. Crow and H.R. Horsman, Corrosion in Lime/Limestone Slurry Scrubbers for Coal-Fired Boiler Flue Gases, *Mater. Perform.*, July 1981, p 35–45
- T.G. Gleason, How to Avoid Scrubber Corrosion, *Chem. Eng. Prog.*, Vol 71 (No. 3), 1975, p 43–47
- E.C. Hoxie and G.W. Tuffnell, A Summary of INCO Corrosion Tests in Power Plant Flue Gas Scrubbing Processes, *Resolving Corrosion Problems in Air Pollution Control Equipment*, National Association of Corrosion Engineers, 1976, p 65–71
- R.D. Kane, "Online Resources Help Plants Fight Corrosion," *Chem. Process.*, May 2001, p 26–27
- T.S. Lee and R.O. Lewis, Evaluation of Corrosion Behavior of Materials in a Model SO_2 Scrubber System, *Mater. Perform.*, May 1985, p 25–32
- T.S. Lee and B.S. Phull, "Use of a Model Limestone SO_2 Scrubber to Evaluate Slurry Chloride Level Effects on Corrosion Behavior," paper presented at the APCA/IGCI/NACE Symposium on Solving Problems in Air Pollution Control Equipment (Orlando, FL), Dec 1984
- B.S. Phull and T.S. Lee, "The Effect of Fly Ash and Fluoride on Corrosion Behavior in a Model SO_2 Scrubber," paper presented at the International Corrosion Forum, National Association of Corrosion Engineers, 25–29 March 1985
- S.L. Sakol and R.A. Schwartz, Construction Materials for Wet Scrubbers, *Chem. Eng. Prog.*, Vol 70 (No. 8), 1974, p 63–68

Fig. 6 A scrubber tower constructed with fiberglass. Courtesy of Branch Environmental Corporation

Corrosion in Recreational Environments

Lorrie A. Krebs, Anderson Materials Evaluation, Inc.
Michael LaPlante, Colt Defense LLC
Jim Langley, USA Cycling Certified Mechanic
Kenneth S. Kutska, Wheaton (IL) Playground District

THE RECREATION INDUSTRY is large and widespread, covering a huge number of sports, games, and activities that take place in environments as diverse as the activities themselves. Also huge is the amount of money spent in the United States alone just on merchandise related to recreational activities. The 1997 Economic Census in the area of Arts, Entertainment, and Recreation reported that Americans collectively spent more than $3 billion on the rental or purchase of equipment and merchandise related to a selected series of activities that included golfing, skiing, racquet sports, and boating among others (Ref 1). Furthermore, this figure reflects only the receipts and revenues of facilities at which people can engage in their activities of choice (golf courses, ski resorts, etc.). The expenditure is significantly greater when one considers the many commercial vendors that supply sporting and recreational equipment without affiliation to a specific recreational facility. For example, a market analysis statement published by Tennis Master Pro Shops, Inc., in the mid-1990s reported overall sales of tennis-related equipment alone to be $2.3 billion annually. According to the National Golf Foundation, golfers spent $4.7 billion on equipment in 2002. These and similar dollar figures for any number of recreational sporting activities clearly indicate that the market for sporting and recreational equipment is a substantial one.

Materials requirements for the equipment needed to participate in these many activities are widely varied as well. As is the case in many industries, strength, size, and weight are often factors when selecting a particular material for a particular purpose. Materials selection for corrosion addresses issues ranging from demanding atmospheric and immersion environments (e.g., marine environments for recreational boating) to small-scale corrosion related to handheld materials that can react to perspiration (e.g., barrel exteriors of some firearms). Despite the wide range of shapes, sizes, and functions covered in this category, some commonalities in materials and material failures can be identified. Aluminum alloys, for example, are used extensively in recreational equipment for a variety of reasons including strength combined with light weight and pleasing appearance. However, improper anodization can cause corrosion-related failures in a range of equipment. Separate incidences of incorrect aluminum anodization have been identified during failure analyses, including the discoloration of knitting needles by handling in one case and the collapse of a residential swimming pool ladder in another. The consequences of such failures range from merely inconvenient to physically harmful and possibly life threatening.

The move from metals to composites in some sports equipment, such as high-end tennis rackets and golf clubs, was motivated more by performance than by corrosion characteristics, although proper cleaning and maintenance will extend the life of metal handheld sports equipment. Other recreational areas are far more sensitive to corrosion issues. Several such activities are explored in this article from the point of view of corrosion in typical environments and the actions taken to improve corrosion performance. The subset of recreational equipment specifically addressed here includes recreational boats, firearms, bicycles, playground equipment, and climbing gear. Corrosive attack may stem from operation and handling of particular items and/or from environmental exposures. The methods chosen to mitigate corrosion while still addressing other functional issues for the given piece of equipment include base materials selection, surface treatments and coatings, and good practice of maintenance procedures.

Corrosion in Boats

There are few environments more conducive to corrosion than that experienced by and on a marine vessel. Boats are composite structures, consisting of a variety of dissimilar metals and other materials, all exposed or immersed to one degree or another in one of the most efficient electrolytes for corrosion, water, and most especially saltwater. Add to this the electrical and electronic systems of the boat, which can result in stray currents, and the potential for corrosion problems increases even more. Metals on boats experience all forms of corrosion—general, pitting, crevice, atmospheric, galvanic, stray-current, fretting fatigue, erosion-corrosion, cavitation, stress corrosion cracking (SCC), and microbiologically influenced corrosion. Consult the next article "Corrosion in Workboats and Recreational Boats" in this Volume for more detailed information on boating corrosion issues. These include corrosion of:

- Hulls, fittings, and fastenings
- Metal deck gear
- Equipment such as winches, lifeline supports, and anchor systems
- Propulsion systems
- Electrical and electronic systems
- Plumbing systems
- Masts, spars, and rigging

Corrosion of Firearms

Corrosion has been the main enemy of firearms since their invention more than 500 years ago. It occurs both on the outside surface of the firearm as well as on the inside surfaces, particularly in the bore of the barrel.

Early Construction and Propellants

The most common form of corrosion in firearms is rust, simple iron oxides, created when chemical iron dissolves in aerated water. This can be caused by a large number of factors such as use in a humid environment or exposure to the elements (rain, snow, etc.). Some people have very acidic sweat and can cause damage merely by handling their firearms. Since none of these factors can be entirely eliminated, owners have always been required to wipe down the exterior surfaces of their firearms after every use to first remove moisture and then to wipe them down again with an oily rag to provide a moisture-proof barrier. More recently, silicone-impregnated cloths have become available to wipe down firearms after use, offering yet another barrier to moisture.

Far more destructive than external surface rust is corrosion in the barrel bore. For hundreds of years the only propellant available for use in firearms was black powder. The residue left after firing is so hygroscopic that corrosion begins almost immediately after firing. Owners were required to wash the barrels out with hot (and preferably soapy) water as soon as possible to prevent corrosion. Modern shooters using replica black powder firearms are still required to do this. Exhaust that emerges near the hammer at firing can also attack the barrel exterior locally, as shown on the replica black powder rifle in Fig. 1.

A little over 100 years ago, manufacturers began to switch to so-called smokeless powders that exhibit this characteristic to a lesser degree. The problem then became one of corrosion caused by chlorates in the priming compounds. After firing they turn into chlorides, which attract moisture. While washing with hot and soapy water would also work with these compounds, chemical cleaning solutions were developed to neutralize these chloride compounds. Even so, cleaning was required as soon as possible after use. Eventually, primers were developed that do not contain these chlorates. Today, in virtually all cases, commercial sporting ammunition is loaded with noncorrosive powders and primers.

Nearly all firearms were originally manufactured from iron, the only available material that was strong enough to contain the pressure of the burning propellant in the barrel. Iron was also used to provide the various other components such as hammers, triggers, and sears. Springs were usually made from steel. In the last half of the 19th century, alloy steel became the preferred material for all components, offering additional strength and allowing reduced weight. Obviously, corrosion was still a problem.

Changes in Construction Materials

Aluminum. In the last 50 years aluminum has been used for components where material strength was less critical and reduced weight offered other advantages. "Rust" is not a problem with aluminum components, although hardcoat anodizing does provide some additional protection from oxidation.

Stainless steel first made a significant appearance in firearms in 1960. The main advantage of stainless, of course, was to reduce corrosion. Even though the materials chosen were usually 400 series stainless or various precipitation-hardening materials, they still provide much more protection against corrosion than alloy steel.

Polymers. In 1980 polymers made their appearance, in much the same way that aluminum had earlier. Polymers are often used for such components as handgun receivers where they serve as a base on which to assemble the more highly stressed components. Obviously, polymers do not corrode and have the added advantage that they can be molded in whatever color the manufacturer desires.

Surface Coatings and Treatments

Plating and Polymer Coatings. Firearms manufacturers realized from the beginning that additional corrosion resistance was required on both interior and exterior surfaces. At first they turned to treatments that were already available. These included plating the exterior surfaces with silver, nickel, and even gold. More recently, manufacturers have used chrome plating, both hard and soft, on the interior and exterior. This type of coating is especially useful in preventing corrosion in barrel bores. Currently, manufacturers are experimenting with advanced plasma vapor deposited coatings such as titanium carbonitride and polyvalent diamond compounds to reduce wear and improve corrosion resistance on both the interior and exterior. In addition, there are now a number of commercially available polymer coatings that users can apply to the exterior of their firearms by such means as painting or spraying. Many of these contain molybdenum disulfide, which acts as a dry-film lubricant in addition to providing corrosion resistance.

Oxidation Treatments. Many early firearms were finished "bright," that is, polished and untreated. This obviously left tremendous opportunity for corrosion and was one of the reasons so much time was spent on keeping firearms clean. Eventually, browning and then bluing processes were developed to protect the interior and exterior surfaces. The processes are similar; both artificially rust the iron surface and leave a pleasing finish. In the first case a thin layer of ferrous hydroxide ($Fe(OH)_2$) is formed on the surface, which gradually converts to brown ferric oxide (Fe_2O_3). The bluing process (originally using steam or boiling water) converted the brown ferric oxide to the much darker ferro-ferric oxide (Fe_3O_4).

Phosphate Treatments. After World War I, additional treatments were developed to provide greater rust protection than browning or bluing. Many of these involved treating the exterior surface with a solution of phosphoric acid and other ingredients, often including manganese phosphate. One of the most popular of these in the United States was the patented Parkerizing process, which utilized zinc phosphate.

Treatments for Stainless Steels. With the use of stainless steels, manufacturers followed two schools of thought. Some manufacturers relied solely on the corrosion resistance of the stainless steel and provided no additional finishing after polishing. Others took the additional step to passivate the components after polishing to remove the free iron particles on the surface and further reduce the possibility of corrosion. The jury is still out as to which approach is superior.

Case Hardening. Some early components such as hammers and triggers were case hardened. This not only increased their hardness and wear resistance, but also provided some corrosion resistance. Only recently have firearms manufacturers gone back to this seemingly simple form of surface treatment, using the Mellonite and Tenifer molten salt processes.

Fig. 1 The red dust shown from two views on the dark barrel exterior of this black powder replica rifle was the result of residue in the exhaust and implies that far worse corrosion is probably present on the interior surface. Courtesy of Daniel Sullivan

Bicycles and Corrosion

In the 138 years that bicycles have existed, the vast majority have been built of alloy steel. While more and more models designed for the enthusiast market of today (2006) feature nearly corrosion-free materials such as aluminum, carbon fiber, composites, and titanium, steel remains the most-used material worldwide because it is affordable, easy to work with, durable and because it makes "nice-riding" bicycles. Consequently, this section focuses primarily on corrosion issues with steel bicycle frames and components because these parts are the most at risk for corrosion and also because there is more historical information regarding their performance. For example, while aluminum was used for bicycle frames as early as 1902, it did not become popular until Cannondale introduced its first bicycles in 1984. Carbon fiber, titanium, and composite bicycles (blends of carbon and aluminum or titanium) have even less history to draw on (Fig. 2). Yet, what is known so far is that they are so much more resistant to corrosion than steel that serious degradation problems are basically unknown.

Unless fastidiously maintained and carefully kept, most steel bicycles succumb to corrosion in one way or another, sooner or later. In time, corrosive attacks occur even on painted and plated steel and can ruin everything from the frame and fork to the drivetrain, wheels, and bearings if ignored. Selected examples are discussed below.

Frames and Forks

Frame and Fork Corrosion. Bicycle frames comprise welded tubular steel of various qualities from high-tensile strength steels to chromium-molybdenum alloys. Weight is always an issue when designing and building bicycle frames. As a result frame tubing is usually as thin as possible, taking into consideration the performance goals of the machine and associated cost. In some cases, the tube wall thickness can be as little as 0.4 mm (0.016 in.). Obviously it does not take much corrosion to weaken tubes this thin.

Manufacturers always protect the steel frame and steel fork exterior surfaces with paint and/or plating, which is adequate for preventing corrosion in most cases. Problems arise when the paint or plating gets chipped, scratched, or worn away, exposing bare metal to a potentially corrosive environment. Corrosion beneath the paint (undercutting) is also possible if the frame was not correctly prepared before painting. If left unchecked, the rust that forms on an exposed surface or beneath the paint will eventually lead to loss of wall thickness and perforation of the tubes, potentially ruining the frame.

Also of concern are the interior surfaces of the tubular frames and forks, which are not painted and are susceptible to rusting once moisture enters the frame. Moisture inside the frame and fork can build up over time due to condensation. Moisture ingress can also occur by seepage past the seatpost from water spray off the back wheel on rainy rides. While most quality frames include venting holes that help some of this moisture evaporate, rust can form any time moisture finds its way into the frame. If left unchecked, as is usually the case, it will eventually weaken the frame by loss of wall thickness and/or by perforation.

The interior surfaces of the tubular steel used for frames and forks can be protected by plating. However, improper plating can cause rapid corrosion of the inside of the frame. This is rarely a concern with production bicycles built by major manufacturers who understand the challenges of proper plating and take precautions to protect the tubing. However, less-established and unprofessional builders occasionally cut corners in the plating process. This also happens when owners, who are repainting their bikes, have them plated by companies that are not experienced with thin-wall tubing. Poor plating practices leave chemicals inside the frame tubing that can actually accelerate the corrosion process. Because this attack takes place hidden from view inside the tubes, a frame can often rust quickly and fail in just a matter of years.

Preventative Measures for Frames and Forks. It is possible to prevent most rusting problems simply by limiting exposure. Avoiding inclement weather whenever possible, cleaning and lubricating the machine immediately after wet rides, and storing the bicycle indoors are all excellent ways to limit moisture ingress to the interior as well as exposure of the exterior surfaces. Another measure that will protect frames and ensure long life is to apply wax. Wax coatings add a protective layer on top of the paint and help to seal any small chips and scratches. Wax coatings are also soft and wear easily, so they should be reapplied roughly every 6 months. It is always best to immediately touch up any chips in the paint and to sand any rust spots to bare metal, refinishing the area appropriately to protect the steel as soon as any problems appear.

Another excellent practice is to protect the inside surfaces of the frame with a rust preventative. Silicone sprays, such as WD-40, can be sprayed through the vent holes and down the tubes. Then by rotating the frame, the liquid can be spread to coat and protect the tubing. An alternative is linseed oil, which dries to form a harder protective coating. There are additional products available through bicycle retailers that are easy to use and very effective.

Corrosion caused by shoddy plating is the result of chemical residue left in the frame during the plating process. This residue accelerates the corrosion process, especially with the ingress of moisture. This is something the plater needs to understand and correct. To prevent this problem it is necessary to bake the frame in an oven at a low temperature (175 °C, or 350 °F) to eliminate the chemicals and protect the tubing. Typically, only experienced professionals perform this step.

Other Components

Component Corrosion. Even on inexpensive bicycles many corrosion-free anodized-aluminum parts are used. However, many of these components include steel subparts, such as axles, nuts, and bolts. There are also many important components, such as chains, cassette cogs, chainrings, spokes, pedals, and derailleurs in which steel is commonly used due to strength requirements. If these steel components are not lubricated regularly, corrosion will quickly cause problems, such as frozen chain links that prevent smooth pedaling and shifting, binding brakes

Fig. 2 This Cannondale composite bicycle has an aluminum frame and a carbon fiber fork, resulting in a more corrosion-resistant bicycle than the more common (worldwide) steel frames. However, important components such as chains, axles, and bearings are still commonly made of steel for strength requirements and must be maintained to prevent corrosion from forming and interfering with performance. Courtesy of Greg Welsh

that will not release or grip, shifting and braking glitches from rusted cables, and bearing damage in the wheels, drivetrain, and steering systems.

These types of corrosion-related failures are all common results of water entering steel components. For instance: water follows exposed cables into the housing, cannot escape, and corrodes the cables, binding them and possibly freezing them inside the housing. Water seeps past seals and gets inside the wheel, steering, and crank bearings where it rusts the balls and races, causing them to grind to dust over time. Water can also enter the chain links freezing the sideplates and rollers, making smooth pedaling and shifting impossible. Corrosion can also cause problems with wheels where it can freeze the nipples to the spokes making it impossible to straighten the wheel should it go out of round or true. If the wheels develop wobbles that cannot be straightened, the braking system is also compromised.

Rubber, nylon, and plastic materials on bicycles, such as tires, brake pads, and seats degrade, too. This is likely if the bicycle is stored outside and exposed to rain, air pollution, and the rays of the sun. It is also caused by riding hard or frequently in bad weather. However, even when carefully cared for, tire sidewalls rot due to the effects of the sun and air, rubber brake pads harden and lose their grip, and synthetic seats may turn brittle and crack.

Another incidence of corrosion that plagues bikes ridden regularly in all weather conditions is certain components freezing in the frame, such as the seatpost and handlebar stem. Sometimes, the materials themselves cause this to occur. For example, the typical setup of a steel frame with an aluminum seatpost has the potential over time of creating electrolysis between the two parts, causing them to almost become one piece and making them immovable (except by a knowledgeable professional mechanic and even then not easily). Even when the materials are the same (steel frame/steel seatpost), they can freeze if corrosion has occurred in the crevice between them.

Preventative Measures for Components. Fortunately, most component corrosion (including frozen components) can be prevented with regular lubrication and maintenance. Components that cannot be protected this way (e.g., tires, brake pads, and seats) can be affordably replaced as needed. By following the industry recommendation of getting a tune-up every six months and a bearing overhaul yearly, cyclists typically will not experience significant component problems due to corrosion. This practice should be coupled with basic home maintenance such as routinely lubricating the chain, derailleur, and brakes to keep a protective coating on most of the corrosion-susceptible parts. It is also possible on many modern bikes to lubricate the cables without tools and without changing any adjustments. If done every six months, the shifting and braking will work optimally and the cables will last almost indefinitely. Another modern trend that is helping to prevent corrosion and related problems is the use of nylon bushings where metal washers were once used, such as in the brakes and derailleurs. These offer a type of nearly maintenance-free permanent lubrication and reduce corrosion-related problems significantly.

Similarly, to prevent corrosion in the wheels, manufacturers are beginning to use stainless steel spokes. If the spoke threads were lubricated during wheel assembly it would prevent the frozen nipples that cause wheel-truing problems. Unfortunately, this is not the common practice unless the cyclists know enough to lubricate the threads occasionally on their own.

Bearing components rely on grease for corrosion prevention. The hubs, crankset, and headset are more challenging to maintain because they are inaccessible without special tools and know-how and because they are not readily apparent and as a result are easily forgotten. As with spoke thread problems, many bearing problems would be solved if manufacturers would, during assembly, simply install more grease in their components. All new bicycles sold by independent bicycle dealers must be assembled before sale, and mechanics often discover underlubricated bearings at that time. Adding lubricant ensures longer life for the bicycle. However, many more bicycles are sold through chain and department stores where such attention and care are rare. Consequently, for bicycles that are ridden regularly or hard or are stored outside, there is usually insufficient grease inside the all-important bearing systems to protect the components for long. This often leads to corrosion, failure, and an expensive repair. For example, if the rear hub bearings and races are damaged by rust, the entire wheel will need to be replaced, an example of a significant expense that could be avoided by proper lubrication at the outset. Generally, all components of a bicycle can be kept free from corrosion and running smoothly with proper lubrication and good general maintenance.

Public Playground Equipment

Metal corrosion is one of many safety concerns with no simple answer regarding public playground equipment. Public leisure industry managers of today (2006) continue to promote and defend their capital replacement, repair, and maintenance budgets, but many times these funding requests are unsuccessful, leaving repair and replacement programs unfunded or filed away in a place called "deferred maintenance." How long can these issues be deferred before they become a real safety issue? There is no definitive answer to this question. Observations throughout the world have shown that there are many public agencies that fail to properly maintain their property. This inaction may be due to a lack of funding, but most often it is due to the lack of knowledge and experience necessary to address the situation before it becomes a serious problem.

The playground equipment industry has grown rapidly over the past 20 years. Once comprising postwar manufacturing companies building one-of-a-kind metal climbers, swings, slides, and other outdoor gymnasia equipment, the industry has evolved dramatically with the development of more sophisticated manufacturing techniques and a wide range of materials. There seems to be no end in sight with regard to the purchasers' ability to design their own elaborate composite play structures utilizing custom computer drafting programs that can, with a touch of the keyboard, add or delete a multitude of play components. Each component may be made in a variety of materials and in almost any color, giving the designer the ability to meet their specific aesthetic needs.

The evolution of playground equipment manufacturing has resulted in the acquisition of many well-known names within the industry by large national and international conglomerates with sales in the billions of dollars. This industry was left unchecked until 1981 when the Consumer Product Safety Commission (CPSC) issued its first federal guideline as a two-part handbook for public playground safety. In 1991, CPSC revised its handbook after conducting an injury analysis using the National Electronic Injury Surveillance System (NEISS) to determine who was being injured on public playgrounds, as well as the type and cause of the injury. In 1990, according to the CPSC, more than 150,000 victims were treated in U.S. hospital emergency rooms for injuries associated with public playground equipment. In 2000, that number had increased to more than 200,000. Some of these injuries, and even deaths, occurred as a result of component and fastener failures. Such failures are often associated with corrosion of a structural member and/or of the fasteners used to support and connect structural or load-bearing components. The ever-increasing number of playground injuries raises serious concerns. While there is no current federal law mandating a standard of care for public playgrounds, there is a voluntary industry consensus-based standard for public playgrounds developed and revised over the past 15 years through the American Society for Testing and Materials (ASTM). ASTM and CPSC along with many other national organizations are continuing to look for answers that will minimize permanently debilitating injuries sustained by children on public playgrounds.

More than 30 years ago, playgrounds were thought to last indefinitely. At that time, manufacturers lacked detailed installation instructions and maintenance instructions were almost nonexistent. Just add a coat of paint and everything would be as good as new. Most equipment was fabricated from steel pipe, and longevity was never considered an issue within the marketplace. Durability of products used in public playgrounds has, until recently, been an unanswered question since no equipment life expectancy was specified at the time of purchase. The marketplace of today (2006) is more sophisticated and still evolving. New designs are made possible by new materials not conventionally used in manufacturing public playground equipment. These rapidly changing innovations in play equipment design, while allowing the creation of ever more imaginative play spaces, make it impossible to forecast a replacement date before a component or fastener fails.

There appears to be no end in sight for the future designs of play equipment with the recent introduction of many other metals, composites, and plastics. With these new materials come problems related to compatibility of adjacent components and fasteners. ASTM F 1487, "Standard Consumer Safety Performance Specification for Playground Equipment for Public Use," section 4, states that playground equipment shall be manufactured and constructed only of materials that have demonstrated durability in the playground or similar outdoor setting (Ref 2). Metals subject to structural degradation such as rust shall be painted, galvanized, or otherwise treated. Woods shall be naturally rot- and insect-resistant or treated to avoid such deterioration. Plastics and other materials that experience ultraviolet (UV) degradation shall be protected against UV light. Regardless of the material or the treatment process, the user of the playground equipment cannot ingest, inhale, or absorb any potentially hazardous amounts of substances through body surfaces as a result of contact with the equipment. Couple these requirements with the need to provide nontoxic environments suitable for people, especially children who are more chemically sensitive, one can see the necessity for a complete, coordinated planning effort throughout all supply lines within the playground industry.

According to the National Recreation and Park Association's (NRPA) National Playground Safety Institute (NPSI) certification program for public playground safety inspectors, one of every three playground injuries is alleged to be caused by improper, or a complete lack of, facility maintenance. As a result of the heightened public awareness of these issues, many agencies have begun to do their part to help reduce preventable injuries by conducting routine inspections, maintenance, and repair of their playgrounds. A thorough facility inspection coupled with timely corrective action by the inspector once a defective or degraded component is identified is the only sure way to eliminate a safety problem before it occurs.

Types of Corrosion Encountered in Public Playground Equipment

Adverse conditions found on playgrounds include prolonged contact with water, contact with excessively high or low soil pH, acid rain, high humidity, salt air, various organisms commonly found in dirt, and human and animal excretions. In addition to water with dissolved ionic components, excretions can corrode metal or can produce waste products that can react with and corrode metal. Therefore, regular cleaning of play equipment not only keeps it attractive and sanitized but also deters corrosion.

Some metals naturally corrode more slowly than others, thereby extending the life and play value of the equipment. The only guidance given through the ASTM F 1487 standard is the reference to ASTM F 1077, "Standard Guide for Selection of Committee F-16 Fastener Specifications" (Ref 3). In addition to ASTM F 1487, there are several other ASTM standards that specify the previously mentioned rather vague compliance standards for materials used throughout the playground equipment industry. These include ASTM F 1148, a standard for home play equipment (Ref 4), ASTM F 1918, a standard for soft-contained play systems (Ref 5), and a soon to be published ASTM standard for public-use play equipment for children 6 to 24 months. These standards require demonstrated durability and state that materials must be corrosion resistant. They do not, however, specify how structural metals are to be protected with barrier coatings that keep the moisture and air away from the metal. Protective, corrosion-resistant coatings typically used on playground equipment components include metal plating, galvanizing, powder coating, liquid vinyl coating, and paints.

The most common form of metal deterioration on the typical public playground is the corrosion of steel, generally referred to as "rust." Any structural metal will corrode over time, however, and without mitigation the structure weakens, deteriorates, and eventually disintegrates as shown in Fig. 3. Corrosion attack can vary, depending on the geographic region. Knowledge of the common everyday environmental factors within regions (e.g., temperature, humidity, rainfall, etc.) can help a planner to make an informed purchasing decision when specifying the type of materials and hardware to be used. Unfortunately, most public officials responsible for the procurement of public play equipment are not well versed in the cause and effect of metal corrosion, the impacts of various environmental factors, or the potential issues associated with adjacent dissimilar materials.

When a component shows signs of failure, it should be replaced with the same component as provided by the original manufacturer unless otherwise authorized by the manufacturer. This replacement needs to take place before the damage becomes serious. Oftentimes, however, when a problem is identified the solution is to run to the local hardware store, lumber yard, or maintenance facility and grab the first thing that appears to fit the need, unaware that this action may cause a much bigger problem. The average person knows little about metal corrosion other than what is referred to as common surface rust resulting from the loss of the protective coating and exposure of the steel to moisture. When these protective coatings are worn away or scratched off during normal use, the unprotected metal underneath is exposed and surface rust begins to form. In its initial stage it is relatively harmless. However, it is good preventive maintenance to routinely inspect all coated metal components to find and repair exposed metal where surface rust is starting before it becomes a larger problem. Removing loose rust with steel wool or a wire brush and applying a rust-neutralizing solution evenly over the affected area followed by an application of the appropriate primer and topcoat recommended by the equipment manufacturer would be the first line of defense before rusting becomes a serious problem as depicted in Fig. 4. The individual equipment manufacturer should always be contacted for proper inspection methods and corrective procedures.

There is a wide variety of types of corrosion that can be found in playground equipment resulting from both man-made and naturally occurring conditions. The forms of corrosion can

(a)

(b)

Fig. 3 Examples of playground and equipment corrosion. (a) When debris and moisture become trapped on surfaces for long periods of time due to lack of maintenance, improper design, or improper installation, corrosion can affect the function and safety of the play component. (b) All materials will corrode over a long period of time, as demonstrated by this stainless steel slide bedway. Courtesy of the NRPA NPSI

include weld decay (Fig. 5), crevice corrosion, pitting or exfoliation, fretting corrosion, under-deposit corrosion, galvanic attack (Fig. 6), white rust corrosion of a galvanized surface (Fig. 7), and soil and/or microbiologically influenced corrosion (Fig. 8). Manufacturers, designers, and engineers working together with an informed and educated marketplace can make great strides in mitigating the inevitable corrosion of playground components by selecting the proper compatible materials for the intended installation environment. Regardless of the type of corrosion, however, taking the necessary precautions of timely inspection, maintenance, and repair of the play equipment can alleviate and/or eliminate corrosive attack. Every owner/operator can maintain and prolong function and play value by keeping their equipment free of foreign substances and by treating rust when it first appears by cleaning, treating, or replacing hardware showing evidence of corrosion and/or by properly cleaning and coating or otherwise encapsulating structural components with appropriate materials.

Unfortunately for older playgrounds, many pieces of play equipment damaged by corrosion have been discontinued or parts are no longer available. When this occurs, the professional judgment, knowledge, and experience of the owner/operator will be challenged to determine the appropriate corrective action. To address this issue, the Park District Risk Management Agency (PDRMA) prepared a comprehensive maintenance training program. These training materials elaborate on all forms of corrosion commonly found in playgrounds of today (2006) (Fig. 7, 8). Detailed explanations of the type and cause of the different forms of corrosion along with instructions on the procedures and materials required to correct each situation are included. Owner/operators can improve conditions by becoming aware of the factors impacting metal corrosion in order to make informed decisions when purchasing, installing, maintaining, and inspecting public play equipment. Understanding the environmental characteristics of a given region, the expected reactions between materials used in fabrication of play equipment, and the appropriate use of corrosion-resistant material or protective coatings that can extend the life expectancy of a product are just a few of the areas addressed. Another factor accelerating the loss of corrosion protective coatings on playground equipment is the amount of use the equipment receives day in and day out. It is critical to understand all aspects of the local playground environment to better maintain and extend the life of a piece of equipment while ultimately reducing the frequency and severity of playground-related injuries.

Free Rock Climbing

The sport of climbing takes many forms, largely dictated by the locale and the desired activity of the climber. These forms include mountain climbing or mountaineering, ice climbing, bouldering, indoor climbing as in a "rock gym," aided rock climbing, and free rock climbing. While there are some commonalities between the equipment used and environments encountered in the outdoor sports of aided and free rock climbing, the focus of this section is on free climbing.

Safety Equipment

While aided ascents require the active use of equipment to make progress along a rock face, free climbing can be done with or without equipment. When equipment is used, it is there to catch a climber in the event of a fall, not to assist in the ascent. The act of using a safety system operated by a climbing partner to catch a falling lead climber is known as belaying. Important metal components of any belay system include carabiners and items that fall under the category of "protection" and include pitons, bolts, chocks and nuts, friends, and variations of these. Some items are carried by the leader. They are placed along the route for the protection of the leader during the initial assent until an anchor is established at the end of a pitch. The belayer then collects the pieces as he/she ascends after the leader. After a careful inspection and cleaning those pieces may be used again on another pitch or climb. Other items are placed and left permanently in the rock face for the use of other climbers who will follow the route at later times. "Permanently" is usually defined as "until it breaks or shows signs of imminent failure." It should be noted that the loading of any of these items can be gradual and relatively static during

Fig. 4 Once protective coatings have been removed from playground equipment, corrosion occurs rather rapidly, accelerated by dirt, moisture, and other environmental factors including human contact. Courtesy of the NRPA NPSI

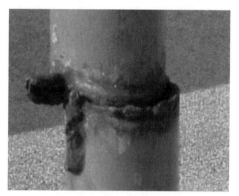

Fig. 5 If a weld joint separates as a result of corrosion, it is difficult to repair to its original structural integrity. Courtesy of the NRPA NPSI

Fig. 6 A noncompatible metal bushing in contact with a stainless steel bolt and other adjacent metal surfaces if left uncorrected may corrode away and weaken the connection. Courtesy of the NRPA NPSI

Fig. 7 White rust on galvanized parts will cause paint to blister off and provides little protection against subsequent rusting of the metal underneath. Courtesy of the PDRMA

Fig. 8 Galvanized post at a concrete footer corrodes from a moist subgrade environment. Chlorides from the fresh concrete can accelerate the corrosion process. Courtesy of the PDRMA

the course of a climb, for example an anchor for a hanging belayer, or it can happen suddenly in the event of a fall of the leader. Longer falls clearly have the capacity to increase the dynamic loading considerably. All of the safety equipment must be capable not only of providing support to static loads, but also of surviving the shock of a fall. The materials used in these devices consist of a variety of alloys, some of which have changed over the years as the sport has progressed. Some of the history and issues pertaining to carabiners and protection are summarized below.

Carabiners are the links between a climber and the rope, the rope and the protection equipment, the rope and a belay anchor, and so forth. Originally fabricated from steel, some pre-WWII European carabiners could bear static loads of no more than 270 kg (600 lb). An increase in weight in the carabiner body improved the load capabilities, but a change in alloy and design undertaken by American manufacturers resulted in far more significant performance gains. The change in design and the use first of 4130 steel (G41300) and then of heat treatable aluminum-copper alloy 24 ST occurred during WWII for the benefit of the American military's Mountain Troops. These improvements ultimately benefited the general climbing community as well (Ref 6).

Today (2006) carabiner bodies are fabricated predominantly from 7xxx-series aluminum alloy bar stock, chosen for their light weight and strength. At least one manufacturer specifies aluminum alloy 7050-T6 (A97075) for their carabiner bodies. The hinge pins may be composed of other alloys. Depending on the length and difficulty of a climb, a leader may carry dozens of carabiners at the outset; consequently, weight continues to be a very important factor. Additional weight can be shaved by reducing the cross section of a carabiner (Fig. 9), but this of course also reduces its rating in terms of the load it can support.

Pitons simply are flat or angled metal blades that can be inserted into a crack in a rock face. The size, style, length, and thickness of the plate can vary greatly, and a wide assortment was necessary in order for a leader to match a piton to the cracks encountered along a route. In the earliest days of the sport, iron pitons were fashioned by local blacksmiths and as a result exhibited dramatic variations in performance, some being too soft to drive reliably into a crack, others being too brittle and suffering fracture at impact. Around the time of WWII, metallurgical studies and advances in design led to the decision to use SAE 1010 low-carbon steel (G10100) with suitable heat treatments, providing a more consistent and reliable product that could be reused provided it could be removed from a crack (Ref 7). Alloy steels have also been introduced in the manufacture of standard pitons providing improved strength and stiffness with little compromise on ductility. Higher production costs for these pitons led to a more expensive final product, and softer steel pitons were still preferred for some time. However, in the case of the very thin and flat "knife blade" pitons, chromium-molybdenum alloy steels such as SAE 4130 (G41300) were essential to meet strength requirements (Ref 8).

By the late 1960s removable protection, such as nuts and chocks, began to replace the piton in some climbing venues (Ref 9). These devices, usually quite simple in design and function, are typically made of aluminum alloys, often anodized, and sometimes combined with stranded steel wire to provide a stiff and strong attachment point for a carabiner. In these cases, the wire is galvanized to protect the strands from corrosion, which could reduce the cross section of the individual strands relatively quickly and severely compromise the usefulness of the protection in the event of a fall. The ease of placement and retrieval during climbing as well as the reduced damage to the rock surface itself helped to popularize this type of protection in many climbing communities. The variety and complexity of available retrievable protection hardware is quite extensive today.

Bolts are placed in a rock face by drilling a suitable opening in the rock. Consequently, a drill, drill bits, and bolts must be carried by the leader for placement during a climb. Some bolts are fully retrievable, while others have collars that are placed in the rock and allow the bolt to be screwed in and out. Still others are placed permanently as complete units. Once a suitable hole is available, the bolt maintains its position either by an expansion mechanism or by gluing. When anchor glue is used, it is squeezed into the hole first. The bolt is then slowly inserted and twisted to maximize coverage and to disrupt and remove trapped air bubbles. The bolt has an exposed threaded end with a nut, to which a variety of hangers can be attached as needed. When tightened, the bolt shaft is kept in tension and the hanger stresses are dependent on the shape of the rock face against which is it pressed. Bolts and nuts are commonly stainless steel, although at least one manufacturer markets mild steel bolts with a thin stainless steel cladding. While stainless steel, and even stainless-steel-clad bolts, may be acceptable in many environments, corrosion issues in some geographic locations have forced recent innovations in materials selection for permanently placed bolts.

Environmental Corrosion Issues

Since most modern protection hardware is retrievable, corrosion issues usually center on maintenance. When gear is used outdoors, particularly in damp or humid locations, and most especially in coastal and tropical locations, proper cleaning is important. Removal of debris, rinsing with fresh water, and allowing pieces to dry is often sufficient to prevent corrosion of retrieved protection and carabiners used in coastal tropical environments as well as less severely corrosive environments. The more serious conditions occur when considering whether or not to rely on permanently placed equipment during the course of a difficult climb.

Equipment that is left behind intentionally or unintentionally should always be approached with caution. "Found" equipment, such as the carabiner shown in Fig. 10, should never be relied upon, even without the dramatically poor appearance of this particular piece. Even if the piece had been cleaned, the poor condition of the hinge and catch pins suggests that it may fail during a fall.

Pitons and bolts that have been left behind as permanently placed protection are another story. While these pieces are intended for the use of later climbers, it would be unwise to assume soundness without inspection. Inspection for a climber typically involves visually inspecting the part for corrosion or other damage, testing it for looseness as appropriate, and finally tapping it with a piton hammer. Motion during tapping would be undesirable, but climbers also listen for

Fig. 9 One way to reduce weight when climbing is to use carabiners with smaller cross sections. The trade-off is in the load-bearing capacity rating. Courtesy of Lorrie Krebs, Anderson Materials Evaluation, Inc.

Fig. 10 The carabiner in the foreground was found in a damp cave in upstate New York. A similar aluminum alloy carabiner is shown for comparison. In addition to white corrosion product generally present over all surfaces, there is red rust at the hinge and catch pins and the carabiner is frozen in the closed position. Courtesy of Lorrie Krebs, Anderson Materials Evaluation, Inc.

a distinctive ringing sound that indicates a piece was well set in hard rock. Many steel pitons of the older variety rust badly in almost any environment prone to regular rain and/or condensing humidity throughout the year because they have little in the way of natural corrosion resistance. Chromium-molybdenum steel pitons can be expected to fare better in similar conditions due to the corrosion resistance afforded by the chromium. Badly rusted pitons can fail in a variety of ways ranging from loosening inside the crack to breaking under impact and should always be avoided.

While rusty pitons have probably been a part of rock climbing for almost as long as pitons have been in use, a corrosion phenomenon relatively new to this sport has become apparent in more recent years. The relatively new popularity of sea-cliff climbing in warm tropical locations such as Sardinia, Thailand, and the Caribbean island of Cayman Brac has revealed a new hazard as documented in part at the website of the American Safe Climbing Association (ASCA, www.safeclimbing.org). Climbers visiting these locations are finding that stainless steel bolts have been fracturing under impact—either the impact of a piton hammer tap or, more seriously, the impact of an actual fall—in as little as 18 months after placement. The cause is chloride-induced SCC. Saltwater present as a spray or mist travels readily in the air, condenses on rock faces, and wicks into the crevice between the hanger and the rock where it attacks the bolt. Failure reports provided by many climbers indicate the bolts typically break in the shaft at the rock face as a result of an impact, large or small, with little or no visual evidence of corrosion observed prior to the break. The stainless steel hangers associated with these bolts typically have a 90° bend at which they are most likely to fail. The stainless steel nuts fitted to the threaded bolts also suffer SCC. In addition, four aluminum alloy carabiners retrieved from rock walls in Cayman Brac after exposures of roughly 6 to 18 months overhanging the ocean were tensile tested to failure. While all four broke near or above their rated strength, the test revealed that the one exposed for 18 months failed due to SCC.

The solution for the bolt problem is a change both in material and in design. New bolts have been developed using titanium to replace stainless steel. These bolts are formed in the shape of a "P," providing both a shaft to be anchored in the rock face and a loop to which a carabiner can be directly attached without threaded parts and associated crevices. The titanium is expected to have a far better survival rate in the aggressive coastal environment, but to add further protection the bolts are placed in the rock face with anchor glue. The adhesive, typically an epoxy, not only securely attaches the bolt to the rock, but also provides a barrier to moisture for the bolt shaft inside the hole. While the cost is substantially greater per bolt, the expected savings in averted injuries and equipment replacements makes the change worthwhile for these climbing sites. Other areas that may benefit from the new bolts are climbing and caving sites in wet areas with rock materials such as limestone, which may have leachable compounds that can dissolve and attack protection hardware. Aluminum alloy carabiners should be cleaned after use to remove potential corrodents. If one is found with a permanent anchor in a tropical coastal location it would be wise to be cautious about its use.

A Swiss-based organization known as the International Mountaineering and Climbing Federation (UIAA—Union Internationale des Associations d'Alpinisme) provides standards for strength requirements of climbing gear that are recognized worldwide. The ASCA reports that a special subcommittee within the UIAA has been formed to investigate marine environment bolt standards and to introduce their first standards specifically addressing corrosion issues.

REFERENCES

1. *Sources of Receipts or Revenue, 1997 Economic Census Arts, Entertainment, and Recreation Subject Series,* U.S. Census Bureau, Aug 2000
2. "Standard Consumer Safety Performance Specification for Playground Equipment for Public Use," F 1487-01e1, ASTM International, 2001
3. "Standard Guide for Selection of Committee F16 Fastener Specifications," F 1077-95a (2002), ASTM International, 2002
4. "Standard Consumer Safety Performance Specification for Home Playground Equipment," F 1148-03, ASTM International, 2003
5. "Standard Safety Performance Specification for Soft Contained Play Equipment," F 1918-04, ASTM International, 2004
6. R.M. Leonard and A. Wexler, Belaying the Leader, *Belaying the Leader, an Omnibus on Climbing Safety,* 6th ed., The Sierra Club, 1964, p 18
7. R.M. Leonard and A. Wexler, Belaying the Leader, *Belaying the Leader, an Omnibus on Climbing Safety,* 6th ed., The Sierra Club, 1964, p 19
8. C. Wilts, The Knifeblade Piton, *Belaying the Leader, an Omnibus on Climbing Safety,* 6th ed., The Sierra Club, 1964, p 50
9. R. Dumais, *Shawangunk Rock Climbing,* Chockstone Press, 1985, p 76

SELECTED REFERENCES

- R.H. Angier, *Firearms Blueing and Browning,* Thomas Samworth, 1936
- A. Gogan, *Fighting Iron—A Metals Handbook for Arms Collectors,* Andrew Mobray Publishers, 1999
- D.G. Wilson, *Bicycling Science,* 3rd ed., MIT Press, 2004

Corrosion in Workboats and Recreational Boats

Everett E. Collier, Consultant

A MARINE VESSEL encounters an environment conducive to corrosion in all its forms: atmospheric, aqueous, galvanic, and microbiologically induced. Boats are composite structures, consisting of a variety of dissimilar metals and other materials, all exposed or immersed in one of the most efficient of electrolytes, water, especially saltwater. Corrosion may be enhanced by stray currents from shipboard electrical and electronic systems. This article focuses on the corrosion and deterioration of components on recreational and small workboats; materials selection and corrosion control also are discussed. An extensive and thorough discussion of the use and protection of both metallic and nonmetallic materials on marine vessels is given in Ref 1.

Hulls, Fittings, and Fastenings

Fasteners and fittings must be compatible with the common marine vessel hull structures of wood, fiberglass composite, steel, aluminum, ferrocement, and sheathed hulls.

Wood Hull Fastenings. Silicon bronze and type 316 stainless steel are preferred above the waterline. Type 304 stainless steel, galvanized steel, and inhibited low-zinc (<15% Zn) brasses are acceptable. Mild steel, the free-machining grades (types 303, 303Se), and the 400-series stainless steels should not be used. Moreover, where the wood may become saturated, such as in hull planking, stainless steel fastenings should not be used. Conditions at or below the waterline (splash zone) are considerably more hostile than they are topside. Constant immersion in seawater or in seawater-saturated wood, or both, is highly conducive to crevice corrosion. These fastenings get no protection from cathodic protection systems such as zinc anodes, so their functional life depends on the corrosion resistance of the material or its protective coating.

Some of the newer superalloys, such as the 6% Mo stainless steels and Ni-Cr-Mo alloys, do much better and are finding increased use under these conditions, but they are expensive. Silicon bronze is preferred, and galvanized steel is still being used.

Galvanized bolts and drifts are commonly found in the most demanding of the below-the-waterline applications: bilge fastenings, floor bolts, and keel bolts. The most corrosive damage occurs at the joint between the members: between the exterior ballast and the keel, between the keel and the floor timbers, between the floor timbers and the frame, and between planking and frames. Corrosion is accelerated in these areas, and the fastenings become wasp-waisted (Fig. 1). Threads should not be exposed in these areas.

The selection, installation, and maintenance of the keel bolt material should be done with consideration of the critical role of these fastenings and the difficulties involved in replacing them. A variety of keel bolt materials are used with cast iron ballast keels. Mild steel is the least expensive and is galvanically compatible with the iron keel; however, it is the most susceptible to accelerated corrosion in the joint area. Hot dip galvanized steel is advised. Type 316 stainless steel is also used, but it is essential to use sealants under the washers at the bolt ends and at the joint to prevent seawater penetration. Silicon bronze, aluminum bronze, nickel-aluminum bronze, and Monel are used, but these alloys are not galvanically compatible with the ballast iron and should be electrically insulated from it. Silicon bronze, aluminum bronze, nickel-aluminum bronze, and Monel are recommended with a lead ballast.

Organic coatings, even on the plank fastenings, are beneficial. A traditional coating choice for wooden hull fastenings is thick or thickened red lead. Some of the newer epoxy-based coatings also work well.

Wood Hull Through-Hull Fittings. On the outside of the hull, through-hull fittings are constantly immersed in seawater. On the inside, they may be immersed intermittently. In the hull, they are exposed to seawater-saturated, oxygen-depleted wood. A combination of corrosion resistance and mechanical strength is needed that exceeds most mild steel and common brass.

Most of the copper alloy seavalves, seacocks, or through-hulls and other underwater fittings are made of cast phosphor bronze (UNS C94300) or a cast leaded tin bronze, which is a highly corrosion-resistant alloy with a minimum yield strength of 90 MPa (13 ksi). This type of bronze corrodes extremely slowly in seawater and wet wood and does not need additional corrosion protection. Providing these parts cathodic protection may supply excess electrons, which would result in the overproduction of hydroxyl ions, the formation of sodium hydroxide in the saltwater-saturated wood around the fitting, and lead to delignification of the wood. Overprotection is generally not a problem where water flows freely over the surface of the fitting, preventing concentration of the caustic alkali (Ref 2).

Composite Hulls. Initially, fiberglass boats were hand lay-up and spray-up boats with successive layers of mat or woven fabric individually applied and impregnated with a liquid resin. In the search for greater strength-to-weight ratios and stiffness, other types of reinforcing fibers, such as other glasses, graphite, and aramid (Kevlar) fibers, and sandwich hulls came into use.

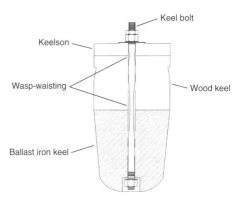

Fig. 1 Fasteners that appear solid at the ends may have hidden corrosion damage, such as the wasp-waisting of bilge fastenings.

Fastenings in Fiberglass. The fastenings most commonly used in fiberglass-reinforced plastic construction are bolts, self-tapping screws, and blind (pop) rivets. Silicon bronze is preferred over galvanized steel; stainless steels should be not used where oxygen is unavailable. There are some situations that are peculiar to fiberglass hulls.

Hard metal fastenings impose high local loads on the relatively thin, brittle laminates, so these loads must be spread through the use of well-bedded wood backing blocks and large (two to three times the bolt diameter) washers. Cathodic overprotection of metal fittings and alkali delignification of the damp backing wood is a potential problem.

Through-bolts are preferable where they can be used. Keep in mind that moisture trapped in the screw hole against the screw can cause crevice corrosion in stainless steel fastenings, and a sealant should be used. Blind rivets are used for deck-to-hull joints and for some small hardware where the backside is inaccessible. Closed-end types (to prevent moisture penetration), in which both the body and the mandrel are type 316 stainless steel or coated (plated or painted) types, should be used.

The use of fiberglass-encased plywood and wood for bulkheads, stringers, and engine beds is common in fiberglass boats. Metal fastenings in these members are susceptible to crevice corrosion, so coatings on the fasteners and sealants under the heads are recommended.

Fastenings in Other Fiber Reinforcements. Carbon fibers are both electrically conductive and highly noble. This has several implications for marine vessels. Less noble fastener metals in contact with carbon fibers and immersed in seawater will corrode. Two dissimilar metals in contact with the same carbon-fiber structure are in electrical contact with each other. If a common electrolyte is present, the less noble metal will corrode. Fasteners should be insulated from contact with the carbon fiber. Thick-wall electrical shrink tubing works well in such applications.

Sandwich core construction increases the rigidity of a structure by increasing its thickness-to-weight ratio. Sandwich core construction consists of thin, resin-impregnated fabric sheets bonded to a thicker lightweight core in between. Honeycomb cores are frequently used in bulkheads, overheads, cabin soles, cabinets, tables, and counters and are available with a variety of core materials, including aluminum. With aluminum cores, there is concern for galvanic corrosion with non-compatible fasteners. The article "Custom Sailing Yacht Design and Manufacture" in *Composites,* Volume 21, of *ASM Handbook,* 2001, provides details of sandwich core construction.

Steel hull boats will inevitably corrode in either fresh or saltwater, and for this reason, they are generally built with heavier-gage steel than is necessary for structural strength in their intended application to compensate for loss of strength as a result of corrosion. The use of a well-planned, carefully applied and maintained epoxy paint system (grit blasting to white metal, anticorrosive primers, and protective topcoats) and the installation of an equally well-planned and maintained cathodic protection system will ensure long service.

Properly designed steel hulls allow rain or seawater coming aboard to drain freely to the bilges. Large limber and lightening holes are provided in floors, frames, struts, and stringers to facilitate both drainage and air circulation. The media used in grit blasting the hull must be free of contaminants. Beach sand will contain unacceptable levels of sodium chloride and other contaminants.

Well-planned cathodic protection requires the correct number and size of zinc anodes to be properly placed on the hull and then to be properly monitored and maintained (Ref 1). Mounting bolts should be welded to the hull to hold the zincs. The hull provides the connection, unlike wood or fiberglass hulls where wires are run from the anode to a common groundpoint.

Aluminum hulls of marine-grade aluminum-magnesium alloys (5000 series) are highly resistant to corrosion in both fresh- and saltwater. A properly designed and built unpainted aluminum hull will last indefinitely in saltwater, despite its reputation to the contrary. A number of small craft that were built from surplus aircraft aluminum-copper alloy (2000 series) after World War II and failed caused that reputation to persist today. The 2000 series is not as resistant to corrosion as other series, so it must be painted or clad for protection (Ref 3).

While it is not necessary to paint aluminum boats above the waterline, hulls must be painted below the waterline to prevent fouling by marine growth. The process consists of grit blasting, chemical cleaning, primer coat, two or more barrier coats, and then the antifouling coat. If the antifouling coat is a copper oxide bottom paint (not recommended), it must be applied over a thick epoxy-based barrier coat. Such a barrier coat is critically important to prevent galvanic interaction between the copper paint and the aluminum.

The interior surfaces of an aluminum hull, especially below the waterline, should also be painted to provide protection from galvanic corrosion caused by dissimilar metal objects falling into the bilge. Most noncommercial owners prefer to paint the topside for aesthetic reasons and to reduce glare and heating effects.

Aluminum is a highly active metal and less noble than any of the other metals commonly used for boats. Galvanic and electrolytic (stray current) corrosion is possible. Metal particles that may be trapped in grinding wheels or hand files—even a copper penny dropped in the bilge—can cause corrosive damage. Any copper bearing alloy, such as brass, bronze, or Monel, will cause corrosive damage unless appropriate precautions are taken. Dissimilar metals must be insulated from contact with the aluminum hull. Insulating pads of neoprene, wood, or some similarly electrically insulating material should be used between the fitting and the aluminum. Fasteners should be coated with a polysulfide sealant or perhaps a zinc-rich primer before installing them to reduce, if not eliminate, any galvanic action between the fastening and the aluminum.

Underwater fittings, through-hulls (with proper backing blocks or built-up hull sections, and properly insulated), propeller and shaft, and rudder and rudder post will require cathodic protection, as with all types of hulls. Aluminum hulls in freshwater use magnesium anodes because of the lower conductivity of the water; in saltwater, zinc anodes are used.

Ferrocement hulls consist of a mild steel and wire mesh armature, plastered with Portland cement. Proper plastering mix and application will avoid future corrosion and structural problems. Galvanizing, professional plastering, moisture-inhibiting coatings, and, more recently, sophisticated concrete additives—plasticizer, pH control, cure-modifying agents, and even epoxy-coated frame and mesh—have overcome early problems of saltwater and oxygen migration through the cement to the armature. This caused extensive rusting. The rust, occupying many times the volume of the unrusted metal, can cause extensive cracking, which further exacerbates the problem.

A robust organic coating system will prevent moisture penetration to the steel armature and will prevent marine growth. The new epoxy-based coating systems have proved useful here; they bond and last well, especially when applied over an epoxy-tar-based primer.

Copper sheathing was once quite commonly used, especially in the tropics where marine borers (teredo worms) attacked wooden bottoms. Thanks to advances in the coating industry, copper sheathing is no longer required for underbody protection; toxic antifoulant coatings have provided a more convenient alternative. Copper sheathing is still being done, however, in restorations of classic vessels and also in special applications, such as for ice protection at the waterline in frigid latitudes.

The copper sheeting is typically quite thin (3 mm, or 0.125 in.) and fastened by copper nails. Galvanic corrosion is a serious consideration. Through-hulls, through-bolts, and other skin fittings must be of bronze or else isolated from contact with the copper sheathing.

Metal Deck Gear

Deck gear is the exterior fittings, fixtures, and structures used in operating or securing the vessel: cleats, mooring bitts, fairleads, pad eyes, lifeline stanchions, pulpits, towers, and ground tackle. Two types of corrosive damage are of concern: that which is primarily cosmetic and proceeds so slowly that sudden catastrophic failure is not likely; and that which can proceed undetected, fail suddenly, and present a safety hazard to the vessel or its crew. A freshwater rinse of deck gear at the end of the day is good practice to reduce the effects of corrosion, as is occasional waxing.

Stainless steels (types 304 and 316) are used extensively in cleats, chocks, samson posts, handrails, davits, and a broad range of other deck hardware and furniture. They are so resistant to atmospheric corrosion that corrosion rates are not considered meaningful. Their greatest susceptibility is to pitting and crevice corrosion. Preventive measures consist of the avoidance of dings, dents, and scratches, which can be the sites for pit initiation.

Stainless steels are also susceptible to crevice corrosion at welds, lap joints, and under bolt heads. Welds, if not properly done—wrong filler metal or excessive heat—are particularly susceptible to intergranular corrosion (weld decay) and degradation of the strength of the joint. With purchased fittings, there is not much one can do about the welds.

Aluminum. Marine-grade aluminum alloys (5xxx and 6xxx series) have excellent resistance to atmospheric corrosion due to their tendency, like the stainless steels, to form a protective oxide film. Marine-grade aluminum, whose exposure is limited to the marine atmosphere, does not require painting if it is kept clean and not abused by repeated scratching, scraping, or chafing.

Most larger aluminum structures—towers, arches, davits, rub rails, handrails, and so on—are bright anodized and require little maintenance. Anodizing is an electrochemical process that produces a surface oxide (Al_2O_3, or alumina) on the metal. This oxide surface is quite hard but can be scratched or rubbed off, eventually exposing bare aluminum that is then subject to pitting corrosion.

Stainless steel fastenings are commonly used with aluminum fittings. The stainless steel fastening is cathodic to the larger anodic aluminum structure. This is acceptable, because it satisfies the rule of cathode-anode area ratios, which states that the area of the cathodic metal must be small compared to that of the anodic metal. The smaller the area of the cathodic metal, the smaller the current draw and hence the smaller the corrosive damage to the anodic metal.

Galvanized steel deck and cockpit fittings are not considered attractive by the boating public, so their use is limited on recreational boats, but they are common on workboats and traditional craft. The service life of a galvanized steel fitting is a direct function of the thickness of the coating, so hot dip galvanizing is favored. Studies have shown that a fitting with a hot dip coating thickness of 0.076 mm (3 mils) may have a useful service life of between 30 and 35 years in a mild marine environment and between 10 and 15 years in a heavy industrial environment.

Bronze. For traditionalists, bronze is the preferred material. The excellent atmospheric corrosion resistance of bronze is due to a protective oxide film (cuprous oxide). This film forms very quickly on exposure to oxygen and continues to build over time, reducing the corrosion rate.

Brass. The brasses are copper-zinc alloys. When the zinc content is greater than 15%, the alloy is susceptible to both dezincification and stress-corrosion cracking. However, when the zinc content is below 15% and when small amounts of tin and arsenic or antimony are added, resistance to both these forms of corrosion increases greatly.

Exposed brass tarnishes rapidly, developing the familiar greenish patina. It must be coated or polished frequently to maintain its characteristic bright, attractive finish. Plain, uncoated brass deck hardware is not very common. Brass is used mostly in light-duty cleats, chocks, vents, and housings and is almost always chrome plated.

Zinc Alloy. Some chrome-plated deck fittings—rail caps, flagstaff sockets, light-duty cleats, and chocks—often have a substrate metal of either zinc or Zamak (a zinc-aluminum alloy). These metals are used because they are inexpensive, are cast at low heat, and are easily die cast, producing parts with fine, sharp detail. They have only moderate strength, hardness, and corrosion resistance and, if the chrome plating should fail, are susceptible to serious degradation of their mechanical properties.

Equipment

Winches are complex assemblies with internals constructed from bronze, stainless steel, and, in some cases, anodized aluminum. These metals are not galvanically compatible. They are not sealed and are susceptible to penetration by saltair and seawater. The inevitable result is corrosion and abrasive wear of critical internal parts. A freshwater rinse will not eliminate the possibility of internal corrosion damage, and periodic servicing is required.

Backing plates are required on any fitting that is subject to large or sudden loads. They are especially important when the fitting is mounted on a relatively thin section, such as a fiberglass deck, cabin top, or coaming (Fig. 2). Backing plates are made of stainless steel, marine-grade aluminum, or plywood.

Susceptibility to atmospheric and galvanic corrosion must be considered. If fasteners are dissimilar metals, the two parts should be electrically insulated. If moisture is allowed to penetrate the space between the fastener threads, there may still be localized crevice corrosion, resulting in "frozen" fastenings. Thus, it is advisable to dip the fastening in some conformable substance, such as silicone or paint.

Lifeline Supports. Lifeline stanchions are usually stainless steel or anodized aluminum. Stanchions are typically inserted into a substantial cast metal base and pinned with a machine screw or cotter pin. Bases, for the most part, are stainless steel, chrome-plated bronze, or chrome-plated Zamak. There are several potential corrosion problems in this arrangement. Because moisture can collect in the socket of the base, there is the potential for galvanic corrosion between the stanchion and the base and between the machine screw or cotter pin and both the stanchion and the base. It is important to be sure that these components can drain freely (Fig. 2) and are galvanically compatible.

Anchor Systems. The elements of the anchoring system are the anchor windlass, bow roller and accessories, chain, end fittings (shackles and thimbles), and the anchor itself.

Windlass. A typical anchor windlass may have parts made of aluminum, bronze, and both carbon and stainless steels. Its location in the bow of the boat subjects it to saltwater spray. Many windlasses have anodized or epoxy-coated aluminum above-deck housings. Corrosion damage to the housing itself consists of paint lifting, chafing, and abrasion on both painted and anodized finishes. Abrasion, mostly by the anchor rode, is the more serious, while galvanic corrosion between the fastenings and the housing must also be considered. Preventive measures consist of keeping the windlass covered when not in use, frequent wash-downs with freshwater, and coating the through-deck fastenings with heavy water-resistant grease or a corrosion inhibitor.

More serious than cosmetic and fastener problems are those that result from internal corrosion damage. Internal corrosion leads to increasingly degraded performance and, ultimately, to seizure of the windlass. If the windlass is electric, seizure can be dangerous, because the nonrotating windlass motor acts essentially as a dead short. Extremely high current can flow, resulting in overheating of the motor and the wires if not properly fused.

Windlasses depend on lubrication for protection from physical wear and corrosion. If left unused for long periods of time, these lubricants drain off metal surfaces, exposing them to the moist internal atmospheres created by condensation, and eventual corrosion. Keep the windlass properly oiled and lubricated, and operate it on a regular basis to keep these lubricants well distributed throughout the internals of the gear box and motor.

Bow Rollers and Accessories. Bow rollers, anchor chocks and brackets, and chain stoppers are exposed to constant saltwater soaking. These items should be of noncorrosive materials, such as stainless steels and Delrin. The bow roller

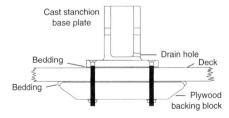

Fig. 2 A backing plate installation distributes stanchion forces. The drain hole in the stanchion base reduces a possible corrosive environment.

should be checked periodically for wear and corrosion of the roller shaft, and all through-deck fastenings should be inspected periodically for corrosion and/or loose nuts.

Chain may be made of low-carbon steel, high-carbon steel, or stainless steel. Hot dip galvanized steels are preferred over electroplated finishes. The actual service life of a hot dipped galvanized chain is dependent on the local conditions but is typically between three and five years.

Stainless steel, typically type 316, has much superior corrosion resistance and is some 10 to 20% stronger than low-carbon steel chains. Corrosion damage to the stainless steel is minimal under normal conditions of immersion in free-flowing seawater. When marine growth is allowed to accumulate on the chain, denying access to the fresh supply of oxygen needed to form a protective coating, pitting and crevice corrosion can result.

The greatest damage to the chain in anchoring or mooring systems results from the wear and abrasion that take place at the bottom end of the mooring or anchor rode. As current and tides change direction and the boat swings on its anchor, the links at the bottom of the chain—and the end fittings—can suffer considerable wear and accelerated corrosion rates. Reversing the chain end-for-end each year, to even out the wear, can significantly extend its useful life.

Thimbles and shackles are available in both galvanized carbon steel and stainless steel (type 18-8 or 316). The greatest wear in the anchor system takes place here, so these are the most common failure points. The physical wear quickly removes the galvanizing, exposing unprotected metal, which quickly forms a soft layer of rust that is rapidly worn away. Little can be done to prevent this, so using larger end fittings will give a longer service life.

Dissimilar metals in ground tackle applications—stainless steel end fittings with galvanized chain and anchors—should be analyzed for cathode-to-anode area ratio, which should be very low (small cathode, large anode). For example, an 8 mm ($5/16$ in.) stainless steel shackle on a 15 m (50 ft), 8 mm ($5/16$ in.) galvanized chain has a ratio of the cathode (stainless shackle) area to anode (galvanized chain) of roughly 1 to 77. Thus, while there may be some localized corrosion at the point of contact, neither the chain nor the shackle will suffer severe corrosive damage, and reasonable service life can be expected.

Anchors are low-maintenance items available in galvanized (hot dipped preferred) steel, stainless steel, aluminum, and titanium.

Propulsion Systems

Propulsion systems include engines, lubrication, cooling, exhaust systems, propellers and propeller shafts, and the rudder. Single-metal uniform corrosion thrives when metals high in the galvanic series, such as mild steel and iron, are exposed to moisture, especially salt-laden moist air. When dissimilar metals, such as aluminum pistons, rods and heads, iron blocks, steel cylinder liners, and copper alloy freeze plugs, are in contact with each other and immersed in the same body of water, galvanic corrosion will also occur. Also, the presence of acids and elevated temperatures can exacerbate and accelerate corrosive damage. All of these conditions exist in internal combustion engines.

Inboard Engines. Engine blocks are large iron castings. Cylinder liners are made of steel with highly polished inner surfaces. Lubricating oil is splashed up onto the inner surfaces of the cylinder liner walls to lubricate the piston travel. As rings wear, some blow-by occurs, resulting in acid contamination of the oil in the oil pan and corrosive attack of cylinder walls and bearings. Pitting of the cylinder walls can also result from cavitation erosion, which can take place on the outer surface of the cylinder liner. This occurs when combustion in the cylinder causes vibration of the cylinder wall (liner), creating vapor bubbles in the coolant fluid. These bubbles then collapse, ultimately removing the protective oxide layer of the metal, leaving it vulnerable to corrosion and pitting of the cylinder liner.

Boats that sit idle for considerable periods of time without running their engines also are subject to corrosive damage. Protective lubricating oil that normally coats metal surfaces gradually drains down, exposing these surfaces to moist, salt-laden air, and the inevitable corrosion begins to take place. This is more likely to occur in diesel engines, because the residual amounts of sulfur are likely to be greater in diesel fuel than in gasoline, especially in certain parts of the world where pollution controls are not stringent. The engines should be run on a regular basis (weekly) to maintain lubrication in the engine and bearings.

Electrical engine components should be mounted as high as possible to avoid water sloshing around in the bilge. Alternators are particularly susceptible to corrosion from moist saltair on terminals and brush springs. A common source of problems is corroded terminals and wire connections in the ignition system. It is a good idea to spray them occasionally with one of the commercially available moisture-displacing lubricants designed for this purpose.

Outboards/Outdrives. The engine blocks in outboards and outdrives are generally of die-cast aluminum. Outboard motor housings, once exclusively aluminum, are now more commonly made of high-impact plastics. The primary form of protection for aluminum outboard motor housings is in the decorative coatings. Typically, there is an epoxy primer undercoating, a decorative epoxy paint, and a clear coating. Chips, scratches, and dents will result in rapid local corrosive attack.

The lower unit is protected by one or more sacrificial anodes most commonly zinc, located below the waterline. Magnesium anodes are also available for motors that are used in freshwater. These must be monitored and replaced when they are no more than 50% depleted.

Frequent flushing of the cooling system is highly recommended, and thorough flushing before storing for the winter is a must.

Lubricating oils reduce wear and also perform functions such as acid neutralization, contaminant suspension, and rust and corrosion inhibition. Additives in the oil enhance these properties. The proper choice and timely replacement of engine oils is important to ensure a reliable and satisfactory life for an engine. In the course of engine operation, products of combustion, wear, and corrosion end up in the oil in the sump. Laboratory spectrographic analysis of a sample of the engine oil can reveal a tremendous amount of information concerning the condition of the engine.

Cooling Systems. There are two basic types of cooling systems in common use in marine engines: once-through systems and closed recirculating systems. In once-through systems, the cooling water passes through the system to be cooled only once and is then discharged. These systems are referred to as direct cooling or raw water systems, because the cooling water is taken from rivers, lakes, or the ocean in which the boat operates. Closed recirculating systems, also referred to as indirect or freshwater cooling systems, continuously recirculate the same water. The heat that is transferred to the recirculated water is subsequently transferred to a second and separate cooling stream through a heat exchanger. Corrosion by-products can accumulate and foul heat-transfer surfaces; however, these systems are amenable to the use of corrosion-inhibiting additives.

To reduce galvanic corrosion, engine manufacturers make provision for the insertion of sacrificial zinc anodes in the cooling system. Figure 3(a) shows a complete engine zinc anode

(a)

(b)

Fig. 3 Zinc sacrificial anodes for an engine cooling system. (a) Engine anode. (b) Replacement pencil anode

assembly. From the outside, the assembly looks like a pipe plug threaded into the block. The replaceable zinc "pencil" (Fig. 3b) threads into the internal end of the anode assembly. Most engines have two or more engine zincs. These should be checked on a regular basis, at least a couple of times a year.

Raw water systems have performed successfully for over 20 years in some very rugged saltwater service, such as commercial fishing boats (Fig. 4a). The engines are designed for raw water cooling, with heavier castings, larger cooling water passages, and thicker wall sections. Failures, when they do occur, are typically in the manifold. The systems are susceptible to clogging and blockage from silt, sand, marine growth, rust, and scale. The silt and sand, if allowed to settle for significant periods of time, can result in poultice corrosion. The sand can cause impingement erosion at elbows or sharp turns in the piping. Raw water systems require regular maintenance; cooling water passages must be kept open, pumps must be inspected for impingement erosion, the impellers must be changed regularly, and filters and thermostats must be kept clean.

The freshwater cooling system is a superior system (Fig. 4b), because the engine operates at a more constant and efficient temperature. Silt, sand, and salt are kept out of the engine. Antifreeze with corrosion-inhibiting additives is used, which makes year-round use of the boat practical. Antifreeze products are based on either ethylene glycol (toxic) or propylene glycol (nontoxic and biodegradable). The coolant fluid in the engine consists of a mixture of 50 to 70% antifreeze and freshwater. Freshwater contains considerable quantities of free oxygen. As the coolant is circulated around the closed, pressurized system and repeatedly heated and cooled, the oxygen content is reduced to negligible levels, which reduces corrosion. To take advantage of this feature of the closed freshwater system, it is a good idea to change the cooling fluid only when necessary, say once a year at winter layup.

The most common form of corrosion in the heat exchanger occurs under sediment deposits. If debris gets by the screens and filters and become lodged in the tubes, it can cause turbulence downstream, which can result in pinholes in the tubes. Water velocities of less than 0.9 m/s (3 ft/s) are not sufficient to prevent sediment buildup, biological fouling, or clogging from small bits of debris collecting in the tubes. Excessive water velocities can cause erosion and impingement damage. Velocities of 1.5 to 1.8 m/s (5 to 6 ft/s) for seawater applications are excessive for copper, but most other commonly used alloys perform well up to 2.7 m/s (9 ft/s). Stainless steels, either type 304 or 316, are not suitable for seawater heat exchanger applications. The alloy of choice for heat exchanger tube applications is 70–30 cupronickel.

Heat exchangers should be periodically flushed out to remove sediment and debris. In brackish, inshore estuarial waters, where there is a lot of silt, mud, and sand, monthly flushings may be required. In clean waters, flushing, disassembly, and brushing at the annual layup is sufficient. Leaving water in a heat exchanger for any significant period of time is highly conducive to corrosion. The water becomes stagnant, bacteria colonies form, microbiologically influenced corrosion results, and crevice corrosion will take place under sediment deposits. Details are found in "Microbiologically Influenced Corrosion" in *ASM Handbook,* Volume 13A, 2003, p 398 to 416.

Heat exchangers are typically fitted inside the boat in the engine room. Another type of heat exchanger, the keel cooler, is mounted outside the boat on the side of the keel or on the underbody (Fig. 5). There is no housing; the tubes themselves are exposed to the water in which the boat is operating.

Exhaust Systems. Exhaust gases are hot, noisy, noxious, toxic, and corrosive. The exhaust must be moved away from the crew, out of the boat, and quieted down in the process. This must be done without restricting the exhaust flow, which would increase the backpressure. The system must be able to withstand the acidic by-products of combustion and the corrosive saltwater coolant at elevated temperatures (Ref 4). There are two basic types of exhaust systems (with some variations): the wet exhaust system and the dry exhaust system.

Wet Exhaust System. Figure 6(a) shows a typical wet exhaust system such as found in a power cruiser where the engine is mounted above the waterline. In this type of system, raw cooling water is injected into the exhaust gas flow after it leaves the manifold and on the downside of the riser. There, it mixes with the exhaust gases, greatly reducing the temperature of the gases. The exhaust stream is then carried through a silencer (or muffler) to the exhaust outlet at the transom and discharged. Sharp and abrupt turns in the system are to be avoided after the cooling water is injected into the system to minimize opportunities for erosion-corrosion damage and backpressure on the engine. The section of the system before the cooling water injection point is hot and must be insulated. If the exhaust drop height is not sufficient to ensure the highest point is above the waterline at all angles of heel, as in the case of sailboats, a riser must be added (Fig. 6b).

Dry exhaust systems are used primarily on commercial and high-performance craft (Fig. 6c). The entire run from the manifold to the outlet is extremely hot; it must be well shielded to prevent burns to both the boat and the crew. When wood is periodically exposed to very hot surfaces (120 °C, or 250 °F) over a long period of time without a significant flow of air, a form of carbon may develop, called pyrophoric carbon, that can be extremely dangerous.

Exhaust Riser. For high temperatures and galvanic compatibility, metallic piping is necessary for components in the dry section between the manifold and the cooling water injection point. Iron is still commonly used for threaded nipples, elbows, and injection elbows.

A siphon break, or vented loop, consists of a tube, connected between the outlet of the raw water pump and the raw water injection point, with an air-admitting valve at the top of the loop. Generally, the valve fails due to corrosion. Clogging by salt crystals and other debris leads to sticking in the open or closed position.

Traditionally, the metallic components in the dry section of the exhaust have been schedule-80 cast iron. This is a robust material that is galvanically compatible with the commonly used cast iron manifolds. It is, however, susceptible to rusting and scaling that can break loose and clog the cooling water injection orifices, resulting in overheating. Alternative alloys have shortcomings in the area of galvanic compatibility or

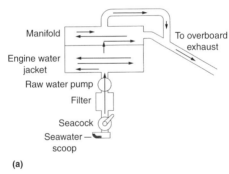

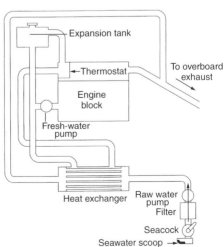

Fig. 4 Engine cooling water systems. (a) Raw water system. (b) Freshwater system

Fig. 5 Keel heat exchanger installation

corrosion resistance. Aluminum, while it would be compatible with engines having aluminum manifolds, is not as hard or as strong as cast iron and poses galvanic compatibility problems with other fittings. Bronze has compatibility problems with both iron and aluminum, is relatively soft, and has a higher susceptibility to erosion- and impingement-corrosion. Also, it is aggressively attacked by the sulfuric acid from residual sulfur in fuels, especially diesel fuels. Most of the common grades of stainless steel (types 18–8, 302, and 304) will corrode if used in wet exhaust systems. Type 316L is sometimes used, but its performance is marginal. Some of the more exotic nickel-base alloys, such as Hastelloy C, Inconel 625, and Incoloy 825, are excellent, but cost limits their use.

Piping. After the point of cooling water injection, wire-reinforced marine exhaust hose is preferable for use in the exhaust line. Hose joints should be double-clamped with all-stainless-steel hose clamps.

Mufflers/Silencers. In the marine industry, mufflers are frequently referred to as silencers, and the terms are used interchangeably. Mufflers are seldom used in dry exhaust systems. When they are used on workboats, because there is no cooling water to bring the temperature down within the range of the nonmetallic materials, stainless steel and hard aluminum are frequently the materials of choice. Corrosion, aside from atmospheric effects, is not a significant problem in these systems.

Propeller, Propeller Shaft, and Rudder. Principal materials used in propellers are Nibral (nickel, aluminum, and bronze alloy) and cast manganese bronze. Nibral is excellent but more expensive. Most propellers are manganese bronze. Manganese bronze (58Cu-39Zn) is really a brass and is subject to dezincification. Popular materials for propeller shafts are stainless steel and Tobin (naval) bronze.

High-performance stainless steels, while more expensive than the conventional austenitics, have found significant use in marine applications. Typical of this group is UNS S20910 (22Cr-13Ni-5Mn), a nitrogen-strengthened grade with superior corrosion resistance. The pitting resistance number (PREN) is 35.7, compared to UNS S31600 with PREN of 32, and it has approximately twice the yield strength. This grade is used in boat shafting, fastenings, and marine hardware. Nitronic 50 (UNS S20910), with a yield strength of 345 MPa (50 ksi), and Aquamet 22, with a yield strength of 896 MPa (130 ksi), are essentially the same alloy, except that the Aquamet grades 17, 18, 19, and 22 are hardened for propeller shaft applications. Nitronic 50 is widely used in sailboat rigging, chains, fasteners, pumps, and other applications.

Another grade of stainless steel that has found wide use in recreational and workboats as propeller and pump shafting, hydrofoil struts, and other marine applications is UNS S17400 (17-4 PH), precipitation-hardening martensitic grade. Corrosion resistance is approximately the same as UNS S30400 (PREN 20); however, yield strength at 1262 MPa (183 ksi) is more than five times that of UNS S30400.

Increasingly stringent environmental and pollution regulations have increased the use of stainless steels. Development efforts have followed two basic approaches. Increased chromium and nitrogen content with a microstructure that is approximately equal parts of austenite and ferrite is referred to as the duplex grade. The other approach was based primarily on increased molybdenum content while retaining the austenite microstructure. This is referred to as the superaustenitic grade. Both approaches depended heavily on increased nitrogen content and control made possible by the development of argon-oxygen decarburization processing.

The duplex grades have better resistance to pitting and crevice corrosion (PREN 39) and to chloride stress-corrosion cracking. They are used primarily in the offshore oil and gas industry. The duplex grades include UNS S31803 (2205), with a yield strength of 517 MPa (75 ksi), and UNS S32550 (255), also called Ferralium 255, with a yield strength of 675 MPa (98 ksi).

The superaustenitics, also refereed to as the 6 Mo stainless steels, are outstanding marine metals. They have the best corrosion resistance of all, with an average PREN of 44.4, and excellent strength. Typical grades are UNS S31254 (254 SMO) and UNS S32654

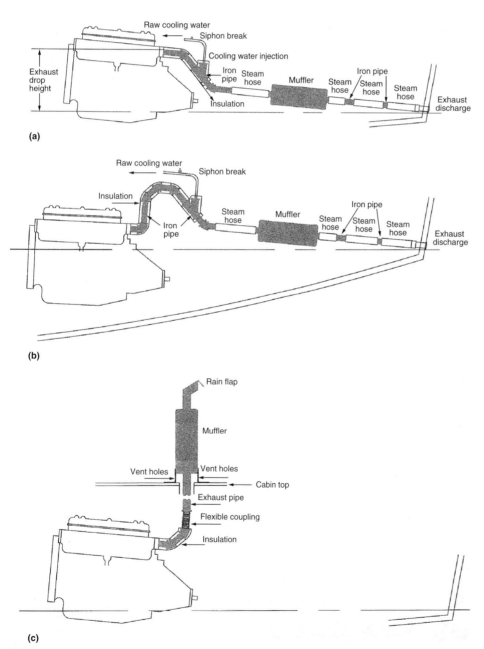

Fig. 6 Exhaust systems. (a) Traditional wet system. (b) Sailboat wet system. (c) Dry system

(654 SMO), with yield strengths of 300 MPa (44 ksi) and 425 MPa (62 ksi), respectively.

While the duplex and superaustenitic alloys have found significant and increasing use in the offshore oil and gas industry, they have thus far made no real penetration into the highly competitive small-boat market because of cost. However, with their superior properties, use is expected to increase on boats.

Tobin bronze, which is roughly 60% Cu and 39% Zn, is, like manganese bronze, subject to dezincification. Bronze propellers on stainless steel shafts require cathodic protection. The propeller, propeller shaft, and rudder assembly can be protected by the use of zinc anodes. The collar zinc, such as that shown in Fig. 7(a), is installed on the propeller shaft. Zinc anodes may also be attached to the struts. The propeller nut zinc, such as that shown in Fig. 7(b), is also sometimes used on the end of the propeller shaft behind the propeller nut.

Alternatively, one or more zinc anodes can be mounted on the outside of the hull in proximity to and in line of sight of the shaft and propeller and then connected electrically to the propeller and shaft (Fig. 8). A wire is run between the zinc mounting bolts on the inside of the hull and the electrical system common ground on the engine block (frequently an engine mounting bolt). This method has a couple of advantages over direct mounting of the zinc on the fitting to be protected. It provides a broader distribution of the ion exchange field and more uniform protection over the surface of the fitting. Also, by electrically connecting other underwater fittings to the common groundpoint, they too will receive cathodic protection from the zincs.

Fig. 7 Zinc sacrificial anodes. (a) Shaft collar zincs are used on immersed exposed propeller and rudder shafts. (b) A propeller nut zinc is threaded onto the end of the propeller shaft

The effectiveness of the cathodic protection system depends on low-resistance electrical paths. Conductors of adequate size and high-conductivity splices and contacts are needed. Connecting the wire to the engine block and relying on the conductivity of the block, transmission housing, and shaft coupling is not satisfactory. The electrical resistance of this path is high and not reliable, especially where a flexible shaft coupling is used. The connection to the shaft should be made through a spring-loaded phosphor bronze shaft brush. If the rudder and rudder stock are to be included in this protection system, the rudder stock must also be connected.

Zinc anodes work best when they are directly attached to the metal they are protecting. In this case, a plate-type zinc, similar to that shown in Fig. 8(a), is commonly used. Single anodes with all of the fittings connected to them provide uniform protection, because the potential is the same. With separate zincs on different fittings, some could be overprotected while others may not be receiving enough protection. Most importantly, the anodes must not be painted. Anodes that have through-holes in them for fastening will fall off as the zinc corrodes and the holes open up. The zinc should have a cast-in

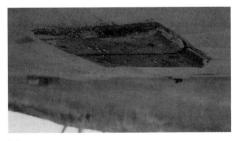

Fig. 8 Zinc sacrificial anodes for the boat underbody. (a) Remote zinc plate. (b) Teardrop zinc. (c) Remote plate installation

plate and fasteners, typically of stainless steel, to prevent this.

The zinc and its through-bolts should be insulated from the wood hull to reduce the possibility of alkali damage to the wood. A teardrop-shaped zinc (Fig. 8b) mounted on, but insulated from, a fairing block reasonably close to the metals it is protecting would be ideal. It should be located so that it will not disturb the smooth flow of water to the propeller, and it should be immersed at all times.

Electrical and Electronic Systems

No other system on the boat is as susceptible to performance-degrading corrosive damage as the electrical and electronic systems and equipment. Components contain many dissimilar alloys and nonmetals in intimate contact. Components are in atmospheres that range from total immersion in the bilge to high humidity, high temperature in the engine room to the salt-laden, spray-soaked flybridge or tuna towers. Because of their increased complexity, smaller size, closer spacings in circuitry and components, and sensitivity to high temperatures, extremely low levels of corrosive contaminants can cause degradation or complete failure in modern electronic equipment.

It is helpful to think about the electrical system in terms of its separate functions: power generation and maintenance, control and distribution, and the loads or devices that it serves.

Power Generation and Maintenance. All loads perform at their best when they have a full 12 V at their terminals. Because there is only 12.5 V on a fully charged battery, minimizing voltage loss by using high-quality copper marine battery cables of 33 mm^2 (0.05 in.^{2}) (American wire gage, or AWG, 2) size is recommended. Automotive or welding cable and clamp-on or solder-on battery clamps are not satisfactory. Preassembled marine cables consist of heavy-duty tinned copper terminal ends and tinned fine-strand copper wire. The ends are molded in an extrathick coating of plastic to prevent water from wicking up into the cable, causing corrosion, increasing resistance, and resulting in cable deterioration and a safety hazard.

Distribution and Control. The distribution panel is the key element in the distribution and control of electrical power to the loads. Typically, it consists of a baked enamel or anodized aluminum panel with circuit protection devices, terminal blocks, and other electrical components mounted on it. The circuit protection devices may be either fuses and toggle switches or circuit breakers. The panel may also contain voltage and current meters. It may be as simple as a six circuit fuse and toggle switch panel or as complex as the 28 branch alternating current (ac) and direct current (dc) instrumented circuit breaker panel shown in Fig. 9(a).

Figure 9(b) shows the back panel wiring for an 11 m (36 ft) inshore commercial workboat, skippered by a hardworking fisherman who did not have time for the electrical niceties. Moisture penetrates into all cracks and crevices—wire splices, butt connectors, screw contacts, fuse blocks, and even nonmarine-grade circuit breakers. Contacts and mating surfaces oxidize, resistance increases, voltage is lost, and heat is generated. At best, performance deteriorates or equipment fails; at worst, the possibility of fire exists. This boat also had a severe stray current corrosion problem.

The distribution panel should be located high above the bilge in a dry, protected location. The enclosure should provide for drainage at the bottom and ventilation at the top. Cable and wire access should be from the side and should lead down and away from the enclosure, forming drip loops (Fig. 10). After all contacts, connections, and metal surfaces are clean and properly made up, a moisture-displacing corrosion-preventive spray should be used.

Wiring and Loads. The only wire that should be used in the boat electrical system is marine-grade multistrand tinned copper wire with a heavy protective sheathing. Consult standards such as Ref 5 or 6. Solid conductor wire simply does not stand up well to the considerable vibration and occasional severe dynamic shock experience on boats, and it also tends to work harden and fracture. This can cause hard-to-find faults. Multistrand wire has more and finer strands and is therefore more flexible and better able to withstand the shock and vibration. Each individual strand is tinned to better withstand the corrosive atmosphere.

If cable ends are unsealed, moisture tends to wick between wire strands under the insulation, causing corrosion and accelerating failure.

Wire sheathing types TW, THW, and THWN (thermoplastic sheathing) are recommended by the American Boat and Yacht Council (ABYC) for general wiring applications, type XHHW (cross-linked polymer sheathing) for high-heat areas, and MTW, which is also oil resistant, for engine room applications. These wires should be type 2 or 3.

Solder Splices. When properly made, soldered splices consisting of two (or more) wires twisted together, soldered, encapsulated in electrician's putty or some other moisture-excluding plastic encapsulation, then covered with either heat-shrink tube or high-dielectric plastic tape, or both, can provide long service. If improperly made, the result is a high-resistance connection; excessive heat results, the solder softens or melts, and the splice comes apart. Soldering, in effect, transforms the flexible-stranded conductor into a solid conductor with its susceptibility to fatigue failure. The ABYC standards do not allow solder to be used as the sole means of mechanical connection of wires.

Crimp connectors, encapsulated to prevent moisture penetration and covered with shrink tube, are preferred. There are three types of crimp terminals: ring, captive spade, and spade terminals. The ring-type terminal is generally recommended, because it cannot come off the stud (screw terminal) if the fastening becomes loose, as the spade terminal may. However, this type of terminal requires the complete removal of the retaining screw or the nut on the stud, and the nut or screw, when removed, can fall into the bilge. For this reason, many prefer the captive spade whose upturned ends assist in preventing this. In-line crimp connectors are also used to connect wires.

There are basically two ways to make a completely airtight and waterproof crimped connection. The first is to use a terminal or connector with an adhesive-lined heat-shrink sleeve. The sleeve should overlap the terminal end, so that when heated, the shrink tube covers the ends of the terminal, making a waterproof fitting. Alternatively, a regular nylon-insulated crimp-type connector can be used but covered with adhesive-lined heat-shrink tubing.

Wire nuts, such as those commonly used in household wiring, should never be used in marine wiring. Typically, the threaded metal insert in these devices is of mild steel and is subject to rusting and a degraded connection.

Electronic equipment includes a wide variety of metals and alloys. Printed circuit boards alone may contain copper, copper alloys, copper-clad materials, electroplating, solder, tin, and lead. Electrical contacts may consist of copper alloys, clad steels, gold, palladium, silver, and tin. Connectors may use beryllium-copper, stainless steels, gold, palladium, silver, and tin. Switches and relays include copper alloys, steels, stainless steels, electroplating, and various contact surface materials. Electronic equipment is highly susceptible to both single-metal (atmospheric) and galvanic corrosion. Many manufacturers of electronic equipment use sealed enclosures to prevent corrosion. Even so, moisture intrusion continues to be a problem. The relative humidity inside the box must remain below 40% to ensure that corrosion problems do not occur. Electronic equipment that uses forced ventilation (fans) from the ambient atmosphere for cooling should be avoided in marine atmospheres. This type of cooling will greatly accelerate corrosion as a result of the moist salt air circulating in intimate contact with the internal circuitry.

Electronic equipment used on boats should be designed for marine applications. Such designs include encapsulated components and protective conformal coatings on printed circuit boards. The printed circuit boards should be mounted vertically above the bottom of the enclosure to prevent moisture accumulation. Cable entries should be on the sides of the box (not the bottom), and cabling should lead down and away from their connectors and be arranged to form drip loops in the cable runs (Fig. 10). Small, sealed packages of desiccant may be placed inside enclosures to absorb moisture during shipping and storage. Often, the desiccant is a salt (lithium chloride) that eventually

Fig. 9 Electrical power distribution. (a) 28 branch alternating current and direct current circuit breaker panel. (b) Wiring for an 11 m (36 ft) workboat without distribution panel

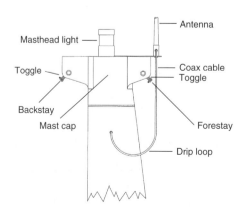

Fig. 10 Typical cable drip loop to prevent water entry into a hollow mast

becomes saturated, leaving a soggy bag of salt inside the equipment, so these should be removed.

Electrical contacts are especially susceptible to corrosive damage. This is due to the low connector contact force and the low voltage, which result in greater sensitivity to tiny amounts of corrosion products and contaminants. Typically, the material in these contact surfaces is tin, gold, or silver plating. Tin plating is susceptible to oxide formation that increases the contact resistance. Most frequently, failure of this type of contact is due to fretting corrosion, that is, repeated oxidation of the contact surfaces due to vibration and minute movement of the contact surfaces. Gold is not subject to atmospheric corrosion. Due to economic and mechanical considerations, the gold plating is usually very thin and has pores at which galvanic corrosion can occur in marine atmospheres. This can seriously degrade the contact surfaces and result in poor or intermittent contact. Silver has, at times, been used as an underplating for the gold. However, it was found that if there were any sulfides in the environment, the contacts deteriorated rapidly. Nickel plating of contacts has proved effective in reducing these effects. Preventive methods are limited to preventing moisture penetration, providing good mechanical support with strain relief at the connector, and ensuring that all coaxial connectors are properly made up and sealed.

Printed circuit board corrosion is minimized with a protective conformal coating, a sprayed plastic type of film that forms a physical barrier against moisture and contaminants.

Bonding, Grounding, and Lightning Protection. Bonding is the practice of electrically connecting the exposed metallic non-current-carrying components on the boat to the common groundpoint of the boat dc electrical system, which is in turn connected to the boat sacrificial anodes. This includes the engine, transmission, propeller and shaft, metal tanks, metal housings of motors, generators and pumps, metal cabinets, and, in short, all large metal objects. The bonding conductor must be of very heavy-gage copper, at least AWG 8, and the sacrificial anodes must be sufficiently massive to protect all of the metals that are immersed—wholly or partially—in seawater. The objective is to eliminate the possibility of any electrical difference of potential between them. There are basically three reasons for bonding: electrical system grounding, lightning protection, and corrosion protection.

The grounding jumper wire provides a low-resistance path to ground, holding the equipment case at ground potential, eliminating the danger of shock and ignition, and reducing the possibility of corrosion due to stray electrical currents.

The lightning protection system consists of an air terminal mounted at the top of the mast and a heavy down conductor—AWG 6 at a minimum—run in as short, straight, and direct a path as possible to a groundplate mounted externally on the vessel bottom. An interconnecting conductor of at least AWG 8 should be connected between the down conductor and large metal objects within 1.8 m (6 ft) of the down conductor to reduce side flashes. This system forms a zone of protection around the boat. The zone of protection has a base radius equal to the height of the air terminal above the water. It must be high enough so that a line drawn from the top of the air terminal down to the water at a 45° angle does not intersect any part of the boat. If this system is properly implemented and the boat is in the water, anyone or anything within the zone of protection will receive some degree of protection. See Ref 7 for details.

Plumbing Systems

In older boats, it was common practice to use steel (carbon or stainless), aluminum or copper tanks, copper or iron piping and fittings, and brass or iron valves. Today, a wide variety of advanced plastic materials have supplanted these corrosion-susceptible metals. Thus, what one can expect in the way of corrosion problems on a boat depends in large measure on the age of the boat and its plumbing.

Freshwater Systems

A typical pressurized freshwater system consists of one or more water storage tanks and a pump—activated by an integral pressure switch—that draws water from the tanks when an opened tap drops the pressure.

Freshwater Tanks. Of the various metals used to construct potable water tanks—carbon steel, stainless steel, aluminum, and copper—none is without definite limitations and shortcomings. All are susceptible to some form of corrosion.

Carbon steel in damp atmospheres is highly susceptible to corrosion, that is, rusting, and requires some sort of specially formulated coating to ensure long-term satisfactory performance.

Stainless steel depends on the availability of a fresh supply of oxygen to sustain the chromium oxide protective film. When held against a moist or damp material, such as support blocks and straps, the necessary supply of oxygen is denied, and the metal is subject to corrosive damage in that area. (Aluminum suffers from this same dependency.) Also, potable water, by definition, is freshwater in which the chloride content is less than 250 ppm by U. S. standards and less than 350 ppm by World Health Organization standards. However, city water may contain chloride levels of 1500 ppm or even higher, depending on the source. This can cause pitting or crevice attack in stainless steels. One should also be concerned about traces of bacteria in the water supply, because stainless steel is susceptible to microbiologically influenced corrosion. Regular flushing of the system a couple of times each season is recommended.

Galvanized Steel. Galvanized tanks have also been used in the past, although their use has been generally discontinued.

Copper was also used quite frequently for water tanks in older boats, because protective films typically form on the surface. However, in soft waters—water with little or no mineral content—these protective films are not able to form, and copper corrosion rates can become significant. A greater concern with the use of copper tanks is the solder used in their fabrication. This can result in dangerously high concentrations of lead in the water, which can represent a serious health hazard. Copper use has fallen quite sharply in recent years, and more attractive alternatives to copper tanks are available.

Aluminum is a popular choice for water tanks. Aluminum and its alloys are lightweight, easy to fabricate, of moderate cost (especially when compared to stainless steel and Monel), and have good corrosion resistance (except for the copper bearing grades, the UNS A92000 and A02000 series) to freshwater. However, much depends on the specific water chemistry. Aluminum is very susceptible to attack by certain heavy metal ions, particularly iron, copper, and lead. Slight concentrations of even a few parts per million will cause severe pitting. Copper tubing upstream of aluminum tanks can easily contain such concentrations. Also, aluminum is highly susceptible to severe pitting resulting from poultice corrosion. This happens when damp organic materials, such as insulation under restraining straps, wood blocking (especially oak), and accumulations of dirt and debris, are in contact with the tank.

Monel is the premium material from which to fabricate water tanks. While more expensive than other metals, it is stronger, far less susceptible to corrosion, and does not impart strange flavors to the water.

Fiberglass. Built-in fiberglass water tanks are frequently used in fiberglass boats.

Polyethylene. Thick-wall high-density polyethylene is an extremely attractive material for water tanks.

Piping. Iron, stainless steel, rubber, clear plastic, or copper were used in older boats. In more recent years, the use of plastic piping is more likely.

Copper. The preceding comments concerning copper tanks also apply to copper tubing. Copper tubing is also very susceptible to impingement corrosion, the erosion that occurs in fast-flowing turbulent water. Furthermore, copper is still subject to corrosion—atmospheric, galvanic, and stray current. There are more attractive alternatives.

Clear plastic vinyl hose is not an attractive alternative, despite the fact that it has been used extensively for quite some time. Clear hose admits light, and this fosters the growth of mold, fungi, and bacteria.

Polybutylene, a plastic resin, was used extensively in the production of water piping from approximately 1979 until approximately

1995. However, oxidants, such as chlorine, react with the polybutylene, causing it to become brittle and develop tiny cracks. The strength of the pipe is seriously diminished and subject to sudden failure. Polybutylene has been abandoned.

Rigid opaque plastic tubing makes excellent water piping. This is the popular polyvinyl chloride and chlorinated polyvinyl chloride plastic tubing and fittings that are increasingly familiar to both homeowners and the boating community. Because it is opaque, mold, fungi, and bacteria are not a problem. It is easier to work with than copper tubing. No solder (or lead) is involved; joints are glued. It does require isolation from shock and vibration.

Flexible opaque plastic tubing suitable for potable water systems is a more recent development. One such hose is referred to as PEX hose, which stands for polyethylene, cross linked. Cross linking refers to the molecular structure of the material and results in considerably increased strength. This is a very attractive alternative to the older materials discussed previously. It is strong, flexible, and easy to lay out and plumb. It will not corrode, and it is opaque, so there are no mold or fungi problems. A similar polyethylene tubing, although not cross linked, is also available commercially. Although not as strong as PEX, its strength is more than adequate for marine freshwater piping applications, and it is more flexible. Both materials are excellent piping materials for marine use in both hot and cold water systems.

Pumps. The pump most commonly used in freshwater systems is the diaphragm pump. Corrosion problems are primarily limited to the housings, which are typically aluminum, and to the fastenings, which are typically stainless steel. This often results in seized fastenings that make the pump difficult to disassemble for simple repairs and cleaning. Also, if wires, contacts, or connections are corroded, voltage at the pump will be lower than it should be, and the pump will run slowly and overheat.

Accumulator tanks may be made of stainless or coated steel, nylon, or acrylonitrile-butadiene-styrene plastic. Corrosion considerations are essentially the same as for water tanks.

Hot Water Tanks. Typical marine water heaters consist of an inner tank, with an inlet fitting at the bottom and an outlet at the top, surrounded by a layer of insulation and all enclosed within a protective shell. An electric heating element is mounted through an access panel on the side of the heater shell. Protective outer shells may be made of stainless steel, coated carbon steel, or polyurethane with a hardened finish coat. Because hot water heaters are usually installed in relatively protected areas, the only concern is atmospheric corrosion.

The tank itself is most often made of stainless steel or glass-lined carbon steel, although marine-grade aluminum is sometimes used. Glass-lined tanks are carbon steel with a porcelain coating. Porcelain is very brittle, and cracking and crazing occur over time. Water inevitably finds its way to the steel substrate. For this reason, glass-lined hot water heaters are usually fitted with a replaceable magnesium sacrificial anode. The anode is frequently an integral part of the hot water discharge fitting. The anode should be checked at least once a year by removing it.

Aside from galvanic corrosion, stainless steels are susceptible to four basic types of corrosion that are significant in this application: pitting, crevice, intergranular, and stress-corrosion cracking. Chloride is the most common agent for initiation of pitting. Freshwater can contain chloride levels as high as 600 ppm, sufficient to cause concern, especially if the water is treated with additional chlorine for purification or flushing. Crevice corrosion can be thought of as a severe form of pitting corrosion, and a well-designed tank will keep crevices to a minimum. Weld decay, a form of intergranular corrosion associated with welds, results from the sensitization developed in the heat-affected zone during welding. An improperly welded tank or the use of improper filler metals or grade of stainless steel can result in susceptibility to crevice corrosion. Stress-corrosion cracking requires residual or applied surface tensile stresses, chlorides, and an electrolyte (water). Most mill products—sheet, plate, pipe, and tubing—contain enough residual tensile stresses from processing to be susceptible without the presence of any external stresses. Cold forming and welding add additional stresses. Laboratory studies have shown that stress-corrosion cracking can occur with chloride concentrations of less than 10 ppm.

Aluminum is sometimes used for both the heating tank and the heat exchanger. The best aluminum is a corrosion-resistant marine grade that is reportedly not as susceptible as stainless steel to leaks at welded seams. Experiential data are lacking, but this is not the material of choice for the top-of-the-line hot water heaters.

Heat exchanger coils should be made of cupronickel. Copper fittings are corrosion resistant and are preferred to galvanized fittings. Copper tubing is frequently used in the hot water circuit and will give good service, keeping in mind problems with erosion-corrosion. Electrical heating elements specifically designed for this application are of tin-plated copper, and they stand up well.

Wastewater Systems. Federal marine sanitation laws require that all boats with installed toilet facilities must also have a marine sanitation device (MSD) that treats and/or stores the waste. The MSDs are Coast Guard-certified types I, II (macerators), and III (holding tanks). The toilet, the chosen MSD, is combined with piping, valves, and perhaps pumps to assemble a sanitation system.

Toilets. There is little to go wrong with a manual marine toilet, as long as it is operated correctly, cared for regularly, and given a periodic rebuild. Keep the pump works lubricated and thus protected from wear and corrosion. Electric macerating toilets are not much more complex than simple manual toilets.

Types I and II MSDs. Where treated discharge is legal, vessels 20 m (65 ft) and under may use either type I or II MSD. Aside from pump impellers, the principal corrosion concern is shaft seals and gaskets damaged by exposure to harsh chemicals. The best preventive procedure is frequent freshwater flushes.

Type III MSD. A type III MSD is a holding tank. Both aluminum and stainless steel have been used for holding tanks. This is not a suitable application for these metals or for any of the common metals, for that matter. Urine is extremely corrosive. It rapidly and aggressively attacks seams, weld laps, and crevices. Typically, metal tanks will begin leaking at the seams and fittings in a year or two. Coatings are not very effective, and repairs are frequent and expensive. The most cost-effective solution for metal holding tank problems is to change them out for a more suitable material. Seamless, thick-walled molded polyethylene is the material of choice.

Vented Loops. When a toilet is installed below the waterline, as it is on most sailboats and in many power boats, there is the risk that water will siphon back into the toilet, perhaps flooding the boat. To prevent this, discharge (and sometimes inlet) hoses lead to and from a U-shaped tube mounted above the waterline at all angles of heel. This device is called a vented loop or siphon break and is available in bronze, Marelon, and plastics. Periodic inspection, disassembly, and cleaning are essential.

Masts, Spars, and Rigging

The mast and its support components—spreaders, stays, shrouds, terminals, toggles, turnbuckles, and chainplates—comprise an integrated, interdependent system that is highly stressed under static conditions and subject to even greater dynamic stresses when sailing. Failure can occur at a number of points in the system, many, or possibly any, of which could bring down the entire rig. The principal components are all susceptible to stress, fatigue, corrosion, and/or rot, and detection of these conditions requires close and regular inspection.

Masts are wood, metal, or composite. Each of these materials presents unique corrosion concerns.

Aluminum spars are generally extruded sections of 6061-T6 (UNS A96061)- or 6063-T6 (UNS A96063)-grade aluminum (Fig. 11). These are excellent corrosion-resistant grades, and properly designed aluminum masts are relatively low-maintenance systems. Problems with aluminum spars are with the fittings rather than the mast extrusion. However, aluminum masts are of relatively thin sections and cannot afford to lose much wall thickness to corrosion without suffering significant strength reduction. They are susceptible to galvanic corrosion, especially with copper, so all copper and copper bearing metals must be isolated from contact with an aluminum mast.

Both UNS A96061 and A96063 are marine-grade alloys with good resistance to atmospheric corrosion. They quickly develop a protective barrier oxide film such that the corrosion rate declines with time. For the most part, the result is a mild roughening of the surface but no significant thinning of the mast section. The consequences of atmospheric corrosion usually have more to do with appearance than mechanical damage, as long as the protective film remains intact. If the film is continually chafed or abraded away, it loses its protection, and significant loss of metal can occur. Some aluminum spars are anodized, an electrochemical bath process done before any fittings are installed. These spars are highly resistant to corrosion and tend to maintain their finish well. A good paint coating can be attractive, protective, and long-lasting if done properly, but painting aluminum requires special care and preparation.

Welds are the first places to look on an aluminum spar for evidence of corrosion. Poor welds will have rough, sharp edges and crevices where water can collect and cause corrosion. Corroded welds should be redone immediately. Rivets are another place that should be checked regularly. Faulty rivets should be drilled out and replaced with the next-larger size. Anywhere water is allowed to collect and stand against the mast, pitting corrosion can be severe. Compare the depth of corrosion pits with the thickness of the mast at normal openings in the mast. If the corrosive penetration approaches 25% of the wall thickness, or if the damage forms a continuous ring around the mast, serious weakening has occurred, and repair or replacement is essential.

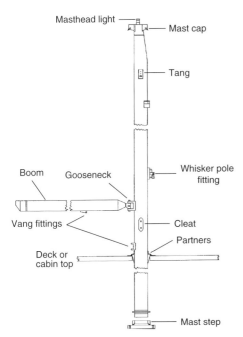

Fig. 11 Typical aluminum mast system

Wood. Most wooden masts are now of hollow-section construction with solid reinforcing blocks where spreaders, tangs, and other components are attached. Sitka spruce is the preferred wood due to its great strength-to-weight ratio. Fastenings in a wooden mast should be of silicon bronze. Mild steel should be avoided because of rusting and deterioration of the wood surrounding the fastening, resulting in a weak and unreliable fastening. Even when galvanized, it is just a matter of time before rusting will occur. Type 316 stainless steel should be used if bronze cannot be used because of galvanic compatibility.

Steel masts are sometimes used in very large sailing vessels, but weight makes steel impractical for most sailboat applications. However, many small commercial boats use steel spars for general lifting or for handling fishing gear. Typically, these are of cold-rolled steel. Almost any low-carbon steel may be used.

With steel spars, any place that water can accumulate is a potential problem area—gaps between adjoining surfaces, pockets, troughs, and corners. All fittings attached to the mast should be continuously welded along their perimeters and the welds ground smooth to avoid water traps and to facilitate painting. It is best if the mast is made air- and watertight, that is, all seams continuously welded and the mast fitted with external halyards and sheaves so that there need be no entry or exit boxes to allow water to get inside. If openings exist through which water can get into the mast, water traps and drains must be provided, and the mast should be painted internally to inhibit corrosion. If electrical wiring is run internally (in polyvinyl chloride tubing), waterproof plugs and connectors or cable gripping wire seals must be used at the entry and exit points. The external surface of the mast must, of course, be painted. Mild steel surfaces subject to wear or abrasion corrode very quickly, which can be dangerous for standing rigging fittings and trawl gear.

Carbon Fiber. Carbon occurs in its natural state as graphite, a highly noble material with a corrosion potential on the order of +0.23 to +0.30 V—at the most noble or cathodic end of the galvanic series. It is highly resistant to atmospheric corrosion. However, the metals typically used in fittings and fastenings—aluminum, at −0.76 to −1.00 V; steel, at −0.60 to −0.71 V; and even bronze, at −0.4 to −0.31 V—are all quite anodic with respect to carbon fiber. Given the necessary electrolyte—saltwater, spray, or dew—the inevitable result is rapid galvanic corrosion of the less noble metal. Titanium, at −0.05 to −0.06 V, and type 316 stainless steel, at 0.00 to −0.10 V, are marginally acceptable. Use of stainless steel fastening galvanic separations in the 200 mV range and an unfavorable cathode-to-anode area ratio are judged based on relative size and duration of localized wetted surfaces. Fastenings should be inspected annually. Alternatively, a fiberglass or epoxy plug for the fastening can be used. Such plugs are often installed by the manufacturer and can be added for aftermarket fittings. Insulating plugs and pads can be Delrin, Teflon, nylon, Mylar, or polyethylene. The edges between the carbon-fiber spar and the fitting should be sealed with a silicone sealant.

Booms and Spreaders. Booms may be made of the materials from which the masts are made; the differences are in the end fittings. Booms normally have an outhaul sheave at the afterend and a gooseneck fitting at the mast. These are, of course, made of metal—bronze or stainless steel in the case of a wood boom, stainless steel or aluminum in the case of aluminum. Bronze or any copper-base alloy cannot be used on an aluminum spar.

If a spreader fails, the mast has lost the support of the shrouds on that side. If the boat is under sail, the mast and, consequently, the boat and crew are significantly endangered. Damage or corrosion at the spreaders is ranked second only to faulty swage fittings as the primary cause of mast failures. For wood spreaders, fastenings at the spreader roots commonly consist of a stainless steel or bronze plate on the top surface and a matching one on the bottom surface, with through-bolts or clevis pins. These should be thoroughly bedded to prevent moisture intrusion. Concerns about aluminum spreaders are essentially the same as those for aluminum masts: Welds, rivets, screws, and through-bolts should show no signs of corrosion. When inspecting fastenings, inspect for stress cracks around the holes. At winter layup, give the spreader tips a generous squirt of a moisture-displacing anti-corrosion spray.

Mast Step. The heel of the mast and the mast step, whether on deck or on the keel, are highly susceptible to corrosion. In the case of wood masts, anticipate rot, deterioration, and corrosion of fastenings in this area. Also, the heel of the mast works when sailing, so there are problems of abrasion and wear to exacerbate deterioration caused by moisture penetration. In the case of aluminum, steel, and carbon-fiber masts stepped on the keel, there is also the potential for galvanic corrosion of bronze or steel keel bolts and floor fastenings, should they come in contact. Avoiding or insulating such contacts and keeping the bilge dry will help to eliminate this source of corrosive damage. Dissimilar metal fasteners in the mast and step must be isolated or insulated. Protective paint coatings and corrosion-inhibiting lubricants are recommended.

Mast Partners. If the mast is stepped on the keel, the place where the mast passes through the deck is referred to as the partners. This opening is typically greater in diameter than the mast to allow for some adjustment. Often, an aluminum mast will be internally reinforced in the area of the partners and provided with drain holes above the boot. Maintenance and prevention consists of keeping drain holes clear and removing the boot periodically for examination of the mast.

Mast Cap. The mast cap, also called the masthead box, is one of the more important of the mast fittings. Mast caps are often cast aluminum alloy on smaller boats and sometimes galvanized steel on older boats, but most are fabricated from stainless steel. Concerns are with galvanic corrosion between the cap and its fastenings and the mast. Corrosion is localized and limited to the interface between the two dissimilar metals.

Sail Track. The extruded luff track in aluminum masts normally does not cause serious problems, as long as nylon slides or bolt ropes are used. On wooden masts, there are basically two types of sail track: the luff groove, a slot routed into the afterside of the mast, and the metal slide track. The luff groove has the virtues of simplicity, immunity to corrosion, and virtual trouble-free service, because it only sees the luff rope. The metal slide track of brass, aluminum, and stainless steel is susceptible to corrosion, which can create friction and jam the slides. Nylon slides help a great deal, but lubricating the slides and track with a moisture-displacing corrosion inhibitor is also very helpful.

On a steel mast, a well-bedded wood spacer is mounted on the afterface of the mast, and the sail track is fastened to this spacer, using machine screws tapped into the mast. One must be concerned with the potential for galvanic corrosion.

Fittings and Fastenings. Some spar fittings are cast aluminum—either 356.0 (UNS A03560) for end caps and other light duties, or 535.0 (UNS A05350) for cleats and other load-bearing fittings. Others, particularly tangs, pad eyes, and the like, are of stainless steel. Even bronze fittings, usually in the form of halyard winches, are sometimes attached to aluminum masts. Here again, all metal fittings that are not aluminum must be insulated from the mast, using pads of neoprene, polyvinyl chloride, Delrin, wood, or some other nonconductive material.

Fastenings are normally stainless steel screws or rivets. Galvanic corrosion will inevitably take place but will be limited to the interface between the fastening and the mast. Look for telltale white powdery residue around the fastening. Preventive maintenance for threaded fastenings consists of annual withdrawal, cleaning of the mating surfaces, and coating with a moisture-displacing anticorrosion lubricant. With rivets, corrosion enlarges the hole, and the rivet becomes loose. The fix consists of drilling out the old rivet and replacing it with the next-larger size, using a closed-end (water- and vaportight) rivet with both the rivet body and mandrel metals of aluminum (5052 or 5056), stainless steel, or Monel. Keep an eye out for cracks and pitting corrosion in mast tangs, chainplates, and other fittings subject to stress and vibration. The part should be replaced if any cracks are found.

Masthead electronics failures are usually the result of faulty connections, most often at the bottom of the mast. The connections at the masthead are also vulnerable to moisture penetration, followed by corrosion, signal degradation, poor performance, and ultimately failure. For electronics, the point of greatest susceptibility is the coaxial cable connector, such as the PL-259 type or BNC types. Also, the coaxial cable itself, by virtue of its construction, has a tendency to wick moisture between the sheath and the tinned-copper braided shield. These connectors have a trouble-free life of approximately three years.

Standing Rigging. There are basically three types of standing rigging in common use: galvanized wire rope, stainless steel wire rope, and rod rigging. Galvanized wire rope is used mostly on commercial vessels for reasons of cost-effectiveness and by traditionalists for authenticity. Stainless wire rope, by far the most popular, is the de facto standard for recreational yachts. Rod rigging is used almost exclusively by high-tech, high-performance sailboats.

Wire rope is made up of multiple strands of thinner wire twisted together into a specific configuration. The most common type for standing rigging is 1 by 19, a central strand with 18 strands of the same gage wire wound around it. Other common types are 7 by 7 and 7 by 19. Because the 1 by 19 is too inflexible to wrap around a thimble or a rigging eye, it requires rigging terminals of another type.

Galvanized and stainless steel wire ropes are both susceptible to several forms of corrosive damage: rusting, pitting, and crevice corrosion. Rod rigging is less susceptible because of the alloy normally used—Nitronic 50—and because rod has a smoother surface and fewer crevices. Both wire and rod are also susceptible to corrosion fatigue and, according to some, stress-corrosion cracking.

The strands at the center of the cables tend to corrode first and most rapidly because of oxygen depletion. The most likely failure point is at the end terminals, but rust and broken strands do occur along the length of the wire rope, especially low on the shrouds and stays. The telltale warning signs are broken strands. Many authorities say that there is no acceptable number of broken strands, and replacement is advised.

Corrosion fatigue can affect the shrouds, stays, and rigging fittings, especially tangs and chainplates. Stress-corrosion cracking as a rigging concern is a bit controversial (Ref 8–13). Failure analysis of riggings is not conclusive, and some controlled experiments on austenitic stainless steel specimens have been unable to replicate stress-corrosion cracking in marine atmospheres. The author's opinion is that stress-corrosion cracking is a possible, even likely, explanation for some otherwise unexplained rig failures. To prevent stress-corrosion cracking, select a metal alloy such as the superaustenitic or 6 Mo (6% Mo) alloys—254 SMO, 20Mo-6, or AL-6XN—or alleviate the cause by reducing residual-stress levels through heat treating, eliminating places where aqueous chlorides can collect or concentrate, lowering the ambient temperature, and applying some form of cathodic protection.

When galvanized wire rope begins to corrode, it rusts, typically in the strands at the core at the lower ends of shrouds. When the rust is very slight, it may be turned end-for-end, getting another season or two out of it. Galvanized wire rope is normally lubricated as part of the manufacturing process, and maintaining the lubrication to help prevent corrosion and reduce wear is a simple matter once a year with an oil-soaked rag.

When stainless steel wire rope begins to corrode, it develops pitting and crevice corrosion in the strands at the core, not visible in the early stages. Pitting and crevice corrosion in stainless steel wire rope are extremely serious and form the initiation sites for stress-corrosion cracking. Stainless steel grades most commonly used in wire rope are the type 18–8's, primarily types 302 and 304. However, type 316 is the standard high-corrosion-resistance grade that is the recommended grade of the Offshore Racing Council.

Wire rope intended for lifelines and steering cables is frequently coated (sheathed) with nylon or vinyl, sealing in cable lubricants and keeping out dirt and moisture, in order to extend its life by protecting it from abrasion. However, moisture can penetrate and wick up under the sheathing if the sheathing becomes cracked or cut, resulting in pitting, crevice corrosion, and broken strands. This can be a serious failing, because the damage is not visible. The Offshore Racing Council regulations do not permit the use of sheathed wire in lifelines.

Rod Rigging. The alloy of choice for rod rigging is a special grade of stainless steel called Nitronic 50 (UNS S20910). This alloy may be more familiar as Aquamet 22, commonly used in propeller shafts. It is a highly corrosion-resistant material that is affected by corrosion fatigue.

Terminals, Turnbuckles, and Chainplates. The most likely cause of rigging system failures is corrosion-induced failure of a terminal fitting, turnbuckle, chainplate, or their fastenings.

Terminal Fittings. One of the oldest methods still in common use is the clamped eye splice. It is simple, quick, inexpensive, and it holds well. It has the advantage of being open to air circulation and to view and the disadvantage of having lots of cracks and crevices to accumulate and concentrate salt deposits, leading to pitting and crevice corrosion. A more popular current option is the Nicopress or, in the United Kingdom, the Talurit fitting. With this type of fitting, a terminal eye is formed by sliding a double-barreled metal sleeve or ferrule over the standing wire rope, bending the rope around the thimble, and passing its bitter end through the other side of the sleeve, then crimping the sleeve with a special tool. Sleeves are available in copper, tin-plated copper, zinc-plated copper, and aluminum. Coating with a dollop of sealant after crimping will also help keep out moisture.

Swage terminals are probably the most widely used of all wire terminal fittings (Fig. 12a). The swage terminal is a stainless steel fitting consisting of an eye, fork (or clevis), or threaded stud on one end and a sleeve on the other. The sleeve is precisely sized to be tight-fitting for the wire rope. The wire is inserted into the sleeve to the full depth, and the sleeve is put through a set of hydraulic rollers and subjected to tremendous pressure. The result is a cold-worked bond between the sleeve and the rope. These fittings have far fewer places for moisture and salt to concentrate, but the stresses introduced by cold working the sleeve, together with the salt-laden atmosphere, set up conditions favorable for stress-corrosion cracking failures. Improper swaging is the major reason these terminal fittings have a poor reputation, especially in the tropics. Attempts are made to seal the opening at the top of the swage with all sorts of things: polysulfide sealants, two-part epoxy adhesives, polyurethanes, and so on. The problem is that moisture is sealed in under the sealant unless the sealant is applied when the terminal is new and in a dry atmosphere.

Another class of terminal fitting is a mechanical assembly that relies on compressive forces applied by the assembly process to hold the wire rope in the fitting (Fig. 12b). Mechanical terminal fittings are quite common on yachts, second only to swage fittings. When properly assembled, mechanical terminals have an excellent reputation for durability and reliability. They are not susceptible to stress-corrosion cracking and, if properly assembled, are highly resistant to moisture penetration and salt concentration. This type of terminal is often referred to by the trade name of one of the best-known manufacturers, Norseman or Sta-Lok.

Another wire terminal end fitting is the spelter fitting, which refers to the molten zinc that is used to hold the end of the wire rope in the terminal fitting. The modern version of the spelter fitting uses fast-setting epoxy in place of the molten zinc.

All of these fittings are continuously exposed to salt-laden moist air, spray, and occasional drenching in saltwater. It is a good idea to seal all openings and crevices on all lower fittings—of any type—with a sealant. However, if the fittings are already in use and have been exposed to salt-laden moisture, the moisture and salt crystals will be sealed in, so thorough flushing and drying before sealing is a must.

Turnbuckles are used to take up slack in the standing rigging. They come in two basic types: open body and closed body. Advocates of closed-body turnbuckles prefer them because they have a tendency to exclude water, and because a grease fitting can be fitted in the barrel and the cavity pumped full of grease periodically to prevent seizing and to inhibit corrosion. Open-body advocates say that this type does not retain water and therefore is less likely to corrode, and they are also easier to inspect. On commercial boats, it is customary to fill the open-body cavity with waterproof grease, wrap the turnbuckle with canvas, and then cover it with a sewn canvas sleeve.

Marine turnbuckles come in four basic materials: galvanized steel, stainless steel, bronze, and a combination of bronze and steel. Galvanized turnbuckles are the most susceptible to corrosion, especially on the threads. Stainless steel does corrode when it is made to stand in oxygen-depleted moisture, which is what happens inside the turnbuckle and around the threads and clevis pins. These are the places where salt crystals concentrate and, given normal loading and some temperature elevation from a tropical sun, the conditions necessary for stress-corrosion cracking are present. Stainless steel is also susceptible to galling. The solution is to keep the thread surfaces clean and well lubricated with a moisture-displacing anticorrosion lubricant. Bronze turnbuckles are resistant to corrosion and galling. There are also stainless-steel-body turnbuckles with bronze threaded inserts and chrome-plated bronze bodies with stainless steel screws. Experience has shown that initial concerns about galvanic reactions between the bronze and the stainless steel were unfounded. This type of turnbuckle has been in wide use for many years with no notable problems. Turnbuckles should receive periodic inspection throughout the sailing season and a thorough examination at least once a year.

Chainplates are the attachment points on the hull for the shrouds and stays. They are typically of steel plate and shaped, formed, and/or welded to the correct angle of alignment with the shroud or stay. Chainplates are under considerable strain, and if they are misaligned, the result is severe working of the fitting and its fastenings, setting up conditions for moisture penetration and several forms of corrosive damage. The more serious failure modes for chainplates are corrosion fatigue and intergranular corrosion (weld decay). Failure usually occurs at the bend or the weld, even though the maximum stress is considerably less than the yield strength. Look for pitting, crevice corrosion, and tiny cracks in these areas. Lifting or bubbling of paint coatings is another indication of pending trouble.

Running Rigging. The running rigging consists of the various lines that control sails and spars: halyards, sheets, downhauls, outhauls, topping lifts, and so on. These lines are natural fiber, synthetic fiber, wire, or some combination of these. Wire rope used in running-rigging applications is typically 7 by 19. It may be galvanized or stainless steel. Sometimes a rope-to-wire splice is used on halyards to provide a fiber "tail" for ease of handling on the hauling end. Like standing rigging, wire running rigging is vulnerable to saltair and spray penetration into the core, resulting in the same corrosion problems and making the same precautionary measures appropriate. Regular inspection for signs of rust and broken strands is important.

Blocks and Sheaves. Old-fashioned wood-cheek blocks are not used much anymore, except by traditionalists. Modern blocks are lighter and more efficient and have aluminum, stainless, plastic, or composite cheeks. Straps are invariably stainless, although titanium is sometimes used in blocks intended for competitive racing. Corrosive damage to the cheeks and straps may be unsightly but is seldom the cause of block failure. Sheaves are most commonly aluminum for wire rope. Bearings may be of bronze, stainless steel, or, in some cases, an advanced plastic. Bushings may be oil-impregnated bronze or, in the more expensive blocks, an advanced plastic. Aside from aesthetic deterioration to the cheeks and straps, corrosion damage results from saltwater, dirt, grit, and salt crystal intrusion into the axle bearing/bushing assembly, the part of the block that is most vulnerable to corrosion. Sheaves should be inspected at least once a year for surface corrosion and wear.

ACKNOWLEDGMENT

Special thanks are expressed to Jon Eaton, Editorial Director, International Marine Publishing, for permitting the author to excerpt so extensively from *Boatowner's Guide to Corrosion: A Complete Reference for Boatowners and Marine Professionals* (Ref 1). Without his generous cooperation, this article would not have been possible.

REFERENCES

1. E. Collier, *Boatowner's Guide to Corrosion: A Complete Reference for Boatowners and Marine Professionals*, International Marine Publishing, 2001
2. "Seacocks, Thru-Hull Connections and Drain Plugs," H-27, American Boat and Yacht Council, July 1997

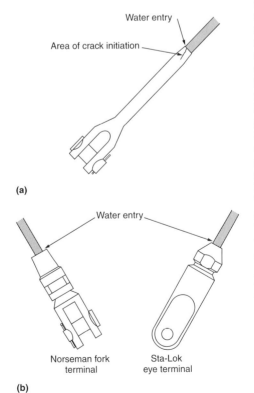

Fig. 12 Terminals for wire rope rigging. (a) Swage terminal fitting. (b) Norseman and Sta-Lok terminal fitting

3. J.G. Kaufman, *Introduction to Aluminum Alloys and Tempers,* ASM International, 2000, p 89
4. "Installation of Exhaust Systems for Propulsion and Auxiliary Engines," P-1, *Standards and Recommended Practices for Small Craft,* American Boat and Yacht Council, July 2002
5. "Conductors, General," Section 425; "Conductors in Circuits of Less Than 50 Volts," Section 430; "Conductors in Circuits of 50 Volts or More," Section 435, Code of Federal Regulations, Title 33, Vol 2, Part 183
6. "Direct Current (dc) Electrical Systems on Boats," E-9, *Standards and Recommended Practices for Small Craft,* American Boat and Yacht Council, 1998
7. "Lightning Protection," E-4, *Standards and Recommended Practices for Small Craft,* American Boat and Yacht Council, 1998
8. C.P. Dillon, Imponderables in Chloride Stress Corrosion Cracking of Stainless Steels, *Mater. Perform.,* Dec 1990, p 66–67
9. "Experience Survey—Stress-Corrosion Cracking of Austenitic Stainless Steels in Water," MTI Publication 27, Materials Technology Institute of the Chemical Processing Industries, 1987
10. J.B. Gnanamoorthy, Stress Corrosion Cracking of Unsensitized Stainless Steels in Ambient-Temperature Coastal Atmosphere, *Mater. Perform.,* Dec 1990, p 63–65
11. R.M. Kain, Marine Atmospheric Stress Corrosion Cracking of Austenitic Stainless Steels, *Mater. Perform.,* Dec 1990, p 60–62
12. K.L. Money, Stress-Corrosion Cracking Behavior of Wrought Fe-Cr-Ni Alloys in Marine Atmospheres, *Mater. Perform.,* 1978
13. J.W. Oldfield and B. Todd, Ambient-Temperature Stress Corrosion Cracking of Austenitic Stainless Steel in Swimming Pools, *Mater. Perform.,* Dec 1990, p 57–58

SELECTED REFERENCES

- D. Ahrens, Engine Advice: What Causes Pitting in Cylinder Liners?, *Mar. Perform. Fish. Prod. News,* Spring 1998, p 27
- C. Cairncross, *Ferrocement Boat Construction,* International Marine Publishing, 1972
- N. Calder, *Marine Diesel Engines,* 2nd ed., International Marine Publishing, 1987
- N. Calder, *Repairs at Sea,* International Marine Publishing, 1988
- N. Calder, Wiring Plans: Tools and Techniques for Ensuring a Reliable Electrical System, *Ocean Nav.,* Nov/Dec 1995, p 83
- N. Calder, *Boatowner's Mechanical and Electrical Manual: How to Maintain, Repair and Improve Your Boat's Essential Systems,* 2nd ed., International Marine Publishing Company, 1996
- J. Clemans, Plumb-It-Yourself, *Motor Boat. Sail.,* March 1999, p 97
- T.E. Colvin, *Steel Boat Building: From Bare Hull to Launching,* Vol 2, 1st ed., International Marine Publishing Company, 1985
- G. Cook, "Ferrocement Boatbuilding," Online Newsgroup rec.boats.building., ⟨graemec@netspace.net.au⟩, Aug 20, 1998
- "Corrosion Resistance of the Austenitic Chromium-Nickel Stainless Steels in Chemical Environments," Publication 2828, Nickel Development Institute
- R.M. Davison et al., Corrosion of Stainless Steels, *Corrosion,* Vol 13, *ASM Handbook,* ASM International, 1987 p 547–565
- C.P. Dillon, *Corrosion Control in the Chemical Process Industries,* 2nd ed., MTI Publication 45, Materials Technology Institute of the Chemical Process Industries, Inc.; NACE International, 1994, p 224
- D. Gerr, Exhaust Installations for Powerboats, *Boatbuilder,* Sept/Oct 1993, p 6
- D.R. Getchell, Sr., *Outboard Boater's Handbook,* International Marine Publishing, 1994
- J.A. Harris, "Failure of Electronic Boards," Newsgroup Corros-L@Listserv.RL.AC.UK, Oct 17, 1997
- G.C. Klingel, *Boatbuilding with Steel,* 1st ed., International Marine Publishing Company, 1973
- J. Klopman, Stainless Steel, *Prof. Boatbuild.,* Aug/Sept 1998, p 70
- I. Nicholson, *Surveying Small Craft,* International Marine Publishing Company, 1974
- R.W. Ross and A.H. Tuthill, "Practical Guide to Using Marine Fasteners," NiDI Technical Series 10 045, The Nickel Development Institute; reprinted from *Mater. Perform.,* April 1990
- "Stainless Steel Fasteners: A Systematic Approach to Their Selection," SS 502-476-18M-CP, American Iron and Steel Institute, April 1976
- R.M. Steward, *Boatbuilding Manual,* International Marine Publishing Company, 1976
- J. Toghill, *The Boat Owner's Maintenance Manual,* John De Graff, Inc., 1979
- A.H. Tuthill, "The Right Metal for Heat Exchanger Tubes," NiDI Technical Series 10 040, The Nickel Development Institute; reprinted from *Chem. Eng.,* Vol 97 (No. 1), Jan 1990
- M. Verney, *Boat Repairs and Conversions,* International Marine Publishing Company, 1977
- G. Virus, "Pop Rivets and Blind Fasteners," Emhart Industries, Aug 1998
- N. Warren, *Metal Corrosion in Boats,* International Marine Publishing Company, 1980
- J.D. Winninghoff, A Few Words of Praise for Aluminum Boats, *Natl. Fisherman,* Sept 1993, p 5
- E.A. Zadig, *The Complete Book of Pleasure Boat Engines,* Prentice-Hall, Inc., 1980

Corrosion of Metal Artifacts and Works of Art in Museum and Collection Environments

Jerry Podany, Paul Getty Museum

BRIDGES, AUTOMOBILES AND SHIPS have little in common with bronze sculptures, silver commemorative plaques, or lead archaeological artifacts, it would seem, other than that they contain metal. Vast resources are spent worldwide to protect the first group from corrosion, despite their exposure to a wide range of corrosive environments. The second group, by contrast, is assumed to be safely sheltered in the protective enclosures of museums or storage facilities where their long-term preservation is assured. However, artifact or bridge, sculpture or ship, all metal corrodes, and the agents of corrosion can be found everywhere, even within the protective walls of a museum. Indeed, the museum walls themselves may be the source of corrosive agents.

Within such protective environments, surprisingly corrosive microclimates can exist. Corrosive pollutants such as ammonia, formaldehyde, acetaldehyde, formic acid, and acetic acid are often found in higher concentrations indoors than outdoors, inducing an array of chemical, electrochemical, and physical corrosion processes. These processes remind us that while the indoor environment can be less aggressive, it is by no means passive. This situation is often due to the fact that the display materials used in museums and for the buildings themselves have dramatically changed over the last century. Cheaper, reformulated products as well as a dramatic increase in synthetic materials have all introduced greater levels of pollutants, most of which are potentially harmful to collections.

There can be an abundance of corrosive agents within storage cabinets, display cases, and galleries. The materials making up display cases, paint on the walls, or the carpets on the floors may all be the sources of an artifact's demise. Indeed, even the materials that were used in the manufacture, cleaning, or repair of an object can contribute to corrosion. Perhaps slower in pace and smaller in scale but destructive nonetheless, these corrosion processes mimic the more dramatic and extensive processes outdoors. It should be noted as well, that buried artifacts can suffer from excavation and subsequent transfer to a seemingly more benign environment. See the article "Corrosion of Metal Artifacts in Buried Environments" in this Volume.

Over the centuries, artists and craftsmen have used an impressive array of alloys. Cast bronze portraits, tin plate, stainless steel and aluminum sculpture, silver and gold jewelry, gilded furniture, and iron decorative plaques are joined in this list by paintings on copper and tin and weavings incorporating metallic threads of gold, copper, and silver. Glass vessels and mosaics shimmer with sandwiched layers of gold, silver, and, in more modern times, even aluminum foil between the layers of glass. The variety of metals and their uses for both artistic expression and industrial design are staggering.

The corrosion processes are detailed elsewhere in this Volume. This article points to the most common of the processes found indoors and focuses on the sources of corrosion in what otherwise would be considered the protective environment of museums and historic collections.

Common Corrosion Processes

The processes and agents of corrosion, like the rusting of iron in the presence of water and oxygen, are just as likely in the museum environment as elsewhere.

Humidity. High relative humidity (RH) is, of course, a leading concern in the preservation of any work of art or historical artifact. Museums or storage facilities lacking adequate climate control in some areas can experience high RH indoors. Collections in tropical zones are particularly challenged, but those in more moderate climates can face 85 to 90% RH for prolonged periods during the summer months.

High RH encourages corrosion in a number of ways. For example, archaeological iron from marine excavations will often have a variety of soluble salts embedded in the corrosion layers. When the humidity is sufficiently high, these salts will attract moisture and dissolve, forming an electrolyte. Iron in this condition exhibits brown droplets of moisture on its surface, giving it the name "sweating" or "weeping" iron.

Temperature fluctuations affect RH. While it is impractical to cool exhibition areas below the comfort level of museum visitors, unnecessary heating of metal objects and their environment by radiant energy can encourage corrosion, because an upward shift in temperature may mean an increase in available water vapor and in the reaction rate. For example, lighting sources within display cases can dramatically increase the interior temperature of the case. The heated air will retain more water vapor. When the lights are turned off at the end of the day, the temperature drops, and the RH rises within the case. As a result, the environment may reach the equilibrium RH of some salts. An electrolyte may form, even in the absence of visible condensation, and corrosion is initiated.

These conditions are relatively straightforward to manage in museums and collections through the strict control of RH by building-wide environmental systems or localized microclimate cases and climate-controlled storage cabinets. Simple preventive efforts may include coatings that protect the metal surface by being somewhat impermeable to moisture. Although direct intervention remains an option, modern conservation practice encourages the control of the environment rather than the imposition of any direct treatment, because such treatments inevitably alter the material nature of the object and are not always effective.

Relative humidity below 40% is an effective way of controlling bronze disease present on archaeological copper alloys that contain layers of reactive cuprous chloride. Bronze disease is essentially a process by which cuprous chloride, nantokite (CuCl), converts to one of the trihydroxychlorides (cupric chloride, $CuCl_2$; atacamite, $Cu_2(OH)_3Cl$; or its dimorph, paratacamite). A dry environment is preferred to direct treatments for bronze disease, such as the application of a water or alcohol solution of benzotriazole, which forms a complex with the copper chlorides but is not always effective over the long term. For these same reasons, the use of effective sulfide scavengers (zinc oxide pellets, for example) and the retention of low RH levels are preferred to the repeated removal of silver tarnish through chemical strippers or abrasive polishes, because both have an irreversible effect on the original material and do nothing to prevent further corrosion.

Temperature. In addition to affecting RH and directly affecting the corrosion kinetics, a temperature increase can also alter the environment by accelerating a variety of chemical processes. An example described at the *Indoor Air Quality in Museums and Archives* Web site (Ref 1) describes a polyvinyl chloride lamp fixture within a museum showcase that may well have released chlorine to form hydrochloric acid vapor into the showcase due to the heat of a light bulb in the case.

Light. One perhaps surprising source of corrosion is radiant energy. In the early 1990s, the role of light in the corrosion of metals was found to be more significant under certain conditions and in the presence of certain pollutant gases than previously assumed, especially in accelerating the corrosion of silver. Oddy and Bradley noted (Ref 2) that when silver coins with a dark-purple patina (typical of silver chloride) were exhibited, their upward-facing surfaces became hazy with a milky-white appearance. By comparison, the opposite sides remained unchanged. Scanning electron microscopy analysis revealed the milky haze to be composed of a white powder that was almost pure silver. Later measurements showed that the light intensity in the location of the exhibition case was as high as 960 lux, with an ultraviolet content of 70 μW/lumen. The researchers postulated that such energy levels were sufficient to reduce the silver chloride on the surface of the coins to pure silver.

Pollutants

The list of aggressive pollutants and their sources in museums and collections is long. Polluting gases such as hydrogen chloride (HCl), ammonia (NH_3), formaldehyde (HCHO), sulfur dioxide (SO_2), hydrogen sulfide (H_2S), carbonyl sulfide (COS), carbon dioxide (CO_2), nitrogen oxides (such as nitrogen dioxide, NO_2), and ozone (O_3) may come into the museum environment from outdoors and be absorbed or adsorbed by metallic surfaces or existing corrosion layers. Once in place, they initiate corrosive processes in the presence of water. For example, adsorbed SO_2 collects sufficient water at 75% RH to undergo a catalytic oxidation to sulfuric acid. Nitrogen oxide pollutants (NO and NO_x) that come predominantly from automobile exhaust or the decomposition of materials are of equal concern, because nitrogen oxides can convert in the presence of water to aggressive acids:

$$2NO_2 + H_2O \rightarrow 2HNO_2 \text{ (nitrous acid)}$$

which then convert to HNO_3 (nitric acid).

Pollutants do not act independently, and their interactions can be quite destructive, although all the interactions with each other and with metal substrates are yet fully understood. Research shows that the reaction of SO_2 with NO_2 on a metal surface can produce nitrous acid (Ref 3):

$$SO_2 + 2NO_2 + 2H_2O \rightarrow 2H^+ + SO_4^{2-} + 2HNO_2$$

The combination of SO_2 and NO_2 can have a corrosive effect on metal a full order of magnitude greater than the effect of SO_2 alone and silver sulfidation caused by hydrogen sulfide increases when NO_2 and ozone are present (Ref 4). Additionally, the combination of NO_2, SO_2, and ozone will tend to inhibit the formation of a protective, oxide layer and encourage the formation of a sulfate layer, which is much more damaging to the metal surface.

Acidic deposition can infiltrate the porous structures of patinas (Ref 5). In museums, although certainly far less dramatic than outdoors, the effect can be significant, because the visual appearance of the patina may be altered, and the analytical/archaeometric value of the patina makeup and structure may be compromised.

While ozone and nitrogen dioxides can be controlled in museum environments through filtration (using activated charcoal, for example), nitrogen monoxide cannot be removed without great difficulty. The presence of nitrogen monoxide is not a major concern, however, unless ozone is present, because ozone will combine with nitrogen monoxide and water to form nitric acid.

Soluble salts that may be present on the metallic surface can act in combination with water as electrolytes. Chlorides, for example, can be present as a residue of previous treatments or repairs. Sodium chloride, which can accelerate the corrosion of metals (especially copper and silver), is often found on archaeological objects excavated from marine burials or from terrestrial sites where the soil includes high concentrations of chlorides. If a collection is housed within a few miles of an oceanfront, salt can also enter as airborne particulates approximately 2 μm (0.08 mils) or larger.

Beyond direct soiling that requires repeated cleaning of the metallic surface, particulates affect corrosion processes due to their hygroscopic, acidic, or alkaline nature. These include sodium chloride (NaCl), ammonium sulfate (($NH_4)_2SO_4$ and ($NH_4)HSO_4$), ammonium chloride (($NH_4)Cl$), and sodium sulfate (Na_2SO_4). A study of relatively polluted urban locations (Ref 6) found such corrosive particulates (especially those less than 0.1 μm, or 0.004 mils) to be present in some museums and historical collections at levels almost as high as in the immediate outdoor environment. While most particulates can be removed by standard filtration systems widely available to museums, some are too small or enter the building through uncontrolled openings, such as windows and doors. These particulates (or aerosols) can be rich in nitrates and sulfates, such as those that are formed by the atmospheric condensation of SO_2 and nitrogen oxides, and as such pose significant corrosion risks to collections. This situation can also be true for a variety of acidic particulates produced in photochemical smog, because anthropogenic particulates of 1 μm (0.04 mils) or smaller are difficult for most general filtration systems to remove. While some of these agents may not corrode metal directly, they can play a catalytic role or may degrade other materials that in turn produce corrosive agents. Additionally, some metals (notably copper, zinc, nickel, and alloys containing these metals) are particularly sensitive to aerosol particulates, such as ammonium sulfate and SO_2, in combination with oxidants, such as NO_2 and ozone (Ref 4).

To the list of corrosive agents must be added the one that walks into the collection on two legs: the visitor. Visitors not only bring considerable amounts of corrosive agents into the museum environment (detritus, fumes from clothing, and human bioeffluent) but also may contribute more directly to the corrosion of the object by improper handling, resulting in physical wear and residues of salts, acids, and oils deposited on the metal by contact with skin.

The Museum as a Source of Corrosion

If one thinks of a museum, gallery, storeroom, or cabinet as an envelope, a number of different corrosion scenarios can be considered. For example, the envelope may not be sufficiently sealed or filtered and may allow outside conditions to influence the interior environment.

Sources for corrosion may come from the interior of the envelope. In this context, even quite small amounts of some pollutants can initiate significant reactions, especially in confined spaces with little or no regular air exchange. In many instances, works of art and artifacts are displayed or stored in exactly such confined spaces (either small storerooms, sealed storage containers, or closed display cases), where deleterious gases and vapors from inappropriate construction or display materials increase to significant concentrations. In wooden storage cabinets, the ratio of wood surface (which may be emitting harmful vapors) to the interior volume is such that even low emission rates lead to dangerous levels of pollutants. The sources for such pollutants in storage facilities and display environments are surprisingly numerous. The processes for manufacturing fabrics used for furniture upholstery, wall coverings, or to cover the interior of display cases may include a host of harmful chemicals, such as reduced sulfides and formaldehyde, that eventually emit gas, either through degradation of the fabric or dye, or because they are present in excess amounts on the fabric itself. Flame retardants, fabric finishes (a formaldehyde-based process that gives us the convenience of permanent press), and water repellents also may emit gases. The effect of such corrosive agents is readily observable. Nails and screws in contact with some fabrics corrode at their contact points after only a few months of exposure (Ref 7). The rapid corrosion of a steel artifact in contact with a dyed yellow fabric was attributed to the inclusion of a halide salt in the dye (Ref 8).

Example 1: Tarnish of Hellenistic Silver Bowls by Use of Sulfur-Laden Velvet. In the mid-1970s, a collection of Hellenistic silver bowls at the J. Paul Getty Museum in California (Fig. 1) was placed in a relatively well-sealed exhibition case that was lined with velvet. Within one month, the surfaces of the bowls began to tarnish noticeably. The case was redesigned, and a more suitable fabric, one not laden with sulfur, was used. Specifically because of such problems, exhibition materials used at the Getty Museum have been regularly tested since 1985 for their potential to damage objects by contact or by proximity.

Plastic and Wood

Plastic products are ubiquitous. Polymers containing hydrolyzable groups such as esters, amides, acetals, and certain ketones can deteriorate and emit by-products such as acetic, sulfuric, formic, and nitric acid, which are harmful to metals. Added opacifiers, plasticizers, fungicides, stabilizers, or excesses of initiators or catalysts may also be a source of corrosion. Polyvinyl-chloride-based products degrade and emit hydrochloric acid through a reaction of the emitted chlorine with water vapor. Some clear plastic sleeves or envelopes commonly used for storing coins provide an example of this problem. Additionally, bacteria can attack some types of vinyl in humid conditions. The result of this attack can be the release of acids that in turn attack coins.

Most wood or wood-based products emit organic acids (formic and acetic acid) that can readily affect many metals, most notably lead. Among the woods most likely to release harmful vapors are oak, Douglas fir, birch, sweet chestnut, and western red cedar. Composite wood products such as plywood, chipboard, and particleboard (particularly if they incorporate urea-formaldehyde adhesives that are ubiquitous in new construction) emit both acetic acid and formaldehyde (and, as a result, formic acid). Kiln-dried woods and those treated to prevent susceptibility to rot are likely to cause greater corrosion of metals than untreated and air-dried woods (Ref 9).

Sulfur

Recently, the most problematic pollutant to attack metals within the museum and collection environments has been sulfur, particularly hydrogen sulfide and carbonyl sulfide. While there has been a large reduction in the problems caused by the burning of sulfur-laden coal and by-products from refineries, which were major contributors to pollution problems of European urban centers during the late 19th and early 20th centuries, some countries still suffer from the effects of this pollutant. Because only minute amounts of sulfur can cause significant visual and physical effects on metal surfaces (particularly silver), sulfur is a major concern for collections. In the past, hydrogen sulfide was considered the most reactive and problematic of sulfur compounds. However, carbonyl sulfide is the most concentrated naturally occurring sulfide in the atmosphere and is equally aggressive.

The sources of sulfur within museums include vulcanized rubber in floor mats and carpet backing, and chemicals used in the various stages of drying and finishing textiles, especially felts and velvets. Wool fabrics not only contain sulfur but also can absorb large amounts that can later be released when the environment is favorable or when the wool ages.

Example 2: Silver Sulfide Dendrites Growing on a Greco-Roman Silver Bowl. Rubber floor mats installed in the 1930s at the Walters Art Museum in Baltimore were meant to protect the objects from breaking should they fall. Sulfurous gases emitted from the mats eventually affected some of the metal artifacts in the storeroom. The objects were undisturbed for a long period of time (many years), and this may have allowed for the dendritic formation of the silver sulfide crystals (Fig. 2). The Walters Art

(a)

(b)

Fig. 1 An exhibition case of Hellenistic silver vessels at the Getty Museum in the late 1970s. (a) Red velvet lines the case walls. (b) One of the ancient Hellenistic silver bowls showing significant tarnishing due to sulfur from the red velvet. Tarnish began to appear on the surface of the silver within one month of the objects being placed in the exhibition case. Courtesy of J. Podany, 1979.

Museum now has very rigorous and effective storage policies and methodologies. All metal objects are stored in a special metal storage vault with low RH, and pollutant scavengers are included to give additional protection to vulnerable objects.

Sulfurous gases can also be released by modeling clays, protein-based glues, and objects that have been repaired with elemental rolled sulfur (resolidified sulfur). Recipes found in old repair and mending manuals called for sulfur in numerous mixtures of fixing cements and glues (Ref 10, 11). Elemental sulfur, sometimes mixed with sand or iron fillings, was often used for fixing iron pins into stone for repair of sculpture or architectural elements. The sulfur was melted and poured into the hole accommodating the pin. Because the sulfur remained rather soft and rubbery until it fully solidified, it gave sufficient time to make adjustments to the pin and the joint. Because many types of repairs to numerous materials were made with elemental sulfur, these have become possible sources of corrosion.

In polluted urban atmospheres, SO_2 can be absorbed by oxides on the metal surface or dissolved in the water film on the surface of metal (Ref 12). It then combines with hydrogen peroxide or ozone to form sulfuric acid that dissolves the protective oxide on the metal, potentially forming oxide-free or less protective anodic (electron-producing) areas adjacent to cathodic (electron-consuming) areas still protected by oxides. The deposition and formation of sulfuric acid in water films on aluminum surfaces has been shown to be a significant cause of corrosion problems in both urban and industrial atmospheres through the dissolution of the oxide film and then the further oxidation of the exposed metal (Ref 13). One must assume that, given high humidity and the presence of sulfuric acid in storerooms or display areas, similar processes occur, albeit slower. Sulfuric acid can initiate the production of salts, such as ammonium sulfate, that attract moisture and function as electrolytes.

A direct role for SO_2 is proposed for local dezincification of brass (Ref 3). Sulfur dioxide reacts with the zinc content of the object, leaving traces of zinc sulfite that rapidly oxidize to zinc sulfate on the surface. This sulfate is highly deliquescent and causes local areas of condensation, even at moderate RH. The zinc sulfate solution thus formed dissolves more SO_2 that in turn oxidizes to sulfur trioxide. An increasingly concentrated solution of sulfuric acid then dissolves more of the copper-zinc alloy, resulting in the production of a mixed copper-zinc basic sulfate.

The effect of sulfur in museum collections has garnered a great deal of attention and research aimed at identifying the corrosion products and their sources. In 1979, visually disfiguring black spots on bronzes in a collection were reported and identified as copper sulfide (Ref 14). The researcher reportedly was successful at culturing bacteria from samples of the corrosion and went on to suggest that the source of the disfiguring corrosion may have been biological. It has since been proven, however, that these areas of corrosion, identified as covellite (CuS), are due to the presence of sulfur in the form of hydrogen sulfide, carbonyl sulfide, and similar pollutants in the air.

Example 3: Copper Sulfide Corrosion ("the Brown Fuzzies") on Copper Alloy Artifacts. In the late 1980s, the museum staff of the Oriental Institute in Chicago noticed a dark-brown, almost velvetlike growth on all of the copper alloy objects within a single display case (Fig. 3). The artifacts, which ranged in style and function, had apparently been mechanically and/or chemically cleaned at an earlier date. Since this incident, the same types of growths have been noticed on copper alloy objects on display in other galleries and in various storage cabinets. Taken at face value, this problem would seem to be a building-wide occurrence and not tied to a specific location.

The largest growths appear cauliflower-like and extend farthest from the surface of the object. Those closest to the surface (at low magnification) seem to have a dendritic structure. Analyses of the corrosion products identified them as djurleite (a copper sulfide, $Cu_{31}S_{16}$), although additional minor phases cannot be ruled out. Remedial measures have included examining all materials within the exhibition and storage environments and replacing all that may have been emitting harmful vapors.

Corrosion from Carbonyl Compounds

The organic acids produced from carbonyl compounds present in the environment can dissolve metal salts and catalyze corrosion processes. In high-humidity conditions, small amounts of volatile ester solvents, such as those retained in paint films or on the surface of metal objects, can hydrolyze and release corrosive lower fatty acids. Drying oil, semidrying oil paints, or paints containing esters of volatile acids are potential sources of corrosive vapors (Ref 15).

Although damage to artifacts was first reported in the late 19th century (Ref 16), it was only in the middle decades of the 20th century that in-depth scientific studies of the effects of these compounds on metal artifacts began to occur (Ref 17). There are still some aspects of the interaction between carbonyl compounds and artifacts or art works, however, that are not well understood.

Enclosed storage and display environments can include low-molecular-weight esters that initiate corrosion by hydrolyzing into their constituents of alcohol and acid (the acid being

(a)

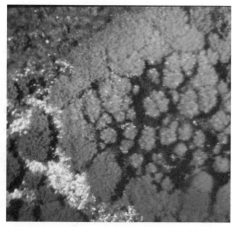

(b)

Fig. 3 Velvetlike corrosion growths on the surface of an ancient copper alloy vessel. (a) The subtle form of the object is all but obscured by corrosion growths. (b) At higher magnification, the velvetlike corrosion begins to reveal its dendritic structure. Original magnification 15×. The corrosion was identified as djurleite, a copper sulfide. The Chicago Oriental Institute.

Fig. 2 Dendritic silver sulfide crystals growing on a Greco-Roman silver bowl. Walters Art Museum, Baltimore. Original magnification 10×. Courtesy of Terry Drayman-Weisser.

the corrosive agent). Their total effect remains less significant, however, than the effects of simple organic acids and aldehydes released from wood, wood products, and a wide variety of commercial synthetic products, such as urea-formaldehyde-containing adhesives, varnishes, paints, composites, and fabrics. Upon degradation, cellulose acetate, for example, can release acetic acid. Degraded cellulose nitrate can release nitric acid.

The organic carbonyl compounds most commonly found in collections are:

- *Acetic acid*: Often the most prevalent pollutant in wooden enclosures such as storage boxes, cabinets, or display cases
- *Acetaldehyde*: A product of incomplete combustion from cigarettes and fireplaces. It is in building products such as polyurethane foam, adhesives, coatings, and inks.
- *Formic acid*: Found in wooden enclosures, especially oak, and in glue
- *Formaldehyde*: Also commonly found in high concentrations within wooden enclosures

These carbonyl compounds can also be found in the outdoor environment of many urban areas. Reported levels in Southern California in a 1983 study were 1 to 29 ppb of formaldehyde, 1 to 13 ppb acetaldehyde, 1 to 8 ppb formic acid, and 2 to 10 ppb acetic acid (Ref 18). It is likely that these levels of formaldehyde and acetyl aldehyde will increase, particularly if the consumption of ethanol-based fuels in urban environments increases. These pollutants often find their way into collection environments. However, major sources of these compounds already exist in the indoor environment.

The pronounced effects of carbonyl compounds on lead were among the first studied. The principal corrosion products of lead found in museum collections are lead formate ($Pb(COOH)_2$), lead acetate ($Pb(CH_3CO)_2$), and basic lead carbonate ($2PbCO_3$). Lead oxide, which readily forms and is more or less protective, will alter to lead carbonate in the presence of organic acids. Lead is more resistant to corrosion when alloyed with as little as 1.5% Sn (Ref 19). Pewter, for example, is often unaffected by conditions that would rapidly corrode objects manufactured of purer lead. The decorative inlays on a number of 19th century Japanese lacquered wooden artifacts experienced a wide range of corrosion damage, based on the percentage of tin alloyed with the lead (which was the predominant metal, in most cases) (Ref 20). The pure-lead inlays showed corrosion so severe that the metal inlays were all but consumed, while lead-tin alloys of up to 80% Sn fared better, some being completely unaffected by the organic acids present in the storage environment. Trace elements in the metal inlays affected the degree of corrosion. Iron and zinc, for example, increased the corrosion damage, while tin, copper, silver, and gold improved the corrosion resistance of the inlay. The corrosion products reported were predominantly lead carbonate with minor amounts of lead formate. This led the researchers to conclude that the corrosion was due to an alteration of the lead oxide by formic acid being released within the storage environment over a period of some 60 years.

Example 4: Corrosion of a Sculpture Made with Incompatible Materials. On occasion, materials used by artists and craftsmen are incompatible. Some materials used for artistic expression may eventually degrade and emit harmful substances that corrode adjacent metal parts of the work of art. Such was the instance when the nitrocellulose segments of an Antoine Pevsner sculpture at the Museum of Modern Art in New York began to degrade and cause the corrosion of those parts of the sculpture made of copper alloy (Fig. 4).

Acetic Acid. The effects of oak planks on lead roofing in the outdoor environment were already being commented on in the 18th century (Ref 21). Today, we understand the mechanism by which acetic acid, evolving from the oak planks, corroded the lead and produced lead acetate ($Pb(CH_3CO)_2$). While one may suspect that such corrosion is only likely outdoors, it is, in fact, also possible within the indoor environments of museums and collections. Acetic acid is released by the hydrolysis of acetyl groups (esters) in the hemicellulose of wood. The rate and amount of this release is dependent on a number of factors, including the ambient RH, temperature, and concentration of esters in the wood. For example, the proportion of acetyl groups giving rise to acetic acid is greater in hardwoods (3 to 5% by weight of the wood) as compared to softwoods (1 to 2% by weight). In many woods, much of the acid content is volatile. Over 90% of the total acid in oak and Douglas fir, the majority of which is acetic acid, is volatile (Ref 22). Corrosion effects should not be surprising when one considers the efficiency by which acetic acid catalyzes the conversion of lead (via its oxide) to the basic carbonate, which was, in fact, the basis for the Dutch process of making lead white pigment. Effects of acetic acid also include the considerable costs faced by commercial wood mills and processing plants due to the corrosion of cutting blades and the corrosion of aluminum coils in dehumidifiers (dryers) used in wood-processing plants.

Wood products have been treated with acetic anhydride to enhance dimensional stability while maintaining other desirable properties. These treated products can release acetic acid. The use of acetylation as an industrial process has been slow to catch on, however, partially due to corrosion of metal fixtures and the entrapment of acetic acid in the wood (Ref 23, 24).

Wood is not the only source of acetic acid in museums and collections. Certain silicone sealants, in which acetic acid is incorporated, can be problematic. Other synthetic products, such as cellulose acetate, release acetic acid as they degrade. Acrylic-based paints often incorporate cellosolve acetate (2-ethoxy ethyl acetate) as a carrier solvent that acts as a source of acetic acid, although the hydrolization is slow. The degradation of widely available polyvinyl acetate adhesives, often used in the construction of furniture and exhibit fixtures, is also a source of acetic acid. Beyond the more common

(a)

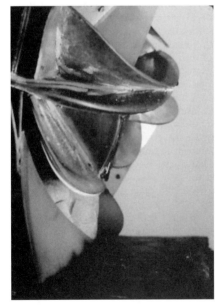

(b)

Fig. 4 *Torso* (1924 to 1926) by Antoine Pevsner constructed in plastic and copper. (a) Full image showing corrosion of the copper segments of the sculpture due to gas emission from and contact with cellulose nitrate segments. (b) A close-up detail of the sculpture. The Museum of Modern Art, New York. Katherine S. Dreier Bequest. ©The Museum of Modern Art, New York, © 2004 Artists Rights Society (ARS), New York/ADAGP, Paris.

release of nitric acid, nitrocellulose-based paints and lacquers can also release acetic acid, which is derived from esters in the paint. Some paints based on drying or semi-drying oils can lose as much as 0.0025 wt% of the paint film per day for some time (Ref 15).

The prevalence of acetic acid in collection environments is so problematic that Ref 9 placed acetic acid in the same corrosive category as SO_2. Significant corrosion occurred at 0.5 ppm concentrations with an ambient RH as low as 70%. Other researchers have found acetic acid active at much lower concentrations. One researcher reported that ancient bronzes corroded relatively rapidly in environments with concentrations of 0.14 ppm (Ref 25). He identified a number of corrosion products on bronze objects stored in wood cabinets, including crystalline and powdery white efflorescence associated with the initial light-blue corrosion products on the surface of the bronzes. This was identified as sodium acetate trihydrate ($NaCH_3COO \cdot 3H_2O$). The formation of this product does not directly involve components of the metal alloy itself but can be disfiguring nonetheless.

It should be noted that the corrosion potential of low concentrations of acetic acid (as well as other corrosive agents) is dramatically enhanced by a rise in RH. While corrosion of lead by acetic acid was found in concentrations above 0.1 ppmv (volume) at 54% RH (Ref 26) and for copper at 50 ppmv (Ref 27), others report acetate corrosion products on copper alloy artifacts housed in storage cabinets where acetic acid was present in concentrations as low as 400 to 500 ppb but where the RH regularly rose to 85% (Ref 28). It was noted that acetic acid can be adsorbed on the surface of metal objects and can thus collect in one environment but react in another if the humidity levels are sufficient.

In the case of lead, acetic acid plays a major corrosive role, far more than formic acid. However, Tétreault et al. have recently shown that in the presence of formic acid, the ability of acetic acid to corrode lead is significantly inhibited, because the corrosion layer is altered and can become increasingly protective (Ref 26).

One process by which acetic acid continues to corrode lead involves acetic acid (CH_3COOH) vapors finding their way into the cracks and fissures of the lead carbonate corrosion product. The resulting lead acetate ($Pb(CH_3COO)_2$) acts as an electrolyte, promoting more corrosion of the metallic lead and formation of additional lead carbonate.

At the tips of the cracks, corrosion occurs by the following process:

$$Pb + 2CH_3COOH + H_2O + 1/2 O_2$$
$$\rightarrow Pb(CH_3COO)_2 + H_2O$$

New acetic acid is then formed by:

$$3Pb(CH_3COO)_2 + 4H_2O + 2CO_2$$
$$\rightarrow Pb_3(OH)_2(CO_3)_2 + 6CH_3COOH$$

The $Pb_3(OH)_2(CO_3)_2$ is hydrocerrusite.

Example 5: Corrosion of Two Chinese Artifacts Possibly Caused by Exposure to Exhibition Case Materials. A Shang dynasty finial (1300 to 1050 B.C.E.) and an Eastern Zhou dynasty ding vessel (500 to 400 B.C.E.) exhibited bright blue corrosion areas while on exhibit at the Asian Art Museum in San Francisco (Fig. 5). Although the exact cause of the corrosion remains unclear, the combination of exhibition case materials, such as wood, fabric, and adhesive, is suspected. All exhibition case materials in the museum have since been changed, and any materials used near art objects must now be tested for corrosive agents. Since that policy has been in place, no similar corrosion problems have been observed.

Example 6: Corrosion of a Japanese Tsuba Stored Near Wooden Boxes. Wood can emit harmful vapors from its structure and from treatments it has received. The vapors, which are often organic acids, can have a significant impact on the preservation of the object. An iron tsuba with inlays of gold, copper, lead, and the traditional Japanese alloys shibuichi (an alloy of silver and copper, often in a ratio of 3 to 1, but recipes are widely diverse) and shakudo (an alloy of copper and 2 to 7% Au) had been stored in the proximity of wooden boxes at the Walters Art Museum in Baltimore and was found to have developed a disfiguring corrosion on its surface (Fig. 6). It was assumed by the conservator who treated the object that the wooden boxes were the source of the organic acids that caused the corrosion. Upon discovery of the condition of the object, it was relocated to a metal storage cabinet coated with baked enamel. The museum now uses Pacific Silvercloth (Wamsutta Industrial Division) over the tsubas, with acid-free tissue between the metal and the cloth (Ref 29). The acid-free tissue was introduced between the object and the silver cloth as a barrier, because the conservator was concerned that silver salts contained in the cloth may adversely affect the copper of the object. The RH within the storage unit is now kept low. Clean lead test coupons have been introduced into the storage environment for several years (starting in 2001), and no corrosion has been observed on the coupon or on the object.

Formaldehyde. Our modern environment is all but permeated with formaldehyde. Worldwide, more than 10 million metric tons are produced annually (Ref 30) for use in a widening array of commercial products. This is an alarming amount, given that concentrations of less than 100 ppb have been shown to cause eye and throat irritation as well as chronic respiratory and allergy problems (Ref 31). With respect to metal corrosion, past research has suggested that concentrations of 0.03 ppm in the atmosphere can initiate corrosion of some metal objects, most specifically lead (Ref 32). Recent studies, however, suggest that formaldehyde has no significant effect on lead or copper (Ref 26). By contrast, the oxidation product of formaldehyde, formic acid, poses a serious problem by corroding lead and copper when the acid is present at 2 ppmv in RH as low as 54%. Corrosive concentrations can be even lower when humidity is higher. For example, 0.4 ppmv of formic acid can initiate corrosion at 75% RH. The reactivity of formic acid may, however, be reduced significantly after the first layer of corrosion product is formed, because that layer can be somewhat protective.

Sources of formaldehyde in the collection environment include urea-formaldehyde insulation,

(a)

(b)

Fig. 5 Two bronze pieces exhibiting corrosion. (a) A finial, approximately 1300 to 1050 B.C.E. from the Shang dynasty (approximately 1600 to 1050 B.C.E.), China. (b) A closeup of the handle of a ritual food vessel (ding) with cover, approximately 550 to 400 B.C.E. Eastern Zhou dynasty (771 to 221 B.C.E.), China. Asian Art Museum of San Francisco, The Avery Brundage Collection. Both images used by permission. Courtesy of Donna Strahan.

chipboard, and urea-formaldehyde adhesives used in construction. While formaldehyde forms durable polymers when combined with phenol and amines, the addition of urea also produces methylol urea compounds. As a result, urea formaldehyde produces ten times more formaldehyde emission than phenol formaldehyde.

Formaldehyde can be in the carpets or tile covering the floor, the plastic containers holding the works of art in storage, or the paints on the gallery wall or display case interior, particularly if the paint is a latex (containing azoni-adamantane chlorides, oxazolidines, and triazines) or an alkyd paint. Foam rubber, fabrics, and fabric finishes (such as flame retardants, or the size often given to yarn in order to reduce dust in the mill during the weaving process) also contribute formaldehyde. A more recent source has been insufficient baking of solvent-based enamel paints used to coat metal storage cabinets. Such cabinets are commonly preferred in museums and historic collections, because wood cabinets and shelves give off gas acetic acid and produce formaldehyde via the degradation of the wood lignin. Formaldehyde is also a by-product of gas and solid fuel combustion as well as cigarette smoke.

Excess formaldehyde is used to speed up the manufacturing of plastics by inducing rapid cross linking of the polymer. Such excess, combined with the emission due to the breakdown of the resins (often through hydrolysis), provides a significant amount of formaldehyde to react with works of art. Formaldehyde is also formed by a reaction of ozone and terpene hydrocarbons present in new constructions (Ref 31).

The formaldehyde released from these products and reactions can produce a number of destructive processes that become especially serious at elevated temperatures. An increase in temperature of 5 to 6 °C (9 to 11 °F) can double the formaldehyde gas concentration in some instances, as can an increase in RH from 30 to 70%. If both temperature and RH increase, the concentration of gas can increase by five times the initial amount. One must also keep in mind that formaldehyde can be released for an extended period of time. It is reported (Ref 33) that a newly constructed house will emit formaldehyde at a fairly steady rate for up to 9 months, and one can assume this would also be true of new construction within a museum. Although formaldehyde itself can catalyze the corrosion process, it is less active in this way than acetic or formic acid. However, formaldehyde can be oxidized in air by peroxides or ultraviolet radiation to formic acid, which is more corrosive.

Past Treatments

Restoration or conservation treatments meant to preserve objects may, if improperly carried out, be the source of significant corrosion. For example, the common use of chemical dips or cleaning polishes (which can contain acids and/or ammonia) can be quite corrosive to silver. Coatings such as linseed oil, dammar varnish, or shellac (often used in the past to protect objects from humidity) age, oxidize, and crack, exposing the previously protected metal surface. These cracks, where fresh metal is exposed, can act anodically to cause corrosion. Likewise, if the coating is not applied properly and evenly over the surface, the uncoated areas of the object can act anodically to the coated cathodic areas. Scratches and chips in the coating behave the same way. Recent studies on the destructive effects of filar corrosion of aluminum and magnesium alloys used for contemporary sculpture (akin to that occurring on aircraft) provide general examples of this process (Ref 34). Neat's-foot oil, linseed oils, and shellac used as protective coatings provide free fatty acids or dicarboxylic acids that form metal soaps with, for example, copper and lead (Ref 35). These copper soaps are especially unstable and promote corrosion of copper and its alloys (Ref 3). In one example, metal soaps formed on a brass wire wound around the socket of an African spear that was coated with animal fat (Ref 36). Paintings on copper that incorporate drying oils, and bronzes that have been coated with drying oils have shown the presence of copper soaps on their surface or within the layers of their structure. Reference 37 states that fatty acids can readily evaporate from oil paint films (palmitic acid evaporating more rapidly than stearic). There are many different fatty acids found in drying oils, especially within the modified oils produced in the last half-century. The four predominant fatty acids, however, are two saturated (palmitic and stearic) acids and two unsaturated (oleic and linoleic) acids. Copper, lead, and zinc palmitate and stearate are most often found as the corrosion products. Palmitic acid released by leather, for example, reacts readily with zinc to form an amorphous precipitate. For this reason, brass attachments, such as studs or clasps, on historic leather jackets, boots, or saddles are often found with a green corrosion product covering them, particularly when the object has been stored in a closed and somewhat humid environment.

It has been pointed out (Ref 38) that oils and fats that cause this type of corrosion on bronzes (as well as on zinc and lead) have three ester linkages that can be hydrolyzed (the addition of water across the bond) to produce free fatty acids that can attack metals, as shown in Fig. 7.

Residues of past cleaning, repair, or restoration efforts can also be sources of corrosive agents. Chlorides remaining from fluxes used for solder repairs are an example. More recently (Ref 28), the presence of sodium-copper carbonate acetate ($NaCu(CO_3)(CH_3COO)$) on copper artifacts in storage at the Athenian Agora site was reported. Some of the sodium compounds formed from such treatments (such as chalconatronate) will readily absorb acetic acid (often from the storage environment) and form sodium-copper carbonate acetate products. When such acetate compounds deliquesce, the resulting salt solution wicks into the pores of the existing corrosion products and into the void or cracks of the metal to cause further corrosion. A strong case was made for the source of the sodium

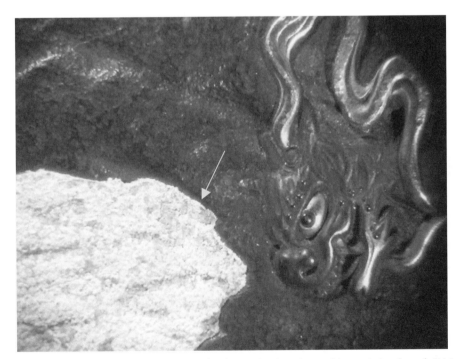

Fig. 6 Japanese tsuba from the Edo/Meiji period with corrosion (arrow) caused by proximity of wood. Original magnification 6×. Courtesy of Terry Drayman-Weisser.

being from previous treatments of the bronzes involving sodium sesquicarbonate, commonly used in the past to stabilize chloride-containing archaeological bronze artifacts. The use of Calgon (Reckitt Benckiser) (previously composed of sodium hexametaphosphate) or the use of sodium hydroxide with Rochelle salts (sodium-potassium tartarate) are also both possible sources of the sodium, because both have been used for cleaning bronze surfaces and sometimes are used with alternating treatments of acetic acid. Patination processes often include ammonia and sodium compounds and involve fuming the object in acetic acid. They may also include a wide range of acetate, chloride, and sulfide compounds (Ref 39).

Example 7: The Corrosion of Copper Alloy Tacks on a Gourd Box from Zaire. Some objects may combine copper alloy elements with a variety of oils and fats applied to the object during manufacture or for restorative or ceremonial purposes. These oils or fats interact with the copper to cause significant corrosion. In this instance, a late-19th century gourd box in the collection of the University of Pennsylvania Museum of Archaeology and Anthropology in Philadelphia showed considerable corrosion of the copper alloy (probably brass) tacks due to the presence of oil (Fig. 8). The corrosion was determined to be either a copper oleate or a copper stearate or a mixture of the two. The corrosion was mechanically cleaned away and the oil on the surface of the object reduced by solvent cleaning.

Preservation

This article has presented a general survey of corrosive agents and processes that exist within what are usually considered the protective environments of museums and collections. While interventive treatments remain a viable option, both passive and active control of the environment to minimize or eliminate corrosive agents are preferred in modern conservation practice.

Environmental control may be as simple as providing a microclimate (usually below 40% RH) or the inclusion of pollution scavengers within display cases or storage cabinets, such as activated charcoal filters, zinc oxide in pellet form, or fabrics/plastics that contain small particles of silver or copper that act as sacrificial reactive sites. Certainly, one method of controlling pollutants in a museum or storage environment is to limit the infiltration of unfiltered outside air. However, if the number of air exchanges is reduced dramatically, a buildup of pollutants can result within the closed environment. While basic filtration of museum or collection spaces is now a common standard, filtration to the level necessary for eliminating all traces of pollution is out of budgetary reach for most small institutions. Small particulates (<1 μm, or 0.04 mils) are difficult to filter, and the removal of nitrous oxides requires specialized equipment. Nonetheless, the inclusion of high-efficiency filters and activated charcoal filters within a heating, ventilation, and air conditioning system can go a long way in protecting collections. Relatively speaking, these can be installed and maintained at a modest cost, given the protection achieved. Simple and more direct methods should, of course, never be overlooked. Eliminating inappropriate materials is a well-documented success (Ref 40). Avoiding the use of alkyd paints, silicone sealants containing acetic acid, polyvinyl chloride or polyvinyl acetate adhesives, urea-formaldehyde adhesives, most wood products, vulcanized rubber products, or any other material that may emit acid, alkaline, or aldehyde by-products is an effective first step. Relatively simple methods of evaluating materials used in construction or finishing of display and storage fixtures and facilities are available and easily applied (Ref 41).

The most well known and widely used of these methods is the Oddy test (Ref 42), involving a sample of the potentially corrosive material in close proximity to or in contact with a copper, a silver, and a lead coupon that is placed in a test tube. The test is undertaken with high RH and elevated temperatures. Although the determination of whether or not the coupons have altered during the 30 day cycle remains somewhat

Fig. 7 The attack on bronzes by oils and fats. (a) Fats with three ester linkages are hydrolyzed, producing free fatty acids. R_1, R_2, and R_3 are long hydrocarbon chains, typically containing 12, 14, 16, 18, or 20 carbon atoms. (b) The free fatty acids attack copper in the presence of water and oxygen, and a fatty-acid salt is formed.

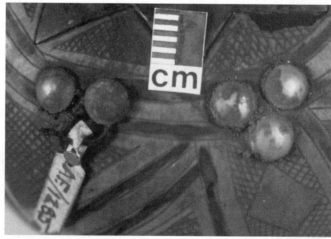

Fig. 8 The interaction of components of an artifact. (a) Gourd box, late 19th or early 20th century, Kasai District, Zaire. (b) Detail shows the corrosion of the copper alloy tacks due to the oil applied to the surface of the gourd during manufacture or use. Original magnification 5×. Conservation Laboratory, University of Pennsylvania Museum of Archaeology and Anthropology

subjective, the test is nonetheless pragmatic, direct, and easy to evaluate. The list and availability of acceptable substitutes for inappropriate materials is continually expanding. Sealing the potentially corrosive material is another option. Plywood used in the construction of exhibition cases is often sealed using aluminum/plastic laminates, such as Marvel Seal (Ludlow Corp.).

Stewardship of collections is recognized as a professional responsibility that includes evaluation and maintenance of the environment where the objects are held. The aim of these efforts is to assure the safest and most benign surroundings possible, thus assuring the future preservation of the collection. An excellent review of air sampling and monitoring approaches can be found in the book *Atmospheric Corrosion* (Ref 4), and an equally excellent review of scavengers and adsorbents is found in Hatchfield's book *Pollutants in the Museum Environment* (Ref 40).

In all the examples noted in this article, appropriate steps were taken by the conservators of the institutions to eliminate the corrosive sources acting on the objects and to avoid future introduction of other potentially damaging materials. These measures, more than any other approach, have been the most effective in preventing future corrosion within museum and collection environments. Each of the examples, the challenges they presented, and the solutions found to resolve the ongoing corrosion threats were generously provided by forward-looking institutions and conservators. Such openness and generosity assures the advance of preservation efforts.

ACKNOWLEDGMENTS

The author would like to thank Terry Drayman-Weisser, Head of Conservation for the Walters Art Museum in Baltimore; Laura D'Alessandro, Objects Conservator for the Oriental Institute Museum in Chicago; Lynda Zycherman, Conservator at the Museum of Modern Art in New York; James Coddington, Conservator at the Museum of Modern Art in New York; Donna Strahan, Head of Conservation for the Asian Art Museum in San Francisco; and Virginia Greene, Senior Conservator at the University of Pennsylvania Museum of Archaeology and Anthropology in Philadelphia, all of whom generously provided case studies; Cecily Grzywacz, Scientist at the Getty Conservation Institute, for her valuable input; and John Twilley, Conservation Scientist, for his generous advice and many insights.

REFERENCES

1. *Indoor Air Quality in Museums and Archives,* http://www.iaq.dk, Morten Ryhl-Svendsen, accessed Jan 2005
2. W.A. Oddy and S.M. Bradley, The Corrosion of Metal Objects in Storage and on Display, *Current Problems in the Conservation of Metal Antiquities,* Proceedings from the International Symposium on the Conservation and Restoration of Cultural Property, Tokyo National Research Institute of Cultural Properties, Japan, 1989, p 225–244
3. D. Scott, *Copper and Bronze in Art: Corrosion, Colorants, Conservation,* Getty Publications, 2002, p 234, 305
4. C. Leygraf and T. Graedel, *Atmospheric Corrosion,* Wiley-Interscience, John Wiley and Sons, 2000, p 103, 116, 199
5. S.D. Cramer, S.A. Matthes, B.S. Covino, Jr., S.J. Bullard, and G.R. Holcomb, Environmental Factors Affecting the Atmospheric Corrosion of Copper, *Outdoor Atmospheric Corrosion,* H.E. Townsend, Ed., ASTM International, 2004, p 245–265
6. W. Nazaroff, M.P. Ligocki, L.G. Salmon, G.R. Cass, T. Fall, M.C. Jones, H. Liu, and T. Ma, *Airborne Particles in Museums,* Research in Conservation Publication 6, Getty Conservation Institute, 1993, p 24–26
7. B. Appelbaum, *Guide to the Environmental Protection of Collections,* Sound View Press, 1991, p 105
8. T. Drayman-Weisser, Some Practical Considerations and Solutions for Preservation in the Museum Environment, *Cultural Property and Its Environment,* Proceedings from the International Symposium on the Conservation and Restoration of Cultural Property, 11-13 Oct 1990, Tokyo National Research Institute of Cultural Properties, Japan, 1990, p 84–86
9. S.G. Clarke and E.E. Longhurst, *J. Appl. Chem.,* Vol 11, 1961, p 435–443
10. H.C. Standage, *Cements, Pastes, Glues and Gums,* Crosby Lockwood and Son, London 1904, p 21, 152
11. C.G. Leland, *A Manual of Mending and Repairing,* Chatto and Windus, London, 1896, p 147, 178
12. J.R. Duncan and D.J. Spedding, *Corros. Sci.,* Vol 14, 1974, p 607–610
13. Q. Dai, A. Freedman, and G.N. Robinson, *J. Electrochem. Soc.,* Vol 142 (No. 12), 1995, p 4063–4069
14. B. Madsen and N. Hjelm-Hansen, Black Spots on Bronzes—A Microbiological or Chemical Attack, *The Conservation and Restoration of Metals,* Proceedings of the Symposium (Edinburgh) Scottish Society for Conservation-Restoration, 1979, p 33–39
15. P.D. Donovan and T.M. Moynehan, *Corros. Sci.,* Vol 5, 1965, p 803–814
16. L. Byne, *J. Conchol.,* Vol 9, 1899, p 172–178, 253–254
17. E. West Fitzhugh and R.J. Gettens, Calcacite and Other Efflorescent Salts on Objects Stored in Wooden Museum Cases, *Science and Archaeology,* R.H. Brill, Ed., MIT Press, 1971, p 91–101
18. C.M. Grzywacz and N.H. Tennent, Pollution Monitoring in Storage and Display Cabinets: Carbonyl Pollutant Levels in Relation to Artifact Deterioration, *Preventive Conservation: Practice, Theory and Research,* (Ottawa), The International Institute for Conservation of Historic and Artistic Works, 1994, p 164–170
19. N.H. Tennent, J. Tate, and L. Cannon, *SSCR J.,* Vol 4 (No.1), 1993, p 8–12
20. D. Heath and G. Martin, The Corrosion of Lead and Lead/Tin Alloys Occurring on Japanese Lacquer Objects, *The Conservation of Far Eastern Art* (Kyoto, Japan), International Inst. for Conservation of Historic and Artistic Works, 1988, p 137–141
21. R. Watson, Essay X: Of Red and White Lead, *Chem. Essays,* Vol 3 (7th ed.), p 337–376
22. B.A. Farmer, *Wood,* Aug 1962, p 443–446
23. S. Kumar, *Wood Fiber Sci.,* Vol 26 (No. 2), 1994, p 270–280
24. M.N. Haque, H.P.S. Abdul Khalil, and C.A.S. Hill, *J. Fibres Compos.,* Vol 1, 2002
25. N. Tennent and T. Baird, *The Conservator,* No. 16, 1992, p 39–47
26. J. Tétreault, E. Cano, M. van Bommel, D. Scott, M. Dennis, M.-G. Barthés-Labrousse, L. Minel, and L. Robbiola, Studies in Conservation, *J. Int. Inst. Conserv. Historic Artistic Works,* Vol 48 (No. 4), 2003, p 237–250
27. J. Tétreault, "Corrosion of Zinc and Copper by Acetic Acid Vapor at 54% RH," Canadian Conservation Institute, Ottawa, 1992
28. A. Paterakis, *J. Am. Inst. Conserv.,* Vol 42 (No. 2), Summer 2003, p 316–321
29. T. Drayman-Weisser, Head of Conservation, Walters Art Museum, Baltimore, MD, personal communication, Jan 2005
30. W. Weirauch, http://www.ub.rug.nl/eldoc/dis/science/j.g.m.winkelman/thesis.pdf,1999, accessed Jan 2005
31. M.H. Sherman and A.T. Hodgson, *Indoor Air,* Vol 14 (No. 1), 2004, p 2–9
32. M.A. Leveque, The Problem of Formaldehyde—A Case Study, *Preprints of Papers Presented at the 14th Annual Meeting,* Proceedings from the American Institute for Conservation of Historic and Artistic Works 14th Annual Meeting, 21-25 May (Chicago, IL), American Institute for Conservation, 1986, p 56–65
33. A.T. Hodgson, A.F. Rudd, D. Beal, and S. Chandra, *Indoor Air,* Vol 10, 2000, p 178–192
34. N. Le Bozec, D. Persson, A. Nazarov, and D. Thierry, *J. Electrochem. Soc.,* Vol 149 (No. 9), 2002, p B403–408
35. J.L. Schrenk, The Royal Art of Benin: Surfaces, Past and Present, *Ancient and Historic Metals: Conservation and Scientific Research,* Proceedings of a symposium organized by the J. Paul Getty Museum and the Getty Conservation Institute, D. Scott, J. Podany, and B. Considine, Ed., Nov 1991, The Getty Conservation Institute, 1994, p 56–59

36. D. Tilbrooke, *ICCM Bull.,* Vol 6 (No. 3–4), 1980, p 46–52
37. L. Robinet and M. Corbeil, *Stud. Conserv.,* Vol 48, 2003, p 23–40
38. J.L. Schrenk, Corrosion and Past "Protective" Treatments of the Benin "Bronzes" in the National Museum of African Art, *Materials Issues in Art and Archeology II,* P.B. Vandiver, J. Druzik, and G. Wheeler, Ed., Materials Research Society, 1991, p 805–812
39. R. Hughes and M. Rowe, *The Coloring, Bronzing and Patination of Metals,* Van Nostrand Reinhold Company, 1983
40. P. Hatchfield, *Pollutants in the Museum Environment,* Archetype Publications, 2002, p 43–54
41. J. Zhang, D. Thickett, and L. Green, *J. Am. Inst. Conserv.,* Vol 33 (No. 1), 1994, p 47–53
42. W.A. Oddy, *Museums J.,* Vol 73, 1973, p 27–28

Corrosion of Metal Artifacts Displayed in Outdoor Environments

L.S. Selwyn, Canadian Conservation Institute, and P.R. Roberge, Royal Military College of Canada

METAL ARTIFACTS are installed in public places to commemorate people, events, and ideas; as public art; or because they are of historical importance. The artifacts can be divided into three broad categories: unique pieces, historical metalwork, and reused pieces. The unique pieces are often custom-built and site-specific; they include works of art, sculptures, statues, monuments, commemorative pieces, war memorials, and fountains. Historical metalwork can be part of a heritage structure or a setting with historical significance; this category includes fences, gates, flagpoles, grave markers, plaques, and decorative elements on historic buildings. The reused pieces can be working artifacts that have been taken out of service and reused as commemorative pieces, such as tanks, cannons, anchors, artillery pieces, trains, aircraft, military vehicles, ships, and rockets. Many of these objects are made in part or totally of metals; copper alloys (bronze, brass) or ferrous metals (wrought iron, cast iron, steels, stainless steels, galvanized steel) are common. Aluminum, Cor-Ten (United States Steel Corp.) weathering steel, lead, zinc, and gold (gilding) are also used.

Being outdoors, cultural artifacts are exposed to uncontrolled conditions of sun, moisture, pollutants, temperature, and winds, as well as microclimates created by the location and influenced by the shape and surface detail of the artifact. They are exposed to a wide range of chemical and physical environments (acid rain, acid particulates, bird droppings), and they are also susceptible to damage by vandalism (graffiti, spray paint, theft). Windblown particles abrade sculptural details. Ultraviolet light and temperature extremes accelerate the breakdown of paints and other coating materials.

Cultural artifacts displayed outdoors in public places are intended to be highly visible. They need to be cared for and maintained; otherwise, the metal artifacts corrode and lose their heritage value and visual appeal. The composition and types of objects, their corrosion problems, and the approaches used to preserve, conserve, and maintain them are described in this article.

Environmental Factors Causing Damage

Some of the general factors are water, temperature fluctuations, freeze-thaw cycles, relative humidity (RH) fluctuations, wind load, particulate matter (sea salts, deicing salts, dirt), pollutants (sulfur dioxide, nitrogen oxides, ozone), local conditions of acidity or alkalinity, vegetation, animals, and people. Also important are design factors that generate local corrosion problems within a monument, such as dissimilar metals in direct contact, pooling water (from lack of drainage), condensed water under overhangs, and trapped water that freezes.

Environments. The location of outdoor metal monuments is classified in one of four general environments: rural, urban, industrial, or marine (Ref 1). Rural locations are relatively clean; urban locations have high levels of car exhaust; industrial locations have high pollutants levels; and marine locations have high levels of salts. Studies of the corrosion rate of small metal testpieces (coupons) in these general environments provide useful information about average corrosion rates due to macroscopic effects such as general climate and pollution conditions (Ref 2). Average corrosion rates are determined by mass-loss measurements, where metal coupons are weighed before and after a period of outdoor exposure; corrosion products are removed before weighing (Ref 3). Typical average corrosion rates for aluminum, copper, iron, lead, and zinc coupons in the four general environments are given in Table 1 (Ref 4–6).

When unprotected bare metal in a metal artifact is exposed outdoors, it interacts with water and environmental chemicals, corrodes, and forms what is referred to as a natural patina. The formation of this patina is influenced by the local microclimate. The overall condition and appearance depends on many factors, including the distance from the ground (including whether it is in contact with the ground), distance from local pollution sources, surface orientation, and degree of shielding from wind, rain, and sunlight. On outdoor bronze statues, for example, the side exposed to strong winds from the sea tends to become covered with lighter-green corrosion products than the sheltered side, which tends to be darker green (Ref 7). Horizontal or inclined surfaces on outdoor bronzes form a patina more rapidly than vertical surfaces do. Laboratory studies of copper exposed to pollutants have demonstrated an enhanced corrosion rate caused by a synergistic effect between sulfur dioxide (SO_2) and nitrogen dioxide (NO_2), and between SO_2 and ozone (Ref 8, 9). These results support the suggestion that outdoor bronzes suffer more corrosion damage when located in areas exposed to heavy traffic (where higher levels of NO_x are expected from car exhaust and possibly the winter use of deicing salt) than when located in more sheltered areas such as in the center of large parks (Ref 10).

Water. In outdoor environments, metals corrode by electrochemical processes that require water as an electrolyte. The corrosion rate of metals is negligible when the RH is low but can be significant when moisture is present to wet the surface. Electrochemical reactions are

Table 1 Corrosion rates for common metals in different kinds of outdoor environments

Metal	Rural		Urban		Industrial		Marine	
	μm/yr	mils/yr	μm/yr	mils/yr	μm/yr	mils/yr	μm/yr	mils/yr
Aluminum	0–0.1	0–0.004	~1	~0.04	...	...	0.4–0.6	0.016–0.024
Copper	~0.5	~0.02	1–2	0.04–0.08	2.5	0.1	~1	~0.04
Iron	4–65	0.16–2.6	23–71	0.92–2.84	26–175	1.04–7	26–104	1.04–4.16
Lead	0.1–1.4	0.004–0.056	1–2	0.04–0.08	0.4–2	0.016–0.08	0.5–2	0.02–0.08
Zinc	0.2–3	0.008–0.12	2–16	0.08–0.64	2–16	0.08–0.64	0.5–8	0.02–0.32

supported when there is enough adsorbed water to approach the behavior of bulk water, which happens on most clean metal surfaces above 65% RH (Ref 11), and when rain, dew, fog, or mist are present. In outdoor corrosion studies, it is often assumed that a metal surface will be wet enough to support corrosion whenever the RH is >80% and the temperature is above 0 °C (32 °F), although historic metals in Antarctica have been found to corrode above 50% RH and above −10 °C (14 °F) (Ref 12). Other factors that contribute to water condensation and metal corrosion are the presence of hygroscopic particles and surface roughness (microscopic cracks, porous particulate matter, corrosion products) that promote capillary condensation (Ref 1).

The metal corrosion rate also depends on the ionic conductivity of water, which is altered by absorption of atmospheric gases, dissolution of surface material (corrosion products), and deposition of airborne dust and salts. Exposure to wind and sun can affect condensation and the rate of drying as well as the amount of contaminants and corrosion products retained on the surface. Rain can help to keep exposed surfaces on outdoor metals less corroded than sheltered areas, where corrosive particles tend to accumulate and drying times are longer. As water flows over a metal surface, it dissolves soluble species and helps to rinse away deposited material. The rate at which metal is flushed from a surface by flowing water is known as the runoff rate and is based on the total metal concentration in collected runoff water (Ref 3, 13). The metal concentration in runoff water reflects the loss of corrosion products and can cause staining of adjacent surfaces. Studies have shown that higher runoff concentrations of copper and zinc are present in the water collected first (first flush) as the easily soluble corrosion products are removed, followed by a lower and reasonably constant concentration in subsequent volumes collected (steady state) as materials with low solubility are slowly dissolved (Ref 14). These studies also determined that the metal concentration in runoff water increased with increasing rain acidity (lower pH) and decreasing rain intensity (drizzle has the longest surface contact period). Other studies have shown that the runoff rates from copper and zinc are higher from surfaces with a low degree of inclination to the horizon (where more pollutants are deposited) and from surfaces oriented in the prevailing wind direction (Ref 15). Comparison studies of the corrosion and runoff rates for copper and zinc have shown that, on an annual basis, runoff rates are lower than corrosion rates, demonstrating that during the corrosion process, corrosion products with relatively low solubility build up on the metal surface (Ref 3).

Pollutants. The key pollutants that interact with outdoor metals and influence their corrosion rate are sulfur dioxide (SO_2), nitrogen oxides (NO_x), ozone (O_3), and chloride ions (Cl^-). Sulfur dioxide and nitrogen oxides form acids that attack the metal directly, whereas ozone and chloride ions enhance corrosion rates (Ref 16, 17). Sulfur dioxide gas is produced by burning fossil fuels (particularly coal), by pulp and paper mills, by smelting of ores, and by natural sources such as volcanoes. Atmospheric interactions convert sulfur dioxide to sulfuric acid (H_2SO_4), which is found in acid precipitation and in atmospheric particulates. The gas-phase nitrogen oxides are by-products of the combustion of fossil fuels, and they show up in acidic rain as nitric acid (HNO_3) and in aerosol particles (Ref 16). Ozone is generated in photochemical smog by the interaction of solar radiation with nitrogen oxide gases and hydrocarbons from vehicle exhaust. Ozone enhances metal corrosion rates because it is a powerful oxidizing agent and, when combined with sulfur dioxide, can have a synergistic effect that increases the corrosion rate of certain metals (Ref 9, 17). Chloride ions are present in precipitation, with the highest concentrations identified in precipitation in coastal regions (Ref 18). The chloride ions are from sea salt particles, winter deicing salts, or atmospheric hydrogen chloride gas (HCl) produced from the combustion of coal and the incineration of waste, especially polyvinyl chloride plastics (Ref 16, 17). Chloride ions increase the corrosion rate of metals because they interfere with the formation of protective surface films and form soluble salts and complex ions with many metal ions.

Outdoor monuments are exposed to pollutants through wet and dry deposition. Metals in unsheltered exposures are affected by both wet and dry deposition, whereas sheltered metals are affected predominantly by dry deposition. Wet deposition transfers atmospheric moisture containing water-soluble species (sulfuric acid, nitric acid) to the surface of outdoor monuments (Ref 16). Dry deposition transfers dry material (gases, microscopic particles, large debris) to outdoor surfaces. Soot, for example, is a microscopic particle produced by incomplete combustion of fossil fuels and often contaminated with sulfate and nitrate ions, absorbed gases, salts, and other material. Particulate material is easily trapped by rough, textured finishes (such as sheet aluminum) or ornate three-dimensional designs (such as pressed zinc) used to decorate historic buildings or in metal sculptures (Ref 19, 20).

Dry deposited and trapped material can damage the underlying metal whenever it gets wet and traps moisture next to the metal surface for long periods of time, promoting local crevice corrosion underneath it, especially if the salts or acids present are hygroscopic. Metal sulfates deliquesce at higher RH values, and these hygroscopic salts retain moisture on the metal surface, increasing the corrosion rate. Zinc sulfate deliquesces at 89% RH; iron(II) sulfate deliquesces at 95% RH; and copper sulfate deliquesces at 97% (Ref 21). Metal chlorides deliquesce at lower RH values: iron(III) chloride and zinc chloride below approximately 10% RH, calcium chloride at 33% RH, iron(II) chloride at 56% RH, and copper(II) chloride at 68% RH (Ref 21, 22). Wet deposited material (along with soluble ionic species from dry deposited material) participates in surface chemical reactions that produce the patina on outdoor monuments.

Sulfate ions participate in the corrosion of outdoor metals by increasing the conductivity of surface water and by interacting with many metals to form both soluble and insoluble corrosion products. The soluble compounds form first; some of these may convert into insoluble ones if they are not washed away by rain. Sulfate-containing corrosion products make up the natural patina on several common outdoor metals. On most outdoor lead, anglesite, a lead sulfate, is one of the main patina components (Ref 23, 24). On outdoor copper roofs and bronze statues, brochantite, a copper hydroxide sulfate, is the main corrosion product in the green patina (Ref 7, 25–29). A series of laboratory experiments have been carried out to study the effect of various pollutants on copper and on copper corrosion products exposed to various pollutants in order to obtain a better understanding of the natural patination process on outdoor copper and bronze monuments (Ref 8–10, 30–33). These studies have demonstrated the importance of the sulfate ion in the development of brochantite in the green patina on outdoor bronzes and its relative stability (Ref 30, 33).

Nitric acid and nitrate ions dissolve in surface water and increase its conductivity. The nitrate ion rarely becomes incorporated into patina layers, presumably because of the high solubility of metal nitrate compounds (Ref 24, 34–36).

Chloride ions can cause serious corrosion problems on many outdoor metals. Not only do chloride ions contribute to the conductivity of surface water, they also promote localized pitting corrosion because they interfere with the formation of protective surface films. Chloride ions form soluble salts and complex ions with many metal ions, thus promoting diffusion away from the metal surface rather than precipitation directly on the surface. Once a pit has initiated at a local surface defect, the dissolution of the metal within the pit causes the local chemical environment to become substantially more acidic than the surrounding bulk environment (Ref 37). Chloride ions from the surrounding electrolyte diffuse into the pit to maintain charge balance with the metal cations. When iron, for example, corrodes in the presence of an acidic solution containing chloride ions, a cyclic corrosion process (a corrosion cycle) is initiated, and this can cause pitting and perforation of the metal (see also the discussion on corrosion of iron alloys in the section "Corrosion of Common Metals used Outdoors" in this article).

Iron and steel objects suffer the worst damage by chloride ions. Rusting, perforation, and component loss have been observed in historic iron fences running alongside busy roads where winter snow, contaminated with deicing salts, is pushed on top of the fence. Paint failure is often the result of inadequate surface preparation that has left hygroscopic salts on the metal surface.

Aluminum artifacts suffer from pitting caused by chloride ions (Ref 36, 38, 39). Aluminum sculptures exposed to chloride ions become covered with spots (white or darkened by dirt), which are gelatinous deposits of aluminum hydroxide covering deep pits (Ref 40). Chloride ions are a contributing factor to the often severe pitting observed on outdoor bronzes, especially when they are located close to roads where deicing salts are used and the surface is frequently splashed with salty water (Ref 41, 42). On outdoor bronzes, the exposure to chloride ions may result in the formation of green copper(II) chloride hydroxides and occasionally the copper(I) chloride; these corrosion products suggest corrosion processes related to bronze disease, a corrosion problem normally associated with archaeological bronzes. For more information on bronze disease, see the article "Corrosion of Metal Artifacts in Buried Environments" in this Volume.

Acid Conditions. The pH of the moisture lying on the metal surface is a crucial factor in the dissolution of any protective surface layer. The pH of clean rain water is controlled by the dissolution of atmospheric carbon dioxide and typically has an equilibrium pH of approximately 5.6 (Ref 24). The presence of carbonate and bicarbonate ions in atmospheric moisture is important for the formation of light-colored patinas on lead and zinc. Hydrocerussite, a lead carbonate hydroxide, forms on outdoor lead, and hydrozincite, a zinc carbonate hydroxide, forms on outdoor zinc.

Acid precipitation (with a pH less than 5.6) develops from the dissolution of water-soluble acids (sulfuric acid, nitric acid, hydrochloric acid). Organic acids (formic, acetic, oxalic) also contribute (Ref 43). Historical assessments of sulfur dioxide concentrations in the United States and Europe show a marked increase between 1890 and the mid-1900s, followed by a leveling off as a result of strict emission controls (Ref 16, 44). In developing countries, sulfur dioxide concentrations may continue to increase because of the continued use of high-sulfur coal (Ref 16). The concentrations of nitrogen oxides and ozone have also increased during industrialization but have not declined during recent decades because of continued burning of fossil fuel in motor vehicles (Ref 33, 45). Since the beginning of the Industrial Revolution in the mid-18th century, the pH of precipitation has decreased (become more acidic) in many places around the world (Ref 16). In Europe, the pH of precipitation is thought to have decreased from approximately 5.4 in 1890 to 4.3 (more acidic) by 1980, while in urban areas of the Northeastern United States, the precipitation pH was estimated at 3.4 for the period 1920 to 1950 but had increased to 4.3 (less acidic) by 1980 (Ref 17). High acidities (pH < 4) can occur on the surface of outdoor metals exposed to acidic fogs or by evaporation of surface electrolytic solutions (Ref 30, 46). Local acidity can damage the surface of metals outdoors by dissolving corrosion products and by attacking the metal. In general, metals experience a thinning or dissolution of protective corrosion-product layers, followed by attack of the bare metal, with exposure to acidic solutions. The solubility of the iron corrosion products that typically form on outdoor iron increases as the pH drops below 6 (Ref 30, 47). For patinated copper, laboratory studies have observed an increase in the copper concentration in runoff water with a decrease in pH (from 4.3 to 3.5); this demonstrates an increased solubility of low-solubility copper compounds with increasing acidity (Ref 14). Other laboratory studies have demonstrated that the patina on copper can be fully dissolved and bare metal exposed if the pH drops below approximately 2.5 (Ref 30, 47). Unsightly grooves develop on lead and zinc roofs as acidic solutions (from acid rain or from organic acids leached from wooden shingles, lichen, or moss) flow over their surfaces. Lead corrodes faster in solutions with pH < 5, and zinc corrodes more rapidly when the pH is < 6 (Ref 24, 48–50). Aluminum corrodes quickly in solutions with pH < 4 (Ref 38, 51, 52).

Alkaline Conditions. Copper alloys and iron alloys are generally not affected by alkaline conditions, but aluminum, lead, and zinc are. These latter three metals corrode faster in alkaline solutions because the corrosion products are more soluble at high pH, and any protective corrosion layer becomes thinner or dissolves (Ref 53). Lead corrosion products are soluble in alkaline solutions with pH > 10 (Ref 24, 48, 49). Zinc oxide and zinc hydroxide are only stable in the pH range of 8 to 12 (Ref 50). The hydrated aluminum oxides responsible for the protection of aluminum become soluble in alkaline solutions with pH > 9 (Ref 38).

Alkaline solutions are present during the curing process in cement, lime-based mortars, and concrete. If these materials remain wet and if the pH is high, contact between them and aluminum, lead, or zinc may cause these metals to corrode or be etched. Lead surfaces, for example, become stained with colored lead monoxides (red litharge, yellow massicot, or alternating layers of red and yellow) when exposed to alkaline environments (Ref 23, 54, 55). It can be dangerous to use alkaline paint strippers on outdoor painted aluminum, lead, or zinc artifacts, because of the possibility that the alkaline stripper will damage the underlying metal.

Temperature. Large sculptures and monuments, particularly those made of zinc, lead, and aluminum, may suffer mechanical failure, with tearing, buckling, and cracking caused by differential thermal expansion and contraction from daily and seasonal temperature fluctuations (Ref 52). Zinc, lead, and aluminum have higher coefficients of linear expansion compared to materials such as brick, concrete, glass, and stone, although there are certain materials (elastomers, polyethylene, ice, epoxy) with much higher values (Table 2). In addition to expansion and contraction problems, zinc and lead also have a tendency to creep (flow slowly due to gravity) under their own weight, and so, large pieces may start to distort, lean, sag, or bulge (Ref 52, 56, 57). In the past, concrete has been used to fill lead and zinc sculpture as a means to provide support against the tendency of these metals to undergo creep. Concrete has also been used to fill the interior of hollow cast iron sculptures (Ref 58). Unfortunately, concrete shrinks on setting, typically leaving narrow gaps between the fill and the metal. Water eventually seeps into the gap, keeps the metal damp, and causes the metal to corrode. When metal, particularly iron, corrodes within a confined space, the increase in volume caused by the formation of corrosion products can place pressure on the surrounding metal and may cause it to crack. Similarly, if water freezes within a confined space, it expands and can also cause the metal to crack (Ref 57, 58).

Vegetation. Historic pieces located close to the ground often suffer corrosion caused by moisture trapped by adjacent shrubs and plants, or when flowers and wreathes are left uncollected and hold moisture next to the metal surface (Ref 59). Fast-growing ivy can encase historic wrought iron fences, disrupt paint, and trap moisture, which promotes rusting. Soil can blow in and fill in the gap between a fence and the ground. A metal sculpture in the shade of trees can remain damp for long periods of time. Trees and shrubs can also cause damage because of aggressive root systems, dropping branches, leaves or seeds, or exuding sticky material. Organic acids (e.g., acetic acid) can be leached from wooden shingles, and oxalic acid can be leached by water running through moss and lichen (Ref 60).

Animals. The complex structures of outdoor metal artifacts often provide convenient nesting grounds or perches for insects, birds, and animals. Nests from wasps, birds, and squirrels have been observed on outdoor sculptures (Ref 28, 61). Birds are often seen sitting on outdoor statues (Fig. 1), where they can scratch the finish and deposit material that causes the metal to

Table 2 Linear expansion coefficient of some common materials

Material	Linear expansivity, $\mu in./in. \cdot °C$
Metals	
Aluminum	23.1
Copper	16.5
Gold	14.2
Iron	11.8
Lead	28.9
Zinc	30.2
Nonmetals	
Brick, building	9
Concrete	10–14
Elastomer, silicone	270
Elastomer, styrene-butadiene	220
Epoxy	81–117
Glass, soda-lime	9
Ice	51
Marble	10
Polyethylene, low density	180–400
Wood, red oak, parallel to grain	5–6
Wood, red oak, perpendicular to grain	31–39

corrode. Nesting material traps moisture next to the metal surface, and degrading organic material introduces additional salts and acidic by-products (such as acetic acid) to the water layer.

Ants and other insects produce formic acid (Ref 51, 52). Bird urine contains varying amounts of ammonia (NH_3), uric acid ($C_5H_4N_4O_3$), and phosphates (PO_4^{3-}), and the pH varies from 5 to 8 (Ref 62, 63). Uric acid, ammonia-containing compounds (e.g., ammonium copper sulfate dihydrate, $Cu(NH_3)_2(SO_4)_2 \cdot 2H_2O$), and phosphates (e.g., libethenite, pyromorphite) have been identified on outdoor bronze statues (Ref 28). Researchers identified pyromorphite, a lead chloride phosphate ($Pb_5Cl(PO_4)_3$), on the gilded bronze *Golden Boy* located on top of the Legislative Building dome in Winnipeg (Ref 64). Bird excrement supports the growth of fungal spores that can lower the pH significantly (e.g., from 7 to less than 4) and damage the underlying metal (Ref 65).

People. Outdoor artifacts are subject to vandalism, especially those situated in isolated locations in parks, gardens, or cemeteries, or if their design and location make them accessible (Ref 42, 59, 66). Graffiti (applied using spray paint, felt-tip markers, paint-filled balloons, or sharp objects) can be a problem; rapid removal of graffiti helps discourage future vandalism. Monuments and their immediate surroundings can be damaged by group activities (e.g., skateboarding, partying) and unsightly garbage (e.g., uneaten food, glass bottles, cigarette butts). Sculptures can be damaged when they are repeatedly touched or climbed on; thin areas of soft metals can be crushed, coatings scratched, bare metal exposed, and small accessories bent or stolen. The bronze sculpture of Denmark's *Little Mermaid* has been decapitated more than once, and statues of political leaders have been pulled down and removed during times of strife. Finally, outdoor artifacts are damaged by neglect; they all need a program that provides some degree of protection against vandalism (e.g., good lighting) as well as regular maintenance.

Design can result in corrosion problems. One common design problem is the direct contact between two dissimilar metals that can lead to accelerated corrosion of the more active metal (galvanic corrosion). Design details that allow water to collect and lie stagnant for long periods of time promote corrosion.

Galvanic corrosion is the accelerated corrosion of a metal in electrical contact with a more noble metal (or a nonmetallic conductor) in an electrolyte. Many outdoor works of art and historically significant artifacts are constructed with two or more metals in direct contact. Artists use combinations of metals to achieve a certain look (e.g., brass on cast iron used in Brancusi's *Infinite Column*, as discussed in Ref 67). Initially, galvanic corrosion is not a problem, because a coating or paint layer isolates the metals from atmospheric moisture. However, over time, if the coating or paint is not maintained and these metals become exposed to water, the more active metals will suffer galvanic corrosion. A number of examples of galvanic corrosion are given subsequently.

Large metal sculptures are usually supported with an internal structure (an armature) made of another metal, often iron. Moisture enters into the interior of these sculptures (through openings or by condensation), and usually, the iron suffers from galvanic corrosion. The corrosion rate of the iron is accelerated when it is in direct contact with more noble metals, such as copper and bronze. The formation of rust exerts huge destructive forces on surrounding metal and can cause it to crack; rust staining of adjacent material is also a problem. Figure 2 shows damage to a cast bronze hand holding a rusting iron staff. Figure 3 illustrates galvanic corrosion

Fig. 1 Bird damage to a 1931 bronze statue located in Kingston, Ontario, that commemorates the 21st Battalion battles in World War I. Courtesy of Pierre Roberge, Royal Military College of Canada. Photograph 2002

Fig. 2 Detail of damage by galvanic corrosion of an iron staff in contact with a cast bronze hand on a statue of *Mercury* (date 1962) located in Kingston, Ontario. Courtesy of Pierre Roberge, Royal Military College of Canada. Photograph 2003

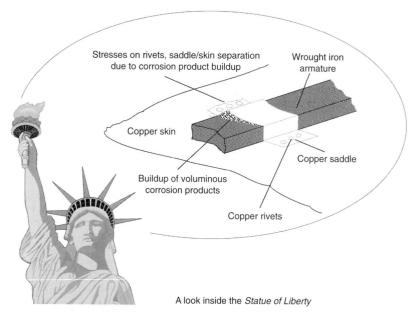

Fig. 3 Galvanic corrosion between copper and wrought iron on the *Statue of Liberty*, causing accelerated corrosion of the iron

on the *Statue of Liberty* (officially titled *Liberty Enlightening the World*) where the wrought iron armature was originally in direct contact with copper (Ref 25). Furthermore, severe rusting of iron bolts can dangerously weaken the attachment points (Ref 68).

Historic iron artifacts (iron cannons, wrought iron gates) suffer accelerated corrosion when brass plaques are directly attached to them. Aluminum suffers galvanic attack when it is in contact with copper or copper-base alloys, such as when brass screws are used to attach lighting elements to an aluminum sculpture (Ref 38). Zinc also suffers galvanic attack when it comes in contact with iron. This is used to advantage in zinc-coated (galvanized) steel but can also cause corrosion problems. If, for example, a pure zinc decorative element or a small piece of galvanized iron is in direct contact with iron (especially a large surface area of iron), then the zinc component suffers accelerated corrosion (Ref 52, 69).

Zinc sculptures that were electroplated with copper (a process that allowed zinc sculptures to be patinated to resemble more expensive bronze sculptures) and then left outdoors suffer from galvanic corrosion. The surface suffers from severe pitting at breaks in the copper where local galvanic cells are created (Fig. 4). The more-active zinc suffers from accelerated corrosion, pits develop and become covered with white zinc corrosion products, and the surface of the sculpture becomes covered with unsightly pimples (Ref 56).

Local galvanic corrosion cells can develop if more-noble metal ions plate from a solution onto the surface of an active metal (Ref 20, 38, 41). This can happen, for example, when rain flows first over a corroding bronze sculpture (where it picks up copper ions in solution) and then flows over and plates onto aluminum letters set in the stone below. The copper ions are reduced to copper metal, and at the same time, the aluminum corrodes and aluminum ions go into solution. Once the copper is plated, it leads to serious attack of the aluminum due to the galvanic cell that is established, and eventually, pitting will occur in the aluminum as it corrodes (Ref 38). Local corrosion cells can also form if steel wool or steel brushes are used to clean artifacts made from other metals. Tiny iron pieces can become trapped in the artifact surface, where they rust and cause staining.

Detailing. The design of a work of art or a heritage object often creates areas where water collects. The expansion force of the corrosion products formed in a confined space (oxide jacking) can cause considerable distortion. Sculptures made of weathering steel can suffer from severe corrosion where water pools because of inadequate drainage (Ref 70, 71). Hollow cast bronzes can crack where trapped water expands during freezing; bronze sculptures usually contain drainage holes (called weep holes) to prevent water from becoming trapped in restricted areas (Ref 42). Heritage vehicles and other complex multiple-part artifacts, especially those containing iron, can suffer severe crevice corrosion where water becomes trapped underneath bolts and gaskets, between unsealed joints (bolted or overlapped seams), or in cracks in poor welds (Ref 72).

Serious corrosion problems, including perforation, can occur where a thin layer of water condenses (often due to poor ventilation) on metals, particularly on the underside of zinc and lead used in historic roofing (Ref 20, 23, 73, 74). This form of corrosion, called underside or condensation corrosion, occurs where water repeatedly condenses as a thin film. Thin water films are nearly saturated with dissolved oxygen, and so, the corrosion rate of a metal is more rapid when covered by a thin film of water compared to a thicker layer. Pollutants or other dissolved materials can increase the acidity of the water and interfere with the development of an adherent corrosion film that usually develops on these metals under conditions of good ventilation. Corrosion products tend to precipitate away from the metal surface, usually as loose white material (called active corrosion on lead and white rust on zinc). More severe corrosion is often encountered where lead and zinc sheets are adjacent to or supported by wood (with oak and cedar being the worst), which releases volatile organic acids such as acetic or formic acids (Ref 51). These acids form soluble compounds with lead and zinc.

Corrosion of Common Metals Used Outdoors

The corrosion processes experienced by copper, iron, lead, zinc, and aluminum—metals commonly used for artifacts in outdoor environments—are discussed in this section. When left uncoated, these metals undergo uniform corrosion and develop a corrosion-product layer. The chemical name, mineral name, and chemical formula for some of the common corrosion products found on metals exposed outdoors are listed in Table 3 (copper), Table 4 (iron), Table 5 (zinc), Table 6 (lead), and Table 7 (aluminum).

Copper Alloys. Relatively pure copper sheets were used to make the outer shell of the 46 m (151 ft) tall *Statue of Liberty* in New York City (Ref 25). Copper sheet is still a popular roofing material, especially on historic buildings (Ref 25, 52, 75). Bronze has been used since antiquity as a casting material for statuary (Ref 76). The large (13 m, or 43 ft, high) bronze of *Great Buddha of Kamakura* in Japan was cast in the 13th century (Ref 77). Another large (11 m, or 36 ft, high) bronze cast in the late 19th century is that of *William Penn*, which sits on top of the Philadelphia City Hall (Ref 52, 68). Although many outdoor bronze statues are unique, there are examples where many copies of the same statue exist. There are some 50 copies of a statue known as the *Hiker* located throughout the United States (Ref 78). Cast bronze statues are hollow, and the larger ones are made of several pieces joined together that may be supported with an internal armature that, in the past, was usually wrought iron. Traditionally, newly cast bronzes are treated with a corrosive chemical solution to produce an intentional patina, called an artificial patina, so the color could be determined by the artist or by the foundry (Ref 79, 80).

Copper alloys generally have good resistance to corrosion in outdoor environments (Table 1). When uncoated copper alloys are exposed outdoors, they undergo relatively uniform corrosion; any intentional patina gradually changes as the surface corrodes. The copper interacts with the environmental species and forms new corrosion products, referred to as a natural patina. Some corrosion products identified on corroded outdoor copper alloys are listed in Table 3 (Ref 7, 26, 28). Most copper(II) compounds are green or blue (Ref 81). The formation of an adherent layer of (usually green) corrosion products on copper helps to slow its corrosion (Ref 80). This naturally formed green patina is often not the artist's original intent, but, because the change from the original artificial patina was slow, the general public tends to forget the original patina color, and many people believe that all bronze sculptures are supposed to be green.

It can take from approximately 8 to 20 years for outdoor copper roofs to develop their characteristic green patina (Ref 82). The rate at which

Fig. 4 Pinhole corrosion on a copper-plated cast zinc statue of *Liberty Enlightening the World* after outdoor exposure. The statue (127 cm, or 50 in., tall) was modeled after Auguste Bartholdi's original model for the *Statue of Liberty*, and many copies were sold to raise money. It was cast in zinc, plated with copper, and given an imitation bronze patina by the foundry Avoiron in Paris sometime between 1878 and 1886. Courtesy of Carol Grissom, Smithsonian Center for Materials Research and Education. Photograph 1981

the green patina forms depends on several factors, including RH, pollution levels, and local acidity. Initially, as the copper corrodes outdoors, it develops a dark layer of cuprite, a copper(I) oxide. Over time, this dark patina gradually changes to a green one as insoluble copper(II) compounds form over the cuprite layer (Ref 83). After decades of exposure, bronze statues can develop a patchy and streaked appearance with alternating black and light-green areas (Ref 28, 84).

The presence of tin in bronze alloys plays a role in determining how the metal corrodes. Tin forms insoluble tin oxides, such as romarchite (SnO) and cassiterite (SnO_2) (Ref 28), which tend to be cathodic relative to the copper alloy, and the subsequent corrosion behavior is the selective dissolution of copper (Ref 27). In exposed areas, a thick layer of corrosion products does not build up, because of periodic washing by rain. These exposed areas tend to be an overall light green, and the main corrosion product is brochantite (Ref 7, 25–29). This surface can develop pits (Fig. 5) as well as small islands of black crusts standing above lower green areas, a process probably related to microsegregation in the alloy (Ref 27, 84, 85). A potential difference of 280 mV has been measured between the anodic green areas and the cathodic black ones (Ref 86). Areas sheltered from rain tend to be black, in part because of the accumulation of debris, soot, pollutants, and other material. These darker areas often contain a complex mixture of dark particulate matter (e.g., quartz, soot), cuprite, green copper(II) hydroxide sulfates (e.g., brochantite, antlerite), and, if chloride ions are present in the environment, green copper(II) chloride hydroxides (e.g., atacamite, clinoatacamite, paratacamite) (Ref 7, 27–29).

Sulfate Formation. Copper(II) hydroxide sulfates form on outdoor copper alloys when sulfate ions (e.g., from acid rain) dissolve in surface water and react with the underlying material. Brochantite is usually detected in the green rain-washed areas (Ref 7, 27–29). Antlerite is usually detected in sheltered areas that favor the accumulation of debris (Ref 7, 27–29). Antlerite precipitates under more acidic conditions than does brochantite (Ref 83, 87). Laboratory studies have shown that once brochantite and antlerite have formed, they remain unaffected by exposure to humid air containing only sulfur dioxide, mixtures of sulfur dioxide and ozone, and mixtures of sulfur dioxide and nitrogen dioxide (Ref 33). However, when copper hydroxide sulfates

Table 3 Some corrosion products identified on copper alloys exposed to outdoor environments

Chemical name	Mineral name	Chemical formula	Color
Oxide			
Copper oxide	Cuprite	Cu_2O	Red
Carbonates			
Copper carbonate hydroxide	Malachite	$Cu_2CO_3(OH)_2$	Green
Copper sodium carbonate trihydrate	Chalconatronite	$CuNa_2(CO_3)_2 \cdot 3H_2O$	Blue
Chlorides			
Copper chloride	Nantokite	$CuCl$	Pale gray
Copper chloride hydroxide	Atacamite	$Cu_2Cl(OH)_3$	Green
Copper chloride hydroxide	Clinoatacamite	$Cu_2Cl(OH)_3$	Green
Copper chloride hydroxide	Paratacamite	$Cu_2Cl(OH)_3$	Green
Sulfides and sulfates			
Copper sulfide	Chalcocite	Cu_2S	Black
Copper sulfide	Geerite	$Cu_{1.6}S$	Black
Copper sulfide	Djurleite	$Cu_{1.96}S$	Black
Copper sulfate pentahydrate	Chalcanthite	$CuSO_4 \cdot 5H_2O$	Blue
Copper hydroxide sulfate	Brochantite	$Cu_4(OH)_6SO_4$	Green
Copper hydroxide sulfate	Antlerite	$Cu_3(OH)_4SO_4$	Green
Copper hydroxide sulfate monohydrate	Posnjakite	$Cu_4(OH)_6SO_4 \cdot H_2O$	Light blue
Copper hydroxide phosphate	Libethenite	$Cu_2(OH)(PO_4)$	Green
Ammonium copper sulfate dihydrate	...	$Cu(NH_4)_2(SO_4)_2 \cdot 2H_2O$	...

Table 4 Some corrosion products identified on ferrous alloys exposed to outdoor environments

Chemical name	Mineral name	Chemical formula	Color
Oxides and hydroxides			
Iron oxide	Magnetite	Fe_3O_4	Black
Iron hydroxide oxide	Goethite	α-FeO(OH)	Yellow-brown
Iron hydroxide oxide	Akaganéite	β-FeO(OH)	Red-brown
Iron hydroxide oxide	Lepidocrocite	γ-FeO(OH)	Orange
Sulfates			
Iron sulfate tetrahydrate	Rozenite	$FeSO_4 \cdot 4H_2O$	Green
Iron sulfate pentahydrate	Siderotil	$FeSO_4 \cdot 5H_2O$	White
Iron sulfate heptahydrate	Melanterite	$FeSO_4 \cdot 7H_2O$	Blue-green
Iron hydroxide sulfate dihydrate	Butlerite	$Fe(OH)SO_4 \cdot 2H_2O$	Orange
Iron potassium hydroxide sulfate	Jarosite	$Fe_3K(OH)_6(SO_4)_2$	Yellow-brown
Iron sodium hydroxide sulfate	Natrojarosite	$Fe_3Na(OH)_6(SO_4)_2$	Yellow-brown

Table 5 Some corrosion products identified on zinc exposed to outdoor environments

Chemical name	Mineral name	Chemical formula	Color
Oxides and hydroxides			
Zinc oxide	Zincite	ZnO	White
Zinc hydroxide	...	$Zn(OH)_2$	White
Carbonates			
Zinc carbonate	Smithsonite	$ZnCO_3$	White
Zinc carbonate hydroxide	Hydrozincite	$Zn_5(CO_3)_2(OH)_6$	White
Zinc carbonate hydroxide monohydrate	...	$Zn_4CO_3(OH)_6 \cdot H_2O$	...
Chlorides			
Zinc chloride hydrate	...	$ZnCl_2 \cdot xH_2O$(a)	White
Zinc chloride hydroxide monohydrate	Simonkolleite	$Zn_5Cl_2(OH)_8 \cdot H_2O$	White
Zinc chloride hydroxide sulfate pentahydrate	...	$Zn_4Cl_2(OH)_4SO_4 \cdot 5H_2O$	...
Sodium zinc chloride hydroxide sulfate hexahydrate	...	$NaZn_4Cl(OH)_6SO_4 \cdot 6H_2O$	...
Sulfates			
Zinc sulfate heptahydrate	Goslarite	$ZnSO_4 \cdot 7H_2O$	White
Zinc hydroxide sulfate hydrate	...	$Zn_4(OH)_6SO_4 \cdot xH_2O$(a)	...

(a) x can vary, depending on the number of waters of hydration.

Table 6 Some corrosion products identified on lead exposed to outdoor environments

Chemical name	Mineral name	Chemical formula	Color
Oxides			
Lead oxide	Litharge	α-PbO	Red
Lead oxide	Massicot	β-PbO	Yellow
Carbonates			
Lead carbonate	Cerussite	$PbCO_3$	White
Lead carbonate hydroxide	Hydrocerussite	$Pb_3(CO_3)_2(OH)_2$	White
Sulfites and sulfates			
Lead sulfite	Scotlandite	$PbSO_3$	White
Lead sulfate	Anglesite	$PbSO_4$	White

are exposed to an aqueous solution of sodium chloride in laboratory studies, they do convert to copper chloride hydroxides (Ref 10). Other copper(II) sulfates (e.g., chalcanthite) and copper(II) hydroxide sulfates (e.g., posnjakite) have occasionally been identified on outdoor copper alloys (Ref 7, 26, 88). These sulfates are considered transitory or intermediate compounds, indicating recent copper corrosion (Ref 26, 30). They can be washed away by rain water or converted to more stable copper(II) hydroxide sulfates, such as brochantite.

Chloride Formation. Copper(II) chloride hydroxides form on outdoor copper alloys when the surface is contaminated by high levels of chloride ions (present in marine environments, nearby use of deicing salts, and air pollution). Insoluble copper(II) chloride hydroxides (e.g., atacamite, clinoatacamite, paratacamite) have been identified on copper alloys exposed outdoors (Ref 7, 27–29). On occasion, copper(I) chloride (nantokite) has also been identified (Ref 27–29). Recent work shows that the x-ray diffraction reference pattern PDF 25-1427 had been incorrectly assigned to paratacamite (rhombohedral) and instead corresponds to a monoclinic crystal structure, newly named clinoatacamite (Ref 89). Therefore, past identification of paratacamite on copper alloys may be clinoatacamite. Laboratory studies have demonstrated that when atacamite is exposed to humid SO_2-containing air, it converts to soluble compounds (e.g., copper(II) sulfate, copper(II) chloride) that are not protective and can stimulate further corrosion of the underlying metal (Ref 10, 31–33). The presence of atacamite or other green copper(II) chloride hydroxides and occasionally the copper(I) chloride nantokite suggests bronze disease, a corrosion process normally associated with archaeological bronzes. More information on bronze disease is in the article "Corrosion of Metal Artifacts in Buried Environments" in this Volume.

Carbonates. Although copper(II) carbonate hydroxides can form on copper corroding in solutions containing carbonate ions, they have only rarely been identified on outdoor copper alloys. Malachite, for example, has been identified on outdoor bronze statues adjacent to damp limestone (calcium carbonate) (Ref 29). It is likely that any copper(II) carbonate hydroxides that do form are gradually converted to copper(II) hydroxide sulfates whenever the surface is exposed to water containing sulfate ions (Ref 90).

Other notable compounds identified on outdoor bronzes include copper sulfides (e.g., chalcocite, djurleite, geerite), copper oxalates, and gypsum (Ref 28, 29). The copper sulfides may be from the reaction of the copper alloy with hydrogen sulfide, especially if the copper alloys are close to pulp and paper mills. Copper oxalates could form because of exposure to oxalic acid, which is secreted by microorganisms such as fungi and lichens and which is a component of acid fog, rain, and mist in urban areas (Ref 43, 60). White accretions of gypsum ($CaSO_4 \cdot 2H_2O$) sometimes develop on the outside of sculptures when residual casting material (e.g., plaster of Paris, $CaSO_4 \cdot 1/2H_2O$) on the inside is dissolved and redeposited on the outside as water seeps through cracks and pores in the casting (Ref 28, 42).

Staining. The corrosion of copper alloys often causes green staining of adjacent material, especially porous stone or concrete (Ref 7, 20). When copper ions are washed off corroding copper, they can become trapped in adjacent material, where they react with other environmental ions and form a green stain of copper(II) compounds (e.g., brochantite, atacamite).

Iron Alloys. Various iron alloys have been used for outdoor works of art or in the construction of heritage buildings and heritage artifacts (Ref 52). In the past, traditional wrought iron (handmade iron with a characteristic fibrous structure due to inclusions of slag) was used for many decorative and functional items, with the peak in wrought iron production occurring in approximately 1850 (Ref 51). Examples include small items such as decorative wrought iron balconies, gates, and fences, and large structures such as the Eiffel Tower in Paris, the internal support structure for the *Statue of Liberty,* and the iron pillar of Delhi (Ref 25, 52, 75, 91). Cast iron is another traditional material, and it is used to make fountains, cannons, building facades, benches, lamps and lamp posts, domes, and even bridges (such as the first European cast iron bridge built in 1781 in Ironbridge, England) (Ref 52, 58). Carbon steels are present in many modern heritage objects and military monuments (Ref 72, 92). Corrosion-resistant stainless steels are popular for polished metal sculptures and as roofing or cladding. In 1930, stainless steel was used to finish the dome on the Chrysler Building in New York City (Ref 93). Weathering steels (Cor-Ten) have been used since the 1960s in architecture and for outdoor sculptures, because this material, when left unpainted, develops a natural dark-brown patina (Ref 71). Weathering steel was used to make the large figure (15 m, or 49 ft) designed by Picasso for the Daley Plaza in Chicago (1967) and known locally as the *Chicago Picasso.*

Most iron alloys rust if they are exposed outdoors without a protective coating (Ref 35, 94). The rust layers that form on corroded iron alloys are porous, poorly adherent, and often cracked. They offer little protection to the underlying metal against the penetration of moisture. The cracking and spalling of rust layers is caused by the formation of new iron corrosion products that have molar volumes approximately three times greater than the molar volume of iron. In general, carbon steels have a higher corrosion rate outdoors than aluminum, copper, lead, or zinc (Table 1), especially in industrial and marine environments (Ref 95).

Some corrosion products that form on iron when it is exposed outdoors are listed in Table 4 (Ref 35, 94, 95). The corrosion of iron is an electrochemical process: the anodic reaction is the oxidation of iron to Fe(II) ions, and the cathodic reaction is usually the reduction of dissolved oxygen. Once Fe(II) ions are present in solution, a variety of corrosion products can form and precipitate. The rust layers generally include an inner layer of black magnetite (plus other amorphous iron corrosion products) and an outer layer of iron hydroxide oxides (e.g., goethite, lepidocrocite), which have the familiar orange, red, and brown colors of rust (Ref 35). Other iron corrosion products can form, depending

Table 7 Some corrosion products identified on aluminum exposed to outdoor environments

Chemical name	Mineral name	Chemical formula	Color
Oxides and hydroxides			
Aluminum oxide (alumina)	...	γ-Al_2O_3	White
Aluminum hydroxide	Bayerite(a)	$Al(OH)_3$	White
Aluminum hydroxide	Gibbsite(a)	$Al(OH)_3$	White
Aluminum hydroxide oxide	Boehmite(b)	$AlO(OH)$	White
Chlorides			
Aluminum chloride hexahydrate	...	$AlCl_3 \cdot 6H_2O$	White
Aluminum chloride hydroxides	...	$Al_x(Cl)_y(OH)_z$	...
Sulfates			
Aluminum sulfate hydrate	Alunogenite	$Al_2(SO_4)_3 \cdot 18H_2O$	...
Aluminum hydroxide sulfates	...	$Al_x(OH)_y(SO_4)_z$	...

(a) The chemical formula for bayerite and gibbsite is sometimes written as $Al_2O_3 \cdot 3H_2O$. (b) The chemical formula for boehmite is sometimes written as $Al_2O_3 \cdot H_2O$.

Fig. 5 Detail of severe pitting on a child's face in the 1887 bronze statue *Stone Age in America* by John J. Boyle located in Philadelphia. Courtesy of the Fairmount Park Art Association. Photograph by Franko Khoury © 1983

on the environmental ions present in surface water.

Sulfate ions accelerate the rusting of iron (Ref 35, 95). When iron rusts in the presence of sulfate ions, soluble iron(II) sulfate forms (Ref 1). Upon evaporation of water, various iron(II) sulfate salts (with different waters of hydration) can crystallize (e.g., rozenite, melanterite, siderotil), but in the presence of moist air, iron(II) sulfate undergoes oxidation and hydrolysis according to:

$$FeSO_4(s) + {}^{1}\!/\!{}_{4}O_2(g) + 1{}^{1}\!/\!{}_{2}H_2O$$
$$\rightarrow FeO(OH)(s) + 2H^+ + SO_4^{2-}$$

Iron oxide hydroxides precipitate (e.g., lepidocrocite, goethite), and the sulfate ion is released as sulfuric acid, making the sulfate ion available to again attack the remaining metal, creating an ongoing corrosion cycle. Some of the sulfate ions are removed from solution as they become incorporated into the rust layer as relatively insoluble iron(III) hydroxide sulfates, such as butlerite, or, if sodium or potassium ions are present, as jarosite or natrojarosite (Ref 35).

Chloride ions are easily transported through a rust layer where they can attack the underlying metal, form soluble iron chloride salts and complexes, promote pitting, and cause rapid, disfiguring corrosion (Ref 35). The hygroscopic nature of iron chloride salts also promotes the formation of blisters (referred to as weeping or sweating). Only rarely do chloride ions become incorporated into the rust layer, but they do promote the formation of iron hydroxide oxides. Laboratory studies have shown that if iron corrodes in solutions containing low concentrations of chloride ions, goethite and lepidocrocite precipitate (Ref 96). However, if the chloride ion concentration is high, another iron hydroxide, known as akaganéite, can precipitate. Akaganéite is normally written β-FeO(OH), but because chlorine atoms stabilize its structure (they are located inside tunnels), a chemical formula more representative of its chlorine content is $FeO_{0.833}(OH)_{1.167}Cl_{0.167}$ (Ref 97–99). For more information on akaganéite formation and weeping on iron, see the article "Corrosion of Metal Artifacts in Buried Environments" in this Volume. Chloride ions accelerate the corrosion rate of iron because they participate in the following corrosion cycle (Ref 100):

$$2Fe(s) + 4H^+(aq) + 4Cl^-(aq) + O_2(g)$$
$$\rightarrow 2Fe^{2+}(aq) + 4Cl^-(aq) + 2H_2O$$

$$2Fe^{2+}(aq) + 4Cl^-(aq) + 3H_2O + {}^{1}\!/\!{}_{2}O_2(g)$$
$$\rightarrow 2FeO(OH)(s) + 4H^+(aq) + 4Cl^-(aq)$$

The crucial factor in this cycle is that the chloride ions form a soluble salt with iron(II) ions, and it is this solubility that allows the cycle to proceed. If the iron chloride salt were insoluble, it would precipitate and the cycle would be broken.

Crevice corrosion can occur on any outdoor metal artifact, but it most often damages iron artifacts, especially large heritage pieces that are not being maintained. Crevices exist at metal-to-metal joints or underneath material held against a metal surface, such as gasketting, biofouling, or external deposits (Ref 37). Because the diffusion of oxygen into a crevice is restricted, an oxygen concentration cell develops, and the region inside the crevice corrodes because it is anodic relative to the region outside the crevice. Both the acidity and the concentration of dissolved metal ions increase inside the crevice, causing a significant increase in the iron corrosion rate within the crevice. Visual evidence of crevice corrosion is often rust leaching out from a seam.

Weathering steels are low-carbon steels with minor additions of other alloying elements (copper, phosphorus, manganese). When exposed outdoors, weathering steels develop a dense adherent oxide film that helps slow the corrosion rate relative to plain carbon steels (Ref 35). The minor alloying elements have a beneficial effect in forming an inner rust layer (consisting mainly of goethite) that is a good barrier to oxygen and water (Ref 101). The formation of a stable patina of rust depends on alternating wet and dry cycles. It can take years for the patina to develop on weathering steel, and during its formation, the runoff can leave stains on adjacent material (Ref 70, 71). The patina can be uneven and streaked (Ref 102). Pooling of water is a major enemy of weathering steels because, if they remain immersed in water, they will corrode with a rate similar to that for plain carbon steel and therefore become susceptible to significant rusting, delamination, and perforation (Ref 70, 71).

Stainless steels, with their chromium-rich oxide film, and other corrosion-resistant specialty iron alloys remain relatively uncorroded in an outdoor environment (Ref 35). Pitting can occur on a microscopic level beneath crevices at inclusions or flaws in the passive layer. A sign of pitting is the formation of small rust spots on the surface of stainless steel. Because chloride ions interfere with the formation of the passive oxide film, they are usually responsible for initiating pitting on stainless steels. Outdoors, where stainless steel is exposed to wind-blown material, dirt can build up and trap moisture and contaminants next to the metal surface. If the oxide film on stainless steels is damaged by chloride ions, oxygen or some other oxidizing agent is needed to reform the film. If dirt restricts oxygen, then the oxide film may not reform, and the stainless steel can then rust. Two popular stainless steels for architectural use outdoors are types 304 (UNS S30400) and 316 (UNS S31600) (Ref 20). In environments contaminated by chloride ions or industrial pollutants, type 316 stainless steel, which contains 2 to 3% Mo, is preferred because it is more corrosion resistant than type 304.

Zinc had its greatest popularity for statuary between 1850 and 1920, when it was used for mass producing low-cost alternatives to bronze or lead. Zinc was used for sculptures, cemetery monuments, architectural elements (roofing, decorative tiles, moldings, animal heads, gargoyles), fountains, and grave markers (Ref 51, 52, 56, 61). In the United States, inexpensive zinc statues and American War memorials (many of which were ordered directly from catalogs) were used to adorn city halls, county courthouses, parks, post offices, and cemeteries (Ref 52, 56, 57). Larger pieces were made by joining together smaller cast or stamped pieces using soft solder. Some zinc sculptures were finished to imitate other (more expensive) materials by painting, electroplating with copper (and patinating to resemble bronze), or even gilding (Ref 56, 57). In the United States, the zinc pieces used in Civil War memorials and cemetery monuments were often left uncoated; they were sandblasted to a matte, stonelike finish and then placed outdoors and allowed to develop a uniform light-gray patina (Ref 52, 56, 57). The use of zinc in sculptures started to decline in approximately 1920, after its poor durability outdoors had been recognized. Zinc is used in modern metal works of art, sometimes as cladding.

In outdoor environments, the corrosion rate of zinc is much lower than iron. This is one of the reasons why zinc is used as a protective coating on iron (galvanized steel). The other reason is that zinc corrodes preferentially to steel because zinc is the more anodic of the two (Ref 50). In general, zinc is more resistant to corrosion in rural atmospheres and less resistant to corrosion in industrial environments polluted with sulfur dioxide. Zinc usually undergoes uniform corrosion when exposed outdoors, and some of the corrosion products that can form are listed in Table 5 (Ref 34, 50). When zinc is placed outdoors, it gradually changes from a bright metallic finish to a matte, gray patina as the zinc undergoes uniform corrosion and forms a somewhat protective film of adherent, relatively insoluble white zinc corrosion products (Ref 34). When zinc corrodes, zinc oxide and zinc hydroxide form first. These corrosion products react with various soluble species (carbon dioxide, chloride ions, sulfate ions) present in surface moisture and convert into other insoluble or soluble zinc corrosion products (Ref 34, 50). In unpolluted environments, zinc reacts with dissolved carbon dioxide gas, and relatively insoluble zinc compounds (e.g., hydrozincite) form in a protective layer (Ref 34, 50, 103). This carbonate layer, however, does not completely protect zinc from further corrosion, especially in marine, urban, and industrial atmospheres contaminated by acid rain, chloride ions, and sulfate ions. Under acidic conditions, the carbonate layer dissolves, and the zinc reacts with soluble contaminants. Zinc reacts with chloride and sulfate ions, at least initially, to form soluble zinc chlorides and zinc sulfates (Ref 34, 50). These soluble salts are readily washed away by rain, which can lead to white staining on adjacent dark building material (Ref 52). Zinc chlorides and zinc sulfates are hygroscopic and, if not washed away, gradually transform into insoluble zinc chloride hydroxides, zinc hydroxide sulfates, and other zinc compounds (containing both chloride and sulfate ions) that become incorporated into the corrosion layer (Ref 104, 105).

Lead is one of the metals used since antiquity (well before 3000 B.C.) (Ref 106). The Romans employed it for water pipes. Cast lead was popular for garden ornaments in England during the late 17th and 18th centuries; many were painted to imitate stone, and some were even gilded (Ref 51). Lead can also be found in roofing, sculptures, fountains, ornate 18th century cisterns, lead cames in stained glass windows (dating as far back as the 12th century), lettering in stone, spacers between stone and as a water barrier under stone, and grout to set iron bars in masonry (Ref 52). It also shows up in soft solder (lead-tin alloys), old pewter (lead is excluded from modern pewter), and in architectural cladding such as lead-coated or terne-coated metals (terne, which means dull, is a lead-tin alloy) (Ref 20). Today, because of health and environmental factors, the use of lead is severely restricted. When used outdoors, for example, the corrosion of lead can result in the amount of lead in precipitation runoff exceeding safe limits (Ref 107).

Lead has a generally good resistance to corrosion in outdoor environments, with its corrosion rate being greater than aluminum, similar to copper (depending on the environment), and smaller than zinc and steel (Table 1). Lead undergoes uniform corrosion when exposed outdoors, and some corrosion products are listed in Table 6 (Ref 23, 107–109).

When lead is first placed outdoors, it forms a thin, almost transparent layer of lead oxides, predominantly the yellow massicot but also some red litharge (Ref 23). Note that the term *litharge* can have two meanings: it can refer specifically to the red lead monoxide (α-PbO) or collectively to lead monoxides that can exist in two polymorphs (α-PbO and β-PbO). The initial patina that develops on outdoor lead artifacts contains white compounds (usually hydrocerussite but sometimes also cerussite) formed by the reaction of the lead oxides with carbon dioxide dissolved in surface water (Ref 23, 107–109). Hydrocerussite forms an adherent, relatively insoluble layer that decreases the corrosion rate of the lead.

Over time, other insoluble lead compounds also form as the hydrocerussite reacts with other dissolved species. Lead sulfite and lead sulfate (anglesite) form on lead in environments polluted with sulfur dioxide (Ref 23). Chloride-containing lead compounds do not appear to be incorporated into the corrosion layer (Ref 24, 107). The most common minerals detected on outdoor lead are anglesite, hydrocerussite, and cerussite (Ref 24).

Aluminum. One of the earliest examples of the use of aluminum is the pyramidal aluminum cap at the top of the Washington Monument (Ref 52). The condition of the aluminum cap in 1999 is shown in Fig. 6; it was cast at great expense in 1884. Aluminum metal was not widely used until after 1886, when it became more affordable as a result of the development of the Hall-Heroult electrolytic process. Another example is the cast aluminum figure of *Eros* on the Shaftesbury Memorial in Piccadilly Circus, London (unveiled in 1893) (Ref 110). Today, aluminum is used for architectural purposes, outdoor sculptures, and airplanes exhibited outside airports or aviation museums (Ref 20, 111).

Aluminum usually corrodes outdoors more slowly than other common metals (Table 1) (Ref 5, 36). The corrosion rate does, however, depend on the specific alloy involved; those alloys containing copper or zinc are much more susceptible to corrosion. In general, when aluminum is exposed outdoors, it undergoes uniform corrosion. Some corrosion products that have been identified on corroded outdoor aluminum are listed in Table 7 (Ref 36, 38).

When aluminum is exposed outdoors, its silvery surface gradually dulls as it weathers and forms a gray surface patina. The patina contains the hydrated aluminum oxides (bayerite, gibbsite, boehmite) mixed with dirt and atmospheric pollutants (Ref 36, 38). In a relatively non-polluted outdoor environment, the hydrated aluminum oxide layer helps to protect aluminum against further uniform corrosion, especially if the pH of surface moisture remains in the range of 4 to 9. Exposure to pollutants, however, can cause local corrosion or increase the overall corrosion rate, especially if the pollutants react with aluminum to form soluble aluminum salts. Sulfate ions, for example, react with aluminum to form soluble aluminum sulfates that, if not washed away by rain, slowly transform into insoluble aluminum hydroxide sulfates (e.g., $Al_x(OH)_y(SO_4)_z$), which are usually amorphous (Ref 36, 39). Sulfate species are abundant constituents of the corrosion layers on outdoor aluminum, especially in marine and industrial environments (Ref 112). Chloride species are also relatively common in aluminum patinas. They are present as soluble aluminum chlorides (that participate in a corrosion cycle causing pitting unless they are washed away by rain), or else they become incorporated into the corrosion layer as insoluble aluminum chloride hydroxides (e.g., $Al_x(Cl)_y(OH)_z$) (Ref 36). Small spots of local corrosion on aluminum usually signal pitting corrosion, caused by the surface becoming contaminated by chloride ions.

Gold (Gilding). The gilding of outdoor statuary and exterior architectural ornaments (such as domes on churches and government buildings) goes back hundreds of years (Ref 113). Some gilded bronzes have survived since antiquity, such as the four fire-gilded bronze horses from the Basilica of St. Mark in Venice and the bronze statue of *Marcus Aurelius* in Rome, gilded by burnishing gold leaf directly onto the surface (Ref 2, 114, 115). Many of the lead garden sculptures at Versailles (built in the 1680s) in France were originally gilded, although the gilding has now been lost. A gilded lead statue that did not survive was that of *King George III* in New York City (unveiled in 1770), which, in 1776, was pulled down and broken up so the lead could be turned into bullets (Ref 52). Today, the most common gilding technique is oil gilding (sheets of gold leaf, less than 1 µm, or 0.04 mil, thick, are adhered to the metal surface using a traditional oil size or a modern adhesive), which allows any metal surface to be gilded (Ref 113). Other examples include the gilded lead in the *Albert Memorial* (1876) in London, the gilded bronze in the *Sherman Monument* (1903) in New York City, and the gilded copper in *Quadriga* (1905) in St. Paul, Minnesota (Ref 116–118).

When inappropriate techniques or primers are used to gild metal surfaces for exterior use, the gilding layer can fail within a few years (Ref 113, 117). This is especially true if galvanic corrosion results from the direct contact between the gold and the underlying metal. As the more anodic base metal corrodes, the corrosion products lift and detach the gilding. When traditional gilding approaches are used, corrosion of the underlying metal is minimized, and the gilding can last for several decades outdoors (Ref 113, 117). The traditional approach to gilding metals for exterior use is to apply a primer (either red lead or zinc chromate) to the underlying metal (prior to applying the size and gilding) to minimize corrosion and provide a solid barrier between the

Fig. 6 The pyramidal cast aluminum cap (14.2 cm, or 5.6 in., square at base; 22.6 cm, or 8.9 in., high) at the apex of the Washington Monument in Washington, D.C., set in place in 1884 and exposed outdoors for 115 years. The darker area shows a region where a gold-plated collar (temporarily removed) serves as a lightning protection system. Courtesy of Judith M. Jacob, National Park Service. Photograph 1999

gold and the underlying metal (Ref 113, 117). Because government regulations are restricting the use of these traditional primers in many countries, new primers are needed, and testing is currently underway to find acceptable replacements (Ref 113).

Preservation Strategies

Whether an artifact is a newly commissioned work of art or one that has been outdoors for many years, it must be cared for. The responsibility belongs to local governments, public and private museums, corporations, or private owners. It must be recognized that maintenance will always be required, that maintenance requires adequate funding, and that lack of maintenance funding will result in neglected artifacts. Furthermore, it is highly recommended that a conservator be consulted when artifacts are newly commissioned or acquired, or already exist but have not been receiving maintenance. Conservators can help identify potential problems and prevent inappropriate (but well-intentioned) attempts to care for artifacts that can damage detailed surfaces and patinas (Ref 102).

It is difficult for a single person to make all the decisions about the conservation, degree of intervention, and requirements for long-term care of an outdoor artifact. Conservators often participate in a multidisciplinary approach to making decisions about one or more outdoor artifacts; such an approach involves the participation of several people with different backgrounds, experience, and knowledge (Ref 102, 119). Depending on the artifacts, this may involve one or more conservators, artists, art historians, curators, fabricators, foundry personnel, past users (of heritage artifacts), local administrators, architects, and engineers.

There is now an increasing public awareness of the need for the preservation of outdoor sculptures, thanks to the Save Outdoor Sculpture! program of Heritage Preservation (www.heritagepreservation.org) that was established in 1989 to develop a national inventory of outdoor sculpture in the United States (Ref 102). The information is stored in the Smithsonian's Inventory of American Sculpture (www.siris.si.edu). Similar programs intended to inventory and preserve outdoor sculpture have been established in other countries.

Preserving Heritage Objects. It can be difficult to maintain an artifact and ensure its stability when it is located in an uncontrolled outdoor environment. An artifact may be so important that any additional corrosion due to outdoor exposure is deemed unacceptable, and a decision is made to move the artifact indoors where the conditions can be better controlled. Such a decision was made with the four gilded bronze horses from the Basilica of St. Mark in Venice; the originals were moved indoors, and reproductions were placed outside (Ref 114).

Applying a protective coating (and then maintaining it) is one of the best ways to protect metal artifacts from corroding in an uncontrolled environment and preserve them for future generations. Of course, certain monuments can be too large or too inaccessible to be coated. The *Statue of Liberty*, for example, remains uncoated and should last for at least 1000 years, assuming the copper corrosion rate remains the same as for the first 100 years (Ref 25).

Conservation. Conservators look for an approach that involves the least possible intervention needed to preserve the artifact and safeguard it for the future. They respect the historical value and past history of the artifact, and try to preserve an artist's original intent by keeping as much original material as possible and, by using techniques that minimize damage to the historic surface, respect the aesthetic qualities of works of art, and are reversible where possible.

Formulation of a conservation treatment begins with conservators examining and documenting the condition of an artifact and, where appropriate, putting forward a treatment proposal. If the treatment proposal is accepted, then conservators carry out the treatment, often with necessary modifications as the treatment proceeds. Before and during the treatment of an outdoor artifact, as much information as possible is researched and recorded (Ref 51). The information collected may include any of the following:

- Object description, location, identification number
- Line drawings, photographs, x-radiographs
- Examination details (e.g., structure, corrosion, associated material)
- Condition before treatment (e.g., evidence of active corrosion)
- Results of any technical analysis (e.g., analysis of metal, repair materials, corrosion products)
- Results of any historical research (e.g., original photographs, technical information)
- Treatment details
- Information from a living artist or research about the artistic intent
- Recommendations for long-term care

Technical analysis and historical research are often required to achieve an understanding of the original aesthetics of the piece (Ref 102). Whenever possible, conservators choose reversible coating materials that can be removed and replaced in the future. When developing a treatment plan for an outdoor artifact, the conservator takes into account the following factors:

- Type of metal, its condition, and the degree of corrosion
- Need to preserve surface details
- Expectations of the owner
- Aesthetics and respect for artist's original intent
- Presence of hazardous materials (e.g., lead paint, asbestos insulation)
- Object location
- Availability of equipment, materials, and funding
- Future maintenance, including maintainability of paint or coating

Repair and Restoration (Reconstruction). Although the main intent of conservation is to preserve the artifact and prevent it from suffering further damage, a heritage object may have deteriorated, and a decision is made for additional intervention. This can involve the repair of damaged parts or the replacement of missing pieces. This can even involve a certain amount of restoration or reconstruction in order to actually return the artifact to a known earlier state or make repairs to the supporting base to ensure the artifact does not fall off. When repairs are made, galvanic corrosion is avoided by using the same metal as in the artifact or by using nonconductive materials (e.g., mastics, nonconducting polymers, washers, synthetic fluorine-containing resin) to separate different metals (Ref 38). When direct contact between different metals cannot be avoided (such as when replacing a rusting internal armature in a copper alloy sculpture), galvanic corrosion is minimized by replacing the iron with a metal as close as possible in the galvanic series to the copper alloy. For example, the corroded wrought iron armature in the *Statue of Liberty* was replaced with type 316L (S31603) stainless steel (Ref 25).

Regular inspections of both an artifact and its immediate surroundings (recommended at least annually) are extremely important to the life expectancy of the artifact, because they allow problems to be identified, regular maintenance to be carried out, and follow-up action to be initiated. During regular inspections, accumulated garbage, leaves, and nests can be removed. Trees can be trimmed and local vegetation cut back to allow for good air circulation and to deter birds. The base of monuments can be checked to see if power mowers have damaged them, if water sprinklers are regularly spraying them, or if deicing salts are accumulating because of nearby use. The artifact can be washed with mild soap and water to remove accumulated dirt, salts, pollution, and bird droppings. Graffiti can be removed and damage identified (and arrangements made for its repair). Reapplication of a wax coating or repair to minor paint flaws can be made. Finally, determination can be made whether any follow-up treatment needs to be carried out.

More frequent inspections will also help in the rapid identification of problems. If, for example, the staff is taught about local cultural artifacts, factors contributing to corrosion and deterioration, conservation, and regular maintenance, then they can participate in the long-term preservation of these artifacts. They can keep a watchful eye on artifacts when working in the vicinity, and they can report any new damage (e.g., vandalism) to an appropriate person so the problem can be addressed as soon as possible.

Conservation Strategies for Specific Metals

In this section, conservation strategies are addressed for the five common metals used in outdoor artifacts: copper alloys, iron alloys, zinc, lead, and aluminum. A brief discussion of gold (gilding) is also included.

Copper Alloys. Most outdoor bronze statues are colored using chemicals to produce an artificial patination, and they are often coated, usually with a wax, before they are placed outdoors (Ref 79). If a sculpture is well maintained, all it needs is regular cleaning to remove soluble material and regular reapplication of a coating. If, on the other hand, bronze sculptures are left outdoors without maintenance, any coating or patina eventually breaks down or wears off, often unevenly. The surface of neglected bronze statuary, after decades of outdoor exposure, becomes covered with copper corrosion products along with dirt, bird droppings, and other debris. Its appearance is disfigured, often patchy and streaked with areas of alternating black and light green (Ref 7, 28, 84). The contrasting light and dark areas obscure the overall sculptural details. The overall approach to the conservation of a neglected and corroded bronze statue is to clean it and then protect it from corrosion with a coating of wax or lacquer (Ref 42, 84, 102, 120, 121). Figure 7 contains an example of the appearance of the bronze statue of Rodin's *The Thinker* before (Fig. 7a) and after (Fig. 7b) receiving conservation treatment. In addition to cleaning and coating, other areas that may need attention include structural repairs, securing supports to the base, replacement of rusting internal armatures or bolts, repairing cracks, replacing lost parts, clearing existing weep holes (or drilling new ones), and repatination.

Cleaning. Conservators have developed a wide range of methods for cleaning corroded bronzes (Ref 122). The first step in removing pollutants, dust, bird droppings, and other surface debris is washing with an anionic or nonionic detergent in tap water, using sponges or soft brushes, and followed by thorough rinsing (Ref 42, 120). Many conservators favor this step as the minimum intervention necessary (Ref 42, 85, 122). If conservators (or others) decide further intervention is needed, then other cleaning techniques are available. However, additional cleaning, especially those methods involving pressure, must be done with caution to avoid damaging the fine detail in artifacts and to minimize the cumulative loss of detail from repeated corrosion removal. The intent of cleaning is usually to even out the patina (by removing the superficial corrosion products but leaving firmly adhered ones) rather than removing all of the patina and exposing bare metal (Ref 84). Solvent cleaning, low-pressure steam cleaning, or blast cleaning with ice or dry ice (frozen CO_2) can be used to remove old coatings (Ref 102). Mechanical cleaning (e.g., fine bronze wool, synthetic abrasive pads) or blast cleaning with soft media (e.g., plastic beads, corncob granules, nut shells) at low pressure can also be used to remove nonadherent material, hard deposits, and some surface corrosion products (Ref 85, 120, 123). Medium-pressure washing can also be used to remove some surface corrosion and readily soluble material (Ref 119, 124).

Repatination may be necessary if the original patina is gone. This may involve repatinating the entire surface or just selectively patinating small areas to visually integrate the color. Repatination is usually carried out using various chemicals to achieve the desired color (Ref 79). Under certain circumstances, the color is altered with pigmented waxes rather than with chemicals. Recipes for green patinas often contain sulfate and chloride ions, and recipes for darker-brown patinas often contain sulfide ions (Ref 79). Many conservators avoid using recipes that contain chloride ions to avoid the possibility of future corrosion problems.

Coating. After cleaning bronze statuary, a protective layer of wax or lacquer is usually applied. Coatings seem to adhere better to a bronze surface after some of the corrosion products have been removed (Ref 84, 85). Conservators sometimes apply benzotriazole, a copper corrosion inhibitor, to the bronze surface before applying a coating, although this may slightly darken the patina, and its effectiveness may be limited (Ref 120, 125). Once the bronze has been coated, it is important that an ongoing program is established to maintain the coating (Ref 126).

Waxes are the most popular coatings used on outdoor bronzes (Ref 120), and they have proven effective in protecting outdoor bronzes from corrosion, as long as the coating is well maintained (Ref 85). Wax coatings usually consist of various mixtures of petroleum-based (e.g., microcrystalline wax) and synthetic (e.g., polyethylene) waxes, although conservators sometimes use commercial paste waxes (Ref 85, 120). In the past, beeswax, lanolin, and other natural-product waxes were used, but today, these are rarely used because of their tendency to form metal soaps from the reaction between the underlying metal and the free fatty acids in the wax (Ref 127). When the wax coating is applied to either a heated (Ref 128) or a cold metal surface, it usually saturates the light-green areas and darkens the overall color, making it easier to read the sculptural form (Ref 42, 85). Waxes are usually reapplied once a year (Ref 85, 120). Old coatings that have become dirty are often cleaned with mineral spirits before more wax is applied (Ref 85). Old waxes are difficult to remove but can often be removed with solvents, steam, or low-pressure blast cleaning (Ref 85, 102). Recently, electrochemical impedance measurements have been used to evaluate various commercial waxes and coatings for possible use on outdoor bronzes (Ref 129–133).

Lacquers have also been used on outdoor bronzes, although, because they tend to be brittle, they are not used on sculptures that the public climbs on. The most popular lacquer is Incralac, which contains the acrylic Acryloid (or Paraloid, Rohm and Haas Co.) B-44 (Ref 102, 120). Conservators are now using thickness gages and perforation detectors to maintain quality control

(a)

(b)

Fig. 7 Detail of the bronze sculpture of *The Thinker* by Auguste Rodin (installed outside the Rodin Museum in Philadelphia in 1929) (a) before treatment and (b) after treatment. Courtesy of the Philadelphia Museum of Art, Conservation Department. Photographs 1992

during lacquer applications, although it can be difficult to measure coating thickness on a heavily textured sculpture (Ref 134). Sometimes, a wax is applied over the lacquer. Lacquer needs regular touchup and periodic replacement, ranging anywhere from 2 to 10 years (Ref 84, 102, 120). Removal of an old Incralac coating is difficult because of its tendency to cross link as it ages (Ref 135). Research is currently underway to identify better clear coatings (ones with good adhesion and good reversibility) for protecting outdoor bronze sculptures (Ref 136–141).

Iron Alloys. Most outdoor artifacts made from iron alloys are protected from corrosion with paint or some other protective coating. Others are left uncoated because they are corrosion resistant (stainless steel art) or they are intended to develop a natural rust-colored patina (weathering steel). Many industrial and military artifacts come painted and well maintained when they are decommissioned and turned into a monument, but others can be seriously rusted and deteriorated when they receive their heritage designation.

Repairs. Although conservators try to save as much original material as possible, they often need to replace missing or severely deteriorated parts (Ref 51, 52). To prevent water from penetrating and being retained in the interior of hollow areas, such as in cast iron balustrades, cracks are repaired (Ref 51, 58). Welding can be used to repair cracks in wrought iron and, under certain conditions, in cast iron as well (Ref 51, 58). Fillers, such as epoxy resins (metal-filled ones) or polyester-based autobody putties, are used to fill small defects and casting flaws (Ref 51, 58).

Cleaning. Uncoated, polished stainless steel sculptures (especially ones sheltered from regular washing by rain) and well-maintained coated iron surfaces all benefit from regular cleaning with a mild detergent and warm water to remove dirt, grease, pollutants, salts, and other accumulated material (Ref 52, 58).

When a coated iron surface is not maintained and the coating has failed to such an extent that serious rusting has become a problem, then the iron is cleaned and a protective coating applied (Ref 142). In the case of historic fences and other architectural pieces, the iron is usually cleaned to bare metal and repainted. For industrial artifacts, it may only be necessary to remove loose surface material and then coat with a wax or a corrosion-prevention product. Old paint is tested before it is removed to determine if it contains lead; stringent safety precautions are needed when removing lead paint to avoid lead poisoning. Hazardous dust and waste must be contained, and all contaminated material must be properly disposed of according to local regulations (Ref 92). Although lead-containing paints are now banned in many countries, red lead was used in the past as an effective corrosion-inhibiting primer on iron, and white lead was used as a pigment in exterior paints.

When an iron surface is to be painted, the standard approach to surface preparation is to remove all dirt, grease, old paint, and existing rust, either by chemical (e.g., phosphoric acid, paint strippers) or mechanical (e.g., abrasive blasting) methods (Ref 52, 58, 142). The drawback with using chemicals (including water) in situ is that they penetrate into hollow interiors, and residues are difficult to fully remove (Ref 58). Chemical removal of rust or paint is usually limited to artifacts that can be fully disassembled, cleaned, rinsed, and reassembled. When abrasive blasting methods are used, various media are tested, starting with the least damaging (i.e., soft abrasives and low pressures), to determine the most appropriate method for removing surface material without damaging (e.g., pitting the surface or perforating the metal) the underlying historic metal. Dry abrasive blasting methods have been used to clean old paint and surface rust from cast and wrought iron artifacts such as fountains, historic fences, and other architectural elements (Ref 51, 58, 143). Wet abrasive blast cleaning is generally not used because of the problem of rapid flash rusting (Ref 58). However, wet-cleaning methods are essential for removing soluble sulfate-containing and chloride-containing contaminants from an iron surface (Ref 51, 144). If these soluble salts are not removed prior to painting, then early paint failure will result (Ref 144).

Another approach to preparing a rusted iron surface for painting is to remove all dirt, grease, and loose material, including loose rust, but leave the existing adherent rust. This rusted surface is then coated with a rust converter primer prior to painting (Ref 69, 145). Rust converters contain a polymeric component (that accepts a topcoat of a compatible paint) and tannic acid (that reacts with rust to form a blue-black iron(III) tannate film) (Ref 145).

Painting iron and steel is the most common and effective method of controlling rusting, and surface preparation is the key to effective paint adhesion (Ref 102). Also important are using a compatible paint system and following manufacturer instructions (Ref 52). Once the iron surface has been suitably cleaned (including removal of contaminating chloride ions) and prepared for painting, it is primed and then topcoated with a compatible paint system. A number of different primers and paints have been used on historic outdoor iron (Ref 51, 52, 58, 143, 144). Alkyd paints are popular because they have proven effective and are relatively easy to maintain (Ref 51).

Waxlike Coatings. The application of a wax or some other corrosion-prevention coating is another method used to control rusting on outdoor iron and steel. With these coatings, it is not necessary to strip off existing rust or paint, only clean by removing dirt, grease, and loose material such as flaking paint and nonadherent rust. Abrasive blasting with soft organic abrasives has proved successful (Ref 92). Wax mixtures, such as those used on outdoor bronzes, have been used on rusted outdoor iron (Ref 70). Commercial corrosion-prevention or rust-preventive products have also been used. Corrosion-prevention products are petroleum-based, and they dry by solvent evaporation, leaving a film (often waxlike) that can provide temporary corrosion protection to outdoor metals. They have been used to protect military monuments (Ref 72), industrial machinery (Ref 92), and cast iron cannons (Ref 146).

Zinc. The problems generally encountered on outdoor architectural and sculptural zinc are largely structural. After years of outdoor exposure without maintenance, the coating fails, and the underlying metal corrodes. Because zinc sculptures were often assembled by soft soldering smaller pieces together, the exposed surfaces can become disfigured, with criss-crossing solder joints standing in relief above the corroded zinc (Ref 56, 61). Frequently encountered on corroded zinc are defective foundry joints, disfiguring repairs, corroded wrought iron armatures, corrosion due to water trapped by concrete in the core, fractures in large zinc pieces unable to support themselves, breakage, and casting flaws (Ref 51).

Conservation of zinc sculptures often involves dismantling them, removing any core material or rusting armatures, and cleaning the surface with relatively neutral solutions (neutral detergents, noncaustic degreasing agents, noncaustic paint strippers) or by using (with extreme caution) low-pressure abrasive blasting with a suitable media soft organic material (Ref 51, 56, 57, 61). Structural repairs are made by soft soldering, welding, or using polyester or epoxy resins reinforced by fiberglass (Ref 52, 56, 57). Corroded wrought iron armatures are usually replaced with stainless steel ones and isolated from the zinc to avoid galvanic corrosion problems (Ref 56, 57). Protective coatings are often applied, but they are dependent on the nature of the original and existing surfaces (Ref 102). Some zinc pieces are left uncoated (especially if they were originally uncoated), some are repainted to resemble the original color, and some are given a protective clear coating of wax or lacquer (Ref 56, 147, 148). It is important to ensure that the intended paint system (especially the primer) has been specifically designed for a zinc surface (whether it is zinc metal, galvanized steel, or a surface covered with a zinc-rich primer). Alkyd or oil-based paints are avoided because zinc soaps form at the interface between the metal and the paint (because zinc reacts with fatty acids in the paint), cause poor adhesion of the paint system, and quite quickly result in peeling paint.

Lead suffers mechanical breakdown with tearing, buckling, and fatigue cracking (Ref 52). Large lead castings of sculpture, garden ornaments, or fountains cannot support themselves; support is provided by leaving the internal core material (plaster and sand) and using wrought iron armatures (Ref 66, 149). Unfortunately, large lead objects often sag or become distorted, because lead is heavy and subject to creep. Ingress of water into the core material or through cracks or porosity in the lead casting causes corrosion of the inner lead surface and rusting of the iron armature (Ref 51). Rusting of the iron armature will cause further cracking of the lead as well as staining.

For outdoor lead, the basic cleaning, coating, and cyclical maintenance strategies for lead are similar to those for bronze (Ref 102). Periodic removal of dirt and corrosive bird droppings is done by washing with a neutral-pH soap in warm water with light brushing (Ref 51). White corrosion (lead carbonate hydroxide) is sometimes removed using a chelating agent such as ethylenediamine tetraacetic acid (Ref 51, 150). Repairs to cracks, tears, and loss are done by lead burning (welding) rather than soldering (Ref 51). Removal of old paint and corrosion products from the surface of lead is done with great care, especially if mechanical methods are used, because lead is soft and easily damaged (Ref 102). Abrasive blasting techniques are avoided for this reason and because they generate toxic lead dust (Ref 51). Corroded wrought iron armatures are replaced with stainless steel ones, a process that usually requires disassembly of the sculpture (Ref 102). To ensure all surfaces receive sufficient support, some conservators fill repaired lead sculptures with closed-cell polyurethane foam, despite the drawback that this foam will be difficult to remove in the future (Ref 51). Lead sculptures can be repainted (if they were originally painted), coated with a protective layer of wax, or left uncoated (Ref 51, 66).

Aluminum. Uncoated and anodized aluminum benefit from regular cleaning to remove airborne deposits, especially chloride ions. Cleaning is usually done with gentle cleaning solutions such as neutral detergents (Ref 151). Aluminum should not be cleaned with strong cleaners such as acids (pH<4), alkaline solutions (pH>9), or chlorinated solvents (Ref 151). Unpainted aluminum and anodized aluminum should be washed regularly (every 1 to 12 months), with the greatest frequency of cleaning in industrial and marine atmospheres and where surfaces are not washed by rain (Ref 151, 152). The least cleaning is needed in rural environments.

The surface oxide on aluminum is relatively thin and easily scratched or damaged by abrasives. Aluminum should be cleaned with nylon brushes, not with steel or brass wool because residual metal fragments will promote galvanic corrosion (Ref 152). When dealing with historic aircraft on exhibit outdoors, regular maintenance, inspection, and repair combined with good ventilation to the interior are needed (Ref 153). A transparent coating will help maintain aluminum that has been finished with a highly reflective surface (Ref 102). When aluminum is painted, the coating needs to be maintained, and certain coatings (e.g., powdered coatings) may be more difficult to maintain than others.

An example of the conservation of an outdoor cast aluminum sculpture (95%Al–2.5%Cu–2.5%Zn) can be found with a figure of *Eros* (Fig. 8) unveiled in Liverpool, England, in 1932 (Ref 154). It had not held up well to outdoor exposure (Ref 110). Over time, the surface became covered with an encrustation of calcium sulfate that had leached from the interior, through metal found to be quite porous. The surface had also been damaged by vandalism and stained with iron corrosion caused by rusting of the internal iron armature. The sculpture has now been moved indoors and the surface encrustation removed, using laser cleaning (Fig. 9) (Ref 110, 155).

Gold (Gilding). Gilded metals have a limited lifetime when exposed outdoors, and regilding is necessary every few decades (Ref 117, 119). This lifetime, however, may be less than a year if the gilding is not done properly or if inappropriate materials are used (Ref 113). It is difficult to implement maintenance for gilded architectural elements, especially domes on churches. Gilded statues, on the other hand, benefit from annual maintenance that can include washing (to remove airborne dirt, salts, and pollutants), local repair of gilding loss, and replenishing of any wax coating if applied (Ref 117).

New Commissions

Another aspect of outdoor artifacts is dealing with new installations, either new commissioned works of art or heritage artifacts no longer in service. The new owner, whether a government or private organization, needs to work with various professionals (conservators, curators, site engineers, architects, the artist, the fabricator, the

Fig. 8 Cast aluminum statue of *Eros* by Alfred Gilbert (cast 1929) in Liverpool, England, before treatment. The damage to wings was caused by corrosion of internal iron armatures. Courtesy of Martin Cooper, National Museums Liverpool. Photograph circa 1994

previous owner) to ensure the new piece can survive outdoors with a minimum of future damage and corrosion problems (Ref 156, 157). It is important that a conservator be consulted to carry out a technical review of the artist's proposals as well as look at site locations to help identify potential problems and make recommendations about maintenance (Ref 158). If an artifact is located at a site where it is easy to maintain, then the chances are good that it will be maintained. The following aspects should be considered when dealing with new commissions:

- Are the materials corrosion resistant?
- Is there potential for galvanic corrosion (are two metals in direct contact)?
- Do both the site and design allow for good water drainage?
- Is the site clear of harmful trees and other vegetation?
- Have all public safety issues been addressed?
- Is the site remote, or will it invite vandalism?
- Is there a rapid response plan for vandalism?
- Are attachment points covered or made tamperproof?
- Can weak areas be reinforced to make them harder to bend or steal?
- Has a maintenance program been recommended?
- Are there funds to implement the maintenance program?
- How easily can the site be accessed for maintenance?

Many communities have legislation requiring a percentage of public construction funds be spent on art in public places, with a provision that part of this money be put aside for future maintenance (Ref 102). In other places, maintenance programs are developed and budgeted for at the time of acquisition. It is hoped that future problems with new commissions (or with newly designated heritage artifacts) can be avoided with the help of the conservation profession, which can identify potential problems and suggest design improvements, including the selection of materials that are durable in an outdoor environment.

Fig. 9 Partially cleaned section of a wing (20 cm, or 7.9 in., long) from *Eros* (see Fig. 8). Laser cleaning has been used to remove the dark-gray corrosion layer from the left half. Courtesy of Martin Cooper, National Museums Liverpool. Photograph 1995–1996

ACKNOWLEDGMENTS

The authors are grateful to the following people for their comments on this article: Bob Barclay, Paul Begin, David Grattan, Michael Harrington, Sandra Lougheed, Susan Maltby, Susan Stock, and Tom Stone. The authors also wish to thank the following people for the use of their figures: Martin Cooper, Laura Griffith, Carol Grissom, Judy Jacob, and Andrew Lins.

REFERENCES

1. M. Tullmin and P.R. Roberge, Atmospheric Corrosion, *Uhlig's Corrosion Handbook,* 2nd ed., R.W. Revie, Ed., John Wiley & Sons, 2000, p 305–321
2. C. Bartuli, R. Cigna, and O. Fumei, Prediction of Durability for Outdoor Exposed Bronzes: Estimation of the Corrosivity of the Atmospheric Environment of the Capitoline Hill in Rome, *Stud. Conserv.,* Vol 44 (No. 4), 1999, p 245–252
3. W. He, I. Odnevall Wallinder, and C. Leygraf, A Comparison Between Corrosion Rates and Runoff Rates from New and Aged Copper and Zinc as Roofing Material, *Water, Air, Soil Pollut.: Focus,* Vol 1 (No. 3–4), 2001, p 67–82
4. A.R. Cook and R. Smith, Atmospheric Corrosion of Lead and Its Alloys, *Atmospheric Corrosion,* W.H. Ailor, Ed., John Wiley & Sons, 1982, p 393–404
5. E. Mattsson, The Atmospheric Corrosion Properties of Some Common Structural Metals: Comparative Study, *Mater. Perform.,* Vol 21 (No. 7), 1982, p 9–19
6. E. Mattsson and R. Holm, Atmospheric Corrosion of Copper and Its Alloys, *Atmospheric Corrosion,* W.H. Ailor, Ed., John Wiley & Sons, 1982, p 365–381
7. H. Strandberg, L.G. Johansson, and J. Rosvall, Outdoor Bronze Sculptures: A Conservation View on the Examination of the State of Preservation, *ICOM Committee for Conservation, 11th Triennial Meeting,* J. Bridgland, Ed., James & James Science Publishers, 1996, p 894–900
8. P. Eriksson, L.G. Johansson, and H. Strandberg, Initial Stages of Copper Corrosion in Humid Air Containing SO_2 and NO_2, *J. Electrochem. Soc.,* Vol 140 (No. 1), 1993, p 53–59
9. H. Strandberg and L.G. Johansson, Role of O_3 in the Atmospheric Corrosion of Copper in the Presence of SO_2, *J. Electrochem. Soc.,* Vol 144 (No. 7), 1997, p 2334–2342
10. H. Strandberg, Reactions of Copper Patina Compounds, Part II: Influence of Sodium Chloride in the Presence of Some Air Pollutants, *Atmospher. Environ.,* Vol 32 (No. 20), 1998, p 3521–3526
11. P.B.P. Phipps and D.W. Rice, The Role of Water in Atmospheric Corrosion, *Corrosion Chemistry, ACS Symposium Series 89,* G.R. Brubaker and P.B.P. Phipps, Ed., American Chemical Society, 1979, p 235–261
12. J. Hughes, G. King, and W. Ganther, Application of Corrosion Data to Develop Conservation Strategies for a Historic Building in Antarctica, *ICOM Committee for Conservation, 13th Triennial Meeting,* R. Vontobel, Ed., James & James Science Publishers, 2002, p 865–870
13. W. He, "Atmospheric Corrosion and Runoff Processes on Copper and Zinc as Roofing Materials," Ph.D. thesis, Department of Materials Science and Engineering, Royal Institute of Technology, Sweden, 2002
14. W. He, I. Odnevall Wallinder, and C. Leygraf, A Laboratory Study of Copper and Zinc Runoff during First Flush and Steady-State Conditions, *Corros. Sci.,* Vol 43 (No. 1), 2001, p 127–146
15. I. Odnevall Wallinder, P. Verbiest, W. He, and C. Leygraf, Effects of Exposure Direction and Inclination on the Runoff Rates of Zinc and Copper Roofs, *Corros. Sci.,* Vol 42 (No. 8), 2000, p 1471–1487
16. T.E. Graedel and P.J. Crutzen, The Changing Atmosphere, *Sci. Am.,* Vol 251 (No. 9), 1989, p 58–68
17. T.E. Graedel, The Corrosivity of the Atmosphere: Past, Present, and Future, *Dialogue/89—The Conservation of Bronze Sculpture in the Outdoor Environment: A Dialogue Among Conservators, Curators, Environmental Scientists, and Corrosion Engineers,* T. Drayman-Weisser, Ed., National Association of Corrosion Engineers, 1992, p 13–31
18. C.E. Junge and R.T. Werby, The Concentration of Chloride, Sodium, Potassium, Calcium, and Sulfate in Rain Water over the United States, *J. Meteorol.,* Vol 15 (No. 5), 1958, p 417–425
19. C. Adams and D. Hallam, Finishes on Aluminium: Conservation Perspective, *Saving the Twentieth Century: The Conservation of Modern Materials,* D.W. Grattan, Ed., Canadian Conservation Institute, 1993, p 273–286
20. L.W. Zahner, *Architectural Metals: A Guide to Selection, Specification, and Performance,* John Wiley & Sons, 1995
21. R. Waller, Temperature- and Humidity-Sensitive Mineralogical and Petrological Specimens, *The Care and Conservation of Geological Material: Minerals, Rocks, Meteorites and Lunar Finds,* F.M. Howie, Ed., Butterworth-Heinemann, 1992, p 25–50
22. G.M. Richardson and R.S. Malthus, Salts for Static Control of Humidity at Relatively Low Levels, *J. Appl. Chem.,* Vol 5, 1955, p 557–567
23. G.C. Allen, L. Black, P.D. Forshaw, and N.J. Seeley, Raindrops Keep Falling on My Lead, *J. Archit. Conserv.,* Vol 9 (No. 1), 2003, p 23–44
24. T.E. Graedel, Chemical Mechanisms for the Atmospheric Corrosion of Lead, *J. Electrochem. Soc.,* Vol 141 (No. 4), 1994, p 922–927
25. R. Baboian, E.B. Cliver, and E.L. Bellante, Ed., *The Statue of Liberty Restoration,* National Association of Corrosion Engineers, 1990
26. K. Nassau, P.K.M. Gallagher, and T.E. Graedel, The Characterization of Patina Components by X-Ray Diffraction and Evolved Gas Analysis, *Corros. Sci.,* Vol 27 (No. 7), 1987, p 669–684
27. L. Robbiola, C. Fiaud, and S. Pennec, New Model of Outdoor Bronze Corrosion and Its Implications for Conservation, *ICOM Committee for Conservation, Tenth Triennial Meeting,* J. Bridgland, Ed., ICOM Committee for Conservation, 1993, p 796–802
28. L.S. Selwyn, N.E. Binnie, J. Poitras, M.E. Laver, and D.A. Downham, Outdoor Bronze Statues: Analysis of Metal and Surface Samples, *Stud. Conserv.,* Vol 41 (No. 4), 1996, p 205–228
29. J. Sirois, E. Moffatt, and M. Singer, "The Analysis of Outdoor Bronze Sculptures from the Assemblée Nationale in Québec City, Canada," Sept 2003 (Ottawa, Canada), NACE Northern Area Eastern Conference
30. K. Nassau, A.E. Miller, and T.E. Graedel, The Reaction of Simulated Rain with Copper, Copper Patina, and Some Copper Compounds, *Corros. Sci.,* Vol 27 (No. 7), 1987, p 703–719
31. H. Strandberg and L.G. Johansson, Some Aspects of the Atmospheric Corrosion of Copper in the Presence of Sodium Chloride, *J. Electrochem. Soc.,* Vol 145 (No. 4), 1998, p 1093–1100
32. H. Strandberg, Perspectives on Bronze Sculpture Conservation: Modelling Corrosion, *Metal 98,* W. Mourey and L. Robbiola, Ed., James & James Science Publishers, 1998, p 297–302
33. H. Strandberg, Reactions of Copper Patina Compounds, Part I: Influence of Some Air Pollutants, *Atmospher. Environ.,* Vol 32 (No. 20), 1998, p 3511–3520
34. T.E. Graedel, Corrosion Mechanisms for Zinc Exposed to the Atmosphere, *J. Electrochem. Soc.,* Vol 136 (No. 4), 1989, p 193C–203C
35. T.E. Graedel and R.P. Frankenthal, Corrosion Mechanisms for Iron and Low Alloy Steels Exposed to the Atmosphere, *J. Electrochem. Soc.,* Vol 137 (No. 8), 1990, p 2385–2394
36. T.E. Graedel, Corrosion Mechanisms for Aluminum Exposed to the Atmosphere, *J. Electrochem. Soc.,* Vol 136 (No. 4), 1989, p 204C–212C
37. M.G. Fontana, *Corrosion Engineering,* 3rd ed., McGraw-Hill, 1986
38. E. Ghali, Aluminum and Aluminum Alloys, *Uhlig's Corrosion Handbook,* 2nd

ed., R.W. Revie, Ed., John Wiley & Sons, 2000, p 677–716
39. N.J.H. Holroyd, Aluminum Alloys, *Environmental Effects on Engineered Materials,* R.H. Jones, Ed., Marcel Dekker, 2001, p 173–251
40. D.A. Scott, *Copper and Bronze in Art: Corrosion, Colorants, Conservation,* Getty Publications, 2002
41. C.A.C. Sequeira, Corrosion of Copper and Copper Alloys, *Uhlig's Corrosion Handbook,* 2nd ed., R.W. Revie, Ed., John Wiley & Sons, 2000, p 729–765
42. S.A. Tatti, Bronze Conservation: Fairmount Park, 1983, *Sculptural Monuments in an Outdoor Environment,* V.N. Naudé, Ed., Pennsylvania Academy of the Fine Arts, 1985, p 58–66
43. S. Steinberg, K. Kawamura, and I.R. Kaplan, The Determination of α-Keto Acids and Oxalic Acid in Rain, Fog and Mist by HPLC, *Int. J. Environ. Anal. Chem.,* Vol 19, 1985, p 251–260
44. H.A. Strandberg, "Perspectives on Bronze Sculpture Conservation: Modelling Copper and Bronze Corrosion," Ph.D. thesis, Department of Inorganic Chemistry, Göteborg University, Sweden, 1997
45. T.E. Graedel, Copper Patinas Formed in the Atmosphere, Part III: A Semi-Quantitative Assessment of Rates and Constraints in the Greater New York Metropolitan Area, *Corros. Sci.,* Vol 27 (No. 7), 1987, p 741–769
46. T.E. Graedel, Corrosion-Related Aspects of the Chemistry and Frequency of Occurrence of Precipitation, *J. Electrochem. Soc.,* Vol 133 (No. 12), 1986, p 2476–2482
47. R.M. Cornell and U. Schwertmann, *The Iron Oxides,* VCH Publishers, 1996
48. F.E. Goodwin, Lead and Lead Alloys, *Uhlig's Corrosion Handbook,* 2nd ed., R.W. Revie, Ed., John Wiley & Sons, 2000, p 767–792
49. D. Greninger, V. Kollonitsch, and C.H. Kline, *Lead Chemicals,* International Lead Zinc Research Organization, 197–
50. X.G. Zhang, Zinc, *Uhlig's Corrosion Handbook,* 2nd ed., R.W. Revie, Ed., John Wiley & Sons, 2000, p 887–904
51. J. Ashurst and N. Ashurst, *Practical Building Conservation: English Heritage Technical Handbook,* Vol 4, *Metals,* Gower Technical Press, 1988
52. M. Gayle, D.W. Look, and J.G. Waite, *Metals in America's Historic Buildings: Uses and Preservation Treatments,* 2nd ed., U.S. National Park Service, 1992
53. M. Pourbaix, *Atlas of Electrochemical Equilibria in Aqueous Solutions,* 2nd ed., National Association of Corrosion Engineers, 1974
54. P.C. Frost, E. Littauer, and H.C. Wesson, Lead and Lead Alloys, *Corrosion,* Vol 1, 3rd ed., L.L. Shreir, R.A. Jarman, and G.T. Burstein, Ed., Butterworth-Heinemann, 1994, p 4:76–4:97
55. W. Hofmann and J. Maatsch, Lead as a Corrosion-Resistant Material, *Lead and Lead Alloys,* Springer-Verlag, 1970, p 268–320
56. C.A. Grissom, The Conservation of Outdoor Zinc Sculpture, *Ancient and Historic Metals,* D.A. Scott, J. Podany, and B.B. Considine, Ed., Getty Conservation Institute, 1994, p 279–304
57. C.A. Grissom and R.S. Harvey, The Conservation of American War Memorials Made of Zinc, *J. Am. Inst. Conserv.,* Vol 42 (No. 1), 2003, p 21–38
58. J.B. Waite, "The Maintenance and Repair of Architectural Cast Iron," Preservation Briefs 27, U.S. National Park Service, 1991
59. M. Belman, Three Case Studies of Outdoor Sculpture with Problematic Intent Issues, *AIC Objects Spec. Group Postpr.,* Vol 9, 2002, p 53–67
60. M. Del Monte, C. Sabbioni, and G. Zappia, The Origin of Calcium Oxalates on Historical Buildings, Monuments and Natural Outcrops, *Sci. Total Environ.,* Vol 67 (No. 1), 1987, p 17–39
61. K. Holm, Production and Restoration of Nineteenth-Century Zinc Sculpture in Denmark, *Ancient and Historic Metals,* D.A. Scott, J. Podany, and B.B. Considine, Ed., Getty Conservation Institute, 1994, p 239–248
62. S. Long and E. Skadhauge, Renal Acid Excretion in the Domestic Fowl, *J. Exp. Biol.,* Vol 104, 1983, p 51–58
63. P.D. Sturkie, Kidneys, Extrarenal Salt Excretion, and Urine, *Avian Physiology,* 4th ed., P.D. Sturkie, Ed., Springer-Verlag, 1986, p 376–377
64. S. Stock and M. Back, Royal Ontario Museum, personal communication, 2004
65. M. Bassi and D. Chiatante, The Role of Pigeon Excrement in Stone Biodeterioration, *Int. Biodeter. Bull.,* Vol 12 (No. 3), 1976, p 73–79
66. K. Blackney and B. Martin, Conservation of the Lead Sphinx at Chiswick House, *Engl. Heritage Res. Trans.,* Vol 1, 1998, p 97–102
67. D.A. Scott, V. Kucera, and B. Rendahl, Infinite Columns and Finite Solutions, *Mortality Immortality?: The Legacy of 20th-Century Art,* M.A. Corzo, Ed., Getty Conservation Institute, 1999, p 106–112
68. A. Lins, Outdoor Bronzes: Some Basic Metallurgical Considerations, *Sculptural Monuments in an Outdoor Environment,* V.N. Naudé, Ed., Pennsylvania Academy of the Fine Arts, 1985, p 8–20
69. S.L. Maltby, "The Conservation and Preservation of Architectural Sheet Metal Work," Sept 2003 (Ottawa, Canada), NACE Northern Area Eastern Conference
70. J. Scott, Conservation of Weathering Steel Sculpture, *Saving the Twentieth Century,* D.W. Grattan, Ed., Canadian Conservation Institute, 1993, p 307–322
71. J.C. Scott and C.L. Searls, Weathering Steel, *Twentieth-Century Building Materials,* T.C. Jester, Ed., McGraw-Hill, 1995, p 73–77
72. B. Hector and P.R. Roberge, "Inhibition of Outdoor Military Monuments with Corrosion Prevention Compounds," Sept 2003 (Ottawa, Canada), NACE Northern Area Eastern Conference
73. W. Bordass, Underside Lead Corrosion, *A Future for the Past,* J.M. Teutonico, Ed., James & James Science Publishers, 1996, p 81–98
74. F. Porter, *Zinc Handbook: Properties, Processing, and Use in Design,* Marcel Dekker, 1991
75. R.S. Hayden and T.W. Despont, *Restoring the Statue of Liberty,* McGraw-Hill, 1986
76. C.C. Mattusch, *Greek Bronze Statuary: From the Beginnings through the Fifth Century B.C.,* Cornell University Press, 1988
77. M. Sekino, Repair of the Great Buddha Statue at Kamakura, *Stud. Conserv.,* Vol 10 (No. 2), 1965, p 39–46
78. J.D. Meakin, D.L. Ames, and D.A. Dolske, Degradation of Monumental Bronzes, *Atmospher. Environ.,* Vol 26B (No. 2), 1992, p 207–215
79. R. Hughes and M. Rowe, *The Colouring, Bronzing and Patination of Metals,* Crafts Council, 1982
80. R. Hughes, Artificial Patination, *Metal Plating and Patination,* S. La Niece and P. Craddock, Ed., Butterworth-Heinemann, 1993, p 1–18
81. H.W. Richardson, Copper Compounds, *Kirk-Othmer Encyclopedia of Chemical Technology,* 4th ed., Vol 7, John Wiley & Sons, 1993, p 505–520
82. T.E. Graedel, K. Nassau, and J.P. Franey, Copper Patinas Formed in the Atmosphere, Part I: Introduction, *Corros. Sci.,* Vol 27 (No. 7), 1987, p 639–657
83. T.E. Graedel, Copper Patinas Formed in the Atmosphere, Part II: A Qualitative Assessment of Mechanisms, *Corros. Sci.,* Vol 27 (No. 7), 1987, p 721–741
84. L. van Zelst and J.L. Lachevre, Outdoor Bronze Sculpture: Problems and Procedures of Protective Treatment, *Technol. Conserv.,* Vol 8 (No. 1), 1983, p 18–24
85. W.T. Chase, Aesthetics, Conservation and Maintenance of Outdoor Bronzes, *AIC Objects Spec. Group Postpr.,* Vol 9, 2002, p 41–52
86. P.D. Weil, L. Gulbransen, R. Lindberg, and D. Zimmerman, The Corrosive Deterioration of Outdoor Bronze Sculpture, *Science and Technology in the Service of Conservation,* N.S. Brommelle, Ed., International Institute for Conservation, 1982, p 130–134
87. A. Lins and T. Power, The Corrosion of Bronze Monuments in Polluted Urban Sites: A Report on the Stability of Copper Mineral Species at Different pH Levels, *Ancient and Historic Metals, Conservation and Scientific Research,* D.A. Scott,

88. M.B. McNeil and D.W. Mohr, Sulfate Formation During Corrosion of Copper Alloy Objects, *Materials Issues in Art and Archaeology III*, P.B. Vandiver, J.R. Druzik, G.S. Wheeler, and I.C. Freestone, Ed., Materials Research Society, 1992, p 1047–1053
89. J.L. Jambor, J.E. Dutrizac, A.C. Roberts, J.D. Grice, and J.T. Szymanski, Clinoatacamite, A New Polymorph of $Cu_2(OH)_3Cl$, and Its Relationship to Paratacamite and Anarakite, *Can. Mineralog.*, Vol 34, 1996, p 61–72
90. T.L. Woods and R.M. Garrels, Use of Oxidized Copper Minerals as Environmental Indicators, *Appl. Geochem.*, Vol 1 (No. 2), 1986, p 181–187
91. F. Habashi, The Iron Pillar of Delhi, *CIM Bull.*, Vol 82 (No. 1), 1989, p 85
92. M. Devine, The Acquisition, Management, and Conservation of Industrial Objects at Parks Canada, *J. Can. Assoc. Conserv.*, Vol 23, 1998, p 3–14
93. R. Score and I.J. Cohen, Stainless Steel, *Twentieth-Century Building Materials*, T.C. Jester, Ed., McGraw-Hill, 1995, p 64–71
94. M. Stratmann, The Atmospheric Corrosion of Iron: Discussion of the Physicochemical Fundamentals of this Omnipresent Corrosion Process, Invited Review, *Ber. Bunsenges. Phys. Chem.*, Vol 94 (No. 6), 1990, p 626–639
95. I. Matsushima, Carbon Steel—Atmospheric Corrosion, *Uhlig's Corrosion Handbook*, 2nd ed., R.W. Revie, Ed., John Wiley & Sons, 2000, p 515–528
96. P. Refait and J.M.R. Génin, The Mechanisms of Oxidation of Ferrous Hydroxychloride β-$Fe_2(OH)_3Cl$ in Aqueous Solution: The Formation of Akaganeite vs. Goethite, *Corros. Sci.*, Vol 39 (No. 3), 1997, p 539–553
97. J.E. Post and V.F. Buchwald, Crystal Structure Refinement of Akaganéite, *Am. Mineralog.*, Vol 76 (No. 1–2), 1991, p 272–277
98. J.E. Post, P.J. Heaney, R.B. Von Dreele, and J.C. Hanson, Neutron and Temperature-Resolved Synchrotron X-Ray Powder Diffraction Study of Akaganéite, *Am. Mineralog.*, Vol 88 (No. 5–6), 2003, p 782–788
99. K. Ståhl, K. Nielsen, J. Jiang, et al., On the Akaganéite Crystal Structure, Phase Transformations and Possible Role in Post-Excavational Corrosion of Iron Artifacts, *Corros. Sci.*, Vol 45 (No. 11), 2003, p 2563–2575
100. A. Askey, S.B. Lyon, G.E. Thompson, et al., The Corrosion of Iron and Zinc by Atmospheric Hydrogen Chloride, *Corros. Sci.*, Vol 34 (No. 2), 1993, p 233–247
101. M. Yamashita, H. Miyuki, Y. Matsuda, H. Nagano, and T. Misawa, The Long Term Growth of the Protective Rust Layer Formed on Weathering Steel by Atmospheric Corrosion During a Quarter of a Century, *Corros. Sci.*, Vol 36 (No. 2), 1994, p 283–299
102. V.N. Naudé and G. Wharton, *Guide to the Maintenance of Outdoor Sculpture*, American Institute for Conservation of Historic and Artistic Works, 1993
103. I. Odnevall and C. Leygraf, The Formation of $Zn_4SO_4(OH)_6 \cdot 4H_2O$ in a Rural Atmosphere, *Corros. Sci.*, Vol 36 (No. 6), 1994, p 1077–1091
104. I. Odnevall and C. Leygraf, Formation of $NaZn_4Cl(OH)_6SO_4 \cdot 6H_2O$ in a Marine Atmosphere, *Corros. Sci.*, Vol 34 (No. 8), 1993, p 1213–1229
105. I. Odnevall and C. Leygraf, The Formation of $Zn_4Cl_2(OH)_4SO_4 \cdot 5H_2O$ in an Urban and an Industrial Atmosphere, *Corros. Sci.*, Vol 36 (No. 9), 1994, p 1551–1567
106. P.M. Sutton-Goold, *Decorative Leadwork*, Shire Publications, 1990
107. S.A. Matthes, S.D. Cramer, B.S. Covino, S.J. Bullard, and G.R. Holcomb, Precipitation Runoff from Lead, *Outdoor Atmospheric Corrosion*, STP 1421, H.E. Townsend, Ed., ASTM International, 2002, p 265–274
108. L. Black and G.C. Allen, Nature of Lead Patination, *Br. Corros. J.*, Vol 34 (No. 3), 1999, p 192–197
109. G.C. Tranter, Patination of Lead: An Infra-Red Spectroscopic Study, *Br. Corros. J.*, Vol 11 (No. 4), 1976, p 222–224
110. J. Larson, Eros: The Laser Cleaning of an Aluminium Sculpture, *From Marble to Chocolate: The Conservation of Modern Sculpture*, J. Heuman, Ed., Archetype Publications, 1995, p 53–58
111. S. Kelley, Aluminum, *Twentieth-Century Building Materials*, T.C. Jester, Ed., McGraw-Hill, 1995, p 47–51
112. J.J. Friel, Atmospheric Corrosion Products on Al, Zn, and AlZn Metallic Coatings, *Corrosion*, Vol 42 (No. 7), 1986, p 422–426
113. M.W. Kramer, Architectural Gilding on Exterior Metal: An Overview of Materials and Methodology, *Gilded Metals: History, Technology and Conservation*, T. Drayman-Weisser, Ed., Archetype Publications, 2000, p 351–361
114. V. Alunno-Rossetti and M. Marabelli, Analyses of the Patinas of a Gilded Horse of St. Mark's Basilica in Venice: Corrosion Mechanisms and Conservation Problems, *Stud. Conserv.*, Vol 21 (No. 4), 1976, p 161–170
115. A.M. Vaccaro, The Equestrian Statue of Marcus Aurelius, *The Art of the Conservator*, A. Oddy, Ed., Smithsonian Institution Press, 1992, p 108–121
116. A. Glass, The Albert Memorial: Saved but Not Restored, *Monuments and the Millennium*, J.M. Teutonico and J. Fidler, Ed., James & James Science Publishers, 2001, p 197–203
117. L. Merk-Gould, Preservation of a Gilded Monumental Sculpture: Research and Treatment of Daniel Chester French's *Quadriga*, *Gilded Metals: History, Technology and Conservation*, T. Drayman-Weisser, Ed., Archetype Publications, 2000, p 337–349
118. D. Wilson, Gilding Conservation in an Architectural Context, *Gilding: Approaches to Treatment*, James & James Science Publishers, 2001, p 27–32
119. P.S. Storch, Ten Years of Sculpture and Monument Conservation on the Minnesota State Capitol Mall, *AIC Objects Spec. Group Postpr.*, Vol 9, 2002, p 14–40
120. T.J. Shedlosky, K.M. Stanek, and G. Bierwagen, On-Line Survey Results of Techniques Used for Outdoor Bronze Conservation, *AIC Objects Spec. Group Postpr.*, Vol 9, 2002, p 3–13
121. L.A. Zycherman and N.F. Veloz, Conservation of a Monumental Outdoor Bronze Sculpture: Theodore Roosevelt by Paul Manship, *J. Am. Inst. Conserv.*, Vol 19 (No. 1), 1980, p 24–33
122. A. Naylor, Modern Conservation of Outdoor Bronze Sculpture, *Monuments and the Millennium*, J.M. Teutonico and J. Fidler, Ed., James & James Science Publishers, 2001, p 91–96
123. N.F. Veloz and W.T. Chase, Airbrasive Cleaning of Statuary and Other Structures, *Technol. Conserv.*, Vol 10 (No. 1), 1989, p 18–27
124. L. Merk-Gould, R. Herskovitz, and C. Wilson, Field Tests on Removing Corrosion from Outdoor Bronze Sculptures Using Medium Pressure Water, *ICOM Committee for Conservation, Tenth Triennial Meeting*, J. Bridgland, Ed., ICOM Committee for Conservation, 1993, p 772–778
125. L.B. Brostoff, T.J. Shedlosky, and E.R. de la Rie, External Reflection Study of Copper-Benzotriazole Films on Bronze in Relation to Pretreatments of Coated Outdoor Bronzes, *Tradition and Innovation*, A. Roy and P. Smith, Ed., International Institute for Conservation, 2000, p 29–33
126. D.R. Montagna, Bronze Conservation in the United States at the Dawn of the New Century, *Monuments and the Millennium*, J.M. Teutonico and J. Fidler, Ed., James & James Science Publishers, 2001, p 128–137
127. A. Burmester and J. Koller, Known and New Corrosion Products on Bronzes: Their Identification and Assessment, Particularly in Relation to Organic Protective Coatings, *Recent Advances in the Conservation and Analysis of Artifacts*, Summer Schools Press, 1987, p 97–103
128. J. Sembrat, The Use of a Thermally Applied Wax Coating on a Large-Scale Outdoor Bronze Monument, *ICOM*

Committee for Conservation, 12th Triennial Meeting, J. Bridgland, Ed., James & James Science Publishers, 1999, p 840–844
129. V. Otieno-Alego, G. Heath, D. Hallam, and D. Creagh, Electrochemical Evaluation of the Anti-Corrosion Performance of Waxy Coatings for Outdoor Bronze Conservation, Metal 98, W. Mourey and L. Robbiola, Ed., James & James Science Publishers, 1998, p 309–314
130. V. Otieno-Alego, D. Hallam, A. Viduka, G. Heath, and D. Creagh, Electrochemical Impedance Studies of the Corrosion Resistance of Wax Coatings on Artificially Patinated Bronze, Metal 98, W. Mourey and L. Robbiola, Ed., James & James Science Publishers, 1998, p 315–319
131. L.A. Ellingson, T.J. Shedlosky, G.P. Bierwagen, E.R. de la Rie, and L.B. Brostoff, The Use of Electrochemical Impedance Spectroscopy in the Evaluation of Coatings for Outdoor Bronze, Stud. Conserv., Vol 49 (No. 1), 2004, p 53–62
132. P. Letardi, Laboratory and Field Tests on Patinas and Protective Coating Systems for Outdoor Bronze Monuments, Metal 2004, J. Ashton and D. Hallam, Ed., National Museum of Australia, 2004, p 379–387
133. P. Letardi, A. Beccaria, M. Marabelli, and G. D'Ercoli, Application of Electrochemical Impedance Measurements as a Tool for the Characterization of the Conservation and Protection State of Bronze Works of Art, Metal 98, W. Mourey and L. Robbiola, Ed., James & James Science Publishers, 1998, p 303–308
134. J. Sembrat, Reliable Methods for the Measurement and Inspection of Protective Barrier Coatings for Outdoor Monuments, Metal 98, W. Mourey and L. Robbiola, Ed., James & James Science Publishers, 1998, p 286–290
135. D. Erhardt, W. Hopwood, T. Padfield, and N.F. Veloz, The Durability of Incralac: Examination of a Ten-Year Old Treatment, ICOM Committee for Conservation, Seventh Triennial Meeting, D. de Froment, Ed., ICOM Committee for Conservation, 1984, p 84.22.1–84.22.3
136. G. Bierwagen, T.J. Shedlosky, and K. Stanek, Developing and Testing a New Generation of Protective Coatings for Outdoor Bronze Sculpture, Prog. Org. Coatings, Vol 48 (No. 2–4), 2003, p 289–296
137. L.B. Brostoff and E.R. de la Rie, Chemical Characterization of Metal/Coating Interfaces from Model Samples for Outdoor Bronzes by Reflection-Absorption Infrared Spectroscopy (RAIR) and Attenuated Total Reflection Spectroscopy (ATR), Metal 98, W. Mourey and L. Robbiola, Ed., James & James Science Publishers, 1998, p 320–328
138. L.B. Brostoff and E.R. de la Rie, Research into Protective Coating Systems for Outdoor Sculpture and Ornamentation, Metal 95, I.D. MacLeod, S.L. Pennec, and L. Robbiola, Ed., James & James Science Publishers, 1997, p 242–244
139. T.J. Shedlosky, A. Huovinen, D. Webster, and G. Bierwagen, Development and Evaluation of Removable Protective Coatings on Bronze, Metal 2004, J. Ashton and D. Hallam, Ed., National Museum of Australia, 2004, p 400–413
140. H. Römich and M. Pilz, Protective Coatings for Outdoor Bronze Sculptures: Available Materials and New Developments, Monuments and the Millennium, J.M. Teutonico and J. Fidler, Ed., English Heritage, 2001, p 120–127
141. M. Pilz and H. Römich, A New Conservation Treatment for Outdoor Bronze Sculptures Based on ORMOCER, Metal 95, I.D. MacLeod, S.L. Pennec, and L. Robbiola, Ed., James & James Science Publishers, 1997, p 245–250
142. G. Prytulak, The Treatment of Rusted Machinery in Preparation for Surface Coatings, J. Int. Inst. Conserv. Can. Group, Vol 16, 1991, p 23–34
143. T.D. Ciampa and N. Goldenberg, Restoration of the Centennial Fence at Washington's Headquarters State Historic Site, APT, Vol 14 (No. 3), 1982, p 27–35
144. K. Blackney and B. Martin, Development and Long-Term Testing of Methods to Clean and Coat Architectural Wrought Ironwork Located in a Marine Environment, Eng. Heritage Res. Trans., Vol 1, 1998, p 103–116
145. N.E. Binnie, L.S. Selwyn, C. Schlichting, and D.A. Rennie-Bisaillion, Corrosion Protection of Outdoor Iron Artifacts Using Commercial Rust Converters, J. Int. Inst. Conserv. Can. Group, Vol 20, 1995, p 26–40
146. E. Busse, The Manitoba North Cannon Stabilization Project, Metal 95, I.D. MacLeod, S.L. Pennec, and L. Robbiola, Ed., James & James Science Publishers, 1997, p 263–268
147. P. Mottner, E. Assfalg, and J. Freitag, Investigations into Intercrystalline Corrosion and Conservation of Zinc, Metal 98, W. Mourey and L. Robbiola, Ed., James & James Science Publishers, 1998, p 329–340
148. P. Mottner, H. Brückner, and J. Freitag, Conservation of Cast Zinc Sculptures in Outdoor Exposure, Metal 95, I.D. MacLeod, S.L. Pennec, and L. Robbiola, Ed., James & James Science Publishers, 1997, p 251–255
149. J. Naylor, Conservation of "Samson Slaying the Philistine," Lead and Tin Studies in Conservation and Technology, Occasional Papers No. 3, G. Miles and S. Pollard, Ed., United Kingdom Institute for Conservation, 1985, p 53–55
150. J. Watson, Conservation of Lead and Lead Alloys Using E.D.T.A. Solutions, Lead and Tin Studies in Conservation and Technology, Occasional Papers No. 3, G. Miles and S. Pollard, Ed., United Kingdom Institute for Conservation, 1985, p 44–45
151. Care of Aluminum, 5th ed., The Aluminum Association, 1983
152. S. Duncan, Aluminium: Its Alloys, Coatings and Corrosion, Modern Metals in Museums, R.E. Child and J.M. Townsend, Ed., Institute of Archaeology, 1988, p 27–32
153. R.C. Mikesh, Preserving Unsheltered Exhibit Aircraft, Yearbook Int. Assoc. Transport Mus., Vol 15/16, 1988/1989, p 45–56
154. M. Cooper, Laser Cleaning in Conservation, Butterworth-Heinemann, 1988
155. M. Cooper, An Introduction to the Laser Cleaning of Metal Surfaces, Back to Basics, H. Moody, Ed., United Kingdom Institute for Conservation Metals Section Press, 2002, p 34–39
156. J. Hughes, Prevention of Conservation Problems of Outdoor Sculpture: The Importance of the Commissioning Documentation, ICOM Committee for Conservation, 11th Triennial Meeting, J. Bridgland, Ed., James & James Science Publishers, 1996, p 876–883
157. J. Hughes, Collection Management of National Memorials and Artworks in Canberra, Australia, Metal 95, I.D. MacLeod, S.L. Pennec, and L. Robbiola, Ed., James & James Science Publishers, 1997, p 310–316
158. S. Lougheed, personal communication, 2004

SELECTED REFERENCES

- L.B. Brostoff, "Coating Strategies for the Protection of Outdoor Bronze Art and Ornamentation," Ph.D. dissertation, University of Amsterdam, The Netherlands, 2003
- Copper Patina Formation (Special Issue), Corros. Sci., Vol 27 (No. 7), 1987, p 639–782
- T. Drayman-Weisser, Ed., Dialogue/89—The Conservation of Bronze Sculpture in the Outdoor Environment: A Dialogue Among Conservators, Curators, Environmental Scientists, and Corrosion Engineers, National Association of Corrosion Engineers, 1992
- C. Leygraf and T. Graedel, Atmospheric Corrosion, Wiley-Interscience, 2000
- V.N. Naudé and G. Wharton, Guide to the Maintenance of Outdoor Sculpture, American Institute for Conservation of Historic and Artistic Works, 1993
- D.A. Scott, Copper and Bronze in Art: Corrosion, Colorants, Conservation, Getty Publications, 2002
- L. Selwyn, Metals and Corrosion: A Handbook for the Conservation Professional, Canadian Conservation Institute, 2004
- H. Strandberg, "Perspectives on Bronze Sculpture Conservation: Modelling Copper and Bronze Corrosion," Ph.D. dissertation, Göteborg University, Sweden, 1997

Corrosion of Metal Artifacts in Buried Environments

Lyndsie S. Selwyn, Canadian Conservation Institute

HUMANS BEGAN USING METALS thousands of years ago to make functional objects such as spear points and rudimentary tools, ornamental objects, and objects of spiritual significance. Metal artifacts from various periods have become buried in underwater locations or in the ground because they were lost, abandoned, or sacrificed. Metal artifacts are often recovered from carefully excavated archaeological sites, while others are discovered by accident as a result of farm work or the construction of roads, railroads, or new buildings (Ref 1). In marine environments, metal artifacts are usually associated with historic shipwrecks located in coastal waters and along maritime trade routes (Ref 2).

The corrosion processes of metals during burial are complex, being affected by environmental pollutants, other archaeological material, geography, microorganisms in the soil, vegetation, land use, soil chemistry, soil physical properties, and the presence or absence of water and air (Ref 3). The burial environment can range from soil through which groundwater flows freely, to seawater with high salt concentrations, or to a polluted harbor contaminated by foul-smelling hydrogen sulfide. The corrosion rate of buried metals is extremely variable, ranging from rapid to negligible, depending on both the metal and its burial environment. During burial, corrosion can penetrate deep into a metal and weaken it, making it extremely fragile and subject to damage during and after excavation. Some objects may be totally corroded and others may still contain a substantial metal core.

Archaeologists use buried cultural artifacts to learn about the past. They need to know the shape of the object in order to recognize its function or interpret its use. The shape of excavated archaeological artifacts is often obscured by corrosion products and other material from the burial environment. Archaeological conservators use various techniques to assist in identifying the original shape of artifacts and to avoid deterioration and possible disintegration after excavation by minimizing further corrosion (e.g., using controlled environmental conditions, carrying out appropriate conservation treatments). Many excavated archaeological artifacts are now preserved within museums and other cultural institutions. Unfortunately, there are also untreated archaeological artifacts that are stored in uncontrolled conditions and are continuing to corrode (particularly iron) because of lack of funds for better storage or for conservation treatments (Ref 4).

The Burial Environment

Artifacts are found buried in soils that can be relatively dry, damp, intermittently wet, or permanently water saturated, or they are found submerged under water in lakes or the ocean (Ref 5). They can be exposed to the atmosphere and surface-weathering processes, or they can be deposited onto lake, estuarine, or marine sediments and eventually become buried within these sediments. Natural burial environments vary from oxidizing to reducing and from acidic to alkaline (Ref 6). If the burial environment changes, corrosion processes and corrosion products can also change. The key environmental variables that affect the corrosion of buried metal artifacts are water (including dissolved salts and gases), sulfate-reducing bacteria, pH (acidity), and potential (oxidizing or reducing capacity) (Ref 7).

Water. The corrosion of buried metal artifacts is electrochemical; water is the electrolyte, and dissolved species (salts, gases) affect the corrosion rate. The amount of water available can range from water vapor at various relative humidities to fluid water under immersed conditions (Ref 8). When metal artifacts sink to the bottom of rivers, lakes, harbors, or the ocean, they are surrounded by water and may become partially or completely buried in sediment. When metal artifacts are buried in soils, they can be located below the water table, where the soil particles are completely surrounded by water, or they can be buried above the water table, where the space between soil particles is filled with moist air and other gases (Ref 9). Variations in the water table can further complicate the corrosion process of metal artifacts. Above the water table, the compactness and water content of soils are important; clays retain moisture within capillaries, whereas sandy soils drain quickly (Ref 8).

Dissolved salts increase the conductivity of water (lower the soil resistivity) and accelerate electrochemical reactions (Ref 9). Consequently, metal artifacts generally corrode faster in water containing high concentrations of dissolved salts; the corrosion rate can, however, become lower if metals react with soluble salts to form an insoluble protective layer. Soluble salts are present in groundwater, lakes, and oceans. The concentration of these dissolved salts is lowest in freshwaters, higher in harbors and coastal waters, higher still in open seawater, and highest in enclosed seas (Red Sea, Mediterranean) that are subject to a high rate of evaporation (Ref 10). Soil solutions associated with calcareous (limestone, dolomite) soils are high in magnesium, calcium, and carbonate ions (Ref 8). In relatively arid regions, the net direction of water movement is upward, and evaporation at the soil surface results in the precipitation of salts containing various cations (e.g., sodium, potassium, magnesium, calcium) and anions (e.g., carbonates, chlorides, sulfates) (Ref 5, 9). Artifacts buried in arid regions are likely to be encrusted with salts, and the surface corrosion products are likely to be metal carbonates and sulfates (Ref 5). In seawater, the major ionic constituents (listed in decreasing order of relative amounts) are chloride, sodium, sulfate, magnesium, calcium, and potassium (Ref 11).

The salts in soil can be markedly altered by human activities such as industrialization, modern agriculture, and changes in land use (Ref 3, 8). The modern burial environment changes as surface soils accumulate a wide range of salts from phosphate fertilizers, waste (human, animal, industrial), deicing salts, weed-killing chemicals, bones, wood ash, and atmospheric pollution (Ref 1, 12). In Sweden, for example, recently excavated bronzes are more deteriorated than earlier finds in the same region (Ref 3). This increased deterioration is attributed to increased atmospheric pollution and acid rain (especially from the Continent and Great Britain) and proximity to roads where deicing salts are used (Ref 3, 13).

Dissolved Gases. Oxygen and carbon dioxide are the principal dissolved gases in surface waters, and they are derived from the atmosphere, biological activity, and the decomposition of organic material. The solubility of these gases decreases with increasing temperature (Ref 14).

Oxygen. Reduction of dissolved oxygen gas is a common cathodic reaction contributing to the corrosion of buried metals (particularly for pH 4 to 9). Metals are more likely to be heavily corroded in aerated damp environments than in low-oxygen environments (with no microbial activity), where the metals are more likely to be covered with only superficial corrosion. Surface waters exposed to the atmosphere generally become saturated with oxygen (solubility 2.7×10^{-4} M) because of local turbulence and where oxygen is released by photosynthesis (Ref 5, 12). In soils where the diffusion of air is restricted, oxygen gas is only slowly incorporated into water. Low oxygen levels exist in stagnant, waterlogged soils (clay, peat bogs) and under marine sediments and concretions. These conditions tend to be the least corrosive to buried metal artifacts, although a corrosive environment may develop under anaerobic conditions if sulfate-reducing bacteria are active (Ref 1, 3, 15). A bronze spearhead, for example, remained well preserved, with the detailed surface still evident, after 3000 years buried in the tidal mud of a river bed in England (Ref 1).

Carbon dioxide (solubility 1.1×10^{-5} M) dissolves in water to produce carbonic acid, H_2CO_3 (Ref 5). Carbon dioxide gas is introduced into water through the dissolution of atmospheric CO_2 and by the aerobic decay of organic matter. Biological activity (supported by sufficient nutrients) removes oxygen and generates carbon dioxide (Ref 9). When the only source of carbon dioxide is atmospheric (e.g., from rainwater), then the water is slightly acidic (pH 5.6). When carbon dioxide is also being generated by the microbial breakdown of organic matter, then the concentration of dissolved carbon dioxide is increased, and the pH becomes more acidic (e.g., pH 4.9) (Ref 5). Seawater is kept in the pH range of 7.5 to 8.3 by the buffering action of the bicarbonate (HCO_3^-)-carbonate (CO_3^{2-}) equilibrium (Ref 14).

Sulfate-Reducing Bacteria (SRB). Buried metals are affected by SRB growing on a metal surface or nearby. These bacteria grow in near-neutral, anaerobic environments containing a source of sulfate ions, such as decomposing organic matter (Ref 15). Sulfate-reducing bacteria grow in waterlogged soils; in mud at the bottom of rivers, lakes, marshes, and estuaries; and under deposits on metals in freshwater and seawater (Ref 15). When SRB grow on a metal surface, they consume hydrogen atoms produced at cathodic sites and reduce sulfate (SO_4^{2-}) ions to sulfide (S^{2-}) ions (Ref 15, 16). If SRB are located on a metal surface, then the sulfide ions react directly with metal ions to form metal sulfides (Ref 16). If SRB are located in close proximity to metals, then the sulfide ions dissolve as HS^- (above approximately pH 7) or H_2S (below approximately pH 7) and diffuse to the metal surface (Ref 17). In marine sediments, the maximum SRB populations are typically located to a depth of approximately 50 cm (20 in.), although SRB can be located at greater depths if decaying organic material is present, such as wooden hulls of shipwrecks (Ref 14). Under certain conditions, SRB remain inactive despite favorable growing conditions. Well-preserved iron, including 2000 year old Roman artifacts, was recovered from such a site in York, England (Ref 18). The inactivity of the SRB was attributed to tannins in the soil (Ref 1).

The best known SRB are the genus *Desulfovibrio*, which thrive at a pH of 4 to 8 and temperatures between 10 and 40 °C (50 and 105 °F) (Ref 15, 16). Active growth of SRB contributes to corrosion on aluminum, copper alloys, iron and steel, stainless steels, and zinc. The corrosion rate can be accelerated by the formation of local deposits of metal sulfides, which generally have good electrical conductivity (Ref 19). Pitting can result when a discontinuous sulfide film is present, because of the galvanic effect created by contact between the cathodic film and the underlying anodic metal (Ref 9, 15). On the other hand, a protective film can also form if the sulfide forms as a continuous, adherent layer over the entire metal surface (Ref 19, 20).

The pH of most environments encountered in nature falls between 4 and 9, although certain environments can have a lower or higher pH (Ref 9, 21). In general, alkaline soils are benign to metals, and acid soils are aggressive (Ref 1). Certain metals, such as aluminum, lead, and zinc, are amphoteric and can experience an increased corrosion rate in both acidic and alkaline conditions (Ref 22).

Acidic environments (pH < 7) develop where organic material is decomposing and large quantities of organic acids are produced (Ref 6, 12). Rainwater, waterlogged soils, and bogs are typically acidic (pH 3 to 6) (Ref 5). In climates where the drainage of water is downward through the soil profile, the environment becomes acidic from the leaching of cations (calcium, magnesium, sodium, potassium) from soil minerals (Ref 5, 9, 12). An acidic environment encourages the oxidation of metals and the dissolution of corrosion products.

Alkaline environments (pH > 7) develop because of the dissolution of inorganic compounds, especially in regions where evaporation exceeds precipitation (Ref 5, 6). Soils containing groundwater that has flowed over limestone are alkaline because of the dissolution of calcium carbonate (Ref 9). In regions where there are calcareous rocks, the soil has good buffering capacity against acidic deposition (Ref 3). The pH of seawater normally falls in the slightly alkaline range between 7.5 and 8.3 (pH > 8 in surface seawater) because of the buffering action of the bicarbonate-carbonate equilibrium (Ref 10, 14). The pH in marine sediment may drop to 7.0 because of the dissolution of hydrogen sulfide produced by SRB.

Potential. The electrode potential (oxidation-reduction potential) of a burial environment is the potential of an inert electrode, such as platinum, placed in that environment and measured with respect to a reference electrode. The measured potential is not a measure of the oxygen concentration but is an indication of the oxidizing or reducing capacity of that environment (Ref 9). When the potential becomes more positive, the conditions are more oxidizing and vice versa (Ref 1). Natural environments can be oxidizing (i.e., able to accept electrons), reducing (i.e., able to donate electrons), or neutral (Ref 5, 23).

Oxidizing environments are typically oxygen rich because they are in contact with the atmosphere, whether by easy access of air (e.g., sandy soils) or by the penetration of water containing dissolved oxygen (e.g., tidal zone, strong wave action) (Ref 6, 24). Reducing environments are oxygen poor. They develop in waterlogged soils, poorly drained swamps, under clay or tightly packed sediment, and under marine concretions where stagnant conditions restrict the replenishment of dissolved oxygen (Ref 6, 24).

Under reducing conditions, SRB activity can cause damage to buried metals (Ref 9, 25). Table 1 lists four ranges of electrode potential (on the standard hydrogen electrode, or SHE, scale corrected to pH 7), categorized according to the corrosivity of burial environments containing SRB (Ref 9, 15, 25). Above 400 mV, the environment is oxidizing and does not support SRB.

Potential versus pH Diagrams for Environments in Nature. One approach to characterizing natural environments is to measure their electrode potential and their pH and then determine which of four general categories they fall into: oxidizing and acidic, oxidizing and alkaline, reducing and acidic, and reducing and alkaline (Ref 22). These categories are shown in Fig. 1 in a diagram of potential versus pH and commonly referred to as a potential-pH or Pourbaix diagram (Ref 23). As mentioned earlier, the pH of most environments in nature falls between 4 and 9. Neutral environments, from an acid-base viewpoint, fall along the vertical dashed line at pH 7.

In natural burial environments under 1 atm pressure, the potential scale is limited by the thermodynamic stability of water (Ref 5, 22).

Table 1 Categories of the corrosivity of burial environments containing sulfate-reducing bacteria according to electrode potential (corrected to pH 7)

Category	Potential, mV (SHE)
Noncorrosive	>400
Slight	400–200
Moderate	200–100
Severe	<100

SHE, standard hydrogen electrode

The region where water is stable lies between the two broken lines labeled "A" and "B" in Fig. 1. The upper limit of stability is given by line B, along which the oxygen gas pressure equals 1 atm. If the potential of the system rises above line B, then the pressure of oxygen gas becomes greater than 1 atm, and water is oxidized with the evolution of gaseous oxygen: $2H_2O \rightarrow O_2(g) + 4H^+ + 4e^-$ (Ref 23). The lower limit of stability is given by line A, along which the hydrogen gas pressure equals 1 atm. If the potential of the system falls below line A, then the pressure of hydrogen gas becomes greater than 1 atm, and water is reduced with the evolution of gaseous hydrogen: $2H_2O + 2e^- \rightarrow H_2(g) + 2OH^-$ (Ref 23). Neutral environments, from an oxidation-reduction viewpoint, have potentials that fall along line C, which is where the hydrogen gas partial pressure equals twice the oxygen gas partial pressure (Ref 23). Oxidizing environments have potentials falling between lines B and C. Reducing environments have potentials falling between lines A and C.

Terrestrial Environments. The corrosion of metals in soils is affected by the availability of oxygen, moisture content, dissolved species, soil conductivity, and biological activity. Soils are complex aggregates of moisture, gases, organic matter, living organisms, precipitated materials, and particles of weathered rock (Ref 8, 9). Sandy soils are almost devoid of organic matter and consist largely of inorganic materials weathered to fairly uniform-sized particles. On the other hand, peat bogs and swamps are primarily water-saturated decaying organic matter mixed with minor amounts of inorganic material (Ref 3). Soils are commonly named according to the general particle size range of inorganic constituents, with, for example, particle diameters being 0.07 to 2 mm (0.003 to 0.08 in.) in sandy soils, 0.005 to 0.07 mm (0.0002 to 0.003 in.) in silts, and less than 0.005 mm (0.0002 in.) in clays (Ref 8).

The aeration of soil is directly related to its pore space and whether this space contains water or moist gases. Water flows quickly through coarse-grained sand, allowing for good drainage and easy access of atmospheric oxygen. Conversely, water has trouble permeating clay soils made up of small particle size (Ref 9). Biological activity tends to decrease the oxygen content and replace the oxygen with gases from metabolic activity, such as carbon dioxide (Ref 8). Anaerobic conditions develop in soils where large amounts of organic matter are decomposing through biological activity or where stagnant, waterlogged conditions significantly decrease oxygen diffusion (Ref 8).

Marine Environments. The corrosion of metal artifacts in a marine environment is affected by the metal composition, water composition, temperature, marine growth, sediment composition, position of artifacts relative to other materials, depth of burial in sediment, and extent of water movement (Ref 14).

A layer of hard concretion usually encrusts shipwrecks, metal artifacts, and other sunken material located in seawater where organisms grow by photosynthesis (Ref 11, 26). The white concretion is mainly calcium carbonate (calcite, aragonite), and it forms a semipermeable barrier between the artifact and the seawater, decreasing the availability of dissolved oxygen and protecting the artifact from physical damage (mechanical abrasion, wave action) (Ref 14, 26, 27). The concretion is usually the result of encapsulation by marine organisms, mainly those such as corals, coralline algae, tubeworms, and shellfish that leave behind a calcium carbonate skeleton after they die (Ref 11). Metals such as copper and lead that release toxic ions discourage the growth of these marine organisms (Ref 27).

Calcium carbonate can also precipitate directly from seawater, particularly in surface waters that can become supersaturated in calcium carbonate because of the dissolution of atmospheric carbon dioxide. Because the solubility of calcium carbonate decreases with increasing pH, significant amounts can precipitate (as calcite and aragonite) if the pH rises above approximately 9 (Ref 11). Precipitation of calcium carbonate occurs in regions where intense photosynthetic activity increases pH (shallow water) as well as on metal surfaces where the cathodic reaction (typically, reduction of dissolved oxygen) increases the local pH (Ref 11). Concretion formation is sometimes prevented by unusual conditions; for example, no concretions formed on artifacts associated with the *RMS Titanic* (1912), which sank off the coast of Newfoundland to a depth of 4 km (2.5 miles), where the water was cold (1 to 2 °C, or 34 to 36 °F) and the hydrostatic pressure high. Instead of concretion, the iron in the *Titanic* became covered with "rusticles" (formed by bacterial action); the rusticles consisted of iron oxide hydroxides, with lepidocrocite making up the exterior shell, and goethite on the inside (Ref 28).

The conditions within the concretion can be significantly different from the surrounding seawater. The semipermeable nature of the concretion can separate anodic and cathodic reactions. At the metal surface beneath the concretion, the metal ions produced by the anodic reaction undergo hydrolysis and increase the acidity. At the same time, the concentration of chloride ions increases underneath the

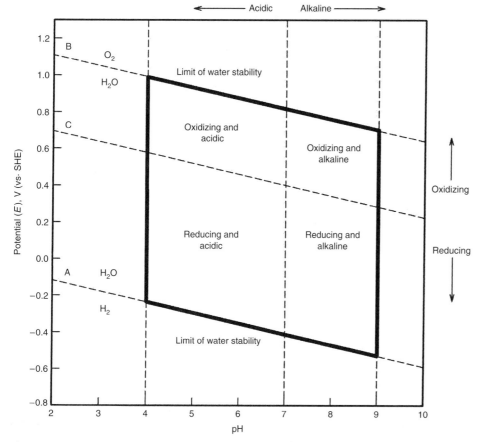

Fig. 1 Schematic diagram of potential versus pH showing the various regions found in nature at a total pressure of 1 atm. The limits of stability for water are given by the lower line A, $E = 0.000 - 0.059$ pH (standard hydrogen electrode, or SHE), and the upper line B, $E = 1.228 - 0.059$ pH (SHE). Line C, $E = 0.813 - 0.059$ pH (SHE), separates oxidizing regions from reducing ones. The dashed vertical line at pH 7 separates acidic and alkaline regions. Adapted from Ref 23

concretion, because these negative anions diffuse inward to maintain charge balance with the positive metal cations (Ref 26, 27). The pH can decrease to approximately 6 under (the relatively thin) concretions on copper alloys (Ref 27) and to approximately 4.8 under thick concretions on iron (Ref 26). The environment can also become anaerobic, with SRB growth supported by a plentiful supply of organic material from the concretion organisms (Ref 14, 29).

The lower pH inside the concretion contributes to the local dissolution of calcium carbonate and the precipitation of metal corrosion products. The interior of the concretion can be stained black or red-brown by iron corrosion products, green by copper chloride hydroxides, or deep blue to black with iron or copper sulfides (Ref 26, 27, 29). In shipwrecks where a collection of artifacts becomes encased with a concretion, several different metals can be in contact or can be corroding in close vicinity. This often results in unusual corrosion product formation. The metal can also completely corrode, leaving powdery corrosion products inside a very detailed natural mold of the object.

Corrosion of Metals during Burial

In antiquity, metals such as gold, copper, and meteoric iron were available in their native form, while other metals, such as tin, lead, and silver, were extracted from their ores. Ancient metals and alloys were made into functional or decorative objects by working or casting (Ref 30). The approximate dates for the first widespread use of various metals are listed in Table 2 (Ref 31). The development of copper smelting led to the Bronze Age (~3000 to 1200 B.C.), and improvements in technology led to the Iron Age (~1200 B.C.) and the Industrial Revolution (A.D. ~1750 to 1850) (Ref 32–36).

Metals are, in general, inherently unstable, and when buried, they react with their environment and convert to more thermodynamically stable corrosion products, such as metal oxides. The reaction products can be solid corrosion products (often with beautiful colors) or dissolved species. Some of the corrosion products identified on archaeological metals are listed in the following tables, along with their mineral name, chemical formula, and the mineral color: Table 3 (copper alloys), Table 4 (silver alloys), Table 5 (iron alloys), Table 6 (lead), and Table 7 (tin and pewter). The outer corrosion layer often incorporates soil minerals; quartz grains, associated burial material; and possibly, organic residues (Ref 37). Depending on the burial environment, metal artifacts can corrode slowly or quickly. The survival of metal artifacts for long periods of time is aided by the formation of adherent corrosion products that slow the corrosion rate, because the adherent material slows the rate at which oxygen and other environmental species reach the underlying metal. The corrosion rate is also influenced by other environmental factors, such as modern pollutants (e.g., deicing salts, fertilizers, acid precipitation), by associated burial material, or by being located in a burial site with human remains.

Table 2 Approximate date of first widespread use of various metals

Metal	Date	Location
Copper		
Native copper	~7000 B.C.	Near East
Smelted copper	~5000 B.C.	Near East
Lead	6000–5000 B.C.	Near East/Balkans
Gold	5000–4000 B.C.	Balkans
Tin	4000–3000 B.C.	Near East
Silver	4000–3000 B.C.	Balkans/Near East
Nickel		
Copper-nickel alloys	2000–1000 B.C.	Near East
Iron	1000–0 B.C.	Near East
Zinc		
Copper-zinc	A.D. 100–200	Rome
Zinc metal	A.D. 900–1000	India
Aluminum	A.D. 1800–1900	Europe/United States

Table 3 Some corrosion products identified on archaeological copper alloys

Chemical name	Mineral name	Chemical formula	Color
Oxides and hydroxides			
Copper oxide	Cuprite	Cu_2O	Red
Carbonates			
Copper carbonate hydroxide	Malachite	$Cu_2CO_3(OH)_2$	Green
Copper carbonate hydroxide	Azurite	$Cu_3(CO_3)_2(OH)_2$	Blue
Copper sodium carbonate trihydrate	Chalconatronite	$CuNa_2(CO_3)_2 \cdot 3H_2O$	Blue
Chlorides			
Copper chloride	Nantokite	$CuCl$	Pale gray
Copper chloride dihydrate	Eriochalcite	$CuCl_2 \cdot 2H_2O$	Blue-green
Copper chloride hydroxide	Atacamite	$Cu_2Cl(OH)_3$	Green
Copper chloride hydroxide	Clinoatacamite	$Cu_2Cl(OH)_3$	Green
Copper chloride hydroxide	Paratacamite	$Cu_2Cl(OH)_3$	Green
Copper chloride hydroxide	Botallackite	$Cu_2Cl(OH)_3$	Green
Phosphates			
Calcium copper sodium chloride phosphate pentahydrate	Sampleite	$CaCu_5NaCl(PO_4)_4 \cdot 5H_2O$	Blue-green
Copper hydroxide phosphate	Libethenite	$Cu_2(OH)(PO_4)$	Green
Sulfides			
Copper sulfide	Chalcocite	Cu_2S	Black
Copper iron sulfide	Chalcopyrite	$CuFeS_2$	Yellow
Copper sulfide	Geerite	$Cu_{1.6}S$	Black
Copper sulfide	Digenite	$Cu_{1.8}S$	Blue-black
Copper sulfide	Djurleite	$Cu_{1.96}S$	Black
Copper sulfide	Covellite	CuS	Dark blue
Sulfates			
Copper hydroxide sulfate	Brochantite	$Cu_4(OH)_6SO_4$	Green
Copper hydroxide sulfate	Antlerite	$Cu_3(OH)_4SO_4$	Green
Copper hydroxide sulfate monohydrate	Posnjakite	$Cu_4(OH)_6SO_4 \cdot H_2O$	Light blue
Copper iron hydroxide sulfate tetrahydrate	Guildite	$CuFe(OH)(SO_4)_2 \cdot 4H_2O$	Yellow-brown
Copper chloride hydroxide sulfate trihydrate	Connellite	$Cu_{19}Cl_4(OH)_{32}(SO_4) \cdot 3H_2O$	Blue

Table 4 Some corrosion products identified on archaeological silver alloys

Chemical name	Mineral name	Chemical formula	Color
Chloride and bromide			
Silver chloride	Chlorargyrite(a)	$AgCl$	White
Silver bromide	Bromargyrite	$AgBr$	Yellow
Silver bromide chloride	Embolite	$Ag(Br, Cl)$	Yellow
Sulfides			
Silver sulfide	Acanthite	α-Ag_2S	Black
Silver sulfide	Argentite	β-Ag_2S	Black
Copper silver sulfide	Jalpaite	Ag_3CuS_2	Black
Copper silver sulfide	Stromeyerite	$AgCuS$	Black
Gold silver sulfide	Uytenbogaardtite	Ag_3AuS_2	Gray-black
Gold silver sulfide	Petrovskaite	$AgAuS$	Black

(a) Formerly known as cerargyrite

Table 5 Some corrosion products identified on archaeological iron alloys

Chemical name	Mineral name	Chemical formula	Color
Oxides and hydroxides			
Iron oxide	Magnetite	Fe_3O_4	Black
Iron oxide	Hematite	$\alpha\text{-}Fe_2O_3$	Red or black
Iron hydroxide oxide	Goethite	$\alpha\text{-}FeO(OH)$	Yellow-brown
Iron hydroxide oxide	Akaganéite	$\beta\text{-}FeO(OH)$	Red-brown
Iron hydroxide oxide	Lepidocrocite	$\gamma\text{-}FeO(OH)$	Orange
Carbonate			
Iron carbonate	Siderite	$FeCO_3$	Yellow-brown
Chlorides			
Iron chloride	...	$FeCl_2$	White
Iron chloride	...	$FeCl_3$	Green
Phosphates			
Iron phosphate octahydrate	Vivianite	$Fe_3(PO_4)_2 \cdot 8H_2O$	Dark blue (or white)
Iron phosphate dihydrate	Strengite	$FePO_4 \cdot 2H_2O$	Pink
Sulfides			
Iron sulfide	Pyrrhotite	$Fe_{1-x}S$ ($x = 0\text{--}0.2$)	Yellow-brown
Iron sulfide	Mackinawite	FeS_{1-x} ($x = 0.01\text{--}0.08$)	Yellow-brown
Iron sulfide	Pyrite	FeS_2	Yellow
Iron sulfide	Greigite	Fe_3S_4	Blue-black
Sulfates			
Iron sulfate tetrahydrate	Rozenite	$FeSO_4 \cdot 4H_2O$	Green
Iron sulfate heptahydrate	Melanterite	$FeSO_4 \cdot 7H_2O$	Blue-green
Iron hydroxide sulfate dihydrate	Butlerite	$Fe(OH)SO_4 \cdot 2H_2O$	Orange
Iron potassium hydroxide sulfate	Jarosite	$Fe_3K(OH)_6(SO_4)_2$	Yellow-brown
Iron sodium hydroxide sulfate	Natrojarosite	$Fe_3Na(OH)_6(SO_4)_2$	Yellow-brown

Table 6 Some corrosion products identified on archaeological lead

Chemical name	Mineral name	Chemical formula	Color
Oxides			
Lead oxide	Litharge	$\alpha\text{-}PbO$	Red
Lead oxide	Massicot	$\beta\text{-}PbO$	Yellow
Carbonates			
Lead carbonate	Cerussite	$PbCO_3$	White
Lead carbonate hydroxide	Hydrocerussite	$Pb_3(CO_3)_2(OH)_2$	White
Chlorides			
Lead chloride	Cotunnite	$PbCl_2$	White
Lead chloride hydroxide	Laurionite	$PbClOH$	White
Lead carbonate chloride	Phosgenite	$Pb_2CO_3Cl_2$	White
Lead chloride phosphate	Pyromorphite	$Pb_5Cl(PO_4)_3$	Yellow
Sulfide and sulfate			
Lead sulfide	Galena	PbS	Black
Lead sulfate	Anglesite	$PbSO_4$	White

Table 7 Some corrosion products on archaeological tin and pewter

Chemical name	Mineral name	Chemical formula	Color
Oxides and hydroxides			
Tin oxide	Romarchite	SnO	Black
Tin hydroxide oxide	Hydroromarchite	$Sn_3O_2(OH)_2$	White
Tin oxide	Cassiterite	SnO_2	White
Magnesium tin hydroxide	Schoenfliesite	$MgSn(OH)_6$	White
Chloride			
Tin chloride hydroxide oxide	Abhurite	$Sn_3Cl_2O(OH)_2$	White
Tin chloride hydroxide	...	$Sn_4Cl_4(OH)_6$	...
Sulfide and sulfate			
Tin sulfide	Ottemannite	Sn_2S_3	Gray

Potential-versus-pH (Pourbaix) Diagrams for Specific Metals. Theoretical potential-versus-pH diagrams (Pourbaix diagrams) can be calculated for metals using thermodynamic data (Ref 17, 22, 38). They provide information about the regions where a metal is corroding and where it can become covered with stable corrosion products. Such calculated diagrams are useful for predicting corrosion behavior. For a given burial site, an accurate Pourbaix diagram can be calculated (using the concentrations of dissolved species in that site) and used as a model to predict the corrosion behavior for metals buried at that site (Ref 39).

Calculated Pourbaix diagrams are useful for interpreting results from measurements of a metal artifact electrode potential (its corrosion potential) and its surface pH, from corrosion product analysis, and from other environmental and electrochemical data. In marine environments, corrosion potential measurements have been used to monitor metals associated with shipwrecks (Ref 40–43). Related studies have been carried out for metal artifacts buried in terrestrial sites (Ref 39, 44, 45).

Patinas. In the context of archaeological metals, particularly copper alloys, the term *patina* is usually associated with the attractive appearance of corrosion products that have formed on metal artifacts after long periods of burial (Ref 37, 46). On high-tin bronzes (i.e., copper alloys containing 20 to 25% Sn), the patinas are often relatively smooth, and they preserve the detail and shape of an artifact, even after thousands of years of burial (Ref 47–49). They can also be highly valued because of their beautiful colors, especially the ones with an overall green color interspersed with splashes of blue (Ref 50). However, the term *patina* is also used in a general sense to describe the surface appearance (attractive or not) of the corrosion products on any archaeological metal artifact (Ref 50).

Copper Alloys. During long periods of burial, the extent of corrosion of copper alloy artifacts ranges from slight (in certain anaerobic sites), with the surface remaining bright, to completely destroyed (in acidic, oxygen-rich deposits), with only a green stain left behind (Ref 12). The extent of corrosion for many other copper alloy artifacts falls in between these two extremes, with some of the metal converted to corrosion products. Copper(I) compounds (cuprite, nantokite, chalcocite) typically form next to the metal surface and, in oxygen-containing environments, green, blue, or blue-green copper(II) compounds form (malachite, atacamite) over the copper(I) compounds. The composition of the copper(II) compounds in the outer layer depends on the relative concentrations of environmental species (Ref 51). Table 3 lists some corrosion products identified on archaeological copper alloys.

Terrestrial Environments. When copper alloys corrode in soil, many of them form a cuprite crust adjacent to the metal and overlaid with green malachite, a copper carbonate hydroxide (Ref 37). Malachite is a common corrosion product because of the reaction between copper and carbon dioxide gas present in humid air or dissolved in groundwaters percolating through damp soil (Ref 1, 13, 37, 50). The uniform growth of malachite produces the attractive green patina on many bronze antiquities (Ref 37). Azurite, a blue copper carbonate hydroxide, is identified less frequently and is less stable than malachite. Azurite is usually located in discrete patches in association with malachite (Ref 37, 50). Chalconatronite, another blue copper compound, has been identified on ancient Egyptian bronzes recovered from soil rich in various salts as a result of evaporation (Ref 52). Chalconatronite can also form on artifacts after treatment in sodium-base solutions (see the section "Treatment of Bronze Disease" in this article). Copper chloride hydroxides are sometimes identified on archaeological copper recovered from terrestrial sites with a high chloride content, such as sites located close to roads where deicing salts are used in the winter (Ref 1, 13). Eriochalcite, a rare blue-green copper chloride, has been identified on a copper alloy object from Egypt (Ref 53).

Nantokite, a copper(I) chloride, can precipitate under or within the cuprite layer when conditions have become sufficiently reducing and acidic because of the inward diffusion of environmental chloride ions toward the anodic regions on corroding copper (Ref 37, 54, 55). Nantokite is soft, has a grayish appearance like paraffin wax, and can be stained green (Ref 50). The accumulation of nantokite (or soluble chloride ions) within the corrosion layers on archaeological copper alloys can result in a corrosion problem after excavation known as bronze disease (see the section "Copper Alloys (Bronze Disease)" in this article).

On bronzes containing a high tin content (e.g., 20 to 25%), two different kinds of patina can form: One type is even and compact, and the other is uneven and rough (Ref 12). High-tin bronzes with an even, compact patina (sometimes called water patina) are highly valued because of their beautiful lustrous appearance, with colors ranging from blue, green to dark green, dark gray, to black (Ref 12, 37, 47–49, 56). The compact patina is thought to form by the selective (and slow) dissolution of copper and the conversion of tin into cassiterite and other tin oxides (Ref 48, 49, 56). Because there is no significant volume change in the outer corrosion layer, the original shape of the artifact is preserved (Ref 47, 49). The uneven, rough, and crusty patinas that form on high-tin bronzes can be quite thick and often contain a mixture of copper chloride hydroxides and malachite over a cuprite layer (Ref 12, 47). These patinas are thought to form under conditions where there is a high corrosion rate, and the original shape of the artifact is usually damaged because of significant volume changes due to corrosion product formation (Ref 47, 49).

Marine Environments. Concretions on copper alloys are usually relatively thin because of the toxic nature of copper corrosion products (Ref 27). Atacamite, a copper chloride hydroxide, usually forms on copper alloys exposed to an aerated marine environment (Ref 57). Various green copper(II) chloride hydroxides (mainly atacamite and paratacamite, rarely botallackite) are often identified on copper alloy artifacts recovered from aerobic marine environments (Ref 37, 50, 57, 58). Recent work by Jambor et al. shows that the x-ray diffraction reference pattern Powder Diffraction File 25-1427 had been incorrectly assigned to paratacamite (rhombohedral) and instead corresponds to a monoclinic crystal structure, newly named clinoatacamite (Ref 59). In the past, clinoatacamite may have been misidentified as paratacamite. Nantokite, the copper(I) chloride, is rarely found on marine copper alloys, presumably because the conditions beneath concretions (pH ~6) are rarely acidic and reducing enough for it to precipitate (Ref 14).

Anaerobic Environments. Dark-colored copper sulfides form on copper artifacts when they are exposed to hydrogen sulfide generated by SRB in anaerobic environments (Ref 19, 24, 50, 57, 60). The corrosion layer can be relatively thick and uneven and can obscure the shape and surface details of an artifact (Ref 60), or it can be relatively thin, uniform, adherent, and protective (Ref 24). Some copper alloy artifacts, protected by copper sulfides, have survived hundreds of years on shipwrecks (Ref 14). Copper sulfides (chalcocite, covellite, digenite, djurleite) have been found on copper alloys recovered from reducing marine environments (Ref 57). Copper sulfides (chalcocite, covellite, djurleite, geerite) have also been identified on copper alloy artifacts recovered from waterlogged sites (Ref 24, 60). Chalcopyrite, a dull gold-colored copper iron sulfide, has been found on copper alloy artifacts, where, because it looks like gilding, it is called pseudogilding (Ref 24, 50, 60–63).

Rarer Corrosion Products. There are rare instances where copper hydroxide sulfates (e.g., brochantite, antlerite, posnjakite) have formed on archaeological copper alloys in soil (contaminated by acid precipitation) (Ref 64) or in freshwater sediments (from the oxidation of copper sulfides) (Ref 24). Bright-blue connellite has also been identified on copper alloys recovered from both marine and terrestrial environments (Ref 37, 57, 65).

The phosphate-containing copper corrosion products, green libethenite and blue-green sampleite, have been identified on copper alloy artifacts (Ref 37, 53, 66). The source of phosphate ions in the burial environment is often from bones or from phosphate-rich soils, such as found in Egypt. Studies of archaeological bronzes recovered in Sweden have shown that phosphate ions in soil (as well as high concentrations of soot) accelerate the corrosion of copper alloys during burial (Ref 3, 13).

Smithsonite (zinc carbonate, $ZnCO_3$) has been identified on copper alloys rich in zinc (Ref 13). Lead compounds, such as cerussite and pyromorphite, have been identified on archaeological copper alloys containing lead and recovered from terrestrial environments (Ref 13, 67, 68). The patina in copper-tin bronzes is often enriched with tin oxides (Ref 48, 49, 56, 67). In marine environments, copper alloyed with tin, zinc, or lead can become covered with various tin, zinc, and lead compounds in addition to copper compounds from the corrosion of these alloying elements (Ref 57).

Silver. Archaeological silver artifacts are typically made from silver-copper alloys. Depending on the aggressiveness of the environment, silver artifacts can develop a thin patina, or all the metal can be converted to silver corrosion products (Ref 12). Table 4 lists some corrosion products identified on archaeological silver alloys.

In well-aerated burial environments, either marine or terrestrial, the silver in these alloys reacts with dissolved halide ions to form various silver halide compounds. Chlorargyrite (silver chloride, also known as horn silver or cerargyrite) is often identified on archaeological silver; less frequently identified are bromargyrite and embolite (Ref 12, 57). During burial, silver-copper alloys often become encrusted with green copper(II) compounds (becoming indistinguishable from corroded copper alloys) as copper in the alloy corrodes, is selectively removed from the alloy, and is deposited as green malachite or atacamite on the outer surface (Ref 12).

In anaerobic environments where SRB are active, silver reacts with hydrogen sulfide to form black sulfides, such as acanthite, or, if copper is also present, argentite, jalpaite, and stromeyerite (Ref 14, 57). In its pure state, argentite exists above 175 °C (350 °F), but if its structure also contains some copper, argentite can exist at room temperature (Ref 19).

During long periods of burial, silver artifacts can become extremely brittle (and fragile) because of microstructural changes or corrosion of the alloy (Ref 69). Microstructurally induced embrittlement is attributed to long-term aging at relatively low temperatures, where precipitation of impurities (e.g., lead) weakens the grain boundaries. Corrosion-induced embrittlement is attributed to segregation of copper at grain boundaries and subsequent intergranular corrosion caused by localized galvanic attack.

Gold and Gilded Metals. Pure gold is rarely corroded by burial environments, although certain organic material (humic material) in reducing environments can dissolve gold (Ref 70). Gold alloys and gilded metals are, however, subject to various forms of corrosion that depend on the local burial conditions. For example, ancient Egyptian gold alloy artifacts, such as the gold recovered from Tutankhamen's tomb, can have a distinctive red or purple surface coloration, identified as gold silver sulfides (uytenbogaardtite, petrovskaite, see Table 4) in a tarnish layer formed in the presence of hydrogen sulfide in the burial environment (Ref 71). During burial in soil or marine environments, the outer surface of gold artifacts can become completely obscured by corrosion products (e.g., green malachite or atacamite, black silver sulfides, light-sensitive silver chloride) coming from the corrosion of the copper or silver in gold alloys or in alloys beneath the gilding (Ref 12, 72). Corrosion beneath the gilding will cause the gilding to become detached from the underlying metal. Finally, gold-silver and gold-copper alloys are also susceptible to intergranular stress-corrosion cracking during burial, which leaves the artifacts in a brittle and fragile state (Ref 72). Galvanic effects contribute to a greater susceptibility to corrosion for gold alloys with higher silver or copper contents.

Iron. Because iron is a relatively active metal, its corrosion rate in many burial environments is quite rapid compared to other common metals of antiquity. The survival rate of iron artifacts is influenced by the burial environment. In aggressive environments, for example, iron objects can be mineralized into solid lumps of iron corrosion products, with little or no metal remaining (Ref 12). On the other hand, iron artifacts buried in anaerobic conditions can survive in a well-preserved state for thousands of years if SRB activity is suppressed (Ref 18). Table 5 lists some corrosion products identified on archaeological iron alloys. The corrosion products that form on iron corroding under reducing conditions are generally iron(II) compounds, such as siderite, vivianite, and pyrite (Ref 73). Under oxidizing conditions the corrosion products on iron are red-brown iron(III) compounds, typically goethite, an iron hydroxide oxide, or black magnetite, an iron(II, III) oxide (Ref 74). Hematite, a red iron oxide, does not normally form as a corrosion product under burial conditions but is sometimes identified on archaeological iron. Because hematite forms when goethite is heated (above approximately 250 °C, or 480 °F), its presence is usually attributed to an object being exposed to fire before burial (Ref 12).

Terrestrial Environments. When iron corrodes during burial in damp, aerated soil, its surface gradually turns into a bulky mass of rust-colored iron corrosion products cemented with soil particles, dirt, clay, and sand (Ref 12). The corrosion products are generally layered, with lower-iron-oxidation-state compounds forming next to the metal surface, and higher-iron-oxidation-state compounds forming an outer corrosion layer. The most common iron compounds identified in the outer corrosion layers are iron hydroxide oxides, mainly goethite and sometimes lepidocrocite (Ref 12, 74). The buildup of corrosion products can lower the corrosion rate of the iron relative to when it was first buried.

Magnetite is the most common iron oxide identified on archaeological iron and is usually located next to the metal surface (Ref 12, 75). When carbonates or phosphates are present in the soil and the conditions are more reducing than are needed for magnetite to form, then protective layers of siderite (an iron carbonate) or vivianite (an iron phosphate) can form (Ref 12, 76). Siderite has been identified on iron artifacts recovered from waterlogged sites in Denmark and England (Ref 39, 77). Vivianite has been identified on iron artifacts recovered from waterlogged sites in England (Ref 77). Although pure vivianite is white, its color on iron artifacts is typically dark blue, because some Fe^{2+} ions have been oxidized to Fe^{3+}; the blue color is from a charge transfer between Fe^{2+} and Fe^{3+} ions on adjacent sites (Ref 78). The source of phosphate ions can be from bones in cemeteries, fish waste, garbage dumps, or (more recently) phosphate fertilizers (Ref 75).

The development of corrosion layers on buried iron promotes pitting and, after extended periods, allows for the separation of anodic and cathodic regions (Ref 21, 76). The anodic half-reaction (metal corrosion) takes place at the interface between the metal and the corrosion products, and the cathodic half-reaction (oxygen reduction) takes place on the outer surface of magnetite, a good electrical conductor. A schematic diagram illustrating this corrosion process is shown in Fig. 2 (Ref 74). At the metal surface, iron(II) ions dissolve, accumulate, and undergo hydrolysis, promoting local acidification. Electrical neutrality is maintained by negative anions diffusing inward from the surrounding burial environment. Chloride ions, in particular, tend to concentrate at the interface because of their high mobility and because they are often the predominant environmental anions (Ref 21, 29). The net result is that, during burial, the cracks, pores, and open spaces within the corrosion layer on archaeological iron become filled with an acidic iron(II) chloride solution (Ref 21, 74). The highly soluble iron(II) chloride remains in solution rather than precipitating as a solid, which can only happen at pH < 2 (Ref 76). The accumulation of this acidic, chloride-rich solution on archaeological iron can result in a serious corrosion problem after excavation (see the section "Iron (Weeping, Akaganéite Formation)" in this article).

Marine Environments. Iron artifacts become covered by massive concretions in most marine conditions (Ref 14). The outer part of concretions is mainly white (calcium carbonate), and the part closest to the iron is red-brown and contains iron corrosion products, such as iron hydroxide oxides, magnetite, siderite, or iron sulfides (Ref 26). The iron beneath the concretion develops an acidic pH of approximately 4.8 (Ref 26). This acidity promotes the dissolution of iron corrosion products, allowing them to diffuse away from the surface until they encounter a more alkaline pH and precipitate.

The shape of a wrought iron artifact is not retained after it has corroded in seawater, but the shape of a cast iron artifact is. Traditional wrought iron (as opposed to modern mild steel) is fairly pure iron that contains glassy inclusions of iron silicate slag and has a characteristic fibrous structure caused by the elongation of the slag during the forging process. As some of the iron dissolves from corroding wrought iron, the slag inclusions are uncovered, and the surface develops a characteristic fibrous or lamellar structure (Fig. 3). Eventually, all of the iron in the wrought iron artifact dissolves, leaving a void in the concretion—a mold of which can often be used to determine the original shape of the artifact (Ref 14).

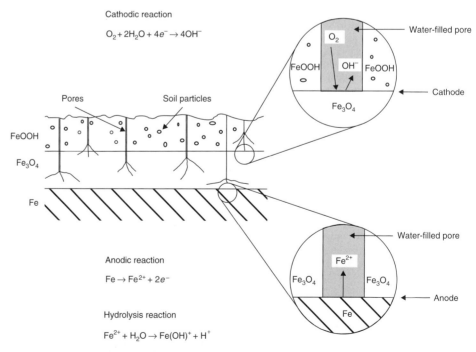

Fig. 2 Schematic diagram showing the corrosion layers on buried iron as well as the separation of anodic and cathodic regions. Adapted from Ref 75

For artifacts made of cast iron (iron alloys containing approximately 2 to 4 wt% C in the form of cementite Fe_3C or graphite), corrosion causes the iron (and eventually the cementite) to dissolve, leaving behind a matrix of soft graphite. This corrosion process is known as graphitic corrosion and is sometimes referred to as graphitization in the conservation literature. The original shape of the artifact (e.g., cast iron cannons and cannon balls) is retained by the graphite matrix, which is usually filled with iron corrosion products (Ref 14).

Anaerobic Environments. Black-colored or gold-colored iron(II) sulfides form on iron artifacts buried in anaerobic environments where SRB are active. The iron sulfides mackinawite, greigite, and pyrite have all been identified on iron artifacts recovered from five different waterlogged land sites in England (Ref 77). Mackinawite initially forms on buried iron because of SRB and then, depending on the oxidizing capacity of the local environment, can transform to greigite, pyrrhotite, or pyrite (Ref 19). Iron sulfides have also been identified in marine concretion (Ref 26) and as precipitates on other archaeological materials: pyrite on waterlogged wood and ceramics, and pyrrhotite on copper alloys (Ref 60, 79). Certain iron sulfides (e.g., pyrite, pyrrhotite) have a metallic luster and yellow-brown color that resemble gold (pyrite is known as "fool's gold"); iron artifacts (or other material) covered with goldlike iron sulfide layers are said to be pseudogilded (Ref 60, 62, 63).

Lead. In most burial environments, lead artifacts have a relatively slow corrosion rate, because many lead compounds that form in natural burial environments are sparingly soluble or insoluble, and the corrosion products can form an adherent, protective layer on lead. The most common lead corrosion products identified on archaeological lead are carbonates, chlorides, phosphates, and sulfates (Ref 80–82). Table 6 lists some corrosion products identified on archaeological lead.

Lead artifacts exposed to alkaline soil often develop bulky corrosion crusts (Ref 12). In the presence of dissolved carbonate ions, the corrosion products are mainly white lead carbonates (cerussite, hydrocerussite) often discolored by oxides, such as red litharge and yellow massicot (Ref 12, 80, 82–84). In acidic soils, lead objects can become very corroded, particularly in the presence of organic acids, because the corrosion products that form are easily soluble in water (especially at pH < 5) and can be leached away (Ref 81, 82). Under certain soil conditions containing chloride ions and a source of phosphate ions (e.g., from decaying bone), pyromorphite can form (Ref 85).

In natural waters, the corrosion of lead depends on the hardness of the water, its pH, and the presence of other dissolved material. When lead artifacts are exposed to hard waters that contain dissolved minerals, such as carbonates, the surface becomes covered with an adherent, protective layer of insoluble salts, usually calcium carbonate along with cerussite and hydrocerussite (Ref 12, 80). In contrast, soft waters do not usually contain enough dissolved material to allow the formation of a protective layer of corrosion products. As a result, the corrosion rate of lead in soft water can be significant—enough to contaminate drinking water (Ref 81). In addition, soft waters that contain organic acids from plants (e.g., peat bogs) promote lead corrosion because of the formation of soluble lead salts. Lead artifacts exposed to such conditions are unlikely to survive for more than half a century (Ref 12, 81).

In marine environments, lead artifacts generally survive extremely well (e.g., several centuries), because they become covered with a protective layer of insoluble lead salts (Ref 12, 82, 86). They may also become covered with a thin layer of concretion (Ref 14). Lead artifacts recovered from an aerobic marine site are normally a light-gray to cream color (Ref 57). The corrosion layer consists of mainly anglesite along with lesser amounts of lead chlorides (e.g., cotunnite, laurionite, phosgenite), hydrocerussite, and the lead monoxides massicot and litharge (Ref 12, 57, 86).

Under anaerobic conditions where SRB are active, lead artifacts become covered with a black layer of galena, because the lead has reacted with hydrogen sulfide (Ref 57, 63). Galena has been found on lead artifacts recovered from anaerobic waterfront sites along the River Thames in London (Ref 60).

Tin and Pewter. Pure tin artifacts are sometimes found at archaeological sites, but it is more common to find tin associated with bronze (a copper-tin alloy), with pewter (either tin-lead alloys or tin-antimony-copper alloys), or as a coating over copper or iron. Leaded tin alloys (e.g., leaded pewter) are relatively resistant to corrosion because they are often protected by lead corrosion products—either lead sulfate (under aerobic conditions) or lead sulfide (under anaerobic conditions) (Ref 14, 57).

In general, when tin corrodes in aerated environments, it forms tin(IV) compounds, typically cassiterite (Ref 57). If oxygen access is restricted, then tin(II) oxides form, such as romarchite and hydroromarchite (Ref 12, 87). On lead-containing pewters, lead corrosion products can form in addition to tin compounds. Table 7 lists some corrosion products identified on archaeological tin and pewter.

In aerated terrestrial environments, tin alloys corrode, sometimes severely, and become covered with tin oxides, and the color can vary from gray to white (Ref 82, 88). The surface can also be stained green if copper is present in the alloy (Ref 12). Tin artifacts can also undergo intergranular corrosion, making them brittle and fragile (Ref 88). Local corrosion on tin is promoted by anions that form soluble tin salts (e.g., chlorides, sulfates), and the corrosion develops as blisters overlying corrosion pits.

In aerated marine environments, tin alloys tend to be extensively corroded, and their surface becomes covered with mainly cassiterite and rarely schoenfliesite, abhurite, and tin chloride hydroxide (Ref 57, 87). Sometimes, there is no residual metal left, and all that remains is a massive pustular outgrowth of tin corrosion products (Ref 14).

Under anaerobic conditions, tin can be well preserved, as demonstrated by the tin ingots and other tin artifacts recovered from the *Ulu Burun*, a 14th century B.C. shipwreck in Turkey (Ref 89). Well-preserved ingots were covered with a thick, hard, dark-brown patina, presumably tin sulfides, beneath a thick concretion layer (Ref 89). To date, ottemannite (Sn_2S_3) is the only tin sulfide that has been identified on a tin alloy recovered from anaerobic conditions (Ref 57). Other stable tin sulfides exist (e.g., SnS_2, SnS) and may be identified in the future (Ref 14).

The term *tin pest* is sometimes used to describe the low-temperature phase change that causes tin to expand and disintegrate into a coarse, friable gray powder (Ref 90). The change is from the β phase (white tin, density 7.3 g/cm^3) to the α phase (gray tin, density 5.8 g/cm^3), and it occurs when tin is cooled below 13 °C (55 °F). The transformation is difficult to initiate, the rate slow, and the change inhibited if small levels of certain impurities (e.g., bismuth, antimony, lead) are present in the tin (Ref 91). Reported cases of tin pest in museum artifacts are generally just normal atmospheric corrosion accelerated by the presence of high relative humidity and pollutants (Ref 90). However, the occurrence of tin pest was recently confirmed on tin ingots recovered from the cold waters surrounding the *Ulu Burun* shipwreck in Turkey (Ref 87, 89).

Aluminum. The use of aluminum only became widespread after 1886 (when an electrolytic process was developed for extracting the metal from its ore), and so the main aluminum artifacts now being recovered from burial environments are aircraft, particularly aluminum aircraft engines. When aluminum corrodes in a burial environment, its surface becomes covered with a gradually thickening layer of aluminum hydroxides ($Al(OH)_3$) (i.e., bayerite, gibbsite) or aluminum hydroxide oxide ($AlO(OH)$) (i.e., boehmite) (Ref 92). The surface of aluminum

Fig. 3 A corroded iron chain with the fibrous characteristic of wrought iron. Courtesy of H. Unglik, Parks Canada

becomes pitted if the environment is contaminated by chloride ions. Aluminum alloys, particularly those that contain copper, can suffer more serious corrosion problems (including perforation) because of galvanic corrosion between intermetallic compounds (e.g., $CuAl_2$) and the aluminum (Ref 14). Because of the relatively few aluminum artifacts recovered from burial environments, little information is available about corrosion products identified on such artifacts. In one instance, aluminum hydroxide (presumably bayerite or gibbsite) was identified on an aluminum engine recovered from a freshwater site (Ref 93).

Corrosion after Excavation

Artifacts may deteriorate rapidly when they are excavated and subjected to a significant change in environment. The form of deterioration depends on the type of metal, its burial environment, and its post-excavation environment (e.g., humidity, temperature, pollutants, handling). Metal artifacts can suffer further deterioration after excavation if they have formed corrosion products that are unstable in damp air, or if they have become contaminated with soluble salts during burial. Salts and other soluble material present in the surface layers of metal artifacts during burial will concentrate and possibly crystallize if the artifact is allowed to dry after excavation. If these salts are hygroscopic, they can promote ongoing corrosion if the surrounding relative humidity is sufficient for them to pick up moisture.

Indoor Environmental Factors. New corrosion problems on excavated metal artifacts can be caused by environmental factors present in a controlled or uncontrolled indoor environment. Exposure of metal artifacts to high relative humidities in an unheated warehouse, for example, will cause most of them to corrode. A well-known example of this is the flash rusting of bare iron; when iron is exposed to damp conditions, it undergoes rapid corrosion and forms bright-orange lepidocrocite. Iron(II) corrosion products identified on archaeological iron appear to be stable if kept dry, but when exposed to damp conditions, they may oxidize. Vivianite may change to strengite, and siderite may convert to iron hydroxide oxides (Ref 39, 76). Although exposure of metal artifacts to light is not usually a problem, color changes have been observed on the light-sensitive silver chloride patina on silver artifacts (Ref 12, 94).

Indoor pollutants can also stimulate new corrosion problems on excavated metals (Ref 95, 96). The key ones are hydrogen sulfide and volatile organic acids, such as acetic (ethanoic) and formic (methanoic) acids (Ref 97, 98). Hydrogen sulfide is released by food, rubber vulcanized with sulfur, protein-based glues, wool, and wooden artifacts excavated from anaerobic burial environments (Ref 99, 100). Hydrogen sulfide can react with silver artifacts to form black whiskers of silver sulfide and with copper artifacts to form black spots of various copper sulfides (Ref 100, 101). Volatile organic acids are given off by wood (with oak and cedar being the worst), poly(vinyl acetate) adhesives, oil-based paints, and certain sealants and plastics (e.g., cellulose acetate) (Ref 95, 96). Acetic acid and other volatile organic acids can cause copper, iron, lead, and zinc to corrode (Ref 95, 96). Lead is particularly susceptible to corrosion when exposed to acetic acid; lead artifacts become covered with powdery white corrosion consisting mainly of hydrocerussite (Ref 102). For more information on the indoor corrosion of metal artifacts, see the article "Corrosion of Metal Artifacts and Works of Art in Museum and Collection Environments" in this Volume.

Specific Corrosion Problems after Excavation

Corrosion products that formed during burial may no longer be stable in a new postexcavation environment. Contamination, especially by chloride ions, may cause continued corrosion of any remaining metal. Initially, no corrosion problem may be apparent, but over time, new corrosion products can form and damage the artifact. New solids that occupy an increased volume will physically stress the artifact, and new soluble material, especially acid reaction products, can chemically attack the remaining metal. Three important corrosion problems (caused by corrosion processes during burial) that can develop after excavation are weeping and akaganéite formation on iron, bronze disease on copper alloys, and the oxidation of metal sulfides.

Iron (Weeping, Akaganéite Formation). During burial, iron artifacts often become contaminated with an acidic iron(II) chloride solution. If these artifacts are left to dry, then any remaining iron will continue to corrode in a process that can eventually destroy the shape of the artifact (Ref 21, 74, 76). An acid regeneration cycle, proposed by Askey et al. for iron contaminated with hydrochloric acid, can be used to explain why archaeological iron disintegrates during storage if left contaminated with chloride ions (Ref 103):

$$2Fe\,(s) + 4H^+\,(aq) + 4Cl^-\,(aq) + O_2\,(g) \rightarrow$$
$$2Fe^{2+}\,(aq) + 4Cl^-\,(aq) + 2H_2O$$

$$2Fe^{2+}\,(aq) + 4Cl^-\,(aq) + 3H_2O + \tfrac{1}{2}O_2\,(g) \rightarrow$$
$$2FeOOH\,(s) + 4H^+\,(aq) + 4Cl^-\,(aq)$$

When an acidic solution of iron(II) chloride is exposed to air, the iron(II) ions in solution can undergo hydrolysis, and oxidation to iron(III) ions and new compounds can form (Ref 74, 76). The crucial factor in this cycle is that the chloride ion forms a soluble salt with iron(II) ions, and it is this solubility that allows the cycle to proceed. If the iron chloride salt were insoluble, it would precipitate, and the cycle would be broken.

The residual soluble ions support ongoing corrosion that causes chemical damage to any remaining iron metal and physical damage to the shape of the object (Ref 74). The chemical damage is caused by the increased acidity that supports the reaction that consumes the remaining iron. The physical damage is caused by the formation of solid iron hydroxide oxides (e.g., goethite, lepidocrocite, akaganéite) within the corrosion layers; these corrosion products have molar volumes approximately three times greater than the molar volume of iron metal (Ref 104, 105). The formation of any of these iron oxyhydroxides within or below the corrosion layer causes stresses, cracks, and damage, which in turn allow easier access for oxygen and moisture, and continued corrosion. Visual symptoms of continuing corrosion induced by chloride contamination on archaeological iron are the formation of akaganéite and weeping.

Akaganéite Formation. The formation of characteristic bright red-brown elongated particles of akaganéite at exposed metal surfaces on excavated archaeological iron is a symptom of active corrosion (Ref 21, 74, 106, 107). Akaganéite particles form at the metal surface underneath the corrosion products and push them off. Figure 4 contains an image of akaganéite (Ref 74).

Laboratory studies of the formation of iron hydroxide oxides from acidic iron(II) chloride solutions have demonstrated that akaganéite precipitates when the chloride ion concentration is high (~2 M) and that goethite and/or lepidocrocite precipitate when the chloride ion concentration is low (Ref 104). Studies of akaganéite have determined that chlorine atoms are present in the crystal structure (approximately 0.2 wt% Cl), located inside tunnels, and a more representative chemical formula for akaganéite is suggested to be $FeO_{0.833}(OH)_{1.167}Cl_{0.167}$ (Ref 108–110). Because akaganéite only forms under conditions of relatively high concentrations of chloride ions, the identification of akaganéite on archaeological iron is an indication

Fig. 4 Scanning electron micrograph of akaganéite on an archaeological iron fragment (gold coated). 600×; picture width, 190 µm. Adapted from Ref 74

that the object is contaminated with high levels of chloride ions (Ref 110).

Weeping. Another symptom of a corrosion problem on excavated iron is the formation of either wet bubbles of acidic liquid (referred to as weeping or sweating iron) or dry, hollow, red spherical shells on an artifact surface (Ref 12, 74, 111). Weeping is attributed to the hygroscopic nature of iron chloride salts (Ref 106). Iron(II) chloride and iron(III) chloride are both hygroscopic and form a series of salts with different waters of hydration, depending on the relative humidity. Iron(II) chloride, for example, deliquesces above a relative humidity of 56% (Ref 112). When the relative humidity is high, these salts absorb water, dissolve, and form wet droplets of orange liquid (Ref 74). Iron hydroxide oxides precipitate around the outside of the droplets (because iron(II) ions in solution undergo oxidation and hydrolysis) and generate a framework for the spherical shells. An example of these hollow shells is shown in Fig. 5 (Ref 74).

Copper Alloys (Bronze Disease). The term *bronze disease* has long been used to describe the appearance of powdery light-green spots on the surface of archaeological copper alloys (Ref 113, 114). Artifacts exhibiting this form of corrosion were once said to be diseased or sick, and the cause was attributed to bacterial or fungal infection (Ref 113). However, it is now known that this corrosion is caused by chloride contamination, either as solid nantokite or soluble chloride ions (Ref 37, 115, 116).

Bronze disease is defined as "the process of interaction of chloride-containing species within the bronze patina with moisture and air" (Ref 116) and also as "a progressive deterioration of ancient copper alloys caused by the existence of cuprous chloride (nantokite) in close proximity to whatever metallic surface may remain" (Ref 37).

The main source of bronze disease on terrestrial archaeological copper alloys is nantokite, usually located adjacent to remaining metal and beneath the outer layers of stable corrosion products. As long as nantokite is isolated from the surrounding air by outer layers of corrosion, it remains unreacted. However, if the protective outer material covering the nantokite is cracked, damaged, or removed, the nantokite is exposed to the surrounding environment. Once exposed, it can react with moisture in the air to form green copper chloride hydroxides (e.g., atacamite, clinoatacamite, paratacamite)—the symptoms of bronze disease. A schematic diagram of bronze disease is shown in Fig. 6 (Ref 114).

Two proposed reactions for the oxidation and hydrolysis of nantokite are (Ref 116, 117):

$$4CuCl(s) + O_2(g) + 4H_2O \rightarrow$$
$$2Cu_2Cl(OH)_3(s) + 2H^+ + 2Cl^-$$

$$6CuCl(s) + 1\tfrac{1}{2}O_2(g) + 3H_2O \rightarrow$$
$$2Cu_2Cl(OH)_3(s) + 2Cu^{2+} + 4Cl^-$$

The reaction products are green copper chloride hydroxides and soluble ions, either hydrochloric acid or copper(II) chloride (both of which promote further corrosion of the remaining metal).

On archaeological copper alloys recovered from marine environments, nantokite rarely precipitates (Ref 14). The artifacts are instead contaminated with chloride ions that have accumulated during burial. These chloride ions, if not removed, can stimulate copper corrosion and generate green copper chloride hydroxides, the same corrosion products that form in bronze disease on terrestrial copper alloys from the reactions of nantokite (Ref 115).

Metal Sulfides. Metals recovered from anaerobic environments where SRB are active can be covered with various metal sulfides, with pyrite being common (Ref 60, 118). Many metal sulfides are unstable in moist air, and they can be oxidized to metal sulfates (Ref 75, 119). The process may even be pyrophoric (Ref 120). Antlerite and guildite, for example, have been identified in spots of green corrosion formed on excavated artifacts covered with copper and iron sulfides (Ref 60).

The problems associated with the oxidation of pyrite are described as pyrite disease when observed in geological and palaeontological collections (Ref 118, 121). Pyrite can be oxidized in moist air (above approximately 60% relative humidity), and the initial reaction is thought to be (Ref 119, 121):

$$2FeS_2(s) + 7O_2(g) + 2H_2O \rightarrow$$
$$2H_2SO_4 + 2FeSO_4(s)$$

The predominant oxidation products are sulfuric acid and various hydrated iron(II) sulfates (e.g., melanterite, rozenite). The iron(II) sulfates can be further oxidized to iron(III) hydroxide sulfates (e.g., butlerite). The volume change associated with the oxidation of the sulfides to sulfates causes mechanical damage to any nearby material. Chemical damage to any underlying material (e.g., iron, wood) is caused by the formation of sulfuric acid, because, besides being acidic, it is also hygroscopic, absorbing moisture at low relative humidity (Ref 122).

Timber and wooden artifacts from ancient shipwrecks can become contaminated by SRB activity. During the 333 years that the 17th century warship *Vasa* (1628) lay at the bottom of Stockholm Harbor, hydrogen sulfide diffused into the timber and transformed into elemental sulfur (Ref 123, 124). The wood in the *Vasa* is now suffering severe damage from sulfuric acid formed by the oxidation of that sulfur (Ref 123, 124). When metals are present in the shipwreck, the wood can become contaminated with metal sulfides. Rozenite, for example, has been identified on wood from Viking ships recovered in Denmark (Ref 125), and melanterite has formed on wood from the *Vasa* (Ref 123). These wooden artifacts can suffer problems after excavation if the humidity is high enough to oxidize the metal sulfides and convert them to sulfates. Butlerite, jarosite, and natrojarosite have formed by this process on wood from the *Batavia* (1629), which sank off the coast of Australia (Ref 79).

Archaeological Conservation

Conservation deals with the preservation of cultural heritage artifacts. Archaeological conservation deals specifically with the preservation of excavated material. The two principal goals of the archaeological conservator are to preserve archaeological material (by stabilizing and protecting it from further damage) and to reveal information about an artifact (shape, material, method of manufacture) and its past (history during use and burial). Many professionals (archaeologists, conservators, scientists, and museum curators) work together to achieve these goals.

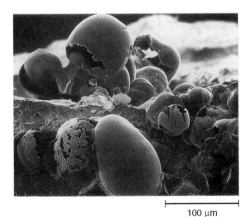

Fig. 5 Scanning electron micrograph of the hollow shells of weeping on archaeological wrought iron. 100×; picture width, 300 μm. Adapted from Ref 74

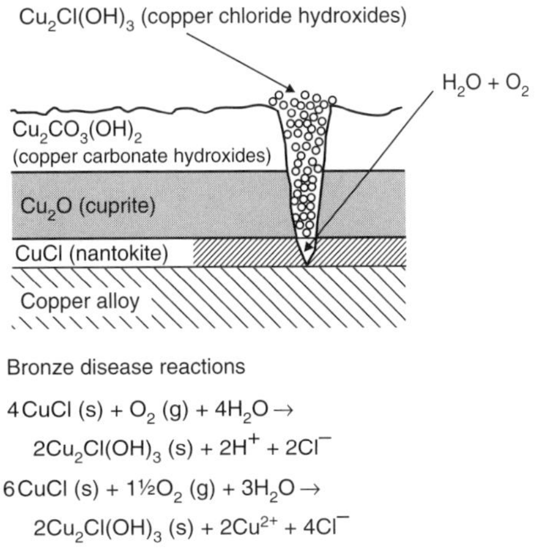

Fig. 6 Schematic diagram of bronze disease on archaeological copper alloys. Adapted from Ref 114

Archaeologists study artifacts recovered from archaeological sites to learn about the past, and they are most interested in the original shape of an artifact. Coins, for example, are useful to archaeologists for dating a site. Archaeological conservators use a variety of techniques to remove surface corrosion and uncover the shape of and inscriptions on an artifact. Conservators preserve metal artifacts by storing them at low relative humidity or by treating them to remove corrosion stimulants (Ref 12). This is particularly important for archaeological iron, because if it is contaminated by chloride ions, it can continue to corrode, and, if left unchecked, this corrosion can completely destroy an artifact. Several books on conservation detail the various options available for treating archaeological metal artifacts (Ref 12, 126–131).

Conservation Strategies

In general, archaeological conservators examine freshly excavated metal artifacts and establish a record for each artifact to track relevant information about the artifact, including treatment details and storage recommendations. Such a record can contain some or all of the following information:

- Description of artifact, where and when it was found, identification number
- Line drawings, photographs, x-radiographs
- Examination details
- Condition before treatment
- Results of any analytical work
- Treatment details
- Recommendations for storage and long-term care

Metal artifacts, particularly iron, are x-radiographed after excavation to record their shape and help with identification (Ref 12). X-radiographs are also useful in detailing the extent of corrosion and the amount of remaining metal. Figure 7 contains photographs of a 17th century cross from an archaeological site in Ferryland, Newfoundland. Figure 7(a) shows the cross before cleaning, Fig. 7(b) shows an x-radiograph, and Fig. 7(c) shows the cross after cleaning.

Not all artifacts receive the same degree of conservation treatment, because of considerations relating to the cost of conservation versus the importance of artifacts. One site, for example, may yield thousands of similar wrought iron nails (Ref 4). A high degree of intervention may be carried out on important objects for display, while only minimal work may be done on artifacts intended for storage.

Conservation treatments are developed to suit individual artifacts, with the extent of intervention kept to a minimum and reversible treatments used where possible. Corrosion products are often removed in order to reveal the shape of the object. Conservators avoid the use of high temperatures to treat a metal artifact so as not to destroy its technological history preserved in the metallurgical structure (Ref 132). When developing a conservation treatment for individual artifacts, the archaeological conservator takes into account the following:

- Type of metal and its condition
- Need to preserve evidence in soil or corrosion layers
- Expectations and intentions of the user (e.g., archaeologist, curator, museum)
- Aesthetics
- Consideration of future location of artifact (e.g., storage or display conditions)
- Availability of equipment, materials, and funding

Storage Prior to Treatment. When metal artifacts are first excavated, they are stored temporarily until they can be looked at in a conservation laboratory and a decision made about long-term storage or treatment. In general, artifacts recovered from a dry environment are left dry (and are often stored in a dry or desiccated environment), and those from a damp site are kept either damp or wet and are sometimes refrigerated or frozen (Ref 12, 131, 133).

Iron artifacts are particularly sensitive to rapid corrosion after exposure to air, especially those from marine environments. There are examples of freshly excavated marine cast iron cannon balls and cannons heating up because of rapid exothermic corrosion reactions; sometimes, the

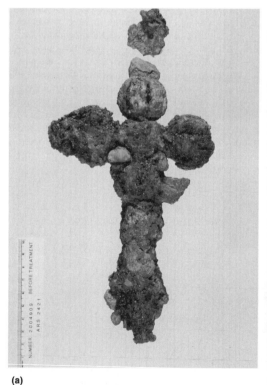 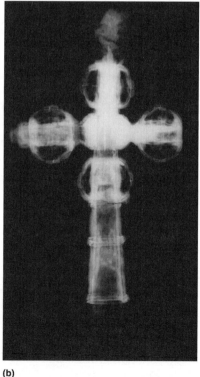 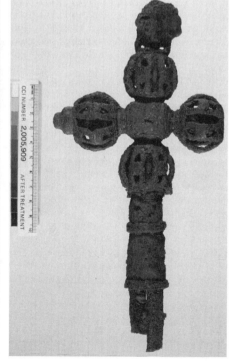

Fig. 7 A 17th century gilded wrought iron cross with brass inlay from an archaeological site in Ferryland, Newfoundland. (a) Before cleaning. (b) X-radiograph. (c) After mechanical cleaning. Courtesy of Lil Hawkins and the Canadian Conservation Institute

surface layers have been observed to literally explode when the cast iron is allowed to dry (Ref 115, 133). Excavated iron artifacts are therefore usually immersed in water temporarily until they can be transferred to a more suitable storage solution prior to conservation treatment. Gilberg studied the storage of cast iron cannon balls in water purged with nitrogen gas and in solutions containing the oxygen scavengers hydrazine or sodium sulfite (Ref 134). The sodium sulfite solution proved the most effective at removing oxygen, but it also contributed to an increased corrosion rate of the iron, because the solution supported the growth of SRB introduced along with the cannon balls. Costain studied alkaline solutions for temporarily storing excavated marine and terrestrial iron and recommended alkaline solutions of either alkaline sulfite (a mixture of sodium hydroxide and sodium sulfite) or sodium hydroxide (Ref 135).

Cathodic protection is now being used to minimize corrosion on iron artifacts awaiting conservation treatment. Sacrificial anode or galvanic cathodic protection is being used to protect marine iron artifacts in seawater (Ref 136, 137). Aluminum anodes, for example, were used on the wrought iron anchor from the *HMS Sirius* (1790), and zinc anodes were used on the cast and wrought iron engine from the *SS Xantho* (1872) (Ref 136). Impressed current cathodic protection is being used (in 2004) to protect the *H.L. Hunley* (1864), a wrought and cast iron submarine that was raised off the coast of Charleston, SC, in 2000 (Ref 138).

Mechanical Cleaning. Conservators often use small hand tools (e.g., scalpels, scrapers, wooden sticks, soft brushes, hammers, chisels) to remove material (e.g., soil, extraneous material, concretions, corrosion products) from archaeological metals (Ref 12, 115). Depending on the nature of this material, they may also need to use power tools (e.g., engraving tools, rotary tools, pneumatic chisels) or air-abrasive units (Ref 139). Such methods provide conservators with a controlled approach to uncovering the shape of an artifact (without adding any additional chemicals) and allow them to search for evidence of plating, inlays, enamels, and other associated materials. They remove material until they reach a level (often a discrete, recognizable boundary) that has preserved the original surface of the artifact, allowing them to determine what the artifact is (Ref 12). The best information, including the shape of the artifact, is often retained in the corrosion layers rather than at the surface of any remaining metal.

Desalination. Archaeological objects recovered from burial environments can be contaminated by salts. The concentration and range of dissolved salts is high in artifacts from marine environments and generally less so for artifacts from terrestrial or freshwater environments. Dissolved salts are located in aqueous solutions filling the porosity of the corrosion products and the cracks, voids, and casting porosity in the remaining metal. If these salts are present in high enough concentration and the artifact is allowed to dry, then crystals of salt precipitate. If seawater is left to evaporate, the most insoluble salts precipitate first and the most soluble ones last, starting with calcium carbonate, followed by gypsum, sodium chloride, magnesium chloride, magnesium sulfate, sodium bromide, and finally potassium chloride (Ref 11). Salt crystallization can cause damage by exerting pressure on surrounding material; more damage can occur if the salts are subjected to fluctuations in relative humidity and go through cycles of deliquescence and crystallization. For some artifacts, soaking or rinsing in water is sufficient to remove soluble salts (Ref 12, 115). For other artifacts, particularly those made from iron alloys or copper alloys, chemicals other than water or a different conservation treatment is needed to specifically remove chloride ion contamination.

Iron artifacts are seriously affected by chloride ion contamination, and research is ongoing about the best approach to remove the contamination (Ref 21). Removal of significant amounts of chloride ions is rarely achieved by soaking in water (Ref 140). One factor contributing to the difficulty in removal of chlorides in terrestrial iron is that they are trapped beneath a hard corrosion crust, making it difficult for any treatment solution to diffuse in or for the chloride ions to diffuse out (Ref 141). Better success has been achieved with alkaline solutions, where various factors contribute to making it easier for the chloride ions to diffuse out of the artifact into the treatment solution (Ref 142). The two alkaline solutions commonly used to treat archaeological iron are sodium hydroxide (Ref 142) and alkaline sulfite (Ref 143).

Copper alloy artifacts recovered from a marine environment are heavily contaminated with chloride ions, and these are removed by extended washing (Ref 37, 115). When the patina needs to be preserved, then washing is done in water or in a sodium sesquicarbonate solution (Ref 144). When the corrosion can be stripped, washing is done using other chemicals, such as sodium dithionite, a strong reducing agent (Ref 145).

Chemical Treatments. In the past, metal artifacts were routinely stripped using chemicals that dissolved all the corrosion products, removed surface detail and evidence of plating, and left only bare metal (Ref 12). The remaining metal remnants were often pitted and bore little resemblance to the shape of the original object (Ref 37). This practice of chemical stripping has now been widely abandoned in favor of other approaches that uncover the original shape of an artifact, which is often preserved in the corrosion products. Chemical treatments are still used in conservation treatments, sometimes to soften concretions or external corrosion crusts, thus making mechanical removal easier, and other times to help in the removal of chloride contamination (Ref 12, 115).

Coins are often recovered as a hoard (a mass of coins) concreted together by corrosion products. Copper alloy coins, including ones with a high silver content, can be encased in green copper corrosion products. Various chemicals have been used to separate the mass of coins into individual ones, which are then treated to remove the corrosion and reveal the inscriptions (Ref 146). A solution of alkaline Rochelle salt (sodium hydroxide plus potassium sodium tartrate) has been found to be effective and cause the least damage (Ref 147).

Sequestering agents are sometimes used to dissolve corrosion products. They are used with caution because they can etch the metal and, if not well rinsed after use, can cause corrosion problems. Two common sequestering agents used in conservation are ethylenediaminetetraacetic acid (EDTA) and diethylenetriaminepentaacetic acid (DTPA); these are most often used to treat lead but have also been used on metals such as copper alloys (Ref 12, 37).

Treatment of Bronze Disease. Bronze disease is usually treated by immersing artifacts in 3 wt/vol% benzotriazole (a copper corrosion inhibitor) in ethanol under vacuum (Ref 148). Small outbreaks can also be sealed under a layer of silver chloride by reacting nantokite with an ethanol paste of silver oxide (Ref 149). In the past, bronze disease was treated by removing all corrosion products by stripping with various chemicals, many of which were sodium base (Ref 150). Another approach was to use chemicals that interacted with nantokite (the main source of bronze disease) but left the patina relatively unaffected. Sodium carbonate (Ref 151) and sodium sesquicarbonate (Ref 152) solutions were used to convert nantokite to cuprite, and acetonitrile solutions were used to dissolve nantokite (Ref 145).

A new corrosion problem is now appearing on copper alloy artifacts that were treated in the past with sodium-base solutions. Fresh growths of blue corrosion, identified as chalconatronite, are forming because of reactions involving the copper alloy and residual sodium ions (Ref 153). These local spots of blue corrosion have been mistaken for bronze disease.

Electrolytic Techniques. Various electrolytic techniques are used for cleaning and stabilizing metal artifacts. These techniques are usually applied to artifacts with a substantial metal core.

Electrolytic cleaning is used to remove corrosion products from metal artifacts. An electrochemical cell is set up, with the artifact as cathode, an appropriate metal as anode, and a suitable electrolyte. Using a battery or other power supply, electrolysis is carried out and hydrogen gas is generated at the surface of the artifact by the reduction of water, $2H_2O + 2e^- \rightarrow 2OH^- + H_2$ (g); these conditions may also cause the corrosion products to be reduced to compounds with a lower oxidation state or to the metallic state (Ref 115). The generation of hydrogen gas at the metal interface is usually effective in mechanically removing the overlying corrosion layers (Ref 12). The disadvantage of this process is that it removes all associated material from the metal surface, including plating, other surface treatments, and information about the original surface of the artifact. Electrolysis can also be used to remove concretions that completely encase a marine artifact (Ref 139) or to locally treat small areas on an artifact surface.

A modification of this cleaning technique is termed galvanic cleaning or electrochemical cleaning (Ref 12, 126). Here, an electrochemical cell is established using an active metal (e.g., aluminum, zinc) as the anode in contact with an artifact (a more noble metal) and immersing them in a suitable electrolyte. Galvanic cleaning was used, for example, to reveal the details of a 7th century A.D. bronze bowl; chalcopyrite (considered dangerous because it may oxidize and form sulfuric acid) was removed by applying a sodium hydroxide solution to a small area and covering it with aluminum foil for short times (Ref 61).

Electrolytic reduction techniques are also used in the removal of chloride ions from contaminated artifacts (Ref 115). Cast iron cannons are often treated using electrolytic reduction (Ref 154). When corroded iron is subjected to electrolysis, the iron hydroxide oxides are thought to be reduced to magnetite, which increases the porosity of the corrosion products and increases the rate at which chloride ions diffuse out of the artifact into the treatment solution (Ref 115, 142, 155, 156). When the removal of corrosion is not wanted, such as in cast iron where the shape of the artifact is maintained by graphite, more controlled reduction techniques (e.g., cathodic reduction at constant potential) are used to avoid the generation of hydrogen gas (Ref 157). Many of the copper alloy artifacts recovered from the *RMS Titanic* (1912) were desalinated using cathodic reduction techniques that avoided the generation of hydrogen bubbles (Ref 158). Recently, cathodic reduction techniques have been used to treat aluminum recovered from marine environments (Ref 93, 159, 160).

Another use of electrolytic reduction techniques is to preserve corroded material by reducing corrosion products back to a cohesive metallic state rather than nonadherent material that is easily brushed away. The reduction process is carried out in a controlled manner that avoids generating hydrogen bubbles and consolidates the corrosion products into a coherent but porous metallic network (Ref 161). There is growing use of cathodic reduction at constant potential for preserving surface details in corroded lead artifacts (Ref 162, 163). A few mineralized archaeological silver artifacts have also been reduced to metallic silver using electrolytic reduction (Ref 164).

Repairs and Reconstruction. Conservators do not try to make the artifact look as it did when new, but they do carry out various repairs or reconstructions to make an artifact more understandable to the viewer and safer to handle. A conservator may reconstruct an artifact by reassembling broken pieces into the shape and form of the original artifact and then filling in missing areas. Filled areas are painted to match the surrounding surface corrosion (e.g., with powdered iron oxide pigments to match iron corrosion). Repair materials or coatings are chosen so they can be removed at a later date, if necessary. Common adhesives are ones containing acrylics (Ref 12). Metal artifacts are sometimes coated with a wax or lacquer after treatment. The popular lacquers are based on stable acrylic resins, such as Acryloid (Paraloid) B-72 or Acryloid B-48 (Rohm and Haas Co.), or Incralac (Albi Manufacturing Division of Stan-Chem Inc.), which contains Acryloid B-44 and benzotriazole (Ref 12). Acryloid B-48 is noted for its adhesion to metal.

Long-Term Storage and Display. Metal artifacts that are considered stable (because they have successfully undergone treatment or are considered relatively inert) are usually stored or displayed along with the general collection inside museums, where the relative humidity is controlled to stay below 60% (Ref 96). Metal artifacts that are considered unstable or are showing signs of instability are usually stored under controlled low relative humidity. Bronze disease on copper alloys can usually be prevented by keeping them at relative humidity below 46% (Ref 116). Active corrosion on archaeological iron contaminated with chlorides is thought to be stopped when artifacts are kept at a relative humidity below 20%, although some artifacts may still continue to corrode (Ref 106, 165). Now being tested for storage of archaeological artifacts are commercially available scavengers that create an environment inside a transparent enclosure (e.g., a heat-sealable barrier film) that is extremely dry and oxygen-free (Ref 166).

In Situ Preservation. Consideration is now being given to in situ preservation of artifacts because of the high cost of conservation, the growing number of archaeological sites, and the need for proper artifact storage after conservation (Ref 167, 168). For a site to be suitable for in situ preservation, it must be safe to keep artifacts in that burial environment and leave them for future generations to excavate. Monitoring of environmental parameters, as is being done at the Nydam Mose site in Denmark (Ref 167, 169), will be necessary to ensure that a burial environment remains safe or to detect changes in the local environment that put metal artifacts at risk of accelerated deterioration (Ref 3, 13).

ACKNOWLEDGMENTS

The author is grateful to the following people for their comments on this article: Bob Barclay, Paul Begin, David Grattan, Judy Logan, Paul Mardikian, Cathy Mathias, Charlotte Newton, and Tom Stone. The author also wishes to thank Lil Hawkins for permission to use photographs of the cross from Ferryland, and Henry Unglik for the use of his photograph.

REFERENCES

1. R.F. Tylecote, The Effect of Soil Conditions on the Long-Term Corrosion of Buried Tin-Bronzes and Copper, *J. Archaeolog. Sci.*, Vol 6 (No. 4), 1979, p 345–368
2. P. Throckmorton, Ed., *The Sea Remembers; Shipwrecks and Archaeology from Homer's Greece to the Rediscovery of the Titanic*, Weidenfeld and Nicolson, 1987
3. M. Fjaestad, I. Ullén, A.G. Nord, K. Tronner, G.C. Borg, and M. Sandberg, Are Recently Excavated Bronze Artifacts More Deteriorated than Earlier Finds?, *Metal 98*, W. Mourey and L. Robbiola, Ed., James & James Science Publishers, 1998, p 71–79
4. C. Mathias, A Conservation Strategy for a Seventeenth Century Archaeological Site at Ferryland, Newfoundland, *J. Int. Inst. Conserv.—Can. Group*, Vol 19, 1994, p 14–23
5. R. Raiswell, Defining the Burial Environment, *Handbook of Archaeological Sciences*, D.R. Brothwell and A.M. Pollard, Ed., John Wiley & Sons, 2001, p 595–604
6. R.M. Garrels and C.L. Christ, *Solutions, Minerals, and Equilibria*, Harper & Row, 1965
7. D.C. Silverman, Aqueous Corrosion, *Corrosion: Fundamentals, Testing, and Protection*, Vol 13A, *ASM Handbook*, ASM International, 2003, p 190–195
8. J.O. Harris and D. Eyre, Soil in the Corrosion Process, *Corrosion*, Vol 1, 3rd ed., L.L. Shreir et al., Ed., Butterworth-Heinemann, 1994, p 2:73–2:86
9. M.J. Wilmott and T.R. Jack, Corrosion by Soils, *Uhlig's Corrosion Handbook*, 2nd ed., R.W. Revie, Ed., John Wiley & Sons, 2000, p 329–348
10. J.C. Rowlands, Sea Water, *Corrosion*, Vol 1, 3rd ed., L.L. Shreir et al., Ed., Butterworth-Heinemann, 1994, p 2:60–2:72
11. M.-L.E. Florian, The Underwater Environment, *Conservation of Marine Archaeological Objects*, C. Pearson, Ed., Butterworths, 1987, p 1–20
12. J.M. Cronyn and W.S. Robinson, *The Elements of Archaeological Conservation*, Routledge, 1990
13. M. Fjaestad, A.G. Nord, K. Tronner, I. Ullén, and A. Lagerlöf, Environmental Threats to Archaeological Artefacts, *ICOM Committee for Conservation, 11th Triennial Meeting*, J. Bridgland, Ed., Sept 1996 (Edinburgh), James & James Science Publishers, 1996, p 870–875
14. N.A. North and I.D. MacLeod, Corrosion of Metals, *Conservation of Marine Archaeological Objects*, C. Pearson, Ed., Butterworths, 1987, p 68–98
15. S.C. Dexter, Microbiologically Influenced Corrosion, *Corrosion: Fundamentals, Testing, and Protection*, Vol 13A, *ASM Handbook*, ASM International, 2003, p 398–416

16. J.-D. Gu, T.E. Ford, and R. Mitchell, Microbiological Corrosion of Metals, *Uhlig's Corrosion Handbook*, 2nd ed., R.W. Revie, Ed., John Wiley & Sons, 2000, p 915–927
17. E. Protopopoff and P. Marcus, Potential versus pH (Pourbaix) Diagrams, *Corrosion: Fundamentals, Testing, and Protection*, Vol 13A, *ASM Handbook*, ASM International, 2003, p 17–30
18. T.W. Farrer, L. Biek, and F. Wormwell, The Role of Tannates and Phosphates in the Preservation of Buried Iron Objects, *J. Appl. Chem.*, Vol 3 (No. 2), 1953, p 80–84
19. M.B. McNeil and B.J. Little, The Use of Mineralogical Data in Interpretation of Long-Term Microbiological Corrosion Processes: Sulfiding Reactions, *J. Am. Inst. Conserv.*, Vol 38 (No. 2), 1999, p 186–199
20. W. Lee, Z. Lewandowski, P.H. Nielsen, and W.A. Hamilton, Role of Sulfate-Reducing Bacteria in Corrosion of Mild Steel: A Review, *Biofouling*, Vol 8, 1995, p 165–194
21. S. Turgoose, Structure, Composition and Deterioration of Unearthed Iron Objects, *Current Problems in the Conservation of Metal Antiquities*, A. Aoki, Ed., Tokyo National Research Institute of Cultural Properties, 1993, p 35–52
22. M. Pourbaix, *Atlas of Electrochemical Equilibria in Aqueous Solutions*, 2nd ed., National Association of Corrosion Engineers, 1974
23. M. Pourbaix, Electrochemical Corrosion and Reduction, *Corrosion and Metal Artifacts: A Dialogue Between Conservators and Archaeologists and Corrosion Scientists*, B.F. Brown et al., Ed., NBS Special Publication 479, National Bureau of Standards, 1977, p 1–16
24. F. Schweizer, Bronze Objects from Lake Sites: From Patina to "Biography," *Ancient and Historic Metals, Conservation and Scientific Research*, D.A. Scott et al., Ed., Getty Conservation Institute, 1994, p 33–50
25. D.E. Hughes, The Microbiology of Corrosion, *Corrosion*, Vol 1, 3rd ed., L.L. Shreir et al., Ed., Butterworth-Heinemann, 1994, p 2:87–2:98
26. N.A. North, Formation of Coral Concretions on Marine Iron, *Int. J. Naut. Archaeol. Underwater Explor.*, Vol 5 (No. 3), 1976, p 253–258
27. I.D. MacLeod, Formation of Marine Concretions on Copper and Its Alloys, *Int. J. Naut. Archaeol. Underwater Explor.*, Vol 11 (No. 4), 1982, p 267–275
28. P. Stoffyn-Egli and D.E. Buckley, The Micro-World of the *Titanic*, *Chem. Br.*, Vol 31 (No. 7), 1995, p 551–553
29. S. Turgoose, The Corrosion of Archaeological Iron during Burial and Treatment, *Stud. Conserv.*, Vol 30 (No. 1), 1985, p 13–18
30. D.A. Scott, *Metallography and Microstructure of Ancient and Historic Metals*, Getty Conservation Institute and the J. Paul Getty Museum, 1991
31. D. Killick, Science, Speculation and the Origins of Extractive Metallurgy, *Handbook of Archaeological Sciences*, D.R. Brothwell and A.M. Pollard, Ed., John Wiley & Sons, 2001, p 483–492
32. J. Day and R.F. Tylecote, Ed., *The Industrial Revolution in Metals*, The Institute of Metals, 1991
33. J.B. Lambert, Metals, *Traces of the Past; Unraveling the Secrets of Archaeology through Chemistry*, Perseus Books, 1997, p 168–213
34. R. Raymond, *Out of the Fiery Furnace; The Impact of Metals on the History of Mankind*, Pennsylvania State University Press, 1984
35. R.F. Tylecote, *A History of Metallurgy*, The Metals Society, 1976
36. R.F. Tylecote, *The Early History of Metallurgy in Europe*, Longman, 1987
37. D.A. Scott, *Copper and Bronze in Art: Corrosion, Colorants, Conservation*, Getty Publications, 2002
38. S.A. Matthes, *Corrosion: Fundamentals, Testing, and Protection*, Vol 13A, *ASM Handbook*, ASM International, 2003, p 34–41
39. H. Matthiesen, L.R. Hilbert, and D.J. Gregory, Siderite as a Corrosion Product on Archaeological Iron from a Waterlogged Environment, *Stud. Conserv.*, Vol 48 (No. 3), 2003, p 183–194
40. D. Gregory, In Situ Corrosion Studies on the Submarine *Resurgam*: A Preliminary Assessment of Her State of Preservation, *Conserv. Manage. Archaeolog. Sites*, Vol 4 (No. 2), 2000, p 93–100
41. I.D. MacLeod, In Situ Corrosion Monitoring of the Iron Shipwreck *City of Launceston* (1865), *ICOM Committee for Conservation, 13th Triennial Meeting*, R. Vontobel, Ed., Sept 2002 (Rio de Janeiro), James & James Science Publishers, 2002, p 871–877
42. I.D. MacLeod, In Situ Corrosion Studies on Iron Shipwrecks and Cannon: The Impact of Water Depth and Archaeological Activities on Corrosion Rates, *Metal 98*, W. Mourey and L. Robbiola, Ed., James & James Science Publishers, 1998, p 116–124
43. I.D. MacLeod, In Situ Corrosion Studies on the Duart Point Wreck, 1994, *Int. J. Naut. Archaeol.*, Vol 24 (No. 1), 1995, p 53–59
44. C. Mathias, Assessment of Corrosion Measurements for Soil Samples Excavated at a Seventeenth-Century Colonial Plantation Site, *Archaeological Conservation and Its Consequences*, A. Roy and P. Smith, Ed., International Institute for Conservation, 1996, p 121–126
45. C. Mathias, Examination of the Interaction between Ferrous Metals and the Archaeological Burial Environment for a Seventeenth-Century Plantation Site, *Sixth International Conference on Non-Destructive Testing and Microanalysis for the Diagnostics and Conservation of the Cultural and Environmental Heritage*, May 1999 (Rome), Euroma, 1999, p 1839–1855
46. P.D. Weil, A Review of the History and Practice of Patination, *Corrosion and Metal Artifacts: A Dialogue between Conservators and Archaeologists and Corrosion Scientists*, B.F. Brown et al., Ed., NBS Special Publication 479, National Bureau of Standards, 1977, p 77–92
47. W.T. Chase, Chinese Bronzes: Casting, Finishing, Patination, and Corrosion, *Ancient and Historic Metals, Conservation and Scientific Research*, D.A. Scott et al., Ed., Getty Conservation Institute, 1994, p 85–117
48. N. Meeks, Patination Phenomena on Roman and Chinese High-Tin Bronze Mirrors and Other Artefacts, *Metal Plating and Patination*, S. La Niece and P. Craddock, Ed., Butterworth-Heinemann, 1993, p 63–84
49. L. Robbiola, J.-M. Blengino, and C. Fiaud, Morphology and Mechanisms of Formation of Natural Patinas on Archaeological Cu-Sn Alloys, *Corros. Sci.*, Vol 40 (No. 12), 1998, p 2083–2111
50. R.J. Gettens, Patina: Noble and Vile, *Art and Technology: A Symposium on Classical Bronzes*, S. Doeringer et al., Ed., MIT Press, 1970, p 57–72
51. T.L. Woods and R.M. Garrels, Use of Oxidized Copper Minerals as Environmental Indicators, *Appl. Geochem.*, Vol 1 (No. 2), 1986, p 181–187
52. R.J. Gettens and C. Frondel, Chalconatronite: An Alteration Product on Some Ancient Egyptian Bronzes, *Stud. Conserv.*, Vol 2 (No. 2), 1955, p 64–75
53. M. Fabrizi and D. Scott, Unusual Copper Corrosion Products and Problems of Identity, *Recent Advances in the Conservation and Analysis of Artifacts*, Summer Schools Press, 1987, p 131–133
54. M.B. McNeil and B.J. Little, Corrosion Products and Mechanisms in Long Term Corrosion of Copper, *Scientific Basis for Nuclear Waste Management XIV*, Vol 212, Materials Research Society, 1991, p 311–316
55. M.B. McNeil and D.W. Mohr, Interpretation of Bronze Disease and Related Copper Corrosion Mechanisms in Terms of Log-Activity Diagrams, *Materials Issues in Art and Archaeology III*, Vol 267, Materials Research Society, 1992, p 1055–1063
56. N. Meeks, Surface Characterization of Tinned Bronze, High-Tin Bronze, Tinned Iron and Arsenical Bronze, *Metal Plating and Patination*, S. La Niece and

P. Craddock, Ed., Butterworth-Heinemann, 1993, p 247–275
57. I.D. MacLeod, Identification of Corrosion Products on Non-Ferrous Metal Artifacts Recovered from Shipwrecks, *Stud. Conserv.,* Vol 36 (No. 4), 1991, p 222–234
58. D.A. Scott, A Review of Copper Chlorides and Related Salts in Bronze Corrosion and as Painting Pigments, *Stud. Conserv.,* Vol 45 (No. 1), 2000, p 39–53
59. J.L. Jambor, J.E. Dutrizac, A.C. Roberts, J.D. Grice, and J.T. Szymanski, Clinoatacamite, a New Polymorph of $Cu_2(OH)_3Cl$, and its Relationship to Paratacamite and "Anarakite," *Can. Mineralog.,* Vol 34, 1996, p 61–72
60. S.J. Duncan and H. Ganiaris, Some Sulphide Corrosion Products on Copper Alloys and Lead Alloys from London Waterfront Sites, *Recent Advances in the Conservation and Analysis of Artifacts,* Summer Schools Press, 1987, p 109–118
61. A. Clydesdale, The Buiston Bowl: Cleaning Trials for Chalcopyrite Removal, *SSCR J.,* Vol 4 (No. 3), 1993, p 12–14
62. G. Eggert and H. Kutzke, All that Glitters is Not Pseudogold: A Study in Pseudo-Pseudogilding, *ICOM Committee for Conservation, 13th Triennial Meeting,* R. Vontobel, Ed., Sept 2002 (Rio de Janeiro), James & James Science Publishers, 2002, p 860–864
63. M.B. McNeil and D.W. Mohr, Formation of Copper-Iron Sulfide Minerals during Corrosion of Artifacts and Possible Implications for Pseudogilding, *Geoarchaeol.,* Vol 8 (No. 1), 1993, p 23–33
64. E. Mattsson, A.G. Nord, K. Tronner, M. Fjaestad, A. Legerlöf, I. Ullén, and G.C. Borg, *Deterioration of Archaeological Material in Soil,* RIK Monograph 10, Central Board of National Antiquities and the National Historical Museums (Stockholm), 1996
65. A. Giumlia-Mair, E.J. Keall, A.N. Shugar, and S. Stock, Investigation of a Copper-Based Hoard from the Megalithic Site of al-Midamman, Yemen: An Interdisciplinary Approach, *J. Archaeolog. Sci.,* Vol 29, 2002, p 195–209
66. M. Fabrizi, H. Ganiaris, S. Tarling, and D.A. Scott, The Occurrence of Sampleite, a Complex Copper Phosphate, as a Corrosion Product on Copper Alloy Objects from Memphis, Egypt, *Stud. Conserv.,* Vol 34 (No. 1), 1989, p 45–51
67. E. Angelini, P. Bianco, and F. Zucchi, On the Corrosion of Bronze Objects of Archaeological Provenance, *Progress in the Understanding and Prevention of Corrosion,* Institute of Materials, 1993, p 14–23
68. X. Zhang, An Unusual Corrosion Product, Pyromorphite, from a Bronze AN: A Technical Note, *Stud. Conserv.,* Vol 47 (No. 1), 2002, p 76–79
69. R.J.H. Wanhill, "Microstructurally-Induced Embrittlement of Archaeological Silver," Report NLR-TP-2001-032, National Aerospace Laboratory (Amsterdam), 2001
70. W.S. Rapson, Effects of Biological Systems on Metallic Gold, *Gold Bull.,* Vol 15 (No. 1), 1982, p 19–20
71. J.H. Frantz and D. Schorsch, Egyptian Red Gold, *Archeomaterials,* Vol 4 (No. 2), 1990, p 133–152
72. D.A. Scott, The Deterioration of Gold Alloys and Some Aspects of Their Conservation, *Stud. Conserv.,* Vol 28 (No. 4), 1983, p 194–203
73. L.S. Selwyn, Corrosion of Archaeological Iron before and after Excavation, Paper 2B.1, *NACE Northern Area Eastern Conference (Ottawa),* NACE International, 1999, p 1–8
74. L.S. Selwyn, P.J. Sirois, and V. Argyropoulos, The Corrosion of Excavated Archaeological Iron with Details on Weeping and Akaganéite, *Stud. Conserv.,* Vol 44 (No. 4), 1999, p 217–232
75. M. McNeil and L.S. Selwyn, Electrochemical Processes in Metallic Corrosion, *Handbook of Archaeological Sciences,* D.R. Brothwell and A.M. Pollard, Ed., John Wiley & Sons, 2001, p 605–614
76. S. Turgoose, The Nature of Surviving Iron Objects, *Conservation of Iron,* Monographs and Reports 53, R.W. Clarke and S.M. Blackshaw, Ed., National Maritime Museum, 1982, p 1–7
77. V. Fell and M. Ward, Iron Sulphides: Corrosion Products on Artifacts from Waterlogged Deposits, *Metal 98,* W. Mourey and L. Robbiola, Ed., James & James Science Publishers, 1998, p 111–115
78. K. Nassau, *The Physics and Chemistry of Color: The Fifteen Causes of Color,* John Wiley & Sons, 1983
79. I.D. MacLeod and C. Kenna, Degradation of Archaeological Timbers by Pyrite: Oxidation of Iron and Sulphur Species, *Fourth ICOM-Group on Wet Organic Archaeological Materials Conference Proceedings,* P. Hoffmann, Ed., Aug 1990, (Bremerhaven), International Council of Museums, 1991, p 133–142
80. P.C. Frost, E. Littauer, and H.C. Wesson, Lead and Lead Alloys, *Corrosion,* Vol 1, 3rd ed., L.L. Shreir et al., Ed., Butterworth-Heinemann, 1994, p 4:76–4:97
81. F.E. Goodwin, Lead and Lead Alloys, *Uhlig's Corrosion Handbook,* 2nd ed., R.W. Revie, Ed., John Wiley & Sons, 2000, p 767–792
82. S. Turgoose, The Corrosion of Lead and Tin: Before and After Excavation, *Lead and Tin Studies in Conservation and Technology,* Occasional Papers, No. 3, G. Miles and S. Pollard, Ed., United Kingdom Institute for Conservation, 1985, p 15–26
83. W. Hofmann and J. Maatsch, Lead as a Corrosion-Resistant Material, *Lead and Lead Alloys,* Springer-Verlag, 1970, p 268–320
84. R.F. Tylecote, The Behaviour of Lead as a Corrosion Resistant Medium Undersea and in Soils, *J. Archaeolog. Sci.,* Vol 10 (No. 4), 1983, p 397–409
85. M. Davis, F. Hunter, and A. Livingstone, The Corrosion, Conservation and Analysis of a Lead and Cannel Coal Necklace from the Early Bronze Age, *Stud. Conserv.,* Vol 40 (No. 4), 1995, p 257–264
86. I.D. MacLeod and R. Wozniak, Corrosion and Conservation of Lead in Sea Water, *ICOM Committee for Conservation, 11th Triennial Meeting,* J. Bridgland, Ed., Sept 1996 (Edinburgh), James & James Science Publishers, 1996, p 884–890
87. L. Lipcsei, A. Murray, R. Smith, and M. Savas, An Examination of Deterioration Products Found on Tin Ingots Excavated from the 14th Century B.C., Late Bronze Age Shipwreck, the *Ulu Burun,* near Kas, Turkey, *Materials Issues in Art and Archaeology VI,* Vol 712, Materials Research Society, 2002, p 451–459
88. H.J. Plenderleith and R.M. Organ, The Decay and Conservation of Museum Objects of Tin, *Stud. Conserv.,* Vol 1 (No. 2), 1953, p 63–72
89. R. Payton, The Conservation of Artefacts from One of the World's Oldest Shipwrecks, the *Ulu Burun,* Kas Shipwreck, Turkey, *Recent Advances in the Conservation and Analysis of Artifacts,* University of London, Institute of Archaeology, Summer Schools Press, 1987, p 41–50
90. M. Gilberg, History of Tin Pest: The Museum Disease, *AICCM Bull.,* Vol 17, 1991, p 3–20
91. R.W. Smith, The α (Semiconductor) ↔ β (Metal) Transition in Tin, *J. Less-Common Met.,* Vol 114, 1986, p 69–80
92. E. Ghali, Aluminum and Aluminum Alloys, *Uhlig's Corrosion Handbook,* 2nd ed., R.W. Revie, Ed., John Wiley & Sons, 2000, p 677–716
93. C. Degrigny, Perfecting an Electrolytic Treatment for the Stabilization of Submerged Aluminum Alloy Aircraft Fragments, *Saving the Twentieth Century: The Conservation of Modern Materials,* D.W. Grattan, Ed., Canadian Conservation Institute, 1993, p 373–380 (in French)
94. A. Oddy and S.M. Bradley, The Corrosion of Metal Objects in Storage and on Display, *Current Problems in the Conservation of Metal Antiquities,* A. Aoki, Ed., Tokyo National Research Institute of Cultural Properties, 1993, p 225–243
95. P.B. Hatchfield, *Pollutants in the Museum Environment,* Archetype Publications, 2002
96. J. Tétreault, *Airborne Pollutants in Museums, Galleries, and Archives: Risk*

Assessment and Control Strategies, Canadian Conservation Institute, 2003
97. S. Bradley and D. Thickett, The Pollution Problem in Perspective, *ICOM Committee for Conservation, 12th Triennial Meeting,* J. Bridgland, Ed., Sept 1999 (Lyon, France), James & James Science Publishers, 1999, p 8–13
98. D. Thickett, S. Bradley, and L. Lee, Assessment of the Risks to Metal Artifacts Posed by Volatile Carbonyl Pollutants, *Metal 98,* W. Mourey and L. Robbiola, Ed., James & James Science Publishers, 1998, p 260–264
99. G. Eggert and U. Sobottka-Braun, Black Spots on Bronzes and Elemental Sulphur, *ICOM Committee for Conservation, 12th Triennial Meeting,* J. Bridgland, Ed., Sept 1999 (Lyon, France), James & James Science Publishers, 1999, p 823–827
100. T.D. Weisser, Some Practical Considerations and Solutions for Preservation in the Museum Environment, *Cultural Property and Its Environment,* Tokyo National Research Institute of Cultural Properties, 1990, p 81–95
101. C. Sease, L.S. Selwyn, S. Zubiate, D.F. Bowers, and D.R. Atkins, Problems with Coated Silver: Whisker Formation and Possible Filiform Corrosion, *Stud. Conserv.,* Vol 42 (No. 1), 1997, p 1–10
102. J. Tétreault, J. Sirois, and E. Stamatopoulou, Studies of Lead Corrosion in Acetic Acid Environments, *Stud. Conserv.,* Vol 43 (No. 1), 1998, p 17–32
103. A. Askey, S.B. Lyon, G.E. Thompson, J.B. Johnson, G.C. Wood, M. Cooke, and P. Sage, The Corrosion of Iron and Zinc by Atmospheric Hydrogen Chloride, *Corros. Sci.,* Vol 34 (No. 2), 1993, p 233–247
104. P. Refait and J.-M.R. Génin, The Mechanisms of Oxidation of Ferrous Hydroxychloride β-$Fe_2(OH)_3Cl$ in Aqueous Solution: The Formation of Akaganeite vs. Goethite, *Corros. Sci.,* Vol 39 (No. 3), 1997, p 539–553
105. D. Watkinson, Degree of Mineralization: Its Significance for the Stability and Treatment of Excavated Ironwork, *Stud. Conserv.,* Vol 28 (No. 2), 1983, p 85–90
106. S. Turgoose, Post-Excavation Changes in Iron Antiquities, *Stud. Conserv.,* Vol 27 (No. 3), 1982, p 97–101
107. F. Zucchi, G. Morigi, and V. Bertolasi, Beta Iron Oxide Hydroxide Formation in Localized Active Corrosion of Iron Artifacts, *Corrosion and Metal Artifacts: A Dialogue Between Conservators and Archaeologists and Corrosion Scientists,* B.F. Brown et al., Ed., NBS Special Publication 479, National Bureau of Standards, 1977, p 103–105
108. J.E. Post and V.F. Buchwald, Crystal Structure Refinement of Akaganéite, *Am. Mineralog.,* Vol 76 (No. 1–2), 1991, p 272–277
109. J.E. Post, P.J. Heaney, R.B. Von Dreele, and J.C. Hanson, Neutron and Temperature-Resolved Synchrotron X-Ray Powder Diffraction Study of Akaganéite, *Am. Mineralog.,* Vol 88 (No. 5–6), 2003, p 782–788
110. K. Ståhl, K. Nielsen, J. Jiang, B. Lebech, J.C. Hanson, P. Norby, and J. Van Lanschot, On the Akaganéite Crystal Structure, Phase Transformations and Possible Role in Post-Excavational Corrosion of Iron Artifacts, *Corros. Sci.,* Vol 45 (No. 11), 2003, p 2563–2575
111. N.A. North, *Corrosion* Products on Marine Iron, *Stud. Conserv.,* Vol 27 (No. 2), 1982, p 75–83
112. G.M. Richardson and R.S. Malthus, Salts for Static Control of Humidity at Relatively Low Levels, *J. Appl. Chem.,* Vol 5, 1955, p 557–567
113. M. Gilberg, History of Bronze Disease and Its Treatment, *Recent Advances in Conservation,* V. Daniels, Ed., British Museum Occasional Paper 65, British Museum Publications Ltd., 1988, p 59–70
114. R.M. Organ, The Conservation of Bronze Objects, *Art and Technology: A Symposium on Classical Bronzes,* S. Doeringer et al., Ed., MIT Press, 1970, p 73–84
115. N.A. North, Conservation of Metals, *Conservation of Marine Archaeological Objects,* C. Pearson, Ed., Butterworths, 1987, p 207–252
116. D.A. Scott, Bronze Disease: A Review of Some Chemical Problems and the Role of Relative Humidity, *J. Am. Inst. Conserv.,* Vol 29 (No. 2), 1990, p 193–206
117. H. Strandberg and L.-G. Johansson, Some Aspects of the Atmospheric Corrosion of Copper in the Presence of Sodium Chloride, *J. Electrochem. Soc.,* Vol 145 (No. 4), 1998, p 1093–1100
118. D. Fellowes and P. Hagan, Pyrite Oxidation: The Conservation of Historic Shipwrecks and Geological and Palaeontological Specimens, *Rev. Conserv.,* Vol 4, 2003, p 26–38
119. R.T. Lowson, Aqueous Oxidation of Pyrite by Molecular Oxygen, *Chem. Rev.,* Vol 82 (No. 5), 1982, p 461–497
120. R. Walker, Instability of Iron Sulfides on Recently Excavated Artifacts, *Stud. Conserv.,* Vol 46 (No. 2), 2001, p 141–152
121. F.M. Howie, *The Care and Conservation of Geological Material: Minerals, Rocks, Meteorites and Lunar Finds,* Butterworth-Heinemann, 1992
122. R. Waller, An Experimental Ammonia Gas Treatment Method for Oxidized Pyritic Mineral Specimens, *ICOM Committee for Conservation, Eighth Triennial Meeting,* K. Grimstad, Ed., Sept 1987 (Sydney, Australia), Getty Conservation Institute, 1987, p 623–630
123. M. Sandström, F. Jalilehvand, I. Persson, U. Gelius, and P. Frank, Acidity and Salt Precipitation on the *Vasa:* The Sulfur Problem, *Eighth ICOM-CC WOAM Conference Proceedings,* P. Hoffmann et al., Ed., June 2001 (Stockholm), Deutsches Schiffahrtsmuseum, 2002, p 67–89
124. M. Sandström, F. Jalilehvand, I. Persson, Y. Fors, E. Damian, U. Gelius, I. Hall-Roth, L. Dal, V. Richards, and I. Godfrey, The Sulphur Threat to Marine Archaeological Artefacts: Acid and Iron Removal from the *Vasa, Conservation Science 2002 Proceedings,* J.H. Townsend et al., Ed., May 2002 (Edinburgh, Scotland), Archetype Publications, 2003, p 79–87
125. K. Jespersen, Precipitation of Iron Corrosion Products on PEG-Treated Wood, *Conservation of Wet Wood and Metal,* I.D. MacLeod, Ed., Western Australian Museum, 1989, p 141–152
126. D.L. Hamilton, *Basic Methods of Conserving Underwater Archaeological Material Culture,* U.S. Department of Defense, 1996
127. C. Pearson, Ed., *Conservation of Marine Archaeological Objects,* Butterworths, 1987
128. W.S. Robinson, *First Aid for Underwater Finds,* 2nd ed., Archetype Publications, 1998
129. C. Sease, *A Conservation Manual for the Field Archaeologist,* UCLA, 1987
130. K. Singley, *The Conservation of Archaeological Artifacts from Freshwater Environments,* Lake Michigan Maritime Museum, 1988
131. D. Watkinson, *First Aid for Finds,* 3rd ed., RESCUE (British Archaeological Trust), 1998
132. R.F. Tylecote and J.W.B. Black, The Effect of Hydrogen Reduction on the Properties of Ferrous Materials, *Stud. Conserv.,* Vol 25 (No. 2), 1980, p 87–96
133. C. Pearson, On-Site Storage and Conservation, *Conservation of Marine Archaeological Objects,* C. Pearson, Ed., Butterworths, 1987, p 105–116
134. M. Gilberg, The Storage of Archaeological Iron in Deoxygenated Aqueous Solutions, *J. Int. Inst. Conserv.-Can. Group,* Vol 12, 1987, p 20–27
135. C.G. Costain, Evaluation of Storage Solutions for Archaeological Iron, *J. Can. Assoc. Conserv.,* Vol 25, 2000, p 11–20
136. I.D. MacLeod, Conservation of Corroded Iron Artefacts—New Methods for On-Site Preservation and Cryogenic Deconcreting, *Int. J. Naut. Archaeol. Underwater Explor.,* Vol 16 (No. 1), 1987, p 49–56
137. D. Gregory, Monitoring the Effect of Sacrificial Anodes on the Large Iron Artefacts on the Duart Point Wreck, 1997, *Int. J. Naut. Archaeol.,* Vol 28 (No. 2), 1999, p 164–173
138. P. Mardikian, Conservation and Management Strategies Applied to Post-Recovery Analysis of the American Civil War Submarine *H.L. Hunley* (1864), *Int. J. Naut. Archaeol.,* Vol 33 (No. 1), 2004, p 137–148

139. J. Carpenter, A Review of Physical Methods Used to Remove Concretions from Artefacts, Mainly Iron, Including Some Recent Ideas, *Bull. Aust. Inst. Maritime Archaeol.*, Vol 14 (No. 1), 1990, p 25–42
140. D. Watkinson, Chloride Extraction from Archaeological Iron: Comparative Treatment Efficiencies, *Archaeological Conservation and Its Consequences*, A. Roy and P. Smith, Ed., International Institute for Conservation, 1996, p 208–212
141. L.S. Selwyn, W.R. McKinnon, and V. Argyropoulos, Models for Chloride Ion Diffusion in Archaeological Iron, *Stud. Conserv.*, Vol 46 (No. 2), 2001, p 109–119
142. N.A. North and C. Pearson, Washing Methods for Chloride Removal from Marine Iron Artifacts, *Stud. Conserv.*, Vol 23 (No. 4), 1978, p 174–186
143. N.A. North and C. Pearson, Alkaline Sulfite Reduction Treatment of Marine Iron, *ICOM Committee for Conservation, Fourth Triennial Meeting*, Oct 1975 (Venice), International Council of Museums, 1975, 75/13/3, p 1–14
144. I.D. MacLeod, Stabilization of Corroded Copper Alloys: A Study of Corrosion and Desalination Mechanisms, *ICOM Committee for Conservation, Eighth Triennial Meeting*, K. Grimstad, Ed., Sept 1987 (Sydney, Australia), Getty Conservation Institute, 1987, p 1079–1085
145. I.D. MacLeod, Conservation of Corroded Copper Alloys: A Comparison of New and Traditional Methods for Removing Chloride Ions, *Stud. Conserv.*, Vol 32 (No. 1), 1987, p 25–40
146. D. Goodburn-Brown and J. Jones, Ed., *Look After the Pennies: Numismatics and Conservation in the 1990s*, Archetype Publications, 1998
147. D. Thickett and C. Enderly, The Cleaning of Coin Hoards: The Benefits of a Collaborative Approach, *The Interface between Science and Conservation*, British Museum Occasional Paper 116, British Museum, 1997
148. C. Sease, Benzotriazole: A Review for Conservators, *Stud. Conserv.*, Vol 23 (No. 2), 1978, p 76–85
149. R.M. Organ, A New Treatment for "Bronze Disease," *Museums J.*, Vol 61 (No. 1), 1961, p 54–56
150. L.E. Merk, A Study of Reagents Used in the Stripping of Bronzes, *Stud. Conserv.*, Vol 23 (No. 1), 1978, p 15–22
151. T.D. Weisser, The Use of Sodium Carbonate as a Pre-Treatment for Difficult-to-Stabilise Bronzes, *Recent Advances in the Conservation and Analysis of Artifacts*, Summer Schools Press, 1987, p 105–108
152. A. Oddy and M.J. Hughes, The Stabilization of "Active" Bronze and Iron Antiquities by the Use of Sodium Sesquicarbonate, *Stud. Conserv.*, Vol 15 (No. 3), 1970, p 183–189
153. C.V. Horie and J.A. Vint, Chalconatronite: A By-Product of Conservation?, *Stud. Conserv.*, Vol 27 (No. 2), 1982, p 185–186
154. C. Pearson, The Preservation of Iron Cannon after 200 Years under the Sea, *Stud. Conserv.*, Vol 17 (No. 3), 1972, p 91–110
155. N.A. North, Electrolysis of Marine Cast Iron, *Papers from the First Southern Hemisphere Conference on Maritime Archaeology*, Australian Sports Publications, 1977, p 145–147
156. C. Pearson, Conservation of the Underwater Heritage, *Protection of the Underwater Heritage*, UNESCO, 1981, p 81–133
157. F. Dalard, Y. Gourbeyre, and C. Degrigny, Chloride Removal from Archaeological Cast Iron by Pulsating Current, *Stud. Conserv.*, Vol 47 (No. 2), 2002, p 117–121
158. S. Païn, R. Bertholon, and N. Lacoudre, The Dechlorination of Copper Alloys with Weak Polarization Electrolysis in Sodium Sesquicarbonate, *Stud. Conserv.*, Vol 36 (No. 1), 1991, p 33–43 (in French)
159. C. Degrigny, The Stabilization of Submerged Aeronautic Remains, *Stud. Conserv.*, Vol 40 (No. 1), 1995, p 10–18 (in French)
160. D.L. Hallam, C.D. Adams, G. Bailey, and G.A. Heath, Redefining the Electrochemical Treatment of Historic Aluminium Objects, *Metal 95*, I.D. MacLeod et al., Ed., James & James Science Publishers, 1997, p 220–222
161. V. Costa, Electrochemistry as a Conservation Tool: An Overview, *Conservation Science 2002 Proceedings*, J.H. Townsend et al., Ed., May 2002 (Edinburgh, Scotland), Archetype Publications, 2003, p 88–95
162. I.A. Carradice and S.A. Campbell, The Conservation of Lead Communion Tokens by Potentiostatic Reduction, *Stud. Conserv.*, Vol 39 (No. 2), 1994, p 100–106
163. C. Degrigny and R. Le Gall, Conservation of Ancient Lead Artifacts Corroded in Organic Acid Environments: Electrolytic Stabilization/Consolidation, *Stud. Conserv.*, Vol 44 (No. 3), 1999, p 157–169
164. R.M. Organ, The Current Status of the Treatment of Corroded Metal Artifacts, *Corrosion and Metal Artifacts: A Dialogue between Conservators and Archaeologists and Corrosion Scientists*, B.F. Brown et al., Ed., NBS Special Publication 479, National Bureau of Standards, 1977, p 107–142
165. S. Keene, Real-Time Survival Rates for Treatments of Archaeological Iron, *Ancient and Historic Metals: Conservation and Scientific Research*, D.A. Scott et al., Ed., Getty Conservation Institute, 1994, p 249–264
166. V. Carrió and S. Stevenson, Assessment of Materials Used for Anoxic Microenvironments, *Conservation Science 2002 Proceedings*, J.H. Townsend et al., Ed., May 2002 (Edinburgh, Scotland), Archetype Publications, 2003, p 32–38
167. H. Matthiesen, D. Gregory, B. Sørensen, T. Alstrøm, and P. Jensen, Monitoring Methods in Mires and Meadows: Five Years of Studies at Nydam Mose, Denmark, *Preserving Archaeological Remains In Situ Proceedings*, Sept 2001 (London), Museum of London, 2001
168. Q. Schiermeier, Undersea Plan Leaves Wrecks to Rest in Peace, *Nature*, Vol 415, 31 Jan 2002, p 460
169. B. Soerensen and D. Gregory, In Situ Preservation of Artifacts in Nydam Mose, *Metal 98*, W. Mourey and L. Robbiola, Ed., James & James Science Publishers, 1998, p 94–99

SELECTED REFERENCES

- D.R. Brothwell and A.M. Pollard, Ed., *Handbook of Archaeological Sciences*, John Wiley & Sons, 2001
- J.M. Cronyn and W.S. Robinson, *The Elements of Archaeological Conservation*, Routledge, 1990
- J.B. Lambert, *Traces of the Past; Unraveling the Secrets of Archaeology through Chemistry*, Perseus Books, 1997
- C. Pearson, Ed., *Conservation of Marine Archaeological Objects*, Butterworths, 1987
- R.W. Revie, Ed., *Uhlig's Corrosion Handbook*, 2nd ed., John Wiley & Sons, 2000
- D.A. Scott, *Copper and Bronze in Art: Corrosion, Colorants, Conservation*, Getty Publications, 2002
- L. Selwyn, *Metals and Corrosion: A Handbook for the Conservation Professional*, Canadian Conservation Institute, 2004

Chemical Cleaning and Cleaning-Related Corrosion of Process Equipment

Revised by Bert Moniz, DuPont Company and Jack W. Horvath, HydroChem Industrial Services, Inc.

CHEMICAL CLEANING is the use of chemicals to dissolve or loosen deposits from process equipment and piping. It offers several advantages over mechanical cleaning, including more uniform removal, no need to dismantle equipment, generally lower overall cost, and longer intervals between cleanings. In some cases, chemical cleaning is the only practical method. Chemical cleaning is usually performed by contractors who specialize in this work. Some cleaning procedures are protected by patents.

The primary disadvantages of chemical cleaning are the possibility of excessive equipment corrosion and the expense of waste disposal. Chemical cleaning solvents must be evaluated in a corrosion test program before acceptance for use.

Chemical Cleaning Methods

There are six major chemical cleaning methods: circulation, fill and soak, cascade, foam, vapor-phase organic, and steam-injected cleaning. A seventh variation is discussed in the section "On-Line Mechanical Cleaning" in this article.

Circulation, the most common method, is applied to columns, heat exchangers, cooling water jackets, and so forth, where the total volume required filling the equipment is not excessive. The equipment is arranged such that it can be completely filled with the cleaning solution and circulated by a pump to maintain flow through the system. Movement of solution through the equipment greatly assists the cleaning action. During circulation, temperature and chemical concentration are measured to monitor the progress. The cleaner may be replenished (sweetened) occasionally to maintain efficiency. Corrosion coupons or on-line monitoring are sometimes used to evaluate the effect of the cleaning chemicals on the equipment materials.

With circulation cleaning, the rate of flow through the equipment is critical. Large-diameter connections are preferred, and a high-capacity pump may be necessary to produce the required circulation flow rate. After cleaning, the equipment is drained, neutralized, flushed, and passivated. When strong acids are used, circulation flow rates may be limited to reduce corrosion.

Fill-and-soak cleaning involves filling the equipment with the cleaner and draining it after a set period of time. This may be repeated several times. The equipment is then water flushed to remove loose insolubles and residual chemicals.

Fill-and-soak cleaning offers limited circulation. The poor access of fresh cleaning solution to the metal, together with the inability to maintain solution temperature, may cause the cleaning action to cease.

The method is limited to relatively small equipment containing light amounts of highly soluble fouling and to equipment in which circulation cannot be properly controlled. Because good agitation is achieved during the flushing stage, flushing should be as thorough as possible. Circulation and fill-and-soak cleaning are sometimes used alternately.

Cascade cleaning, a modification of the circulation method, is usually applied to columns with trays. The column is partially filled, and the liquid is continuously drawn from the reservoir and pumped to the highest point. The liquid then cascades down through the column, cleaning surfaces as it passes over them. The liquid draw-off point must be suitably located to avoid recirculation of loosed foulants, or filtration must be included in the circulation loop. High-capacity pumps and large-diameter piping are required to achieve the necessary transfer of liquid to produce a flow pattern that will contact all fouled surfaces within the column.

The cascade method is primarily used in large columns and is suitable for most types, except for packed columns. Cleaning is not effective in inaccessible areas, such as the underside of trays, due to poor contact with the cleaning solution. Contact may be improved by injecting air or nitrogen at the base of the column. Air should not be used if iron is dissolved in the solvent to avoid causing ferric corrosion (discussed in the section "Ferric Ion Corrosion" in this article). If steam is used to heat the chemicals, the location of the steam-injection point should not lead to localized overheating. Ideally, steam heating of the solvent should occur by steam injection in temporary circulation piping or by use of a temporary heat exchanger. High temperature can also increase corrosion in the vapor space.

Foam cleaning uses a static foam generator that employs air or nitrogen to produce a foamed solvent. As with cascade cleaning, air should not be used if iron is dissolved in the solvent. Foam stabilizers may be added to prolong foam life and increase the effectiveness of the cleaning chemicals. Foam cleaning is used on equipment that cannot support full or partial filling with liquid. Foamed solvent will completely fill fouled equipment, contacting all surfaces. Foam cleaning results in significantly less liquid volume for disposal compared with other methods (Ref 1, 2).

Vapor-phase organic cleaning is used in equipment that is difficult to clean with liquids. For example, vaporized organic solvents are used to remove organic deposits from columns. The organic solvent is vaporized, injected into the top of the column, condensed, collected in a circulation tank, and revaporized.

The principal concerns are the handling and disposal of the solvent and its flammability (when applicable). The circulation tank should be purged and blanketed with nitrogen, fitted with an adequate venting and condensing system, and grounded to prevent accumulation of an electrical charge.

Steam-injected cleaning involves the injection of a concentrated mixture of cleaning chemicals into a stream of fast-moving saturated steam. The steam is injected at one end of the system and condensed at the other. The steam atomizes the chemicals, increasing their effectiveness, and ensures good contact with the metal surface.

Steam-injected cleaning is very effective for critical piping systems. It is also injected into the bottom of columns or other vessels in degassing for later personnel entry. As with foam cleaning, the method produces a relatively low amount of liquid for disposal. Steam-injected cleaning

Chemical Cleaning Solutions

A wide variety of standard chemical cleaning solutions are available (Table 1). Many proprietary solutions are based on these chemicals. Some are patented or involve patented equipment. Chemical cleaning contractors are the best source of information on standard or patented techniques.

For decades, common practice among industrial cleaning contractors and their customers has been to use volume percent (vol%) for surfactants and corrosion inhibitors. Other cleaning chemicals are used on a weight percent (wt%) basis. For acids, the weight percent refers to the concentration of the pure active acid compound in the solution. In this article, concentration is given as weight percent, unless stated otherwise.

For chelants, a sometimes confusing practice used by some contractors is to refer to chelant concentrations as weight percent of the concentrated chelant solution sold by the manufacturer. Ethylenediaminetetraacetic acid (EDTA) salts are commonly sold as approximately 40% active EDTA concentrates. So 15 wt% chelant, by that practice, means 15× 0.40 = 6 wt% as active EDTA. Since other contractors may describe the same concentration solution as 6 wt%, the reader should inquire as to exactly what the active EDTA solids content is for clarity and to be able to compare proposals on a realistic basis.

Chemical cleaning solutions include mineral acids, organic acids, bases, complexing agents, oxidizing agents, reducing agents, and organic solvents. Inhibitors and surfactants are added to reduce corrosion and to improve cleaning efficiency. Following the cleaning cycle, a passivating agent can be introduced to retard further corrosion or to remove trace ion contamination.

Mineral acids are strong scale dissolvers. They include hydrochloric acid (HCl), hydrochloric acid/ammonium bifluoride (HCl/NH_4HF_2), hydrofluoric acid (HF), sulfuric acid (H_2SO_4), nitric acid (HNO_3), phosphoric acid (H_3PO_4), and sulfamic acid (NH_2SO_3H).

Organic acids are much weaker scale dissolvers. They are often used in combination with other chemicals to complex scales. An advantage of organic acids is that they may be disposed of by incineration. They include formic acid (HCOOH), hydroxyacetic/formic acid ($HOCH_2COOH$/HCOOH), acetic acid (CH_3COOH), and citric acid ($HOOCCH_2C(OH)(COOH)CH_2COOH$).

Bases are principally used to remove grease or organic deposits. They include alkaline boilout solutions and emulsions.

Complexing agents are chemicals that combine with metallic ions to form complex ions, which are ions having two or more radicals capable of independent existence. Ferricyanide $[Fe(CN)_6]^{4-}$ is an example of a complex ion. Complexing agents are of two types: chelants and sequestrants. Chelants complex the metallic ion into a ring structure that is water soluble and resistant to precipitation. Sequestrants complex the metallic ion into a structure that is water soluble and resistant to precipitation, but has no ring structure.

Oxidizing agents are used to oxidize compounds present in deposits to make them suitable for dissolution. They include chromic acid (H_2CrO_4), potassium permanganate ($KMnO_4$), and sodium nitrite ($NaNO_2$).

Reducing agents are used to reduce compounds in deposits to a form that makes them suitable for dissolution and to prevent the formation of hazardous by-products. They include sodium hydrosulfite ($NaHSO_2$) and oxalic acid (HOOCCOOH). Reducing agents are also used to prevent ferric iron corrosion. (See the discussion on ferric corrosion found later in this article.)

Inhibitors are specific compounds that are added to cleaning chemicals to diminish their corrosive effect on metals. Most inhibitors are complex proprietary mixtures, and recommendations for their use are available from the supplier.

Surfactants are added to chemical cleaning solutions to improve their wetting characteristics. They are also used to improve the performance of inhibitors, emulsify oils, and for the foaming of solvents. Surfactants are primary ingredients in detergents that are used in neutral, acidic, and alkaline solutions. As with inhibitors, surfactants are trade name products, although

Table 1 Deposit types, solvents, and incompatibilities

Deposits	Commonly used solvents(a)	Temperature °C	Temperature °F	Construction material incompatibilities(b)
Iron oxide: mill scale, or magnetite, Fe_3O_4; red iron oxide, or rust, Fe_2O_3	5–15% HCl	50–80	120–175	Stainless steels, Incoloy 800, soft metals
	2% hydroxyacetic/1% formic	88–96	190–205	Soft metals
	Ammonium bifluoride (ABF additive), $NH_4F \cdot HF$	Depends on acid choice		Titanium, zirconium, stainless steels, Incoloy 800, soft metals
	Chelants: EDTA salts, citric acid salts	65–150	150–300	Soft metals
Copper oxide, CuO copper metal, Cu	HCl + thiourea additive	50–65	120–150	Copper alloys, stainless steels, Incoloy 800, soft metals
	Buffered aqua ammonia + oxidizer; ammoniated citric or EDTA salts+oxidizer	50–65	120–150	Copper alloys, soft metals
Water scale, mainly calcium carbonate, $CaCO_3$	5–15% HCl	50–65	120–150	Stainless steels, Incoloy 800, soft metals
	7–20% sulfamic acid	50–65	120–150	Soft metals
	5–10% formic acid	50–82	120–180	
	Chelants: EDTA salts	65–150	150–300	
Phosphates, or hydroxyapatite, $Ca_{10}(OH)_2(PO_4)_6$	5–15% HCl	65–80	150–175	Stainless steels, Incoloy 800, soft metals
	Sodium EDTA	65–150	150–300	Soft metals
	7–20% sulfamic acid	50–65	120–150	
Calcium sulfate, $CaSO_4$	Potassium hydroxyacetate or potassium acetate, followed by acid stage	20–82	70–180	Depends on the choice of acid
	5–15% HCl	80–88	175–190	Stainless steels, Incoloy 800, soft metals
Silica, SiO_2, or silicate compounds	Long alkaline boilout: followed with acid stage: HCl + NH_4F_2 (ABF)	50–65	120–150	Titanium, zirconium, stainless steels, Incoloy 800, soft metals
Common black iron sulfides, FeS	3–7.5% HCl + H_2S suppressor or use scrubber	50–65	120–150	Stainless steels, Incoloy 800, soft metals
	5–15% sulfuric acid, H_2SO_4 + H_2S suppressor or scrubber	50–65	120–150	Martensitic stainless steels, soft metals
Disulfides, FeS_2	Chromic acid followed by an acid stage	Boiling	Boiling	Depends on the choice of acid
	Alkaline permanganate (AP), followed by an acid stage	>82	>180	Depends on the choice of acid
Carbonaceous deposits, organolignins, other hard organic residues	3% sodium hydroxide + 2% potassium permanganate solution (AP) for thin deposits, followed by acid	82–95	180–205	Soft metals. (Oxalic acid in the rinse between stages prevents chlorine gas generation from HCl + MnO_2.)
Heavy hydrocarbons: tars and asphalts	Alkaline detergents, emulsion solvents	50–95	120–205	Soft metals (if strong alkali is used). (Stay below flash points of solvents.)
Light hydrocarbon: fuels, oils, and greases	Detergents, with or without alkali additives, depending on the deposits	20–82	70–180	Soft metals (if strong alkali is used)

(a) All acids and chelants should be used with corrosion inhibitor additives. (b) Soft metals are low strength metals with low corrosion resistance, such as zinc (galvanizing), aluminum, and magnesium.

some are so common as to approach commodity status.

Hydrochloric acid is the least expensive and most widely used solvent for water-side deposits on steels. The concentration and temperature vary from 5 to 15% and 50 to 80 °C (120 to 175 °F). The acid is generally replenished if the concentration falls below 4%.

Hydrochloric acid must always be used with a filming inhibitor to minimize corrosion. Inhibited HCl is usually suitable for cleaning carbon or alloy steel, cast iron, brasses, bronzes, copper-nickel alloys, and UNS N04400 (Monel 400).

Hydrochloric acid, even inhibited, is not recommended for cleaning stainless steels, UNS N08800 (Incoloy 800), UNS N06600 (Inconel 600), aluminum, or galvanized steel. Its applicability for cleaning titanium or zirconium depends on the contaminants present in the acid. Some technical-grade hydrochloric acid streams are contaminated with HF.

Hydrochloric Acid/Ammonium Bifluoride (ABF). The addition of about 1% ABF (NH_4HF_2) to hydrochloric acid releases hydrofluoric acid (HF) in the solution. This not only acts as an accelerator for the solvent, but more importantly also aids in dissolving silicate scales. A filming inhibitor should be added. Materials compatibility is similar to that of hydrochloric acid except that titanium, zirconium, or tantalum should never be exposed to an acid containing ABF or hydrofluoric acid because they will dissolve rapidly. For calcium water scales, the concentration of ABF additive should be limited to 0.75% or less to prevent the precipitation of calcium fluoride (CaF_2).

Hydrofluoric acid can be used as a preoperational cleaner to remove mill scale, at 1 to 2% and 80 °C (175 °F). It is not widely used in North America, primarily because of safety concerns.

Sulfuric acid is used on a limited scale, from 5 to 10% at temperatures up to 82 °C (180 °F). Although inexpensive, it has several disadvantages, including the extreme care required during handling and the precipitation of calcium sulfate ($CaSO_4$) with deposits containing calcium salts. When inhibited, H_2SO_4 may be used on carbon steel, austenitic stainless steels, copper-nickel alloys, admiralty brass, aluminum bronze, and Monel 400. Sulfuric acid should not be used on aluminum or galvanized steel, and it may be too corrosive on some 400 series stainless steels.

Nitric acid is primarily used for cleaning stainless steel, titanium, or zirconium. It is an extremely strong oxidizer, and its general use is limited because of both corrosion and safety concerns. Conventional inhibitors, detergents, and other additives are not stable in HNO_3. It should not be used on carbon steel, copper alloys, Monel 400, and most other metals. Piping for handling HNO_3 should be austenitic stainless steel or polytetrafluoroethylene (PTFE) lined. A special corrosion inhibitor for nitric acid has been developed. It has been used successfully on several chemical cleaning projects in North America.

Phosphoric acid is sometimes used at 2% and 50 °C (120 °F) for 4 to 6 h to pickle and passivate steel piping. It is not as effective as HCl in removing iron oxide scale, but phosphoric acid has proven effective for cleaning stainless steels.

Phosphoric acid was originally used for removing mill scale from new boilers because it also helped passivate the surface. Another use for it is to brighten aluminum. General use of phosphoric acid is limited by its cost, and by the poor solubility of calcium phosphate at low pH values.

Sulfamic acid is used from 7 to 20% at up to 65 °C (150 °F) to remove calcium and other carbonate scales and also iron oxides. It is not as effective as HCl on iron oxides. If sulfamic acid is heated above 75 to 82 °C (170 to 180 °F), it begins to hydrolyze to sulfuric acid, greatly reducing its effectiveness on calcium deposits, and increasing its corrosiveness on metals. ABF is commonly added to sulfamic acid at 0.25 to 1.0% as an accelerator for iron scale dissolution. Although relatively expensive, sulfamic acid is reasonably safe to handle and may be transported as a solid and diluted on-site. This greatly facilitates handling and makes it preferred for in-house cleaning.

Inhibited sulfamic acid can be used to clean carbon steel, copper, admiralty brass, cast iron, and Monel 400. Without the ABF addition, it is effective on most stainless steels.

Formic acid is generally used as a mixture with other acids (such as citric acid) because alone it is a poor solvent for iron oxide deposits. Accelerators such as erythorbic acid or ABF are commonly added to formic acid solutions to improve iron scale dissolution. Formic acid can be used on stainless steels, is relatively inexpensive, and can be disposed of by incineration.

Hydroxyacetic-formic acid (HAF) is used as a 2%hydroxyacetic-1%formic mixture at 80 to 96 °C (180 to 205 °F) to remove iron oxide and calcium deposits. When added, ABF will also remove silica-containing deposits. The acid must be inhibited for use on carbon steel.

HAF is especially advantageous in nondrainable sections of reactors or boilers, because it may be decomposed to harmless by-products by heating to 150 to 175 °C (300 to 350 °F). The primary use of this cleaner is to remove magnetite (Fe_3O_4) in supercritical boilers and in superheater sections of other boilers. It can be used on stainless steels and is applicable in mixed-metal systems.

Acetic acid is used to clean calcium carbonate scales, but it is ineffective in removing iron oxide deposits. The use of acetic acid in chemical cleaning is limited due to its being weaker than formic acid, necessitating extremely long contact times, and also due to higher cost compared to formic acid.

Citric acid at 3% and 65 °C (150 °F) is used to clean iron oxide deposits from aluminum or titanium. Monoammoniated citric acid is effective in removing iron oxide deposits at pH 3.5. By adjusting the pH to 9 and adding an oxidant, such as sodium nitrite ($NaNO_2$), copper can be removed with the same solution. A patent on this technique has expired.

Where ammonia is a disposal issue, monosodium citrate (MSC) is used in a patented iron scale removal procedure that parallels the monoammonium citrate process, including the passivation. MSC is not effective in copper removal, and great care must be taken to avoid iron salts precipitation.

Citric acid is also used in dilute solution to scavenge residual iron oxide from boilers after HCl cleaning in order to facilitate passivation treatment.

Alkaline boilout solutions are used to emulsify and disperse various oils, greases, and organic contaminants from new steel or stainless steel equipment, such as drum boilers and retubed heat exchangers. Examples are combinations of sodium hydroxide (NaOH, or caustic), trisodium phosphate (Na_3PO_4, or TSP), sodium carbonate (Na_2CO_3, or soda ash), and sodium metasilicate (Na_2SiO_3). These are added in concentrations of 0.5 to 1% and boiled or circulated hot for many hours, often under pressure. A low-sudsing surfactant is usually added. Proprietary compositions are available from suppliers.

Some operators avoid using NaOH because of the possibility of stress-corrosion cracking (SCC, caustic embrittlement) of steel. A controlled amount of sodium nitrate ($NaNO_3$) may be added to the NaOH to inhibit cracking.

Alkaline detergents are used to remove coatings of oily material from inorganic deposits. Oily coatings can prevent dissolution of mineral scales by acids or chelants. For light-bodied oil, a TSP + surfactant combination may be circulated for a few hours at 82 to 95 °C (180 to 200 °F) to remove it. The equipment is then flushed with water.

Chelants will complex such deposit metals as iron or copper. Commonly used chelants include EDTA, citric acid, gluconic acid, and nitrilotriacetic acid (NTA).

Chelating agents must be pH adjusted by using acids or alkalis, depending on the type of chelating agent and the scale constituents to be complexed. For example, under alkaline conditions, the sodium salt of EDTA will complex calcium-containing deposits, but it will not remove iron oxide.

In boiler cleaning, the most common chelating agent is the ammonium salt of EDTA, which is used to remove Fe_3O_4 scale from boilers at 75 to 150 °C (170 to 300 °F). Following this, the solution is oxidized using oxygen gas or hydrogen peroxide to remove copper. This solution must be inhibited to prevent corrosion of the steel.

Some chelating processes are patented, and some have expired patents, such as the Alkaline Copper Removal Process (Dow) which uses tetraammonium EDTA at a pH of 9.0 to 9.5 for removing copper and iron oxides from a boiler while it is intermittently fired. Another chelating agent is the CitroSolv Process, which uses inhibited ammoniated citric acid to dissolve iron oxides and sometimes copper in utility steam generators (patent for Pfizer, also expired.) Chelating agents can often be incinerated, minimizing disposal problems.

Copper complexers such as thiourea are added to HCl when both iron and copper oxide deposits are present. The complexers help dissolve the copper and prevent it from plating out on steel in the system.

Chromic acid has been used at 7 to 10%, primarily to remove iron pyrite and certain carbonaceous deposits that are insoluble in HCl. The iron pyrite is oxidized to the sulfate without the evolution of H_2S. With carbonaceous deposits, H_2CrO_4 degrades the tar that holds the deposits together. The use of H_2CrO_4 in industrial cleaning has become rare due to environmental and safety concerns.

Chromic acid is a strong oxidizing agent and should not be used on copper, brass, bronze, aluminum, zinc, or cast iron. It is acceptable on carbon steel and stainless steel. Steel piping should be used for handling it because nonmetallic hose materials, other than PTFE, are rapidly degraded.

Caustic plus permanganate is widely employed as a pretreatment in refinery-cleaning operations to remove carbonaceous deposits coupled with pyrophoric sulfides. A solution of 1% NaOH and 0.7% $KMnO_4$ is used at 65 to 80 °C (150 to 175 °F). This converts iron sulfide to iron oxide and sulfate, both of which can be removed with acid cleaning. Little or no H_2S is generated. If HCl is used for the acid-cleaning step, oxalic acid should be added to reduce any chlorine gas that might be evolved upon reaction with the insoluble manganese dioxide formed during the first stage of cleaning.

Organic solvents used to dissolve organic deposits include kerosene, xylene, toluene, xylene bottoms, heavy aromatic naphtha, 1,1,1-trichloroethane, N-methyl-2-pyrrolidone, alcohols, glycols, and ortho-dichlorobenzene (Ref 3, 4).

Although very effective in cleaning, the chlorinated solvents have fallen into disuse in recent years due to safety and environmental concerns. In the United States, some solvents are strictly regulated by the Environmental Protection Agency (EPA), Occupational Safety and Health Administration (OSHA), and state and local regulations. The use of any solvents should be carefully reviewed. With few exceptions, rubber hoses or rubber-lined equipment should not be used to handle organic solvents.

Organic emulsions are solvent-based cleaners for heavy organic deposits, such as tar, resid (residual oils), asphalts, and polymers. A typical formulation would consist of an aromatic solvent and emulsifying agents. The mixture should be capable of being heated to 65 to 82 °C (150 to 180 °F). A thorough rinsing step is necessary to remove the sludge left behind when the cleaning agent is drained. Proprietary emulsion solvent blends are also sometimes used.

Passivating solutions are used to prevent carbon steel from rusting after acid cleaning, before it is returned to service. The term passivation is also applied to procedures that are used to remove surface iron contamination from stainless steel and titanium equipment. Most passivating solutions for carbon steel are alkaline and contain an oxidizing agent, even if it is only air. They are applied after rinsing and neutralization.

For the passivation of austenitic stainless steel and welding slag removal, 20% HNO_3 plus 2% HF is commonly used at 50 °C (120 °F) for 6 h. The actual necessary parameters depend on the specific alloy. Titanium can be passivated with nitric acid, but HF should not be used on this metal.

Corrosion inhibitors are added to chemical cleaning agents to permit effective cleaning without excessive corrosion of the equipment. Most inhibitor formulations are organic compounds and function in three ways (Table 2):

- Cathodic inhibitors impede the cathodic half of the corrosion reaction, for example, $2H^+ + 2e^- \rightarrow H_2$.
- Anodic inhibitors limit the anodic half of the corrosion reaction, for example, $Fe \rightarrow Fe^{2+} + 2e^-$.
- Adsorption inhibitors form a physical barrier on the metal surface and prevent corrosion.

Inhibitors must be capable of dispersing or dissolving in the chemical cleaning solution, both fresh and contaminated. They must also be compatible with the equipment materials of construction; for example, low-chloride inhibitors should be used on stainless steels.

The concentration of the inhibitor used is related to the type of solvent, its temperature, the equipment materials of construction, the surface-to-volume ratio in the equipment, and the degree of turbulence. Typical film-forming inhibitor concentrations for mineral acids on steel are of the order of 0.2% (V), more or less.

Most common inhibitors are proprietary formulations, and technical data may be obtained from suppliers. Various inhibitors sometimes contain dyes, which confirm whether or not the inhibitor has been added.

Chemical Cleaning Procedures

The choice of cleaning method, selection of the cleaning solution, and monitoring and documentation requirements should be included in a proposed procedure. Selection of the chemical cleaning contractor is important because the quality of individual crews can vary. Also, the contractor must have the ability and experience to use specific procedures, some of which may require unique steps or patented procedures. The proposed procedure should be written and included in a scope of work that is mutually agreeable to the owner and the contractor. During the project, the key elements must be documented.

Solvent Selection

Proper selection of the solvent system for a chemical cleaning project requires laboratory analysis of a recent sample of the deposit.

Sampling. Great care should be exercised in taking the sample and in shipping it properly, since the analysis is essential to the proper selection of a solvent. For a boiler, the best sample is a tube cut from the water wall at an area that historically had the heaviest deposit, usually the high heat flux area of the boiler. Deposit samples chipped from the metal may suffice in place of a tube cut if no tube section is available and if all deposit layers are included in the sample.

All deposit samples should be labeled with clear distinguishing descriptions to prevent confusion. A sample from before the last cleaning or from the most recent opening of the equipment can be submitted for analysis if a current sample cannot be obtained. Historical information from past cleanings is also helpful, but both of these approaches may fail to account for later episodes of process upsets and changes that can cause layers of deposit which may be more difficult to remove. Water samples, and deposit samples taken from filters, provide only limited information about the deposits on heat-exchanger tubes. Such limited information can be misleading and should not be relied upon as the sole basis for solvent selection.

Laboratories usually require minimum sample sizes. A typical minimum size for a loose deposit sample is 30 g (1 oz). When selecting a tube sample, ASTM D 3483 (Ref 5) specifies a minimum tube length of 0.6 m (2 ft) for removal by dry sawing or grinding. For torch-cut tube sections, laboratories need a minimum tube length of 0.9 m (3 ft). The longer length requirement when the tube section is taken using a cutting torch is because the ends are

Table 2 Classification of corrosion inhibitors

Classification	Example
Cathodic	Alkylamines
	Benzotriazole (L)
	Diphenylamine
	Thiourea (L)
Anodic	Benzotriazole (H)
	Mannich bases
	Pyridines
	Quinolines
	Thiols
	Thioureas (H)
General adsorption	Acetylenic alcohols
	Benzotriazole
	Dibenzlysulfoxide
	Diphenylamine
	Furfuraldehyde

L, at low concentration; H, at high concentration

contaminated by slag and will be cut off and discarded by the lab. Boiler tube sections from tubes that failed should not be used since the violence of the tube blowout will have removed some of the deposit. Samples smaller than the minimums can limit the information that the laboratory can produce.

Lab Report. When the samples are sent, a clear understanding should exist about what will be reported. Allow 2 to 4 weeks for normal handling and reporting. At a minimum, the lab report(s) should provide the following information for each sample:

- Chemical composition of the deposit
- Results of solubility tests
- Descriptions of unusual observations or appearances

Chemical composition of the deposit can be determined by several methods. In one method, the sample is digested in a boiling acid. This is usually followed by spectrometer analysis of the solution.

Results of Solubility Tests. Beaker tests are usually sufficient to indicate which solvents will work in cleaning the equipment, and which is best.

Solvent Choice Based on Deposit Type and Metallurgy. Table 1 summarizes the choices of solvents versus deposits commonly found in boilers and versus construction metals.

In addition to selecting the optimum solvent, other factors essential to successful chemical cleaning are the development of a safe and effective procedure and the prevention of excessive equipment corrosion.

Procedural Stages

The scope of the work should be documented and can be organized in stages.

Planning Stage. The proposed cleaning procedure should be reviewed with the contractor or personnel who will conduct the cleaning. The procedure should outline the specific duties of the contractor and the owner and should include:

- Listing of all chemicals and quantities used, including primary solvent and corrosion inhibitor, surfactants, emulsifiers, neutralizers, passivators, and antifoams
- System volume and solution contact times, temperatures, and pressures
- Method of corrosion control and monitoring
- Special contractor equipment required, such as filter elements, high-volume pumps, reversing manifolds, flow meters, and heat exchangers
- Contractor safety procedures and owner safety requirements, for example, methods of treating hazardous by-products, such as hydrogen sulfide (H_2S) and hydrogen cyanide (HCN), which may be generated during cleaning
- Disposal requirements for waste material, including adherence to federal, state, and local regulations
- Lay-up requirements after cleaning, if equipment is not to be used immediately

When mutually agreed upon, the procedure is developed into a scope of work.

Cleaning Stage. The equipment is readied for cleaning. After isolation, all restrictions, such as orifice plates, should be removed. Tubes of a shell and tube heat exchanger should be cleaned of obstructions before chemical cleaning. Failure to free the tubes could result in solvent remaining in tubes and possible damage at start-up. In general, the chemical cleaning operation consists of:

1. Connecting cleaning system
2. Leak testing of the liquid circulation path, including contractor fittings and temporary piping
3. Establishment of circulation rate and temperature
4. Blending in of chemicals
5. Circulation and monitoring, which include recording of solvent strength and deposit removal; if the solvent strength remains constant for three or more consecutive readings, cleaning is usually considered complete
6. Draining system of solvent
7. Flushing with water
8. Neutralizing, if an acidic solvent was used
9. Passivation, if required
10. Inspection and evaluation of cleaning

Documentation of Cleaning. A simple format will include relevant information on the equipment temperature, materials of construction, chemicals, method of circulation, time of circulation, and evaluation of equipment. The format may be modified for different pieces of equipment.

Corrosion Monitoring in Chemical Cleaning. Chemical cleaning introduces the real possibility of equipment damage from corrosion. Various precautions may be taken to eliminate damage or to reduce the corrosion rate to acceptable levels, such as reducing the cleaning temperature or contact time. Corrosion monitoring during chemical cleaning may consist of on-line electrochemical monitoring, corrosion coupons, or a bypass spool piece containing a sample of the deposit.

On-line electrochemical monitoring provides an instant record of the corrosion rate during cleaning. However, localized corrosion effects may not be indicated.

Corrosion coupons are sometimes used. They should be representative of the materials of construction of the equipment being cleaned, and should be insulated from one another to avoid galvanic effect. If necessary, special coupons, such as U-bends, should be included to check for SCC. The corrosion rates of coupons are generally higher than the actual corrosion rate of the equipment during the cleaning cycle. However, excessively high coupon rates suggest that cleaning has been performed outside of the desired range of temperature, chemical concentration, or circulation rate. Experience dictates that acceptable corrosion rates should be obtained from corrosion coupons. Localized corrosion observed on coupons is unacceptable. A bypass spool piece containing a sample of the deposit is the ideal coupon because it will indicate the effectiveness of the cleaning as well as the condition of the metal after cleaning.

The use of corrosion coupons during chemical cleaning is not recommended, as they may provide inaccurate results because:

- In the early part of the chemical cleaning circulation, the deposit protects the equipment, while the clean metal coupon is exposed to both ferric ion corrosion and acid corrosion. Ferric ion corrosion is rapid, and the coupon can suffer accelerated corrosion during this period
- Sometimes it is not possible or practical to place the coupons in the equipment being cleaned. Corrosion coupons placed in the cleaning contractor's equipment suffer higher corrosion due to high flow rates and turbulence

Postinspection is extremely important to check for damage and to gage the effectiveness of cleaning. Certain scales may increase the corrosiveness of the chemical cleaning solution as they dissolve.

Corrosion Rate. Chemical cleaning contractors measure the general corrosion rate in pounds per square foot per day ($lb/ft^2 \cdot d$). This unit is converted to mils per year (mils/yr) by using a factor that varies with the density of the metal under consideration. To obtain mils per year, multiply pounds per square foot per day by 70.3×10^3/density (grams per cubic centimeter). For steel, 0.01 $lb/ft^2/d = 91$ mils/yr. Further corrosion rate conversion information is found in Reference Information in this Volume.

A significantly higher corrosion rate is tolerable during chemical cleaning than would be allowed in service. For example, 400 mils/yr obtained from coupons may be acceptable for carbon steel. With relatively thick components, such as pump casings, even higher rates may be tolerable. Less than 200 mils/yr should be the goal. Localized forms of corrosion, for example pitting or SCC, are not tolerable.

Localized Corrosion. The principal forms of localized corrosion that occur during chemical cleaning are pitting, SCC, and under-deposit corrosion (crevice corrosion).

Pitting occurs when filming inhibitors break down or there is insufficient inhibitor addition. Pitting will also occur with HCl and incompatible metals, such as aluminum or stainless steels. Oxidizing conditions, such as free Fe^{3+} ion, and the presence of the chloride ion (Cl^-) will also encourage pitting in metals resistant to general corrosion, especially the active-passive metals mentioned. Pitting is especially harmful on thin-wall equipment, for example, heat-exchanger tubes.

Stress-corrosion cracking is likely to occur in specific chemical alloy combinations when the threshold temperature and concentration requirements for cracking are exceeded. This usually occurs in vapor spaces, splash zones, or crevices where the cleaning agent can concentrate. If the chemical cannot be washed out, SCC may occur later, for example, during service. The two most common situations to guard against are the hydroxyl ion (OH^-), for example, from caustic, with carbon steel and the chloride (Cl^-) ion, for example, from HCl, with austenitic stainless steels. In some cases, an inhibitor can be added to minimize cracking. Hydrogen-assisted cracking may also occur in hardenable low-alloy and stainless steels in specific heat treated conditions during acid cleaning.

Cracking of carbon steel (caustic embrittlement) can be inhibited by adding sodium nitrate to the alkaline cleaning solution at a ratio of one part of sodium nitrate per two parts caustic. There is no proven way to inhibit stainless steel from chloride SCC.

Under-deposit corrosion occurs when cleaning is incomplete and the deposits left behind harbor or concentrate corrosives. The corrosion is difficult to see because it is hidden by the deposits.

Temperature, Pressure, and Flow Rate. Of these factors that affect corrosion rate, temperature is the most significant. The higher the temperature of the solvent, the more effective the cleaning generally is. However, corrosion rate increases significantly with temperature and limits the temperature that can be used (Fig. 1). Increasing temperature also increases the amount of inhibitor required.

Each cleaning operation should include a review of the pressure (hydrostatic and gas) that may be experienced. Acceptable limits should not be exceeded. For example, in one operation, impervious graphite heat-exchanger tube failures were significantly reduced when the circulating solvent pressure was decreased from 690 to 207 kPa (100 to 30 psig).

It is extremely difficult to relate flow rate or solution velocity directly to the corrosion rate. This is because the key to increased corrosion rate is turbulence. Experience is the best indicator of velocity or flow limitations during chemical cleaning.

Ferric ion corrosion is a form of general or localized corrosion that occurs on carbon steel, nickel, Monel 400, copper, brass, and zirconium during chemical cleaning. The effect is worst with HCl (because of the presence of the Cl^- ion), but can also occur in other mineral acids, citric acid, EDTA chelant formulations, and formic acid.

Ferric iron corrosion occurs when there are heavy deposits of rust (hematite) or aeration during cleaning. Dissolution of rust or aeration during cleaning generates the Fe^{3+} ion, which corrodes steel as follows:

$$2Fe^{3+} + Fe \rightarrow 3Fe^{2+}$$

Other metals can also be corroded because of the oxidizing nature of the Fe^{3+} ion, especially if the Cl^- ion is present.

Ferric ion corrosion may be prevented with specific inhibitors, for example, stannous chloride ($SnCl_2$). It prevents the formation of Fe^{3+} ion:

$$Sn^{2+} \rightarrow Sn^{4+} + 2e^-$$
$$Sn^{2+} + 2Fe^{3+} \rightarrow Sn^{4+} + 2Fe^{2+}$$

Stannous chloride, though very effective, is rarely used because it introduces a heavy metal into the cleaning waste. The most commonly used ferric iron inhibitor today is erythorbic acid or its less expensive form, sodium salt. Erythorbic acid is the optical isomer of ascorbic acid, which is vitamin C, and it is virtually nontoxic. Proprietary Fe^{3+} ion corrosion inhibitors are also available.

Even if an Fe^{3+} ion inhibitor is added, the total iron content (ferrous and ferric) must be monitored regularly (at least every 30 min) during acid cleaning. Measurement of the Fe^{3+} ion content may be misleading. When the iron content reaches 2 to 3%, the solution should be removed.

Sulfide Corrosion. Sulfides are present in many refinery deposits, and extremely toxic H_2S gas may be released when sulfides are contacted by acid cleaning solutions. Sulfides and H_2S are very corrosive and reduce the effectiveness of corrosion inhibitors. Hydrogen sulfide will also crack hardenable low-alloy steels and blister low-strength carbon steels.

Corrosion problems with sulfides and H_2S may be reduced by lowering the acid concentration and temperature. Inhibitors are effective in preventing cracking or blistering. An aldehyde may be added to the acid cleaning solution to prevent the generation and release of the corrosive and deadly hydrogen sulfide gas. (Ref 6). Glyoxal ($C_2H_2O_2$) is the most commonly used aldehyde in H_2S suppression in chemical cleaning.

Sulfide stringers present in free-machining steels have increased susceptibility toward corrosion during acid cleaning. Selenium-containing free-machining steels or low-sulfur regular grades are less susceptible to attack.

Effectiveness of Corrosion Inhibitor. A useful qualitative test to check whether a filming inhibitor has been added to the system before circulation is the steel wool test. In this test, a small wad of steel wool is completely immersed in a beaker containing the inhibited acid at the strength and temperature to be used during cleaning. If the inhibitor addition is effective, there should be no hydrogen evolution from the beaker for at least 5 min and the wad of steel wool should remain at the bottom of the beaker. Other, more quantitative tests that measure hydrogen generation are also sometimes used.

As noted previously, by-products formed during cleaning can reduce the effectiveness of the corrosion inhibitor. Also, increasing temperature and concentration of cleaning chemicals usually necessitates a greater dose of inhibitor.

"Fish Scaling" of Glass-Lined Equipment. Due to the relatively low heat transfer of glass, the jackets of glass-lined equipment may require frequent cleaning to remove rust, algae, and the like. Hydrochloric acid or any other acid solutions, even if inhibited, should not be used to remove jacket deposits from the jackets of glass-lined equipment. The corrosion rate of the steel jacket is sufficiently high to generate hydrogen. Hydrogen atoms generated by the corrosion reaction diffuse through the steel to the glass/steel interface, where they form hydrogen molecules and accumulate and build up pressure at the glass/steel interface.

The buildup of hydrogen eventually disrupts the bond between the glass and steel and the glass spalls. The initial form of the attack has the appearance of fish scale, but in most cases sufficient hydrogen is generated to result in complete removal glass from a relatively large surface area of the vessel. Reglassing of the entire vessel is then required.

To counter the problem, several proprietary neutral solutions have been developed to remove rust from the jackets of glass-lined vessels. They operate at pH 5 to 7, depending on the formulation, and do not corrode steel. They contain reducing agents, plus a complexing agent, and also surfactants and wetting agents. The reducing agent converts ferric ions in the deposit to ferrous ions, which are then tied up (dissolved) by the complexing agent. Surfactants and wetting agents in the formulations aid dissolution.

For removal of algae alone, a mild solution of sodium hypochlorite may be used.

On-Line Chemical Cleaning

The cleaning methods discussed so far require that the equipment be removed from service and opened to some degree for the cleaning. In some cases, on-line cleaning is effective and more convenient. On-line cleaning can be chemical or mechanical, infrequent or regularly scheduled. Effective on-line cleaning can save time and labor and can prevent a shutdown. Mechanical methods are discussed in a later

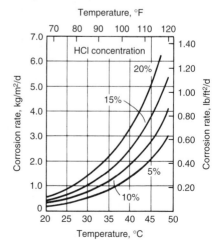

Fig. 1 Effect of temperature on corrosion of mild steel in uninhibited HCl

section. Examples of chemical on-line cleaning include:

- Depressing pH or adding sequestrants and dispersants in a cooling water system
- Increasing chelant or dispersant feed to a boiler
- Injecting inhibited mineral acid into the cooling water stream in front of a heat exchanger for a few minutes and neutralizing it downstream

The first two examples may be executed by a water treatment company. The last example may be executed by a cleaning contractor, but should be coordinated with the water treatment company since it can impact the cooling-water chemistry.

Mechanical Cleaning

Mechanical cleaning is extensively used to clean many types of equipment. There are three broad categories: hydraulic, abrasive, and thermal.

The hydraulic category includes water cleaning, high-pressure water blasting, ultrahigh-pressure water blasting, and ultrasonic cleaning. The abrasive category includes rodding, drilling, sandblasting, pigging and scraping, turbining, and explosive removal. The thermal category includes steam cleaning.

Hydraulic cleaning is carried out at pressures from 2070 kPa to 221 MPa (300 to 32,000 psig). Water cleaning is often used to flush mud and debris from pipelines at pressures to 2070 kPa (300 psig). Slug flow flushing with air injection will obtain better scouring and will conserve water. Water, coupled with a pipeline pig, is often effective in removing scale and corrosion products from lines.

High-pressure water cleaning (hydroblasting), the most common form of mechanical cleaning, is carried out at pressures of 6.9 to 69 MPa (1000 to 10,000 psig). It is used to remove such deposits as light rusting, minerals, and polymers. There are three types of high-pressure water cleaning: tube lancing, line moling, and shotgun jetting.

In tube lancing, a rigid or flexible lance is used to clean the inside of tube bundles, and the operation is repeated for each tube. In line moling, a short self-propelled jet nozzle (mole) and a high-pressure hose are used to clean the inside of piping. In shotgun jetting, a relatively short handheld gun (shotgun) is used to clean surfaces that cannot be cleaned practically by the other means. Shotgun jetting is not effective in cleaning tubes.

High-pressure water cleaning reduces the risk of corrosion damage to equipment, and disposal of the water is less of a problem than with chemical cleaning. Also, the risk of generating hazardous or toxic by-products is eliminated.

The jetted stream is provided by an operator holding a water lance. The lowest pressure with appropriate volume of water is selected, consistent with achieving the required degree of cleaning. This is dictated by experience and risk to personnel.

High-pressure water cleaning must be done by trained personnel using appropriate safety devices and regularly inspected equipment. The area around the project should be suitably barricaded and signs posted.

Line friction leads to a significant loss of pump pressure. To counteract friction loss when pumping through a long hose, organic polymers have been developed that, when added to water in concentrations as low as 0.05 to 0.1%, reduce friction loss as much as 90%.

Ultrahigh-pressure water blasting is conducted at 221 MPa (32,000 psig), using a flow rate of 3.8 to 15 L/min (1 to 4 gal/min). The low flow rate means there is a very low back thrust, allowing the equipment to be operated in congested work spaces with a higher degree of safety. The ultrahigh pressure and low volume of delivery are achieved by using intensifiers rather than the standard plunger pumps. Nozzle configurations are tailored to the application. The nozzle is usually held 19 to 100 mm (0.75 to 4 in.) away from the workpiece. The tooling is specially designed for specific applications, primarily scale, rust, and polymer removal; tube bundle cleaning; and such specialized items as cleaning turbines, diaphragms, and rotors. Cleaning is not effective around more than one 90° turn. The low volume of water minimizes disposal problems.

The technique is particularly advantageous where high-pressure water blasting is ineffective and sandblasting is not desirable, for example, adjacent to pumps, valves, and compressors. Ultrahigh-pressure water blasting has been cost effective in preparing surfaces for inspection, such as the wet fluorescent magnetic-particle inspection of welds. In these cases, an inhibitor is added to the water to prevent flash rusting.

Ultrahigh-pressure water blasting is a proprietary process and is limited to a few vendors who tailor their tools to specific projects. It is more expensive than high-pressure water blasting, but has the advantages of quality and speed.

Hydrokinetic cleaning uses high-frequency vibration or cavitation to remove fouling from the inside of pipes and heat-exchanger tubes. Pulsing a pressurized water column against a plug in the pipe creates the vibration and cavitation. The plug is formed by either the fouling material or by inserting a fabricated "bullet" or "pig" inside the pipe. A small amount of air is also inserted with the water stream to create cavitation. Although the bullet will restrict flow and build pressure in the water column, a small amount of the water is intended to bypass the bullet through the annulus between the bullet and the pipe inside diameter. This creates high-velocity flow at the wall of the pipe and is essential to allow flushing of materials ahead of the bullet as well as to keep the bullet moving through the pipe.

Pulsation of the water stream creates vibration that is transmitted through the water column and to the wall of the pipe and fouling material. The pipe and fouling materials are excited at different frequencies because of their difference in densities. These contrasting vibrations break the cohesion between the fouling material and the pipe wall, and the fouling material is flushed out of the pipe by the water stream. Hydrokinetic cleaning is best when the pipe is partially or completely blocked and the material is polymeric, such as polyethylene.

Aqua milling is a rotary mechanical cutting process for removing scale and fouling from tubing and pipe. The pipe may be coiled, straight, or it may have elbows. Aqua milling can clean pipe with diameters from 25 mm to 3 m (1 in. to 10 ft). Vertical and horizontal vessels up to 7.6 m (25 ft) diam may also be cleaned. Deposits vary from surface scale to full blockage. They may be rubbery, plastic, or brittle.

Unlike high-pressure water cleaning, only one access point is required. Aqua milling is a semiautomatic cleaning system that does not require vessel entry or hands-on lancing. Special jets create hammering impact rather than slicing of scale as seen with high-pressure water cleaning. Aqua milling is not a substitute for high-pressure water cleaning, but may be considered when high-pressure water cleaning takes an extremely long time or when it is necessary to remove personnel from the location of the lance, as in a confined space.

Abrasive cleaning is implemented where hydraulic cleaning may not be practical.

Rodding is used for lightly plugged heat-exchanger tubes where scale has not built up. Care should be taken to prevent scoring of the tubes.

Drilling may be used on tightly plugged heat-exchanger tubes. The drill is hollow, and a fluid connection is used to flush out freed material, eliminating the need to pull the drill out to clear the point and remove loose material.

Sandblasting involves cleaning the internal surface of process lines and equipment by using an abrasive-blasting tool. These are available for different pipe sizes. A standard sandblast nozzle is used that impinges the sand against the conical carbide tip of the nozzle, spraying the sand out radially against the walls of the tube. A circular end plate removes most of the sand as the tool is pulled back out of the pipe.

Piping and Scraping. Pipeline pigs and scrapers are sometimes used, alone or with special chemicals, to clean pipelines. This is a combination of mechanical and chemical cleaning. The pig is a flexible bullet-shaped foam cylinder that is propelled through the pipeline with water. A typical system consists of several pigs, launching and trapping facilities, and a water source. Pigging has been used to clean lines from 25 to 914 mm (1 to 36 in.) in diameter.

Turbining is a tube- or pipe- cleaning method that uses air, steam, or water to drive a motor that turns cutters, brushes, or knockers in order to remove deposits. Many different heads are available for different deposits and equipment, including models for curved tubes. Overzealous turbining may damage pipes because the cutter types will cut into the base metal. Also, turbining is relatively slow and costly compared to other mechanical and chemical methods.

Explosive removal of pipe deposits requires experience and has been used on hard and brittle deposits in piping. Primacord is manufactured in various degrees of explosive power; selection of the proper strength depends on pipe size and type of deposit.

Thermal cleaning is the third major mechanical cleaning method. This method is performed by plant operations personnel. Steam, nitrogen, and oxygen are mixed and injected into the tubing in a furnace section with the fire going for a carefully controlled oxidation of carbon deposits in the tubes.

Steam cleaning can be used to remove mill scale and debris from lines and equipment. When mixed with oxygen, steam cleaning can remove coke and polymers from equipment.

On-Line Mechanical Cleaning

Chemical on-line cleaning was discussed previously. Some examples of mechanical on-line cleaning include:

- Side-stream filtration
- Air bumping of heat exchangers
- Washing soft deposits from steam turbines
- Reversing flow in heat exchangers
- Passing brushes or sponge balls through exchanger tubes

Not all cleaning projects are conducive to on-line cleaning. Hard or insoluble deposits may not be removed effectively. In boilers, if the deposit is heavy, on-line cleaning may cause the deposit to slough and plug tubes—possibly leading to tube failures. Also, some on-line cleaning costs can be high, for example, the initial cost of installing brushes or sponge balls to clean heat exchangers. Each case must be evaluated individually to determine if on-line cleaning will be cost effective.

ACKNOWLEDGMENT

Portions of this article have been adapted from L. Jones, J.D. Haff, and B.J. Moniz, Chemical Cleaning of Process Equipment, *Corrosion*, Vol 13, *Metals Handbook,* 9th ed., ASM International, p 1137–1143.

REFERENCES

1. G.W. Bradley and G.O. Arnold, Foamed Solvents for Cleaning Surface Condensers, IWC-85-52, *Proc. 46th International Water Conference,* Engineer's Society of Western Pennsylvania, 1985
2. C.D. Foster and J. Trimble, Foam Cleaning Boosts Turbine Output, *Power Eng.,* Vol 103 (No. 3), March 1999
3. W.W. Frenier, *Technology for Chemical Cleaning of Industrial Equipment,* NACE International, 2001
4. *Industrial Cleaning Manual,* TPC Publication No. 8, National Association of Corrosion Engineers (NACE), 1982 (out of print)
5. "Test Methods for Accumulated Deposition in a Steam Generator Tube," D 3483, *Annual Book of ASTM Standards,* Vol 11.02, ASTM International
6. W.W. Frenier et al., "Composition and Method for Removing Sulfide-Containing Scale from Metal Surfaces," U.S. Patent No. 4,220,550, 1980

SELECTED REFERENCES

- S.T. Arrington and G.W. Bradley, "Process for Removing Copper and Copper Oxide Deposits from Surfaces," U.S. Patent No. 5,009,714, 1991
- *A Bibliography on Chemical Cleaning of Metal,* TPC Publication No. 6, Vol 1–3, NACE International, 1987
- J.W. Brenner et al., "Foam Cleaning of Surfaces," U.S. Patent 3,037,887, 1962
- C.D. Foster, "Method and Apparatus for Periodic Chemical Cleanings of Turbines," U.S. Patent No. 5,018,355, 1991
- C.D. Foster, "Methods and Apparatus for Periodic Chemical Cleanings of Turbines," U.S. Patent No. 5,090,205, 1992
- C.D. Foster, "Methods and Apparatus for Chemically Cleaning Turbines," U.S. Patent No. 6,311,704, 2001
- C.D. Foster, "Methods for Foam Cleaning Combustion Turbines," U.S. Patent No. 6,478,033, 2002
- C.D. Foster, "Manifold for Use in Cleaning Combustion Turbines," U.S. Patent No. 6,491,048, 2002
- J. Gutzeit, *Cleaning of Process Equipment and Piping,* MTI Publication No. 51, Materials Technology Institute of the Chemical Process Industries, 1997, p 219–239
- J.W. McCoy, *Industrial Chemical Cleaning,* Chemical Publishing Co., 1984, p 153–155
- B.G. Simmons, "Apparatus for Making Foamed Cleaning Solutions and Method of Operation," U.S. Patent 4,133,773, 1979

Corrosion in Specific Industries

Introduction to Corrosion in Specific Industries 337
    Corrosion in the Nuclear Power Industry 337
    Corrosion in Fossil and Alternative Fuel Industries 337
    Corrosion in the Land Transportation Industries 337
    Corrosion in Commercial Aviation 337
    Corrosion in the Microelectronics Industry 338
    Corrosion in the Chemical Process Industry 338
    Corrosion in the Pulp and Paper Industry 338
    Corrosion in the Food and Beverage Industry 338
    Corrosion in the Pharmaceutical and Medical
        Technology Industries 338
    Corrosion in the Petroleum and Petrochemical Industry 338
    Corrosion in the Building Industries 338
    Corrosion in the Mining and Metal Processing
        Industries 338

Corrosion in the Nuclear Power Industry

Introduction to Corrosion in the Nuclear
  Power Industry 339
Corrosion in Boiling Water Reactors 341
    Background to Problem 341
    EAC Analysis 343
    Prediction of EAC in BWRs 350
    Mitigation of EAC in BWRs 354
    Summary of Current Situation and Commentary
        on the Future 356
Corrosion in Pressurized Water Reactors 362
    PWR Materials and Water Chemistry Characteristics 362
    General Corrosion, Crud Release, and Fouling 363
    PWR Primary Side Stress-Corrosion Cracking 367
    Irradiation-Assisted Corrosion Cracking of Austenitic
        Stainless Steels 375
    Steam Generator Secondary Side Corrosion 375
    Intergranular Attack and IGSCC at the Outside Surface 377
    Corrosion Fatigue 380
    External Bolting Corrosion 381
Effect of Irradiation on Stress-Corrosion Cracking and Corrosion in
  Light Water Reactors 386
    Irradiation Effects on SCC 387
        Service Experience 388
        Water Chemistry 391
        Radiation-Induced Segregation 393
        Microstructure, Radiation Hardening, and
            Deformation 398
        Radiation Creep and Stress Relaxation 403
    Mitigation Strategies 404
        Water Chemistry Mitigation—BWRs 404
        Water Chemistry Mitigation—PWR Primary 405

        LWR Operating Guidelines 405
        Design Issues and Stress Mitigation 405
        New Alloys 405
    Irradiation Effects on Corrosion of Zirconium Alloys 406
Corrosion of Zirconium Alloy Components in Light Water
  Reactors 415
    Zirconium Alloys 415
    Corrosion of Zirconium Alloys in Water without
        Irradiation 416
    Heat Flow Conditions 417
    Impact of Metallurgical Parameters on Oxidation
        Resistance 417
    Hydrogen Pickup and Hydriding 417
    Oxide Morphology and Integrity 418
    Effect of Irradiation on Microstructure and Corrosion 418
    LWR Coolant Chemistry 419
    Corrosion of Fuel Rods in Reactors 419
Corrosion of Containment Materials for Radioactive-Waste Isolation 421
    Time Considerations 421
    Environmental and Materials Considerations 422
    Reducing Environments 422
    Oxidizing Environments 425

Corrosion in Fossil and Alternative Fuel Industries

Introduction to Corrosion in Fossil and Alternative Fuel
  Industries 438
    Fuels 438
    Energy Conversion 438
    Efficiency 439
High-Temperature Corrosion in Gasifiers 441
    Corrosion Mechanism and Laboratory Studies 441
    Long-Term Performance of Materials in Service 444
Corrosion in the Condensate-Feedwater System 447
    Corrosion of Condensers 447
        Types of Condensers 447
        Cooling Water Chemistry 448
        Corrosion Mechanisms 448
        Biofouling Control 452
        Other Problems 452
        Corrosion Prevention 452
    Corrosion of Deaerators 452
        Deaerator Designs 452
        Corrosion Problems and Solutions 453
    Corrosion of Feedwater Heaters 456
        Types of Feedwater Heaters 456
        Materials 456
        Water and Steam Chemistry 457
        Corrosion Problems 457

Corrosion of Flue Gas Desulfurization Systems 461
    Flue Gas Desulfurization (FGD) Technology 461
    FGD Corrosion Problem Areas 461
    Materials Selection for FGD Components 462
    Future Air Pollution Control Considerations 463
Corrosion of Steam- and Water-Side of Boilers 466
    Chemistry-Boiler Interactions 466
    Corrosion Control and Prevention 466
    Common Fluid-Side Corrosion Problems 466
Corrosion of Steam Turbines 469
    Steam Turbine Developments 469
    Major Corrosion Problems in Steam Turbines 469
    Turbine Materials 471
    Environment 472
    Design 473
    Solutions to Corrosion Problems 474
    Monitoring 474
    Further Study 474
Fireside Corrosion in Coal- and Oil-Fired Boilers 477
    Waterwall Corrosion 477
    Fuel Ash Corrosion 478
    Prevention of Fireside Corrosion 479
High-Temperature Corrosion in Waste-to-Energy
  Boilers 482
    Corrosion Modes 482
    Corrosion Protection and Alloy Performance 483
Corrosion of Industrial Gas Turbines 486
    Corrosion in the Compressor Section 486
    Corrosion in the Combustor and Turbine Sections 487
Components Susceptible to Dew-Point Corrosion 491
    Dew Point 491
    Components Susceptible to Attack 491
    Mitigation of Dew-Point Corrosion 494
    Guidance for Specific Sections of the Plant 496
Corrosion of Generators 497
    Retaining-Ring Corrosion 497
    Crevice-Corrosion Cracking in Water-Cooled
      Generators 497
Corrosion and Erosion of Ash-Handling Systems 499
    Fly Ash Systems 499
    Wet Bottom Ash Systems 499
    Mitigating the Problems 500
    The Future 500
Corrosion in Portable Energy Sources 501
    Battery Types 501
    Corrosion of Batteries 501
    Corrosion of Fuel Cells 502
Corrosion in Fuel Cells 504
    Fuel Cell Types 504
    Corrosion Processes in Fuel Cell Systems 506
    Materials and Technology Status 511

Corrosion in the Land Transportation Industries

Automotive Body Corrosion 515
    Forms of Corrosion Observed on Automobile
      Bodies 515
    Corrosion-Resistant Sheet Metals 516
    Paint Systems 517
Automotive Exhaust System Corrosion 519
    High-Temperature Corrosion 520
    Cold End Exhaust Corrosion 522

Engine Coolants and Coolant System Corrosion 531
    Antifreeze History 531
    Cooling System Functions 531
    Corrosion 533
    Engine Coolant Base Components and Inhibitors 535
    Engine Coolant Testing 536
    Automotive, Light-Duty versus Heavy-Duty
      Antifreeze/Coolant 536
Automotive Proving Ground Corrosion Testing 538
    When To Use Complete Vehicle Testing 538
    When Complete Vehicle Testing is Less
      Than Adequate 538
    Real-World Conditions the Tests are Aimed
      to Represent 538
    Elements of a Complete Vehicle Corrosion Test 539
    Evaluation of Test Results 539
Corrosion of Aluminum Components in the Automotive Industry 545
    Stress-Induced Corrosion 545
    Cosmetic Corrosion 545
    Crevice Corrosion 545
    Galvanic Corrosion 546
Electric Rail Corrosion and Corrosion Control 548
    Stray-Current Effects 548
    Electric Rail System Design for Corrosion Control 548
    Electric Rail Construction Impacts on Underground
      Utilities 551
    Utility Construction and Funding 553
    Monitoring and Maintenance for Stray-Current Control 556
Corrosion in Bridges and Highways 559
    A Historical Perspective and Current
      Control Strategies 559
        History 559
        Current Corrosion-Control Strategies 560
        Terminology 560
    Concrete: Implications for Corrosion 560
        Cement Chemistry 561
        Additives to Concrete 563
        Aggressive Ions 564
    pH and Corrosion Inhibition 565
        pH of Concrete Pore Water 565
        Chloride Threshold 565
        Applications 566
    Modes of Reinforcement Corrosion 566
        General Corrosion 566
        Localized Corrosion 567
    Prestressing Steel 569
        Types of Prestressed Concrete Construction 569
        Categories of Prestressing Steel 569
        Performance of Prestressing Steel in Concrete
          Highway Structures 569
    Posttensioned Grouted Tendons 570
    Corrosion Inspection 571
        Corrosion Condition Surveys 571
        Assessment of Concrete Quality and Cover 572
        Visual Inspection and Delamination Survey 573
        Reinforcement Potentials 573
        Concrete Resistivity 573
        Chloride and Carbonation Profiles 574
        Corrosion Rate Testing and Other Advanced
          Techniques 575
        Inspection of Steel Elements 575

Corrosion-Resistant Reinforcement 575
        Approaches to Corrosion Resistance 576
        Epoxy-Coated Reinforcement 576
        Stainless Steels and Microcomposite Alloys 578
        Galvanized Reinforcement 580
Performance of Weathering Steel Bridges in
   North America 580
        Weathering Steel as a Material 580
        Rate of Corrosion of Weathering Steel 580
        Performance of Weathering Steel 581
        Recommendations and Considerations
          on the Use of Weathering Steel 581
Coatings 582
        Barrier Coatings for Steel 582
        Concrete Sealers 583
Electrochemical Techniques: Cathodic Protection,
   Chloride Extraction, and Realkalization 583
        Cathodic Protection 584
        Electrochemical Chloride Extraction (ECE)
          and Realkalization 590

Corrosion in the Air Transportation Industry
Corrosion in Commercial Aviation 598
        Corrosion Basics 598
        Commonly Observed Forms of Airplane Corrosion 599
        Factors Influencing Airplane Corrosion 600
        Service-Related Factors 605
        Assessing Fleet Corrosion History 606
        Airworthiness, Corrosion, and Maintenance 607
        New Fleet Design: Establishing Rule-Based
          Corrosion Management Tools 610
        New Airplane Maintenance 611

Corrosion in the Microelectronics Industry
Corrosion in Microelectronics 613
        Characteristics of Corrosion in Microelectronics 613
        Common Sources of Corrosion 614
        Mechanisms of Corrosion in Microelectronics ... 616
        Corrosion Control and Prevention 620
        Corrosion Tests 620
Corrosion in Semiconductor Wafer Fabrication 623
        Corrosion During Fabrication 623
        Corrosion Due to Environmental Effects 626
Corrosion in the Assembly of Semiconductor Integrated
   Circuits 629
        Factors Causing Corrosion 629
        Chip Corrosion 630
        Oxidation of Tin and Tin Lead Alloys (Solders) 630
        Mechanism of Tarnished Leads (Terminations) 630
        Controlling Tarnished Leads at the Assembly 633
Corrosion in Passive Electrical Components 634
        Halide-Induced Corrosion 634
        Organic-Acid-Induced Corrosion 636
        Electrochemical Metal Migration (Dendrite Growth) 638
        Silver Tarnish 640
        Fretting 641
        Metal Whiskers 641
Corrosion and Related Phenomena in Portable Electronic
   Assemblies 643
        Forms of Corrosion Not Unique to Electronics 643
        Forms of Corrosion Unique to Electronics 645
        Corrosion of Some Metals Commonly Found in
          Electronic Packaging 646
        Examples from Electronic Assemblies 647
        Future Trends 650

Corrosion in the Chemical Processing Industry
Effects of Process and Environmental Variables 652
        Plant Environment 652
        Cooling Water 652
        Steam 652
        Startup, Shutdown, and Downtime Conditions 653
        Seasonal Temperature Changes 653
        Variable Process Flow Rates 653
        Impurities 653
Corrosion under Insulation 654
        Corrosion of Steel under Insulation 654
        Corrosion of Stainless Steel under Insulation ... 655
        Prevention of CUI 656
        Inspection for CUI 658
Corrosion by Sulfuric Acid 659
        Carbon Steel 659
        Cast Irons 660
        Austenitic Stainless Steels 660
        Higher Austenitic Stainless Steels 662
        Higher Chromium Fe-Ni-Mo Alloys 662
        High Cr-Fe-Ni Alloy 662
        Nickel-Base Alloys 663
        Other Metals and Alloys 664
        Nonmetals 665
Corrosion by Nitric Acid 668
        Carbon and Alloy Steels 668
        Stainless Steels 668
        Other Austenitic Alloys 670
        Aluminum Alloys 670
        Titanium 671
        Zirconium Alloys 671
        Niobium and Tantalum 672
        Nonmetallic Materials 672
Corrosion by Organic Acids 674
        Corrosion Characteristics 674
        Formic Acid 675
        Acetic Acid 676
        Propionic Acid 678
        Other Organic Acids 679
Corrosion by Hydrogen Chloride and Hydrochloric
   Acid 682
        Effect of Impurities 682
        Corrosion of Metals in HCl 682
        Nonmetallic Materials 686
        Hydrogen Chloride Gas 687
Corrosion by Hydrogen Fluoride and Hydrofluoric
   Acid 690
        Aqueous Hydrofluoric Acid 690
        Anhydrous Hydrogen Fluoride 698
Corrosion by Chlorine 704
        Handling Commercial Chlorine 704
        Dry Chlorine 704
        Refrigerated Liquid Chlorine 706
        High-Temperature Mixed Gases 706
        Moist Chlorine 706
        Chlorine-Water 708

Corrosion by Alkalis 710
    Caustic Soda—Sodium Hydroxide 710
    Corrosion in Contaminated Caustic and Mixtures 721
    Soda Ash 723
    Potassium Hydroxide 723
Corrosion by Ammonia 727
    Aluminum Alloys 727
    Iron and Steel 727
    Stainless Steels 730
    Alloys for Use at Elevated Temperatures 730
    Nickel and Nickel Alloys 731
    Copper and Its Alloys 732
    Titanium and Titanium Alloys 733
    Zirconium and Its Alloys 733
    Niobium and Tantalum 733
    Other Metals and Alloys 733
    Nonmetallic Materials 733
Corrosion by Phosphoric Acid 736
    Corrosion of Metal Alloys in H_3PO_4 736
    Resistance of Nonmetallic Materials 739
Corrosion by Mixed Acids and Salts 742
    Nonoxidizing Mixtures 743
    Oxidizing Acid Mixtures 747
Corrosion by Organic Solvents 750
    Classification of Organic Solvents 750
    Corrosion in Aprotic (Water Insoluble)
        Solvent Systems 750
    Corrosion in Protic (Water Soluble) Solvent Systems 751
    Importance of Conductivity 752
    Corrosion Testing 752
Corrosion in High-Temperature Environments 754
    Oxidation 754
    Carburization 756
    Metal Dusting 757
    Nitridation 758
    High-Temperature Corrosion by Halogen
        and Halides 759
    Sulfidation 759

Corrosion in the Pulp and Paper Industry

Corrosion in the Pulp and Paper Industry 762
    Areas of Major Corrosion Impact 762
    Environmental Issues 763
    Corrosion of Digesters 763
        Batch Digesters 763
        Continuous Digesters 765
        Ancillary Equipment 767
    *Corrosion Control in High-Yield Mechanical
        Pulping* 767
        Materials of Construction and Corrosion
            Problems 768
    Corrosion in the Sulfite Process 768
        The Environment 769
        Construction Materials 769
        Sulfur Dioxide Production 769
        Digesters 770
        Washing and Screening 770
        Chloride Control 770
    *Corrosion Control in Neutral Sulfite
        Semichemical Pulping* 770
        Materials of Construction 770

    Corrosion Control in Bleach Plants 771
        Stages of Chlorine-Based Bleaching 771
        Nonchlorine Bleaching Stages 772
        Process Water Reuse for ECF and Nonchlorine
            Bleaching Stages 772
        Selection of Materials for Bleaching Equipment 773
        Oxygen Bleaching 774
        Pumps, Valves, and the Growing Use of Duplex
            Stainless Steels 774
    Paper Machine Corrosion 775
        Paper Machine Components 775
        White Water 776
        Corrosion Mechanisms 777
    Suction Roll Corrosion 779
        Corrosion 779
        Operating Stresses 779
        Manufacturing Quality 780
        Material Selection 780
        In-Service Inspection 780
    Corrosion Control in Chemical Recovery 780
        Black Liquor 780
        Chemical Recovery Tanks 782
        Additional Considerations for Tanks in Black
            Liquor, Green Liquor, and White
            Liquor Service 783
        Lime Kiln and Lime Kiln Chain 783
    Corrosion Control in Tall Oil Plants 784
    Corrosion in Recovery Boilers 785
        Recovery Boiler Corrosion Problems 786
    Corrosion Control in Air Quality Control 793
        Materials of Construction 793
    *Wastewater Treatment Corrosion in Pulp and
        Paper Mills* 794
        Wastewater System Components and
            Materials of Construction 794
        Parameters Affecting Wastewater
            Corrosivity 794
        Corrosion Mechanisms 795

Corrosion in the Food and Beverage Industries

Corrosion in the Food and Beverage Industries 803
    Corrosion Considerations 803
    Regulations in the United States 803
    Corrosivity of Foodstuffs 804
    Contamination of Food Products by Corrosion 805
    Selection of Stainless Steels as Materials
        of Construction 805
    Avoiding Corrosion Problems in Stainless Steels 807
    Stainless Steel Corrosion Case Studies 807
    Other Materials of Construction 807
    Corrosion in Cleaning and Sanitizing Processes 808

Corrosion in the Pharmaceutical and Medical Technology Industries

Material Issues in the Pharmaceutical Industry 810
    Materials 810
    Passivation 811
    Electropolishing 811
    Rouging 812
Corrosion in the Pharmaceutical Industry 813
    Materials of Construction 813
    Corrosion Failures 815

Corrosion Effects on the Biocompatibility of
    Metallic Materials and Implants 820
        Origins of the Biocompatibility of Metals and
            Metal Alloys 820
        Failure of Metals to Exhibit Expected
            Compatibility 821
        Metal Ion Leaching and Systemic Effects 822
        Possible Cancer-Causing Effects of Metallic
            Biomaterials 823
Mechanically Assisted Corrosion of Metallic
    Biomaterials 826
        Iron-, Cobalt-, and Titanium-Base Biomedical
            Alloys 826
        Surface Characteristics and Electrochemical
            Behavior of Metallic Biomaterials 826
        The Clinical Context for Mechanically
            Assisted Corrosion 827
        Testing of Mechanically Assisted Corrosion 832
Corrosion Performance of Stainless Steels, Cobalt, and
    Titanium Alloys in Biomedical Applications 837
        Chemical Composition and Microstructure of Iron-,
            Cobalt-, and Titanium-Base Alloys 837
        Surface Oxide Morphology and Chemistry 839
        Physiological Environment 840
        Interfacial Interactions between Blood and
            Biomaterials 841
        Coagulation and Thrombogenesis 841
        Inflammatory Response to Biomaterials 841
        General Discussion of Corrosion Behavior of
            Three Groups of Metallic Biomaterials 841
        Corrosion Behavior of Stainless Steel, Cobalt-Base
            Alloy, and Titanium Alloys 844
        Biological Consequences of in vivo Corrosion and
            Biocompatibility 847
Corrosion Fatigue and Stress-Corrosion Cracking
    in Metallic Biomaterials 853
        Background 853
        Metallic Biomaterials 855
        Issues Related to Simulation of the in vivo Environment,
            Service Conditions, and Data Interpretation 861
        Fundamentals of Fatigue and Corrosion Fatigue 863
        Corrosion Fatigue Testing Methodology 867
        Findings from Corrosion Fatigue Laboratory Testing 868
        Findings from in vivo Testing and Retrieval Studies
            Related to Fatigue and Corrosion Fatigue 871
        Fundamentals of Stress-Corrosion Cracking 873
        Stress-Corrosion Cracking Testing Methodology 878
        Findings from SCC Laboratory Testing 880
        Findings from in vivo Testing and Retrieval Studies
            Related to SCC 882
        New Materials and Processing Techniques for CF
            and SCC Prevention 883
Corrosion and Tarnish of Dental Alloys 891
        Dental Alloy Compositions and Properties 891
        Tarnish and Corrosion Resistance 892
        Interstitial versus Oral Fluid Environments and
            Artificial Solutions 894
        Effect of Saliva Composition on Alloy Tarnish and
            Corrosion 896
        Oral Corrosion Pathways and Electrochemical
            Properties 897
        Oral Corrosion Processes 899
        Nature of the Intraoral Surface 902
        Classification and Characterization of
            Dental Alloys 904

Corrosion in the Petroleum and Petrochemical Industry

Corrosion in Petroleum Production Operations 922
    Causes of Corrosion 922
        Oxygen 923
        Hydrogen Sulfide, Polysulfides, and Sulfur 923
        Carbon Dioxide 924
        Strong Acids 925
        Concentrated Brines 926
        Stray-Current Corrosion 926
        Underdeposite (Crevice) Corrosion 926
        Galvanic Corrosion 927
        Biological Effects 927
        Mechanical and Mechanical/Corrosive
            Effects 927
    Corrosion Control Methods 928
        Materials Selection 928
        Coatings 932
        Cathodic Protection 933
        Types of Cathodic Protection Systems 933
        Inhibitors 937
        Nonmetallic Materials 941
        Environmental Control 942
    Problems Encountered and Protective Measures 944
        Drilling Fluid Corrosion 944
        Oil Production 946
        Corrosion in Secondary Recovery
            Operations 953
        Carbon Dioxide Injection 955
        Corrosion of Oil and Gas Offshore Production
            Platforms 956
        Corrosion of Gathering Systems, Tanks, and
            Pipelines 958
        Storage of Tubular Goods 962
Corrosion in Petroleum Refining and Petrochemical Operations 967
    Materials Selection 967
    Corrosion 974
    Environmentally Assisted Cracking (SCC, HEC,
        and Other Mechanisms) 987
    Velocity-Accelerated Corrosion and
        Erosion-Corrosion 999
    Corrosion Control 1002
    Appendix: Industry Standards 1005
External Corrosion of Oil and Natural Gas Pipelines 1015
    Differential Cell Corrosion 1016
    Microbiologically Influenced Corrosion 1017
    Stray Current Corrosion 1017
    Stress-Corrosion Cracking 1018
    Prevention and Mitigation of Corrosion and SCC 1020
    Detection of Corrosion and SCC 1023
    Assessment and Repair of Corrosion and SCC 1023
Natural Gas Internal Pipeline Corrosion 1026
    Background to Internal Corrosion Prediction 1026
    Real-Time Corrosion Measurement and Monitoring 1031
Inspection, Data Collection, and Management 1037
    Inspection 1037
    Noninvasive Inspection 1040

Data Collection and Management 1045
Appendix: Review of Inspection Techniques 1047
Visual Inspection 1047
Ultrasonic Inspection 1047
Radiographic Inspection 1049
Other Commonly Used Inspection Techniques 1050

Corrosion in the Building Industries
Corrosion of Structures 1054
Metal/Environment Interactions 1054
General Considerations in the Corrosion of Structures 1054
Protection Methods for Atmospheric Corrosion 1058
Protection Methods for Cementitious Systems 1060
Case Histories 1062

Corrosion in the Mining and Metal Processing Industries
Corrosion of Metal Processing Equipment 1067
Corrosion of Heat Treating Furnace Equipment 1067
Corrosion of Plating, Anodizing, and
  Pickling Equipment 1071
Corrosion in the Mining and Mineral Industry 1076
Mine Shafts 1077
Wire Rope 1077
Rock Bolts 1078
Pump and Piping Systems 1078
Tanks 1079
Reactor Vessels 1079
Cyclic Loading Machinery 1079
Corrosion of Pressure Leaching Equipment 1079

Introduction to Corrosion in Specific Industries

Hira S. Ahluwalia, Material Selection Resources Inc.

CORROSION-RELATED DIRECT COSTS, such as prevention methods and infrastructure repair and replacement, make up 3.1% of the U.S. gross domestic product—a value of approximately $276 billion annually—according to the 2002 study sponsored by the U.S. Federal Highway Administration (FHWA) (Ref 1), with support from NACE International. The study estimates that 25 to 30% of the cost is entirely preventable. The current state of knowledge contained in the *ASM Corrosion Handbooks* provides valuable information that can be applied to reduce the cost of corrosion. The *ASM Handbook* Volume 13A, *Corrosion: Fundamentals, Testing, and Protection*, 2003, addresses those topics and includes a summary of the FHWA report. The *ASM Handbook* Volume 13B, *Corrosion: Materials*, 2005, addresses the corrosion resistance of materials of construction and includes the article "Global Cost of Corrosion—A Historical Review."

This Section, "Corrosion in Specific Industries," applies the fundamental understanding of corrosion and knowledge of materials of construction to practical applications. The industries addressed are nuclear power, fossil and alternative fuel, land transportation, air transportation, microelectronics, chemical processing, pulp and paper, food and beverage, pharmaceutical and medical technology, petroleum and petrochemical, building, and mining and metal processing. Each specific industry article provides information on the challenges faced in combating corrosion and the appropriate materials of construction and other corrosion control technologies employed to meet these challenges.

Corrosion in the Nuclear Power Industry

The U.S. nuclear industry generates approximately 20% of the electricity needs primarily from reactors designed and built over 30 years ago. Safety concerns continue to plague the industry. Severe cracks found at one nuclear power reactor (ca 2001) and the boric acid corrosion of carbon steel that nearly breached the 152 mm (6 in.) steel dome of another facility (ca 2002) are raising new questions about aging nuclear plants and their inspection methods. By seeking to understand the role of alloys in corrosion, the nuclear industry is in the process of refurbishing and upgrading its facilities in order to extend the design life of plants an additional 20 years. Most nuclear electricity is generated using two types of nuclear reactors that evolved from 1950 designs, namely the boiling water reactor and the pressurized water reactor. The fuel for these types of reactors is similar, consisting of long bundles of 2 to 4% enriched uranium dioxide fuel pellets stacked in zirconium alloy cladding tubes. This Section deals with corrosion issues in light water reactors, research on the corrosion mechanisms, and the development and implementation of engineering solutions.

The burial of spent fuel that has accumulated over the past decades of nuclear power production is another aspect of the industry that has attracted tremendous research and political attention in recent years. Many countries are developing geological repositories for their high-level nuclear waste. These repositories will consist of a stable geological formation within which engineered barriers will be constructed. The engineered barriers are double-walled metallic containers. The article in this Section provides an excellent analysis of the metals and alloys being selected for the metallic cylinders. The corrosion resistance, the effect of metallurgical condition, and the environmental variable that affects the corrosion behavior of the metallic materials are discussed.

Corrosion in Fossil and Alternative Fuel Industries

The three types of fossil fuel power plants primarily used by the electric power industry are coal-fired steam, industrial gas turbine, and combined-cycle power plants. The most common and widely used is the pulverized-coal-fired steam power plant. Because of the complex and corrosive environments in which power plants operate, corrosion has been a serious problem, with a significant impact on reliability of the plants. This Section deals with corrosion of components that continue to have the greatest effect on the reliability of the power plant. Corrosion of tubular components, for example, in condensers and boilers, appears to be the major challenge. Also discussed in this Section are flue gas systems, especially the wet flue gas desulfurization systems. These are strongly affected by corrosion because of the corrosive nature of flue gas impurities such as sulfur dioxide.

Corrosion in the Land Transportation Industries

This section addresses the automotive industry and also discusses issues related to electric rail corrosion and corrosion in bridges and highways. The automobile industry, for the most part, has become one of the major success stories in corrosion management. The use of electrogalvanized steel, aluminum, and polymers has had a major impact on improving the corrosion performance of automotive bodies. Stainless steel use has increased over the past 25 years. Most exhaust systems use stainless steel or aluminized stainless steel for corrosion resistance.

Corrosion in Commercial Aviation

One of the major concerns of the commercial aircraft and airline industry is the aging of aircraft. The risk and cost of corrosion damage is particularly high in aging aircraft. In April 1988, the sudden decompression of a 19 year old Boeing 737 aircraft and the subsequent separation of a major portion of the upper fuselage resulted in federal airworthiness directives, establishing requirements to prevent or control corrosion in aircraft. The Department of Transportation, Federal Aviation Administration (FAA) Title 14 CFR Parts 121, 129, and 135 require that the maintenance and inspection programs include FAA-approved corrosion prevention and control programs (CPCP). This article describes how the implementation of

CPCP programs has dramatically reduced corrosion damage of aircraft structures. Significant improvements have also been made in the corrosion design and manufacturing of new airplanes. Boeing's newest commercial aircraft, the 787, will use approximately 60% less aluminum and significantly more composite materials and increased amounts of titanium. These more corrosion-resistant materials are expected to significantly reduce maintenance costs.

Corrosion in the Microelectronics Industry

Corrosion in electronic components manifests itself in several ways. Computers, integrated circuits, and microchips are now an integral part of all technology-intensive industry products, ranging from aerospace and automotive to medical equipment and consumer products, and are therefore exposed to a variety of environmental conditions. Corrosion in electronic components is insidious and cannot be readily detected. Therefore, when corrosion failure occurs, it is often dismissed as just a failure, and the part or component is replaced. This Section discusses corrosion in passive electronic and semiconductor integrated circuits and components.

Corrosion in the Chemical Process Industry

Before building a chemical plant, or expanding a current plant, one of the first steps to ensuring safety and reliability is to select proper materials of construction. Understanding the importance of materials selection and making an informed decision is critical in achieving safe and reliable operation of the new facility as well as minimizing the capital cost. This Section deals with corrosion by commercially important acids, bases, and solvents. Each article briefly discusses the various grades, methods of manufacture, their effect on most common metals and plastics, and the most common material of construction used in the manufacture and storage of the acids, bases, and solvents. The cross-industry problem of corrosion under insulation is discussed, with a summary of the corrosion mechanism and effective prevention techniques.

Corrosion in the Pulp and Paper Industry

Corrosion in the pulp and paper industry has been a subject of much discussion and research by the Technical Association of the Pulp and Paper Industry and NACE International. The articles in this Section are a summary of the work done by the NACE International committee. Corrosion, materials of construction, and protection methods used in the principal equipment found in 12 different sections of paper mills are discussed.

Corrosion in the Food and Beverage Industry

The food processing industry is one of the largest manufacturing industries in the United States, accounting for approximately 14% of the total U.S. manufacturing output. Stainless steels and aluminum alloys are the primary materials used in the food and beverage industry. The corrosion environment involves moderately to highly concentrated chlorides, with significant concentrations of organic acids. The water side of the processing equipment can range from steam heating to brine cooling. Chemicals used for clean-in-place procedures pose additional corrosion challenges. This article discusses regulations governing materials and the application of stainless steels in the food industry.

Corrosion in the Pharmaceutical and Medical Technology Industries

According to the 2002 corrosion cost study (Ref 1), the total capital expenditures for the pharmaceutical industry are approximately $4.5 billion, with up to $0.5 billion per year in corrosion costs. This figure does not include indirect costs that result because of unexpected loss and downtime of critical equipment and the effect this has on inventories and production scheduling. In the biotechnology sector, the indirect cost can be 10 to 20 times more than the direct cost. The indirect cost is an important consideration when selecting materials of construction for equipment that holds valuable and expensive drugs. Consider a scenario where one batch of final product has to be discarded due to metal-catalyzed damage to the protein. For example, consider a 120 liter final-product vessel constructed of 316 L stainless steel costing $30,000, or if constructed of a Ni-Cr-Mo alloy such as alloy C-22 costing $60,000, and holding bulk product worth $500,000. If one such batch has to be discarded due to the corrosion of 316 L, not only the production cost is lost but the potential revenue and profit. Avoiding the loss of even one single batch easily covers the extra capital expenditure of using more corrosion-resistant materials to ensure no loss of production.

Another incentive to use more corrosion-resistant materials is to ensure compliance to federal regulations, Title 21 CFR Parts 210 and 211, "Current Good Manufacturing Practice in Manufacturing, Processing, Packing, or Holding of Drugs; General and Current Good Manufacturing Practice for Finished Pharmaceuticals." Subpart D-211.65 states "Equipment shall be constructed so that surfaces that contact components, in-process materials, or drug products shall not be reactive, additive, or absorptive so as to alter the safety, identity, strength, quality, or purity of the drug product beyond the official or other established requirements." This regulation clearly states that corrosion of equipment or product contamination of any kind is not acceptable.

Biomedical prosthetic devices are being increasingly used. Any material placed in the body needs to be compatible and not cause adverse reactions in the body. Corrosion effects on the biocompatibility of metallic materials and implants and the corrosion of biomaterials and dental alloys are also discussed in this Section.

Corrosion in the Petroleum and Petrochemical Industry

A large amount of research and industry standards has been published over the last 15 years. The production of oil and gas, its transportation and refining, and its subsequent use as fuel and raw materials for chemicals is a complex and demanding process, where corrosion is an inherent hazard. This Section discusses corrosion problems and methods of control used in petroleum production, in storage and transportation of oil and gas (including external and internal corrosion of pipelines), and in the refining industry. A useful summary of the industry standards is also presented.

Corrosion in the Building Industries

Carbon steel is the main material of construction used in building construction. This Section discusses carbon steel corrosion in structures and methods used to prevent corrosion of structures.

Corrosion in the Mining and Metal Processing Industries

The mining, mineral processing, and extractive metallurgy industries deal with a wide range of corrosive environments and erosive media. In mining and mineral processing, erosion and erosion-corrosion are the major concerns. Extractive metallurgy often involves the use of aggressive acids and other chemicals. This Section addresses the materials selection challenges faced in this industry.

In conclusion, this Section of the *ASM Handbook* provides excellent practical information that can be applied to reduce the significant cost of corrosion to our society. This important Section would not be possible without the dedication of the knowledgeable subject expert authors who devoted considerable time in the preparation of the articles.

REFERENCE

1. G.H. Koch, M.P.H. Brongers, N.A. Thompson, Y.P. Virmani, and J.H. Payer, "Corrosion Costs and Preventive Strategies in the United States," FHWA-RD-01-156, Federal Highway Administration, U.S. Department of Transportation, March 2002

Introduction to Corrosion in the Nuclear Power Industry

Barry M. Gordon, Chair, Structural Integrity Associates, Inc.

THIS SECTION reviews the series of serious corrosion problems that have plagued the light water reactors (LWRs) industry for decades, the complex corrosion mechanisms involved, and the development of practical engineering solutions for their mitigation. However, to understand the nature of LWR corrosion problems, a brief historical perspective of the corrosion design basis of commercial LWRs, that is, boiling water reactors (BWRs) and pressurized water reactors (PWRs), is necessary.

Although corrosion was considered in all plant designs, corrosion was not considered as a serious problem. The major concern was general corrosion, and it was well known at the time of LWR design and construction that the primary structural materials used in the fabrication of the nuclear steam supply system—stainless steels and nickel-base alloys—were characterized by very high general corrosion resistance in high-purity, high-temperature LWR-type environments. The problem was that the "qualifying" laboratory tests did not necessarily reproduce the reactor operating conditions (especially the high residual tensile stresses), and the test times were of short duration relative to the initial plant design lifetime of 40 years, which is currently being extended to 60 years. For example, the initiation time for primary water stress-corrosion cracking (PWSCC) of nickel-base alloys in PWRs can be as long as 25 years! Also, while it was well documented at the time that the presence of chloride in the coolant could cause chloride stress-corrosion cracking (SCC) of stainless steel, there was little evidence that only 0.2 ppm of dissolved oxygen in the highest-purity water could provide the electrochemical driving force for intergranular stress-corrosion cracking (IGSCC) of weld-sensitized austenitic stainless steel.

The initial operation of the early commercial LWRs encountered few corrosion problems. Any problems were quickly repaired and did not have a major impact on plant availability. As more plants entered service and more operating time was accumulated on existing plants, more corrosion-related incidents appeared in the piping, reactor internal components, and other components. Eventually, the corrosion of the plant and fuel materials did seriously impact plant availability, economics, reliability, and, in some cases, plant safety. Table 1 presents a brief summary of the corrosion history of LWRs.

To quantify the fiscal impact of these types of corrosion incidents on the cost of operating nuclear power plants, the Federal Highway Administration, Office of Infrastructure Research and Development, funded an economic evaluation (Ref 1) of the cost of corrosion across segments of industry. Also see the article "Direct Costs of Corrosion in the United States" in *Corrosion: Fundamentals, Testing, and Protection,* Volume 13A, *ASM Handbook,* 2003. The annual nuclear power costs of corrosion were divided into three main categories:

- Corrosion-related causes of partial LWR outages: $5 million/year
- Corrosion-related causes of zero power LWR outages: $665 million/year
- Contribution of corrosion to LWR operation and maintenance (O&M): $2 billion/year

As can be seen in Table 2, which presents the ten most expensive O&M costs of corrosion for one particular reactor site, Oconee units 1, 2, and 3 PWRs, the cost of corrosion in the nuclear power industry is extremely high. Obviously, this type of cost analysis does not include human "costs" for the loss of life, such as occurred in December 1986 and August 2004 at the Surry 2 and Mihama 3 PWRs, respectively, where a total of 9 workers were killed and 10 others were severely injured due to the rupture of a carbon steel piping due to flow-accelerated corrosion in the two plants.

As indicated in Table 1, the major corrosion-related problems addressed by the reactor operators and owners, the Electric Power Research Institute (EPRI), and the nuclear steam system suppliers include the IGSCC of welded austenitic stainless steel pipes in BWRs, PWSCC of nickel-base alloys and steam

Table 1   A brief summary of the corrosion history of light water reactors

Corrosion event(a)	Time of detection
Alloy 600 IGSCC in a laboratory study	Late 1950s
IGSCC/IASCC BWR stainless steel fuel cladding	Late 1950s and early 1960s
BWR IGSCC of type 304 stainless steel during construction	Late 1960s
IGSCC of furnace-sensitized type 304 during BWR operation	Late 1960s
IGSCC in U-bend region of PWR steam generator	Early 1970s
Denting of PWR alloy 600 steam generator tubing	Mid-1970s
PWSCC of PWR alloy 600 steam generator tubing	Mid-1970s
Pellet-cladding interaction failures of BWR zircaloy fuel cladding	Mid-1970s
IGSCC of BWR welded small-diameter type 304 piping	Mid-1970s
IGSCC of BWR large-diameter type 304 piping	Late 1970s
IGSCC of BWR alloy X-750 jet pump beams	Late 1970s
IGSCC of BWR alloy 182/600 in nozzles	Late 1970s
Accelerating occurrence of IGSCC/IASCC of BWR internals	Late 1970s
PWSCC in PWR pressurizer heater sleeves	Early 1980s
Crevice-induced IGSCC of type 304L/316L in BWRs	Mid-1980s
FAC of single-phase carbon steel systems in PWRs	Mid-1980s
PWSCC in PWR pressurizer instrument nozzles	Late 1980s
IGSCC of BWR low-carbon (304L/316L) and stabilized stainless steels (347/321/348) in vessel locations	Late 1980s–present
IGSCC of BWR internal core spray piping	1980s–present
Axial PWSCC of alloy 600 of PWR top head penetration	Early 1900s
Circumferential PWSCC of J-groove welds	Early 1900s
PWSCC of PWR hot leg nozzle alloy 182/82	Early 2000s
PWSCC-induced severe boric acid corrosion of a PWR head	Early 2000s

(a) IGSCC, intergranular stress-corrosion cracking; IASCC, irradiation-assisted stress-corrosion cracking; BWR, boiling water reactor; PWR, pressurized water reactor; PWSCC, primary water stress-corrosion cracking; FAC, flow-accelerated corrosion

Table 2 Ten most expensive operation and maintenance costs of corrosion for the Oconee units 1, 2, and 3 pressurized water reactors

Work activities	Cost	Cost, %	% attributed to corrosion	Cost of corrosion
Steam generators	$22,757,765	8.26	95	$21,619,877
Maintenance engineering support	$13,204,783	4.79	33	$4,357,578
Radiation protection	$12,116,142	4.40	80	$9,692,912
Mechanical components	$10,709,285	3.89	33	$3,534,064
Maintenance function support	$10,675,567	3.87	33	$3,522,937
Work control	$6,073,111	2.20	33	$2,004,127
Chemistry	$5,570,659	2.02	60	$3,342,395
Piping	$2,391,285	0.87	60	$1,434,771
Coatings and painting	$2,279,358	0.83	45	$1,025,771
Decontamination	$1,216,689	0.44	80	$973,351
Remaining activities	$188,590,607	68.43	9	$17,122,624
Total	$275,585,251	100.00	25	$68,896,313

Source: Ref 1

generator corrosion in PWRs, the effect of irradiation on corrosion and SCC, corrosion of zirconium alloy fuel cladding, and the corrosion of containment materials for radioactive waste isolation for both BWRs and PWRs. All of these subjects are addresses in this Section.

REFERENCE

1. G.H. Koch et al., "Corrosion Costs and Preventive Strategies in the United States," Report FHWA-RD-01-156, Federal Highway Administration, Office of Infrastructure Research and Development, Sept 2001

Corrosion in Boiling Water Reactors

F. Peter Ford, General Electric Global Research Center (Retired)
Barry M. Gordon, Structural Integrity Associates, Inc.
Ronald M. Horn, General Electric Nuclear Energy

ENVIRONMENTALLY ASSISTED CRACKING (EAC) of structural materials in the boiling water reactor (BWR) reactor pressure vessel (RPV), core internals, and ancillary piping is the focus of this article. The content is substantially different from that in *Corrosion*, Vol 13, *ASM Handbook* (formerly 9th ed., *Metals Handbook*) published in 1987 to reflect technology advances in the intervening 19 years and their relevance to regulatory requirements and business needs. For instance, this article covers effects of water chemistry on materials degradation, new mitigation approaches and their impact on aging-management programs, plus a discussion of life-prediction capabilities that are having an impact on proactive management of BWR materials degradation.

Boiling water reactor topics dealing with localized corrosion of zirconium alloy fuel cladding, plus a detailed analysis of the effects of irradiation on corrosion are addressed in other articles in this Section, "Corrosion in the Nuclear Power Industry." The corrosion of balance-of-plant structures (e.g., condensers, steam turbines, feedwater heaters, etc.) is covered in the Section "Corrosion in Fossil and Alternative Fuel Industries."

Background to Problem

Most currently operating BWRs are direct-cycle systems. That is, water boils in traversing the fuel channels in the core, with the resultant steam containing approximately 10% moisture above the liquid at the top of the core. This moisture content is reduced to less than 0.1% by steam separators and driers in the top head of the reactor vessel; the steam then leaves the pressure vessel at about 7 MPa (1000 psig) and 288 °C (550 °F) to the steam turbine. Upon exiting the turbine, the steam is condensed and the water purified in demineralizers; this water is then reheated in heat exchangers to approximately 225 °C (435 °F) before returning to the reactor. Although earlier reactor designs relied on natural circulation to transport the water coolant around the reactor internals, the majority of currently operating designs rely on external recirculation pumps and internal jet pumps to control coolant transport and heat transfer within the core. There are differences in specific design details between the various BWR reactor manufacturers (General Electric, ABB-W, KWU-Siemens, Hitachi, and Toshiba), but this general design description is sufficient for the purpose of discussing corrosion issues.

The nuclear steam supply system (NSSS) is very complicated with additional piping associated with, for instance, reactor water cleanup (RWCU), shutdown cooling (SDC), and/or residual heat removal (RHR); these systems ensure core control during normal full-power and transient conditions such as reactor startup and shutdown. A series of emergency core cooling systems (ECCS) ensure core cooling in the event of loss of coolant accidents (LOCA) of varying magnitudes; such piping systems include, for instance, the high- and low-pressure coolant injection (HPCI, LPCI), high- and low-pressure core spray (HPCS, LPCS), and the automatic depressurization system (ADS), which permits effective operation of the low-pressure emergency coolant systems. Although the reliability of these piping systems is critical to the safety of the reactor, there are no corrosion phenomena unique to them and that are not discussed elsewhere in this Volume; thus, for simplicity, discussion of corrosion issues in BWRs is confined to the reactor pressure vessel, its internals, and the external recirculation piping.

Emphasis is placed on materials degradation due to EAC under various combinations of materials, environmental, and stressing conditions. Under static loading conditions EAC is subcategorized as "stress-corrosion cracking" (SCC), or "irradiation-assisted stress-corrosion cracking" (IASCC) for specific irradiated core conditions, and under monotonically increasing strain conditions (for instance, during reactor startup) as "strain-induced corrosion cracking." The detrimental effect of the environment on "dry fatigue" cracking under cyclic loading is subcategorized as "corrosion fatigue." It is important to recognize that these various EAC definitions generally refer to a common environmental degradation process where the stressing mode is changing the kinetics of crack initiation and propagation and, sometimes, the intergranular or transgranular morphology of cracking; the following discussion emphasizes this commonality rather than focusing on the categorizations.

Environmentally assisted cracking has been observed in various BWR structural components over the last 40+ years (Table 1) due to the interactions of specific material, environment, and stress conditions. These occurrences have been described in, for example, the proceedings of a series of biannual symposia and conferences held since 1983 (Ref 1–11), at Electric Power Research Institute (EPRI) sponsored workshops on the development of mitigation approaches

Table 1 Evolution of EAC incidents in boiling water reactors

Component and mode of failure	Alloy	Time period
Fuel cladding, irradiation-assisted SCC	304	1960s
Furnace-sensitized safe ends, IGSCC	304, 182, 600	↓
Weld-sensitized small diameter piping, IGSCC	304	
Weld-sensitized large diameter piping, IGSCC	304	
Furnace-sensitized weldments and safe ends, IGSCC	182/600	
Low-alloy steel nozzles, thermally induced vibration	A508	1980s
Crevice-induced cracking	304L/316L	
Jet pump beams, IGSCC	X750	
Cold work induced IGSCC of "resistant" alloys	304L	
Low-alloy steel pressure vessel, TGSCC	A533B/A508	
Irradiated core internals, IASCC	304, 316	↓
IGSCC/IASCC of low-carbon and stabilized stainless steels	304L, 316L, 321, 347	2000s

SCC, stress-corrosion cracking; IGSCC, intergranular stress-corrosion cracking; TGSCC, transgranular stress-corrosion cracking; IASCC, irradiation-assisted stress-corrosion cracking

(Ref 12, 13) and water chemistry issues (Ref 14), and in information notices published on the U.S. Nuclear Regulatory Commission (USNRC) website (Ref 15).

The earliest incidents of EAC in BWRs occurred during the 1960s and were associated with cracking of type 304 stainless steel fuel cladding (Ref 16); the driving forces for the cracking were the tensile hoop stress in the cladding due to the swelling fuel and the highly oxidizing conditions in the water of the core environment. This problem was effectively mitigated when zirconium alloys were substituted for this component, although, as pointed out in other articles in this Section, other modes of corrosion degradation associated with zirconium alloys became apparent. The earliest indications of cracking in austenitic stainless steel piping were associated with transgranular stress-corrosion cracking (TGSCC) due to chloride contamination encountered either during fabrication or operation (Ref 17); this was again exacerbated by the oxidizing conditions in the water and was effectively managed by appropriate control of the chloride contamination. Subsequently, EAC incidents in BWR subsystems have been dominated by intergranular cracking, and, in some cases, these have been detected after relatively short operating times. For instance, such cracking was observed in late 1965 due to intergranular stress-corrosion cracking (IGSCC) in the weld heat-affected zone (HAZ) of a 150 mm (6 in.) type 304 stainless steel recirculation bypass piping after 68 months of plant operation; this cracking morphology was associated with the thermomechanical fabrication conditions that gave rise to a compositional heterogeneity in the grain boundary (which is discussed later). Subsequently, in 1966 and 1967, there were a number of additional pipe-cracking indications found by ultrasonic examination or by leakage. Numerous indications of IGSCC of welded type 304 stainless steel occurred in the fall of 1974 and early 1975; 64 incidents of cracking were identified during this short period, and all of them occurred in small-diameter lines less than 250 mm (10 in.) (Ref 18). During 1978, incidents of IGSCC were first noticed in a large-diameter (610 mm, or 24 in.) pipe in an overseas BWR (Ref 19). This incident posed additional concern because replacement of such large-diameter lines would be difficult, costly, and would require additional leak-before-break and degraded pipe behavior analysis. In 1982, extensive IGSCC of 710 mm (28 in.) recirculation piping was found in a U.S. BWR, and additional inspections revealed cracking in large-diameter piping of other U.S. BWRs (Ref 20, 21). Since the 1980s, other incidents of IGSCC or interdendritic stress-corrosion cracking (IDSCC) have been observed in nickel-base alloys (alloys 600, 182, and X-750) in, for instance, welded safe ends, access hole covers, shroud head bolts, and jet pump beams, as well as TGSCC in limited regions of stainless steels that had been cold worked (Ref 22–24) by, for instance, severe surface grinding. During this period, the importance of water purity control became increasingly obvious, especially with regard to creviced components where the geometry and oxidizing conditions in the bulk environment could give rise to increased anionic activity in the creviced region (Ref 25, 26) even though the bulk water purity was acceptable at that time. As is discussed later, mitigation of these issues relied on a combination of materials, stress, and environmental modifications, such as the use of lower-carbon (L-grade) stainless steels, which decreased the amount of grain-boundary sensitization and control of the water purity.

Since the mid-1980s, there have been increasing observations of IGSCC in stainless steel core components that have been exposed to neutron irradiation (>1 MeV) beyond a critical fluence level ($>5 \times 10^{20}$ n/cm^2) and that were highly stressed. Initially, these incidents were confined to easily replaced components (e.g., in-core instrumentation tubes) but, more recently, IASCC has been observed at lower fluence levels in larger welded core structures, with the extent of cracking depending on the specific combinations of prior thermal-sensitization due to the initial fabrication procedure, fast neutron flux and fluence, weld residual stress, and coolant purity (Ref 27–30). Thus it is apparent that the initial hypothesis that there is a critical "threshold fluence" required for IASCC depending on whether the component is at "high" or "low" stress is probably in error; this point is discussed later in this article. Although the substitution of type 304 stainless steels with L-grade alloys in these irradiated components has increased the operating time before cracking is observed, incidents can still occur (Ref 31), especially if the component has been severely surface cold worked by grinding, straightening, and so forth during manufacture.

Failure of the ferritic carbon steel and low-alloy steel components (e.g., feedwater piping and unclad regions of the pressure vessel) by EAC in BWRs has been very infrequent compared to the incidence rate observed in the austenitic alloys discussed previously. As is discussed later, there are fundamental reasons why it is remarkably hard to sustain crack propagation in the carbon steel and low-alloy steel/BWR environment systems under good water chemistry and nondynamic loading conditions. The isolated incidents that have been reported have been associated with dynamic loading during plant startup or relatively low-frequency cyclic loading due, for instance, to thermal stratification in piping or inadequate mixing of "hot" and "cold" coolant streams (Ref 32, 33). The morphology of this cracking is generally transgranular, although in a very few isolated incidents the cracking morphology may be intergranular depending on the stressing mode or environmental conditions.

These observations highlight the fact that obvious remedial actions, such as controlling the material sensitivity to grain-boundary compositional heterogeneity (by, for instance, using low-carbon or stabilized stainless steels) to reduce the likelihood of IGSCC, or by improving the water chemistry specifications (by, for instance, controlling the chloride concentration) may not be sufficient for extended operating times. Secondary factors such as surface cold work, radiation effects, or the presence of crevices may well reduce the effectiveness of these "obvious" remedial actions. Thus, long-term management of the materials degradation modes associated with EAC depends on a quantitative understanding of the secondary, as well as primary, interactions among the material, stress, and environmental factors.

Such cracking incidents can be disruptive from an operational viewpoint; for instance, capacity loss due to pipe cracking alone in U.S. BWRs peaked at 16.94% in 1984 before pipe-cracking mitigation actions were widely applied (Ref 34). These incidents are extremely costly in terms of component replacement, lost power availability, and manpower irradiation exposure (Table 2); thus the business driver for the reactor operator and reactor designer to minimize these occurrences and to institute permanent mitigation actions is high. From a regulatory viewpoint there is a requirement that the reactors be periodically inspected and evaluated for continued structural integrity (defined in terms of the risk of structural failure and, ultimately, core damage, and/or unacceptable coolant leakage) during the next operational period before reinspection (normally a fuel reload cycle of 12 to 24 months). These regulatory aspects take on increasing

Table 2 Cost (in millions of U.S. dollars) of repair of boiling water reactor components

Component	Estimated repair cost, $MM	Radiation exposure, man-rems	Repair time, critical path days	Total cost(a), million USD
Shroud support to vessel weld	250	5500	364	487
Control rod drive stub tube to vessel weld	50	700	225	170
In-core housing	50	225	100	102
Recirculation piping	20	1200	125	95
Recirculation inlet and outlet nozzles	25	300	44	50
Jet pump riser brace	10	26	63	42
Core support plate	4	30	50	29
Top guide	8	15	30	23
Instrumentation penetrations	6	200	25	20
Shroud	4	10	12	10
Manway cover (access hole)	2	5	10	7

(a) Assumptions in calculating total cost: (1) Radiation exposure cost; $10K/man-rem, (2) Critical path outage cost; $0.5MM/day, (3) No account taken for "present value analyses." Source: Ref 34

importance as reactor operators in the United States submit license renewal applications to the USNRC to extend the license period from the current 40 years to 60 years. In addition, BWR operators in the United States are now applying to the USNRC for power uprates of up to approximately 20% over the original licensed rating; such increases are accomplished by core redesign and increased coolant flow, thereby raising questions of increased risk of IASCC of austenitic stainless steels, flow-assisted corrosion (FAC) of carbon steels, and (corrosion) fatigue of some components due to flow-induced vibrations. Given the past BWR materials degradation history, there is the expectation from the regulator that the reactor operator (of both BWRs and pressurized water reactors, PWRs) has in place aging-management programs to address the current and future potential degradation modes in his specific plant. Many of these aging-management programs rely on past operating experience, laboratory studies, and analyses by the utilities (e.g., EPRI), reactor manufacturers, national laboratories, and the technical societies who issue USNRC-endorsed manufacturing and inspection codes (e.g., American Society of Mechanical Engineers, ASME). Examples of such aging-management programs relevant to specific components are described in more detail in Ref 35.

To satisfy the safety-driven expectations of the regulator and the business needs of the reactor operator it is necessary to (a) develop inspection techniques that can, with reasonable probability, detect and adequately size defects in sometimes complex geometries under potentially hazardous (e.g., irradiated) conditions, (b) develop algorithms that predict the kinetics of degradation, with defined degrees of uncertainty, in order to either specify the required inspection periodicity, or to define the amount of damage that might occur in a predetermined inspection interval, and, finally, (c) develop cost-effective mitigation actions that will prevent or mitigate the materials degradation.

The interaction between these requirements is shown schematically in Fig. 1. In this illustration it is assumed that a crack was not detected at inspection 1; that is, the crack depth or length is below the nondestructive examination (NDE) resolution limit, and the question is "Can it be expected that a crack will be detected at the planned inspection 2, and if it is, will the crack dimensions be within acceptable values vis-à-vis safety or leakage?" The quantification of the "acceptable value" will depend on either the dimensions of the component (e.g., pipe) to give leakage, or appropriate "margins of safety" on crack dimensions before unstable crack propagation can occur. If the answer to this preliminary question is uncertain then "Can the reactor continue to be operated if a specific mitigation action is introduced at some intermediate time?" It is important to predict and define these events since significant damage can occur relatively quickly under specific operating conditions, especially if the detection techniques have limited resolution. For instance, in 1982 through-wall cracking and leakage was observed in recirculation safe-ends at a domestic plant during a hydrostatic test at a position where nine months previously no cracking had been detected (Ref 20).

While the question of detecting and sizing the defect is beyond the scope of this article, attention is focused on the prediction of the cracking kinetics in BWRs and the methods for quantitatively predicting and mitigating materials degradation.

EAC Analysis

Analysis of EAC phenomena in BWRs since the early 1980s has occurred in two phases. The first and crucial one was to quantify the prime physical parameters that control the cracking process; this was necessary in order to define the mitigation actions. The extent of this understanding developed in the 1970s and early 1980s (Ref 12–14, 17, 18) was not enough, however, to answer the detailed life-management questions posed in Fig. 1. These questions could be answered only with a more quantitative understanding of the relationship between cracking susceptibility and the primary *and secondary* interactions that occur between the various mechanical, environmental, and material parameters controlling the cracking kinetics. This further understanding was developed in the second phase. The following discussion is organized chronologically to mirror this analysis and life-prediction development.

Historically, EAC has been divided into the "initiation" and "propagation" periods. To a large extent this division is arbitrary since, in most engineering situations, "initiation" is defined as the time at which a crack is detected, or when the load has relaxed a specific amount (in, for instance, a strain-controlled test). In both cases, such a definition of initiation corresponds to a crack depth of significant metallurgical dimensions. For instance, detection limits for cracks on the inside surface of a pipe may well be 10% of the pipe wall thickness, that is, approximately 2 mm (80 mils.) if the detection sensor is on the outside surface of the pipe. Similar crack depths correspond to a definition of "initiation" based on a defined amount of load relaxation (say 25%) in a strain-controlled test (upon which current fatigue design codes are based). Thus, for the current objective of lifetime prediction, it is assumed (Fig. 2) that crack initiation (Ref 36) is associated with *microscopic* crack formation, that is, well below a depth of 2 mm (80 mils), at localized corrosion or mechanical defect sites; such sites are related to, for instance, pitting or intergranular attack that may have dimensions of the order of 10 to 100 μm (0.4 to 4 mils), or scratches, weld defects, or design notches. The latter mechanical/design defects will exist at the time of construction, and it is assumed that the former undetectable microscopic initiation sites will exist relatively early in the life of larger-section components. Therefore, the problem of life prediction for larger-section components narrows down to understanding the growth of small cracks from each of these geometrically separated crack-initiation sites, and the coalescence of these small cracks to form a major crack that may then accelerate or arrest depending on the specific material, environment, and stress conditions. For the cracking systems relevant to BWRs this crack coalescence and subsequent growth of a single crack occurs, as shown in Fig. 3 (Ref 37), at crack depths of 50 μm (2 mils), which is well below normal

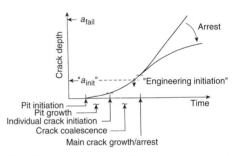

Fig. 2 Sequence of crack initiation, coalescence, and growth during subcritical cracking in aqueous environments. Note that "engineering initiation" corresponds to crack dimensions equal to crack detection capabilities, i.e., function of crack resolution and probability of detection. Source: Ref 36

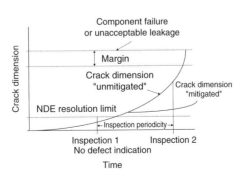

Fig. 1 Evolution of a crack, and the roles of crack dimension resolution and inspection periodicity in managing the structural degradation

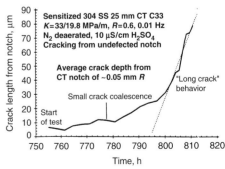

Fig. 3 Observed evolution of intergranular stress-corrosion crack in sensitized stainless steel (SS) compact tension (CT) specimen with single dominant crack being created at ~50 μm depth. Source: Ref 37

commercial crack-detection techniques for systems where the crack detector is well removed from the crack location. Thus, for all practical life-prediction and management purposes, one considers the system conditions that control propagation of an intrinsic single major crack 50 μm (2 mils) deep to a detectable size and the subsequent management of that propagation to ensure maintenance of safety.

There are numerous parameters (Ref 38) that have an effect on EAC propagation in BWR systems (Fig. 4). These initially were characterized (Ref 18) in a macroscopic sense as those that defined the degree of material susceptibility, the "aggressiveness" of the environment, and the magnitude of the tensile stress. This early concept is illustrated in Fig. 5, in which the small region consisting of the overlapping areas of three circles symbolizing the material, environment, and stress factors represents the conditions under which EAC may occur. The presence of these specific factors is a conjoint criterion; that is, all three factors must be present. This explains why the real percentage of, for instance, welded piping exhibiting cracking in a given plant is relatively low. It also indicates that if one factor is missing or reduced sufficiently, then the EAC susceptibility may be significantly reduced. Thus it is evident that there are three obvious methods for mitigating EAC:

- Reducing the tensile stress below some threshold value by new designs or processing techniques
- Changing/eliminating the susceptible material by using an alternate material, alloying addition, surface cladding, material heat treatment, or through processing controls.
- Eliminating or reducing the severity of the corrosive environment below some "threshold" criterion

Tensile Stress Factors

Based on experience with many material/environment cracking systems, the stresses must be tensile to initiate and propagate EAC. Sources of primary and secondary tensile stresses in the BWR reactor components include those produced during both fabrication and reactor pressurization. Fabrication stresses are those introduced during fit-up and assembly in the shop or in the field and include those associated with machining or forming operations (such as surface grinding, welding, or cold straightening). The extent of tensile stress due to some of these sources can be surprisingly large even for those associated with "cosmetic" repair; for example, grinding if done too abusively can introduce surface tensile stresses of the magnitude of the yield point or higher. Weld residual stresses will vary as a function of distance from the weld and through the component thickness (Fig. 6), with the precise variation depending on welding parameters such as heat input, welding speed, number of welding passes, etc. (Ref 18, 39–41). Although these effects are well understood on the basis of data and modeling, the complex parameter interactions can give rise to a scatter in measured residual stresses for a given nominal weld design for which all the relevant welding parameters have not been defined (Fig. 6b). Furthermore, not all sources of tensile stress are immediately obvious; for instance, sufficient tensile stress for EAC may occur due to growth of corrosion product in creviced geometries where there are no other major sources of applied or residual stress (e.g., the crevice regions of certain designs of top guides). Neutron fluence may relax the weld residual stresses in irradiated core structures but, as is discussed later, neutron irradiation may also introduce deleterious (vis-à-vis cracking susceptibility) grain-boundary sensitization and irradiation-induced hardening (Ref 27–30) and such competing effects may give nonmonotonic cracking kinetics.

Finally, cyclic stresses from vibrations or time-dependent straining from changes in reactor operating mode can also add to the total stress effect on cracking. Such varying stresses are of great importance in initiating cracks or in restarting stopped cracks because they provide continuing plastic strain.

The effect of stress level on the time to failure of sensitized austenitic stainless steels in BWR environments has been studied in detail, and a well-defined dependence of stress on time to cracking has been observed (Ref 19). This relationship is shown in Fig. 7 for furnace-sensitized (24 h at 650 °C, or 1200 °F) type 304 stainless steel subjected to various applied tensile stress modes in 0.2 ppm dissolved oxygenated BWR-type water. As expected, the lower the tensile stress the longer the time to component "failure." It is also apparent that the stress-dependency curve is asymptotic to the at-temperature yield strength of the sensitized material. This laboratory observation led to the development at General Electric (Ref 42) of a "stress rule index" (SRI) for evaluating plant

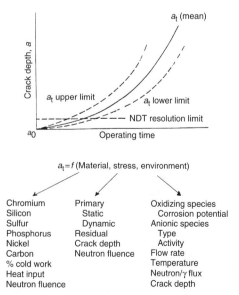

Fig. 4 Materials, stress, and environmental parameters relevant to environmentally assisted cracking in BWRs. NDT, non destructive testing. Source: Ref 38

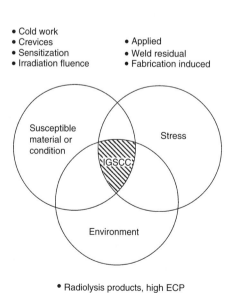

Fig. 5 The three conjoint factors necessary for producing intergranular stress-corrosion cracking (IGSCC). Appropriate changes in the stress, environment, or material conditions make such a Venn diagram applicable to other modes of EAC. ECP, electrochemical corrosion potential. Source: Ref 18

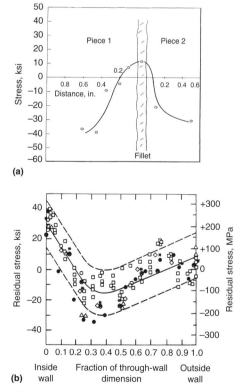

Fig. 6 (a) Variation in longitudinal residual stress adjacent to a weld on the internal pipe surface of a 100 mm (4 in.) pipe. Source: Ref 18. (b) Variation in through-wall residual stress in 700 mm (28 in.) diam schedule 80 piping close to the fusion line. Source: Ref 41

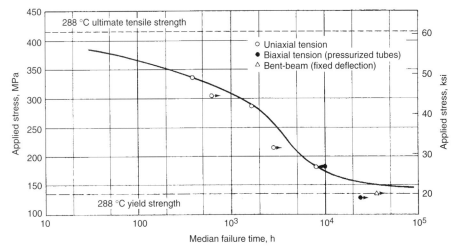

Fig. 7 Stress dependence of intergranular stress-corrosion cracking of a furnace-sensitized type 304 stainless steel in 288 °C (550 °F) water with 0.2 ppm oxygen. Source: Ref 19

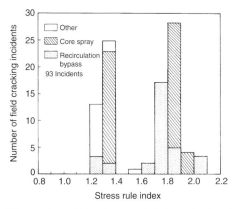

Fig. 8 General Electric stress rule index evaluation of 93 pipe-cracking incidents. Source: Ref 42

performance:

$$\text{SRI} = \left(\frac{(P_M + P_B)}{S_Y}\right) + \left(\frac{(Q + F + \text{Residual})}{(S_Y + 0.002E)}\right) \quad \text{(Eq 1)}$$

where P_M and P_B are the primary membrane and bending stress, respectively, S_Y is the ASME 0.2% offset yield stress at the operating temperature, Q represents the secondary stress (includes thermal), E is the ASME code elastic modulus at applicable temperature, and Residual equals the sum of all sources of residual stress.

Based on observations of numerous cracking incidents in type 304 stainless steel recirculation and core spray lines (Fig. 8), an allowable-stress criterion was defined such that the SRI value must be less than 1.0 to avoid cracking in this class of component. Different limiting SRI values were defined for the onset of cracking in other components, for example, creviced safe ends, and materials of construction.

As discussed earlier, the "failure time" for initially smooth specimens is largely controlled by the propagation rate (Fig. 2) of cracks with dimensions initially well below normally detectable values. An early compilation (Ref 43) of crack-propagation-rate data (da/dt) as a function of stress-intensity factor (K) for sensitized stainless steel in oxygenated water at 288 °C (550 °F) is shown in Fig. 9. Also shown in this figure is the "USNRC disposition curve" (Ref 20) to be used (Ref 44) for disposition of cracks in BWR piping. In Fig. 9:

$$da/dt = 3.590 \times 10^{-8} \, K^{2.161} \text{ in./h} \quad \text{(Eq 2)}$$

where K is the stress-intensity factor in units of ksi$\sqrt{\text{in.}}$ and da/dt is the crack-propagation rate in units of in./h.

It is apparent that there is a significant variance in the data set that may be attributed to the different degrees of the materials grain-boundary sensitization and range of the dissolved oxygen

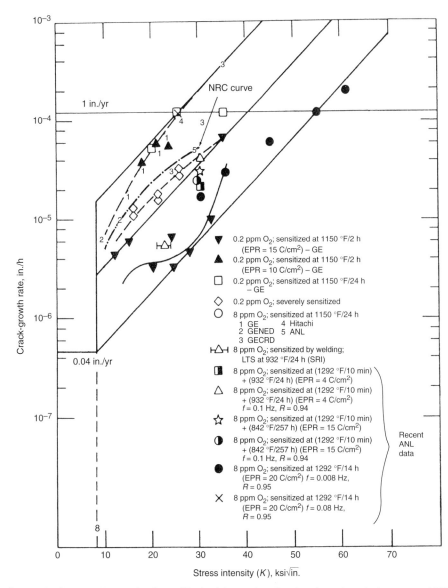

Fig. 9 Crack-propagation-rate data for sensitized stainless steel as a function of stress intensity in oxygenated 288 °C (550 °F) water. Source: Ref 43

content in the water. The corresponding data set for a more controlled oxygen content in the range 0.1 to 0.3 ppm, which is more representative of that in BWR water in the recirculation piping, is shown in Fig. 10 (Ref 38, 45);

the variance is reduced and the USNRC disposition line (Eq 2) approximates the mean of this data set. It is important to point out that the crack-propagation rates are relatively low compared to those in other classical stress-corrosion systems (e.g., copper alloys in ammoniacal solutions and mild steel in carbonate/bicarbonate solutions) and, given the logarithmic scale in Fig. 10, it would take a considerable experimental time to make an unambiguous case for the existence of a "threshold K_{ISCC}" below which the crack-propagation rate is "zero." However, a stress-intensity value can be quoted that corresponds to a very low propagation rate that is "acceptable" over many years of operation. As is discussed later, this "pseudo-K_{ISCC}" value is not a constant since its value will depend on the specific values of other system parameters.

Lower-temperature piping (e.g., the feedwater piping) is generally carbon steel (e.g., SA333 grade 6) whereas the higher-temperature piping comprises welded austenitic stainless steel. The original piping in the United States was predominately type 304 stainless steel (with type 308 weldments), but over the years this has been replaced by lower-carbon L-grade or nuclear-grade (NG) stainless steels that have less susceptibility to grain-boundary sensitization (described below). Core internals such as the core shroud, top guide, and so forth are all wrought and welded austenitic stainless steels of various grades. Stabilized stainless steels such as types 321 or 347 have been used in Germany from the time of construction for the piping and core internals, respectively. Nickel-base alloys are used for nozzle safe ends (e.g., alloy 600) and weld surface preparation butters or weldments (e.g., alloys 82 or 182) for joints between dissimilar ferritic and austenitic stainless steels components.

Material Factors

As indicated in Table 3, the structural materials for the RPV, the reactor internals, and the ancillary piping are ferritic carbon steels or low-alloy steels, and austenitic stainless steels or nickel-base alloys. The RPV is fabricated by welding low-alloy steel plate (e.g., A533 grade B, SA302 grade B), with the nozzle penetrations being forged A508 class 2. Surfaces of the low-alloy steel that are exposed to the water are clad with an austenitic duplex stainless steel (e.g., types 308, 309) for corrosion resistance, although in some of the earlier BWR designs this cladding was absent in the upper-head region on the pressure vessel.

All of the austenitic alloys are susceptible to a phenomenon known as "grain-boundary sensitization" when heated during fabrication in the temperature range approximately 550 to 850 °C (1000 to 1550 °F). This sensitization increases the susceptibility to intergranular localized corrosion and cracking in 288 °C (550 °F) BWR water due to the chromium depletion that occurs adjacent to the grain-boundary chromium-rich carbide ($M_{23}C_6$) precipitates. This phenomenon is understood in terms of the equilibrium relationships between the metal-matrix and carbide precipitates in an 18Cr-8Ni alloy as shown in Fig. 11 (Ref 46). This simplified phase diagram indicates that the carbon should remain in solid

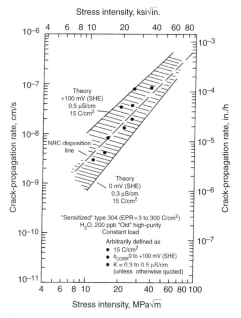

Fig. 10 Crack-propagation-rate/stress-intensity dependence for sensitized stainless steel in 288 °C (550 °F) water with 0.2 ppm oxygen and conductivity of 0.3–0.5 μS/cm. Note the insertion of theoretical relationships (discussed later) and the NRC regulatory relationship used for structural integrity evaluations. NRC, Nuclear Regulatory Commission. Source: Ref 38, 42, 43

Table 3 Compositions of primary materials of boiling water reactors

Alloy	Composition(a), wt%									Typical use
	C	Mn	P	S	Si	Cr	Ni	Fe	Others	
Low-alloy steels										
A533 grade B	0.25	1.15–1.50	0.035	0.035	0.15–0.40	...	0.40–0.70	bal	0.45–0.60Mo	Reactor pressure vessel
SA302 grade B	0.25	1.15–1.50	0.035	0.035	0.15–0.40	...	...	bal	0.45–0.60Mo	Reactor pressure vessel
A508 class 2	0.27	0.5–1.00	0.025	0.025	0.15–0.40	0.25–0.45	0.5–1.0	bal	0.055–0.70Mo 0.05V max	Reactor pressure vessel
SA333 grade 6	0.30	0.29–1.06	0.025	0.025	0.10 min	...	...	bal	...	Feedwater piping
SA193 grade B7	0.37–0.49	0.65–1.10	0.35	0.40	0.15–0.35	0.75–1.20	...	bal	0.15–0.25Mo	Closure studs and bolts
SA320 grade L43	0.38–0.43	0.60–0.85	0.35	0.40	0.15–0.35	0.70–0.90	1.65–2.00	bal	0.20–0.30Mo	Closure studs and bolts
Stainless steels										
304	0.08	2.0	0.045	0.03	1.0	18.0–20.0	8.0–10.5	bal	...	Piping, core internals
304L	0.03	2.0	0.045	0.03	1.0	18.0–20.0	8.0–12.0	bal	...	Piping, core internals
308	0.08	2.0	0.045	0.03	1.0	19.0–21.0	10.0–12.0	bal	...	Weldments, cladding
308L(b)	0.03	1.0–2.5	0.03	0.03	0.3–0.65	19.5–22.0	9.0–11.0	bal	0.75Cu	Weldments, cladding
309	0.20	2.0	0.045	0.03	1.0	22.0–24.0	12.0–16.0	bal	...	Weldments, cladding
309L(b)	0.03	1.0–2.5	0.03	0.03	0.3–0.65	23.0–25.0	12.0–14.0	bal	0.75Cu	Weldments, cladding
316	0.08	2.0	0.045	0.03	1.0	16.0–18.0	10.0–14.0	bal	2.0–3.0Mo	Piping, core internals
316L	0.03	2.0	0.045	0.03	1.0	16.0–18.0	10.0–14.0	bal	2.0–3.0Mo	Piping, core internals
316NG	0.08	2.0	0.045	0.03	1.0	16.0–18.0	10.0–14.0	bal	3.0Mo, 0.16N	Piping
321	0.08	2.0	0.045	0.03	1.0	17.0–19.0	9.0–12.0	bal	Ti: 5×C min	Piping
347	0.08	2.0	0.045	0.03	1.0	17.0–19.0	9.0–13.0	bal	Nb; 20×C min	Core internals
CF-6	0.06	1.5	...	...	2.0	18.0–21.0	8.0–11.0	bal	...	Cast valve fittings
Nickel-base alloys										
600	0.15	1.0	...	0.015	0.5	14.0–17.0	bal	6.0–10.0	0.5Cu	Nozzle safe ends
182(b)	0.10	5.0–9.5	0.03	0.015	1.0	13.0–17.0	bal	10.0	1.0–2.5Nb + Ta	Weldments, buttering
82(b)	0.10	2.5–3.5	0.03	0.015	0.5	18.0–22.0	bal	3.0	2.0–3.0Nb + Ta; 0.75Ti	Weldments
×750	0.08	1.0	...	0.01	0.5	14.0–17.0	bal	...	0.4–1.0Al, 0.5Cu, 0.7–1.2Nb, 2.25–2.75Ti	Jet pump beams

(a) Single values are maximum. (b) Filler metal composition

solution over the temperature range shown provided the carbon is less than approximately 0.03 wt%. Austenite-containing carbon in excess of 0.03 wt% should precipitate $M_{23}C_6$ upon cooling below the solubility line. However, at relatively rapid rates of cooling, this precipitation reaction is partially suppressed, and the room-temperature microstructure is characterized by austenite supersaturated with carbon. If this supersaturated austenite is reheated to elevated temperatures within the (gamma + $M_{23}C_6$) field, further precipitation of the chromium-rich $M_{23}C_6$ will take place preferentially at the austenite grain boundaries. Certain time-temperature combinations will be sufficient to precipitate this chromium-rich carbide, but will be insufficient to rediffuse the larger chromium atoms back from the grain matrix into the austenite near the carbide in the grain boundaries. The net result is the formation of envelopes of chromium-depleted austenite around the carbide (Fig. 12), which in certain environments will not resist intergranular attack and IGSCC. The sensitized regions adjacent to a weld are known as the heat-affected zone (HAZ) (Fig. 13), and it is within this zone where the grain boundaries are in such a chromium-depleted condition that the IGSCC is concentrated (Fig. 14).

This attack, when the chromium depletion is caused by the temperature/time combinations encountered during welding, has been known as "weld decay." However, more modern practice is to use the term "sensitized" to describe chromium depletion, irrespective of whether it has resulted from slow cooling, postweld heat treatment (PWHT), a process designed to reduce weld residual stress in low-alloy steel pressure vessels, elevated-temperature service, or welding. It is important, however, to distinguish between these sensitization definitions because the stress-corrosion response differs with different heat treatments. For example, "furnace-sensitized" type 304 stainless steel imparted by holding at a sensitization temperature is somewhat more susceptible to IGSCC than "weld-sensitized" material in an oxygenated pure water environment because of the higher integrated time at temperature and, hence, the degree of chromium depletion at the grain boundary.

The distance of the crack in the HAZ from the weld fusion line is governed by the material composition and heat input combinations, which maximize the degree of grain-boundary chromium depletion, and the weld joint design, total heat input during welding, and stress relief, which determine the tensile residual stress. In earlier piping systems, for example, where fairly high carbon content steels were common (e.g., 0.05%) the intergranular crack was located approximately 5 mm (200 mils) from the weld fusion line, but with the use of lower-carbon L-grade steels and refined welding practices the corresponding distance is considerably less and is governed by the strain gradient immediately adjacent to the weld fusion line.

From a life-prediction viewpoint, it is important that such sensitization phenomena be quantifiable. This has been the objective of extensive investigations over many decades, with focus on predicting the depleted chromium profile adjacent to the grain boundary as a function of alloy composition, cold work, isothermal temperature/time combinations, and welding history (heat input, component size, number of weld passes, etc.) (Ref 47–51). Alternatively, the extent of chromium depletion may be characterized in terms of the amount of intergranular attack measured by a number of nondestructive electrochemical tests (Ref 52–54).

Grain-boundary chromium depletion may also occur in the grain boundary without the $Cr_{23}C_6$ precipitation that is governed by the thermal equilibrium arguments given previously. This may occur due to the impact of neutron irradiation; the physics of this has also been the object of intense analysis (Ref 27–30) since this depletion, together with irradiation-induced metalloid segregation, can further reduce the IGSCC resistance over that associated solely with thermal sensitization. See the article "Effect of Irradiation on Corrosion and Stress-Corrosion Cracking in Light Water Reactors" in this Volume for additional information on this topic.

As mentioned in the section "Background to Problem" in this article, the ferritic alloys may, under fairly limited environmental, material, and stressing conditions, undergo intergranular cracking down prior austenite grain boundaries. The reason for this preferred cracking path is unknown, but is likely associated with metalloid (e.g., phosphorous, carbon, sulfur) segregation to these grain boundaries during fabrication in a manner akin to that in prior austenite grain boundaries of bainitic turbine rotor and disc NiCrMoV or CrMo steels. In the majority of cases, however, the cracking mode is transgranular with the "material sensitivity" being determined by the MnS precipitate size and distribution (Ref 54), and the Al/N alloying content, which affect dynamic strain aging at a given operating temperature (Ref 32, 55). These issues are expanded upon later, but relate to the activity of S^{2-} anions that can be created and retained at the crack tip and the degree of anelasticity in this region.

A final material parameter of importance vis-à-vis cracking sensitivity of both ferritic and austenitic alloys is the yield stress. A considerable amount of investigation on a variety of materials of construction in light water reactors (LWRs) has indicated that as the yield stress is increased either because of irradiation, cold work, or thermomechanical treatment, the EAC propagation rate in BWR environments increases, with potential changes from an intergranular to a transgranular cracking morphology (Fig. 15) (Ref 56). These effects may be localized. For example, the cracking susceptibility may be increased due to strain mismatch across grain boundaries in the HAZ; in this case the susceptibility may be decreased by appropriate thermomechanical treatment to control such high-angle grain-boundary structures. Another well-recognized example of localized cracking

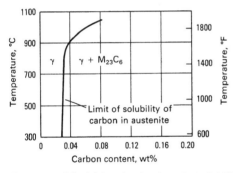

Fig. 11 Solid solubility of carbon in an Fe-18Cr-8Ni alloy. Source: Ref 46

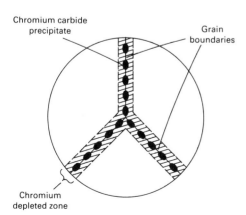

Fig. 12 Grain boundary in sensitized type 304 stainless steel. Source: Ref 19

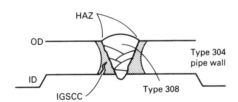

Fig. 13 Location of intergranular stress-corrosion cracking (IGSCC) in heat-affected zone (HAZ) of type 304 pipe. OD, outside diameter; ID, inside diameter. Source: Ref 19

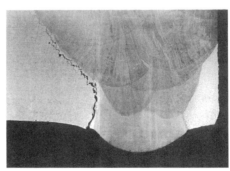

Fig. 14 Pipe test results showing intergranular stress-corrosion cracking in a 400 mm (16 in.) type 304 pipe heat-affected zone. Source: Ref 19

due to yield stress effects is that due to abusive surface grinding that leads to hardened surfaces in which transgranular cracks are easily initiated and propagate approximately 100 μm (4 mils) (depending on the extent of grinding) before transitioning to the normal intergranular mode in the softer underlying matrix. Finally, the straining imparted by weld shrinkage in the HAZs can also produce susceptibility to IGSCC even though there is no sensitization present.

Environmental Factors

The recirculating water coolant in a BWR undergoes radiolysis in the reactor core. The majority of the hydrogen produced in this process is preferentially partitioned to the steam phase, leaving a more-than-stoichiometric amount of dissolved oxygen in the water; this amount is about 0.2 ppm under normal BWR operating conditions. The oxidizing potential of the environment is further increased by the presence of 0.2 to 0.4 ppm dissolved hydrogen peroxide, which is also produced in the core. The corrosion potential of the reactor structural members is controlled by the reduction and oxidation reactions occurring on the metal surface; these reactions are dominated by the reduction of the dissolved oxygen and hydrogen peroxide, and the oxidation of the 0.02 ppm hydrogen that remains in the liquid phase. The specific value of the corrosion potential of the metal surface, which is crucial to the thermodynamics and kinetics of the material degradation modes, is a function of the state of the metal surface (e.g., oxidized versus freshly machined), dissolved gas concentrations, the water flow rate, and the temperature. A representative relationship (Ref 38) between corrosion potential for oxidized stainless steel and the dissolved oxygen content is shown in Fig. 16; these data were obtained from an operating BWR plant and show considerable variability, indicative that the dissolved oxygen content is not the only controlling variable. This factor is emphasized in Fig. 17, showing the significant impact on the corrosion potential of the dissolved hydrogen and hydrogen peroxide concentrations (Ref 57). A further variable of importance is the flow rate of the water past the metal surface, since the reactant concentrations and the resultant reaction rates at that surface are dominated by liquid diffusion control. Very similar corrosion potential versus dissolved oxygen content relationships to those illustrated in Fig. 16 and 17 are observed for low-alloy steels and nickel-base alloys, confirming the common controlling processes of dissolved oxygen and hydrogen peroxide reduction, and of hydrogen oxidation on these structural material surfaces.

The corrosion potential is crucial in determining the resultant EAC susceptibility of austenitic and ferritic alloys in BWR environments; this is illustrated in Fig. 18 and 19 for SCC of sensitized stainless steel and SA333 grade 6 carbon steel, respectively. It is apparent from

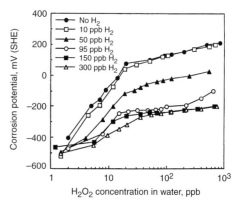

Fig. 17 Laboratory data indicating variation of corrosion potential type 316 of stainless steel with dissolved hydrogen peroxide and hydrogen in 288 °C (500 °F) water. Source: Ref 57

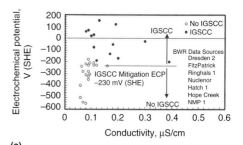

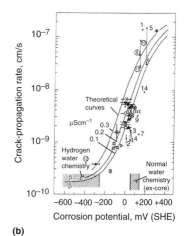

Fig. 18 (a) Corrosion potential and coolant conductivity combinations in which IGSCC is observed in sensitized type 304 stainless during an accelerated slow-strain-rate test conducted at various operating BWRs. Source: Ref 58. (b) Observed and predicted relationships between the crack-propagation rate and corrosion potential for sensitized (15 C/cm^2) type 304 stainless steel in 288 °C (550 °F) water under constant load (27.5 MPa√m, or 25 ksi√in.). Water conductivity in range 0.1 to 0.3 μS/cm. The data point (*) and those that are numbered refer to tests conducted under irradiation conditions in either an operating reactor or a variable energy cyclotron. Source: Ref 38, 59

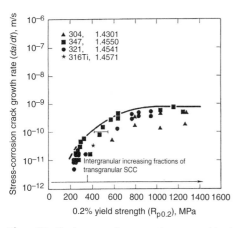

Fig. 15 Crack-propagation rate for nonsensitized stainless steels in simulated boiling water reactor water at 288 °C (550 °F) as a function of yield stress. Source: Ref 56

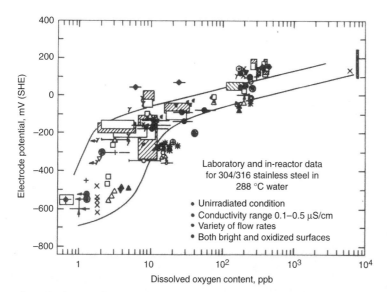

Fig. 16 Relationship between electrochemical corrosion potential (ECP) and dissolved oxygen content. Data from operating reactors, with the different symbols being associated with specific plants. Source: Ref 38

Fig. 18(b) and 19 that there is a very high sensitivity of the crack-propagation rate on the corrosion potential, thereby suggesting a prime mitigating parameter for EAC. Caution should be given, as with the concept of K_{ISCC} discussed earlier, in defining an absolute value of corrosion potential below which EAC "will not occur"; for instance, early investigations (Fig. 18a) ascribed a value of −230 mV standard hydrogen electrode (SHE) as a "threshold potential" for EAC of stainless steels based on data from accelerated slow-strain-rate tests in operating reactors. There are two concerns in this absolute definition. The first is that in many cases (see, for instance, Fig. 18b) cracking can occur, albeit at very slow propagation rates, at corrosion potentials below the "threshold" corrosion potential value; such rates would not be resolvable in an accelerated test such as that used for the data collection shown in Fig. 18(a). The second is that, although under most operating conditions −230 mV (SHE) is a conservative limiting value for constant load/deflection conditions, the absolute "threshold" value will depend on the interactions of other system variables (e.g., stress/strain rate, water purity, alternate alloys, irradiation flux/fluence, etc.) and, in fact, more positive threshold values of corrosion potential than −230 mV (SHE) may be appropriate for many current operating conditions.

Corrosion problems may be encountered on metal surfaces where cation or anion concentration and excessive metal salt/oxide precipitation occurs; these phenomena may be associated with, for instance, the presence of crevices in oxidizing environments, or boiling at heat transfer surfaces. To minimize these potential localized corrosion problems, the BWR water coolant purity is high. For instance in most current operating reactors the coolant purity approaches that of theoretical-purity water, that is, room-temperature conductivity 0.055 μS/cm, with dissolved chloride and sulfate activities maintained below 5 ppb. Early observations (Ref 25, 26) indicated that the incidence of plant component cracking correlated well with the plant reactor water conductivity or coolant impurity level. Subsequent laboratory measurements on the austenitic and ferritic alloys confirmed correlations between the crack-propagation rate and conductivity under both steady-state and transient water purity conditions as shown in Fig. 20 and 21, respectively (Ref 38, 59).

The relationships between the crack-propagation rate and the corrosion potential and conductivity shown in Fig. 18(b) and 19 to 21 indicate that it is not the conductivity of the bulk water that is of fundamental importance, but rather the crack-tip (non-OH⁻) anionic activity. This crack-tip activity will be increased over the bulk value by a factor of 10 to 50 (Ref 38) due to potential-driven diffusion down the crack since the crack mouth corrosion potential under "normal" oxidizing water chemistry conditions may be of the order of +50 mV (SHE), while the crack-tip potential will be of the order of −500 mV (SHE). Such a potential-driven increase in (non-OH⁻) anionic activity at the crack tip will be accompanied by a decrease in pH to maintain local electroneutrality. These combined effects of increased anionic activity and increased acidity at the crack tip will affect the oxidation kinetics occurring in this region and thereby the crack-propagation rate.

Strong acid-forming anions such as sulfate (Ref 60–73) or chloride (Ref 61, 74–76), which originate from resin decomposition in the demineralizer beds, leakage of impurities in defective condenser tubes, and so forth, can have a significant effect on the crack initiation and propagation rate of all the ferritic and austenitic alloys in BWR environments. The theoretical predictions of the effect of corrosion potential and anionic activity on the IGSCC propagation rate of sensitized type 304 stainless steel stressed at 27.5 MPa√m (25 ksi√in.) is indicated in Fig. 22 for sulfate and chloride anions. It is seen that under "normal" water chemistry conditions corresponding to a corrosion potential of approximately +50 mV (SHE) for a recirculation line, increasing the anionic activity from a preferred operating value of <5 to 20 ppb or, in a worst case scenario of 100 ppb, the crack-propagation rate would be predicted to increase by up to two orders of magnitude under oxidizing conditions. Under lower corrosion potential conditions associated with hydrogen water chemistry (HWC) with or without noble metal technology, it is predicted and confirmed (Ref 72) that the increases in propagation rate for stainless steels due to such anionic transients are markedly lower. Preliminary data for nickel-base alloys (Ref 73) indicate that the beneficial effect of lowering the propagation rate in sulfate-containing environments by decreasing the corrosion potential may not be quite as effective as observed in stainless steels, indicating possibly a change in cracking mechanism at low potentials to that pertinent to nickel-base alloys in PWR primary side environments.

Although cationic impurities will not concentrate to the crack tip due to potential-driven mass transport, they can have some effect on crack initiation and shallow crack propagation. For instance, bulk pH changes either side of neutral can be detrimental to crack initiation (Ref 72, 77) with there being more degradation for the acidic shift. However, these changes in cracking susceptibility are likely to be minor in practice since pH changes in the BWR coolant will be limited if the reactor coolant conductivity and anionic activity specification limits are maintained.

Zinc, which may be present in the reactor coolant due to corrosion of the brass condenser tubes or purposely added to control radioactive ⁶⁰Co buildup on the metal surfaces, may increase the cracking resistance of stainless steel, alloy

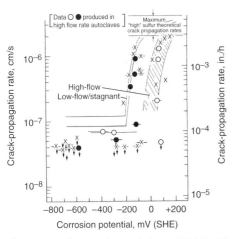

Fig. 19 Predicted (Ref 60) and observed (Ref 61, 62) dependency of the stress-corrosion crack-propagation rate on corrosion potential for A508 steel containing 0.01% S strained at 1 to 1.5×10^{-6} s⁻¹ in 288 °C (550 °F) water with conductivity of 0.2 μS/cm

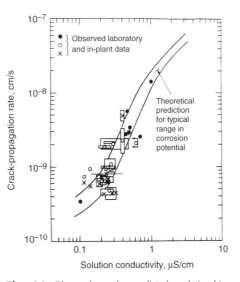

Fig. 20 Observed and predicted relationships between the crack-propagation rate and water conductivity for 316L stainless steel under constant load (27.5 MPa√m, or 25 ksi√in.) in 288 °C (550 °F) water containing 200 ppb oxygen. Source: Ref 38, 59

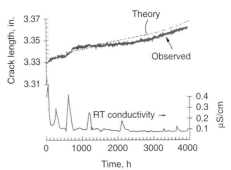

Fig. 21 Observed and predicted (theory) crack depth versus time relationship for sensitized type 304 stainless steel under constant load (27.5 MPa√m, or 25 ksi√in.) during conductivity transients during reactor operation. RT, room temperature. Source: Ref 38, 59

600, and alloy 182 (Ref 78–81) in the doping range 5 to 10 ppb and is effective even in the presence of sulfate and chloride contamination. This beneficial effect is attributed to increases in the fracture strain of the surface oxide and the repassivation rate at the crack tip (Ref 82) which, as is discussed later, are fundamental to the crack-propagation process.

The complex interactions between the cracking susceptibility of a given structural alloy and the dissolved oxygen, hydrogen peroxide and hydrogen concentrations, specific anionic impurity concentrations, temperature, and stressing state, that is, constant versus dynamic loading, are quantifiable and applicable to plant "on-load" operations and, as is discussed later, practical mitigation actions may be defined. Under transient plant operations, such a startup and shutdown, the situation becomes more complicated since, not only are these system parameters (e.g., temperature, dissolved oxygen content, etc.) continuously changing, but there is the possibility that the rate-controlling steps in the cracking mechanism may also change. This created some early concern given the "lore" that EAC problems in operating plant (not just nuclear reactors) can generally be related to such transient operating conditions. This concern was addressed in the early 1980s (Ref 83–90) with the conclusion that, although the EAC susceptibility in BWRs may be increased for both ferritic and austenitic alloys (especially at intermediate oxygen and temperature combinations encountered during plant startup), the cumulative damage during this transient period was minor compared to the damage accumulated during the much longer "on-load" period. One recommendation from these studies, however, which has been widely adopted, is the practice of deaerating the coolant before and during the startup procedure.

Prediction of EAC in BWRs

The previous section discusses the primary materials, environment, and stress parameters that, in isolation, have an effect on the cracking susceptibility of ferritic and austenitic structural alloys in BWRs. This discussion does not lead directly, however, to the quantification of the primary and secondary interactions between these parameters that are necessary for the development of a prediction algorithm of the general form:

Crack dimensions

$= f$ (materials, stress, environment) time

(Eq 3)

where the materials, stress, and environment parameters listed in Fig. 4 may change with structural dimensions and time.

There are two basic approaches that have been taken in deriving such a generic life-prediction algorithm. The first is the classic method of developing a multiparameter correlation function that encompasses all relevant plant or laboratory data. The second is based, first, on developing a quantitative understanding of the mechanism of cracking, followed by qualification of this mechanisms-informed analysis via comparison with the relevant single and multi-effects data sets.

Data-Based Life-Prediction Approaches

These approaches rely on the development of correlation functions based on laboratory or plant operational data. Although sound in principle such approaches are very much limited by the quality, and thereby the variance, of the data upon which the correlation is based. For instance, data for cracking of stainless steel piping as a function of plant operational time (Fig. 23) can exhibit extensive scatter because of an imprecise definition of the system upon which the data was derived. Thus, although analyses of trends in plant behavior as a function of alloy type or water chemistry are useful for predicting the likelihood of a particular failure mode in a given component and for developing risk-informed component inspection procedures (Ref 91), they are insufficient for development of deterministic relationships such as described in Eq 3 or Fig. 4.

Analysis of laboratory data to derive Eq 3 promises greater success than that of plant data since, in principle, such experimental data can be obtained under more stringent system definition and control. Unfortunately, as exemplified by the crack-propagation-rate data for stainless steels in Fig. 9, the expected decrease in data variance is not always achieved, either because at the time that the data was collected the techniques for accurately measuring crack-propagation rates relevant to 40 year component lives had not been developed or because the experimental system was not adequately defined or controlled. For instance, uncontrolled or unmonitored changes in coolant purity can give substantial changes and, thereby, variance in the cracking response of all the BWR structural alloys (Fig. 20). Similarly, the lack of measurement of corrosion potential, plus the practice in the 1970s and 1980s of monitoring solely "dissolved oxygen content" led to inadequate control of the experiment since the corrosion potential could vary by significant amounts and, thereby, lead to large changes in cracking response (Fig. 18 and 19) for relatively small changes in total oxidant concentration and water flow rate. Consequently in more recent times extensive attention has been paid to experimental control and data quality (Ref 92, 93), and old data have been examined and, where necessary, censored with respect to agreed-upon quality-control criteria and adequate system control. Such censoring activities can decrease the data variance at the cost of substantially reducing the number of data points that can be analyzed (Ref 92–94), thereby having an impact on the statistical treatment of these data.

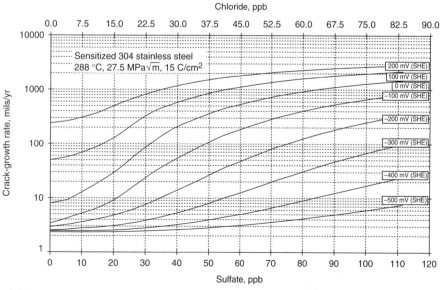

Fig. 22 Crack-growth prediction as a function of sulfate and chloride concentration

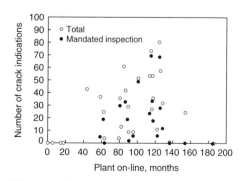

Fig. 23 Relationship between incidences of cracking in BWR piping and operational time. Source: Ref 18

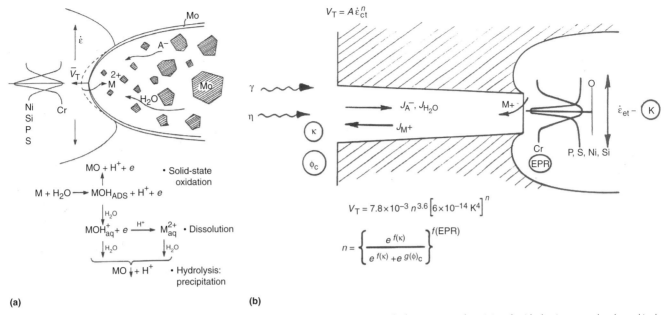

Fig. 24 Relevant phenomena in the crack enclave. (a) Physical processes at the crack tip, illustrating the large amount of precipitated oxide that is expected and noted in the crack enclave. (b) Crack enclave and the relevant phenomena associated with the slip-oxidation mechanism of crack advance. See Eq 5 to 8 for details. Source: Ref 59

Mechanisms-Informed Life-Prediction Approaches

Development of mechanisms-informed models to answer the basic life-management questions posed in Fig. 1 and the related text is viable due to the marked increase in mechanistic understanding of the EAC process that has occurred since the late 1980s and the availability of improved experimental and analytical techniques (e.g., electrochemistry, electron microscopy, material analysis) necessary to quantify the model input and to verify the model predictions.

A requirement at the start of such a mechanisms-informed development is that there be a working hypothesis for crack propagation, which must be then quantitatively validated against well-controlled data. The prime contending hypotheses in the late 1970s were the slip-oxidation, film-induced cleavage, and hydrogen embrittlement models (Ref 95). Although thermodynamic arguments can be made to support all three models for ductile alloys in BWR systems, the soundest kinetic arguments are those based on the slip-oxidation model after considering the range of environments, stressing modes, and materials relevant to BWR operations (Ref 59, 84, 96).

The slip-oxidation model equates, via Faraday's relationship, crack propagation to the oxidation charge density at the crack tip that is associated with the transformation of the metal to the oxidized state (e.g., MO or M^+_{aq}). As shown in Fig. 24, the relevant phenomena in the crack enclave are (Ref 59):

- The creation of a crack-tip environment that may be markedly different from the bulk environment due to the mass transport (controlled by potential drop, convection, etc.) of solvated cations, water, and anions within the crack enclave. A significant driver in this process is the potential drop down the crack length, which exists due to the highly oxidizing conditions at the crack mouth and the deaerated, reducing conditions at the crack tip.
- The metal oxidation rate of the crack-tip alloy whose composition may be significantly different from the bulk alloy and which is being dynamically strained by the residual or applied loads.

There are two basic approaches that have been taken to quantify these processes. The first is that taken by General Electric and formulated in the PLEDGE model (*Plant Life Extension Diagnosis by GE*). This model equates the propagation rate to the oxidation kinetics of the strained crack-tip material in the *localized* crack-tip environment (Fig. 24a) (Ref 38, 59, 84); it emphasizes the importance of the corrosion potential at the crack mouth as being one controlling component in creating a mechanism for anion and acidity concentration at the crack tip (Fig. 24b). The second approach is that of Macdonald et al. (Ref 97–99), which *directly* couples the total oxidation current emanating from the strained crack tip with the reduction rates (involving oxygen and hydrogen peroxide) at the crack mouth. This Coupled *E*nvironment *F*racture *M*odel (CEFM) and the PLEDGE model predict the same general dependencies of SCC for stainless steels on, for instance, corrosion potential, water conductivity, sensitization, and stress intensity, although in several cases there are significant differences in predicted cracking behavior at the extremes of these system parameters.

For brevity, a more detailed description of the prediction capabilities is confined to the PLEDGE model, and its development for EAC in

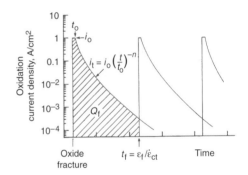

Fig. 25 Oxidation current density versus time relationship for a strained crack tip. Source: Ref 59

the stainless steel/water system and its logical expansion to cover irradiation-assisted cracking (due to changes in material and environmental parameters in the model), and EAC in nickel-base alloys and low-alloy ferritic steels.

The change in oxidation current density with time (and hence oxidation charge density) following the rupture of a protective film at the crack tip is shown in Fig. 25. Initially, the oxidation rate and, hence, crack-propagation rate will be rapid, controlled by activation or diffusion kinetics as the exposed metal dissolves (Ref 59). However, in most of the cracking systems of interest in the BWR, a protective oxide or salt is thermodynamically stable and will reform via solid-state oxidation and/or precipitation from solution on the bare surface. As a result, the oxidation (dissolution) rate, i_t, will fall from the bare surface value, i_o, following a time period, t_o, according to:

$$i_t = i_o(t/t_o)^{-n} \qquad \text{(Eq 4)}$$

Consequently, the rate of total oxidation charge density and, hence, crack-tip advance will slow with time following the initial rupture of the crack-tip oxide. Thus, crack propagation can only be *maintained* if the film rupture process reoccurs. Therefore, for a given crack-tip environment, potential, and material condition, the crack-propagation rate will be controlled by both the change in oxidation charge density with time and the frequency of film rupture at the strained crack tip. This latter parameter will be determined by the fracture strain of the film, ε_f, and the strain rate at the crack tip, $(d\varepsilon/dt)_{ct}$. Thus, by invoking Faraday's law, the average environmentally controlled crack-propagation rate, da/dt, is related (Ref 38, 59, 96) to the strain rate at the crack tip, $(d\varepsilon/dt)_{ct}$, the oxidation rate at the bared crack tip, i_o, and the repassivation rate parameter, n, via:

$$da/dt = (M/z\rho F)(i_o t_o^n/(1-n)\varepsilon_f^n)(d\varepsilon/dt)_{ct}^n \quad \text{(Eq 5)}$$

where M and ρ are the atomic weight and density, respectively, of the crack-tip metal, F is Faraday's constant, and z is the number of electrons involved in the overall oxidation of an atom of metal.

Under constant load or displacement conditions, the crack-tip strain rate will be related to the creep processes at a moving crack tip, and under monotonically or cyclically changing bulk strain conditions it will be related to an applied strain rate. Thus, the slip oxidation mechanism may be applied to not only stress corrosion but also strain-induced cracking and corrosion fatigue.

Recognition of the interacting effects of mass transport within the crack enclave, oxidation or reduction rates at the crack tip and crack mouth, and the crack-tip strain rate has proven of great value in identifying the generic system changes that are likely to reduce the extent of a particular EAC problem (Ref 95). For instance, increasing cracking resistance is to be expected by increasing the passivation rate at the crack tip (by control of the corrosion potential/anionic activity combinations or by material composition changes) and by reduction in crack-tip strain rate (by attention to dislocation morphology, the imposed stress history, etc.) (Ref 84, 95). Thus, there is a mechanisms-informed basis for understanding and developing quantitative predictions of the empirical observations of the effects on EAC susceptibility of, for instance, zinc (effect on oxide fracture strain and passivation rate), chloride and sulfate (effect on passivation rate), and so forth. It is also important to point out that these mechanisms-informed arguments meld into the empirical observations of the factors that influence EAC in BWR systems (Fig. 26).

To advance this concept to application to BWR cracking systems, it is necessary to redefine the fundamental Eq 5 in terms of measurable engineering or operational parameters (Fig. 24b). This involves:

- Defining the crack-tip alloy/environment composition in terms of, for example, bulk alloy composition and heat treatment, anionic concentration or conductivity, dissolved oxygen, hydrogen and hydrogen peroxide contents, or corrosion potential at the crack mouth
- Measuring the reaction rates for the crack-tip alloy/environment system that correspond to the "engineering" system
- Defining the crack-tip strain rate in terms of continuum parameters such as stress, stress intensity, loading frequency, and so forth. There has been extensive work conducted in these areas (Ref 38, 59, 96) for austenitic and ferritic alloys in BWR systems, and the technical details are not reproduced here. Attention is focused on illustrating how these advances have been incorporated into verified, quantitative life-prediction methodologies.

Sensitized Type 304 Stainless Steel

The relevant fundamental reactions pertinent to crack-tip systems have been extensively investigated leading to a quantification of the basic parameters in Eq 5. For stainless steels, this equation for the crack-propagation rate may be simplified to:

$$da/dt = 7.8 \times 10^{-3} n^{3.6} (d\varepsilon/dt)_{ct}^n \quad \text{(Eq 6)}$$

where da/dt is in units of cm/s, n varies between 0.33 and 1.0, depending on factors discussed below, and the crack-tip strain rate is in units of s^{-1}.

As discussed previously, n, the passivation rate parameter in Eq 4 is controlled by the crack-tip environmental (e.g., pH, potential, anionic activity) and material (e.g., percentage of chromium depletion in grain boundary) conditions. To make this practically useful, n has been reformulated in terms of measurable bulk system parameters such as water conductivity (κ), corrosion potential (ϕ_c)—which, in turn, is a function of the dissolved oxygen and hydrogen peroxide content—and the "electrochemical potentiokinetic repassivation" (EPR) parameter,

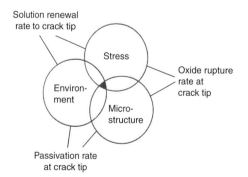

Fig. 26 Relationship between the empirical conjoint factors necessary for EAC: stress, environment, and material conditions, and the crack-tip mechanistic factors, oxide rupture frequency, passivation rate, and solution renewal rate. Source: Ref 38, 59, 96

which is related to the chromium depletion in the grain boundary. Thus, as shown in Fig. 27:

$$n = \left\{ \frac{e^{f(\kappa)}}{e^{f(\kappa)} + e^{g(\phi_c)}} \right\}^{h(EPR)} \quad \text{(Eq 7)}$$

The crack-tip strain rate, $(d\varepsilon/dt)_{ct}$, In Eq 5, may be related to the engineering stress (or stress-intensity) parameters. For instance, under constant load, the crack-tip strain rate is related to the stress-intensity factor, K, via:

$$(d\varepsilon/dt)_{ct} = 6 \times 10^{-14} K^4 \quad \text{(Eq 8)}$$

where the strain rate is in units of s^{-1} and the stress-intensity factor (K) is in units of $ksi\sqrt{in}$.

This formulation of the crack-tip strain rate proved adequate in the early development and application of the model, but it oversimplified the physical fact that as the crack tip advanced the associated stress field activated further dislocation sources. Such activation would increase the crack-tip strain rate by an amount proportional to the crack-propagation rate and the strain profile in front of the crack tip. The resultant crack-tip strain rate would, therefore, be the result of a dynamic equilibrium between this crack-propagation rate-dependent strain rate and the contribution due to exhaustion creep, which would be decreasing with time at the relatively low homologous temperatures relevant to BWR operations, that is:

$$(d\varepsilon/dt)_{ct} = (d\varepsilon/dt)_{creep} + (d\varepsilon/da) \cdot (da/dt) \quad \text{(Eq 9)}$$

where a is the distance in front of original crack tip.

The general validity of the prediction methodology is indicated in Fig. 28, which covers data obtained on sensitized type 304 stainless steel in

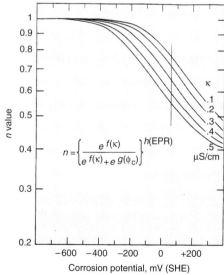

Fig. 27 Relationships between n in Eq 6 and 7 and the corrosion potential and bulk solution conductivity for a sensitized (EPR = 15 C/cm^2) type 304 stainless steel in water at 288 °C (550 °F). EPR, electrochemical potentiokinetic repassivation. Source: Ref 38

8 ppm oxygenated water at 288 °C (550 °F) stressed over a wide range of constant load, monotonically increasing load, and cyclic load conditions. The solid line shows the theoretical relationship from Eq 6 and 7 and the appropriate strain-rate formula for the stressing mode. The agreement between prediction and observation illustrates the applicability of the methodology to the whole stress-corrosion/corrosion-fatigue spectrum. The effect of, for instance, coolant (water) deaeration on the relationship between crack-propagation rate and crack-tip strain-rate is shown in Fig. 29. The degree of the predicted and observed beneficial effect of deaeration is very dependent on the strain rate (i.e., stress level and/or degree of dynamic loading), and this has an impact on evaluating the benefit of deaeration and the choice of the test technique that will be used to quantify the benefit. For instance, the factor of improvement (defined as the ratio of the crack-propagation rates in the two environments at a given crack-tip strain rate) decreases with increasing strain rate: thus factors of improvement determined by an accelerated "slow-strain-rate test" (with crack-tip strain rates in the range 10^{-4} to 10^{-6} s^{-1}) will significantly underestimate the improvement expected in a plant component subject to weld residual stresses (with crack-tip strain rates less than 10^{-8} s^{-1}).

The specific effects of changes in corrosion potential and solution conductivity on the propagation rate under constant load are shown in Fig. 18 and 20; the beneficial effects of improved water chemistry control (by controlling the coolant conductivity or anion content) and coolant deaeration are predicable and quantifiable and thereby lay a solid mechanistic basis for the definition of water chemistry specifications.

Nickel-Base Alloys

The development of prediction models for EAC of nickel-base alloys in BWR systems has been made difficult by the fact that the observed data, against which the prediction models can be validated, either exhibit extreme scatter or were obtained under conditions that do not apply directly to current BWR plant operation (e.g., high conductivity, high stress intensity, corrosion potentials outside range of interest, etc.).

Establishing a mechanistic link between crack-growth rates in stainless steels and ductile nickel-base alloys helps resolve this problem, since it permits the much broader base of data and modeling for stainless steels to provide guidance on the expected cracking susceptibility of nickel-base alloys in BWR systems (Ref 100, 101). Such a mechanistic comparison is reasonable since:

- Intergranular or interdendritic cracking is the dominant failure mode in both the austenitic stainless steels and nickel-base alloys and is associated with chromium depletion at the grain boundary in both alloy classes and hardening or straining. In the case of the stainless steels, the parameter EPR is used to quantify this depletion phenomenon, as described previously. Such a parameter cannot be measured easily in the nickel-base alloys; thus to use the stainless steel prediction algorithms directly, an "equivalent EPR" value has to be introduced.
- Both alloy systems involve ductile face-centered-cubic structures with similar mechanical properties. Thus it is reasonable to use in preliminary model derivations the same crack-tip strain-rate algorithms for ductile nickel-base alloys as for the stainless steels.
- The solubility of the (NiCr) oxides and (FeCr) oxides are very similar in BWR water and thus from an electrochemical viewpoint the crack-tip system (Fig. 24a) and rate-determining steps in the bare surface oxidation kinetics are likely to be similar for the nickel-base and austenitic steel alloy systems. Thus it is to be expected that the i_o, t_o, and n values in Eq 4 will be comparable, and, hence, the general propagation Eq 5 may be used for nickel-base alloys in preliminary analyses.

Despite these assumptions, the predictions are in reasonable agreement with the observations of stress corrosion and corrosion fatigue in the nickel-base alloy/BWR systems. For instance,

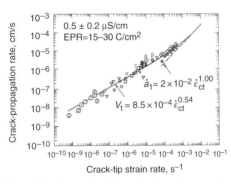

Fig. 28 Observed and theoretical crack-propagation rate versus crack-tip strain-rate relationships for sensitized type 304 stainless steel in oxygenated water at 288 °C (550 °F). EPR, electrochemical potentiokinetic repassivation. Source: Ref 38, 59

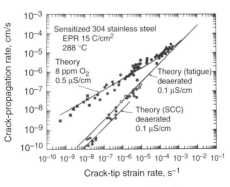

Fig. 29 Observed and theoretical crack-propagation rate versus crack-tip strain-rate relationships for stainless steel in aerated and deaerated water. Source: Ref 38, 59

the crack-propagation rate versus water conductivity relationships observed in operating plant at various corrosion potentials for alloy 182 are in reasonable agreement with the theoretical predictions (Fig. 30). In addition, the variation in cracking of alloy 600 access hole covers and shroud head bolts as a function of the average plant conductivity are also reasonably predicted. The data and predictions for these cases are shown in Fig. 31(a) and (b). In Fig. 31(a), the importance of the detrimental effect of excessive

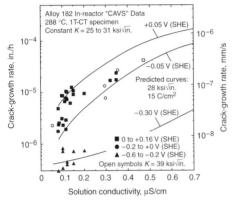

Fig. 30 Comparison between the predicted and observed crack-growth rates versus coolant conductivity for alloy 182 compact tension (CT) specimens exposed to BWR water of varying chemistries

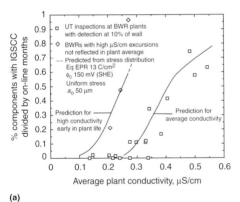

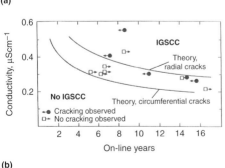

Fig. 31 (a) The predicted and observed effects of average plant water purity on crack incidence in creviced alloy 600 shroud head bolts. (b) Predicted and observed IGSCC on 15 mm (0.600 in.) welded access hole covers

coolant transients is noted even if the average conductivity is subsequently maintained at a low average value.

In conclusion, there is now the methodology to predict the extent of EAC damage for many of the BWR structures, provided the relevant materials, environment, or stressing variables are known, or can be calculated, with sufficient accuracy. This capability offers the opportunity to *proactively* manage the material degradation problems (Fig. 32) by defining cost-effective mitigation actions that may be implemented before costly (Table 2) repairs are necessary.

Mitigation of EAC in BWRs

The following discussion presents several EAC mitigation techniques for BWR components. The mitigation methods are presented under general categories of material solutions, stress solutions, and environmental solutions. A few of the remedies address two of the three necessary conjoint factors illustrated in Fig. 5.

Materials Solutions

The materials solutions to the piping IGSCC concern in the United States consist primarily of replacing the susceptible type 304 and 316 stainless steels with more sensitization-resistant materials, such as types 316 and 304 nuclear grade (NS), redissolving the chromium carbides by solution heat treatment, and cladding with crack-resistant weld metal. Other materials, such as type 347NG, have been used successfully in Europe.

Nuclear-Grade Stainless Steels. The replacement of current piping materials with materials more resistant to sensitization is a straightforward approach to mitigating IGSCC. It is well documented that decreasing the carbon content and increasing the molybdenum content of stainless steel will reduce the kinetics of sensitization (Ref 102–105). However, type 316NG and type 304NG (with no added molybdenum) stainless steels take this theme a step further. Instead of the nominal 0.03% C maximum of the L-grade stainless steels, (Fig. 11) the nuclear grades are characterized by a maximum carbon content of 0.020%. The second important composition characteristic of type 304NG and type 316NG is the specification of 0.060 to 0.100% N. This modification is designed to recover the decrease in alloy strength due to the reduction of the carbon content. Another successful approach that has been used in Germany is the use of low-carbon niobium-stabilized type 347 stainless steel.

For the nuclear-grade materials, full-size pipe tests have shown that factors of improvement over reference type 304 stainless steel performance can be expected to be at least 50 to 100 times in normal BWR operation (Ref 103). The necessary factors for improvement for the current 40 year service life are approximately 20. Therefore, the replacement of type 304 stainless steel piping with type 316NG, type 304NG, or type 347NG provides substantial resistance to IGSCC. It is important to note that, as discussed previously, materials-based resolutions can be rendered insufficient if there are other secondary aggravating effects; for instance creviced design or the introduction of cold-worked microstructures through various fabrication practices. For example, a data-based life-prediction analysis at one Swedish nuclear plant determined that nearly 50% of failures were attributable to cold work fabrication practices (Ref 91). Normally resistant stainless steels can also suffer IGSCC if good-quality water chemistry in the BWR is not maintained.

Solution Heat Treatment. Immunity against IGSCC of type 304 stainless steel can be provided by eliminating weld-sensitized regions. This can be accomplished by solution heat treatment to redissolve the chromium carbides and eliminate chromium depletion around previously sensitized grain boundaries (Fig. 11 and 12). Moreover, solution heat treatment will eliminate detrimental cold work and weld residual stress in the pipe. Following a butt-welding operation, the entire pipe segment is solution annealed at 1040 to 1150 °C (1900 to 2100 °F) for 15 min per 25 mm (1.0 in.) of thickness but not less than 15 min or more than 1 h, regardless of thickness. The pipe segment is then quenched in circulating water to a temperature below 205 °C (400 °F). Solution heat treatment is generally limited to those weld joints made in the shop where heat treatment facilities are available, because of dimensional tolerance consideration, size constraints of the vendor facilities (furnace and quench tank), and cooling rate requirements (dead end legs).

Corrosion-resistant cladding achieves its resistance to IGSCC by using the IGSCC resistance inherent in duplex austenitic-ferritic weld metals (Ref 103). Although the carbide precipitation observed in the HAZ inside surface is also present in the weld metal, the nature of the duplex structure of the weld metal provides resistance to IGSCC in the BWR. In fact, IGSCC propagating from the weld HAZ is generally blunted when it reaches the weld metal if sufficient ferrite is present. As shown in Fig. 14, the cracking will actually curve away from the weld metal. Field experience has indicated that in the as-welded condition very little ferrite is required to prevent IGSCC. These results and numerous laboratory data generated on welded and furnace-sensitized type 308 and type 308L weld metal prompted the conclusion that a minimum amount of ferrite (8%) must be present to provide a high degree of resistance to IGSCC in BWR environments. As with type 304 stainless steel, reducing the carbon level is also beneficial.

In the corrosion-resistant cladding technique, type 308L weld metal is applied using controlled heat input process to the inside surface of the pipe at the pipe weld ends before making the final field weld. This duplex weld metal covers the region that will become sensitized during the final weld process, thus providing IGSCC resistance by maintaining low carbon and a sufficient ferrite level in the region that would normally be sensitized.

Weld overlay repair is similar to corrosion-resistant cladding in that it uses layers of IGSCC resistant duplex weld metal (Ref 104). For cases where alloy 182 is to be overlaid, alloy 82 or 52 weld metal is used. The most significant difference is that the layer of weld metal is placed on the outside surface of the pipe while the pipe is being cooled internally with water and is used to prevent an existing crack from penetrating through the wall. The weld overlay is also applied as a structural reinforcement to restore the original piping safety margins (Fig. 33). An equally important effect of the weld overlay is that it produces a favorable (compressive) residual stress pattern that can retard or arrest crack growth.

The weld overlay technique has the potential for being the most cost-effective method as

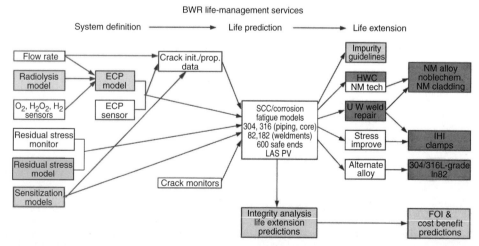

Fig. 32 Proactive management scheme for addressing environmentally assisted cracking (EAC) problems and assessing the cost-effective mitigation action

compared to other repair techniques (pipe replacement, solution heat treatment, corrosion-resistant cladding), which require draining of the system. For additional information, see the article "Hardfacing, Weld Cladding, and Dissimilar Metal Joining" in *Welding, Brazing, and Soldering,* Volume 6 of the *ASM Handbook.*

Tensile Stress Reduction Solutions

The tensile stress solutions primarily affect the weld residual stress profile by placing the inner surface weld residual stress in compression. These solutions, which are discussed below, include heat sink welding, induction heating stress improvement, and last-pass heat sink welding.

Heat Sink Welding. If a pipe can be welded without producing a sensitized structure and high residual tensile stresses in the weld HAZ, the resultant component will be resistant to IGSCC in the BWR environment. The heat sink welding program developed procedures that reduce the sensitization produced on the inside surface of welded pipe and, more important, change the state of surface residual welding stresses from tension to compression. This approach can be used in shop or field applications. Heat sink welding involves water cooling the inside surface of the pipe during all weld passes subsequent to the root pass or first two layers. Water cooling can be applied by using flowing or turbulent water, by spray cooling through a sparger placed inside the pipe, or, in a vertical run, by still water. Laboratory type 304 stainless steel butt welds have been produced to evaluate the inside surface heat sink welding techniques (Ref 105). Residual stresses were measured with strain gages. It was found that in a variety of pipe sizes the inside surface tensile residual stress is reduced substantially or changed from tension to compression as a result of this approach. Heat sink welding, as mentioned previously, has a secondary benefit in that it reduces the time at temperature for sensitization due to the presence of the cooling water heat sink.

Induction Heating Stress Improvement. This technique changes the normally high tensile stress present on the pipe inside surface of weld HAZs to a benign compressive stress (Ref 106, 107). This process involves induction heating of the outer pipe surface of completed girth welds to approximately 400 °C (750 °F) while simultaneously cooling the inside surface, preferably with flowing water (Fig. 34). Thermal expansion caused by the induction heating plastically yields the outside surface in compression, while the cool inside surface plastically yields in tension. After cool down, contraction of the pipe outside surface causes the stress state to reverse, leaving the inner surface in compression and the outside surface in tension (Fig. 35).

Qualification of the effectiveness of induction heating stress improvement has been accomplished by establishing (Ref 104, 107):

- Induction heating stress improvement treatment reliably reduces normally high tensile inside surface residual stresses to a zero compressive state.
- Full-size environmental pipe testing and residual stress tests have demonstrated that these beneficial residual stresses result in a large improvement in IGSCC resistance.
- Metallurgical investigations have shown that an induction heating stress improvement treatment produces no adverse effects. No increase in sensitization is found, and no significant variation in mechanical properties occurs.
- A minimum life of 12 fuel cycles (~216 months) has been measured on induction heated stress improved precracked pipes in the laboratory if the initial cracking does not exceed about 20% of wall thickness.

Last-Pass Heat Sink Welding. As discussed in the previous section, induction heating stress improvement is an IGSCC mitigation technique that favorably alters the weld residual stress pattern. The welding torch is the heat source that initially produces the undesirable residual stress that induction heating stress improvement counterbalances. Analysis was able to establish that this residual stress state could be made compressive by the introduction of inside surface water cooling during the welding operation. Cooling during the entire welding process (heat sink welding) or just during the last pass could be effective in reversing the residual stresses analogous to the induction heating stress improvement process. Qualification of the last-pass heat sink welding process consisted of magnesium chloride ($MgCl_2$) residual stress tests, which verified that the last-pass heat sink welding process does produce compressive residual stresses uniformly around the pipe circumference (Ref 108). Stress-relief strain-gage measurements quantified the compressive axial stress and revealed that the stresses were compressive up to about 50% through-wall.

Finally, intergranular SCC improvement was evaluated by using pipe tests performed at applied stresses above yield in a high-oxygen (8 ppm) high-temperature (288 °C, or 550 °F) water environment. As shown in Fig. 36, at the end of the program period, the pipes had been on test for more than 5500 h, demonstrating a factor of more than 5.5 and more than 6.5 improvement, respectively, at the two test stresses of 193.7 and 211 MPa (28.1 and 30.6 ksi), respectively. This factor of improvement approaches that determined for induction heating stress improvement.

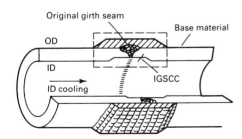

Fig. 33 Weld overlay intergranular stress-corrosion cracking (IGSCC) mitigation technique. ID, inside diameter; OD, outside diameter

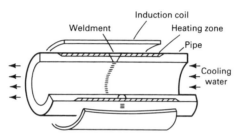

Fig. 34 Heating and cooling process for induction heating stress improvement

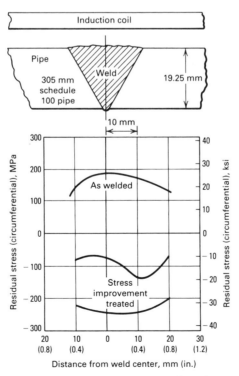

Fig. 35 Residual stress comparison for induction heating stress improvement

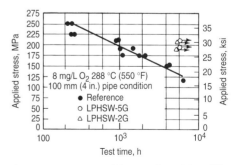

Fig. 36 Comparison of last-pass heat sink welding pipe tests with reference pipe tests

Environmental Solutions

Water Purity Control. As indicated previously in Fig. 20 to 22, 30, and 31, the BWR coolant conductivity or purity has a profound impact on the degree of EAC susceptibility for all the major materials of construction in the BWR system. As a result, water chemistry guidelines recommend maximum limits to the BWR feedwater and recirculation water coolant conductivity and cationic and anionic activities. Action levels (e.g., ultimately reactor shutdown) are prescribed should these limits be exceeded by specific amounts for given time periods. As a result, the purity levels that are maintained in the BWR coolant have markedly improved, with the mean conductivity level of approximately 0.37 µS/cm observed in 1980 decreasing to approximately 0.10 µS/cm in 2003 (Fig. 37).

Hydrogen Water Chemistry and Noble Metal Technology. As indicated in Fig. 18, 19, and 22, there is abundant laboratory and plant experience to indicate that the EAC susceptibility will be markedly reduced by decreasing the corrosion potential of the structural material to below −230 mV (SHE) via control of the reduction kinetics associated with the dissolved oxygen and hydrogen peroxide in the reactor coolant by small hydrogen gas additions. Initial, very short term (~10 h and 4 days) in-reactor hydrogen injection tests were performed as early as 1979 and 1981 (Ref 109, 110). After a somewhat longer (6 weeks) hydrogen demonstration test at the Dresden-2 plant in 1982 (Ref 111), this plant became the first facility to initiate full-time operation on hydrogen injection, that is, hydrogen water chemistry (HWC). The test clearly demonstrated that the addition of hydrogen does indeed reduce the dissolved oxygen content as measured by electrochemical potential techniques plus oxygen analysis, and the beneficial effect of such environmental changes on IGSCC was verified by in-reactor constant extension rate tests at several BWRs (see Fig. 18a and Ref 58 and 112). Crack-growth data versus time and environment on precracked furnace-sensitized type 304 stainless steel obtained at several BWRs revealed that no significant crack extension occurs (Ref 113).

The laboratory hydrogen water chemistry materials program was characterized by an extensive test program (Ref 114). The testing techniques utilized in the program included full-size pipe tests; fracture mechanics studies; electrochemical investigations; constant extension rate tests; straining electrode tests; constant load tests; bent beam tests; fatigue testing; cyclic crack-growth studies; general, galvanic, and crevice corrosion investigations; and corrosion oxide analysis. All of these laboratory studies clearly demonstrated that hydrogen water chemistry mitigates environmental cracking in, for example, BWR recirculation piping.

As shown in Fig. 38, the amount of hydrogen required to achieve this "protection potential" is nominally of the order of 1.0 to 1.5 ppm injected into the feedwater line. However, the kinetics of the H_2/O_2 recombination on the reactor structural surface is catalyzed by gamma radiation primarily in the downcomer region between the RPV wall and the core shroud. Consequently, the actual amount of hydrogen required for protection is plant-geometry specific and, in some cases, considerably more hydrogen is required, especially for protection of core internal structures such as the core shroud and top guides. The consequence of injecting an excessive amount of hydrogen is that volatile nitrogen species (e.g., N_2, NH_3, NO, NO_2) are formed due to homogeneous reduction reactions, and the radiation levels in the main steam line and turbine may increase by a factor of 4 to 5 due to the formation of ^{16}N via a neutron-proton (np) reaction on ^{16}O. As a consequence, hydrogen water chemistry is not effective in protecting reactor core internals without an unacceptable increase in radiation levels outside the reactor containment.

The current strategy for achieving a protection potential condition for all wetted reactor components (including the core internals) is to introduce a noble metal onto the material surface (Ref 115–117); this action effectively increases the exchange current densities for the H_2/H_2O and O_2/H_2O redox reactions and allows recombination of the O_2 and H_2 at much lower hydrogen addition levels. The noble metals favored are primarily platinum and rhodium that may be introduced either as (a) an alloying addition to the structural material, (b) a platinum-containing weld alloy that may be deposited as a cladding, or (c) as an aqueous additive to the feedwater line in the form of $Na_2Pt(OH)_6$ and $Na_3Rh(NO_2)_6$. This latter process, known as noble metal chemical application (NMCA), has found wide application (Fig. 39) in that it effectively lowers the corrosion potential to below −230 mV (SHE) in core structures with less than 0.3 to 0.4 ppm H_2 addition to the feedwater line, that is, injection levels that do not lead to increases in the main steam line radiation level.

Summary of Current Situation and Commentary on the Future

It has been argued that because of defense-in depth design and operating features, materials degradation issues such as SCC of core internals in current BWR designs do not increase significantly a plants' baseline core-damage frequency (CDF) or large energy release frequency (LERF) (Ref 118). However, there is no question that such failures have a marked negative impact on both the economy of plant operation (Table 2) and the public perception of nuclear reactor safety. Thus, the resolution of such problems is essential, especially with changes in the plants' licensing basis associated with, for example, life extension/license renewal and power uprates that may affect EAC susceptibility.

The early empirical understanding of the EAC phenomena (stress-corrosion cracking, corrosion fatigue) and the relative roles of the primary materials, environment, and stress input that define the degree of susceptibility was the basis of many of the EAC mitigation actions currently applied to BWRs. In the last 20 years, this capability has been complemented by a mechanisms-informed life-prediction methodology that accounts for the primary and secondary interactions between these inputs. This allows for a rapid analysis and understanding of the fundamental issues associated with new problems (e.g., cracking of core components) and a definition of appropriate mitigation and system control actions (e.g., development of cracking-resistant alloys and/or environmental modifications) that need to be implemented before costly damage ensues.

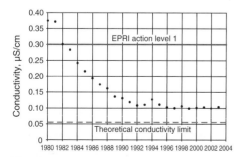

Fig. 37 BWR mean reactor water conductivity history

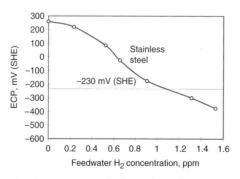

Fig. 38 Variation of electrochemical corrosion potential (ECP) of stainless steel as function of feedwater hydrogen concentration

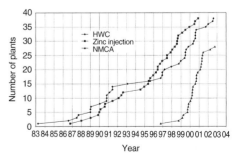

Fig. 39 Chronology of implementation of hydrogen water chemistry (HWC), and zinc injection, and noble metal chemical application (NMCA)

The challenges for the future will be governed by changes in operating and regulatory conditions. For instance, in the United States there is a steady movement toward regulations that are not only performance-based but also risk-informed. This development emphasizes decision-making that takes into account the relative importance of various maintenance, in-service inspection, procurement, and so forth activities in terms of their impact on plant safety (e.g., CDF, LERF) (Ref 119–123). To date, time-dependent materials degradation phenomena have not been included in the probabilistic risk assessments and the associated event- and fault-trees that address, for instance, either the accident initiation frequency or the reliability of components that play a role in subsequent accident mitigation. Thus, the near-term challenge in materials-degradation technology is to extend the current deterministic and mechanisms-informed understanding of the cracking phenomena to cover the uncertainties associated with the random crack-initiation processes, and the uncertainties associated with the propagation rate data and models.

A further operational driver that presents challenges is the constant desire for longer fuel cycles and the resultant increase in time between component inspections. This puts a premium on defining the kinetics of damage accumulation and the development of inspection techniques that have adequate resolution and probability of detection capabilities for the particular system at risk (Fig. 1).

REFERENCES

1. *Proc. First Int. Symposium on Environmental Degradation in Nuclear Power Systems—Water Reactors* (Myrtle Beach, SC), Aug 22–25, 1983, J. Roberts and W. Berry, Ed., National Association of Corrosion Engineers
2. *Proc. Second Int. Symposium on Environmental Degradation in Nuclear Power Systems—Water Reactors* (Monterey, CA), Sept 9–12, 1985, J. Roberts and J. Weeks, Ed., American Nuclear Society
3. *Proc. Third Int. Symposium on Environmental Degradation in Nuclear Power Systems—Water Reactors* (Traverse City, MI), Aug 30–Sept 3, 1987, G.J. Theus and W. Berry, Ed., The Metallurgical Society
4. *Proc. Fourth Int. Symposium on Environmental Degradation in Nuclear Power Systems—Water Reactors* (Jekyll Island, GA), Aug 6–10, 1989, D. Cubicciotti and E. Simonen, Ed., National Association of Corrosion Engineers
5. *Proc. of Fifth Int. Symposium on Environmental Degradation in Nuclear Power Systems—Water Reactors* (Monterey, CA), Aug 25–29, 1991, D. Cubicciotti and E. Simonen, Ed., American Nuclear Society
6. *Proc. Sixth Int. Symposium on Environmental Degradation in Nuclear Power Systems Water Reactors* (San Diego, CA), Aug 1–5, 1993, E. Simonen and R. Gold, Ed., NACE International
7. *Proc. Seventh Int. Symposium on Environmental Degradation in Nuclear Power Systems—Water Reactors* (Breckenridge, CO), Aug 7–10, 1995, R. Gold and A. McIlree, Ed., NACE International
8. *Proc. Eighth Int. Symposium on Environmental Degradation in Nuclear Power Systems—Water Reactors* (Amelia Island, FL), Aug 10–14, 1997, A. McIlree and S. Bruemmer, Ed., American Nuclear Society
9. *Proc. Ninth Int. Symposium on Environmental Degradation in Nuclear Power Systems—Water Reactors* (Newport Beach, CA), Aug 1–5, 1999, S. Bruemmer and F.P. Ford, Ed., The Metallurgical Society
10. *Proc. Tenth Int. Conf. Environmental Degradation in Nuclear Power Systems—Water Reactors* (Lake Tahoe, NV), Aug 5–9, 2001, F.P. Ford and G. Was, Ed., NACE International
11. *Proc. 11th Int. Conf. Environmental Degradation in Nuclear Power Systems—Water Reactors,* Aug 5–9, 2003, G. Was and L. Nelson, Ed., American Nuclear Society
12. *Proc. Seminar on Countermeasures for Pipe Cracking in BWRs,* Vol 1–4, Jan 22–24, 1980, Report WS 79-174, Electric Power Research Institute, May 1980
13. *Proc. Second Seminar on Countermeasures for Pipe Cracking in BWRs,* Vol 1–3, Nov 15–18, 1983, Report NP-3684-SR, Electric Power Research Institute, Sept 1984
14. *Proc. Seminar on BWR Corrosion, Chemistry and Radiation Control,* Oct 8–10, 1984, Electric Power Research Institute
15. http://www.nrc.gov/reading-rm/doc-collections/
16. R.N. Duncan et al., *Nucl. Appl.,* Vol 1 (No. 5), Oct 1965, p 413–418
17. S.H. Bush and R.L. Dillon, Stress Corrosion in Nuclear Systems, *Proc. Conf. Stress Corrosion Cracking and Hydrogen Embrittlement of Iron Base Alloys* (Unieux-Firminy, France), June 12–16, 1973, National Association of Corrosion Engineers, 1977, p 61
18. H.H. Klepfer et al., "Investigation of Cause of Cracking in Austenitic Stainless Steel Piping," NEDO-21000, General Electric, July 1975
19. B.M. Gordon and G.M. Gordon, Corrosion in Boiling Water Reactors, *Corrosion,* Vol 13, 9th ed., *Metals Handbook,* 1987, p 927–937
20. W.S. Hazelton and W.H. Koo, "Technical Report on Materials Selection and Processing Guidelines for BWR Coolant Pressure Boundary Piping," NUREG-0313 Rev. 2, U.S. Nuclear Regulatory Commission, Jan 1988
21. J.C. Danko, Recent Observations of Cracks in Large Diameter BWR Piping; Analysis and Remedial Actions, *Proc. First Int. Symposium on Environmental Degradation in Nuclear Power Systems—Water Reactors* (Myrtle Beach, SC), Aug 22–25, 1983, J. Roberts and W. Berry, Ed., National Association of Corrosion Engineers, p 209–222
22. "Recirculation Inlet Safe End Repair Program," IE Report U.S. Nuclear Regulatory Commission, Dec 8, 1978
23. "Stress Corrosion Cracking in Nonsensitized 316 Stainless Steel," Information Notice 84-39, U.S. Nuclear Regulatory Commission, Dec 7, 1984
24. Y. Kanazawa and M. Tsubota, "Stress Corrosion Cracking of Cold Worked Stainless Steel in High Temperature Water," Paper 237, Corrosion/94 (Baltimore, MD), NACE International, Feb 1994
25. K.S. Brown and G.M. Gordon, Effects of BWR Coolant Chemistry on Propensity for IGSCC Initiation and Growth in Creviced Reactor Internals Components, *Proc. Third Int. Symposium on Environmental Degradation in Nuclear Power Systems—Water Reactors* (Traverse City, MI), Aug 30–Sept 3, 1987, G.J. Theus and W. Berry, Ed., The Metallurgical Society, p 243–248
26. G.M. Gordon and K.S. Ramp, Dependence of Creviced BWR Component IGSCC Behavior on Coolant Chemistry, *Proc. Fourth Int. Symposium on Environmental Degradation in Nuclear Power Systems—Water Reactors* (Jekyll Island, GA), Aug 6–10, 1989, D. Cubicciotti and E. Simonen, Ed., National Association of Corrosion Engineers, p 14.46–14.62
27. P.L. Andresen, F.P. Ford, S.M. Murphy, and J.M. Perks, State of Knowledge of Radiation Effects on EAC in LWR Core Materials, *Proc. Fourth Int. Symposium on Environmental Degradation in Nuclear Power Systems—Water Reactors* (Jekyll Island, GA), Aug 6–10, 1989, D. Cubicciotti and E. Simonen, Ed., National Association of Corrosion Engineers, p 1.83–1.121
28. P.L. Andresen and F.P. Ford, Modeling and Prediction of Irradiation Assisted Cracking, *Proc. Seventh Int. Symposium on Environmental Degradation in Nuclear Power Systems—Water Reactors* (Breckenridge, CO), Aug 7–10, 1995, R. Gold and A. McIlree, Ed., NACE International, p 893–908
29. J.L. Nelson and P.L Andresen, Review of Current Research and Understanding of Irradiation Assisted Stress Corrosion Cracking, *Proc. Fifth Int. Symposium on Environmental Degradation in Nuclear Power Systems—Water Reactors* (Monterey, CA), Aug 25–29, 1991, D. Cubicciotti and E. Simonen, Ed., American Nuclear Society, p 10–26

30. S.M. Bruemmer et al., "Critical Issues Reviews for the Understanding and Evaluation of Irradiation Assisted Stress Corrosion Cracking," TR-107159, Electric Power Research Institute, Nov 1996
31. J. Medoff, "Status Report; Intergranular Stress Corrosion Cracking of BWR Core Shrouds and Other Internals Components," NUREG-1544, U.S. Nuclear Regulatory Commission, March 1996
32. J. Hickling, Strain Induced Corrosion Cracking of Low Alloy Steels under BWR Conditions; Are There Still Open Issues? *Proc. Tenth Int. Conf. Environmental Degradation in Nuclear Power Systems—Water Reactors* (Lake Tahoe, NV), Aug 5–9, 2001, F.P. Ford and G. Was, Ed., NACE International
33. "BWR Feedwater Nozzle and Control Rod Drive Return Line Nozzle Cracking," NUREG-0619, U.S. Nuclear Regulatory Commission, 1980
34. C. Wood, "BWR Water Chemistry Guidelines—2000 Revision," Report TR-103515, Electric Power Research Institute, Feb 2000
35. "Generic Aging Lessons Learned (GALL)," NUREG-1801, Vol 1 and 2, U.S. Nuclear Regulatory Commission, July 2001
36. F.P. Ford, Prediction of Corrosion Fatigue Initiation in Low Alloy Steel and Carbon Steel/Water Systems at 288 °C, *Proc. Sixth Int. Symposium on Environmental Degradation in Nuclear Power Systems—Water Reactors* (San Diego, CA), Aug 1–5, 1993, E. Simonen and R. Gold, Ed., NACE International, p 9–16
37. P.L. Andresen, I.P. Vasatis, and F.P. Ford, "Behavior of Short Cracks in Stainless Steel at 288 °C," Paper 495, Corrosion/90 (Las Vegas, NV), April 1990
38. F.P. Ford, D.F. Taylor, P.L. Andresen, and R.G. Ballinger, "Corrosion-Assisted Cracking of Stainless Steel and Low-Alloy Steels in LWR Environments," Report NP5064S, Electric Power Research Institute, Feb 1987
39. R.M. Chrenko, Residual Stress Measurements on Type 304 Stainless Steel Welded Pipes, *Proc. Seminar on Countermeasures for Pipe Cracking in BWRs,* Vol 2, Jan 22–24, 1980, Report WS 79-174, Electric Power Research Institute, May 1980
40. W.J. Shack and W.A. Ellingson, Measured Residual Stresses in Type 304 Stainless Steel Piping Weldments, *Proc. Seminar on Countermeasures for Pipe Cracking in BWRs,* Vol 2, Jan 22–24, 1980, Report WS 79-174, Electric Power Research Institute, May 1980
41. W.J. Shack, Measurement of Through Wall Residual Stresses in Large Diameter Piping Butt Welds Using Strain Gage Techniques, *Proc. Seminar on Countermeasures for Pipe Cracking in BWRs,* Vol 2, Jan 22–24, 1980, Report WS 79-174, Electric Power Research Institute, May 1980
42. B.M. Gordon, "Corrosion and Corrosion Control in BWRs," NEDO-24819A, Class 1, General Electric Co., Dec 1984
43. W.J. Shack et al., "Environmentally Assisted Cracking in Light Water Reactors," Annual Report, Oct 1981–Sept 1982, NUREG CR3292, U.S. Nuclear Regulatory Commission, Feb 1983
44. "NRC Position on Intergranular Stress Corrosion Cracking in BWR Austenitic Stainless Steel Piping," USNRC Generic Letter 88-01, Jan 25th, 1988
45. F.P. Ford, P.L. Andresen, H.D. Solomon, G.M. Gordon, S. Ranganath, D. Weinstein, and R. Pathania, Application of Water Chemistry Control, On-Line Monitoring and Crack Growth Rate Models for improved BWR Materials Performance, *Proc. Fourth Int. Symposium on Environmental Degradation in Nuclear Power Systems—Water Reactors* (Jekyll Island, GA), Aug 6–10, 1989, D. Cubicciotti and E. Simonen, Ed., National Association of Corrosion Engineers, p 4.26–4.51
46. C. Husen and C.H. Samans, *Chem. Eng.,* Vol 27, Jan 1969
47. S.M. Bruemmer, L.A. Charlot, and B.W. Arey, *Corrosion,* Vol 44, 1988, p 328–333
48. H.D. Solomon and D.C. Lord, *Corrosion,* Vol 36, 1980, p 395–399
49. H.D. Solomon, *Corrosion,* Vol 40, 1984, p 51–60
50. H.D. Solomon, Weld Sensitization, *Proc. Seminar on Countermeasures for Pipe Cracking in BWRs,* Vol 2, Jan 22–24, 1980, Report WS 79-174, Electric Power Research Institute, May 1980
51. S.M. Bruemmer, L.A. Charlot, and D.G. Atteridge, *Corrosion,* Vol 44, 1988, p 427–434
52. W.L. Clarke, "The ECP Method for Detecting Sensitization in Stainless Steels," NUREG/CR-1095, U.S. Nuclear Regulatory Commission, Feb 1980
53. W.L. Clarke, R.L. Cowan, and W.L. Walker, Comparative Methods for Measuring Degree of Sensitization in Stainless Steels, *Intergranular Corrosion of Stainless Alloys,* STP 656, ASTM, 1978, p 79
54. J.H. Bulloch, EAC Phenomena in Reactor Pressure Vessel Steels—Role of MnS Segregation, *Proc. Third Int. Symposium on Environmental Degradation in Nuclear Power Systems—Water Reactors* (Traverse City, MI), Aug 30–Sept 3, 1987, G.J. Theus and W. Berry, Ed., The Metallurgical Society, p 261–268
55. H.D. Solomon and R. Delair, Influence of Dynamic Strain Aging on Low Cycle Fatigue of Low Alloy Steels in High Temperature Water, *Proc. Tenth Int. Conf. Environmental Degradation in Nuclear Power Systems—Water Reactors* (Lake Tahoe, NV), Aug 5–9, 2001, F.P. Ford and G. Was, Ed., NACE International
56. M.O. Speidel and R. Magdowski, Stress Corrosion Cracking of Stabilized Austenitic Stainless Steels in Various Types of Nuclear Power Plants, *Proc. Ninth Int. Symposium on Environmental Degradation in Nuclear Power Systems—Water Reactors* (Newport Beach, CA), Aug 1–5, 1999, S. Bruemmer and F.P. Ford, Ed., The Metallurgical Society, p 325–329
57. Y.J. Kim, *Corrosion,* Vol 52 (No. 8), 1996, p 618–625
58. R.L. Cowan, "Optimizing Water Chemistry in U.S. Boiling Water Reactors," VGB Conference Plant Chemistry (Essen, Germany), Oct 1994
59. F.P. Ford, The Crack Tip System and it's Relevance to the Prediction of Cracking in Aqueous Environments, *Proc. First Int. Conf. Environmentally Assisted Cracking of Metals* (Kohler, Wisconsin), Oct 2–7, 1988, R. Gangloff and B. Ives, Ed., National Association of Corrosion Engineers, p 139–165
60. R.B. Davis and M.E. Indig, "The Effect of Aqueous Impurities on the Stress Corrosion Cracking of Austenitic Stainless Steel in High Temperature Water," paper 128, Corrosion/83 (Anaheim, CA), April 1983, National Association of Corrosion Engineers
61. P.L. Andresen, Effects of Specific Anionic Impurities on Environmental Cracking of Austenitic Materials in 288 Water, *Proc. Fifth Int. Symposium on Environmental Degradation in Nuclear Power Systems—Water Reactors* (Monterey, CA), Aug 25–29, 1991, D. Cubicciotti and E. Simonen, Ed., American Nuclear Society, p 209–218
62. W.E. Ruther and T.F. Kassner, "Effect of Sulfuric Acid, Oxygen and Hydrogen in High Temperature Water on Stress Corrosion Cracking of Sensitized Type 304 Stainless Steel," paper 125, Corrosion/83, (Anaheim, CA), April 1983, National Association of Corrosion Engineers
63. P.L. Andresen, "The Effects of Aqueous Impurities on Intergranular Stress Corrosion Cracking of Sensitized 304 Stainless Steel," report NP-3384, Electric Power Research Institute, Nov 1983
64. M.E. Indig, G.M. Gordon, and R.B. Davis, The Role of Water Purity on Stress Corrosion Cracking, *Proc. First Int. Symposium on Environmental Degradation in Nuclear Power Systems—Water Reactors* (Myrtle Beach, SC), Aug 22–25, 1983, J. Roberts and W. Berry, Ed., National Association of Corrosion Engineers, p 506–531
65. R.J. Kurtz, "Evaluation of BWR Resin Intrusions on Stress Corrosion Cracking of Reactor Structural Materials," report NP-3145, Electric Power Research Institute, June 1983
66. W.J. Shack et al., "Environmentally Assisted Cracking in Light Water

Reactors: Annual Report October 1983–September 1984," NUREG/CR-4287, ANL-85-33, U.S. Nuclear Regulatory Commission, June 1985
67. W.J. Shack et al., "Environmental Assisted Cracking in Light Water Reactors Semi-annual Reports," NUREG/CR-4667, Vol 1, ANL-86-31, U.S. Nuclear Regulatory Commission, June 1986
68. W.E. Ruther et al., "Effect of Temperature and Ionic Impurities at Very Low Concentrations on Stress Corrosion Cracking of Type 304 Stainless Steel," paper 102, Corrosion/85 (Boston, MA), March 1985, *Corrosion*, Vol 44 (No. 11), Nov 1988, p 791
69. L.G. Ljungberg, D. Cubicciotti, and M. Trolle, "The Effect of Sulfate on Environmental Cracking in BWRs Under Constant Load or Fatigue," paper 617, Corrosion/89, (New Orleans, LA), April 1989, National Association of Corrosion Engineers
70. D.A. Hale, "BWR Coolant Impurities Program at Peach Bottom Units 2 and 3," report NP-7310-L, Electric Power Research Institute, May 1991
71. "BWR Coolant Impurity Identification Study," report NP-4156, Electric Power Research Institute, Aug 1985
72. P.L. Andresen, "Specific Anion and Corrosion Potential Effects on Environmentally Assisted Cracking in 288 Water," GE-CRD Report 93CRD215, General Electric Company, Dec 1993
73. L.G. Ljungberg et al., "Stress Corrosion Cracking of Alloys 600 and 182 in BWRs," report TR-104972, Electric Power Research Institute, May 1995
74. B.M. Gordon, The Effect of Chloride and Oxygen on Stress Corrosion Cracking of Stainless Steel; Review of Literature, *Mater. Perform.*, Vol 19 (No. 4), April 1980, p 29
75. M. Hishida and H. Nakada, *Corrosion*, Vol 33 (No. 11), Nov 1977, p 403
76. J. Congleton, W. Zheng, and H. Hua, *Corros. Sci.*, Vol 27 (No. 6/7), 1990, p 555
77. P.L. Andresen, Effect of Material and Environmental Variables on Stress Corrosion Crack Initiation in Slow Strain Rate Tests on Type 304 Stainless Steel, *Symp. Environmental Sensitive Fracture; Evaluation and Comparison of Test Methods*, April 1982, STP 821, American Society for Testing and Materials, 1984, p 271–287
78. P.L. Andresen, "Effects of Zinc Additions on the Crack Growth Rate of Sensitized Stainless Steel and Alloys 600 and 182 in 288 Water," paper 72, Water Chemistry of Nuclear Reactor Systems 6, British Nuclear Energy Society, 1992
79. P.L. Andresen and T.M. Angeliu, "Effects of Zinc Additions on the Stress Corrosion Crack Growth Rate of Sensitized Stainless Steel, Alloy 600 and Alloy 182 Weld Metal in 288 Water," paper 409, Corrosion/95 (Orlando, FL), March 1995, NACE International
80. S. Hettiarachchi, G.P. Wozaldo, and T.P. Diaz, "Influence of Zinc Additions on the Intergranular Stress Corrosion Crack Initiation and Growth of Sensitized Stainless Steel in High Temperature Water," paper 410, Corrosion/95 (Orlando, FL), March 1995, NACE International
81. W.J. Shack et al., "Environmentally Assisted Cracking in Light Water Reactors; Semi Annual Report April-September 1986," NUREG/CR-4667, Vol III, ANL-87-37, U.S. Nuclear Regulatory Commission, Sept 1987
82. T.M. Angeliu, "The Effect of Zinc Additions on the Oxide Rupture Strain and Repassivation Kinetics of Fe-Base Alloys in 288 Water," paper 411, Corrosion/95 (Orlando, FL), March 1995, NACE International
83. F.P. Ford and M.J. Povich, *Corrosion*, Vol 35, 1979, p 569
84. F.P. Ford, "Mechanisms of Environmentally Enhanced Cracking in Alloy/Environment Systems Peculiar to the Power Generation Industries," NP-2589, Sept 1982, Electric Power Research Institute
85. P.L. Andresen, Laboratory Results on Effects of Oxygen Control During a BWR Start up on the IGSCC of 304 Stainless Steel, *Proc. Second Seminar on Countermeasures for Pipe Cracking in BWRs*, Vol 2, Nov 15–18, 1983, Report NP-3684-SR, Electric Power Research Institute, Sept 1984, p 12-1
86. W.L. Pearl, W.R. Kassen, and S.G. Sawochka, Monitoring and Control of Oxygen in Boiling Water Reactors, *Proc. Seminar on Countermeasures for Pipe Cracking in BWRs*, Vol 3, Jan 22–24, 1980, Report WS 79-174, Electric Power Research Institute, May 1980
87. K. Kaikawa et al., Reaerated Startup Operation as a Countermeasure for BWR Pipe Cracking, *Proc. Seminar on Countermeasures for Pipe Cracking in BWRs*, Vol 3, Jan 22–24, 1980, Report WS 79-174, Electric Power Research Institute, May 1980
88. T. Kawakubo, M. Hishida, and M. Arii, A Model for Intergranular Stress Corrosion Cracking Initiation in BWR Pipes and the Virtue of Startup Deaeration, *Proc. Seminar on Countermeasures for Pipe Cracking in BWRs*, Vol 3, Jan 22–24, 1980, Report WS 79-174, Electric Power Research Institute, May 1980
89. P.L. Andresen, The Effects of Surface Preparation, Stress and Deaeration on the Stress Corrosion Cracking of Type 304 Stainless Steel in Simulated BWR Start-Up Cycles, *Proc. Seminar on Countermeasures for Pipe Cracking in BWRs*, Vol 3, Jan 22–24, 1980, Report WS 79-174, Electric Power Research Institute, May 1980
90. M.E. Indig, Controlled Potential Simulation of Deaerated and Non-Dearated Startup, *Proc. Seminar on Countermeasures for Pipe Cracking in BWRs*, Vol 3, Jan 22–24, 1980, Report WS 79-174, Electric Power Research Institute, May 1980
91. K. Gott, Cracking Data Base as a Basis for Risk Informed Inspection, *Proc Tenth Int. Conf. Environmental Degradation in Nuclear Power Systems—Water Reactors* (Lake Tahoe, NV), Aug 5–9, 2001, F.P. Ford and G. Was, Ed., NACE International
92. P.L. Andresen, Stress Corrosion Cracking Testing and Quality Considerations, *Proc. Ninth Int. Symposium on Environmental Degradation in Nuclear Power Systems—Water Reactors* (Newport Beach, CA), Aug 1–5, 1999, S. Bruemmer and F.P. Ford, Ed., The Metallurgical Society, p 411–421
93. P.L. Andresen, K. Gott, and J.L. Nelson, Stress Corrosion Cracking of Sensitized Stainless Steel—A Five Lab Round Robin, *Proc. Ninth Int. Symposium on Environmental Degradation in Nuclear Power Systems—Water Reactors* (Newport Beach, CA), Aug 1–5, 1999, S. Bruemmer and F.P. Ford, Ed., The Metallurgical Society, p 423–433
94. J. Hickling, "Evaluation of Acceptance Criteria for Data on Environmentally Assisted Cracking in Light Water Reactors," SKI Report 94.14, Swedish Nuclear Power Inspectorate, Sept 1994
95. F.P. Ford, Stress Corrosion Cracking, *Corrosion Processes*, R.N. Parkins, Ed., Applied Science, 1982
96. F.P. Ford, *Corrosion*, Vol 52, 1996, p 375–394
97. D.D. Macdonald, P.C. Lu, M. Urquidi-Macdonald, and T.K. Yeh, *Corrosion*, Vol 52, 1996, p 768
98. D.D. Macdonald and M. Urquidi-Macdonald, *Corros. Sci.*, Vol 32, 1991, p 51
99. D.D. Macdonald, *Corros. Sci.*, Vol 38, 1996, p 1033
100. P.L. Andresen, Modeling of Water and Material Chemistry Effects on Crack Tip Chemistry and Resulting Crack Growth Kinetics, *Proc. Third Int. Symposium on Environmental Degradation in Nuclear Power Systems—Water Reactors* (Traverse City, MI), Aug 30–Sept 3, 1987, G.J. Theus and W. Berry, Ed., The Metallurgical Society, p 301–314
101. P.L. Andresen, *Corrosion*, Vol 44, 1988, p 376
102. C. Strawstrom and M. Hillert, *J. Iron Steel Inst.*, Vol 207, 1969, p 77
103. J. Alexander, "Alternative Alloys for BWR Pipe Applications," NP-2671-LD, Electric Power Research Institute, Oct 1982

104. A.E. Pickett, Assessment of Remedies for Degraded Piping, *Seminar on Countermeasures for Pipe Cracking in BWRs*, Electric Power Research Institute, Nov 1986
105. N.R. Hughes, "Evaluation of Near-Term BWR Piping Remedies," NP-1222, Electric Power Research Institute, May 1980
106. Fukushima Daiichi Nuclear Power Station Unit 3, "Primary Loop Recirculation System Work Report for Pipe Branches Replacement," paper presented at BWR Operating Plant Technical Conference No. 8 (Monterey, CA), General Electric Co., Feb 1983
107. H.P. Offer, "Induction Heating Stress Improvement," NP-3375, Electric Power Research Institute, Nov 1983
108. R.M. Horn, "Last Pass Heat Sink Welding," NP-3479-LD, Electric Power Research Institute, March 1984
109. M.E. Indig and A.R. McIlree, *Corrosion*, Vol 35, 1979, p 288–295
110. J. Magdalinski and R. Ivars, "Oxygen Suppression in Oskarshamn-2," Winter Meeting (Washington, D.C.), American Nuclear Society, Nov 1982
111. E.L. Burley, "Oxygen Suppression in Boiling Water Reactors—Phase 2 Final Report," DOE/ET/34203-47, NEDC-23856-7, General Electric Co., Oct 1982
112. B.M. Gordon, R.L. Cowan, C.W. Jewett, and A.E. Pickett, Mitigation of Stress Corrosion Cracking through Suppression of Radiolytic Oxygen, *Proc. First Int. Symposium on Environmental Degradation in Nuclear Power Systems—Water Reactors* (Myrtle Beach, SC), Aug 22–25, 1983, J. Roberts and W. Berry, Ed., National Association of Corrosion Engineers, p 893–932
113. B.M. Gordon, "BWR Material Life Extension Through Hydrogen Water Chemistry," Operability of Nuclear Power Systems in Normal and Adverse Environments (Albuquerque, NM), Sept–Oct 1986, American Nuclear Society
114. B.M. Gordon, "Laboratory Studies of Materials Performance in Hydrogen Water Chemistry," Seminar on BWR Corrosion, Chemistry and Radiation Control (Palo Alto, CA), Oct 1984, Electric Power Research Institute
115. S. Hettiarachchi, G. Wozadlo, P.L. Andresen, T.P. Diaz, and R.L. Cowan, The Concept of Noble Metal Addition Technology for IGSCC Mitigation of Structural Materials, *Proc. Seventh Int. Symposium on Environmental Degradation in Nuclear Power Systems—Water Reactors* (Breckenridge, CO), Aug 7–10, 1995, R. Gold and A. McIlree, Ed., NACE International, p 735–746
116. S. Hettiarachchi, R.J. Law, T.P. Diaz, R.L. Cowan, W. Keith, and R.S. Pathania, The First In-Plant Demonstration of Noble Metal Chemical Addition (NMCA) Technology for IGSCC Mitigation of BWR Internals, *Proc. Eighth Int. Symposium on Environmental Degradation in Nuclear Power Systems—Water Reactors* (Amelia Island, FL), Aug 10–14, 1997, A. McIlree and S. Bruemmer, Ed., American Nuclear Society, p 535–543
117. S. Hettiarachchi, NobleChem from Concept to Operating Commercial Power Plant Application, *Proc. Tenth Int. Conf. Environmental Degradation in Nuclear Power Systems—Water Reactors* (Lake Tahoe, NV), Aug 5–9, 2001, F.P. Ford and G. Was, Ed., NACE International
118. D.G. Ware et al., "Evaluation of Risk Associated with Intergranular Stress Corrosion Cracking in Boiling Water Reactor Internals," NUREG/CR-6677, INEEL/EXT-2000-00888, July 2000
119. "An Approach for Using Probabilistic Risk Assessment in Risk-Informed Decisions on Plant Specific Change to the Licensing Basis," Reg. Guide (RG) 1.174, http://nrc.gov/reading-rm/doc-collections/reg-guides/power-reactors/active/index.html
120. "An Approach for Plant Specific, Risk-Informed Decision-making; In-Service Testing," Reg. Guide (RG) 1.175, http://.nrc.gov/reading-rm/doc-collections/reg-guides/power-reactors/active/index.html
121. "An Approach for Plant Specific, Risk-Informed Decision-making; Graded Quality Assurance," Reg. Guide (RG) 1.176, http://nrc.gov/reading-rm/doc-collections/reg-guides/power-reactors/active/index.html
122. "An Approach for Plant Specific, Risk-Informed Decision-making; Technical Specifications," Reg. Guide (RG) 1.177, http://nrc.gov/reading-rm/doc-collections/reg-guides/power-reactors/active/index.html
123. "An Approach for Plant Specific, Risk-Informed Decision-making for In-service Inspection of Piping," Reg. Guide (RG) 1.178, http://nrc.gov/reading-rm/doc-collections/reg-guides/power-reactors/active/index.html

SELECTED REFERENCES

General Background

- S.H. Bush and R.L. Dillon, Stress Corrosion in Nuclear Systems, *Proc. Conf. Stress Corrosion Cracking and Hydrogen Embrittlement of Iron Base Alloys* (Unieux-Firminy, France), June 12–16, 1973, R.W. Staehle, J. Hochmann, R.D. McCright, and J.E. Slater, Ed., National Association of Corrosion Engineers, 1977, p 61
- R.L. Cowan and G.M. Gordon, Intergranular Stress Corrosion Cracking and Grain Boundary Composition of Fe-Ni-Cr Alloys, *Stress Corrosion Cracking and Hydrogen Embrittlement of Iron-Base Alloys* (Unieux-Firminy, France), June 1973, R.W. Staehle, J. Hochmann, R.D. McCright, and J.E. Slater, Ed., National Association of Corrosion Engineers, 1977, p 1063–1065
- B.M. Gordon and G.M. Gordon, Corrosion in Boiling Water Reactors, *Corrosion*, Vol 13, 9th ed., *Metals Handbook*, 1987, p 929–937
- W.S. Hazelton and W.H. Koo, "Technical Report on Materials Selection and Processing Guidelines for BWR Coolant Pressure Boundary Piping," NUREG-0313 Rev. 2, U.S. Nuclear Regulatory Commission, Jan 1988
- R.M. Horn, G.M. Gordon, F.P. Ford, and R.L. Cowan, Experience and Assessment of Stress Corrosion Cracking in L-Grade Stainless Steel BWR Internals, *Nucl. Eng. Des.*, Vol 174, 1997, p 313–325
- R. Killian et al., Characterization of Sensitization and Stress Corrosion Cracking Behavior of Stabilized Stainless Steels under BWR Conditions, *Seventh Int. Symposium, on Environmental Degradation of Materials in Nuclear Power Systems—Water Reactors* (Breckenridge, CO), NACE International, 1995, p 529–540
- H.H. Klepfer et al., "Investigation of Cause of Cracking in Austenitic Stainless Steel Piping," NEDO-21000, General Electric, July 1975

Modeling and Mechanisms

- P.L. Andresen, Irradiation Assisted Stress Corrosion Cracking, *Stress Corrosion Cracking: Materials Performance and Evaluation*, R.H. Jones, Ed., ASM International, 1992, p 181–210
- P.L. Andresen and F.P. Ford, Fundamental Modeling of Environmental Cracking for Improved Design and Lifetime Evaluations in BWRs, *Int. J. Pressure Vessels Piping*, Vol 59, 1994
- F.P. Ford, Quantitative Prediction of Environmentally Assisted Cracking, *Corrosion*, Vol 52 (No. 5), 1996, p 375–395
- F.P. Ford and P.L. Andresen, Development and Use of a Predictive Model of Crack Propagation in 304/316L, A533B/A508 and Inconel 600/182 in 288 °C Water, *Proc. Third Int. Symposium on Environmental Degradation in Nuclear Power Systems—Water Reactors* (Traverse City, MI), Aug 30–Sept 3, 1987, G.J. Theus and W. Berry, Ed., The Metallurgical Society, 1988, p 789–800
- H.P. Seifert, S. Ritter, and J. Hickling, Environmentally Assisted Cracking of Low-Alloy RPV and Piping Steels under LWR Conditions, *Proc. 11th Int. Conf. Environmental Degradation in Nuclear Power Systems—Water Reactors*, Aug 5–9,

2003, G. Was and L. Nelson, Ed., American Nuclear Society

Mitigation

- B.M. Gordon, "BWR Material Life Extension Through Hydrogen Water Chemistry," paper presented at the Operability of Nuclear Power Systems in Normal and Adverse Environments (Albuquerque, NM), American Nuclear Society, Sept–Oct 1986
- B.M. Gordon, R.L. Cowan, C.W. Jewett, and A.E. Pickett, Mitigation of Stress Corrosion Cracking through Suppression of Radiolytic Oxygen, *Proc. First Int. Symposium on Environmental Degradation in Nuclear Power Systems—Water Reactors* (Myrtle Beach, SC), Aug 22–25, 1983, J. Roberts and W. Berry, Ed., National Association of Corrosion Engineers, p 893–932
- S. Hettiarachchi, NobleChem from Concept to Operating Commercial Power Plant Application, *Proc. Tenth Int. Conf. Environmental Degradation in Nuclear Power Systems—Water Reactors* (Lake Tahoe, NV), Aug 5–9 2001, F.P. Ford and G. Was, Ed., NACE International

Corrosion in Pressurized Water Reactors

Peter M. Scott and Pierre Combrade, Framatome ANP, an AREVA and Siemens company

A PRESSURIZED WATER REACTOR (PWR) is a type of nuclear reactor that uses ordinary light water for both coolant and neutron moderator. In a PWR, the primary coolant loop is pressurized so the water does not achieve bulk boiling, and steam generators are used to transmit heat to a secondary coolant, which is allowed to boil to produce steam either for ship propulsion or for electricity generation. In having this secondary loop, the PWR differs from the boiling water reactor (BWR), in which the primary coolant is allowed to boil in the reactor core and drive a turbine directly. More detailed information on BWRs can be found in the article "Corrosion in Boiling Water Reactors" in this Volume.

PWR Materials and Water Chemistry Characteristics

In order to review PWR materials corrosion issues, it is necessary to start with a description of the main materials and water chemistry characteristics of the primary and secondary water circuits and also to distinguish them carefully from those of BWRs. A schematic diagram of a typical PWR is shown in Fig. 1.

Materials Used. In the PWR, a high pressure of the order of 14 MPa (~2250 psi) is applied to the primary system to prevent general boiling of the primary coolant and ensure that the system is maintained liquid (unlike the BWR where water is allowed to boil as it is heated by the nuclear fuel). The PWR primary water is heated by passage over nuclear fuel that is clad with a zirconium alloy, cooled by its passage through the steam generator tubes, and then recirculated to the nuclear core. The interconnecting primary piping is made of austenitic stainless steel, typically type 304L, while the steam generator tubes are usually fabricated from a nickel-base alloy such as Alloy 600 and more recently, Alloy 690, or in some cases, Alloy 800. The primary pressure vessel is clad with a welded overlay of stainless steel as are the channel heads of the steam generators. The major surfaces in contact with the primary coolant are therefore mainly stainless steels, nickel-base alloys, and zirconium-base alloys.

Secondary-side coolant, on the outside of the steam generator tubes, boils to produce steam that is used to turn the turbines and alternators in order to generate electricity. The pressure vessels of the steam generators are made from low-alloy steel, while the feed water and steam pipe work may be carbon or low-alloy steels, all of which are directly exposed to the secondary water or steam. The remaining components of the secondary circuit are the steam turbines, where various higher-strength low-alloy steels may be used, and the condensers whose tubes may be fabricated from copper-base alloys, titanium, or stainless steel, depending on the age of the plant and available cooling water (seawater, estuarine water, or river water).

A list of common structural steels used in PWRs for pressure vessels and piping is shown in Table 1. Typical austenitic, martensitic, and precipitation-hardened stainless steels are shown in Table 2, and nickel-base alloys in Table 3.

Water Chemistry Characteristics. Typical primary and secondary coolant conditions for PWRs are shown in Table 4. Hydrogen gas is added to the primary circuit coolant at a concentration usually between 20 and 50 mL/kg at standard temperature and pressure (STP) (equivalent to 2–4 ppm by weight). This concentration is sufficient to ensure that any radiolytic decomposition products of water are rapidly scavenged by efficient ion-molecule and radical-molecule reactions. Consequently, corrosion potentials of all relevant structural materials are close to the hydrogen redox potential, as determined by the hydrogen partial pressure and the solution pH (Fig. 2) (Ref 1). By contrast, in direct cycle BWR systems, significant concentrations of oxygen and hydrogen peroxide are present as the end products of water radiolysis (which may be reduced by hydrogen additions in some BWR plants) and these fix the corrosion potential at significantly more positive values (Fig. 2). The other additives to the primary circuit of PWRs are boric acid, which participates with the control rods in moderating the nuclear chain reaction, and lithium hydroxide for pH control

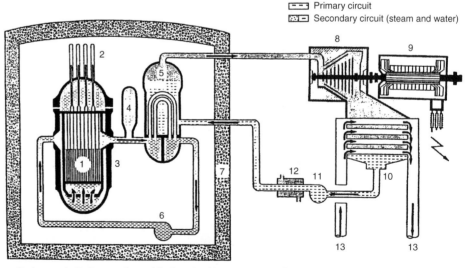

Heat generated in the core (by enriched uranium) is transferred to the primary circuit (pressurized water).

Heat from the primary circuit evaporates water of the secondary circuit.

The turbine generator set converts mechanical power provided by steam into electricity.

1. Reactor core
2. Control rods
3. Reactor vessel
4. Pressurizer
5. Steam generator
6. Primary pump
7. Containment
8. Turbine
9. Generator
10. Condenser
11. Extraction and feedwater pumps
12. Reheater
13. River (or sea) water

Fig. 1 Basic flow diagram of a pressurized water reactor (PWR) nuclear power station

in the range 6.8 to 7.4 (at ~300 °C, or ~570 °F) (Ref 2). This value of the high-temperature pH on the primary side is designed to ensure minimum corrosion rates of the structural materials and minimum solubility of the protective spinel oxides that form on the metal surfaces, as discussed in more detail subsequently. For comparison, no chemical additives are made to BWR water other than hydrogen in some cases.

On the secondary side of PWRs, the objective of water chemistry control is also to ensure that corrosion rates are as low as possible, particularly of nonstainless materials in the feed train, as well as to minimize sludge (magnetite) transfer to heat transfer surfaces. This is achieved by a combination of a suitably alkaline pH (~9–10 as measured at room temperature) using ammonia and/or organic amines such as morpholine or ethanolamine together with hydrazine to reduce oxygen to negligible levels. If copper-base alloys are present in the secondary circuit, the upper limit allowed for pH with ammonia conditioning is 9.2 (at room temperature), which is less than that desired to reduce magnetite solubility to a minimum, and the use of an organic amine is then necessary. Hydrogen from general corrosion and thermal decomposition of hydrazine is stripped preferentially to the steam phase and evacuated with any other noncondensable gases at the condenser. In addition, there is no evidence from in-plant measurements of any significant electrochemical influence of hydrazine itself on corrosion potentials (Ref 3). Consequently, corrosion potentials are somewhat more positive (~200 mV) on the secondary side compared with the primary side (Fig. 2). The most important secondary side corrosion problems are associated with hideout of secondary water impurities in occluded, superheated crevices next to the steam generator tubes where water circulation is restricted by a combination of geometry and sludge deposits.

General Corrosion, Crud Release, and Fouling

The first objective of water chemistry management is to ensure minimum loss of metal due to general corrosion by managing the solution pH, whether on the primary or secondary sides, to coincide as closely as possible with the minimum solubility of spinel oxides such as magnetite. Nevertheless, some loss of metal thickness due to general corrosion does occur (between ~0.2 and 2 µm/yr, or 0.008 and 0.08 mils/yr) depending on the material (Ref 2, 4). While this has negligible influence on the load-bearing capacity of the pressure-retaining components, because of the large surface areas involved, the mass of material converted to oxide can be impressive, being potentially measured in tens of kilograms per year. This can have some important consequences for the operation of PWR systems.

On the primary side, release of cations into solution as well as particles of oxide allows them to pass through the nuclear core leading to transmutation of some elements into radioactive species. These activated species may then be transported out of the core and settle elsewhere, thus giving rise to radiation fields that can impede subsequent maintenance operations. On the secondary side, magnetite released from carbon steel surfaces is transported into the steam generator where it settles and/or precipitates on the steam generator tube surfaces, reducing thermal efficiency. Magnetite will also accumulate in low flow zones such as on the

Table 1 Common pressure-boundary steels used in pressurized water reactor (PWR) pressure vessels and piping (wt%)

Steel designation		Composition, wt%								
UNS No.	ASTM	C	Mn	P	S	Si	Ni	Cr	Mo	V
K12539	A533-B	≤0.25	1.15–1.50	≤0.035	≤0.035	0.15–0.40	0.40–0.70	...	0.45–0.60	...
K12766	A508-2	≤0.27	0.5–1.00	≤0.025	≤0.025	0.15–0.40	0.50–1.00	0.25–0.45	0.55–0.70	≤0.05
K12042	A508-3	≤0.25	1.20–1.50	≤0.025	≤0.025	0.15–0.40	0.40–1.00	≤0.25	0.45–0.60	≤0.05
K03006	A333-6	≤0.30	0.29–1.06	≤0.025	≤0.025	≤0.10	...	...	...	...
(a)	A516	≤0.28	0.60–1.20	≤0.035	≤0.035	0.15–0.40	...	...	...	...

(a) UNS designation is based on yield strength requirement

Table 2 Some stainless steels commonly used in pressurized water reactors (PWRs)

UNS No.	Grade	Composition, wt%								
		C	Mn	Si	Cr	Ni	Mo	P	S	Other
S30403	304L	≤0.03	≤2.0	≤1.0	18–20	8–12	...	≤0.045	≤0.030	...
S31600	316	≤0.08	≤2.0	≤1.0	16–18	10–14	2.0–3.0	≤0.045	≤0.030	...
S32100	321	≤0.08	≤2.0	≤1.0	17–19	9–12	...	≤0.045	≤0.030	Ti>5C
S34700	347	≤0.08	≤2.0	≤1.0	17–19	9–13	...	≤0.045	≤0.030	Nb+Ta>10C
...	308L	≤0.03	≤2.0	≤1.0	19–21	10–12	...	≤0.045	≤0.030	...
...	309L	≤0.03	≤2.0	≤1.0	22–24	12–15	...	≤0.045	≤0.030	...
S66286	A-286	≤0.08	≤2.0	≤1.0	12–15	24–27	1.00–1.50	≤0.040	≤0.030	Ti 1.55–2.00 Al ≤0.35 V 0.10–0.50
S17400	17-4PH	≤0.07	≤1.00	≤1.00	15–17.5	3.0–5.0	...	≤0.040	≤0.030	Cu 3.0–5.0 Nb+Ta 0.15–0.45
S41000	410 ASTM A 240	0.10–0.15	≤1.0	≤1.0	11.50–13.50	...	...	≤0.040	≤0.030	...

Table 3 Nickel-base alloys used in pressurized water reactors (PWRs)

UNS No.	Alloy	Composition, wt%												
		Ni	Cr	Fe	Ti	Al	Nb+Ta	Mo	C	Mn	S	P	Si	Cu	Co
N06600	600	>72.0	14–17	6–10	...	...	...	...	≤0.05	≤1.0	≤0.015	...	≤0.5	≤0.5	≤0.10
W86182	182	Bal	13–17	≤10.0	≤1.0	...	1.0–2.5	...	≤0.10	5.0–9.5	≤0.015	≤0.030	≤1.0	≤0.50	≤0.12
...	82	Bal	18–22	≤3.00	≤0.75	...	2.0–3.0	...	≤0.10	2.5–3.5	≤0.015	≤0.030	≤0.50	≤0.50	≤0.10
N06690	690	>58.0	28–31	7–11	...	...	...	...	≤0.04	≤0.50	≤0.015	...	≤0.50	≤0.5	≤0.10
W86152	152	Bal	28–31.5	8–12	≤0.50	...	1.2–2.2	≤0.50	≤0.045	5.0	≤0.008	≤0.020	≤0.65	≤0.50	≤0.020
N14052	52	Bal	28–31.5	8–12	≤1.0	≤1.10	≤0.10	≤0.05	≤0.040	≤1.0	≤0.008	≤0.020	≤0.50	≤0.30	≤0.020
N08800	800	30–35	19–23	≥39.5	0.15–0.60	0.15–0.60	...	...	≤0.10	≤1.50	≤0.015	...	≤1.0	≤0.75	...
N07750	X750	>70.0	14–17	5–9	2.25–2.75	0.4–1.0	0.7–1.2	...	≤0.08	≤1.0	≤0.010	...	≤0.5	≤0.5	...
N07718	718	50–55	17–21	Bal	0.65–1.15	0.2–0.8	4.75–5.50	2.8–3.3	≤0.08	≤0.35	≤0.010	...	≤0.35	≤0.30	...

tube sheet and in the interstices between the tubes and support plates (see Fig. 5 discussed later in this article) and enhance the phenomenon of hideout of impurity solutes that is responsible for serious corrosion problems on the secondary side of steam generator tubes.

Influence of Corrosion on Primary Circuit Radiation Fields

The primary source of radiation fields from PWR primary circuit components is from corrosion products generated on the surfaces of iron-nickel alloys of the primary pressure boundary. The general corrosion that takes place on these surfaces releases dissolved metallic species, such as cations of iron, nickel, manganese, and cobalt, as well as colloidal oxide particles. Some of these metallic ions are deposited either by particle deposition or by precipitation on hot fuel element surfaces in the core where they are exposed to neutrons and may be activated by nuclear transmutation. The optimum pH for limiting corrosion product production and transport in the primary circuit has been found to be in the range 7.2 to 7.4 (at operating temperature).

Generation of Radioisotopes. The two primary radioactive isotopes generated in the core are ^{60}Co and ^{58}Co (Ref 5). Cobalt-60 arises as a result of neutron capture by naturally occurring ^{59}Co originating as an impurity (with nickel) in iron-nickel alloys used in reactor power plants or from stellites used as hardfacing materials in valves and elsewhere. The high neutron cross section of ^{59}Co and the high energy of the radioactive decay emissions from ^{60}Co cause it to predominate over radioisotopes formed from other elements that are present in higher concentrations. ^{58}Co is formed as a result of a neutron replacing a proton in the nucleus of ^{58}Ni, a naturally occurring isotope of nickel. Because the surface area of nickel-base alloys is usually large in PWRs due to the composition of the steam generator tubes, ^{58}Co is another very significant contributor to radiation fields.

Following the generation of radioisotopes from the deposits present on the surface of the fuel, various processes such as adsorption, ion exchange, dissolution, and erosion can cause the radioactive isotopes to be released from the fuel surface back into the reactor coolant. Transport with the reactor coolant allows some of these to be deposited later and incorporated into the growing oxide films on the surfaces of the reactor coolant system.

The main alloy constituents released by corrosion of the surfaces of the PWR primary circuit are iron and nickel. This fact, combined with the relatively reducing condition present in the primary coolant, results in deposits on the fuel that can be magnetite (in plants with Alloy 800 steam generator tubes), nickel oxide, metallic nickel, or nickel ferrite (in which nickel replaces one of the iron atoms in magnetite, Fe_3O_4, by occupying a fraction of the tetrahedral lattice positions). As a result of the presence of both cobalt and nickel on the fuel, ^{58}Co and ^{60}Co are both significant sources of radiation in PWR systems. ^{58}Co, however, has a shorter half-life of 71 days and consequently is not retained from one refueling outage to the next, as is ^{60}Co, but is freshly generated from fuel cycle to cycle.

One of the most important regions in the PWR primary circuit for radiation control is in the channel heads of the steam generators where many maintenance and nondestructive examinations of steam generators take place.

Table 4 Typical pressurized water reactor (PWR) primary and secondary coolant chemistry conditions

Parameter	Primary water	Secondary water
Pressure, MPa (psi)	14.2 (2060)	5.4–7.2 (783–1044)
Temperature, °C (°F)	286–323 (547–613)	284–305 (543–581)
Conductivity, µS cm^{-1}	1–40	<0.5 (cation conductivity)
Oxygen, ppm	<0.100 (<0.005 normally)	<0.005
Hydrogen, mL/kg at STP	20–50	...
Lithium, ppm (as LiOH)	0.2–3.5	...
Boron, ppm (as H_3BO_3)	0–2300	...
NH_3 or morpholine or ethanolamine, ppm	...	As required for pH
Hydrazine, ppm	...	>[O_2] + 0.005 or >150
Sodium, ppm	...	<0.005
Chloride, ppm	<0.150	<0.03
Fluoride, ppm	<0.150	...
SiO_2, ppm	<0.20	<1.0
pH	6.8–7.4 (at 300 °C)	8.9–10.0 (at 25 °C)

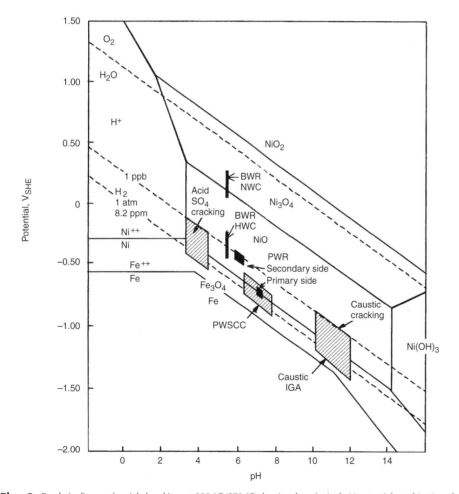

Fig. 2 Pourbaix diagram for nickel and iron at 300 °C (570 °F) showing the principal pH-potential combinations for PWR primary and secondary water, boiling water reactor (BWR), normal water chemistry (NWC), and BWR hydrogen water chemistry (HWC), and the modes of stress-corrosion cracking (SCC) of Alloy 600. IGA, intergranular attack. Source: Ref 1

Consequently, this is a region where much of the radiation exposures associated with primary circuit maintenance occurs.

A second potential source of radioisotopes generated in nuclear power plants is within the fuel itself from nuclear fission and their decay products. These fission products are normally effectively contained within the zirconium-base cladding of each fuel rod although, occasionally, a few rods may develop pinholes or other types of penetrations, thereby permitting the release of fission products into the reactor coolant. Two of the major constituents of fission product radioisotopes are radioactive iodine and cesium.

Iodine, when released to the coolant, will predominantly shift toward volatile gaseous species and will be processed and monitored with the gases that are removed from the reactor coolant. Iodine generally makes an insignificant contribution to the total exposure experienced at nuclear power plants, although there have been occasional situations in which iodine, which has dissolved in the coolant at the time of shutdown, has delayed shutdown operations pending its removal from the coolant systems by ion exchange. Cesium, on the other hand, is readily soluble in reactor water, and therefore is removed from this water by the primary system demineralizers. Thus, cesium ends up in the radioactive waste, where it is processed and packaged on-site and shipped off-site for disposal. For each of these isotopes, and similarly for most fission product isotopes, due to their chemical behavior and/or their low concentration in the coolant water, they contribute relatively little to the radiation fields from components of the primary reactor coolant system.

Another potentially significant source of radioisotopes can arise as a result of failure of the stainless steel cladding of control rods containing silver-indium-cadmium neutron absorber, thus leading to the release of silver and its subsequent activation (Ref 6, 7). Such cladding failures can occur due to wear. Other cladding failures have been caused by undue swelling of the neutron absorber combined with loss of ductility in the cladding due to irradiation damage. The swelling occurs when a second phase is formed following the nuclear transformation reaction of indium into tin.

Radiation Control Techniques

The control of radiation fields has considerable influence on the cost of generation of nuclear power, and the radiation exposure experienced by the personnel who work in nuclear power plants is a key measure of the effectiveness of radiation field control measures (Ref 5). For PWRs, steam generator tube degradation and, more recently, upper head penetrations of the reactor pressure vessel have caused a significant increase in inspection and, in some cases, replacement of steam generators and/or upper heads, with a concomitant increase in personnel radiation exposure. Concurrent with these problems has been the application of a number of techniques for reducing radiation exposure in plants. These have included techniques for reducing the radiation fields such as improved water chemistry control, materials selection, surface treatment, and decontamination (Ref 8–12).

Water Chemistry Control. In PWRs where precipitation and dissolution from fuel surfaces have been shown to be a significant influencing factor in radiation field buildup, the primary water pH can be controlled to reduce the amount of precipitate on the fuel surfaces. Lithium hydroxide (LiOH) concentration decreases as boric acid is reduced through the fuel cycle in order to maintain constant pH throughout most of the cycle. When the pH is high enough, in the region of 7.2 to 7.4 at operating temperature, the thermodynamics of the solubility of nickel ferrite are such that nickel ferrite will not precipitate on the fuel. This reduction in deposition on the fuel reduces the generation of cobalt isotopes and the subsequent buildup of radiation fields in the steam generator and elsewhere. Interest has also been shown in the use of small concentrations of soluble zinc (depleted in the ^{64}Zn isotope to <5% to avoid a zinc activation product) typically at a concentration ~5 to 10 ppb in the primary coolant. Zinc displaces cobalt and other cations from their tetrahedral crystallographic positions in the spinel oxides that form on stainless steels and nickel-base alloys. Cobalt is thus displaced into the primary coolant where it can then be removed by the primary cleanup system demineralizers. Oxide thicknesses and general corrosion rates are also significantly reduced by a factor of three or more by zinc addition (Ref 12). Another technique for reducing the inventory of activated elements in surface oxide films is to add sufficient hydrogen peroxide to eliminate dissolved hydrogen during the last stages of plant cooldown so that the redox potential becomes oxidizing. Cations including cobalt released into solution are then removed by the primary circuit demineralizers.

Materials Selection. High-cobalt alloys (Stellites) are present in most nuclear reactor systems, usually in valves, pumps, and control rod drive mechanisms. New materials containing low amounts of cobalt or no cobalt have been developed that are designed to serve the function of the high-cobalt alloys (Ref 11). Such replacements offer a significant opportunity for radiation field reduction from ^{60}Co together with the reduction of the cobalt source term in structural alloys, particularly nickel-base alloys used for steam generator tubing.

Surface Treatment. Many PWRs experience component replacement, and optimizing the surface condition of these new components can also help with radiation field control. Electropolishing has been shown to reduce the effective surface area by up to a factor of five on pipes and on steam generator channel head surfaces, for example, by removing surface asperities and creating a more uniform surface finish. Electropolishing also removes surface cold work and produces surfaces that are more likely to develop a denser oxide film. Radiation exposures have also been reduced through the use of remote equipment, shielding, and extensive planning and training for high-exposure tasks.

Decontamination solvents that are effective in removing the radioisotopes from the oxide layers have also been developed. These solvents are used particularly when major maintenance and repair work is conducted on equipment that has very high radiation fields. Such decontamination can be combined with procedures such as chemistry control or electropolishing to reduce the subsequent radiation field buildups on these components (Ref 13).

Axial Offset Anomaly

The so-called "axial offset anomaly" is a severe perturbation of the axial neutron flux distribution in the nuclear fuel that is caused by the precipitation of boron-rich minerals on the surfaces of the fuel elements (Ref 14). It tends to be associated with fuel pins having a particularly high thermal flux and incipient local surface boiling so that the boron-rich precipitates tend to occur toward the top of the core. The problem has often been analyzed in terms of the thermodynamic conditions necessary to precipitate lithium metaborate, $LiBO_2$, or even adsorption of nonionized boric acid. Recent chemical analyses of scrapings from fuel have, however, revealed more complex insoluble minerals such as bonaccordite (nickel boroferrite, Ni_2FeBO_5). Such observations emphasize the need for precise thermodynamic data so that the precipitation of such minerals can be predicted and avoided.

Secondary Circuit Corrosion and Fouling

Control of pH. The majority of PWR power plants control the pH of steam generator secondary water using all-volatile treatment (AVT), that is, pH control by ammonia (NH_3) or by amines such as morpholine (C_4H_9NO) and ethanolamine (C_2H_7NO). Many of the early plants used phosphates (Na_3PO_4) whose major advantage is its capacity to buffer against both acidic and basic impurity ingress, a protection that is only weakly afforded by amines. Volatile amines introduce no solids into the system, whereas the salts added in the phosphate method can accumulate and concentrate in superheated occluded areas. For this reason, use of phosphates has been discontinued in most plants and replaced by AVT following observations of caustic cracking of Alloy 600 steam generator tubes and/or wastage caused by concentrated phosphate phases (see the section "Tube Wastage" later in this article).

Control of Oxygen. Oxygen concentrations in power plant secondary feed water are controlled by adding scavenging chemicals (hydrazine, N_2H_4) and/or by mechanical means such as the use of air ejectors on the deaerating section of condensers.

Control of Impurities. Chemical upsets in secondary water can be initiated by condenser tube failures; typical analyses of chemicals that can be introduced are shown in Table 5 (Ref 15). Air in-leakage rates of 2800 L/min (740 gal/min) can be experienced with severe condenser leaks, while a rate of one-tenth of that level has been common in the past. With 500 L/min (130 gal/min) in-leakage, the oxygen concentration in the secondary circuit condensate can increase to 30 ppm. Powder resin filters and deep bed demineralizers are two common condensate polishing systems used in some plants but, if not properly operated/regenerated, these systems can also become sources of ionic impurities.

Effect of Tube Fouling on Steam Pressure. As the PWR fleet has aged, problems have also been experienced with reduced steam pressure as a result of fouling of the secondary side of the tubes of PWR steam generators (Ref 10, 16–18). In a few extreme cases, even the broached holes in the tube support plates of more modern steam generators have become substantially blocked, giving rise to difficulties with secondary water level control. Considerable work has been undertaken to characterize the nature of the deposits responsible and the conditions under which they form. Remedial measures such as pressure pulse cleaning and chemical cleaning and the use of dispersants such as polyacrylic acid to avoid particle adhesion to steam generator tube surfaces have and are being investigated (Ref 17, 18).

Carbon and Low-Alloy Steel Corrosion due to Primary Water Leaks

Numerous incidents of primary water leakage have been reported since 1970 of which some have resulted in significant general corrosion or wastage of nearby carbon or low-alloy steel components (Ref 19). These incidents have been attributed either to boric acid corrosion after concentration of the leaking primary water by evaporation or to the erosion-corrosion effects of high-velocity steam jets. The most serious case of boric acid corrosion was found in 2002 and had culminated in a pineapple-size hole being created in the low-alloy steel of the upper head of a PWR following a leak of primary water through stress-corrosion cracks in an Alloy 600 upper head penetration over a period of about five years.

Canopy seal cracks in the control rod drive mechanism housings above the reactor vessel head are a common source of primary water leaks onto the vessel head, although this has only rarely caused significant corrosion of the low-alloy steel. Another common cause of primary water leaks is from flanged joints ranging from the main vessel head to reactor pressure vessel seal to primary pump and valve flanges. In such cases, high-velocity steam jets have severely damaged and eventually severed high-strength low-alloy steels securing the flanges (Fig. 3) (Ref 20).

Many experimental studies have been carried out to reproduce the corrosion damage of carbon and low-alloy steels observed in PWRs due to primary water leaks (Ref 19–23). Simulations of in-service leaks of primary water using mockups are able to reproduce the high-velocity steam cutting phenomenon as well as the high corrosion rates under accumulated (moist) boric acid deposits. For example, a leak of primary water at 316 °C (601 °F) through a cracked stainless steel tube in a low-alloy steel plate produced a maximum corrosion rate of 55 mm/yr (2.17 in./yr) at the point where the leak escaped from the crevice between the tube and steel plate. Similarly, experiments have been carried out using jets of superheated steam from borated water at 316 °C (601 °F) directed at carbon steel surfaces. Corrosion rates of 15 to 25 mm/yr (0.6 to 1 in./yr) were observed just outside the zone directly impacted by the steam jet, whereas even higher rates were measured in the impact zone where erosion doubtless contributed to metal loss.

Many experiments have also been conducted in aqueous solutions of boric acid as a function of concentration, pH, temperature, and flow velocity in order to quantify corrosion rates of carbon and low-alloy steels (Ref 19–21, 23). Corrosion rates in aerated boric acid solutions with concentrations similar to those of PWR primary water (<2000 ppm boron or 1.1% boric acid) do not exceed 5 mm/yr (0.2 in./yr). High corrosion rates in excess of 10 mm/yr (0.4 in./yr) and up to 120 mm/yr (4.7 in./yr) are observed only in the presence of aerated concentrated boric acid solutions or aerated moist boric acid pastes at intermediate temperatures around 100 °C (212 °F) as shown in Fig. 4 (Ref 23). In static concentrated boric acid solutions, the corrosion rate is observed to decrease rapidly over a period of a few hours, whereas in flowing solutions, such a decrease in the corrosion rate is not observed. This is probably due to the elimination of corrosion products as they are formed, thus avoiding saturation of the solution and consequent slowing of the corrosion kinetics. The weakness of these laboratory corrosion rate studies is that the high rates comparable with those observed in service are

Fig. 3 Reactor coolant pump closure stud degraded by exposure to borated water. Source: Ref 20

Table 5 Typical cooling water analyses

Element	Fresh river water(a)	Fresh lake water(b)	Brackish water(c)	Seawater (d)
Calcium, ppm	58	32	44	400
Magnesium, ppm	15	11	78	1272
Sodium, ppm	13	3.2	603	10,561
Potassium, ppm	(e)	...	20	380
Lead, ppm	(e)	0.004	(e)	0.21
Chloride, ppm	4.8	2.1	1053	18,980
Bicarbonate, ppm	217	149	68	142
Total alkalinity ppm as calcium carbonate, $CaCO_3$	178	(e)	56	(e)
Fluoride, ppm	(e)	0.25	0.08	3.5
Bromide, ppm	(e)	(e)	3.5	65
Sulfate, ppm	45	7	220	2649
Nitrate, ppm	(e)	1.6	1.2	10^{-3} to 7×10^{-1}
Phosphate, ppm	(e)	0.6	...	10^{-3} to 10^{-1}
Silica, ppm	14	5	8.6	0.01 to 7.0
Carbon dioxide, ppm	(e)	3.8	2.9	6
Oxygen, ppm	(e)	(e)	6.2	5
pH	(e)	(e)	6.2	5

(a) Mississippi River. (b) Lake Michigan. (c) Estuary on U.S. East Coast. (d) Typical ocean water. (e) Not determined. Source: Ref 15 and EPRI Report NP 2429, Vol 4, Sludge, Fouling and Sludge Lancing

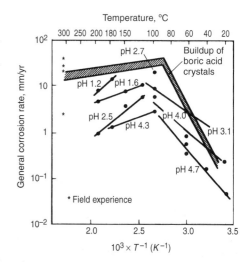

Fig. 4 Variations of iso-pH general corrosion rates of low-alloy steels. Source: Ref 23

reproduced only in aerated conditions, whereas aeration under the thick, moist deposits that form from primary water leaks in service seems rather unlikely.

PWR Primary Side Stress-Corrosion Cracking

There is little doubt that the most severe corrosion problem to affect PWRs is intergranular stress corrosion cracking (IGSCC) of Alloy 600 in PWR primary water. It has become a generic issue rivaling that of IGSCC of sensitized and/or cold-worked stainless steels in BWRs in terms of unanticipated outages and cost of repairs. Rather less-extensive or well known examples of stress corrosion cracking (SCC) in PWRs are those affecting austenitic stainless steels in occluded dead legs in PWR primary loops where it is sometimes difficult to guarantee adequate control of primary water chemistry or proper deoxygenation during plant startup. In addition, high-strength stainless steel and nickel-base alloy fasteners and springs are used extensively in PWR primary circuits, and some service failures of these items have also occurred.

Nickel-Base Alloys

Alloy 600, a nickel-base alloy containing 14 to 17% Cr and 6 to 10% Fe plus various minor elements (Table 3), was initially adopted for use in PWRs for steam generator tubes because of its excellent resistance to chloride cracking (from the secondary side) compared with stainless steel. It was also attractive for primary circuit components because of the closer similarity of its coefficient of thermal expansion to that of the low-alloy steel used to fabricate the reactor pressure vessel, pressurizer, and steam generator shells. A list of PWR components where Alloy 600 and similar nickel-base alloys are used in PWRs is given in Table 6.

The susceptibility of Alloy 600 to IGSCC in operational service in PWR primary water was first revealed in the early 1970s in tight U-bends and in rolled, cold-worked transitions in diameter of steam generator tubes within and/or at the top of tube sheets (Ref 15). This then became a major cause of steam generator tube cracking in the 1980s, and later, premature steam generator retirement and replacement. The IGSCC of pressurizer nozzles and control rod drive mechanism (CRDM) penetrations in the upper heads of PWR reactor pressure vessels followed in the late 1980s and have continued for over a decade (Ref 1, 24). The CRDM penetration cracking appeared first in French PWRs in 1989, but only isolated cases were observed elsewhere until 2000. Apparently interdendritic, but in fact intergranular, SCC of the weld metals, Alloys 182 and 82 (Table 3), the former having a composition similar to Alloy 600, has also been observed recently in major primary circuit welds of several plants, often after very long periods in service ranging between 17 and 27 years (Ref 24). To these can be added the experience of extensive IGSCC in the γ' strengthened analogue of Alloy 600, Alloy X750, which is used for split pins attaching the CRDM guide tubes to the upper core plate. Even Alloy 718, a high-strength nickel-base alloy containing 17 to 21% Cr, which is normally considered a reliable high-strength material in PWR primary water use, has occasionally exhibited IGSCC (Ref 1).

The history of IGSCC in various PWR nickel-base alloy components is summarized subsequently together with the methodologies developed to predict and mitigate cracking until, as sometimes is the case, replacement becomes unavoidable. The main remedy for susceptible Alloy 600 components is replacement, usually by Alloy 690 and its compatible weld metals, Alloys 152 and 52 (Table 3), which have proved to be resistant to IGSCC in PWR primary water both in severe laboratory tests and, to date, after up to 17 years in service.

Alloy 600 Steam Generator Tubes. Most PWR steam generators are of the "recirculating" type, although some are so-called "once-through" where all the secondary water entering the steam generator is transformed into steam; both are illustrated in Fig. 5. Most in-service primary side IGSCC has occurred in recirculating steam generators. An important difference between the two from the point of view of primary side IGSCC is that the once-through steam generators were subjected to a preservice stress relief of the whole steam generator at a temperature of about 610 °C (1130 °F). As well as promoting grain-boundary carbide precipitation in Alloy 600 (discussed in more detail subsequently), some grain-boundary chromium depletion (sensitization) also occurred. In one case, due to accidental ingress of thiosulfate, this led to extensive intergranular attack (IGA) of the sensitized tubes of a plant with once-through steam generators. In general, however,

Table 6 Pressurized water reactor (PWR) components fabricated from nickel-base alloys

PWR components	Nickel-base alloy grades used
Steam generator tubes	600 MA and TT, 690 TT and 800
Steam generator divider plates	600 and 690
Upper head penetrations	600 and 690 (except in Germany)
Lower head penetrations	600
Core supports	600
Pressurizer nozzles	600 and 690
Safe ends	600
Weld metal deposits	82, 182, 52, and 152
Fuel grid and hold-down springs	718
Split pins	X750

MA, mill annealed; TT, thermally treated

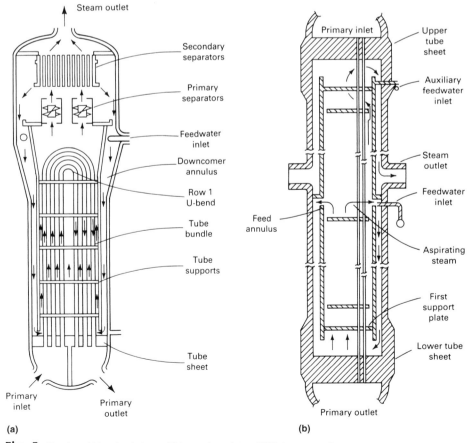

Fig. 5 Sketches of (a) recirculating and (b) once-through type PWR steam generators.

the lower-strength and grain-boundary carbide precipitation in once-through steam generator tubes has proved to be advantageous for resistance to IGSCC on the primary side, despite the sensitization, although even these are now being replaced after typically 20 to 25 years service (Ref 25).

Primary side IGSCC of Alloy 600 steam generator tubing evolved from a laboratory prediction during the 1950s and 1960s to a major degradation mechanism from the 1970s onward. As early as 1957, laboratory studies of cracking of high-nickel alloys in high-purity water at 350 °C (660 °F) were reported (Ref 26), although at that time the importance of the corrosion potential as fixed by the hydrogen partial pressure was not understood. During the following years, numerous laboratory tests were performed in different environments to duplicate and explain these observations. In 1971, the first confirmed primary side cracking of hot-leg roll transition regions at the tube sheet and suspected primary cracking in tight first row U-bends of recirculating steam generators occurred (Ref 27). Leakage at U-bends was experienced in the Obrigheim (Germany) steam generators after only 2 years of operation.

From the mid-1980s to the present day, cracking of Alloy 600 tubes from the primary side became a problem of great importance, as shown as "SCC (ID)" in Fig. 6 (Ref 28). Cracking occurred both in the tight U-bends, mainly on the inner two rows at the apex and at the tangent points with the straight sections, and in the tube sheet at the transition expansion or roll expansion regions of recirculating steam generators. The latter has been responsible for premature steam generator replacement at many plants.

The first roll transitions experiencing IGSCC were located on the hot-leg side where the temperature is typically around 320 °C (610 °F) and is 30 to 40 °C (55 to 70 °F) hotter than the cold-leg inlet at 280 °C (535 °F). Thus, it was clear that temperature had a significant influence on IGSCC initiation, indicating a thermally activated process whose temperature dependence could be fitted to the Arrhenius equation. The apparent activation energy so derived is rather high (~180 kJ/mole for crack initiation) so that a typical temperature difference of 30 °C (55 °F) between hot and cold legs could easily account for a factor of four to five increase in the time to the onset of cracking. Thus, reduction of hot-leg temperature has been one possible mitigating action that has been used. Hot-leg temperature reductions from 4 to even 10 °C (7 to 18 °F) have been applied without necessarily always losing on maximum power output depending on the margins available in the turbine.

The magnitude of the tensile stresses, particularly residual stress from fabrication, also has a major impact on the initiation time for IGSCC so that only the most highly strained regions of steam generator tubing (that is, row one and two U-bends, roll transition regions, expanded regions, and dented areas) have exhibited IGSCC. From laboratory studies of susceptible materials, a threshold stress below which cracking will not occur of ~250 MPa (35 ksi) has been determined, and time-to-failure has been observed to depend approximately on stress^{-4} (Ref 1). Consequently, several stress mitigation techniques have been developed such as local stress relief of first and second row U-bends by resistance or induction heating, and shot peening or rotopeening to induce compressive stresses on the internal surface of roll transitions (Ref 29, 30). While peening helps to prevent initiation of new cracks, it cannot prevent growth of existing cracks whose depths are greater than that of the induced compressive layer, typically 100 to 200 μm (0.004 to 0.008 in.). Thus, peening has been most effective when most tubes have either no cracks or only very small ones, i.e. when practiced before service or very early in life (Ref 30, 31).

Material susceptibility, in combination with the aforementioned factors, is also a major factor affecting the occurrence of IGSCC. Most IGSCC has occurred in mill-annealed tubing. However, it is important to emphasize that there is not a single product called "mill-annealed" Alloy 600 tubing because each tubing manufacturer has employed different production processes. Whereas some mill-annealed tubing has not experienced any IGSCC over extended periods of operation, in other cases it has occurred after only 1 to 2 years of service, particularly at roll transitions. This variability of IGSCC response is even seen between heats from the same manufacturer in the same steam generator, as shown in Fig. 7 (Ref 1). It can be seen in Fig. 7 that susceptibility to IGSCC of the heats of Alloy 600 fits approximately a log-normal distribution so that a rather small fraction of Alloy 600 heats may be responsible for a disproportionately high number of tubes affected by primary side IGSCC. The reasons for such variability are only partly understood.

This microstructural aspect of IGSCC has been observed to be strongly affected by the final mill-annealing temperature and whether precipitation of grain-boundary carbides occurs during this treatment. The most susceptible microstructures are those produced by low mill-annealing temperatures, typically around 980 °C (1800 °F), that produce fine grain sizes (ASTM 9 to 11), copious quantities of intragranular carbides, and, usually, few if any intergranular carbides (Ref 32, 33). Higher mill-annealing temperatures in the range of 1040 to 1070 °C (1900 to 1960 °F) avoid excessive grain growth and leave enough dissolved carbon so that intergranular carbide precipitation occurs more readily during cooling. A further development of this idea was to thermally treat the tubing after mill annealing for ~15 h at 705 °C (1300 °F), which enhances intergranular carbide precipitation and provides enough time for the grain-boundary chromium concentration to recover and avoid sensitization (Ref 33). The beneficial influence of grain-boundary chromium carbides on primary side IGSCC resistance has been extensively evaluated. It has been demonstrated that grain-boundary carbides improve primary-side IGSCC resistance with or without grain-boundary chromium depletion, as also deduced from the generally good operating experience of Alloy 600 tubing of once-through steam generators (Ref 25, 32, 33). However, even thermally treated Alloy 600 tubing has cracked in service, although much less frequently than mill-annealed Alloy 600.

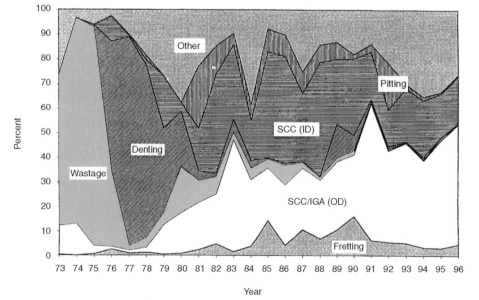

Fig. 6 Worldwide causes of steam generator tube plugging. SCC (ID), primary side SCC; SCC/IGA, outside surface SCC/IGA. Source: Ref 28

This has usually been attributed to a failure of the thermal treatment to produce the desired intergranular carbide microstructure either due to insufficient carbon or factors such as tube straightening prior to thermal treatment, which have favored carbide precipitation on dislocations instead of grain boundaries.

Steam generator tubes with IGSCC detectable by nondestructive testing have usually been preventively plugged either to avoid leakage or before the crack length reaches some predefined conservative fraction of the critical size for ductile rupture. Sleeving has sometimes been deployed as a repair method in operating PWRs to avoid plugging and maintain the affected tubes in service. The sleeves bridge the damaged area and are attached to sound material beyond the damage. The ends of the sleeves may be expanded hydraulically or explosively and are in most cases sealed by rolling, welding, or brazing (Ref 15).

More recent steam generators have been fabricated using Alloy 690 tubes thermally treated for 5 h at 715 °C (1320 °F). As well as being highly resistant in severe laboratory tests to IGSCC in PWR primary water compared with either mill-annealed or thermally treated Alloy 600, the lead steam generators with thermally treated Alloy 690 tube bundles have at the time of writing in early 2004 about 15 years of service with no known tube failures (Ref 34).

Thick Section Alloy 600 Components. Thick section, forged, Alloy 600 components started to crack in the mid 1980s starting with pressurizer nozzles (Ref 1, 35). In France, for example, all pressurizer nozzles were replaced with stainless steel. In 1989, the first cracking of Alloy 600 upper head CRDM nozzles occurred at the Bugey 2 plant in France. At first, it appeared as though this might be a special case because of the combination of a stress concentration due to a counter bore in the nozzles just below the level of the J-groove seal weld with the upper head as well as a relatively high operating temperature that was believed to be close to that of the hot leg in this first-generation French plant. However, the problem spread during the 1990s to CRDM nozzles in other plants with no counter bore, or with a tapered lower section to the CRDM nozzle, and in upper heads where the temperature was the same as the inlet cold-leg temperature (Ref 36, 37).

Three common features of the cracking of upper head CRDM nozzles were the presence of a significant cold-worked layer due to machining or grinding on the internal bore, some distortion or ovalization induced by the fabrication of the J-groove seal welds, and a tendency to occur more frequently in the outer setup circles where the angles between the vertical CRDM nozzle and the domed upper head were greatest. The combination of these three features plus the fact that the upper head is stress relieved before the CRDM nozzles are welded in place pointed to high residual stresses being responsible for these premature failures.

Although the generic problem of Alloy 600 CRDM nozzle cracking appeared first in France, only sporadic instances of similar cracking were observed in other countries until the beginning of the twenty-first century when numerous other incidents were reported. In some cases where cracking was allowed to develop to the point of leaking primary water into the crevice between the CRDM nozzle and the upper head, circumferential cracks initiated on the outer surface of the CRDM nozzle at the root of the J-groove seal weld (Ref 38). This had been observed in 1989 at Bugey 2 but only to a minor extent. No further leaks of primary water due to CRDM nozzle cracking have occurred in France because of an inspection regime adopted to avoid them and a decision taken to replace all upper heads using thermally treated Alloy 690 CRDM nozzles (Ref 36, 37). At the time of writing in early 2004, over 80% of upper heads had been replaced in French PWRs, and the same strategy is slowly being adopted elsewhere as the costs of repeat inspections makes any other policy uneconomic.

Nickel Base Weld Metals. The history of IGSCC in Alloy 600 and similar nickel-base alloys has taken yet another twist in recent years with the discovery of cracked Alloy 182 welds in several PWRs around the world (Ref 38, 39). This has occurred on the primary water side of J-groove welds that seal the CRDM nozzles in the upper head and also in a few cases in the safe-end welds of the reactor pressure vessel or pressurizer. Cracking seems to be significantly exacerbated by the presence of weld repairs made during fabrication, usually to eliminate indications due to hot cracking, thus again emphasizing the role of high residual stress in the failures observed to date. The cracking has often been described as interdendritic, but recent work shows that it is in fact intergranular (Ref 40). Incubation periods seem to be very long, of the order of 20 years, which certainly poses some interesting questions as to the mechanism of SCC.

Life Prediction of Alloy 600 Components. In spite of the improvements available for new plants or for component replacement, many Alloy 600 components, either mill annealed, thermally treated, or forged, remain in service. While most show no sign of cracking in service, it is important to have some tools to assess component life and predict when replacement may become necessary. Prediction methodologies were first developed for steam generator tubes and later extended to pressurizers and upper head CRDM penetrations. Both deterministic and probabilistic methods have been developed and are summarized next (Ref 41, 42).

Modeling of Alloy 600 component life is often based on the following empirical equation:

$$t_f = C \frac{\sigma^{-4}}{I_m} \exp\left(\frac{E}{RT}\right) \quad \text{(Eq 1)}$$

where t_f is the failure time (hours), C is a constant, σ is the applied stress (MPa), I_m is a material susceptibility index (see, Table 7), E is the apparent activation energy (cal/mole or J/mole), R is the universal gas constant (1.987 cal/mole/K or 8.314 J/mole/K), and T is the absolute temperature (in degrees K).

Despite the scatter observed in determinations of apparent activation energy, there is a reasonable consensus that a value of 180 kJ/mole (44 kcal/mole) is adequate for component life

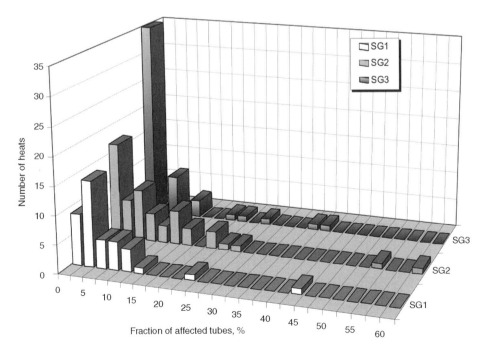

Fig. 7 Distribution of heats of Alloy 600 by susceptibility to intergranular stress-corrosion cracking (IGSCC) on the primary side of steam generator tube roll transitions in three steam generators (SG1, SG2, SG3) of the same plant. Source: Ref 1

estimations and 130 kJ/mole (30 kcal/mole) for crack growth estimations. Establishing the stress including residual fabrication stress on a given component is not trivial, but well tried and proven approaches based on finite element stress analysis or experimental techniques applied to mockups are available. However, dealing with material variability in susceptibility to IGSCC is not so straightforward and, in the case of classification of the susceptibility of CRDM nozzle cracking in U.S. PWRs, material variability has not been taken into account. The strategy for inspection and eventual replacement of upper heads in U.S. plants has been based on an evaluation of equivalent damage years determined only by the Arrhenius part of Eq 1 from the operating time and temperature, as shown in Fig. 8 (Ref 38). Nevertheless, as can be seen in Fig. 8, this approach has apparently correctly identified the plants most at risk from IGSCC of Alloy 600 CRDM nozzles.

A different approach was taken in France to address the same problem. It was observed that even after accounting for differences in stress and temperature, the variability attributed to material response could be considerable; up to a factor of 20 based on French PWR operating experience. One method to account for variability in material resistance to IGSCC has been based on a system of material indices, I_m, in Eq 1 (Ref 43–46). At its simplest, with no direct information about IGSCC susceptibility of individual heats, the guidelines given in Table 7 were adopted. They were based on observations of minimum times-to-failure of plant components or, in cases where no service failures had been observed, of laboratory specimens in accelerated tests of representative plant materials. The constant C in Eq 1 was adjusted so that an index of unity corresponds to a minimum failure time of 10,000 h at a temperature of 325 °C (615 °F) and an applied stress of 450 MPa (65 ksi), as observed in practice in plant and in laboratory tests. In addition, temperature (I_θ) and stress (I_σ) indices were defined relative to the reference conditions of 325 °C (615 °F) and 450 MPa (65 ksi) consistent with Eq 1 as:

$$I_\theta = \exp\left[\left(\frac{-E}{R}\right)\left(\frac{1}{T} - \frac{1}{598}\right)\right]$$

$$I_\sigma = \left(\frac{\sigma}{450}\right)^4$$

(Eq 2)

Thus

$$t_f = \frac{10{,}000}{I_m \cdot I_\theta \cdot I_\sigma}$$

(Eq 3)

In this way, the minimum time to cracking of each generic Alloy 600 primary circuit component was assessed after determining its operating temperature and stress. The results for different generic components of PWR primary circuits are shown in Table 7. Appropriate surveillance strategies were then established.

The quantification of variability of Alloy 600 heat susceptibility to IGSCC has been developed further to assess cracking encountered in the upper head CRDM nozzles of French PWRs and extended to other large Alloy 600 primary circuit components (Ref 42–47). The assessment of heat susceptibility is based on the premise that the distribution of carbides on the grain boundaries and within the grains is the principal parameter controlling susceptibility to IGSCC. Because the carbide microstructure cannot be easily determined by sampling components in service, the approach taken was to study experimentally the microstructure of as many product forms and processing conditions as possible and to categorize them into classes. Three main types of microstructure were recognized and related to the carbon concentration, thermal treatment, especially the temperature at the end of forging or rolling operations, and yield strength after hot working:

- Class A with mainly intergranular carbide precipitates
- Class B recrystallized with carbides mainly on a prior grain-boundary network
- Class C recrystallized with randomized intragranular carbides as well as carbides on prior grain boundaries

These classes were then linked to their IGSCC resistance (i.e., material susceptibility index) as determined from operating experience in the case of upper head penetrations or in accelerated laboratory tests mainly at 360 °C (680 °F). As more product forms associated with other Alloy 600 primary circuit components such as reactor pressure vessel bottom head

Table 7 Minimum failure times for IGSCC of Alloy 600 components in pressurized water reactor (PWR) primary circuits

Alloy 600 parts	Material index (I_m)	Stress index (I_σ)	Temperature index (I_θ)	Overall index product	Time, h	Observation(a)
Hydraulic expansion	0.2	0.4	1	0.08	80,000	NC
Divider plate	0.5	0.3	0.9	0.14	80,000	NC
Hard rolling on cold leg (Ringhals 2)	2	2.2	0.1	0.44	48,000	C
Pressurizer nozzle (San Onofre 3)	0.5	0.1	3.3	0.17	56,000	C
Nozzle (San Onofre)	0.5	0.9	3.3	1.49	8,000	C
Pressurizer nozzle (ANO1)	0.5	0.3	3.3	0.5	84,336	C
Pressurizer nozzle (Palo Verde 1)	0.5	0.4	3.3	0.66	33,320	C
Nozzle (Palo Verde 2)	0.5	1.5	1.1	0.83	25,000	C
Explosive expansion Fessenheim 1	1	0.4	1	0.4	75,000	C
Hard rolling on SG hot leg (Gravelines 6)	0.5	2.2	1	1.1	30,000	C
Hydraulic expansion (Doel 2)	2	0.4	1	0.8	30,000	C
Small U-bends Vallourec	2	2.2	0.3	1.32	30,000	C
Small U-bends Westinghouse	2	1.0	0.3	6	6,000	C
Sensitive hard rolling on SG hot leg	1	2.2	1	2.2	20,000	C
Very sensitive hard rolling on SG hot leg	2	2.2	1	4.4	8,000	C
1300 MW pressurizer nozzle	0.5	3.2	3.3	5.28	8,000	C
Mechanical plugs	0.5	1	1	0.5	40,000	C
French CRDM nozzles	0.5	1.5	0.5	0.4	80,000	C
	0.5	1.5	0.5	0.4	26,800	C
	1.1	2.8	0.08	0.24	72,909	C
	1.1	2.5	0.08	0.22	48,427	C
	1.1	2.5	0.08	0.22	58,868	C
	1.1	2.5	0.08	0.2	90,777	C

(a) NC, noncracked; C, cracked. Source: Ref 43–46

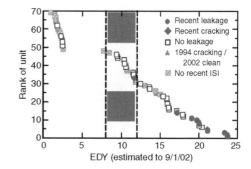

Fig. 8 Equivalent damage years (EDY) for the upper head control rod drive mechanism (CRDM) nozzles of U.S. PWRs in Sept 2002. ISI, in-service inspection. Source: Ref 38

penetrations, steam generator partition plates, and core supports (M-supports) were tested, it became necessary to increase the number of classes of microstructures to seven, for example, to include the influence of twin boundaries, heterogeneous grain sizes, and presence of both intergranular and intragranular carbides (Ref 42, 46).

Inevitably, such an approach to assessing IGSCC susceptibility reveals significant scatter in the susceptibility indices for different heats about the mean associated with each class. That uncertainty has been incorporated into life estimations, as described subsequently. However, practical assessments of the risk of IGSCC of primary circuit components fabricated from Alloy 600 have generally used the best estimate of the susceptibility index based on fabrication data coupled with conservative values of operating stress and best estimates of operating temperature. This method has been used to define appropriate surveillance and maintenance strategies for different generic classes of components in primary circuits of French PWRs.

Application of the Weibull distribution to quantify the dispersion in stress corrosion data is well established and has been successfully applied to IGSCC in Alloy 600 steam generator tubes and upper head penetrations (Ref 41, 42). The dispersion in times to observe detectable cracks arises from the inherent variability in susceptibility of materials to SCC and, in the case of plant components, to uncertainty in the precise operating stress and temperature. The most useful form of the Weibull distribution is the linear transform of the cumulative distribution function:

$$\ln \ln \left(\frac{1}{1-F(t)} \right) = \beta \ln(t-t_0) - \beta \ln(\eta - t_0)$$

(Eq 4)

where $F(t)$ is the fraction of components failed, t is the failure time (hours or years), t_0 is the location parameter or origin of the distribution (hours or years), η is the characteristic life or scale parameter (hours or years), and β is the shape parameter and slope of the linear transform.

The Weibull distribution can be fitted to the early observations of IGSCC as a function of operating time and provides an effective tool for predicting the future development of cracking so that informed inspection and repair plans can be formulated (Ref 41). This approach has also been adapted to French operating experience of IGSCC in the roll transitions of steam generator tubes and incorporating the method of material indices described earlier (Ref 42). For the reference conditions, material index, $I_m = 1$, a stress of 450 MPa (65 ksi), and a temperature of 325 °C (615 °F), the appropriate Weibull distribution parameters were determined to be:

$$\beta = 1.5 \quad \eta = 100,000 \text{ h}$$
$$t_0 = 0.1 \eta = 10,000$$

The slope β is significantly lower than values deduced from U.S. operating experience (Ref 41) probably because the latter did not incorporate an analysis of t_0.

For other Alloy 600 components subject to IGSCC for which estimates of material susceptibility indices were available (e.g., Table 7), the characteristic life and origin of the distribution were easily calculated for each heat using Eq 1 or 3 and the estimated values of operating temperature and stress (Ref 42). The Weibull shape parameter is more difficult to determine, but after examining IGSCC data for other Alloy 600 components and laboratory specimens, the main influence was observed to come from the applied stress, for which an empirical equation was derived:

$$\beta = 0.75 \left(\frac{\sigma - 250}{100} \right)$$

(Eq 5)

Thus, as the stress increases, the slope β increases and the scatter about the mean life decreases, as might be expected.

In order to improve further the stochastic approach to predicting IGSCC in the context of upper head penetration cracking encountered in French PWRs, the inherent dispersion in the input parameters of stress, temperature, activation energy, and material susceptibility index was taken into account. It was clearly not physically realistic to treat heat-to-heat variability as a quantized phenomenon as implied by Table 7, and a log-normal distribution of material susceptibility indices was chosen in accordance with the populations of heats assessed in each class A, B, C, and so forth. The Monte Carlo simulation technique of randomly sampling distributions of the input parameters in Eq 1 to determine a distribution of failure times was adopted, as illustrated in Fig. 9 (Ref 42).

An example of the results using the Monte Carlo approach outlined in Fig. 9 is shown in Fig. 10 in the form of a Weibull distribution comparing the results of these simulations with the inspection results for upper head penetrations in each susceptibility class of Alloy 600 (Ref 42). When the Monte Carlo simulations are repeated many times, the dispersion in the resulting Weibull distribution of failure times is relatively small because the number of penetrations considered for each PWR plant series is over 1000.

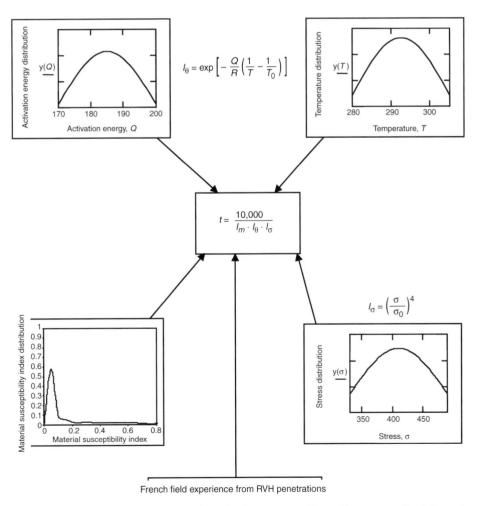

Fig. 9 Illustration of Monte Carlo simulation of upper head penetration cracking. RVH, reactor vessel head. See text for an explanation of other symbols. Source: Ref 42

Thus, the progression of the problem for each design series of PWRs has relatively little inherent uncertainty. On the other hand, if the problem is considered on a plant-by-plant basis, then although the numbers of penetrations in each class of material susceptibility considerably affect the risk of IGSCC occurring, the statistical uncertainty in predictions of the proportion that will crack in a given operating time is much greater because there are less than a hundred CRDM penetrations per upper head. For a given upper head, this statistical uncertainty in the number of cracked penetrations can be of the order of ± 1 to ± 5 on the mean prediction. This uncertainty is easily demonstrated and quantified in a probabilistic sense using the Monte Carlo simulation technique.

It is well known among stress corrosion specialists that surface finish can have a dramatic influence on component susceptibility to cracking in service, but quantifying the effect has not been so obvious until recently (Ref 35, 43, 45, 46). One important consequence of the experience of upper head penetration IGSCC experienced in French PWRs has been a careful characterization and measurement of the thicknesses of cold-worked layers and residual stresses left by different machining techniques as well as a quantitative framework for understanding what is of concern and what is not.

It is observed experimentally that the lifetime to failure by IGSCC of Alloy 600 specimens in PWR primary water depends on the extent of cold work in the surface layer and on its thickness. Most machining methods leave a cold-worked layer on the surface that is normally in compression. If this layer is then placed in tension due to an applied stress exceeding the yield stress of the more ductile substrate, then very high tensile stresses easily up to 1000 MPa (145 ksi) can be generated in the cold-worked layer (Ref 35, 43). Whether this has much effect on the time to failure by IGSCC then depends on the layer thickness. If the compressive strain in the surface cold-worked layer is not reversed, then of course a favorable effect on IGSCC resistance is observed. Thus, the practical problem with remedies such as shot peening, for example, is to ensure that the underlying more ductile substrate is never strained beyond its yield point.

The interpretation and quantification of the influence of surface layers on IGSCC in Alloy 600 has been based on three principles (Ref 43). First, crack growth rates are increased by cold work when the material is loaded above its yield strength. Secondly, the evolution of cracking is characterized by three distinct phases: an incubation period, a slow crack growth period and (when $K_I > K_{ISCC}$) a rapid crack growth period, the difference in growth rates being about an order of magnitude. The third principle used to quantify the effect of cold-worked surface layers on IGSCC life is that both the slow and fast phases of crack growth rate depend primarily on the crack tip strain rate.

The effect of a cold-worked layer with high residual stress caused by yielding of the more ductile substrate on the slow to fast growth rate transition is illustrated in Fig. 11. If the cold-worked surface layer is thin and the transition in growth rates occurs in the underlying material unaffected by cold work, then a smaller effect on lifetime is observed; cold work increases the crack growth rate only over the short distance the crack travels through the cold-worked layer. If the cold-worked and highly stressed layer is thick enough so that the transition in growth rates occurs within it, then the crack length associated with the transition will be much shorter because of the higher stress, and the lifetime to failure will be significantly reduced.

The detailed mathematical treatment of crack growth based on a strain rate dependent model to calculate failure times with various assumptions about the extent and depth of cold work is complex (Ref 43). Nevertheless, the overall effect of surface cold work on IGSCC life that results from this treatment can be presented simply in terms of an effective stress that would induce the same crack initiation time (to reach the rapid crack growth phase) from an electropolished surface compared with the same material with a cold-worked surface layer. Figure 12 shows the effect of thickness of the cold-worked layer with a surface tensile stress of 1000 MPa (145 ksi) overlying a substrate with a yield stress of 300 MPa (44 ksi). Taking one example from Fig. 12, for an applied stress of 500 MPa (73 ksi) and a 200 µm (0.008 in.) thick cold-worked layer characteristic of heavy machining or grinding, the equivalent stress to give the same failure time without the cold-worked layer would be 900 MPa (131 ksi); i.e., a reduction in life due to the cold-worked layer

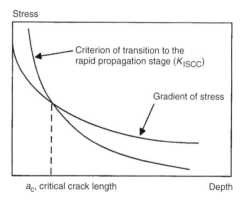

Fig. 11 Illustration of the conditions necessary for transition from slow to rapid intergranular stress-corrosion cracking (IGSCC) propagation. Source: Ref 43

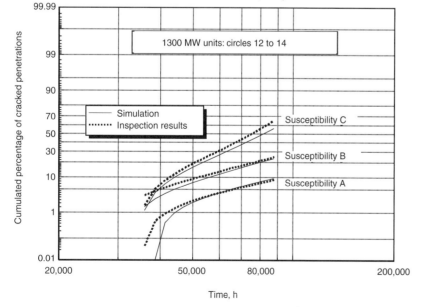

Fig. 10 Results of Monte Carlo simulations of intergranular stress-corrosion cracking (IGSCC) in upper head penetrations of 1300 MW French PWRs and comparison with inspection results for each class of Alloy 600. Source: Ref 42

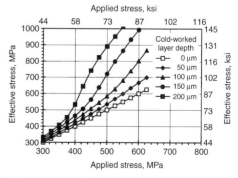

Fig. 12 Equivalent stress for intergranular stress-corrosion cracking (IGSCC) on an electropolished surface as a function of applied stress and thickness of a cold-worked layer on a cold-worked surface. Source: Ref 43

by a factor of $(900/500)^4$ is anticipated, which is equivalent to an order of magnitude.

Crack Propagation of Alloy 600 Components. Once a stress corrosion crack has been detected by nondestructive examination in a PWR primary circuit component, an essential step in the justification of structural integrity and further operation without repair or replacement of the affected component is an assessment of crack growth during the next few operating cycles. Practical approaches to assessing crack growth by IGSCC in Alloy 600 components have relied on empirical measurements of crack growth rates as a function of crack tip stress intensity, K_I, rather than the strain rate based relationships referred to previously that are much more difficult to use. A compilation published in 1999 compared laboratory data with field measurements on upper head penetrations of French 900 and 1300 MW PWRs (Ref 48).

There is fairly general agreement that the most appropriate form of the crack growth equation for IGSCC of Alloy 600 in PWR primary water is (Ref 49):

$$\frac{da}{dt} = C(K_I - 9)^n \quad (K_I \text{ in MPa}\sqrt{m}) \quad \text{(Eq 6)}$$

The values of the coefficient C and exponent n for given practical circumstances vary among different publications, but there is a reasonable consensus that the apparent or effective activation energy to be used for adjusting the coefficient C for temperature is ~130 kJ/mole. Cracking events in Alloy 600 upper head penetrations and weld deposits of Alloys 82 and 182 have once again focused attention on the appropriate values of the coefficients to use for the evaluation of indications found by nondestructive examinations.

Other variables that are known to influence the rate of crack growth in Alloy 600 are cold work, hydrogen overpressure, and possibly pH or lithium hydroxide concentration. Cold work can easily affect the value of the coefficient C by as much as an order of magnitude. Hydrogen overpressure effects are also significant (Ref 50). However, this effect has not been explicitly included in crack growth assessment equations to date, probably because the hydrogen concentration in PWR primary water is controlled within a relatively narrow range. Concerning the possible influence of pH or lithium concentration in PWR primary water on crack growth kinetics, the effect is small within the range of pH or lithium concentrations permitted by the PWR primary water specification (Ref 1, 42).

Summary of Alloy 600 Laboratory Investigations. The generic IGSCC of the nickel-base Alloy 600 and its high-strength analogue, Alloy X750, in PWR primary water has been extensively studied. Despite considerable experimental efforts, no consensus exists as to the nature of the cracking mechanism (Ref 51) and both life modeling and remedial measures have relied on empirical, phenomenological correlations, as described earlier. The essential phenomenological features of primary water IGSCC of Alloy 600 are:

- The profound influence of hydrogen partial pressure (or corrosion potential) and observation of a worst case centered on corrosion potentials near the Ni/NiO equilibrium (Fig. 2)
- The apparently continuous mechanism of failure between 300 °C (570 °F) subcooled water and 400 °C (750 °F) superheated steam
- The high and variable apparent activation energy typically 180 kJ/mole but with a scatter band of 80 to 220 kJ/mole
- The influence of carbon content and microstructure, particularly the favorable influence of grain-boundary carbides and undesirable effect of cold work
- The high stress exponent of ≈4 for lifetime to failure

The mechanism of IGSCC of Alloy 600 in PWR primary water is the subject of intense debate in the technical literature, and it is difficult to find any consensus even on issues as basic as whether cracks advance by an oxidation process at the crack tip or due to embrittlement by hydrogen discharged from the matching cathodic reaction (Ref 51). The magnitude of the activation energy is perhaps one of the most difficult parameters to explain. Physical support for the fourth power dependency on applied stress comes mainly from studies of grain-boundary sliding observed during primary creep in Alloy 600 at temperatures between 325 and 360 °C (615 and 680 °F) (Ref 43, 52). However, the same dependency on stress can also be easily demonstrated if grain-boundary oxidation controlled by parabolic kinetics to some critical depth is required for crack initiation. Grain-boundary sliding rates are observed to depend on grain-boundary carbide coverage, greater coverage being associated with slower grain-boundary sliding rates and higher resistance to IGSCC, as observed in practice. Nevertheless, although grain-boundary carbide morphology is a major reason for heat to heat variability in susceptibility to IGSCC of Alloy 600 in PWR primary water, it is clear that other as yet unidentified metallurgical factors also play a role.

High-Strength Nickel-Base Alloys

Improvements have been made to Alloy X750 for PWR primary water service by increasing the annealing temperature and resultant grain size and by the adoption of single-step aging at 704 °C (1299 °F) for 20 h. Although the main goal of the aging heat treatment is to precipitate the strengthening phase γ', $Ni_3(Ti, Al)$, an added advantage of these heat treatment conditions is a fine, dense $M_{23}C_6$ carbide distribution at grain boundaries (Ref 53).

The need to pay attention to the surface condition of components, particularly with respect to cold work and residual stress, has been emphasized earlier. The example of Alloy X750 is a case in point where the atmosphere used during aging heat treatment was found to influence initiation times for IGSCC in PWR primary water (Ref 53). This was due to oxidation of surface layers that had to be removed by machining after heat treatment in order to ensure optimum performance in service. A more recent similar example is with Alloy 718, a normally highly reliable high-strength alloy for use in PWR primary water. Some studies in the literature have implicated the formation of δ phase during thermal aging as having an aggravating influence on IGSCC susceptibility (Ref 54). Others have not observed a major effect of δ phase on product performance and indeed is a necessary feature to avoid excessive grain growth during solution treatment prior to aging (Ref 1). By comparison, oxidation of the surface during product rolling can have a severe adverse influence on IGSCC initiation in PWR primary water. The reason is not hard to miss in micrographs where intergranular oxidation during product rolling provided a precursor intergranular crack when stressed prior to exposure to PWR primary water. For optimum IGSCC resistance in plant, it was essential to remove the layer affected by the furnace atmosphere, as observed previously for Alloy X750.

Low-Strength Austenitic Stainless Steels

Type 304 and 316 stainless steels are the main materials used for the pressure boundary and other components of PWR primary circuits. Operating experience with these materials since the mid-1960s has generally been excellent. When stress corrosion failures have occurred, they have in most, if not all cases, been due to internal or external surface contamination by chlorides or to out-of-specification chemistry in dead legs or other occluded volumes where primary water chemistry control can be difficult (such as the transient presence of oxygen for significant periods) (Ref 1). Excessive cold work, with the attendant risk of martensite formation in type 304 stainless steel, has also been a contributing factor in some cases. Although low-carbon grades of types 304 and 316 stainless steels have often been used to minimize the risk of sensitization (by grain-boundary chromium depletion) of weld heat-affected zones, there is little doubt that such sensitized materials exist in considerable quantities in many older PWRs. Nevertheless, practical experience shows that deoxygenated, hydrogenated PWR primary water does not cause stress corrosion in such sensitized materials, in contrast to BWR experience with oxygenated water.

The exceptions to this excellent record, aside from inadvertent surface contamination, are associated with the difficulty in maintaining the low concentrations of impurities (<150 ppb Cl^-, for example) required by the primary water specification in areas where water circulation is restricted. In addition, dead legs can trap air bubbles during refueling and maintenance that may then be difficult to eliminate when refilling and pressurizing the primary circuit. Temperatures in such locations are also usually much lower than those in the main circulating part of the primary circuit.

One such location that has experienced problems in PWRs is the canopy seals associated with control rod drive mechanisms. Canopy seals ensure leak tightness of threaded joints that make up the control rod housings above the PWR pressure vessel top head. The threaded joints impede evacuating air during filling of the primary circuit and obviously do not allow free circulation of the primary coolant to the canopy seal. One publication has discussed the origin of canopy seal cracking and leaks in some detail in relation to a few plants (Ref 55). Searches of nuclear industry records of service failures show, however, that this is a widespread generic problem affecting many PWRs worldwide. It divides broadly into two families of failures—those occurring rather rapidly a year or two after entering service and those occurring after more than about 10 years' service (Ref 1). The first group usually has some aggravating mechanical factor associated with poor welding of the canopy seal closure. In the second case, even if welding defects are present, it is far from clear that these are always directly associated with the cracking.

The interpretation of canopy seal cracking has been based on its mainly transgranular morphology, seemingly typical of transgranular stress corrosion cracking (TGSCC) of (non-sensitized) austenitic stainless steels in the presence of chloride and oxygen (Ref 55). Figure 13 shows the classical relationship between oxygen and chloride concentrations that gives rise to TGSCC of nonsensitized stainless steels. It can be seen that any oxygen trapped in this location greatly reduces the tolerance of austenitic stainless steels for chloride contamination. Chloride impurity concentrations acceptable for normal quality primary water would be unacceptable where a trapped air bubble at the high primary circuit pressure is present. Even if the consumption of oxygen by general corrosion is taken into account, such a trapped bubble could be present for a long time, more than sufficient to cause cracking.

The only difficulty with the foregoing interpretation is that the few analyses available of the liquids trapped behind canopy seals hardly ever show any significant chloride, the main impurity being sulfate. The presence of sulfate as the major impurity is not surprising either as a surface impurity on nearby threaded surfaces or from the thermal decomposition of any resin fines that find their way accidentally into the hot parts of the primary circuit. Given that sulfate in combination with oxygen is well known to cause stress corrosion in BWRs, albeit usually intergranular, it is possible that a similar mechanism could be at work in dead legs of PWRs. In addition, recent studies have shown that the combination of very low amounts of chloride and sulfates (of the order of a few ppm) in the presence of dissolved oxygen can promote very fast transgranular cracking of type 304 stainless steel at moderate temperatures of ~200 °C (390 °F) or less (Ref 57).

High-Strength Stainless Steels

Several high-strength stainless steels are used in PWR primary circuits for components such as bolts, springs, and valve stems. The main ones considered here are A-286 precipitation-hardened austenitic stainless steel, type 410 and similar martensitic stainless steels, and 17-4PH precipitation-hardened martensitic stainless steel. Over the years, small numbers of such components have cracked due to stress corrosion or hydrogen embrittlement and, on occasions, loose parts have been generated in the primary circuit.

Alloy A-286 is, an austenitic, precipitation-hardened, stainless steel (Table 2) that is strengthened by γ', $Ni_3(Ti,Al)$, formed during aging at 720 °C (1330 °F). Its use is favored where the expansion coefficient relative to other austenitic stainless steels is an important design factor. Unfortunately, it is susceptible to IGSCC in PWR primary water when loaded at or above the room-temperature yield stress, typically 700 MPa (102 ksi) (Ref 58–63). Cold work prior to aging in combination with the lower of two commonly used solution annealing temperatures of 900 and 980 °C (1650 and 1795 °F) has a particularly adverse effect on resistance to IGSCC (Ref 61). Hot heading of bolts, which can create a heat-affected zone between the head and shank, is another known adverse factor. Nevertheless, even if these metallurgical factors are optimized, immunity from cracking cannot be ensured unless the stresses are maintained below the room-temperature yield stress, which necessitates strictly controlled bolt loading procedures. There is also strong circumstantial evidence that superimposed fatigue stresses can lower the mean threshold stress for IGSCC even further. Finally, the role of impurities, including oxygen introduced during plant shutdown and possibly consumed only slowly in confined crevices, in helping crack initiation is clear from all the evidence available. Once initiated, cracks grow relatively easily even in well-controlled PWR primary water (Ref 62).

Martensitic Steels. Components such as valve stems, bolts, and tie rods requiring rather high strength combined with good corrosion resistance in PWR primary circuit water have been typically fabricated from martensitic stainless steels, particularly type 410 and 17-4 PH (Table 2). A search of the nuclear industry literature reveals a significant number of failures of martensitic stainless steels such as type 410, for example in Nuclear Regulatory Commission (NRC) Information Notices (Ref 64). In most cases, the affected components have usually entered service in an overly hard condition due to tempering at too low a temperature. No in-service aging seems to have been involved, these materials proving susceptible to SCC/hydrogen embrittlement in the as-fabricated condition. An additional problem has been caused by galvanic corrosion with graphite-containing materials in the packing glands of valves, sometimes leading to valve stem seizure. The preferred replacement material has often been 17-4 PH with its higher chromium and molybdenum content no doubt conferring better resistance to crevice corrosion.

A significant number of service failures of 17-4 PH precipitation-hardening stainless steel have also occurred in PWR primary water (Ref 65–67). Initially, intergranular cracking by SCC/hydrogen embrittlement was associated with the lowest temperature aging heat treatment at 480 °C (900 °F) designated H900. This gives a minimum Vickers hardness value of 435 HV, clearly in excess of the limit of 350 HV commonly observed to limit the risk of hydrogen embrittlement. The 593 °C (1100 °F) (H1100) aging heat treatment was subsequently widely adopted and normally yields a hardness value below 350 HV. Nevertheless, a small number of failures, due either to brittle fracture or SCC/hydrogen embrittlement, have continued to occur. The origin of these failures appears to be thermal aging in service.

Two main thermal aging mechanisms of martensitic stainless steels are recognized. The first, "reversible temper embrittlement," is related to the diffusion of phosphorus to grain boundaries at aging temperatures generally above 400 °C (750 °F) and can occur in both type 410 and 17-4 PH stainless steels. The grain boundaries are consequently embrittled and are particularly susceptible to intergranular cracking by hydrogen embrittlement, but no general increase in hardness is observed. The

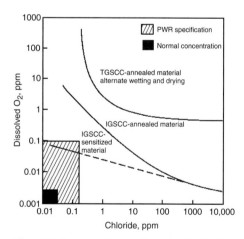

Fig. 13 Stress-corrosion cracking of austenitic stainless steels as a function of chloride and oxygen concentrations in high-temperature water. IGSCC, intergranular stress-corrosion cracking; TGSCC, transgranular stress-corrosion cracking. Source: Ref 56

embrittlement can be reversed by heat treating around 600 °C (1110 °F) and largely avoided by reducing the phosphorous content and by small alloying additions of molybdenum.

The second thermal aging embrittlement mechanism is relevant only to precipitation-hardened stainless steels such as 17-4 PH. It arises from an intragranular decomposition of the martensitic matrix into two phases, α, which is rich in iron, and α′, which is rich in chromium. A generalized increase in hardness is observed with corresponding increases in strength and ductile-brittle transition temperature and loss of fracture toughness. The hardening cannot be reversed without re-solution heating. French studies have shown that this aging mechanism can occur in 17-4 PH steels on time scales relevant to the design lives of PWRs at temperatures exceeding 250 °C (480 °F) and quantitative models for component assessment have been developed (Ref 66, 67). Mechanical fractures occur by cleavage, although those involving corrosion can also be intergranular. Both types of failure have been associated with hardness values after in-service aging significantly exceeding 350 HV. Corrosion-related failures have also been aggravated by impurities from valve packing gland materials.

Irradiation-Assisted Corrosion Cracking of Austenitic Stainless Steels

Irradiation-assisted stress corrosion cracking (IASCC) is a term that defines SCC phenomena in core structural materials of water-cooled and/or moderated nuclear power reactors in which neutron and/or γ irradiation contributes directly to the initiation and propagation of SCC. It is reviewed in detail in the article "Effect of Irradiation on Corrosion and Stress Corrosion Cracking in Light Water Reactors" in this Volume and so is only briefly described here. By implication, in the absence of material damage by fast neutrons and/or modification of the environmental chemistry by ionizing radiations, SCC either does not occur or is significantly less severe. Laboratory and field data show that intergranular SCC of austenitic steels can result from long-term exposure to high-energy neutron radiation (Ref 68–72).

Types 304 and 316 austenitic stainless steels, usually solution annealed and cold worked respectively, are widely used for the fabrication of light water reactor core internals. Unlike fuel elements that are discharged and replaced after a few reactor cycles, the internals are intended to remain for the complete reactor life. Consequently, extremely high neutron doses can be experienced by components in close proximity to the nuclear fuel. Both BWR and PWR core internal structures have experienced cracking attributed to IASCC. An overview of these incidents together with the anticipated end-of-reactor-life doses and characteristic dose ranges for the irradiation-induced physical processes likely to be important in IASCC are shown in Fig. 14 (Ref 70–72).

In the case of the oxygenated coolants of BWRs, the modification of grain-boundary composition due to neutron irradiation and resulting inverse Kirkendall diffusion of point defects, particularly chromium depletion, has been shown to be an important precursor of IASCC. Neutron doses exceeding 5×10^{20} n/cm^2 ($E > 1$ MeV) are associated with the occurrence of IASCC in BWRs, this being the dose required to develop sufficient irradiation-induced chromium depletion at grain boundaries. In addition, the formation of oxidizing species, oxygen and hydrogen peroxide, by radiolysis plays an important role in this manifestation of IASCC.

Pressurized water reactor field experience has also shown that intergranular cracking of highly irradiated core components can occur. Specifically, type 304 cladding of both control rods and (now obsolete) fuel pins as well as cold-worked type 316 core baffle-former bolts of some first generation (CP0 series) 900 MW French PWRs have cracked intergranularly in service (Ref 70, 72). Fast neutron doses of $>2 \times 10^{21}$ n/cm^2 ($E > 1$ MeV) and strains >0.1% have been implicated in the cracking. The same irradiation-induced modification of grain-boundary chemistry accompanied by very significant hardening is observed in stainless steels irradiated in PWRs. Indeed, this occurs to a much greater extent than in BWRs due to the higher neutron doses for equivalent operating times. However, as noted earlier, radiolytic formation of oxygen or hydrogen peroxide is effectively suppressed in PWRs by the addition of 2 to 4 ppm of hydrogen to the primary water coolant. The absence of such oxidizing species is an obvious environmental difference compared with BWRs and renders grain-boundary chromium depletion of no particular consequence in PWR primary water. The specific radiation-induced microstructural and microchemical changes induced by neutron irradiation in PWRs responsible for IASCC susceptibility have not so far been identified, although the considerable irradiation-induced hardening that is observed no doubt plays an important role (Ref 71).

A factor specific to the PWR core baffle-former bolts is the presence of a closed crevice in the particular CP0 series of French 900 MW plants concerned (Ref 70, 72). This has led some investigators to speculate that local radiolytic production of oxidizing species might occur due to limited access of hydrogen. However, no evidence of oxidizing conditions has been found on extracted bolts. Laboratory simulations of crevices irradiated at intensities equivalent to in-core conditions have also failed to reveal the formation of any oxidizing species within the crevice (Ref 68). An alternative hypothesis for baffle bolt cracking is that local boiling due to γ heating could render the confined environment strongly caustic by concentrating lithium hydroxide in the liquid phase. Again, there is no evidence for this in the form of mineral precipitates containing boron that are observed when PWR primary water is allowed to boil (as, for example, in the case of the axial offset anomaly discussed previously in this article).

In addition to the phenomena of radiation-induced hardening and changes to grain-boundary composition, other radiation damage processes could have important influences on the development of IASCC (Ref 70, 72). Irradiation creep can relax residual and applied stresses and is independent of temperature in the range of interest to light water reactors. Swelling, hitherto only considered of importance to fast reactors, could in principle also appear at the high neutron doses associated with the second half of life of PWRs and affect the loads applied to components such as baffle bolts. Thus, although significant advances have been made in the understanding of IASCC, much remains to be learned, and it is today a very active field of research in the context of both BWR and PWR plant aging.

Steam Generator Secondary Side Corrosion

The two types of PWR steam generators in general use are vertical recirculating steam generators (RSGs) and once-through steam generators (OTSGs) (Fig. 5) with tube bundles, depending on age, made from either mill-annealed or thermally treated Alloy 600, or thermally treated Alloy 690, or Alloy 800. Sub-cooled primary water flows through the inside of the tubes and boils secondary water on the shell side of the tubes. The steam quality of the water-steam mixture entering the steam driers of RSGs is typically 10% and the superheat across the tubes may vary from 10 to 40 °C

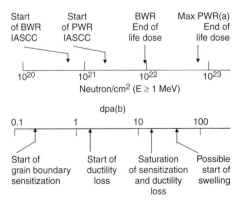

Fig. 14 Summary of irradiation-assisted stress-corrosion cracking (IASCC) in light water reactors and associated metallurgical changes in austenitic stainless steels as a function of neutron irradiation in n/cm^2 ($E \geq 1$ MeV) or displacements per atom (dpa). EFPY, effective full power year. Source: Ref 70–72

(20 to 70 °F). In OTSGs, which of necessity have always used all-volatile water treatment (AVT) water chemistry, the steam quality reaches 100% at the ninth and tenth support plate region, and achieves about 20 to 35 °C (40 to 60 °F) of superheat at the top of the unit. Recirculating steam generators of Russian designed Voda-Vodyanoi Energetichesky Reaktor (VVER) plants are rather different being horizontal with stainless steel tubes running between two large vertical cylindrical collectors for the inlet and outlet primary water.

Vertical PWR steam generators have experienced a variety of corrosion-induced and mechanically induced problems, and many steam generators have been replaced, usually because of SCC of mill-annealed Alloy 600 steam generator tubes (Ref 15, 28, 73). Some have been replaced after only 8 to 12 years of operation, which is far short of the usual initially licensed plant operating period of 40 years. Secondary side corrosion problems include flow-assisted corrosion of carbon steel pipes and denting, wastage, intergranular attack, IGSCC, and pitting on the outside surfaces of the steam generator tubes (Ref 15, 28, 73). The evolution of steam generator tube corrosion with time in terms of relative importance of each damage mechanism is shown in Fig. 6. Mechanical concerns have included water hammer, thermal stratification in secondary feedwater pipes, and fretting and wear of the tubes caused by excessive tube vibration.

As remarked earlier, the vast majority of PWR steam generator corrosion problems, particularly on the secondary side, have concerned mill-annealed Alloy 600. In Germany, Kraftwerk Union AG (KWU) and later Babcock & Wilcox/Atomic Energy of Canada Limited (B&W/AECL) designs for the Canada deuterium uranium (CANDU) series of pressurized heavy water reactors (PHWRs), the tubes were made of Alloy 800. The mill-annealing conditions for Alloy 600 varied among the manufacturers, while B&W/AECL stress-relieved the completed steam generators (605 °C, or 1125 °F, for 8 h) including the Alloy 600 tubing. This latter procedure inevitably "sensitized" the tubing (by precipitating chromium carbides on the grain boundaries and leaving a chromium-depleted band, typically a few tens of micrometers wide, adjacent to grain boundaries). Subsequently, Alloy 600 tubing was thermally treated (705 °C, or 1300 °F, for 15 h) to improve resistance to IGSCC, and to date this tubing has largely resisted secondary corrosion problems. New or replacement RSGs are supplied with thermally treated Alloy 690 or Alloy 800 tubing, which, to date, have also resisted both primary and secondary corrosion problems.

Many secondary side corrosion problems with mill-annealed Alloy 600 tubes have been associated with the interstices between the tubes and the tube supports. The tube support structures for most of the early units were made of carbon steel, while later units switched to types 405, 409, and 410 ferritic stainless steels for additional corrosion resistance (Ref 74). Type 347 austenitic stainless steel has always been used for KWU steam generator tube egg crate (lattice bar) support structures. Tube support structures of early units used plates with drilled holes, then plates with trefoil or quatrefoil broached holes initially with concave lands and then flat lands, or lattice bars (egg crates) (Fig. 15). The objective of the more open tube support designs is to reduce the accumulation of impurities in the interstices by the phenomenon of hideout.

Another corrosion-sensitive zone for steam generator tubes has been in and just above the lower tube sheet. In some of the very early RSG designs, the tubes were only partly expanded just above the seal weld with the lower tube sheet face, thus leaving a crevice between the outside diameter (OD) of the tube and the inside diameter of the hole in the tube sheet. Later, the tubes were expanded into the tube sheet along nearly their full length in order to close all but the last ~4 mm (0.16 in.) of the tube to tube sheet crevice. Tube expansion has been achieved by various methods, mechanical rolling, hydraulic, and explosive. Each expansion method has generated its own characteristic residual stress fields in the tubes that have influenced subsequent stress corrosion behavior if, or when, impurities concentrate by hideout either in the tubesheet crevice or under sludge that accumulates on the upper face of the tube sheet.

The underlying cause of nearly all forms of localized corrosion observed on the secondary side of steam generators is the phenomenon of hideout of low volatility solutes present as impurities, or added deliberately, in superheated crevices with restricted water circulation. Such impurities entering recirculating steam generators in the feed water are relatively insoluble in the steam phase so that they accumulate in the recirculating water. However, their concentration is limited in the recirculating water to about 100 times that of the concentration in the feed water by maintaining a blowdown of recirculating water equivalent to about 1% of the steam mass flow. Significant further concentration of impurities by potentially many orders of magnitude then occurs in occluded superheated crevices by a wick boiling mechanism. Due to the potential variety of impurities entering the steam generators, many complex mixtures of chemicals can be envisaged in these occluded regions where such impurities concentrate by hideout. This severely complicates the task of understanding the mechanisms of tube corrosion and defining adequate remedies. Tube damage such as wastage, pitting, and denting has been attributed to the local formation of strong acids and, evidently, has been largely eliminated by appropriate management of secondary water chemistry (Fig. 6). By contrast, the steadily rising trend in IGA/IGSCC (Fig. 6) suggests that countermeasures have not been completely effective, probably because its

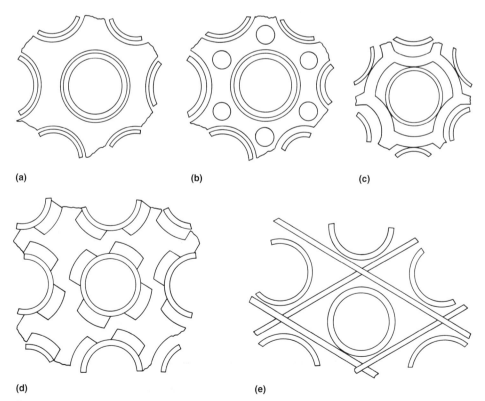

Fig. 15 Steam generator tube support designs. (a) Drilled without flow holes. (b) Drilled with flow holes. (c) Broached trefoil. (d) Broached quatrefoil (concave contact lands). (e) Egg crate (lattice bars)

origin(s) has/have not always been correctly identified. All these different degradation mechanisms are described in more detail hereafter along with the remedies adopted.

Tube Wastage

Tube wastage, or thinning, was one of the first corrosion problems that occurred in RSGs operating with sodium phosphate as a secondary water treatment (Ref 15, 75). It was first observed in commercial PWR plants when phosphate treatment was changed to a low sodium-to-phosphate molar ratio control, in which the molar ratio of Na/PO_4 was maintained at about 2.0. This change was in response to a series of caustic stress corrosion events attributed to operation with uncontrolled high Na/PO_4 ratios (above 2.8), from which free caustic could result. Although the incidence of caustic IGSCC dropped markedly, wastage began to be observed within approximately 1 year after the water chemistry change.

Phosphate wastage in PWR steam generators occurred typically at the interfaces between hot-leg tubes and the tops of sludge piles that accumulated on top of the tube sheet. Where the sludge pile was deep, the zone of wastage extended about 25 mm (1 in.) into the pile, but in most cases, it did not penetrate appreciably into the tube/tube sheet crevice. Wastage was first noted in steam generators of a particular design in the vicinity of antivibration straps, which were relatively wide and oriented in such a way as to define a region of steam blanketing. Corrosion was concentrated at the boundaries of this steam-blanketed region, where a liquid/vapor interface presumably fluctuated over a short length of tubing.

Laboratory work showed that the aggressiveness of concentrated sodium phosphate solutions was related to both concentration and the Na/PO_4 molar ratio and led to recommendations for the Na/PO_4 molar ratio to be controlled in a relatively narrow band from 2.3 to 2.6. The lower limit was related to the rapidly increasing rate of wastage at lower ratios, and the higher limit was selected to avoid free caustic forming. Some plant operators were successful in controlling their steam generator chemistries within this restricted range of compositions, but eddy-current inspections indicated that the rate of attack only slowed but did not stop. Furthermore, a constantly increasing sludge burden, augmented by precipitated phosphate compounds, made such ion ratio control increasingly difficult. One explanation based on laboratory work linked continuing wastage to the formation of an extremely corrosive immiscible liquid phase for Na/PO_4 ratios between 2 and 2.8 at temperatures greater than about 273 °C (523 °F) (Ref 75). In view of the continuing wastage observed in service, the vendors of the affected plants recommended that the plant operators adopt an all-volatile water treatment based on the use of ammonia and hydrazine, and this was almost universally followed with the exception of a few lower temperature plants. However, as described subsequently, this change led to other corrosion problems, namely denting.

Denting

Denting was discovered in 1975 when eddy-current probes were prevented from passing through tube/tube support plate intersections due to tube deformation and associated diameter restrictions (Ref 15, 28, 76). By 1977, denting had become a widespread problem. Denting occurs when the surfaces of the hole in a carbon steel tube support plate corrode to the point at which the iron corrosion products, of volume several times greater than that of the iron from which they came, squeeze and deform the steam generator tubing (and the tube support plate). The cause of denting is best explained by Potter-Mann-type linear accelerated corrosion, in which a nonprotective oxide layer is formed as the corrosion progresses. Sample intersections of tubes and support plates removed from dented steam generators have shown local chloride concentrations of over 4000 ppm accumulated by hideout in the dented region. The source of chloride was generally condenser leakage, particularly at plants cooled by seawater.

Tube deformation has also been reported in some steam generators with carbon steel egg crate supports, implying that concentration of aggressive impurities and carbon steel corrosion may also occur with this design. Similarly, distorted eddy-current signals (dings) have been observed at the ninth and tenth support plates of OTSGs that use a support plate with a trefoil hole design (Ref 15, 25). This is the most likely region for support plate corrosion because it is where dryout and most of the deposition of any low volatility chemicals present in the feedwater occur. However, it is also possible that dings may be caused by tube vibration and impact on the tube support plates. Denting was also observed at the top of the tubesheet crevice in several RSGs but progressed at a slower rate. However, denting has not been reported in steam generators where the support structures are fabricated from type 347 stainless steel or in units that have used lattice bars and broached hole plates made from type 410 stainless steel.

Several major consequences can result from the uncontrolled progression of denting: tube cracking and leaking at U-bends, tube cracking at the constricted crevices, tube support ligament failure, and gross deformation of tube support plates. Laboratory tests have shown that denting was caused by the presence of a high concentration of acidic chlorides in the support plate crevice and the presence of an oxidizing environment outside the crevice (Ref 77). Concentration factors of greater than 20,000 times within crevices have been observed in laboratory tests.

Mitigating Denting Problems. Based on this knowledge of the causes of denting, corrective actions for operating plants were developed to modify the environments and make them less aggressive. These included reducing condenser impurity in-leakage, reducing air in-leakage, producing purer makeup water, and using condensate polishers to purify the feedwater. Another approach employed to modify the environment has been the addition of boric acid, which inhibits acidic chloride attack (Ref 78). The application of these corrective methods greatly reduced denting rates as seen in Fig. 6.

For new plants, the formation of aggressive environments can be minimized by switching from drilled to broached support plates with flat lands with which the concentration of chemicals within the tube/tube support crevice is greatly reduced. For new plants, more corrosion-resistant materials have been used for tube supports such as types 405, 409, and 410 stainless steels (Ref 74, 76).

Pitting

Extensive pitting on the outer surfaces of mill-annealed Alloy 600 tubes has been observed in a few cases. This pitting occurred primarily on the cold leg between the tube sheet and the first support plate in regions where sludge or tube scale was present (Ref 15). Metallographic sections showed a laminar appearance caused by the presence of metallic copper layers. In addition, the corrosion deposit was enriched in chromium and depleted in nickel and iron compared with the base metal. Laboratory tests have shown that pits can be formed above about 150 °C (300 °F) in the presence of high chloride ion concentrations from either seawater or copper chloride (Ref 79). Thus, it was concluded that the pits were caused by chloride, low pH, and an oxidant such as cupric chloride ($CuCl_2$) or oxygen coming from brass condenser tube leaks. Sludge and scale could also have enabled the bulk impurities to be concentrated by the hideout mechanism in addition to the classical electrochemical oxygen concentration cell mechanism.

Mitigating Pitting Problems. The principal corrective action is to modify the environment to make it less corrosive as in the case of wastage. One approach has included reducing the sludge and scale by minimizing the entry of solids (with higher pH and reduced air in-leakage), by sludge lancing, and by chemical cleaning. Another approach has included minimizing the entry of soluble contaminants (principally chlorides and oxidants), by quickly repairing leaky condensers, and by deaerating auxiliary feedwater used for hot standby.

Intergranular Attack and IGSCC at the Outside Surface

This section provides a summary of field observations of intergranular corrosion in its various forms, most notably intergranular attack

(IGA) and intergranular stress corrosion cracking (IGSCC). It also reviews laboratory studies that have resulted in models and computer codes for evaluating and predicting intergranular corrosion, and discusses remedial actions for preventing or arresting intergranular corrosion. Additional information on intergranular corrosion can be found in *Corrosion: Fundamentals, Testing, and Protection,* Vol 13A, of the *ASM Handbook.*

Summary of Field Observations

Intergranular corrosion of mill-annealed Alloy 600 tubes has been experienced in many steam generators operating at seawater- or freshwater-cooled locations (Ref 15, 28). As shown in Fig. 6, this mode of attack has grown in importance, unlike the other modes of attack discussed previously, and is increasingly responsible for premature steam generator replacement. The rate of propagation has been shown to vary widely and, in some cases, has been sufficiently rapid to require mid-cycle inspections and unscheduled outages to plug or repair leaking tubes. The associated economic loss to PWR operators has been significant. Fortunately, most of the corrosion has been confined to crevice locations so that leaks have been small and without risk of a large rupture.

Pressurized water reactor steam generators with mill-annealed Alloy 600 tubes are gradually being replaced due to stress corrosion damage on both primary and secondary sides of the tubes. The newer designs incorporate improvements to limit the fouling of crevices that precedes impurity hideout. The tubes are also fabricated from Alloy 690 or 800 that are more resistant to most of the environments that are hypothesized to exist in superheated crevices and under sludge piles on the secondary side. The population of PWR steam generator tube bundles at immediate risk from IGA/IGSCC is therefore reducing although many mill-annealed Alloy 600 tube bundles still remain in service. In addition, although thermally treated (715 °C, or 1320 °F, for 16 h) Alloy 600 tube bundles are proving very much more resistant to IGA/IGSCC, doubts remain as to their viability over the standard (initial) 40 year licensing period.

Intergranular SCC in mill-annealed Alloy 600 steam generator tubing is illustrated in Fig. 16. The SCC morphology consists of single or multiple major cracks with minor-to-moderate amounts of branching that are essentially 100% intergranular. Experience suggests that IGSCC requires stresses greater than 0.5 yield in order to propagate rapidly. At lower stresses, propagation rates may approach zero, or the corrosion may take the form of intergranular corrosion, which is frequently called "intergranular attack" (IGA).

Intergranular attack is the second generally recognized form of secondary side corrosion attack of mill-annealed Alloy 600 where there is substantial volumetric attack of every grain-boundary, as illustrated in Fig. 17. Stress is not strictly necessary for IGA to occur, which distinguishes it from IGSCC. Nevertheless, the two are closely related, as is apparent in Fig. 17, where stress-assisted fingers of IGA penetrate deeper than the layer of uniform IGA. The microscopic morphology of secondary side cracking observed in service is often that of IGA near the OD surface developing into finger-shaped IGA penetrations and finally into IGSCC toward the primary side. There is a strong presumption that as IGA gradually deepens, its morphology changes to that of IGSCC, although there is no proof that the two forms of attack occur in that order. Figure 18 shows a particularly severe case of IGA/IGSCC from a steam generator that has since been retired and replaced (Ref 80). The detailed morphology of the IGA may vary somewhat from case to case with islands of more resistant grain boundaries sometimes giving a cellular appearance to the IGA.

Combined IGA/IGSCC was first observed in the tube/tube sheet crevice (an annular gap remaining after manufacture of the earliest PWR steam generators) before full depth rolling was generally adopted. However, corrosive attack of mill-annealed Alloy 600 steam generator tubes occurs most commonly in the crevices between the tubes and tube support plates fabricated from carbon steel with cylindrical holes (Ref 28). Stress-corrosion cracking is also observed under the sludge piles that accumulate on tube sheets and can even be located up to several centimeters below the top surface of the tube sheet, despite tube expansion into the tube sheet. There has been a relatively recent trend for circumferential SCC also to occur near the tube sheet surface, although most IGSCC is observed in the axial direction of the tubes. A small number of thermally treated Alloy 600 tubes whose thermal treatment was found to have been unsatisfactory have also suffered from IGA/IGSCC. However, to date (early 2004), neither thermally treated Alloy 690 nor Alloy 800 have been affected. Some IGA/IGSCC has also been reported for plants with quatrefoil broached carbon steel support plates and mill-annealed Alloy 600 tubing but not with quatrefoil broached ferritic stainless steel support plates (Ref 81).

In mill-annealed Alloy 600 tubing, IGA/IGSCC is much more prevalent at the lower levels in recirculating steam generators where

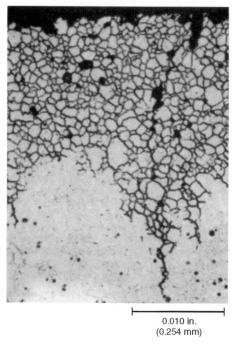

|—————|
0.010 in.
(0.254 mm)

Fig. 17 Intergranular attack of an Alloy 600 steam generator tube (etched sample). Source: Ref 15

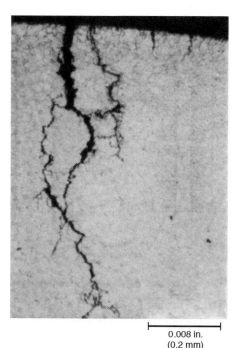

|———|
0.008 in.
(0.2 mm)

Fig. 16 Secondary side intergranular stress-corrosion cracking (IGSCC) in Alloy 600. Source: Ref 15

Fig. 18 An example of severe intergranular attack/intergranular stress-corrosion cracking (IGA/IGSCC) at a tube support location. Source: Ref 80

the temperature difference between the primary and secondary fluids is greatest (Ref 82). This is clearly strong evidence for the importance of impurity hideout, which increases as a function of the available superheat on the secondary side. Broached tube support plates minimize the extent of the narrow gap between the tube and its support plate and hence substantially reduce the tendency for impurity hideout in such locations.

When the fraction of tubes affected by IGA/IGSCC at tube support plate intersections is plotted as a function of time on Weibull coordinates, it is observed that the slopes of the Weibull plots can be rather high, typically between 4 and 9 (Ref 41, 81). This indicates that once IGA/IGSCC starts, its progression to other tubes is quite rapid and relatively consistent among different plants. On the other hand, incubation periods before cracking starts vary considerably. In some cases, IGA/IGSCC has not been observed at all, even on mill-annealed Alloy 600 tube bundles after very long periods of operation. There is a tendency to attribute this variability among different plants mainly to differences in secondary water chemistry and impurities. However, heat-to-heat variability in sensitivity of mill-annealed Alloy 600 to IGA/IGSCC is very important in this respect and the proportion of very sensitive heats varies markedly in different steam generators (Ref 1, 81).

Following the retirement of some steam generators with degraded mill-annealed Alloy 600 tubing, it has been possible to extract and observe metallographically complete tube/tube support plate intersections (Ref 83). These studies have revealed that cylindrical crevices between the tubes and tube support plates are plugged at their entrances and exits with a low porosity (<10%) solid mixture of magnetite and silica. In the center of the crevices, the deposits are mainly magnetite with porosity much higher at around 50%. These features are illustrated in Fig. 19 (Ref 84). Curiously, the tubes appear to be centralized in the crevices by the buildup of deposits and corrosion products. This fouling and plugging of the crevices between tubes and carbon steel tube support plates with cylindrical tube holes is generally acknowledged to be widespread. One consequence is that high forces have become necessary to extract tubes for destructive examination, indicating that the tubes no longer slide easily in the tube support plates as intended by the design.

Deposits found on extracted steam generator tubes have also been examined in detail (Ref 84). On the tube free spans, magnetite deposits are observed overlying a protective nickel/chromium spinel oxide. Within the tube support plate crevices, thin layers rich in alumino-silicates have been observed on the heat transfer surfaces associated with poorly protective oxide films and the presence of IGA/IGSCC.

In some special cases, IGA/IGSCC has also been observed on the free spans of tubes of recirculating steam generators (Ref 28). There are two circumstances where this has occurred. The first is where deposits rich in silica have formed bridges between tubes associated with particularly high and localized heat fluxes. In this case, a crevice is created by the deposits rather than by a design feature so that the fundamental degradation processes occurring under the deposits are probably the same as at tube support plate intersections. The second special case is where lead blankets used for radiological protection have been accidentally left in steam generators after maintenance operations. In these cases, widespread tube free span IGA/IGSCC has been observed. It is well known that lead causes severe IGA/IGSCC of Alloy 600 and other potential tube alloys. What is more difficult to ascertain is whether the traces of lead that are found in nearly every steam generator have a similar bad influence in crevice geometries (Ref 85).

One final feature of steam generator tube degradation that should be mentioned is the relatively recent appearance after about 20 years' service of IGA/IGSCC in the superheated steam zone of OTSG (Ref 25). This tube degradation is associated with scratches on the tube outer surfaces created during tube installation but, unlike most recirculating steam generators, is found on the tube free span between the tube support plates and not in crevices. Nevertheless, the IGA/IGSCC shows strong similarities on the microscopic scale with tube degradation in recirculating steam generators. As noted earlier, a significant difference between OTSG and recirculating steam generators is that the former were stress relieved before entering service and this gave rise to grain-boundary sensitization (i.e., chromium depletion) of the Alloy 600 tubes. The associated presence of chromium carbides on the grain boundaries has been generally regarded as a favorable property improving resistance to IGA/IGSCC (and PWSCC) of these sensitized tubes, whereas the chromium depletion seems to be of no particular significance in deoxygenated secondary (or primary) circuit water.

Laboratory Studies

There is little doubt that the phenomenon of hideout of impurities in crevice geometries of recirculating steam generators lies at the heart of the localized corrosion problems of the tubes that have been encountered. It has been known since the nineteenth century that steam leaks through the riveted joints of steam locomotives and similar boilers caused impurities to concentrate by local evaporation. This usually created a concentrated alkaline liquid at the liquid-vapor interface, which then caused caustic cracking of the boiler shell, sometimes with catastrophic consequences.

The same basic phenomenon has been demonstrated many times in model boiler studies simulating PWR recirculating steam generators (Ref 82). The concentrations of soluble impurities so generated can be amazingly high. By definition, the crevice liquid must have a boiling point equal to the primary side temperature and a vapor pressure equivalent to that of the secondary side pressure. The temperature difference between the primary and secondary sides of PWR steam generator tubes may vary from 45 °C (80 °F) at the base of the hot leg to 22 °C (40 °F) near the upper U-bends. At these degrees of superheat, impurity concentrations of between 25 and 15 M are necessary in order for such solutions to remain in the liquid state at the secondary side pressure. Recent work has shown that only NaOH (and by analogy, KOH) can support superheats >25 °C (45 °F) and remain liquid at the temperature and pressure of PWR steam generators (Ref 86). Sodium chloride can give rise to concentrated solutions with up to 25 °C (45 °F) superheat before precipitation intervenes, while most other impurities such as sulfates and silicates are so insoluble that they can only generate solutions with a few degrees of superheat at most. Thus, of the common impurities that enter steam generators, only NaOH is likely to form a concentrated liquid that can wet the tube OD surfaces (Ref 82), and in other cases, steam blanketing is more likely. However, the presence of silica deposits in the tube/tube support plate crevices, particularly on the heat transfer surface, is incompatible with the formation of highly alkaline crevice environment. Silica is an effective buffering agent in these circumstances that would ensure an upper limit of pH of about 10 at temperature. This may nevertheless be sufficient to cause caustic cracking of Alloy 600 because the minimum pH for this mode of SCC is about 9 at temperature.

Computer codes have been developed to describe the solutions formed by hideout of which the best known for application to PWR

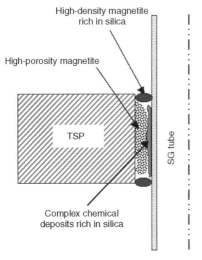

Fig. 19 Sketch of fouled steam generator (SG) tube/tube support plate (TSP) crevices after many years of operation. Source: Ref 84

steam generators is MULTEQ (Ref 87). It has a large thermodynamic database and the calculation routine minimizes the Gibbs free energies of a given chemical system at equilibrium. Various boundary conditions can be imposed such as for static or flowing systems, for evaporation of water at constant temperature or pressure, and removal or retention of vapor and precipitates as they are formed. The code is typically used to calculate the composition of an assumed concentrated fluid present in a superheated crevice. In practice, a concentration factor, which can be as large as 10^6, is applied to the chemical composition of diluted impurities analyzed during hideout return. Hideout return occurs when the steam generator is shut down and the thermal driving force for hideout is removed. The easily soluble impurities then diffuse back into the bulk steam generator water where they can be measured, albeit as very dilute solutions.

Clearly, a code such as MULTEQ can be only as reliable as the thermodynamic database on which it relies and the assumption that the measured hideout return impurities come from a concentrated solution in superheated crevices. The thermodynamic database has been steadily refined over two decades but is still lacking in critical data for the surprising variety of complex minerals found during careful examinations of the deposits formed under heat transfer conditions. Attempts to relate the occurrence of IGA/IGSCC in PWR steam generators with the compositions of the concentrated crevice fluids predicted by MULTEQ have failed to reveal any correlation (Ref 88). Such calculations in recent years have also often predicted neutral to mildly alkaline crevice liquids (assuming such a liquid phase is present) consistent with the deductions from analyses of crevice deposits.

Mathematical models have also been published that incorporate the kinetic aspects of transport of impurity species into tube/tube support plate crevices containing porous deposits (Ref 89, 90). These are very complex calculations including thermal hydraulic, diffusion, and mechanical parameters, and more recently, also electrochemical factors. Hideout due to the wick boiling mechanism is, however, predicted to dominate over any electrochemical contribution due to ion migration. The latter can occur under the influence of any potential difference that might exist between the inside and outside of the crevice if an oxidizing agent is present outside the crevice (such as oxygen or oxidized copper originating from condenser leaks). As an example of the application of such models, assuming a 10^{-7} M solution of NaCl (6 ppm) is present in the recirculating water, approximately one month is predicted to be necessary for the concentration of NaCl to establish its limiting concentration in the crevice solution for the assumed geometry and heat transfer conditions. The crevice is predicted to be half full of this solution of limiting concentration after about 1 year.

Environmental Contaminants. At least seven classes of environmental contaminants have been postulated at various times to explain the occurrence of IGA/IGSCC of mill-annealed Alloy 600 (Ref 73, 91):

- High concentrations of sodium hydroxide (NaOH) and/or potassium hydroxide (KOH)
- The products from the reaction of sulfate ions with hydrazine or hydrogen (reactive low valence sulfur-bearing species are postulated)
- The products of thermal decomposition of ion exchange resins (sulfates and organic residuals)
- Highly concentrated salt solutions at neutral or nearly neutral pH (these salt solutions are the natural consequences of condenser leakage concentrated to high levels by the boiling processes in the steam generator)
- Alkaline carbonates and/or their reaction or hydrolysis products and alumino-silicate deposits (believed to affect the nature of the passive film on the alloy surface)
- Lead contamination
- Polluted steam

All these different modes of SCC of Alloy 600 have been reviewed extensively (Ref 92). Evaluation and modeling of tube damage by IGA/IGSCC in mill-annealed Alloy 600 has, nevertheless, traditionally been based on the assumed formation of solutions in occluded superheated crevices with extreme values of pH less than 5 or greater than 10 at temperature (see Fig. 2) (Ref 82, 93). In practice, most cases have been attributed to caustic cracking and appropriately draconian measures have been taken to restrict as much as possible sodium impurities into steam generators (see Table 4). A few cases of SCC in operating steam generators have been clearly caused by lead, sometimes, but not necessarily, with a marked transgranular component to the cracking. Lead-induced cracking occurs across the whole feasible range of pH. Whether the minor amounts of lead found in practically every steam generator have a critical influence on IGA/IGSCC behavior of Alloy 600 as distinct from aggravating another underlying degradation mechanism remains unresolved (Ref 85). The latter option seems likely in the view of the widely varying and erratic distributions of lead traces found on steam generator tubes.

Remedial Actions

Modification of the crevice environment can appear at first sight to be the most straightforward method of preventing or arresting intergranular corrosion, although implementation can be complicated due to existing deposits impeding access of secondary water to the occluded zone. Attempts to modify the crevice environment have included several factors, such as lowering the temperature, adding a pH neutralizer or buffering agent such as boric acid, removing the aggressive species by flushing or soaking, and changing the concentration and/or anion to cation ratio of bulk water contaminants (Ref 15). Laboratory studies with model boilers have shown the benefit of several of these corrective measures and some have been applied to operating steam generators.

The concentration of contaminants accessible to the crevice may be controlled by eliminating or reducing the entry of contaminants to the steam generator and by controlling the concentrating capability of the sludge pile above the crevice. Reducing the entry of contaminants is best accomplished by preventing condenser leaks, routing drains to the condenser (if condensate is subsequently treated), and properly treating makeup water sources. Use of full-flow condensate polishers for control of ionic species has not been shown to be effective in controlling the species that are probably responsible for IGA and IGSCC, in contrast to the successful control of chloride responsible for denting (Ref 15). Polishers are potential sources of sulfates (SO_4) and sodium (Na^+) if operated or regenerated improperly, and they do not effectively remove silica or organics. However, some plants have used these condensate polishers very effectively to minimize impurity entry into the steam generators.

Control of the sludge pile on the tube sheet, which is an effective concentrating medium, requires three courses of action:

- Effective, periodic sludge lancing
- Minimization of particulate transport by preventing air entry and/or providing for feedwater filtration, such as by powdered resin condensate polishers
- Preventing the entry of chemical species that tend to promote agglomeration

Chemical cleaning has been used to remove the sludge on the tube sheet and more recently, with varying degrees of success, from the interstices between the tubes and tube support plates.

Corrosion Fatigue

Some years ago, cracking of some Alloy 600 OTSG tubes from the outer surface occurred in the upper regions of several OTSGs before the more recent examples of IGA/IGSCC that occurred at scratches described earlier (Ref 15, 25, 73). It was concluded that these cracks were caused by corrosion fatigue resulting from small-amplitude vibration combined with the transport of impurities into the upper regions of the OTSG units, particularly in the open lane. Laboratory tests showed a decrease in Alloy 600 fatigue strength in the presence of chemicals that were judged to be present in these upper regions. A substantial decrease in fatigue resistance of Alloy 600 was observed in acid sulfate/silicate solution. The environment was selected to be consistent with that postulated to exist in the region of interest at the top of an OTSG. The fatigue strength at 10^7 cycles at 289 °C (552 °F) was found to be approximately 56% of the value measured in air at about the same temperature. Aerated acid sulfate/silicate and alkaline sulfate/silicate environments had a less deleterious effect.

Potentially much more serious cracking attributed to a form of corrosion fatigue has been observed in a small number of low-alloy steel steam generator shells and in carbon steel feed water piping. The combination of fabrication and operational factors necessary for such cracking to occur in carbon and low-alloy steels has ensured that it has in reality been highly plant specific, as described subsequently.

Extensive circumferential cracking of the upper shell to cone girth welds of all the Indian Point 3 (Westchester County, NY) steam generators (model 44) was found in 1982 following a steam leak through one of more than a hundred circumferential cracks (Ref 94, 95). Similarly, the steam generator shells of six other plants (both model 44 and 51 steam generators) located in the United States and Europe were also observed to be cracked in the same location (Ref 95–98). In some cases, cracking recurred after local repairs by contour grinding.

The steam generator shell cracking was caused by an environmentally assisted cracking (EAC) mechanism and has been variously called *corrosion fatigue, stress-corrosion cracking,* or *strain-induced corrosion cracking*. In fact, the last seems most appropriate because it recognizes that although the cracking is environmentally controlled, a dynamic strain is necessary to maintain crack propagation (Ref 99). Consequently, crack extension tends to occur intermittently, alternating with pitting at the crack tip during quiescent periods.

This EAC mechanism observed for steam generator shell materials is well known and characterized both for bainitic low-alloy steels as well as for ferritic-pearlitic carbon-manganese steels used extensively in both conventional and nuclear power plants (Ref 99–101). In addition to the dynamic loading requirement usually caused by large thermal transients, cracking has been associated in practice with high residual welding stresses due to poor or nonexistent stress relief (Ref 94). The worst affected plants had weld repairs during fabrication of the final closure weld. In one case the girth weld was remade and stress relieved at a higher temperature of 607 °C (1125 °F) compared with 538 °C (1000 °F) originally. Water chemistry transients, particularly oxygen ingress, occurring at the same time as dynamic loading have also been strongly linked to the observed cracking. The effect of oxygen was exaggerated if copper corrosion products (e.g., from brass condenser tubes) were also present (Ref 95). The only metallurgical factors that appeared to play a role were the sulfur impurity concentration in the steel in the form of manganese sulfide, where the risk of cracking was greater the higher the sulfur content, and possibly also the free nitrogen content via the phenomenon of strain aging (Ref 101). The practical resolution of steam generator shell cracking has been achieved mainly by contour grinding of existing cracks and by ensuring that auxiliary feed water is properly deoxygenated prior to use, particularly during plant startup.

In addition to these reported incidents of steam generator shell cracking, a very large technical literature exists concerning EAC of carbon and low-alloy steels in both nuclear and conventional power plant (Ref 99–101). The incidents range from those where fatigue plays a major part to those where the contribution to damage from mechanical fatigue loading is minor. For example, cracking of carbon steel feed water lines close to the steam generator nozzle observed in twenty-six U.S. and three German PWR plants has generally been attributed to thermal fatigue due to hot/cold water layering at low power, probably assisted by oxygen transients. Deaerator vessels, usually fabricated from carbon steel of a thickness requiring no stress relief following welding, have been the focus of particular attention following several explosions with fatalities (Ref 102). The observed cracking in these cases is usually transgranular cleavage-like in appearance, although it can occasionally be intergranular without any obvious involvement of other chemical pollutants.

External Bolting Corrosion

High-strength martensitic and maraging steels are used in many external fastener applications in nuclear power reactors, and a significant number of failures of this class of component have occurred (Ref 22). Most have been described as corrosion-related failures. The problems encountered with external bolting have affected both support bolting and pressure boundary fasteners.

Failures of low-alloy (4340 and 4140) and maraging steel support bolting have been attributed mainly to hydrogen embrittlement. Steels with ultrahigh yield strengths greater than 1000 MPa (145 ksi) have failed due to a combination of excessive applied stresses and humid or wet environments collecting around the bases of components. Steels with lower yield strengths have also failed due to poor heat treatment or material variability. As a result of these failures, a review of SCC/hydrogen embrittlement properties of high-strength steels exposed to water or saltwater at low temperatures was carried out and regulatory guidelines

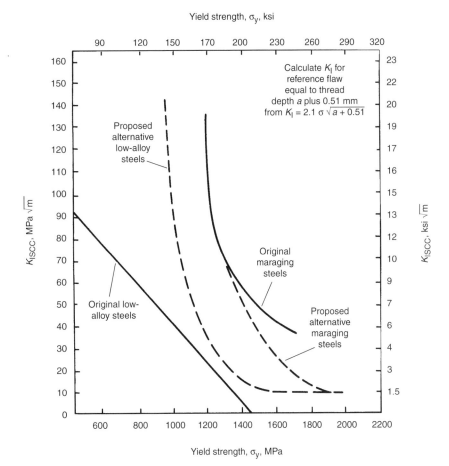

Fig. 20 K_{Iscc} limits for support bolting as a function of yield stress. Source: Ref 103

based on this information were published in the United States (Ref 103). Acceptability of high-strength bolting was based on a lower-bound approach to K_{ISCC} as a function of yield strength as shown in Fig. 20. This fracture mechanics-based approach has some attractions for defining a quality assurance procedure and for defect assessment. However, hydrogen cracks can start from free surfaces, usually in crevices; consequently, it is advisable also to have an upper-bound strength limit (normally defined by a hardness level acceptance criterion such as <350 HV) to guard against this eventuality.

The second category of bolt failures is concerned with the integrity of the primary pressure boundary at locations such as flanges of manway covers and valves. Most of these incidents have been caused by erosion-corrosion in PWR primary water leaks. A small number of failures among this category of bolt have been associated, however, with SCC/hydrogen embrittlement rather than wastage (Ref 22). The ferritic bolting steels involved were not out of specification but had been in contact with molybdenum disulfide lubricants. It has been postulated that the lubricant dissociated on contact with hot water to yield hydrogen sulfide, which is a severe hydrogen embrittling agent for ferritic steels. The remedies for this category of failures are nevertheless clear; avoid leaks at flanges by improved gasket design and avoid sulfide-containing lubricants.

REFERENCES

1. P.M. Scott, Stress Corrosion Cracking in Pressurized Water Reactors: Interpretation, Modeling and Remedies, *Corrosion*, Vol 56 (No. 8), 2000, p 771–782
2. P. Cohen, *Water Coolant Technology of Power Reactors*, 2nd printing, American Nuclear Society, Inc., 1980
3. P.J. Millet and F. Hundley, *Nucl. Energy*, Vol 36, 1997, p 251–256
4. Y. Solomon, An Overview of Water Chemistry for Pressurized Water Reactors, *Proceedings of Water Chemistry of Nuclear Reactor Systems*, British Nuclear Energy Society, 1978, p 101–112
5. R.A. Shaw, Influence of Corrosion on Radiation Fields, *Corrosion*, Vol 13, *ASM Handbook*, ASM International, 1987, p 949–951
6. F. Cattant, C. Lebuffe, J. Dechelotte, A. Raybaud, and M. Ladouceur, Examination of Control Rods in High Activity Cells—Activities Undertaken to Solve the Wear Problem, *Proc. Fontevraud II* (Paris, France), Société Française d'Energie Nucléaire, Sept 10–14, 1990, p 151–159
7. J. Bourgoin, C. Lebuffe, A. Cazus, X. Thibault, M. Monchanin, P.J. Sartor, F. Couvreur, and D. Gosset, Contribution à la connaissance du comportement sous irradiation des crayons de grappes de commande, *Proc. Fontevraud III* (Paris, France), Société Française d'Energie Nucléaire, Sept 12–16, 1994, p 715–725
8. G.C.W. Comley, The Significance of Corrosion Products in Water Reactor Coolant Circuits, *Prog. Nucl. Energy*, Vol 16, 1985, p 41–72
9. C.J. Wood, Reduced Out-of-Core Radiation Eases Maintenance Activities at Nuclear Power Plants, *Power*, Vol 131, 1987, p 29–32
10. F. Nordmann, A. Stutzmann, and J.-L. Bretelle, Overview of PWR Chemistry Options, Paper 64, *Proc. Chimie 2002*, Société Française d'Energie Nucléaire, April 22–26, 2002
11. H. Ocken, Cobalt-Free Hardfacing Alloys Reduce Dose, *Nucl. Eng. Int.*, Jan 1995, p 26–27
12. H. Ocken, K. Fruzzetti, P. Frattini, and C.J. Wood, Recent Developments in PWR Zinc Injection, Paper 41, *Proc. Chimie 2002*, Société Française d'Energie Nucléaire, April 22–26, 2002
13. P. Gosset, M. Dupin, D. Buisine, J.-F. Buet, and V. Brunel, Optimization of the Decontamination in EDF Nuclear Power Plants, Paper 108, *Proc. Chimie 2002*, Société Française, d'Energie Nucléaire April 22–26, 2002
14. J. Blok, S. Chaufriat, and P. Frattini, "Modeling the Axial Offset Anomaly in PWRs," Paper 69, *Proc. Chimie 2002*, Société Française, d'Energie Nucléaire April 22–26, 2002
15. S.J. Green, Steam Generator Failure or Degradation, *Corrosion*, Vol 13, *Metals Handbook*, 9th ed., ASM International, 1987, p 937–945
16. R. Thomson and T. Gaudreau, Modeling and Field Experience of Fouling in Once-Through Steam Generators, *Proc. 7th International Symposium on Environmental Degradation of Materials in Nuclear Power Systems—Water Reactors* (Breckenridge, CO), NACE International, 1995, p 363–372
17. P.J. Millet and F. Hundley, Optimization of Secondary Water Chemistry in U.S. PWRs, *Proc. International Conference on Water Chemistry of Nuclear Reactor Systems*, British Nuclear Energy Society, 1996
18. C.W. Turner, S.J. Klimas, D.A. Guzonas, P.L. Frattini, and K. Fruzzetti, New Insights into Controlling Tube Bundle Fouling Using Alternative Amines, Paper 159, *Proc. Chimie 2002*, Société Française d'Energie Nucléaire, 2002
19. US NRC Information Notice, 86.108, "Degradation of Reactor Coolant System Pressure Boundary Resulting from Boric Acid Corrosion," Dec 29, 1986; Supplement 1, April 20, 1987; Supplement 2, Nov 19, 1987; Supplement 3, Jan 5, 1995. Also refer to: NRC Information Notice 2002-11, "Recent Experience with Degradation of Reactor Pressure Vessel Head," March 12, 2002; and NRC Information Notice 2002-02, "Recent Experience with Reactor Coolant System Leakage and Boric Acid Corrosion," Jan 16, 2003
20. J.H. Hall, R.S. Frisk, A.S. O'Neill, R.S. Pathania, and W.B. Neff, Boric Acid Corrosion of Carbon and Low Alloy Steels, *Proc. 4th Int. Symposium on Environmental Degradation of Materials in Nuclear Power Systems—Water Reactors* (Jekyll Island, GA), NACE 1990, p 9-38 to 9-52
21. J.F. Hall, Corrosion of Low Alloy Steel Fastener Materials Exposed to Borated Water, *Proc. 3rd Int. Symposium on Environmental Degradation of Materials in Nuclear Power Systems—Water Reactors* (Traverse City, MI), The Metallurgical Society, 1988, p 711–722
22. C.J. Czajkowski, Corrosion and Stress Corrosion Cracking of Bolting Materials in Light Water Reactors, *Proc. 1st Int. Symposium on Environmental Degradation of Materials in Nuclear Power Systems—Water Reactors* (Myrtle Beach, SC), NACE, 1984, p 192–208
23. J.M. Gras, D. Noel, B. Prieux, M. Cleurennec, and Y. Rouillon, Corrosion of Low Alloy Steels and Stainless Steels in Concentrated Boric Acid Solutions, Paper OR 20, *Proc. Eurocorr '96*, Session IX Nuclear Corrosion and Protection (Nice, France), 1996
24. W. Bamford and J. Hall, A Review of Alloy 600 Cracking in Operating Nuclear Plants: Historical Experience and Future Trends, *Proc. 11th Int. Conference on Environmental Degradation of Materials in Nuclear Power Systems—Water Reactors* (Stevenson, WA), American Nuclear Society, 2003, p 1071–1079
25. P.A. Sherburne, OTSG Materials Performance—25 Years Later, *Proc. Fontevraud IV International Symposium*, Société Française d'Energie Nucléaire, 1998, p 529–540
26. J. Blanchet, H. Coriou, L. Grall, C. Mahieu, C. Otter, and G. Turluer, Historical Review of the Principal Research Concerning the Phenomena of Cracking of Nickel Base Austenitic Alloys, *Proc. NACE-5 Stress Corrosion Cracking and Hydrogen Embrittlement of Iron Base Alloys*, National Association of Corrosion Engineers, 1977, p 1149–1160
27. H.J. Schenk, Investigation of Tube Failures in Inconel 600 Steam Generator Tubing at KWO Obrigheim, *Mater. Perform.*, Vol 15 (No. 3), 1976, p 25–33
28. D.R. Diercks, W.J. Shack, and J. Muscara, Overview of Steam Generator Tube Degradation and Integrity Issues, *Nucl. Eng. Des.*, Vol 194, 1999, p 19–30
29. G. Frederick and P. Hernalsteen, "Generic Preventive Actions for Mitigating MA Inconel 600 Susceptibility to Pure Water Stress Corrosion Cracking," presented at

The Specialist Meeting on Steam Generators (Stockholm, Sweden), NEA/CSNI-UNIPEDE, Oct 1984
30. P. Saint-Paul and G. Slama, Steam Generator Materials Degradation, *Proc. 5th Int. Symposium on Environmental Degradation of Materials in Nuclear Power Systems—Water Reactors* (Monterey, CA), American Nuclear Society, 1992, p 39–49
31. P. Pitner and T. Riffard, Statistical Evaluation of the Effects of Shot-Peening on Stress Corrosion of Alloy 600 in PWR Steam Generators, *Proc. 6th Int. Symposium on Environmental Degradation of Materials in Nuclear Power Systems—Water Reactors,* The Metallurgical Society, 1993, p 707–712
32. G.P. Airey, The Stress Corrosion Cracking (SCC) Performance of Inconel Alloy 600 in Pure and Primary Water Environments, *Proc. 1st International Symposium on Environmental Degradation of Materials in Nuclear Power Systems—Water Reactors,* NACE, 1984, p 462–476
33. A.A. Stein and A.R. McIlree, Relationship of Annealing Temperature and Microstructure to Primary Side Cracking of Alloy 600 Steam Generator Tubing and the Prediction of Stress Corrosion Cracking in Primary Water, *Proc. 2nd International Symposium on Environmental Degradation of Materials in Nuclear Power Systems—Water Reactors* (Monterey, CA), American Nuclear Society, 1986, p 47–51
34. J.R. Crum and T. Nagashima, Review of Alloy 690 Steam Generator Studies, *Proc. 8th International Symposium on Environmental Degradation of Materials in Nuclear Power Systems—Water Reactors* (Amelia Island, FL), American Nuclear Society, 1997, p 127–137
35. Ph. Berge, Importance of Surface Preparation for Corrosion Control in Nuclear Power Stations, *Mater. Perform.,* Vol 36 (No. 11), 1997, p 56
36. F. Champigny, F. Chapelier, and C. Amzallag, Maintenance Strategy of Inconel Components in PWR Primary Systems in France, *Proc. Conference on Vessel Penetration Inspection, Cracking and Repairs* (Gaithersburg, MD), U.S. Nuclear Regulatory Commission and Argonne National Laboratory, Sept 29–Oct 2, 2003
37. P. Chartier, D. Edmond, and G. Turluer, The French Regulatory Experience and Views on Nickel-Base Alloy PWSCC Prevention, *Proc. Conference on Vessel Penetration Inspection, Cracking and Repairs* (Gaithersburg, MD), U.S. Nuclear Regulatory Commission and Argonne National Laboratory, Sept 29–Oct 2, 2003
38. A. Hiser, U.S. Regulatory Experience and Prognosis with RPV Head Degradation and VHP Nozzle Cracking, *Proc. Conference on Vessel Penetration Inspection, Cracking and Repairs* (Gaithersburg, MD), U.S. Nuclear Regulatory Commission and Argonne National Laboratory, Sept 29–Oct 2, 2003
39. C. Amzallag, J.-M. Boursier, C. Pagès, and C. Gimond, Stress Corrosion Life Assessment of 182 and 82 Welds Used in PWR Components, *Proc. 10th International Conference on Environmental Degradation of Materials in Nuclear Power Systems—Water Reactors* (Lake Tahoe, CA), NACE International, Aug 3–9, 2001
40. L.E. Thomas, J.S. Vetrano, S.M. Bruemmer, P. Efsing, B. Forssgren, G. Embring, and K. Gott, High Resolution Analytical Electron Microscopy Characterization of Environmentally Assisted Cracks in Alloy 182 Weldments, *Proc. 11th International Conference on Environmental Degradation of Materials in Nuclear Power Systems—Water Reactors* (Stevenson, WA), American Nuclear Society, 2003, p 1212–1224
41. R.W. Staehle, J.A. Gorman, K.D. Stavropoulos, and C.S. Welty, Application of Statistical Distributions to Characterizing and Predicting Corrosion of Tubing in Steam Generators of Pressurized Water Reactors, *Proc. Life Prediction of Corrodible Structures* (Houston, TX), R.N. Parkins, Ed., NACE International, 1994, p 1374–1439
42. P. Scott and C. Benhamou, An Overview of Recent Observations and Interpretation of IGSCC in Nickel Base Alloys in PWR Primary Water, *Proc. 10th Int. Symposium on Environmental Degradation of Materials in Nuclear Power Systems—Water Reactors* (Lake Tahoe, CA), NACE International, 2001
43. S. LeHong, C. Amzallag, and A. Gelpi, Modeling of Stress Corrosion Crack Initiation on Alloy 600 in Primary Water of PWRs, *Proc. 9th International Symposium on Environmental Degradation of Materials in Nuclear Power Systems—Water Reactors* (Newport Beach, CA), The Metallurgical Society, 1999, p 115–122
44. C. Pichon, R. Boudot, C. Benhamou, and A. Gelpi, Residual Life Assessment of French PWR Vessel Head Penetrations through Metallurgical Analysis, *Proc. 1994 ASME Pressure Vessels and Piping Conference, PVP 288* (Minneapolis, MN), ASME, 1994, p 41–47
45. C. Amzallag, S. Le Hong, C. Pagès, and A. Gelpi, Stress Corrosion Life Assessment of Alloy 600 Components, *Proc. 9th International Symposium on Environmental Degradation of Materials in Nuclear Power Systems—Water Reactors* (Newport Beach, CA), The Metallurgical Society, 1999, p 243–250
46. C. Amzallag, S. LeHong, C. Benhamou, and A. Gelpi, Methodology Used to Rank the Stress Corrosion Susceptibility of Alloy 600 PWR Components, Pressure Vessel and Piping Conference (Seattle, WA), ASME, 2000
47. C. Pichon, D. Buisine, C. Faidy, A. Gelpi, and M. Vaindirlis, Phenomenon Analysis of Stress Corrosion Cracking in Vessel Head Penetrations of French PWRs, *Proc. 7th International Symposium on Environmental Degradation of Materials in Nuclear Power Systems—Water Reactors* (Breckenridge, CO), NACE International, 1995, p 1–10
48. C. Amzallag and F. Vaillant, Stress Corrosion Crack Propagation Rates in Reactor Vessel Head Penetrations in Alloy 600, *Proc. 9th International Symposium on Environmental Degradation of Materials in Nuclear Power Systems—Water Reactors* (Newport Beach, CA), The Metallurgical Society, 1999, p 235–241
49. G.A. White, J. Hickling, and L.K. Mathews, Crack Growth Rates for Evaluating PWSCC of Thick-Walled Alloy 600 Material, *Proc. 11th International Conference on Environmental Degradation of Materials in Nuclear Power Systems—Water Reactors* (Stevenson, WA), American Nuclear Society, 2003, p 166–179
50. S.A. Attanasio and J.S. Morton, Measurements of Nickel/Nickel Oxide Transition in Ni-Cr-Fe Alloys and Updated Data and Correlations to Quantify the Effect of Aqueous Hydrogen on Primary Water SCC, *Proc. 11th International Conference on Environmental Degradation of Materials in Nuclear Power Systems—Water Reactors* (Stevenson, WA), American Nuclear Society, 2003, p 143–154
51. P.M. Scott and P. Combrade, On the Mechanism of Stress Corrosion Crack Initiation and Growth in Alloy 600 Exposed to PWR Primary Water, *Proc. 11th International Conference on Environmental Degradation of Materials in Nuclear Power Systems—Water Reactors* (Stevenson, WA), American Nuclear Society, 2003, p 29–35
52. F. Vaillant, J.-D. Mithieux, O. de Bouvier, D. Vançon, G. Zacharie, Y. Brechet, and F. Louchet, Influence of Chromium Content and Microstructure on Creep and PWSCC Resistance of Nickel Base Alloys, *Proc. 9th International Symposium on Environmental Degradation of Materials in Nuclear Power Systems—Water Reactors* (Newport Beach, CA), The Metallurgical Society, 1999, p 251–258
53. M. Foucault and C. Benhamou, Influence of Surface Condition on the Susceptibility of Alloy X-750 to Crack Initiation in PWR Primary Water, *Proc. 6th International Symposium on Environmental Degradation of Materials in Nuclear Power Systems—Water Reactors* (San Diego, CA), The Metallurgical Society, 1993, p 791–797

54. M.T. Miglin, J.V. Monter, C.S. Wade, J.K. Tien, and J.L. Nelson, Stress Corrosion Cracking of Chemistry and Heat Treat Variants of Alloy 718, Part 1: Stress Corrosion Test Results, *Proc. 6th International Symposium on Environmental Degradation of Materials in Nuclear Power Systems—Water Reactors* (San Diego, CA), The Metallurgical Society, 1993, p 815–818
55. C.M. Pezze and I.M. Wilson, Transgranular Stress Corrosion Cracking of 304 Stainless Steel Canopy Seal Welds in PWR Systems, *Proc. 4th International Symposium on Environmental Degradation of Materials in Nuclear Reactor Systems—Water Reactors* (Jekyll Island, GA), NACE, 1990, p 4-164 to 4-179
56. B.M. Gordon, The Effect of Chloride and Oxygen on the Stress Corrosion Cracking of Stainless Steel: A Review of the Literature, *Mater. Perform.,* Vol 19 (No. 4), 1980, p 29
57. E. Hachani, P. Combrade, and M. Foucault, unpublished work, 2003
58. G.O. Haynor, R.S. Piascik, D.E. Killian, and K.E. Moore, Babcock & Wilcox Experience with Alloy A-286 Reactor Pressure Vessel Internal Bolting, *Proc. Fontevraud I,* Société Française d'Energie Nucléaire, 1985, p 93–101
59. P.J. Plante and J.P. Mieding, Degradation of Reactor Coolant Pump Impellor-to-Shaft Bolting in a PWR, *Proc. 4th International Symposium on Environmental Degradation of Materials in Nuclear Reactor Systems—Water Reactors* (Jekyll Island, GA), NACE, 1990, p 11-45 to 11-55
60. US NRC Information Notice, "Stress Corrosion Cracking of Reactor Coolant Pump Bolts," 90-68, 1990, and supplement 1, 1994
61. R.S. Piascik, J.V. Monter, G.J. Theus, and M.E. Scott, Stress Corrosion Cracking of Alloy A-286 Bolt Material in Simulated PWR Reactor Environment, *Proc. Second International Symposium on Environmental Degradation of Materials in Nuclear Reactor Systems—Water Reactors* (Monterey, CA), American Nuclear Society, 1986, p 18–25
62. I. Wilson and T. Mager, Stress Corrosion of Age Hardenable NiFeCr Alloys, *Corrosion,* Vol 42 (No. 6), 1986, p 352–360
63. J.B. Hall, S. Fyfitch, and K.E. Moore, Laboratory and Operating Experience with Alloy A286 and Alloy X750 RV Internals Bolting Stress Corrosion Cracking, *Proc. 11th International Conference on Environmental Degradation of Materials in Nuclear Power Systems—Water Reactors* (Stevenson, WA), American Nuclear Society, 2003, p 208–215
64. US NRC Information Notice, "Valve Stem Corrosion Failures," 85-59, 1985
65. J.M. Boursier, D. Buisine, M. Fronteau, Y. Meyzaud, D. Michel, Y. Rouillon, and B. Yrieix, Vieillissement en service des aciers inoxydables, *Proc. Fontevraud IV,* Société Française d'Energie Nucléaire, 1998, p 1123–1134
66. B. Yrieix and M. Guttmann, Aging between 300 and 450 °C of Wrought Martensitic 13-17 wt% Cr Stainless Steels, *Mater. Sci. Technol.,* Vol 9 (No. 2), 1993, p 125–134
67. H. Xu and S. Fyfitch, Aging Embrittlement Modeling of Type 17-4PH at LWR Temperature, *Proc. 10th International Conference on Environmental Degradation of Materials in Nuclear Power Systems—Water Reactors* (Lake Tahoe, CA), Aug 3–9, 2001, NACE International, 2001
68. J.L. Nelson and P.L. Andresen, Review of Current Research and Understanding of Irradiation-Assisted Stress Corrosion Cracking, *Proc. 5th International Symposium on Environmental Degradation of Materials in Nuclear Reactor Systems—Water Reactors* (Monterey, CA), American Nuclear Society, 1992, p 10–26
69. R.M. Horn, G.M. Gordon, F.P. Ford, and R.L. Cowan, Experience and Assessment of Stress Corrosion Cracking in L-Grade Stainless Steel in BWR Internals, *Nucl. Eng. Des.,* Vol 174 (No. 3), 1997, p 313–325
70. P. Scott, Review of Irradiation Assisted Stress Corrosion Cracking, *J. Nucl. Mater.,* Vol 211, 1994, p 101–122
71. S.M. Bruemmer, E.P. Simonen, P.M. Scott, P.L. Andresen, G.S. Was, and J.L. Nelson, Radiation-Induced Material Changes and Susceptibility to Intergranular Failure of Light-Water-Reactor Core Internals, *J. Nucl. Mater.,* Vol 274, 1999, p 299–314
72. P.M. Scott, M.-C. Meunier, D. Deydier, S. Silvestre, and A. Trenty, An Analysis of Baffle/Former Bolt Cracking in French PWRs, *Environmentally Assisted Cracking: Predictive Methods for Risk Assessment and Evaluation of Materials, Equipment and Structures,* STP 1401, R.D. Kane, Ed., ASTM, 2000, p 210–223
73. S.J. Green, Thermal, Hydraulic and Corrosion Aspects of PWR Steam Generator problems, *Heat Transf. Eng.,* Vol 9 (No. 1), 1988, p 19–68
74. C.E. Shoemaker, Selecting Support Structure Alloys for Nuclear Steam Generators, *Proc. 2nd International Symposium on Environmental Degradation of Materials in Nuclear Power Systems—Water Reactors* (Monterey, CA), American Nuclear Society, 1986, p 571–578
75. R. Garnsey, Corrosion of PWR Steam Generators, *Nucl. Energy,* Vol 18, 1979, p 117–132
76. A.R. Vaia, G. Economy, M.J. Wootten, and R.G. Aspden, Denting of Steam Generator Tubes in PWR Plants, *Mater. Perform.,* Vol 19, Feb 1980, p 9
77. P.V. Balakrishnan and R.S. Pathania, Correlation of Tube Support Structure Corrosion Studies, *Proc. 3rd International Symposium on Environmental Degradation of Materials in Nuclear Power Systems—Water Reactors* (Traverse City, MI), The Metallurgical Society, 1988, p 489–499
78. J. Daret and G. Pinard-Legry, Intergranular Attack (IGA) of PWR Steam Generator Tubing: Evaluation of Remedial Properties of Boric Acid, *Proc. 3rd International Symposium on Environmental Degradation of Materials in Nuclear Power Systems—Water Reactors* (Traverse City, MI), The Metallurgical Society, 1988, p 517–523
79. J.F. Sykes and M.J. Angwin, The Causes of Major Pitting of Alloy 600 Steam Generator Tubing in Pressurized Water Reactors, *Proc. 2nd International Symposium on Environmental Degradation of Materials in Nuclear Power Systems—Water Reactors* (Monterey, CA), American Nuclear Society, 1986, p 624–631
80. B. Prieux, F. Vaillant, F. Cattant, A. Stutzmann, and P. Lemaire, Secondary Side Cracking at Saint-Laurent Unit B1: Investigations, Operating Chemistry and Corrosion tests, *Proc. Fontevraud III International Symposium* (Fontevraud, France), Société Française d'Energie Nucléaire, 1994, p 383–393
81. P.M. Scott, A Discussion of Mechanisms and Modeling of Secondary Side Corrosion Cracking in PWR Steam Generators, *Proc. Chemistry and Electrochemistry of Corrosion and Stress Corrosion Cracking: A Symposium Honoring the Contributions of R.W. Staehle,* R.H. Jones, Ed., The Metallurgical Society, 2001, p 107–122
82. A. Baum, Experimental Evaluation of Tube Support Plate Crevice Chemistry, *Proc. 10th Int. Symposium on Environmental Degradation of Materials in Nuclear Power Systems—Water Reactors* (Lake Tahoe, CA), NACE International, 2001
83. L. Albertin, F. Cattant, A. Baum, and P. Kuchirka, Characterization of Deposits in Dampierre-1 Steam Generator Support Plate Crevices, *Proc. 7th International Symposium on Environmental Degradation in Nuclear Power Systems—Water Reactors* (Breckenridge, CO), NACE International, 1995, p 399–408
84. B. Sala, P. Combrade, A. Gelpi, and M. Dupin, The Use of Tube Examinations and Laboratory Simulations to Improve the Knowledge of Local Environments and Surface Reactions in TSPs, *Control of Corrosion on the Secondary Side of Steam Generators,* R.W. Staehle, J.A. Gorman, and A.R. McIlree, Ed., NACE International, 1996, p 483–497

85. L.E. Thomas, V.Y. Gertzman, and S.M. Bruemmer, Crack-Tip Microstructures and Impurities in Stress-Corrosion-Cracked Alloy 600 from Recirculating and Once-through Steam Generators, *Proc. 10th Int. Symposium on Environmental Degradation of Materials in Nuclear Power Systems—Water Reactors* (Lake Tahoe, CA), NACE International, 2001
86. D. You, S. Lefevre, D. Feron, and F. Vaillant, Experimental Study of Concentrated Solutions Containing Sodium and Chloride Pollutants in SG Flow Restricted Areas, Paper 151, *Proc. Chimie* 2002, Société Française d'Energie Nucléaire, 2002
87. W.T. Lindsay, MULTEQ: What It Is and What It Can Do, *Proc. Control of Corrosion on the Secondary Side of Steam Generators*, R.W. Staehle, J.A. Gorman, and A.R. McIlree, Ed., NACE International, 1996, p 567–575
88. A. Stutzmann and F. Nordmann, Studies of IGA Intergranular Attack/Stress Corrosion Cracking Related Parameters and Remedies, *Nucl. Eng. Des.*, Vol 162 (No. 2–3), 1996, p 167–174
89. G.R. Englehardt, D.D. Macdonald, and P.J. Millet, Transport Processes in Steam Generator Crevices, Part I: General Corrosion Model, *Corros. Sci.*, Vol 41, 1999, p 2165–2190
90. G.R. Englehardt, D.D. Macdonald, and P.J. Millet, Transport Processes in Steam Generator Crevices, Part II: A Simplified Method for Estimating Impurity Accumulation Rates, *Corros. Sci.*, Vol 41, 1999, p 2191–2211
91. Q.T. Tran, P.M. Scott, and F. Vaillant, IGA/IGSCC of Alloy 600 in Complex Mixtures of Impurities, *Proc. 10th Int. Symposium on Environmental Degradation of Materials in Nuclear Power Systems—Water Reactors* (Lake Tahoe, CA), NACE International, 2001
92. R. Staehle and J.A. Gorman, Quantitative Assessment of Submodes of Stress Corrosion Cracking on the Secondary Side of Steam Generator Tubing in Pressurized Water Reactors: Parts 1, 2, and 3, *Corrosion*, Vol 59 (No. 11), 2003, p 931–994; Vol 60 (No. 1), 2004, p 5–63; and Vol 60 (No. 2), 2004, p 115–180
93. P.M. Scott and P. Combrade, On the Mechanisms of Secondary Side PWR Steam Generator Tube Cracking, *Proc. 8th International Symposium on Environmental Degradation in Nuclear Power Systems—Water Reactors* (Amelia Island, FL), American Nuclear Society, 1997, p 65–73
94. C.J. Czajkowski, Evaluation of the Transgranular Cracking Phenomenon on the Indian Point No. 3 Steam Generator Vessels, *Int. J. Pressure Vessels Piping*, Vol 26, 1986, p 97–110
95. W.H. Bamford, G.V. Rao, and J.L. Houtman, Investigation of Service-Induced Degradation of Steam Generator Shell Materials, *Proc. 5th International Symposium on Environmental Degradation of Materials in Nuclear Power Systems—Water Reactors* (Monterey, CA), American Nuclear Society, 1992, p 588–595
96. US NRC Information Notice No. 90-04, "Cracking of the Upper Shell to Transition Cone Girth Welds in Steam Generators," 1990
97. C. Tomes, W.H. Bamford, D. Kurek, and G.V. Rao, Steam Generator Shell Cracking: Is It In-Service Degradation, *Service Experience and Life Management in Operating Plants*, ASME, PVP-Vol 240, 1992, p 13–21
98. T.C. Esselman, M. Marina, and S.K. Sinha, Steam Generator Shell Design and Plant Operation to Prevent Cracking, *Pressure Vessel Fracture, Fatigue and Life Management*, ASME, PVP-Vol 233, 1992, p 41–44
99. J. Hickling and D. Blind, Strain-Induced Corrosion Cracking of Low-Alloy Steels in LWR Systems: Case Histories and Identification of Conditions Leading to Susceptibility, *Nucl. Eng. Des.*, Vol 91, 1986, p 305–330
100. P.M. Scott and D.R. Tice, Stress Corrosion in Low Steels, *Nucl. Eng. Des.*, Vol 119, 1990, p 399–413
101. H.P. Seifert, S. Ritter, and J. Hickling, Environmentally-Assisted Cracking of Low-Alloy RPV and Piping Steels under LWR Conditions, *Proc. 11th International Conference on Environmental Degradation of Materials in Nuclear Power Systems—Water Reactors* (Stevenson, WA), American Nuclear Society, 2003, p 73–88
102. "Recommended Practice for Prevention, Detection and Correction of De-aerator Cracking," NACE Standard RP0590-96, NACE, 1996
103. A. Goldberg and M.C. Juhas, "Lower Bound K_{Iscc} Values of Bolting Materials—A Literature Study," NUREG Report CR 2467, 1982

SELECTED REFERENCES

- W. Bamford and J. Hall, A Review of Alloy 600 Cracking in Operating Nuclear Plants: Historical Experience and Future Trends, *Proc. 11th Int. Conference on Environmental Degradation of Materials in Nuclear Power Systems—Water Reactors* (Stevenson, WA), American Nuclear Society, 2003, p 1071–1079
- Ph. Berge, Importance of Surface Preparation for Corrosion Control in Nuclear Power Stations, *Mater. Perform.*, Vol 36 (No. 11), 1997, p 56
- F. Champigny, F. Chapelier, and C. Amzallag, Maintenance Strategy of Inconel Components in PWR Primary Systems in France, *Proc. Conference on Vessel Penetration Inspection, Cracking and Repairs* (Gaithersburg, MD), Sept 29–Oct 2, 2003
- P. Cohen, *Water Coolant Technology of Power Reactors*, 2nd printing, American Nuclear Society, 1980
- C.J. Czajkowski, Corrosion and Stress Corrosion Cracking of Bolting Materials in Light Water Reactors, *Proc. 1st Int. Symposium on Environmental Degradation of Materials in Nuclear Power Systems—Water Reactors* (Myrtle Beach, SC), NACE, 1984, p 192–208
- D.R. Diercks, W.J. Shack, and J. Muscara, Overview of Steam Generator Tube Degradation and Integrity Issues, *Nucl. Eng. Des.*, Vol 194, 1999, p 19–30
- F. Nordmann, A. Stutzmann, and J.-L. Bretelle, Overview of PWR Chemistry Options, Paper *Proc. Chimie* 2002, Société Française d'Energie Nucléaire
- P. Scott, Review of Irradiation Assisted Stress Corrosion Cracking, *J. Nucl. Mater.*, Vol 211, 1994, p 101–122
- H.P. Seifert, S. Ritter, and J. Hickling, Environmentally-Assisted Cracking of Low-Alloy RPV and Piping Steels under LWR Conditions, *Proc. 11th International Conference on Environmental Degradation of Materials in Nuclear Power Systems—Water Reactors* (Stevenson, WA), American Nuclear Society, 2003, p 73–88
- R. Staehle and J.A. Gorman, Quantitative Assessment of Submodes of Stress Corrosion Cracking on the Secondary Side of Steam Generator Tubing in Pressurized Water Reactors: Parts 1, 2, and 3, *Corrosion*, Vol 59 (No. 11), 2003, p 931–994; Vol 60 (No. 1), 2004, p 5–63; and Vol 60 (No. 2), 2004, p 115–180
- G.A. White, J. Hickling, and L.K. Mathews, Crack Growth Rates for Evaluating PWSCC of Thick-Walled Alloy 600 Material, *Proc. 11th International Conference on Environmental Degradation of Materials in Nuclear Power Systems—Water Reactors* (Stevenson, WA), American Nuclear Society, 2003, p 166–179

Effect of Irradiation on Stress-Corrosion Cracking and Corrosion in Light Water Reactors

Gary S. Was and Jeremy Busby, University of Michigan
Peter L. Andresen, General Electric Global Research

A GROWING CONCERN FOR ELECTRIC POWER UTILITIES worldwide has been degradation in core components in nuclear power reactors, which make up approximately 17% of the world's electric power production. Service failures have occurred in boiling water reactor (BWR) core components and, to a somewhat lesser extent, in pressurized water reactor (PWR) core components consisting of iron- and nickel-base stainless alloys that have achieved a significant neutron fluence in environments that span oxygenated to hydrogenated water at 270 to 340 °C (520 to 645 °F). Because cracking susceptibility is a function of radiation, stress, and environment, the failure mechanism has been termed irradiation-assisted stress-corrosion cracking (IASCC). Initially, the affected components have been either relatively small (bolts, springs, etc.) or designed for replacement (fuel rods, control blades, or instrumentation tubes). In the last decade, there have been many more structural components (PWR baffle bolts and BWR core shrouds) that have been identified to be susceptible to IASCC. Recent reviews (Ref 1-5) describe the current knowledge related to IASCC service experience and laboratory investigations and highlight the limited amount of well-controlled experimentation that exists on well-characterized materials. This lack of critical experimentation and the large number of interdependent parameters make it imperative that underpinning science be used to guide mechanistic understanding and quantification of IASCC.

The importance of neutron fluence on IASCC has been well established. As shown in Fig. 1, intergranular SCC (IGSCC) is promoted in austenitic stainless steels when a critical "threshold" fluence is exceeded (such "thresholds" appear only in some tests, as discussed later). Cracking is observed in BWR oxygenated water at fluences above 2 to 5×10^{20} n/cm^2 ($E > 1$ MeV), which corresponds to about 0.3 to 0.7 displacements per atom (dpa). A comparable "threshold" fluence for IASCC susceptibility has been reported for high-stress, in-service BWR component cracking and during ex situ, slow-strain-rate SCC testing of irradiated stainless steels. This indicates "persistent" radiation effects (material changes) are primarily responsible for IASCC susceptibility, although in situ effects such as radiation creep relaxation of weld residual stresses and increased stress from differential swelling can be important.

As with SCC behavior in unirradiated environments, the aqueous environment and stress/strain conditions also strongly influence observed cracking. While some investigators have observed limited intergranular cracking in inert environments, it is clear that the IASCC observations in the laboratory and in plants can only be accounted for by an environmentally assisted cracking mechanism (i.e., IASCC is not a purely mechanical, phenomenon). Water chemistry and electrochemical potential effects are reflected in the sharp increase in IASCC cracking with increasing dissolved oxygen (Ref 1, 5). Cracking is typically not observed in hydrogenated water (i.e., BWR hydrogen water chemistry or PWR water) until a fluence level approximately four times greater than that observed for oxidizing water conditions is attained. Service experience has shown that higher stress components exhibit failures at lower fluence levels.

Recent work has enabled many aspects of IASCC phenomenology to be explained (and predicted) based on the experience with IGSCC of unirradiated stainless steel in reactor water environments. This continuum approach has successfully accounted for radiation effects on water chemistry and its influence on electrochemical corrosion potential. However, all radiation-induced microstructural, and microchemical changes that promote IASCC are not fully known. Well-controlled data from properly

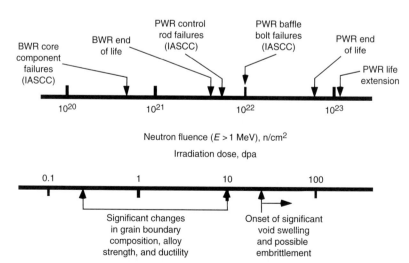

Fig. 1 Neutron fluence effects on irradiation-assisted stress-corrosion cracking susceptibility of type 304 stainless steel in boiling water reactor (BWR) environments. PWR, pressurized water reactor; IASCC, irradiation-assisted stress-corrosion cracking; dpa, displacements per atom. Source: Ref 4

Effect of Irradiation on Stress-Corrosion Cracking and Corrosion in Light Water Reactors

irradiated and properly characterized materials are sorely lacking due to the inherent experimental difficulties and financial limitations. Many of the important metallurgical, mechanical, and environmental aspects that are believed to play a role in the cracking process are illustrated in Fig. 2. Only persistent material changes are required for IASCC to occur, but in-core processes such as radiation creep and radiolysis also have an important effect on IASCC. The following section examines the current understanding of persistent material changes that are produced in stainless alloys during light water reactor (LWR) irradiation based on the fundamentals of radiation damage and existing experimental measurements.

Irradiation Effects on SCC

Gary S. Was, University of Michigan
Peter L. Andresen, General Electric Global Research

Irradiation-assisted SCC can be categorized into radiation effects on the water chemistry (radiolysis) and on the material properties. The cracking response to changes in water chemistry is similar for both irradiated and unirradiated materials, as is discussed later. In both cases, there is a steep increase in environmental cracking kinetics with a rise in the corrosion potential above about 100 mV standard hydrogen electrode (100 mV_{SHE}) (Ref 5–7). At high corrosion potential, the crack growth rate also increases sharply as impurities (especially chloride and sulfate) are added to pure water in either the irradiated or unirradiated cases. In postirradiation tests, the dominant radiation-related factors are microstructural and microchemical changes, which can be responsible for "thresholdlike" behavior in much the same way as corrosion potential, impurities, degree of sensitization, stress, temperature, and so forth. Other radiation phenomena, such as radiation creep relaxation and differential swelling, could also have "persistent" effects if one relies on the sources of stress present during radiation (e.g., weld residual stresses or loading from differential swelling) during postirradiation testing. The effects of radiation rapidly (in seconds) achieve a dynamic equilibrium in water, primarily because of the high mobility of species in water. In metals, dynamic equilibrium is achieved—if ever—only after many dpa, typically requiring years of exposure. While both radiation-induced segregation (RIS) of major elements, and radiation hardening (RH) and the associated microstructural development asymptotically approach a dynamic equilibrium, other factors (e.g., RIS of silicon, or precipitate formation or dissolution) may become important. Yet data on postirradiation slow strain rate tests (SSRT) on stainless steels show that there is a distinct (although not invariant) "threshold" fluence at which IASCC is observed under LWR conditions (Ref 8, p 583). Because this "threshold" occurs at a fraction to several dpa (depending on the alloy, stress, water chemistry, etc.) as shown in Fig. 3, in situ effects (corrosion potential, conductivity, and temperature) may be important, but only "persistent" radiation effects (microstructural and microchemical changes) can be responsible for the "thresholdlike" behavior versus fluence in postirradiation tests.

Having established the importance of persistent changes to the material in postirradiation observations of IASCC, the challenge then becomes one of identifying the specific changes in the material that accelerate SCC. These material changes fall into three categories: (a) microcompositional effects due to radiation-induced segregation of both impurities and major alloying elements, (b) microstructural changes, such as the formation of dislocation loops, voids, precipitates, and the resulting hardening (increase in yield strength), and (c) deformation mode, including formation of dislocation channeling and localized deformation in irradiated materials. As shown by the composite schematic diagram in Fig. 4, all of the observable effects of irradiation on the material increase with dose in much the same manner, making difficult the isolation of and attribution to individual contributions. The dislocation loop microstructure is closely tied to radiation hardening, and both increase with dose until saturation occurs by approximately 5 dpa. Radiation-induced segregation also increases with dose and tends to saturate by approximately 5 dpa. Although dependent on both metallurgical and environmental parameters, IASCC generally occurs at doses between 0.5 dpa (for BWRs) and 2 to 3 dpa (for PWRs) (Ref 3), which encompasses the steeply rising portion of the curves in Fig. 4 that describe the changes in material properties with irradiation. Following a review of service experience, subsequent sections focus on the possible mechanisms by which water chemistry, RIS, microstructure, hardening, deformation mode, and irradiation creep—individually or in concert—may affect IASCC.

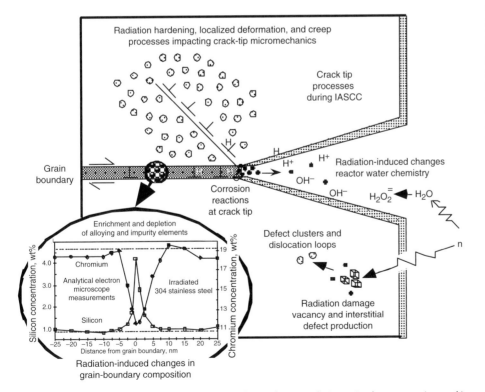

Fig. 2 Mechanistic issues believed to influence crack advance during irradiation-assisted stress-corrosion cracking (IASCC) of austenitic stainless steels.

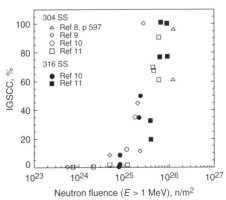

Fig. 3 Dependence of cracking in neutron-irradiated high-purity type 304 stainless steel (SS) on accumulated high-energy neutron fluence. IGSCC, intergranular stress-corrosion cracking

Service Experience

An historical perspective of IASCC service experience is instructive, as the phenomenon extends back many decades, and the early observations and conclusions projected an accurate image of the important characteristics, generic nature, and broad relevance to plant components. As with other instances of environmental cracking, occasional early observations pointed the way toward a growing incidence with time and neutron fluence. First reported in the early 1960s (Ref 1, 5, 12–29), IASCC involved intergranular cracking of stainless steel fuel cladding in a PWR. The findings and conclusions were that intergranular cracking morphology predominated, with initiation of multiple cracks occurring from the waterside. By contrast, only ductile, transgranular cracking was observed in postirradiation mechanical tests performed in inert environments and at various temperatures and strain rates. Grain-boundary carbide precipitation was generally not observed by optical or transmission electron microscopy (although preexisting thermal sensitization was present in some cases). A correlation between time-to-failure and stress level was reported, with failure occurring first in thin-walled rods with small fuel-to-cladding gaps, where swelling strains were largest. The highest incidence of cracking occurred in peak heat flux regions, corresponding to the highest fluence and the greatest fuel-cladding interaction (highest stresses and strains). Similar stainless steel cladding in PWR service exhibited fewer instances of intergranular failure. At that time the PWR failures were attributed to off-chemistry conditions or stress rupture.

Irradiation-assisted SCC has since been observed in a growing number of other stainless steel (and nickel alloy) core components, such as neutron source holders in 1976 and control rod absorber tubes in 1978. Instrument dry tubes and control blade handles and sheaths (Fig. 5), which are subject to very low stresses also cracked, although generally in creviced locations and at higher fluences (Ref 30, 31). Following an initial trickle of failures in the most susceptible components, numerous incidents of IASCC have been observed since the early 1990s, perhaps most notably in BWR core shrouds (Ref 1, 5, 12, 13, 34) and PWR baffle former bolts (Ref 3, 35, 36).

Table 1 presents a broad summary of reported failures of reactor internal components, conclusively showing that IASCC is not confined to a particular reactor design. For example, stainless steel fuel cladding failures were reported years ago in commercial PWRs and in PWR test reactors (Ref 1, 15, 16, 22–29). At the West Milton PWR test loop, intergranular failure of vacuum annealed type 304 stainless steel fuel cladding was observed in 316 °C (601 °F) ammoniated water (pH 10) when the cladding was stressed above yield (Ref 24). Similarly, IASCC was observed in creviced 20Cr/25Ni/Nb stainless steel fuel element ferrules in the Winfrith steam-generating heavy water reactor (SGHWR), a 100 MWe plant in which light water is boiled within pressure tubes, giving rise to a coolant chemistry similar to other boiling water reactor designs (Ref 37). The 20%Cr/25%Ni/Nb stainless steel differs from type 304 primarily in nickel and niobium content, as well as in its lower sulfur (approximately 0.006%) and phosphorus (approximately 0.005%) contents. The ferrules were designed for a 5 year exposure during which the peak fast neutron flux is 2 to 3×10^{13} n/cm$^2 \cdot$s ($E > 1.5$ MeV), yielding a peak fluence over 5 years of 3 to 5×10^{21} n/cm^2.

Reactor type comparisons were also made in swelling tube tests performed in the core of a BWR and a PWR on a variety of commercial and high-purity heats of types 304, 316, and 348 stainless steel and alloys X-750, 718, and 625 (Ref 38, 39). Swelling was controlled by varying the mix of aluminum oxide (Al_2O_3) and boron carbide (B_4C) within the tubes; the latter swells as neutrons transmute boron to helium. Nominally identical strings of specimens were inserted into the core in place of fuel rods. The distinction in the IASCC response between the two reactor types was small.

The oxidizing potential in a PWR core is lower than in BWRs operating in normal water chemistry (currently, many BWRs operate with catalytic surfaces and with some H^2 addition, which dramatically reduces the oxidizing

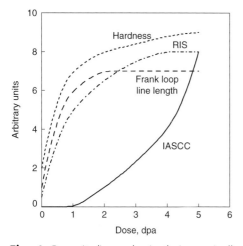

Fig. 4 Composite diagram showing the increase in all parameters (RIS, dislocation loops, hardness) with dose. The IASCC response is shifted by test technique and water chemistry conditions. RIS, radiation-induced segregation; dpa, displacements per atom

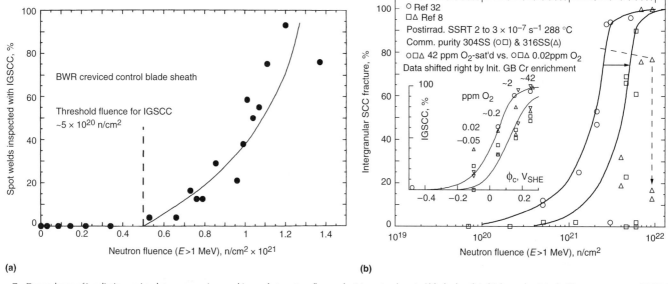

Fig. 5 Dependence of irradiation-assisted stress-corrosion cracking on fast neutron fluence for (a) creviced control blade sheath in high conductivity boiling water reactors (BWRs) (Ref 30) and (b) as measured in slow strain rate tests at 3.7×10^7 s^{-1} on preirradiated type 304 stainless steel (SS) in 288 °C (550 °F) water (Ref 33). The effect of corrosion potential via changes in dissolved oxygen is shown at a fluence of ~2×10^{21} n/cm^2. IGSCC, intergranular stress-corrosion cracking; GB, grain boundary

potential (Ref 40–42). This led some investigators attributed PWR cracking incidents to low-ductility stress rupture (of course, this mechanism would be equally applicable to BWRs). A few laboratory studies reported small amounts of intergranular cracking of irradiated stainless steels in slow strain rate tensile tests in approximately 300 °C (570 °F) inert environments (Ref 43), although in many related experiments (Ref 32, 33) no similar evidence of intergranular failure was found. Small amounts of intergranular cracking in inert tensile tests is not surprising; evidence against it as the sole basis for service failures includes the vastly greater extent of intergranular cracking in tests performed in high-temperature water, the general observation that crack initiation occurs from the water side, and the widespread observation that exposure to hot water produces an environmental enhancement in cracking.

Since the early 1990s, the plant and laboratory evidence of IASCC makes a compelling case that cracking is environmentally assisted and that there is a well-behaved continuum in response over ranges in fluence, corrosion potential, temperature, stress, and so forth (Ref 1, 5, 12, 13). Since there is a consistent trend toward increasing IASCC susceptibility with increasing corrosion potential in BWRs (e.g., Fig. 5b), PWRs should be less susceptible to IASCC. However, other factors distinguish PWRs from BWRs, including their higher temperatures, approximately 10× higher neutron fluence in core structural components, higher hydrogen fugacity, and the borated-lithiated water chemistry (including the possibility of localized boiling and thermal concentration cells in crevices from gamma heating that could lead to aggressive local chemistries). The possible role of RIS of silicon at high fluence may be especially important in accounting for the small effect of corrosion potential on SCC response (Ref 44), which would pose challenges for both BWRs and PWRs.

More detailed and quantitative IASCC field experience was reported by Brown and Gordon (Ref 30, 31), who accumulated and analyzed data for cracking in alloy 600 shroud head bolts (first observed in 1986), and stainless steel safe ends (first observed in 1984) and in-core instrumentation tubes (first observed in 1984), with a focus on components that were creviced, a factor known to exacerbate cracking (Ref 17, 30, 31). The highest radiation exposure occurred for the intermediate range and source range monitor (IRM/SRM) dry tubes, which contain flux monitors housed in thin-walled, annealed stainless steel tubes. Cracking initiated in the crevice between the spring housing tube and the guide plug at fluences between 0.5 and 1.0×10^{22} n/cm^2 ($E > 1$ MeV). Wedging stresses from the thick oxide observed in the crevice were implicated, since other (applied and residual) stresses were negligible. The primary variable from plant-to-plant is the average coolant conductivity, which correlates strongly with cracking incidence (Fig. 6a to c). Each point in Fig. 6 represents inspection results for one BWR plant, and data are normalized using reactor operating time (i.e., "percentage of components with intergranular cracks divided by the on-line exposure time"). The scatter in Fig. 6(a) was attributed to variations in fluence and specific ion chemistry, as well as limitations in the resolution of underwater visual inspection. Scatter can also result from short-term excursions in conductivity, which is not adequately reflected in the average, as identified in Fig. 6(c).

Conductivity correlations were also reported for cracking in shroud head bolts (Fig. 6c) and creviced safe ends (Fig. 6b). The strong influence of conductivity on cracking of stainless steel has also been shown in laboratory tests and plant recirculation piping, where predictive modeling (Ref 1, 5–7, 45, 46) has been compared to field data on the operational time required to achieve a detectable crack (typically 10% of the wall thickness). Preliminary

Table 1 Irradiation-assisted SCC service experience

Component	Material	Reactor type	Possible sources of stress
Fuel cladding	304 SS	BWR	Fuel swelling
	304 SS	PWR	Fuel swelling
Fuel cladding(a)	20%Cr/25%Ni/Nb	AGR	Fuel swelling
Fuel cladding ferrules	20%Cr/25%Ni/Nb	SGHWR	Fabrication
Neutron source holders	304 SS	BWR	Welding and beryllium swelling
Instrument dry tubes	304 SS	BWR	Fabrication
Control rod absorber tubes	304 SS	BWR	B$_4$C swelling
Fuel bundle cap screws	304 SS	BWR	Fabrication
Control rod follower rivets	304 SS	BWR	Fabrication
Control blade handle	304 SS	BWR	Low stress
Control blade sheath	304 SS	BWR	Low stress
Control blades	304 SS	PWR	Low stress
Plate-type control blade	304 SS	BWR	Low stress
Various bolts(b)	A-286	PWR and BWR	Service
Steam separator dryer bolts(b)	A-286	BWR	Service
Shroud head bolts(b)	600	BWR	Service
Various bolts	X-750	BWR and PWR	Service
Guide tube support pins	X-750	PWR	Service
Jet pump beams	X-750	BWR	Service
Various springs	X-750	BWR and PWR	Service
	718	PWR	Service
Baffle former bolts	316 SS cold work	PWR	Torque, differential swelling
Core shroud	304/316/347 L SS	BWR	Weld residual stress
Top guide	304 SS	BWR	Low stress (bending)

(a) Cracking in AGR fuel occurred during storage in spent fuel pond. (b) Cracking of core internals occured away from high neutron and gamma fluxes.
Note: SS, stainless steel; L SS, low-carbon stainless steel; BWR, boiling water reactor; PWR, pressurized water reactor; AGR, advanced gas-cooled reactor; SGHWR, steam-generating heavy water reactor.

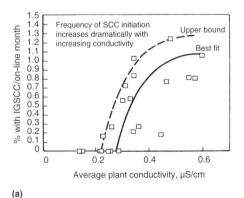
(a)

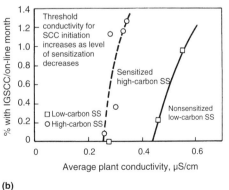
(b)

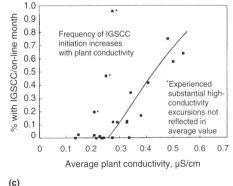
(c)

Fig. 6 The effects of average plant water purity (measured by coolant conductivity) are shown in field correlations of the core component cracking behavior for (a) stainless steel (SS) intermediate range monitor/source range monitor (IRM/SRM) instrumentation dry tubes, (b) creviced stainless steel safe ends, and (c) creviced Inconel 600 shroud head bolts, which also shows the predicted response versus conductivity. SSC, stress-corrosion cracking; IGSSC, intergranular stress-corrosion cracking

prediction of the shroud head bolt cracking also provides reasonable agreement with observation (Fig. 6c).

Cracking has also extensively occurred in high-strength, nickel-base alloy components (Ref 1, 16) as indicated in Table 1. Many incidents occur in lower radiation flux regions (e.g., where the end-of-life fluence is below approximately 5×10^{19} n/cm^2) such as cracking of Inconel X-750 jet pump beams in BWRs. Inconel X-750 cracking has also occurred extensively in PWR fuel hold-down springs, which attain an end-of-life fluence of 1 to 10×10^{21} n/cm^2; this has been attributed in part to vibrational stresses. The effects of irradiation on IASCC in high-strength, precipitation-hardened nickel-base alloy components is not characterized as well as in stainless steels, although in the lower flux/fluence regions the predominant effect of radiation is probably associated with the elevated corrosion potential due to radiolysis.

Two widespread examples of IASCC are in BWR core shrouds (Ref 5, 12, 13, 34) and PWR baffle former bolts (Ref 35, 36), although susceptibility clearly exists in other areas, such as control blade components, fuel components, the BWR top guide, and so forth. Stress-corrosion cracking in the BWR core shroud occurs almost exclusively near the welds (both circumferential and vertical), and initiation is observed from both the inside (ID) and outside diameter (OD) surfaces (the shroud separates the upward core flow from the downward recirculation flow that occurs in the annulus between the shroud and the pressure vessel). This large-diameter welded "pipe" has inherent susceptibility to SCC, related primarily to weld residual stresses and weld shrinkage strains (Ref 47, 48), and cracking is observed in both low-fluence and moderate-fluence areas. The extent of the enhancement in SCC susceptibility by irradiation is limited, because while radiation hardening and radiation-induced segregation occur, radiation creep also relaxes the weld residuals stress. (Of course, radiolysis plays an inherent role, since it is the dominant source of oxidants in the entire BWR circuit, as discussed later.) Predictions indicate that if SCC does not nucleate early in life (e.g., below 0.5 dpa) from high coolant impurity levels or severe surface grinding, susceptibility decreases with fluence in the shroud welds (although many shroud welds are in very low flux areas) because of radiation-induced creep relaxation. The two-sided welding procedure (alternate welding passes on the ID and OD) produces, on average, a reasonably symmetrical residual stress profile that causes the stress-intensity factor to drop below 0 as cracks approach roughly two-thirds of the wall thickness. Thus, in many shrouds cracks do not grow through-wall, although with coincident ID and OD initiation and/or large fit-up stresses, it is possible for through-wall cracks to develop.

Extensive failures of PWR baffle bolts have occurred in the last decade (Ref 35, 36) although large plant-to-plant and heat-to-heat differences are observed. Most baffle bolts are fabricated from type 316 stainless steel cold worked to approximately 15% to increase their yield strength. The complex baffle former structure exists in a PWR because their fuel does not have a surrounding "channel," so the baffle former structure must conform closely to the geometry of the fuel to provide well-distributed water flow. The baffle former plates are usually made from annealed material, typically type 304 stainless steel. Because of their proximity to the fuel, very high fluences can develop—up to ~80 dpa by the end of the original design life. The high gamma flux produces significant heating in the components, in some instances estimated at $+40$ °C ($+104$ °F), especially in designs where the PWR coolant does not have good access to the bolt shank. While the heat-to-heat variations are not understood, it is clear that plants that load-follow (and therefore undergo power level changes and thermal cycles) are much more prone to baffle bolt cracking. Another aggravant is the thermal gradient and possible boiling (resulting in altered water chemistry) in the shank area of the bolt if the coolant access is restricted. However, primary factors must be the very sizeable stress relaxation that occurs early in life (e.g., during the first 5 dpa), followed by preferential radiation swelling of the annealed baffle plates over the cold-worked baffle bolts, which will cause reloading. The dynamic equilibrium between swelling and radiation creep, which determines the "reloading" stress in the bolt, is likely a complex function of local neutron flux, temperature, baffle plate geometry, composition, and other factors.

In the last decade, the number of IASCC incidents has continued to grow, and there can be no question that many components in LWRs are susceptible. Strategies to mitigate IASCC, which are discussed in the section "Mitigation Strategies" in this article, and manage IASCC (e.g., by dispositioning its impact, installing mechanical restraints to mitigate the impact of IASCC in BWR shrouds, or selectively inspecting and replacing baffle bolts) have been successful. The overall trends and correlations for IASCC can be summarized as:

- While intergranular cracks related to radiation effects on solution-annealed stainless steel were once thought to occur only at fluences above approximately 0.3×10^{21} n/cm^2, significant intergranular cracking in BWR core shrouds over a broad range of fluences make it clear that such a distinction (a true fluence threshold) is not justified (Ref 1, 12, 13, 49). Of course, observations of SCC in unsensitized stainless steel (with or without cold work) also render untenable the concept of a threshold fluence, below which no SCC occurs. This also holds for thresholds in corrosion potential, water impurities, and so forth (Ref 5, 44, 49, 50).
- Fluence affects SCC susceptibility, but almost always in a complex fashion. Stress-corrosion cracking in BWR shrouds and PWR baffle bolts does not always correlate strongly with fluence, one important reason for this is that radiation creep produces relaxation of the stresses from welding and in bolts.
- High stresses or dynamic strains were involved in most early incidents; however, cracking has been observed at quite low stresses at high fluences and longer operating exposure. Laboratory and field data indicate that IASCC occurs at stresses below 20% of the irradiated yield stress, and at stress intensities below 10 MPa$\sqrt{m}$ (9.1 ksi$\sqrt{in.}$) (Ref 1, 5, 12, 13, 49, 50).
- A strong effect of corrosion potential is clear from extensive laboratory and field data. Its effect is generally consistent from low to high fluence, although the quantitative change associated with changes in potential vary. Materials prone to high-radiation-induced changes in silicon level may exhibit a very limited effect of corrosion potential (Ref 44). A true threshold potential clearly does not exist, as irradiated materials exhibit IASCC in deaerated water.
- Solution conductivity (i.e., impurities, especially chloride and sulfate) strongly affects cracking propensity in BWR water (Fig. 6). As noted by Brown and Gordon (Ref 31), this correlation applies equally to low and high flux regions and to stainless steels (Fig. 6a and b) and nickel-base alloys (Fig. 6b). Indeed, the correlation closely parallels that from out-of-core (Ref 1, 5–7, 31, 45, 46).
- Crevice geometries exacerbate cracking due primarily to their ability to create a more aggressive crevice chemistry from the gradient in corrosion potential (in BWRs) or in temperature (most relevant to PWRs).
- Cold work often exacerbates cracking (especially abusive surface grinding), although it can also delay the onset of some radiation effects.
- Increasing temperature from approximately 270 to 350 °C (520 to 660 °F) has an important effect on IASCC, enhancing both crack initiation and growth rate (Ref 12, 13, 50).
- Grain-boundary carbides and chromium depletion are not required for susceptibility, although furnace-sensitized stainless steels are clearly highly susceptible to cracking in-core. Chromium depletion remains a primary culprit, although its effect is most pronounced in pH-shifted environments, as can develop when potential or thermal gradients exist. The role of nitrogen, sulfur, and phosphorus, and other grain-boundary segregants is less clear.
- The fluence at which IASCC is observed is dependent on applied stress and strain, corrosion potential, solution conductivity, crevice geometry, and so forth. At sufficiently high conductivities, cracking has been observed in solution-annealed stainless steel in the field (Fig. 6a and b) (Ref 31) and in the laboratory (Ref 1, 5–7, 45, 46). Thus, while convenient in a practical engineering sense, the concept of a "threshold" fluence (or stress, corrosion potential, etc.) is scientifically

misleading (Ref 5, 12, 44, 49, 50); cracking susceptibility and morphology are properly considered an interdependent continuum over many relevant parameters.

The field and laboratory data available in the early 1980s, coupled with broader fundamental understanding of environmental cracking in hot water, led to the hypothesis that—among innumerable possible radiation effects—the most significant factors were RIS at grain boundaries, radiation hardening (elevation of the yield strength), mode of deformation, radiation creep relaxation (of constant displacement stresses, e.g., in welds and bolts), and radiolysis (elevated corrosion potential in BWRs). Other factors could also be important in some instances, such as void formation, which may also affect fracture toughness, and can produce differential swelling that produces reloading of components such as baffle bolts. These factors are addressed in the subsequent sections.

Water Chemistry

Radiolysis and Its Effect on Corrosion Potential. It is widely acknowledged that SCC susceptibility is fundamentally influenced by corrosion potential, not oxidant and reductant concentrations per se (Ref 1, 5, 12, 51). The corrosion potential is a mixed potential formed by a kinetic balance of anodic and cathodic reactions on a metal surface. In the absence of oxidants such as O_2, H_2O_2, and Cu ion, the H_2/H_2O reaction dominates the corrosion potential of most structural materials (and platinum), and the corrosion potential can be calculated from the thermodynamics of the H_2/H_2O reaction, which depends on pH, temperature, and H_2 fugacity.

An important distinction between BWRs and PWRs is the low H_2 concentration in BWRs, which allows the formation of radiolytic species (including oxidants). Above approximately 500 ppb (5.6 cm^3/kg) H_2, radiolytic formation of oxidants is effectively suppressed (Fig. 7a) and the corrosion potential remains close to the thermodynamic minimum as defined by the H_2/H_2O reaction. Boiling water reactors cannot achieve this H_2 level because H_2 partitions to the steam phase, which begins to form about a quarter of the way up the fuel rods. Thus, radiolysis is relevant to BWR water chemistry. There are a number of sequential and nonlinear dependencies that must be considered, for example, radiation flux versus oxidant concentration, oxidant level versus corrosion potential, corrosion potential versus crack chemistry, and crack chemistry versus SCC growth rate (Ref 5, 51).

Water is decomposed by ionizing radiation into various primary species (Ref 52–55) including both radicals (e.g., e_{aq}^-, H, OH, HO_2) and molecules (e.g., H_2O_2, H_2), which can be oxidizing (e.g., H_2O_2, HO_2) or reducing (e.g., e_{aq}^-, H, H_2). The predominant species that are stable after a few seconds are H_2O_2 and H_2, with O_2 forming primarily from the oxidation of H_2O_2 and HO_2 with an OH radical, and from the thermal decomposition of H_2O_2. Because H_2 partitions to the steam phase and H_2O_2 is not volatile, approximately 87% of the water that is recirculated in a BWR (approximately 11 to 14% of the core flow becomes steam) is oxidant rich. Hydrogen (H_2) is introduced in the feedwater, which mixes with the recirculated water near the top of the annulus (the region of down-flow between the core shroud and pressure vessel).

The concentrations of radiolytic species are roughly proportional to the square root of the radiation flux in pure water. The radiation energy-intensity spectrum influences the concentration of each radiolytic specie, which is described in terms of a yield, or G value (molecules produced per 100 eV absorbed by water). In LWRs, the G values for most species are within a factor of approximately 3 for fast neutron versus gamma radiation. Despite this similarity, the influence of fast neutron radiation is much stronger than gamma radiation primarily because the energy deposition rate, or mean linear energy transfer (LET), is greater (40 eV/nm for fast neutrons, versus 0.01 eV/nm for gamma radiation) (Ref 55). Also, the neutron flux in LWRs (e.g., $\sim 1.03 \times 10^9$ rad/h core average and $\sim 1.68 \times 10^9$ rad/h peak in a BWR of 51 W/cm^3 power density) is also higher than the gamma flux ($\sim 0.34 \times 10^9$ rad/h). Indeed, the moderate gamma levels present in the downcomer in the outside annulus of a BWR core actually promote recombination of hydrogen and various oxidants (Ref 53, 56). The contribution of thermal neutrons and beta particles to radiolysis is small in LWRs.

As in many electrochemical processes, the integrated effects of various oxidants and reductants on environmental cracking is best described via changes in corrosion potential, which (at constant pH and temperature) controls the thermodynamic phase stability and influences the kinetics of electrochemical reactions. Since electrochemical potentials are logarithmically dependent on local oxidant, reductant, and ionic concentrations (via the Nernst relationship, $\phi = \phi_o + RT/nF \ln$ [products/reactants]), radiation-induced increases in concentration of various species by many orders of magnitude may have comparatively small effects on the corrosion potential in hot water (note: the Nernst equation is described in detail in *Corrosion: Fundamentals, Testing, and Protection*, Vol 13A, of the *ASM Handbook*). Furthermore, corrosion potentials are mixed potentials involving a balance of anodic and cathodic reactions on the metal surface, which depend on the concentrations of both oxidizing and reducing species. At low oxidant concentrations, the rapid drop in corrosion potential to approximately -0.5 V_{SHE} (Fig. 8a) results from mass transport limited kinetics (e.g., oxygen transport to the metal surface). In this regime, more pronounced shifts in corrosion potential with radiation can occur, presumably from the radiolytic formation of oxidizing species within the mass transport limited stagnant layer.

The relationship between dissolved oxygen and corrosion potential in hot water as a function of radiation type and flux is shown in Fig. 8(a), in which the connected points represent data obtained in controlled, radiation on/off experiments. The data from these latter experiments are shown in Fig. 8(b) in terms of a radiation-induced shift in potential. The curves in Fig. 8(a) represent the scatter band for the data obtained under unirradiated conditions. Similar scatter also exists in the irradiated corrosion potential data in Fig. 8(a) and comprises contributions from both real effects and experimental error. These radiation measurements are quite complex and should involve careful qualification of the radiation resistance and thermodynamic response of reference electrodes. High radiation flux experiments were performed by Taylor (Ref 56) and Miller and Andresen (Ref 57, 58)

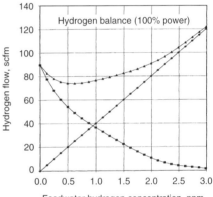

(a)

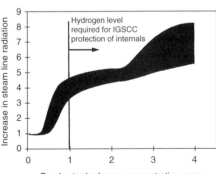

(b)

Fig. 7 (a) The relationship between the H_2 concentration in the feedwater and the H_2 injection flow (straight line, round data points). The descending curve (square data points) is H_2 generated by radiolysis, which decreases as H_2 is injected due to suppression of radiolysis. The curve that decreases and then increases (upper curve, triangle data points) is H_2 in steam. (b) The effect of H_2 injection on the radiation level in the steam lines and turbine from ^{16}N (transmuted from ^{16}O), which results in changes from soluble NO_3^- to volatile NO_x and NH_3. IGSCC, intergranular stress-corrosion cracking

using multiple, fundamentally different types of radiation-hardened reference electrodes. In-reactor measurements have also been performed (Ref 59, 60) using multiple, radiation-qualified silver chloride reference electrodes.

Figures 8(a) and (b) show that little, if any, elevation in corrosion potential results from irradiation sources that do not include neutrons or simulate their contribution (e.g., using high-energy protons as described in Ref 57). Some studies using gamma radiation (Ref 1, 12, 57) showed a significant decrease in corrosion potential, especially in the intermediate (e.g., 10 to 200 ppb) range of dissolved oxygen. This is consistent with enhanced recombination of oxidizing and reducing species, which occurs in the downcomer region of BWRs (Ref 56) and is relied upon to produce SCC mitigation using hydrogen water chemistry (HWC). In instances where neutrons or protons have been used, a consistent, significant elevation in corrosion potential is observed. This is more pronounced in hot water containing low dissolved oxygen concentrations and no dissolved hydrogen (Fig. 8b), where increases of over $+0.25$ V occur. At higher inlet oxygen concentrations (e.g., approximately 200 ppb), the data still show a significant shift (typically $+0.1$ to 0.15 V) in corrosion potential for radiation conditions representative of peak LWR core fluxes (Fig. 8b); less increase is observed for inlet oxygen concentrations associated, for example, with air saturation (approximately 8.8 ppm O_2) or oxygen saturation (approximately 42 ppm O_2 at standard temperature and pressure, STP). A similar elevation in corrosion potential is observed for additions of hydrogen peroxide (200 ppb H_2O_2, Fig. 8a), which suggests that H_2O_2 may be a major factor in increasing the corrosion potential under irradiated conditions.

In-core, in situ measurements in BWRs show that the corrosion potential, which is approximately $+0.2$ to $+0.25$ V_{SHE} in normal water chemistry, can be decreased by >0.5 V by sufficient additions of dissolved hydrogen in a BWR (Ref 60). This is corroborated by other measurements (Ref 56 and Fig. 8b), which show very little radiation-induced elevation in corrosion potential when the fully deaerated inlet water contains moderate dissolved hydrogen (>200 ppb H_2, 0 ppb O_2). However, at high H_2 levels, the core becomes reducing, and the small concentration of ^{16}N (transmuted from ^{16}O) changes from soluble NO_3^- to volatile NO_x and NH_3, causing a large increase in radiation level in the steam lines and turbine (Fig. 7b).

The effect of radiation on the corrosion potential within a crack or crevice has also been of interest, with the possibility that a net oxidizing environment in the crack could be created that could elevate the corrosion potential above the potential at the crack mouth. In the absence of radiation, measurements in high-temperature water in artificial crevices (e.g., tubing) (Ref 61, 62), at the tip of growing cracks (Ref 63), and of short crack growth behavior (Ref 64) show that the corrosion potential remains low (i.e., -0.5 ± 0.1 V_{SHE} in 288 °C, or 550 °F, pure water) for all bulk oxygen concentrations, indicating that complete oxygen consumption occurs within the crack. Measurements of radiation effects in crevices (Ref 57) show that the elevation in corrosion potential is limited to <0.05 V ($\phi_c < -0.45$ V_{SHE}) in-core; this is consistent with interpretation of available corrosion potential data on free surfaces (Ref 1, 5, 12, 13).

These small changes will not significantly affect the approximately 0.75 V ($+0.25$ V_{SHE} {near mouth} minus -0.5 V_{SHE} {in crack}) potential difference in the crack under irradiated normal BWR water chemistry conditions. The potential difference, along with other factors, controls the enhancement mechanism that can lead to an increased anion activity and altered pH at the crack tip (Ref 6, 7, 46, 51). The crack tip pH shifts its corrosion potential (down if alkiline, up if acidic), and thereby changes the potential difference.

Effects of Corrosion Potential on IASCC. Laboratory tests have been conducted (Ref 33, 65) on preirradiated material using SSRT in hot water with additions of oxygen and/or hydrogen peroxide to elevate the corrosion potential to simulate the effect of radiation. Tests by Jacobs et al. (Ref 32) on stainless steel irradiated to $\sim 3 \times 10^{21}$ n/cm^2 showed a strong effect of dissolved oxygen (and, by inference, corrosion potential) on IASCC (Fig. 5b). Similarly, a variety of preirradiated materials were tested by slow strain rate (Fig. 9a), and decreasing average crack growth rates with decreasing corrosion potential were observed (Ref 65, 66). While there might be the suggestion that no IASCC occurs at low corrosion potential, in fact both crack growth rate tests (discussed later) and other tests show that IASCC persists at low corrosion potential. Figure 9(b) shows the effect of fluence and test temperature on %IGSCC of irradiated type 316 stainless steel in PWR primary water (low corrosion potential).

In situ data on fracture mechanics specimens of furnace-sensitized type 304 stainless steel exposed in Nine Mile Point Unit 1 BWR showed

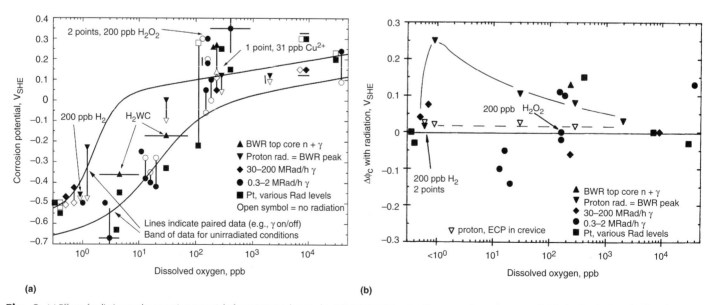

Fig. 8 (a) Effect of radiation on the corrosion potential of type 304 stainless steel in 288 °C (550 °F) water. The curves denote the range of typical values in the unirradiated corrosion potential data (Ref 1). (b) Effect of radiation on the *shift* in corrosion potential from the value under unirradiated conditions for type 304 stainless steel in 288 °C (550 °F) water. With the exception of the boiling water reactor (BWR) measurements, all data were obtained in controlled radiation on/off experiments (Ref 1). Curves in (b) show the trends in the proton-irradiated data, where the effects of radiation (on/off and over a range of fluxes) were evaluated for a variety of dissolved O_2 and H_2 concentrations under otherwise identical conditions.

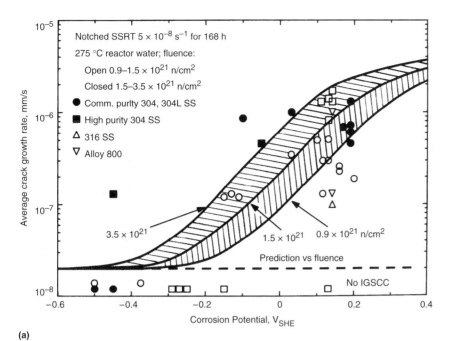

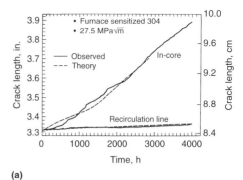

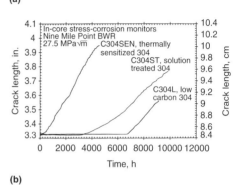

Fig. 10 Data for fracture mechanics specimens of type 304 stainless steel exposed in the high flux region of the core and in the recirculation line of Nine Mile Point Unit 1 Boiling Water Reactor (BWR). All specimens were precracked and wedge loaded to an initial stress-intensity factor of ~27.5 MPa√m (~25 ksi√in.). (a) Comparison of predicted and observed crack length versus time for furnace-sensitized type 304 stainless steel specimens in the core and recirculation line. (b) Crack length versus time for a furnace-sensitized and two annealed specimens of type 304 stainless steel in the core. Source: Ref 1, 6, 12, 13, 59

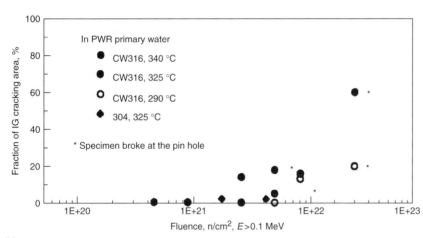

Fig. 9 (a) Comparison of predicted and observed crack growth rates for stainless steels irradiated in a BWR at 288 °C (550 °F) to various fluences. Notched tensile specimens were tested by Ljungberg (Ref 66) at a slow strain rate in 288 °C (550 °F) pure water (Ref 1) and interrupted after a given strain/time. (b) Percentage IGSCC versus fluence for cold-worked type 316 stainless steel (SS) tested at various temperatures. Despite the low potential environment of PWR primary water, at high fluence (especially at higher temperature) there is significant susceptibility to SCC. SSRT, slow strain rate test, IGSCC, intergranular stress-corrosion cracking

that the higher corrosion potentials measured in-core versus in the recirculation piping induced significantly higher measured crack growth rates (Fig. 10). Ex situ crack growth rate testing on irradiated (4 dpa) type 304 stainless steel (Fig. 11a), shows one of many examples of well-behaved crack growth rate at high corrosion potential, along with the strong effect of reduced corrosion potential. One evolving concern for high fluence stainless steels is the prospect of a major reduction in the effect of corrosion potential and stress-intensity factor, and Fig. 11(b) shows that this is indeed observed in unirradiated alloys possessing high silicon. The detrimental effect of silicon is likely associated with its ability to oxidize at all LWR-relevant potentials coupled with its relatively high solubility—that is, it is not a protective oxide like Cr_2O_3 or Fe/Ni oxides/spinels.

These data are compared with other irradiated and unirradiated data in Fig. 12 based on simultaneous measurements of corrosion potential and crack growth rate in fracture mechanics specimens; the accompanying curves represent model predictions (Ref 1, 5–7, 12, 13, 46, 51, 59, 66). Clearly the in situ data compare favorably with the spectrum of unirradiated data and data obtained on a fracture mechanics specimen of furnace-sensitized type 304 stainless steel using high-energy proton irradiation to simulate the mix of neutron and gamma radiation present in power reactors.

Radiation-Induced Segregation

Radiation-induced segregation is a common phenomenon in irradiated alloys and describes the redistribution of major alloying elements and the enrichment or depletion of impurity elements at point defect sinks (Ref 67–76). By virtue of their susceptibility to IGSCC, the sinks of greatest interest are the grain boundaries. Radiation-induced segregation is driven by the flux of radiation-produced defects to sinks and is therefore fundamentally different from thermal segregation or elemental depletion due to grain-boundary precipitation processes. Vacancies and interstitials are the basic defects produced by irradiation and can reach concentrations that are orders of magnitude greater than the thermal equilibrium concentrations. Diffusion of solutes by vacancy or interstitial mechanisms (inverse Kirkendall) is accelerated by the elevated concentration of these defects. If the relative participation of alloying elements in the defect fluxes is not the same as their relative concentration in the alloy, then a net transport of the constituents to or from the grain boundary will occur (Fig. 13). This unequal participation of solutes in the vacancy and/or interstitial fluxes to sinks

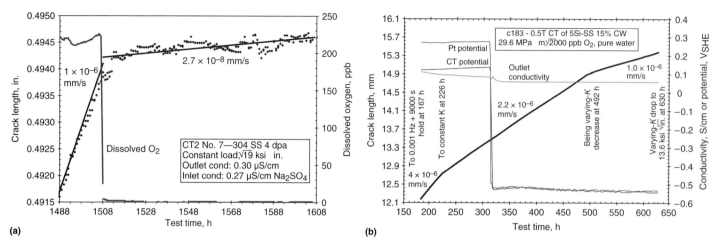

Fig. 11 (a) Crack length versus time for type 304 stainless steel irradiated to 4 dpa in a boiling water reactor showing the elevated crack growth rates at high corrosion potential, but significant decrease in growth rate as the corrosion potential is decreased. These data are plotted as large triangles on Fig. 12. (b) Crack length versus time for a custom "stainless steel" representative of the composition of an irradiated grain boundary, notably containing high silicon. There is no effect of the change in corrosion potential, nor an effect from decreasing stress intensity (K) from 29.6 to 14.3 MPa$\sqrt{m}$ (27 to 13 ksi$\sqrt{in.}$). While the crack growth rate response of such a bulk alloy cannot be presumed to be representative of an irradiated material, these data and other data on type 304L stainless steel possessing 3 or 1.5% Si show broadly similar behavior.

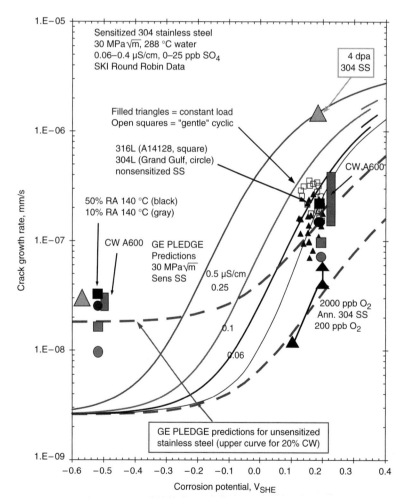

Fig. 12 Observed and predicted relationships of crack growth rate versus corrosion potential for furnace-sensitized type 304 stainless steel (SS) at a constant stress intensity (K) of ≈ 27.5 MPa$\sqrt{m}$ (25 ksi$\sqrt{in.}$). The observed data were obtained in water of conductivity between 0.1 to 0.3 μS/cm. The predicted relationships show the sensitivity of the crack growth rate to changes in combinations of corrosion potential and water purity (0.1 to 0.5 μS/cm) (Ref 5–7, 46, 47). The large triangles were obtained on type 304 stainless steel irradiated to 4 dpa. The larger circles and rectangles represent cold-worked stainless steel and Alloy 600.

results in either enrichment or depletion of an alloying element at the grain boundary. The species that diffuse more slowly by the vacancy diffusion mechanism are enriched, and the faster diffusers become depleted. Enrichment and depletion can also occur by association of the solute with the interstitial flux. The undersized species will enrich, and the oversized species will deplete (Ref 69). The magnitude of the buildup/depletion is dependent on several factors such as whether a constituent migrates more rapidly by one defect mechanism or another, the binding energy between solutes and defects, the dose, dose rate, and the temperature. The RIS profiles are also characterized by their narrowness, often confined to within 5 to 10 nm of the grain boundary, as shown in Fig. 14 for an irradiated stainless steel.

Segregation is a strong function of irradiation temperature, dose, and dose rate. Segregation peaks at intermediate temperatures since a lack of mobility suppresses the process at low temperatures, and back-diffusion of segregants minimize segregation at high temperature (where defect concentrations approach their thermal equilibrium values). For a given dose, lower dose rates result in greater amounts of segregation (at LWR temperatures). At high dose rates, the high defect population results in increased recombination that reduces the number of defects that are available to diffuse to the grain boundary. Figure 15 shows the interplay between temperature and dose rate for an austenitic stainless steel. Radiation-induced segregation occurs in the intermediate temperature range, and this range rises along the temperature scale with increasing dose rate to compensate for the higher recombination rate.

In the Fe-Cr-Ni alloy system, studies have shown that nickel segregates to grain boundaries while chromium and iron deplete. The directions of segregation are consistent with an atomic

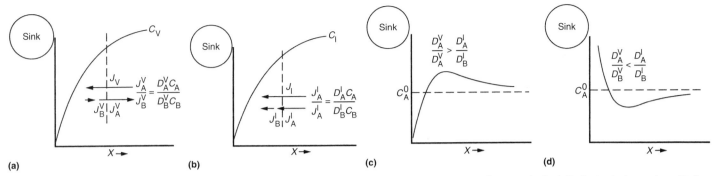

Fig. 13 Inverse Kirkendall effects induced by (a) vacancy flux, (b) interstitial flux, and (c) and (d) the effect of diffusion coefficients on the depth distribution for A atoms in an AB alloy. J, C, and D refer to the flux, concentration, and diffusion coefficient, respectively, of diffusing species defined by the subscript, by way of the complementary species defined by the superscript (V, vacancies, I, interstitials). Source: Ref 68

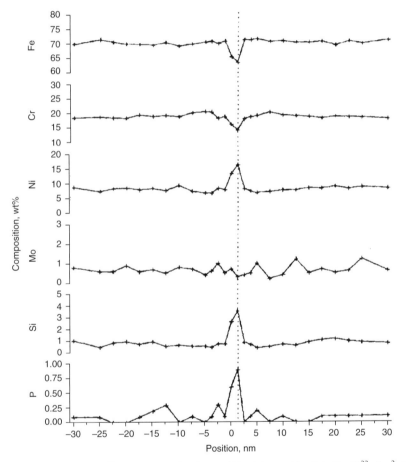

Fig. 14 Typical segregation profiles in commercial purity type 304 stainless steel irradiated to ~10^{22} n/cm^2 at 275 °C (525 °F). Composition profiles were measured using a field emission gun scanning transmission electron microscope (FEGSTEM). Source: Ref 72

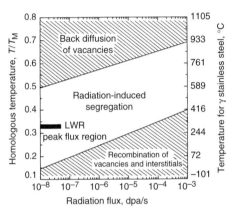

Fig. 15 Dependence of radiation-induced segregation on homologous temperature and dose rate for austenitic stainless steels. LWR, light water reactor

volume effect in which the subsized solute migrates preferentially with the interstitial flux, and the oversized solute participates preferentially in the vacancy flux. The results are also consistent with the diffusivity of the solutes in Fe-Cr-Ni, in which nickel is the slow diffuser, chromium is the fast diffuser, and iron is intermediate. While the inverse Kirkendall mechanism involves preferential association of solutes with both vacancy and interstitials, the observed major element segregation in Fe-Cr-Ni alloys can be qualitatively and quantitatively accounted for by only vacancy exchange, indicating that for this system, preferential participation of solutes by interstitials is not required (Ref 73, 74).

While chromium depletes at grain boundaries and nickel enriches across a wide range of composition in the austenitic Fe-Cr-Ni alloy system, iron can either deplete or enrich according to the magnitude of the diffusion coefficient relative to the other solutes (Ref 77). Allen et al. (Ref 73) showed that iron segregation compensates for the differences in relative diffusivities and abundances of chromium and nickel. Iron is observed to deplete in most austenitic stainless steel alloys and nickel-base alloys, but it can enrich if the ratio of bulk nickel to bulk chromium is very low. Radiation-induced segregation increases with neutron dose in LWRs and saturates after several (~5) dpa in the 300 °C (570 °F) temperature range. Figure 16 shows grain-boundary chromium depletion for austenitic stainless steels as a function of dose. As the slowest diffusing element, nickel becomes enriched at the grain boundary. Since iron depletes in 304 and 316 stainless steels, the nickel enrichment makes up for both chromium and iron depletion and can reach very high levels, up to ~30 wt%. For all of the major elements, the segregation profile is very narrow, generally between 5 and 10 nm full width at half maximum (FWHM). The narrow profile and limitations of the analysis technique makes exact measurements of the grain-boundary values difficult and knowledge of the true grain-boundary compositions is aided by the use of segregation models (Ref 86, 87).

Heat treatment of stainless steels prior to irradiation will often result in an enrichment of chromium at the grain boundary. Subsequent irradiation eventually leads to grain-boundary chromium depletion, but an intermediate step involves the formation of a W-shaped profile

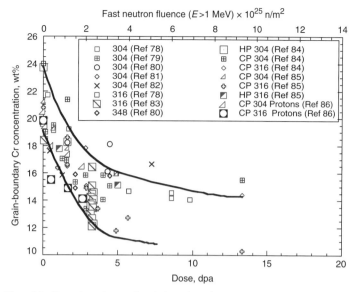

Fig. 16 Dose dependence of grain-boundary chromium concentration for several 300-series austenitic stainless steels irradiated at a temperature of about 300 °C (570 °F). Source: Ref 78–86

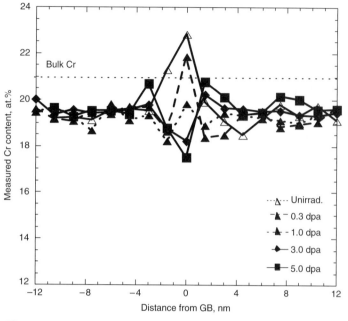

Fig. 17 Variation of the grain-boundary (GB) chromium concentration profile in commercial-purity type 304 stainless steel with dose for 3.2 MeV proton irradiation at 360 °C (680 °F). Source: Ref 86

at low dose. The W-shape is due in part to the initial chromium enrichment. However, inverse Kirkendall-based calculations of grain-boundary composition under irradiation show that the W-shape should transform to a V-shaped depletion profile for doses beyond 0.1 dpa, when in fact, observations show (Ref 86) that it tends to persist beyond 1.0 dpa (Fig. 17).

The source of the tenacity of the W-shape profile may be cosegregation of solutes such as molybdenum or impurities such as carbon or boron. Enrichment of boron, carbon, and phosphorus at grain boundaries in 304 stainless steel has been measured in the unirradiated case (Ref 88). There was a pronounced affinity between molybdenum and nitrogen noted as well, leaving open the possibility that interstitial impurities may play a role in the persistence of the W-shaped profile under irradiation.

Minor alloying elements and impurities also segregate and have been implicated in the IASCC process as well. Manganese and molybdenum strongly deplete at the grain boundary under irradiation (Ref 85). Manganese is a minor alloying element in 304 and 316 stainless steels. As a fast diffuser, it depletes rapidly with dose in preference to chromium. Molybdenum is added to 316 stainless steel to improve pitting resistance, and as an oversized element it also depletes rapidly at the grain boundary. Other minor alloying or impurity elements such as silicon and phosphorus also segregate under irradiation. Silicon strongly enriches at the grain boundary to as much as 10 times the bulk (0.7 to 2.0 at.%) composition in the alloy (Ref 89). However, the γ' phase has not been observed at the grain boundary even in cases where the silicon concentration is well above the solubility limit. Phosphorus is present at much lower concentrations and is only modestly enriched at the grain boundary due to irradiation (Ref 72, 85). Phosphorus tends to segregate to the grain boundary following thermal treatment, which reduces the amount of additional segregation to the grain boundary during irradiation, making the contribution due to irradiation difficult to detect (Ref 85). Undersized solutes such as carbon, boron, and nitrogen should also segregate, but there is little evidence of RIS, due in part to the difficulty of measurement. Another potential segregant is helium, produced by the transmutation of ^{10}B. The mobility of helium is low at LWR core temperatures, but the opportunity for accumulation at the grain boundary is increased by segregation of boron to the boundary. Overall, the behavior of these minor elements under irradiation is not well understood.

Significant work has occurred recently on the effect of oversize solutes on IASCC. These solutes may affect the microchemistry or microstructure of the alloy, thereby altering the IASCC susceptibility. Oversize solutes are believed to affect RIS by acting as vacancy traps, thereby increasing the effective recombination of vacancies and interstitials and thus reducing RIS. Kato et al. (Ref 90) conducted electron irradiations of several stainless steels at temperatures of 400 to 500 °C (750 to 930 °F) up to 10 dpa. Among the solutes added to the reference alloy were titanium, niobium, vanadium, zirconium, tantalum, and hafnium. Results showed that some solutes (zirconium and hafnium) consistently produced a large suppression of radiation-induced chromium depletion, while others resulted in less suppression or suppression at only certain temperatures (Fig. 18). Fournier et al. (Ref 91) conducted irradiation of type 316 stainless steel containing hafnium or platinum using 3 MeV protons (400 °C, or 750 °F) or 5 MeV nickel ions (500 °C, or 930 °F). Nickel irradiations showed little effect of the oversize impurity in reducing grain-boundary chromium depletion (chromium depletion increased in the case of hafnium), but proton irradiation showed a significant suppression of RIS of chromium at low dose (2.5 dpa) with the effect diminishing at higher (5.0 dpa) dose. Platinum had a smaller effect on chromium. Titanium and niobium similarly produced little change in the grain-boundary chromium concentration after irradiation with 3.2 MeV protons to 5.5 dpa at 360 °C (680 °F). In zirconium-doped type 304 stainless steel, there were no consistent results of suppression of grain-boundary chromium after 3.2 MeV proton irradiation to 1.0 dpa at 400 °C (750 °F) (Ref 93). In all, the data on the effect of oversize solutes on RIS of chromium are very inconsistent.

Radiation-induced segregation is often implicated in IASCC of stainless steels, especially in oxidizing environments, due to the wealth of data from laboratory and plant operational experience with sensitized components (Ref 3, 4, 12, 31, 34, 94, 95). As shown in Fig. 16, grain-boundary chromium depletion during irradiation can be severe. Figure 19(a) shows a correlation between grain-boundary chromium level and IGSCC susceptibility in stainless steels where the grain-boundary depletion is due to sensitization (Ref 96). Much data have been accumulated to support the role of chromium depletion as an agent in IGSCC of austenitic alloys in oxidizing conditions. Numerous studies show that as the grain-boundary chromium level decreases, IGSCC increases. Typical chromium-depleted zone widths are of the order of 100 to 300 nm

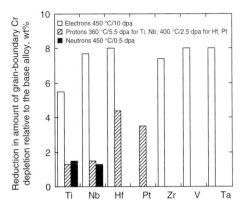

Fig. 18 Effect of oversize solute on the grain-boundary chromium concentration following irradiation with electrons (Ref 90), protons (Ref 91), and neutrons (Ref 92)

FWHM, providing a significant volume of depleted material adjacent to the grain boundary.

Figure 19(b) shows a similar correlation between grain-boundary chromium level and IASCC susceptibility as measured by the percentage of intergranular cracking (%IG) cracking on the fracture surface during SSRT experiments. A major difference between chromium-depletion profiles resulting from RIS and those due to precipitation reactions is that the width of the RIS profiles can be as much as two orders of magnitude smaller, or 5 to 10 nm. There is also a tremendous amount of scatter in the data of Fig. 19(b) that makes a direct correlation difficult to support. In particular, at low grain-boundary chromium levels, the %IG ranges from 0 to 100. However, as the grain-boundary chromium content increases, the data become less "noisy," suggesting that a higher grain-boundary chromium concentration is a strong indicator of resistance to IASCC. Also, there is a trend in the maximum level of %IG with grain-boundary chromium content. Therefore, one interpretation of these data is that grain-boundary chromium depletion is an *important*, but *insufficient* condition for IASCC to occur.

Additional data to support the idea that factors other than grain-boundary chromium depletion are important in IASCC come from the growing number of experiments on postirradiation annealing. Busby et al. (Ref 98) showed that the different rates at which RIS and dislocation loops anneal during a postirradiation heat treatment creates a window in time-temperature space in which annealing of the dislocation loop structure can occur to a great extent with little change in the grain-boundary composition profile. Several studies have been conducted on the postirradiation annealing of irradiated stainless steels and the assessment of the IASCC susceptibility as a function of annealing time and temperature (Ref 99–102) and all show that the grain-boundary chromium concentration remains at the as-irradiated level well beyond the point where the IASCC susceptibility has been substantially reduced. This occurrence is another indication that while RIS may be an important factor in IASCC, it does not alone explain IASCC.

Recent experiments by Allen et al. (Ref 102) and Was et al. (Ref 103) on stainless steels with chromium presegregation provide additional perspective on the role of chromium. In these experiments, several alloys were given a chromium-enrichment heat treatment of 1100 °C (2010 °F) for 20 min followed by air cooling. Grain-boundary chromium enrichment is believed to occur by a thermal, nonequilibrium segregation process (Ref 104). Results of subsequent irradiations showed that despite higher measured grain-boundary chromium concentrations, IASCC susceptibility was no better, and often worse, than when starting without chromium preenrichment (Fig. 20). These data cast further doubt on the sufficiency of grain-boundary chromium depletion as the key factor in IASCC even in oxidizing media.

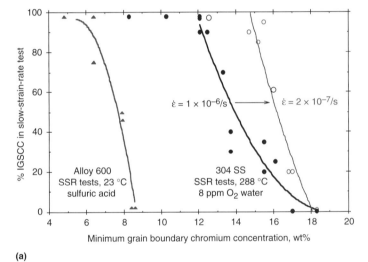

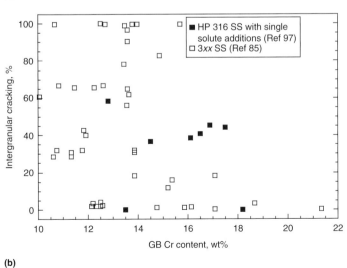

Fig. 19 Effect of grain-boundary (GB) chromium content on intergranular stress-corrosion cracking (IGSCC) for (a) sensitized stainless steel (SS) and alloy 600 (Source Ref 96) and (b) irradiated stainless steels. SSR, slow strain rate.

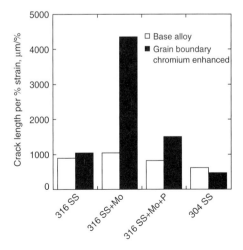

Fig. 20 Effect of chromium enrichment heat treatment on irradiation-assisted stress-corrosion cracking in constant extension rate test in simulated BWR normal water chemistry at 288 °C (550 °F). Source: Ref 102, 103

Under irradiation, both molybdenum and manganese deplete at the grain boundary, but neither is believed to be a significant factor in IASCC. Among the minor alloy elements, only silicon is observed to segregate to high levels and silicon segregation is correlated with IASCC. Experiments by Busby et al. (Ref 97) on a high-purity 316 base alloy doped with 1 wt% Si showed severe IASCC in normal water chemistry and in primary water after proton irradiation to 5.5 dpa at 360 °C (680 °F). Scanning transmission electron microscopy (STEM) measurements of grain-boundary silicon confirm levels up to 6 wt%. Past studies comparing Augel electron spectroscopy (AES) and STEM results have shown that the actual concentration of silicon at the grain-boundary plane may be as high as 15 to 20 wt%. Though the electron beam probe in STEM is very small, the measurement underpredicts the concentration at the grain boundary by as much as a factor of 3. Yonezawa et al. (Ref 105–108) and Li et al. (Ref 109) have provided extensive evidence to show that increased silicon in stainless steel results in increased IGSCC in alloys tailored to imitate the composition of grain boundaries under irradiation. However, it should be noted that these alloys were not irradiated, and this difference may be important in the relevance of such experiments to IASCC. Using 1.5 to 5% Si stainless steels of both standard (e.g., 304L) base composition and synthetic irradiated grain-boundary composition, Andresen has observed (Fig. 11b) significantly increased growth rates, a reduced benefit of a lower corrosion potential, and very little effect of stress-intensity factor between 30 and 14 MPa$\sqrt{m}$ (27 and 13 ksi$\sqrt{in.}$).

The data on impurity segregation effects on IASCC remain inconclusive. Extensive experiments have been conducted to isolate the effect of particular impurities such as silicon, phosphorus, carbon, nitrogen, and boron in IASCC, but none have yielded unambiguous results. Sulfur has not been found to segregate under irradiation, and while phosphorus thermally segregates to a significant extent, irradiation-induced phosphorus segregation is small in comparison. Carbon, nitrogen, and boron cannot be measured in STEM, nitrogen and boron are very difficult to identify in AES, and carbon is a common contaminant. Overall, it has been a challenge to establish a link between impurity-element segregation and IASCC in austenitic stainless steels.

Microstructure, Radiation Hardening, and Deformation

Irradiated Microstructure. The microstructure of austenitic stainless steels under irradiation changes rapidly at LWR service temperatures. Point defect clusters ("black dot damage") form at very low dose, dislocation loops and the network dislocation density evolve with dose over several dpa, and the possibility exists for the formation and growth of helium-filled bubbles, voids, and precipitates in core components in locations exposed to higher dose and temperatures. Typical radiation-induced microstructural features in austenitic stainless steels are dislocation loops, network dislocations, cavities (bubbles and/or voids), and precipitates (Ref 110–116). The microstructure is quite sensitive to temperature with a transition occurring in the 300 °C (570 °F), range. Below 300 °C (570 °F), the microstructure is dominated by small clusters and dislocation loops. Near 300 °C (570 °F), the microstructure contains larger faulted loops plus network dislocations from loop unfaulting and cavities at higher doses. Figure 21 summarizes the types of defect structures reported in 300-series stainless steels as a function of irradiation temperature and dose of relevance to the LWR regime.

Under LWR conditions, the dominant defect structures by far are vacancy and interstitial clusters and Frank dislocation loops. The defect clusters may be of vacancy or interstitial type and are formed as a result of the collapse of a damage cascade during irradiation. The larger, faulted dislocation loops nucleate and grow as a result of the high mobility of interstitials. The loop population grows in size and number density until absorption of vacancies and interstitials equalize, at which point the population has saturated. Figure 22 shows the evolution of loop density and loop size as a function of irradiation dose during LWR irradiation at 280 °C (535 °F). Note that saturation of loop number density occurs very quickly, by ~1 dpa, while loop size continues to evolve up to ~5 dpa. The specific number density and size are dependent on irradiation conditions and alloying elements, but the loop size rarely exceeds 20 nm and densities are of the order of $1 \times 10^{23} \cdot m^{-3}$.

Small defects may be important in the hardening process and in how they affect IASCC. It is generally believed that the small defect clusters are predominantly faulted interstitial loops and vacancy clusters (Ref 117). However, analysis of recent postirradiation annealing experiments by Busby et al. (Ref 118) and Simonen et al. (Ref 119) showed that this description may be inaccurate. Evidence suggests that there are at least two types of defects with different annealing characteristics: vacancy and interstitial faulted loops. Hardness as a function of annealing time shows a stepped behavior that is likely caused by different annealing kinetics of vacancy and interstitial loops. Annealing experiments indicate that the density of vacancy loops is perhaps much higher than previously believed, and higher than the density of interstitial loops (Ref 118).

At temperatures above 300 °C (570 °F), voids and bubbles may begin to form, aided by the increased mobility at the higher temperature. The dislocation structure will evolve into a network structure as larger Frank loops unfault. The reduction in the sink strength of the dislocation loops aids in the growth of voids and bubbles. While their size and number density increase with temperature, the dislocation microstructure continues to be the dominant microstructure component over the temperature range expected for LWR components (<350 °C, or 660 °F).

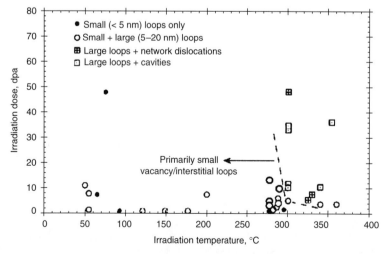

Fig. 21 Summary of reported defect structures in 300-series stainless steels as a function of irradiation dose and temperature. Source: Ref 85

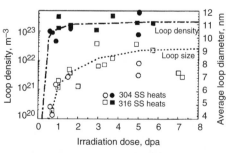

Fig. 22 Measured change in density and size of interstitial loops as a function of dose during LWR irradiation of 300-series stainless steels (SS) at 275 to 290 °C (525 to 555 °F). Source: Ref 85

Irradiation can also accelerate or retard the growth of second phases, modify existing phases, or produce new phases. A key factor in phase formation in austenitic stainless steels under LWR operating conditions is RIS. The RIS of both major and minor elements to sinks under irradiation can induce the formation of phases by exceeding the solubility limit locally. Was et al. (Ref 120) irradiated a high-purity stainless steel containing 1 wt% Si with 3.2 MeV protons to 5.5 dpa at 360 °C (680 °F). They observed the formation of γ' (Ni_3Si) in the matrix but not on the grain boundary. The same was observed in a similar alloy irradiated with neutrons to 7 dpa at 300 °C (570 °F) (Ref 72). Gamma prime is a coherent precipitate that can significantly strengthen the matrix and has the potential to alter the deformation behavior in the unirradiated and irradiated conditions. The occurrence of γ' in the matrix and not on the grain boundary is puzzling since the grain-boundary concentration of silicon is likely greater than 10 or 15% at.% in the boundary plane and the nickel concentration is also enriched. However, most of the effects of irradiation on phase formation and stability do not occur below 370 °C (700 °F). In stainless steels, the principal second phase is chromium carbides, which are stable under irradiation. In high-strength nickel-base alloys, the second phases can undergo several types of transformations. Gamma prime can dissolve, γ'' can dissolve and re-precipitate, and Laves phase can become amorphous under irradiation at LWR operating temperatures. Hence, for these alloys, moderate temperature irradiation can induce important phase changes.

Oversize solutes can also affect the irradiated microstructure by mechanisms similar to those that affect grain-boundary segregation. In both proton and nickel ion irradiations, the addition of hafnium to a type 316 stainless steel base alloy increased loop density, decreased loop size, and eliminated voids (Ref 91). Platinum addition to type 316 resulted in no change in loop density and a small increase in loop size, but increased void size and density. The good agreement between proton and nickel ion irradiation results indicates that the major effect of the oversized solute is not due to the cascade (where there are large differences between proton and nickel ion irradiation), but rather is due to the postcascade defect partitioning in the microstructure evolution. Electron irradiation experiments by Watanabe et al. (Ref 121) and proton irradiation experiments by Was et al. (Ref 120) showed that stainless steel with titanium additions had slightly lower dislocation loop densities and larger sizes compared to the base alloy. Niobium increased loop size only. In contrast to the base alloy, neither the titanium- nor the niobium-doped alloys formed voids under the conditions tested. Zirconium addition to type 304 stainless steel resulted in reduced hardness, decreased loop density, and no change in loop size in proton irradiation to 1.0 dpa at 400 °C (750 °F) when compared to the base alloy (Ref 93). Zirconium-containing samples also had a lower void density with no change in void size as compared to the base alloy.

Irradiation Hardening. A consequence of the formation of the dislocation loop microstructure is hardening of the alloy. The dislocation loops interact elastically with the dislocation network under an applied stress, producing an increase in the yield strength of the alloy. The increase in yield strength is measurable in either a tensile test or by indentation hardness measurement. Accompanying an increase in hardness is a decrease in the ductility and fracture toughness. The yield strength increase with dose in 300-series stainless steels irradiated around 300 °C (570 °F) is plotted in Fig. 23. Note that the yield strength can reach values up to five times the unirradiated value by about 5 dpa. The increase in yield strength follows a square root dependence on dose. Various models have attempted to link the hardening behavior with the dislocation microstructure. Both the source hardening model (Ref 129) and the dispersed barrier hardening model (Ref 130) provide reasonable correlations between hardening and the dislocation loop microstructure. In the dispersed barrier hardening model, the increase in hardness is proportional to $(N_{loop} \times d_{loop})^{1/2}$, where N_{loop} is the loop number density and d_{loop} is the loop diameter.

In addition to hardening with increasing dose, the deformation mode changes dramatically. Homogeneous deformation at low dose is replaced by heterogeneous deformation at higher doses as the defect microstructure begins to impede the motion of dislocations. Plasticity becomes localized to narrow channels that have been cleared of defects by preceding dislocations, providing a path for subsequent dislocation motion. The channels are very narrow (<10 nm) and closely spaced (<1 μm) and typically run the full length of a grain, terminating at the grain boundaries (Ref 131). Dislocation channeling results in intense shear bands that can cause localized necking and a sharp reduction in uniform elongation (Ref 132). As discussed later, these dislocation channels may be an important factor in IASCC.

Hardening has also been cited as a key factor in IASCC susceptibility. Figures 12 and 24(a) show that increasing cold work results in increases in the crack growth rate in nonsensitized austenitic stainless steel tested in 288 °C (550 °F) BWR normal water chemistry (Ref 133). Increasing yield strength resulting from cold work is a fairly good predictor of cracking susceptibility. This dependence also carries over to irradiated material. As shown in Fig. 24(b), increased yield strength due to irradiation results in increased susceptibility to IASCC as measured by the %IG in SSRTs. Note that the correlation is quite good at both high values of yield strength (>800 MPa, or 116 ksi) and at very low values (<400 MPa, or 58 ksi), with more scatter at intermediate values (400 to 800 MPa, or 58 to 116 ksi). The parallel between the effects of cold work and irradiation on IASCC suggests that hardening, irrespective of its nature, is a key factor in IGSCC susceptibility.

Several pieces of evidence suggest that hardening alone is not sufficient to explain IASCC. The annealing experiments shown in Fig. 25 also tracked the change in hardness and dislocation loop microstructure versus annealing condition (Ref 134). Figure 25 shows the change in hardness and the change in the dislocation loop line length

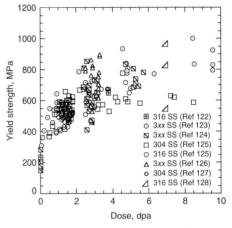

Fig. 23 Irradiation dose effects on measured tensile yield strength for several 300-series stainless steels, irradiated and tested at a temperature of about 300 °C (570 °F). Source: Ref 122–128

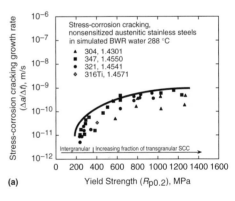

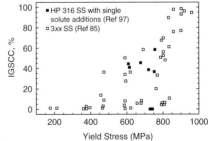

Fig. 24 Effect of yield strength on intergranular stress-corrosion cracking (IGSCC). (a) Crack growth rate of cold-worked, unirradiated 300-series stainless steels (SS) tested in 288 °C (550 °F) simulated BWR water. Source: Ref 131. (b) IGSCC percentage in slow strain rate tests on 300-series stainless steels where hardening is by irradiation. Source: 85, 97, 120

($N_{avg} \times d_{avg}$) along with the change in IASCC susceptibility as annealing progresses, as measured by several authors (Ref 101, 116, 135). Note that except for very short annealing times, the loop line length and the hardness track each other extremely well, as would be expected if the loop structure is controlling the hardening. At very small values of $(Dt)^{1/2}$, the hardening remains flat before proceeding downward with increased annealing. While both hardening and cracking are reduced with increased annealing, the behavior of the hardness cannot fully explain the rapid and complete reversal in IASCC susceptibility with annealing condition. In particular, the cracking susceptibility changes dramatically before the hardness begins to drop from its as-irradiated value, as indicated by the plateau at low $(Dt)^{1/2}$.

Busby et al. (Ref 118) have postulated that the removal of very small defects at short annealing times may be responsible for the change in IASCC behavior. Short times at high temperature (500 °C, or 930 °F) may preferentially remove the small defect clusters either by annihilation or by spontaneous dissociation. The dislocation loops may absorb the free vacancies and interstitials, thus adding to their line length. The loss of the small defect clusters will be offset by the growth of Frank loops producing no net change in measured hardness or yield strength. Despite a lack of hardness change, this process may alter the deformation mode at the local level by removing the small obstacles to dislocation motion, thus changing the character of localized deformation that may affect IASCC.

A set of experiments conducted by Hash et al. (Ref 136) provide supporting evidence that hardening is not the sole factor in IASCC. Hash made a series of samples of commercial purity type 304 stainless steel with nominally the same hardness, but with different contributions from cold work and irradiation. At the extremes were a sample that was cold rolled to a 35% reduction in thickness and no irradiation and one with 1.67 dpa irradiation and no cold work. Three samples had varying amounts of cold work (10%, 20%, and 25%) and corresponding amounts of irradiation dose (0.55, 0.25, and 0.09 dpa) to result in a nominally constant hardness level that varied by no more than 5% over the five samples. Stress-corrosion cracking susceptibility was measured by the amount of IG cracking in a constant extension rate test in 288 °C (550 °F) simulated BWR water with normal water chemistry. Results showed that the IASCC susceptibility was not constant, as would be expected if hardness were the only factor. Rather cracking occurred for only the two highest dose samples (0.55 dpa with 10% cold worked and 1.67 dpa with no cold worked), irrespective of their hardness (Fig. 26). The amount of cracking in the lower dose sample was higher than that in a companion sample at the same dose but without cold work (not shown), indicating that cold work can enhance the IASCC susceptibility. These results also suggest that hardening by irradiation may promote crack initiation. Nevertheless, this result along with the annealing results suggests that the microstructure itself and not just the hardness level play a role in the IASCC process.

Deformation Mode. Results of IASCC experiments on proton-irradiated samples over a wide range of doses and alloys have consistently shown that high-nickel alloys have high IASCC resistance in SSRT. In particular, nickel concentrations ≥18 wt% are highly resistant to IASCC compared to type 304 stainless steel with 8 wt% Ni. Figure 27 shows existing literature data on IASCC versus nickel equivalent (Ni_{Eq}) as determined from correlations by Kodama et al. (Ref 140). It shows that there is a good, but not perfect correlation between Ni_{Eq} and IASCC. Notable exceptions are alloy 800 (Ref 144) and a Fe-20Cr-25Ni-1Nb alloy used in advanced gas-cooled reactors (AGRs) that experienced IGSCC (Ref 145). While nickel may affect IASCC directly through a change in composition, it may also affect IASCC indirectly through a change in the slip character. Higher nickel content in stainless steel increases the stacking fault energy (SFE) significantly. The increased SFE results in a change in the nature of slip from planar to wavy. Swan et al. (Ref 146) conducted a study on the effect of SFE on slip behavior of a series of Fe-18Cr-xNi alloys where $8 < x < 23$. They showed that for the 8% Ni alloy, the slip was entirely planar and as nickel increases, cross slip increases. By 20% Ni, there was no evidence of planar slip and the deformation microstructure consisted of a web of dislocation tangles, evident of wavy slip in a high SFE material. The process is shown in the micrographs and schematics in Fig. 28 in which planar slip is likely to result in greater dislocation interaction with grain boundaries than is wavy slip.

Stacking fault energy has been linked to SCC resistance in stainless steels by Thompson and Bernstein who found that increasing SFE correlates well with increased reduction in area and decreased SCC susceptibility (Ref 147). The IASCC susceptibility of several irradiated stainless steels is plotted as a function of SFE in Fig. 29 and shows that there is a good correlation between SFE and IASCC using Rhodes's (Ref 148) and Schramm's (Ref 149) correlations for SFE. These correlations differ in the elements included and the weights given to each element in terms of Ni_{Eq}. The correlations are also unable to account for several of the minor elements and therefore may deviate substantially from the true SFE in cases where minor elements carry high weight (such as silicon). Note that there is much scatter in both these plots and the plot for Ni_{Eq}

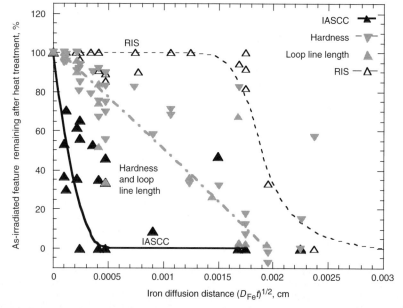

Fig. 25 Removal of radiation-induced segregation (RIS), dislocation microstructure as measured by loop line length and hardness with extent of annealing as measured by $(Dt)^{1/2}$ for iron, to account for annealing at different times and temperatures. IASCC, irradiation-assisted stress-corrosion cracking. Source: Ref 98, 101, 116, 135

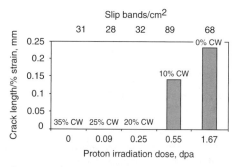

Fig. 26 Degree of irradiation-assisted stress-corrosion cracking in type 304 stainless steel samples with the same hardness but with different combinations of hardening by cold work (CW) and irradiation using 3.2 MeV protons at 360 °C (680 °F). Source: Ref 136

and of a similar degree to that shown in the IGSCC dependence on grain-boundary chromium and yield strength. One potential source of the scatter is the inherent variability in initiation-dominated phenomena. Another is that while grain-boundary chromium and yield strength are measured, Ni_{Eq} and SFE are calculated quantities with considerable uncertainty. So while they may be useful in identifying transitions in behavior, their reliability in quantifying alloy behavior may be much less.

The significance of the SFE is that it controls the deformation mode of the alloy. However, the microstructure can also influence the deformation mode. Farrell et al. (Ref 150) conducted neutron irradiation of type 316 stainless steel at temperatures between 65 and 100 °C (150 and 212 °F) to doses of less than 1 dpa and characterized the deformation behavior as a function of irradiation condition. They found that with increasing dose, the propensity for dislocation channeling increased. A plot of the evolution of the dislocation microstructure and the extent of channeling as determined by the channel area is shown in Fig. 30. Note that the volume of material occupied by channels increases rapidly with dose and saturates at less than 0.5 dpa. These data are important evidence because they show that, in addition to SFE, the defect microstructure created by irradiation can induce planar deformation in the form of narrow dislocation channels. Slip planarity is the common feature between the effect of SFE and irradiation.

The importance of slip localization in IASCC may be in the way in which the dislocations interact with the grain boundary. In planar slip, well-defined and separated slip bands or dislocation channels (for irradiated materials) transmit dislocations during plastic deformation. These slip bands or channels terminate at grain boundaries where dislocations are fed into the grain-boundary region versus the buildup of a tangled dislocation network within the grain. One of two processes could occur to cause grain-boundary cracking. In the more traditional view of dislocation infusion into grain boundaries, the pileup of dislocations in the intersecting channel creates progressively higher stresses at the grain boundary at the head of the pileup. If the stress exceeds a critical value, separation of the grain boundary could occur according to a Stroh (wedge) cracking type of mechanism. This cracking process could occur regardless of the environment and may in fact be the mechanism that occurs in some of the IG cracking observed in very highly irradiated steels in inert environment (Ref 132). At lower fluences, the stress at the grain boundary may cause a rupture of the oxide film, leading to exposure of the metal to the solution and subsequent corrosion and IASCC.

Alternatively, IASCC could occur by deformation in the boundary itself, which would also rupture the oxide film, leading to IASCC. Recent work by Alexandreanu (Ref 151) on unirradiated nickel-base alloys has shown that dislocation absorption by the grain boundaries leads to deformation in the grain boundary, usually by boundary sliding, resulting in IGSCC. He investigated grain boundaries classified according to the misorientation of the two grains and was able to determine that the more random the grain boundary, the greater was the sliding under high-temperature deformation. Boundaries on which sliding occurred were four times more susceptible to subsequent IGSCC in primary water at 360 °C (680 °F) than boundaries that did not slide. In this way, he was able to establish a cause-and-effect between grain-boundary deformation and IGSCC. This same process could explain how dislocation injection into grain boundaries by planar slip or dislocation channeling can result in IASCC. In fact, Dropek and Was et al.

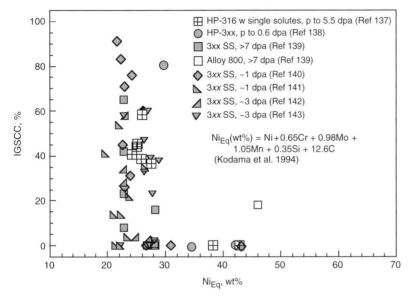

Fig. 27 Intergranular stress-corrosion cracking susceptibility as measured by percentage of intergranular cracking in slow strain rate tests as a function of nickel equivalent (Ni_{Eq}) determined using data from Kodama et al. (Ref 140)

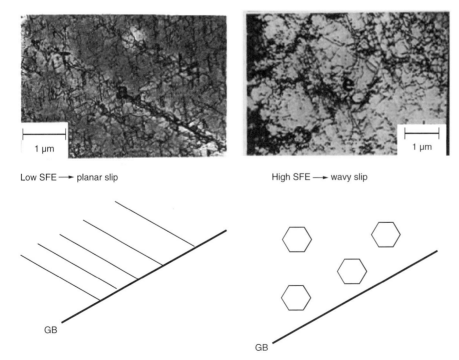

Fig. 28 Micrographs and corresponding illustrations of planar and wavy slip in the vicinity of a grain boundary (GB). SFE, stacking fault energy. Micrographs from Ref 146

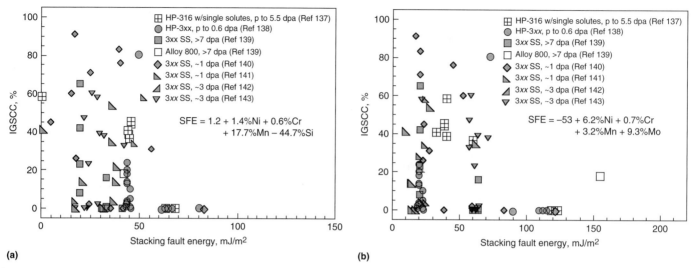

Fig. 29 Irradiation-assisted stress-corrosion cracking susceptibility as measured by percentage of intergranular stress-corrosion cracking (IGSCC) as a function of stacking fault energy determined using (a) Rhodes' correlation (Ref 148) and (b) Schramm's correlation (Ref 149)

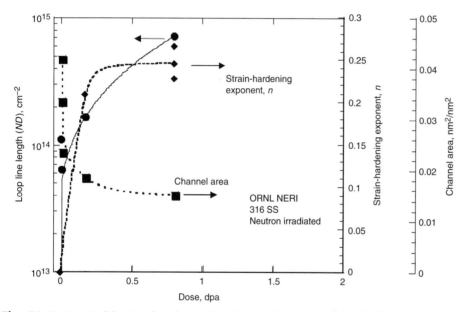

Fig. 30 Variation in dislocation channel area, dislocation loop line length, and strain-hardening exponent as a function of dose for neutron-irradiated type 316 stainless steel (SS). Source: Ref 150

(Ref 152) have reported a beneficial effect of grain-boundary engineering on the IASCC behavior of proton-irradiated, controlled purity type 304 alloys after testing in simulated BWR normal water chemistry at 288 °C (550 °F).

The concept that SFE controls the deformation mode, which then increases IASCC, has support from observations of the relative propensity for IASCC of type 304 and 316 stainless steels. Type 316 has, in some cases (Ref 72), been shown to be more resistant to IASCC than is type 304, which may be explained by its higher SFE. Furthermore, observations by Bailat et al. (Ref 153) on neutron-irradiated samples and by Busby et al. (Ref 118) on proton-irradiated samples both show clear dislocation channel patterns on the surface of type 304 samples that cracked following SSRT, but the absence of those patterns on type 316 samples that did not crack. In addition, SSRT on high-nickel alloys (with SFE around 40 mJ/m^2) showed no IG cracking and also no dislocation channeling pattern. Further support for an SFE-based IASCC mechanism comes from the annealing studies cited earlier. On samples that underwent either a short, low-temperature anneal or were tested in the as-irradiated conditions, the IG cracking was accompanied by a high density of dislocation slip bands on the surface of about 65/mm^2 (42×10^3/in.^{2}). Annealing at 500 °C (930 °F) for 45 min resulted in the elimination of IG cracking and also resulted in a reduction in the surface slip band density to ~25/mm^2 (~16×10^3/in.^{2}).

The experiments on the nature of deformation described earlier (cold worked versus irradiation) also provide supporting evidence for the effectiveness of dislocation channels in IGSCC. In those experiments, hardening by irradiation was much more effective in initiating IGSCC than was cold work, where the dislocation structure will likely be more cellular than that resulting from deformation of an irradiated condition. As shown along the top of Fig. 26, the slip band density correlates well with the amount of hardening introduced by irradiation and also with the amount of IG cracking, providing a tie-in between deformation mode and IASCC initiation.

A final example of the way in which dislocation channels or slip bands may influence cracking is the observation by Bruemmer et al. (Ref 154) of a correlation between slip band intersection with the walls of a growing crack and the accompanying steps on the oxide on the walls. Figure 31(a) shows a transmission electron micrograph (TEM) of an SCC crack in a cold-worked type 316 stainless steel baffle bolt. Note the slip bands intersecting the walls of the narrow crack at 45° to the crack growth direction. The slip band-crack wall intersections are also coincident with steps in the oxide as shown schematically in Fig. 31(b). This image suggests that the flow of dislocations along the slip steps and into the grain boundary may have been responsible for discontinuous growth of the crack along the grain boundary. While the true role of these slip steps and their angle of intersection with the grain boundary are not yet well understood, these observations provide evidence for a dislocation-based mechanism of cracking in irradiated alloys.

The picture that emerges is one where low SFE and irradiation both lead to planar or localized deformation. Deformation bands terminate at grain boundaries, and, as a consequence, large amounts of displacement in the bands must be accommodated by the boundary. This displacement results in shear strain that can rupture the oxide film at the grain-boundary and cause the initiation of an intergranular crack. For an existing grain-boundary crack, localized deformation will supply dislocations to the crack tip resulting in continual rupture of the crack tip oxide, crack extension, and a higher crack growth rate. As such, localized deformation can lead to both crack initiation and enhanced crack growth.

Radiation Creep and Stress Relaxation

Radiation creep at LWR temperatures results from diffusion of the radiation-produced vacancies and interstitial atoms to dislocations, enhancing the climb-to-glide process controlling time-dependent deformation. Radiation creep can produce both beneficial and detrimental effects. Conceptually, its benefits accrue from a steady relaxation of all constant displacement stresses, for example, weld residual stress and in bolts and in some springs. However, under these conditions—and more so under constant load conditions—radiation creep induces elevated creep rates that may help initiate and sustain SCC. This may result from processes such as grain-boundary sliding, which is believed to be an important reason for the preferential growth of intergranular cracks.

Figures 32 to 34 show the load relaxation that occurs under constant displacement conditions. This process is quite reproducible over a wide range of materials and loading modes and generally produces sizeable (>50%) load relaxation within a few dpa. Thus, for example, in areas of the BWR shroud that receive a moderate neutron flux, if SCC initiation does not occur early in life (e.g., by 1 dpa), the relaxation in residual stress should diminish the likelihood of cracking later in life. Similarly, it is not surprising that the incidence of SCC in BWR shroud beltline welds (exposed to the highest neutron fluence) does not show a strong correlation with fluence.

Radiation creep relaxation also affects PWR baffle bolts, which are subject to large variations in fluence and temperature (Ref 35, 36). This suggests that SCC initiates in baffle bolts either early in life (before significant radiation creep relaxation occurs) or later in life when reloading occurs from differential swelling in the (annealed) baffle plates relative to the (cold worked) baffle bolts.

Creep is also effective in reducing the benefit of compressive residual stress provided, for example, by peening. Figure 35 shows the relaxation in surface stress due to neutron or proton irradiation at 288 °C (550 °F) (Ref 159). Between 30 and 60% of the surface residual stress is relaxed by 2 dpa.

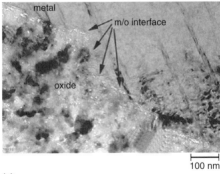

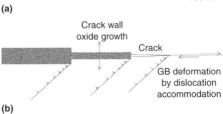

Fig. 31 Transmission electron micrograph of (a) deformation bands intersecting a crack in a baffle bolt (Ref 154) and (b) a schematic of the role that the deformation bands may be playing in IASCC. m, metal; o, oxide; GB, grain boundary

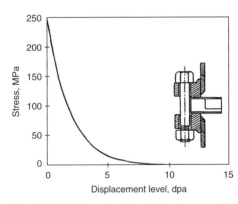

Fig. 32 The effects of radiation-induced creep on load relaxation of stainless steel at 288 °C (550 °F). Source: Ref 155

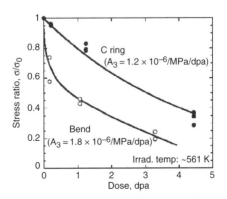

Fig. 34 Stress relaxation of bent beam and C-ring specimens of type 304 stainless steel in Japan Materials Testing Reactor (JMTR) during irradiation at 288 °C (550 °F). Source: Ref 158

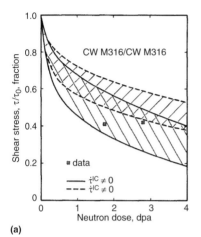

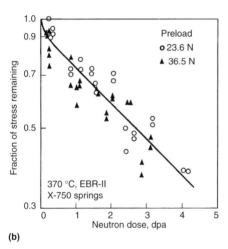

Fig. 33 (a) Radiation creep relaxation of shear stresses in springs of 20% cold-worked (CW) type 316 stainless steel, along with modeling curves showing the effects of radiation-induced creep on load relaxation of stainless steel. Source: Ref 156. (b) Radiation creep relaxation of alloy X-750 springs at 370 °C (700 °F). Source: Ref 157

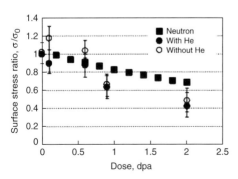

Fig. 35 Comparison of neutron-based prediction with proton-induced creep after removing the effect of thermally induced relaxation. Source: Ref 159

While difficult to prove, the elevated and sustained deformation rates associated with radiation creep can accentuate susceptibility to SCC. Estimates of crack tip deformation rates (Ref 1) indicate that radiation creep is not a large contributor in actively growing cracks, but rather it is expected to promote crack initiation and to sustain crack growth (or promote crack reinitiation, if an existing crack does arrest). One mechanism by which creep can affect crack growth is through its effect on dislocation absorption at grain boundaries, resulting in grain-boundary sliding (Ref 151) as discussed earlier.

It is obviously very important to factor radiation creep relaxation into initial component design and subsequent SCC analysis. Its impact is large and occurs (unsurprisingly) in the same fluence range as RIS and radiation hardening.

Mitigation Strategies

Peter L. Andresen, General Electric Global Research
Gary S. Was, University of Michigan

There are a variety of approaches for mitigating SCC in LWRs, and they are discussed in categories of water chemistry, operating guidelines, new alloys, and design issues and stress mitigation. Since most components in LWRs are not intended to be replaced (and are therefore very expensive to replace), water chemistry is the most attractive mitigation strategy, with operating guidelines and perhaps stress mitigation providing more limited opportunities. While the focus of this article is on IASCC, the primary (especially water chemistry) approaches to SCC mitigation are applicable to both irradiated and unirradiated components.

Water Chemistry Mitigation—BWRs

Because H_2 partitions to the steam phase in BWRs and only about 10 to 15% of the core flow becomes steam (the remainder recirculating in an annulus outside of the core), it is not practical to inject sufficient H_2 to fully suppress radiolysis (which requires about 500 ppb, or 5.7 cm^3/kg), (refer to Fig. 7b). Radiolysis produces substantial concentrations of oxidants, primarily H_2O_2, which is not volatile. Therefore, maintaining a stoichiometric excess of H_2 is impossible once boiling occurs in-core. This produces an elevated corrosion potential on the surfaces of all structural materials, increasing their susceptibility to SCC.

Control of water purity is particularly important in BWRs because the elevated corrosion creates a potential gradient in crevices and cracks that concentrates anionic impurities such as Cl^- and SO_4^{2-} (Ref 5, 52). Until the late 1980s, many BWRs operated at solution conductivities between 0.3 and 0.8 μS/cm, which corresponds to 30 to 90 ppb of (usually acidic) impurities (these were often Cl^- and SO_4^{2-}, which are particularly damaging), that were responsible for extensive cracking. Several plants had severe intrusions of impurities from seawater or release of resins into the reactor water, and experienced severe and immediate SCC. By the early 1990s, the fleet average conductivity had decreased to approximately 0.13 μS/cm, and it is currently approximately 0.11 μS/cm, with much of the residual conductivity (theoretical purity water is 0.055 μS/cm) related to chromate (chromium in the oxide films of structural materials becomes soluble at high potential). A conductivity of 0.10 μS/cm represents only about 6 ppb Cl^- as HCl (1.7×10^{-7} N), and there is a limited value in striving for theoretical purity water because OH^- is present at 2.3×10^{-6} N in pure water; both logic and modern data support the idea that when the "foreign" anions concentration is well below a tenth of the OH^- concentration, their role in carrying ionic current and thereby changing crack chemistry diminishes. However, SCC growth rates in autoclave outlet water of <0.065 μS/cm remain reasonably high at high potential, and crevice chemistry measurements show that a shift from acidic to alkaline crack/crevice chemistry occurs in pure water (Ref 47, 52, 61).

The use of buffering species, such as B/Li, phosphate, ammonia, and so forth, is not effective because of the potent effect of the potential gradient in the crack. Crack growth tests in typical commercial PWR B/Li solutions show that, on changing to high corrosion potential, *very high* growth rates result (Ref 160). "Autobuffering" techniques have been proposed that rely on a change in oxidation state of the species (Ref 52, 161). For example, as NO_3^- is drawn into the crack (which is always at low corrosion potential), it is reduced to NH_4^+, which is rejected from the crack, and also provides some pH buffering capacity. However, while no increase in growth rate results from moderate concentrations of NO_3^- (e.g., up to 100 ppb), it does not appear to reduce the growth rate that is observed in high-purity water at high (or low) corrosion potential. While it may help mitigate the deleterious effects of Cl^- and SO_4^{2-}, these species are kept at very low levels in modern BWRs.

Hydrogen water chemistry (HWC) involves injecting H_2 into the feedwater, which mixes with the recirculating water near the top of the annulus. It is successful in reducing the oxidant level primarily by gamma-enhanced recombination in the annulus (versus suppressing radiolysis in the core). Because the gamma flux is critical, plants with a wide annulus (lower gamma flux at the pressure vessel) are less responsive to HWC. Because residence time in the gamma flux is critical, the water most strongly affected is the recirculation water, which is drawn from the bottom of the annulus, near a (shroud support) ledge that seals the bottom of the jet pumps (note that some BWRs use internal pumps, which avoid external recirculation pumps and piping). The water from the recirculation pump drives the jet pumps (about a 1 to 2 drive-to-suction ratio), the top of which starts in the upper part of the annulus. Thus, two-thirds of the water moves rapidly through most of the annulus region, which reduces the opportunity for gamma recombination. Nonetheless, HWC is used in the majority of worldwide BWRs because it provides significant mitigation in the recirculation piping and varying benefit to the components in the lower part of the BWR (shroud and core support structures, control rod drive housings, etc.). Different H_2 injection rates are used, depending primarily on the plant responsiveness to HWC and their tolerance to increased radiation levels in the steam lines and turbine building. At high H_2 levels, the core becomes reducing and the small amount of radioactive ^{16}N (transmuted from ^{16}O in water) changes from soluble NO_3^- to volatile NO_x or NH_3, as shown in Fig. 7(b). Its half-life is only 7 s, but its presence can raise the radiation level in the steam path by >7 times (Fig. 7). Both the plant design (location and shielding of steam lines and turbine) and the site-boundary limits affect the ability of a given plant to operate at high H_2 injection.

Electrocatalysis employs platinum-group metals to dramatically enhance the efficacy of H_2 injection in lowering the corrosion potential of structural materials (Ref 41, 43). If H_2 is in stoichiometric excess (for forming H_2O), then all of the oxidants (H_2O_2 and O_2) arriving at the surface are consumed catalytically, and the mixed potential for O_2 reduction and H_2 oxidation falls on the H_2/H_2O line at low potential. While making new components out of alloys, or applying weld overlays or thermal spray coatings containing very low levels of platinum-group metals is feasible, the favored approach to creating electrocatalytic surfaces is to introduce soluble platinum-group compounds into the reactor water, for example, at low temperature (e.g., 120 °C, or 250 °F) or during operation to simultaneously catalyze all of the wetted surfaces, a process called NobleChem (Ref 40–42). Provided the distribution of the platinum-group metal on a microscopic scale is relatively uniform (i.e., small platinum particles separated by less than tens of micrometers), catalytic surface loading of well below a monolayer is adequate to achieve low corrosion potentials and excellent SCC mitigation. Deposition can occur either by electroless reduction, or by formation (or introduction) of very fine (nm scale) particles, which stick to the oxidized surfaces of structural materials. Stress-corrosion cracking susceptibility in general, and growth rates in particular, are significantly reduced at low corrosion potential. The extent of the decreases varies with many factors, including yield strength (Ref 1, 5, 6, 12, 13, 40–42, 44, 50). To be effective in mitigating crack growth, the Pt-group metals must be present in the crack to a greater depth than O_2 and H_2O_2, and this requires periodic reapplication. There is isolated evidence that some highly irradiated stainless steels do not show much benefit of low corrosion potential, although a greater body of evidence that shows a strong benefit persists at high fluence (>20 dpa). The differences in response may result from grain-boundary silicon segregation (Ref 44, 50).

Zinc Injection (Ref 162, 163) and Insulated Protective Coatings (Ref 164, 165) are other methods that can mitigate SCC in BWRs. Zinc improves SCC initiation and reduces radiation buildup by improving the quality of the oxide film and substituting for ^{60}Co in the film. It appears to be less effective in cracks, especially at high corrosion potential (where cation transport in cracks is resisted by the potential gradient). However, even at low corrosion potential, providing an adequate and continuing supply of zinc to the tip of a long, growing crack may be difficult.

Insulated protective coatings (IPC) depend on forming a dense or slightly porous layer on the surface that has very low electrical conductivity. The most attractive coating is ZrO_2, which can be deposited by various coating techniques, or formed in situ after creating a metallic zirconium layer on the surface, for example, by wire arc or thermal spray techniques. The IPCs do not rely on H_2 injection or even on the presence of H_2, but it is very unlikely that any technique can be developed to form effective, durable coatings in situ as is done by NobleChem.

Water Chemistry Mitigation—PWR Primary

Commercial PWR primary water uses a coordinated boron (as H_3BO_3) and lithium (as LiOH) chemistry to control reactivity and maintain a near-neutral pH at temperature (typically 6.8 to 7.2 at 300 °C, or 570 °F). The boron level at the beginning of the fuel cycle is typically approximately 1500 ppm (with approximately 2 ppm Li), dropping to 100 to 200 ppm (with approximately 0.5 ppm Li) at the end of cycle. There are some suggestions that the high-temperature pH can be optimized to provide some SCC mitigation, but the evidence is mixed and not well supported by any mechanistic insight. Because PWRs use approximately 30 cm^3/kg (2.7 ppm) H_2, radiolysis is suppressed and the corrosion potential is maintained at its thermodynamic minimum (which varies with temperature, H_2 fugacity, and pH). However, since the relevant metal-oxide phase boundaries have the same pH dependence, there is little hope that small changes in pH will have any effect on SCC. Similarly, while the corrosion potential varies with temperature, H_2 fugacity, and pH, it is close to constant in crevices and cracks, so there is no consequential potential gradient formed.

Zinc Injection and Optimization of H_2 Fugacity. The two most promising water chemistry avenues for mitigating SCC in PWR primary water are Zn injection and H_2 fugacity optimization (for nickel-base alloys). As with BWRs, the case for zinc mitigation of SCC initiation is stronger than crack growth, although its benefit continues to be evaluated. Optimization of H_2 fugacity is relevant only to nickel-base alloys and is based on the observation of a peak in susceptibility to initiation and a peak in crack growth rate near the Ni/NiO phase boundary (Ref 166–168). Because of differences in the temperature dependence on the H_2/H_2O and Ni/NiO boundaries, the corrosion potential (H_2 fugacity) necessary to stay at (or some fixed distance from) the Ni/NiO boundary varies with temperature. Additionally, the fugacity coefficient decreases markedly above 300 °C (570 °F), so more H_2 is needed at high temperature. The peak in growth rate is approximately three times the "background" rate for lower yield strength materials (e.g., annealed alloy 600), and perhaps seven times higher for high yield strength materials (nickel weld metals, alloy X750, etc.). Trying to stay on the low H_2 side of the peak requires very low H_2 levels at 290 °C (555 °F) (the core inlet temperature), which would create the danger of radiolysis and an elevated corrosion potential. High corrosion potential significantly accelerates crack growth in PWR environments. Current programs are focused on evaluating how effective the high H_2 regime could be in mitigating SCC in nickel-base alloys. Some data show increased IASCC in irradiated stainless steels at high H_2 (Ref 169), but it is unclear whether this is a dynamic straining effect on crack initiation (similar to corrosion fatigue crack initiation) or also an effect on crack initiation at constant load or an effect on crack growth rate.

LWR Operating Guidelines

There are limited ways in which plant operation can be adjusted to minimize SCC. Some operating issues, such as optimizing water chemistry, have already been discussed; note, however, that BWRs vary immensely in their practice of maintaining H_2 injection during operation, with the best at ~99% and the worst at <40%. The primary focus of this section relates to load perturbations. It is widely understood that dynamic loading has a prominent role, both in sustaining and accelerating SCC. Thus, avoiding load following and minimizing adjustments that might produce thermal and pressure fluctuations will be beneficial. For example, Scott et al. (Ref 36) showed that the incidence of SCC in PWR baffle bolts was strongly influenced by load following practice. A second element is the number of startups and shutdowns that plants experience. At least under dynamic straining conditions (slow strain rate or slow cyclic loading), most structural materials exhibit a peak in crack growth rate in the vicinity of 175 °C (345 °F), and the startup environment is more oxidizing and often off-chemistry. In BWRs, H_2 injection cannot be started until power is generated (so that there is feedwater flow), and in PWRs, H_2O_2 is injected during cool down. Finally, and potentially very importantly, the dynamic strain associated with pressurization and differential thermal expansion (e.g., in ferrite-containing welds) can be a source of significant dynamic strain that promotes SCC. It may be less important that some growth occurs during the approximately 8 h plant startup transient (when higher growth rates are expected than at 288 °C, or 550 °F), than the fact that the dynamic strains and chemistry conditions would reactivate a crack that became dormant under steady-state operation at 288 °C (550 °F).

Design Issues and Stress Mitigation

There are innumerable design issues related to minimizing the incidence of SCC in components. Obviously, reducing design stresses, mitigating weld residual stresses, eliminating crevices, using resistant materials, avoiding abusive grinding, minimizing neutron dose, and so forth are important. Minimizing or managing thermal gradients is also important. For example, in baffle bolts where significant gamma heating occurs, ensuring good coolant flow into the shank area of the bolt is important. Because above approximately 10 dpa and 300 °C (570 °F), radiation-induced swelling may occur, differential swelling may increase the stress. An example is the use of annealed baffle former plates with cold-worked baffle former bolts; the delayed swelling in the cold-worked bolts makes possible an increase in the loading of the bolts as swelling occurs preferentially in the plates.

Welding always creates vulnerabilities to SCC, and research since the mid-1990s highlights the overlooked importance of weld residual strain, which peaks near the weld fusion line at 15 to 25% equivalent room-temperature strain. This underscores the importance of identifying welding methodologies that minimize both residual strain and residual stress and/or complicate the crack path using an uneven/undulating fusion line during multipass welding.

New Alloys

It is clear that many structural materials used in light water reactors are susceptible to SCC. While improvements such as reduced carbon and added molybdenum (e.g., 316L and 316NG), or stabilizing elements such as titanium or niobium, provide resistance to thermal and weld sensitization, these materials in the unsensitized condition clearly retain significant susceptibility to SCC, especially where weld residual stresses and strains exist. The same broad concerns apply to irradiated conditions, where all variants of commercial stainless steels show marked susceptibility to IASCC. While there remain a few "magic dust" avenues to consider, in general too much time has been expended pursuing subtle variations of the 13-18Cr-8-12Ni nominal stainless steel composition in the hope of achieving dramatic improvements in SCC resistance.

Several known SCC dependencies provide opportunities for improving materials. In both BWRs and PWRs, there is an important role of *yield strength* on crack growth, as shown in Fig. 24 (Ref 5, 45, 51), and the damaging effect of weld residual strains and surface grinding are strongly related to their enhancement in yield strength. Second, chromium depletion makes the grain boundary a highly preferred crack path in pH-shifted crack chemistry that develops at high corrosion potential; at low potential (BWR or PWR) the effect of chromium depletion is not pronounced (Ref 5, 45, 51, 52, 85, 97, 120, 133). Indeed, the presence of grain-boundary carbides retards IGSCC, even at high corrosion potential provided no grain-boundary chromium depletion exists (Ref 44, 50). Materials that have (or achieve by RIS) high silicon levels have high SCC susceptibility (Fig. 11b).

When there is no significant "chemical" preference for crack advance along the grain boundary, it is likely that the intergranular path is preferred because slip accommodation and creep occurs (somewhat) preferentially in the grain boundary (Ref 45, 51). The greater subtlety of the preference for intergranular cracking can be observed by the tendency of the crack to shift to mixed mode or fully transgranular at high "crack tip strain rate," whether imposed by SSRT, high stress-intensity factors, or increased cycling (higher ΔK or frequency, or shorter hold times at K_{max}). Factors that appear to retard these grain-boundary deformation processes include grain-boundary carbides, higher alloying content (e.g., 30% Cr in alloy 690, and even the elevated iron in alloy 800), wavy slip, and lower temperature. One large difference between PWR primary conditions and BWR low potential conditions is temperature, as most structural components in the BWR are at 274 °C (525 °F) after the cooler feedwater mixes in (some, like the top guide and upper inside diameter of the core shroud are at 288 °C, or 550 °F), whereas the PWR primary water ranges from 290 °C (555 °F) (core inlet) to 323 °C (613 °F) (core outlet) to 340 °C (645 °F) (pressurizer).

Radiation-induced segregation occurs by vacancy and interstitial mechanisms, and results in chromium depletion and corresponding nickel enrichment, although iron is generally also depleted in stainless steels. Increasing the chromium level to compensate for RIS is confounded by the need to maintain phase stability; if chromium is increased then nickel must also be increased, and RIS tends to be enhanced at elevated nickel concentrations (other austenite stabilizers such as manganese have also been considered). One alloying approach to reducing the effect of RIS that looks promising involves the use of oversized elements such as hafnium, cerium, zirconium, and yttrium.

Concern for the effect of silicon on SCC has increased, with observations in special heats of stainless steels of high (>1.5%) bulk silicon content of high growth rates and little dependence on corrosion potential or stress-intensity factor as shown in Fig. 11(b) (Ref 45). Since silicon segregates primarily by the interstitial mechanisms, and concentrates by ten times or more at grain boundaries during irradiation even at 1 to 3 dpa, it is appropriate to target new alloys at reasonably low silicon content, for example, <0.1% Si.

Radiation hardening is another major factor in the increased susceptibility to SCC. However, the effect of yield strength on SCC is not linear, and it is unclear that moderate reduction in hardening (e.g., from 900 to 750 MPa, or 130 to 110 ksi) will have a consequential effect—and even more unlikely that alloys can be designed that will not exceed, for example, 400 MPa (58 ksi) in yield strength at high neutron fluence.

Slip localization in the form of dislocation channels is the common mode of plastic deformation for irradiated stainless steels in the 300 °C (570 °F) temperature range. The role of localized deformation in IGSCC and IASCC is not well understood; however, there are parallels between IGSCC and slip planarity in the case of unirradiated materials and between dislocation channeling and IASCC in irradiated alloys. The feature common to high susceptibility for both forms of SCC is localized, planar deformation. These observations suggest that factors that homogenize slip (alloying, second phases, etc.) may also mitigate IGSCC and IASCC.

Finally, altering the character of grain boundaries has proved to be an effective mitigation tool in alloy 600. By applying a processing treatment consisting of deformation followed by annealing at a homologous temperature of about 0.7 T_m, random high angle grain boundaries can be transformed into a network consisting of a high density of low angle and special boundaries. Special boundaries are those that have Σ numbers below 29, where Σ is the reciprocal of the fraction of overlapping lattice sites where the two grains intersect. Preliminary data (Ref 103) show that model type 304 and 316 stainless steel alloys containing a high fraction of special boundaries result in less severe cracking compared to the solution-annealed condition, in which random high-angle boundaries dominate.

Irradiation Effects on Corrosion of Zirconium Alloys

Jeremy Busby, University of Michigan

Zirconium alloys are extensively used in the nuclear industry as nuclear fuel cladding and structural fuel assembly components because they combine a very low neutron absorption cross section with good resistance to high-temperature corrosion and adequate mechanical properties (Ref 170). Nevertheless, uniform waterside corrosion is one of the principal in-reactor degradation mechanisms of zirconium alloys in PWR environments, with localized (nodular) corrosion being of larger concern in BWRs. The intense radiation environment within the reactor core accelerates the degradation of these components, by increasing the rates of corrosion and hydriding, thereby limiting the useful lifetime of the cladding. The integrity of the fuel assembly depends primarily on the performance of the fuel cladding, and its failure is deleterious to the safe and economical operation of nuclear power plants. Therefore, there is great economic incentive to minimize the amount of corrosion that these materials experience in service. The ability of Zircaloy and other zirconium alloys to withstand corrosion and in-reactor degradation is determined by temperature, water chemistry chemical composition, and microstructure (especially the intermetallic precipitates). The key precipitate parameters affecting corrosion are size, morphology, composition, and distribution. Thus the corrosion and hydrogen pickup rates vary not only with alloy composition, but also with the thermomechanical treatment (and resulting microstructure).

Economics have driven the operation of nuclear power plants to more severe duty cycles, including higher coolant temperatures, higher discharge burnups, longer fuel cycles, and longer in-core residence times (Ref 171). In response, fuel vendors have introduced alternate zirconium-base alloys and optimized processing. During the 1990s, small changes in Zircaloy-4 alloy chemistry (nominal composition Zr-1.4Sn-0.2Fe-0.1Cr), such as reduced tin content, have improved in-reactor corrosion performance. For more significant improvements the industry has been replacing the Zircaloy-4 fuel cladding in PWRs with more advanced alloys such as Zirlo (Zr-1.0Nb-1.0Sn-0.1Fe) (Ref 172), achieving significant improvements in performance even in the higher duty cycles. Achieving burnups in excess of 70 GWd/MTU will require further improvements in the corrosion performance of zirconium-base alloys and may require further alloy/processing development.

In the general case, the corrosion kinetics (as reflected by weight gain) usually follow parabolic or cubic behavior (t^n where $n = 1/2$ or $1/3$) until the oxide reaches a thickness of 2 to 3 μm, at which point the oxide undergoes a *transition* to linear kinetics ($n = ~1$). A closer look at the posttransition regime reveals that corrosion still follows a parabolic or cubic relation, however over a much shorter period, which repeats itself throughout the subsequent oxidation process. The pretransition kinetics can be described by the equation, $w(t) = At^n$, where $w(t)$ is the weight gain, t is the exposure time, and A and n are constants.

Irradiation has a considerable influence on the alloy and oxide layer. Irradiation frequently increases the oxidation rate, although the exact mechanism for this increase is not known. In PWR environments, the corrosion rate increases when the oxide thickness exceeds about 5 μm. Under irradiation, the transition to linear kinetics occurs later than that observed in nonirradiated alloys. Neutron or ion irradiation can influence the corrosion behavior of Zircaloy fuel cladding materials by modifying the water chemistry through the creation of free radicals, which are highly mobile and highly reactive. Irradiation can also impact the properties of the growing oxide layer, most notably by increasing diffusion through the oxide layer, which will enhance the corrosion rate. Irradiation can further influence the underlying metal matrix by creating defects in the crystal lattice, causing RIS near the metal/oxide interface, or perhaps most importantly, causing changes in the precipitate structure. Corrosion of Zircaloy is also affected by the water chemistry and cladding temperature, which change when boiling occurs—this begins about 25% of the way along the fuel cladding in a BWR (very limited boiling occurs in the upper part of the core in some PWRs).

The impact of irradiation on diffusion in/through the oxide layer could play a vital role in the corrosion kinetics. Under irradiation, vacancies and interstitials are created in numbers far exceeding thermal equilibrium values. This excess of vacancies and interstitials promotes diffusion processes and thereby corrosion by increasing the transport of oxygen from the water side of the oxide to the oxide/metal interface.

The influence of irradiation on the microstructure of the alloy is one key to understanding corrosion. There is a great deal of empirical knowledge correlating such microstructural parameters as precipitate size, morphology, and distribution (or the annealing parameter) to corrosion performance, hydrogen pickup, and in-reactor performance. The annealing parameter, A, is a simple, single parameter used to describe the combined, accumulated time and temperature of heat treatments. The annealing parameter for any given heat treatment step is usually given as $A = t \exp(-40,000/T)$, where t is the time in hours and T is the temperature in Kelvin. The annealing parameters are then summed for all treatments to gain the cumulative annealing parameter for an alloy (Ref 173). Figure 36 shows the striking correlation of precipitate size with corrosion rate in PWRs and BWRs reported by Garzarolli and Stehle (Ref 173), which is consistent with other studies (Ref 174, 175). In PWRs, large precipitates are desirable to decrease uniform corrosion rates while in BWRs, small precipitates are needed to decrease localized or nodular corrosion. In Zircaloys, these precipitates are typically intermetallics formed when iron and chromium precipitate out of solid solution in the form of $Zr(Fe,Cr)_2$.

In the absence of irradiation, during corrosion of Zircaloy, the zirconia layer closest to the metal/oxide interface forms initially in the tetragonal crystal structure. This dense phase at the metal/oxide interface provides the passive/protective layer during corrosion by slowing the transport of oxygen to the metal/oxide interface. However, it is important to note that the tetragonal phase is not the most stable phase at the temperatures typical of LWR environments. Eventually, as the oxide layer grows thicker, tetragonal zirconia begins to transform into the more stable, monoclinic phase. This phase is not as dense and allows more rapid diffusion of oxygen through the oxide. The transition from tetragonal to monoclinic has been the focus of several recent studies. Yilmazbayhan et al. (Ref 176) measured the fraction of tetragonal phase zirconia using synchrotron radiation. As shown in Fig. 37, the amount of tetragonal phase drops considerably after only a few micrometers of thickness as measured from the oxide/metal interface.

In the initial stages of the oxidation process, the zirconium metal is oxidized preferentially over the precipitates (Ref 177), causing the precipitates to be buried in the oxide layer early in the oxidation process. This is shown schematically in Fig. 38. The precipitates may not begin to oxidize until the surrounding metal phase has been transformed to oxide. During the initial stages of oxidation of a precipitate, the zirconium is oxidized preferentially, releasing chromium and iron to the zirconia matrix (Ref 178). The delayed oxidation of the precipitate is important in the oxidation of Zircaloy as a whole and alters the corrosion process in two ways. First, when the intermetallic undergoes oxidation, the metallic hexagonal close-packed (hcp) phase transforms to the tetragonal oxide, causing an increase in volume. This increase in turn, induces a compressive stress on the oxide layer, which can stabilize the protective tetragonal oxide layer at the metal/oxide interface (Ref 179). The release of chromium and, more significantly, iron, from the second-phase particles, is also important to the uniform corrosion of

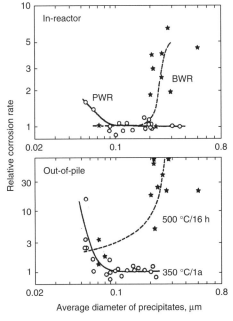

Fig. 36 Relative corrosion rates for cladding with different precipitate sizes in boiling water reactor (BWR)- and pressurized water reactor (PWR)-type environments, when tested in and out of pile. Source: Ref 173

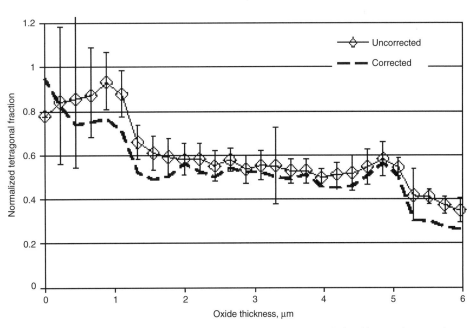

Fig. 37 Tetragonal fraction in oxide formed in Zircaloy-4 tested in pure water, calculated from synchrotron radiation micro-x-ray diffraction, versus distance from the oxide/metal interface. Source: Ref 176

Zircaloy. When released from the precipitates, both chromium and iron will also oxidize (which would also act to slow down the transport of oxygen to the interface). The presence of iron in the form of Fe_3O_4 has been shown to stabilize the tetragonal phase and Fe_3O_4 has been observed in the zirconia matrix. So, the release of iron into the middle of the oxide layer helps create a thicker layer of the protective tetragonal phase, thus reducing corrosion rates.

Under irradiation, the precipitates are affected primarily through amorphization (Ref 180). This process is highly dependent on irradiating particle type, irradiation temperature, and dose rate. The interplay of these variables is shown in Fig. 39 (taken from Ref 180), where the dose required to amorphize precipitates is plotted as a function of irradiation temperature and particle type. Further, postirradiation examination reveals that the precipitates are amorphized from the outside to the inside. That is, the outer shell of the particle is amorphous and depleted in iron while the inner core is of the same structure and composition as that prior to irradiation. An example of this phenomenon is shown in Fig. 40 (taken from Ref 181), which shows a TEM image of a $Zr(Cr,Fe)_2$ precipitate after irradiation with 2 MeV protons at 310 °C (590 °F) to 5 dpa. The energy dispersive spectroscopy (EDS) plot (Fig. 40b) shows that the iron content drops at the edges of the precipitate and rises sharply in the adjacent matrix. The surrounding matrix contains higher levels of chromium and iron, indicating the particles are amorphizing under irradiation.

Since the amorphization of the second-phase particles is not an instantaneous effect, the early stages of corrosion proceed much as they would in the unirradiated case as the precipitates are initially of the same size and composition. However, even as the particles are incorporated into the oxide, irradiation will continue to redistribute iron from the precipitates to the surrounding oxide. Radiation-enhanced diffusion will cause the iron to migrate farther before forming Fe_3O_4. Thus, each precipitate can stabilize the tetragonal phase in a larger volume of oxide matrix. This would provide a more stable and thicker protective layer, consistent with the slight delay in transition observed in LWR conditions. However, this protective oxide remains intact only until it reaches a thickness of approximately 5 μm, at which point, transition occurs and corrosion continues at an accelerated rate, as mentioned previously.

At longer exposures and higher doses, the precipitates entering the oxide at the metal/oxide interface become smaller due to irradiation-induced amorphization. This has several distinct effects on the uniform corrosion rate. First, the smaller particles tend to result in an oxide with smaller, more equiaxed grains, increasing the corrosion rate as discussed previously. Further, since depletion of iron from the precipitates occurs before entering the oxide, there is no supply of iron to contribute to stabilize the tetragonal phase. All these factors combined result in an accelerated corrosion rate in the post-transition regime.

Considerable progress has been made in improving the corrosion behavior of fuel-cladding materials over the last decade (Ref 182). Several new alloys such as Zirlo (Zr-1.0Nb-1.0Sn-0.1Fe), M5 (Zr-1Nb-O), and NDA (Zr-0.1Nb-1.0Sn-0.27Fe-0.16Cr) have been developed to improve the corrosion resistance of Zircaloy. Other solute additions, such as copper, may also improve corrosion resistance. These advanced zirconium-base fuel claddings are currently being tested in reactor. However, a considerable amount of work remains to be done to fully understand the mechanisms of zirconium-base alloy corrosion and develop alloys that are resistant to both irradiation and corrosion in the higher-temperature regimes of advanced reactor designs.

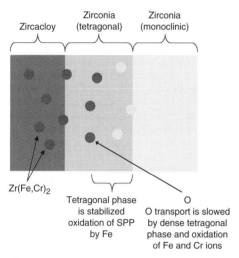

Fig. 38 The role of precipitates in oxidation of Zircaloy-4

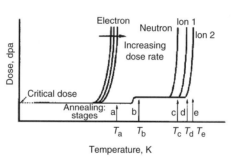

Fig. 39 Dose-to-amorphization versus temperature for different irradiating particles. Source: Ref 180

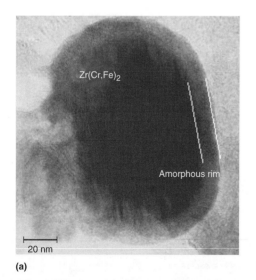

(a)

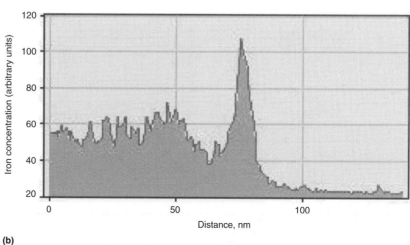

(b)

Fig. 40 (a) A TEM image of a $Zr(Cr,Fe)_2$ precipitate after proton irradiation at 2 MeV, 310 °C (590 °F), to 5 dpa and EDS spectra for the region just outside of the precipitate, showing the concentration of iron in the matrix adjacent to the precipitate. Source: Ref 181

REFERENCES

1. P.L. Andresen, F.P. Ford, S.M. Murphy, and J.M. Perks, *Proc. Fourth International Symposium on Environmental Degradation of Materials in Nuclear Power Systems—Water Reactors,* D. Cubicciotti and G.J. Theus, Ed., National Association of Corrosion Engineers, 1990, p 1–83
2. G.S. Was and P.L. Andresen, *J. Met.,* Vol 44 (No. 4), 1992, p 8
3. P.M. Scott, *J. Nucl. Mater.,* Vol 211, 1994, p 101
4. S.M. Bruemmer, E.P. Simonen, P.M. Scott, P.L. Andresen, G.S. Was, and J.L. Nelson, *J. Nucl. Mater.,* Vol 274, 1999, p 299
5. F.P. Ford and P.L. Andresen, Corrosion in Nuclear Systems: Environmentally Assisted Cracking in Light Water Reactors, *Corrosion Mechanisms,* P. Marcus and J. Ouder, Ed., Marcel Dekker, 1994, p 501–546
6. F.P. Ford and P.L. Andresen, *Proc. Third International Symposium on Environmental Degradation of Materials in Nuclear Power Systems—Water Reactors,* G.J. Theus and J.R. Weeks, Ed., The Metallurgical Society of AIME, 1988, p 789
7. P.L. Andresen and F.P. Ford, *Mater. Sci. Eng.,* Vol A1103, 1988, p 167
8. M. Kodama, R. Katsura, J. Morisawa, S. Nishimura, S. Suzuki, K. Asano, K. Fukuya, and K. Nakata, *Proc. Sixth International Symposium on Environmental Degradation of Materials in Nuclear Power Systems—Water Reactors,* R.E. Gold and E.P. Simonen, Ed., The Minerals, Metals & Materials Society, 1993
9. W.L. Clark and A.J. Jacobs, *First International Symposium on Environmental Degradation of Materials in Nuclear Power Systems—Water Reactors* (Myrtle Beach, SC), NACE, 1983, p 451
10. A.J. Jacobs, *16th International Symposium on Effects of Radiation on Materials,* STP 1175, ASTM International, 1994, p 902
11. M. Kodama, S. Nishimura, J. Morisawa, S. Shims, S. Suzuki, and M. Yamamoto, *Proc. Fifth International Symposium on Environmental Degradation of Materials in Nuclear Power Systems—Water Reactors,* D. Cubicciotti, E.P. Simonen, and R. Gold, Ed., American Nuclear Society, 1992, p 948–954
12. P.L. Andresen, Irradiation Assisted Stress Corrosion Cracking, *Stress Corrosion Cracking: Materials Performance and Evaluation,* R.H. Jones, Ed., ASM International, 1992, p 181–210
13. P.L. Andresen and F.P. Ford, Modeling and Prediction of Irradiation Assisted Stress Corrosion Cracking, *Proc. Seventh International Symposium on Environmental Degradation of Materials in Nuclear Power Systems—Water Reactors,* NACE International, 1995, p 893–908
14. A.J. Jacobs and G.P. Wozadlo, Irradiation Assisted Stress Corrosion Cracking As a Factor in Nuclear Power Plant Aging, *Proc. Int. Conf. on Nuclear Power Plant Aging, Availability Factor and Reliability Analysis* (San Diego, CA), ASM International, 1985, p 173 (A.J. Jacobs, Letter Report and Literature Search, GE Nuclear Energy, San Jose, CA, May 1979)
15. F. Garzarolli, H. Rubel, and E. Steinberg, Behavior of Water Reactor Core Materials with Respect to Corrosion Attack, *Proc. Environmental Degradation of Materials in Nuclear Power Systems—Water Reactors,* National Association of Corrosion Engineers, 1984, p 1–24
16. H. Hanninen and I. Aho-Mantila, Environment-Sensitive Cracking of Reactor Internals, *Proc. Environmental Degradation of Materials in Nuclear Power Systems—Water Reactors,* TMS, 1987, p 77–92
17. R.L. Cowan and G.M. Gordon, Intergranular Stress Corrosion Cracking and Grain Boundary Composition of Fe-Ni-Cr Alloys, *Stress Corrosion Cracking and Hydrogen Embrittlement of Iron-Base Alloys* (Firminy, France), June 1973, R.W. Staehle, J. Hochmann, R.D. McCright, and J.E. Slatern, Ed., National Association of Corrosion Engineers, 1977, p 1063–1065
18. C.F. Cheng, *Corrosion,* Vol 20, 1964, p 341
19. R.N. Duncan, W.H. Arlt, H.E. Williamson, G.S. Baroch, J.P. Hoffman, and T.J. Pashos, *Nucl. Appl.,* Vol 1, 1965, p 413
20. J.S. Armijo, J.R. Low, and U.E. Wolff, *Nucl. Appl.,* Vol 1, 1965, p 462
21. T.J. Pashos et al., Failure Experience with Stainless Steel Clad Fuel Rods in VBWR, *Trans. Am. Nucl. Soc.,* Vol 7 (No. 2), 1964, p 416
22. S. Dumbill, T.M. Williams, J.M. Perks, C.A. English, and W.G. Burns, "Irradiation Assisted Stress Corrosion of Fuel Element Components of a Steam Generating Heavy Water Reactor," NEA/UNIPEDE Specialists Meeting on Life Limiting and Regulatory Aspects of Core Internals and Pressure Vessels (Stockholm, Sweden), Oct 14–16, 1987
23. C.F. Cheng, *J. Nucl. Mater.,* Vol 56, 1975, p 11
24. C.F. Cheng, *Reactor Technol.,* Vol 13, 1970, p 310
25. L.D. Schaffer, Army Reactor Program Progress Report ORNL 3231, Oak Ridge National Laboratory, Jan 1962
26. J.B. Brown, Jr., B.W. Storhok, and J.E. Gates, Post-irradiation Examination of the PM3A Type I Serial 2 Core, *Trans. ANS,* Vol 10, 1967, p 668–669
27. V. Pasupathi and R.W. Klingensmith, "Investigation of Stress Corrosion in Clad Fuel Elements and Fuel Performance in the Connecticut Yankee Reactor," NP-2119, Electric Power Research Institute, Nov 1981
28. I. Multer, "European Operating Experience," Proc. Joint Topical Meeting on Commercial Nuclear Fuel Technology Today (Toronto, Ontario, Canada), 75-CNA/ANS100, April 1975
29. J.T. Storre and D.H. Locke, High Burnup Irradiation Experience in Vulcain, *Nucl. Eng. Int.,* Feb 1970, p 93–99
30. G.M. Gordon and K.S. Brown, Dependence of Creviced BWR Component IGSCC Behavior on Coolant Chemistry, *Proc. Fourth Int. Conf. on Environmental Degradation of Materials in Nuclear Power Systems—Water Reactors,* National Association of Corrosion Engineers, 1990, p 14-46 to 14-62
31. K.S. Brown and G.M. Gordon, Effects of BWR Coolant Chemistry on the Propensity for IGSCC Initiation and Growth in Creviced Reactor Internals Components, *Proc. Environmental Degradation of Materials in Nuclear Power Systems—Water Reactors,* AIME, 1987, p 243–248
32. A.J. Jacobs, D.A. Hale, and M. Siegler, Unpublished Data, GE Nuclear Energy, San Jose, CA, Jan 1986
33. T. Onchi, K. Dohi, N. Soneda, M. Navas, and M. Luisa Castano, Neutron Fluence Dependent Intergranular Cracking of Thermally Sensitized Type 304 Stainless Steel at 290 °C in Inert Gas and in Water, *Proc. 11th International Conf. Environmental Degradation of Materials in Nuclear Power Systems—Water Reactors,* American Nuclear Society, 2003, p 1111
34. R.M. Horn, G.M. Gordon, F.P. Ford, and R.L. Cowan, Experience and Assessment of SCC in L-Grade SS BWR Internals, *Nucl. Eng. Des.,* Vol 174, 1997, p 313–325
35. P.A. Heuze, O. Goltrant, and R. Cauvin, Expertise des vis de liaison cloison-renfort de la centrale de Tihange 1, *Proc. Int. Symp. Fontevraud IV,* Societe Francaise d'Energie Nucleaire, 1998, p 195
36. P.M. Scott, M.-C. Meunier, D. Deydier, S. Silvestre, and A. Trenty, An Analysis of Baffle/Former Bolt Cracking in French PWRs, *Environmentally Assisted Cracking: Predictive Methods for Risk Assessment and Evaluation of Materials, Equipment and Structures,* STP 1401, R.D. Kane, Ed., ASTM, 2000, p 210–223
37. D.I.R. Norris, C. Baker, and J.M. Titchmarsh, Compositional Profiles at Grain Boundaries in 20%Cr/25%Ni/Nb Stainless Steel, *Proc. Radiation Induced Sensitization of Stainless Steels,* Berkeley Nuclear Labs, 1986, p 86–98
38. F. Garzarolli, D. Alter, P. Dewes, and J.L. Nelson, Deformability of Austenitic Stainless Steels and Nickel-Base Alloys in the Core of a Boiling and a Pressurized Water Reactor, *Proc. Third Int. Symp. on Environmental Degradation of Materials in Nuclear Power Systems—Water Reactors,* AIME, 1988, p 657–664

39. F. Garzarolli, D. Alter, and P. Dewes, Deformability of Austenitic Stainless Steels and Nickel-Base Alloys in the Core of a Boiling and a Pressurized Water Reactor, *Proc. Environmental Degradation of Materials in Nuclear Power Systems—Water Reactors,* American Nuclear Society, 1986, p 131–138
40. P.L. Andresen, Application of Noble Metal Technology for Mitigation of Stress Corrosion Cracking in BWRs, *Proc. Seventh International Symposium on Environmental Degradation of Materials in Nuclear Power Systems—Water Reactors,* (Breckenridge, CO), R.E. Gold and E.P. Simonen, Ed., NACE International, 1995, p 563–578
41. S. Hettiarachchi, G.P. Wozadlo, T.P. Diaz, P.L. Andresen, and R.L Cowan, The Concept of Noble Metal Addition Technology for IGSCC Mitigation of Nuclear Materials, *Proc. Seventh Int. Symp. on Environmental Degradation of Materials in Nuclear Power Systems—Water Reactors,* (Breckenridge, CO), R.E. Gold and E.P. Simonen, Ed., NACE International, 1995, p 735–746
42. P.L. Andresen, T.P. Diaz, and S. Hettiarachchi, "Resolving Electrocatalytic SCC Mitigation Issues in High Temperature Water," paper 04668, Corrosion/04, NACE International, 2004
43. M.P. Manahan, J. Santucci, and P. Sipush, A Phenomenological Investigation of In-Reactor Cracking of Type 304 Stainless Steel Control Rod Cladding, *Nucl. Eng. Des.,* Vol 113, 1989, p 297–321
44. P.L. Andresen, M.M. Morra, and W.R. Catlin, "Effects of Yield Strength, Corrosion Potential, Composition and Stress Intensity Factor in SCC of Stainless Steels," paper 04678, Corrosion/04, NACE International, 2004
45. F.P. Ford, P.L. Andresen, H.D. Solomon, G.M. Gordon, S. Ranganath, D. Weinstein, and R. Pathania, Application of Water Chemistry Control, On-Line Monitoring, and Crack Growth Models for Improved BWR Materials Performance, *Proc. Fourth International Symposium on Environmental Degradation of Materials in Nuclear Power Systems,* D. Cubicciotti and G.J. Theus, Ed., National Association of Corrosion Engineers, 1990, p 4-26 to 4-51
46. F.P. Ford, D.F. Taylor, P.L. Andresen, and R.G. Ballinger, "Environmentally Controlled Cracking of Stainless and Low Alloy Steels in LWR Environments," NP-5064M (RP2006-6), Electric Power Research Institute, 1987
47. T.M. Angeliu, P.L. Andresen, J.A. Sutliff, and R.M. Horn, Intergranular Stress Corrosion Cracking of Unsensitized Stainless Steels in BWR Environments, *Proc. Ninth Int. Symp. on Environmental Degradation of Materials in Nuclear Power Systems,* The Metallurgical Society of AIME, 1999
48. T.M. Angeliu, P.L. Andresen, E. Hall, J.A. Sutliff, S. Sitzman, M. Yamamoto, and J. Kuniya, "Strain and Microstructure Characterization of Austenitic Stainless Steel Weld HAZs," paper 00186, Corrosion/2000, NACE International, 2000
49. P.L. Andresen, T.M. Angeliu, and L.M. Young, Immunity, Thresholds, and Other SCC Fiction, *Proc. Staehle Symp. on Chemistry and Electrochemistry of Corrosion and SCC,* TMS, Feb 2001
50. P.L. Andresen, Perspective and Direction of Stress Corrosion Cracking in Hot Water, *Proc. Tenth Int. Symp. on Environmental Degradation of Materials in Nuclear Power Systems—Water Reactors,* NACE International, 2001
51. P.L. Andresen and L.M. Young, Characterization of the Roles of Electrochemistry, Convection and Crack Chemistry in Stress Corrosion Cracking, *Proc. Seventh Int. Symp. on Environmental Degradation of Materials in Nuclear Power Systems—Water Reactors,* NACE International, 1995, p 579–596
52. C.C. Lin, An Overview of Radiation Chemistry in Reactor Coolants, *Proc. Environmental Degradation of Materials in Nuclear Power Systems—Water Reactors,* American Nuclear Society, 1986, p 160–172
53. W.G. Burns and P.B. Moore, *Radiation Effects,* Vol 30, 1976, p 233
54. P. Cohen, *Water Coolant Technology of Power Reactors,* Gordon and Breach Science Publishers, 1969
55. *Proc. of Water Chemistry of Nuclear Reactor Systems,* No. 5, Bournemouth, Oct 23–27, 1989, British Nuclear Energy Society, London, 1989
56. D.F. Taylor, "Response of Electrochemical Sensors to Ionizing Radiation in High Temperature Aqueous Environments," paper 90501, Corrosion/90 (Las Vegas, NV), National Association of Corrosion Engineers, 1990
57. W.D. Miller and P.L. Andresen, unpublished work, GE Nuclear Energy, San Jose, CA, 1989
58. P.L. Andresen, F.P. Ford, J.P. Higgins, I. Suzuki, M. Koyama, M. Akiyama, T. Okubo, Y. Mishima, S. Hattori, H. Anzai, H. Chujo, and Y. Kanazawa, "Life Prediction of Boiling Water Reactor Internals," Proc., ICONE-4 Conference, ASME, 1996
59. R.A. Head, M.E. Indig, and P.L. Andresen, "Measurement of In-core and Recirculation System Responses to Hydrogen Water Chemistry at Nine Mile Point Unit 1 BWR," Contract RP2680-5, Final Report, Electric Power Research Institute, 1989
60. B.M. Gordon, "Hydrogen Water Chemistry for BWR," Task 27 "Materials and Environmental Monitoring with in the Duane Arnold BWR," Contract RP1930-1, Project Manager, J.L. Nelson, Electric Power Research Institute
61. D.F. Taylor and C.A. Caramihas, High Temperature Aqueous Crevice Corrosion in Alloy 600 and 304L Stainless Steel, *Localized Crack Chemistry and Mechanics of Environmentally Assisted Cracking,* AIME, 1983
62. P. Combrade, M. Foucault, and G. Slama, About the Crack Tip Environment Chemistry in Pressure Vessel Steel Exposed to Primary PWR Coolant, *Proc. Atomic Energy Agency Specialists Meeting on Subcritical Crack Growth* (Sendai, Japan), May 1985, NUREG/CP0067, Vol 2, p 201–218
63. G. Gabetta and G. Buzzanaca, Measurement of Corrosion Potential Inside and Outside Growing Crack During Environmental Fatigue Tests, *Proc. Atomic Energy Agency Specialists Meeting on Subcritical Crack Growth* (Sendai, Japan), May 1985, NUREG/CP0067, Vol 2, p 219–230
64. P.L. Andresen, I.P. Vasatis, and F.P. Ford, "Behavior of Short Cracks in Stainless Steels and Inconels in 288 °C Water," paper 495, Corrosion/90 (Las Vegas, NV), National Association of Corrosion Engineers, 1990
65. P.L. Andresen and F.P. Ford, "Modeling of Irradiation Effects on Stress Corrosion Crack Growth Rates," paper 497, Corrosion/89, NACE
66. L.G. Ljungberg, personal communication, ABB Atom, Vasteras, Sweden, March 1991
67. H. Wiedersich, P.R. Okamoto, and N.Q. Lam, *J. Nucl. Mater.,* Vol 83, 1979, p 98
68. P.R. Okamoto and L.E. Rehn, *J. Nucl. Mater.,* Vol 83 1979, p 2
69. P.R. Okamoto and H. Wiedersich, *J. Nucl. Mater.,* Vol 53, 1974, p 336
70. N.Q. Lam, A. Kumar, and H. Wiedersich, *Proc. Effects of Radiation on Materials: Eleventh Conf.,* STP 782, H.R. Brager and J.S. Perrin, Ed., American Society for Testing and Materials, 1982, p 985
71. J.M. Perks, A.D. Marwick, and C.A. English, Fundamental Aspects of Radiation-Induced Segregation in Fe-Cr-Ni Alloys, *Proc. Radiation-Induced Sensitisation of Stainless Steels,* D.I.R. Norris, Ed., Central Electricity Generating Board, Berkeley Nuclear Labs, Berkeley, Gloucestershire, GL13 9PB, 1987, p 15
72. A. Jenssen, L.G. Ljungberg, J. Walmsley, and S. Fisher, *Corrosion,* Vol 54 (No. 1), 1998, p 48
73. T.R. Allen, J.T. Busby, G.S. Was, and E.A. Kenik, *J. Nucl. Mater.,* Vol 255, 1998, p 44
74. G.S. Was, T.R. Allen, J.T. Busby, J. Gan, D. Damcott, D. Carter, and M. Atzmon, Microchemistry of Proton-Irradiated Austenitic Alloys under Conditions Relevant to LWR Core Components, *Proc. Mater.*

75. A.C. Hindmarsh, Lawrence Livermore Laboratory Report UCID-30002, 1974
76. E.P. Simonen, private communication, Battelle Pacific NW Labs, Richland, WA
77. G.S. Was and T.R. Allen, *Mater. Charact.,* Vol 32 (No. 4), 1994, p 239
78. K. Asano, K. Fukuya, K. Nakata, and M. Kodama, Changes in Grain Boundary Composition Induced by Neutron Irradiation on Austenitic Stainless Steels, *Proc. Fifth International Symposium on Environmental Degradation of Materials in Nuclear Power Systems—Water Reactors,* D. Cubicciotti, E.P. Simonen, and R.E. Gold, Ed., American Nuclear Society, 1992, p 838
79. A.J. Jacobs, G.E.C. Bell, C.M. Sheperd, and G.P. Wozadlo, *Proc. Seventh International Symposium on Environmental Degradation of Materials in Nuclear Power Systems—Water Reactors* (Breckenridge, CO), R.E. Gold and E.P. Simonen, Ed., NACE International, 1995, p 917
80. A.J. Jacobs, G.P. Wozadlo, K. Nakata, S. Kasahara, T. Okada, S. Kawano, and S. Suzuki, The Correlation of Grain Boundary Composition in Irradiated Stainless Steel with IASCC Resistance, *Proc. Sixth International Symposium on Environmental Degradation of Materials in Nuclear Power Systems—Water Reactors,* R.E. Gold and E.P. Simonen, Ed., The Minerals, Metals & Materials Society, 1993, p 597
81. E.A. Kenik, *J. Nucl. Mater.,* Vol 187, 1992, p 239
82. S. Nakahigashi, M. Kodama, K. Fukuya, S. Nishimura, S. Yamamoto, K. Saito, and T. Saito, *J. Nucl. Mater.,* Vol 179–181, 1992, p 1061
83. A.J. Jacobs, R.E. Clausing, M.K. Miller, and C.M. Shepherd, Influence of Grain Boundary Composition on the IASCC Susceptibility of Type 348 Stainless Steel, *Proc. Fourth International Symposium on Environmental Degradation of Materials in Nuclear Power Systems—Water Reactors,* D. Cubicciotti and G.J. Theus, Ed., National Association of Corrosion Engineers, 1990, p 14–21
84. J. Walmsley, P. Spellward, S. Fisher, and A. Jenssen, Microchemical Characterization of Grain Boundaries in Irradiated Steels, *Proc. Seventh International Symposium on Environment Degradation of Materials in Nuclear Power System—Water Reactors* (Breckenridge, CO), R.E. Gold and E.P. Simonen, Ed., NACE International, 1997, p 985
85. S.M. Bruemmer, E.P. Simonen, P.M. Scott, P.L. Andresen, G.S. Was, and J.L. Nelson, *J. Nucl. Mater.,* Vol 274, 1999, p 299–314
86. G.S. Was, J.T. Busby, J. Gan, E.A. Kenik. A. Jenssen, S.M. Bruemmer, P.M. Scott, and P.L. Andresen, *J. Nucl. Mater.,* Vol 300, 2002, p 198–216
87. R. Carter, D. Damcott, M. Atzmon, G.S. Was, S.M. Bruemmer, and E.A. Kenik, *J. Nucl. Mater.,* Vol 211, 1994, p 70
88. E.A. Kenik, M.A. Miller, M. Thavander, J.T. Busby, and G.S. Was, Origin and Influence of Pre-existing Segregation on Radiation Induced Segregation in Austenitic Stainless Steel, *Proc. Materials Research Society,* Vol 54, Materials Research Society, 1999, p 445
89. S.M. Bruemmer, New Issues Concerning Radiation-Induced Materials Changes and Irradiation-Assisted Stress Corrosion Cracking in Light Water Reactors, *Proc. Tenth Int. Conf. Environmental Degradation of Materials in Nuclear Power Systems—Water Reactors,* NACE International, 2002
90. T. Kato, H. Takahashi, and M. Izumiya, *J. Nucl. Mater.,* Vol 189, 1992, p 167
91. L. Fournier, B.H. Sencer, G.S. Was, E.P. Simonen, and S.M. Bruemmer, *J. Nucl. Mater.,* Vol 321, (No. 2–3), 2003, p 192–209
92. S. Dumbill and M. Hanks, Strategies for the Moderation of Chromium Depletion at Grain Boundaries in Irradiated Steels, *Proc. Sixth Int. Symp. Environmental Degradation of Materials in Nuclear Power Systems—Water Reactors,* R.E. Gold and E.P. Simonen, Ed., The Minerals, Metals and Materials Society, 1993, p 521
93. J. Gan, E.P. Simonen, S.M. Bruemmer, L. Fournier, B.H. Sencer, and G.S. Was, The Effect of Oversized Solute Additions on the Microstructure of 316SS Irradiated with 5 MeV Ni^{++} Ions or 3.2 MeV Protons, *J. Nucl. Mater.,* Vol 325, 2004, p 94–106
94. J.L. Nelson and P.L. Andresen, Review of Current Research and Understanding of Irradiation-Assisted Stress Corrosion Cracking, *Proc. Fifth International Symposium on Environmental Degradation of Materials in Nuclear Power Systems—Water Reactors,* D. Cubicciotti, E.P. Simonen, and R.E. Gold, Ed., American Nuclear Society, 1992, p 10
95. G.S. Was, T.R. Allen, J.T. Busby, J. Gan, D. Damcott, D. Carter, M. Atzmon, and E.A. Kenik, *J. Nucl. Mater.,* Vol 270, 1999, p 96
96. S.M. Bruemmer and G.S. Was, *J. Nucl. Mater.,* Vol 216, 1994, p 348
97. J.T. Busby and G.S. Was, Irradiation Assisted Stress Corrosion Cracking in Model Austenitic Alloys with Solute Additions, *Proc. 11th Int. Conf. Environmental Degradation of Materials in Nuclear Power Systems—Water Reactors,* American Nuclear Society, 2003, p 995
98. J.T. Busby, G.S. Was, and E.A. Kenik, *J. Nucl. Mater.,* Vol 302, 2002, p 20
99. A.J. Jacobs, G.E.C. Bell, C.M. Shepherd, and G.P. Wozadlo, High-Temperature Solution Annealing as an IASCC Mitigation Technique, *Proc. Fifth International Symposium on Environment Degradation of Materials in Nuclear Power System—Water Reactor,* D. Cubicciotti, E.P. Simonen, and R.E. Told, Ed., American Nuclear Society, 1992, p 917
100. Y. Ishiyama, M. Kodama, N. Yokota, K. Asano, T. Kato, and K. Fukuya, *J. Nucl. Mater.,* Vol 239, 1996, p 90
101. S. Katsura, Y. Ishiyama, N. Yokota, T. Kato, K. Nakata, K. Fukuya, H. Sakamoto, and K. Asano, "Post-Irradiation Annealing Effects of Austenitic Stainless Steel in IASCC," paper 132, Corrosion 98 (Houston, TX), NACE International, 1998
102. J.I. Cole, T.R. Allen, G.S. Was, and E.A. Kenik, The Influence of Pre-irradiation Heat Treatments on Thermal Non-Equilibrium and Radiation-Induced Segregation Behavior in Model Austenitic Stainless Steel Alloys, *Effects of Radiation on Materials: 21st International Symposium,* STP 1447, M.L. Grossbeck, T.R. Allen, R.G. Lott, and A.S. Kumar, Ed., ASTM International, 2004, p 540–552
103. R.B. Dropek, G.S. Was, J. Gan, J.I. Cole, T.R. Allen, and E.A. Kenik, Bulk Composition and Grain Boundary Engineering to Improve Stress Corrosion Cracking Behavior of Proton Irradiated Stainless Steels, *Proc. 11th Int. Conf. Environmental Degradation of Materials in Nuclear Power Systems—Water Reactors,* American Nuclear Society, 2003, p 1132
104. E.P. Simonen and S.M. Bruemmer, Thermally Induced Grain Boundary Composition and Effects on Radiation-Induced Segregation, *Proc. 8th Int. Conf. Environmental Degradation of Materials in Nuclear Power Systems—Water Reactors,* American Nuclear Society, 1997, p 751
105. T. Yonezawa, K. Fujimoto, H. Kanasaki, T. Iwamura, S. Nakada, K. Ajiki, and S. Urata, *Corros, Eng.,* Vol 49, 2000, p 655
106. K. Fujimoto, T. Yonezawa, T. Iwamura, K. Ajiki, and S. Urata, *Corros. Eng.,* Vol 49, 2000, p 701
107. T. Yonezawa, K. Fujimoto, T. Iwamura, and S. Nishida, Improvement of IASCC Resistance for Austenitic Stainless Steels in PWR Environment, *Environmentally Assisted Cracking: Predictive Methods for Risk Assessment and Evaluation of Materials, Equipment, and Structures,* R.D. Kane, Ed., STP 1401, American Society for Testing and Materials, 2000, p 224
108. T. Yonezawa, T. Iwamura, K. Fujimoto, and K. Ajiki, Optimized Chemical Composition and Heat Treatment Conditions of 316 CW and High-Chromium Austenitic Stainless Steels for PWR Baffle Bolts, *Proc. Ninth Int. Conf. Environmental Degradation of Materials in Nuclear*

Power Systems—Water Reactors, The Minerals, Metals & Materials Society, 1999, p 1015
109. G.F. Li, Y. Kaneshima, and T. Shoji, *Corrosion,* Vol 56 (No. 5), 2000, p 540
110. P.J. Maziasz and C.J. McHargue, *Int. Met. Rev.,* Vol 32, 1987, p 190
111. P.J. Maziasz, *J. Nucl. Mater.,* Vol 205, 1993, p 118
112. F.A. Garner, *J. Nucl. Mater.,* Vol 205, 1993, p 98
113. S.J. Zinkle, P.J. Maziasz, and R.E. Stoller, *J. Nucl. Mater.,* Vol 206, 1993, p 266
114. D.J. Edwards, E.P. Simonen, and S.M. Bruemmer, Evolution of Fine-Scale Defects in Stainless Steels Neutron-Irradiated at 275 C, *J. Nucl. Mater.,* Vol 317, 2003, p 13
115. D.J. Edwards, B.A. Oliver, F.A. Garner, and S.M. Bruemmer, Microstructural Evaluation of a Cold-Worked 316SS Baffle Bolt Irradiated in a Commercial PWR, *Proc. Tenth Int. Conf. Environmental Degradation of Materials in Nuclear Power Systems—Water Reactors,* NACE International, 2002
116. D.J. Edwards, E.P. Simonen, F.A. Garner, L.R. Greenwood, B.M. Oliver, and S.M. Bruemmer, *J. Nucl. Mater.,* Vol 317, 2000, p 32
117. R. Stoller, *JOM,* 1996, p 23
118. J.T. Busby, M.M. Sowa, G.S. Was, and E.A. Kenik, The Role of Fine Defect Clusters in Irradiation-Assisted Stress Corrosion Cracking of Proton-Irradiated 304 Stainless Steel, *Proc. 21st International Symposium on Effects of Radiation on Material,* M.R. Grossbeck, T.R. Allen, R.G. Lott, and A.S. Kumar, Ed., STP 1447, ASTM International, 2004, p 78–91
119. E.P. Simonen, D.J. Edwards, B.W. Arey, S.M. Bruemmer, J.T. Busby, and G.S. Was, *Philos. Mag. A,* in press
120. J.T. Busby, E.A. Kenik, and G.S. Was, Effect of Single Solute Additions on Radiation-Induced Segregation and Microstructure of Model Austenitic Alloys, *J. Nucl. Mater.,* in press
121. H. Watanabe, T. Muroga, and N. Yoshida, *J. Nucl. Mater.,* Vol 239, 1996, p 95
122. U. Bergenlid, Y. Haag, and K. Pettersson, "The Studsvik MAT 1 Experiment R2 Irradiations and Post-Irradiation Tensile Test," Studsvik Report STUDSVIK/NS-90/13, 1990
123. G.R. Odette and G. Lucas, *J. Nucl. Mater.,* Vol 179–181, 1991, p 572
124. M. Kodama, S. Suzuki, K. Nakata, S. Nishimura, K. Jukuya, T. Kato, Y. Tanaka, and S. Shima, Mechanical Properties of Various Kinds of Irradiated Austenitic Stainless Steels, *Proc. Eighth International Symposium on Environment Degradation of Materials in Nuclear Power Systems—Water Reactor* (Amelia Island, FL), American Nuclear Society, 1997, p 831
125. A. Jenssen, written communication, Studsvik Nuclear, Sweden, 1998
126. K. Fukuya, S. Shima, K. Nakata, S. Kasahara, A. Jacobs, G. Wozadlo, S. Suzuki, and M. Kitamura, Mechanical Properties and IASCC Susceptibility in Irradiated Stainless Steels, *Sixth International Symposium on Environmental Degradation of Materials in Nuclear Power Systems—Water Reactors,* R.E. Gold and E.P. Simonen, Ed., The Minerals, Metals & Materials Society, 1993
127. T.R. Allen, H. Tsai, J.I. Cole, J. Ohta, K. Dohi, and H. Kusanagi, Properties of 20% Cold-Worked 316 Stainless Steel Irradiated at Low Dose Rate, *Effects of Radiation on Materials: 21st International Symposium,* STP 1447, M.L. Grossbeck, T.R. Allen, R.G. Lott, and A.S. Kumar, Ed., ASTM International, 2003
128. G. Furutani, N. Nakajima, T. Konishi, and M. Kodama, Stress Corrosion Cracking on Irradiated 316 Stainless Steel, *J. Nucl. Mater.,* Vol 288, 2001, p 179
129. B.N. Singh, A.J.E. Foreman, and H. Trinkaus, *J. Nucl. Mater.,* Vol 249, 1997, p 103
130. A. Seeger, On the Theory of Radiation Damage and Radiation Hardening, *Proc. Second United Nations International Conference on Peaceful Uses of Atomic Energy* (Geneva, Switzerland), Vol 6, 1958, p 250
131. S.M. Bruemmer, J. Cole, R. Carter, and G.S. Was, Radiation Hardening Effects on Localized Deformation and Stress Corrosion Cracking of Stainless Steels, *Proc. Sixth International Symposium on Environmental Degradation of Materials in Nuclear Power Systems—Water Reactors,* R.E. Gold and E.P. Simonen, Ed., The Minerals, Metals and Materials Society, 1993, p 537
132. E.E. Bloom, Irradiation Strengthening and Embrittlement, *Radiation Damage in Metals,* N.L. Peterson and S.D. Harkness, Ed., American Society for Metals, 1975, p 295
133. M.O. Speidel and R. Magdowski, Stress Corrosion Cracking of Stabilized Austenitic Stainless Steels in Various Types of Nuclear Power Plants, *Proc. Ninth International Symposium on Environmental Degradation of Materials in Nuclear Power Systems—Water Reactors,* F.P. Ford, S.M. Bruemmer, and G.S. Was, Ed., The Metallurgical Society of AIME, 1999, p 325
134. G.S. Was, Recent Developments in Understanding Irradiation Assisted Stress Corrosion Cracking, *Proc. 11th Int. Conf. Environmental Degradation of Materials in Nuclear Power Systems—Water Reactors,* American Nuclear Society, 2003, p 96
135. A.J. Jacobs, G.P. Wozaldo, and G.M. Gordon, *Corrosion,* Vol 51 (No. 10), 1995, p 731
136. M.C. Hash, J.T. Busby, and G.S. Was, The Effect of Hardening Source in Proton Irradiation-Assisted Stress Corrosion Cracking of Cold-Worked Type 304 Stainless Steel, *Proc. 21st International Symposium on Effects of Radiation on Material,* M.R. Grossbeck, T.R. Allen, R.G. Lott, and A.S. Kumar, Ed., STP 1447, ASTM International, 2004, p 92–104
137. J.T. Busby and G.S. Was, Irradiation-Assisted Stress Corrosion Cracking in Model Austenitic Alloys with Solute Additions, *International Symposium on Environmental Degradation of Materials in Nuclear Power Systems—Water Reactor* (Skamania, WA), Aug 2003, American Nuclear Society, 2003, electronic proceedings only
138. J. Cookson, Ph.D. thesis, University of Michigan, 1996
139. A. Jenssen, L. Ljungberg, J. Walmsley, and S. Fisher, paper 101, Corrosion 96, NACE International, 1996
140. M. Kodama, K. Fukuya, and H. Kayano, Influence of Impurities and Alloying Elements on IASCC in Neutron Irradiated Austenitic Stainless Steels, *Proc. 16th International Symposium on Radiation Effects on Materials,* A.S. Kumar, D.S. Gelles, R.K. Nanstad, and E.A. Little, Ed., STP 1175, American Society for Testing and Materials, 1994, p 889
141. H. Chung et al., IASCC Susceptibility of Model Austenitic Stainless Steels, *Proc. Ninth International Symposium on Environment Degradation of Materials in Nuclear Power Systems—Water Reactors,* TMS, 1999, p 931
142. H. Chung et al., Stress Corrosion Cracking Susceptibility of Irradiated Type 304 Stainless Steels, *16th International Symposium on Effects of Radiation on Materials,* STP 1175, ASTM International, 1994, p 889
143. Y. Tanaka et al., IASCC Susceptibility of Type 304, 304L, and 316L Stainless Steels, *Proc. Eighth International Symposium on Environment Degradation of Materials in Nuclear Power Systems—Water Reactors* (Amelia Island, FL), American Nuclear Society, 1997, p 803
144. A. Jenssen and L. Ljungberg, Irradiation Assisted Stress Corrosion Cracking, Postirradiation CERT Tests of Stainless Steels in a BWR Test Loop, *Proc. Seventh International Symposium on Environmental Degradation of Materials in Nuclear Power Systems—Water Reactors* (Breckenridge, CO), R.E. Gold and E.P. Simonen, Ed., NACE International, 1995, p 1043–1053
145. D.I.R. Norris, C. Baker, C. Taylor, and J.M. Titchmarsh, A Study of Radiation-Induced Sensitization in 20/25/Nb Steel by Compositional Profile Measurements at Grain Boundaries, paper 52, *Materials*

for Nuclear Reactor Core Applications, BNES, London, 1987
146. P.R. Swann, *Corrosion,* Vol 19, 1963, p 102t
147. A.W. Thompson and I.M. Bernstein, The Role of Metallurgical Variables in Hydrogen-Assisted Environmental Fracture, *Advances in Corrosion Science and Technology,* R.W. Staehle and M.G. Fontana, Ed., Vol 7, Plenum Press, 1980, p 53
148. C.G. Rhodes and A.W. Thompson, *Metall. Trans. A,* Vol 8A, 1977, p 1901
149. R.E. Schramm and R.P. Reed, *Metall. Trans. A,* Vol 6A, 1975, p 1345
150. K. Farrell, T.S. Byun, and N. Hashimoto, "Mapping Flow Localization Processes in Deformation of Irradiated Reactor Structural Alloys," report ORNL/TM-2002/66, Oak Ridge National Laboratory, July 2002
151. B. Alexandreanu and G.S. Was, *Corrosion,* Vol 59 (No. 8), 2003, p 705–720
152. R. Dropek, G.S. Was, J. Gan, J.I. Cole, T.R. Allen, and E. A. Kenik, Bulk Composition and Grain Boundary Engineering to Improve Stress Corrosion Cracking Behavior of Proton Irradiated Stainless Steels, *Proc. 11th Int. Conf. Environmental Degradation of Materials in Nuclear Power Systems—Water Reactors,* American Nuclear Society, 2003, p 1132–1143
153. C. Bailat, A. Almazouzi, M. Baluc, R. Schaublin, F. Groschel, and M. Victoria, *J. Nucl. Mater.,* Vol 283–287, 2000, p 446
154. L.E. Thomas and S.M. Bruemmer, Insights into Environmental Degradation Mechanisms from Analytical Transmission Electron Microscopy of SCC Cracks, *Proc. Ninth International Symposium on Environmental Degradation of Materials in Nuclear Power Systems—Water Reactors,* F.P. Ford, S.M. Bruemmer, and G.S. Was, Ed., The Metallurgical Society of AIME, 1999, p 41
155. J.C. Van Duysen, P. Todeschini, and G. Zacharie, Effects of Neutron Irradiations at Temperatures below 550 °C on the Properties of Cold Worked 361 Stainless Steels: A Review, *Proc. Effects of Radiation on Materials: 16th Int. Symposium,* STP 1175, ASTM, 1993, p 747–776
156. J.P. Foster, Analysis of In-Reactor Stress Relaxation Using Irradiation Creep Models, *Proc. Irradiation Effects on the Microstructure and Properties of Metals,* STP 611, ASTM, 1976, p 32
157. L.C. Walters and W.E. Ruther, *J. Nucl. Mater.,* Vol 68, 1977, p 324
158. Y. Ishiyama, K. Nakata, M. Obata, H. Anzai, S. Tanaka, T. Tsukada, and K. Asano, Stress Relaxation Caused by Neutron-Irradiation at 561 K in Austenitic Stainless Steels, *Proc. 11th Int. Conf. Environmental Degradation of Materials in Nuclear Power Systems—Water Reactors,* American Nuclear Society, 2003, p 920
159. B.H. Sencer, G.S. Was, H. Yuya, Y. Isobe, M. Sagasaka, and F.A. Garner, Cross-Sectional TEM and X-Ray Examination of Radiation-Induced Stress Relaxation of Peened Stainless Steel Surfaces, *J. Nucl. Mater.,* Vol 336, 2005, p 314–322
160. P.L. Andresen, SCC Growth Rate Behavior in BWR Water of Increasing Purity, *Proc. Eighth Int. Symp. on Environmental Degradation of Materials in Nuclear Power Systems—Water Reactors* (Amelia Island, FL), American Nuclear Society, 1998, p 675–684
161. P.L. Andresen, The Effects of Nitrate on the Stress Corrosion Cracking of Sensitized Stainless Steel in High Temperature Water, *Proc. Seventh Int. Symp. on Environmental Degradation of Materials in Nuclear Power Systems—Water Reactors* (Breckenridge, CO), R.E. Gold and E.P. Simonen, Ed., NACE International, 1995, p 609–620
162. P.L. Andresen and T.P. Diaz, "Effects of Zinc Additions on the Crack Growth Rate of Sensitized Stainless Steel and Alloys 600 and 182 in 288 °C Water," Bournemouth Conference, BNES, England, Oct 1993
163. P.L. Andresen and T.M. Angeliu, "Effects of Zinc Additions on the Stress Corrosion Crack Growth Rate of Sensitized Stainless Steel, Alloy 600, and Alloy 182 Weld Metal in 288 °C Water," paper 95409, Corrosion/95, NACE International, 1995
164. P.L. Andresen and Y.J. Kim, "Insulated Protective Coating for Mitigation of SCC in Metal Components in High Temperature Water," U.S. patent No. 5,463,281, Nov 7, 1995
165. Y.J. Kim and P.L. Andresen, Application of Insulated Protective Coatings for Reduction of Corrosion Potential of Type 304 Stainless Steel in High Temperature Water, *Corrosion,* Vol 54, 1998, p 1012–1017
166. D.S. Morton, S.A. Attanasio, J.S. Fish, and M.K. Schurman, "Influence of Dissolved Hydrogen on Nickel Alloy SCC in High Temperature Water," paper 99447, Corrosion/99, NACE International, 1999
167. D.S. Morton, S.A. Attanasio, G.A. Young, P.L. Andresen, and T.M. Angeliu, "The Influence of Dissolved Hydrogen on Nickel Alloy SCC: A Window to Fundamental Insight," paper 01117, Corrosion/01, NACE International, 2001
168. D.S. Morton, S.A. Attanasio, and G.A. Young, Primary Water SCC Understanding and Characterization Through Fundamental Testing in the Vicinity of the Ni/NiO Phase Transition, *Proc. Tenth Int. Symp. on Environmental Degradation of Materials in Nuclear Power Systems—Water Reactors,* NACE International, 2002
169. K. Fukuya, K. Fujii, M. Nakano, N. Nakajima, and M. Kodama, Stress Corrosion Cracking on Cold-Worked 316 Stainless Steels Irradiated to High Fluence, *Proc. Tenth Int. Symp. on Environmental Degradation of Materials in Nuclear Power Systems—Water Reactors,* NACE International, 2002
170. C. Lemaignan and A.T. Motta, Zirconium in Nuclear Applications, *Nuclear Materials,* Vol 10B, B.R.T. Frost, Ed., VCH, 1994, p 1–52
171. R.L. Yang, Meeting the Challenge of Managing Nuclear Fuel in a Competitive Environment, *ANS International Topical Meeting on Light Water Reactor Fuel Performance,* American Nuclear Society, 1997, p 3–10
172. G.P. Sabol, G.R. Kilp, M.G. Balfour, and E. Roberts, Development of a Cladding Alloy for High Burnup, *Eighth International Symposium on Zirconium in the Nuclear Industry,* STP 1023, ASTM, 1989, p 227–244
173. F. Garzarolli and H. Stehle, IAEA Symposium on Improvements in Water Reactor Fuel Technology and Utilization, SM 288/24, IAEA, 1986, p 387–407
174. F. Garzarolli and R. Holzer, Waterside Corrosion Performance of Light Water Power Reactor Fuel, *J. Brit. Nucl. Soc.,* Vol 31, 1992, p 65–85
175. D. Charquet, Influence of Precipitate Density on the Nodular Corrosion Resistance of Zr-Sn-Fe-Cr Alloys at 500 °C, *J. Nucl. Mater.,* Vol 288, 2001, p 237–240
176. A. Yilmazbayhan, A.T. Motta, R.J. Comstock, G.P. Sabol, B. Lai, and Z. Cai, Structure of Zirconium Alloy Oxides Formed in Pure Water Studied with Synchrotron Radiation and Optical Microscopy: Relation to Corrosion Rate, *J. Nucl. Mater.,* Vol 324 (No. 1), 2004, p 6–22
177. D. Pecheur, Oxidation of β-Nb and Zr(Fe,V)$_2$ Precipitates in Oxide Films Formed on Advanced Zr-Based Alloys, *J. Nucl. Mater.,* Vol 278, 2000, p 195–201
178. D. Pecheur, F. Lefebvre, A.T. Motta, C. Lemaignan, and J.F. Wadier, Precipitate Evolution in the Zircaloy-4 Oxide Layer, *J. Nucl. Mater.,* Vol 189 (No. 1), 1992, p 318
179. J.M. Leger, P.E. Tomaszewski, A. Atouf, and A.S. Pereira, *Phys. Rev. B,* Vol 47, 1993, p 14075
180. A. Motta, Amorphization of Intermetallic Compounds under Irradiation—A Review, *J. Nucl. Mater.,* Vol 244, 1997, p 227–250
181. X.T. Zu, K. Sun, M. Atzmon, L.M. Wang, L.P. You, F.R. Wan, J.T. Busby, G.S. Was, and R.B. Adamson, Effect of Proton and Ne Irradiation on the

Microstructure of Zircaloy 4, *Philos. Mag. A,* Vol 85 (No. 4–7), 2005, p 649

182. F. Garzarolli, T. Broy, and R.A. Busch, Comparison of the Long-Term Behavior of Certain Zirconium Alloys in PWR, BWR and Laboratory Tests, *Proc. 11th International Symposium on Zirconium in the Nuclear Industry,* STP 1295, Garmisch-Partenkirchen, Germany, ASTM, 1996, p 850–864

SELECTED REFERENCES

- R.H. Jones, Ed., *Stress-Corrosion Cracking: Materials Performance and Evaluation,* ASM International, 1992
- *Proceedings of the International Conference on Environmental Degradation of Materials in Nuclear Power Systems—Water Reactors,* held every 2 years. Most recent is *Proceedings of the 11th International Conference on Environmental Degradation of Materials in Nuclear Power Systems—Water Reactors,* American Nuclear Society, 2003

Corrosion of Zirconium Alloy Components in Light Water Reactors

Clément Lemaignan, CEA France

MOST OF THE POWER REACTORS currently in operation are using natural (light) water as the coolant and moderator and are referred to as light water reactors (LWRs). Higher neutron efficiency may be obtained with heavy water (D_2O) as a substitute for the natural water. However, for all these types of reactors the fuel elements have basically the same design: long tubes filled with fuel oxide pellets—natural or enriched uranium dioxide (UO_2), or a mixed oxide, called MOX, $(U,Pu)O_2$. The tubes may vary in size and geometry. They are all stacked in assemblies for easy handling. For good neutron efficiency the fuel cladding tubes are made of zirconium (Zr) alloys. Typical geometries and operating conditions of the various assembly designs are given in Table 1.

In addition to the cladding, which is a critical component as it acts as the first containment wall for the radioactive fission products, other structural components are made of zirconium-base alloys. Among them, one should consider: the skeleton of the assemblies (the guide tubes—also called thimble tubes—and the spacer grids) and the channel box (in boiling water reactors, or BWRs) or pressure tubes (in CANDU and RBMK as described in Table 1).

All these components are in contact with the primary coolant, whose typical temperatures and pressures are given in Table 1. For all these industrial power plants, the temperatures range between 250 and 310 °C (480 and 590 °F), while high pressure (7 MPa, or 1 ksi, in boiling reactors to 15 MPa, or 2.2 ksi, in pressurized ones) is applied to obtain a high density of the steam or to maintain water in the liquid phase.

This type of environment is highly aggressive for most of the alloys. Claddings and structural components of the early days of the nuclear industry were made of stainless steels. However, the aim of reducing the ^{235}U enrichment by improving the neutron efficiency, and thus the economy of this source of energy, was a strong driving force for the selection of less-neutron-absorbing materials. Zirconium has been chosen as the base component for the development of alloys to be used in power water reactors. Various types of alloys have been studied, and two major families remain: the Zircaloys, that is, Zr-Sn-O (+Fe, Cr, Ni) alloys, and the Zr-Nb-O alloys (with 1 and 2.5 wt% Nb). Advanced ternary and quaternary alloys are currently under development for higher-performance industrial alloys.

Zirconium Alloys

Zirconium, as a metal, is highly reactive with oxygen and this oxide is very stable (much like Al_2O_3 or UO_2). Thus, the extractive metallurgy of zirconium is complex and requires specific procedures (reduction of zirconium tetrachloride, $ZrCl_4$, by liquid magnesium, vacuum arc melting, etc.). In addition, hafnium (Hf), a metal with chemical properties very close to those of zirconium (column IV B of the periodic table of the elements), has to be removed completely from zirconium, since hafnium absorbs neutrons. The corresponding purification process is performed early during the processing, by distillation of molten chlorides (Ref 1).

Pure zirconium exhibits two crystallographic structures, depending on temperature. Below 865 °C (1590 °F), the α structure is hexagonal close-packed (hcp), while above this transition temperature β zirconium is body-centered cubic (bcc). The melting temperature of pure zirconium is close to 1855 °C (3370 °F), classifying it as a refractory metal. In the β phase, all the chemical additions to be discussed later are fully soluble. In the α phase there is only limited solid solution of oxygen, tin, and niobium. The transition metals (iron, chromium, and nickel) are almost nonsoluble in the α phase and form intermetallic precipitates, whose size distribution significantly impacts the corrosion behavior, as is described later in this article. Typical compositions of the zirconium alloys are given in Table 2.

The major alloying elements have the following effects: tin (1.2 to 1.7%) improves the mechanical properties, mainly the creep resistance; niobium (1 or 2.5%) is very effective for corrosion resistance; oxygen (around 1250 ppm) is added voluntarily as a hardening

Table 1 Characteristics of water reactor fuel assemblies

	Fuel rod							Assembly		Coolant			
	Outside diameter		Thickness		Length					Temperature		Pressure	
Reactor type	mm	in.	mm	in.	m	ft	Alloy	Geometry	No. of rods	°C	°F	MPa	ksi
Pressurized water reactor (PWR)	9.5–10.5	0.374–0.413	0.6–0.7	0.024–0.028	3.7–4.3	12.1–14.1	Zry4 (Zircaloy4) Zr-1%Nb (M5®)	Square	240–300	290–335	555–635	15.5	2.25
Boiling water reactor (BWR)	12–13	0.472–0.512	0.8	0.031	3.5–4	11.5–13.1	Zry2	Square with channels	55–65	275–285	525–545	7	1.02
CANDU (CANadian Deuterium Uranium)	13.8	0.543	0.41	0.016	0.5	1.6	Zry4	Cylinder in pressure tube	37	Heavy water 265–310	510–590	10	1.45
WWER (Russian design PWR)	9.1	0.358	0.7	0.028	3.5	11.5	Zr-1%Nb (E110)	Hexagonal	331	290–320	555–610	15.7	2.28
RBMK (graphite-moderated Russian design BWR)	13.6	0.535	0.825	0.032	2–3.5	6.6–11.5	Zr-1%Nb (E110)	Cylinder in pressure tube	2–18	270–284	520–545	6.7	0.97

agent for all the alloys. Transition metals are added to the Zircaloy alloys at low concentrations (below 0.5%). They have been observed to significantly improve the corrosion behavior of the zirconium-tin alloys.

The final heat treatment, after the last cold-rolling pass of the cladding tubes, may be performed within two temperature ranges, corresponding to two types of microstructure:

- The stress-relieved state (SR) is mostly used in PWRs for Zircaloy 4 (Zry4). It corresponds to a final heat treatment at 475 °C (885 °F) for 2 h under vacuum or protective atmosphere, during which the dislocation density is reduced, and without recrystallization of the alloy. The grain shapes remain elongated and highly deformed.
- The recrystallized state (RX) is obtained at higher final heat treatment temperatures of 550 to 600 °C (1020 to 1110 °F). The grains are then equiaxed, with an average grain size in the range of 5 to 7 µm. Due to a much lower dislocation density, the yield strength is reduced, offset by a higher resistance to creep deformation. Zry2 is used in the RX state for BWRs and CANDUs. For the zirconium-niobium alloys, the RX state is generally used. For pressure tubes, specific heat treatments after the extrusion process have been developed. For these alloys, the second-phase particles are niobium-rich bcc precipitates in a Zr (+Nb) matrix.

As zirconium has an hcp structure at temperatures typically used for deformation processing (cold rolling of plates and tubes), the anisotropic deformation mechanisms (dislocation glide on prismatic planes of the hcp crystals and twinning) lead to the development of a strong crystallographic texture of the final products. This texture affects significantly the physical and mechanical behavior (anisotropy in thermal expansion, in mechanical properties, etc.). It does not impact the corrosion itself, but is involved in one by-product of the corrosion: The hydrogen pickup during corrosion leads to the precipitation of brittle hydrides upon cooling, the platelet orientations of which are controlled by this texture.

Corrosion of Zirconium Alloys in Water without Irradiation

As for most of the metallic alloys, the behavior of zirconium in contact with high-temperature water leads to an oxidation reaction, with formation of a solid oxide and release of hydrogen according to:

$$Zr + H_2O \rightarrow ZrO_2 + 4H \quad (Eq\ 1)$$

This leads to the buildup of an oxide layer at the surface of the alloy. The growth of the oxide layer has been shown to occur by diffusion of O^{-2} ions through the oxide and formation of new oxide layer at the metal/oxide interface.

Therefore, this layer acts as a barrier for the development of corrosion. As this layer thickens, the corrosion rate decreases. This protective behavior of the zirconia layer is, however, limited by a transition occurring at a thickness of a few micrometers. The detailed mechanisms of this transformation are still under discussion, but seem to be connected to a phase transformation of the zirconia (ZrO_2) (Ref 2).

Indeed, the crystallographic form of the zirconia at standard temperature and pressure is the monoclinic structure. At high pressure, either the cubic or the tetragonal phase is observed, depending on temperature. During the buildup of the oxide layer, the volume expansion due to the oxide formation at the metal/oxide interface induces compressive stresses in the growing oxide layer to compensate for the local swelling (ZrO_2 has a Pilling-Bedworth factor of 1.56). As the oxide layer grows, these compressive stresses enhance the stability of the tetragonal phase. Due to internal mechanical equilibrium, the compressive stresses cannot increase without limits. For a thickness in the range of 2 to 3 µm, a relaxation process occurs, allowing the zirconia to transform to the monoclinic structure. In an ionic binding material, such as zirconia, this process induces internal cracking. Other phenomena, such as chemical instability of the tetragonal zirconia in contact with high-temperature water, may contribute to the mechanisms of the crystallographic phase change involved in the transition and to the related development of a porous external zirconia layer (Ref 3).

After this transition, the oxide consists of an internal layer, rich in the tetragonal phase and considered to be dense and protective, above which the outer layer is mostly monoclinic and porous. This structure is transferred continuously as the oxide grows, allowing a constant oxidation rate. As a whole, the corrosion kinetics are described in Fig. 1. A pretransition regime, with a parabolic-cubic behavior controlled by the diffusion of oxygen through the entire protective oxide layer, followed by a more linear regime. The later-stage kinetics are controlled by the diffusion of the oxidizing species through the internal oxide layer, the thickness of which could be considered constant after a few damped oscillations. In the case of zirconium-niobium alloys, the transition is much less pronounced and the evolution of the oxidation kinetics with oxide thickness appears to be better described by an interface rate-controlling step. The dissociation of the water molecule at the oxide/coolant interface has indeed been shown to be a rate-controlling step in this type of alloy (Ref 5).

Table 2 Nominal compositions of zirconium alloys used in LWR reactors

Designation		Nominal composition, wt%					
Grade	UNS No.	Sn	Fe	Cr	Ni	Nb	O
Zircaloy 2	R60802	1.2–1.7	0.07–0.2	0.05–0.15	0.03–0.08	...	(a)
Zircaloy 4	R60804	1.2–1.7	0.18–0.24	0.07–0.13	...	...	(a)
Zr-1Nb	...	...	...	...	...	1	(a)
Zr-2.5Nb	R60901	...	...	...	...	2.5	(a)

(a) Oxygen content in all grades ranges from 1100 to 1400 ppm.

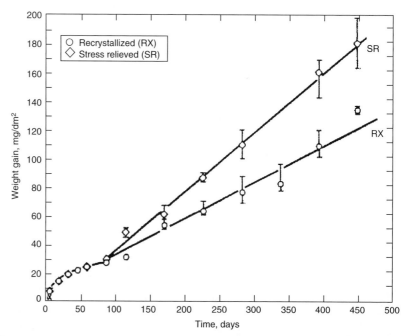

Fig. 1 General oxidation kinetics of Zircaloy in high-temperature water (1 µm ZrO_2 ~15 mg/dm^2). Source Ref 4

Various corrosion kinetics laws have been developed to predict the oxide thickness with time. Detailed modeling procedures separate the pretransition and posttransition phases; the pretransition regime is of the type $\varepsilon_1 = \varepsilon_{1,0} \cdot t^{1/3} \cdot \exp(\Delta H_1/RT)$, and the posttransition regime following the equation $\varepsilon_2 = \varepsilon_{2,0} \cdot t \cdot \exp(\Delta H_2/RT)$. In theses equations, ε_i is the oxide thickness for the ith regime, $\varepsilon_{i,0}$ is an experimental kinetics constant, t is the time, ΔH_i the activation energies of the oxidation process, R is gas constant, and T is absolute temperature. From a practical viewpoint, the pretransition kinetics and thickness can be neglected for oxide thickness typical of claddings after a few years in reactor (larger than 10 μm) (Ref 6).

Typical values of the parameters used in this equation remain within the following ranges: $\varepsilon_{2,0} = 0.5$ to 2×10^8 μm/day, and $\Delta H_2 = 110$ to 120 kJ/mol. For the current operating conditions of the power reactors, a growth of the oxide layer of 5 to 20 μm/yr is observed for standard Zircaloy and much less for the more recent zirconium alloys. The activation energy for corrosion is similar to that for the diffusion of oxygen in zirconia, giving credit to the mechanism of oxide growth by diffusion of O^{-2} ions through the oxide layer. For typical LWR temperature conditions, and due to the high activation energy of the corrosion mechanism, an increase in temperature of 20 K will induce an almost twofold increase in the corrosion rate. Therefore, accurate thermal modeling of the oxide layer is mandatory for a reliable oxidation prediction.

Heat Flow Conditions

During reactor operation, the heat flow through the cladding induces a thermal gradient and the temperature is radial position dependent. Since only the inner part of the oxide is dense and protective, the temperature at this location, that is, at the oxide/metal interface, controls the kinetics of oxide growth. Thus, under conditions of heat flow, such as for fuel cladding, this temperature has to be determined accurately. Usually the coolant temperature and heat flow are well known, and the thermal conductivity of the zirconia is the main parameter controlling the temperature at the metal/oxide interface. The data on the thermal conductivity of such a thin oxide layer, whose structures are complex and may be rather different according to specific corrosion histories, are reported within the range of 1.6 to 2 W/m·K. As the oxide thickens, the thermal barrier due to the oxide layer induces a rise of the temperature at the metal/oxide interface and the corrosion rate increases accordingly. For a typical mean linear heat generation rate of 20 kW/m and an oxide thickness of 40 μm, the oxide/metal interface temperature is higher than the coolant by 20 K, and the corrosion rate increases by a very significant factor of two. In addition to the barrier layer corresponding to the oxide, cruds may be formed during reactor operation and add an additional temperature increase or may change the local chemical conditions, affecting the corrosion rates (Ref 7).

Detailed computations of the metal/oxide interface temperature require accurate thermal-hydraulics and heat transfer modeling codes. Such approaches should consider the bulk coolant temperature, single- and two-phase flow behavior (pressurized and boiling reactors), limited temperature drop in the loose crud layer (frequent in boiling conditions), and the heat transfer in the zirconia. Several thermal-hydraulic codes have been developed for such computations, and their detailed descriptions are beyond the range of this review on corrosion. Typical profiles are illustrated in Fig. 2.

Impact of Metallurgical Parameters on Oxidation Resistance

The Zircaloys have been developed as variants of binary zirconium-tin alloys. Tin has been shown to reduce the detrimental effect of nitrogen contaminations and was added at concentrations of 1.5 to 1.7 wt%. An uncontrolled contamination of a batch of zirconium-tin alloy by a coupon of austenitic stainless steel has been the fortuitous origin of the development of the Zircaloy series. Out of them only the Zry2 and Zry4 variants remain in current use. As described previously, the solubility of the transition metals (Fe, Cr, Ni) is very low in the hcp α-Zr matrix (typically maximum total solubility of 200 ppm for the sum of them), and all these alloying elements form precipitates of $Zr_2(Fe,Ni)$ or $Zr(Fe,Cr)_2$ types. It is therefore not unexpected that the corrosion improvement induced by these transition metals will be strongly dependent of the microstructure of these precipitates (Ref 8).

For pressurized water conditions, large precipitates (larger than 150 nm) are required for the good corrosion resistance, while small ones are preferred for boiling conditions (smaller than 80 nm). The size control of these precipitates is obtained by the successive heat treatments occurring during processing. During the heat treatments in the β-bcc phase, all the alloying elements dissolve and form a solid solution in this high-temperature phase. During quenching, the precipitation of the intermetallics occur, with spatial and size distribution dependent on the cooling rates. The low solubility of these species in the α phase leads to slow coarsening kinetics, and the precipitate sizes remain small unless specific attention is taken in the thermomechanical treatments occurring during tube processing.

In order to control the adjustment of the precipitate sizes of the final products, a cumulative annealing parameter (termed as ΣA or A) is used (Ref 9). It combines all the heat treatments after quenching and is evaluated by:

$$A = \Sigma(A_i) = \Sigma(t_i \cdot \exp(-Q/RT_i)) \quad \text{(Eq 2)}$$

where t_i and T_i are the durations and temperatures of each heat treatment following the β-quench, and Q is defined by the activation temperature, usually taken as $Q/R = 40,000$ K.

Typical values of the A parameter are in the range of 1 to 5×10^{-17} h^{-1} for PWRs and 0.1 to 1×10^{-18} h^{-1} for BWRs, corresponding, respectively, to 200 to 500 nm and 50 to 80 nm for precipitate diameters. An increase in precipitate volume fraction is beneficial for the corrosion resistance (Ref 10). However, the different stoichiometries of the chromium- and nickel-bearing precipitates induce a different effect for increased additions of those elements.

For the zirconium-niobium alloys, the monotectoid transformation at 888 K is to be avoided during heat treatments, as higher temperatures would allow β-Zr precipitates to appear. In order to maintain the microstructure as closely as possible to the equilibrium diagram, heat treatments are therefore performed below this transition temperature, usually at 850 K. The best corrosion resistance is obtained for a microstructure close to the chemical equilibrium. This is indeed obtained with such a thermal history. In addition, during operation in reactors, the irradiation-enhanced diffusion processes allow a further evolution in that direction, improving the corrosion resistance as irradiation proceeds (Ref 11).

For all the alloys, various improvements can be achieved with very specific control of specific minor elements, for example, silicon increased to 150 ppm, carbon limited to 100 ppm, and so forth (Ref 12). Similarly, surface finish also has an impact on the corrosion rates. For instance, it is observed that local impacts or scratches may act as nucleation sites for nodule formation. The usual final surface treatments are either pickling in a nitric acid-hydrofluoric acid bath or finishing (grinding or polishing) with silicon carbide abrasives.

Hydrogen Pickup and Hydriding

The major part of the atomic hydrogen released in the pores of the granular oxide, by the reduction of the water molecules, recombine and form H_2 molecules, remaining in solution in the

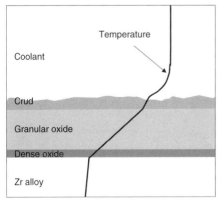

Fig. 2 Temperature profile through the oxide layer in heat flow condition. Temperature decreasing from zirconium alloy to coolant

primary water. However, the metallic zirconium alloys may trap a small part of it. Indeed hydrogen may react with zirconium to form a solid solution at high temperature and will precipitate as zirconium hydrides ($ZrH_{1.6}$) when the temperature is reduced (Ref 13). The diffusion of hydrogen in the zirconia layer is very low ($D_{H\text{-}Zr} < 10^{-15}$ m^2/s at reactor temperatures) and hydrogen ingress cannot occur by bulk diffusion in the oxide layer. Hydrogen uses diffusion short circuits through the metallic precipitates that remain unoxidized in the dense oxide layer. Therefore, they can directly connect the coolant located in the porous oxide to the metallic zirconium alloys.

The fraction of hydrogen trapped in the alloy (pickup fraction) depends on the alloy (composition and microstructure) and also on the oxidation conditions (temperature and water chemistry). Typical values are around 0.15 to 0.25 for Zry4 in PWRs, 0.3 to 0.6 for Zry2 in BWRs, and approximately 0.1 for 1% Nb alloys. The higher value of the hydrogen pickup for Zry2 has been explained by the presence of nickel in this alloy, a noble metal enhancing catalytic reactions with hydrogen. Once trapped in the alloy, the hydrogen cannot escape and its concentration increases with time. It can reach a few hundred ppm at the end of life of the component (Ref 14).

The solubility of hydrogen in zirconium alloys is strongly temperature dependent, with a heat of solution of 64 kJ/mol. At 320 °C (610 °F) the solubility of hydrogen is approximately 90 to 110 ppm, while at room temperature it is below 1 ppm. Thus, after irradiation, the oxidized cladding exhibits precipitation of zirconium hydrides as platelets. The morphology of the hydride platelets is controlled by the cooling rate (hydrogen diffusion kinetics), by the stress state (hydrides expand at formation), and by the crystallographic texture (habit planes of the hydrides close to basal plane of zirconium). These hydrides are brittle at low temperature and the precipitation of hydride platelets during cooling, if their orientations favor a cracking path through them, reduces the ductility of the alloy in case of low-temperature stress or impacts. This could occur during fuel handling in pool and transport. At operating temperatures, the hydrides are ductile and do not affect the mechanical properties of the zirconium alloy components (Ref 15).

Oxide Morphology and Integrity

The oxide layers grown in reactor may exhibit various morphologies:

- In reducing water environments, such as in PWRs, the first oxide layers, before the transition, are black (hypostoichiometric) and of uniform thickness. After the transition, lighter oxides can be observed, with small local variations in thickness. After the transition, layers of internal cracks or porous zones, parallel to the interface, have been reported. They have limited spatial extensions, are separated 3 to 5 μm apart, and may play a role for the spalling of the oxide layer described below (Ref 2).
- In less reducing environments (BWR type), the oxide layers may not be as uniform. At early stages of oxidation, local oxide growth instabilities occur, producing thick oxide spots or nodules. Once nucleated, the nodules grow in diameter and thickness, but their number remains constant. Improvements in alloy compositions and heat treatments reduce the occurrence of this type of corrosion mechanism. Typical aspects of this "nodular" corrosion are given in Fig. 3.

Oxide layers of high thickness have been observed in reactors in several cases of specific conditions, where zirconium alloy components are facing dissimilar metals. This can occur in front of Inconel holding springs or stainless steels parts, for example, control crosses in BWRs. Increases in oxide thickness of five- to tenfold have been reported. Known as "shadow corrosion," this phenomenon only occurs under irradiation and cannot be duplicated without neutron or gamma flux (Ref 16). Local irradiation enhanced radiolysis has been suggested to explain the phenomenon (Ref 17), but local electrochemical conditions under irradiation are now considered as the critical parameter driving "shadow corrosion" (Ref 18).

For high thickness (above 50 to 100 μm), the oxide layer may locally spall out. Flakes of oxide, a few mm^2 in size, are released in the coolant. The effect of this behavior is twofold: due to the thermal barrier made by the zirconia, the local spalling reduces the oxide thickness and, as a consequence, the temperature at the oxide/metal interface is reduced. The oxidation rate should locally decrease. On the other hand, this local "cold spot" is a location for preferential diffusion of the hydrogen in solution. In the worst cases, solid hydrides across the cladding thickness may be observed in spall-out areas (Ref 19). The local massive precipitation of hydrides will increase the local oxidation rate, even if the temperature is lower. In addition, it induces locally a strong embrittlement of the cladding, a matter of concern in the case of low-temperature mechanical loads, as described previously.

Effect of Irradiation on Microstructure and Corrosion

The irradiation damage, induced by the fast neutrons knocking the metal atoms and inducing displacement cascades and point defects, leads to changes in microstructure. Among the various changes induced by irradiation, one should consider those that may affect the corrosion behavior:

- In the Zircaloys, the secondary phases (Zr_2Ni and $ZrCr_2$ types) are ordered phases. During the recombination of vacancies and interstitials, the replacement of a displaced zirconium atom may be realized by an iron, chromium, or a nickel atom and, vice versa, a mechanism that induces a local chemical disorder (antisites). For typical LWR conditions, this is the physical origin of an irradiation-induced crystalline-to-amorphous phase transformation observed within these phases. This occurs especially with the $ZrCr_2$-type precipitates and is linked to significant chemical composition changes (iron is released from the precipitate to the matrix). Since the chemistry and sizes of these phases are critical parameters with respect to corrosion behavior, this irradiation-induced microstructure change has been observed to severely affect the corrosion resistance after irradiation (Ref 20).

 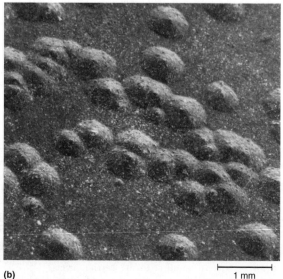

Fig. 3 Nodular corrosion in Zircaloy 2 fuel cladding. (a) Unaided view. (b) Scanning electron micrograph enlargement.

- For the zirconium-niobium alloys, the niobium solubility decreases with temperature and the zirconium matrix is supersaturated with niobium after the final heat treatment at 550 to 580 °C (1020 to 1075 °F). Due to the irradiation-enhanced diffusion of niobium under irradiation, the time under irradiation leads to the chemical equilibrium with additional niobium precipitation as nanometric bcc niobium precipitates. Thus, for these alloys, the irradiation permits a faster return to the chemical equilibrium at the irradiation temperature, a state of higher corrosion resistance (Ref 21).

As an example of the effect of the irradiation-induced microstructure change on oxidation rate, Fig. 4 describes the corrosion behavior of two coupons of Zircaloy (base metal and weld), tested in an autoclave after irradiation, compared with the same alloy not irradiated. The weld had a very fine microstructure due to the β quenching, and the intermetallics dissolved early during irradiation, obliterating the corrosion resistance. For the base metal, a similar behavior is observed, but less pronounced, as the precipitates were larger at the beginning of the irradiation and were only partly transformed during irradiation.

LWR Coolant Chemistry

Various technical constraints impact the water chemistry. In PWRs, soluble neutron absorber is added as boric acid in solution (H_3BO_3) to control the core reactivity. Its concentration decreases as the cycle proceeds, from about 1600 to 500 ppm, at a rate of approximately 3 ppm/day. To counterbalance the low pH, lithium hydroxide (LiOH) is added as an alkalizing agent (2.2 to 0.7 ppm). The normal pH at temperature is 6.9 to 7.2, depending on the time during the cycle. Such a high pH is required to avoid corrosion of stainless steel and the transport of activated chemical species from the core to the cold spots, like the steam generator. However, it has been shown that lithium has a detrimental effect on zirconium alloy corrosion that seems to be mitigated by the presence of the boric acid. A tendency exists to reduce the maximum allowable lithium concentration. In WWER (Russian-designed PWR) the alkalinizing agent is often potassium hydroxide (KOH). In the CANDUs, due to continuous reload, no reactivity compensation is needed in the coolant, and the only chemical addition in water is LiOH to increase the pH to 7.6 in operation.

With respect to the effect of radiolysis, H_2 is added in the primary water of single-phase reactors (PWR, WWER, CANDU) (typically 30 cm^3 STP.kg^{-1}, i.e. 3.5 ppm). It can be shown that H_2 catalyzes the recombination of the radiolysis species, as well as the initial dissolved oxygen, allowing a low oxidation potential in PWR water. In the case of BWRs, the H_2 separates from the water phase and escapes in the steam. In these reactors, a high oxidation potential remains in the coolant outside the core and can lead to intergranular cracking occurring on primary piping. Reducing conditions can, however, be obtained by H_2 additions, and now almost all BWRs operate under hydrogen-controlled water chemistry. This is obtained with continuous H_2 injection in the feedwater of the core (Ref 23). The typical coolant conditions of water reactors are given in Table 3.

Corrosion of Fuel Rods in Reactors

As a whole, the corrosion behaviors in reactors will combine all the effects described previously. Figure 5 shows a plot of the oxide thickness of two fuel rods of the same assembly after 4 years of irradiation in a PWR. Several features can be observed. Oxide thickness generally increases along the fuel rod because the coolant temperature increases as it is heated by the fuel. In the lower part of the core, the corrosion rate is low due to a temperature close to the inlet temperature (see Table 1), while the upper parts of the rods exhibit higher oxide thickness. At a finer scale, large oscillation in oxide thickness can be observed. They correspond to the enhanced heat exchanges induced by the mixing effects of the grids. Downstream of the grids (above them as the coolant flows upward), a more turbulent flow enhances heat exchanges and reduces the temperature of the oxide in contact with the water. Typical distance between grids for PWR and BWR are approximately 0.5 m (1.6 ft). Specific research and development is focused on improving the turbulent efficiency of the grids without increase of the corresponding pressure drops. Finer observation reveals higher-order oscillation related to the fuel pellet/pellet interfaces, where hydride precipitation had occurred, locally increasing the corrosion thickness.

Usually the design rules do not precisely state the maximum oxide thickness allowable during operation. However, other design constraints (remaining cladding mechanical properties, thermal fuel profile, etc.) lead to a maximum oxide thickness of approximately 100 μm. The

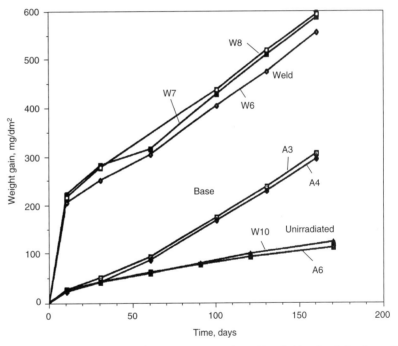

Fig. 4 Corrosion behavior of several Zircaloy specimens in autoclave with and without irradiation. Source Ref 22

Table 3 Typical water chemistry for water reactors

	Reactor type (a)			
	PWR	BWR	CANDU	WWER
H_2, ppm	3–4	...	6 (D_2)	2.6–5.3
O_2, ppb	<5	20–50	<50	<10
pH (25 °C)	10.5–10.9	...	10.5–10.9 (pD)	6–10.2
Conductivity, μS/cm	<1	<0.2	30	...
H_3BO_3, ppm	1800–500	0	0	1350–0
LiOH, ppm	1–2.2	0	...	2.5–7 (KOH)
Cl^-, ppb	<50	<15	<0.2	(Cl + F) < 100 ppb
Fe, ppb	...	<2	<500	<0.2 or <0.02
Cu^+, ppb	...	<0.5	...	<20
SiO_2, ppm	<4	<4	<4	...

(a) See Table 1

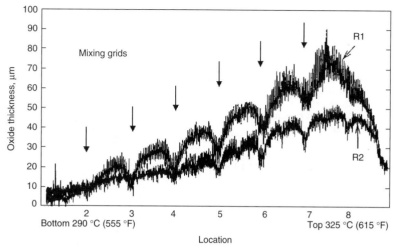

Fig. 5 Oxide thickness along two PWR fuel rods, R1 and R2, of the same assembly. Source Ref 24

oxide thickness observed for the standard alloys irradiated in reactors operating in demanding conditions have been close to this value. This fact has been a strong driving force for the development of new, more corrosion-resistant alloys, such as the low-Sn Zry4 variants, the Zirlo (1-Sn, 1.03-Nb, 0.01-Fe) (Westinghouse Electric Co.), or the M5 (1-Nb, O) (Areva Framatome ANP).

REFERENCES

1. C. Lemaignan and A. Motta, Zr Alloys in Nuclear Applications, *Nuclear Materials*, Vol 10B, BR Frost, Ed., VCH, 1994, p 1–52
2. P. Bossis, G. Lelièvre, P. Barberis, X. Iltis, and F. Lefebvre, Multiscale Characterization of the Metal-Oxide Interface of Zr Alloys, *Zirconium in the Nuclear Industry, 12th Intl. Symp.*, STP 1354, ASTM, 2000, p 918–945
3. X. Iltis, F. Lefebvre, and C. Lemaignan, Microstructure Evolutions and Iron Redistribution in Zircaloy Oxide Layers: Comparative Effects of Irradiation Flux and Irradiation Damages, *Zirconium in the Nuclear Industry, 11th Intl. Symp.*, STP 1295, ASTM, 1996, p 242–264
4. B.C. Cheng, R. Kruger, and R. Adamson, Corrosion Behavior of Irradiated Zircaloy, *Zirconium in the Nuclear Industry, 10th Intl. Symp.*, STP 1245, ASTM, 1994, p 400–418
5. P. Bossis, J. Thomazet, and F. Lefebvre, Study of the Mechanisms Controlling the Oxide Growth under Irradiation: Characterization of Irradiated Zry4 and Zr1Nb-O Oxide Scales, *Zirconium in the Nuclear Industry, 13th Intl. Symp.*, STP 1423, ASTM, 2002, p 190–221
6. B. Cheng, P.M. Gilmore, and H.H. Klepfer, PWR Zry Fuel Cladding Corrosion Performance Mechanisms and Modeling, *Zirconium in the Nuclear Industry, 11th Intl. Symp.*, STP 1295, ASTM, 1996, p 137–160
7. *Waterside Corrosion of Zr Alloys in Nuclear Power Plants*, TecDoc 996, International Atomic Energy Agency, Vienna, Austria, 1998
8. H. Anada and K. Takeda, Microstructures of Oxides on Zircaloy-4, 1.0 Nb Zry4, and Zry2 Formed in 10.3 MPa Steam at 673 K, *Zirconium in the Nuclear Industry, 11th Intl. Symp.*, STP 1295, ASTM, 1996, p 35–54
9. H. Anada, K. Nomot, and Y. Shida, Corrosion Behavior of Zry-4 Sheets Produced under Various Hot Rolling and Annealing Conditions, *Zirconium in the Nuclear Industry, 10th Intl. Symp.*, STP 1245, ASTM, 1994, p 307–327
10. P. Rudling et al., Impact of Second Phase Particles on BWR Zry2 Corrosion and Hydriding Performance, *Zirconium in the Nuclear Industry, 12th Intl. Symp.*, STP 1354, ASTM, 2000, p 678–706
11. O.T. Woo, V.F. Urbanic, R.M. Hutcheon, M. Griffiths, and C.E. Coleman, Corrosion of Electron Irradiated Zr-2.5 Nb and Zry2, *Zirconium in the Nuclear Industry, 12th Intl. Symp.*, STP 1354, ASTM, 2000, p 709–734
12. D. Charquet, Microstructure and Properties of Zr Alloys in the Absence of Irradiation, *Zirconium in the Nuclear Industry, 12th Intl. Symp.*, STP 1354, ASTM, 2000, p 3–14
13. B.F. Kammenzind et al., The Long Range Migration of H Through Zry in Response to Tensile and Compressive Stress Gradients, *Zirconium in the Nuclear Industry, 12th Intl. Symp.*, STP 1354, ASTM, 2000, p 196–233
14. Y. Broy et al., Influence of Transition Elements Fe, Cr and V on Long Time Corrosion in PWRs, *Zirconium in the Nuclear Industry, 12th Intl. Symp.*, STP 1354, ASTM, 2000, p 609–622
15. J.B. Bai, C. Prioul, and D. François, Hydride Embrittlement in Zircaloy-4 Plate, I. Influence of the Microstructure on the Hydride Embrittlement in Zircaloy-4 at 20 and 350 °C, *Metall. Trans.*, Vol 25A, 1994, p 1185–1197
16. R. Adamson, S.P. Lynch, and J.S. Davies, "Hot Cell Observation of Shadow Corrosion Phenomena," Fachtagung des KTG Fachgruppe: Brennelemente und Kernbauteile KFK, FRG, March 2000
17. C. Lemaignan, Impact of β-Radiolysis and Transient Products on Irradiation-Enhanced Corrosion of Zirconium Alloys, *J. Nucl. Mater.*, Vol 187, 1992, p 122–130
18. C. Lemaignan, Physical Phenomena Concerning Corrosion under Irradiation of Zr Alloys, *Zirconium in the Nuclear Industry, 13th International Symposium*, STP 1423, ASTM, 2002, p 20–29
19. A.M. Garde, G.P. Smith, and R.C. Pirek, Effect of Hydride Precipitate Localization and Neutron Fluence on the Ductility of Irradiated Zry4, *Zirconium in the Nuclear Industry, 11th Intl. Symp.*, STP 1295, ASTM, 1996, p 407–430
20. D. Gilbon and C. Simonot, Effect of Irradiation on the Microstructure of Zircaloy-4, *Zirconium in the Nuclear Industry, 10th Intl. Symp.*, STP 1245, ASTM, 1994, p 521–548
21. J.P. Mardon, D. Charquet, and J. Senevat, Influence of Composition and Fabrication Process on Out-of-Pile and In-Pile Properties of M5 Alloy, *Zirconium in the Nuclear Industry, 12th Intl. Symp.*, STP 1354, ASTM, 2000, p 505–524
22. R. Adamson, Effect of Neutron Irradiation on Microstructure and Properties of Zircaloy, *Zirconium in the Nuclear Industry, 12th Intl. Symp.*, STP 1354, ASTM, 2000, p 15–31
23. B.C. Cheng et al., Fuel Performance and Water Chemistry Variables in LWRs, *ANS Light Water Reactor Fuel Performance* (Portland, OR), March 2–6, 1997, American Nuclear Society, p 379–388
24. J.C. Clayton and R.L. Fischer, Corrosion and Hydrating of Zry Fuel Rod Cladding in 633K Water and Reactor Environments, *ANS Light Water Reactor Fuel Performance* (Orlando, FL), April 21–24, 1985, American Nuclear Society, p 3–1 to 3–16

SELECTED REFERENCES

- D. Franklin and P. Lang, Zr Alloy Corrosion: A Review Based on an IAEA Meeting, *Zirconium in the Nuclear Industry, 9th Vol*, STP 1132, ASTM, 1991, p 3–32
- F. Garzarolli, T. Broy, and R.A. Busch, Comparison of the Long-Term Behavior of Certain Zirconium Alloys in PWR, BWR and Laboratory Tests, *Zirconium in the Nuclear Industry, 11th Vol*, STP 1295, ASTM, 1996, p 850–864
- C. Lemaignan, Physical Phenomena Concerning Corrosion under Irradiation of Zr Alloys, *Zirconium in the Nuclear Industry, 13th International Symposium*, STP 1423, ASTM, 2002, p 20–29
- *Waterside Corrosion of Zr Alloys in Nuclear Power Plants*, TecDoc 996, International Atomic Energy Agency, Vienna, Austria, 1998, 313 pages

Corrosion of Containment Materials for Radioactive-Waste Isolation

Raúl B. Rebak and R. Daniel McCright, Lawrence Livermore National Laboratory

RADIOACTIVE MATERIALS are extensively used in a variety of applications such as medical, weapons, and power generation. Once these materials lose their commercial value they are considered radioactive waste. Broadly, the wastes can be separated into defense (weapons) and civilian (power, medical) (Ref 1).

The safe disposal of radioactive waste requires that the waste be isolated from the environment until radioactive decay has reduced its toxicity to innocuous levels for plants, animals, and humans. Many different types of radioactive waste are produced during commercial and defense nuclear fuel cycles. One type of waste, denoted high-level waste (HLW), contains the highest concentration of radiotoxic and heat-generating species. Because of this factor, the most stringent standards for disposing of radioactive wastes are being placed worldwide on HLW, and the majority of the radioactive-waste management effort is being directed toward the HLW problem. One of the most common types of HLW is the spent nuclear fuel (SNF) from commercial nuclear reactors for power generation.

All of the countries currently studying the options for disposing of HLW have selected deep geologic formations to be the primary barrier for accomplishing this isolation. It is postulated that by the own nature of these geological sites, they will contain the waste for long times, avoiding its spread, for example, through water flow. Most of the repository designs also plan to delay the release of radionuclides to the environment by the construction of engineered barrier systems (EBS) between the waste and the geologic formation. The principal engineered component in this multibarrier approach is called the waste package and includes the waste itself, possibly a stabilizing matrix for the waste (together termed the wasteform), and a metallic container that encloses the wasteform. Beyond the metallic containers, other barriers could be added to attenuate the impact of the emplacement environment on the containers. These barriers include, for example, the drip shield proposed in the U.S. design or a bentonite backfill proposed in several European designs (Ref 2, 3).

The waste container may be a single vessel, but current designs generally call for concentric double-walled vessels of dissimilar metals. Each vessel will have a specific function. For example, the container that directly holds the wasteform may be designed to shield radiation and facilitate safe waste handling and emplacement operations. This container may be overpacked with a corrosion-resistant outer layer. This article addresses the long-term corrosion behavior of HLW container materials, more specifically of the outer shell of the containers.

Several countries are currently studying the viability of the geologic disposal of nuclear waste. The design of the different containers for nuclear waste would vary according to the nature of the geologic formation at the site of the repository. That is, the materials (metals or alloys) that may be used for the external wall of the container would depend on, for example, whether the redox potential of the surrounding environment is predicted to be mostly reducing or mostly oxidizing and what major corroding species would be present at the site. In 2001, there were more than 30 nations currently considering the geological disposal of HLW (Ref 3). Table 1 shows a list of nations, the intended geologic formation, the type of environment, and the current proposed design for the containers (Ref 2, 3). The material selection and its behavior in each type of emplacement are discussed separately.

Time Considerations

In the last 15 years, the predicted length of time for the safe survival of the HLW repository sites has increased. In 1987, when the previous version of this article was published, most of the repository designs specified from 300 to 1000 years. In 2004, the minimum length of time specified for many repositories has increased to 10,000, 100,000, and even to 10^6 years (Ref 2, 4–6).

The requirement that a waste container survive intact for thousand of years in an irradiated elevated-temperature geologic environment has created a difficult problem for materials and design engineers to solve. The unique aspect of this problem is associated with making predictions about the corrosion behavior of container materials for these extended periods of time, since many of the alloy systems being considered have been in existence for less than 100 years. The viability of extrapolating corrosion data from short-term testing to long-time performance has been addressed by some investigators (Ref 4, 7, 8). Other researchers proposed models to predict the lifetime performance of container alloys (Ref 9, 10). For example, corrosion-rate values from 5 year tests is expected to be extrapolated to times as long as 10,000 years, that is, more than three orders of magnitude (Ref 11). The problem associated with interpreting results for long-term behavior based on accelerated testing has been addressed by the ASTM Committee C-26 on Nuclear Fuel Cycle, who prepared the standard C 1174 that provides guidelines, among other issues, on the corrosion testing of materials (Ref 12). ASTM C 1174 also recognizes that actual data on the long-term behavior of materials in repositorylike conditions will not be available to be used in the design of the waste package. Therefore, ASTM C 1174 establishes that short-time (10 years) data need to be used to support the development of predictive behavior models for the response of the waste package over times as long as 10,000 years. ASTM C 1174 provides guidelines on how to perform accelerated corrosion tests to compensate for the shorter testing time. These acceleration factors may include the use of more concentrated solutions, higher temperatures, and applied potentials. The accelerating factors described in ASTM C 1174 may be used provided the degradation (alteration) mechanism(s) remains the same. The effects of accelerating conditions should be quantified and

Table 1   Different types of repository design mainly for high-level waste and spent fuel

Country	Potential site	Geologic formation (host rock)	Type of environment	Current container materials and EBS
Argentina	TBD	Clay/granite	Reducing	TBD
Belgium	Boom Clay Formation, Mol-Dessel (URL)	Clay	Backfilled with clay, bentonite, quartz. Strong reducing	Type 316 SS on type 304 SS
Canada	TBD	Plutonic rock (Granitelike)	Bentonite, sand, crushed rock. Reducing	Titanium, copper
China	Beishan (URL)	Granite	Bentonite	TBD
Finland	Olkiluoto	Crystalline	Bentonite, crushed rock. Reducing	Copper on cast iron
France	Meuse/Haute-Marne (URL)	Granite	Bentonite. Reducing	Concrete, steel on stainless steel
Germany	Morsleben, Gorleben (URL)	Salt mine	Salt, concrete	No credit as EBS
India	Rajasthan	Granite	TBD	TBD
Japan	Under study	TBD	Bentonite and sand. Reducing	Carbon steel, Copper overpacks
Korea	TBD	Plutonic rock	Bentonite, sand	Copper or stainless steel on carbon steel
Russia	Several (e.g., Nizhne-Kansky)	Granite	TBD	TBD
Spain	TBD	Granite or clay	Bentonite	Carbon steel
Sweden	Äspö (URL)	Crystalline bedrock	Bentonite. Reducing	Copper on cast iron
Switzerland	Mont Terri, Opalinus Clay, Grimsel (URL)	Crystalline and clay	Bentonite	Steel
United Kingdom	TBD, Cumbria (URL)	TBD(a)	TBD(b)	TBD(c)
United States	Yucca Mountain, Nevada	Unsaturated volcanic tuff	Unsaturated (dry). Oxidizing	Ni-Cr-Mo alloy on type 316 SS, drip shield of Ti alloy

Note: EBS, engineered barrier system; TBD, to be determined; URL, underground research laboratory; SS, stainless steel. Source: Ref 2, 3. (a) Possibly granitic. (b) Possibly reducing. (c) Possibly carbon steel

mechanistically described before they are used for lifetime predictions.

Environmental and Materials Considerations

Many of the countries that are considering the emplacement of HLW in geological repositories have recognized the importance of an underground research laboratory (URL) for the characterization of the rock, the testing of materials for environmental degradation, and other studies. The most common host rocks for nuclear waste repositories in the world are clay and granite (Table 1). The containers are intended to be located in alcoves buried vertically down from the ground level, that is, below the water table. The depth of emplacement may vary from country to country, but it is generally assumed to be on the order of 100 to 500 m (330 to 1640 ft). The only nonsaturated (above the water table) environment for a repository is the one projected for the United States. That is, the Yucca Mountain repository will be emplaced in tunnels buried, still underground, but horizontally into a mountain.

According to the value of the redox potential, an environment can be categorized as mostly reducing or mostly oxidizing. In the case of a reducing redox potential, the cathodic reaction is mostly controlled by the hydrogen discharge reaction. An oxidizing redox potential is established by cathodic reactions other than hydrogen reduction, for example by the reduction of dissolved oxygen. Most of the repository environments in the world will be reducing based on redox potentials, due to the fact of depth (where the solubility of oxygen in water is minimal) and a projected backfill with bentonite (Table 1). The intended function of the backfill is to retard the diffusion of oxygen toward the containers. The repository in the United States will not have restrictions regarding the availability of oxygen to contact the containers; that is, the redox potential will be oxidizing in nature, provided an aqueous solution materializes.

The important environmental parameters affecting corrosion include container temperature, groundwater chemistry (including microbial activity), groundwater flow rate, hydrostatic and lithostatic pressure (which influences water phase and container stress), and radiation flux.

Radiation. Theoretically, radiation can affect the corrosion behavior of the container by affecting both the container environment and its metallurgical properties. The general conclusion reached by most investigators is that the types and dose rates of radiation emitted from decaying wastes are not sufficient to degrade the properties of either the container material or its passivating oxide layer. Only gamma radiation from the HLW can affect the environment, because the other forms of radiation will not penetrate the container walls. It has been suggested that radiation may produce changes in the external environment, for example by controlling microbial activity or radiolysis of ground water. Currently, the United States plans to accept the most active waste for disposal (5 years out of reactor). This waste will produce maximum dose rates at the container surface of 10^3 to 10^5 rads/h.

Water Chemistry. The groundwaters associated with the crystalline-rock formations should all be relatively benign to most materials because of their low ionic strengths, near neutral pH, and low concentrations of halide ions (Table 2) (Ref 6). The corrosivity of these waters could increase if significant groundwater vaporization occurs when high container temperatures exist during the early times following emplacement.

Temperature. The container temperature may be influenced by the design and loading of

Table 2   Groundwater composition for different rock hosts

	Composition, ppm			
Ion	Basalt	Tuff (J-13)	Granite	Clay
Na^+	250	51	0–106	63
K^+	1.9	5	...	7.4
Mg^{2+}	0.4	2	0–6	3.6
Ca^{2+}	1.3	14	10–40	21
Sr^{2+}	...	0.05	...	...
Fe^{2+}	...	0.04	0.02–5	189
NH_4^+	...	...	0.05–0.2	...
Cl^-	148	7.5	4–36	36
SO_4^{2-}	180	22	0.5–24	...
I^-	...	...	...	...
CO_3^{2-}	97	120	90–275	188
Br^-	...	...	...	...
BO_3^{3-}	...	...	...	...
NO_3^-	...	5.6	0.01–0.05	6
HS^-	...	...	0–0.5	...
F^-	37	2.2	0–2	817
$H_3SiO_4^-$	103	61	0–19	8
pH	9–10	7.1	7–9	7.4

the waste package (size, thermal output, radiation output), the rate and density of waste package emplacement, and the thermal properties of the formation. Because heat is a significant by-product of HLW decay, the temperature of all waste containers will initially increase and then decrease as the activity of the waste decays. The predicted temperature history for 3 kW waste packages emplaced in a consolidated volcanic ash (tuff) formation in the United States is not expected to be higher than 160 °C (320 °F) (Ref 6). Typical maximum container temperatures for a number of other repository locations are expected to be lower than 100 °C (212 °F) (Ref 2). The variability in maximum temperature is due primarily to design philosophy. The temperature at a given location can be lowered by longer waste aging before emplacement, lower package loading, and lower overall repository loading. The lower temperatures may enhance the performance of the entire waste package and

Table 3 Approximate chemical composition for candidate alloys

Alloy	UNS No.	ASTM specification	Element, wt%						
			Cr	Cu	Fe	Mo	Ni	Ti	Other
Gray cast iron	F10001–F10012	A319–A159	...	...	~95 (bal)	...	...	...	3–3.5 C, 2–2.4 Si, 0.8 Mn
1018 carbon steel	G10180	A29	...	...	~98 (bal)	...	...	...	0.18 C, 0.5 Mn
4130 alloy steel	G41300	A29	1.0	...	~97 (bal)	0.2	...	...	0.3 C, 0.5 Mn
2.25Cr-1Mo	K30736	A213	2.25	...	bal	1	...	...	0.05 C, 0.4 Mn, 0.2 V
90-10 cupronickel	C70600	B111	...	~88 (bal)	1.3	...	10	...	1 max Mn, 1 max Zn
Type 304	S30400	A182	19	...	~70 (bal)	...	9	...	2 max Mn, 1 max Si
Type 316	S31600	A182	17	...	67 (bal)	2.5	12	...	2 max Mn, 1 max Si
Type 416	S41600	A194	13	...	~85 (bal)	0.6 max	...	...	1.25 max Mn, 1 max Si
Monel 400	N04400	B127	...	~32 (bal)	2.5 max	...	66.5	...	2 max Mn
Incoloy 825	N08825	B163	21.5	2.2	~30 (bal)	3.0	42	0.9	1 max Mn, 0.5 max Si
Inconel 625	N06625	B366	21.5	...	5 max	9.0	~60 (bal)	0.2	4 Nb, 0.5 max Mn
Hastelloy C-276	N10276	B575	16	...	5	16	~60 (bal)	...	4 W, 2.5 max Co
Hastelloy C-4	N06455	B575	16	...	3 max	16	~65 (bal)	...	2 max Co
Hastelloy C-22	N06022	B575	22	...	4	13	~57 (bal)	...	3 W, 2.5 max Co
Ti Gr 2	R50400	B265	...	...	0.3 max	...	...	~99 (bal)	0.25 max O
Ti Gr 7	R52400	B265	...	...	0.3 max	...	...	~98 (bal)	0.2 Pd, 0.25 max O
Ti Gr 16	R52402	B265	...	...	0.3 max	...	...	~98 (bal)	0.06 Pd, 0.25 max O
Ti Gr 12	R53400	B265	...	...	0.3 max	0.3	0.8	~98 (bal)	0.25 max O

decrease the impact of emplacing waste on the geologic formation itself. However, a penalty is incurred for lower temperatures because higher handling and emplacement costs, along with a larger usable area, are required.

Materials. An analysis of Table 1 shows that, except for the United States, most of the materials for the containers will consist of carbon steel, stainless steel, and copper. These metals are not in the high end of the scale of corrosion-resistant alloys. That is, the emphasis of the design is on a controlled environment. Also, the emphasis of the containment of the waste is put on the barriers beyond the outer layer of the metallic container. Contrarily, the materials for the engineered barrier in the U.S. repository are some of the most corrosion-resistant materials available today in the market of industrial alloys. That is, most of the responsibility of containing the waste in the United States is put on the environmental resistance of the alloys of the engineered barriers. The composition of candidate alloys is given in Table 3.

In general, engineers carry out the selection and characterization of the materials for the EBS and project their performance as a function of the emplacement time. However, the fact that one of these repositories is actually going to be built and put in service may depend largely on the public support based on their perception about whether the site would be safe.

Reducing Environments

This section discusses environmental corrosion resistance of materials planned for reducing repositories (Table 1). These containers will generally be surrounded by a backfill of bentonite, which will greatly limit the availability of oxygen to the metal surface. The lack of oxygen (or other oxidizing species) will create a redox potential that will be closer to the hydrogen evolution reaction. Elements such as iron (Fe), nickel (Ni), and copper (Cu) are mostly in the range of corrosion immunity at these reducing potentials in the near-neutral pH range (Ref 13). The most common materials under study in typically reducing environments are carbon steel, stainless steel, copper, and titanium (Tables 1 and 3). For the least corrosive underground waters, carbon steels could be viable materials; however, for the most saline conditions, titanium alloys are also being studied. For example, the groundwater in the crystalline-rock disposal environment in the Canadian shield was expected to be benign (no halide ion, neutral pH, low ion strength) (Ref 14). During a research drilling program, groundwater containing 5.6 g/L of chloride ion was encountered. The existence of isolated pockets of saline ground water in the deep rock formations supports the need for careful site characterization studies (Ref 15).

Carbon Steel and Low-Alloy Steel

Carbon and low-alloy steel have been extensively been tested in groundwater environments for the last 30 years. Corrosion rates measured for carbon steels in granite waters ranged from 3 to 55 μm/yr (0.12 to 2.2 mils/yr), with one study showing that the rate reaches a maximum at around 80 °C (175 °F) (Ref 16). The conditions that would lead to localized corrosion of carbon steels are quite specific and unlikely to be present in typical granitic groundwaters (Ref 17). However, hydrogen embrittlement and hydrogen blistering of low-alloy steels are possible in granitic environments with a high rate of hydrogen production (Ref 17). Carbon steel will also have low corrosion rates in basalt waters (Table 4) (Ref 18). Even in oxygenated solutions at 150 °C (300 °F), the corrosion rate of all tested carbon steels in basalt waters was on the order of 100 μm/yr (4 mils/yr) (Table 4). In another series of tests, 1020 carbon steel, cast carbon steel, cupronickel 90-10, and Fe-9Cr-1Mo steel were tested in anoxic (<0.1 mg/L dissolved oxygen) basalt groundwater at 200 °C (390 °F) for 5 months (Ref 19). The corrosion rates were: 0.9 μm/yr (0.035 mil/yr) for cupronickel 90–10 and Fe-9Cr-1Mo steel, 1.1 μm/yr (0.043 mil/yr) for cast carbon steel and 1.4 μm/yr (0.055 mil/yr) for wrought carbon steel. Under the Swedish program, researchers have studied the anoxic corrosion behavior of carbon steel and cast iron in groundwater at 50 and 85 °C (120 and 185 °F), and the impact of the presence of copper on the type and the mechanical properties of the films formed on the iron alloys (Ref 20, 21). They used a barometric cell filled with simulated groundwater and monitored the redox potential in the cell at 30 °C (85 °F) on a gold electrode. They determined that when steel was introduced to the cell, the redox potential decreased rapidly due to the consumption of the residual oxygen by the corrosion of the steel (Ref 20, 21).

As part of the Japanese program of nuclear waste disposal, the localized corrosion behavior of plain carbon steel was studied at 90 °C (195 °F) using the Tsujikawa-Hisamatsu Electrochemical (THE) method (Ref 22). It was reported that the crevice repassivation potential increased as the chloride concentration decreased and the carbonate/bicarbonate amount

Table 4 General corrosion rates of carbon steels under various conditions in basalt groundwater

Test duration: 2 weeks

Designation		Average corrosion rate, μm/yr (mils/yr)		
		Deoxygenated (<0.1 ppm O$_2$)		Oxygenated (~8 ppm O$_2$)
AISI	UNS	150 °C (300 °F)	250 °C (480 °F)	150 °C (300 °F)
1006	G10060	13 (0.51)	12 (0.47)	118 (4.64)
1020	G10200	13 (0.51)	12 (0.47)	101 (3.97)
1025	G10250	11 (0.43)	13 (0.51)	105 (4.13)

Source: Ref 18

increased; that is, carbonate inhibits the crevice corrosion susceptibility of carbon steel (Ref 22). The passive corrosion behavior of steels was found to be dependent on variables such as groundwater pH, temperature, and available dissolved oxygen (Ref 23). Scientists in the Japanese program have raised the concern that whenever the corrosion of steel decreases due to a decrease in the oxygen content, the alkalinity in the immediacy of the steel increases. This may increase the rate of hydrogen gas production that could be detrimental for the stability of the repository (Ref 24). In another study, it has been shown that the corrosion rate of carbon steel is dependent on the amount of bicarbonate (HCO_3^-) present in the water (Ref 25). At bicarbonate levels of 0.1 M, similar to the geological disposal site, the corrosion of carbon steel is inhibited.

Carbon steel has also been identified as candidate material for rock salt repositories in the German program. Studies were conducted to determine the corrosion response of welded and nonwelded Fe-1.5Mn-0.5Si steel in a magnesium chloride ($MgCl_2$) rich brine (Q-brine) at 150 °C (300 °F) under an irradiation field of 10 gray/hour (Gy/h, one gray equals 100 rads) (Ref 26). Welding was carried out using gas tungsten arc welding (GTAW) and electron beam (EB) welding. The overall corrosion rate of both welded and nonwelded materials was approximately 70 μm/yr (2.8 mils/yr); however, the welded materials experienced some localized attack in the weld seam area. When the material was heat treated for 2 h at 600 °C (1110 °F), the corrosion rate of the welded material increased by approximately 40% (Ref 26).

Carbon steel and low-alloy steel has also been identified as candidate material to contain nuclear waste for an intermediate storage of 100 years in the French program (Ref 27). The dry oxidation testing of carbon steel in dry air (less than 15 ppm water), in air plus 2% water and in air plus 12% water at 300 °C (570 °F) for up to 700 h showed little damage to the tested coupons. When the depth of the oxide layer was extrapolated to 100 years, it resulted in less than 150 μm (6 mils) of damage. The authors also noted little or nil water vapor effect on the oxidation rate at 300 °C (570 °F) (Ref 27).

Copper

The container for the disposition of nuclear waste in Sweden will consist of a 50 mm (2 in.) thick layer of copper over cast ductile iron, which will provide the mechanical strength. Groundwater in granitic rock (as in the Swedish repository) is oxygen free and reducing below a depth of 200 m (655 ft). The redox potential is between −200 and −300 mV in the hydrogen scale and the pH ranges from 7 to 9 (Ref 5). The chloride concentration in the groundwater can vary from 0.15 mM to 1.5 M with an equivalent amount of sodium and less calcium. The corrosion of a copper container in this reducing environment is expected to be less than 5 mm (0.2 in.) in 100,000 years of emplacement (Ref 5). The corrosion of copper is mainly controlled by the availability of oxygen, sulfate, and sulfide in the groundwater. The failure time of the copper layer in the Swedish container has been modeled, and it is predicted that the failure, both by general and pitting corrosion, would be greater than 10^6 years under realistic emplacement conditions (Ref 28).

The anodic behavior of copper was also studied as part of the Japanese nuclear waste disposal program using potentiodynamic polarization tests in simulated groundwater at 30 °C (85 °F) (Ref 29). The amount of dissolved oxygen as well as different additions of chloride, sulfate, and bicarbonate was controlled. The researchers concluded that both sulfate and chloride promote the active dissolution of copper while carbonate is a passivating agent (Ref 29).

In the Canadian design, the thickness of the external layer of the copper layer is 25 mm (1 in.). Scientists have modeled the failure mechanism of copper as a function of the oxygen availability, the temperature, the salinity of the solution, and the redox potential. It is predicted that copper will undergo general corrosion and pitting during the initial warm and oxidizing period, but only general corrosion during the subsequent longer anoxic cooler period. It has been predicted by this model that the Canadian copper container could last more than 10^6 years (Ref 30).

The long-term corrosion rates of many copper-base alloys are also sufficiently low, <20 μm/yr (<0.78 mil/yr) at 200 °C (390 °F), that their use now appears feasible. When a copper container is buried in a mostly reducing environment, the metal will initially be in contact with oxygen, until the oxygen is fully consumed, for example by corrosion (Ref 31–33).

It is known that copper may suffer environmentally assisted cracking (EAC) such as stress-corrosion cracking (SCC) in waters containing, for example, ammonia and nitrite (NO_2^-). It has been shown that copper alloys, candidate materials for the Canadian waste containers, were susceptible to SCC using the slow strain rate technique (Ref 34). It has been reported that the crack growth rate could be as high as 8 nm/s (3×10^{-7} in./s) (Ref 35). However, the conditions under which the damage occurred were extreme and unrepresentative of container emplacement conditions. In the actual container, the general absence of aggressive SCC species, the limited applied strain, and the limited supply of oxygen will limit the susceptibility to environmental cracking. In another study, it has been shown that the minimum stress for crack propagation in copper for the Swedish container was 30 MPa$\sqrt{m}$ (27.3 ksi$\sqrt{in.}$) when tested in a 0.3 M sodium nitrite ($NaNO_2$) solution (Ref 36). A stress intensity of 30 MPa$\sqrt{m}$ (27.3 ksi$\sqrt{in.}$) can be considered high for a statically loaded container having shallow defects on the surface.

Stainless Steel and Nickel Alloys

The cyclic potentiodynamic polarization method (ASTM G 61) was used to evaluate the anodic behavior of corrosion-resistant alloys in oxidized Boom clay water (in Belgium) with varying degrees of added chloride at 90 °C (195 °F) (Ref 37). The original Boom clay water is dominated by chloride and sulfate. The alloys studied included type 316L stainless steel (also with high molybdenum and with titanium) (UNS S31603), alloy 926 (N08926), alloy 904L (N08904), alloy C-4 (N06455), and Ti Gr 7 (R52400) (Table 3). It was found that both R52400 and N06455 resisted pitting corrosion even at added chloride concentrations of 10,000 ppm and N08926 resisted pitting up to 1000 ppm chloride. The other alloys showed minor pitting at 100 ppm chloride and definite pitting corrosion at the higher tested chloride concentrations (Ref 37).

Titanium

Titanium (Ti) alloys are under study as candidate materials for the containers in Canada, Japan, and Germany. The titanium alloys were selected as a potential alternative because of their excellent performance in more aggressive brine solutions compared, for example, to stainless steels. Corrosion rates for Ti Gr 2 and Ti Gr 12 in both oxygenated and irradiated basalt environments are very low—less than 2 μm/yr (0.08 mil/yr). Failure models for the degradation of Ti Gr 2 have been published (Ref 38). The model takes into account the crevice propagation rate as a function of temperature and oxygen availability as well as other factors such as the amount of hydrogen absorbed by the alloy during corrosion before a critical concentration for failure is reached (Ref 38). The localized corrosion resistance of titanium alloys was widely investigated using the THE method (Ref 39, 40). Testing showed that as the temperature and the chloride concentration increased, the repassivation potential (E_{RCREV}) for Ti Gr 1 and Ti Gr 12 decreased to values well below the corrosion potential (E_{corr}) (Ref 39). Ti Gr 12 was more resistant to crevice corrosion than Ti Gr 1. At constant temperature and chloride concentration, E_{RCREV} increased as the palladium (Pd) content in the alloy increased, rapidly up to 0.008% Pd and then slower between 0.008% to 0.062% Pd (Ref 40). In most of the tested electrolytes, the E_{corr} was lower than E_{RCREV} for the palladium-containing titanium alloys, suggesting that crevice corrosion will not initiate in the titanium-palladium alloys at the free corrosion potential (Ref 40).

Titanium alloys were also investigated for their resistance to EAC. One way by which titanium alloys may suffer EAC under reducing conditions is by the formation of hydrides (Ref 41). Slow strain rate testing was conducted using Ti Gr 1 in deaerated 20% sodium chloride (NaCl) at 90 °C (195 °F) at an applied potential of −1.2 V (standard hydrogen electrode, SHE) (Ref 42). It was confirmed that cracks initiated as deep as the presence of hydrides, and it was therefore dismissed that the containers may fail by cathodic EAC (Ref 42).

Oxidizing Environments

As mentioned previously, the United States is currently the only country that has designed and is characterizing a nuclear waste repository with an oxidizing environment surrounding the waste package. This repository is planned to be located in Yucca Mountain, Nevada. The design of the waste package has evolved in the last 10 years (Ref 6, 43). Since 1998, the design specifies a double-walled cylindrical container covered by a titanium alloy drip shield. The outer shell of the container will be a Ni-Cr-Mo alloy (N06022) (Table 3) and an inner shell of nuclear-grade austenitic type 316 stainless steel (S31600). The function of the outer barrier is to resist corrosion, and that of the inner barrier is to provide mechanical strength and mainly shield radiation. The drip shield will be made of Ti Gr 7 (Table 3). The function of the drip shield is to deflect rock fall and early water seepage on the container (Ref 6).

Earlier and Current Design of the Yucca Mountain Waste Package

Waste package designs and selection of materials have changed over the years, and the changes have usually occurred in various design phases associated with the preparation of comprehensive project planning documents. Consideration of Yucca Mountain as a potential repository site began in the late 1970s. At that time, the U.S. Department of Energy was also considering other locations with other types of geological formations. In late 1982, the U.S. Congress passed legislation (Nuclear Waste Policy Act) that codified the process for characterizing prospective geological repository sites. Work continued at the Yucca Mountain site (in tuff rock) and also at specific sites in the state of Washington (in basalt), in Texas (in bedded salt), and at a general site to be selected later in an eastern state (presumably in granite). In late 1987, the Nuclear Waste Policy Act was amended, and work at all the sites except Yucca Mountain was terminated. A comprehensive Site Characterization Plan was formulated for the Yucca Mountain site and issued in early 1989.

The very earliest waste package design was common for all of the U.S. repository sites. This was sometimes called the self-shielded container since it was quite thick (about 300 mm, or 12 in.) and made from steel or cast iron. The thick package could readily withstand high stresses or pressures from the outside, plus the bulk of the waste package reduced the radiation level significantly outside the waste package. In late 1982, it was determined that the repository could be constructed in the unsaturated zone, some 300 m (985 ft) above the permanent water table. This differentiated the Yucca Mountain site from those being studied elsewhere. The early design of the waste package specific for the Yucca Mountain site (often called the Site Characterization Plan design) consisted of a single, relatively thin barrier of approximately 20 mm (0.8 in.) thick. Because the waste packages could be emplaced in drifts in the unsaturated zone, relatively thin barriers would provide sufficient strength against external pressures (this is in contrast with the packages proposed for other countries in the saturated zone that must withstand either hydrostatic or lithostatic pressures). Furthermore, the heat produced by the radioactively decaying waste could be used as a thermal barrier, evaporating water in the vicinity of the waste package and keeping water away for hundreds of years while the temperature on the waste package container surface was well above the boiling point of aqueous solutions relevant to the site.

Early work at Yucca Mountain focused on austenitic stainless steels as the container material, particularly on type 304 stainless steel (S30400) (Table 3). At that time, the waste package was being designed for a 300 to 1000 year lifetime with a fairly large number of breached waste packages permitted in the performance assessment of the system. Because the prevailing environmental conditions at the Yucca Mountain site are oxidizing, materials whose corrosion resistance depends on the formation of protective passive films have been viewed as the most viable materials for construction of the waste package. Over the years, the performance demands on the waste package have increased considerably, and this has resulted in the selection of more and more corrosion-resistant materials for the container. In the 1980s, the container materials work was centered on type 304L stainless steel as the "reference material" with other 300-series austenitic stainless steels and related alloys (such as alloy 825—N08825) as alternatives because of improved resistance against localized corrosion or SCC.

During this same period, some work was performed on copper and copper-base alloys as candidate waste package container materials. Although the prevailing conditions at Yucca Mountain are expected to be oxidizing, some of the copper-bearing alloys, such as aluminum bronze (C61300), copper-nickel (C71500), and nickel-copper alloy 400 (N04400) do perform reasonably well under such conditions. Some of the copper-bearing materials continue in test today as possible alternative materials to those in the present design. These materials generally have corrosion resistance intermediate between corrosion allowance materials, such as carbon steel, and corrosion-resistant materials, such as nickel-base and titanium alloys.

In 1991, the Yucca Mountain Project (YMP) entered into another design phase that included a more robust waste package. The robust feature of the waste package design was achieved in three ways: (a) use of a double container consisting of a corrosion-resistant material for the inner container and a corrosion-allowance material for the outer container, (b) a thicker waste package, and (c) a highly corrosion-resistant material for the inner barrier. The outer barrier would be made of carbon steel, while the inner barrier would be made of a nickel-base alloy. The outer barrier would be about 100 mm (4 in.) in thickness, while the inner barrier would be around 20 mm (0.8 in.) thick. The strategy was that the carbon steel would corrode very slowly in the initially hot and dry atmosphere and then more rapidly as the temperature cooled and aqueous conditions could develop in the repository. It was initially proposed to make the inner container out of alloy 825 (N08825), but because of the need to demonstrate higher corrosion resistance, alloy 625 (N06625) was chosen, only to be finally replaced by alloy 22 (N06022). It was further proposed that the two containers be shrink-fitted together so that a partially corroded steel outer barrier would provide cathodic protection to the inner nickel alloy inner barrier. A further option was added to the design: a thermal sprayed ceramic coating over the carbon steel surface. This design phase ended in 1998 with the issuance of the Viability Assessment Report.

The waste package design was further modified in 1998 for the next phase. The corrosion-resistant barrier became the outer container (alloy 22). The inner container would be fabricated from a nuclear grade of type 316L stainless steel, twice as thick as the external layer. The function of this inner container is to provide added strength and bulk to the waste package. These two container shells will be manufactured separately and assembled as a slip fit with the stainless steel inside the alloy 22 (Ref 6). In addition, a drip shield, made from titanium, will be installed above the waste packages to divert any water seeping and rock falling from the drift wall, away from the waste package. The specific grade of titanium recommended for this design phase was Ti Gr 7 (R52400), containing palladium to provide extremely high corrosion resistance. The strategy of this design was that the highly corrosion-resistant Ti Gr 7 and alloy 22 would provide greater defense in depth since they would be highly unlikely to suffer from the same type of failure. The waste package and drip shield combination is needed to endure for some 10,000 years, with very few breached containers allowed in this time period, in order to achieve a very high standard of protection to the accessible environment around Yucca Mountain. This design remains the current design and was used for the Site Selection phase, which ended in 2002, when Yucca Mountain was selected by the President and approved by the Congress to go forward with a License Application to the U.S. Nuclear Regulatory Commission (NRC). This same design will be used in the License Application process. Since the actual construction of waste packages is still years away, additional testing of alloy 22 is continuing. Also, other Ni-Cr-Mo alloys are currently in test as alternatives to alloy 22. Work is also proceeding on alternative materials to the expensive Ti Gr 7 drip shield, including thermal sprayed ceramics and metal glasses on various substrates.

Effects of Fabrication on Corrosion Behavior. In parallel with the work on investigating the corrosion behavior of the various materials under repository-relevant environmental conditions, efforts have evolved on evaluating processes to fabricate and weld the waste package containers. The corrosion behavior of a material is often related to the metallurgical condition of the material resulting from the processes used in fabrication and welding. In the design proposed for the License Application, the container body will be fabricated from rolled and welded plate. The specified welding process is GTAW with matching filler material (Ref 6). Assembly welds, that is, the welds used to join the pieces of rolled plate together as well as the weld to join the bottom lid to the container shell, will be annealed and quenched. The empty container shells (inner stainless steel and outer alloy 22) will then be filled with the waste, and then the top lids will be welded on. Because of the high radioactivity of the filled waste package, the final closure weld must be made in a hot cell, so that much of the operation will be performed robotically. The GTAW process will be used to make the two closure welds. The annealing and quenching operation on the outer alloy 22 container shell puts the surface in compression. After the top alloy 22 lid is welded onto the now-filled container, the welded area will undergo a stress-relief operation to put down a compressive stress layer. Processes for achieving this include laser shock peening and metal burnishing, both of which can be conducted in a hot cell operation. The strategy is to prevent any possibility of initiating SCC by eliminating sources of tensile stress. Since the first waste package will not be fabricated for several more years, it is expected that changes in the processes described will be made to improve the projected performance of the container material and to ensure a very high degree of quality control. The emplacement of the first waste package at the Yucca Mountain site is planned for 2010.

Carbon and Low-Alloy Steel

Carbon steel was an earlier candidate for the container material in the United States. This is no longer the case; however, the characterization of the corrosion behavior of carbon steels in Yucca Mountain type environments is still important since other components (e.g., rock bolts, tunnel ribs, borehole liners, etc.) in the drifts may still be constructed using carbon steels. Table 5 shows corrosion rates of carbon steels, low-alloy steels, and stainless steels at 100 °C (212 °F) in both steam and water (Ref 44). Table 5 shows that in general all tested materials had lower corrosion rates in steam than in water, especially for the longer testing times. Table 5 also shows that the corrosion rate of stainless steels is almost negligible, even for those with only 13% Cr such as type 416 stainless steel (Table 3). The effect of irradiation on the corrosion rate was also studied (Ref 44). A gamma dose of 3×10^5 rads/h was applied for 3 months in the presence of only J-13 water and in presence of tuff rock plus J-13 water, both at 105 °C (220 °F). J-13 is the name of a well in the Yucca Mountain area from where the groundwater was analyzed. The corrosion rate of carbon steel and alloy steel was higher in the presence of irradiation and also, under irradiated conditions, the corrosion rate was higher when the rock was present in the environment. It was explained that the production of oxygen and hydrogen peroxide may have increased the corrosion rate of the tested materials (Ref 44). Corrosion rates by weight loss were measured for carbon steel, cast iron, and alloy steel in air saturated J-13 water in the temperature range 50 to 100 °C (120 to 212 °F) for 1500 h (Ref 44). Results show that for all the alloys a maximum corrosion rate was reached in the 70 to 80 °C (160 to 175 °F) range and decreased at 100 °C (212 °F). This could be related to the amount of dissolved oxygen in the tested electrolyte. At 100 °C (212 °F), the corrosion rate of 1020 carbon steel, gray cast iron, and 2.25Cr-1Mo alloy steel was approximately 300 μm/yr (11.8 mils/yr); however, for 9Cr-1Mo alloy steel the corrosion rate was only 7 μm/yr (0.28 mil/yr) (Ref 44). Carbon steel 1020 was also investigated regarding the effect of water vapor or relative humidity (RH) on its corrosion behavior at 65 °C (150 °F) (Ref 45). It was determined that a RH between 75 and 85% is necessary before 1020 carbon steel suffered any obvious corrosion damage (Ref 45). The effect of relative humidity on the corrosion susceptibility of carbon steel has also been modeled (Ref 46).

The corrosion rate of 1016 carbon steel was measured in concentrated J-13 type of water (Table 2) using electrochemical techniques (Ref 47). The corrosion rate was always below 10 μm/yr (0.40 mil/yr) for all the tested waters at 25, 50, 70, and 90 °C (75, 120, 160, and 195 °F). Only in dilute aerated water (10 times the concentration of J-13) was the corrosion rate of the carbon steel higher than 10 μm/yr (0.40 mil/yr). A comprehensive review of the corrosion modes of carbon steels regarding nuclear waste disposal evaluated the different degradation modes such as microbiologically influenced corrosion (MIC), general corrosion, localized corrosion, and EAC (Ref 48). The corrosion behavior of A516 carbon steel was studied at 25, 65, and 95 °C (75, 150, and 205 °F) in solutions containing different levels of carbonate/bicarbonate and chloride (Ref 49). It was found that one of the most important parameters controlling the mode and rate of corrosion were pH and the ratio of chloride to total carbonate.

Stainless Steels and Nickel Alloys

The evaluation of the corrosion resistance of type 304L stainless steel and alloy 825 was conducted via electrochemical testing such as the cyclic potentiodynamic polarization (CPP or ASTM G 61) (Ref 50). The testing matrix included 33 different environmental variables including radiation effect. The electrolyte solutions were concentrated versions of J-13 well water (Tuff rock in Table 2). Results showed that both 304L and alloy 825 experienced pitting corrosion under the tested conditions. As expected, alloy 825 was more resistant to localized attack than 304L. Detrimental variables that were identified included chloride and hydrogen peroxide. Beneficial variables included bicarbonate and nitrate (Ref 50). Cyclic potentiodynamic polarization tests were also used to compare the localized corrosion behavior of several resistant alloys, from carbon steel A516 to corrosion-resistant nickel- and titanium-base alloys (Ref 51). The tests were carried out in

Table 5 General corrosion rates of candidate steels in 100 °C (212 °F) tuff-conditioned saturated steam and water

Test duration: ~30 weeks

Material	Corrosion rate in J-13 Water at 100 °C (212 °F)		Corrosion rate in J-13 Steam at 100 °C (212 °F)	
	μm/yr	mils/yr	μm/yr	mils/yr
Carbon steels				
1020	26.8	1.06	12.5	0.492
A36	30.2	1.19	9.74	0.383
A366	28(a)(b)	1.102(a)(b)	52(a)(b)	2.047(a)(b)
Alloy steels				
2.25Cr1Mo	30(a)(b)	1.18(a)(b)	12(a)(b)	0.472(a)(b)
9Cr1Mo	0.70	0.028	−0.55 (mass gain)	−0.022 (mass gain)
Stainless steels				
409	Nil(a)	Nil(a)	Nil(a)	Nil(a)
416	0.3(a)	0.012(a)	0.2(a)	0.008(a)
304L	0.1(a)	0.004(a)	0.1(a)	0.004(a)
316L	0.5(a)	0.020(a)	Nil	Nil
317L	0.5(a)	0.020(a)	0.1(a)	0.004(a)

(a) Data for 6 weeks. (b) Crevice corrosion and uneven surface attack. Source: Ref 44

up to 4 wt% ferric chloride ($FeCl_3$) solution, pH 1.6 at 30, 60, and 90 °C (85, 140, and 195 °F). As expected, the degree of attack increased with the salt concentration and test temperature. Results show that the ranking of the alloys regarding their resistance to corrosion was as follows: K01800 < N08825 ~ N06985 < N06625 < N06030 ~ N06455 < N06022 < R53400. Alloy 22 (N06022) suffered only minor crevice corrosion and Ti Gr 12 (R53400) did not suffer any visible attack. In the creviced areas of tested specimens of alloy 625 and 22, a high concentration of molybdenum oxide was found. It was also reported that, at 90 °C (195 °F), the E_{corr} of Ti Gr 12 was approximately +600 mV (in the saturated silver-silver chloride scale or SSC) and that of A516 carbon steel was approximately −400 mV (SSC), and it did not depend on the concentration of $FeCl_3$ between 1 and 4% (Ref 51). This large difference in the E_{corr} of Ti Gr 12 and A516 could result in galvanic corrosion of A516 if these two metals are put in contact in the studied corroding solution.

Near Field Environment at the Repository

The groundwaters that are associated with Yucca Mountain have been well characterized (Ref 52). In particular, most emphasis has been placed on the composition of water from the J-13 well as being representative of water that would originate from atmospheric precipitation and would infiltrate along fractures in the various tuff layers including the Topopah Spring layer in which the repository would be constructed. Other types of waters, including simple salt solutions, such as concentrated calcium chloride ($CaCl_2$) brines, have also been considered during the testing of waste package (container and drip shield) materials. One such type of water is pore water that would be formed in an upper rock stratum and then find its way also along the fractures leading into the repository drifts. Many of the salts associated with the Yucca Mountain waters are hygroscopic in nature. Over the emplacement years, the temperature decreases and the relative humidity increases in the atmosphere around the container. Eventually a point is reached where an aqueous solution can be maintained because of the deliquescence point of the salt mixture that may be deposited as dust on the surface of the container. Solutions with relatively high concentrations of salts in the water are therefore the most representative of aqueous environmental conditions. Nevertheless, the climate at Yucca Mountain is dry and water quantities reaching the waste package surface are favorably limited. The performance of the material under aqueous conditions may be poorer than those under dry conditions. The dry oxidation rate of alloy 22 at the temperatures of operation is negligible. The enrichment of dilute groundwaters will follow the chemical divide; that is, the nature and amount of each species that could be present in the final drop of water will depend on the relative amount of species in the originating water. Two commonly mentioned waters that may be present at Yucca Mountain are:

- The infiltration water is a bicarbonate-dominated type of water with significant concentrations of sulfate, nitrate, and chloride. Alkali (Na and K) cations dominate over alkaline earth (Ca and Mg). The water also contains some fluoride and is saturated in silica. Its pH is just slightly alkaline in the dilute water, but with concentration of its salts, the pH is raised into the 10 to 11 range, mainly because of the bicarbonate/carbonate buffering effect.
- The pore water is characterized by the absence of any carbonate species. It contains significant amounts of sulfate, nitrate, and chloride, but the alkaline earth cations dominate over the alkali cations. It is slightly acidic when diluted but when concentrated, it remains near neutral, since there is no carbonate to alkalize the solution.

In both cases, a high concentration of nitrate develops, owing to the high solubility of nitrates. Evaporative concentrations of groundwater lead to high nitrate concentrations. This is significant with respect to corrosion performance because the nitrate is an important factor in establishing the overall redox conditions. Nitrate ion also demonstrates inhibiting effects in localized corrosion initiation and probably stress corrosion; thus mitigating the effects of aggressive ions, such as chloride and fluoride. Other oxyanions in solution such as sulfate, carbonate, and silicate also seem to show mitigating or inhibiting effects on localized corrosion.

The peak surface temperature that develops on the waste package container surface depends strongly on how closely the waste packages are placed in the repository drift. Historically, the strategy of the repository design configuration was to elevate the waste package container surface above the boiling point to keep the surface dry and defer any aqueous corrosion effects. In the present design, a maximum peak surface temperature of approximately 160 °C (320 °F) will occur on the waste packages once the repository is closed. The temperature decays slowly, requiring hundreds of years until temperatures below the boiling range are attained. However, more recently the project has also considered and studied an option for keeping the peak surface temperatures at a much lower value (~85 °C, or 185 °F) by using a less compact arrangement of the waste packages in the repository. However, this option means that aqueous corrosion conditions will be encountered at a much earlier time after waste emplacement and puts additional emphasis on the performance of the waste package materials.

Microbial Activity. There is abundance of microbiological life in the repository setting. Microbial activity is expected once sufficient moisture is present (e.g., >90% RH). However, the main concern is their effect on changes in the water chemistry. For instance, types of microbes (bacteria and fungi) producing acids, microbes oxidizing iron (steel is used in the repository construction), and microbes producing strongly reducing conditions (sulfate reducers) could lead to scenarios that might be damaging to alloy 22 and titanium. However, current studies have not found consistent evidence that microbial activity is actually damaging to alloy 22 or Ti Gr 7.

Radiolysis. One further environmental effect is the influence of radiolysis on changes in the water chemistry. The relatively thick two-barrier design attenuates the gamma radiation field by approximately 3 orders of magnitude. The major gamma-producing isotopes have half-lives around 30 years; thus each 100 years in the repository reduces the radiation field by another order of magnitude. The main concern with radiolysis is the combination of a significant gamma field with aqueous conditions, which may produce damaging species such as hydrogen peroxide. This is certainly more of an issue if a lower thermal load of waste packages is used in the repository when water may be present while the waste will be highly radioactive.

Corrosion Behavior of Alloy 22 (N06022)

By designation, the Ni-Cr-Mo alloy 22 (N06022) listed in Table 3 is the corrosion-resistant barrier of the waste containers for the Yucca Mountain Project (Ref 6). When evaluating the corrosion resistance of the container, no credit is given for corrosion resistance to the internal thicker layer of type 316 stainless steel nuclear grade. The container may suffer corrosion if water is present in sufficient amount at the repository site. Dry corrosion of alloy 22 is negligible for the emplacement conditions. There are three main modes of corrosion that the container may suffer during its lifetime of 10,000 or more years. These are: (a) uniform or passive corrosion, (b) localized corrosion (e.g., crevice corrosion), and (c) EAC (e.g., SCC). All types of corrosion will be influenced by the metallurgical condition of the alloy and the type of environment that is present. Metallurgical condition includes, for example, welded versus wrought and annealed versus thermally aged microstructures. The environmental aspect includes temperature, solution composition and pH, redox potential, and effect of radiation or microbial activity. Furthermore, both the metallurgical and environmental conditions will determine the free corrosion potential (E_{corr}) of the container. Since 1998, alloy 22 is being intensively characterized regarding its corrosion resistance in a variety of environmental conditions. The sections that follow describe the results by type of failure mode.

Uniform and Passive Corrosion of Alloy 22. General corrosion (or passive corrosion) is the uniform thinning of a metal (container) at its E_{corr} (Fig. 1). In the presence of aerated multiionic brines, alloy 22 is expected to remain passive at its E_{corr}. The degradation model for the designed container assumes that general corrosion at E_{corr} will progress uniformly over a large surface at a (time-independent) constant rate. This model assumes that the depth of penetration or thinning (x in μm) of the container is equal to the corrosion rate (CR, e.g., in μm/yr) multiplied by the time (t in years) that the container is exposed to an environment under which general corrosion occurs. For example, if the corrosion rate for alloy 22 is 0.1 μm/yr (0.004 mil/yr), the uniform thinning of the container in 10,000 years will be $x = 1$ mm (0.04 in.). General corrosion rates are currently being measured using long-term weight-loss immersion tests and shorter-term electrochemical methods in a variety of environmental conditions. Table 6 shows values of corrosion rates measured under several testing conditions using different testing methods.

After a 2 year immersion of alloy 22 coupons in concentrated aqueous electrolytes from pH 2.8 to 10 in the temperature range between 60 and 90 °C (140 and 195 °F), the average corrosion rate determined by mass loss was approximately 0.02 μm/yr (0.0008 mil/yr) (Ref 53) (Table 6). Electrochemical impedance studies of mill annealed (MA) and aged alloy 22 at E_{corr} in J-13 water at 95 °C (205 °F) showed that after immersion times of less than 3 h in normally aerated solutions, the corrosion rate was approximately 0.2 μm/yr (0.008 mil/yr) (Table 6) (Ref 54). The corrosion rate did not depend on the thermal aging conditions of the tested specimens. Constant potential tests on alloy 22 immersed in deaerated 0.028 M and 0.5 M NaCl solutions of pH 2.7 and 8 were performed at 20 and 95 °C (70 and 205 °F) (Ref 60). After holding the potential constant for 48 h in the range between 0 and +0.4 V (saturated calomel electrode, SCE), Dunn and Brossia reported (Ref 60) passive currents that translated into corrosion rates of less than 0.5 μm/yr (0.02 mil/yr). Constant potential tests were also performed on alloy 22 in deaerated 1 M NaCl + 0.1 M sulfuric acid (H_2SO_4) solution at temperatures between 25 and 85 °C (75 and 185 °F) (Ref 55). At 75 °C (165 °F) and at an applied potential of +200 mV using a silver/silver chloride (Ag/0.1 M AgCl) reference electrode, Lloyd et al. reported a current density of 1.58×10^{-8} A/cm^3 after 10 h of testing (Ref 55). This current density translates into a corrosion rate of 0.138 μm/yr (0.005 mil/yr) (Table 6). Polarization resistance tests were carried out in the vicinity of the E_{corr} of alloy 22 in simulated acidified water (SAW) (Ref 56). The SAW is a multicomponent aqueous solution approximately 1000 times more concentrated than J-13 (Table 2) and acidified to pH near 2.8. After an immersion of 1 h in deaerated conditions, the corrosion rates ranged from 0.480 μm/yr at 30 °C (0.019 mil/yr at 85 °F) to 1.440 μm/yr at 90 °C (0.057 mil/yr at 195 °F). However, after 1 week immersion in aerated SAW, the corrosion rate decreased by more than one order of magnitude to 0.023 μm/yr at 30 °C (0.0009 mil/yr at 85 °F) and to 0.103 μm/yr at 90 °C (0.004 mil/yr at 195 °F) (Table 6) (Ref 56). Constant potential tests were performed in aerated and deaerated SAW and simulated concentrated water (SCW) at 90 °C (195 °F) and at potential values of +0.1 V and +0.4 V (Ag/AgCl) (Ref 57). The SCW has a pH of approximately 8 to 10 and is 1000 times more concentrated than J-13. After 1 day testing in deaerated solutions at +0.1 V, it was reported that the corrosion rates were 0.460 μm/yr (0.018 mil/yr) in SAW and 1.250 μm/yr (0.049 mil/yr) in SCW. After the 1 day current density decay is extrapolated to 1 year, the corrosion rates became 0.021 μm/yr (0.0008 mil/yr) and 0.100 μm/yr (0.004 mil/yr), respectively (Ref 57).

It is difficult to obtain corrosion rates by weight loss of alloy 22 in nonaggressive solutions such as simulated concentrated groundwater, since the mass losses are practically negligible. The corrosion rate for alloy 22 was in general less than 0.01 μm/yr (0.0004 mil/yr) after more than 5 years immersion tests in multiionic solutions (Ref 58, 59). These corrosion rates were obtained after analyzing 122 specimens exposed at two temperatures (60 and 90 °C, or 140 and 195 °F), using two

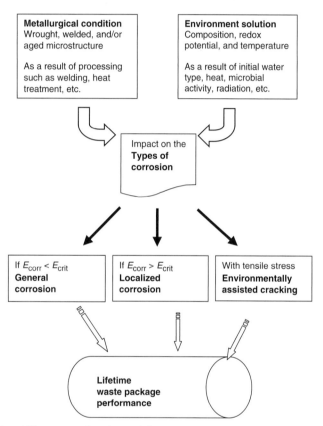

Fig. 1 General model for corrosion degradation of alloy 22

Table 6 Passive corrosion rate of alloy 22 under many conditions

Environment	Measurement method, conditions	Corrosion rate		Ref
		μm/yr	mils/yr	
Concentrated aerated multiionic electrolytes, pH 2.8 to 10, at 60 and 90 °C (140 and 195 °F)	Weight-loss immersion tests for 2 years at E_{corr}	0.02	0.0008	53
Aerated J-13, pH 7.4, at 95 °C (205 °F)	Alternating current impedance at E_{corr} after ~3 h immersion	0.2	0.008	54
Deaerated 1 M NaCl + 0.1 M H_2SO_4 solution at 75 °C (165 °F)	Constant potential +200 mV (Ag/AgCl) for 10 h	0.138	0.005	55
One week in aerated SAW, pH 2.8, at 30 to 90 °C (85 to 195 °F)	Polarization resistance at E_{corr}	0.023–0.103	0.0009–0.004	56
Concentrated deaerated multiionic solutions SAW and SCW at 90 °C (195 °F)	Constant potential +0.1 V (SSC) extrapolated to 1 year current decay	0.02 (SAW) and 0.1 (SCW)	0.0008 (SAW) and 0.004 (SCW)	57
Concentrated aerated multiionic electrolytes, pH 2.8 to 10, at 60 and 90 °C (140 and 195 °F)	Vapor and liquid mass-loss immersion tests for >5 years at E_{corr}	0.01	0.0004	58, 59

SAW, simulated acidified water; SCW, simulated concentrated water; SSC, saturated silver chloride electrode

metallurgical conditions (wrought and welded), three electrolyte solutions (pH 2.8 to 10), and vapor plus liquid phases (Table 7) (Ref 58, 59). Figure 2 shows that the corrosion rate of alloy 22 was practically the same in all the tested conditions in the Long Term Corrosion Test Facility (LTCTF) at Lawrence Livermore National Laboratory (LLNL). Moreover, since the corrosion rate was low and the temperature span of testing was not large, a clear relationship between corrosion rates and temperature could not be obtained. In another immersion test experiment, welded and nonwelded alloy 22 coupons were exposed to basic saturated water (BSW) solution at 105 °C (220 °F) for up to 8 weeks (Ref 61). The reported corrosion rates were on the order of 0.25 to 0.76 μm/yr (0.01 to 0.03 mil/yr) both for welded and nonwelded materials. It was argued that since the mass loss for alloy 22 was the same for the 4 weeks and the 8 weeks exposures, the corrosion rate of alloy 22 would decrease as the testing time is increased (Ref 61).

The low corrosion rates or passive behavior of alloy 22 is believed to be caused by the formation of a protective, chromium-rich oxide film between the alloy (metal) and the surrounding electrolyte. It has been shown that the thickness of this passive film could be only in the range of 5 to 6 nm (Ref 62). The long-term extrapolation of the corrosion rate of alloy 22 has been modeled considering that the dissolution rate is controlled by the injection of oxygen vacancies at the oxide-film/solution interface (Ref 63). It has been concluded that according to the modeling it is unlikely that catastrophic failure of the container may occur due to long-term passive film dissolution (Ref 63).

The available data in the literature reported previously shows that the general corrosion rate of alloy 22 in acidic to alkaline solutions is expected to be well below 0.100 μm/yr (0.004 mil/yr). Therefore, it is concluded that failure of the containers in Yucca Mountain will not occur by passive dissolution.

Localized Corrosion of Alloy 22. Localized corrosion (crevice corrosion) is a type of corrosion in which the attack progresses at discrete sites or in a nonuniform manner. The degradation model (Fig. 1) assumes that localized corrosion will only occur when E_{corr} is equal or greater than a critical potential (E_{crit}) for localized corrosion. That is, if $E_{corr} < E_{crit}$, only general or passive corrosion will occur. The E_{crit} can be defined as a certain potential above which the current density or corrosion rate of alloy 22 increases significantly and irreversibly above the general corrosion rate of the passive metal. In environments that promote localized corrosion, E_{crit} is the lowest potential that would trigger localized (e.g., crevice) corrosion. In environments that are benign toward localized corrosion such as the simulated concentrated groundwaters in Table 2, E_{crit} is of no significance. In these environments, under sufficient polarization, the current density on a metal may increase due to transpassive behavior of the elements in the alloy or to oxygen evolution by the decomposition of water. The margin of safety against localized corrosion will be given by the value of $\Delta E = E_{crit} - E_{corr}$. The higher the value of ΔE, the larger the margin of safety for localized corrosion. It is important to note here that the values of both E_{corr} and E_{crit} depend on the surface condition of alloy 22, the composition of the environment (e.g., chloride concentration), and the temperature. Additionally, the value of E_{crit} may depend on the way (method) it is measured. Researchers commonly use CPP (ASTM G 61) or the THE method (Ref 64) to determine localized corrosion (mostly crevice) repassivation potentials (Ref 65–68). This crevice repassivation potential is generally equated to E_{crit}. In many instances, since the true value of E_{crit} is not known, researchers use values of potential in cyclic polarization curves at which the current density on a test specimen reaches a given value of current density (e.g., the potential to reach 20 μA/cm² in the forward denoted as $E20$ or the potential to reach 1 μA/cm² in the reverse scan denoted as $ER1$) (Ref 56, 60, 68–70). These latter values of potential are for comparative purposes only and may not imply the occurrence of localized corrosion.

Alloy 22 is extremely resistant to localized corrosion such as pitting corrosion and crevice corrosion. Critical temperatures for pitting and crevice corrosion determined through immersion tests in aggressive solutions are always among the highest for nickel alloys (Ref 71, 72). Electrochemical tests also confirmed the resistance of alloy 22 to crevice corrosion. The repassivation potential for crevice corrosion (E_{crit}) of alloy 22 was approximately 300 mV (SCE) in 1 M NaCl at 95 °C (205 °F) (Ref 73–75). Cyclic potentiodynamic polarization tests using seven types of Ni-Cr-Mo alloys showed that the repassivation potential for MA alloy 22 in 1 M NaCl solution at 50 °C (120 °F) was above 400 mV (SCE) (Ref 71). Similar CPP tests in 5 M CaCl$_2$ pH 6.4 solutions showed that the passivity breakdown potential was higher than 800 mV (saturated silver chloride, SSC) at 75 °C (165 °F) and decreased to 195 mV (SSC) at 90 °C (195 °F) (Ref 56). The passivity breakdown potential was taken as the potential for which the current density reached 20 μA/cm² ($E20$). Electrochemical tests show that alloy 22 was susceptible to localized (crevice) corrosion at applied anodic potentials in pure concentrated chloride solutions such as sodium chloride (Ref 65, 73, 75) and calcium chloride (Ref 56, 70, 76). However, when nitrate was added to the chloride containing solution, the susceptibility of alloy 22 to crevice corrosion decreased or disappeared (Ref 56, 60, 69, 70). For example, the crevice repassivation potential of welded alloy 22 in

Table 7 Average corrosion rates for alloy 22 (UNS N06022) in simulated concentrated groundwaters

Determination from weight loss of two-sided specimens exposed for 5 years

Environment(a)	Corrosion rate, nm/yr	
	2.5 × 5 cm specimens(b)	5 × 5 cm specimen(b)
60 °C (140 °F) vapor		
SAW	1.92 ± 3.58	8.69 ± 2.50
SCW	0.37 ± 2.40	4.04 ± 4.56
SDW	0.38 ± 1.08	2.77 ± 3.46
60 °C (140 °F) liquid		
SAW	2.82 ± 2.70	10.27 ± 14.02
SCW	3.23 ± 3.78	12.93 ± 9.60
SDW	1.14 ± 1.08	5.63 ± 4.52
90 °C (195 °F) vapor		
SAW	1.45 ± 2.42	15.13 ± 3.98
SCW	2.12 ± 1.96	5.90 ± 7.08
SDW	1.54 ± 2.20	1.03 ± 1.44
90 °C (195 °F) liquid		
SAW	2.31 ± 2.68	6.14 ± 2.90
SCW	9.50 ± 4.78	8.90 ± 6.64
SDW	0.78 ± 2.20	4.12 ± 6.56
Overall average	2.68 ± 5.76	7.89 ± 14.18

(a) SAW, simulated acidified water; SCW, simulated concentrated water; SDW, simulated dilute water. (b) The 95% confidence ranges are shown as ±2σ. Source: Ref 58

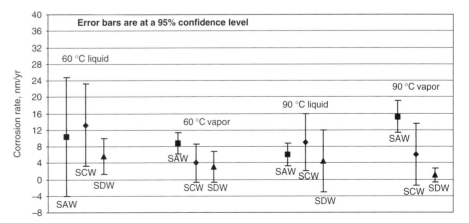

Fig. 2 Corrosion rates for alloy 22 crevice coupons in simulated acidified water (SAW), simulated concentrated water (SCW), and simulated dilute water (SDW). The tests were carried out for 5+ years at the Long Term Corrosion Test Facility at Lawrence Livermore National Laboratory. Source: Ref 58

0.5 M NaCl at 95 °C (205 °F) was 0 V (SCE) (Ref 60). When 0.05 M nitrate was added (Cl$^-$/NO$_3^-$ = 10) the crevice repassivation potential remained unchanged; however, when 0.1 M and higher nitrate concentration was added (Cl$^-$/NO$_3^-$ = 5), the crevice repassivation potential was near 350 mV (SCE) and the alloy was free from crevice corrosion (Ref 60). In another study, the susceptibility of alloy 22 to crevice corrosion was tested as a function of temperature and pH in 5 M lithium chloride (LiCl) containing different amounts of oxyanions (sulfate and nitrate). It was reported that as the ratio Cl$^-$/(NO$_3^-$ + SO$_4^{2-}$) decreased from 100 to 10 to 1, the susceptibility to crevice corrosion decreased (Ref 69). In a third study, it was reported that the breakdown potential of alloy 22 in 5 M CaCl$_2$ solution at 90 °C (195 °F) was 0.195 V (SSC); however, when 0.5 M calcium nitrate, Ca(NO$_3$)$_2$, was added (Cl$^-$/NO$_3^-$ = 10), the breakdown potential increased to 0.76 V (SSC) and the alloy was free from localized corrosion (Ref 56). Figure 3 shows the cyclic potentiodynamic polarization for alloy 22 in 5 M CaCl$_2$ at three different temperatures. As the temperature increases, both the breakdown and repassivation potential decreases. The difference in the breakdown potentials between 105 and 75 °C (220 and 165 °F) is smaller than between 75 and 45 °C (165 and 115 °F) suggesting that there is a threshold temperature below 75 °C (165 °F) on the susceptibility of alloy 22 to localized corrosion as measured using the CPP technique (Fig. 3) (Ref 70). Figure 4 shows the E_{corr} (after 24 h) of alloy 22 both in 5 M CaCl$_2$ and in 5 M CaCl$_2$ + 0.5 M Ca(NO$_3$)$_2$ as a function of the testing temperature (Ref 70). The E_{corr} of alloy 22 in pure CaCl$_2$ solutions does not change with the temperature; however, in the brine that contains nitrate, E_{corr} appears to increase slightly at the highest tested temperature. Figure 4 also shows that the repassivation potential (ER1) of alloy 22 increases when nitrate is added to the solution. For example, at 90 °C (195 °F), for pure chloride solution ER1 is approximately −150 mV (SSC) and for the solution with nitrate approximately +400 mV (SSC). Also, for each solution, there are two different slopes for ER1 depending on the temperature range.

Alloy 22 may be susceptible to crevice corrosion, but only under highly aggressive environmental circumstances such as high chloride concentrations at high temperature and applied anodic potentials. In pure chloride solutions, the crevice repassivation potential (e.g., ER1) decreases as the temperature (T) and the chloride concentration [Cl$^-$] increases following the relationship (Ref 66):

$$ER = A - BT - C \log [Cl^-] \quad (Eq\ 1)$$

where A, B, and C are constants.

Figure 5 shows that the appearance of the corroded metal under the crevice former is with grain-boundary etching, typical of the corrosion of alloy 22 in hot hydrochloric acid (HCl) solutions. Pitting corrosion is seldom observed in alloy 22, and the variables controlling pitting corrosion are unclear. Surface condition could be a factor that may trigger pitting corrosion.

In general, the E_{corr} of alloy 22 is always below the E_{crit} for localized corrosion. For example, Fig. 4 shows that for 24 h tests, E_{crit} (ER1) in each solution was always higher than E_{corr}, that is, alloy 22 would remain free of localized attack at its free corroding potential. In a few circumstances, after long exposure times in aerated electrolytes, E_{corr} of alloy 22 could raise to a potential value comparable to E_{crit}. The fact that E_{corr} may reach the value of E_{crit} is a necessary but not a sufficient condition for crevice corrosion. The occurrence of crevice corrosion will depend not only on the environmental conditions such as chloride to nitrate ratio and temperature but also on the crevice geometry (tightness) and nature of the crevicing material. Crevice corrosion can be initiated in alloy 22 using potentiostatic tests. Results show that crevice corrosion propagated for the first few hours of testing, but after a couple of days crevice corrosion slowed down and seemed to die since the total measured current density reached the passive current density of the alloy in polarization studies. Mechanisms of crevice corrosion initiation, growth, and arrest (stifling) still need to be studied in detail.

Environmentally Assisted Cracking of Alloy 22. Environmentally assisted cracking is a phenomenon by which certain ductile metallic materials lose ductility in the presence of tensile stresses in specific corrosive environments. That is, for EAC to occur, the simultaneous presence of three factors must be present. These are (a) a susceptible microstructure, (b) tensile stresses, and (c) an aggressive environment. If one or more of these variables is eliminated, EAC will not occur. Mill annealed (wrought) alloy 22 is highly resistant to EAC in most environments,

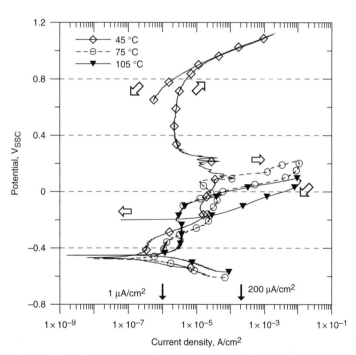

Fig. 3 Cyclic potentiodynamic polarization for alloy 22 (UNS N06022) in deaerated 5 M CaCl$_2$ solutions at different temperatures. The higher the temperature, the lower the breakdown potential (E200) or repassivation potential (ER1)

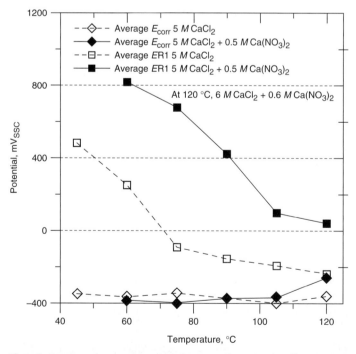

Fig. 4 Parameters from cyclic potentiodynamic polarization for alloy 22 (UNS N06022) in deaerated 5 M CaCl$_2$ and in 5 M CaCl$_2$ + Ca(NO$_3$)$_2$ solutions as a function of temperature. At each temperature (for example at 75 °C, or 165 °F), the repassivation potential (ER1) is raised when nitrate is added to the solution. Source: Ref 70

including acidic concentrated and hot chloride solutions (Ref 74, 77–80). Alloy 22 did not suffer EAC when it was tested in 14 m Cl$^-$ (as magnesium chloride, MgCl$_2$) at 110 °C (230 °F) and 9.1 m LiCl at 95 °C (205 °F) under controlled potential. Wedge opening loaded double cantilever beam (DCB) and compact tension (CT) specimens at stress intensities in the range 32 to 47 MPa$\sqrt{m}$ (29 to 43 ksi$\sqrt{in}$.) were used for times as long as 52 weeks (Ref 75, 77, 80). Similarly, U-bend specimens of alloy 22 did not suffer EAC when exposed to 45% MgCl$_2$ at 154 °C (309 °F) for up to 6 weeks (Ref 78). Slow strain rate tests (SSRT) of alloy 22 at a 1.6×10^{-6} s^{-1} strain rate at the (E_{corr}) in 4 M NaCl at 98 °C (208 °F), saturated CaCl$_2$ (>10 M Cl$^-$) at 120 °C (250 °F) and 1% lead chloride (PbCl$_2$) at 95 °C (205 °F) did not show a loss of ductility or secondary cracking (Ref 79). Welded and wrought U-bend specimens of alloy 22 and other five nickel-base alloys exposed for more than 5 years to multiionic solutions that represent concentrated groundwater of pH 2.8 to 10 at 60 and 90 °C (140 and 195 °F) were free from EAC (Ref 81).

Even though alloy 22 is resistant to EAC in concentrated hot chloride solutions, it may be susceptible under other severe environmental conditions. For example, alloy 22 U-bend specimens suffered transgranular EAC when they were exposed for 336 h to aqueous solutions of 20% hydrogen fluoride (HF) at 93 °C (199 °F) and to its corresponding vapor phase (Ref 82). The liquid phase containing HF was more aggressive than the vapor phase. High concentrations of wet HF is not a representative environment for underground conditions since calcium fluoride (CaF$_2$) would form and this salt is rather insoluble. The susceptibility of alloy 22 to EAC was tested at the E_{corr} in aerated BSW at 110 °C (230 °F) (Ref 83, 84). This BSW multiionic solution is a version of concentrated solutions that might be obtained after evaporative tests of Yucca Mountain groundwaters. Using the reversing direct current potential drop technique, it was reported that the crack growth rate was 5×10^{-13} m/s (16.4×10^{-13} ft/s) in a 20% cold-worked specimen loaded to a stress intensity of 30 MPa$\sqrt{m}$ (27.3 ksi$\sqrt{in}$.). The testing conditions used were highly aggressive and, in spite of that, the measured crack growth rate was near the detection limit of the system (Ref 83, 84). Slow strain rate tests were performed using MA alloy 22 specimens in SCW and other solutions as a function of the temperature and applied potential (Ref 79, 85, 86). Alloy 22 was found susceptible to EAC only in hot SCW solution at anodic applied potentials approximately 300 to 400 mV more positive than E_{corr}. In the anodic region of potentials, the susceptibility to EAC decreased as the temperature decreased. Figure 6 shows the dependence of the time to failure of alloy 22 on temperature when strained at high anodic potentials in SCW solution. Environmentally assisted cracking was not observed for specimens strained in SCW at ambient temperature. Figures 7 and 8 show the influence of temperature for the same applied potential of +400 mV (SSC). While at 86 °C (187 °F), there was secondary cracking and almost no necking (Fig. 7), at 50 °C (120 °F) the secondary cracking disappeared while necking was prominent (Fig. 8). While alloy 22 suffers SCC in hot SCW solution at high anodic potentials using the SSRT technique, cracking did not occur under similar testing conditions using the U-bend or constant deformation technique (Ref 87). Alloy 22 was also slightly susceptible to EAC in SCW at −1000 mV (SSC), especially at ambient temperature. Fluoride seems more detrimental than chloride for the EAC resistance of alloy 22 (Ref 86).

Environmentally assisted cracking is unlikely to initiate and grow in alloy 22 in the repository conditions where the only stresses are residual stresses from fabrication (forming and welding) or from rock fall impact on the container. Nevertheless, it is planned to further mitigate the possibility of EAC by putting compressive stresses on the container surface (Ref 6). Assembly welds made during the container fabrication are expected to be solution annealed to remove residual stress and restore the alloy homogeneity in the welded region. This will help to minimize EAC initiation. Stress mitigation processes around the final closure weld (top lid onto the container body, filled with the waste form) have also been proposed. The two processes under active consideration are laser peening and low plasticity burnishing (Ref 6).

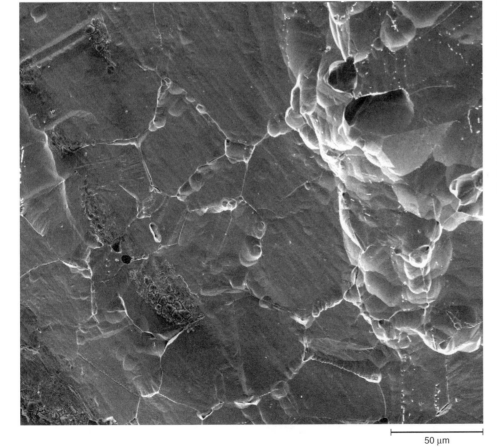

Fig. 5 Typical appearance of crevice corrosion formed on MA alloy 22 when tested using cyclic potentiodynamic polarization in 1 M NaCl at 90 °C (195 °F). The grain-boundary etching appearance is typical of a hot reducing HCl solution that may be forming inside of the crevice. Specimen DEA3129. Original magnification: 1000×. Source: Ref 67

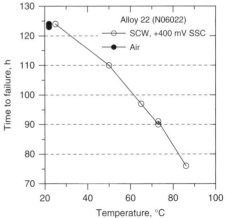

Fig. 6 Time-to-failure for alloy 22 as a function of the testing temperature when slow strain rate test was performed at 1.67×10^{-6}/s at the anodic potential of +400 mV. The higher the temperature, the lower the time to failure. SCW, simulated concentrated water

Laser peening produces a compressive stress layer on the welded waste package surface, and with optimization of the process, this layer may achieve ~3 mm (~0.12 in.) in depth. Burnishing is a mechanical process by which the compressive stresses on the surface are applied via a rolling ball.

Corrosion Behavior of Titanium Alloys

Titanium grade 7 (Ti Gr 7 or R52400) has been chosen to fabricate the detached drip shield for the repository in Yucca Mountain. The presence of the drip shield would deflect early water seepage from the containers. This drip shield would also deflect rock fall from the containers. Ti Gr 7 belongs to a family of commercially pure and modified titanium alloys especially designed to withstand aggressive chemical environments. The superior corrosion resistance of titanium and titanium alloys is due to a thin, stable, and tenacious oxide film that forms rapidly in air and water, especially under oxidizing conditions. Titanium alloys have a broad range of application as corrosion-resistant materials. This includes seawater, wet chlorine, chlorinated organic compounds, and oxidizing acids (e.g., nitric and chromic acids) (Ref 72, 88). Some of the media in which titanium should not be used include hydrofluoric acid, dry chlorine, and hot pure sulfuric acid solutions. The family of titanium corrosion-resistant alloys includes grades from 1 to 34 (Ref 89). Ti Gr 2 (R50400) is commercially one of the more popular grades. Other, more corrosion-resistant grades, which are the focus of the current studies, include Ti Gr 7 (R52400), Ti Gr 16 (R52402), and Ti Gr 12 (R53400) (Table 3). These three grades contain small amounts of alloying elements that improve the corrosion resistance of these alloys. For example, Ti Gr 7 contains 0.12 to 0.25% Pd, Ti Gr 16 contains 0.04 to 0.08% Pd, and Ti Gr 12 contains 0.2 to 0.4 Mo and 0.6 to 0.9% Ni (Ref 89). A detailed review of the general, localized, and environmentally assisted cracking behavior of Ti Gr 7 and other titanium alloys relevant to the application in Yucca Mountain has addressed, among other topics, the effect of alloyed palladium, the properties of the passive films, and the effect of radiation (Ref 41).

General and Localized Corrosion of Titanium Alloys. The general corrosion resistance of Ti Gr 7 is superior to that of Ti Gr 12; however, this effect is more noticeable under reducing conditions due to the beneficial effect of palladium (Ref 88, 90). Corrosion-rate data for Ti Gr 16 are scarce. Some titanium alloys may be susceptible to crevice corrosion under certain conditions; however, these alloys are practically immune to pitting corrosion in halide-containing environments under most practical applications. It has been suggested that the susceptibility to crevice corrosion is due to the formation of a low pH reducing solution under the occluded conditions, where the corrosion rate of titanium is higher than in oxidizing conditions (Ref 91). Halide- and sulfate-containing solutions may induce crevice corrosion in titanium at temperatures higher than 70 °C (160 °F) (Ref 91). Anodic polarization of Ti Gr 7 in chloride- and fluoride-containing solutions at 95 °C (205 °F) have shown that the presence of fluoride produces significantly higher current densities at potentials above the corrosion potential (Ref 92, 93). The presence of fluoride may have also rendered Ti Gr 7 more susceptible to crevice corrosion under anodic polarization (Ref 92).

As with alloy 22, titanium alloys Ti Gr 7, Ti Gr 16, and Ti Gr 12 are being tested in the Long Term Corrosion Test Facility at LLNL. Weight loss, creviced, and U-bend specimens were exposed to three different electrolyte solutions simulating concentrated groundwater for more than 5 years both at 60 and at 90 °C (140 and 195 °F) in the vapor and liquid phases of these solutions. Table 8 shows the corrosion rates calculated by weight loss in the nonstressed specimens (Ref 94). Each titanium alloy exhibited some discoloration (due to the possible formation of an oxide film). Ti Gr 7 generally exhibited the lowest corrosion rates irrespective of temperature or solution type while Ti Gr 12 generally exhibited the highest corrosion rates (Fig. 9) (Ref 94). The specimens immersed in the concentrated alkaline solution (SCW) exhibited the highest corrosion rates regardless of alloy composition. Specimens immersed in the acidified solution and the dilute alkaline solution exhibited significantly lower, and similar, corrosion rates. Ti Gr 12 generally exhibited higher corrosion rates at 90 °C (195 °F) than 60 °C (140 °F). There did not appear to be a temperature dependence on corrosion rate for Ti Gr 7 or Ti Gr 16. For all alloys, there did not appear to be any weld effect on the corrosion rates (Ref 94). In a previous corrosion-rate study by weight loss of Ti Gr 7 in BSW solution at 105 °C (220 °F), no effect was also reported between nonwelded and welded coupons (Ref 61).

Environmentally Assisted Cracking of Titanium Alloys. Titanium and titanium alloys may suffer EAC such as hydrogen embrittlement (HE) and SCC. Embrittlement by hydrogen is a consequence of absorption of atomic hydrogen by the metal to form hydrides (Ref 91, 95, 96). This may happen in service when the titanium alloy is coupled to a more active metal in an acidic solution. A critical concentration of hydrogen in the metal may be needed for HE to occur (Ref 41, 95). The few environments that can induce SCC in titanium are absolute (anhydrous) methanol, red-fuming nitric acid, and nitrogen tetroxide (Ref 91, 97). A few percent

Fig. 7 Mill annealed alloy 22 specimen ARC22-033 strained in simulated concentrated water at 86 °C (187 °F) and +400 mV applied potential. Significant amount of secondary cracking and little necking is observed. Original magnification: 70×

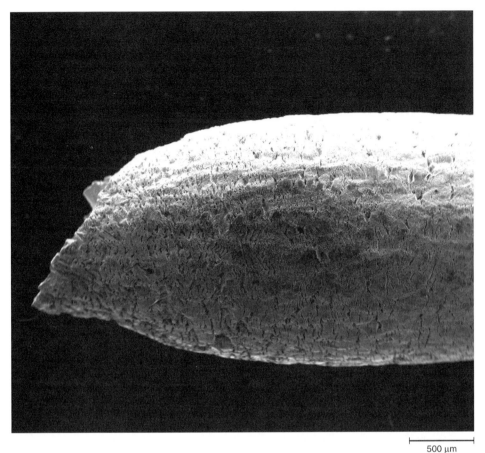

Fig. 8 Mill annealed alloy 22 specimen ARC22-032 strained in simulated corrosion water at 50 °C (120 °F) and +400 mV applied potential. Negligible amount of secondary cracking and obvious necking is observed. Original magnification: 70×

Table 8 Average corrosion rates for titanium alloys in simulated concentrated groundwaters
Determined from weight loss of two-sided specimens. Grade 7 (UNS R 52400) exposed 2.5 years, grade 12 (UNS R53400) and 16 (UNS R52402) exposed for 5 years

	Corrosion rate(b), nm/yr					
	2.5 × 5 cm specimen			5 × 5 cm specimen		
Environment(a)	Ti Gr 7	Ti Gr 16	Ti Gr 12	Ti Gr 7	Ti Gr 16	Ti Gr 12
60 °C (140 °F) vapor						
SAW	0.99 ± 4.8	0.70 ± 2.3	3.32 ± 6.4	2.18 ± 4.2	6.75 ± 10.1	4.75 ± 10.0
SCW	4.40 ± 3.2	1.18 ± 3.3	−2.06 ± 5.9	0.94 ± 5.2	2.65 ± 8.3	7.44 ± 14.7
SDW	0.48 ± 2.3	−2.15 ± 6.1	−1.40 ± 7.9	−0.91 ± 6.0	12.03 ± 16.9	13.80 ± 29.2
60 °C (140 °F) liquid						
SAW	0.49 ± 2.4	6.05 ± 6.8	3.32 ± 5.5	1.25 ± 1.4	14.42 ± 16.6	12.53 ± 14.9
SCW	18.75 ± 11.9	10.60 ± 7.0	26.78 ± 11.2	7.79 ± 5.7	16.47 ± 6.7	71.17 ± 64.5
SDW	1.92 ± 4.7	6.48 ± 10.0	3.58 ± 6.2	−0.61 ± 4.6	15.74 ± 11.2	12.69 ± 20.8
90 °C (195 °F) vapor						
SAW	1.96 ± 8.8	1.69 ± 2.4	5.80 ± 7.8	0.63 ± 4.7	1.64 ± 2.9	9.92 ± 13.9
SCW	0.97 ± 14.9	3.58 ± 11.3	1.37 ± 8.8	−3.07 ± 10.4	1.53 ± 4.2	13.60 ± 25.1
SDW	−2.40 ± 7.7	2.24 ± 10.6	1.44 ± 16.7	−5.53 ± 15.2	5.60 ± 8.0	0.66 ± 1.4
90 °C (195 °F) liquid						
SAW	6.82 ± 14.6	3.49 ± 5.8	3.80 ± 6.4	2.48 ± 1.7	3.45 ± 7.1	−5.40 ± 20.9
SCW	45.97 ± 5.8	18.98 ± 7.1	89.16 ± 72.7	31.47 ± 8.7	70.02 ± 11.1	207.5 ± 108.5
SDW	0.48 ± 2.4	0 ± 0	7.46 ± 21.1	−3.36 ± 5.5	2.80 ± 4.1	3.80 ± 6.3

(a) SAW, simulated acidified water; SCW, simulated concentrated water; SDW, simulated dilute water. (b) The 95% confidence ranges are shown as ±2σ. Source: Ref 94

of water in the environments mentioned previously would inhibit the SCC in titanium.

Regarding the application for the Yucca Mountain repository, slow strain rate tests (3.3×10^{-6} s^{-1}) were performed using smooth specimens of Ti Gr 7 and Ti Gr 12 in 5 wt% NaCl, pH 2.7, at 90 °C (195 °F) at the applied potentials between 0 and −1.2 V (SSC) (Ref 98). It was reported that for Ti Gr 12 (R53400), as the potential decreased the reduction of area at rupture decreased from 40% at 0 V to approximately 15% at −1.2 V (SSC). This behavior was attributed to a HE mechanism due to the formation of hydrides. Under the same tested conditions the reduction of area of Ti Gr 7 (R52400) remained approximately constant at near 50%; that is, Ti Gr 7 was more resistant to EAC than Ti Gr 12. After straining, both Ti Gr 7 and Gr 12 exhibited shallow secondary cracks at all of the tested potentials (Ref 98). The critical concentration of hydrogen to produce HE was calculated to be higher in palladium-containing titanium alloys (such as Ti Gr 7) than for titanium alloys without palladium (such as Ti Gr 12) (Ref 99). It was also anticipated that an enhanced passive corrosion rate of Ti Gr 7 (for example, in an environment containing fluoride ions) could produce enough hydrogen to reach the critical concentration in the metal to cause HE (Ref 99). Intergranular stress-corrosion cracking (IGSCC) changing to transgranular stress-corrosion cracking (TGSCC) was reported on one U-bend specimen of Ti Gr 7 exposed for 155 days in a solution containing 35,500 ppm chloride and 1900 ppm fluoride, pH 6.5, at 105 °C (220 °F) (Ref 100). Stress-corrosion cracking was reported in Ti Gr 7 specimens subjected to constant load tests in a concentrated groundwater solution of pH ~10 at 105 °C (220 °F) (Ref 101). Results from a up to 5 year immersion testing at 60 and 90 °C (140 and 195 °F) of U-bend specimens made of wrought and welded Ti Gr 7 and Ti Gr 16 alloys showed that these alloys were free from EAC in multiionic solutions that could be representative of concentrated groundwater at Yucca Mountain (Ref 90). Welded Ti Gr 12 U-bend specimens suffered EAC in SCW liquid at 90 °C (195 °F). Under the same conditions, nonwelded Ti Gr 12 was free from cracking (Ref 90).

Summary

The following major points are examined in this article:

- Many countries are developing geologic repositories for their high-level nuclear waste.
- Most of the repositories in the world are planned to be in stable rock formations (e.g., granite) below the water table (saturated). The United States is studying a repository above the water table (unsaturated).
- The repositories will consist of a stable geologic formation within which engineered barriers will be constructed. The main part of

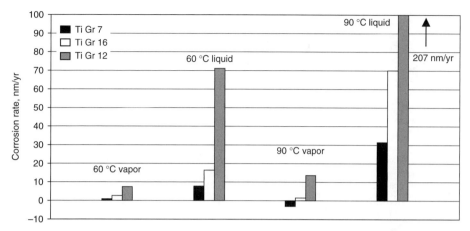

Fig. 9 Corrosion rates for Ti Gr 12, Ti Gr 16, and Ti Gr 7 crevice coupons in simulated concentrated water solution. Ti Gr 12 had the highest corrosion rates. Source: Ref 94

- the engineered barrier is the container of the waste.
- The containers are in general designed as double-walled metallic cylinders.
- From the corrosion point of view, most repositories will deal with reducing redox potentials with the exception of the U.S. repository, which will have a natural oxidizing environment.
- Copper, titanium, stainless steels, and even carbon steels were determined to be suitable materials for the reducing repositories.
- Alloy 22 and Ti Gr 7 are being characterized for the mostly dry oxidizing environment of the U.S. repository.
- Materials for the engineered barriers are constantly being evaluated for general corrosion, localized corrosion, and environmentally assisted cracking resistance.
- The metals are studied under different metallurgical condition, for example, mill annealed, welded, thermally aged, and so forth.
- Environmental variables that affect corrosion behavior include the concentration and type of the aqueous electrolytes at the site (different soluble salts), temperature, redox potential, radiation, microbial activity, and so forth.

ACKNOWLEDGMENT

This work was performed under the auspices of the U.S. Department of Energy by the University of California Lawrence Livermore National Laboratory under contract No. W-7405-Eng-48. This work is supported by the Yucca Mountain Project, LLNL which is part of the Office of Civilian Radioactive Waste Management (OCRWM—DOE).

Reference herein to any specific commercial product, process, or service by trade name, trademark, manufacturer, or otherwise, does not necessarily constitute or imply its endorsement, recommendation, or favoring by the U.S. Government or the University of California. The views and opinions of authors expressed herein do not necessarily state or reflect those of the U.S. Government or the University of California and shall not be used for advertising or product endorsement purposes.

REFERENCES

1. P.A. Baisden and C.E. Atkins-Duffin, Radioactive Waste Management, Chapter 7, Report UCRL-JC-150549, *Handbook of Nuclear Chemistry,* Lawrence Livermore National Laboratory
2. "Engineered Barrier Systems and the Safety of Deep Geological Repositories," Nuclear Energy Agency, Organisation for Economic Co-Operation and Development, Paris, France, 2003
3. P.A. Witherspoon and G.S. Bodvarsson, "Geological Challenges in Radioactive Waste Isolation—Third Worldwide Review," Report LBNL-49767, Lawrence Berkeley Laboratory, 2001
4. D.W. Shoesmith, B.M. Ikeda, F. King, and S. Sunder, Prediction of Long Term Behavior for Radioactive Nuclear Waste Disposal, *Research Topical Symposia—Life Prediction of Structures Subject to Environmental Degradation,* NACE International, 1996, p 101
5. L.O. Werme, Fabrication and Testing of Copper Canister for Long Term Isolation of Spent Nuclear Fuel, *Scientific Basis for Nuclear Waste Management,* Vol 608, Materials Research Society, 2000, p 77
6. G.M. Gordon, *Corrosion,* Vol 58, 2002, p 811
7. A.A. Sagüés, Nuclear Waste Package Corrosion Behavior in the Proposed Yucca Mountain Repository, *Scientific Basis for Nuclear Waste Management,* Vol 556, Materials Research Society, 1999, p 845
8. K. Nuttall, Some Aspects of the Prediction of Long-Term Performance of Fuel Disposal Containers, *Can. Metall. Q.,* Vol 22 (No. 3), 1983, p 404–406
9. D.D. Macdonald, M. Urquidi-Macdonald, and J. Lolcma, Deterministic Predictions of Corrosion Damage to High Level Nuclear Waste Canisters, STP 1194, *Application of Accelerated Corrosion Tests to Service Life Prediction of Materials,* G.A. Cragnolino and N. Sridhar, Ed., ASTM International, 1994, p 143
10. T. Shibata, Statistical and Stochastic Aspects of Corrosion Life Predictions, *A Compilation of Special Topic Reports,* F.M.G. Wong and J.H. Payer, Ed., Waste Package Materials Performance Peer Review, May 31, 2002, p 9-1
11. A.A. Sagüés, Corrosion Performance Projection of Yucca Mountain Waste Packages, *Scientific Basis for Nuclear Waste Management,* Vol 713, Materials Research Society, 2002, p 17
12. "Standard Practice for Prediction of the Long-Term Behavior of Materials, Including Waste Forms, Used in Engineered Barrier Systems (EBS) for Geological Disposal of High-Level Radioactive Waste," C 1174-97, *Annual Book of ASTM Standards,* ASTM International, 1997
13. M. Pourbaix, *Atlas of Electrochemical Equilibria in Aqueous Solutions,* National Association of Corrosion Engineers, 1974
14. D. Cameron et al., The Development of Durable, Man-Made Containment Systems for Fuel Isolation, *Can. Metall. Q.,* Vol 22 (No. 1), 1983, p 89
15. D.W. Shoesmith, F. King, and B.M. Ikeda, The Indefinite Containment of Nuclear Fuel Wastes, *Scientific Basis for Nuclear Waste Management,* Vol 412, Materials Research Society, 1996, p 563–570
16. J.P. Simpson et al., Corrosion Rate of Unalloyed Steels and Cast Irons in Reducing Granitic Groundwaters and Chloride Solutions, *Scientific Basis for Nuclear Waste Management,* Vol 50, Materials Research Society, 1985, p 429–436
17. G.P. Marsh, K.J. Taylor, I.D. Bland, C. Westcott, P.W. Tasker, and S.M. Sharland, Evaluation of the Localized Corrosion of Carbon Steel Overpacks for Nuclear Waste Disposal in Granite Environments, *Scientific Basis for Nuclear Waste Management,* Vol 50, Materials Research Society, 1985, p 421–428
18. R.P. Anantatmula, C.H. Delegard, and R.L. Fish, Corrosion Behavior of Low-Carbon Steels in Grande Ronde Basalt Groundwater in the Presence of Basalt-Bentonite Packing, *Scientific Basis for Nuclear Waste Management,* Vol 26, Materials Research Society, 1984, p 113–120
19. R.P. Anantatmula and R.L. Fish, Corrosion Behavior of Container Materials in Grande Ronde Basalt Groundwater, *Scientific Basis for Nuclear Waste Management,* Vol 84, Materials Research Society, 1987, p 295–306

20. N.R. Smart, A.E. Bond, J.A.A. Crossley, P.C. Lovegrove, and L. Werme, Mechanical Properties of Oxides Formed by Anaerobic Corrosion of Steel, *Scientific Basis for Nuclear Waste Management,* Vol 663, Materials Research Society, 2001, p 477–485
21. N.R. Smart, P.A.H. Fennell, R. Peat, K. Spahiu, and L. Werme, Electrochemical Measurements During the Anaerobic Corrosion of Steel, *Scientific Basis for Nuclear Waste Management,* Vol 663, Materials Research Society, 2001, p 487–495
22. T. Fukuda and M. Akashi, Effect of [HCO3- + CO32-] on Corrosion Crevice Repassivation Potential of Carbon Steel in Simulated Bentonite Environment, *Scientific Basis for Nuclear Waste Management,* Vol 412, Materials Research Society, 1996, p 597–602
23. Y. Fukaya and M. Akashi, Passivation Behavior of Mild Steel Used for Nuclear Waste Disposal Package, *Scientific Basis for Nuclear Waste Management,* Vol 556, Materials Research Society, 1999, p 871
24. A. Fujiwara, I. Yasutomi, K. Fukudome, T. Tateishi, and K. Fujiwara, Influence of Oxygen Concentration and Alkalinity on the Hydrogen Gas Generation by Corrosion of Carbon Steel, *Scientific Basis for Nuclear Waste Management,* Vol 663, Materials Research Society, 2001, p 497–505
25. J. Dong, T. Nishimura, and T. Kodama, Corrosion Behavior of Carbon Steel in Bicarbonate Solutions, *Scientific Basis for Nuclear Waste Management,* Vol 713, Materials Research Society, 2002, p 105
26. E. Smailos, "Influence of Welding and Heat Treatment on Corrosion of the Candidate High-Level Waste Container Material Carbon Steel in Disposal Relevant Salt Brines," paper 00194, Corrosion/2000, NACE International, 2000
27. A. Terlain, C. Desgranges, D. Gauvain, D. Féron, A. Galtayries, and P. Marcus, "Oxidation of Materials for Nuclear Waste Containers under Long Term Disposal," paper 01119, Corrosion/2001, NACE International, 2001
28. K. Worgan, M. Apted, and R. Sjöblom, Performance Analysis of Copper Canister Corrosion under Oxidizing and Reducing Conditions, *Scientific Basis for Nuclear Waste Management,* Vol 353, Materials Research Society, 1995, p 695
29. H. Imai, T. Fukuda, and M. Akashi, Effects of Anionic Species on the Polarization Behavior of Copper for Waste Package Material in Artificial Ground Water, *Scientific Basis for Nuclear Waste Management,* Vol 412, Materials Research Society, 1996, p 589–596
30. F. King, D.M. LeNeveau, and D.J. Jobe, Modelling the Effects of Evolving Redox Conditions on the Corrosion of Copper Containers, *Scientific Basis for Nuclear Waste Management,* Vol 333, Materials Research Society, 1994, p 901
31. F. King, M.J. Quinn, C.D. Litke, and D.M. LeNeveu, *Corros. Sci.,* Vol 37, 1995, p 833
32. F. King and M. Kolář, A Numerical Model for the Corrosion of Copper Nuclear Fuel Waste Containers, *Scientific Basis for Nuclear Waste Management,* Vol 412, Materials Research Society, 1996, p 555–562
33. A. Honda, N. Taniguchi, H. Ishikawa, and M. Kawasaki, A Modeling Study of General Corrosion of Copper Overpack for Geological Isolation of High-Level Radioactive Waste, *Scientific Basis for Nuclear Waste Management,* Vol 556, Materials Research Society, 1999, p 911
34. F. King, C.D. Litke, and B.M. Ikeda, The Stress Corrosion Cracking of Copper Nuclear Waste Containers, *Scientific Basis for Nuclear Waste Management,* Vol 556, Materials Research Society, 1999, p 887
35. F. King, C.D. Litke, and B.M. Ikeda, "The Stress Corrosion Cracking of Copper Containers for the Disposal of High-Level Nuclear Waste," paper 99482, Corrosion/99, NACE International, 1999
36. K. Petterson and M. Oskarsson, Stress Corrosion Crack Growth in Copper for Waste Canister Applications, *Scientific Basis for Nuclear Waste Management,* Vol 608, Materials Research Society, 2000, p 95
37. F. Druyts and B. Kursten, "Influence of Chloride Ions on the Pitting Corrosion of Candidate HLW Overpack Materials in Synthetic Oxidized Boom Clay Water," paper 99472, Corrosion/99, NACE International, 1999
38. D.W. Shoesmith and B.M. Ikeda, Development of Modelling Criteria for Prediction Lifetimes of Titanium Nuclear Waste Containers, *Scientific Basis for Nuclear Waste Management,* Vol 333, Materials Research Society, 1994, p 893
39. M. Akashi, G. Nakayama, and T. Fukuda, "Initiation Criteria for Crevice Corrosion of Titanium Alloys Used for HLW Disposal Overpack," paper 98158, Corrosion/98, NACE International, 1999
40. G. Nakayama, K. Murakami, and M. Akashi, Assessment of Crevice Corrosion and Hydrogen Induced Stress Corrosion Cracks of Ti-Pd Alloys for HLW Overpack in Deep Underground Water Environment, *Scientific Basis for Nuclear Waste Management,* Vol 757, Materials Research Society, 2003, p 771–778
41. F. Hua, K.W. Mon, P. Pasupathi, G.M. Gordon, and D.W. Shoesmith, "Corrosion of Ti Grade 7 and Other Ti Alloys in Nuclear Waste Repository Environments—A Review," paper 04698, Corrosion/2004, NACE International, 2004
42. N. Nakamura, M. Akashi, Y. Fukaya, G. Nakayama, and H. Ueda, "Stress-Corrosion Crack Initiation Behavior in α-Titanium Used for Nuclear Waste Disposal Overpack," paper 00195, Corrosion/2000, NACE International, 2000
43. "Yucca Mountain Science and Engineering Report," DOE/RW-0539, U.S. Department of Energy, Office of Civilian Radioactive Waste Management, May 2001
44. R.D. McCright and H. Weiss, Corrosion Behavior of Carbon Steels under Tuff Repository Environmental Conditions, *Scientific Basis for Nuclear Waste Management,* Vol 44, Materials Research Society, 1985, p 287–294
45. G.E. Gdowski and J.C. Estill, The Effect of Water Vapor on the Corrosion of Carbon Steel at 65 °C, *Scientific Basis for Nuclear Waste Management,* Vol 412, Materials Research Society, 1996, p 533–538
46. J.H. Lee, J.E. Atkins, and R.W. Andrews, Humid-Air and Aqueous Corrosion Models for Corrosion-Allowance Barrier Material, *Scientific Basis for Nuclear Waste Management,* Vol 412, Materials Research Society, 1996, p 571–580
47. T. Lian and D.A. Jones, "Electrochemical Corrosion Behavior of Low Alloy Carbon Steel in Simulated Yucca Mountain Vadose Waters," paper 98163, Corrosion/98, NACE International, 1998
48. G.A. Cragnolino, D.S. Dunn, P. Angell, Y.-M. Pan, and N. Sridhar, "Factors Influencing the Performance of Carbon Steel Overpacks in the Proposed High-Level Nuclear Waste Repository," paper 98147, Corrosion/98, NACE International, 1998
49. C.S. Brossia and G.A. Cragnolino, "Localized Corrosion of Carbon Steel Outer Overpacks for Nuclear Waste Disposal," paper 99468, Corrosion/99, NACE International, 1999
50. J.A. Beavers, N.G. Thompson, and W.V. Harper, Potentiodynamic Polarization Studies of Candidate Container Materials in Simulated Tuff Repository Environments, *Scientific Basis for Nuclear Waste Management,* Vol 176, Materials Research Society, 1990, p 533–540
51. A.K. Roy, D.L. Fleming, and B.Y. Lun, paper 99463, Corrosion 1999, NACE International, 1999
52. J.E. Harrar, J.F. Carley, W.F. Isherwood, and E. Raber, "Report on the Committee to Review the Use of J-13 Well Water in Nevada Nuclear Waste Storage Investigations," LLNL UCID report 21867, University of California, Jan 1990
53. J.C. Farmer, R.D. McCright, G.E. Gdowski, F. Wang, T.S.E. Summers, P. Bedrossian, J.M. Horn, T. Lian, J.C. Estill, A. Lingenfelter, and W. Halsey, General and Localized Corrosion of Outer Barrier of High-Level Waste Container in Yucca Mountain, *Pressure Vessels and Piping Conference,* Vol 408, American Society of Mechanical Engineers, p 53
54. R.B. Rebak, T.S.E. Summers, T. Lian, R.M. Carranza, J.R. Dillman, T. Corbin,

and P. Crook, "Effect of Thermal Aging on the Corrosion Behavior of Wrought and Welded Alloy 22," paper 02542, Corrosion/2002, NACE International, 2002
55. A.C. Lloyd, D.W. Shoesmith, N.S. McIntyre, and J.J. Noël, *J. Electrochem. Soc.,* Vol 150 (No. 4), 2003, p B120
56. K.J. Evans and R.B. Rebak, Passivity of Alloy 22 in Concentrated Electrolytes. Effect of Temperature and Solution Composition, *Corrosion Science—A Retrospective and Current Status in Honor of Robert P. Frankenthal,* PV 2002-13, The Electrochemical Society, 2002, p 344–354
57. T. Lian, J.C. Estill, G.A. Hust, D.V. Fix, and R.B. Rebak, Passive Corrosion Behavior of Alloy 22 in Multi Ionic Aqueous Environments, *Pressure Vessels and Piping Conference* (Vancouver, British Columbia, Canada), Aug 4–8, 2002, Vol 449, American Society of Mechanical Engineers, 2002, p 67
58. L.L. Wong, D.V. Fix, J.C. Estill, R.D. McCright, and R.B. Rebak, Characterization of the Corrosion Behavior of Alloy 22 after Five Years Immersion in Multi-ionic Solutions, *Scientific Basis for Nuclear Waste Management,* Vol 757, Materials Research Society, 2003, p 735–741
59. L.L. Wong, T. Lian, D.V. Fix, M. Sutton, and R.B. Rebak, "Surface Analysis of Alloy 22 Coupons Exposed for Five Years to Concentrated Ground Waters," paper 04701, Corrosion/2004, NACE International, 2004
60. D.S. Dunn, and C.S. Brossia, "Assessment of Passive and Localized Corrosion Processes for Alloy 22 as a High Level Nuclear Waste Container Material," paper 02548, Corrosion/2002, NACE International, 2002
61. F. Hua, J. Sarver, W. Mohn, and G.M. Gordon, "Crevice Corrosion Behavior of Candidate Nuclear Waste Container Materials in a Repository Environment," paper 02529, Corrosion/2002, NACE International, 2002
62. Y.-J. Kim, P.L. Andresen, P.J. Martiniano, J. Chera, M. Larsen, and G.M. Gordon, "Passivity of Nuclear Waste Canister Candidate Materials in Mixed-Salt Environments," paper 02544, Corrosion/2002, NACE International, 2002
63. O. Pensado, D.S. Dunn, and G.A. Cragnolino, Long-Term Extrapolation of Passive Behavior of Alloy 22, *Scientific Basis for Nuclear Waste Management,* Vol 757, Materials Research Society, 2003, p 723–728
64. S. Tsujikawa and Y. Hisamatsu, *Corros. Eng. Jpn.,* Vol 29, 1980, p 37
65. D.S. Dunn, L. Yang, Y.-M. Pan, and G.A. Cragnolino, "Localized Corrosion Susceptibility of Alloy 22," paper 03697, Corrosion/2003, NACE International, 2003
66. J.H. Lee, T. Summers, and R.B. Rebak, "A Performance Assessment Model for Localized Corrosion Susceptibility of Alloy 22 in Chloride Containing Brines for High Level Nuclear Waste Disposal Container," paper 04692, Corrosion/2004, NACE International, 2004
67. K.J. Evans, L.L. Wong, and R.B. Rebak, Transportation, Storage and Disposal of Radioactive Materials, *Pressure Vessels and Piping Conference,* Vol 483, American Society of Mechanical Engineers, 2004, p 137
68. K.J. Evans, A. Yilmaz, S.D. Day, L.L. Wong, J.C. Estill, and R.B. Rebak, Comparison of Electrochemical Methods to Determine Crevice Corrosion Repassivation Potential of Alloy 22 in Chloride Solutions, *JOM,* Jan 2005
69. B.A. Kehler, G.O. Ilevbare, and J.R. Scully, *Corrosion,* Vol 57, 2001, p 1042
70. K.J. Evans, S.D. Day, G.O. Ilevbare, M.T. Whalen, K.J. King, G.A. Hust, L.L. Wong, and R.B. Rebak, Anodic Behavior of Alloy 22 in Calcium Chloride and in Calcium Chloride plus Calcium Nitrate Brines, *Pressure Vessels and Piping Conference,* Vol 467, American Society of Mechanical Engineers, 2003, p 55–62
71. R.B. Rebak and P. Crook, Improved Pitting and Crevice Corrosion Resistance of Nickel and Cobalt Alloys, in *Critical Factors in Localized Corrosion III,* Vol 17, The Electrochemical Society, 1999, p 289
72. R.B. Rebak, Corrosion of Non-Ferrous Alloys Nickel, Cobalt, Copper, Zirconium and Titanium Based Alloys, in *Corrosion and Environmental Degradation,* Vol II, Wiley-VCH, Weinheim, Germany, 2000, p 69–111
73. K.A. Gruss, G.A. Cragnolino, D.S. Dunn, and N. Sridhar, "Repassivation Potential for Localized Corrosion of Alloy 625 and C-22 in Simulated Repository Environments," paper 98149, Corrosion/98, NACE International, 1998
74. D.S. Dunn, G.A. Cragnolino, and N. Sridhar, Passive Dissolution and Localized Corrosion of Alloy 22 High-Level Waste Container Weldments, *Scientific Basis for Nuclear Waste Management,* Vol 608, Materials Research Society, 2000, p 89–94
75. D.S. Dunn and C.S. Brossia, "Long-Term Dissolution Behavior of Alloy 22: Experiments and Modeling," paper 01125, Corrosion/2001, NACE International, 2001
76. S.D. Day, K.J. Evans, and G.O. Ilevbare, "Effect of Temperature and Electrolyte Composition on the Susceptibility of Alloy 22 to Localized Corrosion," PV 2002-24, The Electrochemical Society, 2003, p 534–544
77. Y.-M. Pan, D.S. Dunn, and G.A. Cragnolino, Effects of Environmental Factors and Potential on Stress Corrosion Cracking of Fe-Ni-Cr-Mo Alloys in Chloride Solutions, STP 1401, *Environmentally Assisted Cracking: Predictive Methods for Risk Assessment and Evaluation of Materials, Equipment and Structures,* ASTM International, 2000, p 273
78. R.B. Rebak, Environmentally Assisted Cracking in the Chemical Process Industry, Stress Corrosion Cracking of Iron, Nickel and Cobalt Based Alloys in Chloride and Wet HF Services, STP 1401, *Environmentally Assisted Cracking: Predictive Methods for Risk Assessment and Evaluation of Materials, Equipment and Structures,* ASTM International, 2000, p 289
79. J.C. Estill, K.J. King, D.V. Fix, D.G. Spurlock, G.A. Hust, S.R. Gordon, R.D. McCright, G.M. Gordon, and R.B. Rebak, "Susceptibility of Alloy 22 to Environmentally Assisted Cracking In Yucca Mountain Relevant Environments," paper 02535, Corrosion/2002, NACE International, 2002
80. D.S. Dunn, Y.-M. Pan, and G.A. Cragnolino, "Stress Corrosion Cracking of Nickel-Chromium-Molybdenum Alloys in Chloride Solutions," paper 02425, Corrosion/2002, NACE International, 2002
81. D.V. Fix, J.C. Estill, G.A. Hust, L.L. Wong, and R.B. Rebak, "Environmentally Assisted Cracking Behavior of Nickel Alloys in Simulated Acidic and Alkaline Waters Using U-bend Specimens," paper 04549, Corrosion/2004, NACE International, 2004
82. R.B. Rebak, J.R. Dillman, P. Crook, and C.V.V. Shawber, *Mater. Corros.,* Vol 52, 2001, p 289
83. P.L. Andresen, P.W. Emigh, L.M. Young, and G.M. Gordon, "Stress Corrosion Cracking of Annealed and Cold Worked Titanium Grade 7 and Alloy 22 in 110 °C Concentrated Salt Environments," paper 01130, Corrosion/2001, NACE International, 2001
84. P.L. Andresen, P.W. Emigh, L.M. Young, and G.M. Gordon, "Stress Corrosion Cracking Growth Behavior of Alloy 22 (N06022) in Concentrated Groundwater," paper 03683, Corrosion/2003, NACE International, 2003
85. K.J. King, J.C. Estill, and R.B. Rebak, Characterization of the Resistance of Alloy 22 to Stress Corrosion Cracking, *Pressure Vessels and Piping Conference* (Vancouver, British Columbia, Canada), Aug 4–8, 2002, Vol 449, American Society of Mechanical Engineers, 2002, p 103
86. K.J. King, L.L. Wong, and R.B. Rebak, "Slow Strain Rate Testing of Alloy 22 in Simulated Concentrated Ground Waters," paper 04548, Corrosion/2004, NACE International, 2004
87. D.V. Fix, J.C. Estill, K.J. King, G.A. Hust, S.D. Day, and R.B. Rebak,

"Influence of Environmental Variables on the Susceptibility of Alloy 22 to Environmentally Assisted Cracking," paper 03542, Corrosion/2003, NACE International, 2003
88. R.W. Schutz, *Corrosion,* Vol 59, 2003, p 1043
89. "Standard Specification for Titanium Alloy Strip, Sheet and Plate," B 265, Vol 02.04, *Annual Book of ASTM Standards,* ASTM International, 2002
90. D.V. Fix, J.C. Estill, L.L. Wong, and R.B. Rebak, "Susceptibility of Welded and Non-welded Titanium Alloys to Environmentally Assisted Cracking in Simulated Concentrated Ground Waters," paper 04551, Corrosion/2004, NACE International, 2004
91. R.W. Schutz, Corrosion of Titanium and Titanium Alloys, *Corrosion,* Vol 13, *Metals Handbook,* ASM International, 1987, p 669
92. C.S. Brossia, and G.A. Cragnolino, *Corrosion,* Vol 57, 2001, p 768
93. T. Lian, M.T. Whalen, and L.L. Wong, "Effects of Oxide Film on the Corrosion Resistance of Titanium Grade 7 in Fluoride Containing NaCl Brines," paper 05609, CORROSION/2005, NACE International, 2005
94. L.L. Wong, J.C. Estill, D.V. Fix, and R.B. Rebak, Corrosion Characteristics of Titanium Alloys in Multi-ionic Environments, *Pressure Vessels and Piping Conference,* Vol 467, American Society of Mechanical Engineers, 2003, p 63
95. C.F. Clarke, D. Hardie, and B.M. Ikeda, *Corros. Sci.,* Vol 39, 1997, p 1545
96. C.L. Briant, Z.F. Wang, and N. Chollocoop, *Corros. Sci.,* Vol 44, 2002, p 1875
97. A.C. Hollis and J.C. Scully, *Corros. Sci.,* Vol 34, 1993, p 821
98. A.K. Roy, M.K. Spragge, D.I. Fleming, and B.Y. Lum, *Micron,* Vol 32, 2001, p 211
99. C.A. Greene, A.J. Henry, C.S. Brossia, and T.M. Ahn, Evaluation of the Possible Susceptibility of Titanium Grade 7 to Hydrogen Embrittlement in a Geologic Repository Environment, *Scientific Basis for Nuclear Waste Management,* Vol 663, Materials Research Society, 2001, p 515–523
100. A.L. Pulvirenti, K.M. Needham, M.A. Adel-Hadadi, A. Barkatt, C.R. Marks, and J.A. Gorman, "Corrosion of Titanium Grade 7 in Solutions Containing Fluoride and Chloride Salts," paper 02552, Corrosion/2002, NACE International, 2002
101. L.M. Young, G.M. Catlin, G.M. Gordon, and P.L. Andresen, "Constant Load SCC Initiation Response of Alloy 22 (UNS N06022), Titanium Grade 7 and Stainless Steels at 105 °C," paper 03685, Corrosion/2003, NACE International, 2003

SELECTED REFERENCES

- G.S. Bodvarsson, C.K. Ho, and B.A. Robinson, Ed., *Yucca Mountain Project,* Elsevier Science, Oxford, U.K., 2003
- "Effect of Fabrication Processes on Material Stability—Characterization and Corrosion," Report CNWRA 2004-01, Center for Nuclear Waste Regulatory Analyses, San Antonio, TX, 2004
- D. Féron and D.D. Macdonald, Ed., *Prediction of Long Term Corrosion Behaviour in Nuclear Waste Systems,* No. 36 of the European Federation of Corrosion Publications, The Institute of Materials, Minerals and Mining, London, 2003
- Site of the Nuclear Regulatory Commission Controlling and Authorizing Entity, http://www.nrc.gov, accessed Sept 1, 2005
- Site of the Nuclear Waste Technical Review Board, http://www.nwtrb.gov, accessed Sept 1, 2005
- Site of the Office of Civilian Radioactive Waste Management of the Department of Energy, in Charge of Building the Repository, http://www.ocrwm.doe.gov, accessed Sept 1, 2005
- "Yucca Mountain Science and Engineering Report, Technical Information Supporting Site Recommendation Consideration," report DOE/RW-0539, U.S. Department of Energy, Office of Civilian Radioactive Waste Management, May 2001

Introduction to Corrosion in Fossil and Alternative Fuel Industries

John Stringer

THE PRIMARY FOSSIL FUELS are generally defined as coal, oil, natural gas, tar sands, and shale oil. These can be further subdivided, in terms of the quality of the fuel. For example, coal can be anthracite, bituminous coal, subbituminous coal, or lignite; further classification is possible. Ultimately, a fuel is burnt producing heat, and the heat is then directed to some useful purpose. This purpose might be the generation of electricity, process heat for chemical industry, domestic heating, and so forth.

Fuels

The primary fuels as produced generally contain a number of impurities. In the case of coals, these impurities can be divided in two classes: physical impurities, such as mineral species, which are chemically inert (silica is an example), and chemical impurities, such as sulfides, which undergo modifications during combustion.

Both of these classes of impurities can result in environmental problems. The incombustible physical impurities generate products called ash: the large particle forms of this can be collected and must be disposed of; the small particles may become entrained in the off-gases and may become an environmental hazard with adverse effects on the pulmonary system. The chemical impurities can result in the release of atmospheric impurities such as sulfur oxides and nitrogen oxides, which have been shown to be health hazards, and may also cause damage to other components of the biota, for example trees. In addition, the chemical impurities play a major role in the corrosion of the fuel systems.

The alternative fuels are various; wood, and specifically wood waste from the timber industry, is a major resource. Agricultural wastes are also important, and there are now plants and trees that are being specifically grown for fuel use. Various wastes, such as industrial wastes and municipal solid wastes (MSWs), are increasingly being used as fuels. These fuels also contain impurities of one kind or another; waste wood, for example, can be contaminated with salts from seawater.

The physics and chemistry of these alternative fuels cover a wide spectrum, and even a specific alternative fuel can have a wide range of compositions; MSW is an obvious example. Increasingly, primary fuels are being processed before use. Treatment of coals has been a common practice for many years; for example, many of the coals as mined may contain unacceptably large fractions of mineral species, and this may be reduced at the minehead before shipping to the customer, or at the customer's site before use. In other cases, the final use for a fuel may be better served by a liquid or a gas, and solid fuels may be processed for this purpose. In the case of petroleum-based fuels also, the quality of the required fuel requires adjustment, and this too has been done for many years. The higher boiling-point fractions, which have higher carbon/hydrogen ratios, also have higher viscosities; these are separated by distillation and then hydrogenated to produce a lighter hydrocarbon. The hydrogen for this is generated by steam reformation of the lowest fractions. Coals have been gasified for many years; currently the preferred approach involves entrained gasification at elevated temperatures, with the incombustible fraction forming a slag that is water-quenched in the bottom of the gasifier to allow easy removal by lock-hoppers. The high-temperature gas product is cooled using a heat exchanger within the gasifier, and the gaseous impurities containing sulfur and chlorine are removed from the product stream. The initial product gas may contain hydrogen and carbon monoxide, but this can be shifted to methane to produce a synthetic natural gas. Gasification of the alternative fuels is also possible, with the aim of producing a more practical fuel and controlling harmful impurities. In some cases, a liquid fuel that can be substituted for gasoline is required; biomass conversion to ethanol is an example that has been used for a number of years, and this approach appears to be increasing in importance. Liquefaction of coal has also been a research topic for a long time, and for some years there was an active program in hydrogenation of coal for this purpose. However, although it can be done, the program in the United States was abandoned for economic reasons some years ago.

The control of the emissions to attain regulatory requirements is an important part of the overall systems. In some cases, the modification of the primary fuel mentioned will reduce or even eliminate the impurities, as indicated previously, but often these become concentrated in the by-product, and disposal of this may itself present significant problems.

Energy Conversion

Generally, the chemical energy in a fuel is transformed by oxidation. In the majority of cases, this involves combustion, and the objective of the combustion system is to attain the maximum conversion possible. Typically, this involves attaining the best possible mixing between the fuel and the oxidant (usually air). For the solid fuels, this normally requires reducing the as-received fuel to a particulate form; this is then injected into the combustion chamber through a nozzle in which some or all of the combustion air is used to transport the fuel prior to ignition. The nozzle itself is designed to promote the mixing, and ignition occurs within the combustion chamber at or close to the end of the nozzle. For liquid fuels, the process is much the same, but the reduction in the fuel particle size is usually much simpler. For gaseous fuels, it is still important to achieve good mixing prior to combustion.

There are other approaches. The most significant of these is fluidized bed combustion, in which the solid fuel (or, less commonly, liquid fuel) is injected into a bed of particles through which air is blown from beneath; at a certain flow velocity the bed becomes fluidized and behaves like a liquid; both the chemical mixing and the heat transfer within bed is extremely efficient. The combustion of the fuel takes place within the bed, and part of the heat of combustion may be removed by an in-bed heat exchanger to maintain a constant temperature within the bed. The fuel particle size does not have to be so fine as in the system described previously, and in addition it is possible to achieve the chemical removal of some of the environmentally unacceptable impurities by the use of additives within the bed. There are some corrosion issues specific to this technology; these are not discussed here.

There are other ways of converting the energy, most specifically by using an electrochemical approach. This is the method used in a fuel cell, in which the chemical energy is directly converted to electrical energy. For most of the fuel cells currently being produced, the fuel must be hydrogen, although this can be produced from hydrocarbon fuels by a reformer within the power plant itself. The efficiency of the conversion using this approach is not limited by the thermodynamic Carnot restrictions, and fuel cells have been made to operate at room temperature or temperatures as high as 1100 K. There are a number of corrosion issues in these systems.

Following the combustion, the sensible heat is contained in the combustion gas, and in the majority of cases the useful energy has to be removed from this. In the case of domestic heating, or similar end uses such as product drying, the desired product is a clean, relatively low-temperature heat, and this can be gas (for drying, or forced gas domestic heating systems) or warm water, again for heating systems. The conversion from a relatively hot combustion gas with a number of unacceptable impurities to this clean, lower-temperature product is achieved by a heat exchanger, with the hot dirty gas on one side and the clean, warm product on the other side. The emission from the combustion gas side will be a much cooler gas, with a temperature hopefully only a little above the product temperature. One of the issues with these systems can be related to the dew point on the dirty gas side; generally, it is important to avoid condensation within the heat exchanger itself. The exhaust must also satisfy the environmental requirements and may require treatment to achieve this. The details of the treatment, and the problems involved, will depend very much on the system and particularly on the fuel used.

More generally, the objective is to convert the thermal energy to mechanical energy, and this is done with a heat engine. In most cases, this involves expanding a hot, high-pressure gas through an engine to produce a mechanical result that is most often the rotation of a shaft. The most common example is the automobile engine, in which gasoline and air are drawn through a valve to a mixing nozzle into a cylinder by a piston; the mixture is then compressed by the piston rising in the cylinder. The compressed mixed gases are then ignited by a spark and explode, driving the piston down. The piston is attached to a crankshaft, rotating a shaft; as the shaft rotates a valve opens, allowing the combustion gas exhaust to escape. Here, the combustion gas is the "working fluid," and the compression is achieved by the piston in the cylinder. This is called the Otto cycle.

In the case of a steam turbine system, the combustion gas is used to boil water inside a heat exchanger; the compression is achieved by pumping the inlet water. The compressed steam is then allowed to expand through a turbine, and it is condensed by a condenser at the turbine exhaust; the condensate is then recycled to the boiler. This called the Rankine cycle. The steam is the working fluid, and this is an example of an indirect working fluid.

A gas turbine (also called a combustion turbine) is generally a direct cycle; air is compressed by a turbine; at the exit of this, fuel is injected into the compressed air in a combustion chamber, and the hot gas is expanded through another turbine. This drives the compressor, and this may also power the driveshaft. The working fluid in this case is the combustion gas. This is called the Brayton cycle.

Efficiency

The efficiency of a heat engine, μ, is restricted by thermodynamic principles originally outlined by Carnot, and it is determined by the maximum temperature in the thermodynamic cycle T_{max}, and the minimum temperature, T_{min}. Here, both temperatures are absolute. The ideal Carnot efficiency μ_C is:

$$\mu_C = (T_{max} - T_{min})/T_{max}$$

The maximum and minimum temperatures in a cycle are determined at the entrance and exit of the working component: thus, in the case of the ranking cycle it is the steam temperature at the entrance to the turbine and the temperature at the exit of the turbine, which is close to the condenser entry temperature. For the majority of the steam turbines used to generate electricity in the United States at the moment, these temperatures are 811 and 300 K; μ_C is thus 0.63. However, a real system is not ideal, and the actual efficiency in a practical system is closer to 0.38. In the case of a combustion turbine the maximum temperature is considerably higher, but the minimum cycle temperature, which is that of the turbine exhaust, is also considerably higher. The cycle efficiency turns out to be more or less the same. However, if the exhaust gas from a combustion turbine is used to raise steam to power a steam turbine, the overall efficiency of the system is considerably higher. This is called a combined cycle, and efficiencies as high as 0.60 have been achieved.

Since 1987 there have been a number of significant changes in the system requirements for the fossil and alternative fuel industries. These essentially fall into three groups: the first is a desire to improve the overall efficiencies of the overall energy conversion, the second is a need to accept a wider range of fuels, and the third is to achieve much improved control over the environmental implications.

The first issue essentially requires increasing the temperature difference across the cycle. For a simple Rankine cycle, there is usually virtually no possibility of reducing the minimum cycle temperature, so this really means increasing the steam temperature at the turbine inlet. Achieving this results in a need for a different suite of materials for the high-temperature parts of the system; this is particularly true for the Rankine cycle machines, since the Brayton cycle turbines used for power generation have inlet temperatures at the power turbine still less than those in aircraft engines, so there are materials capable in principle of meeting these requirements. It is not, of course, quite as simple as this, because the component sizes are larger, and this means that some of the materials used in aviation engines cannot be used for the land-based engines. These new materials requirements arise primarily for mechanical property reasons, and the issues that arise for achieving adequate corrosion resistance combined with the mechanical criteria present further problems. In some cases, this may require coatings or claddings. The requirements of most concern are on surfaces in contact with the combustion gases, but it can be expected that the steam-side surfaces may also experience enhanced attack; issues such as the exfoliation of steam-side oxide may become more important.

The second issue has two aspects. One is blending of fuels: a high sulfur coal with a low sulfur coal, for example, or municipal solid waste with coal. The second is the ability to switch fuels: having a gas turbine able to switch from natural gas to oil, for example.

The third issue involves a number of aspects. One of particular importance at the moment is controlling combustion to reduce NO_x emissions. In the case of a large, coal-fired utility boiler, for example, if the coal is combusted in the boiler at low excess air, or even at substoichiometric air in the vicinity of the fuel nozzles, with the balance of the air added higher in the boiler, the emissions can be reduced considerably. However, the consequence is that the lower water-wall tubes in the boiler experience a significantly more corrosive environment, associated with the impact of uncombusted coal particles and unoxidized pyrite particles. Another approach is to replace the direct firing of coal by the use of a gasifier before the combustion; as indicated previously, the removal of impurities, including ash, is easier, but the corrosion problems in the boiler are replaced by the corrosion issues in the gasifier. The corrosion issues in the systems used to remove the impurities are also different.

The high-temperature corrosion problems in most of these systems involve two principal mechanisms: mixed-gas corrosion—typically involving conditions that may be sulfidizing and oxidizing, or carburizing and oxidizing—and molten salt accelerated corrosion. In the latter case, alkali metals—sodium, and potassium—in the impurities are often vectors in the attack. In the cases of petroleum-based fuels, vanadium species may be important.

All of the heat engines are designed to make the most use of the heat, and this means that there are several "warm-water" streams in the systems. Some of these are very pure waters, as is the case in the recycle of the boiler condensate; some of them are more or less impure "gray" waters. There are important corrosion issues in these locations also. The condensers are typically cooled with water that may come from local rivers or the ocean, and the flows are considerable; it is generally impossible to undertake any treatment of these waters. The problem of process condensates on the combustion gas side has been mentioned previously, and one obvious and important example is condensation in the exhaust stack or immediately downstream of the stack. These can result in both environmental issues and corrosion damage.

It is apparent that there are a wide range of corrosion-related issues in the utilization of fossil fuels and alternative fuels, and recent studies have given some good estimates of the economic impact of these issues. The following articles describe a number of these problems in more detail.

High-Temperature Corrosion in Gasifiers

Wate Bakker, Electric Power Research Institute

INTENSIVE DEVELOPMENT of gasification technology has been carried out since the early 1970s (Ref 1–3). At present (2005), several commercial- or nearly commercial-scale coal-gasification combined-cycle (CGCC) plants are in operation worldwide. These include plants based on the Texaco and Dow processes in the United States, a plant based on the Prenflo (Krupp-Uhde) process in Spain, and a plant based on the Shell process in the Netherlands. All present plants include entrained slagging gasifiers and cold-gas cleanup equipment. In this type of gasifier, the fuel—coal, petroleum coke, or heavy oil bottoms—are pulverized and fed to the gasifier in dry or slurry form. Gasification is carried out at very high temperatures, above the melting point of the mineral matter in the fuel, usually at 1300 to 1600 °C (2370 to 2910 °F). The raw hot gas, consisting mainly of CO and H_2, with H_2S as the major corrosive impurity, is cooled in syngas (synthetic gas) coolers and then water quenched to remove particulates and water-soluble impurities such as NH_3 and chlorides. After quenching, H_2S is removed in one of several commercially available gas-purification systems, for instance, the Selexol process. The purified gas is then fed to a gas turbine combined-cycle plant for combustion and electricity generation. Figure 1 shows a schematic of such a plant, which is quite complex.

Several corrosion and erosion problems have occurred in various plants, mostly at the pilot plant stage. These include coal slurry erosion and corrosion in feed systems, refractory corrosion and erosion in gasifiers, high-temperature corrosion in syngas coolers, and aqueous corrosion in downstream and gas-purification equipment. Most of the problems found are not unique to gasifiers, but have been encountered elsewhere and have been resolved using existing technology. However, high-temperature corrosion in syngas coolers is unique to gasifiers; in these systems there are interactions that occur between the materials of construction and various impurities present in coal as well as other fuels. Therefore, this article concentrates on high-temperature corrosion in syngas coolers.

Details on operating conditions in syngas coolers are provided in Refs 1 and 4. The composition of the raw syngas depends on the gasification process used and the fuel composition, especially its sulfur and chlorine content. The syngas composition of processes with water-coal slurry systems is typically (vol%) 35 to 45 CO, 10 to 15 CO_2, 27 to 30 H_2, 15 to 25 H_2O, 0.2 to 1.2 H_2S, and 50 to 500 ppm HCl; for a dry-fed gasifier, the CO content of the gas is considerably higher (62 to 64%), while the CO_2 and H_2O contents are generally less than 4%. Thus the partial oxygen pressure (p_{O_2}) of a syngas from slurry-fed gasifier is considerably higher than that from a dry-fed gasifier. The syngas composition reflects the high-temperature equilibrium composition at 1300 to 1500 °C (2370 to 2730 °F), while the heat exchangers operate at 300 to 600 °C (570 to 1110 °F). This makes it difficult to calculate p_{O_2} and p_{S_2} values, since the gases are not in equilibrium at the metal temperature (Ref 1). Heat-exchanger syngas operating temperatures in present gasifiers are generally less than 450 °C (842 °F), partly due to high-temperature corrosion issues that indicate that the use of higher heat-exchanger temperatures with current designs are not cost effective.

Corrosion Mechanism and Laboratory Studies

The major corrosion mechanism in syngas coolers is sulfidation for low-alloy steels and sulfidation/oxidation for more highly alloyed chromium-containing materials. Due to the low p_{O_2} and relatively high p_{S_2} of the raw syngas, the sulfidation rate of carbon and low-alloy steels is too high to allow the use of these low-cost materials. This is shown in Fig. 2, which contains information from laboratory as well as plant studies, with good agreement between the two. It is further observed that the corrosion rate of low-alloy steels appears largely independent of the chlorine content of the coal. The plant data in Fig. 2 were obtained from coals or coal-derived fuels with chlorine contents ranging from 0.05 to

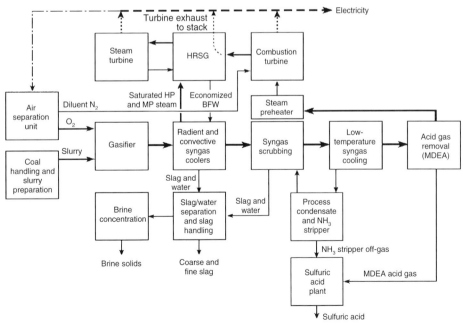

Fig. 1 Coal-gasification combined-cycle CGCC power plant. HRSG, heat recovery steam generator. Acid gas removal uses MDEA, methyldiethanolamine

0.3%, resulting in 100 to 600 ppm HCl in the syngas (Ref 1). Laboratory tests confirm that HCl contents in this range do not increase corrosion rates, at least not under the highly reducing conditions present in gasifiers.

Due to the unacceptably high corrosion rates of low-alloy boiler steels, stainless steel heat-exchanger tubes, coatings, or claddings must be used to provide low corrosion rates. A complicating design factor is that some construction codes (such as those issued by the American Society for Mechanical Engineers) limit the use of certain stainless steels in water wetted service. Extensive laboratory corrosion studies on both model and commercial alloys have been reported earlier (Ref 1, 5) and are summarized here. All experiments were done under isothermal conditions. It was found that the corrosion rate of stainless steels containing 20% or more chromium with or without various minor alloying elements, such as silicon, aluminum, and titanium, did not simply increase with increasing p_{S_2}/p_{O_2} ratio of the syngas, but also depended on the HCl content of the syngas, the presence of chlorides in deposits, and alloy composition.

In the absence of HCl in the syngas, the corrosion loss indeed increases as a function of the p_{S_2}/p_{O_2} ratio as shown in Fig. 3 for alloys 800 (UNS N08800, Fe-20Cr-32Ni-0.35Al-0.35Ti) and 310 (UNS S31000, Fe-25Cr-20Ni). Corrosion rates are relatively low and parabolic reaction kinetics are usually observed (Ref 6, 7). Scanning electron microscopy/energy-dispersive x-ray analysis (SEM/EDAX) indicates that at high p_{S_2}/p_{O_2} ratios, a somewhat protective $FeCr_2S_4$ spinel is formed. At lower p_{S_2}/p_{O_2} ratios, this layer gradually changes to $FeCr_2O_4$. Below the $FeCr_2S_6$ layer, an inward-growing scale/precipitation zone is formed consisting of $FeCr_2(S,O)_4$ spinel, Cr_2O_3, and iron-rich alloy remnants. This corrosion mechanism is labeled type A corrosion and is schematically shown in Fig. 4.

When HCl is present in the syngas, the corrosion mechanism changes considerably, as is shown by a comparison of Fig. 5 to Fig. 3 (MA 956 is Fe-20Cr-4.5Al-Ti-Y, UNS S67956). Alloys with a relatively low chromium content such as alloy 800 now show a maximum corrosion rate at a p_{S_2}/p_{O_2} ratio of about 18, while alloys with a higher chromium content or similar oxide-forming alloying elements such as aluminum or silicon are relatively less affected by the addition of HCl to the gas. SEM/EDAX analysis

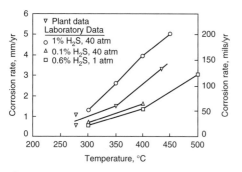

Fig. 2 Corrosion of low-alloy steels in syngas environments. Plant data are compared to laboratory data for various hydrogen sulfide atmospheres.

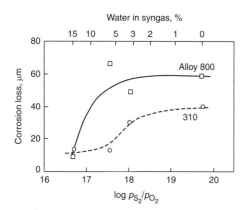

Fig. 3 Corrosion losses of stainless steels as a function of partial pressure ratio log p_{S_2}/p_{O_2} without HCl. Laboratory test 540 °C (1000 °F), 600 h. Solid line, nickel alloy 800 (UNS N08800); broken line, type 310 stainless steel (UNS S31000)

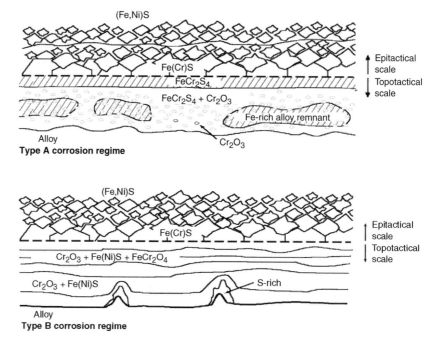

Fig. 4 Scale morphology for type A and B corrosion

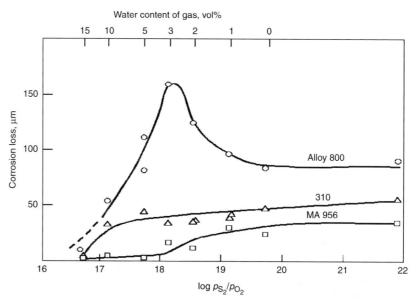

Fig. 5 Corrosion loss of three stainless steels as a function of partial pressure ratio log p_{S_2}/p_{O_2} with 400 ppm HCl. Laboratory test 600 h, 540 °C (1000 °F)

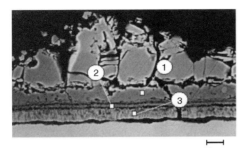
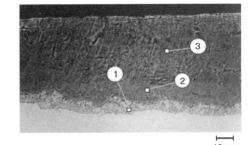

Area	Main species present	Composition, at.%						
		O	Si	S	Cr	Mn	Fe	Ni
1	S-rich spinel	0	nd	58	28	3	11	0
2	(S,O)-rich spinel	30	2	29	34	1	5	1
3	Cr_2O_3 + Fe	29	2	3	30	0	33	2

nd, not determined

(a)

Area	Main species present	Composition, at.%						
		O	Si	S	Cr	Mn	Fe	Ni
1	Cr_2O_3 + Fe	37	2	4	33	<1	23	1
2	Cr_2O_3 + FeS + Fe	49	2	13	33	1	3	0
3	Cr_2O_3 + FeS	49	2	10	31	1	6	1

(b)

Fig. 6 Corrosion of type 310 (UNS S31000) stainless steel in syngas with 1200 ppm HCl. The electron dispersive spectroscopy (EDS) analysis is given for indicated areas along with the main species present. (a) Without deposit of 5% chloride, corrosion loss is 15 μm (0.6 mils). (b) With deposit of 5% chloride, corrosion loss is 86 μm. (a) and (b) Original magnification: 750×

of alloy 800 after exposure to a syngas with a p_{S_2}/p_{O_2} ratio of about 18 and HCl content of 400 ppm indicated the absence of a protective $FeCr_2S_4$ scale. An outward-growing FeS-rich scale is present on top of an inward-growing porous Cr_2O_3-rich scale containing Fe(Ni)S, which appears nonprotective. This corrosion mechanism is labeled type B corrosion and is also schematically shown in Fig. 4.

The main difference between type A and B corrosion is that a protective $FeCr_2S_4$ is formed in type A. The absence of this protection in type B leads to higher corrosion rates. Further studies have shown that the onset of the less desirable type B corrosion depends on the amount of oxide-forming alloying elements in the alloy, the amount of HCl in the syngas, and the presence of chlorides in deposits, which probably increases the p_{Cl_2} near the metal surface. For instance, the corrosion loss of stainless steel 310 after 164 h exposure to syngas with a p_{S_2}/p_{O_2} ratio of 18.5 and 1200 ppm is about 15 μm. However, when a deposit containing 5% chloride is present the corrosion loss increases to 86 μm and the corrosion mechanism changes from type A to type B (Fig. 6). Type A attack is shown in Fig. 6(a); here there is a sulfur-rich $FeCr_2S_4$ outer scale and a uniform inner-scale layer. Type B attack is shown in Fig. 6(b); here the outer scale is a mixture of sulfides and oxides and sulfur-rich nodules are observed at the metal/scale interface.

Aqueous corrosion during downtime can further increase overall corrosion rates in the presence of chloride-containing deposits. During downtime, the hygroscopic chlorides can migrate through cracks in the scale to the scale/metal interface and form an $FeCl_3$-rich layer beneath the scale. During heatup, the $FeCl_3$ layer causes spalling of the protective Cr_2O_3-rich scale and a new outward-growing FeS-rich scale and inward-growing Cr_2O_3-rich scale is formed. This sequence repeats itself during each thermal cycle. Figure 7 shows the scale microstructure after three 200 h high-temperature cycles and three 24 h downtime periods in humid air at 400 °C (752 °F). During the first 200 h, the very thin protective layer 1b is formed, which spalls during downtime. During the next 200 h, a faster-growing, less protective layer (1) forms, which also spalls during downtime. Similarly, layer 1a forms during the third exposure. Layer 2 is the oxide- and chloride-rich layer formed below the protective oxide-rich scale during downtime. Layers 3 and 3a are the nonprotective outward-growing FeS scales formed during the last two exposure periods.

From the laboratory experiments described previously, it is clear that the high-temperature corrosion of stainless steels in syngas environments is very complex and not readily predictable from simple gas/solid reaction mechanisms, especially when HCl is present in the gas and/or chlorides are in the deposit.

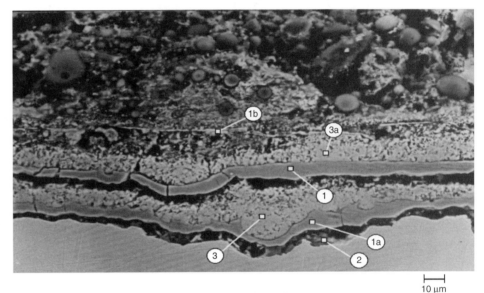

Area	O	Si	S	Cr	Fe	Ni	Mn	Cl
1	43	2.0	8	34	8	2	2.0	1.3
2	56	0.6	3	7	23	5	0.7	5.0
3	0	0.4	49	2	38	10	0.4	0.7

Fig. 7 Scanning electron microscopy image of type 310 (UNS S31000) stainless steel after 3 high temperature/downtime corrosion cycles. Details of the labeled layers are given in the text. Original magnification: 500×

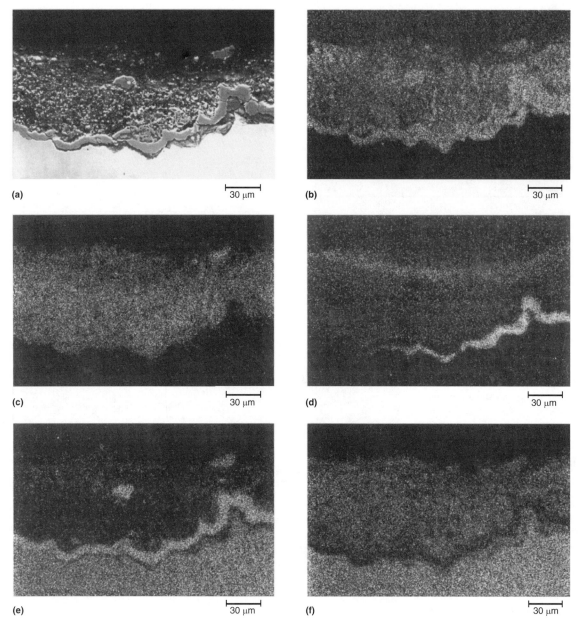

Fig. 8 Scanning electron microscopy image and elemental maps of scale on 310 (UNS S31000) stainless steel after 17,000 h exposure in a syngas cooler. (a) Backscattered electron image. (b) Oxygen map. (c) Sulfur map. (d) Chlorine map. (e) Chromium map. (f) Iron map distribution

Long-Term Performance of Materials in Service

Laboratory tests are usually conducted at atmospheric pressure under isothermal conditions for a few hundred to a few thousand hours. Materials and components of electric power plants should last at least 100,000 h. In syngas coolers, components are also exposed to elevated pressures, 4 to 6 MPa (40 to 60 atm), with significant thermal gradients. Thus, laboratory tests are useful to determine corrosion mechanisms and guide alloy selection, but are generally not suitable to predict actual corrosion rates. For this reason, the Electric Power Research Institute (EPRI) has cooperated with several gasifier operators to obtain long-term corrosion rate data in various pilot and commercial-size gasification plants. Several materials have been exposed in various gasifiers for more than 50,000 h (Ref 4, 8–10). Individual tests have ranged from 2000 to 17,000 h, at temperatures ranging from 300 to 550 °C (572 to 1022 °F).

In general, the corrosion mechanisms found in laboratory tests are also found in service, such as type A corrosion under highly sulfidizing conditions at relatively low HCl levels and type B corrosion under less sulfidizing conditions, at somewhat higher HCl levels and high rates of corrosion due to scale spallation when chlorides are present in the deposits. Figure 8 shows an example of pitting and scale spallation of 310 stainless steel after 17,000 h exposure in a Texaco process gasification plant. The distribution of elements is useful in identifying corrosion products.

In the absence of scale spallation, stainless steels containing at least 20% Cr generally provide a satisfactory corrosion rate of about 0.1 mm/yr or less in the 300 to 400 °C (572 to 472 °F), the temperature range at which most heat exchangers in syngas coolers operate (Fig. 9). HR-160 (UNS N12160) shown in Fig. 9 is nominally 37Ni-29Co-28Cr-2.75Si. Corrosion mechanisms ranged from type B corrosion for the lower-chromium steels to type A corrosion for steels with a higher-chromium content such as 310 stainless steel.

When scale spallation due to chlorine migration to scale/metal interface occurs, corrosion rates increase considerably even at relatively low temperatures. Figure 10 shows the corrosion loss of the 310 stainless steel of Fig. 8, as a function of time. The rate is linear and is about 0.25 mm/yr. Here a molybdenum-containing alloy such as alloy 28 (27Cr-31Ni-3.5Mo-Cu, UNS N08028) is a better choice because it has a parabolic corrosion rate, with less than 0.25 mm loss after 17,000 h exposure. When the gasified fuel has very high chlorine levels, extremely high corrosion rates are possible. This is illustrated in Fig. 11, which gives the results of a 5500 h exposure test in a gasifier using a fuel containing 0.2 to 0.5% Cl. Here stainless steels have completely unacceptable corrosion rates, and molybdenum-containing alloys, preferably with a high nickel content such as alloy 625, are the only acceptable materials. It is interesting to note that alloy 625 is also the preferred alloy for waste-to-energy plants, where chlorine is the major corrodent under oxidizing conditions (Ref 11).

Material Selection. Field corrosion data from various gasifiers indicate that metal wastage rates experienced so far will not lead to catastrophically high corrosion rates. When sulfidation/oxidation is the only or major corrosion mechanism (type A corrosion), corrosion rates less than 0.1 mm/yr are generally experienced for steels with >20% Cr at temperatures below 400 °C (572 °F). Thus alloy 800 (20Cr-32Ni-Al-Ti), which is code approved for boiler service, would be an economical choice and is indeed frequently used. However, when the chlorine content of the fuel is somewhat higher and the syngas is less sulfidizing, type B corrosion can lead to highly increased corrosion rates. Under these conditions, alloy 800 and the weld filler metal FM82 (20Cr-72Ni-Nb, UNS W86082) are less suitable and alloys with a higher chromium content are needed. Type 310 stainless steel (Fe-25Cr-20Ni, UNS S31000) has adequate corrosion resistance, but is not approved for boiler service due to potential waterside corrosion problems. Alloy 28 is code approved and is also more resistant to chlorine-induced scale spallation (Ref 12). It is therefore the logical choice for application where type B corrosion or scale spallation is likely.

In isothermal laboratory corrosion tests, scale spallation under chlorine-rich deposits is observed only in tests combining exposure at high temperature, with shorter periods of exposure to humid air (downtime corrosion). Similar scale spallation is also frequently found in service, but it also occurs when exposure to aqueous corrosion conditions is infrequent, that is, when the plant operates continuously with only one or two shutdowns per year. Recently, corrosion tests were run under thermal gradients to study waterwall corrosion in boilers (Ref 13). It was found that chloride species in deposits can also move to the metal/scale interface and cause scale spallation when a thermal gradient is present, especially when the thermal gradient is steep

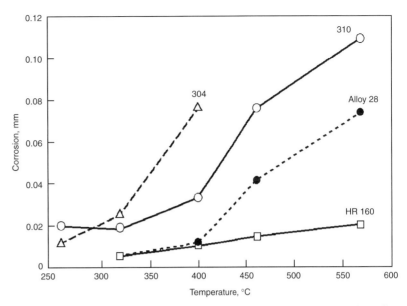

Fig. 9 Corrosion loss of stainless steels after 2000 h exposure in the Prenflo demonstration plant. Alloy 28 (UNS N08028), HR 160 (UNS N12160)

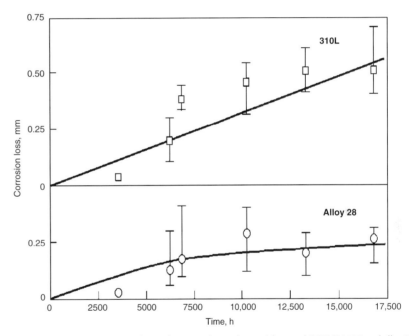

Fig. 10 Corrosion loss as a function of time of type 310 low-carbon stainless steel (UNS S31002) and alloy 28 (UNS N08028), during exposure in a syngas cooled at 390 to 420 °C (734 to 788 °F)

enough to cause melting of the chloride mixture. Thus, the authors now believe that aqueous corrosion during downtime is not the only cause of scale spallation, but that thermal gradients in deposits may be the main culprits, especially in commercial gasifiers, where downtime is infrequent.

Fuels used for present and future gasification plants can have dangerous chlorine levels. Refinery wastes, petroleum coke, and heavy oil bottoms have nominally low chlorine levels. However "phantom chlorides" have become a problem in refineries (Ref 14). In one exposure test using heavy fuel oil, nominally free of chlorine, spallation due to chlorine corrosion was also experienced (Ref 14). The chlorine content of most coal is relatively low, 0.1% or less. However, the resulting HCl levels during gasification are much higher than in boilers, and chlorides are frequently found in heat-exchanger deposits, when gasifying coals with chlorine contents as low as 0.1%.

Biomass can also have surprisingly high chloride levels, although the chlorine level is

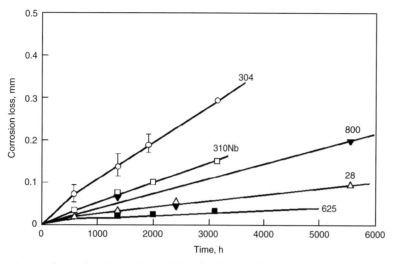

Fig. 11 Corrosion losses of stainless steels and nickel alloys from a coal-gasification plant, gasifying coal residues with 0.2 to 0.5% Cl. 304 (UNS S30400), 310Nb (UNS S31040), 800 (UNS N08800), 28 (UNS N08028), 625 (UNS N06625)

very source dependent. Thus, the chlorine content of the biomass should be an important selection criterion.

Conclusions on corrosion performance of CGCC are:

- Commercial-scale gasification power plants have operated trouble free for long periods of time and are commercially available.
- Corrosion losses due to high-temperature sulfidation alone are generally low enough to provide acceptable long-term service, especially when the chromium content of the alloy is 20% or higher.
- The presence of chloride-rich deposits can cause scale spallation, resulting in accelerated corrosion. When this occurs, common stainless steels have linear corrosion rates. Molybdenum-containing steels, which have more adherent scales and exhibit parabolic corrosion rates, are preferred.

ACKNOWLEDGMENT

This material by Wate Bakker has appeared in *Materials Research in Brazil*, Vol 7, (No. 1), 2004.

REFERENCES

1. W.T. Bakker, Effect of Gasifier Environment on Materials Performance, *Mater. High Temp.*, Vol 11 (No. 1–4), 1993, p 90
2. J. van Liere, Present Status of Advanced Power Plants. *Mater. High Temp.*, Vol 14 (No. 2, 3), 1997, p 7
3. W.T. Bakker, Present Status of Coal Gasification Technology, *Proc. Materials for Coal Gasification*, V. Hill, S. Dapkunas, and W. Bakker, Ed., Oct 1987, ASM International, 1988
4. W.T. Bakker and J. Stringer, Mixed Oxidant High Temperature Corrosion in Gasifiers and Power Plants, *Mater. High Temp.*, Vol 14 (No. 2), 1997, p 101
5. W.T. Bakker, The Effect of Chlorine on Mixed Oxidant Corrosion of Stainless Steels, *Mater. High Temp.*, Vol 143 (No. 3), 1997, p 197
6. J. Norton et al., *Proc. Heat Resistant Materials II*, K. Natesan, P. Ganesan, and G. Lai, Ed., ASM International, 1995, p 111
7. R.C. John et al., *Mater. High Temp.*, Vol 11 (No. 1–4), 1993, p 124
8. W.T. Bakker, J.B.M. Kip, and H.P. Schmitz, *Mater. High Temp.*, Vol 11 (No. 1–4), 1993, p 133
9. W. Schellberg et al., *Mater. High Temp.*, Vol 14 (No. 2), 1997, p 159
10. M.E. Fahrion, *Mater. High Temp.*, Vol 11 (No. 1–4), 1993, p 107
11. G. Sorell, *Mater. High Temp.*, Vol 14 (No. 2, 3), 1997, p 137
12. A.M. Lancha, M. Alvarez de Lara, D. Gomez-Briceño, and P. Coca, *Mater. High Temp.*, Vol 20 (No. 1), 2003, p 75–83
13. W.T. Bakker, J.E. Blough, S. Warner, and S.C. Kung, The Effect of Deposits on Waterwall Corrosion, *Proc. Materials for Advanced Power Engineering* (Liege, Belgium), J. Lecomte-Beckers, M. Carton, F. Schubert, and P.J. Ennis, Ed., European Commission/University of Liege, 2002
14. M.V. Veasey, Phantom Chlorides Create Real Problems for Refineries, *Mater. Perform.*, May 2002, p 17–19

Corrosion in the Condensate-Feedwater System

Barry C. Syrett, Electric Power Research Institute
Otakar Jonas and Joyce M. Mancini, Jonas, Inc.

IN STEAM TURBINE ELECTRICAL POWER GENERATION PLANTS, turbine exhaust steam is condensed, and the condensate is reheated for return as feedwater to the boiler. Water from the condenser to the feedwater pump is called condensate, and from the pump to the boiler it is feedwater. This article addresses the major heat-transfer components in this part of the water-steam loop of a power plant: the condenser and the feedwater heater. These components are characterized by providing heat-transfer through metal walls between moving fluids. Both open-type feedwater heaters and condensers can function as deaerators. Deaerators may be separate units that also provide feedwater storage.

Corrosion of Condensers

Barry C. Syrett, Electric Power Research Institute
Otakar Jonas, Jonas, Inc.

The functions of a condenser include condensing turbine exhaust steam, generating a vacuum (which has a large effect on turbine efficiency), and deaerating the condensing steam. The condenser is a particularly critical component in a power plant, because condenser tube leaks and air inleakage affect many other components in the steam-water cycle (Ref 1–35). The root causes of many of the corrosion problems in fossil fuel boilers, nuclear steam generators, low-pressure steam turbines, and feedwater heaters have been traced to condensers that have leaked and allowed contamination of the steam condensate with raw cooling water and air. Most tube leaks are caused by corrosion, but some tube failures are purely mechanical, such as those caused by steam impingement (erosion), tube-to-tubesheet joint leaks, mechanical ruptures from foreign object impact, and tube vibration resulting in fretting wear and fatigue (Ref 1–3). Purely mechanical failures are not discussed further.

Various methods are used to locate the sites of air and cooling water inleakage (Ref 34, 35), with helium leak detection being the best. Also, the presence of cooling water leaks can be detected by monitoring the condensate for sodium or cation conductivity (Ref 11).

Types of Condensers

Condensers may be cooled by water, air, or water spray.

Water-Cooled Condensers. A typical large utility condenser is a shell and tube heat exchanger comprising a carbon steel shell, two tubesheets, and thousands of straight tubes running between the tubesheets. It is usually located under the low-pressure (LP) turbine base, connected by a flexible joint, although smaller condensers may be connected to the LP turbine by a pipe. Heat is transferred from steam on the outside of the condenser tubes (the shell side or steam side) to water on the inside (the tube side or water side). The cooling water passing through the tubes typically is pumped from a natural body of water, such as a river, lake, or ocean, or from various types of cooling towers. A schematic of a typical electric power plant condenser is shown in Fig. 1.

The lowest-pressure feedwater heaters may be located in the condenser neck. Condenser tubes are arranged to facilitate proper flow of wet steam through the array of tubes toward the air-removal region, from which the noncondensable gases are removed through a vent collector (Fig. 1) by external air ejectors or vacuum pumps (Ref 1–9). Makeup water for the main steam cycle is usually added into the condenser through a sparger (perforated pipe). The condensate, collected in the hotwell at the base of the condenser, is removed by one or more condensate pumps.

Air-cooled condensers are used where cooling water is not available or in environmentally

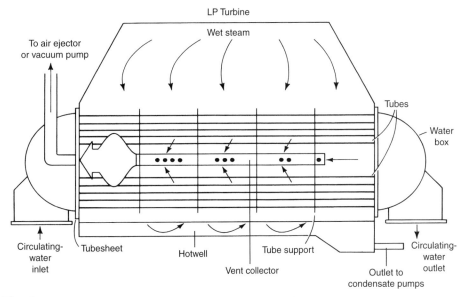

Fig. 1 Schematic of a typical condenser in an electric power plant. LP, low pressure

sensitive areas (Ref 2, 10). All components, including panels of finned tubes, are usually made of carbon steel. Here, the wet steam is on the inside of the tubes, and the coolant (air) is on the outside. The components are usually adequately protected against corrosion by normal steam and condensate chemistry (Ref 11). However, when the chemistry is out of control (low pH, high O_2), large quantities of iron corrosion products are generated, which can be transported into the feedwater and become harmful to boilers and turbines. Some designs of air-cooled condensers are subject to flow-accelerated corrosion (FAC) on the steam side, particularly at the transitions from headers to condenser tubes. During startups early in the life of the condenser, the condensate may be contaminated with organic preservatives, silica, and other impurities picked up from the tube walls. It is recommended that the condensate be filtered to minimize these effects.

The external surfaces of the finned tubes are prone to fouling from pollen, dust, insects, leaves, plastic bags, bird carcasses, and so on. Not only is the air flow affected but also the heat-transfer coefficient; the deterioration in performance increases unit operating costs.

To improve the heat-removal capacity of an air-cooled condenser under conditions of high ambient air temperature, operators sometimes spray water on the external surfaces of the heat-exchanger tubes to reduce the surface temperature. Unfortunately, depending on the hardness of the water used, this sometimes leads to scale formation on the tube fins and reduces the heat-transfer rate if the deposits are allowed to accumulate.

There have been a few air-cooled condensers made using aluminum tubes, but the experience with these has not been very good.

Other Types of Condensers. In smaller utility and industrial systems, the condenser can be part of a dry-wet cooling tower. In these designs, galvanized carbon steel condenser tubes are located at the bottom of the cooling tower and cooled on their outer surfaces by water spray. Typical problems include FAC on the steam side at the tube inlets and corrosion and erosion of the zinc plating on the tube outer surfaces.

Cooling Water Chemistry

The goal of cooling water chemistry control is to avoid formation of mineral scale, organic growth, and corrosion (Ref 28–32). However, the overall control is also significantly influenced by environmental regulations.

Cooling water chemistry control is governed by the chemical composition of the available cooling water sources, which range from low-conductivity, low-hardness freshwater to high-hardness freshwater to sea and brackish water and cooling tower water. All of these cooling waters can have various tendencies to corrode condenser materials and to introduce organic fouling and mineral scaling.

The cooling water system treatment program actually chosen is often determined by the cost of the treatment relative to the costs of the corrosion, deposition, and scaling that would result in the absence of the treatment. Control of deposition, scaling, and microbiological growth in recirculating systems is achieved through the addition of chemicals that adjust the pH and alkalinity of the cooling water, inhibit scale growth, act as dispersants or complexing agents, or inhibit microbiological growth.

In once-through cooling water systems, the same approach cannot be taken, due to cost and environmental regulations. The water chemistry typically is not controlled, except for filtration of the water and, quite frequently, some type of biofouling control. In both once-through and recirculating systems, the addition of chlorine or nonoxidizing biocides is common.

Corrosion Mechanisms

Data collected by the North American Electric Reliability Council from 1476 fossil units in 168 utilities between 1996 and 2000 show that condensers were responsible for 25,955 forced and scheduled outages and deratings, costing the utilities 53,869 GWh (Ref 36). Table 1 lists several of the components that were responsible for the failures as well as the associated losses. The component failure statistics shown in Table 1 do not specify the damage mechanism or cause of the outage, but the main component problems, such as those in tubes, tubesheets, expansion joints, vacuum pumps, or air-inleakage-related problems, are frequently caused by corrosion or erosion. The leaks in these components allow corrosive impurity ingress, which promotes corrosion in other steam cycle components.

Corrosion mechanisms active in condensers and giving rise to known service problems are summarized in Table 2 (Ref 1–3, 11–25, 37). Information on tubesheet materials requires special explanation. Because tubesheets are very thick (>25 mm, or >1 in.), corrosion rates can be 15 to 50 times higher than in condenser tubes and still be considered acceptable in most cases. Furthermore, even if corrosion is a serious problem in a tubesheet, inspections are usually scheduled frequently enough, and the tubesheet is thick enough, that suitable repairs can be made or corrosion protection procedures can be instituted long before the tubesheet is penetrated. Thus, although Table 2 indicates that copper alloy tubesheets have suffered significant (often severe) galvanic corrosion under certain conditions, leakage of cooling water through the tubesheet from the water side to the steam side has rarely occurred. Each of the corrosion mechanisms found in condensers is reviewed as follows.

Erosion-Corrosion (Water Side). Erosion-corrosion is a relatively common water-side phenomenon that is only a problem in copper alloy condenser tubes (Ref 2, 3, 5, 14, 22). It occurs in areas where the turbulence intensity at the metal surface is high enough to cause mechanical or electrochemical disruption of the protective oxide film. In these turbulent regions, pitlike scalloped features develop. Turbulence increases as the flow velocity increases, and it is greatly influenced by geometry. For example, turbulence intensity is much higher at tube inlets than it is several feet down the tubes, so, if erosion-corrosion is a problem, it is commonly found in the first 25 cm (10 in.) of copper alloy tubes. Tube inserts and tube end coatings have been used to circumvent this problem. A tube insert is a tightly fitting internal sleeve, typically 150 to 300 mm (6 to 12 in.) long, made from an erosion-corrosion-resistant material that shields the susceptible tube ends. However, unless there is a smooth transition (feathered edge) between the end of the insert or coating and the tube, the transition can create turbulent conditions and promote erosion-corrosion further down the tube (Fig. 2). Electric Power Research Institute (EPRI)-funded experiments have demonstrated that inlet-end erosion-corrosion can also be prevented by installing a cathodic protection system in the water box region (Ref 38). Reducing the cooling water flow velocity and distributing flow

Table 1 Generating capacity lost by forced and scheduled outages and deratings caused by condensers from 1996 to 2000

Component or activity	Total occurrences	Outages per unit-year(a)	Total losses for all units, MWh	Loss per unit-year, MWh/unit-year(a)	Loss per outage, MWh/outage
Tubes (leaks)	4569	0.675	1.4×10^7	2081.4	3084
Tubes (fouling—shell side)	91	0.013	8.2×10^4	12.1	899
Tubes (fouling—tube side)	4340	0.641	4.1×10^6	600.4	936
Tube and water box (cleaning)	9530	1.408	8.6×10^6	1275.7	906
Other tube casing or shell incidents	224	0.033	2.1×10^6	308.7	9330
Tubesheets	61	0.009	2.3×10^6	334.0	37,073
Expansion joints	124	0.018	2.8×10^6	409.2	22,340
Inspection	113	0.017	5.0×10^5	73.5	4402
Major overhaul	43	0.006	8.4×10^6	1234	194,271
Air inleakage	147	0.022	7.3×10^5	108.5	4996
Miscellaneous	1446	0.214	3.1×10^6	458	2144

MWh, megawatt hours. (a) Total number of unit-years = 6771. Source: Ref 36

uniformly across the tubesheet also reduce erosion-corrosion. Some suggested maximum cooling water flow velocities are shown in Table 3.

Water box and water intake design can be the source of inlet end erosion-corrosion problems. The preferred arrangement is a hemispherical inlet chamber with the inlet pipe centered in the hemispherical head. This arrangement results in reasonably uniform flow to all tubes and minimizes tube inlet end erosion-corrosion. Other arrangements (inlet perpendicular to the water box, flat rather than hemispherical water boxes, and other variations due to space limitations) often lead to turbulence and erosion-corrosion at the inlet end of nonferrous-tubed condensers.

Erosion-corrosion may also occur when marine life or debris in a tube creates a partial blockage, resulting in locally high velocities through the restricted opening. In these cases, the best solution is to keep the tubes clean, using one or more of the following methods:

- Install or improve intake screens
- Install on-line sponge ball cleaning
- Periodically reverse flow (backwash)
- Manually clean with brushes, balls, scrapers, and so on (off line)
- Prevent biofouling by chlorination or thermal shock
- Decrease intake screen openings to one-half the tube diameter

Alternatively, some copper alloys—such as aluminum brass in saline waters—benefit from periodic dosing of the water with ferrous ions (Fe^{2+}), which are usually added as ferrous sulfate ($FeSO_4$) solution. The Fe^{2+} ions deposit as a protective lepidocrocite [FeO(OH)] layer on the copper alloy surface.

Sulfide Attack (Water Side). This form of attack affects only copper alloys and occurs when the cooling water, most often brackish water or seawater, is polluted with sulfides, polysulfides, or elemental sulfur (Ref 2, 3, 5, 12, 39). As little as 10 mg/m^3 (10 ppb) of sulfide in the cooling water can have a detrimental effect, and concentrations far greater than this are often measured in polluted harbors and estuaries. Sulfide attack manifests itself in many ways. It can greatly increase general corrosion rates, and it can induce or accelerate dealloying, pitting, erosion-corrosion, intergranular attack, and galvanic corrosion. Penetration rates in polluted waters can be extraordinarily high, sometimes as high as 20 mm/yr (790 mils/yr). No copper alloy is resistant to sulfide attack, and the relative performance of copper alloys in polluted or brackish waters seems to depend on the precise environmental conditions (Ref 14, 20–22, 38–43).

If the incoming water contains sulfide and there is no obvious method of eliminating the source, one possible method of reducing or preventing the problem is to dose the water periodically or continuously with $FeSO_4$ or some other source of Fe^{2+} ions (Ref 44). If this approach is used, compliance with environmental discharge requirements for ferrous ions should be checked. However, sulfide attack can also occur in condensers cooled with nominally unpolluted water if marine organisms trapped within the condenser during downtime are allowed to die and putrefy to produce sulfides.

This can be prevented by maintaining circulation of the cooling water during downtime or, for longer outages, by draining the tubes and drying them with warm, dehumidified air. In addition, sulfate-reducing bacteria can produce sulfides under debris and deposits where the oxygen content is low. Thus, the risk of sulfide attack is greatly reduced if the copper alloy tubes are regularly cleaned (Ref 44).

Dealloying (Water Side). Dealloying is the selective corrosion of one or more components of a solid-solution alloy and is also known as parting. The more active (reactive) components of the alloy corrode preferentially, leaving behind a surface layer that is rich in the more noble alloying elements. Dealloying has been reported in condenser tubes, tubesheets, and water boxes (Ref 2, 3, 5, 15). Dezincification may occur in brass tubes and tubesheets, denickelification in copper-nickel alloy tubes, dealuminification in aluminum bronze tubesheets, and graphitic corrosion in cast iron water boxes.

Dealloying is rarely the cause of condenser tube failures, but when it does occur, it is generally found beneath deposits and at hot spots and is promoted by stagnant conditions. This results in plug-type dealloying. A much less localized form of dealloying, termed layer-type dealloying, is rare in tubes but has occasionally occurred in brass tubesheets, particularly in conjunction with galvanic corrosion induced by titanium or stainless steel condenser tubes (Ref 19).

The occurrence of dealloying in copper-base alloys is critically dependent on the major and minor alloying elements present. Copper-zinc

Table 2 Corrosion mechanisms that have caused problems in power plant condensers under certain conditions

Alloy	Erosion-corrosion	Sulfide attack	Dealloying	Pitting/crevice corrosion	Galvanic corrosion	Environmental cracking(b)	NH$_3$ attack
Carbon steel							
Steel condenser shell, tubesheets, spargers	S	N	N	W	W	N	N
Copper alloys							
Muntz metal tubesheets	N	w	w	N	W	N	N
Aluminum bronze tubesheets	N	w	N	N	w	N	N
Aluminum bronze	W	W	w	w(c)	w	s	s
90Cu-10Ni	W	W	w	w(c)	w(d)	N	s
70Cu-30Ni	W	W	N	w(c)	w(d)	N	s
Aluminum brass	W	W	w	W(c)	W	WS	S
Admiralty brass	W	W	w	W(c)	W	WS	S
Stainless steels							
304 and 316	N	N	N	W	N(e)	w	N
AL6X (UNS N08366)	N	N	N	w	N(e)	N	N
AL29-4C (UNS S44735)	N	N	N	N	N(e)	w	N
Sea-cure (UNS S44660)	N	N	N	w(f)	N(e)	w	N
Titanium alloys							
Commercial-purity titanium	N	N	N	N	N(e)	N(g)	N

(a) W, water-side problem; S, steam-side problem; WS, both sides; N, not a problem; w, small sensitivity on water side; s, small sensitivity on steam side. (b) Includes stress-corrosion cracking and hydrogen embrittlement. (c) Possible problem only if sulfide present. (d) Problems have occurred in similar alloys. (e) Induced in adjacent copper alloys, iron, and carbon steels when used in seawater or other highly conductive waters. (f) Problem in heats containing only 25.5% Cr and 3% Mo. (g) Brittle titanium hydride may form and crack if excessively high cathodic protection currents are applied.

Fig. 2 Erosion-corrosion occurring immediately downstream of a nylon insert in an aluminum brass condenser tube cooled by seawater

Table 3 Recommended maximum velocity limit for condenser tube alloys in seawater

	Maximum velocity	
Material	m/s	ft/s
Copper	0.9	3
Admiralty brass	1.5	5
Aluminum brass	2.4	8
90Cu-10Ni	3.0	10
70Cu-30Ni	3.7	12
Stainless steels and titanium	No limit	

Source: Ref 5

alloys are highly susceptible to attack, while copper-nickel alloys are generally considered to be more resistant. In copper-zinc alloys, dealloying (dezincification) is inhibited by the addition of arsenic, antimony, and phosphorus. Dealloying attack is frequently associated with improper heat treatment of the multiphase alloys that are used for tubesheets. However, heat treatment has little effect on dealloying of single-phase materials such as admiralty brass, 90Cu-10Ni, and 70Cu-30Ni.

Graphitic corrosion, which occurs occasionally in cast iron condenser water boxes, has also been labeled a form of dealloying, even though, in this case, selective corrosion does not occur in a solid-solution alloy. Graphitic corrosion results from the deterioration of gray cast iron in which the metallic constituents are selectively leached or converted to corrosion products, leaving behind a surface profile that is largely unchanged but is composed of a weak, porous mass of graphite.

A cathodic protection system installed in the water box will control both galvanic corrosion and dealloying. Where dealloying problems have been encountered in condenser tubes, frequent tube cleaning and minimizing the occurrence of stagnant conditions (such as wet lay-up) can reduce the rate of attack. Alternatively, tube materials that are resistant to dealloying, such as titanium, stainless steel, and inhibited copper-base alloys, can replace the dealloying-sensitive alloys.

Crevice Corrosion and Pitting (Water Side). Brass and austenitic stainless steel condenser tubes are known to have failed by cooling-water-side pitting and crevice corrosion (Ref 2, 3, 5, 14, 16–18). There is limited evidence that copper alloys have adequate resistance to these forms of corrosion if the cooling water is completely free of sulfide. Certainly, the susceptibility seems to be greatly increased when sulfide is present.

Pitting and crevice corrosion of stainless steels are more dependent on the chloride content of the cooling water than on the sulfide content, although laboratory data have demonstrated that the detrimental effects of chloride are accentuated in the presence of sulfide. Some alloys, such as AISI type 304 and 316 stainless steels, which generally perform well in freshwaters or slightly brackish waters, suffer rapid pitting and crevice corrosion in seawater (Fig. 3). Stagnant seawater left in austenitic stainless steel tubes can perforate the tubes in a few weeks. The more highly alloyed stainless steels, including AL6X (UNS N08366), AL29-4C (Fe-29Cr-4Mo-0.35Si-0.02C-0.02N-0.24Ti), and Sea-cure (Fe-27.5Cr-3.4Mo-1.7Ni-0.4Mn-0.4Si-0.02C-0.5Ti + Nb), generally perform well even in seawater. However, a few failures have been reported for AL6X and for some of the early heats of Sea-cure.

The initiation of pitting in stainless steels may be stimulated by the dissolution of inclusions at the metal surface. Manganese sulfide (MnS) inclusions, for instance, are formed in stainless steels when the manganese content approaches 2%. These inclusions are soluble in water and, depending on the size of the inclusion, could result in the formation of a corrosion cell or just dissolve and leave a pitlike hole.

On the cooling water side, crevice corrosion occurs where surfaces are in close proximity and are not boldly exposed to the water environment, such as at a poorly formed tube-to-tubesheet joint. Imperfect construction methods that result in tubesheet hole ovality, tube-to-tubesheet joint underrolling, and inadequate flaring have been identified as potentially leading to crevices that can stimulate crevice corrosion.

Tube cleanliness is a critical issue, because debris, biofilms, and deposits promote the formation of concentration cells (the precursor to crevice corrosion) and the production of sulfides. The tube-cleaning techniques summarized earlier in this section are therefore equally useful in preventing crevice corrosion and pitting in copper alloys and stainless steels. Other solutions include:

- Eliminating tube-to-tubesheet crevices by seal welding the joints
- Applying cathodic protection or protective coatings
- Avoiding lay-up periods with stagnant water left in the condenser tubes, particularly sea and brackish water in austenitic stainless steel tubes
- Flushing during outages
- Chemically treating cooling water

Galvanic Corrosion (Water Side). Galvanic corrosion is corrosion associated with the current resulting from the electrical coupling of dissimilar metals (electrodes) in an electrolyte. Galvanic corrosion is not a problem in low-conductivity waters, such as those normally found on the steam side of condensers. However, it can be a water-side problem in condensers cooled with seawater or with medium- or high-conductivity fresh and brackish waters.

In seawater, tube materials, such as copper-nickel alloys, stainless steels, and titanium, are more noble than tubesheet materials, such as Muntz metal and aluminum bronze. Consequently, the tubesheet may suffer galvanic attack when fitted with these more noble tubes (Fig. 4) (Ref 2, 3, 19). Laboratory tests have demonstrated that the rate of galvanic corrosion of a Muntz metal tubesheet fitted with titanium or stainless steel tubes can exceed 5 mm/yr (200 mils/yr) in seawater (Table 4) (Ref 19). Similarly, if stainless steel inserts are installed in copper alloy tubes to prevent inlet-end

Fig. 3 Example of pitting in type 316 stainless steel in seawater service

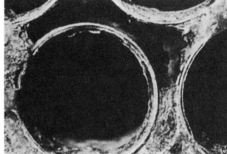

Fig. 4 Galvanic corrosion of a Muntz metal tubesheet, fitted with AL6X stainless steel tubes, after 1 year of seawater service

Table 4 Tubesheet corrosion rates for tubesheet/tube galvanic couples exposed under simulated condenser conditions

Tubesheet/tube couple	Tube end	Corrosion rate(a)							
		Seawater, 22 °C (72 °F)		Brackish water, 22 °C (72 °F)		Seawater, 6 °C (43 °F)		Brackish water, 11 °C (52 °F)	
		mm/yr	mils/yr	mm/yr	mils/yr	mm/yr	mils/yr	mm/yr	mils/yr
Aluminum bronze/Al6X	Inlet	0.63	25	0.18	7	0.05	2	0.10	4
	Outlet	0.63	25	0.15	6	0.09	3	0.22	9
Aluminum bronze/Ti 50A	Inlet	1.9	73	0.20	8	0.06	2	0.20	8
	Outlet	3.0	120	0.37	15	0.05	2	0.23	9
Muntz/Ti 50A	Inlet	9.2	360	2.3	90	0.76	30	0.63	25
	Outlet	6.3	250	0.83	32	0.36	14	1.0	39
Muntz/Al6X	Inlet	4.0	160	1.6	60	0.70	28	0.71	28
	Outlet	3.8	150	0.98	39	0.36	14	0.37	14
Muntz/Al6X with cathodic protection	Inlet	0.83	33	0.21	8	0.40	16	0.15	6
	Outlet	0.52	20	0.24	9	0.13	5	0.07	3

(a) Corrosion rates determined by weight loss. Area ratio, 1:1. Source: Ref 19 erosion-corrosion, rapid galvanic corrosion can be promoted in the tube close to the insert/tube interface.

Other galvanic couples can exist in a condenser. In each case, the recommended method of alleviating the problem is to install a cathodic protection system in the water box. Electrical current requirements can be reduced by coating the tubesheet and water box with a nonconducting material.

Environmental Cracking (Water Side and Steam Side). Stress-corrosion cracking (SCC) and hydrogen embrittlement cracking are forms of environmental cracking that can affect condenser tubes (Ref 2, 3, 5, 14, 15). Hydrogen embrittlement cracking was identified as a problem on the water side of ferritic stainless steel tubes in a couple of condensers fitted with cathodic protection systems. It is believed that hydrogen was generated on the surface of the tubes by the passage of too high a cathodic protection current and that this hydrogen promoted slow crack growth and failures at the ends of the tubes. The tube ends were particularly susceptible to hydrogen embrittlement cracking, because roller expansion during fabrication introduced higher-than-normal residual stresses in these zones.

Entry of hydrogen into cathodically protected titanium tubes is also possible if too high a cathodic protection current is delivered. In such cases, the absorbed hydrogen can react with the metal to form a brittle titanium hydride phase, which could conceivably crack and lead to premature failure. However, so far, no titanium condenser tube failures have been reported. One electric power producer that grossly overprotected the water box region of a titanium-tubed condenser discovered that the ends of the tubes were severely hydrided, but even here the affected tubes did not leak.

Apart from the rather unusual failures in ferritic stainless steels, environmental cracking is only a problem in copper alloys, specifically the brasses. Here, SCC (not hydrogen embrittlement cracking) is the mechanism of failure. Most SCC failures initiate on the steam side of the tubes when the steam condensate contains high concentrations of NH_3 and oxygen (Ref 11). The NH_3 is derived from the chemicals added for boiler feedwater chemistry control, and oxygen originates from air that leaks into the system through imperfectly maintained turbine glands, expansion joints, valve packing glands, and so on. The NH_3 and oxygen concentrations are particularly high in the air-removal section, and it is here that SCC occurs most frequently. Stainless steel or copper-nickel is often used in this section, rather than brass, to avoid SCC due to NH_3.

Steam-side SCC can be controlled by ensuring that the condensate on the tubes has a low oxygen concentration. The maintenance of an airtight system requires continuing attention to all seals, glands, and joints that are subjected to internal pressures less than atmospheric during start-up, normal operation, or shutdown. Helium tracer and similar techniques allow leaks to be detected with moderate ease.

Water-side SCC of brasses has occurred less frequently than steam-side attack, and in most cases, the chemicals responsible for the failure were not positively identified. However, NH_3 and its derivatives (nitrates and nitrites) are often suspected of promoting SCC. Possible sources of these chemicals are farm fertilizers (runoff) and decaying organisms in polluted water. Water-side failures frequently initiate beneath surface deposits, probably because the deleterious species can concentrate beneath the deposit to levels that favor SCC. Thus, once again, tube cleanliness is important, and cleaning techniques can be used to minimize water-side SCC.

No matter which side of the tube is susceptible to environmental cracking, the cracking can be controlled by reducing or eliminating residual tensile stresses. Roller expansion of the tubes during installation will always introduce some residual stresses, but care should be taken to avoid expansion beyond the back of the tubesheet, which can lead to particularly high residual stresses. In addition, fully stress-relieved tubes should be used, and during installation, they should not be bent or mechanically abused.

In addition to the condenser tubes, the condenser shell, support structure, makeup inlet sparger, and drain baffles, all typically made of welded carbon steel, can also be subject to environmental cracking. A combination of residual welding stresses, vibration caused by steam flow and flashing of various drains, and wet steam and condensate can lead to SCC and corrosion fatigue. Improvement of weld quality, control of vibration by design, and avoidance of critical transonic flow in drains will help to prevent these problems.

Condensate Corrosion (Steam Side). Condensate corrosion may occur on the steam side of copper alloy condenser tubes either in a relatively uniform manner (i.e., general corrosion) or preferentially at certain locations. The latter form of attack is known as ammonia attack, ammonia grooving, or condensate grooving. Both types of condensate corrosion occur in locations where the steam condensate contains high levels of ammonia, carbonate, and dissolved oxygen (Fig. 5) (Ref 20–23). Consequently, condensate corrosion, like SCC, is most prevalent in the air-removal section. Ammonia is a component of the all-volatile treatment of feedwater, and carbonate (carbon dioxide) and oxygen come from air that leaks into the condenser. Other sources of CO_2 are decomposed organic impurities and organic water treatment chemicals and aerated makeup water.

General Corrosion. This type of corrosion may occur during operation or during shutdown periods. General corrosion during operation is reduced by controlling air inleakage or by ensuring—through good design—that air is removed effectively from the condenser. General corrosion during extended shutdown periods can be minimized by eliminating moisture from the steam side by flowing either filtered dehumidified air or nitrogen through the system. Condensers may also be retrofitted with stainless steel or titanium tubes to avoid condensate corrosion and the subsequent transport of copper corrosion products. Copper can play a critical role in the corrosion of other components in the main steam cycle.

Ammonia Grooving. This form of condensate corrosion is localized, because the corrosive environment is localized (Ref 2, 5, 14, 23) (Fig. 6). For example, the slight tilt routinely

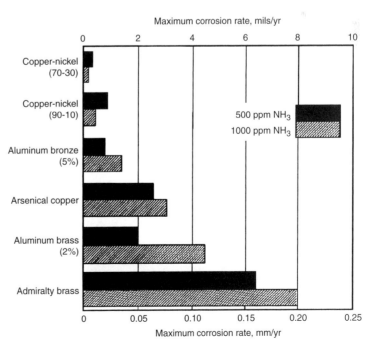

Fig. 5 Corrosion rates of copper alloys in aerated ammonia drip solutions. Source: Ref 26

given to condenser tubes may promote flow of some of the steam condensate toward one side of a tube support plate. There, the flow from a large number of tubes can collect and run down the plate surface. Such localized flow can create deep circumferential corrosion grooves in the tubes immediately next to the support plate (Fig. 6b) if the condensate contains high levels of ammonia and oxygen. Ammonia grooving can be controlled by reducing the oxygen concentration in the condensate, as discussed previously for steam-side SCC, or by selecting more resistant alloys, such as copper-nickel alloys, or completely resistant alloys, such as stainless steels and titanium.

Biofouling Control

Minimizing biofouling on the cooling water side of a condenser will reduce corrosion (Ref 24–26). In some instances, microbial activity does nothing more than create localized geometric conditions, for example, a crevice beneath a microbial growth, that stimulate crevice corrosion. In other circumstances, microbes produce metabolites, such as organic or mineral acids, ammonia, or hydrogen sulfide, that are corrosive to structural materials. A few microbes can concentrate chlorides and other halides, which can result in severe, localized corrosion of passive materials, such as stainless steels and nickel alloys.

Furthermore, detached biofouling organisms can clog or partially plug tubes, possibly causing accelerated attack immediately downstream of the entrapped organisms. During shutdowns or in stagnant water conditions, the organisms may die and putrefy, producing trace quantities of sulfide and nitrogen compounds that may cause localized attack or cracking of some alloys.

The two most common methods used to control biological fouling in surface condensers are chlorination and thermal shock. Chlorination is usually achieved in one of three ways: injection of chlorine gas through an appropriate regulator, injection of sodium hypochlorite, or the generation of chlorine in situ using a duel-electrode generator. Typically, the cooling water is dosed with chlorine sufficient to produce a small residual amount of free chlorine (~0.5 ppm). The frequency, duration, and amount of chlorine dosage vary from plant to plant and can be quite dependent on the time of year.

To a lesser degree, thermal shock is employed as a method for controlling biological fouling. Recirculating the cooling water flow through the condenser so that the temperature increases to 49 °C (120 °F) will kill off most of the organisms. Thermal shock is effective against larger organisms that have a higher tolerance for chemical biocides.

Other Problems

Other problems include:

- Erosion and erosion-corrosion of the carbon steel condenser shell and the structural elements that generate iron oxides, which are transported into the steam cycle
- General corrosion and pitting corrosion of condenser shell bottoms that sit on a concrete slab
- Air ejector erosion
- Corrosion fatigue of carbon steel baffle and sparger welds

Corrosion Prevention

Many corrosion mechanisms active in condensers can be prevented if only two maintenance procedures are followed. First, if air is eliminated from the steam, condensate corrosion and SCC of copper alloy tubes can be minimized. Second, if condenser tubes are kept clean and free of deposits, debris, and biofouling on the water side, sulfide attack, dealloying, erosion-corrosion, crevice corrosion, pitting, and SCC can be prevented or minimized.

Corrosion of Deaerators

Otakar Jonas and Joyce M. Mancini, Jonas, Inc.

Deaerators have been used for more than 50 years in fossil utilities and industrial cycles to remove dissolved gases (mostly oxygen, carbon dioxide, and nitrogen) from feedwater through mechanical deaeration (Fig. 7). By removing dissolved gases from the droplets of feedwater formed in spray nozzles and trays, deaerators help to reduce corrosion in the preboiler system and the boiler. Deaerators also provide water storage capacity to supply boiler feed pumps and allow chemical oxygen scavengers to have a residence time at appropriate elevated temperatures. In addition, the hydrostatic head provides the proper flow conditions and pressure for the boiler feed pump suction. The design, material of construction, and environment affect the corrosion of deaerators (Ref 45–72).

Deaerator Designs

There are many deaerator design configurations, both horizontal and vertical, with different types of internals where feedwater and steam interact. Typical designs incorporate spray, tray, or combination spray/tray deaerating/heating sections. In electric utility deaerators, turbine extraction steam is used. The deaerator is located between the low-pressure and high-pressure feedwater heaters. In industrial steam cycles, main or hot reheat steam is used for deaeration. In combined cycle units, low-pressure boiler steam is often used for deaeration, and a deaerator may be combined with the low-pressure drum.

Spray Deaerators. Feedwater sprays into a steam-filled space where it is heated and scrubbed to release gas. As fresh steam enters the unit, it passes through the sprayed water, agitating the water a second time and liberating most of the remaining volatile impurities.

Tray Deaerators. Feedwater is directed onto a series of cascading trays. Water falls from tray to tray by overflowing or passing through small holes. Steam, which engulfs the trays, heats and deaerates water as it falls.

Combination Heaters. Feedwater first sprays into a steam-filled space, then rains down on a series of trays through which it passes for further agitation and scrubbing.

Both spray- and tray-type deaerators should be able to remove oxygen to less than 7 ppb, and the deaeration efficiency should be good even at low loads. Incomplete deaeration occurs when trays

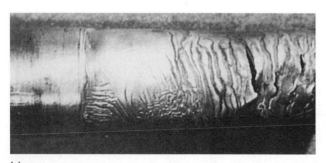
(a)

(b)

Fig. 6 Examples of ammonia grooving on admiralty brass. (a) The unattacked tube end (left) was protected by the tubesheet. (b) Attack occurred at one side of a tube support plate

are not properly aligned, spray nozzles plug or fail, or when the deaerator steam temperature or flow are low.

Corrosion Problems and Solutions

Stress-corrosion cracking (SCC) and corrosion fatigue (CF) of the deaerator storage tank and deaerating vessel welds is an industrywide (utility and industrial steam plant), generic problem. It has led to catastrophic failures and fatalities (Ref 45, 50, 51, 54, 55, 58–61) and is still a major maintenance and inspection burden, with cracking in more than 30% of utility and industrial plant deaerators. NACE Task Group T-7-H7 on Deaerator Cracking identified seven catastrophic failures of deaerator storage and hot water tanks due to stress-corrosion cracking and corrosion fatigue (Ref 52). Similar cracking occurs in blowdown flash tanks and other welded carbon steel pressure vessels and piping. Other deaerator damage mechanisms include general corrosion, pitting, and flow-accelerated corrosion (Table 5).

Weld Cracking (Stress-Corrosion and Corrosion Fatigue). Deaerator cracking has been predominantly found in the welds of the vessel shell (Fig. 7) in either the heat-affected zones (HAZs) or in the welds (Ref 45–49, 53). Weld cracking most frequently forms perpendicular to the welds, but cracking parallel to the weld, especially at the toe, is also common (Fig. 8). In many vessels, the most serious cracking has been found in the head-to-shell circumferential weld. Welds of various internal attachments also crack.

In a paper mill, cracking at a head-to-shell circumferential weld resulted in a catastrophic failure and fatalities. The break before leak occurred during normal operation after the slow-growing corrosion crack reached approximately 50% of the wall thickness over approximately 280° of the circumference. At that point, the stress-intensity factor reached the material fracture toughness value and the head separated. The hot pressurized water flashed into steam, propelling the deaerator through the mill.

Cracks perpendicular to the welds are believed to be SCC cracks, and cracks parallel with the welds, CF cracks. There can be an interaction of both cracking mechanisms. Crack initiation has often been attributed to pitting. Stress-induced pitting (Ref 69) can also be an initiation mechanism.

There is a good correlation between the crack propagation rate and the time in service (Fig. 9a), which may indicate that the older deaerators were fabricated with lower residual welding stresses. Increases in welding productivity by faster, higher-energy input welding can produce higher residual stresses. There has been little correlation found between cracking rate and either operating pressure or the design-to-operation pressure ratio (Fig. 9b, c) (Ref 50, 54).

A survey of deaerators inspected by the wet fluorescent magnetic particle method indicated that the presence of "water hammer" (caused by water entering a steam line or steam-filled space) is a possible cause of deaerator cracking (Ref 50, 53). Of the four deaerators reported to have water hammer but no cracks, three were in service less than 4 years at the time of inspection. This limited survey could not draw any conclusions about the effect of basic structural design, materials of construction, deaerator type, operating pressure, operating history, or water chemistry on cracking.

Internal surfaces of the vessel where attachments are welded to the exterior surfaces have also shown cracking. Cracking is perpendicular to the surface and primarily in a straight line, although some have shown limited branching.

Causes and Prevention of Weld Cracking. There are no confirmed root causes of deaerator weld cracking, and there has not been any comprehensive research to determine root causes and verify corrective measures. This is in addition to the general ignorance about stress-corrosion cracking and corrosion fatigue mechanisms.

The most likely root causes of deaerator weld cracking (based on hundreds of cases) are listed in Table 6, along with possible preventive measures.

Corrosion cracks can initiate at welds because of high residual tensile stresses, weld defects,

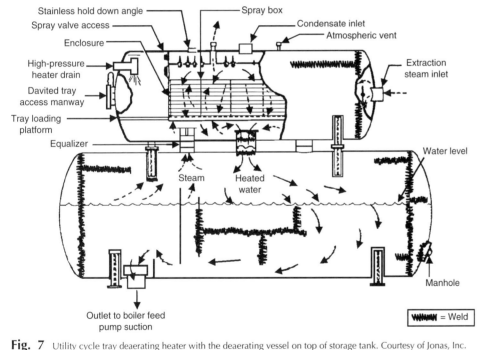

Fig. 7 Utility cycle tray deaerating heater with the deaerating vessel on top of storage tank. Courtesy of Jonas, Inc.

Table 5 Typical deaerator corrosion problems

Problem	Damage mechanism(a)	Root cause(b)	Inspection and detection(c)	Possible safety issue	Cost per event(d) ($1000)
Corrosion cracking of welds in horizontal deaerators	SCC and CF, P	Unknown; could include residual welding stresses, water piston, water hammer, CH	MP, UI, AE, V	Catastrophic failure, steam leaks	100–10,000
Flow-accelerated corrosion wall thinning of vessels and piping	FAC, possibly cavitation	High flow velocity, turbulence, low oxygen, excess oxygen scavenger, material, CH	V, UI	Leaks	100–1000
Cracking of internal supports and spray nozzles	CF	O, D	V, poor deaeration	None	50–200
General corrosion of vessels and piping	GC	Layup, CH	V, high iron in feedwater	None	N/A

(a) CF, corrosion fatigue; SCC, stress-corrosion cracking; P, pitting; FAC, flow-accelerated corrosion; GC, general corrosion. (b) D, design and material selection; CH, water chemistry; O, operation. (c) V, visual; UI, ultrasonic inspection; MP, magnetic particle; AE, acoustic emissions. (d) Lost production and repairs per one event. The cost of lost production is typically much higher than the loss from repairs with a ratio up to 10 to 1. Source: Jonas, Inc.

increased corrosion susceptibility of weld and HAZ, and stress concentrations occurring in these locations. However, both stress-relieved and nonrelieved welds experience the cracking, and, in a few cases, residual stresses after stress relief were still close to yield strength. In many cases, the cracking mechanism appears to be low-cycle corrosion fatigue (Ref 49), and it occurs over a long period. Applied stresses contributing to the cracking are likely caused by water piston (Ref 45, 50, 54, 64), water hammer, and by internal pressurization.

Water piston is an unstable water surface waviness (Kelvin-Helmholtz instability) occurring in partially filled horizontal liquid storage vessels, such as in horizontal deaerator storage tanks (Ref 45, 50, 54, 64). The inflow of water can cause the standing waves to close the whole cross section (a piston), and when there is a pressure differential along the vessel, the piston is propelled toward the end of the vessel. This can occur repeatedly, leading to a generation of alternating stresses and to corrosion fatigue of the head-to-shell welds. It is recommended that designers perform an evaluation for water piston conditions (Ref 45, 50, 54, 64).

Weld Inspection. It is recommended to periodically inspect deaerator welds, to determine crack depth, crack propagation rate under corrosion fatigue, and stress-corrosion conditions, and to calculate residual life and a safe inspection interval.

For new deaerators, stress relief of welds and radiographic weld inspection are recommended but not mandated in the ASME Boiler Pressure Vessel Code.

Deaerators should be inspected during commissioning of new units and about every 5 years thereafter, or when operation has changed (e.g., from base load to cycling operation). Deaerators should be monitored for water piston, water hammer, and other unsteady phenomena that can generate fatigue stresses; operation should be adjusted to eliminate such conditions. The NACE Recommended Practice for Prevention, Detection, and Correction of Deaerator Cracking (Ref 46) provides guidance on proper inspection techniques. It recommends wet fluorescent magnetic particle inspection after light grinding of the welds inside the vessel. Acoustic emission and ultrasonic inspection methods have also been used.

An engineering evaluation of inspection results should be done with considerations of corrosion fatigue and stress-corrosion crack propagation rates and service experience. The remaining life from the time of inspection to a leak or break should be calculated and repair or next, nondestructive test (NDT) interval recommended (Ref 50). The most important NDT information is crack depth. A block diagram of the life-evaluation procedure is shown in Fig. 10. In this procedure, data on operating history, stress analysis, and results of NDT are used to determine the maximum stress intensity for SCC and CF crack growth calculations in which the material properties for specific water chemistry and operating temperature are used. The time for SCC and CF cracks to lead to a leak or final fracture are calculated (residual life), and the NDT interval is determined, usually as 50% of the residual life.

Preferential Corrosion in the HAZ. During welding, the base metal is rapidly heated and cooled with filler material being added to the joint. Temperatures experienced in the HAZ during welding range from ambient at a distance away from the weld to the melting point. As a result, the microstructure across the HAZ and weld metal will vary from that of the base metal.

Although not widely recognized, the HAZ can be subject to preferential corrosion (Fig. 11) (Ref 63). The phenomenon has been observed in a wide range of environments with the common links being an elevated conductivity and pH values normally ranging from below 7 to about 8.

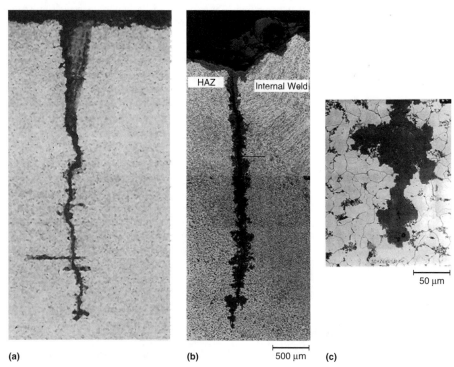

Fig. 8 Cracks associated with carbon steel deaerator welds, (a) transverse in the weld and (b) longitudinal. (c) Closeup of crack tip in (b). Courtesy of Jonas, Inc.

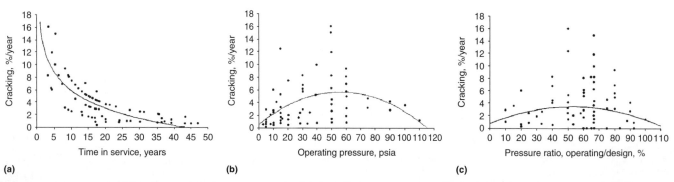

Fig. 9 Deaerator storage tank weld cracking: crack depth as a percentage of wall thickness per year versus (a) time in service, (b) operating pressure, and (c) operating/design pressure ratio. Source: Ref 50

The reasons for localized weld attack are not fully known. There is a microstructural dependence, and there are differences in weld and base metal composition and strength. Corrosion has been shown to be more severe when the material composition and welding are such that hardened structures are formed (Ref 63). It is known that hardened steel may corrode more rapidly under acidic conditions than tempered material. On this basis, water treatments ensuring alkaline conditions should be less likely to induce HAZ corrosion. However, even at pH = 8, hydrogen ion reduction accounts for approximately 20% of the total corrosion (Ref 63). Therefore, a pH well above 8 would be required to completely suppress the pH effect.

Flow-accelerated corrosion (FAC) can be found in selected sections of deaerators and piping where there is high velocity or turbulent flow, typically at the outlet of the deaerating vessel and other piping connections (Ref 70–72). In vertical deaerators, where there is tangential flow between the deaerating and storage section, FAC is found more frequently than in horizontal deaerators. Because the water in the deaerator is at saturation conditions, any changes in the static pressure can also lead to evolution of steam and gas bubbles and cavitation. Erosion and FAC of the impingement baffles near the steam inlet connection has been a problem in some spray-type deaerators.

The parameters that govern FAC include: flow velocity and turbulence, component geometry, temperature, feedwater chemistry (pH, oxygen, and oxygen scavenger), and material composition (chromium, copper, molybdenum) (Ref 70–72). Flow-accelerated corrosion is well understood, and inspection methods and engineering solutions have been developed. Besides the effect on integrity, the main effect of FAC is the generation of iron oxides and their deposition in the boiler and turbine (via attemperating sprays).

Deaerator and associated piping design should be evaluated for susceptibility to FAC and cavitation using commercially available software (Ref 72). The areas identified as susceptible to FAC should be inspected visually and for thinning by ultrasonic thickness testing. Maintaining feedwater pH above 9.5 with residual concentration of oxygen and minimizing the concentration of oxygen scavenger can significantly reduce FAC. In once-through and drum boiler cycles with condensate polishing and good feedwater, FAC can be reduced by oxygenated feedwater treatment. Components with a high thinning rate identified by wall thickness measurements should be replaced with alloy steel components.

General Corrosion. There is often some general corrosion of deaerators, particularly during unprotected layup with aerated water left in the storage tank or by humid air. The main effect of general corrosion of deaerators is an increase in the iron concentration in feedwater leading to scale formation in the boiler. In cases of poor feedwater chemistry with high cation conductivity, low pH, and high oxygen concentration, there can be severe general corrosion. General corrosion can be controlled to acceptable levels by maintaining feedwater chemistry within recommended guidelines (Ref 66, 67) and using proper layup procedures (Ref 67, 68).

Pitting. A deaerator cracking survey (Ref 53) found that 30% of deaerators with pits also had cracks in the weld areas. However, pitting in deaerators is not a major problem. It sometimes occurs when feedwater contains chloride, oxygen, CO_2, and other corrosive impurities. Pitting can be active during both operation and layup (Ref 45, 46, 50, 54). Metal oxide and mineral scale, which is often thicker around welds, can concentrate impurities and accelerate pitting. The main effect of pits in the weld area is the initiation of cracking.

Pitting in deaerators can be avoided by maintaining good feedwater chemistry and protection during layup (Ref 67, 68). To minimize pitting and its effect on crack initiation, the deaerator scale and deposits should be removed during inspections and pits in weld and other areas susceptible to cracking should be removed by grinding; however, the minimum allowable wall thickness should be preserved.

Tray Enclosure Box in Vertical Deaerators. General corrosion, pitting, and SCC resulting in massive cracking of 304L stainless

Table 6 Probable root causes and preventive measures for deaerator weld cracking

Probable root cause	Prevention
Residual welding stresses	Qualified welding procedure, stress relief, control of fatigue stresses (water piston and water hammer)
Generic problem of SCC and CF of carbon steel welds, even in high-purity water	Reduction of residual stresses and control of transient stresses, water piston, and water hammer
Rapid load and pressure changes producing condensation and evaporation (flashing) shocks	Adjust operating procedures, test during commissioning
Water piston (Ref 45, 50, 54, 64) and water hammer	Evaluate deaerator design, adjust operation, test during commissioning
Weld defects	Qualified welding process and welders, radiography of welds
Water chemistry (a secondary effect)	Follow recommended guidelines

SCC, stress-corrosion cracking; CF, corrosion fatigue. Source: Jonas, Inc.

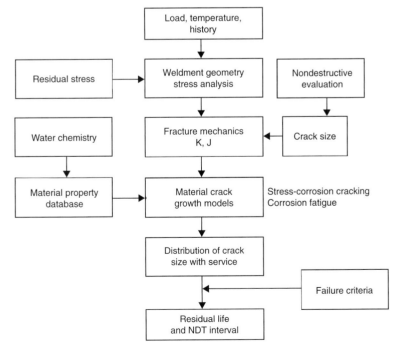

Fig. 10 Life assessment of deaerators. Source: Ref 50

Fig. 11 Preferential heat-affected zone corrosion in carbon steel from service in aqueous conditions. Original magnification: 5×. Source: Ref 63

steel enclosure boxes have been experienced in several vertical deaerators. The corrosive environment was not positively identified, but there is a suspicion of the presence of organic acids and elevated oxygen.

Spray Valve Spring SCC and CF. Stress-corrosion cracking and CF of spray valve springs, usually made of higher-strength stainless steels, have been experienced in numerous deaerators. Cracks are detected during routine inspections and replaced.

Corrosion of Feedwater Heaters

Otakar Jonas and Joyce M. Mancini, Jonas, Inc.

Feedwater heaters (FWH) increase overall cycle efficiency by heating feedwater using low-pressure (LP) and high-pressure (HP) steam from the turbines. Design, materials of construction, and environment affect the corrosion of feedwater heaters (Ref 73–96).

Types of Feedwater Heaters

Channel Feedwater Heaters. This design consists of a flat tubesheet that holds the straight or U-shaped tubes in a shell (Ref 80). Feedwater heaters are typically designed with condensing and subcooling zones (two-zone) or with desuperheating, condensing, and subcooling zones (three-zone). A typical example of a three-zone heater is shown in Fig. 12.

Header Feedwater Heaters. This design is a header-type construction, patterned after boiler header applications (Ref 74, 76, 81, 82). The feedwater inlet and outlet headers have tubes and nipples attached to each header. Between the headers, the tubes are in a W-shape in the heater. The header-type heaters are much larger, more expensive, and more difficult to maintain than the tubesheet-type heaters. The main advantage of the header-type heaters is better thermal design for cyclic service.

Materials

The frequency of failures and failure modes for feedwater heaters vary considerably from one class of materials to another (Ref 73, 76, 77, 79, 80, 82, 83, 88–92). All feedwater heater parts (e.g., shell, head, internals, and piping) except the tubes are carbon steel. Commonly used FWH tube materials are admiralty brass and other brasses, copper-nickel alloys, Monel, carbon steel, and stainless steels (Table 7).

Admiralty and Other Brasses. The major corrosion problems associated with admiralty and other brasses are general corrosion, flow-accelerated corrosion (FAC), and stress-corrosion cracking (SCC) (Ref 79, 88–92). In most instances, general corrosion of these tubes is not associated with premature tube failure, but the copper released in the system has been implicated in corrosion failures of other power-plant components in the steam/water cycle.

Copper-Nickel Alloys. The major corrosion problems associated with copper-nickel alloy tubes are general corrosion (copper and nickel release), exfoliation, and stress-corrosion cracking (Ref 79, 83, 88–92). Exfoliation tends to be most severe in units that operate in peaking service, presumably because O_2 levels are higher; the tubes most affected are those made of 70Cu-30Ni, while 90Cu-10Ni are reported to be least attacked. In this type of service, SCC failures have been reported in 90Cu-10Ni and 80Cu-20Ni tubes. Most frequently, these failures have occurred on the steam side in locations where high residual stresses were present.

Monel Alloy 400. Few failures of Monel alloy 400 tubes have been reported in feedwater heater applications (Ref 88, 92). Because of the excellent corrosion resistance and high allowable stresses of Monel alloy 400, it is most frequently used in high-pressure feedwater heaters where its high cost can be justified. In general, the environments found in HP FWHs are more aggressive than those found in the lower-pressure units. Those Monel failures that have occurred generally have been attributed to secondary causes such as tube-vibration-induced wear, mechanical damage during installation, and steam impingement as a result of improper FWH design. As is the case with copper-base alloys, corrosion product release is of concern with Monel. A few SCC failures also have been reported. In most cases, these have been intergranular in nature and have been associated with high residual stresses. Initiation of SCC has been reported on both the steam side and water side.

Carbon Steel. The major problems associated with carbon steel in FWHs are general corrosion, water-side flow-accelerated corrosion (FAC), and steam-side erosion (Ref 79, 88–92). The water-side FAC failures have occurred primarily at the tube inlets and generally are found

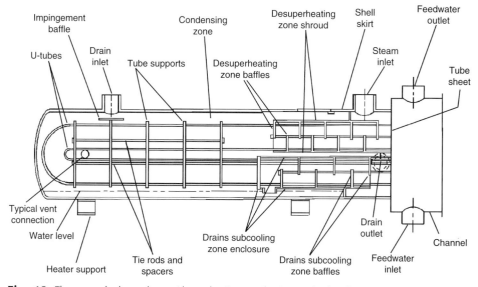

Fig. 12 Three-zone feedwater heater (desuperheating, condensing, and subcooling zones). Courtesy of Heat Exchange Institute

Table 7 Feedwater heater tube materials susceptibility to corrosion

Material	Susceptibility to corrosion(a)						
	General corrosion	Pitting	Crevice corrosion	FAC(b)	SCC(c)	Dealloying	Intergranular corrosion
Arsenical admiralty	H	L	L	H	H	M	H
Arsenical copper	M	M	M	H	M	L	H
Carbon steel	H	M	M	H	H	NA	H
Copper-nickel (70/30)	M	M	L	M	M	L	H
Copper-nickel (80/20)	M	M	L	M	M	L	H
Copper-nickel (90/10)	M	M	L	M	M	M	H
Monel (alloy 400)	M	M	M	M	M	L	M
Stainless steel (304)	L	H	H	L	H	NA	H

(a) Susceptibility scale: L, low; M, moderate; H, high; NA, not applicable. (b) FAC, flow-accelerated corrosion; (c) SCC, stress-corrosion cracking

in the first few heaters in the high-pressure units. On the steam side, erosion failures have occurred on the peripheral tubes because of direct steam impingement and in the drain cooling section because of condensate flashing and steam in-leakage. Unlike erosion and FAC, general corrosion of carbon steel FWHs usually is not associated with actual failures, but the corrosion product generation and transport that occurs because of general corrosion is of concern for other cycle components.

Stainless Steels. The most common stainless steel used for LP FWH tubes is type 304, which is highly resistant to general corrosion and FAC (Ref 79, 88–92). Under some circumstances, 304 stainless steel can be vulnerable to SCC, pitting, and crevice corrosion. Infrequent service failures of type 304 stainless steel in FWHs suggest that the conditions required for their occurrence are rather unlikely under normal operating conditions. There have been cases where large manganese sulfide (MnS) inclusions dissolved and caused perforation of 304 stainless steel tubes. Other stainless steels used for feedwater heaters include 304N, AL-6X, AL-6XN, type 329, type 439, and Sea-Cure.

Water and Steam Chemistry

The water chemistry factors that influence feedwater heater tube corrosion are water and steam contaminants (chloride, dissolved oxygen, etc.), fluid temperature, and the pH achieved by adjustment with ammonia (Ref 77, 79, 80, 94, 96). Systems with all steel and stainless steel components generally operate with a pH between 9.2 and 9.6 to minimize carbon steel corrosion. This limits the entrainment of corrosion products in the condensate. Problems are encountered with mixed metal systems containing both copper alloys and carbon steel. These systems are generally operated at a pH of 9.0 to 9.3. However, this range is only a compromise and is not an optimum range for the control of corrosion of either material. Recommended guidelines are available that specify control parameters and monitoring locations (Ref 94, 96).

Turbine extraction steam chemistry is, for the superheated steam extractions, the same as the turbine steam chemistry (Ref 94). The concentrations of impurities in this steam are in the low ppb range. During the heat-transfer process in feedwater heaters, the steam is desuperheated and the impurities redistribute between moisture droplets and liquid film on tube surfaces and the steam. Similarly as in the turbine, the liquid phase will have higher concentrations of low-volatility impurities, such as salts, and lower pH, because ammonia will be mostly in the gaseous phase. This will influence tube corrosion and FAC as well as FAC of the heater shell. In the zone just before condensation (slight superheats), there will be a "salt zone" in which salts and acids concentrate to percent concentrations (see the article "Corrosion of Steam Turbines" in this Volume). Corrosion is the most severe in the salt zone.

Corrosion Problems

Corrosion mechanisms that have led to failures or serious problems in feedwater heaters include: general corrosion, flow-accelerated corrosion, stress-corrosion cracking, crevice corrosion, and pitting. Each form of failure occurs only under specific environmental and metallurgical conditions. Corrosion problems associated with common feedwater heater tube materials can be summarized:

- Flow-accelerated corrosion and erosion of tube inlets
- Cavitation
- Debris or "clamshell" pits
- Support wear
- Fatigue cracks (near supports and midspan)
- Tube-to-tube wear
- Dealloying of copper alloys
- Overrolling (at tubesheets or supports) or freeze bulges
- Stress-corrosion cracking
- Steam moisture erosion
- Corrosion, wear, and flow-accelerated corrosion at supports
- Underdeposit pitting of stainless steel
- Manufacturing defects including MnS inclusions in austenitic stainless steel

Feedwater heater corrosion and other problems are:

- Feedwater heater tube leaks, see Fig. 13
- Thinning of heater shell and piping by flow-accelerated corrosion
- Erosion of support plates and drain
- Deposition of copper alloys (snakeskin) in high-pressure heater tubes caused by general corrosion elsewhere in the system
- Loose tube plugs
- General corrosion of copper alloy tubes with redeposition in the HP turbine leading to a loss of efficiency
- Water induction and water hammer
- Phosphate hideout and boiler tube corrosion due to copper alloy tubing corrosion and deposition of zinc and nickel in boiler tubes

Tubes

The most frequent problems with feedwater heaters involve tube leaks. In spite of many possible damage locations, it is often possible to minimize any downtime by plugging a leaking or corroded tube (Ref 74).

Flow-Accelerated Corrosion. Experience has demonstrated that carbon steel tube bundles are among the components most vulnerable to flow-accelerated corrosion (FAC) (Fig. 13), although copper alloy tubes are also susceptible (Table 7) (Ref 77, 80, 83, 86, 87, 89, 92). Feedwater heater tubes are typically designed with the minimum wall thickness necessary to support the mechanical loads because a thinner wall promotes more efficient heat transfer. Therefore, even a small reduction in wall thickness due to corrosion can result in failure. The primary causes of FAC are excessive water velocities, either by poor design or by abnormal operation of the feedwater heaters, and channel geometries that result in local zones of turbulence at the tubesheet. Low feedwater pH, low oxygen concentration, and high oxygen scavenger concentration accelerate FAC.

The inlet-end of the tubes is the most susceptible. Flow-accelerated corrosion can be minimized by the use of flow diffusers in the inlet channel to reduce inlet turbulence and ensure even distribution of the flow to all tubes. Tube inserts may also be used; however, care should be taken to avoid galvanic corrosion. Other measures for preventing flow-accelerated corrosion in feedwater heater tubes include:

- Maintaining feedwater velocity within recommended limits (e.g., 1.7 to 2.4 m/s, or 5.5 to 8 ft/s, for carbon steel depending on temperature) (Ref 75)
- Evaluating design changes which could reduce or prevent FAC

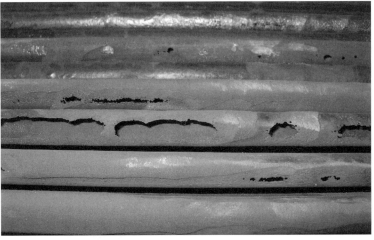

Fig. 13 Flow-accelerated corrosion of carbon steel feedwater heater tubes. Courtesy of Jonas, Inc.

- Substituting more resistant tube alloys
- Maintaining water chemistry within recommended limits (Ref 94)

General corrosion occurs in all feedwater heater tube alloys (Ref 77, 80, 87–93). Corrosion rate, however, varies substantially among the tube materials. Passive materials, such as the austenitic stainless steels, usually demonstrate negligible uniform corrosion rates and are therefore considered almost immune. In general, susceptibility to general corrosion decreases as alloy selection changes from carbon steel to brass, copper-nickel, Monel 400, and stainless steel (immune). The nature of the corrosion product film is the primary controlling variable.

The variables affecting general corrosion, and therefore the corrosion product film, are pH, temperature, fluid velocity, and concentration of corrodent (e.g., O_2, Cl, CO_2, organic acids). The influence exerted by each variable greatly depends on the alloy selected.

Fossil plant cycling can accelerate the corrosion of carbon steel and copper alloy tubes. Typically, the highest concentrations of oxygen and CO_2 are present during low loads and during outages. Because the corrosion rate of carbon steel is directly proportional to the oxygen and carbon dioxide concentrations, greater corrosion is likely with cycling service. Cycling service also creates thermal expansion-contraction problems that arise from the difference in the thermal expansion coefficients of the carbon steel tube and its corrosion product film. This leads to repeated fracturing of the protective corrosion product film.

Deposition in Other Parts of the Cycle. General corrosion of copper alloy tubes can lead to copper transport and deposition in boilers and turbines, resulting in a loss of efficiency and megawatt (MW) generating capacity.

Snakeskin. Corrosion of admiralty brass low-pressure feedwater heater tubes can lead to a redeposition of copper on high-pressure feedwater heater tubes (Ref 87, 88). The deposition results in a thin flaky film, which resembles a snakeskin, that can cause significant reduction in the heat transfer rate. These deposits can then be released during load changes and start-up and redeposit downstream in the boiler, superheater, and HP turbine.

Maintaining water chemistry within recommended limits (Ref 94) and proper layup (Ref 95) will minimize the rate at which general corrosion occurs.

Pitting of feedwater heater tubes is generally limited to austenitic stainless steels and carbon steels (Ref 77, 80, 87, 88, 92, 93). Pitting of austenitic stainless steel alloys is caused by the contamination of feedwater by chlorides. Pitting typically occurs when chlorides are concentrated under deposits, within crevices, and through evaporation, for example, in the desuperheating zone, shell side. The few reported incidents of chloride-induced pitting in austenitic stainless steel (type 304) are generally associated with excessive seawater intrusion permitted by condenser tube leaks. Failure to control feedwater oxygen content has also resulted in incidents of pitting in carbon steel tubing. As temperature increases, pitting resistance decreases.

Dissolution of inclusions can result in the initiation of pitting and perforation of stainless steel tubes. Manganese sulfide (MnS) inclusions are formed in stainless steels when the manganese content approaches 2%. These inclusions are soluble in water and, depending on the size of the inclusion, could result in the formation of an active site for pit initiation. Using stainless steels with low manganese concentrations reduces the size and number of inclusions.

Fouling. Poor water chemistry, corrosion products generated upstream, and condenser leakage are the primarily causes of tube fouling (Ref 79). It may also be a result of inadequate initial cleaning of the system or from leaks of turbine oil into the condensate.

Corrosion of LP FWHs can cause severe tube deposits inside tubes and at the tubesheet of the HP heaters. The reason is that the concentration of corrosion products (magnetite, copper oxides) can be higher than their solubility in feedwater. To prevent fouling, feedwater chemistry should be maintained within recommended limits (Ref 94), and the feedwater heater should be cleaned as necessary.

Stress-corrosion cracking (SCC) occurs when an alloy under a tensile stress (applied and/or residual from manufacture or welding) is exposed to a specific corrosive environment (Ref 77, 80, 87–93). Failures due to SCC in feedwater heaters have not been significant in number, and gross deviation from normal feedwater chemistry is the universal cause. The potential for concentration of corrosive chlorides occurs mainly on the shell side in the desuperheating zone in the event of a small leak. In the feedwater heater desuperheating zone, salts and acids can be present as very corrosive concentrated aqueous solutions. This is a similar situation to the "salt zone" corrosion occurring in turbines.

Copper alloys are susceptible to NH_3-induced SCC, with brass (copper-zinc) exhibiting the highest stress-corrosion crack growth rates. Stress-corrosion cracking can be minimized by reducing the residual stresses in the material, for example, by stress relieving tube U-bends. Because of the susceptibility of copper alloys to other corrosion mechanisms and because of the overall need to reduce copper within the system, the use of copper alloys has been greatly diminished.

The limits of chloride and oxygen in water for avoiding stress-corrosion cracking of type 304 stainless steel are well known, and the limits for avoidance of pitting are similar. There have been a few reported SCC failures in Monel 400 tubes, primarily at the U-bend.

Exfoliation. Typically, exfoliation occurs preferentially along the boundaries of grains that were elongated during tube fabrication (Ref 80, 87, 88). The water insoluble corrosion products that form produce expansion forces sufficient to forcibly separate the elongated grains. The 70Cu-30Ni and 80Cu-20Ni alloys are susceptible to exfoliation, and susceptibility increases with increasing alloy nickel content. The 90Cu-10Ni alloy is immune to exfoliation (Ref 80).

Tubes that have experienced this form of corrosion exhibit heavy scaling, which has a distinct "leafy" (laminated) appearance. The scale material is composed of copper and nickel oxides in approximately the same proportion as the copper and nickel content of the specific alloy.

Oxygen is a critical corrodent in the exfoliation process, and peaking/cycling service increases the rate of exfoliation due to greater and more frequent exposure to oxygen.

Tube-to-Tubesheet Joint

Crevice corrosion often starts at the back of a tube-to-tubesheet joint weld where the manufacturer did not adequately expand the tube into the tubesheet after welding (Ref 77, 80, 92, 93). Stagnant water and crud that can be in constant contact with the weld nugget may eventually corrode the weld, especially in a vertical channel down design. When this happens, a worm-hole condition is established that can work its way into the face of the tube inlet. Once this penetration is made, it is aggravated by the pressure differential from the channel to the shell; rapid erosion occurs, resulting in catastrophic tube-to-tubesheet failure. In some cases, the tube-to-tubesheet joint weld failed around the complete circumference, and the separated tube end was found by boroscopic examination beyond the back of the tubesheet. Failures of this kind allow large volumes of water to enter the shell and possibly the turbine before being detected.

Heater Shell

The shell of a feedwater heater is normally carbon steel and is susceptible to general corrosion and flow-accelerated corrosion.

Flow-Accelerated Corrosion. Thinning from flow-accelerated corrosion (FAC) has been detected on the carbon steel shells of feedwater heaters and is caused by the two-phase flow inside the shell (Ref 80, 85). Most feedwater heaters also use carbon steel tube support plates and flow baffles, which are also subject to FAC. Some long-term erosion in the condensing zone is unavoidable.

ACKNOWLEDGMENT

The authors of "Corrosion of Condensers" wish to acknowledge Roland Coit who co-authored an earlier version of the section "Corrosion of Condensers" in the article "Corrosion in Fossil Fuel Power Plants," *Corrosion,* Volume 13, *ASM Handbook,* 1987, p 986–989.

REFERENCES

1. R. Coit, Condensers, *The ASME Handbook on Water Technology for Thermal Power Systems,* American Society of Mechanical Engineers, 1989
2. O. Jonas and J. Mancini, Condensers, Chapter 11, *Low Temperature Corrosion Problems in Fossil Power Plants—State of Knowledge,* Electric Power Research Institute, Dec 2003
3. *Productivity Improvement Handbook for Fossil Steam Power Plants,* 2nd ed., Electric Power Research Institute, June 2000
4. R. Putnam, *Steam Surface Condensers: Basic Principles, Performance Monitoring and Maintenance,* American Society of Mechanical Engineers, 2001
5. *Condenser Application and Maintenance Guide,* Electric Power Research Institute, Aug 2001
6. *Standards for Steam Surface Condensers,* 9th ed., Heat Exchange Institute, Cleveland, OH, 1995
7. *Addendum 1 to Standards for Steam Surface Condensers,* 9th ed., Heat Exchange Institute, Cleveland, OH, 2002
8. *Standards for Direct Contact Barometric and Low Level Condensers,* 6th ed., Heat Exchange Institute, Cleveland, OH, 1995
9. *ANSI/ASME Performance Test Code for Steam Condensers,* ANSI/ASME PTC.12.2-1983, ASME, American Society of Mechanical Engineers, 1983
10. R. Putnam and D. Jaresch, The Impact of Air Cooled Condensers on Plant Design and Operation, *Condenser Technology: Seminar and Conference,* Electric Power Research Institute, Sept 2002
11. "Interim Consensus Guidelines on Fossil Plant Chemistry," EPRI CS-4629, Electric Power Research Institute, June 1986
12. *Power and Energy: Corrosion in Condensers and Feedwater Heaters,* Corrosion Information Compilation Series, National Association of Corrosion Engineers, 1996
13. B. Buecker and E. Loper, Selecting Condenser Replacement Tubes, *Power Eng.,* March 1996
14. R. Muller and H. Seipp, Condensers and Other Heat Exchangers, *Corrosion in Power Generating Equipment,* Plenum Press, 1984
15. J. Beavers, A. Agrawal, and J. Payer, Dealloying and Stress-Corrosion Cracking (Waterside), *Prevention of Condenser Failures—The State of the Art,* Electric Power Research Institute, March 1982
16. C. Kovach, Report on Twenty-Five Years Experience with High Performance Stainless Steel Tubing in Power Plant Steam Condensers, *Energy-Tech.,* Aug 2002
17. J. Tverberg and E. Blessman, The Performance of Superferritic Stainless Steels in High Chloride Waters, *Condenser Technology: Seminar and Conference,* Electric Power Research Institute, Sept 2002
18. H. Khatak and B. Raj, Ed., *Corrosion of Austenitic Stainless Steels—Mechanisms, Mitigation, and Monitoring,* ASM International, 2002
19. G. Gehring and R. Kyle, "Galvanic Corrosion in Steam Surface Condensers Tubed with Either Stainless Steel or Titanium," Corrosion 82, National Association of Corrosion Engineers, 1982
20. "Corrosion of Cu-Ni-Zn Alloys in Water-Ammonia Power Plant: Development of High Temperature Potential-pH (Pourbaix) Diagrams," TR-113697, Electric Power Research Institute, Nov 1999
21. "State-of-Knowledge of Copper in Fossil Plant Cycles," TR-108460, Electric Power Research Institute, Sept 1997
22. B. Boffardi, Control the Deterioration of Copper-Based Surface Condensers, *Power,* 1985
23. N. Polan, G. Sheldon, and J. Popplewell, "The Effect of NH_3 and O_2 Levels on the Corrosion Characteristics and Copper Release Rates of Copper Base Condenser Tube Alloys under Simulated Steam-Side Conditions," Paper 81-JPGC-Pwr-9, Joint ASME/IEEE Power Generation Conference, American Society of Mechanical Engineers/Institute of Electrical and Electronics Engineers, Oct 1981
24. *Condenser Microbiofouling Control Handbook,* Electric Power Research Institute, April 1993
25. W. Characklis et al., Biofouling Film Development and Its Effects on Energy Losses: A Laboratory Study, *Condenser Biofouling Control—Symposium Proceedings,* Electric Power Research Institute, 1980
26. H. Michels, W. Kirk, and A. Tuthill, The Influence of Corrosion and Fouling on Steam Condenser Performance, *J. Mater. Energy Syst.,* Dec 1979
27. *Consensus for the Lay-up of Boilers, Turbines, Turbine Condensers, and Auxiliary Equipment,* ASME Research Report, CRTD Vol 66, American Society of Mechanical Engineers, 2002
28. B. Buecker, "Condenser Chemistry and Performance Monitoring: A Critical Necessity for Reliable Steam Plant Operation," Paper IWC-99-10, Proceedings of the 60th Annual International Water Conference, Engineers' Society of Western Pennsylvania, 1999
29. P. Puckorius and G. Loretitsch, "Cooling Water Scale and Scaling Indices: What They Mean—How to Use Them Effectively—How They Can Cut Treatment Costs," Paper IWC-99-47, Proceedings of the 60th Annual International Water Conference, Engineers' Society of Western Pennsylvania, Oct 1999
30. "Effect of Targeted Chlorination on the Corrosion Behavior of Copper-Nickel Condenser Tubing," TR-101405, Electric Power Research Institute, Oct 1992
31. "Effects of Selected Water Treatments and Cathodic Protection, Corrosion and Embrittlement of Condenser Tubes," EPRI CS-5589, Electric Power Research Institute, Jan 1988
32. "Failure Cause Analysis—Condenser and Associated Systems," EPRI CS-2378, Electric Power Research Institute, June 1982
33. "Steam Plant Surface Condenser Leakage Study," NP-481, Electric Power Research Institute, March 1977
34. E. Szymanski, Troubleshooting of Condenser Tube Leaks, *Ultrapure Water,* March 2002
35. "Condenser Leak-Detection Guidelines Using Sulfur Hexafluoride as a Tracer Gas," EPRI CS CS-6014, Electric Power Research Institute, Sept 1988
36. "Generating Availability Report, Statistical Data 1982–2000," pc-GAR, North American Electric Reliability Council, 2002
37. B.C. Syrett and R.L. Coit, Materials Degradation in Condensers and Feedwater Heaters, *Proc. of AIME/ANS/NACE International Symposium on Environmental Degradation of Materials in Nuclear Systems—Water Reactors,* Aug 1983 (Myrtle Beach, SC), National Association of Corrosion Engineers, 1984, p 87
38. "Effects of Sulfide, Sand, Temperature, and Cathodic Protection on Corrosion of Condensers," EPRI CS-4562, Electric Power Research Institute, May 1986
39. B.C. Syrett, D.D. Macdonald, and S.S. Wing, Corrosion of Copper-Nickel Alloys in Seawater Polluted with Sulfide and Sulfide Oxidation Products, *Corrosion,* Vol 35, 1979, p 409
40. D.D. Macdonald, B.C. Syrett, and S.S. Wing, The Corrosion of Copper-Nickel Alloys 706 and 715 in Flowing Seawater, Part II: Effect of Dissolved Sulfide, *Corrosion,* Vol 35, 1979, p 367
41. B.C. Syrett and S.S. Wing, Effect of Flow on Corrosion of Copper-Nickel Alloys in Aerated Seawater and in Sulfide-Polluted Seawater, *Corrosion,* Vol 36, 1980, p 73
42. B.C. Syrett, The Mechanism of Accelerated Corrosion of Copper-Nickel Alloys in Sulphide-Polluted Seawater, *Corros. Sci.,* Vol 21, 1981, p 187
43. L.E. Eiselstein, B.C. Syrett, S.S. Wing, and R.D. Caligiuri, Accelerated Corrosion of Copper-Nickel Alloys in Sulphide-Polluted Seawater—Mechanism No. 2, *Corros. Sci.,* Vol 23, 1983, p 223
44. B.C. Syrett, Protection of Copper Alloys from Corrosion in Sulfide-Polluted Seawater, *Mater. Perform.,* Vol 20 (No. 5), 1981, p 50
45. O. Jonas and J. Mancini, Chapter 14—Deaerators, *Low Temperature Corrosion Problems in Fossil Power Plants—State of Knowledge,* Technical Report 1004924, Electric Power Research Institute, 2003

46. *Recommended Practice for Prevention, Detection, and Correction of Deaerator Cracking,* Standard Recommended Practice RP0590-96, NACE International, 1996
47. J. Kelly et al., An Overview of Deaerator Cracking, *Mater. Perform.,* Sept 1988 (Part 1) and Oct 1988 (Part 2)
48. H. Herro, "Corrosion Fatigue and Deaerator Cracking," Paper 213, Corrosion 87, National Association of Corrosion Engineers, March 1987
49. J. Copeland, A. Eastman, and C. Schmidt, "Fatigue and Stress Corrosion Cracking Evaluations in Deaerators," Paper 216, Corrosion 87, National Association of Corrosion Engineers, March 1987
50. O. Jonas, Deaerator Corrosion, An Overview of Design, Operation, Experience, and R&D, *Proc. American Power Conference,* Vol 49, A. 979, Illinois Institute of Technology, 1987
51. R. Beckwith et al., System Design, Specifications, Operation, and Inspection of Deaerators, *TAPPI J.,* July 1987
52. D. McIntyre, "A Review of Corrosion and Cracking Mechanisms in Boiler Feedwater Deaerators," Paper 309, Corrosion 86, National Association of Corrosion Engineers, March 1986
53. J. Robinson, "Deaerator Cracking Survey: Basic Design, Operating History, and Water Chemistry Survey," Paper 305, Corrosion 86, National Association of Corrosion Engineers, March 1986
54. O. Jonas "Deaerator, Design and Operation, Effects on Water Chemistry, Corrosion Problems," EPRI Symposium on Fossil Utility Water Chemistry (Atlanta, GA), June 11–13, 1985, Electric Power Research Institute
55. National Board of Boiler and Pressure Vessel Inspectors, More on Those Dangerous Deaerators, *National Board Bull.,* Vol 42 (No. 1), 1984
56. NACE Activity Coordinates the State-of-the-Art Concerns about Deaerator (Feedwater Tank) Integrity, *Mater. Perform.,* Nov 1984
57. S. Strauss, Clues Offered on Deaerator Problems, *Power,* April 1984
58. National Board of Boiler and Pressure Vessel Inspectors, Dangerous Deterioration of Deaerators, *National Board Bull.,* Oct 1983
59. TAPPI Engineering Division—Deaerator Advisory, *TAPPI J.,* 1983
60. Warning Issued on Cracking in Deaerators and Deaerator Storage Vessels, *Mater. Perform.,* Vol 22 (No. 11), 1983
61. S. Strauss, Concern Rises for Safety of Feedwater Deaerators, *Power,* Nov 1983
62. Recommended Practices for Prevention, Detection, and Correction of Deaerator Cracking, *TAPPI J.,* Sept 1991
63. T. Gooch and P. Hart, "Review of Welding Practice for Carbon Steel Deaerator Vessels," Paper 303, Corrosion 86, National Association of Corrosion Engineers, March 1986
64. R. Cranfield, L. Dartnell, and D. Wilkinson, Solving Fluid-Flow Problems in Deaerators Using Scale Models, *CEGB Res.,* June 1982
65. G. Bohnsack, "German Views on the Influence of Feedwater Chemistry on Crack Damage in Deaerators," Paper 301, Corrosion 86, National Association of Corrosion Engineers, March 1986
66. *Interim Consensus Guidelines on Fossil Plant Chemistry,* Technical Report CS-4629, Electric Power Research Institute, June 1986, and other EPRI guidelines
67. *Cycling, Startup, Shutdown, and Layup—Fossil Plant Cycle Chemistry Guidelines for Operators and Chemists,* Technical Report 107754, Aug 1998
68. *Consensus for the Lay-up of Boilers, Turbines, Turbine Condensers, and Auxiliary Equipment,* Research Report, CRTD—Vol 66, ASME 2002
69. R. Parkins, Stress Corrosion Spectrum, *Br. Corr. J.,* Vol 7, Jan 1972
70. *Flow-Accelerated Corrosion in Power Plants,* Technical Report 106611-R1, Electric Power Research Institute, 1998
71. O. Jonas, Alert: Erosion-Corrosion of Feedwater and Wet Steam Piping, *Power,* Feb 1996
72. *CHECWORKS: Flow Accelerated Corrosion Application,* Version 1.0G, SW-103198-P1R1DK, Electric Power Research Institute, Aug 1998
73. O. Jonas and J. Mancini, Chapter 13—LP Feedwater Heaters, *Low Temperature Corrosion Problems in Fossil Power Plants—State of Knowledge,* Technical Report 1004924, Electric Power Research Institute, Dec 2003
74. *Productivity Improvement Handbook for Fossil Steam Power Plants,* Technical Report 114910, 2nd ed., Electric Power Research Institute, June 2000
75. *Standards for Closed Feedwater Heaters,* 7th ed., Heat Exchanger Institute, 2004
76. *Practical Aspects and Performance of Heat Exchanger Components and Materials,* International Joint Power Generation Conference, PWR—Vol 19, ASME, 1992
77. *Recommended Guidelines for the Operation and Maintenance of Feedwater Heaters,* Technical Report CS-3239, Electric Power Research Institute, Sept 1983
78. L. Neel, Cycling to Failure—Feedwater Heaters, *Power Eng.,* Nov 1997
79. *High-Reliability Feedwater Heater Study,* Technical Report CS-5856, Electric Power Research Institute, June 1988
80. *Feedwater Heater Maintenance Guide,* Technical Report 1003470, Electric Power Research Institute, May 2002
81. *Feedwater Heater Technology Seminar and Symposium,* Technical Report 1004022, Electric Power Research Institute, Sept 2001
82. *Proceedings: 1992 Feedwater Heater Technology Symposium,* Technical Report 102923, Electric Power Research Institute, Sept 1993
83. *State-of-Knowledge of Copper in Fossil Plant Cycles,* Technical Report 108460, Electric Power Research Institute, Sept 1997
84. *Flow-Accelerated Corrosion in Power Plants,* Technical Report 106611, Electric Power Research Institute, 1996
85. CHECWORKS Users Group, "Recommendations for Inspecting Feedwater Heater Shells for Possible Flow-Accelerated Corrosion Damage," Electric Power Research Institute, Feb 2000
86. *Corrosion in Condensers and Feedwater Heaters,* NACE International, 1996
87. *Manual for Investigation and Correction of Feedwater Heater Failures,* Technical Report GS-7390, Electric Power Research Institute, Nov 1991
88. R. Bell, Corrosion of Deaerators and Feedwater Heaters, *Corrosion,* Vol 13, *Metals Handbook,* ASM International, 1987
89. J. Beavers et al., "Corrosion Related Failures in Feedwater Heaters," Corrosion/84, National Association of Corrosion Engineers, April 1984
90. E. Brush and W. Pearl, Corrosion and Corrosion Product Release in Neutral Feedwater, *Corrosion,* Vol 28 (No. 4), April 1972
91. E. Brush and W. Pearl, Corrosion Behavior of Nonferrous Alloy Feedwater Heater Materials in Neutral Water with Low Oxygen Contents, *Corrosion,* Vol 25 (No. 3), March 1969
92. *Corrosion-Related Failures in Feedwater Heaters,* Technical Report CS-3184, Electric Power Research Institute, July 1983
93. *Failure Cause Analysis—Feedwater Heaters,* Technical Report CS-1776, Electric Power Research Institute, April 1981
94. *Interim Consensus Guidelines on Fossil Plant Chemistry,* Technical Report CS-4629, Electric Power Research Institute, June 1986
95. *Cycling, Startup, Shutdown, and Layup—Fossil Plant Cycle Chemistry Guidelines for Operators and Chemists,* Technical Report 107754, Electric Power Research Institute, Aug 1998
96. Consensus for the Lay-up of Boilers, Turbines, Turbine Condensers, and Auxiliary Equipment, *ASME Research Report,* CRTD—Vol 66, 2002

SELECTED REFERENCE

- N. Lieberman, *Working Guide to Process Equipment,* McGraw-Hill, 2003

Corrosion of Flue Gas Desulfurization Systems

W.L. Mathay, Nickel Institute

AIR POLLUTION is a world-wide concern that can adversely affect the environment, structures, and human health. It is a major contributor to acid rain, which occurs as a result of sulfur oxide emissions being converted into acids in the atmosphere. These emissions can be carried long distances by the prevailing winds and can fall to earth with rain, creating very acidic conditions at ground level. It has long been recognized that coal-burning power plants are the major source of sulfur oxides, and along with vehicles and other combustion units, a major source of nitrogen oxides, another significant pollutant.

Recognizing the need to control acid rain, the U.S. Environmental Protection Agency (EPA) began establishing sulfur dioxide (SO_2) emission restrictions and time frames for compliance beginning in the 1960s and continuing through the 1990s. During the latter period, restrictions on emissions of nitrogen oxides (NO_x) were also established. These restrictions and others dealing with toxic metals and particulates continue to be adjusted as environmental conditions dictate. In recent years, other countries with coal-burning power plants have adopted SO_2 and NO_x emission restrictions as well.

Flue Gas Desulfurization (FGD) Technology

Numerous methods for controlling SO_2 emissions have been investigated, with most of the emphasis being placed on systems for the removal of SO_2 from power-plant flue gases. In these systems, combustion gases generated by the firing of coal are sent to scrubbing equipment where SO_2 is removed by reaction with an absorbing slurry, solution, or reactant. At one time there were over 250 flue gas desulfurization (FGD) processes under consideration, but wet lime/limestone scrubbing with the production of gypsum as a by-product has become the most widely used process because of its high removal efficiency and operational reliability. These efficiencies have been improved to the point where SO_2 removal rates of over 95% have become common. In addition, forced-air oxidation has been used to control the oxidation of reaction products, decrease scaling, and aid in the production of commercial-grade gypsum for wallboard production.

Inasmuch as wet scrubbers have become the predominant method of SO_2 emissions control, their use in NO_x control also has been considered. However, they are ineffective in reducing NO_x emissions because of the relative insolubility of nitrogen oxides in water. Thus, power plant operators turned to the use of low-nitrogen-oxide burners, changes in firing practice, and the installation of selective catalytic reduction (SCR) systems. The SCR system uses a catalyst that promotes reaction of the nitrogen oxides with ammonia to produce nitrogen and water from the nitrogen oxides (Ref 1). However, the use of SCR systems can lead to the formation of additional sulfur trioxide (SO_3) beyond that which is formed during the coal combustion process. Ultimately, the SO_3 can react with water to form sulfuric-acid-mist aerosols that escape the scrubber as fine particulates. Scrubbers are highly efficient in absorbing SO_2 but are inadequate in capturing the SO_3 aerosols that can result in the formation of a very visible acid plume exiting the chimney.

FGD Corrosion Problem Areas

A number of years ago, in order to understand the corrosion problems encountered in wet scrubbing, the American Society for Testing and Materials (now ASTM International) prepared a special technical publication, STP 837, which presented a simplistic generic FGD system and attempted to define the associated corrosion problem areas (Ref 2). Initially, a prescrubber leading into the main scrubber or absorber was used primarily for prequenching and particulate control. However, later designs eliminated the prescrubber and brought the hot flue gases from the electrostatic precipitator directly into the absorber, as shown in Fig. 1 (Ref 3).

There are a number of FGD designs involving different flue gas/scrubbing medium contact methods, but in the commonly used spray tower, similar to the generic scrubber, the flue gas is brought into the lower part of the absorber at temperatures ranging from 149 to 182 °C (300 to 360 °F) and is quenched by counter-current sprays of the scrubbing medium. Corrosion can be quite severe in the quench area because wet-dry zones occur near the flue gas inlet, causing scale deposition. Crevice corrosion can occur beneath these deposits because of the presence of chlorides and possibly fluorides from the coal and the low pH values that develop as corrosion proceeds in an autocatalytic manner. The potential severity of the problem is best shown by the fact that chloride concentrations as high as 100,000 ppm have been found beneath the scale deposits. Scale deposition and corrosion also can extend into the inlet duct because of the carry-back of the scrubbing medium and its reaction with the hot, incoming flue gas. Usually the most corrosion-resistant materials are required in this area.

In the absorber sump or recycle tank below the liquid level, there may be concerns about erosion because of the solids content of the scrubber slurry. In some cases FGD designers have chosen to use thicker alloys to avoid the problem.

As the flue gas leaves the wet-dry or quench zone and travels up through the absorber and the mist eliminator area, usually much less corrosive conditions are encountered. The gas temperature drops to about 50 to 65 °C (122 to 149 °F) and the pH value of the environment generally rises, allowing the use of less corrosion-resistant materials. However, under certain circumstances improved boiler efficiencies may lead to greater flue gas velocities, which can result in "blow-through" of the corrosives from the lower part of the absorber. Thus, the environment approaching the absorber outlet may remain quite aggressive and dictate the use of materials similar to those used in the lower part of the absorber above the wet-dry zone.

In the outlet duct and into the chimney liner, the environment again can become quite corrosive because the scrubbed gas is saturated with

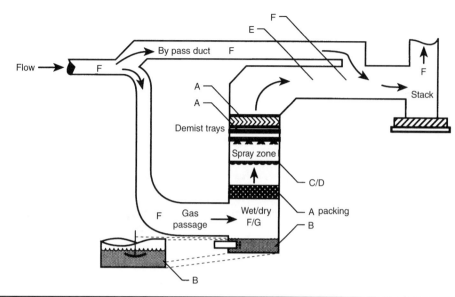

Fig. 1 Operating zones for generic FGD system, ASTM D-33.09 modified. Source: Ref 3

	Qualitative definition of operating zones		
Code	Chemistry	Mechanical environment	Temperature
A	Mild corrosive (vapor)	Mild	Mild
B	Moderate (immersion)	Mild	Mild
C	Moderate	Moderate	Mild
D	Moderate	Severe	Mild
E	Severe	Mild	Moderate
F	Severe	Mild	Severe
G	Severe	Severe	Severe

moisture and contains sulfur acids of varying concentrations as well as possibly hydrochloric and hydrofluoric acids. These acids may condense in the outlet duct and in the chimney liner, causing serious corrosion problems. Currently there are a number of FGD installations that have bypass ducts that can introduce hot, unscrubbed flue gas into the scrubbed gas stream and increase the severity of the acid attack. However, newer designs generally have eliminated the bypass ducts because of SO_2 emission restrictions.

In the past, several types of reheaters have been used in the outlet duct, but their use has been abandoned because of the many corrosion problems encountered with the reheaters. Thus, the use of the so-called 'wet-stack' has become common practice. In a wet-stack, materials that are resistant to acid attack at the temperatures involved must be used.

If a plant has installed an SCR system for NO_x capture, it may need to install a wet electrostatic precipitator (WESP) with several electrical fields following the absorber to capture the SO_x aerosols that are not removed in the absorber (Ref 4). One plant has integrated the absorber and the WESP, with the flue gas flowing up through the absorber and then through the WESP before passing through a final mist eliminator. After leaving the final mist eliminator, the flue gas passes through an outlet hood and duct and then enters the chimney liner.

Materials Selection for FGD Components

To comply with environmental emission restrictions, air pollution control systems must utilize corrosion-resistant construction to ensure availability and reliability. The corrosion problem areas have been determined and the major forms of corrosive attack encountered; crevice corrosion, pitting corrosion, and acid attack have been identified. Design features that minimize the possibility of corrosion in certain components also have been recognized. These include the sloping of ducts to avoid puddling of liquids, and the use of welding procedures that provide low profile, smooth welds to avoid scale build-up and crevice corrosion.

A number of materials have been used to minimize corrosion problems in FGD systems, but initial cost is often a determining factor. For example, flake-glass-filled organic coatings and linings on carbon steel have been used for absorbers and ductwork because of low initial costs. However, inadequate surface preparation and improper application can result in short service lives and lead to costly touch-ups and/or recoating.

If properly applied and cured, rubber linings on carbon steel have performed well in absorbers, slurry piping, and pumps, but some can become brittle and can blister from water absorption. Periodic repairs and maintenance may be required, resulting in costly outages.

Acid-resisting bricks and mortar have been used successfully for many years as chimney liners. Problems with mortar permeability and substrate corrosion, as well as the need for increased foundation support, can be deterrents to usage.

In recent years, borosilicate-glass blocks have found application in outlet ducts and as chimney liners because of their corrosion-resisting and insulating properties. However, care must be taken to avoid surface damage to the blocks that can lead to liquid penetration and deterioration of the membrane backing.

Fiberglass reinforced plastic (FRP) laminates can be cost-effective materials for absorbers, ductwork, slurry piping, and chimney liners. For satisfactory performance they require good quality control procedures throughout fabrication and installation, and adequate support after fabrication.

Titanium-sheet linings and intermittently bonded cladding on carbon steel have been used for outlet ductwork and chimney liners. Again, proper fabrication and installation procedures are necessary. Titanium has generally good corrosion resistance but can be severely attacked by fluorides.

Although each of the foregoing materials may offer certain advantages, nickel-containing alloys are increasingly preferred for FGD service. They are readily fabricated and installed and, when properly selected, offer excellent corrosion resistance along with long-term reliability at low life-cycle costs.

The nominal chemical compositions of the various alloys considered for use in FGD systems are shown in Table 1. Included in generally increasing corrosion-resistance order are the austenitic stainless steels, types 316L (UNS S31603) and 317L (S31703), and the 4% molybdenum stainless steels, types 317LM (S31725), 317LMN (S31726), and 904L (N08904). (For the most part, alloy 904L has been replaced by type 317LMN.)

Next are the austenitic/ferritic duplex and super duplex stainless steels, typified by alloys 2205 (S31803 and S32205) and 2507 (S32750), respectively, the 6 to 7% Mo super austenitic stainless steels, a 9% Mo-Ni, alloy 625 (N06625), and the 13 to 16% Mo-Ni alloys. Of this latter group, alloy C-276 (N10276) and C-22 (N06022) were the most corrosion-resistant FGD alloys available until alloy 59 (N06059), alloy 686 (N06686), and alloy C-2000 (N06200) were developed to provide greater corrosion resistance.

The user has the option of using the stainless steels and nickel alloys as sheet linings on carbon steel (often called wallpapering), as alloy-clad carbon-steel plate, or as solid alloys, with the selection being determined by cost and overall project considerations. NACE International has developed detailed instructions for the installation of metallic wallpaper sheet linings (Ref 5) and alloy-clad plate (Ref 6) in air pollution control equipment. Usually there is no cost advantage to the use of the lower-alloyed stainless steels as sheet linings or clad plate over solid plate. However, the higher alloys, particularly those containing 13 to 16% Mo, when used as 1.6 mm (0.0625 in.) thick sheet linings or cladding can be economically attractive (Ref 7).

As for the specific applications for the foregoing constructional materials, types 316L and 317L stainless steels are used for some low-chloride absorber applications. However, increased chloride levels are being encountered in the scrubbing slurries as plants have moved toward zero-discharge of liquids. This increase in chloride levels has dictated the application of more corrosion-resistant absorber materials. The 4% and 6 to 7% Mo stainless steels and alloys C-276 and C-22 have been used depending on pH and chloride levels. The duplex and super duplex stainless steels are also finding increased usage in place of the 300-series stainless steels.

For the very corrosive wet-dry zones in the absorber inlet and inlet duct, the high Ni-Cr-Mo alloys such as alloy C-276 are usually specified. By contrast, at the top of the absorber in the mist eliminator area, types 316L and 317L stainless steels have been used successfully. If "blow-through" conditions exist, higher alloys such as the 4 to 6% molybdenum alloys should be considered.

In the outlet duct and in the chimney liner, alloys resistant to the acid concentrations and temperatures that can occur are shown in Fig. 2 (Ref 3). In many instances, alloy C-276 or a higher alloy has been used, usually as sheet linings or clad plate, to provide the desired corrosion resistance.

Often, spray piping in the absorber is fabricated of FRP, rubber-lined or covered-steel, or a high-alloy material such as alloy C-276. Absorber internals and supports usually are fabricated of the same alloy as the absorber, although some use has been made of duplex stainless steels. Pumps and other items of equipment also make use of rubber-covered components and duplex and super duplex stainless steels.

In those installations where there is a WESP following the absorber, the 4 to 6% Mo stainless steels have been used for the casing and most of the internal structures and components. However, because the flue gas entering the WESP contains high levels of sulfuric acid and chlorides, materials such as alloy C-276 may be required. Alternatively, FRP has been used.

Numerous laboratory and field corrosion tests as well as service performance studies have been used to evaluate alloys and to develop alloy comparisons for use in FGD systems. In this regard, comparisons of alloys have been aided by the use of Pitting Resistance Equivalent Numbers (PREN). The PREN values are calculated using equations involving alloying elements such as chromium (Cr), molybdenum (Mo), and nitrogen (N) that are important to providing pitting corrosion resistance. The higher the PREN value, the greater the corrosion resistance of the alloy. One of the earliest equations used was $PREN = \%Cr + 3.3\ (\%Mo) + 16(\%N)$. Although initially applied to stainless steels, the equation has been extended to include nickel-base alloys, and in many instances the nitrogen multiplier has been increased from 16 to 30. Alloy producers regularly provide PREN equations and values in their literature, including both stainless steels and nickel-base alloys.

More definitive alloy selection assistance is available from the use of charts such as Fig. 3 (Ref 1). This chart offers suggested guidelines for the selection of stainless steels and nickel alloys for FGD equipment based on chloride levels and pH values at the absorber operating temperatures of 50 to 65 °C (122 to 149 °F). However, when using this type of information, it is important to remember that each FGD system is somewhat unique, no matter how similar in design.

Future Air Pollution Control Considerations

For the foreseeable future, coal will continue to be a major fuel for power generation. New and retrofitted FGD systems will continue to be installed. Governmental regulations will continue to call for further reductions in emissions of SO_x and NO_x, as well as reductions in particulates and toxic metals such as mercury.

Table 1 Nominal chemical compositions of metals considered for flue gas desulfurization service

Common name	UNS No.	Composition, wt%								
		Cr	Ni	Mo	Fe	Co	W	Cu	C	Other
316L	S31603	17	12	2.2	bal	...	...	...	0.03	0.05 N
317L	S31703	19	13	3.2	bal	...	...	...	0.03	0.05 N
317LM	S31725	19	15	4.2	bal	...	...	...	0.03	0.05 N
317LMN	S31726	19	15	4.2	bal	...	...	...	0.03	0.15 N
904L	N08904	21	25	4.2	bal	...	1.5	0.02	...	
Alloy 2205	S31803/S32205	22	5.5	3	bal	...	...	...	0.03	0.15 N
Alloy 2507	S32750	25	7.0	4	bal	...	...	0.5	0.03	0.28 N
6–7% Mo stainless steels	...	20–24	18–25	6–7	bal	...	...	0–1.00	0.02	0.15–0.5 N
Alloy 625	N06625	21.5	61	9	4	...	...	...	0.05	3.65 Cb + Ta
C-276	N10276	16	57	16	5	1.3	3.8	...	0.01	...
Alloy C-22	N06022	22	59	13	3	...	3.0	...	0.01	...
Alloy C-2000	N06200	23	59	16	<1	...	...	1.6	0.01	...
Alloy 59	N06059	23	59	16	1	...	...	...	0.01	...
Alloy 686	N06686	20.5	57	16.3	1	...	4.0	...	0.005	...

Alloys C-22 and C-2000 are trademarks of Haynes International. Alloy 686 is a trademark of Special Metals. Alloy 59 is a trademark of Krupp-VDM. Alloy 2507 is a trademark of Sandvik Metals.

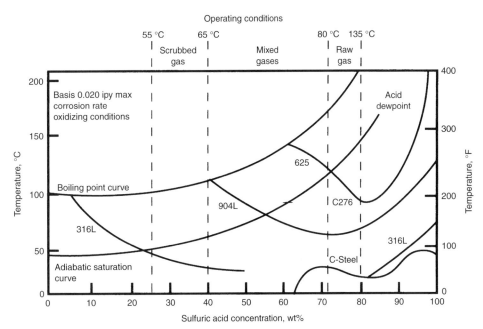

Fig. 2 Suggested alloys to resist sulfuric acid corrosion in FGD systems. Source: Ref 3

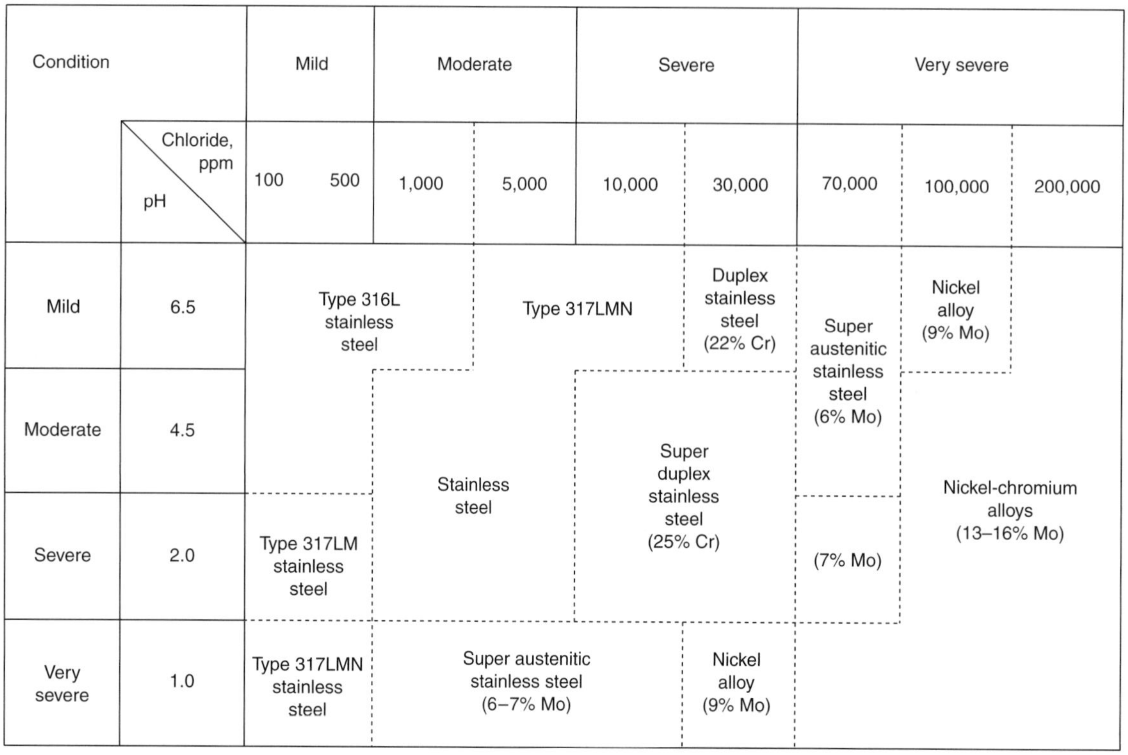

Condition		Mild		Moderate		Severe		Very severe		
pH \ Chloride, ppm		100	500	1,000	5,000	10,000	30,000	70,000	100,000	200,000
Mild	6.5	Type 316L stainless steel		Type 317LMN		Duplex stainless steel (22% Cr)		Super austenitic stainless steel (6% Mo)	Nickel alloy (9% Mo)	
Moderate	4.5			Stainless steel		Super duplex stainless steel (25% Cr)			Nickel-chromium alloys (13–16% Mo)	
Severe	2.0	Type 317LM stainless steel						(7% Mo)		
Very severe	1.0	Type 317LMN stainless steel		Super austenitic stainless steel (6–7% Mo)		Nickel alloy (9% Mo)				

Fig. 3 Guidelines for selection of stainless steels and nickel alloys for flue gas desulfurization equipment at 50 to 65 °C (122 to 149 °F). Source: Ref 3

Both wet and dry FGD scrubbing systems will be used for SO_2 emission control, depending upon the type of coal burned. Systems for reducing NO_x, sulfuric acid aerosols, fine particulates, and toxic metal emissions will be required. Thus in the United States, because existing and proposed federal emission standards involve multiple pollutants, multi-pollutant control technologies must be considered from an economic standpoint.

The U.S. Environmental Protection Agency has recognized the fact that there is a significant relationship between SO_2, particulates, and toxic metals. Power plants burning high-sulfur coals and equipped with wet scrubbers are said to be able to remove up to 90% of particulate heavy metals (Ref 8). The addition of a WESP to the scrubbing system will result in the capture of sulfuric acid aerosols that escape the scrubber and will aid further in the removal of fine particulates and metals.

The removal of mercury poses a problem because its capture requires that it be in the ionic or oxidized form. Wet scrubbers at plants burning high-sulfur coals apparently can remove most of the oxidized mercury present in the flue gas reaching the scrubber. Research indicates that the catalysts in SCR installations that are ahead of the scrubber can oxidize the elemental mercury with the gas, thus increasing mercury removal by the scrubbers. However, plants burning low-sulfur coal appear to emit mercury that is mostly elemental. Dry scrubbers and activated carbon injections with fabric filters for particulate and mercury collection are being considered for low-sulfur coal-burning plants.

As yet, the integration of processes to control multiple-pollutant emissions has not made significant progress because of the lack of information on the economics involved. Also, integration with existing pollution control equipment must be considered if retrofit installations are to be made.

REFERENCES

1. B. Buecher, Emission Control: SCR Design and Operation, *Power Engineering*, August 2002, p 24–26
2. "Manual of Protective Linings for Flue Gas Desulfurization Systems," STP 837, ASTM, 1984
3. W.H.D. Plant and W.L. Mathay, "Nickel-Containing Materials in Flue Gas Desulfurization Equipment," NiDI Technical Series No. 10 072, Nickel Development Institute, June 1999
4. R. Blagg and R. Keeth, "Wet Electrostatic Precipitators for Power Boilers—New and Retrofit Applications," Combined Power Plant Air Pollution Control Mega Symposium (Chicago, IL), A&WMA, EPRI, EPA, and NETL, August 2001
5. "Installation of Thin Metallic Wallpaper Lining in Air Pollution Control and Other Process Equipment," NACE Standard Recommended Practice RP0292-2003, NACE International, 2003
6. "Installation of Stainless Chromium-Nickel Steel and Nickel-Alloy Roll-Bonded and Explosion-Bonded Clad Plate in Air Pollution Control Equipment," NACE Standard Recommended Practice RP0199-2004, NACE International, 2004
7. R.E. Avery, "Utilization of Nickel-Containing Alloys For FGD Systems in North America and Europe," Paper 455, presented at Corrosion '96, National Association of Corrosion Engineers, 1996
8. R.W. McIlvane, "Voluntary Early Reduction Option For Utility Air Toxics," distributed by the McIlvane Company, Northbrook, IL

SELECTED REFERENCES

- D.C. Agarwal and J. Kloewer, "Nickel-Base Alloys: Corrosion Challenges in the New Millenium," Paper 01325, presented at Corrosion 2001, NACE International, 2001
- R. Francis et al., "Applications and Experiences with Super Duplex Stainless Steels in Wet FGD Scrubber Systems," Paper 98479, presented at Corrosion '98, NACE International, 1998
- J.F. Grubb, Selection of Alloys for Air Pollution Control Equipment: Stainless Steels, *Proceedings of the Mega Symposium*, (Chicago, IL), Aug 20–23, 2001

- J. F. Grubb et al., "A 6% Mo Stainless Steel for Flue Gas Desulfurization," Paper 00583, presented at Corrosion 2000, NACE International, March 2000
- C.V. Mathai and E.D. Elliott, "The Clear Skies Initiative," EM, Air & Waste Management Association, May 2002, p 25–34
- B.S. Phull et al., "Corrosion Resistance of Duplex and 4–6% Mo-Containing Stainless Steels in FGD Scrubber Absorber Slurry Environments," Paper 00578, presented at Corrosion 2000, NACE International, March 2000
- A. Saleem, "Design and Operation of Single Train Spray Tower FGD Systems," *The 1991 SO_2 Control Symposium,* Vol 3, (Washington, D.C.) 1991, p 329–349

Corrosion of Steam- and Water-Side of Boilers

Phillip Daniel, Babcock & Wilcox Company

HIGH-TEMPERATURE WATER AND STEAM react with boiler steel to form oxides, hydroxides, hydrates, and hydrogen, but formation of a protective oxide layer, such as magnetite (Fe_3O_4), on the metal surface generally impedes further reaction. Such protective oxide films are maintained and corrosion prevented by appropriate control of boiler water chemistry. However, excessive impurities in boiler feedwater can prevent effective control, form harmful deposits, and aggravate corrosion. Although deposition, corrosion, and chemistry control processes vary with boiler pressure and cycle design, similar processes are involved for a wide range of power boilers and for many industrial boilers and heat-recovery steam generators as well.

Chemistry-Boiler Interactions

As steam forms, dissolved solids concentrate in the boiler water. Where the solubility of an impurity is exceeded, steam formation may leave behind precipitated deposits. These deposits in turn provide a sheltered environment that further increases chemical concentrations and deposition rates. In a clean boiler tube, concentration of chemicals at the tube surface is limited by free exchange of fluid between the surface and boiler water flowing through the tube. Even where there are thin deposits, boiling generally produces sufficient flow within the deposits to limit the degree of concentration. However, as deposits accumulate, they restrict flow to the surface.

Typical boiler deposits are largely hardness precipitates (calcium and magnesium salts) and metal oxides. The nature of these deposits is typically a function of the makeup water purification system. Hardness deposits are most common with softened makeup water, while metal oxides are more likely with demineralized or reverse-osmosis-type makeup water. Most hardness salts are easily precipitated due to their retrograde solubility. Hardness salts can enter the cycle as impurities in makeup water and in cooling water from condenser leaks. Metal oxides are largely from corrosion of preboiler cycle components. Scaling occurs when these minerals and oxides precipitate and adhere to boiler internal surfaces, where they impede heat transfer. The result can be overheating of tubes, sometimes followed by failure of tubes if corrective measures are not taken. Deposits can also increase boiler circuit pressure drop in forced circulation boiler systems.

As they concentrate, some chemicals become corrosive. Corrosion can occur even in a clean boiler, but the likelihood of substantial corrosion is much greater beneath thick porous deposits that facilitate the concentration process. The concentration beneath deposits can be more than 1000 times higher than that in the bulk boiler water, and the temperature at the base of these deposits can substantially exceed the saturation temperature. As deposits accumulate, control of boiler water chemistry to avoid the formation of corrosive concentrates becomes increasingly important and difficult. Hence, operation of a boiler with excessively thick deposits must be avoided.

Although separation devices remove most water droplets carried by steam, some residual droplets containing small amounts of dissolved solids always carry through with the steam. Also, at higher pressures, there is some vaporous carryover. Excessive carryover of impurities can damage superheaters, steam turbines, or downstream process equipment.

Corrosion Control and Prevention

Deposition and corrosion in boilers are controlled by water purification and chemical treatment. Impurities are removed by purification of makeup water, polishing of returned condensate, deaeration, and blowdown. Chemicals are then added to control pH, electrochemical potential, and dissolved oxygen concentration. Care must also be taken to prevent corrosion when boilers are out of service, on standby or for maintenance. Proper water chemistry control improves boiler reliability and reduces maintenance and component replacement costs. It also improves performance and life of heat exchangers, pumps, turbines, and piping throughout the steam generation, use, and condensation cycle.

General water chemistry control limits and guidelines have been developed and issued by various groups of boiler owners and operators (e.g., American Society of Mechanical Engineers, Ref 1–3; Electric Power Research Institute, Ref 4; and Vereinigung der Grosskraftwerkbetreiber, Ref 5), water treatment specialists, utilities, and industries. Also, manufacturers provide recommended chemistry control limits for each boiler and for other major cycle components. However, optimal water and steam chemistry limits for boilers, turbines, and other cycle components depend on equipment design and materials of construction for the combination of equipment employed. Hence, for each boiler system, boiler-specific water chemistry limits and treatment practices must be developed and tailored by competent specialists familiar with the specific boiler and its operating environment.

In addition to customized boiler-specific guidelines and procedures, qualified operators are essential, and vigilance is required to detect early signs of chemistry upsets. Operators responsible for plant cycle chemistry must understand boiler water chemistry guidelines and how they are derived and customized. They must also understand how water impurities, treatment chemicals, and boiler components interact. Training must therefore be an integral, ongoing part of operations and should include management, control room operators, chemists, and laboratory staff. Training and plant operating procedures must also address safety and health hazards associated with boiler operation, with steam and water sampling, and with chemicals used for water purification and treatment. If not handled properly, many of these operations and chemicals can pose severe safety and health hazards.

Common Fluid-Side Corrosion Problems

With adequate control of water and steam chemistry, internal corrosion of boiler circuitry can be avoided. Yet, chemistry upsets (transient losses of control) do occur. Where preventative measures fail and corrosion occurs, good monitoring and documentation of system chemistry can facilitate identification of the root cause. Identification of the root cause is an essential step toward avoidance of further corrosion. Where corrosion occurs and the cause is unknown, the documented water chemistry, location of the corrosion, appearance of the corrosion, and chemistry of localized deposits and corrosion products often suggest the cause. Common causes are flow-accelerated corrosion, oxygen pitting, chelant corrosion, caustic corrosion, acid corrosion, organic corrosion, phosphate corrosion, hydrogen damage, and corrosion-assisted cracking. Additionally, pitting and microbiological corrosion can occur under poor lay-up conditions. Further discussion of corrosion and failure mechanisms is provided in Ref 6 to 9.

One distinguishing feature of corrosion is its appearance. Metal loss may be uniform, so the surface appears smooth. Conversely, the surface may be gouged, scalloped, or pitted. Other forms of corrosion are microscopic in breadth, and subsurface, so they are not initially discernable. Subsurface forms of corrosion include intergranular corrosion, corrosion fatigue, stress-corrosion cracking, and hydrogen damage. Such corrosion can occur alone or in combination with surface wastage. In the absence of component failure, detection of subsurface corrosion often requires ultrasonic, dye penetrant, or magnetic-particle inspection. These forms of corrosion are best diagnosed with destructive cross-sectional metallography.

Another distinguishing feature is the chemical composition of associated surface deposits and corrosion products. Deposits may contain residual corrosives such as caustic or acid. Magnesium hydroxide in deposits can suggest the presence of an acid-forming precipitation process. Ferrates and sodium carbonate indicate caustic conditions. Sodium-iron phosphate indicates phosphate wastage. Organic deposits suggest corrosion by organics, and excessive amounts of ferric oxide or hydroxide with pitting suggest oxygen attack.

Flow-accelerated corrosion is the localized dissolution of feedwater piping in areas of flow impingement. It occurs where metal dissolution dominates over protective oxide scale formation. For example, localized conditions are sufficiently oxidizing to form soluble Fe^{2+} ions but not sufficiently oxidizing to form Fe^{3+} ions needed for protective oxide formation. Conditions known to accelerate thinning include flow impingement on pipe walls, low pH, excessive reducing agent (oxygen scavenger) concentrations, chemicals that increase iron solubility, and thermal degradation of organic chemicals. Thinning can occur at any feedwater temperature, but it is most rapid in the range of 120 to 205 °C (250 to 400 °F). Thinned areas often have a scalloped or pitted appearance. Feedwater piping failures, such as that shown in Fig. 1, can occur unexpectedly and close to work areas and walkways. To ensure continued integrity of boiler feedwater piping, it must be periodically inspected for internal corrosion and wall thinning. Any thinned areas must be identified and replaced before they become a safety hazard. The affected piping should be replaced with low-alloy chromium-bearing steel piping, and the water chemistry control should be appropriately altered.

Oxygen pitting and corrosion during boiler operation largely occur in preboiler feedwater heaters and economizers, where oxygen from poorly deaerated feedwater is consumed by corrosion before it reaches the boiler. However, oxygen pitting within boilers occurs when poorly deaerated water is used for startup or for accelerated cooling of a boiler. It also occurs in feedwater piping, drums, and downcomers in some low-pressure boilers that have no feedwater heaters or economizer. Oxygen pitting is often characterized by well-formed, relatively deep and rounded pits surrounded by passive zones of little or no corrosion. Because increasing scavenger concentrations to eliminate residual traces of oxygen can aggravate flow-accelerated corrosion, care must be taken to distinguish between oxygen pitting and flow-accelerated corrosion, which generally occurs only where all traces of oxygen have been eliminated.

Complexing agent corrosion occurs where appropriate feedwater and boiler water chemistries for chelant or some polymer treatments used in some low-pressure boilers are not maintained. Potentially corrosive conditions include low pH or excessive concentration of free chelant and dissolved oxygen or polymer. Especially susceptible surfaces include flow impingement areas of feedwater piping, riser tubes, and cyclone steam/water separators. Affected areas are often dark colored and have the appearance of uniform thinning or flow-accelerated corrosion.

Corrosion fatigue is cracking well below the yield strength of a material by the combined action of corrosion and alternating stresses. Stress may be of mechanical or thermal origin. In boilers, corrosion fatigue also occurs in ferritic and austenitic superheaters, but it is most common in water-wetted surfaces where there is a mechanical constraint on the tubing. For example, corrosion fatigue occurs in furnace wall tubes adjacent to windbox, buckstay, and other welded attachments. Failures are thick lipped, with little or no reduction in wall thickness. On examination of the internal tube surface, multiple initiation sites are evident. Cracking is transgranular. Environmental conditions facilitate fatigue cracking where it would not otherwise occur in a benign environment. One such factor is low-pH transients associated with, for example, cyclic operation, condenser leaks, and phosphate hideout and hideout return.

Phosphate corrosion occurs on the inner steam-forming side of boiler tubes by reaction of the steel with phosphate to form maricite ($NaFePO_4$). Figure 2 shows ribbed tubing that has suffered this type of wastage. The affected surface has a gouged appearance, with maricite, hematite, and magnetite deposits. Phosphate corrosion occurs predominantly in boilers experiencing phosphate hideout. Although not always apparent, common signs of phosphate corrosion include difficulty maintaining target phosphate concentrations, phosphate hideout and pH increase with increasing boiler load or pressure, phosphate hideout return and decreasing pH with decreasing load or pressure, and periods of high iron concentration in boiler water. Potential for phosphate corrosion increases with increasing internal deposit loading, low effective sodium-to-phosphate molar ratios, increasing phosphate concentration, inclusion of low-molar-ratio phosphates (disodium and especially monosodium phosphate) in phosphate feed solution, and increasing boiler pressure. To avoid acid phosphate corrosion,

Fig. 1 Feedwater pipe thinning and rupture

Fig. 2 Acid phosphate corrosion of ribbed tubing

operators should monitor boiler water conditions closely, ensure accuracy of pH and phosphate measurements, ensure purity and reliability of chemical feed solutions, ensure that target boiler water chemistry parameters are appropriate and are attained throughout boiler load changes, and watch for signs of phosphate corrosion. Phosphate corrosion in high-pressure utility boilers may be addressed with equilibrium phosphate chemistry control.

Underdeposit acid corrosion and hydrogen damage occur where boiler water acidifies as it concentrates beneath deposits on steam-generating surfaces. Hydrogen from acid corrosion diffuses into the steel, where it reacts with carbon to form methane. The resultant decarburization and methane formation weakens the steel and creates microfissures. Thick-lipped failures occur when the degraded steel no longer has sufficient strength to hold the internal tube pressure. Signs of hydrogen damage include underdeposit corrosion, thick-lipped failure, and steel decarburization and microfissures. The corrosion product from acid corrosion is mostly magnetite. Affected tubing, which may extend far beyond the failure, must be replaced. The boiler must be chemically cleaned to remove internal tube deposits, and boiler water chemistry must be altered or better controlled to prevent acid formation as the water concentrates. Operators should reduce acid-forming impurities by improving makeup water, reducing condenser leakage, or using condensate polishing. For drum boilers, operators should use phosphate treatment with an effective sodium-to-phosphate molar ratio of 2.8 or greater.

Caustic corrosion, gouging, and grooving occur where boiler water forms a caustic residue as it evaporates. In vertical furnace wall tubes, this occurs beneath deposits that facilitate a high degree of concentration, and the corroded surface has a gouged appearance. In inclined tubes where the heat flux is directed through the upper half of the tube, caustic concentrates by evaporation of boiler water in the steam space on the upper tube surface. Resulting corrosion is in the form of a groove that is generally free of deposits and centered on the crown of the tube. Deposits associated with caustic gouging often include Na_2FeO_4. To prevent reoccurrence of caustic gouging, operators should prevent accumulation of excessive deposits and should control water chemistry so that concentration of boiler water does not form caustic. The latter can generally be achieved by ensuring appropriate feedwater chemistry with coordinated or congruent phosphate boiler water treatment, taking care to control the effective sodium-to-phosphate molar ratio as appropriate for the specific boiler and the specific chemical and operating conditions. In some instances, where caustic grooving along the top of a sloped tube is associated with steam/water separation, better mixing can be maintained by the use of ribbed tubing.

Caustic stress-corrosion cracking can occur where caustic concentrates in contact with steel that is highly stressed. Caustic cracking is rare in boilers with all welded connections. This generally occurs in boilers using a high-alkalinity caustic boiler water treatment, and it is normally associated with unwelded rolled joints and welds that are not stress relieved. Caustic stress-corrosion cracking at a rolled tube joint generally requires a roll leak that produces very high caustic concentrations at this location when the boiler water flashes to steam. On metallographic examination, caustic cracking is intergranular and has the branched appearance characteristic of stress-corrosion cracking. It can generally be avoided by use of coordinated phosphate treatment. Where a high-alkalinity caustic phosphate boiler water treatment is used for low-pressure boilers, nitrate is often added to inhibit caustic cracking.

Stress-corrosion cracking also occurs in superheaters and reheaters, usually in austenitic stainless steel elements where contaminants concentrate as standing water evaporates during boiler startups. Cracking occurs in areas of high stress associated with bends, welds, and attachments. Potential for stress-corrosion cracking can be minimized by control of carryover, attemperation water, chemical cleaning procedures, and out-of-service conditions to minimize contamination, especially contamination with hydroxide, chloride, or sulfide.

Out-of-service corrosion is predominantly oxygen pitting. Pitting attributed to out-of-service corrosion occurs during outages but also as aerated water is heated when boilers return to service. Especially common locations include the waterline in steam drums, areas where water stands along the bottom of horizontal pipe and tube runs, and lower bends of pendant superheaters and reheaters. Pinhole failures are more common in thinner-walled reheater and economizer tubing. Such corrosion can be minimized by following appropriate lay-up procedures for boiler outages and by improving oxygen control during boiler startups.

ACKNOWLEDGMENT

The material presented here is abridged from "Water Chemistry, Water Treatment, and Corrosion," chapter 43, *Steam, Its Generation and Use,* 41st ed., The Babcock & Wilcox Company, Barberton, OH, 2004.

REFERENCES

1. H.A. Klein and J.K. Rice, A Research Study on Internal Corrosion in High Pressure Boilers, *J. Eng. Power (Trans. ASME),* Vol 88 (No. 3), July 1966, p 232–242
2. P. Cohen, Ed., *The ASME Handbook on Water Technology in Thermal Power Systems,* The American Society of Mechanical Engineers, 1989
3. *Consensus on Operating Practices for the Control of Feedwater and Boiler Water Quality in Modern Industrial Boilers,* The American Society of Mechanical Engineers, 1994
4. Numerous guidelines for control of boiler water chemistry, Electric Power Research Institute
5. *VGB Guidelines for Feedwater, Boiler Water and Steam for Pressure above 68 Bar,* VGB Kraftwerkstechnik, 1988
6. H.H. Uhlig, *Corrosion and Corrosion Control,* Wiley, 1985
7. G.A. Lamping and R.M. Arrowood, Jr., "Manual for Investigation and Correction of Boiler Tube Failures," Report CS-3945, Electric Power Research Institute, 1985
8. D.N. French, *Metallurgical Failures in Fossil Fired Boilers,* Wiley, 1983
9. R.D. Port and H.M. Herro, *The NALCO Guide to Boiler Failure Analysis,* McGraw-Hill, 1991

SELECTED REFERENCES

- *Power Plant Chemistry* (journal of all power plant chemistry areas), Power Plant Chemistry GmbH, Neulussheim, Germany
- Water Chemistry, Water Treatment, and Corrosion, Chapter 43, *Steam, Its Generation and Use,* 41st ed., The Babcock & Wilcox Company, 2004

Corrosion of Steam Turbines

Otakar Jonas, Jonas, Inc.

THE STEAM TURBINE (see Fig. 1) is the simplest and most efficient engine for converting large amounts of heat energy into mechanical work. As the steam expands, it acquires high velocity and exerts force on the turbine blades. Turbines range in size from a few kilowatts for one-stage units to 1500 MW for multiple-stage multiple-component units comprising high-pressure, intermediate-pressure, and up to three double-flow low-pressure turbines. For mechanical drives, single-, double-, and multiple-stage turbines are generally used. Most larger modern turbines are multiple-stage axial-flow units. Turbine inlet steam pressure and temperature conditions are governed by the boiler and range from less than 1.4 MPa (200 psi) saturated to more than 35 MPa (5000 psi) at 650 °C (1200 °F) superheated and supercritical. See the article "Corrosion in Supercritical Water—Ultrasupercritical Environments for Power Production" in this Volume. Steam is often extracted from the turbine for heating of feedwater and used in industrial processes. It is exhausted as wet subatmospheric steam, with up to 12% moisture (Fig. 7), into a condenser or at higher than atmospheric pressure for use as process steam.

Turbines are typically designed for a 25 to 40 year life. During this long life, corrosion and other material damage can accumulate and lead to premature failures. Corrosion usually results from a combination of water chemistry, design, and materials selection problems (Ref 1–67).

The cost of corrosion to U.S. fossil power plants in 1998 was estimated to be $4.6 billion, which was equivalent to $9 million/GW of installed capacity. Of that, corrosion fatigue (CF) and stress-corrosion cracking (SCC) in low-pressure (LP) utility turbines accounted for over $600 million (Ref 2). The cost of industrial steam turbine corrosion is estimated to be similar. Data collected by the North American Electric Reliability Council for 1476 fossil units between 1996 and 2000 show that LP turbines were responsible for 818 forced and scheduled outages and deratings, costing the utilities 39,574 GWh (Ref 68). In the list of steam cycle component contribution to forced and planned outages, steam turbines have been in the top ten for 30 years. These outages are often characterized as low-frequency, high-impact events. For example, LP turbine disc SCC can result in a 6 month long outage for a new or weld-repaired rotor, costing over $100 million, mostly for lost production.

Steam Turbine Developments

To reduce LP turbine SCC problems, most shrunk-on discs have been redesigned and replaced (Ref 13–21). Many LP rotors with shrunk-on discs have been replaced by integral or welded rotors. Weld repair techniques for the disc-blade attachment areas were developed, including weld repair of the low-alloy steel disc with 12Cr stainless steel (Ref 16). Marginally designed LP turbine blades have been replaced by better designs with reduced vibration excitation and with blades without tenon-to-shroud crevices (with integral shroud sections) (Ref 2, 3, 6). Mixed-tuned blade rows are being used for longer blades where neighboring blades do not have similar resonance frequencies. Computational flow dynamics and viscous flow modeling are used for the blade path design and development of new blade shapes (banana blades) and profiles. Better steam chemistry and cycle chemistry monitoring are resulting in better control of local corrosive environments on blade and disc surfaces. Extending turbine inspection intervals up to 10 years is bringing higher demands on steam purity and on the evaluation of inspection results, such as pitting and cracking.

Major Corrosion Problems in Steam Turbines

Corrosion fatigue, SCC, pitting, and erosion-corrosion are the primary corrosion mechanisms in steam turbines (Ref 1–37). In addition, there is water droplet erosion of last rows of LP turbine blades (Ref 2, 3, 6, 29, 38) and solid-particle erosion in the high-pressure and intermediate-pressure turbines and turbine valves caused by exfoliation of oxides in superheaters, reheaters, and steam piping (Ref 3–6, 8). There have also been cases of SCC of the crossover pipe stainless steel and Inconel expansion bellows (Ref 1, 2, 5–7, 26). Figure 1 and Table 1 indicate the distribution of corrosion within the turbine. Pitting and CF of turbine blades and SCC of discs are currently the two costliest problems. Much research has been devoted to these two problems (Ref 1–10, 13–24), and progress has been made in both operation (better steam chemistry) and design (lower stresses, no tenon crevices, limit on maximum strength to reduce susceptibility to SCC).

It is estimated that inadequate mechanical design (use of high-strength materials susceptible to environmental cracking, high steady and vibratory stresses, stress concentration, and vibration) is responsible for approximately 50% of the problems, inadequate steam chemistry for approximately 20%, and marginal thermodynamic and flow design for approximately 20%. Poor manufacturing and maintenance practices account for the remaining 10% of the problems. The root cause of most corrosion-related outages is design, with high local stresses in discs and blades and flow excitation of blades playing the major role. Stress-corrosion cracking of blade attachments (Ref 1–3, 5–7, 13–16) and erosion-corrosion of wet-steam piping (Ref 1–3, 5, 28) seem to be generic problems, particularly in nuclear turbines.

Blade failures (Ref 1–7, 22, 23, Fig. 1) are most frequent (in fossil utility turbines) in the last minus-one row (L-1), which is at or immediately before the saturation line (Ref 1–3, 9–12, 22, 23). In industrial turbines, blade failures are also most frequent in the stages in which first moisture occurs (Ref 51). Blade pitting and cracking had been more frequent in once-through boiler units (these units used all-volatile water treatment and condensate polishers) than in drum boiler units. This was attributed to better neutralization of acids and salts by sodium phosphate in the drum boilers (Ref 35). Statistics from Germany, the United Kingdom, and Japan indicate a very low incidence of blade failures. Typical locations of CF cracks are shown in Fig. 1. A cracked L-1 blade is shown in Fig. 2, and a CF (Goodman) diagram for three stainless steel blade alloys is shown in Fig. 3.

Cyclic stresses are caused by turbine startups and shutdowns (low number of cycles, high cyclic stresses), by the turbine and blades ramping through critical speeds at which some components are in resonance (high amplitude, high frequency), and by over 15 flow-induced blade excitation mechanisms during normal operation that include:

- Synchronous resonance of the blades at a harmonic of the unit running speed
- Nonuniform flows
- Blade vibration induced from a vibrating rotor or disc
- Self-excitation, such as flutter
- Random excitation—resonance with adjacent blades
- Shock waves in the transonic flow region and shock wave—condensation interaction
- Wrong incidence flow angle and flow separation

There are pronounced thermodynamic effects that cause the concentration of impurities at surfaces and turbine component corrosion. They are most frequent where the metal surface temperature is slightly above the saturation temperature of steam. For the low-pressure fossil utility turbines, this usually occurs on the L-1 blades and on various surfaces of the last two discs. It is one of the reasons the L-1 blades in fossil utility turbines have higher failure rates than any other blade row. Another reason for the high L-1 blade failure rate is the effects of transonic flow (shock waves) and an interaction of the shock waves with the Wilson (delayed droplet nucleation) line (start of condensation; periodic destruction of the Wilson line), resulting in high vibratory stresses (Ref 10, 52, 62–65).

Stress corrosion of low-pressure turbine discs is specifically distributed for each type of turbine and each disc location, indicating a correlation with surface stress, temperature, and steam conditions (Fig. 1) (Ref 1–3, 13–15, 48, 49). This correlation is often related to the impurity concentration by evaporation of moisture. The tendency toward cracking increases with yield strength, stress, and operating temperature. Figure 4 shows massive SCC, caused by high concentration of NaOH in steam, that was discovered during a routine inspection. Figure 5 illustrates the stress-corrosion behavior of NiCrMoV disc steel in "good" water and steam. The effects that yield strength and temperature have on the average crack growth rate of NiCrMoV disc steels are shown in Fig. 6.

There have been catastrophic failures of shrunk-on discs caused by SCC from the keyways, and hundreds of discs have been redesigned and replaced (Ref 1–7, 9, 13, 17–21). This problem was solved by eliminating the keyways and replacing discs with the lower-yield-strength material, typically less than 900 MPa (130 ksi). The current generic problem is SCC of blade attachments in both fossil and nuclear LP turbines (Ref 1–3, 14, 15).

Corrosion of other turbine parts is due to one or more of the same causes described previously (Table 1). There is a strong SCC/low-cycle fatigue interaction in many turbine components (well recognized in piping) that may be important in stationary blade, bolt, expansion bellows, and turbine cylinder failures. The highest incidence of turbine bolting failures has been in high- and intermediate-pressure cylinders and nozzle blocks, most of it due to the presence of NaOH. Stress-corrosion cracking of expansion bellows in steam pipes was a problem in fossil units with high concentrations of NaOH (Ref 26). To find the true causes of corrosion, it is essential to analyze the local temperature, pressure, chemistry, moisture velocity, and stress and material conditions.

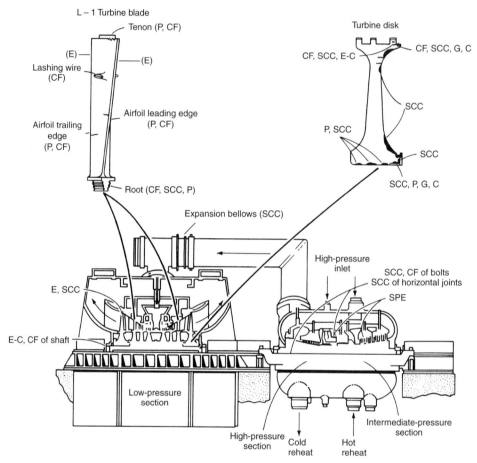

Fig. 1 Medium-sized utility turbine with locations of corrosion and erosion of steam turbine components. P, pitting; CF, corrosion fatigue; SCC, stress-corrosion cracking; C, crevice corrosion; G, galvanic corrosion; E, erosion; E-C, erosion-corrosion; SPE, solid-particle erosion

Table 1 Corrosion mechanisms in steam turbine components

Component	Material	Corrosion mechanisms(a)
Rotor	Forged Cr-Mo-V or Ni-Cr-Mo-V low-alloy steel	P, SCC, CF, E
Shell	Cast carbon or Cr-Mo-V low-alloy steel, fabricated carbon steel for low-pressure turbines	SCC, E-C
Discs, bucket wheels	Forged Cr-Mo-V, Ni-Cr-Mo-V or Ni-Cr-Mo low-alloy steel, 12Cr weld repair	P, SCC, CF, E-C
Dovetail pins	Cr-Mo low-alloy steels; 5CrMoV, similar to ASTM A681 grade H-11	SCC
Blades, buckets	12Cr stainless steels, 15-5PH, 17-4PH, Ti6-4, PH13-8Mo, Fe-26Cr-2Mo	P, CF, SCC, E
Bucket tie wires	12Cr stainless steels (ferritic and martensitic)	SCC, P, CF
Shrouds, bucket covers	12Cr stainless steels, 15-5PH, 17-4PH, Ti6-4, PH13-8Mo, Fe-26Cr-2Mo	P, SCC
Stationary blades	Type 304 stainless steel, other stainless steels	SCC, SCC-LCF
Expansion bellows	Types 321 or 304 stainless steels, Inconel 600	SCC, SCC-LCF
Erosion shields	Weld-deposited or soldered Stellite type 6B; hardened blade materials	SCC, E
Bolts	Incoloy 901, Refractalloy 25, Pyromet 860, . . .	SCC, SCC-LCF
Wet-steam piping	Carbon steel	E-C
Valve bushings and stems	13Cr-Mo and other stainless steels	P, OX

(a) P, pitting; SCC, stress-corrosion cracking; CF, corrosion fatigue; E, erosion; E-C, erosion-corrosion; LCF, low-cycle fatigue; OX, oxidation in steam. General corrosion is experienced by all carbon and low-alloy steel components. Solid-particle erosion is experienced in high- and intermediate-pressure inlets (nozzle blocks, stationary and rotating blades, and valves). It is caused by exfoliation of steam-grown oxides in superheater tubes and in steam pipes.

Pitting (Ref 1–3, 9, 10) is often the initiating corrosion mechanism for SCC and CF. Pit growth seems to decrease with time (parabolic under constant environmental conditions), and pitting is often much faster during unprotected layup (maintenance shut-down) than during normal operation. The main environmental factor causing pitting is the high concentration of chloride in steam and deposits; however, other chemicals can also cause pitting. The presence of pits significantly reduces fatigue strength of turbine materials (Ref 9).

Erosion-corrosion (also called flow-accelerated corrosion) is most pronounced in carbon steel pipes with high-velocity turbulent flow and low-pH moisture containing high concentrations of CO_2 or other acid-forming anions (Ref 2, 27, 28, 30–33). In fossil units, LP extraction pipes can suffer severe erosion-corrosion.

Water droplet and solid-particle erosion (Ref 24, 28, 29) are not believed to be related to corrosion; however, some studies have shown that water droplet erosion is accelerated by the presence of corrosive impurities. Water droplet erosion can initiate fatigue failures of blades by causing surface notches and even changing resonant frequencies of vibration. There have been fatigue failures of last row blade airfoils initiated by water droplet erosion of the trailing edges caused by reversed wet-steam flow during low-load operation or by excessive use of hood sprays.

Turbine Materials

There is little worldwide variation in materials for blades, discs, rotors, and turbine cylinders, and only a few major changes have been introduced in the last two decades (Ref 2, 6). These materials are listed in Table 1. Titanium alloy blades are slowly being introduced for the last low-pressure stages. Also, improved melting practices, control of inclusions and tramp elements, and weld repair techniques are being used for discs and rotors.

Considering the typical design life of 25 to 40 years and the relatively high stresses, turbine materials perform remarkably well. Turbine steels are susceptible to SCC and CF in numerous environments, such as caustic, chlorides, hydrogen, carbonate-bicarbonate, carbonate-CO_2, acids, and, at higher stresses and strength levels, pure water and steam (Ref 2, 6, 13–23).

No turbine can permanently tolerate concentrated caustic, sodium chloride (NaCl), or acids. Under such conditions, disc and rotor materials would crack by SCC in a few hundred hours at stresses as low as 10% of the yield strength (Ref 1–21). Chromium steel (12% Cr) blades may tolerate NaOH but would pit and crack by CF at very low vibratory stresses (as low as 6.9 MPa, or 1 ksi) in many other corrodents. With high concentrations of impurities for long periods of time, there would always be a weak link in the turbine or within the power cycle that would fail prematurely. Although titanium could be used for blades and would tolerate most of the impurities (except hydroxides), corrosion-resistant materials for the large forgings of rotors and discs are difficult to find because of the cost, forgeability, banding of alloying elements and impurities, and other problems. Plating and surface coatings appear to be only marginal temporary solutions.

In addition to the susceptibility to localized corrosion in concentrated impurities, the low-alloy and carbon steels used extensively in steam turbines have rather narrow ranges of passivity (Ref 32, 33). This makes these steels vulnerable

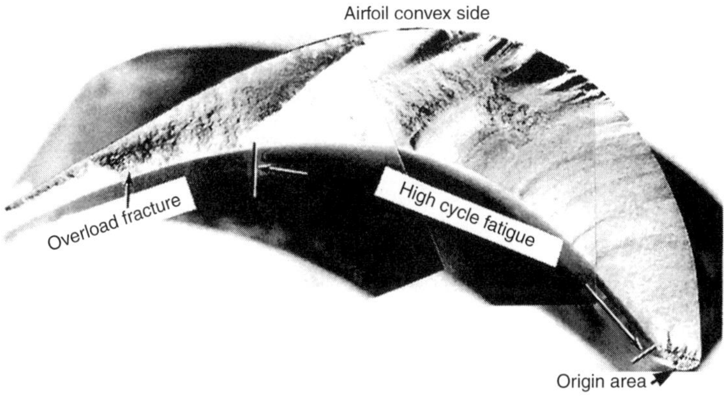

Fig. 2 Corrosion fatigue of an L-1 blade airfoil. Courtesy of O. Jonas, Jonas, Inc.

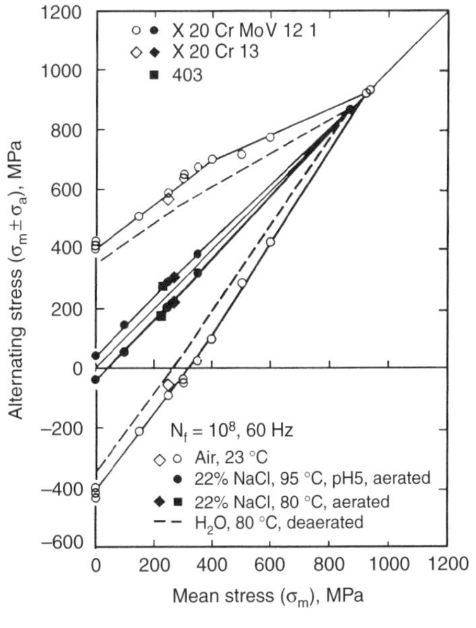

Fig. 3 Corrosion fatigue (Goodman) diagram for three stainless steel blading alloys in various environments. Based on 10^8 cycles to failure at 60 Hz. Source: Ref 23

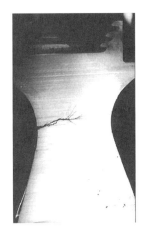
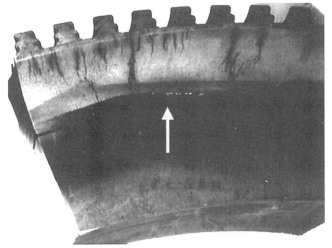

Fig. 4 Massive stress-corrosion cracking of L-1 disc caused by high concentration of NaOH in steam. Courtesy of O. Jonas, Jonas, Inc.

to pH and oxygen excursions and consequent pitting and other forms of localized corrosion during operation and layup. In chloride solutions, the pH region for passivation becomes even narrower. The material properties needed for designing against corrosion and for failure analysis and evaluation of residual life include fracture toughness (K_{Ic}), stress-corrosion threshold stress (σ_{SCC}), threshold stress intensity (K_{ISCC}), crack propagation rate ((da/dt)$_{SCC}$), CF limit, CF threshold stress intensity (ΔK_{th}), CF crack propagation rate (da/dN), pitting rate, and a pit depth limit, all for a range of temperatures and environments (Ref 1, 2, 24, 32, 33).

Environment

Normally, utility turbine steam is very pure, with the total concentration of impurities less than 50 ppb and the limit for Na, Cl^-, SO_4^{2-}, Fe, and Cu in the 3 to 5 ppb range (Ref 39–45). The normal limit for cation conductivity of condensed steam is 0.1 to 0.3 µS/cm. Corrosive impurities are transported into the turbine steam from the preboiler cycle (feedwater and attemperating water) and the boiler as a total (mechanical plus vaporous) carryover. Their major sources include condenser leaks, air in-leakage, makeup water, and improperly operated condensate polishers. Water treatment chemicals, such as NaOH, used for boiler water treatment and amines, oxygen scavengers, dispersants, and chelants can also generate corrosive conditions in the turbine (Ref 50–55). The organics generate acidic conditions by decomposition and formation of acids. The turbine environment is controlled through a control of impurity ingress and various feedwater, boiler water, and steam chemistry limits (Ref 39–46). The corrosiveness of the steam turbine environments (on the surface, at temperature) is caused by one or more of the following:

- Concentration of impurities from low part per billion levels in steam to percent levels on turbine surfaces and the formation of concentrated aqueous solutions (concentration by deposition or evaporation of moisture)
- Insufficient pH control (in both acid and alkaline regions)
- High-velocity, high-turbulence, and low-pH moisture

The situation is illustrated in Fig. 7 and 8, which shows a Mollier diagram with a typical LP turbine steam expansion line, the thermodynamic regions of impurity concentration (NaOH and salts), and the resulting corrosion. The conditions on hot turbine surfaces (in relation to the steam saturation temperature) can shift from the wet-steam region into the salt zone and above by evaporation of moisture. Condensation shock and the transonic flow shock waves can also evaporate moisture and cause impurity concentration. This is why SCC of discs often occurs in the wet-steam regions, and it emphasizes the need to consider local surface temperatures in design and failure analysis. The surfaces may be hot because of heat transfer through the metal (Ref 47, 48) or because of the stagnation temperature effect (zero flow velocity at the surface and conversion of kinetic energy of steam into heat) (Ref 54). Depending on their vapor pressures, near the saturation line, impurities can be present as a dry salt or as a concentrated aqueous solution. In the wet-steam region, they are either diluted by moisture or could concentrate by evaporation on hot surfaces (Fig. 7, 8) (Ref 47, 48).

The steam impurities that are of most concern include chlorides, sulfates, fluorides, carbonates, hydroxides, organic and inorganic acids, oxygen, and CO_2. Their behavior in turbine steam and deposits is well documented (Ref 1–12, 49–59). There are strong synergistic effects and interactions with metal oxides. However, pure water and wet steam can also cause SCC and CF of turbine blade and disc materials.

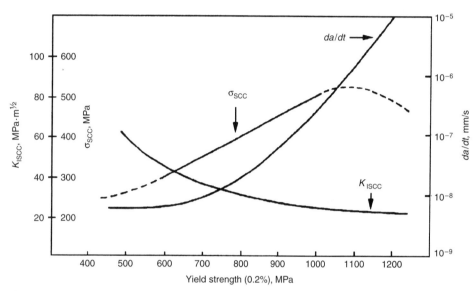

Fig. 5 Stress-corrosion behavior of NiCrMoV disc steel vs. yield strength for "good" water and steam. K_{ISCC}, threshold stress intensity; σ_{SCC}, threshold stress; da/dt, stage 2 crack growth rate. Courtesy of O. Jonas, Jonas, Inc.

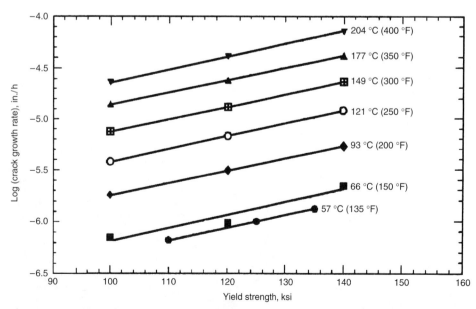

Fig. 6 Average crack growth rates vs. yield strength for several operating temperatures for NiCrMoV disc steel. Log (da/dt) = −2.093 −3218.9 (1/T) + 0.01207 (YS). da/dt, crack growth rate (in./h); YS, yield strength; T, temperature. Source: Ref 34

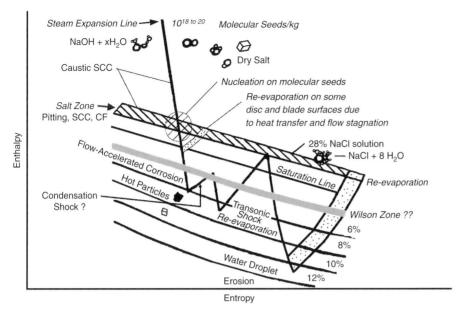

Fig. 7 Mollier diagram showing the low-pressure turbine steam expansion line, thermodynamic regions of impurity concentration, and corrosion mechanisms. Percent moisture lines are given. SCC, stress-corrosion cracking; CF, corrosion fatigue. Courtesy of O. Jonas, Jonas, Inc.

In addition to corrosion during operation, turbines can corrode during:

- Manufacture (machining fluids and lubricants)
- Storage (airborne impurities and preservatives)
- Erection (airborne impurities, preservatives, and cleaning fluids)
- Chemical cleaning of boiler (storage of acid in condenser hotwells)
- Nondestructive testing (cleaning and testing fluids)
- Layup (deposits plus humid air)

Many of the aforementioned substances may contain high concentration of sulfur and chlorine, which could form acids on decomposition. Decomposition of typical organics—for example, carbon tetrachloride (CCl_4)—occurs above 150 °C (300 °F). Therefore, the composition of all of these substances should be controlled (maximum of 50 to 100 ppm sulfur and 50 to 100 ppm chlorine has been recommended), and most of them should be removed before operation. Molybdenum disulfide (MoS_2) has been implicated as a stress corrodent in power system applications (Ref 1, 2, 60, 61).

Layup corrosion increases rapidly when the relative humidity of the air reaches approximately 60%. When deposits are present, layup corrosion can be in the form of pitting or SCC.

Design

Design disciplines that affect turbine corrosion can be separated into four parts:

- Mechanical design (stresses, stress concentrations, and stress intensity, or K_I)
- Heat transfer (surface temperatures and heated crevices)
- Flow and thermodynamics (moisture velocity, location of the salt zone, stagnation temperature, and interaction of shock wave with condensation, that is, the Wilson line)
- Physical shape (crevices, obstacles to flow, and surface finish)

The various aspects of design against environment-induced cracking are well documented (Ref 1–12, 18–21, 24, 34), but only a few of these references deal with the complexity of the problem (Ref 18, 20). Probabilistic approaches are discussed in Ref 34. Problems with such approaches include the lack of a large number of statistical data on corrosion properties of materials, service stresses, and environments. The mechanical design concepts for avoiding turbine corrosion should include evaluation of safety factors against σ_{SCC}, K_{ISCC}, $(da/dt)_{SCC}$ (Fig. 5), (da/dN), ΔK_{th}, CF limit, pitting rate, and a pit depth limit.

True residual stresses (welding, overspeed effects) should also be considered. Implementation of the aforementioned concepts requires that

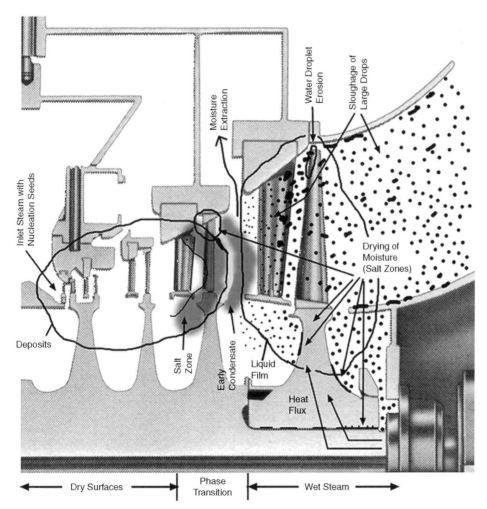

Fig. 8 Cross section of a low-pressure turbine illustrating the locations where impurities can concentrate by precipitation, deposition, and by evaporation of moisture to percent concentrations, thus becoming very corrosive. Courtesy of O. Jonas, Jonas, Inc.

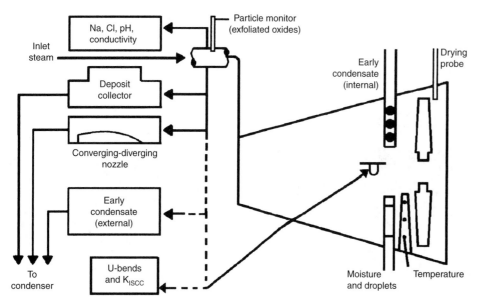

Fig. 9 Low-pressure turbine troubleshooting instrumentation can identify specific corrosive conditions. K_{ISCC}, threshold stress intensity. Courtesy of O. Jonas, Jonas, Inc.

corrosion testing generates quantitative data to be used by the designer.

Solutions to Corrosion Problems

Control of the steam and surface chemistry to minimize turbine corrosion includes (Ref 2, 5, 9, 10, 39–46):

- Decreasing concentration of corrosive impurities in makeup and feedwater, low air inleakage and condenser leakage, and so on
- Oxygenated water treatment for once-through fossil boiler units for excellent feedwater chemistry and clean boilers
- Layup protection
- Turbine washing after chemical upsets to remove deposited impurities
- Reduction or elimination of copper and its oxides and their synergistic corrosion effects by reducing oxygen concentration, operating with a reducing (negative oxidation-reduction potential) environment and a low ammonia concentration, or by replacing copper alloys with steel or titanium

Design and material improvements that reduce turbine corrosion include (Ref 2, 6, 18, 24, 32, 33):

- Welded rotors, large integral rotors, and discs without keyways—eliminate high stress concentrations due to disc keyways
- Replacement of higher-strength NiCrMoV discs with lower-strength (yield < 900 MPa, or 130 ksi) discs and avoiding high-strength blades
- Repair welding of discs and rotors, also with 12% Cr stainless steel weld metal
- Mixed-tuned blade rows to reduce random excitation
- Free-standing and integrally shrouded LP blades without tenon crevices and with lower stresses, as well as frictional damping
- Titanium LP blades—corrosion resistant in turbine environments except for NaOH
- Lower stress and stress concentrations—increasing resistance to SCC and CF
- Flow path design using computerized flow dynamics and viscous flow—lower flow-induced vibration, which reduces susceptibility to CF
- Curved (banana) stationary blades, which reduce nozzle-passing excitation
- New materials for blade pins and bolting that are resistant against SCC
- Flow guides and double-ply expansion bellows—reduce impurity concentration, better SCC resistance
- Moisture extraction to improve efficiency and reduce erosion-corrosion and water droplet erosion, as well as use of alloy steels to reduce flow-accelerated corrosion

Monitoring

The field monitoring equipment shown in Fig. 9 can be used to diagnose and prevent many common LP turbine corrosion and deposition problems (Ref 68). In addition, there are monitors available to detect vibration and provide vibration signatures of the whole machine, blade and rotor cracking, steam leaks, air inleakage, rotor position, and wear of bearings.

Further Study

It is estimated by the author that 70% of knowledge to solve and prevent corrosion problems in steam turbines is available. The percentage of available knowledge for understanding the effects of stress and environment is much lower, approximately 40%. The knowledge that is missing or needs improvement includes:

- Interaction of SCC and CF
- Threshold stress required to initiate SCC in blade attachments
- Effects of steeple geometry (stress concentrations and size) on SCC and CF. The steeple is the disc rim blade attachment that matches the blade root (Fig. 4)
- Effects of overloads during heater box and overspeed tests on stress redistribution and SCC in steeples, blade roots, and disc keyways
- Effectiveness of grinding out SCC and CF cracks as a corrective measure
- Effects of organic water treatment chemicals, boric acid, and organic impurities on composition of water droplets, SCC, CF, and pitting
- Effects of electrical charges carried by water droplets on corrosion
- Effects of galvanic coupling of dissimilar materials, such as the blade-steeple materials, on corrosion
- Effects of residues of preservatives and anaerobic adhesives on SCC, CF, and pitting of blade attachments
- Effects of blade trailing edge erosion on cracking
- Accelerated stress-corrosion testing
- Effects of variable amplitude loading on CF crack initiation and propagation
- Effects of water droplet pH and composition on erosion, and how to predict erosion
- Existence of Wilson delayed droplet nucleation zone in turbines
- Effects of shot peening to reduce stresses, SCC, and CF of blade steeples and blades
- Understanding the basic mechanisms of stress corrosion, CF, fatigue, and stress-induced pitting

REFERENCES

1. O. Jonas, Steam Turbine Corrosion: Stress—Environment—Material, *Proceedings of Int. Conference on Materials and Corrosion Experience in Fossil Power Plants,* Nov 2003, Electric Power Research Institute, 2003
2. O. Jonas, Low Pressure Steam Turbines, *Low Temperature Corrosion Problems in Fossil Power Plants—State of Knowledge,* 1004924, Electric Power Research Institute, Dec 2003
3. T. McCloskey, Troubleshooting Turbine Steam Path Damage, *Advances in Life Assessment and Optimization of Fossil Power Plants—Proceedings,* 1006965, Electric Power Research Institute, June 2002
4. W. Sanders, *Steam Turbine Path Damage and Maintenance,* Vol 1 (Feb 2001) and Vol 2 (July 2002), Pennwell Press

5. O. Jonas, Corrosion and Water Chemistry Problems in Steam Systems—Root Causes and Solutions, *Mater. Perform.*, Dec 2001
6. *Turbine Steam Path Damage: Theory and Practice*, AP-108943, Electric Power Research Institute, June 1998
7. O. Jonas and B. Dooley, Major Turbine Problems Related to Steam Chemistry: R&D, Root Causes, and Solutions, *Proceedings: Fifth International Conference on Cycle Chemistry in Fossil Plants*, TR-108459, Electric Power Research Institute, Nov 1997
8. K. Cotton, *Evaluating and Improving Steam Turbine Performance*, Cotton Fact, Inc., New York, 1993
9. O. Jonas, Steam Turbine Corrosion, *Mater. Perform.*, Feb 1985
10. O. Jonas, Understanding Steam Turbine Corrosion, *Corrosion/84 Proceedings*, National Association of Corrosion Engineers, 1984
11. V. Scegljajev, *Parni Turbiny (Steam Turbines)*, SNTL, Prague, 1983
12. O. Jonas, Steam, *Kirk-Othmer Encyclopedia of Chemical Technology*, John Wiley & Sons, 1983
13. *Proceedings: Workshop on Corrosion of Steam Turbine Blading and Discs in the Phase Transition Zone*, TR-111340, Electric Power Research Institute, Sept 1998
14. *Low-Pressure Rotor Rim Attachment Cracking Survey of Utility Experience*, TR-107088, Electric Power Research Institute, March 1997
15. E. Nowak, Low Pressure Turbine Stress Corrosion Cracking Investigation at the Navajo Generating Station, *Proceedings: Steam Turbine Stress Corrosion Workshop*, TR-108982, Electric Power Research Institute, Sept 1997
16. R. Kilroy et al., A 12% Chrome Weld Repair Increases Stress Corrosion Cracking Resistance of LP Finger Type Rotor Dovetails, *Proceedings of the International Joint Power Generation Conference: Volume 2: Power*, American Society of Mechanical Engineers, Nov 1997
17. N. Cheruvu and B. Seth, Key Variables Affecting the Susceptibility of Shrunk-On Discs to Stress Corrosion Cracking, *The Steam Turbine Generator Today: Materials, Flow Path Design, Repair, and Refurbishment*, American Society of Mechanical Engineers, 1993
18. W. Engelke et al., Design, Operating and Inspection Considerations to Control Steam Corrosion of LP Turbine Discs, *American Power Conference Proceedings* (Chicago, IL), Illinois Institute of Technology, April 1983
19. F.F. Lyle, Jr. and H.C. Burghard, Jr., *Steam Turbine Disc Cracking Experience*, NP-2429-LD, Electric Power Research Institute, June 1982
20. J.M. Hodge and I.L. Mogford, *Proc. Inst. Mech. Eng.*, Vol 193, 1979, p 93
21. J.L. Gray, *Proc. Inst. Mech. Eng.*, Vol 186 (No. 32), 1972, p 379
22. R.I. Jaffee, Ed., *Corrosion Fatigue of Steam Turbine Blade Materials*, Pergamon Press, 1983
23. A. Atrens et al., Steam Turbine Blades, *Corrosion in Power Generating Equipment*, Plenum Press, 1984
24. F.J. Heymann et al., "Steam Turbine Blades: Considerations in Design and a Survey of Blade Failures," EPRI CS-1967, Electric Power Research Institute, Aug 1981
25. H. Haas, Major Damage Caused by Turbine or Generator Rotor Failures in the Range of the Tripping Speed, *Der Maschinenschaden*, Vol 50, 1977, p 6
26. L.D. Kramer et al., *Mater. Perform.*, Vol 14, 1975, p 15
27. R. Svoboda and G. Faber, Erosion-Corrosion of Steam Turbine Components, *Corrosion in Power Generating Equipment*, Plenum Publishing Corporation, 1984
28. J.P. Cerdan et al., Erosion Corrosion in Wet Steam: Impacts of Variables and Possible Remedies, *Water Chemistry and Corrosion in Steam Water Loops of Nuclear Power Stations Symposium Proceedings*, ADERP, March 1980
29. V.A. Pryakhin et al., Problems of Erosion of the Rotating Blades of Steam Turbines, *Teploenergetika*, Oct 1984
30. *Flow-Accelerated Corrosion in Power Plants*, TR-106611, Electric Power Research Institute, 1996
31. O. Jonas, Control Erosion/Corrosion of Steels in Wet Steam, *Power*, March 1985
32. O. Jonas, Guidelines Help Designers Protect Against Localized Corrosion, *Power*, Vol 130 (No. 8), Aug 1986, p 35–38
33. O. Jonas, Design Against Localized Corrosion, *Second International Symposium on Environmental Degradation of Materials in Nuclear Power Systems—Water Reactors, Proceedings* (Monterey, CA), 9–12 Sept 1985, American Nuclear Society, Metallurgical Society of AIME, and NACE
34. W.G. Clark et al., "Procedures for Estimating the Probability of Steam Turbine Disc Rupture from Stress Corrosion Cracking," Paper 81-JPGC-PWR-31, ASME/IEEE Power Generation Conference (New York), American Society of Mechanical Engineers/Institute of Electrical and Electronics Engineers, Oct 1981
35. B.W. Bussert, R.M. Curran, and G.C. Gould, Paper 78-JPGC-Pwr-9, *Joint Power Generation Conference Proceedings*, American Society of Mechanical Engineers, 1978
36. O. Jonas, Tapered Tensile Specimen for Stress Corrosion Threshold Stress Testing, *J. Test. Eval.*, Vol 6 (No. 1), 1978, p 40
37. O. Jonas, Steam Generation, *Corrosion Tests and Standards—Application and Interpretation*, ASTM, 1995
38. F. Heymann, Liquid Impingement Erosion, *Friction, Lubrication, and Wear Technology*, Vol 18, *ASM Handbook*, ASM International, 1992
39. A.F. Aschoff, Y.H. Lee, D.M. Sopocy, and O. Jonas, *Interim Consensus Guidelines on Fossil Plant Cycle Chemistry*, CS-4629 and several other EPRI guidelines, Electric Power Research Institute, June 1986
40. *Boiler Water Limits and Achievable Steam Purity for Water-Tubed Boilers*, American Boiler Manufacturers Association, 1995
41. *Guideline for Boiler Feedwater, Boiler Water and Steam of Steam Generators with a Permissible Operating Pressure > 68 Bar*, VGB Technischen Vereinigung, 1988
42. "Recommendations for Treatment of Water for Steam Boilers and Water Heaters," BS2486, British Standards Institute, 1997
43. "Water Quality of the Feedwater and the Boiler Water for Recirculating Boilers," B8223, Japanese Institute of Standards
44. "Feedwater Quality for Once-Through Boilers," B8224, Japanese Institute of Standards
45. *Consensus for the Lay-Up of Boilers, Turbines, Turbine Condensers, and Auxiliary Equipment*, Research Report, CRTD-Vol 66, American Society of Mechanical Engineers, 2002
46. O. Jonas and B. Dooley, Impurity Concentration Processes in Steam Turbines, *Third International VGB/EPRI Conference on Steam Chemistry, Proceedings* (Freiburg, Germany), VGB/Electric Power Research Institute, June 22–25, 1999
47. O.I. Martynova, *Transport and Concentration Processes of Steam and Water Impurities in Steam Generating Systems*, Moscow Power Institute, 1962, p 547–562
48. O. Jonas and B. Dooley, Steam Chemistry and Its Effects on Turbine Deposits and Corrosion, *57th International Water Conference Proceedings* (Pittsburgh, PA), Engineers' Society of Western Pennsylvania, Oct 1996
49. O. Jonas, B. Dooley, and N. Rieger, Steam Chemistry and Turbine Corrosion—State-of-Knowledge, *54th Int. Water Conference Proceedings* (Pittsburgh, PA), Engineers' Society of Western Pennsylvania, Oct 1993
50. O. Jonas and B.C. Syrett, Chemical Transport and Turbine Corrosion in Phosphate Treated Drum Boiler Units, *Int. Water Conf. Proceedings* (Pittsburgh, PA), Engineers' Society of Western Pennsylvania, Nov 1987
51. A. Whitehead and G.F. Wolfe, Steam Purity for Industrial Turbines, *Eighth ASME Industrial Power Conference Proceedings* (Houston, TX), American Society of Mechanical Engineers, Oct 1980
52. O. Jonas and A. Pebler, *Characterization of Operational Environment for Steam Turbine-Blading Alloys*, CS-2931, Electric Power Research Institute, 1984
53. O. Jonas, M. Roidt, and A.S. Manocha, Dynamic Deposition and Solubility of NaCl in Superheated Steam, *44th International Water Conference Proceedings* (Pittsburgh,

PA), Engineers' Society of Western Pennsylvania, Oct 1983
54. O. Jonas, Beware of Organic Impurities in Steam Power Systems, *Power,* Vol 126 (No. 9), Sept 1982
55. W.T. Lindsay, Jr., *Power Eng.,* May 1979, p 68
56. O. Jonas, Turbine Steam Purity, *Combustion,* Vol 50 (No. 6), Dec 1978
57. M.A. Styrikovich, O.I. Martynova, and Z.S. Belova, *Therm. Eng. (USSR),* Vol 12 (No. 9), 1965, p 115
58. F.F. Straus, "Steam Turbine Blade Deposits," Bulletin 59, University of Illinois, June 1946
59. D.J. Turner, "SCC of LP Turbines: The Generation of Potentially Hazardous Environments from Molybdenum Compounds," RD/L/N 204/74, Central Electricity Research Laboratory, 1974
60. J.F. Newman, "The SCC of Turbine Disc Steels in Dilute Molybdate Solutions and Stagnant Water," RD/L/N 215/74, Central Electricity Research Laboratory, 1974
61. *Turbine Steam, Chemistry and Corrosion-Generation of Early Liquid Films in Turbines,* TR-113090, Electric Power Research Institute, May 1999
62. O. Jonas, Condensation in Steam Turbines—New Theory and Data, *International Joint Power Generation Conference Proceedings* (Baltimore, MD), Aug 23–26, 1998, American Society of Mechanical Engineers
63. *Moisture Nucleation in Steam Turbines,* TR-108942, Electric Power Research Institute, Oct 1997
64. O. Jonas, Effects of Steam Chemistry on Moisture Nucleation, *Moisture Nucleation in Steam Turbines,* TR-108942, Electric Power Research Institute, Oct 1997
65. M. Stastny, O. Jonas, and M. Sejna, Numerical Analysis of the Flow with Condensation in a Turbine Cascade, *Moisture Nucleation in Steam Turbines,* TR-108942, Electric Power Research Institute, Oct 1997
66. O. Jonas, On-Line Diagnosis of Turbine Deposits and First Condensate, *55th Annual Int. Water Conf. Proceedings* (Pittsburgh, PA), Engineers' Society of Western Pennsylvania, 1994
67. O. Jonas and J. Mancini, Steam Turbine Problems and Their Field Monitoring, *Mater. Perform.,* March 2001
68. "Generating Availability Report, Statistical Data 1982–2000," pc-GAR, North American Electric Reliability Council, 2002

Fireside Corrosion in Coal- and Oil-Fired Boilers

Steven C. Kung, The Babcock & Wilcox Company

THE PRESENCE OF CERTAIN IMPURITIES in coal and oil is responsible for the majority of fireside corrosion experienced in utility boilers. In coal, the primary impurities are sulfur, alkali metals, and chlorine. The most detrimental impurities in fuel oil are vanadium, sodium, sulfur, and chlorine. During combustion, various species containing these impurity elements are produced in the form of vapors and condensed phases. For example, the combustion products from sulfur are predominately SO_2 or H_2S in the gas phase, depending on the stoichiometry of the air/fuel mixture. Vapor species such as alkali sulfates, alkali vanadates, and vanadium sulfates are also formed and subsequently condense on the boiler tubes where the temperatures are lower. The lower metal temperatures inevitably allow continuing accumulation of the deposit, thus creating local environments on the tubes that are significantly different from the bulk of the combustion products.

In general, fireside corrosion in boilers can be loosely divided into two categories by location (i.e., waterwall corrosion in the lower furnace and fuel ash corrosion of superheaters and reheaters in the upper furnace). Different corrosion mechanisms operate on these tube surfaces, as dictated by the local chemistry of combustion gases and deposits, the boiler tube compositions, and the gas and metal temperatures.

Waterwall Corrosion

The lower furnace of a coal- or oil-fired boiler is essentially a large enclosed volume where the combustion of fossil fuel, as well as the cooling of the combustion products, takes place. The furnace enclosures are made of water-cooled tubes, generally in a welded membrane construction. In some cases, these enclosures comprise tangential tube construction or closely spaced tubes with an exterior gas-tight seal. By design, the tube wall and membrane surfaces are exposed to the combustion gases on the fireside, while exterior insulation and lagging prevent heat loss and provide personnel safety.

The majority of construction materials used for furnace walls is carbon steel due primarily to its lower cost. The use of low- and medium-alloy steels, characterized by the addition of a small amount of chromium, molybdenum, and other elements for improved mechanical properties, is also common. For example, ASME SA213-T2 (0.5Cr-0.5Mo, UNS K11547), SA213-T11 (1Cr-1Mo, UNS K11597), and SA213-T22 (2.25Cr-1Mo, UNS K21590) are alloy steels widely used in boilers based on their exceptional high-temperature strength and resistance to graphitization.

Conventional Burners. In boilers equipped with conventional burners, severe fireside corrosion on furnace walls is typically found in small areas, mostly adjacent to the burners where high heat flux is encountered. On these tube wall surfaces, a thick scale is produced that often displays circumferential grooving resembling an alligator hide, as illustrated in Fig. 1. The appearance of these grooves is a result of pointed cracks filled with a mixture of iron oxide and iron sulfide corrosion products. Severe corrosion found in the conventional boilers is attributed to localized reducing conditions caused by improper mixing of air and fuel, along with a high local heat flux leading to elevated thermal stress.

Low-NO_x Burners. To significantly reduce nitrogen oxide (NO_x) formation in utility boilers, a number of low-NO_x burner technologies have been developed in the last decade. Although different in design, all low-NO_x burner technologies implement the strategy of staged combustion that creates substoichiometric combustion zones in the lower furnace. As a result, the fuel is only partially oxidized in the primary combustion zone to suppress NO_x formation. The remaining fuel is combusted with additional air supplied through secondary air ports. Under the substoichiometric combustion conditions, sulfur from the fuel exists primarily as hydrogen sulfide (H_2S) in the flue gas and metal sulfides in the ash (Ref 1). When in direct contact with the furnace walls, such reducing and sulfidizing combustion products cause rapid attack of the boiler tubes through sulfidation.

Consequently, many utility boilers retrofitted with low-NO_x burners have experienced accelerated metal wastage on the lower furnace walls. The wastage is most severe in supercritical boilers burning high-sulfur fuel due to higher metal temperatures and H_2S concentrations. However, subcritical units burning low- and medium-sulfur fuel are not totally immune to the problem. The wastage rates of furnace walls are typically up to 0.5 mm/yr (20 mils/yr) in subcritical boilers and 1 to 2 mm/yr (40 to 80 mils/yr) in supercritical units (Ref 2).

Extensive laboratory studies have investigated the effect of low-NO_x corrosion on furnace walls (Ref 1, 3, and 4). Various alloys and coatings

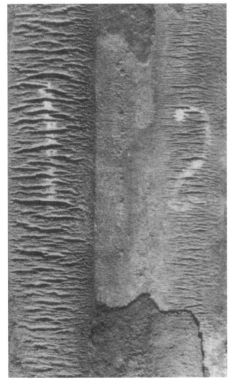

Fig. 1 Surface of a waterwall showing circumferential grooving resembling alligator hide

were exposed to simulated low-NO_x combustion environments in a high-temperature retort. The laboratory environments consisted of 5 to 12% CO, 500 to 5000 ppm H_2S, approximately 5% H_2O, and the balance CO_2, N_2, and H_2. The test temperatures ranged from 260 to 482 °C (500 to 900 °F), typical metal temperatures for waterwalls. From the corrosion data generated from these studies, a corrosion model was developed that can be used to estimate the corrosion rates of waterwall alloys (carbon and low-alloy steels) as a function of metal temperature, H_2S concentration in the flue gas, and chromium content in the alloy (Ref 4). Figure 2 shows an example of the corrosion rates predicted for SA213-T2 under different H_2S concentrations as a function of metal temperature. Accuracy of the corrosion model was validated by comparing the predicted value to the actual corrosion rate of SA213-T2 on a test panel installed in a coal-fired utility boiler (Ref 4, 5). During the test, the metal temperature and gas composition adjacent to the test panel were carefully monitored. For a long-term exposure, good agreement was obtained between the predicted and actual corrosion rates. This corrosion model is applicable for conditions where the corrosion mechanism is dominated by sulfidation due to H_2S vapor in the gas phase.

Additional studies identified the root cause for the extremely high wastage found on the furnace walls of some coal-fired boilers retrofitted with low-NO_x burners. Severe sulfidation attack can occur on these boiler tubes when covered with an FeS-rich ash deposit under oxidizing or alternating oxidizing/reducing combustion gases (Ref 6, 7). The primary source for FeS is pyrite (FeS_2) present in coal as a natural mineral. In boilers equipped with conventional burners, the pyrite is fully oxidized to iron oxide prior to depositing on the boiler tubes. However, during staged combustion for low-NO_x burners, the pyrite is only partially oxidized to FeS before condensing on certain regions of the furnace walls along with unburned carbon. Laboratory studies demonstrated that subsequent exposure of the FeS-rich ash to oxidizing gases can produce elemental sulfur in the deposit at a concentration up to several percent along with H_2S, COS, and CH_3SH. Depending on the partial pressure of oxygen, the elemental sulfur is formed via:

$$3FeS(s) + 2O_2(g) \rightarrow Fe_3O_4(s) + 3S(l, g)$$
(Eq 1)

$$2FeS(s) + 3/2O_2(g) \rightarrow Fe_2O_3(s) + 2S(l, g)$$
(Eq 2)

The physical state of the elemental sulfur produced is determined by the local temperature. Above the boiling point of 444 °C (832 °F), sulfur vaporizes, and at lower temperature liquid sulfur is stable. The formation of elemental sulfur adjacent to the metal surfaces makes the local environment far more corrosive than expected for gas metal attack under reducing conditions where the corrosion is dominated by H_2S.

Following the discovery of this sulfidation mechanism, another corrosion model was developed using the corrosion data generated from laboratory studies (Ref 6). This model focuses on prediction of severe corrosion rates of alloys covered with FeS-rich ash as a function of O_2 and CO concentrations and metal temperature. Figure 3 gives an example of the average and maximum corrosion rates estimated for SA213-T2 as a function of temperature. This corrosion model was also validated by a field study in which a test panel was installed in a coal-fired utility boiler experiencing severe metal wastage on the furnace walls after low-NO_x burner retrofit. For the long-term field exposure, good agreement was observed between the predicted and actual wastage rates (Ref 8).

Chlorine in coal can also play an important role in waterwall corrosion under reducing combustion conditions. The chlorine concentration of coal produced in the United States typically varies from 0.01 to 0.5%, with the majority falling within 0.05 to 0.2%. A concentration of 0.2% or higher is considered high-chlorine coal and 0.05% or lower is low-chlorine coal. During combustion, chlorine is primarily converted to HCl in the gas phase and alkali chloride in the ash. Laboratory and field data indicate that the corrosion rates of carbon and low-alloy steels are not significantly affected by the presence of a small amount of HCl or alkali chloride from burning low-chlorine coal. However, when the HCl and chloride concentrations are sufficiently high from burning coal containing $\geq 0.2\%$ Cl, sulfidation attack can be accelerated under substoichiometric combustion conditions (Ref 9, 10). It has been proposed that the presence of chloride in the ash in direct contact with the scale can alter the growth mechanism of FeS formed on the boiler tubes, thus allowing rapid outward diffusion of iron through the scale (Ref 11, 12). In extreme cases, iron chloride is formed at the metal/scale interface, which then leads to severe metal loss via vaporization of $FeCl_x$, a mechanism similar to that found in municipal solid waste boilers. A direct correlation of corrosion rate with the coal chlorine content has been reported by researchers in the United Kingdom, where the chlorine concentration available in coal has increased steadily from 0.35 to 0.65% over the last two decades (Ref 10, 13).

Mechanisms for the formation of iron chloride at the scale/metal interface have been proposed (Ref 13, 14). Chlorine can react with iron to form iron chlorides underneath the scale during boiler operation (Ref 13, 15). The condensation of moisture can also occur during boiler downtime when the ambient temperature falls below the dew point (Ref 12). As a result, water dissolves chlorides in the ash and enters the scale through its pores and cracks. Such a mechanism implies that downtime corrosion can contribute to the overall corrosion wastage observed on the boiler tubes.

Fuel Ash Corrosion

In the upper furnace, fireside corrosion is often found on the superheaters and reheaters, in particular, in the steam outlet sections where the metal temperatures are the highest. The combustion gas is typically oxidizing, containing about 3% excess oxygen. As a result, sulfur is present in the flue gas primarily as SO_2 and SO_3. The tubes suffering severe wastage are generally covered with a thick layer of ash deposit that becomes partially molten at the metal temperatures. The formation of a molten phase at the tube surfaces allows a "hot corrosion" attack, which causes dissolution and fluxing of the otherwise protective oxide scales formed on the alloys (Ref 16).

Coal ash corrosion was extensively studied in the 1980s and 1990s. In coal-fired boilers, accelerated corrosion is related to the formation of a fused complex sulfate, alkali iron trisulfate $(Na,K)_3Fe(SO_4)_3$, which exists at a metal temperature of 593 to 732 °C (1100 to 1350 °F). The melting point of synthesized alkali iron trisulfate varies with the Na-to-K ratio, with the lowest temperature of 552 °C (1025 °F) occurring at the ratio of 2/3 (Ref 17). Because the corrosion mechanism requires the formation of a molten phase, variation of the Na-to-K ratio in the coal inevitably affects the corrosivity of the ash. The melting point can be further reduced

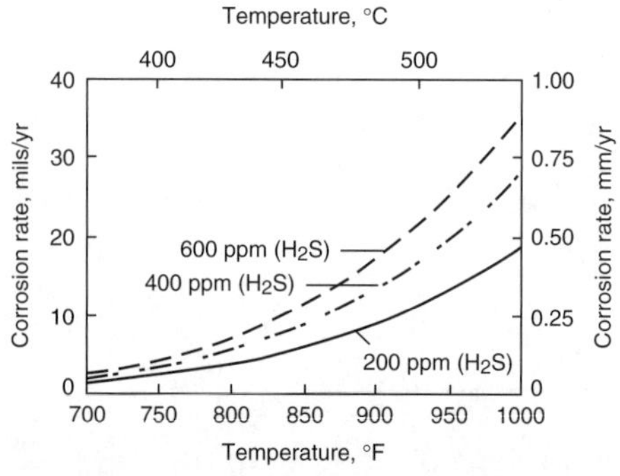

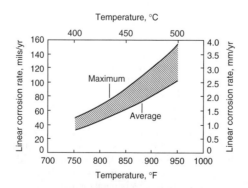

Fig. 2 Predicted corrosion rates of SA213-T2 (UNS K11547) alloy steel under different H_2S concentrations as a function of metal temperature

Fig. 3 Maximum and average corrosion rates predicted for SA213-T2 (UNS K11547) alloy steel covered with FeS-rich ash in combustion gas containing 1% O_2 as a function of metal temperature

when additional compounds dissolve in the trisulfate formed in the ash. The complex sulfate has been shown to exhibit an acid-base behavior that can be defined and measured quantitatively by electrochemical probes (Ref 18, 19). This fused salt acts as an effective solvent for protective scales, aggravating the hot corrosion attack.

Coal ash corrosion is found on superheaters and reheaters covered with a deposit that is strongly bonded to the tube surfaces. The deposit consists of at least three distinct layers (Ref 20). The outer layer constitutes the bulk of the deposit and has an elemental composition similar to that of fly ash. This hard and brittle layer is typically porous, allowing the gas species to diffuse through easily. Underneath the outer layer, a white to yellow-colored layer is present. This intermediate layer, often referred to as the "white layer," can melt as low as 538 °C (1000 °F) upon heating in air. This layer is rich in sulfur and alkali metals due to the presence of a complex sulfate that is the essential ingredient for fuel ash corrosion. The inner layer usually is thin and consists of corrosion products produced from oxidation and sulfidation of the base metal.

The greatest metal loss on superheaters and reheaters occurs on the upstream side of the boiler tubes at the 2 and 10 o'clock positions, as illustrated in Fig. 4 (Ref 21). At these locations, the outer ash layer is relatively thin because of gas stream erosion and the gas temperature is high. Consequently, a high heat flux is created locally, which leads to a high metal temperature and thus the formation of a molten intermediate ash layer. The corrosion wastage is less severe at the 12 o'clock position because sufficient thermal insulation is provided by the thicker deposit layer. The metal wastage tapers off to little or none on the back side of the tubes away from the hot gas stream.

The corrosion rate is not a linear function of metal temperature. Instead, a bell-shape curve is often used to describe the behavior of fuel ash corrosion. The relative weight losses of ASME SA213-T22 (2.5% Cr) and 304 stainless steel (18Cr-8Ni, UNS S30400) in a simulated coal-fired combustion environment are shown in Fig. 5. The wastage rates exhibit a sharp increase above the metal temperature of 538 °C (1000 °F), reach the maximum at 677 to 732 °C (1250 to 1350 °F), and decrease rapidly at higher temperatures. Such a corrosion curve derives from the melting temperatures of the complex sulfates as well as the thermodynamic stability of these compounds. In a temperature range from as low as 538 °C (1000 °F) to the maximum, the corrosion mechanism is dominated by the formation of a molten phase in the intermediate ash layer. With the active fuel ash corrosion taking place, the rate increases greatly with the metal temperature. Beyond the maximum, the complex sulfates are no longer stable. As a result, active fuel ash corrosion ceases and the metal loss is significantly reduced.

In oil-fired boilers, vanadium plays the most important role in fuel ash corrosion. During combustion, vapors of vanadium pentoxide (V_2O_5) and alkali sulfates are formed. These vapors subsequently condense onto the cooler tube surfaces along with other ash constituents. Vanadium pentoxide reacts with alkali sulfates in the deposit to form low-melting temperature alkali vanadate compounds and/or their solutions (Ref 22). Once formed, the molten phase can attack the superheaters and reheaters via the hot corrosion mechanism (Ref 16). Although the concentration of vanadium in fuel oil is relatively low, typically less than 300 ppm, it tends to accumulate on the tube surfaces and can become 80% of the deposit constituents.

Because alkali sulfates play an important role in lowering the liquidus temperature of a deposit, the severity of oil ash corrosion also depends on the sodium and sulfur concentrations in the oil. In general, the corrosion attack is enhanced by increasing the sodium and sulfur contents. Furthermore, chlorine is generally present in fuel oil at a concentration up to 100 ppm. As a result, chloride is commonly found in the oil ash deposit as NaCl. Laboratory studies indicate that the addition of sufficient NaCl to V_2O_5 can increase the corrosion rates of boiler tubes significantly (Ref 23). Surprisingly, adding a low level of HCl to the flue gas has no effect on oil ash corrosion (Ref 24).

Prevention of Fireside Corrosion

Means for addressing fireside corrosion range from fuel additives to changes in boiler operations to changes in tube materials. In the lower furnace where severe corrosion is caused by the presence of reducing/sulfidizing conditions, the problem can sometimes be minimized by adjusting the air and fuel distribution to the burners. Other remedies include air blanketing, burner modifications, and refined specifications on coal fineness. To reduce the amount of impurities introduced to the boilers, fuel switching or blending to lower the overall sulfur and alkali contents is sometimes recommended. Furthermore, a significant amount of sulfur and alkali metals in coal can be removed by coal washing. However, washing does not remove chlorine and can therefore inadvertently amplify the problems associated with chlorine. Of course, modification of any of the boiler operations can lead to a higher cost in plant operation. Therefore, any benefit gained from the increased tube life must be carefully weighed against the increased cost.

From the material selection perspective, the sulfidation resistance of waterwall alloys has been shown to increase with their chromium content (Ref 3, 4). Tube alloys containing 9 to 12% Cr exhibit satisfactory resistance when exposed to low-NO_x combustion gases. Furthermore, a higher chromium content is needed when the corrosion mechanism involves an FeS-rich deposit along with an oxidizing or alternating reducing/oxidizing environment. Because it is not practical to construct the waterwalls with high chromium content alloy steel or stainless steel tubing, application of a protective clad or coating may be necessary to combat fireside corrosion in the lower furnace.

Various coatings are available for waterwall protection; these can be applied either in the shop or directly in the field. The primary techniques used for field-applied coatings are metal spray and weld overlay. The most common metal spray processes consist of wire arc spray and high-velocity oxyfuel (HVOF). Both spray processes typically achieve a coating layer of high-chromium material 0.4 to 0.75 mm (15 to 30 mils) thick. The arc spray process allows low energy input and can achieve a high spray rate. Consequently, this process is relatively inexpensive and fast. The HVOF process generally produces a denser coating with less oxide inclusion and porosity. Therefore, the HVOF-applied coatings are less permeable to the corrosive gases and can provide better protection to the furnace walls. However, HVOF is more expensive than arc spray. Common alloy powders selected for the sprayed coatings on waterwalls include 309, 310 stainless steel, and 50Ni-50Cr.

Weld overlay produces a coating of high-chromium alloy composition on the tube surfaces at a typical thickness of 1.5 to 2.0 mm (60 to 80 mils). The common materials of selection are alloys 622 (UNS N06022), 625 (UNS N06625), 309 (UNS S30900), and 312 (UNS S31380). These coatings are relatively dense,

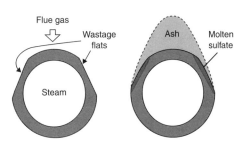

Fig. 4 Fuel ash corrosion on superheater and reheater tubes showing the maximum metal loss at the 2 and 10 o'clock positions

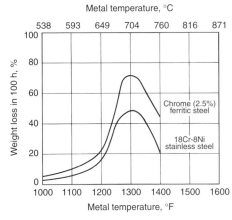

Fig. 5 Corrosion rates of chrome (2.5%) ferritic steel, SA213-T22 (UNS K21590), and 18Cr-8Ni stainless steel, type 304 (UNS S30400), exhibiting a bell-shape curve as a function of temperature

impermeable to the combustion gases, and well bonded to the metal substrate. Therefore, the long-term performance of weld overlay is likely to be better than that of metal spray. However, the major drawback of weld overlay is its high cost and sensitivity to coefficient of thermal expansion (CTE) mismatch between the weld metal and substrate. A high CTE mismatch often leads to distortion and cracking of the waterwalls.

While the coatings mentioned previously can also be applied in the field or shop, alternative processes lack such flexibility. In particular, coatings applied by laser cladding and diffusion processes must be done in the shop on replacement panels. Following the coating applications, these panels are then installed on the furnace walls in areas where severe corrosion is experienced. Similar to weld overlay, laser cladding can produce a dense and impermeable coating of high-chromium alloy composition. However, because of its low heat input, the laser-clad coating suffers little from chromium dilution caused by interdiffusion with iron in the substrate. The most common alloy compositions applied by laser cladding are alloys 622, 625, and 309. Due to CTE mismatch, distortion of the laser-clad panels can also occur. Therefore, panel straightening is usually required after such coating applications.

Diffusion coatings involve alloying of protective elements, such as chromium, aluminum, and silicon, into the substrate surface. The coating is commonly applied by the pack cementation technique or its derivatives, which involve exposure of the substrate to a pack mix heated to high temperatures under an inert or reducing environment. The pack mix consists of powders of active metal(s), a halide salt activator, and an inert oxide powder as the matrix. For boiler applications, chromizing, which produces a higher chromium concentration at the tube surfaces, is the most widely accepted process. Co-diffusion coatings that introduce two or more protective elements simultaneously, such as chromizing/aluminizing and chromizing/siliconizing, have also been developed (Ref 25). See the article "Pack Cementation Coatings" in *ASM Handbook*, Volume 13A, 2003.

In the upper furnace of coal-fired utility boilers, where molten ash is the dominant cause of corrosion, different methods have been implemented to protect the superheaters and reheaters. These include the use of heat shields to guard the most vulnerable tubes and the use of soot blowers to minimize the deposit buildup. Fuel switching and blending can also be effective when cost benefit prevails. Other approaches consist of changes to furnace geometry, burner configuration, tube bundle arrangement, and the use of gas tempering. All of these approaches aim at reducing the gas and metal temperatures as well as minimizing the temperature imbalance. In terms of alloy selection, the allowable stress and cost, as well as oxidation and corrosion resistance, determine the materials prescribed for superheaters and reheaters. Whenever possible, the use of carbon steel is always maximized due to its lower cost. However, careful selection of alloy steels and stainless steels is necessary for tube elements where the gas and metal temperatures are the highest.

In units where coal ash corrosion occurs, economical heat-resistant steels, such as 304, are often used in the outlet sections of superheaters and reheaters. For extreme cases found in boilers burning coal containing high sulfur and alkali contents, selection of a higher corrosion-resistant alloy (alloy 800H clad with alloy 671, for example) becomes necessary. The use of sprayed metal coatings and ceramic coatings to combat coal ash corrosion is generally ineffective due to coating spallation, erosion, and rapid deterioration.

The steam outlet temperature of a supercritical boiler is typically designed to be less than 566 °C (1050 °F). However, to further increase plant efficiency and reduce emissions, ultra-supercritical coal-fired utility boilers are being developed. The steam outlet temperature and pressure in the proposed ultrasupercritical units can be as high as 750 °C (1380 °F) and 100 MPa (14.5 ksi), respectively. Aside from having adequate creep and oxidation resistance, the superheater and reheater materials must withstand the coal ash corrosion that occurs at a rate near the peak of the bell-shape curve (Fig. 5). Under these conditions, few commercially available alloys can meet the stringent requirements. See the article "Corrosion in Supercritical Water—Ultrasupercritical Environments for Power Production" in this Volume for more details on these units.

To select candidate superheater and reheater materials, test loops consisting of various high-performance alloys and coatings were installed in a coal-fired utility boiler burning high-sulfur coal as part of a long-term field study (Ref 26). The test loops were designed to simulate the steam conditions expected in an ultra-supercritical unit. The test results reveal a strong correlation between the corrosion resistance of alloy and its chromium concentration (Fig. 6). Only a chromium content of 40% or greater in the alloy can provide adequate resistance to such coal ash corrosion. An even better correlation is found between the corrosion resistance and the sum of chromium and nickel contents among the alloys investigated (Fig. 7). Such a correlation suggests that the use of high-chromium nickel-base alloys, such as 671 (50Ni-50Cr) and 740 (Ni-25Cr-20Co-Nb-Ti-Al), may be necessary for the superheaters and reheaters of ultra-supercritical boilers (Ref 27).

At present, no alloys are immune to fuel ash corrosion in oil-fired boilers. In general, the higher the chromium content, the more corrosion resistant the alloy is. Therefore, high-chromium alloys and coatings are commonly used in oil-fired boilers to combat the corrosion attack. Ferritic steels are more corrosion resistant to oil ash corrosion than nickel-containing austenitic alloys (Ref 28). Among them, SA213-T9 and SA268-TP410 are most widely used. However, on components such as uncooled tube supports where the metal temperatures are high, the use of a higher-chromium alloy or coating, such as 50Ni-50Cr, is necessary.

Other remedies for oil-fired boilers include reducing the tube metal temperature by the use of

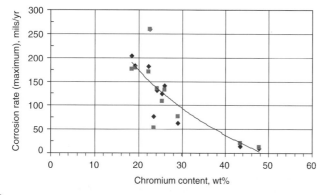

Fig. 6 Maximum coal ash corrosion rates of alloys in ultrasupercritical boilers as a function of chromium concentration. Squares and diamonds are data from different test loops.

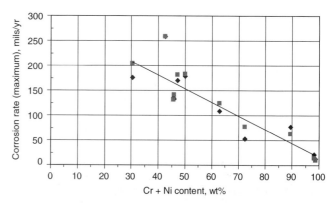

Fig. 7 Maximum coal ash corrosion rates of alloys in ultrasupercritical boilers as a function of Cr + Ni concentration. Squares and diamonds are data from different test loops.

a parallel steam flow or decreasing the steam outlet temperature. Operating oil-fired boilers with low excess air can also decrease the metal loss by altering the dominant vanadium compounds formed. At a low partial pressure of oxygen, vanadium exists predominantly as noncorrosive high-melting V_2O_4 instead of corrosive low-melting V_2O_5. Inhibitors and additives can be used to reduce oil ash corrosion with varying success. MgO is the most effective inhibitor that reacts with vanadium to form a compound having a melting point above the tube metal temperatures (Ref 29). Without the presence of a molten phase, the oil ash corrosion is significantly impeded.

REFERENCES

1. P.L. Daniel, S.F. Chou, L.W. Rogers, and P.S. Ayres, Jr., "Fireside Corrosion in Low NO_x Combustion Systems," Report No. GS-6339, Electric Power Research Institute, April 1989
2. W. Bakker, "Waterwall Wastage in Low NO_x Boilers—Root Cause and Remedies," Topical Report TR-111155, Electric Power Research Institute, Oct 1998
3. S.C. Kung and C.F. Eckhart, "Corrosion of Iron-Base Alloys in Reducing Combustion Gases," Paper No. 242, CORROSION/93 (New Orleans, LA), NACE International, March 1993
4. S.C. Kung, Prediction of Corrosion Rate for Alloys Exposed to Reducing/Sulfidizing Combustion Gases, *Mater. Perform.,* Vol 36 (No. 12), Dec 1997, p 36–40
5. C.F. Eckhart, R.F. DeVault, and S.C. Kung, "Full-Scale Demonstration of Low-NO_x Cell Burner Retrofit Long-Term Testing," Program Task Report DOE/PC/90545-1, Department of Energy, Jan 1994
6. S.C. Kung, "Effect of Iron Sulfide on Furnace Wall Corrosion," Report TR-111152, Electric Power Research Institute, Nov 1998
7. W.T. Bakker and S.C. Kung, "Waterwall Corrosion in Coal-Fired Boilers: a New Culprit-FeS," Paper No. 246, CORROSION/2000 (Orlando, FL), NACE International, March 2000
8. W.T. Bakker, J.L. Blough, S.C. Kung, T.L. Banfield, and P. Cunningham, "Long-Term Testing of Protective Coatings and Weld Overlays in a Supercritical Boiler Retrofitted with Low-NO_x Burners," Paper No. 02384, CORROSION/2002 (Denver, CO), NACE International, April 2002
9. P.L. Daniel, L.D. Paul, J.M. Tanzosh, and R. Hubinger, Paper No. 138, CORROSION/88 (Houston, TX), NACE International, 1988
10. D.J. Leeds, "A Summary of Observations Relating Furnace Wall Fireside Corrosion to Chlorine Content of Coal," SSD/MID/M26/79, 1979
11. S.C. Kung, "Corrosion Studies for Low-NO_x Burner Technology," Report TR-108750, Electric Power Research Institute, Sept 1997
12. S.C. Kung and W.T. Bakker, "Effect of Sulfur and Chlorine on Furnace Wall Corrosion," Paper No. 180, CORROSION/98, NACE International, March 1998
13. P.J. James and L.W. Pinder, "The Impact of Fuel Chlorine on the Fireside Corrosion Behavior of Boiler Tubing: A UK Perspective," CORROSION/97, NACE International, March 1997
14. L. Darken and R. Gurry, *Physical Chemistry of Metals,* International Student Edition, McGraw-Hill, Tokyo, 1953
15. M.E. Whitehead, "Consideration of a Mechanism of Furnace Wall Corrosion," SSD/MID/M7/79, 1979
16. R.A. Rapp and Y.S. Zhang, Hot Corrosion of Materials: Fundamental Studies, *J. Met.,* Dec 1994, p 47
17. C. Cain, Jr. and W. Nelson, *J. Eng. Power, Trans. ASME,* Oct 1961, p 468
18. P.P. Lebianc and R.A. Rapp, *J. Electrochem. Soc.,* Vol 139 (No. 3), March 1992, p L31
19. P.P. Lebianc and R.A. Rapp, *J. Electrochem. Soc.,* Vol 140 (No. 3), March 1993, p L41
20. S.C. Stultz and J.B. Kitto, Ed., *Steam/Its Generation and Use,* 40th ed., The Babcock & Wilcox Co., 1992
21. A.J.B. Cutler, T. Flatley, and K.A. Hay, *Metall. & Mater. Technol.,* Feb 1981, p 69
22. S.C. Stultz and J.B. Kitto, Ed., *Steam/Its Generation and Use,* 40th ed., The Babcock & Wilcox Co., 1992, p 20–23
23. T. Kawamura and Y. Harada, "Control of Gas Side Corrosion in Oil Fired Boilers," Technical Bulletin No. 139, Mitsubishi Heavy Industries, Ltd. May 1980
24. W.D. Halstead, Progress Review No. 60: Some Chemical Aspects of Fireside Corrosion in Oil Fired Boilers, *J. Inst. Fuel,* July 1970
25. R. Bianco, M.A. Harper, and R.A. Rapp, Codeposition of Elements in Diffusion Coatings by the Halide-Activated Pack Cementation Method, *J. Met.,* Nov 1991, p 68–73
26. D.K. McDonald, P.L. Daniel, D.J. DeVault, and E.S. Robitz, Jr., "Coal Ash Corrosion Resistant Materials Testing," Topical Report DE-FC26–99FT40525, U.S. Department of Energy, June 2004
27. G. Smith and L. Shoemaker, Advanced Nickel Alloys for Coal-Fired Boiler Tubing, *Adv. Mater. Process.,* Vol 162 (No. 7), July 2004, p 23
28. P.A. Alexander, R.A. Marsden, J.M. Nelson-Allen, and W.A. Stewart, Operational Trials of Superheater Steels in a C.E.G.B. Oil-Fired Boiler at Bromborough Power Station, *J. Inst. Fuel,* July 1963
29. J.H. Swisher and S. Shankarnarayan, Inhibiting Vanadium-Induced Corrosion, *Mater. Perform.,* Sept 1944, p 49

SELECTED REFERENCES

- N. Birks and G.H. Meier, *Introduction to High Temperature Oxidation of Metals,* Edward Arnold, London, 1983
- P. Kofstad, *High Temperature Corrosion,* Elsevier Applied Science, London and New York, 1988
- G.Y. Lai, *High-Temperature Corrosion of Engineering Alloys,* ASM International, 1990
- R.A. Rapp, Ed., *High Temperature Corrosion, NACE-6,* National Association of Corrosion Engineers, 1983

High-Temperature Corrosion in Waste-to-Energy Boilers

George Y. Lai, Consultant

Corrosion Modes

THE COMBUSTION OF MUNICIPAL SOLID WASTE in a waste-to-energy (WTE) boiler for power generation produces a very corrosive environment for the boiler tube materials. Municipal solid waste typically contains plastic materials, textile, leathers, batteries, food waste, and other miscellaneous materials. These constituents are the source of chlorine, sulfur, sodium, potassium, cadmium, zinc, lead, and other heavy metals that form corrosive vapors of various metal chlorides and sulfates during combustion. These chloride and sulfate vapors, along with fly ash, condense and deposit on the cooler surfaces, such as the waterwalls that surround the combustion zone, and heat-exchanger surfaces in the convection path in the upper furnace. Vulnerable heat-exchanger parts in the convection path are screen tubes, superheater tubes, and generating banks. The waterwall, screen tubes, and generating bank tubes are typically made of carbon steels, while superheater tubes are made of carbon steels or chromium-molybdenum steels. The corrosion problems with these boiler tubes in WTE boilers is discussed in detail in Ref 1 to 4, and a review on WTE boilers and their associated corrosion and materials issues is found in Ref 5.

Corrosion Modes

Typical gases and combustion products found in the flue gas stream consist of O_2, CO_2, H_2O, SO_2, HCl, HF, and N_2. Among these gaseous constituents, SO_2, HCl, and HF are considered to be corrosive at high temperatures. Reference 4 reports typical ranges of concentrations in mass-burning units as 100 to 200 ppm SO_2, 400 to 600 ppm HCl, and 5 to 20 ppm HF; in refuse-derived-fuel units (RDF) 200 to 400 ppm SO_2, 600 to 800 HCl, and 10 to 30 ppm HF. These levels of SO_2, HCl, and HF are not likely to be responsible for high wastage rates observed for nickel-base alloy 625 (Ni-22Cr-9Mo-3.5Nb, UNS N06625) in superheaters in some boilers. As shown in Fig. 1, for example, alloy 625 overlay cladding (approximately 2 mm thick) was completely corroded after 15 months of service in a superheater (Ref 6). Laboratory tests (Ref 7) simulated conditions in N_2 with 10% O_2, 50 ppm SO_2 with 500 ppm HCl in one environment and with 4% HCl in another environment. Tests were conducted at 593 °C (1100 °F) for 1008 h. This temperature is much higher than superheater metal temperatures in WTE boilers. It was found that alloy 625 showed very little corrosion attack in both 500 ppm and 4% HCl environments. The extrapolated corrosion rates were 2.79 μm/yr (0.1 mils/yr) and 15.75 μm/yr (0.6 mils/yr) in 500 ppm HCl and 4% HCl environments, respectively. The level of 4% HCl is significantly higher than the level expected in the combustion of municipal waste. From these data, it can be concluded that gaseous HCl plays no significant role in causing significant wastage rates observed for both waterwalls and superheaters.

The metal chloride vapors or deposits that are not usually measured in the combustion flue gas stream could be responsible for causing high corrosion rates for metallic components. Some of the metal chlorides exhibit high vapor pressures. For example, $ZnCl_2$ exhibits 10 Pa (10^{-4} atm) of vapor pressure when the temperature reaches 349 °C (660 °F). This vapor pressure is generally considered to be high enough to cause high-temperature corrosion. Studies of the corrosion of alloys by $ZnCl_2$ vapor at 540 °C (1000 °F) found alloy 625 suffering an extrapolated corrosion rate of 3.07 mm/yr (121 mils/yr) (Ref 8). Some of the salt vapor might have deposited on the test specimen as well in the test. Thus, the corrosion attack could be due to both $ZnCl_2$ vapor and liquid deposit. Nevertheless, the extrapolated corrosion rate was in the same order of magnitude as the wastage rates observed for alloy 625 cladding in coextruded composite or weld overlay superheater tubes in some aggressive WTE boilers (Ref 6, 9). Many chlorides exhibit low melting points. Figure 2 shows the

Fig. 1 The overlay cladding, approximately 2 mm (80 mils) thick of alloy 625 (UNS N06625) overlay superheater tube, was corroded away in 15 months in a waste-to-energy boiler.

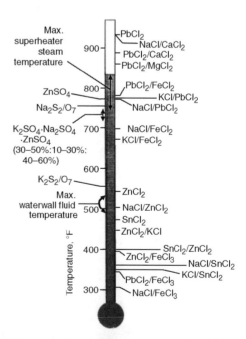

Fig. 2 Melting points of various metal chlorides including eutectics and sulfur-containing compounds that might be present in waste-to-energy combustion environments (Ref 10)

melting points of those metal chlorides including eutectics and sulfur-containing products that might be present in WTE combustion environments (Ref 10).

Analyses using scanning electron microscopy with energy dispersive x-ray spectroscopy (SEM/EDS) performed on the ash deposits and corrosion products formed on the waterwall tube or superheater tube showed the presence of chlorine, sodium, potassium, zinc, lead, and other elements. The corrosion products at the corrosion front were found to contain chlorine, sodium, potassium, zinc, and lead, as well as the elements from the tube metal. Figure 3 shows an example of one analysis on the deposits and corrosion products formed on a steel tube in the waterwall of a WTE boiler (Ref 6). The corrosion products at the corrosion front (location numbers 5, 6, and 7 in Fig. 3) were found to contain iron, chlorine, zinc, cadmium, and sodium. It implies that zinc, cadmium, and sodium along with chlorine in this case were participating in the corrosion of the steel tube. Thus, it is believed that the molten chloride salts involving zinc, lead, sodium, potassium, and other metals are primarily responsible for corrosion attack on the waterwall. In the superheater area, because of higher temperatures, some chlorides, such as $ZnCl_2$, may exhibit high vapor pressures. In this case, corrosion attack may involve chloride attack by both liquid and vapor phases.

Corrosion Protection and Alloy Performance

The discussion on the corrosion protection and alloy performance will focus on two main areas of the boiler: furnace water walls and super heaters.

Furnace Waterwalls. Due to aggressive environments in both the lower furnace and the convection path, the conventional tube materials, such as carbon steels and chromium-molybdenum steels, for waterwalls and superheaters require some sort of corrosion protection. The current prevailing method of protection for the furnace waterwalls is the use of alloy 625 overlay cladding applied by automatic gas metal arc welding (GMAW) process (Ref 6, 11). The overlay cladding is typically performed on-site in the boiler. Figure 4 shows a schematic for the application of the overlay cladding for the waterwall, which consists of membranes and tubes. The performance of the alloy 625 overlay has been found to be satisfactory in both mass-burning and RDF units. Corrosion attack tends to be of pitting type, as shown in Fig. 5. The area that has suffered corrosion attack can be readily repaired by overlay welding. This repair is performed by first grinding out the corroded or pitted area, and then overlay welding.

In general, the tube metal wastage rate for alloy 625 overlay on the waterwalls has been observed to be about 0.13 mm/yr (5 mils/yr) for many boilers (Ref 12). Localized or pitting corrosion attack with higher corrosion rates has also been encountered. These 625 overlays in WTE boilers in general exhibit a dilution of 10% or less. This means about 10% or less Fe is observed in the alloy 625 overlay, which typically originates as < 1.0% Fe in the weld wire. It is believed that excessive dilution (15 to 20% Fe) might significantly enhance the wastage rate. High iron content in the overlay, which results from high dilution, could increase the probability of forming low-melting-point iron-containing chlorides, such as $FeCl_2$ and/or $FeCl_3$ and their eutectics, thus promoting higher wastage rates.

Superheaters. One of the very common corrosion protection methods for superheaters is the use of bimetallic tubes with alloy 625 cladding on the outer diameter of the carbon or chromium-molybdenum steel tube. The 625 cladding can be fabricated by overlay welding (Ref 13) or by coextrusion manufacturing (Ref 9). Figure 6 shows an example where alloy 625 overlay tubes have been performing well in the superheater in a WTE boiler (Ref 6). There was no evidence of corrosion attack after 4 1/2 years of service in a superheater producing 405 °C (761 °F) and 42 bars (609 psi) superheated steam. Table 1 summarizes the comparative performance between bare carbon steel tube and alloy 625 overlay tube in a side-by-side test in another boiler. The superheated steam temperature and pressure for this boiler were 399 °C (750 °F) and 52 bars (750 psi), respectively. Wastage rates were observed to be about 2.8 mm/yr (110 mils/yr) for carbon steel and 0.46 mm/yr (18.3 mils/yr) for alloy 625 overlay (Ref 6). Based on these data, the replacement intervals for carbon steel tubes are expected to be 1.4 years and alloy 625 overlay tubes 5.8 years in this boiler, which is a service life improvement of slightly more than fourfold. For some boilers, the environments have been found to be much more aggressive such that alloy 625 cladding in either coextruded tubes or weld overlay tubes exhibited wastage rates exceeding 1.27 mm/yr (50 mil/yr) (Ref 9, 12). The overlays made from other nickel alloys, such as alloy 622 (Ni-21Cr-14Mo-3W), C276 (Ni-16Cr-16Mo-4W, UNS N10276), 52 (Ni-30Cr-10Fe, UNS N06052) and

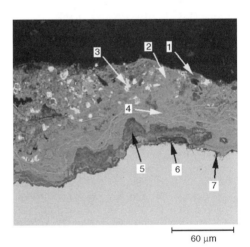

Location	Composition, %					
	Fe	Cl	Zn	S	Na	Other(a)
1	15	...	2	9	...	Ca 31, Si 29, Mg 14
2	63	16	9	2	...	Pb 4
3	20	13	3	11	4	K 3, Ca 2
4	67	12	7	4	6	...
5	72	6	7	4	4	K 2
6	88	2	2	...	1	Cd 2
7	76	8	5	2	2	Cd 2

(a) Trace elements not reported here

Fig. 3 Scanning electron micrograph (backscattered electron image) showing the deposits and corrosion scales formed on a carbon steel superheater tube suffering a severe tube wall wastage. Chemical compositions at different locations were analyzed by energy dispersive x-ray analysis as indicated.

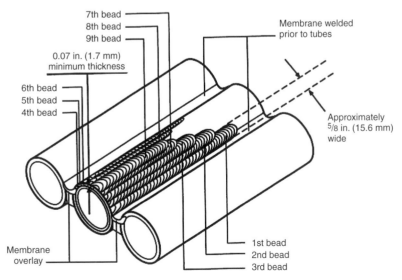

Fig. 4 Weld overlay cladding of the furnace waterwall for corrosion protection of carbon steel boiler tubes and membranes in WTE boilers. Source: Welding Services Inc.

Fig. 5 Alloy 625 overlay on the waterwall revealing typical pitting type corrosion attack

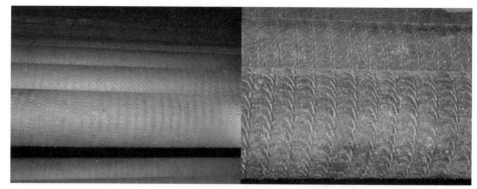

Fig. 6 Overview (left) and close-up view (right) of alloy 625 overlay tubes after 4½ years of service in a superheater in a WTE boiler. Weld bead ripples are still clearly visible.

Table 1 Wastage rates of carbon steel tubes and 625 overlay tubes in a finishing superheater in a waste-to-energy boiler in New York

	Wastage rate	
Tube construction	mm/yr	mils/yr
Carbon steel	2.8	110
625 overlay	0.46	18.3

Source: Ref 6

Fig. 7 Carbon steel superheater tubes protected by metallic tube shields awaiting installation at one waste-to-energy plant

72 (Ni-44Cr, UNS N06072), were tested and found to be not as good as the alloy 625 overlay (Ref 12).

The wastage rates observed for alloy 625 cladding in either coextruded tubes or weld overlay tubes were found to range from less than 0.25 to more than 1.25 mm/y (less than 10 to more than 50 mils/yr) depending on the boiler (Ref 9, 12). There have been significant variations in wastage rates from boiler to boiler. The possible reason could be that the superheater metal temperatures are much higher than those of waterwalls. This may make the superheater much more sensitive to waste constituents, particularly the concentrations of chlorine, zinc, lead, and other heavy metals. Other factors that may also be important in affecting the wastage rates include boiler design, temperatures and velocities (also local velocities) of the flue gas stream entering into the superheater bundle, and steam temperature and pressure.

Another common protection method is the use of tube shields, which are typically made of type 309 (UNS S30900), 310 (UNS S31000), 253MA (UNS S30815) stainless steel, or other austenitic alloys. Figure 7 shows carbon steel superheater tubes protected by metallic tube shields awaiting installation at a WTE plant. The tube shields are generally exposed to the temperatures as high as those of flue gas stream entering into the superheater tube bundle, which can vary from 650 to 900 °C (1200 to 1650 °F). As a result, the tube shields typically suffer chloride attack and sulfidation attack. Furthermore, the shields are typically attached to the tube by mechanical clamps and fillet welds. In addition to high-temperature corrosion, the shields also suffer warping, distortion, and creep damage. Common problems associated with tube shields include:

- Accumulation of ash/salt deposits behind the shields causing accelerated corrosion on the tube
- Impeding gas flow by fallen shields

Tube shields are generally considered to be a "sacrificial" part and are to be replaced regularly during the plant maintenance shutdown in order to avoid the aforementioned problems.

REFERENCES

1. E. Haggblom and J. Mayrhuber, Materials Problems in Energy Utilisation from Waste, *High Temperature Materials for Power Engineering 1990*, Part I, E. Bachelet et al., Ed., Kluwer Academic Publishers, Dordrecht, 1990, p 91
2. H.H. Krause, "Historical Perspective of Fireside Corrosion Problems in Refuse-Fired Boilers," Paper No. 200, Corrosion 93, NACE International, 1993
3. H.H. Krause and I.G. Wright, "Boiler Tube Failures in Municipal Waste-To-Energy Plants: Case Histories," Paper No. 561, Corrosion 95, NACE International, 1995
4. P.Z. Kubin, "Materials Performance and Corrosion Control in Modern Waste-To-Energy Boilers Applications and Experience," Paper No. 90, Corrosion 99, NACE International, 1999
5. G. Sorell, The Role of Chlorine in High Temperature Corrosion in Waste-To-Energy Plants, *Corrosion in Advanced Power Plants*, W.T. Bakker et al., Ed., Proc. Second International Workshop on Corrosion in Advanced Power Plants (Tampa, FL), March 3–5, 1997
6. G.Y. Lai, "Corrosion Mechanisms and Alloy Performance in Waste-To-Energy Boilers Combustion Environments," Presented at North America Waste-To-Energy Conference (NAWTEC 12), (Savannah, GA), May 17–19, 2004
7. G.D. Smith and P. Ganesan, Metallic Corrosion in Waste Incineration: A Look at

Selected Environmental and Alloy Fundamentals, *Heat-Resistant Materials II,* K. Natesan, P. Ganesan, and G. Lai, Ed., ASM International, Materials Park, Ohio, 1995, p 631
8. B. Gleeson, J.E. Barnes, and M.A. Harper, "Corrosion Behavior of Various Commercial Alloys in A Simulated Combustion Environment Containing $ZnCl_2$," Paper No. 196, Corrosion 98, NACE International, 1998
9. A. Wilson et al., Composite Tubes in Waste Incineration Boilers, *Stainless Steel World 1999 Conference,* Book 2, KCI Publishing BV, The Netherlands, 1999, p 669
10. I.G. Wright, V. Nagarajan, and H.H. Krause, "Mechanisms of Fireside Corrosion by Chlorine and Sulfur in Refuse-Firing," Paper No. 201, Corrosion 93, NACE International, 1993
11. P. Hulsizer, Paper No. 246, Corrosion 91, NACE International, 1991
12. Welding Services Inc. unpublished data
13. P. Hulsizer, Dual Pass Weld Overlay Method and Apparatus, U.S. Patent No. 6,013,890, Jan 11, 2000

Corrosion of Industrial Gas Turbines

Henry L. Bernstein and Ronald L. McAlpin, Gas Turbine Materials Associates

INDUSTRIAL GAS TURBINES are used in a variety of industries to generate electricity, produce steam, and provide mechanical shaft power. The utility industry often refers to them as combustion turbines so as to distinguish them from steam turbines (since both use gases to generate energy). They are also used on ships for propulsion and to generate electricity, in which case they are called marine gas turbines.

The corrosion issues for industrial gas turbines are similar for all applications and depend more on the quality of the fuel, air, and water used in the engine than on the specific industrial application. In the compressor section, aqueous corrosion is the principal concern. In the combustor and turbine sections, high-temperature environmental attack in the form of high-temperature oxidation and hot corrosion are the most important.

Corrosion in the Compressor Section

Compressor blades and vanes are usually made from martensitic or precipitation-hardening stainless steels and can either be coated or bare. Compressor disks, and many turbine disks as well, are made from low-alloy steels and, sometimes, martensitic stainless steels. Humidity in the atmosphere can be sufficiently high that water condenses on the forward compressor components during operation. Inlet air water fogging and evaporative cooling are sometimes used to increase power, especially in newer installations. In addition, water vapor can collect in compressor and turbine sections of peaking units that are idle. Some utilities heat or dehumidify their turbines when they are not in use to avoid this problem or cover the inlet and/or outlet to prevent air circulation through the turbine.

Forms of Corrosion. General corrosion and pitting can result from the accumulation of water on compressor blades, stators and disks. If corrosion is extensive, parts may need to be replaced. Pitting on blades and vanes can increase the likelihood of fatigue crack initiation. Corrosion fatigue can dramatically reduce the life of steels used for compressor blades, as shown in Fig. 1.

Fouling can result from carryover of fines from the inlet air filters as well as the buildup of corrosion products. Fouling and corrosion can be exacerbating problems, as the presence of deposits and moisture increases the likelihood of pitting in the blade alloys. Fouling of compressor blades also reduces the efficiency of the compressor, increasing the heat rate and reducing the output of the engine. Corrosion problems are compounded when aggressive environments are present, such as salt air or airborne contaminants.

Corrosion from condensation during outages is often encountered in low-alloy steel disks used in the hot section of the turbines. Exposure to moisture during outage periods can increase the fretting wear of the blade attachments on the turbine disks.

Periodic cleaning of the compressor can restore much, but not all, of the efficiency and power lost because of fouling (Ref 2). On older units with uncooled turbin blades and vanes, abrasive blasting using crushed nutshells or other media is sometimes used. Water washing is the primary method of compressor cleaning in most industrial gas turbines and can be performed either off-line or online. Online washing usually employs just water, and allows operators of turbines used in intermediate- or base-load service to clean without interrupting operation. Care must be exercised with online washing because of the risk of water-droplet erosion to the compressor blades. Original equipment manufacturers (OEMs) may limit the frequency and duration of online water washing cycles. Off-line water washing solutions can include detergents, dispersants, and organic solvents, with detergents being the most common due to environmental considerations. Control of the wash-water quality is very important to prevent corrosion, especially for online washing. (Water containing small amounts of sodium or potassium can cause hot corrosion in the turbine section of the engine.) Thorough rinsing after the off-line washing is required to remove all detergent.

Off-line water washing provides a more thorough cleaning of the compressor. Online washing is only effective on the first few compressor stages because the water evaporates as the air is heated. The effectiveness and benefits of online washing are debated and probably depend on a variety of factors, including the location of the site, air filtration, frequency of washing, and the design of the washing system.

Coating Systems. Nickel-cadmium coatings can be used to protect the blades and vanes from corrosion and erosion and to improve the surface finish and compressor efficiency. This coating can pit in acidic conditions and can degrade by cadmium vaporization in the latter compressor stages. Since the nickel-cadmium coating is cathodic to the base metal, it can increase the rate of corrosion at pits and coating defects. Because cadmium is toxic, nickel-cadmium coatings have been discontinued.

Current (2006) coating systems consist of metallic-ceramic coatings, an example of which is shown in Fig. 2. These coatings are used on the

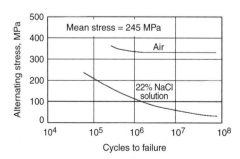

Fig. 1 Reduction of fatigue life of a martensitic stainless steel in a saltwater environment. Source: Adapted from Ref 1

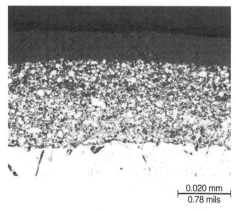

Fig. 2 Compressor coating. The coating is composed of an aluminum-filled inorganic phosphate with a polytetrafluoroethylene (PTFE) layer on top. Beneath the PTFE layer is an ion-reactive primer that inhibits corrosion, which otherwise would cause delamination of this top layer. As-polished. Courtesy of Sermatech International

blades, vanes, and disks. These coatings are anodic, which will limit base-metal corrosion even when defects or erosion damage is present. The base coat is protected by a seal coat, which can also be chemically active to promote passivation of the galvanic layer. Seal coats can include fillers and pigments for improved toughness or can be smooth to improve aerodynamics. Organic seal coats can be used to improve surface slickness, limit fouling tendency, and facilitate cleaning of the blades for improved efficiency. Organic coatings are typically limited to 260 °C (500 °F) operation.

Corrosion in the Combustor and Turbine Sections

The materials commonly found in the combustor and turbine sections are superalloys and stainless steels. Nickel-base superalloys are typically used for combustor liners, transitions, blades, vanes, shrouds, and sometimes disks because of their high strength at high temperatures. Cobalt-base superalloys are typically used for vanes because of their good resistance to hot corrosion and good weldability. Iron-base superalloys are typically used for disks and lower-temperature blades, vanes, and shrouds. Stainless steels, both austenitic and martensitic, are used for lower-temperature static parts, fuel nozzles, and sometimes disks. Low-alloy steels are sometimes used for disks. While corrosion issues can occur for all of these materials and components, these issues are most often encountered with the blades and vanes.

When the alloys do not have sufficient resistance to corrosion by themselves, high-temperature coatings are used. These coatings are a separate metallurgical system that is applied over the surface of the part. These coatings are discussed in a separate section of this article.

The main corrosion issues in the combustor and turbine sections are high-temperature environmental attack in the form of high-temperature oxidation and hot corrosion. Hot corrosion is usually subdivided into two types—high-temperature hot corrosion (type I) and low-temperature hot corrosion (type II)—although additional categories have been used. The approximate temperature regimes over which these forms of attack occur, and their severity, are shown in Fig. 3.

Further information about high-temperature environmental attack and high-temperature coatings can be found in the article "Design for Oxidation" in Volume 20 of the *ASM Handbook*, in Part 4 of the book, *Superalloys II* (Ref 3), and in review papers (Ref 4, 5). There is a voluminous and ever-expanding collection of literature in this area, and these sources contain references to some of this literature. While much of the basic knowledge of this area has been well established, there are still aspects that are not well understood.

High-Temperature Oxidation

High-temperature oxidation is the oxidation of the metal and its alloying elements. It results in the formation of an external oxide scale that may or may not be protective. A protective scale grows slowly (due to a slow rate of diffusion of metal and oxygen ions through the scale) and is adherent to the metal substrate. For nonprotective scales, an internally oxidized region and an alloy-depleted layer occur beneath the metal surface. This region often contains nitrides as well as oxides. Micrographs of external and internal oxidation are shown in Fig. 4. High-temperature oxidation accelerates with higher temperatures.

High-temperature oxidation is typically not a major problem in industrial gas turbines. It usually becomes a problem when temperatures are hotter than expected, due to the prediction of too low of a temperature during the design process, or the operation of the engines at too high of a temperature. Sometimes it is a problem because the base metal or coating selected has insufficient oxidation resistance.

High-temperature oxidation can usually be controlled by the use of base metals having sufficient oxidation resistance by themselves, or by the use of a protective coating. (It should be noted that the base metal has to have sufficient resistance to oxidation by itself so that if portions of the coating are lost, the component can successfully operate to the next inspection interval.) Occasionally, the temperature of the component is so high that additional cooling of the component is needed and/or a thermal barrier coating is needed to reduce the metal temperature.

Hot Corrosion

Hot corrosion can be very destructive, as shown in Fig. 5.

High-temperature hot corrosion, also called type I hot corrosion, occurs at metal temperatures between 815 and 955 °C (1500 and

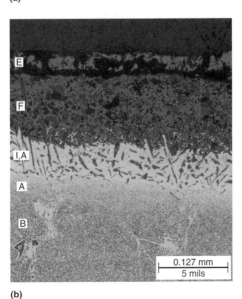

Fig. 4 Micrographs of external and internal oxidation. (a) High-temperature oxidation of the tip of an industrial gas turbine blade. Below the tip, a coating is protecting the base metal. (b) Micrograph of the oxidation shown in (a). There is an external oxide E; a layer of fully oxidized base metal, F; an internally oxidized layer, I; and an alloy-depleted layer, A. The alloy-depleted layer includes the internally oxidized layer. The base metal is shown at B. Kallings etch

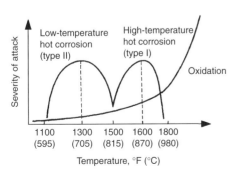

Fig. 3 Regimes of high-temperature attack. Temperatures are approximate.

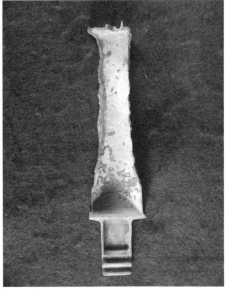

Fig. 5 Severe attack of an aeroderivative gas turbine blade by hot corrosion

1750 °F) when sulfur is present along with potassium or sodium. It is most severe at temperatures between 845 and 925 °C (1550 and 1700 °F). High-temperature hot corrosion is caused by the formation of alkali metal salts, such as sodium sulfate (Na_2SO_4) and potassium sulfate (K_2SO_4), which melt and form a liquid on the surface. This molten salt attacks the oxide scale, destroying the protection normally afforded by this scale. Once the oxide scale is breached, sulfidation and oxidation of the metal occurs, which results in the rapid destruction of the metal surface.

A micrograph of high-temperature hot corrosion is shown in Fig. 6. A broad, relatively uniform attack is present that consists of sulfides at the tip of the attack and internal oxidation behind it. The attack is usually deeper down the grain boundaries. The internal attack of the metal is led by sulfur, which diffuses into the metal faster than oxygen. As oxygen diffuses behind the sulfur, it oxidizes the sulfide particles that have formed because the oxides are more stable than the sulfides. The oxidation of the sulfides releases elemental sulfur, which then diffuses deeper into the metal, producing more sulfides, and the process repeats itself.

Low-temperature hot corrosion, also called type II hot corrosion, occurs at temperatures between 595 and 815 °C (1100 and 1500 °F) with a maximum rate of attack between 675 and 730 °C (1250 and 1350 °F). It is caused by the mixture of alkali sulfates with alloy metal sulfates, such as cobalt sulfate ($CoSO_4$) or nickel sulfate ($NiSO_4$), and requires a significant partial pressure of sulfur trioxide (SO_3). This sulfate mixture destroys the oxide scale and its protective ability. The rate of attack can be as great as that for high-temperature hot corrosion.

A micrograph of low-temperature hot corrosion is shown in Fig. 7. Low-temperature hot corrosion results in pitting of the metal surface, a porous layered scale, little base-metal depletion, and few internal sulfides.

Transition Hot Corrosion. Between about 760 and 815 °C (1400 and 1500 °F) there is a form of hot corrosion that has been called transition hot corrosion. This attack is characterized by the formation of discrete internal sulfides beneath the surface and an oxide scale on the surface. This form of attack is not very aggressive and has not received very much study. It can also occur at temperatures much lower than 760 °C (1400 °F).

Hot Corrosion versus Sulfidation. Hot corrosion is often termed sulfidation because sulfides are formed in the metal or in the scale. However, sulfidation can also refer to the gaseous reaction of sulfur with a metal in a manner similar to oxidation, whereas hot corrosion refers to attack of the metal by molten sulfates.

Elements Influencing Hot Corrosion. Hot corrosion in industrial environments is modified by the presence of other elements, such as calcium, chlorine, unburned carbon, and so forth. Exactly how these other elements affect hot corrosion is not known, but they typically make the hot corrosion more aggressive. As such, one can consider hot corrosion of industrial gas turbines to be caused by a chemical soup, with sulfur and either sodium or potassium as necessary ingredients.

Base Metals and Coatings. Hot corrosion is a serious problem for industrial gas turbines. Base metals have varying degrees of resistance to hot corrosion, ranging from no resistance to good resistance. However, there are no base metals that are so resistant that they can successfully operate continuously in a hot corrosion environment. All base metals will eventually be badly attacked when operated in hot corrosion environments. Base metals for industrial gas turbines are usually selected so that they have some resistance to hot corrosion, although this is not always the case. These alloys have a composition that prevents them from achieving the strength levels of aircraft alloys, which usually have much less resistance to hot corrosion.

Similarly, protective coatings have varying degrees of resistance to hot corrosion, ranging from limited resistance to excellent resistance. However, they also will eventually be compromised and attacked by the hot corrosion environment.

Mitigating Hot Corrosion. The best solution to hot corrosion is to avoid it by removing the contaminants that cause it. Hot corrosion is caused by sulfur plus the presence of alkali metals, principally sodium and potassium. Usually, the sulfur cannot be economically removed, but sodium and potassium can be. These elements can enter from the air, fuel, or water that is used in the engine. Proper filtration of the air and control of the quality of the fuel and water will eliminate hot corrosion conditions. The amount of allowable sodium and potassium from all sources taken together varies by engine model and typically varies from 0.1 to 2 ppm, with a value of 0.5 ppm being most typical.

Hot Corrosion Caused by Lead and Vanadium. Two other forms of hot corrosion not discussed previously are occasionally seen. These are hot corrosion by lead and hot corrosion by vanadium. Lead hot corrosion is caused by the formation of lead oxide, which melts at 888 °C (1630 °F). Vanadium hot corrosion is caused by vanadium pentoxide, which melts at 690 °C (1275 °F). Neither base metals nor coatings can tolerate these forms of hot corrosion, which are very aggressive. Lead hot corrosion is controlled by ensuring that there is no lead in either the fuel or the air (typically a maximum of 0.1 to 2 ppm is allowed). Vanadium hot corrosion is controlled in the same manner. Vanadium in liquid fuel can be removed by the use of magnesium, which ties up the vanadium.

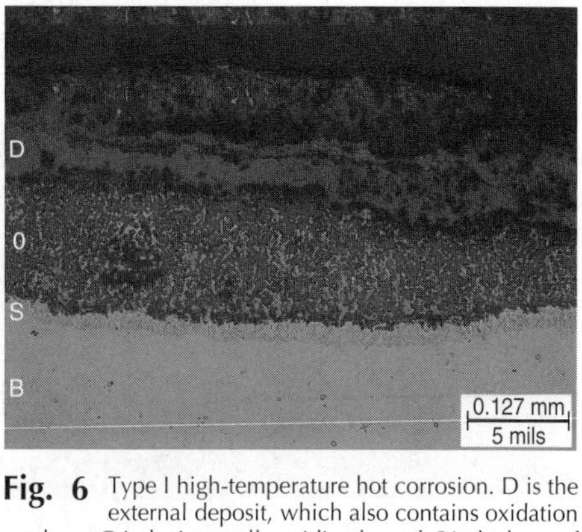

Effect of High-Temperature Corrosion on Mechanical Properties

The effect of high-temperature oxidation and hot corrosion on the mechanical properties of superalloys is generally negative. Both forms of environmental attack reduce the cross-sectional area of the part, thereby increasing the stresses. More importantly, these forms of attack reduce creep and fatigue life by forming notches in the metal that are either the precursors to fatigue cracks or assist in the growth of existing cracks.

An example of the reduction of creep life is shown in Fig. 8 for several nickel-base superalloys. The rupture life in salt environments that cause hot corrosion can be greatly degraded as compared to the life in air. The effect of the hot corrosion is reduced at low stresses, which indicates that a threshold stress level may exist for each alloy, below which rupture life may become insensitive to hot corrosion.

High-Temperature Coatings

The use of high-temperature coatings is the principal method to provide protection from high-temperature oxidation and hot corrosion in industrial gas turbines. These coatings work by forming a thin oxide barrier on the surface, as shown schematically in Fig. 9. (The oxidation protection of base metals works in the same manner.) This barrier separates the base metal and the coating from the reactive gases and is extremely thin, on the order of 2 μm (100 μin.). Without this barrier, the reactive gases would oxidize and corrode the base metal, as well as the coating itself. This oxide barrier is primarily

Fig. 6 Type I high-temperature hot corrosion. D is the external deposit, which also contains oxidation products. O is the internally oxidized metal. S is the layer of sulfides. B is the base metal. As-polished

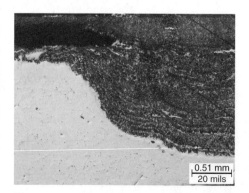

Fig. 7 Type II low-temperature hot corrosion, showing the layered appearance of the scale. As-polished

alumina (Al_2O_3) for high temperatures and chromia (Cr_2O_3) for lower temperatures.

The role of the coating is to form an oxide barrier. Ideally, this barrier is perfectly adherent to the substrate, has no imperfections, and grows very slowly after first forming. Since one cannot achieve ideal oxide barriers, the coating must be able to reform the oxide barrier should some of it become damaged due to cracking, spallation, and so forth. The role of the coating is to act as a reservoir of the oxide-forming elements, which are primarily aluminum and chromium. Thus, these coatings contain as much aluminum and chromium as possible, without making the coating too brittle. As the protective oxide is removed and reformed, the aluminum and chromium content of the coating is reduced. Eventually, the aluminum and chromium levels are too low to reform this protective oxide layer. At this point, the coating itself is attacked by oxidation. As such, the coating is expected to wear out and be replaced by a fresh coating. This is why coatings are called sacrificial.

For hot corrosion conditions, the protective oxide layer is attacked by the molten corrosive on its surface. After an incubation period, the oxide is removed and the coating is then attacked by the molten corrosive. The role of the coating is to form an oxide that maximizes the length of this incubation period.

Types of Protective Coatings. There are two categories of coatings used—diffusion coatings and overlay coatings. Diffusion coatings are formed by diffusing aluminum or chromium into the surface of the material to be coated. Overlay coatings are made by depositing a coating over the surface of the material to be coated.

Thermal barrier coatings (TBCs) are a third type of coating that is used in industrial gas turbines. This coating consists of a ceramic layer on top of either a diffusion or overlay coating. The ceramic lowers the temperature of the metal by acting as an insulator, as long as the opposite surface of the part is cooled by some means. If there is no cooling of the opposite surface of the metal, then both the metal and the TBC will eventually reach the same temperature because there is no way to remove the heat that the metal picks up. The diffusion or overlay coating that the ceramic is applied over provides corrosion protection because the ceramic layer is porous to oxygen and molten corrosives. However, the ceramic is not tolerant of molten corrosives. When the engine is cooled down, the molten corrosives inside the ceramic solidify and cause the ceramic to spall off. Because TBCs are designed for thermal protection, as opposed to protection from high-temperature oxidation and hot corrosion, they will not be dealt with further. Their resistance to these forms of environmental attack is the same as the diffusion and overlay coatings discussed below.

The protective elements in coatings, aluminum and chromium, form the protective oxide scale. Nickel and/or cobalt form the matrix of the coatings. Other elements used in coatings, but in smaller quantities, are silicon, platinum, rhodium, palladium, yttrium, hafnium, and tantalum. Silicon improves the hot corrosion resistance. The other elements improve both the oxidation and hot corrosion resistance of the coating.

Yttrium and other rare earth elements improve the adherence of the oxide scale to the substrate by reducing the amount of the oxide that spalls off during a thermal cycle. Platinum, rhodium, and palladium also reduce oxide spalling, due to reasons that are not understood. These elements are sometimes called "active elements" because of their affinity for oxygen. It has been shown that yttrium forms sulfides with sulfur in the coating. If there is free sulfur in the coating, it migrates to the oxide-coating interface, weakening this interface, and increasing the amount of oxide spalling.

Diffusion aluminide coatings are nickel aluminide (NiAl) and cobalt aluminide (CoAl). NiAl forms on nickel-base superalloys and CoAl forms on cobalt-base superalloys. Diffusion chrome coatings are also used. There are various processes used to make these coatings,

including pack diffusion, chemical vapor deposition, and slurries.

A micrograph of a nickel-aluminide coating is shown in Fig. 10. The coating consists of the protective portion and an interdiffusion zone. The interdiffusion zone contains various intermetallic compounds, including sigma phase, due to the loss of elements from the base metal to form the coating. (This sigma phase is not detrimental.)

Aluminide coatings for industrial gas turbines are usually modified by the addition of chromium (chromium usually comes from the superalloy), silicon, platinum, or other noble metals. Chromium and silicon impart resistance to hot corrosion. Platinum significantly improves the resistance to high-temperature oxidation and high-temperature hot corrosion.

Overlay coatings are a layer of a special alloy selected for resistance to environmental attack and is deposited on the metal surface. Overlay coatings can be made thicker than diffusion coatings, which may provide more protection to the base metal. The microstructure of an overlay coating is shown by the middle portion of the coating in Fig. 11. It consists of intermetallic

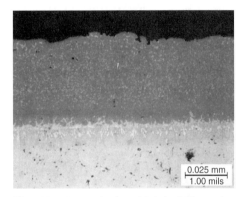

Fig. 10 Micrograph of a nickel-aluminide coating. As-polished

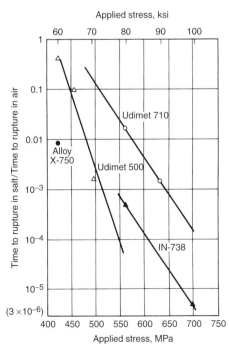

Fig. 8 Degradation in rupture life for various superalloys due to hot corrosion at 705 °C (1300 °F)

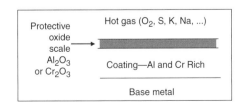

Fig. 9 High-temperature coatings protect the base metal by forming a protective oxide, which acts as a barrier to the corrosive gases.

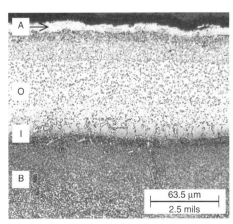

Fig. 11 Duplex coating consisting of an overlay coating (O) with an aluminide layer (A) on top. The interdiffusion zone with the base metal (B) is indicated by the letter I. There is another interdiffusion zone between A and O. Lactic acid etch

compounds in a matrix. These compounds are aluminides—NiAl, Ni$_3$Al, or CoAl. The matrix consists of nickel, cobalt, chromium, and yttrium.

There are a wide variety of overlay coating compositions. These coatings are usually called MCrAlY (pronounced "M crawl-ee") coatings, since chromium, aluminum, and yttrium are almost always present. The "M" in these coatings is either nickel, cobalt, or a mixture of these elements. The concentrations of the coating elements depend on the intended use of the coating. Other elements such as silicon, hafnium, and tantalum are sometimes added to the coating to improve the performance. Overlay coatings, such as NiCrSi, also can be made without aluminum or yttrium and can provide protection at lower temperatures.

Chromium and aluminum provide the oxidation and hot corrosion resistance of the coating. Their levels are adjusted to provide maximum resistance to one or another form of attack or are balanced to provide protection from several types of attack. Higher aluminum (>8 wt%) is used for oxidation protection, and higher chromium (>23 wt%) is used for protection from low-temperature hot corrosion. In order to be fatigue resistant, the aluminum and chromium levels should be kept low because higher levels reduce the ductility of the coating. However, if these levels are too low, there will be inadequate protection from high-temperature attack.

The majority of the overlay coating is composed of cobalt, nickel, or a combination of the two. Cobalt is better for hot corrosion, and nickel is better for high-temperature oxidation. A mixture of both cobalt and nickel is a compromise for protection from both types of attack. NiCrAlY coatings are more ductile than CoCrAlY coatings, but NiCoCrAlY coatings can be more ductile than either.

Overlay coatings are generally applied by plasma spray, high-velocity oxyfuel (HVOF) spray, or electron beam physical vapor deposition (EBPVD). Frequently, a diffusion anneal treatment is carried out after coating to obtain diffusion between coating and substrate, which gives good bonding, and homogenizes the as-sprayed microstructure.

Sometimes a diffusion aluminide coating is applied over an overlay coating. These duplex coatings provide increased protection by substantially increasing the aluminum content of the outer layer. However, this increased aluminum level makes these coatings more prone to cracking on the outer layer. The microstructure of a duplex coating is shown in Fig. 11.

The effect of coatings on the mechanical properties of turbine components ranges from beneficial to harmful. If the coating is properly applied, there is usually no effect on the tensile or creep strength of the metal. In high-cycle fatigue, coatings are usually considered to reduce the fatigue life because they crack at higher stresses. In low-cycle fatigue, or thermal mechanical fatigue, coatings will reduce the fatigue life if the stresses or strains are high enough to cause the coating to crack, because the coatings are not as ductile as the metal substrate. However, coatings may prolong the fatigue life if they do not crack, because they prevent the oxidation of the metal substrate, which can be the precursor to the formation of a crack.

REFERENCES

1. B. Becker and D. Bohn, "Operating Experience with Compressors of Large Heavy-Duty Gas Turbines," Paper 84-GT-133, American Society of Mechanical Engineers, 1984
2. I.S. Diakunchak, "Performance Deterioration in Industrial Gas Turbines," Paper 91-GT-228, American Society of Mechanical Engineers, 1991
3. C.T. Sims, N.S. Stoloff, and W.C. Hagel, *Superalloys II: High-Temperature Materials for Aerospace and Industrial Power,* John Wiley & Sons, 1987
4. H.L. Bernstein, High Temperature Coatings for Industrial Gas Turbine Users, *Proc. 28th Turbomachinery Symposium* (College Station, TX), 1999
5. H.L. Bernstein and J.M. Allen, A Review of High Temperature Coatings for Combustion Turbine Blades, *Proc. Steam and Combustion Turbine-Blading Conference and Workshop—1992,* TR-102061, Research Project 1856-09, Electric Power Research Institute, April 1993, p 6-19 to 6-47

SELECTED REFERENCES

- A.K. Koul, V.R. Parameswaran, J.-P. Immarigeon, and W. Wallace, Ed., *Advances in High Temperature Structural Materials and Protective Coatings,* National Research Council of Canada, 1994
- Protective Coatings for Superalloys, *ASM Specialty Handbook: Heat-Resistant Materials,* J.R. Davis, Ed., ASM International, 1997, p 335–344

Components Susceptible to Dew-Point Corrosion

William Cox, Corrosion Management Ltd.
Wally Huijbregts, Huijbregts Corrosion Consultancy
René Leferink, KEMA

DEW-POINT CORROSION occurs when gas is cooled below the saturation temperature pertinent to the concentration of condensable species contained by a gas. In the context of this article, it is the attack in the low-temperature section of combustion equipment resulting from acidic flue gas vapors that condense and cause corrosion damage to the plant materials. Waste flue gas produced by the combustion of fossil fuels may contain several components, such as sulfur trioxide (SO_3), hydrogen chloride (HCl), nitrogen dioxide (NO_2), carbon dioxide (CO_2), and water (H_2O), and therefore may display several dew-point temperatures at which the various species begin to condense.

Dew Point

The dew point is referred to as the temperature below which a gas must be cooled at constant pressure and constant vapor content in order for saturation to occur. Any additional cooling will result in the formation of a liquid or, at low temperatures, a solid.

In the flue gas of an industrial process, gaseous compounds such as SO_3, HCl, and NO_2 can be present. These compounds can react easily with water to form solutions of sulfuric acid, hydrochloric acid, and nitric acid, respectively. The dew points of all these solutions are higher than are the dew points of water, which means that these acids will deposit well in advance of water in a flue gas that is being cooled.

Apart from the general corrosion that is caused by the deposition of these acids, other more specific types of corrosion may occur. For example, the presence of hydrochloric acid can produce stress-corrosion cracking (SCC) in austenitic-type stainless steels and nitric acid can cause nitrate SCC in carbon steels.

The dew points of the various gases can be calculated from the formulas of:

- Verhoff (Ref 1) for SO_3 (Fig. 1)
- Kiang (Ref 2) for SO_3 and HCl (Fig. 2)
- Perry (Ref 3) for NO_2 (Fig. 3)

The droplets that condense are often highly concentrated, and acid concentrations in excess of 10% can be expected. At temperatures below the dew point, the acid concentration will decrease according to the boiling line of the acid (for sulfuric acid, see Fig. 4). A gas containing 10 vppm SO_3 and 4% water has a dew point of ~130 °C (265 °F) as shown in Fig. 1. At this temperature, the first condensed droplets will have a concentration of 85% sulfuric acid (Fig. 4). At a lower temperature of about 40 °C (105 °F), the concentration will still be approximately 40%.

Components Susceptible to Attack

Corrosion-rate peaks related to individual condensation processes may appear, but the formation of protective corrosion products and the deposition of soot ash can moderate the corrosive effects of deposited acids. In conventional boiler plants, the risk areas normally include the posteconomizer flue gas handling sections such as air heaters, ducting and precipitators, induced-draft fans, and chimney stacks.

Dew-point corrosion problems in nominally dry flue gas handling systems are discussed in this section. Dew-point corrosion caused by SO_3 and HCl in fossil-fired power plants and the secondary factors affecting it are thoroughly discussed in Ref 4 and 5. Many of these cases are also valid for waste incinerators. Nitrate SCC in heat-recovery steam generators (HRSGs) is a more recent problem area (Ref 6, 7). Problems in wet flue gas desulfurization (FGD) systems are covered in the article "Corrosion of Flue Gas Desulfurization Systems" in this Volume.

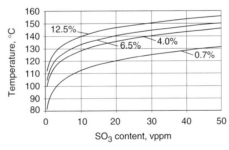

Fig. 1 Dew-point behavior of SO_3 at various water contents in the gas. Source: Ref 1

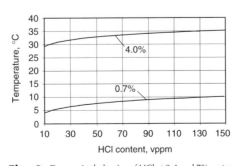

Fig. 2 Dew-point behavior of HCl at 0.4 and 7% water contents in the gas. Source: Ref 2

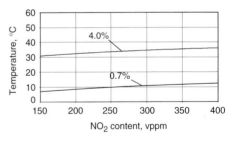

Fig. 3 Dew-point behavior of NO_2 at 0.7 and 4% water content in the gas. Source: Ref 3

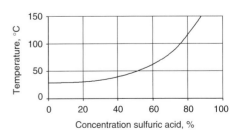

Fig. 4 Concentration of condensed sulfuric acid from an SO_3 gas containing 4% water

Most fossil-fired power plants are constructed of carbon steel, and past work largely was conducted on this material. Extensive dew-point corrosion work is reported in Ref 4 and 5. In specialized applications, low-alloy steels, stainless steels, nickel-base alloys, and organic and inorganic coatings are also used. The latest developments are condensing heat-recovery systems. In an effort to extract as much energy as possible from the flue gas, it is cooled to temperatures well below the dew point. This has the benefit that the energy that is released by condensation of water also can be utilized. In addition to water, the acid will be stripped from the flue gas. Due to the excess amount of condensing water, the concentration of the acids will not be so high as is the case where condensation occurs immediately below the dew point. Nevertheless, unprotected carbon steel cannot be used for long periods in condensing heat-recovery systems.

There are several choices for material in a condensing heat-recovery system. These include:

- The regular replacement of carbon steel tubes
- The use of more highly alloyed material
- The application of polymers as listed in Table 1
- The coating of carbon steel tubes with polytetrafluoroethylene (PTFE)

The practice from condensing heat-recovery boilers shows that the equipment and designs are maturing, but problems may still arise. For example, in a PTFE coating, cracks can be formed that result in crevice corrosion of the steel substrate. Additionally, HCl can diffuse through a PTFE coating over time and can result in corrosion of the underlying steel. Net present value calculations can be a powerful method to identify which material is the optimal choice. In 1997 calculations were made by KEMA (the Dutch Electricity Research Institute) for heat-exchanger bundles. The design conditions were: mass flow 700 kg/s (1540 lb/s), temperature 170 °C (340 °F), and pressure 25 bar (300 psig). The materials involved were carbon steel, stainless steel, and polyvinylidene fluoride (PVDF). At that time, regular replacement of the carbon steel tube bundles was found to be the most cost-effective option.

In the early 1990s, several HRSGs of combined cycle power plants in the Netherlands suffered failures in carbon steel tubes. Cracks were mostly found in the low-temperature heat exchangers, which typically operate at temperatures between 70 and 90 °C (160 and 195 °F), where nitrate condenses from the flue gas.

In some HRSGs, thermal insulation material is mounted on the inside of a closed casing. The advantage is a lower metal temperature of the casing. This will reduce the risk of creep damage to casing steel, and therefore less expensive materials can be used. However, there is a temperature gradient between the inside and outside of the insulation where sulfuric acid, nitric acid, and water can be condensed during startup and shutdown, and even during normal operation conditions. Thus, there is a risk of nitrate SCC not only in the low-temperature heat exchangers, but also in the hot boiler casings.

Dry Flue Gas Handling Systems in Conventional Combustion Systems

A conventional coal-fired power generation boiler is illustrated in Fig. 5; the locations of items prone to dew-point corrosion are indicated. The most susceptible areas are discussed in this section.

Penthouse Casing and Hanger Bars. The penthouse casing encloses the tube header pipework above the boiler furnace roof. Flue gas leakage from the furnace can cause corrosion of the casing and furnace support steelwork. The hanger bars holding the furnace tube bundles are particularly susceptible to attack near the seals retaining the bars passing from the casing to the external environment (Fig. 6). Bars can corrode such that insufficient cross section remains to support the load, and failure allows collapse of the furnace roof and superheater tube bundles.

Air Heater Cold Ends. Air heaters are commonly either of honeycomb matrix (Ljungström or Rothemühle) or shell and tube construction. In both Ljungström and Rothemühle air heaters, energy from the exhaust flue gas is absorbed by metal plates. When the plates then come in contact with the cold intake air, the heat energy is released. For both heat-exchanger types, damage is usually most severe at the cold end. A Ljungström rotating-matrix air heater is shown schematically in Fig. 7. Rothemühle air heaters are similar except that the matrix is stationary and the flue gas/inlet air is supplied by an arrangement of rotating hoods. In the Ljungström type, the ductwork is stationary, and the heat-exchanger matrix rotates in either a horizontal or a vertical plane.

Corrosion damage is sustained by the thin steel heat-exchanger elements and by support steelwork and air seal materials. It is caused partly by cold air leaking from the inlet to the outlet duct (bypassing the air heater matrix), but damage is often worsened by mechanical interaction between the rotating and fixed components as well as by displaced heat-exchanger elements that have fallen from their baskets. The highest dew-point corrosion rates are often associated with operation of a cold-end soot-blower, which removes fouling deposits by steam or air jets.

Increased dew-point attack seems linked to moisture droplets entrained in the blowing medium impinging directly on the lower edges of the air heater elements, causing dissolution of aggressive salts, removal of protective bonded deposits, mechanical abrasion or erosion of the element surface, and fatigue cracking of the element plates. The local injection of considerable moisture increases the acid dew-point temperature of the flue gas in a region where the metal temperature may already approach this dew point.

In tube-type air heaters, corrosion is again normally found at the cold end. The same temperature and sootblowing parameters described for matrix air heaters also apply to tube-type heaters, but bypass air leakage is not a problem until tubes have been perforated. However, tube air heaters are often prone to poor gas distribution across the heat-exchanger surface. This leads to localized cool spots where dew-point corrosion takes place (Ref 6).

Table 1 Polymers used in condensing heat-recovery systems

Type	Maximum temperature °C	Maximum temperature °F	Cost ratio
Polypropylene (PP)	80(a)	175(a)	1
Polyvinylidene fluoride (PVDF)	150(b)	300(b)	25
Perfluoralkoxy copolymer (PFA)	240	465	55
Polyetherketone based (PEEK)	300–330	570–625	95

(a) At 2 to 3 atm. (b) At 5 to 6 atm

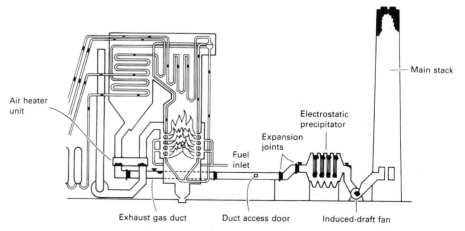

Fig. 5 Fossil-fired power generation boiler showing areas susceptible to dew-point corrosion (black areas)

Ductwork, Expansion Joints, Inspection or Sampling Ports, and Access Doors. Damage at the locations shown in Fig. 5 is often caused by constant low operating temperatures, which normally are related to air entry at fabrication faults, leaking expansion joints and door seals, or careless replacement of sampling port covers. Such leaks can cause an appreciable reduction in duct metal temperature due to stratification effects within the gas stream. Attack rates of 5 mm/yr (200 mils/yr) have been reported in precipitator outlet manifold ductwork. Such damage substantially affects ductwork integrity and frequently leads to increased attack rates on plant components downstream, particularly on the electrostatic precipitator housing and fittings and on the chimney stack.

Electrostatic Precipitators and Filter Bag Houses. Fly ash precipitators are not normally used on oil- or gas-fired equipment, but are usually present on coal-fired systems or incinerator exhaust streams. Filter bags are sometimes used instead; however, both methods of dust collection are located in large insulated housings, and their internal components are expensive to maintain and replace.

Precipitator housings are normally constructed of low-carbon steel and concrete. Corrosion attack of the housing is often associated with high levels of air in-leakage upstream due to inadequate maintenance or poor housekeeping. Occasionally, poor internal gas flow distribution may allow cool zones to develop, especially under low-flow conditions. Damage appears as casing perforation and sulfation or spalling of concrete, corrosion of rebar, and loss of structural integrity.

Damage to precipitator components is very expensive and necessitates plant shutdown for repair or replacement. Low-carbon steel collector plates and emitter wires may sustain attack during in-service periods of low flue gas temperature, which result from excessive air entry upstream of the precipitators or from inadequate sealing of the dust discharge doors and plate suspension masts (Fig. 8). Off-line attack resulting from the sweating of acid-saturated pulverized fuel ash may also be important. Precipitator plate attack is more marked because the thin-gage plates are attacked simultaneously from both sides, effectively doubling the plate thinning rate.

Filter bag installations normally are less expensive than are precipitators, but require more maintenance. Dew-point corrosion frequently is displayed on the housings because repeated access to replace holed or detached bags tends to cause wear to access door seals. Low-temperature operation or frequent startup routines may subject the bag house, bag supports, and rapping equipment to extended periods of corrosion attack; this causes blinding of bags and premature holing of the bag material due to chafing during operation or rapping cycles (Ref 8).

Induced-Draft Fan Seals. Corrosion damage to the induced-draft fan housings, control vanes, and impeller seals again is due to air in-leakage. Severe damage normally is associated with the neglect of seal condition, and attack of the fan housing itself rapidly can become severe if neglected. Fan housing washing drains should not be left open when the fan is in service. Impeller damage is less common because this component contacts the bulk flue gas.

Chimney Stacks. The stack is often the most vulnerable unit in the gas-handling system because it experiences the results of poorly controlled operation of the furnace and other components of the gas train. Damage normally is due to continuous condensation during operation and the buildup of agglomerated deposits that insulate the stack inner surface. Stack temperature monitoring often is ignored or neglected. Stacks frequently suffer severe attack before the damage is evident, and inconvenient emergency repairs, necessitating unscheduled unit shutdown, must be made. Damage can occur to liners and/or to the body of the structure, whether it is low-carbon steel, nickel alloy, or concrete (Fig. 9). In prefabricated sectional steel stacks, corrosion occurs preferentially on the inner surface near the external flanges or ladder attachments because of local cooling effects.

Heat-Recovery Steam Generators in Gas Turbines

In recent years gas turbines fueled on natural gas have become increasingly popular. To in-

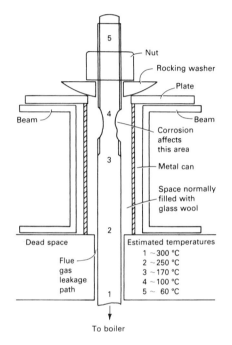

Fig. 6 Hangar bar packing box showing areas prone to corrosion. Source: Ref 4

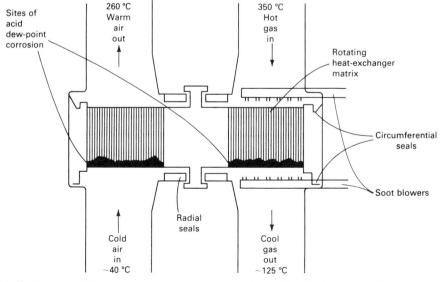

Fig. 7 Ljungström air heater showing the locations of the air seal and cold end basket corrosion sites

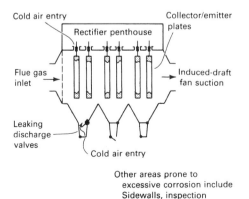

Fig. 8 Electrostatic precipitator showing areas prone to excessive corrosion

crease the total efficiency and the versatility of gas turbine powered installations, the hot exhaust gases from the combustion turbine, at temperatures in excess of 500 °C (930 °F), are cooled in a HRSG. Experience of nitrate SCC in HRSGs has been described in detail in Ref 6 and 7.

Low-Temperature Heat Exchangers. In general, the heat exchangers are fabricated from a standard low-carbon steel used for construction purposes. In some cases soon after going into service (i.e., within one year) nitrate-induced SCC can occur in the HRSGs. Most observed cracking occurs in bends and finned tubes where mechanical stresses were relatively high. Microscopic analyses of samples revealed that intergranular corrosion had occurred, and it was frequently reported that complete grains of material had become detached. A cross section of a cracked tube is depicted in Fig. 10.

It was discovered that in deposits found near cracked pipes on failed heat exchangers, both sulfates and nitrates were present. The large amount of sulfate in the deposits was explained by the presence of sulfur salts in the intake air. If air filters are used, these salts can be removed from the air. However, during conditions of atmospheric fog, small moisture droplets precipitate onto the filters. Salts in the air filter dissolve in these droplets. If the fog conditions persist for a longer period, the air filter will become saturated with water and salt-containing droplets can be sucked into the combustion air, thereby entering the gas turbine.

Ammonium nitrate deposits have been found in the cooler parts of such waste-heat-recovery boilers, close to the stack. Under normal conditions, the flue gas will be cooled to about 80 °C (175 °F). However, below 170 °C (340 °F), ammonium nitrate is solid and is most likely to be deposited on the tube walls. Because ammonium nitrate will not leave the boiler in a vapor state, it accumulates over time on the heat-exchanger surfaces near to the stack. Ammonium nitrate is hygroscopic and is easily dissolved in water. During startup and shutdown periods, when the local temperature is below the water dew point a liquid film containing up to 75% ammonium nitrate can be expected on the wall. At 60 °C (140 °F), the ammonium nitrate concentration of the liquid can increase to as much as 80% (10 mol/L). In these environments intergranular corrosion of carbon steel and low-alloy steel will occur easily, initiating the stress-assisted intergranular corrosion or SCC.

Cold Casings. In some HRSGs, thermal insulation material is mounted on the inside of the casing. Thus, there is a temperature gradient between the inside and outside of the insulation, where sulfuric acid, nitric acid, and water can be condensed during startup and shutdown periods, and even during normal operation. Sulfuric and nitric acid will react with the insulation forming calcium nitrates. Calcium and ammonium nitrates are both environments where intergranular SCC in carbon steel will be initiated very easily. Calcium and ammonium nitrate both are hygroscopic materials that can take up water even above the dew point of a gas. Up to a temperature of 120 °C (250 °F), this can create a very corrosive environment and may result in a severe risk of nitrate SCC in susceptible materials.

Mitigation of Dew-Point Corrosion

In boiler technology, the "acid dew point" usually means the sulfuric acid dew-point temperature, because this is the highest temperature at which acid condensation occurs. Earlier studies showed that, in dust-free flue gases, low-carbon steel corrosion reaches a maximum at about 25 to 50 °C (45 to 90 °F) below the acid dew-point temperature. More recently (Ref 4), it has been seen that at much lower temperatures (below 60 °C, or 140 °F) the corrosion rate of low-carbon steel increases rapidly to at least twice the high-temperature maximum, and to at least three times the high-temperature maximum if HCl is present (Fig. 11).

During the period from 1942 to 1955, it was found difficult to define a precise temperature below which rapid corrosion would start and continue; therefore, a threshold value was sought above which the plant could be operated confidently. A dew-point probe and meter was developed to identify the dew-point temperature, and maintenance of the flue gas temperature at 20 to 30 °C (36 to 54 °F) higher than the dew-point temperature was recommended to avoid condensation and corrosion.

Increasing fuel costs, higher overall efficiencies, and environmental considerations have focused renewed effort on this subject and the re-examination of earlier philosophies. The drive toward less expensive low-grade (more sulfur-rich) fuels, together with improved heat extraction processes, has produced a move toward lower back-end temperatures with better combustion control, materials selection, and use of additives to minimize dew-point corrosion. A further refinement arises from the development of continuous in-plant monitors of dew-point corrosion that can identify harmful operating regions directly.

It is difficult to give blanket advice on lessening dew-point corrosion problems because much depends on the precise plant configuration and service environment. However, general comments on materials selection, plant operation, use of neutralizing additives, maintenance, good housekeeping, and lagging (insulation) are offered in the sections that follow.

Materials Selection

Sulfuric Acid Dew Point. Most power boiler exhaust components are made of carbon steel. The material can last indefinitely in conventional plants if metal temperature is maintained above the acid dew-point temperature when the plant is in service and if reasonable care is taken to prevent excessive off-line attack during shutdown periods. In marginal conditions, cast iron may extend service life, but normally this is simply due to its greater cross section.

A few low-alloy steels have shown superior performance to low-carbon steel in particular applications. Notably, Cor-Ten, a high-strength low-alloy weathering steel, has been used with fair success for cold-end air heater elements. The improved corrosion performance seems related to the formation of a thin protective layer of corrosion product. Care must be taken to ensure that a cyclic wet/dry operating regime occurs, especially during the early period of exposure. Continuous exposure in the wet condition can result in higher corrosion rates than may occur on plain carbon steel.

Expensive alloy materials have been used where plain-carbon steel corrosion is severe, but they have not always been successful. The good but sometimes unreliable performance of such materials is due to the presence of a preformed

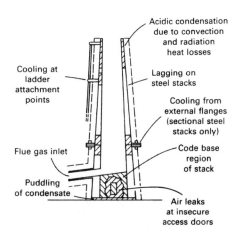

Fig. 9 Chimney stack showing sites prone to corrosion

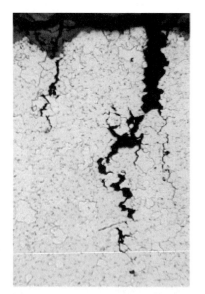

Fig. 10 Intergranular nitrate SCC in a low-temperature heat exchanger

protective oxide. Failure is associated with deposit formation, which causes differential aeration and concentration cells, producing pitting attack.

Coating use is hazardous because of difficulties in surface preparation, quality of application, physical durability, adhesion under cycling temperature, and dust abrasion. However, in certain practical application tests, good performance has been reported. For concrete stack lining in power stations, an expensive fluoroelastomer has given outstanding protection against acid dew-point attack, but the material is permeable to moisture and hence is unsuitable for use on a carbon steel substrate. Less expensive modified coal tar epoxy has performed well in some applications, but good quality surface preparation is critical. Glass flake polyester coatings can be satisfactory provided that the substrate cannot flex, especially during cold weather, because such coatings can be very brittle at low temperatures. An isocyanate-cured epoxy material has also been successful in less troublesome plant conditions. For air heater elements, double-dipped enamel coatings are likely to be most effective, though such coatings add significantly to the capital cost. Major difficulties in ensuring defect-free adherent coatings have occurred with arc sprayed aluminum, chromia, and alumina materials, but arc sprayed aluminum has been reported to be generally satisfactory for high-temperature gas turbine stack applications. However, in these stacks, severe thermal conditions can lead to exfoliation and acidic rust flaking.

Nitric Acid Dew Point. The use of stressed carbon steels in conditions where there is a risk of condensation in the presence of nitrates or NO_x means that there is a substantial probability that SCC is likely to occur, sooner or later. Stress-corrosion cracking due to nitrates is nearly always associated with intergranular corrosion. Leferink (Ref 7) studied the resistance of several ferritic steels to intergranular corrosion in ammonium nitrate solutions in the range of 2 to 35 wt% at 90 °C (195 °F) for a period of 100 h. A clear correlation was observed between the steel composition and the concentrations of ammonium nitrate that were necessary to support intergranular corrosion. For all of the tested steels: carbon steel, 15Mo3 steel, 1Cr-0.4Mo steel, 2.25Cr-1Mo steel, and 12% Cr steel, the correlation was:

Critical percentage ammonium nitrate (wt%)
$= 6Mo + 2Cr + 1Mn + 8 - 12Si - 8Cu - 1C$

As nitrate-induced intergranular corrosion is often a precursor to intergranular SCC, SCC is particularly influenced by chemical corrosion. Annealing a material, or the selection of a higher-strength material, will be of minimal help in preventing SCC on the longer term. The selection of a low-alloyed 2.25Cr-1Mo steel as a replacement for carbon steel in critical locations, in general, is a better choice for the prevention of intergranular corrosion in dissolved nitrate environments.

Plant Operation

Sulfuric Acid Dew Point. Although materials selection can be pertinent in certain plant components, careful plant operation, good housekeeping, and comprehensive maintenance represent a major and more general route to technically efficient, fuel economical, and cost-effective operation. Classical dew-point conditions do not necessarily yield unacceptably high corrosion rates in practice. Control of the secondary factors alone can lead to significant savings by minimizing replacement costs in key areas. Improved on-line monitoring capabilities allow the onset of corrosion to be detected and avoided, often by minor operational refinement, and the possibility of lowering back-end temperatures without risk to the plant can be explored confidently. Precise operational control is important. The inherent variability of excess oxygen, furnace temperature, combustion method, and fuel ashing characteristics can affect low-temperature corrosion more than changes in fuel sulfur content.

Nitric Acid Dew Point. Even though knowledge of the condensation of corrosive gases is available in the literature, condensation of acids other than sulfuric acid frequently is overlooked at the design stage. This leads to many failures in technical installations. The operation of equipment that can generate condensing gases makes it necessary to think carefully about the operational regimes (both on- and off-line) and of the possible risk of acid condensation attack. The condensation of nitric acid can cause SCC of carbon steel. However, reaction products of nitric acid with the steel or insulation also can result in the formation of corrosive ammonium nitrate or calcium nitrate. As mentioned previously, calcium and ammonium nitrate both are hygroscopic materials that can take up water even above the dew point of a gas and up to a temperature of 120 °C (250 °F). This can create a corrosive environment that may result in nitrate SCC in susceptible materials. Cold boiler/combustion plant casings can be used only at very low NO_2 levels. Even at NO_2 levels below 20 vppm, there is a danger that nitrates will accumulate in the insulation after prolonged periods of operation.

Maintenance, Good Housekeeping, and Lagging

A good housekeeping list can be established by adequate planned maintenance procedures introduced for any particular installation because practical dew-point factors are now better understood. Poorly designed or maintained thermal insulation, badly sealed inspection doors, leaking expansion joints, and so on, are unacceptable if future installations are to operate at optimum levels.

Additives

Sulfuric Acid Dew Point. Neutralizing additives, usually calcium or magnesium oxide/hydroxide, tend to have an ad hoc usage, especially in oil-fired situations. Their benefit in preventing acid smuts is proven, but their ability

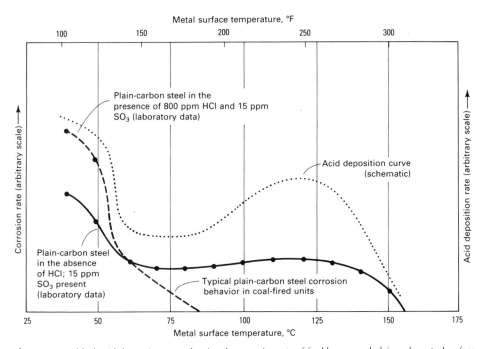

Fig. 11 Modified acid dew-point curve showing the corrosion rate of freshly prepared plain-carbon steel surface exposed to dust-free flue gas in a laboratory test rig. The increase in corrosion rate at lower temperatures in the presence of HCl is shown. The modification of classic dew-point corrosion behavior in coal-fired power plants also is indicated.

to reduce corrosion rates consistently is less certain. Modern electrochemical monitoring techniques can provide a continuous means of assessing additive performance, although in many cases fluctuations in fuel composition and operational control may mask the contribution of the additive. Direct injection of ammonia has been employed to reduce flue gas acidity, but the resultant formation of sticky bisulfite deposits can lead to severe fouling.

Nitric Acid Dew Point. No additives are available to prevent nitrate SCC.

Guidance for Specific Sections of the Plant

Penthouse Casing and Hanger Bars. The casing and hanger bars should be maintained during operation at a temperature in excess of the acid dew-point temperature. This is achieved by external insulation of all items. Hanger bar packing boxes should be routinely checked to ensure that excessive clearance has not developed between the packing and the bar, allowing flue gas leakage. This is especially important above the superheater pendant tubes, which tend to require more maintenance than do other tube bundles (Ref 4).

Air Heater Cold Ends. In matrix-type air heaters, air leakage should be controlled, possibly by allowing only a predetermined acceptable leakage rate, and condensate injection should be prevented, especially that due to leaking sootblower steam isolation valves. Compressed air is preferred as the blower medium, and slight preheating to dry the air should be considered. Adequate condensate drainage facilities must be maintained for all types of blowers. Air heaters should not be water washed on-line unless absolutely necessary, but thorough water washing off-line is recommended. In many cases, the need to do this may point to inadequate operational control, for example, during low-road conditions. For lower outlet flue gas temperature units, the use of Cor-Ten steel or double-dipped enameled elements should be considered for severe service. Adequate specification and coating quality control must be ensured when using enameled elements. Good gas distribution across the air heater minimizes cold spots. The use of fuel additives should be considered for combating unavoidable corrosion attack.

Special care should be taken in shell and tube air heaters to ensure good temperature distribution across the unit. Buildup of gas-side deposits, which may allow sub-dew-point metal temperatures to occur beneath, must not be permitted. Deposits (iron sulfates, ash, fly-ash) are rather hygroscopic and corrosive during shutdown situations. Sulfur dioxide will be absorbed and oxidized to SO_3 in presence of ash and increase the corrosiveness. Leakage from perforated cold-end tubes must be minimized (Ref 5).

Ductwork, Expansion Joints, Sampling Ports, and Access Doors. All air in-leakage must be prevented. Door seals should be replaced routinely, and a gas-tight fit ensured. Thermal insulation should be maintained in good condition, leaking expansion joints replaced, and sampling port covers properly refitted after use.

Electrostatic Precipitators and Filter Bag Houses. Adequate maintenance can prevent cold air entry, especially at discharge doors and electrode support masts and at access door seals. External lagging must be in good condition to maintain housing temperature above the acid dew point. Air entry upstream of the precipitator or bag house should be minimized. If possible, the unit should be kept warm during hot-standby conditions. Operation periods in low gas flow or startup conditions should be minimized.

Induced-Draft Fan Seals. These seals should be maintained in good condition on a routine basis. Fan housing washing drains should not be left uncovered.

Stacks. Adequate temperature monitoring should be ensured at top and bottom. Stack operating temperature should be maintained within design limits at all times. For low-carbon steel stacks, adequate insulation and lining conditions must be ensured. Stacks should not be run wet or cold unless specifically designed to do so. Routine inspection programs, for monitoring stack lining and structural integrity, should be initiated.

Low-Temperature Heat Exchangers in HRSG of Gas Turbines. The gas cannot be cooled too much because of the danger of nitric acid and ammonium nitrate formation. Regular inspection of the tube surfaces is necessary. A continuous monitoring system is advised.

Cold Casing in HRSG of Gas Turbines. Cold HRSG casings are especially vulnerable to dew-point attack because the condensation occurs in the insulation and near the carbon steel casing. A hot (closed) casing with insulation on the exterior is considered the preferred choice.

ACKNOWLEDGMENT

Portions of this article were adapted from W.M. Cox, D. Gearey, and G.C. Wood, Components Susceptible to Dew-Point Corrosion, *Corrosion,* Vol 13, *ASM Handbook* (formerly 9th ed. *Metals Handbook*), ASM International, 1987, p 1001–1004.

REFERENCES

1. F.H. Verhoff and J. Banchero, Predicting Dew Points of Flue Gases, *Chem. Eng. Prog.,* Aug 1974
2. Y.H. Kiang, Predicting Dew Points of Acid Gases, *Chem. Eng.,* Feb 1981, p 127
3. R.H. Perry and C.H. Chilton, *Chemical Engineers Handbook,* 5th ed., McGraw Hill, 1973
4. D.R. Holmes, Ed., *Dewpoint Corrosion,* Ellis Horwood Ltd., 1985
5. "Final Report of the Dewpoint Investigation," U.K. Department of Trade and Industry, June 1986
6. W.M.M. Huijbregts and R.G.I. Leferink, "Latest Advances in the Understanding of Acid Dewpoint Corrosion: Corrosion and Stress Corrosion Cracking in Combustion Gas Condensates," *Anti Corros. Methods Mater.,* Vol 51 (No. 3), 2004, p 173–188
7. R.G.I. Leferink and W.M.M. Huijbregts, Nitrate Stress Corrosion Cracking in Waste Heat Recovery Boilers, *Anti-Corros. Methods Mater.,* Vol 49 (No. 2), 2002, p 118–126
8. E. Land, "The Theory of Acid Deposition and Its Application to the Dew Point Meter, *J. Inst. Fuel,* June 1977

Corrosion of Generators

William G. Moore, National Electric Coil

A SIGNIFICANT DISCOVERY in the corrosion of generators occurred in the 1990s—crevice-corrosion cracking in water-cooled generators. Before delving into crevice-corrosion cracking, however, this article reviews the generator industry experience with stress-corrosion cracking (SCC) of 18Mn-5Cr alloy retaining rings. This alloy was primarily 18% Mn and 5% Cr. Chemical composition details of this nonmagnetic alloy, used for generator retaining rings, can be found in ASTM A 289 (Class B).

Retaining-Ring Corrosion

Retaining rings support the generator rotor end turn coils during rotation and are under residual stresses and severe operating stresses. In the presence of moisture or other corrodents, and this high operating stress, this 18Mn-5Cr material has been shown to be highly susceptible to SCC. Figure 1 shows fluorescent dye penetrant indications of SCC on the inner retaining ring surface. This ring had been part of a rotor stored in uncontrolled conditions for about 5 years.

Failures in the industry prior to 1987 had been documented in published papers (Ref 1, 2). Of course, failure of a retaining ring inside an operating generator can cause extensive damage and is a threat to the safety of those nearby. As a minimum, failure of a ring requires replacement of the stator winding, stator core, and the rotor winding, and could possibly extend to damage to the generator frame and rotor forging. The loss of generating capacity is a significant cost.

Fortunately, from 1987 to the present, original equipment manufacturers and independent repair shops have pursued an aggressive approach of nondestructive examination of the 18Mn-5Cr rings and replacement with a superior alloy resistant to SCC. This superior alloy is made from 18% Mn and 18% Cr (18Mn-18Cr). This material substitution has kept catastrophic failures to a minimum, but has cost the industry millions of dollars. One industry failure of the 18Mn-18Cr material has been documented, but the root cause is suspected to be primarily due to heavy concentration of chlorides, rather than moisture alone causing SCC. The 18Mn-18Cr composition is now the only composition that satisfies ASTM A 289/289M.

Many 18Mn-5Cr rings are still in operation on generator rotors today. It is important to minimize moisture in the generator at all times and to inspect the rings if exposed to moisture. In operation, moisture can be minimized by maintaining a high level of hydrogen purity through a hydrogen gas dryer and filter. There is a higher probability of SCC damage occurring to 18Mn-5Cr rings during the shutdown process. If the coolant gas dew point is higher than the retaining-ring temperature, moisture will condense on the rings, which can lead to SCC (Ref 3). Hydrogen coolers should be inspected at major outages for leaks, and tubes must be plugged as necessary.

A higher probability of SCC can occur if the rotor is pulled out of the stator and left sitting on the turbine deck or shop floor. The rings are still under high stress at this time due to the designed prestressing of large shrink fits. Condensation and resulting moisture on the rings should be prevented by turning on the strip heaters in air-cooled machines, or if the rotor is out of the stator, enclosing it in a tarp with some heaters or lamps underneath sufficient to maintain the rotor above the dew point.

Corrosion of magnetic retaining rings is also a concern (Ref 4). Magnetic retaining rings were mostly used on older generators manufactured prior to about 1965. These units tend to be smaller, but in the presence of moisture can rust and pit as shown in Fig. 2.

As with nonmagnetic rings, pits due to corrosion in magnetic rings can initiate cracking. Especially susceptible are rings that operate in hydrogen, are made of very hard materials (>38 HRC), and were manufactured prior to about 1960. Magnetic rings are more easily repairable than nonmagnetic.

Crevice-Corrosion Cracking in Water-Cooled Generators

Another serious corrosion issue that is unique to water-cooled generators has developed on a specific class and age of units. The problem was first recognized and publicized in the very early 1990s and has become a major concern due to the costs related to solving the problem. The problem occurs only on generators with stator windings that are water cooled. Units

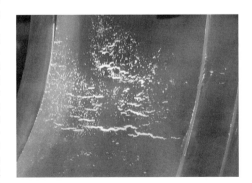

Fig. 1 Fluorescent dye penetrant examination showing linear indications of stress-corrosion cracking in an 18Mn-5Cr retaining ring

Fig. 2 Corrosion-related pitting on surface of magnetic retaining ring

manufactured from about 1970 to 1986, estimated to be about 350 in total, can develop water leaks (Ref 5). These leaks allow cooling water to penetrate the coil insulation, reducing the dielectric strength of the insulation, eventually causing a phase-to-ground electrical fault in the generator. This will trip the unit off-line. Consequential damage to the stator winding and stator core can also occur, often requiring a stator rewind and core repair.

The problem occurs specifically at a brazed connection at the end of the stator coil. The braze material, which joins the copper clip or sleeve to the hollow strand bundle, is relatively porous in nature. Stator coil cooling water can become trapped in these porous openings. The water corrodes the phosphorus rich braze material, eventually finding its way through the hollow copper strand, creating a leak path.

Figures 3 and 4 show the location of this crevice-corrosion cracking in generator water-cooled stator coils.

Research is ongoing about the corrosion mechanism, levels of dissolved oxygen, and various alternative repair mechanisms. There are three primary repair alternatives to address corrosion in water cooled coils. The first repair option, pioneered by the OEM, is to inject epoxy resin into the "strand package" area. The intent is to allow the epoxy resin to fill existing crevices and pockets in the porous braze material, preventing water from entering these areas in the future and causing a reinitiation of the corrosion process.

The second repair option is to completely remove the clip assembly at the end of the stator coil, replacing it with a new clip-to-strand connection. A patented, two-piece clip design (Ref 6) allows two separate brazes to be made at two different temperatures. This option allows better access to the strand package during brazing and ultimately allows better control of the brazing process. This process, in conjunction with improved braze material, eliminates the porosity in the braze material, eliminating pockets where cooling water can collect and stagnate. This prevents initiation of crevice-corrosion cracking.

The third alternative is to rewind the stator with completely new coils, utilizing an improved clip-to-strand design, new materials and braze alloys, and better controlled processes as described previously. The coil rewind option is often necessary if the crevice-corrosion cracking on the existing coils has progressed substantially to the point that a single coil, or more than one coil, has developed leaks, ultimately damaging the coil insulation.

All three alternatives, along with improved efforts to monitor and control oxygen content in the cooling water, have been in practice for a few years and proven to be successful in addressing the issue of corrosion in water-cooled generators.

REFERENCES

1. R.T. Hagaman, "Failure Experience with Generator Rotors," Proc. EPRI Workshop on Rotor Forgings for Turbines and Generators, (Palo Alto, CA), Sept 14, 1980, Electric Power Research Institute
2. R. Viswanathan, "Retaining Ring Failures," Workshop Proceedings: Retaining Rings for Electric Generators, EPRI EL-3209, Electric Power Research Institute, Aug 1983
3. J.D. Albright and D.R. Albright, "Generator Field Winding Shorted Turns: Moisture Effects," Steam Turbine Generator Workshop (Nashville, TN), Aug 25–27, 2003, Electric Power Research Institute
4. W.G. Moore, "Evaluation, Repair and Restoration of Generator Magnetic Retaining Rings," Rotating Electric Machinery Colloquium (Lake Buena Vista, FL), Sept 8, 1999, IEEE
5. A.M. Iversen, "GE TIL 1098 Update," Maintaining the Integrity of Water-Cooled Generator Stator Windings Conference (Tampa, FL), Nov 18, 1996, Electric Power Research Institute
6. "MD & A Two-Piece Clip," brochure, Mechanical Dynamics and Analysis, St. Louis, MO, 2001

Fig. 3 End of a water-cooled coil. This "bottle clip," made of copper, allows the coil to be electrically connected to the next coil by means of the laminated flexible connector and also provides a means of cooling water to pass through this coil and out the machine.

Fig. 4 Borescopic view through the bottle clip opening into the strand package, which consists of hollow strands that carry cooling water and solid strands that carry the majority of the current. This bundle of copper strands (both the hollow and solid) are brazed together into a single, consolidated "strand package." Braze material, which is porous, covers the surface of the strands, and extends axially back into the coil between strands. It is this material that is susceptible to crevice-corrosion cracking as cooling water is trapped in the porous openings and initiates the corrosion process.

Corrosion and Erosion of Ash-Handling Systems

ASH HANDLING is a major challenge for utilities and industries using coal as a primary fuel. The combustion process—cyclone, pulverized-coal, or fluidized-bed firing—determines the type and characteristics of the ash. Waste ash, fly ash, and bottom ash are generated in large volumes and must be dealt with in an environmentally acceptable manner. Fly ash comprises small dustlike particles (100 µm or less in diameter); bottom ash consists of much larger slag particles removed at furnace bottoms.

Conventional ash-handling systems collect, convey, and dispose of waste ash by methods that depend on site-specific considerations, government regulations, and economic considerations. Generally, the ash will be handled wet or dry, but the handling system may include combinations of wet and dry conveying. Fly ash is collected dry and is typically conveyed dry for storage or disposal. Wet conveying and storage of fly ash has become environmentally more difficult, and regulations now favor dry conveying systems. Bottom ash is collected in a water bath and is almost always handled wet. Most ash-handling systems convey bottom ash and fly ash separately to storage or disposal, but conveyance in the same pipeline is also acceptable.

Typical operating problems associated with conventional fly ash/bottom ash handling systems may be grouped into two categories:

- Corrosion and erosion (material deterioration)
- Scaling and plugging (material buildup)

These problem areas must be addressed by designers of ash-handling systems and plant operators, who should consider factors such as equipment selection, materials applications, maintenance policy, and system concept. The handling system concepts for fly ash/bottom ash are dry/dry, dry/wet, wet/dry, or wet/wet. The subject of ash handling is discussed in detail in Ref 1.

Fly Ash Systems

Corrosion and corrosion-related problems are more prevalent in wet systems than they are in dry systems. Both systems are reviewed subsequently.

Dry Fly Ash Systems. The pneumatic conveying of coal ash particles can produce plugging of conveying lines, sticking of the ash to chutes and hoppers, and erosion of internal parts of ash-handling equipment. Corrosion of internal components is normally small to nonexistent. Only the introduction of unwanted moisture would create an environment that produces corrosion severe enough to cause system outages or to require frequent maintenance.

Wet Fly Ash Systems. The transport of fly ash with water as the conveying medium is normally achieved by using a slurry of 10 to 15% ash by weight. The introduction of conveying water sets up an array of water chemistry situations that can lead to scaling and/or corrosion. The tendency for scaling and corrosion depends in part on the chemical characteristics of the ash sluice water and in part on the composition of coal ash.

Table 1 shows the variations in coal ash composition with coal type. The soluble ash species of elements such as iron, calcium, sodium, magnesium, potassium, and a variety of trace elements produce a wide range of pH levels in the ash slurry. For example, a dramatic change in coal supply can result in changes of 1 to 2 pH units. The dominant alkaline constituents are Fe_2O_3, CaO, MgO, Na_2O, and K_2O; the acid constituents are SiO_2, Al_2O_3, and TiO_2 (Ref 2). This interaction of water and ash is the cause of most corrosion problems.

Scale formation on equipment internals and pipeline walls is always a frequent maintenance problem. The composition of typical scale indicates that compounds of calcium, magnesium, sodium, and silica are available to go into solution, concentrate, and precipitate out on internal surfaces. The reduction in pipeline diameter caused by heavy scale buildup, the reduction in pump and valve internal clearance by scaling, and the plugging of small control lines often requires dismantling and descaling, or eventual replacement. Most corrosion and erosion-related failures are complicated by scale formation.

Wet Bottom Ash Systems

Hot ash (clinkers or agglomerated slag) deposits of varying sizes are quenched and collected for conveying in the bottom ash system. Bottom ash systems are generally conveyed wet from collection to final disposal. However, some systems incorporate dewatering equipment and haul the ash by mobile equipment or belt conveyors to suitable disposal sites. The pumping and pipeline conveying of bottom ash slurries (normally 15 to 20% ash by weight) generally results in a highly abrasive but only moderately corrosive condition. Agglomerated granular (fused) bottom ash, with its irregular shape and large size, gives the slurry its abrasive nature. Compared with fly ash, bottom ash is inert and insoluble; therefore, chemical interaction between the bottom ash and conveying water is less worrisome.

Current environmental regulations have resulted in the development of the zero-discharge (closed-loop) system. The reuse of conveying water in either bottom ash or fly ash systems concentrates soluble salts and will usually result in more corrosive water chemistry. While most failures result from erosion, scaling,

Table 1 Variations in coal ash composition by coal type

	Composition, %									
Coal type	SiO_2	Al_2O_3	Fe_2O_3	TiO_2	CaO	MgO	Na_2O	K_2O	SO_3	Ash
Anthracite	48–68	25–44	2–10	1–2	0.2–4	0.2–1	...	...	0.1–1	4–19
Bituminous	7–68	4–39	2–44	0.5–4	0.7–36	0.1–4	0.2–3	0.2–4	0.1–32	3–32
Subbituminous	17–58	4–35	3–19	0.6–2	2.2–52	0.5–8	...	...	3–16	3–16
Lignite	6–40	4–26	1–34	0–8	12.4–52	2.8–14	0.2–28	0.1–1.3	8.3–32	4–19

Source: Ref 1

or plugging, corrosion may become a problem if soluble salts are allowed to accumulate too long in closed-loop systems.

Mitigating the Problems

Operating problems due to corrosion, erosion, scaling, and plugging can be minimized by consideration of the following:

- Piping systems should have gradual bends, turns, and transitions and should be without sharp angles to reduce erosion and/or plugging.
- Materials of construction (metals, nonmetals, and coatings) that are resistant to acid attack should be selected.
- All storage vessels should have constant agitation to keep solids in suspension to facilitate pumping.
- Pipe diameters and line velocity must be such that suspended solids do not cause plugging.
- Piping systems should be flushed during shutdown to remove solids before solids cementation reactions can occur.
- Abrasive-resistant liners (replaceable metal, plastic, ceramic, and elastomers) or surface treatments should be selected for system components (pump impellers and casings, agitator blades, and pipe liners) for better control of erosion and corrosion of wetted parts.

It is not possible to recommend a single approach or a material of construction that will work for every ash-handling application. Corrosion problems are reduced and equipment lifetime is extended by good case-by-case materials selection, the existence of adequate quality control, operational simplicity, and the use of effective inspection/maintenance practices.

The Future

Present and pending regulations (mainly federal in the United States) tend to restrict the wet-dry options for utility and industry owners. Future systems for fly ash will be dry processes and will require suitable reuse or environmentally safe disposal of the ash. Wet systems for both bottom ash and combined fly ash/bottom ash will drift to the zero-discharge concept and recycle their conveying water.

The article "Environmental Performance of Concrete" in *ASM Handbook,* Vol 13B discusses the advantages of using fly ash as a cementitious material. In the United States in 2000, approximately one-third of the more than 100 million tons of coal combustion products were recycled (Ref 3). In order to meet specifications for use in concrete, the fly ash requires further processing. This equipment is subject to the corrosion and erosion or fly ash in a more agressive environment.

ACKNOWLEDGMENT

This article is adapted from the section "Corrosion of Ash-Handling Systems," by L.D. Fox, in the article "Corrosion in Fossil Fuel Power Plants," in *Corrosion,* Volume 13, *ASM Handbook,* p 985–1010.

REFERENCES

1. *Coal Ash Disposal Manual,* EPRI CS-2049, Electric Power Research Institute, Oct 1981
2. *Steam: Its Generation and Use,* 39th ed., The Babcock & Wilcox Company, 1978, p 15-4
3. K. Bargaheiser and T.S. Butalia, Use of High Volume Fly Ash Mixes in Preventing Corrosion in Concrete, *International Ash Utilization Symposium,* University of Kentucky, accessed at http://www.flyash.org, January 2005

SELECTED REFERENCES

- A.C. Cuddon and C. Allen, The Wear of Tungsten Carbide—Cobalt Cemented Carbides in a Coal Ash Conditioner, *Wear,* Vol 153 (No. 2), 1992, p 375–385
- Conveyors Used to Haul Ash to Waste-to-Energy Facility, *Civic Public Works* (Canada), Vol 41 (No. 10), 1989, p 20–22
- J. Stringer and I.G. Wright, Materials Issues in Fluidized Abed Combustion, *Proceedings of Materials for Future Energy Systems 1985,* American Society for Metals, 1984, p 123–126
- *Improving the Convenience of Solid Fuel in the Domestic Market Sector,* Commission of the European Communities (Luxembourg), 1997, p 117
- R.O. Heckroodt and A.N. Robinson, Erosion by Pulverized Fuel Ash in Power Generating Units, *Proceedings of the International Symposium on Ash,* Council for Scientific and Technical Research, Pretoria, South Africa, 1987, p 15
- J.T. Ellis and P.A. Cobham, Abrasion Tetrotechnology, *Soc. Symp. Proc.,* Vol 14 (No. 9), Materials Research Society, 1985, p 32–35
- R.L. Mace, Specialty Coating Applications in the Power Industry, *1983 Plastics Seminar,* NACE, Houston, TX, 1983, p 12.1–12.9
- J.H. Mallinson, Abrasion of Fiber-Reinforced Plastics in Corrosion Environments, *Chem. Eng.,* Vol 89 (No. 11), 1982, p 143–144

Corrosion in Portable Energy Sources

Chester M. Dacres, DACCO SCI, Inc.

BATTERIES AND FUEL CELLS are popular forms of portable electrical energy sources. A battery is a device that uses electrochemical processes to convert chemical energy into electrical energy. Batteries are classified into two groups: primary or nonrechargeable batteries, which are discarded after a single use, and secondary or rechargeable batteries, which can be recharged by reversing the flow of current through the electrodes to reconstitute the chemical reactants to their original state (Ref 1).

Batteries and fuel cells make constructive use of electrochemical reactions by converting chemical energy to electrical energy. The design and materials used to construct these devices must address the destructive aspect of reactions present in order to assure useful service life for the devices. This article provides a look at the operation and corrosion problems inherent in batteries and fuel cells. The articles "Anodes for Batteries" and "Fuel Cells" in *Corrosion: Fundamentals, Testing, and Protection*, Volume 13A of *ASM Handbook*, 2003, address the constructive aspects of corrosion mechanisms. The article "Corrosion in Fuel Cells" in this Volume provides details of research and development of fuel cells.

Fuel cells are devices similar in purpose and construction to storage batteries; however, they continuously convert energy from various chemicals (fuels) directly into energy. In a fuel cell, the electrochemical reaction of a fuel and an oxidant directly generates a flow of electric current when power is demanded of the cell by an external load. There is no intermediary stage where energy must be expended to produce heat before producing power, and there are no moving parts. Fuel cells are highly efficient, noiseless, and generate relatively less heat and toxic fumes than do internal combustion engines. Fuel cells can be thought of as batteries that do not wear out, do not need recharging, and are lighter in weight (Ref 2).

Battery Types

A battery consists of a negative electrode (anode) from which electrons flow into the external circuit. The reactions at the negative electrodes usually involve strong reductants, which tend to give up electrons while forming positive ions or cations. A battery also has a positive electrode (cathode) into which electrons flow from the external circuit. The reactants at the positive electrode are usually strong oxidizing agents that accept electrons with ease, while forming negative ions or anions. The third component of a battery is the electrolyte that allows ionic conduction between the positive electrode and the negative electrode. The electrolyte in this type of galvanic cell need not be limited to a liquid solution but may in some cases be a solid ionic conductor. A battery can also have a separator, which is an inert porous insulating medium or an ion-selective permeable membrane that permits the movement of ions between the electrodes while physically separating the electrodes. There are a number of galvanic cells that could be classified as batteries; however, the ones listed in Table 1 are commercially available nonrechargeable and rechargeable batteries (Ref 1).

Corrosion of Batteries

The tendency of metals to oxidize during battery discharge is reflected in their negative Gibbs free energy change. Metals that oxidize more readily have more negative Gibbs free energy change and are considered better anodes for batteries. Battery performance depends not only on this free energy change but on how much of this energy can be extracted from a given mass of metal during the corrosion reaction when the battery discharges. Some batteries are comprised of many components that are susceptible to corrosion. For example, lithium-sulfur dioxide ($LiSO_2$) batteries have several parts that corrode and can contribute to a short battery life. The electrolyte in the $LiSO_2$ batteries contains LiBr as the electrolyte salt and SO_2 as the cathode reactant. The electrolyte dissolves in acetonitrile and sometimes mixes with propylene carbonate. Hardware corrosion problems in $LiSO_2$ cells include glass corrosion, tantalum corrosion in welded regions, corrosion of lithium in contact with nickel, and stress-corrosion cracking of the battery case. Figure 1 shows each of these sites on a simplified battery schematic (Ref 3–5). Glass corrosion is primarily the result of attack by lithium metal to produce lithium oxide:

$$4Li + SiO_2 \leftrightarrow 2Li_2O + Si \quad \text{(Eq 1)}$$

The corrosion product Li_2O further corrodes the glass:

$$Li_2O + SiO_2 \leftrightarrow Li_2SiO_3 \quad \text{(Eq 2)}$$

$$2Li_2O + SiO_2 \leftrightarrow Li_2SiO_4 \quad \text{(Eq 3)}$$

The mechanism of glass corrosion is common to all lithium cells in which a liquid electrolyte containing lithium ions (Li^+) wets the glass, and

Table 1 Commercial primary and secondary batteries

Battery system	Net electrochemical reaction
Secondary (rechargeable)	
Lead-acid	$PbO_2 + Pb + 2H_2SO_4 \rightleftharpoons 2PbSO_4 + 2H_2O$
Nickel-cadmium	$2NiOOH + Cd + 2H_2O \rightleftharpoons 2Ni(OH)_2 + Cd(OH)_2$
Nickel-hydrogen	$2NiOOH + H_2 \rightleftharpoons 2Ni(OH)_2$
Primary (nonrechargeable)	
Leclanche dry cell	$Zn + 2MnO_2 \rightarrow ZnO \cdot Mn_2O_3$
Alkaline	$Zn + 2MnO_2 \rightarrow ZnO + Mn_2O_3$
Silver-zinc	$Ag_2O_2 + 2Zn + 2H_2O \rightarrow 2Ag + 2Zn(OH)_2$
Reuben mercury cell	$HgO + Zn + H_2O \rightarrow Hg + Zn(OH)_2$
Zinc-air	$Zn + O_2 + 2H_2O \rightarrow 2Zn(OH)_2$
Lithium-iodine	$2Li + I_2 \rightarrow 2LiI$
Lithium-sulfur dioxide	$2Li + 2SO_2 \rightarrow Li_2S_2O_4$
Lithium-thionyl chloride	$4Li + 2SOCl_2 \rightarrow 4LiCl + S + SO_2$
Lithium-manganese dioxide	$Li + Mn(IV)O_2 \rightarrow LiMn(IV)O_2$
Lithium-carbon monofluoride	$nLi + (CF)_n \rightarrow nLiF + nC$

the phenomenon has been noted in Li/V$_2$O$_5$, Li/SOCl$_2$, and other systems (Ref 6).

Secondary rechargeable batteries usually have considerably longer service lives than primary batteries. A much-used secondary battery is the lead-acid storage battery. A lead-acid battery consists of groups of positive and negative plates immersed in a concentrated electrolyte solution of sulfuric acid (H$_2$SO$_4$). The positive plates contain a web of lead alloy grids coated with lead dioxide (PbO$_2$). Negative plates contain a web of lead alloy grids coated with pure lead, usually known as sponge lead because of its porous nature. The following reactions occur during charging, and the reverse of these occur upon discharge (Ref 7–11):

Positive

$$PbSO_4 + 2H_2O \leftrightarrow PbO_2 + 3H^+ + HSO_4^- + 2e^-$$

(Eq 4)

Negative

$$PbSO_4 + H^+ + 2e^- \leftrightarrow Pb + HSO_4^-$$ (Eq 5)

During initial charging (first time), on the positive grid, a lead sulfate/lead oxide paste and a very small amount of the lead of the grid surface are converted to PbO$_2$, providing good bonding to the grid. On the negative grid, the paste is merely converted to a pure lead sponge. The major corrosion problem in lead-acid batteries occurs during subsequent cyclic discharging and recharging. Overcharging can lead to further conversion of lead to PbO$_2$ on the positive grid, according to the reaction:

$$Pb + 2O^{2-} \leftrightarrow PbO_2 + 4e^-$$ (Eq 6)

In general, the rate of this process is higher at lower acid concentrations (for example, 2.17 N, or 1.065 specific gravity) than at higher concentrations. It is necessary to design the grids such that corrosion of the positive grid is uniform rather than intergranular (Ref 12–19). Electric and hybrid vehicles use valve-regulated lead-acid cells that are subjected to dynamic operation with charge, rest, and discharge periods on the order of seconds. Such operation requires more sophisticated models that incorporate the electrochemical double layer to retard the corrosion process (Ref 20).

Alloys used to make grids include pure lead, lead-antimony with antimony levels from 0.5 to 6%, and, more recently, lead-calcium alloys. Pure lead has limited application as a grid material because it is extremely soft and prone to severe corrosion creep. However, to take advantage of the good corrosion resistance of the pure metal (better than the commonly used alloys), batteries for certain standby power applications have been designed to allow for creep (Ref 21, 22).

Corrosion of Fuel Cells

Corrosion of components in a fuel cell system is known to be a key factor in the reliability and service life of these systems. Fuel cells constitute a particular example of a battery where the two electrodes are not consumed during discharge but merely act as reaction sites for the reactants (fuel and oxidant), which are stored externally to the cell. The most common fuel is hydrogen, while air or pure oxygen is the most common oxidant. Many types of fuel cells have been developed, usually classified according to the type of electrolyte used. The five main types are:

- The phosphoric acid fuel cell (PAFC)
- The solid polymer electrolyte fuel cell (SPFC), where the electrolyte is a good proton-conducting fluropolymer at 100 °C (212 °F)
- The alkaline electrolyte fuel cell (AFC), a relatively low-temperature system (80 °C, or 175 °F), which is used on the space shuttle as the principal power source
- The molten carbonate fuel cell (MCFC), which can operate on virtually any fuel due to its very high operating temperature of 1000 °C (1830 °F) (Ref 1)
- The solid oxide fuel cell (SOFC)

Two types of fuel cell technologies, PAFC and MCFC, are discussed here; the others are addressed in the article "Corrosion in Fuel Cells" in this Volume. Figure 2 shows a schematic of the repeating components of the PAFC power generation section (Ref 23). The PAFC operates under pressurized conditions that create a corrosive environment because of high operating temperatures and high water activity (Ref 24). The primary corrosion reactions are the anodic oxidation of the various allotropic forms of carbon and the dissolution of platinum (probably to form complexed platinum ion, Pt^{2+}, in solution).

The following generic cell reaction is representative of the carbon oxidation:

$$C + 2H_2O + 4S \leftrightarrow CO_2 + 4(HS)^+ + 4e^-$$

(Eq 7)

Extensive postmortem analyses have documented the corrosion-susceptible regions. Material loss, as a fraction of the material initially present, is greatest in the catalyst layer, followed by the losses in the substrate and then loss from the bipolar plate. This ranking is what would be expected based on relative surface area and extent of electrolyte wetting of the different components, assuming that the intrinsic corrosion resistance of each component is the same (Ref 25).

The MCFC component assembly follows the bipolar principle of the PAFC stack but uses completely different materials. Figure 3 shows the basic repeating elements in a MCFC stack. The electrolyte is a molten lithium carbonate (Li$_2$CO$_3$), potassium carbonate (K$_2$CO$_3$), and/or sodium carbonate (Na$_2$CO$_3$) mixture, with the most commonly chosen composition being 62Li$_2$CO$_3$-38K$_2$CO$_3$ (usually referred to as a tile, because the carbonate salts and acuminate powder are hot pressed to form a solid piece). The ionic current in the cell is carried by the carbonate ion (CO$_3^{2-}$); therefore, the half-cell reactions at the anode and cathode are different from their PAFC counterparts. In this case, the cathode reaction is the reduction of oxygen by reaction with carbon dioxide (CO$_2$) to form CO$_3^{2-}$:

$$\tfrac{1}{2}O_2 + CO_2 + 2e^- \leftrightarrow CO_3^{2-}$$ (Eq 8)

At the anode, hydrogen reacts with the CO$^{2-}{}_3$ ion to form CO$_2$ and water:

$$H_2 + CO_3^{2-} \leftrightarrow H_2O + CO_2 + 2e^-$$ (Eq 9)

The components subject to corrosion and the nature of the corrosion processes are reasonably well understood from postmortem analyses of tested cells. With type 316 stainless steel as the current collector, moderate-to-severe attack of the anode current collector was observed after a few thousand hours of operation, with much less attack observed at the cathode. The region of the current collector that was most severely attacked was the point of contact with the anode material

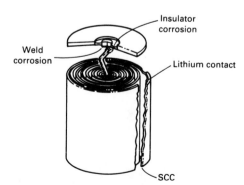

Fig. 1 Sites of corrosion in LiSO$_2$ batteries. SCC, stress-corrosion cracking

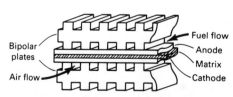

Fig. 2 Schematic of the repeating components of the phosphoric acid fuel cell power generation section

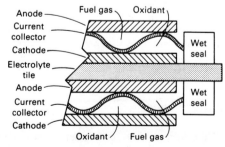

Fig. 3 Schematic of the repeating components of the molten carbonate fuel cell power generation section

(and thus in contact with the carbonate melt) and areas near the point of contact, probably owing to creep of carbonate onto the current collector (Ref 26).

A more serious and as yet unsolved corrosion process is the finding of significant solubility of NiO in the carbonate melts. Early theoretical analysis of metal corrosion in molten alkali carbonates presumed that NiO dissolution occurs by the reaction (Ref 27):

$$NiO + CO_2 \leftrightarrow Ni^{2+} + CO_3^{2-} \quad (Eq\ 10)$$

Work that addresses corrosion phenomena specific to MCFC technology is much less plentiful. The theoretical overview of the requirements for corrosion resistance in MCFC applications has served as a guide for much of the recent work (Ref 28). Recent work has been done to enhance the understanding of the predictions of the distributions of current density, water flux, corrosion, and membrane conductivity in fuel cells (Ref 29).

REFERENCES

1. A.I. Ittia, *What is Electrochemistry*, 4th ed., The Electrochemical Society, 1997, p 3–5
2. *Fuel Cells: Power for the Future,* Fuel Cell Research Associates, 1960, p 9
3. P. Bro and S.C. Levy, Lithium Sulfur Dioxide Batteries, *Lithium Battery Technology,* H.V. Venkatasetty, Ed., John Wiley & Sons, 1984, p 79–126
4. S.C. Levy, Modified Li/SO$_2$ Cells for Long-Life Applications, *Proceedings of the 29th Power Sources Symposium,* The Electrochemical Society, June 1980, p 96–109
5. S.C. Levy, Corrosion Reactions in Lithium-Sulfur Dioxide Cells, *Corrosion in Batteries and Fuel Cells and Corrosion in Solar Energy Systems,* C.J. Johnson and S.L. Pohlman, Ed., The Electrochemical Society, 1983, p 9–16
6. C.R. Walk, Lithium-Sulfur Dioxide Cells, *Lithium Battery Technology,* H.V. Venkatasetty, Ed., John Wiley & Sons, 1984, p 281–302
7. J. Burbank, A.C. Simon, and E. Willihnganz, The Lead-Acid Cell, *Advances in Electrochemistry and Electrochemical Engineering,* Vol 8, P. Delahay and C.W. Tobias, Ed., John Wiley & Sons, 1971
8. H. Bodie, *Lead-Acid Batteries,* R.J. Brodd and D.V. Kordesch, Trans., John Wiley & Sons, 1977
9. A.T. Kuhn, Ed., *Electrochemistry of Lead,* Academic Press, 1979
10. B.D. McNicholl and D.A. Rand, Ed., *Power Sources for Electric Vehicles,* Elsevier, 1984
11. D. Pavlov, Ed., *Advances in Lead-Acid Batteries,* Vol 84–14, The Electrochemical Society, 1984
12. J.J. Lander, Anodic Corrosion of Lead in Sulfuric Acid Solutions, *J. Electrochem. Soc.,* Vol 98, 1951, p 213–219; Vol 103, 1956, p 1–8
13. V.K. Dantam, "Effect of Tin on the Corrosion Properties of Wrought Lead-Calcium Alloys," Report V223022, General Motors Corporation, Delco Remy Division, April 1984
14. J.J. Lander, Effect of Corrosion and Growth on the Life of Positive Grids in the Lead-Acid Cell, *J. Electrochem. Soc.,* Vol 99, 1952, p 467–473
15. A.G. Cannone, D.O. Feder, and R.V. Biagetti, Positive Grid Design Principles, *Bell Syst., Tech. J.,* Sept 1970, p 1279–1303
16. M. Torralba, Present Trends in Lead Alloys for the Manufacture of Battery Grids—A Review, *J. Power Sources,* 1976–1977, p 301–310
17. V.K. Dantam, "Testing Low Antimony Alloy Grids in a F-11 Design Battery," Report V223128, General Motors Corporation, Delco Remy Division, July 1986
18. V.K. Dantam, "Effect of Aluminum on Microstructural Properties of Wrought Lead-Calcium-Tin Alloys," Report V223049, General Motors Corporation, Delco Remy Division, June 1984
19. D. Marshall and W. Tiedeman, Microstructural Aspects of Grid Corrosion in the PbO$_2$ Electrode, *J. Electrochem. Soc.,* Vol 123, 1976, p 1849–1855
20. V. Srinivasan, G.Q. Wang, and C.Y. Wang, *J. Electrocem. Soc.,* Vol 150 (No. 3), 2003, p A316–A325
21. C.M. Dacres, S.M. Reamer, R.A. Sutula, and I.A. Angres, *J. Electrochem. Soc.,* Vol 128, 1981, p 2060
22. C.M. Dacres, R.A. Sutula, and B.F. Larrick, *J. Electrochem. Soc.,* Vol 130, 1983, p 981
23. A.J. Appleby, Carbon Components in the PAFC—An Overview, *Proc. Workshop on the Electrochemistry of Carbon,* Vol 84-5, S. Sarangapani, J. Akridge, and B. Schumm, Ed., The Electrochemical Society, 1984, p 251–273
24. W.A. Nystrom, Raw Material and Processing Effects on the Electrochemical Wear of Carbon Composite Materials, *Proc. Workshop on the Electrochemistry of Carbon,* Vol 84-5, S. Sarangapani, J. Akridge, and B. Schumm, Ed., The Electrochemical Society, 1984, p 363–387
25. P. Stonehart and J. MacDonald, "Stability of Acid Fuel Cell Cathode Materials," EPRIEM-1664, Electric Power Research Institute, 1981
26. N.S. Choudhury, "Development of MCFCs for Power Generation," final report to the United States Department of Energy, Contract DE-ACO-77ET11319, General Electric Company, 1980
27. M. Ingram and G. Janz, The Thermodynamics of Corrosion in Molten Carbonates. Applications of E/T CO$_2$ Diagrams, *Electrochim. Acta,* Vol 10, 1965, p 783
28. R. Rapp, Materials Selection and Problems the MCFC, *Proceedings of the DOE/EPRI Workshop on MCFS,* Oak Ridge National Laboratory, 1979, p 230
29. W.-K. Lee, S. Shimpalee, and J.W. Van Zee, *J. Electrochem. Soc.,* Vol 150 (No. 3), 2003, p A341–A348

Corrosion in Fuel Cells

Prabhakar Singh and Zhenguo Yang, Pacific Northwest National Laboratory

AMONG ENERGY CONVERSION SYSTEMS being developed for using commonly available fossil and hydrocarbon-based fuels, fuel cells have attracted the most attention from utilities, automotive manufacturers, and military hardware designers. Desirable characteristics are (Ref 1, 2):

- Silent operation
- Ecological soundness
- Significantly high electrical conversion efficiency (up to 70% chemical to electrical)
- Modularity of construction (from a few watts to megawatts)
- Multifuel capability (coal-derived syngas, gaseous and liquid hydrocarbons)

Fuel cells also offer the ability to hybridize with gas turbines as well as the potential to develop near-zero-emissions power plants and to capture greenhouse gas from the exhaust, ultimately leading to a hydrogen economy and infrastructure. At present, widespread use of fuel cells is limited by high cost and their inability to achieve stable electrical performance and longer operational life. The cost per kilowatt from fuel cells can be five to six times that from conventional natural-gas-fueled turbines. To be commercially viable, fuel cells must also provide a lifetime of 5 to 10 years for stationary and automotive applications. This section addresses issues related to fuel cell electrical performance degradation and processes limiting the overall life of cells and stacks. Corrosion processes operating in a variety of fuel cell systems are presented and discussed, and corrosion mitigation schemes are highlighted. The understanding of corrosion processes in fuel cells is also expected to help in developing robust materials and a knowledge base for technologies such as electrolysis for the production of hydrogen, energy storage, and gas separation membranes.

Although the underlying science behind the fuel cell technology is well over a century old (invented by Sir William Grove in 1839), most technical and engineering progress in the development of integrated power systems is only a couple of decades old. Aggressive research and engineering development programs (R&D) worldwide have resulted in the development of commercial and prototypical precommercial demonstration units for stationary, automotive, and military applications. The 100 to 250 kWe (electrical kW) class fuel cell power plants are currently being deployed for the generation of both heat and power at industrial facilities. Smaller power plants are increasingly finding applications in automotive prime propulsion, auxiliary power, portable electronic devices, telecommunications, military hardware, and residential applications as a preferred power source and replacement for conventional internal combustion engines and batteries.

Fuel cells convert the chemical energy of gaseous or liquid fuels directly into electrical energy and heat via an electrochemical oxidation process. The absence of high-temperature combustion processes and an open flame front eliminates the formation of pollutants such as NO_x, volatile hydrocarbons, and particulate matter. Not limited by the Carnot cycle, fuel cells offer high chemical-to-electrical conversion efficiency. A fuel cell device is comprised of an anode electrode (exposed to fuel), a cathode electrode (exposed to oxidant), and a gas-impermeable electrolyte that separates the anode and cathode electrodes (Fig. 1). Electrolyte allows the passage of ions participating in the electrochemical oxidation. Fuel is oxidized at the anode electrode (production of electrons), whereas the oxidant is reduced (consumption of electrons) at the cathode electrode. While electrolyte allows for the transport of ions, electrons flow in the external circuit (load), producing electrical power. The fuel cell produces direct current electricity as long as fuel and oxidants are supplied. To build up a useful voltage, a series of cells are electrically connected and integrated into a stack. Commonly used fuels in conventional fuel cells are reformed gaseous and liquid hydrocarbons, hydrogen, coal-derived syngas, and liquid methanol. The commonly used oxidant is air. The theoretical cell voltage, called the open circuit voltage or Nernst voltage (V_{Nernst}), is governed by the chemical potential gradient in oxygen across the electrolyte and is given by:

$$V_{Nernst} = RT/4\,F \ln(PO_{2\,oxidant}/PO_{2\,fuel}) \quad (Eq\ 1)$$

where R is the gas constant, T is the absolute temperature (K), F is the Faraday constant, and $PO_{2\,oxidant}$ and $PO_{2\,fuel}$ are partial pressures.

The actual operating cell voltage (V_{cell}) during a given current flow is reduced by ohmic losses ($V_{ohmic} = IR$), due to cell internal resistance (R), and by polarization losses ($V_{polarization}$), due to activation and mass transport:

$$V_{cell} = V_{Nernst} - V_{ohmic} - V_{polarization} \quad (Eq\ 2)$$

The electrical output of a fuel cell is conventionally represented by a current-voltage relationship, shown in Fig. 2. At lower current densities, cell voltage drop is mostly associated with the activation or electron exchange process limitations (region 1), whereas at higher current densities, ohmic resistance (region 2) and mass transport (region 3) processes limit the overall performance. Degradation in the electrical performance and cell voltage loss with time are often attributed to corrosion and corrosion product formation and buildup, resulting in the poisoning of the electrode, localized metal loss, and development of resistive interfaces.

Fuel Cell Types

Review of recent technical literature indicates an explosive growth in the development and demonstration of fuel cell technologies, ranging from unconventional biological, direct carbon, and single-chamber fuel cells to engineering and scaleup of more conventional fuel cells. This section focuses on the "mature technologies," which, depending on the operating temperature and type of electrolytes used, can be grouped into five major types (Fig. 3):

- Alkaline fuel cells (AFCs)
- Phosphoric acid fuel cells (PAFCs)
- Polymer electrolyte membrane fuel cells (PEMFCs)
- Molten carbonate fuel cells (MCFCs)
- Solid oxide fuel cells (SOFCs)

Alkaline fuel cells use a liquid alkaline electrolyte, such as KOH, and operate in the 100 to 250 °C (212 to 480 °F) temperature range. The liquid electrolyte transports OH^-, and the cell reactions are:

$$2H_2 + 4OH^- \rightarrow 4H_2O + 4e^- \quad (anode) \quad (Eq\ 3)$$

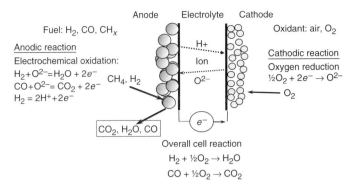

Fig. 1 Fuel cell. Fuel introduced at the anode is oxidized, and liberated electrons produce useful power in the external circuit.

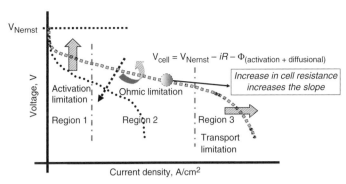

Fig. 2 Current-voltage characteristic of a fuel cell. Operating cell voltage is less than theoretical (Nernst voltage) due to activation (region 1), resistance (region 2), and mass transport (region 3). To improve output: in region 1, reducing activation polarization will move V-I curve upward; in region 2, reducing ohmic polarization reduces slope of V-I curve; in region 3, reducing diffusional polarization moves bend in V-I curve to the right. Degradation in electrical performance with time is attributed to electrode deactivation, electrode poisoning, and increase in cell resistance.

$$O_2 + 2H_2O + 4e^- \rightarrow 4OH^- \text{ (cathode)} \quad \text{(Eq 4)}$$

Alkaline fuel cells have been successfully used in the National Aeronautics and Space Administration (NASA) Apollo and Orbiter space missions. When operated on pure H_2 and O_2 gases, AFCs show stable performance and relatively high power densities. However, cell performance significantly degrades during their operation on fuels and oxidants containing CO_2 (present as impurities in commercial reformed fuels and ambient air), due to the formation of solid carbonates (K_2CO_3). Presence of corrosive liquid hydroxides, electrolyte transport, and internal shorting also remain a concern for accelerated corrosion of conventional cell, stack, and the balance of plant components fabricated from low-cost commercial materials. For successful long-term operation of cells and stacks, it has been recommended that CO_2 must be scrubbed from both fuel and oxidant gas streams (Ref 3).

Phosphoric Acid Fuel Cells. As the name implies, PAFCs use proton-conducting phosphoric acid (H_3PO_4) as an electrolyte and operate in the 150 to 200 °C (300 to 390 °F) temperature range. Electrode reactions are:

$$H_2 \rightarrow 2H^+ + 2e^- \text{ (anode)} \quad \text{(Eq 5)}$$

$$\tfrac{1}{2}O_2 + 2H^+ + 2e^- \rightarrow H_2O \text{ (cathode)} \quad \text{(Eq 6)}$$

The PAFCs most commonly use platinum or platinum alloy catalysts for electrode reactions. Highly dispersed catalysts are supported on carbon substrates that also act as a gas diffusion layer. The liquid electrolyte is constrained in a porous ceramic matrix. Bipolar (dual-atmosphere) gas separators and current collectors are fabricated from graphitic carbon with embossed patterns for fuel and oxidant flow. Stack designs incorporate provisions for cooling to remove heat produced during power generation. Although large commercial power plants employing PAFC stacks have been installed worldwide and have successfully operated on a variety of reformed hydrocarbon fuels, mass commercialization has not been achieved due to overall high system construction cost (>$3000/kW). Limited life and performance degradation observed in the PAFC systems are attributed to electrode poisoning and catalyst deactivation due to gas-phase impurities (CO and H_2S), sintering and agglomeration of platinum, migration and evaporation of electrolyte, and corrosion of cell and stack component materials. Corrosion resulting in softening and large volume increases has been observed on bipolar graphite current collectors nominally prepared by the low-temperature carbonization (<1000 °C, or 1830 °F) process. Such bipolar plates, when carbonized at higher temperatures (>2500 °C, or 4530 °F), remain corrosion tolerant and retain their shape during long-term operation. Cell performance degradation mitigation strategies have also been developed and implemented in commercial power plants.

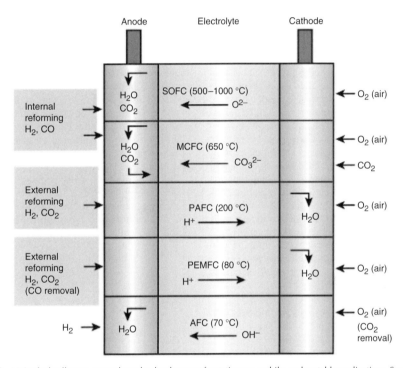

Fig. 3 Major fuel cell types currently under development for stationary, mobile, and portable applications. See text for fuel cell types. Source: Adapted from Ref 3

Polymer electrolyte membrane fuel cells, also known as solid polymer fuel cells, have attracted attention in recent years due to increased interest in hydrogen and its potential application in automobiles as the prime propulsion power system (Ref 1–4). Automotive manufacturers in North America, Europe, and Japan have invested heavily in this technology to replace the internal combustion engine, with the intent of developing a hydrogen-base fuel infrastructure. The 50 kW and larger automotive power plants have been developed and installed in passenger cars and large public transport buses. Similar power plants are also being field tested in stationary industrial power plants where hydrogen gas is readily available as a by-product of chemical processes (such as in the chlorine-alkali industry). The PEMFCs use a polymeric proton-conducting electrolyte and carbon-base electrodes containing noble metal and alloy catalysts for electrode reactions. Electrode processes for both oxidation of fuel and reduction of the oxidant are similar to the PAFC electrode processes. The PEMFCs nominally operate at 60 to 90 °C (140 to 194 °F) on high-purity hydrogen gas, because trace amounts of gas-phase contaminants such as CO, NH_3, and H_2S severely degrade the electrical performance due to the poisoning of cell electrodes. The presence of ammonia in the fuel stream and the formation of peroxides have been reported to react adversely with the membrane, limiting the overall life of the fuel cell. Due to lower operating temperatures, the cell electrodes and stack components remain prone to flooding with water and susceptible to galvanic corrosion. Conventional fuel cell stacks use graphitic molded bipolar gas separators and current collectors. Due to lower exposure temperature, such current collectors have been reported to resist corrosion. Metallic current collectors, coated with polymeric or metallic conducting coatings, have also been used in cell stacks for lightweighting and volume reduction. Care must be taken in developing crack-free or pinhole-free coatings to minimize localized corrosion (galvanic cell formation) and the release of metal ions that can contaminate the electrolyte membrane and poison the electrode catalysts.

Molten carbonate fuel cells operate in the 600 to 750 °C (1110 to 1380 °F) temperature range and use a variety of fuels, ranging from hydrocarbons to reformed hydrocarbon fuels (H_2, CO, CH_4, etc.), to produce electricity. Electrode reactions are:

$$H_2 + CO_3^{2-} = H_2O + CO_2 + 2e^- \text{ (anode)} \quad \text{(Eq 7)}$$

$$CO + CO_3^{2-} = 2CO_2 + 2e^- \text{ (anode)} \quad \text{(Eq 8)}$$

$$CO_2 + \tfrac{1}{2}O_2 + 2e^- = CO_3^{2-} \text{ (cathode)} \quad \text{(Eq 9)}$$

It is noted that the cathodic reaction in MCFCs involves the presence of CO_2 gas. This is normally accomplished by taking the CO_2 from the used fuel stream. The MCFCs use a molten mixture of alkali metal carbonates as electrolyte for the transport of CO_3^{2-} ions participating in the anodic oxidation of fuel and the cathodic reduction of oxidant. The molten electrolyte, usually a binary mixture of lithium and potassium carbonate, is retained in a porous matrix sandwiched between an anode (nickel alloys usually containing chromium or aluminum to form respective internal oxides) and cathode (lithiated nickel oxide) electrodes. Gas separation and cell-to-cell connection is provided by conductive metal current collectors. Nickel-clad stainless steels (such as types 316, 310, Crutemp, or 446) have commonly been used as the current collector where nickel is exposed to the fuel gas and stainless steel is exposed to the oxidant gas. Electrically insulating gas seals, called wet seals, are formed by aluminizing the metal substrates. Because of their higher operating temperature, MCFCs offer the potential for hybridization and higher electrical efficiency, high-quality product heat for combined heat and power applications, and fuel flexibility, including the use of hydrocarbons and internal reformation. The 250 kWe and MWe class power plants are currently being developed for distributed power generation, including industrial, military, and telecommunication remote power systems.

One of the primary shortcomings of the current technology is its lack of durability. High-temperature exposure, complex fuel and oxidant gas atmospheres, and the presence of molten carbonate electrolyte accelerate the corrosion and dissolution of oxides formed on the metallic current collectors and cathode electrode. The presence of bipolar exposure conditions also leads to a localized form of metal attack. Carburization and sensitization of the metallic substrates have also been observed during long-term tests in a carbonaceous gas environment. To enhance component life, the current R&D focus is on the selection of advanced materials compositions as well as the engineering design of components and the selection of operating conditions without compromising the electrical performance.

Solid oxide fuel cells operate at 600 to 1000 °C (1100 to 1830 °F) and use a solid-state electrolyte membrane for the transport of oxygen ion and electrochemical oxidation of fuels over a porous anode electrode:

$$H_2 + O^{2-} \rightarrow H_2O + 2e^- \text{ (anode)} \quad \text{(Eq 10)}$$

$$CO + O^{2-} \rightarrow CO_2 + 2e^- \text{ (anode)} \quad \text{(Eq 11)}$$

$$CH_4 + 2O^{2-} \rightarrow CO_2 + 2H_2O + 4e^- \text{ (anode)} \quad \text{(Eq 12)}$$

$$\tfrac{1}{2}O_2 + 2e^- \rightarrow O^{2-} \text{ (cathode)} \quad \text{(Eq 13)}$$

In its most common form, SOFCs consist of a thin, dense film of doped zirconia electrolyte (such as yttria-stabilized zirconia) between a porous nickel cermet anode (Ni-ZrO_2) and a ceramic perovskite cathode (doped ABO_3, where A is a lanthanide group metal, and B is a transition metal) (Ref 5). Reactant flow and cell-to-cell electrical connection is achieved through metallic current collectors. High-temperature operation of the cells provides high-quality product heat, tolerance to poisons (sulfur, CO, and NH_3), ability to internally reform hydrocarbons on the cell electrode, hybridization with gas turbines, and significantly higher electrical conversion efficiency. The SOFC power generation systems ranging from several watts to several hundred kilowatts in size have been successfully assembled and field-tested.

Technology issues are currently being addressed by large industrial research projects as well as by a number of U.S. government programs, including:

- Department of Energy: Solid-State Energy Conversion Alliance and Coal-Based Fuel Cell Systems programs
- Defense Advanced Research Projects Agency: Palm Power program
- Department of Defense: Tactical Quiet Generator and Unmanned Underwater and Aerial Vehicles programs

Some of the important development areas being addressed include identification and testing of advanced cell and stack component materials for improved electrical performance and performance stability. Although ferritic stainless steels and electronically conducting doped lanthanum chromites have been commonly used as current collectors, due to their matched coefficient of thermal expansion in intermediate- and high-temperature SOFCs, long-term stability issues remain. Accelerated corrosion, oxide evaporation, and contamination of electrodes are key development areas currently being addressed. Ferritic and austenitic stainless steel interconnects exposed to the bipolar environment show accelerated corrosion and localized metal loss. Chromia scale evaporation and subsequent condensation/interaction with cathode electrodes have also been observed. At 1000 °C (1830 °F), significant interaction of the anode electrode with silica vapor (originating from ceramic alumina-base thermal insulation) has been observed. Several advanced coatings, alloy formulations, and design modifications have been developed to mitigate such degradation.

Corrosion Processes in Fuel Cell Systems

The heart of the fuel cell power generation system is the fuel cell stack, where chemical energy of the fuel is efficiently converted into electrical power by the electrochemical oxidation process. To allow for the use of conventional fuels in cell stacks, efficient use of process heat, and conversion of direct current power into alternating current, the balance of the fuel cell power plant consists of fuel processor, heat exchangers, combustors, and power conditioner (Fig. 4a). The operating conditions in fuel cell power generation subsystems vary widely, depending on the nature of the fuel and the type of fuel cells used. Figure 4(b) shows the operating temperature range and exposure conditions for the fuel processor, fuel cell stack, and combustor. The introduction and presence of hydrocarbon fuel gas mixtures near the inlet of the fuel processor develops higher carbon activity in the fuel processor. As the hydrocarbons reform to produce H_2 and CO mixtures along the flow direction, hydrocarbon and steam partial pressures reduce. Exposure conditions in the cell stacks, on the other hand, lead to the consumption of H_2, CO, and CH_x fuel species and the production of H_2O and CO_2 along the fuel flow direction. Formation of H_2O and CO_2 in the gas stream and reduction in the H_2/H_2O and CO/CO_2 ratio also gradually increase the oxygen partial pressure in the fuel stream near the fuel outlet. The exiting fuel from the cell stack is conventionally recirculated and fed to the reformer or fully combusted to produce thermal energy before its safe release into the environment.

Corrosion processes in fuel cells are divided into three major groups:

- Corrosion due to solid-gas interactions
- Corrosion due to solid-liquid interactions
- Corrosion due to solid-solid interactions

Corrosion due to solid-gas interactions represents a wide variety of corrosion processes that range from oxidation, carburization, and metal dusting, to surface oxide evaporation and bipolar corrosion during exposure to gas-phase reactants and simultaneous exposure to an oxidizing (oxidant) and a reducing environment (fuel). Oxidation of metals and alloys in simple and multiconstituent gas atmospheres is well studied and documented. The classic metal oxidation and surface oxide buildup mechanism developed initially by Wagner (Ref 6) obeys the parabolic time law, and the rate-limiting steps are controlled by the diffusion of ions in the oxide lattice. Electrical neutrality in the growing scale is maintained by the diffusion of electrons in the same direction as the cations, and local equilibrium is maintained throughout the scale. The oxidation mechanism has been further extended (Ref 7–9) to account for the oxidation of alloys, internal oxide formation, transition from internal to external oxidation, complex oxide buildup in multioxidant systems, and oxide growth due to the molecular transport of reactants within the growing scale. The oxidation of iron- and nickel-base chromia- and alumina-forming alloys and the role of alloying constituents on the development of protective oxides have been studied (Ref 9–13). Oxide scale morphology predominantly depends on the exposure temperature and chromium level. At lower chromium levels (<5%), both iron and nickel alloys form mixed oxides and often show acceleration in the oxidation rate, whereas at 10 to 12% Cr level, alloys initially show the formation of protective surface oxides. Long-duration exposures of such alloys lead to internal oxidation and formation of a chromium-depleted zone near the surface. At chromium levels >15 to 18%, the alloys form a protective chromia scale and also are capable of rehealing (due to a higher chromium reservoir) in the event of scale spallation and cracking.

The addition of rare earth elements, such as zirconium, yttrium, and hafnium, to bulk alloys has proven beneficial in improving the scale adherence under thermal cyclic conditions. Rare earth additions facilitate the formation of anchors at the metal-oxide interface and also provide vacancy sink (annihilation and prevention of pores) for improving the adherence of the oxide with the metal substrate (Ref 14, 15). For applications where electrically conductive oxides are required in order to minimize the resistive loss, alloys are tailored to reduce the concentrations of elements such as silicon and aluminum. Silicon present in the alloy has a tendency to segregate and form continuous SiO_2 at the scale-metal interface and increase the electrical resistance of the scale. At certain levels, aluminum also tends to form alumina at the metal-oxide interface, increasing the overall scale resistance. It should, however, be noted that the additions of aluminum and silicon improve the scaling resistance and hence are selectively used where the electrical conduction in the scale is not an important requirement. Oxidation of metals and alloys in a complex gas environment has been extensively studied for its importance in understanding the scaling process and materials selection for applications in cell and stack components and fuel processors exposed to a reducing gas environment. The presence of an environment consisting of H_2, CO, CH_4, CO_2, H_2O, and so on establishes scaling processes very different from simple oxidation in oxygen or air. Such exposures have the potential to form condensed products from reactions with carbon, oxygen, and so on. Metals such as nickel and copper remain stable in most fuel environments and up to 99% fuel utilization. Experiments conducted on commercial Ni-200 foil samples in N_2-3%H_2-3%H_2-3%H_2O for 200 h showed thermal grooving of the metal surface at 800 and 1000 °C (1470 and 1830 °F). Cross-sectional analysis of the metal showed no surface oxide formation. Grain boundaries of the samples were, however, found to be decorated with SiO_2. Molecular transport of gaseous species and the establishment of redox (H_2-H_2O and CO-CO_2) reactions within the oxide scale enhance the scaling and localized metal loss. It is also observed that the presence of CO and CO-CO_2 redox couples facilitates the carbide formation at the metal-scale interface.

Long-duration exposures of both austenitic and ferritic stainless steels during the operation of molten carbonate fuel cells have confirmed such observations, showing significant

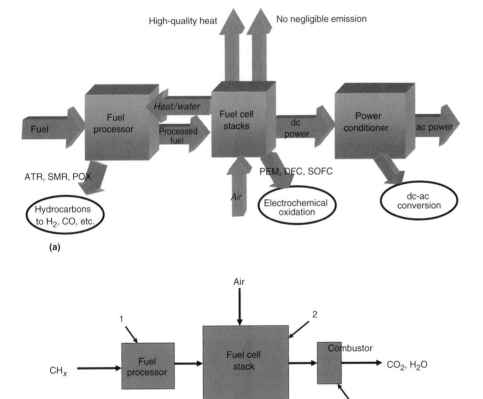

Fig. 4 Fuel cell system including auxiliary equipment. (a) Fuel cell power generation system incorporating fuel processor, fuel cell stack, and power conditioner. ATR, autothermal reforming; SMR, steam methane reforming; POX, partial oxidation; PEM, polymer electrolyte membrane; DFC, direct fuel cell; and SDFC, solid oxide fuel cell. dc, direct current; ac, alternating current. (b) Respective operating temperatures and exposure environments. Area 1: High carbon activity; partial pressure, $P_{H_2O\ in} > P_{H_2O\ out}$; temperature ($T$), >600 °C (1100 °F); there can be contaminate inclusions. Area 2: $P_{H_2O\ out} > P_{H_2O\ in}$; $P_{H_2\ in} > P_{H_2\ out}$; T, room temperature to 1000 °C (1830 °F). Area 3: $P_{H_2\ out} = 0$ $P_{H_2O\ out} + P_{N_2\ out} + P_{CO_2\ out} = 1$; $T > 800$ °C (1470 °F)

chromium carbide formation at the metal interface (Ref 16). At 600 to 700 °C (1110 to 1290 °F), carbide formation within the bulk metal, especially along the grain boundaries, also occurs due to sensitization. Carburization of stainless steels in high- ($a_c = 10^{-1}$) and low- ($a_c = 10^{-20}$) carbon atmospheres was examined in the 600 to 700 °C (1110 to 1290 °F) temperature range, and it was found that carbides can form by both molecular transport and solution diffusion processes in high-carbon-activity environments, whereas carbide formation is expected only by molecular transport in low-carbon environments. Establishment of a redox gas atmosphere within the oxide enhances the metal ion transport by short circuit diffusion, leading to thicker scale growth and underlying metal loss. Stainless steels (types 304 and 316) form a multilayer scale, where the outer scale is comprised of iron oxide and the inner scale predominantly consists of porous chromia and nickel. At higher concentrations of chromium (24 to 26%) and nickel (>20%) in type 310, the scale morphology changes considerably, forming uniform outer and inner oxide layers, followed by a nickel-enrichment layer in the metal near the metal-oxide interface. To minimize carbide formation and sensitization, common practice involves the use of low-carbon bulk alloys. Ingress of carbon through the scale is also minimized by reducing the molecular transport and forming defect-free oxides at the surface. In the presence of hydrocarbons, both nickel- and iron-base alloys remain susceptible to carbon deposition and metal dusting. Nickel-base chromia-forming alloys perform better than iron-base alloys under SOFC conditions, because the oxide scale, consisting of Cr_2O_3 (formed on nickel-base alloys), provides better protection than spinel-containing scales formed on iron-base alloys. Metal dusting is accelerated in low-humidity and high-pressure environments and can occur more frequently near the fuel inlet. Carbon formation in the nickel anode and catalyst also remains a concern during fuel cell operation. Gaseous reactions that lead to metal dusting are:

$$CO + H_2 \rightarrow C + H_2O \quad \text{(Eq 14)}$$

$$2CO \rightarrow C + CO_2 \quad \text{(Eq 15)}$$

$$CH_4 \rightarrow C + 2H_2 \quad \text{(Eq 16)}$$

The mechanisms of metal dusting have been studied and discussed (Ref 17–19). Factors influencing metal dusting include carbon activity in the gas stream, oxide morphology and composition, temperature, and exposure time. Degradation of nickel due to carburization and metal dusting has been frequently observed in the reformer and also on the cell anode under upset operating conditions, where filamentary carbon forms at the substrate and breaks the integrity of the support. Large volume changes associated with the carbon formation result in pressure buildup in the reformer and cell stack. At elevated temperatures (600 to 1000 °C, or 1110 to 1830 °F), representative of high-temperature solid oxide and molten carbonate fuel cells, oxide evaporation also becomes a concern for long-term metal substrate protection. Evaporation of chromia and silica are especially of interest, because they not only relate to the oxide loss from the scale surface but also the contamination of electrodes and electrical performance loss. Evaporation of chromia and formation of CrO_3 vapor phase above 1000 °C (1830 °F) in air is well documented in superalloy literature (Ref 10). At intermediate temperatures and in the presence of water vapor, chromia also forms hydrated oxides that have higher vapor pressure than CrO_3 (Ref 20, 21):

$$Cr_2O_3 + \tfrac{1}{2}O_2 \rightarrow 2CrO_3 \quad \text{(Eq 17)}$$

$$Cr_2O_3 + 2H_2O + \tfrac{3}{2}O_2 \rightarrow 2CrO_2(OH)_2$$

(most predominant) (Eq 18)

Evaporation of chromia results in the deviation of the scale growth kinetics from parabolic to paralinear rates. Evaporation reactions and vapor pressures of chromium oxyhydroxides are shown in Fig. 5. Vapor species formed at the scale surface migrate along the reactant flow direction and have the potential to react and poison the contacting electrode. Like the accelerated evaporation of chromia in an oxidant gas environment, silica evaporation has also been reported in a SOFC reducing (fuel) environment containing steam. At 1000 °C (1830 °F), gaseous hydrated silica vapors form according to the following reaction:

$$SiO_2 + 2H_2O \rightarrow Si(OH)_4 \quad \text{(Eq 19)}$$

The silica vapor entrains in the fuel gas deposits, poisoning the nickel electrode and hydrocarbon reforming catalyst (Ref 22). Silica transport in SOFCs has been studied (Ref 23), and mechanisms of vaporization, transport, and deposition over nickel reforming substrates have been identified. A bipolar exposure (air/metal/ H_2-H_2O) condition results from simultaneous exposure to an oxidizing (cathodic) and reducing (anodic) environment and is typical of cell interconnects and gas separator exposure. Long-term electrical tests in MCFC and SOFC power generation systems showed accelerated corrosion, metal loss, and bulk structural changes on interconnects and gas separators. A few citations exist in the literature concerning corrosion under bipolar exposure. One of the earlier reports (Ref 24) indicated unusual oxidation and metal loss during the oxidation of several stainless steels under MCFC operation conditions. Simple oxidation in oxidizing or reducing gases was not able to explain such degradation. Recently, the bipolar exposure conditions were simulated, and the corrosion behavior of candidate current collector materials silver, iron alloys, and nickel alloys were investigated (Ref 24–27).

Although silver remains stable in oxidant and fuel gas atmospheres, bipolar exposure resulted in bulk porosity formation and cracking. Figure 6 shows the micrographs of silver exposed to flowing air and fuel (H_2-3%H_2O). Dissolution of both hydrogen and oxygen in the bulk metal and their subsequent reaction form high-pressure water vapor at locations such as grain boundaries and other nucleation sites. This mechanism results in the formation of interconnected porosity and cracks (Fig. 7). The thermodynamic models (Ref 25) of the reactions and associated Gibbs free energies (ΔG_n) are:

$$H_2(g) \leftrightarrow 2[H]_{Ag} \quad \Delta G_1 \quad \text{(Eq 20)}$$

$$O_2(g) \leftrightarrow 2[O]_{Ag} \quad \Delta G_2 \quad \text{(Eq 21)}$$

$$H_2(g) + \tfrac{1}{2}(O_2) \leftrightarrow H_2O(g) \quad \Delta G_3 \quad \text{(Eq 22)}$$

$$2[H]_{Ag} + [O]_{Ag} \leftrightarrow H_2O(g) \quad \Delta G_4 \quad \text{(Eq 22a)}$$

The net energy balance is:

$$\Delta G_4 = \Delta G_1 + \Delta G_{2v} + \Delta G_3$$

$$\log K_4 = \log(P_{H_2O}) - 2\log(aH) - \log(aO)$$

$$\log(P_{H_2O}) = -\Delta G_4/4.575T + 2\log(aH) + \log(aO)$$

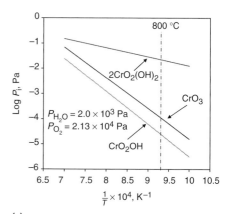

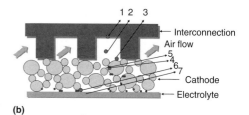

Fig. 5 Chromia scale evaporation in an oxidizing environment. (a) Partial pressure of various gaseous species formed in humidified air. Chromia evaporation is dominated by $CrO_2(OH)_2$ formation, especially at low temperatures (note inverse temperature scale). At 800 °C (1470 °F), for example, partial pressure of $CrO_2(OH)_2$ is 220 times greater than CrO_3. The level of $CrO_2(OH)_2$ gas is still significant at 800 °C (1470 °F) and 0.022 Pa (3.2×10^{-6} psi), which is 52% of the pressure at 900 °C (1650 °F). (b) Schematic presentation of possible formation, transport, and interactions of chromium species during solid oxide fuel cell operation near the cathode. 1, chromium evaporation from interconnection surface; 2, gas-phase transport of chromia vapor; 3, contact with cathode surface; 4, reaction with cathode; 5, diffusion into cathode; 6, reduction and deposition at cathode/electrolyte interface; 7, diffusion into electrolyte/barrier layer

where K_4 is the equilibrium constant, P_{H_2O} is the partial pressure of H_2O, T is absolute temperature (K), and aH and aO are activities of hydrogen and oxygen. The relation of temperature to the partial pressure of H_2O formed in bulk silver is:

Temperature, K	$\log(P_{H_2O})$, atm
600	17.05
700	14.15
800	11.97
900	10.29
1000	8.92

These values of steam pressure remain high in the operating temperature range.

Similar experiments conducted on types 304 and 430 stainless steels showed localized scale growth and underlying metal loss (Fig. 8, 9) (Ref 26–28). Similar scale morphology has been reported to develop during the oxidation.

Similar scale morphology has been reported to develop during the oxidation of stainless steels in steam-containing environments (Ref 29, 30). Exposure of nickel and nickel alloys to similar bipolar exposure conditions only slightly modified the scale morphology and the scaling rate. Although the exact mechanism of the acceleration in corrosion observed on stainless steels under bipolar exposure is not fully understood, it is hypothesized by the authors that the diffusion of hydrogen through the metals from the fuel side to the oxidant side could result in the H_2-H_2O redox reaction at the metal-oxide interface at the oxidant side, leading to accelerated outward diffusion of cations in the oxide scale and localized growth of oxide nodules. Schematic presentation of the reaction processes is shown in Fig. 10. In the presence of steam, it is postulated that the localized defects present in the scale lead to the inward diffusion of H_2O, the possible development of H_2-H_2O redox atmosphere, and the outward migration of cations. The observed acceleration in the oxidation can also occur due to the modification of oxide stoichiometry and changes in the scale defect structure or the transport of protons through the scale. Further work is needed to understand the role of the aforementioned processes and bipolar exposure on the scaling behavior.

Corrosion due to Solid-Liquid Interactions. The presence of reaction product, such as liquid water in PEMFCs and liquid electrolytes in PAFCs, AFCs, and MCFCs, creates the potential for enhanced corrosion and for corrosion processes different than the solid-gas interactions previously presented. Corrosion processes in the presence of liquids vary from simple galvanic, to complex oxide dissolution and metal pitting, to hot corrosion where accelerated metal corrosion is controlled by the basicity or acidity of the melt. All forms of corrosion tend to accelerate the localized metal loss. At lower temperatures in PEMFCs, for example, the formation of liquid water corrodes the metal bipolar plates and contaminates the electrodes and catalysts. In PAFCs, the liquid phosphoric acid tends to corrode the carbon electrodes and current collectors, whereas in MCFCs, the molten carbonate electrolyte dissolves oxides formed on both current collectors and seals and develops nonprotective scale.

Corrosion in the presence of water formed by the oxidation of hydrogen has been studied in PEMFCs (Ref 31–33). The PEMFC environments comprise weakly acidic media and are often found to be corrosive to the stack components, in particular, cell-to-cell interconnects, that are made from either graphitic carbon-base materials or stainless steels. The carbon/graphite-base materials show promise in the chemical stability required to withstand the high-humidity and low-pH environment of the PEMFC stack, whereas metal bipolar plates suffer from surface oxide formation. Metal ions (e.g., Fe^{n+}) leaching from the metallic interconnects into the polymer membrane block the sulfonic acid sites and reduce the ionic conductivity of the membrane. Passivating oxides or oxyhydroxide layers formed on the surface of the metallic interconnects also increase the contact resistance between the bipolar plate and electrodes, resulting in long-term performance loss. The metal ions also poison the electrode catalysts, leading to degradation in cell performance. To protect the metallic interconnects and prevent metal ion poisoning to the polymer membrane, a thin, inert, yet electrically conductive coating is often applied onto the metallic interconnects, using various surface-coating techniques. Both precious metals (e.g., gold) and conductive oxides (e.g., SnO_2) have been successfully used as the coating materials. A great level of success has been reported with the precious metal coatings. Overlay oxide coatings appear to be acceptable in terms of cost but have thus far not proven sufficiently viable due to the formation of pinhole defects, which can result in local galvanic corrosion and metal ion contamination to the membrane (Ref 34). The presence of pinholes has been reported to be mitigated via the application of an interlayer between the metal and an overlay coat (Ref 35, 36). For example, the combination of an electroless nickel interlayer and a physical vapor deposited CrN outer layer

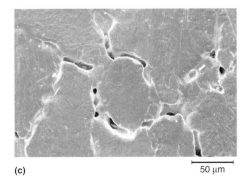
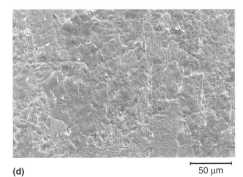

Fig. 6 (a) and (b) Microstructural cross sections of silver tubular sections, and (c) and (d) surfaces. (a) and (c) were exposed to a bipolar condition (H_2 + 3%H_2O/silver/air) at 700 °C (1290 °F) for 700 h. (b) and (c) were exposed to air at the same temperature and duration. (c) Grain-boundary porosity is present. (d) No grain-boundary porosity

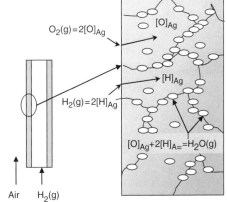

Fig. 7 Schematic presentation of porosity formation and structural degradation in silver exposed to hydrogen/air dual environment. Partial pressure of steam formed within the bulk metal remains very high at all exposure temperatures.

exhibited a better corrosion resistance than the CrN layer only (Ref 35). In addition to CrN, other transition metal nitrides as well as their carbides appear to offer a combination of high electrical conductivity and good corrosion resistance that makes them attractive for potential use as a protective surface layer in this application. Recent work has shown promise with thermally nitrided metal surfaces (Ref 37, 38).

Corrosion of metallic components in the presence of molten carbonate electrolyte mixtures (usually lithium and potassium carbonate eutectic) has been the subject of intense study during the development of materials for MCFCs. Scaling of both iron- and nickel-base alloys has been extensively studied in both anode and cathode gas atmospheres (Ref 39). Experiments in the anode gas showed the precipitation of iron oxides at the electrolyte-gas interface, indicating scale dissolution and precipitation-type fluxing reaction. The molten electrolyte does not show any presence of chromium. In the case of chromium-plated steels, the highest protection is achieved due to the formation of reaction products consisting of Cr_2O_3 and $LiCrO_2$. The scale formed is 20 to 30 times thinner than those formed on the stainless steels. Chromium plating also offers a barrier to carbon diffusion and carburization of the metal substrate. The oxidation resistance of Al_2O_3-forming alloys remains excellent in the presence of molten electrolyte; however, the alloys provide highly electrically resistive scale not suitable for electrical conduction. Nickel-clad or copper-plated stainless steels showed the stability of nickel and copper in the fuel gas. The process of oxide scale fluxing in the molten electrolyte mixture in the fuel gas is considered identical to the hot corrosion fluxing mechanisms proposed (Ref 40).

In the cathode gas atmosphere, stainless steels do not show fluxing of the oxides. The electrolyte, however, dissolves surface Cr_2O_3 and other oxides to form dissolved species such as CrO_4^{2-}, FeO_2^{2-} and NiO_2^{2-}. Figures 11 and 12 present the corrosion processes operating in the anode and cathode gas atmospheres. The relative amount of dissolved species depends on the oxide melt equilibrium. Thermodynamic calculations indicate that the ratio of CrO_4^{2-} to FeO_2^{2-} is very high in the cathode gas, indicating preferential dissolution of Cr_2O_3 from the scale. The Cr_2O_3 dissolution and formation of chromates is:

$$2Li_2CO_3 + Cr_2O_3 + 3/2 O_2 = 2Li_2CrO_4 + 2CO_2$$
(Eq 23)

$$2K_2CO_3 + Cr_2O_3 + 3/2 O_2 = 2K_2CrO_4 + 2CO_2$$
(Eq 24)

Dissolution of Cr_2O_3 and its higher solubility in the electrolyte make the chromium plating a poor choice for cathode-side application. Alloys forming exclusive corrosion products, such as SiO_2, MnO, (Mo, V, or W)$_xO_y$, also show excessive dissolution in the electrolyte. Dissociation and consumption of carbonate melts lead to permanent loss of the electrolyte from the fuel cell system. Formation of hydroxides (reactions with H_2O) also increases the loss of the electrolyte. Loss of the electrolyte from the cell increases cell resistance and electrical performance degradation. The MCFC wet seal area that separates the fuel, air, and ambient atmosphere experiences accelerated corrosion in the presence of molten electrolyte because of the development of localized corrosion cells during simultaneous exposure to reducing and oxidizing atmospheres. Aluminizing of the metal surface has demonstrated corrosion prevention due to the formation of a passivating alumina scale (Ref 41).

Air/air
(a)

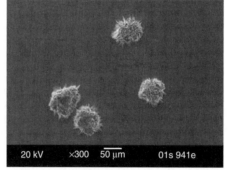

Air/fuel
(b)

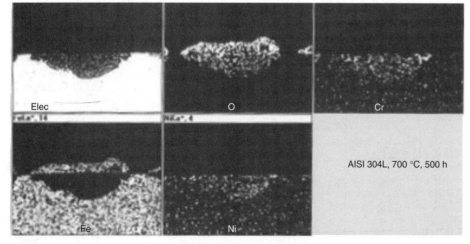

(c)

Fig. 8 Surface morphology and elemental distribution in scales formed on type 304 stainless steel during exposure to single and bipolar exposure conditions. (a) Formation of uniform surface oxide layer in air. (b) Development of local iron-oxide-rich nodules during exposure to bipolar condition. (c) Elemental distribution in the scale formed during bipolar exposure. Note iron enrichment in the outer oxide; chromium and nickel within the inner scale

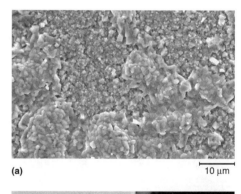

(a)

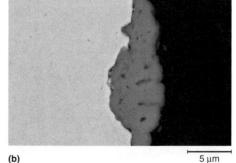

(b)

Fig. 9 SEM observation of the scale on type 430 stainless steel after isothermal oxidation at 800 °C (1470 °F) for 300 h. (a) Surface and (b) cross-sectional microstructures of the airside scale on the coupon that was simultaneously exposed to air on one side and moist hydrogen on the other. Results of the energy-dispersive spectroscopy linear analysis on cross sections are also included. Source: Ref 26

Corrosion due to Solid-Solid Interactions. Solid-solid interactions in fuel cells have been most frequently associated with high-temperature solid-state reactions where adjacent components of the cell stack or subsystems react to form undesirable corrosion products and structures that limit the performance and life of the fuel cell system. Such interactions vary from Kirkendall void formation due to interdiffusion of metals to compound formation near the electrode-electrolyte interfaces. Interdiffusion of metals leading to Kirkendall void formation has been observed during the corrosion evaluation of copper plating in the fuel gas environment of MCFCs (Ref 42). Copper plating that is thermodynamically stable under the anode environment shows extensive diffusion in the contacting nickel anode, resulting in the formation of large porosity in copper and swelling in nickel. Higher diffusion of copper in nickel leads to net positive flow and transport of copper and formation of voids in the coating layer. Diffusing copper in nickel forms a nickel-copper alloy. Solid-state interdiffusion has also been observed in nickel plating and aluminized coatings formed over stainless steels. In SOFCs, the most commonly observed solid-solid interactions represent reactions between interconnects and electrodes. One such reaction is the formation of resistive pyrochlore ($La_2Zr_2O_7$) at the perovskite (doped $LaMnO_3$) cathode electrode and yttria-stabilized zirconia electrolyte interface:

$$2LaMnO_3 + 2ZrO_2 \rightarrow La_2Zr_2O_7 + 2MnO + \tfrac{1}{2}O_2$$

(Eq 25)

Reduced oxygen partial pressure favors the forward reaction, whereas the presence of MnO favors the backward reaction and dissolution and removal of the pyrochlore. Reactions of the perovskite cathodes with the interconnect surface oxides, such as Cr_2O_3, Al_2O_3, SiO_2, and FeO, also result in the formation of discrete compounds such as $LaCrO_3$ and $LaSiO_3$. Substitution of the cations present in the surface oxides also modifies the electrical properties of the cathode. For ceramic chromite interconnects, prolonged exposure to elevated temperatures shows interdiffusion of manganese from the cathode electrode and the formation of interface porosity. Optimization of the interconnection chemistry (doping of both A and B sites) and lowering of the fabrication and cell operation temperatures reduce such interactions. In the SOFC seal area, similar solid-state reactions involving surface chromium oxide (formed on the current collectors) and glass result in the formation of chromates (Ref 43).

Materials and Technology Status

Global interest in the development and deployment of fuel cell power generation systems has considerably increased in the last decade. As the major fuel cell technologies (PEMFCs, MCFCs, and SOFCs) mature and transition from laboratory to commercial products, R&D efforts focus on the long-term performance stability, development of chemically and structurally compatible component materials, and understanding of long-term degradation processes. Efforts range from the optimization of engineering design that can allow for the mitigation of hot spots and thermal stresses, to the development of bulk materials and surface coatings for minimizing corrosion, to the development of operating conditions that successfully eliminate or reduce undesirable transients (changes in temperature, environment). To

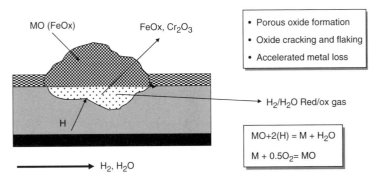

Fig. 10 Mechanism for the development of scale under the bipolar condition

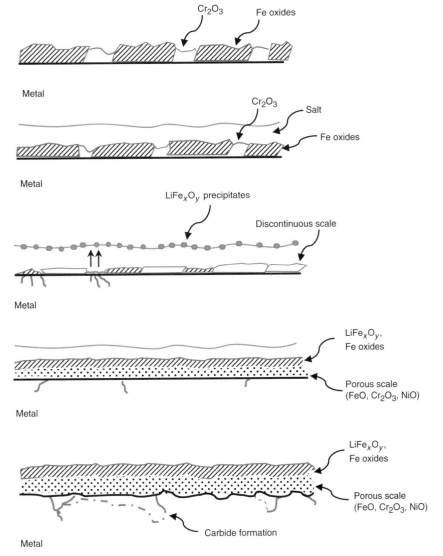

Fig. 11 Representation of scale growth in the anode atmosphere of the molten carbonate fuel cell. Time progresses from top to bottom. Oxide fluxing is observed during the oxidation of stainless steels.

remain cost-competitive, the development activity also focuses on lower-cost materials and the fabrication process.

The PEMFCs are widely recognized as the most viable fuel cell technology for vehicle prime propulsion. While significant advancements have been made over the past several years in improving the electrical performance of the PEMFC, the current focus on technology development is geared toward improvement in system life, performance stability, and cost reduction. From a corrosion perspective, the key development areas that are worthy of mention are the development of bipolar plate materials, corrosion-resistant coatings, and corrosion inhibitors for the coolant systems. Because the state-of-the-art bipolar plate is the most bulky component in the PEMFC stack and one of the most expensive to manufacture, a significant effort is underway in the development of lightweight and chemically stable materials and coatings. Overall, the bipolar plates are made from either a carbon- or a metal-base material. While the carbon-base materials, particularly carbon-carbon and carbon-polymer composites, offer the chemical stability required to survive the high-humidity and low-pH environment of the PEMFC stack, these materials usually require complex manufacturing processes and are therefore quite expensive to manufacture. On the other hand, metallic bipolar plates can be manufactured in very thin form (150 to 250 μm, or 6 to 10 mils) to reduce weight and volume in the overall stack and can be mass-produced in the desired shape, using inexpensive stamping or embossing processes. The key challenge with metal interconnects is surface corrosion. A number of approaches employing surface coatings are currently being studied. Although the greatest level of success has been achieved with noble metal coatings such as gold and palladium, the coating process remains prohibitively expensive. Transition metal nitrides and carbides offer a combination of high electrical conductivity and good corrosion resistance that make them attractive for potential use as a protective surface layer in this application (Ref 38). While recent work has shown promise with thermally nitrided metal surfaces and the development of pinhole-free coatings, the research is focused on developing alloys that not only can be uniformly nitrided for corrosion resistance and electrical conduction but also exhibit mechanical properties suitable for net shape forming. Because the exposed metal surfaces in the cell and stack cooling systems consist of a variety of ferrous and nonferrous alloys, such as stainless steel, aluminum, copper, and so on, there is a need for the development of inhibited coolant that minimizes corrosion and also remains compatible with the electrical requirements. Coolants with both metallic and nonmetallic additives are also being developed to meet the electrical conductivity (lower conductivity and hence lower shunt current), heat capacity (higher heat capacity), and thermal conductivity (higher thermal conductivity) requirements.

The MCFCs. Recent R&D focus centers around the development of cost-effective and corrosion-tolerant cell, stack, and balance of plant materials to improve the overall system life. Component designs, chemistry, and fabrication processes are some of the key areas currently being addressed. Cell operating conditions, including nominal and off-design conditions, are shown in Table 1. Because on-anode use of hydrocarbon fuels has proven highly effective in improving the electrical efficiency and thermal management of the stack, there remains significant interest in assessing the material stability under hydrocarbon exposure

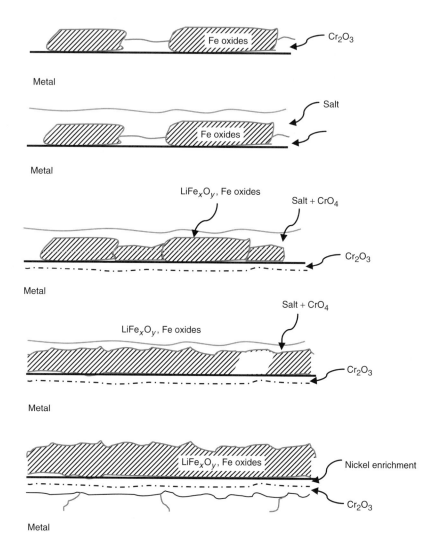

Fig. 12 Representation of scale growth in the cathode atmosphere of the molten carbonate fuel cell. Time progresses from top to bottom. Oxide dissolution is observed during the oxidation of stainless steels.

Table 1 Operation parameters of molten carbonate fuel cells and their impact on corrosion

Operation parameters	Operating range	Impact on corrosion
Normal		
Fuel gas	Low- and medium-Btu coal gas	Affects hydrogen diffusion rate and extent of corrosion
Oxidant gas	Air, 5–10% CO_2	May affect soluble corrosion products formation
Contaminants	1–3 ppmv H_2S	May cause intergranular corrosion, enhanced corrosion
	20–40 ppbv SO_2	...
Temperature	600–750 °C (1110–1380 °F)	High-temperature-enhanced corrosion
Pressure	1–10 atm	May lead to increased carburization
Fuel utilization	0–75%	Higher utilization may result in higher corrosion.
Oxidant utilization	0–50%	Minor effects expected but not characterized
Thermal cycling	>10 cycles	Severe corrosion due to cracking
Off-design conditions		
Fuel utilization	~100%	May result in cracking of protective scale, severe corrosion
Oxidant utilization	~100%	Not known
Electrolyte	Flooded	Excessive fluxing of corrosion layer

Table 2 Comparison of nickel-base alloy and ferritic stainless steel properties relevant for solid oxide fuel cell interconnection

Property	Nickel-base 230(a)	Stainless steel(b)
Composition	Ni-22Cr-14W-2Mo-0.5Mn	Fe-22Cr-0.5Mn
UNS No.	N06230	...
Coefficient of thermal expansion, 10^{-6}/K(c)	15.2	12.2
Oxidation resistance, $10^{-13} g^2/cm^4 \cdot s$		
At 700 °C (1290 °F)	0.05	2.0
At 800 °C (1470 °F)	0.36	7.0
Surface resistivity, $m\Omega cm^2$		
At 700 °C (1290 °F)	5.0	4.0
At 800 °C (1470 °F)	10.0	9.0
Ultimate tensile strength, MPa (ksi)		
At room temperature	865 (125)	443 (64)
At 760 °C (1400 °F)	605 (88)	<100 (15)
Corrosion resistance, dual atmosphere	Enhanced formation of uniform chromia scale	Grow hematite to accelerate attack
Manufacturability	Easy	Easy
Raw material cost	Fairly expensive	Fairly expensive

(a) Haynes 230 superalloy. (b) Crofer 22 APU. (c) Applicable, room temperature to 800 °C (1470 °F)

conditions. Processes such as carburization and metal dusting are being studied in depth to understand long-term impact on the breakdown of the protection offered by growing surface scale. The role of claddings and alloy compositions is being examined to minimize carbon ingress in the bulk metal. The sensitivity of nickel cladding toward substrate carburization is also a topic of research. Another area of research includes electrolyte management and protection of metals under the wet seal. Means of controlling the amount of the electrolyte in contact with exposed metal surfaces are being examined as a means of controlling the hot corrosion fluxing of the oxide. Oxide dissolution and contamination of the lithiated nickel oxide also depend on the extent and type of oxide dissolved in the molten electrolyte. At the systems level, R&D focus areas cover high-temperature heat exchanger and reformer materials.

The SOFCs. Materials research in SOFCs has steadily progressed in the areas of electrodics, electrolyte materials development, and optimization of ceramic processing techniques. A thrust toward the development of intermediate-temperature SOFCs has resulted in expanded efforts in the areas of metallic current collectors and protective coatings development (Ref 44, 45). Table 2 shows the properties of a typical iron- or nickel-base alloy of interest to SOFC developers. Until recently, the leading material for interconnect was doped lanthanum chromite ($LaCrO_3$), a ceramic that could easily withstand the traditional 900 to 1000 °C (1650 to 1830 °F) operating temperature. The high cost of raw materials and difficulties in obtaining high-density chromite parts at reasonable sintering temperatures has led to efforts toward the development of low-cost iron-base alloys and appropriate coatings for applications in the 600 to 800 °C (1110 to 1470 °F) temperature range. Ferritic alloys, such as chromia-forming ferritic stainless steels, demonstrate a close thermal expansion matching to ceramic electrodes. Research efforts on advanced alloy formulations as well as simple metallic systems such as nickel and silver are being pursued to understand oxidation processes in anode and cathode gas and also under bipolar exposure conditions. Corrosion in seals and optimization of sealing materials have also attracted considerable attention in recent years. A wide variety of seal designs (compressive, compliant, rigid, self-healing) and materials are currently being investigated. At the system level, the need to develop corrosion-tolerant heat exchanger and reformer materials has been identified.

REFERENCES

1. A.J. Appleby, *Sci. Am.,* July 1999, p 58
2. B.C.H. Steele and A. Heinzel, *Nature,* Vol 414, 2001, p 345
3. K. Kordesch et al., *J. Power Sources,* Vol 86, 2000, p 162
4. S.G. Chalk, J.F. Miller, and F.W. Wagner, *J. Power Sources,* Vol 86, 2000, p 40
5. N.Q. Minh, *J. Am. Ceram. Soc.,* Vol 76, 1993, p 563
6. C. Wagner, *Z. Elektrochem.,* Vol 63, 1959, p 772
7. R.A. Rapp, *Metall. Trans. A,* Vol 15, 1984 p 765
8. N. Birks and G.H. Meier, *Introduction to High Temperature Oxidation of Metals,* E. Arnold, London, 1983
9. P. Kofstad, *High Temperature Corrosion,* Elsevier Applied Science Publishers Ltd., London, 1988
10. G.E. Wasielewski and R.A. Rapp, *High Temperature Oxidation in the Superalloys,* C.S. Sims and W. Hagel, Ed., John Wiley & Sons, Inc., 1972, p 287
11. G.S. Giggins and F.S. Pettit, *Trans. Met. Soc. AIME,* Vol 245, 1969, p 2495
12. N. Birks and H. Rickert, *J. Inst. Met.,* Vol 91, 1961, p 308
13. W.J. Quadakkers, J. Piron-Abellan, U. Flesch, V. Shemet, and L. Singheiser, *Mater. High Temp.,* Vol 20, 2003, p 115
14. P.Y. Hou and J. Stringer, *Mater. Sci. Eng. A,* Vol 202, 1995, p 1–10
15. J. Jedlinski and G. Borchardt, in *Adv. Mater. Process., Proc. Eur. Conf., 1st,* Euromat '89 (Aachen, FRG), H. Eckart and V. Schumacher, Ed., 1990, p 601–606
16. P. Singh and N. Birks, *Oxid. Met.,* Vol 13 (No. 5), 1979, p 457
17. Z. Zemg and K. Natesan, *Solid State Ionics,* Vol 167, 2004, p 9
18. H.J. Grabke, *Mater. High Temp.,* Vol 7, 2000, p 483
19. C.H. Toh, P.R. Munroe, and D.J. Young, *Mater. High Temp.,* Vol 20, 2003, p 527
20. C. Gindorf, L. Singheiser, and K. Hilpert, *Steel Res.,* Vol 72, 2001, p 528
21. K. Hilpert, D. Das, M. Miller, D.H. Peck, and R. Weiß, *J. Electrochem. Soc.,* Vol 143, 1996, p 3642
22. J.R. Rostrup-Nielsen, *Catalysis,* Vol 5, J.R. Anderson and M. Boudart, Ed., Springer Verlag, 1984
23. P. Singh and S.D. Vora, *Proc. 29th International Conference on Advanced Ceramics and Composites,* The American Ceramic Society, 2005, p 99–104
24. P. Singh, L. Paetsch, and H.C. Maru, in *Corrosion '86,* National Association of Corrosion Engineers, 1986, p 86–1
25. P. Singh, Z. Yang, V. Viswanathan, and J.W. Stevenson, *J. Mater. Perform. Eng.,* Vol 13, 2004, p 287
26. Z. Yang, M.S. Walker, P. Singh, and J.W. Stevenson, *Electrochem. Solid-State Lett.,* Vol 6, 2003, p B35
27. Z. Yang, G. Xia, P. Singh, and J.W. Stevenson, *Solid State Ionics,* Vol 176, 2005, p 1495
28. Z. Yang, M.S. Walker, P. Singh, J.W. Stevenson, and T. Norby, *J. Electrochem. Soc.,* Vol 15, 2004, p B669
29. G.C. Wood, I.G. Wright, T. Hodgkiess, and D.P. Whittle, *Werkst. Korros.,* Vol 21, 1970, p 900
30. D.L. Douglass, P. Kofstad, A. Rahmel, and G.C. Wood, *Oxid. Met.,* Vol 45, 1996, p 529
31. J.A. Turner, Proc. 2000 Hydrogen Program Preview, NREL/CP-570-28890 US/DoE
32. J. Sholta, B. Rohland, and J. Garche, in *New Materials for Fuel Cell and Modern Battery Systems II,* O. Savadogo and P.R. Roberge, Ed., Editions de l'Ecole Polytechnique de Montreal, Quebec, Canada, 1997, p 330
33. M.C. Li, C.L. Zeng, S.Z. Luo, J.N. Shen, H.C. Lin, and C.N. Cao, *Electrochim. Acta,* Vol 48, 2003, p 1735
34. C.L. Ma, S. Warthesen, and D.A. Shores, *J. New Mater. Electrochem. Syst.,* Vol 3, 2000, p 221
35. W. Brandl and C. Gendig, *Thin Solid Films,* Vol 290–291, 1996, p 343
36. N. Cunningham, D. Guay, J.P. Dodelet, Y. Meng, A.R. Hill, and A.S. Hay, *J. Electrochem. Soc.,* Vol 149, 2002, p 905

37. M.P. Brady, K. Weisbrod, I. Paulauskas, R.A. Buchanan, K.L. More, H. Wang, M. Wilson, F. Garzon, and L.R. Walker, *Scr. Mater.,* Vol 20, 2004, p 1017
38. H. Wang, M.P. Brady, K.L. More, H.M. Meger III, and J.A. Turner, *J. Power Sources,* Vol 138, 2004, p 79
39. P. Singh and H.C. Maru, Paper 344, Corrosion '85, National Association of Corrosion Engineers, 1985
40. F.S. Pettit and C.S. Giggins, *Superalloys II,* C.T. Simms, N.S. Stoloff, and W.C. Hagel, Ed., Wiley, 1987, p 327
41. C.Y. Yuh, P. Singh, L. Paetsch, and H.C. Maru, *Corrosion '87,* National Association of Corrosion Engineers, 1987, p 276–1
42. P. Singh, *Proc. Symposia on Corrosion in Batteries and Fuel Cells and Corrosion in Solar Energy Systems,* C.J. Johnson and S.L. Pohlman, Ed., The Electrochemical Society, 1983, p 124
43. Z. Yang, K.S. Weil, D.M. Paxton, and J.W. Stevenson, *J. Electrochem. Soc.,* Vol 150, 2003, p A1188
44. Z. Yang, K.D. Meinhardt, and J.W. Stevenson, *J. Electrochem. Soc.,* Vol 150, 2003, p A1095
45. W.Z. Zhu and S.C. Deevi, *Mater. Res. Bull.,* Vol 38, 2003 p 957

Automotive Body Corrosion

D.L. Jordan and J.L. Tardiff, Ford Motor Company

CORROSION is a process defined as the reaction of a material with its environment. Corrosion behavior is controlled by the susceptibility of the material, the aggressiveness of the environment, and the effectiveness of any intervening barrier meant to keep the material and environment from interacting.

The materials used in as-produced automobile bodies are relatively straightforward to identify or characterize. Steel, with zinc and zinc alloys applied as sacrificial coatings, currently is the predominant metallic material for automotive body panels and structures. Aluminum, stainless steel, metallic and metallic-coated fasteners, and metallized plastic trim materials all are common and must be considered when developing the overall corrosion protection strategy. A variety of highly engineered and well-characterized paints and organic sealants are used as barrier coatings.

On the other hand, the automotive operating environment is not so simple to characterize. The use of road deicing salts, mainly sodium chloride, in North America has increased tenfold since the 1950s and varies dramatically among different geographical regions and within different state road jurisdictions. Some locales use combinations of sodium chloride, calcium chloride, magnesium chloride, and other potentially corrosive chemical deicers (Ref 1). Dust-control practices in rural areas frequently include the use of corrosive hygroscopic chemicals such as calcium chloride. The type and amount of air pollution varies with time and location. Driving and maintenance behaviors vary dramatically among the driving populace, and commercial car washes introduce a variety of chemicals that can beneficially or detrimentally affect the corrosion performance of the vehicle. Indeed, even the atmospheric component of the automotive environment varies dramatically with geographical location. Consequently, the environment to which a vehicle is exposed varies widely and can be expected to engender a variety of corrosion types and corrosion rates.

Exposure of known materials to an ever-changing environment results in the formation of possibly unique soluble and insoluble corrosion products that can influence subsequent corrosion reactions and rates. Automotive engineers are charged with providing vehicles that are resistant to the corrosive effects of all commonly encountered environmental input, even if a given vehicle may never be exposed. In view of the likelihood that little can be done by the vehicle manufacturer to control the environment in which a motor vehicle is operated, the general strategy is to provide a cost-effective and production-friendly combination of corrosion-resistant materials and barrier coatings.

The likely futility of forecasting corrosion lifetimes of materials in such a dynamic and operator-specific environment was described more than a generation ago when LaQue wrote, "There is no chance whatsoever of coming up with a test by which so many hours of exposure to the test can be established as being equivalent to so many months, or years, of exposure to 'atmospheric corrosion,' whatever that might be" (Ref 2). On the other hand, the economic importance of lifetime prediction, improved understanding of corrosion mechanisms, and the development of advanced material characterization tools have contributed to the continued scientific pursuit of corrosion forecasting methods. A universally acceptable test to predict automotive body corrosion lifetimes has yet to be developed.

During the late 1970s through the early 1990s, the Automotive Corrosion and Prevention Committee of the Society of Automotive Engineers (now SAE International) held a biannual series of well-attended symposia that were dedicated to the understanding and solution of automotive body corrosion issues. By the late 1990s, technological solutions were available and, when properly executed, have provided the long-term corrosion protection desired by the consumers and manufacturers. Since that time, the industrywide effort has given way in large part to scattered individual or small group efforts to reduce cost without sacrificing quality.

This article discusses the commonly encountered forms of automotive body corrosion, corrosion-resistant sheet metals, and paint and sealant systems for corrosion control in automotive body applications. Corrosion problems encountered in automotive exhaust systems and automotive engine coolant systems are described in subsequent articles in this section of the Handbook.

Forms of Corrosion Observed on Automobile Bodies

The complexity of the automotive environment, coupled with unique and sometimes incompatible materials combinations that are necessitated by increasingly dynamic and demanding vehicle functionality requirements, leads to the occurrence of a number of forms of corrosion. While the basic fundamentals of electrochemical corrosion (i.e., liberation of metal ions at the anode and reduction of a species such as dissolved oxygen or the hydronium ion at the cathode) are well established and apply regardless of the particular materials application, corrosion control continues to evolve in response to the specific corrosion environment, economics, and available technology.

The primary types of corrosion observed on automobile bodies are classified as (a) cosmetic or underfilm corrosion, where the combination of compromised paint adherence and the formation of corrosion products leads to increasing amounts of unsightly paint loss, and (b) perforation, where body panels thin to the point of developing holes and losing mechanical functionality. Perforation may be caused by a variety of corrosion types, including general or uniform corrosion, galvanic corrosion, crevice corrosion, poultice or underdeposit corrosion, and pitting.

General or uniform corrosion occurs over the entire exposed surface of a component and is manifested by a generalized thinning of the material. Because the attack occurs uniformly on the surface of the component, it is the least likely to result in significant mechanical property loss.

Cosmetic or underfilm corrosion is electrochemical in nature and results in paint delamination and exposure of unprotected metal to corrosive attack. Successive exposure to cyclic conditions results in further delamination and exposure of fresh metal. Several phenomenological models to explain cosmetic corrosion have been proposed, but a consensus has not been reached (Ref 3, 4).

Galvanic corrosion results from the contact of two dissimilar metals, both of which must be immersed in the same corrosive electrolyte at the same time. The more active metal or alloy becomes the anode of the couple and may be subject to accelerated attack. This type of corrosion is becoming a significant concern in the automotive industry since weight-reduction efforts often include the use of aluminum and magnesium, both of which may suffer galvanic attack when in contact with steel components. Careful application of electrical isolation and sealants can disrupt the electrochemical circuit and limit the amount of galvanic attack in such cases. Figure 1 is a schematic illustration of the mechanism of galvanic corrosion.

Crevice corrosion is a severe form of localized attack that is normally associated with small volumes of stagnant electrolyte that can form in occluded areas at joints or under fasteners. Concentration gradients between the stagnant electrolyte and the adjoining exposed electrolyte drive the electrochemical cell. Crevice attack can be very rapid and usually gives little warning because it occurs in generally nonvisible areas. Figure 2 shows a simplified schematic diagram of one model of crevice corrosion.

Poultice or underdeposit corrosion is a special form of crevice corrosion that occurs under deposits of road debris, such as mud that can deposit inside fenders and other partially enclosed areas of vehicles. The poultice can accumulate and concentrate road salts and other potentially corrosive substances. Oxygen concentration gradients under and adjacent to the poultice exacerbate attack under the poultice. Figure 3 shows the mechanism of poultice corrosion.

Pitting is similar to crevice corrosion in that it is a localized attack. It occurs most often in areas of low pH that are depleted in oxygen, but have high chloride content. Once pitting is initiated, the mechanism is similar to that of crevice corrosion, with the pit acting as the crevice. Pits can initiate at metal inhomogeneities, discontinuities in passive films or other protective coatings, under surface deposits, or at other defects. Figure 4 shows a schematic diagram of the mechanism of pitting corrosion.

All of these forms of corrosion are the result of an interaction of a metal with its environment, and all require the presence of an electrolyte (in most cases water) to facilitate charge transfer. These and other forms of corrosion that occur in the automotive environment are discussed in greater detail in *ASM Handbook*, Volume 13A, 2003.

Corrosion-Resistant Sheet Metals

Beginning in the 1950s, automakers responded to the challenge of increased corrosion of body panels by introducing precoated sheet steels, in particular galvanized (zinc-coated) steels. The term precoated is meant to denote the application of a corrosion-resistant metallic coating to the steel while still in coil form. After fabrication into a part, all metal surfaces (except for cut edges) contain the corrosion-protective sacrificial coating.

In the 1980s, numerous zinc-base sacrificial coatings for steel were developed using existing steel manufacturing infrastructure and popular accelerated corrosion tests of the day. It became apparent that the performance of different coating systems was highly dependent on the nature of the accelerated corrosion test, and correlation with service conditions was poor. Time and experience provided in-service corrosion data that helped lead to the current dominance of electrogalvanized steel (EG or EL), hot dip galvanized steel (GI or HD), and hot dip galvannealed steel (GA) for corrosion-resistant automotive body panels. Other more complicated and less robust coating candidates were developed and vigorously marketed, but vanished with increasing understanding of corrosion mechanisms and the development of rational accelerated corrosion test methods. Coating weights are typically 60 to 70 g/m^2 (0.2 to 0.23 oz/ft^2) for EG and GI and 35 to 50 g/m^2 (0.15 to 0.16 oz/ft^2) for GA. The selection of one coating over the other is based on a balance of economics with the infrastructure of the manufacturing facility, as described below.

Practically all steel automotive skin panels (door outers, floor pans, wheelhouse inners) and structural parts (cross members, rails, beams) thinner than approximately 2 mm (0.08 in.) have a precoating of zinc or zinc alloy for corrosion protection. The incidence of cosmetic corrosion and perforation corrosion of car bodies decreased dramatically during the 1990s to the point where corrosion warranties approaching and even surpassing 10 years are common.

Electrogalvanized steel is recognized as the most robust solution to the manufacturing/functionality needs of corrosion-resistant coated sheet steel for exposed automotive body parts, despite the comparatively higher cost in most markets. The smooth, uniform, and spangle-free surface that results from electroplating zinc onto meticulously cleaned steel strip is very friendly to stamping operations and provides an excellent base for paint. In addition, resistance spot weldability is good. Finally, the relatively low-temperature electroplating process has minimal effect on the mechanical properties of the steel, allowing greater flexibility in the application of increasingly popular high-strength, paint-bake hardenable steel substrates.

Hot dip galvanized steel is produced by passing a freshly cleaned and annealed steel strip though a bath of molten zinc. The corrosion resistance of the resultant zinc layer is comparable to that obtained by electrogalvanizing. The presence of a small amount (0.20 to 0.50 wt%) of aluminum in the hot dip coating can complicate zinc phosphate pretreatment (Ref 6) and subsequent paint adhesion. In addition, hot dip is less robust than electrogalvanized

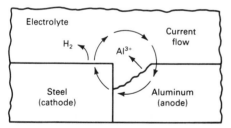

Fig. 1 Mechanism of galvanic corrosion, which occurs when dissimilar metals are placed in contact with one another and are exposed to a common electrolyte. The more electrochemically active metal will act as the anode and will corrode preferentially (often at an accelerated rate), and the less active metal acts as a cathodic surface for the reduction of an oxidizing species.

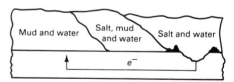

Fig. 3 Mechanism of poultice corrosion. The most common cause of this type of corrosion is thought to be electrolyte composition gradients. In the example shown, clumps of mud and water have collected, and the varying concentrations of salt and water within the clump encourage corrosion.

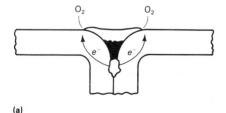

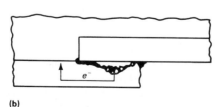

Fig. 2 Mechanism of crevice corrosion at a joint. Crevice corrosion is common at weldments or sheet metal joints (a) and can occur in apparently sealed lap joints (b).

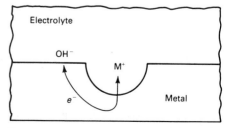

Fig. 4 Mechanism of pitting corrosion. As with crevice corrosion, pitting occurs in localized areas that are depleted of oxygen, low in pH, and high in chlorides.

steel to variation in typical resistance spot welding processes, due in part to the effects of aluminum in the hot dip coating and in part to comparatively higher coating weight variability.

Hot dip galvannealed steel has a coating that consists of a series of zinc-iron intermetallic compounds. It is produced by heating hot dip galvanized steel to facilitate interdiffusion of the iron and zinc. Galvannealed steel has excellent resistance spot welding characteristics and good corrosion resistance. Due to the brittle nature of the zinc-iron intermetallic coating, stamping operations often result in the localized removal of small coating particles, a process called "powdering" (when the particles result from fracture within the intermetallic layer) or "flaking" (when the coating delaminates from the steel surface). Furthermore, impingement by stones or road debris may cause chips that remove paint as well as the metallic coating, resulting in severe localized corrosion.

Paint Systems

Automotive paint systems have two primary functions—to provide a protective barrier against corrosion and to be aesthetically pleasing to the consumer. A typical automotive paint system consists of five layers—pretreatment, electrocoat primer, spray primer, basecoat, and clearcoat. Although the corrosion performance is dependent on the entire paint system, the key components of the paint system for corrosion protection are the pretreatment and the electrocoat primer. The character of the metallic substrate is of utmost importance in the pretreatment and electrocoat primer steps. After electrocoating (a process more correctly termed electrophoretic deposition or cataphoretic deposition), subsequent spray painting steps are minimally affected by the character of the substrate.

The pretreatment for steel and galvanized steel is typically zinc phosphate, while the pretreatment for aluminum is either zinc phosphate or a zirconium polymeric layer, depending on the mix of substrates in the phosphate system. The zinc phosphate pretreatment layer is usually no more than 2 µm thick. It promotes adhesion of the paint layer to the metal substrate. The electrocoat primer is most frequently a cathodically deposited epoxy resin (Ref 7). Electrocoat primer is approximately 20 to 30 µm thick and is the primary layer for paint system adhesion and subsequent corrosion protection. The spray primer layer is approximately 20 to 30 µm thick and is important for stone chip protection and adhesion of the topcoat (basecoat and clearcoat.) The basecoat is typically 15 to 25 µm thick and provides the color of the vehicle. The clearcoat is typically 40 to 50 µm thick and provides gloss and ultraviolet light protection to the paint system. In general, the pretreatment and electrocoat primer thicknesses are consistent among automotive manufacturers, but the spray primer, basecoat, and clearcoat thicknesses vary depending on the particular chemistry being used by the manufacturer. Some manufacturers use solventborne systems, while others use waterborne systems, powder paint systems, or a combination of the three for the spray primer, basecoat, and clearcoat layers. The use of different paint systems is an attempt by manufacturers to meet increasingly stringent environmental regulations in North America and Europe.

Recent environmental regulations have had a profound effect on the pretreatment and electrocoat primer systems that are now allowable in manufacturing facilities. Postphosphate treatments that contain hexavalent chromium compounds were used for many years to improve the underfilm corrosion performance of painted sheet metal, but hexavalent chromium is no longer allowed in automotive manufacturing facilities in North America and Europe. Similarly, and specifically for environmental reasons, lead has been removed from electrocoat primer formulations. Some tests suggest that the elimination of hexavalent chromium posttreatments and lead in electrocoat primer have had a detrimental effect on the long-term underfilm corrosion performance of vehicles, but continuing advances in other parts of the paint system are closing the performance gap.

Both hexavalent chromium and lead apparently played a beneficial role on metal substrates that had been sanded after the pretreatment process. It is known that sanding (i.e., metal finishing) of the steel, galvanized steel, or aluminum after pretreatment and electrocoat has a detrimental effect on underfilm corrosion performance. Sanding prior to the paint process also can adversely affect corrosion performance, particularly on aluminum. Therefore, care must be taken to minimize defects in the substrate and the paint coating that require sanding.

Surface Preparation for Paint Processes

Metal Cleaning. Preparing the surface for painting begins with cleaning oil and grease from the surface and removing solid contaminants such as fibers and weld balls (spatter). The cleaning system consists of alkaline cleaners and water neutralizers/rinses. If solid contaminants are not removed from the metal surface, a variety of defects that may necessitate sanding or other dirt-producing rework of the paint system may result. If oils and grease are not removed, they can inhibit the nucleation and growth of phosphate crystals on the metal surface and provide sites for adhesion loss and subsequent corrosion.

The phosphate portion of the pretreatment process consists of a rinse conditioner, phosphate stage, rinses, and posttreatment. The rinse conditioner provides sites where phosphate crystals can nucleate on the metal surface. In the phosphate stage, zinc phosphate crystals deposit at the nucleation sites. Excess process chemicals are removed, and the remaining chemicals are neutralized in the rinse stages. The posttreatment typically is a zirconium polymeric or similar rinse that prohibits further crystal nucleation and seals any surface that is not covered by phosphate crystals. The final rinse prior to the electrocoat primer process is critical, as unreacted pretreatment chemicals can adversely affect the success of the electrocoat primer bath.

Critical characteristics of the pretreatment system are phosphate crystal size and phosphate coating weights. It should be noted that aluminum parts are frequently processed concurrently with steel and galvanized steel parts, often as a closure panel on a mixed-metal vehicle. The inclusion of aluminum substrates in the pretreatment process presents some challenges to the system. In particular, the outermost aluminum oxide layer is etched in the phosphate stage and significantly increases the amount of sludge produced in the bath. If the liberated soluble aluminum ions are not removed from the bath as sludge, they will inhibit deposition of phosphate on the metal. Similar concerns are valid for magnesium body panels. For aluminum-intensive vehicles, dedicated pretreatment processes that do not have to satisfy the needs of mixed-metal vehicles may be used.

The electrocoat primer process immediately follows the pretreatment process. Almost all automotive manufacturers now use electrodeposition to apply the primary corrosion protection layer to the vehicle body. In the cathodic electrodeposition process, the vehicle is negatively charged and immersed into a tank containing positively charged electrocoat primer. The electrocoat primer is attracted to the metal and deposits uniformly on the pretreated metallic surfaces. Both exterior and interior surfaces are coated, although the coating on interior surfaces may be thinner than that on exterior surfaces due to throwing power limitations. The thickness of the electrocoat primer can be varied by changing the voltage and by operating auxiliary anodes in shielded areas.

After the vehicle is coated with electrocoat primer, the body is rinsed to remove excess solids that could result in surface defects. The excess solids are recovered and returned to the electrocoating tank. Vehicle bodies typically are baked in an oven for 30 to 45 min to cure the electrocoat primer for optimum performance. It is important for all parts of the vehicle to attain an adequate level of cure, which can be challenging considering the varying thicknesses of the steel and the numerous enclosed sections on the vehicle. Inadequate cure significantly reduces the corrosion protection on the vehicle. Overcure can reduce the adhesion of the spray primer to the electrocoat primer and can cause components in the coating to bake out, both of which are detrimental to long-term corrosion performance.

Spray primer or primer-surfacer provides several functions to the paint system, the most important being protection from stone chipping. Stone chips that penetrate through the spray primer to the electrocoat or to the metal surface provide sites for corrosion initiation. The goal of the spray primer layer is to dissipate energy from impinging stones, thus minimizing damage to the underlying corrosion preventive layers. Other important functions of the spray primer layer are to provide adhesion between the topcoat (basecoat and clearcoat) and the electrocoat primer and to keep ultraviolet (UV) light from reaching the UV-sensitive electrocoat primer layer.

There are several different spray primer paint systems being used by automotive manufacturers, depending on the company strategy for meeting environmental regulations. Solventborne, waterborne, and powder spray primers are electrostatically sprayed onto the vehicle. The vehicle is grounded, while the paint is positively charged. This process provides the most efficient transfer of paint to the vehicle body. Only the exterior surface, door openings, and underhood areas are sprayed with spray primer.

After application of the spray primer, the paint is cured for a prescribed time/temperature cycle in an oven. Cure of the paint is extremely important for the performance of the coating. Undercure (too little time or too low temperature) prevents the primer from adequately protecting the underlying layers from UV light and stone chips. Overcure (too much time or too high temperature) causes the paint to become brittle, resulting in degraded chipping performance.

The topcoat consists of basecoat and clearcoat that are applied wet-on-wet, or without a full bake between layers. The primary purposes of these two layers are to provide aesthetics (color, gloss) and UV protection. Basecoat layers are typically solventborne or waterborne. Clearcoat layers are usually solventborne, although there has been progress in the application of powder clearcoat technologies.

Automotive topcoat is applied electrostatically, in the same manner as the primer layer. The applied thickness of each layer is critical—in basecoat to attain the desired color and in clearcoat for protection from UV light. Curing of this layer is also important to attain the appropriate performance characteristics. A 30 to 40 min bake is typical for topcoat systems.

Coil Coating

Some automotive manufacturers use coil-coated or preprimed steel as a means to improve corrosion protection at areas where it is difficult to apply electrocoat primer. This process provides a primer layer with an extremely consistent thickness. In the coil-coating process, a coil of steel typically is pretreated with zinc phosphate and is continuously coated with a polymeric film, which is cured before the coil is rewound for shipment to the user. The coil-coated material is then subjected to the usual stamping and welding processes that comprise automotive body construction.

The coil-coated parts are then processed through the spray primer, topcoat, and clearcoat processes by the normal methods outlined previously. Note that the coil-coated parts are subjected to the same cleaning and pretreatment processes as noncoil-coated parts, but no phosphate crystals deposit onto the primed surface.

REFERENCES

1. J. Starling, "Are Ice-Clearing Chemicals Killing Your Trucks?" Maintenance & Technology Council—The American Trucking Association, 2001
2. F.L. LaQue, Corrosion Tests and Service Performance, *Relation of Testing and Service Performance,* STP 423, American Society for Testing and Materials, 1967, p 61
3. A. Sabata, "Study of the Corrosion Mechanisms in Modern Painted Precoated Automotive Sheet Steels," Ph.D. dissertation T-3777, Colorado School of Mines, 1989
4. D.L. Jordan, "Location and Identity of the Cathodic Reaction During Underfilm Corrosion of Painted Galvanized Steel," Ph.D. dissertation, Illinois Institute of Technology, 1996
5. J.C. Bittence, Waging War on Rust, Part I: Understanding Rust, *Mach. Des.,* Oct 7, 1976, p 108–113; Part II: Resisting Rust, *Mach. Des.,* Nov 11, 1976, p 146–152
6. J.A. Kargol et al., The Influence of High Strength Cold Rolled Steel and Zinc Coated Steel Surface Characteristics on Phosphate Pretreatment, *Corrosion,* Vol 39 (No. 6), June 1983, p 213–217
7. J. Gezo, E-Coat Technologies and Where They Fit, *Ind. Paint Powder,* Vol 80 (No. 3), March 2004, p 14–20

SELECTED REFERENCES

- R. Baboian, The Automotive Environment, *Automotive Corrosion by Deicing Salts,* R. Baboian, Ed., National Association of Corrosion Engineers, 1981, p 3–12
- R. Baboian, "Chemistry and Corrosivity of the Automotive Environment," paper 371, CORROSION/91, National Association of Corrosion Engineers, 1991
- R. Baboian, Ed., *Corrosion Testing and Standards: Application and Interpretation,* Manual Series MNL 20, ASTM International, 1995, p 561–573
- S.W. Dean, Automobile Atmospheric Corrosion—Testing and Evaluation, *Mater. Perform.,* March 1991, p 48
- M.G. Fontana, *Corrosion Engineering,* McGraw-Hill, 1985
- F.L. LaQue, A Critical Look at Salt Spray Tests, *Mater. Meth.,* Vol 35 (No. 2), 1952, p 77–81
- Light Truck Frame Project Team, "A Guide to Corrosion Protection for Passenger Car and Light Truck Underbody Structural Components," Auto/Steel Partnership, 1999
- M.R. Ostermiller and H.E. Townsend, "On-Vehicle Cosmetic Corrosion Evaluations of Coated and Cold Rolled Steel Sheet," Technical paper 932335, Society of Automotive Engineers, 1993
- R.W. Revie, Ed., *Uhlig's Corrosion Handbook,* 2nd ed., John Wiley & Sons, 2000
- L.A. Roudabush and T.E. Dorsett, "A Review of Perforation Corrosion Testing—1980 to 1990," Technical paper 912285, Society of Automotive Engineers
- L.C. Rowe, "The Application of Corrosion Principles to Engineering Design," Technical paper 770292, Society of Automotive Engineers, 1977
- E.N. Soepenberg, "A History of Materials Technology for the Automotive Industry—A Continuing Environment-Materials-Energy Debate," paper 546, CORROSION/93, National Association of Corrosion Engineers, 1993
- H.E. Townsend et al., Hot-Dip Coated Sheet Steels—A Review, *Mater. Perform.,* Vol 25 (No. 8), Aug 1986, p 36–46
- R.R. Wiggle, A.G. Smith, and J.V. Petrocelli, Paint Adhesion Failure Mechanisms on Steel in Corrosion Environments, *J. Paint Technol.,* Vol 40 (No. 519), 1968, p 174–186

Automotive Exhaust System Corrosion

Joseph Douthett, AK Steel Corporation

CORROSION ENGINEERS looking for the optimal real-world environment to evaluate stainless and coated steel alloys need to look no further than the modern automobile and truck exhaust system. Imagine a test sequence that includes an aqueous environment with a pH that varies from basic 8 to 9 down to acidic 2, with dissolved chloride levels from 50 to 2000 ppm. Now, include cyclic humidity conditions where components experience temperatures from ambient up to 870 °C (1600 °F). Finally, add crevices and welds, mix the various materials, shake with a combination of low- and high-cycle fatigue stresses, and one has the modern automotive exhaust system. Steel automotive exhaust systems suffer from at least eight common forms of corrosion, including general cosmetic attack, chloride pitting, crevice, galvanic and intergranular corrosion, hot salt attack, oxidation, and corrosion fatigue. The general cosmetic attack is basically a form of superficial pitting discoloration, as opposed to the more traditional carbon steel general attack where base metal is removed in measurably thick layers.

Since the early 1970s, automobile exhaust systems have gone through a gradual but steady evolution of design and materials. From the days of aluminized carbon steel mufflers, often with bare carbon steel connecting and internal tubes, today's (2006) exhaust system is a high-technology sound-deadening, pollution-removing system with one or more catalytic converters, mufflers, resonators, and flexible joints, all constructed of stainless steel. With an average North American annual production of 16 million light trucks and cars, the exhaust industry yearly consumes large volumes of stainless alloys, including many miles of fabricated tubing. There are close to 200 million vehicles on North American roads, which means that a new production vehicle can expect to avoid recycling and the crusher for some 13 years after first being sold. With initial car loans close to 60 months, it is understandable why car builders are pushing the original equipment exhaust producers to develop systems with 7 to 10 year durability and warranties of 3 to 4 years.

Traditionally, the normal exhaust system has been divided into two parts: a hot end and a cold end (Fig. 1). The hot end, closest to the engine, includes the exhaust manifold, downpipe, catalytic converter, and often a flexible coupling. On some vehicles, the main underbody catalytic converter is supplemented by a smaller close-coupled (next to the manifold outlet) light-off converter designed to reduce pollution on vehicle start-up. The exhaust hot end is considered a part of the vehicle emission control system and therefore comes under government-mandated warranties, which, in 2004, became 10 years and 193,000 km (120,000 miles). The cold end of the exhaust includes the resonator, intermediate pipe, muffler, and tail pipe. Its durability target or warranty is chosen by the carmaker based on competitive pressures and consumer complaints. Today (2006), cold end exhaust targets are to have the original equipment manufacturers (OEM) systems survive on the vehicle a minimum of 7 to 10 years. Cold end warranties are generally 3 to 4 years and 58,000 to 77,000 km (36,000 to 48,000 miles).

Figure 1 is a schematic of the standard OEM exhaust system and lists the stainless alloys most often used to construct these components. For each component in Fig. 1, an attempt was made to add a typical in-service maximum metal skin temperature. Obviously, that temperature varies with engine, driving cycle, component design, and degree of insulation and heat shielding. Table 1 shows the compositions of the most common alloys used in exhaust system construction. The heavy use of ferritic 400-series stainless alloys is immediately obvious. The 300-series stainless steels in North America have limited use, most often as tail pipes or fabricated manifolds. Carbon steels, normally as aluminized (type 1) steel, find their way into some heat shield applications, as occasionally do aluminum alloys. Bare carbon steels still are used as 9.53 to 12.7 mm (0.375 to 0.500 in.) thick flat flanges or as 12.7 mm (0.5 in.) diameter rod hangers to attach systems to the car frames. In normal exhaust construction, stainless components will range in thickness from 0.5 mm (0.020 in.) muffler wraps up to 2 mm (0.080 in.) wall thickness downpipes. Some fabricated stainless steel flanges will measure 3.1 to 4.7 mm (0.125

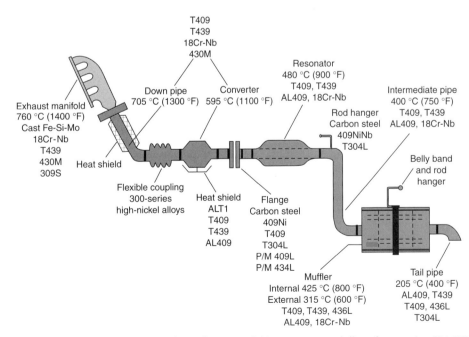

Fig. 1 Exhaust system components with typical upper metal skin temperatures and alloys of construction. P/M 409L and P/M 434L produced by powder metallurgy (P/M) processing

to 0.187 in.) thick, with flat flanges up to 9.5 mm (0.375 in.). This is the thickest stainless coil plate that is made in North America. Currently, approximately 60% of North American exhaust system components are constructed from bare 11% Cr T409, with another 20% upgraded to hot dip aluminum-coated T409 (referred to here as AL409). The 17 to 18% Cr stabilized ferritics (T439, 18Cr-Nb, 441, and 436L) and the austenitic alloys 304L and 309S make up most of the remaining tonnage, except for an 11 to 12% Cr ferritic alloy, 409Ni, used in some heavier flange stampings. A Japanese-developed hot end high-chromium alloy referred to as 430M has seen limited use, mostly on vehicles brought to the United States from Japan.

High-Temperature Corrosion

Oxidation. With exhaust metal skin temperatures of 870 °C (1600 °F) or higher, catastrophic oxidation is a concern with exhaust components. Experience coupled with laboratory cyclic oxidation testing has generally helped avoid such failures. Tables 2 and 3 show the results of cyclic oxidation (25 min of heat followed by 5 min of ambient air cooling) exposure at 815 °C (1500 °F) and 900 °C (1650 °F) in a natural-gas-fired atmosphere on several of the more standard exhaust hot end materials. The 316Ti alloy is a flexible coupling alloy, while AL439 is an aluminum-coated 439 long-life material, developed for some luxury car muffler applications. Weight changes were recorded during and after 260 to 350 h of testing at temperature. The ferritic stainless alloys showed weight increases as oxide scales thickened over time. The austenitic alloys (304L, 316Ti, and 309S), however, tended to show weight decreases. This was due to the higher coefficients of thermal expansion, which caused protective scales to spall over time, leading to weight loss and sample thinning. While stronger and more formable, the thinning and loss of scale characteristic of austenitic stainless alloys in cyclic heating has proven to be the "Achilles heel" of this class of exhaust materials in hot end exhaust applications.

In cyclic tests at 900 °C (1650 °F), 409 alloy had a thick scale characteristic of breakaway catastrophic oxidation. The 17% Cr titanium-stabilized 439LT coupon showed the beginning of excessive scaling. These laboratory-tested coupons are shown in Fig. 2. The 18Cr-Nb, (designated 18CR-CB in Fig. 2), AL409, and AL439 samples had stable scales, although the AL409 developed heavy oxidation at sheared edges and at the punched hole not protected by a coating of aluminum. Of the austenitic alloys, 309S was the best, although scale losses were high. The 304L and 316Ti alloys lost so much weight that they structurally fell apart. Aluminum-coated stainless alloys benefited at elevated temperatures (>540 °C, or 1000 °F) by having the aluminum coating diffuse into the base alloy to form a scale-resistant iron-aluminum intermetallic layer that then formed a thin scale of Al_2O_3 on the sample surface.

From an oxidation standpoint, Table 4 shows what is considered to be a conservative upper temperature limit for standard exhaust alloys in cyclic oxidation service. This temperature is generally not exceeded in continuous service. However, brief exposures to temperatures 28 to 55 °C (50 to 100 °F) above the cyclic service limit can be tolerated with little likelihood of seeing the onset of catastrophic oxidation.

Hot Salt Attack. When salt crystals are added to the high-temperature oxidizing environment, the phenomenon known as hot salt attack can occur. This form of attack is more rapid and catastrophic than aqueous chloride pitting on a scaled stainless surface. Hot salt attack occurs when a quantity of salt is trapped against a hot surface. In normal converter/downpipe applications, salt-laden water splashes against the hot exhaust components, the water evaporates, and salt crystals can accumulate on the steel surface, leading to hot salt attack. Crevices, insulation, or depressions in the formed steel surfaces provide collecting basins for salt solutions to be trapped.

Table 1 Typical compositions for exhaust system alloys of construction

	Composition, wt%									
Alloy	Cr	Ni	Mo	Si	Mn	Ti	Nb	Cu	C	N
T409(a)	11.2	0.25	0.20	0.50	0.25	0.20	...	...	0.01	0.015
T439(a)	17.3	0.25	0.20	0.30	0.25	0.30	...	...	0.015	0.015
18Cr-Nb	18.0	0.25	0.20	0.40	0.25	0.20	0.55	...	0.015	0.015
430M	19.2	0.20	0.20	0.40	0.30	...	0.60	0.50	...	0.015
436L	17.2	0.25	1.3	0.30	0.30	0.30	...	...	0.010	0.015
409Ni	11.0	0.85	0.20	0.35	0.75	0.20	...	...	0.015	0.015
T304L(a)	18.1	8.5	0.20	0.30	1.70	...	...	0.20	0.025	0.03
309S	22.2	12.5	0.20	0.30	1.80	...	...	0.20	0.050	0.10
FeSiMo(b)	...	...	0.8	4.5	0.5	...	...	...	3.00	...
ALT1(c)	...	...	...	...	...	...	...	...	...	...
AL409(d)	...	...	...	...	...	...	...	...	...	...

(a) T is tubular material. (b) High-temperature cast iron. (c) Hot dip type 1 (90%Al/10%Si) coating on a low-carbon steel. (d) Hot dip type 1 (90%Al/10%Si) coating on 409

Fig. 2 Cyclic oxidation exposure coupons after 1050 cycles (unless otherwise noted) at 900 °C (1650 °F). Each cycle: 25 min heat exposure and 5 min cooling

Table 2 Results of cyclic oxidation test exposure at 815 °C (1500 °F) for selected exhaust stainless steel alloys

The weight change values given in the table are in mg/cm².

	Oxidation cycles(a)					
Alloy	50	100	150	200	350	520
T409	1.66	2.09	2.42	2.51	2.64	2.71
439LT	0.06	0.16	0.12	0.16	0.23	0.25
18Cr-Nb	0.05	0.09	0.12	0.12	0.17	0.23
AL409	0.34	0.43	0.50	0.53	0.64	0.74
AL439	0.33	0.40	0.43	0.45	0.54	0.59
T304L	−3.92	−6.85	−8.59	−12.2	−19.4	−31.0
316Ti	−5.24	−9.66	−11.9	−13.2	−18.1	−24.8

(a) Each cycle consists of 25 min heating and 5 min air cool.

Table 3 Results of cyclic oxidation test exposure at 900 °C (1650 °F) for selected exhaust stainless steel alloys

The weight change values given in the table are in mg/cm².

	Oxidation cycles(a)					
Alloy	51	100	200	350	500	700
T409	1.84	4.00	10.4	19.8	63.1	205
439LT	0.29	0.36	0.50	0.79	1.94	9.35
18Cr-Nb	0.26	0.42	0.59	0.76	1.30	5.24
AL409	0.81	1.21	1.74	2.43	3.94	3.72
AL439	0.51	0.62	0.73	0.78	0.98	1.16
T304L	−27.6	−57.0	−112	−193	(b)	(b)
316Ti	−5.29	−15.7	−42.6	−110	−173	(b)
309S	0.17	0.31	0.37	−0.17	−14.1	−74.2

(a) Each cycle consists of 25 min heating and 5 min air cool. (b) Discontinued

Table 4 Maximum sustained service oxidation temperature for exhaust stainless steel alloys

	Temperature	
Alloy	°C	°F
Bare carbon steel	425	800
ALT1(a)	675	1250
T409	815	1500
T304L	815	1500
AL409(b)	845	1550
T439	885	1625
436L	885	1625
18Cr-Nb	900	1650
309S	925	1700

(a) ALT1, aluminum-coated low-carbon steel. (b) AL409, aluminum-coated 409, T1 alloy.

Water will evaporate, leaving behind salt deposits. With high-enough temperatures, such deposits can melt and act as a fluxing agent capable of removing the stainless alloy protective chromium oxide scale. This fluxing reaction occurs according to the equation:

$$Cr_2O_3 + 4NaCl + \frac{5}{2}O_2 = 2Na_2CrO_4 + 4Cl$$

The sodium chromate is water soluble and easily removed. With the protective scale gone, the salt and free chloride then react with chromium in the base steel according to the following equations:

$$Cr + 2NaCl + 2O_2 = Na_2CrO_4 + Cl_2$$

$$Cr + \frac{3}{2}Cl_2 = CrCl_3$$

The chromium chloride is volatile, and the sodium chromate is water soluble. The result is intergranular attack.

An example of hot salt attack is shown in Fig. 3. Two layers of 0.203 mm (0.008 in.) thick T304 were sandwiched together with an internal layer of fiber insulation to form a catalytic converter heat shield. The semicircular ends of the laminate heat shield were not watertight. Saltwater soaked the fiber mat, putting salt crystals in intimate contact with the 304 surfaces. The 304 layer closest to the hot converter shell experienced temperatures high enough to melt the salt. The subsequent intergranular attack is shown in the metallographic cross section (Fig. 3). Hot salt attack is a more rapid form of corrosion than chloride pitting. The accompanying intergranular corrosion occurs over a wide area, often removing layers of metal, much like catastrophic oxidation.

Laboratory screening tests to evaluate the hot salt corrosion resistance of exhaust alloys is fairly simple. Coupons are immersed in a concentrated sodium chloride (NaCl) salt solution for 5 min, placed in a metal rack, and heated in air for 2 h. At the end of the heating cycle, the samples are cooled for 5 min and again immersed in salt solution. Ten salt exposures followed by heat are carried out at each temperature. Samples are weighed before and after testing, and weight loss per unit of sample surface area is calculated. A battery of hot salt tests generally consists of 10-cycle evaluations at each of five different exhaust temperatures from 400 to 800 °C (750 to 1470 °F). This temperature range covers hot salt environments from the intermediate pipe to the manifold. Weight loss versus temperature plots are shown in Fig. 4 and 5. Figure 4 compares the resistance of five bare exhaust alloys, while Fig. 5 shows results for standard exhaust alloys 409 and 439 and compares them to hot dip aluminum-coated carbon steel (ALT1) and coated 409 and 439. Figure 5 also includes the aluminum-bearing (1.7%) stainless heat-resistant alloy AK 18SR, which contains 18% Cr.

The hot salt tests on standard uncoated stainless exhaust alloys in Fig. 4 show that weight losses increase with test temperature. The high corrosion rates at 800 °C (1470 °F) make sense, given that 800 °C (1470 °F) is near the melting point of pure NaCl. Increasing the chromium content proved beneficial, as did raising the molybdenum content. Alloy T304 with 18%Cr-8%Ni had approximately the same hot salt resistance as the 17 to 18% Cr ferritic alloys (439, 18Cr-Nb, and 436L).

Figure 5 shows how adding aluminum, either as a coating or alloy addition, affected hot salt resistance. Adding aluminum to the base alloy was beneficial, but there was not as much improvement as when it is added as a coating. Even carbon steel after aluminizing resisted hot salt attack better than the bare stainless alloys. At most of the higher temperatures used in the hot salt evaluations, the aluminum coating alloyed with the substrate, forming an iron-aluminum intermetallic with an outer scale of Al_2O_3.

Thermal Fatigue. A final form of environment-assisted hot end exhaust system failure is thermal fatigue. This type of component failure is oxidation-assisted mechanical fatigue and generally occurs only in hot end applications. Figure 6 shows the exterior of a truck T409 stainless steel catalytic converter shell that failed in service, developing a through-thickness fracture at the deepest midlength rib on the washboardlike converter outer surface. Also in Fig. 6 is a 50× cross section of the rib with a

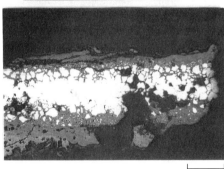

Fig. 3 Laminated T304 heat shield where insulation became soaked with salt solution, leading to hot salt intergranular attack. At bottom, cross section. 200×

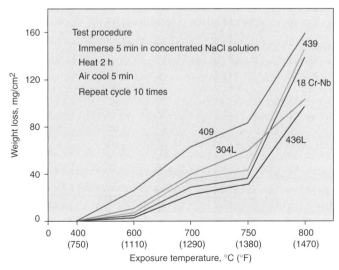

Fig. 4 Hot salt comparisons of standard exhaust alloy stainless steels with weight losses after 10 cycles versus exposure temperature

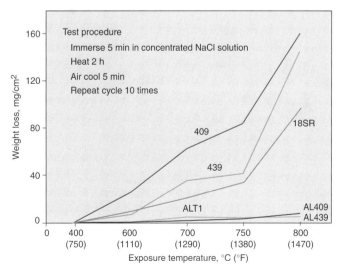

Fig. 5 Hot salt comparisons of standard and aluminum-coated exhaust stainless steels with weight losses after 10 cycles versus exposure temperature

series of shallow notches in the metal, characteristic of fatigue. Cyclic stresses on the hot shell surface lead to small surface cracks that, due to the elevated service temperature, are oxidized. The oxide within the shallow cracks wedged open the steel surface so that, on subsequent heating cycles, a notch was formed that cracked further and again oxidized. With time, one of these fissures oxidized more rapidly, and eventually, so much of the steel cross section was cracked and weakened that the next heatup led to a ductile rupture failure of the shell. The rib that fractured in this component happened to be the deepest stamping in the converter shell and was situated in a location that was exposed to high heat from internal exhaust gases. Had the rib been shallower or possibly insulated from the internal gases, the temperature and stress concentrations would have been less and fatigue would have been avoided. Without changing shell design, the way to eliminate the problem is to move to a stronger, more oxidation-resistant alloy or thicken the shell wall.

Thermal fatigue is obviously design-critical. Designs that concentrate in-service stresses or lead to hot spots promote such failures. Experience has shown that failed or too few hanger assemblies are a major cause of thermal fatigue. Heat-affected zones (HAZs) are sensitive to such vibrational failures, particularly a ferritic 400-series alloy HAZ if the metal on the other side of the weld is a stronger, higher coefficient of thermal expansion 300-series austenitic stainless steel. As with the converter shell described previously, sometimes a design showing fatigue problems must remain, and material change must be considered. In these instances, an alloy with greater strength at temperature or more oxidation resistance is the best solution. Table 5 is a comparison of exhaust alloy strength at 815 °C (1500 °F). The 300-series alloys offer the better strength but not necessarily the better resistance to oxidation. Among the 400-series alloys, the addition of niobium, Nb (or columbium, Cb) improves elevated-temperature strength. Alloys such as T441 or 18Cr-Nb have frequently found favor as a way to upgrade the elevated-temperature strength and oxidation resistance of T409.

Cold End Exhaust Corrosion

Exhaust components further from the engine experience lower gas temperatures and conditions where condensate forms in exhaust service. Added to these condensates are the exhaust gas ions and compounds formed from combustion of fuel. These include carbon dioxide (CO_2), sulfur trioxide (SO_3), ammonia (NH_3), and water (H_2O). They leave the catalytic converter and form droplets of condensate when dewpoints of 120 to 175 °C (250 to 350 °F) are encountered on cooler intermediate pipe and muffler walls. Add to this mix gaseous hydrogen chloride (HCl) from chlorides left on converter ceramic monoliths (when precious metals were deposited from a chloride-rich solution), and the potential exists for chloride pitting in an acid environment. New converters can produce condensates with 1000 to 2000 ppm of chlorides from the water-soluble chlorides left on ceramic honeycomb biscuits after manufacturing. These chlorides (as gaseous HCl) are absorbed in water vapor from the combusted hydrocarbons. In several hundred miles of driving, the chloride level of new car condensates will drop below 100 ppm and approach the 50 ppm of chloride in normal gasoline. As the chloride level returns to 50 ppm, the pitting tendency decreases, but still the potential exists for pitting, particularly with dried chloride deposits that are rewetted with each new driving cycle.

Condensate Pitting Corrosion. As the dewpoint on cool component walls is reached, droplets of condensate laden with chlorides, sulfur compounds, and ammonia ions form. The amount of diluting water present is dependent on heat-up rate and climatic conditions. Initially, the condensate can be neutral or basic (pH ≥ 7), depending on the amount of dissolved ammonia. As the ammonia boils off and the droplets shrink in size, the pH tends to become more acidic. Some pH values have been measured as low as 1 to 2. The slower the rate of evaporation, the longer the time that chloride-laden acidic droplets have to initiate and propagate pits. For the same mileage, short driving trips where mufflers never get hot enough to completely dry out are more detrimental to muffler longevity than extended interstate trips.

Drain holes in muffler end plates are important as a means to flush out condensates and accumulating salt deposits. A double-walled T409 muffler, shown in Fig. 7, was constructed with no drain hole and was removed after 4000 km (2500 miles) of driving. Internal condensate puddled in an area adjacent to a spot-welded internal baffle and pitted through both the inner and outer shell. Numerous small through-thickness pits created by the trapped condensate formed on the inner shell. A drain hole put in the bottom of the single-thickness muffler end plate and mounted on the vehicle with an angle to ensure drainage would have prevented such rapid pitting. On some vehicles, a drain hole is replaced by a siphon tube designed to use venturi-like air flow to evacuate condensate. Such devices only function if the vehicle is in motion and must be mounted with care for proper operation.

Laboratory tests to evaluate condensate resistance of stainless exhaust alloys consist of a three-step procedure. Coupons 75 by 100 mm (3 by 4 in.) are heated in air for 1 h at 260 °C (500 °F) to simulate heat conditions inside the muffler. At this temperature, a light straw-colored heat tint develops and slightly deteriorates surface corrosion resistance. After the heat cycle, the samples are placed in a beaker partially filled with synthetic condensate solution (Fig. 8a). This aqueous solution contains 5000 ppm of sulfate ion (SO_4^{3-}), 100 ppm of chloride ion (Cl^-), 100 ppm of nitrate ion (NO_3^-), 100 ppm of formic acid (HCOOH), and initially has a pH of 5.3 to 5.5. The solution is boiled 16 h, with no liquid being added and no condenser on top of the beaker to maintain the volume. As the solution evaporates, the pH decreases to 3.3 to 3.5 by the conclusion of the test cycle. The evaporating condensate creates a

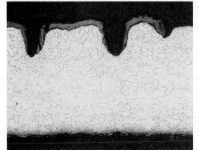

Fig. 6 Catalytic converter shell with thermal fatigue cracks in formed exterior surface rib. In cross section at bottom, note numerous outer-diameter surface cracks. 50×

Table 5 Comparison of elevated-temperature strengths for commonly used exhaust stainless steel alloys

Alloy	Short-time 0.2% yield strength		Short-time ultimate tensile strength		Rupture strength(a)		Fatigue endurance(b)	
	MPa	ksi	MPa	ksi	MPa	ksi	MPa	ksi
T409	20.7	3.0	29.0	4.2	6.2	0.9	10.3	1.5
T439	23.4	3.4	30.3	4.4	6.9	1.0	9.6	1.4
18Cr-Nb	40.0	5.8	50.3	7.3	12.4	1.8	27.6	4.0
T304L	64.8	9.4	109	15.8	22.1	3.2	...	...
309S	131	19.0	179	26.0	29.6	4.3	...	...

(a) 1000 h stress. (b) Limit at 10^7 cycles

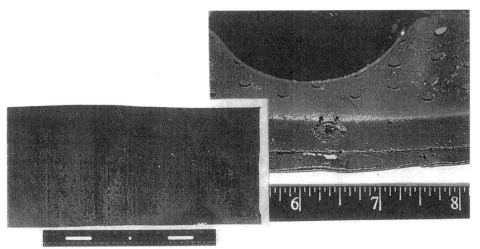

Fig. 7 South American muffler (409 shell and baffles) designed without drain holes in end caps or baffles (4000 km, or 2500 miles, of driving). Lack of condensate drainage has led to premature pitting perforations.

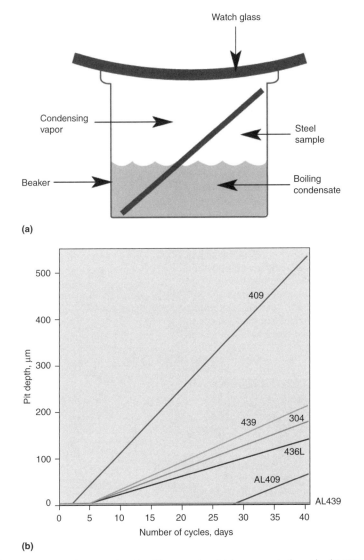

Fig. 8 Laboratory test to evaluate condensate pitting corrosion. (a) Arrangement of sample. (b) Pitting condensate corrosion results for six exhaust alloys. Pit depths represent the average of the ten deepest pits. Details of the test procedure are given in text.

Automotive Exhaust System Corrosion / 523

pitting condition similar to that which occurs in a muffler. After boildown, the coupons are exposed for 6 h in a humidity cabinet at 50 °C (125 °F) and 85% relative humidity. The humidity cabinet rewets dried salts and allows pitting to continue. One test cycle consists of the daily three-part routine. Total test duration is generally 8 to 10 weeks (40 to 50 daily cycles), with weekends spent in the humidity cabinet. Samples are periodically removed from the test for pit depth measurements.

Pit depths were measured by one of two methods. After a light glass bead blasting or peening to remove built-up corrosion deposits, pointed micrometers were used to measure the deeper-appearing pits, and an average of the ten deepest pits was calculated. The second method involved chemically cleaning corrosion debris with a solution of inhibited 1 to 1 HCl/H_2O in accordance with ASTM G 1 (Ref 1). The ten deepest pits were then circled, and a calibrated microscopic eyepiece was focused on the bottom of the pit and then the surface to estimate the pit depths. Agreement between the two depth-measuring procedures was fairly good.

Figure 8(b) shows the boiling beaker condensate corrosion results for six exhaust stainless alloys, four bare and two aluminum coated. On this graph, 500 μm represents approximately 0.508 mm (0.020 in.) of pit depth. Bare 409 (11% Cr) suffered the deepest pitting attack, followed by 17%Cr 439, 18%Cr-8%Ni T304, and 17%Cr-1%Mo 436L. The aluminum-coated 409 outperformed all the bare stainless alloys, while aluminized 439 was not exposed long enough to show any appreciable pits. The aluminized 409 showed no pitting for 28 of the 40 test cycles, the time for the acid chloride solution to corrode away the galvanic protection offered by the aluminum outer layer and the underlying iron-aluminum intermetallic diffusion coating. Unlike coated carbon steel, stainless alloys with hot dip aluminum coatings have a diffusion layer of iron-aluminum under the thicker aluminum coating that also offers galvanic sacrificial protection for the stainless substrate.

Exterior Salt Pitting. On cold end exhaust components, where skin temperatures do not quickly boil away aqueous solutions, exterior stainless surfaces undergo wet chloride pitting. This classic form of pitting corrosion occurs because cold end exhaust components experience more wet time due to their location under the vehicle, and the lower skin temperatures extend drying times. Unlike internal condensate pitting (discussed earlier), exterior salt corrosion does not involve acidic pH levels. Exterior surfaces experience a variety of chloride compounds. Sodium chloride (rock salt), the mainstay of road deicing compounds for the last 50 years, has recently been joined by calcium and magnesium chloride in either solid or liquid form as a pretreatment prior to snowfall or when low temperatures are anticipated. For traction control, many states use sand or fine aggregate (crushed rock) mixed with the salt. Poulticelike mixtures of mud and chlorides coated on exhaust component surfaces promote pitting and form crevice-corrosion conditions.

The more recent use of calcium chloride, not only as a winter antiicing compound but as a sprayed-on summer road dust deterrent, could be introducing a second pitting problem for exhaust components as well as automotive exterior trim. Calcium chloride ($CaCl_2$) crystals are hygroscopic. When deposited on metallic surfaces, such crystals absorb water vapor, become more adherent, and, with the presence of chlorides, continue the pitting process.

Exhaust gas temperatures lead to exterior heat tinting of cold end components and outright oxidation and scale formation on hot end components. Heat tints on intermediate pipes, mufflers, and tail pipes are generally thin interference oxide films that vary in color from straw yellow at low temperatures (200 to 260 °C, or 400 to 500 °F) to darkening shades of brown and even blue as temperatures increase toward 540 °C (1000 °F). Above approximately 650 °C (1200 °F), the scales thicken and take on the dark-brown or gray coloration generally associated with oxidation. Experience has shown that the presence of scale and heat tint lead to increased pitting attack. On cold end components, this increase normally takes the form of many shallow pits rather than a few deep pits. The red iron oxide formed within and around the rim of these shallow pits gives the tinted surface the appearance of having undergone significant corrosion, albeit cosmetic or superficial in nature. The greater the degree of heat tinting, the more rapid and severe the resultant cosmetic attack. The increased propensity to pit on a tinted surface may be related to the rougher surface being a collection for chlorides. Another theory suggests that the increased chromium level in the oxide film causes a corresponding lower chromium level in the substrate immediately under the heat tint layer. Wet chlorides penetrating cracks or voids in the heat tint find an active, lower-chromium-content alloy underneath.

Figure 9 shows a single-shell 17%Cr T439 muffler with 106,500 km (66,000 miles) of exposure during driving. Note the band of cosmetic rust running circumferentially around the midlength of the muffler. Sitting between two rows of spot welds, this area is probably an internal tuning chamber that ran hotter and caused the external surface to tint and cosmetically corrode in the presence of road salt and moisture. Heat tint and exterior salt cosmetic attack were the main reasons behind the development of hot dip aluminized 409 muffler steels. Being anodic to the steel substrate, the aluminum coating prevents surface red rust from forming on 409 until the exterior aluminum coating and alloy layer of iron-aluminum intermetallic underneath are consumed. Aluminum corrosion product is white and not nearly as visible from a distance. Aluminum or the Al-10wt%Si alloy used on these coated stainless alloys will not heat tint, retaining a silver metallic appearance until it is consumed in the iron-aluminum intermetallic layer and as Al_2O_3. After the aluminum is consumed, which generally requires temperatures above 427 °C (800 °F), the coated T409 becomes dark gray in surface appearance and is prone to red rusting. To further illustrate the cosmetic corrosion benefits of aluminum coatings, Fig. 10 shows two exhaust systems from the same-make vehicle. One is completely constructed of aluminized 409, except for flanges and hangers. The all-AL409 unit saw 201,600 km (125,000 miles) of driving. The second system has bare 409 for the exposed pipes and saw only 56,500 km (35,000 miles) of service. While both systems are of sound integrity, the cosmetic attack of the bare 409 pipes is obvious. North American cold end exhaust components, at least muffler wraps and tail pipes, visible to the public have seen significant growth in the use of aluminum-coated 409 stainless, depending on the 11% Cr base alloy for durability and the thin exterior aluminum surface (coating weights generally are applied at a minimum of 85 g/m^2, or 0.25 oz/ft^2, total for both sides) for cosmetic corrosion resistance. The protective nature of the aluminum coating, galvanic even to uncoated adjacent filler welds, has been found to be a more effective way to avoid cosmetic rust than sprayed-on black paints applied after component manufacturing. Such paints are often seen to peel and flake in a year or less and can actually make corrosion worse due to their tendency to trap moisture and salt against the underlying steel surfaces.

To be realistic, laboratory external salt pitting tests need to combine simulated road salt splash with cyclic heating and exposure to high but noncondensing humidity. The samples should be wet but not washed. Coupons can be flat, rectangular panels or have 90° bends and even 25 mm (1 in.) diameter stretch domes pushed to a height of 5 mm (0.2 in.) added to better evaluate the effect of forming, particularly on coated alloys. Road deicing salts are often mixtures of chlorides, but most laboratory tests use only NaCl. As described previously, sets of test coupons were removed on a periodic basis, cleaned, and average pit depths measured. Heat, salt, and humidity can be combined into endless test possibilities, with the heat-cycle temperature chosen based on which exhaust component skin temperature (Fig. 1) is being simulated.

External pitting test A (muffler shell temperatures) was carried out in a controlled relative humidity (RH) cabinet. Coupons received a weekly 260 °C (500 °F) 1 h heat exposure to form a heat tint. The samples were then placed in the cabinet and given a repeating 6 h cycle of:

- 15 min immersion in 5% NaCl solution
- 20 min dry in 40 °C (104 °F)/35% RH air
- Raise conditions to 60 °C (140 °F)/85% RH and hold balance of cycle

Figure 11 shows average pit depths after 1176 h of testing with the procedure outlined previously. Results show the benefit of increasing chromium content and, at longer test times, of increasing the molybdenum (1.3% in 436M2) content. The aluminum coating was also beneficial until the galvanic protection of the free-aluminum and alloy layer was eliminated. Then, the underlying 409 base alloy corroded at a rate approaching that of the bare base alloy.

Test A requires a sophisticated programmable humidity cabinet to conduct. A second, simpler test, external salt pitting test B, was developed and has been proven to correlate well with field survey results as well as on-car component durability tests. Test B uses a weekly heat treat cycle of 1 h, but only a once-per-day salt dip and air dry, and spends the balance of each day at 85% RH and 60 °C (140 °F). Figures 12 through 14 show the results of conducting test B at varying

Fig. 9 T439 muffler shell after 106,500 km (66,000 miles) of driving. Cosmetic superficial pitting seen in band above a tuning chamber area that saw higher temperatures and heat tinting

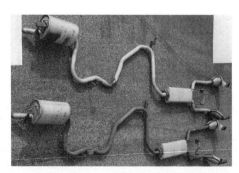

Fig. 10 Two identical exhaust systems collected from the field. The upper system is AL409 with 201,600 km (125,000 miles) of service. The lower system has bare T409 tubes with 56,500 km (35,000 miles) of service.

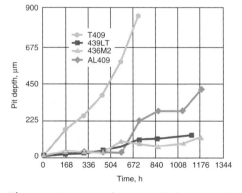

Fig. 11 Comparison of average pit depths over time for four exhaust stainless steel alloys exposed to cyclic exterior salt humidity cabinet pitting test A. See text for test details.

heat treat temperatures, namely 315, 425, and 540 °C (600, 800, and 1000 °F), respectively. The 315 °C (600 °F) test results resemble test A at 260 °C (500 °F). At these hotter muffler temperatures, 17%Cr T439 and aluminum-coated 439 (a premium durability product used on some luxury vehicles) show essentially no attack after 12 weeks of testing. At 425 °C (800 °F) intermediate pipe temperatures, the depth of pitting increased on all the alloys evaluated. Finally, with a converter can 540 °C (1000 °F) heat cycle, bare 409 began rapid pitting, which appears to cease between 2 to 4 weeks, only to reinitiate after 4 weeks. Visual coupon observations indicated the 540 °C (1000 °F) temperatures caused 409 to form a thicker scale jacket than on the higher-chromium alloys. This thicker oxide film filled active pits with debris that shut down the pitting mechanism for a time. The decrease in pitting corrosion of AL409 at this highest temperature was associated with the total alloying of the free-aluminum layer. While no longer bright cosmetically, the thick alloy layer does provide galvanic and barrier coating resistance to chloride attack. Results shown here with a 540 °C (1000 °F) heat treatment do appear to correlate with converter exhaust systems in the field. One week in this test B cycle appears equivalent to 1 year of service in the severe salt belt of North America.

Crevice corrosion occurs when liquids can penetrate enclosed areas with a limited supply of oxygen and set up a concentration cell. With stagnant solutions that have limited supplies of oxygen, the stainless alloys cannot maintain their passive protective films. The presence of chlorides only accelerates the reaction. Figure 15 is a schematic of such a crevice. In cold exhaust applications, there are three areas particularly prone to crevice attack. One is between the wrap and shell of double-walled mufflers, and the other is on external muffler surfaces when belly bands are used for underbody attachment. Welded-on rod or sheet-metal hangers can also cause small crevices where they attach to the muffler end caps or exhaust tubes. A third area of potential crevice attack is the muffler internal surface where baffles are spot welded. Small gaps between the baffle bottom surface and inner muffler shell can create a crevice where stagnant liquid can collect. To combat baffle crevice corrosion, the baffles frequently have a drain hole on their bottom edges to permit liquids to pass through on their way to drain holes in the end plates.

Crevice attack between the wrap and shell of mufflers is primarily the result of leakage from internal condensate and intrusion of external road salt solution through the muffler shell lock seams and rolled-on end plates. Figure 16 shows half of a T409 double-walled conventional muffler with the wrap (right side) and shell (left side, with baffles and internal tubes attached) separated. The component experienced 132,500 km (82,000 miles) of driving service. In this figure, the two single-wall end plates have been removed by sawing. The insert in Fig. 16 shows an end view of the muffler and the lock seam used to attach the wrap to the shell. There is another lock seam on the other side of the

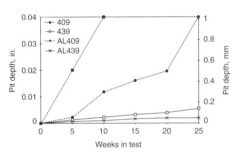

Fig. 12 Exterior cyclic salt pitting test B after 25 weeks with a weekly 1 h 315 °C (600 °F) heat exposure and daily salt solution dip and humidity exposure

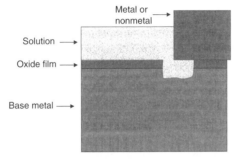

Fig. 15 Schematic of crevice-corrosion condition where stagnant liquid with a limited supply of oxygen leads to breakdown of the stainless alloy passive film

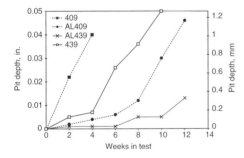

Fig. 13 Exterior cyclic salt pitting test B after 12 weeks with a weekly 1 h 425 °C (800 °F) heat exposure and daily salt solution dip and humidity exposure

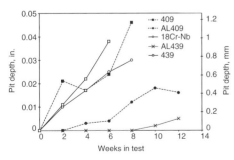

Fig. 14 Exterior cyclic salt pitting test B after 12 weeks with a weekly 1 hr 540 °C (1000 °F) heat exposure and daily salt solution dip and humidity exposure

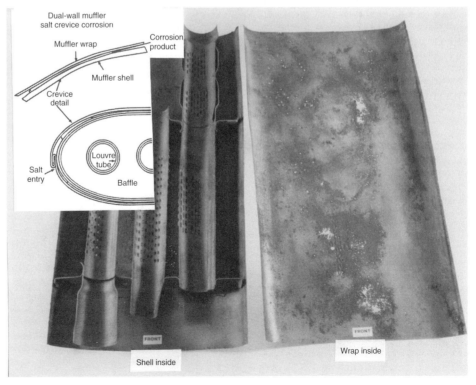

Fig. 16 Dual-wall 409 muffler after 132,500 km (82,000 miles) with crevice corrosion between the outer wrap and inner shell

muffler, and this one joins the inner shell oval. Springback of the 180° lock seam bends prevents an airtight seal and allows acid condensate from inside the muffler and road salt solution from outside the muffler to enter the narrow space between the shell and wrap. This interwrap area has no drainage, and salt levels can build up, with resultant corrosion on the inside of the outer wrap and outside of the inner shell. Because muffler exterior wraps are often the thinner of the two layers, perforation generally occurs first to this layer. In Fig. 16, the pinholes of perforation appear in greater number on the thinner 409 wrap. Single-shell mufflers do not have such crevices and offer longer corrosion durability, often at the expense of increased noise-level concerns.

To combat double-shell interwrap crevice corrosion, muffler builders have begun adopting a double-lock design where the wrap and shell are simultaneously roll locked together on the same side of the muffler. Figure 17 shows a cross section of this form of muffler seal. While this forming technique requires more power, it does retard liquid intrusion into interwrap areas.

Another source of interwrap leakage in double-wall mufflers is rolled-on end caps. As shown in Fig. 18, if the outer wrap is tucked into the head and not bent 180° with the end cap and inner shell, salt solution intrusion from outside surfaces is possible. The wrap needs to be extended into the shell/end cap seam and the entire thickness rerolled to seal off leaks. Figure 19 shows an AL409 muffler wrap after 27,400 km (17,000 miles) of driving, with internal leakage around the rolled head causing exterior corrosion of the protective aluminum coating. The aluminum-coated stainless alloys, when used in double-wall muffler construction, have been found to improve leak tightness. The softer free-aluminum exterior coating acts similar to a gasket in crimping operations when compared to the smoother, harder mating surfaces presented when bare stainless shells are joined.

Exterior wrapped belly bands for muffler attachment to the car frame also create life-shortening crevices. Figure 20 shows a muffler shell constructed of T409 after 114,500 km (71,000 miles) of driving in Montreal, Canada. The belly band has been removed and the entire surface glass bead blasted to expose corrosion damage. The crevice under the band has perforated the 409 shell, structurally damaging the shell integrity.

Evaluating crevice-corrosion resistance in the laboratory required the development of a suitable coupon configuration. Figure 21 shows a test coupon made up of two sheet pieces held together with a hem flange 180° bend, similar to the design used to seal the bottom of car door skin panels. The two-piece test panel is exposed to internal condensate solutions or external road salt mixtures and, after the appropriate cycles, can be pulled apart, cleaned as described previously, and pit depths measured by a pointed micrometer or calibrated microscope. The two sheet samples making up the crevice test panel can be of the same alloy or of mixed materials, which adds the possibility of a galvanic corrosion contribution.

Table 6 shows crevice-corrosion pit depths after 10 weeks of cyclic condensate exposure. The cyclic test incorporated a weekly 1 h heating of 315 °C (600 °F), followed by daily spraying with the synthetic condensate (450 ppm chloride, 2000 ppm sulfate, and pH 2.5) and a 90 min drying cycle. After the dry period, the hem flange coupons were put in a humidity cabinet for 22.5 h of 60 °C (140 °F)/85% RH. Each weekend, samples remained in the humidity. After 10 weeks of this routine, the hem flange pieces were pulled apart, cleaned, and the average of the ten deepest pits determined. For each alloy, pits were

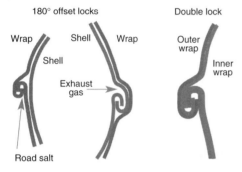

Fig. 17 Muffler lock seam constructions. The double-lock seam construction (right) helps prevent liquid penetration between the wraps.

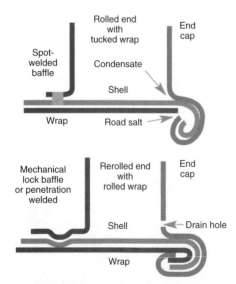

Fig. 18 Muffler end cap constructions. The more desirable corrosion-resistant muffler design is shown at bottom.

Fig. 19 Muffler with 27,400 km (17,000 miles) of service and AL409 wrap and end cap. Condensate leakage through the rolled head has attacked the aluminum coating.

Fig. 20 Crevice corrosion of a 409 muffler shell under a belly band (removed from photo) on a vehicle driven 114,500 km (71,000 miles) in Montreal, Canada.

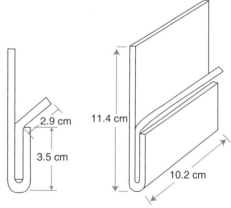

Fig. 21 Sheet-metal crevice-corrosion hem flange coupon designed for laboratory corrosion tests where both crevice and boldly exposed surface areas are created. With two samples, the test permits the mixing of base metal alloys.

Table 6 Synthetic muffler condensate pit depth after 10 weeks

Alloy	Boldly exposed area(a)		Crevice area(a)	
	mm	in.	mm	in.
T409	0.43	0.017	0.25	0.010
T439	0.08	0.003	0.18	0.007
430M	0.13	0.005	0.13	0.005
18Cr-Nb	0.03	<0.001	0.36	0.014
AL409	0.08	0.003	0.13	0.005

(a) Average depth of ten deepest pits

measured within the crevice area and on the boldly exposed surfaces. Comparing pit depths to muffler surveys from the salt belt of the U.S. Northeast revealed that approximately 1 week of this cyclic test was the equivalent of 6 months service in the field.

A comparison of Table 6 bold exposure and crevice pit depths shows the crevice depths to be deeper than or equal to the boldly exposed areas for all alloys except 11%Cr 409. The lower general corrosion resistance of this 11%Cr alloy appeared to cause the crevice opening to fill with oxide debris and choke off the corrosion reaction. While T409 showed the deepest pitting, the 17%Cr T439, the 19%Cr 430M, and the aluminum-coated 409 were more resistant and essentially equivalent to each other.

Intergranular corrosion (IGC), while infrequent, has been found in exhaust applications, generally in the cold end of the system where corrosive liquids are more prone to interact with weld HAZ areas. Ferritic stainless alloys, like their austenitic 300-series counterparts, can be sensitized from improper heating that results in the formation of chromium-rich carbides ($M_{23}C_6$) along grain boundaries and the creation of chromium-depleted base metal areas adjacent to the grain boundaries. These lower-chromium grain-boundary areas are prone to intergranular attack from corrosive liquids. With a lower solubility for carbon compared to 300-series austenitics, the ferritic stainless alloys can be sensitized at lower levels of dissolved carbon as well as with dissolved nitrogen, forming Cr_2N particles. In most exhaust ferritic alloys, titanium or titanium along with niobium are added to reduce soluble carbon and nitrogen contents, making the base alloy and welds more resistant to IGC attack and reducing chances of forming chromium compounds if components are exposed to temperatures in excess of 400 °C (750 °F) (Ref 2). The chromium carbides and nitrides are thought to precipitate up to 900 to 925 °C (1650 to 1700 °F) and will dissolve if heated above 1000 °C (1830 °F) (Ref 3). Unfortunately, the 1000 °C (1830 °F) temperature needed to dissolve chromium precipitates is also high enough to dissolve some (titanium, niobium) carbonitrides, putting interstitial elements back into solution. This soluble carbon and nitrogen can reprecipitate as chromium compounds if, after high-temperature exposure, cooling rates through the red heat zone are too fast for more stable titanium and niobium precipitates to form. With soluble carbon and nitrogen present, exposure temperatures of 705 to 900 °C (1300 to 1650 °F) will lead to grain-boundary chromium precipitate formation. However, above 705 °C (1300 °F), the diffusion rate of chromium in the ferritic stainless base alloy is high enough to replenish chromium-depleted regions adjacent to grain boundaries. No IGC would be expected. Exposure temperatures between 400 and 705 °C (750 and 1300 °F) will also lead to chromium carbide and nitride formation, but chromium-depleted base metal grain-boundary areas will not be replenished, and IGC is possible if such regions are exposed to moisture and chlorides.

With ferritic stainless alloys, the greatest concern of IGC comes during welding. Stoichiometric calculations predict that the minimum stabilizer needed to remove soluble carbon and nitrogen contents is:

$$\%Ti\ min = 4(\%C) + 3.5(\%N)$$

$$\%Nb\ min = 7.75(\%C) + 6.65(\%N)$$

Hydrocarbons on the steel surface due to lubricants used in component production can provide a source of potential carbon during welding operations. Austenitic filler metals in gas metal arc (GMA) (nonstandard term is metal inert gas) welds can bring higher soluble carbon and nitrogen levels into the weld pool and mix with the ferritic base alloy. The ferritic weld pool is at risk for nitrogen pickup if shielding gas coverage terminates while the weld pool is still at high temperatures. With these added risks, stabilizer levels slightly in excess of minimums have been found desirable. In marginally stabilized ferritic welds, there is the added concern that rapid cooling of the weld may not give enough time for the carbon and nitrogen to be completely combined with titanium/niobium before some chromium compounds begin forming at the lower temperatures.

With the ultralow levels of carbon and nitrogen possible with vacuum oxygen decarburization and some argon oxygen decarburization melting practices, some stabilization ratios even call for a minimum amount of stabilizer regardless of interstitial levels. While some excess stabilizer is desirable, excessive levels of titanium or niobium are felt to be detrimental to weld toughness. A balance is needed. The dual use of titanium and niobium is considered by some to be a way to obtain such a balance. Weld ductility with titanium- or titanium + niobium-stabilized base alloy ferritics is normally considered acceptable, although not as good as the base metal. Experience with niobium-only stabilized melts has shown that autogenous gas tungsten arc welds tend to have a more dendritic cast structure, with reduced formability and toughness. Titanium nitride (TiN) forms at higher temperatures than either titanium carbide (TiC) or niobium compounds. It is thought that these nitrides form in the melt and act as nucleation sites for ferritic grains. If a ferritic stainless were to have only one stabilizer to remove carbon and nitrogen content, titanium would be the best choice. Recent advances in stainless melting technology have largely overcome earlier concerns that titanium stabilization will lead to unsightly and abrasive surface streaks of TiN.

Laboratory tests to evaluate ferritic stainless alloys for IGC are covered by ASTM A 763 (Ref 4). As a screening tool to look for susceptibility of intergranular attack with stabilized ferritic alloys, the electrolytic 10 wt% oxalic acid test (practice W) can be used. If one or more grains are found completely surrounded by ditched (corroded) grain boundaries, the procedure then calls for additional testing using the more definitive practice Z. This practice calls for 24 h of boiling in copper-copper sulfate (Cu-$CuSO_4$) solution acidified with 16% sulfuric acid (H_2SO_4). After boiling, the samples are bent 180° around a diameter equal to $4t$, where t is the sample thickness. Examined at 5 to 20×, the bent outer-diameter surfaces must not show cracks in order to be considered resistant to IGC. It bears mentioning that the practice Z test is designed for alloys with a minimum of 17 to 18% Cr. T409, with its nominal 11% Cr addition, will be heavily attacked by the Cu-$CuSO_4$ solution and, at times, dissolved in 24 h, depending on initial sample thickness. To evaluate IGC in stabilized 11 to 12% Cr alloys, several modified practice Z procedures have been used but are not incorporated into the ASTM standard. In one practice Z modification, the standard solution was used, but the test cycle was reduced to 16 h of exposure in a 30 °C (85 °F) solution. A second, modified cycle exposed the samples to boiling solution for 20 h but lowered the H_2SO_4 solution strength to 0.5 wt%. Reducing the aggressiveness of the test solution permits comparisons of the 11 to 12% Cr alloys without generally attacking and dissolving the base metal.

When T409 became the material of choice for catalytic converter shells in the mid-1970s, the 11% Cr ferritic was melted to a stabilization specification of 6×C minimum. In those days, the carbon and nitrogen levels were each close to 0.03%, and titanium levels ranged from 0.3 to 0.5%. With advances in melting, the carbon levels were reduced to typically less than 0.015%, and nitrogen levels fell to 0.02%. With lower carbon content, titanium could be reduced and meet the 6×C minimum. Unfortunately, with no nitrogen factor in the specification and TiN forming in the melt, marginally stabilized 409 melts were produced. Failures such as the one shown in Fig. 22, while rare, were encountered. Here, a 409 tail pipe GMA welded to a 409 rear muffler head fractured in the HAZ after several years of driving service in Canada. A micrograph of the weld break in cross section shows the intergranular nature of the failure.

In response to failures like the one in Fig. 22, ASTM International and the stainless steel producers redefined the minimum stabilization in 409 to include both carbon and nitrogen. ASTM International recognizes three T409 stabilization specifications (Ref 5):

$$Ti\ min = 6(C+N)$$

$$Ti\ min = 8(C+N)\ and\ 0.15\%\ min$$

$$(Ti + Nb)\ min = 0.08 + 8(C+N)$$

Since adopting the higher stabilization minimums, no major IGC exhaust failure incidences have been reported.

Galvanic corrosion due to the coupling of dissimilar metals in the presence of a corrosive solution is the last form of corrosion found to have occurred in exhaust system components. Fortunately, like intergranular attack, exhaust designers have learned from experience, and premature failures from this type of attack have been eliminated.

In galvanic or bimetal corrosion, the two dissimilar metals have different corrosion potentials for the solution involved. When electrically coupled in the presence of the conducting solution, current flows between the metal pair. The more noble or corrosion-resistant material will experience a reduced rate of attack and become the cathode. The less noble alloy becomes the anode and undergoes more rapid corrosion, actually sacrificing itself to protect the more noble material of the pair. The rate of weight loss or pitting for the anodic metal in the couple depends on how far apart the two materials are in the electromotive force series for the corroding liquid involved, as well as the relative size difference of the anode and cathode. A small anode area coupled to a large-area cathode is the worst possible combination, leading to rapid attack of the less noble alloy. Earlier in this article, one form of galvanic corrosion was alluded to in the description of the galvanic protection that aluminum coatings provide when hot dip coated onto carbon and stainless steel substrates. Here, the aluminum outer layer sacrifices itself for the steel substrate if a scratch or cut edge is encountered. In this type of application, the area of the aluminum anode is large and the steel cathode small. Protection is then provided with a minimum of attack to the less noble aluminum layer.

To see the worst form of exhaust system galvanic corrosion, one must go back to the days (prior to 1995) when exhaust designers mixed the stainless 11%Cr T409 alloys with straight aluminized type 1 carbon steel (ALT1). One scenario involved upgrading the corrosion resistance of mufflers by specifying 409 while continuing to call for ALT1 for the attached intermediate pipe. The two dissimilar metals were joined with a GMA weld using a stainless 409Nb weld filler. Exposure to corrosive road salt solutions on one surface and muffler condensate on the other created a galvanic cell. The aluminum coating on the carbon steel substrate sacrificed itself to protect the 409 muffler shell and weld (Fig. 23). With time, the aluminum was stripped away, revealing the underlying carbon steel base metal. With extended exposure times, the loss of aluminum became great enough that a carbon-steel-to-409 couple was established. The less noble carbon steel now attempted to protect its stainless material mate. A corrosion failure next to the weld soon ensued. A rough rule of thumb would predict that 409 pipes to 409 muffler shells should last a minimum of 5 to 7 years without perforation. The ALT1 tubes and shells, depending on the type of protection used on welds, could survive 3 years or more under the car. The aforementioned dissimilar-metal galvanic couple was predicted to fail in the coated carbon steel in 18 to 24 months of service.

In a similar problem, Fig. 24 shows an internal view of an exhaust muffler taken off of a vehicle with less than 16,000 km (10,000 miles) of driving. The muffler designer mixed 409 internal baffles and tubes with an ALT1 exterior shell to take advantage of the coated carbon steel cosmetic resistance to early red rusting. Unfortunately, the galvanic cell created by the stainless baffles contacting the inner surface of the ALT1 shell in the presence of corrosive condensate liquids led to accelerated galvanic attack of the shell and a linear perforation of the shell under the baffle in a very short driving time.

Failures similar to the two previously described examples occurred in the 1980s and early 1990s as car makers and OEM designers struggled to upgrade cold end exhaust system durability by selectively mixing stainless components with the old standard ALT1 carbon steel. However, with the arrival in 1995 of the 3 year/58,000 km (36,000 mile) warranties, the replacement of coated carbon steels in OEM exhausts accelerated, with the final replacement occurring in one major truck line in model year 1996.

Since the switch to all stainless, catastrophic galvanic corrosion failures have essentially disappeared, although dissimilar-metal couples still exist. The use of aluminum-coated 409 for cold end cosmetic applications has led to the frequent coupling of a bare stainless component to an aluminized one. Once again, the aluminum layer sacrifices itself to protect its bare stainless neighbor. Fortunately, in this instance, as the aluminum and underlying iron-aluminum intermetallic layer are removed, a stainless substrate is revealed with far better durability than was found with coated carbon steel.

Early in the development of aluminum-coated 409, corrosion studies found that the intermetallic iron-aluminum diffusion layer that forms when the hot dip Al-10%Si layer is applied to the 409 substrate is anodic or protective to the base alloy. On aluminized carbon steels, the alloy layer, once the free aluminum is removed, offers no additional corrosion protection. Regardless of the substrate composition, red rust will appear once the free aluminum is gone and the iron of the alloy layer is exposed to moisture.

Figure 25 shows a muffler with 27,400 km (17,000 miles) of exposure. The muffler shell or

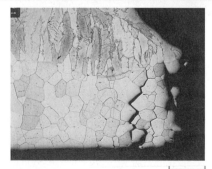

Fig. 22 Tail pipe and muffler end cap, both of 409, that failed by intergranular corrosion after driving service in Canada. Cross-sectional photo at bottom is 50×. Base alloy had Ti/(C + N) ratio of 5.8.

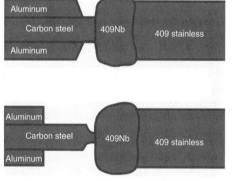

Fig. 23 Galvanic corrosion when mixing muffler materials. Aluminum sacrifices itself to protect the 409 stainless steel, exposing the carbon steel substrate.

Fig. 24 Internal view of muffler with 14,500 km (9,000 miles) of service. Condensate galvanic corrosion between the 409 stainless baffles and ALT1 shell has destroyed the wrap.

Fig. 25 Muffler with AL409 wrap and bare stainless belly band and end cap (right side of image). After 27,400 km (17,000 miles) of driving service, the band and muffler end plate have led to galvanic attack on the aluminum coating of the wrap.

wrap was made of AL409, while the end cap (on the right) and the belly band were bare stainless steel. Note that the band and cap are not heavily rusted or discolored (the benefit of galvanic protection), while the AL409 shell areas adjacent to the cap and belly band show dark-gray stripes where free aluminum has been galvanically removed, exposing the underlying alloy layer. With additional time, those exposed alloy layers will begin showing red rust. In a similar vein, the muffler and tail pipes shown in Fig. 26 were half constructed of bare 409 and half AL409. The muffler/pipes were installed on a vehicle and driven 6 months, with daily spraying with a salt solution. Cosmetically, the aluminum coating made a great difference. The circumferential GMA weld (409Nb) joining the muffler halves further demonstrates the galvanic benefit of the aluminum coating. On the aluminized muffler half, a band of aluminum has been galvanically removed, sacrificing itself to protect the weld bead and bare 409 shell on the other side of the weld. On the bare 409 muffler half, a strip of 409 with far less superficial red rust is apparent, the benefit of the protective nature of aluminum.

From a galvanic standpoint, bare stainless alloys can be coupled to each other with but a minimal dissimilar cell being established. One exception to this is the muffler section shown in Fig. 27. Here, a double-wall muffler used a thick inner shell of 409 for support and a thinner outer layer of 439 for better cosmetic corrosion resistance. Intrawrap seepage of condensate and road salts caused the 409 to corrode, probably a little faster than normal, because it was the less noble member of the 409/439 couple. With time (103,000 km, or 64,000 miles, of driving), the corrosion product on the 409 surface grew in volume, exerted pressure on the thinner 439 outer wrap, and forced a mechanical fatigue tear. In Fig. 27, a portion of the 439 skin has been cut away to reveal the underlying corrosion debris, but some of the 439 tear is still visible running parallel to the muffler long dimension.

Fig. 27 Galvanic corrosion between a T439 outer muffler wrap and inner T409 shell caused corrosion debris to build up between the layers and split the outer wrap. Crack visible in wrap despite section removal to show corrosion debris

In concluding this section on dissimilar-metal galvanic couples, the story would not be complete without discussing the two areas of OEM exhaust systems that still use some bare carbon steel. Solid rod hangers 10 to 15 mm (0.39 to 0.59 in.) in diameter, to attach exhaust systems to the underbodies of vehicles, and connecting flat flanges 9.5 to 12.7 mm (0.375 to 0.500 in.), used mainly between hot end exhaust components, are still frequently made of low-carbon steels occasionally sprayed with glass film paints to impart some barrier corrosion protection. While such carbon steel components would be anodic when mated to stainless materials, exhaust designers have felt the heavy bare steel thicknesses would allow these parts to meet the warranty targets. If not carbon steels, such parts have been made from wrought austenitic stainless alloys such as 304L, a nickel-modified 409 (409Ni) for improved cold weather toughness, or

Fig. 28 Galvanic corrosion of solid carbon steel hanger rods after 1.5 years of driving service in salt belt. Muffler end plate and sheet-metal hanger are 18Cr-Nb.

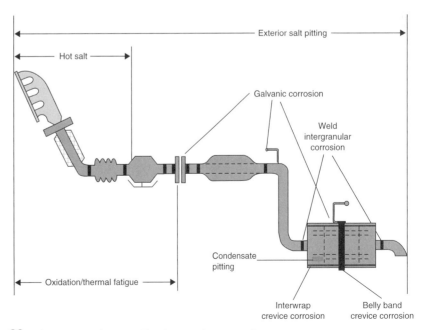

Fig. 26 Muffler with wrap and heads half made of bare 409 (bottom portion) and half of AL409 (top portion) and test driven 27,400 km (17,000 miles). Bare 409 gas metal arc weld and band of bare 409 adjacent to the weld galvanically protected by aluminum coating

Fig. 29 Exhaust system schematic with major corrosion concerns by component

from powder metallurgy stainless components. Standard exhaust ferritic alloys, such as 409 or 439, have been found to suffer from lower-temperature brittle impact fractures when produced to these heavier flange/hanger rod thicknesses.

Even assuming the bare carbon steel flanges or rod hangers have the corrosion durability to meet warranty targets, galvanic corrosion of these components can lead to cosmetic corrosion concerns. Figure 28 shows the muffler end plate and tail pipe of an 18%Cr exhaust section. For attachment purposes, a stainless sheet-metal bracket has been welded to the muffler cap and attached to rubber grommets with bare carbon steel rod hangers. After 1.5 years of service, the attack on the carbon steel rods is obvious. When and if this system fails, the hanger will probably be the most likely point. However, even prior to an actual system failure, concern could be expressed that iron oxide red rust from the hanger may cover the attachment bracket and even bleed down onto the muffler face or tail pipe.

Conclusion

In concluding this survey of corrosion in modern automobile exhaust systems, a schematic (Fig. 29) of a typical exhaust system is once more reproduced, and identifying arrows are added where one would expect to find the various forms of corrosion discussed previously. While today's (2006) typical system is easily capable of 7 to 10 years of durability from a corrosion standpoint, the materials needed for a 15 year exhaust life are probably already available, if engineers can optimize designs to minimize the corrosion mechanisms discussed in this article.

ACKNOWLEDGMENT

Appreciation is expressed to Richard Strait, Principal Marketing Engineer at AK Steel, who developed many of the corrosion tests and conducted most of the failure analyses and field surveys reported or referred to in this article.

REFERENCES

1. "Standard Practice for Preparing, Cleaning, and Evaluating Corrosion Test Specimens," G 1, *Annual Book of ASTM Standards,* Vol 03.02, ASTM International
2. I. Franson and J. Fritz, Stabilization Requirements for T409 (UNS S40900) Ferritic Stainless Steel, *SAE Technical Paper Series 971005,* International Congress and Exposition, Society of Automotive Engineers, Feb 1997, p 155–161
3. J. Gates and R. Jago, Breakdown of Stabilization in Type 444 Ferritic Stainless Steel, *Met. Forum,* Vol 7 (No. 1), 1984, p 38–45
4. "Standard Practices for Detecting Susceptibility to Intergranular Attack in Ferritic Stainless Steels," A 763, *Annual Book of ASTM Standards,* Vol 01.03, ASTM International
5. "Standard Specification for Chromium and Chromium-Nickel Stainless Steel Plate, Sheet, and Strip for Pressure Vessels and for General Application," A 240/A 240M, *Annual Book of ASTM Standards,* Vol 01.03, ASTM International

SELECTED REFERENCES

- A. Ando, N. Hatanaka, and T. Kittaka, "Development of Al Coated Stainless Steel Sheet Material for Automotive Exhaust Systems," Paper 384, Corrosion 91, March 1991 (Cincinnati, OH), National Association of Corrosion Engineers
- A. Sabata, C.S. Brossia, and M. Behling, "Localized Corrosion Resistance of Automotive Exhaust Alloys," Paper 549, Corrosion 98, NACE International, 1998

Engine Coolants and Coolant System Corrosion

Aleksei V Gershun and Peter M Woyciesjes, Prestone Products

A GOOD ENGINE COOLANT must minimize the degradation of nonmetals and prevent the corrosion of the metals in the cooling system. Excessive deterioration of any material can lead to a loss of coolant from the cooling system, to a reduction in the strength of a material or component, and to the subsequent malfunction of a part leading to engine failure. Even minor deterioration can cause deposits to form on heat exchanger surfaces, reducing heat transfer and/or restricting coolant flow in the radiator and heater core tubes and ultimately to a plugged heater core or to overheating of the engine from a plugged radiator.

Antifreeze History

As automobiles developed and liquid cooled engines became more commonplace in the period 1910 to 1925, one of the first cooling system problems to develop was leakage. Roadways were extremely rough and leaks caused by vibration were relatively common, particularly at radiators.

In the early days, car use was generally confined to periods when temperatures remained consistently above freezing because of difficulties with cold weather starting and the lack of heat in the passenger area. Cars did not have heaters and often were equipped only with side curtains, making driving very uncomfortable. In cold climates, cars were usually stored in winter. As engines and lubricants improved, heaters were developed and glass side windows became the rule and more winter driving was attempted. This required an engine coolant that would not freeze in the cold months. Many compounds such as honey, sugar, molasses, and ethyl alcohol were used. Alcohol became the most satisfactory engine coolant with "antifreeze" properties. Despite good cooling ability, alcohol had significant drawbacks of boilaway, odor, corrosion, and flammability. The car owner was perpetually uncertain as to the level of freezing protection.

The antifreeze business, as we know it today, began with the marketing of ethylene glycol antifreeze in 1927. It provided consumers with the first true engine coolant/antifreeze. It provided the engine with the necessary boil-over protection in the hot months and freeze-up protection in the colder months.

Over the next decades, ethylene glycol coolants saw many advances in inhibitor technologies. Glycol coolants continued to provide better and longer-lasting protection to the cooling system.

United States automakers, Ford Motor Company, General Motors, and DaimlerChrysler, began installing a 50% water and 50% ethylene glycol coolant solution into their new vehicles. This led to the emergence of engine coolant/antifreeze as a year-round functional fluid, making it as important as engine oil or automatic transmission fluid.

The early 1970s marked a significant turning point in engine design and inhibitor technology for engine coolants. Engine coolants were reformulated to incorporate the unique, patented silicone-silicate copolymer, which greatly enhanced inhibitor effectiveness, particularly for aluminum cooling system components. Lighter-weight aluminum components were starting to become more prevalent in vehicle design, particularly in the engine cooling system. The new silicate-silicone coolant technology was ideally suited to protect engines from the increased operating temperatures created by higher-opening temperature thermostats, exhaust emission controls, trailer hauling, air conditioning, and other power options.

In the late 1980s, the need for globalization of coolant technologies to meet a single performance standard was addressed. With concerns of using phosphates in Europe and concerns of using borate and silicate in engine coolants in Asia, the direction of coolant technology started to shift to a coolant that did not contain any phosphate, silicate, or borate. Organic acid-based inhibitor technologies filled this niche nicely. In 1995 General Motors was the first U.S. manufacturer to switch to organic acid coolant technology as its original factory fill coolant. Since then, Ford and Chrysler have changed to hybrid organic acid technology, which combines silicate technology with organic acids.

What do we have in store for the future? Engine coolant technologies still face many challenges. With the desire to continue to go to lighter materials in the cooling system such as magnesium components, inhibitor technology will change to address the potential corrosion issues of these new materials. Furthermore, advances in vehicle design and technology, such as new brazing technologies for heat exchangers (radiators and heater cores) or new engine/motor designs (fuel cell technology or heavy-duty engines with new exhaust gas recirculation (ERG) systems), will require engine coolant technology to continue to change to keep up with these improvements.

Cooling System Functions

The cooling system has several functions, but the most important is controlling the temperature of the internal combustion engine. Under load, an engine can theoretically produce enough heat to melt an average 90 kg (200 lb) engine block in 20 minutes. Even in normal driving conditions, combustion gas temperature may be as high as 2500 °C, (4500 °F), and lubricated parts, like pistons, may run 110 °C (200 °F) or more above the boil point of water. No matter what time of year, there is a safe high-operating temperature that helps provide maximum engine performance with the least engine wear. It is the function of the cooling system to maintain this operating temperature and reach it as quickly as possible during summer and winter.

The cooling system provides heat to the passenger compartment because the hot coolant passes through the heater core. As fresh air comes in, the heater core warms it before it enters the passenger compartment. If the vehicle uses the same ducts and controls for the heater, warm air from the heater can also be used to regulate the temperature of the chilled air released by the air conditioner.

Another function of the cooling system, especially in vehicles with automatic transmissions, is to control the temperature of the transmission fluid carried to and from an oil cooler in the radiator by oil lines connected to the transmission.

The primary function of the coolant is to maintain the optimum operating temperatures of the engine by preventing the engine from boiling over by carrying excess heat away from the engine and by preventing freezing in the winter. Vehicles today have higher operating temperatures because of air conditioners, antipollution devices, automatic transmissions, new under-the-hood materials, and increased engine cooling loads. Smaller vehicles have higher cooling loads because the engine works at a higher percentage of its capacity to power the vehicle. These higher temperatures demand a coolant with a boiling point higher than water, so vehicles come from the factory with a mixture of coolant concentrate and water generally 44 to 55% coolant concentrate. The engine coolant water mixture should be maintained within this range, with 50% being optimal.

A good coolant also has inhibitors that fight rust, corrosion and other cooling system problems, which can result in clogged radiator and heater core passages that eventually reduce the efficiency of the cooling system. This can lead to overheating of the engine and engine failure. The correct mixture of coolant and corrosion inhibitors will prevent this.

The cooling system must be able to adequately cool a vehicle traveling at 125 mph in 50 °C (120 °F) temperatures in Arizona. The same vehicle should not experience thermostat cycling problems when driving in −29 °C (−20 °F) weather in Minnesota.

Achieving optimal engine performance requires careful balance between the coolant and the various cooling system components. The coolant and cooling system work together to control the temperature of the engine. Cool metal temperatures are good for horsepower, but bad for hydrocarbon emissions. Higher temperatures are good for reducing hydrocarbons, but bad for nitrogen oxide emission (NOx) and may also cause spark knock. High temperatures are also good for reducing oil viscosity, lowering friction, and reducing fuel consumption, up to a point. Very high temperatures quickly degrade oil life, and temperature gradients must also be controlled. High gradients can lead to thermal distortion, thereby degrading performance and reducing engine life.

Optimizing coolant temperature requires knowledge of the engine operating conditions. Some factors that affect metal operating temperatures include:

- Coolant velocity
- Coolant temperature
- Coolant pressure
- Coolant composition
- Coolant concentration
- Engine load
- Amount of entrapped air in the system
- Water pump size
- Flow restrictions
- Heat exchanger size and design

All of these factors must be considered. Every component in the cooling system affects other component performance. For example, a change to the fan clutch can cause a water pump failure; a change in the size of the radiator tube can correct an overheating engine.

The cooling system needs to be designed, constructed, and maintained as a system. Individual components cannot be developed independently of each other. The system must operate under the worst conditions the customer will subject it to.

Cooling System Operation. When a gasoline engine is started, fuel and air are pumped into the combustion chamber and ignited by the spark plug, and the coolant pump begins circulating coolant through the engine block. The combustion process generates heat while the coolant system removes heat and maintains the optimum engine temperature.

Initially, coolant goes through the cylinder block from the front to the rear (Fig. 1), circulating around the cylinders as it passes through the cylinder block. At the rear of the cylinder block it passes up to the cylinder head through openings in the cylinder head gasket that seals the cylinder block and cylinder head together.

The coolant then circulates from the rear to the front of the cylinder head through passages around combustion chambers. These passages permit extra cooling of high-heat areas such as the spark plugs and exhaust valve seals.

As the coolant leaves the cylinder head it comes in contact with the thermostat. Until the coolant bulk temperature reaches ~82 °C (180 °F) the coolant will continue to circulate through the engine block. When the coolant reaches this temperature, the thermostat will open and send hot coolant to the radiator. The coolant enters the radiator at the tank inlet at the top and flows across or down through the tubes in the core, where it is cooled, and then flows to a tank on the other side of the core. From there it returns to the engine via the coolant pump.

Materials. Alloys of cast iron or cast aluminum are used for cylinder blocks, cylinder heads, intake manifolds, coolant pumps, and coolant outlets. Wrought brass or wrought aluminum are used for radiators and heater cores. For brass radiators and heater cores, high or low lead solders are used to join the components. Stamped steel is frequently used for small components such as core hole plugs, coolant pump housing closures, and coolant pump impellers.

Synthetic elastomers, such as ethylene propylene diene monomer (EPDM), are used for hoses to provide flexible connections from the engine to the radiator and heater core. Rigid plastics, such as high-density polyethylene and reinforced nylon, are typically used for coolant recovery reservoirs, and plastic radiator or heater tanks, respectively.

Cylinder head gaskets are embossed steel shim construction or silicone elastomer composition overlaid on a steel core and reinforced with steel grommets. Occasionally, stainless steel may be used for embossed shim cylinder head gaskets where extra corrosion resistance is desired.

The coolant comes in direct contact with all of these construction materials, and the combination of materials and coolant is frequently associated with corrosion problems or failures with cooling system components. In this respect, the coolant should be recognized as a critical material and a full-time functional fluid.

Components. The major components of a cooling system are as follows.

Heat Exchanger. The radiator and heater core consists basically of a series of small tubes and fins surrounded by air passages. Hot coolant from the engine flows through the tubes and is cooled by air flowing around them. Heater cores are very similar to radiators in their design and construction. To prevent the heater core from experiencing excessive flows, some vehicle manufacturers place a variable restriction in the coolant circuit just ahead of the heater core.

The coolant fan draws cooling air in from the outside through the radiator core air passages to cool the radiator during idling and low-speed operation.

The coolant pump (often referred to as the water pump) forces coolant through the engine block and the engine coolant components. To provide adequate engine cooling for a typical

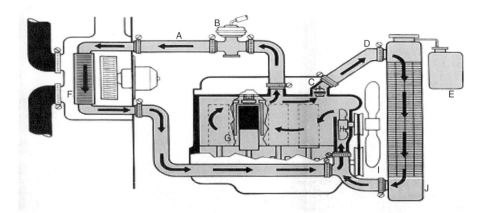

Fig. 1 Circulation and components of the cooling system. A, heater hose; B, heater control valve; C, thermostat; D, radiator hose; E, overflow reservoir; F, heater core; G, engine; H, coolant pump; I, fan; J, radiator

passenger car, the pump may have to circulate as much as 28,000 L (7397 gal) per hour.

The fan belt is extremely important to the operation of the vehicle cooling system because it drives both the fan and coolant pump.

The thermostat is a heat-operated valve that controls the amount of coolant that flows from the engine to the radiator. In cool weather or for light driving when maximum cooling is not needed, the thermostat is nearly closed.

Pressure Cap. Cooling systems are built to function at high temperature and pressure. The radiator pressure cap controls the pressure that builds up in the cooling system at high operating temperatures. This pressure, in turn, raises the boiling point of the coolant. A pressure cap that's working properly reduces the chance of boiling and loss of coolant through overflows. The correct pressure cap is determined by car model and year and can be found in the owner's manual.

Radiator hoses and coolant hoses are flexible tubes that are used to connect the engine components. The basic hose design consists of three elements. There is an inner hose (tube), reinforcement knit (yarn), and a protective outer layer (cover). The coolant hoses are made of elastomers that must be designed to resist heat and coolant; most hoses use EPDM, the most cost-effective option. For high-heat applications, silicon hoses are available. There are several materials that are commonly used to reinforce hoses: nomex, kevlar, rayon, polyester, or nylon. The yarn that is used in the reinforcing knit can be wound in several different ways: spiral wrap, plain stitch, plain radial, lock stitch, and braid.

Corrosion

Corrosion is the destructive electrochemical reaction between a metal or metal alloy and its environment. Vehicle manufacturers around the world install inhibited engine coolant in new cars and trucks to protect the cooling system from corrosion. They recommend all subsequent coolant changes or cooling system "topoffs" be done with a well-inhibited engine coolant rather than just water or water with an added rust inhibitor.

Drivers are frequently interested in how much freezing protection they are getting. Most vehicle manufacturers are not as concerned with freeze point because most coolants on the market are glycol based and provide adequate protection. Manufacturers are more concerned with the degree of corrosion protection for the various metals found in automotive cooling systems: aluminum alloys, cast iron, copper, brass, steel, and various brazes and solders.

The basic mechanisms involved in cooling system corrosion are no different from those in other systems and, therefore, methods of control are similar. Metals are corroded by direct chemical attack as a result of electrochemical processes. When certain conditions exist, cavitation-erosion-corrosion can also occur.

Various forms of corrosion that occur in cooling systems are:

- Uniform corrosion
- Galvanic corrosion
- Crevice corrosion
- Solder bloom
- Pitting corrosion
- Intergranular corrosion
- Erosion corrosion
- Cavitation corrosion

Each form of corrosion is described as it affects cooling system corrosion failures.

Uniform corrosion of a cast iron block in an acidic, used coolant environment is an example. Uniform corrosion is preferred from a technical viewpoint because it is predictable and spread over a large surface (not localized). Cooling system components can be designed to withstand general surface corrosion.

Galvanic corrosion will occur if two different metals are immersed in the conducting coolant and are connected by an external circuit to permit a flow of current. In this event, the more active metal, based on its position in the electromotive force series, will usually act as an anode and dissolve as the more noble cathode accepts electrons through the external circuit. The extent of galvanic corrosion depends on many things, such as the magnitude of galvanic potential, conductivity of the fluid (coolant), area ratio of the anode and cathode, surface conditions of the coupled metals, and coolant flow.

Galvanic potential is related to the magnitude of the current produced by coupling dissimilar

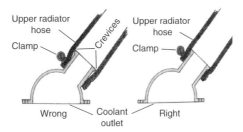

materials exposed to a common electrolyte. The potential depends on the metals that are coupled and on the characteristics of the electrolyte to which the metals are exposed. A galvanic series has been developed experimentally that lists many commonly used metals according to their galvanic potential in flowing seawater (see "Reference Information" in volume 13B, *Corrosion: Materials*). Although this series is specific to aerated seawater, differences with other aqueous solutions are often minor, i.e., depleted engine coolants, coolants that have been overdiluted with water, and coolants that have degraded.

Crevice Corrosion. The corrosion rate of an alloy is often greater in the small sheltered volume of a crevice. This crevice may be created by the form of the fluid channel, a mechanical seam, a fastener (bolt, rivet, washer) of the same or different alloy, a deposit (sand, corrosion product, or other insoluble solid), or a nonmetallic gasket or packing. If the crevice is made up of differing alloys or if the deposit is conductive, crevice corrosion may be further complicated by galvanic effects.

Severe crevice corrosion damage has been observed at hose connections with aluminum coolant outlets, pump inlets, and pump bypass fittings. In most cases, the problem is caused by fittings, where the hose and hose clamp are unnecessarily too far down on the connection, as shown in Fig. 2. Even if perforation does not occur, the corrosion products formed in the crevice can interfere with proper sealing of the hose, causing coolant leakage. Positioning the hose clamp tightly against the bead usually prevents such failures.

Engine block core hole plugs fail from perforation caused by crevice corrosion if they are not installed properly (Fig. 3). If the core-hole plugs located on the outer walls of the cylinder block are not inserted flush to the coolant jacket, deposits build on the core-hole plug, setting up crevice corrosion. Perforation of the thin steel plug, followed by coolant leakage, is the inevitable result. Prevention is achieved simply by locating the inner face of the plug flush with the coolant jacket wall.

In certain radiator and heater core designs, a turbulator is inserted in the tubes to enhance heat transfer. If the turbulator is located in tight

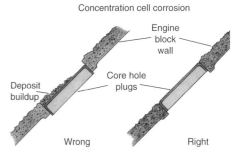

Fig. 2 Crevice corrosion can occur at hose connections. (a) The right and wrong way to position a hose clamp on a coolant outlet. (b) Severe crevice attack of an aluminum coolant outlet due to incorrect hose positioning

Fig. 3 An improperly installed steel core hole in an engine block leads to deposit buildup that can form a corrosion cell.

contact with the inside diameter of the tube, crevice corrosion can be initiated at the fixed contact sites. This can be avoided by ensuring that design-in minimum clearances are met to ensure against crevice sites.

With recently developed aluminum radiators and heater cores, there is a trend to employ plastic inlet and outlet tanks that are gasketed and clamped to the heater plates. Crevice corrosion can ensue if deep crevices are formed at the interface of the gasket and the aluminum heater plate.

Solder Bloom. Brass radiator soldering techniques and materials selection have a direct bearing on solder corrosion and clogging at the tube-to-heater plate joint. Lead solders are used in manufacturing cores and complete radiators because of their high temperature strength and melting point. The lead content makes the alloy difficult to inhibit from forming voluminous corrosion products, i.e., solder bloom. If excessive amounts of solder are deposited around the tube joint, and if the interior of the tube opening is wetted by the solder, there is a better chance that partial clogging of the tube opening will occur after 3 or 4 years of service (Fig. 4). Extended use and/or dilution of the coolant increases the probability of clogging.

Pitting Corrosion. There are a number of design and material factors with both brass and aluminum radiators that affect inherent resistance to pitting corrosion. High flow rates through the tubes can generate impingement erosion forces sufficient to erode passive or protective films and cause pitting perforation. This type of attack has been observed for downflow radiators just below the entrance of the tube in both aluminum and brass radiators.

Contamination of the coolant with aggressive ions such as chloride and sulfate tend to exacerbate the problem by breaking down protective films or by preventing their formation. Direct attack can also occur by the deposition of a more noble metal on one lower in the electromotive series. For instance, the deposition of copper on the more electronegative aluminum can initiate pitting corrosion that results in perforation of the radiator tubes.

Pitting corrosion by definition is confined to small areas; the highly localized corrosion results in pits that can lead to early failures in the thin metal sections of radiator tubes and heater cores. Similar attack in the engine block would not be as critical considering the greater thickness of metal (Fig. 5).

Intergranular Corrosion. Reactive impurities may segregate, or passivating elements such as silicon (in aluminum alloy) may be depleted at the grain boundaries. As a result, the grain boundary or adjacent regions are often less corrosion resistant. Preferential corrosion at the grain boundary may be severe enough to drop grains out of the surface. Intergranular corrosion (IGC), sometimes called intergranular attack, is a common problem in many alloy systems, including aluminum alloy with cladding materials found in cooling systems.

Cladding materials consisting of aluminum-silicon alloy have a noble electrode potential and are severely attacked at the aluminum and silicon boundaries. The IGC is caused by cathodic-type of galvanic action based on silicon diffusion or precipitation along the boundaries. Relatively shallow silicon diffusion tends to produce IGC that spreads only on the surface of materials. The deeper silicon diffusion produces more severe intergranular corrosion in the material thickness direction.

Erosion Corrosion. High velocity fluid flow can result in erosion-corrosion (Fig. 6). Design factors, materials of construction, and a depleted coolant can further aggravate this damage. A coolant flowing slowly through a component may show low or modest corrosion rate, but rapid movement of the same fluid can physically erode and remove the protective film, exposing reactive alloy beneath, and accelerating attack. Low-strength alloys that depend on a surface oxide layer for corrosion resistance are most susceptible. The attack generally follows the directions of localized flow and turbulence around surface irregularities.

Cavitation corrosion is a complex phenomenon involving mechanical damage from the pressure shock waves or jets, created by collapsing vapor bubbles in the coolant. The cavitation process begins with the coolant being converted to vapor at a high temperature or a low pressure point. Subsequently, when temperature

Fig. 5 Pitting corrosion in aluminum cylinder head port

(a)

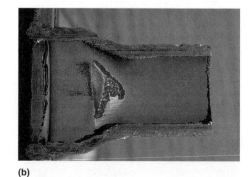

(b)

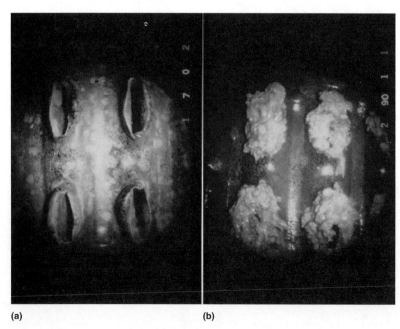

(a)   (b)

Fig. 4 Solder bloom formation can block and restrict the flow of coolant in the radiator. (a) New radiator core. (b) Solder bloom after just 22,000 miles of normal highway operation

Fig. 6 Erosion-corrosion related to high coolant flow. (a) Radiator tank erosion on wall opposite inlet. (b) Tube narrowing causes increased velocity and turbulent flow.

decreases or the pressure increases, the vapor bubbles collapse at very high rate and produce intense shock waves or jets. These waves radiate out pounding the nearby metal surface and causing the surface to erode or pit. Cavitation can also cause the fracture of brittle materials and fatigue failure of ductile metals.

Aluminum water pumps are vulnerable to cavitation-erosion-corrosion in the presence of improperly inhibited, depleted, or diluted engine coolant. The extent of damage and tendency to perforate are also influenced by the cooling system and coolant pump design. Internal configurations that concentrate large pressure gradients (sudden pressure changes) can cause localized pitting and erosion, which can result in decreased pump performance and potential perforation failure of the pump body or cover (Fig. 7).

Aluminum cylinder heads are exposed to potential cavitation corrosion conditions in the high-heat flux areas where nucleate boiling is generated (Fig. 8). Here again, coolant inhibitor type, glycol concentration, degree of inhibitor depletion, materials selection, and component design have a major effect on the extent of attack. Design deficiencies in terms of flow rate, flow distribution, and surface-to-volume ratios in the coolant jacket passages can increase the intensity of the corrosion damage (Fig. 9).

Further improvements in engine cooling system performance and reliability can be obtained by increased awareness of the following factors:

- Design and materials application should reflect thorough study of specific corrosion mechanisms.
- Material changes, particularly toward light alloys, should be accompanied by full consideration of differences in properties and design requirements.
- The role of the coolant as a full-time, high-performance functional fluid should be more critically evaluated.
- Consumer and service industry practices should be studied further to assess the opportunity to reduce cooling system failures and increase overall engine performance and service life by using good quality coolant.

Engine Coolant Base Components and Inhibitors

No single component or inhibitor will satisfactorily protect all of the metals in an automotive cooling system. The proper selection of an inhibitor combination depends on such things as the metal used, the composition of the coolant base, flow conditions, environmental temperatures, federal and local regulations, and heat flux at the heat transfer surfaces. Ethylene and propylene glycol are the most commonly used base fluids to provide the necessary boiling and freezing point protection required in vehicles. To these base fluids, an assortment of organic and inorganic inhibitors are added to provide protection against corrosion and to maximize the performance of the cooling system.

Glycols are the principal component of all modern coolants and comprise ~95% of the engine coolant because of their unique ability, when diluted with water, to elevate the boiling point and depress the freeze point of the engine coolant-water solution. Ethylene glycol is the primary glycol used. When diluted to a 50 vol% solution, it will provide boiling point protection to 130 °C (265 °F) (in the cooling system, under a 0.10 mPa, or 15 psi, pressure cap) and freeze-point protection to -37 °C (-34 °F).

This increased boiling point allows a car engine to run hotter while still avoiding the catastrophic coolant loss and potential engine damage that would accompany violent boiling. The decreased freezing point provides the cooling system protection against freeze up and damage to the components due to thermal expansion of the coolant during the cold winter months.

Buffers. The pH of a coolant depends on the nature of the metal to be protected in the cooling system, as well as the selection of the inhibitors used in the inhibitor package. Some inhibitors initially prefer a high-pH (~10) environment. Ideally, the pH of the coolant should be between 7 and 9 while in use. The buffer prevents the pH of the coolant from drifting up or down from the desired set point. The buffer must neutralize any acids generated by the degradation of glycol or from corrosion processes and consume free hydroxide formed by the corrosion of aluminum alloys. Some common buffers used over the years include borates, phosphate, benzoate, and triethanolamine salts. These buffers have been used by themselves and in combination with each other.

Borates are used in the coolant to provide a buffer solution (pH ~8–9) to prevent the pH from dropping to a range where the corrosion of ferrous metals, such as cast iron and steel, is unacceptably high.

Phosphates (di and mono salts) serve two functions. Their principal function is to buffer the pH of the fluid from dropping to a value that would be corrosive to ferrous metals. It has a wide buffering range, pH 7 to 10. Hydrogen phosphate ion also serves as a primary corrosion inhibitor for aluminum and provides some additional protection for solders and ferrous metals.

Sodium benzoate also has some buffering capability; however, it buffers in the range pH 4 to 6 in contrast to borate and phosphate, which buffer in the range pH 7 to 10. In addition, sodium benzoate serves as a corrosion inhibitor for the ferrous metals, cast iron and steel.

Triethanolamine is used most often to neutralize phosphoric acid to give the triethanolammonium salt in much the same way that sodium hydroxide is used to neutralize phosphoric acid to give the sodium phosphate salts in other formulas. The triethanolammonium ion formed in this neutralization acts as a buffer to prevent the pH of the antifreeze from increasing to a range that would be corrosive to aluminum. The phosphate ion formed in this neutralization acts as a buffer to prevent the pH of the antifreeze from dropping to a range that would be corrosive

(a)

(b)

Fig. 7 Corrosion perforation of (a) aluminum impeller and (b) aluminum cover of a coolant pump resulting from a used and degraded engine coolant

Fig. 8 Nucleate boiling-induced cavitation corrosion caused perforation at the exhaust valve port of this aluminum cylinder head.

Fig. 9 Severe cavitation corrosion in an aluminum cylinder head exhaust port manifold

to ferrous metals. Triethanolamine also serves as a primary corrosion inhibitor for ferrous metals in this formula.

Inhibitors. ASTM D 4725, "Standard Terminology for Engine Coolants," defines an inhibitor as a chemical compound added to engine coolant to mitigate cooling system degradation. Following are some common coolant corrosion inhibitors and their functions.

Silicate and Silicones. Silicate is one of the most effective corrosion inhibitors. It is used primarily to provide protection of the aluminum components in a car's cooling system. Silicates also provide protection to other metals, including cast iron and steel. Care has to be taken when using silicate chemistry because the failure to incorporate it properly in a stable form can lead to dramatic corrosion of aluminum components, such as engine blocks, heads, and radiators.

Silicone is one of the most important components of fluids based on silicate technology because it provides for the proper stabilization of inorganic silicate. Failure to add adequate amounts of an effective silicone stabilizer will result in the formation of silicate gels in the coolant concentrate or in the cooling system. This can lead to severe corrosion and flow-related problems due to blockage of narrow cooling system passages in the heater core and radiator tubes.

Triazoles and Thiazoles. These groups of inhibitors are typically used to protect copper and brass components. In engine coolants they are used both by themselves and in combination with each other.

Benzotriazole, commonly abbreviated BZT, is the parent compound of the more-familiar homologue, tolyltriazole. Benzotriazole is classically used to protect copper and brass by forming a tough, protective film over the surface of the metals. It may also protect solder and aluminum, but these mechanisms are not yet well understood.

Tolyltriazole (TTZ) primarily protects copper and brass and may also provide some protection for aluminum and solder.

Mercaptobenzothiazole (MBT) serves, along with benzotriazole and tolyltriazole, as a corrosion inhibitor for copper and brass. One aspect of the chemistry of mercaptobenzothiazole that should be recognized and understood is its sensitivity to light. Mercaptobenzothiazole undergoes an oxidative coupling reaction upon exposure to light to produce a compound that is insoluble in ethylene glycol.

Organic acids encompass a very large group of materials. Organic acids are typically added to the engine as the alkali or alkaline metal, aliphatic and/or aromatic, mono and/or di carboxylic acid salts. The major purpose of these inhibitors is to replace traditional inhibitors such as silicate-silicone, nitrites, nitrate, phosphates, polyphosphates, amines, and borates.

Organic acid technology has shown long-term stability, good multimetal protection, and long-term protection of the cooling system.

Other Common Corrosion Inhibitors. Sodium molybdate's primary function is to protect ferrous metals, cast iron, and steel from corrosion. However, with the right combination of other inhibitors, sodium molybdate may also provide some protection for aluminum and solder.

Nitrate (as sodium or potassium salts) is a very good inhibitor for aluminum, usually added to a coolant to provide protection against pitting and crevice attack.

Nitrite (as sodium or potassium salts) is generally used as an inhibitor for cast iron to prevent cylinder liner pitting in heavy-duty engine applications. Concerns have been raised over the use of nitrite in conjunction with certain amines because of the potential of forming nitrosamines (a potential carcinogen). To avoid exposure to the consumer, nitrite has traditionally been used only in heavy-duty fleet coolant applications.

Antifoam is added to the coolant to control (minimize the amount and speed up the break time) the foam formed when the coolant is mixed with water. Failure to add it can lead to foaming problems when the customer installs the coolant in the car.

Dyes are added to the coolant to give it a characteristic color that allows the consumer to differentiate it from other automotive fluids (e.g., transmission fluid, windshield washer fluid). They come in a number of colors including yellow, green, red, blue, and orange and serve only to color the fluids; they have no corrosion-inhibiting properties.

Engine Coolant Testing

Certain chemical, physical, and performance properties of an engine coolant must be checked to ensure that it meets the specific requirements of the cooling system. ASTM, SAE, and vehicle manufacturers have developed standards and methods for this purpose (Table 1). ASTM standards are designed to check the physical and chemical performance of the engine coolant but are not designed to analyze the composition of the coolant.

These standards and test methods are used to control the quality and to ensure that a coolant complies with specified performance limits and to engineering specifications. Some methods may be used in test programs to determine the changes to coolant after use in a vehicle. The absolute composition of the coolant concentrate can be obtained only by chemical analyses.

Specific tests are used to evaluate the effect of coolant on nonmetallic materials, such as radiators or heater hoses. Standard test methods are used to measure the hardness of elastomeric materials and determine changes in material (e.g., swelling, hardness, tensile strength, weight change) after exposure to coolant solutions under stressed conditions.

In the past, the corrosion of metals in the cooling system has received more attention than the degradation of nonmetals. Most tests look at corrosion of metal coupons that have been exposed to control test conditions.

The coolant performance tests can be divided into three categories:

Laboratory tests

- Physical properties tests
- Glassware corrosion test
- Foam tests
- Electrochemical tests
- Simulated service tests
- Heat rejection tests
- Stability tests

Special tests

- Pump cavitation tests
- Aluminum transport tests
- Engine dynamometer tests
- Compatibility tests

Field service tests

- Fleet proving ground tests
- Controlled vehicle fleet tests

Automotive, Light-Duty versus Heavy-Duty Antifreeze/Coolant

It is helpful to describe the differences between light-duty automotive and heavy-duty diesel engine coolants. Diesel engines are similar to gasoline engines in many ways. Both are internal combustion engines and most versions use a four-stroke cycle, but there are fundamental differences that are responsible for the development of different coolant formulations for each application.

Table 1 ASTM standard and test methods to determine physical, chemical, and performance properties of engine coolant concentrate

ASTM designation	Title
D 1119	Ash content of engine antifreezes, antirusts, and coolants
D 1120	Boiling point of engine coolants
D 1121	Reserve alkalinity of engine antifreezes, antirusts, and coolants
D 1122	Specific gravity of engine antifreezes by the hydrometer
D 1123	Water in engine coolant concentrate by Karl Fishier
D 1177/D 6660	Freezing point of aqueous engine coolant solution
D 1287	pH of engine antifreeze, antirusts, and coolants
D 1881	Foaming tendencies of engine coolants in glassware
D 1882	Effect of cooling system chemical solutions on organic finishes for automotive vehicles
D 3634	Trace chloride ion in engine antifreeze/coolants

The need for unique coolant chemistries is a function of the system operational design, the demands made upon the system by these conditions, and the typical cooling system maintenance program. Table 2 contrasts differences between automobile and heavy-duty diesel service. The expected life to overhaul for a heavy-duty engine is four to five times that of an automobile gasoline engine, despite the fact that the load factor is over twice as much and the gross vehicle weight per horsepower is five to eight times as high. For the diesel to work over twice as hard and still last four times as long, it is essential that the cooling system is well designed and maintained, filled with a quality, fully formulated heavy-duty coolant and periodically supplemented with supplement coolant additive or inhibitor extending packages (extender).

On the other hand, new gasoline engines use more lightweight metals (aluminum alloys and magnesium) as compared with diesel engines, which use cast iron and steel. Aluminum and cast iron require different inhibitor packages for protection against corrosion.

In addition to differences in design and cooling system conditions, use of heavy-duty coolant in automotive application is restricted by toxicity of some chemicals that are required to be used in heavy-duty vehicles. For example, the nitrite required in coolant formulation to protect replaceable or wet liners in wet sleeve heavy-duty engine cylinders might produce a nitrosoamine that is considered a carcinogenic agent and could be harmful.

For these reasons, it is recommended that automotive coolant be used in automotive and light-duty applications, and fully formulated heavy-duty coolant be used in heavy-duty vehicles powered by diesel engines.

SELECTED REFERENCES

- *ASTM Standards on Engine Coolants,* Vol 15.05, ASTM International, 2004
- *Engine Cooling Testing: State of the Art,* STP 705, Roy E. Beal, Ed., ASTM International, 1980
- *Engine Cooling Testing: Second Symposium,* STP 887, Roy E. Beal, Ed., ASTM International, 1986
- *Engine Cooling Testing: Third Symposium,* STP 1192, Roy E. Beal, Ed., ASTM International, 1991
- *Engine Cooling Testing: Fourth Symposium,* STP 1335, Roy E. Beal, Ed., ASTM International, 1997
- *Worldwide Trends in Engine Coolants, Cooling System Materials and Testing,* SAE SP-811, Society of Automotive Engineers, Feb 1990
- *Engine Coolants and Cooling System Components,* SAE SP-1162, Society of Automotive Engineers, Feb 1996
- *Engine Coolants and Cooling Systems,* SAE SP-1456, Society of Automotive Engineers, March 1999
- *Engine Coolant Technology,* SAE SP-1612, Society of Automotive Engineers, March 2001

Table 2 Differences in automotive and heavy-duty diesel engine service

Parameter	Light-duty automobile	Heavy-duty diesel
Expected life to first overhaul, miles	70,000–100,000	300,000–500,000
Total expected life, miles	150,000	1,000,000
Usage rate, miles/yr	10,000–15,000	120,000–180,000
Load factor, %	Less than 30	~70
Gross vehicle weight per horsepower, lb/HP	30	200
Engine block material	Cast iron or cast aluminum	Cast iron
Cylinder head	Cast aluminum	Cast iron
Uses wet sleeve cylinders	No	Yes
Radiator/heater core	Aluminum alloy	Brass
Uses of nitrite	No	Yes

Modified version of R.D. Hudgens and R.D. Hercamp, "Test Methods for the Development of Supplemental Additives for Heavy-Duty Diesel Engine Coolants," ASTM STP 887, ASTM, 1986, p 189–215

Automotive Proving Ground Corrosion Testing

Mats Ström, Volvo Car Corporation

COMPLETE VEHICLE ACCELERATED CORROSION TESTING as performed on the proving ground is a "mandatory" testing tool among vehicle manufacturers around the globe. It serves to verify the total quality of the corrosion protection of new models close to the start of production. This activity encompasses cars, sport utility vehicles (SUVs), trucks, and other commercial vehicles. In the automotive industry, complete vehicle testing is often regarded as an acceptable substitute for real service exposure, with the drawbacks of high cost and long lead time. It serves as a decision-making tool for issues ranging from a cost cut of cents per unit, to million-dollar investment issues. The test shall reproduce potential upcoming customer issues to the right proportions in a fraction of the time it takes outdoors. The determined failure modes should preferably be unambiguous and well documented, to let a nonexpert be able to evaluate and report the test outcome.

Apart from presenting and discussing the typical element of complete vehicle corrosion testing, an aim of this article is to create some awareness of the difficulty in applying reliable corrosion tests for everything at once and how this difficulty can be handled. A spectrum of physical, chemical, and electrochemical mechanisms are involved in a single atmospheric corrosion case. It is a challenge to simulate such a complex event to proceed in the similar manner in a typical 25-fold increase in rate, at which an accelerated test is expected to deliver reliable results. Taking an entire vehicle into account, one has to multiply this achievement for dozens of combinations of coatings, metallic materials, and designs, with the final goal of a balanced test outcome. To accomplish this within the same test regime, defined by a limited number of accelerating tools, is a formidable task that will at best, result in a reasonable compromise. This suggests that the test results should be treated as risk indicators that have to be assessed and weighed, rather than as precise predictions. Considered in this article is the extent that the high expectations can be met and the alternative evaluation routes at hand when the lack of realistic results becomes too obvious. A number of practical examples regarding both corrosion cases and test equipment are discussed.

When To Use Complete Vehicle Testing

The primary value of complete vehicle testing is the evaluation of how technical solutions function in interaction with their specific vehicle environments, for example, considering exposure to wear from road grit, dirt, and salt load, elevated temperatures, and mechanical effects such as chafing and fretting.

The second value of a complete vehicle test is the ability to verify everything in one comprehensive test. Regarding new models to be launched, complete vehicle testing is preferably used when the quality of a test object on most levels has reached a productionlike status. The test is then used to verify and finalize the performance of most components and systems. The results may indicate the need for upgrades as well as suggest product rationalizations.

For models already in production, complete vehicle testing offers the opportunity to ensure that all parts are produced according to specifications. All production processes tend to drift with time. When implemented, product rationalizations have a tendency to result in inadvertent changes to other attributes than the intended ones. The very long warranty commitments offered on body perforations and the potential for causing future associated out-of-control costs is a strong driver to run more corrosion tests of products already in production.

Proving ground testing offers the most convenient way of comprehensive comparisons of quality outcome when vehicle manufacturers produce the same models in more than one plant. However, when only one or a few units are tested, it is unlikely to detect issues that come and go depending on stochastic or drifting variations in the manufacturing processes.

When Complete Vehicle Testing is Less Than Adequate

In spite of good intentions to detect and counteract issues at an early stage in a product program, it is of limited value to test very early prototypes because:

- Test objects lack, to a large extent, the detailed relevance regarding finishes and geometries that often are key variables in their corrosion performances. They do not represent the outcome from the actual production equipment.
- A significant time of the product program has elapsed before test results are reported and a wide span of changes has already affected the model. Because of this, negative test results are easily disregarded and decisions postponed, even in cases when actions would have been called for.
- Early prototypes are very expensive test objects.

Real-World Conditions the Tests are Aimed to Represent

The automotive corrosive environment is formed by the sum of a very large number of stochastic events on a time scale ranging from hours up to the time scale represented by averaged climate condition for a given global location (Ref 1). Humidity and its variation are key variables for corrosion propagation. Prevailing high humidities are typical for northern locations with maritime influence in winter season, exemplified by the Canadian east coast and the northwest areas of Europe, but also around the North American Great Lakes region. To generate high corrosive conditions that become a concern for vehicles, contaminants that promote corrosion have to be added. The wintertime use of deicing salts in cold and temperate areas is a major seasonal corrosion contributor in this context. Airborne saline load from the sea is another source, affecting many coastal areas. Subtropical sites where both high humidity and temperature boost the corrosive effect of such a salt deposition are the worst of this kind. A third corrosive source is gaseous pollutants, especially sulfur dioxide, which still is a factor to consider in so-called emerging markets.

The salt type most widely used for wintertime deicing in the cold markets is sodium chloride. Some calcium chloride is often added to improve wet sticking to the roads (North America). In areas frequently affected by temperatures far below freezing, calcium chloride has found extensive use, but seems to have been gradually replaced by magnesium chloride in recent years. The use of calcium chloride for dust control of rural roads can also be mentioned in this context.

The salt load on a vehicle over time varies from saturated conditions to infinite dilution. The concentration of salt on the road will be dependent on the degree of drying or dilution, and the salt concentration of the wet films on the vehicle will be determined by the relative humidity. Temperature increase is often associated with rainfall, and "pure" water road splash will sometimes wash away salt, which limits the time with prevailing high salt loads at high outdoor temperatures. This goes for well-exposed areas not affected by established corrosion. In pockets and semi-open cavities, wash-off will seldom occur, if at all. Here, road dirt and soluble salts slowly form accretions by airborne transport. These areas often pose the worst corrosion risk on a vehicle. In established rust on mild steel, chloride ions act as a dissolution catalyst deep down at the rust/steel interface, and the corrosion process needs only infrequent and low-level wetting to continue almost year round. To the basic environmental corrosion factors is then added the complexities of how the vehicle is used by the customer in terms of:

- Road surfaces and the associated mechanical impact on vehicle components, including abrasive wear from stones and grits picked up and released from the tires
- Exposure to dirt and frequency and type of car washing
- Different driving modes often associated with local market conditions, which may cause high heat loads on one extreme and insufficient warming up on the other
- The way and frequency with which different vehicle systems are operated by the customer, for example, door locks and hand brake

In a representative accelerated test, these factors must be compressed into a standardized, comprehensive scheme of recurring events, suitable to run on durability tracks at a proving ground, combined with forced climatic conditions in a static chamber exposure.

Elements of a Complete Vehicle Corrosion Test

Proving ground corrosion tests typically consist of a repeated cycle of elements, where the four main test blocks are: (1) a driving sequence on various proving ground tracks, (2) exposing the vehicle in one or more ways to road deicing salt(s), (3) static exposure to forced climatic conditions, and (4) additional elements such as operating the mechanical systems of the vehicle, car washes, and so forth. Based on the content of about 10 car manufacturer's complete vehicle corrosion tests (Ref 2), the following representative test conditions can be identified:

Test duration varies from 10 weeks to more than 20 weeks. Depending on test length and, sometimes, corrosion type, the test is interpreted to correspond to a service in the field ranging from 3 to 10 years in a corrosive reference market.

Use of Tracks. The common procedure is to have the vehicle running on tracks according to a specified sequence of events during workdays. This takes a few hours, depending on the brand-specific content. Otherwise, the vehicle is exposed in a climatized garage for most of the time. The durability tracks at the proving ground usually have their core function in vehicle programs for pure mechanical endurance testing. For corrosion testing, the tracks allow field-representative effects of twisting, vibrations, chafing, and fretting to interact with the effects of the corrosive environment. Among the elements of the tracks are cobblestone pavings or Belgian blocks, washboards with various wavelengths, potholes, and railroad crossings. Handling roads (curves, hills) accentuate side forces and chassis dynamics. The SUVs and heavy-duty vehicles have more demanding rough roads included in the driving schemes.

Another regular element of driving is mileage accumulation, usually performed on the high-speed track. Typical speeds range from 80 to 130 km/h (50 to 80 mph). The purpose is general aging by driving, drying up of wet films of salt and other contaminants, and in some tests the generation of high peak temperatures in subsequent parking.

A mandatory testing component is driving on a gravel road in order to generate self-induced abrasive wear from grit picked up and released from the tires of the test vehicle. Grit size used in these tests usually ranges from 4 to 10 mm (0.16 to 0.4 in.) in diameter. The road surface under the gravel varies from asphalt to dirt beds in different tests and the depth of the gravel from scattered to thick beds. Some tests also have a follower procedure with a vehicle running ahead of the test object in order to generate stone pecking on the test vehicle, usually implemented at the test start.

Dust tracks and mud pits are used to build up dirt accumulation from high-speed and low-speed conditions, respectively.

There are still some vehicle manufacturers that do not test exclusively under proving ground conditions, that is, road elements of the tests are simulated in static rigs or performed on public roads.

Three methods are used to contaminate the vehicles with salt solution:

- Highway simulating splash/spray at medium speed (60 to 90 km/h, or 35 to 55 mph). The length of these passages range from 50 to 400 m (165 to 1310 ft), sometimes combined with washboard road disturbances (Fig. 1).
- Low-speed passages through troughs with typically 50 to 80 mm (2 to 3 in.) depth of saltwater
- Exposure to spray or salt fog solution on the static vehicle

Sodium chloride dominates as the dissolved corrosive agent. In some tests, calcium chloride and other contaminants are added. The concentration of salt in solution has typically been 3 to 5 wt%, but lower concentrations at the 0.5 to 1% level are found in more recently upgraded test procedures. No vehicle manufacturer, to the author's knowledge, applies gaseous corrosive agents such as sulfur dioxide in complete vehicle test application.

The major part of corrosion takes place under immobile conditions in special garages at elevated temperature and humidity. The exposure to forced climatic conditions is often programmable with respect to temperature and humidity cycling (Fig. 2, 3). The percentage of a full test the vehicle spends under climatized conditions ranges from 30 to 85%. It is important to note that the lower end represents very high humidity conditions only, whereas extensive periods of semidry conditions are included in the climate programs in the latter case.

About half of the vehicle manufacturers use some kind of cold room to reach vehicle temperatures far below freezing. In one known case, this was combined with stone-pecking exposure. Rig testing that combines climatic exposure with mechanical loads is seldom used. One case of salt spraying on the static test vehicle in wind-tunnel-like conditions is reported.

Evaluation of Test Results

Corrosion inspections during the test are usually conducted weekly or biweekly. The final inspection after test completion often takes substantial time. In many cases it comprises a total dismantling of the vehicle, including opening of joints and cavities in the sheet metal structure. Detailed analysis is usually performed and reported in-house, but some companies let main system suppliers take care of substantial parts of the test evaluation.

Corrosion requirements are classed into either of two major attributes: appearance and function. Requirements on appearance can stretch from what the customer finds on the vehicle in the retail showroom to penetrative corrosion in closures after several years. In terms of test inspections, this corresponds to results at a fraction of the test time and at the completed test, respectively. The latter may even involve projections extending far beyond the final result. Long antiperforation warranties on body and

Fig. 1 Proving ground highway simulating splash/spray during high wintertime conditions with deicing salt use. Note the shallow splash groves in the highway.

Fig. 2 Garaging of a test car in a proving ground humidity chamber after a daily test track driving sequence

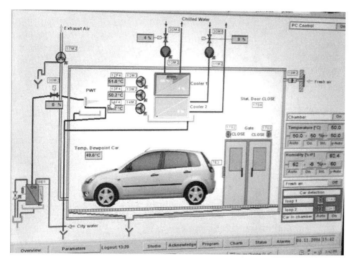

Fig. 3 Computer-controlled monitoring of corrosive conditions in a modern programmable complete vehicle humidity chamber

closures are strong drivers for such extrapolations.

Requirements on function usually address mechanical or electrical functionality when affected by corrosion, ranging from poor serviceability to structural disintegration. A major part of these demands are of the full-service-life type.

There are two approaches to proving ground corrosion tests for systems that are expected to function throughout the full vehicle service life:

- Full-service-life test: Test is run for at least 20 weeks, interpreted to directly represent full service life, meant to correspond to at least 10 years-in-service (YiS). In its most simplified approach, components that pass the test without actual breakdown are checked off as okay without further analysis. Systems with lower requirements are then checked for function at specified times during the exposure.
- Projection: Tests are run for typically 12 weeks, often interpreted to correspond to 6 YiS in a certain corrosive reference market. Damage to parts expected to function the full vehicle service life is extrapolated to their critical time regarding required customer YiS. The parts are then assessed as to whether they are likely to pass the requirement or not.

In both approaches there are different views on how to design the test, all from pushing very hard to acting more restrictively and selectively on the parameter settings.

Simulation and Acceleration

Which test is the most severe? How many years does the test represent? Can one test more quickly? These are common questions from the engineering community, which typically is not aware of the complications discussed in this section.

As already indicated in the introduction, for a complex product it is not possible to acquire such a balanced outcome from an accelerated test that it can be unambiguously and clearly translated to "x" years in customer service. Increased acceleration of a test usually leads to decreased correlation with the field tests (Ref 3, 4) for the same reason as any remodeling of reality inevitably introduces scaling errors. With the good intention of designing a tough and quick test, conditions for creating nonrelevant phenomena are introduced (Ref 3, 5). If results generated in such tests are not modulated through field references, misleading conclusions can be converted into automotive manufacturing practice.

Effects of Humidity and Salt. The exposure to humidity, wetness, and associated dissolved aggressive ions is the standard cause of corrosion on automotive products. However, too much of either wetness or salt load in a test may counteract its intended purpose.

Some corrosion mechanisms require that the electrolytic activity be somewhat restricted, otherwise their typical mode will hardly manifest at all. A typical example is filiform corrosion underpaint "creepage" on aluminum (Ref 6). Another example is galvanic corrosion localized to a bimetallic interface, and especially under crevice conditions. Under realistic conditions, deep pitting in the less noble metal may lead to localized penetration at the bimetallic contact. A high salt load in a test may not be able to mimic this behavior. The overall corrosion rate will be high, but the associated high electrolytic activity that leads to the anodic reaction becomes "smeared out" over a large surface. Under these conditions, the local maximum penetration depth will in effect be smaller than in a case with a lower electrolytic activity.

Zinc coatings will dissolve very actively when affected by long exposure to wetness and/or high salt loads, resulting in ample formation of white corrosion products that generally differs from those formed in the field. Painted or passivated zinc-coated steels become "prematurely" rusty, especially at sheet metal edges. The reason is that the basic carbonates and basic chloride of zinc are outside their thermodynamic stability ranges and cannot act as passive films. Instead, a self-promoting active mode with zinc oxide formation will dominate (Ref 7, 8).

With high salt loads, the rust formed on poorly treated steel/cast iron components develops into test-specific morphologies (Fig. 4, 5). Openly exposed surfaces quickly become heavily rusted since steel/iron readily responds to a wide range of nominally accelerated conditions, whereas the same material in confined mode is easily made nonreactive by excessive electrolytic activity (Fig. 6). Motor vehicles in the field display more selective rust locations and have almost exclusively rust penetrations located in or at crevices such as spot-weld overlaps and hem flanges. Unfortunately, crevice corrosion seems to be the most difficult type to accelerate and to keep to scale with other corrosion types. Therefore, in accelerated testing under harsh conditions, rust penetrations, if occurring, are located in open and worn positions rather than in crevices.

A special manifestation of the problem described previously is often encountered in standard scribe tests for evaluation of adhesion/corrosion properties of paint systems applied on a steel base (Fig. 4). Already a moderately aggressive combination of salt and wetness causes inhibition of paint creepage if sufficient cyclic drying is not introduced in the test procedure.

Figure 7 illustrates the acceleration dilemma, showing typical corrosion rate responses for steel and zinc as a function of "nominal corrosivity" (qualitative, collective term for degree of humidity exposure/time of wetness, extent of drying, and load of deliquescent salts). In this simple representation there is no accelerating condition to be found that results in balanced scaling between the presented corrosion modes. To make the problem worse, there seem to be corrosion mechanisms that require a high and extended wet load in order to be initiated. Based on field experience, the hydrolysis of polymers and their adhesion interfaces to metals are likely to be precursors to the onset of corrosion.

Effects of High Test Temperatures. Raising the temperature is the standard acceleration tool, as this increases the probability to exceed activation barriers for single chemical steps (Arrhenius equation). However, activation energies and temperature dependencies differ for different types of reactions and hence their temperature dependence. Furthermore, a corrosion process usually comprises a number of parallel and consecutive subreactions where feedback is common, and a temperature increase may facilitate passivation or activate corrosion. There is also a thermodynamic displacement factor that affects the stability ranges of different participating species when changing the temperature. One example is the Schikorr reaction $(3\ Fe(OH)_2 \rightarrow Fe_3O_4 + 2H_2O + H_2)$ (Ref 9), which is barely detectable below 50 °C (120 °F) but may produce hydrogen in the corrosion cells at the steel interface at higher temperatures (the probable cause of the blisters in Fig. 5).

Zinc readily forms protective layers of basic salts of carbonate and chloride in the high-temperature regime if there is access of atmospheric carbon dioxide (CO_2), which is the key passivation promoter (Ref 7). This is the probable cause of the lack of positive temperature dependence for the atmospheric corrosion of zinc found under moderate salt exposures (Ref 8). However, long periods of wetness and high loads of deposited salt counteract the formation of protective films and promotes continuing high activity, as discussed previously. Hence, zinc becomes sensitive to the testing conditions, and consequently the results become sluggish. An example is zinc corrosion associated with interfacial undermining of paints and sealants, which is rather frequently encountered on automobiles in corrosive regions. This mode of corrosion is difficult to simulate in a high-temperature regime unless high loads of salt are used to counteract the enhanced tendency for passivation.

Corrosion Effects of Driving Events. The durability test track driving effects, such as twisting, shear forces, and vibrations, have limited impact on most corrosion results. However, when mechanical forces trigger corrosion, the alarm bell should sound, since many findings such as this have shown to reflect highly significant customer quality issues. Examples are wear on paint from mounted components or between sheet metal parts. Electrical cables and wire harnesses can be routed so that they inadvertently rub on distribution pipes. Fine particles from dust roads can contribute to fretting wear between components, which are not sufficiently fixed to each other, eventually resulting in issues affecting either appearance or function. On the other hand, the mechanical impact of stone pecking from rough road conditions can result in exaggerated damages. Most proving ground tests have a constant interaction between a corrosive environment and gravel roads, which causes many painted steel components to become quickly stripped of paint and exposed to corrosive conditions sooner than what is representative for most customer vehicles.

Compromising Corrosive Conditions. In spite of identified drawbacks, an accelerating parameter set characterized by a high exposure temperature combined with a low salt load results in a better balance for most corrosion types than vice versa (Ref 3). However, using high-humidity exposure conditions at temperatures above 50 °C (120 °F) seems dubious, due to thermodynamic and kinetic considerations. The relative humidity conditions should be cycled between wet and semidry conditions, where slow drying in the intermediate humidity range seems to yield the most fieldlike results. The design of the humidity cycle has dramatic

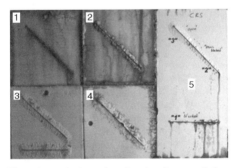

Fig. 4 Monitoring of underpaint "creepage" from scribe line(s) in the same automotive paint system on cold-rolled steel test panels. Panels 1–3: 12 week results from proving ground tests with decreased nominal corrosion load (panel No. 1 most severe) showing "blocked," "semiblocked," and "open" creepage modes, respectively. Panel 4: typical result from 3 year on-vehicle exposure in corrosive panel position/environment. Panel 5: 12 week laboratory test showing all three creepage modes on the same panel due to different salt accumulation caused by scribe orientation and panel inclination

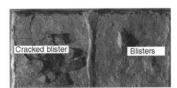

Long-term high humidity and 5% NaCl, at 50 °C, 12 weeks, 380 μm average corrosion depth

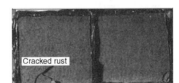

Cyclic humidity conditions, 1% NaCl at 40 °C, 12 weeks, 380 μm average corrosion depth

Fig. 5 Rust morphology under different degrees of corrosivity. Upper panels show gas blisters in the diffusion-tight rust. Lower panels show a more open rust morphology at nominally the same average corrosion depth

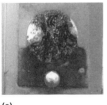 

Fig. 6 Monitoring of crevice corrosion (confined mode) in electrocoated and primed mild steel sheet panels. A 25 mm (1 in.) circular portion of exposed metal (lid removed) is shown. (a) shallow 6 week attack and magnetite slurry after nominally hard exposure conditions resulting in almost constant wetness. (b) 2 year on-vehicle exposure

influence on the corrosion results, especially on the relation between that of open exposure and confined mode.

Plain sodium chloride solution is found to be an appropriate corrosion promoter and is used in most test procedures. The possible benefits of more complex blends have not been sufficiently demonstrated. The strong corrosive effect of the sodium ion, which has not been recognized in the past (Ref 10), is one good reason for taking this position. Much larger effects are found both from the salt load level and from how the humidity exposure is designed.

Aspects of Test Equipment and Track Conditions

The climate chamber (garage) is where most of the actual corrosion takes place even in the proving ground case. There are some very important factors to consider in order to have defined conditions. As described in this section, these include the level of relative humidity (RH), exposure to road salt, and particulate contaminations.

High humidity exposure at saturated conditions (100% RH) is rather widely used in cyclic corrosion tests utilizing deliquescent salt(s). However, equilibrium salt concentrations are not defined at 100% RH. Any salt film, irrespective of amount or type, will dilute infinitely as water is gradually absorbed on the exterior of the test object. Runoff will inevitably result until, eventually, there is practically no salt load left on the surfaces. This may seem relevant enough, since undeniably this occurs under real field conditions. However, there is an associated reproducibility (and perhaps a repeatability) problem. The dilution rate is dependent on the mass transport of water to the surfaces of the test object, and accordingly to the speed of the air (steam) flow in the climate chamber. The dilution rate on a vehicle will not be uniform even within the same chamber, and of course, much less so for different types of chambers (and vehicles). Completely unacceptable is a tolerance span of the type "95 to 100% RH" as seen in some test specifications, which means that quite different climatic conditions are accepted within the same test specification.

A related phenomenon is when a cool vehicle is parked in the chamber after driving on the tracks. Outdoor conditions are often below freezing in temperate areas, whereas the dew point in the test chamber may be close to 50 °C (120 °F). Under such a dramatic transition the vehicle immediately becomes soaked, and run-off of the salt film will quickly occur, which is probably not intended. A remedy is to have a program that restarts each time the test vehicle is garaged and then starts with a controlled ramping up to set temperature before the humidification is initiated.

Dilution of the salt film under climatic exposure can, however, be beneficial in one important aspect. It counteracts the previously discussed negative effects of very high salt loads used in many tests.

Exposure to Road Salt and Conditions for a Run-Through Spray/Splash Section. The single most important factor that decides what will corrode and to what extent the corrosion will proceed is how the salt solution is distributed over the vehicle. In areas using road-deicing agents on the highways in wintertime, the vehicle will typically be exposed to salty aerosols for miles. Spray and splashes will come from passing vehicles, but the effect on the undercarriage is mainly self-induced, predominantly by high-speed driving through shallow grooves on normal paving. On the proving ground, these conditions are best reproduced in a passage with spraying and splash conditions. However, there are some aspects to consider when the described highway elements are translated into a test section. Spraying nozzles must be able to generate a representative pattern and spectrum of droplet sizes, to represent both structural ingress and sufficient exterior wetting. A proving-ground passage must necessarily be very short compared to any real highway condition, but must generate sufficient speed, and both needs are difficult to satisfy. A scaling error that very easily occurs is the gradual buildup of salt depositions in positions "semiaccessible" to the spray, or affected by secondary runs. In highway conditions these positions will be rinsed by the same kind of road spray, containing only rainwater during most of the year. A remedy to this unbalance is either to decrease salt concentration or to alternate with pure water in the spray passage.

The main alternatives to a spray/splash passage—static spraying or ploughing through a trench of saltwater at low speed—will bias some vehicle areas but leave others unaffected. With these less field-representative salt applications there is an obvious risk that the resulting corrosion will lack correlation with what is found in the field.

Particulate Contaminations. Another problem is the restricted ability to generate fieldlike depositions of mud and dirt in some "semi-accessible" positions. Dirt accumulations can have considerable long-term impact on various surface treatments. The dirt accumulation on most field vehicles is a slow but continuous process where road particles (ranging from mm to µm) from all kind of pavings settles in aerodynamic "backflow" areas. In an accelerated mode, it is difficult to keep up with this process even if passing dust-generating test sections. Frequent underbody splashes may also counteract dirt accumulation. The use of slow driving in so-called mud pits will generate dirt accumulations, but is less representative of the accumulations found on most customer vehicles.

Monitoring the Corrosivity

A traditional way of monitoring the corrosivity of proving ground corrosion environments is to assemble a number of standardized mild steel plates in a standard position and to evaluate them on a number of occasions during or after the corrosion test, usually as metal loss by gravimetry after removing the corrosion products. Ironically, mild steel in open corrosion mode shows a high and surprisingly stable corrosion rate under a very broad range of nominally differing corrosive conditions (Fig. 7). For reasons discussed previously, zinc provides a more sensitive choice in this respect. A more sophisticated way is to monitor the corrosion environment itself in field and proving grounds, respectively, which is treated in Ref 11 to 13.

An Evaluation Strategy that Complies with Constraints in Accelerated Testing

The major advantage of the full-service-life test approach is that it offers a straightforward evaluation and that the results are easily comprehended when communicated to the engineering staff and its management. Both these factors also save manpower and do not require expertise involvement. On the other hand, the obvious drawbacks can be summarized:

- *Long test:* Within the time constraints of contemporary new automotive brand model programs there is no time for long-term testing. Waiting for a reasonably relevant prototype to be tested means that start of production will have passed when the test results are at hand. The inevitable imbalance in acceleration of different corrosion types results in gradual loss of correlation to field as the test progresses, rather than in otherwise undetected relevant failures emerging with time.
- *Severe test:* Running a more severe test with the aim to cover full-service-life conditions will seldom reach the intended target. More individual issues will be found, but a considerable share of those will be test artifacts

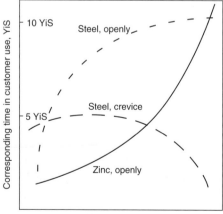

Fig. 7 Typical corrosion rate response (YiS, years in service) for steel and zinc during a 12 week test as a function of increased "nominal corrosivity" (a qualitative term for degree of humidity exposure/time of wetness, extent of drying, and salt load)

seldom (or never) encountered in customer use. However, the test protocol calls for cost-driving actions. Eventually this may take another turn, and the test outcome may be questioned as a whole, including the relevant issues.
- *Evaluation:* From a statistical standpoint, the least improvement for a single (or few) sample test is to consider not only explicit failures. Damages that represent obvious risks (e.g., base metal corrosion to considerable depth) must also be treated as deviations since they indicate that a significant fraction of the customer fleet is likely to violate the quality requirement unless countermeasures are taken.

This brings to consideration the alternative, the comparatively shorter "projection" type of test. Figure 8 illustrates how this procedure is aimed to function at its most favorable conditions. To use and interpret comparatively short tests suiting the product programs requires support from corrosion experience originating from the company's technical organization. Individuals with experience must be at hand and responsible for collecting data and building knowledge and be able to correctly interpret indications of corrosion in a short test. With the often lean staffing of many base organizations, this may be problematic.

In order to limit the need for expertise involvement on every issue, the following strategy can be implemented:

- Determine test acceleration factors for typical, recurring classes of material/designs based on gathered knowledge from field surveys, on-vehicle testing, quality issues, warranty databases, etc.
- Create real-time scales (rulers) of corrosive degeneration, backed up by photos, physical components, and so forth. Define where on the scale a given test delivers. Designed correlation test panels, exposed both in field and tests can facilitate this process.
- Extrapolate the test result for a certain corrosion type/component to the time specified by the service-life requirement (linearly often sufficient if data are lacking).
- Formalize and document such a process; that is, there should be system requirements that explicitly state the acceptable outcome of the test, based on the projection procedure (numbered circles in Fig. 8).
- In such a document, be very restrictive in the acceptance of corrosion modes that are known to be associated with a lack of robustness, such as adhesion failures.

New materials, surface treatments, or their new combinations/implementations should first have been investigated in laboratory testing by designed experiments in order to understand their operating limits and degree of robustness. Verifying solely on a test vehicle late in a product program will not deliver the sufficient degree of quality assurance. The incorporation of new knowledge and field experiences must be an ongoing process resulting in requirements and test conditions being revised and updated on a regular basis.

Conclusions

The aim of this article has been to create some awareness of the difficulty in applying reliable corrosion tests for all corrosion-sensitive automotive systems at once. To put it frankly, a proving ground test only partly proves things. A main message is: do not push too hard or kinetic backlashes may be misleading, for example, steel corrosion becomes diffusion-blocked by amorphous rust and crevice corrosion becomes impeded by constant wetness. Complexity and strong feedback mechanisms make sophisticated predictability and ranking of different corrosion protection systems impractical under accelerated conditions. The climatic accelerators must be applied with caution. The exposure temperature is preferably used as the main accelerator, although 50 °C (120 °F) seems an appropriate upper limit. Frequent and long drying cycles at intermediate relative humidities followed by rewetting is recommended. Avoid excessive salt loads, both regarding frequency and concentration. Try to apply the salt load as fieldlike as possible in order to reach good results. Increase the probability to discover potentially nonrobust behavior by widening the "spectrum" of the test, especially regarding the humidity conditions.

Even so, a well-designed test of 10 to 12 weeks duration cannot be driven so hard that extrapolations of the test results can be avoided for vehicle systems that are expected to function throughout the whole service life, or a test must be very long (>20 weeks). For such projections, use a formalized and documented procedure based on field data, which explicitly states the acceptable outcome of the test for a certain corrosion mode.

Finally, there is somewhat of a contradiction in the whole concept outlined previously. By the time one has gathered sufficient knowledge to perform good testing, so much will be known about what is tested that testing will be less important. At this stage, one will be able to specify the most cost-effective solutions at the early design phase of a product program, and the complete vehicle test will be the final sign-off verification tool. In conclusion, there is no way to escape the constant need of "one foot in the field" for highprecision quality assurance.

REFERENCES

1. M. Ström, G. Ström, W.J. van Ooij, A. Sabata, R.A. Evans, and A.C. Ramamurthy, A Statistically Designed Study of Atmospheric Corrosion Simulating Automotive Field Conditions Using a High Performance Climate Chamber—Status Report of Work in Progress, SAE 912282, *Fifth Automotive Corrosion and Prevention Conference and Exposition*, Oct 21–23, 1991, (Dearborn, MI), SAE International
2. Internal company information, collected by visits, surveys with interviews, in unofficial automotive conferences, etc.
3. M. Ström and G. Ström, A Statistically Designed Study of Atmospheric Corrosion Simulating Automotive Field Conditions under Laboratory Conditions—Final Volvo Report on the AISI Cosmetic Corrosion Set of Materials, SAE 932338, *Sixth Automotive Corrosion and Prevention Conference and Exposition*, Oct 4–6, 1993, (Dearborn, MI), SAE International
4. D. Davidson, C. Meade, L. Thompson, T. Mackie, F. Lutze, D. McCune, B. Tiburcio, H. Townsend, K. Smith, and R. Tuszynski, Perforation Corrosion Performance of Autobody Steel Sheet in On-Vehicle and Accelerated Tests, SAE 2003-01-1238, *World Congress*, March 3–6, 2003, (Detroit, MI), SAE International
5. F. Zhu, "Atmospheric Corrosion of Precoated Steel in a Confined Environment," Doctorial Thesis, Division of Corrosion Science, Dept. of Materials and Engineering, Royal Institute of Technology, Stockholm, 2000
6. N. 'Le Bozec, D. Persson, D. Thierry, and S.B. Axelsen, Effect of Climatic Parameters on Filiform Corrosion of Coated Aluminium Alloys, *Corrosion*, Vol 60 (No. 6), 2004, p 584–593
7. J.-E. Svensson and L.-G. Johansson, *J. Electrochem. Soc.*, Vol 140, 1993, p 2210

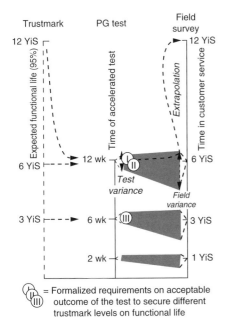

Fig. 8 Example of how functional requirements on components (involving a specific corrosion mode) are verified by formalized proving ground (PG) test outcome criteria (YiS, years in service). The most benign case for assessing long-term requirements is a "short" test based on previous field survey results.

8. R. Lindström, J.-E. Svensson, and L.-G. Johansson, The Atmospheric Corrosion of Zinc in Presence of NaCl; The Influence of Carbon Dioxide and Temperature, *J. Electrochem Soc.,* Vol 147, 2000, p 1751
9. G. Schikorr, *Z. Anorg. Allg. Chem.,* Vol 212, 1933, p 33
10. R. Lindström, J.-E. Svensson, and L.-G. Johansson, The Influence of Salt Deposits on the Atmospheric Corrosion of Zinc; The Important Role of the Sodium Ion, *J. Electrochem Soc.,* Vol 149, 2002, p 641
11. T. Wang, G. Gao, J. Bomback, and M. Ricketts, A Vehicle Micro Corrosion Environment Study, SAE 2000-01-1194, *SAE 2000 World Congress,* March 6–9, 2000, (Detroit, MI), SAE International
12. T. Wang, G. Gao, J. Bomback, and M. Ricketts, "A Vehicle Micro Corrosion Environmental Study of Field and Proving Ground Tests," SAE 2001-01-0646, *SAE 2001 World Congress,* March 5–8, 2001, (Detroit, MI), SAE International
13. T. Wang, D. Nymberg, and M. Ström, A Comparative Study of Corrosion Test Environments at Three Proving Grounds, SAE 2003-01-1240, *SAE 2003 World Congress,* March 3–6, 2003, (Detroit, MI), SAE International

Corrosion of Aluminum Components in the Automotive Industry

Greg Courval, Alcan International Limited

ALUMINUM COMPONENTS and assembled structures generally show very high resistance to corrosive environments when properly designed and manufactured with appropriate alloys. This article provides an overview of the principle forms of corrosion that can occur on automotive aluminum components and offers general guidelines on how best to avoid these situations. To this end, it is more informative to examine the mechanisms by which corrosion can occur, rather than to focus on specific components. This article does not cover the fundamental principles of aluminum corrosion as this can be found in other parts of the Handbook. In addition, it is generally known that specific corrosion issues have to be examined on a case-by-case basis. There are many unique features of any particular component or assembled module that dictate the methods that can be used to provide good corrosion protection. The guidelines offered are therefore of a general nature, and great care to detail and field experience must be applied to any specific problem.

Only the most common forms of aluminum corrosion are covered:

- Stress-induced corrosion
- Cosmetic corrosion
- Crevice corrosion
- Galvanic corrosion

In addition, pitting and intergranular corrosion can occur on bare aluminum components such as heat exchangers or suspension parts. This type of corrosion may lead to a deterioration in surface appearance, but in some cases also affects the service life of the part. Corrosion fatigue can be an issue affecting dynamically loaded aluminum structures, and care must be taken to avoid sharp notches and fine surface cracks. The best way to avoid corrosion fatigue is through proper design and the use of suitable protective coatings. A good review of pitting, intergranular, and corrosion fatigue is available in the *Aluminium Automotive Manual* (Ref 1).

Stress-Induced Corrosion

Aluminum Association (AA) $5xxx$ series alloys and certain AlMg casting alloys, commonly used for structural components, with magnesium content above about 3% can be susceptible to stress-corrosion cracking after extended exposure to elevated temperatures. Some $2xxx$, $7xxx$, and $6xxx$ with high copper additions are also susceptible to stress-corrosion cracking, but these are generally not used in automotive applications. For this reason, the AA guidelines (Ref 2, 3) strongly recommend alloys with a maximum of 3% Mg for structural components in automotive applications, where exposure for long periods to temperatures in excess of about 75 °C (167 °F) occurs. When the use of AlMg alloys with higher magnesium content is desired, consultation with the material producer is recommended and their applicability must be evaluated in detail. The thermal exposure of the part during its lifetime should be known, and preferably a realistic, full component test should be performed. These guidelines are generally followed, and the author and colleagues are not aware of any in-service failures of components due to stress-corrosion cracking. For example, the Honda NSX (Ref 4), which has used 5182, a 4.5% Mg alloy in a number of structural components for almost 15 years, has no known record of field failures due to stress-corrosion cracking. Therefore, this issue is not covered any further in this article.

Cosmetic Corrosion

Cosmetic corrosion generally refers to paint performance of class "A" outer body panels. To fully understand and evaluate the reasons related to cosmetic corrosion issues, the performance of the complete original equipment manufacturer cleaning, phosphating (or alternative chemical pretreatment), and multilayer paint system must be taken into consideration, as well as the characteristics of the alloy and its surface condition. A complete review of all these factors is well beyond the scope of this article. For a more detailed analysis, the reader is referred to the work underway on correlating lab tests of cosmetic corrosion with in-service performance (Ref 5). From general experience, it can be stated that, provided all the necessary coating system preparation and application precautions are taken, service performance of painted aluminum closures has been excellent.

With respect to the painting of individual components, surface preparation prior to paint application is absolutely critical in order to achieve optimum corrosion performance. This is particularly the case for components such as cast aluminum wheels.

Crevice Corrosion

Crevice corrosion on aluminum components can occur whenever moisture enters the crevice area between two adjoining surfaces, especially if salts are also present. Often, the two adjoining surfaces would be aluminum but, as is shown in some examples, one of the surfaces can also be of another material. It should be noted that once the corrosion starts, it can propagate intergranularly. Crevice corrosion is most critical for situations where the crevice is formed by two conducting materials with different electrochemical potentials leading to an additional galvanic effect. The next section covers galvanic corrosion.

Figure 1 shows an example of crevice corrosion underneath a carpet that covered aluminum steps on a vehicle. Moisture and salts had penetrated into the crevice under a fastener between the carpet and the aluminum surface. The carpet served as a reservoir for electrolyte. In order to rectify this situation, a sealant was applied over the entire area between the carpet and the step, forming a barrier to penetration of moisture and salts. This example was selected to demonstrate how the use of sealants can be very effective in eliminating crevice corrosion issues.

In addition to the use of sealants, design of joints effectively eliminates the potential of crevice corrosion. In essence, components should be constructed in such a manner that entrapment areas for moisture, salts, and dirt are avoided as much as possible (Fig. 2). The joints are made to allow good drainage and prevent the buildup of poultices that can trap salts and moisture. More complete guidelines for the design of components to avoid or reduce corrosion issues can be found in the Aluminum Association literature (Ref 2) and in the article "Designing to Minimize Corrosion" in Volume 13A, *ASM Handbook*.

Consideration should also be given to allowing paint or other corrosion protective coatings access into crevice areas. This is particularly important whenever electrocoating techniques are applied. If a joint area is very tight, the paint cannot penetrate the crevice. However, a small gap between adjoining surfaces sufficient to allow access provides a barrier to corrosion activity. A large gap that would allow pooling of the paint beyond its recommended thickness should be avoided.

Another example of crevice corrosion is shown in Fig. 3. In this case, the rubber gasket around a bare aluminum windshield frame resulted in minor corrosion-related discoloration of the metal. Please note that the small hole in the middle of the component indicates the position of a screw and is not a result of any corrosive activity. Application of a sealant between the gasket and the aluminum or prior painting of the aluminum surface would have prevented corrosion from occurring. Whenever a sealant is used in any of these applications, it is important to ensure the sealant remains pliable. Sealants that dry out and crack allow access of moisture into crevice areas and can even accelerate the rate of crevice corrosion. The example shown in Fig. 3 crosses the boundary between crevice and galvanic corrosion in that rubber or other plastics containing carbon black can show galvanic corrosion in contact with aluminum. Selection of the proper grade and level of carbon black in the rubber must be considered to avoid this issue (Ref 6).

Another form of crevice corrosion originates from the use of certain foam materials occasionally used for sound-dampening purposes. Some of these foams can retain moisture and salts and may lead to corrosion if there is no barrier material between the foam and the aluminum component. It is best to avoid the use of such materials and select a product that does not act to retain moisture.

In summary, guidelines for the prevention of crevice corrosion effects include:

- Design of the component to allow good drainage and prevention of moisture, salt, and dirt retention in crevice areas is critical.
- Wherever possible, sufficient gaps should be left between components on a vehicle body to allow ingress of electrocoat primer into crevice areas.
- The use of paint and sealers between adjoining surfaces can be very helpful in preventing most crevice corrosion issues.

Galvanic Corrosion

Galvanic corrosion can occur whenever two dissimilar metals (or more generally electrically conductive materials) are in contact in the presence of an electrolyte. A common incidence of galvanic corrosion is caused by the use of steel fasteners. Often, the aluminum component is fastened to a steel component, which is then another source of galvanic action. The primary prevention method is the use of barrier materials to isolate steel from aluminum, thereby eliminating electrical contact. An example of such a system is shown in Fig. 4. The choice of the fastener protection material is critical. There are a number of sacrificial coatings that provide excellent protection against galvanic corrosion effects between aluminum and steel; most important are zinc powder and/or aluminum flake in an organic binder. Various grades are available, and detailed discussion with the manufacturer is required to be sure the coating meets the requirements for the specific application. It is also necessary to ensure the coating is properly applied, otherwise much of the protection could be lost prior to or during assembly. Other types of protective coatings on fasteners are also available, including aluminum and tin-plated layers. Stainless steel fasteners can be used without any protective coating, but it should be noted that their use does not necessarily prevent galvanic corrosion from occurring. Experience has shown that properly galvanized steel fasteners are very effective in providing excellent galvanic corrosion prevention.

The nonconductive barrier material between the steel and aluminum components may be a paint, sealer, or polymeric material such as a polyester tape. There is a wide choice of barrier materials, and consideration must be given to cost and ease of manufacture as well as the degree of corrosion prevention required based on the location of the joint. The requirement for a sealer to remain flexible and not to dry out or become brittle applies in this use as well. In any situation where moisture and salts can accumulate, additional sealant should be applied outside of the immediate joint area to isolate the seam as much as possible.

It is important to note, as shown in Fig. 4, that the barrier material should extend beyond the immediate contact area between the dissimilar

Fig. 1 Crevice corrosion under a floor cover in the proximity of a fastener

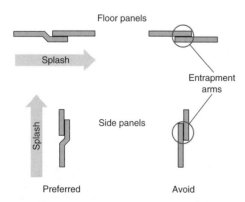

Fig. 2 Design guidelines for corrosion prevention. The component is fabricated so that retention of moisture, salts, and dirt in crevice areas is avoided.

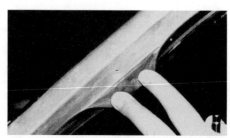

Fig. 3 Crevice corrosion under a windshield gasket

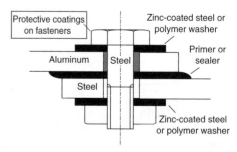

Fig. 4 Methods of sealing crevices between steel and aluminum surfaces

metals. A distance of about 10 to 20 mm (0.375 to 0.75 in.) is usually adequate. If the barrier material covers only the adjoining surfaces, electrolyte can form a bridge between the two metals and corrosion can occur.

Prepainting of the steel component prior to joining to the aluminum sections is often a very effective method of preventing galvanic corrosion. Again, consideration must be given to the overall requirements for manufacturing and cost.

Although most galvanic corrosion concerns are associated with steel-to-aluminum contact, other metals such as copper can also result in galvanic corrosion with aluminum if joints are not properly protected. One particular example is the use of copper grounding wire in contact with an aluminum structural component. An isolated joint between the copper and an aluminum wire, with the aluminum wire subsequently joined to the structural component has been proposed (Ref 6) as a means of avoiding these situations.

These measures provide excellent corrosion performance in service. However, there are alternative joining technologies such as adhesive bonding to replace fasteners if concerns over galvanic corrosion remain.

In summary, these guidelines can help prevent galvanic and crevice corrosion of aluminum components:

- Avoid entrapment areas where moisture, salts, and poultices can accumulate in or near a galvanic couple. Sufficient space should be left between steel and aluminum components to allow electrocoat paint to enter the crevice and form a barrier between the dissimilar metals.
- Use properly protected fasteners and consider the material selected for the fastener. Primers, sealers, and other barrier materials should be applied appropriately to fasteners and the adjoining surfaces of a galvanic couple. The barrier material should extend beyond the immediate contact area of the joint.
- Use of prepainted components prior to assembly should be considered where possible. If applicable, also the use of alternative joining methods such as adhesive bonding should be considered.

REFERENCES

1. European Aluminium Association, *The Aluminium Automotive Manual*, www.aluminiumtechnologie.de/aam/, accessed March 2006
2. "Aluminum: The Corrosion Resistant Automotive Material," publication AT7, Aluminum Association, 2001
3. "Aluminum for Automotive Body Sheet Panels," publication AT3, Aluminum Association, 1998
4. Y. Komatsu et al., "Application of All Aluminum Automotive Body for Honda NSX," technical paper 910548, Society of Automotive Engineers, 1991
5. G. Courval et al., "Development of an Improved Cosmetic Corrosion Test by the Automotive and Aluminum Industries for Finished Aluminum Autobody Panels," technical paper 2003-01-1235, SAE International, 2003
6. Y. Komatsu et al., "New Pretreatment and Painting Technology for All-Aluminum Automotive Body," technical paper 910887, Society of Automotive Engineers, 1991

Electric Rail Corrosion and Corrosion Control

Stuart Greenberger and Teresa Elliott, City of Portland, Oregon

ELECTRIC STREETCAR TROLLEYS AND LIGHT RAILS are once again popular, and their resurgence has a large impact on the city street—above and below ground. City streets are both traffic rights-of-ways and utility corridors. On downtown streets, electric rail transit shares the road with water, sewer, natural gas and fuel lines, electric power, communication, and traffic utilities. This article presents the fundamentals of stray-current corrosion caused by electric rail transit systems, electric rail system design for corrosion control, and the electric rail impacts to utilities that neighbor the transit system. Additional information can be found in the article "Stray-Current Corrosion" in Volume 13A.

Corrosion-control design elements for electric rail include substation spacing and grounding, track and track slab design, and construction acceptance criteria. The impacts of electric rail construction on underground utilities are discussed as direct physical interferences, maintenance access encroachments, stray-current effects, and utility relocation design considerations. Construction issues include construction funding, construction sequencing, and post-construction monitoring and maintenance by the transit agency and by the utilities.

Stray-Current Effects

Streetcar and light rail trains are powered by electric motors. Electric current flows from a direct current substation to the train through an overhead wire, and the current returns from the train to the substation through the rails. Because the earth acts as a parallel conductor to the rail, a portion of the current will return to the substation through the soil. This current returning through the soil is referred to as "stray current" because it follows a path other than the intended circuit. The engineering convention of electric-current flow from positive to negative, through metal conductors and the soil, is used throughout this article.

Stray currents cause electrolytic corrosion of both the transit system and neighboring utilities. The rail and rail fasteners corrode where stray current leaves the rail and enters the soil. Stray-current flow in the earth is through a large, three-dimensional space, through both the soil and metallic underground utilities such as pipes and cables. The current follows multiple paths, which can be quite a maze as the current transfers from one underground structure to another. Corrosion will occur where the current leaves neighboring underground utility structures on its return route to the transit system.

Corrosion occurs at those locations where the current leaves a structure because, in the transfer of current across a metal/soil interface, there is electron transfer and associated electrochemical reactions. Metal atoms lose electrons and go into solution. The amount of metal loss is proportional to the total amount of direct current. Therefore, metal loss (corrosion) is related to the magnitude and the duration of the current. Metal loss from current discharge is quantified in Faraday's law of electrolysis:

$$m = \frac{Ita}{nF} \qquad \text{(Eq 1)}$$

where m is the metal loss (lb), I is the current (A, or 1 C/s), t is the time (s), a is the atomic weight (lb), n is the charge equivalents exchanged, and F is Faraday's constant (96,500 C/charge equivalent).

The metal losses that result from one ampere per year of stray current for some common construction metals are 6.5 lb (2.95 kg) for aluminum, 20 lb (9.1 kg) for steel or iron, 24.5 lb (11.1 kg) for zinc, 45 lb (20.4 kg) for copper, and 74.5 lb (33.8 kg) for lead. This metal loss can also be expressed as a surface corrosion or penetration rate in mils per year by considering the mass loss per unit area and the density of the material. This proportionality is shown in Eq 2, with the conversion factor of 12,000 mils/ft:

$$p = \frac{12,000\, ir}{d} \qquad \text{(Eq 2)}$$

where p is the corrosion (penetration) rate (mils/yr), i is the current density (A/ft^2), r is the metal loss rate (lb/A-yr), and d is the metal density (lb/ft^3).

The corrosion rate of steel and iron is 488 mils/yr (12.4 mm/yr) at a current density of 1 A/ft^2 (10.8 A/m^2) or 0.5 mils/yr (12.7 µm/yr) at a current density of 1 mA/ft^2 (10.8 mA/m^2).

Estimating the corrosion effects on a structure requires engineering judgment. For example, in the case of a small uniform perforation of a pipe, it might be appropriate to estimate pipe life by simply dividing the pipe wall thickness by the corrosion rate. More typically, the corrosion rate would be multiplied by a factor. In the case of pipe life, the corrosion rate factor seldom would be less than four, even in fairly uniform corrosion conditions. This assumes a multiplier of two for electrolytic current density distribution across the corroding area and a multiplier of two for a hoop-stress safety factor. Hence, the estimated pipe life would be one-quarter the previous case of a small, uniform perforation of a pipe.

In addition to metal loss from stray-current discharge, stray current can be deleterious in areas where current flows onto a structure. Electrochemical reactions at the metal/soil interface where current flows onto a structure can result in high pH (caustic) buildup and also hydrogen evolution. Amphoteric metals such as lead and aluminum dissolve in high-pH/caustic environments. Hydrogen evolution causes embrittlement of high-strength steel and disbondment of protective coatings.

Electric Rail System Design for Corrosion Control

Stray current can be minimized by transit system designs that encourage current flow through the rail and discourage current flow from the rail to earth. In the past, electric transit systems were constructed with poorly bonded rail, noninsulated embedded track, and directly grounded substations. As a consequence, large magnitude stray currents occurred along the length of the transit system. Today stray current is mitigated, in part, by construction with continuously welded and cross-bonded rail, and high resistance rail-to-earth track, referred to as track-to-earth resistance since the rails are cross bonded. Substations are closely spaced and are either diode grounded or ungrounded. These design elements are illustrated in the electrical schematic of a single train powered by a single substation in Fig. 1. Current flow is from the substation, through the overhead wire catenary system (OCS), and back to the substation through both the track and the earth.

The quantitative effects of transit system design elements are demonstrated by considering earth as a parallel conductor to the track. The traction current will flow in the two parallel paths, the track and the earth. The magnitude of the stray current is the traction current times the ratio of the track resistance to the overall circuit resistance:

$$I_{UG} = I_{TOTAL} \frac{R_{TRACK}}{(R_{TRACK} + 2R_{TTE\ UG})} \quad (\text{Eq 3a})$$

$$I_{DG} = I_{TOTAL} \frac{R_{TRACK}}{(R_{TRACK} + R_{GM} + R_{TTE\ DG})} \quad (\text{Eq 3b})$$

where I_{UG} is the stray current in an ungrounded system, I_{DG} is the stray current in a diode-grounded system, I_{Total} is the total traction current, R_{Track} is the track resistance, R_{GM} is the ground mat resistance, $R_{TTE\ UG}$ is the track-to-earth resistance in an ungrounded system, and $R_{TTE\ DG}$ is the track-to-earth resistance in a diode-grounded system.

Stray current is minimized by a low track resistance and a high track-to-earth resistance. Note that for the ungrounded system, stray-current discharge is shown, in Fig. 1(a), to occur over one-half the track length, and stray-current return is shown to occur over the other half of the track length. For the diode-grounded circuit, the stray-current discharge is shown in Fig. 1(b) to occur over the entire length of the track, and the stray-current return is shown to occur at the diode ground mat. The diode ground mat is an engineered low-resistance ground return.

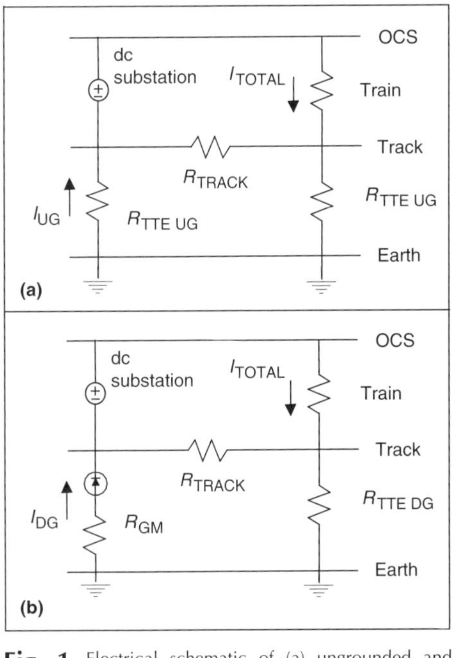

Track (Conductor) Resistance. Rails are continuously welded and cross bonded to minimize the electrical resistance of the track. Rails are cross bonded with electric cables so that the four rails of two-train track (typical of bidirection traffic) are electrically continuous. The resistance of a single rail is approximately 0.01 Ω/1000 ft or 0.0025 Ω/1000 foot of four-rail track that is cross bonded.

Track-to-earth resistance is historically expressed in units of Ω/1000 ft (1 Ω/1000 ft is 0.3 Ω/km). Noninsulated tie-and-ballast track with wooden ties, steel spikes, and clean rock ballast has a resistance on the order of 10 to 100 Ω/1000 ft of four-rail track. Insulated tie-and-ballast track with concrete ties, insulated fastener clips, and clean rock ballast has a resistance on the order of 100 to 500 Ω/1000 ft of four-rail track. Insulated direct fixation track (track anchored onto a concrete slab) also has a resistance on the order of 100 to 500 Ω/1000 ft of four-rail track.

Figure 2 shows noninsulated tie-and-ballast track with wooden ties, steel spikes, and rock ballast that is fouled with soil fines and moss. This track might have a resistance less than 10 Ω/1000 ft of four-rail track. Figure 3 shows insulated tie-and-ballast track with concrete ties. Note the clean, well-drained ballast and insulating clips between the ties and rails. This track might have a resistance on the order of 500 Ω/1000 ft of four-rail track.

Insulated embedded track is claimed to provide exceedingly high track-to-earth resistance. Indeed, sections of embedded track with boot-type insulation have been measured at 10,000 and even 100,000 Ω/1000 ft of four-rail track. However, low-resistance point source grounds are possible because of construction defects. Sensitivity analysis, by a calculation of electrode resistance, shows that a single flaw, or several small flaws, will lower the track section resistance to 100 Ω/1000 ft. Direct shorts can result in a point source short of 10 Ω or less.

For tie-and-ballast and direct fixation tracks, there is fairly uniform current discharge along the track given the presumed condition of stray current due to uniform leakage at each fastener. Directly shorted fasteners are point source shorts. These shorts are exposed and can be readily located and repaired. In contrast, a short in an insulated embedded track system is more difficult to locate and repair because the insulated rail is cast into the concrete slab. Typical defects in the installation of insulated embedded track include damage to the insulating elastomeric boot, inadequate boot splice connections, and shorts to the gauge bar and surrounding track slab rebar.

Figure 4 shows the complexity of embedded track construction. The photograph shows a rail with an insulating boot, a gauge bar, and surrounding track slab rebar. The boot must be protected during storage, handling, positioning, and welding of the rail and during rebar installation and placing of concrete.

Figure 5 shows an insulated embedded rail constructed with a potting compound that is now degraded from environmental and mechanical damage. Figure 6 shows construction of a track slab with an insulating elastomeric boot. Although the elastomeric boot is a promising construction technique, the remaining concern is

Fig. 3 Insulated tie-and-ballast track with concrete ties and insulated clips. The ballast rock is clean and well-drained; however, leaves are accumulating.

Fig. 1 Electrical schematic of (a) ungrounded and (b) diode-grounded rail systems. Refer to the corresponding text, and in particular Eq 3(a) and (b), for an explanation of the abbreviations shown in this figure.

Fig. 2 Noninsulated tie-and-ballast, with wooden ties, metal spikes, and ballast rock that is fouled with soil fines and moss

Fig. 4 Embedded track slab construction, rail with insulating boot held by gage bar and surrounded by the slab reinforcing steel.

that the product does not have a long performance history, and repair of a material failure may be costly and disruptive. Figure 7 shows a boot with separation of the splice connection and damage to the running surface.

Substation Spacing and Grounding. The spacing of substations along a proposed electric rail line is a key element that determines the stray current in a system. Closely spaced substations lessen the length of track between stations, thereby reducing the track circuit resistance and increasing the track-to-earth resistance. This lowers the magnitude of stray current. Substations can be ungrounded, diode grounded, or solidly grounded.

The diodes of diode-grounded substations prevent current discharge from the ground mat and the direction of stray-current flow is from the track, through the earth, to the ground mat. The diodes also have a forward bias that prevents current flow until a threshold voltage is exceeded. However, the threshold voltage can be regularly exceeded, causing the diodes to conduct often. When the diodes are conducting, there is an additional concern that stray current can be concentrated around the substation ground mat and cause corrosion at utilities neighboring and serving the substation.

Systems with ungrounded substations can have a greater track-to-earth circuit resistance than systems with diode-grounded substations when the diode is conducting. As illustrated in Fig. 1(a), in an ungrounded system, half of the track length contributes to current discharge, and half of the track length contributes to current return, so the track resistance is $2\,R_{TTE\ UG}$. In contrast, diode-grounded systems have a relatively low resistance ground mat ($R_{GM} \ll R_{TTE\ DG}$), and when the diode is conducting, current discharges from the full length of track to the ground mat. Typical resistances of diode system ground mats vary from 0.01 to 0.1 Ω where the direct current (dc) ground mat is cross bonded with the substation alternating current (ac) ground mat, 0.1 to 1.0 Ω for a low-resistance dc ground mat, and 1.0 to 10 Ω or greater for a high-resistance dc ground mat.

Considering the case of an electric rail system of 2 miles of track at 200 Ω/1000 ft ($R_{TTE\ DG} \approx 20\ \Omega$, $R_{TTE\ UG} \approx 40\ \Omega$), and a diode ground mat less than 1 Ω ($R_{GM} \ll R_{TTE\ DG}$), the stray current would be approximately fourfold greater when the diode is conducting ($I_{DG} \approx 1/R_{TTE\ DG} \approx 1/20$) than when the system is ungrounded ($I_{UG} \approx 1/(2\,R_{TTE\ UG}) \approx 1/80$). An electric rail system with a high resistance diode-ground mat and/or a low track-to-earth resistance would be, operationally, intermediate between a diode-grounded and ungrounded system. The actual difference in overall stray current between an ungrounded and a diode-grounded system is dependent on all elements of system design and construction.

The principles presented for the simple case of a single train and a single substation, shown in Eq 3(a) and 3(b), are fundamentally correct and instructive regarding corrosion-control design elements. In practice, electric rail transit systems are modeled to simulate the more realistic case of multiple moving trains and multiple substations, where long sections of track can discharge to distant track and to distant substation ground mats. These simulations are used for both the design of the electric rail system and to determine the stray-current exposure to neighboring utilities.

Another corrosion consideration regarding ungrounded versus diode-grounded electric rail systems is the direction of current flow between the track and the earth. In an ungrounded system there is current reversal to and from the track at any location because, as the train travels, the location of the electric load changes. In a diode-grounded system, the current flow is principally from the track to the earth and back through the diode-ground mat because the track has a much greater resistance than the diode-ground mat.

Figure 8 shows pipe-to-soil measurements made during the monitoring of a cathodically protected pipeline at a light rail crossing. During the monitoring shown in Fig. 8(a), the transit system was operated diode grounded, and stray-current flow was always from the track and toward the pipe in the immediate area of the crossing. The pipe-to-soil potential was always more negative than –1.0 V.

During monitoring shown in Fig. 8(b), the transit system was operated ungrounded, and current flow was both to and from the track. The centerline measurement of pipe-to-soil potential is more positive than –1.0 V. Stray-current flow reversed with train movement, and since there is the possibility of current flow from the pipe to the track, there is an increased prospect of corrosion of the pipe at the track crossing. This corrosion mechanism of closely coupled stray current from the pipe to the track can concentrate corrosion at this location. It is of particular concern because there is a multitude of utility crossings along an electric rail system. The location noted in Fig. 8 had been operated for 20 years as a diode-grounded system with an insulated embedded track. The rail had appreciable corrosion where installation of the elastomeric boot was inadequate.

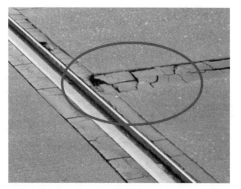

Fig. 5 Embedded track insulated with potting compound, now degraded from mechanical and environmental damage.

Fig. 6 Embedded track insulated with a surrounding elastomeric boot that is visible at a construction joint in the track slab.

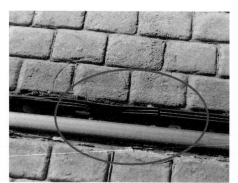

Fig. 7 Embedded track with boot damage. Boot shows surface raveling and an open gap at a splice connection.

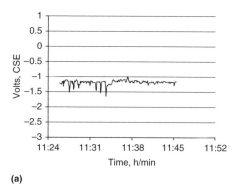

(a)

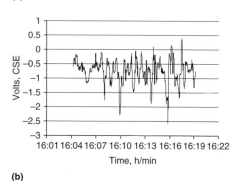

(b)

Fig. 8 Substations operated diode-grounded (a) and operated ungrounded (b). CSE, copper/copper sulfate reference electrode

There are some advantages to electric rail system operations having substations ungrounded (Ref 1, 2). When operated ungrounded, the rail voltage varies both positive and negative, and absolute voltages are lower. In some circumstances, it may be possible to better detect ground fault conditions if the system is ungrounded rather than diode grounded. In respect to corrosion control, overall stray current could be reduced when the system is ungrounded because of the increase in the overall circuit resistance. Also, the corrosion of rail is more evenly distributed because the location of the current discharge and return changes as the train moves.

In the case of a system with solidly grounded substations, the overall circuit resistance will be particularly low and the magnitude of stray current appreciably higher than either a diode-grounded or ungrounded system. If the substation is grounded directly or indirectly to neighboring utilities, stray current will occur over a greatly expanded area. Critical facilities are sometimes protected in this manner by providing an interference bond. However, this is of necessity and not favored in new construction.

Trains and Station Platforms. The total traction current shown in Eq 3(a) and 3(b) is largely determined by train operation. Two-car trains (called "consists") have twice the current load of a single-car train, and the heavier the train and passenger weight the greater the traction current. Streetcars and light rails require lower peak traction currents in downtown areas where speed and acceleration are curtailed. The peak traction current of a quickly accelerating train is more than an order of magnitude greater than the average running load.

Underground utilities should be considered when installing station platforms because stray current is likely to be greater at these locations. Acceleration, deceleration, and the associated increase in traction current occur near station platforms, as well as at traffic signals and hills. Deceleration affects traction current when trains use regenerative breaking, and the motor is used to generate power that is fed back into the system. In downtown areas there will likely be a platform every few blocks in contrast to suburban areas where platforms might be miles apart.

Track Slab Design. The track slab of embedded rail is designed to support the train and to bridge a distance that accommodates minor underground utility work, often a span of less than 10 ft (3 m). Figure 9 shows utility installation under an existing track slab.

To protect the track slab reinforcing steel from stray-current corrosion, the rebar may be coated and installed electrically discontinuous, or the rail insulation may provide sufficient stray-current control and standard bare rebar construction is used.

Alternately, rebar mats can be installed bare with welded rebar to provide electrical continuity, though wire-tie construction is incidentally electrically continuous. Sometimes rebar mats are characterized as "collector mats," believed to direct the preponderance of stray current back to the rail. This characterization does not consider that the rebar mat is a parallel conductor in equilibrium with the soil electric field strength along its length. In response to the corollary concern of a rebar mat concentrating ground current effects at utility crossings, the track slab design can include dielectric insulation, such as a dielectric membrane between the slab and the earth. Dielectric insulation of the track slab also increases the track-to-earth resistance at that location and provides redundancy at critical utility crossings.

Construction Quality Acceptance. To avoid low-resistance point shorts, the acceptance criterion for track-to-earth resistance should require resistance testing on suitably short sections of track or require alternate methods that detect low-resistance shorts. Often, test measurements are made on long sections of track and are prorated to uniform values specified as $\Omega/1000$ ft. This can be misleading because the corrosion caused by uniformly distributed stray current is typically less a concern than the situation where there is actually large magnitude stray current from a point source. A criteria of 100 $\Omega/1000$ ft that is prorated could result in acceptance of ten 1000 Ω shorts in a 1000 ft section, or one 100 Ω short in a 1000 ft section, or one 20 Ω short in a 5000 ft section. All these equate to a prorated 100 $\Omega/1000$ ft. Specifying an actual test length limits the possible magnitude of a short that can go undetected. If the actual test section length were required to be 1000 ft, then the lowest resistance single short accepted would be a minimum of 100 Ω.

In the case of embedded track, locating shorts and flaws in insulation requires considerable field testing. Repairs require jackhammer excavation in the track slab. Track construction methods should include flaw (holiday) detection just prior to placing concrete. At that time, an insulating boot can be tested with a high-voltage holiday detector. After concrete encasement, flaws can be detected by techniques similar to those used to find flaws in buried pipeline coatings, typically, radio frequency holiday detectors and voltage gradient measurements.

Electric Rail Construction Impacts on Underground Utilities

Electric streetcar trolleys and light rails are intended to serve urban areas. These are often downtown areas with congested traffic, street-level businesses, and myriad large-sized underground utilities serving the area. To minimize project cost and disruption to the public, transit authorities endeavor to minimize utility relocations. This consideration should be a balance between initial capital construction and the continuing cost and disruption caused by utility maintenance.

Despite the best transit design, utilities are impacted by the project. Some underground utility relocation will be necessary to accommodate electric rail transit construction. Electric rail transit construction impacts to underground utility systems can be characterized as direct physical interferences, maintenance access encroachments, and stray-current corrosion. These impacts to the utilities are evaluated together to ensure uniform integrity and to balance risk and cost.

Direct Physical Interferences. Construction of electric rail tracks, catenary poles, and duct banks obviously requires utility relocations where there is a direct physical interference. There may be additional physical interferences from street improvements caused by, or in conjunction with, the transit project. These could be from road alignment and grade changes, construction of retaining walls and bridges, sidewalks, drainage, streetlights, and traffic signals. Moreover, there may be direct impacts of other utilities relocating to accommodate construction. Typical utilities that relocate and compete for space in the public right-of-way include water, sewer, natural gas and fuel lines, electric power, communications, and traffic utilities. Finding space for every utility in the congested right-of-way becomes a challenge.

Maintenance Access Encroachments. Utilities located directly underneath or in close proximity to the proposed tracks are inaccessible for routine maintenance, utility operations, and future improvements. Electric rail systems are usually considered regional facilities with uninterruptible service during revenue hours. Trains normally run every 5 to 15 min during operating hours (which are often called the revenue hours), 7 days a week. Nonrevenue hours are generally 2:30 to 4 AM and are the only hours when trains are not running that utility work can occur adjacent to tracks without impeding the trains.

Fig. 9 Utility work at existing track slab. Tracks remain operational during excavation; however, utility work stops and moves out of the way to allow train to pass unimpeded.

This limits utility operations, maintenance, and construction to a restrictive transit schedule unless the utilities are located a sufficient distance from the tracks such that work can occur without interrupting the transit operations.

Safety requirements limit crews from working close to moving trains or close to the high-voltage overhead catenary power line. Excavation under or adjacent to the track is limited to prevent undermining the rail system, to minimize weight load on the existing utility, and to protect the workers in the trench. Excavation parallel to a track can be more problematic than utility crossings under a track, since work parallel to the track would require shoring systems to provide geotechnical support to the trench in addition to protecting the workers. Shoring can be a formidable task given the proximity of neighboring utilities. In addition, trench work parallel to the track is problematic in the physical constraints of the heavy equipment needed to install, maintain, and replace utilities. The construction equipment straddles the utility trench and can encroach on the track right-of-way and tangle with the OCS or impede the movement of the train.

Stray-Current Corrosion. Many existing buried utilities are bare metal. These include concentric neutrals on power lines, lead sheathed communication cables, steel fuel lines, and water lines that are typically gray and ductile cast iron with copper service lines. Portions of the water lines can be electrically continuous because of standard construction methods, that is, leaded joints for cast iron and mechanically restrained joints for ductile iron. Fuel lines are usually coated steel with cathodic protection. The concerns for electric rail construction are stray-current corrosion of the bare metal utilities and electric rail interference to cathodically protected utilities.

Figures 10(a) and (b) show recording chart data of track-to-earth currents and the corresponding earth gradients, resulting from an intentional test short on an ungrounded rail system. The earth gradients are potential (voltage) measurements between a copper/copper sulfate electrode reference cell (CSE) at a remote location and a CSE reference cell in an area of interest. The data are also shown in a cumulative frequency graph used to determine time average values (Fig. 10c and d). The chart profiles are characteristic signatures of multiple train movements. The change in the readings through time is associated with train location and acceleration.

Measurements of current flow through the test short (Fig. 10a), show reversals with both discharge and pickup at the short. The time average current discharge of the short was approximately 20% of peak. Computations of electrolytic corrosion effects are based on time average current.

The earth gradient profile (Fig. 10b), corresponds directly to the current flow shown in Fig. 10(a). Using Ohm's law, the test short is computed to have a constant track-to-earth short resistance of approximately 100 Ω:

$$E = IR \qquad \text{(Eq 4)}$$

where E is the track-to-earth voltage at a given time, (V), I is the track-to-earth current at a given time, (A), and R is the track-to-earth resistance constant (Ω).

The expression track-to-earth resistance is somewhat misleading in that the resistance is not a contact resistance, but is the resistance of current flow between remote earth and the structure. The resistance is a consequence of soil resistivity and current density. The current density is minimal in remote earth, but streamlines of current (current density) converge in the vicinity of the electrode.

The resulting gradient is paramount in stray-current control and is shown in Fig. 11. Figure 11 shows two curves—the current density in soil resulting from a hemispherical field such as a discrete short, and a hemicylindrical field representing uniform discharge along the track. The figure compares the current density at 1 ft (0.3 m) distance from a current source through a 25 ft (7.6 m) distance. These are common distances for utility clearances in street construction. As a useful field approximation, note that the geometric relationship between current density in the soil and distance from a current source is such that current density is reduced roughly half or more with each doubling of distance from the current source. The change in current density through distance is characteristically a rectangular hyperbola ($y = c/x$) as shown in both the hemispherical and hemicylindrical case.

This geometric relationship between current density and distance can be seen in the comparison on the chart recording of Fig. 10(b) with the simultaneous chart recording of Fig. 12. The earth gradient between remote earth and the immediate vicinity of the short, Fig. 10(b), is roughly an order of magnitude greater than the gradient between remote earth and a location

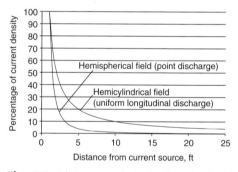

Fig. 11 Relative current density in soil as a function of distance from the current source, with the maximum current density of interest 1 ft (0.3 m) from the source.

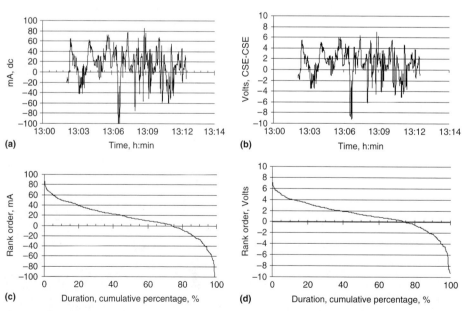

Fig. 10 Track-to-earth current (a) and earth gradient (b) for a 100 Ω test short on an ungrounded embedded track system and their respective time durations (c) and (d). The CSE reference cells were located at the short and remote (175 ft, or 53.3 m).

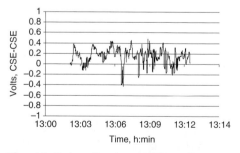

Fig. 12 Earth gradient for a 100 Ω test short on an ungrounded embedded track system. The CSE reference cells were located 25 ft (7.6 m) from the short and remote (175 ft, or 53.3 m). Refer to Fig. 10(b). See text for details.

25 ft from the short (Fig. 12). Note that Fig. 12 voltage is graphed at 1/10 scale of Fig. 10(b) and that the gradient profiles are analogous.

An engineering approach to electric rail stray current can be developed from "Grounding Principles and Practice—Fundamental Considerations on Ground Currents" (Ref 3), which presents the case of a short pipe in an extended field, a long pipe in a limited field, and a long pipe in an extended field. Ground currents are addressed in terms of current density (current per cross-sectional area), electric field strength (as the product of current density and soil resistivity), and voltage (as the line integral of field strength).

Equation 5(a) given below can be used to calculate stray current for the case of a short bare pipe in an extended field. Only current density in the soil and the pipe dimensions are necessary to estimate stray-current corrosion. Electrolytic current density, as the average entrance and exit current density over the surface of a pipe, is derived as a function of electric field strength and the ground resistance of the structure. The calculation is independent of soil resistivity because soil resistivity cancels in the relationship of electric field strength and structure-to-earth resistance. Equation 5(b) calculates the case of a long pipe in a limited ground field. Again, the distance between the pipe and the current source is a determinant of corrosion. Equation 5(c) calculates the case of a long pipe in an extended field. The distance between the pipe and current source is not predominant. The ratio of the soil/metal resistivity is so great (roughly 10^9) that in the equilibrium of electric field strength between metal and soil, large cross sections of soil are involved. In this case, electric rail influence can extend over wide areas and long distances. In urban areas, the effects might be spread over a multitude of structures including building foundations (most often electrically continuous through utility connections). In suburban and rural areas, the effect might be concentrated on the few parallel utility structures. Equations 5(a) to (c) are presented as:

$$i_{\text{avg pipe}} = i_{\text{soil}} \frac{\frac{l}{a}}{\ln\left(2\frac{l}{a}\right)} \qquad \text{(Eq 5a)}$$

$$i_{\text{max pipe}} = \frac{I_{\text{soil}}}{\pi a S} \frac{1}{\ln\left(4\frac{S}{a}\right)} \qquad \text{(Eq 5b)}$$

$$\frac{i_{\text{pipe}}}{i_{\text{soil}}} = \frac{e_{\text{pipe}}/\rho_{\text{pipe}}}{e_{\text{soil}}/\rho_{\text{soil}}} = \frac{\rho_{\text{soil}}}{\rho_{\text{pipe}}} \qquad \text{(Eq 5c)}$$

where a is the pipe radius (ft), e_{pipe} is the electric field strength of pipe (V/ft), e_{soil} is the electric field strength of soil (V/ft), i_{pipe} is the current density in pipe (A/ft^2), $i_{\text{avg pipe}}$ is the average surface current density over pipe half length (A/ft^2), $i_{\text{max pipe}}$ is the maximum surface current density over pipe half length (A/ft^2), i_{soil} is the current density in soil (A/ft^2), I_{soil} is the current in soil from electrode (A), l is the half length pipe (ft), ρ_{pipe} is the resistivity of pipe ($\Omega \cdot$ ft), ρ_{soil} is the resistivity of soil ($\Omega \cdot$ ft), and S is the distance between pipe and current source (ft).

The principals presented in Ref 3 have been adapted to utility work in electric rail corridors (Ref 4). Graphical methods are provided to quantitatively assess stray-current effects on existing bare utilities for several scenarios, including uniform conditions along the track, areas of failed track insulation, and point source track shorts. The approach is to elect an acceptable corrosion rate and compare the associated ground current to anticipated rail system characteristics. Dynamic models of track potentials and stray current can be utilized but are not necessary.

The graphical approach provides for sensitivity analysis by demonstrating the effects of changed rail conditions and also suggests reasonable values to assume in solving otherwise indeterminate problems.

Matrix tables are then used to assess risk. For example, a design that is contingent on installing and maintaining an exceedingly high track-to-earth resistance can be scrutinized to show the effects of a compromised or degraded installation. The matrix tables are also used to find a convergence of design criteria that balances risk. Presumably, in cases where there is adequate maintenance access and/or interruption of transit is a viable option, a greater corrosion risk might be acceptable. Alternately, where utility access is problematic and/or disruption of transit service is not an option, more stringent corrosion control would be required.

Table 1 shows the application of a matrix table used in a project to evaluate possible relocation of waterlines proximate to a light rail (Ref 4). The designer may consider that a 100 Ω short is a likely possibility. Using the values presented in the table, a long pipe located within 3 ft (0.9 m) of a 100 Ω short will corrode at an unacceptable rate. However, at 10 ft (3 m) distance, a 20 Ω short would result in a moderate corrosion rate and a 20 Ω short may be considered unlikely. Note that in this case, stray current from low-resistance shorts governs, and that track with a uniform resistance is less of a concern regarding stray current. To maximize the benefit to project cost, the corrosion risk associated with distance from the track is then compared to the distance criteria for access and maintenance requirements. This approach to engineering uncertainty, the use of sensitivity studies, graphical solutions, and decision matrices are discussed in Ref 5.

Electric rail construction is used most often on shared transportation right-of-ways that have buried utilities. The utility engineers must elect to either leave the utilities that neighbor electric rail in place or relocate the utilities. If the utilities are to be relocated in place or at a further distance, mitigation can include the use of alternate nonmetallic materials, casings, and cathodic protection. Corrosion-control measures can also include mitigation bonds and galvanic drains; however, there are common misunderstandings in their use, and the reader's review of a primer on the subject is suggested (Ref 6).

Utility Construction and Funding

By their nature, large electric rail projects are complicated and challenging. There are a number of issues and criteria to be considered while resolving conflicts between existing utilities and electric rails (Ref 7, 8).

Early involvement by utility owners in the design phase of the transit system is essential to assess the need and cost of relocations. Utility relocation and coordination can be critical variables affecting transit project schedule, scope, and budget. Early involvement allows utilities and project designers to proactively identify and resolve conflicts that could otherwise delay design and construction.

Construction issues involve staging and sequencing of work and mitigating traffic impacts and impacts to adjacent property owners, neighborhoods, and businesses. Construction sequencing and staging are interrelated. Usually, utility work is done in advance of the main civil track construction. Utilities must provide uninterrupted service to customers, and that might

Table 1 Risk assessment matrix

Based on a selected current density of 0.001 A/ft^2 for an acceptable corrosion rate of the pipe. See text for details.

Design separation pipe-to-rail, ft	Maximum allowable stray current, mA	Equivalent short resistance, Ω	Risk
Long pipe, limited ground field			
3	22	100	High/likely
10	200	20	Moderate/unlikely
50	670	3	Low/unlikely
Short pipe, limited ground field—hemispherical current source field			
3	15	100	High/likely
10	170	10	Moderate/unlikely
50	4300	0.5	Low/unlikely
Short pipe, limited ground field—hemicylindrical current source field			
3	50 mA/10 ft	0.5 Ω/1000 ft = 50 Ω/10 ft	Low/unlikely
10	170 mA/10 ft	0.1 Ω/1000 ft = 10 Ω/10 ft	Low/unlikely
50	850 mA/10 ft	0.02 Ω/1000 ft = 2 Ω/10 ft	Low/unlikely

Units used in this table are traditional for the industry. Source: Ref 4

require alternate service arrangements during construction. There are usually operational and seasonal restrictions that limit when work can occur on the utility system. For instance, water-supply systems often cannot be taken out of service during peak season, which is most often May to October. For a natural gas utility, the peak season is during the winter and the utility tries to limit their downtime to the summer months. Often each utility must work in turn, rather than concurrently. For safety reasons, no utility can work in the area while the power poles and electrical lines are being relocated. In underground work, typically, the deepest utility relocates first and shallowest relocates last. However, sometimes, other sequencing arrangements are necessary. With limited routes for relocation available, there may also be an issue of where utilities can physically locate.

Construction and maintenance limitations arise in accommodating electric rail corridors (Ref 7). Figures 13 and 14 illustrate the challenge of installing mains or relocating pipelines to accommodate the electric rail. Figure 13 shows a typical choke point where several utilities converge to make way for the new electric rail next to an existing viaduct. Since the viaduct supports half of the roadway, only half of the right-of-way is available for utilities. With the proposed light rail over the centerline of the roadway and an existing 96 in. (2.44 m) sewer occupying the travel lane, there is little room for remaining utilities. Figure 14 shows the vertical offset crossing of a 24 in. (0.6 m) ductile iron main crossing a 12 in. (0.3 m) main. Utility relocations often require vertical offsets to route under and over existing structures.

Working Clearances. Utility relocations are designed to meet current federal, state, and local regulations and industry standards. Regulations cover safety clearances that are required for permanent and temporary obstructions. Designs typically satisfy both conditions. When it is not possible, designs address permanent safety requirements, and requirements for temporary safe working conditions are satisfied as they arise during construction, operations, and maintenance.

A design envelope for clearances around an electric rail is shown in Fig. 15. Some of the safe clearance requirements affecting utilities in electric rail corridors are presented. There may be other clearance requirements—the listing is not all-inclusive and requirements may differ by jurisdiction. Aboveground clearances include the permanent and temporary clearances shown from overhead power lines and working clearances from moving trains. Underground clearances include the vertical distance to the track and a corresponding horizontal separation from the track, depending on depth of excavation. These clearances affect the decisions by utilities to relocate facilities and affect the decisions of where to relocate or install new facilities.

Safety clearances are governed by the Occupational Health and Safety Administration (OSHA) and by various organizations responsible for industry standards. Notes 1 and 2 in Fig. 15 illustrate clearances from the overhead power lines, which are governed by OSHA, National Electrical Safety Code (NESC), and the U.S. Department of Transportation. Oregon OSHA (Ref 9) and NESC (Ref 10) require a minimum 10 ft working clearance between overhead electrical lines (50 kV or less) and personnel or equipment (Fig. 15, note 1).

Fig. 13 Utility relocation encumbered by congestion. Utilities are squeezed together on one side of a proposed light rail. The other side of the road is a bridge viaduct.

Fig. 14 Vertical offset of a 24 in. (0.6 m) ductile iron water main to accommodate an existing 12 in. (0.3 m) main.

The NESC (Ref 11) requires an 18 ft (5.5 m) minimum vertical clearance over roadways for electrical lines 22 kV or less (Fig. 15, note 2). This same regulation requires a minimum of 22 ft (6.7 m) vertical clearance for electrical lines crossing railroads.

Electric transit systems that operate on fixed tracks are subject to regulations similar to heavy rail systems and governed by the Federal Transit Administration (FTA). Federal Rail Administration safety program (Ref 12) requires a working clearance between moving trains for road workers, rail workers, and their equipment (Fig. 15, note 3). This clearance, which is based on speed of the train, results in a 13 ft (4 m) worker clearance (Ref 12). The consequence of this federal regulation is that anyone working within this "fouling zone" is required to have flaggers, stop working, and move their equipment out of the way when a train is passing the work zone. During train operating hours (revenue hours), trains could be passing every 7 min. Oregon adopted the FTA regulations and applies them to all transit rail systems operating in Oregon (Ref 13 and 14). Typically, as a condition of the right-of-way permit to allow utilities to occupy a jurisdictions right-of-way, utilities and public agencies are required to conform to the rail safety standards.

There are national railroad standards (Ref 15 and 16) for cased crossings for flammable and nonflammable substances crossing railroads (Fig. 15, note 4). Gas and petroleum lines are subject to the requirements for flammable substances, while water lines are subject to the requirements for nonflammable substances. Casing lengths for nonflammable substances are required to extend at least two times the depth of the bottom of the casing below the railroad subgrade or extend at least 5 ft (1.5 m) beyond a 2-to-1 maximum excavation slope on each side of the track centerline, whichever is greater. Heavy rail companies may require the casings to extend across the entire length of the railroad right-of-way. Excavation by utilities is not allowed in the ballast or subballast zones (Ref 15, 16).

OSHA also restricts the maximum excavation slopes without shoring (Ref 17–19). A 1.5-to-1 is the steepest slope allowed, though a 2-to-1 maximum slope is preferred (Fig. 15, note 5).

Designing relocations to incorporate safety clearances provides access and continued utility operations and, more appropriately, assigns the cost to the capital project rather than shifting the cost to utility operations and maintenance.

Track Design Envelopes and Utility Design. Figure 16 illustrates differences in the track design envelope and its impacts on utilities. Figure 16(a) is the tie-and-ballast track section with a short ballast wall. This design is commonly used when a dedicated right-of-way can be provided adjacent to a roadway. Figure 16(b) is embedded track and is common when the electric rail transit shares the right-of-way with motorists. Figure 16(c) is the tie-and-ballast scenario often seen where the track system can be located in a dedicated right-of-way, similar to the heavy rail systems. Figure 16(d) is a variation of Fig. 16(a) and (c), used when the track system is built at a different elevation from the adjacent ground and there is insufficient room for side slopes.

Tie-and-ballast systems have deep design envelopes of 5 ft (1.5 m) or more, while

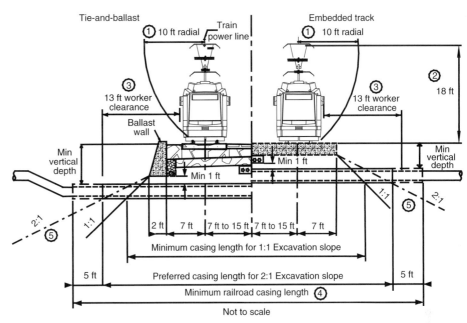

Fig. 15 Work clearances for utilities adjacent to and crossing under electric rail systems. Circled numbers are referenced in the text as notes. Units of measure given in feet.

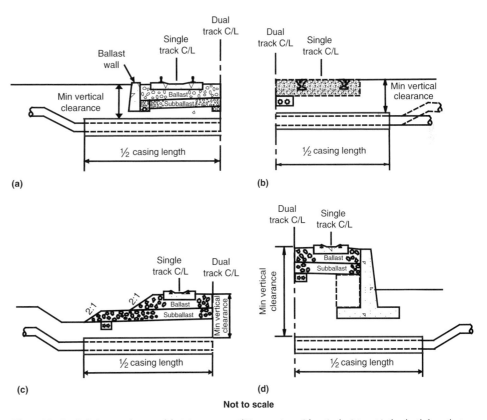

Fig. 16 Track design envelopes and their impacts on utilities crossings. Of particular interest is the depth from the top of the rail to the top of the casing, the length of the casing, and the vertical offset of the carrier pipe outside of the casing. (a) Tie-and-ballast adjacent to roadway. (b) Embedded track. (c) Tie-and-ballast in dedicated right-of-way. (d) Tie-and-ballast retaining wall. C/L, centerline

embedded track systems can have shallower envelopes, about 2 to 3 ft (0.6 to 0.9 m) in depth. Underground utilities are typically 2 or 3 ft (0.6 or 0.9 m) minimum depth. Most utilities will be physically impacted by tie-and-ballast construction because of the depth of the envelope. With embedded track, fewer utilities will be in direct physical conflict with the envelope construction.

The track slab design can be a critical cost factor because it can govern utility relocations. A shallow track slab design that incorporates the electric rail "utilidor" (utility corridor for the track system duct banks and track drainage) near the surface might allow for utility crossings in the same trench configuration without elevation changes. Elevation changes entail vertical offsets and often require a deep excavation to clear other neighboring utilities. Figure 16(b) illustrates the benefit of replacing a pipe in the existing configuration without elevation changes. Note, an offset is required if there is an elevation change, and often the elevation change must be particularly deep to go under conflicting interferences. The deeper the excavation, the greater the cost and disruption of construction and the more difficult it is to later access the pipeline.

The Portland Water Bureau developed a 10 ft (3 m) separation criterion between the electric rail transit and the water facilities. This is considered a convergence in the design requirements for access, corrosion control, and cost/benefit. The clearance entails relocating parallel waterlines 10 ft (3 m) from the track edge and crossing pipelines are cased under the track with the casing extending 10 ft (3 m) from each side of the track edge.

Construction limitations often necessitate exceptions to this criterion. In downtown streets, sometimes there simply is not enough unoccupied space to install a utility pipe parallel or extend a casing the specified distance from the track. Then the requirement is augmented with secondary insulation of the track slab and/or upgrade of the pipeline materials and construction. This addresses corrosion control; however, access is compromised.

At times, construction must be site specific. Where there is a pipeline that cannot be taken out of service, a casing such as a box culvert can be built around the existing pipe. If there is insufficient space for the casing to extend beyond the permitted excavation slope, a retaining wall can be provided to permit access for maintenance without undermining the track. This provides access for maintenance and construction, but again access is compromised.

Utility Relocation Funding Issues. Utility relocations can be eligible for up to 100% reimbursement to a utility owner in accord with U.S. Department of Transportation rules, provided there are no state or local regulations prohibiting reimbursement. The term "relocation" includes "removing and reinstalling facilities, including necessary temporary facilities, acquiring necessary right-of-way for the new location; moving, rearranging or changing the type of existing facilities; and taking any necessary safety and protective measures" (Ref 20, 21). The intent is to "restore the utility to a functional capacity equivalent to that in place." Often, municipally owned utilities are reimbursed for relocating facilities located in the public rights-of-way or in private easements. However, costs for relocating public nonmunicipal and private utilities might not be reimbursed. Nonreimbursed expenditures are passed on to the ratepayers in the cost of services.

There is limited funding for electric rail construction, and reimbursement funds are not available for electric rail expansion or other amenities, such as station enhancements. Electric rail projects are often coupled with urban development and city street improvement projects that are outside of transit funding, so there is an impetus to minimize utility reimbursement to make these amenities available.

After electric rail construction is complete, there continue to be long-range impacts on the utility system. There should be long-term agreements regarding any additional cost of operations, repairs, and damages to the utility and limited liability of the utility to the transit system.

Monitoring and Maintenance for Stray-Current Control

Stray currents affect both the electric rail facilities and neighboring utilities. Effective stray-current corrosion control requires ongoing monitoring and maintenance after construction.

Monitoring and Maintenance by the Transit Authority. Though corrosion control is inherent in the design of the electric rail, the transit authority must prevent upset operating conditions. If an inoperative substation is allowed to remain off-line, or if there is imbalance among substations, then these upset conditions essentially increase the distance between substations and, thereby, increase stray currents.

Transit systems are designed to shut down if there is a critical short of the high-voltage supply. However, there can be high-resistance shorts that allow continued operation but result in stray current. This is often caused by current flow around fouled insulators. Transit authorities routinely visually inspect for this during regular maintenance.

The track-to-earth resistance can be greatly lowered by shorts, inadvertent or from construction, or from fouling of the ballast or fouling around insulators on tie-and-ballast, direct fixation, or embedded track. Transit authorities have wash-down maintenance programs to clean fouled ballast and insulators.

Those systems with diode grounding can be monitored for stray current by measuring ground return through a return shunt. This can be used to immediately detect upset conditions and to monitor for long-term, system-wide, degradation in stray-current control. Systems that are ungrounded can be monitored with propriety stray-current monitoring based on track voltage signatures. However, presently these methods are sensitive only to very low-resistance shorts.

Radio frequency holiday short detection surveys are practical for rail line surveys. Earth gradient measurements are practical at fixed critical locations, such as utility crossings, but because gradients vary with train movements they are less practical for line surveys. This difficulty can be overcome in a track maintenance survey by impressing a fixed current and monitoring the gradient. Often both radio frequency and impressed gradients are used to locate and confirm a short.

Track slab design can include electrically segmented track sections with current shunts to monitor stray current in the steel reinforcement.

Actual corrosion damage to the transit system is monitored by inspection and maintenance operations, with particular attention to track, fasteners, and supports for the OCS.

Monitoring by the Utilities. Utilities can monitor earth gradients at critical locations, such as crossings. Earth gradient measurements between a reference cell proximate to the track and a remote reference cell produce a profile with a characteristic signature that correlates with train movement. All of the earth gradient charts presented in this article show the characteristic profile of the change in stray current through time when trains are moving past a fixed location.

This method is a measure of the rail system conditions, and it does not require test stations. Nor does this method require a connection to the electric rail system or a neighboring structure. For safety, work in the electric rail corridor requires the cooperation of the transit authority. A test connection to the track can cause a dangerous malfunction of the train signal, and personnel are exposed to train traffic and possibly high-voltage environments.

Figure 17 shows typical gradient measurements of tracks at noninsulated tie-and-ballast track (Fig. 17a), insulated tie-and-ballast track (Fig. 17b), embedded track insulated with potting compound (Fig. 17c), and embedded track insulated with an elastomeric boot (Fig. 17d). Note that the scale of Fig. 17(a) is tenfold greater than the others. The gradient for the noninsulated track is an order of magnitude greater than the gradient of the insulated track. This is because the track-to-earth resistance of the noninsulated track is much less than the insulated track systems. The gradient for the track insulated with a potting compound is appreciably greater than the gradient of the other examples of insulated track. The potting compound is failing and the track-to-earth resistance of this track is less than the other insulated track sections. The small bias around the baseline of gradient measurements can be attributed to natural variations in soil; however, an offset of 100 mV or greater might indicate a steady-state source of stray current.

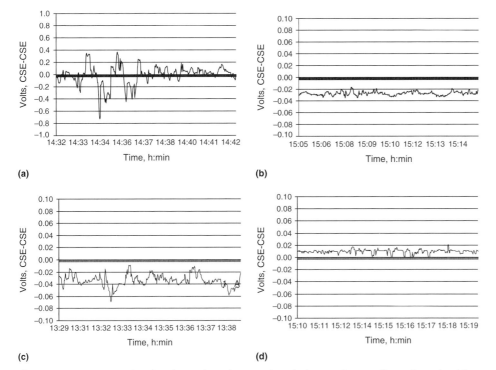

Fig. 17 Comparative track earth gradients. The tracks are not shorted. The CSE reference cells were located 15 ft from the track and remote (175 ft). See text for details.

With the cooperation of the transit authority, a calibrated track-to-earth test short can be used to provide a qualitative and semiquantitative measure of site-specific conditions. Subsequent profiles can be readily compared for changes and to verify that the original design assumptions are maintained. Figure 10(b) shows the measured effect of a 100 Ω short in an embedded track compared to the properly insulated section of embedded track in Fig. 17.

Utilities can install corrosion coupon monitoring test stations in critical locations such as crossings. Coupons are dimensioned bare areas made of the same material as the structure and electrically connected to the structure. Coupons are intended to represent the surface of the structure, and the coupon test stations are wired to measure polarized potentials and current flow (Ref 22).

Traditional corrosion test stations can also be installed. These test stations provide electrical connections to the utility and locations for reference cells and can include calibrated current spans or isolation joints with calibrated shunts for current measurement.

Test stations also facilitate utility line surveys. Line surveys include stray-current "exposure" and "mutual" surveys (Ref 23, 24). A method for locating high-exposure areas in electric rail corridors is to plot a profile of time-variant structure-to-soil potentials and simultaneous measurements of pipeline current (if available). This is done at several pipeline locations. In this way, areas of current pickup, current discharge, and locations of maximum exposure can be determined.

A mutual survey consists of simultaneous measurements of the source of stray current and structure-to-earth potentials. The data are graphed to verify a correlation between the transit operation and effects on the structure. The graph is commonly referred to as a beta curve when the data are a plot of the voltage between the structure and the transit system versus the structure-to-soil potential. A profile drawing of the pipeline locations versus the slopes of the beta curves indicates the locations of maximum exposure. The magnitude of the interference can be estimated by providing a calibrated test current that approximates the slope of the line at that location.

ACKNOWLEDGMENT

The authors wish to thank T. Heilig and K. Looijenga, Tri-County Metropolitan Transportation District of Oregon (Tri-Met), Portland, OR, for their cooperative stray-current testing and review.

REFERENCES

1. K. Pham, T. Heilig, K. Looijenga, and X. Ramirez, "DC Frame Fault & Ground Fault Field Testing On TriMet Portland Light Rail System," 2005 ASME/IEEE Joint Rail Conference (Pueblo, CO), March 2005
2. T. Heilig, Memo—The Grounding Diode in Tri-Met's System, Tri-Country Metropolitan Transportation District of Oregon (Tri-Met), Portland, Oregon, June 30, 2002
3. R. Rudenberg, Grounding Principles and Practice—Fundamental Considerations on Ground Currents, *Elect. Eng.*, Vol 64 (No. 1), 1945, p 1–13
4. S. Greenberger, "Parametric Stray Current Monitoring and Mitigation for Electric Rail Stray Current," paper No. 05249, Corrosion 2005, NACE International, 2005
5. S. Vick, *Degrees of Belief—Subjective Probability and Engineering Judgment*, ASCE Press, 2002, p 293
6. R. Steifert, Do these Corrosion Mitigation Processes Mystify You? *Pipeline Eng.*, June/August 1968, p 65–73
7. T. Elliott, S. Greenberger, and M. Collentine, "Considerations and Criteria for Relocating Utilities in Electric Rail Corridors," paper No. 05240, Corrosion 2005, NACE International, 2005
8. S. Greenberger, Underground Water Utilities—Crowded and Complex, *Mater. Perform.*, Vol 41 (No. 7), July 2002, p 8
9. Oregon Occupational Health and Safety Administration, Rule 437-83-2813, Min 10 ft working clearance
10. National Electrical Safety Code (NESC) ANSI C2, rules 234B, 2002, Min 10 ft working clearance
11. National Electrical Safety Code (NESC) ANSI C2, rules 232A, 2002, Vertical Clearances above Roads and Railroads
12. "Roadway Worker Protection," Title 49, Code of Federal Regulations, Part 214, U.S. Department of Transportation, Federal Rail Administration, Dec 16, 1996
13. Oregon Department of Transportation, (ODOT) Rail Section, ORS 267.230 applicability of law to mass transit
14. Oregon Department of Transportation, (ODOT) Rail Section, OAR 824 definition of railroad includes lightrails, streetcars, and trolleys on fixed rails
15. "Track Safety Standards," Title 49, Code of Federal Regulations, Part 213, U.S. Department of Transportation, Federal Rail Administration, Dec 16, 1996
16. "Pipeline Installation Procedures for Crossings" and "UPRR Pipeline Installation Engineering Specifications," Union Pacific Railroad, http://www.uprr.com/reus/pipeline/pipe/procedur.shtml, revised Aug 2003
17. "Title 29, Code of Federal Regulations, Part 1926," U.S. Department of Transportation, Maximum excavation slope requirements
18. Oregon Occupational Health and Safety Administration, Rule 1926-650, Maximum excavation slope requirements
19. Oregon Occupational Health and Safety Administration, Rule, OAR 437-83-2813, Maximum excavation slope requirements
20. "Title 23, Code of Federal Regulations, Part 645, Utility Relocations," Federal Highway Administration, April 2003

21. "Highway/Utility Guide," Federal Highway Administration, 1993
22. "The Use Of Coupons for Cathodic Protection Monitoring Applications," ANSI/NACE Standard RP0104, NACE International, 2004
23. "Control of External Corrosion on Underground of Submerged Metallic Piping Systems," ANSI/NACE Standard RP0169, NACE International, 2002
24. A. Peabody, *Peabody's Control of Pipeline Corrosion,* R. Bianchetti, Ed., 2nd ed., NACE International, 2001

SELECTED REFERENCES

- American Public Transportation Association Standards, http://www.aptastandards.com
- Appalachian Underground Corrosion Short Course—Basic, Intermediate, Advanced Course Books, College of Engineering, West Virginia University http://www.aucsc.com/bookstore.htm
- T. Barlo and D. Zdunek, "Stray Current Corrosion In Electrified Rail Systems" Northwestern University Infrastructure Technology Institute, May 1995, http://www.iti.northwestern.edu/projects/stray2.html
- "Practical Guide to Railway Engineering," The American Railway Engineering and Maintenance of Way Association. http://www.arema.org
- M. Szeliga, Ed., Stray Current Corrosion, NACE International, 1995
- "Grant Management Guidelines," FTA C5 010.1C, U.S. Department of Transportation, Federal Transit Administration, http://www.fta.dot.gov/legal/guidance/circulars/5000/324_288_ENG_HTML.htm

Corrosion in Bridges and Highways

Jack Tinnea, Tinnea & Associates, LLC
Lianfang Li, W.R. Grace & Co.
William H. Hartt, Florida Atlantic University
Alberto A. Sagüés, University of South Florida
Frank Pianca, Ontario Ministry of Transportation
Bryant "Web" Chandler, Greenman Pedersen, Inc.

BRIDGES AND HIGHWAYS are a core component of any country's or region's transportation system. In the United States, they comprise the largest public infrastructure system, consisting of approximately 583,000 bridges and 6.3 million kilometers (3.9 million miles) of public roads (Ref 1). Approximately 200,000 of those bridges are constructed of steel, 235,000 of conventionally reinforced concrete, 108,000 of prestressed concrete, and the balance constructed with other materials. These roads and highways range from quite primitive pavements with earth, gravel, or stone covered by a thin bituminous surface course to a continually reinforced portland cement concrete (PCC) roadway with or without a bituminous wear course.

This article is concerned with bridges and dowels and reinforcement used in PCC roadways. It is not that other types of roads and highways do not have maintenance issues. They do, but they do not suffer from corrosion problems. Bridges and reinforced PCC roadways do suffer from corrosion, and the cost of that corrosion is staggering. In the United States alone, a 2002 study reported annual direct corrosion costs of $8.3 billion (Ref 2). Those costs included $3.8 billion for bridge replacement, $2 billion for bridge decks, $2 billion to repair concrete substructures, and $0.5 billion for maintenance coating of steel elements. Estimates of indirect costs to the bridge user, from traffic delay, detours, and lost productivity, were ten times the direct costs.

Highways and bridges are a ubiquitous component of the infrastructure, and the factors affecting their corrosion are varied and at times subtly complex. A brief overview is provided of the rise in awareness of the corrosion issues affecting bridges and highways and of the terminology used in following sections on highways and bridges. Much of the bridge and highway infrastructure is comprised of reinforced concrete, so a discussion of concrete and its role as an electrolyte follows. The next section addresses reinforcement, including conventional, prestressed, cable stays, and corrosion-resistant reinforcement. Following this is a discussion of electrochemical methods for inspection and corrosion control for embedded reinforcement. The final section discusses the corrosion of metal bridges and corrosion control, including the use of weathering steels and coating systems.

A Historical Perspective and Current Control Strategies

Jack Tinnea, Tinnea & Associates, LLC

Over the past thirty years, both the popular media and professional published record have seen a somewhat explosive growth of articles concerning the decay and aging of public works facilities and buildings. Roads and bridges, with their potholes and rusted trusses, became a most visible icon for what is now collectively known as "the crumbling infrastructure," with much of that "crumbling" the result of corrosion processes.

History

The maintenance and repair of public works is an ancient art. It has come to prominence in the last quarter of the 20th century for several reasons. The first half of the 20th century saw structural engineers embrace reinforced concrete as a core building material. The second half of the 20th century saw an explosion in the building of highways and bridges and other infrastructure items in North America, in war-ravaged Europe and Asia, and in many developing countries. A review of the published literature over this latter period shows an amazing growth of interest in corrosion of metals embedded in concrete (Ref 3).

It has long been known that seawater aggravates the corrosion of coastal bridges. People have attempted to address the corrosion of metals in marine environments since antiquity. Although engineers from the Roman Empire may have discussed the corrosion of iron bars grouted in stone or set in natural cement structures, published modern discussion of concrete-reinforcement corrosion dates to the early 20th century (Ref 4, 5). Much of the early work on reinforcing-steel corrosion addressed reinforced concrete in marine environments (Ref 5–7). In time, field experience and research showed that the processes that attack concrete, such as carbonation (Ref 8), alkali soils (Ref 9), and cracks (Ref 10), also can contribute to the corrosion of embedded reinforcement.

In addition to a rapid growth in the building of highways and bridges, the 1950s saw substantial growth in the use of roadway deicing salts in areas where snow and ice on the bridge decks adversely affected traffic safety. Because most of the early highway deicers contained chloride ions, this safety measure brought the ocean and its corrosion problems to inland areas. The earliest signs that the application of deicers caused corrosion problems appeared on metal elements at or above bridge decks. Included were steel items such as guardrails and truss elements that passing traffic would spray with slush and water containing chloride ions. Typically, these steel elements were protected by paint or galvanizing. Often, these coatings were not what today (2006) would be considered an integral component of a maintenance program. Rather, they were merely thin coats of paint applied over mill scale, with little or no surface preparation (Ref 2). The application of deicers produced a winter salt spray that greatly accelerated the rate of corrosion and subsequent coating failure.

In addition to causing corrosion problems for atmospherically exposed metals, chloride ions from deicers slowly diffuse through the concrete to the surface of underlying steel reinforcement. When the concentration of chloride ions reaches a threshold level, corrosion commences. In time, as this corrosion proceeds, the buildup of corrosion products at the rebar-concrete interface generates sufficient pressure to crack the concrete. By the mid-1960s, highway engineers and scientists recognized that the marinelike problems with reinforcement corrosion had reached inland areas (Ref 11).

Changes in environmental regulations required owners and designers to look at highway and bridge corrosion in new ways. Lead- and chromate-bearing paints were the coating of choice for steel bridges during the North American interstate building boom from the early 1950s through the 1960s. In the United States, it is estimated that 80 to 90% of the over 200,000 steel bridges have lead-base coating systems (Ref 12, 13). Recoating of those structures requires integral lead abatement to provide protection to both the surrounding environment and the workers. Globally, there are over a half-million steel bridges, so the scale of maintenance painting is clearly enormous (Ref 14).

Environmental considerations carry over to today's (2006) concrete mix designs. Fly ash, a by-product from burning coal, is a pozzolanic material that can be used to replace cement. Pozzolans are finely divided mineral additions, or mineral admixtures, that, when mixed with calcium hydroxide, alkalis, and water, produce a cementitious material. Good mix designs, which incorporate fly ash or similar mineral admixtures, can accomplish several things. First, they find a productive use for a material that may otherwise be sent to a landfill. Second, their use reduces greenhouse gas emissions by allowing the cement content to be reduced. The U.S. Environmental Protection Agency (EPA) estimates that fly ash cement replacement in the United States reduced annual greenhouse emissions by approximately 2.7 million metric ton carbon equivalents (Ref 15). The EPA goes on to say that if all the fly ash produced in the United States were employed as a cement replacement in concrete mixes or other similar uses, such as stabilizing road beds, it would reduce greenhouse emissions by the equivalent of removing 13 million cars from the road for a year. In addition to environmental benefits, fly ash and its pozzolanic cousin, silica fume, produce concretes that are less permeable to chloride ions than comparable mixes that employ only portland cement.

Current Corrosion-Control Strategies

Strategies to combat corrosion in highways and bridges approach the problem from several directions. For reinforced concrete these include:

- Improvements in concrete and grout mix designs, including high-performance concretes
- Improvements in structure design and detailing
- Use of corrosion-resistant alloys or nonmetallics as reinforcement
- Electrochemical techniques, including cathodic protection (CP), chloride extraction, and realkalization

For steel bridges and steel substructure components, the corrosion-control approaches include:

- Designs with structural elements out of the path of salt water that is sprayed by passing traffic
- Designing drains to avoid run-off water discharging onto substructure elements
- Development of nonhazardous high-performance coatings and primers
- Metallizing and thermal sprays
- Use of weathering steels

In addition to controlling corrosion, maintenance of bridges and highways must address the economic, health, and safety role these structures play. Some bridges are historic, so corrosion control and maintenance must also address this special issue. Fortunately, asset management has evolved considerably, and systems exist that assist owners in managing often complex infrastructures (Ref 16–19). Computerized bridge management systems available in the United States today (2006) include PONTIS and BRIDGIT. In Europe, the highways are managed today (2006) on an individual country basis. The Bridge Management in Europe (BRIME) project was undertaken to provide a framework for European highway network management that eventually could save billions of euros (Ref 20).

Many of the methods used to control corrosion in highways and bridges, such as coatings, corrosion-resistant metals, and CP, are discussed at length in these three volumes of Volume 13 of the *ASM Handbook*. Unique issues germane to the application of such methods to highways and bridges are detailed in the following sections. Because concrete is ubiquitous in highways and bridges, the properties of concrete that play a significant role in the corrosion of embedded reinforcement are also discussed.

Terminology

Corrosion practitioners and metallurgists new to the area of corrosion in highways and bridges can find themselves somewhat overwhelmed by terminology employed to describe the structures and their elements. Fortunately, there are excellent glossaries available, including some on the internet. Appendix III to the BRIME project report (Ref 20), "Glossary of Terms Used in Bridge Engineering," provides an extensive glossary of bridge engineering terminology and includes excellent graphics. For terms specific to concrete, refer to Ref 21 and the glossary in Ref 22.

Concrete: Implications for Corrosion

Jack Tinnea, Tinnea & Associates, LLC

Reinforced concrete is a heterogeneous composite material. Concrete affords corrosion protection for reinforcing steels by providing a physical barrier to the surrounding environment. In acting as a barrier coating, concrete provides shielding from aggressive ions such as chlorides. Concrete also acts as a barrier to oxygen, reducing its availability at the surface of the metal. However, the cement paste that binds concrete is a porous barrier. Actions taken that impede the ability of a concrete to transport ions will improve the ability of the concrete to protect embedded reinforcement.

Beyond being a barrier, concrete also provides embedded metals an alkaline environment. If other factors remain constant, the corrosion rate of steel in concrete is inversely proportional to the pore water pH and directly proportional to the oxygen and chloride ion content at the surface of the embedded steel. Movements of ions and gases in concrete are many orders of magnitude slower than that observed in seawater. Often, corrosion of metal embedded in concrete is in cathodic control, with the rate of corrosion limited by oxygen availability.

Although ubiquitous, concrete is a complex mixture of solid materials and voids whose chemistry and geometry greatly influence the corrosion behavior of embedded metals. It is comprised of sand and gravel bound in a cementitious matrix. The sand and gravel contained in concrete are known respectively as fine and coarse aggregate. The cement binder can be portland cement, activated blast furnace slag cement, or blends of portland cement and slag cements. Mineral admixtures may also be added to the cementitious matrix to improve corrosion resistance and as sustainable engineering practice.

In concrete, transport mechanisms frequently dominate the corrosion behavior of embedded metals. The observed corrosion rate is often under cathodic control governed by oxygen migration from the concrete surface to the surface of the embedded reinforcement. Figure 1 (Ref 23) shows key factors that impact the corrosion of metals embedded in concrete. Only the microclimate is included in this graphic; for large structures, macroclimate effects impact the entire structure, not just different structural elements or elements with isolated conditions.

The observed corrosion of steel in concrete often derives from the influence of multiple factors, as shown in Fig. 1, with different processes dominating at different scales. Transport in concrete principally occurs in pores and voids contained within the concrete. The scale of these voids runs from a few nanometers for interparticle spacing to a few millimeters for entrapped air voids. That is a range of almost 7 orders of magnitude and is comparable in range to comparing an object the size of a basketball to another the size of the moon. That range of scales impacts reactant transport and hence corrosion. To properly resolve corrosion issues of reinforcement requires learning why concrete is more than an inert sponge that simply surrounds the metal.

Cement Chemistry

Portland cement is a mixture of silicates and aluminates formed at high temperature in a kiln and collectively known as clinker. The clinker is finely ground with a relatively small amount of a calcium sulfate known as gypsum. The resulting mixture is cementitious and reacts with water through a process known as hydration. Cements employed in concrete are known as hydraulic cements. The reaction products that result from several simultaneous hydration processes form the microstructure of the cement binder in hardened concrete. Blast furnace slags are also used as cements. However, to achieve strength in a reasonable time, they must be activated. Slags can also be blended with portland cement. For a general introduction to concrete, there are several texts available that can be quite helpful to those dealing with corrosion of metals embedded in concrete (Ref 24, 25). For those who wish a more thorough discussion of the chemistries involved, Ref 26 provides a good start. Because the cement microstructure often dominates the transport of corrosion reactants and products, a thorough discussion of structure may be found in Ref 27.

Cement Compounds. The hardened cement paste (HCP) is comprised of three principal solid phases: a calcium-silicate hydrate (C-S-H), calcium hydroxide, and calcium sulfoaluminate hydrate. The C-S-H is an amorphous hydrate with a variable composition. The C-S-H makes up more than 50% of the hydration product solid volume and is the greatest contributor to the strength of the cement. Calcium hydroxide comprises 20 to 25% of the paste solid volume, while the calcium aluminate hydrate volume is 10 to 15%. Because iron can substitute for aluminum, and SO_4^{2-} can be partially or fully replaced by other anions, such as $2OH^-$, CO_3^{2-}, $2Cl^-$, $2NO_3^-$, and so on, the designation AFm is often used to refer to all HCP compositions with the same basic structure. The largest portion of the remaining solid volume comes from unhydrated cement residues (Ref 24).

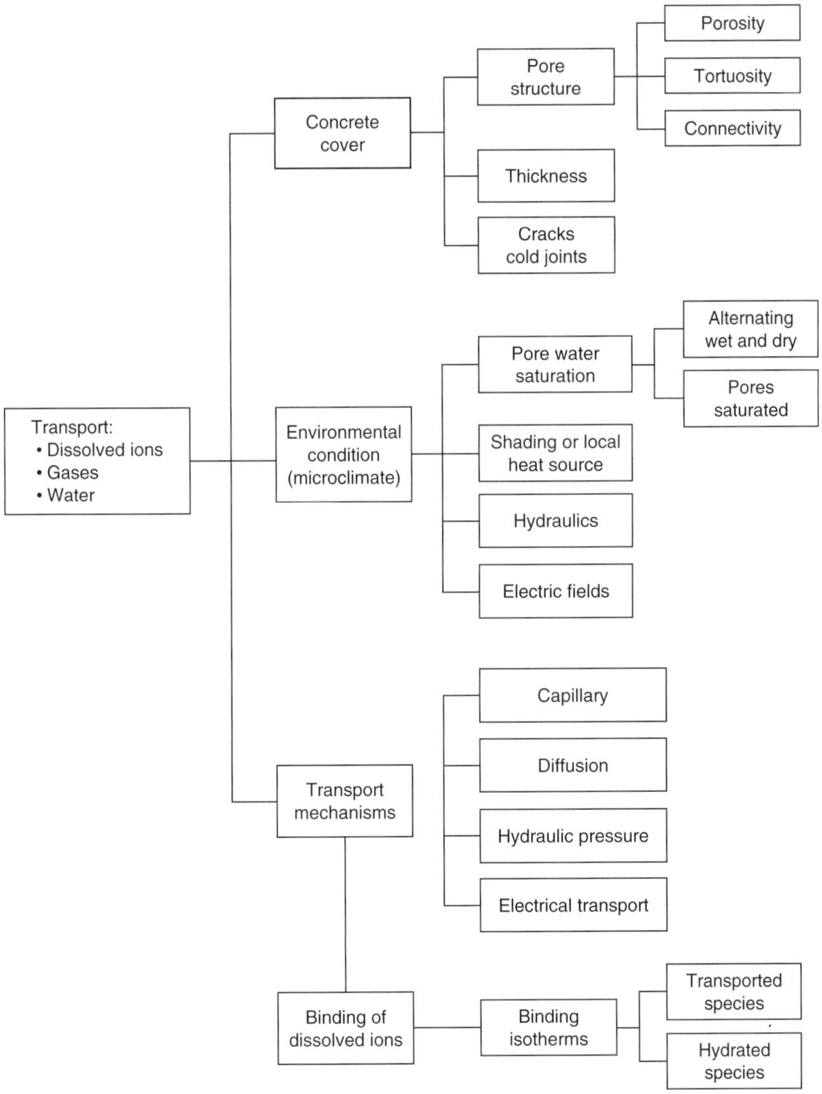

Cement Paste Microstructure. Figures 2(a) through (c) show the HCP microstructure in successively greater magnification. This series of photographs covers 3 orders of magnitude in magnification. As can be seen, pore structure is a major component of HCP microstructure. The approximate range for pore sizes is from 0.1 to 1000 nm. Because most corrosion-related transport in sound concrete occurs in the capillary pores, it can be helpful to make a distinction between capillary pores and gel pores that are present in the C-S-H. One approach is to make a clear break at 10 nm (0.01 μm) (Ref 24, 27, 28), with the capillary pores being those with the larger effective radii. A second approach allows a bit of overlay, with small capillaries overlapping the range from 2.5 to 10 nm (0.0025 to 0.01 μm). The capillary pores are voids in the cement matrix that are remnants of water-filled space or relics from the hydration process. They exist as a random network of irregular pores that are interconnected by varying degrees.

Transport behavior in large capillaries, from 10,000 to 50 nm (10 to 0.05 μm), is not too different than that of bulk water (Ref 24, 25). From 50 to 10 nm (0.05 to 0.01 μm), surface forces within the capillaries start to play a role, and as their size diminishes, these forces can effectively increase water viscosity (Ref 28). Liquid-based transport in the gel pores of the C-S-H, those smaller than 2.5 nm (0.0025 μm), is felt to play only a minor role in corrosion (Ref 29), because the rate of liquid transport is approximately 1000 times slower in gel pores than that occurring in the capillary pores (Ref 30). From a corrosion standpoint, the capillary pores have historically received the greatest interest. They are most involved in the transport of chloride ions and oxygen and therefore have the greatest direct influence on the ability of a concrete to protect embedded steel reinforcement. However, water loss from gel pores can affect shrinkage and cracking. Cracking provides reactants with much wider pathways, which facilitate transport and accelerate corrosion.

The size and distribution of the capillary pores in concrete can change with time and variation in environmental conditions. If the hydration process can be continued, the formation of reaction products can reduce the effective pore size. The relative volume of water used to charge the concrete mix also impacts pore size and

Fig. 1 Factors involved in corrosion of metals embedded in concrete. Adapted from Ref 23

distribution. A common term used to express that volume of water is the water-cement (w/c) ratio. Figure 3 shows the effect of w/c ratio on capillary pore size distribution as measured through mercury intrusion porosimetry (MIP) (Ref 31). Although MIP is not able to provide an absolute measure of pore size, Fig. 3 illustrates that reducing the w/c ratio yields smaller capillaries. A reduction of HCP capillary size will restrict transport of chloride ions and oxygen and thus impede corrosion of embedded steel.

The average size of the capillary pores is but one structural factor that influences the reactant transport of a given concrete. Three other factors are porosity, tortuosity, and connectivity. Porosity is a measure of how much of the nominal volume of a concrete is occupied by the pores and entrained voids. As noted previously, increasing the w/c ratio will increase pore size, hence increase the porosity, and make the concrete more permeable. Capillary pores in concrete do not run in straight lines. Tortuosity describes the convolutions in these pathways and the impact of capillary narrowing. Increasing tortuosity slows the transport of ions and gases and hence can delay the onset of corrosion and slow its progress when initiated. Connectivity is a term that describes how well the pores are interconnected. Low connectivity means the pores are poorly interconnected, which will impede the transport of dissolved gases and ions and slow the onset and rate of corrosion. Conversely, high connectivity means it is easier for reactants to move, and corrosion will initiate sooner and with a higher subsequent rate. In summary, better control of reinforcement corrosion results by lowering both the porosity and connectivity of the concrete and by increasing the tortuosity of the pores.

Interfacial Transition Zone. The microstructure of HCP is modified near inclusions such as aggregates and reinforcing steel (Ref 32–34). This modification extends 20 to 50 μm (0.8 to 2.0 mils) away from the inclusion surface and is called the interfacial transition zone (ITZ). The origin of the ITZ is the wall effect, which reduces the density of small particles near much larger particles and is further enhanced by the effects of localized bleeding. This creates more space adjacent to the inclusion (aggregate, reinforcing steel) that must be filled by hydration products. The ITZ is characterized by enhanced porosity and a greater amount of cement hydration (Fig. 4). Computer simulations have shown that the ITZ around the aggregates can form a percolating network of porosity (Ref 34). The impact of ITZ and percolation should be considered when formulating concrete mix designs for reinforced structures in aggressive environments or where long life is desired.

Reactant Transport in Concrete. As discussed previously, the initiation of corrosion of steel reinforcement in concrete and its subsequent corrosion rate often depend on the transport rates of aggressive ions and oxygen within the concrete matrix. If the concrete contains large voids, rock pockets, or is cracked or delaminated, then capillary transport becomes much less an issue, because the movement of reactants within large defects is orders of magnitude more rapid than in sound concrete. Table 1 shows the transport modes common in concrete.

Depending on the situation, one or more modes may play a role. Simple ionic diffusion is usually the dominant mechanism of transport in the core of concrete structures. In the case of a retaining wall, where the soil behind the wall has a high chloride or sulfate concentration and hydrostatic pressure is present, permeability may be the larger transport mode. Several transport mechanisms may be at work within a vertical piling where lower areas are fully submerged in seawater while upper areas are exposed to the atmosphere. In this case, ionic diffusion may dominate in submerged areas below the tidal zone. Capillary suction may dominate locations above high water, where evaporation of pore

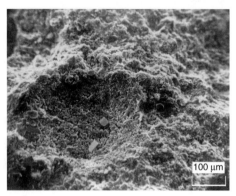

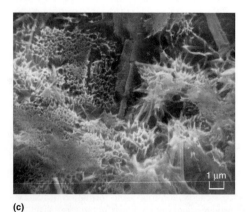

Fig. 2 (a), (b), (c) Fractured surface of hardened cement paste, with increasing magnification

Fig. 3 Influence of water-cement (w/c) ratio on effective pore diameter

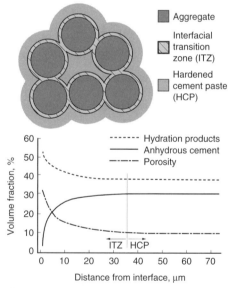

Fig. 4 Schematic view of interfacial transition zone

Table 1 Modes of transport in concrete

Mode of transport	Basic parameter	Description
Permeability	K_p	Bulk flow of water or solution under a pressure head
Ionic diffusion	D	Diffusion of ions under a concentration gradient
Capillary suction	$\gamma/r(a)$	Spontaneous filling of very small pores (wicking)
Ionic mobility	u	Movement of ions under a potential gradient

(a) γ, surface tension of water; r, radius of the pore

water wicks water up from submerged areas or pulls seawater into the concrete after surface splashing. Within the tidal zone, which is between these two areas, a combination of modes is often at work.

pH. It is the high pH of concrete that provides a great deal of the corrosion protection to embedded steels. Figure 5 shows the inverse relationship between the corrosion rate of steel in concrete and pH (Ref 31). The liquid in the capillary pores is not pure water. Low-alkali cements yield a pH of approximately 13, while high-alkali cements can raise the pH to above 13.5 (Ref 24). At the elevated pH present in concrete, traditional low-alloy reinforcing steels, such as ASTM A706 (Ref 35), will develop a dense oxide coating that exhibits passive behavior similar to stainless steels. In fact, the corrosion potential of passive low-alloy reinforcements may be slightly more noble than stainless steel bars embedded in the same concrete. The corrosion inhibition by concrete is discussed in detail in the next section of this article.

Exclusive of its impact on corrosion rates, high alkali content can cause adverse reactions affecting concrete durability. Some aggregates used in concrete are sensitive to alkali content. This is known as alkali-aggregate reaction and can lead to cracking of the concrete, popping, or spalling. High alkali content can also lead to aesthetic issues such as staining of limestones and the formation of powdery deposits on the concrete surface (known as efflorescence).

Additives to Concrete

Although standard concrete mixes provide corrosion protection to concrete, materials are added to concrete to improve its handling characteristics during construction. Other added materials are actual by-products of waste generated in industry or electrical power production and are included as part of sustainable construction efforts. Still other materials are added to improve the ability of a concrete to protect the embedded reinforcement. Often, these additives fill more than one of the aforementioned roles.

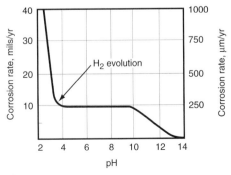

Fig. 5 Influence of pH on corrosion rate

Chemical admixtures are often dissolved or dispersed in the mix water. In the United States, materials are typically selected that meet ASTM C 494 or C 1017 (Ref 36, 37). Included are the following.

Accelerators. These are materials that accelerate setting and strength development in concrete. Included are calcium and sodium chloride, triethanolamine, soluble formats, nitrates, nitrites, and thiocyanates. It is noteworthy from a corrosion standpoint that chloride-bearing accelerators are often used during winter concreting. If chloride ions are included in the initial mix, a substantial fraction will become incorporated into the gel phase of the cement paste and may not be as problematic toward corrosion as are free chloride ions dissolved in the pore water. Also, when chloride ions are included in the initial mix, they will be uniformly distributed throughout the matrix. From a corrosion standpoint, uniform chloride content is typically less problematic than situations where the chloride ion content at the reinforcement surface varies at adjacent locations due to variation in depth of concrete cover or to the presence of voids and cracks.

Air-Entraining Agents. These compounds promote formation of small bubbles in the concrete mix. Generally, these entrained air voids are roughly spherical, with diameters ranging from 0.05 to 1.25 mm (0.002 to 0.05 in.). Air entrainment improves the resistance of a concrete to degradation from freeze-thaw cycles, deicers, sulfate attack, and alkali-aggregate reaction. As little as 1.5 vol% of entrained air can provide protection against freeze-thaw weathering. Air entrainment also provides a low-cost method to improve concrete workability. Improving concrete durability will also enhance the ability of the concrete to protect embedded reinforcement. Improving workability will increase the likelihood that the as-constructed concrete is placed properly, without large voids or rock pockets, and hence improves corrosion protection.

Alkali-Reactivity Reducers. While the high pH of concrete provides protection to embedded ferrous reinforcement, that high alkalinity can attack some aggregates. The presence of alkali oxides, such as Na_2O and K_2O, in cement are what elevate the typical concrete pH above 12.4, which is the value that would be expected for a saturated lime solution, $Ca(OH)_2$. As mentioned, air entrainment can help with alkali reactivity. The hydration processes of some mineral admixtures will reduce the pH. Lithium and barium salts are also employed. Lithium is also used as a humectant in the application of cathodic protection to avoid drying at the anode-concrete interface (see the section "Electrochemical Techniques: Cathodic Protection, Chloride Extraction, and Realkalization" in this article).

Corrosion inhibitors reduce steel corrosion activity. Inhibitors are most commonly introduced to the concrete in the mix water prior to placement. Nitrite salts, particularly calcium nitrite, are overwhelmingly the most widely used and have the largest body of work reported in the technical literature. Other compounds added to the mix water as corrosion inhibitors include sodium benzoate, phosphates, fluosilicates, and fluoaluminates. Topical inhibitors applied to the concrete surface, such as proprietary amines, may provide some corrosion reduction at crack locations. However, in uncracked concrete, it is likely their transport to the steel surface is too slow to provide significant corrosion inhibition within a reasonable time interval. Manufacturers of inhibitors applied to the surface of corrosion-distressed concrete report effective diffusion coefficients for their products that are almost 1 order of magnitude smaller than those observed for chloride ions in diffusion-resistant latex-modified concrete. With normal concretes, the reported effective topical inhibitor diffusion coefficients indicate that the inhibitors would take over ten times as long to reach the surface of the embedded steel as the chloride ions that initiated the corrosion. If it took 20 years for chloride ions to reach the surface of reinforcing steel embedded in normal concrete, it may take almost 200 years for topically applied inhibitors to reach that same steel surface in sound concrete.

Crack Reducers. Concretes with low or ultralow w/c ratios may suffer from cracking if too much water is lost to the surrounding environment. Such high-performance concretes are less readily penetrated by aggressive ions in the bulk than standard concrete. However, if cracks are present, that benefit may be lost, and the resulting cracked structure may be more prone to corrosion. Proper curing can help to reduce cracking. To provide greater protection with crack-prone concretes or installation conditions, crack reducers may provide benefit. Some crack-reducing admixtures lower the surface tension of the pore water. Small, nonmetallic fibers or microfibers can be added to the mix to provide restraint against cracking.

Permeability Reducers. Polymers can be added to the mix water to reduce permeability and otherwise impede the transport of aggressive ions. The most common of these are latex polymers that are used in latex-modified concretes. Water reducers and mineral admixtures can also reduce permeability.

Retarders. These materials retard the setting time of concrete. Included are sugars, citric acid, tartaric acid, borax, and lignins. From a corrosion standpoint, the chief benefit from using retarders is to secure well-placed and uncracked concrete.

Superplasticizers. These products allow the w/c to be reduced by approximately 12% or more while maintaining a flowable concrete. Reducing water reduces porosity, which, if no cracks or placement-related voids occur, is beneficial in controlling corrosion.

Water Reducers. These products differ from superplasticizers in the degree of water reduction allowed. They reduce the water demand by 5% or more and, like superplasticizers, reduce the concrete porosity, thereby impeding transport of aggressive ions, oxygen, and reaction products.

Mineral Admixtures. Several mineral admixtures are added to concrete. They can react with water on their own or may require the action of lime derived from the cement. Concretes made with some of these materials can be both stronger and provide greater resistance to corrosion than normal concretes. In addition, the use of some of these materials can remove what would otherwise be a waste stream from industry and provide cement replacement that reduces greenhouse gas emissions.

Cementitious. These materials have hydraulic properties and react with water on their own to form a binder. They can also be blended with portland cement. Included here are blast furnace slags and natural cements.

Pozzolans are siliceous or aluminosilicate compounds that, on their own, do not have cementing properties. The ancient Greeks and Romans knew about natural pozzolans that, when mixed with hydraulic limes, could be used in building (Ref 24). These naturally occurring pozzolans were volcanic materials. Today (2006), pozzolans added to concrete mixes include diatomaceous earths, opaline cherts, calcined clays, and industrial by-products such as fly ash and silica fume. Pozzolans react with calcium hydroxide to form C-S-H.

Pozzolanic and Cementitious. These combine the properties of the first two mineral additives. Included in this group are high-calcium fly ash and blast furnace slag.

Normally Inert Fillers. These materials are added to improve workability or aesthetics and normally provide little benefit to corrosion control. Included here are marble, dolomite, quartz, and granite.

Pozzolan additions can improve the strength of the concrete. Compared to portland cement, pozzolanic reactions proceed slowly. The slow reaction rate can lead to partial filling of capillary pores. The reaction with lime reduces both the hydroxide ion concentration and the ionic strength of the pore solution. Pozzolan additions, particularly silica fume, can produce a capillary pore network with lowered connectivity and increased tortuosity even if the porosity is not markedly reduced. These pore structure changes impede the ability of aggressive ions, such as chloride, from penetrating the high-performance concrete protecting the reinforcing steel. On the other hand, reducing the number of hydroxide ions will lower the pH and thereby lower the threshold chloride ion concentration required to initiate corrosion (Ref 38).

Aggressive Ions

Several aggressive ions are involved in reinforcement corrosion. Most significant in terms of widely observed damage are chlorides, carbon dioxide, and sulfates. These ions can directly destabilize the protective passive oxide layer on the surface of the steel, or indirectly destabilize that layer by reducing the pore water pH or by attacking the protective concrete and thereby exposing the embedded reinforcement to a more corrosive environment.

Chloride Ions. The passive layer that protects steel in concrete can be destabilized by chloride ions. The threshold at which normal low-alloy steel reinforcement, such as ASTM A706 (Ref 35), will lose passivity depends on pH. Once passivity is lost, the observed corrosion rate is typically most influenced by the permeability of the surrounding concrete to oxygen, the moisture content of the concrete (resistivity), and the ambient temperature. Because of variations in cements, concretes, and local conditions, it is difficult to provide a single, specific chloride ion concentration at the rebar surface that is the global corrosion threshold. Further discussion of pH impact on the influence of chloride ions on reinforcement corrosion is provided in the sections "pH and Corrosion Inhibition" and "Corrosion-Resistant Reinforcement" in this article.

Chloride ion contamination of concrete characteristically follows one or more of four paths. First is contamination from saltwater for reinforced concrete structures exposed to or submerged in marine environments. Second is contamination from chloride ions present in the soils or fill surrounding buried or partially buried structures. Third is from deicers used on highways and bridges. The fourth is from the initial concrete mix, where chloride-bearing accelerators and chloride-contaminated aggregate and/or mix water is used.

For the first three paths, chloride ions enter the concrete at an exposed surface. Depending on the exposure, the chloride ion transport may occur through capillary action, diffusion, or hydraulic effects. For atmospherically exposed concrete, the pores of concrete within 12 to 25 mm (0.5 to 1.0 in.) of the surface may not be saturated with water. It is common for this area to be exposed to alternating periods of wetting and drying. Capillary action is often the dominant transport mechanism in this near-surface area. Deeper into the concrete, the moisture content is generally not as dynamic, so diffusion is normally the dominant transport mechanism. Figure 6 shows a typical profile of chloride ion concentration as a function of depth into atmospherically exposed concrete. Note how capillary action dominates near the surface and, as shown, how washing by the rain can sometimes lead to the highest chloride concentration being a short distance into the concrete rather than at the surface (closest to the chloride ion source).

At depth, diffusion is shown to dominate in Fig. 6. Binding between the chloride ions and cement hydration products can affect the rate of chloride ion transport. That said, observations by numerous works show that after the first 12 to 25 mm (0.5 to 1.0 in.), the chloride profiles fit well with diffusion models. In such efforts, an effective diffusion coefficient is determined, rather than deriving the actual diffusion coefficient and making allowance for sorption-desorption isotherms.

Static conditions are rarely observed in reinforced-concrete structures. Most efforts to fit profiles of chloride content versus depth from the concrete surface employ Fick's second law:

$$\frac{\partial C}{\partial t} = D \frac{\partial^2 C}{\partial x^2} \qquad (Eq\ 1)$$

where C is the chloride ion concentration at some distance x below the concrete or boundary surface at some time t, and D is the diffusion coefficient. In actual practice, the effective diffusion coefficient, D_c, would replace D. Integral solutions to Eq 1 that employ the error function are commonly used to calculate the effective diffusion coefficient, D_c (Ref 39). However, if the calculated diffusion coefficient is used to estimate time to corrosion, models should allow for the fact that the embedded reinforcing steel is a nonporous barrier in a porous matrix and will block diffusional flow. Under such restraint, the chloride ion concentration at the reinforcement surface will increase more rapidly than would be the case with unimpeded diffusion at similar depth in the bulk concrete (Ref 40).

Carbonation. Carbon dioxide is a normal constituent of the air surrounding reinforced-concrete structures. It readily dissolves in water to produce carbonic acid. Rain water typically contains sufficient carbon dioxide to reduce its pH to approximately 5.6 (Ref 41). In some areas, industrial sources or burning of vegetation or agriculture can introduce pollutants such as SO_2 and NO_x, which can reduce the pH even further. In 1997, the average field-measured precipitation pH in the Northeast United States was 4.4 (Ref 42), although wet-deposition pH levels in the Northeast, by 2004 averages, continue to be less than 4.7, with occasional readings dropping to under 4 (Ref 43–46).

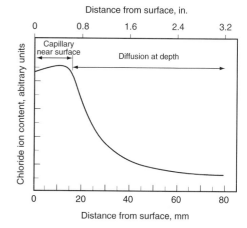

Fig. 6 Chloride profile showing capillary effects near the surface and diffusion at depth

Carbonation can influence corrosion of steel reinforcement in several ways. First, carbon dioxide will react with lime, forming insoluble calcium carbonate. This reaction lowers the pore water pH, and that reduction can destabilize passivation. Precipitation of calcium carbonate can self-limit carbonation by effectively blocking pores in the cement. In newer concretes with relatively lower w/c ratios and pores with smaller diameters, carbonation may be limited to just a few millimeters from the concrete surface. In older, high-w/c-ratio concretes with relatively larger pore diameters, calcium carbonate precipitation may not be able to limit subsequent carbonation penetration.

Sulfate Reactions. Trisulfoaluminate (AFt) is formed by the reaction of calcium aluminate with gypsum during early hydration. As with AFm, iron can substitute for aluminum in AFt, also known as ettringite. When the sulfate present during initial hydration is exhausted, further hydration of the aluminates/ferrites converts AFt to AFm. If, after the cement hardening is complete, there is additional reaction with calcium sulfate, AFm can convert back to AFt. An increase in solid volume of approximately 55% attends the conversion of AFm to Aft, a volume change that produces internal stresses that can eventually crack the concrete (Ref 24).

In some areas, the soil or water has a high sulfate concentration. This can cause problems if reinforced concrete abuts such soils and if conditions are favorable to wick water through the overlying concrete. Delayed ettringite formation is another process that can also crack the concrete. The resulting cracks typically provide easier access of corrodants to the reinforcement than sound concrete. If degradation continues, the embedded steel may become directly exposed to the surrounding environment. Increases in corrosion would be expected in both situations.

Compared to submerged or buried exposure, there appears significantly less published literature discussion of aerosol sulfate attack on concrete for atmospheric SO_2 (Ref 47). In boldly exposed areas, it appears that reaction products from SO_2 attack will wash away (Ref 48). However, it has been shown that atmospheric SO_2 can lead to the formation of expansive AFt reaction products and degradation of portland cement paste (Ref 47). The degradation of the cement and concrete can lead to increased embedded reinforcement corrosion.

Electrical Resistivity. Corrosion of steel in concrete often involves discrete anodic and cathodic areas. Given that situation, the ability of a concrete to conduct electricity can play a significant role in regulating the rate of embedded reinforcement corrosion. Table 2 outlines a general relationship between concrete resistivity and the corrosion rate of embedded steel.

Additions of some mineral admixtures, such as silica fume, can produce severalfold increases in the resistivity of the resulting concrete. Reducing the w/c ratio also tends to increase concrete resistivity.

pH and Corrosion Inhibition

Lianfang Li, W.R. Grace & Co.

The corrosion rate of steel in concrete is greatly influenced by the pH of the pore water. The impact of pH on the corrosion process is discussed as follows in some detail.

pH of Concrete Pore Water

Concrete is a porous composite material. Its capillary pores, depending on the environmental moisture condition, are partially or fully filled with a liquid called pore water or pore solution. Interestingly, the pH of concrete pore water most often is not controlled by the calcium hydroxide that is abundant in cement paste (the pH of a saturated solution of $Ca(OH)_2$ being ~12.6 at room temperature) but by the small quantity of alkalis present in the cement clinker (normally less than 1% by weight of cement). Those alkalis are expressed analytically in terms of the equivalent K_2O and Na_2O but are actually present either as alkali sulfates or alkali-calcium sulfates or in solid solution within certain compounds of the cement. Much of the alkali content, especially that present as alkali sulfate, is readily soluble, following a solubility process that can be expressed as (Ref 50).

$$A_2SO_4 + Ca(OH)_2 \rightarrow CaSO_4 + 2AOH \quad (Eq\ 2)$$

where A represents either a sodium (Na) or potassium (K) ion. As a result of the aforementioned reaction, a relatively large amount of Na^+, K^+, and OH^- ions in the hydrated cement paste is released into the pore water. Figure 7 (Ref 51–60) plots the pore water pH as a function of the cement equivalent Na_2O_e content ($Na_2O_e\% = Na_2O\% + 0.66K_2O\%$). It is obvious that pore water pH increases as cement $Na_2O_e\%$ increases. For a cement with Na_2O_e of 0.6%, the limit specified by ASTM C 150 (Ref 61), the expected pore water pH is approximately 13.5. Nevertheless, in addition to cement alkali content, other factors, such as concrete w/c ratio, pozzolanic additions, concrete admixtures, and so on, may also strongly affect pore water pH.

Table 2 Concrete resistivity versus relative corrosion rate

Concrete resistivity, $\Omega \cdot cm$	Relative corrosion rate
<5000	Very high
5000–10,000	High
10,000–20,000	Moderate
>20,000	Low

Source: Ref 49

Chloride Threshold

It is well known that reinforcing steel (rebar) embedded in alkaline, chloride-free concrete exhibits passivity. Under this passive condition, the corrosion rate of steel is negligible. However, chlorides can penetrate into concrete structures that are exposed to deicing salts or a marine environment. When the chlorides at the steel reach a threshold level (often called the chloride threshold), the passive film on the steel surface is disrupted, and active corrosion initiates. The chloride threshold level is commonly presented as a concrete total chloride content, expressed either as a percentage of the weight of cement or concrete, or in kilograms of chlorides per cubic meter of concrete. Nevertheless, a unique value of the chloride threshold for steel in concrete does not exist. Glass and Buenfeld (Ref 62) reported threshold values for carbon steel in the range of 0.17 to 2.5% by weight of cement for actual concrete structures. This large variability can be attributed to many factors, some of which are difficult to identify. Stainless steel reinforcement has substantially higher chloride thresholds (Ref 63, 64).

Effect of pH on Chloride Threshold. The beneficial effect of hydroxide ions in enhancing the resistance to chloride-induced corrosion of carbon steel in alkaline media is widely understood. Based on several published studies, the relationship between the threshold chloride concentration and the pH of alkaline solutions simulating concrete pore water is presented in Fig. 8 (Ref 65–68). It is apparent that the threshold chloride level increased monotonically with the increase of solution pH for all of these cases, irrespective of differences in the solution

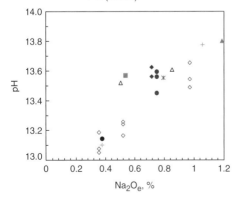

Fig. 7 Relationship between pore water pH and cement equivalent, Na_2O_e, content

and carbon steel chemical composition, specimen size and surface finish, testing procedures, and criteria employed to define the threshold. For that reason, the following equation has been proposed to relate the carbon steel threshold chloride level to the solution pH:

$$pH = n \log[Cl^-] + K \quad (Eq\ 3)$$

where n and K are constants, and [Cl$^-$] is the carbon steel corrosion threshold chloride concentration. Typical values of n are ~0.5 to 1.0; however, values much higher than 1.0 have also been reported (Ref 62).

In comparison, studies addressing the relationship between chloride threshold and concrete pore water pH are scarce, partly because of the technical difficulty involved in accurately measuring the pore water pH at the rebar/concrete interface, especially at the time of corrosion initiation. A recent study conducted at Florida Atlantic University (Ref 69) compared the time to corrosion of ASTM G 109-type concrete specimens (Ref 70) using cements of different alkali contents (Fig. 9). The average bulk concrete pore water pH, determined by an ex situ leaching method (Ref 60), was approximately 13.6 for Na$_2$O$_e$ of 0.97% and 13.2 for Na$_2$O$_e$ of 0.52%. It can be seen that at w/c = 0.41, the time for 50% (two out of four) of the carbon steel specimens to start active corrosion was ~1300 days for pH = 13.6 concrete but only approximately 200 days for pH = 13.2 concrete, suggesting the beneficial effect of using high-alkalinity cement.

Applications

In view of the existing evidence, it can be concluded that the chloride threshold of carbon steel rebar in concrete can be elevated by raising the pH of concrete pore water. In other words, the corrosion initiation time of steel in concrete can be extended by increasing the concrete alkalinity. This can be achieved with minimum cost by either using cements of high alkalinity or externally adding sodium or potassium hydroxide into the mixing water while making the concrete. Approximately 0.5 kg NaOH is needed to raise cement alkalinity by 0.1% for concrete with a cement factor of 400 kg/m^3. Consider the case where concrete has the following properties: surface chloride content is 5% by weight of cement, the effective diffusion coefficient is 10^{-8} cm^2/s, and the threshold chloride is 0.5% by weight of cement. If it is assumed that the chloride threshold is doubled when the cement alkalinity is doubled, calculation based on Fick's diffusion law suggests that the corrosion initiation time for carbon steel with a 7.5 cm (3.0 in.) concrete cover will be extended by approximately 20 years as a result of cement alkalinity increase.

The previous estimation does not take into account the possible loss of hydroxide ions due to leaching and/or carbonation. Values of diffusion coefficients for OH$^-$ ions in concrete are scarce, but it is expected to be on the same order of magnitude as that of chlorides for the same type of concrete (Ref 71). Therefore, it is recommended that concrete should be of high quality whenever high-alkali cement is selected for corrosion purposes. As an example, comparing the time-to-corrosion values shown in Fig. 9, concrete with Na$_2$O$_e$ of 0.97% will outperform that with Na$_2$O$_e$ of 0.52% at w/c = 0.41 but not at w/c = 0.50. Another issue related to the use of high-alkali cement is that it increases the risk of concrete expansion due to alkali-silica reaction (ASR) (Ref 72). Widespread occurrence of ASR problems has been reported within the United States (Ref 73). Therefore, before high-alkali cement is used, both fine and coarse aggregates should be carefully examined for alkali-silica reactivity by following testing procedures as specified in ASTM C 1293 (Ref 74) and/or ASTM C 1260 (Ref 75). Other issues related to the use of high-alkali cement are that it may affect the behavior of concrete admixtures and may also undesirably change the properties of fresh concrete.

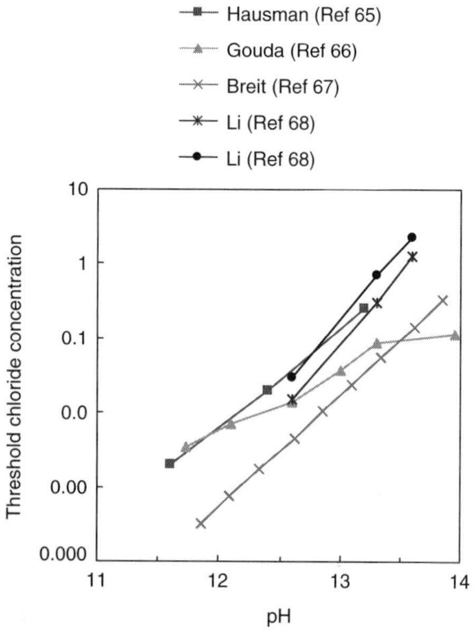

Modes of Reinforcement Corrosion

Jack Tinnea, Tinnea & Associates, LLC

Corrosion in concrete is often visualized as a pothole on the deck of a bridge. This is a common highway corrosion failure mode in areas with frequent application of roadway deicers. Spalls and delaminations are also commonly encountered in marine areas. Localized corrosion of steel in concrete can also occur and is often not recognized prior to failure. Spalls, potholes, and delaminations are not a trivial maintenance issue, but they rarely lead to sudden failure. In part, this is due to the fact that they are easily identified, either visually or through sounding techniques. Localized corrosion can lead to catastrophic failure. Localized corrosion of reinforcing steels typically occurs at locations with abrupt changes in the environment surrounding the rebar. This can occur at cracks in the concrete, within posttensioning ducts, at locations where reinforcement exits the concrete, or in areas of high stress.

General Corrosion

When aggressive ions reach the surface of the reinforcement, or if pH is reduced, the steel loses its passive protection and corrodes. The resulting corrosion product occupies greater volume than the parent metal, with variations due to oxidation state and amount of water included in the final product. The corrosion in concrete generally follows a two-step process, shown in Eq 4 and 5:

$$Fe^0 \rightarrow Fe^{2+} + 2e^- \text{(ferrous)} \quad (Eq\ 4)$$

$$Fe^{2+} \rightarrow Fe^{3+} + 1e^- \text{(ferric)} \quad (Eq\ 5)$$

In oxygenated concrete, the insoluble corrosion products collect at the reinforcement-concrete surface. Hydration also often occurs. As Fig. 10 shows, these hydrated corrosion products occupy significantly more volume than the parent metal. This increase in volume, which is restrained by the presence of the rebar, leads to the development of tensile hoop stresses in the surrounding concrete (Fig. 11a). Concrete is quite strong in compression but has tensile strength that typically runs only 10 to 15% of its compressive strength. Because of this limited

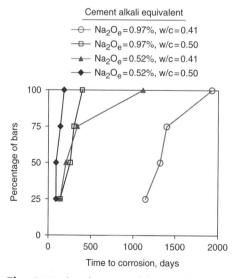

Fig. 8 Relationship between threshold chloride concentration and pH of alkaline solution

Fig. 9 Number of specimens failed as a function of chloride ponding time. w/c, water-cement ratio. Adapted from Ref 66

tensile strength, the concrete cracks to relieve the tensile hoop stresses generated by the expansive corrosion products. The failure mode depends on the strength of the concrete, the depth of concrete cover over the reinforcement, and the spacing between the bars. If there is shallow concrete cover and/or wide spacing of the reinforcement, then vertical cracks form to relieve the corrosion-initiated stress (Fig. 11b). When the reinforcement is more closely spaced or is overlain with a greater depth of concrete cover, the stress-relieving cracks will run through the concrete from bar to bar, forming what is known as a delamination plane (Fig. 11c). Continued general corrosion of the reinforcement, or mechanical action such as from traffic, can dislodge some or all of the delaminated concrete, creating a spall or pothole.

Often, section loss of the steel reinforcement is quite modest until the loss of concrete bond and/or cover. Therefore, the initial problems associated with general corrosion of steel embedded in concrete derive from damage of the concrete rather than a reduction of reinforcement cross section. Some corrosion-related problems, such as cracking, rust staining, or spalls, are visually obvious. Delaminations, which may not be visually apparent, are typically identified through traditional sounding techniques.

Localized Corrosion

Routine inspections of reinforced concrete structures employ visual inspection and/or sounding techniques to identify reinforcement corrosion. These techniques are good at identifying situations where stress from expansive corrosion products has cracked, delaminated, or spalled the concrete covering the distressed reinforcement. These standard techniques are not similarly effective with situations where localized corrosion has severely damaged reinforcing steel without spalling or delaminating the concrete. Localized corrosion can occur with both conventional reinforcement and prestressing steel. This section addresses localized corrosion involving conventional reinforcement. Localized corrosion is also discussed in the sections "Prestressing Steel" and "Posttensioned Grouted Tendons" in this article.

Cracks in concrete can come from a number of sources other than corrosion. Early loss of moisture can lead to shrinkage cracking. Differential settlement of piling can crack pile caps and beams. The presence of such cracks can disrupt the normal protection afforded by concrete and allow corrosion to occur where the crack intersects the reinforcement. Here, the corrosion activity is localized and involves what are known as macrocells. Other cracklike crevices, transitions, or cold joints can also lead to localized corrosion of embedded reinforcement. Cold joints are the result of common construction practice where the concrete is placed in immediately adjacent areas. Transport of chlorides or other aggressive ions to the steel surface occurs much faster at such locations than in the bulk concrete.

Several authors have looked at localized/macrocell/crevice corrosion from laboratory, theoretical, and modeling paradigms (Ref 76–81). Localized corrosion of concrete reinforcement has frequently been observed in field situations (Ref 82). Several examples of those situations are discussed as follows.

Localized Corrosion—Adjacent Cathode. The substructure and supports of a bridge were suffering significant corrosion damage as the result of chloride ion contamination from both deicing salts and the close proximity to the ocean. The continuing chloride ion exposure eventually led to corrosion of the reinforcing steel. In most areas, the chloride ions diffused from the surface to the bar, and corrosion activity was related to depth of cover and proximity to the ocean. In these cases, the damage that resulted was spalling and delamination of the concrete cover.

Figure 12 shows an elevation view of a beam supporting a bridge where differential settlement had generated a narrow (<0.5 mm, or 0.02 in., wide) crack. It appears that the concrete had cracked early in the service life of the bridge. The bridge is located in an area with significant tidal variation (>3 m, or 10 ft). At extreme high tides or during storms, the crack would be wetted with saltwater splash. The narrow crack wicked chlorides rapidly to the level of the reinforcement. Although there was a rust stain on the concrete surface, sounding the concrete with a hammer did not indicate a problem, that is, a hollow sound indicating delamination, so initially no repairs were scheduled. The rust staining was attributed to a tie wire or other trivial surface artifact. Measurements of the corrosion potential of embedded steel in the vicinity of the crack showed rapid change of potential (>150 mV) within a short distance (<150 mm, or 6 in.). An abrupt change in the corrosion potential of reinforcement in the vicinity of cracks, cold joints, or even in apparently sound concrete is often indicative of severe corrosion of embedded reinforcing steel. The area was excavated, and all of the lower bars in the pile cap supporting the deck, and shown in Fig. 12, were severely corroded (>50% loss of cross section) or had fully failed. Figure 13 shows a closeup of one of the bars. Figure 14 shows a similar failure of a stirrup at a location where early cracking of the beam occurred. All the stirrups on one side

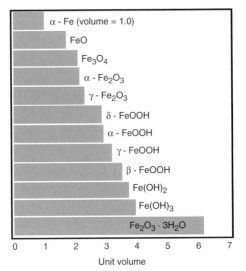

Fig. 10 Comparison of the unit volumes of steel and its corrosion products

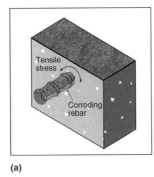
(a)

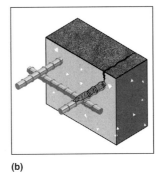
(b)

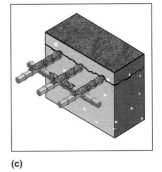
(c)

Fig. 11 Expansive corrosion products from rebar generate stress in concrete. (a) Tensile hoop stresses develop adjacent to the corroded steel. (b) Wide bar spacing and/or shallow cover favor formation of vertical cracks to relieve the stress created by the expansive corrosion products. (c) Close bar spacing and/or deep cover favor formation of a delamination plane to relieve the stress created by expansive corrosion products.

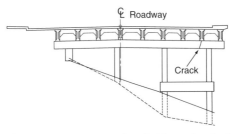

Fig. 12 Elevation view of cracked beam. Localized corrosion cut all the rebars crossing the crack, including those at the beam bottom.

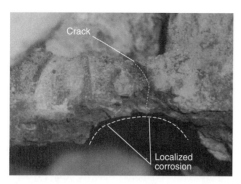

Fig. 13 Closeup of reinforcing bar that suffered full-section failure. Approximately three-fourths of this bar was fully corroded, with the remaining one-fourth cracked.

Fig. 14 Closeup of a fully failed pile cap stirrup. Note that immediately above the severely corroded area, the bar shows no sign of corrosion. The upper area was in a passive condition and served as the cathode for corrosion of the area just below. Close interval mapping of reinforcement corrosion potential can identify these situations.

of the beam suffered >50% section loss due to localized corrosion.

In both cases, the pile cap and the beam stirrups, the cracks had high average times of wetness. This fact kept the chloride ions available for corrosion activation, iron transport, and hydrolysis. As a consequence, soluble corrosion ferrous ions (Eq 4) were transported from the steel surface. A schematic of the corrosion process is shown in Fig. 15. Here, iron from the steel reinforcement goes into solution as Fe^{2+}. The ferrous ion is able to move away from the surface of the parent steel as a soluble hydroxide or as a chloride complex. Completion of the oxidation process, and formation of voluminous red rust corrosion products (Eq 5), occurs either at the concrete surface (the crack opening) or in concrete pores ($Fe^{2+} \rightarrow Fe^{3+}$). The differential in chloride concentrations allows steel exposed to the crack to be the anode and metal exposed to the adjacent chloride-free concrete to be the cathode.

Being able to transport reduced iron, or its chloride complexes, from the corroding surface allows entrained air and voids in the concrete to provide locations that can accommodate the formation of relatively large red rust hydrated corrosion product. This dispersal of the corrosion products avoids developing tensile hoop stresses at the reinforcement surface and subsequent cracking, spalling, and delamination of the concrete.

In both of the cases shown in Fig. 14 and 15, traditional sounding techniques did not indicate a corrosion problem. Close-interval (150 mm, or 6 in.) measurement of the embedded reinforcement corrosion potential clearly indicated abrupt changes in potential and provided a good indicator that localized corrosion was present.

Localized Corrosion—Remote Cathode. A situation similar to the bridge cap beam can occur on retaining walls and tunnels. Figure 16 shows a localized corrosion failure that occurred at a cold joint in a retaining wall with a double mat of reinforcement. One face of the retaining wall was exposed to chloride-contaminated fill, while the other face was exposed to the atmosphere. Chloride ions migrated along the cold joint to the layer of reinforcement closest to the fill and became the anode in the distinct macrocell. The cathode was the mat of reinforcing steel that was closest to the atmospherically exposed surface and had a higher oxygen concentration.

A concrete construction cold joint may be thought of as a very narrow crack. Although it is a narrow crack, it is a much less encumbered chemical transport path than is the much more tortuous passageway through the cement pore network provided by an equal thickness of bulk concrete. Generally, it is desirable to avoid construction joints in areas where corrosion accelerators are present (Ref 83). However, budgets and other considerations may result in the presence of construction joints where corrosive materials, such as chloride- or high-sulfate-bearing soils, can enter the joint.

In cases of tunnels or retaining walls, the cathode can be at a distance from the anode in a macrocell configuration. With both adjacent and remote cathode configurations, time of wetness at the anode significantly affects the observed corrosion. In the confined area, particularly cracks or cold joints, aggressive ions can build up, and hydrolysis reactions can produce a significant drop in pH. With a remote cathode, the effective cathode area is much larger than the anode, and oxygen transport to that area may play less of a role in controlling the corrosion rate than concrete resistivity. The resistivity of the concrete can attenuate the effective cathode, and a relatively low-resistivity concrete can, under comparable circumstances,

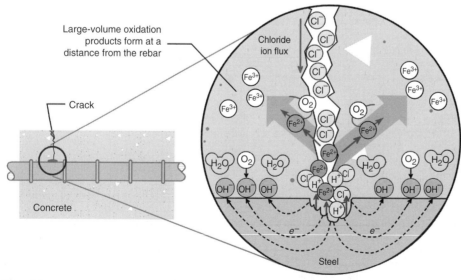

Fig. 15 Schematic of localized corrosion with adjacent cathode

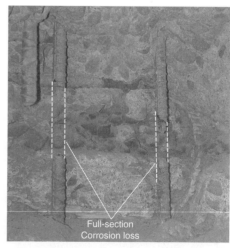

Fig. 16 Localized full-section corrosion failure with remote cathode

provide a larger cathode than a high-resistivity concrete.

Prestressing Steel

William H. Hartt, Florida Atlantic University

Prestressed concrete bridge construction uses high-strength steel as either solid bars or, more commonly, seven-wire spirally wound strand. The underlying principle is that tensile forces in the steel, which are applied either before or after concrete placement and setting, depending on the prestressing method, impart compressive stresses to the concrete upon force release. Presence of a compressive stress state in the concrete, which is relatively brittle, enhances tensile load-bearing capacity. Because of reduced material cost, structural efficiency, and a more streamlined appearance, more than 50% of all new bridge construction now uses prestressing. Additional insight into the principles of prestressed concrete construction can be obtained from standard texts in the field (Ref 84).

Types of Prestressed Concrete Construction

There are two general categories of prestressed concrete construction: pre- and posttensioned. For the first (pretensioned), the steel is tensioned prior to concrete pouring, with the steel ends being released subsequent to concrete setting. As such, the prestressing is bonded directly to the concrete. Production of pretensioned concrete takes place at a central fabrication yard and involves standardized components such as pilings and I-beams that are subsequently transported to the jobsite. Posttensioning, on the other hand, requires that the steel is contained within plastic ducts that are positioned within the concrete formwork at the construction site. (Historically, metallic and even paper ducts have been employed in posttensioned construction; however, these have been discontinued because of corrosion in the former case and inadequate protection from inwardly migrating aggressive species in the latter.) These ducts are tensioned and anchored subsequent to concrete setting. As such, it is the duct and not the posttensioned steel that is bonded to the concrete. Ducts are grouted subsequent to tensioning to provide added corrosion protection to the strand. Earlier bridge construction methods employed an inhibited grease.

Within the past two decades, segmental posttensioning, which is a modification to conventional posttensioning, has increasingly been employed. Here, multiple strands are contained within a duct, and the system is extended through multiple precast hollow roadway or column concrete sections. Upon tensioning, the segments are squeezed together, providing structural integrity to the assembly.

Categories of Prestressing Steel

Prestressing steel strand normally conforms to ASTM standard specification A 416/A 416M (Ref 85). This specification covers two types and two grades of uncoated strand. The types are normal and low relaxation, where only thermal treatment (stress relief) is employed for the former and continuous thermal-mechanical treatment for the latter. Low-relaxation strand, for which a maximum relaxation loss of 2.5% is specified during 1000 h at a load of 70% of the specified minimum breaking strength, is the industry standard, because any relaxation after tensioning reduces compressive stress in the concrete. The grades are grade 1725 (250) and grade 1860 (270), where the numerical designation is the minimum ultimate tensile strength in Système Internationale (SI), 1725 and 1860 MPa and customary units (250 and 270 ksi). No upper strength limit is specified; while there are no reports of this being a problem for strand, excessive strength has led to premature failures of prestressed concrete cylinder pipe, which is wire-wound (Ref 86). (The failures were attributed to dynamic strain aging. This was apparent as wire splitting, which led to loss of prestress and pipe rupture and reflected the fact that brittle fracture and environmental cracking tendency generally increase with increasing strength.) Strands must necessarily be of relatively high strength and support high loads if meaningful compressive stresses are to be imparted to the concrete, because cross-sectional area for the strand is small compared to that of the concrete.

The ASTM A 416/A 416M specification (Ref 85) requires that the wire from which strand is fabricated be drawn carbon steel and that the final product (strand) possess specified dimensions and a requisite minimum strength. As such, steel composition is not addressed; however, attainment of the required mechanical properties using a carbon steel is most reliably achieved with a near-eutectoid composition (0.77 wt% C). Quenched and tempered steel is precluded because of its relatively brittle nature and susceptibility to environmental cracking.

A recent specification, ASTM A 882, has been published for filled, fusion-bonded, epoxy-coated prestressing strand (Ref 87) that otherwise conforms to ASTM A 416. The coating is applied by the electrostatic deposition method and is provided with or without embedded grit to enhance the bond to concrete. Because of the inherent viscoelastic nature of epoxies, a maximum relaxation loss of 6.5% is allowed, compared to 2.5% for the uncoated (bare) strand. Caution must be exercised, because elevated temperatures could cause coating softening and result in loss of prestress transfer. The specification requires the epoxy to be capable of reaching 66 °C (150 °F) without reducing transfer.

Performance of Prestressing Steel in Concrete Highway Structures

Corrosion of reinforcing steel, both conventional and prestressed, as a consequence of either marine exposure or application in northern climates of chloride-based deicing salts (or both) has evolved to become a formidable technical, societal, and economic problem in North America and elsewhere. Such damage is more critical for prestressing than conventional reinforcement, because higher stresses are present in the former. Consequently, a relatively small amount of corrosion on a single wire can cause its fracture. This then transfers load to the remaining six wires of the strand, which, in combination with further corrosion, can lead to fracture of the strand and load transference to adjacent strands.

Corrosion of pretensioned strand in concrete normally proceeds by the same mechanism as for conventional reinforcement and involves progressive migration of chlorides into the concrete until a critical concentration is reached at the steel depth. At this time, passivity is compromised, and corrosion initiates. This is followed by accumulation of solid corrosion products about the steel, which leads to concrete cracking and spalling. As for conventionally reinforced steel, methods for corrosion assessment, such as measurement of potential and polarization resistance, apply. In the case of posttensioning, corrosion damage results from water intrusion into the duct or end anchorage, irrespective of the presence of chlorides. Such intrusion can occur during construction when the posttensioning is exposed or subsequently as a consequence of leakage. For this reason, sound grouting materials and practices and protection of end anchorages are critical. Tendon failures on segmental bridges as a consequence of strand corrosion from excessive bleed water have recently been reported (Ref 88). This, in turn, has led to specification of thixotropic grouts. A point of concern for posttensioning is that the conventional corrosion assessment methods for pretensioning are not applicable, because the strands are electrically shielded by the duct. Techniques based on impulse radar, high-energy linear accelerator, impact echo (Ref 89, 90), magnetic flux leakage (Ref 91), and vibration testing (Ref 90) (the latter is applicable to external tendons only) can be employed for situations where access is restricted by expense or design.

The possibility of hydrogen embrittlement as a consequence of corrosion is also a point of concern. Hydrogen ions are produced by the reaction:

$$Fe + 2H_2O \rightarrow Fe(OH)_2 + 2H^+ \qquad (Eq\ 6)$$

with subsequent formation of adsorbed atomic hydrogen, H_{ads}, according to the reduction reaction:

$$H^+ + e^- \rightarrow H_{ads} \quad \text{(Eq 7)}$$

Alternatively, in deaerated regions or with excessive cathodic polarization, the reaction:

$$H_2O + e^- \rightarrow H_{ads} + OH^- \quad \text{(Eq 8)}$$

occurs. The H_{ads} product has two reaction paths available:

$$H_{ads} + H_{ads} \rightarrow H_2$$

or

$$H_{ads} \rightarrow H_{dis} \quad \text{(Eq 9)}$$

where H_{dis} is hydrogen dissolved in the steel that contributes to embrittlement.

Cathodic protection (CP) is the only proven long-term service method controlling ongoing corrosion of steel in chloride-contaminated concrete. Because of the relatively high resistivity of concrete and the need for a high driving voltage, most CP systems for reinforcement have been of the impressed-current type. However, concern with overprotection and hydrogen generation (Eq 8) in some areas has largely precluded application of this technology to prestressed concrete bridge structures. With this in mind, a recent state-of-the-art report (Ref 92) has addressed protection criteria for prestressed concrete. Galvanic anode (sacrificial) CP has been successfully employed in warm marine applications, but performance is often insufficient in colder, drier climates. Improved anodes (Ref 93) and application of humectants (Ref 94) are being developed to overcome this poor performance.

Posttensioned Grouted Tendons

Alberto A. Sagüés,
University of South Florida

Posttensioned (PT) tendons are placed through a concrete structure after the concrete is hardened, and then stretched to provide compressive forces that hold the structure together. The ends of the cables are held by anchorages (Fig. 17) that are usually embedded in the concrete. The anchorages typically have a wedge plate with wedges that hold the stressed cables in place after tension is applied by a hydraulic jack. A tendon usually has multiple (often 19) strands. Each strand is typically made of seven 5 mm (0.2 in.) diameter wires, with a central straight wire and the remaining six in a helical pattern around the center wire. The strands themselves run nearly parallel to each other over most of the length of the tendon, with little or no braiding effect. The wires are usually made of a near-eutectoid composition low-alloy carbon steel, heat treated and cold worked (later stress relieved) to achieve a very high tensile strength (1.86 GPa, or 270 ksi). The microstructure resembles fine pearlite, with grains very elongated in the drawing direction. The wire surface usually has only a thin mill scale. Anchorage bodies are commonly made of cast iron, while the wedge plate and wedges are typically carbon steel.

Tendons are placed in ducts that may be embedded in concrete (internal tendons) or have portions of the tendon in air space between concrete components (external tendons). Ducts can be metallic (galvanized, corrugated, or straight pipe), as is often the case for internal tendons, or polymeric (such as high-density polyethylene (HDPE), commonly used for external tendons. The ducts and anchorages can be filled with cementitious grout or grease. In case of failure of one or more strands, the grout serves a mechanical function by frictionally transferring load to the remaining unbroken strands or the anchorage, as long as there is a sufficient transfer length. Such mechanical transfer is not effective in the case of grease-filled tendons. Staehle and Little (Ref 95) documented corrosion development by fungal action in a greased PT assembly. The focus of this section is on corrosion of PT grouted tendons, in particular related to recent tendon failures in bridge applications.

The tendons concerned are critical structural components carrying forces often greater than 10^6 N (225,000 lbf), and failure of more than one or two nearby tendons in a structure can be catastrophic. Given the intrinsically low corrosion resistance of the tendon steel, control of its immediate environment is essential. Hardened cementitious grout should have high pore water pH (typically >13), very low chloride content (e.g., <100 ppm of the grout mass) to ensure stable steel passivity (Ref 67), and high electric resistivity (e.g., well above 10 k$\Omega \cdot$cm) to minimize adverse galvanic coupling should active corrosion develop.

Instances of PT strand corrosion in segmental bridges took place in the 1980s in Europe (Ref 96), leading to an application moratorium in the United Kingdom for a time. Corrosion of PT strands has received renewed attention since a series of failed tendon incidents occurred in the late 1990s in Florida. Those occurred around the (then) 18 year old Niles Channel Bridge in the Florida Keys (Ref 97, 98), the 7 year old Mid Bay Bridge in the Western panhandle (Ref 99), and the 15 year old Sunshine Skyway Bridge over Tampa Bay (Ref 100). In each of these bridges, the observed damage consisted of one completely cut tendon plus one to several partially detensioned tendons. The completely cut tendons had failed in the anchorage region following severe corrosion of many of the strands, as illustrated in Fig. 18.

Mechanical failure appeared to be the result of simple overload of locally reduced cross section, usually in the form of generalized corrosion along several centimeters of strand length, with occasional pitting. In the partially detensioned tendons, severe corrosion and mechanical failure had affected some but not all of the strands. In one instance, the area of severe corrosion was approximately 2 to 3 m (7 to 10 ft) away from the anchorage. All cases of severe corrosion were associated with large void areas that should have been filled with grout. Significant chloride contamination of the remaining grout was observed at some of the corroded regions. In many cases, the affected anchorages were at bridge expansion joints or underneath horizontal surfaces, where salt and moisture availability was significant. Additionally, HDPE duct cracks were commonly encountered in the Mid Bay Bridge. Corrosion was not unique to any particular anchorage manufacturer.

The available evidence suggests that the key deterioration factor in the Florida incidents was the formation of large grout voids at or near the anchorages. The most likely cause of voids is the development of large amounts of bleed water during the grouting process at the time of construction of the affected bridges (Ref 101, 102). Bleeding is assisted by wicking of water along the strands and vibration resulting from the grout-pumping operations. The bleed water may be subsequently reabsorbed into the grout or lost by evaporation through incomplete anchorage sealing that may result from deficiencies at the pourback on the wedge plate or elsewhere (e.g., duct connections or duct cracking). Thus, some of the steel is not in contact with the bulk of the protective highly alkaline grout that is intended to promote passivity of the steel surface. The region where the steel emerges from the grout-void interface is particularly vulnerable because

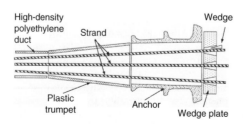

Fig. 17 Idealized schematic of a typical anchorage system. Only three strands shown for clarity. Anchor, wedge plate, and wedges are ferrous alloys.

Fig. 18 Failed strands from corrosion in grout void area of an anchorage. Courtesy of Florida Department of Transportation

of a combination of detrimental factors. The pH of the grout pore water there could be lowered by carbonation from atmospheric carbon dioxide. At the same time, evaporative chloride ion enrichment may occur, elevating the chloride content (that could be either the small native content of the grout or some larger amount from external contamination) to a level sufficient to cause passivity breakdown at the local pH and potential conditions. After initiation, further corrosion may be accelerated through galvanic coupling of the relatively small active region with a larger passive surface consisting of the rest of the strand assembly in the grout, plus other metallic parts of the anchorage. In the absence of free bleed water, the extent of the corrosion macrocell action is expected to depend on the polarization characteristics of the anodic and cathodic zones and on the resistivity of the intervening hardened grout.

The aforementioned corrosion scenario has been supported by preliminary laboratory experiments that showed lowered pH and enhanced chloride content (as well as increased grout porosity) at the grout-void interface of a model anchorage assembly (Ref 102). Adverse galvanic coupling with the anchorage and severe active corrosion were documented for that region. The local corrosion rates were on the order of 10^2 μA/cm^2, sufficient to cause mechanical failure of at least some wires after short periods, for example, 1 year. Fast corrosion failure of an experimental anchorage has been reported elsewhere (Ref 103). Additional evidence of corrosion while bleed water was still present is detailed in Ref 104. These cases suggest that some of the damage in the affected bridges may have been quite advanced early in the service life of the affected bridges. Newly introduced low-bleed, high-w/c-ratio grouts (Ref 105, 106) have shown encouraging performance in preventing void formation. Significant self-desiccation appears to take place during extended curing of those grouts, as indicated by terminal electric resistivity and internal relative humidity values $>10^4$ Ω·cm and $<80\%$, respectively (Ref 107). Active corrosion is difficult to sustain in cementitious media under those conditions. However, initial modeling calculations indicate that if a small active anodic zone still existed at the grout-void interface, macrocell coupling with extended cathodes elsewhere on the strand and anchorage body could still be substantial at moderate-to-high resistivities, given enough oxygen availability (Ref 107–109).

Early detection of PT corrosion is highly desirable but not easy. Even in the case of essentially external tendons, much of the damage can happen within anchorage areas that are normally encased in thick concrete bulkheads. Direct inspection of anchorages can be made by removing the mortar pourback covering the wedge plate and inserting a boroscope through an unused wedge seat into grout voids, should those become observable. This procedure was used extensively to assess the condition of the surface strands exposed at voids (Ref 99) in the Mid Bay Bridge. Open-circuit potential measurements using a reference electrode inserted in the assembly to reveal possible active corrosion, as in ASTM C 876 (Ref 110), are feasible and have been used routinely as well. However, the corrosion may have existed early in the life of the structure, and potentials at the time of inspection may be in the passive range, even if the damage was extensive earlier. A similar situation exists for polarization resistance or electrochemical impedance measurements. Continuing electrochemical testing (either by static potential, electrochemical noise, or transient polarization methods) is, in principle, feasible but has not yet been practically implemented. Nondestructive magnetic assessment of the external portion of tendons using a traveling probe has been conducted routinely in several bridges, but the method is limited to regions away from the anchorage, where severe damage is less common (Ref 111). Radiographic methods to evaluate anchorages are feasible, to some extent, but generally slow and expensive. Pulse-echo and related ultrasonic assessment, with acoustic excitation at individual strands exposed at the back of the wedge plate, has been attempted but with limited success. Macroscopic acoustic methods exist based on monitoring to record individual strand breakages (Ref 111, 112); however, these procedures require continuous data acquisition at fast rates and over the entire service life of the structure, and a dedicated data processing service with consequent cost issues.

Electrical resistance measurements of tendons (end-to-end or at intermediate positions, using a traveling potential contact) have been performed in external tendons, using highly sensitive electric bridge circuitry to reveal partial discontinuities at wire breaks, including those in the hidden anchorage area (Ref 98). This method has shown promising results, but it is time-consuming and requires disruptive placement of electric contacts through the tendon duct. Vibration testing (using either natural traffic excitation or controlled impact) has been used extensively in recent years to determine the state of tension of external tendons. Tendons showing unusually low tension (or unusual disparities of tension at opposite ends of the same tendon) are suspected of having experienced breakage of one or more strands somewhere along the tendon, including end anchorages. The procedure is relatively inexpensive and fast and has been used together with boroscope examination of anchorages to identify suspect units in bridges with large tendon inventories (Ref 98, 113). Vibration tests have correctly identified some tendons with damage in progress, but not always. Corrosion can still remain undetected if grout and friction permit efficient transfer of force to adjacent strands, so that the tension at the external portion remains essentially unaffected. Due to the shortcomings of individual inspection methods, a reasonable chance of detecting PT corrosion before complete tendon failure requires a combination of various techniques at present.

The corrosion incidents in Florida highlight the sensitivity of PT systems to design weaknesses and construction quality. As a result of those incidents, the Florida Department of Transportation has developed an extensive set of new design and quality guidelines aimed at control of PT corrosion in new structures (Ref 114). Those guidelines include the use of low-bleed grouts, grout-pumping procedures for prevention of voids, redundant anchorage capping and improved polymer ducts for isolation from the external environment to prevent moisture ingress, and construction detailing to avoid chloride contamination of regions around anchorages.

Related structural components experience some corrosion modalities that are common with those of grouted tendons. A detailed treatment of cable corrosion in bridges and other structures is found in Ref 115. Issues on corrosion of prestressing steel are summarized in Ref 116.

Corrosion Inspection

Jack Tinnea, Tinnea & Associates, LLC

Corrosion inspection of bridges and highways requires an understanding of both civil and corrosion engineering. In the United States, the ownership of bridges is diverse and includes private and government agencies, including groups from local, state, and federal government. Although there is local variation in inspection requirements, at the top of the inspection protocol are the National Bridge Inspection Standards. Inspection typically defers to the American Association of State and Highway Transportation Officials' (AASHTO) *Manual for Condition Evaluation of Bridges* (Ref 117). For structures that include underwater reinforced-concrete components, the book *Underwater Investigations—Standard Practice Manual* (Ref 118) is very helpful. An excellent guide written in Spanish is also available (Ref 119). NACE International is working on a recommended practice for conducting corrosion inspections of reinforced-concrete structures that should be completed in 2006.

Corrosion Condition Surveys

Corrosion condition surveys are a part of standard bridge and highway inspections that should be conducted at regular intervals. Bridges are typically inspected in 24 month intervals and underwater inspections in 60 month intervals. If premature deterioration is suspected, then the time interval may be shortened as required. In some cases, it may be necessary to employ sophisticated equipment and techniques to determine the extent of damage. This discussion highlights corrosion issues that may be present in bridges being inspected. It is important to recognize that proper bridge inspection should be performed by qualified individuals with specific training and professional licensing. As required, corrosion specialists may be involved to assist in defining corrosion conditions and to provide input on appropriate action to resolve corrosion problems.

Bridges and highways include a number of elements that may encounter corrosion problems. Included are:

- Bridge decks and reinforced slabs
- Superstructure and substructure members
- Deck joint elements
- Bearings
- Railings
- Culverts

Metal elements can be painted, galvanized, thermal sprayed, or left bare, such as weathering steels. Reinforced concrete can include reinforcement that is conventional or prestressed. The reinforcement can be bare, galvanized, epoxy coated, cathodically protected, or fabricated from corrosion-resistant materials.

Elements to consider in a corrosion inspection of reinforced-concrete highways and bridge members include:

- Assessment of concrete strength and condition
- Cover thickness survey
- Visual inspection techniques for cracks and rust staining
- Delamination survey
- Corrosion potential measurements and mapping
- Concrete resistivity determination
- Chloride profile determination
- Carbonation testing
- Corrosion rate monitoring
- Other advanced techniques

Elements to consider in a corrosion inspection of the structural steel elements of a bridge include:

- Visual inspection techniques for rust staining
- Assessment of coating condition
- If present, identification of crevice corrosion on bolted joints or hinges

There is such a very wide range of bridge types, structures, and environments that it is not possible to provide a single strategy. However, in all cases, it is important that individuals with extensive experience in the area of highway-related corrosion review the corrosion survey findings and that they discuss the implications of the results with an appropriate structural engineer.

Assessment of Concrete Quality and Cover

An accurate assessment of the concrete of a reinforced-concrete structure and the typical depth by which it covers the reinforcement can provide important information for a corrosion inspection. Generally speaking, good-quality concrete allows slower diffusion of aggressive ions and also slower degradation if corrosion initiates. Figure 19 shows how the quality of concrete can influence the corrosion threshold (Ref 120). Here, two curves are shown that delineate good-quality and bad-quality concrete. Most concretes encountered in field situations fall between these two curves. As can be seen, the chloride content threshold required to initiate corrosion varies both by concrete quality and the humidity of the local environment or microenvironment.

Figure 20 shows a variation on what has become the classical representation of cumulative corrosion damage versus time (Ref 121). A good understanding of this behavior is necessary when relating the findings of a corrosion inspection to the expected service life of a structure. The corrosion initiation phase is the time before the onset of significant reinforcement corrosion. This is the period when aggressive ions, such as chlorides, penetrate the concrete cover and reach the level of the reinforcement, or when the carbonation front reaches the embedded metal surface. This period is shown ending at point t_0, when conditions reach the local corrosion threshold. Note that there may be some initial cracking early in the corrosion initiation phase. These may derive from processes other than corrosion, such as drying shrinkage. If such cracks reach the steel surface, their presence may greatly compress the corrosion initiation phase.

The next time interval shown in Fig. 20 is that of the corrosion propagation phase. Note that after a relatively short nonlinear period immediately after t_0, there is a relatively linear increase in damage with time during most of the remaining propagation phase. Shortly before the point noted as "Corrosion failure," the rate of corrosion damage typically shows a dramatic increase. "Corrosion failure" is assumed to be the point of maximum allowable corrosion damage and is also noted by the time t_F. The point of corrosion failure, t_F, would be the end of the service life with no remediation or repairs.

If maintenance is applied to extend the life of the structure, that maintenance could come either before or after the onset of corrosion, t_0. The point t_{RM} and the damage-versus-time plot that follows reflect reactive maintenance, or that

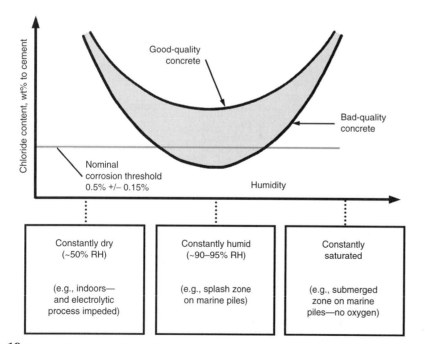

Fig. 19 Impact of concrete quality on chloride content (wt% to cement) corrosion threshold. RH, relative humidity

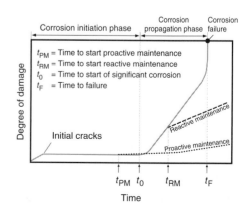

Fig. 20 Plot of cumulative damage versus time, with several maintenance options

occurring after the onset of significant corrosion. The point t_{PM} and the following plot reflect proactive maintenance. For example, if chloride ion contamination was the defining corrosion issue for a structure, electrochemical chloride extraction could be applied either before or after the onset of significant corrosion activity and damage. Proactive maintenance typically will minimize incurred damage.

Visual Inspection and Delamination Survey

A visual inspection can help to identify problem areas. Sounding techniques can help to identify areas of cracked concrete that have not yet spalled. ASTM method D 4580 covers measurement of delaminations by sounding. Visual inspection and sounding techniques can help the corrosion inspection team in directing their efforts. These techniques should be thought of as complementary, with sounding being able to help in identifying distress that is not immediately apparent visually. However, if there are visual indications of corrosion, for example, rust staining, but sounding does not indicate delamination, further examination should be considered. For example, potential mapping may help to clearly understand the situation. If the cracks that make up the delamination are filled with water, sounding techniques may not indicate a problem. Likewise, if the rust stains are indicative of localized corrosion, there may be severe reinforcement section loss but without development of tensile hoop stresses that crack and spall the concrete. In both of these cases, potential mapping may help to more clearly define the situation.

Reinforcement Potentials

Corrosion potentials are employed in evaluating the corrosion condition of a reinforced structure. Figure 21 shows the standard setup used to obtain corrosion potentials of reinforcement in concrete. Figure 21(a) shows a passive noncorroding situation. Figure 21(b) shows chloride ion contamination reaching the level of the reinforcement, and the reading indicates that corrosion is occurring. The absolute value of these potentials can provide some guidance on corrosion activity. Table 3 presents guidelines from ASTM standard test method C 876 on interpreting corrosion potentials of steel in concrete (Ref 110).

In addition to absolute corrosion potentials, patterns seen in maps of close-interval potential readings can reflect structural activity or localized corrosion. A close-interval map obtained on the surface of a floor beam is shown in Fig. 22 (Ref 122). Note that the most negative readings, that is, the values indicative of corrosion, correspond to areas where bending moment, fatigue, or load transfer are occurring. The potential map that appears at the bottom of Fig. 22 essentially reflects a stress map of the floor beam.

Oxygen is a cathodic depolarizer. For reinforcement potentials, this means that if other conditions such as pH, chloride content, and concrete quality remain consistent, then the availability of oxygen will affect the measured potential. In dry concrete, where oxygen has relatively easy access to the steel surface, potentials will be more positive than if that same specimen concrete was fully saturated with water. Full saturation impedes oxygen transport to the steel surface, and, in time, reduced oxygen at the metal surface will shift corrosion potential readings in a negative direction. In tunnels, for example, it is not unusual to encounter readings that ASTM C 876 would indicate reflect a high corrosion probability. However, these relatively negative readings really reflect an absence of oxygen. Figure 23(a) and (b) show how it may be necessary to look at the variation in potential in addition to its absolute value (Ref 123). In Fig. 23(a), the three-dimensional (3-D) map shows readings from a marine piling that was submerged almost all of the time. At this location, sea life and microbes on the pile surface consume essentially all available oxygen. The resulting corrosion potentials, although fairly negative, also show little variation. This is a situation that is characteristic of low oxygen/no oxygen available at the steel surface. The corrosion rate for reinforcing steel in such anaerobic conditions is low, and no corrosion activity or damage was noted in this area of the pile. In Fig. 23(b), the corrosion potentials are much more electropositive, but the 3-D map shows an abrupt change in potential at appproximately the 61 cm (24 in.) location on the length axis. Note in the grid above the 3-D map that the area with the negative potentials is coincident with an area with delaminated concrete cover, indicating problems with reinforcement corrosion. If localized corrosion is suspected, it is helpful to excavate a worst-case area for visual inspection.

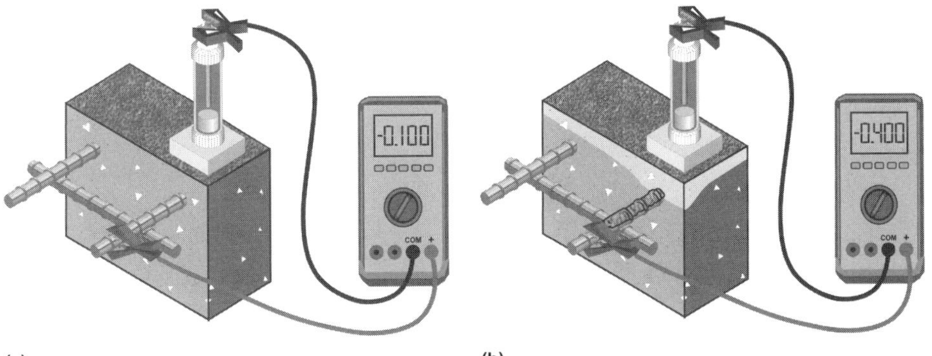

Concrete Resistivity

Electrical resistance plays a major role in corrosion in concrete and in the effectiveness of cathodic protection (CP) if applied in a restoration. The corrosion rate of a corrosion macrocell, or output of a CP system, follows Ohm's law, $E = IR$, where E is output voltage, I is CP current, and R is the total circuit resistance. The total circuit resistance comes from several sources, including:

- Concrete resistivity
- Concrete-CP anode interfacial resistance
- Concrete-reinforcement interfacial resistance
- Resistance between individual elements of the reinforcement matrix

Concrete resistivity, the measure of unit resistance of the concrete, is dynamic and plays a large role in the operation of CP systems applied to reinforced-concrete structures. In addition to providing a means of estimating CP current output, concrete resistivity can provide information on the following properties:

- The amount of moisture contained in the cement pores
- The size and distribution of the cement pores (and microcracks)
- How complex a path the pores follow (tortuosity)
- How well the pores are interconnected

Concrete resistivity is strongly influenced by the quality of the concrete and hence reflects the concrete w/c ratio, cement content, use of mineral and other admixtures, and curing conditions.

Of all the factors that affect concrete resistivity, moisture influence is greatest. A high w/c

Fig. 21 Schematic showing the layout for potential testing. (a) The meter reading of −0.100 V to a copper-copper sulfate electrode reflects a noncorroding condition. (b) The −0.400 V to a copper-copper sulfate electrode reading reflects a corroding condition.

Table 3 Interpretation of corrosion potential data

Reinforcement potential (CSE), V	Corrosion condition
> −0.20	>90% probability of no corrosion
−0.20 to −0.35	Corrosion activity uncertain
< −0.35	>90% probability of corrosion

CSE, copper-copper sulfate electrode

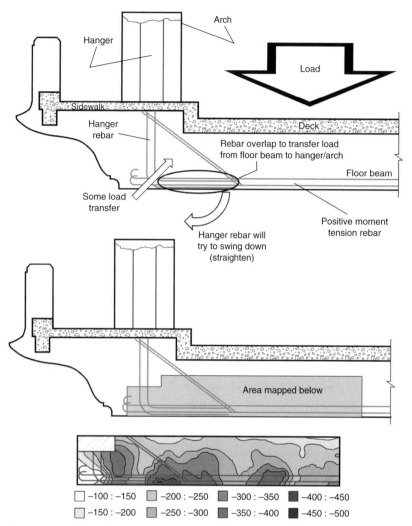

Fig. 22 Close-interval potential map showing the effect of loading and load-transfer fatigue on floor beam reinforcement corrosion

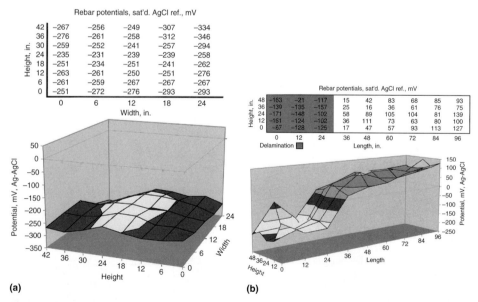

Fig. 23 Map of potential variations. (a) Low potential variation with relatively negative potentials indicates very low oxygen availability and little corrosion. (b) Abrupt variation in potential indicates corrosion activity.

ratio means more and larger pores and a less tortuous path (Ref 31). This, in turn, affords additional room to hold water and generally produces lower resistivities than are observed with low-w/c-ratio mixes. The presence of mobile ions, such as chloride, may have some direct impact on concrete resistivity by increasing the electrical conductivity of the pore water. However, chloride ions also tend to be hygroscopic and improve the capacity of concrete to retain moisture (Ref 124, 125). If other factors are held constant, the addition of chloride ions will reduce hydroxide ion concentration. Given that the specific ion equivalent mobility and conductivity of the replaced hydroxide ion are approximately twice that of the chloride ion, from a conductivity standpoint alone, chloride ion additions would be expected to increase concrete resistivity, not decrease it. Because that is not the generally observed situation, it seems reasonable to expect that hygroscopic action, which increases water retention, more greatly influences measured resistivity than changes in pore water conductivity that result from the addition of chloride.

Electrical resistivity of the concrete is strongly influenced by cement pore structure. It provides a good measure of the permeability and corrosion condition of a concrete. Several authors have examined concrete resistivity as a measure of the corrosion condition (Ref 126–128). Table 4 provides an overview of those suggested relationships.

It must be understood, however, that simply because a concrete resistivity is low does not mean that the reinforcement corrosion rate is high. Following Feliu's (Ref 127) rating system, a low concrete resistivity simply means that concrete resistivity does not seriously impede corrosion. For example, take a situation where reinforcement is embedded in uncontaminated concrete, and the cement pores are fully saturated with water. Because of the moisture content, the concrete resistivity will be low. Further, because of limited oxygen availability, the potential of the reinforcement will be shifted in a negative direction, perhaps to indeterminate values. This does not reflect a corrosive environment but is a reflection of environmental conditions. If this concrete dries, its resistivity will increase, and the potential of the embedded reinforcement will become more positive because of cathodic depolarization from increased oxygen availability.

Chloride and Carbonation Profiles

Concrete samples can be obtained to determine chloride ion penetration or carbonation, if that is suspected. Chloride samples may be obtained as powder through rotary hammer drilling, by cutting cores, or otherwise obtaining samples. If cores or other solid samples are used, the samples are ground prior to testing. Several test methods are available. AASHTO method T260, "Sampling and Testing for Chloride Ion in Concrete and Concrete Raw Materials," is often used. AASHTO method T260 procedure A provides for both acid-soluble and water-soluble chloride content analysis, either by potentiometric titration or ion-selective electrode. Procedure B of T260 employs the atomic adsorption method for determination of acid-soluble chloride content. ASTM International has two methods: C 1152 for acid-soluble chloride extraction, and C 1218 for water-soluble chloride extraction. For both methods, the extract is subsequently tested for chloride content using ASTM method C 114. Another method for testing for water-soluble chlorides that uses whole pieces of concrete rather than powdered samples is the American Concrete Institute (ACI) soxhlet extraction process, test method 222.1.

Acid-soluble chloride ion testing is also known as total chloride testing. If chloride ions are contained in the aggregate, acid-soluble testing will include those ions, even if they normally would not be a significant contributor to in situ corrosion. If chloride-bearing aggregate or chloride ions were otherwise present in the initial concrete mix, conducting water-soluble chloride and/or soxhlet tests can aid in data discrimination. For example, water-soluble testing of some central North American limestone coarse aggregates may show chloride ion contents up to 0.1%, yet no corrosion problems with these aggregates has been reported (Ref 129). The ACI 222.1 soxhlet method has shown the ability to remove chloride ions from the concrete while not extracting significant amounts of chloride from the coarse aggregate.

As shown in Fig. 19, chloride threshold values depend on the quality of the concrete and local moisture conditions. Because chloride ions may effectively be bound in aggregate or in the cement paste, the threshold also depends on the analytical method. For a thorough discussion of these issues refer to Ref 129.

Carbonation test samples are conducted on freshly fractured concrete specimens obtained by coring or other means. The depth of carbonation is determined by applying a fine mist of phenolphthalein solution employed as a pH indicator. An aqueous solution of 1% phenolphthalein dissolved in 70% ethyl alcohol will provide a good pH indicator (Ref 130). Noncarbonated areas will turn magenta on application of the phenolphthalein solution, whereas carbonated areas will not.

Corrosion Rate Testing and Other Advanced Techniques

Corrosion rate testing has been applied in the field to investigate corrosion of reinforced-concrete bridges and highways. Most work has involved the linear polarization technique, typically employing a guard ring to contain the applied current. The galvanostatic pulse technique has also been employed. Obtaining meaningful results with these testing approaches requires more than casual familiarity with the techniques. Acoustic testing, such as pulse echo, and ground-penetrating radar have been employed with some success to identify corrosion problems in reinforced concrete. Several variations in radiography have also been tried.

Inspection of Steel Elements

Characteristics to be noted depend on the component of the structure.

Painted steel inspection includes defining the condition of the coating. This can range from:

- Little evidence of corrosion, suggesting that the coating is sound
- Some chalking or peeling but no exposed metal
- Moderate to heavy surface rust
- Coating system has failed
- Corrosion has caused sufficient section loss to warrant a structural evaluation.

Weathering steel inspection would look to identify:

- Tightly adhering oxide layer with little or no section loss
- Some surface rust or pitting with a granular or dusty oxide layer
- Measurable section loss due to corrosion, some flaking
- Corrosion is advanced, with laminar sheets of rust present.

Steel pin or hanger inspection would include similar inspection breakdowns, as given previously for painted steel and weathering steel. Additional attention should be given to crevice areas, particularly if salt spray or drainage problems are an issue.

Steel culvert inspection would look to identify that:

- The culvert shows little deterioration, if galvanizing is present and if it is in good condition.
- There is slight to moderate corrosion, particularly on the invert; there is minor damage or separation of seams.
- There is significant corrosion and some through-pits on the invert; there is minor to moderate distortion, deflection, or separation of the seams.
- Major corrosion is present, with obvious distortion, deflection, or settlement.

Bearing inspection would look to identify that:

- The bearing shows minimal signs of corrosion.
- There is moderate to heavy corrosion, with perhaps some pitting, but the bearing is still functioning.
- Heavy corrosion has occurred with section loss; shear keys may have failed, or corrosion may restrict motion.

Table 4 Resistivity and possible corrosion rates

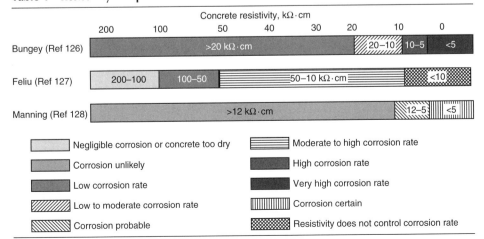

Corrosion-Resistant Reinforcement

Jack Tinnea, Tinnea & Associates, LLC

Corrosion-resistant reinforcement (CRR) is employed in bridges and highways in efforts to help reduce the staggering maintenance costs related to infrastructure corrosion. The CRR has taken many forms, often novel, to extend the service life of reinforced-concrete highway structures that range from simple reinforced-concrete culverts to bridges soaring over wide rivers.

Approaches to Corrosion Resistance

The CRR has taken three major paths to achieve its goal of reducing vulnerability to corrosion. These include:

- Coated metal reinforcement
- Use of corrosion-resistant alloys
- Replacement of metal reinforcement with nonmetallic products

A problem in the development of CRR stems from the fact that the extended life conditions, which reinforced-concrete structures experience in the as-constructed environment, are not easily replicated in accelerated testing protocols. Furthermore, new high-performance concretes (HPC) and a general move toward placing greater concrete cover over reinforcement greatly impede the influx of corrosive ions such as chlorides. This in turn means that the CRR will exist in a high-pH environment for decades before it becomes necessary for the CRR to provide improved corrosion resistance.

Selecting the appropriate design approach for corrosion control in a reinforced-concrete structure requires a clear estimate of the desired service life. That seems simple but often is problematic in the public works area. For example, if it is likely that a highway bridge will become functionally obsolete in 40 years due to expected traffic volume increases, it may not make economic sense to specify 100 year service life materials by selecting an HPC mix, which provides exceptional resistance to chloride ion penetration, and coupling that with 316LN stainless steel reinforcement. However, many reinforced-concrete bridges remain in use well beyond their planned service lives and continue as important parts of local infrastructures. Evaluation of the potential for functional obsolescence requires input from planners and traffic engineers. In short, making a proper estimate of a reinforced-concrete highway structure requires a multidisciplinary team. Working with a multidisciplinary team makes unique demands on management, but the savings that result, particularly in large projects, can yield handsome returns.

Economics and Life Cycle. Growth in China, India, and other countries has led to volatility in the costs of construction materials. In 2003, the price of plain reinforcing bar increased 55%; in 2004, the increase was only 1.2%, and in 2005, the price dropped 7.4% (Ref 131). Such economic pressures are not likely to abate in the foreseeable future, so inclusion of specific costs in this article could lead to serious error. However, given the general trends over the past decade, it is possible to understand when it is reasonable in highway and bridge design to proceed with one type of material over another.

Figure 24 compares relative costs for several types of reinforcing bar versus the age of the structure. The microcomposite alloy is a proprietary steel, MMFX-II, that meets ASTM A 1035 (Ref 132). It is an alloy that is almost free of interlath carbides, with a microstructure that consists of nanosheets of austenite between laths of dislocated martensite or ferrite (Ref 133). For this representation, it is assumed that the structure is located in an aggressive environment with chloride ion exposure. It is also assumed that the concrete cover will provide 25 years of protection before the chloride ion concentration reaches the threshold necessary for corrosion of carbon steel. The onset of corrosion for both epoxy-coated reinforcement (ECR) and carbon steel is shown to occur at the same time, because the base metals are the same. A 5 year propagation period is assumed for carbon steel, while a 10 year propagation period is assumed for ECR. The increase in total cost during those two propagation periods is from either rehabilitation or replacement. It is assumed that the chloride ion threshold for the microcomposite alloy is twice that of carbon steel. After corrosion commences, it is assumed that the microcomposite will behave in a manner similar to carbon steel and show a 5 year propagation period. The two cost lines for 304 and 316 stainless steel are not expected to increase during the 60+ years shown in Fig. 24.

It appears reasonable to assume that with improved concretes and increased cover, today's (2006) bridge in an aggressive environment can expect to provide 20 to 30 years' service or more before uncoated ASTM A615 (Ref 134) black steel reinforcement will start to show signs of corrosion. Work performed by the Virginia Department of Transportation found that with newer construction techniques, such as low-w/c concrete and adequate concrete cover, less than 25% of all Virginia bridge decks would require rehabilitation within 100 years (Ref 135). The studied Virginia bridges include those exposed to marine environments, high deicing salt applications, and traffic, and others, typically in rural areas, that see no marine exposure, less traffic, and hence less frequent deicer applications.

Options. There are other options beyond materials selection that can dramatically extend the service life of a highway structure. Some are the simple steps of paying attention to detailing. Run-off water from a bridge deck often contains chloride-bearing deicing salts. This can be particularly problematic on curved or skewed bridge decks. If only scuppers are used that are fitted with short drain pipes, the wind may carry saltwater that drains to support piles and causes corrosion in an area where it could easily and very inexpensively be avoided. If a 100 year design life is desired and posttensioning (PT) is to be part of the design, it is reasonable to use the best technology available to avoid premature corrosion of the PT strands. It is also relatively inexpensive insurance to include spare empty PT strand ducts that, although sealed during construction, are readily available for installation of supplementary PT strands, should loss of prestressing occur. Figure 25 is a picture of a bridge where this technique was employed.

Epoxy-Coated Reinforcement

The ECR provides protection to carbon steel reinforcement by providing a barrier to potentially corrosive conditions, such as exposure to chloride ions. The coating is applied by fusion bonding, where a 100% dry powder mixture of epoxy resins, curing agents, pigments, fillers, and other additives is electrostatically applied to a preheated steel surface. The applied heat melts the powder constituents, which then react to form a cross-linked polymer coating. The first field installation of ECR was in 1973 on a bridge deck near Philadelphia, Pennsylvania.

Effectiveness of ECR. Since 1973, the presence of ECR has become commonplace on highway and bridge construction sites. Many early studies indicated that the technology appeared capable of providing effective corrosion control for reinforced-concrete structures in the highway and bridge arena (Ref 136–138). In 1992, an article was published by an early proponent of the technology that questioned ECR effectiveness (Ref 139). A year later, a news article (Ref 140) documented the debate. At the 85th meeting of the Transportation Research

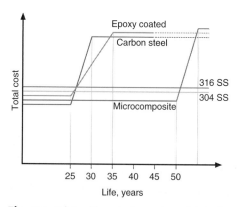

Fig. 24 Relative life costs of corrosion-resistant reinforcement

Fig. 25 Alsea Bay Bridge, Waldport, Oregon. Courtesy of Oregon Department of Transportation

Board (TRB) in 2006, a panel discussion was held on the pros and cons of using ECR (Ref 141); it showed that the debate has not been resolved.

Between the start of the debate in 1992 and the TRB panel discussion in 2006, a number of reports, papers, and articles were prepared on ECR. Some of these discussed problems with ECR not providing significant corrosion control (Ref 142–144), while others reported good performance with the technique (Ref 145–147). Between these two extremes were researchers who reported mixed results on the level of corrosion protection provided by ECR (Ref 83, 148–150). To better understand the performance of ECR, a brief discussion of several issues from the literature follows.

Coating Disbondment. NACE International defines disbondment as "the loss of adhesion between a coating and the substrate." It is not unusual for organic coatings to take up water when they are exposed to wet, damp, or humid environments (Ref 151). Water moves through the coating either at capillaries or pores (Ref 152) or by attacking relatively hydrophilic areas of low-molecular-weight/poorly cross-linked material (Ref 153). The water produces a multimolecular layer that moves laterally along the epoxy-steel interface. The formation of this water layer produces a loss in adhesion between the epoxy and the steel. For purposes of discussion, this is called wet adhesion loss.

A second form of coating disbondment can result from corrosion processes (Ref 149, 154). This can occur from either undercutting at the boundary of a corroding site or through cathodic disbondment, which occurs at a distance and often generates blisters rather than multimolecular layer separations.

Wet loss of adhesion can occur within four years of installation and is not necessarily associated with visible corrosion (Ref 143, 154). The wet adhesion loss can be substantially recovered if the reinforcing bars that have suffered that loss are dried (Ref 155). However, although adhesion recovery is possible for ECR dried in a laboratory, the concrete is likely to retain a high relative humidity at reinforcement depth in most highways and bridges, and such self-healing may not occur. Disbondment that results from corrosion undercutting or blisters formed during cathodic processes will never recover lost adhesion.

Coating Protection and Corrosion. Nicks, cuts, and other defects in the epoxy coating can result in the underlying carbon steel being directly exposed to the surrounding environment. For this reason, most specifications that include ECR discuss proper handling, placement, and coating repair prior to the placement of concrete. Even with the inclusion of special handling provisions, researchers in Texas reported that "experience has shown that epoxy-coated steel is treated like uncoated steel at the project site" (Ref 156). Figures 26 and 27 illustrate failures to observe proper handling at a major Midwestern construction site.

Coating with Defects. Figure 28 shows a schematic drawing of coating degradation that occurs as the result of a coating defect. The figure and discussion follow a model outlined by Nguyen and his colleagues at the National Institute of Standards and Technology (NIST) (Ref 153). In Fig. 28(a), step 1 is the formation of the coating defect, and step 2 is the transport of water and other reactants, such as oxygen and chloride ions, to the metal surface. Steps 3 and 4 are the formation of the anode (corrosion site) and cathode(s).

In Fig. 28(b), note that the cathodic reaction is not benign, because it leads to delamination of the coating. Here, the sodium ion is part of a hydrolysis reaction that produces an alkaline solution at the metal-coating interface:

$$\tfrac{1}{2}O_2 + H_2O + 2e^- + 2Na^+ \rightarrow 2(Na^+OH^-) \quad \text{(Eq 10)}$$

Sodium ions diffuse along the metal-coating interface (step 5) and initiate blisters at a distance (step 6). In Fig. 28(c), this alkaline solution then creates an osmotic pressure gradient (step 7), the main force behind blister growth (step 8). Step 9 is the anodic hydrolysis reaction:

$$Fe^{2+} \; 2Cl^- + 2H_2O \rightarrow Fe(OH)_2 \downarrow + 2(H^+Cl^-) \quad \text{(Eq 11)}$$

The reaction shown in Eq 11 produces hydrochloric acid that will reduce the local pH and accelerate corrosion. It is to avoid this corrosion process that strict handling and placement specifications typically are employed for structures with ECR bars. Problems with coating defects can greatly increase subsequent corrosion.

Coating without Defects. However, even without field-detectable defects, epoxy coatings have pathways or pores that can be involved in absorption and diffusion of water and sodium and chloride ions. Figure 29 shows a schematic of the NIST model for coating degradation without pre-existing defects (Ref 153). In Fig. 29(a), the gray hydrophilic regions are areas with low-molecular-weight materials that have relatively little cross linking. As such, they can take up water and ions and are subject to hydrolysis.

The hydrolysis and dissolution of the hydrophilic regions are, in fact, an attack by water that eventually leads to the formation of pathways through the epoxy coating to the steel surface

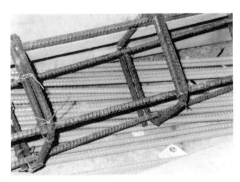

Fig. 26 Bare reinforcement cage resting on a loose pile of epoxy-coated reinforcement

Fig. 27 Epoxy-coated reinforcement (ECR) loosely stacked on in-place ECR

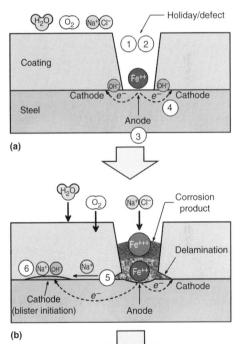

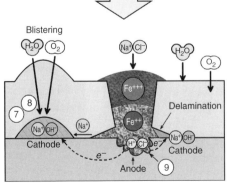

Fig. 28 Failure at a coating defect

(step 1 in Fig. 29b). As in the case of ECR with a defect, water and reactants reach the steel (step 2), leading to the formation of anodes and cathodes (steps 3 and 4, respectively).

In Fig. 29(c), the small and mobile sodium ion diffuses along the steel-coating interface in step 5. Through hydrolysis (step 6), alkaline solutions develop. This leads to delamination of the coating adjacent to the conductive pathway.

In Fig. 29(d) (step 7), the alkaline pools generate osmotic pressure gradients that transport water to the reaction site. This inflow of water and expansion due to the formation of corrosion products causes swelling and blister formation in step 8. Generally speaking, the corrosion observed on initially undamaged ECR is much less than that seen with damaged bar (Ref 150).

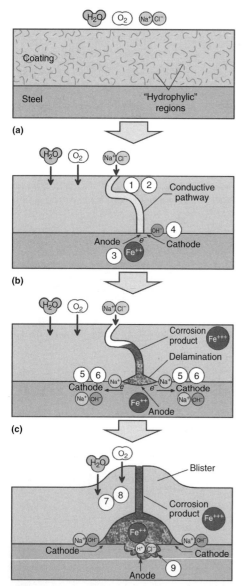

Fig. 29 Failure without a coating defect

Hydrolysis. For both initially undamaged and damaged coatings (Fig. 28c and 29d), step 9 shows pits forming in the steel as the result of hydrolysis reactions. These reactions are aggravated by the presence of chloride ions, because iron will form soluble complexes with chlorides. These complexes aid in the removal of reaction products from the active site. Hydrolysis also will reduce the local pH.

Control of the Cathodic Reaction. The ECR was initially employed to shield steel reinforcement from a corrosive environment. To many, this meant ECR protected steel from aggressive ions such as chlorides. Given this view, many structures were built with both ECR and uncoated steel. In particular, it is the practice of some agencies to use ECR for the top mat of a bridge deck and uncoated black steel on the bottom mat. The logic is that the top mat is protected from chloride-bearing deicers by the epoxy coating, while the bottom mat has such a deep concrete cover that corrosion is not likely to occur.

This thinking does not account for the fact that if ECR were used in the bottom mat, the epoxy coating would shield that steel from oxygen, which is necessary for the corrosion process. Corrosion of steel in concrete is typically in cathodic control, being limited by the diffusion rate of oxygen through the concrete. The presence of an epoxy coating greatly diminishes oxygen availability at the steel surface (Ref 157, 158). Even with special chairs and ties, it is unlikely that an ECR top mat will be completely electrically isolated from a bare steel bottom mat. When chloride ions reach threshold level at a defect in the ECR top mat, this will create a small anode connected to a very large uncoated carbon steel bottom mat cathode (Ref 159). These small anode/large cathode situations can lead to localized corrosion of reinforcement with significant section loss that does not generate tensile hoop stresses, cracking, or delamination.

Modified ECR Techniques. Additional protection has been attempted to provide integral protection for the almost unavoidable holidays and coating defects. These attempts include the use of inhibitors in the concrete and the coating. An arc-sprayed zinc coating was recently applied to reinforcing bar before applying the epoxy coating (Ref 64). Self-healing resins are also being investigated.

Manufacture and Construction. The coating application, fabrication, handling, and installation of ECR are covered under the standards and practices listed in Table 5.

Stainless Steels and Microcomposite Alloys

The rationale for using stainless steels and microcomposite alloys in highways and bridges is quite simple. These alloys are more resistant to chloride contamination than are carbon steels, and they have substantially higher chloride ion thresholds. A higher chloride ion threshold means that if other factors remain constant, it will take a longer time to reach the corrosion threshold, so a longer maintenance-free service life is enjoyed.

An early example of the use of stainless steel reinforcement is in a marine pier constructed between 1937 and 1941 in Progreso, Mexico. A recent inspection of that pier showed that even with chloride concentrations of approximately 1.2 wt% of dry concrete measured at reinforcement depth, the structure was showing no signs of general corrosion problems after 60 years service (Ref 63). The chemistry of the alloy used in the Progreso pier is similar to type 304.

Stainless steels are iron alloys that produce a corrosion-resistant chromium oxide layer on their surfaces. This generally requires a minimum of 11 to 12% Cr. Other materials, such as nickel, molybdenum, nitrogen, and others, are

Table 5 Standards related to epoxy-coated reinforcement

Designation	Title	Process covered	Reference
AASHTO M284-05	Epoxy-Coated Reinforcing Bars: Materials and Coating Requirements(a)	Coating application Fabrication (bending) Field handling, coating repair, and installation	160
ASTM A 775/A 775M	Standard Specification for Epoxy-Coated Reinforcing Steel Bars	Coating application Fabrication (bending) Field handling, coating repair, and installation	161
ASTM A 934/A 934M	Standard Specification for Prefabricated Epoxy-Coated Reinforcing Steel Bars	Coating application Fabrication (bending) Field handling, coating repair, and installation	162
ASTM A 884/A 884M	Standard Specification for Epoxy-Coated Steel Wire and Welded Wire Reinforcement	Coating application	163
ASTM D 3963/D 3963M	Standard Specification for the Fabrication and Job Site Handling of Epoxy-Coated Reinforcing Steel Bars	Fabrication (bending) Field handling, coating repair, and installation	164

(a) This AASHTO standard is identical to ASTM A 775/A 775M-04.

added to improve mechanical properties, provide resistance to chloride-environment pitting, and improve strength. For example, by lowering carbon content, the L-grades improve corrosion resistance but reduce the yield strength. By adding nitrogen, the N-grades raise the strength of L-grades back toward that of standard grades and provide added corrosion protection. Stainless steel reinforcement is available at a cost premium compared to carbon steel alternatives.

Stainless Steel Alloys. Alloys used in bridges and highways include the austenitic alloys 304 (S30400), 316 (S31600), and XM19 (S20910) and the duplex alloys 2101 and 2205 (S31803). Ferritic stainless alloys, such as 3Cr12 (S41003) and CRS 100, are also being used as reinforcement. Specialty alloys are also employed, such as Nitronic 60 (S21800) for galling situations and XM29 (S24000) where nonmagnetic steels may be required.

Microcomposite Alloy. ASTM A 1035 describes a Fe-C-Cr-Mn alloy that has an average chromium content of ~9%. Although high, that chromium content does not reach the threshold of 10.5% used in defining stainless steels. The corrosion resistance derives from a morphology where the nanometer structure consists of dislocated martensite or ferrite phases separated by austenite sheets. In addition to corrosion resistance, this structure provides increased strength, qualifying as ASTM A615 grade 75.

Corrosion Resistance. Table 6 is a matrix showing several corrosion-resistance indexes for carbon steel, several stainless steels, and the microcomposite alloy.

The pitting resistance equivalent number (PREN) shown in Table 6 is a number that gives a general guideline in evaluating pitting resistance. For stainless steels in concrete, there is a general correlation between PREN and the chloride ion threshold (Ref 167). The chloride ion threshold "Carbon steel index" sets carbon steel as the base, with a value of 1, and expresses the other alloy chloride ion corrosion threshold as a multiple of that observed for carbon steel. The data were obtained from outdoor southern exposure protocol (Ref 168) specimens that followed a weekly regimen of ponding with a saturated NaCl solution for three days and drying for four days (Ref 64). The chloride ion threshold "Cl^-/OH^- ratio" expresses the chloride ion corrosion threshold as the molar ratio of the chloride and hydroxyl ions.

In an effort to reduce costs, stainless-steel-clad conventional reinforcements have been produced. Although stainless steel has a relatively high hydrogen overvoltage and as such does not make as good a cathode as, for example, passive carbon steel in concrete, problems have been observed when defects are present either at the ends, bends, or at the seams of stainless-clad reinforcement (Ref 64, 166, 169). Welding clad reinforcement in the field could also be problematic. The data in Table 6 also suggest potential problems with cladding defects or where seams have separated during fabrication.

To place these corrosion-resistant reinforcement (CRR) alloys in an exact order according to corrosion resistance when used as reinforcement in concrete is difficult through accelerated laboratory testing. Generally, as stated previously, there is reasonable correlation between the PREN number and corrosion resistance. Again, accelerated testing can be helpful, but long-term behavior may differ. For example, in the over 60 year old Progresso pier, the chloride ion concentration of 1.2 wt% at the level of the stainless steel bar exceeded literature-reported lab study thresholds for stainless steels of between 0.7 and 1.0 (Ref 63).

Applications. An extensive study of corrosion-resistant reinforcements stated that stainless steel reinforcing bar should provide a service life in excess of 100 years, even with concrete cracking (Ref 170). This study also recommended the use of 316 alloys over 304 alloys for conventional reinforcement in marine structures and bridges where repair closure is particularly problematic. Stainless alloys can also be beneficial for doweling applications.

In addition to reinforcement, stainless steels may be used for hinged arches (Ref 171). Figure 30 shows an arch hinge used on a coastal bridge. As can be seen, the tide sometimes covers parts of the hinge. The hinge plates were fabricated from duplex stainless steel 2205, while antigalling Nitronic 60 was used for the pins. Figure 31 shows the 316LN studs used to connect the hinges to the concrete arch and abutment.

Costs. Although more expensive than plain carbon steel or ECR, life-cycle cost analysis clearly shows stainless steel is preferable for bridges in high-traffic areas subject to deicing salts or in marine environments (Ref 142, 171). Cost comparisons between ECR and stainless need to look at installation costs in addition to materials costs. For example, as was discussed earlier, ECR requires special handling to avoid incidental damage during construction. Figure 32 shows that with stainless steel reinforcement, workers can walk on placed mats without damaging the system or inducing future corrosion (Ref 171). This can substantially reduce construction costs. With CRR, it is also good practice to specify use of stainless steel chairs, tie wires, and incidental hardware that workers find easier to use than similar specialty hardware employed to place ECR. On three new

Fig. 30 Stainless steel arch hinge. Note marine growth

Fig. 31 316LN anchor studs used to connect stainless steel arch hinge to concrete

Table 6 Corrosion-resistance indexes

Alloy	PREN(a)	Chloride ion threshold	
		Carbon steel index(b)	Cl^-/OH^- ratio(c)
ASTM A615 carbon steel	0	1.0	0.25–0.34
ASTM A1035 microcomposite	10	4.7–6.0	4.9
304 austenitic stainless(d)	19	>11.2–14.2(e)	...
304LN	21	...	...
316L-clad with 1×25 mm (0.04×1.0 in.) cut	25	6.5–8.5(e)	...
316L-clad with drilled holes	25	>11.2–14.2(e)	0.25(e)
316L-clad	25	>11.2–14.2(e)	4.9(e)
316/316L austenitic stainless	25	...	...
316LN	27	>11.2–14.2(e)	20(e)
XM19	33	...	...
XM29	23	...	...
2101 duplex stainless	26	2.6–3.4	9.7(e)
2205 duplex stainless	34	...	...

(a) Pitting resistance equivalent: PREN = %Cr + 3.3 × %Mo + 16 × %N (Ref 165). (b) Carbon steel index = [Cl^-] threshold observed ÷ [Cl^-] threshold carbon steel. The range of values results from test variation: bent or straight bar, top test bar coupled with similar material or with carbon steel (Ref 64). (c) Ref 166. (d) Bars formed by welding 304 wire to 304 rods to form the ribs "positive-machined." (e) Bars were pickled.

coast bridges in Oregon, the cost premium for using stainless steel reinforcement compared to black iron bars was only 10% of the total project cost (Ref 172).

As mentioned previously, some CRR metals, such as A1035, have higher strength than carbon steel. Because A1035 also qualifies as ASTM A615 grade 75, designers can reduce the amount of steel required with a similar design using ASTM A615 grade 60 reinforcement. Reducing the amount of steel required helps reduce the typical CRR materials cost premium. Further, reductions in the amount of steel required in congested areas will also reduce labor costs.

Galvanized Reinforcement

Galvanized reinforcement is another option to control corrosion in reinforced-concrete highway and bridge structures. Typically, the coating is applied by the hot dip process, where the reinforcement is immersed in a molten zinc bath. The coatings that form include several iron-zinc intermetallic layers covered with essentially pure zinc as the outer layer. This zinc coating provides both a barrier coating on the carbon steel and sacrificial cathodic protection (Ref 173). Zinc is protected by a passive surface film at pH values between approximately 6 and 12.5 (Ref 174).

Fresh concrete is quite alkaline, with the pH typically well above the 12.5 pH passive upper limit. When zinc is first exposed to this environment, the reaction is vigorous, producing calcium hydroxyzincate precipitate and generating hydrogen gas. This process stops after the final set of the concrete. If local sources of portland cement are particularly alkaline, it is suggested that some trial mixes be made that employ mineral admixtures such as fly ash or silica fume. These additions can help moderate excessively high pH during early hydration and, if properly used, will also provide a concrete that is less permeable to chloride ion penetration. Secondary mineralization, where the zinc replaces calcium in the cement-phase matrix, continues at a much reduced rate, forming and maintaining a good barrier coating to the reinforcement.

Passivation of carbon steel reinforcement will be lost if the pore water pH drops below approximately 11.5. Also, the ratio of chloride ions to hydroxyl ions impacts corrosion, so if chloride ions are present, passivation may be lost at higher pH values. Zinc-coated steel has been shown to remain passive to approximately pH 9.5. Further, galvanized bars have been shown to resist corrosion at much higher chloride levels, at least 2.5 times higher, than bare carbon steel (Ref 175).

In-depth inspection of two bridge decks fabricated with galvanized reinforcement showed little evidence of corrosion after almost 30 years in service (Ref 173). In fact, metallographic examination of rebar samples obtained from the two bridges indicated that the remaining zinc coating thickness exceeded ASTM A 767 class II requirements for new bars.

Performance of Weathering Steel Bridges in North America

Frank Pianca,
Ontario Ministry of Transportation

The first major step in obtaining public acceptance of weathering steel in North America occurred in 1963 when Eero Saarinen, a well-known architect, selected this steel to construct a building in Moline, Illinois. The first use of weathering steel in a highway bridge occurred in 1964. The market for weathering steel began to expand rapidly in the 1960s as the unique combination of corrosion resistance and strength properties caught the eye of the design engineer, along with the potential advantages of low maintenance, freedom from painting, and weight savings.

In contrast to the introduction of most new materials and products, which follow a lengthy procedure of laboratory testing and field trials, there was a total commitment from many transportation agencies to the use of weathering steel.

Steel industry records indicate that the use of weathering steel in bridges increased each year to a high in 1980, when it accounted for approximately 12% of the total steel awarded for bridges. In that same year, the state of Michigan instituted a complete moratorium on the use of uncoated weathering steel in its highway program, because of reported problems with some of its weathering steel bridges.

Thereafter, the use of weathering steel in bridges declined to a low of approximately 10% in 1987, but, as painting costs continued to rise and a better understanding of the nature of the corrosion problems developed, the use of this steel again began to climb. By the end of 1989, the use of weathering steel had risen to a new high level of approximately 15%. In 1990, the Michigan Department of Transportation (DOT) rescinded the moratorium and agreed once again to consider the use of weathering steel for new bridges.

Weathering Steel as a Material

Weathering steel is a low-alloy steel containing up to 3% of alloying elements such as copper, chromium, nickel, vanadium, phosphorus, silicon, and manganese. Of all the elements listed, copper, nickel, chromium, and silicon contribute most to improving the atmospheric corrosion resistance. Weathering steel is also known as atmospheric corrosion-resistant steel and is marketed under the trade names of several steel companies. In Canada, the applicable material specification is CSA G40.21 type A, and in the United States, ASTM A588 (the grade most commonly used on bridges).

On exposure to the atmosphere and under suitable conditions, the corrosion rates of weathering steels are considerably less than ordinary mild steel in the same environment. As the steel weathers, it forms an oxide that is light colored, loose, and fine grained. Further oxidation results in a tightly adhering stable oxide film, or patina, that is rough, dark brown in color, and inhibits further corrosion. The rapidity with which weathering steel develops a stable patina and its characteristic color is determined largely by the atmospheric conditions of exposure. The time required has been quoted as generally more than 18 months (but less than 3 years) to many years.

It is instructive to examine the conditions most favorable for the development of a stable patina. They are exposure to intermittent wetting and drying, such as is provided by direct exposure to rain, wind, and sun, and the absence of corrosive pollutants, particularly salt. Experience has shown that the presence of salt prevents the formation of a stable oxide, and corrosion proceeds unchecked. The design, fabrication, erection, and maintenance of weathering steel structures play a major role in assuring satisfactory performance.

Rate of Corrosion of Weathering Steel

A number of studies have been carried out to measure the rate of corrosion of weathering steel. Many of the studies have included a comparison with other steels. A direct comparison with mild steel is difficult, because the inclusion of scrap metal in the production of mild steel can result in composition variation that results in differing corrosion performance. A typical figure is that weathering steel has an atmospheric corrosion resistance of approximately 1.5 to 4 times that of carbon steel containing no alloyed

Fig. 32 Workers can walk on stainless steel bar, where they should not with epoxy-coated reinforcement.

copper. However, it is the actual corrosion rates that are of interest, rather than whether or not weathering steel performs well or badly in relation to a material, carbon or mild steels, that would not be used unprotected in outdoor environments.

The corrosion penetration rate found for weathering steel in a variety of bridges that are performing satisfactorily ranges from 3 to 7.5 μm/year/surface (0.12 to 0.30 mil/year/surface). Weathering steel corroding at a rate higher than 7.5 μm/year/surface (0.30 mil/year/surface) cannot be expected to develop a protective oxide coating.

Performance of Weathering Steel

A study undertaken by the Michigan DOT in the late 1970s showed bridge corrosion problems associated with the contamination of weathering steel surfaces by deicing salts. Some design details that permitted crevice corrosion, galvanic corrosion, or the accumulation of deicers resulted in serious corrosion damage. These findings led to a decision to impose a partial moratorium on the use of weathering steel in Michigan bridges. The moratorium, instituted May 2, 1979, prohibited the use of unpainted steel in depressed roadway sections (where the most serious corrosion problems were observed) and in urban and industrial areas where heavy salting and industrial pollution created a corrosive environment. On February 6, 1980, the moratorium was made statewide.

In response to the Michigan moratorium, a study was launched under the auspices of the American Iron and Steel Institute (AISI) that included the examination of 49 weathering steel bridges in seven states. An analysis of the 49 bridges indicated the following:

- Thirty percent of the bridges showed good performance in all areas.
- Fifty-eight percent of the bridges showed good overall performance, with moderate corrosion in some areas.
- Twelve percent of the bridges showed good overall performance, with heavy localized corrosion in some areas.

The areas that exhibited moderate-to-heavy corrosion were found to be those most frequently subjected to contact with salt-laden road runoff passing through leaking and open expansion joints. Tunnellike conditions, particularly where the underclearance was less than 6 m (20 ft), also were detrimental, because they resulted in splashing and deposition of salt-laden snow onto the bottom of the beams.

The inspection results did indicate the need for remedial painting of certain portions of bridges exhibiting excessive corrosive attack. The report also stated that there was a need for additional research on the development of suitable painting systems for weathering steel subject to corrosive attack.

At approximately the same time, the National Cooperative Highway Research Program (NCHRP) initiated a study to build on the AISI findings. The major findings of the NCHRP investigation were "... that weathering steel can provide a satisfactory service life with limited maintenance if the structural details are designed in a manner that prevents accelerated attack, vulnerable areas are painted, and contamination with chlorides is avoided."

Michigan rescinded the moratorium in 1990, and the Michigan DOT decided to consider the use of weathering steel whenever a proposed bridge location was within the guidelines of the Federal Highway Administration Technical Advisory, October 1989.

Figure 33 shows a protective patina tightly adhering to weathering steel on the exterior of a cable-stay-supported bridge. The relative humidity at the bridge location averages over 80% for four months and over 70% for the remainder of the year. Figure 34 shows a light-colored sheet of weathering steel corrosion product peeling from the interior of the cable stay support pier adjacent to the exterior view shown in Fig. 33. This wallpaper-like corrosion product has a thickness of approximately 2.5 mm (0.1 in.). Although near the coast, corrosion product samples from the interior show no signs that chloride ions aggravated the corrosion. The clear difference in behavior between interior and exterior exposure appears to be the result of the high time of wetness that occurs within the piers. Thus, the interior environment minimizes the alternating wet-dry cycle necessary to produce a protective patina, and instead of protection, the interior conditions produce exfoliation of sheets of corrosion product.

Recommendations and Considerations on the Use of Weathering Steel

The most important considerations in designing a weathering steel bridge are preventing water ponding, diverting the flow of runoff water away from the steel superstructure, preventing the accumulation of debris that traps moisture, and avoiding environments in which the bridge would be contaminated with salt.

Leaking joints are the most serious and common cause of corrosion problems in weathering steel bridges. Runoff water leaking through the deck joints persistently wets the bearings, flanges, webs, stiffeners, and diaphragms in the vicinity of the joint, where it can migrate for long distances along the bottom flange and wick up the web.

To avoid the problems created by leaking expansion joints, bridges should have a continuous superstructure, fixed or integral bearings at piers or abutments, and no bridge deck expansion joints unless absolutely necessary. When expansion joints cannot be avoided, they should be provided only at abutments.

It has been suggested that weathering steel coupons be used for up to two years to determine if corrosion rates are acceptable. It is desirable to monitor the atmosphere for constituents such as chlorides, sulfur dioxide, and moisture. Although no specific data define acceptable times of wetness for weathering steel, times greater than 60% are usually considered very damp for corrosion purposes. This can include low-level water crossings that may lead to high humidity and frequently wet conditions. Over stagnant or sheltered water, vertical clearance to the bottom flange of the bridge girder on bridges should be at least 3 m (9.8 ft), and over moving water at least 2.5 m (8.2 ft).

For aesthetic reasons, every effort should be made to minimize unsightly rust staining of concrete supports visible to the public by preventing runoff water from draining over the weathering steel and onto the concrete.

The color and texture of mill scale do not match those of corroded weathering steel surfaces. Unless mill scale is removed, the steel surface will appear matted, flaky, and not uniform for several years, depending on the degree of exposure and the aggressiveness of the local

Fig. 33 Protective patina tightly adhering to weathering steel on bridge exterior

Fig. 34 Wallpaper-like corrosion product peeling from weathering steel on bridge interior

environment. Therefore, all surfaces visible to the public should be blast cleaned. Near-white blast cleaning is recommended for weathering steel surfaces exposed to public view, and commercial blast cleaning for surfaces not visible to the public.

Combinations of high traffic speed, low clearance, and heavy use of deicing salt on the lower roadway pose a severe corrosion hazard. The spray plume kicked up by passing trucks is approximately twice as high as the truck and settles on the vulnerable overhead low points of the structure.

Inspection and Maintenance. An effective inspection and maintenance program is essential to ensuring that a weathering steel bridge reaches its design life. Weathering steel is not a maintenance-free material. The following examples illustrate the type of periodic maintenance that may be needed:

- Removing loose debris with a jet of compressed air or vacuum-cleaning equipment
- Scraping off sheets of delaminated rust
- High-pressure hosing of wet debris and aggressive agents from the steel surfaces
- Tracing leaks to their sources
- Repairing leaky joints
- Installing drainage systems, drip plates, and deflector plates that divert runoff water away from the superstructure and abutments
- Cleaning drains and downspouts
- Cleaning and caulking all crevices
- Remedially painting areas of excessive corrosion

Rehabilitation. Weathering steel bridges that are undergoing a high degree of uniform or local corrosion must be remedially painted to protect the affected areas against further section loss. However, little information is available on the long-term performance of coating systems over weathered A588 steel substrates. In the absence of such information, coating systems are often recommended based on their performance over uncoated weathering steel or carbon steel. This area has been identified in a number of studies as requiring further research work.

Conclusion

All available data indicate that good performance of uncoated weathering steel bridges is attainable. With the proper application of guidelines, bridge owners and consultants can be assured of obtaining consistently good performance with their weathering steel bridges. However, weathering steel does not provide a service life for the superstructure that is entirely free from maintenance. Proper detailing and choice of a suitable environment can reduce the need for some, but not all, maintenance. An effective inspection and maintenance program is essential to ensuring that a weathering steel bridge reaches its design life.

Coatings

Bryant "Web" Chandler,
Greenman Pedersen, Inc.

According to the U.S. Federal Highway Administration, there are just under 600,000 bridges (6 m, or 20 ft, and longer) in the national highway system, of which 190,000 to 210,000 are steel structures. They all require protection from corrosion in one way or another.

Barrier Coatings for Steel

The best way to protect steel bridges is by the use of coatings. Coatings provide protection as a barrier coat and/or for cathodic protection (CP). As a barrier, the coating isolates the metallic substrate from the electrolyte, thus eliminating one of the essential ingredients for corrosion.

Electrolytes such as liquids that include deicing salts and chemicals, salt air, acid rain, industrial pollutants, and engine exhaust gases, will promote corrosion. A good barrier coat will provide the needed isolation of the electrolyte. However, all coatings are permeable to moisture, and thus, the electrolytes or soluble salts (chlorides, sulfates, and nitrates) must be removed from the steel before coating application. If the soluble salts are not removed, then moisture is drawn through the coating, creating a condition called osmotic blistering. These blisters will break, exposing the bare steel and resulting in corrosion. There are several different testing procedures available to determine the presence of soluble salts.

Barrier coats, prior to the banning of coatings in the 1990s that contained a high volume of volatile organic compounds, included vinyl, chlorinated rubber, alkyds, low-solids epoxy, epoxy phenolics, urethanes, and other miscellaneous coatings. They have been replaced with acrylics, high-solids zinc-rich primers, high-solids epoxy, epoxy mastic, urethanes, moisture-cured urethanes, and siloxanes. Low-viscosity 100% solids (no solvent) penetrating primers are available where proper surface preparation is not possible. Advanced-technology epoxies, moisture-cured urethanes, and siloxanes can be applied and will cure below 0 °C (32 °F). This feature extends the traditional painting season.

The coatings that protect the metallic bridges by CP include zinc-rich primers and thermal spray materials. The most common thermal spray coating (TSC) used on bridges in the United States is an alloy of zinc and aluminum (85%Zn-15%Al) applied by the electric arc or flame spray process. The thermal spray process melts the wire or powder and sprays the molten droplets onto the substrate with the aid of compressed air. The TSC bond to the steel structure is not metallurgical but mechanical; thus, a very clean abrasive-blast profile is required to obtain a good adhesion strength, something greater than 5 MPa (700 psi). The process of TSC application leaves a slightly porous coating and should be sealed within 8 to 12 h after application, before oxides have a chance to form. The viscosity of the sealer should be relatively thin so as to penetrate the pores of the TSC and only add 13 to 38 μm (0.5 to 1.5 mils) of coating thickness. A color coat can be added for aesthetic reasons and will yield longer corrosion protection to the TSC, much the same as painting galvanized steel does.

The zinc-rich primers that protect by CP include inorganic and organic binders. The binder holds the zinc particles in place and provides the intimate contact with the steel. The binder for inorganic zinc is typically a silicate. Binders for organic zinc coatings include epoxies and urethanes. The application of inorganic zincs requires an experienced applicator and a cleaner blast surface than organic zincs. Inorganic zinc is most often applied in a shop environment. Field touchup of inorganic zinc is normally done with organic zinc.

Bridges erected or painted prior to the mid-1970s were commonly painted with a red lead primer, an excellent low-cost corrosion-inhibiting pigment but containing a hazardous heavy metal and often including cadmium and chromium in the topcoats. These coatings were tolerant of minimal surface preparation. Those old bridges are now requiring complete repainting, considerable maintenance painting, or overcoating. Maintenance painting and/or overcoating requires close evaluation of the old coating to determine if it is justifiable and can support additional paint that will increase the stresses already present in the old coating. An evaluation would include the extent of failed paint or corrosion, the adhesion, the paint thickness, and the budget. Environmental or human factors may enter the evaluation process as well if the location of the bridge is close to schools, playgrounds, or other sensitive receptors to hazardous heavy metals.

Bridge design plays a big part in corrosion prevention. Complex truss bridge design is more prone to corrosion than the girder design. The decks of girder bridges are supported underneath by transverse floor beams and longitudinal steel girders, all relying on piers to support the structure—a fairly simple structure with few corrosion-prone areas. Conversely, truss bridges, both through-truss and under-the-deck truss, have many locations that are prone to entrapment of dirt, debris, and moisture, precursors to corrosion. The entrapment areas are often locations of high structural stresses, all the more important to prevent corrosion with the use of coatings.

Weathering steel does not normally require a protective coating. However, deicing salts can have a very adverse effect. As a result, weathering steel that is subject to frequent wetting from the bridge deck runoff should be protected with multiple barrier coats. Those areas to be protected include the steel immediately under bridge deck expansion joints, abutments, deck drains, and so on.

Concrete Sealers

Despite its durability, concrete is a porous material, which means it can absorb water, chlorides, stains, and other water- or oil-based materials that it comes in contact with. It is necessary to inhibit the transfer of chlorides or other soluble salts into the concrete, where they can attack bare carbon steel reinforcing steel (rebar) embedded within the concrete structure. Rebar, subject to corrosion, will cause spalling and cracking of the concrete due to the seven to ten times volume expansion with the formation of rust.

Sealers are the most widely used materials to restrict or prevent the penetration of soluble salts or other contaminants while allowing the surface to breathe, so moisture within the concrete does not become trapped.

Generally, there are two broad categories of sealers:

- Film formers block the penetration of water and contaminants by forming a barrier on the concrete surface. They may also impart a gloss or sheen, depending on their chemical makeup.
- Penetrants actually penetrate into the concrete surface to increase water repellency and resist stains on absorbent concrete. Usually, they provide invisible protection without changing the surface appearance.

The penetrant type of sealer is the most widely used on concrete bridges, especially on their decks to inhibit the intrusion of chlorides and other deicing chemicals. Boiled linseed oil thinned with mineral spirits had been the sealer of choice for many highway departments for many years, because it is relatively cheap and abundant but not without its faults. Linseed oil will run off or leach out from the concrete and pollute the environment.

Other types of sealers, both penetrants and film forming, include silanes, siloxanes and siliconates, epoxies, gum resins and mineral gums, stearates, acrylics, silicates and fluorosilicates, urethanes, polyesters, chlorinated rubber, silicones, and vinyls.

Important considerations and properties for sealers include penetration depth, ultraviolet resistance, reactivity of concrete materials, service life, chloride and water absorption, water vapor transmission, crack bridging, and deicer scaling resistance. These properties delay or prevent the onset of rebar corrosion and deicer scaling.

Testing by various state transportation departments indicates that penetrating sealers work most effectively if they are applied within three to six months after bridge construction or concrete overlay, with repeated applications approximately every five years.

One of the popular generic penetrating sealers uses silicates that react with lime, alkali, and available moisture to seal the capillaries within the concrete. The sealant properties that control the reduction of concrete permeability are molecular size, viscosity, contact angle (wetting ability), and surface tension. Penetrating sealers do not have the ability to seal cracks and thus must be used in conjunction with a high-molecular-weight material to make an effective deck sealant that will prevent or slow down the migration of chloride ions to the rebar mat.

Electrochemical Techniques: Cathodic Protection, Chloride Extraction, and Realkalization

Jack Tinnea, Tinnea & Associates, LLC

Reinforcement in concrete does not normally corrode. This is because the concrete provides a high-pH environment, typically in excess of 12.4. The corrosion rate of mild and low-alloy steels typically employed as reinforcement in concrete is very low at high pH, because the metals form stable oxide layers and exhibit behavior similar to that of stainless steels. This behavior is known as passivation.

If the normal high pH of the concrete is lost, then corrosion can occur. Carbon dioxide in the atmosphere is able to do this through a process known as carbonation. Certain ions, such as chloride, are aggressive toward steel. If the concentration of an aggressive ion at the surface of the steel reaches a threshold level, corrosion will commence. The exact threshold level concentration is a function of the ion concentrations, pH, and oxygen availability.

A typical corroding area on a reinforcing bar is shown in Fig. 35(a). Corrosion occurs at the anode. The metal does not corrode at cathodic areas. A cross section of Fig. 35(a) is shown in Fig. 35(b). Note that corrosion is typically observed in the 10 to 2 o'clock position relative to the surface, with the rebar areas at greater depth continuing to manifest passive behavior. An electrical circuit analog of this corroding area is shown in Fig. 35(c).

The corrosion potential of steel embedded in concrete reflects both the anodic and cathodic reactions and is therefore a mixed potential. The rebar corrosion potentials are usually obtained with reference electrodes placed on the concrete surface. The corrosion potential readings often vary from location to location on the reinforced-concrete surface, sometimes 200 mV or more within 300 mm (1 ft).

Cathodic protection provides an electrical current to the reinforcement that polarizes the steel surface. Ideally, the application of CP makes all areas cathodic and thereby reduces the corrosion rate that would normally occur. With a successful CP system, the corrosion rate is reduced to levels that are acceptable or even to levels that have no engineering significance.

Electrochemical chloride extraction (ECE) can be looked upon as a high-current CP system that is applied on a temporary basis. The ECE process has been shown able to extract a significant quantity of chloride ions from contaminated reinforced concrete. The basic ECE processes can be modified to allow for realkalization of carbonated concrete.

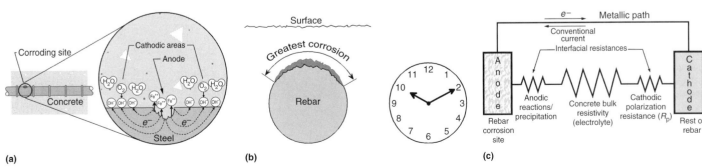

Fig. 35 Corrosion in rebar. (a) Flow of corrosion current in the reinforcement rod. (b) Cross section of corroding rebar. (c) Electrical circuit analog of corroding rebar

Cathodic Protection

History of Cathodic Protection. Cathodic protection has been used for over 100 years to provide corrosion protection to metals exposed to waters and soils. The initial application of CP to steel embedded in concrete dates to laboratory work performed by Richard Stratfull and the California Department of Transportation in 1959 (Ref 176). It was not until the mid-to-late 1970s that full-scale demonstration projects were implemented (Ref 177, 178), and not until the late 1980s that application of such systems became routine.

It has been over 30 years since the first full-scale installation of a CP system to control corrosion of a reinforced-concrete structure. Since that time, acceptance of the process remains sporadic. Some large property owners, such as transportation ministries and departments, accept CP as an integral component of their restoration and maintenance strategy. However, they are industry exceptions. Even endorsement by the Administrator of the United States Department of Transportation's Federal Highway Administration resulted in acceptance of the technology by only a few of the 50 different state departments of transportation.

Why were owners reluctant to embrace CP? There are several reasons for this hesitation. Lack of familiarity with the technology is clearly one factor. Corroding and deteriorating concrete structures tend to be the province of civil engineers. Most early approaches to controlling reinforcement corrosion were simply to remove the unsound concrete and patch the structure. This approach involved technology familiar to the owners and engineers. In time, it became apparent that a "chip-and-patch" solution leads to chipping and patching greater areas each year.

Retaining specialty engineers to design CP systems did not appear to be the principal impediment to acceptance of the technology. Hesitation over embracing CP included concerns over life-cycle costs of a cathodic system versus the life-cycle costs of installing a new bridge deck. Another issue is costs for long-term maintenance of the systems by specialty contractors. There is often significant variation in the relative restoration costs, depending on the agency involved and the design style employed for the bridges and bridge decks.

Elements of a CP System. There are four elements to a CP system. These are the anode, cathode, electrolyte, and metallic path, shown in Fig. 36(a). In a concrete CP system, the cathode is the reinforcement or other embedded metals. The anode is either embedded in the concrete or affixed to the structure surface. The metal path is the reinforcement, reinforcement tie wires, and electrical cables installed as part of the CP system installation. The electrolyte is the concrete.

The electric current in the anode, cathode, and metallic path is carried by electrons, while in the electrolyte the current is carried by ions. Although concrete is not thought of as a conductor of electricity, if it were unable to conduct electricity, there would not be a corrosion problem. As stated earlier, concrete consists of coarse and fine aggregate bonded by hydrated cement. Within this matrix, the conduction of electricity typically occurs in the cement phase.

The anode(s) can be embedded in the concrete, surface-applied, or remote to the reinforced concrete. Embedded anodes include anodes covered with concrete overlays or covered with shotcrete; anodes embedded in grouted jacket systems; anodes installed in slots sawn or holes drilled in the concrete; anodes installed in repair excavations; and anodes embedded in the original concrete at the time of construction. The last example, where CP anodes are installed during initial construction, is also known as cathodic prevention (Ref 179).

Anodes can be applied to the surface of the concrete as coatings or with conductive adhesives. Coatings include carbon-loaded organic coatings and thermal-spray-applied metal coatings. If the reinforced-concrete is partially submerged or buried, it is possible to use submerged or buried anodes that are located remote from the reinforced-concrete structure. Figure 36(b) shows a remote CP configuration. Note that a second electrolyte (i.e., soil or water) is shown in series between the anodic reactions and the bulk concrete resistivity. For applications in seawater, the level of the tide must be included in design considerations. Remote submerged anodes are only able to provide protective current to areas that are submerged. In remote submerged anode systems, the resistivity of the bulk concrete rapidly attenuates the protective current flowing to embedded reinforcement located above the waterline.

Note that in Fig. 36(a) and (b), protective current should flow to all of the reinforcement. The goal of CP is to polarize all portions of the reinforcement to a cathodic potential that suppresses corrosion even in crevices or at the base of pits. That is achieved by polarizing the structure to a potential equivalent to, or more negative than, the corrosion potential of the most anodic site. The geometry and variations in electrolyte resistivity of the structure can result in uneven CP current densities collecting on the reinforcement surface at different locations in the area being protected, with some areas receiving too much current while others receive currents that are insufficient to provide complete protection. Situations of adverse current-density disparity can be avoided with proper attention during the CP system design.

Types of CP Systems. Cathodic protection systems derive the necessary electric potential from an impressed-current source or from the galvanic action of sacrificial anodes.

Impressed-Current Systems. The driving voltage necessary to power CP systems can come from external direct current (dc) power supplies. To provide external power, a transformer is coupled with a rectifier to convert available alternating current (ac) to dc of an appropriate voltage range, as shown in Fig. 37(a). This approach is known as impressed-current cathodic protection (ICCP). Anodes employed in ICCP systems include inert metals such as titanium with metal oxide coatings, carbon-loaded coatings and fillers, conductive ceramics, and soluble metallic anodes such as zinc. The rectifier is placed in the metallic path, shown in Fig. 37(a), with its positive terminal connected to the anode and its negative terminal to the reinforcement, the cathode.

With an ICCP system, the output voltage can be adjusted. With this feature, if the concrete resistivity increases, for example, from drying during the summer, the output voltage can be increased to overcome the increased circuit resistance. Conversely, if seasonal wetting reduces the concrete resistivity, the rectifier can be adjusted to decrease the current output. Some commercially available rectifiers have the capacity to automatically adjust the output to respond to preset voltage, current, or reinforcement potential parameters.

Galvanic Anode Systems. Galvanic anodes used in sacrificial cathodic protection (SACP) provide a potential that does not require dc or rectified ac power. The SACP employs a metal that is more electrochemically active than the one to be protected. Active metals such as aluminum, magnesium, zinc, and their alloys are typically used as galvanic anodes for mild steel. In SACP, a direct electrical connection is made between the galvanic anode and the metal to be protected, as shown in Fig. 37(b). This is a

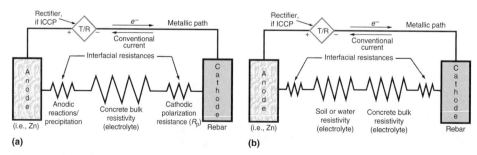

Fig. 36 Electrical circuit analog of reinforced-concrete cathodic protection system components. (a) Anode is embedded or affixed to structure. (b) Anodes installed remote from the reinforced-concrete structure. ICCP, impressed-current cathodic protection; T/R, transformer/rectifier

sacrificial process whereby the galvanic anode corrodes, thus generating the electric current that cathodically protects the other metal.

The SACP provides a simple system that will not lead to conditions of overprotection. However, the output of SACP systems cannot be easily adjusted. The inability to readily adjust the galvanic output presents the designer with limitations on where SACP may be applied. It may be possible to extend the range of SACP systems by augmenting the initial output through the temporary use of rectifiers or zinc-air batteries.

How CP Controls Corrosion. Figure 38 is an Evans diagram of SACP applied in a neutral or alkaline environment. Because cement is alkaline, this diagram is appropriate to discuss using CP to control the corrosion of steel in concrete. The ordinate in Fig. 38 reflects the steel (reinforcement) potential measured with respect to a reference electrode, and the abscissa is the log of the corrosion current. The rate of corrosion is directly proportional to the corrosion current.

Three anodic curves and one cathodic curve are shown. The curves are labeled with the dominant reaction. The anodic curves are the three straight lines that angle up to the right at approximately 30°. The cathodic curve starts angling down to the right, then plunges in an almost vertical direction and finally returns to angling down to the right. The line labeled with the reaction $Fe \rightarrow Fe^{2+} + 2e^-$ is the reaction for the corrosion of iron. Note that this curve crosses the heavier cathodic reaction curve at the point labeled "Corr." This location (potential) represents a condition where steel reinforcement is freely corroding in an alkaline concrete environment. On the ordinate, this point is identified as $E_{corr, S}$ (the corrosion potential for the reinforcement), and as I_A on the abscissa (the corrosion current).

Parallel to and below the iron anodic curve is a curve for zinc corrosion in a similar environment. Zinc is more electronegative than steel. If the zinc is connected to the steel, the steel will be polarized in a negative direction (potential) to the point labeled "CP." This is the mixed potential for the steel-zinc electrochemical couple. Note the marked reduction in the iron corrosion rate that is attained by polarizing to this potential. The Evans diagram in Fig. 38 provides a brief introduction to the traditional basis for CP. In concrete, however, the opportunity exists to alter the environment surrounding the reinforcement. This is a situation that is not typically possible in applications of CP in natural waters or even soils.

CP Design Overview. Most of the early applications of CP to reinforced-concrete structures were ICCP systems, and ICCP requires routine monitoring by specialty contractors. The SACP is an inherently simpler technology than ICCP and requires far less monitoring. Although there were several installations of SACP systems early in the development of CP for reinforced structures, there was little reporting of their success. Part of that absence from the literature derived from early papers reporting that high current densities, ranging from 10 to over 30 mA/m^2, were required for effective CP (Ref 180, 181).

General Design Considerations. Design of CP systems is beyond the scope of this article. See the article "Cathodic Protection" in *Corrosion: Fundamentals, Testing, and Protection*, Volume 13A of *ASM Handbook*, 2003. Several texts discuss CP of concrete reinforcement (Ref 38, 182, 183). NACE International has published several relevant documents (Ref 184–189).

Other standards that address CP of reinforced concrete include the European Union's EN 12696:2000, "Cathodic Protection of Steel in Concrete, Part 1: Atmospherically Exposed Concrete" (Ref 190), Australian standard AS 2832.5, "Cathodic Protection of Metals—Steel in Concrete Structure" (Ref 191), and the Norwegian NORSOK standard M-503 (Ref 192).

NORSOK-503 only briefly addresses CP of concrete.

Specific Design Considerations. Generally, CP is applied to bare black steel reinforced-concrete structures. It can also be applied to prestressing steels, but special care is required to avoid excessive polarization, hydrogen evolution, and embrittlement of the high-strength steels employed (Ref 186). Cathodic protection of epoxy-coated reinforcement can be problematic because of electrical continuity problems introduced by the presence of the epoxy coating and the likelihood of significant portions of the reinforcement network being electrically isolated from each other. Beyond the basic corrosion issues that affect the application of CP, there are other factors that impact the restoration decision process. These include:

- Availability of alternate vehicle traffic routes during restoration or replacement
- Availability of ac electrical power
- Historic value of the structure
- Use of stay-in-place formwork
- Use of epoxy-coated reinforcement
- Use of resin injection, waterproofing, or other electrically nonconductive materials
- Size of the structure
- Presence of prestressing elements
- Shoreline considerations

Current Requirement. A portion of the basis for early practice to employ relatively high current densities came from attempts to polarize the reinforcement potential to levels used to protect pipes in soil and marine environments. Because these current densities typically were beyond the

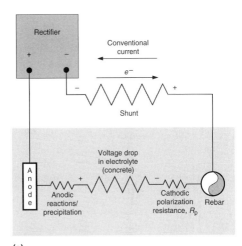
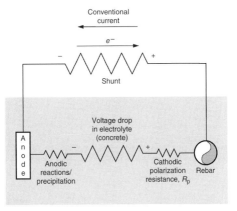

Fig. 37 System components for cathodic protection. (a) Impressed current. (b) Sacrificial or galvanic system

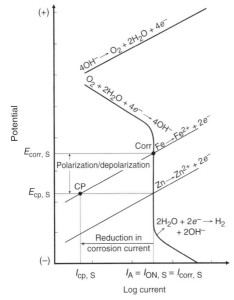

Fig. 38 Evans diagram for the application of a galvanic cathodic protection (CP) system. Note: Log current is the absolute value of the current density, where both the anodic and cathodic currents are plotted as positive values. See the text for description.

capacity of most galvanic systems, development of SACP was impeded.

In time, it was noted that substantially less current was required for effective CP (Ref 193). A system that employed perforated zinc sheets on a bridge deck with a porous asphalt overlay, which was initially believed to provide insufficient current output, proved successful in effective long-term corrosion control (Ref 194). Success with SACP was also reported with similar perforated zinc sheets in protecting marine piling (Ref 195). The perforated zinc sheets for both installations came from scrap metal left over from the manufacture of U.S. pennies.

Effective protection of bare steel in quiet seawater requires CP currents of approximately 20 to 30 mA/m^2 of uncoated steel (Ref 196). Cathodic protection currents in the range of 0.5 to 1 mA/m^2 are required to provide similar protection to bare steel buried in many natural soils with resistivities of 5 to 15 $\Omega \cdot$m, while resistivities of 15 to 400 $\Omega \cdot$m for a similar level of protection will require currents of only 0.1 to 0.5 mA/m^2 (Ref 196).

Currents on the order of 1 to 7 mA/m^2 may be required to fully polarize actively corroding areas when CP is first applied to corroding steel in concrete (Ref 197). Subsequent to polarization, CP may typically be maintained with significantly smaller currents that range from 0.25 to 1.0 mA/m^2 (Ref 198–200). Protection of piling in subtropical marine splash zones may require higher currents—5 mA/m^2 to protect epoxy-coated reinforcement, and 10 mA/m^2 to protect bare reinforcement (Ref 201).

It is not surprising that less current is required to protect steel reinforcement in concrete than steel in seawater. Several factors are responsible. It is reasonable to expect seawater to have a much higher concentration of aggressive ions, such as chloride or sulfate, than high-resistivity natural soils. In addition, oxygen availability is expected to be less for steel buried in soil than steel exposed to seawater. Both of these factors lead to expectations that less CP current will be required to protect a structure buried in the soil than the current needed to protect that same structure if it were submerged in seawater.

Resistance in Circuit. Electrical resistance plays a major role in the effectiveness of CP of reinforced-concrete structures. The output of a CP system follows Ohm's law, $E = IR$, where E is output voltage, I is CP current, and R is the total circuit resistance. Resistance impedes the current output of the CP system and, depending on where that resistance occurs, can influence its effectiveness. An increase in resistance at the reinforcement surface reflects reduced corrosion and is desirable. Variation in resistance in other areas produces variation in the CP system output. Typical approaches to CP design attempt to address these factors. The relatively fixed output voltage of SACP systems means that they are more impacted by system resistance than are ICCP systems.

The overall CP system circuit resistance as applied to concrete structures comes from several sources, including:

- Concrete resistivity
- Concrete-anode interfacial resistance
- Concrete-reinforcement interfacial resistance
- Resistance between individual elements of the reinforcement network

Concrete resistivity depends on several factors. These include the initial concrete mix design, the presence of mobile soluble salts in the pore water, and the moisture content of the cement. Addition of mineral admixtures, such as silica fume or fly ash, can increase a concrete resistivity. The presence of chloride ions can reduce a concrete resistivity. However, the largest single contributor to the resistivity of a concrete is the concrete moisture content. The resistivity of any concrete is much lower when its moisture content is high than when it is dry.

Cathodic protection systems are broken down into zones to assist in output control. If adjacent areas vary in initial bulk concrete resistivity, a higher current will flow to the area with lower concrete resistivity. Designing zone breaks to separate those areas of different concrete resistivity can provide better current control and direct how much current is applied to the different areas.

Interfacial resistance at the anode, however, does not improve corrosion protection. In fact, efforts are undertaken to reduce anode-concrete interfacial resistance (ACIR) and thereby increase current output of both ICCP and SACP systems. Several processes involved may act to produce increases in ACIR. One is simply moisture content. This is particularly true with atmospherically exposed CP systems where the anode is applied at or near the concrete surface and where normal variations in local rainfall and humidity produce variations in the level of concrete pore saturation. The ACIR will drop on cool, wet, and humid days and increase on hot and dry days (Ref 172).

Another source of ACIR increase derives from anode material precipitating in the cement pores, reducing their effective hydraulic radius. This process is known as secondary mineralization and has been observed on CP systems that employ metallized zinc-bearing coatings (Ref 202). Zinc is able to replace calcium in the cement matrix and can produce almost a waterproofing effect.

Humectants, which are hygroscopic and attract water, have been employed to counteract drying in the cement pores surrounding CP anodes. Included in these attempts are lithium salts and calcium chloride (Ref 203, 204). Lithium salts also help to avoid alkali-silica reactions that can be an issue with some aggregates employed in concrete.

Polarization resistance, R_p, is the resistance at the concrete-reinforcement interface. Changes in R_p reflect the likelihood of corrosion. Initially, in uncontaminated concrete where the pH is high (~13), the reinforcement R_p value is relatively high, because the steel is passive and the corrosion rate is low (Fig. 39a). The R_p is reflected in the slope of the line identified with the reaction label Fe→Fe^{2+} + 2e^-, and the anodic Tafel constant, β_a, is shown as >700 mV per decade. Note that the corrosion potentials are relatively electropositive and tightly clustered, and the slope of the iron oxidation line is steep.

Contamination by chloride ions or loss of pH through carbonation can initiate corrosion. With the onset of corrosion, the observed R_p becomes less. In Fig. 39(b), a second iron oxidation line is added, showing active corrosion and an anodic Tafel constant β_a < 150 mV per decade. The slope flattens for active corrosion iron oxidation reaction, and reinforcement potentials become more electronegative and dispersed. On corroding reinforced structures, a family of Fe→Fe^{2+} + 2e^- curves would fill between the two lines shown in Fig. 39(b). The different iron oxidation reaction slopes would reflect the local conditions, such as the ratio of chloride ions to hydroxyl ions ([Cl$^-$]/[OH$^-$]), pH, and oxygen availability.

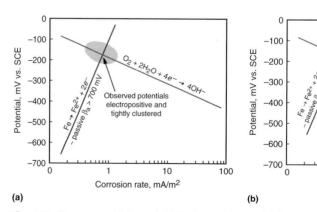

Fig. 39 Current-potential diagrams. (a) Passive conditions. Tafel slope, β_a > 700 mV. (b) Corroding conditions. Tafel slope, β_a < 150 mV. Corrosion occurs as passivity is lost and active corrosion proceeds. SCE, saturated calomel electrode

The application of CP changes the reaction at the reinforcement surface from iron oxidation, $Fe \rightarrow Fe^{2+} + 2e^-$, to oxygen reduction $1/2 O_2 + H_2O + 2e^- \rightarrow 2OH^-$. The passage of CP current will move chloride ions away from the reinforcement, and soluble alkali ions, such as sodium and potassium, toward the rebar. With time, CP affects chloride removal and increases the pH (realkalization). These processes return the environment surrounding the reinforcement to one that supports passive behavior of the embedded steel. If a properly operated CP system is turned off and allowed to depolarize, the observed steel oxidation R_p increases (Fig. 40; also Ref 199). Note that in Fig. 40, the reinforcement corrosion potentials become more electropositive and more tightly clustered. If CP was not able to change the environment surrounding the embedded reinforcement, it would be much less effective. In the lower right of Fig. 40, the application of 40 mA/m² of CP current is shown. Following the actively corroding iron oxidation line, $\beta_a < 150$ mV, the application of CP reduces the pre-CP corrosion current of 25 mA/m² to 18 mA/m², a corrosion rate too high to sustain structure durability. However, through the removal of chloride ions, realkalization, and a return to a passive iron oxidation line, $\beta_a > 700$ mV, the resulting repassivation produces a far greater corrosion rate reduction.

Figure 41 shows a Pourbaix diagram of the repassivation process. When a reinforced-concrete structure is first constructed, the embedded steel is in the passive regime, as indicated by point "P." In this example, carbonation of the concrete reduces the pH, initiating active corrosion (as show by the arrow from "P" to point "A"). The initial application of CP polarizes the steel to point "CP$_i$." With time, realkalization increases the pH, moving the steel to point "CP$_f$." If the CP systems were turned off, the steel would then return to point "P."

Figure 42(a) shows the polarization decay of a CP system zone that had been in operation for over five years (Ref 194). The CP anode system was a saw-slot type installed on a bridge deck. This installation configuration allowed measuring reinforcing-steel corrosion potentials on a square grid of 1.5 m (5 ft). The ordinate of Fig. 42(a) is the average of 200 individual corrosion potential readings relative to a copper-copper sulfate electrode (CSE). The abscissa is the time that the 200 readings were obtained. Note that depolarization continues throughout the 90 day testing period. Also note that after approximately 30 days, the average corrosion potential reflects a return to passive conditions. The depolarization testing was terminated after 90 days.

Table 7 lists statistics for the pre-CP and 90 day depolarized (post-CP) corrosion potentials shown in Fig. 42(a). The post-CP mean corrosion potential for the depolarized rebar is almost 100 mV less active (more electropositive) than the pre-CP value. The most active post-CP reading, −219 mV, is over 300 mV more positive (less active) than the most active pre-CP reading. Pre-CP readings reflect multimodal behavior, with some areas actively corroding and others not. The post-CP data more closely follow standard distribution, as reflected by its lower standard deviation, skewness, and kurtosis. Standard deviation is a metric of the spread or dispersion of a data set. Skewness describes the degree of asymmetry of the sample distribution. Positive skewness indicates that the distribution tails toward more positive values, while negative skewness indicates that the distribution tails in a

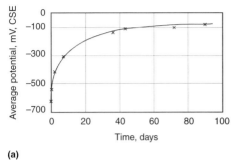

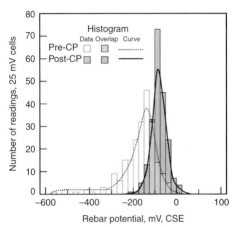

Fig. 42 (a) Decay of average corrosion potential over a 90 day period. Each data point is the average of 200 individual readings obtained on a square grid of 1.5 m (5 ft) on center over an entire cathodic protection (CP) zone (saw-slot-style anode system). No additional readings were obtained after 90 days. If the CP system were left turned off indefinitely, chloride ions would diffuse back to the steel surface, and corrosion would again commence. (b) Histograms for pre- and post-CP potential surveys of the same CP zone. Post-CP survey potentials were obtained after 90 day depolarization. The abscissa shows the rebar potential observed. Each histogram cell (the rectangular columns) is 25 mV wide. The height of the histogram cell, corresponding to the ordinate value, reflects the number of readings observed within each 25 mV cell. Curves for the two data sets smooth the histogram stair steps. For both data sets, individual readings were obtained on a square grid of 1.5 m (5 ft) on center over an entire CP zone (saw-slot-style anode system). CSE, copper-copper sulfate electrode

Table 7 Pre- and post-cathodic protection (CP) rebar potential data

Variable	Pre-CP	90 day depolarization
Sample size	220	206
Mean, mV, CSE	−170	−82
Median, mV	−152	−83
Maximum, mV	−26	21
Minimum, mV	−570	−219
Range, mV	544	240
Standard deviation	±83	±36
Skewness	−1.76	−0.11
Kurtosis	5.10	1.19

CSE, copper-copper sulfate electrode. Source: Ref 194

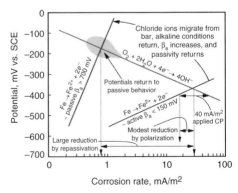

Fig. 40 Cathodic protection (CP) of reinforcement can lead to repassivation of the rebar when chloride ions migrate from rebar and/or realkalization of the concrete surrounding the rebar occurs. Under these conditions, the CP can be removed. The rebar will depolarize and return to passive conditions, as shown. When chloride ions return to the rebar surface or pH is again lost, corrosion reoccurs.

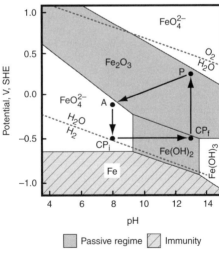

Fig. 41 Path of cathodically protected steel through an Fe-H$_2$O Pourbaix diagram. Using cathodic protection (CP), the corrosion rate is substantially reduced. Under favorable conditions, chloride ions migrate from the rebar surface, alkalinity is reestablished in the concrete surrounding the rebar, and the reinforcement then will reflect passive behavior when the CP current is removed. SHE, standard hydrogen electrode

negative direction. The kurtosis of a data set characterizes vertical asymmetry from normal distribution. Positive kurtosis reflects peaked distribution, while negative kurtosis indicates flattened distribution. The data in Table 7 reflect that this application of CP current was able to return the reinforcement to a passive environment, with the observed corrosion potentials more positive and tightly clustered, following the pattern shown in Fig. 40.

Figure 42(b) shows histograms of the pre-CP and 90 day depolarization post-CP data sets. The abscissa in Fig. 42(b) reflects the measured reinforcement corrosion potential (mV, CSE). For both sets, the histogram cell width was 25 mV. The ordinate for Fig. 42(b) is the reading observed in each 25 mV wide cell. Smoothed curves overlay their respective histogram. The pre-CP readings show a much broader data set that tails to the left (more negative/more corroding corrosion potentials). The tail of the pre-CP readings is reflected in the -1.76 skewness of the pre-CP data set compared to almost no skew with the post-CP data. Both data sets are more peaked than a normal distribution curve, with the post-CP kurtosis much less than that of the pre-CP. The histograms in Fig. 42(b) and the statistics in Table 7 show that post-CP readings are much more tightly clustered and better follow a normal distribution curve than do the pre-CP readings. The pre-CP readings include areas of active corrosion, while the 90 day depolarized post-CP readings reflect a return of passive conditions. This observed statistical behavior of the corroding pre-CP data follows previous observations and discussions that abrupt changes in rebar potential are indicative of corrosion activity. The use of statistics is then a valuable tool to use with absolute corrosion potential values in interpreting bridge data.

Resistance in the reinforcement network can lead to some elements being effectively isolated from the CP system. Generally, when reinforced-concrete structures are fabricated, electrical continuity between the reinforcement bars is not specifically addressed. The standard practice of overlapped bars, multiple crossings between longitudinal and transverse bars, and the use of metal tie wires to secure reinforcement prior to placing concrete provides reasonably good electrical continuity. However, prior to the design of a CP system, it is important to have a sense of the degree of electrical continuity within the reinforcement network.

Effect of Environmental Factors on CP. For SACP systems, environmental factors may require greater attention than is necessary for ICCP systems. If the concrete dries or there is a substantial increase in the ACIR, one may simply increase the output voltage of an ICCP system to sustain the desired output of CP current. With SACP systems, it is not possible to make such simple adjustments.

Temperature impact on SACP current output has its foundation in the Arrhenius equation. The output from sacrificial systems depends on the corrosion rate of the galvanic anode material. As with most chemical reactions, the rate of corrosion can be expressed as follows:

$$\text{rate} = k[A]^a[B]^b \quad \text{(Eq 12)}$$

where the rate is in mol $dm^{-3}s^{-1}$, k is the rate constant, a and b are the order of reaction with respect to the reactants A and B, respectively, and [A] and [B] are concentrations in mol dm^{-3}.

The rate constant is determined from the Arrhenius equation:

$$k = Ae^{-\frac{E_A}{RT}} \quad \text{(Eq 13)}$$

where k is the rate constant, A is the frequency factor (derives from frequency of collisions and their orientation and is roughly constant over small temperature changes), e is the base for natural logarithms, E_A is the activation energy, R is the gas constant, and T is the absolute temperature in K.

If the impact of Arrhenius on the rate of reaction is examined, one can note a general guideline that an increase of 10 °C doubles the reaction rate. For example, assume a reaction at 20 °C is increased to 30 °C and that the activation energy is 50 kJ mol^{-1}:

$$\text{At } 20\,°C: \; e^{-\frac{E_A}{RT}} = e^{-\frac{50,000}{8.31 \times 293}} = 1.21 \times 10^{-9} \quad \text{(Eq 14)}$$

$$\text{At } 30\,°C: \; e^{-\frac{E_A}{RT}} = e^{-\frac{50,000}{8.31 \times 303}} = 2.38 \times 10^{-9} \quad \text{(Eq 15)}$$

From this example, it can be seen that for reactions at approximately 20 °C, which have activation energies of approximately 50 kJ mol^{-1}, the rate of reaction will approximately double for each 10 °C increase in temperature.

For SACP systems applied as thermal-sprayed coatings, it is reasonable to assume that temperature will also affect metal loss at the surface. The literature reports significant variation in observed service lives for metallized CP systems. In Florida, the estimated service lives of metallized SACP in subtropical marine environments range from 5 to 10 years (Ref 205). Service lives of over 25 years based on bond strength are observed for metallized-zinc CP systems installed in temperate marine environments along the Oregon coast (Ref 206). In both examples, the majority of the bridges are not exposed to serious air pollution. The rainfall along the Oregon coast is higher than the Florida Keys, where many of the CP-protected structures are located. Also, the average relative humidity is slightly higher along the Oregon coast than in the Keys. Higher rainfall and humidity should make for a more aggressive environment and hence shorter service life for CP systems along the Oregon coast. One corrosion rate factor where Florida clearly exceeds Oregon is average annual temperature. The annual average temperature in Key West, Florida, is approximately 26 °C (79 °F), while in Newport, Oregon, it is only 10 °C (50 °F). If those temperatures are inserted into Eq 13, with E_a of 50 kJ mol^{-1}, the rate constants are 1.78×10^{-9} and 5.69×10^{-10} for Key West and Newport, respectively. The "rule-of-thumb" approach in this example would lead one to predict that the corrosion rate of metallized-zinc coatings on Florida bridges may be expected to be approximately three times greater ($1.78 \times 10^{-9} \div 5.69 \times 10^{-10} = 3.1$) than that of Oregon coast bridges. That ratio approximates the observed service lives seen with field applications in the two areas.

Moisture is another environmental factor that can influence CP systems protecting concrete structures. The moisture content greatly influences the bulk resistivity of concrete and hence the output of any CP system. Rain, sunshine, and humidity can have large effects on reinforcement CP systems. The atmospheric exposure is typically much more dynamic than traditional buried and submerged environments for CP systems. Marine applications that are submerged do not see such variations, but areas in the tidal and splash zones can see significant moisture effects.

An example is a reinforced-concrete pile that supports a coastal bridge (Fig. 43). The lowest portions of the pile are buried in the mud or sand. Above that is a section of pile that is continually submerged. Above the submerged area is the tidal zone. The next area is what is known as the splash zone, and the last is the atmospheric zone. Adjacent to the zone identifications are typical relative rates of corrosion observed.

In areas just above high tide, the drying action of air and/or sun exposure can create a wicking action, where salt water that has entered the concrete evaporates. When the water evaporates, the sea salts remain behind. Repeated cycling of tides and evaporation can significantly concentrate chloride ions in this area.

Because many reinforced-concrete structures are directly exposed to the weather, any corrosion control efforts, including CP, should

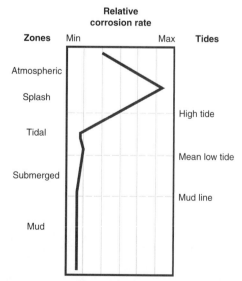

Fig. 43 Marine tidal zones

examine drainage. Ponding of rainwater or splashed seawater on the concrete can accelerate corrosion. Local ponding, by increasing the local moisture content, can lead to localized areas where CP systems have unusually high current output. If attention to drainage and ponding is not included in the design, excessive currents can develop, leading to localized failure of the CP anodes (Ref 194, 202).

CP Criteria. Concrete, absent contamination by aggressive ions or the loss of pH through carbonation, provides a much different environment to embedded reinforcement than natural soils or waters do to buried or submerged steel. Typically, natural soils or waters do not provide an environment that passivates steel, as does concrete. Passivating conditions will be restored to the steel reinforcement if the application of CP can cause migration of chloride or other contaminating ions from the steel surface and/or return the high pH to the steel surface. In fact, it has been noted for some years that the operation of CP systems will dramatically reduce the chloride content in previously contaminated structures, as seen in Fig. 44 (Ref 194) and elsewhere (Ref 207, 208).

If structures that previously enjoyed long-term CP were allowed to depolarize over extended periods, it was noted that reinforcement in those previously protected structures exhibited characteristics similar to noncorroding, noncontaminated structures. If sufficient current can be applied to provide initial corrosion control, then with time, the chloride concentration at the level of the reinforcement should diminish (Fig. 44), and reduced current will be required to sustain protection (Fig. 39a, 40).

A minimum decay of reinforcement polarization of 100 mV is a generally accepted criterion for CP of steel in concrete (Ref 184, 209). This criterion is applicable for systems when the cathodic reaction is under mass-transfer control, while the anodic reaction is under activation control, and also when the cathodic reaction kinetics are under activation control (Ref 209). The former case reflects high corrosion rates or those where the concrete pore structure is near saturation and the rate of oxygen diffusion controls the rate of corrosion. The latter case is typical of low current densities, and the basis for protection derives from the return of a high-pH/low-chloride environment around the reinforcement that restores passivation. Further discussion of CP criteria as applied to reinforced concrete appears in several NACE International publications (Ref 184–186).

CP Anode Configurations. In general terms, CP systems involve either distributed anodes or discrete anodes. Distributed anodes conform, more or less, to the surface of the concrete being covered. Discrete anodes are isolated individual installations and are typically located to address a specific corrosion area. It is possible to install discrete anodes as networked groups so they become more distributed in application. Before addressing current practice for reinforced concrete, the following brief discussion of earlier approaches, some of which are, at the time of this writing, still providing complete protection to the reinforcement, is included.

Conductive asphalt overlays applied to bridge decks were the original applications of CP to steel in concrete (Ref 176–178, 210). Conductivity was gained by adding carbon coke to the asphalt. Connection to the asphalt was typically made through the use of high-silicon cast iron "pancake" anodes. In a nomenclature that continues to this day, the cast iron anodes were called primary anodes, while the conductive overlay was called a secondary anode. These systems were effective from a corrosion standpoint. However, if the bridge did not have a previous overlay, the anodes added a significant dead load. Another problem with the coke addition is that it softened the asphalt and often would require a second course of nonmodified asphalt as a riding surface.

Cast-in-place conductive polymer secondary anodes have been installed in several forms. Some of the earliest involved a Federal Highway Administration (FHWA)-developed cast-in-place polymer. This polymer is a polyester type that was loaded with conductive carbon filler. Electrical connection and continuity of the polymer was achieved with graphite-fiber and/or platinized niobium-copper-cored primary anodes. Early attempts involved installation of the cast-in-place conductive anode material into slots cut into the riding surface of bridge decks. Although this approach avoided the dead-load issues of the conductive asphalt overlays, there were problems with wear of the polymer and having the polymer pop out of the slots.

To address anode wear and pop-out problems, a mound installation was developed. Here, the cast-in-place polymer secondary anode was installed as a mound on top of a scarified deck surface. Again, carbon-fiber or platinized primary anodes were employed to make connections and provide longitudinal continuity. After the conductive polymer cured, the deck was overlaid with a wearing course. The addition of the wearing course again involves significant addition to dead weight.

Precast and extruded polymer secondary anodes were also used to apply CP to reinforced-concrete structures. The precast polymer anodes were most often fabricated using carbon-filled polyester resins. As with the cast-in-place installations, carbon-fiber or platinized primary anodes were run down the core of the precast anode to provide electrical connection and continuity should the resin crack. Some of these precast installations are still in operation.

The extruded polymer anodes were manufactured with copper core wires with a cross-linked carbon-filled outer polymer coating. These anodes tended not to weather well in bridge installations. The outer polymer tended to become brittle and crack. The copper core wire then failed at the cracks, and electrical continuity was lost (Ref 194).

Current practice ICCP systems include distributed and discrete anodes.

Distributed anode configurations include carbon-filled organic coatings, thermal-sprayed metal coatings, and rare earth catalyzed-titanium expanded mesh as secondary anodes. Primary anode materials and installations vary considerably. The two coating approaches are typically installed directly on the surface of reinforced sub- and superstructure elements. Zinc and catalyzed titanium are the most frequently used metals in the thermal spray installations. The catalyzed-titanium expanded mesh can be installed on decks that are overlain with a sufficiently conductive concrete or on sub- and/or superstructure elements by covering with a sufficiently conductive shotcrete.

Discrete anode configurations generally involve the use of catalyzed-titanium inserts or conductive ceramics. The conductive ceramics are a titanium suboxide material, usually in a tube form. These anodes are installed in holes drilled or cut into the concrete.

Current practice SACP systems likewise include distributed and discrete anodes.

Distributed anode configurations include thermal-sprayed metal coatings, expanded mesh, and perforated scrap material. Successful thermal spray metals include low-iron high-purity zinc (e.g., ASTM B418, type II), an 85-15 zinc-aluminum alloy, and a proprietary 88Al-12Zn-0.02 In alloy. High-purity zinc expanded mesh and scrap (Ref 195) has also been used successfully, particularly in wrapped or jacketed installations on concrete piles in marine environments (Ref 123). The SACP systems that use zinc sheets attached to the concrete surface with a proprietary hydrogel adhesive are still in operation but, at the time of this writing, are no longer being manufactured.

Discrete anodes are fabricated from small ingots of galvanic anode material. The SACP anodes provide protection by corroding. Because discrete anodes are typically embedded into the concrete, the volume of corrosion product generated by SACP anodes must be considered. For example, aluminum produces

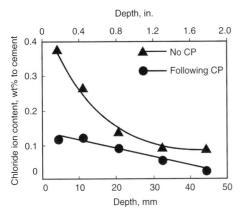

Fig. 44 Chloride migration resulting from cathodic protection (CP)

high-volume corrosion products, and early attempts to use aluminum anodes generated cracks and delaminations of the concrete. Zinc corrosion products are not so voluminous and can actually replace calcium in the cement phase through a secondary mineralization process (Ref 202, 211). Because zinc is an amphoteric material, it is necessary to keep the pH in the bedding material high. Likewise, keeping the grout surrounding the discrete SACP anodes relatively moist will improve output by lowering the grout resistivity. To accomplish these ends, proprietary additives, such as lithium hydroxide and lithium halides, are employed in the fabrication and/or installation of discrete anodes.

CP Installation. Installation of CP systems involves several common issues, regardless of whether SACP or ICCP is employed. The following are brief discussions of these shared factors.

Repairs. It is important to make repairs that are compatible with CP. If nonconductive materials are used, such as injected epoxy, care should be exercised in their application. This is to avoid having the nonconductive material inadvertently shield reinforcement from the CP current source. Secondly, it is necessary to confirm electrical continuity of the reinforcement. Repairs typically afford the opportunity to perform this work, because reinforcement is usually exposed during this process.

Short Circuits. With ICCP, and sometimes with SACP, it is important to avoid creating short circuits between the impressed-current anode and the embedded reinforcement. With ICCP, short circuits can seriously cripple the system and adversely affect current distribution and the effectiveness of the system. With SACP systems, short circuits can limit the ability to monitor or control the system. Metal at or near the surface, such as tie wires and support chairs used to position reinforcement prior to concrete placement, can be a problem on soffits.

In many SACP systems, short circuits are not necessarily a problem. Some SACP systems are designed so that the impact of short circuits limiting system control is accepted and special short-free areas are created for monitoring. If reinforcement is exposed, direct connection between the bar and the anode is possible. For example, Fig. 45 shows a thermal-sprayed zinc coating protecting both exposed and embedded reinforcement on a marine pile. Here, no primary anode was used. Rather, the zinc was directly sprayed on both the exposed reinforcement and over the surface of the adjacent sound concrete, protecting both exposed and buried steel.

Surface Preparation. When employing surface-applied coatings for either ICCP or SACP, care should be exercised during surface preparation. As with all coating projects, success in applying CP coatings starts with good surface preparation. Abrasive blasting is frequently employed, but care must be exercised to avoid excessive removal of the high-cement-content surface layer and exposing too much coarse aggregate. In concrete, it is the cement phase that is responsible for conducting electricity.

Figure 46(a) shows a proper abrasive blast that cleans the surface and provides a good profile to improve coating adhesion. Figure 46(b) shows a surface where too much cement-rich material has been removed. Note that a significant portion of the metallized coating is applied directly to coarse aggregate. Typically, coarse aggregate has a very much higher resistivity than the cement phase of concrete and also forms a weaker bond with the coating. The effective electrical contact area between the metallized coating and the underlying concrete is much less in Fig. 46(b) than in the situation shown in Fig. 46(a), because the gravel is such a poor electrical conductor. A CP system installed with a proper surface preparation, as shown in Fig. 46(a), will have far better current output than a system installed with a surface prepared as shown in Fig. 46(b).

CP System Energizing. After the system has been installed, it must be energized. Energizing and the initial evaluation of CP systems for reinforced-concrete structures require the input of and testing by specialists. The most commonly employed criterion of CP is a potential decay of 100 mV. In unsaturated atmospherically-exposed concrete, where oxygen is readily available but the concrete is not too dry, polarization decay can occur within 4 h (Ref 184). For concrete with a high moisture content, which impedes oxygen transport, or very dry concrete, depolarization can take a longer time, 10 to 24 h or more. Australian standard AS 2832.5, "Cathodic Protection of Metals—Steel in Concrete Structure," allows extended depolarization for up to 72 h (Ref 191). Observation of several extended interruptions of reinforced-concrete CP systems has shown that depolarization continues for over 90 days (Fig. 42a; Ref 194).

Electrochemical Chloride Extraction (ECE) and Realkalization

Because the contamination of concrete by chloride ions is often the cause of corrosion in reinforced structures, it is natural that solutions were sought to prevent chloride ion contamination from occurring. For structures that were already contaminated with chloride ions, it was also reasonable that researchers would investigate how to remove these troublesome ions. In the 1970s, the FHWA funded two studies on removing chloride ions from chloride-contaminated reinforced-concrete structures (Ref 212, 213). Additional research by a private Norwegian company led to several patents, including ones in North America (Ref 214–216). The Strategic Highway Research Program (SHRP) also investigated the ECE technique (Ref 217). Field data indicate the technique is able to stop corrosion for 5 to 10 years.

Fig. 45 Directly applied thermal-sprayed zinc sacrificial cathodic protection system protecting both exposed and embedded reinforcement

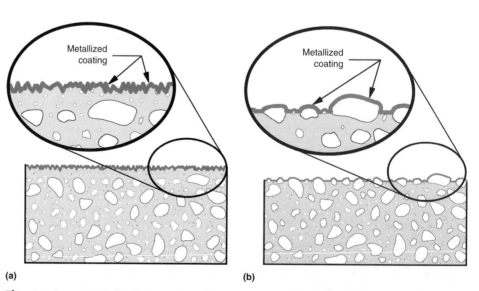

Fig. 46 Cross sections of coated concrete. (a) Proper surface preparation where most of the contact between the coating and the concrete is with the cement phase. (b) Improper surface preparation with too much contact between the coating and the nonconductive coarse aggregate

ECE and Realkalization Theory. As discussed earlier, chloride ions migrate toward the anode. The ECE process is based on the fact that anions, like chloride, migrate to the anode when a direct current is passed through an electrolyte. Hydroxyl ions are the only other mobile anion present in quantity in chloride-contaminated concrete. To balance the charge transfer occurring in ECE (and in CP), cations migrate to the cathode, the reinforcement. In chloride-contaminated concrete, these will be sodium, potassium, and calcium. The amount of current that each ion carries depends on its charge, its concentration, and its mobility in the cement pore structure. Temperature will also play a role.

The ECE works by removal of chloride ions. How alkalinity is restored, although essentially the same process, is a bit more complex. At the surface of the steel (the cathode), two reduction reactions typically take place:

$$O_2 + 2H_2O + 4e^- \rightarrow 4(OH^-) \quad \text{(Eq 16)}$$

$$2H_2O + 2e^- \rightarrow H_2 + 2(OH^-) \quad \text{(Eq 17)}$$

Both of these reactions produce hydroxyl ions, OH^-, which will increase the pH at the reinforcement surface. Oxygen availability is limited in concrete, which slows the progress of Eq 16. The current densities employed with ECE/realkalization are 100 to 500 times the current densities applied in CP systems. Most of the current collected on the reinforcement at ECE/realkalization current densities does so through the reduction of water (Eq 17). This process generates hydrogen gas and must be carefully considered if high-strength steels, such as those commonly used in prestressed concrete, are present in the candidate structure.

Evaluation of Candidate Structures. Successful application of ECE or realkalization depends on an accurate understanding of several concrete characteristics. Overall circuit resistance impacts the current output through Ohm's law, $I = E/R$. The concrete resistivity and cover depth over the reinforcement impact circuit resistance. A high degree of electrical continuity throughout the reinforcement matrix should also be confirmed.

Another concern with the application of ECE or realkalization is that the process can cause reaction with alkali-sensitive aggregate when present. Use of additives to the electrolyte, such as lithium borate, can reduce or avoid problems with alkali-reactive aggregate.

Preliminary investigations should confirm the availability of utilities (power and water) at the site. Also, because the temporary anodes may be in place for periods of four to eight weeks, the impact of that presence on other activities in the vicinity and the security of the system should be considered.

Application of ECE/Realkalization. If the reader is interested in applying ECE and realkalization technology, it is suggested that they review the SHRP publication (Ref 217) and the NACE International state-of-the-art report on ECE (Ref 218).

Repairs. As with CP, necessary repairs are made to the structure prior to commencing the ECE/realkalization process. Likewise, repair materials employed are cementitious, and shrinkage and/or resistivity requirements should be specified as is appropriate.

Anode Electrolyte. The anodes used for ECE and realkalization are bedded in a temporary electrolyte. Numerous electrolytic solutions have been used. The most commonly used electrolytes are potable water and saturated lime water (calcium hydroxide). In situations where reaction with alkali-sensitive aggregate is possible, lithium bromate can be added to the solution. Lithium electrolytes are, however, expensive both to purchase and to dispose.

Anode reactions that can occur include:

$$2H_2O \rightarrow O_2 + 4H^+ + 4e^- \quad \text{(Eq 18)}$$

$$2Cl^- \rightarrow Cl_2 + 2e^- \quad \text{(Eq 19)}$$

Equation 18 will tend to make the electrolyte acidic. The reaction in Eq 19 shows the evolution of chlorine gas. If the electrolyte pH is allowed to drop below 4, significant amounts of chlorine gas can be generated. This could present health and safety issues. It is common practice to maintain the electrolyte pH to levels above 7 to minimize chlorine generation.

Catalyzed titanium is often used as an anode material for ECE and realkalization. It has the advantage of not corroding.

Steel anodes have a price advantage, but they are not inert. The corrosion of steel anodes minimizes electrolyte acidification and chlorine generation. However, the steel corrosion can produce rust stains on the concrete surface.

Electrolyte Medium. Given that the chloride-contaminated structure likely includes decks, soffits, and vertical surfaces, it is necessary to confine the electrolyte in a different way. For the topside of a deck, a simple dam can confine the electrolyte. For columns, a surface-mounted tank can be used. For vertical surfaces, cellulose fiber (e.g., used newspapers) can be used to formulate a spray-on electrolyte. Felt mats are also employed. After the electrolyte medium is placed, it is charged with the selected anode electrolyte.

Current Application. After repairs, verification of reinforcement continuity, and installation of the temporary anodes, anode electrolyte, and electrolyte medium, the transformer/rectifier that powers the system is energized. The initial voltage typically is slowly brought up to produce an initial current level of 1.0 A/m^2 (0.1 A/ft^2). The maximum system output voltage is kept below 50 V for safety reasons. The concrete resistivity often will drop as the concrete gets wet. Maximum current outputs are generally kept at values less than 2.0 A/m^2 (0.2 A/ft^2) of concrete surface, or less than 5.0 A/m^2 (0.5 A/ft^2) of steel surface. The transformer/rectifier units employed often provide both voltage and current regulation.

REFERENCES

1. "The Federal Role in Highway Research and Technology," Special Report 261, Transportation Research Board, National Research Council, Washington, D.C., 2001
2. G.H. Koch, M.P.H. Brongers, N.G. Thompson, Y.P. Virmani, and J.H. Payer, "Corrosion Costs and Preventive Strategies in the U.S.," FHWA-RD-01-156, Federal Highway Administration, Washington, D.C., 2002
3. K.M. Al-Haimus and G.T. Halvorsen, "Annotated Bibliography on Corrosion in Reinforced Concrete," Department of Civil Engineering, West Virginia University, Morgantown, WV, 1987
4. M. Toch, Electrolytic Corrosion of Structural Steel, *Trans. Am. Electrochem. Soc.*, Vol 9, 1906
5. E.B. Rosa, B. McCollum, and O.S. Peters, "Electrolysis in Concrete," U.S. Bureau of Standards Technologic Paper 18, U.S. Department of Commerce, Washington, D.C., 1913, p 1–137
6. H.K.G. Bamber, C.R. Marsh, and F.S. Wentworth-Shields, A British Example of Electrolytic Corrosion of Steel in a Reinforced Concrete Structure, *Eng. News-Rec.*, Vol 66 (No. 7), Aug 17, 1911, p 207
7. R.J. Wig and L.R. Ferguson, Reinforced Concrete in Seawater Fails from Corroded Steel, *Eng. News-Rec.*, Vol 79 (No. 15), Oct 11, 1917, p 689–693
8. J.R. Baylis, Corrosion of Concrete, *Trans. ASCE*, Vol 90, 1927, p 791
9. T.E. Stanton, Durability of Concrete Exposed to Sea Water and Alkali Soils—California Experience, *Proc. Am. Concr. Inst.*, Vol 44, 1948, p 821
10. N.J. Rengers, Influence of the Width of Cracks on Corrosive Attack, *Beton Eisen*, Vol 34, March 20, 1935, p 161–162
11. B. Ost and E.E. Monfore, Penetration of Chloride into Concrete, *J. PCA Res. Develop. Lab.*, Vol 8 (No. 1), Jan 1966, p 46–52
12. L.M. Smith and G.L. Ticklenberg, "Lead-Containing Paint Removal, Containment, and Disposal," FHWA-RD-94-100, Federal Highway Administration, McLean, VA, 1995
13. B.R. Appelman, Removing Lead Paint from Bridges: Costs and Practices, *J. Prot. Coatings Linings*, Vol 15 (No. 8), 1998, p 52–60
14. *Bridge Management*, Organization for Economic Co-operation and Development (OECD), Paris, France, 1992, p 26–28
15. "Life-Cycle Greenhouse Gas Emission Factors for Fly Ash Used as a Cement Replacement in Concrete," EPA530-R-03-016, United States Environmental Protection Agency, Washington, D.C., 2003
16. "Transportation Asset Management Today," American Association of State

Highway Transportation Officials, http://assetmanagement.transportation.org/tam/aashto.nsf/home, 2004 (accessed May 2006)
17. "Asset Management for the Roads Sector," ITRD E108508, Organization for Economic Co-operation and Development, Paris, France, 2001
18. "Highway Asset Management Systems, A Primer," Transportation Association of Canada, Ottawa, ON, Canada, 1999
19. "Bridge Management Systems—The State of the Art," Austroads Inc., Sydney, NSW, Australia, 2000
20. R.J. Woodward et al., "BRIME—Bridge Management in Europe," Final Report Contract RO-97-SC.2220, Forum of European National Highway Research Laboratories, Brussels, Belgium, 2001, p 1–6
21. ACI Committee 116, "Cement and Concrete Terminology (ACI 116R-00)," American Concrete Institute, Farmington Hills, MI, 2005
22. T.P. Fahl, *Concrete Principles,* American Technical Publishers, 2001
23. L. Bertolini, P. Pedeferri, and R.P. Polder, *Corrosion of Steel in Concrete,* Wiley-VCH, Weinheim, 2004
24. S. Mindess, J.F. Young, and D. Darwin, *Concrete,* 2nd ed., Prentice-Hall, 2003
25. A.M. Neville, *Properties of Concrete,* 4th ed., Longman Scientific, New York, NY, 1996
26. P.C. Hewlett, Ed., *Lea's Chemistry of Cement and Concrete,* 4th ed., Butterworth-Heinemann Ltd., Oxford, U.K., 2001
27. K.K. Aligizaki, *Pore Structure of Cement-Based Materials,* 1st ed., Taylor & Francis, New York, NY, 2006
28. J.F. Young, A Review of the Pore Structure of Cement Paste and Concrete and Its Influence on Permeability, *Concrete Permeability,* D. Whiting and A. Wallitt, Ed., SP-108, American Concrete Institute, Farmington Hills, MI, 1988, p 1–18
29. D.M. Roy, P.W. Brown, D. Shi, B.E. Scheetz, and W. May, Concrete Microstructure Porosity and Permeability, *Strategic Highway Research Program Report SHRP-C-628,* National Research Council, Washington, D.C., 1993, p 45
30. V.C. Johansen, W.A. Klemm, and P.C. Taylor, Why Chemistry Matters in Concrete, *Concr. Int.,* Vol 24 (No. 3), 2002, p 84–89
31. J. Tinnea and J.F. Young, The Chemistry and Microstructure of Concrete: Its Effects on Corrosion Testing, *Corrosion and Prevention—2000 Proceedings,* Australasian Corrosion Association, Auckland, New Zealand, 2000
32. K.L. Scrivener, The Microstructure of Concrete, *Materials Science of Concrete I,* J.P. Skalny, Ed., American Ceramic Society, Westerville, OH, 1989, p 129–162
33. D.P. Bentz, E. Schlangen, and E.J. Garboczi, Computer Simulation of Interfacial Zone Microstructure and Its Effect on the Properties of Cement-Based Materials, *Materials Science of Concrete IV,* J.P. Skalny and S. Mindess, Ed., American Ceramic Society, Westerville, OH, 1995, p 155–200
34. E.J. Garboczi and D.P. Bentz, Fundamental Computer Simulation Models for Cement-Based Materials, *Materials Science of Concrete II,* J.P. Skalny, Ed., American Ceramic Society, Westerville, OH, 1989, p 249–278
35. "Standard Specification for Low-Alloy Steel Deformed and Plain Bars for Concrete Reinforcement," A 706/A 706M, ASTM
36. "Standard Specification for Chemical Admixtures for Concrete," C 494/C 494M, ASTM
37. "Standard Specification for Chemical Admixtures for Use in Producing Flowing Concrete," C 1017/C 1017M, ASTM
38. A. Bentur, S. Diamond, and N.S. Berke, *Steel Corrosion in Concrete,* E & FN Spon, London, U.K., 1997
39. R.E. West and W.G. Hiem, Chloride Profiles in Salty Concrete, *Mater. Perform.,* Vol 24 (No. 7), 1985, p 29–36
40. S.C. Kranc, A.A. Sagüés, and F.J. Presuel-Moreno, Decreased Corrosion Initiation Time of Steel in Concrete due to Reinforcing Bar Obstruction of Diffusional Flow, *ACI Mater. J.,* Vol 99 (No. 1), 2002
41. J.H. Seinfeld and S.N. Pandis, *Atmospheric Chemistry and Physics—From Air Pollution to Climate Change,* John Wiley and Sons, New York, NY, 1998
42. C.T. Driscoll et al., "Acid Rain Revisited: Advances in Scientific Understanding Since the Passage of the 1970 and 1990 Clean Air Act Amendments," Hubbard Brook Research Foundation, Hanover, NH, 2001
43. "National Atmospheric Deposition Program 2004 Annual Summary," NADP Data Report 2005-01, National Atmospheric Deposition Program Office, Champaign, IL, 2005
44. "2004 Annual and Seasonal Data Summary for Site NY68, Part 2: Statistical Summary of Precipitation Chemistry for Valid Samples," National Atmospheric Deposition Program Office, Champaign, IL, Aug 15, 2005
45. "2004 Annual and Seasonal Data Summary for Site NY99, Part 2: Statistical Summary of Precipitation Chemistry for Valid Samples," National Atmospheric Deposition Program Office, Champaign, IL, Aug 15, 2005
46. "2004 Annual and Seasonal Data Summary for Site PA29, Part 2: Statistical Summary of Precipitation Chemistry for Valid Samples," National Atmospheric Deposition Program Office, Champaign, IL, Aug 15, 2005
47. K. Van Balen et al., "Environmental Deterioration of Ancient and Modern Hydraulic Mortar," European Commission Research Report XX, Luxembourg, Luxembourg, 1999
48. WBK and Associates, *Sulphur Dioxide: Environmental Effects, Fate and Behavior,* Alberta Ministry of Environment, Edmonton, AB, Canada, 2003, p 25–27
49. P. Langford and J. Broomfield, Monitoring the Corrosion of Reinforcing Steel, *Construct. Repair,* Vol 1 (No. 2), 1987, p 32–36
50. R.F.M. Bakker, Chapter Three: Initiation Period, *Corrosion of Steel in Concrete,* Report of the Technical Committee 60-CSC, RILEM, P. Schiessl, Ed., Chapman and Hall, London, 1988, p 23
51. S. Diamond, Effects of Two Danish Flyashes on Alkali Contents of Pore Solutions of Cement-Flyash Pastes, *Cem. Concr. Res.,* Vol 11 (No. 3), 1981, p 383–394
52. C. Arya and Y. Xu, Effect of Cement Type on Chloride Binding and Corrosion of Steel in Concrete, *Cem. Concr. Res.,* Vol 25 (No. 4), 1995, p 893–902
53. C.L. Page and Ø. Vennesland, Pore Solution Composition and Chloride Binding Capacity of Silica-Fume Cement Pastes, *Mater. Struct.,* Vol 16 (No. 91), 1983, p 19–25
54. D. Constantiner and S. Diamond, Pore Solution Analysis: Are There Pressure Effects?, *Mechanisms of Chemical Degradation of Cement-Based Systems,* K.L. Scrivener and J.F. Young, Ed., E & FN Spon, London, 1997, p 22–29
55. J. Tritthart, Chloride Binding in Concrete, Part II: The Influence of the Hydroxide Concentration in the Pore Solution of Hardened Cement Paste on Chloride Binding, *Cem. Concr. Res.,* Vol 19 (No. 5), 1989, p 683–691
56. M. Kawamura, O.A. Kayyali, and M.N. Haque, "Effects of a Flyash on Pore Solution Composition in Calcium and Sodium Chloride-Bearing Mortars, *Cem. Concr. Res.,* Vol 18 (No. 5), 1988, p 763–773
57. O.A. Kayyali and M.N. Haque, Environmental Factor and Concentration of Cl^- and OH^- in Mortars, *J. Mater. Civil Eng.,* Vol 2 (No. 1), 1990, p 24–34
58. J.A. Larbi, A.L. Fraay, and J.M. Bijen, The Chemistry of the Pore Fluid of Silica Fume-Blended Cement Systems, *Cem. Concr. Res.,* Vol 20 (No. 4), 1990, p 506–516
59. J. Duchesne and M.A. Berube, Evaluation of the Validity of the Pore Solution Expression Method from Hardened Cement Pastes and Mortars, *Cem. Concr. Res.,* Vol 24 (No. 3), 1994, p 456–462
60. L. Li, J. Nam, and W.H. Hartt, Ex-situ Leaching Measurement of Concrete Alkalinity, *Cem. Concr. Res.,* Vol 35 (No. 2), 2005, p 277–283

61. "Standard Specification for Portland Cement," C 150, ASTM
62. G.K. Glass and N.R. Buenfeld, The Presentation of the Chloride Threshold Level for Corrosion of Steel in Concrete, *Corros. Sci.*, Vol 39 (No. 5), 1997, p 1001–1013
63. P. Castro-Borges et al., Performance of a 60-Year-Old Concrete Pier with Stainless Steel Reinforcement, *Mater. Perform.*, Vol 41 (No. 10), 2002, p 50–55
64. G.G. Clemeña and Y.P. Virmani, Comparing the Chloride Resistances of Reinforcing Bars, *Concr. Int.*, Vol 26 (No. 11), 2004, p 39–49
65. D.A. Hausmann, Steel Corrosion in Concrete—How Does It Occur?, *Mater. Prot.*, Vol 6 (No. 11), 1967, p 19–23
66. V.K. Gouda, Corrosion and Corrosion Inhibition of Reinforcing Steel, Part I: Immersed in Alkaline Solutions, *Br. Corros. J.*, Vol 5 (No. 9), 1970, p 198–208
67. W. Breit, Critical Chloride Content—Investigation of Steel in Alkaline Chloride Solutions, *Mater. Corros.*, Vol 49, 1998, p 539–550
68. L. Li and A.A. Sagüés, Chloride Corrosion Threshold of Reinforcing Steel in Alkaline Solutions—Open-Circuit Immersion Tests, *Corrosion*, Vol 57 (No. 1), 2001, p 19–28
69. J. Nam, W. Hartt, K. Kim, and L. Li, "Effect of Cement Alkalinity upon Time-to-Corrosion of Reinforcing Steel in Concrete Undergoing Chloride Exposure," Paper 03290, Corrosion/2003, NACE International, 2003
70. "Standard Test Method for Determining the Effects of Chemical Admixtures on the Corrosion of Embedded Steel Reinforcement in Concrete Exposed to Chloride Environments," G 109, ASTM
71. J. Marchand, E. Samson, Y. Maltais, R.J. Lee, and S. Sahu, Predicting the Performance of Concrete Structures Exposed to Chemically Aggressive Environment—Field Validation, *Mater. Struct.*, Vol 35, Dec 2002, p 623–631
72. M.H. Shehata and M.D.A. Thomas, The Effect of Fly Ash Composition on the Expansion of Concrete due to Alkali-Silica Reaction, *Cem. Concr. Res.*, Vol 30, 2000, p 1063–1072
73. "Guidelines for the Use of Lithium to Mitigate or Prevent ASR," Publication FHWA-RD-03-047, Federal Highway Administration, Jan 2003
74. "Standard Test Method for Determination of Length Change of Concrete Due to Alkali-Silica Reaction," C 1293, ASTM
75. "Standard Test Method for Potential Alkali Reactivity of Aggregates (Mortar-Bar Method)," C 1260, ASTM
76. Z.T. Chang, B. Cherry, and M. Marosszeky, *The Proceedings of Australasian Corrosion Association Inc. Corrosion and Prevention—2001* (Newcastle, Australia), 2001, p 035
77. S. Jäggi, H. Böhni, and B. Elsener, *Proceedings of Eurocorr 2001* (Riva del Garda, Italy), 2001, p 045
78. A.A. Sagüés and H.W. Pickering, Technical Note: Crevice Effect on Corrosion of Steel in Simulated Concrete Pore Solutions, *Corrosion*, Vol 56 (No. 10), 2000
79. Z.T. Chang, B. Cherry, and M. Marosszeky, *The Proceedings of Australasian Corrosion Association Inc. Corrosion and Prevention—2000* (Auckland, New Zealand), 2000, p 029
80. K. Cho and H.W. Pickering, *J. Electrochem. Soc.*, Vol 137 (No. 10), 1990, p L3313
81. E.A. Nystrom, J.B. Lee, A.A. Sagüés, and H.W. Pickering, *J. Electrochem. Soc.*, Vol 141 (No. 2), 1994, p 358
82. J.S. Tinnea, Paper 706, *Proceedings of the 13th International Corrosion Conference* (Granada, Spain), 2002
83. J.L. Smith and Y.P. Virmani, *Materials and Methods for Corrosion Control of Reinforced and Prestressed Concrete Structures in New Construction*, FHWA Publication 00-081, Federal Highway Administration, McLean, VA, 2000
84. E.G. Nawy, *Prestressed Concrete: A Fundamental Approach*, Prentice Hall, 1989
85. "Standard Specification for Steel Strand, Uncoated Seven-Wire for Prestressed Concrete," A 416, ASTM
86. R.O. Lewis, "Quality and Performance of Prestressing Wire for Prestressed Concrete Cylinder Pile," American Concrete Pressure Pipe Association, Reston, VA, July 2002
87. "Standard Specification for Filled Epoxy-Coated Seven-Wire Prestressing Steel Strand," A 882, ASTM
88. R.G. Powers, A.A. Sagüés, and Y.P. Virmani, Corrosion of Post-Tensioned Tendons in Florida Bridges, *Proceedings 17th U.S.-Japan Bridge Engineering Workshop*, Public Works Research Institute, Tskuba, Japan, 2001
89. A. Ghorbanporr, "Evaluation of Post-Tensioned Concrete Bridge Structures by the Impact-Echo Technique," Report FHWA-RD-90-096, Federal Highway Administration, Washington, D.C., 1993
90. A.A. Sagüés, Improved Vibration Techniques and On-Going Research Efforts, *Proceedings Symposia on New Directions for Florida Post-Tensioned Bridges*, Florida Department of Transportation, 2003, p 151
91. A. Ghorbanporr and S. Shi, Assessment of Corrosion of Steel in Concrete Structures by Magnetic Based NDE Techniques, *Techniques to Assess the Corrosion Activity of Steel Reinforced Concrete Structures*, STP 1276, N.S. Berke, E. Escalante, C. Nmai, and D. Whiting, Ed., ASTM, 1995
92. "State-of-the-Art Report: Criteria for Cathodic Protection of Prestressed Concrete Structures," Technical Committee Report 01102, NACE International, 2001
93. S.F. Daily, "Galvanic Cathodic Protection of Reinforced and Prestressed Concrete Structures Using a Thermally Sprayed Aluminum Alloy," presented at NACE International Northern Area Eastern Regional Conference, Aug 26–29, 2001 (Halifax, NS)
94. J.E. Bennett, "Chemical Enhancement of Metallized Zinc Anode Performance," Paper 98640, Corrosion/98, NACE International, 1998
95. R. Staehle and B. Little, Corrosion and Stress Corrosion Cracking of Post Tension Cables Associated with Fungal Action, *Microbiological Influenced Corrosion, Proc. of the Corrosion/2002 Research Topical Symposium*, NACE Press, 2002, p 33
96. R. Woodward and F. Williams, Collapse of Ynys-Y-Gwas Bridge, West Glamorgan, *Proc. Inst. Civ. Eng.*, Part 1, Vol 84, Aug 1988, p 635–669
97. R.G. Powers, "Corrosion Evaluation of Post-Tensioned Tendons on the Niles Channel Bridge," Florida Department of Transportation, Gainesville, FL, June 1999
98. A.A. Sagüés, S.C. Kranc, and R.H. Hoehne, "Initial Development of Methods for Assessing Condition of Post-Tensioned Tendons of Segmental Bridges," Final Report BC374, May 17, 2000 (available through the Florida Department of Transportation at http://www.dot.state.fl.us/) (accessed May 2006)
99. Corven Engineering, Inc., "Mid-Bay Bridge Post-Tensioning Evaluation Final Report," Final Report to Florida Department of Transportation District 3, Oct 10, 2001 (available through the Florida Department of Transportation at http://www.dot.state.fl.us/) (accessed May 2006)
100. Parsons Brinckerhoff Quade & Douglas, Inc. in association with Concorr Florida, Inc. and Kisinger Campo & Associates Corp., "Sunshine Skyway Bridge Post-Tensioned Tendon Investigation," FPN: 411135 1 32 01, Report to Florida Department of Transportation District Seven, Tampa, FL, Feb 6, 2002
101. A. Ghorbanpoor and S.C. Madathanapalli, "Performance of Grouts for Post-Tensioned Bridge Structures," Report FHWA-RD-92-095, National Technical Information Service, Springfield, VA, 1993
102. R.G. Powers, A.A. Sagüés, and Y.P. Virmani, Corrosion of Post-Tensioned Tendons in Florida Bridges, *Proceedings, 17th U.S.-Japan Bridge Engineering Workshop*, Nov 12–14, 2001, Public

Works Research Institute, Technical Memorandum of PWRI 3843, H. Sato, Ed., 2002, p 579–594
103. H. Tabatabai, A.T. Ciolko, and T.J. Dickson, Implications of Test Results from Full-Scale Fatigue Tests of Stay Cables Composed of Seven-Wire Prestressing Strands, *Proceedings of the Fourth International Bridge Engineering Conference,* Vol 1, Transportation Research Board, Washington, D.C., Aug 28–30, 1995, p 266
104. M.D. Bricker and A.J. Schokker, "Corrosion from Bleed Water in Grouted Post-Tensioned Tendons," Research and Development Bulletin RD137, Portland Cement Association, Skokie, IL, 2005
105. D.P. Bentz, K.H. Hansen, and M.R. Geiker, Shrinkage-Reducing Admixture and Early Age Desiccation in Cement Pastes and Mortars, *Cem. Concr. Res.,* Vol 31 (No. 7), 2001, p 1075
106. P. Harkins, "Product Evaluation and Qualified Products List J-Grout/Mortar," Florida Department of Transportation, Tallahassee, FL, Aug 2002
107. H. Wang, A.A. Sagüés, and R.G. Powers, "Corrosion of the Strand-Anchorage System in Post-Tensioned Grouted Assemblies," Paper 05266, Corrosion/2005, NACE International, 2005
108. A.A. Sagüés, R.G. Powers, and H.B. Wang, "Mechanism of Corrosion of Steel Strands in Post-Tensioned Grouted Assemblies," Paper 03312, Corrosion/2003, NACE International, 2003
109. H. Wang and A.A. Sagüés, "Corrosion of Post-Tensioning Strands," Final Report BC353-33, Nov 1, 2005 (available through the Florida Department of Transportation at http://www.dot.state.fl.us/) (accessed May 2006)
110. "Standard Test Method for Half-Cell Potentials of Uncoated Reinforcing Steel in Concrete," C 876, ASTM
111. A.T. Ciolko and H. Tabatabai, "Nondestructive Methods for Condition Evaluation of Prestressing Steel Strands in Concrete Bridges, Final Report, Phase I: Technology Review," NCHRP Web Document 23 (Project 10-53), National Cooperative Highway Research Program, Transportation Research Board, Washington, D.C., March 1999
112. J.F. Elliott, Continuous Acoustic Monitoring of Bridges, *Corros. Prevent. Control,* Sept 2000, p 67
113. A.A. Sagüés, S.C. Kranc, and T.G. Eason, Vibrational Tension Measurement of External Tendons in Segmental Post-Tensioned Bridges, *ASCE J. Bridge Eng.,* 2006, in press
114. "New Directions for Florida Post-Tensioning Bridges," Vol 1–10B, Structures Design Office, Florida Department of Transportation, Tallahassee, FL, 2006 (available through the Florida Department of Transportation at http://www.dot.state.fl.us/) (accessed May 2006)
115. F.L. Stahl and C.P. Gagnon, *Cable Corrosion in Bridges and Other Structures,* ASCE Press, New York, NY, 1995
116. ACI Committee 222, "Corrosion of Prestressing Steels," ACI 222.2R-01, American Concrete Institute, Farmington Hills, MI, May 2001
117. *Manual for Condition Evaluation of Bridges,* 2nd ed. with 2001 and 2003 revisions, American Association of State Highway and Transportation Officials, Washington, D.C., 2003
118. K.M. Childs, Ed., *Underwater Investigations—Standard Practice Manual,* American Society of Civil Engineers, Reston, VA, 2001
119. L. Uller and O.T. de Rincón, "Manual of Inspection, Evaluation, and Diagnosis of Corrosion in Reinforced Concrete Structures," CYTED, Maracaibo, Venezuela, 1998 (in Spanish)
120. News Bulletin 166, Euro-International Concrete Committee, Lausanne, Switzerland, 1985 (in French)
121. K. Tuutti, "Corrosion of Steel in Concrete," Report 4, Swedish Cement and Concrete Research Institute, Stockholm, 1982
122. J.S. Tinnea and N.J. Feuer, "Evaluation of Structural Fatigue and Reinforcement Corrosion Interrelationships Using Close Grid Computer Generated Equipotential Mapping," Paper 259, Corrosion/85, NACE International, 1985
123. J.S. Tinnea, K.M. Howell, and M. Figley, "Triple System Galvanic Protection of Reinforced Concrete: Energizing and Operation," Paper 337, Corrosion/2004, NACE International, 2004
124. J. Broomfield and S. Millard, Measuring Concrete Resistivity to Assess Corrosion Rates, *Concrete,* Vol 36 (No. 2), 2002, p 37–39
125. W.H. Hartt and S. Venugopalan, *Corrosion Evaluation of Post-Tensioned Tendons on the Mid Bay Bridge in Destin, Florida—Final Report,* Florida Department of Transportation Research Center, 2002, p 11
126. J.H. Bungey, *Testing of Concrete in Structures,* 2nd ed., Chapman & Hall, 1989
127. S. Feliu, J.A. Gonzalez, and C. Andrade, Electrochemical Techniques for On-Site Determination of Corrosion Rates of Rebars, *Techniques to Assess the Corrosion Activity of Steel Reinforced Concrete Structures,* STP 1276, N. Berke, E. Escalante, C. Nmai, and D. Whiting, Ed., ASTM, 1996, p 107–118
128. D.G. Manning, "Detecting Defects and Deterioration in Highway Structures," National Cooperative Highway Research Program Synthesis of Highway Practice 118, 1995
129. P.C. Taylor, M.A. Nagi, and D.A. Whiting, "Threshold Chloride Content for Corrosion of Steel in Concrete: A Literature Review," Portland Cement Association R&D Serial 2169, Skokie, IL, 1999
130. M.A. Sanuan, C. Andrade, and M. Cheyrezy, Concrete Carbonation Tests in Natural and Accelerated Conditions, *Adv. Cem. Res.,* Vol 15 (No. 4), 2003, p 173
131. Fourth Quarterly Cost Report, *Eng. News-Rec.,* Vol 255 (No. 5), Dec 19, 2005, p 26
132. "Standard Specification for Deformed and Plain, Low-Carbon, Chromium, Steel Bars for Concrete Reinforcement," A 1035/A 1035M, ASTM
133. G. Kusinski and G. Thomas, High Strength, Toughness and Nanostructure Considerations for Fe/Cr/Mn/C Lath Martensitic Steels, *Proceedings: First International Conference—Super High Strength Steels,* Associazione Italiana di Metallurgia, Rome, Italy, 2005
134. "Standard Specification for Deformed and Plain Carbon-Steel Bars for Concrete Reinforcement," A 615/A 615M, ASTM
135. M.C. Brown and R.E. Weyers, "Corrosion Protection Service Life of Epoxy-Coated Reinforcing Steel in Virginia Decks—Final Report," Virginia Transportation Research Council Report VTRC 04-CR7, 2003, p iii
136. J.R. Clifton, H.F. Beeghly, and R.G. Mathey, "Nonmetallic Coatings for Concrete Reinforcing Bars," Federal Highway Administration Report FHWA-RD-74-18, 1974
137. W.R. Baldwin, An Update on Epoxy Coated Reinforcing Steel, *Solving Rebar Corrosion Problems in Concrete* (Chicago, IL), 2002, Seminar reprints, NACE International, 1983
138. Y.P. Virmani and K.C. Clear, "Time-to-Corrosion of Reinforcing Steel in Concrete Slabs," FHWA/RD-80/012, Federal Highway Administration, McLean, VA, 1983
139. K.C. Clear, Effectiveness of Epoxy-Coated Reinforcing Steel, *Concr. Int.,* Vol 14 (No. 5), 1992, p 58–62
140. Epoxy-Coated Rebar Debate Still Going, *Civ. Eng., ASCE,* Vol 63 (No. 3), 1993, p 22–27
141. "Use of Epoxy Coated Rebar in Concrete Bridge Components: Pros and Cons," Session 210, Y.P. Virmani, presiding, 85th Annual Meeting of the U.S. Transportation Research Board, 2006
142. C.M. Hansson et al., "Corrosion Protection Strategies for Ministry Bridges—Final Report," University of Waterloo, MTO Consulting Assignment 9015-A-000034, 2000
143. W.A. Pyć et al., "Field Performance of Epoxy-Coated Reinforcing Steel in Virginia Bridge Decks," Virginia Transportation Research Council Report VTCR 00-R16, 2000

144. B.S. Covino et al., "Performance of Epoxy-Coated Steel Reinforcement in the Deck of the Perley Bridge," Ministry of Transportation Report MI-180, 2000
145. H.J. Gillis and M.G. Hagen, "Field Examination of Epoxy-Coated Rebars in Concrete Bridge Decks," Minnesota Department of Transportation Report 94-14, 1994
146. "Evaluation of Bridge Decks Using Epoxy Coated Reinforcement," West Virginia Department of Transportation Report MIR 1261603, 1994, and http://www.wvdot.com//10_contractors/10f6a_epoxy.htm, 2006 (accessed May 2006)
147. T.M. Adams, J.A. Pincheira, and Y.H. Huang, "Assessment and Rehabilitation Strategies/Guidelines to Maximize the Service Life of Concrete Structures," Wisconsin Department of Transportation Report WHRP 02-003, 2001
148. A. Sagüés, "Corrosion of Epoxy Coated Rebar in Florida Bridges," Florida Department of Transportation WPI 0510603, 1994
149. K.C. Clear, W.H. Hartt, J. McIntyre, and S.K. Lee, "Performance of Epoxy Coated Steel Reinforcing Steel in Highway Bridge," NCHRP Report 370, National Academy Press, Washington, D.C., 1995
150. S.K. Lee and P.D. Krauss, "Long-Term Performance of Epoxy-Coated Reinforcing Steel in Heavy Salt-Contaminated Concrete," Federal Highway Administration Report FHWA-HRT-04-090, 2004
151. H. Leidheiser, Jr., Whitney Award Lecture—1983, Towards a Better Understanding of Corrosion Beneath Organic Coatings, *Corrosion*, Vol 39 (No. 5), 1983, p 197
152. H. Leidheiser and W. Funke, Water Disbondment and Wet Adhesion of Organic Coatings on Metals: A Review and Interpretation, *J. Oil Color Chem. Assoc.*, Vol 70 (No. 5), 1987, p 121–132
153. T. Nguyen, J.B. Hubbard, and J.M. Pommersheim, Unified Model for the Degradation of Organic Coatings on Steel in a Neutral Electrolyte, *J. Prot. Coatings,* Vol 68 (No. 855), 1996, p 45–56
154. A. Sagüés, "Coating Disbondment in Epoxy-Coated Reinforcing Steel in Concrete—Field Observations," Paper 325, Corrosion/1996, NACE International, 1996
155. J.W. Martin et al., "Degradation of Powder Epoxy-Coated Panels Immersed in Saturated Calcium Hydroxide Solution Containing Sodium Chloride," Federal Highway Administration Report FHWA-RD-94-174, 1995
156. L.M. Wolf and R.L. Sarcinella, Use of Epoxy-Coated Reinforcing Steel and Other Corrosion Protection Philosophy in Texas: The 'Belts and Suspenders' Approach, *International Conference on the Corrosion and Rehabilitation of Reinforced Concrete Structures,* Federal Highway Administration Publication FHWA-SA-99-014, 1998
157. H. Chen and H.G. Wheat, "Evaluation of Selected Epoxy-Coated Reinforcing Steels," Paper 329, Corrosion/1996, NACE International, 1996, p 5
158. P. Montes, T.W. Bremner, and I. Kondratova, Eighteen-Year Performance of Epoxy-Coated Rebar in a Tunnel Structure Subjected to a Very Aggressive Chloride-Contaminated Environment, *Corrosion,* Vol 60 (No. 10), 2004, p 974–981
159. L.M. Samples and J.A. Ramerez, "Methods of Corrosion Protection and Durability of Concrete Bridge Decks Reinforced with Epoxy-Coated Bars—Phase I," Federal Highway Administration Report FHWA/IN/JTRP-98/15, 1999
160. "Epoxy-Coated Reinforcing Bars: Materials and Coating Requirements," AASHTO M 284-05, American Association of State Highway Transportation Officials, Washington, D.C., 2005
161. "Standard Specification for Epoxy-Coated Reinforcing Steel Bars," A 775/A 775M, ASTM
162. "Standard Specification for Prefabricated Epoxy-Coated Reinforcing Steel Bars," A 934/A 934M, ASTM
163. "Standard Specification for Epoxy-Coated Steel Wire and Welded Wire Reinforcement," A 884/A 884M, ASTM
164. "Standard Specification for the Fabrication and Job Site Handling of Epoxy-Coated Reinforcing Steel Bars," D 3963/D 3963M, ASTM
165. J. Sedricks, *Corrosion of Stainless Steels,* 2nd ed., John Wiley & Sons, Inc., 1996
166. M.F. Hurley and J.R. Scully, "Threshold Chloride Concentrations of Selected Corrosion Resistant Rebar Materials Compared to Carbon Steel," Paper 259, Corrosion/2005, NACE International, 2005
167. L. Bertolini et al., Behavior of Stainless Steel in Simulated Concrete Pore Solution, *Br. Corros. J.,* Vol 31 (No. 3), 1996 p 218–222
168. D.W. Pfeifer and M.J. Scali, "Concrete Sealers for Protection of Bridge Substructures," National Cooperative Highway Research Program Report 244, Washington, D.C., 1981
169. F. Cui and A.A. Sagüés, "Corrosion Behavior of Stainless Steel Clad Rebar," Paper 645, Corrosion/2001, NACE International, 2001
170. D.B. McDonald, D.W. Pfeifer, and M.R. Sherman, "Corrosion Evaluation of Epoxy-Coated, Metallic-Clad and Solid Metallic Reinforcing Bars in Concrete," Federal Highway Administration Report FHWA-RD-98-153, 1998, p 73–75
171. F.J. Nelson, "Use of Stainless Steel Reinforcing in Coastal Bridges," International Bridge Conference Paper IBC-04-40, Engineers' Society of Western Pennsylvania, Pittsburg, PA, 2005
172. S.D. Cramer et al., Prevention of Chloride-Induced Corrosion Damage to Bridges, *ISIJ Int.,* Vol 42 (No. 12), 2002, p 1376–1385
173. M. Nagi, "Long Term Performance of Galvanized Reinforcing Steel in Concrete Bridges—Case Studies," Paper 264, Corrosion/2005, NACE International, 2005
174. M.C. Andrade and A. Marcais, Galvanized Reinforcements in Concrete, *Surface Coatings-2,* A. Wilson and H. Posser, Ed., Elsevier Applied Sciences, London, 1988
175. S.R. Yoemans, Performance of Black, Galvanized, and Epoxy-Coated Reinforcing Steels in Chloride-Contaminated Concrete, *Corros. J.,* Vol 50 (No. 1), 1994, p 80
176. R.F. Stratfull, Progress Report on Inhibiting the Corrosion of Steel in a Reinforced Concrete Bridge, *Corrosion,* Vol 15 (No. 6), 1959
177. R.F. Stratfull, "Experimental Cathodic Protection of a Bridge Deck," Transportation Research Record 500, Transportation Research Board, Washington, D.C., 1974
178. R.F. Stratfull, Cathodic Protection of a Bridge Deck: Preliminary Investigation, *Mater. Perform.,* Vol 13 (No. 4), 1974
179. S.F. Daily and K. Kendell, Cathodic Protection of New Reinforced Concrete Structures in Aggressive Environments, *Mater. Perform.,* Vol 37 (No. 10), 1998, p 19–25
180. G.H.C. Chang, J.A. Apsotolos, and F.A. Myhres, "Cathodic Protection Studies on Reinforced Concrete," California Department of Transportation Report FHWA/CA/TL-81/02, 1981
181. P.M. Ward, Cathodic Protection: A User's Perspective, *Chloride Corrosion of Steel in Concrete,* STP 629, ASTM, 1977, p 150–163
182. J.P. Broomfield, *Corrosion of Steel in Concrete,* E & FN Spon, London, U.K., 1996
183. P.M. Chess, *Cathodic Protection of Steel in Concrete,* E & FN Spon, London, U.K., 1998
184. "Impressed Current Cathodic Protection of Reinforcing Steel in Atmospherically Exposed Concrete Structures," RP0200-2000, NACE International, 2000
185. "Sacrificial Cathodic Protection of Reinforced Concrete Elements—A State-of-the-Art Report," NACE International Publication 01105, NACE International, 2005
186. "State-of-the-Art Report: Criteria for Cathodic Protection of Prestressed Concrete Structures," NACE International Publication 01102, NACE International, 2002
187. "Use of Reference Electrodes for Atmospherically Exposed Reinforced Concrete

188. "Standard Test Method: Testing of Embeddable Impressed Current Anodes for Use in Cathodic Protection of Atmospherically Exposed Steel-Reinforced Concrete," TM0294-2001, NACE International, 2001
189. "Standard Test Method: Testing Procedures for Organic-Based Conductive Coating Anodes for Use on Concrete Structures," TM0105-2005, NACE International, 2005
190. "Cathodic Protection of Steel in Concrete, Part 1: Atmospherically Exposed Concrete," BS/EN 12696:2000, CEN, Brussels, 2000
191. "Cathodic Protection of Metals, Part 5: Steel in Concrete Structures," AS 2832.5, Standards Australia, Sydney, 2002
192. "Cathodic Protection," M-503, Rev. 2, NORSOK Standard, 1997
193. H.M. Laylor, "Demonstration Project Soffit Cathodic Protection System in a Coastal Environment," Oregon Department of Transportation Report FHWA/OR/RD-87/04, 1987
194. J.P. Broomfield and J.S. Tinnea, "Cathodic Protection of Reinforced Concrete Bridge Components," National Research Council Report SHRP-C/UWP-92-618, 1992
195. R.J. Kessler, R.G. Powers, and I.R. Lasa, Cathodic Protection Using Scrap and Recycled Materials, *Mater. Perform.,* Vol 30 (No. 6), June 1991, p 29–31
196. R. Baboian, Ed., *NACE Corrosion Engineer's Reference Book,* 3rd ed., NACE Press, Houston, TX, 2002, p 162
197. G.K. Glass, A.M. Hassanein, and N.R. Buenfeld, Cathodic Protection Afforded by an Intermittent Current Applied to Reinforced Concrete, *Corros. Sci.,* Vol 43 (No. 6), 2001, p 1111–1131
198. L. Bertolini et al., Cathodic Protection and Cathodic Prevention in Concrete: Principals and Applications, *J. Appl. Electrochem.,* Vol 28 (No. 12), 1998, p 1321–1331
199. G.K. Glass and A.M. Hassanein, Surprisingly Effective Cathodic Protection, *J. Corros. Sci. Eng.,* Vol 4 (Paper 7), 2003, http:www.jcse.org (accessed May 2006)
200. P. Pedeferri, Cathodic Protection and Cathodic Prevention, *Construct. Build. Mater.,* Vol 10 (No. 5), 1996, p 391–402
201. A.A. Sagüés and R.O. Powers, "Sprayed Zinc Galvanic Anodes for Concrete Marine Bridge Structures," Strategic Highway Research Project Report SHRP-S-405, 1994
202. R.P. Brown and J.S. Tinnea, Cathodic Protection Design Problems for Reinforced Concrete, *Mater. Perform.,* Vol 30 (No. 8), 1991, p 28–31
203. G.R. Holcomb et al., Humectant Use in Cathodic Protection of Reinforced Concrete, *Corrosion,* Vol 56 (No. 11), 2000, p 1141–1157
204. K.C. Clear, "A 'New,' High Current Output, Galvanic Sacrificial Anode, Electrochemical Rehabilitation System for Reinforced and Prestressed Concrete Structures," Paper 556, Corrosion/99, NACE International, 1999
205. R.J. Kessler, R.G. Powers, and I.R. Lasa, "An Update on the Long Term Use of Cathodic Protection of Steel Reinforced Concrete Marine Structures," Paper 254, Corrosion/2002, NACE International, 2004
206. S.J. Bullard et al., "Intermittent Cathodic Protection for Steel Reinforced Concrete Bridges," Paper 086, 15th International Corrosion Conference (Granada, Spain), 2002
207. S.D. Cramer et al., "Carbon Paint Anode for Reinforced Concrete Bridges in Coastal Environments," Paper 265, Corrosion/2002, NACE International, 2002
208. S.D. Cramer et al., "Evaluation of Rocky Point Viaduct Concrete Beam—Final Report," Project SPR 381, Oregon Department of Transportation, Salem, OR, and Federal Highway Administration, Washington, D.C., FHWA-OR-RD-00-18, June 2000
209. G.K. Glass, Technical Note: The 100-mV Potential Cathodic Protection Criterion, *Corrosion,* Vol 55 (No. 3), 1999, p 286–290
210. H. Schell, D.G. Manning, and F. Pianca, "A Decade of Bridge Deck Cathodic Protection in Ontario," Ministry of Transportation, Research and Development Branch, Toronto, 1987
211. B.S. Covino et al., "Performance of Zinc Anodes for Cathodic Protection of Reinforced Concrete Bridges—Final Report," Oregon Department of Transportation Special Report SPR 364, 2002
212. J.E. Slater, D.R. Lankard, and P.J. Moreland, "Electrochemical Removal of Chlorides from Concrete Bridge Decks," Transportation Research Record 604, 1976
213. G.L. Morrison et al., "Chloride Removal and Monomer Impregnation of Bridge Deck Concrete by Electro-Osmosis," Report FHWA-KS-RD 74-1, Kansas Department of Transportation, 1976
214. Method and Apparatus for Removal of Chlorides from Steel Reinforced Concrete Structures, U.S. Patent 5,141,607, Aug 25, 1992
215. Process for Rehabilitating Internally Reinforced Concrete by Removal of Chlorides, U.S. Patent 5,228,959, July 20, 1993
216. Apparatus for the Removal of Chloride from Reinforced Concrete Structures, U.S. Patent 5,296,120, April 17, 1994
217. J.E. Bennett, T.J. Schue, K.C. Clear, D.L. Lankard, W.H. Hartt, and W.J. Swiat, "Electrochemical Chloride Removal and Protection of Concrete Bridge Components: Laboratory Studies," Report SHRP-S-657, National Research Council, 1993
218. "Electrochemical Chloride Extraction from Steel Reinforced Concrete—A State-of-the-Art Report," NACE International Publication 01101, NACE International, 2001

SELECTED REFERENCES

- AISI Task Group on Weathering Steel Bridges, "Performance of Weathering Steel in Highway Bridges—A First Phase Report," American Iron and Steel Institute, Washington, D.C., 1982
- P. Albrecht et al., "Guidelines for the Use of Weathering Steel in Bridges," NCHRP Report 314, U.S. Department of Transportation, Federal Highway Administration, Washington, D.C., June 1989
- B.R. Appleman, "Lead-Based Paint Removal for Steel Highway Bridge," National Cooperative Highway Research Program Synthesis 251, Transportation Research Board, Washington, D.C., 1997
- S.-L. Chong and Y. Yao, "Laboratory Evaluation of Waterborne Coatings on Steel," Report FHWA-RD-03-032, U.S. Federal Highway Administration, McLean, VA, 2003
- D.C. Cook and A.C. Van Orden, "The Luling Bridge: An Inside Story," Paper 00449, Corrosion/2000, NACE International, 2000
- J.D. Culp and G.L. Tinklenberg, "Corrosion of Unpainted Weathering Steel: Causes and Cure," paper presented at the Second World Congress of Coating Systems for Bridges and Steel Structures, Oct 26–27, 1982 (New York, NY)
- "Design Division Informational Memorandum 263-B," Michigan Department of Transportation, Lansing, MI, June 5, 1979
- "Design Division Informational Memorandum 271-B," Michigan Department of Transportation, Lansing, MI, March 7, 1980
- C.L. Farschon, R.A. Kogler, and J.P. Ault, "Guidelines for Repair and Maintenance of Bridge Coatings: Overcoating," Report FHWA-RD-97-092, U.S. Federal Highway Administration, McLean, VA, 1997
- "Guide Specification for Coating Systems with Inorganic Zinc-Rich Primer—S 8.1," American Association of State Highway and Transportation Officials and Nations Steel Bridge Alliance, Washington, D.C., 2002
- M.B. Kilcullen and M. McKenzie, Weathering Steels, *Corrosion in Civil Engineering,* Institution of Civil Engineers, London, U.K., 1979, p 95–105
- R.A. Kogler, Painting Highway Bridges and Structures, *Steel Structures Painting Manual,* Vol 1, *Good Painting Practice,* 4th ed., Steel Structures Painting Council, Pittsburgh, PA, 2002

- R.A. Kogler, J.P. Ault, and C.L. Farschon, "Environmentally Acceptable Materials for the Corrosion Protection of Steel Bridges," Report FHWA-RD-96-058, U.S. Federal Highway Administration, McLean, VA, 1997
- D.G. Manning, "Accelerated Corrosion in Weathering Steel Bridges," ME-84-03, Ontario Ministry of Transportation and Communications, Downsview, ON, Canada, 1984
- D.G. Manning, "Accelerated Corrosion in Weathering Steel Bridges—An Update," ME-84-04, Ministry of Transportation and Communications, Downsview, ON, Canada, 1984
- W.L. Mathay, Uncoated Weathering Steel Bridges, *Highway Structures Design Handbook,* Vol 1, American Institute of Steel Construction, Pittsburgh, PA, 1993
- F. Pianca, "Weathering Steel—Performance of Test Coupons Over a Ten-Year Period of Environmental Exposure," MAT-96-02, Ontario Ministry of Transportation, Downsview, ON, Canada, 1997
- "Uncoated Weathering Steel in Structures," FHWA Technical Advisory T 5140.22, U.S. Department of Transportation, Washington, D.C., Oct 3, 1989

Corrosion in Commercial Aviation

Alain Adjorlolo, The Boeing Company

THE COMBINED SHARE of composite materials and titanium alloys in commercial aircraft structure has steadily increased at the expense of aluminum, the major material in use (by weight) and one that is corrosion susceptible (Fig. 1). The susceptibility of aluminum to corrosion and its wide application in the airplane structure have driven incremental and cumulative corrosion-control improvements over many years of airplane design. Boeing's newest commercial aircraft, the 787, will use significantly less aluminum. This choice is already causing changes in design and manufacture for corrosion prevention.

Corrosion control has been an integral part of the design and manufacture of commercial aircraft ever since corrosion-susceptible, high-strength or high-toughness aluminum alloys and steels became the design materials of choice. As commercial aircraft age in service, the structures experience many forms of corrosion, dependent on the alloy and temper, the finish, the assembly design and location on the airplane, the drainage, and the internal and external environments. Unattended, these may become serious airworthiness issues. In the United States, the Federal Aviation Administration (FAA) requires that the manufacturer "... show that catastrophic failure due to fatigue, corrosion or accidental damage, will be avoided throughout the operational life of the airplane ..." [Title 14—Code of Federal Regulations, Section 25.571], or that the airline operator have a plan to address corrosion problems and reduce their severity between inspections. "Each certificate holder is primarily responsible for the airworthiness of its aircraft ... the performance of the maintenance ... in accordance with its manual and the regulations ..." [Title 14—Code of Federal Regulations, Section 121.363]. The manual refers to the maintenance plan for the airplane, which is approved by the regulatory authority to ensure compliance with the "regulations." The manufacturer, the airline industry, and the regulatory authority act together to create conditions for continued airworthiness of all airplanes.

This article surveys forms of corrosion encountered in commercial airplanes, the factors that are important to design for corrosion prevention, some important lessons learned from 40 years of commercial jet airplane service history, and how corrosion data are collected. This article also provides an overview of the implementation and evolution of airline corrosion prevention and control programs (CPCPs) and directions being considered in the design for corrosion prevention of newer airplanes.

Corrosion Basics

When aluminum corrodes in an aqueous environment, its ions are liberated into the environment at anodic sites and it is said to be oxidized.

At anodic sites (dissolution or oxidation):

$$Al = Al^{3+} + 3e^- \quad \text{(Eq 1)}$$

Simultaneously, excess electrons left behind at anodes travel to adjacent cathodic sites, where they are absorbed by some species in the environment. The species of interest in corrosion in natural environments is oxygen, which combines with water in acquiring the electrons generated in the reaction shown in Eq 1 through the oxygen reduction reaction.

At the cathodic sites (reduction in aerated environments):

$$O_2 + 2H_2O + 4e^- = 4OH^- \quad \text{(Eq 2a)}$$

However, in the absence of oxygen water reduction takes place (reduction in nonaerated environment):

$$2H_2O + 2e^- = 2OH^- + H_2 \quad \text{(Eq 2b)}$$

The corroding metal itself provides a path for electron transfer from anodes to cathodes, while the moisture on the surface provides the medium for the migration of ionic charges (from cathodes to anodes for negative ions, and anode to cathode for positive ions), completing the electrical circuit.

Whether or not oxygen is present, the net effect is that cathodic locations become increasingly alkaline. The presence of soluble salts introduces negative ions, which increase the ionic conductivity of the moist environment and thus allow reduction reactions to occur further away from the anode. In addition, they may render the environment more aggressive toward aluminum in specific ways. For example, the effect of ever-present chloride ions is to attract positive hydrogen ions at the anode. Hydrolysis of metal chlorides thus formed increases the acidity of the anode environment. Because aluminum and its alloys corrode at high rates in both acid and alkaline environments, the simultaneous increase in acidity at the anode and corresponding increase in alkalinity at the cathode lead progressively to more corrosive conditions alloys. On mating surfaces, evidence suggests that the presence of carbonate ions affect the morphology of the corroded surface (Ref 1).

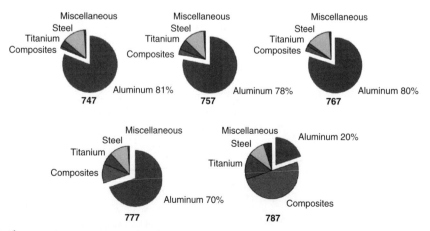

Fig. 1 Material usage in commercial airplane construction

Commonly Observed Forms of Airplane Corrosion

The general mechanism described previously gives rise to these most commonly observed forms of corrosion in commercial aviation (Ref 2):

- General corrosion
- Exfoliation corrosion
- Pitting corrosion
- Microbiologically induced corrosion
- Galvanic corrosion
- Filiform corrosion
- Crevice corrosion
- Stress-corrosion cracking (SCC)
- Fretting and other types of corrosion wear

General corrosion is corrosion that is upon large surfaces of the material and that results in the overall thinning of the structure. It is the result of shifting anodes and cathodes on a metal surface. This form of corrosion has been observed particularly between disbonded aluminum skins and the interior or exterior fuselage belly skins.

Exfoliation corrosion is the leafing of rolled or extruded forms of aluminum alloys (Fig. 2a). Corrosion starts at an exposed end grain and proceeds through multiple parallel paths perpendicular to the dimension that is reduced in the forming process (short transverse direction). This form of corrosion is observed wherever the aluminum product forms listed previously are used, that is, wing and empennage (tail) skins, and wing, fuselage, and empennage stringers and chords.

Pitting is an acute perforation of the alloy. This form of corrosion is observed in all aluminum product forms used for fuselage skin, webs, and bulkheads. It is also observed on steels used in landing-gear and flight-control components. Pits are stress raisers and initiation sites for fatigue or stress-corrosion crack propagation.

Microbiologically induced corrosion manifests itself as pitting or intergranular corrosion. It is promoted by the presence of fungi (*Cladosporium resinae*) brought in by jet fuel. Figure 2(b) shows fungi responsible for microbial corrosion on a fuel tank access door. This form of corrosion is observed inboard of the wings, and is frequently observed on airplanes operated in warmer climates.

Crevice corrosion is corrosion that occurs inside crevices, tight joints, under absorbent materials, debris, or any recessed areas where stagnant conditions exist. Local oxygen depletion and acidification of the crevice environment give rise to an electrical potential difference with the region outside the crevice and leads to severe corrosion in the crevice. This form of corrosion is observed at unsealed joints and at joints where the fillet seal is broken. It is also observed on fuselage structures held in intimate contact with wet insulation blankets for extended periods of time. Crevice corrosion may be observed on aluminum or stainless steel tubing under rubber clamps.

Stress-corrosion cracking is a brittle fracture of susceptible alloys occurring under load at stresses below the yield strength of the alloy. It results from the combination of tensile stress and environmental factors specific to the alloy. In commercial airplane operating environments, both aluminum and steels are susceptible. In aluminum alloys, cracking may result from wedging, that is, from tensile stresses caused by the expanded volume of corrosion products along grain boundaries, or from installation stresses. Alloy steels that are heat treated to strengths greater than 1520 MPa (220 ksi) are the most susceptible. Fracture occurs by a hydrogen embrittlement mechanism; that is, hydrogen produced through the reduction reaction (Eq 2b) diffuses to regions where subsurface residual stresses are present from service conditions. No SCC fractures have been observed with titanium components in service, although some alloy heat treat combinations of Ti-6Al-4V are susceptible to this form of damage.

Galvanic corrosion is the accelerated corrosion of a metal in electrical contact with another material that is more noble in the galvanic series in the presence of moisture. Galvanic couples frequently encountered in commercial aviation are: titanium-aluminum, carbon fiber reinforced plastic (CFRP)-aluminum, aluminum-nickel bronze (Fig. 2c, 2d) and corrosion-resistant steel (CRES)-aluminum. Galvanic couples may exist between exposed cut edges of CFRP and aluminum. Figure 3 shows major galvanic groupings of alloys used in airplane design. Galvanic corrosion is also observed on stainless steel blind rivets, and alloy steel fasteners and clipnuts after their cadmium plating has worn away.

Filiform corrosion is surface corrosion that develops as wormlike filaments. These filaments grow under protective coatings and typically radiate out from fastener heads and skin edges. The head of the filament is the active corroding end and is acidic, while the tail is alkaline and filled with hydrated aluminum oxide.

Fretting and other wear-related forms of corrosion damage result from the synergistic effects between corrosion and wear. Fretting corrosion occurs when corrosion is associated with relative motion of small amplitude between two metallic surfaces. Hard abrasive particles break off from oxide corrosion products and

Fig. 2 Forms of corrosion in aircraft. (a) Exfoliation corrosion. (b) Microbiologically induced corrosion on fuel tank access door. (c) (d) Galvanic corrosion under aluminum-nickel bronze bushing

abrade the metal, maintaining fresh surface for further corrosion. Fretting is often found on fuel tank access doors.

Contact fatigue is an important form of damage on surfaces that experience moderate to high bearing stresses: rolling surfaces of flap tracks, flap carriage components journals, pins, and bearings. Undercutting of the plating by contact fatigue is often seen on chrome plated surfaces resulting in steel corrosion, as seen on the aft journal of a 747-400 flap carriage (Fig. 4).

Factors Influencing Airplane Corrosion

Each of the corrosion forms described can be the cause of major structural airworthiness issues. Each one of the effects is in direct relation to the structural design, to corrosion susceptibilities introduced during manufacturing, to operational environments (both internal and external), to aircraft use, and to maintenance practices, which vary from operator to operator. Figure 5 summarizes these factors from the manufacturing perspective.

Design Factors

Improving the corrosion resistance of commercial aircraft structures has resulted from improvements made in these areas of design:

- Materials selection
- Protective finish systems selection
- Drainage
- Sealing improvements to eliminate crevices
- Use of corrosion inhibiting compounds
- Maintenance access

Materials Selection. An important goal of material selection is to ensure the use of inherently more corrosion-resistant materials.

Aluminum Alloys. For aluminum alloys, which are continuously developed to improve structural performance and life-cycle cost (Ref 3), the trend has been to increase the use of overaged rather than peak-aged tempers. Several stress-corrosion fractures of 7079-T6 aluminum alloy in the late 1960s to early 1970s led to their replacement with 7075-T73, which has better corrosion resistance and a crack growth rate at least 2 orders of magnitude lower. Alloy 7178-T6 was also replaced because of low stress-corrosion resistance. The current choices in aluminum alloys for applications requiring higher strength with improved stress-corrosion and exfoliation resistance are overaged 7050, 7150, and 7055 plates and extrusions for upper wing skin, and 2324-T39 or 2024-T39 plates alloys for the lower wing skins. Clad 2024-T3/2524-T3 sheets or 2024-T351/2524-T351 plates are used on the pressurized fuselage. Cladding on the exterior skin of fuselage 2xxx series and 7xxx series alloys, which consists of 1230 and 7032 respectively, is provided for increased corrosion resistance. 2xxx-T3x alloys have relatively low exfoliation and stress-corrosion cracking resistance and therefore must be very well protected. The chemical-milled or machined surfaces of 2024-T351 clad aluminum plate used for bilge skin has proven to be a corrosion concern, regardless whether the milled cladding is internal, as in the 747-400 (Fig. 6), or external, as in the 777 (Fig. 7). Figure 7 shows how susceptibility to corrosion increases once the pure aluminum cladding is removed.

Steels and Other Metallic Materials. During the late 1960s to early 1970s, multiple failures in service of H-11 alloy steel bolts by stress-corrosion cracking led to their replacement throughout the entire Boeing fleet. Other steels no longer used because of their SCC susceptibility include PH 13-8Mo at strengths above 1520 MPa (220 ksi), 17-7PH, and maraging steels. For high-strength steels (1520 MPa and above), current designs use vacuum arc remelted and grain-refined AISI 4340M or 300M material for reduced susceptibility to hydrogen embrittlement in plating solutions, and to stress-corrosion cracking by the hydrogen embrittlement mechanism in service. The improved

	Class	Metals
Reactive (Anodic)	I	Magnesium and its alloys
	II	Cadmium, zinc, aluminum, and their alloys
	III	Iron, steels (except corrosion-resistant steels), tin, and their alloys
Passive (Cathodic)	IV	Copper, brass, bronze, chromium, nickel and nickel-base alloys, corrosion-resistant steels, titanium and their alloys, graphite

Fig. 3 Galvanic grouping of metals and alloys commonly used in commercial airplane design

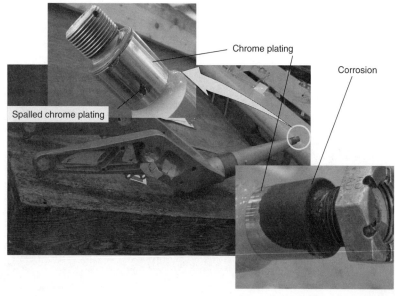

Fig. 4 Chrome plating wear and corrosion of steel on a 747 flap carriage journal. The bottom view shows missing chromium plating and corroded journal.

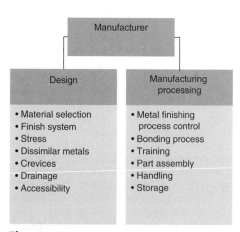

Fig. 5 Manufacturing factors contributing to corrosion

performance is the result of improved transverse ductility and toughness, which raises SCC resistance. Nevertheless, the alloy remains susceptible to corrosion and pitting. Corrosion pits can act as stress risers and lead to fractures. Some stress-corrosion fractures of high strength alloy steels have been observed in a few instances, including the bolt holes of the 767 main landing gear aft trunnion (Fig. 8). Alloy 9Ni-4Co-0.3C steels were used for engine mounts because of their good heat resistance combined with good stress-corrosion resistance. However, because of repeated corrosion from chipped finish, nickel alloy 718 or titanium is now preferred. The general trend is to eliminate corrosion-prone materials from design, such as increasing the use of corrosion-resistant steels (CRES) to replace low-alloy steels. To that end, nonferrous materials such as magnesium and its alloys are no longer used despite their weight benefit. Also, interest is being shown in Custom 465, a high-strength stainless steel. Despite its high cost, it is finding applications in trailing-edge links and flap track, and torque tube fittings in future programs, including the 787.

Protective Finishes—Organic and Inorganic. Protective finishes provide corrosion resistance, wear and abrasion resistance, and improved appearance. In combination with selected alloys, protective finish systems are subjected to extensive testing using industry standards. A list of standards and approaches to testing for corrosion protection for application to commercial aircraft design are described elsewhere (Ref 4).

For corrosion protection, the system consists of an inorganic coating, which serves as surface preparation, and one or more coats of an organic finish. For all exterior aerodynamic aluminum surfaces, the finish systems include chromate conversion coating, filiform corrosion-resistant primer, and polyurethane enamel as topcoat. For interior aluminum structures, it consists of chromate conversion coating or anodize, solvent-resistant chromate epoxy primer prior to assembly, and an additional coat of primer after assembly (Fig. 9). An integral fuel tank primer designed to resist the growth of fungi is used inside the wing. For aluminum parts that are to be bonded, the system consists of phosphoric acid anodize and a chromate bonding primer prior to assembly bonding. Corrosion protection of alloy steels components is provided by cadmium plating and chromate primer. Additionally, these components are topcoated with polyurethane enamel paint when used externally. Corrosion-resistant steels alloys are also cadmium plated prior to priming when galvanic compatibility with aluminum is required. Other corrosion-resistant materials such as titanium and nickel alloys may be left bare or painted to improve the appearance of the part. However, they are always painted to prevent galvanic corrosion resulting from contact with dissimilar metals when they are part of an assembly that includes aluminum alloys. To prevent galvanic corrosion of aluminum in assemblies with CFRP, the surfaces must

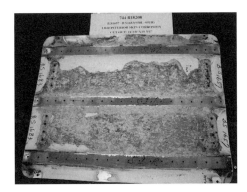

Fig. 6 Corrosion of 747-400 interior 2024-T351 aluminum belly skin machined interior aluminum cladding

Fig. 7 Corrosion of 777-300 exterior 2024-T351 aluminum belly skin machined exterior aluminum cladding

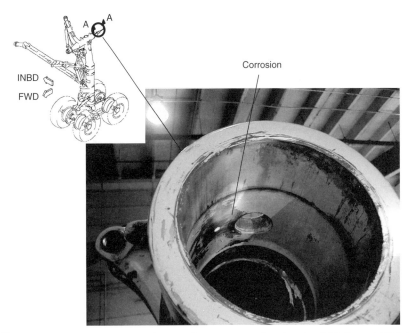

Fig. 8 Corrosion of the 767 steel main landing gear aft trunnion. Insert shows location. FWD, forward; INBD, inboard

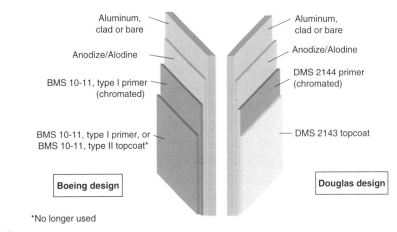

Fig. 9 Design finishes for all interior aluminum structures

have a cocured fiberglass ply, polyvinyl fluoride film, or be primed and topcoated as shown in Fig. 16.

Wear and abrasion resistance is provided to aluminum by hard anodize coatings. On steels, chromium plating is used. Macrocracking of chromium plating, and undercutting from corrosion in service (Fig. 4) is leading to gradual replacement with thermal spray coatings in selected applications. Unlike chromium plating, these may be applied to titanium alloys, allowing their use in wear applications. Initially, thermal spray WC-Co coatings were used. These have been replaced with more corrosion- and wear-resistant WC-Co-Cr coatings because of fretting and sliding wear issues experienced by WC-Co on seal surfaces.

Environmental Challenges. The aerospace finishing industry is increasingly being challenged by environmental pressures to limit the use of cadmium and chromium, materials on which current corrosion performance standards are based. Progress has been made in finding equivalent replacements in some areas. At Boeing, thermal sprayed WC-Co-Cr is gradually replacing chromium plating on all steels, and zinc-nickel alloy plating is replacing cadmium on all steels except on those with strength greater than 1520 MPa (220 ksi). However, proposed OSHA rules for an 8 h time-weighted average permissible exposure limit of one microgram of Cr(VI) per cubic meter of air (1 $\mu g/m^3$) for all Cr(VI) compounds (Ref 5) represent a significant challenge since chromates, which are used as corrosion inhibitors in primers, remain unsurpassed in their ability to protect aluminum against corrosion. Passage of this rule will require considerable adjustments industrywide.

Airplane Drainage. Fleet history indicates that corrosion is most often found at joints and pockets, under debris, wherever moisture traps exist, as shown in Fig. 10. This may be exacerbated by a design that does not provide adequate drain holes, or by maintenance practices that block drain holes or fail to open clogged ones as shown in Fig. 11. A complete and effective drainage system consisting of drain holes, tubes, where necessary, and drain valves must be part of the design. The current standard is to provide drain paths through $^3/_8$ in. minimum diam or 0.11 in.^{2} minimum area holes, observing airplane angle with the horizontal when on the ground, to allow cross drainage. Holes or drain valves must also be provided at low points of the fuselage and wing.

Pressurized fuselage drain valves are spring loaded to close in flight and open on the ground.

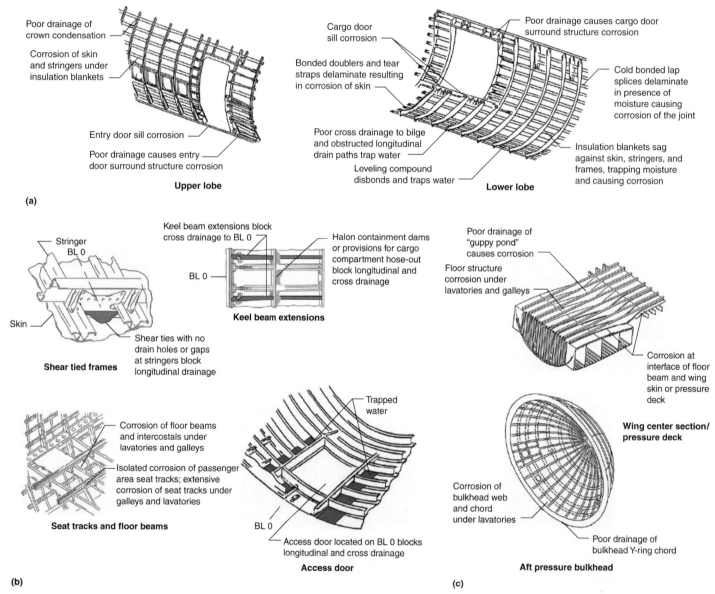

Fig. 10 Fuselage corrosion locations learned from airplanes in service. (a) Upper and lower lobe. (b) Lower section of fuselage. BL 0 is the lowest longitudinal line of the fuselage (butt line). (c) Aft and wing

Enough of these drain valves must be provided to drain an estimated maximum amount of condensed moisture between landing and takeoff. Figure 12 shows typical locations of drain holes and a sketch of a fuselage drain valve. The drainage system is typically validated with the first airplane built and appropriate modifications made to address areas where moisture accumulation is found.

Sealants and Corrosion-Inhibiting Compounds (CICs). The use of sealants and corrosion-inhibiting compounds to keep moisture out of metallic joints has increased dramatically in recent years. Polysulfide sealants applied to mating (faying) surfaces keep moisture out of major assembly joints, such as wing chord to spar (Fig. 13), fuselage lap joints, and fuselage skin to stringer. In fact, this use of faying surface sealant has dramatically reduced corrosion of the outer wing at wing front and rear spar locations. The seriousness of such findings cannot be overstated as the corrosion may be impossible to detect until it exceeds limits allowed by design (Fig. 14). On similar metal contacts, for example, aluminum-aluminum or aluminum-cadmium plated steel, sealants prevent moisture entry and thus prevent crevice corrosion. Fastener holes are a major moisture path in permanently fastened installations. All exterior fasteners through structures required to sustain flight and ground loads and installations involving the stacking of different alloys are installed wet with sealant in order to block moisture path to mating surfaces.

Organic corrosion-inhibiting or moisture-displacing compounds are sucked by capillary force into unsealed joints to displace any moisture that may be present. Originally applied to structural joints, their use has expanded to open surfaces, driven in large part by airline maintenance, because they do not require disassembly of structural joints and often protect the finish against rapid deterioration from environmental exposure.

Sealant and CIC Use. The passenger floor structures have been chronic corrosion areas for Boeing 7*xx* series airplanes at "wet" locations, such as under galleys, lavatory, and entryways (Fig. 15). Here the improved use of sealant and CICs is preventing corrosion. Today (2006), areas under passenger entryways, and areas extending 20 in. beyond galley and lavatory footprints, floor structures are covered with CIC. Floor panels are attached to the structure with plastic clipnuts, and the panel joints are sealed and overlaid with durable polyurethane seam tape. The entire floor in then covered with larger sections of polyurethane tape (Fig. 16). This design effectively prevents moisture ingress to the floor structure. The improved sealing avoids the more costly, but certainly more durable solution, implemented on the 777 airplanes with the use CFRP floor beams and titanium seat tracks, materials that do not corrode.

Special Assembly Consideration: CFRP-Aluminum Joints. Nowhere is the risk of corrosion greater than with assemblies involving CFRP and aluminum, materials at opposite poles of the galvanic chart (Fig. 3). Graphite fibers are excellent electrical conductors, do not corrode, and act as cathodes when attached to aluminum. Preventing severe corrosion of any attached aluminum structure involves the use of sealant to block all moisture paths, including fastener holes. It is also as important to ensure that cut edges, where graphite fibers are exposed, are sealed (Fig. 17).

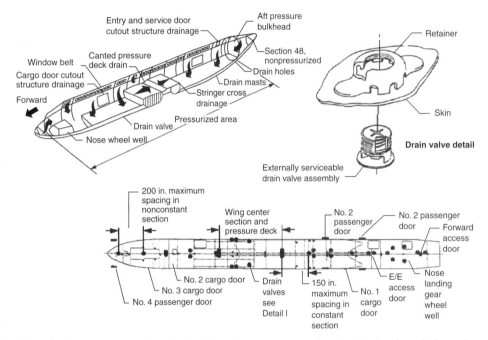

Fig. 12 Fuselage drainage and drain valve design. Drain locations are indicated as closed circles in the lower plan view

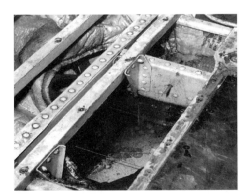

Fig. 11 Moisture accumulation from blocked drain

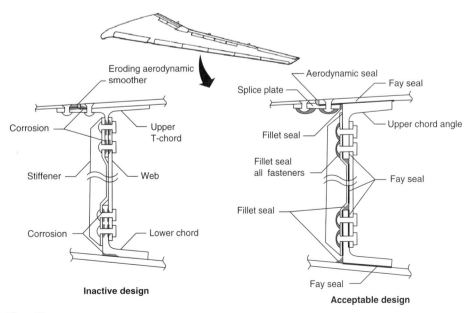

Fig. 13 Wing front spar with corrosion sites. Sealant locations in acceptable design

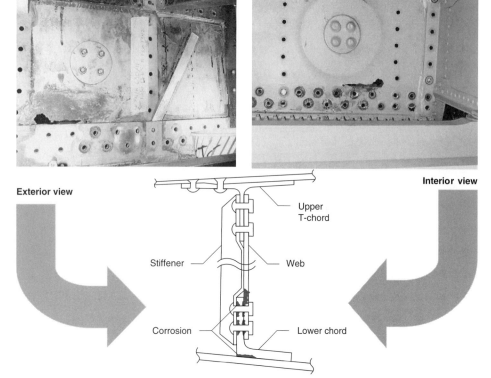

Fig. 14 Corrosion between wing front spar lower chord and web of 747

Fig. 15 Aluminum floor structure corrosion in a 767 at forward entry door

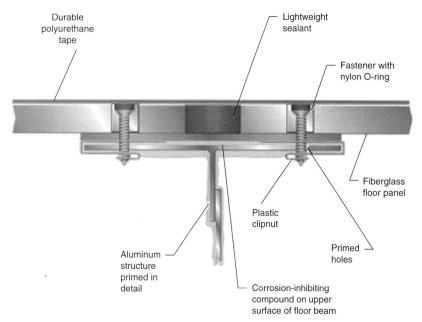

Fig. 16 Typical floor structure design for corrosion protection

Accessibility. Designing readily accessible structures is an important part of corrosion prevention. Accessibility is provided to allow easy visual inspection and repair of major structural components. The design must provide removal panels and access doors to serve that purpose.

Manufacturing Factors

Minimizing defects, and finding and correcting any that occur during airplane manufacture, is an important step in corrosion control.

Assembly Defect Control. Defects introduced during the assembly of the airplane, may become initiation sites for corrosion, and must be avoided. If this is not possible, mitigating measures are often part of the design in the form of drawing notes, for example:

- Drain holes provided between detail parts of an assembly may occasionally be accidentally plugged during manufacturing with sealants. A note to remove excess sealant to avoid blocking a moisture path may be needed.
- Scratches may be introduced in the protective finish during fastener installation. Finish touch-up after assembly is required per design.
- Loss of paint adhesion due to contamination of surfaces to be painted. This discrepancy may be detected during fastener assembly, requiring a local refinishing. In the example shown in Fig. 18, undetected silicone contamination during the fabrication process and constant moisture exposure in service caused a loss of primer adhesion in a 747-400 airplane bilge.

Process Controls. Testing and process control are used to assess and avoid delayed fractures that can be caused by the manufacturing process. Cold bonding of lap joints was eliminated as a process in 1972 following significant bondline corrosion problems observed on Boeing airplanes built in the early 1970s. These can now be avoided with strict controls of the bonding process, with the development of phosphoric acid anodizing, and strict process controls including a wedge test (Ref 4) to assess bond quality. Stress-corrosion cracking or hydrogen embrittlement in high-strength steels are routinely avoided with process controls that periodically assess embrittling conditions in process solutions. Part handling and assembly may create conditions where corrosion may be initiated in service. These must be recognized and addressed with adequate training/certification of inspection personnel and quality-control procedures.

In order to increase airplane structure durability and avoid catastrophic accident with corrosion as its root cause, commercial airplanes design and manufacturing have exploited improvements in materials and finish systems and sealing methods, expanded CIC use, improved access, and implemented better manufacturing controls. The resulting increase in value achieved with a more corrosion-resistant airframe can be maintained if operational factors leading to the deterioration of the structure are understood and adequately addressed.

Service-Related Factors

This section surveys factors responsible for airplane corrosion in service. The deterioration of the airplane structure from corrosion in service is the result of interaction with the airplane operating environment. Two types of aircraft environments should be distinguished: internal and external environments. Internal environments affect the interior fuselage, while the exterior environments affect all exterior structures including wing leading and trailing edge cavities, nose and main landing gear cavities, and nonpressurized fuselage tail section.

Internal Environments

The internal environment of the airplane fuselage contains various fluids from these sources:

- Condensation on internal surfaces
- Fluid spills in cargo compartments
- Toilet and galley leakage
- Wet insulation blankets in contact with structure
- Rainwater

Condensation. Moisture is generated inside the airplane cabin because of passenger respiration and perspiration. Condensation is heavily influenced by aircraft loads factors (the percent of seats occupied on any flight), and airplane utilization rates (the number of flight cycles or hours accumulated per day). Most of this humidity collects on the fuselage skin as ice during cruise and thaws during descent. The time of wetness on long-range flights differs from short-stop flights. On longer-range flights moisture exists as ice much longer than as liquid. Because the converse is true for short-range flights, corrosion may be more of a problem on airplanes operated on short-haul flights at high utilization rates.

Cargo and Other Accidental Spills. In addition to being the collection zone for all condensation on the airplane, the structure below the cargo floor, or bilge, may be exposed to a variety of fluid spills with different levels of aggressiveness. They are direct consequence of transported cargo: live animals, fish, flowers, and chemicals. Despite measures put in place to control spills (e.g., special containers for the transport of fish), such spills do occur. On occasion, an airplane may be accidentally exposed to unique environments, such as hangar fire suppression systems, deluging the aircraft with fire-extinguishing foam while all airplane doors are open.

Lavatories and Galley Leakage. Older model lavatories were bolted directly to the floor structure. More recent airplane models have adopted vacuum toilets and self-contained lavatories. While this has reduced the amount of water spills from lavatories, leakage still occurs and seeps through floor panel joints onto the floor structure. These spills are often the source of corrosion of the floor structure under galleys and lavatories

Rainwater. Entry door locations are exposed to rain. Entry doors, similarly to galleys and lavatories, are also designated as "wet areas" for the purpose of designing the passenger floor.

External Operating Environment and Utilization

Factors that affect corrosion in external environment are:

- Geographic areas at or near coastal or industrial environment
- Flight regime
- Volcanic ash
- Airplane utilization
- Runway deicers

Significant amounts of moisture can accumulate on external surfaces of airplanes. Of particular importance, external surfaces accumulate aggressive species found in industrial pollutants, salt aerosols, volcanic ash (SO_2, Cl^-, H_2SO_4 droplets, etc.).

While the corrosion concerns due to high humidity, temperatures, and pollutants have always been known to airline operators, the increasing use of potassium or sodium formate/acetate as runway deicers at airports in countries with colder climates is more recent as some airport authorities have replaced glycol or urea-based runway deicers with potassium formate or acetate for environmental reasons. This concern has been greatest for Boeing 737-600/700/800/900 operators with airplanes flying into airports where these deicers are used. In contrast with other airplane models, electrical connectors located in the landing gear wheel well may be splashed by runway fluids because of their proximity to the ground. Corrosion of connector pins and the resulting electrical connector failure was found serious enough to warrant regulatory action by the FAA (Ref 6).

Each airplane may be subject to a unique combination of the factors listed previously, which may lead to considerable variability in the onset and severity of corrosion depending where in the world it normally travels. However, among factors influencing corrosion of commercial airplanes one must also include maintenance practices and in-service finish deterioration. Figure 19 summarizes these factors from the operator perspective.

Maintenance Practices

Maintenance practices have a major impact on corrosion of the airplane structure. The goal of maintenance related to corrosion is to restore the level of protection to that of the as-designed airplane. However, the quality of corrosion-control measures implemented during maintenance can vary widely. This can range from

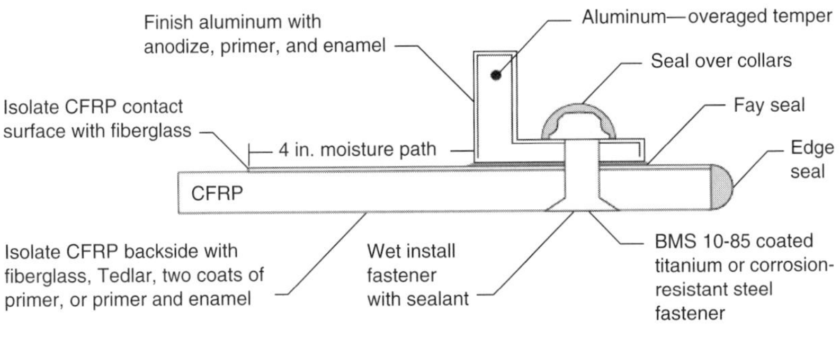

Fig. 17 Protecting aluminum against galvanic corrosion from contact with carbon fiber reinforced plastic (CFRP)

Fig. 18 Loss of paint adhesion caused by silicone contamination

barely adequate in some countries, to excellent. The training of maintenance workers may also be inadequate and the restoration of finishes may be left to low-skilled workers. In other instances, the level of cleanliness required of surfaces prior to the application of finishes may be inadequate. Maintenance tasks designed to reduce or prevent corrosion may not be adequately understood. On occasion, drain holes have been plugged with sealant during repairs.

Organic finish deterioration invariably occurs over the life of an airplane due to environmental exposure (Fig. 18). On exterior coatings, this deterioration manifests itself in the loss of flexibility, color, and gloss and requires repainting after 4 to 8 years. Deterioration of corrosion protective paints used in interior structures is very slow and may not be apparent over the entire life of the airplane. It is a general practice in the airline industry to apply corrosion-inhibiting compounds on most surfaces as frequently as necessary to increase the life of the paint.

Additionally, incidental finish damage and the restoration practices impact the appearance of corrosion. Deterioration that includes scratches and gouges in the protective finish may occur in service from debris. Scratches or gouges may be introduced in areas adjacent to repairs. It may be difficult, if not impossible, for maintenance crews to match the level of cleanliness for painting and sealing during maintenance repair that is achieved during manufacturing.

Areas of the airplane that are more likely to suffer from corrosion are often those that are more frequently accessed for inspection during maintenance. For instance, the floor structure in wet areas, which has benefited from improved sealing to make it more resistant to corrosion, is rarely found corroded during the first inspection, but more often during the second, suggesting that the floor was not reinstalled as carefully as during original manufacture. While floor panel sealing has improved, inspection of the floor support structure has remained more frequent than may be necessary because of past corrosion problems.

Assessing Fleet Corrosion History

Most of the improvements made in the design and manufacture for corrosion prevention of commercial airplane are the result of corrosion data gathered from the in-service fleet. Several data-gathering tools have been used to assess the condition of airplanes in service:

- Aging fleet surveys
- Operator assistance requests messages (also known as "telex")
- Corrosion Prevention and Control Program (CPCP) reports

Aging Fleet Surveys. Aircraft surveys are visits undertaken by the manufacturer at the operator's maintenance base to assess the condition of the airplane structure with respect to corrosion and fatigue deterioration and to review maintenance practices for the fleet of similar models owned by the operator. These surveys provide valuable lessons about areas of chronic corrosion on the aging fleet or the efficacy of retrofit actions. Sometimes, survey trips are scheduled specifically to draw lessons to be used in the design of new or derivative models. Figure 20 shows a list of major corrosion locations discovered on the 767 airplanes during several 767 fleet surveys.

Operator and Maintenance Repair Organization (MRO) Messages. The repair of corrosion damage that exceeds limits specified in the Structural Repair Manual (SRM) must be approved by the regulatory authority (FAA for U.S. registered airplanes). Airlines often direct requests to the manufacturer for assistance on engineered repairs. These requests provide detailed corrosion location and residual thicknesses after corrosion removal. For some locations, the number of such messages has grown steadily with the age of the fleet. Fuselage Sections 44 (over the wing) and 46 (aft fuselage) have been particularly hard hit on the first generation, or "classic," fleet of 747 (Fig. 21a). In contrast, the request for corrosion repair assistance has remained constant for Sections 42 and 46 (Fig. 21b). As a result of this service history, drainage of the sloping pressure deck has been improved on the 747-400 and on the 777 airplanes. Sealing of the floor in areas in the wet areas of the 747-400 has also been improved as a result.

Corrosion Prevention and Control Program (CPCP) Reporting. The CPCP is one of several programs mandated by the FAA on all airplanes operated in the United States. It defines minimum requirements to address corrosion before it becomes an airworthiness concern. Under this program, operators are required to submit to the regulatory authority and the manufacturer a report for each severe corrosion finding (called a level 2 finding). The report contains the exact location of the corrosion, the probable causes of corrosion, and whether corrosion was caused by a unique event, such as an accidental spill, or by a condition that might exist on other airplanes in the operator's fleet. The intent of the additional reporting to the manufacturer is to assess the effectiveness of the baseline CPCP in order to adjust it as necessary. Figure 22 shows a worldwide distribution of level 2 reports. This distribution may be inaccurate in describing the worldwide fleet status with respect to severe corrosion: they indicate a larger percentage of "level 2" corrosion in developed countries for all models. However, these reports have been valuable in helping the manufacturer identify chronic corrosion problems areas and underlying design weaknesses. It should be noted since 2002 that reporting to the manufacturer is no longer required. However, reporting to the FAA remains in effect as defined under Title 14—CFR Section 121.703.

Information from the sources listed previously has contributed greatly to increasing Boeing's understanding of regions with chronic corrosion problems and their severity. This information has driven cumulative design improvements (Fig. 23). These have been incorporated into "Book 4—Design for Corrosion Prevention" used by Boeing-Puget Sound engineers as a guide for the design of new 7*xx* series airplane models or derivatives. Similarly, lessons learned for Douglas models and derivatives were incorporated into the Type Finish Specifications TF-109 (DC-9, MD80, MD90, Boeing 717) and TF-110 (DC-10/MD-11).

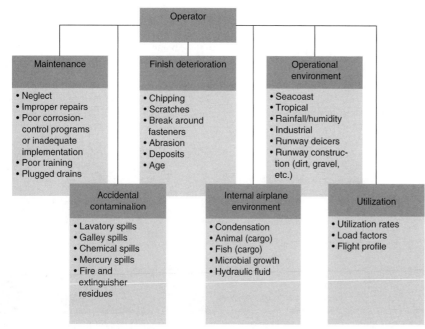

Fig. 19 Operating factors contributing to corrosion

Location of corrosion	Airplane delivery date												
	June '83	Mar '85	Aug '82	Oct '82	Dec '82	May '86	Jun '83	Apr '93	Feb '83	Sep '84	Jun '83	Aug '82	Jun '83
	Airplane survey date												
	May '90	Jun '90	Nov '90	Jun '91	Sep '91	Feb '92	Jun '92	Nov '92	Dec '92	Oct '93	Sep '94	May '95	Apr '98
Keel beam chords	■	■	■	■		■		■	■	■	■		
Floor beams at fwd entry door	■	■	■	■		■		■	■	■	■		
Station 434 stringer splices		■	■	■	■				■	■	■		
Forward cargo lower sill failsafe strap						■							
Cargo door beams nutplate attachments	■						■			■		■	
Aft cargo compartment frame upper chord								■	■	■	■		
Seat tracks		■	■										
Floor beams under galley complex							■	■					
Lower skin antenna locations				■	■						■		
Aft entry door									■	■			
Sill inboard chord cargo door pan strap		■						■	■				
Station 1480 floor beam													
Trailing edge flap rear spar upper chord													
Horizontal stabilizer Z stringer													■

■ Corrosion found during survey

Fig. 20 Aging survey results for model 767

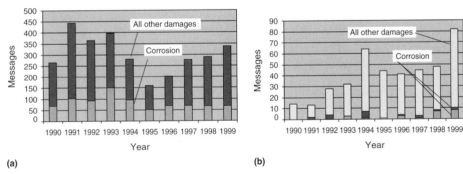

Fig. 21 Pattern of corrosion-related messages. (a) 747 Section 46 monocoque structure. (b) 747-400 Section 46 including monocoque and floor structure corrosion

Airworthiness, Corrosion, and Maintenance

Many commercial aircraft are being operated well beyond their original economic design service objective (DSO), a minimum period of service during which the airplane structure is essentially free of detectable fatigue cracks. This practice will continue as long it is profitable to do so. However, since continued safe operation and prevention of accidental hull loss is the prime consideration, how corrosion is addressed as the airplane structure ages is of paramount importance.

Airworthiness defines the ability of an airplane to sustain all fail-safe design loads (limit loads) throughout its operational life without catastrophic failure. For continued airworthiness, it is now recognized that fail-safe design loads must be sustained in the presence of all damages that may occur in the airplane operating life, that is, accidental, fatigue, or corrosion damage, until such time when the damage is detected through inspection and repaired. This concept, known as damage tolerance, became the standard principle for establishing maintenance programs for all airplane models since 1979. All post-1979 models have included damage-tolerance principles in their design and incorporated them into their baseline maintenance programs.

The maintenance of earlier, fail-safe-designed, airplanes did not include provisions for multiple inspections of an airplane structure near or above the DSO before a crack reached its critical length. The application of damage-tolerance principles to these older airplanes relied instead on data from extensive structural testing and tear-down. The required inspection methods and frequency were determined using linear elastic fracture mechanics analysis techniques relating detectable damage, damage growth, and critical damage size. These were published in a supplemental structural inspection document (SSID) for the detection of fatigue damage for each affected model.

Effect of Corrosion on Airworthiness. It was recognized at the outset of the SSID program development that uncontrolled corrosion could act synergistically with fatigue, resulting in faster crack initiation, unpredictable crack growth, and significant reduction in the damage detection period. Thus, for fatigue inspection frequencies to be valid, corrosion had to be detected early and treated quickly. Nevertheless, no industry-wide action was taken to mandate timely corrosion detection and repair, and it was left to the discretion of each airline operator to develop its own corrosion-prevention program.

The severe impact of unattended corrosion on airworthiness of a structure designed to carry a significant portion of pressure loads was dramatically demonstrated to the world with the explosive decompression of a 737 fuselage in 1988 (Ref 7). Several small fatigue cracks had grown and linked up in a disbonded and corroded fuselage lap joint. This event led the FAA to mandate aging airplane programs that included CPCP, SSID, mandatory service bulletin modifications, and repair assessment programs for all older airplanes. Several airworthiness directives (ADs) were issued in 1990 for each older airplane model CPCP.

The ADs required that operators of older airplanes that did not have an effective corrosion-prevention program adopt a baseline CPCP as a supplement to their maintenance program. To ensure commonality of approach between airplane manufacturers, the baseline programs for each airplane were developed under an Airworthiness Assurance Working Group (AAWG) steering committee, and through structure task groups (STG) that included airplane manufacturers, operators, and the FAA. For Boeing aircraft, the mandate was applicable to pre-1979 airplanes and their derivatives, that is, 747-400 and 737-300/400/500. Because post-1979, "damage-tolerant" airplanes (757 and 767) incorporated into their maintenance program all recommended inspection tasks and intervals for the aircraft components and structure, including corrosion tasks, an AD for implementation was not required. However, their CPCP remained separate from the structural inspection programs.

Maintaining airworthiness for older airplanes and earlier damage-tolerant ones meant conducting a variety of independently developed programs requiring multiple access and ground time for inspections that targeted the same structures. Airplanes certified in 1993 and later have a consolidated structural maintenance program, which provides a single opportunity to inspect for accidental, fatigue, and corrosion damage. This results in significant cost savings to operators. These consolidated programs were developed under a Maintenance Steering Group (MSG), using a methodology called MSG3-Rev 2 (Fig. 24). This methodology can be used for analysis of airplanes certified before 1993 (Ref 8) at the request of the airplane operator.

The approach and evolution of the CPCP program for all airplanes is reflected in Fig. 25.

See the article "Predictive Modeling of Structure Service Life," *ASM Handbook*, Volume 13A, 2003, for more on these design philosophies and fleet maintenance.

Corrosion prevention during maintenance is accomplished via observance of practices contained in CPCP documents for pre-1993 airplanes and consolidated maintenance documents for newer airplanes. The airplane structure is subdivided into primary structure, that is, structure required to sustain ground and flight loads, and secondary or principal structural element (PSEs). Principal elements are structural details or assemblies judged significant because their failure would result in the reduction of the airplane residual strength. Most primary structural elements and PSEs are in similar airplane zones and are grouped into corrosion tasks.

Inspection Methodology and Maintenance Decisions—Corrosion Prevention and Control Program. Maintaining the airworthiness of all aging aircraft structures regardless of when they were designed means adopting good maintenance practice and developing skills in the areas of corrosion detection and repair. The baseline CPCP contains minimum standards to maintain safety. The program defines:

- Corrosion levels
- Basic tasks
- Implementation age (I)
- Repeat interval (R)
- Mandatory reporting requirements for all severe corrosion (level 2 and 3) findings
- Periodic review of these reports by the STG to adjust the program, as necessary

Corrosion Level Determination. Three corrosion levels are defined, based on allowable damage and whether the corrosion is local or widespread. Allowable limits for each structure, that is, the maximum amount of material that may be removed from the structure by blend-out before its strength falls below ultimate design loads, are set during design and found in the Structural Repair Manual (SRM) for each airplane model. Local or widespread corrosion indicates whether corrosion is restricted to a single frame chord, stringer, stiffener, and skin not exceeding one stringer, frame or stiffener bay, or whether the corrosion involves two or more of these structures in adjacent locations (Fig. 26). A more complete description of the corrosion level assignments to corrosion findings is shown in Fig. 27:

- Level 1: Corrosion found between successive inspections that is within allowable limits
- Level 2: Corrosion found between successive inspections that exceeds allowable limits
- Level 3: Corrosion found during first or successive inspections that is determined by the operator to be an urgent airworthiness concern

Basic Tasks. The assignment of corrosion level is made following corrosion removal

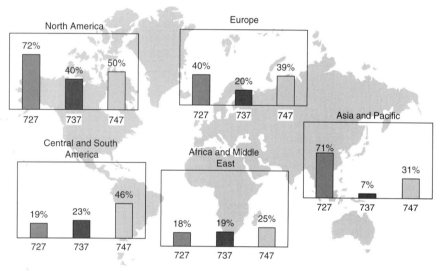

Fig. 22 Percentage of in-service airplanes with CPCP reports of level 2 corrosion. Refer to Fig. 27

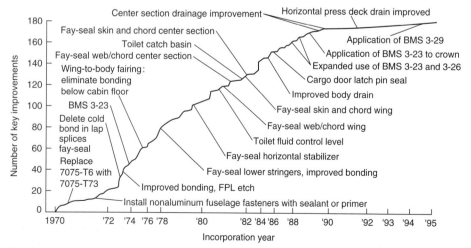

Fig. 23 Cumulative number of corrosion resistance improvements on the 747

and cleanup, or corrosion blend-out, which is necessary when corrosion is found. Most importantly, it is intended to trigger certain actions on the part of the operator to ensure that any subsequent finding on the same structure will be at level 1 or better, and to demonstrate compliance with the CPCP program. The set of required actions includes six tasks in each PSE zone:

- Ensuring adequate access through the removal of all systems equipment and interior furnishing
- Ensuring cleanliness of the structures to allow inspections
- Visual inspections, recognizing corrosion types and symptoms
- Performing the repairs, including complete corrosion removal and adequately restoring all protective finishes
- Clearing of blocked drainage paths
- Application of suitable water-displacing and corrosion-inhibiting compounds

Implementation Age and Repeat Inspection Interval. The basic tasks are to be accomplished at the implementation age (I) and each repeat interval (R). Values of I and R vary with the structure location, the airplane model, and are based on previous service history. For example, all 747 fuselage bilge structure must be inspected first after 6 years, with a repeat inspection every 4 years. By contrast, for the less corrosion susceptible fuselage crown I is 15 years and R is 8 years. The interior of the main wing inspection requirements are 20 years for I and 10 years for R. In contrast, the more corrosion-susceptible exterior leading edge wing are 6 years and 2 years, respectively. Figure 28 contrasts I and R values for the original 737 and 747, or "classic" airplanes, with their more modern derivatives showing that improved design of the derivatives airplanes often leads to relaxed initial inspection requirements.

Inspection Methods. Step 3 of the basic tasks instructs maintenance personnel to "*visually* inspect all PSEs and other structures listed in the baseline program from a distance considered necessary to detect early stages of corrosion, or indication of other discrepancies such as cracking." Successful implementation of the program thus requires effective visual corrosion recognition and detection, in addition to repair and other prevention efforts.

If severe corrosion is found on a structure during CPCP inspections (namely, level 2 or 3), the operator is required to take appropriate measures. The CPCP program must be adjusted to reduce the severity of corrosion at the following inspection. Thus, an effective CPCP is one that controls corrosion of any PSE to level 1 or better between successive inspections. Figure 29 illustrates the fleet damage rate with and without an effective corrosion-prevention program. Without an effective corrosion program, maintenance costs resulting from the repairs of severe corrosion and fatigue damage escalate sooner in the life of the aircraft than DSO. With an effective corrosion program, the design economic life of the airplane may be fully realized. Program adjustments may be made to the basic tasks or to the repeat intervals to maintain airworthiness. Operators may escalate their programs by reducing the baseline inspection interval if unable to control corrosion. Conversely, they may request approval from the regulatory authority for "de-escalating" specific tasks if the corrosion history justifies it.

It has become clear in recent years that the CPCP may not have the full capability to detect corrosion that occurs between mating surfaces until it is made visible in its advanced stage through a severely bulging structure or hole punctured through the structure. Figure 14 shows corrosion through the wing front spar lower chord and web of a 747. In these cases, low-frequency eddy-current (LFEC) and ultrasonic inspection methods must supplement visual inspection, and additional adjustment to supplemental fatigue inspection must be made to account for cracking scenarios that

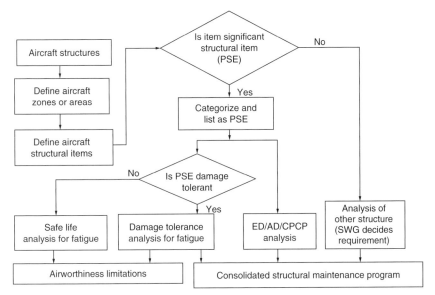

Fig. 24 Logic diagram for developing consolidated airplanes maintenance programs. MSG, Maintenance Steering Group; SWG refers to Structures Working Group. See text for other acronyms.

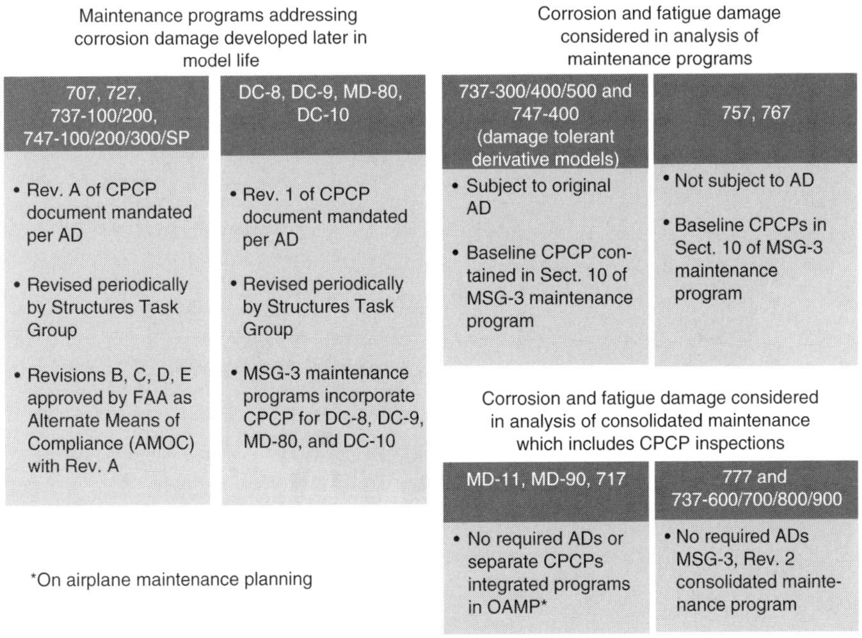

Fig. 25 Summary and evolution of Corrosion Prevention and Control Programs (CPCP) programs for several groups of airplane models. See text for acronyms.

result from hidden corrosion (Ref 9). It should be noted that application of faying surface sealant, now required between major joints as shown in Fig. 13 for the wing chord to web attachment, has significantly reduced these occurrences.

New Fleet Design: Establishing Rule-Based Corrosion Management Tools

The advent of airplanes in which the majority of structure is composites has created a need for a more methodical approach to manage galvanic

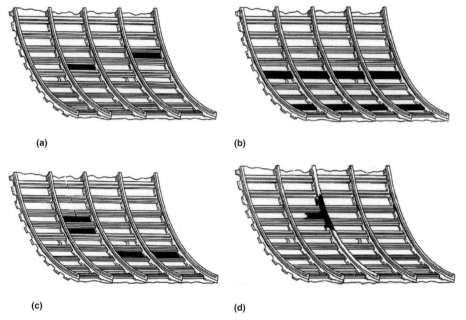

Corrosion severity relative to SRM allowable blend-out limits		Corrosion level for given extent and number of inspection intervals to reach given severity			
		Local		Widespread	
		Single interval	Multiple interval	Single interval	Multiple interval
Well below		Level 1	Level 1	Level 1	Level 1
Approaching		Level 1	Level 1	Level 2	Level 1
At limit		Level 1	Level 1	Level 2	Level 1
Above limit	(a)	Level 2	Level 1(c)	Level 2	Level 1(c)
	(b)	Level 3	Level 1(d)	Level 3	Level 1(d)

Fig. 27 Corrosion level determination per Corrosion Prevention and Control Programs. Structural Repair Manual, SRM. (a) Corrosion is not an urgent airworthiness concern. (b) Corrosion is an urgent airworthiness concern. (c) Operator experience over several years has demonstrated only light corrosion between successive inspections, and cumulative lend-out now exceeds allowable limit. (d) Highly unlikely event following multiple applications of preventive measures and corrosion-inhibitive compounds

Fig. 26 Examples of local and widespread corrosion on fuselage structures. (a) Local corrosion involves nonadjacent frame bays. (b) Local corrosion—involves nonadjacent stringers. (c) Widespread corrosion—involves adjacent stringer frame bays. (d) Widespread corrosion—involves adjacent frame and skin

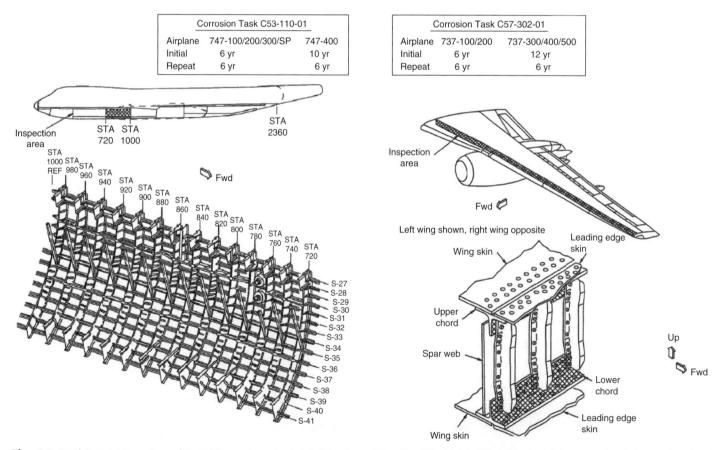

Fig. 28 Initial Corrosion Prevention and Control Program inspection task. Initial and repeat interval for "Classic" 747, 737 airplanes and their more modern derivatives. Location of the area of inspection is given.

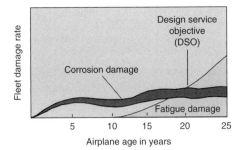

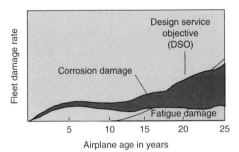

Fig. 29 Fleet damage rate under (a) an effective and (b) an ineffective corrosion-control program

coupling and assess the strategic value of eliminating aluminum in large areas to gain significant maintenance breakthroughs. Recent efforts at Boeing on the 787 have focused on each of these areas.

Managing Galvanic Coupling: The Corrosion Risk Assessment (CRA). Galvanic coupling of aluminum with CFRP across a broad range of parts and airplane locations is inevitable. On the 777, this was managed through the use of isolation practices summarized in Fig. 17 and required extreme attention to details by specialists. Carbon fiber reinforced plastic was primarily used on the horizontal and vertical stabilizers. This approach worked well because the applications were sufficiently limited to allow review of design details by these specialists. However, with airplanes such as the 787, the extent of galvanic CFRP-aluminum coupling use is such that reliance on specialists of different airframe sections to review all specific details is impractical in terms of cost and speed. Fortunately, efforts on the 777 have enabled the creation of CRA, an airplanewide evaluation process that can be used by designers to assess and manage risk.

The CRA considers parameters such as environment severity (determined by the location of the assembly on the airplane), the part protection scheme (applied finish, fasteners installation, and graphite cut-edge sealing, and isolation material), and the susceptibility of the protection scheme to damage. Taken together, these parameters define a corrosion risk for the assembly.

Assessing Strategic Value: The Rectification Impact Assessment (RIA). The ability to further reduce maintenance costs on past airplanes was limited by their use of aluminum. Federal Aviation Administration mandated corrosion prevention and control programs (CPCPs) are required if any structurally significant items are made of aluminum. The frequency and costs of this inspection depend on the environment and service history of the airline. The expanded use of CFRP on airplanes such as the 787 now enables designers to consider the benefit of eliminating the remaining aluminum structure to achieve further savings in maintenance (CPCP), nonroutine inspections, and corrosion repair. The need for a system by which these decisions can be managed and uniformly made is addressed by the rectification impact assessment (RIA).

The RIA, in contrast to the CRA, recognizes that corrosion may nonetheless occur on any aluminum structure, including one that is a primary load path, due to a potential manufacturing defect, and such corrosion may result in maintenance repairs of great difficulty. The relative "pain" of having to perform these repairs is assessed by considering RIA parameters: corrosion detectability, repairability, and the impact or structural significance of the part. Naturally, a structurally significant assembly that is difficult to detect and repair increases the "pain and cost" of a nonroutine maintenance repair or rectification. A high corrosion-rectification risk indicates a "painful" repair.

To design for corrosion resistance and reduce corrosion-related maintenance costs, minimum thresholds of CRA and RIA are chosen based on the 40 year cumulative corrosion history of mostly aluminum airplanes. In order to use aluminum in a new design, these minimum threshold scores must be achieved.

New Airplane Maintenance

The expanded use of composite material in airplane design (Fig. 1) also provides opportunities for additional changes in approach to maintenance. A consolidated structural maintenance program covering accidental, fatigue, and corrosion damage will still exist for all metallic structures. However, with fatigue inspections of metallic materials starting at 100% of DSO instead of 75% of DSO as for previous airplane models, and the use of titanium alloys for improved corrosion resistance where minimum CRA and RIA thresholds are not met, airplanes such as the 787 are expected to significantly reduce maintenance costs compared to the aluminum airplane fleet.

Additional savings may be achieved with structural health maintenance using strain gages to monitor structural damage and determine which structures to inspect in the event of hard landing. Although it is farther in the future, in situ corrosion and humidity monitoring with optical fiber-based sensors may offer additional benefits (Ref 10). A system for the detection of humidity has been recently field tested by a major U.S. airline in the bilge of a 767 airplane.

Conclusions. With more than 40 years of corrosion information accumulated on commercial airplanes in service, significant lessons have been learned. These lessons are being applied in the design of new and the next generation of commercial airplanes. Significant reduction in corrosion has been achieved with careful material selection, enhanced and strategically placed drainage, state-of-the-art primer, coatings, and enamels, systematic use of sealants to eliminate the presence of moisture entry into structural joints, and electrical isolation for CFRP-aluminum assemblies. Additionally, implementation of a self-adjusting CPCP allows operators to keep corrosion within the limits of applicability of damage-tolerant designs. For the next generation of carbon fiber reinforced plastic airplanes, a disciplined corrosion assessment purports to remove aluminum from areas where corrosion risk is too high and to improve the protection during design, further reducing maintenance costs.

REFERENCES

1. K. Ferrer and R. Kelly, "The Impact of the Carbon Dioxide System on the Cathodic Kinetics within Aircraft Lap Splice Joints—(University of Virginia)," 2002 Meeting of the Electrochemical Society (Salt Lake City, UT), 2002
2. D. Banis, J.A. Marceau, and M. Mohaghegh, Design for Corrosion, *Aero*, No. 07, 2000 http://www.boeing.com/commercial/aeromagazine/aero_07/index.html
3. M.V. Hyatt and E.S. Axter, Aluminum Alloy Development for Subsonic Aircraft, *Proc. International Conference on Recent Advances in Science and Engineering of Light Metals*, K. Herano et al., Ed., Institute of Light Metals, Tokyo, 1991, p 273–280
4. A. Adjorlolo and J.A. Marceau, Chapter 61, Commercial Aircraft, *Corrosion Tests and Standards: Applications and Interpretation*, R. Baboian, Ed., MNL20, 2nd ed., ASTM International, 2005, p 687–692
5. Occupational Safety and Health Administration (OSHA) "Occupational Exposure to Hexavalent Chromium," Federal Register, Proposed Rules, Vol 69 (No. 191), Oct 4, 2004, p 59305–59474
6. Federal Aviation Administration, 14 CFR Part 39—Airworthiness Directive 2005-18-23; Boeing Model 737-600, -700, -700C, -800, and -900 Series Airplane, Department of Transportation, Federal Register, Vol 70 (No. 177), Sept 14, 2005, p 54253–54258
7. "Aircraft Accident Report—Aloha Airlines Flight 243, Boeing 737-200 N73711, Near Maui, Hawaii, April 28, 1988," NTSB/AAR89-03, National Transportation Safety Board, U.S. Government, June 14, 1989
8. A. Akdeniz and G.K. Das, Integrated Maintenance of Commercial Fleet for

Economic Management of Inspection and Repair, *FAA/NASA/DOD Conference on Aging Aircraft* (Albuquerque, NM, Sept 1999, p 29–38

9. A. Akdeniz and G.K. Das, Influence of Undetected Hidden Corrosion on Structural Airworthiness of Aging Jet Transports, *The Second Joint ASA/FAA/DoD Conference on Aging Aircraft* (Williamsburg, VA), Aug 31 to Sept 3, 1998

10. J.L. Elster, A. Trego et al., Corrosion Monitoring in Aging Aircraft Using Optical Fiber-Based Chemical Sensors, *DOD/NASA/AFRL Aging Aircraft Conference* (St. Louis, KS), May 15–18, 2000, p 13–20

SELECTED REFERENCES

Title 14—Code of Federal Regulations (also known as Federal Aviation Regulations), U.S. Government Printing Office, www.gpoaccess.gov, accessed, 2006

- 14 CFR Section 25.571, "Damage Tolerance and Fatigue Evaluation of Structure"
- 14 CFR Section 25.1529, "Instructions for Continued Airworthiness"
- 14 CFR Part 121 Subpart C, "Structures"

Federal Aviation Administration, U.S. Department of Transportation, www.faa.gov, accessed 2006

- Order 8300.10, "Airworthiness Inspector Handbook," change 22, 11/30/05
- Order 8300.12, "Corrosion Prevention and Control Programs," www.airweb.faa.gov

FAA Advisory Circulars, www.airweb.faa.gov/Regulatory_and_Guidance_Library

- AC 43-4A, "Corrosion Control for Aircraft"
- AC 91-56, "Supplemental Structural Inspection Program for Large Transport Category Airplanes"
- AC 91-60, "Continued Airworthiness of Older Airplanes"
- AC 120-16C, "Continued Airworthiness Maintenance Programs"

Corrosion in Microelectronics

Jianhai Qiu, Nanyang Technological University, Singapore

TREMENDOUS TECHNOLOGICAL ADVANCES have been made in the microelectronics industry. From industrial control and telecommunications equipment to consumer digital electronics, the industry is driven by ever-increasing demands for performance, miniaturization, and low-cost reliability. The continued shrinkage of integrated circuit (IC) design, coupled with the need to integrate new materials for high-speed performance at low cost, continues to pose new challenges to reliability engineers.

In the early 1970s, IC failure rate in the range of 1000 to 2000 FIT (1 FIT is 1 in 1,000,000 failures per 1000 h) was considered acceptable. This was reduced to 100 FIT in the 1990s. By the year 2007, a failure rate of 0.1 FIT is expected (Ref 1). Increasingly, corrosion has become a significant reliability concern and has received considerable attention in the literature (Ref 2–15).

Common problems encountered with ICs and supporting microelectronic devices include (Ref 16):

- Galvanic corrosion
- Electrical short due to corrosion
- Increased contact resistance due to corrosion
- Degradation of organic surface coatings
- Swelling of materials due to moisture pickup
- Degradation of electrical and thermal properties in insulating materials
- Changes in mechanical properties (loss of strength)

Electronic materials are selected and optimized for their electronic, magnetic, and optical properties. Devices and components inevitably use dissimilar metals, for example, aluminum and gold at a bonding pad on an IC. Electrical bias is often present during operation. In the presence of moisture and contaminants, corrosion ensues. The extremely small physical dimensions make the devices and components even more susceptible to corrosion. A heat exchanger can still function with kilograms of metal loss due to corrosion, but an electronic device may fail after the loss of only one picogram. In heat exchangers, corrosion allowance or tolerance are built into the equipment, whereas in microelectronic devices and components, there is no corrosion allowance at all.

Common components susceptible to corrosion attack include IC chips, circuit boards, and associated components such as connectors, switches, and transformers. A video games vendor spent several thousand dollars per month on replacing corroded circuit boards. Consumer digital electronics such as digital cameras and hand phones may also suffer from severe corrosion attacks after use in outdoor environments. The accumulated corrosion at the connector of a digital camera over a few months' usage was sufficient to cause inconsistent battery operation and inaccurate indication of the need for battery recharging. Even the equipment built for corrosion testing and measurement is not immune to corrosion in the laboratory atmosphere. The electrode cable connecting a potentiostat became loose due to the corrosion, and alternating current impedance response and direct current polarization resistance were affected, with an error of 43% (Ref 17).

The fundamental principles of corrosion are detailed in *Corrosion: Fundamentals, Testing, and Protection*, Volume 13A of *ASM Handbook*, 2003. Corrosion in microelectronics, as in other industries, is governed by the electrochemical principles for aqueous corrosion. There are, however, some characteristics unique to corrosion in microelectronics.

Characteristics of Corrosion in Microelectronics

The materials, the design, the package type, and the environment influence corrosion in microelectronics.

Materials selected for use in microelectronics are primarily based on their electrical, magnetic, and optical properties, ranging from noble metals such as gold and silver to more reactive metals such as aluminum and its alloys. Table 1 shows typical applications of conductors and contact materials (Ref 18). The functionality of devices requires extensive use of dissimilar-metal contacts. Galvanic corrosion, pitting in electrodeposits (Fig. 1), anodic corrosion (oxidation of a metal, such as aluminum, at the anode sites), and cathodic corrosion (attack on cathodically biased aluminum by a hydroxyl ion from cathodic reduction of oxygen or water) are direct consequences of the inherent multimetal system designs.

Design and Packaging. The submicron separation of metal lines on IC devices, the extremely thin and often porous gold and silver plating on substrates, and the extensive use of dissimilar metals make microelectronic devices and components extremely susceptible to

Table 1 Typical applications of conductors and contact materials

Material group	Alloys	Typical applications
Copper alloys	C11100, C19400, C10100, C19500, C10400	Lead materials, leadframes
Electroplated copper	...	Printed wiring
Beryllium-copper alloys	C17000, C17200, C17400	Contact springs
Brass alloys	C36000, C27000, C26000	Contact pins
Phosphor-bronze alloys	C51000, C52400	Contact pins
Electroplated gold	99.9Au-0.1Co; 99Au-1Co	Finish for contact surfaces
Gold alloy	99.99Au	Bond wires
Silver alloy	...	Contact materials in relays
Electroplated silver	...	Finish for high-frequency components
Tungsten	Tungsten powder/binder(a)	High-temperature cobalt-fired metallization
Molybdenum	Molybdenum or molybdenum/manganese powder/binder(a)	Co-fired metallization
Electroplated rhodium	...	Finish for mating contact surfaces; high-wear applications
Aluminum	Al-1Si; Al-1Mg	Bond wires, metal lines
Kovar	29Ni-17Co-54Fe	Pins for glass-sealed feed-through leadframes; glass-sealing applications
Alloy 42	42Ni-58Fe	Leadframes; brazed leads

(a) Powder with glass/ceramic binder in as-fired condition. Source: Ref 18

corrosion. In order to minimize manufacturing cost, the design may use alternative processes and materials (e.g., gold interconnect layers replaced with aluminum alloy), with limited understanding of the possibility of corrosion during manufacturing processes and in service. Although rare, under favorable conditions, including the presence of an electrical bias, even noble metals such as gold will corrode actively. Polymers such as epoxy and silicone used in plastic packages are relatively permeable to water or moisture and other corrosive chemicals. Virtually all plastic-encapsulated packages are therefore susceptible to corrosion. Some flame retardants such as bromine added to epoxy encapsulant can significantly accelerate corrosion (Ref 19). For hermetically sealed packages, the sealed-in moisture and contaminants are sometimes sufficient to cause corrosion of the exposed metals. The seals may also leak at edges or cracks.

The Environment. Microelectronic components and systems traditionally were found mostly indoors, and within packages or cabinets when used outdoors, and so they were sheltered from direct weathering agents such as sunlight and precipitation. Now, more are used outdoors, so if the enclosure is improperly designed, the corrosivity can increase due to elevated temperatures or humidity. The atmosphere to which a microelectronic component and package is exposed can vary from a purified and filtered indoor to an industrial or marine atmosphere in outdoor exposure.

Common Sources of Corrosion

Corrosion is defined as "the chemical or electrochemical reaction between a material, usually a metal, and its environment that leads to the deterioration of the material and its property" (Ref 20). Under this definition, reliability issues pertaining to humidity, contamination, temperature, and electrical bias are all within the domain of corrosion. Corrosion can originate from many sources during manufacturing, storage, shipping, and service. Moisture, temperature, contaminants, and electrical bias, either alone or in combination, often initiate or accelerate corrosion of many metals. The form and rate of corrosion are dependent on the materials, the fabrication and assembly processes, the design, the package type, and the environmental conditions. Figure 2 shows the common sources of corrosion in microelectronics.

Relative Humidity and Moisture. Air contains water vapor in addition to its major constituents of nitrogen and oxygen gases. At any given temperature, the amount of water contained in a unit volume of air (g/m^3) is referred to as the absolute humidity. The higher the air temperature, the more water it can contain. Relative humidity (RH) is the amount of moisture in the air (absolute humidity) compared with the maximum amount of moisture that the air can hold at the same temperature. Because warm air can hold more water than cool

Fig. 1 Pitting of electrodeposit on a copper leadframe

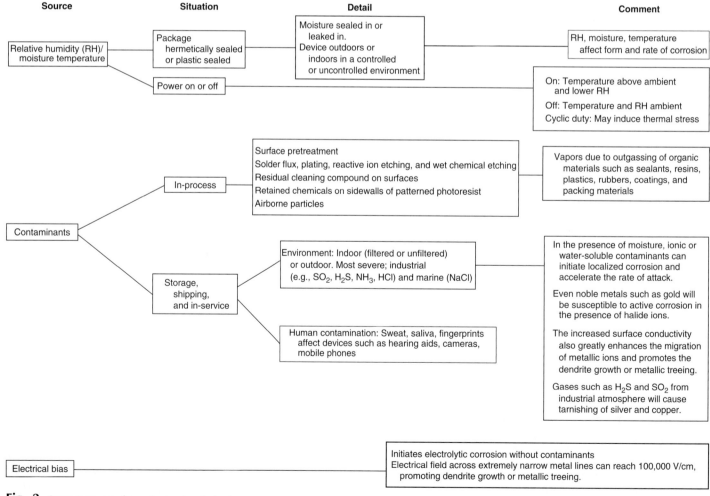

Fig. 2 Common sources of corrosion in microelectronics

air, there is less water in 20 °C (68 °F) air compared to 25 °C (77 °F) when they are both at 60% RH. There exists a critical RH above which significant corrosion takes place. Research (Ref 21) indicates that three monolayers of condensed water molecules are the absolute minimum amount of moisture that can raise the surface conductance to a significant level. As the thickness of the surface moisture film increases, the corrosion rate for most metals rises sharply, and a peak is observed at approximately 1 μm (0.04 mil) (Fig. 3). Further increases in moisture film thickness result in a reduction in the corrosion rate due to diffusion-controlled oxygen reduction. The corrosion rate levels off at a moisture film thickness above 1 mm (0.04 in.), which is equivalent to the total immersion condition.

In hermetically sealed packages, the sources of moisture include the sealed-in moisture and that which leaked in through cracks or edges. Polymer-encapsulated packages are permeable to moisture and other corrosive chemicals, such as SO_2, H_2S, and NaCl, depending on the location and the nature of the exposure site. These corrosive chemicals, when dissolved in the moisture film, can increase the surface conductance, modify the pH of the electrolyte, and lead to accelerated corrosion attack on most metals.

Temperature. Dewpoint temperature, the temperature at which moisture will condense on a surface, is an important concept in corrosion. At the dewpoint temperature, the air immediately next to the surface is at 100% RH, and moisture cannot evaporate from the surface. When electronics are using power, the heat generated may be beneficial and raise the surface temperature (possibly above the dewpoint temperature) and lower the surface RH below the critical RH. This can reduce the corrosion rate. However, as the combination of ambient conditions and internal heating increases the surface temperature, the corrosion rate roughly doubles for every 10 °C (18 °F) rise in temperature. It was calculated that at 70 °C (158 °F), without packaging, a bare aluminum bond pad (100 by 100 by 1 μm, or 4 by 4 by 0.04 mils) would be completely corroded within 67 h, and an interconnect segment (5 by 5 by 1 μm, or 0.2 by 0.2 by 0.04 mil) would be completely corroded within 10 min (Ref 23). The effect of temperature and RH on the mean time to failure has received considerable attention. Table 2 shows models proposed in the literature (Ref 24).

Contaminants that are of concern to microelectronics include ionic contaminants, corrosive gases, and vapors (Ref 6). Ionic contaminants or water-soluble salts have long been recognized to cause corrosion, particularly in the protective coatings industry. In microelectronics, the reliability hazard of ionic contaminants increases dramatically as the device size shrinks and the density increases. For example, a current of 10^{-13} A has no significant effect on standard circuit board conductor stripes operating for many years but will cause failure of a 0.1 μm (0.004 mil) wide conductor in less than 1 min. The ionic contaminants can either directly induce or indirectly assist the corrosion of metallic materials. For example, chloride is well recognized to cause breakdown of normally protective oxide films, leading to localized corrosion of both positively biased and negatively biased aluminum (Ref 30–32). Chloride is also involved in the formation of gold complex ions and can cause corrosion of gold conductors (Ref 33). Other ionic contaminants, when dissolved in the moisture film, increase the conductivity of the surface electrolyte and hence accelerate corrosion. Contamination may come from several sources, such as in-process contamination due to the presence of residual chemicals, storage- and shipping-related contamination, in-service contamination, and contamination due to human contact (Fig. 2).

In-Process Sources. Microelectronic systems are fabricated by a number of complex and often proprietary processes. Residual chemicals from processes can introduce significant amounts of ionic contaminants. Solder fluxes, plating, and various etching processes may leave residual chemicals on the surfaces of metals or sidewalls of patterned photoresist. Microelectronic devices are particularly susceptible to fine airborne particles with aerodynamic diameters of 0.05~2 μm (0.002~0.08 mil) (Ref 34–36). These fine particles are rich in ammonium acid sulfate and are difficult to remove by filtration. See the article "Corrosion in Semiconductor Wafer Fabrication" in this Volume.

Storage and shipping of components, semifinished and finished devices, can lead to contamination by exposure to unregulated atmospheres. The devices are also subject to the fluctuation of ambient RH and temperature (Ref 8). Condensation of moisture and deposition of airborne particles can lead to localized corrosion, such as pitting and galvanic corrosion. Tarnishing of noble metals by corrosive gases, such as NH_3, SO_x, and H_2S, are also encountered.

In-service microelectronic devices must withstand normal and abnormal conditions. Airborne particles have caused field failures in facilities such as telecommunication centers (Ref 37). Fires involving polyvinyl chloride cables and pipes release hydrogen chloride vapor that can pose serious reliability hazards. Graphitic carbon formed in fires can significantly increase the leakage current, because graphitic carbon is conductive even at low RH. In locations within 1 km (0.6 mile) from the coast, airborne chloride concentrations are sufficiently high to cause corrosion failures of electronic components. Circuit boards in radar systems, video game machines, and entertainment equipment are particularly susceptible to corrosion in the marine atmosphere, largely due to the high chloride contents and the high RH. For industrial-control electronics, the common atmospheric contaminants corrosive to components and ancillary equipment are given in Table 3 (Ref 38).

Human contamination is known to be a source of IC failures (Ref 11, 39). Throughout materials processing, device fabrication, assembly, shipping, installation, and operation, human contact, either direct or indirect, is inevitable. Body effluvia from perspiration, skin oil secretions, and coughs and sneezes contain significant amounts of chlorides and alkali ions that can induce or assist corrosion of most metals, including gold. Consider how a hearing aid is in direct contact with the human body. Perspiration (a source of chloride and humidity) and water splash can cause failures of the microelectronics if contamination is present and the packaging is not properly designed.

Electrical bias is present in all microelectronic active devices, and it is also one of the most important factors in corrosion failures. At the present time (2006), the conductor width and separation on ICs has reached submicron level. The electrical fields across these extremely narrow metal lines can exceed 100,000 V/cm. A few monolayers of adsorbed water may increase the surface conductivity by several orders of

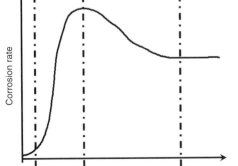

Fig. 3 Dependence of corrosion rate on moisture film thickness. Adapted from Ref 22

Table 2 Models of the effect of relative humidity and temperature on mean time to failure (MTTF)

MTTF	Constants	Ref
$Ae^{B(T+RH)}$	$B \approx -0.06$ to -0.09	25
$A(RH)^n e^{\frac{E_a}{k_B T}}$	$E_a \approx 0.77$ to 0.81 eV; $n \approx -2.5$ to -3.0	26
$Ae^{\left[\frac{E_a}{k_B T} + B(RH)^2\right]}$	$E_a \approx 0.70$ to 0.95 eV; $B = -0.0004$	27
$Ae^{\left(\frac{E_a}{k_B T} + \frac{B}{RH}\right)}$	$E_a = 0.65$ eV; $B = 304$	28
$Ae^{\left[\frac{E_a}{k_B T} + \frac{B(RH)}{k_B T} + C(RH)\right]}$	$E_a \approx 1.0$ to 1.1 eV; $B \approx -4.4 \times 10^{-3}$ to -7.7×10^{-3}; $C \approx 0.076$ to 0.13	29

A, empirical constant; B, C, n, constants; T, temperature; RH, relative humidity; k_B, Boltzman's constant; E_a, activation energy. Adapted from Ref 24

Table 3 Common industrial pollutants corrosive to microelectronics

Pollutant	Sources	Susceptible metals
Sulfur dioxide (SO_2)	Fossil fuel combustion, petrochemical industries, pulp and paper industry, metal-producing industry	Most metals
Nitrogen dioxide (NO_2)	Auto and truck emissions, fossil fuel combustion, various industries	Copper, brass, synergistic with SO_2
Hydrogen sulfide (H_2S)	Pulp and paper industries, chemical industry, sewage plants, landfills, oil refineries, animal shelters, volcanic activity, swamp areas, marine tidal areas	All copper- and silver-base metals
Chlorine (Cl_2) (gas most important)	Bleaching plants in industries, metal production, polyvinyl chloride plants, cleaning agents	Most metals synergistic with other pollutants
Ammonia and its salts (NH_3 and NH_4^+)	Fertilizer, animal and human activity, detergents	All copper-base alloys, nickel, silver
Chloride (Cl^-)	Sea salt mist, road salt areas	Most metals
Soot (C)	Combustion, auto and truck emissions, steel production	Synergistic with other pollutants; provides cathodic sites for most metals
Ozone (O_3)	Formed in polluted areas; highest concentrations in smog	Strong oxidant to produce acids that attack most metals
Mineral acids (H_2SO_4, HCl, HF, HNO_3)	Pickling industry, chemical industry, metals production, semiconductor industry	Most metals, glass, ceramics
Organic acids	Wood, packing material, animals, preservatives	Long-term effects on some metals

Adapted from Ref 38

Fig. 4 Example of dendrite growth on the surface of an integrated circuit, in this case, silver, which is no longer used. Source: Ref 42

magnitude. Under these high-voltage gradients and high-conductivity conditions, most metals, including noble metals, will corrode actively through the electrolytic process, even without the presence of contaminants. The metal ions will diffuse through the surface electrolyte film toward the cathode, where reduction reactions lead to the formation of metal dendrites. The dendrites grow toward the anode and eventually short the circuits (Ref 40). This phenomenon is sometimes referred to as metallic treeing (Ref 41). Figure 4 shows the silver dendrite growth on the surface of an IC (Ref 42). Silver is not used in microelectronics for this reason.

Mechanisms of Corrosion in Microelectronics

In microelectronics, reliability hazards are most often caused by aqueous corrosion. Gaseous corrosion, which involves chemical reactions of metals at higher temperatures in the absence of moisture, is also possible, but it is generally less of a threat to reliability than aqueous corrosion. The mechanisms of corrosion in microelectronics and ancillary components include:

- Anodic, cathodic, and electrolytic reactions resulting in uniform corrosion

Table 4 Electrochemical series for reactions important in corrosion

Standard conditions: 25 °C (77 °F); 1 atm; H_2O; unit activities, metals present in their standard states

Electrode reaction	Standard potential, V (SHE)(a)
$Au^{3+} + 3e = Au$	+1.50
$2O_2 + 4H^+ + 4e = 2H_2O$	+1.23
$Pt^{2+} + 2e = Pt$	+1.20
$Ag^+ + e = Ag$	+0.80
$Cu^{2+} + 2e = Cu$	+0.34
$2H^+ + 2e = H_2$	0.00
$Pb^{2+} + 2e = Pb$	−0.13
$Sn^{2+} + 2e = Sn$	−0.14
$Mo^{3+} + 3e = Mo$	−0.20(b)
$Ni^{2+} + 2e = Ni$	−0.25
$Co^{2+} + 2e = Co$	−0.28
$Fe^{2+} + 2e = Fe$	−0.44
$Cr^{3+} + 3e = Cr$	−0.74
$Zn^{2+} + 2e = Zn$	−0.76
$Al^{3+} + 3e = Al$	−1.66

(a) SHE, standard hydrogen electrode. Source: Ref 43 except (b) Ref 44

- Galvanic corrosion
- Pitting
- Creep corrosion
- Dendrite growth
- Fretting
- Stress-corrosion cracking
- Hydrogen embrittlement
- Whisker growth

The form and rate of corrosion are influenced by the nature of the materials involved, the design, the packaging type, and the environmental conditions.

Uniform corrosion often occurs in the presence of a moisture film and ionic contaminants. If the electrical bias is sufficiently high, even noble metals will undergo electrolytic corrosion. The tendency of a metal to undergo oxidation is indicated by the electromotive force (emf) series, which is an orderly list of the standard reduction potentials of pure metals in equilibrium with 1 M aqueous solution of their own ions at 25 °C (77 °F). Table 4 gives the standard reduction potentials for some of the metals (and reactions involving oxygen and hydrogen) relevant to corrosion in microelectronics. A more extensive list can be found in *Corrosion:*

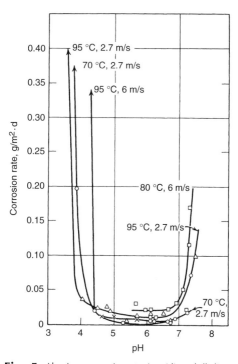

Fig. 5 Aluminum corrosion rates in acidic and alkaline environments at various temperatures and flow rates. Source: Ref 46

Materials, Volume 13B of *ASM Handbook,* 2005. The more negative the potential, the greater the tendency for the metal to undergo oxidation (corrosion). Aluminum is a reactive metal and is located at the lower end of the emf series. Reactive ion etching of aluminum metallization in plasma containing chlorine-base gases such as Cl_2, CCl_4, $CHCl_3$, and BCl_3 can result in rapid corrosion in the presence of moisture (Ref 24, 45). Other ionic contaminants may induce either an acidic or alkaline condition to the moisture film, causing accelerated corrosion of aluminum. Figure 5 shows that the corrosion rate of aluminum increases drastically outside the pH range of 4.5 to 7 (Ref 46). The potential-pH diagram at 25 °C (77 °F) for aluminum (Fig. 6) also indicates that active corrosion occurs outside the pH range of 4 to 8

(Ref 47). The presence of ionic contaminants such as Na^+ and K^+ can increase the conductivity of the electrolyte and shift the pH of the moisture film to the threshold value of 8 and above. The amount of sodium ions required is calculated to be 1.15×10^{-5} μg for a hermetically sealed package with an internal volume of 0.1 cm^3 containing 5000 ppm_v of water vapor, of which 1% condenses on the chip (Ref 23).

Anodic Corrosion. The corrosion process, as in any other electrochemical system, consists of two half-cell reactions: anodic reaction and cathodic reaction. For example, the anodic reaction is simply the oxidation of aluminum:

$$Al = Al^{3+} + 3e^- \quad (Eq\ 1)$$

This oxidation process occurs at the metal/oxide interface and involves diffusion of both Al^{3+} and O^{2-} or OH^-.

There are several possibilities for the cathodic reactions, depending on the pH of the electrolyte. Hydrogen evolution occurs in acidic environments where pH < 7:

$$2H^+ + 2e^- = H_2 \quad (Eq\ 2)$$

In neutral or alkaline environment where pH ≥ 7, oxygen reduction takes place:

$$O_2 + 2H_2O + 4e^- = 4OH^- \quad (Eq\ 3)$$

If oxygen is limited, direct reduction of water will be predominant:

$$2H_2O + 4e^- = H_2 + 2OH^- \quad (Eq\ 4)$$

The overall reaction in the acidic environment is obtained by combining Eq 1 and Eq 2:

$$2Al + 6H^+ = 2Al^{3+} + 3H_2 \quad (Eq\ 5)$$

Equations 1, 3, and 4 can be combined to give the overall reactions in neutral and alkaline environments, where oxygen reduction is predominant:

$$4Al + 3O_2 + 6H_2O = 4Al(OH)_3 \quad (Eq\ 6)$$

Where reduction of water is predominant:

$$2Al + 6H_2O = 2Al(OH)_3 + 3H_2 \quad (Eq\ 7)$$

Cathodic Corrosion. The corrosion of aluminum discussed previously is anodic oxidation and, in the absence of ionic contaminants or electrical bias, is sometimes referred to as anodic corrosion. When aluminum metallization is cathodically biased, cathodic reduction takes place in accordance with Eq 2, 3, and 4, depending on the pH of the moisture film. In both Eq 3 and 4, hydroxyl ions (OH^-) are produced, leading to an increase in pH. Another source of alkalinity is the ionic contaminants, such as alkali ions of Na^+ and K^+, that hydrolyze in the presence of moisture and increase the pH. Aluminum is amphoteric; that is, when the pH of the electrolyte is above 8 (Fig. 6), aluminum oxide is no longer stable and will undergo active dissolution (Ref 44, 46, 47), resulting in opens at the cathodic line. This phenomenon is often referred to as cathodic corrosion. The hydroxyl ions attack the aluminum hydroxide (Eq 5) at the metal/electrolyte interface, resulting in the formation of soluble aluminate ions and hydrogen gas. The reaction is:

$$Al(OH)_3 + OH^- = Al(OH)_4^- \quad (Eq\ 8)$$

where oxygen reduction is predominant. The reaction is:

$$Al(OH)_3 + OH^- = Al(OH)_4^- \quad (Eq\ 9)$$

where reduction of water is predominant.

Research on the corrosion of pure aluminum in NaOH solution, with the concentration ranging from 0.01 to 1 M, showed that the predominant cathodic reaction is the direct reduction of water rather than the reduction of oxygen. Hence, the hydroxyl attack on aluminum oxide during cathodic corrosion is by the mechanism in Eq 9, with the release of hydrogen gas (Ref 48). The activation energy for cathodic corrosion was reported to be similar to that for the dissolution of aluminum in NaOH solution (Ref 49).

Electrolytic Corrosion. Noble metals are generally more resistant to corrosion than aluminum. For example, gold has a standard potential of 1.5 V (Table 4) and will not undergo oxidation under normal circumstances. However, when the electrical bias is higher than 1.5 V, gold can still undergo electrolytic corrosion, leading to the formation of $Au(OH)_3$ (Ref 33). In the presence of halide ions, complex ion formation lowers the standard reduction potentials of gold from 1.5 to 0.96 V (Table 5), causing active corrosion.

The anodic reaction is:

$$4Au + 16Cl^- = 4AuCl_4^- + 12e^- \quad (Eq\ 10)$$

The cathodic reaction is:

$$3O_2 + 12H^+ + 12e^- = 6H_2O \quad (Eq\ 11)$$

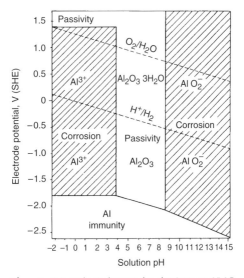

Fig. 6 Potential-pH diagram for aluminum at 25 °C (77 °F). SHE, standard hydrogen electrode. Source: Ref 47

The overall reaction is spontaneous, with a release of free energy of 22 kJ/mol (Ref 43):

$$4Au + 3O_2 + 16Cl^- + 12H^+ = 4AuCl_4^- + 6H_2O \quad (Eq\ 12)$$

For copper, the following reactions are involved.
The anodic reaction is:

$$Cu = Cu^{2+} + 2e^- \quad (Eq\ 13)$$

The cathodic reaction is:

$$O_2 + 4H^+ + 4e^- = 2H_2O \quad (Eq\ 14)$$

The overall reaction is also spontaneous, with a release of free energy of 172 kJ/mol:

$$O_2 + 4H^+ + 2Cu = 2H_2O + 2Cu^{2+} \quad (Eq\ 15)$$

Both ionic contaminants, such as chloride and gaseous contaminants such as ammonia and hydrogen sulfide, can significantly decrease the reduction potentials of the copper and silver systems (Table 5) to values as low as −0.62 V for copper and −0.45 V for silver, causing accelerated corrosion even in the absence of an electrical bias. This enhanced reactivity is responsible for the tarnishing of copper and silver in humid environments containing NH_3 and H_2S gases (Ref 51).

Dendrite growth, or electrolytic migration, refers to the reduction of metal ions in the form of metal dendrites along the diffusion path between conductors. The process involves the following steps:

1. Electrolytic dissolution of the metal (usually noble metals such as gold, silver, copper)
2. Migration of metal ions from anode to cathode through the surface electrolyte film
3. Electrodeposition of metal ions at the cathode in the form of dendrites
4. Propagation of dendrites toward the anode
5. Shorting of the anode to cathode across the dielectric

Noble metal ions, such as gold and silver ions, have a greater tendency to be reduced in the presence of moisture (Table 4). The source of metal ions mainly comes from the oxidation reaction at the anode. Under the electrical field, the positively charged metal ions migrate

Table 5 Standard electrode potentials for some noble metal systems

System	Redox potential, V	Ref
$Au^{3+} + 2e = Au^+$	1.41	50
$Au^{3+} + e = Au^{2+}$	<1.29	50
$Au^{2+} + e = Au^+$	<1.29	50
$AuCl_2^- + e = Au + 2Cl^-$	1.11	47
$AuCl_4^- + e = Au + 4Cl^-$	1.00	50
$AuBr_2^- + e = Au + 2Br^-$	0.96	50
$AgCl(s) + e = Ag + Cl^-$	0.22	47
$Ag_2S + 2H^+ + e = 2Ag + H_2S(g)$	−0.45	47
$CuCl_2^- + e = Cu + Cl^-$	0.14	47
$[Cu(NH_3)_4]^{2+} + 2e = Cu + 4NH_3$	−0.05	47
$CuS(s) + 2H^+ + e = Cu + H_2S(g)$	−0.62	47

through the surface electrolyte film toward the cathode where they are deposited (Fig. 4). It is believed that this diffusion-controlled process enhances the stability and propagation of dendrites. The phenomenon is sometimes also referred to as metallic treeing (Ref 41). Gold, silver, copper, and tin are some of the common electronic materials susceptible to dendrite growth on devices such as ICs and circuit boards (Ref 33, 52–59). Moisture and contaminants are known to accelerate dendrite growth.

Galvanic corrosion arises due to the difference in the electrochemical potentials of dissimilar metals in electrical contact, resulting in the accelerated attack on the anode, the metal with a more negative potential. This is probably the greatest single source of all corrosion failures. Galvanic effects are also responsible, in most cases, for the initiation and propagation of pitting corrosion on noble metal electrodeposits and on aluminum and gold metallization. In microelectronic components and devices, it is almost mandatory to use many different metals, either in direct physical or electrical contact. Metal structures such as molybdenum-gold metallization and the aluminum metallization/gold bond wire interfaces are typical examples. In the presence of moisture and contaminants, a galvanic cell will form. The metal with a more positive potential acts as the cathode, and the one with a less positive potential acts as the anode. The driving force of the galvanic corrosion is the potential difference between the two dissimilar metals. For the aluminum-and-gold galvanic couple at the interface of the aluminum metallization/gold bond wire, the galvanic cell potential under standard conditions can be calculated from the standard reduction potentials (Table 4):

$$E_{cell} = E_{cathode} - E_{anode} = E_{Au} - E_{Al}$$
$$= +1.50 \text{ V} - (-1.66 \text{ V}) = +3.16 \text{ V}$$

This voltage is over twice the voltage in AA alkaline batteries. The anode will experience accelerated attack.

For the aluminum-copper galvanic couple in the aluminum-copper bonding pads (Ref 60), the galvanic cell potential calculated from the standard reduction potentials is 2.00 V. For the molybdenum-gold couple in molybdenum-gold metallization, the cell potential is 1.70 V. It is noted that the emf series assumes a standard equilibrium condition, where the metal is in contact with a $1\ M$ aqueous solution of its own ions. This metal ion concentration is too far from actual concentrations in the surface electrolyte film. A more widely used series is the one measured in natural seawater, such as is found in the article "Galvanic Series of Metals and Alloys in Seawater" in *Corrosion: Materials*, Volume 13B of *ASM Handbook*, 2005. The aluminum-gold couple in seawater exposures produces roughly 1.2 V, which is still a significant driving force for accelerated attack at the anode.

Microelectronic components and devices are rarely exposed to an immersion condition, and corrosion of the various metal structures is mainly atmospheric in nature. Therefore, the galvanic series for metals exposed to atmosphere (Table 6) seems to most closely match the

Table 6 Metals and alloys compatible in dissimilar-metal couples

Group number	Metallurgical category	Electromotive force (emf), V	Anodic index(a), V	Compatible couples(b)
1	Gold, solid and plated; gold-platinum alloys; wrought platinum	+0.15	0	
2	Rhodium plated on silver-plated copper	+0.05	0.10	
3	Silver, solid or plated; high-silver alloys	0	0.15	
4	Nickel, solid or plated; monel metal, high-nickel-copper alloys	−0.15	0.30	
5	Copper, solid or plated; low brasses or bronzes; silver solder; German silvery high copper-nickel alloys; nickel-chromium alloys; austenitic corrosion-resistant steels	−0.20	0.35	
6	Commercial yellow brasses and bronzes	−0.25	0.40	
7	High brasses and bronzes; naval brass; Muntz metal	−0.30	0.45	
8	18% Cr type corrosion-resistant steels	−0.35	0.50	
9	Chromium plated; tin plated; 12% Cr type corrosion-resistant steels	−0.45	0.60	
10	Tin-plate; terneplate; tin-lead solder	−0.50	0.65	
11	Lead, solid or plated; high-lead alloys	−0.55	0.70	
12	Aluminum, wrought alloys of the 2000-series	−0.60	0.75	
13	Iron, wrought, gray or malleable; plain carbon and low-alloy-steels; armco iron	−0.70	0.85	
14	Aluminum, wrought alloys other than 2000-series aluminum, cast alloys of the silicon type	−0.75	0.90	
15	Aluminum, cast alloys other than silicon type; cadmium, plated and chromated	−0.80	0.95	
16	Hot-dip zinc plate; galvanized steel	−1.05	1.20	
17	Zinc, wrought; zinc-base die-casting alloys; zinc plated	−1.10	1.25	
18	Magnesium and magnesium-base alloys, cast or wrought	−1.60	1.75	

(a) Anodic index is the absolute value of the potential difference between the most noble (cathodic) metals listed and the metal or alloy in question. For example, the emf of gold (group1) is +0.15 V, and the emf of wrought 2000-series aluminum alloys (group 12) is −0.60 V. Thus, the anodic index of wrought 2000-series aluminum alloys is 0.75 V. (b) "Compatible" means the potential difference of the metals in question, which are connected by lines, is not more than 0.25 V. An open circle indicates the most cathodic members of a series; a closed circle indicates an anodic member. Arrows indicate the anodic direction

exposure conditions encountered in microelectronics (Ref 61, 62). The greater the anodic index (the potential difference between the anode and cathode members of a couple), the higher the galvanic corrosion risk. The conductivity of the surface electrolyte film will affect the rate and distribution of galvanic attack. Moisture accumulation when the power is off and contamination from both ionic and gaseous sources will increase the surface conductivity and hence enhance galvanic corrosion.

Pitting is localized attack on the metal structure, resulting in holes or cavities. In microelectronics, it is sometimes referred to as pore corrosion. It is commonly associated with noble metal coatings such as gold and silver on copper and nickel. The ultrathin and often porous noble metal coatings can leave the substrate exposed to the surface electrolyte film. The galvanic cell potential between the noble coatings and the less-noble substrate will initiate and accelerate the attack on the anodic substrate. The large cathode-to-anode surface area ratio also increases the acceleration factor, leading to rapid pitting of the underlying substrate. Halide ions such as chlorides are known to cause breakdown of oxide films and lead to the initiation of pitting (Ref 47, 63). On aluminum, the aluminum oxide is cathodic to the metal substrate. Ionic contaminants such as chloride can cause localized acidification due to the hydrolysis of aluminum chloride:

$$Al^{3+} + 3Cl^- = AlCl_3 \quad \text{(Eq 16)}$$

$$AlCl_3 + 3H_2O = Al(OH)_3 + 3HCl \quad \text{(Eq 17)}$$

In reaction Eq 17, hydrogen chloride is produced, and this lowers the pH and accelerates the corrosion of aluminum in reaction Eq 1. As more Al^{3+} ions accumulate at the anode, electrostatic attraction tends to cause more Cl^- to diffuse into the pit bottom. This will accelerate reactions, leading to more HCl formation, which, in turn, causes more dissolution of Al^{3+} ions in the pit. This cycle of $Al^{3+} \rightarrow Cl^- \rightarrow HCl \rightarrow Al^{3+}$ repeats itself, and the process is self-accelerating, or autocatalytic, in nature.

Pitting on aluminum can also be initiated by the presence of more-noble metal ions such as copper. Copper ions from corrosion products can be readily reduced on aluminum. The galvanic effect between copper and aluminum is sufficient to initiate pitting on aluminum. Intermetallics in aluminum alloys such as $FeAl_3$ and $CuAl_2$ are cathodic precipitates that can also initiate pitting due to the same galvanic principle. Figure 7 shows schematically the mechanism of pitting on aluminum and the common factors involved in the process.

Creep corrosion is a mass-transport process during which solid corrosion products migrate over a surface away from the source metal. It is a known failure mechanism in electrical contacts and connectors. For example, tarnish films formed on copper and silver in humid environments containing NH_3 or H_2S can creep over adjoining gold surfaces, leading to increased contact resistance of the gold. The driving force for the gold/copper system is most likely associated with the local galvanic cell between the dissimilar metals. For components with noble metal preplated leadframes, creep corrosion is considered to be a potential reliability hazard for long-term field applications. Field failure due to creep corrosion in telecommunication electronics is reported in the literature. The source, process, and possible products of creep corrosion on IC packages are discussed in Ref 2.

Fretting is a deterioration at the interface between contacting surfaces as the result of corrosion and slight oscillatory slip between the two surfaces. Associated components such as connectors, switches, and relays must maintain good conductive contact surfaces for thousands or millions of operation cycles. The effect of fretting can seriously affect the contact resistance (Ref 64–66). Even noble metals are susceptible to fretting (Ref 67).

Stress Corrosion Cracking (SCC) is a cracking process that requires the simultaneous action of a corrodent and sustained tensile stress. It has been known for almost a century that copper and its alloys are susceptible to SCC in humid atmospheres containing ammonium compounds (Ref 68). Other environments that can cause cracking of copper and its alloys include moist SO_2, acetates, citrates, formates, tartrates, nitrites, and NaOH solutions. Aluminum and its alloys are susceptible to cracking in moist air and halide solutions. Precipitates along the grain boundaries in some aluminum alloys can result in intergranular corrosion. A quantitative model describing SCC of wire microjoints in microelectronics is given in Ref 69.

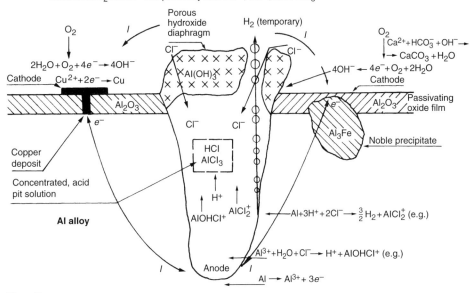

Fig. 7 Electrochemical mechanism of pitting on aluminum. Source: Ref 47

Table 7 Designing for corrosion-resistant finishes

Problem	Solution	Example
Dissimilar metals	Select metals from appropriate table of permissible couples (Table 6).	Use nickel or rhodium, not brass or bronze, next to silver.
	Plate with compatible metal to reduce potential difference.	Tinplate aluminum and bronze used together.
	Keep affected area of less-noble metal as large as possible.	Stainless steel hardware in sheet aluminum may be satisfactory because of large area of aluminum (but not the reverse).
Contact	Apply corrosion inhibitors.	Vapor-phase corrosion inhibitors
	Interpose inert barrier or gaskets to prevent contact (extend 6.4 mm, or 1/4 in., beyond joint).	Vinyl tape, rubber gasket, dielectric washer
	Paint both metals (or cathode, at least) with alkali-resistant organic coating.	MIL-P-52192 or MIL-P-15930
Electrolyte	Avoid designs where moisture can be trapped.	Use sealant bead on crimped, spot-welded, and threaded joints.
	Use desiccant.	Useful only in hermetically sealed compartment MIL-S-7124
	Seal joint with organic insulation.	
	Seal metal faces against contact with electrolyte.	Primer, paint, or sealant
General	Where possible, avoid use of magnesium.	Protection of magnesium requires very special attention.

Source: Adapted from Ref 62

Hydrogen embrittlement, or hydrogen-induced cracking, refers to the severe loss of ductility caused by the presence of hydrogen in the metal. In a corrosion process, both the reduction of hydrogen ions and the reduction of water will generate hydrogen atoms on the surface of the cathode. Processes such as electrocleaning, pickling, and plating can also produce hydrogen atoms. Absorbed hydrogen atoms can diffuse into the metal and cause the embrittlement.

Whisker growth refers to the filamentary growth on metallic materials that can cause short circuits. It is not a corrosion process but a mechanism of stress relief. The whiskers are not considered to be true corrosion products. Metal whiskers can grow even in hermetically sealed packages during storage. Common materials susceptible to whisker growth include electroplated tin, zinc, cadmium, copper, and silver. Failures caused by whisker growth in missile and aircraft radar systems are reported (Ref 70, 71). The process is retarded under low-humidity and low-temperature conditions.

Corrosion Control and Prevention

Control and prevention measures are naturally centered on the factors of materials selection, design, package type, and environmental conditions.

It is possible to minimize the corrosion effect by selection of compatible materials based on the design requirements. For avoiding galvanic corrosion, Table 6 provides guidelines on the compatibility of different metals. Table 7 provides guidelines on designing for corrosion-resistant finishes. Many metals used in microelectronic components and devices require protection from general atmospheric exposure (Table 8). The package type will play a vital role in preventing or reducing moisture ingress. Without the presence of an electrolyte, the corrosion process cannot take place.

Hermetic packages made of ceramics, glass, and metals are more resistant to moisture ingress than any plastic or polymeric packages by several orders of magnitude. The water molecules are small in size (with a diameter of 0.34 nm) and can diffuse through very small openings such as pores and cracks commonly found in plastic and polymeric materials. Figure 8 illustrates the time scale for moisture penetration for some common packaging materials. The following are essential and effective measures for minimizing the moisture retention in hermetic packages and/or moisture ingress in plastic packages:

- Proper selection and design of packaging materials
- Baking
- Dry sealing ambient
- Low leak rate for hermetic packages
- Low moisture permeation for plastic packages
- Wafer coating with good adhesion to the die surface

Moisture initiates and ionic contaminants assist and accelerate the electrochemical processes. Strict control of the sources of ionic contamination is essential (Fig. 2) in keeping the corrosion risks to the minimum. This includes processing equipment and materials (such as furnaces, etchants, and cleaning chemicals), packaging materials, fabrication environment, and human contact during assembly and handling. In addition to ionic contaminants, corrosive gases and vapors such as H_2S, SO_x, and NH_3 also require special attention, particularly the organic vapors from polymeric materials used in packaging (Table 9).

Corrosion inhibitors have been widely used in other industries to control corrosion processes. For multimetal systems, as encountered in microelectronic components and devices, effective inhibition of corrosion requires the inhibitor to simultaneously retard the corrosion of different metals, such as aluminum, copper, and tin. Inhibitors can be added to the polymeric encapsulant, typically epoxy resin, to retard the transport of ionic contaminants through the package as well as to prevent the formation of corrosive electrolytes in the presence of moisture. Recent advances in corrosion-inhibitor formulation have resulted in an increased use of vapor-phase corrosion inhibitors in microelectronic control systems in the military and aerospace industries. The inhibitors are produced in a variety of formats, such as emitting devices, tablets, powder packets, stretch films, paper, and electronic spray. Within an enclosure, as is most often the case with microelectronic components and devices, the vapor-phase corrosion inhibitor can be an effective means of reducing the corrosion risks associated with moisture and contaminants during storage, shipping, and operation (Ref 73).

Corrosion Tests

The rate and forms of corrosion are affected by variables in both the materials and the environments. Hence, both materials tests and environmental tests are performed on microelectronic packages to detect flaws or defects that may lead to device failures during fabrication or field operation. Various electrochemical techniques based on direct current (dc) and alternating current (ac) are discussed in

Table 8 Metals requiring protection from atmospheric exposure

Metals	Withstands exposure (including salt air and high relative humidity)	Requires protection
Aluminum	...	X
Aluminum alloys	...	X
Brasses	...	X
Copper	...	X
Copper alloys	...	X
Kovar	...	X
Alloy 42 (Ni-Fe)	...	X
Gold plating	X	...
Molybdenum	...	X
Nickel alloys	X	...
Nickel plating	X	...
Rhodium plating	X	...
Silver plating	...	X
Tin plating	X	...
Tin-lead alloys	X	...
Tungsten	...	X
Low-carbon steels	...	X
18-8 stainless steels	X	...

Source: Ref 21

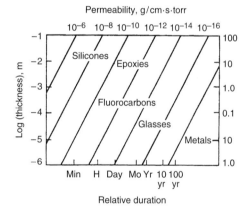

Fig. 8 Time for moisture to permeate various packaging materials. Adapted from Ref 72

Table 9 Ranking of organic vapor sources corrosive to metals

Material	Severely corrosive	Somewhat corrosive	Noncorrosive
Adhesive	Urea-formaldehyde	Phenol-formaldehyde	Epoxy
Gasket	Neoprene(a)-asbestos	Nitrile-asbestos	...
	Resin-cork	Glue-cellulose	...
Insulation (wire)	Vinyl	Teflon(a)	Polyurethane
	Polyvinyl chloride	Nylon	Polycarbonate
	Vinylidene fluoride	Polyimide	...
Sealer	Polysulfide	Epoxy	Silicone
Sleeving	Vinyl	Silicone	...
	Polyvinyl chloride	...	...
Tubing	Neoprene(a), heat-shrinkable tubing	...	...
Plastics	Melamine	Polyester	Silicone
	ABS(b)	Diallyl phthalate	Epoxy
	Phenolic	...	Polyurethane
Varnish	Vinyl	Alkyd	...

Note: Extent of attack is worse when plastic is incompletely cured; at high temperatures of operation, it may vary with the plastic used. (a) Trademark of E.I. du Pont de Nemours & Co., Wilmington, DE. (b) ABS, acrylonitrile-butadiene-styrene. Source: Ref 62

Corrosion: Fundamentals, Testing, and Protection, Volume 13A of *ASM Handbook,* 2003. Direct current electrochemical techniques, such as potential and polarization measurements, are used to determine porosity in gold plating on copper-nickel substrates. These techniques also provide rapid and effective methods for the evaluation and selection of processing liquids, such as the etchants and detergents used in fabrication. Electrochemical impedance spectroscopy (EIS) is a proven powerful, nondestructive technique for the evaluation of polymer coatings. In encapsulated triple-track test structures, for example, EIS can be used to monitor the degradation of plastic packaging material and to determine the moisture pickup. The corrosion resistance and/or the effect of moisture ingress on the metal lines can also be monitored with EIS. Electrochemical techniques (ac and dc) have the advantages of high sensitivity and fast response, with typical test duration of less than 1 h. This is in sharp contrast with the various "accelerated" salt fog and environmental (temperature, humidity, and bias, or THB) tests that usually require 1000 h or more (Ref 74, 75). Microelectronic components and devices intended for exposure to corrosive environments are often subjected to environmental tests such as salt fog tests. These tests are similar to the original salt fog test proposed in 1914 and later standardized by ASTM International (Ref 76). Experience has shown that there is little correlation between the salt fog test results and the actual service life of electronic components, because the salt fog tests do not reproduce the actual service condition. Nevertheless, the tests are still widely used today, primarily as an arbitrary performance test for closely related materials. In microelectronics, the most commonly used accelerated tests are THB tests at 85% RH and 85 °C (185 °F), pressure-cooker tests at 100% RH and temperature above 100 °C (212 °F), and the highly accelerated stress test at RH below 100% and temperature up to 150 °C (300 °F).

The service life can be predicted based on the measured mean time to failure (MTTF) under accelerated test conditions:

$$\mathrm{MTTF}_{field} = \mathrm{AF} \times \mathrm{MTTF}_{test} \quad (Eq\ 18)$$

where MTTF_{field} and MTTF_{test} are the mean time to fail under field and test conditions, and AF is the acceleration factor. The AF can be expressed as:

$$\mathrm{AF} = e^{\frac{E_a}{k_B}\left(\frac{1}{T_{field}} - \frac{1}{T_{test}}\right)} \quad (Eq\ 19)$$

where E_a is activation energy, k_B is Boltzman's constant, and T is temperature in Kelvin.

Studies have shown that accelerated-temperature stress testing cannot be used to extrapolate to lower-temperature use conditions (Ref 77, 78), because the failure mechanisms introduced at the higher stress levels are not representative of actual use conditions. A more popular expression for the AF includes the humidity and temperature:

$$\mathrm{AF} = \left(\frac{\mathrm{RH}_1}{\mathrm{RH}_2}\right)^n e^{\frac{E_a}{k_B}\left(\frac{1}{T_1} - \frac{1}{T_2}\right)} \quad (Eq\ 20)$$

where RH_1, RH_2, T_1, and T_2 are relative humidity and temperature, respectively, at two stress conditions.

Because corrosion is also influenced by contaminants, the commonly used THB tests have a problem with the omission of contaminants that may become a controlling factor in a failure mechanism. Without contaminants in the stressing environment, the THB testing is essentially a measure of the cleanliness of the part prior to testing. In order to improve the correlation of the environmental test results with actual field performance, controlled contamination can be introduced into the test environment.

REFERENCES

1. E.A. Amerasekera and F.N. Najm, Ed., *Failure Mechanisms in Semiconductor Devices,* John Wiley & Sons Ltd., 1997
2. P. Zhao and M. Pecht, *Microelectron. Reliab.,* Vol 43 (No. 5), 2003, p 775–783
3. G.M. Whelan, M. Kinsella, L. Carbonell, H.M. Ho, and K. Maex, *Microelectron. Eng.,* Vol 70, 2003, p 551–557
4. D. Rocak, K. Bukat, M. Zupan, J. Fajfar-Plut, and V. Tadic, *Microelectron. J.,* Vol 30, 1999, p 887–893
5. D. Ernur, V. Terzieva, W. Wu, S.H. Brongersman, and K. Maex, *J. Electrochem. Soc.,* Vol 151, 2004, p 636–643
6. C. Hillman, B. Castillo, and M. Pecht, *Microelectron. Reliab.,* Vol 43 (No. 4), 2003, p 635–643
7. A. Scandurra, G. Curro, F. Frisina, and S. Pignataro, *J. Electrochem. Soc.,* Vol 148, 2001, p 289–292
8. Y. Zhang, M. Pecht, and L. Lantz, *Microelectron. Reliab.,* Vol 38, 1998, p 1811–1816
9. G.S. Frankel, *IEEE Trans. Compon., Packag. Manuf. Technol. B,* Vol 18 (No. 4), 1995, p 709
10. F.W. Ragay, J.A. Pol, and J. Naderman, *Microelectron. Reliab.,* Vol 36, 1996, p 1931–1934
11. M. Brenman and J. Mejerovich, *Microelectron. J.,* Vol 20, 1989, p 43–47
12. E. Brambilla, P. Brambilla, C. Canali, F. Fantini, and M. Vanzi, *Microelectron. Reliab.,* Vol 23, 1983, p 577
13. D. Ernur, S. Kondo, D. Shamiryan, and K. Maex, *Microelectron. Eng.,* Vol 64, 2002, p 117–124
14. K.W. Rosengarth, Jr., *Solid State Technol.,* June 1984, p 191
15. H.M. Berg and W.M. Paulson, *Microelectron. Reliab.,* Vol 20, 1980, p 247–263
16. R.V. Nash, "MIL-STD-810D and Humidity," Proc. Institute of Environmental Sciences, 1988
17. J.H. Qiu, *Mater. Perform.,* Vol 44 (No. 3), 2005, p 22–23
18. J.Y. Evans and J.W. Evans, *Handbook of Electronic Package Design,* M. Pecht, Ed., Marcel Dekker Inc., 1991, p 773
19. K.N. Ritz, W.T. Stacy, and E.K. Broadbent, The Microstructure of Ball Bond Failures, Proc. 25th IRPS, IEEE, 1987, p 28–33
20. "Standard Terminology Relating to Corrosion and Corrosion Testing," G 15, ASTM International
21. G.B. Cvijanovich, Conductivities and Electrolytic Properties of Adsorbed Layers of Water, *Proc. NBS/RADC Workshop, Moisture Measurement Technology for Hermetic Semiconductor Devices, II* (Gaithersburg, MD), E.C. Cohen and S. Ruthberg, Ed., National Bureau of Standards, 1980, p 149–164
22. N.D. Tomashov, *Theory of Corrosion and Protection of Metals,* The MacMillan Company, 1966, p 368
23. R.K. Lowry, *Microcontam.,* May 1985, p 63–100
24. P. Marcus and J. Oudar, Ed., *Corrosion Mechanisms in Theory and Practice,* Marcel Dekker, Inc., 1995
25. B. Reich and E.B. Hakim, *Solid State Technol.,* Sept 1972, p 65
26. D.S. Peck, *Ann. Proc. Reliab. Phys.,* Vol 24, 1986, p 44
27. S.P. Sim and R.W. Lawson, *IEEE 17th Ann. Proc. Reliab. Phys.,* Vol 17, 1979, p 103
28. N. Lycoudes, *Solid State Technol.,* Oct 1978, p 53
29. N.L. Sbar and R.P. Kozackiewicz, *IEEE Trans. Electron. Dev.,* ED-26, 1979, p 56
30. S.C. Koelsar, *Ann. Proc. Reliab. Phys.,* Vol 12, 1974, p 155
31. W.M. Paulson and R.W. Kirk, *Ann. Proc. Reliab. Phys.,* Vol 12, 1974, p 172
32. R.B. Comizzoli, *RCA Rev.,* Vol 37, 1976, p 483
33. R.P. Frankenthal and W.H. Becker, *J. Electrochem. Soc.,* Vol 126, 1979, p 1718
34. J.D. Sinclair, *J. Electrochem. Soc.,* Vol 135, 1988, p 89
35. R.P. Frankenthal, R. Lobnig, D.J. Siconolfi, and J.D. Sinclair, *J. Electrochem. Soc.,* Vol 140, 1993, p 1902
36. J.D. Sinclair, L.A. Psota-Kelty, C.J. Weschler, and H.C. Shields, *J. Electrochem. Soc.,* Vol 137, 1990, p 1200
37. W.C. Ko and M. Pecht, *Handbook of Electronic Package Design,* M. Pecht, Ed., Marcel Dekker Inc., 1991, p 579–616
38. R.B. Comizzoli, *J. Electrochem. Soc.,* Vol 139, 1992, p 2058
39. R. Baboian, *Corrosion Tests and Standards,* R. Baboian, Ed., ASTM, 1995, p 638
40. R.W. Thomas and D.W. Calabrese, The Identification and Elimination of Human Contamination in the Manufacture of ICs, *Proceedings of the 23rd Annual International Reliability Physics Symposium* (Orlando, FL), IEEE, 1985, p 139

41. W.R. Bratschun and J.L. Wallner, *Electronics Packaging and Corrosion in Microelectronics*, M.E. Nicholson, Ed., ASM International, 1987, p 41–47
42. J.R. Devany, *Electronics Packaging and Corrosion in Microelectronics*, M.E. Nicholson, Ed., ASM International, 1987, p 287–293
43. D. Pletcher and F.C. Walsh, *Industrial Electrochemistry*, 2nd ed., Chapman & Hall, 1993, p 491
44. M. Pourbaix, *Atlas of Electrochemical Equilibria in Aqueous Solutions*, NACE, 1974
45. W.-Y. Lee, J.M. Eldridge, and G.C. Schwartz, *J. Appl. Phys.*, Vol 52, 1981, p 2994
46. J. Draley and W. Ruther, *J. Electrochem. Soc.*, Vol 104, 1957, p 329
47. G. Wranglen, *An Introduction to Corrosion and Protection of Metals*, Chapman & Hall, 1985, p 11–13, 96, 251
48. S. Pyun and S.M. Moon, *J. Solid State Electrochem.*, Vol 4, 2000, p 267–272
49. E.P.G.T. van de Ven and H. Keolmans, *J. Electrochem. Soc.*, Vol 123, 1976, p 143
50. L.L. Shreir, R.A. Jarman, and G.T. Burstein, *Corrosion*, Vol 1, 3rd ed., Butterworth-Heinemann, 1994, p 6:8–11
51. J.H. Qiu, "Tarnishing of Silver: Causes and Prevention," Coinformation, MITA (P) 098/04/95, 1995
52. B.D. Yan, G.W. Warren, and P. Wynblatt, *Corrosion*, Vol 43, 1987, p 118
53. F. Chen and A.J. Osteraas, *Electronics Packaging and Corrosion in Microelectronics*, M.E. Nicholson, Ed., ASM International, 1987, p 175–184
54. J.J. Steppan, J.A. Roth, L.C. Hall, D.A. Jeannotte, and P.S. Carbone, *J. Electrochem. Soc.*, Vol 134, 1987, p 175
55. S.L. Meilink, M. Zamanzadeh, G.W. Warren, and P. Wynblatt, *Corrosion*, Vol 44, 1988, p 644
56. M. Zamanzadeh, Y.S. Liu, P. Wynblatt, and G.W. Warren, *Corrosion*, Vol 45, 1989, p 643
57. M. Zamanzadeh, S.L. Meilink, G.W. Warren, P. Wynblatt, and B.D. Yan, *Corrosion*, Vol 46, 1990, p 665
58. G.W. Warren, P. Wynblatt, and M. Zamanzadeh, *J. Electron. Mater.*, Vol 18, 1989, p 339
59. A.D. Marderosian, *IEEE 15th Ann. Proc. Reliab. Phys.*, Vol 15, 1977, p 92
60. S. Thomas and H.M. Berg, *Microelectron. Reliab.*, Vol 28, 1988, p 827
61. "Finishes for Ground-Based Electronic Equipment," MIL-F-14072D, U.S. Department of Defense, 1990
62. C.A. Harper and R.M. Sampson, *Electronic Materials and Processes Handbook*, 2nd ed., McGraw-Hill, 1994
63. J.H. Qiu, *Surf. Interface Anal.*, Vol 33, 2002, p 830–833
64. J.H.A. Glashorster, *IEEE Trans. Compon. Hybrids Manuf. Technol.*, CHMT-10, 1987, p 68
65. J.J. Mottine and B.T. Reagor, *Microelectron. Reliab.*, Vol 25, 1985, p 386
66. M. Antler, *IEEE Trans. Compon. Hybrids Manuf. Technol.*, CHMT-7, 1984, p 129
67. M. Sun, M. Pecht, and M.A.E. Natishan, *Microelectron. J.*, Vol 30, 1999, p 217–222
68. H. Moore, S. Beckinsale, and C.E. Mallinson, *J. Inst. Met.*, Vol 25, 1921, p 35–152
69. A.H. Rawicz, *Microelectron. Reliab.*, Vol 34, 1994, p 875–882
70. M. Moore, "Navy Details Flaws in New Version of Phoenix Missile," *Los Angeles Times*, Sept 4, 1986
71. B. Nordwwall, Air Force Links Radar Problems to Growth of Tin Whiskers, *Aviat. Week Space Technol.*, June 30, 1986, p 65–69
72. K.B. Doyle, *Microelectron. Reliab.*, Vol 14, 1977, p 303–307
73. B. Miksic and A.M. Vignetti, *Suppl. Mater. Perform.*, Jan 2001, p 18–21
74. "Test Methods and Procedures for Microelectronics," MIL-STD-883, 1998
75. "Test Method Standard for Semiconductor Devices," MIL-STD-750, 1995
76. J.A. Capp, A Rationale Test for Metal Protective Coatings, *Proc. ASTM*, Vol 14 (No. II), 1914, p 474
77. P. Lall, M. Pecht, and E.B. Hakim, *Influence of Temperature on Microelectronics and System Reliability*, CRC Press, 1997
78. J. Kopanski, D.L. Blackburn, G.G. Harman, and D.W. Berning, *Transactions of the First International High Temperature Electronics Conference* (Albuquerque, NM), Sandia National Laboratories, 1991, p 137–142

SELECTED REFERENCES

- R.B. Comizzoli, R.P. Frankenthal, and J.D. Sinclair, Ed., *Corrosion and Reliability of Electronic Materials and Devices*, The Electrochemical Society Inc., 1994
- G. Di Giacomo, *Reliability of Electronic Packages and Semiconductor Devices*, McGraw-Hill, 1996

Corrosion in Semiconductor Wafer Fabrication

Mercy Thomas, Gary Hanvy, and Khuzema Sulemanji, Texas Instruments

THE MICROELECTRONICS INDUSTRY is one of the fastest growing industries in the United States due to the constant development and application of new products in data storage, communication, and computing. Constantly shrinking device dimensions enable integrated circuits to become more powerful while the cost of manufacturing continues to drop. At the heart of the industry is the ability at the wafer-fabrication plant to implement the complicated processes needed to deposit, pattern, and etch circuits in semiconductor materials.

One key to the fabrication of integrated circuits is the method of removing selected areas of a conductive-deposited aluminum-copper film. The remaining metal lines are interconnects that join the active components in the integrated circuit. The copper addition to the aluminum increases the reliability of interconnects by preventing the failure mode of electromigration. Electromigration is the unwanted movement of metal atoms in the conductor influenced by current density and temperature that can result in short circuits or open circuits. Metal plasma etch is one of these techniques in which metal lines (or leads) are formed by removing surrounding metal in a chemically reactive low-pressure plasma.

Corrosion is one of the major problems associated with the metal plasma etch process. Dry etching in a chlorine-containing plasma is a potentially important technique for patterning aluminum alloy lines for very large scale integration (VLSI) circuits. However, aluminum alloy lines formed by dry etching are prone to corrode upon atmospheric exposure. The inclusion of copper in the alloy may result in accelerated corrosion of aluminum in the presence of chlorine and moisture during the plasma etching. This is not only an in-process corrosion problem, but also a reliability issue for the metallization of the circuits. Corrosion is typically detected by either manual or more commonly by automated microscopic inspections. The experience of one wafer-fabrication facility to combat the problem is outlined in this article.

Corrosion During Fabrication

In wafer fabrication, much effort has been made to anticipate and avoid corrosion because failure to do so can result in unexpected and catastrophic failure of semiconductor parts either during quality-assurance testing during production or in the field. The impact of corrosion is severe on processing yields and product quality.

Metallization and Chlorine-Induced Corrosion. Corrosion most commonly appears within the metallization process levels between metal deposition and the formation of the oxide layer following metal lead etch and solvent cleanup. A typical corrosion defect is shown in Fig. 1. The mechanism of corrosion is:

$$Al + (BCl_3 \text{ or } Cl_2) \rightarrow AlCl_3$$

With an atmosphere of 30 to 60% relative humidity (RH),

$$AlCl_3 + 3 H_2O \rightarrow Al(OH)_3 + 3 HCl$$
$$2 AlCl_3 + 6 H_2O \rightarrow Al_2O_3 + 3 H_2O + 6 HCl$$
$$2 Al + 6 HCl \rightarrow 2 AlCl_3 + 3 H_2$$

The source of the chlorine could be the sidewall polymer where it can be trapped and then can attack the aluminum leads if they are not put in the solvent stripper. Trapped chlorine in thin-etch residue causes severe pitting corrosion. The etch residue can consist of organic, inorganic (from etch species of Cl and F), metal-organic, and metalinorganic species. Combinations of these make them more difficult to remove. Aluminum oxychloride (AlClO) is formed due to time delay between etch and wet strip. AlClO is soluble in water and stripper. Chloride corrosion at the plasma step is not seen until the wet strip is complete; hence passivating the sidewall is critical to prevent latent corrosion due to residual halides from etch residues.

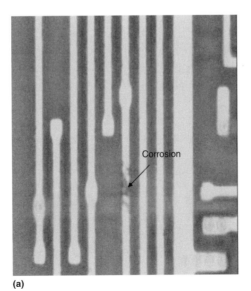
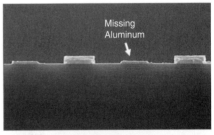

Fig. 1 Corrosion of the aluminum alloy metal lines on a wafer seen (a) in plan view with a light microscope and (b) in profile with a scanning electron microscope (SEM)

The corrosion-prevention effort timeline (Fig. 2) summarizes the activities at the fabrication plant to identify corrosion sources and implement the corrective actions. Each stage emphasized specific areas of the wafer-fabrication facility.

Fabrication Stages. The physical locations of stations in a typical wafer-fabrication processing unit are given in Fig. 3. The occurrence and reasons for corrosion are summarized.

Prior to Metal Etch. Corrosion is due to supercarrier or boat contamination (Fig. 4).

Metal Lead Etch in Batch Etchers. Batch etchers (Fig. 5) are identified as 8330 on the fabrication unit layout (Fig. 3). Corrosion is due to:

- Presence of photoresist remaining after the metal lead etch process
- Lack of a fluoride substitution process in the metal-etch recipe
- Lack of a organic polymer passivation in the metal-etch recipe

Metal Lead Etch Stage in Single-Wafer Etchers. Single-wafer etchers are identified as 9600 in the fabrication unit layout (Fig. 3). Corrosion is due to:

- Improper control of atmospheric passivation module (APM) process
- Decoupled source quartz (DSQ) photoresist-removal process
- Polymer corrosion due to mid-chamber wear
- Environmental effects due to proximity of etchers to each other or due to foreline maintenance

Solvent Cleanup after Metal Etch. Corrosion is due to:

- Solvent hood setup (dispersion plates) and filtration system
- Failure to maintain product time limits between metal lead etch and solvent cleanup
- Failure to use nitrogen cabinets for storage between metal lead etch and solvent cleanup
- High RH levels in the fabrication area
- Failure to strip all polymer from the wafer in the solvent cleanup process

Environmental effects that could occur at various production stages include:

- Opening of the foreline to chlorine or etch polymer contamination
- Metal-etch tool cleanup affecting other nearby tools
- Poor airflow

Corrosion at these stages is examined in more detail in the sections that follow.

Corrosion Prior to the Metal Lead Etch Process. The source of this defect is identified by observing the pattern of the corrosion on the wafers. Corrosion occurred only around the extreme edge of the wafer. Little or none was observed around the wafer flat or at the wafer edge opposite the flat. When stored in a supercarrier, the wafer flat is always up in the boat and the wafer round (edge opposite the flat) is always down. The areas clear of corrosion are never in contact with the interior surface of the boat slots. Further testing with controlled conditions duplicated the corrosion defect, and the source was determined to be from the supercarrier boats used when transporting wafers from one processing stage to another.

Stage 1	Stage 2	Stage 3	Stage 4	Stage 5
Focused on optimizing recipes and implementing N$_2$ storage stations	Focused on single-wafer etcher (Model TCP 9600)	Focused on identifying sources of Cl$_2$/HCl contamination in the metal-loop process	Focused on environmental HCl/Cl$_2$ contamination sources	Focused on improvements in the photoresist removal process/tool

Corrosion prevention efforts

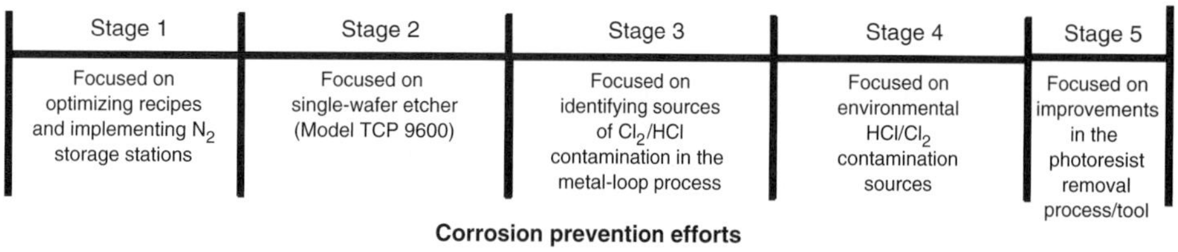

Fig. 2 Timeline of the corrosion-prevention project from one fabrication unit showing stages of work

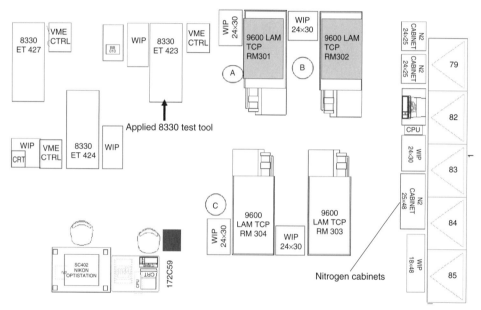

Fig. 3 Floor plan of a fabrication unit with single (9600) and batch (8300) etchers. Applied test tool and sample points (A, B, and C) were used in corrosion study.

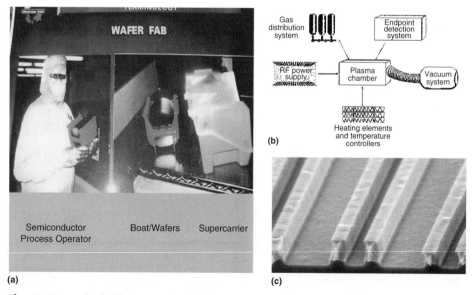

Fig. 4 Aspects of wafer fabrication and metal etching. (a) Operator with supercarrier and boat fixtures. (b) Diagram of etcher-system components. (c) SEM image of a metal-etch feature

Other wafer-fabrication units within the company were polled, and the consensus was that they did not have the same issue. The units without the incoming corrosion problems managed their supercarrier boats in a manner that made sure boats were switched either before or just after metal deposition.

In order to implement a boat switch at metal deposition, more boats would have to be purchased. To avoid this additional cost, a novel deionized water boat wash procedure was implemented whereby supercarrier boats were collected at the metal-deposition process and washed. Now wafers completing the metal-deposition process are placed into a freshly cleaned boat.

Corrosion Issues Related to Batch Metal-Etch Systems. Batch metal-etch systems, such as the 8330, can be considered a source of corrosion since the photoresist cannot easily be immediately stripped following etch as can be done in single-wafer metal-etch systems. Figure 6 shows that batch systems are more likely to produce parts with corrosion defects than are the three single-wafer etchers.

Corrosion in the batch metal-etch system stems from the presence of aluminum chloride residue. The residue is produced from photoresist and other foreign materials present. The aluminum lead sidewall will trap chloride. The more porous the sidewall, the heavier the corrosion.

Corrosion-prevention measures in the batch systems include:

- Ensure a residue-free etch as much as possible.
- Perform a fluoride-substitution treatment by converting the chloride residue into fluoride that is inert to moisture. This step is called the "fluorine/chlorine exchange step." The step consists of washing the parts with 100 standard cubic centimeters per minute (sccm) CF_4 for 2 min.
- Perform an organic polymer passivation step using trifluoromethane (CHF_3) to form a strippable protective coating. This step is called the "anticorrosion step." The step consists of coating the parts with 60 sccm CHF_3 for 6 min.
- Strip the photoresist as soon as possible after etch. This requires tight control of sit times between etch and cleanup.

Of these items, three are key: the fluoride substitution, the organic polymer passivation, and the control sit time between etch and cleanup (Ref 1).

Corrosion Issues Related to Single-Wafer Metal-Etch Systems. Single-wafer systems, such as the 9600, have an advantage over batch systems in the prevention of corrosion by the sequence of process modules. The main etch chamber is followed by the DSQ photoresist strip chamber, and then by the APM rinse. Even so, the most common corrosion causes can be grouped under the two categories of APM and DSQ setup and maintenance. When either fails, the typical concentration of the corrosion is dense across the wafers as in the wafer map (Fig. 7).

Atmospheric passivation module processing has critical parameters that need to be controlled and monitored closely: water temperature and pressure, wafer chuck spin speed, exhaust, nitrogen nozzle pressure and positioning, and nitrogen curtain flow. Of these, water temperature has the largest impact on corrosion based on test results. The defects for wafers processed at various APM water temperature set points and spun at 500 and 1000 rpm are shown in Fig. 8. Corrosion defects reached a first maximum at the 55 °C (130 °F) set point at 1000 rpm and the 60 °C (140 °F) set point at 500 rpm. Spin speed did not have much impact until the 80 °C (176 °F) set point was reached. Choosing the correct water-temperature set point and preventing drift of all the critical APM parameters through proper adjustment and maintenance of the equipment is key to controlling APM process corrosion.

Decoupled source quartz photoresist removal process relies on these critical parameters that must be monitored: photoresist strip rate, photoresist strip uniformity, paddle temperature, and water vapor concentration. If the DSQ setup is optimum and is operating correctly, the water vapor introduced in the initial step is the key parameter from the list. It neutralizes or removes the residual chloride surrounding the etched metal leads. Next in importance is the photoresist strip rate and uniformity. It is essential that the photoresist is removed completely, leaving no residual on the wafer for electronic as well as corrosion reasons. Proper monitoring and maintenance of the DSQ chamber is needed to ensure corrosion is kept to a minimum (Ref 2, 3).

Mid-chamber Wear. Wafer corrosion on the single-wafer system can occur due to mid-chamber wear in the following manner. As a result of process etch gases eroding the anodized aluminum walls of the plenum, a corrosion by-product is formed. This by-product will then be deposited on the wafer surface, making the photoresist very hard to remove in the DSQ and even in the following solvent cleanup steps. The residual photoresist will result in metal lead corrosion on the wafer.

To prevent this, the machine manufacturer developed an upgrade whereby they lined the gas plenum with ceramic. This extends the top plate

Fig. 5 A batch metal etcher, model 8330, installed in the fabrication unit

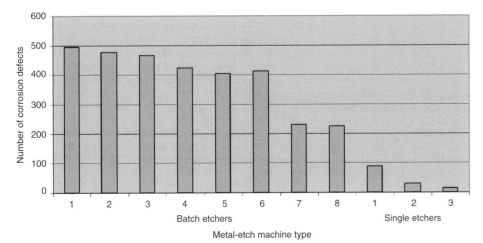

Fig. 6 Data showing batch etchers are more prone to corrosion than single-wafer etchers

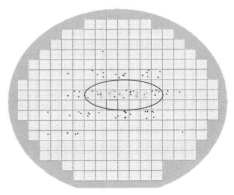

Fig. 7 Wafer map of typical corrosion defects. Black spots are random corrosion. Gray (or green) spots in circle are clustered corrosion.

over the anodized aluminum middle chamber, fully enclosing the gas plenum with ceramic (Fig. 9). It prevents exposure of the chamber walls to process gases and extends the mid-chamber life indefinitely, as well as helping to prevent corrosion to the product (Ref 4).

Moisture from the fabrication-room environment is a corrosion concern. When wafers are sitting in the clean room after metal lead etch, they are at a high risk to be corroded by exposure to the atmosphere. To prevent moisture attacking the metal leads, wafers are stored in nitrogen cabinets, where nitrogen is continuously circulated. Nitrogen serves as a corrosion prevention, but does not guarantee the absence of corrosion. Time limits are set for the sit time, the time between metal lead etch and solvent cleanup, to reduce the chance of corrosion. The sit-time limit is approximately 1 to 2 h after metal lead etch. Wafers are stored in a nitrogen environment and can be held up to the sit time limit before processing in solvent cleanup. Chloride residue from the photoresist attacks metal leads and needs to be removed before the waiting period.

The oven-bake process is introduced to dry out residual chloride from photoresist. Wafers are put in ovens as soon as they come from the batch metal etchers. Ovens operate at 150 to 250 °C (300 to 480 °F), which dries out the chlorine before the ash and solvent strip. However, the oven-baking step hardens the photoresist, and it is then more difficult to remove. The ash and solvent-strip removal parameters must be optimized.

Solvent Cleanup after Metal Etch. The composition, flow characteristics, and maintenance of the solvent will impact the possibility of degradation to the wafers.

Solvent Deionize Water Content. Solvents are used at the stripping step. The cleaning effectiveness of many of solvents is sensitive to the water content of the solvent. A low deionized water content would not provide an efficient cleaning, but a high deionized water content would cause corrosion.

Dispersion Plates. A dynamic flow in the solvent tank is essential to provide a good mechanical motion between solvent and wafers. Dispersion plates should be installed at the bottom of solvent tanks. These are stainless steel plates with small holes (Fig. 10) that restrict overly concentrated flow in the wafer zone and help to deliver an even flow to the wafers. A good circulation of solvent takes the polymer and photoresist from the wafers to the bottom of the tank where they are trapped by the filters. If wafers are processed in a tank of dirty solvent, corrosion appears as in Fig. 11. A good filtration system will keep the solvent clean.

Dual-Filtration for Solvent Bench. As line widths of conductors on the wafers decrease, defects that were previously not a concern become more important. Flow rate is a key parameter to monitor in a solvent-bath filter system. A single-filter system may not be sufficient on a dirty solvent-strip process. For an application where the photoresist remains after the etch, a dual-filter system consisting of a coarse polyguard filter plus a fine microguard filter works well filtering solvent containing both large and small photoresist particles. Improperly sized filters can leave a residue in the solvent that results in wafer corrosion. Also, solvent-bath change and filter change at rigid frequencies are required due to loading effects. The frequency of change can be time based or based on the number of wafers processed.

Fluorine corrosion occurs when an excessive amount of water remains on the wafer surface (Fig. 12). Heater failure, poor nitrogen flow, or a bad door seal of a spin-rinse dryer can cause the wafers not to dry completely after a rinse cycle. Corrosion occurs if wafers are not dried immediately to minimize the wafer contact to moisture.

Material Segregation. Metal leads with chloride residue should be processed separately from other material in solvent cleanup to avoid cross-contamination. The residual chloride may corrode wafers without photoresist. For instance, wafers from a batch-metal etcher and wafers from a single-wafer etcher must be run independently in the solvent because residual chloride is present on batch wafers, but not on single wafers. Chloride residue attacks the single wafers and results in corrosion damage.

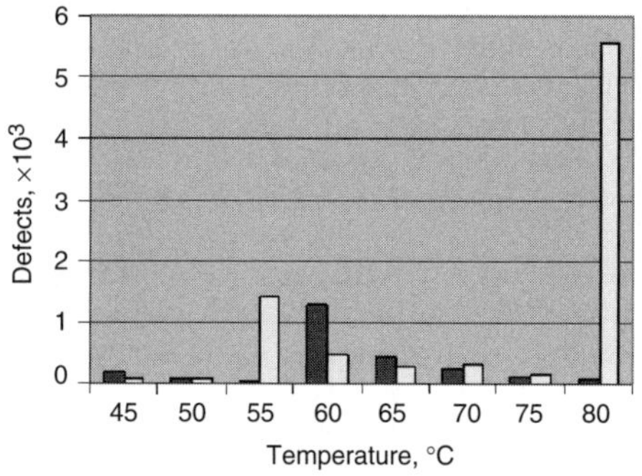

Corrosion Due to Environmental Effects

Environmental factors within a wafer-fabrication unit (fab), such as acid vapors released into the air, are potential sources for contamination. A study focusing on the release of chlorine vapors into the fab environment during a tool maintenance activity and its effects on wafer integrity was conducted.

This study considered a ballroom-type wafer fab using an 8330 batch etcher (Fig. 5) as the test tool that was the source of the chlorine vapor. The 8330 etcher requires preventative maintenance (PM) to remove polymer or etch residue buildup within the process chamber.

At PM intervals, the process chamber is vented to the fab room where low concentrations of chlorine within the polymer are released. The chlorine interacts with the water vapor in the air and forms low-grade hydrochloric acid (HCl) vapors.

The first phase of the study involved placing preinspected wafers with 2.0% aluminum copper deposition (witness wafers) in locations near the 8330 metal etcher before opening the process chamber to atmosphere. The locations are identified around the test tool as A, B, and C in the Fig. 3.

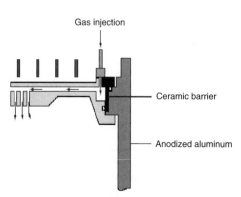

Fig. 8 Correlation of corrosion defects to atmospheric passivation module water temperature and wafer spin speed. Dark gray bars, 500 rpm; light gray bars, 1000 rpm

Fig. 10 Dispersion plates used in solvent tanks to control flow of liquid

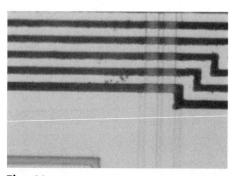

Fig. 9 Mid-chamber ceramic barrier. Courtesy of LAM Research Corporation

Fig. 11 Corrosion on wafers from solvent tanks

The chamber was then opened and a normal PM was performed. At the conclusion of the PM, the wafers were inspected for visual defects (Table 1). The wafers were then processed on a TCP 9600 metal etcher using a standard recipe to determine if the postetch process enhances the effects of corrosion.

The data indicate that wafers with metal deposition, which are exposed to HCl vapors or byproducts, have the potential to corrode. This is confirmed by the incomplete etch defects viewed after metal-etch processing. Of the three sets of wafers placed around the metal etcher, only the wafers placed in a location A showed metal corrosion. This indicates that there are other fab conditions that affect corrosion such as residence time and air-flow dynamics.

The second phase of the study focused on the allowable residence time of HCl vapors and its impact on wafer integrity. A repeat of the first phase of the study was performed using only location A (Fig. 3). Witness wafers were set out at the initial opening of the process chamber and at 10 to 30 min intervals thereafter. A visual inspection of each set of wafers was then made and is shown in relation to the PM steps (Fig. 13).

Figure 13 shows that there is some amount of residence time before the corrosion can be seen. No corrosion was seen during the initial inspection, although the wafer had been exposed to a burst of vapors as the chamber was opened. It was approximately 40 min later before the corrosion could be seen on the original witness wafer. The wafers that were switched out at specific intervals continued to show corrosion as the PM continued. Corrosion did not stop until the chamber cleaning maintenance was complete.

The results of both the first and second phases of this case study raised the question of the importance of air-flow dynamics. A third phase of this case study was initiated focusing on air-flow dynamics in and around the original test tool (Fig. 14). A fogger-type device was used to view the flow dynamics at different locations throughout this area.

The results of the test indicated that good vertical laminar flow existed in most areas. Location points A and B (Fig. 14), however, showed an air-flow drift toward location C (which is location A test point used in the first phase (Fig. 3). The air-flow dynamics of the room guided the HCl vapor toward this location,

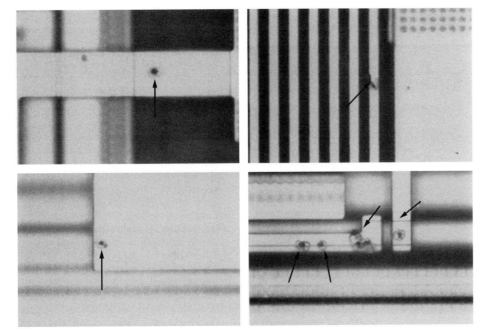

Fig. 12 Fluorine-induced corrosion indicated by arrows on metal leads

Table 1 Results of witness wafer test

Wafer location(a)	Wafer No.	Post PM inspection(b)	Inspection after wafer processing(c)
A, RM301 wafer-staging rack	2	Heavy corrosion	Incomplete etch
	3	Heavy corrosion	Incomplete etch
B, Between RM301/RM302 wafer-staging rack	6	No defects	No defects
	7	No defects	No defects
C, RM304 wafer-staging rack	4	No defects	No defects
	5	No defects	No defects

(a) Refer to Fig. 3. (b) Preventative maintenance performed on 8330 ET 423 etcher. (c) Processed on 9600 TCP single-wafer etcher

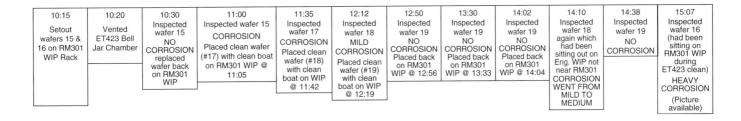

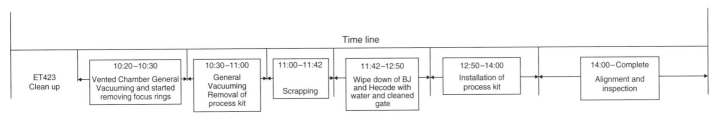

Fig. 13 Residence time study of HCl vapors in a wafer-fabrication unit environment to track the appearance of corrosion on wafers. Parallel maintenance activities are tracked on time line. WIP is work in progress rack.

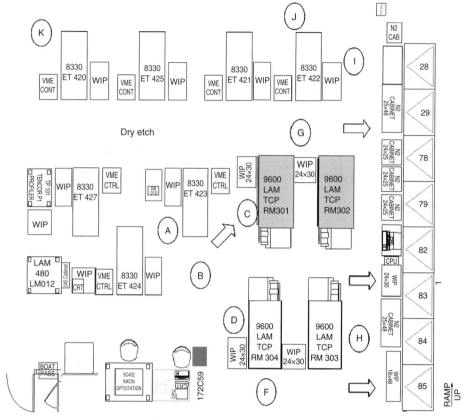

Fig. 14 Air-flow dynamics in the fabrication unit. Arrows indicate horizontal flow. Circled letters are test locations. Shaded objects are the single-batch etchers.

and only wafers in this vicinity were affected. The area, in general, was observed, and it was noticed that variation in floor tile perforation and the bulkiness of the tools contributed to this air-flow drift.

This case study showed that factors outside the normal processing of wafers (example foreline maintenance and other PM) or tool-specific problems can contribute to metal-line corrosion. The best solution is to remove the possibility of HCl exposure altogether. Unfortunately, that is not always possible. Measures should be taken to mitigate the amount and frequency of HCl exposure. In certain scenarios, consideration should be given to stop wafer processing and remove the wafers from the area until normal fab unit conditions return.

ACKNOWLEDGMENT

Special thanks to Martha Adams, Plasma PE senior technician who helped set up and run some of the case study evaluations.

REFERENCES

1. *8100/8300 Basic Etch Process Course Manual,* module 4, Applied Materials, Inc. 1994
2. "Elimination of Post-Etch Aluminum Corrosion on the Lam Research Transformer Coupled Plasma (TCP) 9600 Metal Etcher," Doc ID No. 93112092A-ENG, Sematech
3. S.G. Bradley and W.-B. Chou, Lam Research Transformer-Coupled Plasma (TCP) 9600 Downstream Quartz (DSQ) Hardware Evaluation Report, Doc ID No. 95072915A-TR, Sematech, Aug 31, 1995
4. "Ceramic Lined Gas Plenum Upgrade," Product Bulletin, LAM Research Corp., 1993

SELECTED REFERENCES

- "DSQ Stripper, TCP 9600 Operation and Maintenance," Rev B, *Lam Research (LAM) Corporation Manuals,* Feb 1995
- "TCP 9600, System Operation," Rev C, *Lam Research (LAM) Corporation Manuals,* April 1993

Corrosion in the Assembly of Semiconductor Integrated Circuits

A.C. Tan, Micron Semiconductor Asia

METALS USED IN SEMICONDUCTOR INTEGRATED CIRCUITS (SICs) serve several functions. Within the chip, metal lines provide local interconnection to join a collection of circuit elements globally to various areas of the chips, and they also provide input/output signals. Aluminum is currently the most commonly used metal for integrated circuits (metallization and bonding pads), but copper is increasingly being used for the high-end (ultralarge-scale integrated circuitry) products because it exhibits a lower electrical resistivity and better electromigration resistance than that of aluminum. The chip is connected to the outside world by a wire bonding (or other forms of interconnects) and a leadframe (which is the metallic portion of an assembled IC package and is used to complete the electrical connection path from the die to defined circuit elements of a printed board assembly), or a patterned laminate (also known as a plastic interposer, commonly with plated copper traces). The terminations of the leadframes are plated with a solderable metal or alloy, such as tin and tin-lead alloys. These metals are susceptible to corrosion.

In a typical IC component, corrosion may be observed at the chip level and at the termination (also known as leads) area, as illustrated in Fig. 1. In this figure, a thin-small-outline package (TSOP) is shown along with the various areas where corrosion can occur. For chip scale package (CSP) type of packages with solder balls as interconnects between the package and a printed circuit board, similar corrosion sites may occur, except that the entry path may be different. Generally, it is rare to see a solder ball corroded. Chip corrosion refers to either the corrosion at the bond pad, or corrosion at the metallization, with thickness usually around 0.5 to 1 μm. Device terminations refer to the leadframe, commonly Alloy 42 (ASTM F 30, 42% Ni, 58% Fe) or copper, usually coated with either solder or tin, with typical thickness ranging from about 6 to 12 μm. In some components, the copper leadframes are coated with a solderable palladium or gold flash (typical thickness, <0.125 μm).

Chip corrosion could cause a reliability (functional) failure in an electronic device. Corrosion of the device terminations (known as macrointerconnects) results in lead (termination) tarnishing, which is, in a mild form, an aesthetic issue, and in the worst form, a contact resistance problem or a solderability problem at the board assembly. Due to the corrosion, the metals are said to be corroded, oxidized, or tarnished (Ref 1–5). Generally, the macrointerconnects are more exposed to the ambient environments, while the chips are protected by the encapsulants (compounds for a plastic encapsulated device). However, the dimension of the macroconnects is generally in mils (1 mil = 25 μm), while the metallization of the chips is about 1 μm. Failure due to the corrosion of the macrointerconnects is not as catastrophic as corrosion at the chip. Generally, corrosion of the macrointerconnects may be "reworked," while chip corrosion would mean that the chips are to be scrapped, not reworkable at all.

Factors Causing Corrosion

Corrosion of metals has been extensively studied with volumes of monographs and articles (Ref 2–4), and the key factors contributing to corrosion of electronic components are:

- Chemicals (salts containing halides, sulfides, acids, and alkalis)
- Temperature
- Air (polluted air)
- Moisture
- Contact between dissimilar metals in a wet condition
- Applied potential differences
- Stress

The most common factors are the combination of all of these factors, except the stress factor, which occurs less frequently.

Routes of Corrosion. Semiconductor components are subjected to a variety of exposures during the manufacturing of the components and when assembled on the board. Thus, corrosion may be routed differently. In general, two types are observed: corrosion driven by the application of an applied potential, and corrosion occurring under open circuit (static) conditions. The focus of this article is on the latter. For corrosion driven by an applied voltage, readers may refer to Ref 6 to 8.

Chemicals are found practically everywhere in typical device fabrication (known as front-end) and assembly (known as back-end) processes and can contaminate the device when care is not taken to minimize the exposure. It is known that air contains airborne particles (especially in a place with heavy industries or heavy vehicular flow). Airborne particles are, basically, a mixture of hydrocarbons, silica, magnesium oxide, sulfides, and iron oxides. The actual composition varies with different air environments and sources.

Wet processes involving solvents, plating baths, and fluxing agents are used in the IC assembly and chemicals are also used in making the packaging materials (known in the electronics industry as piece parts). Examples of the piece parts used in the assembly of SIC are leadframes, die-attach (adhesives), wire (about 1 mil in diameter), and molding compounds. Chemicals are used in the plating process and in the post-plating processes (e.g., burn-in, cleaning of the units after a burn-in process). The most corrosive chemicals used in the assembly are the plating chemicals (precleaning chemicals and plating chemicals) or fluxing. In the early days prior to the banning of ozone-depleting chemicals, a chlorinated solvent, 1,1,1-trichloroethane was a common solvent. If the solvent is dispensed at the vicinity of a furnace,

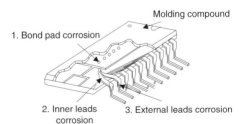

Fig. 1 Areas where corrosion can be observed in a typical IC component. 1, Bonding pad corrosion; 2, Inner lead corrosion; 3, Leadframe (external leads) corrosion due to environmental impact. The IC package shown here is a partial thin small outline package (TSOP) with molding compound removed showing the silicon die, bonding pad, and the external and inner leads. Source: Ref 1

decomposition of the 1,1,1-trichloroethane is expected, giving rise to a corrosive hydrochloric (HCl) gas, which could subsequently corrode the terminations if the components are stored in the vicinity.

The chemical ingredients used by the piece parts (materials for the packaging and assembly of silicon dies) are generally inert (noncorrosive); however, ionic impurities (due to the incoming materials and processing) may be present, and hence it is imperative that the level of the ionic impurities is controlled, usually to less than 10 ppm. One of the important ingredients used in the molding compounds is a releasing agent, carnauba wax, which is basically a mixture of long-chain hydrocarbons, ester, amides, and organic acids. Outgassing of these compounds occurs when the encapsulant is heated. This may cause tarnishing of the leads and an increase in contact resistance. Additionally, the outgassing of the wax materials is also detrimental to the wire bond integrity when the SIC is undergoing a high-temperature storage (HTS) test because some of the outgassing products are corrosive.

Process Flow of IC Packaging. The assembly of electronics parts follows a series of processes, as illustrated in Fig. 2, where some corrosive chemicals and elevated temperatures are involved. Corrosion (chip corrosion, tarnished terminations, and rusty discoloration with iron rust or rings of multiple color) has been observed since the beginning of semiconductor

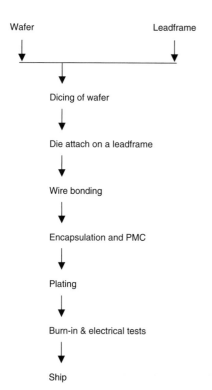

Fig. 2 Generic process flow of a semiconductor device. Key areas where corrosion could occur are shown. PMC refers to post mold cure, a process whereby the encapsulant (molding compound) is further cured.

device manufacturing. With a better understanding and lessons learned from operation in the field, its occurrence is more manageable now. Let us examine the assembly processes where corrosion may take place when process or material control is lacking. Generally, the ionic contents of the packaging materials are to be controlled, at about 10 ppm or lower. Each of the steps has potential area for corrosion if not properly controlled. Among the processing steps, the plating process is the most corrosive. Surface finishing (plating) is a process (an integral part of the assembly) using a series of corrosive chemicals for etching, plating, and neutralization. If residues are left on the package (plastic body and leadframes), the reliability of the device can be potentially affected.

In a burn-in process, packages are subjected to a long duration (hours) of heat treatment (for static condition, about 125 °C, or 255 °F, for a dynamic condition, between cold and hot). Tarnishing of the leads can be observed if no proper care is taken to manage the testing conditions. Solderability problems are likely to occur when the leads are severely tarnished.

Chip Corrosion

Chip corrosion (bond pads and metallization) has been extensively documented and cited in the literature (Ref 1, 6–10). Fluorine is commonly found on the bond pads due to the fluorine-containing etching chemicals not being completely removed. Hence, the patterned wafers are tightly packed so that moisture is kept away from the wafers. The bond pads, when exposed to the moist air, are susceptible to corrosion, as it is known via auger analysis that traces of fluorine are usually present. Corrosion products are the mixture of aluminum oxides and aluminum oxyfluoride salts. Bonding with these corroded pads would encounter "nonstick," or weak adhesion wire bonding, hence creating a serious reliability problem where the aluminum pad and gold wire are separated. Piece parts used in the assembly, such as die-attach and molding compounds, may contribute to the corrosion if the impurities levels (particularly halides and salts) of the piece parts are not controlled.

Brominated resins are commonly used as a flame retardant in a compound formulation and provide a source of halide ion (e.g., bromide) to cause corrosion of the chips. Surface finishing processes, such as solder dipping or plating (tin or solder), are another source of corrosive chemicals. When left uncleaned, the chemicals may diffuse into the compound via crevices and result in poor adhesion between the compound and the leadframe.

Oxidation of Tin and Tin Lead Alloys (Solders)

Oxidation refers to a reaction whereby the metal is transformed to a metal oxide upon exposure to air and temperature. It is a corrosion reaction. For example, like most metals, tin or solder has a native oxide layer on the metal surface. Tin oxide (SnO or SnO_2, or a mixture of the two oxidation states of tin) is a passivating film with thickness approximately a few angstroms. In the presence of wet conditions, temperature, and ionic species, further oxidation may occur, producing more tin oxides. The color of the oxides depends on the thickness. It ranges from tinge-blue or yellow, yellow, purplish blue, and black, and all of these are tarnishing (Ref 2, 5). Oxidation of solders is similarly dependent on the aging condition (O_2-rich or N_2-rich), ionic species, temperature, and moisture level (Ref 11, 12).

Generally, solders are also passivated with a layer of a mixture of tin and lead oxides. Tin is slightly more anodic than lead. The standard electrode potentials for tin and lead are -136 and -126 mV, respectively, thus tin oxide is preferentially formed. However, the composition of the oxides involves more than the electrode potentials; kinetics plays an important role as well (Ref 11, 12). Reflowed solders have two distinguishable phases, tin-rich and lead-rich. Electroplated solder (in various compositions) contains tin and lead and has a well-defined grain structure. The reflowed solder and electroplated solder contain mainly tin and lead, but they are different structurally. Hence, the nature of the oxides formed for the reflowed solder and electroplated solder would depend on the surface composition and the environment. When there is a limited supply of oxygen (e.g., in a N_2 purged oven), there is a competition between the surface composition and structure, as described in Ref 13 and 14. In a normal condition (air-filled burn-in oven), a mixture of the oxides is formed. The composition of the solder oxide depends on the surface composition and the environment.

Tin oxide is a passivating film, but, in the presence of air (O_2), corrosive chemicals (e.g., Cl^- ion), and moisture, the oxide is attacked, converting tin oxide to a stannous chloride ($SnCl_2$) salt. The $SnCl_2$ salt is a loosely bonded salt; in the presence of water (moist air) and carbon dioxide, it may be changed to a tin carbonate. The corrosion reactions make the surface tarnished or colored, depending on the extent of the oxidation. For long-term storage of IC units with a solder finishing, the solder oxides formed are a mixture of tin and lead oxides, and the salts, commonly lumped as "oxides."

Mechanism of Tarnished Leads (Terminations)

Tarnish of leads has been observed in several forms:

- All leads are tarnished (various colors and shades).
- Tarnish occurs at the interface between compound and leadframe (usually rusty color).

- One or two pins are tarnished while the rest of the device leads are not tarnished.
- Leads are rusty with multiple color on the plastic body and leadframe.

The mechanisms are corrosion (aqueous and dry corrosion) and galvanic corrosion. The ingredients that trigger the corrosion are the usual factors: chemicals, moisture, and temperature. The place of occurrence could be found after plating and post-plating processes such as trimform, burn-in, and solder reflow.

Tarnished Leads Caused by a Galvanic Corrosion. A classic example of a galvanic corrosion is the aluminum bond pad corrosion (Ref 1, 5–10) where the gold acts as the cathode, and aluminum acts as the anode. The corrosion leads to a continuity failure when the aluminum pad is depleted. The attributes of this chip corrosion are the ingress of moisture and ionic contaminants. Two major sources of ionic contaminants are observed: piece part (molding compounds and die attach) and the residues from the flux or plating chemicals. The ionic impurities of the die-attach and molding compounds of the early years (1970s and 1980s) were not properly controlled, and corrosion of the chips was commonly encountered. However, most current die-attach and compounds have the chloride level and other aggressive ions controlled at about 10 ppm. Similarly, if the HCl acid and other corrosive chemicals remain on the plastic body, the chemicals could ingress into the plastic compound and corrode the bond pad. Eventually, a discontinuity problem can be observed. The aluminum bonding pad is eventually corroded, forming aluminum trihydrate ($Al(OH)_3$) as a corrosion product. The rinsing and drying parts of the plating and soldering process need to be optimized such that the package, after plating or soldering, is optimally cleaned. Rinse effectiveness of the process may be measured by the use of a resistivity measurement (Ref 1). The industry has learned from the harms caused by the corrosive residues, and actions have been taken so that the history of corrosion is not repeated. With better management of the wet processes, leadframe design, and materials selection, bond pad corrosion is now rarely experienced.

Galvanic corrosion can also be observed in the assembly. One case is illustrated in Fig. 3, where only a few pins of the entire package are corroded (in this case, two pins of the entire package). In another similar case, an imprint of terminations was seen on a gold lid of a ceramic package as shown in Fig. 4. These two cases are the result of a mishandling arising from a wet process (see also the subsequent discussion on oxide wash and ink symbol rework).

The burn-in process, if not controlled properly, can result in severely tarnished leads. Tarnished leads found during this burn-in process are caused by factors such as temperature, bad door seal of the burn-in oven, humidity, plating chemical residues remaining on the terminations, and the cleanliness of the burn-in sockets. The outgassing of polymeric materials used in a burn-in board and encapsulants (e.g., epoxy molding compounds used in many IC devices) could also contribute to the tarnishing of the device leads. Traditionally, a subsequent "oxide wash" is the usual remedy to salvage the affected lot. Typically in an oxide wash process, an acid or alkaline solution is used to clean off the oxides. By mistake or improper design of a loading and unloading tool, a galvanic couple is created. A stainless steel jig or handler is commonly used in the industry because it is robust and can be easily used in a wet process. However, the package terminations (pins) coated with solder/tin may be in contact with a stainless steel jig during the wet process. Corrosion of the solder (tin) takes place when there is a contact, and in the presence of a wet condition (e.g., an acidified solution). Figure 5 schematically explains the galvanic action taking place. The etched tin or solder surface, when subjected to drying after the wet process, is further oxidized, resulting in a darker appearance as compared with the noncorroded pins (refer to Fig. 3). This mechanism explains the appearance of one or two pins being tarnished, while the remaining pins are unchanged. Galvanic corrosion does not take place when there is no contact.

The imprint of leads on a gold lid of a ceramic package (a high-reliability product), as illustrated in Fig. 4, shows another example of galvanic corrosion due to dissimilar electrodes in contact. In the presence of an electrolyte and temperature, galvanic corrosion occurs. In the days of the ink symbolization process, mistakes could be made (e.g., wrong symbol or poor symbol). Traditionally, a rework of the cured ink symbol was done by soaking the gold lid package in a sodium hydroxide solution at an elevated temperature. If the cleaning was properly carried out, the cured ink would be removed and the package reused. However, if the ceramic

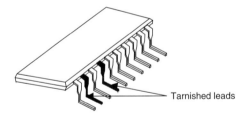

Fig. 3 Diagram showing a partial thin-small-outline package with tarnished leads caused by galvanic corrosion. Notice that two pins, out of ten, are tarnished. In some cases, only one pin is tarnished. The tarnished leads are seen in a wet process (e.g., oxide wash process).

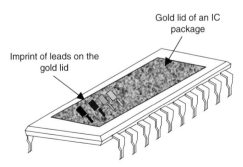

Fig. 4 Pictorial illustration of an imprint of device terminations on the gold lid of a ceramic package. The imprints are formed when the units are soaked in a sodium-hydroxide solution at an elevated temperature.

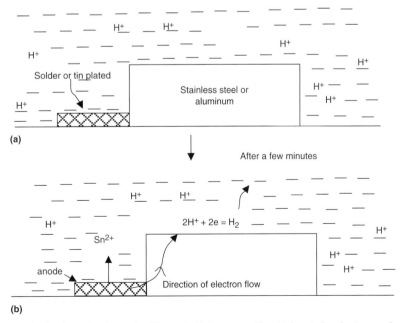

Fig. 5 Mechanism for a galvanic reaction. The terminal is in contact with a stainless steel or aluminum surface seen in (a). The presence of an acidified electrolyte completes the path that allows the charges to be transported from one dissimilar metal to the other. In this case, the solder (or tin) is the anode, which dissolves anodically with eventual consumption of the solder (seen in b). This results in a tarnished lead.

units were thrown into the sodium hydroxide solution, contacts between the terminations and the gold lid were made unintentionally. If there is a contact between the gold lid and the terminations, galvanic corrosion occurs where the tin of the lead surface is transferred onto the gold surface.

In a sodium hydroxide solution, the reaction product is $Sn(OH)_2$ or $Sn(OH)_4^{2-}$. The products float around the vicinity of the reaction site. If the stannous hydroxide particles are on the cathode side (gold lid), the stannous salts may be redeposited at the area of contact on the cathode (the gold lid of a ceramic device) as follows:

$$Sn(OH)_2 + 2e^- = Sn$$

(redeposited at the area of the gold lid)

$$Sn(OH)_4^{2-} + 2e^- = Sn + 4OH^-$$

(redeposited at the area of the gold lid)

The ceramic body is heavy enough to make a very tight contact between the leads and the gold lids, as units are stacked randomly on top of each other. The end result is a spongy deposit of tin, which appears as a gray spot. When the soaking is further prolonged, an imprint of the terminations can be seen, as shown in Fig. 4. The concentration of the sodium hydroxide solution is 1 mol/dm^3, and the temperature is kept at 50 to 60 °C (120 to 140 °F). The main application of the solution is to strip off the ink symbols. However, due to improper handling (making a contact between gold and solder), galvanic corrosion occurs.

Tarnishing due to Chemical Residues. Tarnishing of the leads occurs for the entire terminations and with different color appearance. Plating, soldering, and other wet processes tend to leave residues on the surface when the rinsing and drying are not optimized. The chemical residues are corrosive (acids or alkalis), and in the presence of temperature and high humidity, corrosion of the solder or tin surfaces is expected. There are two common types of burn-in processes: static, where the temperature is fixed (usually about 125 °C, or 255 °F), or a dynamic burn-in, where the devices are subjected to a low temperature (commonly below 25 °C, or 75 °F), then to a high temperature (about 125 °C, or 255 °F). In the static burn-in process, the chemical residues react with the solder or tin, forming oxides, with an appearance of a tarnished surface. Though a N_2-rich or a dry-air purging may be engaged, if the amount of residues is large, corrosion could still occur. In a dynamic burn-in, care should be taken to minimize the amount of moisture (the level is influenced by the percentage of relative humidity) being trapped by freezing during the cold cycle. The common method is to purge with a stream of inert gases (N_2 or clean dry air) and optimize air leakage during the burn-in process.

Non-optimization of the rinsing steps in a plating process could also result in another form of tarnishing—tinged yellow tarnished leads—after a long burn-in. Solderability is not affected, but production flow is slowed down by the time spent deciding whether to accept the tinged appearance. It is believed that adding a neutralization step in a plating process (Ref 5) would help to minimize this tinged appearance. It is believed that the neutralization step immediately after plating helps to fill the pores of the plating layer. This means that no plating residues are trapped during the water rinsing step.

Tarnishing Caused by Base Metal Migration. Plating layers, especially those with low thicknesses, are generally porous. This means the protective efficiency is limited when subjecting the devices to a prolonged temperature treatment (e.g., burn-in). An example is copper leadframes plated with tin or a solder coating. The copper base metal migrated and reached the surface of the thin plated layer and then oxidized to a copper oxide, with its characteristic copper color. Apart from the thickness factor, the rinsing process in plating is also an important parameter to take care of. If the thin plating layer is immersed in the rinsing process for too long and at too high a rinsing temperature, the plated layer is believed to become more permeable. The copper from the base metal can easily and similarly migrate upward. Upon a thermal treatment (drying, burn-in, or other heat treatment), the copper at the plated surface gives rise to a copper color. For Alloy 42-based leadframe, the migration of nickel and iron is probable; you may see a rusty color (due to the formation of an iron oxide) on the plated layer.

Tarnishing by Plating Additives. Electroplated metals are known to be porous and may contain occluded additives, as compared with metals made by casting. If there is a breakdown in the additives, or improper current density, the plated layer may be more porous, and hence, more prone to oxidation (corrosion). The O_2 may diffuse into the porous layer, causing more oxidation, and hence, tarnishing. Occluded additives may be quantified in terms of "carbon content" and may be controlled at less than 10 ppm. High carbon content is usually found in bright plating. For semibright plating, the carbon content is relatively lower. Tarnishing is due to the gradual outgassing of the additives (moisture as a by-product) when the IC devices are aged. Thus, one needs to optimize the plating bath composition, rinsing steps, and conduct a series of stringent tests to ensure that the plated layer has a consistent built–in tarnish-resistance capability. An example of the test is to bake the plated sample at 150 °C (300 °F) for 120 h without nitrogen purge. If no tarnishing of the leads is seen, it is said that the plating baseline is optimized with strong resistance to tarnishing. Another test is subjecting the plated sample to a solder reflow process. Tarnishing of the plated layer (purple or brown appearance) after exposure to a reflow process is an indication that the plating process is not optimized.

Rusty and Multicolor Seen on the Leadframes and Package Body. This is characteristic of Alloy 42 leadframes (42% Ni and 58% Fe), which is anodic in nature (easily corroded when wet). When the Alloy 42 leadframes are bundled together prior to plating, care is taken not to have compound and the leadframe in contact because such packing would result in an imprint of the leadframe fingers, with appearance ranging from a light gray imprint to the extreme rust color on the package body (molding compound).

In an assembly plant, one may observe multicolor rust and various shades of blue (in some assembly sites, a term referred to as "rainbow" color) on the plated terminations, plastic body of a device, or at the interface between the plastic compound and the terminations. Such rusty stains were seen occasionally after an autoclave test or at end of the line. The stains were attributed to:

- Air (with the active corrosion ingredient oxygen) not fully pumped out (due to a faulty valve in the pressure cooker test chamber)
- Chemical residues remained on the plated terminations.
- The leadframes were not annealed properly during the manufacturing of the leadframes. Stressed leadframes are more prone to corrosion, compared with annealed leadframes.

The rainbow stains or rust is a mixture of nickel and iron, based on the energy-dispersive x-ray analyses done on the stain area. The corrosion products are a mixture of nickel hydroxide and iron hydroxide.

Palladium-coated Alloy 42 leadframes are seen with rainbow-colored stain on the body and the leadframes after a 96 h pressure cooker (autoclave) test. Palladium plated on Alloy 42 is an example of a dissimilar metal pair coupled with stress generated by the trim-form process. Corrosion of the alloy occurs, giving rise to a mixture of iron and nickel oxides. Hence, it is not recommended to have the palladium on Alloy 42 for this reason. Such corrosion is not seen on the palladium-plated copper leadframes because the electrode potentials of copper and palladium are both on the noble end of the electromotive force (emf) series. The addition of another noble layer (e.g., gold) may remedy the

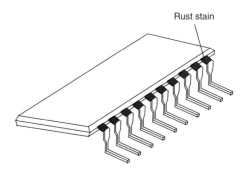

Fig. 6 Tarnished area of interface between the molding compound and the leadframe of a partially drawn thin-small-outline package. The color of the stain is a typical rust color (for Alloy 42 leadframe) or green color (for copper leadframe).

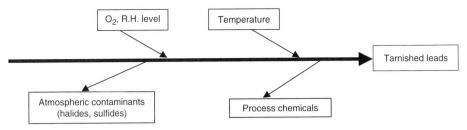

Fig. 7 A "fish-bone" diagram showing the causes of tarnishing of leads.

corrosion seen at the palladium-coated Alloy 42 leadframes.

Rust Stain Seen at the Interface between Leadframe and Compound. An example of corroded inner leads (Alloy 42) is illustrated in Fig. 1 (see point 2). This is due to poor adhesion between the compound and the leadframe. The chemicals (acids or alkalis) can diffuse into the package and react with the Alloy 42. The corroded products, oxides of iron and nickel, are illustrated in Fig. 6 where the stain mark is seen.

Controlling Tarnished Leads at the Assembly

Is it possible to have zero tarnish at the end of the line? Ideally, one should aim for "zero defect" tarnished leads, but practically it is difficult to achieve this without proper planning of the plating process and post-plating handling such as burn-in, oxide wash of the leads, and the washing of the burn-in board. The essential process is summarized by the "fish-bone" diagram in Fig. 7. One common factor overlooked is the oven: air leakage is often observed as doors are not properly sealed (sealant wear off) and at outlets for piping connections. In tropical countries the air is moist with a certain high relative humidity value. If not properly controlled, tarnishing of the leads may result. Generally, the most serious cause of corrosion of the terminations seen in this industry is the plating (or flux) residues, acids, and alkalis. With time and temperature, severely tarnished leads can be observed. Generally, SIC devices are stored in a warehouse where the air humidity should be controlled at about 55 to 60% relative humidity. If there is a chemical spillage (e.g., acid spillage, acid rains, or release of hydrogen sulfide due to some industrial accident), actions need to be taken to protect the SIC devices by relocation or proper sealing of the windows and rooms.

REFERENCES

1. A.C. Tan, *Lead Finishing in Semiconductor Devices: Soldering,* World Scientific, 1990
2. U.R. Evans, *The Corrosion of Metals,* Edward Arnold, 1981, London
3. G. Wranglen, *An Introduction to Corrosion and Protection of Metals,* Chapman and Hall, 1985
4. J.N. West, *Basic Corrosion and Oxidation,* John Wiley & Sons, 1980
5. A.C. Tan, Chapter 7 (Corrosion), *Tin and Solder Plating in the Semiconductor Industry, a Technical Guide,* Chapman and Hall, 1993
6. R.B. Camizzoli, *Materials Developments in Microelectronics Packaging Conference Proceedings,* P.J. Singh, Ed., ASM International, 1991, p 311
7. J.W. Osenbach, *Semicond. Sci. Technol.,* Vol 11, 1996, p 155
8. J.J. Steppan, J.A. Roth, L.C. Hall, D.A. Jeanotte, and S.P. Carbone, *J. Electrochem. Soc.* Vol 134, 1987, p 175
9. W.M. Paulson and R.P. Lorigan, *Reliability Physics Symposium Proceedings,* Vol 14, 1976, p 42–47
10. S.C. Kolesar, *Reliability Physics Symposium Proceedings,* 1974, p 155
11. A.J. Bevelo, J.D. Verhoeven, M. Noack, and A. Leels, Auger Study of the Oxidation of Liquid and Solid Tin, *Surf. Sci.,* Vol 134, 1983, p 499
12. R.A. Konetzki, T.A. Chang, and V.C. Marcottee, Oxidation Kinetics of Pb-Sn Alloys, *J. Mater. Res.,* Vol 4, 1989, p 1421
13. R.P. Frankenthal and D.J. Sincondolfi, AES Study of Tin Lead Alloys: Effects of Ion Sputtering and Oxidation on the Surface Composition and Structure, *J. Vac. Sci. Technol.,* Vol 17, 1980, p 1315
14. R.P. Frankenthal and D.J. Sincondolfi, *Surf. Sci.,* Vol 104, 1981, p 205

Corrosion in Passive Electrical Components

Stan Silvus, Southwest Research Institute

THE BROAD SPECTRUM OF ELECTRICAL AND ELECTRONIC COMPONENTS can be divided into two classes: semiconductor devices (i.e., active devices) and passive components. This article concentrates on passive components such as resistors, capacitors, wound components, sensors and transducers, relays and switches, connectors, printed circuit boards, and hardware. Various types of corrosion-related failure mechanisms and their effects on electronic and electrical components are illustrated by actual field-failure examples.

Corrosion and related phenomena cause a variety of failures in electronic and electrical passive components. Although the most pervasive factor in corrosion-related failures is solder-flux residue, environmental factors play important roles also. Thus, for long-term reliability, electronic and electrical components should be kept clean (i.e., free of ionic contaminants), and they should be protected from environmental contaminants that cause corrosion or tarnish. Some corrosion mechanisms are either driven by or accelerated by electrical potential differences, but in such cases, it is actually the electric-field strength (i.e., the ratio of potential difference to physical separation) that is the ultimate driving force; accordingly, during circuit-board design, spacing between adjacent electrical conductors must be given due consideration.

Halide-Induced Corrosion

Many types of solder fluxes contain halides. Frequently, the halide constituent is the chloride ion, but in some solder-flux formulations, bromide and fluoride ions may be present alone or in combination with chloride ions. Of course, solder flux is a necessity in assembly of electronic components into systems, so the effects of halides from this source are constant considerations. Moreover, the service environment to which electrical and electronic components are exposed may contain halides. Near a seacoast, for example, the air may contain salt, or a process fluid to which a component is exposed may contain a halide.

Wound components (e.g., relay and solenoid-valve coils, transformers, inductors, motors, etc.) contain copper wire that is susceptible to halide-induced corrosion (Ref 1). Usually, in such devices there is also an electrical potential that accelerates corrosion, but the potential is not a requirement for halide-induced corrosion of copper wire. One of the worst cases for electrical-potential acceleration of copper corrosion is the use of a wound component in a low-side-switched circuit (Fig. 1). In this frequently used circuit, one end of the coil is permanently connected to the positive side of a direct-current (dc) power supply, and the opposite end of the coil is switched to ground whenever it is desired to actuate the device. Because of the permanent connection to the power source, a low current may pass continuously through leakage-resistance paths from the copper wire to the grounded magnetic-core structure (Fig. 2), and this current causes corrosion of the copper magnet wire and attendant removal of copper ions from the wire; the ultimate result is an open circuit in the coil.

Necessary conditions for halide-induced corrosion of the copper magnet wire in a coil include (a) a pinhole or crack in the thin polymeric insulation on the wire, (b) a halide-containing material, and (c) moisture. Typical specifications for polymer-insulated magnet wire allow as many as 15 pinholes per 100 ft of wire, so there are plenty of pinholes to satisfy one of the conditions. Because relative humidity greater than about 45% provides enough moisture to support corrosion, sufficient moisture is present except in unusually dry environments.

The most common source of halide contamination in coils is solder-flux residue. Typically, ends of the fine magnet wire in a coil are soldered to larger wires that form the coil leads. After soldering has been completed, residual flux is usually removed by brushing or spraying the joint with alcohol or other suitable solvent. In many cases the coil is not protected from flux-containing solvent drops or overspray during the cleaning operation. When no protection is provided, flux-containing liquid tends

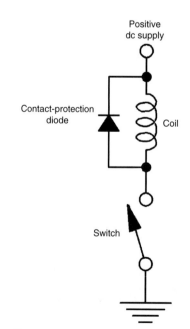

Fig. 1 Coil in low-side-switched circuit

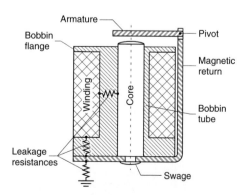

Fig. 2 Electrical-leakage paths in a typical relay structure (not to scale; proportions distorted for clarity)

to wick into the coil winding. As suggested in Fig. 2, the most critical locations are near grounded metallic structures, that is, inside surfaces of the coil-bobbin flanges and the bottom layer of the winding. If an insulation pinhole is present at one of these critical locations, then corrosion of the wire is likely to occur. Additionally, if an electrical potential difference exists (e.g., as in a low-side-switched circuit), then the rate of corrosion increases. Figure 3 shows a typical example of corrosion in the bottom layer of a low-side-switched relay coil. The magnet wire had been severed by corrosion at the indicated site, and corrosion products at the site contained chloride ions. The source of chloride ions was the solder flux that had been used in terminating the coil ends.

Two examples of chloride-induced corrosion of very fine magnet wire in the coil of a small magnetic sensor are shown in Fig. 4 and 5. These photographs illustrate the two common locations for coil corrosion: wires in the bottom layer of the coil (Fig. 4) and wires that touch a bobbin flange (Fig. 5). The source of the chloride ion in this case was the flux that was used when the sensor terminals were soldered to a printed circuit board. The circuit board had been cleaned after the soldering operation, but the sensor package had relatively large openings around its terminals, and flux-containing cleaning solvent entered through these openings and wicked into the coil.

Ends of corroded wires typically have irregular shapes as illustrated by the examples shown in Fig. 6 through 8. Dangling insulation fragments (Fig. 6), hollow ends leaving insulation sheaths (Fig. 7), and very unusual shapes (Fig. 8) are commonly observed.

When coil failure is suspected, the first electrical check should be made by connecting a high-input-resistance (1 to 10 GΩ) dc voltmeter across the coil terminals. If a voltage on the order of a few millivolts or higher is present, then it is very likely that corrosion was the cause of failure. After the terminal-voltage measurement has been made, it is permissible to check coil resistance with an ohmmeter to confirm that the coil is an open circuit.

Preventive measures include: (a) careful cleaning of solder joints with protection of the coil from flux-laden solvent drops and overspray, (b) protecting the coil from human spittle, which contains chloride ions, and (c) sealing the coil against entry of flux and other contaminants during subsequent soldering and cleaning operations and field service. Additionally, in low-side-switched circuits, use of a negative power supply can eliminate acceleration of corrosion; however, chemical corrosion may still occur if the coil is contaminated, so cleanliness remains the key to protecting coils from corrosion.

Sometimes, corrosion failures in coils are initiated by other failures. Figure 9 is an as-received view of the inside of an encapsulated coil; a package crack caused by elevated-temperature aging of the bobbin material is pointed out. Voltage measurement suggested that the coil had corroded internally, and a subsequent resistance measurement verified that the coil was an open circuit. Because corrosion was suspected, the bobbin material was removed in the vicinity of the crack to facilitate inspection of the magnet wire. A corrosion site was found against the inside of the bobbin flange (Fig. 10, 11). Microprobing of the wire showed that the site pointed out in Fig. 11 was the only break in the coil. Energy-dispersive x-ray spectroscopy (EDS) of the corrosion product detected chlorine. The failed coil had been in service near a seacoast, and it appeared that the detected chloride ions and the moisture required to support corrosion were airborne. Location of the corrosion site suggested that contaminants entered the coil through the crack in the bobbin.

Although this article deals primarily with passive devices, there are some exceptions to the division between nonsemiconductor devices and semiconductor devices. An example of a fringe-area device is a pressure transducer that

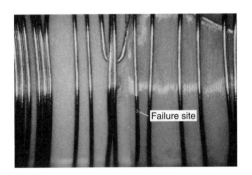

Fig. 3 Corrosion failure site in bottom layer of relay-coil winding

Fig. 4 Corrosion failure site in bottom layer of coil in magnetic sensor

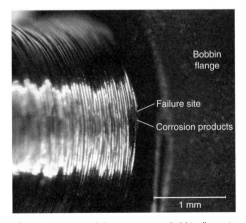

Fig. 5 Corrosion failure site against bobbin flange in magnetic-sensor coil

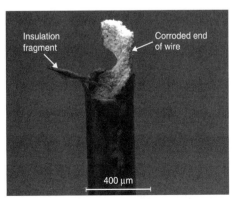

Fig. 6 Scanning electron micrograph showing end of corroded copper magnet wire

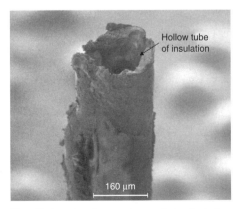

Fig. 7 Scanning electron micrograph showing hollow insulation sheath at end of corroded copper magnet wire

incorporates a silicon strain-gage bridge. There are four diffused resistors on a silicon strain-gage die. Two of these resistors are in a high-stress region of a thin silicon diaphragm, and the other two are in a low-stress region. Resistances of the highly stressed resistors change when pressure is applied to the thin diaphragm, but resistances of resistors in the low-stress region change very little. The function of the latter two resistors is to provide temperature compensation of the bridge. The aluminum bond pads and interconnecting metallization on a silicon strain-gage die removed from a failed pressure transducer were corroded (Fig. 12). Corrosion products, shown in detail in Fig. 13, contained bromine. Further investigation revealed that the silicone-rubber seal between the process fluid and the silicon die had a leak, and this leak permitted the process fluid, which contained a water-soluble bromide compound, to contact the aluminum metallization and cause corrosion.

The connector shown in Fig. 14 had tin-plated carbon steel pins. It is evident that the pins were rusty and that some of the corrosion products had made their way to the interior walls of the connector shell. An EDS analysis of the corrosion products detected iron, chlorine, and oxygen, suggesting that both iron oxide and iron chloride were present. Information provided by the user of the connector revealed that the component had been used in a sheltered outdoor location near a seacoast; hence, it appeared that the corrosion-causing chloride ions were airborne.

Organic-Acid-Induced Corrosion

Some solder fluxes contain organic acids, and residues of this type of solder flux are probably the most common sources of organic-acid contamination in electronic components. However, other materials contain organic acids or related substances that sometimes cause corrosion in electronic components.

The inside of a good aluminum electrolytic capacitor is shown in Fig. 15. A roll, comprising two specially treated aluminum-foil plates and two interleaved electrolyte-saturated paper separators, is housed in a drawn-aluminum can that has a plastic cover sealed into one end. Two aluminum terminals penetrate the cover to provide connections to external circuitry, and the plate foils are connected to the inside ends of the terminals by aluminum ribbons or tabs that are spot welded or crimped at their ends. Because the capacitor contains an electrolyte, all of the metals inside the device are aluminum so that there is no support for galvanic corrosion. Despite this single-metal construction, the capacitor shown in Fig. 16 failed because its positive tab or ribbon had corroded. The negative tab was undamaged, suggesting that the dc potential impressed across the capacitor during normal service played a key role in the corrosion process.

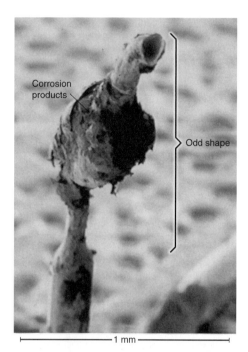

Fig. 8 Scanning electron micrograph showing unusually shaped end of corroded copper magnet wire

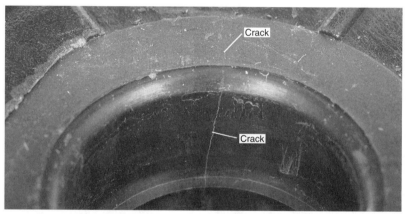

Fig. 9 Inside of failed coil bobbin showing crack (millimeter scale)

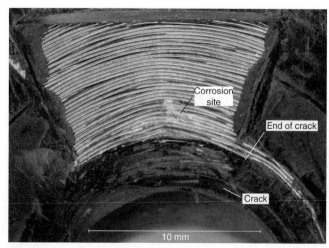

Fig. 10 Overall view of corrosion failure site in failed coil

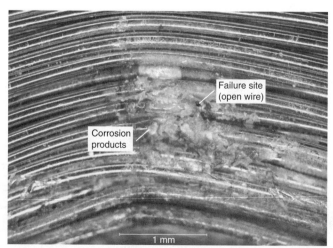

Fig. 11 Detail view of corrosion failure site in failed coil

Considering that the electrolyte does not attack aluminum, it was apparent that a foreign substance was involved in the failure shown in Fig. 16. Chlorine-containing solvents are notorious for penetrating the seals of aluminum electrolytic capacitors. When such materials get inside the capacitor, they decompose, yielding chloride ions that attack the aluminum. However, in the failed capacitor, no chlorine was detected in the corrosion products. Normally, the insides of aluminum electrolytic capacitors have an aminelike (i.e., fishy) odor, but the failed capacitor had a very strong phenolic odor. It was determined that the cover was made of phenolic plastic, and the particular variety of plastic used in the cover had a phenolic constituent that was leached by the electrolyte (Ref 2). The phenolic leachant was the foreign material that facilitated electrochemical corrosion of the positive tab. Use of a different cover material was the preventive measure employed in subsequently produced capacitors.

Capacitors intended for continuous across-the-power-line service in alternating-current (ac) circuits are frequently of the metallized-polypropylene-film type. In addition to having very low power loss, these capacitors have a desirable self-healing capability if a small short circuit (e.g., at a pinhole or inclusion in the polypropylene film) occurs between opposite-polarity plates. A self-healing event, also called a clearing event, causes the thin deposited-metal coating on the polypropylene film to evaporate in a small area around the defect, thereby effectively isolating the short circuit from the rest of the capacitor. A typical cleared site is small and approximately circular. A few cleared sites are expected in a capacitor that has been in service.

A metallized-polypropylene-film capacitor is made of two long, narrow strips of thin polypropylene film, each of which is coated on one side with a thin layer of metal (the plate or electrode). A margin along one longitudinal edge of each strip of film is left without metallization. Two such strips, aligned longitudinally and stacked, are wound tightly into a roll with the metallized side of one strip of film in contact with the nonmetallized side of the second strip and with the unmetallized margins on opposing edges of the roll. When the film is unrolled, as in a failure analysis, the two layers of film usually do not separate; that is, they remain stacked. An additional operation is required to separate the two layers (Fig. 17).

Because the very thin deposited-metal plates are translucent, transmitted-light examination of the two stacked film layers readily reveals thin spots in the metallization. Such thin spots may include normal clearing sites and areas in which metallization has been destroyed by corrosion or another mechanism. When the two stacked layers of metallized film are backlit, areas in which the metallization is intact on both layers appear dark. However, areas in which the metallization is missing from only one layer have intermediate brightness, and areas in which the metallization is missing from both film layers are bright.

Two backlit stacked layers of metallized polypropylene film removed from a good

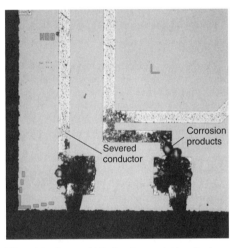

Fig. 12 Corner of failed strain-gage-bridge die showing corroded metallization

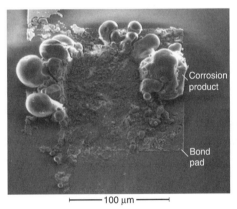

Fig. 13 Scanning electron micrograph of corrosion products around bond pad on failed strain-gage-bridge die

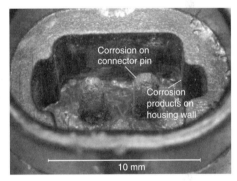

Fig. 14 Interior of failed electrical connector showing corroded pins

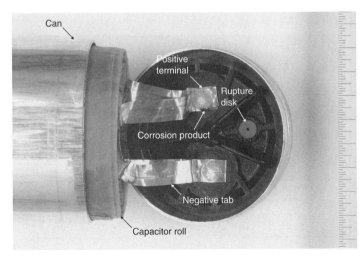

Fig. 15 Interior of good aluminum electrolytic capacitor (millimeter scale)

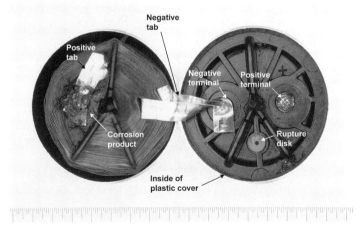

Fig. 16 Interior of failed aluminum electrolytic capacitor showing effects of corrosion (millimeter scale)

capacitor are shown in Fig. 18. Only a few small areas of missing metallization are apparent. In contrast, the backlit film layers shown in Fig. 19 were removed from a failed capacitor that had only two-thirds of its original capacitance. In the darkest areas in Fig. 19, there was metallization on both layers of film; however, in the gray areas, there was metallization on only one of the film layers, and in the brightest areas, the metallization was missing from both layers. Large sizes and irregular shapes of the missing metallization areas and absence of film defects in these areas indicated that clearing was not the failure mechanism; corrosion of the zinc-aluminum metallization had occurred instead. Elemental analysis of areas in which the metallization was missing did not detect halides, but did detect constituents of organic acids. Moreover, there was evidence that the spot-welded wire-lead attachments to the terminations at the ends of the metallized-film roll had been manually reworked with a soldering iron, an operation that required use of flux. Although an attempt had been made to clean the solder joints, some solder-flux residue remained. Elemental analysis indicated that the flux residue was of the organic-acid type. Leaving potentially corrosive solder-flux residue on the ends of the roll was inadvisable, but cleaning the joints with a solvent and allowing the flux-laden solvent to wick into the roll greatly accelerated the significant metallization damage that subsequently occurred.

Electrochemical Metal Migration (Dendrite Growth)

To some extent, all metals are susceptible to electrochemical metal migration. Conditions necessary for metal migration include (a) a migration-susceptible metal, (b) an ionic contaminant, (c) an electrical potential difference, and (d) moisture. Among the metals commonly used in electronic assemblies, silver, lead, tin, and copper are particularly susceptible to migration, with silver being the most susceptible (Ref 3). The most common ionic contaminant is solder-flux residue, either the halide-containing type or the organic-acid type. A potential difference is required for electronic circuits to function, so one of the conditions is always met in operating circuit boards; however, under some conditions, galvanic potentials developed between adjacent dissimilar metals may provide sufficient driving force for metal migration to occur on a nonoperating circuit board. Ditz (Ref 4) has shown that a relative humidity higher than about 43% provides enough moisture to support silver migration, and it is likely that only slightly higher humidity levels will support migration of the other metals widely used in electronic assemblies.

In the past, printed circuit boards assembled with halide-containing fluxes were cleaned after soldering had been completed. The cleaning process may have included washing the board in a suitable solvent or in deionized water, perhaps with a nonionic detergent added. Recently, however, many printed circuit board assemblers are using "no-clean" solder fluxes that are based on organic acids. The idea behind "no-clean" fluxes is that during the high-temperature reflow step of the soldering process, the corrosive constituents are either evaporated, decomposed into benign substances that may be left in place, or encapsulated in a hard shell that effectively passivates them. In the latter case, the hard shell should never by damaged; if it is damaged, the corrosive substances will be released, and corrosion may follow. Although the "no-clean" process normally goes as intended, sometimes

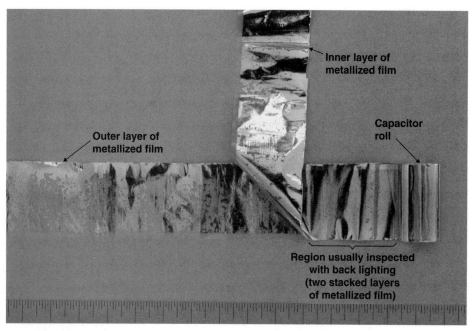

Fig. 17 Partially unrolled metallized-film capacitor showing individual film layers and stacked film layers (millimeter scale)

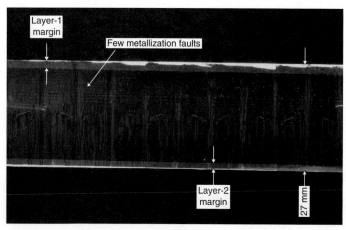

Fig. 18 Backlit view of two stacked film layers removed from a good metallized-polypropylene-film capacitor

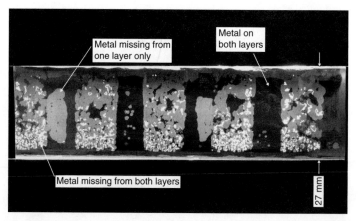

Fig. 19 Backlit view of two stacked film layers removed from a failed metallized-polypropylene-film capacitor

solder flux wicks into the small clearances under surface-mounted components where some of the flux may be protected from heat. In this case, evaporation, decomposition, and encapsulation of the acid constituents do not necessarily occur, and the result is that a potentially corrosive material may be left in critical locations on the circuit board.

Many surface-mounted electronic components have a layer of silver in their solderable terminations. An electrolytic contaminant, such as solder-flux residue, an electrical potential difference, and moisture combine to support electrochemical migration of silver, a metal highly susceptible to this phenomenon. The resulting dendritic growth usually leads to a short circuit in parallel with the component. Figure 20 shows the back side of a surface-mounted chip resistor that failed in this manner.

As mentioned earlier, silver is not the only electromigration-susceptible metal widely used in the electronics industry. Figure 21 is a scanning electron micrograph of the area between two adjacent solder joints on the ceramic substrate of a thick-film-hybrid resistor network. The dendrites pointed out in this micrograph are lead (Pb) that was leached from the adjacent solder joint. An EDS analysis of the substrate in the vicinities of the dendrites detected the electrolyte-forming elements sodium and chlorine. Interestingly, the dendrites did not complete a metallic short circuit between adjacent solder joints, but instead shortened the insulating path length enough to produce an unacceptably low resistance through the sodium-and-chlorine-based electrolyte film that formed on the ceramic-substrate surface.

The failing resistor-network substrate had been encapsulated in a transfer-molded plastic that did not adhere well to the metal terminals of the device. Lack of adhesion caused thin gaps at the terminal entries (Fig. 22). Moreover, the resistor network was mounted on the wave-soldered side of the printed circuit board and, hence, was immersed in chlorine-containing flux just before exposure to the thermal shock of the high-temperature solder wave. Following the soldering operation, the circuit board was cleaned in softened (i.e., sodium-containing) water. The result of this process was that chlorine-containing solder flux and sodium-containing water wicked into the areas between solder joints by way of the terminal-entry gaps. The resulting electrolyte then supported electrochemical migration of the lead constituent of the solder.

Figure 23 is a detail view of the area between two adjacent solder joints on a printed circuit board. Tin dendrites growing between the solder joints are clearly visible. The circuit board had been exposed to a chlorine-containing flux during the soldering operation and had not been cleaned properly. Subsequently, the board was conformally coated with a clear polymeric material. Many conformal-coating materials do not adhere well to solder-flux residue, and they eventually delaminate from the circuit-board surface forming thin pockets in which moisture accumulates. The environment in these pockets becomes ideal for metal migration, and dendrites, such as those shown in Fig. 23, grow. The eventual result is a short circuit between the adjacent solder joints.

Less organized metal migration is shown in Fig. 24. In this case, the migrating metal was

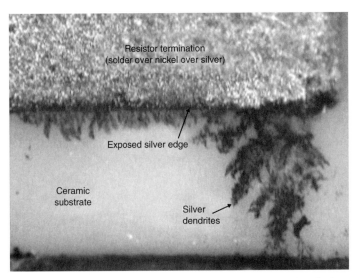

Fig. 20 Back side of failed surface-mounted chip resistor showing silver dendrites. Courtesy of Pat Kader, ENI

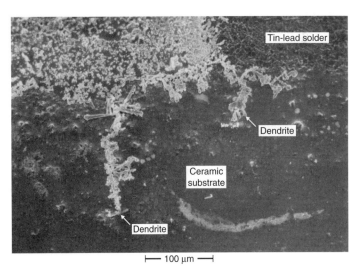

Fig. 21 Scanning electron micrograph of lead dendrites on substrate of failed thick-film-hybrid resistor network

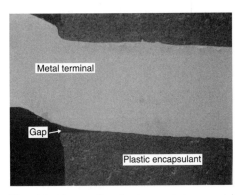

Fig. 22 Cross section showing gap between metal terminal and encapsulant at terminal entry of thick-film-hybrid resistor network

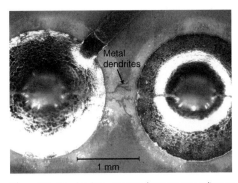

Fig. 23 Tin dendrites growing between two adjacent solder joints on failed printed circuit board

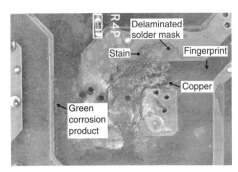

Fig. 24 Copper migration and corrosion on surface of failed printed circuit board

copper, and the ionic contamination was from the organic-acid solder flux used in circuit-board assembly. There was plenty of moisture present because failure occurred during a temperature-humidity-bias (THB) test in which the relative humidity was maintained at 85%. Whether the fingerprint pointed out in Fig. 24 was present during the test was unknown. In addition to metal migration, there was a small amount of copper corrosion.

Three of the four conditions for metal migration are almost always present on electronic circuit boards, leaving ionic contamination as the only condition that may be practically controlled. For best reliability, circuit boards should be thoroughly cleaned after soldering, and cleaning is essential if boards are to be conformally coated. "No-clean" solder fluxes work satisfactorily in relatively benign operating environments, but in critical applications in which long-term reliability under adverse conditions is required, other types of flux should be used, and the circuit boards should be cleaned after soldering has been completed. In cleaning, particular attention should be given to removal of solder flux trapped under components, particularly surface-mounted devices.

Silver Tarnish

Silver is frequently used in electronic devices because of its high electrical and thermal conductivities. However, silver is highly susceptible to tarnishing in the presence of sulfur-containing materials such as airborne hydrogen sulfide and sulfur dioxide. The tarnish product, silver sulfide, has poor electrical conductivity. As the tarnishing process progresses, electrical resistance of the conductor or contact surface increases until an open-circuit failure occurs. The only practical preventive measures are to avoid use of silver in sulfur-containing environments or hermetically seal silver in an inert atmosphere.

A common source of sulfur-containing gases is vulcanized rubber, which is sometimes used for insulating electrical wires. Other sources of sulfur-containing gases include brown paper bags and cardboard boxes. If rubber objects are enclosed in the same housing with silver-containing electronic components, failure is almost inevitable. A typical example (Ref 5) is illustrated in Fig. 25, which shows a corner of a surface-mounted chip resistor that has a layer of silver in each of its terminations. The silver layer had been consumed by tarnish, and the resistor had become an open circuit. The circuit board on which this resistor and others like it were mounted had been enclosed in a poorly ventilated housing along with a vulcanized-rubber conduit that contained some of the system interconnection wiring. A sulfur-containing gas evolved from the rubber and reacted with the silver terminations of the resistor to yield silver sulfide. A scanning electron micrograph of a cross section through the failed resistor termination (Fig. 26) shows the silver-depleted region that broke the circuit between the resistive element and the termination.

A similar failure occurred in a trimming potentiometer that was installed in a control system located in a tire-manufacturing plant where substantial concentrations of sulfur-containing gases were present. The effect of these gases was to tarnish the silver terminations at the ends of the potentiometer resistance element (Fig. 27). The silver was consumed with the result that the potentiometer became an open circuit. Figure 28 is a scanning electron micrograph showing a detail view of one of the tarnished terminations; the grasslike appearance is typical of tarnished thin films of silver.

Many relay and switch contacts are made of silver or high-silver-content alloys. In environments that are contaminated with sulfur-containing gases, surfaces of such contacts tarnish, and the contacts become incapable of completing an electrical circuit when called upon to do so. Figure 29 shows a typical heavily tarnished silver alloy relay contact. Even though the mechanism in this relay provided desirable wiping action, as evidenced by the elongated

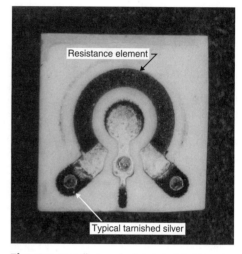

Fig. 27 Overall view of substrate in failed trimming potentiometer showing tarnished silver terminations

Fig. 25 Corner of failed surface-mounted thick-film chip resistor showing tarnished silver. Courtesy of Steve Axtell, Vishay

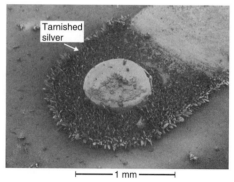

Fig. 28 Scanning electron micrograph of tarnished silver termination on substrate in failed trimming potentiometer

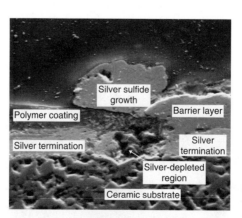

Fig. 26 Scanning electron micrograph of cross section through termination region of failed surface-mounted thick-film chip resistor. Courtesy of Steve Axtell, Vishay

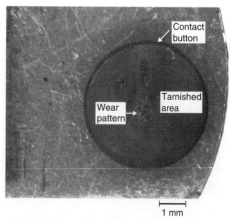

Fig. 29 Heavily tarnished silver alloy relay contact

wear pattern, the wiping action was not sufficiently aggressive to break through the tarnish film that had built up over the long periods of time during which the contact had been held in the open state. Failure of this contact was caused by sulfur-containing gases that evolved from the rubber insulation on wires that were housed in the same enclosure.

Fretting

The rotor of a motor-operated switch was bound so tightly that the mechanism could no longer rotate. Inspection of the drive mechanism showed that grease in the sleeve bearings had become hard (Fig. 30), and subsequent EDS analysis detected the sleeve-material elements in the hardened grease. Disassembling the mechanism revealed additional hardened grease on the corresponding journal (Fig. 31). Cleaning the sleeve and journal, applying fresh grease, and reassembling the device restored normal operation.

The motor in the switch was operated from a 60 Hz ac source. As a result, low-level 120 Hz vibration was present while the coil was energized continuously for months at a time. The almost continuous vibration caused fretting of the sleeve bearing, and the resulting fine metallic particles led to hardening of the grease and consequent binding of the mechanism. Using dc excitation would have eliminated the vibration and minimized the possibility of fretting of the sleeve-bearing surface.

Metal Whiskers

Of the metals frequently used in electronic devices, tin is the most likely to form whiskers. Such whiskers may be several millimeters long, and, of course, they are electrically conductive. Examples of tin whiskers projecting from the tin-plated base of an insulated standoff terminal are shown in Fig. 32. One of the whiskers on this terminal contacted an adjacent resistor lead, thereby causing a short circuit to ground. In this situation, there was not enough energy available to evaporate the whisker, so circuit malfunction ensued. However, in many circuit applications, there is enough energy available to evaporate a small-diameter metal whisker, and in such cases, whiskers usually are nuisances rather than causes of hard failures. Tin whiskers are of great concern in outer-space applications in which evaporation of a metal whisker may trigger catastrophic arcing.

It is generally thought that metal whiskers form as the result of residual stresses in plated layers, but the underlying base material also has an influence. One often suggested preventive measure is to anneal or reflow the plating. Brusse et al. (Ref 6) have published an extensive treatise and bibliography on metal whiskers.

There is a modern environmentally driven trend toward lead-free electronic components and equipment. The lead in conventional tin-lead solder alloys appears to suppress tin-whisker formation, so eliminating lead from solder takes away this desirable feature. In this regard, the National Aeronautics and Space Administration (NASA) prohibits use of pure tin and requires at least 3 wt% Pb in tin-lead alloys to suppress tin-whisker formation (Ref 7). There is evidence that other alloying elements (e.g., silver, indium, antimony, and bismuth) suppress tin-whisker growth, but the referenced NASA document does not mention these as acceptable alternatives.

Most of the proposed lead-free solder formulations have high tin content, and the solderable surfaces of many lead-free components are plated with pure tin. Thus, tin-whisker formation is resurfacing as a potential reliability concern. The IPC (formerly the Institute for Printed Circuits) Solder Products Value Council has recently published a report (Ref 8) that recommends the 96.5Sn-3.0Ag-0.5Cu composition (designated SAC305) as the alloy of choice for lead-free applications. Supporting this recommendation was an extensive round-robin evaluation that included determination of assembly-line performance characteristics, metallurgical cross sectioning, metallographic analysis, thermal cycling, thermal shock, and other tests. Apparently, however, the potential for tin-whisker growth was not assessed. At this writing, experimental lead-free solders have not been in field service long enough for accumulation of reliable statistical data on tin-whisker formation, so only time will tell whether the proposed lead-free solder alloys are really trouble free.

ACKNOWLEDGMENT

The author thanks Steve Axtell and Pat Kader, who supplied some of the key illustrations used in this article, and Ken Cook who took the remainder of the photographs. The author also thanks the peer reviewers for their many helpful suggestions.

REFERENCES

1. S. Silvus, Failure Analysis of Relay and Solenoid-Valve Coils, *Electron. Dev. Fail. An. News,* May 2001, p 4–8
2. R. Alwait and Y. Liu, Electrolytes for High Voltage Aluminum Electrolytic Capacitors, *Proc. Capacitor and Resistor Technology Symposium,* Components Technology Institute, 1996, p 32–38
3. S. Silvus, The Wonders and Wanderings of Silver, *Electron. Dev. Fail. An.,* Nov 2001, p 37–39
4. M. Ditz, Determination of the Moisture Threshold for Silver Migration, *Proc. International Symposium for Testing and Failure Analysis,* ASM International, Nov 1994, p 181–187
5. S. Axtell, Failure Analysis of Thick-Film Resistors in Sulfur-Containing Environments, *Microelectronic Failure Analysis Desk Reference, 2002 Supplement,* T. Kane, Ed., Electronic Device Failure Analysis Society, affiliate of ASM International, Nov 2002, p 161–173
6. J. Brusse, G. Ewell, and J. Siplon, Tin Whiskers: Attributes and Mitigation, *Proc. Capacitor and Resistor Technology Symposium,* March 2002. Components Technology Institute, p 67–80
7. M.J. Sampson, "Tin Whiskers," Parts Advisory NA-044, NASA Goddard Space Flight Center, Oct 1998

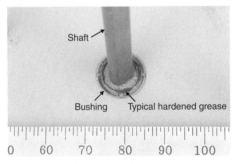

Fig. 30 Exterior view of shaft and sleeve bearing on failed motor-operated switch (millimeter scale)

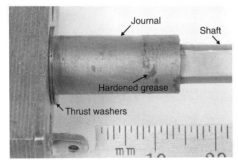

Fig. 31 Hardened grease on rotor journal of failed motor-operated switch (millimeter scale)

Fig. 32 Scanning electron micrograph of tin whiskers on base of insulated standoff terminal

8. "Round Robin Testing and Analysis of Lead Free Solder Pastes with Alloys of Tin, Silver and Copper—Final Report," SPVC2005-CD, IPC Solder Products Value Council, IPC Association Connecting Electronic Industries, Bannockburn, IL, 2005

SELECTED REFERENCES

- S. Chang, Silver Migration in Lead Borosilicate Glass on Soda Lime Glass Substrate, *Proc. International Symposium for Testing and Failure Analysis—1980* (Los Angeles, CA), ASM International, Oct 1980, p 153–157
- J. Devaney, G. Hill, and R. Seippel, *Failure Analysis Mechanisms, Techniques, and Photo Atlas,* Failure Recognition and Training Services, Monrovia, CA, 1983
- S. Doss and G. Condas, Corrosion Mechanisms of Thin Film Disk Magnetic Recording Structures, *Proc. International Symposium for Testing and Failure Analysis—1986* (Los Angeles, CA), ASM International, Oct 1986, p 35–38
- M. Jacques and D. Knauss, Crystalline Growth on RF Tuning Screws, *Proc. International Symposium for Testing and Failure Analysis—1985* (Long Beach, CA), ASM International, Oct 1985, p 10–15
- M. Johnson and S. Smith, Failure Modes and Mechanisms of Non-Semiconductor Electronic Components, *Microelectronic Failure Analysis Desk Reference,* 3rd ed., T. Lee and S. Pabbisetty, Ed., ASM International, 1993, p 303–320
- P. Martin, *Electronic Failure Analysis Handbook,* McGraw-Hill, 1999
- B. Melody, The Potential for Positive Tab Corrosion in High Voltage Aluminum Electrolytic Capacitors Caused by Electrolyte Decomposition Products, *Proc. Capacitor and Resistor Technology Symposium,* Components Technology Institute, 1993, p 199–205
- M. Minges, Ed., *Electronic Materials Handbook,* Vol 1, *Packaging,* ASM International, 1989
- C. Murphy and L. Kashar, Analysis of Failures in Half-Size Crystal Can Relay Moveable Contact Arm, *Proc. International Symposium for Testing and Failure Analysis* (Los Angeles, CA), ASM International, Oct 1984, p 196–200
- S. Silvus, Failure Analysis of Passive Components, *Microelectronic Failure Analysis Desk Reference,* 4th ed., R. Ross, C. Boit, and D. Staab, Ed., ASM International, 1999, p 379–399
- S. Silvus, Failure Analysis of Passive Components, *Microelectronic Failure Analysis Desk Reference,* 5th ed., B. Holdford, Ed., Electronic Device Failure Analysis Society, affiliate of ASM International, 2004, p 219–228
- J. Siplon, G. Ewell, E. Fasco, J. Brusse, and T. Gibson, Tin Whiskers on Discrete Components: The Problem, *Proc. International Symposium for Testing and Failure Analysis* (Phoenix, AZ), ASM International, Nov 2002, p 421–434
- Y. Takeda, S. Sakamoto, S. Sumi, and K. Torazawa, Pin-Hole Failure Analysis of Plastic-Based Magneto-Optical Disk, *Proc. International Symposium for Testing and Failure Analysis* (Los Angeles, CA), ASM International, Nov 1989, p 223–229
- D. Van Westerhuyzen, P. Backes, J. Linder, S. Merrell, and R. Poeschel, Tin Whisker Induced Failure in Vacuum, *Proc. International Symposium for Testing and Failure Analysis* (Los Angeles, CA), ASM International, Oct 1992, p 407–412
- F. Wang, H. Zhang, P. Ronkainen, and P. Välimäki, Tantalum Capacitor Failure Caused by Silver Migration, *Proc. International Symposium for Testing and Failure Analysis* (Santa Clara, CA), ASM International, Nov 2001, p 451–455
- E. Williams, Tin Whiskers on Flat Pack Lead Plating between Solder Dip and Sealing Glass, *Proc. International Symposium for Testing and Failure Analysis* (Long Beach, CA), ASM International, Oct 1985, p 16–21

Corrosion and Related Phenomena in Portable Electronic Assemblies

Puligandla Viswanadham, Nokia Research Center
Sridhar Canumalla, Nokia Enterprise Systems

AS INFORMATION PROCESSING ELECTRONICS migrated to homes and outdoor environments, an entirely new set of performance and reliability expectations were imposed on the products. Portable electronics comprise a host of products such as personal digital assistants, camcorders, mobile phones, external medical electronics such as hearing aids and self-diagnostic or monitoring tools, and human-portable military hardware. These products are low-cost, high-volume devices with short product development cycle as well as product life. The use profile is considerably diverse and variable. Not only do these products experience wide variations in temperature and humidity, they are also prone to exposure to condensing atmospheres and a variety of hostile gaseous and particulate contamination, particularly in urban industrial environments. As a consequence, corrosion-related exposures tend to be more prevalent in consumer and portable electronics. In addition, these products are subject to a multitude of mechanical stresses such as shock, vibration, and bend. A discussion of these mechanical aspects is considered outside the scope of this article. Nevertheless, it is important to recognize that consumer expectations of portable electronic product quality and reliability remain high.

Portable communication and entertainment are becoming the growth segments in the consumer electronics industry, and these portable electronic devices are sometimes also viewed as fashion accessories with opportunities for personalization. These devices are miniaturized, lighter, and relatively affordable at the same time. Even in the military electronics domain, the human-portable military electronic hardware has to be highly flexible, robust, and usable in a variety of environmental conditions. In general, people carry these personal portable electronic devices wherever they go, and the devices are exposed to the same environments as humans. Traditional heavy, brittle, hermetic ceramic packages that are prone to failure due to mechanical shock, vibration, and drop have given way to lighter, low-profile, plastic packages, which are not hermetically sealed. Instead, the isolation from the environment arises from the use of primarily epoxy-based molding compounds, underfills, and so forth. Therefore, although the integrated circuits (ICs) in the electronic packages are relatively safe from moisture ingress, most portable electronic devices (appliances) are not very tolerant of environmental conditions such as condensed water, humidity, and temperature fluctuations. In addition, even low levels of corrosive substances such as chlorinated water can be lethal to electronic circuits. In this context, corrosion and related phenomena in portable electronic products are extremely important to the reliable operation of a portable consumer electronic product, and some of the salient aspects are discussed in the following sections.

Some of the metallic materials that are exposed to corrosion in electronic assemblies are:

- At the package level, materials such as aluminum, alloy-42, copper alloys, tin and tin-lead alloys, nickel, gold, palladium, and palladium-silver are employed in lead frame materials, pad metallurgies, lead finishes, etc.
- At the second level assembly where the packages are assembled onto the printed wiring board, the materials include copper, immersion tin, immersion silver, tin-lead solders, nickel, and gold.

Table 1 shows a list of typical metallurgies (and their nominal compositions) used in the various aspects of electronic packaging. Some of the salient corrosion aspects of these materials applicable to electronic packaging are discussed subsequently with emphasis placed on copper and copper alloys, tin and tin-lead alloys, and aluminum.

Owing to their portability, the electronic products considered in this article are more vulnerable to atmospheric corrosion than are traditional business and office machines. The effects of humidity and prolonged exposure to various atmospheric contaminants are discussed in the section "Forms of Corrosion Not Unique to Electronics." Corrosion and related phenomena such as electrochemical migration and conductive anode filament formation are also important in portable electronic assemblies. These subjects are addressed in the section "Forms of Corrosion Unique to Electronics." Examples of electrochemical migration involving copper, tin, and silver are also presented.

Forms of Corrosion Not Unique to Electronics

There are a number of basic forms of corrosion that materials experience based on their nature and the environment. The various forms of corrosion, which are described in detail in *Corrosion: Fundamentals, Testing, and Protection,* Volume 13A of the *ASM Handbook,* are:

- Uniform corrosion
- Galvanic corrosion
- Crevice corrosion
- Intergranular corrosion
- Selective leaching (dealloying)
- Environmentally induced cracking
- Fretting corrosion
- Hydrogen embrittlement
- Atmospheric corrosion

Depending on the severity, corrosion in electronic assemblies results in the following failure pathways:

- Oxidative materials degradation resulting in loss of electrical continuity
- Partial materials degradation accompanied by the formation of a conductive oxidation product, such as a salt, which could result in lower surface insulation resistance (SIR)
- Electrical shorts between adjacent conductive features
- Intermittent shorts or opens depending on the humidity levels and the ionic nature of the corrosion product

Relevant Forms of Corrosion

Of particular interest in electronic packaging are uniform corrosion, pitting corrosion, environmentally induced cracking, galvanic corrosion, and atmospheric corrosion (Ref 1). Atmospheric corrosion is discussed in the section "Condensing and Noncondensing Humidity."

Uniform corrosion is evenly distributed over the surface, and the rate of corrosion is the same over the entire surface. A measure of the severity is the corrosion product film thickness or the average penetration of the metal. A common example is aluminum bond pad corrosion on nonpassivated chip metallization. A metallization is defined as a single or multilayer film pattern of conductive material on a substrate to facilitate interconnection.

Pitting and Crevice Corrosion. Localized corrosion appears as pits or crevices in the metal. The bulk of the material remains passive but suffers localized and rapid surface degradation. In particular, chloride ions are notorious for inducing pitting corrosion. Once a pit is formed, the environmental attack is locally autocatalytic. Generally, pitting corrosion on the surface results in an array of pits with varying depths.

Environmentally induced cracking occurs under the combined influence of a corrosive environment and static or cyclic stress. Cracking under static loading conditions is known as stress-corrosion cracking (SCC) and cracking under cyclic loading conditions is known as corrosion fatigue. Residual stresses in electronic package termination leads from lead forming and bending operations were observed to cause SCC failures in the presence of moisture (Ref 2). Stress-corrosion cracking of package leads was also reported in the presence of solder flux residues (Ref 3).

Galvanic corrosion is driven by the electrode potential differences between two dissimilar metals coupled electrically. The result is an accelerated corrosive attack of the less noble material. Galvanic corrosion tends to be particularly severe if the anodic surface is small compared with that of the more noble cathode, or in cases where a more noble metal is coated onto a less noble one. For instance, when porous gold plating over a nickel substrate is exposed to a corrosive environment, the gold coating acts as a large cathode relative to the small area of exposed nickel. This sets up a galvanic cell at the exposed substrate, which experiences intense anodic dissolution. Galvanic corrosion has been observed to enhance pore corrosion when the substrate metal is less noble than the coating, and vice versa (Ref 4).

Condensing and Noncondensing Humidity

In contrast to most business and office machines, portable consumer electronic hardware is subject to a much wider range of humidity exposures owing to their portability and are therefore more prone to condensed moisture and liquid damage. The nature and amount of contamination present can influence the failure mechanisms as well. In general, dry contamination does not result in an increase in the leakage current.

While all polymeric materials are permeable to moisture and water vapor, the rate of moisture ingress depends on such factors as the nature of resin, filler material, curing agent, hydrophobic nature of filler particles, and their volume fraction, size distribution, and so forth. An important aspect in corrosion related failure is the level of extractable ionic contaminants such as chloride, bromide, and so on. Typically they are in the 5 to 20 parts per million (ppm) range. When correctly implemented, polymeric encapsulant and packaging materials have been shown to provide thousands of hours of reliable performance under biased humidity conditions. Strong interfacial adhesion, low ionic contamination in the encapsulant, and good cleanliness ensure good corrosion resistance.

Condensed moisture is likely to be acidic in nature due to the dissolution of ambient gases such as carbon dioxide, oxides of sulfur, and oxides of nitrogen. The surface morphology of the area affected will generally contain reaction products due to the corrosion and the formation of unique metal migration species.

Atmospheric Corrosion. Extended exposure of portable electronic hardware to urban atmospheres is a concern of considerable importance in terms of product performance and reliability. Reaction products of pollutant species such as sulfur dioxide (SO_2), nitrogen compounds (NO_x), hydrogen sulfide (H_2S), chlorine-containing species, mercaptans, and organic sulfur compounds can be detrimental. Some airborne nitrates tend to be hygroscopic at humidity levels above 50% relative humidity (RH) and are known to cause electrolytic SCC of nickel-brass (Ref 5). While electrical performance degradation can be caused by the electrical leakage through moisture films in the conductor interspaces, concomitant corrosion of the anodic conductor also disrupts the current flow, which can sometimes be accompanied by metal dendrite growth. The chlorine, NO_3, and SO_4 anions are often associated with corrosion susceptibility, leading to increased leakage currents causing open circuits with thin film conductors (Ref 6). Temperature, relative humidity, pollutant and particle concentration, and sample orientation are perhaps the most important aspects in the evaluation of portable electronic hardware for corrosion in an urban environment. Researchers (Ref 7) evaluated the indoor corrosion of copper, nickel, cobalt, and iron in urban atmospheric conditions. Outdoor corrosion products are generally complex mixtures of hydroxides, sulfates, carbonates, and chlorides in the case of copper metallurgy. In the case of silver metallurgy, the products are silver sulfide, hydroxide, chloride and nitrate. Nitrogen dioxide (NO_2) is a precursor to nitrous and nitric acid and

Table 1 Typical metallurgies used in electronic packaging

Category	Metallurgy	Typical composition
Wire bonding	Aluminum	Al with 1% Si
		Al with 1% Mg
	Gold	Au with 5 to 10 ppm Be
		Au with 30 to 100 ppm Cu
Die attach materials	Tin alloys	92 Sn-8Sb
	Lead alloys	95Pb-5Sn
		97.5Pb-1.5Ag-1Sn
	Gold alloys	98Au-2Si
		8Au-12Ge
		80Au-20Sn
Lead frame materials	Copper alloys	Cu-0.1Zr
		Cu-2.35Fe-0.03P-0.12Zn
		(Cu + Ag)99.8-0.11Mg-0.6P
	Alloy 42	58Fe-42Ni
	Kovar	53Fe-17Co-29Ni
Solders	Tin-lead alloys	63Sn-37Pb
		90Pb-10Sn
		95Pb-5Sn
	Lead-free alloys	Sn-Bi
		Sn-3.8Ag-0.7Cu and other variations with minor elements such as Sb, Bi
PWB surface finishes	Solders	63Sn-37Pb
	Tin	Matte Sn or immersion Sn
	Lead-free alloys	Sn-Ag-Cu
	Silver	Immersion Ag with additives
	Nickel-gold	Electroless Ni-immersion Au
Component lead terminations	Nickel-gold	Plated Au
		Electroless Ni-immersion Au
	Tin	Pure Sn plating
	Silver-palladium	Ag-Pd
	Nickel-palladium	Ni-Pd
	Nickel-palladium-gold	Ni-Pd-Au
		Ni-Pd-Ni-Ni-Pd-Au
Connector metallurgies	Copper alloys	98Cu-2Be
		Cu-(3.5–10)Sn-1P

PWB, printed wiring board

its concentration is generally high in urban atmospheres. Table 2 compares the pollutant concentrations in indoor and urban atmospheres (Ref 7).

Hostile environmental test conditions can consist of H_2S, NO_2, SO_2, S_x, and Cl_2 species in the concentrations indicated subsequently. While most component tests are done at 70 °C (160 °F) and 70% RH, machine testing is done at 25 °C (75 °F) and 70% RH with the pollutant concentrations in parts per billion (ppb) of 350 ppb of SO_2, 3 ppb of Cl_2, 500 ppb of NO_2, 0.5 ppb of S_x, and 44 ppb of H_2S (Ref 8).

Forms of Corrosion Unique to Electronics

Electronic packaging is a hierarchical structure that allows a central processing unit to communicate with memory and input/output units such as display, keyboard, and data storage devices. The packaging integrates:

- Chips into single chip modules (SCM) or multichip modules (MCM), which are sometimes known as first-level package
- Components (SCMs, MCMs, connectors, discretes such as resistors, capacitors, inductances, etc.) on a printing wiring board (PWB) that may be referred to as second-level package
- PWB assemblies, cables, power supplies, cooling systems, and peripherals into a frame or a box and termed a third-level package

The chip or integrated circuit (IC) device is connected to its carrier in one of many ways. In one scheme, the back side of the chip is bonded to the substrate with an adhesive or a solder, and the bond pads on the active side of the chip are connected to the bond pads on the carrier with a gold or aluminum wire with ultrasonic or thermocompression techniques. In another scheme, called flip chip attach, the chip is placed upside down with the active side facing the carrier and is attached to the carrier with either solder balls or gold bumps. This scheme provides the highest interconnection density for a given area and electrical performance. The entire interconnection structure, except for the termination leads, is either encapsulated with a molding compound or encased in a metal. The leads are formed in such a manner as to facilitate either insertion into the corresponding holes or surface mounted onto the corresponding foot prints.

The first-level packages are assembled onto the PWB either by insertion mounting technology with a wave solder machine or a surface mount technology with a solder reflow oven. The reader is referred to the excellent resources on this subject in the selected references at the end of this article.

In addition to the universally observed forms of corrosion that are not specific to electronics applications, there are two other corrosion-related phenomena that are found only in electronics, namely, electrochemical migration (ECM) and conductive anodic filament formation (CAF).

Electrochemical Migration (ECM)

In ECM, which is generally observed under the influence of applied direct current bias voltage, oxidation of the metal occurs, as in the case of corrosion, at the anode. The positively charged ion, under the influence of the applied potential, travels through the electrolyte medium between the conductors toward the negative electrode (i.e., at the cathode). At the cathode the positively charged ions are reduced to neutral metal atoms. As the reduction process at the cathode continues, filamentary metal dendrites form, extending back toward the source anode. The filament growth continues until the anode is depleted, or until an electrical short occurs when the dendrites reach the anode. The following conditions need to be met for ECM to occur: (a) the presence of exposed metal, (b) sufficient moisture to form an electrolyte, (c) presence of ionic contamination for a conductive path between the anode and cathode, and (d) an applied potential.

Dendritic growth can occur between any two conductors on or in a PWB. It could be on the surface or internal to the PWB. The risk or failure locations can be at any of the following locations (see Fig. 1 for an explanation of these terms):

- Trace to trace
- Trace to pad
- Pad to pad

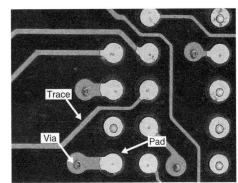

Fig. 1 Schematic of a printed wiring board explaining the different components that can be affected by electrochemical migration

- Plated through hole (via) to inner plane
- Plated through hole to plated through hole (via to via)

The amount of moisture required to facilitate ECM depends on the nature of the substrate, its ability to absorb moisture, and the surface topography and roughness. At the same RH, hydrophobic materials acquire fewer monolayers of moisture than hydrophilic materials. Hence, more moisture collects on hydrophilic materials and they tend to promote ECM. Exposed epoxy-glass is much more hydrophilic than a PWB surface covered with protective coatings. Exactly how many monolayers are needed to promote ECM is a matter of considerable debate. A few layers to several hundred layers have been proposed. While a few monolayers are adequate to promote ion transport on certain surfaces, others might require more monolayers. It is generally believed that the behavior of 200 monolayers of moisture closely matches the bulk behavior. Both adsorption, which is a surface phenomenon, and absorption, which is a bulk phenomenon, play crucial roles. Surface defects, such as cracks and crevices, modify the ECM due to capillary action.

If the PWB material under consideration is porous, moisture penetrates into the bulk and can lead to vertical migration. Nucleation sites are required for dendrites to initiate and grow. Surface roughnesses, presence of sharp corners, kinks, and so forth provide such nucleation sites.

Applied voltage, while providing the necessary potential for the anodic and cathodic reactions, also provides the driving force for the ion transport to their respective electrodes. Most consumer electronics operate in the 3 to 15 V range. Higher voltages accelerate the ion transport. Dendrite formation as a function of voltage was studied (Ref 9). At low voltages of 2 to 4 V, the time to failure versus voltage followed the log-linear form of the Butler-Volmer equation (Ref 10). At higher voltages the time to failure saturates at a specific value and becomes transport limited.

As consumer electronic products migrate toward finer pitch packages, the spacing between conductors becomes smaller, approaching 50 to 75 μm (2 to 3 mils) wide lines and spaces. Thus, the voltage gradients become steeper compared with coarse pitch circuitries. In the presence of steeper voltage gradients, ionic species migrate faster to their respective electrode, hastening the times to cause potential product failure by ECM. As the thrust toward miniaturization continues and finer pitch packages come into use, the surface cleanliness of the PWB assemblies will become an important challenge. Assemblies thus become less tolerant to even minute amounts of residues containing ionic contaminants.

Dendrite formation as a function of conductor spacing also was studied using water drop test (Ref 9). Time to failure was studied with lead spacing in the range of 62 to 325 μm (2.5 to 13 mils). Time to dendrite growth was found to increase monotonically with lead spacing. The

Table 2 Comparison of typical indoor and outdoor pollutant concentrations

Species	Indoor, μg/m³	Outdoor, μg/m³
SO_2	12.2	17.6
NO_2	26.0	43.5
NH_3	14.9	10.0
Reduced S	0.48	0.48
Cl gases	0.2	1.5
Airborne dust	16.0	61

Source: Ref 7

nature of the ionic contaminant has by far the most significant impact on dendrite growth. The severity depends on the particular ionic species involved, its mobility or transport number, ionic radius, electronegativity, electron affinity, charge to size ratio, and so forth. In PWB assemblies, the source of ionic contaminants that make up the conductive medium is generally traced to activators in the flux. These may be acid functional groups such as those found in abeitic, tartaric, succinic acids, and so forth or halides such as chlorides, bromides, and the like. Tall oil water white gum resins contain diterpenic acids, esters, and monoether alcohols. During the package-to-board interconnection reflow process, autooxidative polymerization and/or degradation, disproportionation, and/or free radical chain reactions are known to take place, giving rise to complex reaction products. Halide ions such as chloride, bromides, organic acid groups, and so on can remain entrapped on the PWB due to inadequate cleaning. Also, some of the chloro-fluorocarbon-alcohol mixtures that are used as cleaning solvents decompose to give rise to hydrogen chloride at elevated temperatures. Active metals such as zinc and magnesium are known to catalyze the decomposition reactions. Subsequently, the hydrogen chloride attacks the metals with the liberation of the corresponding salts and hydrogen.

Conductive Anodic Filaments (CAF)

Another corrosion-related aspect unique to electronics is conductive anodic filament (CAF) formation. As in the case of ECM, anodic oxidation is a precursor to this phenomenon. A current leakage is observed between conductors on the same layer of the PWB or between conductors of adjacent layers. Several factors contribute to the leakage current between conductors separated by a dielectric medium. Because the PWB is made up of epoxy resin, glass fibers, and plated and etched copper traces, the final leakage current is made up of several contributions and aided by environmental factors such as temperature, humidity, and nature and extent of contamination. The degree of cure of the polymer and the porosity of the material can have significant impact on the leakage current. As discussed earlier, ECM can occur either along the surface of a PWB between any two biased conductors, referred to as surface migration, or it can occur between conductors separated by a dielectric medium, known as vertical migration. Both phenomena are observed. The moisture trapped in the pores of the surface in the presence of ionic contamination promotes migration. In a normal FR-4 printed circuit board, several interfaces exist. There is a copper-resin interface, a glass-resin interface, and a glass-silane interface if silane-coupling agents are used as glass-epoxy adhesion promoters. These interfaces can degrade under mechanical stresses, temperature, and/or humidity. Inherent interfacial mechanical stresses may be relieved under temperature and humidity conditions resulting in interfacial debonding. Interfacial degradation provides opportunity for moisture entrapment and formation of electrochemical cells.

The anodic oxidative dissolution and the cathodic reduction reactions also induce an acidity or alkalinity gradient in the migration path. In other words, the pH (a measure of the hydrogen ion concentration) continually changes along the path. At the anode, an oxidation reaction takes place. In cases where sufficient moisture is present and is trapped along the defect interfaces, electrolysis of water occurs at the anode, generating hydrogen ions and oxygen evolution according to the equation:

$$2H_2O = 4H^+ + O_2(g) + 4e^- \quad (Eq\ 1)$$

A reduction reaction occurs at the cathode accompanied by hydrogen evolution and hydroxyl ion generation according to the equation:

$$2H_2O(l) + 2e^- = H_2(g) + 2OH^- \quad (Eq\ 2)$$

Thus, there exists a pH gradient between the two electrodes. The metal in question will undergo oxidation to give a positively charged ion M^{n+} that migrates toward the cathode, according to the equation:

$$M = M^{n+} + ne^- \quad (Eq\ 3)$$

As the metal ions migrate toward the cathode, they precipitate at the point in the pH gradient where the pH promotes the precipitation of insoluble corrosion products. This point is determined by the thermodynamics of the corrosion product solubility reactions. The filament grows toward the anode as time progresses. These filamentary growths are termed CAFs when the corrosion product is conductive and is formed due to the anodic reaction. These have been known to form at the fiber and the epoxy interfaces (Ref 11). Conductive anodic filaments are generally formed in glass fiber-reinforced composites, which have propensity for interfacial damage due to temperature and humidity.

Thus, the conditions necessary for the CAF formation are:

- High temperature
- High humidity
- Interfacial degradation in the PWB (copper/epoxy, epoxy/glass)
- Applied voltage
- An acidity (pH) gradient between the two electrodes
- Precipitation of the reaction product leading to filamentary growth

Corrosion of Some Metals Commonly Found in Electronic Packaging

Copper. In most instances, the copper surface on the PWB is protected with organic solder preservatives such as benzimidazoles, benzotriazoles, polyalkylbenzimdazoles, and so forth, until component assembly is performed. These surface coatings are extremely thin and are prone to damage due to mechanical abrasion and to instability in thermal exposures. After the interconnection process, exposed copper surfaces can experience corrosion and have the potential for product failures.

Copper and its alloys react readily with ambient air (Ref 12, 13). In the absence of moisture, a thin layer of thermodynamically stable cuprous oxide grows on copper according to the equation:

$$2Cu(s) + 1/2 O_2(g) = Cu_2O(s) \quad (Eq\ 4)$$

The growth kinetics are parabolic at ambient temperature for about 100 h and linear thereafter. A layer of 10 nm is formed in about 10 days and has 5×10^{-8} moles of Cu_2O/cm^2.

In the presence of moisture and CO_2, a green patina is formed on copper according to the equation:

$$2Cu(s) + O_2(g) + CO_2(g) + H_2O(l)$$
$$= Cu_2CO_3(OH)_2(s) \quad (Eq\ 5)$$

This green patina can form an impervious barrier to further ingress of moisture and CO_2 and provides a permanent light green coating on copper. This patina sometimes provides an architectural appeal for buildings with copper sheet roofing and is sometimes referred to as architectural weathering (Ref 14).

Tin and Tin-Lead Alloys. Until recently, the most prevalent interconnection alloy in the electronic industry has been the eutectic tin-lead alloy with 72 at.% Sn and 38 at.% Pb that melts at 183 °C (361 °F). The corrosion reactions of solder alloys with ambient air and moisture are governed by thermodynamic and kinetic factors. Solder quickly reacts to form a tin-oxide layer when exposed to low oxygen partial pressures of 0.1 torr. At relatively higher oxygen pressures of 1 torr, a lead-rich oxide exists initially and quickly becomes tin-rich in the early stages of oxidation. Subsequently, stannic oxide (SnO_2) is the primary component with the remaining tin evenly distributed as tin and stannous oxide (SnO). The lead species are distributed evenly as lead and lead-oxides. Thus:

$$Pb(s) + O_2(g) = PbO_2(s) \quad (Eq\ 6)$$

and

$$PbO_2(s) + SnO(s) = PbO(s) + SnO_2(s) \quad (Eq\ 7)$$

The enthalpy, ΔH^0, for reaction in Eq 7, is -56.7 kcal/mole and accounts for diffusion of tin to the surface and results in the formation of layers that are alternatively rich in tin and lead. Differences in oxidation behavior between lower and higher oxygen partial pressure environments are kinetic. At higher oxygen partial pressures, surface tin is rapidly consumed followed by oxidation of lead. At lower oxygen partial pressures, the tin oxidation rate is lower than the tin diffusion rate from the bulk to the surface (Ref 15, 16). In an Auger electron spectroscopic study of oxidation of molten solder, de Kluizenaar reported that a complex oxide layer is formed on clean eutectic Sn/Pb solder surface (Ref 17). It consisted of (a) a thin layer of SnO_2 about 2 nm thick, (b) a layer of SnO mixed with finely dispersed metallic lead, and (c) a transition layer of tin and metallic tin and lead. The tin oxide concentration decreased with depth into the solder while the tin metal concentration increased. Below the oxide layer was the underlying solder alloy (Ref 17).

The following reactions are possible with moisture and the solder constituents:

$$Pb(s) + 2H_2O(l) = Pb(OH)_2(s) + H_2(g) \quad (Eq\ 8)$$

$$Sn(s) + 2H_2O(l) = Sn(OH)_2(s) + H_2(g) \quad (Eq\ 9)$$

with ΔH^0 values of -10.6 and -22.4 kcal/mole, respectively. It can be seen by the larger negative value of the enthalpy that the tin reaction is favored and, hence, the OH species is associated with tin rather than lead.

Almost all portable electronic appliances are exposed to outdoor atmospheres including the urban industrial environments in metropolitan areas. The atmospheric pollutants include such gases as oxides of nitrogen, oxides of sulfur, hydrogen chloride, hydrogen sulfide, and so forth. The corrosive reactions encountered can be significant in the performance of portable electronic appliances.

Dry $NO_2(g)$ promotes oxidation of the solder surface, without attendant nitrogen species on the solder surface, according to the equations:

$$Pb(s) + NO_2(g) = PbO(s) + NO(g) \quad (Eq\ 10)$$

$$2Pb(s) + NO_2(g) = 2PbO(s) + \tfrac{1}{2}N_2(g) \quad (Eq\ 11)$$

$$PbO(s) + NO_2(g) = PbO_2(s) + NO(g) \quad (Eq\ 12)$$

and

$$SnO + NO_2(g) = SnO_2(s) + NO(g) \quad (Eq\ 13)$$

with the corresponding enthalpies of, -38.9, -74.2, -0.2, and -56.9 kcal/mole, respectively. In the presence of moisture, however, both NO^- and NO_2^- are formed according to the equations:

$$2[O^{2-}]\ \text{lattice} \xrightarrow{H_2O} 2\square\ O + 4e^- + O_2(g) \quad (Eq\ 14)$$

$$NO_2(g) + e^- \xrightarrow{H_2O} NO_2^- \quad (Eq\ 15)$$

$$NO(g) + e^- \xrightarrow{H_2O} NO^- \quad (Eq\ 16)$$

with lead nitrate as the primary product (Ref 18, 19). The square in Eq 14 represents vacancy in the lattice.

Metallic lead on the surface of solder alloys is readily covered by a tenacious adherent layer of oxide according to the equation:

$$Pb(s) + \tfrac{1}{2}O_2(g) = PbO(s) \quad (Eq\ 17)$$

However, in the presence of moisture and chlorides, lead oxide is converted into loosely adhering $PbCl_2(s)$ as follows:

$$PbO(s) + 2HCl(\tfrac{l}{g}) = PbCl_2(s) + H_2O(l) \quad (Eq\ 18)$$

When $H_2O(l)$ and $CO_2(g)$ are present, lead chloride is converted to porous lead carbonate according to:

$$PbCl_2(s) + H_2O(l) + CO_2(g)$$
$$= PbCO_3(s) + 2HCl(l) \quad (Eq\ 19)$$

The porosity of lead carbonate facilitates further oxidation of lead according to Eq 17 until all the lead is consumed.

Aluminum. In first-level packaging, failures due to corrosion can result in electrical opens in lead wires, electrical shorts due to dendrite growths, metal migration, peeling, as well as assembly parametric changes or degradation. As indicated earlier, chloride and bromide ion contamination is not uncommon in plastic encapsulated microelectronics. In the presence of moisture and applied potential, the aluminum pad metallization can become a hydrated aluminum hydroxide. Chloride ions, hydroxyl ions, and water molecules compete for available bonding sites on the hydrated aluminum hydroxide surface. Chloride ions react with aluminum hydroxide according to the equation:

$$Cl^- + Al(OH)_3 = Al(OH)_2Cl + OH^- \quad (Eq\ 20)$$

Subsequently, the chloride ion reacts with the exposed aluminum as:

$$4Cl^- + Al = AlCl_4^- + 3e^- \quad (Eq\ 21)$$

The tetrachloro-aluminum ion reacts with water liberating hydrogen and chloride ions and aluminum hydroxide as follows:

$$2AlCl_4^- + 6H_2O(l) = 2Al(OH)_3 + 6H^+ + 8Cl^- \quad (Eq\ 22)$$

Thus, the chloride ion is regenerated, and the aluminum oxidation process continues until all the aluminum is consumed.

A case of galvanic corrosion is encountered sometimes with aluminum and gold components. Aluminum, being the less noble metal, corrodes more readily, forming aluminum oxide.

Ball bond corrosion at elevated temperatures is sometimes encountered in aluminum and gold at the package level due to the presence of bromine in the flame retardant of the polymer molding compound. At temperatures above 150 °C (300 °F), the brominated compounds undergo decomposition and disproportionation reactions giving rise to a variety of corrosive species such as hydrogen bromide and aliphatic alkyl bromides.

Researchers (Ref 20) proposed the following reactions:

$$CH_3Br = CH_3^+ + Br^- \quad (Eq\ 23)$$

$$4HBr + 2O^{2-} = 4Br^- + 2H_2O \quad (Eq\ 24)$$

$$Au_4Al + 3Br^- = 4Au + AlBr_3 + 3e^- \quad (Eq\ 25)$$

and

$$2AlBr_3 + 3O^{2-} = Al_2O_3 + 6Br^- \quad (Eq\ 26)$$

The time to failure due to these corrosion reactions depends on the time it takes for the moisture and the corrosive species to penetrate to the interfaces of importance. Defects such as cracks, crevices, voids, and so forth, already present in the structure due to other process deficiencies, will accelerate the rate of penetration.

Examples from Electronic Assemblies

Corrosion of Gold-Plated Connectors. Gold plating of connectors is a common practice designed to protect the underlying copper and nickel layers from corrosive attack and also promote good electrical contact. However, under the action of friction, the relatively thin and inert gold coating can be removed locally, thereby exposing the copper and nickel layer underneath. In such cases, fretting corrosion, pitting corrosion, and localized galvanic corrosion can occur simultaneously, especially in the presence of ionic species such as chlorides. The resulting corrosion product, which is usually nonconductive, can cause electrical failure due to open or intermittently open contact. An example of gold-plated connector corrosion from a device exposed to field conditions is shown in Fig. 2. The energy-dispersive x-ray analysis (EDX) elemental map for gold indicates that the gold coating is intact over the major portion of the surface. However, in the central portion of the image, the gold and nickel coatings have been removed completely, and the underlying copper is exposed, identified as the bright and gray areas in the copper elemental map. The shape and size of the gray area in the center of the exposed copper area coincides with the gray and bright areas seen in the maps for oxygen and chlorine, respectively. It can be surmised that the central area of the exposed copper substrate (gray area in copper EDX map) is covered with corrosion products containing oxides and chlorides of copper, which attenuates the copper signal in the EDX map. The absence of any areas with high concentrations of nickel indicates that the mating surface of the connector has probably worn through the nickel layer in the area of contact.

Electrochemical Migration. In several studies, the propensity for ECM of different metallization systems can be ranked as follows: $Ag > Pb > Cu > Sn$ (Ref 21). Although ECM phenomena have been observed with many metals, only silver (Ref 21, 22), and copper to a limited extent (Ref 23), and perhaps tin (Ref 24), have been found to exhibit this behavior in the presence of noncondensing but humid conditions. Indeed, it was concluded (Ref 25) that silver migration presents the greatest risk because dendritic growth can occur on ceramic and plastic substrates whether silver is outside the package and fully exposed to humid air or inside and only partly exposed. Although researchers (Ref 25) suggested that copper migration and tin migration did not pose as large a risk, in mobile electronic products, which see a wide range of corrosive species during their lifetime, ECM of copper and tin can be as prevalent as silver migration. In addition, residues on the substrates that originate from the PWB assembly process play an important role through water adsorption and conductivity behavior modification.

It was noted (Ref 26) that in practice, ECM could manifest itself as two separate, though not always distinct, effects that lead to impairement of the circuit's electrical integrity. The first kind, dendritic or filamentary bridging between the anode and cathode, has been discussed at length before. The second kind of ECM is colloidal staining, which can also cause a short. Deposits of colloidal silver, copper, or tin have been observed to originate at the anode without necessarily remaining in contact with it. One example each is provided next for ECM phenomena involving copper, tin, and silver.

Copper ECM. Copper forms complex species such as $CuCl_4^{2-}$, $CuCl_2(H_2O)$, $Cu(H_2O)^{2+}$, and so forth in the presence of halide-containing species and moisture. An example of copper ECM resulting in copper dendrite formation is shown in Fig. 3 for a device exposed to temperatures ranging from 10 to 55 °C (50 to 130 °F) in a humid atmosphere of 95% RH. If copper plated through-hole vias or copper-conductor pads are too close, ECM can occur when the assembly is exposed to humid environments in the presence of an ionic contaminant.

The tin ECM mechanism is similar to that of copper, but it is much more prevalent because tin constitutes a major portion of several commercial solder compositions such as 62Sn36Pb2Ag, 90Pb10Sn, and Sn3.5Ag0.7Cu (see also Table 1). In addition, exposed tin is more widespread on an assembled PWB as compared with copper. The example shown in Fig. 4 is from a test vehicle that failed upon exposure to a similar temperature-humidity environment as in the copper ECM example. In this case, the potential difference between the terminals of a capacitor with tin terminations resulted in the migration of tin from the anode toward the cathode across the dielectric material (gray area) between the tin terminations (bright areas) in the left side of Fig. 4. The right side of Fig. 4 shows a higher magnification view of the tin dendrites at the cathode end of the termination.

The composition of the ECM products can be revealed by elemental analysis mapping data of the surface of the same capacitor as shown in

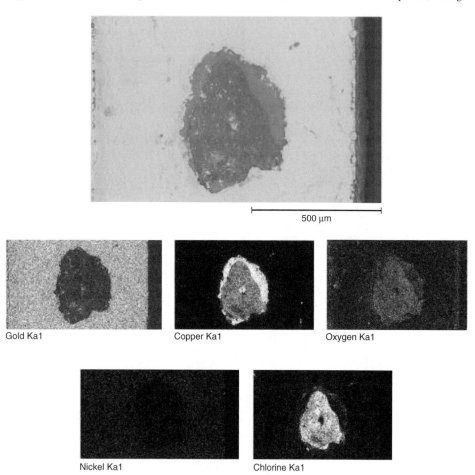

Fig. 2 Corrosion of gold-plated connector along with energy-dispersive x-ray analysis elemental maps of Au, Cu, O, Ni, and Cl

Fig. 3 Copper electrochemical migration on a printed wiring board subjected to high-humidity heat exposure

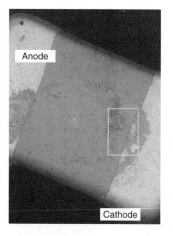 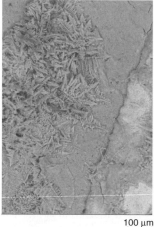

Fig. 4 Electrochemical migration on a capacitor with tin termination. The right side of the figure shows a higher magnification view of the dendrites at the cathode end of the termination.

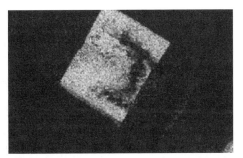

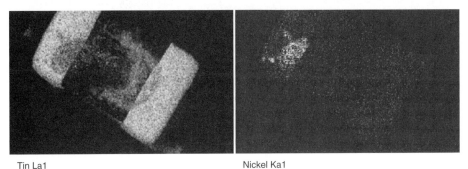

Fig. 5 Composition of the electrochemical migration product of the capacitor shown in Fig. 4. Energy-dispersive x-ray analysis elemental maps are Ba, Ti, O, Sn, and Ni. These maps show the presence of tin electrochemical migration between the terminations and exposure of the nickel barrier layer under the consumed tin surface at the anode.

Fig. 5. The capacitor dielectric material is composed primarily of barium titanate, and the barium and titanium maps show the extent of the exposed dielectric surface. For the capacitor to function, it is necessary that the cathode and anode terminations are electrically insulated from each other. Therefore, in a functional capacitor, one would not find any tin covering the dielectric material between the terminations (gray area in Fig. 4). However, in this failed capacitor, in addition to the tin at the terminations, which is normal, the tin map also shows the presence of tin between the terminations, where there should be none.

In several passive components (capacitor, resistors, etc.), nickel is used as a barrier layer between the silver adjacent to the dielectric and the tin termination. In this particular case, the nickel barrier layer at the anode is visible in areas where the tin from the surface has been consumed by the ECM process, exposing the nickel barrier layer underneath the tin. Another example of tin ECM is shown in Fig. 6, where a colloidal form of ECM can be observed in addition to dendrite formation.

Silver ECM can occur on the PWB if there is exposed metal in the termination or pad finish, or it can occur on the surface of passive devices separate from the surface of the PWB. The occurrence of ECM on the surface of passive devices can potentially be a more serious reliability risk because of the current trend toward smaller-size passives, which provides a ready site for ECM. Furthermore, silver commonly coats the ends of the passive device to ensure that there is a good contact between the electrodes in a capacitor and the termination, which is often made of tin. Therefore, nickel is used as a barrier layer between the silver base and the tin outer layers. If the nickel layer is not continuous or if there are gaps through which silver can be exposed to the environment, silver ECM can occur, as shown in Fig. 7. Here, dendrites of

Fig. 6 Tin electrochemical migration (ECM) involving both formation of dendrites and colloidal form of ECM on a resistor with pure tin termination.

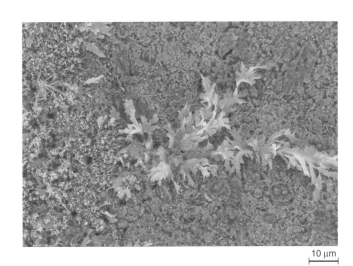

Fig. 7 Silver electrochemical migration on a resistor with tin termination. Inadequate protection due to poor quality nickel barrier layers enabled the silver to exhibit electrochemical migration.

silver can be seen growing on the surface of the passive component after temperature-humidity tests.

Future Trends

As portable consumer electronic hardware becomes more complex with multitudes of functions and increased data handling capacity, further miniaturization and higher levels of integration at all levels of packaging are a natural trend. More functions will be integrated into the device. The silicon device thickness will be in the range of 40 to 50 μm. Stacked devices and folded and stacked packages will be more prevalent with a combination of multiple levels of wire bonding and/or flip chip interconnection. Another emerging trend in packaging is the three-dimensional integration at the wafer level. New materials that will have better mechanical properties and moisture resistance will be developed. More functions will be embedded into the PWB and these may include active, passive, and optical devices, attendant with new embedded interconnection schemes. The PWB technology itself will witness revolutionary changes with thinner and improved materials capable of 10 to 25 μm (0.4 to 1 mils) vias, 10 to 20 μm (0.4 to 0.8 mils) lines and spaces, and structures involving several layers of stacked vias. Consequently, hitherto unknown failure mechanisms are likely to be encountered. As the feature sizes diminish, the distinction between first- and second-level packaging becomes less clear. Failure analysis, even at the PWB assemblies, will be a formidable challenge. With shorter product development cycles and faster-to-market business environments, the need for more automated analytical tools with minimal operator intervention for rapid and repeatable root cause analysis will increase. The reliability demands will be higher to ensure customer satisfaction and product acceptance. Innovative reliability test practices will be needed to shorten the test durations to accommodate faster development schedules. The implications for reliability, failure, and root cause analysis will be significant.

As miniaturization and integration continue in portable electronic appliances, there will be increasing emphasis on new materials developments to meet performance demands. As the products become smaller and lighter, the need for thinner and lighter materials will increase. Simultaneously, better high-strength plastic composite materials will be required in order to meet the form factor and robustness requirements. As portable electronic hardware is more likely to be exposed to hostile environments, such as condensing and noncondensing humidity, and corrosive gases, than traditional business and office machines, better hydrophobic materials, and more chemically inert materials will be in demand. The electronic components and packages inside the mechanics will be smaller, lighter, and low profile. This implies that the amount of molding compound surrounding the devices will be thinner and could be less impervious to corrosive elements. Thus, the need for better encapsulating molding compounds that are more impervious to corrosive elements will be felt. Molecular level coatings may be required for protection against corrosion and ECM on the PWBs.

Static charge and the propensity for particulate dust attraction could also be a concern in product performance. Particulate materials can act as nucleation sites for aiding corrosion reactions. New antistatic materials may be needed for adequate product protection.

The recent enormous interest and developmental activities in the area of nanomaterials and technology show considerable promise and application potential in portable electronics. Developments in nanotechnology are occurring along various fronts. These include development of high-strength composites, abrasion-resistant materials, displays, semiconductors, fuel cells, and so forth. It can be expected that the developments in nanotechnology will be extensively applied in addressing corrosion-related issues in portable electronic assemblies.

ACKNOWLEDGMENTS

The authors acknowledge the valuable assistance of their colleagues Steve Dunford, Sesil Mathew, and Murali Hanabe and appreciate the management support of Timothy Fitzgerald and Ramin Vatanparast.

REFERENCES

1. M. Tullmin and P.R. Roberge, Corrosion of Metallic Materials, *IEEE Trans on Reliability*, Vol 44 (No. 2), p 271–278, 1995
2. J.D. Guttenplan, Corrosion in the Electronics Industry, *Corrosion*, Vol 13, *ASM Metals Handbook*, 9th ed., ASM International, 1987
3. A.J. Raffalovich, Corrosive Effects of Solder Fluxes on Printed Circuit Boards, *IEEE Trans. Components, Hybrids, Mfg. Technol.*, Vol 7 (No. 4), 1971, p 155–162
4. K. Yasuda, S. Umemura, and T. Aoki, Degradation Mechanisms in Tin and Gold Plated Connector Contacts, *IEEE Trans. Components, Hybrids, Mfg. Technol.*, Vol 7 (No. 3), 1987, p 4556–4462
5. H.W. Hermance, C.A. Russell, E.J. Bauer, T.F. Eagan, and H.V. Wadlow, *Environ. Sci. Technol.*, Vol 5, 1971, p 781
6. G.M. Munier, L.A. Psoto, B.T. Reagor, B. Russiello, and J.O. Sinclair, *J. Electrochem. Soc.* and *Electrochem. Sci. and Technol.*, Vol 127 (No. 2), 1980, p 265
7. D.W. Rice, R.J. Cappell, W. Kinsolving, and J.J. Laskowski, Indoor Corrosion of Metals, *J. Electrochem. Soc.*, Vol 127 (No. 4), p 891
8. R. Dreikorn and L.L. Marsh, *IBM Technical Disclosure Bulletin*, Vol 22 (No. 9) 1980
9. L.C. Mathew and D.L. Roth, The Water Drop Test – Highly Accelerated Migration Testing, *Fourth Electronic Materials and Processing Conference* (Montreal, Canada), P.J. Singh, Ed., ASM International, 1991
10. J.O.M. Bockris and A.K.N. Reddy, *Modern Electrochemistry*, Plenum Press, New York, c 1970
11. A.A. Shukla, T.J. Dishongh, M. Pecht, D. Jennings, Hollow Fibers in Woven Laminates, *Printed Circuit Fabrication*, Vol 20 (No. 1), Jan 1997, p 30–32
12. J. Ambrose, R.G. Barradas, and D.W. Smith, *Electroanal. Chem.* and *Interfacial Electrochem.* Vol 47, 1973, p 47–64
13. G. Tessier, C. LeGressius, and J. Bouygues, *IEEE Trans.*, Vol CHMT-5, (No. 2), 1982, p 217–224
14. H.H. Manko, *Solders and Soldering*, 2nd ed., McGraw-Hill, 1979, p 37
15. R.P. Frankenthal and D.J. Siconolf, *J. Vac. Sci. Technol.*, Vol 17 (No. 6), 1980, p 1315
16. W.J. Carter, and D.M. Hercules, *Appl. Spectrosc.*, Vol 33, 1979, p 287
17. E.E. de Kluizenaar, *J. Vac. Sci. Technol.*, Vol A1 (No. 3), 1983, p 1480
18. H.G. Tompkins, *J. Electrochem. Soc.*, Vol 32, 1972, p 269
19. H.G. Tompkins, *J. Electrochem. Soc.*, Vol 33, 1973, p 651
20. K.V. Ritz, W.T. Stacy, and E.K. Broadbent, Microstructure of Ball Bond Corrosion Failures, *Proc. 25th Annual International Reliability Physics Symposium*, IEEE International, 1987, p 28–33
21. G. Harsanyi, and G. Inzelt, Comparing Migratory Resistive Short Formation Abilities of Conductor Systems Applied in Advanced Interconnection Systems, *Microelectronics Reliability*, Vol 41, 2001, p 229–237
22. G.T. Kohman, H.W. Hermance, and G.H Downes, Silver Migration in Electrical Insulation, *Bell Syst. Tech. J.*, Vol 34, 1955, p 1115
23. A. Dermarderosian, The Electrochemical Migration of Metals, *Proc. International Society of Hybrid Microelectronics*, 1978, p 134
24. J.N. Lahti, R.H. Delaney, and J.N. Hines, The Characteristic Wearout Process in Epoxy-Glass Printed Circuits in High Density Electronic Packaging, *Proc. 17th Annual Reliability Physics Symposium*, IEEE International, 1979, p 39
25. P. Dumoulin, J.-P. Seurin, and P. Marce, Metal Migration outside the Package during Accelerated Life Tests, *IEEE Trans. Components, Packaging, and Manufacturing Technol.*, Vol 5 (No. 4), 1982, p 479–486
26. S.J. Krumbein, Metallic Electromigration Phenomena, *IEEE Trans. Components,*

SELECTED REFERENCES

- J.H. Lau, Ricky S.W. Lee, and Rickey S. Lee, Chip Scale Package: Design, Materials, Process, Reliability, and Applications, McGraw-Hill, 1999. Handbook of Area Array Packaging, P. Totta and K. Puttlitz, Ed., Kluwer Academic, 2001
- M.G. Pecht, L.T. Nguyen, E.B. Hakim, Plastic Encapsulated Microelectronics, John Wiley and Sons, 1995
- R.R. Tummala, E.J. Rymaszewski, and A.G. Klopfenstein, Ed., Microelectronics Packaging Handbook, Parts I, II, and III, Kluwer Academic, 1997
- P. Viswanadham and P. Singh, Failure Modes and Mechanisms in Electronic Packages, Chapman and Hall Publications, New York, 1998

Hybrids, and Mfg. Technol., Vol 11 (No. 1), 1988, p 5–15

Effects of Process and Environmental Variables

Revised by Bernard S. Covino, Jr., National Energy Technology Laboratory

TWO CHEMICAL PROCESSING PLANTS making the same product and using the same or a similar process will sometimes have different experiences with corrosion. At one plant, a steel pipeline may last for many years in a given service, yet a seemingly identical pipeline may fail within weeks or months in the same service at another plant. A major piece of equipment may suddenly fail after 15 or 20 years of service as a result of less than 1 ppm of metal ion contamination in a new source of raw material.

In designing a chemical processing plant, considerable attention is paid to fluid flows, sizing of lines, pumps and processing equipment, and temperatures and pressures. Materials are selected on the basis of past experience, corrosion tests, corrosion literature, and the recommendations of materials suppliers. The nominal compositions of process streams and raw materials are known with some accuracy. Occasionally, however, other variables exist that can cause corrosion failures. The purpose of this article is to bring those variables and their effects to the attention of the corrosionist. As such, Table 1 lists some of the variables that can affect the performance of metals in corrosive environments and gives an indication of their effect on seven different metal groups: carbon steels, austenitic stainless steels, nickel and nickel-copper alloys, nickel-chromium-molybdenum alloys, titanium, zirconium, and copper and copper alloys.

Plant Environment

One factor that is occasionally overlooked is the surrounding environment. This includes ambient temperature and composition. Temperature can be affected by geographic location (see the section "Seasonal Temperature Changes" in this article), location next to a hot or cold process, and so on. The composition of the environment can be affected not only by what is released in the plant but also by what is released by other plants in close proximity. Temperature and atmospheric composition primarily affect the external surfaces of the plant and plant equipment, causing corrosion that must be repaired periodically.

Cooling Water

The quality of water used to cool shell and tube heat exchangers can vary from season to season and year to year, particularly if the water is taken from a river that discharges into seawater. Water composition, especially chloride, metal impurities, and dissolved oxygen, can greatly affect the corrosion of equipment.

Steam

The corrosive properties of the steam used in a plant are seldom considered, particularly if the steam will be supplied by a public utility or power plant that has good boiler water treatment that will prevent condensate corrosion. However, impurities in steam can cause corrosion problems.

Water Content. One of the most important process variables, from the corrosion viewpoint, is the amount of water in a process stream. The water may be present as water vapor in a gas

Table 1 Effects of process variables on corrosion of metals in aggressive environments

Process variable	Carbon steels	Austenitic stainless steels	Nickel, nickel-copper alloys	Nickel-chromium-molybdenum alloys	Titanium	Zirconium	Copper, copper alloys
Velocity							
Increasing	>(a)	0 to >	0 to >	0 to >	0(b)	0	>
Decreasing	<L(c)(d)	<L	0 to <	0 to <	0	0	<
Aeration	>	0 to <	>	V(e)	0 to <	0	>
Galvanic couple	>	0	0	0	HE(f)	HE	0
Impurities							
Nonchloride oxidants	V	<	>	V	0	0	>
FeCl$_3$, CuCl$_2$	>	>L, SCC(g)	>	0 to >	0	>E(h)	>
Neutral chlorides	>	>L, SCC	0	0 to >	0, unless H(j)	0	>
Ammonia	0	0	>	0	0	0	>SCC
Sodium, potassium	0, SCC if H	0, SCC if H	0	0	0, unless H	0	0, unless H
Fluorides	0	0	0	0	>	>	0
Nitrites	V	V	>	V	0(k)	0	>SCC
Nitrates	<, SCC	0 to <	0	0	0	0	0
HCl in chlorine	>	>	>	>	>(m)	>	>
Mercury	0	0	0, SCC if H	0	0, unless H	0	>SCC
Sulfur compounds	V	0, SCC if S(n)	>	0, SCC if S	0	0	>

(a) >, generally increases attack. (b) 0, generally has no effect on metal in question. (c) <, generally decreases attack. (d) L, local attack or crevice corrosion is possible. (e) V, may increase or decrease attack, depending on specific alloy and environment. (f) HE, hydrogen embrittlement is possible. (g) SCC, stress-corrosion cracking is possible. (h) E, embrittlement. (j) H, hot (from 80 to 150 °C, or 175 to 300 °F, depending on variable and alloy). (k) Except in fuming HNO$_3$. (m) Except for dilute solutions. (n) S, sensitized by heat treatment

stream or may be soluble. Water is most likely to contribute to corrosion when it is present as a separate liquid phase. When the water is present as vapor, it is likely to condense at cold spots, such as pipe supports welded directly to process piping or nozzles in the top or vapor phase of reactors, even though the lines and nozzles are insulated.

Aeration. The design of the plant and, in particular, equipment selection can influence the amount of air introduced into a process stream, which in turn may profoundly affect corrosion. This can account for different corrosion experiences in plants having otherwise identical chemistries and products. In different cases, the quantity of introduced air can be either beneficial or detrimental. Increasing the amount of air can increase the corrosion rate of mild steel components, but reducing the amount of air may transform a normally passive stainless steel piece of equipment to a state of corrosion.

Startup, Shutdown, and Downtime Conditions

Startups and shutdowns are variables that can easily be overlooked in the design stage or conducted improperly by operations. If improperly conducted, corrosion can occur due to dewpoint corrosion or due to process fluids left in the equipment. Inert gas blanketing may be needed to reduce these types of corrosion problems. Problems with the cleaning of process equipment generally occur during downtimes.

Seasonal Temperature Changes

Ambient temperature is an important variable that can vary widely. Metal parts exposed to the sun may reach 60 °C (140 °F) in the summer. During the winter, colder cooling water can cause a lower process exit temperature than desired. To correct this problem, plant operators will frequently decrease the water flow. All too often, the water flow is throttled at the water inlet. This can result in a fluctuating liquid level, with buildup of solids, pitting, and/or stress-corrosion cracking.

Variable Process Flow Rates

Plants are frequently operated at flow rates other than design capacity. They may be turned down to considerably less than capacity or pushed above it. Either condition can cause corrosion problems.

Impurities

A principal cause of unexpected corrosion failures is the presence of small amounts of impurities in the chemical environment. A number of common impurities affecting corrosion are presented in Table 1.

SELECTED REFERENCES

- S.A. Abdul-Wahab, Statistical Prediction of Atmospheric Corrosion from Atmospheric-Pollution Parameters, *Pract. Period. Hazard., Toxic, Radioac. Waste Mgmt.*, Vol 7 (No. 3), July 2003, p 190–200
- M.G. Fontana and N.D. Greene, *Corrosion Engineering*, McGraw-Hill Book Company, 1978
- L.L. Shreir, R.A. Jarman, and G.T. Burstein, Ed., *Corrosion*, 3rd ed., Vol 1–2, Elsevier, 1994

Corrosion under Insulation

Hira S. Ahluwalia, Material Selection Resources, Inc.

CORROSION UNDER INSULATION (CUI) is a well-understood problem and mitigation methods are well established; however, it is pervasive and continues to cost the process industry many millions of dollars annually. In 2002, a large chemical company spent more than $5 million dollars replacing type 304 stainless steel equipment because of chloride stress-corrosion cracking (CSCC) under insulation (Ref 1). The cost of the downtime due to the loss of production was even more significant. In another example, one petrochemical company estimates that CUI accounts for as high as 40 to 60% of the company's piping maintenance costs (Ref 2). Similar case histories are commonplace within the process industries. An effective CUI prevention strategy based on life-cycle costs can significantly reduce costs due to downtime, maintenance repair, and inspection. The corrosion of carbon steel under wet insulation is nonuniform general corrosion and/or highly localized pitting. In austenitic stainless steels, the main forms of corrosion are pitting and CSCC.

Corrosion of Steel under Insulation

The corrosion of carbon steel under wet insulation is nonuniform general corrosion and/or highly localized pitting. Figure 1 shows a section of a large carbon steel storage tank that has undergone corrosion in a localized region leading to a through wall hole. The corrosion occurred on the sidewall near the tank bottom where the coating failed, exposing the carbon steel to wet corrosive conditions under the insulation.

The Mechanism

Carbon steel does not corrode simply because it is covered with insulation, but because it is contacted by aerated water. The primary role of insulation in corrosion is to provide an annular space or crevice for the retention of water with full access to oxygen (air) and other corrosive media. In addition, the insulation provides a material that may wick or absorb and may contribute contaminants that increase or accelerate the corrosion rate. The corrosion rate of carbon steel is principally controlled by the temperature of the steel surface, availability of oxygen and water, and the presence of corrosive contaminant species in the water.

Water Sources. There are two primary water sources involved in CUI of carbon steel. First, breaks in the weatherproofing can lead to infiltration of water to the metal surface from external sources such as rainfall, drift from cooling towers, condensate falling from cold service equipment, steam discharge, process liquid spillage, spray from fire sprinklers, deluge systems, washrooms, and from condensation on cold surfaces after vapor barrier damage. Second, a major corrosion problem develops in situations where there are cycling temperatures that vary from below the dew point to above ambient temperatures. In this case, the classic wet/dry cycle occurs when the cold metal develops water condensation that is then baked off during the hot/dry cycle. The transition from cold/wet to hot/dry includes an interim period of damp/warm conditions with attendant high corrosion rates.

Contaminants. Chlorides and sulfates are the principal contaminants found under insulation. These may be leached from the insulation materials or from external waterborne or airborne sources. Chlorides and sulfates are particularly detrimental because their respective metal salts are highly soluble in water, and these aqueous solutions have high electrical conductivity. Furthermore, hydrolysis of the metal salts can create acidic conditions leading to localized corrosion.

Temperature. It is generally accepted that carbon steel operating in the temperature range of −4 to 150 °C (25 to 300 °F) is at the greatest risk from CUI. Equipment that operates

Fig. 1 Through wall corrosion under insulation of a large coated carbon steel storage tank. Through hole at arrow. Courtesy of Paul Powers, GE Inspection

continuously below −4 °C (25 °F) usually remains free of corrosion. Corrosion of equipment above 150 °C (300 °F), above the boiling point of water, is reduced because the carbon steel surface remains essentially dry. Corrosion tends to occur at those points of water entry into the insulation system where the temperature is below 150 °C (300 °F) and when the equipment is idle.

Figure 2 shows the corrosiveness of water versus temperature. The problem of steel corrosion under insulation can be classified as equivalent to corrosion in a closed hot-water system. In an open system, the oxygen content decreases with increasing temperature to the point where corrosion decreases even though the temperature continues to increase (Ref 4). In a closed system, the corrosion rate of carbon steel in water continues to increase as the water temperature increases. Estimated corrosion rate data of carbon steel under insulation plotted in Fig. 2 that were obtained from actual plant case histories confirm that the rate increases with temperature in a manner similar to that of a closed system (Ref 3). It is inferred that the same oxygen cell corrosion mechanism is taking place as in a closed system. The corrosion rates from field measurements are shown to be greater than laboratory tests, due to the presence of salts in the field. Salts increase the conductivity of the water film and thereby influence the corrosion rate.

Insulation. Corrosion under insulation of carbon steel is possible under all types of insulation. The rate of corrosion may vary depending on the characteristics of the insulation material. Some insulation materials contain water-leachable salts that may contribute to corrosion, and some foams may contain residual compounds that react with water to form an acidic environment. The water retention, permeability, and wettability properties of the insulation material also influence the corrosion of carbon steel.

Corrosion of Stainless Steel under Insulation

Corrosion under insulation in austenitic stainless steel is manifested by chloride-induced stress-corrosion cracking (CISCC), commonly referred to as external stress-corrosion cracking (ESCC), because the source of chlorides is external to the process environment. Figure 3 shows ESCC of a 100 mm (4 in.) type 304 stainless steel pipe that operated in the 50 to 100 °C (120 to 212 °F) range. Figure 4 shows the typical transgranular lightning strike appearance of ESCC in the pipe. External SCC of austenitic stainless steel is possible when the equipment is contacted by aerated water, and chlorides or contaminants are present in the temperature range 50 to 150 °C (120 to 300 °F) in the presence of tensile stresses.

The Mechanism

A detailed discussion on the mechanism of SCC can be found in a number of publications (Ref 5, 6). The mode of cracking is normally transgranular. It is well established that the propensity for ESCC is greatest when the following conditions are present:

- A susceptible 300 series austenitic stainless steel
- The presence of residual or applied surface tensile stresses
- The presence of chlorides, bromide (Br^-), and fluoride (F^-) ions may also be involved
- Metal temperature in the range 50 to 150 °C (120 to 300 °F)
- The presence of an electrolyte (water)

Alloys. The stainless steels that are commonly affected by ESCC in the chemical process industries are the 300 series stainless steels, including type 304 (UNS S30400 and S30403), type 316 (UNS S31600 and S31603), type 317L (UNS S31700), type 321 (UNS S32100), and type 347 (UNS S34700). It should be noted that other stainless steels could also undergo ESCC under specific corrosive conditions.

Role of Stress. For ESCC to develop, sufficient tensile stress must be present in the material. If the tensile stress is eliminated or greatly reduced, cracking will not occur. The threshold stress required to develop cracking depends somewhat on the severity of the cracking medium. Most mill products, such as sheet, plate, pipe, and tubing, contain enough residual tensile stresses from processing to develop cracks without external stresses. When the austenitic stainless steels are cold formed and welded, additional stresses are imposed. The incidence of ESCC is greater in process piping because of the high hoop stresses normally present in piping systems. As the total stress increases, the potential for ESCC increases.

Chlorides. The Cl^- ion is damaging to the passive protective layer on 18–8 stainless steels. Once the passive layer is penetrated, localized corrosion cells become active. Under the proper set of circumstances, SCC can lead to failure in only a few days or weeks. Sodium chloride, because of its high solubility and widespread presence, is the most common corrosive species (Ref 5). This neutral salt is the most common, but

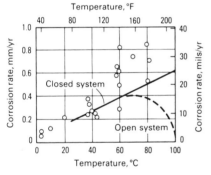

Fig. 2 Effect of temperature on corrosion of steel in water. Data points are from plant measurements of corrosion under insulation (Ref 3). Closed system, oxygen held in system; open system, oxygen free to escape (Ref 4)

Fig. 3 External stress-corrosion cracking of a type 304 stainless steel, 100 mm (4 in.) schedule 40 pipe. The piping system was insulated with calcium silicate insulation and operated at temperatures between 50 and 100 °C (120 and 212 °F).

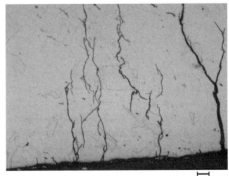

Fig. 4 Cross section through type 304 stainless steel pipe showing transgranular external stress-corrosion cracking

not the most aggressive. Chloride salts of the weak bases and light metals, such as lithium chloride (LiCl), magnesium chloride ($MgCl_2$), and aluminum chloride ($AlCl_3$) can even more rapidly crack the 18–8 stainless steels under the right conditions of temperature and moisture content.

The sources of chlorides in ESCC are from insulating materials and external sources. The insulating materials include insulation, mastics, sealants, adhesives, and cements. Experience has shown that insulating materials with chloride content as little as 350 ppm can contribute to ESCC. Typically, if the insulating material is the source of leachable chlorides, failure occurs only after a few years of operation. However, external sources of chloride account for most of the ESCC failures. The sources include rain, coastal fog, wash water, fire and deluge system testing, and process leaks or spills. Other sources of chloride ions known to be aggressive include chlorine, hydrogen chloride gas, hydrochloric acid, and hydrolyzed organic chlorides. Clearly, the presence of chlorides under acidic conditions is more aggressive than neutral or basic conditions. Failures due to chlorides from external sources tend to occur after 5 or more years of service.

The concentration of chlorides necessary to initiate SCC is difficult to ascertain. Researchers have developed cracking in solutions with remarkably low levels of chlorides (<10 ppm). The situation of chlorides under insulation is unique and ultimately depends on the concentration of chlorides deposited on the external surface of the metal. Deposits near ESCC failures have been found with as little as 1000 ppm chloride. If chlorides are detected, there will probably be some localized sites of high concentration.

Temperature. The most important condition affecting chloride concentration is the temperature of the metal surface. Temperature has a dual effect; first, elevated temperatures will cause water evaporation on the metal surface, which results in chloride concentration. Second, as the temperature increases the susceptibility to initiation and propagation of ESCC increases. External SCC occurs more frequently in the range 50 to 150 °C (120 to 300 °F). Below 50 °C (120 °F), chlorides do not concentrate to levels that cause ESCC. Above 150 °C (300 F), water is not normally present on the metal surface, and failures are uncommon. Equipment that cycles through the water dew point is particularly susceptible because during each temperature cycle the chloride salts in the water concentrate on the surface.

Electrolyte. Water is the fourth necessary condition for ESCC. Since SCC involves an electrochemical reaction, it requires an electrolyte. As water penetrates the insulation system, it plays a key role at the metal surface, depending on the equipment operating conditions. Examination of the phenomenon of corrosion of steel under insulation provides a better appreciation of the widespread intrusion of water (Ref 7–9). In effect, water may be expected to enter the metal/insulation annulus at joints or breaks in the insulation and its protective coating. The water then condenses or wets the metal surface, or if it is too hot, the water is vaporized (Ref 7).

This water vapor (steam) penetrates the entire insulation system and settles into places where it can recondense. Because the outer surface of the insulation is designed to keep water out, it also serves to keep water in. The thermal insulation does not have to be in poor condition or constantly water soaked. A common practice in chemical plants is to turn on the fire protection water systems on a regular basis. This deluges the equipment with water. Some seacoast locations use seawater for the fire protection water. Hot food-processing equipment is regularly washed with tap water, which contains chlorides. All insulation system water barriers eventually develop defects. As the vessel/insulation system breathes, moist air contacts the metal surface. From the insulation standpoint, the outer covering acts as a weather barrier to protect the physical integrity of the insulation material. The outer coverings are not intended, nor can they be expected, to maintain an airtight and watertight system.

Prevention of CUI

Design of Insulation System. When designing an insulation system the goal is to prevent ingress of moisture. Poorly designed or applied insulation and protrusions through thermal insulation permit water to bypass the insulation, thereby corroding the substrate material. References 10 and 11 provide detailed information on the mechanical design of insulation systems. In the author's experience, attachments to vessels and piping stems are common locations that allow water to bypass the insulation and to concentrate at the attachment point. Examples of such attachments are shown in Fig. 5 and 6. Attention to details such as these is important in order to produce a high-quality insulation system. Although the design of the insulation systems is important, prevention methodologies based on design alone are not advisable or practical in a chemical plant.

The physical characteristics of thermal insulation materials can vary widely. Some insulation materials contain a leachable inhibitor to neutralize the pH of the water in contact with the metal surface. The degree of water absorbency can also vary. For some systems, the coefficient of thermal expansion will influence the system design. For example, cellular glass insulation expands about the same as carbon steel, whereas cellular foam expands nine times more than carbon steel and therefore require expansion joints. General industry experience over the last 20 years indicates that corrosion is possible under all types of insulation. The common types of insulation materials and their recommended service temperatures are listed in Table 1.

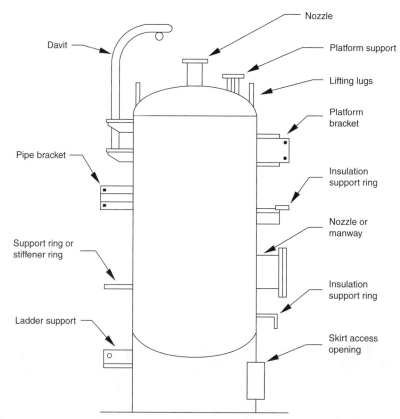

Fig. 5 Typical vessel attachments where water may bypass insulation. Source: Ref 11

Selecting and specifying the correct insulation material can reduce corrosion of both carbon steel and stainless steels.

Organic Coating System. The application of organic coatings on both carbon steel and stainless steel equipment beneath insulation is an effective method of having a physical barrier to the corrosive electrolytes and thereby prevent corrosion. This method is effective only if a holiday-free coated surface is obtained. In the chemical plant environment, the average life cycle of a coating system is 5 to 13 years (Ref 12). In some cases when a correctly selected and correctly applied coating system is used, a 20 year service life can be achieved. Some of the parameters that need to be considered when selecting a coating system include: coating selection, surface preparation requirements, environmental requirements, compatibility with insulating material, coating tests, coating vendor selection, specifications, inspection, and selection of coating applicator.

Coating systems that have been shown to have been used successfully in the process industries include liquid-applied coatings such as epoxies, urethanes, and polyurethanes; fusion-bonded coatings; brushable coal tar or asphalt-base coatings; mineralization coatings; and tapes. More information on the selection of protective coating is available from coating manufacturers' literature and in Ref 10. Information pertaining to organic coatings can also be found in "Organic Coatings and Linings." Vol 13A, of the *ASM Handbook*.

Personnel Protection Cages. In many instances thermal insulation is used for personnel protection from hot surfaces. The unnecessary use of thermal insulation creates a location for potential corrosion. In these cases wire "stand-off" cages should be used instead. These cages are simple in design, low in cost, and eliminate the concerns with CUI.

Thermal Spray Aluminum (TSA). For services too severe for organic coatings, such as temperature cycling above and below 150 °C (300 °F), TSA provides the best choice for corrosion protection beneath insulation. The TSA protects equipment by acting as a barrier coating and serves as a sacrificial anode protecting the substrate at the sites of any chips or breaks in the coating. The U.S. Navy has demonstrated that the use of TSA has resulted in substantial reduction in cost of its corrosion-control effort aboard its ships (Ref 13). A large petrochemical company has increased the use of TSA in its plant and has shown that large savings can be obtained based on life-cycle cost (Ref 12). Over a 20 year cost analysis, the replacement of an existing carbon steel pipe with TSA-coated carbon steel pipe compared with replacement with painted carbon steel that needs to be painted at least once during this period was shown to result in a savings of more than 100% (Ref 12). The development of more mobile thermal spray equipment with high deposition efficiency is likely to increase the use of TSA in the chemical process industry.

Aluminum Foil Wrapping of Stainless Steel Pipe. This technique is widely used in Europe by end users and engineering companies, but has not been widely accepted in North America. Aluminum foil wrapping has been used successfully for more than 30 years in preventing ESCC by chemical companies in Europe. The aluminum foil provides electrochemical protection by preferentially undergoing corrosion and maintaining a safe potential for stainless steel. The system relies on good weatherproofing and the prevention of immersion conditions. The system can be applied by the insulation contractor and, furthermore, it takes less time to apply than a coating and has minimum substrate preparation.

Wrapping pipe with 46 standard wire gage (SWG) 0.1 mm (0.004 in.) aluminum foil can prevent CISCC of stainless steel pipe operating continuously between 60 and 500 °C (140 and 930 °F). The pipe should be wrapped with 50 mm (2 in.) overlap, formed to shed water on the vertical line, and held with aluminum or stainless wire. The foil should be molded around flanges and fittings. Steam-traced lines should be double wrapped, with the first layer applied directly onto the pipe, followed by the steam tracing and then more foil over the top. On vessels, the aluminum foil is applied in bands held by insulation clips and insulation support rings (Ref 14).

Use of Higher Alloyed Material. To eliminate ESCC higher nickel-, chromium-, and molybdenum-containing alloys (superaustenitic stainless steels), and the lower-nickel, higher-chromium duplex alloys can be used. These

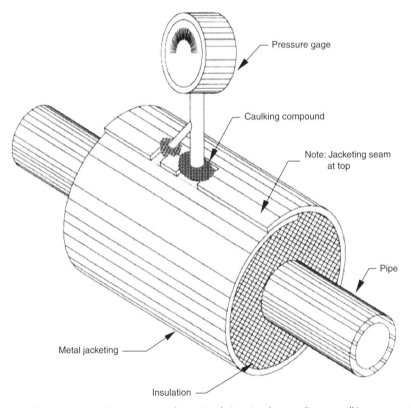

Fig. 6 Attachment to piping where water may bypass insulation. Attachment relies on caulking compound only. Source: Ref 11

Table 1 Typical service temperatures for thermal insulation materials

Thermal insulation material	Recommended service temperature	
	°C	°F
Polystyrene foam	−73 to +60	−99 to +140
Polyurethane foam—rigid	−73 to +82	−99 to +180
Polyisocyanurate foam—rigid	−73 to +120(a)	−99 to +248(a)
Flexible foamed elastomer	−40 to +104	−40 to +219
Cellular glass	−240 to +121	−400 to +250
Glass fiber	27 to 343	81 to 649
Mineral wool	27 to 982	81 to 1800
Calcium silicate	27 to 649	81 to 1200
Pearlite-silicate	27 to 593	81 to 1099

(a) Although some manufacturer's literature shows upper temperature limits approaching 149 °C (300 °F), experience indicates that polyisocyanurate foam begins to degrade at about 93 °C (199 °F) in the presence of moisture. A suggested practical upper limit is 66 °C (151 °F).

alloys are more resistant to SCC and have been found to be resistant to ESCC under insulation. The higher cost of some of these materials makes this option unattractive. However, in some applications the lean duplex stainless steel alloys may offer a low life-cycle cost alternative.

Inspection for CUI

Inspection of insulated piping, vessels, and other components is a major challenge and can be costly and time consuming. The goal for the process industries should be to move toward an inspection-free and maintenance-free philosophy by using the appropriate corrosion-prevention methods described in the previous section with a focus on life-cycle cost.

The American Petroleum Institute code, API 570, "Inspection, Repair, Alteration, and Re-Rating of In-Service Piping Systems" (Ref 15), identifies corrosion under insulation as a special concern and requires that an appropriate amount of external visual inspection be conducted on piping systems within susceptible temperature ranges. The use of risk-based inspection (RBI) assessment conducted in accordance with API RP 580 (Ref 16) provides a methodology for prioritizing CUI-related maintenance and inspection activities. The intent of using RBI is to manage the probability of failure in piping and vessels while establishing an optimum inspection program. At the same time, a significant portion of the risk in the plant can be addressed by focusing on a relatively few items in the unit. The factors that are usually considered in a RBI analysis include: location of equipment, temperature, materials of construction, age of the equipment, the type and condition of the coating system, insulation type, and risk potential in terms of process, business, environment, and safety. Guidelines on how to conduct a visual inspection to detect signs of CUI are detailed in Ref 10 and 17.

Quantifying CUI in piping in most cases requires the removal of insulation and surface preparation prior to inspection. The cost of insulation removal, inspection, and reinstallation can be very high. A number of nondestructive evaluation (NDE) methods that do not require removal of insulation have been developed to inspect for CUI. The Materials Technology Institute (MTI) sponsored a project to identify and evaluate the effectiveness of several NDE methods (Ref 18). The NDE methods evaluated were neutron backscatter, tangential radioscopy, through transmission radioscopy, pulsed eddy current, electromagnetic encircling coils, and three types of ultrasonic guided wave methods. The study concluded that the NDE methods could detect CUI; however, no technique is suitable for every application. The techniques vary widely in several ways such as speed, ease of inspecting piping, detectability of defects, and safety. The factors that influence the ease of inspecting piping include: pipe orientation; number of obstacles such as hangers and valve tees; proximity to the large metal masses; and insulation tie wires and jacket straps. Detectability of defects is influenced by the orientation, size, and type of defects. It should be noted that the CUI pattern may be nonuniform and spot nondestructive evaluation may be misleading.

REFERENCES

1. K. Bartlett, GE Advanced Materials, Evansville, IN, private communications
2. B. Fitzgerald et al., STG 36 Oral Presentation, CORROSION, NACE, March 2004
3. W.G. Ashbaugh, Corrosion of Steel and Stainless Steel under Thermal Insulation, *Process Industries Corrosion,* B.J. Moniz and W.I. Pollock, Ed., NACE, 1986, p 761
4. F.N. Speller, *Corrosion—Causes and Prevention,* 2nd ed., McGraw-Hill, 1935, p 153, Fig. 25
5. R.H. Jones, Ed., *Stress-Corrosion Cracking—Materials Performance and Evaluation,* ASM International, 1992
6. R.H. Jones, *Stress-Corrosion Cracking, Corrosion: Fundamentals, Testing, and Protection,* Vol 13A, *ASM Handbook,* ASM International, 2003, p 346–366
7. P. Lazar, Factors Affecting Corrosion of Carbon Steel under Insulation, *Corrosion of Metals under Thermal Insulation,* STP 880, American Society for Testing and Materials, 1980
8. T. Sandberg, Experience with Corrosion Beneath Thermal Insulation in a Petrochemical Plant, *Corrosion of Metals under Thermal Insulation,* STP 880, American Society for Testing and Materials, 1980
9. V.C. Long and P.G. Crawley, Recent Experiences with Corrosion Beneath Thermal Insulation in a Chemical Plant, *Corrosion of Metals under Thermal Insulation,* STP 880, American Society for Testing and Materials, 1980
10. "The Control of Corrosion under Thermal Insulation and Fireproofing Materials—A Systems Approach," RP 0198-98, NACE International
11. J.B. Bhavsar, Insulation Design Practices for Mitigation of Pipe and Equipment Corrosion, *Corrosion under Wet Thermal Insulation,* CORROSION 1989 Symposium, p 15–32, NACE, 1990
12. B.J. Fitzgerald et al., No. 03029, CORROSION 2003, NACE
13. R. Parks and R. Kogler, U.S. Navy Experience with High Temperature Corrosion Control of Lagged Piping System Components Using Sprayed Aluminum Coatings, *Corrosion under Wet Thermal Insulation,* CORROSION 1989 Symposium, NACE, 1990, p 71–76
14. R. Smith, Eutech, Oral Presentation at Stainless Steel World Conference, Netherlands 1999
15. "Inspection, Repair, Alteration and Re-Rating of In-Service Piping Systems," API 570, American Petroleum Institute
16. Risk Based Inspection, API 580, American Petroleum Institute
17. J.W. Kalis, "Insulation Outlook," National Insulation Association, Feb 2002
18. "Detection of Corrosion through Insulation," project 118, Materials Technology Institute, 1998

Corrosion by Sulfuric Acid

S.K. Brubaker, E.I. Du Pont de Nemours & Company, Inc.

SULFURIC ACID PRODUCTION AND ITS USE is critical to modern industry. Its use plays some part in the production of nearly all manufactured goods. More of this acid is used each year than any other manufactured chemical.

Today, most sulfuric acid is made by the contact process, in which elemental sulfur or sulfur-containing waste is burned to form sulfur dioxide (SO_2). Sulfur dioxide is converted to sulfur trioxide (SO_3) by contact with a vanadium catalyst. The SO_3 is absorbed in oleum (fuming H_2SO_4) or H_2SO_4 in a series of towers.

The basic materials of construction for the production and handling of concentrated sulfuric acid during the early 1900s were lead, steel, cast iron, and brick. In the mid-1900s stainless steels and the "20" type alloys came into use. During the later 1900s, super stainless steels, high-nickel alloys, and fluoropolymers were developed and found to be useful for higher-temperature acid handling and processing.

The corrosiveness of H_2SO_4 depends on many factors, particularly temperature and concentration. Strong, hot conditions present the greatest problems, and few materials except platinum, tantalum, fluorocarbon plastics, and brick-lined steel will resist 60 to 98% H_2SO_4 at 120 °C (250 °F). Other variables also influence the resistance of materials to H_2SO_4. The presence of oxidizing or reducing contaminants, velocity effects, solids in suspension, and galvanic effects can alter the serviceability of a particular material of construction.

It is unwise to select materials of construction for equipment that will handle H_2SO_4 solely on the basis of published corrosion data unless the conditions are adequately and specifically covered by the reference data or plant experience. The corrosion of metals in H_2SO_4 is complex, and an understanding of electrochemical theory is useful (Ref 1–3). Seemingly minor differences in impurities, velocity, or concentration may significantly impact service corrosion rates. Impurities such as halides generally increase corrosion. Aeration or the presence of oxidizing agents generally accelerates corrosion of nonferrous metals and reduces corrosion of stainless alloys, but the extent of these effects depends on specific conditions.

It is advisable, therefore, to consider all general corrosion data only as an indicator of relative resistance and as a guide for further review. Final selection of materials for specific equipment depends, of course, on such factors as allowable corrosion rate, desired mechanical and physical properties, fabrication requirements, availability, and cost (Ref 4).

Carbon Steel

General Resistance. Steel has been used for more than 100 years for handling concentrated H_2SO_4 at ambient temperatures under static and low-velocity conditions (<0.9 m/s, or 3 ft/s). A soft sulfate film forms that is highly protective unless physically disturbed. The actual corrosion rate of steel depends on temperature, acid concentration, iron content, and flow, because these parameters determine the dissolution rate of the protective sulfate film. Figure 1 shows the corrosion resistance of steel as a function of temperature for nonflowing conditions at concentrations above 65%. In the concentration range of 65 to 100%, steel is applicable for ambient-temperatures storage tanks and piping at low velocities. However, the corrosion rate of carbon steel may be excessive in the 80 to 90% acid concentration range for general applications. Within the 100 to 101% range, steel exhibits potentially high corrosion rates as the temperature exceeds 25 °C (75 °F) and is not recommended, except where low temperatures and velocities can be ensured. Steel is used in oleum service above 101% acid concentration at ambient and moderate temperatures.

Localized attack can occur even at flow velocities within the prescribed limits. Discontinuities such as short-radius elbows, excessive penetration of welds, and pipe mismatch may cause sufficient downstream turbulence to disturb the protective sulfate film, resulting in high corrosion rates. Weldments must be thoroughly inspected to ensure that they contain no slag, surface porosity, laps, excessive penetration, or other welding defects that might encourage accelerated corrosion. In addition, steel vessels and piping should be free of mill scale, or nonuniform corrosion may occur.

Hydrogen grooving is another form of localized attack that occurs on vertical or inclined surfaces exposed to the liquid phase. During the corrosion of steel by H_2SO_4, hydrogen is evolved. If produced in sufficient quantities, the hydrogen forms small bubbles that stream along preferred paths on vertical and inclined surfaces, disrupting the soft protective iron sulfate film. Channels and deep grooves may eventually form. Grooving is commonly observed in the tops of horizontal manways on the side of vertical storage tanks and on the top 180° of horizontal pipe runs (Fig. 2). In piping, stagnant conditions promote grooving; therefore, a minimum velocity of 0.3 m/s (1 ft/s) is often recommended.

Grooving, combined with erosion-corrosion, also occurs on the sidewalls of tanks (Fig. 3). Location of the liquid inlet in the roof near the shell has, in at least two cases, resulted in combined erosion-corrosion and hydrogen grooving, which caused catastrophic rupture of large storage tanks (Ref 5).

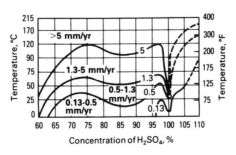

Fig. 1 Corrosion of steel by H_2SO_4 as a function of temperature and acid concentration. Source: Ref 3

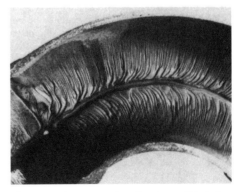

Fig. 2 Hydrogen grooving of a 75 mm (3 in.) diam steel elbow. The elbow was sectioned; the top half is shown.

Influence of Copper. The addition of copper to steels for improved resistance to H_2SO_4 has been studied with varied results (Ref 6, 7). It is generally believed that the advantages, if any, of the use of copper-bearing steels for handling strong H_2SO_4 are not sufficient to warrant the additional expense.

Anodic protection has been used to lower corrosion rates on steel tanks storing 93 to 99% H_2SO_4. The intent is to prolong tank life and minimize iron pickup. Typically, corrosion rates can be lowered by up to 80%. In this process, the steel is electrochemically driven from the active to the passive region by an applied potential between a cathode and the steel. If proper passivity is achieved, the corrosion rate decreases to less than 0.1 mm/yr (4 mils/yr) (Ref 8). The use of stainless steel and alloy 20Cb-3 in tanks being protected must be carefully analyzed because the passive protection potential for steel may overlap active regions for these alloys. With improvements in the H_2SO_4 production process, acid with lower iron content is being produced. This acid is more corrosive to steel than the high-iron acid. Thus, the use of anodic protection has become increasingly important.

Mechanism of Corrosion Protection. Steel corrodes in the active state at all concentrations up to 100% H_2SO_4. At concentrations below 60 to 65%, the iron sulfate corrosion product readily goes into solution, and corrosion rates are high. At higher concentrations, the initial corrosion rate is high, but is quickly reduced by the accumulation of iron sulfate corrosion product on the surface. It has been shown that the dissolution and diffusion of ferrous sulfate from the surface is the rate-limiting step (Ref 9). The effects of velocity, concentration, and temperature on the corrosion process in piping and storage tanks have been modeled (Ref 10, 11). At concentrations from 65 to 100% at ambient temperatures, the sulfate layer diffuses into solution sufficiently slowly for the use of steel as a material of construction. Corrosion rates are typically 0.15 to 1.0 mm/yr (6 to 40 mils/yr). In the oleum range above 101% H_2SO_4 concentration (4% free SO_3), steel corrodes in the passive region because of the oxidizing effect of SO_3. Corrosion rates are typically <0.1 mm/yr (<4 mils/yr) at ambient temperature. In the 100 to 101% concentration range, the steel goes through an active-passive transition state, and measured corrosion rates are erratic and excessive above 25 °C (75 °F).

Cast Irons

Gray cast iron has been used for piping and coolers for H_2SO_4 since acid plants were first built. Due to the brittle nature of gray cast iron and the availability of alternate materials, it has fallen out of favor over the last 20 years. Gray cast iron is more resistant to corrosion than steel in the 65 to 100% acid concentration range at ambient temperature. The higher carbon and silicon content of cast irons leads to conditions favoring superior resistance at higher velocities and elevated temperatures, at least in the concentration range of 90 to 100%. Gray cast iron is less sensitive to velocity than steel and is frequently used up to 1.7 m/s (6 ft/s) in larger-diameter piping. The superior resistance of cast iron may be due to interference of the graphite flake network with the reaction between the acid and the metallic matrix (Ref 12). Graphite flakes may act as cathodic areas and shift the corrosion potential to a more favorable region (Ref 13).

Gray cast iron normally should not be used in the oleum range because of a tendency toward cracking. Free SO_3 is thought to attack the graphite flakes. Corrosion products form in the voids and strain the structure by volumetric expansion. Cracking of gray cast iron has also occurred in concentrated H_2SO_4 service, in which the cast iron was attached to stainless steel coolers protected by anodic protection. It is theorized that the anodic current causes the cast iron to behave as though it is in oleum service. This problem has been solved by placing 3 m (10 ft) of ductile iron piping between the cooler and the gray cast iron pipe. Ductile cast iron is not subject to the same phenomena as gray cast iron, because the graphite is in the form of isolated nodules and the metal matrix is substantially stronger.

Gray cast iron is brittle and can rupture without warning. Care must be taken to support cast iron pipe properly and to replace it before it thins excessively because of corrosion. For replacement or new piping, ductile cast iron, stainless steels with or without anodic protection, and plastic-lined pipe are now favored.

Ductile cast iron performs nearly as well as gray cast iron in concentrated H_2SO_4 at ambient temperatures. At moderate to higher temperatures, gray cast iron has lower corrosion rates, but ductile iron is often used as it is less brittle. Laboratory testing comparing the materials has been inconclusive, probably because chemical compositions of the iron vary considerably. Actual plant experiences with piping tend to indicate up to 50% higher corrosion rates for ductile cast iron versus gray cast iron, but this can be compensated for by the use of additional corrosion allowance. Ductile iron clearly has lower temperature limits than gray cast iron, and its use at higher temperatures needs to be carefully monitored.

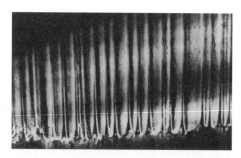

Fig. 3 Hydrogen grooving on the sidewall of an H_2SO_4 storage tank

A modified ductile cast iron, with the trade name of "MONDI," was introduced in the 1980s and reportedly has better corrosion resistance than conventional ductile cast iron. This alloy has about 3.5% Si versus the 1.8 to 2.8% for standard ductile iron. This ductile iron has been used at 65 to 70 °C (150 to 160 °F) in 93% acid and 110 to 120 °C (230 to 250 °F) in 98 to 99% acid.

High-Silicon Cast Iron. Another material that has long been used for handling H_2SO_4 is high-silicon iron. Iron with 14.5% Si has exceptional resistance to H_2SO_4 in all concentrations to 100% up to the atmospheric boiling points. Corrosion rates are normally less than 0.12 mm/yr (5 mils/yr), as shown in Fig. 4 (Ref 14).

The corrosion resistance of high-silicon iron is due to the formation of a strong silicon-rich abrasion-resistant film. Even the most severely abrasive slurries can be handled. High-silicon irons are rapidly attacked in oleum or other services containing free SO_3. Only SO_3, SO_2, and fluorine/fluoride contaminants are known to drastically alter the corrosion resistance of high-silicon iron to H_2SO_4. The high-silicon irons are available only in the cast form. In addition, they have low tensile strength and virtually no ductility. High-silicon irons are susceptible to thermal and mechanical shock failures.

Austenitic Stainless Steels

The resistance of austenitic stainless steels to H_2SO_4 is complex due to the active-passive nature of the alloys. An excellent summary that includes corrosion rate data is provided elsewhere (Ref 15).

Mechanism of Protection. Stainless steels depend on electrochemical passivity for resistance to corrosion in H_2SO_4 solutions. Stable passivity is achieved at ambient temperatures in the very low and very high acid concentrations and in oleum.

Corrosion Resistance. At ambient temperatures, austenitic stainless steels, for example, type 304 and 316, exhibit stable passivity in H_2SO_4 above 93% concentration and are frequently used for piping and tankage where product purity is desirable. The corrosion rates are essentially nil. Molybdenum stretches the

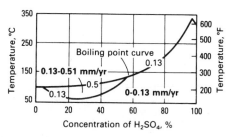

Fig. 4 Corrosion of high-silicon cast iron in H_2SO_4 as a function of temperature and acid concentration. Source: Ref 14

passive region, making type 316 and 317 acceptable for service in 90 to 93% H$_2$SO$_4$ at ambient temperature. Type 316 is the preferred choice for ambient temperature 93% service as type 304 may be active-passive at slightly lower concentrations. The upper temperature limit for stable passivity for types 304 and 316 in 93% H$_2$SO$_4$ is believed to be around 40 °C (105 °F). For 98.5% H$_2$SO$_4$, the upper stable passive limit is believed to be above 70 °C (160 °F) (Ref 16). As concentration increases above 99%, corrosivity decreases rapidly, allowing the use of stainless steels above 100 °C (212 °F).

In dilute acid, the molybdenum grades, such as type 316 and 317, are useful, although type 304 may be used when only a trace of acid is present. Figure 5 shows corrosion data for these alloys in as-mixed and refluxed (aerated) H$_2$SO$_4$ solutions (Ref 17). Stainless steels have poor resistance to deaerated dilute solutions. Higher chromium is beneficial for oxidizing conditions. Type 310 stainless steel with 25% Cr and no intentionally added molybdenum is more resistant than the lower-chromium molybdenum-bearing grades when oxidizing agents are present (Ref 15).

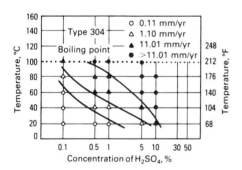

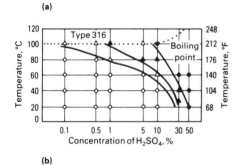

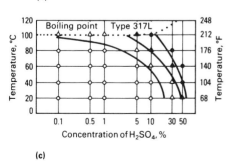

Fig. 5 Isocorrosion diagrams for (a) type 304, (b) type 316, and (c) type 317L stainless steels in aerated H$_2$SO$_4$ up to 50% concentration. Source: Ref 17

Effect of Velocity. If a stainless steel is in solidly passive behavior, velocity appears to have little effect on corrosion rate. Laboratory tests of type 304L with velocities to 6 m/s (20 ft/s) in 93% H$_2$SO$_4$ at ambient temperature have shown passive behavior (Ref 16). However, once the alloy drops to active-passive behavior, usually because of increasing temperature, velocity has a major effect. Under abrasive conditions, cast stainless steels have shown active-passive behavior in 96% H$_2$SO$_4$ even at ambient temperatures (Ref 18).

Effect of Aeration and Oxidants. Highly aerated solutions are much more suitable for stainless steels than air-free ones. Similarly, the presence of oxidizing impurities stabilizes the passive film, and the resistance to H$_2$SO$_4$ of austenitic stainless steels improves markedly. Cations that are easily reducible, such as Fe^{3+}, Cu^{2+}, stannic (Sn^{4+}), and cerric (Ce^{4+}) ions, are oxidizing agents that can inhibit the attack of stainless steels in H$_2$SO$_4$ solutions. It was found that 0.19 g/L of Fe^{3+} ion was sufficient to cause passivity and low corrosion rates in boiling 10% H$_2$SO$_4$, but 0.115 g/L did not give inhibition (Ref 19).

Other oxidizing agents, such as chromic acid (H$_2$CrO$_4$) and nitric acid (HNO$_3$), were shown to be effective in reducing corrosion rates (Ref 20). Nitric acid concentrations as low as 1.5% mixed into H$_2$SO$_4$ were found to inhibit the corrosion of stainless steel over a wide range of H$_2$SO$_4$ concentrations at ambient and elevated temperatures (Ref 21). Oxidants in sufficient quantities were shown to reduce the corrosivity of H$_2$SO$_4$ on stainless steel by shifting the corrosion potential from an active to a passive state (Ref 22).

Effect of High Concentration and SO$_3$. In strong H$_2$SO$_4$ (above 98% concentration) and in oleum, the oxidizing power of the acid increases dramatically and corrosion rates are reduced (Ref 23, 24). Figure 6 shows an isocorrosion

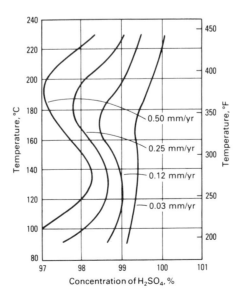

Fig. 6 Isocorrosion diagram of type 304L stainless steel in highly concentrated H$_2$SO$_4$. Source: Ref 24

diagram generated for type 304L stainless steel in an absorption tower environment. Most stainless steels and nickel-base alloys have similar reductions in corrosion rates with increasing concentration. However, the higher molybdenum-containing alloys, such as alloys B-2, C-276, 904L, and 20Cb-3, are distinctly inferior (Table 1).

Extreme care must be taken when using stainless steels in the 98 to 100% concentration range at high temperatures as velocity conditions, reductions in acid concentration, or changes in oxidant levels may initiate high corrosion rates. For contrast, compare the higher corrosion rates generated in flowing 98.7% H$_2$SO$_4$ at 100 °C (212 °F) shown in Table 2 (Ref 25) with the data for generally static conditions in Table 1. In the oleum range, stainless steels with higher chromium and low-to-nil molybdenum are free from concerns about minor acid concentration variations, and corrosion resistance is extended well in excess of 100 °C (212 °F).

Anodic protection is a practical method of extending the useful temperature and concentration range for stainless steels in H$_2$SO$_4$. Nearly all new acid coolers are built with anodic protection. With anodic protection, a stainless steel component (anode) is held in the passive condition by an impressed current from a cathode. Fortunately, H$_2$SO$_4$ is a good conductor of electricity and therefore has good throwing power. Complex shell sides of stainless steel shell and tube coolers handling concentrated H$_2$SO$_4$ can be easily protected using alloy C-276 cathodes that extend the full length of the tube bundle. In these units, the tubes joints are welded, not rolled, so that anodic current can protect the tube to tubesheet crevice. Piping is more difficult to protect because the current must be thrown a greater distance from the cathodes,

Table 1 Corrosion rates of various metals in 99% H$_2$SO$_4$ at 100–120 °C (212–250 °F)

Alloy	Corrosion rate	
	mm/yr	mils/yr
Steel	>2.4	94.5
Cast iron	0.12	4.7
Ductile iron	0.25	9.8
Type 304L	0.02	0.8
Type 316L	0.06	2.4
Alloy 904L	0.19	7.5
Alloy 20Cb-3	0.08	3.1
Alloy C-276	0.33	13.0
Alloy B-2	2.3	90.6
A-611	0.04	1.6
E-Brite 26-1	<0.01	0.4

Table 2 Corrosion test in flowing 98.7% H$_2$SO$_4$ at 100 °C (212 °F)

Alloy	Corrosion rate	
	mm/yr	mils/yr
Type 304	0.5	19.7
Type 316	3.44	135.0
Alloy 904L	2.3	90.6

Source: Ref 25

which are typically located every 4.5 to 6 m (15 to 20 ft). Stainless steel may be protected in 93% H_2SO_4 up to 60 to 70 °C (140 to 158 °F) and in 98 to 99% H_2SO_4 up to 120 °C (248 °F). Corrosion rates can be reduced to 0.01 to 0.1 mm/yr (0.4 to 4 mils/yr).

Silicon Stainless Steels. Austenitic stainless steels containing 5 to 6% Si have been developed for concentrated H_2SO_4 service. Cast and wrought versions are available. The cast version, UNS J93900, has a typical composition of Fe-5Si-21Cr-16Ni-0.02C (Ref 26). The wrought version of this alloy UNS S30601 has a typical composition of Fe-5.3Si-18Cr-18Ni-0.02C. The wrought alloy has shown useful corrosion resistance in 93% H_2SO_4 up to 70 °C (158 °F) and in 99% H_2SO_4 up to 120 °C (248 °F) without anodic protection. Anodic protection extends the temperature and concentration limits. UNS S32615 and UNS S38815 are other high-silicon stainless steels reported to have resistance to hot concentrated acid. Over the last 20 years, piping, distributors, pump tanks, coolers, and absorbing towers handling hot 93 to 99% H_2SO_4 have been made from silicon stainless steels.

Corrosion protection is obtained by the formation of a tenacious silicon-rich film formed on the surface during the initial days of corrosion. The silicon-rich film is attacked by oleum and the corrosion resistance is destroyed. Severe corrosion has occurred in absorbing towers where stagnant acid touching the wall in and below the packing area was locally enriched in SO_3, allowing oleum formation. The silicon stainless steel alloy manufacturer should be consulted for the specifics of the alloy performance.

Cast stainless steels have essentially the same corrosion resistance to H_2SO_4 as their wrought counterparts. Because the cast versions contain second-phase ferrite for castability, care must be taken that proper heat treatment is performed for maximum corrosion resistance. Preferential corrosion as the result of the duplex structure has been shown to be a problem. However, properly cast and heat treated duplex materials perform well (Ref 15).

One cast alloy that does not have a close wrought counterpart is ACI CD-4MCu (UNS J93370). Its corrosion resistance lies between alloy 20Cb-3 and type 316. Figure 7 shows the isocorrosion diagram for this cast alloy. Useful corrosion resistance, less than 0.5 mm/yr (20 mils/yr), of CD-4MCu in H_2SO_4 extends over the entire concentration range at ambient temperatures, and the alloy is used well above 100 °C (212 °F) in oleum service.

Higher Austenitic Stainless Steels

Like the austenitic stainless steels, the corrosion resistance of the higher austenitic stainless steels is also complex. However, the range of passivity and corrosion resistance is extended because of the higher alloy content. Like the stainless steels, active and active-passive electrochemical behavior are the modes of corrosion. Resistance is achieved in the passive state. A summary that includes corrosion data is provided in Ref 15.

The 20-type alloys are usually the first considered when an H_2SO_4 environment is too corrosive for the use of steel, 300 series stainless steels, or cast iron. This group contains both wrought alloy 20Cb-3 (N08020) and cast alloy ACI CN-7M (N08007) that are roughly equivalent in resistance to H_2SO_4. These alloys contain 20Cr-30/34Ni-2.5Mo-2.5Cu-bal Fe.

Cast ACI CN-7M. Figure 8 shows an isocorrosion diagram for cast ACI CN-7M. This alloy is generally suitable to 80 °C (175 °F) at concentrations to 50%. For higher concentrations, good corrosion resistance is expected to 65 °C (150 °F).

Wrought Alloy 20Cb-3. The wrought counterpart to cast ACI CN-7M was developed in 1947. In 1948, niobium was added to this alloy for stabilization against sensitization and intergranular attack. In 1963, the nickel content was raised to 33 to 35% in order to give greater resistance to chloride stress-corrosion cracking (SCC) and to improve resistance to boiling H_2SO_4 under heat-transfer conditions. Minor changes were subsequently made to impart greater resistance to intergranular corrosion. This alloy, now known as alloy 20Cb-3, has a typical composition of Fe-34Ni-20Cr-3.3Cu-2.5Mo. Corrosion resistance of alloy 20Cb-3 is similar to that of CN-7M. Figure 9 shows an isocorrosion diagram for alloy 20Cb-3.

Iron-Base Ni-Cr-Mo-Cu Alloys. Figure 10 shows an isocorrosion diagram for a 25Ni-20Cr-4.5Mo-1.5Cu alloy such as alloy 904L (UNS N08904). The copper addition of 1.5% makes the alloy suitable for the entire concentration range at ambient temperature. Other alloys in this family (UNS N08926 and N031254) with less copper and more molybdenum have similar resistance.

Higher Chromium Fe-Ni-Mo Alloys

Alloy 31 (UNS N08031), also known as Nicrofer 3127 hMo, was developed in the late 1980s and was designed for increased temperature limits in H_2SO_4 over the 20 to 80% composition range versus other stainless steels and nickel-base alloys. The typical composition of this alloy is 32Fe-31Ni-27Cr-6.5Mo-1.2Cu. It shows useful resistance in 0 to 60% acid at 85 °C (185 °F) and in 60 to 80% acid at 70 °C (160 °F). Its short-term isocorrosion diagram is shown in Fig. 11 (Ref 27). Similar alloys with less chromium (N08026) or less molybdenum (N08028) have slightly less resistance.

High Cr-Fe-Ni Alloy

Alloy 33 (R20033) was specifically designed for high-temperature, concentrated sulfuric acid in 1995. The typical composition is 33Cr-32Fe-31Ni-1.6Mo-0.4N. The high chromium and low molybdenum content provides excellent resistance in hot concentrated oxidizing acid. Corrosion data summarizing alloy 33 versus other alloys is shown in Tables 3 and 4 (Ref 28, 29).

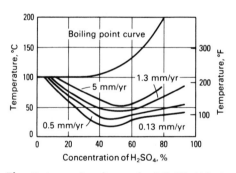

Fig. 7 Isocorrosion diagram for ACI CD-4MCu in H_2SO_4. Source: Ref 15

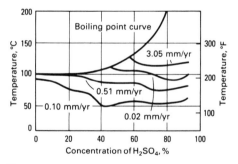

Fig. 9 Isocorrosion diagram for alloy 20Cb-3 in H_2SO_4. Source: Ref 15

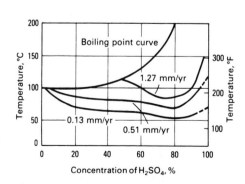

Fig. 8 Isocorrosion diagram for ACI CN-7M in H_2SO_4. Source: Ref 15

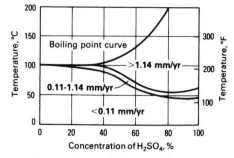

Fig. 10 Isocorrosion diagram for alloy 904L in H_2SO_4. Source: Ref 15

Nickel-Base Alloys

The nickel-base alloys have good resistance to corrosion in H_2SO_4 up to 95% concentration because of their high alloy content. Frequently, low corrosion rates occur in both the active and passive corrosion states. Thus, reliable corrosion behavior is achieved over a wide range of concentrations, temperatures, and impurity levels. Extensive discussion, data, and references are given in Ref 15 and 30 to 32.

Nickel-Base Fe-Cr-Mo-Cu Alloy. Alloy 825 (N08825) has useful resistance to H_2SO_4 up to 40% concentration and in concentrated acid (Fig. 12). The higher nickel content of this alloy versus the 20-type alloys offers slightly improved corrosion resistance at high temperatures in low and high concentrations of H_2SO_4.

Nickel-Base Cr-Fe-Mo-Cu Alloys. The nickel-base "G" alloys (G, G-3, and G-30) are modifications of the now obsolete Hastelloy F. A typical composition of G-30 (N06030) is 43Ni-30Cr-15Fe-5.5Mo-5Co-2.7W-1.7Cu. Alloys G-3 and G-30 are modified versions of alloy G with the same general corrosion resistance. It was modified to resist formation of grain-boundary precipitates during prolonged heating, such as stress relief. These alloys are promoted for their resistance to halide-contaminated H_2SO_4 environments. Figures 13 and 14 show isocorrosion diagrams for alloy G.

Nickel-Copper Alloys. Alloy 400 is used for handling H_2SO_4 under reducing conditions. Thus, this alloy offers an alternative to stainless steels and other alloys exhibiting active-passive behavior when H_2SO_4 solutions are not oxidizing, such as deaerated dilute acid. Alloy 400 exhibits reasonably low corrosion rates in air-free H_2SO_4 up to 85% concentration at 30 °C (86 °F) and up to 60% concentration at 95 °C (203 °F).

Nickel-Base Molybdenum-Chromium Alloys. The nickel-base "C" alloys, containing about 16% Mo and 16 to 22% Cr are available in wrought (alloy C-276) (UNS N10276) and cast forms (ASTM A494, grades CW-12MW, CW-6M, CW-2M, CX2MW). In general, the corrosion resistance is excellent (Fig. 15, 16). At room temperature, the corrosion rate for all concentrations is less than 0.1 mm/yr (4.0 mils/yr). Other wrought alloys in the "C" family with similar resistance are N06686, N06059, and N06022. Because of the higher chromium content, these "C" alloys are more resistant to H_2SO_4 containing oxidizing contaminants than alloy 400 or the "B" alloys. Another material in this group, with about 9% Mo, is alloy 625 (Fig. 17) (Ref 30).

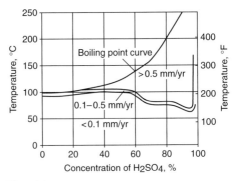

Fig. 11 Isocorrosion diagram for alloy 31 (Nicrofer 3127 hMo) in H_2SO_4. Source: Ref 27

Table 3 Corrosion test in a sulfuric acid plant, 99.1% H_2SO_4, >1.2 m/s (>3.9 ft/s) velocity, 150 °C (300 °F), 134 days

Alloy	Corrosion rate	
	mm/yr	mils/yr
Alloy 825	1.46	57.5
Type 316 Ti	0.81	31.9
Alloy 690	0.09	3.5
Alloy 33	<0.01	<0.4

Source: Ref 28, 29

Table 4 Corrosion test in a sulfuric acid plant, 96 to 98.5% H_2SO_4, >1 m/s (>3.3 ft/s), 135 to 140 °C (275 to 285 °F), 14 days

Alloy	Corrosion rate	
	mm/yr	mils/yr
Type 316 Ti	0.24	9.5
Type 304	0.16	6.3
Alloy G-30	0.08	3.1
Alloy A-611	0.03	1.2
Alloy 33	<0.01	<0.4

Source: Ref 28, 29

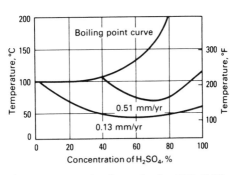

Fig. 12 Isocorrosion diagram for alloy 825 in H_2SO_4. Source: Ref 15

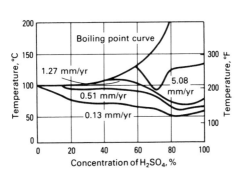

Fig. 13 Isocorrosion diagram for alloy G in H_2SO_4. Source: Ref 15

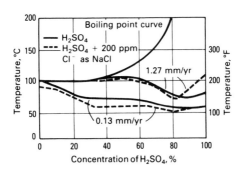

Fig. 14 Isocorrosion diagram for alloy G in H_2SO_4 solutions contaminated with Cl^- ion. Source: Ref 15

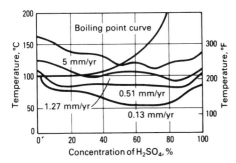

Fig. 15 Isocorrosion diagram for alloy C-276 in H_2SO_4. Source: Ref 15

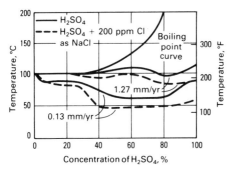

Fig. 16 Isocorrosion diagram for alloy C-276 in H_2SO_4 and in H_2SO_4 contaminated with Cl^- ion. Source: Ref 15

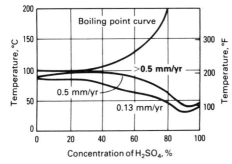

Fig. 17 Isocorrosion diagram for alloy 625 in H_2SO_4. Source: Ref 30

Nickel-Base Molybdenum Alloys. The "B" alloys contain about 28% Mo and 1% Cr and are available in the cast and wrought forms. The corrosion resistance of these alloys is excellent in pure H_2SO_4 over a wide range of temperatures and concentrations. However, oxidizing contaminants, such as Fe^{3+} ions, increase corrosion rates considerably and have caused premature failure. Chlorides also increase corrosion rates. Alloys B-2 (N10665) and B-3 (N10675) are the wrought materials in this alloy group (Fig. 18). The cast version with similar corrosion resistance is ASTM A494, grade N-12MV.

Nickel-base Cr-Fe-Co-Si alloys were developed specifically for hot concentrated H_2SO_4 pump and valve applications. Proprietary alloy 55 is one such cast alloy (Fig. 19). Figure 20 shows the effect of temperature in 98% H_2SO_4 as determined by laboratory tests. Alloy 66 is a ductile cast alloy that can also be made in the wrought form. Both forms have useful corrosion resistance in the 0 to 60% and 80 to 99% ranges, but performance is erratic in the 60 to 80% range (Fig. 21) for alloy 66.

Nickel-base Cr-Mo-Cu alloys are proprietary alloys designed to resist H_2SO_4 concentrations to 98% at temperatures to 100 °C (212 °F). Illium 98 is a weldable, machinable cast alloy. Illium B, also a cast alloy, is a version of 98 that has been modified for enhanced corrosion resistance; however, Illium B is not easily welded (Fig. 22).

Nickel-Base Chromium-Silicon Alloy. The D-205 (Haynes International) nickel-base alloy was first marketed in the 1990s to provide outstanding resistance to high-temperature concentrated sulfuric acid with the added capability of fabrication in thin sections and complex shapes. Its composition is 65Ni-20Cr-6Fe-5Si-2.5Mo-2Cu-0.03C. Its advantage over the high-silicon stainless steel is formability in thin sections. It is ideal for plate and frame heat exchangers. Corrosion resistance at less than <0.1 mm/yr (<4 mils/yr) is reported for 93% acid at 90 °C (195 °F) and for 98 to 99% acid at 130 °C (265 °F) (Ref 32). It is also reported to have excellent resistance to concentrated acid with oxidizing impurities.

Other Metals and Alloys

Zirconium. The corrosion resistance of zirconium (R60702) in H_2SO_4 depends on the formation of a highly ordered and corrosion-resistant passive film. However, as temperature and/or concentration increases, the transpassive (breakdown) potential decreases, and there is less tolerance for oxidizing agents, such Fe^{3+}, Cu^{2+}, and nitrate (NO_3^-) ions. Thus, zirconium has excellent resistance to H_2SO_4 up to 50% concentration at temperatures to boiling and above. From 50 to 65% concentration, resistance is generally excellent at elevated temperatures, but the passive film is a less effective barrier. Experience has shown that welding and the presence of oxidizing species in more than 50% H_2SO_4 can encourage selective attack. In concentrated H_2SO_4 above 70%, the corrosion rate of zirconium increases rapidly with increasing concentration (Ref 33). Zirconium is generally not used above 55% concentration in the welded condition unless it has been heat treated to maximize corrosion resistance (Ref 34). Figure 23 shows the corrosion rates of zirconium as a function of temperature and H_2SO_4 concentration. Control of tin content helps control harmful second-phase particles and improves corrosion resistance in the welded condition (Ref 35). Stress cracking has been reported in the base alloy in the 64 to 69% concentration range (Ref 36).

Tantalum and Ta-2.5W alloys have resistance to H_2SO_4 over the entire range of concentration and temperature except for very strong and exceptionally hot conditions. Figure 24 shows the corrosion rates of tantalum in 98% acid and oleum. A common use is for reboilers used for the concentration of H_2SO_4 in the 70% concentration range. Recent long-term testing has shown that the Ta-2.5W alloy performs at least as well as unalloyed tantalum in hot concentrated sulfuric acid service. In addition, the Ta-2.5W alloy shows clear superiority in hot concentrated acid containing oxidizing impurities such as nitrates (Ref 37, 38).

Unalloyed titanium is rapidly attacked by all concentrations of H_2SO_4 except very dilute

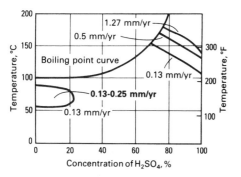

Fig. 18 Isocorrosion diagram for alloy B-2 in H_2SO_4. Source: Ref 15

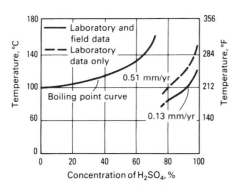

Fig. 19 Isocorrosion diagram for alloy 55 in H_2SO_4. Source: Ref 15

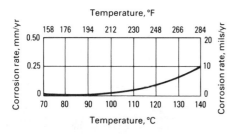

Fig. 20 Corrosion of alloy 55 in 98% H_2SO_4. Source: Ref 15

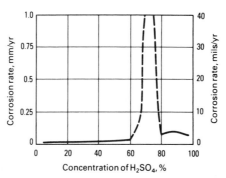

Fig. 21 Corrosion rates of alloy 66 in H_2SO_4 solutions of 100 °C (212 °F). Source: Ref 15

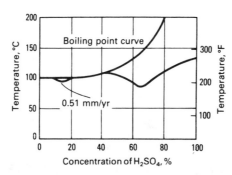

Fig. 22 Isocorrosion diagram for Illium B in H_2SO_4. Source: Ref 15

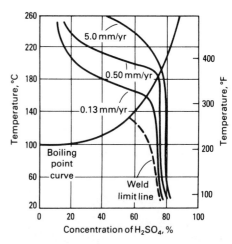

Fig. 23 Corrosion of zirconium by H_2SO_4 as a function of temperature and acid concentration. Source: Ref 33, 34

solutions. Impurities in the form of oxidizing agents may act as inhibitors; for example, titanium resists 0 to 50% concentrations of H_2SO_4 saturated with chlorine (Ref 39, 40). Titanium has also been used in the H_2SO_4 leaching of nickel ores. Resistance is attributed to the presence of heavy-metal oxidizing agents, such as Fe^{3+} and Cu^{2+} ions (Ref 41, 42).

Lead resists H_2SO_4, but its protective sulfate film is increasingly soluble above 90% concentration. The film is easily damaged by erosion or abrasion even at low velocities, with the rate of attack increasing rapidly with concentration. Lead is useful as a pan material to catch acid drippings. It also has been used as a membrane behind brick in H_2SO_4 concentrating units and scrubbing towers.

Copper and copper-base alloys are not widely used in relatively pure H_2SO_4 because of the drastic effect of oxidizing conditions. The resultant Cu^{2+} ions also cause autocatalytic corrosion. Brass is not often used because of the potential for dezincification. Bronzes have the greatest applicability, with the normal copper-tin bronzes showing acceptable service below 60% concentration at 79 °C (174 °F). The silicon bronzes extend the suitable range to about 70% concentration. The aluminum bronzes have resistance approaching that of alloy 400; however, dealuminification can be a problem, particularly with alloys containing more than 8% Al (Ref 4).

Nonmetals

Nonmetallic materials of construction have wide application in H_2SO_4. Most of these materials have good corrosion resistance to the pure acid, particularly in dilute concentrations. They are relatively unaffected by most inorganic contaminants, except for strongly oxidizing agents such as HNO_3, peroxides, and dichromates. Materials selection and design are normally done in collaboration with knowledgeable materials suppliers and fabricators.

Brick linings have been used for the most severe H_2SO_4 conditions. Under most conditions, acid-resistant fire clay brick can be used. Bricks and ceramics can be attacked at extremely high temperatures, and performance should be monitored or testing done. At concentrations below 70% lead, plastic, or elastomeric corrosion-resistant membranes are required behind the brick. The brick serves as a thermal and abrasion barrier. At higher concentrations, brick can be used directly against steel, although polytetrafluoroethylene (PTFE) or mortar barriers are often used. For moderate concentrations and temperatures, plastic mortars, such as furans and phenolics, are used. For high temperatures and concentrations, the silicate cements are required. Potassium silicate cement is now preferred instead of sodium silicate because its sulfate salt does not hydrate. Mortars based on pure colloidal silica are also available for extreme conditions.

Polyvinyl chloride (PVC) has excellent resistance to H_2SO_4 up to at least 93% concentration at ambient temperature. Chlorinated PVC resists slightly higher concentrations and temperatures. These materials have been used for piping, but are not popular for longer runs, because they have high coefficients of expansion and require essentially continuous support. Some companies prohibit the use of solid nonreinforced plastic piping because of the potential for breakage.

Lined Pipe. Polyvinylidene chloride (PVDC), polypropylene (PP), polyvinylidene fluoride (PVDF), PTFE, and perfluoroalkoxy (PFA) are used for linings in pipe for handling H_2SO_4. The temperature range of chemical resistance is shown in Table 5. At elevated temperature, performance is also limited by mechanical properties and permeation.

Polyethylene is reported to be resistant to attack by H_2SO_4 up to 98% concentration at ambient temperature for short duration service. Polyethylene is widely used in laboratories for beakers, bottles, and so on, which do not require heating over direct flames. The use of polyethylene equipment for industrial handling of H_2SO_4 has mixed reviews. Tanks constructed of molded, seamless, high-density, cross-linked polyethylene (XLPE) have been used to store relatively small volumes of concentrated H_2SO_4 up to 96%. However, failures due to acid attack have been reported, and some vendors restrict life guarantees for high concentrations. Some operators limit the use of solid nonreinforced plastic tankage due to performance concerns.

Fluoroplastics. Fully fluorinated plastics, such as PTFE, fluorinated ethylene propylene (FEP), and PFA, are unattacked by H_2SO_4 and oleum at all concentrations. Other fluoroplastics, including PVDF, ethylene-chlorotrifluoroethylene (ECTFE), and ethylene-tetrafluoroethylene (ETFE), are resistant to acid up to 98% concentration, but have less resistance to oleum and SO_3. Table 6 lists temperature limitations for chemical attack. At elevated temperature, fluoroplastic linings are limited by mechanical properties, permeation, and adhesives.

Fluoroplastics are being increasingly used for H_2SO_4 applications, such as linings in pumps and valves. Fiberglass-backed sheet linings of FEP, PTFE, and PVDF permit fabrication of large components, such as tanks. Rotomolding and dispersion powder linings of PVDF, ECTFE, ETFE, and PFA, with or without fiber reinforcement, are applied to a broad range of steel equipment (Ref 45). Equipment size is limited only by available ovens.

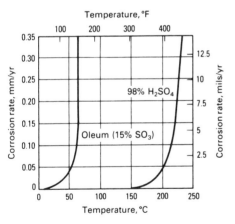

Fig. 24 Corrosion of tantalum in 98% H_2SO_4 and in oleum

Table 5 Chemical resistance of lined-pipe systems to sulfuric acid

	Maximum temperature					
	PP		PVDF		PTFE/PFA(a)	
Corrosive stream, sulfuric acid, %	°C	°F	°C	°F	°C	°F
10%	110	225	120	250	230	450
16–60%	95	200	120	250	230	450
60% saturated with Cl_2	25	75	95	200	230	450
78%	52	125	95	200	230	450
93–98%	NR	NR	65	150	230	450
>98% and oleum	NR	NR	NR	NR	230	450

(a) PTFE/PFA is chemically resistant to at least 230 °C (450 °F); however, at temperatures above 130 °C (265 °F) life may be limited due to permeation and mechanical properties. NR, not resistant. Source: Ref 43, 44

Table 6 Chemical resistance temperature limits for fluoroplastic linings in sulfuric acid

	Maximum temperature(a)		
Material	°C	°F	Concentration
PTFE	230	450	All
PFA	230	450	All
FEP	205	400	All
ECTFE	150	300	<98
ETFE	150	300	<98(b)
PVDF	120	250	<16
	95	200	30–60
	65	150	85–93
	50	120	94–98(b)

PTFE, polytetrafluoroethylene; PFA, perfluoroalkoxy; ECTFE, ethylene chlorotrifluoroethylene; FEP, fluorinated ethylene propylene copolymer resin with TFE; ETFE, modified ethylene-tetrafluoroethylene; PVDF, polyvinylidene fluoride. (a) Above 110 °C (230 °F), service as a lining may be limited by permeation, mechanical properties, thermal expansion, and adhesives. (b) PVDF and ETFE may be attacked by free SO_3 present at concentrations slightly higher than 98%. Source: Ref 43, 44

Polyester Resins. Polyester and vinyl-ester resins, normally used to form a fiberglass-reinforced plastic (FRP) laminate, are available to resist dilute H_2SO_4 at 93 °C (200 °F). At 50 to 70% concentration, FRP tanks and vessels with proper resins can be safely used to 65 °C (150 °F) and can even find application in special cases where the temperature approaches 80 °C (175 °F). At 70 to 75% acid, the temperature should be restricted to less than 40 °C (104 °F). Above 80% concentration, most polyester and vinyl-ester resins are rapidly degraded even at ambient temperatures.

Carbon and Impervious Graphite. Impervious graphite has excellent corrosion resistance to all but highly oxidizing concentrations of H_2SO_4. Impervious graphite coolers are used in 93% H_2SO_4 (Ref 46). Phenolic and PTFE impregnants are readily available. Carbon is somewhat superior to graphite in the higher concentrations and is generally preferred for hot acid at concentrations above 60%. Carbon is less conductive and lacks strength and is therefore unsuitable for cooler construction. Both carbon and graphite are, of course, brittle (Ref 47, 48).

Phenolic Linings. Heat-cured phenolic linings on steel are often used to prevent iron contamination in 93% H_2SO_4 and have reasonable life in 98% H_2SO_4 at ambient temperature. These thin-film linings should not be used for concentrations below 70% because of rapid corrosion of exposed steel substrate at expected pinholes.

Butyl rubber and neoprene exhibit good resistance to 50% H_2SO_4 at modest temperatures and will resist 75% acid under ambient conditions.

Glass and glass-lined equipment is widely used in H_2SO_4 service. For severe H_2SO_4 applications, glass-lined steel is a popular material of construction.

REFERENCES

1. O.W. Siebert, *Mater. Perform.*, Vol 20 (No. 2), Feb 1981, p 38
2. H.S. Tong, "Corrosion and Electrochemical Behavior of Fe-Cr-Ni Alloys in Concentrated Sulfuric Acid Solutions," paper presented at Symposium on Progress in Electrochemical Corrosion Testing, American Society for Testing and Materials, May 20–25, 1979
3. M.G. Fontana, *Ind. Eng. Chem.*, Vol 43, Aug 1951, p 65a
4. "Materials of Construction for Handling Sulfuric Acid," Technical Committee Report 5A151, National Association of Corrosion Engineers, 1985
5. M. Tiivel and F. McGlynn, "Avoiding Problems in Sulfuric Acid Storage," paper presented at AIChE Meeting (New Orleans, LA), American Institute of Chemical Engineers, April 1986
6. E. Williams and M.E. Komp, *Corrosion*, Vol 21 (No. 1), Jan 1965, p 9–14
7. H. Endo and S. Morioka, "Dissolution Phenomenon of Copper-Containing Steels in Aqueous Sulfuric Acid Solutions of Various Concentrations," paper presented at the Japanese Metal Association Third Symposium, April 1938
8. D. Fyfe et al., *Chem. Eng. Prog.*, March 1977
9. B.T. Ellison and W.R. Schmeal, *Elec. Soc.*, Vol 125, 1978, p 524
10. S.W. Dean Jr. and G.D. Grab, "Corrosion of Carbon Steel by Concentrated Sulfuric Acid," paper 147, presented at Corrosion/84, National Association of Corrosion Engineers, 1984
11. S.W. Dean and G.D. Grab, "Corrosion of Carbon Steel Tanks in Concentrated Sulfuric Acid Service," paper 298, presented at Corrosion/85, National Association of Corrosion Engineers, 1985
12. M.G. Fontana and N.D. Greene, *Corrosion Engineering*, McGraw Hill, 1967
13. E. Maahn, *Br. Corros. J.*, Vol 1, Nov 1966
14. M.G. Fontana and N.D. Greene, *Corrosion Engineering*, McGraw Hill, 1967
15. "The Corrosion Resistance of Nickel-Containing Alloys in Sulfuric Acid and Related Compounds," Corrosion Engineering Bulletin 1, The International Nickel Company, 1983
16. J.E. Strutt, "Corrosion Resistance of Stainless Steels in 93% and 98.5% Sulfuric Acid," Materials Technology Institute, Sept 1985
17. H. Abo, M. Ueda, and S. Noguchi, *Boshoku Gijutso*, Vol 23, 1974, p 341–346 (in Japanese)
18. P.F. Wieser et al., *Mater. Protec.*, Vol 12 (No. 7), July 1973, p 34–38
19. M.A. Streicher, *Corrosion*, Vol 14 (No. 2), Feb 1958, p 59t–70t
20. G.C. Kiefer and W.G. Renshaw, *Corrosion*, Vol 8, Aug 1950, p 235
21. J.R. Auld, Effect of Heat Treatment and Welding on Corrosion Resistance of Austenitic Stainless Steels, *Proc. Second International Conference on Metallic Corrosion*, National Association of Corrosion Engineers, 1967
22. T.N. Anderson et al., *Metall. Trans.*, Vol IIA, Aug 1980
23. D.R. McAlister et al., "Heat Recovery From Concentrated Sulfuric Acid," U.S. patent 4,576,813, granted March 1986
24. D.R. McAlister et al., "A Major Breakthrough in Sulfuric Acid," paper presented at AIChE 1986 Annual Meeting (New Orleans, LA), American Institute of Chemical Engineers, April 1986
25. M. Renner et al., "Corrosion Resistance of Stainless Steels and Nickel Alloys in Concentrated Sulfuric Acid," paper 189, presented at Corrosion/86, National Association of Corrosion Engineers, 1986
26. D.J. Chronister and T.C. Spence, Influence of Higher Silicon Levels on the Corrosion Resistance of Modified CF-Type Cast Stainless Steels, *Proc. NACE Corrosion/85 Symposium on Corrosion in Sulfuric Acid*, National Association of Corrosion Engineers, 1985, p 75
27. Materials data sheet No. 4031, Thyssen-Krupp VDM, Aug 2002
28. F. White, M. Kohler, and M. Renner, "Alloy 33, An Optimized Material for Sulphuric Acid Service," presented at 1996 Sulfur Conference (Vancouver), Oct 1996
29. M.H. Renner and D. Michalski-Vollmer, "Corrosion Behavior of Alloy 33 in Concentrated Sulfuric Acid," paper presented at Stainless Steel World 1999 (La Haye, The Netherlands), Stainless Steel World
30. J.R. Crum and M.E. Adkins, Correlation of Alloy 625 Electrochemical Behavior with the Sulfuric Acid Isocorrosion Chart, *Proc. NACE Corrosion/85 Symposium on Corrosion in Sulfuric Acid*, National Association of Corrosion Engineers, 1985, p 23
31. N. Sridhar, "Mechanisms of Corrosion in Concentrated Sulfuric Acid," paper presented at Sulfur '85 International Conference (London), Nov 10–13, 1985
32. "Guide to Corrosion Resistance Alloys," H-2114B, Haynes International, 2001
33. R.T. Webster and T.L. Yau, Zirconium in Sulfuric Acid Applications, *Proc. NACE Corrosion/85 Symposium on Corrosion in Sulfuric Acid*, National Association of Corrosion Engineers, 1985, p 69
34. M.A. Maguire and T.L. Yau, "Corrosion-Electrochemical Properties of Zirconium on Mineral Acids," paper 265, presented at Corrosion/86, National Association of Corrosion Engineers, 1986
35. D. Holmes, "Effect of Heat Treatment and Tin Content on the Corrosion of Zirconium 702 in Sulfuric Acid," paper 04222, Corrosion/2004, NACE International
36. T. Yau, "Understanding Corrosion Behavior from Electrochemical Measurements," paper 04227, Corrosion/2004, NACE International
37. M. Coscia and M. Renner, Corrosion of Tantalum and Tantalum-2.5% Tungsten in Highly Concentrated Sulfuric Acids, *Mater. Perform.*, Vol 37, 1998, p 52
38. M. Renner et al., Application Limits of Tantalum and Ta-2.5%W for Sulfuric Acid Handling, *Mater. Corros.*, Vol 49, 1998, p 877–887
39. L.W. Gleekman, *Corrosion*, Vol 14, Sept 1958
40. P.J. Gegner and W.L. Wilson, *Corrosion*, Vol 15 (No. 7), 1959
41. J.P. Cotton, *Chem. Eng. Prog.*, Vol 66 (No. 10), Oct 1970
42. N.G. Feige, "The Industrial Applications of Titanium in the Chemical Industry," paper presented at the Symposium on Titanium-Zirconium for the Chemical Process Industries (New Orleans, LA), Nov 1975

43. Chemical Resistance Chart, Edlon-PSI, www.edlon.com/pdf/chemresi.pdf, 1998
44. Chemical Resistance Ratings, Crane Resistoflex, www.resistoflex.com/chooseliner.asp, May 2004
45. R.E. Tatnall and D.J. Kratzer, The Use of Fluoroplastics in Sulfuric Acid Service, *Proc. NACE Corrosion/85 Symposium on Corrosion in Sulfuric Acid,* National Association of Corrosion Engineers, 1985, p 85
46. J.R. Schley, *Chem. Eng.,* Feb 18, 1974
47. J.R. Schley, *Chem. Eng.,* March 18, 1974
48. E. Shields and W.J. Dessert, *Pollut. Eng.,* Dec 1981

SELECTED REFERENCES

- M. Davies, "Materials Selection for Sulfuric Acid," publication MS-1, Materials Technology Institute, 2005
- *DECHEMA Corrosion Handbook,* Vol 8, *Sulfuric Acid,* D. Behrens, Ed., John Wiley & Sons, 1991, 284 pages

Corrosion by Nitric Acid

Hira S. Ahluwalia, Material Selection Resources Inc.
Paul Eyre, Dupont
Michael Davies, Cariad Consultants
Te-Lin Yau, Yau Consultancy

NITRIC ACID (HNO_3) is one of the most widely used acids in the chemical processing industry. In the United States alone, there are more than 65 HNO_3 manufacturing facilities with a total capacity of more than 11 million tons/yr. Plants range in size from 6,000 to 700,000 tons/yr. Approximately 70% of all HNO_3 produced is used in the manufacturing of ammonium nitrate that is then used in fertilizers. Nitric acid is also a key component in the manufacturing of adipic acid and terephthalic acid. Other uses include the production of industrial explosives such as nitroglycerin and trinitrotoluene (TNT), dyes, plastics, synthetic fibers, and in metal pickling and the recovery of uranium.

The basic HNO_3 process involves the oxidation of a preheated air-ammonia mixture in the presence of a catalyst at approximately 925 °C (1700 °F). The resulting gas, consisting of various oxides of nitrogen, water vapor, and air, passes through heat-recovery equipment, a cooler condenser, and an absorption tower. Condensed acid in the concentration range 60 to 65% is drawn from the cooler condenser and absorption tower. Uncondensed vapor and air are recirculated. The higher the pressure in the condensing and absorption stage, the greater the process efficiency. This process can be performed at one or multiple pressures. Newer processes typically operate at a low and high pressure to favor the reactions. The azeotrope of HNO_3 and water at atmospheric pressure is 67% HNO_3. Higher concentrations of acid in the range 90 to 100% are produced by extractive distillation using concentrated sulfuric acid as a dehydrating agent. The terminology used to describe HNO_3 mixtures that are generally accepted is shown in Table 1 (Ref 1).

Nitric acid is a strongly oxidizing acid that is aggressively corrosive to many metals. Acid temperature and concentration affect corrosion rate as does the composition of the alloy exposed to it. General attack, crevice attack, and intergranular attack are common in HNO_3. Stress-corrosion cracking (SCC) can occur in zirconium alloys exposed to hot >70% HNO_3. Velocity, aeration, and the presence of impurities such as chlorine and fluorine in the acid all affect corrosion rate. Stainless steels, chromium-containing alloys, aluminum, and—to a lesser degree—titanium and zirconium alloys are useful in HNO_3. Carbon steel and copper alloys are not. The materials of construction used for handling HNO_3 will depend on the application and the type of equipment, acid concentration, and temperature. Table 2 provides a summary of the common materials used at various acid concentration ranges and temperatures (Ref 1).

Carbon and Alloy Steels

The ferrous alloys, consisting of iron plus carbon or other elements in small amounts, are passivated upon immersion in concentrated HNO_3, but the protective film is so easily damaged that they are not useful in HNO_3 applications. In practice, both wrought and cast ferrous alloys corrode very rapidly. The application of anodic protection to the corrosion of carbon steel has been studied and found to be effective in reducing the corrosion rate in various strengths of HNO_3 (Ref 2). The technique, however, has not found any industrial applications.

Stainless Steels

The presence of many of the common alloying elements in stainless steels can be either detrimental or dramatically beneficial to the resistance of an alloy in HNO_3. Chromium is considered the most beneficial element in providing corrosion resistance of stainless steels in HNO_3. If carbon, phosphorus, sulfur, titanium, or molybdenum is present, the general corrosion resistance of the alloy system will decrease with increasing concentration of that element. These elements either form separate phases that are susceptible to selective attack by HNO_3 or tie up chromium and increase the sensitivity of the alloy to selective attack at the grain boundaries. Addition of silicon to stainless steels increases their corrosion resistance at high acid concentrations.

The most widely used grade of stainless steels for HNO_3 is the low-carbon grade UNS S30403 (304L) and to a lesser extent the stabilized austenitic grades UNS S32100 (321) and UNS S34700 (347). The isocorrosion diagram for type 304 stainless steel in HNO_3 is shown in Fig. 1. This diagram shows the effect of temperature and HNO_3 concentration on the corrosion of type 304 stainless steel. Increasing either or both raises the corrosion rate; nevertheless, there is a large useful area extending from 0 to 90% concentration and up to the boiling point below 50% concentration in which the predicted corrosion rate is less than 0.13 mm/yr (5 mils/yr) (Ref 3). The corrosion of austenitic stainless steels in HNO_3 is accompanied by the formation of hexavalent chromium (Cr^{6+}), a complex chromium compound that increases the corrosivity of HNO_3 solutions. The effect of Cr^{6+} buildup is shown in Fig. 2 to be clearly detrimental (Ref 4).

Table 1 Nitric acid nomenclature

Name	Composition, wt%
Weak or tower acid	50–65% (nominally 57%)
Commercial 42° baumé	67% (American Chemical Society, ACS, grade)
Chemically pure (CP) or reagent grade 43° baumé	70.3%
Strong or concentrated acid	95–99%
White fuming nitric acid (WFNA)	95–99% <0.5% NO_x
Red fuming nitric acid (RFNA)	95–99% + NO_x 6–30%
Inhibited RFNA	98% plus 15% H_2SO_4

Source: Ref 1

Table 2 Common materials of construction for equipment in nitric acid service

Acid conc, %	Temperature range, °C (°F)	Storage tank	Pressure vessel	Heat exchangers Heaters	Heat exchangers Coolers	Pipe fittings	Pumps	Valves	Gasket seals
<40	Ambient	304L	304L 310 (sp)	NA	NA	PE/PP/Fe 304L	CF-8M CF-8	CF-8M PVDC/Fe	PTFE/SS PE/PP
	52 (125) to BP	304L 316L	304L 310 (sp)	Alloy 20 310 (sp)	304L Alloy 20	304L PTFE/Fe	CF-8M CD-4MCu	CF-8M CF-3M	PTFE/SS PTFE
	>BP	NA	304L 310 (sp) Zr	310 (sp) Zr	Alloy 20 310 (sp) Zr	304L PTFE/Fe Zr	CF-8M CD-4MCu Zr	CF-3M Zr	PTFE/SS PTFE
40–70	Ambient	304L 316L	304L 310 (sp)	NA	NA	304L PVC/Fe	CF-8M CF-3M	CF-8M CF-3M	PTFE/SS PTFE
	52 (125) to BP	304L 316L	304L 310 (sp)	Alloy 20	Alloy 20 310 (sp)	304L PTFE/Fe	CF-8M CF-3M	CF-8M CF-3M	PTFE/SS PTFE
	>BP	NA	310 (sp)	Zr	310 (sp) Zr	Ti Gr 2 Zr	CD-4MCu Ti Gr 2 Zr	CF-8M CD-4MCu Zr	PTFE
70–93	Ambient	304L 316L	304L 310 (sp)	NA	NA	304L Ti Gr 2	CF-8M CF-3M	CF-8M CF-3M	PTFE/SS
	52 (125) to BP	304L	304L (NAG)	Ti Gr 2	Ti Gr 2	310L Ti Gr 2	CF-8M CD-4MCu		PTFE
	>BP	NA	Ti Gr 2	4% Si SS Ti Gr 2	Zr 702	Ti Gr 2 310L	DMET5 Ti Gr 2	DMET5 Ti Gr 2	PTFE/SS PTFE
93–100	Ambient	A93003 A95056	A95056 A95454	NA	NA	A93003 4% Si SS	CD-4MCu DMET5	DMET5	PTFE A91100
	52 (125) to BP	4% Si SS 310 (sp)	4% Si SS	Ti Gr 2 Ti Gr 12	Ti Gr 2 Ti Gr 12	4% Si SS Ti Gr 2	DMET5 14.5% Si Fe	DMET5	PTFE/SS PTFE
	>BP	NA	4% Si SS	Ti Gr 2 Ti Gr 12	Ti Gr 2 Ti Gr 12	4% Si SS Ti Gr 2	14.5% Si Fe	DMET5	PTFE/SS PTFE

BP, boiling point; conc, concentration; DMET5, Durcomet 5 (UNS J93900); Fe, iron; Gr, grade; NA, not applicable; NAG, nitric acid grade; PE, polyethylene; PP, polypropylene; PTFE, polytetrafluoroethylene; PVC, polyvinyl chloride; PVDC, polyvinylidiene chloride; sp, special grade for HNO_3 service; SS, stainless steel; Ti, titanium; Zr, zirconium. Source: Ref 1

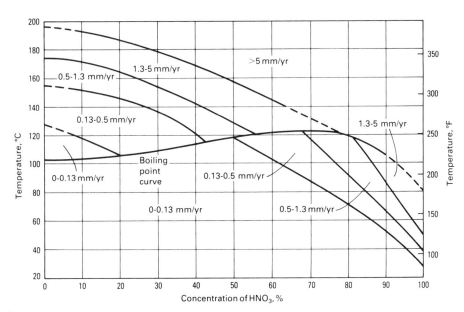

Fig. 1 Isocorrosion diagram for annealed type 304 stainless steel in HNO_3. Source: Ref 3

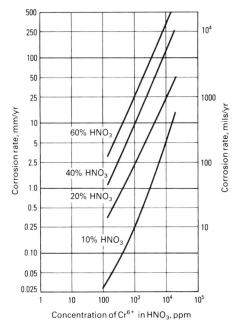

Fig. 2 Effect of hexavalent Cr^{6+} contamination on the corrosion rate of type 304 stainless steel in HNO_3. Source: Ref 4

The problem of intergranular corrosion in weldments of the high-carbon UNS S30400 (304) has been minimized by the use of type 304L grades. Modern steelmaking practice makes it possible to make steels with precise element control resulting in the 300 series stainless steels having elements such as chromium produced to the low end of the specification. Reduction in the chromium level may result in increased corrosion rate of the stainless steels.

The difference in the corrosion behavior between type 304L and 316L in HNO_3 is often not substantial and can sometimes be attributed to differences in chromium content rather than the presence or absence of molybdenum. Since type 304L is less expensive than type 316L, it is preferred for HNO_3 applications. The exception would be if substantial levels of impurities are present in the acid that would benefit from type 316L molybdenum grade. The molybdenum grade might also be better in heat-exchanger applications where the resistance to pitting on the water side would be beneficial. Welds of type 304L welded with E308L, in the "as welded" condition and welded and sensitized condition show no preferential attack in boiling HNO_3. Welded and sensitized stabilized grades, such as types 347 and 321, can suffer on occasion from "knife-line" attack in boiling HNO_3. Sensitized welds in type 316L are subject to accelerated attack in boiling HNO_3 (Ref 1).

Nitric Acid Grade (NAG) Wrought Stainless Steels. In HNO_3 up to 67% concentration at elevated temperature, type 304L stainless steels are in the transition region between passive and transpassive behavior and can therefore lead to intergranular corrosion at heat-affected zones (HAZ) near welds. To further reduce intergranular corrosion, type 304L stainless steels can be produced using eletroslag refining techniques with extremely low carbon, silicon, phosphorous, and sulfur levels with optimized microstructures. These stainless steels are referred to as types 304 Special or 304 NAG (UNS S30403 NAG). Type 316L NAG is also produced by control of composition to minimize second-phase formation (ferrite and sigma), together with low residuals of silicon, phosphorous, and sulfur.

The beneficial effects of increasing chromium additions to an alloy system exposed to HNO_3 have been universally accepted. Type 310L (UNS S31050), an austenitic alloy that contains nominally 25% Cr, 20% Ni, and very low carbon (0.02%) as well as being low in other interstitial elements offers significant advantages over type 304L at high temperatures and is often used for condensers in concentrating of the acid. The excellent performance of type 310L depends on obtaining a fine-grained microstructure (ASTM 4 to 6) and ensuring that it is free of carbide precipitation (Ref 1).

The other alloying element that improves the corrosion resistance of stainless steels in HNO_3 environments is silicon. With levels above 2% Si in the alloy, a thick silica layer forms on top of the normal chromium oxide film present on stainless steels. The two most common austenitic stainless steel alloys containing silicon are A610 (UNS S30600), which contains nominally 18% Cr, 15% Ni, and 4% Si, and A611 (UNS S30601), which contains nominally 18% Cr, 17% Ni, and 5.3% Si. Both these alloys have excellent resistance to highly concentrated HNO_3; however, they are less corrosion resistant than types 304L and 310L at lower concentration. Figure 3 shows the corrosion rate of 304L, 310L (2521LC), and A610 (1815LC Si) in boiling HNO_3 (Ref 5).

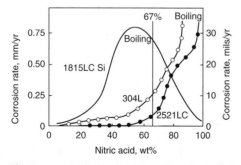

Fig. 3 Corrosion rates of 304L, 310L (2521LC), and A610 (1815LC Si) in boiling HNO_3. Source: Ref 5

Vapor/Liquid. Most corrosion data are derived from laboratory tests made in the liquid phase; however, in actual plant equipment there is often a vapor space above the liquid acid where acid vapor composition may vary from the bulk liquid. This acid vapor may condense on the equipment inner surface and be more or less corrosive than the bulk liquid. Strong acid vapors will generally be more corrosive than the bulk liquid. Vapor-phase corrosion is a potential problem for stainless steel tanks handling concentrations greater than 90%. In this case, the vapors contain much stronger HNO_3 than the liquid because of the higher volatility of HNO_3 over strong acid, and the condensate can be corrosive to these stainless steels. Vents on high HNO_3 concentration aluminum storage tanks can cause accelerated corrosion in adjacent areas because atmospheric moisture enters the tank and dilutes the acid into the range where it is corrosive to the aluminum.

Cast Stainless Steels. The equivalent cast version of types 304 and 304L, ACI CF-8 (J92600) and ACI CF-3 (J92500), respectively, have approximately the same corrosion resistance to HNO_3 as the wrought alloys. The molybdenum-bearing cast ACI grades CF-8M (J92900) and CF-3M (J92800) are more available and are used for pumps and valves in HNO_3 service. The cast version of alloy 20, ACI CN-7M (N08007), is also commonly used in pumps and valves in HNO_3 service at elevated temperatures up to 70% concentration. Cast low-carbon, high-silicon stainless steels have been developed for HNO_3 service based on 20% Cr and 13% Ni with varying amounts of silicon up to 6.5%. A cast version of S30600 (J93900), containing 21% Cr, 16% Ni, and 5% Si, known as Durcomet 5, is used for pumps and valves in concentrated HNO_3 service. Silicon cast irons containing 14.5% Si (e.g., UNS F47003) exhibit excellent corrosion resistance to HNO_3 above 45% at temperatures up to the boiling point. The resistance increases with increasing concentration. Corrosion rate becomes nil in strong acid at high temperatures. The corrosion resistance can be attributed to the formation of an adherent siliceous film (Ref 1).

Other Austenitic Alloys

Because of the generally good performance of the stainless steels with a wide range of acid concentrations and temperatures, the more costly and more highly alloyed stainless steels and Ni-Cr-Mo alloys are not widely used in the handling of straight HNO_3 solutions. These latter materials are more likely to be used with mixtures of HNO_3 with other acids, such as sulfuric (H_2SO_4) or hydrofluoric (HF) acids, or where some chlorides may be present to cause SCC or pitting of the austenitic stainless steels. Some of the alloys with high chromium content that have been used in various HNO_3 applications include alloy 20 (N08020), G30 (N06030), alloy 33 (R20033), alloy 28 (N08028), alloy 22 (N06022), and alloy 625 (N06625). The nickel-base alloys that are essentially chromium-free, such as alloys 200 (N02200), 400 (N04400), B2 (N10665), and their variants, suffer immediate attack in HNO_3.

Aluminum Alloys

All aluminum alloys, welded and unwelded, have good resistance to uninhibited fuming HNO_3 (red and white) up to 50 °C (122 °F). Above this temperature, most aluminum alloys exhibit knife-line attack adjacent to the welds in uninhibited acid. Above 50 °C (122 °F), the depth of knife-line attack increased markedly with temperature. One exception was in the case of a fusion-welded 1060 (UNS A91060) alloy in which no knife-line attack was observed even at temperatures as high as 70 °C (160 °F). In inhibited fuming HNO_3 containing at least 0.1% HF, no knife-line attack has been observed for any commercial aluminum alloy or weldment even at 70 °C (160 °F) (Ref 6).

Aluminum alloys are good only for very high concentrations, for example, greater than 80% at room temperature and greater than 93% at 43 °C (110 °F). Aluminum alloys commonly used are UNS A91100, A93003, A95052, and A95454. The first two alloys must be welded with A91100 or A91060 rod, the second two with A95356. Welds made with A91100 may suffer intergranular attack if the iron content is too high, and the high-silicon rods, for example, A94043, are subject to accelerated preferential attack. UNS A91100 is a commercial grade of aluminum containing 98.85% Al. The material, although structurally weak, has excellent resistance to corrosion in 95%+ acid at ambient temperatures. The isocorrosion diagram for A 91100 is shown in Fig. 4 (Ref 3). The alloy should be fusion welded using only matching 1100 filler metal to maintain the corrosion resistance. UNS A93003 is a commercial grade of aluminum containing 1.25% Si, 0.7% Mn, and 0.2% Fe. This alloy offers improved strength properties over A91100 but at some loss of corrosion resistance. This alloy can replace A91100 when a little more strength is required and should be fusion welded using 1100 filler metal to avoid weld corrosion. UNS A95454 is an aluminum alloy containing 2.7% Mg for additional strength. Its corrosion resistance in HNO_3 is slightly reduced from A91100 or A93003. Its primary application is found in rail and highway equipment tankage. UNS A95454 should be fusion welded using A95356 filler metal. UNS A95056 is an alloy containing 5% Mg that is still stronger than A95454 and thus offers a higher-strength alternative at some loss of corrosion resistance. Primary usage is in transportation tankage. For fusion welds, filler metal A95356 should be used. Other aluminum alloys of the 5000 series designation may also be used in HNO_3 applications. Their corrosion resistance

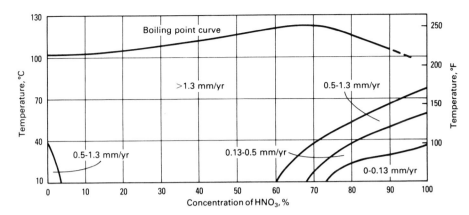

Fig. 4 Isocorrosion diagram for aluminum alloy A91100 in HNO_3. Source: Ref 3

Table 3 Effect of added Ti^{4+} on the corrosion of unalloyed titanium

Ti ion Added, mg/L	Corrosion rate			
	for 40% HNO_3		for 68% HNO_3	
	mm/yr	mils/yr	mm/yr	mils/yr
0	0.75	29.5	0.81	31.9
10	...	...	0.02	0.79
20	0.22	8.7	0.06	2.36
40	0.05	1.97	0.01	0.40
80	0.02	0.79	0.01	0.40

Source: Ref 8

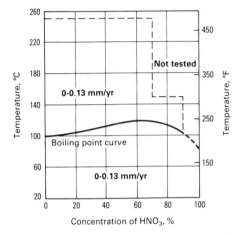

Fig. 6 Isocorrosion curve for zirconium in HNO_3. Source: Ref 8

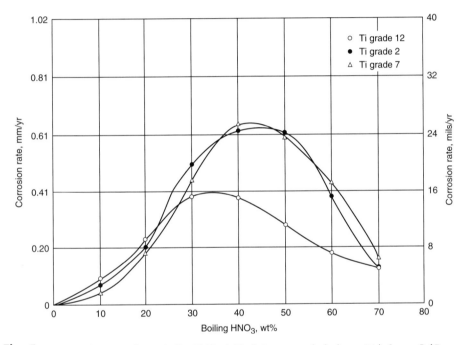

Fig. 5 Corrosion of titanium alloys in boiling HNO_3. Acid solutions were refreshed every 24 h. Source: Ref 7

should be comparable to the specific alloys mentioned, but there is no test data. Alloy A96061 should not be used in HNO_3, as erratic corrosion has been observed.

Titanium

Unalloyed titanium has been extensively utilized for handling and producing HNO_3 in applications where stainless steels have exhibited significant uniform or intergranular attack. Titanium offers excellent resistance over the full concentration range at subboiling temperatures. The corrosion resistance up to 70% concentration at 70 °C (160 °F) is below 0.04 mm/yr (1.6 mils/yr). As temperature exceeds 80 °C (175 °F), however, the corrosion resistance becomes highly dependent on acid purity. The corrosion resistance of titanium in boiling HNO_3 over a range of strengths is shown in Fig. 5. Ti grade 12 (R53400), which contains molybdenum and nickel, is a higher-strength alloy with improved corrosion resistance in HNO_3 (Ref 7).

In hot, very pure solutions or vapor condensates of HNO_3, significant general corrosion (and trickling acid condensate attack) may occur in the 20 to 70 wt% range. Under marginal high-temperature conditions, higher-purity unalloyed grades of titanium (i.e., grade 1) are preferred for curtailing accelerated corrosion of weldments. In hot pure acid, the resistance of titanium improves dramatically when there are trace amounts of oxidizing ions, in particular, Ti^{4+}. That is, the initial high corrosion rate of titanium will quickly inhibit its corrosion as illustrated in Table 3 (Ref 8). Often, oxidizing impurities, such as Si^{4+}, Fe^{3+}, Cr^{6+}, and Pt^{2+}, are present in the HNO_3 stream. These impurities are harmful to stainless steels but are very beneficial to titanium. Consequently, titanium becomes more suitable than stainless steels in applications such as reactors, stripper reboiler loops, spent HNO_3 recovery systems, and evaporator reboilers. For cooler condenser applications, titanium is vulnerable to corrosion if significant condensation occurs. Similarly, titanium may be attacked in the vapor phase of strong HNO_3.

Titanium alloys have good corrosion resistance in white fuming HNO_3. However, pyrophoric reactions may occur when they are exposed to red fuming HNO_3 or to nitrogen tetroxide gas. The attack is intergranular and results in breaking down the surface into fine particles. Fine titanium particles are highly reactive. The critical variables are the nitrogen dioxide and water contents of the acid. Fuming HNO_3, containing less than 2% water or more than 6% nitrogen dioxide, may cause this pyrophoric reaction to occur. Both water and nitrogen monoxide are effective inhibitors for this reaction, but increasing oxygen and nitrogen dioxide are detrimental in this situation (Ref 7).

Zirconium Alloys

The excellent corrosion resistance of zirconium in HNO_3 has been recognized for more than 30 years. Zirconium alloys have excellent corrosion resistance under the most severe conditions of concentration and temperature (Fig. 6). The corrosion resistance of zirconium is less than 0.13 mm/yr (5 mils/yr) below 250 °C (480 °F) up to 70% concentration and below 150 °C (300 °F) between 70 to 90% concentration. Zirconium alloys are not normally used above 90% acid concentration.

Zirconium is normally susceptible to pitting in acidic oxidizing chloride solutions. However, nitrate ion is an inhibitor for the pitting of zirconium because of its passivating power. The minimum [NO^{3-}]/[Cl^-] molar ratio required to inhibit pitting of zirconium is between 1 and 5 (Ref 9–11). Results of tests indicate that the resistance of zirconium is not degraded in up to 70% HNO_3 with the addition of 1% ferric chloride, 1% sodium chloride, 1% seawater, 1% ferric ion, or 1% stainless steel at 204 °C (399 °F). Still, the presence of an appreciable amount of hydrochloric acid (HCl) should be avoided since zirconium is not resistant to aqua regia.

In the production of HNO_3, ammonia is oxidized with air over platinum catalysts. The resulting nitric oxide (NO) is further oxidized into nitrogen dioxide (NO_2), then absorbed in water to form HNO_3. Acid of up to 70% concentration is produced at temperatures up to 204 °C (399 °F) by the process. Zirconium is one of very few materials that are suitable for this process when the temperature is near the high end.

The polarization curves of zirconium in HNO_3 are shown in Fig. 7. Zirconium has the passive-to-active transition similar to that which occurs in H_2SO_4. However, corrosion potentials are very noble because of the oxidizing nature of HNO_3. Stress-corrosion cracking occurs in zirconium when it is above the transition region between passivity and transpassivity (Ref 12, 13). The relationship between HNO_3 concentration, potential, and susceptibility to SCC is shown in Fig. 8 (Ref 1, 14). The addition of titanium to zirconium increases the transition potential and thus resistance to SCC. Stress-corrosion cracking of zirconium equipment can be reduced by a stress-relieving treatment.

Additional concerns include the accumulation of chlorine gas in the vapor phase and the presence of fluorides. Chlorine gas may be generated by the oxidation of chlorides by HNO_3. Areas that can trap chlorine gas should be avoided for zirconium equipment when chlorides are present in HNO_3. The corrosion of fluoride-containing HNO_3 solutions can be controlled by adding an inhibitor, such as zirconium sponge or zirconium nitrate, to convert fluoride ions into noncorrosive complex ions (Ref 8).

Niobium and Tantalum

Niobium and tantalum are among the metals most corrosion resistant to HNO_3. They exhibit resistance in the acid at all concentrations to temperatures as high as 250 °C (480 °F). The presence of chloride impurities does not degrade their resistance. They are not vulnerable to pyrophoric reaction in red fuming HNO_3 either.

Nevertheless, there are differences among these metals. Tantalum can tolerate more H_2SO_4 contamination than niobium can. This is an important consideration when H_2SO_4 is used as the dehydrating agent in making concentrated HNO_3. On the other hand, niobium can tolerate more fluoride contamination than tantalum can (Ref 8).

Nonmetallic Materials

Plastics. The use of plastics, mainly fluoroplastic materials, is confined to linings for pipe and vessels, rather than as solid construction, because of the hazardous nature of HNO_3 (see Table 2). Polytetrafluoroethylene (PTFE) will resist up to 70% acid to about 230 °C (450 °F), but a limit of about 25 °C (75 °F) is suggested above that concentration, primarily because the vapors tend to permeate the plastic. Fluorinated ethylene propylene (FEP) will tolerate up to 100% acid to 200 °C (390 °F), while polyvinylidene fluoride (PVDF) is recommended to about 50% concentration to 52 °C (125 °F) (Ref 1).

The chlorinated and vinyl ester thermosetting resins have been used in combination with fluoroplastics in dual-laminate construction storage tank applications. If used without a fluoroplastic liner the vinyl ester resins have satisfactory resistance up to 20% acid to about 65 °C (150 °F) and up to 35 °C (95 °F) at 40% concentration. The vinyl ester resin is not suitable above the 40% acid concentration.

Elastomers. Elastomeric products can be successfully used as sealing components at various temperatures and concentrations of HNO_3. Selection of the proper elastomeric material is critical for ensuring long-term, reliable service. Variations in elastomer compound formulations and cure chemistry can result in performance differences. Therefore, testing of elastomeric seals is always recommended before use. The perfluoroelastomers (FFKM) and PTFE-based elastomers are the only elastomers that are useful over a wide range of concentration and temperature. Rubber and other synthetic elastomers can be used only in very dilute HNO_3 at ambient temperatures (Ref 1). Due to the wide variation in types and quality levels of elastomers, it is important that compatibility testing is done before the appropriate selection is made.

Glass-Lined Equipment. Glass-lined steel pipe, fittings, and other equipment are available and widely used in HNO_3 (Ref 1). Glass-lined steel equipment can normally be used at any strength of HNO_3 up to the thermal limits of the glass lining. These thermal limits depend on the design of the equipment and the possibilities and extent of thermal shock, and they are aimed at keeping the glass under compressive stress. A typical maximum operation temperature is around 230 °C (450 °F). Glass-coated steel equipment is fully resistant to HNO_3 up to 70% concentration and to 125 °C (260 °F). Exposure to conditions beyond these limits will reduce the life of glass-coated equipment. Fluorides at all concentrations should be avoided since they attack the silica network, and this attack can be very severe at concentrations of HF <10 ppm. If any detectable fluorides are present, attack is likely to be accelerated. Glass-lined reactors are used by all large chemical companies. In some cases they are being used with highly concentrated HNO_3 with reaction temperatures up to 160 °C (320 °F).

REFERENCES

1. M. Davies et al., MTI publication, MS-5 Material Selector for Nitric Acid, Materials Technology Institute (MTI), 2004
2. T.P. Sastry and V.V. Rao, Anodic Protection of Mild Steel in Nitric Acid, *Corrosion*, Vol 39 (No. 2), 1983, p 55–60
3. M.G. Fontana and N.D. Greene, *Corrosion Engineering*, McGraw Hill, 1967
4. M.W. Wilding and B.E. Paige, "Idaho National Engineering Laboratory Survey on Corrosion of Metals in Solutions Containing

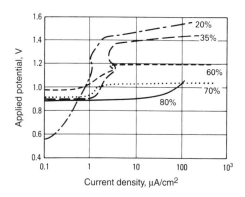

Fig. 7 Anodic polarization curves for unalloyed zirconium in near-boiling HNO_3 as a function of concentration (wt%). Applied potential is given in volts versus the saturated calomel electrode (SCE). Source: Ref 8

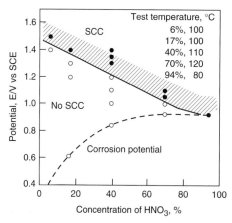

Fig. 8 Relationship between potential and HNO_3 concentration on the susceptibility of zirconium to SCC. Source: Ref 14

Nitric Acid," report N77-32302, National Technical Information Service, Dec 1976
5. R.R. Kirchheiner, U. Heubner, and F. Hoffman, "Increasing the Lifetime of Nitric Acid Equipment Using Improved Stainless Steels and Nickel Alloys," paper 318, Corrosion 1988, NACE International, p 22
6. P. Eyre, Du Pont, e-mail communication, 2004
7. R. Schutz and D. Thomas, Corrosion of Titanium and Titanium Alloys, *Corrosion,* Vol 13, 9th ed., *Metals Handbook,* ASM International, 1987, p 677–679
8. T.L. Yau, "Metallic Materials for Nitric Acid Service," paper No. 02121, CORROSION 2002, NACE International
9. V.V. Andreeva and A.I. Glukhova, *J. Appl. Chem.,* Vol 11, 1961, p 390
10. G. Jangg, R.T. Webster, and M. Simon, *Werkst. Korros.,* Vol 29, 1978, p 16
11. M. Maraghini et al., *J. Electrochem. Soc.,* Vol 101, 1954, p 400
12. T.L. Yau, *Corrosion,* Vol 39, 1983, p 167
13. J.A. Beavers, J.C. Griess, and W.K. Boyd, *Corrosion,* Vol 36, 1981, p 292
14. H. Nagano and H. Kajimura, *Corros. Sci.,* Vol 38 (No. 5), 1996, p 781–791

Corrosion by Organic Acids

Revised by L.A. Scribner, Becht Engineering

ORGANIC ACIDS represent a key group of industrial chemicals. They are often used in their pure form, and they are used as an intermediate in a wide variety of chemical reactions to make products ranging from polyester clothing to amino acids used in vitamins. Acetic acid is synthetically produced in the largest volume of all of the carboxylic or organic acids and is best known by the general public as the weak aqueous solution, "vinegar." The simple, straight chain aliphatic acids are discussed in this article. They are often called "fatty acids" because those containing an even number of carbon atoms (four or greater) exist in a combined form with glycerol as fats and oils.

The subject of corrosion by organic acids is complicated not only by the numerous acids to be considered, but also because the acids typically are not handled as a pure chemical but as process mixtures, with inorganic acids, organic solvents, salts, and as mixtures of several organic acids (Ref 1). They are even used as solvents in chemical reactions. The corrosion of materials by organic acids is also complicated by the virtually unlimited number of possible compounds. The corrosion of metals by organic acid is often confounded by trace impurities such as oxygen and metallic salts. This article concentrates on corrosion by acetic (CH_3COOH), formic (HCOOH), and propionic (CH_3CH_2COOH) acids and gives some information on longer-chain organic acids.

Corrosion Characteristics

Organic acids are weak acids when compared to the common inorganic acids such as hydrochloric (HCl) or sulfuric (H_2SO_4), but still hydrolyze well enough to act as true acids toward most metals. Aliphatic organic acids are usually considered to be slightly reducing. They are often handled in copper, which does not directly displace hydrogen from acids. The 400-series stainless steels exhibit borderline passivity and thus are seldom selected, whereas the 300-series stainless steels are the materials of choice. The grades in the 300 series of stainless steels require oxidizing conditions to maintain their passivity, especially at high temperatures. The reversal of corrosion resistance as the environment changes from oxidizing to reducing characteristics makes contaminants extremely important because they tend to shift the oxidizing capacity of the acid mixture. Aeration (i.e., dissolved oxygen or DO), ferric ions, peracids, or peroxides will cause rapid attack of copper and copper alloys, while the presence of chlorides, which are reducing, can have disastrous effects on stainless steels. Additional information on the role of contaminants in acetic acid corrosion is provided in Ref 2. Corrosion testing can be difficult in organic acid media. Electrochemical measurements are most successful in dilute aqueous solutions of the acids because the conductance is very low in high concentrations or in solutions of nonaqueous solvents. The addition of sodium or chloride salts improves the ease in making electrochemical measurements. Reported electrochemical data obtained in strong acetic acid, acetic acid anhydride, and formic acid solutions showed active-passive behavior of stainless steels that is consistent with field experience (Ref 3).

Figure 1, an electronically produced electrochemical stability diagram (Pourbaix diagram) of copper in acetic acid, shows results that are similar to those that have been shown by testing. The light shaded fields are those that delineate regions where copper is corroded by acetic acid. The addition of oxidizing impurities, such as ferric ions or dissolved oxygen, will serve to raise the electrochemical potential into the regions where copper forms stable soluble species and corrosion proceeds. By reference to this Pourbaix diagram, copper does not show passivity in acetic acid, but instead shows either immunity or active corrosion. One can surmise that the same electrochemical response will be found for formic and other similar organic acids.

Data obtained by simple immersion testing in the laboratory often show erroneous results unless the atmosphere is carefully controlled.

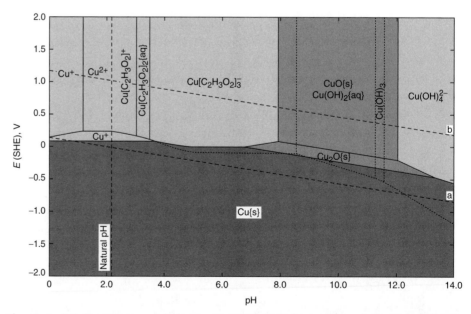

Fig. 1 Pourbaix diagram for copper in 5 M acetic acid at 80 °C (175 °F) showing the natural pH and the possible effects of oxidizing impurities. In the diagram, the darker gray area shows the region of immunity or no corrosion, the lighter gray area shows the stability of copper oxides or regions of possible passivity where low corrosion rates are expected, and the light areas show the regions of soluble copper salts and where high corrosion rates are expected. E(SHE), electrode potential measured by the standard hydrogen electrode

Without very good atmospheric control, the solution will be saturated with air at the beginning of the test, but will lose air as the temperature is increased until, at boiling conditions, almost all of the air will be removed. This situation can lead to results that vary widely, depending on the length of the test. Short tests of metals that exhibit active-passive behavior, such as the 300-grade stainless steels, can also be misleading, because the metal may remain passive for a short time and then may start to corrode actively after longer exposures.

Truly anhydrous organic acids are usually much more corrosive to stainless steels than organic acids containing even traces of water. Corrosion rates reported for glacial (concentrated) acids often reflect this effect, because the acids may be truly anhydrous or may contain small amounts of water.

Formic Acid

Formic acid is the most corrosive of the common organic acids because it is the most highly ionized, being about 10 times more ionized or acidic as acetic acid. Formic acid acts as both an organic acid and as an aldehyde and hence has reducing properties. Formic acid is somewhat unstable as the concentration approaches 100% and decomposes to carbon monoxide and water—especially when heated above 35 °C (95 °F). The following discussion is divided into metal and alloy groups used to handle formic acid.

Steel is attacked by formic acid at all concentrations and temperatures and is normally not considered suitable for formic service. Steels that have been hardened to high strengths can fail by hydrogen cracking when exposed to organic acids. Stressed 18% Ni maraging steel cracks when exposed at ambient temperatures to 10% aqueous formic acid, but does not crack in 91% formic acid, or in either glacial acid or 10% acetic acid (Ref 4).

Aluminum shows good resistance to formic acid at any concentration at ambient temperature (Fig. 2). Although pure aluminum and pure formic acid of all concentrations appears to be compatible from a corrosion viewpoint, the contamination of the acid by various impurities, such as heavy metal salts, limits the use of aluminum to shipment and storage of formic acid with a concentration greater than 95% (Ref 5).

Alloy 5086 (A95086) is an often-used grade of aluminum that is used to make storage tanks and, as shown in Fig. 3, shows acceptable corrosion rates to the higher concentrations of formic acid. Because the temperature of chemicals in shipment can get as warm as 45 °C (115 °F), testing was done at this temperature. The rate of attack on alloy 5086 decreases rapidly with increasing concentration of acid. Importantly, the acid used in the tests became turbid with aluminum salts, which lowered the acid purity. If short exposure times and/or lower temperatures during storage are forecast, or if a lower-quality formic acid that contains aluminum salts is acceptable, this aluminum can be used for the storage of 95 to 99% acid.

Copper and its alloys, except yellow brasses, which may dezincify, respond somewhat differently than aluminum on exposure to formic acid. Table 1 shows typical corrosion rates for copper (C10300) and 90Cu-10Ni copper-nickel (C70600) in various concentrations of formic acid.

Because the corrosion resistance of copper to attack by reducing acids such as formic is determined by the presence of oxidizing agents, the control of dissolved oxygen and other oxidizing species is critical. The corrosion of copper can be autocatalytic if oxygen is present since the cupric ions thus formed are in themselves oxidizing agents (Ref 6). If air or other oxidants are present, high corrosion rates are encountered; if the acid is free of air and other oxidants, copper often provides good resistance to formic acid of all concentrations to the atmospheric boiling point and even at higher temperatures. The anomalies in the data shown in Table 1, such as the higher rate of attack in 50 and 70% formic acid, are probably caused by incomplete deaeration or cupric ion accumulation during laboratory tests.

Stainless Steels. The corrosion resistance of the various grades of stainless steels is strongly influenced by the impurities present in the acid—oxidizing or reducing. The 400-series stainless steels are usually not resistant to formic acid. However, if a PH grade were needed, type 15-7Mo (S15700) would be a candidate based on its similarity to type 316L (S31603).

Type 304 (S30400) stainless steel and its similar nonmolybdenum bearing alloys such as types 321 (S32100) and 347 (S34700) have excellent resistance to formic acid and at all concentrations, but only at ambient temperatures, and they are the preferred, low-cost material of construction for storage of the acid. However, type 304 stainless steel should not be considered for elevated-temperature use. Table 2 shows typical rates of attack on various stainless steels in several concentrations of formic acid at the atmospheric boiling temperature.

Because of intergranular corrosion, the L-grade (low-carbon) versions of stainless steels should be used in any application where the temperature exceeds ambient. Type 316L is the minimum alloy of choice of the common 300 grades of stainless steel when the temperature exceeds ambient and shows excellent resistance to formic acid in all concentrations at ambient temperatures and is resistant to at least 5% formic acid at the atmospheric boiling temperature. However, type 316 (S31600) stainless steel can be seriously attacked by intermediate strengths of formic at higher temperatures, and corrosion tests are advisable. The azeotropic concentration (78% formic acid) dripping condensate appears to cause high rates of attack on type 316 stainless, whereas those concentrations of acid above 90% are less corrosive. The control of corrosion on the 300 grades of stainless steels is a strong function of the molybdenum content. Type 317L (S31703) is the high-temperature alloy that is often

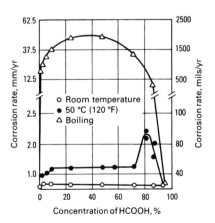

Fig. 2 Rate of corrosion for 1100-H14 (A91100) aluminum in aqueous reagent grade formic acid

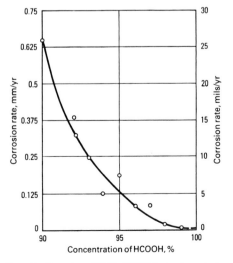

Fig. 3 Corrosion of 5086 (A95086) aluminum in formic acid at 45 °C (115 °F)

Table 1 Corrosion of copper and 90-10 copper-nickel by formic acid

Laboratory tests in deaerated acid at atmospheric boiling temperature; test duration; 96 h

Acid concentration, %	Corrosion rate			
	Copper C10300		Copper-nickel C70600	
	mm/yr	mils/yr	mm/yr	mils/yr
1.0	0.02	0.8	0.02	0.9
5.0	0.02	0.7	0.02	0.9
10.0	0.02	0.6	0.02	0.7
20.0	0.20	7.8	0.40	15.7
40.0	0.14	5.5	0.34	13.3
50.0	0.26	10.2	0.54	21.1
60.0	0.05	2.0	0.03	1.3
70.0	0.76	30.0	0.76	30.0
80.0	0.20	7.8	0.13	5.0
90.0	0.22	8.7	0.19	7.6

selected for formic acid service because of its higher molybdenum content compared to type 316L.

Duplex Alloys. Laboratory corrosion tests of the newer grades of molybdenum-containing duplex stainless steels show corrosion resistance that exceeds that of type 317L. Alloys 2205 (S31803 and S32205) and 2507 (S32750) have corrosion rates that are lower than type 317L stainless steel. As shown in Fig. 4, alloy 2507 has outstanding resistance to hot formic acid throughout the full concentration range at 100 °C (212 °F) (Ref 7).

The 20-type alloys are more resistant to formic acid than type 316 stainless steel, and its use should be considered in higher concentrations at higher temperatures. Other alloys with chromium, nickel, and molybdenum contents higher than type 316 stainless steel, such as alloys 28 (N08028) and 904L (N08904), also show superior resistance to mixtures of formic and acetic acid and would be expected to be better in formic acid itself. Under gaskets and in other occluded areas (e.g., threaded connections), type 316L stainless steel sometimes undergoes crevice corrosion. Weld overlays of alloy 20 have proven partially successful for solving this problem. The superaustenitic alloys containing 6% Mo (e.g., alloy 6XN; N08367) should be even better than the alloy 20, which contains only 2 to 3% Mo. A 6% Mo version of the 20-type alloy, alloy 20Mo-6 (N08026), is now available.

Titanium. While titanium will resist acetic acid under almost all conditions to the boiling point, it will resist formic acid, only under strongly oxidizing conditions. Titanium exhibits borderline passivity in formic acid, and under normal conditions pure formic will corrode titanium at the boiling point in all concentrations above 10% (Ref 8). A titanium heat exchanger disappeared in less than one day when strong conditions, approaching the anhydrous state, had been encountered in a distillation system. When titanium does corrode, even slightly, hydriding of the metal occurs and loss of ductility results.

Nickel Alloys. Several of the high nickel alloys such as alloys B-2 (N10665), C-276 (N10276), and C-4 (N06455) have shown outstanding resistance to formic acid in process equipment and are reported to have very good resistance even at temperatures above the atmospheric boiling point.

Corrosion data for alloy C (N10002) are shown in Fig. 5 (Ref 9). Note that Fig. 5 also shows a zone of higher corrosion for alloy B-2. In all other areas, alloy B corroded at lower rates. Alloy B (N10001) sometimes exhibits less corrosion resistance in intermediate strengths than alloy C but, because alloy B usually costs more, it is not often selected. However, in those services that are very reducing, such as sulfuric acid catalyzed esterification reactions, alloy B is often the better choice. In services where contamination from sodium chloride or other halogens is possible, alloy B should be superior to alloy C. As is to be expected, alloy B is not better than alloy C types when strongly oxidizing conditions prevail.

The Cr-Mo-Ni alloys are all similar to the C and B alloy types. Alloy 625 (N06625) contains only 9% Mo and should be expected to be somewhat less corrosion resistant than alloy C-276, but is so similar in resistance as to be usable in all but the most severe of services.

Other Metals. Lead is substantially non-resistant to formic acid and other organic acids. Commercially pure zirconium exhibits very good resistance to boiling formic acid with rates below 0.02 mm/yr (0.8 mil/yr), even when the acid is contaminated with metal salts or iodine. This suggests that zirconium is a good candidate material for the distillation portions of formic acid plants, in which the acid concentration and the temperatures are both high (Ref 10).

Nonmetals. Formic acid is an excellent solvent, which makes it very destructive to most organic materials and coatings. Therefore, plastics are not normally considered for this service. An exception is polyethylene, which is good with all concentrations to about 35 °C (95 °F). The fluorocarbon plastics are resistant to their normal temperature limitations, except as coatings, which have a temperature limitation of about 93 °C (200 °F).

Hard rubber, neoprene, and butyl rubbers are resistant at ambient temperatures and fluoroelastomers to about 50 °C (120 °F). Rubber linings can be used for storage if discoloration of the acid is not a matter of concern.

Acetic Acid

Acetic acid is made by a variety of processes, the best known being the butane oxidation and the Monsanto low-pressure processes. Historically, acetic acid was produced by fermentation of grain and then by oxidation of acetaldehyde. The fermentation process was usually carried out in copper equipment or wooden tanks, and combinations of copper and stainless steel equipment were used for acetaldehyde oxidation.

The Monsanto process is based on the carbonylation of methanol (with carbon monoxide) using a rhodium catalyst and an alkyl halide such as methyl iodide. This process has all but replaced the now-antiquated Wacker Process (Ref 9). The Monsanto process (and various improvements thereof) is the most popular method for the production of acetic acid. As can be surmised, the addition of the halogen reaction promoter to the reaction cycle requires the use

Table 2 Corrosion of stainless steel by formic acid
96 h laboratory tests at atmospheric boiling temperature

Acid concentration	Corrosion rate									
	304(a)		316(a)		316(b)		Alloy 20Cb-3(c)		26 Cr-1Mo(a)	
	mm/yr	mils/yr	mm/yr	mils/yr	mm/yr	mils/yr	mm/yr	mils/yr	mm/yr	mils/yr
1.0	0.17	6.8	0.09	3.6	...	...	...	...	<0.02	<0.8
5.0	0.77	30.8	0.04	1.6	...	...	...	...	...	...
10.0	1.33	53.2	0.26	10.4	...	...	...	...	...	...
20.0	1.89	75.6	0.27	10.8	...	...	...	...	...	...
40.0	3.40	136	0.20	8	...	...	...	...	...	...
50.0	4.20	168	0.50	20	0.46	18.4	0.03	1.2	...	...
60.0	3.40	136	0.46	18.4	...	...	...	...	...	...
70.0	3.97	159	0.48	19.2	0.64	25.6	...	...	<0.02	<0.8
80.0	4.20	168	0.47	18.8	...	...	...	...	...	...
90.0	3.23	129	0.41	16.4	0.61	24.4	...	...	0.10	4
100.0	...	...	...	...	0.25	10	...	...	...	...

(a) Oxygen not controlled. (b) Deaerated, coupons exposed to the refluxed condensate. (c) 48 h exposure

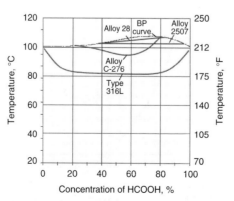

Fig. 4 Isocorrosion diagram at 0.1 mm/yr (4 mils/yr) in formic acid for S32750 (alloy 2507), N08028 (alloy 28), N10276 (alloy C-276) and S31603 (type 316L) stainless steel. BP curve, boiling point curve

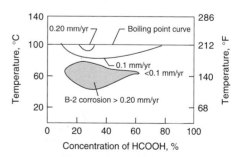

Fig. 5 Isocorrosion curves for alloys B-2 and C-276 in formic acid. In the gray shaded area, alloy B-2 corrodes at rates that exceed 0.2 mm/yr (8 mils/yr). In all other areas, the rate for alloy B-2 is lower. Alloy C-276 data are shown unshaded.

of very corrosion-resistant alloys and spreads trace halides throughout the rest of the plant. Various patents have been filed to overcome the corrosion caused by the halogens that are in the reaction cycle. These have varied from the use of silver compounds to form insoluble silver salts to distillation schemes to minimize carrying bromides into other parts of the plant. The bromide addition, if carried downstream in sufficient quantities, certainly has the potential to pit, crack, and corrode the stainless steel used in the downstream equipment. Data derived directly from the field exposure of alloys are not available, but the problems facing the corrosion engineer who selects materials for these processes have been outlined in Ref 11. The reaction system contains the halogen promoter, and it has been suggested that alloy B should be tested along with titanium, zirconium, and tantalum for this high-pressure, high-temperature system (Ref 11). Nickel-molybdenum and nickel-copper alloys are considered attractive for the recovery system that still contains the halide, using the more conventional materials only to handle the acid after the halide has been removed (Ref 11). Excellent corrosion resistance of commercially pure zirconium in the reaction system has been reported in Ref 12.

Steel is attacked quite rapidly, with hydrogen evolution, at concentrations stronger than 1×10^{-3} molar, even at room temperature. Glacial acetic at room temperature is less aggressive than aqueous solutions of the acid, but still causes a corrosion rate of attack of 0.8 to 1.3 mm/yr (30 to 50 mils/yr). Therefore, steel is normally unacceptable for use in acetic acid service.

Aluminum shows good resistance to nearly all concentrations of acetic acid at temperatures up to 50 °C (120 °F). Alloys 1100 (A91100), 3003 (A93003), and those from the 5000 series such as 5052 (A95052) are used for storage and shipment of the acid. It is rapidly attacked below about 95% acetic acid at the boiling point and is again attacked very rapidly in concentrations near 100% or in those mixtures containing excess acetic anhydride. Aluminum again becomes resistant to pure acetic anhydride, although it causes contamination of the anhydride due to formation of a white crystalline solid, aluminum triacetate, which precipitates in the liquid. The copper-bearing 2000-series, alloys of aluminum are not suitable for acetic acid service because of high corrosion caused by the copper-rich precipitates.

An excellent summary of the use of aluminum in acetic acid and anhydride has been published in Ref 13. Figures 6 and 7 show the resistance of aluminum in acetic acid and acetic anhydride, $(CH_3CO)_2O$. The data are for aluminum 1100, but similar rates would be expected for alloys such as 3003, 6063 (A96063), or 5086. The corrosion resistance of aluminum is strongly affected by contaminants in acetic acid, and aluminum can corrode in almost any concentration of acetic acid at any temperature if the acid is contaminated with the proper species.

Copper and Copper Alloys. With the exception of the alloys that contain more than 15% Zn, copper (C12000) and all of its alloys show good resistance to all concentrations of acetic acid up to and even above the atmospheric boiling temperature in the absence of oxygen or other oxidants. This good performance is predicted by the previous discussion on corrosion characteristics of the electrochemical stability of copper in organic acids. Copper was used almost exclusively to handle acetic acid until the advent of stainless steels; today type 316L stainless steel and higher alloys are often used. Type 316L stainless is not always the best solution. A case was reported in which type 316L was badly attacked in air-free hot acetic acid service, whereas copper under the same conditions was acceptable (Ref 14). The absence of oxidizing agents is a requirement for copper to be usable in acetic acid solutions, as well as other organic acids. Slight contamination of acetic acid with air, through storage under an air atmosphere or by ingress of air through a vacuum leak, can increase the corrosion to rates that are unsuitable. Copper is nearly immune to attack by pure, uncontaminated acetic acid.

Laboratory tests that introduced oxygen into room-temperature 50% acetic acid gave corrosion rates of 1.9 mm/yr (76 mils/yr) compared to only 0.08 mm/yr (3.2 mils/yr) in nitrogen blanketed tests. Similar data in 6% acetic acid showed a corrosion rate of only 0.03 mm/yr (1.2 mils/yr) when passing hydrogen over the solution, whereas when oxygen is passed over the solution, the rate rose to 0.58 mm/yr (23 mils/yr) (Ref 1). Copper alloyed with increasing quantities of nickel showed increased corrosion resistance to acetic acid. The nickel addition also increased the resistance to the effect of oxidants. Tests in air sparged with 50% aqueous acetic acid at the boiling point gave rates of 7.75 mm/yr (310 mils/yr) for copper, 4.7 mm/yr (188 mils/yr) for copper containing 30% Ni, and 2.1 mm/yr (84 mils/yr) for copper containing 70% Ni. Similar reductions in corrosion rates with increasing nickel content were noted when ferric ion was added to the solution. However, the rates still are too high for economical use even though the nickel addition was beneficial—again pointing out the harmful effects of oxidizing impurities. A 10% aluminum bronze alloy (C61800) is reported to be a superior copper alloy for handling acetic acid with rates of 0.05 to 0.07 mm/yr (2 to 3 mils/yr) in boiling glacial acid (Ref 15).

Stainless Steels. Low corrosion rates for the straight chromium 400-series stainless steels in dilute acetic acid solutions can often be shown in laboratory tests. However, these alloys exhibit very tenuous passivity, and field experience with these materials indicates a susceptibility to high corrosion rates and pitting attack. An exception is a high-purity chromium-molybdenum ferritic stainless steel such as alloy 26-1 (S44627) that shows good resistance.

Type 304L (S30403) stainless steel is the lowest grade commonly used. Type 304 (S30400) stainless steel finds wide use in dilute acetic acid solutions and in the shipment and storage of concentrated acetic acid. Previously published data (Ref 4) show that glacial acetic acid can be handled in type 304 to a temperature of about 80 °C (175 °F), and it has been satisfactory for lower concentrations to the boiling point of the acid. Intergranular corrosion of sensitized type 304 stainless steel will occur in 60 °C (140 °F) and hotter acetic acid. To prevent this intergranular attack, the use of the low-carbon grade 304L is recommended for welded construction. The effect of oxidizing impurities cannot be overlooked when using type 304 stainless steel. The hotter the solution, the more critical is the presence of adequate quantities of oxygen or other oxidizing impurities. If oxidizing conditions are lost, rapid corrosion can ensue.

Type 316L stainless steel is the alloy most commonly used in equipment processing acetic

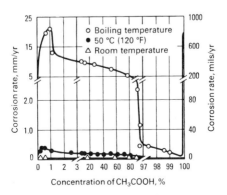

Fig. 6 Effect of concentration and temperature on corrosion of 1100-H14 (A91100) aluminum alloy in acetic acid

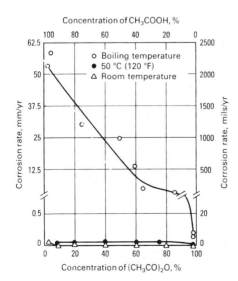

Fig. 7 Metal loss for 1100-H14 (A91100) aluminum alloy in acetic-acid/acetic-anhydride solutions at atmospheric boiling temperature

acid. Even glacial acid at temperatures above the atmospheric boiling point can be handled if the impurities are held within proper levels. The low-carbon grade 316L is required for the higher-temperature applications where welding is required. Type 316L, rather than type 304L, is most often the required alloy for tankers shipping commercially pure acetic acid because of the lower metal pickup from corrosion.

The old acetaldehyde oxidation process for manufacturing acetic acid also produced acetic anhydride as a coproduct, and it is often found in acetic acid streams. If acetic acid contains only small quantities of acetic anhydride or if the acid is truly anhydrous, the rate of attack on type 316L is very high. The introduction of just a few tenths percent of water will reduce the corrosion back to normal rates. Figure 8 illustrates the reduction in corrosion rate as more acetic anhydride is added.

Leakage of chloride-bearing water to acetic acid process streams (e.g., from leaking condensers) results in contamination of the acid with sodium chloride, with subsequent formation of hydrochloric acid. The hydrochloric acid can then move through the system causing its own problems. Corrosion of stainless steel equipment by the hydrochloric acid, for example, produces ferric chloride (along with chromium and nickel chlorides) which, because of its volatility and strong oxidizing nature, is particularly pernicious. The volatility of ferric chloride allows it to move freely through a processing system, and, because it is a strong oxidant, it promotes stress-corrosion cracking (SCC). Besides SCC, a condition of accelerated corrosion and pitting results from chloride contamination. Reportedly, slightly less than 20 ppm of chloride in the organic acid stream can be tolerated, but higher concentrations are likely to cause rapid equipment failure (Ref 16).

Heat transfer, as in heat exchangers, can alter the corrosion mechanism and corrosion rate.

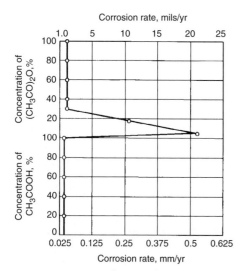

Fig. 8 Laboratory corrosion rates of type 316 stainless steel (S31600) in acetic-acid/acetic-anhydride mixtures

A method to test metals under heat-transfer conditions has been described in Ref 17, and data on stainless steels and some nickel alloys under heat transfer conditions have been developed. Higher alloys such as alloys 20 (N08020), 825 (N08825), 904L (N08904), and the 6% Mo superaustenitic stainless steels show better resistance to acetic acid than does type 316 stainless steel—especially when being used to heat the acid, for example, in evaporators. Alloy 20 weld overlay has been used successfully to combat crevice corrosion of type 316 stainless steel in areas such as flange faces.

The duplex stainless steels that contain molybdenum are very good in boiling acetic acid. Among the duplex grades, alloy 2205 offers an advantage over type 316L in the presence of chlorides. In 80% acid containing 2000 ppm chlorides at 90 °C (195 °F), it showed a rate of less than 0.05 mm/yr (2 mils/yr) versus greater than 1 mm/yr (40 mils/yr) for type 316L. However, in boiling 99.5% acid with 200 ppm chlorides, alloy 2205 exceeded 1 mm/yr (40 mils/yr), whereas alloy 2507 corroded at less than 0.02 mm/yr (0.8 mil/yr) (Ref 6).

Titanium is resistant to all concentrations of acetic acid up to the atmospheric boiling point. Electrochemical corrosion studies in acetic acid solutions suggest that it is possible to attack titanium in anhydrous acetic acid, although titanium has been successfully used in commercial practice. However, some failures of titanium in hot strong acetic acid have been reported with most failures associated with hydrogen embrittlement. Laboratory tests confirmed hydrogen absorption in 95% acid at 210 °C (410 °F) with 1000 ppm hydrobromic acid. Titanium is often used for the reactors in the production of terephthalic acid by the air oxidation of xylene using cobalt, manganese, and hydrogen bromide (HBr) catalysts. In these processes, acetic acid is used as the reaction solvent. Even when chlorinated feedstocks such as mono- and dichloro-terephthalic acid are used, titanium can be made to work if the HCl that is generated by decomposition is prevented from condensing on unwashed surfaces. The high-strength titanium alloys should not be used because of their propensity for SCC.

Zirconium is resistant to all of the concentrations of acetic acid that titanium is resistant to but does not appear to be subject to corrosion in anhydrous conditions or to corrosion from condensation on unwashed surfaces. It is used in the iodide-catalyzed methanol/carbon-monoxide production process for acetic acid. For reasons not fully understood, zirconium is not good if copper ions, especially cupric chloride, are present in hot strong acetic acid/anhydride mixtures (Ref 12).

Nickel Alloys. Alloys C-276 and B-2 are resistant to acetic acid solutions at all normal concentrations and temperatures. These materials are sometimes used where the acetic acid is mixed with inorganic acids and salts, which limits the use of stainless steel or copper alloys. Alloy B-2 is used under reducing conditions, such as with combinations of acetic acid and sulfuric acid, while alloy C-276 is commonly used in acetic acid solutions, which are highly oxidizing in nature. Alloy C-276 has been used for glacial acetic acid evaporators.

Other Alloys. Silver has been used frequently in European practice to handle acetic acid, and it is quite resistant to all concentrations at normal temperatures. Because silver has a higher electrochemical potential than copper, it is much more resistant to acids contaminated with metals ions such as iron and copper. Because of cost considerations, silver has found very little continued use.

Nonmetallic Materials. In the food industry, wooden stave storage tanks have been used for the storage of dilute acetic acid, and many years of service are common with wood tanks in 3 to 4% acid. However, stainless steel tanks are now widely used with the exception of pickle production, wherein copious amounts of salt are dissolved in the weak acid.

Successful use of both hard and soft rubber linings has been experienced in the storage of acetic acid, provided that discolorization of the product is not objectionable. Soft rubbers tend to swell. Butyl rubber will withstand glacial acetic acid to about 80 °C (175 °F). Ethylene-propylene rubber is satisfactory above 5% concentration at room temperature.

Of the plastic materials, polyethylene drums are used to handle chemically pure glacial acetic acid. The fluorinated plastics are completely resistant to their normal temperature limits. Many plastics tend to be susceptible to solvent action by the acid. Of the thermosetting resins, chlorinated polyethers and furanes have been used to about 100 °C (212 °F). Reinforced phenolics are suitable to about 120 °C (250 °F). Glass-reinforced plastic construction using a bisphenol polyester and vinyl esters has been successfully employed.

Carbon and graphite are resistant to the boiling point, and impervious graphite heat exchangers have been used in many demanding acetic acid services.

Glass linings have been used to handle acetic acid. No attack on borosilicate glass is reported below about 149 to 177 °C (300 to 350 °F). When the glass is attacked, it becomes porous and fades from a dark blue to a pale blue via ion exchange. This is easily detected by simple wetting of the surface to detect a color change.

Propionic Acid

Propionic acid or methyl acetic acid boils at 141.4 °C (286.5 °F) and is water soluble. It is a weaker acid, but otherwise very similar to acetic acid.

Steel is attacked at rates of about 0.6 mm/yr (24 mils/yr) in pure propionic acid at room temperature and at much higher rates in aqueous solutions of the acid. Steel has very limited usefulness in propionic acid.

Aluminum will resist 100% propionic acid at room temperature, but becomes unacceptable at about 80 °C (175 °F). The corrosion diagram displayed in Fig. 9 shows that the attack is very similar to that in acetic acid (Ref 4). Like acetic acid, if the propionic acid is contaminated with anhydride or heavy metal ions, high rates are possible.

Copper and copper alloys containing not more than 15% Zn will handle propionic acid in all concentrations. The data shown in Fig. 10 indicate attack on copper in boiling 100% propionic acid, but this is believed to be an anomaly caused by incomplete deaeration of the solution since propionic acid been routinely handled in copper-lined equipment. As with formic and acetic acid, copper and its alloys are satisfactory only if the solutions are completely deaerated and do not contain other oxidizing agents (Ref 1).

Stainless Steels. The 400-series stainless steels will pit in propionic acid solutions, probably because of borderline passivity, and are therefore unreliable. They should not be used unless they are tested in the exact environment.

Type 304 stainless steel shows good resistance to propionic acid at room temperature and to aqueous solutions up to about 50% concentration at the atmospheric boiling point. Field experience has been that type 304 demonstrates borderline passivity above 80% and is not suitable for such service.

Figure 10 shows the resistance of various materials, including type 316 stainless, to boiling propionic acid solutions. These were short-term tests in which the gaseous atmosphere above the boiling solutions was not controlled. It is likely that this resulted in two erroneous results (Ref 1). The rate of copper corrosion was found to increase rapidly above 65% concentration but copper would have very low corrosion rates at all concentrations if the atmosphere above the solution was oxygen free. Secondly, type 304 stainless steel was shown to have decreasing corrosion rates between 80 and 100% concentration which is not borne out by field experience. Field experience shows that type 304 stainless steel exhibits border-line passivity in this range of concentrations at the boiling temperature and should not be considered for such service.

Type 316L stainless steel and the molybdenum-containing duplex stainless steels, such as alloy 2205, are the preferred material for handling hot concentrated solutions of propionic acid. The low-carbon grades should be utilized to avoid possible intergranular attack. They are quite suitable for all ranges of concentration, noting that oxidizing conditions are beneficial and that the rate can be somewhat high around 65% concentration as show in Fig. 10 (Ref 21).

Nickel Alloys. Alloys B-2 and C-276 show excellent resistance to propionic acid solutions under reducing and oxidizing conditions, respectively. Other nickel alloys of similar composition show similar good resistance to propionic acid, but are not normally used since type 316L is suitable for most concentrations at high temperatures and is much less expensive.

Other Organic Acids

The solubility in water decreases with increasing molecular weight of the aliphatic mono acids (e.g., butyric, pentanoic, etc.). Such acids are usually noncorrosive to type 304L, for example, until the temperature becomes high enough to promote dissociation. This critical temperature is usually close to the atmospheric boiling point, at which temperature type 316L is usually required.

It is impossible to cover here the corrosion characteristics of the many different organic acids. However, a sampling of corrosion rates of various metals in several of the longer chain aliphatic acids, aromatic acids, and some dicarboxylic acids is shown in Tables 3 and 4 (Ref 9). Table 4 shows data on the corrosion of some less familiar alloys in various organic acids.

Steel is usable at ambient temperatures in the higher molecular weight acids, and it is used conventionally to store many of the acids and their corresponding anhydrides.

Aluminum shows good resistance to the acids at room temperature and is widely used for their handling. Some of the higher molecular weight acids cause severe attack of aluminum at highly elevated temperatures; therefore, the use of aluminum must be considered for the specific acid and temperature desired.

Copper and copper alloys exhibit good resistance to all of the higher molecular weight acids and can be used quite widely to handle the acids, even at elevated temperatures in the absence of oxidants.

Stainless Steels. Type 304 stainless steel has excellent resistance to the higher molecular weight organic acids at room temperature and at lower concentrations at high temperatures. With the concentrated acids, type 304 is sometimes severely corroded. Type 316 stainless steel is then required and is usable in almost all of the acids, even at elevated temperatures.

Nickel Alloys. The Ni-Mo and Ni-Mo-Cr alloys show excellent resistance to the higher molecular weight acids, but the expense of these alloys is rarely justified unless other contaminants, such as inorganic acids, are also present. The nickel alloys, particularly nickel-copper, have been used to process the higher molecular weight acids at elevated temperatures. They can be particularly useful when contamination of the acids prohibits the use of type 316 stainless steel.

ACKNOWLEDGMENT

This article has been adapted from George Elder, Corrosion by Organic Acids, *Corrosion*, Volume 13, *ASM Handbook* (formerly 9th ed. *Metals Handbook*), ASM International, 1987, p 1157 to 1160.

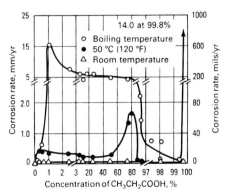

Fig. 9 Resistance to corrosion of 1100-H14 (A91100) aluminum alloy in propionic acid solutions at various temperatures

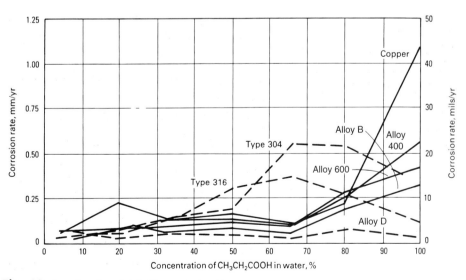

Fig. 10 Corrosion of metals in boiling propionic acid

Table 3 Corrosion of metals in refined organic acids

Acid	Steel mm/yr	Steel mils/yr	Copper mm/yr	Copper mils/yr	Silicon-bronze mm/yr	Silicon-bronze mils/yr	Type 304 stainless steel mm/yr	Type 304 stainless steel mils/yr	Type 316 stainless steel mm/yr	Type 316 stainless steel mils/yr
50% acrylic in an ether at 88 °C (190 °F)	...	...	...	...	...	...	<0.02	<1	<0.02	<1
90% benzoic at 138 °C (280 °F)	...	...	...	...	...	...	0.38	15	0.13	5
Butyric										
Room temperature	0.15	6	0.05	2	0.05	2	<0.02	<1	<0.02	<1
115 °C (240 °F)	...	...	...	...	...	...	0.08	3	0.08	3
Boiling (163 °C, or 325 °F)	...	...	...	...	...	...	1.42	56	0.12	5
Crotonic (crude product), 92 °C (200 °F)	...	...	...	...	...	...	<0.02	1	<0.02	1
2-ethylbutyric										
Room temperature	0.18	7	0.02	1	0.02	1	<0.02	<1	<0.02	<1
150 °C (300 °F)	0.86	34	0.41	16	0.23	9	0.53	21	<0.02	<1
2-ethylhexoic										
Room temperature	0.02	1	<0.02	<1	<0.02	<1	<0.00	<1	<0.02	<1
190 °C (375 °F)	1.27	50	<0.02	<1	<0.02	<1	0.20	8	<0.02	<1
Heptanedionic (pimelic), 225 °C (435 °F)	...	...	...	...	...	...	0.94	37	0.18	7
Hexadienoic (sorbic) as water slurry, 88 °C (190 °F)	...	...	...	...	...	...	<0.02	<1	<0.02	<1
Iso-octanoic										
Room temperature	<0.02	<1	<0.02	<1	<0.02	<1	<0.02	<1	<0.02	<1
190 °C (375 °F)	0.89	35	<0.02	<1	<0.02	<1	0.20	8	<0.02	<1
Iso-decanic										
Room temperature	<0.02	<1	<0.02	<1	<0.02	<1	<0.02	<1	<0.02	<1
190 °C (375 °F)	0.84	33	<0.02	<1	<0.02	<1	0.20	8	<0.02	<1
2-methylpentanoic										
Room temperature	0.02	1	0.08	3	0.10	4	<0.02	<1	<0.02	<1
150 °C (300 °F)	0.53	21	0.30	12	0.08	3	<0.02	<1	<0.02	<1
Pentanedioic (gluloric), 210 °C (410 °F)	...	...	...	...	...	...	0.68	27	0.20	<1
Pentanoic (valeric)										
Room temperature	0.05	2	0.05	2	0.05	2	<0.02	<1	<0.02	<1
114 °C (237 °F)	1.37	54	0.68	27	0.13	5	<0.02	<1	<0.02	<1

Note: Room-temperature values at 26 °C (79 °F)

Table 4 Corrosion of miscellaneous alloys by organic acids

48 h tests of atmospheric boiling temperature; atmosphere not controlled

Test medium	Type 329 mm/yr	Type 329 mils/yr	Tantalum mm/yr	Tantalum mils/yr	Titanium mm/yr	Titanium mils/yr	Zirconium mm/yr	Zirconium mils/yr	Alloy 26-1 (S44626) mm/yr	Alloy 26-1 (S44626) mils/yr	MP35N (R30035) mm/yr	MP35N (R30035) mils/yr
Glacial CH_3COOH	<0.02	<1	nil	nil	<0.02	<1	nil	nil	<0.02	<1	0.05	2
99% CH_3COOH, 1% $(CH_3CO)_2O$	...	...	nil	nil	<0.02	<1	<0.02	<1	...	...	...	...
90% CH_3COOH, 10% $(CH_3CO)_2O$	...	...	nil	nil	<0.02	<1	<0.02	<1	...	...	...	...
50% CH_3COOH, 50% $(CH_3CO)_2O$	0.58	23	nil	nil	0.18	7	<0.02	<1	<0.20	8	<0.02	<1
90% CH_3COOH, 10% HCOOH	0.71	28	nil	nil	<0.02	<1	<0.02	<1	...	...	...	...
70% HCOOH	1.27	50	<0.02	<1	...	...	<0.02	<1	<0.02	<1	0.13	5
20% HCOOH	...	...	<0.02	<1	...	...	<0.02	<1	<0.02	<1	...	...
2-ethylbutyric acid	...	...	nil	nil	<0.02	<1	<0.02	<1	<0.02	<1	...	...
10% aqueous oxalic acid	0.36	14	nil	nil	...	...	<0.02	<1	0.36	14	0.10	4

REFERENCES

1. G.B. Elder, Corrosion by Organic Acids, *Metals Handbook,* Vol 13, 9th ed., American Society for Metals, 1987, p 1157–1160
2. C.P. Dillon, *Mater. Protect.,* Vol 21 (No. 9), 1965, p 4
3. P. Kangas and M. Newman, "Performance of Duplex Stainless Steels in Organic Acids," s-52-89-ENG, AB Sandvik Steel brochure, Oct 1997
4. G.B. Elder, Corrosion by Organic Acid, *Process Industries Corrosion,* National Association of Corrosion Engineers, 1975, p 247–254
5. T-5A-7c Work Group Report, *Mater. Perform.,* Vol 13 (No. 7), 1974, p 1310

6. C.P. Dillon, *Materials Selector for Hazardous Chemicals,* Vol 2, *Formic, Acetic and Other Organic Acids,* Materials Technology Institute of the Chemical Process Industries, 1997
7. P. Kangas and M. Newman, "Performance of Duplex Stainless Steels in Organic Acids," AB Sandvik Steel, Nov 1997
8. R.L. Kane et al., *The Corrosion of Light Metals*, John Wiley & Sons, 1967, p 337
9. L.A. Scribner, "Corrosion by Organic Acids" paper 01343, Corrosion 2001, NACE International
10. Kemira Specifies Zircadyne 702 for Use in a Formic Acid Application, *Outlook,* Vol 11 (No. 1), Teledyne Wah Chang, Albany, OR, 1990
11. H. Togano et al., Corrosion Tests on Materials Used in the Synthesis of Acetic Acid from Methanol and Carbon Dioxide. I. Examination in Acetic Acid Solutions at Increased Temperature, *Tokyo Kogyo Shikensho Hokoku,* Vol 57, 1962, p 342–350, part II in Vol 60 (No. 6), 1965, p 221–231
12. T.-L. Yau, Zircadyne Improves Organics Production, *Outlook,* Vol 16 (No. 1), Teledyne Wah Chang, Albany, OR, 1995, p 3–4
13. A.B. McKee et al., *Corrosion,* Vol 10 (No. 1), 1954, p 786t
14. M.G. Fontana, Some Unusual Corrosion Problems in the Chemical Process Industries, *First International Congress on Metallic Corrosion,* Butterworth, 1961, p 587
15. H. Leidheiser, Jr., *The Corrosion of Copper, Tin and Their Alloys,* John Wiley & Sons, 1971, p 142
16. "Corrosion Resistance of Nickel-Containing Alloys in Organic Acid and Related Compounds," CEB-6, The International Nickel Company, 1979, p 11
17. N.D. Groves et al., *Corrosion,* Vol 17 (No. 4), 1961, p 173t

SELECTED REFERENCES

- C.P. Dillon, *Materials Selector for Hazardous Chemicals,* Vol 2, *Formic, Acetic and Other Organic Acids,* Materials Technology Institute of the Chemical Process Industries, 1997
- *Materials Selection for the Chemical Process Industries,* 2nd ed., C.P. Dillon, Ed., S. Dean, Ed. of 2nd ed., Pub 45, Materials Technology Institute, 2004

Corrosion by Hydrogen Chloride and Hydrochloric Acid

Revised by J.R. Crum, Special Metals Corporation

HYDROCHLORIC ACID (HCl) is an important mineral acid with many uses, including acid pickling of steel, acid treatment of oil wells, chemical cleaning, and chemical processing. It is made by absorbing hydrogen chloride in water. Most acid is the by-product of chlorinations. Pure acid is produced by burning chlorine and hydrogen. Hydrochloric acid is available in technical, recovered, food-processing, and reagent grades. Reagent grade is normally 37.1% (Ref 1). Hydrochloric acid is a corrosive, hazardous liquid that reacts with most metals to form explosive hydrogen gas and causes severe burns and irritation of the eyes and mucous membranes. Safe handling procedures for HCl are described in Ref 2. Additional safety information is available from the manufacturer.

Concentrated HCl is transported and stored in rubber-lined or fiberglass-reinforced plastic (FRP) tanks, although plastic-fabricated polyester-reinforced thermoset plastic storage tanks have also been used. Pipelines are usually plastic-polypropylene-lined (PP) steel or FRP. Processes involving aqueous acid are commonly carried out in glass-lined steel equipment. Nonmetallic materials are normally preferred, when possible, because of the corrosive action of this strongest of acids on most metals.

Candidate metals and alloys for handling HCl are shown in Fig. 1. Materials are judged to be suitable if they have a corrosion rate under 0.5 mm/yr (20 mils/yr) when exposed to uncontaminated HCl. Several metals or alloys have useful corrosion resistance in low concentrations of HCl. As the concentration and temperature increase, the corrosivity increases dramatically. As a result, the selection of metals and alloys becomes more limited, as shown in Fig. 1. In practice, contamination is not uncommon and can be catastrophic. Selection of a candidate metal should be based on extensive corrosion testing or, preferably, field experience, using the grade of acid that will be available.

Effect of Impurities

Hydrochloric acid may contain trace amounts of impurities that will change the aggressiveness of the solution.

Fluorides. Acid recovered from the manufacture of fluorocarbons may contain trace amounts of HF. It has been reported that such acids may contain more than 0.5% HF (Ref 4). Commercial suppliers remove most of the fluoride from HCl by selective absorption, and it is unlikely that an unsuspecting customer would receive acid containing as much as 0.5%. However, glass-lined steel and the refractory metals, such as zirconium, niobium, tantalum, and titanium (but not molybdenum), have very low tolerance levels for fluorides. Zirconium is reported to tolerate less than 10 ppm (Ref 5). Tantalum may tolerate 10 ppm or more. The limits are undefined except for a few specific cases, and it is best to consider all of these metals and glass-lined steel to be essentially nonresistant to fluorides and to know the source and specification limits of the acid if these materials are used.

Ferric Salts. The presence of Fe^{3+} ions in HCl has a profound effect on the corrosion of many metals and alloys otherwise resistant to HCl. Nickel-base alloys, including alloy B-2 (UNS N10665), the copper alloys, and unalloyed zirconium, are affected. Although the acid specification may be low in iron, acid can easily become contaminated during shipment and handling.

Cupric Salts. Cupric ions have an accelerating effect on the corrosion of many metals that is similar to that of Fe^{3+} ions. Like Fe^{3+} ions, Cu^{2+} ions can cause pitting and stress-corrosion cracking (SCC) of zirconium. The presence of Cu^{2+} ions can also lead to the autocatalytic acceleration of nickel-copper and copper-nickel alloys. It is unlikely that commercial acid would contain Cu^{2+} salts. This is more of a problem of in-process contamination by exposure of copper-containing metals or introduction as an impurity in a chemical or raw material used in the process.

Aeration, although less damaging than the presence of oxidizing metal salts, accelerates the corrosion of many metals. Figure 2 shows the effect of aeration on the corrosion rate of Nickel 200 (UNS N02200) and alloy 400 (UNS N04400).

Chlorine contamination accelerates the corrosion of all metals except unalloyed tantalum and noble metals; however, unalloyed titanium can be protected by the presence of chlorine in dilute HCl (Ref 6). Chlorine may be present in acid recovered from a chlorination process but would be removed before sale.

Organics. Hydrochloric acid can become contaminated with organic solvents, such as carbon tetrachloride (CCl_4) or chlorobenzene (C_6H_5Cl), when recovered as a by-product of a chlorination process. Even a few parts per million of organic contaminants can, over a period of time, destroy rubber linings, elastomer membranes behind brick linings, and certain plastics and elastomers. It is unlikely that such acid would be shipped, although organic contamination is often encountered in plant processes.

Corrosion of Metals in HCl

Most corrosion data and graphical information published by metal suppliers and others are based on tests conducted in reagent-grade HCl. In some cases, as in Fig. 1, limitations will be noted regarding the presence of aeration or impurities. As mentioned previously, these data should be used only as a guide in selecting metals for further testing and evaluation.

Carbon and alloy steels are unsuitable for exposure to HCl except during acid cleaning.

Austenitic Stainless Steels. The commonly used austenitic stainless steels, such as types 304 and 316, are nonresistant to HCl at any concentration and temperature. At ambient temperatures and above, corrosion rates are high. Nickel, molybdenum, and, to a lesser extent, copper impart some resistance to dilute acid, but pitting, local attack, and SCC may result (Ref 7). Subambient temperatures will slow the corrosion rate but will invite SCC. Type 316 stainless steel has been known to crack in 5% HCl at 0 °C (32 °F) (Ref 8). At high corrosion rates (>0.25 mm/yr, or 10 mils/yr), SCC is unlikely to occur. However, the corrosion products, particularly $FeCl_3$, will cause cracking. Chlorides can penetrate and destroy the passivity (oxide film) that is responsible for the corrosion resistance of stainless steels, and the corrosion engineer should resist every attempt to use stainless steels in environments containing chlorides.

Superaustenitic stainless steels and higher-nickel alloys possess useful resistance to HCl at lower concentrations, as shown in Fig. 3. Stainless steel 316 shows very little resistance, as mentioned previously, while resistance increases in alloys containing higher nickel and molybdenum contents. Some other higher-nickel alloys, such as 20Cb-3 (UNS N08020) and 825 (UNS N08825), with their high nickel content (32 to 42%), 2 to 3% Mo, and 3 to 4% Cu, resist HCl at all concentrations below approximately 40 °C (100 °F).

The standard ferritic stainless steels, such as types 410 and 430, should not be considered, because their corrosion resistance to HCl is lower than that of carbon steel. An exception is 29-4-2 stainless steel, which reportedly resists up to 1.5% HCl to the boiling point and remains passive (Ref 10). However, it is not suitable at higher concentrations, and the alloy is susceptible to SCC, although its resistance is reported to be high.

Nickel and Nickel Alloys. Nickel 200 and alloy 400 have good resistance (<0.25 mm/yr, or 10 mils/yr) to dilute (<10%) HCl in the absence of air or oxidizing agents at ambient temperatures. Alloy 400 has been used at concentrations below 20% at ambient temperatures under air-free conditions and under 10% concentration aerated, but penetration rates generally exceed 0.25 mm/yr (10 mils/yr) and may approach 1 mm/yr (40 mils/yr).

Increasing the temperature affects the corrosion rate of Nickel 200 more than that of alloy

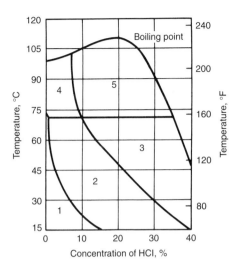

Metals in each zone

Zone 1

Alloy 400(a)(b)(c)
Copper(a)(b)(c)
Nickel 200(a)(b)(c)
Silicon bronze(a)(b)(c)
Silicon cast iron(b)(d)
Alloy B-2,3,4(b)(c)
Alloys C-276, 22, 686, 59, 825, 2000
6 Mo superaustenitic stainless steel
Titanium grade 7
Titanium grade 2(e)
Platinum
Tantalum
Zirconium(a)(c)

Zone 2

Silicon bronze(a)(c)
Silicon cast iron(b)(d)
Alloy B-2,3,4(b)(c)
Alloys C-276, 686(f)
Alloys 22, 59, 2000(g)
Platinum
Tantalum
Zirconium(a)(c)

Zone 3

Silicon cast iron(b)(d)
Alloy B-2,3,4(b)(c)
Alloys C-276, 686(f)
Alloys 22, 59, 2000(g)
Platinum
Tantalum
Zirconium(a)(c)

Zone 4

Tungsten
Titanium grade 7(h)
Alloy B-2,3,4(b)(c)
Platinum
Tantalum
Zirconium(a)(c)

All zones (including 5)

Platinum
Tantalum
Zirconium(a)(c)
Alloy B-2,3,4(b)(c)

(a) No aeration. (b) No $FeCl_2$ or $CuCl_2$ contamination (c) No free chlorine. (d) Contains Cr, Mo, and Ni. (e) <10% HCl at 25 °C (75 °F). (f) 65 °C (149 °F). (g) 55 °C (131 °F). (h) <5% HCl at boiling temperature

Fig. 1 Metals with reported corrosion rates <0.5 mm/yr (<20 mils/yr) in HCl. Source: Ref 3

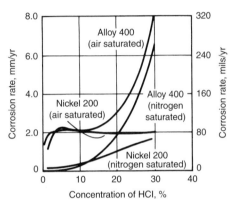

Fig. 2 Corrosion of Nickel 200 and alloy 400 in HCl solutions at 30 °C (85 °F). Source: Ref 5

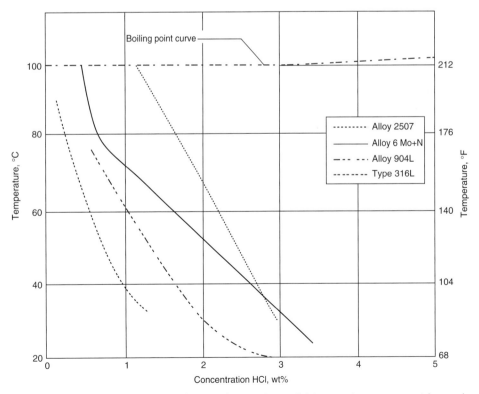

Fig. 3 Hydrochloric acid isocorrosion diagram of 0.1 mm/yr (4 mils/yr) comparing common stainless steels. Source: Ref 9

400 in 5% HCl (Fig. 4). When HCl is formed by hydrolysis of chlorinated hydrocarbons, acid concentrations are often less than 0.5%. Under these conditions, Nickel 200 and alloy 400 have found application at temperatures below 200 °C (390 °F) (Ref 5).

Alloy 600 exhibits corrosion resistance that is inferior to that of Nickel 200 and alloy 400 but has little, if any, catalytic effect on hydrolysis.

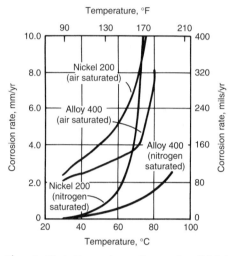

Fig. 4 Effect of temperature on the corrosion of Nickel 200 and alloy 400 in 5% HCl. Source: Ref 5

Many nickel alloys containing chromium and molybdenum have useful resistance to HCl in all concentrations up to approximately, 65 °C (150 °F) and in very dilute acid up to the boiling point (Fig. 5). Corrosion resistance increases with molybdenum content. The most resistant of these alloys, such as alloy C-276 (UNS N10276), also contain tungsten. The materials shown in Fig. 5 are useful for handling hot concentrated HCl, while the other materials discussed previously can only be relied on to resist minor HCl contamination in other chemical processes. These nickel-base, chromium- and molybdenum-containing alloys are also very resistant to chloride SCC. Resistance to pitting and crevice corrosion in acid chloride solutions improves with increasing chromium and molybdenum contents.

The most corrosion resistant of the nickel-base alloys to HCl are the B alloys (Ni-28Mo). Alloy B-2 (a more stable version of the original alloy B, (or UNS N10001) is one of the few metals with a corrosion rate under 0.5 mm/yr (20 mils/yr) in all concentrations and temperatures up to the atmospheric boiling point in nonaerated acid in the absence of oxidizing agents (Fig. 6). Solution heat treatment of alloy B-2 after welding is required for maximum corrosion resistance. Alloys B-3 (UNS N10675) and B-4 (UNS N10629) are variations of this alloy, containing small amounts of chromium, that are more metallurgically stable and, as a result, more weldable and resistant to intergranular attack and SCC. However, small amounts of oxidizers, such as ferric ion, can destroy passivity of alloy B-2 and lead to high corrosion rates (Fig. 7). The presence of approximately 21% Cr in other high-molybdenum alloys, such as alloys C-22 (UNS N06022, 59Ni-21Cr-13Mo-3W-4Fe), 686 (UNS N06686, 59Ni-21Cr-16Mo-3.8W), and 59 (UNS N06059, 59Ni-23Cr-15Mo), promotes passivation in the presence of ferric ion contamination and leads to improved corrosion resistance (Fig. 8).

Alloy C-276 is another nickel alloy that has seen wide use in HCl service. It has excellent corrosion resistance (<0.13 mm/yr, or 5 mils/yr) in all concentrations of HCl at room temperature and good resistance (<0.5 mm/yr, or 20 mils/yr) to all concentrations up to 50 °C (120 °F). At concentrations under 10%, its resistance often exceeds that of B-2. Oxygen and strong oxidizing agents accelerate corrosion, although markedly less than for alloy B-2.

Copper and copper alloys have limited utility in HCl service because they are so sensitive to velocity, aeration, and oxidizing impurities. However, copper is one of the few

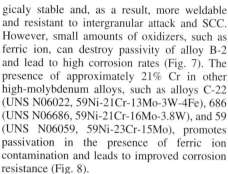

Fig. 6 Isocorrosion diagram for alloy B-2 in HCl

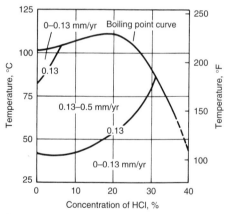

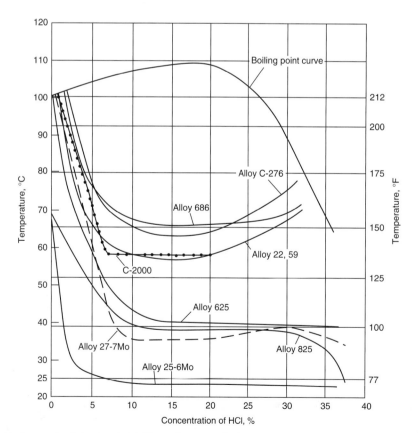

Fig. 5 Comparative behavior of nickel-base alloys in HCl. The isocorrosion lines indicate a corrosion rate of 0.51 mm/yr (20 mils/yr). Source: Ref 11

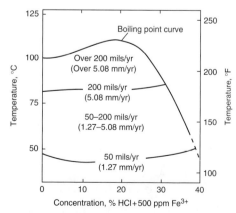

Fig. 7 Isocorrosion diagram for alloy B-2 to HCl with 500 ppm ferric ions. Source: Ref 12

common metals above hydrogen in the electromotive force (emf) series, and under reducing conditions, it can have low corrosion rates. An example is the use of silicon-bronze (UNS C65500) agitators and hardware in the manufacture of $ZnCl_2$ by dissolving zinc in HCl (Ref 8), with the evolution of hydrogen.

Typical corrosion rates for silicon bronze in HCl under nonaerated nonoxidizing conditions are 0.08 to 0.1 mm/yr (3 to 4 mils/yr) in up to 20% acid and 0.5 mm/yr (20 mils/yr) in concentrated acid at 25 °C (75 °F). At 70 °C (160 °F), rates are approximately 1 mm/yr (40 mils/yr) in up to 20% acid and over 6.4 mm/yr (250 mils/yr) in concentrated (35 to 37%) acid.

Corrosion-Resistant Cast Iron. A high-silicon iron alloyed with small amounts of molybdenum, chromium, and copper has good resistance to all concentrations of HCl to temperatures as high as 95 °C (200 °F). It is one of the few metals commonly used for pumps and valves in handling commercial grades of acid (Ref 7). Fluoride impurities are damaging.

Zirconium. The corrosion resistance of zirconium to HCl exceeds that of most metals, except tantalum and such noble metals as gold and platinum. Aeration does not have an appreciable effect. Corrosion rates are less than 0.13 mm/yr (5 mils/yr) at all concentrations to the atmospheric boiling point and above (Ref 13). However, chlorine and relatively small amounts of Cu^{2+} and Fe^{3+} ions accelerate corrosion and can cause pitting and embrittlement. Figure 9 shows an isocorrosion diagram of zirconium in HCl.

It is important to avoid galvanic effects when connecting zirconium to other metals immersed in an electrolyte, because zirconium, like all the reactive metals, is sensitive to hydrogen embrittlement when it is the cathode of an electrochemical cell. Applications of zirconium equipment in HCl include pumps, valves, piping, and heat exchangers (Ref 13).

Titanium and Titanium Alloys. Although titanium has limited resistance to HCl, it is, unlike other metals, passivated rather than corroded by the presence of dissolved oxygen, Fe^{3+} and Cu^{2+} ions, nitrates, chromates, chlorine, and other oxidizing impurities (Ref 6). In commercial applications involving hot, dilute acid, enough impurities are often present to provide a high degree of protection. For example, the corrosion rate of unalloyed titanium in boiling 4% HCl was lowered from 21.4 mm/yr (843 mils/yr) to 0.01 mm/yr (0.4 mils/yr) by the addition of 0.2% $FeCl_3$ (Ref 6).

An isocorrosion diagram for unalloyed titanium and two titanium alloys is shown in Fig. 10. No titanium alloy has resistance to concentrated grades of HCl.

Tantalum and its alloys are the most resistant to HCl. Tantalum resists concentrations below 25% up to 190 °C (375 °F) and concentrations to 37% at temperatures to 150 °C (300 °F) (Ref 14). It was found that tantalum is embrittled by concentrations of 25% or higher at 190 °C (375 °F) (Ref 14). Corrosion rates at that temperature were less than 0.025 mm/yr (<1 mil/yr) at 25% concentration or less, 0.1 mm/yr (3.9 mils/yr) at 30%, and 0.29 mm/yr (11.6 mils/yr) at 37%. Embrittlement was most pronounced in 37% acid. It also was found that corrosion and embrittlement could be avoided by coupling tantalum with platinum. This discovery formed the basis for the later development of the titanium-palladium alloy. Tantalum, like zirconium and titanium, is embrittled by the absorption of atomic hydrogen. This occurs if the metal is corroded in a nonoxidizing chemical or if it becomes the cathode of an electrolytic cell. For example, it is possible to embrittle tantalum plugs in a glass-lined vessel, if the vessel is equipped with an agitator of a metal lower in the emf series.

A comparison of reactive metals is made in Fig. 11. Nickel-molybdenum alloy B-2 and high-silicon steel, at the bottom of the figure, form a baseline for comparison. At temperatures above the HCl boiling point, tantalum and zirconium and their alloys and, in some cases, niobium are among the few metals or alloys that are resistant to attack.

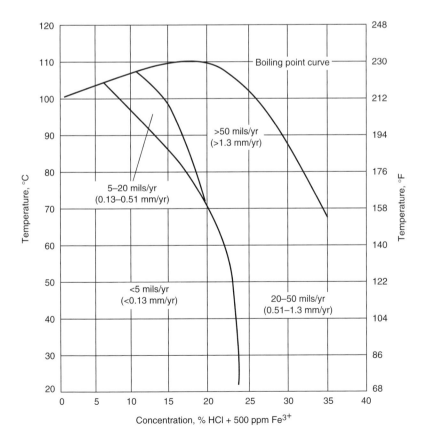

Fig. 8 Isocorrosion diagram for alloy 686 in HCl with 500 ppm ferric ions. Source: Ref 11

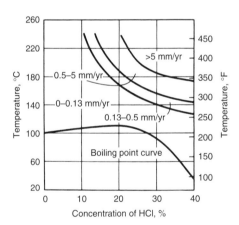

Fig. 9 Isocorrosion diagram for zirconium in HCl solutions. Source: Ref 13

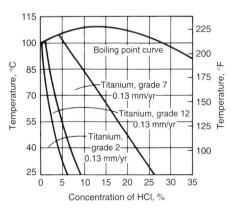

Fig. 10 Isocorrosion diagram for titanium in naturally aerated HCl solutions. Source: Ref 5

Noble Metals. Silver forms a protective chloride film in HCl and is resistant as long as the film is not dissolved or disturbed. At ambient temperatures, in the absence of oxidizing agents, silver resists all concentrations, with a corrosion rate of less than 0.025 mm/yr (<1 mil/yr) at concentrations of 20% or lower. At the boiling point, corrosion rates vary from 0.025 mm/yr (1 mil/yr) at 5% to 0.5 mm/yr (20 mils/yr) at 20%. Increasing the temperatures, aeration, and velocities will increase corrosion rates (Ref 7).

Refined gold is extremely resistant to all concentrations of HCl to and above the atmospheric boiling point. Oxidizing agents such as $FeCl_3$ and HNO_3 cause attack (Ref 7).

Refined platinum is almost as resistant as gold. It is slightly attacked by 36% acid at 100 °C (212 °F), the point at which gold shows no appreciable attack (Ref 7).

Nonmetallic Materials

Elastomers used in HCl service include natural rubber, neoprene, and others.

Natural rubber is used to line steel tanks, tank cars, and piping for handling commercial, uncontaminated, concentrated HCl (Ref 1). Natural rubber forms a protective rubber hydrochloride surface that slows further penetration, and a life expectancy of 20 years at ambient temperature is not uncommon. The lining must be properly compounded for the service, must be of sufficient thickness (a minimum of 6.4 mm, or 0.25 in.), and must be properly applied and inspected. Soft rubber linings are recommended to 40 to 45 °C (100 to 110 °F), depending on the manufacturer (Ref 16, 17). A three-ply construction (soft-hard-soft) is often used to take advantage of the impermeability of the harder rubber and the resistance to damage, thermal shock, and abrasion of the soft rubber. Three-ply construction is recommended to temperatures of 55 °C (130 °F) (Ref 16). For higher temperatures, semihard rubber is recommended to 70 °C (160 °F), and hard rubber or flexible ebonite to a maximum temperature of 95 °C (200 °F) (Ref 18). The latter must be protected against exposure to sunlight and thermal and mechanical shock.

Natural rubber is severely affected by many organic contaminants, some of which may be present in by-product acid. Because the effect is cumulative, even trace amounts will eventually cause failure.

Neoprene. Although relatively unaffected by ambient-temperature HCl, neoprene is not recommended for linings, because of its permeability to concentrated acid (Ref 16, 17).

Other Elastomers. A number of elastomers have resistance to HCl that is as good as or better than that of rubber, but they are seldom used as linings for this service because of economics and difficulty of application. However, they should be considered for exposure to sunlight, organics, higher temperatures, or flexing (the hard rubber hydrochloride surface cracks when flexed). These include butyl rubber, nitrite butyl rubber, ethylene propylene diene monomer, and chlorosulfonated polyethylene elastomer (Ref 18, 19). Fluoroelastomers have good resistance to hot acid, but because of high cost, are limited to small parts, such as gaskets and O-rings.

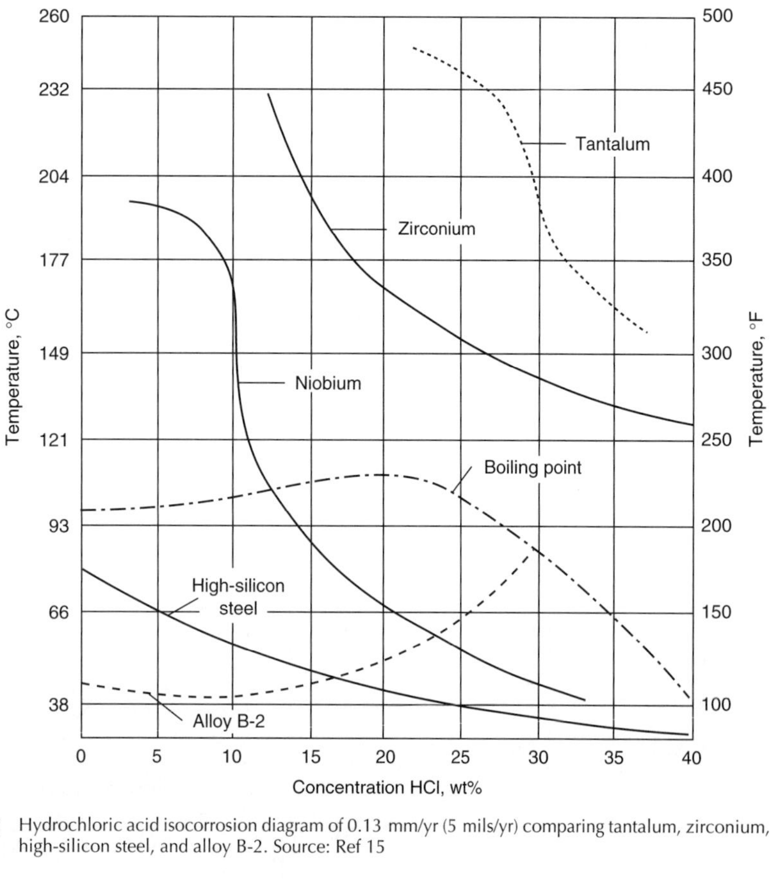

Thermoplastics. A large number of thermoplastics are suitable for handling all concentrations of HCl. Some of the most commercially important are described as follows.

Polypropylene. Polypropylene-lined steel pipe has largely replaced rubber-lined pipe for commercial HCl service, primarily because of economics and ease of field fit-up. Lined pipe is recommended for all concentrations of HCl to temperatures of 110 °C (230 °F) (Ref 20). Polyproplyene is also available in small-to-moderate sized tanks and various molded shapes.

Polyethylene is unaffected by ambient-temperature acid, but its lack of stiffness, high thermal coefficient of expansion, and rapid falloff of strength with increasing temperature limit its use to underground or continuously supported pipelines. High-density cross-linked polyethylene is stronger and stiffer and is available in tanks up to 3785 L (1000 gal).

Polyvinyl Chloride (PVC). Both plasticized and unplasticized PVC have good resistance to HCl at ambient temperature. Unplasticized PVC piping systems are practical for complex systems that can be properly supported.

Polyvinylidine chloride has been used for many years as a lining for HCl pipelines. It is recommended for all concentrations to 80 °C (175 °F) (Ref 20). It has better resistance than rubber or polypropylene to certain hydrocarbons that may be present in process streams.

Polyvinylidine Fluoride (PVDF). Pipe lined with PVDF is recommended for all concentrations of HCl to 135 °C (275 °F) (Ref 20). Polyvinylidine fluoride is also available in lined valves and pumps. It resists a wide variety of other chemicals and solvents that may be present in process streams or in recovered acid. Linings, frequently reinforced with graphite, glasscloth, or veil, have been applied with multiple thermally fused layers with varying degrees of success. Each case should be approached with caution.

Fluoroplastics. In addition to PVDF, a number of other fluorocarbon plastics resist all concentrations of HCl at temperatures as high as 260 °C (500 °F) (Ref 20). The fully fluorinated polytetrafluoroethylene (PTFE) and perfluoroalkoxy have the highest temperature resistance and are the most expensive. Fluorinated ethylene propylene is limited to approximately 200 °C (400 °F) but is more easily molded than PTFE. Ethylene-chlorotrifluoroethylene and ethylene-tetrafluoroethylene are limited to 150 °C (300 °F) but have superior

Fig. 11 Hydrochloric acid isocorrosion diagram of 0.13 mm/yr (5 mils/yr) comparing tantalum, zirconium, niobium, high-silicon steel, and alloy B-2. Source: Ref 15

abrasion and permeability resistance (Ref 21). Most fluoroplastics are available as lined pipe, valves, and pumps (Ref 21).

Other Thermoplastics. Other plastics, such as acrylonitrile-butadiene-styrene, polysulfone, polyphenylene sulfide, and polyphenylsulfone, are resistant to HCl. Nonresistant thermoplastics include such polyimides as nylon and methacrylates.

Reinforced thermoset plastics for HCl service include polyester, epoxy, phenolic, and furan resins.

Polyester Resins. There have been a number of failures of glass-reinforced polyester resin storage tanks in concentrated HCl service. A life expectancy of 5 to 10 years has been reported (Ref 22). However, there are custom-built glass-reinforced plastic storage tanks that have given satisfactory service in concentrated HCl for 10 to 20 years and longer (Ref 8).

Standard, filament-wound storage tanks are not suitable for concentrated acid. Vessels should be fabricated in accordance with the standards detailed in Ref 23. The inner resin-rich layer should be thicker than 0.8 mm (1/32 in.), with two layers of veil (Ref 8). Bubbles and porosity are not acceptable, because they will cause blistering, which will lead to degradation. The novalac epoxy vinyl-ester resins resist blistering better than other resins, although none are recommended for storage of concentrated acid (Ref 22).

In Europe, glass-reinforced plastic tanks for storing concentrated acid are frequently lined with PVC sheet. The novalac resins have superior resistance to some chlorinated hydrocarbons (Ref 29). Custom-built storage tanks using this resin have been in service for over 10 years in concentrated acid containing CCl_4 in excess of solubility (Ref 8). Pumps and piping systems of polyester glass-reinforced plastic are also used for moving concentrated acid.

The epoxy resins are generally less resistant to HCl than the resistant grades of polyesters. Manufacturers of glass-reinforced plastic pipe limit recommendations for their epoxy pipe to concentrations of 10 or 20% at ambient temperatures, while the polyester pipe is recommended for all concentrations at higher than ambient temperatures. Recommendations by a leading manufacturer are given in Table 1. It should be noted, however, that a noted pump manufacturer makes a silica-filled epoxy pump that is widely used for pumping concentrated HCl.

Phenolic Resins. Vessels, ducts, and piping systems fabricated of suitably filled phenolic resins will resist all concentrations of acid to temperatures as high as 175 to 200 °C (350 to 390 °F) (Ref 26). The phenolics are also resistant to many solvents. Because phenolic resins are weak in tension, equipment and pipe are often armored with epoxy-impregnated glass cloth or are filament wound. Piping systems must be carefully designed and installed with adequate support to minimize tensile stresses. Phenolic resins are often used as cements for brick linings.

Furan resins are also used to construct vessel duct and piping systems that are resistant to all concentrations of wet HCl (Ref 26). The solvent and alkali resistance of furan resins exceeds that of phenolics, and they are easier to handle during fabrication. Thus, furan glass-reinforced plastic vessels are available, and furan-glass membranes are used in brick-lined equipment. Temperature resistance is somewhat lower than that of the phenolics. Furans are the most common of the plastic cements used for brick linings.

Other nonmetallic materials for HCl service include impervious graphite, carbon, glass, glass-lined steel, and wood.

Impervious graphite is the most commonly used material of construction for HCl absorption and refining equipment. Resin-impregnated graphite equipment resists all concentrations of HCl to temperatures of 165 to 185 °C (330 to 365 °F), depending on impregnating resin and manufacturer. Polytetrafluoroethylene-impregnated graphite is reported to be resistant to a maximum temperature of 230 °C (450 °F) (Ref 27). By using graphite impregnated with pure carbon, it is possible to increase the maximum-use temperature in air to 400 °C (750 °F) (Ref 27). Graphite has excellent thermal conductivity and chemical resistance (except to strong oxidants and concentrated H_2SO_4). However, it is extremely weak in tension and is seldom used for piping. Heat exchangers are available in block and crossbore construction to obtain more rugged units than shell and tube designs.

Carbon has poor heat transfer but excellent abrasion resistance. It is available as bricks for brick-lined construction. Like graphite, carbon resists all concentrations of HCl to 400 °C (750 °F) in air.

Glass. Complete small chemical plants are available in glassware. Piping systems can be obtained armored with glass fabric. Piping systems must be properly supported, because tensile stresses can result in delayed fracture.

Glass-Lined Steel. The corrosion resistance of a proprietary glass lining in HCl is shown in Fig. 12. The minimum thickness of high-voltage-tested glass linings is approximately 1 mm (40 mils). The calculated life expectancy may occasionally be somewhat greater than actual experience because of localized chains of bubbles. However, these data provide a reasonably realistic design basis.

Wood is an economical material for handling dilute acids. Tanks and large-diameter piping cost appreciably less than rubber-lined steel or thermosetting plastic construction. Although unsuitable for concentrated HCl, wood has excellent resistance up to 10% concentration. The National Wood Tank Institute recommends wood in 2% HCl to temperatures of 60 °C (140 °F) and 10% HCl to 70 °C (160 °F) (Ref 29). Cypress and fir were unaffected after 31 days of exposure in both 5 and 10% acid at room temperature and showed slight fiber disintegration and slight embrittlement and softening after 8 h in boiling 10% HCl (Ref 30).

For storage and handling of HCl, the following materials are usually considered adequate (Ref 31):

Application	Material
Tanks	FRP or rubber-lined steel
Tank trucks	Rubber-lined steel
Railroad cars	Rubber-lined steel
Piping	FRP or plastic-lined steel (e.g., PP)
Valves	PTFE or fluorinated ethylene propylene (FEP)-lined, cast alloy B (UNS N10001), N-12M(a)
Pumps	PTFE-lined, impervious graphite, cast alloy B(a)
Gaskets	Rubber, felted PTFE, FEP envelope, flexible graphite

(a) Requires acid free of iron or other oxidizing contaminants

Hydrogen Chloride Gas

Corrosion rates determined in dry hydrogen chloride and chlorine gases are listed in Table 2. Based on these results, the recommended upper

Table 1 Recommended maximum temperature limits for fiberglass-reinforced plastic piping in HCl

Acid concentration, wt%	Maximum recommended temperature			
	Epoxy resin		Vinyl-ester resin	
	°C	°F	°C	°F
1	93	200	93	200
10.5	66	150	93	200
20	24	75	93	200
36.5 (conc)	NR	NR	66	150(a)

NR, Not recommended. (a) Maximum temperature tested; could be suitable at higher temperature. Source: Ref 25

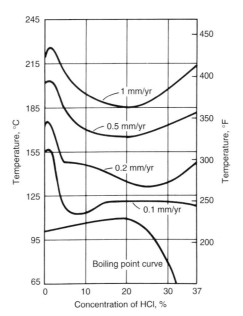

Fig. 12 Isocorrosion diagram for Glasteel 5000 in HCl. Source: Ref 28

Table 2 Corrosion of metals in dry hydrogen chloride

	Temperature that corrosion rate is exceeded(a)									
	0.0635 mm/mo (2.5 mils/mo)		0.127 mm/mo (5 mils/mo)		0.254 mm/mo (10 mils/mo)		1.27 mm/mo (50 mils/mo)		2.54 mm/mo (100 mils/mo)	
Metal	°C	°F	°C	°F	°C	°F	°C	°F	°C	°F
Platinum	1260	2300	...	...	...	...	...	...	...	...
Gold	980	1800	...	...	...	...	...	...	...	...
Nickel	455	850	510	950	565	1050	675	1250	705	1300
Alloy 600 (UNS N06600)	425	800	480	900	540	1000	675	1250	730	1350
Alloy B (UNS N10001)	370	700	425	800	480	900	650	1200	705	1300
Alloy C (UNS N10002)	370	700	425	800	480	900	620	1150	675	1250
18-8-Mo (UNS S31600)	370	700	370	700	480	900	595	1100	650	1200
25-12-Cb	340	650	400	750	455	850	565	1050	620	1150
18-8 (UNS S30400)	340	650	400	750	455	850	595	1100	650	1200
Carbon steel	260	500	315	600	400	750	565	1050	620	1150
Ni-Resist (type 1)	260	500	315	600	370	700	540	1000	595	1100
Alloy 400 (UNS N04400)	230	450	260	500	340	650	480	900	565	1050
Silver	230	450	290	550	340	650	455	850	...	...
Cast iron	205	400	260	500	315	600	455	850	510	950
Copper	95	200	570	300	205	400	315	600	370	700

Note: These values are based on short-time laboratory tests. They should be interpreted only as being indicative of the limitations of the materials and should not be used for estimation of the service life of equipment. (a) Temperature is approximate. Original data in customary units. Source: Ref 32

Table 3 Suggested upper temperature limits for continuous service in dry HCl gas

	Temperature	
Material	°C	°F
Platinum	1200	2190
Gold	870	1600
Nickel 201	510	950
Alloy 600	480	900
Alloy B	450	840
Alloy C	450	840
Type 316 stainless steel	430	805
Type 310Cb stainless steel	430	805
Type 304 stainless steel	400	750
Carbon steel	260	500
Alloy 400	230	445
Silver	230	445
Cast iron	200	390
Copper C11000	90	195

Source: Ref 33

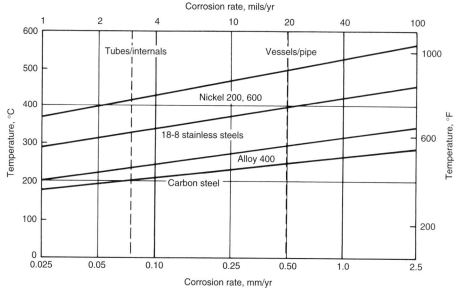

Fig. 13 Maximum temperature limits for various alloys in HCl service. Source: Ref 34

temperature limits for various metals in HCl gas are shown in Table 3. These guidelines were developed from short-term tests. Because corrosion rates generally decrease with time as films and scales form, these limits have proved to be conservative. Temperature-versus-design corrosion rates for steel and several alloys are shown in Fig. 13.

The addition of water vapor to HCl gas has little effect on upper temperature limits. The addition of 0.2% water vapor by weight has been found to have no effect (Ref 33). However, water vapor has a severe effect at lower temperatures. In a study (Ref 35), the suggested lower limit for carbon steel in a 60% H_2O, 40% HCl environment was 200 °C (390 °F) or slightly lower. The lower temperature limits for nickel alloys were 115 °C (240 °F) for C-276, 140 °C (285 °F) for alloy 625, and 260 °C (500 °F) for Nickel 200, alloy B-2, and alloy 600. Grade 7 (0.2% Pd) titanium had satisfactory resistance up to 120 °C (250 °F). None of the materials tested, including Nickel 200, titanium, zirconium, and cobalt-base alloys, was suitable at or slightly below the dewpoint. Minimum temperatures for metals ranged from 30 to 100 °C (85 to 210 °F) above the dewpoint, depending on the hygroscopicity of the corrosion products.

ACKNOWLEDGMENT

This article has been adapted from Thomas F. Degnan, "Corrosion by Hydrogen Chloride and Hydrochloric Acid," *Corrosion*, Volume 13, *ASM Handbook* (formerly *Metals Handbook*, 9th ed.), ASM International, 1987, pages 1160–1166.

REFERENCES

1. "Storage and Handling of Hydrochloric Acid," Technical Bulletin, E.I. DuPont de Nemours & Company, 1972
2. "Manufacturing Chemists Association Chemical Safety Data," Sheet SD-39, Manufacturing Chemists Association, 1970
3. *Corrosion Data Survey—Part I, Metals*, National Association of Corrosion Engineers, 1974

4. L.W. Gleekman, *Chemical Process Industries Symposium*, National Association of Corrosion Engineers, 1975, p 225
5. *Resistance to Corrosion*, 4th ed., Inco Alloys International, 1985, p 21–24
6. "Corrosion Resistance of Titanium," Titanium Metals Corporation of America
7. "Engineering Guide to DuPont Elastomers," E.I. DuPont de Nemours & Company
8. E.I. DuPont de Nemours & Company, private communication
9. J.M. Nicholls, Corrosion Properties of Duplex Stainless Steels: General Corrosion, Pitting and Crevice Corrosion, *Duplex 94 Stainless Steels*, Vol 1, The Welding Institute, Cambridge, U.K., 1994
10. *Properties of a High Purity 29%Cr-4%Mo-2%Ni Ferritic Alloy for Aggressive Environments*, Allegheny Ludlum Steel Corporation, 1982, p 9
11. "High Performance Alloys for Resistance to Aqueous Corrosion," Publication SMC-026, Special Metals Corporation, 2000
12. "Hastelloy B-2 Alloy," Publication H-2006D, Haynes International, 1997
13. "Zircadyne Corrosion Properties," Teledyne Wah Chang Albany, 1981
14. C.R. Bishop and M. Stern, *Corrosion*, Vol 17 (No. 8), 1961, p 379t
15. J.J. Hunkeler, Tantalum and Niobium, *Process Industry Corrosion*, B.J. Bomiz and W.I. Pollock, Ed., NACE International, 1986, p 545
16. "Uniroyal Rubber Linings," Uniroyal, Inc., 1968
17. "B.F. Goodrich Tank Linings," B.F. Goodrich, Inc., 1974
18. NACE Technical Committee, *Corrosion*, Vol 17 (No. 9), 1961, p 453t
19. "The General Chemical Resistance of Various Elastomers," The Los Angeles Rubber Group Inc., 1970
20. "Chemical Resistance Guide for Systems Using Dow Plastic-Lined Piping Products," The Dow Chemical Company, 1978
21. R.E. Tatnall and D.J. Kratzer, Paper 307, presented at Corrosion/85, National Association of Corrosion Engineers, 1985
22. J.E. Niesse, *Mater. Perform.*, Vol. 21 (No. 1), 1982, p 25
23. T.G. Priest and O.W. Seibert, *Mater. Perform.*, Vol 20 (No. 10), 1981, p 38
24. W.W. McClellan, T.F. Anderson, and R.F. Stavinoha, 30th Anniversary Technical Conference, Section 6A, Reinforced Plastics/Composites Institute, The Society of the Plastics Industry, Inc., 1975
25. "Chemical Resistance Red Thread, Green Thread, Polythread, Chemline Piping Systems for the Chemical Process Industries," Product Bulletin 9002, A.O. Smith-Inland Inc.
26. 21st Biennial Report—Materials of Construction—Section 1, *Chem. Eng.*, Vol 71 (No. 24), 9 Nov 1964, p 170
27. "Graphitar, A Challenge to Corrosion," Carbone-Lorraine Industries Corporation
28. "Pfaudler 5000 Glassteel Technical Data," The Pfaudler Company Inc., 1979
29. "Wood Tanks for Corrosive Applications," Technical Bulletin 758, National Wood Tank Institute, 1975
30. A.P. Pfeil, *Des. Eng.*, June 1961, p 114
31. C.P. Dillon, *Materials Selection for the Chemical Process Industries*, McGraw-Hill, Inc., 1991
32. "Corrosion Engineering Bulletin CEB-3," The International Nickel Company, Inc.
33. M.H. Brown, W.B. DeLong, and J.R. Auld, *Ind. Eng. Chem.*, Vol 39 (No. 7), 1947, p 839–844
34. C.M. Schillmoller, *Chem. Eng.*, Vol 87 (No. 5), 10 March 1980, p 161
35. J.P. Carter, B.S. Covino, Jr., T.J. Driscoll, W.D. Riley, and M. Rosen, *Corrosion*, Vol 40 (No. 5), 1984, p 205

Corrosion by Hydrogen Fluoride and Hydrofluoric Acid

Herbert S. Jennings, DuPont Fluoroproducts

COMMERCIAL AQUEOUS HYDROFLUORIC ACID (HF) concentrations are provided in bulk at 49 and 70% by weight. By definition, anhydrous hydrogen fluoride (AHF) has less than 400 ppm water by weight. Both HF and AHF are of great industrial importance. Hydrofluoric acid is used for many industrial applications, such as etching and cleaning of metals and ceramics, acid treating oil and gas wells, and metallurgical ore and refining processes. Anhydrous hydrogen fluoride is consumed in the manufacture of fluorine and the production of fluorochemicals that are used to produce lubricants, plastics, and elastomers. Anhydrous hydrogen fluoride is also a catalyst for petroleum and chemical reactions.

The Hydrogen Fluoride Industry Practices Institute has guidelines for materials of construction for handling HF (Ref 1) and AHF (Ref 2). NACE International T5A171 (Ref 3) and Materials Technology Institute publication MS-4 (Ref 4) provide materials and corrosion data for critical design situations. More information can be obtained from HF and AHF producers, materials suppliers, and equipment fabricators.

Much of the data presented are from tests under ideal conditions with no impurities. Even in ideal conditions, corrosion rates may vary significantly from test to test. Test duration is important, because most metals form protective scales, and initial high corrosion rates will decrease with time to the rates found in commercial use (Ref 5). Corrosion rates of metals in HF and AHF are a function of concentration, velocity, temperature, impurities, and the nature of the corrosion product film. Corrosion reduces metal thickness and generates metal fluorides and atomic hydrogen.

The first part of this article presents corrosion data for materials in HF. The second presents data for AHF and includes discussions of hydrogen blistering and cracking of carbon steels and data for materials in high-temperature HF and AHF.

Aqueous Hydrofluoric Acid

Information on suitable materials of construction for handling HF of various strengths and at various temperatures is given in Ref 3 and in alloy producers' literature. Figure 1 (Ref 3), an isocorrosion diagram for various metals, indicates temperature and concentration conditions where metals have corrosion rates of 0.5 mm/yr (20 mils/yr) or less. This figure is generated mostly from laboratory tests. Additional corrosion data for various metals in selected HF concentrations are shown in Tables 1 and 2 (Ref 6).

Carbon and Low-Alloy Steels

Carbon steels have useful corrosion resistance from 70 to 100% HF. In liquid concentrations below 70% HF, steel has a significant corrosion rate. Corrosion rates are reduced when HF solution temperatures are less than 20 °C (70 °F). Carbon steel in continuous exposure to 70% HF must be limited at temperatures no greater than 30 °C (90 °F). Under the influence of velocity, turbulence, or impingement, carbon steel is susceptible to erosion-corrosion; therefore, in 70% HF, it is a good practice to limit velocity to 0.6 m/s (2 ft/s). Increasing temperature, velocity, and aeration or lowering concentration can increase the corrosion rates of steels in HF.

Hydrofluoric acid vapors will condense at 38% concentration on surfaces that are at the dewpoint of 111.4 °C (232 °F). The concentration of condensing HF depends on the concentration of liquid HF. When the liquid is lower than 38% concentration and the temperature is below the dewpoint, the condensing vapor will be water enriched. However, above a liquid phase with higher than 38% HF concentration, the vapor phase condensate will be HF enriched. The HF-water/liquid-vapor equilibrium exhibits a constant boiling mixture or azeotrope at approximately 38 wt% HF. At 100 kPa (1 atm) or less pressure, there will be no condensation on surfaces that are above 111.4 °C (232 °F). Carbon steel corrosion data in HF liquid and vapor at 38 °C (100 °F) (Ref 7) are presented in Fig. 2. The maximum corrosion rate in liquid is at approximately 35% HF, 20 mm/yr (800 mils/yr). This high rate is indicative of the poor adherence and high solubility for ferrous fluoride (FeF_2) and ferric fluoride (FeF_3) in 35 to 38 wt% HF. Steel in HF vapor exhibits a maximum corrosion rate at approximately 20% HF, 2.6 mm/yr (105 mils/yr).

Impurities in HF. Generally, carbon steel has higher corrosion rates in HF with air than without air. Notably, steel corrosion rates may be lowered by increased total acidity in HF by fluorosilicic and sulfuric acids. These impurities have a protective effect for steels in HF (Ref 8). Commercial HF quality has improved over time, so now these impurities are present in HF at only very low levels, and recent corrosion rates on steels tend to be higher than that reported in the 1950s.

Residual Elements in Carbon Steels. Parallel with changes in HF quality are changes in global steel-making practices, where more recycled metals are used. Some carbon steel products have slightly more residual elements (RE), such as copper, nickel, or chromium, than previous decades. These slight changes are a cause for higher steel corrosion rates in HF. Also, certain alloy steels exhibit high corrosion rates in HF. Dilute acids corrode iron and carbon steel at similar rates; compare the rates of iron and steel to those of alloy steels in Table 3. The HF causes exceptionally high corrosion rates on alloy steels with 1.35 to 17% Cr (ASTM A335 grades and type 430 stainless steels) (Ref 9).

Several HF alkylation units have reported corrosion of selective carbon steel piping or fittings that have slightly higher RE. These failures were associated with steels with slightly higher levels of copper, nickel, or chromium. The RE for all of the components was within nominal specifications for the pipe and fittings (Ref 10–14). The high-RE-component corrosion product layer was loose, porous, and nonprotective. Corrosion test data on low- and high-RE steels supported the relationship with corrosion performance (Ref 10). A 0.2% limit on RE for the sum of copper, nickel, and chromium (Σ_{RE}) was proposed for carbon steels in HF alkylation services. Electrochemical corrosion studies on high-Σ_{RE} samples indicated a diffusion-controlled anodic process (Ref 15). Low-carbon steel had lower corrosion rates than higher-alloyed steels in concentrated HF. Some RE were beneficial or detrimental to corrosion resistance, depending on HF concentration and the presence of oxygen. The quantity and form of carbon in steel influenced corrosion in HF.

Another electrochemical study proposed reducing the 0.2% Σ_{RE} guideline to 0.15% Σ_{RE} (Ref 16). This study was performed with 60% HF at 71 °C (160 °F) and 1% HF at 40 °C (105 °F). An observation was that >0.18% C was beneficial for both low- and high-Σ_{RE} steel resistance. Copper and nickel were detrimental to corrosion resistance in HF. Chromium was beneficial for steels with >0.18% C but detrimental when carbon was lower. Tests in 1% HF found that this environment enabled galvanic corrosion. The low-Σ_{RE} steels were preferentially corroded, while high-Σ_{RE} steels exhibited lower corrosion rates. Figure 3, from this study, presents normalized corrosion data from 60% HF at 71 °C (160 °F); the corrosion rates were lowest when carbon > 0.18% and Σ_{RE} = 0.1%.

Galvanic corrosion is reported in 70% HF at 52 °C (125 °F) for carbon steels welded to higher alloys (Ref 17). Compared to 70% HF, the galvanic corrosion rates dropped significantly in 82 and 100% HF.

Austenitic Stainless Steels

Austenitic stainless steels have resistance to dilute HF. Type 304 stainless steel (UNS S30400) has very poor resistance to greater than 1% HF concentration. In concentrations over 5% HF, type 304 exhibits higher corrosion rates than carbon steels at similar concentrations and temperatures. Type 316 stainless steel (UNS S31600) has useful resistance only at 23 °C (73 °F) or lower temperatures, at 10% HF concentrations or less, and this alloy may pit or crack under these conditions (Ref 18). Corrosion rates of type 316 are reported in Table 4 at 24 °C (75 °F) for 1 and 10% HF (Ref 19). Table 5 presents corrosion rates for type 316 exposed to 5% HF at 52 and 79 °C (125 and 175 °F) (Ref 20). Table 6 presents corrosion data for types 304, 316, and 309Cb (UNS S30900) (Ref 8). Weld fusion line corrosion has been observed on type 304 in dilute HF (<1%) and attributed to preferential corrosion of ferrite in the weld heat-affected zones (HAZs). Ferritic and martensitic stainless steels are subject to high corrosion rates and cracking in HF.

Cracking. Chloride stress-corrosion cracking (SCC) of sensitized type 304 is greatly aggravated by the simultaneous presence of acid fluorides (from weld electrode coating) and chlorides. Sensitized type 304 had SCC susceptibility to NaF over the range of 50 to 80 °C (120 to 175 °F) to 40 to 400 ppm F$^-$, but with both Cl$^-$ and F$^-$, the SCC threshold dropped to 10 to 100 ppm F$^-$ (Ref 21).

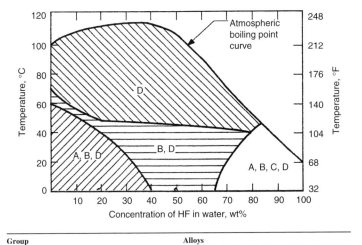

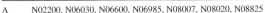

Group	Alloys
A	N02200, N06030, N06600, N06985, N08007, N08020, N08825
B	N06022, N06059, N06200, N06686, N10276, N10665, N10675, N26455, N26022, N30007, N30107
C	Carbon steel(a)
D	C70600, C71500, N04400, N24135, P00020, P04995, P07015, R03600

(a) Carbon steel may suffer hydrogen stress cracking, hydrogen embrittlement, or hydrogen-induced cracking.

Fig. 1 Isocorrosion diagram for metals and alloys in hydrofluoric acid (HF). Regions indicate where corrosion is less than 0.5 mm/yr (20 mils/yr). This information is for guidance only. It represents low-flow, oxygen-free, uncontaminated conditions. Velocity and/or impurities can make these materials unsuitable. See text for information about stress-corrosion cracking in nickel-rich and nickel-base alloys. Source: Ref 3

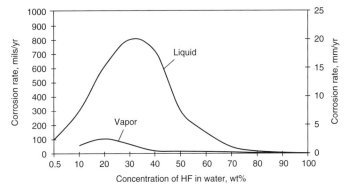

Fig. 2 Carbon steel corrosion in hydrofluoric acid (HF) liquid and vapor at 38 °C (100 °F). These data represent corrosion rates on carbon steel in HF that is free of contamination and at low- or no-flow conditions. Rates shown for 70 to 100% HF combine laboratory and service experience data. Service experience data reflect ambient temperatures of 21 to 38 °C (70 to 100 °F). Source: Ref 7

Table 1 Corrosion of selected metals and alloys in boiling 48% hydrofluoric acid

Exposure for 40 days; vapor phase purged with nitrogen

	Corrosion rate			
	Immersed		Vapor	
Metal	mm/yr	mils/yr	mm/yr	mils/yr
Platinum	nil		nil	
Silver	nil		nil	
Copper	0.05	2	0.06	2
90-10 copper-nickel	0.07	3	0.02	1
80-20 copper-nickel	0.12	5	0.02	1
70-30 copper-nickel	0.08	3	0.02	1
Monel 400	0.46	18	0.85	34
Lead	5	200	3.8	150

Source: Ref 6

Table 2 Corrosion of selected metals and alloys in commercial hydrofluoric acid solutions

Exposure for 35 days at 60 °C (140 °F); vapor phase purged with nitrogen

	Corrosion rate in various acid concentrations											
	50%				65%				70%			
	Liquid		Vapor		Liquid		Vapor		Liquid		Vapor	
Metal	mm/yr	mils/yr	mm/yr	mils/yr	mm/yr	mils/yr	mm/yr	mils/yr	mm/yr	mils/yr	mm/yr	mils/yr
Platinum	nil		nil		nil		nil		...		...	
Silver	0.01	0.4	nil		0.02	0.7	0.0005	0.02	0.02	0.7	0.003	0.01
Monel 400	0.46	18	0.12	5	0.13	5	0.06	2	0.14	5	0.05	2
Magnesium	0.21	8	0.03	1	0.06	2	0.06	2	...		...	
Hastelloy C	0.74	29	0.6	24	0.19	8	0.24	10	...		...	

Source: Ref 6

Nickel-Rich Austenitic Stainless Steels

In HF service, nickel-rich (nickel > 30%) austenitic stainless steels with chromium, molybdenum, and copper, such as alloys G3 (N06985), G30 (N06030), CN-7M (N08007), 20 (N08020), and 825 (N08825), have better resistance than other austenitic stainless steels but only over a slightly greater range of concentration and temperature, as indicated in Fig. 1. These alloys have good resistance to all liquid concentrations of HF at ambient temperatures and up to 10% HF at 70 °C (160 °F) (Ref 8). Corrosion data for alloys 20 and 825 in 1 and 10% HF are presented in Table 4. Also, data for alloy 825 are presented in Tables 5 and 6. The isocorrosion diagram (Fig. 4) includes data on alloys 825 and G30.

Weld Corrosion. In HF, the welds in alloys 20, 825, and G30 are more susceptible to corrosion than wrought metal. Weld attack by HF-HNO_3 is reported for alloys 825 (Ref 23) and G30 welded with electrode ERNiCrMo-11 (Ref 24). Weld attack is due to alloy segregation and the detrimental effect of reactive metals (tantalum, niobium, and titanium) alloyed in the base and weld metals.

Oxygen Influence. Alloys 20, 825, G3, and G30 have higher corrosion rates when oxygen is present in HF liquid (Ref 25). In aerated HF, the alloys with better corrosion resistance have higher chromium and molybdenum, with nickel adding some limited benefits. In N_2-purged HF, the more resistant alloys contain nickel, molybdenum, and copper (G3 and G30).

Cracking of alloys 20 and 825 is reported in 10% HF vapors at 76 °C (169 °F) (Ref 19). The cracked samples had either a thin layer of copper deposit or significant staining, suggesting vapor-phase condensation corrosion, with accumulation and concentration of CuF on the sample surface. The CuF is oxidized to CuF_2 in the presence of oxygen. This may not occur when the metal is submerged in liquid HF, due to limited exposure to oxygen or dilution of corrosion products that reduce CuF on the metal surface. These alloys may experience SCC if exposed to low levels of fluorides with chlorides (Ref 26).

Nickel and Nickel-Base Alloys

Nickel and nickel alloys are among the more corrosion-resistant materials for HF. Figure 1 shows that nickel alloys from groups B and D are useful across the range of HF concentrations. In general, the corrosion rates of nickel alloys increase with HF concentration (up to the azeotrope of HF) and temperature. Nickel alloys and welds containing niobium tend to have elevated corrosion rates in HF above 40 °C (105 °F). Hydrofluoric acid liquid or vapor with oxygen increases corrosion of these alloys. The presence of oxidizing salts in HF may be beneficial to nickel-chromium alloys and detrimental to nickel-copper and nickel-molybdenum alloys. Corrosion data for various nickel alloys are presented in Fig. 4 to 6 as well as Tables 1, 2, 4, 5,

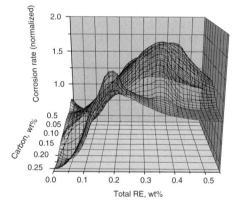

Fig. 3 Three-dimensional graph showing trends with respect to composition of carbon and residual elements (RE) in steels. The lowest corrosion rates were observed when carbon > 0.18% and Σ_{RE} = 0.1%. Source: Ref 12

Table 3 Corrosion rates of iron, carbon steel, and alloy steels in dilute acids
Exposure for 3 h at 45 °C (115 °F)

Metal	1.7 N HF mm/yr	1.7 N HF in./yr	1.5 N H_2SO_4 mm/yr	1.5 N H_2SO_4 in./yr	1.4 N HCl mm/yr	1.4 N HCl in./yr
Armco iron	14	0.550	22	0.866	13	0.512
ASTM A106 carbon steel	26	1.02	20	0.787	23	0.906
ASTM A335						
Grade P11 (1.5Cr-0.25Mo)	290	11.4	41	1.6	4	0.157
Grade P22 (2Cr-1Mo)	640	25.2	330	13	10	0.394
Grade P5 (5Cr-0.5Mo)	>290	>11.4	94	3.7	3	0.118
Grade P9 (10Cr-1Mo)	330	13	83	3.27	6	0.236

Source: Ref 9

Table 4 Corrosion of various alloys in hydrofluoric acid immersion and vapor exposures
72 h tests

		24 (75) Immersed mm/yr	24 (75) Immersed mils/yr	24 (75) Vapor mm/yr	24 (75) Vapor mils/yr	50 (122) Immersed mm/yr	50 (122) Immersed mils/yr	50 (122) Vapor mm/yr	50 (122) Vapor mils/yr	76 (169) Immersed mm/yr	76 (169) Immersed mils/yr	76 (169) Vapor mm/yr	76 (169) Vapor mils/yr
Alloy	%HF												
316L	1.0	0.56	22	0.38	15	2.90	114	1.12	44	7.04	277	1.68	66
	10.0	8.46	333	6.25	246	...	...	...	...	...	...	...	...
20	1.0	0.04	1.6	0.17	6.7	0.24	9.4	1.04	41	0.30	12	0.50(a)	20(a)
	10.0	0.06	2.4	4.22	166	0.35	14	2.97(a)	117(a)	0.55	22	1.42(a)	56(a)
825	1.0	0.01	0.4	0.01	0.4	0.09	3.5	0.08	3.1	0.29	11	2.10	83
	10.0	0.07	2.8	2.02	80	0.45	18	4.22(a)	166(a)	0.88	35	6.12(a)	241(a)
Ni 200	1.0	0.06	2.4	1.11(b)	44(b)	0.20	7.9	1.85(a,b)	73(a,b)	0.18	7	3.53(a,b)	139(a,b)
	10.0	0.07	2.8	1.87(b)	74(b)	...	...	...	...	0.27	11	4.47(a,b)	176(a,b)
400	1.0	0.04	1.6	0.91	36	0.18	7.1	3.38	133	0.21	8.3	2.27(a)	89(a)
	10.0	0.05	2.0	2.24(b)	88(b)	0.15	6.0	5.87(b)	231(b)	0.30	12	4.39	173
600	1.0	0.05	2.0	1.06	42	0.33	13	3.36(a)	132(a)	0.22	8.7	2.59(a)	102(a)
	10.0	0.07	2.8	8.36(b)	329(b)	0.36	14	10.7(a)	421(a)	0.33	13	5.00(a)	197(a)
C22	1.0	0.01	0.4	0.005	0.2	0.05	2.0	0.04	1.6	0.19	7.5	0.59	23
	10.0	0.03	1.2	0.09	3.5	0.19	7.5	1.25	49	0.39	15	2.33	92
C276	1.0	0.02	0.8	0.04	1.6	0.12	4.7	0.32	13	0.21	8.3	1.25	49
	10.0	0.06	2.4	0.35	14	0.21	8.3	2.35	93	0.34	13	2.57(a)	101(a)

Note: Rates are calculated assuming uniform weight loss. (a) Adherent corrosion products on sample leading to underestimation of corrosion rate. (b) Indicates extreme attack at vapor-solution interface. Source: Ref 19

Table 5 Wrought alloys exposed to 5% hydrofluoric acid

96 h tests

Alloy	Corrosion rate at temperature, °C (°F)			
	52 (125)		79 (175)	
	mm/yr	mils/yr	mm/yr	mils/yr
316L	15.4	607	98.5	3877
904L	2.2	85	11.6	457
825	0.27	11	0.81	32
625	0.25	10	1.14	45
C276	0.24	10	0.40	16
C22	0.22	9	0.63	25
C2000	0.06	2	0.27	11

Source: Ref 20

Table 6 Corrosion of austenitic stainless steels and alloy 825 in hydrofluoric acid

Laboratory tests with no aeration or agitation (except boiling tests)

Concentration HF, %	Temperature		Test duration, days	Corrosion rate							
				Type 304		Type 316		Type 309 Cb		Incoloy 825	
	°C	°F		mm/yr	mils/yr	mm/yr	mils/yr	mm/yr	mils/yr	mm/yr	mils/yr
0.05	60	140	10	0.3	12	...	...	0.25	10	...	...
0.1	60	140	10	0.64	25	...	...	0.69	27	...	...
0.15	60	140	10	1.2	47	...	...	1.1	44	...	...
0.2	60	140	10	1.6	62	...	...	1.4	54	...	...
10	16	60	30	0.01	0.4	<0.002	<0.1	...	...	...	...
20	102	215	3	...	...	...	...	...	...	1.04	41
38	110	230	2	...	...	51	2000	...	...	...	...
38	Boiling		4	...	...	...	...	...	...	0.25	10
48	Boiling		4	...	...	...	...	...	...	0.23	9
50	60	140	35	...	...	...	...	...	...	0.05	2
65	60	140	35	...	...	...	...	...	...	0.13	5
70	60	140	35	...	...	...	...	...	...	0.13	5
70	21	70	42	...	...	1.24	49	...	...	0.35	14
90	4	40	0.2	0.9	35	...	...	...	...	...	...
90	21	70	1	0.76	30	...	...	...	...	...	...
90	21	70	1	0.28	11(a)	...	...	...	...	...	...
98	34–44	95–110	3.5	...	...	0.05	2	...	...	...	...

(a) Velocity: 0.14 to 0.43 m/s (0.4 to 1.4 ft/s). Source: Ref 8

and 7 to 11 permitting comparison among these alloys and welds over a range of conditions.

Concentration. A general trend is that increasing HF concentration increases corrosion of nickel alloys at least up to the azeotrope. In contrast, Table 2 data on nickel alloys 400 and C trend to lower rates as concentration increases from 50 to 65% HF. The trend to higher corrosion is noticeable in Tables 4, 6, 7, and 8 as concentration increases from 1 to 48% HF. Similarly, this trend is noticeable in Fig. 1 (among group A alloys), Fig. 4 (for 10 nickel alloys), and in Fig. 5 (for 5 Ni-Cr-Mo alloys in 1 to 40% HF). In Fig. 5, several alloys have acceptable corrosion resistance at 10 to 20% HF at 79 °C (175 °F); however, at 20% and higher concentrations, there is divergence in corrosion resistance. Corrosion data for wrought nickel alloys should be considered with the weld data in Tables 7 and 8.

Temperature. Increasing temperature generally increases corrosion of nickel alloys in HF liquid and vapor (Tables 4, 5, 7, and 9). Figure 4 indicates that in liquid HF, only alloy 400 (UNS N04400) has acceptable corrosion at higher temperatures, while other nickel alloys are limited to lower concentrations with increasing temperature. Figure 6 presents corrosion data on alloys 400 and C2000 (UNS N06200) in 20% HF liquid and vapor over a range of 38 to 93 °C (100 to 200 °F) (Ref 20). One of the conclusions from this study was that nickel alloys containing high chromium and exposed in HF vapors had decreased corrosion rates with time, but when immersed, the corrosion rates were not dependent on time. The corrosion rates for alloys 400 and C2000 increased with HF concentration and temperature. Table 9 presents data from this same study for alloys 400, C276 (UNS N10276), C2000, 242 (UNS N10242) and B3 (UNS N10675) in liquid 20% HF and vapor (Ref 27). At this concentration, liquid- and vapor-exposed alloys 400, C276, C2000, and 242 showed lower corrosion rates at 93 °C (200 °F) than at 66 °C (150 °F). This may indicate that during the 240 h test period, the corrosion product layer became more protective, or the oxygen in the system was reduced.

Alloys Containing Niobium. Nickel alloys containing niobium are prone to higher corrosion rates in HF. Table 5 includes corrosion data on wrought nickel alloys 625 (UNS N06625), C276, C22 (UNS N06022), and C2000 in 5% HF. Notably, from among these alloys, 625 had the highest corrosion rate at 79 °C (175 °F). This alloy has nominal 3.7% Nb. In Fig. 4, it is noticeable that among the Ni-Cr-Mo alloys, alloy 625 has the lower useful temperature (<40 °C, or 105 °F) in HF.

Weld Corrosion. Selective weld corrosion is common in HF. There are two primary causes for

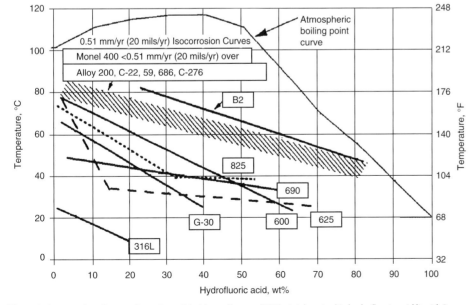

Fig. 4 Isocorrosion diagrams for various nickel-base alloys and 316L stainless steel in hydrofluoric acid liquid. Source: Ref 22

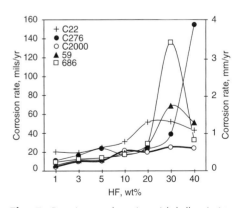

Fig. 5 Corrosion rates for various nickel alloys in 1 to 40% hydrofluoric acid (HF) at 79 °C (175 °F) for 24 h. Rates increase with HF concentration up to 20% HF, then corrosion behavior diverges. The x-axis is nonlinear. Source: Ref 20

weld corrosion. The first cause is segregation of certain elements in the cooling weld puddle. This is typical among nickel alloy welds and is notable by comparing the corrosion rates of welds to the base metal in Tables 7 and 8. Each weld has higher corrosion rates in comparison to the like base metal in corresponding test conditions. Alloy C22 (UNS N06022) weld samples had acceptable rates in both immersion and vapor exposure in 24 and 48% HF at 24 °C (75 °F). Alloy C22 welds had unacceptably high rates in liquid and vapors at both concentrations at 50 and 76 °C (122 and 169 °F). Although high, the alloy C22 weld samples had lower rates than alloy 400 weld samples in vapor exposure. In immersion conditions, alloy 400 weld corrosion rates generally were lower than rates of welded alloy C22 in similar conditions.

The second cause of selective weld corrosion is the alloying of niobium or other refractory metals in weld or base metals. Niobium affects the HF corrosion resistance of cast nickel-copper alloys M35-1 (N024135) and M30C (N024130) (Ref 28). Table 10 is HF corrosion data for these cast alloys and weld materials with various amounts of niobium. The cast alloy M35-1, with less than 0.5% Nb, had lower corrosion rates than M30C (1.3% Nb) in 49% HF. Weld deposits with 2.08% Nb had higher corrosion in aerated and nitrogen-purged 40% HF than those without niobium.

Hydrofluoric Acid with Oxygen: Liquid and Vapor Corrosion. Alloy 400 is used extensively in HF alkylation units and in the handling of HF. It has excellent resistance to liquid HF over the entire concentration range in the absence of oxygen to at least 150 °C (300 °F) (Ref 29). Alloy 400 has high corrosion rates in aerated 30 to 75% HF and even higher rates in HF vapor with air. Furthermore, denickelification of alloy 400 has been reported in aerated HF (Ref 5). Figure 6 shows that corrosion rates for alloy 400 in HF vapors can be higher than in liquid, and in these conditions, Ni-Cr-Mo alloys such as C2000 have better resistance. Oxygen increases the corrosiveness of HF for most nickel alloys but particularly for nickel-copper and nickel-molybdenum alloys. Data on a number of nickel alloys in HF liquid and vapor are presented in Tables 4 and 7. In Table 4, each wrought nickel alloy has acceptable corrosion rates when immersed in 1 and 10% HF during 72 h tests. However, alloys 200, 400, and 600 had higher corrosion rates in HF vapor (in air) than immersed samples and higher than (Ni-Cr-Mo) alloys C22 and C276. Table 7 presents data on wrought alloys C22 and 400 exposed to 24 and 48% HF in 72 h tests. Immersion samples had lower corrosion rates than respective vapor

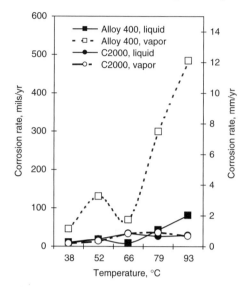

Fig. 6 Corrosion rates for alloys 400 and C2000 in 20% hydrofluoric acid liquid and vapors, 38 to 93 °C (100 to 200 °F), 240 h, showing comparable liquid rates but diverging vapor rates. Source: Ref 20

Table 7 Corrosion rates of alloy C22 and 400 in 24 and 48% hydrofluoric acid
72 h immersion tests

			Corrosion rate										
			Test temperature, °C (°F)										
			24 (75)				50 (122)				76 (169)			
			Immersed		Vapor		Immersed		Vapor		Immersed		Vapor	
Alloy	Condition	%HF	mm/yr	mils/yr	mm/yr	mils/yr	mm/yr	mils/yr	mm/yr	mils/yr	mm/yr	mils/yr	mm/yr	mils/yr
C22	Unwelded	24	0.06	2	0.48	19	0.29	11	1.51	59	1.42	56	3.81	150
		48	0.06	2	0.50	20	0.19	7	0.95	37	0.43	17	3.81	150
	Welded	24	0.38	15	0.43	17	0.68	27	1.90	75	1.63	64	3.30	130
		48	0.34	13	0.60	24	1.65	65	1.14	45	3.00	118	2.34	92
400	Unwelded	24	0.08	3	1.67	66	0.28	11	5.26	207	0.36	14	5.21	205
		48	0.06	2	1.94	76	0.34	13	6.17	243	0.44	17	5.61	220
	Welded	24	0.08	3	1.55	61	0.34	13	5.72	225	0.36	14	5.21	205
		48	0.44	17	2.02	80	0.33	13	5.41	213	0.69	27	5.28	208

Source: Ref 19

Table 8 Corrosion of wrought nickel alloys and welds in 20 and 48% hydrofluoric acid
96 h exposure (solution changed at 48 h) at 79 °C (175 °F)

		Concentration HF, %			
		20		48	
Alloy	Site	mm/yr	mils/yr	mm/yr	mils/yr
C276	Base metal	0.59	23	0.98	38
	Weld	1.17	46	1.38	55
C22	Base metal	1.31	52	0.69	27
	Weld	3.85	152	16.2	637
C2000	Base metal	0.51	20	0.49	19
	Weld	0.58	23	0.67	26
59	Base metal	0.70	28	...	...
	Weld	3.35	132	8.28	326
686	Base metal	0.73	29	...	...
	Weld	1.06	42	6.08	240

Source: Ref 20

Table 9 Corrosion rates on nickel alloys (U-bends)
Exposed to 20% hydrofluoric acid, 240 h

	Corrosion rate at temperature, °C (°F)											
	66 (150)				79 (175)				93 (200)			
	Immersed		Vapor		Immersed		Vapor		Immersed		Vapor	
Alloy	mm/yr	mils/yr	mm/yr	mils/yr	mm/yr	mils/yr	mm/yr	mils/yr	mm/yr	mils/yr	mm/yr	mils/yr
400	0.165	6.5	6.48	255	0.003	<0.1	6.78	267	0.003	<0.1	3.81	150
C276	2.95	116	1.02	43	6.07	239	2.39	94	0.732	29	0.798	31
C2000	0.856	34	0.864	34	0.411	16	0.363	14	0.353	14	0.376	15
242	1.82	72	0.683	27	1.66	65.4	0.503	20	0.434	17	0.615	24
B3	2.84	112	1.87	74	9.98	393	4.55	179	2.62	103	2.39	94

Source: Ref 27

samples. Alloy 400 (wrought) immersed samples had acceptable rates in all conditions but unacceptable rates in all vapor-exposed conditions. Alloy C22 (wrought) immersed samples had acceptable rates up to 50 °C (122 °F). Alloy C22 (wrought) vapor samples had unacceptable rates at higher temperatures, but each vapor sample had lower corrosion rates than alloy 400 in comparable exposures. Table 9 presents corrosion data on nickel alloys in 240 h exposures to 20% HF liquid and vapor. High corrosion rates are noticeable for vapor-exposed alloys 400 and B3 versus liquid exposures; however, (Ni-Cr-Mo) alloys C276, C2000, and 242 had equal or lower corrosion rates in vapor versus those in liquid HF.

Corrosion test data from condensing conditions at 50 and 70 °C (122 and 158 °F) with 5% HF, 15% H_2O, and 8% O_2 are presented in Table 11. Corrosion rates at 50 °C (122 °F) on nickel-copper and iron-base alloys 400 and 25-6Mo (UNS N08926) were unacceptably high, but rates for Ni-Cr and Ni-Cr-Mo alloys 600, 622 (UNS N06022), 686 (UNS N06686), 825, and alloy C276 were low to moderate. The chromium-bearing nickel alloys were more resistant in the oxygen-rich HF. Corrosion rates at 70 °C (158 °F) were negligible due to little condensation.

Cracking. Alloy 400 samples, after exposure to vapors over 5 and 48% HF, had SCC in the base and weld metal of welded samples (Ref 30). Welded and stress-relieved samples made by the shielded metal arc weld process had SCC. Oxygen and stress level were factors affecting the severity of SCC. No SCC was found on submerged samples; wrought, solution-annealed vapor samples; or welded and stress-relieved samples made by the gas tungsten arc weld process. Data from another investigation of nickel alloy SCC in HF are presented in Table 12. Alloys 400 and 500 (UNS N05500) in vapors of 48% HF at 60 °C (140 °F) had intergranular and transgranular SCC. Alloy 500, in the hot rolled or aged condition, was susceptible to SCC. Annealed or cold rolled alloy 400 U-bends cracked. Welded alloy 400 cracked in the base and weld metal. There was no SCC in air-free 48% HF after 30 days. However, alloy 500 cracked after 15 days in aerated immersion tests in 48% HF.

Alloy 400 cracking mechanism was unknown until it was found that aqueous HF solutions with cupric fluoride (CuF_2) would cause rapid SCC (Ref 32). Investigations on nickel-copper and copper-nickel alloys with HF and CuF_2 at 23 °C (73 °F) led to Fig. 7 and 8. Figure 7 indicates that when stressed to 80% of the ultimate tensile strength, alloys with 65 to 80% Ni required the least time to crack. The most SCC-susceptible composition was an alloy with approximately 69% Ni. Alloy 400 contains 63 to 70% Ni. Alloys with 70% Cu were not susceptible to SCC. Both increasing CuF_2 concentration in HF and tensile stresses reduced time for SCC. Figure 8 indicates that the time to crack alloy 400 in HF with CuF_2 is reduced by increasing stress and temperature.

Usually, SCC is found in the vapor phase due to condensation of HF vapor. A thin liquid film forms that is gradually enriched with CuF (cuprous fluoride) corrosion product as well as oxygen from the atmosphere, and CuF is oxidized to CuF_2. The bulk liquid phase provides much greater dilution of corrosion products, and oxygen absorption is confined to the liquid-vapor interface, thus limiting CuF_2 in the bulk solution. Preventing SCC requires total immersion and purging vapor spaces with nitrogen. Thermal stress relief is beneficial but has not always been successful in preventing SCC. As shown in Fig. 8, with time, SCC can occur at relatively low stress levels. Oxygen may enter when systems are opened for maintenance, in feed materials, or formed as a reaction product. General practice in HF alkylation service is to stress relieve cold-worked or welded alloy 400 at 621 °C (1150 °F).

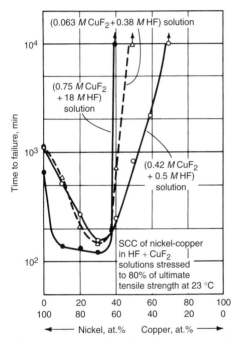

Fig. 7 Relationship of time for stress-corrosion cracking (SCC) failure for various nickel-copper alloys exposed to hydrofluoric acid (HF) and CuF_2 solutions while stressed at 80% ultimate tensile strength. Source: Ref 32

Table 10 Effect of niobium (Nb) in nickel-copper exposed to hydrofluoric acid

	Corrosion rate at given exposure								
	49% HF, 49 °C (120 °F), 120 h		40% HF, 70 °C (158 °F), 16.5 days, N_2 purge		40% HF, 70 °C (158 °F), 16.5 days, aerated		40% HF, 70 °C (158 °F), 12 days, N_2 purge		40% HF, 70 °C (158 °F), 12 days, aerated	
Metal (%Nb)	mm/yr	mils/yr	mm/yr	mils/yr	mm/yr	mils/yr	mm/yr	mils/yr	mm/yr	mils/yr
M35-1 (0.5)	0.35	14	...	...	...	...	...	...	...	...
M30C (1.3)	0.62(a)	24.5(a)	...	...	...	...	...	...	...	...
Weld filler (0%)	...	...	0.09	3.6	1.7	67	0.05	2	1.07	42
Weld filler (2.08%)	...	...	0.71	28	2.64	104	0.05	2	1.32(a)	52(a)

(a) Localized attack. Source: Ref 28

Table 11 Corrosion data evidence of stress-corrosion cracking (SCC) for various alloy coupons

Immersed in condensing 5% HF, 15% H_2O, and 8% O_2 at 50 and 70 °C (122 and 158 °F). SCC determined with U-bends

	Corrosion rate at conditions				SCC	
	50 °C (122 °F), 55 h		70 °C (158 °F), 100 h			
Alloy	mm/yr	mils/yr	mm/yr	mils/yr	50 °C (122 °F), 55 h	70 °C (158 °F), 100 h
600	0.23	9	0.02	1	No	No
400	1.9	75	0.15	6	Yes	No
825	0.62	25	0.04	2	No	No
C276	0.39	15	0.03	1	No	No
686	0.15	6	0	0	No	No
622	0.66	26	0.02	1	No	No
25-6 Mo	0.94	37	0.09	4	No	No

Source: Ref 22

Table 12 Results from nickel alloy U-bends exposed in 48% hydrofluoric acid vapors at 60 °C (140 °F)

Alloy	Condition	Hardness(a), HV	First observed cracking, days
500	Hot rolled	148–210	2–6
	Aged	32(b)	5
400	Annealed	109–136	4–15
	Cold rolled	250	6–15
	Welded	...	4–14
400 + 1% Si	Cold rolled	293	15
85 Ni, 15 Cu	Annealed	147	5
Ni 200	Cold rolled	232	14
600	Cold rolled	295	14–15

(a) Vickers hardness, test force 30 kgf. (b) HRC value approximately equal to 309 HV, 150 kgf test force. Source: Ref 31

However, nickel alloy and HF manufacturers recommend stress relief at 705 °C (1300 °F) for 2 h to avoid SCC of alloy 400. This is a prudent practice in systems where HF-exposed metal is rarely exposed to oxygen. However, stress-relieved alloy 400 exposed to HF, SO_2, and air has failed by intergranular corrosion after several years of service.

Cold rolled alloys 200 and 600 experienced SCC when exposed in vapors of 48% HF at 60 °C (140 °F), as shown in Table 12. Alloy 200 had intergranular cracking, while alloy 600 had transgranular cracking on the tension side. There are reports of SCC of alloys 200, C276, and B2 in HF with CuF_2 (Ref 32) and when located downstream from copper-bearing alloys.

Copper Alloys

Corrosion data on wrought and welded 70-30 copper-nickel (UNS C71500) in 5 and 48% HF are presented in Table 13. Higher corrosion rates occurred for vapor-exposed samples. Heat treatment had only a slight effect on corrosion rates. Although intergranular corrosion was noted, there was no SCC of 70-30 copper-nickel. The liquid-vapor interface and vapor-exposed samples had heavy corrosion. Copper alloys resist HF at very low velocities in the absence of air, oxygen, and other oxidizing contaminants. Copper tubing has been used for short-term services at ambient temperature in 70% HF free of oxygen. Copper sheets have been applied inside distillation columns in areas that are subject to HF accumulation.

Denickelification of 90-10 and 70-30 copper-nickel wrought and weld metals in a $HF-H_2SO_4$ mixture of 5 to 15% acid in water at 150 to 170 °C (300 to 340 °F) is reported (Ref 33). Severe denickelification was observed on 70-30 copper-nickel equipment, while 90-10 copper-nickel had only minor effects. Crevice areas had notable denickelification for both alloys. Welded 70-30 copper-nickel base metals had uniform corrosion, but welds had denickelification; corrosion data are presented in Table 14. The weld metal denickelification was two to eight times higher than the base metal. Denickelification was also reported on copper-nickel alloys in HF pressurized with air (Ref 5); however, denickelification was severe for alloy 400. Less nickel in the nickel-copper/copper-nickel alloy system reduces the susceptibility for denickelification in HF service.

Aluminum bronze and aluminum-silicon bronze dealloy and crack in HF exposure. Aluminum bronze (Ref 34) and aluminum-silicon bronze (Ref 35) are prone to dealuminification and intergranular SCC in HF. Copper-zinc alloys have not been used for HF services. Copper-zinc and some other copper alloys are subject to embrittlement or are sensitive to velocity.

Precious and Other Metals

Silver (UNS P07020), gold (UNS P00020), platinum (UNS P04995), tungsten (UNS R07005), and molybdenum (UNS R03600) are resistant to corrosion in HF of any concentration at ambient temperatures and, in most cases, to the boiling point or higher (Ref 36, 37). Corrosion data for platinum and silver are presented in Tables 1 and 2. Also, iridium (UNS P06100), osmium, ruthenium, rhodium (UNS P05981), and palladium (UNS P03995) resist 40% HF at room temperature (Ref 38). Palladium may have higher corrosion rates in aerated HF or in HF with concentrated sulfuric acid.

Silver has low corrosion rates up to and above the HF boiling point and has been used in HF services such as stills, heating coils, condensers, rupture disks, and pressure diaphragms. However, silver in HF has high corrosion rates when also exposed to oxygen, hydrogen sulfide, or sulfur contamination (Ref 37). Sulfur compounds may be present in commercial HF.

Molybdenum has good resistance to HF. The corrosion rates of molybdenum in 25 to 50% HF at 100 °C (210 °F) are 0.4 to 0.5 mm/yr (16 to 20 mils/yr) in aerated acid. Rates in the absence of air are negligible (Ref 39). Molybdenum is used only for small parts that do not require welding.

Magnesium has good resistance up to 40% HF at 25 °C (77 °F), because the protective film formed is insoluble in aqueous acid. Corrosion data for magnesium are presented in Table 2. However, magnesium is infrequently used in HF service, because it has unacceptable corrosion in HF at higher concentrations and temperatures.

Reactive Metals. Hydrofluoric acid rapidly attacks reactive metals, niobium, tantalum, zirconium, and titanium, either as pure or alloyed metal. Even a percent or two of reactive metals in stainless or nickel alloys reduces their corrosion resistance to HF.

Nonmetals

Thermoplastics. Several thermoplastics are resistant to HF. Most have service limitations, but their lower costs and minimal contamination of HF make them candidates for consideration. Some thermoplastics change physical properties with exposure to HF. Also, when HF concentration is higher than the azeotrope, there will be AHF permeation through the plastic. Contamination, temperature conditions, mechanical damage, and ultraviolet light have caused thermoplastics to fail in service. Thermoplastics used in hazardous services such as HF are mainly applied as liners inside a metal or reinforced

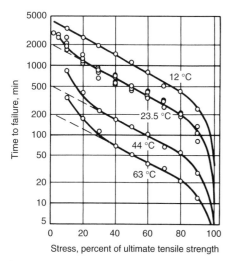

Fig. 8 Relationship of time for stress-corrosion cracking failure in a hydrofluoric acid and CuF_2 solution versus stress at various temperatures on alloy 400. Source: Ref 32

Table 13 Corrosion rates of UNS alloy C71500 from exposure to 5 and 48% HF
Tested at 66 °C (150 °F)

	Corrosion rate at concentration							
	5% HF				48% HF			
	Partially immersed		Immersed		Partially immersed		Immersed	
Condition	mm/yr	mils/yr	mm/yr	mils/yr	mm/yr	mils/yr	mm/yr	mils/yr
As received	0.74	29	0.08	3	1.93	76	0.13	5
As welded	0.86	34	0.10	4	2.16	85	0.18	7
Weld, stress relieved (482 °C, or 900 °F)	0.84	33	0.15	6	2.31	91	0.18	7
Weld, annealed (816 °C, or 1500 °F)	0.69	27	0.10	4	2.03	80	0.15	6
Weld, annealed (927 °C, or 1700 °F)	0.81	32	0.10	4	2.74	108	0.18	7
Base metal, annealed (816 °C, or 1500 °F)	0.51	20	0.05	2	1.68	66	0.10	4

Source: Ref 30

Table 14 General and dealloying corrosion of welded 70-30 copper-nickel
Samples exposed in process unit to 5 to 15% hydrofluoric acid with H_2SO_4 in water at 150 to 170 °C (300 to 340 °F)

	General corrosion rate		Base metal denickelification	
Weld process, filler	mm/yr	mils/yr	mm/yr	mils/yr
GTAW, ERCuNi	1.07	42	2.87	113
SMAW, ECuNi	1.30	51	9.53	375
GTAW, ERNiCu-7	2.84	112	5.23	206

Note: GTAW, gas tungsten arc welding; SMAW, shielded metal arc welding. Source: Ref 33

thermosetting plastic vessel or pipe. Figure 9 presents maximum HF concentration and temperature for continuous use of plastics and elastomers in HF service (Ref 3), with the understanding that maintenance or replacement will be required after some time. Fully fluorinated thermoplastics are resistant to all HF concentrations up to and above the boiling point. Fully fluorinated plastics, such as polytetrafluoroethylene (PTFE), perfluoroalkoxy (PFA), and fluorinated ethylene propylene (FEP) do not lose strength, swell, crack, embrittle, and are not chemically attacked by HF. A chemically modified PTFE fluoropolymer, MFA (perfluoromethyl vinyl ether and tetrafluoroethylene copolymer), is weldable (Ref 40) and has increased permeation resistance (Ref 41) and improved deformation resistance under load and creep conditions. It is good practice to take care when installing fluoroplastic linings to avoid excess stress that can lead to internal liner damage and promote HF permeation. The industry trend is toward fluoropolymer-lined equipment to minimize metal ion contamination of HF.

Other thermoplastic materials, such as polyethylene, polypropylene, chlorinated polyvinyl chloride (CPVC), and polyvinyl chloride (PVC), are less costly but are limited to lower temperatures and concentrations below 70% HF, as seen in Fig. 9. Tanks fabricated from high-density polyethylene (HDPE) ribbon as helical weld seams have provided over 15 years of satisfactory service in concentrations of 5 to 70% HF at ambient temperatures and atmospheric pressure, with no measurable deterioration of properties (Ref 42). Cross-linked polyethylene tanks have provided over 15 years of service in HF service at ambient temperatures. Small midwall blisters, containing HF, have formed in low-density polyethylene and HDPE after a year exposure to concentrations over 60% HF at 30 °C (85 °F) but with no deterioration of mechanical properties. However, a rotational-molded HDPE tank cracked through the wall at the bottom knuckle by environmental stress cracking after two years of service (Ref 43). Unplasticized PVC and CPVC have resistance to HF solutions and AHF vapors; however, PVC and CPVC plasticized with vinyl acetate or vinyl chloride is not resistant.

Reinforced Thermosetting Plastics. Reinforced thermosetting plastic resins commonly recommended for HF services are epoxy vinyl esters (Ref 44). However, these have been limited to 10% HF concentrations below 65 °C (150 °F). The corrosion-resistant barrier typically has at least a double layer of apertured synthetic polyester, apertured ethylene chlorotrifluorethylene, or carbon-reinforcing veil materials. Additional information may be obtained from manufacturers and fabricators.

Elastomers. Carbon-filled and peroxide-cured fluoroelastomers and perfluoroelastomers have been used as O-rings in all HF concentrations at ambient temperatures and above. Generally, elastomers with higher fluorine content have better HF resistance (Ref 45). Consult manufacturers for guidance on the best compounds, fillers, and cures for a specific HF service.

As seen in Fig. 9, several rubber types are useful in HF concentrations below 60% at temperatures below 70 °C (160 °F). Consultation with the rubber supplier is essential to assure that available rubbers will have low silica or magnesia compounding (below 0.3 wt%) for HF service. Some rubber suppliers may be unaware that clays they use as additives contain silica and magnesia compounds. Specially formulated chlorobutyl rubber is used for lining containers for 70% HF. In 70% HF, the useful life of chlorobutyl in warm climates is approximately three years, while in cool climates it is four to six years. Longer chlorobutyl service life is experienced with lower HF concentrations. Bromobutyl rubber is used in HF service in Europe. Table 15 presents 50% HF elastomeric exposure data at 22 °C (72 °F) (Ref 46).

Carbon and graphite are resistant to HF. Impervious graphite impregnated with PTFE fluoropolymer has provided satisfactory service to at least 80% HF at 150 °C (300 °F) but may be prone to AHF permeation. Impervious graphite impregnated with phenolics has provided satisfactory service to 110 °C (230 °F) for 48% HF and to 85 °C (185 °F) for 60% HF. Impervious graphite impregnated with furan has cracked in

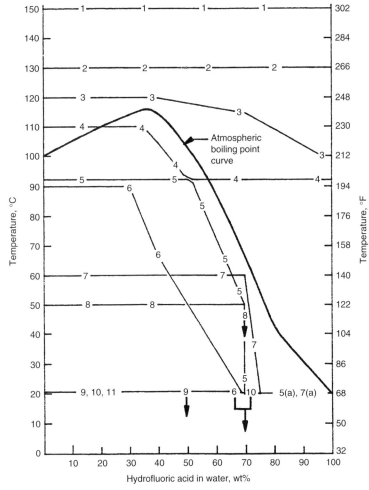

Fig. 9 Guidance for use of plastics and elastomers in hydrofluoric acid (HF). Regions below and left of lines are suitable in uncontaminated HF. The down arrows (↓) are the upper limit of acid concentration for the material. (a) Vapor only. Materials: 1, PTFE, perfluoroalkoxy (PFA); 2, fluorinated ethylene propylene (FEP); 3, ethylene tetrafluoroethylene (ETFE); 4, polyvinylidene fluoride (PVDF); 5, polypropylene; 6, chlorobutyl, bromobutyl, ethylene chlorotrifluoroethylene (ECTFE); 7, cross-linked, high-density, low-density, or copolymer polyethylene; 8, chlorosulfonated polyethylene; 9, soft natural rubber; 10, polychloroprene, butyl rubber, ethylene propylene diene monomer; 11, polyvinyl chloride (PVC), unplasticized. Source: Ref 3

Table 15 Polymer exposures in 50% hydrofluoric acid at 22 °C (72 °F) for 30 days

Polymer	Weight change, %	Volume change, %
Chlorobutyl	5.24	6.02
Chlorobutyl/natural blend	2.57	2.57
Bromobutyl	2.03	2.02
Chlorinated polyethylene	40.86	38.19

Source: Ref 46

HF services (Ref 47). Carbon brick linings (ash content <1%) with phenolic mortar and fluorinated plastic membranes have been applied in carbon steel vessels for high-temperature HF-water scrubbers. Flexible graphite has been used for gaskets and packing materials in HF services. Carbon-carbon composites have excellent resistance to HF (Ref 48); however, their use is limited to internal components, because they are permeable.

Alumina, Sapphire, and Carbides. Alumina (>99.7% purity) and sapphire have resistance to HF. Sintered silicon carbides (100% alpha grade) have been used for pump bearings in HF service. Sintered silicon carbide fibers had essentially no mass loss in 53% HF at 25 °C (77 °F) in 125 h (Ref 49).

Commercially available polycrystalline alumina and sapphire (single-crystal alumina) were tested in 38% (20 M) HF at 90 °C (195 °F) for two weeks (Ref 50). All of the polycrystalline alumina had high corrosion rates, and most samples were consumed during the exposure. The silicate-based, glassy grain-boundary films were attacked. Sapphire had a 7 μm/yr (0.27 mil/yr) corrosion rate. High-purity alumina, doped with 500 ppm MgO and exposed in the same environment, had corrosion rates of less than 25 μm/yr (1 mil/yr) compared to samples without MgO that had 0.25 to 7.6 mm/yr (10 to 300 mils/yr) corrosion rates. The MgO acted to disperse silica and silicate into alumina grains so that HF could not attack them.

Various commercially available ceramic oxides, carbides, nitrides, and borides were exposed to 38% (20 M) HF at 90 °C (195 °F) for two weeks (Ref 51). The polycrystalline oxide ceramics, such as Al_2O_3, TiO_2, and ZrO_2, were severely attacked, but polycrystalline MgO corroded at only 10 μm/yr (0.4 mil/yr). The ceramics with the best resistance were sapphire and pure polycrystalline carbides such as SiC, B_4C, WC, and TiC. Notably, the pure SiC and alpha silicon carbides with no free silica or silicon ranged from no attack up to 5 μm/yr (0.2 mil/yr). However, commercial reaction-bonded and sintered SiC had corrosion rates of 0.18 to 3.8 mm/yr (7 to 150 mils/yr). Sapphire and pure polycrystalline carbides had the lowest corrosion rates.

Anhydrous Hydrogen Fluoride

Many metals and alloys have resistance to AHF, and most have high initial corrosion rates that diminish with the development of a surface film or scale of corrosion product. Therefore, the validity of short-term rates should be confirmed by longer tests. Table 16 lists data on the corrosion resistance of various metals in liquid AHF and AHF-hydrocarbon mixtures at temperatures of 15 to 90 °C (60 to 190 °F). Alloys have better resistance to AHF than HF, because AHF is less corrosive.

Carbon and Low-Alloy Steels

Liquid AHF is commonly handled in carbon steel railcars, storage tanks, and piping at temperatures up to 65 °C (150 °F) with velocity limited to 1.8 m/s (6 ft/s). At lower temperatures, steels can withstand higher liquid velocities. Gaseous AHF is less corrosive and is handled successfully at temperatures up to 200 °C (390 °F) with velocities up to 10 m/s (33 ft/s) and up to 300 °C (570 °F) in low-flow conditions. Figure 10 presents corrosion rates versus temperature for carbon steels exposed to AHF liquid and vapor. These data were generated in stagnant 100 h (average) tests. Steel vessels and piping can be passivated by exposure to AHF in stagnant vapor or liquid for approximately 24 h prior to velocity exposure. Carbon steels used for AHF storage for 40 years, at ambient conditions of 0 to 40 °C (30 to 105 °F), have lost less than 0.013 mm/yr (0.5 mil/yr). Even though metal loss is very low, there are two other problems caused by the reaction of AHF with iron. First, the corrosion product is approximately two times more voluminous than the corroded iron. The iron fluoride scale is hard, abrasive, and wedges between mating surfaces of pipe, flanges, and valve parts, which may lead to pipe joint leaks and inoperable valves. The fluoride scale also may flake off of surfaces and be carried into valves and pumps, causing wear. Second is that atomic hydrogen is generated and could lead to hydrogen damage of the carbon steel.

Residual Elements in Carbon Steels. Alloy steels often have higher corrosion than carbon steels in HF and AHF (Ref 18). Steels with high residual elements can have high corrosion rates in AHF, because the corrosion product layer is loose, porous, and nonprotective. Steels containing a sum of copper, nickel, and chromium, $\Sigma_{RE} > 0.2\%$, sometimes have increasingly high corrosion rates with increasing temperature. A fluorochemical plant had corrosion of approximately 25.4 mm/yr (1000 mils/yr) on pipe, while carbon steel fittings on either side had only 0.89 to 1.27 mm/yr (35 to 50 mils/yr). The conditions were approximately 10% AHF and 90% organic at 80 °C (175 °F). Turbulent areas had higher corrosion loss. The pipe was ASTM A333 grade 9 with 0.81% Cu and 1.86% Ni. The fittings had no copper, nickel, or chromium. Table 17 data are from in-plant and lab corrosion studies in AHF with steels of varied compositions. This study indicates that steels with $\Sigma_{RE} > 0.2\%$ are susceptible to higher corrosion, and that rates increase with temperature and amount of AHF in the process.

Blistering and Cracking of Steels. Atomic hydrogen is formed when HF corrodes steel. Atomic hydrogen may either combine to form molecular hydrogen or be absorbed into the steel. Sulfides and arsenic in HF inhibit the formation of molecular hydrogen, promoting the entry of atomic hydrogen into the steel, where it may recombine at inclusion sites or laminations to form blisters or may migrate to dislocations in hardened steel to cause hydrogen embrittlement (Ref 8). Hydrogen may accumulate in closed vessels and cause unexpected leaks, ruptures, or reactions due to the increase in pressure (Ref 54). Hydrogen permeating through steel has accumulated in crevices created by support saddles, external attachments, slip-on flanges, reinforcing pads, and legs, causing bulges and cracking. Therefore, designs and procedures should consider ways of safely dealing with hydrogen. Carbon steel AHF vessels have experienced hydrogen blistering (Ref 55), hydrogen-induced cracking (HIC) (Ref 56), and hydrogen embrittlement. Hydrogen stress cracking has also occurred at hardened welds.

Steel Blistering. Hydrogen blistering is associated with oxide/sulfide inclusions that can mostly be prevented by specifying HIC-resistant steels. These are produced with low sulfur, calcium treatment, and vacuum degassing.

Table 16 Corrosion of metals and alloys in anhydrous hydrogen fluoride
Exposure for 6 to 40 days

Metal	Corrosion rate at temperature, °C (°F)							
	15–25 (60–80)		24–40 (80–100)		55 (130)		80–90 (180–190)	
	mm/yr	mils/yr	mm/yr	mils/yr	mm/yr	mils/yr	mm/yr	mils/yr
Carbon steel	0.07	3	0.16	6	0.35	14	2.3	89
Low-alloy steel	0.15	6	0.14	6	...	...	2	78
Austenitic stainless steel	0.16	6	0.12	5	...	...	0.06	2
Monel 400	0.08	3	0.02	1	...	...	...	...
Copper	0.33	13	...	...	...	...	...	...
Nickel 200	0.06	2	...	...	...	...	...	...
70-30 copper-nickel	...	...	0.05	2	0.008	0.3	0.25	10
80-20 copper-nickel	...	...	0.13	5	...	...	...	...
Red brass	0.76	30	0.4	16	...	...	1.3	50
Admiralty brass	0.25	10	0.33	13	0.01	0.4	0.5	20
Aluminum-bronze	0.37	14	...	...	...	...	...	...
Phosphorus-bronze	0.5	20	0.48	19	...	...	1.5	60
Inconel 600	...	...	0.07	3	...	...	...	...
Duriron	...	...	1.1	45	...	...	...	...
Aluminum	0.52	20	...	...	...	...	24.8	976
Magnesium	0.13	5	0.43	17	nil		nil	

Source: Ref 52

Fine-grain practice (killed) steels are often preferred for lower-temperature service because of their improved notch toughness, particularly when normalized or quenched and tempered. Blistering occurs more readily in fine-grain practice steels, because the added aluminum forms clusters of Al_2O_3 particles that segregate in the center of the plate and form sites for hydrogen accumulation. Type II manganese sulfide inclusions, which predominate in killed steels, flatten out and also serve as sites for hydrogen accumulation. Steels for blistering environments should be hot rolled or annealed rather than cold rolled (Ref 57). Steels produced with only low sulfur are more resistant but not completely immune to hydrogen blistering. Both oxygen and sulfur are determinants for the distribution of oxide and sulfide inclusions in steels (Ref 58). NACE standard TM0284 (Ref 59) defines a test to determine steel resistance to hydrogen blistering and step-wise cracking. The HIC-resistant steels have good results in this test. NACE report 5A171 (Ref 3) has a discussion of the specifications and testing that may be applied to these types of steels.

Steel Cracking. Another effect of atomic hydrogen is cracking of carbon and alloy steels with hardness over HRC 22, or approximate tensile strength of 793 MPa (115 ksi) or higher. Susceptibility to cracking increases with residual and applied stresses. In AHF, cold-worked carbon steels are susceptible to cracking as well as stress-oriented corrosion (Ref 53). The American Petroleum Institute's RP 751 (Ref 60) discusses recommendations for carbon equivalency and postweld heat treatment (PWHT) for steel less than or equal to 38 mm (1.5 in.) thick. A more detailed discussion on steps to minimize carbon steel susceptibility to cracking can be found in NACE standard RP0472 (Ref 61). As pointed out in the standard, stress-relief heat treatment, although desirable, will not necessarily reduce hardness, particularly that of high-manganese welds, and does not avoid the necessity for testing. Thick steel plates may cool rapidly when welded and may harden welds or HAZs. Higher-strength steels tend to have a harder weld HAZ that is typically where cracking is found after exposure to HF and AHF. Proper weld techniques, such as preheat, uphill welding of vertical seams, use of temper beads, and PWHT, are useful to reduce hardness of weld HAZ.

Austenitic Stainless Steels

Austenitic stainless steels have good resistance to liquid and vapor AHF up to 100 °C (210 °F). In pure liquid AHF at ambient temperatures, austenitic stainless steel types 304, 304L, 316, 316L, CF8M, and CF3M have been used for pumps, valves, piping, storage tanks, and transportation equipment. However, both type 304 stainless steel and carbon steel have been known to fail by direct impingement of AHF steams. Corrosion rates for type 304 stainless steel and carbon steel exposed in AHF vapor are shown in Fig. 11. Type 304 stainless steel corrosion rates increase with temperature above 100 °C (210 °F), and the corrosion rates are more erratic above 100 °C (210 °F) in tests and service.

Cracking. A cold-worked type 303 stainless steel fastener failed rapidly in a HF plant. Cold-worked type 304 stainless steel fasteners (ASTM A193, grade B8, class 2) had few failures. The failed fasteners were magnetic. Alpha-martensite is formed when alloys such as types 301, 303, and 304 stainless steels are cold worked (Ref 8), thus increasing their magnetic permeability. Increasing α-martensite content increases susceptibility to hydrogen embrittlement (Ref 62) and corrosion in nonoxidizing acids (Ref 63). Magnetic permeability may be tested by Severin gage (12.5 maximum) to sort those that are magnetic. Some users specify type 316 stainless steel that is metallurgically stable and does not develop α-martensite.

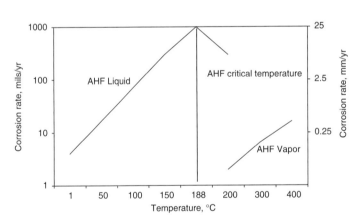

Fig. 10 Corrosion rates of carbon steels in static anhydrous hydrogen fluoride (AHF) liquid and vapor. These data emphasize that liquid AHF is more corrosive than vapor, and that carbon steels have acceptable corrosion in static AHF at 300 °C (570 °F). The critical point occurs at 188 °C (370 °F) and 64.88 bars (941 psia). Source: Ref 53

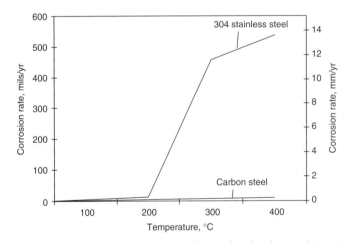

Fig. 11 Corrosion rates for type 304 stainless steel and carbon steel in static anhydrous hydrogen fluoride vapor. Type 304 corrosion rates are erratic above 100 °C (210 °F).

Table 17 Carbon and alloy steels in anhydrous hydrogen fluoride (AHF) liquid

	Composition, %					Corrosion rate at condition(b)							
						A		B		C		D	
ASTM material	C	Cu	Ni	Cr	Σ_{RE}(a)	mm/yr	mils/yr	mm/yr	mils/yr	mm/yr	mils/yr	mm/yr	mils/yr
A516-55	0.14	0.03	0.02	0.02	0.07	1.19	47.0	0.55	21.5	0.12	4.7	0.84	33.0
A516-55	0.15	0.09	0.07	0.06	0.22	1.88	74.0	...	...	...	...	0.89	35.0
A106-B	0.17	0.28	0.11	0.10	0.49	2.79	110.0	0.77	30.4	0.11	4.4	1.22	48.0
A588-A	0.13	0.35	0.61	0.20	1.16	...	...	>0.94(c)	>37.0(c)	0.14	5.4	1.63	64.0
A203-E	0.15	...	3.47	...	3.47	4.09	161.0	>1.14(c)	>45.0(c)	0.17	6.6	1.40	55.0
A710-A	0.05	1.05	0.77	0.68	2.50	5.54	218.0	>1.27(c)	>50.0(c)	0.19	7.3	1.02	40.0

(a) Σ_{RE}, sum of residual elements. (b) A, 14 day lab test, AHF, 90 °C (195 °F); B, 1.3 yr in-plant, 100% AHF, 80 °C (175 °F); C, 1.3 yr in-plant, 100% AHF, 45 °C (115 °F); D, 6 mo in-plant, 10% AHF + 1.4% Cl_2 organic, 90 °C (195 °F). (c) Estimated, coupon missing. Source: Ref 53

Cold-worked type 316 stainless steel tubing cracked in AHF at 40 °C (105 °F) (Ref 53). These were transgranular and intergranular cracks, with initiation on the AHF side. This was classified as AHF environmental stress cracking. Another case of AHF environmental cracking occurred in type 316L stainless steel cold bent piping after two years. The process contained organics, AHF, and trace anhydrous HCl (AHCl) at −29 °C (−20 °F). Intergranular cracking was observed in elbows but not in piping or welds. There was no sensitization. Cold-worked type 316L stainless steel is susceptible to intergranular SCC in AHF with traces of AHCl (Ref 64).

Nickel-Rich Austenitic Stainless Steels

In liquid AHF service, high-nickel austenitic stainless steels with chromium, molybdenum, and copper, such as alloys G3, G30, 20, and 825, have excellent corrosion resistance up to at least 125 °C (255 °F). Their resistance to AHF in flowing conditions makes these alloys useful for applications such as pumps, valves, and heat-exchanger tubing. The preferred material for cast pumps and valves for AHF service is CN-7M casting alloy (Ref 18).

Weld Corrosion. Alloy 20, 825, G, and G30 welds are susceptible to corrosion when exposed to AHF. Autogenous welds in alloy 20 tubing had a rate of 0.254 mm/yr (10 mils/yr) versus 0.0254 mm/yr (1 mil/yr) on the base metal when exposed to 10% AHF, 1% Cl_2 and chlorofluorocarbons at 95 °C (205 °F) (Ref 53). Weld attack of these alloys is due to weld alloy segregation as well as the detrimental effect of reactive metals such as niobium, tantalum, and titanium. These reactive metals may be alloyed in the base metal as well as welding consumables.

Cracking. Alloys 20, 825, and G30 have given satisfactory service in AHF environments that have SO_2; air or oxygen has caused SCC of alloy 400. However, there have been cases of in-service SCC of these nickel-rich austenitic stainless steel alloys when exposed to AHF and cupric fluoride. These were downstream of copper-bearing alloys.

An alloy 20 cold-formed dished head on a pressure vessel cracked within a week of exposure to AHF. The cracking was transgranular and concentrated at the knuckle radius where the cold deformation was greatest. A replacement head that was annealed after forming gave good service.

Nickel and Nickel Alloys

Corrosion data for alloys 200, 400, and 600 in AHF are presented in Table 16. Nickel alloys have excellent corrosion resistance to AHF. Alloy 200 is used for rupture discs in AHF storage tanks and railcars at ambient temperatures and for pipe and equipment to handle AHF liquid and vapor from −40 to 600 °C (−40 to 1110 °F). Alloys 600, C22, and C276 are used in valves, piping, and tubing for AHF up to at least 149 °C (300 °F). Alloy 600 has been used in valves and other equipment in place of alloy 400 to avoid possible SCC. It is the most widely used alloy for hot HF vapors, combining excellent chemical resistance and good metallurgical stability. Alloy 400 is used up to at least 149 °C (300 °F) in AHF alkylation service where process conditions are air-free.

Cracking. Alloys 400 and 500 fail by SCC in AHF after opening systems to air for maintenance or inspections. General practice in AHF alkylation service is to stress relieve alloy 400 at 621 °C (1150 °F). However, nickel alloy and HF manufacturers recommend stress relief at 705 °C (1300 °F) for 2 h to avoid SCC of alloy 400. This is a prudent practice in systems where HF-exposed metal is rarely exposed to oxygen. However, for alloy 400 equipment that will be exposed to AHF and oxygen, stress relief may reduce the potential for SCC, but it will not prevent intergranular corrosion (Ref 53). In such cases, other alloys should be considered. Nickel alloys may be susceptible to SCC when CuF_2 is present in AHF. There are reports of SCC in AHF for various nickel alloys that were downstream from copper-containing alloys.

Environmental stress cracking of alloy 600 has occurred in AHF at 135 °C (275 °F) and in AHF with 5% Cl_2 at 100 °C (210 °F). In both cases, corrosion loss was nil, and transgranular cracking occurred within days of initial service. Cracking was found only at the knuckle radius of cold-formed, dished heads (Ref 53). Associated equipment with annealed alloy 600 had no cracking in the same environments. To prevent cracking in AHF, alloy 600 should be solution annealed after cold forming.

Nickel and Nickel Alloys in High-Temperature AHF and HF

Resistance of nickel alloys in high-temperature HF and AHF is dependent on the formation of thin, adherent, protective metal fluoride films. Nickel fluoride and molybdenum fluoride films are relatively stable compared to other nickel halides and are protective to the extent that their vapor pressure is less than or equal to 10 Pa (10^{-4} atm), and their melting point is higher than the service temperature (Ref 65). Corrosion is only one mode of high-temperature degradation in AHF/HF. Internal degradation also occurs, due to permeation of high-temperature AHF/HF, which affects corrosion behavior and mechanical properties. Permeability is a product of fluoride diffusivity and solubility in metal. Chromium, niobium, tantalum, titanium, and iron are detrimental to fluorination resistance because they form high-vapor-pressure and low-melting-point metal fluorides that increase corrosion susceptibility of nickel-base alloys. These elements also increase internal damage. Corrosion data presented for metals in high-temperature AHF/HF are from tests of a few hours to a few days, which is appropriate for ranking relative corrosion resistance and observation of short-term physical changes in the metal. Above 400 °C (750 °F), some high-strength nickel alloys undergo long-term metallurgical changes, such as aging or long-range ordering, thus increasing hardness, reducing fracture toughness and creep ductility, and increasing susceptibility to hydrogen embrittlement or corrosion. Alloys 200, 201, and 600 are relatively unaffected.

Alloying Element Influence on Corrosion in High-Temperature AHF. Corrosion data for various metals in AHF at 500 to 600 °C (932 to 1112 °F) from 4 to 15 h tests are found in Table 18, and from 36 h tests in Table 19. Corrosion data for pure nickel in Ar-5% AHF at 1000 °C (1832 °F) from 15 h tests are given in Table 20, and from 15 h tests in Table 21. From Table 18, the lowest corrosion rates were observed for copper and nickel alloys 200, 400, and 600, while the highest corrosion rates were for iron-base alloys. The highest corrosion rates were associated with high chromium and niobium content (309, 310, and 347). Table 19 presents some of the same alloys but with longer exposure and lower corrosion rates; note that alloys C and B (UNS N10002 and N10001) with molybdenum had low corrosion rates. Exposed nickel alloys in AHF at 650 °C (1200 °F) with more chromium had higher corrosion rates and higher internal fluoride formation in AHF (Ref 69). In 10 h exposures to AHF at 850 °C

Table 18 Corrosion of metals and alloys by gaseous anhydrous hydrogen fluoride at elevated temperatures
Tested for a duration of 4 to 15 h

	Corrosion rate at temperature, °C (°F)			
	500 (932)		600 (1112)	
Metal	mm/yr	mils/yr	mm/yr	mils/yr
Ni 200	0.91	36	0.91	36
400	1.22	48	1.83	72
600	1.52	60	1.52	60
Copper	1.52	60	1.22	48
1020 steel	15.54	612	7.62	300
Type 304	...	...	13.41	528
Type 309	5.79	228	168	6600
Type 310	12.2	480	305	12,000
Type 347	183	7200	177	6960
Type 430	1.52	60	11.6	456

Source: Ref 66

Table 19 Corrosion tests in anhydrous hydrogen fluoride for nickel and nickel alloys
Tested at 500 to 600 °C (932 to 1112 °F), 36 h duration

	Corrosion rate		
Metal	mm/yr	mils/yr	Comment
Alloy C	0.01	0.3	Iridescent tarnish film
Alloy 600	0.02	0.7	Iridescent tarnish film
Alloy B	0.05	2	Black film
Ni 200	0.23	9	Black film
Ni 201	0.36	14	Black film
Alloy 400	0.33	13	Adherent dark film
Alloy 500	0.41	16	Adherent dark film

Source: Ref 67

(1560 °F), nickel, cobalt, copper, and molybdenum had lower corrosion rates than iron, chromium, niobium, and tantalum (Ref 70). In Ar-5% AHF exposures at 1000 °C (1830 °F), high-purity nickel alloy 227 (UNS N02270) had the lowest corrosion rates and had no measurable internal fluoridation (Ref 68). Alloys based on nickel and cobalt had higher corrosion rates than pure nickel and exhibited internal fluoride formation. Nickel-molybdenum alloys had lower corrosion rates and internal fluoride formation than other alloys among these observations. Extensive internal fluorides were found in alloys 600, 617 (UNS N06617), and 602CA (UNS N06025).

Water and Oxygen Influence on High-Temperature AHF and HF Corrosion. For some metals, HF and water is less corrosive than AHF at high temperatures. Corrosion data are presented in Table 22 for various alloys in 5% HF, 8% O_2, 15% H_2O (balance as CO_2 and N_2) for 100 and 155 h exposures at 450 °C (842 °F) (Ref 22). These low corrosion rates may be due to water vapor and oxygen promoting a protective metal oxide film. Nickel-chromium alloys exposed to HF with steam at 650 °C (1200 °F) produced a chromium-oxide-rich surface scale with fluorides in the underlying alloy; also, there was an internal corrosion product network with Cr_2O_3 and chromium fluorides (CrF_2 and CrF_3) (Ref 69). Table 23 data are for alloys 200 and 400 exposed in 50% HF and 50% steam for 4 to 15 h from 550 to 750 °C (1022 to 1382 °F). The trend indicates increasing corrosion with temperature. From a comparison of data in Table 18 to AHF data in Table 23, steam in HF increases corrosion of Ni 200 but decreases corrosion of alloy 400 at 600 °C (1112 °F). From 10 h exposures to HF-50% H_2O at 850 °C (1560 °F), nickel, cobalt, copper, iron, chromium, and tantalum had lower corrosion rates than in AHF, while molybdenum had higher rates (due to oxide formation) (Ref 70). These alloys were examined after 50 h exposures to 90% HF and 10% steam at 850 °C (1560 °F), and types 304, 310, and alloy 800 (UNS N08800) had corrosion rates of 100 to 130 mm/yr (4000 to 5000 mils/yr); thick, brittle corrosion products; and a porous metal matrix. Alloy 600 had a corrosion rate of 23 mm/yr (900 mils/yr), with external iron oxide scaling and internal precipitation of chromium fluorides. Alloy 625 had a corrosion rate of 6.7 mm/yr (265 mils/yr), with an external scale rich in niobium and oxygen and internal chromium fluoride precipitation.

Welds. Weld deposits of alloys 62 (1.5 to 3% Nb; UNS N06062), 82 (2 to 3% Nb; UNS N06082), 132 (1.5 to 4% Nb; UNS W86132), and 182 (1 to 2.5% Nb; UNS W86182) were exposed to AHF at 350 to 400 °C (660 to 750 °F) for two weeks. These welds had corrosion rates of 0.25 to 0.58 mm/yr (10 to 23 mils/yr), while alloy 600 was 0.05 mm/yr (2 mils/yr) (Ref 53). In-service weld attack was noted in AHF, starting at 150 °C (300 °F) and increasing with temperature. At 350 °C (660 °F), weld corrosion rates were 0.4 to 0.6 mm/yr (16 to 24 mils/yr) versus rates on alloy 600 of 0.025 mm/yr (1 mil/yr). Weld corrosion deposits were higher in niobium, tantalum, and titanium than the weld metal.

Cracking. Table 22 includes a report of SCC of alloy 600 in 5% HF, 8% O_2, 15% H_2O (balance as CO_2 and N_2) at 450 °C (842 °F). This was a very low-corrosion-rate environment, and cracking occurred within 100 h on U-bend samples that were not initially cold worked but subject to strain when bent.

Table 20 Corrosion rates and observations on pure metals exposed in Ar-5% anhydrous hydrogen fluoride
Tested at 1000 °C (1832 °F)

Material	Weight change, mg/cm²	Corrosion rate mm/yr	mils/yr	Appearance	Comment
Cr(a)	−35.25	...	...	Green	Extensive internal fluorides to a depth of 0.21 mm (8.3 mils)/side; assuming linear rates 36.7 mm/yr (1445 mils/yr)
Ni 200(b)	−0.1	0.036	1.4	Unchanged	No evidence of internal fluorides
Ni 201(b)	−0.1	0.036	1.4	Unchanged	No evidence of internal fluorides
Ni 227(b)	0	0	0	Unchanged	No evidence of internal fluorides

(a) Cr exposed for 50 h. (b) Ni exposed for 15 h. Source: Ref 68

Table 21 Metallographic measurements and observations on alloys exposed in Ar-5% anhydrous hydrogen fluoride (AHF)
Tested at 1000 °C (1832 °F) for 15 h

Alloy	Maximum depth of affected metal mm/side	mils/side	Appearance	Comment
242	0.015	0.6	Surface scale	Also resistant in Ar-35% AHF
B3	0.020	0.8	Surface scale	...
Nickel aluminide	0.025	1.0	Surface scale	Extensive corrosion in Ar-35% AHF
188	0.051	2.0	Surface scale	...
617	0.13	5.3	Nodules on surface	Extensive internal fluorides
600	0.15	5.9	Nodules on surface	Extensive internal fluorides
602CA	0.18	7.1	Surface scale: molten corrosion product	Extensive internal fluorides

Source: Ref 68

Table 22 Corrosion rates and stress-corrosion cracking (SCC) evidence for various alloys
Exposed in 5% hydrofluoric acid, 8% O_2, 15% H_2O, and balance as CO_2 and N_2 at 450 °C (842 °F); SCC determined with U-bends

Alloy	Corrosion rate 100 h mm/yr	mils/yr	155 h mm/yr	mils/yr	SCC 100 h	155 h
600	0.01	0.4	0	0	Yes	Yes
400	0	0	0.002	0.1	No	No
825	0.01	0.4	0	0	No	No
C276	0	0	0	0	No	No
686	0	0	0	0	No	No
C22	0	0	0	0	No	No
25-6Mo	0	0	0	0	No	No

Same specimens were evaluated at 100 h and after an additional 55 h exposure. Source: Ref 22

Table 23 Corrosion of alloys 200 and 400 in 50% hydrofluoric acid and 50% steam
Tests were 4 to 15 h

Alloy	Corrosion rate at temperature 550 °C (1022 °F) mm/yr	mils/yr	600 °C (1112 °F) mm/yr	mils/yr	650 °C (1202 °F) mm/yr	mils/yr	700 °C (1292 °F) mm/yr	mils/yr	750 °C (1382 °F) mm/yr	mils/yr
200	0.79	31	1.83	72	2.74	108	3.66	144	3.05	120
400	...	...	0.61	24	1.52	60	3.96	156	5.18	204

Source: Ref 70

Copper Alloys

Table 16 is static-condition corrosion data for copper, copper-nickel, brass, and bronze alloys in AHF. Copper-nickel alloys 90-10 (UNS C70600) and 70-30 have been successfully employed for heat-exchanger tubing because of their corrosion resistance in flowing AHF (Ref 3). Copper-nickel alloys have also been used for bolting in AHF in lieu of alloy 400 because of their SCC resistance. Copper tubing has been used at ambient temperature for AHF liquid and vapors at low velocities.

Precious and Other Metals

Silver, gold, platinum, tungsten, and molybdenum resist corrosion in AHF. Tungsten and molybdenum are used for special parts in valves and equipment that may be fabricated without welding. Silver, despite its higher cost compared to other materials, has been used in AHF services such as stills, heating coils, condensers, rupture disks, and pressure diaphragms. However, silver in AHF service has high corrosion rates when also exposed to oxygen, hydrogen sulfide, or sulfur contamination. Gold, platinum, and palladium were exposed to Ar-5% AHF at 1000 °C (1830 °F) for 15 h (Ref 68). Gold had no corrosion, platinum had a corrosion rate of 0.03 mm/yr (1.1 mil/yr), and palladium had a weight gain, but the appearance was unchanged.

Aluminum and its alloys have not been employed in AHF service, even though they have some resistance, as noted in Table 16, because they experience very high corrosion rates in HF. Hydrofluoric acid solutions are formed during water exposure, which would be required when washing out equipment for inspections and maintenance.

Nonmetals

Thermoplastics. Fully fluorinated plastics are resistant to AHF up to and above the boiling point. Fully fluorinated plastics, such as PFA, PTFE, FEP, and MFA, do not lose strength, swell, crack, embrittle, or undergo chemical attack by AHF. However, AHF will permeate through fully fluorinated plastics, and the rate of permeation increases with increasing temperature, concentration, vapor pressure, and with decreasing liner thickness. More information is presented in the HF section about thermoplastics in this article.

Elastomers. Manufacturers should be consulted for guidance on the best compounds, fillers, and cures for specific AHF service conditions. Only a limited group of fluoroelastomers, designated FKM, in accordance with ASTM D 1418, have resistance to AHF, and these generally are carbon-filled and peroxide-cured. Perfluoroelastomers designated FFKM, in accordance with ASTM D 1418, generally have resistance to AHF at ambient temperatures and above. There are, however, performance differences among the various grades of FFKM (Ref 2). The higher the fluorine content of the elastomers, generally the better the chemical resistance in AHF.

Carbon and graphite (both with low ash) resist AHF. Flexible graphite (low ash or nuclear grade) has been used for gaskets and packing materials in AHF services. Carbon-carbon composites have excellent resistance to AHF; however, their use is limited to internal components because they are permeable. Two grades of graphite (with <5 and 700 ppm total ash) were exposed to Ar-5% AHF for 15 h at 1000 °C (1830 °F) (Ref 68). Both samples had no weight change and appeared unchanged.

Alumina (>99.7% purity), sapphire, and ruby have been used in instruments and analyzers in AHF services. Alpha silicon carbide (fully reaction bonded with no free silicon) has been used in AHF for pump bearings and heat-exchanger tubing.

ACKNOWLEDGMENTS

The author acknowledges E.I. DuPont for the opportunities to learn about corrosion by HF and AHF and to prepare this article. Appreciation goes to Paul Crook of Haynes International for being so responsive in providing corrosion data on nickel alloys. Thanks to Jeffrey M. Chandler, Amy Anderson-Maifeld, Brian J. Saldanha, and Larry D. Schwarz for their comments and suggestions.

REFERENCES

1. "Materials of Construction Guideline for Aqueous Hydrofluoric Acid," Hydrogen Fluoride Industry Practices Institute, revised May 1996
2. Materials of Construction Guideline for Anhydrous Hydrogen Fluoride, *Recommended Practices for the Hydrogen Fluoride Industry,* Vol 1, Hydrogen Fluoride Industry Practices Institute, Jan 2005
3. "Materials for Receiving, Handling, and Storing Hydrofluoric Acid," Publication 5A171, NACE International, revised Sept 2001
4. T.F. Degnan, *Materials Selector for Hazardous Chemicals, MS-4: Hydrogen Fluoride and Hydrofluoric Acid,* H.S. Ahluwalia, Ed., Materials Technology Institute of the Chemical Process Industries, 2003
5. M.E. Holmberg and F.A. Prange, *Ind. Eng. Chem.,* Vol 37, 1945, p 1030
6. H.A. Pray, F.W. Fink, B.E. Friedl, and W.J. Brain, Report 268, Battelle Memorial Institute, 1953
7. M. Howells, correspondence to NACE Task Group T-5A-36, March 8, 1994
8. T.F. Degnan, Corrosion by Hydrogen Fluoride and Hydrofluoric Acid, *Corrosion,* Vol 13, *Metals Handbook,* 9th ed., ASM International, 1987, p 1166
9. G. Trabanelli, A. Frignani, G. Brunoro, C. Monticelli, and F. Zucchi, *Mater. Perform.,* Vol 24 (No. 6), 1985, p 33
10. H.H. Hashim and W.L. Valerioti, Corrosion Resistance of Carbon Steel in Hydrofluoric Acid Alkylation Service, *Mater. Perform.,* Vol 32 (No. 11), 1993, p 50
11. M. Baker, "1997 Turnaround Findings," presentation to NACE Committee T5A-36 (San Diego, CA), 1998
12. A. Gysbers, "HF Alkylation Unit Experiences with Localized Corrosion," presentation to NACE Committee T5A-36 (San Diego, CA), 1998
13. O. Forsen, J. Aromaa, and M. Somervuori, "Materials Performance in HF-Alkylation Units," Paper 342, Corrosion '95 (Orlando, FL), National Association of Corrosion Engineers, 1995
14. J.D. Dobis, D.R. Clarida, and J.P. Richert, "A Survey of Plant Practices and Experiences in HF Alkylation Units," Paper 511, Corrosion '94 (Baltimore, MD), National Association of Corrosion Engineers, 1994
15. G. Chirinos, S. Turgoose, and R.C. Newman, "Effects of Residual Elements on the Corrosion Resistance of Steels in HF," Paper 513, Corrosion '97 (New Orleons, LA), National Association of Corrosion Engineers, 1997
16. A. Gysbers, H.H. Hashim, D.R. Clarida, G. Chirinos, J. Marsh, and J. Palmer, "Specification for Carbon Steel Materials for Hydrofluoric Acid Units," Paper 651, Corrosion/2003, NACE International, 2003
17. W.K. Blanchard, Jr. and N.C. Mack, "Corrosion Results of Alloys and Welded Couples over a Range of Hydrofluoric Acid Concentrations at 125 °F," Paper 452, Corrosion '92, National Association of Corrosion Engineers, 1992
18. T.F. Degnan, Materials of Construction for Hydrofluoric Acid and Hydrogen Fluoride, *Process Industries Corrosion: Theory and Practice,* B.J. Moniz and W.I. Pollock, Ed., National Association of Corrosion Engineers, 1986, p 275
19. S.J. Pawel, Corrosion of High-Alloy Materials in Aqueous Hydrofluoric Acid Environments, *Corrosion,* Vol 50 (No. 12), 1994, p 963
20. P. Crook, "Internal Corrosion Data," Haynes International, Jan 12, 2004
21. M. Takemoto, T. Shonohara, M. Shirai, and T. Shinogays, "Case Study of External Stress Corrosion Cracking (ESCC) of Austenitic Stainless Steels, and a Simulation Test," Paper 143, Corrosion '84 (New Orleans, LA), National Association of Corrosion Engineers, 1984
22. J.R. Crum, G.D. Smith, M.J. McNallan, and S. Hirnyj, "Characterization of Corrosion Resistant Materials in Low and High Temperature HF Environments," Paper 382, Corrosion '99 (San Antonio, TX), National Association of Corrosion Engineers, 1999

23. B.E. Paige, "Evaluation of Welds and Stresses in High Nickel Alloys in Fluoride Solutions," Paper 80, Corrosion/77, National Association of Corrosion Engineers, 1977
24. B.C. Norby, "Improved Corrosion Resistance on UNS N06030 in Nitric/Hydrofluoric Acid Solutions by Welding with a Dissimilar Weld Wire," Paper 107, Corrosion '92, National Association of Corrosion Engineers, 1992
25. G. Schmitt, and S. Losaker, Performance of CRA in Hydrofluoric Acid, *EUROCORR '97 Proceedings,* Vol 1, European Federation of Corrosion, Trondheim, Norway, 1997, p 677
26. C. Haver, H.J. Choi, and R. Bhavsar, "Halogen Ion Stress Corrosion Cracking of an Encapsulated UNS N08825 Control Line," Paper 00131, Corrosion/2000, National Association of Corrosion Engineers, 2000
27. R.B. Rebak, J.R. Dillman, P. Crook, and C.V.V. Shawber, Corrosion Behavior of Nickel Alloys in Wet Hydrofluoric Acid, *Mater. Corros.,* Vol 52, 2000, p 289
28. T.C. Spence, "Cast Ni-Cu Alloys in HF," presentation to NACE International Committee T5A-36 in Corrosion/97, March 11, 1997
29. *Special Metals High-Performance Alloys for Resistance to Aqueous Corrosion,* SMC-026, Special Metals Corporation, 2000
30. M. Schussler, Metal Materials for Handling Aqueous Hydrofluoric Acid, *Ind. Eng. Chem.,* Vol 47, 1955, p 135
31. H.R. Copson and C.F. Cheng, Stress Corrosion Cracking of Monel in Hydrofluoric Acid, *Corrosion,* Vol 12 (No. 12), Dec 1956, p 647–653
32. L. Graf and W. Wittich, *Werkst. Korros.,* Vol 17, 1966, p 385
33. B. Saldanha, "Denickelification of Cu-Ni Alloys," presentation to NACE International Committee T5A-36 in Corrosion/2000, 27 March 2000
34. R.J. Tzou and H.C. Shih, *Corrosion,* Vol 45 (No. 4), April 1989, p 328–333
35. J.J. Hoffman, J.W. Slusser, and J.L. O'Leary, "Stress Corrosion Cracking Susceptibility of Aluminum-Silicon-Bronze," Paper 03510, Corrosion/2003, NACE International, 2003
36. H.H. Uhlig, *Corrosion Handbook,* John Wiley & Sons, 1948, p 115, 303, 315, 769
37. F.L. LaQue and H.R. Copson, *Corrosion Resistance of Metals and Alloys,* 2nd ed., A.C.S. Monograph 158, Reinhold, 1963
38. G. Smith and E. Zysk, Corrosion of the Noble Metals, *Corrosion,* Vol 13, *ASM Handbook,* ASM International, 1987, p 793–807
39. "Technical Notes," Climax Molybdenum Company, 1959
40. M. Bunner, Fluoropolymer Lining Systems: An Economical Alternative, *Mater. Perform.,* Vol 39 (No. 1), 2000, p 38
41. M. Conde and C. Taxen, "Hydrochloric Acid and Water Permeability in Fluoropolymer Tubes," Paper 00572, Corrosion/2000, National Association of Corrosion Engineers, 2000
42. J.C. Stonehill, presentation to NACE International Committee T5A-36, 28 March 1995
43. J.H. Van Sciver, Chemical Resistance, *Mater. Perform.,* Vol 27 (No. 12), 1988, p 49
44. C. Gleditsch, "Corrosion Barrier Compositions of FRP Chemical Process Equipment," Paper 03619, Corrosion/2003, NACE International, 2003
45. E.P. Ferber and J.E. Alexander, "Fluoroelastomers for Harsh Environments," Paper 408, Corrosion '96, National Association of Corrosion Engineers, 1996
46. L. DeLashmit, "Rubber Linings for Hydrofluoric Acid," presentation to NACE Committee T5A-36, March 27, 2000
47. T.F. Degnan, Materials for Handling Hydrofluoric, Nitric, and Sulfuric Acids, *Process Industries Corrosion,* National Association of Corrosion Engineers, 1975, p 229
48. W. Darden and G. Taccini, "The Use of Carbon/Carbon Composites and Graphite in the New Millennium," Paper 01339, Corrosion/2001, National Association of Corrosion Engineers, 2001
49. J.K. Wessel and W.G. Long, "Application of Continuous Fiber Reinforced Ceramic Composites (CFCC) in Corrosive/Erosive Environments," Paper 00563, Corrosion/2000, National Association of Corrosion Engineers, 2000
50. K.R. Mikeska and S.J. Bennison, Corrosion of Alumina in Aqueous Hydrofluoric Acid, *J. Am. Ceram. Soc.,* Vol 82 (No. 12), 1999, p 3561
51. K.R. Mikeska, S.J. Bennison, and S.L. Grise, Corrosion of Ceramics in Aqueous Hydrofluoric Acid, *J. Am. Ceram. Soc.,* Vol 83 (No. 5), 2000, p 1160
52. "Hydrofluoric Acid Alkylation," Phillips Petroleum Company, 1946
53. H.S. Jennings, "Materials for Hydrofluoric Acid Service in the New Millennium," Paper 01345, Corrosion/2001, National Association of Corrosion Engineers, 2001
54. C.C. Seastrom, "Pressure Increase during Storage of Anhydrous Hydrogen Fluoride," presentation to NACE T5A-36, April 27, 1999
55. R.L. Schuyler, Hydrogen Blistering of Steel in Anhydrous Hydrofluoric Acid, *Mater. Perform.,* Vol 18 (No. 8), 1979, p 9
56. R.D. Merrick, An Overview of Hydrogen Damage to Steels at Low Temperatures, *Mater. Perform.,* Vol 28 (No. 2), 1989, p 53
57. D. Warren, Hydrogen Effects on Steel, *Mater. Perform.,* Vol 26 (No. 1), 1987, p 38–48
58. C.C. Seastrom, Minimizing Hydrogen Damage in Carbon Steel Vessels Exposed to Anhydrous Hydrogen Fluoride, *Mechanical Working and Steel Processing Proceedings,* Iron and Steel Society, 1990, p 507–514
59. "Evaluation of Pipeline and Pressure Vessel Steels for Resistance to Hydrogen-Induced Cracking," ANSI/NACE standard TM0284, American National Standard Institute and NACE International, revised Jan 2003
60. "Safe Operation of Hydrofluoric Acid Alkylation Units," Publication RP 751, American Petroleum Institute, Feb 1999
61. "Methods and Controls to Prevent In-Service Environmental Cracking of Carbon Steel Weldments in Corrosive Petroleum Refining Environments," NACE standard RP0472, National Association of Corrosion Engineers, Sept 2000
62. H. Hanninen and T. Hakarainen, On the Effects of α-Martensite in Hydrogen Embrittlement of Cathodically Charged AISI Type 304 Austenitic Stainless Steel, *Corrosion,* Vol 36 (No. 1), 1980, p 47
63. J.D. Fritz and B.W. Parks, "The Influence of Martensite on the Corrosion Resistance of Austenitic Stainless Steels," Paper 00507, Corrosion/2000, NACE International, 2000
64. J. Ely, "Environmental Cracking of Type 316L Stainless Steel in AHF Service," presentation to NACE TEG 119X, 2002
65. P. Elliott, Practical Guide to High-Temperature Alloys, *Mater. Perform.,* Vol 28 (No. 4), 1989, p 57
66. W.R. Myers and W.B. Delong, Fluorine Corrosion, *Chem. Eng. Progr.,* Vol 44 (No. 5), 1948, p 359
67. "Corrosion Resistance of Nickel-Containing Alloys in Hydrofluoric Acid, Hydrogen Fluoride and Fluorine," Publication CEB-5, INCO, 1968
68. J.J. Barnes, "Materials Behavior in High-Temperature HF-Containing Environments," Paper 518, Corrosion/2000, National Association of Corrosion Engineers, 2000
69. G. Marsh and P. Elliott, *High Temperature Corrosion in Energy Systems,* M.F. Rothman, Ed., The Metallurgical Society of AIME, 1985, p 467–481
70. C.J. Tyreman and P. Elliott, "High Temperature Metallic Corrosion in HF-Containing Gases," Paper 135, Corrosion '88, National Association of Corrosion Engineers, 1988

Corrosion by Chlorine

E.L. Liening, The Dow Chemical Company

CORROSION OF METALS by dry liquid chlorine, dry gaseous chlorine, moist chlorine, selected mixed gases with chlorine, and chlorine-water is considered in this article. Also considered are nonmetals used with chlorine and those materials of construction frequently used in handling chlorine.

Chlorine is a yellow-green halogen gas with an atmospheric boiling point of −24 °C (−11 °F) and a liquid density of 1.468 g/cm^3 at 0 °C (32 °F). Chlorine is soluble in water, forming a highly oxidizing hypochlorous acid solution. Solubility of chlorine ranges from 10.46 g/L of water at 0 °C (32 °F) to 6.4 g/L at 25 °C (75 °F).

Materials (including nonmetals) used in chlorine manufacturing are reviewed in Ref 1 and 2. Reference 3 is a good source for references on corrosion by chlorine published prior to 1976. The Chlorine Institute (Ref 4), a trade association of chlorine manufacturers, also provides information on handling chlorine.

Handling Commercial Chlorine

Commercial chlorine is widely used for bleaching and water treatment. It is typically handled as a dry liquid (less than 40 ppm water) under pressure in carbon steel tanks. Piping is commonly carbon steel. Cast iron is not considered suitable for handling commercial chlorine because of its susceptibility to brittle failure. Equipment containing cast components, such as pumps and valves, is typically ductile iron or alloy 400 (UNS N04400). Valve trim commonly includes alloy 400 seats and alloy C276 stems. Gaskets are often spiral-wound alloy 400 with polytetrafluoroethylene (PTFE) filler (Ref 1–3).

Equipment handling commercial chlorine must be free of water, organic contamination, and metal shavings to avoid unwanted exotherms and possible ignition events. Cleaning guidelines for piping systems handling commercial chlorine are given by The Chlorine Institute (Ref 4). Ultraviolet lights are sometimes used to inspect for organic contamination before exposing equipment to liquid chlorine.

Chlorine lines that are refrigerated below 0 °C (32 °F) for extended times may become covered in ice. Alloy 20 (UNS N08320) valves are used in such conditions, and valve bonnet bolting should be adequately corrosion resistant to withstand moist chlorine gas that accumulates under the ice as the result of fugitive emissions from the valve. This may require alloy C276 or similar nickel-chrome-molybdenum alloy bolting.

Dry Chlorine

Dry chlorine is not corrosive to steels, stainless steels, or nickel alloys at ambient temperatures. It is commonly shipped and handled in carbon steel equipment, with higher-alloy materials such as nickel, Monel 400, and Hastelloy C usually used for critical parts. Valves are frequently steel, with Monel 400 or Hastelloy C trim and stem. Steel is usable up to approximately 150 °C (300 °F) and possibly higher under certain conditions. Stainless steels are usable up to approximately 300 °C (570 °F), and nickel is commonly used up to approximately 500 °C (930 °F). Alloy 600 (UNS N06600) or low-carbon nickel is often used at temperatures where graphitization may occur. Moisture will greatly accelerate attack on any of these materials, with the additional danger of stress-corrosion cracking (SCC) of stainless steels.

Iron and Steel. Eight-hour tests showed corrosion rates for steel to be below 0.0025 mm/yr (0.1 mil/yr) at up to 250 °C (480 °F), but ignition occurred at 250 °C (480 °F) (Ref 5). A slightly lower ignition temperature was reported for 16 and 20 h tests (Ref 6). Ignition of steel wool (grade 00) occurred at temperatures as low as 185 °C (365 °F) (Ref 5).

A maximum-use temperature for steel is given as 205 °C (400 °F) in Ref 6 and 7, but the discussion in Ref 7 suggests that a prudent maximum temperature may be nearer 150 °C (300 °F) because of exotherms from reaction with grease-contaminated equipment. For grease-free and properly cleaned equipment, however, a maximum-use temperature of 200 °C (390 °F) may be acceptable.

Tests with various irons and steels in flowing chlorine indicated that ignition occurs at lower temperatures for steels with higher alloy contents, especially carbon (Ref 8, 9). Ignition at temperatures as low as 220 °C (430 °F) was found for iron containing 0.3% C and 6% other alloy content. This probably accounts for the lower use temperature for cast iron versus steel given in Ref 6. Iron alloys with silicon contents in the 10 to 15% range, however, reportedly resisted attack by dry chlorine (Ref 10, 11), but their poor impact properties limit practical applications in such critical service.

Below 250 °C (480 °F), the presence of oxygen and moisture had little effect on iron chlorination rates (Ref 12). Because iron is attacked less rapidly by hydrogen chloride than chlorine, the presence of hydrogen chloride in dry chlorine should have little effect on iron chlorination rates. In practice, however, the use of steel is avoided where moisture may be present.

Table 1 indicates corrosion rates and suggests upper temperature limits for use of metals in dry chlorine, based on 2 to 20 h tests (Ref 6). The temperature limits for many of the alloys correspond to higher corrosion rates than may usually be tolerated for expensive alloys. The rates shown, however, are in most cases higher than those that would occur in prolonged exposures. This is because, for most of the alloys, passivation occurs by the formation of a metal chloride film, after which corrosion decreases rapidly. In many cases, corrosion at higher temperatures is roughly proportional to the vapor pressure of the particular metal chlorides formed. Suggested upper design limits for steel and some other alloys in dry chlorine are shown in Fig. 1, based on corrosion rates of approximately 0.08 mm/yr (3 mils/yr) for tubes and internals and 0.5 mm/yr (20 mils/yr) for vessels and pipes. Similar design guidelines are discussed in Ref 14 and 15.

Aluminum. Care must be taken when evaluating results for aluminum, because the protective aluminum oxide film may delay the onset of corrosion. In one study, there was a 5 h delay before reaction in dry chlorine at 500 °C (930 °F) (Ref 8). Aluminum was reported to be usable up to approximately 120 °C (250 °F) (Ref 6), and moisture in dry chlorine at room temperature increased attack (Ref 16). This is supported by the work documented in Ref 17 and is attributable to condensation. At higher temperatures (130 to 630 °C, or 265 to 1165 °F), the presence of water was reported to decrease attack on aluminum (Ref 18).

Copper. A maximum-use temperature of 205 °C (400 °F) was suggested for copper in dry chlorine (Ref 6). Ignition of copper was observed at temperatures as low as 260 to 300 °C (500 to 570 °F) at high velocities (Ref 19). Ignition occurred at 290 to 310 °C (555 to 590 °F) at a velocity of 250 mL/min, and no ignition occurred at 40 mL/min (Ref 12). The same researchers reported that below 200 °C (390 °F), the presence of oxygen and water vapor accelerates attack on copper, while above 250 °C (480 °F), they reduce the rate of attack and move the ignition temperature to near 350 °C (660 °F) (Ref 12, 20).

Stainless Steels. Austenitic stainless steels were shown to be significantly more resistant to dry chlorine than steel, aluminum, or copper (Ref 6, 8). Type 304 and type 316 stainless steels may be used up to 300 °C (570 °F). Moisture was reported to accelerate attack below 370 °C (700 °F) but to exert little effect or even decreased attack above the temperature (Ref 6, 18). The presence of moisture also increases the possibility of SCC.

Nickel and Nickel Alloys. Nickel 200 and nickel-base alloys show excellent resistance to dry chlorine, as indicated in Table 1. The good performance of nickel was supported by data in Ref 21 for tests in chlorine at a pressure of 13.2 kPa (0.13 atm) and in Ref 8 in flowing chlorine. However, the reports indicated higher corrosion rates (2.4 and 2 mm/yr, or 95 and 81 mils/yr) for nickel at 525 to 540 °C (975 to 1000 °F) than the study described in Ref 6. A temperature of 500 °C (930 °F) seems a wise upper limit for routine use of nickel in dry chlorine.

Water vapor was found to increase attack on nickel below 550 °C (1020 °F) but had little effect above that temperature (Ref 18). Between 425 and 760 °C (795 and 1400 °F), graphitization of Nickel 200 may occur, and a low-carbon version of nickel with a maximum of 0.02% C is normally used. When sulfur compounds are present, alloy 600 is often substituted for nickel to avoid intergranular attack.

The data given in Table 1 indicate that Inconel 600 and nickel-molybdenum alloy B perform nearly as well as nickel in dry chlorine, and Ni-Cr-Mo alloy C somewhat less well. Chromel A and alloy 400 perform much better than stainless steels but not as well as the other alloys mentioned previously. Cast Alloy Casting Institute (ACI) alloy CW-12M (Ni-18Cr-18Mo) was reported to corrode in dry chlorine at 0 to 60 °C (32 to 140 °F) at approximately the same rate as Ni-Cr-Mo alloy C (Ref 22, 23).

Monel 400 is commonly used as trim on valves, but it should be used with care in refrigerated systems. Water vapor in chlorine at temperatures below the dewpoint is corrosive to a wide variety of nickel, nickel-copper, nickel-chromium-iron, and nickel-chromium-molybdenum alloys (Ref 24).

Other Metals. Limited information on the performance of various other alloys in dry chlorine is also available. Magnesium was found to perform as well as Chromel A and actually better than alloy 400 (Ref 6). A maximum-use temperature of 455 °C (850 °F) was suggested, but use of magnesium in chlorine is not widespread.

Lead was found to be resistant in dry chlorine up to 275 °C (525 °F) (Ref 25). Average corrosion rates for lead in dry flowing chlorine were reported as 0.06 mm/yr (2.4 mils/yr) at 200 °C (390 °F), 0.13 mm/yr (5.1 mils/yr) at 250 °C (480 °F), 0.14 mm/yr (5.5 mils/yr) at 275 °C (525 °F), 1.5 mm/yr (59 mils/yr) at 295 °C (565 °F), and 2.5 mm/yr (98 mils/yr) at 310 °C (590 °F) (Ref 9, 18).

Unalloyed zirconium corrodes at less than 0.13 mm/yr (5 mils/yr) in dry chlorine near room temperature but is not resistant in wet chlorine (Ref 22, 23, 26–28). Reactor-grade zirconium tubing was reported to stress-corrosion crack in 0.01 mg/cm³ of chlorine gas at 360 to 400 °C (680 to 750 °F) (Ref 29).

Titanium was found to ignite in dry chlorine at temperatures as low as −18 °C (0 °F) (Ref 22, 23, 30, 31). However, small amounts of moisture in chlorine can passivate titanium.

Niobium suffered no attack in dry chlorine up to approximately 200 °C (390 °F) (Ref 32).

Tantalum performs well in dry chlorine up to 250 °C (480 °F) (Ref 33, 34). Attack on tantalum was shown to begin at 250 °C (480 °F), to be

Table 1 Approximate temperatures at which the indicated corrosion rates occur in dry chlorine

	Temperature for corrosion rate, mm/yr (mils/yr)											
	0.76 (30)		1.5 (60)		3 (120)		15 (600)		30 (1200)		Maximum(a)	
Alloy	°C	°F	°C	°F	°C	°F	°C	°F	°C	°F	°C	°F
Nickel 201	510	950	540	1000	590	1100	650	1200	680	1250	540	1000
Inconel 600	510	950	540	1000	565	1050	650	1200	680	1250	540	1000
Hastelloy B	510	950	540	1000	590	1100	650	1200	...	...	540	1000
Hastelloy C	480	900	540	1000	560	1050	650	1200	...	...	510	950
Magnesium	450	850	480	900	510	950	540	1000	565	1050	450	850
Ni-20Cr-1Si	425	800	480	900	540	1000	620	1150	...	...	450	850
Monel 400	400	750	450	850	480	900	540	1000	540	1000	420	800
Type 316 stainless steel	310	600	345	650	400	750	450	850	480	900	340	650
Type 304 stainless steel	290	550	315	600	340	650	400	750	450	850	310	600
Platinum	480	900	510	950	540	1000	560	1050	560	1050	260	500
Hastelloy D	205	400	230	450	290	550	...	...	...	...	205	400
Deoxidized copper	180	350	230	450	260	500	260	500	290	550	205	400
Carbon steel	120	250	180	350	205	400	230	450	230	450	205	400
Cast iron	90	200	120	250	180	350	230	450	230	450	180	350
Aluminum alloy 1100	120	250	150	300	150	300	180	350	180	350	120	250
Gold	120	250	150	300	180	350	200	400	200	400	...	...
Silver	40	100	65	150	120	250	230	450	260	500	...	...

(a) Suggested upper temperature limit for continuous service. Source: Ref 6

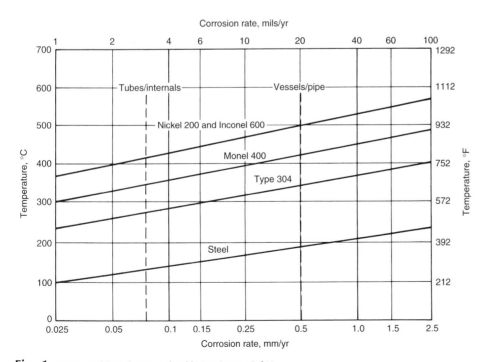

Fig. 1 Design guidelines for use in dry chlorine. Source: Ref 13

violent after 35 min at 450 °C (840 °F), and to be instantaneous at 500 °C (930 °F) (Ref 35). This is consistent with the data given in Ref 18. Pitting of tantalum was reported in a mixture of dry chlorine and anhydrous methanol at 65 °C (150 °F), presumably caused by the presence of halogenated HCOOH contamination (Ref 29).

Nonmetals. The use of carbon steel in dry chlorine generally makes it unnecessary to use the more costly nonmetallic alternatives. Keeping that in mind, the following comments can be made about nonmetallics that sometimes find applications in dry chlorine service.

Glass-lined steel is resistant to dry chlorine as either the liquid or gas, although gaskets can be short-lived. The common PTFE envelope-style gasket for glass-lined steel is susceptible to permeation of chlorine through the PTFE envelope, followed by degradation of the filler material.

PTFE equipment can be used at temperatures up to 150 °C (300 °F) if adequate consideration is given to chlorine permeation effects. Examples are PTFE-lined pipe and vessels and wire-reinforced PTFE hose. Particularly for the hose, care should be taken to limit exposure so that the hose burst pressure is not reduced by corrosion of the metal reinforcement caused by permeated chlorine. Such hoses should be removed from service before burst pressure is affected. Note that this may be before visible indications of degradation are apparent. The practical problem of knowing when to decommission a hose adversely affects the practicality of this option.

Fiber-reinforced pipe and vessels can be used in dry chlorine gas if the appropriate resins, reinforcements, and designs are used. Polyester resins are generally not suitable (Ref 2), but some formulations of vinyl-ester resins are usable up to 121 °C (250 °F) (Ref 36). Consult the manufacturers for design and fabrication guidelines for these materials in this service.

Polyvinylidene fluoride (PVDF)-lined equipment can be used up to 93 °C (199 °F), with appropriate care taken to account for permeation. Permeation is significantly less than for PTFE, but it cannot be ignored. Monolithic PVDF equipment is available but should be used only with careful consideration of piping and other secondary stresses that could break nozzles and joints. Most users opt for more robust systems because of the hazard associated with chlorine releases.

Polypropylene, another common lining material for pipe and vessels, is not reliably resistant to dry chlorine and should not be used in this service.

Polyethylene, a common monolithic pipe and vessel material, is only marginally resistant to dry chlorine and is generally avoided.

Graphite is a common gasket material. It is not suitable in liquid chlorine, because it reacts directly with it to form an interstitial compound (Ref 2). However, so-called flexible graphite is free of fillers and impregnates and can be used in wet and dry chlorine gas.

Elastomer performance is summarized in Ref 2 for both dry and wet chlorine. Natural, butyl, silicone, and nitrile elastomers are reported as not suitable for service. Ethylene propylene rubber, ethylene-propylene-diene monomer, and chlorosulfonated polyethylene (Hypalon, Dupont Dow Elastomers) are reported as suitable up to 65 °C (150 °F). Fluoroelastomers and perfluoroelastomers are generally suitable to higher temperatures. As with most elastomeric applications, actual performance is highly dependent on the circumstances of use. Dynamic versus static conditions and trapped versus open-seating geometries are particularly important.

Refrigerated Liquid Chlorine

Refrigerated liquid chlorine can be handled in steel, but special care should be taken at potential leak sites, such as at valves and nonwelded fittings. Because the chlorine is refrigerated, pipelines and associated equipment become encased in ice formed from the moisture in the air. Chlorine from even small leaks is then trapped beneath the ice, forming wet chlorine gas that is corrosive even at low temperatures. Therefore, alloy 20 materials are often used for valves and other fittings in refrigerated liquid chlorine equipment.

Similarly, ordinarily nonwetted parts, such as bonnet bolts in valves, are typically replaced by more resistant alloys. Nickel-copper alloys, such as alloy 400, have been used, but they are sensitive to oxidizing corrosives. Figure 2 shows an example of corrosion of alloy 400 bonnet bolts. Chlorine had leaked along the valve stem and was trapped beneath the ice, where it became wet and very corrosive. Nickel-chromium-molybdenum alloys, such as alloy C276, are used in order to avoid this type of attack on critical components such as bonnet bolts.

High-Temperature Mixed Gases

Several investigations were conducted on alloys in high-temperature mixed gases of chlorine, oxygen, and argon. Various alloys were tested in an atmosphere of argon, 20% O_2, and 2% Cl_2 at 900 °C (1650 °F) for 8 h (Ref 37). Alloy 214 performed the best, with metal loss of less than 0.025 mm/yr (1 mil/yr). Alloy R-41 (UNS N07041) was the next best performer, followed by Inconel 601 and 600, type 310 stainless steel, and alloy 625. Hastelloy S was less resistant than the aforementioned group, followed by alloy 800H, alloy X, alloy C276, and alloy 188. Longer-term tests in a similar environment yielded the following in order of performance: alloys 214, 601, R-41, X, C276, and 625.

Corrosion in argon-oxygen-chlorine mixtures was investigated on a more basic level (Ref 38). Work was performed in environments containing argon, 20% O_2, and 0.25% Cl_2 at 900 to 1000 °C (1650 to 1830 °F) and in air and 2% Cl_2 at the same temperatures (Ref 39, 40). A report on the corrosion at 260 to 425 °C (500 to 795 °F) of a variety of engineering alloys considered for use in a waste heat recovery system for a chlorinated hydrocarbon incinerator is given in Ref 41.

Moist Chlorine

Many industrial chlorine environments contain substantial water, particularly those in chlorine manufacture prior to the drying operation. Wet chlorine gas is extremely corrosive at temperatures below the dewpoint, because the condensate is a very acidic and oxidizing mixture. One of the most commonly used metals for wet chlorine service is titanium, especially for wet chlorine compressors. Titanium is perfectly passive if there is enough water in the chlorine, but it ignites if there is not enough. With sufficient water, titanium is resistant up to at least 175 °C (345 °F) and probably higher. Wet chlorine gas can be handled at room temperatures with stainless steels similar to alloy 20 if condensation is not too great and if corrosion

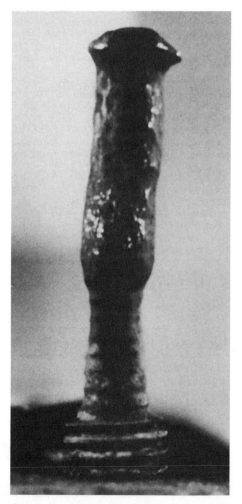

Fig. 2 Monel 400 bonnet bolt extensively corroded by chlorine gas trapped beneath ice covering a valve in liquid chlorine service

rates up to 1.3 mm/yr (50 mils/yr) are tolerable. Alloy C276 is suitable to somewhat higher temperatures, and niobium or tantalum is required at even higher temperatures.

Iron and Steel. The response of iron-base alloys to moisture in chlorine depends on the temperature range and particular class of alloys. At practical-use temperatures, small amounts of moisture and oxygen were reported to have little effect on iron chlorination rates; these contaminants were found to reduce chlorination rates above 300 °C (570 °F) (Ref 8, 20, 42). In practice, steel is generally avoided in any service where chlorine gas may collect water vapor.

Stainless Steels. Moisture has a significant effect on the corrosion of type 304 stainless steel in chlorine, as indicated in Ref 6, which documents tests conducted on type 304 stainless steel in chlorine containing 0.4% H_2O. Rates of approximately 30.5 mm/yr (1200 mils/yr) were found at 40 °C (105 °F), versus an estimated 0.3 mm/yr (12 mils/yr) at 100 °C (210 °F) in dry chlorine. Corrosion in wet chlorine decreased with increasing temperature until approximately 370 °C (700 °F), at which point the corrosion was approximately 4.6 mm/yr (180 mils/yr) and the effect of the moisture disappeared. The detrimental effect of moisture at low temperatures is believed to exist for chromium and austenitic stainless steels in general. Several alloys were tested in water-saturated chlorine gas for 6 weeks at room temperature (Ref 43). The results (Table 2) show substantially higher rates than in dry chlorine. The data support the view that moisture in room-temperature chlorine gas increases corrosion rates of stainless steels, including cast stainless steels.

Aluminum. Aluminum alloy 1100 is more readily attacked by wet than dry chlorine at room temperature, especially if condensation is present. At temperatures above 130 °C (265 °F), however, the presence of moisture greatly reduces corrosion on aluminum. Data from Ref 18 (Table 3) show that larger amounts of water more effectively reduce corrosion on aluminum. Aluminum appears to be usable to 130 °C (265 °F) at 0.06% H_2O, 200 °C (390 °F) at 1.5% H_2O, and 545 °C (1015 °F) at 30% H_2O.

Copper and copper alloys were found to suffer accelerated attack in chlorine gas saturated with water vapor (Ref 44). Copper alloys do not have adequate corrosion resistance for practical use in moist chlorine at temperatures above 200 °C (390 °F). Water vapor at room temperature was shown to accelerate attack on copper and copper alloys (Ref 45). The same effect was reported on copper at temperatures below 200 °C (390 °F) (Ref 12). Above 250 °C (480 °F), however, water vapor and oxygen in chlorine was reported to reduce attack of copper and to move its ignition temperature from approximately 300 °C (570 °F) to approximately 350 °C (660 °F) (Ref 12, 20).

Nickel and nickel alloys are adversely affected by the presence of moisture in chlorine at temperatures up to their maximum-use temperatures in dry chlorine. Water vapor at 1.5% was found to double the reaction rate between chlorine and nickel, while 30% H_2O increased the rate from 2 to 20 times (Ref 12, 18). Above 550 °C (1020 °F), moisture was reported to have little effect, but the rates at that temperature even in dry chlorine make nickel marginal.

Alloy C276 corrodes 2 to 1000 times faster, and Chlorimet 3 (62Ni-18Cr-18Mo) 100 to 1000 times faster, in wet chlorine than in dry. Numerous other nickel, nickel-copper, nickel-chromium-iron, and nickel-chromium-molybdenum alloys are also reported to suffer greatly accelerated attack by wet chlorine at such temperatures.

Titanium is well known for its resistance to corrosion by wet chlorine and is widely used in various chlorine-manufacturing equipment, such as compressors that are exposed to wet chlorine. However, a minimum quantity of water is required to maintain the passivity of titanium. The amount of water required at temperatures between 25 and 175 °C (75 and 350 °F) was found to depend on chlorine pressure, temperature, flow rate, purity, and degree of surface abrasion of the titanium (Ref 30). Figure 3 shows a general guideline for water content needed to maintain passivity of commercial-purity titanium in chlorine at temperatures to 105 °C (220 °F).

Crude cell chlorine tested under static conditions required approximately 0.5% H_2O at 125 °C (225 °F) and 1.2% H_2O at 175 °C (350 °F) (Ref 30). Less water was required at flow rates above 0.15 m/s (0.5 ft/s). Pure (99.5%) chlorine requires relatively more water: approximately 0.93% at room temperature and 1.5% at 200 °C (390 °F) under static conditions.

Other Metals. Relatively limited information is available for other materials. Unalloyed zirconium was found to corrode in wet chlorine at a rate of 2 mm/yr (80 mils/yr) at 15 °C (60 °F) (Ref 23) and 4.9 mm/yr (192 mils/yr) at 25 °C (75 °F) (Ref 27). It also corroded at over 1.3 mm/yr (50 mils/yr) in room-temperature chlorine containing 0.3% H_2O (Ref 28).

Niobium was found to be resistant to wet chlorine at up to 100 °C (210 °F) and tantalum at up to 150 °C (300 °F) (Ref 11, 34, 46, 47).

Tantalum reportedly performed well in chlorine plus 1.5% H_2O at up to 375 °C (705 °F) and in chlorine plus 30% H_2O at up to 400 °C (750 °F) (Ref 35). However, other studies indicated higher rates under these conditions (Ref 18). Another source reported corrosion of tantalum in wet chlorine at temperatures above 350 °C (660 °F), but the amount of water was not specified (Ref 46).

Nonmetals are frequently used in wet chlorine gas because of its corrosivity to most metals, particularly to carbon steel equipment that is widely used in the dry gas. Caution is warranted because, as with many metals, many nonmetals work in either wet or dry chlorine but not both.

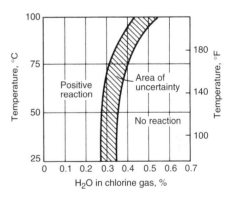

Fig. 3 Estimated water required to passivate unalloyed titanium in chlorine gas

Table 2 Corrosion of cast alloys by water-saturated chlorine gas
Exposure for 6 weeks at room temperature

Alloy	Wrought version	Corrosion rate mm/yr	mils/yr
ACI CF-8M	Type 316	0.79	31
ACI CN-7M	...	1.04	41
ACI CD-4MCu	Duplex stainless steel	1.24	49
ACI CW-12M	Hastelloy C	0.056	2.2

ACI, Alloy Casting Institute. Source: Ref 43

Table 3 Effect of water in chlorine gas on corrosion of aluminum alloy 1100

Temperature		Corrosion rate for amount of water in chlorine, wt%					
		0.06		1.5		30	
°C	°F	mm/yr	mils/yr	mm/yr	mils/yr	mm/yr	mils/yr
130	265	<0.5	<20	<0.5	<20	...	...
140	285	...	...	...	...	<0.5	<20
170	340	3.6	140	...	...	0.53	21
200	390	...	...	<0.5	<20	...	...
290	555	...	...	1.04	41	<0.5	<20
320	610	...	...	1.04	41	...	...
350	660	...	...	5.3 m/yr	210 in./yr	...	...
400	750	...	...	21 m/yr	830 in./yr	...	...
545	1015	...	...	...	...	<0.5	<20
615	1140	...	...	...	...	5.3	210
630	1165	...	...	...	...	7.9 m/yr	311 in./yr

Source: Ref 18

Table 4 Corrosion in chlorine-saturated water
Temperature: 25 °C (75 °F)

Alloy	Test duration, days	Corrosion rate mm/yr	Corrosion rate mils/yr
ACI CF-8M	42	0.013	0.50(a)
Type 316	56	0.008	0.30
ACI CN-7M	42	0.05	1.8(a)
20Cb-3	56	0.008	0.30
ACI CD-4MCu	42	0.06	2.5(a)
ACI CW-12M	42	0.023	0.90
Hastelloy C276	56	0.0025	0.10
Monel 400	56	24	948
Titanium, grade 2	56	0.0005	0.02

ACI, Alloy Casting Institute. (a) Crevice corrosion. Source: Ref 43

Table 5 Corrosion of alloys in chlorine-ice
147 day tests at −20 °C (−4 °F)

Alloy	Corrosion rate μm/yr	Corrosion rate mils/yr
Steel	38	1.5
ACI CF-8	0.25	0.01
ACI CF-8M	0.25	0.01
ACI CN-7M	<0.25	<0.01
ACI CD-4MCu	0.25	0.01
Alloy 255	<0.25	<0.01
N-12M	76	3.01
ACI CW-12M	3.8	0.15
M-35	29	1.15

ACI, Alloy Casting Institute. Source: Ref 43

The materials used in equipment for the production of chlorine are reviewed in Ref 48.

Glass-lined steel, fiber-reinforced plastic, PTFE equipment, PVDF equipment, and selected elastomers are resistant to moist chlorine under the same conditions as described previously for dry chlorine. The same caveats apply regarding permeation, following manufacturers' recommendations, and properly accounting for secondary stresses on monolithic nonmetallic equipment.

Polypropylene, polyethylene, and graphite also have the same limitations in moist chlorine as described for dry chlorine.

Chlorine-Water

Chlorine dissolved in water forms a mixture of HCl and hypochlorous acid (HClO). See the article "Corrosion by Hydrogen Chloride and Hydrochloric Acid" in this Volume. Hypochlorous acid is very oxidizing, which makes the mixture extremely corrosive. Relatively little information is available on corrosion in water containing substantial levels of chlorine, especially near saturation. Aluminum was reported to be unsuitable in HClO and to be attacked with extensive pitting in chlorine-water environments (Ref 17). Zirconium was found to corrode at less than 0.025 mm/yr (1 mil/yr) in chlorine-saturated water (Ref 26).

Corrosion of several alloys in chlorine-saturated water at 25 °C (75 °F) was investigated (Ref 43), and the results, presented in Table 4, show that rates are generally low for the chromium-containing alloys tested and for titanium. Only Monel 400 showed a very high rate (24 mm/yr, or 948 mils/yr), which is to be expected because of its sensitivity to oxidants. No large difference in performance between cast alloys and their wrought equivalents was found, except for crevice attack on ACI CF-8M, CN-7M, and CD-4MCu.

Related tests were performed in chlorine ice at −20 °C (−4 °F) (Ref 43). The results, given in Table 5, show that corrosion rates for chromium-containing alloys are below 0.0025 mm/yr (0.1 mil/yr), except for ACI CW-12M at 0.0038 mm/yr (0.15 mil/yr). Higher, but not unacceptable, rates were found for steel, N-12M, and M-35. These alloys are sensitive to oxidants in acidic environments, which explains their poorer performance.

REFERENCES

1. C.P. Dillon, *Corrosion Control in the Chemical Process Industries,* McGraw-Hill Book Company, 1986, p 209–212
2. "Materials Selector for Hazardous Chemicals. Volume 3: Hydrochloric Acid, Hydrogen Chloride and Chlorine," MTI Publication MS-3, Materials Technology Institute, Inc., 1999
3. "Bibliography of Corrosion by Chlorine," TPC-4, National Association of Corrosion Engineers, 1976
4. The Chlorine Institute, Inc., www.chlorineinstitute.org, accessed Feb 2005
5. G. Heinemann, F.G. Garrison, and P.A. Haber, *Ind. Eng. Chem.,* Vol 38 (No. 5), 1946, p 497
6. M.H. Brown, W.B. DeLong, and J.R. Auld, *Ind. Eng. Chem.,* Vol 39 (No. 7), 1947, p 839
7. W.Z. Friend and B.B. Knapp, *Trans. AIChE,* Section A, 25 Feb 1943, p 731
8. K.L. Tseitlin and J.A. Strunkin, *J. Appl. Chem. (USSR),* Vol 31 (No. 12), 1958, p 1832
9. K.L. Tseitlin, *J. Appl. Chem. (USSR),* Vol 28 (No. 5), 1955, p 467
10. W.A. Luce and R.B. Seymour, *Chem. Eng.,* Vol 57 (No. 10), 1950, p 217
11. S.D. Kirkpatrick and J.R. Callahan, *Chem. Eng.,* Vol 57 (No. 11), 1950, p 107
12. P.L. Daniel and R.A. Rapp, *Halogen Corrosion of Metals,* Vol 5, *Advances in Corrosion Science and Technology,* M.G. Fontana and R.W. Staehle, Ed., Plenum Press, 1976, p 55
13. C.M. Schillmoller, *Chem. Eng.,* Vol 87 (No. 5), March, 1980, p 161
14. G.N. Kirby, *Chem. Eng.,* Vol 87 (No. 23), 1980, p 86
15. N.C. Horowitz, *Chem. Eng.,* Vol 88 (No. 7), 1981, p 105
16. E. Rabald, *Corrosion Guide,* American Elsevier, 1968
17. P. Juniere and M. Sigwalt, *Aluminum—Its Application in the Chemical and Food Industries,* Crosby Lockwood & Son Ltd., 1964
18. K.L. Tseitlin and J.A. Strunkin, *J. Appl. Chem. (USSR),* Vol 29 (No. 11), 1956, p 1793
19. K.L. Tseitlin, *J. Appl. Chem. (USSR),* Vol 27 (No. 9), 1954, p 889
20. K.L. Tseitlin, *J. Appl. Chem. (USSR),* Vol 29 (No. 2), 1956, p 253
21. B.J. Downey, J.C. Bernel, and P.J. Zimmer, *Corrosion,* Vol 25 (No. 12), 1969, p 502
22. R.S. Sheppard, D.R. Hise, P.J. Gegner, and W.L. Wilson, *Corrosion,* Vol 18 (No. 6), 1962, p 211t
23. P.J. Gegner and W.L. Wilson, *Corrosion,* Vol 15 (No. 7), 1959
24. "Huntington Alloys—Resistance to Corrosion," Publication 25M(11-70)S-37, Huntington Alloys, Inc., 1970
25. *Lead for Corrosion Resistant Applications,* Lead Industries Association, 1974
26. "Zirconium and Hafnium," Publication 10M-101570, Amax
27. L.B. Golden, I.R. Lane, Jr., and W.L. Acherman, *Ind. Eng. Chem.,* Vol 44 (No. 8), 1952, p 1930
28. "Zircadyne Corrosion Data," TWCA-8101Zr19, Teledyne Wah Chang, 1981
29. E. Rabald, *Werkst. Korros.,* Vol 12 (No. 11), 1961, p 695
30. E.E. Millaway and M.H. Kleinman, *Corrosion,* Vol 23 (No. 4), 1967, p 88
31. G.E. Hutchinson and P.H. Permar, *Corrosion,* Vol 5 (No. 10), 1949, p 319
32. "Columbium," Publication 313-PD1, KBI Division of Cabot Corporation, 1985
33. "Tantalum," Publication 312-PD1, KBI Division of Cabot Corporation, 1985
34. "Corrosion Resistant Materials," Bulletin 104 PD1, Kawecki Berylco Industries, Inc., 1977
35. M. Schussler, *Corrosion Data Survey on Tantalum,* Fansteel Inc., 1972
36. "Chemical Resistance and Engineering Guide, DERAKANE Epoxy Vinyl Ester Resins," The Dow Chemical Company, March 1999
37. S. Baranow, G.Y. Lai, and M.F. Rothman, "Materials Performance in High Temperature, Halogen-Bearing Environments," Paper 16, presented at Corrosion/84, National Association of Corrosion Engineers, 1984
38. M.J. McNallan, J.M. Oh, and W.W. Liang, "High-Temperature Corrosion of Metals in Argon-Oxygen-Chlorine Mixtures," DOE/ER-12093-T1, Gas Research Institute, 1982
39. M.J. McNallan, M.H. Rhee, S. Thongtem, and T. Hansler, "The Effect of Temperature on the High-Temperature Corrosion of

Superalloys in Argon-20% Oxygen-0.25% Chlorine," Paper 11, presented at Corrosion/85, National Association of Corrosion Engineers, 1985
40. P. Elliot, A.A. Ansari, R. Prescott, and M.F. Rothman, "Behavior of Selected Commercial-Base Alloys during High Temperature Oxychlorination," Paper 13, presented at Corrosion/85, National Association of Corrosion Engineers, 1985
41. W.C. Fort III and W.R. Dicks, *Mater. Perform.,* Vol 25 (No. 3), 1986, p 9
42. E.E. Millaway and L.C. Covington, "Resistance of Titanium to Gaseous and Liquid Fluorine," Titanium Metals Corporation of America, 1959
43. E.L. Liening, Report ME-4242, The Dow Chemical Company, April 1980
44. V. Pershke and L. Pecherkin, *Khimistroi,* Vol 6, 1934, p 140
45. The American Brass Company, *Chem. Eng.,* Vol 58 (No. 1), 1951, p 108
46. "Corrosion Resistance of Tantalum and Niobium Metals," Bulletin 3000, NRC Inc.
47. W.E. Bratt, L.R. Scribner, and C.G. Chisholm, *Chem. Eng.,* Vol 54 (No. 2), 1947, p 219
48. W.H. Shearon, Jr., F. Chrencik, and C.L. Dickinson, *Ind. Eng. Chem.,* Vol 40 (No. 11), 1948, p 2002

Corrosion by Alkalis

Michael Davies, CARIAD Consultants

TRUE ALKALINE CHEMICALS include caustic soda, or sodium hydroxide (NaOH), caustic potash, or potassium hydroxide (KOH), and soda ash, or sodium carbonate (Na_2CO_3). They are reviewed in this article and provide a basis for general discussion of various alkaline exposures.

Caustic Soda—Sodium Hydroxide

Caustic soda is one of the three most prominent products of the chemical industry, the other two being sulfuric acid and soda ash. By the 1990s, more than 13 million tons were used annually in the United States alone. Major West European caustic soda capacity in 2003 was estimated to be 11.3 million tons/yr (Ref 1). The annual worldwide production of caustic soda is of the order of 45 million tons (Ref 2). As of July 2002, more than 500 companies produced chlor-alkali (a general term to cover the coproduction of chlorine and caustic soda) at more than 650 sites worldwide, with a total annual capacity of more than 51 million metric tons of chlorine. About half of all plants are located in Asia, but many of these are relatively small (Ref 3).

Caustic soda is the most important commercial caustic chemical and is used in a variety of processes, for example, plastics (notably polyvinyl chloride, PVC), pulp and paper, soap, glass, aluminum, and so forth. It is also used for acid waste neutralization, although soda ash is equally effective and less expensive. In waste-management programs, caustic soda is observed to be about 100 times more soluble than lime, a less expensive substitute for such applications.

There are a number of processes that can be used to make caustic soda, but the majority is produced as a by-product of the electrolysis of brine to produce chlorine. The three major electrolysis processes used in the manufacture of chlorine and caustic soda are:

- Diaphragm cell
- Mercury cell
- Membrane cell

These three processes produce caustic soda of differing strengths and purities, and most of the diaphragm and membrane cell caustic must be further purified and then concentrated by evaporation.

Caustic potash or potassium hydroxide is an analogous chemical, sometimes used as a substitute in papermaking processes, but it is more expensive and of lesser commercial importance. The term "caustic" is often used to describe both caustic soda and caustic potash.

Aluminum and Aluminum Alloys

Aluminum is rapidly attacked by even dilute solutions of caustic soda at all temperatures. The aluminum ion (Al^{3+}) is readily complexed by hydroxyl ions (OH^-). Dilute solutions ($<1\ N$) can be inhibited by saturating with potassium dichromate. Aluminum should not normally be considered for service above about pH 8.5 (Ref 4). The corrosion of aluminum in caustic is controlled by the competing processes of film growth and dissolution. The film formed consists of an inner compact layer and an outer crystalline one. At pH around 9 or less corrosion rate is low, but at higher pH cavities form in the outer layer permitting access of the fluid to the surface and increasing corrosion rate. At pH 12 these cavities are more prominent and severe localized corrosion occurs. The corrosive effect of increasing solution pH for aluminum at different temperatures is clearly seen in Fig. 1 (Ref 5). The addition of 1000 ppm chloride ions decreased the corrosion rate at 60 °C (140 °F), but had no appreciable effect at 30 °C (85 °F).

Impurities present in NaOH can have a strong effect on the corrosion of aluminum. Impurities such as Fe^{2+} can be reduced on the metal surface forming preferential sites for hydrogen evolution and greatly increasing corrosion current. Aluminates in solution contribute to a slight decrease in corrosion current (Ref 6).

Iron and Steel

Ferrous alloys are very commonly used in caustic soda service, provided iron contamination is not objectionable and provided certain restrictions are imposed on service conditions to avoid stress-corrosion cracking (SCC). At ambient temperatures, iron and steel are protected by a passive layer of magnetite. Magnetite is the least soluble iron oxide in alkaline solutions, so corrosion is largely prevented under these conditions (Ref 7). At elevated temperature, magnetite has been found to be either protective or nonprotective depending on the growth conditions. A protective film grows at the metal/oxide interface with "excess" iron ions forming soluble ferroates while the nonprotective magnetite film becomes highly stressed as it grows at the oxide/solution interface. The breakdown of the protective film also occurs at more elevated temperatures and under turbulent or erosive conditions (Ref 8).

Cast Irons. Gray cast iron is resistant to caustic but is usually not used because of safety

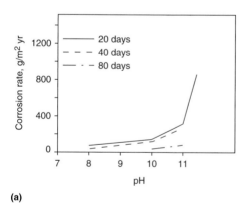

(a)

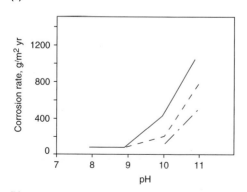

(b)

Fig. 1 Effect of pH on the corrosion rate of aluminum in NaOH at 30 °C (85 °F) (a) and 60 °C (140 °F) (b). Source: Ref 5

problems associated with its brittle nature. At one time, caustic soda solutions were concentrated in cast iron evaporators.

Hot alkalis at >30% concentration attack unalloyed irons. Temperature should be <80 °C (<175 °F) in concentrations up to 70% if the corrosion rate is not to exceed 0.25 mm/yr (10 mils/yr). The corrosion rate of ductile cast iron is similar to that of gray cast iron, but ductile iron can be susceptible to cracking in highly alkaline solutions while gray cast iron is not (Ref 9). Ductile cast iron is, however, sometimes used for specific items, such as valves or pumps.

The addition of nickel greatly reduces the corrosion rate of cast iron in boiling 50 to 65% caustic, as shown in Table 1 (Ref 10). This was an 81 day test under vacuum of 660 mm (26 in.) mercury.

The austenitic nickel cast irons of greater than about 15% Ni content, for example, NiResist (International Nickel Co.) types 1 (F41000), 2 (F41002) and the ductile NiResist type D2 (F43000) are much more resistant than unalloyed cast iron in caustic solutions up to about 70%. Corrosion rate should be <0.25 mm/yr (<10 mils/yr) in solutions up to 70% NaOH at temperatures approaching boiling (Ref 9). The resistance is roughly proportional to the nickel content unless sulfur or sulfur compounds are present. The nickel-containing cast irons may be susceptible to SCC, especially in the presence of high chlorides, so it is considered a reasonable precaution to stress relieve these alloys before use in hot caustic soda solutions. Detailed corrosion data for NiResist is compared with that of unalloyed cast iron in Table 2 (Ref 11).

The resistance of unalloyed and alloyed cast irons in molten caustic soda is shown in Table 3 (Ref 12). These data show that nickel has superior resistance to all these cast irons, including the highly alloyed ones.

High-silicon cast irons have good resistance to relatively dilute caustic soda solutions at moderate temperatures. This type of iron, for example, the 14.5% Si cast iron (F47003) is not resistant at higher strengths of caustic at elevated temperature because NaOH reacts with the siliceous film from which it derives its acid resistance. High-chromium cast irons are also not resistant to strong alkaline solutions.

Carbon and Low-Alloy Steels. Some early data on the corrosion of steel in ambient temperature is shown in Table 4. These data show that corrosion rate decreases with increase of caustic concentration at room temperature (Ref 13). At ambient temperature, steels are only slightly attacked by caustic soda with solution strength having little effect on rate. Unalloyed and low-alloy steel corrode at <0.005 mm/yr (<0.2 mil/yr) in up to 20% NaOH at ambient temperature. In stronger solutions, 20 to 50%, the corrosion rate will be <0.01 mm/yr (<0.4 mil/yr), and the steel is still usually protected by the presence of a passive layer of magnetite (Ref 7). At higher strengths and temperatures, this oxide layer no longer provides effective protection and corrosion rates increase.

The stated limits for the use of carbon and low-alloy steels in caustic soda solution at temperatures above ambient vary widely. Typical suggested limits on the basis of metal loss by corrosion are 50% concentration at temperature up to 85 to 90 °C (185 to 195 °F) (Ref 14) and 50% at up to 60 °C (140 °F) (Ref 15). One source says that the corrosion rate in 50% will be <0.025 mm/yr (<1 mil/yr) at 40 °C (105 °F), 0.13 mm/yr (5 mils/yr) at 60 °C (140 °F), and 0.2 mm/yr (8 mils/yr) at 55 to 75 °C (130 to 165 °F) (Ref 16), while another source says that the corrosion rate in 50% will be 0.06 mm/yr (2.5 mils/yr) or less at 21 to 38 °C (70 to 100 °F) (Ref 15). Some of the discrepancy between various stated safe limits is caused by other factors such as purity of the caustic soda, velocity, oxygen, concentration effects leading to caustic SCC, and so forth. In other strengths of caustic there is less discrepancy, and typically carbon steel is said to be satisfactory up to 70% at up to 80 °C (175 °F) (Ref 17) and in 75% at up to 100 °C (212 °F), assuming iron contamination is acceptable (Ref 18).

These limits of operating conditions are based on reasonable rates of metal loss and assume that iron contamination of the solution is acceptable. If this is not the case, then a more resistant material must be employed or the vessel must be lined. Storage tanks are often lined with neoprene, phenolic-epoxy, or other resistant coating or lining. In the absence of concerns about iron contamination, bare steel tanks are used effectively to store caustic solutions up to about 50% concentration and up to about 65 °C (150 °F) (Ref 9).

If steel is under tensile stress, as from welding or cold work (e.g., field-flaring of pipe flanges) and is exposed to caustic soda, caustic SCC is

Table 2 Corrosion data for NiResist and unalloyed cast iron in caustic soda

| | | Temperature | | Average corrosion rate | | | |
| | | | | of cast iron | | of NiResist | |
Medium	Location	°C	°F	mm/yr	mils/yr	mm/yr	mils/yr
50% caustic soda plus suspended salt	Salt tank	82	180	0.16	6.3	<0.01	0.01
50% caustic soda	Distributor box to settler	Hot	Hot	0.51	20	0.10	4
50% caustic soda	Evaporator	Hot	Hot	0.76	30	0.15	6
50% caustic soda	High-concentration evaporator	Hot	Hot	1.02	40	0.51	20
Anhydrous sodium hydroxide	Flaker pan	371	700	12.95	510	0.33	13
75% caustic soda	Storage tank after vacuum evaporator	135	275	1.78	70	0.10	4
NaOH and KOH each 90%	Flaker pan	371	700	12.7	500	0.33	13
Caustic soda and dissolved silicates	Drop kettle in metal cleaner manufacture	49	120	0.76	30	0.13	5

Source: Ref 11

Table 1 Effect of nickel additions on corrosion of cast irons in boiling 50–65% NaOH

	Corrosion rate	
Nickel, %	mm/yr	mils/yr
0	1.9	73
0	2.3	91
0	2.2	86
3.5	1.2	47
5	1.24	49
15	0.8	30
20	0.08	3.3
20 (plus 2% Cr)	0.15	6
30	0.01	0.4

Source: Ref 10

Table 3 Corrosion of cast irons by molten NaOH at 510 °C (950 °F)

| | Corrosion rate | | Pit depth | |
Material	mm/yr	mils/yr	mm	mils
Gray iron	2.5–3.4	97–135	0.13	5
Ductile iron	5.3	207	...	...
White iron	3.8	151	0.5	20
3% Ni-Fe	1.8	71	...	...
Austenitic, type 1	15.9	628	1.5	60
Austenitic, type 2	24.2	954	1.8	70
Ductile austenitic, type 2	11.8	466	1.5	60
Austenitic, type 3	2.2	87	0	0
Austenitic, type 4	13.6	534	1.0	40
Wrought nickel	0.23	9	...	...

Source: Ref 12

Table 4 Corrosion of steel in NaOH solutions (22 day laboratory test at room temperature)

	Corrosion rate	
Concentration of NaOH, g/L	mm/yr	mils/yr
0	0.05	2
0.001	0.05	2
0.01	0.05	2
0.1	0.05	2
1.0	0.018	0.7
10	0	0
100	0.0025	0.1
540	0	0

Source: Ref 13

possible. Under these conditions a safe operating limit might be up to 50% caustic at up to 65 °C (150 °F), although cracking has occurred at temperatures as low as 48 °C (118 °F) (Ref 19). A survey of field experience was used to produce a curve that indicated temperatures and caustic strengths above which cracking was possible. These data were thought to be conservative but were based on many years of practical experience (Ref 20). Another curve depicting the parameters of caustic concentration and service temperature above which caustic cracking may be a problem was produced from short-term (up to 62 days) laboratory tests (Fig. 2) (Ref 21). This curve is also thought to be conservative but is widely used and reproduced in various forms. Failures that occasionally occur in the safe zone are likely to be associated with other factors such as localized overheating or contaminants present in the caustic.

Figure 3 shows the suggested regions, as defined by temperature and concentration, in which postfabrication thermal stress relief (or an alternative alloy) is recommended to avoid SCC in steels (Ref 22).

An iron alloy with 3% Ni was found to be immune to caustic cracking, and the addition of 0.5% Mo had little effect on that SCC resistance. However, segregation of phosphorus at grain boundaries significantly reduced resistance. Similarly, the presence of carbon or carbides increases the risk of this type of intergranular SCC. These conclusions are probably applicable to other low-alloy steels (Ref 23).

Stainless Steels

All of the standard stainless steels are resistant to general corrosion by all concentrations of caustic soda up to about 65 °C (150 °F). If low levels of iron are required in the product then more resistant alloys or coatings may be needed. The resistance to general corrosion in caustic is almost directly proportional to nickel content. Stainless steels, however, are subject to SCC at certain concentrations and temperatures.

Ferritic Grades. The standard ferritic grades of stainless steel such as types 409 through 430 (S40900 to S43000) and 434 and 444 (S43400 and S44400) contain 11 to 17% Cr with carbon and nitrogen levels kept low to avoid embrittlement. These standard ferritic grades are not generally as resistant in caustic as the standard austenitic stainless steels. For example, in 60% NaOH at 100 °C (212 °F) type 444 corroded at 0.61 mm/yr (24 mils/yr) while type 304 austenitic had a corrosion rate of 0.07 mm/yr (3.0 mils/yr) (Ref 24). Some of the standard ferritic grades are compared with the standard austenitic stainless steels in Table 5 (Ref 12).

There are now much more highly alloyed and corrosion-resistant ferritic stainless steels, commonly known as superferritics. The first of these superferritic stainless steels was based on 26% Cr, 1% Mo (S44627), and the niobium-stabilized XM-27 (S44627). These were developed to provide better resistance to chloride SCC than the austenitic 300 series. These ferritic steels have low interstitial content with high chromium and very low carbon levels. Other superferritic stainless steels are based on a 29% Cr, 4% Mo alloy, and they need low C + N levels, that is, less than 0.025%, to avoid intergranular corrosion caused by chromium depletion from precipitation of carbides and nitrides. Some of the current ferritics contain higher levels of C + N and have additions of titanium or niobium as carbon/nitrogen stabilizers. Superferritic steels include AL 29-4C (Allegheny Ludlum) (S44735), AL 29-4-2 (Allegheny Ludlum) (S44800), Sea-Cure (Crucible, Inc.) (S44660), E-Brite (Allegheny Ludlum) (S44627), and Monit (Uddeholm, now Outokumpu Stainless) (S44635).

While standard austenitic stainless steels show high rates in boiling 50% caustic, the higher alloy 20 and the superferritic grades tend to be resistant, as shown in Table 6 (Ref 25) and Table 7 (Ref 26–29). The corrosion rates in all

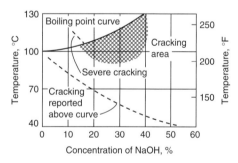

Fig. 2 Temperature and concentration of caustic soda that can cause SCC of carbon steels. Source: Ref 21

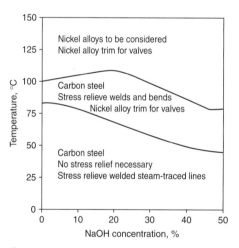

Fig. 3 Temperature and concentrations of caustic soda that require stress relief to prevent SCC of carbon steel. Source: Ref 22

Table 6 Corrosion rates of stainless steels in 50% NaOH at 140 °C (290 °F)

	Corrosion rate	
Type	mm/yr	mils/yr
304 (S30400)	>4.57	>180
316 (S31600)	>3.05	>120
439 (S43035)	>7.62	>300
444 (S44400)	>7.37	>290
XM-27 (S44627)	<0.025	<1
29-4C (S44735)	<0.025	<1
29-4-2 (S44800)	<0.025	<1

Source: Ref 25

Table 5 Corrosion of stainless steel in NaOH solutions

	Concentration of NaOH, %	Temperature		Test duration, days	Corrosion rate	
Type		°C	°F		mm/yr	mils/yr
302	20	50–60	120–140	134	<0.0025	<0.1
304	22	50–60	120–140	133	<0.0025	<0.1
309	20	50–60	120–140	134	<0.0025	<0.1
310	20	50–60	120–140	134	<0.0025	<0.1
410	20	50–60	120–140	134	0.0025	0.1
430	20	50–60	120–140	134	0.0025	0.1
304	72(a)	120–125	245–255	119	0.09	3.7
316	72(a)	120–125	245–255	119	0.08	3.1
329	72(a)	120–125	245–255	119	0.0025	0.1
21Cr-4Ni-0.5Cu	72(a)	120–125	245–255	119	0.15	6
410	72(a)	120–125	245–255	119	0.8	32
302	73(b)	100–120	210–245	88	0.97	38
304	73(b)	100–120	212–245	88	1.1	45

(a) Solution moderately agitated (b) No aeration. Source: Ref 12

Table 7 Corrosion rates of austenitic and ferritic stainless steels in boiling 50% NaOH solution

	Corrosion rate			
	of base metal		of weld	
Alloy	mm/yr	mils/yr	mm/yr	mils/yr
304 (S30400)	3.0	118	3.3	130
304L (S30403)	1.8	71	2.2	87
316 (S31600)	3.1	123.6	3.5	136.8
316L (S31603)	1.97	77.6	2.17	85.4
317L (S31703)	0.83	32.8	0.81	31.9
20 (N08020)	0.18	7.2	0.15	6.0
AL 29-4C (S44735)	0.01	0.4	...	...
AL 29-4-2 (S44800)	<0.1	0.1	...	...
E-Brite (S44627)	0.003	0.11	...	...

Source: Ref 26–29

cases are unacceptably high, except for alloy 20 (N08020) and the superferritic stainless steels. In the standard 300 series stainless steels the weld is less resistant than the base metal and the low-carbon L grade more resistant than the normal carbon grades.

A high-purity 30% Cr, 2% Mo (Shomac 30-2) alloy was studied in hot, concentrated caustic and was found to be resistant to caustic strengths up to 50% at temperatures below 120 °C (250 °F) under isothermal conditions. In more concentrated solutions at higher temperatures most specimens suffered from intergranular attack (IGA). Under heat-transfer conditions, corrosion was more severe and IGA was seen even at lower caustic concentrations. The cause of IGA was found to be carbide or nitride precipitation at grain boundaries leading to localized chromium depletion. This mechanism was analogous to that occurring in austenitic stainless steels so the same test methods for susceptibility can be applied. Galvanic coupling to more active metals prevented IGA, but potential must be strictly controlled to avoid active dissolution (Ref 30).

Ferritic alloys E-Brite (S44627), 2¼Cr-1Mo (K21950), type 446 (S44600), and 26-1S (S44626) were included in tests to determine corrosion and caustic SCC in high-temperature NaOH solutions (Ref 31). It was found that in deaerated 50% NaOH at temperatures in the range 284 to 332 °C (543 to 630 °F) these ferritic alloys were severely corroded and their susceptibility to caustic SCC was dependent on their heat treatment.

Superferritic alloys are sometimes used for heat-exchanger tubing. They have also been successfully used in caustic evaporators, but require a sustained supply of oxidizing contaminants, such as chlorates, to maintain passivity at these higher concentrations and temperatures. The high-purity 26-1 grade (S44626) containing niobium and molybdenum is useful up to about 175 °C (350 °F) in caustic evaporators, depending on the caustic, chloride, and other contaminant concentration (Ref 9). This type of ferritic stainless steel has been generally successful in evaporators, but failures have occurred from either localized or general corrosion. These failures have been associated with one or more of the following factors (Ref 14):

- Increase in carbon levels in the alloy due to pickup from oil, grease, or other hydrocarbon contamination during tube preparation
- Temperature in the first-effect evaporator >150 °C (>300 °F)
- High local tube temperatures caused by blockages with insoluble salts

It has been shown that environmentally assisted cracking (EAC) in caustic is related to hydrogen formation so that conditions that encourage hydrogen permeation also encourage caustic cracking. Type 410 martensitic steel was found to be immune to cracking in concentrated, deaerated caustic soda solution up to 90 °C (195 °F) (Ref 32).

The precipitation-hardening (PH) stainless steels see limited service in caustic environments. These alloys exhibit very high strengths combined with good notch toughness and corrosion resistance. For this reason, they are often the preferred material for such parts as valve stems and certain critical fasteners. Some of these alloys are susceptible to hydrogen embrittlement in corrosive or hydrogen-rich environments.

Corrosion rates are acceptable in concentrated caustic at moderate temperatures, as shown in Table 8 (Ref 33). At boiling point in 50% caustic this martensitic alloy 17-4PH (S17400) in condition H1075 was severely corroded. Tests were carried out for five 48 h periods.

Duplex stainless steels are not commonly used in caustic soda applications except where their resistance to chloride SCC is useful from the water side of heat exchangers. The modern grades, for example, alloy 2205 (S31803), typically also contain molybdenum to achieve a composition of about 22% Cr, 5.5% Ni, 3% Mo, and 0.03% C max, strengthened and stabilized by nitrogen additions. The mixed austenite-ferrite structure imparts strength and resistance (but not immunity) to chloride pitting and SCC. They have a higher strength than the lower austenitic grades such as type 304 (S30400) but are subject to temper embrittlement at about 475 °C (885 °F).

Corrosion testing in a range of caustic solutions at boiling point showed that duplex alloys can be used in boiling solutions up to at least 30% with negligible corrosion (Fig. 4) (Ref 34). The tests also showed that these duplex alloys were not susceptible to SCC in boiling caustic solutions from 20 to 70%.

Corrosion rates of duplex stainless steels Uranus 50 (Creusot-Loire) (21Cr, 6.6Ni, 2.3Mo) and Ferralium 255 (Langley Alloys) (26Cr, 5.6Ni, and 3.2Mo) in boiling caustic soda up to 30 mass% was less than 0.01 mm/yr (0.39 mil/yr) with uniform corrosion rates. The rate increased with caustic concentration, reaching a maximum at 60%, and in these more concentrated solutions the austenitic phase was slightly more attacked than the ferritic phase. Small flow rates (e.g., 0.08 m/s, or 2.6 ft/s, rotating disc) increased the corrosion rate by a factor of six compared with stagnant conditions. Further increases in flow rate up to 1 m/s (3.3 ft/s) had no further effect on corrosion rate (Ref 35). E-Brite and 7-Mo (Carpenter

Table 8 Corrosion of 17-4PH (S17400) in hot, concentrated caustic

Concentration, %	Temperature		Corrosion rate	
	°C	°F	mm/yr	mils/yr
30	80	176	0.18	7.1
50	80	176	0.10	3.9
30	Boiling	Boiling	0.28	11.0
50	Boiling	Boiling	14.2	559

Source: Ref 33

Technology) stainless steels showed good resistance to SCC and corrosion in 50% caustic at 135 °C (275 °F) (Ref 9).

Welding tends to lead to variations (20 to 80%) in the austenite/ferrite ratio in the as-cast weld bead and fusion zone. To minimize this effect, they are welded with a nickel-rich, overmatching rod whose higher nickel content is beneficial in caustic service.

Recent corrosion testing of various duplex stainless steels in 30 to 70% NaOH solutions found that their resistance was on the order of:

2304 < 2205 < 2507 < S32906 (Safurex)

< alloy X (a proprietary duplex alloy)

Alloy 200 (N02200) had a lower corrosion rate than any of the duplex alloys in pure caustic solutions at any test temperature. Statistical analysis of the results showed that the most important alloying elements to produce resistance to caustic corrosion in the duplex alloys were chromium, nitrogen, and nickel. The effect of chromium in the austenite phase was investigated in a set of experimental duplex alloys. It was found that corrosion resistance was directly proportional to the chromium content. The range of chromium contents was from 26.54 to 29.04% with corrosion rates in boiling 60% NaOH of 0.39 to 0.05 mm/yr (15.4 to 1.97 mils/yr), respectively. Stress-corrosion cracking tests in 50% NaOH at 137 °C (279 °F) showed that S32906 was immune to cracking in this environment (Ref 36).

Because of the uncertainties about the amount of chlorides and oxidizing salts such as hypochlorites and chlorates, extensive field corrosion tests should be made before selecting duplex stainless steels for caustic soda service.

Austenitic Grades. There is little difference in the corrosion resistance of type 304L (S30403) and 316L (S31603) in 50 or 73% caustic solutions. No difference was found between the corrosivity of diaphragm cell caustic and mercury cell caustic in an extensive survey of caustic soda producers (Ref 37).

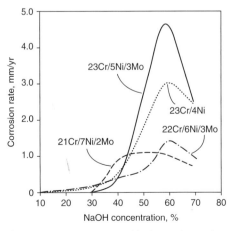

Fig. 4 Corrosion resistance of duplex stainless steels in boiling NaOH solutions. Source: Ref 34

Corrosion of 18-8 steels in hot, concentrated caustic solutions was found to increase under heat-transfer conditions, possibly due to localized evaporation and erosion of the passive film by gas bubbles. The molybdenum-containing type 316 (S31600) was found to be less resistant than type 304 (S30400). Annealed and sensitized specimens were subject to IGA when held in the passive region. When in the active region, attack was general and transgranular (Ref 38).

All austenitic stainless steels resist general corrosion by all concentrations of caustic soda up to about 65 °C (150 °F). Other authorities give higher temperature limits for 50% caustic, but these are probably influenced by other factors such as presence of impurities in the caustic. Type 304 and 316 show low corrosion rates in boiling caustic at concentration up to nearly 20%. Type 316 has a better resistance to pitting than type 304 in caustic solutions, especially if chlorides are present. The low-carbon grades perform marginally better than the high-carbon grades because of their resistance to sensitization. This means that type 316L is a good choice for caustic solutions as long as operating conditions are such that caustic SCC is not a problem (Ref 9). Stress-corrosion cracking of 304 and 316 grades occurs at around 100 °C (212 °F). An isocorrosion curve in mm/yr for these grades is shown in Fig. 5 (Ref 39). This figure also shows the temperature and concentration region where caustic SCC of these steels is likely.

The resistance to caustic SCC in 30% NaOH at 200 °C (390 °F) increased in the order: N08904 (least resistant) < S31603~F138 < S31803 (most resistant). (F138 is a special grade of S31603 with lower levels of inclusions.) In this environment the addition of sulfide (20 g/L $Na_2S \cdot 9H_2O$) dramatically increased the susceptibility of all these alloys to SCC (Ref 40). Slow-strain-rate tests in the same environment determined that the order of resistance for the alloys tested was:

904L > 825 > alloy 28 > alloy 33

Resistance increased as the chromium content of the alloy increased. When sulfide ions were added, alloy 33 (R20033) showed excellent corrosion resistance to pure caustic soda but was very prone to SCC even at very low strain rates (Ref 41). The zone of concentration and temperatures in which SCC is possible for 904L (N80904) is shown in Fig. 6 (Ref 42). This figure also shows the isocorrosion curve at 0.1 mm/yr (4 mils/yr) corrosion rate for this and other stainless steels and for titanium.

Caustic cracking can also be caused by potassium hydroxide. A type 304L (S30403) bypass line in a steam methane reforming unit failed by SCC in KOH that was formed from a potassium-promoted catalyst. Cracking at relatively low temperatures was possible only in the presence of hydrogen and/or carbon monoxide. This study found that NaOH caused cracking at lower temperatures than did KOH (Ref 43).

Cast stainless steels perform well in caustic soda, and they are frequently used in pumps and valves. While corrosion rates are similar to those of their wrought equivalents, the nature of the cast surface means that caustic SCC is not a problem. Cast stainless steels are often used at operating conditions well in excess of those of the wrought version (Ref 9).

The equivalent cast version of types 304 (S30400) and 304L (S30403) are CF-8 (J92600) and CF-3 (J92500), respectively, and they exhibit approximately the same corrosion response as the wrought alloys (Ref 44).

Castings can, however, have surface layers containing more than the maximum allowable carbon content of 0.08%, which can significantly reduce corrosion resistance of the surface. In the cast form, the difference between CF-8 (J92600) or CF-3 (J92500) and the corresponding molybdenum-bearing grades CF-8M (J92900) and CF-3M (J92800) is insignificant. CF-3 or CF-8 are not particularly common, and manufacturers of cast pumps and valves tend to standardize on CF-8M, which has a broader range of applications. The molybdenum-grade castings are often more available and less expensive than the 304 or 304L grade equivalents. Since the cast version of these alloys is unlikely to be welded, there is rarely a justification for specifying the low-carbon grades in this case. This assumes that the valves or pumps, if weld repaired by the manufacturer, are properly reheat-treated (solution annealed) to restore optimum corrosion resistance. Availability and price are likely to favor the non L grade, and a properly heat treated casting in CF-8 or CF-8M is likely to be as corrosion resistant as its low-carbon cast or wrought equivalent.

High-Performance Austenitic Alloys

There is a group of high-performance austenitic alloys, sometimes called corrosion-resistant alloys (CRAs) or superaustenitic alloys, which consist of both stainless steels (i.e., containing >50% Fe) with high chromium and nickel content and alloys with nickel as the predominant alloying element. The latter group are not true nickel-base alloys because they contain <50% Ni. The original high-performance austenitic alloys were the molybdenum-free alloy 800 (20Cr, 30Ni; N08800) and the molybdenum-containing alloy 20Cb3 (Carpenter Technology) (20Cr, 33Ni, 2.5Mo; N08020). Molybdenum is now a common alloying element in this group of superaustenitic alloys that started with the stainless steel 254SMO (20Cr, 18Ni, 6Mo; S31254). The molybdenum in this type of alloy gives resistance to chloride pitting, crevice corrosion, and chloride SCC.

Alloy 800 (N08800) has been used in up to 73% caustic at 120 °C (250 °F), but is susceptible to caustic SCC above 150 °C (300 °F). Alloy 825 (N08825), which has 3% Mo, is slightly more resistant than alloy 800 (N08800). These nickel-chromium-iron alloys with and without molybdenum (e.g., N08800, N08825, N08020) have useful resistance up to 73% caustic at temperatures up to about 120 °C (250 °F). The beneficial effects of nickel and molybdenum in nickel-containing and nickel-base alloys is shown in Fig. 7 (Ref 45). These data show that both these alloying elements improve corrosion resistance in this caustic solution. The effect of molybdenum is minimal above about 3 to 4% Mo.

In studies of austenitic and duplex alloys in 30 wt% NaOH at 150 °C (300 °F), it was found that under deaerated conditions chromium and nickel passivate the alloys and iron is dissolved. Under very oxidizing conditions chromium was dissolved, probably as hexavalent chromium (Cr^{VI}), and nickel and iron form a thicker, less protective oxide film. Caustic SCC cracking occurred only in alloy 800 (N08800) at high anodic potentials where chromium is being dissolved from the oxide (Ref 46).

In oxygenated 50% caustic at 300 °C (570 °F) an increase in nickel content between 15 and

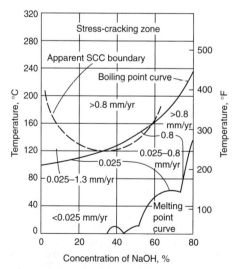

Fig. 5 Isocorrosion curves for type 304 and 316 stainless steels in caustic soda also showing limits of SCC. Source: Ref 39

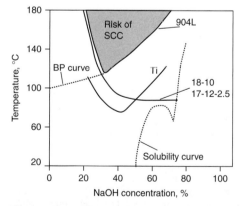

Fig. 6 Isocorrosion curve for a corrosion rate of 0.1 mm/yr (0.4 mil/yr) for 904L (N80904), other stainless steels, and titanium. BP, boiling point; SCC, stress-corrosion cracking. Source: Ref 42

45% was found to improve the caustic SCC resistance of Fe-Ni-Cr alloys containing 10 to 15% Cr, but had little effect on alloys with 20 to 25% Cr (Ref 47). Nickel also has a beneficial effect on the corrosion resistance of iron-base alloys in hot caustic solutions. The higher the nickel content, the lower is the corrosion rate as shown in Fig. 8 (Ref 48). The maximum benefit is reached at about 20% Ni with little further benefit from further increases in nickel into the nickel-base alloys. These data also show that when the caustic was deaerated (with hydrogen) the corrosion rates for all of these metals and alloys were reduced.

Comparative tests of various stainless steels in 50% NaOH are shown in Table 9 (Ref 49). This table shows the lowest temperature at which the corrosion rate exceeds 0.127 mm/yr (5 mils/yr). These same data are plotted for the three alloying elements in Fig. 9. Nickel has a strong beneficial effect on corrosion resistance while there is no apparent benefit of additional molybdenum or chromium in this environment, within the range of compositions represented by these steels.

Data on the corrosion of the 6% Mo austenitic stainless steel, AL-6XN (N08367), and other steels and alloys in boiling 50% NaOH are shown in Table 10 (Ref 50). These data show that none of these materials is really acceptable for this use.

For pumps and valves, a cast version of the "20" type alloy CN 7M (20Cr, 29Ni, 3Mo; N08007) is quite often employed (Ref 51, 52). Corrosion rates are below 0.13 mm/yr (5 mils/yr) for the complete range of NaOH concentration at temperatures up to the atmospheric boiling point.

Alloy 28 (N08028) is another Fe-Ni-Cr alloy that has useful resistance to caustic solutions, particularly if contaminated. This alloy is compared with alloy 800 (N08800) and 904L (N08904) in NaOH with and without chlorides in Table 11 (Ref 53). Tests were carried out for an initial 1 day period, followed by two other test periods of 3 days each. These data show that the presence of chlorides do not significantly increase corrosion in these alloys; in fact, they decrease the corrosion of alloy 28 and alloy 800 in the 43% caustic.

The effect of nickel on the corrosion resistance of Fe-Cr-Ni alloys is shown in Fig. 10 (Ref 54). The results of tests in 28% NaOH with 8% NaCl at 99 °C (210 °F) show that nickel-free and high-nickel alloys are resistant with a maximum corrosion rate at about 5% Ni. The solid data points in this figure relate to alloy 28 (N08028) that has good resistance in this chloride-contaminated environment. This alloy has been used for the evaporation of diaphragm cell NaOH. In this application erosion by sodium chloride crystals can be a problem, and this alloy showed considerably better performance than pure nickel.

Alloy 33 (stellite) (R20033) is an austenitic alloy based on chromium with a nominal analysis of 33% Cr, 32% Fe, 31% Ni, 1.6% Mo, 0.6% Cu, and 0.4% N. Testing in caustic soda has shown that it could be used to replace standard austenitic stainless steels in, for example, pipes or vessels for hot caustic solution where type 316Ti is limited to 90 °C (190 °F). Alloy 33 could be acceptable in 25 and 50% NaOH up to boiling point based on 28 day laboratory tests (Table 12) (Ref 55).

Krupp VDM alloys were exposed in caustic evaporators in various Russian plants and subject to laboratory evaluation. This study concluded that (Ref 56):

- Alloy 201 and its welded joints were resistant to corrosion in evaporation of electrolytic caustic soda (from both mercury and diaphragm cells) up to 60% NaOH at temperatures up to 170 °C (340 °F); increased grain growth (index 2 to 3) was seen in the heat-affected zone (HAZ) of the welded joints

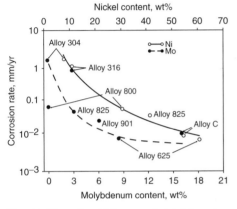

Fig. 7 Effect of molybdenum and nickel in various alloys on corrosion rate in NaOH solutions. Source: Ref 45

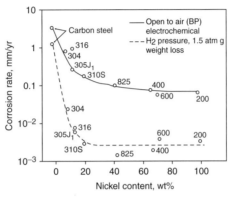

Fig. 8 Effect of nickel content on corrosion in NaOH of various alloys with and without oxygen in the atmosphere. Source: Ref 48

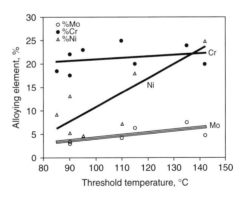

Fig. 9 Effect of molybdenum, nickel, and chromium on the threshold temperature at which the corrosion rate exceeds 0.127 mm/yr (5 mils/yr) in 50% NaOH. Source: Ref 49

Table 9 Lowest temperature at which corrosion rate of stainless steels exceeds 0.127 mm/yr (5 mils/yr) in 50% NaOH

Alloy	Composition, %			Temperature	
	Ni	Cr	Mo	°C	°F
304	9	18.5	...	85	185
316L	13	17.5	2.7	90	194
2205 Code Plus Two	5	22	3.2	90	194
SAF 2304 (S32304)	4.5	23	...	95	203
SAF 2507	7	25	4.0	110	230
254 SMO (S31254)	18	20	6.1	115	239
654 SMO (S32654)	22	24	7.3	135	275
904L (N08904)	25	20	4.5	(a)	(a)

(a) Boiling (~142 °C, or 288 °F). Source: Ref 49

Table 10 Corrosion of various alloys in boiling 50% NaOH

Alloy	Corrosion rate	
	mm/yr	mils/yr
316L	1.92	77.59
317L	0.83	32.78
904L	0.24	9.61
AL-6XN	0.29	11.42
276	0.45	17.77

Source: Ref 50

Table 11 Corrosion rates of various alloys in NaOH and NaOH/NaCl

	Corrosion rate							
	in 28% NaOH at 99 °C (210 °F)		in 28% NaOH/8% NaCl at 99 °C (210 °F)		in 43% NaOH at 135 °C (275 °F)		in 43% NaOH/6.7% NaCl at 135 °C (275 °F)	
Alloy	mm/yr	mils/yr	mm/yr	mils/yr	mm/yr	mils/yr	mm/yr	mils/yr
Alloy 28	0.008	0.31	0.008	0.31	0.074	2.91	0.045	1.77
Alloy 800	0.011	0.43	0.013	0.51	0.397	15.6	0.283	11.1
904L	0.013	0.51	0.018	0.71	0.301	11.9	0.349	13.7

Source: Ref 53

of alloy 201, which may be a contributing factor to the failure of the welded joint of the alloy 201 in diaphragm caustic soda in more severe conditions (65 to 70% NaOH, 180 to 195 °C, or 355 to 385 °F). Intergranular and knifeline (IGA) corrosion was seen in the welded joint, and the weld metal suffered selective corrosion. The overall corrosion rate of the welded joint of alloy 201 was 0.8 mm/yr (31.5 mils/yr).

- The base metal and welded joints of alloy 201 were fairly corrosion resistant when concentrating to more than 60% NaOH of mercury cell caustic soda; alloy 33 and its welded joints possess good resistance under the conditions of evaporation concentration of diaphragm cell caustic soda (up to 65% NaOH), but, as the concentration increased to 70%, the corrosion rate of the welded joint increased to 0.3 mm/yr (11.8 mils/yr) with general corrosion and marked etching of the weld metal; at higher concentrations of the diaphragm cell caustic soda the alloy suffered pitting corrosion.
- Alloy 33 was resistant in concentration from 45 to 60% of mercury cell caustic soda (free from chlorides), but not as resistant as alloy 201 (0.019 and 0.0065 mm/yr, or 0.75 and 0.26 mil/yr, respectively). At caustic soda concentrations that increased from 60 to 97.5%, the corrosion rate of alloy 201 was about three times that of alloy 33 (0.336 and 0.125 mm/yr, or 13.2 and 4.9 mils/yr, respectively). Upon further concentration of the caustic soda to 99.5%, the corrosion rate of alloy 201 was about 20 times higher than that of alloy 33 (0.0485 and 0.0022 mm/yr, or 1.9 and 0.09 mils/yr, respectively).

At temperatures of the order of 300 °C (575 °F), 25% caustic will cause SCC of the CRAs. It is very important when both chlorides and caustic are present in an environment known to cause SCC to know which is the active species. If it is the chloride, a CRA will suffice to overcome the problem; whereas, if caustic is the cause of the cracking then a high-nickel alloy will be needed.

Nickel and Nickel Alloys

The nickel-base alloys are generally preferred for caustic soda service, but are not without their problems. They may be discussed under separate groupings that distinguish between the chromium-free alloys (resistant to reducing corrosive conditions) and the chromium-bearing alloys, which are more resistant under oxidizing conditions. While nickel and nickel alloys can suffer from caustic SCC in very hot, concentrated caustic, alloy 600 is more resistant than unalloyed nickel (Ref 57). Alloy 400, alloy 600, and alloy 690 are all susceptible to SCC over a wide range of caustic concentrations at temperatures above 290 °C (550 °F) (Ref 58).

Chromium-Free Alloys. Commercially pure nickel, alloy 200 (N02200), is one of the best metals for resisting corrosion while simultaneously avoiding unacceptable metal contamination. Alloy 200 has a corrosion rate of <0.005 mm/yr (<0.2 mil/yr) below 50% NaOH at up to boiling temperature. Above 315 °C (600 °F), the low-carbon variant, alloy 201 (N02201), must be used to avoid embrittlement by graphitization and attendant intergranular attack (Ref 59). The mechanical strength of nickel also diminishes at elevated temperatures, encouraging the use of chromium-bearing nickel alloys.

Nickel has corrosion rates of <0.0025 mm/yr (<0.1 mil/yr) up to 73% NaOH at 115 °C (240 °F) and only 0.025 mm/yr (1 mil/yr) at 130 °C (265 °F). Caustic soda is usually produced at 11 to 15% solution and concentrated up to 50% or higher by evaporation in alloy 200 vessels. It has outstanding corrosion resistance to caustic soda at all concentrations up to anhydrous at boiling or molten temperatures (Ref 9).

Nickel is more severely attacked in hot, concentrated caustic under heat-transfer conditions and in the presence of chlorates in solution (Ref 60). The presence of oxidizable sulfur compounds also increases the corrosion of nickel in caustic. The effect is more pronounced with sulfides such as hydrogen sulfide, mercaptans, or sodium sulfide and, to a lesser extent, with partly oxidized compounds such as thiosulfates and sulfites (Ref 59).

When nickel ion contamination must be minimized, cathodic protection may be applied in concentrators and storage tanks, with effective current densities as low as 0.11 A/m^2 (0.01 A/ft^2) (Ref 16). The corrosion rate increases linearly with increasing concentration of chlorate and other oxidizing contaminants.

The behavior of alloy 201 (N02201) is compared with other nickel alloys in molten caustic in Table 13 (Ref 61). It can be seen that nickel is more resistant to corrosion in molten caustic than the more highly alloyed materials even at elevated temperature.

In molten caustic containing 0.5% sodium chloride, 0.5% sodium carbonate, and 0.03% sodium sulfate at 510 °C (950 °F), the corrosion rate of wrought nickel was 0.23 mm/yr (9 mils/yr). This high rate was still much lower than rates measured for other alloys tested (Ref 12). Rates of over 1.27 mm/yr (50 mils/yr) were measured for nickel in laboratory tests in 75% caustic concentrated to the anhydrous product at 480 °C (900 °F). This rate was also much lower than the other alloys tested apart from silver, which corroded at a rate of 0.135 mm/yr (5.3 mils/yr) (Ref 62). In plant exposures at up to 540 °C (1000 °F) in 73% caustic soda, the corrosion rate of nickel was 6.60 mm/yr (260 mils/yr). This unexpectedly high corrosion rate was probably due to sulfur and nitrate contamination (Ref 63).

The least expensive chromium-free, nickel-base alloy is alloy 400 (N04400), which is stronger than nickel itself. It is used in caustic soda service up to about 73% but, having a

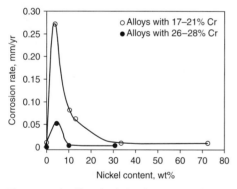

Fig. 10 The effect of nickel on the corrosion of Fe-Ni-Cr alloys in caustic soda. Source: Ref 54

Table 12 Corrosion rates of stainless steels in sodium hydroxide at various conditions

	Corrosion rate													
	in 25% NaOH						in 50% NaOH							
	at 75 °C (167 °F)		at 100 °C (212 °F)		at boiling point, 104 °C (219 °F)		at 75 °C (167 °F)		at 100 °C (212 °F)		at 125 °C (257 °F)		at boiling point 146 °C (295 °F)	
Alloy	mm/yr	mils/yr	mm/yr	mils/yr	mm/yr	mils/yr	mm/yr	mils/yr	mm/yr	mils/yr	mm/yr	mils/yr	mm/yr	mils/yr
316Ti	<0.01	<0.39	0.12	4.72	0.63	24.8	0.08	3.15	0.35	13.8	1.60	63	7.99	315
X1CrNiMoN 25-25-2	<0.01	<0.39	0.03	1.18	0.02	0.79	<0.01	<0.39	<0.01	<0.39	0.26	10.2	1.35	53
Alloy 33	<0.01	<0.39	<0.01	<0.39	<0.01	<0.39	<0.01	<0.39	<0.01	<0.39	<0.01	<0.39	<0.01	<0.39

Source: Ref 55

copper content of around 25 to 35% may be objectionable from the contamination standpoint in some applications where it might be suitable from the corrosion viewpoint. It is less resistant than nickel at concentrations above 73% at the atmospheric boiling point and can suffer SCC in hot caustic. It is subject to SCC at higher temperatures, for example, in 2 to 2.7 MPa (300 to 400 psi) steam contaminated with alkaline boiler-treating chemicals. Also, liquid metal embrittlement (LME) has occurred with alloy 400 (N04400) components handling mercury cell caustic.

Alloy K-500 (N05500), a tougher and more abrasion-resistant variant that undergoes precipitation hardening, has substantially the same corrosion characteristics as alloy 400.

Results of plant tests of various nickel-base alloys and other materials in caustic evaporators are shown in Table 14 (Ref 59). These data show the superiority of nickel over the other alloys, but also show that alloy 400 and alloy 600 have acceptable corrosion resistance.

Nickel-clad steel plate is often used to simultaneously provide the corrosion resistance of nickel and the strength of steel at moderately elevated temperatures. Of course, to prevent corrosion of the substrate and iron contamination, special precautions must be taken in fabricating and welding.

The molybdenum-rich alloy B2 (N10665) has good corrosion resistance in the absence of oxidizing contaminants, but finds no application because of the excellent resistance of nickel itself.

Although electroless nickel plating (ENP) has a corrosion resistance superior to pure nickel 200 (N02200), such coatings are used only to minimize iron contamination since they contain "holidays" that allow the caustic solution to access the steel surface. The resistance of ENP in caustic depends on its phosphorus content. The phosphorus is present since ENP is produced by the autocatalytic reduction of nickel in the presence of sodium hypophosphite. In tests of ENP samples with various phosphorus contents (from 0 to 18.2 at.%) in 50% NaOH at room temperature it was found that (Ref 64):

- All coatings had corrosion rates of ≤ 0.002 mm/yr (≤ 0.08 mil/yr).
- High-phosphorus electroless nickel (HPEN) was less resistant than low phosphorus (LPEN) or medium-phosphorus electroless nickel (MPEN) coatings in hot, concentrated NaOH. The LPEN and MPEN coatings had corrosion rates similar to pure nickel in this environment.
- This inverse effect of phosphorus on corrosion resistance was said to be due to the formation of soluble nickel complexes in the presence of high phosphorus leading to the removal of nickel from the film instead of the formation of the protective nickel hydroxide/nickel oxide film.

Chromium-Bearing Alloys. The Ni-Fe-Cr alloys can be subject to caustic SCC in hot, concentrated caustic. In deaerated 50% NaOH the presence of nickel improves the resistance to SCC. In aerated 50% NaOH both chromium and nickel need to be present to resist caustic SCC. The presence of molybdenum in these alloys was found to have no effect on the cracking resistance in deaerated 50% NaOH. Similarly, no particular heat treatment improved the resistance of these alloys to caustic SCC. The effect of temperature in the range tested, 284 to 332 °C (543 to 630 °F), was found to have only a minor effect on the extent of cracking (Ref 31). The beneficial effect of nickel in weaker deaerated caustic (10%) at 316 °C (601 °F) was found to be less pronounced with alloy 800, alloy 600, and alloy 690, all showing similar cracking resistance (Ref 65).

The most commonly used chromium-bearing nickel alloy is the basic Ni-Cr-Fe alloy 600 (76Ni, 16Cr, 8Fe; N06600). This alloy combines corrosion and heat resistance with excellent mechanical strength and workability. The high nickel content provides resistance to caustic corrosion. It is also resistant to chloride-induced SCC and to corrosion by many organic and inorganic acids and water should equipment be exposed to these environments external to the caustic application.

Alloy 600 (N06600) is the preferred alloy for NaOH service at higher temperatures, especially if sulfur compounds are present. The high chromium content of alloy 600 confers resistance to sulfur compounds and oxidizing environments in caustic. Alloy 600 (N06600) has performed well in alkaline sulfur solutions, such as those encountered in the manufacture of sulfate or kraft paper. However, sulfur contamination of nickel alloys can cause a form of LME-induced cracking at welding temperatures. This is not, in fact, true LME, but a form of intergranular penetration caused by the formation of a low-melting-point nickel/nickel sulfide eutectic. Alloy 600 is more resistant to this type of attack than is alloy 201 (Ref 14).

Table 13 Corrosion rates of nickel and nickel alloys in molten caustic soda

	Corrosion rate							
	at 400 °C (750 °F)		at 500 °C (932 °F)		at 580 °C (1076 °F)		at 680 °C (1256 °F)	
Alloy	mm/yr	mils/yr	mm/yr	mils/yr	mm/yr	mils/yr	mm/yr	mils/yr
Alloy 201 (N02201)	0.022	0.9	0.033	1.3	0.064	2.5	0.96	37.8
Alloy 400 (N04400)	0.046	1.8	0.13	5.1	0.45	17.6	...	...
Alloy 600 (N06600)	0.028	1.1	0.061	2.4	0.13	5.1	1.69	66.4

Source: Ref 61

Table 14 Corrosion rates of nickel and other materials in caustic evaporation plants

	Corrosion rate							
	in 14% NaOH in first effect multieffect evaporator at average temperature 88 °C (190 °F)		in 23% NaOH in tank receiving liquor from evaporator at average temperature 104 °C (220 °F)		in 30–50% NaOH in single effect evaporator at average temperature 81.7 °C (179 °F)		in up to 50% NaOH in evaporator	
Alloy	mm/yr	mils/yr	mm/yr	mils/yr	mm/yr	mils/yr	mm/yr	mils/yr
Alloy 200	<0.001	0.02	0.004	0.16	0.003	0.10	0.003	0.1
Alloy 400	0.001	0.05	0.005	0.20	0.005	0.19	...	...
Alloy 600	<0.001	0.03	0.004	0.17	...	...	0.008	0.3
Mild steel	0.21	8.20	...	...	0.09	3.70	...	...
Cast iron	0.21	8.20	...	...	0.18	7.00	0.56	22
NiResist type 1	0.07	2.90	...	...	...	...	0.10	4
NiResist type 2	...	...	...	...	...	...	0.07	2.6
3% Ni cast iron	...	...	...	...	...	...	0.23	9
Copper	...	...	...	...	0.06	2.30	...	...
75Cu-20Ni-5Zn	...	...	...	...	0.01	0.50	...	...
14% Cr steel	...	...	...	...	0.84	33.0	...	...

Source: Ref 59

Minimal corrosion rates for alloy 600 are found in boiling NaOH up to about 50% concentration, as shown in Fig. 11 in comparison with alloy 201 (N02201) (Ref 66, 67). However, this alloy is subject to caustic SCC above about 190 °C (375 °F) in strong caustic, although short-term U-bend tests showed no cracks after 1 month in various solutions of caustic up to 70% at 184 °C (363 °F).

The possibility of SCC can be lessened if the alloy is fully stress relieved for 1 h at 900 °C (1650 °F) or for 4 h at 790 °C (1450 °F). In tests to assess the effects of heat treatment on SCC resistance, alloy 600 (N06600) and alloy 690 (N06690) were compared in 40% NaOH at 315 °C (600 °F). It was found that sensitized alloy 600 was more resistant to SCC than fully annealed alloy 600. The explanation given is that the chromium carbides present in the sensitized alloy have a more beneficial effect on SCC than the detrimental effect of chromium depletion. Effects of carbides and chromium depletion on IGA were not included in these tests. Resistance to SCC also increased with an increase in grain size and with chromium content (Ref 68).

Alloy 600 is used worldwide as a steam generator tube material in pressurized light water reactor nuclear plants where caustic SCC is also encountered. In this case, the environment is typically 10% NaOH at around 300 °C (570 °F) (Ref 69). Detailed information on the corrosion behavior of alloy 600 in nuclear reactor environments can be found in the articles "Corrosion in Pressurized Water Reactors" and "Effect of Irradiation on Corrosion and Stress-Corrosion Cracking in Light Water Reactors" in this Volume.

A variant of alloy 600 (N06600) is alloy 601 (N06601). This solid-solution-strengthened higher-chromium (23%) alloy was developed primarily for high-temperature applications. However, it too shows excellent corrosion resistance up to 98% caustic because of the high nickel content.

The Ni-Cr-Mo alloys, such as alloy 625 (N06625) and alloy C-276 (N10276) are rarely employed in caustic service because the conventional nickel alloys are usually suitable. The improvement in corrosion resistance afforded by the high molybdenum content is minimal for all practical purposes in this service. Although not usually associated with alkaline service, these more expensive alloys may be usefully employed or encountered in some special applications. In 10% NaOH at 93 °C (199 °F), there was no measurable corrosion on alloy 625 (N06625). In boiling 50% NaOH at 143 °C (289 °F), a corrosion rate of 0.061 mm/yr (2.4 mils/yr) was measured on alloy 625, while on alloy C-276 the corrosion rate was 0.452 mm/yr (17.8 mils/yr) (Ref 61). The other chromium-containing nickel-base alloy that is sometimes used in caustic is alloy 690 (N06690).

The corrosion rate of VDM alloy 6030 (N06690) in 70% NaOH at 170 °C (340 °F) was 0.03 mm/yr (1.18 mils/yr). In the same tests alloy 400 (N04400) and alloy 33 (R20033) had the same corrosion rate, that is, 0.03 mm/yr (1.2 mils/yr), while alloy 59 (N06059) corroded at 0.48 mm/yr (18.9 mils/yr) and alloy C-22 (N06022) at 0.51 mm/yr (20.1 mils/yr) (Ref 70).

Clamps made from a proprietary nickel-base alloy with 16% Mo, 15% Cr, and W, holding internal steam coils in a reactor, failed by caustic SCC after three years service. The reactor environment consisted of 20% organics, 8% NaCl, 4% NaOH, balance water and the steam in the coils was at 180 °C (355 °F). It was postulated that the caustic must have become more concentrated under the clamps. An extensive investigation found that while the 22% Cr and 25% Cr duplex stainless steels were not immune to caustic SCC, they were much more resistant than the original nickel-base alloy. The failed clamps were replaced with a duplex stainless steel that is still in service after more than 12 years (Ref 71).

Copper and Copper Alloys

There is very little published corrosion data on copper alloys in caustic soda. This is because the major industries that use large amounts of caustic find copper contamination intolerable. For example, copper causes rancidity in soap and discoloration in rayon.

Copper, gunmetal (C90550), and bronze have limited resistance to caustic soda and potash as do copper-nickel alloys and high-tin aluminum bronze (Ref 72). High-zinc brasses suffer dezincification and should be avoided. However, conventional bronze castings, such as the 85Cu-5Sn-5Zn-5Pb (C83600) can be used for valves or pumps for 25% caustic solutions, for example. Bronzes solution annealed at 850 °C (1560 °F) can show SCC at pH 12.3 (about 500 ppm NaOH).

Corrosion resistance does increase with nickel content and the 90-10 cupronickel (C70600) and 70-30 grade (C71500) are sometimes used for caustic service. Because of improved heat-transfer properties and strength, the 70-30 (C71500) grade has been used in evaporators up to 50% concentration in applications in which copper contamination is acceptable.

Copper piping has sometimes been used for caustic soda solutions in situations in which thermal stress relief of steel piping was impractical (Ref 17). In the absence of oxidizing agents (e.g., chlorites and chlorates), copper may be used up to 73% NaOH to 100 °C (212 °F). In molten caustic, however, copper is much less resistant than iron or nickel, being attacked at about 12.7 mm/yr (500 mils/yr) as compared with 0.51 mm/yr (20 mils/yr) and 0.18 mm/yr (7 mils/yr), respectively, for the alternative materials.

Titanium and Titanium Alloys

Titanium (e.g., grade 2, R50400) is a reactive metal that forms a thin, tenacious, self-healing oxide film in oxygen-containing environments. It resists alkaline media, including caustic solutions at subboiling conditions, but is not recommended for boiling concentrated solutions. Temperature and concentration limits for titanium are 80 °C (175 °F) and 50% NaOH. Titanium is at "high" risk for LME in caustic contaminated by mercury. Titanium is also rapidly attacked by hot caustic containing powerful oxidizing agents, for example, permanganate in pulp and paper applications.

Titanium and titanium alloys experience excessive pickup of hydrogen from corrosion reactions, causing hydride formation and possible hydrogen embrittlement in NaOH above 80 °C (175 °F) at pH of 12 or more (>400 ppm NaOH) (Ref 9). Corrosion of titanium increases rapidly as temperature increases, for example, the corrosion rate is 0.18 mm/yr (7 mils/yr) in 73% NaOH at 130 °C (265 °F), but is >1 mm/yr (>39.4 mils/yr) at 190 °C (375 °F).

Titanium finds application (subject to the limitations noted) chiefly where the surrounding media external to the caustic vessel or assembly contains impurities such as chloride, chlorate, hypochlorite, and wet chlorine. For example, titanium was not corroded in 10% NaOH + 15% NaCl at 82 °C (180 °F) or in 60% NaOH + 2% sodium hypochlorite + trace ammonia. In 50% NaOH containing free chlorine at 38 °C (100 °F), the corrosion rate was 0.023 mm/yr (0.9 mil/yr). The corrosion behavior of commercially pure titanium in various caustic solutions is shown in Table 15 (Ref 73, 74).

In a highly alkaline mixture of 52% NaOH and 16% NH$_3$ titanium corroded at >0.9 mm/yr (>35.4 mils/yr) in long-term tests at about 140 °C (285 °F). In 73% NaOH without ammonia the corrosion rate of titanium was only about 0.18 mm/yr (7.1 mils/yr) at about 130 °C (265 °F) (Ref 75).

In tests to assess the resistance of titanium and other alloys to caustic SCC at elevated

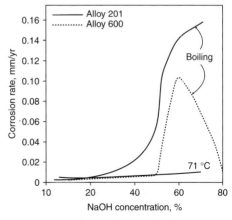

Fig. 11 Isocorrosion curves for alloy 600 (N06600) and alloy 201 (N02201) in caustic soda. Source: Ref 66, 67

temperature, it was found that the corrosion rate was too high to detect discrete cracks in the specimens. These tests were carried out in 50% NaOH at temperatures in the range of 284 to 332 °C (543 to 630 °F) (Ref 31).

Zirconium and Zirconium Alloys

Zirconium (e.g., R60702) is a reactive metal, forming a very resilient oxide film (zirconia) that provides resistance to almost all alkalis, fused or in solution. The oxide film is self-healing and is protective up to 300 °C (570 °F). Zirconium has outstanding resistance to sodium hydroxide, tending to mirror the performance of nickel. Zirconium is considered useful in 73% NaOH at up to 138 °C (280 °F). Like titanium, zirconium is considered at high risk for LME in caustic contaminated with mercury. Zirconium is not attacked by oxidizing media unless chlorides are present (Ref 16).

Zirconium is very resistant to molten sodium hydroxide at temperatures above 1000 °C (1830 °F). It is resistant to SCC in boiling caustic solutions (Ref 9). Corrosion rates of zirconium Zr702 (R60702) in pure and contaminated caustic soda show that it is resistant to all conditions except very hot, very strong solutions (Table 16) (Ref 76).

Cast pumps and valves utilize the benefits of zirconium where service conditions are too severe for stainless steels, nickel alloys, or titanium.

Other Metals and Alloys

The common nonferrous metals, such as zinc, tin, and lead, are amphoteric and find no application in caustic service.

In ambient aqueous alkaline solutions, niobium has corrosion rates of less than 0.025 mm/yr (1 mil/yr). At higher temperatures, even though the corrosion rate does not seem excessive, niobium is embrittled even at low concentrations (5%) of sodium hydroxide and potassium hydroxide. Like tantalum, niobium is embrittled in salts that hydrolyze to form alkaline solutions. These salts include sodium and potassium carbonates and phosphates (Ref 77).

Tantalum is destroyed in caustic solutions by the formation of successive surface oxide layers. The rate of attack increases with increasing concentration or temperature, and tantalum is dissolved in molten alkalis. Tantalum has, however, been used in dilute alkaline solutions (pH 10) in a paper mill (Ref 78). Ta-10W alloy showed passive behavior in 5, 10, and 15% NaOH at 50 °C (120 °F) and in 5% NaOH at 75 °C (165 °F). This Ta-10W alloy was found to be slightly more corrosion resistant than tantalum when both materials were in the passive state but less resistant under active corrosion conditions. The Ta-10W alloy can be used in caustic solutions only over a very limited range of concentrations at low temperatures (Ref 79).

Silver is probably the most resistant metal in caustic soda and is not attacked by sodium hydroxides at temperatures below 500 °C (930 °F). At one time it was used in the evaporation of anhydrous caustic.

Nonmetallic Materials

Elastomers. Some suggested temperature limits for rubbers and elastomers for immersion service in caustic soda solutions are given in Table 17. These data are a compilation from various sources.

Plastics. There are areas of use in which both thermoplastic and thermosetting resins find application in caustic service. They also find some application as immersion coatings, for example, modified epoxy-phenolics.

From a practical viewpoint, the majority of polymeric materials can handle caustic soda up to about 80 °C (175 °F). Careful attention needs to be given to the coating techniques or to fabrication involving polymeric materials. Factory-applied systems can usually be expected to be more reliable than those applied on site. Preparation of vessels to be coated should be carried out by competent companies to a recognized standard, such as NACE RP0178, "Fabrication Details, Surface Finish Requirements, and Proper Design Considerations for Tanks and Vessels to Be Lined for Immersion Service." Transport of coated components should be in accordance with the supplier's directives.

Some plastics, for example, hydrolyzable esters, are easily permeated by dilute caustic and are therefore more resistant in more concentrated (40 to 50% NaOH) solutions.

Thermoplastics. Polyolefins are inherently resistant to caustic, within their temperature limits. High-density polyethylene (HDPE) and polypropylene (PP) generally show better chemical resistance than low density PE. The HDPE

Table 15 Corrosion of commercially pure titanium in various solutions of NaOH

Concentration, %	Temperature °C	Temperature °F	Corrosion rate mm/yr	Corrosion rate mils/yr
5–10	21	70	0.001	0.04
10	Boiling	Boiling	0.02	0.79
28	Boiling	Boiling	0.0025	0.10
40	RT	RT	0.13	5.1
	93	199	0.064	2.5
	121	250	0.127	5.0
50	38–57	100–135	0.013	0.5
	66	151	0.018	0.7
	80	176	0.0025–0.013	0.1–5.0
	121	250	0.033	1.3
50–73	190	374	>43	>1.09
73	60	140	0.18	7.1
	110	230	0.051	2.0
	116	241	0.127	5.0
	129	264	0.178	7.0
75	121	250	0.033	1.3
Saturated	RT	RT	0	0

RT, room temperature. Source: Ref 73, 74

Table 16 Corrosion rates of zirconium (R60702) in sodium hydroxide solutions and mixtures

Environment	Temperature °C	Temperature °F	Corrosion rate mm/yr	Corrosion rate mpy
5–10% NaOH	21	70	<0.025	<1
28% NaOH	RT	RT	<0.025	<1
10–25% NaOH	B	B	<0.025	<1
40% NaOH	100	212	<0.025	<1
50% NaOH	38–57	100–135	<0.025	<1
50–73% NaOH	188	370	0.508–1.27	20–50
73% NaOH	110–129	230–264	<0.05	<2
73% NaOH to anhydrous	212–538	414–1000	0.508–1.27	20–50
9–11% NaOH, 15% NaCl	82	180	<0.025	<1
10% NaOH, 10% NaCl, plus wet cobalt chloride	10–32	50–90	<0.025	<1
0.6% NaOH, 2% NaCl, trace of NH_3	129	264	<0.025	<1
7% NaOH, 53% NaCl, 7% $NaClO_3$, 80–100 ppm NH_3	191	376	<0.025	<1
52% NaOH, 16% NH_3	138	280	<0.127	<5
20% NaOH with suspended salts, violent boiling	60	140	0.254–0.508	10–20
50% NaOH, 750 ppm free Cl_2	38–57	100–135	<0.025	<1

RT, room temperature; B, boiling. Source: Ref 76

resists all concentrations of NaOH to 60 °C (140 °F). It should be noted that some copolymers are subject to environmental stress cracking (ESC) and that others, for example, PE/vinyl acetate (PVA) copolymers, are nonresistant because the PVA is subject to saponification. ASTM D 1693-05 provides a standard test method for evaluating ESC of ethylene plastics.

Generally, PVC, in rigid form, resists alkaline solutions to about 50 °C (122 °F). PVC-lined fiberglass-reinforced plastic (FRP) piping is used for caustic service, but unsupported PVC is not safe. For safety, the chlorinated polymer CPVC, which can tolerate 10 to 70% caustic at up to about 80 °C (175 °F) is often preferred. This limit for CPVC is not universally accepted, and some suppliers suggest much lower limits. Similarly, there are a range of limits suggested for polyvinylidene fluoride (PVDF) and some other polymers. When using any polymeric materials, it is advisable to consult the supplier and best to test the polymer under actual service conditions.

Table 18 shows some suggested temperature limits for thermoplastic materials in caustic service, as a compilation of several sources. Use of many of these materials is confined to linings for pipe and vessels, rather than as solid construction items, because of their lack of strength and the hazardous nature of caustic soda. Fluoroplastic materials are to be preferred, although PP-lined steel pipe has also been used to handle 50% NaOH at 90 °C (195 °F). Recommended temperature limits for plastic-lined steel pipes are given in Table 19 (Ref 80).

Thermoplastics are also used as the resistant liner in dual-laminate construction in which FRP is used as the reinforcing, structural element. The corrosion resistance of the FRP depends on the resistance of the thermoplastic liner, although resistant resins are often used in the construction of the FRP reinforcement in case of permeation and leaks in the thermoplastic liner. Most common thermoplastics are used in this type of construction, and many of them are suitable for use in caustic applications (Table 20) (Ref 81).

Thermosetting Resin Materials. The thermosetting resins are used almost entirely in reinforced construction, with the exception of modified epoxy coatings. Fiberglass-reinforced plastic is used for piping and equipment in caustic service, but not usually for high-concentration, high-temperature service (Ref 16, 82).

Such applications are contradictory to some extent in that the glass reinforcement is subject to attack by even quite weak alkalis. The use of FRP is effective because of the incorporation of special interior surface treatment, that is, the use of synthetic surfacing veils (polyesters and acrylics are common). The selection and protection of a highly resistant inner surface (the corrosion barrier) is critical for satisfactory service. The temperature limitations vary with the type of resin employed, and it should be noted that lower NaOH concentration applications may

Table 17 Rubber/elastomers maximum temperature in various strength solutions of NaOH

| Elastomer | Maximum temperature ||||||||
| | in 10% NaOH || in 30% NaOH || in 50% NaOH || in 70% NaOH ||
	°C	°F	°C	°F	°C	°F	°C	°F
Butyl rubber, grade 1	100	212	100	212	100	212	100	212
Hard natural rubber	100	212	100	212	100	212	100	212
Soft natural rubber	60	140	60	140	60	140	60	140
Hypalon (chlorsulfonated polyethylene)	120	248	120	248	120	248	120	248
Nitrile-butadiene rubber	90	194	90	194	90	194	70	158
Butadiene	60	140	60	140	60	140	60	140
Silicone	25	77	25	77	25	77	25	77
Polysulfide	25	77	25	77	25	77	25	77

Table 18 Thermoplastics maximum temperature in various strengths of NaOH

| Plastic | Maximum temperature ||||||||
| | in 10% NaOH || in 30% NaOH || in 50% NaOH || in 70% NaOH ||
	°C	°F	°C	°F	°C	°F	°C	°F
Acrylic	25	77	25	77	25	77	25	77
Acrylonitrile-butadiene-styrene (ABS)	40	104	40	104	40	104	40	104
Polyethylene (PE)	60	140	60	140	60	140	60	140
Polypropylene (PP)	100	212	100	212	100	212	100	212
Chlorinated polyether	60	140	120	248	120	248	100	212
Polystyrene	60	140	60	140	60	140	60	140
Polycarbonate	25	77	25	77	NR	NR	NR	NR
Polyvinyl chloride (PVC)	60	140	40	104	40	104	40	104
Chorinated polyvinyl chloride (CPVC)	80	176	80	176	80	176	80	176
Polyvinylidene chloride (PVDC)	25	77	25	77	40	104	40	104
Polyvinylidene fluoride (PVDF)	60	140	60	140	60	140	60	140
Fluorinated ethylene-propylene (FEP)	200	392	200	392	200	392	200	392
Polytetrafluorethylene (PTFE)(a)	220	428	220	428	220	428	220	428

(a) Limit to 150 °C (300 °F). NR, not resistant

Table 19 Temperature limits for plastic-lined pipe in caustic soda

| NaOH concentration | Temperature limit ||||||||
| | for PVDC Saran || for PP || for PVDF || for PTFE ||
	°C	°F	°C	°F	°C	°F	°C	°F
<10%	65	150	93	200	79(a)	175	230	450
10–50%	24	75	93	200	52(a)	125	230	450
50%	24	75	93	200	NR	NR	230	450
>50%	...	...	65	150	NR	NR	230	450

PVDC, polyvinylidene chloride; PP, polypropylene; PVDF, polyvinylidene fluoride; PTFE, polytetrafluoroethylene. (a) If mercury amalgam is present rating drops to not resistant (NR). Source: Ref 80

Table 20 Temperature limits for various plastics in dual-laminate construction in up to 10% caustic soda solutions

| Plastic | Temperature limit ||
	°C	°F
Polyvinyl chloride (PVC)	40, 60(a)	104, 140(a)
Chlorinated PVC (CPVC)	80	176
Polyethylene (PE)	60	140
Polypropylene (PP)	100	212
Polyvinylidene fluoride (PVDF)	...	...
Ethylene chlorotrifluoroethylene (ECTFE)	100	212
Ethylene trifluoro ethylene (ETFE)	100	212
Fluorinated ethylene propylene (FEP)	150	302
Perfluoro alkoxy (PFA)	150	302

(a) Conditional at this temperature, can attack or cause swelling. The medium can attack the material or cause swelling. Restrictions must be made in regard to pressure and/or temperature, taking the expected service life into account. The service life of the installation can be noticeably shortened. Source: Ref 81

be more aggressive to FRP than the stronger alkalis because of hydrolysis effects.

There are a great many resins used in conventional FRP construction. Conventional isophthalates and phenolic resins are not resistant to caustic and should not be considered. Some special formulations, such as bisphenol-A fumarate (Bis-A), are very good up to 50% caustic at about 80 °C (175 °F). The hydrogenated Bis-A is limited to dilute caustic at room temperature. Vinyl esters are also useful, although only in the lower concentrations at moderate temperatures. It must be emphasized that the chlorinated polyester (chlorendic anhydride) should not be used in NaOH.

Polyurethane shows variable performance in caustic limited by the permeability of the material; the lower the permeability of the polyurethane, the better the performance. This material is limited to 10 to 15% NaOH concentration. Epoxy resin formulations are widely used up to about 90 °C (195 °F). Furane (furfural-furfuryl alcohol) resins can be used both as reinforced materials (e.g., with graphite) and as coatings for immersion service in intermediate concentrations. They resist boiling 20% NaOH but not 5%. Cold-setting resins (as coatings for steel) have shown suitable resistance for up to 20% concentration NaOH even at the boiling point.

Temperature limits for applicable thermosetting resins in immersion service at various caustic concentrations are shown in Table 21 (data derived from various sources). The supplier should always be consulted in selecting thermoplastic or thermosetting construction for caustic service, and adequate consideration must be given to the mechanical and physical properties, particularly thermal expansion.

Carbon and Graphite

Corrosion resistance of carbon and graphite in caustic is good; impervious graphite, that is, graphite impregnated with carbon, not organic binders, is suitable for 80% NaOH up to 80 °C (175 °F) (Ref 83). The resistance of resin-impregnated graphite depends on the resins used. Graphite impregnated with phenolic resin is marginally resistant to 10 to 15% NaOH at up to 100 °C (212 °F). The furane-impregnated grade is rated at up to 50% NaOH at 130 °C (265 °F), and the grade that is impregnated with fluorocarbon is claimed to be acceptable at up to 80% or even 100% at up to 230 °C (446 °F) (Ref 84). Graphite impregnated with other resins are only slightly, or not at all resistant to >10% NaOH (Ref 83).

Impregnated raw graphite uses modified phenolic or furan resins under high heat and pressure. The phenolic binders and cements are attacked by caustic. Impervious graphite tube-and-shell heat exchangers contain cemented joints, which are traditionally made with phenolic resin for other services. It is essential that joints in equipment intended for caustic service be made with an epoxy or furane resin cement.

Impervious graphite is used for seals and gaskets because it has dimensional stability combined with self-lubricating, nonfatiguing, and noncontaminating properties.

Ceramics

Ceramics find practically no application in caustic service. Glass-lined vessels are only suitable for very low concentrations of caustic at very low temperatures, 1% or less at around 60 °C (140 °F) (Ref 85).

Rates of corrosion of borosilicate glass pass through a maximum at around 20% concentration at temperatures up to about 60 °C (140 °F). At higher temperatures, the maximum is at higher concentrations and corrosion rates at all concentrations increase dramatically.

Certain grades of silicon carbide can be used in caustic soda service. The rate of attack of silicon carbide (Hexoloy, Saint-Gobain Advanced Ceramics) and other ceramics are shown in Table 22. The weight loss rate is given in $mg/cm^2 \cdot yr^{-1}$, and a material with a rate of 0.3 to 9.9 $mg/cm^2 \cdot yr^{-1}$ is considered suitable for long-term service (Ref 86).

Corrosion in Contaminated Caustic and Mixtures

There are various kinds and degrees of contamination encountered in caustic soda production and usage, and these may profoundly affect the corrosion resistance of the materials of construction commonly used in this service. There are some impurities inherent in the manufacturing process that are substantially removed during the refining process. In other instances, the caustic soda may be mixed with other process chemicals, either deliberately or inadvertently. This section discusses the effects of contamination of and by caustic and of admixtures of caustic with other chemicals.

Contaminants in Caustic Soda

Contaminants in caustic soda solutions derive from impurities in the feed brine, reactions occurring in the chlorine cell, or corrosion products formed during production, processing, shipping, or storing. Since the source of the caustic influences the purity of the product it could also affect the corrosiveness of the solution. A survey of producers, however, carried out by the National Association of Corrosion Engineers, found no substantial difference in the corrosion of steels or nickel alloys in 50 or 73% caustic soda derived from either mercury or diaphragm cells. The conclusion was that differences in concentration and temperature were more important than the manufacturing method used (Ref 87). This does not mean that all caustic soda solutions are equal in terms of their corrosion behavior, and the presence of impurities does influence the attack suffered by metals and alloys. Producers go to great lengths to either reduce aggressive impurities and/or to use materials that are more resistant to the type of caustic that they produce and concentrate.

Chlorates are produced at the anodes of the electrolytic cells and are strong oxidants. They are usually present at >100 ppm and have a beneficial effect on stainless steels and an adverse effect on nickel. It has been found that nickel pickup (from corrosion in the first effect evaporator tubing) was directly proportional to the chlorate content of the cell liquor, within the 120 to 200 ppm range encountered (Ref 88). Some plants add a continuous feed of 10% sucrose solution to react with this chlorate. No residues are left from this reaction, and product quality is not compromised (Ref 89). More commonly, the chlorate is removed before or during evaporation by proprietary treatments or by extraction with ammonia, which also reduces the dissolved salt (Ref 14).

Chlorate ions can increase the rate of corrosion of carbon steel. For 48% NaOH with 0.5% $NaClO_3$, carbon steel showed a tenfold increase

Table 21 Temperature limits for thermosets used in fiber-reinforced plastics construction in various strengths of NaOH

	Temperature limit							
	in 10% NaOH		in 30% NaOH		in 50% NaOH		in 70% NaOH	
Resin	°C	°F	°C	°F	°C	°F	°C	°F
Epoxy	60	140	60	140	80	176	120	248
Furane	60	140	120	248	120	248	120	248
Bisphenol-A fumarate	65	149	65	149	80	176	NR	NR
Hydrogenated Bis-A polyester	25	77	NR	NR	...	...	...	...
Chlorinated polyester	25	77	NR	NR	...	...	...	...
Vinyl ester	60	140	60	140	100	212	NR	NR
Polyurethane	60	140	NR	NR	...	...	...	...

NR, not resistant

Table 22 Corrosive weight loss of various ceramics in 50% NaOH at 100 °C (212 °F)

Material	Weight loss, $mg/cm^2 \cdot yr^{-1}$
Hexoloy SA (no free Si)	2.5
Hexoloy KT (12% free Si)	>1000
Tungsten carbide (WC-6Co)	5.0
Aluminum oxide (99.0%)	75

in corrosion rate compared with caustic that did not contain chlorate ions (Ref 90). Chlorates also increased the corrosion rate of nickel alloys (Ref 14).

Corrosion of nickel in the caustic plant is ascribed either to chlorate ions (ClO_3^-) or to hypochlorites (OCl^-). General corrosion rates can exceed 0.5 mm/yr (20 mils/yr). Corrosion is aggravated by high velocity, as over welds on pump discharge piping. Velocity and increased temperature are thought to be the most probable cause of attack of nickel in caustic service, rather than chlorate or hypochlorite levels. In laboratory tests at 185 °C (365 °F) in actual first effect caustic liquor (43% NaOH, 0.15% ClO_3^-) and in a synthetic mix (51% NaOH; 0.20% ClO_3^-) corrosion and SCC resistance of alloys 200 and 26-1 were both superior to any of the other alloys tested. The relative corrosion resistance of the other alloys in static and circulating simulated and actual liquors was:

alloy 600 ≫ alloy 825 ≅ alloy 800 ≅ alloy 28

Some of these data are summarized in Fig. 12, which shows that chlorates increase the corrosion of nickel and nickel alloys, although the effect is not sufficient to cause failures of nickel in evaporators without the presence of additional factors, such as temperature or velocity. Increased nickel content in the alloys increases the corrosion resistance, particularly in caustic without chlorates. Alloy 26-1, which has no nickel, has good corrosion resistance but can suffer from SCC in boiling solutions of >30% NaOH and can also be subject to IGA. The presence of hypochlorites was found to increase the corrosion rates of all the alloys to a lesser or greater extent (Ref 91).

The effect of sodium chloride and sodium chlorate additions to hot caustic soda solutions on the corrosion of the ferritic stainless steel, E-Brite (S44627), is shown in Table 23 (Ref 29). The corrosion rate of this alloy is little affected by these contaminants, common in cell liquors, unlike that of nickel in which the corrosion rate is accelerated. This alloy is attacked in hotter, stronger caustic solutions as shown in the data for 70% NaOH at 188 °C (370 °F) that is included in this table (Ref 92). Corrosion is decreased in caustic by the addition of sodium chlorate and hypochlorite up to 1000 ppm, but IGA occurred when these additions were at 10,000 ppm.

The effect of chlorides and chlorates on the corrosion resistance of another ferritic alloy, Sea-Cure (S44660), in hot caustic soda solutions is shown in Table 24 (Ref 93). These data compare the behavior of this alloy with that of type 316, ferritic stainless steel 26-1S, and pure nickel. These data show that Sea-Cure is as resistant to these conditions as is alloy 200, except for the tendency to IGA in this environment.

The effects of the presence of chlorides and chlorates was investigated for a duplex stainless steel (S32906) and alloy 200 (N02200). These data show that when higher levels of chloride and chlorate are present, simulating membrane and diaphragm caustic, the duplex alloy is much more resistant than nickel (Table 25) (Ref 36).

Chlorides. Because NaOH is produced by electrolysis of brine, chlorides may be present from about 20 to 5000 ppm, the minimum concentration being typical of membrane cell caustic. Chlorides in caustic soda solution do not cause chloride SCC of austenitic stainless steels. A solution of 0.5 g/L NaOH at pH 12 is sufficiently alkaline to avoid this type of cracking, which is common in neutral or acidic solutions (Ref 9). It has also been claimed that chlorides inhibit caustic SCC. Mercury cell caustic typically contains 20 to 30 ppm chlorides, while diaphragm cell caustic is more likely to contain up to 1% chlorides. While chloride-contaminated caustic soda solution may not cause SCC of austenitic stainless steel if the caustic soda is consumed, for example in a neutralizing process, then the residual neutral chloride-containing solution may then initiate cracking (Ref 14). See Table 23 for the effect of chlorides with and without chlorates on the ferritic E-Brite stainless steel (S44627), Table 24 for Sea-Cure (S44600), and Table 11 for the effect of chlorides on alloy 28.

Zirconium is resistant to sodium hydroxide solutions even when high levels of chlorides are present (see Table 16).

Alloy 800 (N08800) was tested in a NaOH cooling tank at 70 to 105 °C (160 to 220 °F) for 119 days. The environment was 50% NaOH with 10 to 15% NaCl, and the corrosion rate measured was negligible at <0.003 mm/yr (<0.12 mil/yr) (Ref 94).

Chlorine/Hypochlorite. Hypochlorites are formed when chlorine is introduced into water or alkaline solutions:

$$Cl_2 + 2NaOH \rightarrow NaCl + NaOCl$$

Sodium hypochlorite (NaOCl) is liable to initiate pitting in otherwise passive metals because of the concentration cells between freely exposed and occluded surfaces, analogous to an oxygen concentration cell.

The corrosion resistance of various alloys in caustic soda containing chlorine was tested in field trials. The results of these tests showed that alloy 33 (R20033) performed better than the other alloys tested (Table 26) (Ref 95). All of the alloys showed acceptable corrosion rates in the liquid and the vapor, although pitting and crevice corrosion, severe in cases, was observed.

Mercury can be entrained from the production source, while mercuric ions can be reduced to metallic mercury at local cathode sites. Metallic mercury contamination can lead to LME or liquid metal cracking (LMC) in some alloy systems. Titanium and zirconium and their alloys, copper and copper alloys, aluminum and aluminum alloys, alloy 400, and alloy 200 at elevated temperatures are known to be at risk from this form of attack (Ref 96).

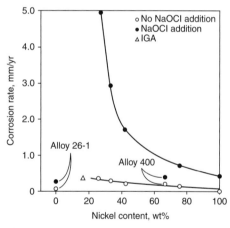

Fig. 12 Effect of nickel content on corrosion of various materials in simulated first effect liquor at 185 °C (365 °F). Source: Ref 91

Table 23 Resistance of E-Brite alloy to caustic solutions containing NaCl and NaClO$_3$

% NaOH	% NaCl	% NaClO$_3$	°C	°F	mm/yr	mils/yr
20	10	...	104	219	0.015	0.59
45	5	...	143	289	0.041	1.61
50	...	...	143	289	0.003	0.12
50	5	...	152	306	0.076	2.99
50	5	0.1	152	306	0.069	2.72
50	5	0.2	152	306	0.028	1.10
50	5	0.4	152	306	0.028	1.10
70	5	0.15	188	370	0.110	4.33

Source: Ref 29

Table 24 Corrosion rates of various alloys in hot caustic solutions

	Corrosion rate					
	in 55% NaOH + 8% NaCl + 0.3% NaClO$_3$ at 99 °C (210 °F)		in 50% NaOH at 143 °C (290 °F)		in 55% NaOH + 8% NaCl + 0.3% NaClO$_3$ at 158 °C (315 °F)	
Alloy	mm/yr	mils/yr	mm/yr	mils/yr	mm/yr	mils/yr
316	0.15	5.9	0.38	15.0	Very high	Very high
26-1S	<0.0025	0.1	0.015	0.59	0.02	0.79
Alloy 200	<0.0025	0.1	0.023	0.91	0.07	2.8
Sea-Cure	<0.0025	0.1	0.025	1.0	IGA	IGA

IGA, intergranular attack. Source: Ref 93

Sulfur. Sulfides, sulfite, or sulfate contaminants in caustic can increase the rate of corrosion of carbon steel if the passive oxide film on the steel is damaged, scratched, defective, and so forth. In considering low-alloy steels for service in the cellulose and paper industry, where strongly alkaline solutions often contain sulfide, it has been found that sodium sulfide and sodium thiosulfate exacerbate damage, whereas sodium sulfite/sulfate have little effect (Ref 97).

Sulfur species can also accelerate the corrosion of nickel alloys, especially at elevated temperatures.

Iron. Because much of the processing, storage, and transportation of caustic soda is effected in steel, iron contamination is often present, either as the soluble ferroite or ferroate, or as fines of colloidal ferric oxide. The resultant discoloration is objectionable for many applications, notably the soap/detergent, pulp/paper, and rayon/textile industries. More resistant materials of construction are often selected just to prevent this undesirable contaminant.

Soda Ash

Soda ash is anhydrous sodium carbonate (Na_2CO_3) chemical abstracting service (CAS) number 497-19-8. It exists as a naturally occurring mineral in parts of Africa and in the United States, notably in Wyoming. Native soda ash is found in the form of trona, which is sodium carbonate and bicarbonate, or as brines that are mixtures of sodium carbonate, sulfate, and sulfite (Ref 98). Soda ash can be produced by the reaction of caustic soda and carbon dioxide or in the Solvay (ammonia soda) process that reacts limestone, ammoniated brine, and coke.

Soda ash is used in household cleaners (washing soda), glass making, water treatment, chemical processing, and so forth. In some pulp and paper applications, soda ash is reacted with limestone to produce caustic soda, used in neutralizing and other processes.

Pure soda ash is a hygroscopic white powder whose molecular weight is 105.99 and specific gravity 2.533. It has a melting point of 851 °C (1564 °F) and generates heat of solution in water. It has a strong degree of alkalinity.

Sodium carbonate decahydrate ($Na_2CO_3 \cdot 10 H_2O$) melts at 34 °C (93 °F) to form a solution of about 37% concentration. Corrosion problems may develop because it can either lose carbon dioxide to release free caustic or further absorb CO_2 to form the bicarbonate ($NaHCO_3$).

Sodium carbonate at elevated temperature will cause caustic corrosion in a manner similar to caustic soda corrosion. The propensity to caustic SCC is less with soda ash, and carbon steel is routinely used to handle boiling solutions. The alkalinity of soda ash is less than that of NaOH, and the carbonate ions have an inhibiting effect on many metals. Soda ash can form bicarbonate, and this may corrode steel because of its lower pH (Ref 16).

A typical application is the use of sodium carbonate solutions for the absorption and release of carbon dioxide from hydrocarbon streams, in a manner analogous to alkanolamine acid gas scrubbing systems. The carbonate reacts selectively with CO_2 to form bicarbonate, which is reheated in the separation column to release the acid gas, returning the regenerated lean Na_2CO_3 solution to the absorber. This process is inherently corrosive to steel, particularly in the regeneration step.

Potassium Hydroxide

Potassium hydroxide (KOH), CAS 1310-58-3, is very similar to sodium hydroxide in its chemical and corrosion characteristics, but it is much less common. Caustic potash is produced in a similar manner to caustic soda by the electrolysis of potassium chloride. The equipment for evaporation is largely based on nickel and nickel alloys with cathodic protection being employed for >50% KOH solutions, which have a higher boiling point than NaOH.

Corrosion data are limited, but KOH corrosion behavior is similar to that of NaOH, and can be extrapolated from the NaOH data and experience. Caustic SCC can occur in caustic potash solutions at elevated temperatures, but cracking is not so severe in the potassium salt for alloy 400 and alloy 600. The failure of a low-carbon steel (ASTM A516, grade 70) pressure vessel after 10 years service in a hydrogen sulfide absorber was diagnosed as due to caustic SCC from KOH. The environment was 20% aqueous KOH, potassium carbonate, and arsenic at 33 °C (91 °F). Although the operating temperature was low for SCC, the vessel had been exposed to a fire that may have taken the temperature above the induction temperature. It may also have increased the local concentration of KOH and added stress to the existing residual and operating stresses (Ref 99).

In a problem analogous to that which occurs with caustic carryover in steam, SCC of type 347 (S34700) by potassium hydroxide has been reported. In one plant, a catalyst containing potassium oxide was to be used in a hydrogenator. When the plant was started up with 300 °C (570 °F) steam after a catalyst change, sufficient KOH was formed and entrained to cause failure of the stainless equipment downstream.

Some data are available for alloy 200 in caustic potash under velocity conditions and when mixed with potassium chloride (KCl) and potassium chlorate ($KClO_3$) (Table 27) (Ref 100).

Data for various metals and alloys in KOH are shown in Table 28 (Ref 101). This table shows that KOH at these temperatures and concentrations is not very corrosive even to low-carbon steel.

Zirconium has a corrosion rate of <0.025 mm/yr (<1 mil/yr) in 0 to 50% KOH

Table 25 Corrosion rates of S32906 and N02200 in boiling NaOH solutions simulating membrane and diaphragm cell liquors

	Corrosion rate			
	in 50% NaOH + 30 ppm Cl^-, 20 ppm ClO_3^-		in 50% NaOH + 7% NaCl, 800 ppm ClO_3^-	
Alloy	mm/yr	mils/yr	mm/yr	mils/yr
S32906	0.001	0.04	0.016	0.63
N02200	0.001	0.04	0.15	5.9

Source: Ref 36

Table 26 Corrosion rates of various alloys in NaOH/NaOCl exposed to liquid and vapor

	Corrosion rate				
	in 50% NaOH, 110 °C (230 °F), 180 days, liquid, mm/yr (mils/yr)	in 20% NaOH + NaOCl (80–100 g Cl_2/L), 30 °C (85 °F), 195 days		12.5% NaOH + NaOCl (130 g Cl_2/L), 30 °C (85 °F), 225 days	
Alloy		Liquid, mm/yr (mils/yr)	Vapor, mm/yr (mils/yr)	Liquid, mm/yr (mils/yr)	Vapor, mm/yr (mils/yr)
316Ti	≤0.01 (≤0.39)	0.01 (0.39), pitting and severe crevice corrosion	0.02 (0.79), pitting	...	≤0.01 (≤0.39), severe pitting
926	...	≤0.01 (≤0.39), pitting and severe crevice corrosion	≤0.01 (≤0.39)	≤0.01 (≤0.39)	≤0.01 (≤0.39), some uniform corrosion
654 SMO	≤0.01 (≤0.39)	≤0.01 (≤0.39), some uniform corrosion	...	≤0.01 (≤0.39)	≤0.01 (≤0.39)
C-4	...	≤0.01 (≤0.39), pitting	...	≤0.01 (≤0.39)	≤0.01 (≤0.39), some crevice and uniform corrosion
33	≤0.01 (≤0.39)	≤0.01 (≤0.39)	≤0.01 (≤0.39), some crevice corrosion	≤0.01 (≤0.39), some crevice corrosion	≤0.01 (≤0.39), some crevice corrosion

Source: Ref 95

Table 27 Laboratory corrosion tests of alloy 200 in potassium hydroxide

Environment (solution composition and velocity)		Temperature		Corrosion rate	
		°C	°F	mm/yr	mils/yr
30% NaOH, saturated with KCl + 0.05% KClO₃	Liquid	Boiling	Boiling	0.005	0.20
	Vapor			0.008	0.31
47% NaOH saturated with KCl + 0.078% KClO₃	Liquid	Boiling	Boiling	0.002	0.08
	Vapor			0.008	0.31
50% KOH	6.6 m/min (21.7 ft/min)	150	300	WG	WG
	106 m/min (348 ft/min)			WG	WG
70% KOH	6.6 m/min (21.7 ft/min)	150	300	0.01	0.39
	106 m/min (348 ft/min)			0.04	1.57

WG, weight gain. Source: Ref 100

Table 28 Corrosion rate of metals and alloys in KOH solutions

Alloy	in 13% KOH at 30 °C (85 °F), 13% KCl added		in 50% KOH at 25 °C (80 °F)	
	mm/yr	mils/yr	mm/yr	mils/yr
Titanium	0.023	0.9	0.01	0.4
Zirconium	0.005	0.2	0.0015	0.06
Nickel	Nil	Nil	0.00008	0.003
Monel	Nil	Nil	0.00005	0.002
Inconel	Nil	Nil	Nil	Nil
Low-carbon steel	0.013(a)	0.5(a)	0.0013	0.05

(a) Slight attack under spacer. Source: Ref 101

at temperatures up to boiling (Ref 102). Zirconium in 50% KOH at 27 °C (81 °F) corroded 0.0015 mm/yr (0.06 mil/yr), and in short-term tests in 50% to anhydrous KOH at 241 to 377 °C (466 to 711 °F), the corrosion rate was <0.001 mm/yr (<0.03 mil/yr) (Ref 103).

REFERENCES

1. Caustic Soda, *Eur. Chem. News*, Vol 78, June 2003, p 18
2. *Caustic Soda (Sodium Hydroxide)*, ChemLink Australasia, http://www.chemlink.com.au/caustic.htm, 1997
3. E. Linak, *CEH Report—Chlorine/Sodium Hydroxide*, SRI Consulting, http://ceh.sric.sri.com/Enframe/Report.html, 2002
4. *Aluminium with Food and Chemicals*, Alcan Industries Ltd., London, U.K. 1966, p 91
5. M.R. Tabrizi, S.B. Lyon, G.E. Thompson, and J.M. Ferguson, The Long Term Corrosion of Aluminium in Alkaline Media, *Corros. Sci.*, Vol 32 (No. 7), 1991, p 733–742
6. M.L. Doche, J. Rameau, R. Durand, and F. Novel-Cattin, Electrochemical Behaviour of Aluminium in Concentrated NaOH Solutions, *Corros. Sci.*, Vol 41 (No. 4), 1999, p 805–826
7. M. Hagen, Corrosion of Steels, *Corrosion and Environmental Degradation*, Vol II, M. Schutze, Ed., Wiley-VCH, Weinheim, Germany, 2000, p 1–68
8. S. Giddey, B. Cherry, F. Lawson, and M. Forsyth, Stability of Oxide Films Formed on Mild Steel in Turbulent Flow Conditions of Alkaline Solutions at Elevated Temperatures, *Corros. Sci.*, 43 (No. 8), 2001, p 1497–1517
9. B.D. Craig and D.B. Anderson, Ed., *Handbook of Corrosion Data*, ASM International, 1997, p 761–790
10. "Corrosion Resistance of Nickel and Nickel-Containing Alloys in Caustic Soda and Other Alkalis," CEB-2, International Nickel Company, 1973, p 17
11. R. Covert, J. Morrison, K. Rohrig, and W. Spear, *Ni-Resist and Ductile Ni-Resist Alloys*, Reference book No. 11018, Nickel Development Institute, Toronto, Canada, 1998, 42 pages
12. F.L. LaQue and H.R. Copson, *Corrosion Resistance of Metals and Alloys*, Reinhold Publishers, 1963, 365 pages
13. E. Heyn and D. Bauer, in P.J. Gegner, *Corrosion Resistance of Materials in Alkalis and Hypochlorites*, Process Industries Corrosion, National Association of Corrosion Engineers, 1975, p 296–305
14. C.M. Schillmoller, Select the Right Alloys for Caustic Soda Service, *Chem. Eng. Prog.*, May 1996, p 48–55
15. NACE Network, reported in *Mater. Perform.*, Vol 42 (No. 2), 2003, p 82–83
16. J.K. Nelson, Corrosion by Alkalis and Hypochlorites, *Corrosion*, Vol 13, 9th ed., *Metals Handbook*, ASM International, 1987, p 1174–1180
17. C.P. Dillon, *Corrosion Control in the Chemical Process Industries*, 2nd ed., MTI, 1994, 424 pages
18. R.B. Norden, Materials of Construction, *Chemical Engineers Handbook*, 5th ed., J.J. Carberry, M.S. Peters, W.R. Schowalter, and J. Wei, Ed., McGraw-Hill, 1973, p 23-14
19. *Corrosion by Caustic Solutions*, The Hendrix Group Inc., http://www.hghouston.com/naoh.html, 2002
20. H.W. Schmidt, P.J. Gegner, G. Heinemann, C.F. Pogacar, and E.H. Wyche, Stress Corrosion Cracking in Alkaline Solutions, *Corrosion*, Vol 7, 1951, p 295–302
21. A.A. Berk and W.F. Waldeck, Caustic Danger Zone, *Chem. Eng.*, Vol 57 (No. 6), 1950, p 235–237
22. *Corrosion Data Survey*, National Association Corrosion Engineers, 1985, p 176
23. N. Bandyopadhyay and C.L. Briant, Caustic Stress Corrosion Cracking of Low Alloy Iron Base Materials, *Corrosion*, Vol 41 (No. 5), 1985, p 274–280
24. A. Sabata and W.J. Schumacher, Martensitic and Ferritic Stainless Steels, *CASTI Handbook of Stainless Steels and Nickel Alloys*, S. Lamb, Ed., CASTI Publishing Inc., Edmonton, Canada, 2000, p 144
25. H.E. Deverell and I.A. Franson, Practical Guide to Newer Ferritic Stainless Steels, *Mater. Perform.*, Vol 28 (No. 9), 1989, p 52–57
26. *ATI Technical Data Sheets*, on CD, Allegheny Ludlum Corp., 2001
27. "AL 29-4C," technical data sheet No. B-51-Ed5/7.5M/793/GP, Allegheny Ludlum Corp., 1982, p 11
28. "AL 29-4-2," technical data sheet No. B-153-Ed 1-10M-582P, Allegheny Ludlum Corp., 1993, p 3
29. "E-Brite Alloy," technical data sheet No. B-150-Ed1-10M-181P, Allegheny Ludlum Corp., 1980, p 7
30. M. Yasuda, F. Takeya, S. Tokunaga, and F. Hine, Corrosion Behavior of a Ferritic Stainless Steel in Hot Concentrated NaOH Solutions, *Mater. Perform.*, Vol 23 (No. 7), 1984, p 44–49
31. A.R. McIlree and H.T. Michels, Stress Corrosion Behavior of Fe-Cr-Ni and Other Alloys in High Temperature Caustic Solutions, *Corrosion*, Vol 33 (No. 2), 1977, p 60–67
32. J.G. Gonzalez-Rodriguez, V.M. Salinas-Bravo, and A. Martinez-Villafane, Hydrogen Embrittlement of Type 410 Stainless Steel in Sodium Chloride, Sodium Sulfate and Sodium Hydroxide Environments at 90 °C, *Corrosion*, Vol 53 (No. 6), 1997, p 499–504
33. "Armco17PH Precipitation-Hardening Stainless Steel," Bulletin No. FS-11, Armco Advanced Materials Co., 1994
34. E.-M. Horn et al, *Corrosion and Environmental Degradation*, Vol II, M. Schutze, Ed., Wiley-VCH, Weinheim, Germany, 2000, p 69–111
35. E.-M. Horn, S. Savakis, G. Schmitt, and I. Lewandowski, Performance of Duplex Steels in Caustic Solutions, *Duplex Stainless Steels '91*, Les Editions de Physique, Les Ulis, France, 1991, p 1111–1119
36. D. Leander, Corrosion Characteristics of Different Stainless Steels, Austenitic and Duplex, in NaOH Environment, *Stainless Steel World Conference* (Maastricht, Netherlands), Stainless Steel World, 2003, 9 pages
37. C.M. Schillmoller, "Alloy Selection for Caustic Soda Service," Technical Series No. 10019, The Nickel Development Institute, Toronto, Ontario, Canada, March 1988, 9 pages
38. M. Yasuda, S. Tokunaga, T. Taga, and F. Hine, Corrosion Behavior of 18-8 Stainless Steels in Hot Concentrated Caustic Soda

Solutions under Heat-transfer Conditions, *Corrosion,* Vol 41 (No. 12), 1985, p 720–727
39. F.L. LaQue and H.R. Copson (1963), in Sodium Hydroxide Advisor, *ChemCor 6,* MTI/NACE/ NiDI/NIST, 1992
40. G. Rondelli, B. Vincenti, and E. Sivieri, Stress Corrosion Cracking of Stainless Steel in High Temperature Caustic Solutions, *Corros. Sci.,* Vol 39 (No. 6), 1997, p 1037–1049
41. G. Rondelli and B. Vincenti, Susceptibility of Highly Alloyed Austenitic Stainless Steels to Caustic Stress Corrosion Cracking, *Mater. Corros.,* Vol 53, 2002, p 813–819
42. *Corrosion Handbook for Stainless Steels,* Sandvik Steel, Sandviken, Sweden, 1999, p II:54
43. S.W. Dean, Caustic Cracking from Potassium Hydroxide in Syngas, *Mater. Perform.,* Vol 38 (No. 1), 1999, p 73–76
44. R.W. Monroe and S.J. Pawel, Corrosion of Cast Steels, *Corrosion,* Vol 13, 9th ed., *Metals Handbook,* ASM International, 1987, p 580
45. R.C. Scarberry et al. (1967) in Sodium Hydroxide Advisor, *ChemCor 6,* MTI/NACE/ NiDI/NIST, 1992
46. T. Cassagne and P. Combrade, "Stress Corrosion Cracking of Stainless Steels in Caustic Environments: A Laboratory Study," Innovation Stainless Steel, Florence, Italy, 1993, p 3.215–3.220
47. J.E. Truman and R. Perry, The Resistance to Stress Corrosion Cracking of Some Cr-Ni-Fe Austenitic Steels and Alloys, *Brit. Corros. J.,* Vol 1, 1965, p 60–66
48. M. Yasuda, K. Fukumoto, H. Koizumi, Y. Ogata, and F. Hine, On the Active Dissolution of Metals and Alloys in Hot Concentrated Caustic Soda, *Corrosion,* Vol 43 (No. 8), 1987, p 492–498
49. "SAF Type 2507," booklet No. AP-26-12/01, Avesta Polarit, 2001, p 3
50. J.F. Grubb, Ed. *AL-6XN Alloy,* Allegheny Ludlum Corp., 1995, 38 pages
51. "Durcomet 100," bulletin A/7d, The Duriron Co. Inc., 1981, 8 pages
52. "Durimet 20," bulletin A/1f, The Duriron Co. Inc., 1981, 6 pages
53. S. Bernhardsson, S. Lagerberg, C. Martensson, and M. Tynell, "Corrosion Performance of a High-Nickel Stainless Alloy in the Process Industry," Sandvik, Sandviken, Sweden, undated, 14 pages
54. S. Bernhardsson and J. leGrand, "Special Stainless Steels for the Chemical and Petrochemical Industry," lecture No. S-52-65-ENG, Sandvik, Sandviken, Sweden, 1981, 15 pages
55. M. Kohler, M. Heubner, K.W. Eichenhofer, and M. Renner, "Alloy 33, A New Corrosion Resistant Austenitic Material for the Refinery Industry and Related Applications," paper No. 338, Corrosion 95, NACE International, 1995, p 14
56. Yu.B. Danilov, V.A. Kachanov, E.K. Gvozdikova, T.É. Shepil', V.N. Khil', T.A. Balak, and V.S. Gorlova, Corrosion Studies of Alloys from Krupp VDM in Caustic Soda Production, *Chem. Petroleum Eng.,* Vol 37 (No. 3–4), 2001, p 253–258
57. M. Hagen, Corrosion of Steels, *Corrosion and Environmental Degradation,* Vol II, M. Schutze, Ed., Wiley-VCH, Weinheim, Germany, 2000, p 69–111
58. G. Kobrin, Materials Selection, *Corrosion,* Vol 13, 9th ed., *Metals Handbook,* ASM International, 1987, p 328
59. "Resistance to Corrosion," publication No. 3M8-88 S-37, Inco Alloys Int., 1985, p 33
60. M. Yasuda, F. Takeya, and F. Hine, Corrosion Behavior of Nickel in Concentrated NaOH Solutions under Heat Transfer Conditions, *Corrosion,* Vol 39 (No. 10), 1983, p 399–406
61. "Corrosion Resistant Alloys—Specifications and Operating Data," Rolled Alloys Inc., 2002, p 11
62. CEB-2 (International Nickel Company Inc., 1969) in P.J. Gegner, Corrosion Resistance of Materials in Alkalis and Hypochlorites, *Process Industries Corrosion,* National Association of Corrosion Engineer, 1975, p 298
63. R.S. Sheppard et al. (1962) in P.J. Gegner, Corrosion Resistance of Materials in Alkalis and Hypochlorites, *Process Industries Corrosion,* National Association of Corrosion Engineer, 1975, p 298
64. R.L. Zeller and L. Salvati, Effects of Phosphorus on Corrosion Resistance of Electroless Nickel in 50% Sodium Hydroxide, *Corrosion,* Vol 50 (No. 6), 1994, p 457–467
65. A.J. Sedriks et al. (1979) in A.J. Sedriks, *Corrosion of Stainless Steels,* John Wiley & Sons, 1979, p 175
66. "Nickel Alloys 200, 2201, 270, 301," brochure 10M-22-79 T-15, Inco Alloys International, 1979, p 17
67. "Inconel Alloy 600," No. SMC 027, Inco Alloys International, 2000, p 15
68. H.P. Kim, S.S. Hwang, Y.S. Lim, I.H. Kuk, and J.S. Kim, Effect of Heat Treatment and Chemical Composition on Caustic Stress Corrosion Cracking of Alloy 600 and Alloy 690, *Key Eng. Mater.,* Vol 183–187, 2000, p 707–712
69. J.S. Baek, J.G. Kim, D.H. Hur, and J.S. Kim, Anodic Film Properties Determined by EIS and their Relationship with Caustic Stress Corrosion Cracking of Alloy 600, *Corros. Sci.,* Vol 45 (No. 5), 2003, p 983–984
70. U. Heubner and M. Kohler, "High-Alloy Materials for Aggressive Environments," report No. 26, ThyssenKrupp VDM GmbH, Werdol, Germany, 1998, p 122
71. G. Notten, E. van den Heuvel, and H. Verhoef, "Stress Corrosion Cracking Resistance of Duplex Stainless Steels in Caustic Environments," paper no. DA2_023, Stainless Steel World, Duplex America 2000, 2000, p 53–59
72. "Corrosion Resistance of Copper and Copper Alloys," No. 106, Copper Development Association, London, U.K.
73. "Corrosion Resistance of Titanium," ref 1431531969, IMI Kynoch Ltd., Birmingham, U.K., 1969, p 28
74. "Corrosion Resistance of Titanium," brochure No. TMC-0105, Timet, 1999, p 17
75. Ammonia and Ammonium Hydroxide section, *Dechema Corrosion Handbook,* CD, Dechema aV, Frankfurt, Germany, 2001
76. "Zircadyne Corrosion Data," bulletin No. TWCA-8101ZR, Teledyne Wah Chang, 1987, 25 pages
77. "Niobium," ref NioNio-056 on CD, Allegheny Ludlum Corp., 2001, 42 pages
78. M. Schussler, "Corrosion Data Survey on Tantalum," Fansteel Inc., 1972, p 50
79. A. Robin, Corrosion Behaviour of Ta-10 wt% W Alloy in Sodium Hydroxide Solutions, *Corros. Eng. Sci. Technol.,* Vol 38 (No. 3), 2003, p 211–217
80. "Chemical Resistance Guide," Dow Chemical, 1991, 20 pages
81. "Chemical Resistance of Thermoplastics used in Dual Laminate Constructions," Dual Laminate Fabrication Association, 143 pages, http://www.dual-laminate.org/html/corrosion_guide.html, 2002
82. P.J. Gegner, Corrosion Resistance of Materials in Alkalis and Hypochlorites, *Process Industries Corrosion,* National Association of Corrosion Engineers, 1975, p 296–305
83. "The Chemical Industry Build on Graphite," Brochure PE 200/07, SGL Carbon Group, Meitingen, Germany, 2001, 24 pages
84. "Corrosion Chart for Grafilor," brochure GC 5 FED 7821, Le Carbone-Lorraine, Moselle, France, undated, 22 pages
85. "Worldwide GLASTEEL 9100," brochure No. SB95-910-5, Pfaudler Reactor Systems, 2000, p 5
86. "Hexoloy SA Corrosion Test in Liquids," Saint-Gobain Advanced Ceramics, http://www.carbo.com/datasheets/corrosiontest.html, 2003
87. NACE Task Group T5-A report, *Mater. Protect. Perform.,* Vol 10 (No. 7), 1971, p 39
88. B.M. Barkel, "Accelerated Corrosion of Nickel Tubes in Caustic Evaporation Service," paper No. 13, Corrosion/79, National Association of Corrosion Engineers, 1979, 9 pages
89. M.P. Sukumaran Nair, Stress Corrosion Cracking—A Caustic Experience, *Chem. Eng.,* Jan 2003, p 1–3
90. K. Hauffe (1986) in NACE Network, reported in *Mater. Perform.,* Vol 42 (No. 2), 2003, p 82–83

91. J.R. Crum and W.G. Lipscomb, "Correlation between Laboratory Tests and Field Experience for Nickel 200 and 26-1 Stainless Steel in Caustic Service," Corrosion '83, paper No. 23, National Association of Corrosion Engineers, 1983, 18 pages
92. J.R. Kearns, M.J. Johnson, and I.A. Franson, "The Corrosion of Stainless Steels and Nickel Alloys in Caustic Solutions," Corrosion '84, paper No. 146, National Association of Corrosion Engineers, 1984, 18 pages
93. "Trent SEA-CURE Stainless Steel for Power Generation and Chemical Processing," brochure No. A18-7/00-5000, Trent Tube, 2000, 20 pages
94. "Incoloy Alloys 800, 800H and 800HT," brochure No. 1A1 172/7M, Inco Alloys International, 1997, 28 pages
95. M. Kohler, U. Heubner, K.W. Eichenhofer, and M. Renner, "Progress with Alloy 33 (UNS R20033), a New Corrosion Resistant Chromium Based Austenitic Material," paper No. 428, Corrosion/96, NACE International, 1996, 18 pages
96. J.R. Davis, Ed., *Corrosion—Understanding the Basics,* ASM International, 2000, p 191
97. D.A. Wensley and R.S. Charlton (1980) in Sodium Hydroxide Advisor, *ChemCor 6,* MTI/NACE/ NiDI/NIST, 1992
98. "Lake Natron Soda Ash Project," National Development Corp., Dar es Salaam, Tanzania, http://www.ndctz.com/sodaash.htm, 1997
99. R.T. King, Failure of Pressure Vessels, *Failure Analysis and Prevention,* Vol 11, 9th ed., *Metals Handbook,* American Society for Metals, 1986, p 658
100. "Wiggin Corrosion Resisting Alloys," Wiggin Alloys Ltd., Hereford, U.K., 1983, p 18
101. P.J. Gegner and W.L. Wilson *Corrosion,* Vol 15 (No. 7), 1959
102. K. Bird, The Caustic Truth about Zircadyne, *Outlook,* Vol 15 (No. 3), 1994, p 6–7
103. D.R. Knittel and R.T. Webster, "Corrosion Resistance of Zirconium and Zirconium Alloys in Inorganic Acids and Alkalies," Symp on Industrial Applications of Zirconium and Titanium, ASTM, 1979

SELECTED REFERENCES

- "Corrosion Resistance of Nickel and Nickel-Containing Alloys in Caustic Soda and Other Alkalis," CEB-2, International Nickel Company Inc., 1973, p 17
- B.D. Craig and D.B. Anderson, Ed., *Handbook of Corrosion Data,* ASM International, 1997
- M. Davies, "Materials Selection for Ammonia and Caustic Soda," No. MS-6, Materials Technology Institute of the Chemical Process Industries, in press

Corrosion by Ammonia

Michael Davies, CARIAD Consultants

ANHYDROUS AMMONIA (NH_3) is a major commercial chemical that is used in the manufacture of fertilizers, nitric acid (HNO_3), acrylonitrile, and other products. The world total production of NH_3 in 1996 was 93,500 thousand metric tons of nitrogen. In 2002, according to the International Fertilizer Association, the world production was 108,320 thousand metric tons (89,025 thousand metric tons of nitrogen). Approximately 80% of the U.S. apparent domestic NH_3 consumption was for fertilizer use, including anhydrous NH_3 for direct application, urea, ammonium nitrate (NH_4NO_3), ammonium phosphates, and other nitrogen compounds.

Ammonia is made by a number of processes; the two principal ones are:

- Steam reforming of natural gas or other light hydrocarbons (natural gas liquids, liquefied petroleum gas, naphtha). Modified steam reforming using excess air in the secondary reformer and heat-exchange autothermal reforming are also being used to some extent.
- Partial oxidation of heavy fuel oil, coal, or vacuum residue is used.

The hydrogen-containing gas from the primary reformer is mixed with nitrogen in the secondary reformer to produce a gas containing CO_2, H_2, and N_2. The CO_2 is stripped off, and the remaining synthesis gas (syngas) is reacted over an iron catalyst to produce NH_3 that is condensed and separated from the gas stream. Ammonia that leaks into cooling towers can be converted to HNO_3 by nitrifying bacteria and pose a corrosion threat throughout the plant. Since much of the NH_3 production process operates at elevated temperatures, materials resistant to high-temperature corrosion and with sufficient mechanical strength are needed. The anhydrous NH_3 is purified and shipped or stored, usually in steel tanks.

Liquid anhydrous NH_3 is stored in one of three ways, selected on economic grounds (Ref 1, 2):

- Flat-bottomed steel storage tanks are used for storing very large volumes of NH_3, with typical capacity of 10,000 to 30,000 tons (up to 50,000). These tanks are designed for low pressure so the anhydrous NH_3 must be fully refrigerated at atmospheric pressure and at the atmospheric boiling point of -33 °C (-28 °F).
- Semirefrigerated tanks or spheres are used where larger volumes of NH_3 must be stored. These are held at some intermediate temperature, for example, -12 °C (10 °F), between ambient and fully refrigerated -33 °C (-28 °F) conditions. Cylindrical storage vessels (bullets) are also commonly employed.
- Pressurized tanks (around 2 MPa, or 300 psig) at ambient temperature in spheres or horizontal cylinders up to about 1700 tons are also used.

Ammonia and ammonium hydroxide (NH_4OH) are not particularly corrosive in themselves, but corrosion problems can arise with specific materials, particularly when contaminants are present. Air or oxygen contamination is a factor in many instances, causing general corrosion of some materials and localized corrosion, specifically stress-corrosion cracking (SCC), in others. Contamination with carbon dioxide can lead to corrosion due to carbamates, which are sometimes encountered in NH_3 recovery systems. High-temperature corrosion will occur in hot dissociated NH_3.

Aluminum Alloys

Aluminum and its copper-free alloys show good resistance to dry, gaseous NH_3 at ambient or elevated temperatures. Corrosion rates of <0.025 mm/yr (<1 mil/yr) at 21 °C (70 °F) and <0.05 mm/yr (<2 mils/yr) at 100 °C (212 °F) are typical (Ref 3).

Aluminum may be used for cargo tanks for the anhydrous product. If moisture is present there is some attack on aluminum, but a protective film soon forms and corrosion stops. Aluminum tubing is used in NH_3 refrigeration units operating in liquid NH_3 containing 5% water. In moist NH_3 vapor, corrosion is low (<0.1 mm/yr, or 3.9 mils/yr) below about 50 °C (120 °F). Under condensing conditions of steam and NH_3, aluminum can be attacked and the rate does not decrease with time. Attack is prevented if the CO_2-to-NH_3 ratio is at least 2.5 to 1. This high level of carbon dioxide can be achieved, for example, in NH_3 recovery plants by the decomposition of alcohols or other organic chemicals. Hydrogen sulfide also inhibits corrosion under condensing conditions. Aluminum is used for compressors, heat exchangers, evaporators, condensers, and piping in the production of NH_3. Aluminum pressure vessels are used in the storage and transport of NH_3 (Ref 4).

There is mild action on aluminum in NH_4OH solutions at temperatures below about 50 °C (120 °F). The greatest attack occurs in concentrated solutions (around 25%) and at about 5% concentration. For aluminum to perform well in NH_4OH the solution must be free from heavy metals and halogen ions. Attack is limited if the solution is saturated with aluminum ions before exposure takes place. As with anhydrous NH_3, aluminum is attacked at higher temperature, but this attack stops as a protective film forms (Ref 5, 6). Even at ambient temperature, corrosion rates decrease with time of exposure (Ref 7).

The initial attack of aluminum by dilute NH_3 solutions (up to ~10%) is controlled by the diffusion of OH^- ions to the surface and is a function of pH. The surface is passivated once sufficient corrosion product has been produced to form a protective film. Under some exposure conditions, the corrosion product may continue to dissolve rather than form a protective film (Ref 4).

Iron and Steel

Ferrous alloys are generally not corroded by NH_3 or NH_4OH at ambient or elevated temperatures. Corrosion can, however, occur in the presence of contaminants and ferrous alloys can be subject to SCC.

Cast Irons

Cast iron has been widely used in NH_3 production and handling. Since it is generally thick walled, somewhat higher corrosion rates can be tolerated as long as the iron in solution is acceptable. Cast iron strippers have been used to concentrate crude gas liquors with about 1 to 2% NH_3 up to 25% NH_4OH. A corrosion rate of 1 mm/yr (40 mils/yr) has been found in the vapor space of a ductile cast iron NH_3 pressure distillation plant operating at about 88 °C (190 °F). White cast iron with >18% Cr content (ASTM A532 grade IID) has been used in some applications.

High-nickel cast irons have been used in valves, pumps, and so forth, in NH_4OH solutions. Austenitic nickel cast iron NiResist cooling coils corroded at about 0.005 mm/yr (0.1 mil/yr) at 25 °C (75 °F) and 0.15 mm/yr (5.9 mils/yr) at 70 °C (160 °F) in flowing NH_3 gas (Ref 3). Laboratory testing of NiResist (International Nickel Company) cast irons and unalloyed cast irons gave the results shown in Table 1 (Ref 8). This table also shows data from NH_3-containing environments in industrial applications. In most applications unalloyed cast iron is at least as good as the NiResist irons, but in the concentrated and contaminated solutions the alloyed irons are superior. In most cases in this table, the recommended NiResist irons are types 1 (F41000) and 2 (F41002) with type 3 (F41004) also appropriate for the dilute solutions.

Carbon Steels

Ammonia is essentially noncorrosive to steels at ambient temperatures, which accounts for their widespread use. Carbon steels (commonly referred to as "steels" in many standards and regulations pertaining to NH_3 storage and transport) include both carbon steels and carbon-manganese steels and are commonly specified for process equipment. These alloys are by far the most commonly used materials for the storage and handling of NH_3. The standard used to specify the steel is selected based on the mechanical properties required by the application, with consideration of environmental factors.

Even at elevated temperatures, corrosion rates in anhydrous NH_3 are low, less than 0.05 mm/yr (1.9 mils/yr) in the range from 297 to 589 K (24 to 316 °C, or 75 to 600 °F) (Ref 9). One very important exception relative to corrosion resistance of steels is the tendency of anhydrous NH_3 to cause SCC (see below).

Ordinary carbon and alloy steels are satisfactory in NH_4OH service, although a superficial rusting will occur in the vapor space.

Stress-Corrosion Cracking of Steels. Ammonia SCC in carbon steel vessels was first reported in the mid-1950s in agricultural service tanks. Cracking occurred in areas of high residual stress, such as welds and cold-formed dished heads. Hot forming or stress relieving the heads considerably reduced the occurrence of cracking, as did the addition of a small amount of water to the NH_3.

Throughout the 1960s and early 1970s, cracking problems appeared to be mainly associated with high-strength quenched-and-tempered steels. Later, there were reports of cracking occurring in spheres containing anhydrous NH_3 with water additions and also in spheres that had been stress relieved after finding and repairing cracks.

The cause of this cracking is now accepted to be high local stresses and the presence of air contamination, although nitrogen and carbon dioxide are also thought to play a role. Cracking is accelerated by the use of high-strength steels, the presence of hard welds and air contamination (Ref 4, 10). The highest susceptibility to SCC has been found to be in liquid NH_3 with 3 to 10 ppm oxygen and a water content <100 ppm. Stress-corrosion cracking can occur, however, in NH_3 with an oxygen content down to 0.5 ppm when the water content is very low (Ref 11).

Possible ways to control or reduce liquid NH_3 SCC in carbon steels include (Ref 12):

- Eliminate oxygen
- Add around 0.2% water to the NH_3
- Use steel with an actual yield strength less than 300 MPa (44 ksi)
- Reduce residual stress by stress relieving
- Inspect often enough to detect cracks before they grow to dangerous proportions
- Use a sacrificial anode, such as a thermal sprayed zinc coating
- Cathodically protect the steel
- Use a different material such as stainless steel
- Paint the steel with a suitable protective coating

Some of these options are impractical at the present time (2005) since:

- There is no commercially available paint suitable for NH_3 submersion duty.
- Using stainless steel would increase the capital cost of vessel construction by at least 100% and increase steel mass.
- Cathodic protection has not yet been shown to be reliable in refrigerated installations and would not protect in vapor spaces, for example, in road tankers.
- Zinc thermal spraying of NH_3 tank internals has been used, largely on a trial basis. More experience is required.

The remaining options that are currently used for storage vessels or road tankers are evaluated in Table 2 (Ref 12).

Stress-corrosion cracking of carbon steel occurred in NH_3 receiver tanks used for recycling NH_3 in a urea plant. These tanks (SA 516 grade 70) suffered extensive cracking in welds and HAZ in spite of the addition of 0.2% water to the system. It was concluded that the water was not uniformly distributed throughout these vessels and was, probably inadequate to

Table 1   Corrosion data for NiResist and unalloyed cast iron in NH_3 solutions and environments

| | Temperature | | Average corrosion rate for: | | | |
| | | | Cast iron | | NiResist | |
Medium	°C	°F	mm/yr	mils/yr	mm/yr	mils/yr
5% NH_4OH	15.6	60	Nil	Nil	<0.01	0.01
10% NH_4OH	15.6	60	Nil	Nil	<0.01	0.2
25% NH_4OH	15.6	60	Nil	Nil	<0.01	0.18
50% NH_4OH	15.6	60	Nil	Nil	Nil	Nil
75% NH_4OH	15.6	60	Nil	Nil	Nil	Nil
Concentrated NH_4OH	15.6	60	0.05	2	Nil	Nil
5–6 vol% NH_3; 150 ppm phenol in H_2O vapor	102	215.6	0.05	2	0.02	0.9
NH_3 liquor separator tank	...	...	<0.01	0.09	<0.01	0.05
NH_3 liquor; 6.5 g/L NH_3	102	215.6	0.08	3	0.015	0.6
NH_3 liquor with sulfates, sulfides, etc.	37.8	100	<0.01	0.1	<0.01	0.01

Source: Ref 8

Table 2   Ammonia SCC mitigation measures

Corrosion mitigation scheme	Refrigerated tanks	Road tankers
Eliminate oxygen	This is a normal commissioning procedure, but regular inspection reintroduces oxygen. The number of inspections should be minimized, on-line inspection is desirable.	Not usually practical
Passivate ammonia with 0.2% water	This is now normal practice.	Vapor space not protected
Use steel with <300 MPa (<44 ksi) actual yield strength	Most tanks are constructed of grade 490 steel (ASTM A515 or equivalent) so actual yield easily exceeds 300 MPa (44 ksi).	Tankers in Australia are often made of quenched and tempered steel where yield >600 MPa (>87 ksi) to reduce tare weight.
Stress relieve	Not practical	Mandatory, especially where yield >300 MPa (>44 ksi) which is almost always.
Inspect regularly	Traditionally, tanks were shut down to inspect. Inspection on-line now becoming more possible and used.	Inspection is mandatory, usually every one or two years.

Source: Ref 12

protect the vapor space in the condensing NH_3 (Ref 13).

Nitriding of Steels. At high temperatures, NH_3 may dissociate into hydrogen and nascent nitrogen. The nascent nitrogen has a high affinity for iron and reacts to form a very hard, brittle metal nitride. This is sometimes a desirable reaction, and wear-resistant surfaces are commonly produced on steel parts by nitriding in an NH_3 atmosphere.

Although commercial nitriding is performed at temperatures above the normal service temperature for steels (495 to 565 °C, or 925 to 1050 °F), significant nitriding can occur at lower temperatures, resulting in loss of ductility. Ammonia dissociation is catalyzed by iron, which also contributes to the damage potential. For these reasons, steels are restricted to use at temperatures below 300 °C (600 °F) in NH_3 service. In NH_3 converters, nitriding layers can develop over time to a depth of several millimeters and these hard layers can cause brittle, surface cracks to form. Austenitic steels, in converter baskets for example, develop thin, hard nitrided layers that tend to flake off (Ref 1). Nitriding of pipes, grid supports, and so forth in the NH_3 converter is limited by using stainless steels with a higher nickel content, for example, type 347.

Hydrogen Attack. At high temperatures and pressures, hydrogen that is present in NH_3 synthesis can dissociate; the atomic hydrogen entering the steel lattice reacts with carbon to form methane (CH_4). This weakens the steel in two ways, by removing carbon from the steel (decarburization) and by forming blisters or fissures in so doing. This can eventually cause the vessel or pipe to rupture, often without any obvious prior deformation. Areas around weld seams are particularly prone to this phenomenon.

The risk of attack may exist at temperatures as low as 200 °C (390 °F) and hydrogen partial pressure as low as 7 bar (700 kPa). Selection of appropriately resistant materials can largely eliminate this problem. The classic Nelson curves described in the American Petroleum Institute (API) document 941 (Ref 14) give guidance on the stability limits of various alloys in terms of temperature and hydrogen partial pressure. These original guidelines have been modified in recent years on the basis of ongoing experience. For example, the low-alloy steels with 0.25 and 0.5% Mo are now classed as unalloyed steels from the viewpoint of resistance to hydrogen attack (Ref 1).

Steels with low levels of molybdenum and no chromium failed in a number of cases in catalytic reforming service, so care was recommended in the use of these alloys for that application. However, failures caused by hydrogen also occurred in hydrodesulfurization, NH_3 synthesis, and other parts of NH_3 production plants, leading to their removal from these safe operating charts (Ref 15). Curves (modified from Nelson curves) that incorporate findings from the extensive operating experience are available that provide safe operating conditions for carbon and low-alloy steels (Ref 14, 16). An example of a modified Nelson curve is shown in Fig. 1.

The standard chromium-molybdenum steels are being modified and improved so that they can be used at higher temperatures. A modified $2\frac{1}{4}$Cr-1Mo has been developed that is usable at temperatures up to 484 °C (903 °F) instead of being limited to 454 °C (849 °F) as are conventional steels of this type. This higher-temperature operation is permitted under ASME II rules. The modification included the addition of $\frac{1}{4}$% V. The API 941 standard has placed this modified steel at the same limits in terms of temperature and hydrogen as the 3Cr-1Mo steel. The first vessel made from this steel was an NH_3 converter fabricated in 1995. The vanadium-modified steel also has a better resistance to temper embrittlement than does the standard, unmodified version (Ref 16).

Hydrogen damage can also occur from the dissociation of molecular hydrogen into atomic hydrogen at elevated temperatures. Atomic hydrogen can diffuse into metal structures and recombine into molecular hydrogen at defects or discontinuities within the metal. This is also known as hydrogen embrittlement and is usually associated with welds that have not received correct postweld heat treatment (PWHT). In contrast to the phenomenon described above, this effect of hydrogen is reversible and hydrogen can diffuse back out of the structure if held at atmospheric pressure at around 300 °C (570 °F). Slow cooling from elevated-temperature and pressure operation is often recommended to permit this diffusion process to occur.

Temper Embrittlement. If heat-resistant steels are held at tempering temperatures, that is, above about 400 °C (750 °F), for long periods their impact properties can decline. The transition temperature between ductile and brittle behavior can be elevated to 60 °C (140 °F) from the normal 0 °C (32 °F) or below. This tendency to embrittle can be reduced by controlling the level of trace elements (Si, P, Mn, Sn) in the steel. In this respect, modern steels are generally much cleaner than some of the older steels and are less prone to this phenomenon. Vessels or pipes in which temper embrittlement is anticipated should not be pressurized at low temperatures.

Hydrogen Sulfide Attack. Most high-temperature steels are attacked by hydrogen sulfide (H_2S) in the gas stream in partial oxidation plants. The use of austenitic stainless steels eliminates this problem, but stress relief of welds is advised in these plants to avoid SCC by chlorides sometimes present in the feed oil (Ref 1).

Corrosion can also occur from the condensation of sulfur acids at the colder ends of flue gas ducts. Type 304 stainless steel is used at the cold ends. Carbon steel air preheater tubes in this area can corrode. Alternative materials include cast iron, glass, or coated steel.

Alloy Steels

Steels alloyed with small amounts of chromium, molybdenum, nickel, or other elements are referred to as low-alloy steels, and they can provide an economical alternative to high-alloy steels at both high and low temperatures. Low-alloy steels have been used effectively in NH_3 storage vessels. These steels are also subject to SCC with high stresses and air contamination being the main factors causing cracking. The addition of 0.1 to 0.2% water inhibits this attack (Ref 4). Chromium-molybdenum steels are the most common steels of this type used in NH_3 applications.

Chromium-molybdenum steels are commonly used at elevated temperatures where the alloying additions provide resistance to hydrogen attack and increase the strength of the alloy. Typical alloys are 1.25Cr-0.5Mo (K11597), 2.25Cr-1Mo (K21950), 5Cr-0.5Mo (K41545),

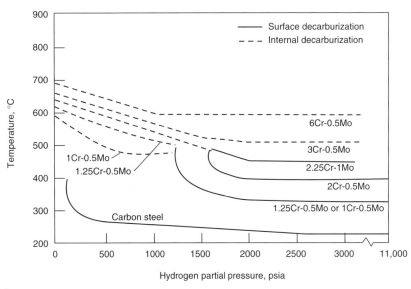

Fig. 1 Modified Nelson curve showing the operating limits for steels in hydrogen service. Source: Ref 14

7Cr-0.5Mo (S50300), and 9Cr-0.5Mo (S50400). Their use in NH_3 service is chiefly in NH_3 synthesis, the alloy selection being based on API Publication 941 (Ref 14). Old editions of this standard have curves showing the resistance of low-alloy steels, with 0.5% Mo without chromium. As mentioned previously in the section "Hydrogen Attack," service experience has shown that this alloy has little more resistance to hydrogen attack than carbon steel so the API publication was updated in 1991 to reflect this experience. Other uses for these chromium-molybdenum steels are in high-strength parts such as fasteners.

Nickel Alloy Steels. Addition of nickel to steel greatly enhances the low-temperature toughness (impact properties). For this reason, these materials are sometimes specified for low-temperature NH_3 service, especially as weld filler metals. Typically, the nickel-alloy steels contain 3.5% (K32025), 6%, or 9% Ni (K81340).

Stainless Steels

From the point of view of corrosion, stainless steels are not normally required in NH_3, but they do find many applications, particularly for low temperatures and to avoid iron contamination. All grades of stainless steel are resistant to NH_4OH solutions at up to the atmospheric boiling point.

Ferritic Grades. The standard ferritic grades are rarely used in NH_3 service, but the modern superferritic steels find occasional applications. The first superferritic steels were based on 26% Cr, 1% Mo (S44627), and the niobium-stabilized XM-27 (S44627). These were developed to provide better resistance to chloride SCC than the austenitic 300 series. Superferritic steels include AL 29-4C (Allegheny Ludlum Corporation) (S44735), AL 29-4-2 (Allegheny Ludlum Corporation) (S44800), Sea-Cure (Plymouth Tube Co.) (S44660), and Monit (Uddeholm, originally; now, Outokumpu Stainless) (S44635). The superferritic AL 29-4C (S44735) has been used successfully in an NH_3 stripper reboiler handling water, NH_3, H_2S, and steam (Ref 17).

Precipitation-Hardening Grades. The PH stainless steels see limited service in NH_3 environments, but play an important role nonetheless. These alloys exhibit very high strengths combined with good notch toughness and corrosion–resistance properties. For this reason, they are often the preferred material for such parts as valve stems and certain critical fasteners. Some of these alloys are susceptible to hydrogen embrittlement in corrosive or hydrogen-rich environments.

Duplex stainless steels are not used in NH_3 service except where their resistance to chloride SCC is useful from the water side of heat exchangers. The modern grades, for example, alloy 2205 (S31803), typically also contain molybdenum to achieve a composition of about 22% Cr, 5.5% Ni, 3% Mo, and 0.03% C max. They are strengthened and stabilized by nitrogen. The mixed austenite-ferrite structure imparts strength and resistance (but not immunity) to chloride pitting and SCC. The duplex 3RE60 (S31500) has been used to replace standard austenitic stainless steels in this type of situation. The duplex stainless steels have higher strength than the lower austenitic grades such as type 304 (S30400), but are subject to temper embrittlement at about 475 °C (885 °F).

Welding tends to lead to variations (20 to 80%) in the austenite/ferrite ratio in the as-cast weld bead and fusion zone. To minimize this effect, they are welded with a nickel-rich, overmatching rod.

Austenitic Stainless Steels. The standard austenitic grades, for example, types 304 (S30400) and 316 (S31600), find widespread use in NH_3 service since they are resistant to general corrosion and NH_3 SCC. Stabilized and low-carbon grades are not generally needed in NH_3 applications. Type 304 (S30400) stainless steel was found to be resistant to intergranular attack (IGA) in 28% NH_3 solution at room temperature even after sensitizing for 1 h at 677 °C (1250 °F) (Ref 18). Stainless steels can, however, fail by SCC that is initiated by sensitization formed during heating, for example when welding. Such a failure in a reformer tube occurred at the welded junction between the type 321 flange and HK-40 catalyst tube. The weld and HAZ were sensitized when joined using a high-carbon welding rod. The high-nickel alloy weld that was intended to protect the cast steel (HK-40) from corrosion had inadvertently been machined off, thus exacerbating the corrosion (Ref 19). Fuel gas lines that contain H_2S (stripped from feedstock naphtha) are subject to corrosion. The use of steam-traced type 304 limits this problem.

The high cost of stainless steels relative to carbon and low-alloy steels often precludes their use for large equipment, but their many advantages lead to their selection for smaller parts and components. They are used in low-temperature service since they have an extremely low nil-ductility transition temperature (NDTT) and exhibit excellent notch toughness at temperatures far below the atmospheric boiling temperature of NH_3. In elevated-temperature services in NH_3 synthesis, they are used because of their resistance to hydrogen attack and nitriding. Start-up heater coils in the NH_3 synthesis section are vulnerable to high-temperature corrosion and fatigue. Usually, type 321 performs well here.

Some shallow nitriding does occur after years of service in aggressive environments, but due to the excellent notch toughness of the base material, this effect does not reduce the structural integrity. Another common area of application of the basic grades is in heat-exchanger tubes, where the material selection is influenced by the fluid used to heat or cool the NH_3.

These standard grades can suffer from chloride SCC, particularly in partial oxidation plants in which the feed oil contains chlorides. Chloride SCC failures can also occur under upset condition. An example of such a failure occurred in type 321 and 310S tubes in a waste heat boiler in an ammonium synthesis converter. The direct cause of this SCC was the failure of a feedwater boiler pump that was not repaired or replaced and permitted a buildup of deposits in the system. General and intergranular corrosion was also observed on these tubes (Ref 20).

There are many more specialty grades of stainless steels that see service where some factor limits the use of other materials. A very common example is the use of special ferritic grades or duplex grades of stainless steels as heat-exchanger tubes where austenitic stainless steels are subject to pitting or chloride SCC from cooling waters. The martensitic grades of stainless steel are used where high strength is required, especially at elevated temperatures, as in compressor rotors.

Cast Stainless Steels. The equivalent cast version of types 304 (S30400) and 304L (S30403) are CF-8 (J92600) and CF-3 (J92500), respectively, and they exhibit approximately the same corrosion response as the wrought alloys. However, castings can have surface layers containing more than the maximum allowable carbon content of 0.08%, which can significantly reduce corrosion resistance of the surface.

In the cast form, the difference between CF-8 or CF-3 and the corresponding molybdenum-bearing grades CF-8M (J92900) and CF-3M (J92800) is insignificant. The CF-3 or CF-8 cast steels are not particularly common, and manufacturers of cast pumps and valves tend to standardize on CF-8M, which has a broader range of applications. The molybdenum grade castings are often more available and less expensive than the 304 or 304L wrought grade equivalents. Since the cast version of these alloys is unlikely to be welded, there is rarely a justification for specifying the low-carbon (L) grades in this case. This assumes that the valves or pumps, if weld repaired by the manufacturer, are properly reheat treated (solution annealed) to restore optimum corrosion resistance. Availability and price are likely to favor the non-L grade, and properly heat treated castings in CF-8 or CF-8M are likely to be as corrosion resistant as their low-carbon cast or wrought equivalents.

Alloys for Use at Elevated Temperatures

Many of the operations in NH_3 production take place at elevated temperatures. Some of the materials used for these applications are conventional iron-base alloys that have been discussed previously or nickel-base alloys. However, specialized elevated-temperature duties in NH_3 production, petroleum refining, and so forth, have generated a demand for ever better materials specific to these duties. Many of the materials used here are proprietary and have been developed to have high strength and good creep-resistant properties in aggressive, gaseous environments.

The standard stainless steels—types 304, 310, and 347—were commonly used but often failed by cracking at elevated temperatures. A better alloy, HK-40 (J94204, 25Cr-20Ni), was developed, and this became the industry standard (Ref 21).

A later material development produced the HP (N08705, 26Cr-35Ni) alloy with 26% Cr and 35% Ni, which was modified by the addition of alloy elements such as molybdenum, niobium, or tungsten. These modified alloys have improved creep resistance, but still possess good ductility and weldability. Various HP-modified alloys are used in NH_3 applications, one of the most common being HP-45Nb (Ref 22). A more recent development is the production of microalloyed HP alloys in which trace quantities of titanium, zirconium, and rare earths are added during casting. These alloys have better resistance to carburization and better high-temperature creep-rupture properties (Ref 21).

There are new alloys, many of them proprietary such as 25Cr-35Ni-15Co-5W, 28Cr-48Ni-5W, and 35Cr-45Ni plus additional elements, and these offer higher strength and good high-temperature properties (Ref 22).

There has also been some development of coatings to resist high-temperature attack by NH_3. Uncoated type 304 (S30400) and 316 (S31600) and the same steels coated with silica applied by a sol-gel procedure were tested in anhydrous NH_3 at high temperature. The uncoated samples were attacked and formed a nitride scale that embrittled the metal. After 115 h of testing at 500 °C (932 °F) uncoated samples were completely degraded, while the coated samples were only lightly attacked. Multilayer coatings were most effective, and the stainless steel substrates were not sensitized to intergranular attack by the high-temperature coating process (Ref 23).

Metal Dusting

The phenomenon of metal dusting occurs during high-temperature operation, for example, in steam superheaters downstream of the secondary reformer. It is related to the process of carburization in which carbon migrates into the structure forming hard carbides. Carburization occurs above about 800 °C (1470 °F) in the presence of hydrocarbons that crack to provide the carbon.

Metal dusting occurs at 500 to 800 °C (930 to 1470 °F) on iron-nickel or iron-cobalt alloys in gases containing carbon monoxide. Carbon formation is catalyzed by iron, nickel, or cobalt, and the effect is to produce a surface dust layer consisting of a mixture of metal, oxides, and carbon. This dusting is usually observed as pitting or general corrosion attack. Theoretically, alloys that form protective films of the oxides of chromium, silicon, or aluminum should be more resistant. Virtually all high-temperature alloys can be prone to this attack, but higher steam/CO levels help as does coating with aluminum. In nickel-base alloys such as 601, 601H, 625, and so forth, the attack is tolerable in normal operation (Ref 1).

Hydrogen sulfide in the gas offers some protection from metal dusting since the adsorbed sulfur blocks the surface for the adsorption of CO or CH_4 and other hydrocarbons, and the molecules cannot adsorb and dissociate if their adsorption sites are occupied by sulfur. Similarly, dense oxide scales can prevent ingress of carbon into the structure. If sulfur cannot be tolerated in the process, nickel alloys with high chromium and aluminum or silicon additions are the best materials to resist metal dusting (Ref 24).

Alloy resistance to metal dusting is dependent on its ability to form a protective chromium oxide scale and can be ranked according to its chromium equivalence:

$$\text{Cr equivalent} = \text{Cr}\% + 3 \times (\text{Si}\% + \text{Al}\%)$$

Values of Cr equivalent for alloys (some of which are proprietary) commonly used in high-temperature applications in ammonia plants are shown in Table 3 (Ref 22).

There are also nickel alloys that are said to have exceptional resistance to metal dusting and elevated-temperature corrosion. The standard alloys such as alloy 600 and alloy 601 have been used in these environments, but other alloys have been developed with superior high-temperature behavior. Alloy 693 (N06693) was tested for a year in CO and 20% hydrogen at 621 °C (1150 °F). Pit depths measured on this alloy were only 0.031 mm (1.2 mils). Pit depths for other alloys tested under the same conditions were (Ref 25):

Alloy	Pit depth	
	mm	mils
800 (N08800)	8.164	321
DS	3.451	136
601 (N06601)	0.033	1.3
620	0.34	13.4
690 (N06690)	0.293	11.5

An example of the failure of alloy 601 occurred in a heat exchanger on an NH_3 plant after only 2½ years service. This failure was found to be caused by contamination of the process side by steam. This caused mineral deposition that destroyed the protective film within about 6 months after the steam ingress into the process side (Ref 26).

Another nickel alloy developed for resistance to metal dusting and carburizing is alloy 602CA (N06025). The metal wastage rate of this alloy is compared with other nickel alloys in a strongly carburizing $CO-H_2-H_2O$ gas in Fig. 2 (Ref 27). This alloy has been tested in an NH_3 plant in Europe and showed no metal dusting at temperatures of 450 to 850 °C (840 to 1560 °F). Alloy 601 had some attack, and alloy 800H was severely attacked.

There are also coatings being developed to resist metal dusting. Diffusion coatings based on oxide formers such as silicon, titanium, chromium, and aluminum have been tested and found to show good potential to extend the lifetime of iron-base and nickel-base alloys under metal dusting conditions (Ref 28).

Nickel and Nickel Alloys

Nickel alloys are seldom used in NH_3 service except at elevated temperatures. They are very

Table 3 Chromium equivalents for alloys commonly used in NH_3 plants

Alloy	Nominal composition, wt%				Chromium equivalent	Expected performance
	Cr	Ni	Si	Al		
Wrought alloys						
304	18	10	...	...	18	Poor
800/800H	20	32	0.3	0.3	22	Poor
803	25	35	0.3	0.3	27	Fair
310	25	20	0.3	...	26	Fair
600	15	72	...	...	15	Fair
601	22	60	...	1.5	27	Good
617	22	52	...	1.2	26	Good
214	16	76	...	4.5	30	Good
APM	22	...	...	6.0	40	Best
Cast alloys						
HK-40	25	20	1.0	...	28	Good
HP-Mod	26	35	1.5	...	30	Good
XTM	35	48	1.5	...	40	Best

Source: Ref 22

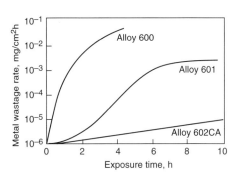

Fig. 2 Metal wastage rates of nickel-base alloys in a strongly carburizing atmosphere at elevated temperatures. Source: Ref 27

resistant to dry NH_3, but can be attacked by gaseous NH_3 if more than about 1% water is present. They are resistant to anhydrous NH_3 and exhibit good resistance to nitriding (Ref 29).

Data in Table 4 show the effect of temperature on nitrogen absorption and nitriding depth for nickel alloys compared with a type 310 stainless steel (Ref 30). Samples were exposed to pure NH_3 for 168 h.

Nickel alloy 200 (N02200) will resist NH_4OH only up to about 1% concentration. Dissolved oxygen (DO) may maintain passivation up to about 10% concentration. Higher concentrations are highly corrosive to nickel even in the presence of air (Ref 31).

The corrosion rate of alloy 200 in agitated solutions of various strengths of NH_4OH at room temperature is shown in Table 5 (Ref 31). These data show that even with agitation nickel is attacked by moderate concentrations of NH_4OH at room temperature.

Alloy 400, with about 30% Cu content, is more resistant than alloy 200 as shown in Table 6 (Ref 31). In solutions of >3% NH_4OH the corrosion rate is increased considerably by aeration and agitation.

The corrosion rate of alloy 800 in 5% and 10% NH_4OH at 80 °C (176 °F) was <0.003 mm/yr (0.12 mil/yr) in 7 day laboratory tests (Ref 32).

The Ni-Cr-Mo alloys such as alloy 625 (N06625) and alloy C-276 (N10276) are resistant but find no application because of the adequate resistance of lesser alloys.

Copper and Its Alloys

Copper alloys are generally to be avoided in NH_3 service. Although resistant to pure, dry NH_3, contamination by water and oxygen will cause SCC and general corrosion. Corrosion of various alloys in deaerated NH_3 are shown in Fig. 3(a) and are compared with the behavior of the same alloys in aerated ammonia in Fig. 3(b) (Ref 33). These data clearly show the corrosive effect of the presence of air in the solution. Carbon steel (A-285) is included in these data for comparison, and its corrosion behavior is also adversely affected by the presence of oxygen at lower oxygen levels.

Ammonia and copper typically react to form an intensely blue copper/ammonium complex. All copper-base alloys can be made to crack in NH_3 vapor, NH_3 solutions, ammonium ion solutions, and environments in which NH_3 is formed. It is generally true that any metal with a small grain size is more resistant to SCC, irrespective of whether the cracking is transgranular or intergranular. The effect of grain size on the time to cracking of yellow brass (C26800) in NH_3 is shown in Fig. 4 (Ref 34).

While some copper-base alloys are far superior to others in their resistance, and dry anaerobic NH_3 does not cause corrosion, it is general industry practice to avoid all use of copper-base alloys in NH_3 and related services. It can be noted in passing that copper in solid solution in ferrous metals (generally less than 3%) added to attain certain physical, mechanical, or corrosion properties does not pose a problem in NH_3 applications.

Copper alloys C11200 and C26000 were penetrated at a rate of 5 μm/yr (0.2 mil/yr) in anhydrous NH_3 at atmospheric temperature and pressure. Corrosion rates were also low if small amounts of water were present, but oxygen was also probably excluded (Ref 4).

Table 4 Effect of temperature on nitriding depth in various nickel alloys

Alloy	Nitrogen absorption, mg/cm²			Nitriding depth			
	at 650 °C (1200 °F)	at 980 °C (1800 °F)	at 1090 °C (2000 °F)	at 980 °C (1800 °F)		at 1090 °C (2000 °F)	
				μm	mils	μm	mils
214	1.5	0.3	0.2	35.6	1.4	18	0.7
230	0.7	1.4	1.5	124	4.9	389	15.3
617	1.3	1.5	1.9	381	15.0	>559	>22
601	1.1	1.2	2.6	168	6.6	>584	>23
X	1.7	3.2	3.7	188	7.4	>584	>23
556	4.9	6.7	4.2	373	14.7	>508	>20
800H	4.3	4.0	5.5	282	11.1	>762	>30
310	7.4	7.7	9.5	384	15.1	>787	>31

Source: Ref 30

Table 5 Corrosion of alloy 200 in agitated NH_4OH solutions at room temperature

NH_4OH concentration, %	Corrosion rate	
	mm/yr	mils/yr
1.1	0	0
12.9	14.2	559
20.2	9.4	370
27.1	4.6	181

Source: Ref 31

Table 6 Corrosion of alloy 400 in agitated NH_4OH solutions at room temperature

NH_4OH concentration, %	Corrosion rate	
	mm/yr	mils/yr
2.7	0	0
3.6	1.8	71
5.5	7.6	299
8.2	8.1	319
11.1	8.3	327
18.3	5.9	232
25.8	0.9	35

Source: Ref 31

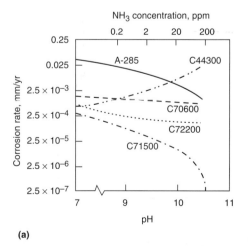
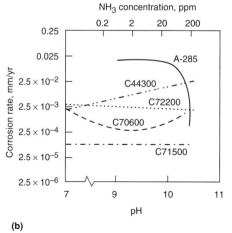

Fig. 3 Corrosion of various copper alloys in (a) deaerated and (b) aerated NH_3. A carbon steel (A-285) is included for comparison. Source: Ref 33

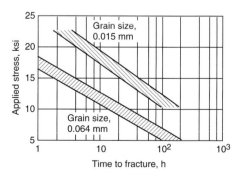

Fig. 4 The effect of grain size on the time to fracture of yellow brass (C26800) in NH_3. Source: Ref 34

Copper alloy tubes in utility condensers have often been attacked by NH_3 on the steam side. The NH_3 in this case comes from decomposition of amines or hydrazine added to the boiler feed to control oxygen and passivate the boiler surface. Admiralty brass is commonly used to tube such condensers and has been subject to this attack by NH_3. Copper alloy condensers form a surface layer of cuprous oxide when placed in service. If excess oxygen is present, this oxide layer is converted to cupric oxide, which is readily complexed by NH_3. Attack by NH_3 in this type of utility condenser is accelerated by air in-leakage. In cases where boiler chemistry cannot be altered or controlled to avoid NH_3 formation, a more resistant alloy, such as copper-nickel, is used (Ref 35, 36).

All copper-base alloys are attacked by NH_4OH unless air is rigorously excluded, which is not feasible in plant practice. The deep royal blue of the copper/ammonium complex is immediately obvious.

Dealloying can occur also in some copper alloys in NH_3 solutions. It was found that dezincification occurred in 70Cu-30Zn brass in 10 N NH_4OH solution at 32 °C (90 °F). Corrosion and dezincification were increased by increasing stress applied to the alloy. The mechanism of this increased attack was shown to be due to an increase in open circuit potential and a shift in the polarization curve under the influence of applied stress (Ref 37).

Titanium and Titanium Alloys

Titanium is not attacked by atmospheres containing NH_3, but can be corroded at elevated temperatures. The protective oxide film is effective in NH_3 up to at least 300 °C (570 °F) (Ref 4). At higher temperatures, NH_3 will decompose into nitrogen and hydrogen that may cause hydrogen embrittlement of titanium. Titanium corroded at 11.2 mm/yr (440 mils/yr) in an NH_3-steam mixture at 221 °C (430 °F). This high corrosion rate was thought to be associated with hydriding. Titanium shows excellent resistance to corrosion in up to 70% NH_4OH up to the boiling point (Ref 38). The corrosion rate of titanium in 100% anhydrous NH_3 is <0.13 mm/yr (5 mils/yr) at 40 °C (105 °F). In 28% NH_4OH solution at room temperature, the rate is 0.0025 mm/yr (~0.1 mil/yr) (Ref 39).

Zirconium and Its Alloys

Zirconium is resistant to NH_3 even at elevated temperatures. The corrosion rate of zirconium Zr702 (R60702) in wet NH_3 at 38 °C (100 °F) is less than <0.127 mm/yr (<5 mils/yr) and in 28% NH_4OH at up to 100 °C (212 °F) is <0.025 mm/yr (<1 mil/yr) (Ref 40). Zirconium is stable in NH_3 up to about 1000 °C (1830 °F) (Ref 4).

Niobium and Tantalum

Niobium is not attacked by 13% and 25% NH_4OH solutions at 20 to 100 °C (68 to 212 °F) (Ref 3).

Tantalum is resistant to anhydrous liquid NH_3, but should not be exposed to the gaseous mixtures encountered in NH_3 synthesis at elevated temperature. Above about 250 °C (480 °F), it reacts rapidly with hydrogen to form brittle hydrides. Tantalum is not corroded by 10% aqueous NH_4OH solutions up to 100 °C (212 °F), but is attacked by hot, concentrated NH_4OH solutions (Ref 3).

Other Metals and Alloys

Lead is resistant in NH_3 at temperatures up to 60 °C (140 °F) and hard lead (alloyed with antimony) is satisfactory up to 100 °C (212 °F) in dry NH_3. Lead is also resistant to NH_4OH at room temperature (Ref 41). The corrosion rate is very sensitive to the presence of air and agitation. In 27% NH_3 solution at 20 °C (68 °F), lead had a corrosion rate of 110 g/m^2/d (3.3 mm/yr) with rapid agitation and only 21 g/m^2/d (0.63 mm/yr) without agitation (Ref 7).

Tin is resistant to dry NH_3 and saturated NH_3 solutions, but dilute NH_3 solutions corrode tin.

Zinc is not resistant to NH_4OH.

Magnesium is not attacked by wet or dry NH_3 at ordinary temperatures, but attack may occur if water vapor is present.

The precious metals are also not used. In fact, there is a potential hazard if silver is exposed under some conditions because explosive azides may be formed. Precious metals are not employed, and silver or silver-rich alloys are not to be employed in NH_3 or NH_4OH.

Nonmetallic Materials

Although ferrous and nonferrous alloys are the most commonly used materials for the manufacture, handling, and storage of NH_3 and NH_4OH, nonmetallic materials are also used for some applications. These include:

- Elastomers
- Plastics, both thermoplastics and thermosetting resins
- Carbon and graphite
- Glass

Elastomers. Some elastomeric materials are attacked by NH_3, for example, butyl, Viton A (E.I. Du Pont Co.), hard rubber, isoprene, and natural rubber. Satisfactory materials include Chemraz (Greene, Tweed), Kalrez (E.I. Du Pont Co.), chloroprene rubber, acrylonitrile rubber (Buna N), and butadiene-styrene rubber (Buna S). Ethylene-propylene(-diene) rubber (EPDM) is resistant to NH_3, but may be attacked by oils present in compressed gas systems. Buna N elastomer used in pumps is rated as suitable for cold NH_3 gas but fair to poor in hot gas. Neoprene is suitable in cold NH_3 while polytetrafluoroethylene (PTFE) is suitable in cold or hot NH_3 (Ref 42).

Satisfactory materials in NH_4OH include Chemraz, Kalrez, and chloroprene rubber. Some elastomers that are attacked and are unsatisfactory are butyl, Viton A, hard rubber, isoprene, and natural rubber.

Temperature limits for various elastomers are given in Table 7. These data are a compilation from various sources.

Plastics. Most plastics are chemically resistant to NH_3 and NH_4OH at ambient temperatures. However, due to the hazardous nature of NH_3, the use of plastics is generally not recommended except in seals and gaskets, where fluoropolymers are typically used.

Thermoplastics. Resistant grades of fluoropolymers are ethylene chlorotrifluoroethylene (ECTFE), for example, Halar (Allied Corp.), ethylene trifluoroethylene (ETFE), for example, Tefzel (E.I. Du Pont Co.), polyvinylidene fluoride (PVDF), for example, Kynar (Arkema), fluorinated ethylene propylene (FEP), perfluoro alkoxy (PFA), and polytetrafluoroethylene (PTFE). Some plastics that are generally unsatisfactory in NH_3 are acrylonitrile-butadiene-styrene (ABS), epoxy, certain polyesters, polyisobutylene, and polystyrene. Data on temperature limits for various common thermoplastics are shown in Table 8. These data are a compilation from various sources. Various manufacturers list elevated-temperature limits for plastics in liquid NH_3, and these are included in this table. They are not, however, of very practical interest since it is highly unlikely that plastics would be used in elevated-temperature (therefore, also high-pressure) liquid NH_3. In the unlikely event that an application for a plastic

Table 7 Temperature limits for various elastomers used in NH_3 service

Plastic	Gaseous NH_3 °C	°F	Liquid NH_3 °C	°F	Saturated NH_3 solution °C	°F
Natural rubber (NR)	60	140	20	68	60	140
Nitrile butyl rubber (NBR)	40	105	20	68	60	140
Ethylene propylene(-diene) monomer (EPDM)	60	140	20	68	80	175
Chloroprene (CR)	60	140	20	68	80	175
Chlorosulfonated polyethylene (CSM) Hypalon	60	140	20	68	...	...
Fluoroelastomers Viton	NR	NR	NR	NR	...	...

NR, not resistant. Hypalon and Viton are registered trademarks of E.I. Du Pont Co.

was being considered in liquid NH_3, the subzero properties would be more relevant and should be investigated.

One area of interest is the difference in behavior of polyvinyl chloride (PVC) and chlorinated polyvinyl chloride (CPVC). In many environments CPVC is more resistant than PVC and can withstand higher temperatures. This is not the case, however, in NH_3 and amines. Polyvinyl chloride has generally good resistance to NH_3 and some amines, even at somewhat elevated temperatures, while CPVC has extremely poor resistance to NH_3 or NH_4OH, and limited resistance to most amines, even at ambient temperatures. This is due to the extremely high reactivity of amines and chlorine, the higher availability of chlorine in CPVC, and the lower bond strength of CPVC versus PVC. Even at fairly low concentrations and temperatures, NH_3 and many amines are capable of rapid dehydrochlorination of CPVC. One CPVC pipe handling 28% NH_4OH at ambient temperature failed after only 1 year in service (Ref 43).

If polyethylene is used to store concentrated aqueous ammonium solutions, there is a weight loss due to outward diffusion through the plastic. A solution of 27% NH_4OH solution kept in a 500 mL bottle (1 mm (0.04 in.) wall thickness) at 20 °C (68 °F) for 54 days lost 3.6% of its weight (Ref 3).

Use of many of these thermoplastics is confined to linings for pipe and vessels, rather than as solid construction items, because of the hazardous nature of NH_3. Recommended temperature limits for plastic-lined steel pipes are given in Table 9 (Ref 44).

Thermoplastics are also used as the resistant liner in dual-laminate construction in which fiber-reinforced plastic (FRP) is used as the reinforcing, structural element. The corrosion resistance depends on the resistance of the thermoplastic liner, although resistant resins are often used in the FRP reinforcement in case of permeation and leaks in the thermoplastic liner. Most common thermoplastics are used in this type of construction, and many of them are suitable for use in NH_3 applications. Data on temperature limits for various thermoplastics used for dual-laminate construction for NH_3 service are given in Table 10 (Ref 45).

Thermosetting Resins. The fiberglass-reinforced thermosetting composites, commonly called FRPs, have a resistance determined by the polymer used. Some suggested concentration and temperature limitations of FRP are shown in Table 11. Data are a compilation from various sources.

Carbon and graphite are resistant to NH_4OH. However, with impervious graphite heat exchangers, an epoxy resin should be specified rather than phenolic, which is attacked by alkaline chemicals. Carbon and graphite are resistant to all solutions of NH_3 up to their limiting temperature, which depends on the individual grade and formulation. Graphite is resistant to anhydrous NH_3 over the full range of concentration up to the temperature limit of the graphite. Fluorocarbon-bonded graphite, Diabon F100 (SEL Carbon Group), is resistant in 20% NH_3/caustic NH_3 at up to 40 °C (105 °F) (Ref 46).

Glass is resistant to about 60 °C (140 °F) in very dilute solutions of NH_4OH (perhaps 1%), but will withstand solutions to pH 14 at room temperature in lined steel (Ref 47).

REFERENCES

1. M. Appl, *Ammonia: Principles and Industrial Practice,* Wiley-VCH, Weinheim, Germany, 1999, p 209–221

Table 8 Temperature limits for various plastics used in NH_3 service

	Gaseous NH_3		Liquid NH_3		10% NH_3 solution	
Plastic	°C	°F	°C	°F	°C	°F
Polyethylene (PE)	60	140	60	140	60	140
Polybutylene (PB)	60	140	20	68	60	140
Polypropylene (PP)	60	140	60	140	60	140
Polyvinyl chloride (PVC)	60	140	60	140	60	140
Chlorinated PVC (CPVC)	40 fair	105 fair	...	...	40 fair	105 fair
Polyvinylidene fluoride (PVDF)	80	175	40	105	100	212
Polytetrafluoroethylene (PTFE)	120	250	120	250	120	250

Table 9 Temperature limits for plastic-lined pipe used in NH_3 service

	PP		PVDF		PTFE	
Ammonia or hydroxide	°C	°F	°C	°F	°C	°F
Anhydrous gas	65	150	NR	NR	230	450
Anhydrous liquid	110	225	NR	NR	230	450
1% NH_4OH	110	225	110	225	230	450
10% NH_4OH	110	225	110	225	230	450
Concentrated NH_4OH	110	225	110	225	230	450

NR, not resistant. Source: Ref 44

Table 10 Temperature limits for various plastics used in dual-laminate construction for NH_3 service

	Gaseous NH_3 (technically pure)		Aqueous NH_4OH (~25%)	
Plastic	°C	°F	°C	°F
Polyvinyl chloride (PVC)	60	140	40, 60(a)	105, 140(a)
Chlorinated PVC (CPVC)	...	...	40(a)	105(a)
Polyethylene (PE)	60	140	60	140
Polypropylene (PP)	60	140	60	140
Polyvinylidene fluoride (PVDF)	40, 100(a)	105, 212(a)	...	...
Ethylene chlorotrifluoroethylene (ECTFE)	20	68	100	212
Ethylene trifluoroethylene (ETFE)	150	300	150	300
Fluorinated ethylene propylene (FEP)	150	300	150	300
Perfluoro alkoxy (PFA)	150	300	150	300

(a) Conditionally resistant at this temperature. The medium can attack the material or cause swelling. Restrictions must be made in regard to pressure and/or temperature, taking the expected service life into account. The service life of the installation can be noticeably shortened. Source: Ref 45

Table 11 Suggested temperature limits for fiberglass-reinforced plastic used in NH_3 service

	Bisphenol A fumurate		Vinyl ester		Epoxy		Furane	
NH_3 or NH_4OH(a)	°C	°F	°C	°F	°C	°F	°C	°F
Liquified NH_3	NR	NR	NR	NR	NR	NR	NR	NR
NH_3 gas, dry	60	140	93	200	...	...	...	...
NH_3 gas, wet	93	200	93	200	...	...	...	...
5% NH_4OH	82	180	82	180	82	180	38	100
10% NH_4OH	60	140	60	140	71	160	38	100
20% NH_4OH	60	140	60	140	66	151	NR	NR
29% NH_4OH	38	100	38	100	52	126	NR	NR

(a) Limiting allowable concentration. NR, not resistant

2. *Production of Ammonia,* Vol 1, EFMA, European Fertilizer Manufacturers' Association, Brussels, Belgium, 2000, 44 pages
3. Ammonia and Ammonium Hydroxide section, *Dechema Corrosion Handbook,* CD, Dechema aV, Frankfurt, Germany, 2001
4. B.D. Craig and D.B. Anderson, Ed., *Handbook of Corrosion Data,* ASM International, 1997, p 128–135
5. *Aluminium with Food and Chemicals,* Alcan Industries Ltd, London, U.K., 1966, p 18
6. F.L. LaQue and H.R. Copson, *Corrosion Resistance of Metals and Alloys,* 2nd ed., Reinhold Publishing Corp., 1963
7. E. Rabald, *Corrosion Guide,* Elsevier Scientific Publishing Co., Amsterdam, Netherlands, 1968, p 46–50
8. R. Covert, J. Morrison, K. Rohrig, and W. Spear, *Ni-Resist and Ductile Ni-Resist Alloys,* Reference book No. 11018, Nickel Development Institute, 1998, 42 pages
9. P. Ludwigsen and H. Arup, Stress Corrosion Cracking of Mild Steel in Ammonia Vapor above Liquid Ammonia, *Corrosion,* Vol 32 (No. 11), 1976, p 430–431
10. A.W. Loginow, Stress-Corrosion Cracking of Steel in Liquefied Ammonia Service, *Mater. Perform.,* Vol 25 (No. 12), 1986, p 18–22
11. L. Lunde and R. Nyborg, Stress Corrosion Cracking of Carbon Steel Storage Tanks for Anhydrous Ammonia, *Proc. International Fertiliser Society,* Proceeding No. 307, 1991
12. P. McGowan, "Managing Ammonia Stress Corrosion Cracking," Plenary Address to the Australasian Corrosion Association Conference, Auckland, 2000
13. K.C. Pattnaik and M.P. Gupta, Stress Corrosion Cracking of Ammonia Receiver Tank, *Br. Corros. J.,* Vol 30 (No. 1), 1995, p 80
14. "Steels for Hydrogen Service at Elevated Temperatures and Pressures in Petroleum Refineries and Petrochemical Plants," API 941, American Petroleum Institute
15. K.L. Baumert, G.V. Krishna, and D.P. Bucci, Hydrogen Attack of Carbon-0.5 Molybdenum Piping in Ammonia Synthesis, *Mater. Perform.,* Vol 25 (No. 7), 1986, p 34–37
16. L.P. Antalffy and G.T. West, "The New Generation Vanadium Modified Steels," minutes of EFC WP15 meeting, Total Fina Elf, Paris, France, 2002
17. "AL 29-4C," Technical Data Sheet B-151-Ed5/7.5M/793/GP, Allegheny Ludlum Steel Corp., 1993, 8 pages
18. J.M. Stone, in *Corrosion Resistance of Nickel and Nickel-Containing Alloys in Caustic Soda and Other Alkalies,* CEB-2, International Nickel Co., 1973, p 20
19. S.K. Bhaumik, R. Rangaraju, M.A. Parameswara, T.A. Bhaskaran, M.A. Venkataswamy, A.C. Raghuram, and R.V. Krishnan, Failure of Reformer Tube of an Ammonia Plant, *Eng. Fail. Anal.,* Vol 9, 2002, p 553–561
20. K.M. Verma, H. Ghosh, and K.C. Pattnaik, Stress Corrosion Failure of Waste Heat Boiler Tubes in an Ammonia Synthesis Converter, *Br. Corros. J.,* Vol 15 (No. 4), 1980, p 175–178
21. M.P. Sukumaran Nair, Tackling Corrosion in Ammonia Plants—Selecting the Proper Materials, *Chemical Processing,* Dec 2001
22. S.B. Parks and C.M. Schillmoller, Improve Alloy Selection for Ammonia Furnaces, *Hydrocarbon Proc.* (Int. ed.), Vol 76 (No. 10), 1997, p 93–98
23. O. de Sanctis, L. Gómez, N. Pellegri, and A. Durfi, Behaviour in Hot Ammonia Atmosphere of SiO_2-Coated Stainless Steels Produced by a Sol-Gel Procedure, *Surf. Coat. Technol.,* Vol 70, 1995, p 251–255
24. H.J. Grabke, Metal Dusting, *Mater. Corros.,* Vol 54 (No. 10), 2003, p 736–746
25. S. McCoy, "Inconel Alloy 693 Exceptional Metal Dusting and High Temperature Corrosion Resistance," minutes of EFC WP15 meeting, Total Fina Elf, Paris, France, 2002
26. H.J. Grabke and M. Spiegel, Occurrence of Metal Dusting—Referring to Failure Cases, *Mater. Corros.,* Vol 54 (No. 10), 2003, p 799–804
27. D.C. Agarwal, L. Stewart, and M. McAllister, "Alloy 602CA (UNS N06025) Solves Pig Tail Corrosion Problems in Refineries," paper No. 03495, Corrosion 2003, NACE International, 2003, 17 pages
28. C. Rosado and M. Schutze, Protective Behaviour of Newly Developed Coatings Against Metal Dusting, *Mater. Corros.,* Vol 54 (No. 11), 2003, p 831–853
29. "Haynes 230 Alloy for Industrial Heating Applications—Data Summary," Brochure H-3033G Haynes International, 2002, p 19
30. F.G. Hodge, "High Performance Alloys Solve Problems in the Process Industries," Corrosion/91, paper No. 173, NACE, 1991, p 9
31. "Corrosion Resistance of Nickel and Nickel-Containing Alloys in Caustic Soda and Other Alkalies," CEB-2, International Nickel Company, 1973, p 21
32. *Solutions to Materials Problems,* CD Inco Alloys International, 1997
33. N.W. Polan, G.P. Sheldon, and J.M. Popplewell, "The Effect of NH_3 and O_2 Levels on the Corrosion Characteristics and Copper Release Rates of Copper Base Condenser Tube Alloys under Simulated Steam Side Conditions," paper No. 81-JPGC-Pwr-9, Joint ASME/IEEE Power Generation Conference, Oct 1981
34. H.H. Uhlig, *Corrosion and Corrosion Control,* John Wiley & Sons, 1971
35. B.J. Buecker, Watch Out for Steamside Corrosion in Utility Condensers, *Mater. Perform.,* Vol 31 (No. 9), 1992, p 68–70
36. B.J. Buecker and E. Loper, Steam Surface Condenser Tubes: Watch Out for Sneaky Corrosion, *Mater. Perform.,* Vol 39 (No. 5), 2000, p 60–64
37. T.K.G. Namboodhiri and R.S. Tripathi, The Stress-Assisted Dezincification of 70/30 Brass in Ammonia, *Corros. Sci.,* Vol 26 (No. 10), 1986, p 745–756
38. "Corrosion Resistance of Titanium," brochure No. TMC-0105, Timet, 1999, p 22
39. "Corrosion Resistance of Titanium," Ref 1431531969, IMI Kynoch Ltd., Birmingham, U.K., 1969, p 28, 36
40. "Zircadyne Corrosion Data," bulletin No. TWCA-8101ZR, Teledyne Wah Chang, 1987, 25 pages
41. *Corrosion of Lead,* Lead Development Association, London, U.K., 1971
42. *Corrosion Resistance Guide,* Yamada Pumps, 2001, p 6
43. M.L. Knight, "Failure Analysis of PVC and CPVC Piping Materials," paper No. 03606, Corrosion 2003, NACE International, 2003, 7 pages
44. *Chemical Resistance Guide,* Dow Chemical, 1991, 20 pages
45. "Chemical Resistance of Thermoplastics used in Dual Laminate Constructions," Dual Laminate Fabrication Association, 2002, 143 pages; http://www.dual-laminate.org/html/corrosion_guide.html, accessed Jan 2003
46. "The Chemical Industry Builds on Graphite," brochure PE 200/07, SGL Carbon Group, Meitingen, Germany, 2001, 24 pages
47. "Worldwide GLASTEEL 9100," brochure No. SB95-910-5, Pfaudler Reactor Systems, 2000, p 5

Corrosion by Phosphoric Acid

Ralph (Bud) W. Ross, Jr., Consultant

PHOSPHORIC ACID (H_3PO_4) is a major industrial chemical. It follows sulfuric and nitric acids in importance and tonnage produced. Phosphoric acid is made from phosphate rock by two methods: the wet process and electric furnace. Phosphoric acid is not a particularly strong acid. It is a stronger acid than acetic acid but weaker than sulfuric acid and hydrochloric acid. Phosphoric acid is incompatible with strong caustics and readily reacts with metals to form flammable hydrogen gas. Synonyms for phosphoric acid are orthophosphoric acid, metaphosphoric acid, and white phosphoric acid. Industrial or merchant grades include superphosphoric, defluorinated, and technical-grade phosphoric acid. Food-grade phosphoric acid is primarily produced by the electric furnace method to eliminate impurities in the acid.

Phosphoric acid is less corrosive than sulfuric and hydrochloric acids. Pure phosphoric acid has no effective oxidizing power and is classified as a nonoxidizing acid, much like dilute sulfuric acid. Commercial phosphoric acid made by wet process/acid digestion of phosphate rock usually contains impurities such as fluorides, sulfates, fluorosilicic acid, metal ions, and chlorides that markedly increase its corrosivity. Oxidizing compounds such as ferric salts may also be present to influence corrosion.

Phosphoric acid is used in nearly every major industry. In the chemical industry, it is used in the manufacture of phosphates and as a catalyst for other chemical reactions, such as ethanol production. The electronics industry uses H_3PO_4 in the preparation of semiconductors and printed circuit boards. Pure H_3PO_4 is used in the food industry as an acidulant in cola-type beverages and in jams and jellies. Combined with ammonia, it produces nutrients for yeast production. It is also used in sugar refining and production of pet foods. In water treatment, H_3PO_4 is used to remove mineral deposits from process equipment and boilers. It acts as a binder for aluminum refractories. In metal processing, H_3PO_4 is the main ingredient in phosphating solutions for ferrous metals. It is used to clean aluminum in bright dip baths. Phosphoric acid is also used as a nutrient for biological processes in secondary wastewater treatment plants. Its major use is in the agriculture industry, where it is used to produce fertilizer solutions and as a mineral supplement in cattle feed. Other uses are the production of pharmaceuticals, flame retardants, electrolytic fuel cells, cement processing, tanning of leather, firebrick and silica brick, crucibles and molds, toilet preparations, varnish, and synthetic rubber.

Corrosion of Metal Alloys in H_3PO_4

A survey of metal groups reveals a range of suitability for use in the manufacture, transport, and handling of H_3PO_4.

Aluminum shows no useful corrosion resistance in pure phosphoric acid. It is not used in the manufacturing or handling of phosphoric acid because of its high corrosion rates and the evolution of dangerous hydrogen gas from these reactions.

Carbon Steel and Cast Irons. Wrought and cast carbon and low-alloy steels show very high rates of corrosion in all concentrations of phosphoric acid, even at room temperature, and are not used in this service. Gray cast irons and nickel cast irons are not suitable, but high-silicon iron ($\approx 14\%$ Si), although brittle, shows good to excellent resistance in all concentrations of pure phosphoric acid, although there is measurable attack above 100 °C (212 °F). However, they do not resist acid contaminated with fluorides.

Stainless Steels. The utility of stainless steels in phosphoric acid varies with the particular family of alloys and improves with higher alloy content. The 11 to 13% Cr martensitic grades and related low-chromium ferritic grades, such as types 405 (UNS S40500) and 409 (UNS S40900), find no application in this service.

Ferritic Stainless Steels. The 17% Cr stainless steels, such as the prototypical 430 (UNS S43000), show very high rates of corrosion in phosphoric acid in all concentrations at room temperature. The extralow-interstitial-grade 26-1 (UNS S44626) ferritic stainless steel shows active/passive behavior, as does 29-4-2 (UNS S44800). Preliminary corrosion testing by the Tennessee Valley Authority Fertilizer Division has shown 29-4-2 to have acceptable corrosion resistance in some phosphoric acid solutions. More corrosion data are needed for these modern ferritic stainless steels before they can be used in such service. Also, because of high nil ductility transition temperatures, their use is usually limited to thin-walled equipment such as heat-exchanger tubes.

Austenitic Stainless Steels. Conventional stainless steels are the prime materials for phosphoric acid service.

Type 304L (UNS S30403) stainless steel has been used in some phosphoric acid solutions, because it has good general corrosion resistance up to 80% acid below 74 °C (165 °F). Because the alloy is subject to stress-corrosion cracking in the presence of chlorides and is very susceptible to pitting and crevice corrosion from contamination by chlorides or fluorides, very little is used for handling, storage, or manufacturing of H_3PO_4. In the active state, it will be rapidly corroded by phosphoric acid, with the release of hydrogen gas (Table 1) (Ref 1) that may produce flammable and explosive hydrogen/air mixtures.

For comparison, at 196 ppm Cl^-, type 316 (UNS S31600) corrosion rate is 0.03 mm/yr (1 mil/yr). Type 316L stainless steel is the preferred material in the phosphoric acid industry for acid storage, handling, and transport (Ref 2). It has excellent corrosion resistance: less than 0.25 mm/yr (10 mils/yr) from <1 to 60% phosphoric acid (44% P_2O_5) up to the boiling point. From 60% (44% P_2O_5) to approximately 90% acid (66% P_2O_5), 316L has good resistance up to approximately 115 °C (240 °F). In superphosphoric acid (typically, 69% P_2O_5), 316L has excellent resistance at even higher temperatures, 120 to 185 °C (250 to 365 °F). The isocorrosion chart for 316L in phosphoric acid is shown in Fig. 1 (Ref 1).

Table 1 Effect of chloride on corrosion of type 304 stainless steel in 75% phosphoric acid

Chlorides, ppm	Corrosion rate	
	mm/yr	mils/yr
8	0.04	1.4
15	0.28	11.0
23	12.9(a)	508(a)
39	23.1(a)	910(a)
117	24.7(a)	973(a)
196	27.5(a)	1081(a)

Temperature was 29.4 to 37.8 °C (85 to 100 °F) for 23 h. (a) Hydrogen gas evolved, and acid turned green.

Type 316L and most of the austenitic stainless steels are very susceptible to the effects of increasing temperature and acid concentration. Maximum corrosion of 316 stainless steel occurs at 79 to 91% acid concentration (Fig. 2) (Ref 3).

Type 317L (UNS S31703) stainless steel behaves similarly to 316L and can be used interchangeably up to approximately 60% acid. At this point, 316L apparently exhibits better corrosion resistance than 317L in pure acid (Fig. 1) (Ref 1, 4). Note the significantly better performance of 316L than 317L between 60 and 100% H_3PO_4 (44 and 73% P_2O_5). Above 105% H_3PO_4 (76% P_2O_5), the two stainless steels show similar corrosion resistance.

Modern duplex (austenitic/ferritic) stainless steels are exemplified by alloys 255 (UNS S32550) and 2205 (UNS S31803). Figure 1 also presents corrosion data for the duplex alloy 255. Surprisingly, this duplex exhibits corrosion resistance similar to 317L but inferior to 316L. This relationship would probably change in contaminated phosphoric acid, where chlorides and fluorides result in accelerated corrosion of 316 stainless steel and may attack the ferrite phase in duplex grades. Duplex alloy 2205 has been introduced in Europe as a practical material for marine cargo applications, having been installed in 14 cargo ships since 1987 (Ref 5).

Figure 3 compares the corrosion resistance of 316, 317, and cast alloy 20 (UNS N08007) in 105% super-phosphoric acid (76% P_2O_5) (Ref 6). Note that all of the alloys show less than 0.25 mm/yr (10 mils/yr) corrosion rate between 90 and 150 °C (195 and 300 °F).

Superaustenitic Stainless Steels. The alloy 20Cb-3 (UNS N08020), 20Mo-4 (UNS N08024), and 20Mo-6 (UNS N08026) stainless steels show no better corrosion resistance than type 316L up to approximately 60% pure phosphoric acid (44% P_2O_5) at boiling temperatures (Fig. 4) (Ref 7). Between 70 and 90% acid, alloy 27-7MO (UNS S31277) shows improved corrosion resistance over 316L, 317L, and alloy 825 (UNS N08825) below the boiling point (Fig. 5) (Ref 8, 9). At higher concentrations and temperatures, these alloys would have very high corrosion rates. The 6 to 7% Mo superaustenitic stainless steels, alloys AL-6X (UNS N08366), AL-6XN (UNS N08367), 254 SMO (UNS S31254), 654 SMO (UNS S32654), 1925hMo (UNS N08926), 20Mo-6 (UNS N08026), and 27-7MO (UNS S31277), may also be useful in contaminated phosphoric acid because of their higher molybdenum content. These alloys are usually referred to as 6% Mo alloys.

Alloys 28 (UNS N08028), also known as 3127LC, UR SB8, and alloy 31 (UNS N08031), are similar in composition and have the highest phosphoric acid corrosion resistance of the superaustenitic stainless steel grades because of their increased chromium and molybdenum content. These alloys can be used in pure phosphoric acid up to 80% acid (58% P_2O_5) to the boiling point (Fig. 5) (Ref 4). Their corrosion resistance falls as acid concentration increases

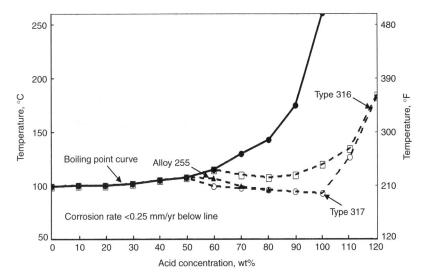

Fig. 1 Isocorrosion diagram for various stainless steels in phosphoric acid

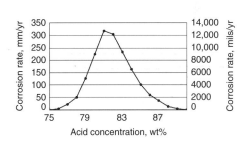

Fig. 2 Effect of acid concentration on corrosion rate of type 316L stainless steel in phosphoric acid at 163 °C (325 °F)

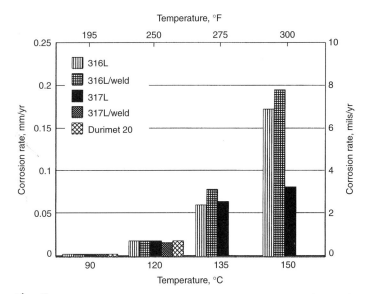

Fig. 3 Corrosion behavior of various stainless steels in 105% superphosphoric acid

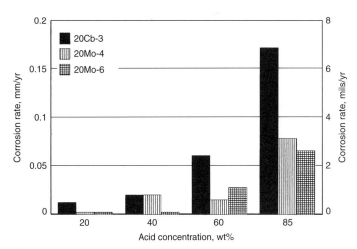

Fig. 4 Corrosion of alloy 20 stainless steels in boiling phosphoric acid

from 80 to 90%. The corrosion resistance of these alloys is quite good up to 260 °C (500 °F) at 110% acid concentration.

Pure superphosphoric acid is highly corrosive toward both stainless steels and nickel-base alloys at 200 °C (390 °F). However, its corrosiveness declines when it is combined with various metallic ions. Magnesium has been shown to be very effective in reducing corrosion of these alloys in hot superphosphoric acid (Ref 10). The test results have shown that an addition of approximately 0.25% Mg^{2+} can drastically reduce corrosion rates in hot superphosphoric acid containing minor contaminants. Magnesium salts inhibit corrosion by forming a protective layer on the surface of the metal. Magnesium occurs naturally in many phosphates and can be added to the acid in order to reduce corrosion.

Alloy 825 (UNS N08825) is quite similar to alloy 20Cb-3 and is sometimes employed in contaminated acid.

Nickel-Rich G-Type Alloys. A family of chromium-bearing nickel-rich alloys has been developed specifically for wet-process phosphoric acid service where chloride and fluoride contamination is present. The prototype was alloy F (UNS N06001) with a nominal composition of 22Cr-45Ni-22Fe-6Mo-Cb,Si,W. This alloy evolved into superior grades based on alloy G (UNS N06007).

The modern G-type alloys, G-3 (UNS N06985), G-30 (UNS N06030), and G-35 (UNS N06035), show excellent corrosion resistance in wet-process phosphoric acid (Ref 11). However, they are not cost-effective in pure phosphoric acid contrasted to the less expensive, conventional types 316L and 317L. They would show resistance comparable to alloy 28 in pure phosphoric acid. The order of resistance between the G-type alloys and alloy 28 (UNS N08028) appears to vary with the specific phosphate ore processed, which is to say that the kind and concentration of contaminants vary in an unpredictable manner in various geographic locations. Alloy G-35 showed somewhat better corrosion resistance than alloys G-30 or 31 (UNS N08031) in wet-process phosphoric acid at 121 °C (250 °F), which suggests the alloy may offer better performance in evaporator tubing applications (Fig. 6) (Ref 12).

Copper and copper alloys have been used in the past in heat exchangers for phosphoric acid. However, the corrosion rates of the copper alloys are generally very high in phosphoric acid, due to aeration and oxidizing contaminants in the acid. Copper alloys display high corrosion rates, greater than 1.27 mm/yr (50 mils/yr) in phosphoric acid from 5 to 90% (4 to 65% P_2O_5) at 25 °C (75 °F), depending largely on the presence of oxidizing species in the acid. Higher temperatures cause very significant increases in corrosion rates (Ref 13).

Contamination of phosphoric acid containing sulfate, sulfite, chloride, and fluoride ions by dissolved oxygen (DO) can rapidly increase the corrosion rate of copper alloys (as high as 150 times the rates reported previously). Ferric ions cause accelerated corrosion even in the absence of DO. This limits the usefulness of copper alloys in phosphoric acid service. Generally, copper and copper alloys will form voluminous and porous films and scales that result in pitting underneath. There is little evidence of the use of copper and copper alloys in phosphoric acid service in the published literature.

Nickel Alloys. The nickel-base (>50% Ni) alloys comprise two groups: with and without chromium additions. Only the second group demonstrates significant passivation effects.

Chromium-Free Alloys. Pure nickel alloys, such as alloys 200 (UNS N02200) and 201 (UNS N02201), have no practical applications in phosphoric acid. High rates of corrosion, between 0.51 and 1.27 mm/yr (20 and 50 mils/yr), can be experienced, especially in aerated solutions or solutions containing oxidizing ions (e.g., Fe^{3+}, Cu^{2+}). Also, the 68%Ni-32%Cu-alloy 400 (UNS N04400) is not generally used in phosphoric acid service. Alloy 400 does exhibit excellent corrosion resistance in deaerated phosphoric acid, with corrosion rates of less than 0.05 mm/yr (2 mils/yr) between 10 and 100% acid up to 80 °C (175 °F) (Ref 14).

The nickel-molybdenum alloy B-2 (UNS N10665) contains 28% Mo. It shows excellent corrosion resistance up to 50% acid up to 175 °C (350 °F). Both wrought and cast versions of alloy B-2 are available for pump and valve components. Because of its 28% Mo content, it is quite an expensive nickel-base alloy. It is also subject to accelerated corrosion when oxidizing ions (e.g., Fe^{3+}, Cu^{2+}, etc.) are added to the acid, and it has little to offer in phosphoric acid service.

Chromium-Bearing Alloys. The basic nickel-chromium alloy 600 (UNS N06600) is of limited use in phosphoric acid and is not generally used. In Fig. 5, a wide range of conditions for use of alloy 825 is shown. Alloy 625 (UNS N06625) and the C-family of alloys are more expensive materials available for pure phosphoric acid service, but they are not more resistant than alloy 825 (Ref 14). Consequently, they are used infrequently. The C-family includes C-4 (UNS N06455), C-276 (UNS N10276), C-22 (UNS N06022), C-2000 (UNS N06200), 59 (UNS N06059), and 686 (UNS N06686). They are similar in composition and corrosion characteristics and may also be useful in contaminated phosphoric acid because of their higher molybdenum content. These alloys also exhibit active/passive behavior near the boiling point and should not be used under these conditions. In fact, the Ni-Cr-Mo grades often show less resistance than alloy G-30 (UNS N06985) under wet-process conditions.

Lead shows excellent resistance to pure phosphoric acid up to 85% acid (58% P_2O_5) to 95 °C (205 °F) and reasonably good resistance to 207 °C (405 °F). The corrosion rates in the latter case are 0.51 to 1.27 mm/yr (20 to 50 mils/yr) under static conditions. Lead-lined vessels have been used in the past in the manufacture and storage of phosphoric acid. Lead is significantly affected by velocity conditions, because the protective phosphate film is easily removed. Because of environmental and health concerns, lead has no use in today's modern production and handling of phosphoric acid.

Titanium Alloys. Titanium grade 2 (UNS R50400) shows good corrosion resistance to aerated pure solutions of phosphoric acid up to 30% (22% P_2O_5) at room temperature (Fig. 7) (Ref 15). Titanium grade 12 (UNS R53400)

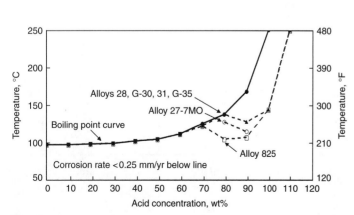

Fig. 5 Isocorrosion diagram for alloy 28 and various nickel alloys in phosphoric acid

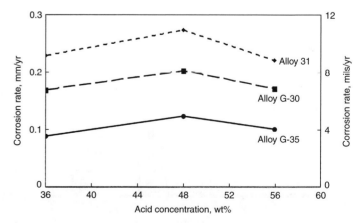

Fig. 6 Corrosion of alloys G-35, G-30, and 31 in wet-process phosphoric acid at 121 °C (250 °F)

offers no particular advantage. Increasing phosphoric acid concentration and/or temperature causes a rapid decrease in corrosion resistance of titanium. Titanium grades 7 (UNS R52400) and 11 (UNS R52250) offer better resistance than pure titanium by virtue of their palladium additions. Contamination of phosphoric acid with free fluoride ion would also accelerate corrosion of titanium alloys. Because of costs and marginal corrosion resistance, titanium alloys are not used in phosphoric acid service.

Zirconium alloys resist attack in phosphoric acid at concentrations up to 55% (40% P_2O_5) and temperatures exceeding the boiling point. Corrosion rates of zirconium alloys increase rapidly, as the temperature increases, for concentrations above 55% acid. Unalloyed zirconium (UNS R60702) and its alloys with tin and niobium, respectively (UNS R60704, UNS R60705), resist dilute phosphoric acid solutions at elevated temperatures, as shown in Table 2 (Ref 16).

Zirconium alloys exhibit better corrosion resistance to boiling phosphoric acid solutions than titanium alloys (Fig. 8). Fluorides, even at very low impurity levels, will attack zirconium. Because of costs and marginal corrosion resistance, zirconium alloys are not used in phosphoric acid service.

Precious Metals. The only metals that are corrosion resistant to pure phosphoric acid for all concentrations and temperatures are gold, platinum, platinum plus iridium, and platinum-rhodium. Silver was found to be satisfactory for all concentrations up to approximately 185 °C (365 °F).

Corrosion of Metals in Phosphoric Acid Vapors. As shown in Table 3, only the most corrosion-resistant alloys showed acceptable resistance to phosphoric acid vapors: alloys 20Cb-3, 825, G, G-3, and B-2 (Ref 17). The newer nickel-rich, highly-alloyed stainless steels (alloy 28) and other nickel-base alloys (625, C-276, G-30) would be expected to show acceptable resistance in phosphoric acid vapors, although there is no other information available.

Resistance of Nonmetallic Materials

Nonmetallic materials may be chemically attacked in some corrosive environments, which can result in swelling, hardening, or softening phenomena (if the material is a plastic or elastomer); extraction of ingredients; chemical conversion of the nonmetallic constituents (molecular changes); cross-linking oxidation; and/or substitution reactions. The general properties of the nonmetallic can be greatly changed by these chemical reactions. Other environmental effects can adversely affect a nonmetallic, such as stress, change in weight, change in volume, and change in mechanical properties. A change in hardness is very indicative of attack on organic materials. The nonmetallic material can also undergo discoloration.

Quantitative corrosion or environmental degradation data for nonmetallic materials of construction used in phosphoric acid applications are not as extensive as they are for metals. The reason is that nonmetallics do not deteriorate or corrode by the specific forms of corrosion experienced with metals. Therefore, when corrosion tests are conducted on nonmetallics, the data usually show a "go" or "no-go" result.

Rubber and elastomeric materials can be successfully used as sealing components at various temperatures and concentrations of phosphoric acid. Selection of the proper elastomeric material is critical for assuring long-term, reliable service. Although Table 4 contains information on the performance of elastomer families, variations in elastomer compound formulations and cure chemistry can result in performance differences (Ref 18–21). Therefore, testing of elastomeric seals is always recommended before use.

Most standard elastomeric materials are not suggested for use in superphosphoric acid, except chlorbutyl rubber. High-performance fluoroelastomers and perfluoroelastomers should perform well in superphosphoric acid, but testing is advised. Rubber linings are used widely in acid evaporator bodies and may also be used for scrubbers, storage tanks, and so on in the manufacture and transportation of phosphoric acid. Rubber-lined steel pipe, pumps, and valves are also used, as well as gaskets. In conjunction with acid-resistant brick as an outer lining, rubber linings can be used at higher temperatures, up to approximately 115 °C (240 °F) (Ref 19).

Elastomers are combustible materials that may evolve toxic, corrosive, and otherwise dangerous fumes. Extreme caution should be followed when welding or cutting near elastomer sealing components. One should follow industry safety standards and good fire-prevention techniques at all times.

Plastics. Both thermoplastic and thermosetting polymers find application in phosphoric acid under specific conditions.

Table 2 Corrosion of zirconium alloys in phosphoric acid

| | Temperature | | Corrosion, rate | | | | | |
| | | | Zr702 | | Zr704 | | Zr705 | |
Acid concentration, %	°C	°F	mm/yr	mils/yr	mm/yr	mils/yr	mm/yr	mils/yr
5–30	25	75	<0.13	<5	...	...	...	...
5–35	60	140	<0.13	<5	...	...	...	...
5–50	100	210	<0.13	<5	...	...	...	...
35–50	25	75	<0.13	<5	...	...	...	...
45	Boiling		<0.13	<5	...	...	...	...
50	Boiling		<0.13	<5	0.13–0.25	5–10	0.25–0.38	10–15
65	100	210	0.13–0.25	5–10	...	...	<0.51	<20
70	Boiling		>1.27	>50	...	...	>1.27	>50
85	38	100	0.13–0.51	5–20	...	...	...	...
85	Boiling		>1.27	>50	...	...	>1.27	>50
20	150	300	<0.03	<1	...	...	...	...

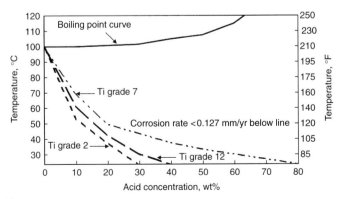

Fig. 7 Isocorrosion diagram for titanium alloys in phosphoric acid

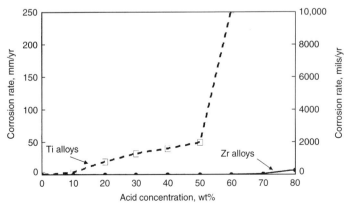

Fig. 8 Corrosion behavior of titanium and zirconium alloys in boiling phosphoric acid

Thermoplastics are materials that can be softened or fused when heated, then they reharden when cooled. Major thermoplastics used today are:

- *Polyolefins:* polyethylene, polypropylene, polybutylene, and polymethylpentene
- *Vinyls:* PVC and CPVC
- *Fluoropolymers:* PTFE, FEP, PFA, CTFE, ECTFE, ETFE, and PVF
- *Crystalline polymers:* nylons, polyesters, acetals, and polyphenylene sulfide
- *Amorphous polymers:* ABS, SAN, polycarbonate, PSO, PES, and PPO.

The unfluorinated thermoplastics generally have strict individual temperature limits that vary from as low as 4.4 to 15.6 °C (40 to 60 °F) to 82 to 107 °C (180 to 224 °F).

Maximum corrosion resistance of the thermoplastics is achieved by replacing hydrogen with chlorine and/or fluorine in the molecular structure of the plastic (fluoroplastic). Thermal resistance and strength are thereby increased, with a useful maximum temperature range of the order of 121 to 260 °C (250 to 500 °F). The fluoroplastics find major applications in liners for pipe, valves, pumps, vessels, tower packing, seals, and packing. The most resistant plastics are fluorinated: PVDF, FEP and PTFE or PFA. The nonfluorinated thermoplastics are generally not used in the manufacture of phosphoric acid but may be used in the storage of the acid.

Plastics, in general, offer good to excellent corrosion resistance in phosphoric acid, ease in fabrication, low cost, smooth surfaces (which results in less scale buildup), absence of corrosion deposits, light weight, and ease of installation. They also provide good thermal insulation, require low tooling costs, are non-conductive, and do not contaminate the acid with metal ions. Their disadvantages are degradation in hot, strong acid solutions; flammability (producing deadly and corrosive fumes); poor mechanical and fatigue strengths at elevated temperatures; a tendency to leak at joints; low temperature limits; and poor abrasion resistance. Quality control during manufacture and fabrication of the plastic components is critical. There may be permeation, absorption, and swelling and leaching problems in some cases, and they may suffer from ultraviolet degradation or delamination. Table 5 presents chemical resistance data on many of the common plastics used in phosphoric acid service (Ref 22, 23).

Thermoset plastics are materials that become permanently rigid when heated or cured. Reinforced thermosetting plastics are produced by combining a fiber (glass, wood, paper, cotton, metal, Kevlar, or carbon) and a resin (polyester, epoxy, furan). In some applications, phenol-formaldehyde resins have been used to approximately 95% acid at 60 °C (140 °F) (Ref 24).

The primary product used in the chemical processing industry is fiberglass-reinforced plastic (FRP), which is used in the handling, storage, and transportation of phosphoric acid solutions because of its excellent corrosion resistance in a wide variety of acid conditions. The FRP is resistant to all phosphoric acid contaminants at room temperature. It may have a temperature limit of approximately 60 °C (140 °F); however, some suppliers have reported that several resins may be available that will offer higher operating temperatures. Therefore, the manufacturer of the equipment and the resin supplier should be consulted to ensure that the operating temperature and the phosphoric acid concentration are compatible for the composition of the FRP selected for the application. However, FRP is subject to impact damage. Glass fibers in the FRP will be attacked by fluoride ions in the phosphoric acid or by hot concentrated H_3PO_4 if the outer laminants are damaged, exposing the glass fibers.

Carbon and graphite are chemically identical but have different physical structure and physical properties. They are resistant to all

Table 3 Corrosion of alloys in phosphoric acid vapors

Material (UNS)	Temperature		Corrosion rate	
	°C	°F	mm/yr	mils/yr
Type 304 (S30400)	10–93	50–200	0.51–1.27	20–50
Type 304 (S30400)	120–150	250–300	>1.27	>50
Type 316 (S31600)	10–93	50–200	0.51–1.27	20–50
Type 317 (S31700)	10–93	50–200	0.51–1.27	20–50
20Cr-25Ni-4.5Mo (N08904)	10–93	50–200	0.51–1.27	20–50
Copper alloys	93	200	>1.27	>50
Nickel (N02200), Ni-Cu (N04400)	93	200	0.51–1.27	20–50
Nickel (N02200), Ni-Cu (N04400)	120–150	250–300	>1.27	>50
Alloy 600 (N06600)	93	200	>1.27	>50
Super SS (N08020, N08825)	10–93	50–200	>1.27	>50
Alloy G (N06007), G-3 (N06985)	10–93	50–200	>1.27	>50
Alloy B-2 (N10665)	10–93	50–200	>1.27	>50
Lead (L50045)	93	200	0.51–1.27	20–50
Lead (L50045)	120–150	250–300	>1.27	>50

Table 4 Compatibility of rubber and other elastomeric materials with phosphoric acid

Elastomer family (class)	Immersion test temperature		Performance rating(a)
	°C	°F	
Natural rubber, soft (NR)	≤60	≤140	A
Natural rubber, semihard (NR)	≤82	≤180	A
Butyl (IIR)	≤66	≤150	A
Chlorobutyl	≤66	≤150	A
Fluoroelastomer (FKM)	≤100	≤210	A
Neoprene (CR)	≤66	≤150	A
Nitrile, soft (NBR)	≤38	≤100	A
Nitrile, hard (NBR)	≤66	≤150	A
Perfluoroelastomer (FFKM)	≤260	≤500	A
	260–300	500–570	B

(a) A, Suitable up to service temperature of elastomer; B, satisfactory but testing recommended

Table 5 Chemical resistance of plastics in phosphoric acid

Material (class)	Maximum concentration, % acid	Maximum temperature	
		°C	°F
Polyethylene (PE)	50	60	140
	80–100	23	73
Polypropylene (PP)	50–100	100	210
Polybutylene (PB)	50	60	140
	75	23	73
Polystyrene	60	60	140
Acrylo-butadiene-styrene (ABS)	50	23	73
Polyvinyl chloride (PVC 2110)	100	23	73
(PVC 1120)	100	60	140
Chlorinated PVC (CPVC)	70	99	210
	85	23	73
Polycarbonate (PC)	95(a)	60	140
Polyvinylidene fluoride (PVDF)	<85	107	225
	100	135	275
Polymonochlortrifluoroethylene (PCTFE)	95	100	210
Polyfluoroethylenepropylene (FEP)	95	205	400
Synthetic fluorine-containing resin (PTFE)(b)	95	260	500

(a) Subject to environmental stress cracking. (b) PTFE, polytetrafluoroethylene

concentrations of phosphoric acid up to and including the boiling point and are used extensively in the production of phosphoric acid. Carbon is used primarily in carbon brick and ring packing for towers. Graphite is used in vessels, piping, and pumps.

Graphite is used in fibrous form as gasket material. Impervious graphite is impregnated with organic resins to fill the pores inherent in the original structure. Impervious graphite is used in heat exchangers because of its good heat-transfer rate (equivalent to copper alloys). Impervious graphite is brittle and must be protected from mechanical shock or severe tensile stress, such as by water freezing in exchanger tubes or mechanical abuse during cleaning. The temperature limit is approximately 260 °C (500 °F) (Ref 25).

Ceramic Materials. Many of the ceramic materials, such as glass, stoneware, and acid-resistant brick, can be used in the production and handling of phosphoric acid. Cement and concrete structures are nonresistant, however.

Glass. Borosilicate glass can be used in phosphoric acid in all concentrations up to 95% acid (69% P_2O_5) and 100 °C (212 °F) temperature (Ref 26). Glass is severely limited by fluoride contamination and should not be used in such phosphoric acid solutions. Glass is used for piping, pumps, and flow meters. Very high concentrations of phosphoric acid and temperatures above 100 °C (212 °F) should be avoided because of chemical attack. Glass is also prone to mechanical damage, and appropriate armor or other protective systems should be employed.

Porcelain and stoneware can be used in phosphoric acid up to 95% phosphoric acid (69% P_2O_5) at 60 °C (140 °F) maximum temperature or up to 100 °C (212 °F) at concentrations less than 60%. Stoneware and porcelain can be made into sinks, pipes, valves, pumps, insulators, and acid nozzles. Both materials are subject to mechanical damage and have low tensile strengths.

Acid-resistant brick is used extensively in the manufacture of phosphoric acid. It is common to use membrane linings under acid-resistant brick in tanks and other process vessels to minimize the possibility of corrosion of the steel substrate. Acid-resistant brick-lined steel components have been used in conjunction with lead, rubber, or plastic membranes. Areas subject to acid spillage are also made of acid-resistant brick. The limiting factor in using acid-resistant brick would be the cements and mortars. Manufacturers should be consulted for a particular cement or mortar, especially if fluoride contamination exists in the phosphoric acid.

Because there are some inconsistencies in the published data purporting to indicate where various non-metallics can be used safely in phosphoric acid, it is strongly recommended that either the end user conduct corrosion tests or have previous experience with a particular non-metallic before selecting one for use in the manufacture, transportation, or storage of phosphoric acid. Nearly all of the information supplied previously on nonmetallics was given as general guidance only, and very little referred to actual service experience or case histories.

Another problem area with the use of some of the nonmetallics is that fabrication of equipment from nonmetallics is quite variable in quality. The reason for this is that many of the initial ingredients in the plastic are supplied to others who combine, mix, and formulate the materials before fabricating the final product. Therefore, many processing variables can occur that can lower the mechanical and chemical properties of the nonmetallic. Excellent service with nonmetallics can be achieved if qualified component fabricators are selected to manufacture the needed equipment.

ACKNOWLEDGMENT

This article has been adapted from the MS Series publication *Corrosion by Phosphoric Acid,* published and reprinted by permission of the Materials Technology Institute (MTI) 2004 for the Chemical Process Industries Inc., St. Louis, MO.

REFERENCES

1. "Corrosion Resistance of Nickel-Containing Alloys in Phosphoric Acid," Corrosion Engineering Bulletin CEB-4, The International Nickel Company, Inc., New York, 1966, p A-415
2. "Phosphoric Acid Handling, Storage Procedures and Precautions," Publication 9109, Monsanto Company, St. Louis, MO
3. F.H. Haynie, P.J. Moreland, W.K. Boyd, and W.B. Krauskopf, *Mater. Prot. Perform.,* Vol 9 (No. 12), Dec 1970, p 35
4. C.N. Schillmoller, "Alloys Selection in Wet Process Phosphoric Acid Plants," Technical Series 1010015, Nickel Development Institute, Toronto, Canada
5. B. Leffler, Alloy 2205 for Marine Chemical Tankers, *Mater. Perform.,* April 1990
6. "Superphosphoric and Polyphosphoric Acids," Technical Bulletin 202, FMC Corporation, Philadelphia, 1982
7. "Alloy Data Publications 20Mo-4 Stainless," Technical Bulletin Stainless Steel 58, No 2-87/6.5M, Carpenter Technology Corporation, 1987
8. F.A. Hendershot, W.G. Lipscomb, and R.W. Ross, *The Use of Nickel Alloys in the Chemical Process Industry,* INCO Alloys International, Huntington, WV, 1987
9. Special Metals Corporation, Huntington, WV, private communication, 2004
10. G. Berglund and S. Bernhardsson, "Metallic Tubular Phosphoric Acid Evaporators—Corrosion Performance and Plant Experience," AICHE Spring Meeting for Area 13E, American Institute of Chemical Engineers, April 1986
11. A.I. Asphahani, D.C. Agarwal, and P.E. Manning, "Hastelloy Alloy G-30: A Specialty Corrosion Resistant Alloy for Applications in the Fertilizer Industry," Paper T.F.12513-Doc. 3325B, ACS 194th National Meeting, Sept 1987 (New Orleans, LA), American Chemical Society
12. P. Crook and M.L. Caruso, *A New Nickel Alloy for Use in the Agrichemical Industries,* Haynes International, Kokomo, IN, May 2003
13. *Corrosion,* Vol 13, *Metals Handbook,* 9th ed., ASM International, 1987, p 627
14. "High-Performance Alloys for Resistance to Aqueous Corrosion," Technical Bulletin SMC026/9M, Special Metals Corporation, 2000, p 22
15. "Corrosion Resistance of Titanium," Timet Publication M2-5M-588 Rev., Timet Corporation, Pittsburgh, PA, p 17
16. *Zirconium Data Chart,* Teledyne Wah Chang Albany, Albany, OR
17. *Corrosion Data Survey, Metals Section,* 5th ed., NACE, 1974
18. *Corrosion Chart,* IPC Industrial Press Limited, London, Dec 1975
19. DuPont Dow Elastomers, private communication, 2004
20. *Permobond Rubber Linings,* Catalog 580, Uniroyal, 1984
21. "Protective Linings: Rubber, Synthetic, PVC Resist Corrosion and Abrasion," Bulletin SR-87-TL-001, B.F. Goodrich, Akron, OH, 1987
22. "Thermoplastic Piping for the Transport of Chemicals," Report TR19/10-84, Technical Report of the Plastics Pipe Institute, Wayne, NJ, 1984
23. "Chemical Resistance Chart—Kynar Polyvinylidene Fluoride," Publication TR-15M-1-89-Ruv, Pennwalt Corporation, Philadelphia, PA, 1989
24. "Chemical Resistance Guide for Red Thread II, Green Thread and Poly Thread Products," Bulletin E5600, Smith Fiberglass Products, Inc., Little Rock, AR, April 15, 1989
25. M.G. Fontana, *Corrosion Engineering,* 3rd ed., McGraw-Hill Book Co., 1986, p 358
26. "Process Pipe Fittings and Hardware," PFH-6/87-CP, Corning Glassworks, Big Flats, NY, 1987

Corrosion by Mixed Acids and Salts

Narasi Sridhar, Southwest Research Institute

PROCESS FLUIDS are often found to contain various salts from catalysts, impurities from mixing water, or corrosion products. Therefore, estimating the performance of materials in the chemical process industry (CPI) requires an understanding of the chemistry of these environments and its effects on corrosion. The corrosion rate in many acid or acid-plus-salt mixtures is a complex function of solution and alloy compositions, rendering interpolation of data from pure acids difficult (Ref 1). For example, fluoride acts as an aggressive species in dilute sulfuric acid but as a corrosion inhibitor in concentrated sulfuric acid (Ref 2).

The inability to predict the corrosion behavior of materials in complex CPI streams has considerable economic and safety impacts. Mixtures of acids or acids and salts are of great importance to the CPI for use in digestion of solids, as a promoter in reactions, as a scale remover, and as a complexant, for example. Often, when a pure acid is thought to have been used, minor dissolved species from catalysts or the corrosion of upstream components can complicate materials selection. Corrosion data in pure acids are more readily available than in acid mixtures. However, the corrosion behavior of materials in acid mixtures cannot always be interpolated from a knowledge of their corrosion behavior in individual acids. The object of this article is to assess the performance of Ni-Fe-Cr-Mo alloys in mixed acids and salts in an objective manner. The nominal composition of these alloys is given in Table 1. The table lists a range of commercially available alloys, although a few representative alloys are discussed with respect to their corrosion behavior. However, caution must be used in interpreting the data for materials selection:

- Data are based on short-term immersion tests and may not accurately reproduce plant conditions.
- Weight-loss data are not reliable indicators of the susceptibility or rate of localized corrosion.
- Corrosion phenomena, such as stress-corrosion cracking, galvanic effects, and intergranular corrosion, are not considered.
- Other factors, such as heat transfer, flow effects, presence of minor constituents, vapor-liquid interfaces, galvanic effects,

microstructure (welding, other fabrication), and corrosion test methods, may play a role in corrosion processes.

The information presented should be used as guidance for further process-specific testing. The focus is on Ni-Fe-Cr-Mo-type alloys because these alloys constitute the bulk of the corrosion-resistant alloys used in the CPI, and considerable data are available related to these alloys. Data for other alloys sometimes used in the CPI, such as titanium and zirconium, are presented elsewhere in this Volume.

For the nonoxidizing acids, the corrosion rate (CR) of Ni-Fe-Cr-Mo alloys may be described by an equation of the form:

$$\ln(CR) = -A \times (Mo + 0.5W) + B \quad (Eq\ 1)$$

where both A and B depend on the acid mixture. It is possible that this behavior is determined by the relationship between the (Mo + 0.5W) content and the exchange current density for the anodic dissolution and proton reduction reaction in the nonoxidizing acids. The decrease in anodic exchange current density by molybdenum results in an increase in the corrosion potential as a function of increasing molybdenum content. For the oxidizing acids, the effect of increasing chromium and molybdenum are both important, because the anodic curves are affected by two different but related mechanisms. The passive current density is reduced by increasing chromium and molybdenum, and the critical potential (the upper inflection point of the anodic curve) is increased by increasing chromium and molybdenum. The corrosion potential is determined by the intersection of the cathodic curve with the anodic curves. With

Table 1 Nominal compositions of pertinent Ni-Fe-Cr-(Mo) corrosion-resistant alloys

		Composition, wt%									
UNS No.	Name	Ni	Cr	Fe	Mo	W	Cu	C	Nb	Ti	Al	Others
...	27-7Mo	27	22	42	7.2	...	1	...	...	...	...	0.34N
S34700	347	11	18	69	...	...	...	0.08	0.5	...	...	2Mn
S32550	Ferralium 255	5.5	25	64	3	...	1.7	...	...	...	...	0.17N
S32100	321	10.5	18	69	...	...	...	0.08	...	0.3	...	2Mn
S31603	316L	12	17	65	2.5	...	...	0.03	...	...	...	...
S31600	316	12	17	65	2.5	...	...	0.06	...	...	...	...
S31254	254 SMO	18	20	55	6.25	...	0.75	0.02	...	...	...	0.20N
S30403	304L	10	19	70	...	...	...	0.03	...	...	...	2Mn
S30400	304	9.25	19	70	...	...	...	0.08	...	...	...	2Mn
R20033	33	31	33	32	1.6	...	0.6	0.01	...	...	...	0.4N
N26022	Cast C (CX2MW)	56	21.5	4	13.5	3	...	...	...	...	...	...
N10665	B-2	67	1	2	28	...	...	...	...	...	...	...
N10276	C-276	54	15.5	6.5	16	3.75	...	...	...	...	...	...
N10001	B	59	1	6	29.5	...	...	...	...	...	...	...
N08926	25-6Mo, 1925hMo	25	20	47.5	6.5	...	1	0.02	...	...	...	0.2N
N08904	904L	25.5	21	44	4.5	...	1.5	...	...	...	...	...
N08825	825	42	21.5	28.5	3	...	2.25	...	...	...	...	...
N08367	AL 6XN	24.5	21	45	6.5	...	...	...	...	...	...	...
N08031	31	31	27	32	6.5	...	...	...	...	...	...	0.2N
N08028	Sanicro 28	31	27	34	3.5	...	1	...	...	...	...	...
N08020	20Cb-3	35	20	35	2.5	...	3.5	...	...	...	...	...
N06985	G-3	49.5	22.25	18	6	...	0.95	0.015	...	...	...	...
N06920	H-9M	42	21.75	18.5	9	1.5	...	...	...	...	...	...
N06693	693	62	29	4.25	...	...	...	...	1.5	...	3.25	...
N06690	690	58	29	9	...	...	0.5	0.05	...	...	...	...
N06686	686	57	21	2	16	3.7	...	0.02	...	0.14	...	...
N06625	625	59	21.5	5	9	...	...	0.05	...	...	...	...
N06455	C-4	65	16	3	15.5	...	...	0.01	...	0.7	...	...
N06200	2000	57	23	2	16	...	1.6	...	...	...	...	...
N06059	59	59	23	1	16	...	...	0.01	...	...	...	...
N06030	G-30	37.5	29.75	15	5	2.75	1.7	0.03	...	...	...	...
N06022	22, 622	56	22	3	13.75	3	...	0.015	...	...	...	...

increasing chromium, the corrosion current density is reduced, because the passive current density is reduced. Assuming that the cathodic polarization is unaffected by chromium and molybdenum, the corrosion potential increases with an increase in chromium and molybdenum. The critical potential is also increased by molybdenum and chromium. The increase in critical potential by chromium and molybdenum would depend on factors such as temperature and concentrations of aggressive/inhibitive anionic species. Therefore, the effect of alloying elements in oxidizing solutions would be expected to be more complex.

The acid and acid-plus-salt mixtures can be classified into the following general categories, with an example of each:

- Nonoxidizing acid mixtures (H_2SO_4 + H_3PO_4)
- Nonoxidizing acids with halides (H_2SO_4 + HCl)
- Oxidizing acid mixtures without halides (H_2SO_4 + HNO_3)
- Oxidizing acid mixtures with halides (HNO_3 + HF)

Nonoxidizing solutions are commonly referred to as reducing solutions. However, the term *reducing acid* is not used here, because it implies that the acid reduces the metal, which is clearly not the case for metallic corrosion. In nonoxidizing acids, the main cathodic reaction is the reduction of proton, H^+ (or hydronium ion, H_3O^+, for aqueous solutions), and the redox potential (open-circuit potential of a noncorroding electrode) generally follows the H^+/H_2 reduction equilibrium and kinetics. The presence of halides in these solutions generally does not lead to localized corrosion but can lead to high rates of active corrosion. While alloys that rely on passive film for corrosion resistance can perform adequately in nonoxidizing environments, the alloys that perform best in these environments also have low exchange current densities for anodic dissolution as well as cathodic reduction of hydrogen ions. In oxidizing acids or acid-salt mixtures, the cathodic reactions generate a redox potential higher than that of the proton reduction. In these environments, corrosion protection is afforded primarily by the passive film. The alloys that do not readily form a passive film (e.g., nickel-molybdenum and nickel-copper alloys) generally corrode at a high rate in oxidizing aqueous environments. The presence of halides can lead to localized corrosion of the passive alloys, depending on the alloy composition, the halides (Ref 3), and the other species in solution. In the absence of halides, the corrosion rates can be high if the redox potential is such that the transpassive regime is reached (e.g., chromic acid).

Nonoxidizing Mixtures

Reagent-Grade Phosphoric Acid Mixtures. The wet-process phosphoric acid is made by reacting fluorapatite rock with sulfuric acid and then evaporating the dilute phosphoric acid in successive stages (Ref 4, 5). The resultant acid, usually referred to in P_2O_5 equivalents, is evaporated from 28% to as high as 54.9% P_2O_5. The phosphoric acid contains varying concentrations of residual sulfuric acid, hydrofluoric acid, and fluosilicic acid. Chlorides, aluminum ions, ferrous/ferric ions, and other organic species also are present from the mineral and water used in the reactions (Ref 4–7). It is important to note that because of the presence of oxidizing impurities (especially Fe^{3+} and Al^{3+} ions) in the real-world wet-process acid, it can be an oxidizing acid. On the other hand, the presence of organic species can reduce the redox potential substantially. The corrosion performance of alloys in wet-process acid acquired from manufacturers is therefore highly variable but can be characterized as oxidizing (Ref 8).

Pyrolytic process (furnace-grade acid) is generally pure and nonoxidizing. The corrosion behavior of alloys in wet-process acids is described in the article "Corrosion by Phosphoric Acid" in this Volume. Several investigators (Ref 6, 7) have used reagent-grade phosphoric acid to understand the role of various residual impurities in phosphoric acid on corrosion. In these investigations, reagent-grade phosphoric acid was mixed with various concentrations of sulfuric, hydrofluoric, and fluosilicic acids. These mixtures are all nonoxidizing, as described in this section. Some claim that the synthetic acid mixtures adequately represent the commercial wet-process acid (Ref 5). These studies are typically based on a limited number of acids and alloys.

The effect of molybdenum in Ni-Fe-Cr-Mo alloys on the corrosion rate in reagent-grade mixtures of 75% H_3PO_4 (Fig. 1) is that:

- The corrosion rate decreases with an increase in molybdenum concentration, including alloys without any chromium, symptomatic of nonoxidizing environments.
- Addition of H_2SO_4 increases the corrosion rate slightly.
- Addition of hydrofluoric and fluosilicic acids significantly increases the corrosion rate.
- The effect of fluosilicic acid is much more significant than that of hydrofluoric acid (HF).

The last observation seems to be in apparent contradiction with the results (Ref 5) that indicated that HF is more aggressive than fluosilicic acid. Here, the effect of fluosilicic acid was measured in terms of the change in the passive potential range in a polarization curve. Such a change may not be directly correlated to corrosion rate.

Anodic polarization curves of type 317L stainless steel in reagent-grade phosphoric acid with additions of sulfate, chloride, and fluoride, among other cationic species, showed that chloride had a more pronounced effect than fluoride on the active region of the polarization curve, but fluoride had a more pronounced effect in the pseudo-passive region (Ref 7). The effect of a number of halides on the corrosion rate of 304 and 316 stainless steel in phosphoric acid is shown in Fig. 2. The rate for the molybdenum-containing 316 is lower than that for the non-molybdenum 304. Fluoride seems to have a much greater effect on corrosion than chloride. Bromide is considerably less aggressive than the

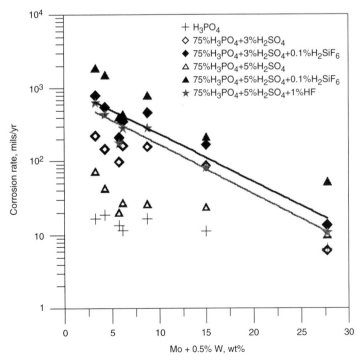

Fig. 1 Effect of molybdenum on corrosion rates in reagent-grade phosphoric acid mixtures at 149 °C (300 °F)

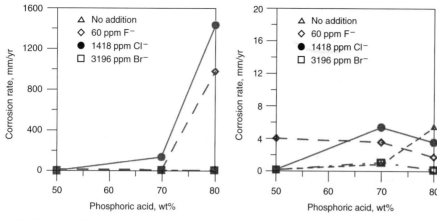
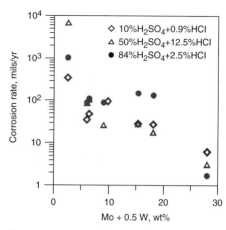

Fig. 2 Effect of F⁻, Cl⁻, and Br⁻ on corrosion rate of stainless steel in reagent-grade phosphoric acid at 80 °C (176 °F). (a) Type 304. (b) Type 316

Fig. 3 Effect of alloying elements on corrosion rate of Ni-Fe-Cr-Mo alloys in H_2SO_4 + HCl mixtures at 79 °C (175 °F)

other two halides. Such effects may be related to the stability of halide complexes in the acid.

The effect of alloying elements on corrosion can be fitted to a logarithmic relationship (Eq 1), where CR is corrosion rate in mils/yr, A is 0.15, B is 7 for the 75% H_3PO_4 + 3% H_2SO_4 + 0.1% H_2SiF_6 mixture, and B is 6.5 for the 75% H_3PO_4 + 5% H_2SO_4 + 1% HF mixture.

H_2SO_4 and HCl Mixtures. The effect of molybdenum plus tungsten on the corrosion rates of alloys in H_2SO_4 + HCl mixtures (Fig. 3) shows that increasing concentration of molybdenum results in a reduction in corrosion rate. For the chromium-containing alloys, which exhibit passive behavior, the effect of molybdenum may be attributed to its role in increasing the critical potential for localized corrosion for the chromium-containing alloys. For some chromium-containing alloys, only active behavior is seen in these acids. This is the case for type 316L stainless steel. The nickel-molybdenum alloys, such as alloy B-2, also do not exhibit significant passivity in these environments. In these cases, the beneficial effect of molybdenum may be attributed to a decrease in the exchange current density for anodic dissolution (Ref 9). It should be noted that even for the case of passive behavior and localized corrosion, the beneficial effect of molybdenum may be attributed to its effect on the active corrosion in the pit electrolyte (Ref 10). It should be cautioned that the use of alloy B-2 in these mixtures is predicated on the absence of oxidizing agents, such as dissolved iron. Based on the data (Fig. 3), the effect of alloying elements on the corrosion rate can be written in the form of Eq 1, where CR is corrosion rate in mils/yr, and A and B depend on the acid concentration. For 84% H_2SO_4 + 2.5% HCl mixture, A is 0.18 and B is 6.6.

The isocorrosion curves from data from two different sources for alloy G-30 are shown in Fig. 4. While the two sources tested at different temperatures and times, there is general agreement in the data trends. Certain regions of the data field are blocked because of lack of data for developing contours.

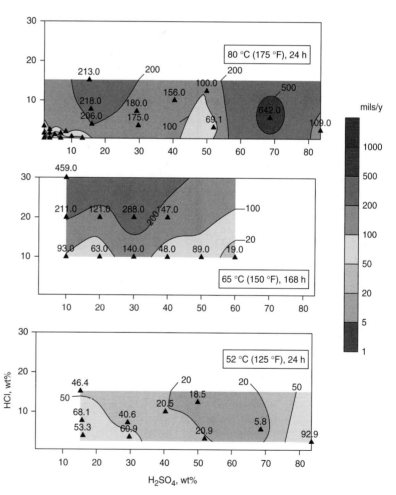

Fig. 4 Isocorrosion contours for alloy G-30 based on data from two sources. The temperatures are indicated within the figures. The numbers next to the symbols are test data. Contour values and test data are both mils/yr. Courtesy of Haynes International and Special Metals

One important aspect of corrosion behavior in mixed acids is the role of microstructural features of the alloy. This is especially true for duplex stainless steels. Generally, corrosion of duplex stainless steels in sulfuric acid occurs preferentially in the austenite phase (Ref 11). However, when significant concentrations of chloride are present in sulfuric acid through the addition of HCl, corrosion occurs preferentially in the ferrite phase (Ref 11). If the ferrite phase is

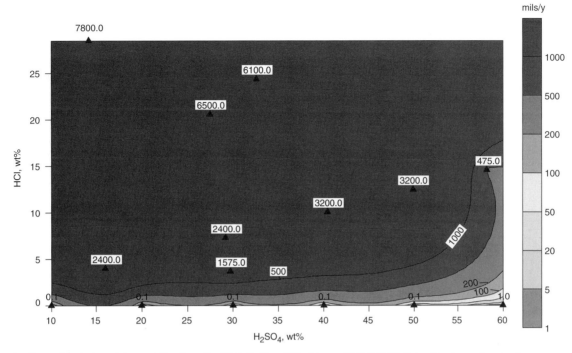

Fig. 5 Isocorrosion diagram for a duplex stainless steel, Ferralium alloy 255, in H_2SO_4 + HCl mixtures at 52 °C (125 °F). Data point values and contour lines are corrosion rate in mils/yr.

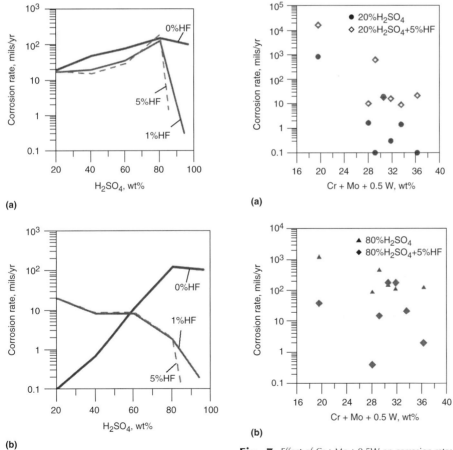

Fig. 6 Effect of HF addition to mixture at 79 °C (175 °F) on corrosion rates of alloys (a) 625 and (b) G-30

Fig. 7 Effect of Cr + Mo + 0.5W on corrosion rates in different concentration ranges of sulfuric acid with and without HF at 80 °C (175 °F). (a) 20% H_2SO_4. (b) 80% H_2SO_4

continuous, high corrosion rates may be observed through grain dropping. An isocorrosion curve for a duplex stainless steel is given in Fig. 5.

H_2SO_4 and HF Mixtures. Unlike chloride ions, fluoride seldom promotes localized corrosion. Typically, increased active dissolution is observed in the presence of fluoride (Ref 12). The effect of fluoride ion on corrosion in a broad range of sulfuric acid concentrations was studied (Ref 2). Two examples of results for alloys 625 and G-30 are shown in Fig. 6. At concentrations below approximately 60 wt% H_2SO_4, the addition of HF increases the corrosion rate of alloy G-30, whereas above 60 wt% H_2SO_4, the addition of HF decreases the corrosion rate. Higher concentrations of HF lead to a greater decrease in corrosion rates. For the case of alloy 625, the change in the effect of HF occurs at a higher concentration of H_2SO_4. It was found that in dilute sulfuric acid (10 wt%), the addition of HF resulted in complete depassivation and active corrosion. On the other hand, in 90 wt% sulfuric acid, the addition of HF resulted in a 3 order of magnitude decrease in current density and a broad passive region extending over 1 V. The increase in rotation speed of a rotating cylinder electrode served to increase passive potential range as well as decrease passive current density (Ref 2).

The effect of alloying elements Cr-Mo-W on corrosion rate in sulfuric acid with and without HF is seen in Fig. 7. In 20% sulfuric acid, the corrosion rate generally decreases with an increase in the level of chromium and molybdenum. It has also been shown previously that copper and nickel contents play a crucial role

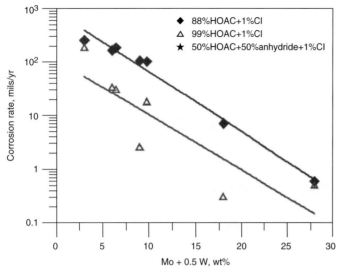

Fig. 8 Effect of alloying elements on corrosion rates in acetic acid and NaCl mixtures. Boiling point of 100% acetic acid is 117.9 °C (244.2 °F). Boiling point of 100% acetic anhydride is 139.6 °C (283.3 °F).

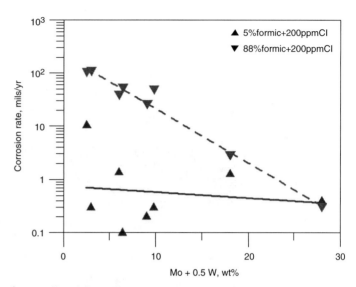

Fig. 9 Effect of alloying elements on corrosion rates in formic acid and NaCl mixtures, boiling, 24 h

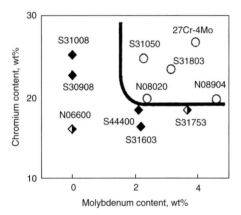

Fig. 10 Effect of chromium and molybdenum content on corrosion in simulated wet-process phosphoric acid (50 wt% P_2O_5, 3% H_2SO_4, 0.5% HF, and 0.5% $Fe_2(SO_4)_3$) at 50 °C (120 °F) for 120 h. Corrosion rates: open circle < 46 mils/yr; half-diamond < 460 mils/yr; solid diamond > 460 mils/yr

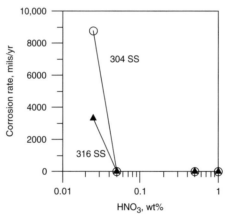

Fig. 11 Effect of nitric acid on corrosion rate in 50% sulfuric acid at 60 °C (140 °F). Source: Ref 14

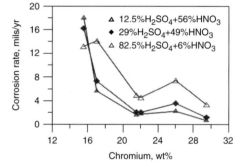

Fig. 12 Effect of alloying element on corrosion rate in sulfuric and nitric acid mixtures

(Ref 13). Alloys containing nickel greater than approximately 34% perform significantly better than alloys with nickel less than this level. Within each of these two alloy classes, the corrosion rate seems to decrease with an increase in Cr + Mo + 0.5W. The correlation with just (Mo + 0.5W) was poor for both acid concentration ranges. Copper has a smaller effect, and the effect reaches a plateau at approximately 1% Cu. As shown in Fig. 7, Cr + Mo + 0.5W has a more significant effect on corrosion rates in dilute acid than in concentrated acid. The trend with alloying elements seems to be the same in H_2SO_4 + HF mixtures, although HF increases the corrosion rates in dilute acids and decreases it in concentrated acids.

Acetic Acid Mixtures. In the absence of oxidizing salts, mixtures of acetic acid with anhydride, NaCl, $FeCl_3$, $CuCl_2$, and $Fe_2(SO_4)_3$ exhibit the same relationship as other nonoxidizing acids. However, because of the low dielectric constant of the acid and the resultant low ion dissociation, the corrosion rates are lower than in comparable inorganic acids.

Figure 8 shows increasing molybdenum results in decreasing corrosion rates, symptomatic of nonoxidizing acids. The effect of alloying elements can be fit to a logarithmic relationship where CR is in mils/yr.

For the 88% acetic acid + Cl solutions, the effect of alloying elements is given by the following regression fit (coefficient of determination, $R^2 = 0.98$):

$$\ln(CR) = -0.26 \times (Mo + 0.5W) + 6.8 \quad (Eq\ 2)$$

For the 99% acetic acid + Cl, the effect of alloying elements may be given as (coefficient of determination, $R^2 = 0.74$):

$$\ln(CR) = -0.24 \times (Mo + 0.5W) + 4.7 \quad (Eq\ 3)$$

As expected, for the acetic acid and anhydride mixture, the regression fit gives a similar slope as the more concentrated acetic acid (coefficient of determination, $R^2 = 0.72$):

$$\ln(CR) = -0.21 \times (Mo + 0.5W) + 4.5 \quad (Eq\ 4)$$

Formic Acid Mixtures. Formic acid can be regarded as a strong acid because of its relatively high dielectric constant. Therefore, the corrosion behavior of austenitic alloys in formic acid mixtures should resemble that in inorganic nonoxidizing acids. This is illustrated for two formic acid and NaCl mixtures in Fig. 9. For the 5% formic acid and 200 ppm Cl^- mixture, the corrosion rates are relatively low for most alloys, and there is significant scatter in the corrosion rates. However, for the 88 wt% acid, the corrosion rates decrease with (Mo + 0.5W), similar to other nonoxidizing acids. The corrosion rate in this acid can be given by the regression expression:

$$\ln(CR) = -0.23 \times (\text{Mo} + 0.5\text{W}) + 5.5 \quad (\text{Eq 5})$$

Oxidizing Acid Mixtures

The effect of alloying elements in oxidizing acid mixtures is more complex than for non-oxidizing mixtures.

Wet-process phosphoric acid can range in its redox potential corrosivity, depending on the source of the rock. With a laboratory-simulated acid (50% P_2O_5 containing 4% SO_4^{2-}, 1% F^-, and 0.02% Cl^-), the addition of Fe^{3+} decreased the corrosion rate of type 317L stainless steel dramatically, because the potential shifted from the active to the passive regime (Ref 7). On the other hand, if there is sufficient chloride present, then the presence of Fe^{3+} can increase corrosion rates. In wet-process phosphoric acid, this showed that the corrosion rates of stainless steels increased in the order 904L > 317L > 316L, indicating that both chromium and molybdenum played a role. A number of stainless steels in both plant wet-process acid and laboratory mixture simulating the plant acid were examined (Ref 6). The test results are plotted in terms of chromium and molybdenum contents of various alloys in Fig. 10.

Sulfuric Acid and Nitric Acid Mixtures. Small additions of nitric acid or nitrates can decrease the corrosion rates of chromium-containing alloys in sulfuric acid (Fig. 11). The effect of nitrate generally depends on the chromium content of the alloy (Fig. 12). With the exception of the 26-Cr alloy, there is a decrease in corrosion rate with chromium content. The 26-Cr alloy is a duplex stainless steel with low nickel (5Ni), which may be the reason for the difference. Another possibility is the effect of microstructure in inducing interphase corrosion in the duplex stainless steel. No correlation with molybdenum or chromium and molybdenum contents was observed with these data.

The effects of temperature and acid mixture compositions on the corrosion rate of type 316L stainless steel are shown in isocorrosion diagrams (Fig. 13).

Nitric and Hydrofluoric Acid Mixtures. These mixtures are commonly found in descaling baths and fluoropolymer processing. For the chromium-containing alloys, the corrosion rate appears to be a function of chromium, with molybdenum and tungsten playing a relatively minor role (Fig. 14). However, the nickel-molybdenum alloy (alloy B-2) was not as high as expected, although it is still such that its use is prohibited in this environment.

Nitric and Hydrochloric Acid Mixtures. The effect of alloying elements on corrosion depends on the acid composition. This is illustrated in terms of isocorrosion diagrams as functions of chromium and molybdenum contents of the alloy (Fig. 15). Because some of the alloys have tungsten, the effect of tungsten is included in terms of its atomic concentration with respect to molybdenum. For the low nitric acid concentration solutions (<4% HNO_3), the corrosion rate decreases with an increase in molybdenum content. For the high nitric acid concentration, the corrosion rate increases with molybdenum concentration. Thus, high-molybdenum alloys may be detrimental in high nitric acid solutions containing chloride. Nitric acid additions to HCl cause two opposing effects that depend on the concentrations of both constituents as well as the alloy composition. Nitric acid (nitrate) helps in the repassivation of localized

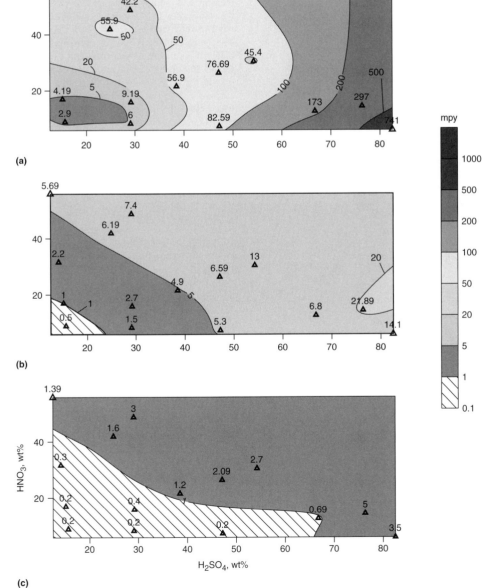

Fig. 13 Isocorrosion diagrams for type 316L in sulfuric acid and nitric acid mixtures. (a) Boiling. (b) 79 °C (175 °F). (c) 52 °C (125 °F)

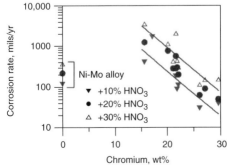

Fig. 14 Effect of alloying elements on corrosion rate of Ni-Fe-Cr-Mo-W alloys in HNO_3 + 3% HF at 79 °C (175 °F)

corrosion and therefore inhibits corrosion. On the other hand, nitric acid also increases the redox potential of the solution and therefore increases the susceptibility to localized corrosion. This results in a complex behavior of alloys in these mixtures.

Acetic Acid and Oxidizing Agents. Oxidizing agents containing chloride or other halides have been known to exacerbate corrosion in acetic acid. The effect of $FeCl_3$ and $CuCl_2$ in glacial acetic acid on corrosion is shown as a function of (Mo + 0.5W) (Fig. 16). The corrosion rate in the same acetic acid in the presence of nonhalide ferric salt is quite low. Note that corrosion rate is not a good parameter for evaluating the behavior of chromium-molybdenum alloys because of localized corrosion. Nevertheless, because localized corrosion propagation was considerable, corrosion rate is used as a measure of relative susceptibility rather than actual penetration rate. In the 0.1% $FeCl_3$-containing acid, all alloys exhibited significant corrosion. The B-2 nickel-molybdenum alloy shows uniform corrosion, because it does not form a passive film. In aqueous solutions with similar concentrations of Fe^{3+}, the corrosion rate of alloy B-2 would be significantly higher.

Although the redox potentials were not measured, it is likely that the redox potentials of the solutions with 0.03% $FeCl_3$ and 0.02% $CuCl_2$ were lower than the solution with 0.1% $FeCl_3$. This may explain the lower corrosion rates of most alloys in the former mixtures. Also, alloy C-276 showed a significantly lower corrosion rate in the $CuCl_2$ solution, suggesting that the corrosion potential of this alloy in this solution was below the critical potential for localized corrosion. It can be seen (Fig. 17) that the effect of $FeCl_3$ on corrosion in acetic anhydride is less than in acetic acid but is nevertheless significant. Further characterization of the dissociation and redox kinetics is necessary in order to better interpret these results.

ACKNOWLEDGMENTS

The author thanks the contributors of data: Dr. Paul Manning (Haynes International), Mr. James Crum (Special Metals), Dr. Hira Ahluwalia (Materials Resources), Dr. Michael Renner (Bayer), and Mr. Steven Grise (DuPont). The project to compile the data into a database was sponsored by the Materials Technology

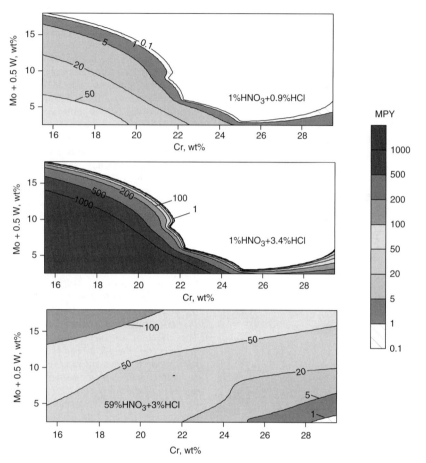

Fig. 15 Isocorrosion diagrams as functions of chromium and molybdenum contents of alloys in three mixtures of nitric and hydrochloric acid

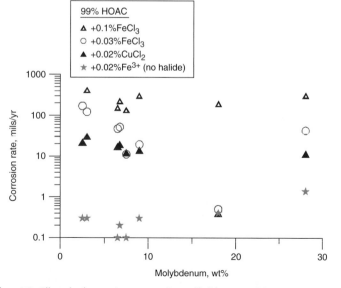

Fig. 16 Effect of redox species concentration and halide concentration on corrosion rate in 99% acetic acid. Boiling point is 117.9 °C (244.2 °F). 24 h

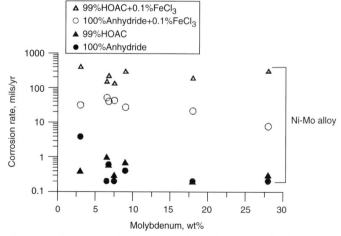

Fig. 17 Effect of anhydride on corrosion rate in the presence of $FeCl_3$. Boiling, 24 h

Institute of the Chemical Process Industry. The data used in this section arise from this database. The author acknowledges the guidance and support provided by Dr. F. Galen Hodge during the course of the project.

REFERENCES

1. N. Sridhar, J.B.C. Wu, and S.M. Corey, Corrosion of Ni-Base Alloys in Acid Mixtures, *Mater. Perform.,* Vol 24 (No. 10), 1987, p 17
2. N. Sridhar, "The Effect of Velocity on Corrosion in H_2SO_4 + HF Mixtures," Paper 19, Corrosion/90, NACE International, 1990
3. G.H. Koch, "Localized Corrosion in Halides Other than Chlorides," MTI Publication 41, Materials Technology Institute of the Chemical Process Industries, Inc., 1995
4. *Sulfuric/Phosphoric Acid Plant Operations,* Chemical Engineering Progress Technical Manual, American Institute of Chemical Engineers, 1982
5. J.P. Audouard, D. Catelin, and P. Soulignac, Application of Stainless Steels in Wet Process Phosphoric Acid, *Corrosion Prevention in Process Industries,* R.N. Parkins, Ed., NACE International, 1990, p 71–82
6. M. Honda and Y. Kobayashi, Stainless Steels for Wet Process Phosphoric Acid, *Mater. Perform.,* Vol 33 (No. 9), 1994, p 66–69
7. T. Sydberger and S. Nordin, Corrosivity of Wet Process Phosphoric Acid, *Corrosion,* Vol 34 (No. 1), 1978, p 16–22
8. N. Sridhar, J.B.C. Wu, and P.E. Manning, Corrosion Resistant Ni-Cr-Mo Alloys, *J. Met.,* Vol 37 (No. 12), 1985
9. C.R. Brooks, J.E. Spruiell, and E.E. Stansbury, Physical Metallurgy of Nickel-Molybdenum Effects, *Int. Met. Rev.,* Vol 29 (No. 3), 1984
10. R.C. Newman, Understanding the Corrosion of Stainless Steel, *Corrosion,* Vol 57 (No. 12), 2001, p 1030–1041
11. N. Sridhar, J. Kolts, and M.V. Zeller, Correlation of Microstructure and Selective Dissolution in Ferralium Alloy 255, *Metallography and Corrosion,* J.L. McCall et al., Ed., NACE International, 1986, p 105
12. H.-H. Strehblow, B. Titze, and B.P. Loechel, *Corros. Sci.,* Vol 19, 1979, p 1047–1057
13. N. Sridhar, Behavior of High Performance Alloys in Sulfuric Acid, *Mater. Perform.,* Vol 25 (No. 3), 1988, p 40
14. L.L. Faingold and T.Y. Shatova, *Zashch. Met.,* Vol 17 (No. 3), 1981, p 312–315

Corrosion by Organic Solvents

Hira S. Ahluwalia, Material Selection Resources Inc.
Ramgopal Thodla, General Electric Company

CORROSION is much less predictable in organic or mixed-solvent environments than in aqueous process environments. As a result, many chemical companies face greater uncertainty when selecting process equipment materials to manufacture chemical products using organic or mixed solvents than when the process environments are only aqueous. Chemical companies handle this uncertainty by overdesigning the equipment (wasting money and energy), rather than by accepting increased risks of corrosion failure (personnel hazards and environmental releases). Of particular interest to the chemical industry is the corrosion resistance of metals in organic solutions containing low concentrations of water and acid contamination. The corrosion of metals by organic solvents relevant to the chemical industry has received little systematic study; however, Ref 1 and 2 provide excellent information on this subject and are the basis for this article.

Classification of Organic Solvents

The structure and type of the organic functional group influences the reactivity of metals in organic solvents. The role of the organic solvent in many cases is to transport corrosive reactants and the corrosion precipitates formed during the corrosion process. The aggressiveness of the corrosive species is a function of the type of solvent used and their solvation properties. The various physical and chemical interaction forces acting during solvation can affect the corrosion behavior of various alloys in a given organic solvent or mixtures of solvents.

Aprotic and Protic Solvents. Organic solvents can be divided according to their properties and corrosivity into two general groups. These groups are referred to as aprotic and protic solvents. Aprotic solvents are water insoluble, whereas protic solvents are water soluble. The aprotic solvents can be further divided into two classes, nonpolar and dipolar. The nonpolar aprotic solvents include aromatic and aliphatic hydrocarbons and symmetrical halogenated hydrocarbons. Solvation in these systems occurs by relatively weak van der Waals-London forces. The dipolar aprotic solvents include dimethyl formamide, propylene carbonate, acetonitrile, ketones, aldehydes, esters, and asymmetric halogenated hydrocarbons. The dipolar aprotic system displays electrostatic forces due to ion-dipole and dipole-dipole interactions. The protic solvents include organic acids, alcohols, amines, and amides. Solvation in protic solvents occurs through dipole-dipole interaction, ion-dipole interaction, and hydrogen bonding. The protic or aprotic character of the solvent is determined by the ability to provide protons. Protic media contain active hydrogen protons and have high conductivity and dielectric constants. The aprotic solvents do not contain active hydrogen protons and have very low dielectric constants and low conductivity. Table 1 shows examples of protic and aprotic systems with some of their characteristic physical properties (Ref 2).

One-Component and Multicomponent Systems. The classification of organic solvents into one-component and multicomponent systems provides a more phenomenological way of systematization (Ref 1). Pure organic solutions are considered one-component, whereas homogeneous mixtures with predominant organic components are considered as multicomponent. In one-component solvent systems, an oxidizing group within the molecule structure is responsible for the corrosion process and the solvent and aggressive reagent are the same. Multicomponent systems generally contain the oxidizing species as solutes, and often the solvent and aggressive species are different. Tables 2 and 3 list examples of both classes. The corrosive species in one-component systems tends to consist of acidic hydrogen atoms of hydroxyl or carboxy groups as well as halogen atoms of halogenated hydrocarbons. These hydrocarbons have a strong affinity to electronegative metals. In multicomponent systems, the corrosive species are similar to those in aqueous solutions and include oxygen, solvated protons, halogens, metal ions of higher valence, and other oxidizing compounds.

Corrosion in Aprotic (Water Insoluble) Solvent Systems

Aprotic solvents are essentially nonconductive and have very low solubility for water. The chlorinated solvent systems are of particular interest to the chemical industry

Table 1 Examples of aprotic and protic solvents and their physical properties

Solvent	Electrical conductivity, S/m	Solubility of water in solvent at 25 °C (75 °F), ppm	Dielectric constant	Boiling point °C	Boiling point °F
Aprotic solvents					
Toluene	8×10^{-14}	334	2.38	110	230
Trichloroethylene	8×10^{-10}	320	3.42	87	189
Methylene chloride	4.3×10^{-9}	1700	9.1	40	104
0-dichloro benzene	3×10^{-9}	309	2.27	180	356
1,1,1 trichloroethylene	7.3×10^{-8}	340	7.5	74	165
Protic solvents					
Methanol	1.5×10^{-7}	Infinite	33	65	149
Dimethyl formamide	6×10^{-6}	Infinite	37	153	307
Dimethyl acetamide	2×10^{-5}	Infinite	38	166	331
Propylene carbonate	4×10^{-5}	8.3 wt%	69	242	468
Dimethyl sulfoxide	2×10^{-7}	High	47	189	372
Water					
Water	5.5×10^{-6}	...	80	100	212

Source: Ref 1

because hydrochloric acid (HCl) can be generated by thermal or hydrolytic degradation.

In dry aprotic solvent systems, even in the presence of acid constituents, corrosion generally does not occur because the solutions are nonconductive and no electrochemical processes can take place. For example, bubbling up to 3 wt% HCl gas into dry trichloroethylene does not corrode carbon steel or stainless steels in contrast to high corrosion rates that would be expected if the acid was in a water solution.

In the presence of low concentration of acids or water, corrosion in trichloroethylene becomes significant. If excess water beyond the solubility is present, two phases are formed resulting in the water phase becoming very corrosive because it extracts the acid from the solvent phase. If the water is dissolved, but less than 100 ppm in the presence of acid, little or no corrosion will be observed in the solution, condensing, or vapor zones. However, if the dissolved water is more than 200 ppm in the presence of acid, or if there is a small amount of phase water, minor corrosion would be expected in the solution, but severe corrosion may occur in the condensing zone. If the dissolved water exceeds about 600 ppm in the organic solutions, corrosion would be expected in the boiling solution and the condensing zone (Ref 2).

The corrosion resistance of several metallic materials of construction exposed to contaminated boiling trichloroethylene is tabulated in Table 4. The pH of the environment ranged from 0.5 to 2.5, and 200 to 600 ppm soluble water was present. The corrosion resistance of all the alloys tested was good when exposed to the liquid and vapor phase; however, significant corrosion was observed in the condensing zones. A nickel-molybdenum alloy (UNS N10001) and tantalum that normally have good corrosion resistance in hydrochloric solutions were the only alloys to show excellent performance in the condensing zone (Ref 2).

In aprotic solvent systems with chlorinated hydrocarbons, HCl, and H_2O as components, the degree of corrosion of unalloyed metals, stainless steels, and nickel-base alloys depends on the presence of critical concentrations of dissolved water and HCl of 200 and 40 ppm, respectively. The HCl and H_2O content should be kept below these critical values in order to avoid the two-phase state and maintain a low corrosion rate both in solution and in condensing areas. In the presence of water, the solvents trichloroethylene, 1,1,2-trichloroethane, and methylene chloride hydrolyze with a resulting buildup of acidity.

Table 2 Examples of one-component systems

Class of solvent	Example	Formula	Corrosive group
Carboxylic acids	Acetic acid	CH_3COOH	—COOH and/or H^+_{solv}
Alcohols, aromatics, or aliphatic	Phenol, methanol, ethanol	C_6H_5OH	—C—OH and/or H^+_{solv}
Halogenated hydrocarbons	Trichloroethylene	$CHClCl_2$	—C—Cl

Source: Ref 1

Table 3 Examples of multicomponent systems

Class of solvent	Example (formula)	Components	Corrosive group or components
Alcohols, aromatics, or aliphatic	Ethanol (CH_3CH_2OH)	H_2O, O_2, inorganic or organic acids	H^+_{solv}, O_2, —COOH
Esters	Ethyl acetate ($CH_3COOCH_2CH_3$), propylene carbonate ($CH_3CHCH_2CO_3$)	H_2O, acids, O_2	H^+_{solv}, O_2, —COOH
Aldehydes	Formaldehyde (HCHO)	H_2O, formic acid, O_2	H^+_{solv}, O_2, —COOH
Halogenated hydrocarbons	Ethylene chloride (CH_2=CHCl)	H_2O, HCl, O_2	H^+_{solv}, O_2, —Cl
Hydrocarbons	Heptane (C_7H_{16})	H_2O, O_2, halogens, HX	Halogens, O_2, H^+_{solv}

Table 4 General corrosion rate of various materials exposed to boiling trichloroethylene 200–600 ppm soluble water, pH 0.5–2.5

		Corrosion rate					
		of solution		of vapor		of condensate	
UNS No.	Common name	mm/yr	mils/yr	mm/yr	mils/yr	mm/yr	mils/yr
G10080	1008 carbon steel	0.025	1	0.025	1	>25	>1000
S30400	304 SS annealed	<0.025	<1	<0.025	<1	>2.5	>100
	304 SS sensitized	0.1	4	<0.025	<1	>25	>1000
N10276	Alloy C-276	<0.025	<1	<0.025	<1	1–2.25	40–90
N10001	Alloy B	<0.025	<10	...	...	0.05	2
N02200	Nickel 200	...	...	...	...	>2.5	>100
N04400	Alloy 400	<0.025	<1	...	...	>2.5	>100
R05200	Tantalum	...	...	...	...	0.075	3

Source: Ref 2

Table 5 Effect of water content on the general corrosion rate of various materials exposed to boiling 0.5 wt% HCl in dimethyl formamide

	Corrosion rate							
	of 1008 carbon steel		of type 304		of nickel 200		of alloy 800	
Water content, wt%	mm/yr	mils/yr	mm/yr	mils/yr	mm/yr	mils/yr	mm/yr	mils/yr
0.0075	43	1720	8.7	348	0.95	38	4.08	163
0.85	36.25	1450	4.13	165	0.75	30	1.65	66
1.4	42.5	1700	1.35	54	0.625	25	1.38	55
2.0	46.37	1855	0.015	0.6	0.45	18	0.83	33
5.0	43.02	1721	0.005	0.2	0.05	2	0.013	0.5
20.0	576	14.4	0.008	0.3	0.83	33	...	...

Source: Ref 2

Corrosion in Protic (Water Soluble) Solvent Systems

Protic solvents have high dielectric constants and electrical conductivity and dissolve high levels of water. In contrast to aprotic systems, corrosion rate for many materials can be high in essentially water-free organic systems containing acids. The effect of water and acids has been studied in several protic organic solvents in several different metal systems (Ref 2). Table 5 shows the effect of water content on the corrosion of materials exposed to 0.5% wt HCl in dimethyl formamide (DMF). The data show that a large reduction in corrosion rate occurs when the water content in HCl-acidified organic solvents exceeds 1 wt%. The amount of water required to inhibit corrosion depends on the alloy and on the level of acid present. The role of the water in these systems in forming and maintaining the passive film is thought to be important.

For example in case of tantalum, which is an extremely corrosion-resistant material, the presence of water in methanol solutions plays a strong role in corrosion resistance. The pitting and repassivation potential of tantalum are a strong function of water concentration as seen in Fig. 1. The pitting and repassivation potential increase sharply as the water concentration increases from 0.03 to 4 wt% (Ref 3).

Aeration and higher temperatures increase the corrosion rates in some cases even higher than in aqueous solutions containing the same level and type of acidity (Ref 4).

Importance of Conductivity

The difference in the corrosion characteristic of protic and aprotic solvents is not necessarily dependent on water content, but rather on the solution conductivity. Increasing the acidity or the soluble water content are merely ways to increase the solution conductivity. The corrosion data developed by Demo (Ref 2) correlating the corrosion rate with solution conductivities, acid content, and water content show quite clearly the importance of conductivity as the primary parameter determining the corrosivity of a organic solvent. Table 6 presents the relationship among solution conductivity, acidity, and corrosion rates of type 304 stainless steel in protic and aprotic solvents.

Based on the data in Table 6, corrosion will not occur in organic solvent solutions having a conductivity of less than about 10^{-5} S/m even with acid levels up to 1 to 3 wt%. If solution conductivities are equal to or higher than 10^{-5} S/m, corrosion will be increasingly severe as the HCl content exceeds 40 ppm. For other acids the minimum acid content is higher. In low-conductivity solvent systems, the conductivity increases as the soluble water content increases. The conductivities approach 10^{-5} S/m when the water content is 200 ppm, thus leading to corrosion under appropriate acidic conditions.

Corrosion Testing

The basic types of corrosion processes are similar in both aqueous and organic liquid environments; however, special consideration must be given to the use of testing techniques to organic systems due to the complexity of the environment. The wide variety of organic liquids and mixtures increases the number of experimental variables that need to considered and controlled. A review by Brossia et al. (Ref 5) discusses the important variables and parameters that require consideration when testing in organic systems.

Some of the important environmental variables that influence corrosion testing in organic liquids include:

- Metal, alloy, or nonmetallic material in contact with the organic solution
- Solution conductivity
- Solution acidity
- Water content of the solution
- Presence and stability of oxide or other preexisting films
- Type and concentration of surface contaminants or inclusions
- Functional group and concentration of organic solvent
- Solvent oxidation or reduction products
- Type and concentration of supporting electrolyte
- Applied potential
- Possible mechanistic paths for passivation
- Temperature

As in all corrosion testing, matching the test environment to the actual service environment is critical for obtaining useful information. In addition, by studying the effects of these variables on the corrosion processes, better insights into the controlling mechanisms, as well as the sensitivity of the corrosion processes to changes in the service environment, can be gained (Ref 5).

There are several aspects to conducting electrochemical tests in organic liquids that are often not encountered or important while testing in aqueous solutions. These include effects due to low solution conductivities, the importance of the water concentration in the solution, the existence of an extremely wide variety of liquid compositions, the complexity of products from electroactive organic liquids, and a lack of thermodynamic data (Ref 5).

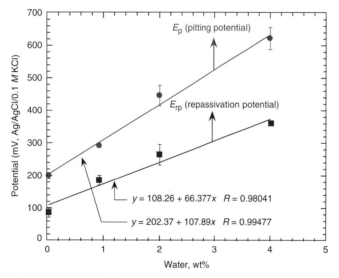

Fig. 1 Pitting and repassivation potential of pure tantalum as a function of water concentration (0–4 wt%) in methanol solutions. 99.5–95.5 wt% MeOH. 0.5 wt% HCl. Room temperature, N_2 purge

Table 6 Relationship among solvent conductivity, acidity (HCl), and corrosion rates of type 304 stainless steel in trichloroethylene (aprotic) and dimethyl formamide (protic) boiling solutions

Solvent	Acidity as HCl, ppm	Water, ppm	Conductivity at boiling point, S/m	Corrosion rate in solution	
				mm/yr	mils/yr
Trichloroethylene	<10	70	$<10^{-9}$	0	0
	600	70	2.4×10^{-8}	0	0
	1450	127	1.0×10^{-8}	0	0
	1450	900	2.0×10^{-3}	6.25	250
				(Condensate)	
DMF	<10	75	1.2×10^{-4}	0	0
	20	75	7.1×10^{-4}	0.025	1
	100	75	2.1×10^{-2}	0.3	12
	5000	8500	0.34	4.13	165
Water	<10	...	7.2×10^{-4}	0	0
	5000	...	6.4	3.43	137

DMF, dimethyl formamide. Source: Ref 2

REFERENCES

1. E. Heitz, Corrosion of Metals in Organic Solvents, *Advances in Corrosion Science and Technology,* Vol 4, M.G. Fontana and R.W. Staehle, Ed., Plenum Publishing, 1974
2. J.J. Demo, paper No. 175, *Corrosion 91,* NACE International, 1991
3. T. Ramgopal, unpublished results, private communication
4. J.J. Demo, *Chem. Eng. World,* Vol 7, 1972, p 115–124
5. C.S. Brossia and D.A. Shifler, Organic Liquids, Chapter 38, *Corrosion Tests and Standards: Application and Interpretation,* MNL 20, R. Baboian, Ed., 2nd ed., ASTM International, 2004

SELECTED REFERENCES

- P.L. DeAnna, *Corros. Sci.,* Vol 25, 1985, p 43–53
- P. Hronsky, *Corrosion,* Vol 137, 1981, p 161
- R.G. Kelly and P.J. Moran, *Corros. Sci.,* Vol 30, 1990, p 495
- Z. Szklarska-Smialowska and J. Mankowski, *Corros. Sci.,* Vol 22, 1982, p 1105

Corrosion in High-Temperature Environments

George Y. Lai, Consultant

SOME MANUFACTURING PROCESSES require chemical reactions to proceed at high temperatures. High-temperature corrosion of the processing equipment by the processing stream can be a critical issue when making an alloy selection. High-temperature corrosion can result in a premature failure of the component. In some cases, contamination of the product by the corrosion products can also present a serious manufacturing problem. Thus, understanding the high-temperature corrosion behavior of alloys is an important step toward the selection of appropriate alloys for process equipment. Representative high-temperature equipment includes calciners, reactor vessels, pyrolysis furnace tubes, fired and process heater tubes, heat exchangers, waste heat boilers, and so on. Selection of a suitable alloy for the construction of the high-temperature equipment depends on a number of factors, such as mode of corrosion, type of equipment, operating temperatures, stresses, thermal cycling, materials properties, thermal stability of the alloy, corrosion allowable, and component design life and cost, among other factors. This article briefly describes the high-temperature corrosion modes that are frequently encountered in the chemical process industry. These corrosion modes include oxidation, carburization, metal dusting, nitridation, halogen corrosion, and sulfidation. More detailed treatments of these corrosion modes as well as other corrosion modes, such as ash/salt deposit corrosion, molten salt corrosion, and molten metal corrosion, can be found in Ref 1 and the articles "Molten Salt Corrosion," "Liquid Metal Corrosion," and "High-Temperature Gaseous Corrosion" in *Corrosion: Fundamentals, Testing, and Protection*, Volume 13A of *ASM Handbook*, 2003.

Oxidation

The reactor (or calciner), where the chemical reaction is taking place in an oxidizing environment, can suffer oxidation attack in its internal diameter. Furthermore, the reactor (or calciner) is also subject to oxidation attack on the outer diameter when heated from the outer diameter to allow the chemical process reaction to take place inside of the reactor (or calciner). Oxidation is also a critical issue in process heater or pyrolysis tubes, heat exchangers, piping, boilers, and numerous process equipment.

Carbon and chromium-molybdenum (Cr-Mo) steels are widely used for construction of high-temperature components, such as reactor vessels, calciners, heat exchangers, piping, boilers, and so on. Iron oxidizes to form magnetite (Fe_3O_4) and hematite or ferric oxide (Fe_2O_3) at temperatures up to 570 °C (1060 °F) and forms ferrous oxide (FeO), Fe_3O_4, and Fe_2O_3 at temperatures above 570 °C (1060 °F). Carbon steels are generally not considered for applications above approximately 540 °C (1000 °F), due to decreased mechanical strength and increased scaling (accelerated oxidation). An early laboratory oxidation study (Ref 2) on carbon steel in air showed a wastage rate of approximately 0.3 mm/yr (12 mils/yr) at 595 °C (1100 °F), as shown in Fig. 1. This was confirmed by a recent laboratory oxidation study in air (Ref 3) that showed the oxidation attack of 0.25 mm (10 mils) after one year at 595 °C (1100 °F).

Most oxidation data arise from laboratory testing that is commonly short term and isothermal. In actual plant services, components are subject to more complex service conditions, such as thermal cycling, overheating, heat fluxes, and the presence of various combustion species. Any of these conditions can accelerate oxidation rates. In addition, the high-temperature strength of the material decreases rapidly with increasing temperature. Furthermore, the alloy can undergo microstructural changes under long-term exposure to high temperatures. These microstructural changes can significantly degrade the material strength and properties. For example, carbon steel can suffer graphitization after exposure at 595 °C (1100 °F) in less than 1000 h (Ref 4). Chromium addition can improve the mechanical properties of steel and its resistance to

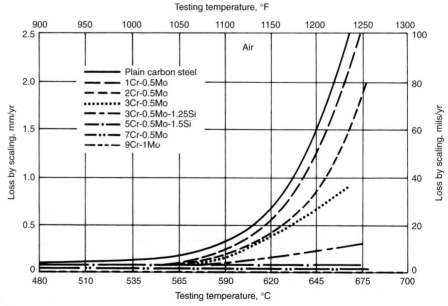

Fig. 1 Effect of chromium on the oxidation resistance of steels in air. Source: Ref 2

microstructural changes and oxidation attack (Fig. 1).

Stainless Steels. For components operating at higher temperatures, stainless steels that form a continuous chromium oxide scale (Cr_2O_3) offer good service. Because of their higher elevated-temperature strengths and their ease of fabrication and welding, austenitic stainless steels are much more widely used for the construction of high-temperature components than ferritic stainless steels. Chromium dictates oxidation resistance, with higher chromium content generally improving oxidation resistance for the alloy. The ASTM (ASME) specification range for chromium is 16 to 18% for type 316, 17 to 20% for type 321 and 347, 18 to 20% for type 304, 22 to 24% for type 309, and 24 to 26% for type 310. Type 310 is generally considered to be the best in the group, followed by type 309 and then those with lower chromium ranges. The benefit of chromium is dramatically illustrated in Fig. 2 (Ref 5). Testing was conducted at a very high temperature of 980 °C (1800 °F) in order to discriminate these stainless steels. The figure also shows that the stainless steels with lower chromium contents exhibited a shorter incubation time to breakaway oxidation than those with higher chromium contents. Breakaway oxidation occurs when the chromium oxide scale can no longer reform on the metal surface, thus causing the stainless steel to suffer rapid growth of iron oxides. As the chromium oxide scale grows with time, the chromium concentration in the matrix underneath the oxide scale decreases with increasing service time. The chromium concentration can be reduced to a level that is too low to maintain a protective chromium oxide scale. As a result, formation of nonprotective iron oxides and iron-chromium oxides is initiated, causing breakaway oxidation.

Alloy producers tend to manufacture stainless steels to the bottom of the specification range for key alloying elements, such as chromium, to reduce materials cost in today's business climate. This is illustrated for type 304 (Ref 6):

	Cr, wt%	Ni, wt%
ASTM A 240, type 304	18.00–20.00	8.00–10.50
Pre-1965 heats	18.7	9.9
Current production	18.3	9.0

As a result, the chromium content can be insufficient to maintain a continuous chromium oxide scale during prolonged service or when subjected to thermal cycling or overheating conditions, thus promoting breakaway oxidation. These "lean" stainless steels can be further aggravated by the surface depletion of chromium resulting from the manufacturing process that may involve excessive pickling after black annealing (annealing in air or combustion atmosphere) during successive reductions in cold rolling in flat product manufacturing or pilgering in tubular manufacturing. In thin-gage sheet or tubular products, the reduced chromium concentration at the surface can be too low to form or maintain a continuous chromium oxide scale during service. As a result, iron oxides and isolated nonprotective iron-chromium oxide nodules develop on the metal surface, resulting in breakaway oxidation and unsatisfactory service.

Iron-Nickel-Chromium (Fe-Ni-Cr) and Nickel-Base Alloys. As the temperature increases to the temperature range where stainless steels can no longer meet both strength and oxidation requirements, it may be necessary to consider Fe-Ni-Cr and nickel-base alloys. Figure 2 shows significantly better cyclic oxidation resistance for nickel-base alloys in comparison with some stainless steels. Compositions of some of these alloys are tabulated in Table 1. In addition to increases in nickel and chromium, in some cases many alloys have used silicon or aluminum along with rare earth elements, such as cerium, lanthanum, and yttrium, or the reactive element zirconium for improving the resistance to oxidation and other high-temperature corrosion attack. All the alloys listed in Table 1, except alloys 214 and MA 956, are chromia formers (alloys that form chromium oxide scales when heated to elevated temperatures). The small additions of these alloying elements enhance the protective nature of the chromium oxide scale. Addition of alloying elements, such as molybdenum and tungsten, increase elevated-temperature strength, which may be as important a design consideration as oxidation resistance as the service temperature increases.

At very high temperatures, 1100 °C (2010 °F) and higher, chromium oxide scales can suffer rapid growth, rendering them nonprotective. In addition, internal oxidation attack becomes severe at these temperatures. This is illustrated in Fig. 3 (Ref 7). The figure shows the oxidation morphology of alloy 601 after testing in air for 1056 h (44 days) at 850, 1000, 1100, and 1200 °C (1560, 1830, 2010, and 2190 °F). It is clear from the micrographs that the chromium oxide scales were no longer protective at 1100 and 1200 °C (2010 and 2190 °F). Alternate alloys for these applications will be alumina formers that form a protective aluminum oxide scale. Table 2 indicates oxidation data generated in air for 360 days at 1200 °C (2190 °F) in comparing an alumina former (alloy 214) with several chromia formers (Ref 8). Commercial alloys that are applicable at these temperatures are very limited. Wrought nickel alloy 214, when used at this temperature range, has adequate oxidation resistance; however, grain coarsening and lower creep-rupture strength can be significant issues. Oxide-dispersion-strengthened alloy MA 956, which is produced by a mechanical alloying powder process, exhibits adequate oxidation resistance and creep-rupture strengths. However, fabrication and joining of this alloy, as

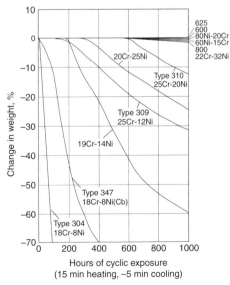

Fig. 2 Cyclic oxidation resistance of several stainless steels and nickel-base alloys in air at 980 °C (1800 °F). Source: Ref 5

Table 1 Compositions of selected wrought iron-nickel-chromium and nickel-base alloys for high-temperature chemical processing environments

		Composition, wt%				
Alloy	UNS No.	Ni	Fe	Cr	Al	Others
253MA	S30815	11	bal	21	...	1.6Si, 0.05Ce
800H/HT	N08800	32	bal	21	0.4	0.4Ti
330	N08330	35	bal	19	...	1.3Si
353MA	S35315	35	bal	25	...	1.6Si, 0.05Ce, 0.15N
803	S35045	35	bal	27	0.3	0.4Ti
HR-120	N08120	37	bal	25	...	0.7Nb, 0.2N
556	R30556	20	bal	22	...	18Co, 0.6Ta, 0.2N
45TM	N06045	bal	23	27	...	2.7Si, 0.06Ce
HR-160	N12160	bal	3.5(a)	28	...	29Co, 2.75Si, 0.5Ti, 0.5Nb
601	N06601	bal	14	23	1.4	...
602CA	N06025	bal	9	25	2.1	0.08Y, 0.05Zr
X	N06002	bal	18	22	...	9Mo, 0.6W
333	N06333	bal	18	25	...	3Mo, 3W
617	N06617	bal	1.5	22	1.2	12.5Co, 9Mo
230	N06230	bal	3(a)	22	...	14W, 5Co max, 2Mo, 0.02La
214	N07214	bal	3	16	4.5	0.01Y, 0.05Zr max
MA 956(b)	S67956	...	bal	20	4.5	$0.5Y_2O_3$

(a) Maximum. (b) MA 956 is an oxide-dispersion-strengthened alloy produced by mechanical alloying powder process, no longer active. Listed for reference

well as cost, can be significant issues. Comparison oxidation data for some of the heat-resistant alloys are also available in technical publications (Ref 1, 3, 8) and alloy producers' technical data brochures (Ref 9–17).

In order to deal with the grain coarsening and relatively lower creep-rupture strengths that are associated with wrought alloy products at very high temperatures, engineers have the option of using castings. Because the manufacturing process does not involve forging and hot or cold rolling, casting alloys can be designed to contain large amounts of stable carbides and solid-solution-strengthening alloying elements for improving creep-rupture strengths. Furthermore, the cast product can be made with thicker walls to prolong service life. Many cast alloys contain large amounts of carbon along with tungsten, niobium, titanium, and rare earth elements. Cast alloys involving the addition of titanium and rare earth elements are commonly referred to as microalloyed products. Many reformers and pyrolysis furnace tubes are made of casting alloys. Table 3 lists some of the casting alloys.

Carburization

Metals and alloys are susceptible to carburization when exposed to environments containing methane (CH_4), carbon monoxide (CO), carbon dioxide (CO_2), hydrogen (H_2), or other hydrocarbon gases at elevated temperatures. Metals and alloys can also be carburized when in contact with graphite at elevated temperatures. An alloy is likely to be carburized when the carbon activity (a_c) of the environment is greater than that of the alloy. Thermodynamic considerations for carburization are discussed in detail elsewhere (Ref 18, 19) and in the articles "Thermodynamics of Gaseous Corrosion" and "Kinetics of Gaseous Corrosion Processes" in *Corrosion: Fundamentals, Testing, and Protection*, Volume 13A of *ASM Handbook*, 2003. The carburization reaction generally follows the following steps (Ref 19):

1. Transport in the gas atmosphere by gas flow and diffusion in the boundary layer
2. Carbon transfer to the metal phase by phase-boundary reactions
3. Inward diffusion of carbon
4. Reaction of carbon with carbide-forming alloying elements in alloy interior

For step 2, addition of a small amount of sulfur to the process environment can significantly reduce the rate of carburization (Ref 19). For step 3, the composition of the alloy can be adjusted to decrease the diffusivity of carbon in the alloy interior, thus resulting in better resistance to carburization. Nickel reduces the diffusivity of carbon in Fe-Ni-Cr alloys (Ref 20). Accordingly, increasing nickel in Fe-Ni-Cr alloys improves the alloy carburization resistance, as shown in Fig. 4 (Ref 21).

Most high-temperature alloys form an adherent chromium oxide (Cr_2O_3) scale, which is considered to be an effective barrier against carbon ingress. Nevertheless, carburization has been encountered for chromia formers. Carbon permeation through the oxide scale occurs through pores and fissures in the oxide scale (Ref 22). Adding silicon to chromia formers can significantly increase the alloy carburization resistance, as shown in Fig. 5 (Ref 23). Furthermore, aluminum oxide (Al_2O_3) scale is a significantly better barrier against carburization. This is illustrated in Fig. 6, which shows nickel-base alloy 214 (N07214) exhibiting essentially no evidence of carburization attack under the severe carburizing test condition (Ref 24). The figure also shows excellent carburization

Table 2 Long-term oxidation tests at 1200 °C (2190 °F) for some alumina and chromia formers exposed to still air for 360 days

Specimens were cycled to room temperature once every 30 days

Alloy	Metal loss		Average metal affected(a)	
	mm	mils	mm	mils
214	0.01	0.39	0.02	0.79
HR-160	0.42	16.5	2.62	103.1
230	1.51	59.4	2.64	103.9
601	0.70	27.6	2.95	116.1
617	2.23	87.8	>6.36	>250.4
RA85H	0.78	30.7	>6.39	>251.6
HR-120	Consumed	Consumed	>6.37	>250.8
800H	Consumed	Consumed	>6.35	>250.0

(a) Metal loss plus average internal penetration. Source: Ref 8

Table 3 Compositions of selected heat-resistant casting alloys for chemical processing environments

Alloy	Composition, wt%					
	C	Si	Cr	Ni	Fe	Others(a)
HK40	0.4	2 max	26	20	bal	...
HP40	0.4	2 max	26	35	bal	...
Thermax 63	0.45	2 max	23	35	bal	0.5W
Thermax 63W	0.45	2 max	25	35	bal	15Co, 5W
Manaurite 36X	0.4	2 max	25	34	bal	1.2Nb
Manaurite 36XS	0.4	2 max	25	34	bal	1.5Nb, 1.5W
Manaurite XT	0.4	2 max	35	44	bal	1.5Nb, 1.5W
More 10	0.4	2 max	25	35	bal	1.2Nb
More 10MA	0.45	1.8	25	35	bal	1.2Nb, Ti, RE
More 40MA	0.4	1.8	35	45	bal	1.2Nb, Ti, RE

(a) RE, rare earth element(s)

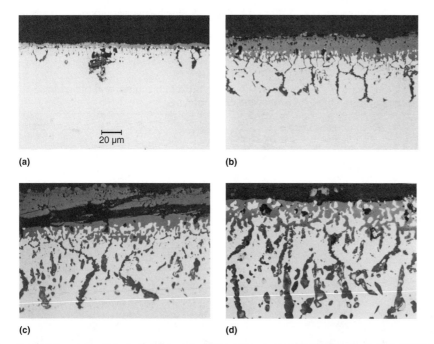

Fig. 3 Oxidation of alloy 601 after testing in air for 1056 h (44 days) at (a) 850 °C (1560 °F), (b) 1000 °C (1830 °F), (c) 1100 °C (2010 °F), and (d) 1200 °C (2190 °F). Courtesy of ThyssenKrupp VDM

Fig. 4 Effect of nickel on the carburization resistance of cast Fe-Ni-Cr and Ni-Cr alloys. Data were generated in pack carburization tests at 1100 °C (2010 °F). Source: Ref 21

resistance for alloy 602CA (N06025), which forms duplex Cr_2O_3/Al_2O_3 oxide scales.

Metal Dusting

Metal dusting typically occurs in carburizing environments containing primarily H_2 and CO. The corrosion attack is localized in the form of pitting. The dustlike corrosion products consist of carbon soot and powders of metals, oxides, and carbides. The temperature range for metal dusting attack is usually between 480 and 900 °C (900 and 1650 °F), with the maximum rate of attack occurring at 600 to 700 °C (1100 to 1300 °F), depending on the alloy and environment.

Significant advancement in understanding metal dusting behavior of alloys has been achieved in the past decade, with most of the contribution being made by Grabke and his coworkers (Ref 25–31). The thermodynamics and mechanisms of metal dusting were summarized in his latest paper (Ref 31). Metal dusting can occur when the carbon activity (a_c) of the carburizing environment is greater than 1. The carbon transfer from the environment to the metal (iron and nickel and their alloys) results in oversaturation and disintegration into fine metal particles and graphite (Ref 31). The resistance to metal dusting attack of many high-temperature alloys has also been extensively investigated (Ref 25–37). Figure 7 shows an example of metal dusting attack for alloy 800 (Ref 32).

Researchers (Ref 32) reported metal dusting behavior of many commercial alloys, including Fe-Ni-Cr alloys and nickel-base alloys. The Fe-Ni-Cr alloys, such as 800H, HK40, and HP40, were found to suffer rapid metal dusting attack. Nickel-base alloys were found to be significantly better. Among nickel-base alloys, high-chromium alloys appeared to be much more resistant. Furthermore, aluminum addition to high-chromium nickel-base alloys appeared to be beneficial also. This is illustrated in Fig. 8 (Ref 32). Alloy 602CA (2.1% Al) was found to be much more resistant to metal dusting than alloy 601, (1.4% Al), which, in turn, was much better than alloy 600 (0% Al). In this test, however, the 602CA test specimen was black annealed (pre-heat treated in air), while alloys 601 and 600 were in as-ground conditions. Preoxidation treatment of 602CA in air may have produced a duplex Cr_2O_3/Al_2O_3 oxide scale prior to the testing in the metal dusting environment. The effect of the surface condition of specimens on resistance to metal dusting has been investigated by several authors (Ref 27, 34). It has been found that the surface deformation due to grinding in the as-ground surface condition could significantly improve the resistance to metal dusting, because a better chromium oxide scale is formed. Figure 9 shows the metal dusting behavior of alloy 601 with various surface finishes, including a typical production product of the annealed and pickled condition (Ref 34). The specimen with as-ground surface finish exhibited the best resistance. The production sheet material with an annealed and pickled

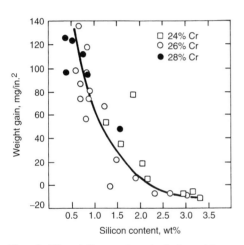

Fig. 5 Effect of silicon on the carburization resistance of cast Fe-20Ni-Cr alloys tested at 1090 °C (2000 °F) for 24 h in ethane. Source: Ref 23

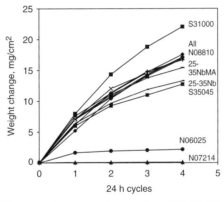

Fig. 6 Carburization resistance of type 310 (S31000), 800H (N08810), 25–35NbMA (HP microalloyed), 25–35Nb (HP alloy), 803 (S35045), 602CA (N06025), and 214 (N07214) at 980 °C (1800 °F) for 96 h in H_2-2%CH_4 with cycling to room temperature and data collection every 24 h. Source: Ref 24

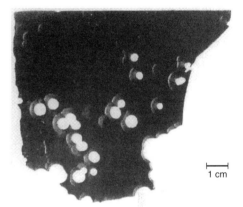

Fig. 7 An alloy 800H component that suffered metal dusting in a synthesis gas at 550 °C (1020 °F). Source: Ref 31

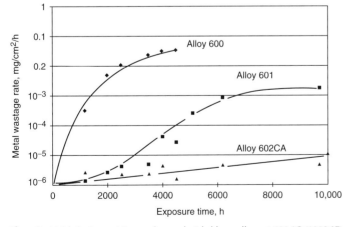

Fig. 8 Metal dusting resistance of several nickel-base alloys at 650 °C (1200 °F) in H_2-24%CO-2%H_2O during the first 5000 h of exposure and in H_2-40%CO-2%H_2O during the last 5000 h of exposure. Source: Ref 32

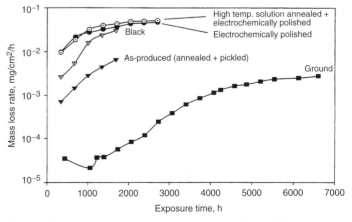

Fig. 9 Effect of surface finishes of the test specimens of alloy 601 on the metal dusting resistance in laboratory testing at 620 °C (1150 °F) in H_2-80%CO. Source: Ref 34

condition was found to perform poorly. The process of black annealing and pickling could produce the surface condition with depleted chromium concentration at and near the surface. Equally unfavorable surface conditions were produced by black annealing, electrochemically polishing, or high-temperature solution annealing plus electrochemically polishing. A black-annealed surface for a chromia former, such as alloy 601, is not beneficial in developing favorable oxide scales to resist metal dusting. On the other hand, for an alumina former or an alloy containing both chromium and aluminum, a duplex Cr_2O_3/Al_2O_3 oxide scale may develop as a result of black annealing (preoxidation in air). Figure 10 summarizes the metal dusting test data up to 16,000 h of exposure for more than 30 commercial alloys (Ref 35). All the specimen surface conditions were in as-ground condition. The best alloy was found to be alloy 693, a newly developed nickel-base alloy containing 29% Cr and 3.1% Al with small amounts of niobium and zirconium.

It has been demonstrated that the as-ground surface provides additional improvement in resisting metal dusting attack. However, in terms of actual applications, it is unlikely that the tubing or the fabricated component is provided to the user for applications in the as-ground condition. However, one can take advantage of the favorable surface condition of the metal product if the general manufacturing processes are understood for different products in nickel-base alloys. For plate and some sheet products as well as some tubular products, the manufacturing process typically involves black annealing (i.e., annealing in air) and pickling. This type of surface finish is not desirable in terms of metal dusting resistance. Some sheet and tubular products are cold rolled (cold pilgered for tubular products) and bright annealed (i.e., annealing in an inert environment, typically H_2 atmosphere). For chromia formers, the metal surface typically exhibits a shiny appearance with no preformed chromium oxide scale. Nevertheless, these products are expected to perform better than those processed by black annealing and pickling. For alumina formers as well as high-silicon alloys, however, the bright annealing tends to preform an aluminum-rich oxide scale or silicon-rich oxide scale for high-aluminum- and silicon-containing alloys, respectively. This is because the dewpoint for most of the bright annealing atmospheres is low enough to prevent the formation of Cr_2O_3 but high enough to form Al_2O_3 and SiO_2. The bright-annealed surface condition for these two types of alloys can significantly improve the alloy resistance to metal dusting. This is illustrated in Fig. 11. HR-160 alloy (Ni-29Co-28Cr-2.75Si-0.5Ti) was shown to exhibit no metal dusting attack after testing for 10,000 h in H_2-49%CO-2%H_2O at 650 °C (1200 °F). On the other hand, alloy 601 (Ni-23Cr-14Fe-1.4Al) was found to have suffered severe metal dusting attack. Test specimens of both alloys were obtained from bright-annealed sheet products. The bright-annealed sheet specimen of alloy 601 is believed to contain no preformed Cr_2O_3 oxide, while HR-160 bright-annealed sheet specimen most likely contained a preformed SiO_2 oxide film.

Nitridation

Metals and alloys are generally susceptible to nitridation when exposed to ammonia-bearing or nitrogen-base atmospheres at elevated temperatures. Nitridation can cause the alloy to become embrittled. Many austenitic stainless steels have been used for processing equipment in ammonia-bearing environments. If the conditions are too hostile for stainless steels because of higher temperatures, higher ammonia concentration, or both, nickel-base alloys should be considered. Increasing nickel content in Fe-Ni-Cr alloys or Ni-Cr-Fe alloys increases the alloy nitridation resistance, as illustrated in Fig. 12 (Ref 1). High temperatures, 980 °C (1800 °F), for example, tend to result in the formation of internal nitrides, while low temperatures, 650 °C (1200 °F), for example,

Fig. 11 Optical micrographs showing the cross section of (a) alloy 601 and (b) alloy HR-160 specimens after laboratory testing at 650 °C (1200 °F) for 10,000 h in H_2-49%CO-2%H_2O. Both specimens were in a bright-annealed finish condition (no surface grinding) prior to the exposure test. Source: Ref 36

Fig. 10 Metal dusting behavior of a large number of commercial alloys at 620 °C (1150 °F) in H_2-80%CO. Source: Ref 35

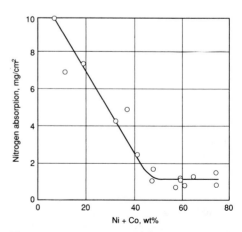

Fig. 12 Resistance to nitridation improves with increasing nickel and cobalt in iron-, nickel-, and cobalt-base alloys at 650 °C (1200 °F) for 168 h in ammonia (100% NH_3 in the inlet gas and 30% NH_3 in the exhaust). Source: Ref 1

cause nitride layer to form on the surface. Figure 13 shows a nitrided layer formed on the metal surface for type 304 stainless steel and alloy X (Ni-22Cr-9Mo-18.5Fe) when exposed to ammonia environment at 650 °C (1200 °F).

High-Temperature Corrosion by Halogen and Halides

Many metals and alloys are susceptible to severe corrosion attack when exposed to halogen gases at elevated temperatures. Halogen gases are very mobile species with high solubility and diffusivity that react readily with many metals and alloys to form metal halides, many of which exhibit high vapor pressures and/or low melting points. As a result, the metal-halogen reaction can proceed at a fairly rapid rate. Figure 14 provides a very general guideline for several alloys in dry chlorine environment (Ref 37). In chlorination reactions (i.e., the environment consisting of hydrogen chloride, HCl, with no oxygen present), nickel and nickel-base alloys, including molybdenum-containing nickel-base alloys, are much more resistant than iron-base alloys, as illustrated in Fig. 15 (Ref 38). For oxidizing environments containing Cl_2 (O_2-Cl_2 gas mixtures), molybdenum- (as well as tungsten-) containing nickel-base alloys were found to suffer higher corrosion attack, particularly at higher temperatures, as illustrated in Fig. 16 (Ref 39). When tested in Ar-20%O_2-2%Cl_2 at 900 °C (1650 °F), alloys 625 (9% Mo), X (9% Mo), 188 (14% W), C-276 (16% Mo, 4% W), and S (14.5% Mo) suffered higher corrosion rates than some simple Fe-Ni-Cr alloys, such as alloy 800H (which contains neither molybdenum nor tungsten). This was attributed (Ref 39) to the formation of oxychlorides of molybdenum and tungsten, which have very high vapor pressures. Also shown in Fig. 16, alumina-forming nickel-base alloys, such as alloy 214 (4.5% Al), performed very well in O_2-Cl_2 environments.

In F_2-containing environments without the presence of O_2, nickel is probably the best engineering material in resisting corrosion attack to fluorine gas corrosion, attributable to the formation of adherent nickel fluoride scale (Ref 40). Nickel is also the most resistant engineering material in hydrogen fluoride (HF) environments. Precious metals, such as gold and platinum, were found to be very resistant to high-temperature corrosion in HF environments (Ref 41). Nickel-molybdenum alloys, such as alloy B-3, or Ni-Mo-Cr alloys containing low chromium, such as alloy 242, are much more resistant to high-temperature HF corrosion than nickel-chromium alloys, such as alloys 600, 617, and 602CA (Ref 41). Chromium in nickel-chromium alloys tends to cause internal attack by the formation of internal fluorides (Ref 42).

Sulfidation

Sulfidizing environments can best be categorized into three types: sulfur vapor and H_2-H_2S environments, oxidizing/sulfidizing environments, and SO_2-bearing environments. In sulfur vapor and H_2-H_2S environments, the oxygen potentials (P_{O_2}) are so low that iron and chromium oxides are not thermodynamically stable. Under these conditions, alloys form sulfide scales, and the alloys that form chromium sulfide scales are more resistant than those forming iron sulfide scales. Thus, increasing chromium generally improves the alloy sulfidation resistance. Some of the sulfidation data generated from these environments are available elsewhere (Ref 43–45).

In oxidizing/sulfidizing environments (also referred to as reducing, mixed-gas environments), which typically consist of H_2, CO, CO_2, H_2O, H_2S, and so on, there are generally sufficient oxygen and sulfur potentials (P_{O_2}, P_{S_2}) to form oxides and sulfides for most high-temperature alloys, particularly chromia formers. The corrosion reaction in these environments generally involves oxidation and sulfidation. In most cases, the alloy remains protected during the initial oxidation period until breakaway corrosion is initiated, which is then followed by rapid sulfidation attack.

Higher-nickel-content alloys are most susceptible to sulfidation attack. Decreasing nickel content in nickel-base alloys generally reduces

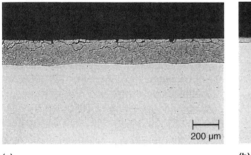
(a)

(b)

Fig. 13 Typical morphology of the nitride layer formed on the metal surface at 650 °C (1200 °F) in ammonia for 168 h for (a) type 310 stainless steel and (b) alloy X. Source: Ref 1

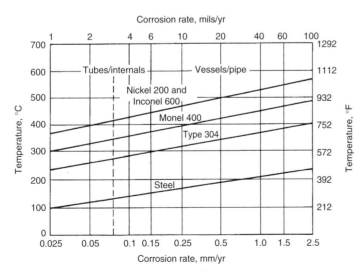

Fig. 14 General guidelines for several alloys for use in dry chlorine environment. Source: Ref 37

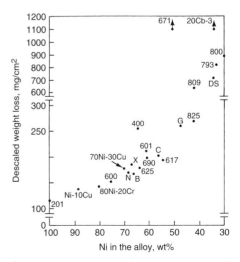

Fig. 15 Effect of nickel on the corrosion resistance of alloys in Ar-30%Cl_2 at 705 °C (1300 °F) for 24 h. Source: Ref 38

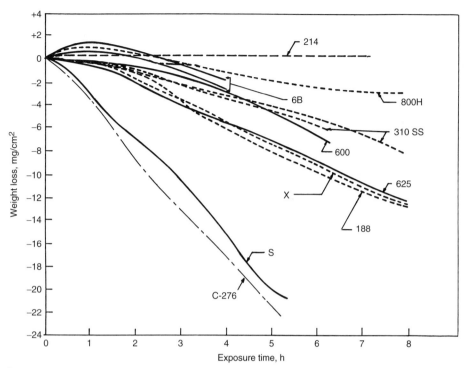

Fig. 16 Corrosion resistance of various alloys in Ar-20%O$_2$-2%Cl$_2$ at 900 °C (1650 °F). Source: Ref 39

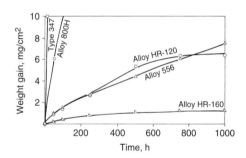

Fig. 17 Sulfidation behavior of type 347 stainless steel, alloy 800H, alloy HR-120, alloy 556, and alloy HR-160 in H$_2$-7%CO-1.5%H$_2$O-0.6%H$_2$S (oxygen potential, P$_{O_2}$ = 10^{-23} atm; sulfur potential, P$_{S_2}$ = 10^{-9} atm; carbon activity, a_c = 0.3–0.4). Source: Ref 45

susceptibility to sulfidation attack. Chromium oxide scale is generally protective against sulfidation. Increasing chromium content in iron-, nickel-, and cobalt-base alloys improves sulfidation resistance. For severe sulfidizing environments (i.e., high temperatures, high P$_{S_2}$, low P$_{O_2}$), chromium content should preferably be higher than 25%. Other beneficial alloying elements include cobalt, silicon, aluminum, and titanium. Formation of a duplex Cr$_2$O$_3$/SiO$_2$ oxide scale offers a very effective barrier to sulfidation attack in nickel-base alloy HR-160, as illustrated in Fig. 17 (Ref 45). Another commercial nickel-base alloy containing high chromium and silicon is 45TM, which is also reported to provide excellent sulfidation resistance (Ref 46).

Environments containing SO$_2$ are generally highly oxidizing. However, sulfides tend to form internally underneath the chromium oxide scales. A test was performed (Ref 47) to examine the corrosion behavior of cobalt-bearing alloy 556 and type 304 stainless steel in Ar-5%O$_2$-5%CO$_2$ and Ar-5%O$_2$-5%CO$_2$-10%SO$_2$ environments at 980 °C (1800 °F) for 550 h. Test results suggested that an alloy that formed a protective oxide scale in a purely oxidizing environment would form a similarly protective oxide scale in SO$_2$-O$_2$ environments. Chromium, again, is a very effective alloying element in improving the alloy sulfidation resistance.

REFERENCES

1. G.Y. Lai, *High Temperature Corrosion of Engineering Alloys*, ASM International, 1990
2. A.W. Zeuthen, *Heat./Piping/Air Cond.*, Vol 42 (No. 1), 1970, p 152
3. R.C. John, Paper 73, Corrosion 99, NACE International, 1999
4. *Failure Analysis and Prevention*, Vol 10, Metals Handbook, 8th ed., American Society for Metals, 1975, p 533
5. H.E. Eiselstein and E.N. Skinner, in *STP 165*, ASTM, 1954, p 162
6. J.C. Kelly, Today's versus Yesterday's Stainless Steels, *Mater. Perform.*, March 1992
7. ThyssenKrupp VDM, unpublished data
8. M.A. Harper and G.Y. Lai, in *Environmental Effects on Engineered Materials*, R.H. Jones, Ed., Marcel Dekker, Inc., 2001, p 75
9. "RA 353MA Alloy Data Sheet," Rolled Alloys, Temperance, MI
10. "Haynes 214 Alloy," H-3008B, Haynes International, Kokomo, IN
11. "Haynes 230 Alloy," H-3000, Haynes International, Kokomo, IN
12. "Haynes HR-120 Alloy," H-3125A, Haynes International, Kokomo, IN
13. "Haynes HR-160 Alloy," H-3129A, Haynes International, Kokomo, IN
14. "INCONEL Alloy 601," Special Metals Corporation, Huntington, WV
15. "INCONEL Alloy 617," Special Metals Corporation, Huntington, WV
16. "Nicrofer 6025HT—Alloy 602CA," Material Data Sheet 4037, ThyssenKrupp VDM, Germany
17. "Nicrofer 45TM—Alloy 45TM," Material Data Sheet 4039, ThyssenKrupp VDM, Germany
18. G.Y. Lai, *High Temperature Corrosion of Engineering Alloys*, ASM International, 1990, p 47
19. H.J. Grabke, "Carburization—A High Temperature Corrosion Phenomenon," MTI Publications 52, Materials Technology Institute of the Chemical Process Industries, Inc., St. Louis, MO, 1998
20. O. Demel, E. Keil, and P. Kostecki, SGAW Report 2538, Osterreichische, Association for Studies of Atomic Energy, Vienna, Seibersdorf Research Center, Institute for Metallurgy
21. H.J. Grabke, U. Gravenhorst, and W. Steinkusch, *Werkst. Korros.*, Vol 27, 1976, p 291
22. I. Wolf and H.J. Grabke, *Solid State Commun.*, Vol 54, 1985, p 5
23. L.H. Wolfe, *Mater. Perform.*, April 1978, p 38
24. G.Y. Lai et al., Paper 3473, *Corrosion 2003*, NACE International, 2003
25. H.J. Grabke et al., Metal Dusting of Nickel-Based Alloys, *Mater. Perform.*, July 1998, p 58
26. J.C. Nava Paz and H.J. Grabke, Metal Dusting, *Oxid. Met.*, Vol 39 (No. 5/6), 1993, p 437
27. H.J. Grabke, R. Krajak, and E.M. Muller-Lorenz, Metal Dusting of High Temperature Alloys, *Werkst. Korros.*, Vol 44, 1993, p 89
28. H.J. Grabke, C.B. Bracho-Troconis, and E.M. Muller-Lorenz, Metal Dusting of Low Alloy Steels, *Werkst. Korros.*, Vol 45, 1994, p 215
29. H.J. Grabke, Metal Dusting of Low- and High-Alloy Steels, *Corrosion*, Vol 51 (No. 9), 1995, p 711
30. H.J. Grabke, Nickel-Based Alloys in Carbonaceous Gases, *Corrosion*, Vol 56 (No. 8), 2000, p 801
31. H.J. Grabke, Metal Dusting, *Mater. Corros.*, Vol 54 (No. 10), 2003, p 736
32. J. Klower et al., "Metal Dusting and Carburization Resistance of Nickel-Base

Alloys," Paper 139, Corrosion 97, NACE International, 1997
33. B.A. Baker and G.D. Smith, "Alloy Selection for Environments Which Promote Metal Dusting," Paper 257, Corrosion 2000, NACE International, 2000
34. B.A. Baker and G.D. Smith, "Alloy Solutions to Metal Dusting Problems in the Petrochemical Industry," presented at International Workshop on Metal Dusting, Sept 26–28, 2001 (Argonne, IL), Department of Energy, 2001
35. B.A. Baker et al., "Nickel-Base Material Solutions to Metal Dusting Problems," Paper 2394, Corrosion 2002, NACE International, 2002
36. D.L. Klarstrom and H.J. Grabke, "The Metal Dusting Behavior of Several High Temperature Alloys," Paper 1379, Corrosion 2001, NACE International, 2001
37. "Materials Selector for Hazardous Chemicals, Hydrochloric Acid, Hydrogen Chloride and Chlorine," Vol 3, MTI Publications MS-3, Materials Technology Institute of the Chemical Process Industries, Inc., 1990
38. R.H. Kane, in *Process Industries Corrosion,* B.J. Moritz and W.I. Pollock, Ed., National Association of Corrosion Engineers, 1986, p 45
39. J.M. Oh, M.J. McNallan, G.Y. Lai, and M.F. Rothman, *Metall. Trans. A,* Vol 17, June 1986, p 1087
40. R.L. Jarry, J. Fischer, and W.H. Gunther, *J. Electrochem. Soc.,* Vol 110 (No. 4), 1963, p 346
41. J.J. Barnes, "Materials Behavior in High Temperature HF-Containing Environments," Paper 518, Corrosion 2000, NACE International, 2000
42. G. Marsh and P. Elliott, in *High Temperature Corrosion in Energy Systems,* M.F. Rothman, Ed., The Metallurgical Society of AIME, 1985, p 467
43. G. Sorell, "Compilation and Correlation of High Temperature Catalytic Reformer Corrosion Data," NACE Technical Committee Report, Publication 58-2, National Association of Corrosion Engineers, 1957
44. E.B. Backensto and J.W. Sjoberg, "Iso-Corrosion Rate Curves for High Temperature Hydrogen-Hydrogen Sulfide," NACE Technical Committee Report, Publication 59-10, National Association of Corrosion Engineers, 1958
45. J.F. Norton, Joint Research Centre, Petten Establishment, Petten, The Netherlands, unpublished data, 1989
46. D.C. Agarwal et al., Paper 471, Corrosion95, NACE International, 1995
47. J.J. Barnes and G.Y. Lai, Paper 90276, Corrosion/90, National Association of Corrosion Engineers, 1990

Corrosion in the Pulp and Paper Industry

Harry Dykstra, Acuren*

THE $165 BILLION Pulp, paper and allied products industry supplies the United States with approximately 300 kg (660 lb) of paper per person every year. More than 300 pulp mills and 550 paper mills support its production. The total direct annual cost of corrosion is estimated at U.S. $6 billion and these costs will likely increase. The infrastructure of North American producers has not been replaced because most new production has been located to South America and South East Asia. The North American mills face tough competition from lower cost areas of the world with the result that extending the life of existing mill equipment has offset funding for capital projects. Increased efficiencies are often realized through process changes that require modification of existing equipment to function under environments that it was not specifically designed for. An understanding of the corrosion mechanisms active in the pulp and paper industry is essential for successful maintenance and operation of these plants as the operating conditions evolve and the infrastructure continues to age.

The pulp and paper processes require a myriad of materials and fabrication methods to provide service in a mixture of organic and inorganic environments that range from atmospheric to boilers producing molten smelt under controlled reducing conditions. The pH ranges from near 1 in acidic, oxidizing bleach environments to 14 in hot digester liquors. Often, mills using the same process can have completely different corrosion issues due to site-specific conditions that develop based on the type of wood being processed, on the chemicals used in different parts of the process, corrosion control measures that have been introduced and the age of the equipment. The operating conditions at one location can be altered due to process modifications that are made at other locations in the mill. As a result, unlike many other chemical industries, historical data of process equipment in the pulp and paper industry is no predictor of future materials performance.

Areas of Major Corrosion Impact

Corrosive environments are found in a number of process stages in the pulp and paper industry including:

- Pulp production
- Pulp processing and chemical recovery
- Pulp bleaching
- Paper manufacturing

Each manufacturing step has its own corrosion problems related to the size and quality of the wood fibers, the amount of and temperature of the process water, the concentration of the treatment chemicals, and the materials used for machinery construction.

Pulp Production

There are several different methods of pulp production to make different strengths and grades of paper. The most common classifications are chemical, mechanical, and semichemical pulping techniques.

Chemical pulping uses various chemicals to produce long, strong, and stable fibers and to remove the lignin that bonds the fibers together. The chemicals used will vary depending on the type of chemical pulping used. In North America, there are two main types of chemical pulping performed: kraft pulping, which is also referred to as sulfate pulping, and sulfite pulping.

The kraft process starts with feeding wood chips into large pressure vessels called digesters (see the section "Corrosion of Digesters" for details). Chips are cooked under pressure in a steam heated aqueous solution of sodium hydroxide (NaOH) and sodium sulfide (NaS_2), also known as white liquor. The spent pulping liquor, called black liquor, is concentrated by evaporation and burned to generate high-pressure steam for the mill processes. The inorganic portion of the black liquor is then used to regenerate the NaOH and NaS_2 needed for pulping.

The kraft process is the predominant pulping process used in North America to extract fibers from wood for use in the manufacture of paper, tissue, and paperboard (approximately 80% of pulp is produced by the kraft process). The term "kraft" is derived from the German word for strong, which reflects the high strength of paper products derived from kraft pulp. The high strength, together with effective methods of recovering pulping chemicals, explains the popularity of kraft pulping. Emphasis in this article has been placed on the corrosion problems associated with the kraft process.

The sulfite pulping process uses cooking liquors of sodium bisulfite or magnesium bisulfite in a pulp digester with a pH of 3. It is generally considered more corrosive than the kraft process environment. See the section "Corrosion in the Sulfite Process" for details.

Mechanical Pulping. In mechanical pulping, wood fibers are separated from each other by rolling, rubbing, or teasing, either against themselves, or against a harder, specially designed surface. Chemical additions or the application of heat and pressure can be used to increase the efficiency of the process, but the predominant pulping forces are mechanical. See the section "Corrosion Control in High-Yield Mechanical Pulping" for details.

Semichemical pulping techniques use weak chemical solutions of sodium sulfite (Na_2SO_3) and sodium carbonate (Na_2CO_3) to help digest the lignin in the pulp. See the section "Corrosion Control in Neutral Sulfite Semichemical Pulping" for details.

Pulp Processing and Chemical Recovery

To recover chemicals from the black liquor, the slurry goes through a chemical recovery process. The liquor passes through evaporators, recovery boilers, and causticizers to eventually produce white liquor that is recycled back to the digester. See the sections "Corrosion Control in Chemical Recovery" and "Corrosion in Recovery Boilers" for details. A by-product of the kraft black liquor process is tall oil, which is refined and used in paint varnish, linoleum, drying oils, emulsions, lubricants, and soaps.

---

*Harry Dykstra, Chair, NACE International Task Group. Contributors: Angela Wensley, Angela Wensley Engineering; Chris Thompson, Paprican; Max D. Moskal, Mechanical and Materials Engineering; Donald E. Bardsley and William Miller, Sulzer Process Pumps Inc.; David Bennett, Corrosion Probe Inc.; Craig Reid, Acuren; Arthur H. Tuthill, Tuthill Associates; Douglas Singbeil, Paprican; Randy Nixon, Corrosion Probe, Inc.

See the section "Corrosion Control in Tall Oil Plants" for details.

Pulp Bleaching

Pulp bleaching is performed on the pulp in order to increase its brightness. Bleaching is an extremely corrosive process that is carried out under acidic conditions with strong oxidants such as elemental chlorine (Cl_2), chlorine dioxide (ClO_2), ozone (O_3), sodium hypochlorite (NaClO), oxygen (O_2), and hydrogen peroxide (H_2O_2). In recent years, chlorine-free bleaching operations have been developed. This has resulted in major changes in materials selection for bleach plants. See the section "Corrosion Control in Bleach Plants" for details.

Paper Manufacturing

Using a paper production machine, the processed pulp is converted into a paper product. At the beginning of this stage, the water content of the paper is greater than 99%. In the wet-end operation, the slurry of pulp is deposited onto a continuously moving belt that suctions the water from the slurry using gravity, vacuum chambers, and vacuum rolls. The continuous sheet then moves through additional rollers that compress the fibers and remove the residual water. See the sections "Paper Machine Corrosion" and "Suction Roll Corrosion" for details of the corrosion problems encountered in the wet end of a paper machine.

Following the pressing of the wet-end operations, the continuous sheet is compressed by steam-heated rollers to allow the fibers to begin bonding together. Coatings are then applied to enhance the surface appearance before the sheet is spooled for storage. Because corrosion is not a serious concern with these dry-end operations, this subject is not addressed further in this article.

Environmental Issues

Environmental issues involving the pulp and paper industry require that water usage by pulp and paper mills be reduced because they are among the largest industrial process water users in North America. Today, the pulp and paper industry uses a lower volume of process water, recycles and reuses more water, and cleans water before releasing it, all in an effort to reduce costs as well as responding to increasingly strict environmental regulations. Such efforts, unfortunately, can sometimes create additional corrosion-related problem. See, for example, the section "Wastewater Treatment Corrosion in Pulp and Paper Mills."

Similarly, air quality emission standards have been made more strict and pulp mills are now required to reduce the amount of sulfur compounds released to the atmosphere. This has also led to corrosion problems as discussed in the section "Corrosion Control in Air Quality Control."

Corrosion of Digesters

Angela Wensley,
Angela Wensley Engineering

In the kraft process, hot alkaline (sodium sulfide and sodium hydroxide) liquid is used to dissolve the lignin from wood chips and to separate individual wood fibers for use in papermaking. Wood chips are exposed to cooking liquors for several hours at elevated temperature and pressure in a process called digestion. Digestion may occur by repetitive batch processes in batch digesters, or the process may occur continuously in continuous digesters. This section describes both types of digesters, their materials of construction, corrosion problems encountered, and methods to protect digesters from corrosion. Ancillary equipment to digesters is also discussed.

Batch Digesters

A typical batch digester consists of a vertical cylindrical vessel with a hemispherical or ellipsoidal top head and a conical bottom, as shown in cross-sectional view in Fig. 1. Batch digesters are typically 2.4 to 4.0 m (8 to 13 ft) in diameter and up to 18.3 m (60 ft) high. Soft or hard wood chips are fed into the top of the vessel, along with hot cooking liquor, which helps pack the chips in the vessel. The liquor consists of a mixture of white and black liquors in various volume ratios depending on the pulp product being manufactured.

After filling with wood chips and liquor, the vessel is closed and cooking begins, with heat supplied by direct injection of steam (Fig. 1a) or by indirect steam heating (Fig. 1b) in an external heat exchanger. A typical batch cook lasts about 2 h. The cooking temperature of approximately 170 °C (338 °F) is reached after about 1 h. At this time, direct steaming is usually stopped. Some facilities remove the liquors and pulp by displacement instead of blowing, but this is not a common practice.

At the end of the cook, the pulp is blown from the bottom of the vessel into a blow tank. From there the pulp goes to brown stock washers where the spent cooking liquor is separated from the pulp. Steam from the blow tank is removed for heat recovery and condensed in brown stock wash water.

Over the years there has been a trend to increase production by decreasing batch cook times. This requires the use of higher ratios of white-to-black liquor and higher temperatures. Both these practices cause increased corrosion rates in both carbon steel and stainless steel digesters.

Materials of Construction

Although there is a trend to construct batch digesters from solid duplex stainless steels, most batch digesters have been constructed from carbon steel with generous corrosion allowances ($\geq$19 mm, or 0.75 in.), such that they can remain in service for perhaps 10 years before some means of protection must be used. In the 1950s and 1960s, digesters in North America were constructed using a modified low-silicon (0.02% Si max) grade of ASTM A 285 carbon steel, with low-silicon welds on the process side. Today (2006) most new carbon steel batch digesters are made from ASTM A 516 grade 70, a higher-strength, pressure vessel steel in which the silicon content is controlled in the range 0.15 to 0.30% Si, and without low-silicon weld caps. Higher-silicon steels corrode more rapidly in alkaline pulping liquors.

Numerous batch digesters have been constructed from clad plate (either roll- or explosion-bonded) with stainless steel on the inside and carbon steel on the outside. Types 304L (S30403) and 316L (S31603) stainless steels have been most commonly selected for clad plate (although these experience corrosion). Some batch digesters have been constructed with a stainless steel weld overlay lining, although this practice is not common. Stainless steel weld overlays are discussed in some depth in this section under "Protection of Batch Digesters." Some have been constructed of cold stretched type 304 (S30400) stainless steel in accordance with the Swedish cold stretching code.

Duplex stainless steels in either solid or clad form have been used for several years for construction of new batch digesters worldwide.

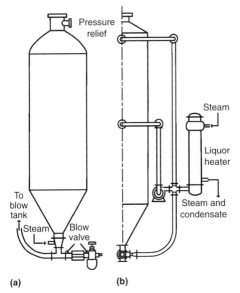

Fig. 1 Two types of batch digesters. (a) Direct-heating digester. (b) Indirect-heating digester

North America has been slow to adopt these materials, but the number of new duplex stainless steel batch digesters is expected to increase. The most common duplex alloy used for duplex digester construction is S32205 (formerly known as S31803 and commonly known as "alloy 2205"). Due to their higher strength, duplex stainless steel digesters may be significantly thinner than carbon steel digesters designed to hold the same pressure.

Corrosion

Corrosion of carbon steel kraft batch digesters has been a known problem for more than 50 years. Pioneering work (Ref 1–6) revealed that the silicon content of the steel controlled the corrosion rate of digester steels, with higher contents being increasingly susceptible. Carbon steel batch digesters do not corrode uniformly and often experience the most corrosion in an inverted horseshoe-shaped pattern where the liquor contacts the wall during filling (Ref 7). The zone of most severe corrosion varies from mill to mill, and perhaps from digester to digester. In most digesters the corrosion is most pronounced in the cylindrical section. In other digesters it is worse in the bottom cone; in yet others, in the top dome.

Corrosion of stainless steels (both in wrought form and weld overlays) is primarily a function of the chromium content of the overlay. Austenitic stainless steel grades such as type 316L (16 to 18% Cr) and type 304L (18 to 20% Cr) can experience rapid corrosion, up to 1 mm/yr (40 mils/yr). With its higher chromium content, type 304L performs better in digesters than type 316L. Work by Audouard to investigate corrosion during "hot plate boiling" has revealed that duplex stainless steel with even higher chromium content (22 to 27% Cr) resists corrosion better than conventional austenitic grades (Ref 8–12). Corrosion testing has also shown that molybdenum is not a beneficial alloying addition for corrosion resistance of stainless steels in digester liquors.

Stainless steel weld overlays with low chromium content can also experience rapid corrosion. Conventional type 309L (S30980) stainless steel weld overlays may have an as-deposited chromium content of 20%, which is higher than for type 304L but is still insufficient for corrosion resistance in aggressive batch digester environments. Corrosion testing in several batch digester liquors has revealed that at least 25% Cr is required for stainless steel weld overlays to have best corrosion resistance in aggressive digester environments (Ref 13). Duplex stainless steel weld overlays such as type 312 (S31200) can give 22% to 28% Cr when applied over a carbon steel substrate, depending on the welding process and mode employed.

Early work (Ref 14) suggested that lower-than-intended chromium content resulted in poor weld microstructures, which in turn corroded (Fig. 2). More recent work has focused on the role of chromium content on overlay corrosion. Technical Association of the Pulp and Paper Industry (TAPPI) guidelines give minimum as-deposited chemistry requirements for austenitic stainless steel weld overlays, along with criteria for soundness and structural uniformity. However, even weld overlays that meet the minimum of 18% Cr in the TAPPI guideline can experience rapid corrosion in many batch digesters (Ref 15). Another problem with stainless steel weld overlays is that once the overlay is penetrated (e.g., at a pinhole), the underlying carbon steel can corrode at a great rate, producing a large cavity that can grow completely through the digester wall.

Although duplex stainless steels resist corrosion in digesters better than conventional austenitic stainless steels (Ref 8–13), duplex stainless steels can experience selective corrosion of the austenite phase in the microstructure because this phase is lower in chromium content than the ferrite phase (Ref 16). Olsson reports very low corrosion rate—approximately 0.1 mm (5 mils) after 3 years in service—where selective phase corrosion has occurred (Ref 17). For particularly aggressive batch digester environments, superduplex stainless steels (such as S32750) have superior corrosion resistance because the ferrite phase has a higher chromium content than that in conventional grades (such as S32205). The modern trend is to construct new batch digesters from duplex stainless steels (Ref 18, 19). The higher strength and low corrosion allowance for duplex materials allow thinner walls to offset much of the higher cost compared to carbon steel.

Protection of Batch Digesters

Thinned carbon steel digesters are most commonly protected by application of a layer of stainless steel weld overlay (Ref 14, 20–24). Weld overlay has also been applied to extend the service life of digesters with corroded stainless steel cladding or with corroded overlay. Other protective measures include application of thermal spray coatings (Ref 16) and anodic protection (Ref 25–27). Buildup with carbon steel weld metal is not considered to be a protective option, as such buildup (e.g., E7018) characteristically has high silicon content and typically corrodes much faster than the original digester wall. However, buildup of very thin sections with carbon steel before applying stainless steel overlay or thermal spray coating is good practice and normally done.

Stainless steel weld overlays are best applied before the corrosion allowance has been completely consumed. TAPPI TIP 0402-03 "Guidelines for Corrosion Resistant Weld Overlays in Sulphate and Soda Digester Vessels" provides much useful information (Ref 28). These overlays are applied automatically using either the submerged arc welding (SAW) process or the gas metal arc welding (GMAW) process. Other welding processes may be used for pickup repairs or for smaller areas of overlay, such as around projecting nozzles where automatic equipment does not work.

A SAW overlay is typically applied horizontally, with twin electrodes traveling around the circumference of the digester; the second electrode follows behind the first, completely remelting the deposit. A GMAW overlay can be applied either horizontally ("conventional" overlay with a single electrode) or vertically over lengths up to 4 m (13 ft), with single or dual torches.

Horizontal weld overlay has been the "conventional" overlay mode for more than 40 years and typically gives an overlay thickness of 6 mm (0.25 in.), for either the SAW or GMAW processes, with a minimum of 4.8 mm (0.188 in.) overlay thickness. The vertical down mode typically gives an overlay with a nominal thickness of 4.8 mm (0.188 in.) and a minimum thickness of 2.5 mm (0.100 in.). Vertical overlay may be suitable in certain batch digesters if it can be established (either by corrosion testing or by service experience) that the overlay alloy does not corrode rapidly in the particular liquor environment.

The as-deposited composition of a weld overlay is a result of the dilution of the filler metal with the substrate. For overlay on carbon steel, dilution with the carbon steel results in a lower alloy content than that of the wire or electrode. For overlay on stainless steel (or for two-layer overlay), the as-deposited composition may be close to that of the wire or electrode.

In the past, most stainless steel weld overlays in digesters were type 309 with 20 to 23% Cr and 10 to 12% Ni in the weld deposit. For SAW overlay, the wire and flux are specially manufactured to provide the desired as-deposited chemistry. For GMAW overlay, type ER309LSi (S30988) wire has been widely used.

In recent years there has been an increasing interest in application of type 312 duplex stainless steel weld overlays. The SAW wire and flux chemistries that give as-deposited compositions resembling type 312 stainless steel are available.

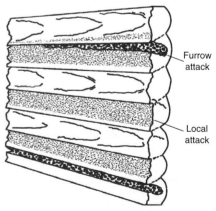

Fig. 2 Typical problems with digester weld overlays. Accelerated corrosive attack on a stainless steel weld overlay lining in a kraft pulp digester vessel after 18 months of service. Localized furrow attack occurs between weld beads displaying rapid-etching (lean alloy) structure. Deposited by submerged arc process (E-310 electrode).

A GMAW overlay can be done with ER312 wire. Internal microcracking has occurred in some type 312 overlays. Provided the internal microcracks are not exposed to the surface, there is little driving force for their growth and no way for them to act as paths for liquor to gain access to the carbon steel substrate.

Overlays made with either type 309 or 312 stainless steels have microstructures consisting of a mixture of austenite and ferrite. Conventionally, however, the type 309 overlays have been called "austenitic" because they have lower ferrite than their type 312 counterparts, which are termed "duplex." The SAW overlays tend to have higher ferrite contents than their GMAW counterparts. The presence of ferrite in the microstructure of stainless steel weld overlays is highly desirable. Stainless steel overlays with less than 3% ferrite are prone to hot cracking, particularly if sulfur is present as a contaminant.

Thermal Spray Coatings. Many batch digesters have been protected from corrosion by the application of thermal spray coatings. Most coating alloys are either alloy 625 (N06625) or similar alloy, but with somewhat lower molybdenum content. The predominant process for applying these coatings has been twin-wire arc spray (TWAS). In recent years, coatings have also been applied by the high-velocity oxygen fuel (HVOF) process.

Twin-wire arc spray coatings are typically applied 2 mm (80 mils) thick to overcome the porosity inherent in that process and to prevent liquor access to the carbon steel substrate. High-velocity oxyfuel coatings are characteristically much less porous and are applied thinner (e.g., 0.6 mm, or 25 mils).

Thermal spray coatings are maintenance coatings inasmuch as they typically have a service life of up to 8 years before recoating is required. The main advantage of thermal spray coatings is that they introduce no significant heating, and thus there are no heat-affected zones (HAZs) as with welds and no distortion or delamination of the vessel (as sometimes happens with weld overlays).

In-service problems with thermal spray coatings include disbonding and blistering. The bond to the carbon steel substrate is mechanical, and poor surface preparation can lead to poor bonding. The blistering is believed to be caused by osmotic pressure buildup when liquor permeates the coating.

Anodic Protection. Monitoring corrosion potential in carbon steel batch digesters has revealed that the potential cycles change from active to passive over the course of a cook (Ref 29). With anodic protection, an external rectifier and internal cathodes are used to supply current to the digester wall, thus raising the corrosion potential more rapidly to the passive zone. Although the concept of anodic protection of batch digesters has been understood for many years, there are as yet no anodically protected batch digesters in North America. However, there are numerous anodically protected continuous digesters in Finland.

Anodic protection is only able to protect the digester wall below the liquor level. For those digesters where corrosion of the top dome is the problem, anodic protection would be of no benefit.

Continuous Digesters

Continuous digesters first appeared commercially in the late 1950s. Figure 3 shows a schematic flow diagram for a typical continuous digester, the majority of which are of a "Kamyr" design. A prominent feature of most Kamyr systems is a cylindrical digester shell, having a vertical axis and a length-to-diameter ratio ranging from about 5-to-1 to 15-to-1. The shell diameter typically decreases from bottom to top with a series of conical transitions. The top and bottom heads are ellipsoidal.

As the chips descend through the digester, they are impregnated with liquor, cooked, washed, and discharged into a blow tank. The spent liquor is extracted through extraction screens. Heat is supplied by indirect heating of the cooking liquor in external heat exchangers. Continuous digesters can be either of "hydraulic" (filled to the top with cooking liquor) or "vapor phase" (with steam injection in the top) design. The hydraulic digester is by far the predominant type.

Digesters can also be single-vessel or two-vessel systems. In two-vessel systems, there is a separate impregnation vessel, with cooking, extraction, and washing done in the digester. For this section, the term "digester vessel" refers to either the single-vessel or two-vessel systems, including the impregnation vessel.

In conventional operation the cooking liquor (a mixture of white and black liquors) is added in the top of either the digester (for single-vessel systems) or impregnation vessel (for two-vessel systems). In extended delignification processes such as modified continuous cooking (MCC), extended modified continuous cooking (EMCC), isothermal cooking (ITC), or low-solids cooking, white liquor is added lower in the digester. The extended delignification processes are also characterized by higher temperatures in the bottom of the digester (e.g., below the extraction screens).

Materials of Construction

The pressure shell of the earliest continuous digesters were built of low-silicon ASTM A 285 grade C carbon steel "modified for digester service," together with low-silicon caps for the process-side welds. In the late 1960s, medium-silicon steels such as A 516 grade 70 became the predominant material of construction for continuous digesters, and the use of low-silicon weld metal was discontinued.

Some of the nonpressurized internal equipment in continuous digesters have traditionally been constructed from type 304L stainless steel. This includes the central pipes, screens, and internal cone. Many of the nozzles in carbon steel digesters are also type 304L stainless steel. There is a recent trend to replace corroded carbon steel blank plates with type 304L stainless steel blank plates and type 304L standoff rods on the back.

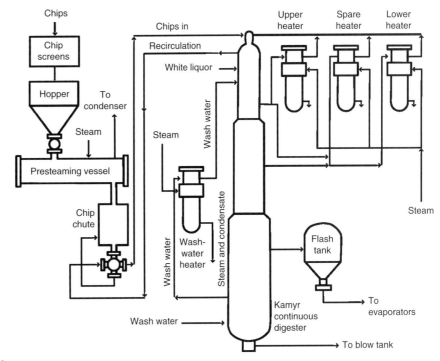

Fig. 3 Kamyr continuous digester

Several continuous digesters have been constructed from roll-clad plate with type 304L stainless steel on the process side and A 516 grade 70 carbon steel on the outside. Indeed, most digesters have also been constructed with type 304L stainless-clad top and bottom heads, either as a loose liner or with roll-clad plate.

Some new continuous digesters have been built from duplex stainless steel (S31803). This alloy is quite resistant to corrosion in continuous digester environments (Ref 13, 16). Duplex stainless steel has also been used for the replacement of one digester top.

Corrosion

The most serious corrosion problem with carbon steel continuous digesters has been caustic stress-corrosion cracking (SCC) of un-stress-relieved seam welds in the impregnation zone or in the impregnation vessel for two-vessel systems (Ref 30–40). The ASME Boiler and Pressure Vessel Code (Ref 41) does not require postweld stress relief treatment for wall thicknesses less than 32 mm (1.25 in.), which often is the case at the top of continuous digesters. In 1980 there was catastrophic caustic SCC failure of an un-stress-relieved top section of a continuous digester. The combination of high tensile stress (from residual welding stresses) and corrosion potential in a critical range is a prerequisite for caustic SCC. Since the early 1980s most if not all continuous digesters have been fully postweld heat treated, even though stress relief was not mandated by the ASME Code for wall thicknesses less than 32 mm (1.25 in.). Carbon steel welds and weld buildup made in the impregnation zone of digester vessels and not subsequently stress relieved are susceptible to caustic SCC. Below the cooking screens, caustic SCC has not been reported.

Until the late 1980s corrosion thinning was not a serious problem with continuous digesters, except in unusual cases where corrosive wood species such as western red cedar were being pulped. Earlier research had identified wood extractives such as catechols as very corrosive to steel under conditions of alkaline pulping (Ref 42). In more recent years there have been many cases of rapid thinning at rates approaching 6 mm/yr (250 mils/yr) of carbon steel continuous digesters, around and below the extraction screens and in the wash zone (Ref 43). High corrosion rates observed in conjunction with new pulping technologies such as EMCC and ITC are believed to be caused by a combination of higher temperatures and low hydroxide concentration that contributes to a loss of passivation of carbon steel (Ref 13, 44).

Many digesters have experienced extensive metal loss from cleaning with hydrochloric acid (HCl). Even when properly inhibited, corrosion damage (usually in the form of pitting) can occur if the temperature is above 70 °C (160 °F), which is often the case. Acid cleaning done at ≤50 °C (≤120 °F) is not considered to be corrosive. Pitting of the carbon steel wall in the impregnation zone is usually a sign of acid cleaning damage, as continuous digesters do not usually experience corrosion thinning or pitting in this area. Alternative acids for cleaning, such as sulfamic acid and formic acid, are less corrosive to digesters (Ref 45).

Preferential weld corrosion is often observed in continuous digesters and is the result of the poorer corrosion resistance of weld metal (which has a coarse-grain structure similar to a casting), compared with the parent metal plate (which is typically lower in silicon content). The common practice of restoring corroded weld seams without subsequent stress relief results in welds with high residual stresses that may make the digester more susceptible to caustic SCC.

Stainless steels in digester vessels can experience corrosion as a result of HCl cleaning, which preferentially attacks the ferrite phase in welds, but can also cause widespread pitting if the temperature is high enough. Attack of circumferential welds in type 304L central pipes has resulted in central pipe failures. The welds often have incomplete penetration, which contributes to failure. Stainless steel top and bottom dome liners may experience SCC or intergranular attack (IGA) if they are heat treated with the digester. This practice can result in sensitization through chromium carbide precipitation at the grain boundaries. Continuous digester vessels constructed from roll-clad austenitic stainless steel can also undergo IGA if they become sensitized during postweld heat treatment, which is mandatory for the carbon steel digester shell (Ref 46).

Protection of Continuous Digesters

As corrosion rates have been observed to increase, protective measures such as corrosion-resistant weld overlay, thermal spray coating, and anodic protection have been increasingly employed. Measures such as anodic protection, taken to prevent corrosion by the pulping liquor, will not protect the digester from acid damage because the anodic protection system must be turned off during acid cleaning.

As with batch digesters, carbon steel weld buildup is not considered a permanent protective measure because it is susceptible to high corrosion rates, which can expose welding defects such as porosity. It is also susceptible to caustic SCC if in the impregnation zone.

Stainless steel weld overlays are being increasingly used in continuous digesters for protection of the carbon steel shell from corrosion thinning. The overlay properties in continuous digesters are essentially the same as those discussed in the section on stainless steel weld overlay of batch digesters. Service experience and corrosion testing (Ref 13, 16, 47) have shown that type 309 stainless steel weld overlays (applied by either the SAW or GMAW processes) have good corrosion resistance, even under nonconventional cooking operation. Indeed, the 18% minimum Cr level recommended in the TAPPI guidelines (Ref 28) is likely adequate for good corrosion resistance. A minimum of 20% Cr is recommended for aggressive continuous digester environments (Ref 13).

Since continuous digesters are so large, it is often not practical to overlay large areas in one shutdown, so overlay is typically done over a period of years. For digesters thinned to near minimum, stainless steel weld overlay of digester walls is sometimes done in combination with anodic protection.

Preferential corrosion or "fingernailing" is often seen in the carbon steel adjacent to stainless steel weld overlay, most often in the impregnation and cooking zones. While fingernailing resembles galvanic corrosion, it is simply the preferential corrosion of the HAZ in the carbon steel, which has poorer corrosion resistance than the parent metal. Caustic SCC often begins at the bottom of a fingernailing crevice.

Protection of carbon steel weld seams susceptible to caustic SCC (in the impregnation zone) with stainless weld overlay bands has not been successful. The residual tensile stresses in the carbon steel at the termination of the overlay are high enough to promote caustic SCC (Ref 48, 49). However, anodic protection can prevent caustic SCC in these HAZs adjacent to the overlay.

Alloy 625 (ERNiCrMo3) and alloy 82 (ERNiCr3) nickel-base weld overlays have good corrosion resistance in continuous digester liquors and were widely used in the 1980s. However, most overlay being applied in continuous digesters today (2006) is type 309 stainless steel.

Thermal spray coatings, both TWAS and HVOF, have been applied in continuous digesters to protect large areas just above or below the extraction screens from corrosion thinning. Because thermal sprays do not produce a HAZ, they can also protect weld seams in the impregnation zone from caustic SCC. The thermal spray coatings applied in continuous digesters are the same as those applied in batch digesters, that is, predominantly alloy 625 and similar nickel-base alloys.

Although laboratory testing has indicated thermal spray coatings can protect against both caustic SCC (Ref 32) and thinning (Ref 50), service experience has been mixed. With good surface preparation, coatings adhere well and provide several years of corrosion protection. However, there have been problems with blistering and disbonding.

Anodic Protection. Since the early 1980s anodic protection has been successfully used to protect partially stress-relieved continuous digesters from both caustic SCC and corrosion thinning (Ref 51–57). Anodic protection of continuous digesters requires multiple external rectifiers and internal cathodes. There are two main cathode designs: centrally mounted (on standoffs from the central pipe) and wall-mounted. Anodic protection for thinning can reduce corrosion rates, but may not necessarily reduce them to zero. Anodic protection can, however, extend the life of the corrosion allowance so that more permanent protective measures (such as stainless steel weld overlay) can be carried out later.

Ancillary Equipment

Stainless steels have been used as materials of construction for much of the equipment ancillary to digesters. This includes piping, valves, and pumps. Some of the major ancillary equipment is discussed in this section.

Liquor Heaters. External heat exchangers are used for indirect heating of the digester, most of which today (2006) are a two-pass shell and tube construction (Fig. 4). Batch digesters usually have one heat exchanger, while Kamyr units usually have three (see Fig. 3). With the continuous cook, two exchangers are in service while the third is being cleaned or in a standby mode. Tubing is 25 to 37 mm (1 to 1.5 in.) in outer diameter (OD), and from 3 to 4.6 m (10 to 15 ft) in length. Cooking liquor circulates through the tubes, with saturated steam on the shell side. Shell side temperature is approximately 200 °C (390 °F), while the liquor is 150 to 170 °C (300 to 340 °F).

For many years, welded type 304L stainless steel tubes have been the "standard" material of construction in liquor heaters. Unfortunately, austenitic stainless steels types 304L and 316L are susceptible to both chloride and caustic SCC, which has caused many tube failures. Stress-corrosion cracking of liquor heater tubes can occur from either the steam side or the liquor side (Ref 58). Inadvertent introduction of superheated steam has caused rapid SCC of type 304 tubing. Type 304L stainless steel tubes are also susceptible to rapid liquor-side thinning, which eventually leads to tube rupture. Thinning in batch digester liquor heaters is believed to be due to high-temperature operation. In continuous digester liquor heaters, thinning may be due to HCl cleaning. Hydrochloric acid cleaning is detrimental to the welds in type 304L stainless steel welded tubing, unless the manufacturer of the welded tubing has processed the tubing to reduce the normal ferritic content of the welds.

Type 304L tubes are normally used for new construction—where cost often controls material selection—and they are replaced when SCC or thinning causes unacceptable amounts of downtime. Duplex stainless steels such as 3RE60 (S31500), alloy 2205 (S32205), and alloy 2507 (S32750) are resistant to SCC in liquor heater service but are also susceptible to thinning, especially at higher temperatures. High-nickel alloys such as alloy 600 (N06600) and alloy 800 (N08800) are resistant to SCC and have improved resistance to acid cleaning damage.

Chip conveyors, which bring chips from the wood yard to the chip feeders, use type 304 for bends and other components subject to chip abrasion. Chip feeders for continuous digesters are typically made from centrifugally cast, precipitation-hardened stainless steel alloy CB-7Cu-1 (J92180), in the solution annealed and aged (H925) condition, for best abrasion resistance. Rotors are manufactured from cast martensitic alloy CA-6NM (J91540) or from alloy CB-6 (J91804). Rotors are quenched and tempered to 240 to 302 HB. Modified versions of alloy CA-6NM have been used to enhance weldability. Rotor cracking problems have been experienced and have been largely due to casting shrinkage. Manufacturers and users have begun to specify radiographic testing of rotors to ensure quality of the casting. Worn rotors are typically rebuilt by welding with modified type 410 (S41000) stainless steel applied by the submerged arc welding method. Corrosion of carbon steel feeder housings beneath the liner is a common problem and can result in cracking of the liner. Significant corrosion must be repaired by removal of the liner and welding a stainless steel overlay onto the housing. This is then precision machined to accept the liner.

Steaming Vessels. Chips are usually presteamed in a steaming vessel prior to introduction into the cooking vessel through a rotary-type high-pressure feeder. The steaming vessel is a horizontal cylindrical vessel that has conventionally been constructed from carbon steel with a partial cladding of type 304L stainless steel on the inside. A wear plate of type 304L or type 316L stainless steel is usually installed along the bottom of the vessel to protect the carbon steel wall from wear by the chips as they pass through the vessel. The wear plate usually corrodes rapidly and needs to be replaced every few years. The steaming nozzles, if constructed from type 304L or type 316L, may also experience SCC from the inside.

In the 1980s several steaming vessels were constructed from solid type 304L stainless steel. Most of these vessels experienced external SCC beneath the insulation when the insulation became wet from liquor spills. These vessels also had internal SCC in the steaming nozzles.

Duplex stainless steels such as alloy 2304 (S32304) and alloy 2205 have superior resistance to SCC and wear and are preferred for the internal lining, particularly the wear plate. Some steaming vessels have been constructed from clad duplex stainless steel (roll-clad or explosion clad) and are relatively maintenance-free.

Flash Tanks, Blow Tanks, Valves, and Pumps. In a continuous digester there are typically two flash tanks for the liquor extracted from the digester. The flash tanks were typically made from carbon steel, but there have been many reports of severe corrosion or erosion-corrosion of these vessels. High rates of flash tank corrosion usually occur when the digester is also experiencing rapid corrosion thinning. Types 304L and 316L were rated as marginal and duplex alloy 2205 is preferred (Ref 59). Corrosion was attributed to the presence of organic acids in the flash tank environment.

Thermal spray coating or lining with type 304L stainless steel has extended the life of corroding flash tanks. Replacement of flash tanks with solid duplex alloy 2205 is a solution to the corrosion problem.

The blow tank for batch digesters may be of carbon steel, type 304L, or for larger tanks, alloy 2205 construction. Blow valves are usually CF-3 (J92500) cast stainless steel. Cast duplex stainless steels such as CD-4MCuN (J93372) or CD-6MN (J93371) are preferred for pumps due to abrasion from sand and grit loadings.

Corrosion Control in High-Yield Mechanical Pulping

Chris Thompson, Paprican

High-yield mechanical pulping is the generic name for producing pulps mechanically using either disk refiners or grindstones (also

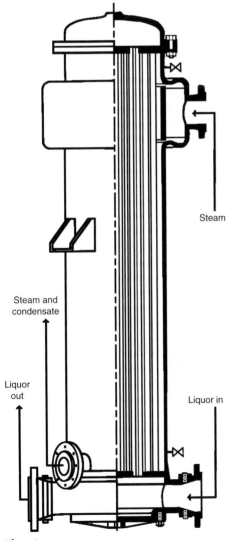

Fig. 4 Digester liquor heater

called "pulpstones"). Today (2006), high-yield mechanical pulping is dominated by disk refiner technology and, in North America, stone groundwood technology is rapidly disappearing from the scene. Newsprint and groundwood specialties typically contain high amounts of refiner pulps, sometimes as much as 100%.

A large number of refiner pulping processes have been developed, primarily as a result of the need to produce tailor-made pulps and to decrease energy requirements. The simplest process is refiner mechanical pulping (RMP), in which the chips are refined at atmospheric pressure with no prior use of heat or chemical action. For newsprint production, the dominant methods of refiner pulping are thermomechanical pulping (TMP), in which the chips are softened by heating in a steaming tube before entry into the primary refiner, and chemi-thermomechanical pulping (CTMP), in which the chips are also impregnated with a cooking chemical such as sodium sulfite. In both TMP and CTMP systems, chip steaming and refining take place at temperatures above 100 °C (212 °F), so that the steaming vessel and the first-stage refiner are pressurized. In the secondary refiner, the pulp is usually discharged at atmospheric pressure. The distinction between TMP and CTMP can be blurred, as TMP facilities sometimes add sulfite solution to the chips before refining. A schematic of a TMP system is shown in Fig. 5.

Materials of Construction and Corrosion Problems

Modern refiners are large and powerful, with disk diameters up to 1.8 m (70 in.) and motor powers of 11 MW (15,000 hp) and higher. In operation, pulping is achieved by passing wood chips between two or more closely spaced sets of serrated refiner plates, at least one set of which is rotating. Each set of plates is assembled by bolting individual plates onto a backing disk. Refiner plates are made from abrasion-resistant white cast irons, such as 25% Cr, Ni-HiCr type D (ASTM specification A 532) or high-carbon cast stainless steels, and are treated as consumables. For more information on the wear and corrosion of refiner plates, see Ref 59.

Relatively little information exists in the literature regarding corrosion in high-yield pulping environments. A recent review is given in Ref 60. A survey of materials problems and inspection procedures for refiner systems is reported in Ref 61. A nagging concern is stress-corrosion cracking (SCC) and corrosion fatigue, particularly with regard to pressurized steaming tubes, refiner disks, and refiner drive shafts. Many of the most severe problems reported to date have occurred in coastal mills, where high chloride levels have been encountered. However, chlorides are often a problem for inland mills as well. Chlorides can be introduced from deicing salts spilled on wood in the yard. Caustic used in the mill often contains chlorides. Even if chlorides are low, they can concentrate in crevices, under deposits, and just above the waterline in the vapor phase (Ref 60). Mills reducing their freshwater usage can also be vulnerable to unexpected attack, as contaminated white water from the paper mill may be fed back to the mechanical pulp mill.

The combination of the presence of chloride ions, the use of chemical pulping or bleaching agents, and the release of volatile wood acids can result in the localized buildup of highly aggressive conditions, even when nominal white-water conditions are considered to be relatively innocuous. In some of the cases mentioned in Ref 60, corrosion was initiated under deposits or in dead zones where the local environment was considerably more aggressive than the nominal environment. For example, deposit analyses from a coastal presteaming vessel showed pH values of 1.9 and chloride levels of about 13,000 ppm (compared to a nominal chloride level in the TMP white water of about 200 ppm). One mill has reported replacement of two alloy 904L (N08904) steaming vessels with alloy C-276 (N10276) and alloy C-22 (N06022), due to serious chloride SCC of 904L, an alloy that is normally resistant to chloride SCC (Ref 62).

Type 316L (S31603) is the standard alloy used by manufacturers for the major components of refiner systems (Ref 60). Exceptions are where high strength is required (e.g., for drive shafts and refiner backing disks) and in areas where corrosion is considered to be a particular concern. Upgrading to alloy 904L or alloy 2205 (S32205) has been suggested for high chloride environments. Alloys C-276 and C-22 have been used for particularly aggressive and acidic conditions in steaming tubes. For cavitation resistance, alloy 2205 clad plate has been suggested for some applications. The metallurgy used for refiner backing disks, drive shafts, and casings varies with the manufacturer and specific process conditions. Alloys that have been used include forged 15-5PH (S15500), CA6-NM (J91540), and CB-6 (J91804). Refiner casing alloys include 316L, 904L, CA-15 (J91150), CA6-NM, and 2205 clad carbon steel. The exact choice depends on the anticipated service conditions. Chloride SCC is reported to be such a serious problem that the use of type 316L for pressurized components has been discontinued (Ref 62). Chloride SCC of CB-6 refiner cases in less than two years has also been reported (Ref 62). Driveshafts are not designed to be exposed to the process and have typically been made from alloy steels such as 4140 or 4340; some use of martensitic stainless steel and the duplex stainless steel 2507 (S32750) has been reported.

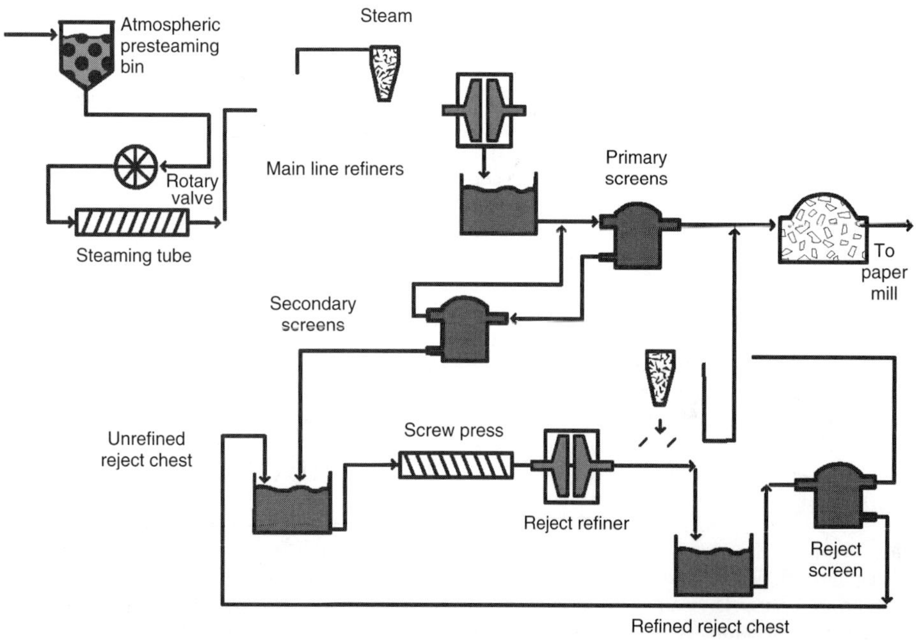

Corrosion in the Sulfite Process

Max D. Moskal, Mechanical and Materials Engineering

Once the principal method for producing chemical pulp, the sulfite process has lost ground to the kraft, mechanical, and semichemical pulping methods. Nevertheless, sulfite pulping remains an important process.

Many similarities can be seen between kraft and sulfite mills, although chemistries are quite different. Each has batch or continuous

Fig. 5 Flow diagram of a thermomechanical pulping mill

digestion, and some sulfite operations have chemical recovery. Following digestion, processes for washing, bleaching, and papermaking can be identical. Consequently, the following discussion is limited to liquor preparation, digestion, and chemical recovery. Figure 6 is a flow diagram of a typical magnesium-base pulping and recovery operation.

The Environment

Sulfite cooking liquor consists of free sulfur dioxide dissolved in water together with sulfur dioxide (SO_2) in the form of bisulfite. Until about 1950, sulfite pulping was an acid process using calcium as the base. Because the acid process was limited to only certain wood species, other base materials were introduced. These soluble bases are sodium, magnesium, and ammonium, and their use has broadened sulfite pulping for all wood species. Subsequent to the introduction and application of magnesium and sodium bases, several two-stage sulfite pulping methods using these bases had been developed and utilized commercially. These methods are generally characterized by a dramatic change in cooking pH between the initial stage (for penetration and sulfonation) and the second stage (for dissolution and removal of the lignin) (Ref 63).

Depending on the process, the pH of the cooking liquor can range from acid to neutral to alkaline. In acid bisulfite pulping, any pH within the range of 1.5 to 4.0 can be achieved by controlling the ratio of free and combined SO_2. True bisulfite pulping, with equal amounts of free and combined SO_2, would be carried out at a pH of 4.0 to 5.0. Another important sulfite pulping process is neutral sulfite semichemical pulping, which is described in the next major section of this article.

Sulfuric acid is the principal aggressor in an acid sulfite mill. It can be generated in any number of unit operations. The environment usually contains chlorides that are often present in other chemicals used and are always present in logs that have been floated to coastal mills. Chlorides increase the aggressiveness of sulfuric acid.

Construction Materials

It can be said that stainless steels and sulfite pulping grew up together. In fact, many new alloys were tested in acid sulfite mills. Type 316 (S31600) stainless steel became the principal material of construction for sulfite mills, although more highly alloyed materials are required in a number of applications. Type 317L (S31703) and the higher molybdenum stainless steels are finding wider application, especially in the higher chloride environments. More recently, duplex stainless steels have been specified, due to their excellent resistance to pitting and stress-corrosion cracking. There are also a number of applications where specialty alloys such as alloy 20 (N08020), 6% Mo superaustenitic alloys, G-30 (N06030), and C-276 (N10276) are advantageous.

Sulfur Dioxide Production

The principal method for producing SO_2 is the burning of elemental sulfur, although some Canadian mills roast iron pyrites. The steps for preparing a sulfite cooking liquor are:

1. Burning of sulfur
2. Rapid cooling (to prevent sulfur trioxide formation)
3. Absorption of sulfur dioxide in a weak alkaline solution
4. Fortifying the raw acid

Mild steel is satisfactory for sulfur burners and combustion chambers (Ref 64). This is not a good environment for austenitic stainless steels because of the formation of low-melting-point nickel sulfide scale.

The hot SO_2 is cooled rapidly in either heat exchangers or direct-contact water spray towers. The heat exchangers, or pond coolers, are lead-lined steel pipe. Cooling towers are commonly acid-brick lined, but are occasionally constructed of 316L stainless steel. When stainless steel is used, walls should receive a constant supply of wash water. Figure 7 shows one design for a SO_2 cooler. In this case, the first section is lead or acid-brick lined, and the second section is stainless steel. Emergency water is available in case there is an excursion in temperature.

Piping from the cooling tower to the absorption tower can be exposed to sulfuric acid. Piping has been made from 316L or 317L stainless steel, although some mills use fiberglass-reinforced plastic (FRP).

The SO_2 compressors can be lined with lead or rubber, or more commonly 316L or 317L. Sometimes a more acid-resistant, copper alloy (alloy 20) stainless steel would be preferred (Ref 65).

The SO_2 fans and compressors can be lined with lead or rubber, or more commonly, with 316L or 317L construction. A better choice

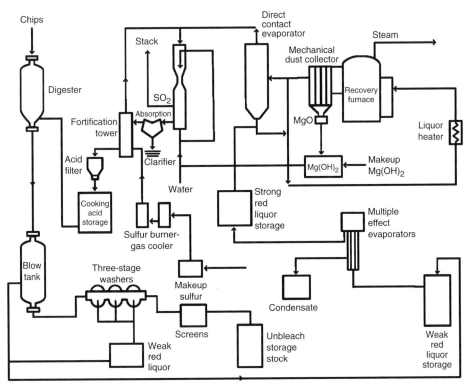

Fig. 6 Flow diagram of a magnesium-base (sulfite) mill and recovery system

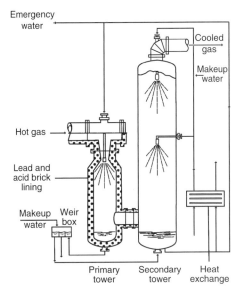

Fig. 7 Spray-type sulfur dioxide cooler

would be alloys 20 or 904L (N08904), as this equipment is prone to accumulation of carryover sulfur and attack by the small amounts of sulfuric acid that are inadvertently formed in the burner (Ref 65).

Atmospheric absorption towers can be 316L or 317L or FRP. With FRP construction, all internals would be 316L or 317L stainless. Both the body and internals of pressure towers can be of stainless steel construction.

Acid tanks are usually acid-brick-lined, although 316L and 2205 (S32205) duplex stainless steel have been used. However, sulfuric acid can be present, and if the temperature exceeds 85 °C (185 °F), acid-brick-lined tanks are preferred.

Digesters

Older batch digesters were acid-brick-lined. However, digesters have more recently been constructed from 317L, 904L, 317LMN (S31726), and 2205 duplex stainless steels. Duplex stainless steel alloys offer better protection against chloride stress-corrosion cracking compared to austenitic stainless steels. In brick-lined digesters, most fittings, screens, sleeves, and man-way hatches are 316L or 317L stainless steel.

Continuous digesters for sulfite pulping are very similar to kraft continuous digesters in design. For the sulfite process, continuous digesters are all stainless steel construction using 316L (with a minimum of 2.75% Mo) or 317L (Ref 66). Duplex stainless steel has also been used in several continuous digesters.

Washing and Screening

Pumps, piping, screens, knotters, washers, and deckers may all be exposed to small amounts of bisulfite in the spent liquor, so stainless steel construction is still necessary for corrosion protection. Type 316L stainless is almost standard. However, crevice corrosion may occur under pulp deposits; so the choice may be for 317L, 2205 duplex or higher alloyed materials.

A number of sulfite mills operate liquor recovery systems. However, there appears to be considerable variation in operating conditions. Sulfite recovery boilers are primarily of carbon steel construction, with significant corrosion allowances in critical areas such as screens, superheaters, and waterwall tubes. More recent construction has incorporated composite 304L with SA210 steel tubes in the lower furnace waterwalls. Carbon steel remains the choice for floor tubes. The SO_2 content of sulfite recovery flue gas is much higher than in kraft systems, increasing the potential for sulfur trioxide (SO_3) formation.

Scrubbing systems are usually of FRP and/or 316L/317L stainless steel. Where high levels of chlorides and/or SO_3 are present, high-molybdenum stainless steels or nickel-base alloys give better service (Ref 65). The magnesium sulfite operation presents a special case in scrubber corrosion, where the magnesium oxide (MgO) slurry in water is combined with hot SO_2 gas. In this case, the SO_2 gas enters the top of one leg of the scrubber and is combined with the MgO slurry shower. Problems of severe stress-corrosion cracking and crevice corrosion under wet salt have occurred in 317L stainless steel scrubber walls. One solution has been to replace the affected components with Ni-Cr-Mo-W (e.g., C-276) alloys. Consideration could be given to the use of duplex stainless steels in this application given the excellent resistance to pitting and stress-corrosion cracking of these alloys.

Jonsson reported that a steel corresponding to type 316L is suitable material for all components in the evaporators such as tubes, vapor bodies, liquid flash tanks, vapor pipes, condenser pipes, liquor pipes, valves, pumps, and condensers where calcium bisulfite is evaporated (Ref 65). For evaporation of magnesium-base cooking liquor, 316L with 2.75% Mo or 317L is often prescribed, since the corrosive conditions are considered to be somewhat more severe than for calcium bisulfite. For sodium-base cooking liquor, the conditions are not so corrosive, and therefore 316L is used for the first effect and 304L or mild steel for the weak liquor effects in the evaporator train. Carbon steels in the weak liquor stages have sometimes been replaced by stainless steels, primarily to reduce fouling and secondarily to reduce corrosion. The corrosion problems are less severe for ammonium-base cooking liquor.

Chloride Control

Chloride ions entering the pulp mill cycle from saltwater wood storage can cause serious corrosion of stainless steel equipment throughout the liquor cycle. A control system conceived for the MgO recovery process can greatly reduce corrosion by maintaining a low chloride ion level. The system is based on the knowledge that the chlorides leave the recovery boiler as hydrochloric acid and can be readily removed from the flue gas by a venturi absorber (Ref 67).

Corrosion Control in Neutral Sulfite Semichemical Pulping

Chris Thompson, Paprican

In neutral sulfite semichemical (NSSC) pulping, hardwood chips are cooked in a liquor solution consisting of sodium sulfite (Na_2SO_3) with a small quantity of sodium carbonate (Na_2CO_3), sodium bicarbonate ($NaHCO_3$), or sodium hydroxide (NaOH). A typical strong liquor contains 120 to 200 g/L of Na_2SO_3 and 30 to 50 g/L of Na_2CO_3. The liquor is generally prepared by burning sulfur to form sulfur dioxide (SO_2), and then absorbing the gas in a solution of Na_2CO_3 (soda ash), although small mills may purchase chemicals if readily available from a nearby source.

Cooking is done in digesters similar to those used in the kraft process, except that the continuous digesters are of a different design. Typical continuous digesters include horizontal multiple-tube units or inclined-tube digesters. The cooking occurs in either the liquid or the vapor phase. For vapor-phase cooking, the chips must pass through a press impregnator before entering the digester. Figure 8 is a flow diagram for a typical NSSC mill (Ref 68). Cooking temperatures range from approximately 180 to 195 °C (355 to 380 °F) and the pH range is from 7.2 to 9. Heating is either directly by steam injection or indirectly with an external heat exchanger.

A number of mills are combined kraft/NSSC operations because both are used in the manufacture of corrugated cardboard. Consequently, the sulfur and soda obtained from NSSC spent cooking liquors are introduced into the kraft chemical recovery system to replace the salt cake lost in the kraft mill. During carbonation of the smelt, hydrogen sulfide is released. This is burned to SO_2 in a separate reactor or in the oxidation zone of the recovery boiler.

Materials of Construction

Because NSSC spent liquors are more corrosive than kraft liquors, stainless steels are generally used in all equipment. While type 304 (S30400) has been used successfully, type 316 (S31600) is generally preferred because of its better resistance to corrosion. When welding is involved, the low-carbon variations, types 304L (S30403) and 316L (S31603), are used to prevent problems with intergranular corrosion. While NSSC pulp can be bleached, using chlorine, caustic soda, or hypochlorite, bleaching is not usually required.

In a related process, pulping is done at more acidic pH levels, using bisulfite rather than sulfite as the active chemical. In some instances, this has caused rapid attack of the vapor-phase region of the digester, due to the formation of sulfuric acid vapor (Ref 69). The conditions that promote the formation of acid vapors are the presence of air and a low cooking liquor pH. Digesters for use in semichemical bisulfite pulping have been made from a range of materials, including carbon steel and austenitic stainless steels. The accelerated corrosion of a type 317L (S31703) continuous bisulfite semichemical digester has been described by Murarka (Ref 69). Formation of acid vapors was minimized by raising the liquor pH from 4 to 6 and by using a plug screw feeder to reduce the entry of oxygen with the incoming chips. Additional precautions taken were feeding

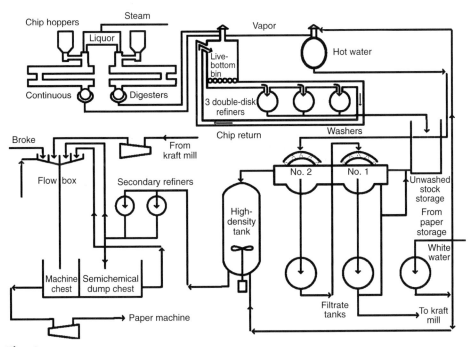

Fig. 8 Flow diagram of a typical neutral sulfite semichemical mill

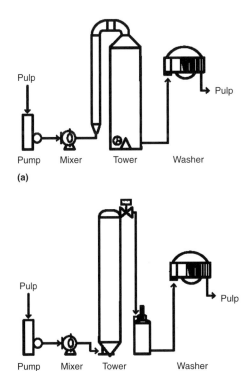

Fig. 9 Bleaching stage components. (a) Atmospheric (upflow/downflow). (b) Pressurized (upflow)

Table 1 Bleach stage nomenclature

Symbol	Process/processing chemical
C	Chlorine (Cl_2)
D	Chlorine dioxide (ClO_2)
D_0	Full chlorine dioxide substitution for chlorine
D_c	Partial chlorine dioxide substitution for chlorine
D_n	Neutralized with caustic
E	Caustic extraction
E_o	Oxidative caustic extraction
E_{op}	Peroxide/oxygen caustic extraction
H	Hypochlorite
O	Oxygen (O_2)
Z	Ozone (O_3)
P	Peroxide (H_2O_2)
P_{HT} or P_O	Peroxide, hot or pressurized

the chips into the digester below the liquor level, and preheating the water in the blow tank to lower its dissolved oxygen content. The vapor space of the digester was clad in alloy 904L (N08904) stainless steel. Presteaming the chips and replacing the air in the digester with steam prior to startup have also been advocated to minimize the possibility of acidic corrosion (Ref 59, 70).

Corrosion Control in Bleach Plants

Donald E. Bardsley and William Miller, Sulzer Process Pumps Inc.

The oxygen delignification and bleach plant sequence is a continuation of lignin removal following the digester cooking process described in the section "Corrosion of Digesters." This section describes the various stages of chlorine-based and nonchlorine bleaching, process water reuse for elemental chlorine-free and nonchlorine bleaching stages, selection of material for bleaching equipment, recent developments in oxygen bleaching, and the use of more highly corrosion-resistant materials for bleach plant equipment.

Stages of Chlorine-Based Bleaching

Bleaching takes place in the oxygen delignification and bleach plant sequences in 3 to 5 steps, each step referred to as a stage (refer to Table 1 for bleach stage nomenclature). Each stage comprises four basic equipment components:

- A pump to move the pulp through the stage at the desired consistency
- A mixer to blend the pulp, chemicals, and steam
- A reaction tower or vessel (atmospheric or pressurized) with dilution and agitation equipment for discharge consistency control
- A washer to separate residual chemicals and reaction by-products from the pulp

These four components are shown in Fig. 9(a) for an atmospheric bleaching stage (Cl_2, ClO_2, E_o) and in Fig. 9(b) for a pressurized stage (O_2, O_3, P_O). The upflow section of the atmospheric stage can also be pressurized (E_o, E_{op}, P_{HT}). These stages are divided into two process categories within each bleaching sequence: delignification and brightening. There are a large number of different bleaching stages used in various combinations.

Delignification. Immediately following the cooking process, the delignification stages selectively continue the bulk removal of lignin from the fiber. Inclusive in this group of bleaching stages are oxygen delignification (O), chlorination (C) or chlorine dioxide (D) stages, and oxidative extraction (E_o) or peroxide-reinforced extraction stages (E_{op}). In a bleaching sequence the delignification stages are typically grouped as D_cE_o, D_0E_{op}, and OD_0E_{op}.

Brightening. Following the delignification stages are the brightening stages. These stages are most effective when the lignin level in the pulp has been reduced to less than 1% by weight. Permanent brightness development occurs as this remaining lignin is removed or modified in the brightening stages. These are typically chlorine dioxide stages (D), with intermediate caustic neutralization steps (D_n), or mild caustic extraction stages (E). In a bleaching sequence, the brightening stages are typically grouped as D, DED, DD, or D_nD.

Bleaching Sequences. The final International Organization for Standardization (ISO) brightness target and environmental constraints define the combination of delignification and brightening stages in a bleach plant design. The most typical sequences are:

- $D_cE_oD_nD$ to achieve 88 to 90% target brightness
- $ODE_{op}D$ to achieve 85 to 88% target brightness

These sequences were selected to encompass the scope of bleach stage design parameters currently utilized in bleach plants. The ODE_{op}D sequence, with no elemental chlorine used, is referred to as an elemental chlorine-free (ECF) sequence. Figure 10 schematically outlines the $D_cE_oD_nD$ bleaching sequence.

Table 2 summarizes the typical operating and design parameters for $D_cE_oD_nD$ and ODE_{op}D sequences to achieve 88 to 90% and 85 to 88% ISO target brightness, respectively, on softwood pulp. The O_2 delignification stage in the ODE_{op}D sequence is designed to reduce the brown stock kappa number to 18. The kappa number measures the amount of lignin present in a kraft pulp:

kappa number $\times$ 0.15% = % lignin in pulp

For example, if the kappa number of a conventionally cooked pulp after O_2 delignification is 18, this kappa number corresponds to a lignin content of 2.7%.

Nonchlorine Bleaching Stages

Nonchlorine bleaching stages, utilized in both the delignification and brightening stages of a bleaching sequence, are designed to reduce or eliminate the use of chlorine-based (ClO_2) bleaching chemicals. These stages are the building blocks for the totally chlorine-free (TCF) bleaching sequences.

Delignification. The function of the delignification stages is identical to those described in "Stages of Chlorine-Based Bleaching." Inclusive in this category for nonchlorine bleaching stages, the bleaching stages utilized are oxygen delignification (O), ozone (Z), and pressurized hydrogen peroxide reinforced, oxidative extraction stages (E_{op}).

Brightening. Following the delignification stages are the brightening stages, as previously described. For nonchlorine bleaching the brightening stages are typically pressurized, oxidative hydrogen peroxide stages (P_{HT}, P_O, P).

Bleaching Sequences. Nonchlorine bleaching stages in ECF sequences are used primarily to maintain final brightness targets while lowering active chlorine demand. The nonchlorine stages can be utilized to reduce active chlorine in both the delignification and brightening stages.

The final brightness target and environmental constraints define the combination of nonchlorine delignification and brightening stages in an ECF bleach plant design. For example purposes only, a proposed $D_0(ZE_{op})DP_{HT}$ bleaching sequence is shown schematically in Fig. 11. The purpose of this figure is to illustrate how a plant might apply an ozone (Z) delignification stage and/or a P_{HT} brightening stage.

Bleaching Stages. The range of operating conditions for nonchlorine bleaching stages are outlined in Table 3. These conditions are used to optimize the displacement of active chlorine in the delignification and brightening stages while maintaining pulp quality.

Process Water Reuse for ECF and Nonchlorine Bleaching Stages

Many bleach plants today recycle filtrates countercurrently to accomplish at least partial closure (see Fig. 12, 13). The bleach plant effluent, which flows to the waste treatment plant, is primarily overflow from the D_c/D_{100} and E_{op} stages.

There are several benefits to recycling filtrates, such as:

- Reduction in water consumption and effluent flow
- Reduction in chemical consumption (NaOH and H_2SO_4)
- Reduction in heating requirements

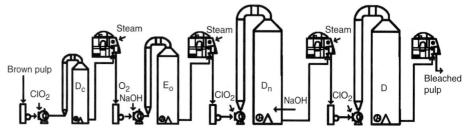

Fig. 10 Flow diagram of a conventional $D_cE_oD_nD$ bleaching sequence

Table 2 Typical operating and design parameters for conventional and ECF bleaching sequences

	Conventional	ECF(a)
Bleaching sequence	$D_cE_oD_nD$	OD(E_{op})D
Brown stock kappa No.	30	30
Target Brightness (% ISO)	88–90	85–88
Stage	NA	O
% NaOH (on pulp)	NA	1.5–2.0
% O_2	NA	1.5–2.0
Temperature	NA	85–100 °C (185–210 °F)
Final pH	NA	10.0–12.0
Bleach tower	NA	Cone top and bottom (100 psig at bottom, pressurized)
Retention time, min	NA	60
Stage	D_c	D
% ClO_2 substitution	50	100
% active Cl_2 (on pulp)	6.0–7.8	3.6–4.7
Temperature	30–50 °C (90–120 °F)	50–65 °C (120–150 °F)
Final pH	1.5–2.0	2.5–3.5
Consistency, %	3–4	3–12
Bleach tower	Upflow (atm)/downflow	Upflow (atm)/downflow
Retention time, min	60 (15/45)	45 (15/30)
Stage	E_o	E_{op}
% O_2 (on pulp)	0.5	0.5
% H_2O_2 (on pulp)	NA	0.5
Final pH	10.5–12.0	10.5–12.0
Temperature	75–85 °C (165–185 °F)	75–90 °C (165–195 °F)
Bleach tower	Upflow (atm)/downflow	Upflow (pressurized)/downflow
Retention time, min	45 (5 min at 20 psig/40 min)	45 (15 min at 60 psig/30 min)
Stage	D_n	D
% ClO_2 (on pulp)	0.8–1.2	1.0–1.4
pH	3.5–4.5 (dilution zone is neutralized to 7.0–9.0 pH)	3.0–4.5
Consistency, %	10–12	10–12
Temperature	70–80 °C (160–175 °F)	70–80 °C (160–175 °F)
Bleach tower	Upflow (atm)/downflow	Upflow (atm)/downflow
Retention time, min	180 (45/135)	240 (45/195)
Brightness, % ISO	87.0–88.0	85.0–88.0
Stage	D	
% ClO_2 (on pulp)	0.1–0.3	...
pH	4.5–5.0	...
Temperature	70–80 °C (160–175 °F)	...
Consistency, %	10–12	...
Bleach tower	Upflow (atm)/downflow	...
Retention time, min	180 (45/135)	...
Brightness, % ISO	88–90.0	...

NA, not applicable; atm, atmospheric pressure. (a) Totally chlorine-free, ECF

In cases where bleach stage residual is high, filtrate recycling can also lead to corrosion in certain parts of the bleach plant.

The effluents from the bleaching operation contribute a significant portion of the chemical oxygen demand (COD), color, and absorbable organic halide (AOX) to the overall effluent stream. Environmental regulations have put limitations on the levels of these components discharged with bleach plant mill effluent. This has led to a decrease in the use of chlorine and chlorine-containing compounds and to the trend toward ECF and nonchlorine bleaching sequences. Recent bleach plant design is also focusing on reducing the effluent flows by increasing the level of filtrate recycling.

Reevaluation and increased monitoring of the construction materials used throughout the bleach plant are becoming more critical as new bleaching processes and increased filtrate recycling are utilized in bleach sequences (Ref 71).

Selection of Materials for Bleaching Equipment

Materials for bleach plant equipment have undergone major changes over the past 30 years. In the 1970s recycling of wash water and approaches to closure of the wash-water cycle increased corrosivity. Changes in bleaching processes and reduction in time available for maintenance, together with recycling of wash water, required upgrading to more highly alloyed, more corrosion-resistant materials in the late 1970s and 1980s. More recently, the need to reduce dioxins has resulted in replacing chlorine with chlorine dioxide (ECF bleaching). Substituting oxygen and hydrogen peroxide for both chlorine and chlorine dioxide, as some mills have done, is known as TCF bleaching. Totally chlorine-free bleaching requires less highly alloyed materials than ECF bleaching. Type 316L (S31603), CF-3M (J92800), and CF-8M (J92900) are quite adequate for most applications in TCF bleaching.

For environmental reasons, a number of mills have eliminated the chlorination stage. Only a few continue their use of chlorine, and these are expected to discontinue use of chlorine within the next few years. In ECF bleaching where elemental chlorine has been eliminated, the bleaching sequence normally comprises two to three chlorine dioxide stages. The use of ECF bleaching is expected to continue to grow in North American mills. Only a few mills have adopted TCF bleaching because it lacks economic appeal. The hypochlorite stage, which was very popular in the 1950s, has been largely abandoned in North America and Europe, but is still in use in South America and the Far East.

Some of the more common materials used in the older chlorine (C), chlorine dioxide (D), and hypochlorite (H) stages are shown in the first three columns of Table 4. Piping that leads to the chlorine or chlorine dioxide tower is fiberglass-reinforced plastic (FRP). The tower itself is steel with a membrane, acid-brick lining, or solid FRP. Piping from the towers is either 6% Mo stainless steel or FRP. Cast alloys CW-2M (N26455) and CX-2MW (N26022) performed well in chlorine-stage chemical mixers. In both the chlorine stage with chlorine dioxide substitution and the chlorine dioxide stage, these Ni-Cr-Mo alloys (equivalent to wrought alloys C-4 and C-22, respectively) have suffered substantial corrosion. Titanium is the preferred material for both the chemical and high shear mixers in the D stage washer. In the old C stage, precautions must be taken to prevent dry chlorine from entering the titanium mixer in order to avoid a pyrophoric reaction. Superaustenitic 6% Mo stainless steels have become standard for the C and D stage washers.

During ECF bleaching when countercurrent washing is utilized in the bleach plant, caustic filtrate is applied as shower water on the first-stage chlorine dioxide bleach dioxide filter to displace the chlorine dioxide in the pulp mat. Residual chlorine dioxide in the pulp mat combines with the caustic, resulting in the formation of sodium chlorate and sodium chlorite in:

$$2ClO_2 + 2NaOH \rightarrow NaClO_3 + NaClO_2 + H_2O$$

The resulting condition is known as near-neutral chlorine dioxide with a pH of 6 to 7.

Superaustenitic, 6% Mo stainless steels have adequate resistance to near-neutral chlorine dioxide. However, the nickel-base alloy filler metals, AWS ERNiCrMo-4 (N10276) and AWS ERNiCrMo-10 (N06022), normally used for fabrication, are subject to what has been termed transpassive corrosion in this near-neutral chlorine environment. These nickel-base filler metals undergo molybdenum dissolution, which renders them unsuitable for this near-neutral chlorine dioxide environment. After some investigation, it was found that AWS ERNiCrMo-11 (N06030) with higher chromium and lower molybdenum had much better corrosion resistance in the near-neutral chlorine dioxide environment while still maintaining good resistance to acid chloride environments. ERNiCrMo-11 has become the

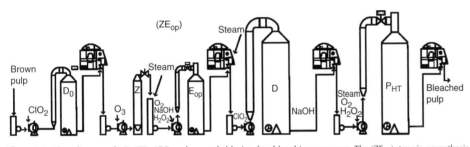

Fig. 11 Flow diagram of a $D_0(ZE_{op})DP_{HT}$ elemental chlorine-free bleaching sequence. The (ZE_{op}) stage in parenthesis denotes no wash between stages. In some instances, a chelation or acid-rinse step followed by a wash may be required prior to the P_{HT} and P_o stages.

Table 3   Typical operating and design parameters for nonchlorine bleaching stages

Bleach stage	Reinforced oxidative extraction, E_{op}	Ozone, Z	Pressurized perixode, P_{HT}, P_O
Chemical (on pulp)	0.5–1.0 (H_2O_2), 0.5 (O_2)	0.2–0.5 (O_3)	1.0–3.0 (H_2O_2), 0.5 (O_2)
Temperature, °C (°F)	85–105 (185–200)	50–60 (120–140)	85–100 (185–210)
Final pH	10.0–10.5	2.5–3.0	10.0–10.5
Consistency, %	12–14	12–14	12–14
Bleach tower	Upflow (press)/downflow	Upflow (press)	Upflow (press)/downflow or pressurized upflow
Retention time(a), min	60 (15 at 0.4 MPa (60 psig)/45 at atm)	5–10 at 1.0 MPa (150 psig)	180 (15 at 0.4 MPa (60 psig)/165 atm), 120 at 0.4–0.7 MPa (60–100 psig)

(a) atm, Atmospheric pressure

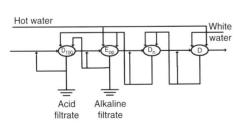

Fig. 12 Process water reuse for $D_{100}E_{op}D_nD$ bleaching sequence

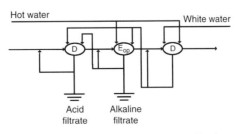

Fig. 13 Process water reuse for $D(E_{op})D$ bleaching sequence

preferred filler metal for fabricating 6% Mo stainless steel washers for near-neutral chlorine dioxide stage service.

For the newer, less corrosive, oxygen, peroxide, and caustic stages (shown in the remaining columns of Table 4), 316L and its cast counterpart CF-3M are the principal materials used for the towers, pumps, and mixers. For 316L towers, it is necessary to protect the outside from chloride stress-corrosion cracking (SCC) beneath the insulation, as described in this article and in the article "Corrosion Under Insulation" in this Volume. Some mills have selected duplex stainless steel for the oxygen towers because of its better resistance to chloride SCC. Zirconium is used for the high-shear mixer in the hot alkaline peroxide stage.

For bleach plant service, it is desirable that all wetted surfaces be clean and free of embedded iron and other fabrication-related defects. It is essential that type 316L hydrogen peroxide towers and other peroxide wetted surfaces are free of embedded iron, as embedded iron degrades hydrogen peroxide. ASTM A 380 and A 967 provide the guidelines for detection and removal of embedded iron and other fabrication defects using mechanical, chemical, and electrochemical methods.

Oxygen Bleaching

Oxygen has become a principal technology to replace chlorine-based chemicals for bleaching pulp. Oxygen usage is expected to continue growing as chlorine usage declines. Initially, oxygen bleaching was done on high-consistency pulp, leaving a gas phase at the top of the vessel. Pressures were as high as 0.8 MPa (120 psig) and temperatures as high as 135 °C (275 °F), conditions that are conducive to internal chloride SCC. The interface between liquid and vapor in high consistency bleaching provided a "hot wall" effect, where chlorides could concentrate just above the liquid level, increasing the probability of internal chloride SCC.

In order to prevent internal chloride SCC of 316L when gas-phase oxygen bleaching began in the early 1970s, Scandinavian mills "cold stretched" 316L oxygen reactors. Cold stretching was achieved by overpressuring the vessel sufficiently so that, at operating pressures, the inside surface remained in compression and resistant to chloride SCC. Cold stretching also allowed a reduction in wall thickness under Scandinavian pressure vessel codes. In North America, where codes did not allow a reduction in wall thickness for cold stretching, upgrading to solid or clad alloy 20 (N08020) became the practice to resist chloride SCC in oxygen reactors.

The initial process was changed to treat medium consistency pulp in an upflow mode instead of downflow mode, eliminating the vapor space to the top of the vessel, where chlorides could concentrate and initiate chloride SCC. This reduced the potential for internal chloride SCC of 316L without the need for cold stretching. Most of the oxygen reactors treating medium-consistency pulp in the upflow mode have been 316L. Recently, some have been fabricated from alloy 2205 (S32205), a duplex stainless steel which is resistant to chloride SCC.

Type 316L is quite susceptible to external SCC under insulation when the insulation becomes wet from rain or even high humidity. NACE Publication RP 0198 "Standard Recommended Practice—The Control of Corrosion under Thermal Insulation and Fireproofing Materials" provides the necessary information for keeping the outside surface of 316L reactors, and other 304L (S30403) and 316L vessels and towers, dry and resistant to SCC. Although internal chloride SCC has not been reported in vessels where pulp is treated with oxygen in the upflow mode with no vapor space at the top, a number of mills have elected to upgrade to alloy 2205 for better resistance to both internal and external chloride SCC. Type 316L and CF-3M are satisfactory for other components in the oxygen and hydrogen peroxide bleaching stages.

Pumps, Valves, and the Growing Use of Duplex Stainless Steels

Since conditions vary considerably from mill to mill, individual mill site experience generally provides the best guide for alloy selection and upgrading. The pitting resistance equivalent

Table 4 Materials selection guide for bleach plant equipment

Bleaching stage	Chlorination, C	Chlorine dioxide, D	Hypochlorite, H	Caustic extraction, E	Caustic/ Oxygen, E_o	Peroxide, P	Ozone, Z	Oxygen, O
Tower								
Shell	Mild steel	Mild steel	316L(a)	316L(a)	316L(a)	316L(a)(b), 2205(b)	316L(a)	316L(a), alloy 20, 904L, 2205
Lining	Membrane/tile	Membrane/tile	...	...	...	...	...	...
Nozzles	Membrane/tile	Membrane/tile	316L	316L	316L	316L	316L	as above
Washer								
Vat	6%Mo	6%Mo	316L/317L	317L/6%Mo	317L/6%Mo	316L	316L	316L
Drum	6%Mo	6%Mo	316L/317L	317L/6%Mo	317L/6%Mo	316L/2205	316L/2205	316L/2205
Piping								
Before tower	FRP/Ti(c)	FRP/Ti	316L	316L	316L	316L	316L	316L
After tower	6%Mo	6%Mo	316L	316L	316L	316L	316L	316L
Filtrate	6%Mo	6%Mo	316L	316L	316L	316L	316L	316L
Top scraper	C-276/CW-2M	Titanium	C-276/CW-2M	316L	316L	316L	316L	316L
Tower agitators								
Casing	C-276/CW-2M	6%Mo	CF-3M/G-30	CF-3M	CF-3M	CF-3M	CF-3M	CF-3M
Propeller	C-276/CW-2M	6%Mo	CF-3M/G-30	CF-3M	CF-3M	CF-3M	CF-3M	CF-3M
Line or radial chemical mixers								
Casing	C-276/CW-2M	Titanium	CF-3M	CF-3M	CF-3M	316L/Zr(d)	CF-3M	CF-3M
Impeller	C-276/CW-2M	Titanium	CF-3M	CF-3M	CF-3M	316L/Zr(d)	CF-3M	CF-3M
Steam mixers								
Casing	316L	316L	316L	316L	316L	316L	316L	316L
Impeller	316L	316L	316L	316L	316L	316L	316L	316L
High-shear mixers								
Casing	Titanium(c)/CW-2M	Titanium	CF-3M	CF-3M	CF-3M	316L/Zr(d)	CF-3M	CF-3M
Impeller	Titanium(c)/CW-2M	Titanium	CF-3M	CF-3M	CF-3M	316L/Zr(d)	CF-3M	CF-3M
Pumps								
Low-density stock	CG-3M	CG-3M	CF-3M	CF-3M	CF-3M	CF-3M	CF-3M	CF-3M
Filtrate	CG-3M	CG-3M	CF-3M	CF-3M	CF-3M	CF-3M	CF-3M	CF-3M
Filtrate tanks	Tile or FRP	Tile or FRP	316L	316L	316L	316L	316L	316L
Sewer lines	FRP	FRP	316L	316L	316L	316L	316L	316L

Note: 6%Mo denotes superaustenitic stainless steels. (a) To prevent external SCC, special care of type 316 in insulation is required. (b) Heat tint and embedded iron must be removed from all wetted surfaces to reduce decomposition of hydrogen peroxide. (c) Titanium must be protected from contact with dry chlorine by interlocks to avoid pyrophoric reactions. (d) Zirconium is used for the high-temperature (>100 °C, or 212 °F), more aggressive applications.

nitrogen number (PREN) provides very useful guidelines for upgrading bleach plant materials when upgrading is necessary. The PREN is a calculated parameter used to estimate the expected resistance to localized corrosion by chlorides. For stainless steels, it is calculated from composition using the empirical formula:

$$PREN = \%Cr + 3.3 \times \%Mo + 16 \times \%N$$

where chromium, molybdenum, and nitrogen are the respective concentrations of these elements in the alloy expressed in percentages by weight. In general, the larger the numerical value of PREN is, the higher the pitting resistance will be, although a high numerical value of PREN should not be viewed as an absolute guarantee of freedom from localized attack.

The earlier practice of upgrading to 317L (S31703, PREN = 30 to 35) in North America, and to alloy 904L (N08904, PREN = 36) in Europe, when 316L (PREM = 24 to 30) proved marginal or inadequate, is being supplemented, and to an extent superseded, by the current trend of upgrading to a duplex alloy, usually alloy 2205 (PREN = 36). Alloy 2205 is becoming the preferred upgrade whenever higher strength, better resistance to chloride SCC, or better resistance to abrasion is needed. Alloy 2205 is also often the choice for large field-erected vessels, where its higher strength can be used to offset much of the higher material cost.

The cast version of 317L, CG-3M (J92999), and 6% Mo alloys, CN-3MN (J94651), and CK-3MCuN (J93254), have become standard for filtrate pumps in the chlorine and chlorine dioxide stages. CF-3M, CG-3M, and the duplex grade CD-3MN (J92205) are used for pulp stock feed pumps in these two stages. CF-3M and CG-3M and the cast duplex grades are the principal stock pump materials in the six less corrosive bleaching stages. Duplex is preferred wherever there is abrasive wear. CK-3MCuN and CN-3MN are preferred for valves in the chlorine stages. CG-3M, CG-8M (J93000), and CN-7M (N08007) are used for valves in the six less corrosive services. CN-7M is preferred for control valves.

Titanium is the only material that has been used for continuous bleaching towers, despite the considerable maintenance required.

Paper Machine Corrosion

Angela Wensley,
Angela Wensley Engineering

MOST OF THE CORROSION PROBLEMS IN PAPER MACHINES occur in the wet end (Fig. 14) and in ancillary equipment handling white water. Areas susceptible to corrosion are those exposed to the white-water environment by immersion, splashing, vapor mist, or in the damp crevices beneath pulp pads or other deposits. Materials of construction susceptible to corrosion include stainless steels, carbon steel, copper alloys, and cast irons.

This section is divided into three parts. The first part describes the materials of construction of the paper machine components and the specific corrosion problems affecting them; the second part deals with the composition and corrosive nature of white water; the third part is a more detailed discussion of specific corrosion mechanisms.

Paper Machine Components

Good introductions to the corrosion problems in paper machine wet ends and dryer sections have been published by Bennett and Sharp (Ref 72, 73). Most wetted paper machine surfaces are constructed using austenitic stainless steels such as type 304L (S30403) and type 316L (S31603) (Ref 74, 75). Type 304L stainless steel was traditionally used in white waters of lower chloride content, but is not commonly used in new paper machines. Type 316L stainless steel is the standard stainless steel used in modern paper machines, particularly those where chloride ions are present in significant concentration (>100 ppm). Type 316L stainless steel contains molybdenum that confers a resistance to chloride and thiosulfate pitting and crevice corrosion. Stainless steels having even higher molybdenum contents have been used as materials of construction in particularly corrosive paper machine white waters that result from high degrees of recycled water or system closure (Ref 76, 77). Specification of a corrosion-resistant grade of stainless steel is not in itself sufficient action to ensure that the paper machine is free from corrosion. Cleaning of the stainless steel during fabrication is essential to avoid contamination of surfaces. Iron oxides, heat tint, weld spatter, and slag from welding can all impair the corrosion resistance of the underlying stainless steel. Embedded dirt or iron, from inappropriate use of carbon steel wire brushes for weld cleanup, may provide sites for pitting to initiate on the stainless steel surfaces. It may be necessary to passivate the stainless steel surface with a nitric acid solution while in service in order to dissolve embedded iron while leaving the stainless steel intact. Pickling solutions are much more aggressive than passivating solutions, since they typically contain fluoride in addition to the nitric acid, and are used to remove both heat tint and surface iron contamination. However, pickling solutions can also dissolve the stainless steel metal surface leaving it slightly rough.

Stock Piping. Paper machine stock piping is typically thin-gage (schedule 10) type 304L or 316L austenitic stainless steel. Corrosion is generally not a problem in stock lines if the correct stainless alloy, weld filler metal, welding process, and cleaning procedure have been chosen. Corrosion problems may occur at weld joints with incomplete penetration or mismatch, where surfaces have become roughened, or where weld projections exist that provide sites for stock hang-up. Crevice corrosion or microbiologic attack is liable to occur under deposits (Ref 78). Fatigue cracking problems are often associated with welding defects such as incomplete penetration in stock piping. Great care must be taken to ensure that good quality full-penetration welds are realized. For root passes, inert gas purging is preferred over so-called antioxidation fluxes.

Headboxes may also experience internal mechanical damage (e.g., from rectifier roll failure), and there is always the possibility of stray-current damage in addition to being susceptible to corrosion damage (Ref 79). Stock hang-up on rough or corroded surfaces may result in the frequent release of pulp pads causing product quality problems or even sheet breakage. Clean, smooth, surfaces inside the headbox are essential. Pickling, passivation, and buffing with abrasives are commonly used to provide a smooth surface finish. Electropolishing improves the finish for surfaces of 0.4 μm (15 μin.) and above (Ref 80). Headboxes have typically been constructed using type 316L stainless steel, although stainless steels with higher molybdenum content such as type 317L have also been used where corrosion resistance is a concern.

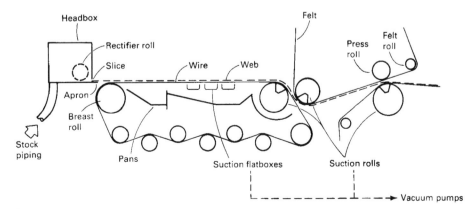

Fig. 14 Wet end of a fourdrinier paper machine

A calcium deoxidized machinable version of type 316L stainless steel is sometimes used in headboxes because of its superior machinability and electropolishing properties. Machinable stainless steels have greater freedom from macroinclusions than do conventional grades, but the microinclusions provide more sites for initiation of pitting corrosion. Machinable stainless steels have less corrosion resistance than conventional grades of stainless steel with an equivalent molybdenum alloying content.

Headbox apron and slice lips are particularly prone to pitting corrosion attack. Corrosion pits that initiate outside the liquid flow may grow in size and eventually reach the edge where they will have a detrimental effect on sheet formation. The use of highly pitting-resistant stainless steel, such as the 6% Mo grades, has been successful in preventing pitting of slice and apron lips.

Paper Machine Wires. Before the advent of synthetic plastic mesh fabrics, paper machine wires were typically phosphor bronze screens. Bronze screens are still used on a few older machines. Bronze wire life is limited by wear and corrosion attack (Ref 81). Corrosion inhibitor additions can increase the service life of bronze wires (Ref 78, 82). Sacrificial galvanic corrosion of the wires can result when the bronze screens come in contact with a bronze couch roll, effectively providing cathodic protection to the couch roll. The change to synthetic fabrics has resulted in the loss of this "cathodic protection" to bronze couch rolls (Ref 83).

Wet-End Structures. Wet-end carbon steel structural members are typically protected by using either paint coatings or stainless steel cladding (Ref 84). Where coatings fail or where white water can gain access beneath stainless steel cladding (at incomplete seal welds or perforations), the underlying structural steel may corrode away rapidly. Industry experience with wet-end coating systems involving inorganic zinc-rich primers and epoxy-polyamide top coats has not been good. Immersion service coating systems such as organic zinc/amine adduct-cured epoxy and inhibited epoxy/amine adduct-cured epoxy systems both appear to offer a longer service life (Ref 85, 86).

Vacuum Pumps. Liquid ring vacuum pumps typically have ductile cast iron rotors and gray cast iron casings. Corrosion of the rotors and casings occurs when corrosive seal waters are used. Ideally, the seal water should be cool freshwater. Paper machine white water is considered unsuitable for use as seal water due to its high temperature and low pH. Even freshwater may become corrosive if the flow rate is so low that the temperature of the water increases inside the pump or if it becomes contaminated with white-water carryover. A discharge seal water temperature of greater than 50 °C (120 °F) and/or a pH of less than 4.5 can cause accelerated cast iron corrosion. Corrosion rates may be lowered by adding a small amount of caustic to the seal water. High-discharge seal water conductivity (in excess of 200 μS) may indicate excessive white water-carryover. Accelerated impeller corrosion can occur at rotor tip speeds above 27 m/s (80 ft/s) (Ref 87). The presence of sand or grit in the water will cause erosion that shortens the life of liquid ring pumps.

A residue of graphite is left behind as cast irons corrode. Pearlitic ductile iron forms a more protective graphite surface layer than ferritic ductile iron does since the three-dimensional carbonaceous residue from pearlite dissolution acts as a binder for the graphite. Rapid liquid ring vacuum pump corrosion can be expected if environmental, velocity, or erosion conditions are such that the protective graphite film is removed. Liquid ring vacuum pumps have been built using stainless-lined bodies and stainless steel rotors for service in particularly corrosive white waters.

Roll Journals. Corrosion fatigue cracking results from the combination of cyclic stresses and a corrosive environment. Corrosion fatigue cracking failures occur after fewer stress cycles and at lower stress levels than failures due to pure mechanical fatigue. Small-diameter journals for felt rolls in wet-end environments are susceptible to corrosion fatigue, particularly if the machine speed has been increased significantly from its original design speed. Corrosion fatigue cracking may also occur at abrupt changes in the section size of journals. The environment promoting corrosion fatigue may be produced by splashing white water, the formation of damp pads of pulp on the journals, or even the humid atmosphere of the machine. Felt roll journals may also be exposed to acidic felt cleaning chemicals.

Improved corrosion fatigue performance can be achieved by protecting journals from their environment, by incorporating a redesign to increase their diameter, by removing step changes in section size, and by polishing to remove surface flaws (Ref 88).

Proper materials selection can ensure better resistance to corrosion fatigue cracking. Roll journals have been constructed using medium- or high-carbon steels, low-alloy steels, cast irons, and various grades of stainless steel. Fatigue strength is generally proportional to tensile strength. The most commonly used materials for roll journals are high-strength low-alloy steels such as G41400 (SAE 4140) and G43400 (SAE 4340). Austenitic stainless steels have relatively poor corrosion fatigue resistance and are not commonly used for roll journals in wet-end environments. Duplex stainless steels such as alloy 2205 (S32205) and precipitation-hardened stainless steels such as 17-4 PH (S17400) (in condition H1100) offer better corrosion fatigue resistance than more conventional journal materials. The use of integral (one-piece) ductile cast iron roll heads and journals having a generous radius between the head and journal may also improve resistance to corrosion fatigue. No roll journal material is immune to corrosion fatigue since there is no fatigue limit below which an indefinite service life could be expected.

Wet-end roll journals can be physically protected from exposure to white water. Stainless steel shields, paint coatings, and thermal spray coatings have all been used for journal protection. The disadvantage of physical barriers is that they hinder the inspection of the roll journal surfaces for cracking. Dry-end (reel spool) journals are usually made from 4140 or 4340 low-alloy steels in the quenched-and-tempered condition. Type 4340 is often preferred for larger section sizes since it is hardened deeper.

Suction roll shells operate in a corrosion fatigue environment without a safe fatigue limit. All suction roll shells will have a finite service life regardless of the material of construction. Corrosion-related failures of suction roll shells are discussed in the section "Suction Roll Corrosion" in this article.

White Water

White water is the aqueous medium in which the cellulose fibers are suspended. White waters contain appreciable amounts of dissolved organic and inorganic compounds, some of which contribute significantly to white-water corrosivity.

Closure of papermaking systems occurs when white water is recycled instead of being discharged to the environment. Closure results in increased concentrations of dissolved inorganic and organic solids, a decrease on the pH, and an increase in temperature. All these factors are known to increase corrosivity. The degree of closure can be expressed as the volume of water consumed per ton of paper produced (Ref 89). Closed-system paper mills may discharge less than 5 m^3/tonne (195 ft^3/ton), whereas an open-system paper mill may discharge more than 40 m^3/tonne (1560 ft^3/ton).

Closure is considered detrimental for corrosion. However, there is no clear parallel between increasing closure and increasing paper machine corrosion (Ref 89, 90). Accelerated corrosion may be expected for materials that undergo general corrosion attack, such as carbon steels, cast irons, and copper alloys. Increasing closure may decrease the "margin of safety" for stainless steels. The margin of safety for a stainless steel is the difference between the pitting corrosion breakdown potential and the free corrosion potential of the stainless steel. The grades of stainless steel having higher contents of molybdenum have larger margins of safety and therefore are more resistant to pitting corrosion. The molybdenum content of modern type 316L stainless steel is often only 2.01% Mo, which is at the low end of the specified range of 2% to 3% Mo for this alloy (Ref 91). Older versions of type 316L stainless steel tended to contain molybdenum toward the higher end of this range, typically at least 2.6%. New paper machines made using modern type 316L stainless steel may be less corrosion resistant than paper machines made using the older type 316L.

Composition. Paper machine white waters can vary widely in composition. The corrosivity of white water depends primarily on such parameters as the pH, temperature, and the concentrations of inorganic anions such as chloride and thiosulfate. The inorganic compounds found in white waters come from the wood, the pulping process, bleaching or brightening processes, and the chemicals used in the wet end of the machine, such as fillers, sizing agents, retention aids, defoamers, slimicides, and dyes. Seawater absorbed by logs during ocean transport in floating rafts will result in high concentrations of chloride ions. Kraft pulping will introduce sulfates. Chemi-thermomechanical pulping may add sulfites. Chlorine dioxide and hypochlorite bleaching will introduce chloride ions. Unreacted oxidants such as chlorine dioxide can be carried over onto the paper machine with the bleached pulp. White waters may also contain decomposition products of sodium hydrosulfite used in the brightening of mechanical and thermomechanical pulps. The average composition of the white water in a West Coast newsprint mill using semibleached kraft, chemi-thermomechanical pulp, and groundwood is:

Temperature, °C (°F)	48 (118)
pH (initial)	4.6
pH (on standing)	3.0
Sodium chloride (NaCl), ppm	311
Sodium sulfate, ppm	855
Sodium thiosulfate, ppm	34
Sodium sulfite, ppm	75
Conductivity, μS	1749

The contribution of each of the aforementioned parameters on white-water corrosivity is discussed below.

Temperature. Increased temperature due to closure contributes to energy savings and also improves drainage in the wet end of a paper machine that permits an increase in machine speed. However, higher temperatures are detrimental for corrosion. As a rule of thumb, the rates of most chemical reactions (including corrosion reactions) double for each 10 °C (18 °F) increase in temperature. Increases in corrosion rates of this magnitude probably apply to general corrosion of the carbon steels, cast irons, and copper alloys. Stainless steels have a "critical pitting temperature." Pitting corrosion of stainless steels will initiate above this temperature. This temperature is dependent on the alloy and on the chloride concentration in the white water. Stainless steels with higher molybdenum content have higher critical pitting temperatures. A higher chloride concentration lowers the critical pitting temperature.

pH. In acid papermaking the white-water pH is typically in the range from 4 to 6. The pH is controlled on the acid side by alum additions and/or by souring the white water with sulfur dioxide, sulfite, or sulfuric acid.

Below a pH of about 4 the general corrosion rates of many nonpassive metals are accelerated. This is due to hydrogen ions becoming so plentiful that the reduction of the hydrogen ion to form hydrogen gas replaces the reduction of dissolved oxygen molecules as the predominant cathodic half reaction in the corrosion process. Under acid conditions, corrosion is no longer limited by the rate of diffusion of dissolved oxygen into the metal. This decreased pH also lowers the stability of passive films on stainless steels, making them more susceptible to the initiation of pitting. Low pH conditions also increase the probability that once a pit is initiated it will continue to grow.

If white water is left standing, as may occur during a shutdown, the white-water pH may decrease from 0.5 to 2.0 units within approximately 24 h. There are many reasons for this. One possible mechanism is the continued oxidation of hydrosulfite and bisulfite that generates free hydrogen ions.

Chlorides. Transporting logs in seawater to coastal mills may induce chloride concentrations in the process streams that are orders of magnitude higher than those found in inland mills. Since chloride ions are very soluble, they tend to persist through the pulping process and eventually end up in the white waters. Increasing chloride concentrations increase the general corrosion rate of carbon steels and may also increase the corrosion rate of copper-base alloys. Chlorides are also agents for pitting corrosion, crevice corrosion, and stress-corrosion cracking (SCC) of austenitic stainless steels. Stress-corrosion cracking of conventional austenitic stainless steels usually requires temperatures above 60 °C (140 °F). Stress-corrosion cracking increasingly becomes a problem in mills with a high degrees of closure and where white-water temperatures have crept upward.

Sulfates. Sulfate ions (SO_4^{2-}) appear in white water as a residual chemical from the kraft chemical pulping process. Sulfates can also come from seawater, alum, sulfuric acid additions, and the oxidation of other sulfur-containing ions, such as bisulfite (HSO_3^-), thiosulfate ($S_2O_3^{2-}$), and hydrosulfite ($S_2O_4^{2-}$). Sulfate is the thermodynamically stable oxidation state of sulfur in white waters. The prolonged contact of white water with air will result in the conversion of all lower-oxidation state forms of sulfur to the stable sulfate form.

Sulfate is not an aggressive anion, and it can act as an inhibitor for chloride-related pitting. The inhibiting effect is particularly pronounced if the molar concentration of sulfate is in excess over the molar concentration of chloride ions (Ref 91, 92). On the other hand, sulfate ions make a significant contribution to the white-water conductivity. Sulfate tends to increase the rate of electrochemical corrosion reactions by facilitating charge transfer electrolyte in electrochemical corrosion reactions. Sulfate ions are also consumed by sulfate-reducing bacteria that result in microbiologically influenced corrosion (MIC) of paper machine equipment. Sulfate can also enhance thiosulfate pitting in white waters containing thiosulfate ions (Ref 93).

Thiosulfates. Mills that use hydrosulfite brightening have appreciable amounts of corrosive thiosulfate ions in their white waters as the result of anaerobic decomposition of warm hydrosulfite solutions during storage. Hydrosulfite is very rapidly and stoichiometrically converted to thiosulfate under nonoxidizing conditions. By contrast, the oxidation of hydrosulfite in air does not produce thiosulfate. In practice, adding hydrosulfite to meet target brightness levels serves only to introduce even more thiosulfate into the white-water system (Ref 94). The kinetics of the oxidation of thiosulfate to sulfate is so slow that appreciable thiosulfate concentrations can build up.

Thiosulfate concentrations in white water can be reduced by ensuring that the concentration in the original source is as low as possible. Hydrosulfite solutions stored at 2 to 9 °C (36 to 48 °F) and/or the addition of an alkaline stabilizer to maintain a pH of 10 or more minimizes thiosulfate formation (Ref 95). Brightening at a pH level over 5 will also help to keep thiosulfate levels down. Thiosulfate ions have been found to be particularly aggressive pitting agents for austenitic stainless steels (Ref 93, 96). Thiosulfate ions also contribute to rapid general corrosion of copper-base alloys (Ref 97, 98).

Sulfites. Sulfite additions are used in chemi-thermomechanical pulping and are also made elsewhere to control pH or to brighten stock. Bisulfite (the stable form of sulfite in white water) is also produced during the decomposition of hydrosulfite brightening solutions. Sulfites are readily oxidized to sulfate in the presence of dissolved oxygen. Sulfite is not considered to be aggressive; in fact, sulfites are often used as oxygen-scavenging inhibitors in other aqueous systems as in boiler feedwater systems. Sulfites should have inhibitive properties in paper machine white water. Very high concentrations of bisulfite (>200 ppm SO_3^{2-}) in white water may result in sulfur dioxide evolution that in turn can cause atmospheric (dew-point) corrosion problems on aluminum paper machine dryer hoods.

Conductivity. Higher conductivity is generally an indication of increased corrosivity in white waters.

Corrosion Mechanisms

Corrosion is an electrochemical process. Corrosion reactions are the net sum of anodic (oxidation) and cathodic (reduction) partial reactions. The corrosion of iron in waters of near-neutral pH is the sum of an anodic dissolution of iron and a cathodic reduction of dissolved oxygen:

$$2Fe \rightarrow 2Fe^{2+} + 4e^- \text{ (oxidation reaction)}$$

$$O_2 + 2H_2O + 4e^- \rightarrow 4OH^- \text{ (reduction reaction)}$$

$$2Fe + O_2 + 2H_2O \rightarrow 2Fe^{2+} + 4OH^-$$
(net reaction)

The corrosion rate of iron in near-neutral waters is limited by the solubility of oxygen in water and the rate of diffusion of dissolved oxygen to the metal surface.

The corrosion of iron in more acidic waters (pH < 4) produces a reduction in hydrogen ions:

$$Fe \rightarrow Fe^{2+} + 2e^- \text{ (oxidation reaction)}$$

$$2H^+ + 2e^- \rightarrow H_2 \text{ (reduction reaction)}$$

$$2Fe + 2H^+ \rightarrow Fe^{2+} + H_2 \text{ (net reaction)}$$

The rate of the corrosion reaction in acid solutions is not diffusion limited and can proceed at a rapid rate.

Chloride Pitting and Crevice Corrosion. Stainless steels rely on the formation of a stable passive film for immunity to corrosion in paper machine white waters. The passive film is thin, is highly resistant to dissolution, and protects the underlying stainless steel from corrosion. Oxidizing conditions must be present for the passive film to form and be maintained. Dissolved oxygen from the air is usually sufficient to maintain stable passivation. Stainless steels can withstand slightly reducing conditions in flowing white waters without suffering appreciable corrosion attack. Excessively oxidizing conditions, resulting from the presence of oxidizing agents such as chlorine dioxide, may result in the corrosion potential of the stainless steel exceeding its breakdown potential resulting in a localized breakdown of the passive film. The passive film breakdown occurs only in isolated locations such as weak spots in the film due to the presence of inclusions or other defects in the underlying metal. These isolated locations of breakdown develop as pits. Pitting attack is favored by certain environmental conditions:

- High chloride concentration
- High temperature
- Oxidizing conditions
- Low pH
- Stagnant or low-velocity conditions

A newly initiated corrosion pit tends to repassivate unless the conditions for its initiation are maintained (corrosion potential above the breakdown potential). Figure 15 is a schematic diagram of the pitting process.

If the pits do not repassivate they will become anodic with respect to the surrounding metal. Once a pit has become established, the driving force for its growth is the development of an oxygen concentration cell. The surrounding metal surface remains passive because it has ready access to dissolved oxygen. Reduction of dissolved oxygen on the surrounding metal is the cathodic half of the net corrosion reaction. Anodic dissolution of metal in a pit can occur at a very high rate since the surface area of the pit is much smaller than that of the surrounding metal surface. Subsequent hydrolysis of dissolved metal complexes inside the pit results in the generation of free hydrogen cations. Anions diffuse into the pit from the outside by the process of electromigration to maintain electron neutrality of the solution in the pit. Of the available anions in the white-water solution, chlorides are the most mobile and can enter the pit more readily than larger anions such as sulfate. Generation of hydrogen cations due to metal ion hydrolysis and the electromigration of chloride anions into the pit can produce an environment inside the pit that resembles hydrochloric acid more than it does white water. Pit growth may then accelerate as acid corrosion.

Once pitting corrosion has become established, it is difficult to stop. Lowering the corrosion potential below the breakdown potential by eliminating the presence of residual bleach plant oxidants may not stop pitting corrosion until a substantially lower potential value is reached (the repassivation potential).

Crevice corrosion is a form of localized corrosion that occurs beneath deposits or in other areas shielded from direct contact with dissolved oxygen in the white water. Crevice corrosion resembles pitting corrosion (both are sustained by oxygen concentration cells), but crevice corrosion initiates more readily than pitting corrosion because the conditions for an oxygen concentration cell already exist with a crevice. The metal inside a crevice has difficulty maintaining passivation due to restricted access to dissolved oxygen. The inside of the crevice becomes anodic (metal dissolution) while the surrounding metal becomes cathodic (oxygen reduction).

White-water system closure can result in chloride concentrations and temperatures exceeding the values required for localized corrosion breakdown. Resistance to pitting corrosion of new or replacement equipment can be accomplished by selecting stainless steel having higher molybdenum content. Type 304L stainless steel contains no appreciable molybdenum and is highly susceptible to localized corrosion in closed white-water systems. Type 316L stainless steel has a specified molybdenum range of 2 to 3%, although modern type 316L typically contains molybdenum at the very bottom end of this range. Type 317L stainless steel has a specified molybdenum content of 3 to 4% and is more resistant to localized corrosion than type 316L stainless steel.

Thiosulfate Pitting. Thiosulfate is an aggressive pitting agent especially for type 304L stainless steel. Thiosulfate pitting can only occur below the potential for thiosulfate reduction, unlike chloride pitting that occurs only above the passive film breakdown potential (Ref 93, 96). A reduction of thiosulfate in the presence of hydrogen ions produces an adsorbed sulfur monolayer on the metal surface. The adsorbed sulfur activates the anodic dissolution of the metal and hinders repassivation. Excess hydrogen ions must be present for acidification of the pit; further, there must also be a larger amount of inert ions (sulfate and chloride) that can be transported into the pit to meet charge transfer requirements and take part in metal-complexing hydrolysis reactions. The worst case of thiosulfate pitting occurs within the molar concentration ratio:

$$((Na_2SO_4 + NaCl)/Na_2S_2O_3) = 10 \text{ to } 20$$

Above this range (ratio > 20) there is insufficient thiosulfate to reach the pit nucleus. Below this range (ratio < 10) there is too much thiosulfate reduction to bisulfite, which prevents acidification of the pit.

Once formed, thiosulfate pits are very stable and not subject to spontaneous repassivation. Scratches encourage the initiation of pits. Sensitization makes the heat-affected zones of welds particularly susceptible to thiosulfate pitting. A few large pits tend to form in thiosulfate pitting rather than many small pits that form in chloride pitting. Type 316L stainless steel is the minimum grade that should be used for white-water service where high thiosulfate levels may exist. In practice, it appears that corrosion problems can be mitigated if thiosulfate levels are controlled below 5 and 10 ppm for equipment made using types 304L and 316L stainless steel, respectively.

Microbiological Corrosion. Paper machine white waters contain nutrients that can sustain bacterial growth particularly if the pH is near neutral. White-water temperatures within the range of 40 to 50 °C (105 to 120 °F) also favor microbiological growth. Higher temperatures may prevent the growth of some forms of bacteria, although the increased temperatures can increase the metabolism of those bacteria that can adapt to heat. The result is the formation of slimes (Ref 78, 99).

Stock and white-water flow systems are designed to minimize slime accumulations. Surfaces are polished and weld projections removed to prevent pulp hang-ups. MIC can occur wherever slime deposits develop (Ref 100). Once a slime deposit has grown to a sufficient thickness to exclude oxygen, a colony of sulfate-reducing bacteria (*Desulfovibrio desulfuricans*) can become established. Enzymes produced by these anaerobic bacteria catalyze the reduction of sulfates resulting in the formation of free sulfide ions. Chemically reducing conditions quickly develop that result in the depassivation of the stainless steel surface beneath the deposit. Active corrosion, in the form of large shallow pitting, then takes place. The pits are characteristically covered with a black crust. Pitting perforations through stainless steel equipment due to MIC are usually small

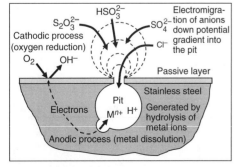

Fig. 15 Pitting corrosion process in paper machine white water

since the entry of oxygen at a leak will stop the activity of sulfate-reducing bacteria.

Suction Roll Corrosion

Max D. Moskal, Mechanical and Materials Engineering

Suction rolls are used to remove water from paper at the wet end of the paper machine (Fig. 14). The paper web is passed through a roll nip, one roll of which is the suction roll. The suction roll is drilled with a pattern engineered to provide the required water removal and suction roll shell strength. A vacuum is applied to the inside of the roll to draw the water through the web and into the roll. Rubber or plastic coatings are used in either the suction roll or the mating roll to give the desired nip pressure for optimized water removal. Different paper machine designers will use suction rolls in different locations with water-removal schemes designed for their specific machine configuration.

Suction rolls are part of the paper machine and are usually drilled to about 20% open area. They are used to remove water from paper at the wet end of the paper machine. Prior to 1950, suction rolls were cast from copper alloy C83600 (85Cu-5Sn-5Pb-5Zn). As paper machine width, speed, and nip pressure increased, stronger alloys were introduced: nickel aluminum bronze, forged stainless steel type 410 (S41000), and centrifugally cast stainless steels CA-15 (J91150), type CF-3M (J92800), and CF-8M (J92900).

Suction roll speeds and nip pressure continued to increase, and failures of these roll materials led to the introduction of special grades of cast and wrought duplex stainless steels. The composition of suction roll alloys is given in Tables 5 to 7. Tables 5 and 6 show currently available suction roll alloys. Table 7 shows alloys, many of which are still in use, that have been discontinued by producers in favor of the alloys shown in Table 6. In the selection and performance of suction roll materials, corrosion resistance and fatigue strength in paper machine white water, nip pressure, roll width, and residual stress level all play important roles.

Corrosion

Corrosivity of the white-water environment is largely predetermined. However, the papermaker can still exert some control. Following the practices outlined in Ref 101 can reduce sulfate ion contamination and residual hydrosulfite, which can be especially damaging.

Freshwater showers of the suction rolls provide more effective cleaning and also reduce corrosive effects. Good maintenance of the roll is an important factor in minimizing corrosion. Drilled holes must be kept clean of deposits by the use of needle showers and periodic off-machine cleaning. A careful program of biological control should be maintained. The papermaker should also avoid the use of cleaning chemicals (such as muriatic acid) that are known to damage shell alloys and roll covers.

Operating Stresses

Both the applied stress and internal residual stresses (inherent in the shell material) contribute to the stress in operation of the suction roll. The responsibility for the level of applied stress rests chiefly with the roll designer. Each shell material will have an upper limit of applied stress that will give long life in a particular environment. The magnitude of this stress is a function of roll configuration and applied loads.

Drilled hole patterns are designed to reduce the effects of applied stress. The most important factor influencing the applied stress is shell thickness. Applied stress decreases as the shell

Table 5 Nominal composition of currently available copper-base suction roll alloys

Material designation(s)	UNS	Composition(a), wt%							
		Cu	Sn	Pb	Zn	Al	Ni	Fe	Mn
1N Bronze(b) GC-CuSn5ZnPb	C83600	84–86	4–6	4–6	4–6	...	1.0 max	...	...
GC-CuAl9.5Ni(c)(d)	C95810 (mod)	81.5–83.5	...	...	...	9.0–9.5	4.0–4.5	3.0–3.5	0.5–1.0
GC-CuSn10Zn(c)	C90500	86–89	9–11	1.5 max	1–3	...	1.0 max	...	...
GC-CuSn10(e)	...	88–90	9–11	1.0 max	...	...	...	...	...

(a) Composition range; additional elements may be added or present in minor amounts. (b) Trade name Sandusky International. (c) Trade name Kabelmetre Alloys. (d) Size currently limited to 3.5 m (11.5 ft) maximum length. (e) Discontinued

Table 6 Nominal composition of currently available stainless steel suction roll alloys

Material designation(s)	Composition(a), wt%						
	C	Cr	Ni	Mo	Cu	Mn	Si
Austenitic							
CF-3M (J92800)	0.02	17.7	13.8	2.3	...	1.3	0.8
Martensitic							
CA-15, C-169	0.07	12.4	0.6	0.5	...	0.5	0.6
PM-4 1300M (J91150)	0.1	13.0	1.0	2.1	...	0.9	0.7
Duplex—centrifugally cast							
Alloy 86, alloy EPV	0.02	26.0	6.8	...	2.0	0.8	0.7
ACX-100(b)	0.02	24.0	5.7	2.4	0.5	0.8	0.8
ACL-105, KRC-A894	0.02	22.5	4.5	1.5	...	0.6	0.6
KCR-110	0.02	21.0	3.2	0.7	...	0.8	1.0
Duplex—rolled and welded							
3RE60 SRG(c)	0.02	18.5	5.0	2.8	...	1.5	1.5
2205 SRG Plus(c)	0.02	22.0	5.2	2.9	...	1.5	0.7
2304 AVS(c)	0.02	22.7	4.7	0.3	...	1.5	0.8
Duplex—powder metallurgy							
Duplok 27(d)	0.02	26.5	6.5	3.0	...	0.7	0.4
Duplex—forged							
PM-2-21-6MC	0.06	21.5	5.0	0.5	1.0	0.5	0.6

(a) Typical composition; balance of composition is iron; other elements may be added or present in minor amounts. (b) Contains nitrogen and cobalt. (c) Contains nitrogen. (d) Hot isostatic pressed; contains 0.3% N

Table 7 Nominal composition of discontinued stainless steel suction roll alloys

Material designation(s)	Composition(a), wt%						
	C	Cr	Ni	Mo	Cu	Mn	Si
Austenitic							
CF-8M (J9290)	0.05	17.7	13.8	2.3	...	1.3	0.8
PM-3-1811-MN	0.015	16.5	13.5	2.0	...	1.6	0.5
Martensitic							
DSS-69	0.04	12.4	4.0	0.7	...	0.7	0.6
A-70	0.03	11.9	4.0	1.5	...	0.8	0.5
Duplex—centrifugally cast							
A-63	0.05	21.8	9.4	2.7	...	0.8	1.3
A-75	0.02	26.0	6.8	...	...	0.8	0.5
VK-A170	0.07	23.3	10.7	2.1	...	0.7	1.5
VK-A171	0.07	22.2	8.3	1.2	...	0.8	1.1
VK-A271	0.06	24.6	4.3	0.7	...	0.7	1.3
VK-A378(b)	0.05	20.0	5.0	2.0	3.0	0.7	1.0
KCR-A682(c)	0.06	18.0	5.5	2.3	3.2	0.7	0.7
Duplex—forged							
PM-3-1804M	0.06	17.9	4.0	2.0	...	0.6	0.6
PM-3-1808N	0.08	18.0	9.0	...	...	2.0	1.0
PM-2-22-5	0.07	26.0	4.0	0.8	...	1.2	1.3

(a) Typical composition; balance of composition is iron; other elements may be added or present in minor amounts. (b) Contains nitrogen and tungsten. (c) Contains niobium

thickness increases. A stress calculation procedure has been developed by the Technical Association for the Pulp and Paper Industry (TAPPI) to use as a guide for determining stresses in suction rolls (Ref 102), although there is little agreement among roll manufacturers as to how to make this calculation. The TAPPI procedure was recently modified so that stress calculations can be made on a personal computer. While the TAPPI procedure still needs to be refined, it provides mill personnel with a uniform method for determining minimum thickness requirements and drill patterns and for evaluating roll failures regardless of the roll manufacturer.

Gun-drilled or reamed holes for improved surface finish also reduces corrosion and improves the cleanliness of drilled holes in service. Gun-drilling or reaming is usually not performed on bronze shells.

Residual stress that originates in the alloy and from the heat treatment is also an important factor contributing to roll life. Shells made from materials with known high residual stresses should be designed more conservatively or, preferably, not used at all.

Manufacturing Quality

There is little published information relating to suction roll manufacturing quality. Rolls should be inspected using liquid penetrant prior to drilling to detect flaws such as porosity, slag, and cracks (Ref 103). Skillful control of melting and casting procedures can minimize these defects. Weld repairs should be used with discretion and only before shell heat treatment. Type 1N bronze should not be weld repaired because of the hot shortness that develops in the alloy. Careful attention should be given to drilling and machining quality and dimensional control. Industry-wide standards giving acceptance criteria for casting flaws and drilling quality have not been developed.

Material Selection

When a new or replacement suction roll is contemplated, the question of material selection is of primary importance. Selection of a new alloy should involve an evaluation of the existing roll condition and the factors contributing to its deterioration. Changes in the corrosion environment that have occurred on the paper machine should also be evaluated. The temptation is to select an identical alloy if the previous material served for 15 or 20 years. However, it is likely that changes in the machine environment have occurred during that period and that the life of a new roll of the same alloy could be substantially reduced. Furthermore, the roll material in use may have been superseded by an improved alloy and no longer be in production. Table 7 lists suction roll alloys that have been discontinued by manufacturers.

Most suction roll vacuum boxes on older paper machines were manufactured using epoxy painted cast iron. More recently, manufacturers have used weld fabricated type 316L (S31603) stainless steel suction boxes to better resist corrosion damage.

In-Service Inspection

Periodic inspection of the suction roll is a critical step that helps extend roll life and minimize unexpected shutdowns. Rolls should be removed from the machine, disassembled, and thoroughly cleaned, then inspected. The inspection interval is determined by local experience and is normally about once a year. Inspection should include an evaluation of drilled hole cleanliness, fatigue cracking of the shell, and rubber cover deterioration or detachment. A careful visual examination is made for corrosion and cracks. Water-washable liquid penetrant has been successfully used to detect cracks on the inside and outside surfaces.

When cracking is observed during an annual inspection, immediate consideration must be given to replacement because of the long lead time required for shell manufacture. Consideration should also be given to inspecting the roll more often than once per year to observe the rate of crack growth. A careful record must be made whenever cracks are observed. TAPPI TIP 0402-19 (Ref 104) provides guidelines for suction roll inspection. TAPPI TIP 0402-01 (Ref 105) provides a standard form for documenting inspection results and roll failures.

Corrosion Control in Chemical Recovery

David Bennett, Corrosion Probe Inc.
Craig Reid, Acuren

This section addresses corrosion and chemical recovery associated with kraft pulping liquors. As mentioned previously in this article, the kraft process is the most commonly used pulping process used in North America.

Black Liquor

Black Liquor Evaporation. The flow diagram of the chemical recovery system of a typical kraft mill is shown in Fig. 16. The cyclic recovery processes recover the inorganic chemicals (white liquor) used to cook or digest the wood chips. First the black liquor is evaporated so that it is dry enough to be burned. Second, the reduced, molten smelt produced in the recovery boiler is dissolved in water to make green liquor. This is then causticized to produce the white liquor. Organic chemicals dissolved out of the wood are the fuel portion of the black liquor fired in the recovery furnace. This combustion also raises steam for energy generation and for process use.

After separation from the pulp in brown stock washers, weak black liquor at a concentration of about 15% solids is pumped to a multiple-effect evaporator (MEE) in which solids are concentrated to at least 50%. Further evaporation of the black liquor in a concentrator or direct contact evaporator increases solids content still higher so it has enough heat value to be the primary fuel in the recovery furnace.

Black liquor inorganic constituents are mostly sodium carbonate and sodium sulfide with other constituents as listed for a typical composition for a southern pine sulfate kraft black liquor leaving the digester. The pH is 11.6 and the concentration is 23% solid:

Component	Composition, gm/L
Sodium sulfide	4.9
Sodium hydroxide	6.8
Sodium sulfate	2.0
Sodium carbonate	55.9
Sodium triosulfate	14.3
Sodium chloride	0.17

Pulp mills with a chemical recovery process utilize a series of MEEs similar to that shown in Fig. 17 to make the liquor concentrated enough for combustion in the recovery boiler. Multiple-effect evaporator systems vary as to the number of evaporators, concentrator systems, capacity, and so forth. Most evaporators are steam heated, using either shell-and-tube or falling-film heat-exchanger designs. Figure 18 shows a schematic of a falling-film unit.

Some kraft mills concentrate the black liquor to around 80% solids, but most mills concentrate the liquor to somewhere in the 65 to 75% solids range. The high temperatures associated with evaporating the black liquor above about 40% solids make stainless steels essential for the liquor-wetted surfaces. Stainless steel heat-exchanger surfaces are more or less universal in all evaporation stages.

Carbon steel may still provide satisfactory corrosion resistance in the bodies of the first two or three evaporation effects (these usually are the highest-numbered effects), especially where liquor temperatures are below 95 °C (200 °F). Although the exact mechanism for corrosion of carbon steel in black liquor is not fully understood, the tenacity of the self-protective passive film is a crucial factor.

Stainless steels, with their much more tenacious passivity, have a favorable history of resisting corrosion in all black liquor environments. As a result, stainless steels are used both to repair existing equipment where carbon steel has corroded and in new equipment where corrosion of carbon steel might be expected.

The MEEs from the third or fourth to the first effects and the concentrator are always made completely of stainless steel, most often type 304L (S30403). Many mills have used thermal

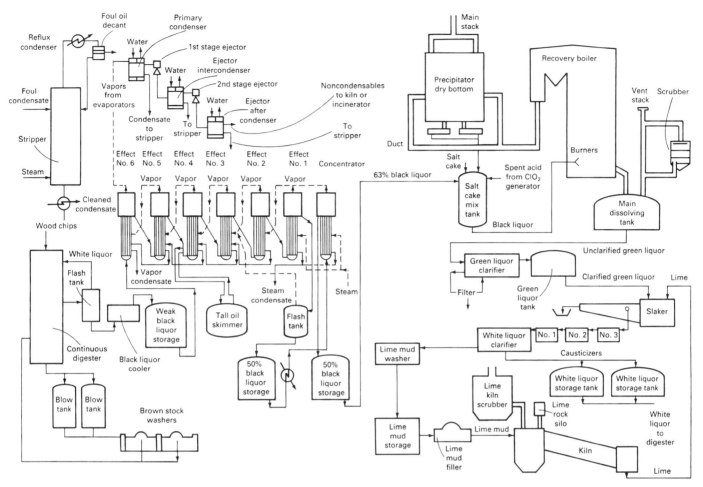

Fig. 16 Flow diagram of kraft pulping and chemical recovery

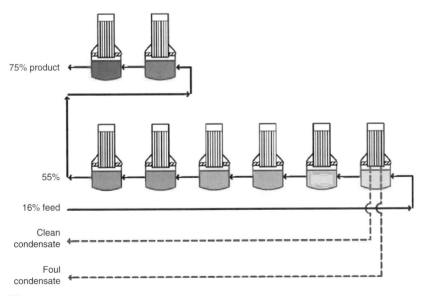

Fig. 17 Multiple-effect evaporator/concentrator design

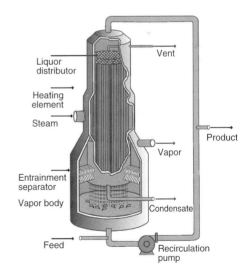

Fig. 18 Cross-sectional view of a falling-film black liquor evaporator and concentrator design

spray coatings or sheet linings of duplex or austenitic stainless steel to protect carbon steel components against liquor corrosion. It is sometimes necessary to protect the carbon steel tube sheets and vapor zone target areas in MEEs with corrosion barriers.

In liquors concentrated to higher than about 70% dry solids, type 304 and type 316 concentrator tubing is subject to both significant general corrosion and to caustic stress-corrosion cracking. Rapid, localized corrosion can develop in areas of higher flow velocity or increased turbulence. Caustic cracking is believed to occur because at these high dry solids environments, the temperatures, and caustic concentrations are entering ranges where both types 304L and 316L are subject to stress-corrosion cracking (SCC). The solution is to either use more highly alloyed austenitic stainless steels or to use duplex stainless steels that are more resistant to both caustic cracking and to corrosion in high turbulence areas.

Waste streams, such as sulfuric acid effluent from the tall oil plant and spent acid from chlorine dioxide operations—both are sources of sulfates for maintaining the sulfur balance—may be added to the evaporator black liquor stream before the first effect or concentrator. This sometimes makes liquors too corrosive for type 304L stainless steel, especially if the residual alkalinity in the liquor is eliminated by these additions. Higher alloyed austenitic steels are generally adequate for handling black liquors.

Stainless steels perform well as pumps, piping, tube sheets, liquor boxes, and surfaces subject to evaporator-related black liquor flow turbulence or splashing. Type 304L stainless steel is also widely used throughout the steam and condensate piping systems. This includes all vapor bodies, separators, condenser tubes, pumps, and piping.

This discussion has concerned sulfate (kraft) and soda (i.e., alkaline) spent liquor evaporators. Evaporators for acid (i.e., sulfite) spent liquors require type 316L (S31603) or higher molybdenum content stainless alloys for all liquor and vapor contacted surfaces. Evaporators for neutral pulping (sulfite semichemical), in which the liquors have lower but still alkaline pH and a higher concentration of sulfite ions, normally use type 316 (S31600) stainless steels for heat-exchanger tubing and other liquor contacted surfaces.

Black liquor oxidation is carried out in many mills as an odor-control measure. By oxidizing the liquor, hydrogen sulfide (H_2S) and mercaptan levels in evaporator condensate and in the noncondensable phases are reduced. Oxidation of sulfides in the liquor to thiosulfate also reduces the hydrogen sulfide content of the boiler flue gas.

As with all liquor handling equipment at elevated temperatures, tanks, piping and other components for black liquor oxidation are most often made of type 304L stainless steel.

Black Liquor Tanks. Black liquor may be stored at 30% solids, which is a typical liquor concentration in soap separation tanks, as well as in the more concentrated "heavy black liquor" form, which may be stored before or after the liquor concentrator.

Black liquors must be kept hot for easier pumping. This makes corrosion of carbon steel in black liquor service unpredictable. Stainless steel construction, most commonly with type 304L, can minimize corrosion in black liquor tanks and is essential to avoid flow-induced corrosion around agitators and in pumps and piping. There is also one report of SCC in the tidal zone of type 304L tanks storing 72% solids black liquor. The tanks are located in a mill processing seawater floated logs. In this same mill type 304L tanks storing 50% solids black liquor have shown no signs of corrosion or SCC after more than 15 years.

Stainless steel sheet linings can protect corroding surfaces in carbon steel tanks, provided the lining is properly designed to remain intact. Stainless steel linings also are used in carbon steel flash tanks on continuous digester systems.

A significant amount of welding is involved in properly installing a stainless steel lining because the dimensions of each lining panel must be small enough to minimize thermal expansion effects and avoid subsequent weld cracking. Another drawback of internal linings is the challenge of inspecting for corrosion behind the lining.

For new black liquor tanks, it is generally more economical to specify solid stainless steel construction instead of using clad material or lining a carbon steel tank. Clad material at the thicknesses involved (<15 mm, or 0.6 in.) is usually more expensive than solid stainless steel plate.

Storage tanks in a mill with high chloride levels in the heavy black liquor have experienced internal SCC and pitting in the wet-dry region of the walls. Testing showed that 6% Mo stainless steel adequately resisted the unusually aggressive environment in these tanks.

Appropriate inorganic corrosion-resistant linings can provide good chemical resistance to black liquor at all concentrations and temperatures.

Chemical Recovery Tanks

Smelt Dissolving Tank. The molten smelt is quenched and dissolved in a large tank with internal agitators—the smelt dissolving tank—to produce green liquor. This tank rumbles continuously, and the heat produced by heavy smelt inflows can cause significant hydraulic pulses. Most dissolving tanks have extra-thick walls and are made of carbon steel.

Internal corrosion protection is usually required because the agitated contents have a high content of abrasive, suspended solids. The continuous flows around agitators and the hydraulic pulsing help the heavy suspended solids content cause erratic, localized erosion-corrosion of carbon steel. Erosion-corrosion of carbon steel can also occur behind linings that are breached and are not sufficiently tightly attached.

There are three general ways to prevent corrosion in dissolving tanks:

- Stainless steel construction with weldable 12% Cr steel, type 304, or alloy 2205 (S32205) stainless steel
- Sheet lining with type 430 (S43000), type 304, or alloy 2205 stainless steel and proper lining design
- Lining with a reinforced, inorganic cementitious material

All stainless steels have good corrosion resistance in green liquor immersion. Duplex and ferritic stainless steels resist SCC above the liquid level better than types 304L or 316L. Solid stainless steel dissolving tanks are increasingly common. Pumps, agitators, and nozzles that regularly see flow conditions are made of stainless steel. Duplex stainless steels are more resistant than austenitic grades to abrasive erosion-corrosion, such as experienced by agitator blades, due in part to the higher hardness of duplex stainless steel.

Green Liquor Tanks. Carbon steel green liquor tankage corrodes most rapidly where air dissolved in the green liquor produces thiosulfate ions. This usually occurs—and limits the worst corrosion to—half a meter or so immediately below the liquor level. Fluctuating levels reduce thinning rates by spreading the corrosion over a wider area.

Stainless steels such as austenitic type 304L and duplex grades such as alloy 2304 (S32304) or alloy 2205 have excellent corrosion resistance in all green liquor service conditions, including pumps and piping. Newer duplex stainless steels with lower molybdenum and nickel contents are also likely to perform well in green liquor service. A 12% Cr stainless steel also is suitable for green liquor tank construction. Special care should be taken when using hydrochloric acid to remove carbonate scale buildup since corrosion of these stainless steels can occur even if the acid is inhibited.

Many carbon steel clarifiers and tanks in green liquor service have internal stainless steel linings or solid plate inserts to prevent corrosion. Solid type 304L stainless steel also is widely used for rakes, internal piping, and nozzles.

Slakers and Causticizers. Slakers are tanks in which calcined lime is converted to calcium hydroxide, or a lime solution. The lime solution is added to the green liquor in the causticizer tanks to convert the sodium carbonate to sodium hydroxide, making white liquor from green liquor.

Slaker and causticizer environments are strongly alkaline and have high levels of suspended particles. The abrasion-erosion effects of the suspended materials can cause rapid corrosion of carbon steel where protective scales or deposits do not form. The abrasion from these suspended materials normally has an aggressive "polishing" effect on stainless steels. Harder stainless steels such as the duplex grades or work-hardened austenitic grades are significantly more resistant to this abrasive "polishing" mechanism.

Slaking and causticizing tanks are typically constructed of stainless steel or of carbon steel with an inorganic lining, commonly a gunned, sulfate-resistant, portland cement-based material. Pumps and piping are the standard grades of stainless steel or duplex grades to better resist the abrasive conditions.

White Liquor Tanks. White liquor is clarified or filtered and then stored in tanks. Onset and rates of carbon steel corrosion are very unpredictable in white liquors. Aeration of white liquor produces intermediate, oxidized sulfur species such as thiosulfate or polysulfide. These chemical species affect the passivity of carbon steel and can increase corrosion, although polysulfide levels higher than about 2% help stabilize the passive surface layer, thereby reducing carbon steel corrosion.

Stainless steel is fully corrosion resistant to white liquor due to its significantly more tenacious passive film. Types 304 and 316 are subject to intergranular cracking in with liquor at temperatures above 70 °C (160 °F) when sensitized by welding. Sensitization can be avoided by specifying the low-carbon grades, for example, type 304L, or a duplex stainless steel.

Molybdenum-containing stainless steels such as types 316L and 317L (S31703) have measurably poorer corrosion resistance than 304L in white liquor environments. Duplex stainless steels, with their higher chromium content, have better corrosion resistance than 304L. They also perform better under erosion-corrosion conditions, where the higher hardness of the duplex stainless steels contributes to their superiority. There are also recently commercialized grades of duplex stainless steel with lower alloying content to reduce material costs but good corrosion resistance in white liquor service.

For liquor heaters, tubes of alloy 2205 stainless steel, 26% Cr ferritic stainless steel (e.g., S44660), and nickel (or nickel alloys) all outperform type 304, which can experience internal corrosion at temperatures above 150 °C (300 °F) and also experience SCC, especially from the steam side.

As mentioned previously for green liquor piping, acid cleaning austenitic stainless steel tubes in white liquor heaters with hydrochloric acid can damage the tubes. Pitting and SCC can occur with even short exposure to hot hydrochloric acid, even if the acid is inhibited.

Clarifier rakes, piping, pumps, valves, nozzles, and so forth for white liquor service are type 304L or alloy 2205 wrought stainless steel, or CF-3 (J92500) or CD-4MCuN (J93372) castings. As with wrought materials, duplex stainless steel castings have higher resistance to erosion-corrosion than austenitic stainless steel castings.

Additional Considerations for Tanks in Black Liquor, Green Liquor, and White Liquor Service

As stainless steel replaces carbon steel for these chemical recovery tanks, two important precautions are sometimes overlooked. First, when hydrotesting newly installed stainless steel tanks, it is necessary to drain the water used for hydrotesting or place the tank in service promptly to avoid microbiologically influenced corrosion (MIC). Microbiologically influenced corrosion has been reported in type 304L (and 316L) tanks in a number of cases where water used for hydrotesting was left in the tank for a month or more. It is important to drain the water completely since MIC can also occur in water left in the bottom of incompletely drained tanks. Microbiologically influenced corrosion has not been reported when tanks have been placed in service within a few days of hydrotesting.

Second, the exterior of type 304L tanks is subject to SCC under the insulation, especially when the insulation is regularly wetted, for example, by rain or leakage from nearby lines. Underinsulation SCC has been such a frequent occurrence that NACE International developed a Standard Recommended Practice: "The Control of Corrosion under Thermal Insulation and Fireproofing Materials" (RP 0198). This document outlines good practices to prevent under-insulation SCC. Duplex and more highly alloyed stainless steels are far less susceptible to MIC and chloride SCC. However, both draining promptly after hydrotesting and keeping the insulation dry are good anticorrosion practices to follow, regardless of the material of construction.

Lime Kiln and Lime Kiln Chain

Lime Kiln. Lime mud from the white liquor clarifier is washed and fed to the lime kiln, where moisture is driven off and calcium oxide is formed. This is added to the green liquor to make white liquor. The lime mud is conveyed into the chain section of the kiln by a short section of spiral flights welded to the kiln shell. Progressing into the kiln from the feed end, four zones of the chain section can be defined (Table 8). The carbon steel shell in the hot end of the kiln is protected with heat-resisting brick, but the kiln internals can be subject to corrosion and mechanically enhanced corrosion and cracking.

In the first zone, the kiln shell and the flights in the conveying section are subject to moderate abrasive corrosion and alloy 2205 duplex stainless steel has been used for the flights and to line the kiln shell to better resist the abrasion in zones 1 and 2.

Lime Kiln Chain. Heavy chains are used to collect heat from the hot exit gases and transfer the heat to the lime mud to evaporate the moisture from the mud. The chain section in the kiln can be 15 m (50 ft) or more long, so the chain nearest the burner is subject to both convective and radiative heat transfer.

Carbon steel and low-alloy steels can be used for the first two zones of chain if the metal temperature does not exceed 500 °C (930 °F). Cast or wrought stainless steel chain is required in the hotter zones where chain life is determined by scaling rather than by wear. Both austenitic and ferritic stainless steels are used. Selection is based on experience in a particular kiln and cost.

Austenitic stainless steels are considerably stronger than the ferritic stainless steels above about 540 °C (1000 °F). Type 304 stainless steel or its cast equivalent CF-8 (J92600) can be used in zones 2, 3, or 4. Special heat-resisting grades are also available for zone 4.

Type 304 stainless steel chain:

- Has reasonable oxidation resistance up to 815 °C (1500 °F)
- Can be exposed for 10,000 h at 650 °C (1200 °F) and still retain adequate room-temperature ductility
- Has a reasonable creep rate at 650 °C (1200 °F)

While type 304 is the basic alloy used for lime kiln chain, other alloys with better resistance to high temperatures are also used. Table 9 shows some aluminum- and silicon-alloyed ferritic stainless steels used for lime kiln chain.

The calcined lime discharged from the kiln passes through lump-crushers and grizzlies and is conveyed away from the lime kiln by a drag chain. Typically, HH (J93503) and HN (J94213) cast stainless steels are used for grizzlies and chain links, while type 310 austenitic stainless steel is used for chain pins. Table 10 gives the composition of these stainless steels.

Table 9 Typical compositions of aluminum- and silicon-alloyed ferritic stainless steels used in lime kilns

		Composition, wt%		
Zone	DIN	Cr	Al	Si
2/3	1.4713	6.0–8.0	0.7–1.2	0.7–1.2
3/4	1.4724	12.0–14.0	0.7–1.2	0.7–1.2
4	1.4742	17.0–19.0	0.7–1.2	1.0–1.5
	1.4762	23.0–25.0	1.2–1.7	1.0–1.5

Table 10 Composition of high-temperature alloys used in lime kilns

		Composition, wt%			
	UNS	C	Cr	Ni	Other
Wrought					
309	S30900	0.20	23.0	13.5	...
310	S31000	0.25	25.0	20.5	...
330	N08330	0.08	18.5	35.0	1.0 Cu
Cast					
310	J94302	0.25	25.0	20.5	...
HH	J93503	0.35	26.0	12.5	...
HN	J94213	0.035	21.0	25.0	...

Table 8 Lime kiln heat-transfer chain zones and conditions

		Gas temperature		Lime temperature		Mud moisture content, wt%
Zone	Function	°C	°F	°C	°F	
1	Dust curtain	200–300	390–570	...	...	20–40
2	Cold evaporation	...	...	100–170	210–340	10
3	Hot evaporation	...	...	150–260	300–500	0
4	Heating/radiation curtain	650–760	1200–1400	540–650	1000–1200	0

At the service temperatures encountered, the HH and HN cast steels are subject to precipitation of secondary phases (sigma phase) that can lower toughness. Drag chain failures have occurred during start-up when the chain is cold and jammed with hard deposits of calcined lime. Failures can be avoided by appropriate design and start-up procedures.

The higher chromium content of the cast heat-resistant alloys provides better oxidation resistance at temperatures above 650 °C (1200 °F). However, the higher chromium content makes these alloys more susceptible to sigma phase formation than type 304.

Corrosion Control in Tall Oil Plants

Max D. Moskal, Mechanical and Materials Engineering
Arthur H. Tuthill, Tuthill Associates

Tall oil is a mixture of rosin and fatty acids with unsaponifiables in the black liquor from the digesters. It is removed from the black liquor by skimmers and sent to a storage tank for processing with sulfuric acid (H_2SO_4) in the tall oil plant. Mappin provides flow diagrams for the batch and continuous acidulation processes (Ref 106). Conventional materials are noted in Fig. 19.

Batch acidulation is the most prevalent process method. Most batch reactors are constructed of carbon steel with membrane and acid-brick linings. However, Ketchum reported in 1990 that a batch reactor constructed of alloy 20 (N08020) was free of significant corrosion wastage after 24 months in service (Ref 107, 108). Evidence that corrosion-resistant alloys can provide low maintenance in the long term makes these alloys cost-effective alternatives for reactor construction. The reactor for the continuous process is either fiberglass-reinforced plastic (FRP), alloy 20, or alloy 904L (N08904). Older lead-lined reactors have been discontinued.

Sulfuric acid is the principal corrodent in tall oil plants. Carbon steel has good resistance to strong 98% sulfuric acid and is the principal material used for strong acid storage tanks. Flow disturbs the protective sulfate film that protects carbon steel; therefore, type 316L (S31603) is normally used for piping. Type 316L is also useful in weak sulfuric acid at low temperatures. Copper additions to stainless steel improve corrosion resistance to sulfuric acid. Alloy 20, which was developed for sulfuric acid service and has 3.5% Cu, is the principal upgrade in North America. Alloy 904L, which has 1.5% Cu, is the principal upgrade in Europe when concentrations and temperatures are too high for satisfactory performance for type 316L. In the mid-range of concentrations at elevated temperatures, the nickel-molybdenum alloy B (N10665) is one of the few metallic materials—other than tantalum and 12% Si cast iron—that has useful corrosion resistance.

CN-7M (N08007) is the basic cast alloy used for pumps, valves, and other cast components in sulfuric acid service. Alloy 20 is its wrought counterpart.

Isocorrosion charts provide useful estimates of probable performance of the large number of alloys used (Ref 109), and promoted for use, in sulfuric acid. Two isocorrosion charts shown in Fig. 20 and 21 (Ref 110) compare the isocorrosion curves for a number of common alloys, including 904L and 6% Mo (254 SMO) in chloride-free sulfuric acid and acid containing a low concentration of chloride (200 ppm).

Chemical species, other than chlorides, and including organic species, often present in sulfuric acid, can have a significant effect on corrosivity. The positive or negative impact of other species present on corrosivity, and the nature of sulfuric acid itself, tend to make alloy selection site specific. The article "Corrosion by Sulfuric Acid" in this Volume should be consulted for additional information.

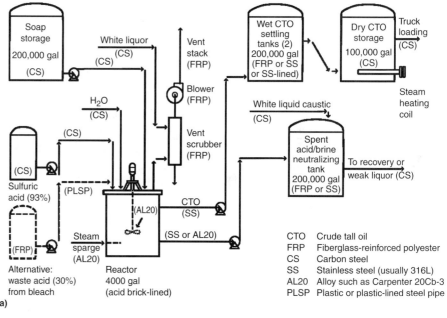

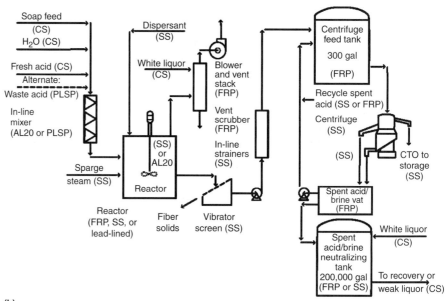

Fig. 19 Flow diagrams and typical materials of construction for the (a) batch and (b) continuous acidulation processes.

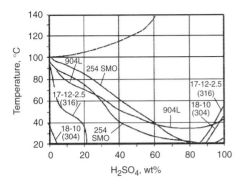

Fig. 20 Isocorrosion diagram, 0.1 mm/yr (4 mils/yr), of austenitic stainless steels in naturally aerated sulfuric acid of chemical purity. The dashed line represents the boiling point.

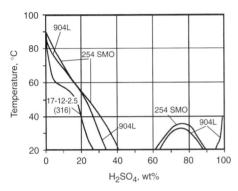

Fig. 21 Isocorrosion diagram, 0.1 mm/yr (4 mils/yr), of austenitic stainless steels in naturally aerated sulfuric acid with 200 ppm chloride addition

Corrosion in Recovery Boilers

Douglas Singbeil, Paprican

Recovery boilers in the pulp and paper industry are large, fuel-to-energy boilers designed to combust organic waste from the pulping process (Ref 111). They are similar in many respects to other types of boilers, including those that burn coal, municipal waste, and woodwaste or bark, and are prone to many of the same corrosion and materials issues. What differentiates recovery boilers is their function as a chemical reactor (Ref 112). Their role of recovering the spent inorganic cooking chemicals and converting them into forms suitable for reuse in the pulping process is paramount over steam and energy production.

Most recovery boilers in the pulp and paper industry burn kraft process liquor (black liquor) as a fuel, although sulfite and neutral carbonate liquors are also burned. As a consequence, the Tomlinson furnace, or kraft recovery boiler, is the focus of the balance of this section. The first Tomlinson furnace entered service in the late 1930s, but the design has evolved significantly over the decades and modern kraft recovery boilers bear little resemblance to the original boiler. The basic components of the modern recovery boiler have been in place since the late 1960s, and in North America many boilers still date from this era, although most have been upgraded or substantially rebuilt.

A typical modern recovery boiler (Fig. 22) built in the period 1995 to 2005 processes upward of 3000 tonnes (3300 tons) dry solids fuel/day, operates at a steam pressure of between 8.3 and 12.4 MPa (1200 and 1800 psi), and produces as much as 150 kg/s (330 lb/s) of steam. Older recovery boilers are generally smaller, and many operate at lower steam pressures, some as low as 4.1 MPa (600 psi). Water-cooled tubes make up the floors and walls of the furnace cavity, while the heat from combustion gases is extracted by superheaters, a boiler bank, and economizers before the gases exit the boiler. In modern boilers, wall and floor panels are made with sealed membranes between tubes, although sealed and unsealed tangent tube construction has been used in the past. Black liquor is delivered to the boiler as a hot, aqueous slurry at between 63 and 85% solids content. The black liquor is fed into the boiler from all four walls through liquor guns, which are tubes fitted with a nozzle, most commonly just a splashplate designed to break up the spray of liquor into fine droplets as it enters the furnace (Ref 113). Depending on the boiler design and size,

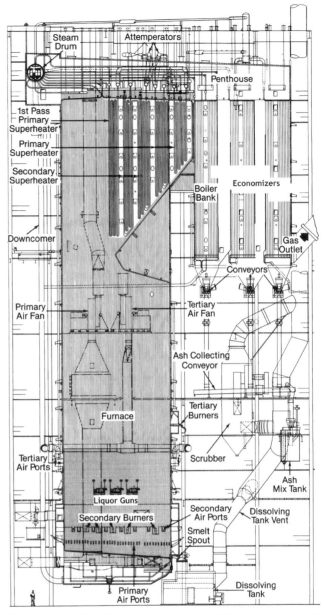

Fig. 22 Typical kraft recovery boiler used in the wood pulp industry. This is a modern, single-drum design, with the steam drum located outside the gas passage. Most boilers built prior to 1990 incorporated a generating bank, with an upper steam drum and lower mud drum, in place of the boiler bank and external steam drum. Courtesy of the Babcock & Wilcox Company

anywhere from 4 to 12 liquor guns might be employed. Combustion air is typically added at three levels in the furnace (older designs may use only two levels, and some new designs incorporate as many as 4 to 6 levels). A less than stoichiometric amount of air is added through the lower, primary air ports, with the balance being added through secondary and tertiary air levels.

Approximately 20 to 25% of the dry solids in black liquor are inorganic chemicals (Table 11) (Ref 114). It is critical to the process that the sulfur and soda content from the inorganic compounds be recovered and, at the same time, that the sulfur-containing salts are reduced back to the sulfide form necessary to catalyze the dissolution of lignin from wood chips. These latter reactions are endothermic and only occur at high temperatures in the presence of carbon char, or other reducing compounds—hence the need for the recovery boiler.

As the liquor burns, it goes through a sequence of discrete steps that include drying (evaporation of water), pyrolysis, and devolatilization (release of organic and inorganic volatile gases), char burning (production of CO and CO_2), and finally reactions between remaining molten salts (Ref 115). The hot combustion gases that are produced carry with them a fraction of the sodium (in the form of carbonate) and sulfur (as sulfate). The balance of the inorganic salts flow down the walls or fall to the bottom of the boiler to form a molten pool of smelt that is decanted off in a continuous stream through spouts into a dissolving tank. Depending on the boiler manufacturer, the floor might be flat (fully decanting) or sloped at a small angle outward toward the spout openings or inward toward a common center header. The top of this molten smelt pool is usually covered by a porous layer of char, in which the bulk of the reduction reactions occur as the molten salts pass through it.

Recovery Boiler Corrosion Problems

Corrosion of the component parts in a kraft recovery boiler has been a major challenge for the industry. It limits the overall energy efficiency of the boiler, largely due to restrictions placed on maximum steam temperatures to prevent molten salt corrosion of the superheater tubes. In addition, the hot, molten pool of inorganic salts (smelt) at the bottom of the boiler can react with water to create a very energetic, and often catastrophic, explosion. Consequently, unlike most other types of boilers, kraft recovery boilers cannot tolerate even the smallest leak of water into the furnace cavity. Complete destruction of the boiler, and corresponding risks to operating personnel, could result.

All kraft recovery boilers are subjected to a strict regimen of inspection and repair designed to identify and rectify problems before they compromise the safe operation of the boiler (Ref 116). Nonetheless, tube leaks do occur. Historically, about 40% of smelt/water explosions in kraft recovery boilers have been a consequence of a pressure part failure, although the modern record is much improved (Ref 117).

Fire-side corrosion problems to which kraft recovery boilers are susceptible include:

- General corrosion
- Fatigue
- Corrosion fatigue
- Stress-corrosion cracking

Water-side corrosion problems include:

- Overheating
- Stress-assisted corrosion
- Oxygen pitting
- Flow-assisted corrosion

As might be expected for a reactor design that has been in a steady state of evolution for nearly 80 years, and with models in operation from numerous manufacturers, many corrosion problems have been specific to certain designs or methods of manufacture (Ref 118). Many of these historical problems no longer affect modern recovery boilers and are not mentioned here. The water-side corrosion problems to which recovery boilers are subjected are similar to those found in other boilers, and reference will be made to these only in passing. Readers are referred to other articles in this Volume for more information on these types of problems. See, for example, the article "Corrosion of Steam-land Water-side of Boilers" in this Volume.

Corrosion in the Lower Furnace

The following paragraphs describe the general fire-side environment responsible for corrosion in each section of a recovery boiler, providing comments on common problems prevalent in modern recovery boilers, as well as ways to resolve them. Most of the discussion is focused on issues that affect pressure parts, rather than attachments and other noncritical parts of the boiler. For reference throughout this section, a list of alloys that are commonly used in recovery boilers is shown in Table 12.

The environment in the lower furnace of a recovery boiler is a maelstrom of liquor, char, particulate, combustion gases, and molten salts. The liquor entering the boiler is not homogeneous; composition and calorific value change rapidly, which affects both the combustion and distribution of liquor into the furnace. The lower waterwalls of the furnace are always

Table 11 Composition of typical North American as-fired kraft liquors

Element/ compound	Concentration, % of dry solids		
	Coastal softwood	Interior softwood	Interior hardwood
Na^+	18	18	19
K^+	1.5	1.2	3.4
Other cations	0.08	0.06	0.08
OH^-	0.9	0.5	2.1
CO_3^{2-}	5.9	4.2	4.4
Cl^-	2.9	1.0	0.5
Other anions	0.07	0.03	0.11
S	4.9	5.4	5.0
Combustible organics	~66	~70	~65

Source: Ref 114

Table 12 Alloys used in kraft recovery boilers

Alloy	Composition, wt%						Comments
	Fe	Cr	Ni	Mo	C	Other	
Furnace							
A210-A1	bal	...	...	...	0.27	...	Also used as core for composite tubes
A178-A	bal	...	...	...	0.18	...	...
304L	bal	18	8	...	0.03	...	As a coextruded tube
Sanicro 38(a)	bal	20	38	2.5	0.025	Cu = 1.7	As a coextruded tube
HR11N(a)	bal	27	38	0.5	...	...	As a coextruded tube
Sanicro 63(a)	3	21	bal	8.5	...	Nb = 3.4	As a coextruded tube
Super 625(a)	16	21	bal	8.5	...	Nb = 0.6	As a coextruded tube
WSI Unifuse 625(a)	5	21	61	8.5	0.1	Nb = 3.5	As a weld overlaid tube
18% Cr ferritic weld overlay	bal	17	...	...	0.1	Nb = 0.85, Ti, Al	Used in Japan, South America
25% Cr ferritic weld overlay	bal	25	...	...	0.09	Nb = 0.95, Ti, Al	Used in Japan, South America
45CT(a)	0.1	43.5	bal	...	0.04	Ti	Twin-wire arc spray
Alloy 625	<5	20	bal	8	<0.10	...	HVOF, HVAF coating
Superheaters							
T2	bal	0.05	...	0.44	0.20	...	...
T11	bal	1.0	...	0.44	0.15	...	...
T22	bal	1.9	...	0.87	0.15	...	...
T91	bal	8.0	...	0.85	0.12	...	...
310	bal	24	19	...	<0.08	...	Also as coextruded tube
347H	bal	17	9	...	<0.1	Nb = 8×C-1.0	...
800H	bal	19	30	...	0.05	...	...
825	bal	19.5	38	2.5	<0.05	Cu = 1.5	...
YUS170(a)	bal	23	12	0.50	<0.06	...	Used in Japan
MN25R(a)	bal	23	13	0.50	...	...	Used in Japan
Boiler bank and economizers							
A210-A1	bal	...	...	...	0.27	...	...
A178-A	bal	...	...	...	0.18	...	...

(a) Proprietary alloy names trademarks of their respective manufacturers

covered by layers of both molten and solid salts, as well as residual liquor and char. The depth of the molten smelt/char bed on the floor of the boiler can vary from as low as 100 to 200 mm (4 to 8 in.) around the edges, up to more than 2 m (6 ft) high in the center of the boiler. The molten smelt on the floor of the furnace is typically at a temperature of 750 to 850 °C (1400 to 1560 °F), while the combustion gases above the bed are generally hotter than 1000 °C (1800 °F). The combustion gases in this portion of the boiler are reducing, due to the substoichiometric supply of primary air, and have been analyzed to contain as much as 5 vol% hydrogen sulfide (H_2S) as well as significant amounts of carbon monoxide (CO) and other reducing gases (Table 13) (Ref 119).

Corrosion of Carbon Steel Tubes. Surprisingly, the water-cooled tubes that make up the walls and floor of the boiler were made from ASTM A 210 grade A1 carbon steel (CS) or equivalent until the early 1970s, and this material is still in use today in many boiler floors and in the walls of older, low-pressure recovery boilers. The reason is simple—the first melting point of most deposits, including the smelt, in recovery boilers is never lower than about 500 °C (930 °F), while the tube surface temperature of a typical low-pressure (<6.2 MPa, or 900 psi) recovery boiler is not much greater than the water temperature in the tubes—about 300 °C (570 °F). A layer of insulating frozen smelt quickly forms on the tubes and acts as a barrier to corrosion. To aid in formation of the frozen smelt layer on the tube surface, and to keep it from sloughing off, carbon steel studs about 10 mm diameter by 25 mm long (0.5 in. diameter by 1 in. long) are generally resistance welded in a dense pattern to the surface of the tubes (Ref 120).

Studded carbon steel floor tubes have generally been free of corrosion problems (Ref 121). The most rapid corrosion of carbon steel wall tubes in the lower furnace occurs where the liquor lands on the walls in a zone from the top of the smelt bed to just below the secondary air-port openings (Ref 122). Estimates for average rates of corrosion on carbon steel tubes in the lower furnace vary from between 0.05 and 0.2 mm/yr (2 and 8 mils/yr) up to a maximum of 0.4 to 0.8 mm/yr (16 to 32 mils/yr). Much higher short-term wastage rates—as much as 5 mm/yr (200 mils/yr) are occasionally reported (Ref 123). The corrosion is manifested as general wastage of the tube surface, and analyses of corrosion products invariably show they are composed of iron sulfide, with little or no iron oxide present. While wastage patterns of approximately 1 m² (11 ft²) or more are not unusual, small intense areas of localized corrosion are also found. These latter areas are often much smaller than the inspection grid spacing and present considerable challenges to locate before a tube rupture occurs (Ref 124). Overheating due to internal scale formation or nonuniform water circulation in a tube is sometimes, but not always, a factor that contributes to corrosion of carbon steel tubes.

The mechanisms responsible for corrosion of carbon steel tubes in the lower furnace are well understood. Due to the large temperature difference between the "cold" tube surfaces and the first melting point of the salts in the boiler, fluxing from molten salts is unlikely to contribute significantly to corrosion, except under unusual circumstances. Some corrosion may be the result of exposure to wet liquor, or hydrated sodium sulfide, but it is reactions between the steel and pyrolysis gases that circulate through the porous deposits adhering to the surface of the walls that are likely responsible for most of the observed corrosion. Long-term average corrosion rates can be readily accounted for by isothermal reactions between carbon steel, H_2O, O_2, and gaseous H_2S (along with other sulfur species) in the reducing atmosphere of the lower furnace (Ref 123). Additional factors have been identified as being particularly likely to contribute to accelerated corrosion (Fig. 23) (Ref 125):

- Organic sulfur compounds released from the liquor in large quantities during the early stages of pyrolysis. These gases (principally methyl mercaptan—CH_3SH) contribute to higher sulfur activities than H_2S and on a equal volume basis, CH_3SH is as much as 22 times more corrosive than H_2S at temperatures between 300 and 400 °C (570 and 750 °F).
- Increased temperature combined with temperature cycling. Carbon steel corrosion rates increase rapidly as temperatures approach 400 °C (750 °F), but cycling over a limited range of only 300 to 400 °C (570 to 750 °F) will increase corrosion rates by a factor of 3 times or more over the highest isothermal rate.
- Cycling between oxidizing and reducing conditions, which accelerates corrosion rates of carbon steel by a factor of 2 to 3 times.

In practical terms, liquor droplets that hit the surface of the tubes are quickly covered by newly arriving material, including molten smelt running down the wall. As these droplets heat, pyrolysis gases, including H_2S and CH_3SH, are released to diffuse through a porous inner layer and react with the tube surface. Variations in the flux of droplets to the wall, and in the energy reaching the tube surface, account for the wide variability in corrosion rate that can be observed. Operating practices for the boilers and overall boiler design play a large role in determining the fraction of liquor droplets that reach the furnace walls before pyrolysis occurs. Controlling this problem, while meeting the process needs of the boiler as a chemical reactor, is a complex issue. Generally speaking, operating practices that promote liquor burning on the walls will also increase carbon steel corrosion rates.

Corrosion Control of Carbon Steel Tubes. A number of strategies have been employed to mitigate corrosion of carbon steel tubes in the lower furnace. These include "studding" the tubes, applying protective coatings, or replacing carbon steel with more resistant alloys. Each of these options is discussed briefly below.

Studded tubes are still found in some low-pressure recovery boiler walls, and in the floors of many boilers, including higher-pressure units. On occasion, chromized studs (carbon steel with a thin chromium-rich surface layer applied by pack cementation) have been used instead of plain carbon steel. The cooling of the studs is not as efficient as on the tube surface, and over time the studs will corrode back to nubs, particularly in areas of the boiler subject to high local heat fluxes (Fig. 24). These high heat fluxes promote more rapid degradation of the studs and also encourage a wash of molten smelt over their surface that contributes to corrosion. Studding is a proven long-standing method of reducing the rate of corrosion on carbon steel tubes, but studded tubes are sometimes subject to the formation of numerous shallow, parallel notches in the tube between studs (often known as "elephant hide"), as well as deeper circumferential cracks at the base of the studs—both likely caused by intense cyclic thermal fluxes (Ref 126). The process of applying studs requires careful quality control, particularly when new studs are applied in the field over old, worn-down studs. Lack of fusion at the edge of the weld, or formation of other discontinuities can act as stress raisers that progress into thermal fatigue cracks (Fig. 25). Studs have been burned through the tube during manual application, and the formation of solidification cracks due to the presence of sulfur in the weld pool is also possible when applying studs to a poorly cleaned tube.

Table 13 Lower furnace gas compositions in four recovery boilers

Gas	Avg. composition, vol%	Min. composition, vol%	Max. composition, vol%
N_2	bal	bal	bal
H_2	2–15	0–3	4–22
O_2	5–8	1–4	9–16
CO	2–7	0–7	4–23
CO_2	12–15	5–13	14–19
H_2S	0.02–0.8	0.001–0.3	0.3–5
SO_2	0.005–0.3	0.0001–0.001	0.05–5

Source: Ref 119

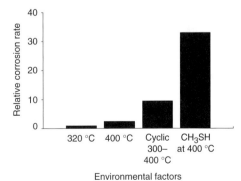

Fig. 23 Laboratory test results showing relative effect of environmental factors on corrosion of carbon steel in lower furnace. Corrosion rates of alloys are normalized to that of A210-A1 at 320 °C (610 °F).

Protective Coatings for Carbon Steel Wall Tubes. Some manufacturers and end-users opt to apply a protective coating over the fire-side surface of the carbon steel wall tubes in the lower furnace. A range of alloys and application methods have been used for this purpose, most of which would now be considered obsolete. These coatings fail over time because the porous coating structure allows gases to penetrate to the carbon steel substrate. Newer coatings are generally specified on the basis of as-applied coating density and coating thickness. They are often applied with an automated high-velocity oxyfuel (HVOF) or equivalent technique. Alloy 625 or nominal 50%Cr-50%Ni material are common choices for recovery boiler coatings, but the alloy choice is generally less important than the method of application and level of care taken prior to and during spraying. For the most part, these coatings are applied as an alternative to complete wall replacement where rapid corrosion has occurred. Since they are almost always field-applied under adverse working conditions, coating quality can be variable, and spalling is frequently encountered (Fig. 26). With stringent quality assurance/quality control in place, and the use of fully automated application, coating lifetimes of 5 to 10 years have occasionally been achieved with only minor annual touch-ups required.

Pack cementation coatings have also been applied to large boiler wall panels (Ref 127). This process leaves a thin, corrosion-resistant layer of nearly pure chromium on the surface of the tubes (Fig. 27). Several boilers have been built with chromized lower furnace walls with favorable, long-term performance reported by the mills. As with other coatings, quality control and assurance are critical to obtaining a consistently uniform product, particularly when dealing with large panels, or panels that include complex shapes and bends. As one example, the chromizing process by nature leaves porosity in the surface known as Kirkendall voids. Under ideal conditions, these voids are found near the surface and do not affect the integrity of the chromized coating. However, improper processing can cause these voids to form as low as the interface with the carbon steel, leaving a potential migration path for sulfur to the underlying carbon steel. Uneven coating thicknesses and poor adhesion between the chromized layer and the underlying steel are other problems that can arise during the manufacturing process.

Composite Tube Construction. Beginning in the mid-1970s, the industry began a transition toward the use of composite tubes in the lower furnace as an alternative to carbon steel tubes. This change has been extraordinarily successful and was the key technological innovation that permitted operation of kraft recovery boilers at steam pressures greater than 8.3 MPa (1200 psi). Composite tubes are made with a corrosion-resistant outer layer metallurgically bonded to an inner core of carbon steel. The vast majority of composite tubes in service are coextruded, with an outer alloy layer formed over an inner core of A210-A1 carbon steel (CS), but some are produced via a weld overlay process—either spirally welded around the entire tube circumference, or applied as a one-sided weld overlay on carbon steel panels. Stainless steel type 304L/CS (30403) was the first alloy chosen for use as a composite tube in kraft recovery boilers, and 304L has largely been the material of choice since then, although other alloys are also now used (Table 13). These include proprietary variants of alloy 825 and alloy 625. Ferritic stainless steel weld overlays with between 18 and 25 wt% Cr are used in Japan (Ref 128) and in South America, but have no extensive history in North America or Europe.

Composite tube panels are usually built with a membrane construction, much as for carbon steel walls and floors. Depending on the boiler manufacturer and the date of construction, the membranes between tubes can be solid stainless alloy, roll clad alloy on carbon steel, or weld overlay on carbon steel.

While composite tubes solved nearly all of the problems that affected carbon steel tubes in the lower furnace, particularly those related to gas-phase corrosion, two new issues were introduced when they came into service: preferential corrosion of the external alloy layer and cracking that originates from the fire-side surface. In both cases, the introduction of composite tubes revealed aspects of the corrosiveness of the lower furnace boiler environment that had been completely unsuspected when carbon steel was the material of choice.

Preferential corrosion of the external layer is a particular problem with 304L/CS composite tubes, but has also been observed for composite tubes made from most other alloys, including alloy 625/CS. Surprisingly, once the underlying carbon steel is exposed, it is common for very little corrosion of exposed carbon steel to take

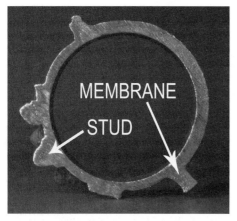

Fig. 24 Cross section through a studded carbon steel boiler tube, showing reduction in dimensions of the studs that occurs in operation. Note the loss of wall thickness in the tube around the entire fireside circumference, including the crotch of the tube near the membranes.

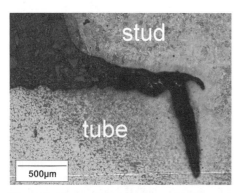

Fig. 25 Micrograph showing a thermal fatigue crack that originated at a discontinuity in the base of a stud on a carbon steel tube. The crack has turned and begun to propagate into the thickness of the tube.

Fig. 26 Spalling of a plasma sprayed coating as a result of poor surface preparation prior to the application of the coating

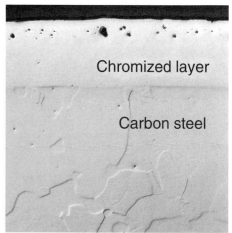

Fig. 27 Cross-section photograph through a chromized tube. The chromized layer (between 0.2 and 0.4 mm thick) contains about 40 wt% Cr at the surface and about 16 wt% Cr at the diffusion boundary with the carbon steel. The voids near the surface are normal artifacts of the chromizing process, and do not affect the corrosion resistance of the coating unless they provide a migration path to the carbon steel surface.

place until the exposed area is quite large. "Balding" was first observed at the corners of air-port openings, on the cold side of the tubes in hidden areas behind flat studs and cast inserts (Ref 129). Subsequently, similar preferential corrosion has been observed at primary air-port openings further around the tube circumference toward the fire-side crown (Fig. 28), as well as on the cold side of the tubes behind insert plates in liquor gun, secondary air port, and burner openings, and on the cold side of the tubes underneath the smelt spout openings, where they are shielded by the smelt hoods.

Neither field observations nor laboratory experiments have clearly established the corrosion mechanism, but this may simply be because there is more than one mechanism operating, depending on the location in the boiler. The generally accepted mechanism for the cold-side corrosion is condensation of sodium hydroxide fumes from combustion gases (Ref 129). Preferential corrosion toward the fire side of the tubes is more likely to be caused by exposure to wet liquor or hydrated sodium sulfide, as formation of NaOH is not thermodynamically favored and alloys that should be resistant to attack by NaOH corrode in these locations (Ref 130). Laboratory corrosion tests in hydrated sodium sulfide/sodium hydroxide produce alloy rankings consistent with most of the field observations (Fig. 29). One effective solution to "balding" at the corners of air-port openings is to ensure that cast inserts are well-sealed with refractory when they are first installed, and to follow up regularly on maintenance outages with restoration of degraded refractory seals. A variety of weld repair methods have been employed, with caution, to restore the protective alloy cover when concerns existed about corrosion of exposed carbon steel. For the most part, either 309 stainless steel or alloy 625 has been chosen as the filler metal, with mixed success reported for each.

Occasionally, corrosion at the crown of composite tubes has been observed around the level of the primary air ports, and in the lower parts of the spout openings that are subject to washing by the molten smelt leaving the boiler. In the latter case, 304L/CS composite tubes that form spout openings are often replaced with ones made with alloy 825/CS for improved resistance to both corrosion and cracking.

The second issue that affects the use of composite tubes in recovery boilers is cracking of the external corrosion-resistant layer and propagation of these cracks into the underlying carbon steel. Cracking in composite tube panels occurs in many locations, including on the crown of tubes, near the membrane-to-tube weld, in the membrane itself, and at membrane terminations for port openings (Ref 131). The mismatch in thermal expansion coefficients between the alloys used to make composite tubes can be significant, particularly when 304L is used as the outer layer. Consequently, stresses induced in the tubes due to thermal fluctuations were originally considered to be the root cause of cracking in composite tubes. The true situation is actually much more complex, and a number of mechanisms are now known to contribute to the problem, depending on where the cracks form (Ref 132, 133).

Cracking at membrane terminations and along welds for sealed air port inserts in composite wall panels is a common problem in recovery boilers. Thermal stresses undoubtedly play a role at these locations. Many of these cracks remain in the membrane, or propagate along the weld and do not progress into the pressure part, but there is always a concern that they might do so (Fig. 30). Cracking in these locations has been minimized by very careful joint and weld design and attention to detail in fabrication. Recent practice has been to replace seal-welded air-port inserts with bolt-on cast inserts to avoid thermal cracking at the welds. This avoids one problem at the risk of introducing preferential corrosion at the corners of the openings, as described previously.

Cracking of Composite Floor Tubes. Cracking of 304L/CS composite floor tubes is common in nearly every boiler that uses these tubes, to the extent that this alloy is no longer specified for this application. Portions of wall tubes that are exposed to a similar environment (i.e., at or below the normal level of the smelt bed) are also sometimes susceptible to cracking. Cracking often appears as "craze" cracking on the tube crowns, but is also found along tube/membrane welds and in the membranes themselves (Fig. 31). Fortunately, these cracks almost invariably stop when they reach the stainless/carbon steel interface or turn and propagate along it. Over time, several of these latter cracks may join and cause spalling of the stainless layer. Very few, out of many thousands, of such cracks that have been examined, have ever been documented to penetrate into the carbon steel, and all of these were atypical in cause.

The mechanism of cracking remained controversial for many years, but it is now generally agreed that the root cause of most composite floor tube cracks is stress-corrosion cracking (SCC) that occurs underneath piles of residual smelt during boiler water washes (Ref 132).

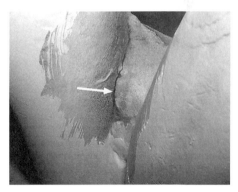

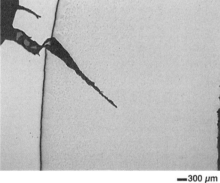

Fig. 30 Recovery boiler sealed air ports. (a) Crack running along the heat-affected zone of a weld between a composite tube and a crotch plate in a spout opening. (b) Subsequent failure analysis found the crack penetrating into the carbon steel layer of the composite tube.

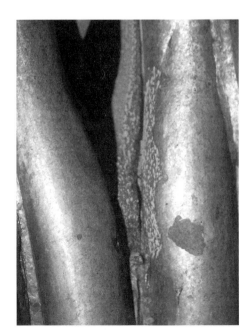

Fig. 28 Preferential corrosion of the 304L stainless steel layer in a composite tube. Corrosion occurs at the edge of the bolt-on casting in a primary air port and extends out toward the fire-side crown of the tube

Fig. 29 Plot comparing average corrosion rate of alloys in 75%Na_2S-25%NaOH (dry wt%) salt hydrate mixture at 180 °C (355 °F) for a period of 24 h. Error bars show maximum and minimum values obtained in triplicate experiments.

Boiler water washes are performed periodically to clear the superheaters and boiler banks of deposits that plug the gas passages. Depending on the needs of a particular boiler, they may occur anywhere from once annually to six or more times a year. Although the smelt bed is burned down and allowed to cool before washing the boiler, large piles of frozen, but still very hot, smelt are often left in place on the floor. The residual heat in these piles is sufficient to maintain the underlying tubes in a temperature range found appropriate for SCC. Small amounts of water flowing underneath these piles of smelt will quickly form concentrated "soups" or poultices of hydrated salts on the surface of the tubes.

Key observations that have led to this conclusion are:

- Measurements of tube surface temperatures during operation revealed that thermal cycling of the tubes does not correlate with the location of cracking, nor was the magnitude and frequency of thermal cycling sufficient to induce fatigue cracks in 304L.
- Modeling and measurement of residual stresses in composite tubes show that the outer stainless layer is in compression during normal operation, but surface stresses become tensile at about 200 °C (390 °F) as the tube is cooled toward ambient.
- It was found that 304L and other alloys are very susceptible to SCC when exposed to hydrated salt mixtures (Na_2S, $NaOH$, and Na_2CO_3) that simulate the poultices present on the tubes at temperatures in the range of 160 to 200 °C (320 to 390 °F) (Fig. 32).

Remedial measures that have been adopted for floor tube cracking include monitoring the cracks over time and then grinding the cracks out and repair welding as necessary, or by replacing 304L/CS composite tubes with either studded carbon steel tubes or composite tubes made with alloy 825 or alloy 625 external layers. A very few mills report no cracking of the 304L/CS composite floor tubes in their boiler, but these mills invariably follow stringent procedures designed to completely burn down the remnant bed on the floor before beginning the water wash or never water wash the boiler.

Cracking of Composite Wall Tubes at Openings. Tubes that form port openings in composite tube walls are also susceptible to cracking in-service, especially at the primary air port and spout openings (Ref 134). Cracks found in these locations include the familiar craze cracks found in floor tubes, as well as single, circumferentially oriented cracks. Unlike the situation for floor tubes, several cracks in tubes that form primary air-port openings have propagated into the carbon steel and tube leaks have resulted from cracking. Cracking at primary air-port openings is generally confined to the two tubes that form either side of the opening, and cracks are nearly always found only in the lower half of the opening. Most cracks are found on the tube bends, but some appear 150 mm (6 in.) or more below the level of the membrane termination (Fig. 33). A strong correlation has been made between frequent, high-magnitude thermal cycles on the air-port opening tubes and cracking, leading to the conclusion that corrosion fatigue is an operative mechanism along with SCC.

Substantial evidence has been accumulated to show that cracking of composite tubes at primary air-port openings occurs as a result of a complex interplay between air-port design, construction practices, boiler operation, and tube metallurgy. Consequently, although boilers from nearly all manufacturers have had cracked tubes at primary air-port openings, not all boilers are affected. Long-term boiler trials have demonstrated that the thermal fluctuations at port openings can be calmed by careful optimization of boiler operation and that the severity and number of cracks is substantially diminished as a result (Ref 135). From a metallurgical viewpoint, laboratory tests have shown that cold-worked alloys appear much more susceptible to cracking than those that have been annealed (Fig. 34).

Cracking at primary air-port openings has been controlled by a combination of optimizing boiler operation, replacing 304L/SC composite tubes with either alloy 825/CS co-extruded or 625/CS weld overlay tubes and changing the air-port design to reduce stresses induced by manufacture and optimize air flow through the opening (Ref 136). Optimized boiler operation appears to be a critical step in preventing

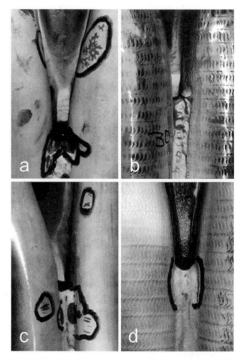

Fig. 33 Cracks revealed by visible dye penetrant testing on 304L and weld overlay (WO) 625 composite tubes that form primary air-port openings. (a) Craze cracks on 304L. (b) Membrane cracks on WO625. (c), Circumferential cracks on 304L. (d) Tube-membrane weld cracks on WO625

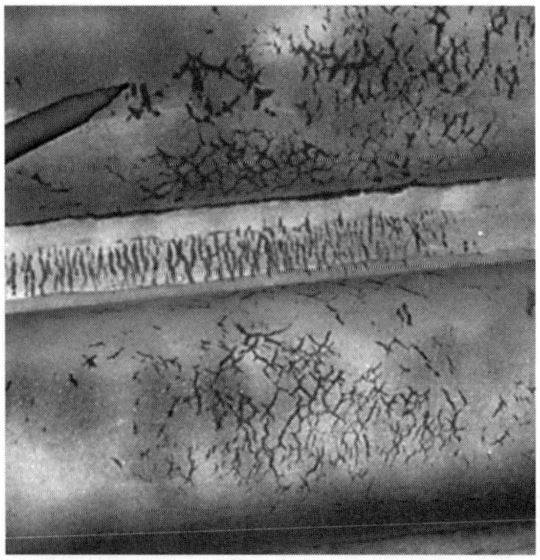

Fig. 31 Cracks revealed by visible dye penetrant testing in a 304L composite floor tube. Note also the cracking along the tube/membrane interface and in the membrane.

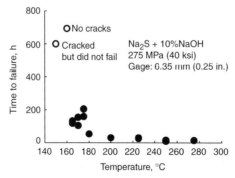

Fig. 32 Results from constant stress SCC tests of 304L shown as a function of temperature in a hydrated mixture of Na_2S and NaOH. Solid circles are data points for specimens that failed during the test. Open circles are for specimens that either cracked, but did not fail during the test, or did not crack at all.

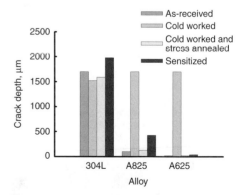

Fig. 34 Resistance of alloys to stress-corrosion cracking in hydrated 75% Na_2S-25%NaOH (dry wt%) hydrated salt mixture at 180 °C (355 °F) for a period of 48 h

cracking. None of the other options by themselves have proved sufficient to ensure that cracking will not occur. Alloy 825/CS wall tubes have provided exceptional service in many mills, although cracking of this alloy in tubes that form primary air-port openings has occurred in at least one boiler. Coextruded alloy 625/CS tubes have experienced extensive cracking in a boiler when used as air-port opening tubes.

Cracking of 304L/CS composite tubes in spout openings has generally been controlled by replacing the tubes that form the openings with ones made from alloy 825/CS composite tubes. Small incipient cracking is still often found in the alloy 825/CS tubes, but has been easily controlled by inspection and light grinding as part of regular maintenance. Corrosion of the alloy 825 also occurs where it is exposed to flowing smelt, but at a lower rate than for 304L. Coextruded alloy 625/CS tubes have experienced extensive cracking when used as a spout opening tube.

Smelt Spouts and Air-Port Castings. Smelt spouts guide the molten smelt out of the spout opening in the boiler wall and let it drop into a dissolving tank located underneath the boiler. They are water-cooled, due to the intense heat to which they are subjected from the flowing smelt. In some designs, the end of the spout is butted against the outside of the boiler wall, while in other designs, the spout intrudes into the spout opening to protect the tubes from flowing smelt. Although not a pressure part, corrosion of spouts has contributed to leaks of water into the furnace cavity (Ref 118). Consequently, operating guidelines for kraft recovery boilers generally call for the spouts to be replaced during the annual maintenance shutdown, regardless of their physical condition. For this reason they are generally built from carbon steel. In most boilers, this material provides adequate service. In order to extend the lifetime of spouts, and to prevent premature failure, water-cooled spouts are sometimes clad with corrosion-resistant alloys such as alloy 625 or have had chromized coatings applied, but there has been little systematic study of the value of these approaches to life extension. Proprietary dry spouts, with no water cooling, have been cast from different alloys, including variants of a 50Cr-50Ni alloy.

Corrosion of the cast inserts in primary airport openings is a maintenance problem in that the flow of primary air into the boiler can be adversely affected by damaged castings, and localized corrosion of composite tubes at the corners of the air-port openings has been linked to poor sealing between the casting and the tube that forms the opening (Fig. 28). Most often these castings are made from ductile iron, but occasional attempts have been made to use different alloys, also including a 50Cr-50Ni alloy.

Mid-Furnace Corrosion

The mid-furnace is a transition zone between the reducing conditions found in the lower furnace and oxidizing conditions more typical of boiler flue gases. Traditionally, construction of the boiler walls has been of carbon steel, and few corrosion problems were reported. When composite tubes were introduced, the height of the cut-line between the composite walls and the carbon steel walls was established at a level above the secondary air ports. This decision was based on historical corrosion patterns of the time. Over the years, the height of the cut-line has been raised and was eventually "standardized" at about 1 m (3 ft) above the tertiary air-port openings. This has now proved to be insufficient to cope with modern boiler design and operating practices. Many boilers now experience what is known as "cut-line corrosion," or corrosion of the carbon steel tubes beginning at or just above the butt weld between the two types of tube (Fig. 35). In some boilers, the corrosion has been localized to within a few inches of the butt weld, while in others, corrosion of the carbon steel tubes has continued for some distance upward from the weld. Investigation has shown that the environment close to the wall is rich in H_2S, as well as organic sulfur species such as CH_3SH, and that the gas phase cycles continuously between oxidizing and reducing conditions, both of which are known to significantly increase corrosion rates of carbon steel (Ref 137). While the conditions necessary to promote accelerated fire-side corrosion clearly exist in mid-furnace regions, it is still not understood why corrosion would be concentrated so close to the butt weld in some cases and go well up the wall in others. Solutions to mid-furnace corrosion are relatively straightforward—use tubes with a thicker corrosion allowance, apply protective coatings over the corroded carbon steel, or extend the composite tube cut-line to higher elevations in the boiler.

Corrosion of Superheaters

Corrosion of superheater tubes is the Achilles heel of a recovery boiler and limits the overall energy efficiency that can be achieved. This is because the surface temperature of the tubes begins to come very close to the first melting point of deposits that form on the surface of the tubes (Ref 138). Most recovery boilers are designed for flue gas furnace exit temperatures of 950 to 980 °C (1740 to 1800 °F), but actual operational values may be as much as 200 °C (360 °F) lower. By the time the flue gas reaches the boiler bank, nominal temperatures are 600 to 650 °C (1110 to 1200 °F). Depending on the composition of the deposits, the first melting point will vary from about 500 to 600 °C (930 to 1110 °F), with a typical value of about 560 °C (1040 °F) in boilers with limited chloride and potassium inputs to the liquor. Both of these elements are commonly found in superheater deposits. They each have strong influences on the melting behavior of salts in the superheater, although their individual effect is quite different (Ref 139). When chloride is present in a deposit, a further increase in concentration does not change the first melting point of the deposit—only the proportion of liquid phase in the deposit at the first melting point (Fig. 36). Conversely, the proportion of liquid phase in the deposit at the first melting point does not change at all with increasing potassium content, but the first melting point will decrease substantially (Fig. 37). To ensure that molten deposits are not in contact with the tube surfaces, current boiler designs typically limit superheater steam temperatures in high-pressure recovery boilers to a maximum of

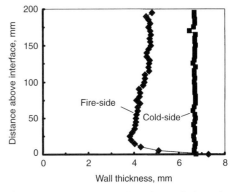

Fig. 35 Thickness profile of a carbon steel tube at the cut-line with the composite tube lower furnace walls

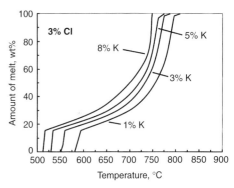

Fig. 36 Influence of chloride (at fixed potassium content) on deposit first melting temperature (Ref 139)

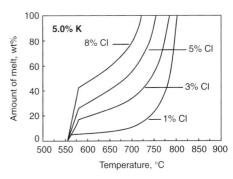

Fig. 37 Influence of potassium (at fixed chloride content) on deposit first melting temperature (Ref 139)

about 480 °C (900 °F) to allow for temperature gradients across the tube thickness. A few boilers already operate at superheater steam temperatures above 510 °C (950 °F), and it is expected that more will be built as companies attempt to maximize energy production (Ref 128).

Relative to other boilers, kraft recovery boilers carry a much higher flux of particulate through the superheater, boiler bank, and economizer sections of the boiler (Ref 112). The composition of the deposits changes substantially as the flue gas passes through the boiler (Fig. 38), and also as a function of how the boiler is operated and the overall dimensions of the boiler. Recovery boilers are specifically built with a tall furnace to increase residence time and ensure complete combustion of the fuel, but entrainment of unburned liquor droplets or droplets in various stages of combustion in the superheaters is a common occurrence, especially in older, overloaded boilers. Plugging of the gas passages in the superheaters and boiler banks is a major operational problem for pulp mills and requires constant sootblowing with steam lances, as well as frequent shutdowns to allow the gas passages to be washed clean with hot water. Modern, well-controlled boilers are generally able to run between annual maintenance shutdowns without needing a water wash, but in older boilers, or boilers with high fuel loads, as many as five or six water washes might be required annually. Any of these operational issues can influence the location and severity of corrosion in the superheaters.

Dramatic changes have been made to boiler operation since 1980 to meet the demand for greater efficiency, higher throughputs, and reduced stack emissions, and these have had a major effect on the environment in the superheater. Total reduced sulfur (TRS) and SO_2 concentrations in the flue gas are now much lower than in the past, while the amount of condensed sodium carbonate fume has significantly increased. Altogether, these changes make the superheater environment much less corrosive. In the past, many superheater corrosion problems were often caused by acidic sulfate deposits containing sodium and potassium bisulfate and pyrosulfate salts—all of which are liquid at superheater tube temperatures (Ref 118). Under modern operating conditions, these acidic sulfates no longer form in sufficient concentrations to cause corrosion, due to the low SO_x concentration in the flue gas and because any that are formed are quickly neutralized by the available sodium carbonate. Locally reducing conditions—formed under droplets of unburned liquor that are carried into the superheaters and impact the tube surfaces—have also been responsible for corrosion of the hotter superheater bends, but this problem is also largely of historical interest, except perhaps for older recovery boilers.

Materials used for superheater tubes in kraft recovery boilers are generally governed by the same design constraints as for other types of boilers, including the need for strength at temperature and good creep resistance. Ferritic alloys T11 (K11597) and T22 (K21590) are commonly used for superheater tubes in lower pressure recovery boilers, while T91 (K90901) sees service in higher pressure boilers (see Table 12). In boilers with particularly aggressive environments, the hottest superheater bends have been specified in corrosion-resistant austenitic alloys, including 347H (S34709), 310H (S31009), alloy 800 (N08800), and alloy 825 (see Table 12). Coextruded TP310H/T22 tubes have also been used in some boilers. For high-pressure recovery boilers, with superheater tube temperatures significantly higher than 500 °C (930 °F), chloride and potassium removal from the process becomes a critical part of corrosion control. Even so, more corrosion-resistant alloys might be required. Experience reported from Japanese boilers operating with a superheater steam temperature of about 500 °C (930 °F) has

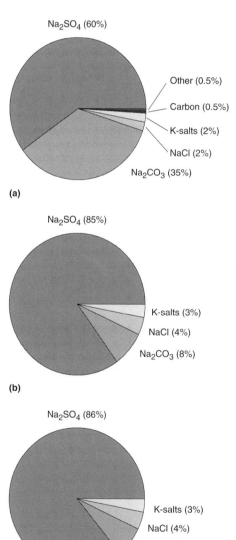

Fig. 38 Changes that occur in deposit chemistry as the flue gases pass through the (a) superheaters, (b) boiler bank and, (c) economizer

been positive with a proprietary alloy that contains about 25 wt% Cr, except when high sulfur auxiliary fuel was burned in the boiler (Ref 128).

Boiler Bank and Economizers

Flue gas temperatures in the boiler bank quickly fall below the first melting point of the deposits, and consequently the environment is generally not very corrosive. Carbon steel tubes are standard in this part of the boiler and lifetimes of 25 to 30 years for the tubes are common. However, this part of the boiler has not been entirely trouble-free. Throughout the 1980s, many two-drum recovery boilers experienced a particular form of localized corrosion called "near-drum thinning" (Ref 140). This phenomenon involved wastage of the tubes adjacent to the tube seat in the mud drum (Fig. 39). Typically, only small areas around the tube circumference were affected, usually at ten and two o'clock with respect to the direction from which the sootblower passed by the drum at any given tube location. Corrosion rates typically varied between 0.13 and 0.51 mm/yr (5 and 20 mils/yr), although some boilers experienced much more rapid corrosion. The mechanism was determined to be formation of liquid sodium bisulfate underneath deposits on the tube surface. The corrosion process was cyclic—every time the sootblower closest to the drum operated, deposits and corrosion product were removed from the surface of the tube, leaving the process to start over again. As in the superheater sections, process conditions no longer favor the formation of sodium bisulfate, and this type of corrosion is now only rarely encountered and then only in older boilers that still fire low-solids liquor and cannot maintain hot enough beds to generate the necessary sodium carbonate fume in the flue gas. Other design-specific problems that have been encountered in this area of the boiler are corrosion fatigue at header welds and dew-point corrosion of economizer tubes.

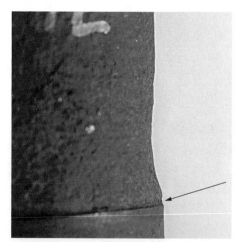

Fig. 39 Generating bank tube showing near-drum thinning just above the tube seat for the mud drum (arrow points to location where the tube enters the drum)

Corrosion Control in Air Quality Control

Craig Reid, Acuren

For many years kraft pulp mills have collected and disposed of "foul" condensates and odorous noncondensable gases (NCGs) containing hydrogen sulfide and reduced organic sulfur compounds in order to reduce the impact of emissions on neighboring communities and the environment. These compounds are collectively described as total reduced sulfur (TRS). Typically, low-volume high-concentration (LVHC) gases and off-gases from the stripping of foul condensates are collected at concentrations above their upper explosive limit (UEL) and incinerated in the lime kiln, recovery boiler, power boiler, or dedicated thermal oxidizer. The stripped condensate is recycled within the process and/or sent to the effluent treatment system of the mill. Heat recovered from the gases and condensates is used to heat process water. Some high-volume low-concentration (HVLC) gases are collected below their lower explosive limit (LEL) and incinerated, while others are released to the atmosphere.

In recent years, environmental regulations have progressively restricted the release of TRS and other compounds in NCGs and foul condensates. Accordingly, there has been continuous evolution of processes and process equipment for collection and disposal of NCGs and foul condensates (Ref 141–154).

On Nov 14, 1997 the U.S. Environmental Protection Agency (U.S. EPA) issued the National Emissions Standards for Hazardous Air Pollutants (NESHAP) for selected pulp and paper categories. These new regulations required the collection and disposal of dilute NCGs and condensates from sources that previously were released to the environment by most mills. Depending on the age and process configuration of a mill, these sources may include:

- LVHC vents
- Knotting and screening
- Brown stock washing
- Brown stock decker
- Oxygen delignification
- Weak black liquor storage tank vents

Existing mills can thus end up with two or more parallel systems to collect and dispose of NCGs and strip and recycle foul condensates. It is also apparent that process evolution will continue. Table 14 lists sources of odors and odor-removal uses.

Materials of Construction

Carbon steel is not suitable for NCG collection systems. Concentrated NCGs are saturated in water, and thus condensation occurs in the collection system. Dissolved NCGs in this condensate make it corrosive to carbon steel. Fiberglass-reinforced plastic (FRP) is generally not recommended for NCG collection systems because of attack on the resin by turpentine and methanol in the NCG system. The FRP resins are also subject to failure should NCG ignite.

Types 304L (S30403) and 316L (S31603) stainless steels have been the materials of choice for process equipment and piping handling NCGs and foul condensates. They do not experience general corrosion and remain free of fouling, which is beneficial for heat-transfer equipment.

Design of collection systems is important. Pipe networks should be properly grounded and potential sparking sources eliminated. Provisions should be made to prevent carryover of black liquor droplets into the NCG piping, especially where the black liquor has significant chloride ion content. Pipe slope should be adequate to ensure rapid condensate flow, and condensate separators should be installed at the low points of the lines (Ref 150). It may also be necessary to purge some lines and equipment on start-up and/or shutdown.

The pH of foul condensates is generally above 8, but in vapor space condensates the pH can be acidic due to dissolved NCGs. Depending on the mill and processes involved, chloride ions may be present. Preferential corrosion of weld metal may occur in the presence of some condensates if they are allowed to collect in "dead" zones. For example, Fig. 40 and 41 show corrosion of ferrite in type 316L weld metal after 20 years of service. Depending on temperature, stress-corrosion cracking of types 304L and 316L has also occurred.

Table 14 Sources of odors and removal units

Removal units	Sources of kraft mill odors
Lime kiln	Digesters
Power boiler	Accumulators
	Electrostatic precipitators
	Contaminated hot water
	Dissolving tanks
	Bleach plant towers
	Washers
	Tailings
	Oxygen delignification
Recovery boiler	Digester relief
Thermal oxidizer	Multieffect evaporators
	Makeup
	Sewer
	Causticizing
	Slaking
	Screens
	Foam tanks
	Brown stock washers and decker
Wet scrubber	Digester blow tanks
	Direct contact evaporators
	Sewer outfall
	Mud filters
	Knotters
	Driers
	LVHC vents
	Liquor storage tanks

It is not known if sulfur compounds have played a role in the instances of weld metal corrosion or stress-corrosion cracking. Further, the collection and analysis of NCGs and condensates for diagnosis of corrosion problems is not straightforward, and results may vary depending on sampling and analysis procedures. Thus it is not generally possible to predict which areas of process equipment, if any, may experience corrosion or stress-corrosion cracking. Duplex stainless steels, higher alloy austenitic stainless steels, and nickel-base alloys have been used in response to local occurrences of stress-corrosion cracking.

Stainless steels and nickel-base alloys suitable for high-temperature service are commonly used for burner components for firing NCGs in lime kilns and for some components of dedicated thermal oxidizers (lime kiln materials are discussed in the section "Corrosion Control in Chemical Recovery"). Very aggressive environments from sulfur compounds in the NCGs can develop in thermal oxidizers, which incorporate a quench chamber and scrubber for the combustion gases. Extensive use of nickel-base alloys and type 317L (S31703) stainless steel has been reported.

Because of the wide variety of processes and environments involved in collecting, processing, and incinerating NCGs, it is important when selecting equipment to review past performance and determine if process conditions will be outside the envelope of experience. In all cases, it is desirable to ensure quality in welding to maximize the margin against corrosion or fatigue.

Fig. 40 Preferential corrosion of type 316L weld metal where aggressive condensate collected in a vapor line

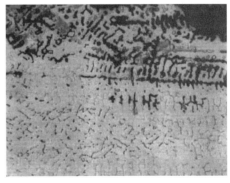

Fig. 41 Closer view of corrosion of ferrite in weld metal

Wastewater Treatment Corrosion in Pulp and Paper Mills

Randy Nixon, Corrosion Probe, Inc.

Corrosion problems are typically more severe in pulp mill wastewaters, especially for bleached pulp mills, than in paper mill wastewaters. These problems are present in wastewater from kraft pulping, sulfite pulping, and thermomechanical pulping (TMP) mill processes. Each of these is discussed in earlier sections of this article. The corrosion problems encountered include oxygen-driven pitting corrosion of submerged carbon steel, vapor phase acidic attack of concrete, microbiologically influenced corrosion (MIC) of carbon and austenitic stainless steels (liquid-phase), liquid-phase chemical attack of concrete, and several other corrosion mechanisms. These mechanisms affect the performance of various components of the wastewater treatment plant process as discussed in this section.

Wastewater System Components and Materials of Construction

The wastewater treatment process in most pulp and/or paper mills consists of two treatment stages:

1. Primary treatment stage, which includes neutralization or equalization and primary clarification
2. Secondary treatment stage, which includes aeration and secondary clarification

Neutralization. When the mill has both acidic and alkaline sewers, it will generally have a neutralization or equalization tank or chamber (generally lined concrete) where the two wastewaters mix and are adjusted for near-neutral pH. This usually includes the addition of sulfuric acid and/or sodium hydroxide. If pH swings are not problematic, it is also common to let the wastewaters mix in the primary clarifiers. The piping materials used for collecting wastewater throughout mills varies widely. Most mills in the United States built their treatment plants as mandated by law in the mid- to late-1970s. At that time, welded carbon steel, fiberglass-reinforced plastic (FRP), precast concrete cylinder pipe, and reinforced-concrete pipe were used. Since that time, much of the collection piping for bleached pulp mills has been replaced or sliplined with FRP or high-density polyethylene (HDPE), respectively, which have performed extremely well.

The primary clarifier provides quiescent flow conditions and retention times where the settlement and removal of suspended solids is accomplished. Most primary clarifiers in mills are circular concrete tanks with sloped bottoms. A center-supported, motor-driven rake mechanism as shown in Fig. 42 circulates around the tank to push the settled sludge into troughs in the bottom of the tank for conveyance to a sludge thickener or to sludge dewatering equipment. The center inlet column, feed well, baffles, rake arms, and other metallic components are generally constructed from carbon steel coated with a polymer-based coating system, usually epoxy. Type 316 stainless steel has been used in some rake-mechanism replacement projects. This has generally been very successful.

From the primary clarifiers, the clarified wastewater flows through launder troughs and piping to an aeration tank of some kind. The secondary treatment stage now begins.

Aeration. Bacteria are used to cleanse the wastewater biologically by consuming the harmful organic nutrients under oxidizing conditions. This process is largely focused on biological oxygen demand (BOD) reduction and removal. Aeration ensures the health and efficiency of the aerobic bacteria used to metabolize organic compounds and, in the process, converts dissolved solids into suspended solids.

Several types of aeration treatment are used in pulp and paper mills. Trickling filters allow wastewater to be aerated by ambient air while it flows through synthetic media with numerous small openings. Ambient aeration tanks are also used. These large open-topped concrete tanks have piping manifolds in their bottoms that convey compressed ambient air into the wastewater as bubbles. Agitators are used to keep the wastewater well mixed with bacteria and aerated. A third option (used by many Canadian mills) is pure oxygen reactors. These covered concrete tanks provide serpentine-type flow through the use of baffles and include agitators for mixing. Pure oxygen is introduced into the wastewater through piping jets. Since pure oxygen is more efficient than ambient air (20% oxygen), less overall gas flow is required into the wastewater.

Secondary Clarifiers. The overflow from the aeration tanks or chambers goes to the secondary clarifiers. Here the suspended solids created by bacterial action are settled out and removed as sludge. These concrete tanks with metallic rake mechanisms or collectors are generally similar in construction to the primary clarifiers.

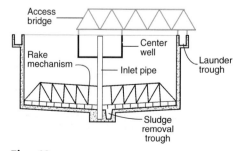

Fig. 42 Primary clarifier

The concentration of bacteria in the aeration tanks is maintained by the continuous return of settled biological floc to the aeration tanks from the secondary clarifiers. The return activated sludge (RAS) is a key factor in enhancing BOD removal in the secondary stage of treatment.

The clarified effluent from the secondary clarifiers overflows into the launder troughs and is then commonly discharged. The sludge is either fed back to the aeration tank as RAS or is pumped to a dewatering stage prior to being landfilled or burned in a biomass boiler. Unlike municipal wastewater treatment plants, most pulp and paper mill treatment plants do not use anaerobic digesters to further treat sludge nor do most mills chlorinate the treated wastewater before discharge.

Parameters Affecting Wastewater Corrosivity

Many parameters affect wastewater corrosivity. The most important parameter is pH as it impacts both the corrosion of metals and concrete.

pH. Table 15 (Ref 155) shows different mill wastewater pH ranges for different types of pulp mill and paper mill effluents. Most pulp mills today (2006) neutralize wastewater pH such that the wastewater is less aggressive based solely on pH once the primary clarifiers are reached.

Wastewaters with pH values below about 4 are aggressive to concrete structures and promote higher corrosion rates in carbon steel components. Low pH wastewaters will also affect the corrosion rates of austenitic stainless steels (types 304 and 316) provided high chlorides or other aggressive ions are present. pH values of 6 to 12 are generally not aggressive toward concrete. Rebar corrosion can be sustained in concrete where the pH of the surrounding concrete has been reduced from 12.5 (original) to 9.0, chloride is present, and oxygen diffusion is permitted (Fig. 43).

Chlorides in mill wastewaters generally increase the conductivity of the electrolyte and thus accelerate the corrosion of carbon steels. Higher chloride concentrations promote localized corrosion (pitting and crevice corrosion) of austenitic stainless steels. Localized corrosion of stainless steels is generally limited to collection system (types 304L or 316L) piping and is not found in clarifier components due to the large degree of dilution that occurs. However, chloride

Table 15 Wastewater pH ranges for various pulp and paper effluents

Effluent type	pH range
Kraft pulping (bleached)	2.0–12.0
Sulfite pulping	1.0–4.0
Thermomechanical pulping	4.0–5.0
Alkaline papermaking	7.0–10.0
Acidic papermaking	3.0–6.0

Source: Ref 155

ion ingress into reinforced concrete structures, especially precast, prestressed reinforced concrete piping and inlet structures to treatment plants has caused substantial reinforcing steel corrosion problems in some mills.

Microbiologically influenced corrosion (MIC) occurs commonly in RAS systems causing underdeposit corrosion of carbon steels and stainless steels due to the action of sulfate reducing bacteria (SRB). Uncoated carbon steel and 316L stainless steel have suffered severe corrosion losses due to SRB action in RAS piping systems and in clear effluent piping, respectively (Ref 156). Figures 44 and 45 show carbon steel and 316L stainless steel, respectively, attacked by SRB action.

Where hydrogen sulfide (H_2S) is present in aerobic headspaces or vapor-phase areas in covered wastewater tanks, sulfur-oxidizing bacteria (SOB) metabolize the H_2S and other reduced sulfur compounds to produce sulfuric acid. This biogenic acid formation, common to municipal sewers, causes aggressive corrosion of carbon steel, austenitic stainless steels, and hydrated portland cement. The acid generated by the MIC process causes rapid dissolution of the normally alkaline cement paste that holds concrete together.

Temperature in wastewater systems is typically in the range of 25 to 45 °C (80 to 110 °F) and is not a significant factor contributing to corrosion. When wastewater temperatures exceed this range, however, there can be increased corrosion of carbon steel and stainless steel. With a high chloride concentration and a wastewater temperature above 60 °C (140 °F), stress-corrosion cracking of austenitic stainless steels can occur. This has seldom been observed in mill wastewater except in some collection piping applications near the source of bleach plant effluents.

Sulfate increases wastewater conductivity and provides nutrition for SRB. Both of those factors contribute to corrosion of metal components. Sulfate attack of portland cement concrete piping and tankage can occur, especially where wetting and drying of the concrete with wastewater is common. Sulfate attack may be found in concert with sulfuric acid attack.

Carbon dioxide (dissolved CO_2) concentrations can be high in mill wastewaters particularly in the covered, slightly pressurized headspaces of pure oxygen reactors. A great deal of CO_2 evolves from the bacterial metabolism of organic compounds during secondary treatment. When aeration tanks are open to the atmosphere, the CO_2 gas largely dissipates into the air. In a covered, slightly pressurized headspace, more of the CO_2 remains dissolved as carbonic acid and the wastewater pH is slightly depressed. This increases the rate of aqueous carbonation attack of hydrated portland cement paste. The dissolved CO_2 concentration appears to have little impact on metal corrosion in mill wastewater systems.

Conductivity clearly contributes to the corrosivity of wastewater as it does in most aqueous environments. For carbon steel or other iron-base alloys, the higher the conductivity, the higher the corrosion rates. This is not necessarily true for stainless steels provided they remain passive and are not subjected to some form of localized corrosion.

Differential aeration influences corrosion in wastewater. For example, aggressive cells are established between the rake arm surfaces within the sludge layer and those rake arm surfaces above the sludge. The surfaces in the sludge corrode faster than the surfaces above the sludge because of differential aeration and/or concentration effects. These effects increase the open-circuit potential difference between the two regions, one cathodic and the other anodic (Ref 156).

Electrolytic couples (galvanic couples) can also occur between the rake mechanism and the reinforcing steel in the concrete walls and floor of clarifiers. With the two metals in a common electrolyte, the steel (rebars) within the alkaline concrete remain cathodic to the submerged steel rake mechanism and center supports, and so forth. Because there is a large surface area of reinforcing steel, corrosion of the immersed-steel clarifier can be significantly accelerated (Ref 156). Figure 46 shows a rake mechanism that suffered severe corrosive attack.

Velocity also affects wastewater corrosivity. Increased velocity typically increases the corrosion rate of carbon steel and iron-base alloys. An increase in velocity from 0 to 0.08 m/s (0 to 0.25 ft/s) can double the corrosion rate of carbon steel. This effect can occur between the tip of the rake mechanism and the center column of a clarifier.

In summary, many factors contribute to the corrosivity of wastewater, but the most influential parameters appear to be pH, reduced sulfur compound concentration, and velocity.

Corrosion Mechanisms

Numerous corrosion mechanisms occur in pulp and paper mill wastewater systems. The most common mechanisms are discussed separately for metal and concrete substrates.

Metal Corrosion

Corrosion of metals takes several forms in mill wastewater piping and equipment. General corrosion is influenced by pH, dissolved oxygen concentration at the corroding surfaces, conductivity, and velocity. Corrosion rates are

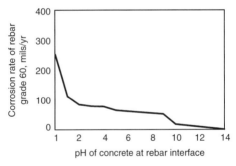

Fig. 43 Effect of pH on corrosion rate of carbon steel rebar in concrete

Fig. 44 Typical carbon steel corrosion morphology due to SRB action

Fig. 45 Resultant leak of 316L stainless steel due to internal SRB activity

Fig. 46 Badly corroded rake mechanism in a primary clarifier

often greater at the center well in clarifiers where dissolved oxygen levels are higher and at rake arm tips where velocities are higher. Experience has also shown that with lower pH wastewater (as in sulfite pulp mills), the corrosion rates of immersed carbon steel components are greater than in more neutral TMP or neutralized kraft pulp mill effluents.

Electrolytic (galvanic) corrosion of immersed carbon steel can occur in areas exposed by local failure of coatings. With typical anodic areas (metal exposed by coating breakdown) being smaller than the cathodic area on coated rake mechanisms, the corrosion rate of the exposed metal can be quite high. Proper recoating of the carbon steel combined with cathodic protection has been very effective in prevention of this rather typical electrolytic corrosion mechanism. Also, replacement of rake-mechanism parts, weirs, baffles, and other components with FRP has been very successful and cost effective.

Microbiological Corrosion. Pulp mill wastewaters often contain high concentrations of sulfur compounds including sulfates. Sulfates are metabolized under anaerobic conditions in wastewater piping to produce dissolved sulfides eventually forming hydrogen sulfide. The SRB involved cause corrosion of carbon steel waste water piping in the collection system. This same mechanism occurs in carbon steel and stainless steel RAS piping systems and in final effluent (clear effluent) piping where biofilm deposits develop and create an anaerobic environment. Once the biofilm prevents oxygen from being present, SRB colonize. As they metabolize the sulfates, enzymes are produced that create acidic conditions beneath the biodeposits and corrosion proceeds. The same corrosion problem is common in carbon steel sludge piping used to convey waste-activated sludge (WAS) from the clarifiers to the dewatering presses or filters.

Replacement of carbon steel wastewater collection and RAS piping with FRP and HDPE has been very successful. Reasonable performance has been attained in valves and pumps for these same systems using type 317 stainless steel; however, other stainless steels more resistant to SRB associated corrosion are being studied. Promising results have been reported for alloy 2205 duplex stainless steel (Ref 157).

Another microbial corrosion problem related to sulfur species involves the metabolism of hydrogen sulfide by SOB in aerobic headspaces of pipes and tanks. Hydrogen sulfide gas evolved from anaerobic sludge layers dissolves in the headspace condensates, and the *thiobacillus* bacteria metabolize the H_2S, ultimately producing sulfuric acid. The sulfuric acid can produce pH levels as low as 1 to 2, resulting in aggressive corrosion of carbon steel, 304 stainless steel, and concrete. This problem has been identified in stainless steel WAS piping, wastewater collection piping, covered tanks, pure oxygen reactor concrete, and in stainless steel agitator shafts in pure oxygen reactors (Ref 157).

Other forms of metallic corrosion mechanisms are rare in mill wastewater systems.

Concrete Corrosion

There are four basic mechanisms of concrete corrosion in mill wastewater systems: acid attack, aqueous or liquid phase carbonation, chloride-related deterioration, and sulfate attack.

Acid Attack. Acids react with concrete in a standard, acid-base neutralization reaction. The most common acid environment in mill wastewater systems is the sulfuric acid formed by bacterial action in aerated headspaces. This biogenic acid attack occurs in vapor spaces and has been discussed previously. Acid attack of concrete also occurs in the liquid phase where manufactured acids depress the pH of mill wastewaters. The rate of attack depends on the types of acids present and their concentrations.

The hardened cement paste in concrete is very alkaline, usually starting out with a pH above 12.5. The cement reacts with acidic solutions to produce soluble salts. These are mostly calcium salts because the hydrated cement is about 25% calcium hydroxide. Acid attack dissolves the cement phase of the concrete away, exposing the less soluble aggregate materials (Fig. 47).

In mill wastewater streams, the typical acids or acid sources are: sulfuric and hydrochloric acids from upsets in ClO_2 production, D-stage bleaching filtrates, and sulfite pulping process streams. Prevention of aqueous acidic attack of concrete has largely been achieved through the use of neutralization tanks and pH adjustment systems for the mill effluents. However, organic linings such as epoxies, polyurethanes, and vinyl esters for tanks have also been effective. Similarly, the biogenic headspace acidic attack has been averted through the use of anchored thermoplastic sheet linings or liquid applied organic linings.

Carbonation of Concrete (Liquid Phase). Carbonation occurs naturally in all concrete exposed to the atmosphere or to carbonated waters. It involves the reaction of carbon dioxide, CO_2, with the hydrated constituents of portland cement paste, especially calcium hydroxide, $Ca(OH)_2$. In the case of $Ca(OH)_2$, the reaction is:

$$Ca(OH)_2 + CO_2 \rightarrow CaCO_3 + 2H_2O$$

The reactions are actually more complex, with several reactions occurring between CO_2 and the other constituents of the hydrated cement paste. Regardless, all of these reactions produce carbonates, and they invariably result in shrinkage of the cement paste. Carbonation also occurs in concrete where CO_2 has been absorbed by water, producing carbonic acid.

The rate of carbonation in hardened concrete depends strongly on the permeability of the concrete, which in turn depends on the water-to-cement ratio, wastewater temperature, the concentration of CO_2 (carbonic acid) in the water, and the hardness and alkalinity of the water. Permeable concrete shows a greater rate of carbonation.

Liquid-phase carbonation of concrete occurs widely in mill wastewater treatment plant structures primarily in aeration tanks, oxygen reactors, and primary and secondary clarifiers. It is more common in pure oxygen reactors where greater dissolved CO_2 concentrations are possible due to the pressurized headspaces in these structures (Fig. 48). The CO_2 is generated by the decomposition of organics by aerobic bacteria in the reactors or aerators. Carbonation is also more aggressive where source waters are lower in hardness and alkalinity.

The rate of cement paste attack by liquid-phase carbonation is generally very slow. Usually the carbonation rate is approximately 2 to 3 mm ($1/16$ to $1/18$ in.) of depth loss in the first 1 to 2 years of exposure. Due to the pore-blocking effects of calcium carbonate and water, this rate slows down significantly after the first couple of years, depending on the extent of turbulent flow present.

Cement paste depth losses are typically 13 to 19 mm ($1/2$ to $3/4$ in.) over 15 or so years of exposure. The rate of cement paste loss is strongly influenced by flow conditions. Under quiescent conditions, the rate of attack is extremely low. The high solubility of calcium carbonate makes carbonation especially influenced by flow conditions. In turbulent flow conditions, the rates are much higher as the reaction products are washed away exposing fresh alkaline surfaces for further attack.

Carbonation can be reduced by ensuring high-quality, low-density concrete is used through lower water to cement ratios and proper curing

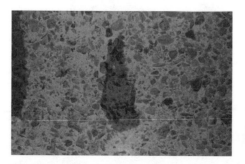

Fig. 47 Typical acidic attack of concrete in a neutralization chamber

Fig. 48 Cement paste losses due to liquid-phase carbonation in a pure oxygen reactor for a pulp and paper mill

practices. Protective organic or inorganic linings and sealers are also effective.

Chloride-Induced Concrete Deterioration (Vapor Phase). When a wet concrete mix is placed around steel reinforcing bars, a corrosion reaction occurs between the steel and the alkaline paste to produce a tightly adherent, protective oxide film on the steel surface. This passive film is stable at pH above 11 and prevents corrosion of the steel if the film is not damaged or removed.

Passive protection of the steel can be lost when the pH at the rebar/concrete interface decreases because moisture and oxygen penetrate the concrete and reach the steel. Chloride ions are particularly harmful because they disrupt the passive oxide film on the reinforcing steel (Fig. 49). Penetration can occur slowly by permeation of the concrete or rapidly through crack paths. Penetration usually occurs at different times in different locations to form local corrosion sites (anodes) on the reinforcing steel. The uncorroded, still-passive areas serve as the cathodes.

As the steel corrodes, expansion forces occur within the concrete because the steel corrosion products (rust) have a much greater volume than the steel that corroded to produce them. These expansion forces are much greater than the tensile strength of the concrete cover over the reinforcing steel. The concrete therefore cracks and spalls allowing further ingress of the corrosive environment to the steel to cause further corrosion.

The concrete matrix is not chemically influenced by chlorides. However, concrete exposed to an external source of chlorides can still deteriorate surprisingly rapidly because the presence of chloride ions significantly accelerates the corrosion of reinforcing steel in concrete.

Chloride ions are sufficiently small and mobile to readily enter the concrete and penetrate and undermine the stability of the passive film. This makes it easier for corrosion of the steel reinforcement to start and increases the conductivity of the environment around the steel. This latter effect increases the corrosion rates by enlarging the effective size of the cathodic zone around the corroding, anodic sites and generally increasing the chemical activity of the corrosion reactions.

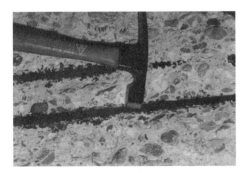

Fig. 49 Chloride ion induced rebar corrosion in a concrete splitter box that directed flow to two primary clarifiers in a bleached sulfite mill

Once rebar corrosion has started, chloride ions also play a pivotal role in acidifying the local anodic zone environment including the surrounding concrete. This is because iron cations react with water (hydrolyze) to form a solid precipitate of iron hydroxide or rust and produce hydrogen cations and chloride anions—that is, hydrochloric acid. A simplified overall reaction is:

$$2H^+ + 2OH^- + Fe^{2+} + 2Cl^-$$
$$\text{(Water)}$$
$$= Fe(OH)_2 + 2H^+ + 2Cl^-$$
$$\text{(Rust)} \quad \text{(Acid)}$$

Increased corrosion of the steel due to loss of passivity produces more ferrous cations that eventually form more rust and generate more acid. This lowers the pH and increases the concentration of chloride ions at the steel surface, promoting more corrosion and continuing the destructive cycle.

Chloride-induced reinforcing steel corrosion in wastewater treatment systems occurs in vapor phase areas that are subjected to cyclic wetting and drying conditions. Below the water line, sulfates and other precipitates commonly have a pore blocking effect that retards chloride intrusion and oxygen diffusion into the concrete. In lift stations, collection pipes, and in other chambers that experience varying flow elevations, chloride-induced rebar corrosion can cause severe degradation in a few years. The source of the chlorides can be chloride-rich accelerating admixtures used when the concrete was placed. This is, however, unusual in structures built since 1975 when use of such admixtures were generally stopped based on revised concrete design practices (see American Concrete Institute, ACI, documents 318 and 222R). The usual source of chloride ions in wastewater streams today (2006) is bleach plant effluents. Chlorides may also enter the wastewater stream in coastal mills that process sea floated logs. Mill closure to reduce water use may increase the chloride ion concentration in the mill process streams and thus in the wastewater to be treated.

Chloride ion related concrete degradation can be prevented by the use of dense concrete with adequate concrete cover thickness over reinforcing steel bars, through HDPE sliplining of concrete piping, and through the use of protective organic linings over concrete structures.

Sulfate Attack (Vapor Phase). Sulfate attack is the mechanism where sulfates (SO_4^{2-}, commonly from H_2SO_4) react with certain constituents in hydrated portland cement paste to form reaction products larger in volume than the reactant compounds. This increase in volume produces expansive forces, which can exceed the tensile strength of the concrete and promote microcracking and disintegration of the concrete matrix. When accompanied by water or water vapor allowing penetration into concrete capillary porosity, the sulfate attack mechanism is generally accelerated. Evaporation cycles can increase the sulfate reactions through the concentration of sulfates in the pores of the concrete.

Sulfate attack appears to be a two-stage chemical reaction. First, the sulfate ion reacts with the calcium ions in hydrated portland cement paste, mainly with calcium hydroxide. This reaction forms calcium sulfate or gypsum:

$$H_2SO_4 + Ca(OH)_2 \rightarrow CaSO_4 + 2H_2O$$

The second reaction occurs between the sulfate ion of the gypsum and hydrated calcium aluminate (mainly tricalcium aluminate or C_3A) to form calcium sulfoaluminates, mainly tricalcium sulfoaluminate or ettringite:

$$CaSO_4 + 2H_2O + C_3A$$
$$\rightarrow 3CaO \cdot Al_2O_3 \cdot 3CaSO_4 \cdot 3H_2O$$

The most destructive expansion occurs with the second reaction (Ref 158).

Sulfate attack of concrete occurs in sulfite pulp mill effluents and in the equalization tanks where sulfuric acid is used for wastewater pH adjustment prior to primary clarification. It does not appear to be a problem in most TMP or kraft pulp mill wastewaters except where upset conditions have occurred, resulting in higher-than-usual sulfate concentrations in structures where level changes are frequent. Prevention is again achieved through the design and construction of dense concrete, the use of organic linings, or through proper process control upstream of the wastewater treatment plant.

ACKNOWLEDGMENTS

This article was produced by the NACE International Task Group 283 as a revision to the article "Corrosion in the Pulp and Paper Industry," *Corrosion*, Vol 13, *Metals Handbook*, 9th ed., ASM International, 1987, p 1186–1220. We thank NACE International and the Chair of the Task Group, Harry Dykstra. NACE International standards and committee reports dealing with corrosion control are available from NACE International, 1440 South Creek Dr., Houston, Texas, 77084 (www.nace.org).

REFERENCES

1. C.-G. Von Essen, Corrosion Problems in Sulphate Pulp Mills, *TAPPI J.*, Vol 33 (No. 7), 1950, p 14A–32A
2. R.A. Huseby and M.A. Scheil, Corrosion of Mild Steel and Mild Steel Welds in Sulphate Digesters, *TAPPI J.*, Vol 34 (No. 4), 1951, p 202–209
3. B. Roald, Corrosion in Sulphate Digesters, *Norsk. Skogind.*, Vol 7 (No. 11), 1953, p 382–388
4. E.W. Hopper, Some Observations on the Causes and Prevention of Kraft Digester Corrosion, *TAPPI J.*, Vol 36 (No. 8), 1953, p 345–352
5. L. Ruus and L. Stockman, Investigations on Corrosion in Sulfate Digesters, *TAPPI J.*, Vol 38 (No. 3), 1955, p 156A–161A

6. B. Roald, The Effect of Steel Quality on Sulphate Digester Corrosion, *Norsk Skogind.*, Vol 11 (No. 11), 1957, p 446–450
7. J.J. Wegerif, Alkaline Digester Corrosion, *Pulp Paper Can.*, Vol 59, 1958, p 104–108
8. J.-P. Audouard, A. Desestret, G. Vallier, J. Chevassut, and J.-P. Mader, Study and Development of Special Austenitic-Ferritic Stainless Steel Linings for Kraft Pulp Batch Digesters, *Proc. Third Int. Symp. Corrosion in the Pulp and Paper Industry*, National Association of Corrosion Engineer, 1980, p 30–39
9. J.-P. Audouard, A New Special Austenitic-Ferritic Stainless Steel for Kraft Pulp Batch Digesters, *Proc. Fourth Int. Symp. Corrosion in the Pulp and Paper Industry*, National Association of Corrosion Engineer, 1983, p 43–47
10. J.-P. Audouard, Duplex Stainless Steels for Kraft Pulp Batch Digesters—Welding and Corrosion Performance, *Proc. Fifth Int. Symp. Corrosion in the Pulp and Paper Industry* (Vancouver, BC, Canada), Canadian Pulp and Paper Association, June 1986, p 193–199
11. J.-P. Audouard, F. Dupoiron, and D. Jobard, Stainless Steels for Kraft Digesters in New Pollution Free Mills, *Proc. Sixth Int. Symp. Corrosion in the Pulp and Paper Industry* (Helsinki, Finland), 1989
12. J.-P. Audouard, Corrosion Performance of Duplex Stainless Steels for Kraft Digester Applications, *Proc. Fifth World Congress and Exposition on Duplex Stainless Steels* (Massricht, Netherlands), Oct 21–23, 1997
13. A. Wensley, Corrosion of Batch and Continuous Digesters, *Proc. Ninth Int. Symp. Corrosion in the Pulp and Paper Industry*, Canadian Pulp and Paper Association, May 1998, p 27–37
14. K.L. Crooks and G.E. Linnert, Stainless Steel Weld Metal Overlay Linings for Pulp Digester Vessels, *TAPPI J.*, Vol 44 (No. 8), 1961, p 544–554
15. A. Wensley, Corrosion of Weld Overlay in Batch Digesters, *Proc. TAPPI Eng. Conf.*, TAPPI Press, 1998, p 1193–1200
16. A. Wensley, M. Moskal, and W. Wilton, Materials Selection for Batch Digesters, paper No. 378, *Corrosion 97 Conf.*, NACE International, 1997
17. J. Olsson, B. Leffler, and C. Jorgensen, Experiences of 2205 for Pulp Digesters and Other Pressure Vessels, *Proc. Fifth World Congress and Exposition on Duplex Stainless Steels* (Maastricht, Netherlands), Oct 21–23, 1997
18. P.H. Thorpe, Duplex Stainless Steel Pulp Digesters—Fabrication and User Experience in Australia and New Zealand, *Proc. Eighth Int. Symp. Corrosion in the Pulp and Paper Industry* (Stockholm, Sweden), TAPPI, NACE, May 1995, p 20–25
19. M. Moskal, G. Cheetham, J. Paultre, and W. Wilton, Quality Requirements for Duplex Stainless Steel Digester Fabrication, *Proc. Ninth Int. Symp. Corrosion in the Pulp and Paper Industry*, Canadian Pulp and Paper Association, May 1998, p 67–73
20. M.F. Frombach, Stainless Steel Overlay of Alkaline Digesters, *Proc. Tech. Section CPPA*, 1962, p D108-D117
21. P.A. Kammer and H.F. Reid, Welded Overlays for Sulfate Digesters, *Proc. First Int. Symp. Corrosion in the Pulp and Paper Industry*, 1974, p 144–149
22. T. Erikkson, E. Borjesson, and L. Johansson, Experience from Overlay Welded Lining in Digesters, *Proc. Eighth Int. Symp. Corrosion in the Pulp and Paper Industry* (Stockholm, Sweden), TAPPI, NACE, May 1995, p 58–64
23. R. Workman, Advances in Weld Overlay Technology Provide Corrosion Protection in Kraft Digesters, *Proc. TAPPI Eng. Conf.*, TAPPI Press, 1995, p 173–176
24. P. Hulsizer, Expedited Weld Overlay of Continuous Digesters, *Proc. TAPPI Eng. Conf.*, TAPPI Press, 1995, p 193–202
25. W.A. Mueller, Anodic Protection of Alkaline Pulping Digesters, *Pulp. Paper Can.*, Vol 60, 1959, p T3–T8
26. T.R.B. Watson, Anodic Protection of Kraft Digesters, *Mater. Perform.*, Vol 3 (No. 6), 1964, p 54–56
27. W.P. Banks, M. Hutchison, and R.M. Hurd, Anodic Protection of Carbon Steel Alkaline Sulfide Pulp Digesters, *TAPPI J.*, Vol 50 (No. 2), 1967, p 49–55
28. Guidelines for Corrosion Resistant Weld Overlay in Sulphate and Soda Digesters, TIP 0403-02, Technical Information Paper, TAPPI Press, 1998
29. S. Östlund, B. Ernerfeldt, B. Sandberg, and M. Linder, Investigations Regarding Corrosion in Batch Digesters, *Proc. Sixth Int. Symp. Corrosion in the Pulp and Paper Industry* (Helsinki, Finland), 1989, p 48–59
30. D. Singbeil and D. Tromans, Stress Corrosion Cracking of Mild Steel in Alkaline Sulfide Solutions, *Proc. Third Int. Symp. Corrosion in the Pulp and Paper Industry*, National Association of Corrosion Engineers, 1980, p 40–46
31. D.C. Bennett, Cracking in Continuous Digesters, *TAPPI J.*, Vol 65 (No. 12), 1982, p 43–45
32. R. Yeske, "Stress Corrosion Cracking of Continuous Digesters for Kraft Pulping—Final Report to the Digester Cracking Research Committee," Project 3544, Institute of Paper Chemistry, 1983
33. D.C. Bennett, Cracking of Continuous Digesters—Review of History, Corrosion Engineering, Aspects and Factors Affecting Cracking, *Proc. Fourth Int. Symp. Corrosion in the Pulp and Paper Industry* (Stockholm, Sweden), National Association of Corrosion Engineers, 1983, p 2–7
34. O. Pichler, Stress Corrosion in the Pre-Impregnation Vessel of a Continuous Kraft Digester, *Proc. Fourth Int. Symp. Corrosion in the Pulp and Paper Industry* (Stockholm, Sweden), National Association of Corrosion Engineers, 1983, p 15–20
35. N.J. Olson, R.H. Kilgore, M.J. Danielson, and K.H. Pool, Continuous Digester Stress Corrosion Cracking Results from a Two-Vessel Kraft System, *Proc. Fourth Int. Symp. Corrosion in the Pulp and Paper Industry* (Stockholm, Sweden), National Association of Corrosion Engineers, 1983, p 21–26
36. P.E. Thomas and D.M. Whyte, Stress Corrosion Cracking in a Kraft Continuous Digester, *Appita*, Vol 37 (No. 9), 1984, p 748–752
37. D. Singbeil, Kraft Continuous Digester Cracking: Causes and Prevention, *Pulp Paper Can.*, Vol 87 (No. 3), 1986, p T93–T97
38. D. Singbeil, Kraft Continuous Digester Cracking: A Critical Review, *Proc. Fifth Int. Symp. Corrosion in the Pulp and Paper Industry* (Vancouver, B.C., Canada), Canadian Pulp and Paper Association, June 1986, p 267–271
39. R. Yeske, In-Situ Studies of Stress Corrosion Cracking in Continuous Digesters, *TAPPI J.*, Vol 69 (No. 5), 1986, p 104–108
40. D. Crowe, Stress Corrosion Cracking of Carbon Steel in Kraft Digester Liquors, *Proc. Sixth Int. Symp. Corrosion in the Pulp and Paper Industry* (Helsinki, Finland), 1989, p 35–47
41. Pressure Vessels, *American Society of Mechanical Engineers (ASME) Boiler and Pressure Vessel Code*, Section VIII, Division 1
42. C.B. Christiansen and J.B. Lathrop, Field Investigation of Corrosion in Alkaline Pulping Equipment, *Pulp Paper Can.*, Vol 55, 1954, p 113–119
43. A. Wensley, Corrosion and Protection of Kraft Digesters, *TAPPI J.*, Vol 79 (No. 10), 1996, p 153–160
44. L. Kiessling, A Study of the Influence of Modified Continuous Cooking Processes on the Corrosion of Continuous Digester Shells, *Proc. Eighth Int. Symp. Corrosion in the Pulp and Paper Industry* (Stockholm, Sweden), TAPPI, NACE, 1995, p 12–19
45. D.C. Crowe, Corrosion in Acid Cleaning Solutions for Kraft Digesters, *Proc. Seventh Int. Symp. Corrosion in the Pulp and Paper Industry* (Orlando, FL), TAPPI, NACE, Nov 1992, p 33–40
46. A. Wensley, Intergranular Attack of Stainless Steels in Kraft Digester Liquors, paper No. 465, *Corrosion 96 Conf.*, NACE International, 1996
47. A. Wensley, Corrosion Testing in Digester Liquors, *Proc. TAPPI Eng. Conf.*, TAPPI Press, 1996, p 291–298
48. M.D. Moskal, Residual Stresses in Digester Weld Overlays, *Proc. Fifth Int. Symp. Corrosion in the Pulp and Paper Industry*

49. A. Wensley, Corrosion Control in a Kraft Continuous Digester, *Proc. Sixth Int. Symp. Corrosion in the Pulp and Paper Industry* (Helsinki, Finland), 1989, p 7–19
50. A. Wensley, Corrosion in Digester Liquors, *Proc. Eighth Int. Symp. Corrosion in the Pulp and Paper Industry* (Stockholm, Sweden), TAPPI, NACE, May 1995, p 26–37
51. J.I. Munro, Anodic Protection of Continuous Digesters to Prevent Corrosion and Cracking, *Proc. TAPPI Eng. Conf.,* TAPPI Press, 1983, p 181–185
52. H. Savisalo, M. Pullianen, and T. Kerola, Corrosion of Kraft Continuous Digesting: The Results of In-situ Electrochemical Testing, *Proc. Fifth Int. Symp. Corrosion in the Pulp and Paper Industry* (Vancouver, B.C., Canada), Canadian Pulp and Paper Association, June 1986
53. D. Singbeil and A. Garner, Anodic Protection to Prevent the Stress-Corrosion Cracking of Pressure Vessel Steels in Alkaline Sulfide Solutions, *Mater. Perform.,* Vol 26 (No. 4), 1987, p 31–36
54. D. Singbeil, Does Anodic Protection Stop Digester Cracking? *Proc. Sixth Int. Symp. Corrosion in the Pulp and Paper Industry* (Helsinki, Finland), 1989, p 109–116
55. L. Keissling, Application of an Anodic Protection System to a Preimpregnation Vessel, *Proc. TAPPI Eng. Conf.,* TAPPI Press, 1989, p 659–666
56. L.H. Boulton, G. Wallace, B.J. Barry, and D. Hodder, The Effects of Anodic Protection on Carbon Steel Corrosion Rates in a Kraft Pulp Digester as Measured by TLA, *Proc. Sixth Int. Symp. Corrosion in the Pulp and Paper Industry* (Helsinki, Finland), 1989, p 67–83
57. M. Pullianen and H. Savisalo, Experiences of Corrosion Control in Continuous Hydraulic and Two Vessel Digesters by Anodic Protection, *Proc. Sixth Int. Symp. Corrosion in the Pulp and Paper Industry* (Helsinki, Finland), 1989
58. D.A. Wensley, Corrosion of Liquor Heater Tubes, *Proc. TAPPI Eng. Conf.,* TAPPI Press, 1990, p 219–230
59. C.B. Thompson and D.A. Wensley, Corrosion of Mechanical Pulping Equipment, *Corrosion,* Vol 13, *ASM Handbook* (formerly 9th ed., *Metals, Handbook*), ASM International, 1987, p 1215–1218
60. P.H. Clarke, D.L. Singbeil, and C.B. Thompson, System Closure and Corrosion in High Yield Pulping Processes, *Ninth Int. Symp. Corrosion in the Pulp and Paper Industry,* Technical Section, Canadian Pulp and Paper Association, May 1998, p 131
61. C.B. Thompson, A Survey of Inspection Procedures and Materials Problems in Chip Refiner Systems, *Pulp Paper Canada,* Vol 94 (No. 7), 1993, T199–T203
62. C. Reid and S. Lawn, Macmillan Bloedell Experience with Stress Corrosion Cracking in CTMP Equipment, *TAPPI Eng. Conf. Proc.,* Book 3, TAPPI Press, 1992, p 1001–1014
63. G.A. Smook, *Handbook for Pulp and Paper Technologists,* TAPPI, Canadian Pulp and Paper Association, 1982
64. S.D. McGovern, A Review of Corrosion in the Sulphite Pulping Industry, *Pulp and Paper Industry Corrosion Problems,* Vol 3, National Association of Corrosion Engineers, 1980
65. K.-E. Jonsson, Stainless Steels in the Pulp and Paper Industry, *Handbook of Stainless Steels,* McGraw-Hill, 1977, p 43–1 to 43–12
66. *Pulp and Paper Manufacture,* Vol 1, 2nd ed., McGraw-Hill, 1969
67. R.A. McIlroy, The Control of Chloride Ions in the MgO Recovery System, *TAPPI J.,* Vol 56 (No. 9), 1973
68. *TAPPI J.,* Vol 41 (No. 12), 1958
69. S.K. Murarka, W.T.A. Dwars, and J.F. Langlois, Potential Formation of Corrosive Acidic Condensate and Its Monitoring in a Continuous Vapour Phase Bisulphite Chemi-Mechanical (BCMP) Process, *Pulp and Paper Industry Corrosion Problems,* Vol 5, Canadian Pulp and Paper Association, Montreal, Canada, 1986, p 187
70. A. Garner, et al., Corrosion in the Pulp and Paper Industry, *Corrosion,* Vol 13, *ASM Handbook* (formerly 9th ed., *Metals Handbook*), ASM International, 1987, p 1198
71. D.E. Bardsley, Compatible Metallurgies for Today's New Bleach Plant Washing Processes, *Eighth Int. Symp. Corrosion in the Pulp and Paper Industry* (Stockholm, Sweden), TAPPI, NACE, May 1995
72. D.C. Bennett and W.B.A. Sharp, Corrosion in Paper Machine Wet Ends, *TAPPI Eng. Conf.,* TAPPI Press, 1997, p 481
73. D.C. Bennett and W.B.A. Sharp, Corrosion Concerns in Dryer Sections, *TAPPI Eng. Conf.,* TAPPI Press, 1997, p 501
74. J.M. Muhonen, Corrosion of Stainless Steels in Whitewater, *First Int. Symp. Corrosion in the Pulp and Paper Industry,* 1974, p 75
75. M. Kurkela, N. Suutala, and J. Kemppainen, On the Selection of Stainless Steels in Bleach Plants and White Water Systems, *Fifth Int. Symp. Corrosion in the Pulp and Paper Industry* (Vancouver, B.C., Canada), Canadian Pulp and Paper Association, June 1986, p 127
76. J.D. Rushton and P.A. Kelly, White Water Corrosion Data from an Integrated Newsprint Operation, *TAPPI Eng. Conf.,* TAPPI Press, 1984, p 41
77. C.B. Thompson and A. Garner, Paper Machine Corrosion and Progressive Closure of White Water Systems, *TAPPI Eng. Conf.,* TAPPI Press, 1996, p 545
78. A.J. Piluso and C.C. Nathan, Chemical Treatment to Control Corrosion in the Wet-End Operations of Pulp and Paper Mills, *First Int. Symp. Corrosion in the Pulp and Paper Industry,* 1974, p 14
79. W.A. Mueller and J.M. Muhonen, Pitting Corrosion of Stainless Steels in Six Paper Machine Headboxes: Mechanism and Prevention, *TAPPI J.,* Vol 55 (No. 4), 1972, p 589
80. J.L. Ewald and D.P. Hundley, Surface Finish Requirements for the Headbox, *TAPPI J.,* Vol 63 (No. 11), 1980, p 121
81. J.P. Gerhauser, Corrosion of Fourdrinier Wires, *TAPPI J.,* Vol 43 (No. 4), 1960, p 207A
82. I.J. Magar, Laboratory Evaluation of Inhibitor Performance on Bronze in Synthetic White Water, *TAPPI J.,* Vol 75 (No. 3) 1992, p 167
83. C.C. Nathan and A.J. Piluso, Wet End Corrosion Problems in Paper Mills, *Second Int. Symp. Corrosion in the Pulp and Paper Industry,* 1977, p 126
84. E. Danielsson, Corrosion and Corrosion Control in the Paper Machine Wet End, *Third Int. Symp. Corrosion in the Pulp and Paper Industry,* National Association of Corrosion Engineers, 1980, p 191
85. D.C. Bennett and I.J. Magar, Protective Coatings for Wet Ends of Paper Machines, *Sixth Int. Symp. Corrosion in the Pulp and Paper Industry* (Helsinki, Finland), 1989, p 348
86. D.C. Bennett and I.J. Magar, Strategies for Coating Paper Machine Wet Ends, *TAPPI Eng. Conf.,* TAPPI Press, 1997, p 497
87. P.H. Thorpe, Corrosion in Paper Machines—An Overview, *Third Int. Symp. Corrosion in the Pulp and Paper Industry,* National Association of Corrosion Engineers, 1980, p 184
88. B.M. Blakey and P.H. Thorpe, Failure and Redesign of Press Roll Ends in Paperboard Machines, *Appita,* Vol 33, 1979, p 45
89. G. Sund and S. Strom, The Consequences of System Closure for Corrosion in Swedish Pulp and Paper Mills, *Fifth Int. Symp. Corrosion in the Pulp and Paper Industry* (Vancouver, B.C., Canada), Canadian Pulp and Paper Association, June 1986, p 51
90. D.F. Bowers, Changes in Water Properties and Corrosivity with Closure, *TAPPI J.,* Vol 66 (No. 9), 1983, p 103
91. D.A. Wensley, Localized Corrosion of Stainless Steels in White Waters, *Mater. Perform.,* Vol 28 (No. 11), 1989, p 68
92. D.C. Bennett and C.J. Federowicz, Prediction of Localized Corrosion of Stainless Steels in White Water, *Mater. Perform.,* Vol 21 (No. 4), 1982, p 39
93. R.C. Newman, Pitting of Stainless Alloys in Sulfate Solutions Containing Thiosulfate Ions, *Corrosion,* Vol 41 (No. 8), 1985, p 450
94. R.W. Barton, Thiosulfate Generated by Excess Application of Sodium Hydrosulfite,

Pulp Paper, Vol 59 (No. 6), 1985, p 108

95. A. Garner, Thiosulfate Corrosion in Paper Machine White Water, *Corrosion,* Vol 41 (No. 10), 1985, p 587
96. R.C. Newman, W.P. Wong, H. Ezubier, and A. Garner, Pitting of Stainless Steels by Thiosulfate Ions, *Corrosion,* Vol 45 (No. 4), 1989, p 282
97. W.S. Butterfield and P.E. Glogowski, Determination of the Corrosivity of Thiosulfate Ion in Paper Mill Systems, *TAPPI Eng. Conf.,* TAPPI Press, 1984, p 61
98. A. Wensley, H. Dykstra, and A. Augustyn, Corrosion Monitoring in Paper Machine White Waters, *TAPPI Eng. Conf.,* TAPPI Press, 1997, p 489
99. V.V. Gorelov and A.K. Talybly, Corrosion Protection of the Equipment in Closed and Reduced Water Recycling Systems, *Fourth Int. Symp. Corrosion in the Pulp and Paper Industry* (Stockholm, Sweden), National Association of Corrosion Engineers, 1983, p 113
100. P.H. Thorpe, Microbiological Corrosion of Stainless Steel in Paper Machines and Its Causes, *Fifth Int. Symp. Corrosion in the Pulp and Paper Industry* (Vancouver, B.C., Canada), Canadian Pulp and Paper Association, June 1986, p 169
101. A. Garner, Suction Roll Failures in Canada, *Pulp Paper Canada,* Vol 86 (No. 3), 1985, p 81–89
102. "Guide for Evaluation of Paper Machine Suction Roll Shells," TIP 0402-10, Technical Information Papers, TAPPI Press
103. "Liquid Penetrant Testing of New Suction Roll Shells," TIP 0402-11, Technical Information Papers, TAPPI Press
104. "Guidelines for Nondestructive Examination of Suction Roll Shells," TIP 0402-19, Technical Information Papers, TAPPI Press
105. "Suction Roll Shell Report," TIP 0402-01, Technical Information Papers, TAPPI Press
106. F.H. Mappin, Construction Materials and Corrosion in Tall Oil Production, *TAPPI J.,* Vol 70 (No. 1), 1987, p 59
107. M. Ketchum, Black Liquor Soap Separation and Acidulation: Changing Some Old Habits, *TAPPI J.,* Vol 73 (No. 2), 1990, p 107
108. M. Ketchum, personal communication, May 1999
109. "The Corrosion Resistance of Nickel-Containing Alloys in Sulfuric Acid and Related Compounds," INCO CEB No. 1, Bulletin No. 1318, available from the Nickel Development Institute, Toronto, Ontario, Canada, 1983
110. *Corrosion Handbook,* Avesta Sheffield, Stockholm, Sweden, 1994
111. S.C. Stultz and J.B. Kitto, Ed., *Steam—Its Generation and Use,* 40th ed., Babcock & Wilcox, 1992
112. T.N. Adams, W.J. Frederick, T.M. Grace, M. Hupa, K. Iisa, A.K. Jones, and H. Tran, Ed., *Kraft Recovery Boilers,* TAPPI Press, 1997
113. D. Levesque, M.P. Fard, and S. Morrison, BL Spray: Understanding the Effect of Black Liquor Properties and Splash Plate Nozzle Configuration on Spray Characteristics, *Proc. 2004 Int. Chemical Recovery Conf.* (Charleston, SC), June 2004, TAPPI Press, CD-ROM, 2004
114. M. Towers, V. Uloth, R. Hogikyan et al., "Mill-Wide Sampling at Four Canadian Kraft Mills. Part I: Process Stream Analysis," PPR1278, Paprican, Pointe Claire, PQ, 1997
115. R.P. Green and G. Hough, Ed., *Chemical Recovery in the Alkaline Pulping Processes,* 3rd ed., TAPPI Press, 1992
116. "Guidelines for Nondestructive Thickness Measurement of Black Liquor Recovery Boiler Tubes," TIP 0402-18, Technical Information Papers, TAPPI Press, 1999
117. T. Grace, Smelt Water Explosions, *Kraft Recovery Short Course* (Orlando, FL), Jan 7–10, 2002, TAPPI Press, 2002, p 405–418
118. J.L. Barna, R.J. Mattie, J.B. Rogan, and S.F. Allison, Fireside Corrosion Inspections of Black Liquor Recovery Boilers, *Kraft Recovery Short Course* (Orlando, FL), Jan 7–10, 2002, TAPPI Press, 2002, p 419–435
119. O. Stelling and A. Vegeby, Corrosion on Tubes in Black Liquor Recovery Boilers, *Pulp Paper Canada,* Vol 70 (No. 8), 1969, p 51–77
120. J.A. Dickinson, M.E. Murphy, and W.C. Wolfe, Kraft Recovery Boiler Furnace Corrosion Protection, *TAPPI Eng. Conf.,* TAPPI Press, 1981, p 607–612
121. L Clement and J.D. Blue, Recovery Boiler Floor Design and Alternative Materials, *Proc. Tenth Latin American Recovery Congress* (Concepcion, Chile), Aug 26–30, 1996
122. R.G. Tallent and A.L. Plumley, Recent Research on External Corrosion of Waterwall Tubes in Kraft Recovery Furnaces, *TAPPI J.,* Vol 52 (No. 10), 1969, p 1955–1959
123. D.L. Singbeil and A. Garner, Lower Furnace Corrosion in Kraft Recovery Boilers, Part I: A Review of Research Findings to Date, *TAPPI J.,* Vol 72 (No. 6), 1989, p 136–141
124. D.G. Bauer and W.B.A. Sharp, The Inspection of Recovery Boilers to Detect Factors that Cause Critical Leaks, *TAPPI J.,* Vol 74 (No. 9), 1991, p 92–100
125. D.L. Singbeil, L. Frederick, N. Stead, J. Colwell, and G. Fonder, Testing the Effects of Operating Conditions on Corrosion of Water Wall Materials in Kraft Recovery Boilers, *Proc. TAPPI Eng. Conf.,* TAPPI Press, 1996, p 647–680
126. *The Recovery Boiler Reference Manual for Owners and Operators of Kraft Recovery Boilers,* Vol II, American Paper Institute, 1991
127. A. Plumley, J. Henry, and A. Jones, Chromium Diffusion Coating—Definitive Protection for Papermill Recovery Boilers, *TAPPI Eng. Conf.,* TAPPI Press, 1997, p 1073–1078
128. Y. Arakawa, Y. Taguchi, T. Maeda, and Y. Baba, Experience with High Pressure and High Temperature Recovery Boilers for Two Decades, *Pulp Paper Canada,* Vol 106 (No. 12), 2005, p 89–92
129. T. Odelstam, H.N. Tran, D. Barham, D.W. Reeve, M. Hupa, and R. Backman, Primary Air Port Corrosion of Composite Tubes in Kraft Recovery Boilers, *Proc. TAPPI Eng. Conf.,* TAPPI Press, 1987, p 585–590
130. J.R. Kish, D.L. Singbeil, J.R. Keiser, and F.R. Jette, "Assessment of Molten $Na_2S \cdot xH_2O$/NaOH Solutions as a Cause of Composite Tube Corrosion in Kraft Recovery Boilers," paper No. 06236, Corrosion 2006, NACE International, 2006
131. H. Dykstra, N. Riseborough, and A. Wensley, Corrosion and Cracking of Lower Furnace Wall Tubes in Recovery Boilers, *TAPPI Eng. Conf.,* TAPPI Press, 1999, p 1071–1090
132. J.R. Keiser, D.L. Singbeil, P.M. Singh, G.B. Sarma, X.L. Wang, C.R. Hubbard, and R.W. Swindeman, Why Do Kraft Recovery Boiler Composite Floor Tubes Crack? *TAPPI Eng. Conf.,* TAPPI Press, CD-ROM, 2000
133. J.R. Keiser, D.L. Singbeil, G.B. Sarma, J.R. Kish, et al., Causes and Solutions for Recovery Boiler Primary Air Port Composite Tube Cracking, *11th Int. Symp. Corrosion in the Pulp and Paper Industry* (Charleston, SC), June 2004, TAPPI Press, 2004, p 107–122
134. J.R. Kish, D.L. Singbeil, and J.R. Keiser, North American Experience with Composite Tubes in Kraft Recovery Boilers, *Pulp Paper Canada,* Vol 106 (No. 4), 2005, p T75
135. J.R. Kish, I. Karidio, D.L. Singbeil, L.A. Frederick, J.R. Keiser, K.A. Choudhury, and F.R. Jetté, Reducing Primary Air Port Composite Tube Cracking in a Kraft Recovery Boiler through Changes in Operation, *Proc. TAPPI Engineering, Pulping and Environmental Conference,* TAPPI Press, CD-ROM, 2005
136. J.R. Kish, D.L. Singbeil, J.R. Keiser, D.A. Wensley, and F. Jette, Cracking and Performance of Composite Tubes and Air Port Designs in a Kraft Recovery Boiler, *11th Int. Symp. Corrosion in the Pulp and Paper Industry* (Charleston, SC), June 2004, TAPPI Press, 2004, p 135–154
137. V.R. Behrani, J. Mahmood and P.M. Singh, "Mid-Furnace Corrosion in Kraft Recovery Boilers and Its Control," paper No. 06238, *Corrosion 2006,* NACE International, CD-ROM, 2006

138. M. Makipaa, E. Kauppinen, et al., Superheater Tube Corrosion in Recovery Boilers, *Tenth Int. Symp. Corrosion in the Pulp and Paper Industry* (Helsinki, Finland), Aug 21–24, 2001, Vol 1, VTT Manufacturing Technology, Helsinki, 2001, p 157–180

139. K. Salmenoja, "Field and Laboratory Studies on Chlorine-Induced Superheater Corrosion in Boilers Fired with Biofuels," Report 00-1, Faculty of Chemical Engineering, Abo Akademi, Abo, Finland, 2000

140. R. Thompson, D.L. Singbeil, C.E. Guzi, and D. Streit, Corrosion of Generating Bank Tubes at the Mud Drum Interface in Kraft Recovery Boilers, *Proc. Seventh Int. Symp. Corrosion in the Pulp and Paper Industry* (Orlando, FL), Nov 1992, TAPPI, NACE, 1992, p 309–318

141. P. Hynninen, Ed., Environmental Control, *Papermaking Science and Technology,* Book 19, Finnish Paper Engineers' Association and TAPPI, Helsinki, Fapet Oy, 1998, p 102–107

142. M. Kouris, and M.J. Kocurek, Mill Control and Control Systems: Quality Testing, Environmental, Corrosion, Electrical Systems, *Pulp and Paper Manufacture Series,* Vol 9, TAPPI Press, 1992, p 235–341

143. A.F. Jagel, et al., Louisiana-Pacific: Approaching the Odor Free Kraft Mill, *Proc. TAPPI Int. Environmental Conf.,* TAPPI, 1998, p 151–156

144. D.C. Meissner, et al., Incinerating NCG in Recovery Boilers, *Proc. TAPPI Int. Environmental Conf.,* TAPPI, 1998, p 157–167

145. E. Sebbas, Reuse of Kraft Mill Secondary Condensates, *TAPPI J.,* Vol 81 (No. 7), 1998, p 53–58

146. T. Burgess, "The Basics of Foul Condensate Stripping," TAPPI Kraft Recovery Operations Short Course, TAPPI, 1998, p 4.1-1 to 4.1-12

147. T. Burgess, "Collecting and Burning Non-Condensable Gases," TAPPI Kraft Recovery Operations Short Course, TAPPI, 1998, p 4.2-1 to 4.2-18

148. B.R. Blackwell, et al., Review of Kraft Foul Condensates: Sources, Quantities, Chemical Composition, and Environmental Effects, *TAPPI J.,* Vol 62 (No. 10), 1979, p 33–37

149. B.R. Blackwell, Review of Methods for Treating Kraft Foul Condensates, *Proc. 72nd Annual Meeting of the Canadian Pulp and Paper Assoc.* (Montreal, Quebec, Canada), 1986, p B59–B74

150. K.L. Nguyen, Design Criteria for Kraft Non-Condensables Thermal Oxidation Systems, *Appita,* Vol 44 (No. 6), 1991, p 371–372

151. M. Bell, Dilute Non-Condensable Gas Systems are Designed for Greater Efficiency, *Pulp Paper,* June 1996, p 127–129

152. J.F. Straitz, Capture Heat from Air Pollution Control, *Chem. Eng.,* Oct 1993, p 6–14

153. R.D. Tembreull, A.L. Farr, and G. Lloyd, Design and Installation of a Replacement Thermal Oxidizer for Odor Abatement, *TAPPI J.,* Vol 82 (No. 4), 1999, p 157–164

154. S. Al-Hassan, et al., "Corrosion of Carbon and Stainless Steels in Condensed Black Liquor Vapors," paper No. 00594, Corrosion/2000, NACE International, 2000

155. C. Reid and D. Christie, "Corrosion in Secondary Effluent Treatment Systems," Corrosion/2004 (New Orleans, LA), NACE International, 2004

156. J. I. Munro, Corrosion Mechanisms and Mitigation Methods for Effluent Clarifiers, *Third Int. Symp. Corrosion in the Pulp and Paper Industry,* TAPPI, NACE, 1980, p 204–209

157. A. Iversen, MIC on Stainless Steels in Wastewater Treatment Plants, *ACOM,* Vol 1, 2004, p 2–20

158. R.A. Nixon, Basics of Corrosion of Concrete in Wastewater, *Water Environment Federation Proceedings,* WEFTEC Conference, 2001

SELECTED REFERENCES

Kraft Digester Corrosion

- S.J. Clarke and N.J. Stead, Corrosion Testing in Flash Tanks of Kraft Pulp Mills, *Mater. Perform.,* Vol 38 (No. 11), 1999, p 62–66
- F. Dupoiron, J.-P. Audouard, G. Schweitzer, and J. Charles, "Performances of Special Stainless Steels and Alloys in Pulp and Paper Industries: Industrial References and Filed Test Results," paper No. 420, Corrosion/94, NACE International, 1994
- H. Leinonen, "Corrosive Environments in Different Stages of Modern Kraft Batch Cooking," paper No. 590, Corrosion/2000, NACE International, 2000
- A. Wensley, Corrosion of Carbon and Stainless Steels in Kraft Digesters, paper No. 589, Corrosion/2000, NACE International, 2000

Bleach Plant Corrosion

- D.E. Bardsley, The Effect of Chlorine Dioxide Substitution on Various Metals Specifically for Mixer Applications, *Proc. Seventh Int. Symp. Corrosion in the Pulp and Paper Industry* (Orlando, FL), TAPPI, NACE, Nov 1992
- D.E. Bardsley, Selection and Performance Materials for Bleach Plant Washers and Auxiliary Equipment, *Pulp Washing Conf. Proc.,* CPPA/TAPPI, 1996, p 127–132
- S. Clarke and D.E. Bardsley, "Weld Consumables for Superaustenitic Stainless Steels in Bleach Plant Service," report PPR 1291, Pulp and Paper Research Institute of Canada, June 1997
- G. Coates and D.E. Bardsley, "North American Pulp Mill Applications of Duplex Stainless Steels, *Proc. Fifth World Conf. on Duplex Stainless Steels* (Maastricht, Netherlands), Oct 21–23, 1997, Stainless Steel World, 1997, p 857–864
- D.C. Crowe and A.H. Tuthill, Corrosion of Metallic and Nonmetallic Piping for Bleach Plant D Stage Filtrate, *Eng. Conf. Proc.,* Sept 25, 1990, TAPPI Press, available as NiDI reprint No. 14013
- D.W. Reeve, "Bleaching Chemicals and Sequences," Bleach Plant Seminar, Section 1–3, TAPPI, 1998
- B. Roy, G. DosSantos, D.E. Bardsley, B. van Lierop, and R. Berry, "Material Selection for High Temperature Peroxide Bleaching," report PPR 1292, Pulp and Paper Research Inst. of Canada, June 1997
- R.W. Schutz and M. Xiao, "Practical Windows and Inhibitors for Titanium Use in Alkaline Peroxide Bleach Solutions," paper No. 427, Corrosion/94, NACE International, 1994
- "GTAW Root Pass Welding of 6% Molybdenum Austenitic Stainless Steel Pipe—Open Root Joint with Hand Fed Filler," TIP 0402-20, Technical Information Papers, TAPPI Press
- A.H. Tuthill and D.E. Bardsley, Performance of Highly Alloyed Materials in Chlorine Dioxide Bleaching, *Eng. Conf. Proc.,* Sept 25, 1990, TAPPI Press, available as NiDI reprint No. 14014
- A.H. Tuthill and R.E. Avery, Specifying Stainless Steel Surface Treatments, *Adv. Mater. Process.,* Vol 142 (No. 6), 1992, available as NiDI reprint No. 10068
- A.H. Tuthill, R.E. Avery, and A. Garner, Corrosion Behavior of Stainless Steel, Nickel Base Alloy and Titanium Weldments in Chlorination and Chlorine Dioxide Bleaching, *Proc. Seventh Int. Symp. Corrosion in the Pulp and Paper Industry* (Orlando, FL), Nov 1992, TAPPI, NACE, available as NiDI reprint No. 14026
- O.A. Varjonen and T. Hakkarainen, "Corrosion of Titanium (Grade 5) in Alkaline Hydrogen Peroxide Bleaching Solution—Laboratory Experiments," paper No. 425, Corrosion/94, NACE International, 1994
- A. Wensley, et al., Corrosion of Stainless Steel Alloys in a Chlorine Dioxide Bleached Pulp Washer, *Engineering Conf. Proc.,* 1991, TAPPI Press, p 499–509

Corrosion in Chemical Recovery

- J. Aromaa and A. Klarin, Materials, Corrosion Prevention and Maintenance, *Papermaking Science and Technology,* Book 15, Fapet Oy, Helsinki, Finland, 1999, available from TAPPI Press
- D.C. Crowe, On-Line Corrosion Monitoring in Kraft White Liquor Clarifiers, *Proc Eng. Conf.* (Dallas, TX), Sept 11–14, 1995, TAPPI, p 203
- S. Meiley and D. Bennett, Corrosion Monitoring and Condition Evaluation in

Liquor Tanks, *Proc. Seventh Int. Symp. Corrosion in the Pulp and Paper Industry* (Orlando, FL), Nov 1992, TAPPI, NACE
- T. Odelstam, H.N. Tran, D. Barham, D.W. Reeve, M. Hupa, and R. Backman, Air Port Corrosion of Composite Tubing in Kraft Recovery Boilers, *Proc. Eng. Conf.* (New Orleans, LA), Sept 1987, TAPPI
- D. Singbeil, R. Prescott, J. Keiser, and R. Swinderman, Composite Tube Cracking in Kraft Recovery Boilers—A State of the Art Review, *Proc. Eng. Conf.* (Nashville, TN), Oct 6–9, 1997, TAPPI, p 1001
- H.N. Tran, N. Mapara, and D. Barham, The Effect of H_2O on Acidic Sulfate Corrosion in Kraft Recovery Boilers, *TAPPI J.,* Vol 79 (No. 11), 1996, p 155–160
- A. Wensley and P. Champagne, Effect of Sulfidity on the Corrosivity of White, Green and Black Liquors, paper No. 281, Corrosion/99 (San Antonio, TX), April 26–29, 1999, NACE International

Recovery Boiler Corrosion

- J.L. Barna, R.J. Mattie, J.B. Rogan, and S.F. Allison, Recovery Boiler Corrosion is Due to Complex Mechanisms and Conditions, *Pulp Paper,* Vol 63 (No. 6), 1989, p 90–98
- J.R. Kish, D.L. Singbeil, and J.R. Keiser, "Stress Corrosion Cracking Susceptibility of Alternative Composite Tube Alloys in Simulated Kraft Recovery Boiler Environment," paper No. 04244, Corrosion/2004, NACE International, 2004
- M. Makipaa, E. Kauppinen, T. Lind, J. Pyykonen, J. Jokiniemi, P. McKeough, M. Oksa, T.H. Malkow, R.J. Fromdham, D. Baxter, L. Koivisto, K. Saviharju, and E. Vakkilainen, Superheater Tube Corrosion in Recovery Boilers, *Proc. Fall Technical Conference,* (Charleston, SC), June 6–10, 2004, TAPPI Press, CD-ROM, 2004
- D.L. Singbeil, R. Prescott, J.R. Keiser, and R.W. Swindeman, Composite Tube Cracking in Kraft Recovery Boilers—State of the Art Review, *Proc. Eng. Conf.* (Atlanta, GA), TAPPI Press, 1997, p 1001–1024
- H.N. Tran, D. Barham, and M. Hupa, Fireside Corrosion in Kraft Recovery Boilers—An Overview, *Mater. Perform.,* Vol 37 (No. 7), 1988, p 40–46

Corrosion in the Food and Beverage Industries

Shi Hua Zhang and Bert Moniz, DuPont Company
Michael Meyer, The Solae Company

A RECENT SURVEY estimated the total annual direct cost of corrosion in the food-processing industry as $2.1 billion dollars. This cost assumes the stainless steel consumption and cost in this industry are entirely attributed to corrosion (Ref 1). The total estimated stainless steel cost for the food-processing industry is $1.8 billion per year, accounting for about 86% of the total annual corrosion cost. This cost includes stainless steel utilized in beverage production, food machinery, cutlery and utensils, commercial and restaurant equipment, and appliances. Other annual corrosion costs include $250 million for the cost of aluminum cans in the beverage industry and $50 million for corrosion inhibitors used in the food-processing industry (Ref 1). It can be seen from these costs that the appropriate stainless steel selection is the primary method for addressing corrosion problems in the food and beverage industries. As a result, this article focuses primarily on the applications of stainless steels.

Corrosion Considerations

There are several distinct aspects of corrosion that should be taken into account when solving the corrosion problems in the food and beverage industries:

- Stainless steels are the primary material of construction for equipment used in the food and beverage industries due to their high corrosion resistance, cleanability, and sanitization properties. The majority of corrosion problems faced in the food and beverage industries are related to stainless steel. It is essential to understand the corrosion properties and corrosion mechanisms of stainless steels to more efficiently solve corrosion problems in the food and beverage industries.
- Although in general the corrosivity of foodstuffs is less than that of most aggressive chemicals, some foodstuffs are quite corrosive. Factors attributing to the corrosivity of foods and beverages include organic acids with low pH, moderately to highly concentrated chloride from salt, and the high temperature at which some foods are processed. When selecting materials of construction for food- and beverage-processing equipment, all these factors should be considered and evaluated.
- In the food and beverage industries, small amounts of corrosion that do not detrimentally affect equipment integrity and safety cannot be tolerated due to concern for potential contamination and food spoilage of products by metallic ions.
- Due to the hygiene requirement in food-processing equipment, cleaning and sanitizing are necessary processes frequently performed in any food-manufacturing process. Since some cleaning and sanitizing media are quite corrosive to metals, the selection of proper cleaning and sanitizing agents and the strict control of the cleaning and sanitizing process parameters become some of the critical parts of an overall corrosion-control strategy.
- Some of the processes used in the food and beverage industries, such as cooling water treatments, are similar to those used in other industries, such as in the chemical processing industry. General practices used in these industries for attacking similar corrosion problems may be adopted for use in the food and beverage industries.
- When materials and chemicals (such as corrosion inhibitors) are in direct contact with foods and beverages during manufacturing, processing, and handling, they are regulated and must comply with the applicable regulations.

Regulations in the United States

Materials used in food processing must be certified as "food grade," meaning it is safe for use in food. With polymers and elastomers, food grade implies the product is made from virgin material; that is, it has not been contaminated or reprocessed.

3-A Sanitary Standards Inc. (3-A SSI) formulates standards and practices for the sanitary design, fabrication, installation, and cleanability of dairy and food equipment/systems that are used to handle, process, and package consumable products where a high degree of sanitation is required (Ref 2).

The dairy industry first developed standards in the 1920s to define its sanitation requirements. 3-A standards exist for 70 different types of equipment, such as heat exchangers, holding tanks, homogenizers, evaporators, gaskets, pumps, and so forth. Conforming equipment may display the 3-A symbol, which certifies the design meets 3-A standards in hygienic design and cleanability. 3-A standards have often been incorporated de facto by other segments of the food industry.

It is difficult to obtain a waiver from 3-A standards. For example, 3-A sanitary piping and pump standards are dominant, because of the sheer weight of experience contained in them. There currently are no 3-A standards addressing equipment or piping corrosion.

The American Society of Mechanical Engineers (ASME) Bioprocessing Equipment (BPE) Code has increasing influence within the industry in aspects of equipment design (Ref 3). It addresses:

- The need for equipment designs that can be cleaned and sterilized
- Special emphasis on the quality of weld surfaces once the required joint strength is obtained
- Standardized definitions used by the entire community from fabricators to end users
- The need to accommodate and integrate other relevant standards into the Code

Bioprocessing equipment covers design for cleanability and sterility, dimensions and tolerances for stainless steel tube welding, hygienic clamp fittings, material joining guidance, stainless and higher alloy materials surface finishes, and sanitary equipment sealing. The BPE Code follows ASME B31.3 for general piping requirements.

The BPE Code systems may be used for services that contact water, such as water for injection (WFI), purified water, clean steam, and other products.

The American Welding Society (AWS) is becoming increasingly involved in regulating fabrication practices in food and pharmaceuticals, but has developed relatively few standards (Ref 4). When a few large shops used to control the industry there was no need to capture the nuances of fabrication. Smaller, less experienced shops require more guidance, and the AWS standards help with welding procedure guidance. Among these standards are:

- 18.1 Specification for Welding of Austenitic Stainless Steel Tube and Pipe Systems in Sanitary (Hygienic) Applications. It outlines welding standards for use in the manufacture and construction of dairy- and food-product processing plants.
- 18.2 Guide to Weld Discoloration Levels on Inside of Austenitic Stainless Steel Tube. It shows the degrees of coloration on the inside of an austenitic stainless steel tube with increasing amounts of oxygen in the backing shield gas.

The Food and Drug Administration (FDA) through hazard analysis and critical control point (HACCP) follows seven principles (Ref 5):

- Analyze for potential hazards. They could be biological, such as a microbe; chemical, such as a toxin; or physical, such as ground glass or metal fragments.
- Identify critical control points. Indicate places in the production line, from raw state through processing and shipping to assimilation by the consumer, where a potential hazard can be controlled or eliminated. For example, cooking, cooling, packaging, and metal detection.
- Establish preventive measures. Select critical limits for each control point. For example, with cooked food it might include designating the minimum cooking temperature/time to ensure microbe elimination.
- Create procedures to monitor the critical control points. They might include determining how and by whom time and temperature should be monitored.
- Decide on required corrective actions when monitoring indicates failure to meet a critical limit. For example, reprocess or dispose of food if the minimum cooking temperature is not met.
- Write procedures to verify that the system is working properly. For example, testing time-and-temperature recording devices to verify that a cooking unit is performing as required.
- Install effective record-keeping practices, backed by sound scientific knowledge to document the HACCP system. This would include records of hazards and their control methods, the monitoring of safety requirements, and action taken to correct potential problems. An example of sound scientific knowledge includes published microbiological studies on time and temperature factors for controlling food-borne pathogens.

The HACCP system is based on sound science and focuses on identifying and preventing hazards from contaminating food and permits more efficient and effective oversight. Record keeping allows investigators to assess long-term compliance rather than on the day of an inspection. If there is a potential corrosion problem, HACCP would require a prescription to address it.

The FDA also takes responsibility for determining whether and how manufactured materials may be used in contact with food products. Definitions for proper use are found in a series of regulations published annually under Government Regulations CFR 21 (see discussion below). The FDA provides certain specifications regarding composition, additives, and properties. A material that meets these standards can then be stated as FDA compliant. It is the responsibility of the end user to use the product in a manner compatible with FDA guidelines (Ref 6).

Code of Federal Regulations (CFR) Title 21, Subpart B—Substances utilized to control the growth of microorganisms, Section 178.1001 Sanitizing Solutions, describes sanitizing solutions that may be safely used on food-processing equipment and utensils, and on other food-contact articles as specified in this section, within prescribed conditions.

United States Department of Agriculture (USDA). The USDA has jurisdiction over equipment used in federally inspected meat and poultry processing plants and over packaging materials used for such products. Materials used in this equipment are approved on an individual basis. Determining suitability for use of components and the materials from which they are made is the responsibility of the equipment manufacturer.

Corrosivity of Foodstuffs

The corrosivity of a foodstuff is determined primarily by its fluid pH and the chloride ion concentration. The pH ranges of a number of foodstuffs are shown in Table 1 (Ref 7, 8). Many foods are acidic in nature, and the chloride ion concentration of foods can be as high as a few thousand ppm. Foods such as liquid sugars and gelatin may contain chloride ion on the order of 2000 ppm (Ref 9), while some dairy whey solutions may contain levels very much greater than this (Ref 7). In general, the most corrosive foodstuffs and processes are those containing high chloride levels, high oxygen content, at a low pH, and operating at relatively high temperatures. Most stainless steels provide good corrosion resistance by forming a robust passive film on the metal surface. High chloride ion and low pH conditions tend to damage the passive film of the stainless steels and initiate localized corrosion, such as pitting. Three factors, pH, chloride, and temperature play an even more important role in determining the corrosion severity when 300 series stainless steels (austenitic stainless steels) are used as the materials of construction. Currently, 300 series stainless steels are widely used in food and brewery industries due to their good corrosion performance. However, two concerns when selecting these stainless steels are their susceptibility to stress-corrosion cracking (SCC) and pitting. Although the SCC of the austenitic stainless steels may be affected by many factors, in general the prerequisite conditions for the initiation

Table 1 Approximate pH values of various foods

Food	pH
Apples	2.9–3.3
Apricots	3.6–4.0
Asparagus	5.4–5.8
Bananas	4.5–4.7
Beans	5.0–6.0
Beers	4.0–5.0
Beets	4.9–5.5
Blackberries	3.2–3.6
Bread, white	5.0–6.0
Butter	6.1–6.4
Cabbage	5.2–5.4
Carrots	4.9–5.3
Cherries	3.2–4.0
Cider	2.9–3.3
Corn	6.0–6.5
Crackers	6.5–8.5
Dates	6.2–6.4
Eggs, fresh white	7.6–8.0
Flour, wheat	5.5–6.5
Gooseberries	2.8–3.0
Grapefruit	3.0–3.3
Grapes	3.5–4.5
Hominy (lye)	6.8–8.0
Jams, fruit	3.5–4.0
Jellies, fruit	2.8–3.4
Lemons	2.2–2.4
Limes	1.8–2.0
Maple syrup	6.5–7.0
Milk, cows	6.3–6.6
Olives	3.6–3.8
Oranges	3.0–4.0
Oysters	6.1–6.6
Peaches	3.4–3.6
Pears	3.6–4.0
Pickles, dill	3.2–3.6
Pickles, sour	3.0–3.4
Pimento	4.6–5.2
Plums	2.8–3.0
Potatoes	5.6–6.0
Pumpkin	4.8–5.2
Raspberries	3.2–3.6
Rhubarb	3.1–3.2
Salmon	6.1–6.3
Sauerkraut	3.4–3.6
Shrimp	6.8–7.0
Soft drinks	2.0–4.0
Spinach	5.1–5.7
Squash	5.0–5.4
Strawberries	3.0–3.5
Sweet potatoes	5.3–5.6
Tomatoes	4.0–4.4
Tuna	5.9–6.1
Turnips	5.2–5.6
Vinegar	2.4–3.4
Water, drinking	6.5–8.0

Source: Ref 7, 8

of SCC are: (a) a mechanism for concentrating chloride, (b) temperatures above 60 °C (140 °F), and (c) a tensile stress within the material, such as residual stress at welds. Pitting may occur in stagnant conditions in waters and when oxidizing chloride-containing solutions are used to clean equipment.

Corrosion in Organic Acids. When organic acids are present, molybdenum-containing type 316 stainless steel offers better resistance to corrosion than molybdenum-free type 304 stainless steel. In general, the corrosivity of organic acids decreases as the carbon chain length of the acid increases. The most corrosive organic acid is formic, followed by acetic and oxalic. Elevated temperatures and the presence of chloride increase attack. Table 2 contains the corrosion-rate data of types 304 and 316L stainless steels in various organic acids (Ref 10).

Some of the other chemical constituents of food fluid can act as corrosion inhibitors. For example, phosphates, polypeptides (proteins), and some anions including hydroxyl, carbonate, sulfate, and nitrate are known to inhibit pitting corrosion of stainless steel in chloride-containing solutions (Ref 7, 11). A recent study (2003) has shown that some of the amino acids have certain inhibition effects on the corrosion of carbon steels in acid environments (Ref 12, 13).

Specific Considerations for Beer and Wine. Beer has a pH of about 4 when fresh, but the pH can drop to 3.5 or below if the beer is exposed to oxygen such that it sours (Ref 14). Stainless steel is generally capable of handling these acidity levels.

In the wine industry, sulfur dioxide is widely used to suppress biochemical action. Sulfur dioxide is the most corrosive agent found in wines as it forms sulfurous acid (Ref 15). In addition, tartaric, acetic, tannic, malic, and citric acids are also found in wines to a greater or lesser extent. Stainless steel is resistant to all these weak acids in the concentration in which they may occur (Ref 15). The selection of the exact stainless steel to use is really dependent on the free sulfur dioxide content in the wine. For example, in Germany, where the free sulfur dioxide content is limited by law to a maximum of about 50 ppm, type 304 stainless steel has been widely used for many years with excellent results. However, it is generally accepted that if the free sulfur dioxide content exceeds about 75 ppm, it is safer to use type 316L stainless steel (Ref 15). When wine tanks are only partially filled, sulfurous acid droplet condensation may occur on the tank roof and walls in the vapor phase above the wine level at much lower free sulfur dioxide contents and cause pitting at these vapor locations. To minimize the risk associated with these phenomena, it is preferable to fabricate the tank out of type 316L stainless steel.

Contamination of Food Products by Corrosion

Since stainless steel—particularly type 304 stainless steel, which contains 19% Cr and 9% Ni—is widely used both in industry and domestically for the processing and preparation of food, the potential contamination of food products by corrosion of stainless steel has been a concern for a long time. This potential food contamination issue had been addressed as early as 1961 (Ref 15, 16). It was found that when an 18Cr-15Ni stainless steel utensil was used for cooking fruits, vegetables, poultry, rice, and custard for 1 h, the chromium and nickel contents of the foods were found to be 0.12 and 0.13 ppm, respectively. When a stainless steel containing 3.7% Mo was used in another test, the residual molybdenum content was 0.05 ppm. It was concluded, therefore, that the quantities of these elements likely to be found in foods as a result of contact with stainless steel are of no pharmacological significance and that stainless steel alloys most likely to be used with foods or by the food-processing industry are safe.

In a 1997 study (Ref 17), lemon marmalade, green tomato chutney, potatoes, rhubarb, and apricots were cooked from 15 min to 2 h at normal domestic cooking temperature using 12 stainless steel saucepans from four manufacturers. The foods were purposely selected because their acidity or chloride content made them more likely than other foods to be aggressive against stainless steel. The study results showed that using new pans caused a higher nickel and chromium release. For example, the original mean chromium content of rhubarb before cooking was <30 ppb, while the chromium content increased to 202 ppb after cooking using a new stainless steel saucepan. The nickel content of the rhubarb increased from 39 to 296 ppb in the same case. After the stainless steel saucepans were used many times, the chromium and nickel release rate was decreased drastically, probably due to the passive oxide film formed on the surface. For instance, the chromium content was reduced to 49 ppb after the new saucepan was used for five cooking operations. The nickel content was reduced to 66 ppb in the same case. The surface treatment conditions of the new pans have a significant effect on the chromium or nickel release rate. The new stainless steel saucepans with four kinds of surface treatments including (a) mild abrasion using a stainless steel reviver, (b) treatment with nitric/hydrofluoric acid (HNO_3/HF), (c) electropolishing, and (d) ultrasonic cleaning were used in the test for evaluating surface finish effects. It was found that the electropolished saucepans had the lowest chromium and nickel release rate. The chromium release was reduced about five times by electropolishing treatment, from 202 to 40 ppb. The HNO_3/HF treatment resulted in the smallest improvement, as the results were unclear (appeared to be random in nature). The chromium content was increased from 202 to 250 ppb, while the nickel content was reduced from 296 to 90 ppb after the HNO_3/HF treatment. In all experiments the ratio of chromium to nickel in the food product was less than 2 to 1, the ratio of the concentration of these metals in the stainless steel. This suggests the acids in fruits react preferentially with nickel. The study showed that although chromium and nickel pickup due to the use of stainless steel utensils was evident, the amount of the chromium and nickel pickup may not be significantly high compared to the daily intake from the diet and water. The daily intake of chromium from the diet and water has been was reported as 50 to 200 µg/day (Ref 17, 18), and average human daily intake of nickel is approximately 200 µg/day (Ref 17, 19).

In the wine industry, the potential contamination of wines, particularly by iron ions, through the corrosion of stainless steels had been extensively studied before stainless steels were popularly used in the industry. Tests made by the Services Technique du Comite Interprofessional du Vin de Champagne showed that over a 4 year period, there was no difference in iron content between wine stored in a stainless steel tank and the same wine stored in corked glass magnums (Ref 15, 20). A number of tests carried out in several countries including Germany, France, and Switzerland showed that there was no difference in the iron content and the turbidity between wine in contact with stainless steel and wine in contact with the more traditional equipment, such as concrete and glass-lined tanks (Ref 15). The good performance of stainless steels in the wine industry has been confirmed by experience.

Selection of Stainless Steels as Materials of Construction

Selecting the most appropriate materials of construction for food-processing equipment is

Table 2   Comparative corrosion resistance of type 304 and type 316L stainless steel

	Temperature		Corrosion rate			
			304		316L	
Solution	°C	°F	mm/yr	mils/yr	mm/yr	mils/yr
20% acetic acid	Boiling	Boiling	0.76	30	0.007	0.3
10% citric acid	99	210	0.21	8.3	0.13	0.5
45% formic acid	Boiling	Boiling	1.22	48	0.28	11
65% nitric acid	Boiling	Boiling	0.18	7	0.20	8
10% oxalic acid	Boiling	Boiling	1.22	48	0.91	36
0.4% hydrochloric acid	27	80	0.25	10	0.130	5
40% sodium bisulfide	Boiling	Boiling	0.041	1.6	0.028	1.1

Source: Ref 10

the most effective method for addressing/preventing potential corrosion problems in the food and beverage industries. The processing equipment must have good corrosion resistance to foodstuff without contamination, and it must survive cyclic cleaning and sanitizing processes. Stainless steels have been the most popular and widely used materials of construction in the food and beverage industries due to their excellent general corrosion resistance, reasonably good pitting and crevice corrosion resistance, cleaning ease, and appealing appearance. A discussion of specific stainless steels, their corrosion resistance, fabrication and applications is presented in this section, and compared to other potential materials of construction.

Stainless Steel Types. Although there are several families of stainless steels (Ref 21), the austenitic family is by far the most widely used and the ferritic family is sometimes used. Chromium and nickel are the predominant alloying elements. For example, the most popular austenitic stainless steel, type 304 (UNS S30400), contains about 18% Cr and 8 to 10% Ni. The alloying elements play an important role in offering different corrosion resistances to stainless steels. In general, chromium will promote the passivation of stainless steels and increase the corrosion resistance of steels in oxidizing environments. Nickel usually promotes corrosion resistance of steels in reducing environments. Molybdenum, an element sometimes added to stainless steels, increases resistance to localized corrosion, such as pitting and crevice corrosion. An example of a molybdenum-containing stainless steel is type 316 (S31600).

The austenitic stainless steels offer excellent general corrosion resistance, good pitting, and crevice corrosion resistance with ease of fabrication and weldability (Ref 22). Although all kinds of localized corrosion, such as pitting, crevice corrosion, and intergranular corrosion could occur under certain environmental conditions, the most insidious corrosion form that austenitic stainless steels are extremely susceptible to is SCC. Stress-corrosion cracking is a cracking process that requires the simultaneous action of tensile stress and a specific corrosive environment and leads to an early failure of the steel (Ref 21, 23). In the case of 300 series stainless steels, practical experiences and experimental tests show the prerequisite conditions for the initiation of SCC include:

- The existence of some tensile stress. This can be caused by either a residual stress from the manufacturing processes or stress developed during normal operating conditions.
- Corrosive environment. Chlorides are present in tens of ppm. However, any conditions that generate local concentrations of chloride, such as condensation and alternate wet-dry areas, make the situation even more severe.
- Temperature. The threshold temperature above which SCC may occur is normally 50 to 60 °C (120 to 140 °F) (Ref 22, 24).

Although the elimination of one of these conditions would minimize or stop SCC, it is not practical to do so in most engineering applications. A more practical option for solving SCC is to select or change the material of construction so that is not susceptible to SCC.

A typical ferritic stainless steel is type 444 (S44400), which contains 18% Cr, 0.5% Ni, 2% Mo, and carbon as low as 0.025%. One of the advantages of ferritic stainless steel is immunity to SCC, while the resistance to pitting and crevice corrosion is comparable to that of type 316 stainless steel in most environments. Due to a relatively high ductile-to-brittle transition temperature, it is recommended that the maximum thickness of type 444 ferritic stainless steel that can be satisfactorily welded is 3 mm (0.125 in.).

A third family of stainless steel that finds increasing favor in food processing are the duplex stainless steels, which contain approximately equal amounts of ferrite and austenite in their microstructure. The advantages of duplex stainless steels include a high strength (stronger than carbon steels and austenitic stainless steels), good general corrosion resistance and pitting/crevice corrosion resistance, and very good SCC resistance compared with austenitic stainless steels. An example of a duplex stainless steel is 2205 (S31803), which contains 22% Cr, 5% Ni, 3% Mo, and a maximum carbon content of 0.03%. One concern with duplex grades is that they very easily form brittle intermetallic phases. For this reason, the maximum recommended service temperature for duplex stainless steels is about 315 °C (600 °F).

Basics of Stainless Steel Selection. Selecting materials of construction for food and brewery processing equipment normally begins with type 304L (S30403) stainless steel. When an environment is more corrosive than 304 stainless steel can withstand, the material is usually upgraded to type 316L (S31603) stainless steel to improve its pitting and crevice corrosion resistance. When an environment becomes extremely aggressive, the material may need to be upgraded to a more corrosion-resistant superaustenitic stainless steel, such as AL-6XN (N08367), which contains high level of chromium (20%), nickel (24%), and molybdenum (6%). If chloride SCC of a 300 series stainless steel occurs, other stainless steels, such as ferritic or duplex stainless steels, may be chosen to reduce or eliminate SCC. In some cases, the selection of more expensive nickel-base alloys is necessary if the processing conditions are extremely corrosive.

Joining Stainless Steels. The general rules for fabricating austenitic stainless steel equipment include (Ref 25):

- Always use high-purity inert welding gases and cover gases to protect the welding zone from oxidation.
- If two surfaces are tack welded to hold them in place prior to making the primary weld, ensure the tack welds are well purged with inert gas and free from oxidation. Oxidation along the tack weld edges can lead to a leak path in the weld.
- Always clean the surface prior to welding to remove any organic materials, moisture, and dirt. A dirty weld may result in chromium reacting to produce chromium carbide or may decompose during welding to create hydrogen gas and a porous weld.
- Always use aluminum oxide grinding wheels, not silicon carbide, for dressing of weld surfaces. The carbide may react with the chromium and decrease the corrosion resistance of the weld metal. Grinding wheels used on stainless steel should always be based on iron-free abrasives to avoid surface contamination by iron with associated possibilities of rust spots. If grinding wheels have been used on other ferrous metals, they should not be used on stainless steels (Ref 26).
- Because stainless steel has lower conductivity than carbon steel, 30% less heat input is required and the welds require longer cooling times.
- Since the coefficient of thermal expansion for austenitic stainless steel is higher than carbon steel or ferritic or martensitic stainless steels, low heat input should be used to prevent distortion and reduce residual stress.
- If multiple weld passes are required, maintain the interpass temperature at less than 100 °C (200 °F) to reduce the opportunity for cracking and distortion.
- Avoid crater cracks by controlling the size of the termination weld pool. If crater cracks occur, remove them by grinding the weld with an aluminum oxide wheel before proceeding.

The general rules for joining ferritic stainless steel equipment include:

- Low heat input should be used in the welding process to prevent excessive grain growth.
- Surface contamination with oil or other extraneous materials must be avoided to prevent the pickup of carbon during welding.
- To prevent nitrogen pickup during welding, the weld pool must be fully shielded with an inert gas. The actual gas used will depend on the welding process employed, but for the gas tungsten arc welding (GTAW) process, an argon or argon/helium mixture is recommended. Argon + 1 to 2% oxygen or argon/helium are the preferred gases for the gas metal arc welding (GMAW) process (Ref 24).
- In general, the maximum thickness of ferritic plates that can be welded is (3 mm) (0.125 in.) without structural integrity problems.

The general rules for joining duplex stainless steel equipment include:

- The weld procedure should be selected to produce a weld with good phase balance of ferrite and austenite. Rapid cooling rate would promote too much ferrite formation, which may lead to a ductility reduction, while slow cooling rate can result in too much austenite formation. Too much austenite formation may reduce SCC resistance.

- Typically 100% argon is used as a shielding gas. However, for welding of thin sections to each other or thin sections to thick sections, 1 to 3% nitrogen should be added to promote austenite formation because of the faster cooling rates of thin sections.
- Autogenous welding (welding without filler metal) is not recommended because it may promote too much ferrite formation and reduce ductility and corrosion resistance. Gas tungsten arc welding is the recommended weld process. Filler metal (usually nickel overalloyed filler) is mandatory.
- Preheat and postweld heat treatment are not normally required or recommended. Preheat would slow the cooling rate, which may lead to too much austenite formation and reduce the resistance to SCC.
- As with ferritic stainless steels, heat input should be minimized through low-diameter welding wire/electrodes and low welding currents avoided (Ref 24).

Avoiding Corrosion Problems in Stainless Steels

Liquid Flow Rate and Stagnant or Dead Leg Areas. Stainless steels maintain their corrosion resistance due to the passive film formed on the surface. Pitting corrosion occurs due to the film breakdown at locations where locally concentrated pockets of chloride or other halide ions occur. Pitting corrosion occurs more readily under any dirt deposits such as "milkstone" (a deposit found in the dairy industry originating from milk and hard water consisting primarily of protein and calcium phosphate) or in stagnant areas. Stainless steel pitting risk can be significantly reduced by maintaining high fluid flow rates and/or eliminating stagnant areas. Higher flow rates not only scour the surface of deposits, but also provide more oxygen to the surface, which will result in the formation or reformation of a more stable passive film.

Avoid Crevices and Discontinuities. Crevices or discontinuities can trap fluids that harbor bacteria growth and encourage solids buildup on metal surfaces (Ref 26). Such crevices and discontinuities will increase the possibility of crevice corrosion and localized corrosion such as pitting.

Select Appropriate Surface Finish. Although surface finish is important for equipment cleanability, smooth surface finish also reduces the susceptibility to corrosion. Since stainless steel corrosion usually occurs as localized corrosion, a smoother surface that has less adherence of milkstone, "beerstone" (the deposition of hardness salts and proteinaceous matter in the brewing industry), and "winestone" has a reduced possibility of suffering localized corrosion. In general, a 150 grit finish (3A finish from 3A Dairy Standard (Ref 27) and No. 4 finish from AISI classification) is adequate for most food and brewery services. Table 3 lists various finishes classified by the AISI (American Iron and Steel Institute).

Hydraulic Pressure Testing (Hydrotesting). If possible, use deionized (DI) water for any hydraulic pressure test. If other water types are used, completely drain and dry the system immediately after the test to prevent pitting corrosion, which may otherwise occur in times as short as overnight.

Stainless Steel Corrosion Case Studies

Stress Corrosion Cracking of Type 304 Stainless Steel Vessel. In a U.K. sweets-manufacturing company, a 570 L (150 gal) type 304 stainless steel jacket vessel containing a glucose and citric acid mixture at 100 °C (212 °F) with chloride present had severe SCC after only 18 months of service. Cracking started on the product side and completely penetrated the vessel wall while there was no sign of general corrosion. Similar corrosion problems were also found in welded pipework handling the same type of food product. The solution to this SCC problem was installing alloy 825 (N08825), a nickel alloy containing 40% Ni, 20% Cr, 3% Mo, and 2% Cu (Ref 15).

Pitting of Type 304 Stainless Steel Plate and Frame Heat Exchanger. In a French beet sugar mill, severe pitting was found on the type 304 stainless steel plates in a plate-type heat exchanger. Pitting occurred after only 12 months of service. Failure analysis tests showed that deposits in the pits contained 500 ppm chloride. Type 316 stainless steel plates were evaluated and, although better than type 304, were still pitted to some extent (Ref 15). Titanium would provide a better solution for this application.

Stress-Corrosion Cracking of Type 304 Stainless Steel Conveyor-Belt Components. Conveyor belts used by a food company in preparing hot dogs were made of type 304 stainless steel. Hot dogs contain salt and are cured at 80 to 82 °C (175 to 180 °F). The presence of chloride and elevated temperature combined with the residual stress in the conveyor belt side links caused SCC failure in about 2 years. To solve this problem, all the conveyors were converted to type 444 ferritic stainless steel and have been functioning without SCC for more than 5 years (Ref 22).

Stress-Corrosion Cracking of Type 304 Pressure Cooker. A major manufacturer of restaurant pressure cookers for chicken experienced SCC failure of these units at the flanges due to the combined effect of high temperature and salt in the cooking oil. Conversion to type 444 ferritic stainless steel solved the problem (Ref 22).

Stress-Corrosion Cracking of Type 304 Stainless Steel Glucose Tank. A major brewery in Canada suffered SCC of a type 304 glucose tank. The temperature and the chlorides in the glucose provided the necessary ingredients for SCC. The problem was solved by substituting a type 444 ferritic stainless steel tank (Ref 22).

Other Materials of Construction

The materials described in this section offer alternative solutions to stainless steels, but often do not compete because of the wide-ranging beneficial properties of stainless steels.

Aluminum is still used extensively in certain areas of food manufacture and distribution because of its good corrosion resistance, cleaning ease, material of construction cost, material weight, and high thermal conductivity (Ref 28). Examples of aluminum applications include beer kegs, beer cans, soft drink cans, and numerous small-scale equipment uses in the brewing industry (Ref 28). In other food manufacturing and preparation areas, aluminum is used extensively to process butter, margarine, table oils, edible fats, meat and meat products, fish and shellfish, and certain sorts of vinegar, mustards, and spices (Ref 28). The primary factor limiting the widespread use of aluminum in the food and beverage industries is its susceptibility to galvanic corrosion due to its very negative open circuit potential. Aluminum also cannot survive the harsh cleaning chemicals and conditions required for clean-in-place applications, because the pH range over which it is resistant is relatively narrow.

Copper has been the traditional metal used in breweries for centuries due to its excellent thermal conductivity, ductility, fabricability, and reasonably good corrosion resistance. However, with the use of new alkaline cleaners, some corrosion problems have occurred. Conventional brazed joints in large-diameter kettles have been preferentially corroded and required silver braze repair. Metal toxicity and its catalytic activity in the development of oxidative rancidity in fats and oils have limited the use of copper as a material of construction in the food industry (Ref 28). Even at the sub-part-per-million level, copper in vegetable oils and animal fats rapidly causes the development of off-flavors (Ref 28).

Table 3 Surface finishes on stainless steel sheet

Finish	Surface description
No. 1	Hot-rolled, annealed, and descaled
No. 2B	Cold-rolled, annealed, and descaled plus light planishing pass between polished rolls (bright finish)
No. 2D	Cold-rolled, annealed, and descaled (dull finish)
No. 3	Polished finish (100 mesh grit)
No. 4	Polished finish (120–150 mesh grit), general-purpose finish for food equipment
No. 6	Dull satin polish finish
No. 7	Bright polish finish
No. 8	Mirror polish finish

Source: American Iron and Steel Institute

Titanium. There are certain areas of the food industry, especially in equipment involving heat transfer, where stainless steels just cannot withstand the corrosive effects of salty, low-pH environments (Ref 28). Titanium is increasingly being accepted as an alternative in such cases. It is believed titanium is the only answer to address corrosion under the most aggressive conditions. For example, a well-known food processor has been using titanium for many years for processing hot salted cream cheese and salted dressings containing vinegar and salt. The titanium performed well and solved a pitting problem by replacing the stainless steels (Ref 29).

It was reported in 1981 that several hundreds of plate-type titanium heat exchangers were employed by the American food industry, and many of them had been in service for ten years. These heat exchangers were employed very successfully in corn products, barbecue sauce, pickles, salad dressings, and other foods with temperature ranging up to about 120 °C (250 °F) (Ref 29). One company found that titanium equipment lasted three times longer than stainless steel equipment in a hot bean sauce (Ref 29).

Titanium is a light metal with a high strength/weight ratio. Its density is almost half of stainless steels. Although relatively expensive, the low density of titanium can offset the price differential for the raw material by almost 50%.

Corrosion in Cleaning and Sanitizing Processes

An important activity in any food-processing system is the elimination of microbial growth or contamination since it directly affects food product safety, quality, and shelf life. Clean processing systems are usually characterized by frequently executed cleaning and sanitizing processes. In these processes, cleaning chemicals are used to remove various inorganic and organic soils, and sanitizing chemicals are used to disinfect the equipment or piping surface. Although stainless steels can survive most of the cleaning and sanitizing environments created by these corrosive chemicals, corrosion can still occur if the cleaning and sanitizing process parameters are not properly controlled. The critical parameters of the cleaning and sanitizing process include the chemical concentration, temperature, and duration of the cleaning and sanitizing process.

Sodium hypochlorite sanitizing compounds are normally supplied in alkaline solution with 3 to 15% available chlorine. These solutions function most efficiently in the pH range 8 to 9. Values higher than pH 9 result in slower germicidal action and lower corrosion rates, whereas values below pH 8 provide rapid germicidal action but excessive corrosion rates (Ref 15). To ensure the best sanitizing conditions while avoiding stainless steel pitting, process parameters such as the sodium hypochlorite concentration, temperature, and duration of contact time, should be controlled in the sanitizing process. The recommended concentration of sodium hypochlorite in the sanitizing solution is 200 ppm. The contact time with stainless steel surfaces should not be more than 30 min. Solution temperature greater than 49 °C (120 °F) can increase the corrosive effect and eventually may cause pitting (Ref 30). The maximum recommended temperature for a 200 ppm sodium hypochlorite is 49 °C (120 °F) to prevent pitting.

Gluconic acid solutions are effective in removing beerstone deposits, and they do not corrode the stainless steel equipment. Stainless steel brewery fermenting tanks with aluminum fittings have been successfully cleaned with a 4 to 5% solution of gluconic acid (Ref 15, 31) at temperatures in the range of 15 to 35 °C (60 to 95 °F). The formulation shown in Table 4 has been found to be useful in removing beerstone from stainless steel beer barrels (Ref 15).

Phosphoric Acid. Milkstone deposits can be removed conveniently from milking machines and other equipment with 30% phosphoric acid solution containing 2 to 5% of a nonionic wetting agent, used at the rate of 6 to 30 g/L (1 to 5 oz/gal) of hot water at 70 to 80 °C (160 to 180 °F) (Ref 15, 32).

Prerinsing of the milking system is accomplished using clean warm water (38 to 43 °C, or 100 to 110 °F) to remove residual fat, protein, and sugars. The rinse temperature should not exceed 46 °C (115 °F) since the soil can be "cooked" onto the equipment or piping surface and become harder to remove (Ref 30).

Milk and wash lines should be looped to avoid "dead ends" in the piping system (Ref 30). These dead ends not only create poor cleaning, but also generate an environment that facilitates corrosion due to chemical accumulation and/or oxygen depletion.

Mechanically Assisted Cleaning and Clean-in-Place (CIP). Effective cleaning requires both chemical and mechanical action. In a manual operation, the mechanical action should come from the use of the appropriate brushes and physical effort. Metal scouring pads, wire bristle brushes, or other abrasive materials should be avoided, as these cleaning tools can damage the surface being cleaned (Ref 30). For CIP systems, mechanical action is achieved from the solution turbulence, which is a result of the fluid velocity. The minimal acceptable fluid velocity flow for the CIP solutions through the system should be 1.5 m/s (5 ft/s) (Ref 30), but a higher velocity of 2 m/s (7 ft/s) is more desirable.

Table 4   Solution for removing beerstone from stainless steel barrels

Chemical	Quantity
Caustic soda, kg (lb)	9–11 (20–25)
Sodium gluconate, kg (lb)	4.5–9 (10–20)
Sodium oleate, kg (lb)	0.045–0.09 (0.1–0.2)
Water, L (gal)	45 (12)

Source: Ref 15

REFERENCES

1. "Corrosion Costs and Preventive Strategies in the United States," FHWA-RD-01-156, Federal Highway Administration, Sept 2001
2. 3-A Sanitary Standards, Inc., www.3-a.org, accessed July 2004
3. "ASME Bioprocessing Equipment," BPE-2002, American Society of Mechanical Engineers, July 1, 2002
4. American Welding Society, www.aws.org, accessed July 2004
5. HACCP, http://vm.cfsan.fda.gov/~lrd/haccp.html, accessed July 2004
6. "Current Good Manufacturing Practices in Manufacturing, Packaging or Holding Human Food," Code of Federal Regulations 21 CFR 110, U.S. Food and Drug Administration, April 1, 1999
7. A.J. Betts and L.H. Boulton, Paper 32/17, 32nd Conference of the Australasian Corrosion Association, 1998, p 1–14
8. *CRC Handbook of Chemistry and Physics*, CRC Press, 1998, p D-146
9. C.T. Cowan, *Food Eng. Int.*, Vol 2, Sept 1977, p 34
10. Stainless Steels in Corrosion Service, *Properties and Selection: Stainless Steels, Tool Materials, and Special-Purpose Metals*, Vol 3, 9th ed., Metals Handbook, American Society for Metals, 1980, p 56
11. Z. Szklarka-Smialowska, *Pitting Corrosion of Metals*, National Association of Corrosion Engineers, 1986, p 281
12. S. Lyon, *Nature*, Vol 427, Jan 20, 2004, p 406
13. H. Ashassi-Sorkhabi, M.R. Majidi, and K. Seyeddi, *Appl. Surf. Sci.*, doi: 10.10/j.apsusc 2003.10.007, 2003
14. Corrosion-doctors organization, http://www.corrosion-doctors.org/foodindustry/beer-corrosion.htm, accessed July 2004
15. B.J. Connolly, *Austral. Corros. Eng.*, Vol 15 (No. 11), 1971, p 23
16. A.J. Lehman, *Culinary Stainless Steels*, Vol 25 (No. 3), July 1961, p 123
17. G.N. Flint and S. Packirisamy, *Food Additives Contamin.*, Vol 14 (No. 2), 1997, p 115–126
18. "Chromium," Environmental Health Criteria 61, World Health Organization, Geneva, Switzerland, p 17–21
19. B. Thomas, J.A. Roughan, and E.D. Watters, *J. Sci. Food Agriculture*, Vol 25, 1974, p 771–776
20. P. Geoffroy and J. Perin, Les Aciers Inoxydables-Leur Utilisation en Oenologie, Reprint from article published in *Le Vigneron Champenois*, No. 7/8, July-Aug 1968
21. R.M. Davison, T. DeBold, and M.J. Johnson, Corrosion of Stainless Steels, *Corrosion Metals Handbook*, Vol 13, 9th ed., ASM International, 1987, p 547–565
22. J.D. Redmond, Solving Brewery Stress Corrosion Cracking Problems, *Technical Quarterly—Master Brewers Association for the Americas*, Vol 21 (No. 1), 1984, p 1–7

23. S.W. Jr. Dean, Review of Recent Studies on the Mechanism of Stress Corrosion Cracking in Austenitic Stainless Steels, *Stress Corrosion New Approaches,* STP 610, ASTM, 1976, p 308
24. G. Daufin and J. Pagetti, New Stainless Steels, *IDF Bull.,* Vol 189, 1985, p 13–23
25. J.C. Tverberg, Stainless Steel in the Brewery, *Technical Quarterly—Master Brewers Association for the Americas,* Vol 38 (No. 2), 2001, p 67–82
26. E.C. Hale, Welded Stainless Steel Tube in Food and Drink Processing, *Anti-Corros. Meth. Mater.,* Vol 22 (No. 8), 1975, p 7–10
27. "3-A, Accepted Practices for Permanently Installed Sanitary Product-Pipelines and Cleaning Systems," No. 605-02, International Association of Milk, Food, and Environmental Sanitarians, United States Public Health Service, The Dairy Industry Committee
28. *Wiley Encyclopedia of Food Science and Technology,* F.J. Francis, Ed., Vol 1–4, 2nd ed., John Wiley & Sons, 1999
29. H.B. Bomberger, Titanium for Food and Drug Processing, *Titanium for Energy and Industrial Applications,* D. Eylon, Ed., AIME, 1981, p 277–283
30. P.T. Tybor and W.D. Gilson, "Cleaning and Sanitizing in Milking Operations," Bulletin 1025, the University of Georgia and Ft. Valley State College, the U.S. Department of Agriculture and counties of the state cooperating, Revised 1989
31. D.C. Horner, The Use of Gluconates in Metal Cleaning Solutions, *Met. Finish. J.,* Vol 12, March 1969, p 99–103
32. A. Twomoy and H.R. Prenter, *New Zealand J. Agriculture,* Vol 17 (No. 1), July 1968, p 44–47

Material Issues in the Pharmaceutical Industry

Paul K. Whitcraft, Rolled Alloys

THE OVERVIEW of materials of construction corrosion failure mechanisms in the article "Corrosion in the Pharmaceutical Industry" in this Volume continues to be a valuable resource for information on this topic. Some trends in the availability and application of materials in this industry, and current issues relative to pharmaceutical equipment construction, warrant closer examination.

Materials

In an industry conservative by nature, the change in material use is slow to occur and driven primarily by the failure of existing material systems to resist corrosion from new processes. Equipment involved in product contact has been constructed largely from austenitic stainless steel 304L (UNS S30403) and 316L (UNS S31603). The satisfactory performance of these materials in most applications, combined with good material availability at acceptable price levels, produces little incentive to change. Historically, in areas where 316L was not adequate, the high-Ni-Cr-Mo alloy C-276 (UNS N10276) was the alternate choice.

In the 1970s and 1980s, numerous alloys intermediate to 316L and C-276 were developed that offered the chemical and corrosion resistance necessary for the pharmaceutical industry. A few of these alloys are popular in the sense of market availability and use, and while alloy development continues, it is at a much slower pace. These alloys can generally be characterized into two families, identified as superaustenitic and duplex stainless steels.

Superaustenitic Stainless. Of the superaustenitic stainless alloys, the AL-6XN (Allegheny Ludlum) alloy (UNS N08367) has been the most widely used. This alloy is nominally 25Ni-20Cr-6Mo, with the balance essentially iron. See the article "Corrosion of Nickel and Nickel-Base Alloys" in *Corrosion: Materials, Volume 13B of ASM Handbook,* 2005, for information on other Ni-Cr- Mo alloys and aqueous and high-temperature corrosion. The high-chromium and molybdenum additions result in an alloy that is very resistant to pitting attack, but more importantly, resistant to crevice corrosion. Laboratory tests for evaluating resistance to localized corrosion have been developed and are useful in determining the relative resistance of various materials to acid chloride environments. These tests should not be used to predict precise behavior in other environments, and comparison of one data set to another must be done with caution because there can be variability in results. However, materials showing resistance in these tests have demonstrated an ability to perform well under conditions where 316L has failed due to localized corrosion, such as crevice corrosion.

Crevice corrosion occurs in 304L and 316L stainless steels when the conditions in a tight crevice become more severe than the surrounding environment. Stainless steels gain their corrosion resistance by forming a protective oxide surface layer. This layer, composed primarily of chromium oxides, is readily formed in the presence of oxygen. This protective layer is susceptible to damage from halogens, but if oxygen is present in the environment, repair of the film is often rapid enough to prevent any significant damage. In the absence of a source to replenish oxygen, such as in a tight crevice, corrosion can accelerate as the available oxygen is consumed and the pH drops as a result of the corrosion process.

Another location where depleted oxygen can lead to corrosion of stainless steel is at low points of piping systems without adequate drains.

Duplex Stainless Steels. A number of duplex stainless steels exist. Alloy 2205 (UNS S32205) contains 22Cr-5Ni-3Mo and is widely used in industrial and chemical process applications. To date, duplex alloys have seen little application in pharmaceutical equipment, but their relative cost and properties make them suitable candidates to replace 304L and 316L. The chromium and molybdenum contents result in a material with pitting and crevice corrosion resistance superior to that of 316L stainless and that is also resistant to chloride-induced stress-corrosion cracking because of the austenite-ferrite (duplex) microstructure. In general, duplex alloys are magnetic, slightly less formable, and require more care in fabrication than the austenitic grades.

The correlation between resistance to chloride pitting and crevice corrosion and alloy content is shown in Table 1. Nitrogen has become an important alloy addition in these alloys as well. A pitting index, called the pitting resistance equivalent number (PREN) with nitrogen, has been developed to predict resistance of an alloy based on its composition. The PREN is defined as $\%Cr + 3.3 \times \%Mo + 30 \times \%N$. Other PREN formulas use $\%Cr + 3.3 \times \%Mo + 16 \times \%N$, so it is important to know the basis of the value when making comparisons. There is a term for tungsten in other PREN formulas. The level of localized corrosion resistance offered by the duplex and superaustenitic stainless steel materials fills the large gap between 316L and C-276. The superaustenitic stainless grade should be

Table 1 Typical corrosion characteristic values for materials of construction in the pharmaceutical industry

Alloy	UNS No.	Pitting resistance equivalent number (PREN)(a)	Critical pitting temperature(b), °C	Critical crevice temperature(c), °C	Stress-corrosion cracking, boiling 25% NaCl
304L	S30403	19.8	0(d)	<−3	Cracks in <72 h
316L	S31603	24.9	15	−3	Cracks in <72 h
2205	S32205	37.6	40	17	Resists
AL-6XN	N08367	47.2	80	43	Resists
C-276	N10276	65.0	>110	>75	Resists

(a) Using PREN = %Cr +3.3× %Mo + 30× %N. (b) ASTM standard G 48, method C, critical pitting temperature test. (c) ASTM standard G 48, method D, critical crevice temperature test. (d) Estimated

seriously considered as a less expensive alternative to the higher-nickel alloy. Also see the articles "Corrosion Resistance of Stainless Steels and Nickel Alloys" and "Effects of Metallurgical Variables on the Corrosion of Stainless Steels" in *Corrosion: Fundamentals, Testing, and Protection*, Volume 13A of *ASM Handbook*, 2003.

Material Replacement. With advances in melting technology, primarily the argon-oxygen decarburization refining process that facilitated the development of these higher alloys, came the ability to control alloy content to more precise levels. One result of this advancement to the end user is that the alloy content of grades such as 316L is actually lower than it was 25 years ago. Given the ability to control molybdenum levels to tighter ranges, it is not surprising that the molybdenum content in 316L today (2006) is typically below 2.1%, while 25 years ago it probably averaged nearer 2.5%. This has resulted in a significant cost-savings to users of these alloys in most applications. In those areas where the alloy offered corrosion resistance that was barely adequate, the subtle change in chemistry may mean that a replacement 316L component no longer performs well in areas where it was once satisfactory.

Passivation

Stainless steels offer useful corrosion resistance because they exhibit passive corrosion behavior as a result of the formation of protective oxide films on the exposed surfaces. Under normal circumstances, stainless steels will readily form this protective layer immediately on exposure to oxygen. This oxygen may be from any source, including air, dissolved oxygen in water, or other oxidizing media. When this protective film is violated or fails to form, active corrosion is likely to occur. Some fabrication processes can impede the reformation of the passive layer, and, to ensure that it is formed, stainless steels are subjected to passivation treatment. The common concern in this regard is contamination of the surface by carbon steel. This local contamination may result from a variety of causes, including handling equipment, forming tools, fixtures, and clamps.

Treatments. The production of stainless steels commonly involves the use of strong oxidizing acids such as nitric and nitric-hydrofluoric acid mixtures to remove the oxide scales formed during thermal treatment. This pickling process provides two benefits. First, it removes the oxide scale and passivates the underlying metal surface. Second, due to its aggressive nature, the process will remove any chromium-depleted layer that may have formed as a result of the scale formation.

For passivation treatments other than scale removal following thermal treatment, less aggressive acid solutions are usually employed. The primary purpose of these treatments is to remove contaminants that may be on the component surface and could prevent the formation of the oxide layer locally. The most common contaminant is embedded or free iron particles from forming or machining tools. Mechanical polishing can be employed to provide a uniform surface finish and remove these contaminants. The polishing materials used should be devoted to stainless steel use only, because they can carry over small particles from one part to the next. In addition, these fine work-hardened particles, even from a stainless vessel, can have a lower threshold for corrosion and act as an initiation site if not removed. A dilute (10%) solution of nitric acid is effective at removing free iron or similar contaminants. For ferritic, martensitic, or precipitation-hardening grades, a nitric acid solution inhibited with sodium dichromate is used so as not to attack the stainless too aggressively. Phosphoric acid at 1% concentration and citric acid at up to 20% concentration are also effective for the more resistant stainless alloys. Other commercially available chelating agents can be employed. The use of these mild acids or chelating agents can also represent a significant advantage in terms of relief from environmental issues.

Welding. A passivation treatment is also advisable following welding. Welding processes, even with proper gas shielding, may result in some oxidation (heat tint) on or adjacent to the weld. Under severe corrosion conditions, these areas will be more likely to initiate corrosion. However, the magnitude of the increased tendency to corrode is difficult to quantify, and in many instances, this condition may have no impact on actual service performance. Only extensive corrosion testing or trial experience should be used to justify the serviceability of material exhibiting heat tint. Cleanup of localized areas is best addressed through conventional weld-cleaning methods, such as pickling pastes. Electrolytic pickling and descaling processes may also be effective. Stainless wire brushing is also an acceptable means of heat tint removal. The conventional postfabrication passivation techniques, such as dilute phosphoric, nitric, or citric acid, will not remove the heat tint from welding. However, such treatments will ensure that no free iron or active areas remain.

Further information on pickling, passivation, and cleaning treatments may be found in ASTM International standards A 380 (Ref 1) and A 967 (Ref 2).

Electropolishing

Electropolishing is a controlled corrosion process, resulting in the uniform removal of metal from the surface. See the article "Electropolishing" in *Corrosion: Fundamentals, Testing, and Protection*, Volume 13A of *ASM Handbook*, 2003. Electropolishing is not a passivation treatment, although the proper execution of the process will result in a passive surface. Proper electropolishing technique maintains the part in the electrochemically passive range, while the passivated layer is only allowed to grow several atoms thick, at most. The electrolyte simultaneously promotes dissolution of this layer. The electropolishing process does remove surface impurities, as is accomplished with passivation. During the cleaning and rinsing process following electropolishing, the material does passivate naturally upon exposure to oxygen-containing rinse water or air. Ultimate passivation of the surface is assured, because any contaminants that may have been on the surface have also been removed by the polishing process. No additional passivation procedure is required. In fact, exposure to some passivation treatments, such as dilute nitric acid, may be detrimental to the intended result by dulling the luster of the polish.

Electrolytes used for electropolishing are usually proprietary mixtures with contents that are not quantitatively revealed. The electrolyte will typically have the ingredients to facilitate three different actions of the polishing process:

- An etchant that facilitates the breakdown of a passive film
- An oxidizer that helps form a passivating film
- A highly viscous constituent that promotes the formation of a diffusion layer

The oxidizer and the etchant assist in maintaining the part in a pseudopassive state, while the diffusion layer control is necessary to promote uniform metal loss. For electropolishing 316L stainless, the electrolyte will often contain perchloric acid, which can provide both oxidizing power and a halide etchant. Acetic anhydride may be used to control the diffusion layer. Stainless and higher-nickel alloys may also use nitric acid, sulfuric acid, phosphoric acid, hydrogen peroxide, and methyl alcohol. Determining the ideal electrolyte ingredients is an important part of the polishing process. Because electropolishing is an electrochemical process, other variables to be controlled include the voltage, current density, and solution temperature. Voltages that are too low can cause etching of the surface due to more rapid general corrosion. Voltages that are too high will result in pitting. Part geometry and cathode design are critical, because it is imperative that the current be distributed as uniformly as possible over the part surface. High current densities will also result in pitting.

The large majority of stainless steel production today (2006) is by the argon-oxygen decarburization (AOD) process, followed by continuous casting. The AOD process, along with other controls in the steel-making process, has resulted in a more consistent and uniform product. Continuous casting does create the opportunity for types of surface or near-surface imperfections that can have an impact on electropolishing behavior. Static casting of ingots can result in similar problems with cleanliness. Although electropolishing imperfections on AOD stainless products are rare, the use of a double-melted product will ensure an essentially imperfection-free electropolished

Rouging

Rouging is a phenomenon of particular interest to the pharmaceutical industry. It is the presence of a surface layer of oxide on stainless equipment or piping typically handling high-purity water at temperatures above ambient. This includes stills, steam systems, purified water, and water for injection. The oxides can vary in composition, degree of oxidation, color, texture, and adherence. Although generally shown to be innocuous, the mere presence of these deposits can raise concern.

The rouge itself is typically composed primarily of iron oxides or iron hydroxides, but because these are developing on stainless surfaces, they may also contain oxides of chromium, nickel, and molybdenum. There is empirical data indicating that resistance to rouging increases with increasing chromium-iron ratios in the passive layer and/or the thickness of the passive layer itself. Because both electropolishing and passivation increase the chromium-iron ratio, application of these processes can increase resistance to rouging. Even with such treatments, the passive layer can break down due to the ionizing effect of high-purity water. The low oxygen content of these waters also slows the rate of repassivation and may cause the layer to linger in intermediate states of oxidation. Repeated cycles of this process result in the entrapment of various oxides in the passive layer, hence the wide range of colors.

Rouging has also been observed to result from deposits of corrosion products from upstream equipment. Such deposits can simply be wiped from the surface and reveal an unaffected electropolished surface underneath. Upstream equipment potentially identified as the source has included stills (often with carbon steel components), lower-alloy stainless steel piping (304L/316L) welds, and stainless pump components. Stainless pump housings are often produced from as-cast products. Foundry castings often contain higher ferrite contents than wrought stainless steels to facilitate castability. Ferrite contents in cast 304 or 316 products (CF-8, CF-8M) can exceed 20%, and these may be less corrosion resistant than their wrought counterparts that typically contain no more than 5% ferrite.

Removal of rouging can be accomplished mechanically but is usually addressed by chemical cleaning. Repassivation treatments with nitric, phosphoric, citric, or other oxidizing acid solutions have been effective in removing or fully reoxidizing this layer. As with any chemical reaction, the process is time-dependent and can be influenced by temperature. For more resistant rouge patterns, reducing acids such as hydrofluoric or hydrochloric may be used in combination with a passivation treatment. The use of these acids in strong concentrations, however, may etch the surface.

Potentiodynamic polarization studies have been conducted to measure the efficacy of passivation treatments. It has been shown that the breakdown (pitting) potential is raised by passivation or electropolishing techniques that result in higher chromium-iron ratios and increased thickness of the passive layer. These potentials can be increased by as much as 50 to 100 mV over mechanically polished or pickled surfaces and have been equated to increased resistance to rouging. Additionally, the breakdown potential of the molybdenum-bearing superaustenitic stainless alloy N08367 was shown to be in excess of 400 mV higher than 316L prior to enhancing passivation treatments, and another 50 mV higher following such treatments. Such studies would suggest that higher alloys, such as N08367 or the C-276-type alloy, are highly resistant to rouging.

Issues of rouging have recently attracted the attention of at least two organizations concerned with materials performance. ASTM International Committee G-1 on corrosion is considering the development of a test method that could be used to evaluate a material resistance to rouging. The existence of such a test method could help in understanding the conditions under which rouging will occur, as well as provide an indication of the materials that will resist its formation. The Materials Technology Institute of the Chemical Process Industries is developing a project designed to better define the sources and mechanism of the formation of rouge, and this program anticipates funding work to identify the mechanism and ways to prevent its formation. See the article "Rouging of Stainless Steels in High-Purity Waters" in this Volume for details on the water chemistry, materials of construction, and classification of rouge.

REFERENCES

1. "Standard Practice for Cleaning, Descaling, and Passivation of Stainless Steel Parts, Equipment, and Systems," A 380, ASTM International
2. "Standard Specification for Chemical Passivation Treatments for Stainless Steel Parts," A 967, ASTM International

SELECTED REFERENCES

- A. Grant, "CIP and Passivation," ASME Bioprocess Seminar (San Juan, PR), Jan 2003
- "MTI Projects," Materials Performance Institute of the Chemical Process Industries, http://www.MTI-Global.org, 2004 (accessed April 17, 2006)
- P. Schweitzer, *Corrosion and Corrosion Protection Handbook,* Marcel Dekker, 1996
- A. Sedriks, *Corrosion of Stainless Steels,* John Wiley & Sons, 1979
- J. West, *Electrodeposition and Corrosion Processes,* Van Nostrand Reinhold, London, U.K., 1971
- P. Whitcraft, Corrosion of Pharmaceutical Equipment, *Encyclopedia of Pharmaceutical Technology,* Marcel Dekker, Inc., 2003

Corrosion in the Pharmaceutical Industry

THE PREVENTION AND MITIGATION of corrosion in the pharmaceutical industry presents a demanding challenge to materials engineers. In most cases, the processes require equipment and piping systems to be fabricated from material having extremely low corrosion rates when exposed to a wide variety of corrosive media and operating conditions. Any substance produced by corrosion reactions could contaminate the product being manufactured. This contamination must be removed in a subsequent process step so that the product is in compliance with the stringent purity and quality demands established by the applicable government regulatory agencies for the good of the consumer.

Materials of Construction

The materials of construction found in pharmaceutical production facilities include:

- Metal alloys, both solid and clad
- Thermosetting plastics and thermoplastics, both solid and as linings
- Ceramics and glass, both solid and as linings
- Impregnated carbon

Stainless Steels

The articles "Corrosion of Wrought Stainless Steels" and "Corrosion of Cast Stainless Steels" in *Corrosion: Materials,* Volume 13B of *ASM Handbook,* 2005, provide details on the general use of stainless steels. The article "Material Issues in the Pharamacuetical Industry" in this Volume provides further information.

Austenitic stainless steels, especially low-carbon and stabilized grades, have been the dominant alloys in the pharmaceutical industry. These alloys exhibit good corrosion resistance in many media, are easily fabricated, have excellent strength over a wide temperature range, are readily available, and are relatively inexpensive. The surface condition of the austenitic stainless steel is critical where it is in contact with the pharmaceutical product and where the stainless steel is required to resist an aggressive environment. The highly protective chromium oxide film that gives stainless steel its corrosion resistance is tenacious, durable, and self-healing in the presence of oxygen; however, this film can be damaged during fabrication and by cleanup practices after fabrication. Fortunately, problems can be minimized by following good design, procurement, fabrication, handling, and cleanup practices.

Austenitic stainless steels are widely used in oxidizing environments, high-purity water service, and in fine chemical and pharmaceutical production equipment and piping. They are not suitable for use in chloride-containing environments, particularly at high chloride concentrations and at high temperatures.

Stainless steel producers have developed more highly alloyed superaustenitic stainless steels having excellent resistance to general corrosion, pitting, and crevice attack in chloride-containing environments. These alloys are highly resistant to intergranular corrosion and stress-corrosion cracking (SCC); this makes them useful in oxidizing chloride solutions, oxidizing acids, and brines. The high molybdenum (2.5 to 6.5%) content and increased chromium and nitrogen give the superaustenitics good resistance to pitting and crevice corrosion. The relatively high nickel content (18 to 31%) and the high chromium and molybdenum levels give the alloys excellent SCC resistance. The presence of copper in the alloys improves resistance to sulfuric (H_2SO_4), phosphoric (H_3PO_4), and acetic acids.

The superstainless alloys are less costly than the nickel-base alloys and are readily available in a wide range of product forms, such as pipe and tubing, sheet, plate, and forgings, as well as a full range of welding consumables. In addition, the superaustenitic stainless steels are more workable and have better weldability than high-alloy ferritic steels. When the superaustenitics are specified, the welding specifications of alloy producers should be followed explicitly so that the full chemical and cracking resistance of the alloy is maintained.

Duplex stainless steels are duplex in structure; at room temperature, their equilibrium structure is a mixture of austenite and ferrite phases. The compositions of these alloys are carefully controlled to maintain the proper balance of austenite to ferrite. Most of the physical properties of the duplex stainless steels are between those of the austenitic and ferritic stainless alloys. The thermal conductivity of the duplex stainless alloys is less than half that of carbon steel but approximately 25% higher than that of the austenitic stainless steels.

The coefficient of thermal expansion of carbon steel is similar to that of the stainless duplex alloys and is approximately 40% less than that of the austenitic stainless alloys. Duplexes have excellent toughness as well as high strength. Compared with the ferritic stainless steels, the ductile-to-brittle transition of the duplex alloys is more gradual and occurs at a lower temperature, thus allowing the production of a wide range of product forms.

The duplex alloys are not suitable for cryogenic service; the austenitic grades are preferred in this application. The duplex grades have good resistance of chloride SCC; however, the various alloys can show variability in pitting and crevice-corrosion susceptibility because of the segregation of the ferrite and austenite phases. The high chromium content of the duplex stainless steels makes them strongly resistant to oxidation. However, prolonged exposure at elevated temperatures (above 345 to 370 °C, or 650 to 700 °F) can affect toughness and corrosion resistance in aqueous media.

All commonly used duplex stainless steels are included in ASTM International specifications for sheet, strip, plate, and seamless and welded tubing and pipe. Many duplex stainless steels are also included in Section VIII, Division 1, of the American Society of Mechanical Engineers (ASME) Boiler and Pressure Vessel Code. These steels are called austenitic/ferritic corrosion-resistant steels in international standards such as EN 10213-4.

The best corrosion resistance and mechanical properties in welded duplex stainless steels are achieved when a 50-50 austenite-ferrite phase is formed in both the weld metal and the heat-affected zone (HAZ). Welding practices that emphasize cleanliness, avoid carbon contamination, and use dry inert gas shielding are required. Because of the sensitivity of ferrite to hydrogen embrittlement, shielding gases containing hydrogen should not be used. Duplex alloys include alloy 2205 (UNS S31803), 44LN (UNS S31200), and Ferralium 255 (UNS S32550).

High-purity ferritic stainless steels were introduced in the United States in the 1970s. The oldest, E-Brite 26-1 (26Cr-1Mo, UNS S44627), was first electron beam refined for low carbon and gas content and is now vacuum melted. Other high-purity ferritic alloys with similar chromium and molybdenum contents and stabilizing additions of titanium have been introduced. The 26-1 alloy generally has corrosion resistance equal to or better than types 304 and

316 stainless steel. In addition, the 26-1 alloy is resistant to chloride-induced SCC, a major shortcoming of the austenitic stainless steels. With the introduction of argon oxygen decarburization, a variety of new alloys were possible.

Welded ferritic stainless steels typically have poor weld zone ductility and are notch sensitive in the HAZ. If these alloys are heated to between 400 and 480 °C (750 and 900 °F) for prolonged times or are slowly cooled within this temperature range, notch toughness is further reduced, and the material becomes brittle. The use of the ferritics at low temperatures is limited because of their high ductile-to-brittle transition temperatures. If the corrosion resistance of the ferritic alloys is to be maintained, extreme care must be taken to avoid contamination with nitrogen, oxygen, hydrogen, or carbon during welding. Detailed tungsten inert gas welding procedures developed by the alloy producer should be used to maintain ductility and corrosion resistance.

Nickel and Nickel-Base Alloys

See the article "Corrosion of Nickel and Nickel-Base Alloys" in *Corrosion: Materials,* Volume 13B of *ASM Handbook,* 2005, for additional details.

Commercially pure nickel, nickel 200 (UNS N02200), is highly resistant to many corrosive media. It is most useful in reducing environments, and it can be used under oxidizing conditions that cause the development of a passive oxide film. Nickel has poor corrosion resistance in H_2SO_4, HCl, HNO_3, and H_3PO_4. Pure nickel has outstanding resistance to alkalis, the exception being ammonium hydroxide (NH_4OH), which rapidly corrodes nickel. Oxidizing acid chlorides such as ferric, cupric, and mercuric are very corrosive. Nickel is used for containing very reactive chlorides, such as phosphorus oxychloride, phosphorus trichloride, nitrosyl chloride, benzyl chloride, and benzoyl chloride. Pure nickel resists anhydrous chlorine, anhydrous hydrogen chloride, phenol, and bromine.

Nickel-Copper Alloys. Alloy 400 (UNS N04400) is more resistant than nickel to corrosion under reducing conditions and more resistant than copper to corrosion under oxidizing conditions. As a solid-solution alloy 400 is free from the corrosion that can result from local galvanic action between the phases of multiphase alloys. Alloy N04400 is generally resistant to SCC; exceptions are mercury and solutions of its salts, fluorosilicates, concentrated caustic soda (NaOH), and potassium hydroxide (KOH). Alloy N04400 is resistant to all common dry gases at room temperatures. It is not resistant to chlorine, bromine, nitric oxides, ammonia, sulfur dioxide, and hydrogen sulfide in the presence of moisture. It is useful in handling H_2SO_4 under air-free reducing conditions. Aeration causes a sharp increase in the corrosion rate. Alloy N04400 can handle aerated HCl at 10% concentration at room temperature; above room temperature, applications are usually limited to 3 to 4% HCl. In unaerated hydrofluoric acid (HF), N04400 is resistant to all concentrations up to the boiling point. It has very poor corrosion resistance in HNO_3 but has good resistance to NaOH up to approximately 50% concentration and to NH_4OH up to 3% concentration.

Nickel-Chromium Alloys. The high nickel content of these, such as alloy 625 (UNS N06625), gives them considerable resistance to corrosion under reducing conditions and in strong alkaline environments. Also, because of the high nickel content, the alloys are virtually immune to chloride SCC. However, at high temperature and in contact with concentrated alkalis, they are subject to SCC. Mercury will also cause SCC at elevated temperatures.

Nickel-molybdenum alloys were developed to be resistant to HCl at all temperatures and concentrations. The nickel-molybdenum alloys have good corrosion resistance to other nonoxidizing environments, including boiling 60% H_2SO_4, pure H_3PO_4 at most concentrations and temperatures, wet hydrogen chloride gas, hydrogen chloride to 455 °C (850 °F), and wet halogenated organics. The presence of ferric or cupric salts or other oxidizing agents will cause rapid corrosion of these alloys.

Nickel-Chromium-Molybdenum Alloys. The addition of chromium to the nickel-molybdenum alloys increases the resistance to oxidizing environments, giving them good resistance to HNO_3 and H_3PO_4, as well as to most chloride salts. The alloys are very resistant to pitting and crevice corrosion because of the molybdenum content. The nickel-chromium-molybdenum alloys are the most versatile corrosion-resistant alloys available. They are some of the few alloys that are resistant to wet chlorine gas, hypochlorite, and chlorine dioxide solutions. They have good resistance to ferric and cupric chlorides and other oxidizing salts, alkalis, and acids.

Nickel-chromium-molybdenum-copper alloys were developed to resist H_2SO_4 and HNO_3 over a wide range of concentrations and temperatures. They are resistant to H_3PO_4, even when the acid contains fluorides or oxidizing compounds.

Titanium

Titanium forms a tight, adherent oxide film that makes it resistant to many oxidizing reagents, including HNO_3 and chromic acid (H_2CrO_4). Titanium is attacked by reducing acids such as H_2SO_4 and H_3PO_4. It is useful in moist chlorine gas and hypochlorite at ambient and elevated temperatures; most organic acids; and water, seawater, and brine solutions at temperatures to the boiling point. At elevated temperatures, titanium is subject to pitting and crevice corrosion in a chloride environment. Titanium will not resist red fuming HNO_3 or HF in any concentration or temperature and will ignite at very low temperatures in dry chlorine gas. More information on the corrosion of titanium and its alloys is available in the article "Corrosion of Titanium and Titanium Alloys" in *Corrosion: Materials,* Volume 13B of *ASM Handbook,* 2005.

Zirconium

The two zirconium-base alloys used most frequently in the pharmaceutical and chemical-processing industries contain some hafnium, which is metallurgically and chemically similar to zirconium and does not reduce its corrosion resistance. The alloy UNS R60702 contains a minimum of 99.2% (Zr + Hf), with a maximum hafnium content of 4.5%. The second alloy, UNS R60705, contains a minimum of (95.5% Zr + Hf), with a maximum of 4.5% Hf and 2.0 to 3.0% Nb. Both are approved for use in the construction of pressure vessels according to the ASME Boiler and Pressure Vessel Code, Section VIII. The hafnium content is limited for the nuclear grades of zirconium.

Zirconium has excellent resistance to HCl at all concentrations up to temperatures of 120 °C (250 °F). It resists HNO_3 in all concentrations up to 90% and temperatures to 150 °C (300 °F); however, SCC may occur above 70% concentration if high tensile stresses are present. Zirconium is corrosion resistant in H_3PO_4 in concentrations up to 55% at temperatures of 175 °C (350 °F). Above 55% concentration, the corrosion rate increases with temperature, but even in 85% H_3PO_4 at 60 °C (140 °F), the corrosion rate is still less than 0.13 mm/yr (5 mils/yr). If there are fluoride ion impurities present at any concentration of H_3PO_4, zirconium may be subject to rapid corrosion attack. Zirconium alloys are resistant to H_2SO_4 up to 75% concentration and at temperatures to boiling. Ferric, cupric, and nitrate ion impurities cause corrosion of zirconium in H_2SO_4 concentrations above 65%. Fluoride ion concentrations as low as 1 ppm in 50% H_2SO_4 will cause corrosion of zirconium alloys. Zirconium has no corrosion resistance to HF and is rapidly attacked at concentrations as low as 0.001%.

Zirconium is resistant to virtually all alkaline solutions, either fused or in solution to boiling temperature. Zirconium has excellent resistance to corrosion in most organics and organic acids. It should be noted that SCC of zirconium occurs in ferric and cupric chloride solutions, concentrated HNO_3, methanol-hydrochloric acid and methanol-iodine solutions, and liquid mercury or cesium. Detailed information on the corrosion of zirconium and its alloys is available in the article "Corrosion of Zirconium and Zirconium Alloys" in *Corrosion: Materials,* Volume 13B of *ASM Handbook,* 2005.

Impervious Graphite

Impervious graphite is made of raw graphite impregnated under pressure with phenolic, furan, or fluorocarbon resins. The resulting nonporous graphite is impermeable to gases and liquids and is highly corrosion resistant in acids and many solvents. Impervious graphite is dimensionally stable, does not fatigue, and will withstand thermal shock. The phenolic and furan impregnants leach out from the graphite when exposed to ammoniacal compounds, NaOH and KOH, wet halogens, hydrogen peroxide (H_2O_2), strong HNO_3, hypochlorites, and some solvents.

The introduction of the fluorocarbon impregnants to graphite has markedly increased corrosion resistance in solvents, acids, and alkalis. The maximum service temperature for the phenolic and furan impregnants is 170 °C (340 °F); for the synthetic fluorine-containing resin impregnant, 205 °C (400 °F).

Fluoropolymers

All fluorocarbons have high molecular weights, high melting points, and excellent chemical resistance. They have found wide application in chemical and pharmaceutical plants as pipe liners, nozzle liners, gaskets, expansion joints, valve liners, diaphragms for valves and pumps, seals and seal components, and barrier linings for vessels.

Polytetrafluoroethylene (PTFE) has a service temperature of 245 to 260 °C (475 to 500 °F) and is immune to most corrosive environments. Among the materials that attack PTFE are molten alkali metals and free fluorine. This material is rapidly permeated by bromine and oxides of nitrogen, and low-molecular-weight amines tend to plasticize the polymer. It can also be used at cryogenic temperatures, giving it the widest temperature range of any polymer.

Perfluoroalkoxytetrafluoroethylene (PFA) is a copolymer of tetrafluoroethylene and a perfluorovinyl ether. It has high molecular weight, making it suitable for extrusion and molding of parts. Ultrahigh-purity PFA has been developed for the pharamaceutical and semiconductor industries, where chemical inertness and freedom from contamination are needed. This complies with Food and Drug Administration (FDA) regulations as a coating that can be used in contact with food. The upper service temperature is 260 °C (500 °F), higher than fluorinated ethylene propylene. It is nonflammable and has high dielectric properties, low permeability, which is always helpful for lining and barrier applications, and possesses high tensile properties at elevated temperatures.

Fluorinated ethylene propylene (FEP) is a copolymer of tetrafluoroethylene and hexafluoropropylene. It is a fully fluorinated thermoplastic with a service temperature of 175 °C (350 °F). Like PTFE and PFA, this material is chemically inert, with a slightly lower permeability than that of PTFE. The uses for FEP are similar to those for PFA and PTFE.

Ethylene-chlorotrifluoroethylene (ECTFE) is the result of a 1:1 alternating copolymer of ethylene and chlorotrifluoroethylene, having a service temperature of 160 °C (320 °F). Because ECTFE is not completely fluorinated, there are sites along the polymer chain at which chemical attack may occur. This material is attacked by aromatic solvents above 120 °C (250 °F), chlorinated hydrocarbons, ethers, methanol, butanol, and ketones above 65 °C (150 °F), acetic acid above 95 °C (200 °F), and H_2SO_4 and aromatic amines above 65 °C (150 °F). It resists mineral acids up to 120 °C (250 °F), inorganic alkalis, inorganic salts, and oxidizing acids at room temperatures.

Polyvinylidene fluoride (PVDF) is a crystalline, high-molecular-weight, partially fluorinated polymer of vinylidenedifluoride, having a service temperature of 135 °C (275 °F). It is attacked by hot alkalis, hot H_2SO_4, solvents, and warm organic acids. This material has good resistance to chlorine, bromine, and their compounds; however, it will exhibit cracking when subjected to stress in the presence of nascent halogens. Virgin unplasticized PVDF has been successfully used for high-purity water piping as a replacement for electropolished type 316L stainless steel.

Glass-Lined Steel

This material offers the corrosion resistance of glass combined with the strength of steel, making it useful for process equipment operating at elevated pressure and temperature. Glass-lined steel has excellent resistance to corrosion over a wide range of pH and environments. Most glass-lined steel applications will not adversely affect product purity, flavor, or color. Glass-lined steel has an extremely smooth surface that resists fouling and is easily cleaned, which makes it attractive for use in pharmaceutical and fine chemical manufacture.

Corrosion Failures

The types of failures experienced in the pharmaceutical industry are similar in many ways to those seen in the chemical-processing industries (see the articles in the Section "Corrosion in the Chemical Processing Industry" in this Volume). Three primary causes of failure in the manufacture of pharmaceuticals—embedded iron, failures of glass linings, and corrosion under thermal insulation—are discussed.

Embedded Iron

A common problem during the fabrication of stainless steel equipment is the embedding of iron in the stainless steel surface. The iron corrodes when exposed to moist air or when wetted, leaving rust streaks. Larger embedded iron particles can also initiate crevice-corrosion attack in the stainless steel. Embedded iron cannot be tolerated in a fabrication destined for an application in the pharmaceutical industry in which the stainless steel is used to prevent contamination.

Fabrication Practices. The following practices are recommended for minimizing embedded iron in fabrication. First, sheet, strip, and pipe are usually purchased in a surface finish called 2B (a bright, cold-rolled finish; see the article "Surface Engineering of Stainless Steels" in *Surface Engineering,* Volume 5 of *ASM Handbook,* 1994). Plate is normally hot rolled, annealed, and pickled and is furnished with a 2B mill finish. If a plate having a better surface finish is required, it should be specified during the procurement stage. When cleanliness is very important, sheet and plate can be ordered with a protective adhesive paper that can be left in place during storage and fabrication. Pipe and tubing can be ordered with protective end covers, especially if the pipe and tubing is to be stored outdoors.

Sheet and plates should be stored indoors and upright in racks, not horizontally on the floor. The dragging of sheets and plates over each other and worker foot traffic are often primary causes of embedded iron and deep surface scratches.

Care should be exercised in handling the sheet and plate on layout tables, forming roll aprons, and benches. This will minimize iron contamination.

Design plays an important role in iron contamination. Equipment and piping should be free draining. If internal attachments are needed, they should not interfere with free drainage. Bottom connections should be completely free draining. Vessel bottoms used as a work area during construction collect debris, and the foot traffic grinds the debris into the surface. It is suggested that the vessel bottom be flushed down and drained completely at the end of the workday to remove collected dirt and debris. If the vessel is a large flat-bottomed unit, a slatted wood floor should be installed to reduce the grinding of contaminants into the vessel bottom by foot traffic.

Testing new fabrications for embedded iron is relatively easy. The surfaces should be washed with clean water, drained completely, and after a 24 h waiting period, inspected for rust streaks on the surface. Water testing of the fabrication as a minimum should definitely be part of the equipment purchase order. For items to be used in a pharmaceutical plant, a precision inspection—the ferroxyl test for free iron (Ref 1)—should be specified in purchase documentation. This test can be easily performed in the field as well as in the fabricator shop.

Removal of Embedded Iron. Pickling is the most effective method of removing embedded iron. The surfaces must be cleaned of all surface oil, grease, and other organic materials so that the surface is wetted by the pickling solution. The pickling solution is a mixture of HNO_3 and HF at 50 °C (120 °F). The solution removes the embedded iron and other metallic contaminants and leaves the surface clean and in its most corrosion-resistant condition. It should be noted that HNO_3 alone will remove only superficial iron contamination and will leave the deeply embedded particles. Small items are usually pickled by immersion. Piping and vessels that are too large for immersion can be pickled by circulating the pickling solution through them. It is recommended that a competent chemical cleaning contractor be employed for the pickling operation. If the ferroxyl test shows only spotty patches of iron contamination, then it is recommended that these be removed by the use of an HNO_3-HF pickling paste, rather than a complete pickling bath.

Another method of cleaning the stainless steel surfaces is the use of glass bead blasting. The beads should be clean and of a proper size to abrade the surface slightly and remove the contamination. Gritblasting and sandblasting are not recommended; they leave a rough profile that makes the stainless steel prone to crevice corrosion.

Organic contamination on stainless steel surfaces increases crevice corrosion. Contaminants include grease, oil, marking crayons, paint, and adhesive tape. Removal of organic contaminants is best accomplished by the use of a nonchlorinated solvent. It is important that nonchlorinated solvents be used. If a proprietary degreasing solvent is used, it should be tested to ensure that it does not contain chlorides. Residual chlorides remain in crevices and cause chloride SCC of austenitic stainless steels.

Weld Defects. Austenitic stainless steel surfaces can be affected by slag from coated welding electrodes, arc strikes, welding stop points, grinding marks, and weld spatter. These factors have initiated corrosion in aggressive environments that normally do not attack stainless steels. Arc strikes damage the protective oxide film of the stainless steel and create crevicelike imperfections in or near the HAZ. Weld stops create pinpoint defects in the weld metal. Arc strikes and weld stop points are actually more damaging than embedded iron, because they occur where the protective film has been weakened by the heat of welding.

Weld stop defects can be avoided by using runout tabs, by beginning the arc immediately ahead of the stop point, and by welding over each intermediate stop point. Arc strikes are more difficult to eliminate. Initially, the arc can be struck on a runout tab. It can also be struck on the weld metal when the filler metal will tolerate arc strikes. If the metal will not tolerate arc strikes, the arc must be struck alongside the filler metal in or adjacent to the HAZ.

Weld spatter creates a small weld in which the molten glob of metal touches and adheres to the surface. The protective oxide film is penetrated, and small crevices or pits are formed where the film has been weakened.

Heat tint formation also weakens the oxide film. The weakening is greater for some degrees of heating than for others, as indicated by the extent of color change. The necessity of removing heat tint is greatest where the environment is very aggressive and the stainless steel approaches the limit of its corrosion resistance.

Pickling by immersion in the standard HNO_3-HF solution is the simplest and preferred method of heat tint removal when size permits. Glass bead blasting, using beads that are clean and of proper size in order to prevent overroughening of the surface, can also be employed to remove the heat tint.

Small slag particles from coated electrodes resist cleaning and tend to collect in slight undercuts or other irregularities. To remove slag from 300-series stainless steel, wire brushes fabricated from 300-series stainless steel should be used. For critical service, brushing should be followed by local pickling or glass bead blasting. Grinding is frequently used to remove slag, arc strikes, weld spatter, and other imperfections. Grinding wheels and continuous-belt grinders can overheat the surface and reduce corrosion resistance; therefore, they have limited usefulness. Abrasive disks and flapper wheels are not as harmful to the metal surface. Disks must be kept clean and replaced frequently. These procedures are good commercial fabrication practices and should be specified during the bidding and procurement stages in an effort to eliminate cost overruns and poor service performance.

Failures of Glass-Lined Steel Equipment

Most glass-lined equipment failures are not related to chemical deterioration of the lining but rather to mechanical and thermal influences. The typical failures encountered in the use of this type of equipment in the pharmaceutical and fine chemical industries usually involve mechanical shock, corrosion, abrasion, thermal shock, and thermal stress.

Mechanical shock is the most common cause of glass failure in pharmaceutical production facilities. It accounts for approximately 70% of the failures in glass-lined steel process equipment and is frequently the result of human error.

The most obvious cause of glass failure due to mechanical shock is objects falling on either the exterior or interior of the vessel. Care must be observed at all times when working near glass-lined equipment, because a shock to the outside of a vessel may cause damage to the glass lining.

Lifting lugs are normally supplied on glass-lining equipment and should be used for lifting the equipment and setting it in place. The lugs are specifically designed for this purpose and should be used in accordance with manufacturer's recommended procedures for handling and rigging. Shortcuts in rigging, such as using a nozzle as a lifting lug, can easily subject the glass lining to undue stress and cause damage to the lining. If mechanical shock has occurred or is suspected, the equipment interior should be inspected immediately and, if necessary, repaired.

Entering the interior of glass-lined steel equipment always creates a potential for mechanical damage. In addition to compliance with the routine safety precautions, the mechanic should wear clean, soft rubber-soled shoes or sneakers to prevent scratching of the lining and should remove all loose objects from pockets prior to entering the lined equipment. Even metal belt buckles should be removed to prevent accidental scratching of the glass. Tools can be lowered to the mechanic when he is safely inside.

If it becomes necessary to remove product or by-products from the walls of glass-lined steel equipment, care must be taken to avoid scratching of the glazed surface. Metal tools should never be used. Plastic or wood scrapers can be used; high-pressure water jets are preferable.

The introduction of nucleated glass linings by fabricators of glass-lined steel equipment has reduced the damage caused by impact and, to some extent, has lessened its occurrence. Nucleated or partially nucleated glass linings have higher tensile strength and fracture energy than conventional glassed steel linings. These properties have lessened the tendency toward spalling from mechanical shock, the releasing of internal stresses, or thermal shock. Nucleated glass linings cannot prevent cracking entirely, but they tend to limit the extent of the damage by restricting it to a small area, usually requiring only a tantalum plug for repair.

Corrosion Attack of Glass Linings. Glass-lined steel is not completely inert and is constantly undergoing local chemical reactions at the glass surface. Glass-lined steel can be used with corrosive materials because of the low rate of reaction; the slower the rate, the longer the useful life of the glass lining.

Acids. Except for HF, concentrated H_3PO_4, and phosphorus acid (H_3PO_3) above 85%, glass-lined steel is resistant to corrosion by acids. Generally, the corrosion rate decreases with concentration but accelerates with increasing temperature. Acid attack is more severe in the vapor phase than in the liquid phase, especially in more dilute solutions, because of the water vapor in the vapor phase. Usually, acid attack will result in a gradual loss of the fire-polished surface, but the lining will generally retain a dull, smooth finish.

Hydrofluoric acid will completely destroy glass-lined steel equipment. Even with concentrations as low as 20 ppm, fluorides in an acid environment corrode glass severely, especially in continuous reactions in which the fluorides are repeatedly replaced. Hydrofluoric acid reacts with silicon dioxide, the primary ingredient in glass, and destroys the silicon dioxide structure, forming silicon tetrafluoride and water vapor. The silicon tetrafluoride then hydrolyzes into silicon dioxide and hydrofluorine silicic acid, which are absorbed by the condensing water vapor. When this contaminated condensate reaches the heated vessel wall in the vapor space, it evaporates, depositing silicon dioxide and liberating the hydrofluorine silicic acid, which then disintegrates into silicon tetrafluoride and HF. The chain reaction keeps repeating itself with continuous replenishment of HF. The corrosion occurs both in the liquid phase and in the cooler areas of the vapor space in which the fluoride vapor can condense.

Glass is not attacked by fluorine and its compounds in an alkaline environment, nor is it attacked by anhydrous hydrogen fluoride gas. The prerequisite for HF attack in the vapor space is the formation of water; thus, the corrosion rate will be greatly reduced if the vapor area is heated. In the liquid phase, fluorides will severely etch the glass and produce a roughened surface, with a complete loss of the fire-polished surface. In the vapor phase, the attack is more localized and concentrated; chipping and pinholes will be seen but with considerably less loss of the fire-polished glass surface.

Reagents that contain fluoride impurities must be carefully analyzed to determine the fluoride level before they are used. Frequently, technical-grade H_3PO_4 and its salts are contaminated with fluoride, as are other mineral acids. It is important to realize this when using these chemicals in recovery operations.

Alkaline attack of glass linings is much more severe than acid attack. The attack takes place only in the liquid phase in the case of nonvolatile alkalis. The greater the concentration and pH of the alkali, the greater the amount of corrosion. As the operating temperature and the pH increase, the corrosion resistance will decrease. Corrosion by alkalis is evidenced by pinholes, chipping, and a severe loss of the fire-polished surface.

Many glass-lined steel reactors have been lost prematurely in service by improper charging of reactants into the vessels. Caustic reactants charged into a vessel should always be fed directly into the liquid phase. If fed through a nozzle, the alkali will run down the sidewall of the reactor in the vapor space and cause severe alkaline attack, especially if the reactor is being heated.

Water can cause severe corrosion of glass-lined steel, and the severity increases with water purity and temperature, becoming greatest above the boiling point. When water droplets condense on the relatively cool surface of the glass-lined equipment in the vapor space, they leach out alkali ions from the glass and form an alkaline solution that attacks the glass. A small amount of acid added to the water usually slows the corrosion caused by condensation in the vapor space. This addition of acid is frequently useful in steam distillations. See the article "Rouging of Stainless Steels in High-Purity Waters" in this Volume.

Abrasion Failures. Failure of glass linings by abrasion alone is not very common. It is evidenced by a loss of fire polish of the glazed surface and results in a rough, sandpaper-like finish. Abrasion in conjunction with acid corrosion results in severe failure. The abrasive action weakens the silica network mechanically, allowing acid corrosion to accelerate rapidly.

Thermal shock failure occurs because of abrupt changes in the temperature of the glass lining and results in relatively small but thick pieces of glass spalling off in rigid fractures. There are four operations in which sudden temperature variations can cause thermal shock:

- Sudden cooling of a glass-lined surface by subjecting a preheated surface to a cold liquid
- Sudden heating of a glass-lined steel wall by rapidly circulating a very hot fluid through the jacket of a cold vessel
- Sudden heating of the glass-lined steel surface by introducing a hot fluid into a cold vessel
- Sudden cooling of the vessel wall by rapid circulation of a cold fluid through the jacket of a preheated vessel

Thermal shock is strictly an operational problem and can be eliminated by adequate process controls. Unlike failure due to mechanical shock, thermal shock usually damages the glass lining so that repairs with tantalum plugs are not practical or possible. Reglassing of the vessel is then required.

Failure due to thermal stress is caused by differential heating or cooling that is not instantaneous. Thermal stress may occur on the vessel wall just below the area at the top jacket closure ring or in the area where the bottom jacket closure ring is welded to the vessel. In either case, the inside of the jacket can be heated or cooled, while the unjacketed area is not. At temperatures approaching 205 °C (400 °F), sufficient strain can be developed in the glass lining at these areas to cause the glass to crack. Thermal stress can be reduced by careful control of heating or cooling operations. Also, insulation of the unjacketed areas will help to reduce the extreme temperature variations.

Water hammer has also been known to cause shock waves that add to the thermal stresses at the area of the jacket closure. If these stresses are allowed to persist for long periods of time, cracks can develop with or without chipping.

Overstressing of Nozzles. Glass is strong in compression, but a concentrated point load can damage a glass-lined nozzle. The convex radius at the top of the nozzle also makes it susceptible to damage due to overstressing. The two situations that can cause overstressing of the nozzle and lead to possible failure are overtorquing and overstressing by the attached piping.

Overtorquing of bolts or clamps used to secure nozzles on glass-lined vessels to piping can cause glass lining to spall in large segments that may include a considerable area of the nozzle. Overstressing of nozzles through external piping can break the glass lining on the nozzle face. The design of external piping systems should include expansion joints or bellows as close as possible to the nozzles to minimize eccentric loads and moments and to allow for thermal expansion of the piping and the vessel. Cold springing when connecting pipe to a nozzle cannot be tolerated. Supporting heavy gear drives on a nozzle can also create an unsafe condition with regard to the glass lining.

Failures of Repair Plugs. Tantalum is the most commonly used repair material for glass linings. Its corrosion resistance is very close to that of glass, except in fuming H_2SO_4 above 65 °C (150 °F), H_2SO_4 above 98%, free sulfur trioxide (SO_3) above 65 °C (150 °F), or nascent hydrogen.

A repair made with tantalum will protect a glass-lined vessel for its useful life if the repair plug is installed in accordance with the recommendations of the manufacturer. However, the repair plug must be inspected periodically to ensure that it is secure.

Galvanic corrosion can cause hydrogen embrittlement of tantalum repair plugs, because the tantalum is in contact with the steel substrate. If a second dissimilar metal is present in the vessel, such as a metallic dip tube fabricated from a material other than tantalum, a galvanic cell is produced. Tantalum, being more noble, will generally act as the cathode where hydrogen is liberated.

Failure Due to Hydrogen Damage. Acids on the exterior steel surfaces of glass-lined steel equipment will, in time, react with the steel, forming nascent hydrogen. This hydrogen diffuses through the steel behind the glass and causes the glass to spall because of pressure buildup.

To avoid this type of failure, all acid spills on the outside of a glass-lined steel vessel should be washed off immediately to avoid acid attack. The so-called acid-resistant coatings give some degree of protection, but none is completely effective against acid attack on the steels.

Jacket cleaning can create another source of nascent hydrogen. The cleaning solutions recommended by the manufacturer should be used for cleaning the inside jacket of a glass-lined vessel. Strong acid solutions, inhibited or otherwise, should never be used for descaling the jacket, in order to avoid any possibility of nascent hydrogen formation.

Corrosion Beneath Thermal Insulation

Serious corrosion problems frequently occur under thermal insulation applied to vessels and piping components in pharmaceutical plants when the insulation becomes wet. Corrosion beneath insulation is an insidious problem that is also discussed in the article "Corrosion Under Insulation" in this Volume. The insulation usually conceals the corroding metal, and the situation can go undetected until metal failure occurs. The corrosion of metal components beneath insulation has led to very high maintenance costs and lost production time and has frequently required complete replacement of major components. In addition, operator and plant safety may be jeopardized.

Thermal insulation received from manufacturers and distributors is dry, or nearly so. Therefore, if the insulation remains dry, there is no corrosion problem. Possible solutions to the problem of metallic corrosion beneath wet insulation include keeping the insulation dry or protecting the metal.

Unfortunately, the application of this solution is not that simple. Insulation can become wet in storage and during field erection. Moisture or weather barriers are not always installed correctly, or they are not effective in preventing water entry. Weather barriers and protective coatings can become damaged and are often not maintained and repaired.

The problem is further complicated by the fact that the degree of corrosion by wet insulation appears to be dependent on the insulation material as well as the atmospheric contaminants and moisture entering from external sources. Water extracts from calcium-silicate-base insulation, fiberglass, cellular glass, and ceramic fiber are generally neutral to alkaline, with pH values in the range of 7 to 11. Cellular glass is free of soluble chloride, while calcium silicate, fiberglass, and some ceramic fibers contain chlorides. Mineral wool gives a neutral environment when wet, usually a pH value of 6 to 7 with a low chloride content (2 to 3 ppm). Water extracts from organic foams can be quite acidic, with pH values of 2 or 3. In addition, where halogenated fire retardants have been added to the foam, water extracts show high levels of free halide, depending on the degree of hydrolysis achieved.

Carbon and low-alloy steels are normally passive in alkaline environments and have minimal corrosion rates. However, chloride ions (Cl^-), either from the insulation material itself or from airborne or waterborne contaminants, tend to break down the passivity locally and initiate pitting corrosion. If penetration by acidic airborne or waterborne contaminants of sulfur or nitrogen oxides is possible or if water extracts from the insulation are acidic, such as from organic foams, then general corrosion occurs. Occasionally, airborne or waterborne contaminants, notably the nitrate anion (NO_3^-), cause external SCC of nonstress-relieved carbon and low-alloy steel systems, especially if a cyclic wetting and drying concentration mechanism is present. Generally, plant facilities operating continuously or intermittently between 65 and 205 °C (150 and 400 °F) are subject to corrosive attack.

The most significant corrosion problem that occurs when insulated austenitic stainless steels are subjected to moisture is external SCC. The problem occurs because chlorides tend to concentrate under the insulation at the surface of the metal when the insulation becomes wet. The moisture can leach soluble chlorides out of the insulation, or the entering moisture may already contain chloride from the environment. At the warm metal surface, the moisture is vaporized, leaving behind an increasing concentration of chlorides. During operation, when the equipment or piping is in the susceptible temperature range, chloride SCC can then occur rapidly. The four factors necessary for SCC are an austenitic stainless steel, Cl^- ions, tensile stress on the metal, and temperatures between 50 and 230 °C (120 and 450 °F).

Unfortunately, thermal stress relief, which is usually effective in preventing SCC of carbon and low-alloy steels, is normally not practical for austenitic stainless steels. However, there are numerous controllable factors in the design, construction, and maintenance of insulated equipment that have a marked effect on the amount of damage caused by corrosion under thermal insulation.

Equipment Design. The design of pressure vessels, tanks, and piping generally includes numerous details for support, reinforcement, and connection to other equipment. These details may include stiffening rings, insulation support rings, gussets, brackets, reinforcing pads, ladder brackets, flanges, and hangers. The design of equipment, including these details, is the responsibility of engineers/designers using construction codes to ensure reliable designs for both insulated and uninsulated equipment. Unfortunately, consideration of the problem of insulating these details and of leaving adequate room for the insulation is completely lacking in these codes and in the instructions to the engineers/designers. As a result, the items are designed as though they will not be insulated.

Undesirable geometries and design features include:

- Flat horizontal surfaces (such as vacuum rings)
- Structural shapes that trap water (H-beams, channels)
- Shapes or configurations that are impossible to weatherproof properly (structural members, gussets)
- Shapes that lead moisture and contaminants into the insulation hanger rods (angle iron brackets)
- Inadequate spacing that causes interruption of the vapor or weather barrier (nozzle extensions, ladder brackets, deck or grating supports). The weather barrier on such designs is frequently broken because of inappropriate details for insulated equipment or the lack of space for the specified insulation thickness.

The consequence of an incomplete moisture barrier is that more water and contaminants get into the insulation at each exposure cycle; this increases the time required for drying, cools the insulated equipment to temperatures at which corrosion is possible, and thus increases the damage. Some equipment details, such as gussets, brackets, and hangers, actually funnel water and contaminants into the insulation. Another consideration is the increased cost of insulating equipment that was not designed for insulation. In such cases, the insulation and jacketing must be cut and fitted by installers; thus, needless man-hours are spent insulating a complicated detail, with the result being an installation that is doomed to failure. The solution to such problems is to specify the type of insulation, thickness, and weather jacketing in the design stage.

The service or operating temperature of the equipment is very important in corrosion beneath thermal insulation. Higher temperatures make water more corrosive, and paints and caulking will fail prematurely.

Generally, corrosion associated with equipment operating below freezing temperatures is corrosion outside of, not under, the insulation. Equipment operating between freezing and the atmospheric dewpoint is subject to continuous corrosion, and damage can occur as quickly as it does under warm insulation. However, corrosion beneath warm insulation is more difficult to control, because of the drying out of water entering the insulation and the concentration of contaminants carried in with the water drying out repeatedly in the same location.

Insulation Materials. Corrosion is possible beneath all types of insulation. The insulation material is only a contributing factor. The insulation characteristics that are most influential in the corrosion of metal beneath insulation include water absorbency, chemical contributions to the water phase (not only from the insulation but also from external sources), and service temperature. Some of the more widely used insulation materials are discussed as follows.

Polyurethane foam is primarily used for cold and antisweat service. It does not absorb or wick water as long as the cell structure remains intact. It is permeable to water vapor in cold service when the required vapor barrier fails. Vapor diffuses through cell walls to the temperature zone in which it condenses and diffuses further to the point at which it freezes. The maximum service temperature is 80 °C (180 °F). If used in continuously cold service, it does not corrode unprotected metal surfaces. If in intermittent service to its maximum service temperature, it can cause corrosion of unprotected wet metal surfaces from released chlorides in fire retardants and blowing agents. The ultraviolet rays from the sun will decompose this insulation.

Polyisocyanurate foam is a fire-resistant organic foam having a low flame propagation rate. It does not absorb and wick water as long as the cell structure remains intact. It is permeable to water vapor in cold service when the required vapor barrier fails. The maximum service temperature is 120 °C (250 °F).

When polyisocyanurate is exposed to heat and moisture, the cell structure in the heated zone breaks down. The decomposition products may contain chlorides from the fire retardant and blowing agent and may aggressively corrode unprotected metal surfaces.

Flexible foamed elastomer does not readily absorb or wick water, and it has a maximum service temperature of 80 °C (180 °F). Although not corrosive by itself, it will support the corrosion of unprotected metal surfaces when water is present, especially when the water contains chlorides from an external source.

Cellular glass is a rigid glass foam whose blowing agent contains carbon dioxide and hydrogen sulfide. It does not absorb and wick water. The maximum service temperature is 480 °C (900 °F). When water is present and the cell structure is damaged, release of the foam blowing agent may cause corrosion on unprotected carbon steel surfaces.

Glass fiber for insulation is usually a pure glass fiber containing various types of binders. Fiberglass will absorb and wick water; however, it drains excess moisture better than other types of insulation. The maximum service temperature is 230 °C (450 °F), with special formulations to 455 °C (850 °F).

Mineral wool is a natural mineral (rock wool) or metal slag fiber that is basically an impure glass. It readily absorbs and wicks water. Maximum service temperatures range from 650 to 980 °C (1200 to 1800 °F), depending on type and manufacturer. The fact that mineral wool will wick and hold water makes it a contributory factor in the corrosion of unprotected wet metal surfaces.

Calcium silicate insulation is a cementitious mixture and readily absorbs and wicks water. Calcium silicate insulation can hold up to 400% of its own weight in water without dripping. The maximum service temperature is 650 °C (1200 °F). Although its pH is initially high (10 average), it is aggressive in supporting corrosion on unprotected wet metal surfaces because of its moisture retention, particularly when the moisture contains chlorides from an external source.

Protective coatings are extremely important in preventing the corrosion of metal surfaces beneath thermal insulation. The assumption that a single coat of primer is adequate, based on the assumption that the weatherproofing never allows water to penetrate into the insulation system, is not valid.

Basically, service under thermal insulation is virtually an immersion service. Once the weather- or vaporproofing barrier is broken, metal surfaces under isolation are wet longer than the surfaces of most uninsulated equipment. Under warm insulation, the coating is subject to higher temperatures than most coated, uninsulated equipment. The coatings beneath thermal insulation fail because of chemical degradation and the permeability of the coating. Highly permeable coatings allow corrosion to initiate behind the coating film, even in the absence of breaks or pinholes in the coating.

In selecting a protective coating for use beneath insulation, consideration should be given to its abrasion resistance, temperature resistance, chemical resistance, and its ability to resist hot water vapors. Organic coating materials and surface preparation are discussed in details in the articles "Organic Coatings and Linings" and "Paint Systems" in *Corrosion: Fundamentals, Testing, and Protection*, Volume 13A of *ASM Handbook*, 2003.

In general, for carbon steel and stainless steel piping and equipment operating at −45 to 120 °C (−50 to 250 °F) under insulation, it is recommended that the surface be abrasive blast cleaned to NACE International No. 2 or SSPC: The Society for Protective Coatings SP10 near-white metal standards having a surface profile of 50 to 75 μm (2 to 3 mils). The freshly cleaned surfaces should then be coated with 305 μm (12 mils) (total dry-film thickness) of an epoxy phenolic coating applied in two or more coats or with 305 μm (12 mils) (total dry-film thickness) of a high-melting-point, amine-cured, coal tar epoxy applied in one or two coats.

For carbon steel piping and equipment operating at temperatures of 120 to 260 °C (250 to 500 °F) with intermittent cycling service into the hot water range under insulation, the same abrasive cleaning is recommended. The freshly cleaned surfaces should then be coated with 150 μm (6 mils) (total dry-film thickness) of a copylymerized silicone resin applied in two or more coats.

Carbon steel piping and equipment operating continuously above 260 °C (500 °F) under insulation should have the surface abrasive blast cleaned to NACE International No. 3 or SSPC: The Society for Protective Coatings SP6 commercial blast standards having a surface profile of 25 to 50 μm (1 to 2 mils). The freshly cleaned surfaces should then be coated with 100 μm (4 mils) (total dry-film thickness) of a black zinc-free modified silicone coating applied in two coats.

For stainless steel piping and equipment operating at temperatures from 120 to 370 °C (250 to 700 °F) under insulation, it is recommended that the surface be cleaned according to SSPC: The Society for Protective Coatings SP1 solvent cleaning standards. The freshly cleaned surfaces should be coated with 100 μm (4 mils) (dry-film thickness) of a black zinc-free modified silicone coating applied in two coats.

Inorganic and organic zinc-rich primers have given very poor performance under thermal insulation and should not be used. Possible reasons for this poor performance include:

- There is the likelihood of reversal of polarity of galvanic couples with increasing temperature.
- Chemical salts that are carried in and deposited with the water interfere with or destroy the effectiveness of the coating.
- The environment beneath the insulation is not freely ventilated and may not have adequate oxygen or carbon dioxide for film-forming reactions to occur.

Weatherproofing and Vaporproofing. The outer covering of the insulation system is critical. It is the principal barrier to the water necessary for the corrosion of metals beneath thermal insulation. Also, it is the only part of the insulation system that can be quickly inspected and economically repaired.

The purpose of the weather barrier, which should be used on warm or hot equipment, is to keep liquid water out but to permit the evaporation of any moisture that gets in. The purpose of a vapor barrier is to keep both liquid and vapor out of the insulation system. Vapor barriers should be used on all cold and dual-service insulation systems.

ACKNOWLEDGMENT

This article is adapted from "Corrosion in the Pharmaceutical Industry" by Ralph J. Valentine, VAL-CORR, in *Corrosion*, Volume 13 of *ASM Handbook*, 1987, p 1226 to 1231.

REFERENCE

1. "Cleaning, Descaling, and Passivation of Stainless Steel Parts, Equipment, and Systems," A 380, *Annual Book of ASTM Standards*, Vol 01.03, ASTM International

SELECTED REFERENCES

- D.H. DeClerck and A.J. Patarcity, 32nd Report on Materials of Construction, *Chem. Eng.*, Vol 93 (No. 22), p 46–63
- A. Garner, "Corrosion of High Alloy Austenitic Stainless Steel Weldments in Oxidizing Environments," Paper presented at Corrosion/82, National Association of Corrosion Engineers, 1982
- J. Kearns and G.E. Moller, Reducing Heat Tint Effects on the Corrosion Resistance of Stainless Steels, *Mater. Perform.*, May 1994
- S. Lederman, "Glass Lined Equipment—Typical and Atypical Failures Encountered in the Field," Paper presented at Corrosion/82, National Association of Corrosion Engineers, 1982
- W.I. Pollock and J.M. Barnhart, Ed., *Corrosion of Metals Under Thermal Insulation*, STP 880, American Society for Testing and Materials, 1985

Corrosion Effects on the Biocompatibility of Metallic Materials and Implants

Kenneth R. St. John, The University of Mississippi Medical Center

IN THE FIELD OF MEDICAL DEVICE DEVELOPMENT AND TESTING, corrosion of metallic parts can lead to significant adverse effects on the biocompatibility of the device. As corrosion occurs, the products of corrosion may accumulate in adjacent tissues; ionic species released may participate in metabolic processes as a substituent for the normal metallic ions in the processes and may impact the overall function of the device in its intended environment. Organometallic species may be formed by the reaction of proteins with the metals, inhibiting the normal function of the proteins. In addition, some individuals have hypersensitivity to certain metal ions that may produce localized or systemic immune responses.

In general, metals and metal alloys that have been qualified for use in medical devices achieve their compatibility through the formation of continuous passive films that prevent or significantly limit the corrosion rates encountered in the physiological environment. One mechanism of biocompatibility failure is the occurrence of conditions that damage the passive film or prevent its formation. Changes in environmental conditions surrounding the device, wear of moving surfaces, and fretting between components may all contribute to the loss of corrosion protection. In some cases, the increased corrosion may itself contribute to changes in the physiological environment and further corrosion problems.

All implanted metals release metal ions into the surrounding tissue and the tissues must then respond in some fashion. In general, the response is relatively innocuous and the metals are said to be biocompatible.

Origins of the Biocompatibility of Metals and Metal Alloys

As surgery for the repair of anatomical structures and the replacement of damaged organs and organ components have developed, a concern for the long-term survival of metallic devices and device components in the body has focused on both the damage to the components themselves and injury to surrounding tissues and the body as a whole as a result of the presence of the metals in the physiological environment. Some of the concerns have included toxic responses to metallic ions released from the material, the long-term possibility of neoplastic transformations to cells and the development of cancer in response to the products of metal corrosion, and changes in the function of the material in the physiological environment as a result of the material/tissue interaction. As metals became more widely used in surgery, surgeons also wondered whether the appropriateness of certain metals could be determined in advance of their use and the long-term success of implantation of metals improved.

Mechanisms of Metal and Alloy Biocompatibility. It is almost universally true that metallic materials that are considered to be biocompatible or suitable for use in human implantation or tissue-contact device use derive their compatibility from the fact that they have a nonporous stable passive film on the surface that minimizes the diffusion of metal ions from the bulk material and prevents corrosion of the material in contact with human tissues. Based on testing results (Ref 1), it is clear that most of the alloying elements used in the production of the alloys used in surgery are not intrinsically inert, but the formation of the passive film protects the alloy from the corrosive physiological environment. Stainless steel and cobalt-chromium alloys rely on the presence of chromium in the alloy for their passivity. Titanium and its alloys rely on the titanium passive layer, and alloys containing zirconium rely on the zirconium passive layer. The biocompatibility of these materials is determined in vitro (outside the body) and in vivo (in the living body) under standardized and controlled conditions. In most cases, the breakdown of the passive layer on the metal surface is a major factor in biocompatibility problems with metallic devices. The absence or interruption of the passive layer removes the corrosion protection, at least temporarily, and leads to increased corrosion.

The usual response to the presence of metallic materials in living tissue is the production of a fibrous tissue capsule that surrounds the implant and provides a partial barrier to physiological interaction with the metal and creates a microenvironment around the implant. In testing for acceptable biocompatibility, the thickness of the fibrous tissue capsule and the quantity and identity of the cells present in the surrounding tissues are compared with those for alloys with a long history of successful use with an acceptable tissue response. Some materials, most notably titanium and tantalum, can be implanted into tissue without the formation of a reactive fibrous tissue capsule (Ref 2–4). This characteristic is used in the implantation of titanium and tantalum into bone because, in the absence of micromotion, the bone will grow back to the surface without an intervening soft tissue capsule, giving the potential for better mechanical stability of the device.

Early Studies of Metal Biocompatibility. Among the pioneers in the study of the tissue response to metals (*circa* 1930–1960) were Venable and his coworkers (Ref 5) and Laing (Ref 1, 6). Venable based his conclusions on observations from an animal study and concluded that pure metals were inert and, unless two dissimilar metals were connected, any reaction seen in tissues was purely a chemical reaction to body acids and had no electrolytic significance. He and his coworkers had observed extremely variable clinical results in the wide variety of metals and alloys being used at the time in surgery and were attempting to perform a scientifically based study to confirm for themselves which metals should be used in their patients. Their understanding of the electrochemical reactions occurring in their experimental animals was minimal, but it should be recognized that, in 1937, the general state of knowledge about corrosion and corrosion processes was much less understood. Venable did identify the need for consideration of the corrosion resistance of implant materials during the materials selection process.

Laing tested 42 alloys and 13 pure metals in rabbits to characterize the tissue reaction (thickness of reactive pseudomembrane) and the metal content in tissues surrounding the implant specimens. His results showed that some stainless steels (particularly type 316L and precipitation-hardening type 17-7 PH) caused much less tissue response than the remainder of the stainless steels he tested. He recognized that physiological environments differ from the environments for which stainless steels were developed and were being used at the time. Therefore, test methods specific to suitability for use in the body would need to be developed. His observations of commercially pure titanium and titanium-aluminum-vanadium alloys gave evidence of the minimal response of the body to titanium and some of its alloys and, in 1967, preceded most of the research that has led to recognition of titanium alloys as particularly suitable for use in contact with bone. He found that the metals being used at the time for implant manufacture elicited a relatively minor response by soft tissues. Many of these same materials are still used in medical device manufacture. His observations of the response to pure titanium, zirconium, and niobium led him to suggest that they be considered for future development as alloy components, as occurred in the 1990s.

Failure of Metals to Exhibit Expected Compatibility

Metallic materials present in the human body are not subjected to uniform environmental conditions in all circumstances. The normal physiological environment may be generalized as pH 7.4, 0.9% (by weight) saline, at 37 °C (98.6 °F). These conditions vary in certain organs for metabolic or ion transport reasons (Ref 7), with the pH of gastric fluid as low as 0.8. The pH of urine ranges from 4.5 to 8.0 because of the role of the kidney in maintaining systemic pH. The pH in the microenvironment around dental caries-causing bacteria may be as low as 2.2 (Ref 8). During normal fracture healing, the pH of the local tissues has been found to decrease to 5.5 in the absence of infection (Ref 6). The same author showed a decrease to 5.5 and then an increase to 9.0 in the presence of infection. Others (Ref 9) state that infection results in decreased pH and decreased oxygen tension. A microelectrode study in 1988 (Ref 10) showed that pH under osteoclasts (the cells responsible for bone resorption) could be as low as 3.0 and under activated macrophages (cells involved in the response to bacteria and foreign materials), as low as 3.6. When an infection is present at the surgical site, the physiological response to the infection may include fever (a change in temperature systemically); a change in concentration of oxygen in the surrounding tissues; and inflammation, which will include a major local increase in defensive cells and a local increase in temperature. Because sodium chloride concentration, temperature, oxygen concentration, and pH can vary due to physiological needs, infection, and/or inflammation, the corrosive environment around the implanted metal may be significantly different from that under standardized testing conditions. One report (Ref 11) suggests that the increased corrosion in the presence of infection may be self-perpetuating because the change in pH and the inflammatory response to the infection is a defensive reaction. In some cases, the increased levels of metal ions due to the higher rate of corrosion may lead to an immune response to the metal, and the initial inflammatory response to the infection may then become an inflammatory response to the corrosion products and metal ion concentrations, perpetuating the low pH and oxygen concentration conditions. Additionally, another study (Ref 12) showed that the presence of low levels of metal ions significantly reduced the ability of phagocytes to attack and kill bacteria, meaning that the infection can increase the corrosion rate and the corrosion products can serve to perpetuate the infection.

The details of device design (such as size, shape, and degree of conformity to surrounding tissues) and mechanical loading conditions introduce factors that may lead to physical breakdown of the passive film, such as fretting, erosion, or wear. Both particulate debris and metal ions are released into the surrounding tissues as a result of these processes. The tissue response to the products of corrosion may then exacerbate the conditions that initially contributed to the corrosion.

Dissimilar metals may be combined into a single tissue site either by design or, inadvertently, by the surgeon in placing devices composed of more than one component. Additionally, devices placed in close proximity to each other may become part of a complete circuit, even though it may not be immediately apparent that the necessary physical or electrical contact is formed.

Response to Severe Corrosion of Implant Metals. In the early days of the use of metals and metal alloys in surgery, selection of constituent materials was not necessarily based on their corrosion properties in 0.9 wt% sodium chloride solution. Surgeons or machinists with whom they were associated selected materials of construction and produced many of their own designs for solving clinical problems. In some cases, a great deal of corrosion occurred after implantation (Ref 13–15) and, in some cases, necessitated device removal and debridement (surgical removal) of surrounding tissues. In the first study (Ref 13), a type 316 stainless steel nail was attached to a type 303 stainless steel plate and the plate was then affixed to the bone with screws, some of which were 303 and some 316. Plates in the other two studies (Ref 14, 15) were both Lane plates, but after 29 years' and 64 years' implantation, respectively, the original alloy composition was not available and was difficult to determine on the highly corroded plates. The metallurgists who studied these plates found them to be similar to type 4068, a molybdenum-bearing low-alloy steel (Ref 14), or type 1060, a high-carbon steel (Ref 15), although compositional analysis of a severely corroded part is likely to be highly subject to error. The surrounding tissues were found to be filled with considerable amounts of corrosion product, and there was a chronic inflammatory response to the corrosion products around the nail/plate and around the plate implanted for 29 years, but not the one implanted for 64 years. There was partial destruction of bone near the plate implanted for 29 years, which had been infected for 26 years. The plate left in place for 64 years was removed for reasons unrelated to the corrosion of the plate; the patient had been asymptomatic, and there had been no problems reported with the plate. Another example of tissue destruction due to the use of an alloy without the required physiological corrosion resistance is found in a case study from 1960 (Ref 16) in which the pitting corrosion of a fracture fixation pin (type 420 stainless steel), originally implanted in 1956, caused bone destruction and pain at the site of the deposition of corrosion deposits.

A cobalt alloy plate and screws (Ref 17) removed from a patient due to persistent pain were found to have failed due to crevice corrosion between the plate and bone. The composition of the plate was similar to that of Stellite 25 (Co-20Cr-15W-10Ni), although only 12.8% tungsten was detected in the alloy, while the screws were similar in composition to Stellite 21 (Co-27Cr-5.5Mo). The authors noted that the corrosion and corrosion deposits were associated with the plate and not the screws and that the patient's chronic pain was no longer present after surgery to remove the implants. Similar materials are still specified for use in medical devices and standardized through material specifications (Ref 18–20). Reports of corrosion problems and/or adverse tissue response with these alloys has been extremely rare and, while the cited case study serves to describe how tissues respond when corrosion occurs, it does not indicate a problem with the use of the alloys, but possibly with mixing the alloys.

In all except one of the cases cited, the surgeons reoperated on the patients because of chronic or intermittent pain associated with the corroded device and the pain was alleviated after removal surgery. It appears that one response to corrosion product is a painful inflammatory response at the site of the corrosion product deposits in the bone or other tissue.

Response to Particulate Materials. It is well known in the medical device literature that the response of tissue to the presence of particulate debris is increased levels of macrophages and other inflammatory cells in the area as the body attempts to fight off the foreign substance. When this response is due to bacteria, the response will result in the death of the bacteria; the death of the defensive cells that engulfed the bacteria and; over a period of time, a reduction in the inflammatory response to preinsult levels. While most of the studies of the tissue response to particles relate to the presence of wear debris around devices that have moving parts, particulate corrosion product may also be released from an implant site. When the defensive response is to foreign particles, the initial response is the same but the particles remain after the reactive cells themselves have died and the tissue response is repeated or escalated. There appears to be a concentration of particles in the tissue that can be handled without long-term problems, but there appears to be a threshold concentration, which will vary from patient to patient, above which the tissue destruction due to the inflammatory response will make the process self-perpetuating. The response to the products of corrosion is the same as to wear particles and, in sufficient concentrations, will lead to tissue destruction around the corroding part. This type of response was reported in 1994 (Ref 21) in a study in which 15 explanted hip prostheses (ten at revision, five at autopsy) that had macroscopic evidence of corrosion were further evaluated, along with surrounding tissues. All had cobalt-chromium femoral heads and most of the stems were manufactured from the same alloy. It was found that most of the tissues contained corrosion product whose principal component was chromium and which was identified as chromium (III) orthophosphate hexahydrate. Lysis (destruction) of surrounding bone was seen in four of the cases and was the reason for device removal. The osteonecrotic (pertaining to dead bone) lytic lesions were inspected histologically and deposits of the corrosion product were found in all. It was not possible to say how much the corrosion products contributed to the failures because polyethylene wear particles from the acetabular cup bearing surface were also present. The corrosion particles were found to have been engulfed in phagocytic cells, suggesting that at least part of the osteolytic response was due to the corrosion product.

The presence of particulate corrosion products in tissues surrounding implanted devices has the potential to cause osteolytic problems, and some problems, which were originally considered to have been due to metal sensitivity (Ref 22), now appear to have been particulate related.

Metal Ion Leaching and Systemic Effects

There appears to be little question that serum and tissue metal ion levels experience an increase in the early postoperative period, particularly for total joint replacements, but most investigators have shown a decrease in metal ion levels after a period of time, although not always to presurgical levels. It also appears that, when the bearing surfaces are both metallic, instead of one of them being polyethylene, the metal ion increases are both greater and more sustained (Ref 23, 24). These studies showed that the metal ion levels in patients with metal-on-metal bearings were much higher than normal baseline levels of the metal ions measured before surgery or in patients without implants. Despite these findings, there is little evidence of long-term adverse effects due to the levels of metal ions. Many of the original case reports describing elevated levels of metal ions and corrosion products around prostheses that failed due to loosening attribute the loosening to the metal ion levels and corrosion product in tissues immediately surrounding the prostheses. These studies were conducted before the wear of polyethylene was identified as the principal contributing factor to bone loss around implants, and before the presence or absence of polymeric particles was routinely reported. Given the black or gray color of metal products in the tissue, it is likely that polyethylene particles could be completely missed in these tissue samples. For this reason, case reports published before the 1980s must be considered with this fact in mind.

Metal Binding and Effects on Metabolic Processes. Most of the metal ions that may potentially be released from implanted devices are present at low levels in the normal physiology. In fact, many of these metals are essential for health. The normal physiological levels of metal ions in the body exist because of their role in many metabolic processes, and changes in their levels or specific chemical forms may adversely impact these normal functions, including the immune system and the functions of essential enzymes. Iron is a necessary part of hemoglobin, without which oxygen transport in red blood cells cannot occur. Vitamin B_{12} is a cobalt compound and cannot exist without the presence of trace amounts of cobalt in the body. Cr^{3+} is essential to metabolic function, but, if Cr^{6+} becomes substituted in the organometallic compound, the compounds become nonfunctional. It was reported (Ref 21) that the corrosion products found were composed of trivalent chromium. The passive film on stainless steel and cobalt-chromium alloys is also trivalent chromium. One study (Ref 25) concluded, after laboratory and animal studies of DNA mutation, that trivalent chromium compounds are "relatively nontoxic and beneficial for human health." Another study (Ref 26), looking at laboratory studies of DNA damage due to trivalent and hexavalent chromium compounds, determined that Cr^{3+} was cytotoxic while Cr^{6+} was genotoxic. The authors further reported that Cr^{6+} is converted to Cr^{3+} in a three-step reduction in charge and suggested that the toxicity of the two forms are related. In most cases, the body can control the physiological levels of essential metals through ingestion and excretion, but at high enough levels, the metal ions can be toxic. There do not appear to be reports in the literature of systemic disease due to metal ion release from medical implants and chemical interaction with metabolic processes, but the acceptable levels of many of these metals are based on experience with industrial exposures through pathways other than direct absorption in the tissues.

Hypersensitivity to Metal Ions. One of the consequences of the release of metal ions from an implanted device is the potential for an allergic response to the metal. One of the most common examples of hypersensitivity is the body's reaction to nickel. Nickel is present in the stainless steel commonly used in fracture fixation devices and other implantable devices (for example, 13 to 15% Ni in type 316L implant-grade stainless steel). Many persons, particularly women, are sensitive to nickel ions and can develop a reaction to the presence of stainless steel devices under the skin, as commonly occurs in devices used for the stabilization of fractured bones while awaiting healing. This type of response does not usually occur at the first exposure to the allergen but rather is a response that increases as a result of repeated exposures to the metal ions. Women are much more commonly exposed to stainless steel in jewelry such as earrings for pierced ears, and thus the likelihood of a response to an implanted alloy containing nickel is much greater. Some patients who might be susceptible to an immune response to nickel may previously have experienced inflammation around earrings or under wristwatches. Sweating around and under stainless steel jewelry will create a corrosive sodium chloride environment that may be conducive to corrosion of the alloy in the jewelry. The reaction to nickel has been considered enough of a concern that stainless steel alloys with little or no nickel content have been developed for implant use and are being used in some applications (Ref 27, 28). Table 1 lists chemical composition requirements for low-nickel implant-grade stainless steels as well as conventional (higher-nickel) grades used for fracture fixation devices. Reactions to other metal ions are reported less frequently, but there have been reports of sensitivity to chromium (Ref 29), cobalt (Ref 29, 30), and gold (Ref 31–33); however, these seem to be of lesser concern.

A recent clinical study (Ref 34) documented preoperative metal sensitivity to at least one component of a commonly used cobalt-chromium alloy (UNS R30075, Ref 19) in 26% of 92 patients before surgery to implant a total knee replacement. Five of the patients developed eczema at the surgical site or extending over the whole body. Two of those patients had their implants replaced with devices made of ceramic and experienced resolution of symptoms, providing strong evidence for the implant being the source of the problem. While patients are not routinely screened for allergies to the metals present in medical alloys, these authors recommended such screening (Ref 34). Five out of the 24 patients who showed a metal sensitivity actually experienced reaction to the metal ions after surgery. The remaining 19 patients, with equivalent scores in the sensitivity testing, did not exhibit a response after surgery. One of the problems to be addressed in this area is the fact that clinical decision making based on the results of testing that is only about 20% predictive could lead to the needless elimination of patients from consideration for certain types of medical devices that have the potential for improving quality of life.

Table 1 Chemical compositions of austenitic stainless steels used for implantable fracture fixation devices

ASTM designation	UNS No.	Composition(a), %										
		C	Mn	P	S	Si	Cr	Ni	Mo	N	Cu	Others
F 138(b)	S31673(b)	0.03	2.00	0.025	0.010	0.75	17.00–19.00	13.00–15.00	2.25–3.00	0.10	0.50	
F 1314	S20910	0.03	4.00–6.00	0.025	0.010	0.75	20.50–23.50	11.50–13.50	2.00–3.00	0.20–0.40	0.50	0.10–0.30 Nb; 0.10–0.30 V
F 1586	S31675	0.08	2.00–4.25	0.025	0.010	0.75	19.50–22.00	9.00–11.00	2.00–3.00	0.25–0.50	0.25	0.25–0.80 Nb
F2229	S29108	0.08	21.00–24.00	0.03	0.010	0.75	19.00–23.00	0.10	0.50–1.50	0.90 min	0.25	

UNS, Unified Numbering System. (a) Single values are maximum values unless otherwise indicated. The balance is Fe. (b) Also commonly referred to as type 316L or 316LVM (vacuum melted type 316L)

Possible Cancer-Causing Effects of Metallic Biomaterials

There has been a concern that the release of metal ions or corrosion products in the tissues surrounding medical devices might induce mutations in the cellular DNA and eventually lead to the development of malignant tumors. Considering the large number of metallic devices that have been implanted over the last 75 years, there does not appear to be an epidemic of malignant neoplastic transformations around medical devices, but individual reports of tumors adjacent to metallic implants have been published.

Studies have shown that sufficient quantities of some metal ions will produce changes in cells that could lead to the development of cancer tumors. An in vitro study of cobalt in fibroblast cell culture (Ref 35) found changes in nuclei and cells division in daughter generations of rat fibroblasts exposed to cobalt chloride as compared with daughter generations of cells not exposed to cobalt. The large quantity of metal that may be necessary to elicit cellular changes does not provide useful information about the possibility of cellular changes at the much lower concentrations and dosages that are likely around implants manufactured from commonly accepted alloys. The suitability of commonly accepted alloys is primarily due to their corrosion resistance in the physiological environment.

Recently, the resurgence in the use of metal-on-metal articulations in total hip prostheses has revived questions about the potential impact of metal ions on cells at the chromosomal level. The very low wear rate of these articulations (less than 1 mg per million cycles) means that the wear primarily results in the removal of the passive layer from the polished surfaces, followed immediately by repassivation, a process similar in mechanism to fretting corrosion. A clinical trial in Italy (Ref 36) showed no correlation between chromosomal changes in peripheral lymphocytes and metal ion concentrations in 30 patients six months after surgery. Another study (Ref 37) showed chromosomal changes in 31 total hip replacement patients at the time of removal of a failed prosthesis, as compared with 30 primary hip replacement patients. Neither study reported the development of any malignancies in study patients but both recommended continued study because the types of chromosomal changes being reported have the tendency to be cumulative and have long-term effects.

Tumors have been found in close proximity to implanted metallic devices as reported in papers published between 1967 and 1997 (Ref 38–51) in which tumors were suspected to have a relationship to metal ions or corrosion product released from devices. Each report discusses only one or two patients and represents a very small percentage of all devices placed into patients during the time periods in question. Only two of these reports (Ref 47, 48) appear to deal with prostheses in which there was deliberate movement of two metallic surfaces against each other (metal-on-metal hip prostheses), which holds the potential for increased release of metal ions with repetitive repassivation as wear occurs. Additionally, in a study of 1358 patients for up to ten years, cancers of the lymphatic and blood cell production tissues were significantly higher in patients with implants than those without implants (Ref 52). Tumors of the breast, colon, and rectum were significantly less than expected. The conclusions suggest that there may be an association with the prostheses but that other factors, such as drug therapies used, should also be considered before a direct correlation is assumed.

Neoplastic and carcinogenic effects of chemicals in the body are frequently a response to long-term exposures to low levels of chemicals in the body, and any determination of these effects requires long-term epidemiological studies to confirm a direct relationship. Except for tissue destruction in response to particulate corrosion products and allergic responses to metal ions, tumor formation in association with metallic medical devices appears to be one of the few theoretical adverse effects of corrosion that has had reports of its actual occurrence in human patients. If this problem were common and widespread, there should be a great deal more reports of the problem than currently exist. Since these first reports appeared, there have been hundreds of thousands of metallic device implantations with very few additional reports. In addition, many devices already implanted at that time continue to be present in living patients, extending the period of exposure to the patients.

Summary

The metallic materials used in surgery currently derive their biocompatibility from the alloying elements responsible for the development of continuous stable passive layers on the surfaces. There is a very low level of release of metal ions even under the ideal conditions of passivity and when there is no damage to the surfaces. Corrosion of these metals and alloys may occur when environmental conditions change, when mechanisms for repetitive surface damage occur, or in rare situations in which dissimilar metals are used in an inappropriate combination. Case reports and data from the use of other alloys in the period before about 1965 show the possible adverse tissue effects when corrosion products are released in the surrounding tissues. These early results were the basis for controlled animal studies to try to identify which alloys were the most appropriate for implantation based on the tissue response to the corrosion products from the alloys. As standardization has occurred and government regulatory bodies have begun to regulate these devices and the materials of manufacture, the experience of previous generations has been brought to bear, and the corrosion of metallic medical devices has become more uncommon.

The potential adverse effects of metal ion release into living tissues can be described based on information from industrial exposure, pollution, and other exposures to high or extended doses of metal particles and dissolved metal ions. Testing in the laboratory and in animals has occasionally produced results that, while of concern in the initial reporting, do not appear to be borne out based on clinical experience. The results of laboratory tissue and cell culture testing as well as the testing of simulated failure products in vivo do not necessarily take into account all of the protection mechanisms and physiological response characteristics of the actual implant usage situation. Potential mechanisms of long-term adverse responses do exist and the continued monitoring of clinical results is occurring. While some proposed problems, such as tumor formation around devices or toxic responses to metal wear particles, have not been seen to have occurred in sufficient numbers to be proved to be anything other than coincidental, longer-term data collection on the increasing numbers of devices in use may provide additional data to prove or disprove the theories.

It is clear that when corrosion occurs, the consequences to the surrounding tissues can be quite severe, but research, government regulation, and consensus standards for both biocompatibility testing methods and for the composition of metals and alloys suitable for use in surgery have improved the likelihood that metallic medical devices will be present in the physiological environment without significant corrosion and adverse consequences.

REFERENCES

1. P.G. Laing, A.B. Ferguson, and E.S. Hodge, Tissue Reaction in Rabbit Muscle Exposed to Metallic Implants, *J. Biomed. Mater. Res.*, Vol 1, 1967, p 135–149
2. S.G. Steinemann, J. Eulenberger, P.A. Maeusli, and A. Schroeder, Adhesion of Bone to Titanium, *Biological and Biomechanical Performance of Biomaterials*, P. Christel, A. Meunier, and A.J.C. Lee, Ed., Elsevier, 1986, p 409–414
3. J.E. Sundgren, P. Bodö, and I. Lindström, Auger Electron Spectroscopic Studies of the Interface between Human Tissue and Implants of Titanium and Stainless Steel, *J. Colloid Interf. Sci*, Vol 110, 1986, p 9–20
4. J. Black, Biological Performance of Tantalum, *Clin. Mater.*, Vol 16, 1994, p 167–173
5. C.S. Venable, W.G. Stuck, and A. Beach, The Effects on Bone of the Presence of Metals; Based upon Electrolysis, *Ann. Surg.*, Vol 105, 1937, p 917–938
6. P.G. Laing, Compatibility of Biomaterials, *Orthop. Clin. N. Amer.*, Vol 4 (No. 2), 1973, p 249–273
7. A.C. Guyton and J.E. Hall, *Textbook of Medical Physiology*, 9th ed., W.B. Saunders, 1996, p 386
8. M.C. Badet, B. Richard, and G. Dorignac, An In Vitro Study of the pH-Lowering Potential of Salivary Lactobacilli Associated With Dental Caries, *J. Appl. Microbiol.*, Vol 90, 2001, p 1015–1018
9. J.T. Mader and J. Calhoun, Osteomyelitis, *Principles and Practice of Infectious Diseases*, 5th ed., G.L. Mandell, J.E. Bennett, and R. Dolan, Ed., Churchill Livingston, 2000, p 1183
10. I.A. Silver, R.J. Murrills, and D.J. Etherington, Microelectrode Studies on the Microenvironment beneath Adherent Macrophages and Osteoclasts, *Exp. Cell Res.*, Vol 175, 1988, p 266–276
11. S. Hierholzer, G. Heirholzer, K.H. Sauer, and R.S. Paterson, Increased Corrosion of Stainless Steel Implants in Infected Plated Fractures, *Arch. Orthop. Trauma. Surg.*, Vol 102, 1984, p 198–200
12. T. Rae, The Action of Cobalt, Nickel, and Chromium on Phagocytosis and Bacterial Killing by Human Polymorphonuclear Leucocytes; Its Relevance to Infection after Total Joint Arthroplasty, *Biomaterials*, Vol 4, 1983, p 175–180
13. J. Cohen and G. Hammond, Corrosion in a Device for Fracture Fixation, *J. Bone Joint Surg.*, Vol 41A, 1959, p 524–534
14. J.R. Cahoon, On the Corrosion Products of Orthopedic Implants, *J. Biomed. Mater. Res.*, Vol 7 (No. 4), 1973, p 375–383
15. J.P. McAuley, K.V. Gow, A. Covert, A.G. McDermott, and R.H. Yabsley, Analysis of a Lane-Plate Internal Fixation Device after 64 Years In Vivo, *Can. J. Surg.*, Vol 30, 1987, p 424–427
16. J. Cohen and W.S. Foultz, Failure of a Steinmann Pin Used for Intramedullary Fixation, *J. Bone Joint Surg.*, Vol 42A, 1960, p 1201–1206
17. J. Cohen and J. Wulff, Clinical Failure Caused by Corrosion of a Vitallium Plate, *J. Bone Joint Surg.*, Vol 54A, 1972, p 617–628
18. "Standard Specification for Wrought Cobalt-20 Chromium-15 Tungsten-10 Nickel Alloy for Surgical Implant Applications (UNS R30605)," ASTM F 90, *Annual Book of ASTM Standards*, ASTM International
19. "Standard Specification for Cobalt-28 Chromium-6 Molybdenum Casting Alloy and Cast Products for Surgical Implants (UNS R30075)," ASTM F 75, *Annual Book of ASTM Standards*, ASTM International
20. "Standard Specification for Wrought Cobalt-28 Chromium-6 Molybdenum Alloys for Surgical Implants (UNS R31537, UNS R31538, and UNS R31539)," ASTM F 1537, *Annual Book of ASTM Standards*, ASTM International
21. R.M. Urban, J.J. Jacobs, J.L. Gilbert, and J.O. Galante, Migration of Corrosion Products from Modular Hip Prostheses, *J. Bone Joint Surg.*, Vol 76A, 1994, p 1345–1359
22. E.M. Evans, M.A.R. Freeman, A.J. Miller, and B. Vernon-Roberts, Metal Sensitivity as a Cause of Bone Necrosis and Loosening of the Prosthesis on Total Joint Replacement, *J. Bone Joint Surg.*, Vol 56B, 1974, p 626–642
23. M.T. Clarke, P.T.H. Lee, A. Arora, and R.N. Villar, Levels of Metal Ions after Small- and Large-Diameter Metal-on-Metal Hip Arthroplasty, *J. Bone Joint Surg.*, Vol 85B, 2003, p 913–917
24. L. Savarino, D. Granchi, G. Ciapetti, E. Cenni, M. Greco, R. Rotinin, C.A. Veronesi, N. Baldini, and A. Giunti, Ion Release in Stable Hip Arthroplasties Using Metal-on-Metal Articulating Surfaces: A Comparison between Short- and Medium-Term Results, *J. Biomed. Mater. Res.*, Vol 66A, 2003, p 450–456
25. D. Bagchi, S.J. Stohs, B.W. Downs, M. Bagchi, and H.G. Preuss, Cytotoxicty and Oxidative Mechanisms of Different Forms of Chromium, *Toxicol.*, Vol 180, 2002, p 5–22
26. J. Blasak and J. Kowalik, A Comparison of the In Vitro Genotoxicity of Tri- and Hexavalent Chromium, *Mut. Res.*, Vol 469, 2000, p 135–145
27. L.D. Zardiackas, S. Williamson, M. Roach, and J.-A. Bogan, Comparison of Anodic Polarization and Galvanic Corrosion of a Low-Nickel Stainless Steel to 316LS and 22-Cr-13Ni-5Mn Stainless Steels, *Stainless Steels for Medical and Surgical Applications*, G.L. Winters and M.J. Nutt, Ed., 2003, ASTM International, p107–118
28. "Standard Specification for Wrought, Nitrogen Strengthened 23Manganese-21 Chromium-1Molybdenum Low-Nickel Stainless Steel Alloy Bar and Wire for Surgical Implants (UNS S29108)," ASTM F 2229, *Annual Book of ASTM Standards*, ASTM International
29. C. Szliska and J. Raskoski, Sensitization to Nickel, Cobalt, and Chromium in Surgical Patients, *Contact Dermatitis*, Vol 23, 1990, p 278–279
30. F.C. Antony, W. Dudley, R. Field, and C.A. Holden, Metal Allergy Resurfaces in Failed Hip Endoprostheses, *Contact Dermatitis*, Vol 48, 2003, p 49–50
31. W.T. Trathen and R.J. Stanley, Allergic Reaction to Hulka Clips, *Obstet. Gynecol.*, Vol 66, 1985, p 743–744
32. H. Petros and A.L. MacMillan, Allergic Contact Sensitivity to Gold with Unusual Features, *Br. J. Dermatol.*, Vol 88, 1973, p 505–508
33. S. Comaish, A Case of Contact Hypersensitivity to Metallic Gold, *Arch. Dermatol.*, Vol 99, 1969, p 720–723
34. Y. Niki, H. Matsumoto, T. Otani, T. Yatabe, M. Kondo, F. Yoshimine, and Y. Toyama, Screening for Symptomatic Metal Sensitivity: A Prospective Study of 92 Patients Undergoing Total Knee Arthroplasty, *Biomaterials*, Vol 26, 2005, p 1020–1026
35. M. Daniel, J.T. Dingle, M. Webb, and J.C. Heath, The Biological Action of Cobalt and Other Metals, Part I: The Effect of Cobalt on the Morphology and Metabolism of Rat Fibroblasts In Vitro, *Br. J. Exp. Pathol.*, Vol 44, 1963, p 163–175
36. A. Massè, M. Bosetti, C. Buratti, O. Visentin, D. Bergadano, and M. Cannas, Ion Release and Chromosomal Damage from Total Hip Prostheses with Metal-on-Metal Articulation, *J. Biomed. Mater. Res.*, Vol 67B, 2003, p 750–757
37. A.T. Doherty, R.T. Howell, L.A. Ellis, I. Bisbinas, I.D. Learmouth, R. Newson, and C.P. Case, Increased Chromosome Translocations and Aneuploidy in Peripheral Blood Lymphocytes of Patients Having Revision Arthroplasty of the Hip, *J. Bone Joint Surg.*, Vol 83B, 2001, p 1075–1081
38. J.W. Harrison, D.L. McLain, R.B. Hohn, G.P. Wilson III, J.A. Chalman, and K.N. MacGowan, Osteosarcoma Associated with

Metallic Implants: Report of Two Cases in Dogs, *Clin. Orthop. Rel. Res.*, Vol 116, 1976, p 253–257
39. K. Sinibaldi, H. Rosen, S.-K. Liu, and M. DeAngelis, Tumors Associated with Metallic Implants in Animals, *Clin. Orthop. Rel. Res.*, Vol 118, 1976, p 257–266
40. P.V. Hautamaa, D.W. Gaither, and R.C. Thompson, Malignant Fibrous Histiocytoma Arising in the Region of a Femoral Fracture, *J. Bone Joint Surg.*, Vol 74A, 1992, p 777–780
41. Y.-S. Lee, R.W.H. Pho, and A. Nather, Malignant Fibrous Histiocytoma at Site of Metal Implant, *Cancer*, Vol 54, 1984, p 2286–2289
42. A.W. Hughes, D.A. Sherlock, D.L. Hamblen, and R. Reid, Sarcoma at the Site of a Single Hip Screw: A Case Report, *J. Bone Joint Surg.*, Vol 69B, 1987, p 470–472
43. K.J.J. Tayton, Ewing's Sarcoma at the Site of a Metal Plate, *Cancer*, Vol 45, 1980, p 413–415
44. J.J. Ward, D.D. Thornbury, J.E. Lemons, and W.K. Dunham, Metal-Induced Sarcoma: A Case Report and Literature Review, *Clin. Orthop. Rel. Res.*, Vol 252, 1990, p 299–306
45. A. Martin, T.W. Bauer, M.T. Manley, and K.E. Marks, Osteosarcoma at the Site of Total Hip Replacement, *J. Bone Joint Surg.*, Vol 70A, 1988, p 1561–1567
46. M. Haag and C.P. Adler, Malignant Fibrous Histiocytoma in Association with Hip Replacement, *J. Bone Joint Surg.*, Vol 71B, 1989, p 701
47. M. Swann, Malignant Soft-Tissue Tumour at the Site of a Total Hip Replacement, *J. Bone Joint Surg.*, Vol 66B, 1984, p 629–631
48. H.G. Penman and P.A. Ring, Osteosarcoma in Association with Total Hip Replacement, *J. Bone Joint Surg.*, Vol 66B, 1984, p 632–634
49. J. Bago-Granell, M. Aguirre-Canyadell, J. Nardi, and N. Tallada, Malignant Fibrous Histiocytoma at the Site of a Total Hip Arthroplasty, *J. Bone Joint Surg.*, Vol 66B, 1984, p 38–40
50. J.P. Nelson and P.H. Phillips, Malignant Fibrous Histiocytoma Associated with Total Hip Replacement, *Orthop. Rev.*, Vol 19, 1990, p 1078–1080
51. R.S. Bell, S. Hopyan, A.M. Davis, and A.E. Gross, Sarcoma of Bone-Cement Membrane: A Case Report and Review of the Literature, *Can. J. Surg.*, Vol 40, 1997, p 51–55
52. W.J. Gillespie, C.M.A. Frampton, R.J. Henderson, and P.M. Ryan, The Incidence of Cancer Following Total Hip Replacement, *J. Bone Joint Surg.*, Vol 70B, 1988, p 542

SELECTED REFERENCES

- J.A. Helsen and H.J. Breme, Ed., *Metals as Biomaterials*, John Wiley & Sons, West Sussex, UK, 1998
- H. Sigel, Ed., *Carcinogenicity and Metal Ions (Metal Ions in Biological Systems, Vol 10)*, Marcel Dekker, 1980
- H. Sigel, Ed., *Concepts on Metal Ion Toxicity (Metal Ions in Biological Systems, Vol 20)*, Marcel Dekker, 1986

Mechanically Assisted Corrosion of Metallic Biomaterials

Jeremy L. Gilbert, Syracuse University

METALLIC BIOMATERIALS have been used for the past several decades in a wide array of applications spanning several medical and dental fields. Much early work in metallic biomaterials was performed in the dental community where gold alloys, dental amalgams, and base metal alloys of cobalt, nickel, and iron have been used. More recently, the use of titanium has increased dramatically in dental applications such as dental implants. Metal alloys have been used in orthopedic applications in greatest amounts related to fracture fixation devices and total joint arthroplasties. Alloys of cobalt, titanium, and iron also have been used in cardiovascular applications such as heart valves, stents, and pacemakers. In spinal applications, metallic alloys are also used in significant quantities where adaptations and innovations from the fracture fixation field have evolved into a set of highly specialized designs for spinal instrumentation (e.g., pedicle screws, fusion instrumentation, etc.) Most of the earlier designs and devices used in spinal instrumentation were made from iron-base alloys (type 316L stainless steel, ASTM F138) (Ref 1), but more recently other alloys are being explored, alone and in combination, due to the current understanding of each material's behavior and interaction with the biological system. There is an increasing complexity in designs being developed where multiple screw-countersink combinations, rod-set screw designs and other crevicelike and restricted environments are developed. Here, restricted means locations where small volumes of fluid can be relatively isolated from the bulk solution environment and where possible chemical reactions may alter the chemical makeup of this fluid (e.g., make it more concentrated in metal ions, lower pH, etc.).

Questions may arise as to what combinations of metal alloys can be used and in what types of designs such that there are no untoward effects on the patient. In particular, there are concerns that combinations of dissimilar metals may result in adverse corrosion reactions, leading to negative consequences.

The goal of this article is to review the current understanding of the corrosion interactions between alloys, in particular iron-base, titanium-base, and cobalt-base alloys, in complex geometries and in applications where there are significant cyclic stresses and potential for wear and fretting motion. This review attempts to frame the nature of these metal surfaces, their propensity for corrosion reactions when combined with similar or different alloys in complex restrictive (e.g., crevicelike) environments within the human body, and under loading conditions.

Iron-, Cobalt-, and Titanium-Base Biomedical Alloys

Stainless Steels. The main alloys used in orthopedics and spinal instrumentation today are the iron-base alloys (low-carbon type 316L stainless steel; ASTM F 138, Ref 1), as well as newer alloys with low nickel and high nitrogen- and manganese-containing alloys, e.g., ASTM F 1586 (Ref 2), and F 2229 (Ref 3), respectively. These alloys (primarily vacuum melted type 316L stainless steel) are austenitic, single-phase alloys and have served as plates, screws, nails, rods, and wires in a wide array of applications. Their strength and corrosion resistance are very good, although alloys based on titanium and cobalt-chromium have better corrosion and mechanical properties and have been making inroads into uses where type 316L had typically been used.

Cobalt-Chromium Alloys. These alloys come with a variety of chemistries that fall under the range of cobalt-base alloys. There are several ASTM designations for these alloys, including F 75 (Ref 4), F 799 (Ref 5), F 1537 (Ref 6), and so forth, which include Co-Cr-Mo, Co-Cr-Ni-W-Mo, and Co-Cr-Ni-Mo alloy systems. These have been used in cast, wrought, and forged conditions and, more recently, as powder metallurgy products that are sintered into net shape. These alloys have a complex metallurgy (Ref 7), with multiple phases possible and present depending on the fabrication process, including complex carbides. The chemistry may vary somewhat within the microstructure as well as due to chemical heterogeneity during solidification or processing.

Titanium and its alloys have enjoyed widespread use in biomedical applications either as commercially pure (CP) titanium, ASTM F 67 (Ref 8), F 1341 (Ref 9), as well as alloys including Ti-6Al-4V ELI, ASTM F 136 (Ref 10) and others. More recently, alloys of nickel and titanium have been developed to take advantage of the super elastic and shape memory characteristics of this alloy (ASTM F 2063) (Ref 11). Titanium has very good mechanical properties and corrosion properties (in the absence of surface mechanical abrasion) and is known to be highly tolerated in the body environment. In fact, titanium has taken on the aura of being the "most biocompatible" metal in use today. There are a variety of rationales for this primarily linked to the relative inertness of the alloy surface. However, recent work indicates that there may be other factors that influence its behavior, including semiconducting properties, catalytic activity, and the relatively high electric field conditions that exist at its thin oxide film surface (Ref 12–14). More details are presented in a companion article, "Corrosion Performance of Stainless Steels, Cobalt and Titanium Alloys in Biomedical Applications," in this Volume.

Surface Characteristics and Electrochemical Behavior of Metallic Biomaterials

The surfaces of iron-, cobalt-, and titanium-base alloys have some similarities (and some differences) that are worthy of mention. One of the main reasons that each of these alloy systems is in such significant use today is due to their corrosion resistance within the environment of the human body.

The main factor governing the corrosion resistance of medical alloys is the presence of a nanometer-scale oxide film that spontaneously forms on the surface. These oxide films provide a kinetic barrier to corrosion. That is, each of these alloys has a high free energy driving oxidation of the metal components and each would corrode at extremely high rates if the oxide film were not present to limit specific steps in the corrosion reaction (Ref 15, 16). However, the oxide films that form act to limit access of the metal to the environment and the environment to the metal such that continued oxidation/corrosion is greatly reduced. The importance of these oxides to the continuing resistance to corrosion cannot be understated. It is also important to understand that high thermodynamic driving forces for corrosion are present and remain even in the presence of these oxide films and these forces, which manifest primarily as an extremely high electric field across the oxide, continue to attempt to drive corrosion and assist in several processes, including ion migration (chemical segregation) and electronic charge transport.

One of the most important aspects of the corrosion resistance of each of these alloy systems is the integrity of the oxide film. If, for whatever reason (mechanical abrasion, chemical effects, biological effects), the oxide films are compromised, then the high driving forces for corrosion will take over and severe, high-rate corrosion can and will take place. This has been documented in several ways in the literature and is discussed subsequently. Oxide films are not, in a literal sense, films per se, but rather they can take on a variety of morphologies including domes, overlapping domes, oriented elongated domes, and film-dome combinations. The details of the oxide morphology on the surface is also affected by surface treatments, prior electrochemical history, and immersion time. These effects are also reviewed in the article, "Corrosion Performance of Stainless Steels, Cobalt and Titanium Alloys in Biomedical Applications," in this Volume.

A large body of literature exists on the electrochemical behavior of these alloys in physiologically representative solutions. Corrosion testing has been performed in 0.9% sodium chloride (NaCl), phosphate buffered saline (PBS), Ringer's solution, Hanks' balanced salt solution, and a variety of other solutions that vary in complexity (e.g., protein additions, aerated/deaerated, etc., as discussed in the aforementioned companion article in this Volume). All of these alloys show fairly high corrosion resistance over a range of electrochemical potentials when performing polarization testing. These types of corrosion tests are helpful in understanding the corrosion behavior of alloys in the absence of mechanical factors (i.e., where the oxide films are not stressed or abraded) but have a limited value in understanding the true nature of the surface behavior. Typical results from these tests cannot always tell the range of potential where the oxide is stable, whether oxide reduction is possible, what type of breakdown occurs at anodic or cathodic potentials (e.g., pitting vs. generalized corrosion, corrosion inside a crevice, etc.), and what level of corrosion is obtained at these surfaces in the absence of mechanical factors. Standard electrochemical testing (anodic polarization testing, cyclic polarization testing, etc.) is also not able to provide insight into the effects of combining components (or different alloys) or understanding how surface abrasion or fretting-crevice corrosion may impact the corrosion stability.

Standard electrochemical tests are also limited in providing insight and knowledge about medical devices that may be made of various alloys, combined and connected with potential crevices, and the possibility that they may be cyclically loaded and stressed. Under these types of geometries and mechanical loading states, corrosion becomes a much more complicated process. This requires that we develop new ways of thinking about medical device corrosion in the presence of mechanical factors.

It should also be mentioned that most electrochemical testing of biomedical alloys has been done in environments that, at best, represent the body environment in a highly simplified manner. Most testing is done at neutral pH with normal physiological ionic concentrations of inorganic salts (e.g., PBS, NaCl, etc.). Little work has been done to investigate the complex environments that arise either as a result of local biochemical changes (e.g., local inflammatory reactions at a metal surface due to the presence of inflammatory cells and wound healing) or the restricted environments that may arise in the geometrically complex implants that are being designed. In this context, solution chemistry can be highly different from the pH 7.4 saline solutions typically used in corrosion experiments including superoxide radicals and their by-products (e.g., peroxynitrite, hydrogen peroxide), pH as low as 1, high local levels of metal ions, proteins, and lytic enzymes whose effect is unknown.

The Clinical Context for Mechanically Assisted Corrosion

Since the early 1980s there has been an increasing awareness of the development of mechanically assisted corrosion in orthopedic and other medical device constructs. Early work in fretting corrosion (Ref 17–19) and fretting corrosion fatigue (Ref 20) have focused on nails, plates, screws, and screw-hole countersink interactions. Since the late 1980s, the development of implant modularity has been the primary reason why mechanically assisted corrosion processes have become a significant concern. Implant modularity is where different components of a total joint prosthesis are combined, for example, at the time of surgery using metal-metal conical taper connections that the surgeon assembles. These conical tapers are sometimes referred to as Morse, or modular tapers (see Fig. 1 for an example). Many advantages have been identified for the use of modularity that makes this design approach an important advance. Modularity provides surgeons with greater flexibility intraoperatively to make the right choice of components. Different alloys can be used for different aspects of the prosthesis (e.g., titanium alloys for the stem of the hip replacement, due to better bone biocompatibility, and a cobalt-chromium alloy for the head, due to better wear resistance). Also, modularity allows for a large number of combinations of designs to be available with much fewer parts needed for manufacture and hospital inventory. Finally, modularity provides greater flexibility during revision surgery to replace portions of a device without having to remove and replace the entire prosthesis.

Early studies of corrosion in conditions similar to modular implants included investigations into the corrosion behavior of assemblies of alloys and mixed alloys with crevices (Ref 21–26). These studies showed little evidence that mixing alloys or the presence of crevices would have a deleterious effect on the corrosion behavior. However, early reports (Ref 27–29) of femoral stems of total hip prostheses that were retrieved during revision surgery (i.e., after implantation, used for a period of time, and then removed due to such complications as loosening, infection, pain, and other causes) showed some evidence that significant corrosion attack was taking place at the modular junctions of these devices. These early reports focused mostly on the idea that mixed alloys were causing a galvanic attack since the combinations reported on were Ti-6Al-4V stems connected to cobalt-chromium-molybdenum heads. However, other studies (Ref 30–36) around the same time began to show that corrosion within these modular tapers was not limited to dissimilar metal combinations (Fig. 2). Corrosion was also observed in cases of cobalt-chromium stems with cobalt-chromium heads, and titanium stems with titanium heads (Ref 30, 33).

More recent work has shown that modular connections in a wide variety of devices, for example, intramedullary rods (i.e., rods that are placed in the center canal of long bones) and revision hip stems with modular connections, are susceptible to corrosion attack (Ref 37–39) and that severe corrosion can be observed even with type 316L stainless steel or titanium-titanium interfaces (Fig. 3).

Retrieval studies (Ref 40) have shown that modular taper corrosion appears to be linked to the mechanics of the taper. Tapers with low flexural rigidity (small diameter tapers made from low modulus materials, e.g., titanium) show a higher probability of being corroded at the same time after implantation than high rigidity materials and designs. Also, the materials combination appears to play a role, with titanium stems in contact with cobalt-chromium heads having a higher extent of corrosion compared with cobalt-chromium/cobalt-chromium combinations. Stainless steel alloys, as well as titanium-titanium junctions have also been shown to be susceptible to corrosion at the modular taper (Ref 38) as shown in Fig. 3.

Thus, the findings from these studies have demonstrated that when devices are implanted with modular connections, there is a significant probability that corrosion will occur at the taper interface. The probability of corrosion in modular connections in hip prostheses is between 28 and 42% in retrieved cases with factors affecting this rate being materials combination, flexural rigidity, and time of implantation (Ref 40). How these percentages relate to the overall population of femoral hip prostheses is not known. However, there are two possibilities: either the corrosion process has affected the reason for retrieval, and hence the rates seen are higher than the overall population, or the corrosion process has not influenced the rate of revision. If it is the first case, then this is clear evidence that corrosion is a clinically significant process (having raised the revision rate). If it is the latter case, then the percentages seen in the retrieval studies are a reflection of the overall rate of corrosion in the population of modular devices. This would imply that the corrosion process occurs at very high rates in modular connections. It is important to note that this corrosion process occurs in all passivating alloys used in orthopedics and that there are mechanical factors that influence its occurrence.

The main reason why corrosion in modular connections was not foreseen prior to their introduction is that virtually all corrosion testing of these alloys had not included mechanical loading as part of the test. These devices experience high cyclic mechanical loads and stresses that are complex and variable. Typical loading forces in total joints can be several times the weight of the patient, and normally active patients will load their legs on the order of 1 to 2 million cycles per year. When these mechanical effects are included, then corrosion of passivating alloys becomes very much a reaction to electrochemical driving forces in conjunction with mechanically driven oxide disruption processes. Thus, to understand the electrochemical behavior of metallic medical devices, the mechanical environment should be included in any evaluation of their performance.

Other reports of corrosion attack include evidence of a mechanically assisted crevice corrosion attack of cemented stems of total hip prostheses (Ref 38, 41, 42). It was found (Ref 41) that upon revision surgery, the pH of the solution present at the interface between the stem and the orthopedic bone cement holding it in place was significantly lower (around 3) than the physiological pH normally assumed (pH 7.4). The researchers also noted that pain was associated with severely corroding and low pH cases. Pitting and crevice attack of cemented stainless steel hip stems has also been reported (Ref 38).

There are other clinical ramifications to this corrosion process. For instance, there have been reports of fatigue failures emanating from corroded taper connections (Ref 32). Also, corrosion by-products, which include particles and ions, can be released into the local tissue and can cause osteolysis (a cellular-driven process of "bone destruction" that leads to lesions in the bone), and can be systemically distributed via the bloodstream. Particles have been found in lymph nodes removed from the implant site as well as in such organs as the liver, spleen, and kidney. Blood and urine levels of corrosion by-products (e.g., Co ions and Cr ions) have been identified in patients with modular tapers that are corroded and the level is correlated to the severity of corrosion (Ref 15, 37, 42–47).

There are other devices that are susceptible to mechanically assisted corrosion, for instance, vascular stents, which are small-scale wire mesh (much like chicken wire) tubes typically made from type 316L stainless steel or NiTi shape memory alloys. These stents are inserted on a catheter to the intrasvascular site (inner lumen) to provide structural support to prevent restenosis (renarrowing) of arteries with atherosclerotic plaques. These stents are inserted in a compressed (small diameter) state and then deployed by expanding the stent and deforming the wires. This deformation results in breaches on the oxide film and transient high corrosion rate processes that cause significant local currents and large drops in open circuit potential (OCP) of the stent. An example of the OCP drop during deployment of a type 316L stainless steel stent is shown in Fig. 4.

Mechanically Assisted Crevice Corrosion

From these clinical and retrieval reports a better understanding of the specific factors that give rise to the observed corrosion behavior has been obtained. Specifically, questions arose about the possibility of galvanic attack and about what role mechanical factors had on the process. Because evidence of corrosion was found even where the alloys at the taper interface were chemically identical, galvanic effects, while still a potential contributor, could not entirely explain the presence of this corrosion process. Thus, mechanical factors were explored in more detail.

Over the past decade, ideas related to corrosion of medical devices in biological

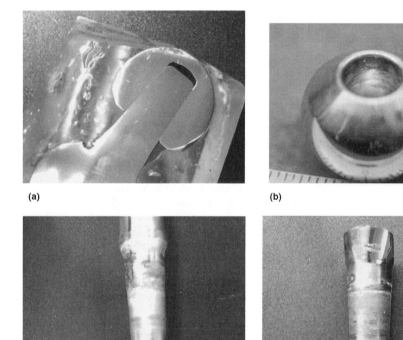

Fig. 1 Optical photographs of modular tapers from a variety of prostheses showing evidence of corrosion. (a) Cross section of a retrieved modular femoral component for a total hip replacement. The head comprises Co-Cr-Mo alloy and the stem is a Ti-6Al-4V alloy. The taper interface is revealed by sectioning. (b) Retrieved Co-Cr-Mo modular head that had sat on a Ti-6Al-4V stem. Note the dark and discolored region inside the taper recess and signs of etching. (c) Taper junction (both male and female portions) of a 316L stainless steel intramedullary rod after retrieval. (d) Retrieved Ti-6Al-4V neck and Ti-6Al-4V "thimble" (sectioned in half) used to attach a ceramic head to the femoral stem. Two modular interfaces are established in this design, one comprising the Ti-Ti junction and the other the Ti-ceramic junction. Note the interface regions are highly corroded on both Ti interfaces and a purple color was present (lighter gray in black-and-white, indicative of Ti^{3+} ions). Also found at these interfaces is evidence of Ti-phosphate particles, which indicate that the pH had dropped below 1 inside the taper.

environments in the presence of mechanical factors have been clarified with a series of articles (Ref 16, 48–59). These articles have investigated the clinical aspects of mechanically assisted corrosion, experimental testing of the mechanism in vitro, and testing of the basic aspects of oxide disruption and reformation. Mechanically assisted corrosion as a term is not widely used; however, it captures a range of possible mechanisms that are considered by the general corrosion community. Mechanically assisted corrosion encompasses all mechanisms that combine mechanical factors with electrochemical factors to increase the rate of corrosion. Effects such as fretting corrosion, corrosion fatigue, wear-assisted corrosion, stress-assisted corrosion, stress-corrosion cracking fall under this more general term. Also part of this model is the restricted (or crevicelike) environment that can serve as the location for significant changes in local solution chemistry. Thus, mechanically assisted corrosion can and often does occur in crevicelike environments. The essential feature of mechanically assisted corrosion processes is that alloys with oxide films that serve as kinetic barriers to corrosion (i.e., are passive films) will experience dramatic increases in corrosion rate when exposed to mechanical factors that can disrupt (or distort) these oxide films. With increased corrosion rates can come other changes in terms of the stability of the oxide film, shift in potential, local consumption of oxygen, decreasing pH, and so forth, all of which can affect (or feed back) to the mechanically assisted corrosion process.

The mechanism of mechanically assisted corrosion comes about from the fact, as mentioned previously, that oxides provide a barrier to corrosion and that disruption of this barrier will significantly alter the corrosion behavior. The features needed for mechanically assisted corrosion are: (1) a passive film-covered alloy (e.g., type 316, cobalt-chromium, titanium, etc.), (2) mechanical stress and/or abrasion sufficient to disrupt the oxide film typically in a cyclic fashion, and (3) an aqueous electrolyte. Additionally, if there are restricted geometries where fluid ingress and egress are limited (e.g., crevices, screw-countersink interfaces, implant-bone cement interface, etc.), then these crevice-like environments can provide a set of additional factors that may predispose these alloys to aggressive and severe attack.

The mechanism of mechanically assisted corrosion is as follows. Figure 5 gives a model of the femoral component of a modular total hip replacement head-neck junction (at left). At higher magnification, the head and neck come together at the conical taper interface (center). However, the contact surfaces are not perfectly matched and are sometimes machined to give high and low points in the cross section and, therefore, engage in asperity-asperity contact. These geometric gaps provide locations for fluid ingress into the interface region between the two surfaces. During loading, small-scale cyclic motion (fretting) can take place at this interface that, if the contact stresses are large enough, can result in disruption of the oxide film by the counterface (right).

Once the oxide film is breached or disrupted (or stressed), dramatically larger corrosion currents are generated as this breached film area reforms new oxide film. There are two major sources of Faradaic (corrosion-related) currents (or current densities, current per unit area) during this disruption event (Fig. 5, lower right). One gives rise to ionic dissolution currents ($i_{dissolution}$), and the other to film formation currents (i_{film}). The current density during these periods of oxide film disruption can be many orders of magnitude higher than the currents measured at nondisrupted oxide film covered surfaces. For example, typical current densities for oxide film covered alloys of titanium, cobalt-chromium, and type 316L are in the range of 0.1 to 5 $\mu A/cm^2$ when there is no mechanical disruption. However, currents across disrupted oxide film surfaces can reach up to 5×10^5 to 1.5×10^8 $\mu A/cm^2$ (Ref 49, 51, 56, 60–66). That is a roughly 10^6 to 10^8 increase in current density that results when an oxide film is disrupted.

Some of this current goes to form new oxide film that then acts to shut down the corrosion reactions. Figure 5 shows a differential equation that describes the relationship between the total current transient, i_{total}, and the film and dissolution currents. This model, adapted from Ref 67, assumes that at any instant after disruption, there is a fraction of the volume of oxide removed that

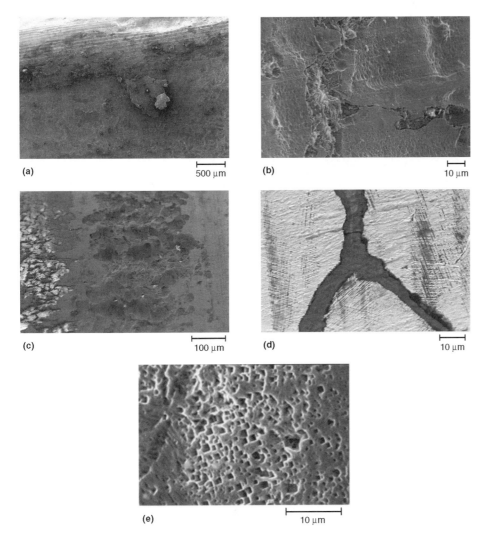

Fig. 2 Selected scanning electron micrography (SEM) micrographs of retrieved modular connections from different orthopedic implants showing evidence of corrosion attack. (a) SEM of head taper near the free surface (top). Note the machining lines at the top near the exterior of the device and the corrosion evident within about 500 μm (20 mils) from the free surface. (b) Higher-magnification SEM of the cast Co-Cr-Mo (ASTM F 75) head inside the taper showing the corrosion attack (pitting) and etching of the grain-boundary carbides. (c) SEM micrograph of a region of a retrieved cast Co-Cr-Mo head taper showing severe corrosion attack of the interdendritic regions of the microstructure (center) by body fluids and the presence of corrosion debris (at left). This is at the deepest-most region of the taper crevice. (d) Backscattered electron micrograph of a grain-boundary triple point of a cast Co-Cr-Mo femoral stem after retrieval. Intergranular carbides are evident and etching by body fluids to reveal the crystallographic orientation of the grains can be seen. (e) Example of a retrieved wrought Co-Cr-Mo alloy head of total hip replacement showing cuboidal pitting attack within the taper region. Source: C.A. Buckley, Ph.D. thesis, Northwestern University, 1994

has reoxidized from the bare metal (θ), and a fraction that has not ($1-\theta$). The ionic dissolution currents are assumed to occur only in the non-reoxidized regions and to obey a Tafel-like behavior. In Fig. 5, ρ is the oxide film density, V is the volume of film removed, n is the charge per cation, F is Faraday's constant (96,480 C/mol), Mw is the molecular weight of the oxide, i_{odiss} is the dissolution exchange current density, η is the overpotential for active dissolution, A_o is the area scratched, b_a is the anodic Tafel slope, and τ is a time constant for repassivation. One can readily solve this equation if one assumes that the total current behaves according to an exponential decay process (Ref 49). Experiments have shown this to be a good approximation (Fig. 6). Film reformation can take place quickly or it can take a significant amount of time depending on the specific circumstances. In a typical environment and at a "normal" fixed potential, repassivation rates are on the range of a few milliseconds; however, this can be affected by pH, potential, and the presence of proteins (Ref 49, 56, 65). If the material is not potentiostatically held, then oxide disruption results in a negative excursion in potential and a much longer repassivation rate (Ref 68, 69).

Thus, the mechanism of mechanically assisted corrosion has at its core a mechanical disruption of the oxide. Therefore, mechanical factors related to oxide film stability are important. How well adhered the oxide is to the substrate; how hard the substrate is; the work-hardening behavior of the substrate, oxide, and metal modulus; the fracture strain/stress of the oxide; and the extent of residual stress in the oxide (resulting from the lattice mismatch between the oxide and the metal) are all important mechanical factors that affect the stability of the film. Some of these properties may be affected by immersion into physiological solution (Ref 70). Also, the mechanical process that may give rise to the oxide disruption is important. This can include wear (fretting or otherwise), surface deformation, or bulk deformation that causes dislocation motion up to the surface. It should be noted that the fretting resistance of titanium alloy surfaces is relatively poor compared with cobalt-chromium and stainless steel alloys. That is, it is easier to abrade the oxide of titanium than it is cobalt-chromium or stainless steel alloy surfaces (Ref 16). Thus, one factor in the use of these materials is that titanium may have a greater susceptibility to corrosion events during mechanically assisted corrosion, which is supported in the retrieval studies.

In the instant of time immediately following oxide disruption, very high rates of oxidation (corrosion) take place. Some of this oxidation goes to form new oxide film that, once reestablished, reduces the rate of oxidation until the surface oxide is reformed and returns to its resting condition. The electrochemistry that results in the instant following disruption has several aspects to it that should be understood.

First, both dissolution and film formation reactions result in the generation of free electrons (see equations, Fig. 5). Here, free electrons refer to those electrons liberated in the oxidation reactions that remain (for a time) in the metal. The film formation reaction takes oxygen from water and generates free electrons and hydrogen ions, while metal atoms oxidize to metal ions and leave electrons behind in the metal as they dissolve into solution or go to form new oxide film. Thus, both reactions create a significant excess of electrons that remain (at least for a period of time) in the metal. This has the effect of lowering (making more negative) the overall potential of the metal being corroded. Thus, if one were to monitor the potential of a metal being subjected to mechanically assisted corrosion, one would see that it would decrease (become more negative) with increasing amounts of oxide film disruption per unit time (Ref 50, 68, 69). The

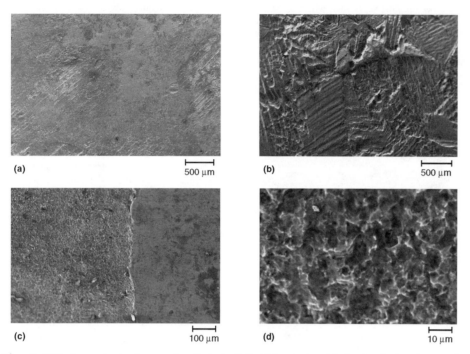

Fig. 3 SEM micrographs showing corrosion attack of a 316L stainless steel modular junction (both surfaces are the same alloy) from a retrieved intramedullary rod (e.g., see Fig. 1c) and from a retrieved SROM modular total hip replacement where the body of the femoral component has a major tapered interface between two Ti-6Al-4V alloy surfaces. (a) Low-magnification SEM micrograph of cone surface of the 316L intramedullary rod showing corrosion attack. (~35×). (b) Higher magnification of 316L stainless steel surface showing the mechanically assisted corrosion attack at a severe stage. The surface has been etched and pitted by the body fluids such that any sign of fretting has been eliminated. (c) Low-magnification SEM of a Ti-6Al-4V/Ti-6Al-4V modular interface for a SROM modular total hip stem. The modular interface in this case is between the stem and the proximal shell, which is designed to provide a site for bony ingrowth. The interface between the stem of the in-growth pad has severe corrosion. The left side of the micrograph shows severe corrosion and the right side does not (and was outside of the crevice zone). (d) Higher magnification of the stem side of the SROM interface showing attack of the Ti-6Al-4V microstructure

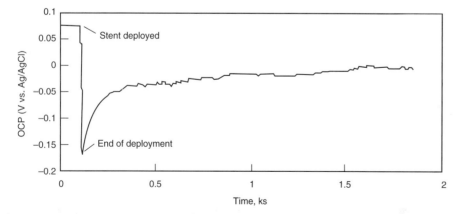

Fig. 4 Plot of the open circuit potential (OCP) shift that takes place during deformation associated with the deployment of a 316L stainless steel vascular stent. The angular wire bends that make up the stent geometry are plastically deformed during deployment, which disrupts the oxide film on the surface and results in a sudden generation of electrons (from the oxidation process), which causes the OCP to drop.

extent of potential excursion depends on the rate of excess electron generation (the corrosion rates), the relative surface area abraded to the total surface area of the metal exposed to solution, the rate at which the electrons are consumed by reduction reactions (e.g., $H_2O + 1/2 O_2 + 2e^- \rightarrow 2OH^-$), and the area available for the consumption of electrons by reduction reactions.

For example, if the fraction of a titanium sample area, being subjected to mechanically assisted corrosion is a significant percentage (greater than 1%) of the total area, potential drops of about -200 to -400 mV can be observed. If the area abraded rises to 10 to 30%, the drop can be up to -600 to -800 mV, and if the entire area is abraded, potential drops of up to -1 V can be observed (Ref 68, 69).

This has implications for the subsequent corrosion behavior. For instance, with a drop in potential, the electrochemical driving force for continued corrosion and/or oxide film formation is decreased. Therefore, with abrasion, the voltage drops and the amount of film that grows and the amount of ions dissolved per unit area may be decreased as the potential gets more negative. Also, it takes a significant amount of time (e.g., several hundred seconds) for the voltage to recover after an abrasion event when reduction reactions alone are responsible for the process (Ref 71). Thus, when abrasion takes place on a small relative area, there is a small voltage drop (because the majority of the area is unabraded and therefore potential controlling), and the repassivation of the abraded area is relatively fast. When the area fraction abraded increases, there is a greater potential drop and it takes longer for the repassivation process to take place.

It should be noted that oxide film formation does not require the presence of dissolved oxygen. Hydrolysis of water is all that is needed. In fact, organic solutions with as little as 1% water present can repassivate titanium (Ref 64, 65).

Second, sudden disruption of a surface and the electrochemical reaction that ensues can and does give rise to a transient electric field that can propagate away from the surface. This field results in an imbalance of charge at the surface, which causes a flux of ions in the solution. This effect can be sensed up to 5 to 7 mm (0.20 to 0.28 in.) away from a 50×10 μm (0.4 mil) scratch imparted in 1 ms (Ref 72). There are bioelectric ramifications of this process, including nerve stimulation (e.g., pain) near the site of an abrasion.

Third, it should be noted in the preceding electrochemical reaction for repassivation (film growth) that another by-product of repassivation is the generation of hydrogen ions at the site of abrasion. These hydrogen ions may remain in the solution and result in a significant drop in the pH in the vicinity of the mechanically assisted corrosion. This is especially true if the local environment is restricted, with limited transport of fluid into and out of the crevice, as is the case in modular tapers. In these cases, the pH of the solution near the abrasion, along with the concentration of metal ions in the solution can be dramatically altered from the bulk solution (body environment). Ionic concentrations have been measured within the modular connections of total hip replacements with upward of 3 to 6 ppm of Co ions during laboratory testing of modular components (Ref 50). Measured pH within these modular tapers has been observed to be as low as 3 (Ref 73). There is other evidence that the pH in modular tapers can achieve a pH level below 1 in some cases. The evidence for this (Fig. 7) is that in some retrieved and highly corroded titanium-titanium modular components, highly corroded interfaces of Ti-6Al-4V couples retrieved have been observed where the attack started as mechanically assisted corrosion but clearly became an electrochemical attack (evidence of fretting was corroded away), as shown in Fig. 7(a), (b), and (c). Also observed in these retrieved devices was the precipitation of titanium phosphate particles in the inner region of these modular pieces (Fig. 7d). Severe corrosion of Ti-6Al-4V of the type observed can occur only when the pH of the solution is approximately 1 or less (Ref 74). Titanium phosphate can occur in this state only if the titanium is present as ions and subsequently reacts with phosphate ions at pHs below about 1. This can be shown by placing titanium particles into an aqueous solution of phosphoric acid (pH = 0). A purple color will ensue over time that indicates Ti^{3+} ions were going into solution (Ref 75). With time, the hydrogen ions reacted with the titanium particle surfaces to form hydrogen gas which raised the pH. As the pH reached a level above 1, the titanium ions precipitated out as titanium phosphate particles identical to that seen in the retrieved implant case (Fig. 1d and 7).

Drops in pH and negative excursions in potential can combine to result in placing a metal surface in an active condition. That is, there are ranges of pH and potential where the oxide film is no longer stable and it will not reform once abraded. The underlying metal will still be subject to corrosion and will corrode actively under these conditions. This means that, once the conditions for active corrosion are established, mechanical abrasion is no longer a necessary condition for continued high levels of corrosion within the crevice. This has been observed to be the case in testing of modular tapers where electrochemical evidence of continued corrosion after 1 million cycles of fatigue loading

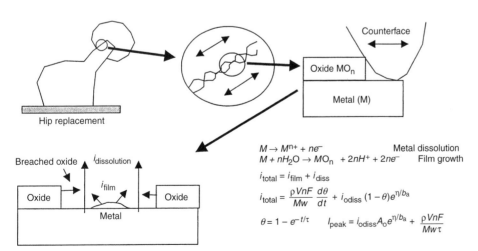

Fig. 5 Schematics across spatial scale showing aspects of mechanically assisted corrosion process for a modular connection in a total hip replacement. The contact region within the taper is rough with regions where solution can gain access. The counterfaces can slide relative to one another due to either rigid body motion or elastic deformation. Sliding with sufficient surface contact stress will cause oxide abrasion and high-current densities associated with repassivation and ionic dissolution. The total current transient then can be thought of as the sum of these two contributions. If one assumes θ to be the fraction of abraded oxide volume that has reformed at any time, and $(1-\theta)$ to be the fraction not yet reformed, then the above differential equation can be used to model the transient response. Source: Ref 49, 52

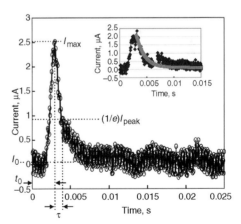

Fig. 6 Typical current transient that results during an 80 μm scratch test with a diamond stylus (radius of 20 μm, or 0.8 mil) on a Co-Cr-Mo surface in phosphate buffered saline potentiostatically held at a passivating potential (0 mV SCE) and loaded with a 1.5 GPa contact stress. The scratch requires about 1 ms to impart and the current transient rise and decay occur within about 2 to 5 ms. The scratch is not complete before the beginning of repassivation so the peak current underestimates the real bare surface current density. Based on the geometry of the scratch (80×10 μm^2), the current density is (2.5 μA/800 μm^2 = 0.31 A/cm^2). Inset is another example where data are fit with an exponential equation of the form $I = 0.1 + 2.15 \text{EXP} -(t - 0.003)/0.0015)$. See text for details.

remained for up to 72 h after the end of loading (Ref 50) and in retrieved implants where pitting and grain-boundary attack are seen without signs of fretting (i.e., the fretting scars were corroded away) (Ref 29, 37, 39). Low pH may also promote hydrogen embrittlement of the metals; however, this has not been investigated for biomaterials.

Fourth, mechanically assisted corrosion can occur on any metal surface where an oxide is present. That is, mechanically assisted corrosion has been reported on stainless steel-stainless steel couples (intramedullary rods in vivo (Ref 37), cobalt-chromium to cobalt-chromium couples (modular head-neck junctions in total hips) (Ref 29, 32, 40), titanium alloy-titanium alloy couples (Ref 29, 39), as well as with any combination of these alloys including cobalt-chromium to stainless steel (Ref 58) and cobalt-chromium to titanium. The necessary conditions for this to occur include sufficient mechanical factors to repeatedly disrupt oxide films, and restricted environments that can result in large changes in local chemistry.

Testing of Mechanically Assisted Corrosion

There have been several tests developed to investigate aspects of mechanically assisted corrosion. These include device tests of geometries that represent modular hip connections and their loading and electrochemical environment (Ref 50–54, 58, 76). Also, high-speed scratch testing has been performed where controlled contact stresses, voltage, solution conditions, and scratch area are maintained and the current transient associated with oxide film disruption is monitored (Ref 49, 55, 56, 60–65). While there are some limitations to these tests, they are able to provide a significant amount of insight into the processes occurring.

Scratch Test. In particular, the electrochemical scratch test method (Ref 49, 55, 56) has been able to show many of the important mechanical and electrochemical factors that influence the behavior of these oxide film-covered alloys during mechanical abrasion. During scratch testing, a well-characterized stylus (diamond) is brought into contact, with a known force, with an oxide film-covered surface immersed in a physiologic solution and potentiostatically held. Then, a high-speed piezoelectric actuator applies a scratch of known length (e.g., 80 μm, or 3 mils) in 0.5 ms. The current response (Fig. 6) to this abrasion event is then measured and used to analyze the behavior. Note that the transient response appears to be an exponential decay process (i.e., $I = I_0 + I_{peak}e^{-(t-t_0)/\tau}$, where I_0 is the baseline current, I_{peak} is the peak current away from the baseline, t_0 is the time to reach the peak, and τ is the time constant for the decay). The magnitude of the current spike (above the background current level) depends on the scratch length, scratch speed, the contact load, the potential, and the solution conditions (Ref 49, 55, 57, 77).

For example, the effects of contact stress and potential on the scratch behavior of cobalt-chromium is summarized in Fig. 8. These plots show how the peak current in the transient varies with applied load at a fixed potential (Fig. 8a) and how the peak current varies with applied potential at a fixed load (Fig. 8b). From Fig. 8(a), it can be seen that the load (and therefore the contact stress) to cause oxide film disruption is about 0.05 N (or about 1 GPa) (Ref 49, 55, 56) and that below this load, no current transient is observed. Increases in the peak transient with load above this level indicates greater amount of oxide area abraded with the higher loads due to the load dependence of the contact area. These observations are consistent with Hertzian contact stress analysis, where the area of contact varies with load. Figure 8(b) shows how the peak current varies with potential during a scratch test in a phosphate buffered saline. Here, one can see that below approximately −500 mV vs. SCE (i.e., the passivating potential, E_{pp}), there is no measurable current transient, while above this voltage up to about +500 mV (i.e., the breakdown potential, E_b), the peak current increases with potential. Above +500 mV SCE, the peak current drops as the material enters the transpassive region and the oxide film is no longer as good a barrier to corrosion. Thus, this plot clearly demonstrates the potential range wherein the oxide film reforms after mechanical disruption.

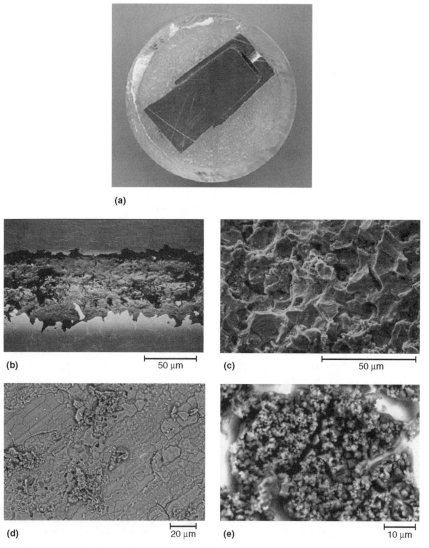

Fig. 7 SEM micrographs of Ti-6Al-4V/Ti-6Al-4V modular connections for a total hip replacement device from a retrieved device (due to ceramic head fracture). This modular connection consisted of a femoral neck taper onto which was placed a Ti-6Al-4V "thimble" to allow a ceramic head to attach to a titanium stem (see Fig. 1d). (a) Optical micrograph of cross section of thimble-neck taper junction intact. (b) SEM micrograph of the interface in (a) showing the Ti-6Al-4V microstructures for both sides of the interface and the gap between the two filled with corrosion debris. Note that there is an approximate 50 μm (2 mils) gap between the two sides and apparent corrosion attack at the grain boundaries of the titanium microstructure. (c) SEM micrograph of the neck surface in the junction region after separation of the interface. Note the penetrating corrosion attack of the grains in this micrograph. (d) Backscattered electron micrograph of the thimble portion of the device (Fig. 1d) showing the corrosion debris accumulation in this region. (e) Backscattered electron micrograph of Ti-phosphate precipitates within the thimble region (near the hole at the proximal most part of the device). These Ti-phosphate particles are indicative of a very low pH condition within these taper environments (below 1).

Below −500 mV, no oxide reforms (or there is no oxide film present due to reduction of the oxide), and above +500 mV, the Cr_2O_3-rich film is no longer thermodynamically stable as the Cr increases its valence from +3 to +6. These observations have been confirmed by electrochemical atomic force microscopy (AFM) tests of cobalt-chromium-molybdenum and direct observation of the film formation process at this voltage (Ref 78–81). The time constants for repassivation are also dependent on potential (Fig. 8c). Here, as the potential falls below −500 mV, the time constant drops to zero, and above +500 mV it rises rapidly as the oxide becomes transpassive.

Scratch testing, therefore, is a highly systematic and sensitive means for assessing the details of the mechanically assisted corrosion process where mechanical, electrical, and chemical aspects of the corrosion process can be evaluated. There are some limitations to the scratch test. Primarily, the rate of scratching may not be high enough to fully abrade the surface before some repassivation takes place. Thus, peak current densities may be underestimated by this technique.

In Vitro Fretting Corrosion Test. Another testing method used to evaluate medical devices, specifically modular connections, is the in vitro fretting corrosion test method (Ref 50–54). In these tests, the sample where the fretting occurs (i.e., the taper region, see Fig. 9) is separated spatially from the rest of the component. An ammeter is placed between the two portions and a voltmeter with a reference electrode is used to monitor the overall potential of the test system during fretting crevice corrosion testing. The sample is then immersed in saline and testing is begun by applying cyclic mechanical loads to the head of the device and monitoring the electrochemical response. When fretting begins, one can observe a current waveform that is in phase with the loading cycle (Fig. 10a, b, c). One can use this test setup to evaluate materials and design variations, as well as test specific effects of the solution on the performance of these tapers.

One such test is where the cyclic load magnitude is incrementally increased and the increase in current and decrease in potential can be monitored (Fig. 10b and c, respectively). By monitoring the current and potential changes, the mechanical conditions whereby fretting crevice corrosion starts can be determined. Also, long-term cyclic tests can be performed with periodic sampling of current and potential (Ref 50). These measurements (along with post-test optical and SEM analysis of the taper surfaces, and particle and ion analysis of the solution) can provide detailed understanding of the long-term behavior of these taper systems. This approach is also easily adapted to the study of other medical devices that contain metal-metal contact points that are potentially susceptible to mechanically assisted corrosion. Testing of other devices, including spinal constructs, vascular stents, bearing surfaces, and so forth, are all possible with this test methodology.

Summary

Oxide film covered alloy systems used in medical applications are all susceptible to increased corrosion when mechanical aspects are incorporated into the overall behavior of the

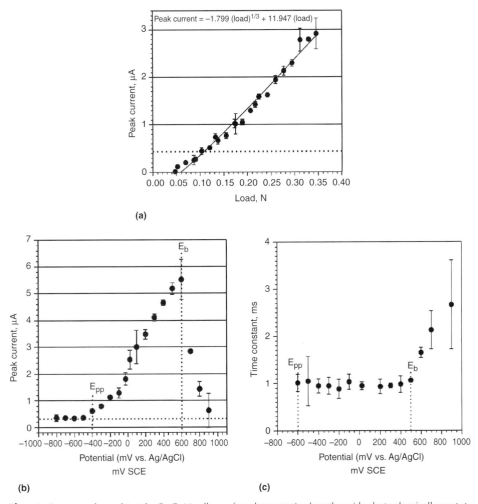

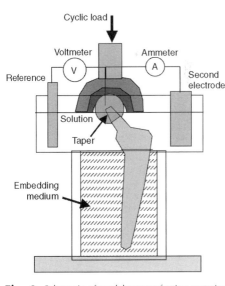

Fig. 8 Summary of scratch test for Co-Cr-Mo alloy surface demonstrating how the oxide electrochemically reacts to disruption while under potentiostatic control. (a) Peak current variation with contact load (or contact stress). (b) Peak current as a function of applied potential at a fixed contact load and scratch length. (c) The time constants for recovery as a function of applied potential. Note that the response depends on the applied load with a lower limit load indicating where the oxide does not disrupt. The potential effects on peak current and time constants show the range of potentials where the oxide film is stable and where it either does not reform (at low potentials) or has lost its passivity (at high potentials). Note that the time constants increase significantly at transpassive potentials (above about 500 mV).

Fig. 9 Schematic of modular taper fretting corrosion test setup used to evaluate the susceptibility of total hip replacement designs to mechanically assisted corrosion. Monitoring of OCP of the test setup and current passed between the working electrode and a second electrode meant to represent the remaining stem portion allows for quantitative assessment of the conditions needed to cause fretting crevice corrosion. Also, solution ion levels, particulate levels, and pH can be monitored in this test during and after either short-term and/or long-term cyclic loading.

device. Because most metallic medical devices are used specifically to withstand the mechanical environment, the corrosion behavior of these materials cannot be separated easily from their mechanical behavior. Little is known at present about the mechanical behavior of nanometer thick films in aqueous environments. However, it is clear that there is a significant probability for oxides to have variable mechanical properties, including adhesive strength, modulus, fracture strain, and so forth when exposed to potentials and solution conditions at the surface of the metal alloy. Also, the high electric field conditions of the oxide in solution can affect the adsorption of proteins to the surface and segregation of metal atoms and ions near the surface. Much is yet to be understood about metallic surface oxides in physiological solutions and the advent of the electrochemical AFM is providing a significant new capability (Ref 78–81) to understand and study these films in situ.

From this analysis of mechanically assisted corrosion of implant alloys, there are several factors that should be considered in the design and use of these materials. When crevices are expected that will experience significant cyclic loading, care must be taken to minimize one of several aspects of this overall process. One can design out the possibility of mechanically assisted corrosion by either eliminating the crevice altogether (eliminate modularity), minimize the ability of fluid to ingress into the crevice site, or, if this is not possible, design the crevice geometry to allow for fluid recycling so that low pH, high ion conditions do not develop. Surface treatments that harden the surface and prevent oxide disruption may also provide some benefit. However, these approaches need to be taken with caution as the typically hard materials that make up these coatings (physical vapor deposition or chemical vapor deposition type coatings, or ion implantation, for example), may abrade themselves with time and result in an aggravated condition for continued wear. Designs that minimize the potential for fretting motion (either by elastic deformation or rigid body motion) also should be considered.

It is clear that metallic biomaterials will continue to be used in medical devices for the foreseeable future and that the mechanical and electrochemical performance of these surfaces will be an important part of the success or failure of these materials and the devices they comprise. The complex, interdependent nature of mechanically assisted corrosion mechanisms is evident in this article. Understanding of these processes requires in-depth knowledge of surface contact mechanics, surface structure and chemistry, electrical phenomena, and corrosion behavior. The biological context and how it may affect these processes is also very poorly understood. Most tests have not included the complexity and local heterogeneity that may arise when a metal is implanted into a living system (inflammation, proteins, enzymes, cells, etc.), which may affect the overall process in vivo. Much is still to be learned about how these biological species and processes interact with the highly energetic surfaces of metallic biomaterials.

ACKNOWLEDGMENTS

I would like to acknowledge several of my students who contributed to this work: Christine Buckley, Jay Goldberg, Saryn Goldberg, and Spiro Megremis. Also, discussions with Dr. Zhijun Bai were particularly helpful in the preparation of this manuscript.

REFERENCES

1. "Standard Specification for Wrought 18 Chromium-14 Nickel-2.5 Molybdenum Stainless Steel Bar and Wire for Surgical Implants (UNS S31673)," F 138, *Annual Book of ASTM Standards,* Vol 13, ASTM International, 1996
2. "Standard Specification for Wrought Nitrogen Strengthened 21 Chromium-10 Nickel-3 Manganese-2.5 Molybdenum Stainless Steel Bar for Surgical Implants (UNS S31675)," F 1586, *Annual Book of ASTM Standards,* Vol 13, ASTM International, 1996
3. "Standard Specification for Wrought Nitrogen Strengthened 23 Manganese-21 Chromium-1 Molybdenum Low-Nickel Stainless Steel Alloy Bar and Wire for Surgical Implants (UNS S29108)," F 2229, *Annual Book of ASTM Standards,* Vol 13, ASTM International, 1996
4. "Standard Specification for Cobalt-28 Chromium-6 Molybdenum Castings and Casting Alloy for Surgical Implants (UNS R300075)," F 75, *Annual Book of ASTM Standards,* Vol 13, ASTM International, 1996
5. "Standard Specification for Cobalt-28 Chromium-6 Molybdenum Alloy Forgings for Surgical Implants (UNS R31537, R31538, R31539)," F 799, *Annual Book of ASTM Standards,* Vol 13, ASTM International, 1996
6. "Standard Specification for Wrought Cobalt-28 Chromium-6 Molybdenum Alloy for Surgical Implants (UNS R31537, R31538, R31539)," F1537, *Annual Book of ASTM Standards,* Vol 13, ASTM International, 1996
7. T. Kilner, R.M. Pilliar, and G.C. Weatherly, *J. Biomed. Mater. Res.,* Vol 16 (No. 1), 1982, p 63–79
8. "Standard Specification for Unalloyed Titanium for Surgical Implant Applications (UNS R50250, R50400, R50550, R50700)," F 67, *Annual Book of ASTM Standards,* Vol 13, ASTM International, 1996
9. "Standard Specification for Unalloyed Titanium Wire for Surgical Implant Applications (UNS R50250, R50400, R50550, R50700)," F 1341, *Annual Book of ASTM Standards,* Vol 13, ASTM International, 1996
10. "Standard Specification for Wrought Titanium-6 Aluminum-4 Vanadium ELI (Extra Low Interstitial) Alloy for Surgical

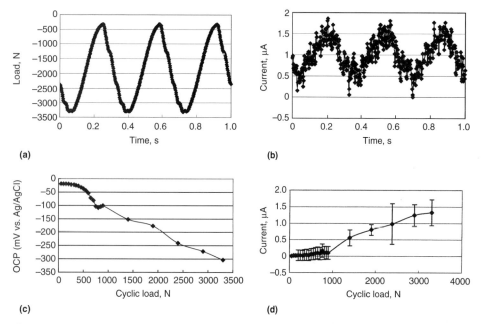

Fig. 10 Summary of results of one of the mechanically assisted corrosion tests of a modular femoral hip taper interface. (a) Load vs. time showing a compressive cyclic load to 3300 N applied at 3 Hz. (b) The corresponding fretting current measured between the working and second electrodes (see Fig. 9). Note that the current is correlated with the loading. (c) Summary of the OCP vs. applied cyclic load. (d) Mean current (and rms current—error bars) vs. applied cyclic load. Note that there is a load at onset where fretting corrosion can be detected (~500 N) in both the currents and potential shifts observed. Also, note the increased corrosion currents measured with increasing load and the larger drop in OCP indicating increased corrosion reactions.

Implant Applications (UNS R56401)," F 136, *Annual Book of ASTM Standards,* Vol 13, ASTM International, 1996
11. "Standard Specification for Wrought Nickel-Titanium Shape Memory Alloys for Medical Devices and Surgical Implants," F 2063, *Annual Book of ASTM Standards,* Vol 13, ASTM International, 1996
12. A. Boltz and M. Schaldach, *Med. Biolog. Eng. Comput.,* Vol 31, 2003, p 123–130
13. A. Bolz, M. Amon, and C. Ozbek, *Texas Heart Inst. J.,* Vol 23 (No. 2), 1996, p 162–166
14. D. Scharnweber, R. Beutner, S. Robles, and H.J. Worch, *Mat. Sci: Mater. in Med.,* Vol 13, 2002, p 1215–1220
15. J.J. Jacobs, J.L. Gilbert, and R.M. Urban, *Current Concepts Review, J. Bone and Joint Surgery,* Vol 80-A (No. 2), Feb, 1998, p 268–282
16. J.L. Gilbert and J.J. Jacobs, "Modularity of Orthopedic Implants," STP 1301, J.E. Parr, M.B. Mayor, and D.E. Marlow, Ed., ASTM International, 1997, p 45–59
17. S.A. Brown and J.P. Simpson, *J. Biomed. Mater. Res.,* Vol 15, 1981, p 867–878
18. S.A. Brown and K. Merritt, *J. Biomed. Mater. Res.,* Vol 15, 1981, p 479–488
19. R.M. Rose, A.L. Schiller, and E.L. Radin, *J. Bone and Joint Surg.,* Vol 54-A (No. 4), 1972, p 854–862
20. H.R. Piehler, M.A. Portnoff, L.E. Sloter, E.J. Vegdahl, J.L. Gilbert, and M.J. Weber, *Corrosion and Degradation of Implant Materials: 2nd Symp.,* ASTM Special Technical Publication 859, A.C. Fraker and C.D. Griffin, Ed., 1985, p 93–104
21. L.C. Lucas, R.A. Buchannan, and J.E. Lemons, *J. Biomed. Mater. Res.,* Vol 15, 1981, p 731–747
22. C.D. Griffin, R.A. Buchanan, and J.E. Lemons, *J. Biomed. Mater. Res.,* Vol 17, 1983, p 489–500
23. W. Rostoker, *J. Biomed. Mater. Res.,* Vol 12, 1978, p 823–829
24. W. Rostoker, C.W. Pretzel, and J.O. Galante, *J. Biomed. Mater. Res.,* Vol 8, 1974, p 407–419
25. D.C. Mears, *J. Biomed. Mater. Res. Symp.,* No. 6, 1975, p 133–148
26. T.P. Hoar and D.C. Mears, *Roc. Roy. Soc., Ser. A,* Vol 294, 1966, p 486–510
27. J.P. Collier, V.A. Surprenanyt, R.E. Jensen, and M.B. Mayor, *Clin. Orthop.,* Vol 271, 1991, p 305–312
28. J.P. Collier, M.B. Mayor, R.E. Jensen, V.A. Surprenant, H.P. Surprenant, J.L. McNamara, and L. Belec, *Clin. Orthop.,* Vol 285, 1992, p 129
29. J.P. Collier, V.A. Surprenant, R.E. Jensen, M.B. Mayor, and H.P. Surprenant, *J. Bone and Joint Surgery,* Vol 74-B, 1992, p 511–517
30. J.L. Gilbert, C.A. Buckley, and J.J. Jacobs, *J. Biomed. Mater. Res.,* Vol 27, 1993, p 1533–1544
31. J.R. Lieberman, C.M. Rimnac, K.L. Garvin, R.W. Klein, and E.A. Salvati, *Clin. Orthop.,* Vol 200, 1994, p 162–167
32. J.L. Gilbert, C.A. Buckley, J.J. Jacobs, K.C. Bertin, and M.R. Zernich, *J. Bone and Joint Surgery,* Vol 76-A, 1994, p 110–115
33. E.B. Matheisen, J.U. Lindgren, G.G. Blomgren, and F.P. Reinholt, *J. Bone and Joint Surgery,* Vol 73-B, 1991, p 569–575
34. S.D. Cook, R.L. Barrack, and A.J.T. Clemow, *J. Bone and Joint Surgery,* Vol 76-B, 1994, p 68–72
35. J.P. Collier, M.B. Mayor, I.R. Williams, B.A. Surprenant, H.P. Surprenant, and B.H. Currier, *Clin. Orthop.,* Vol 311, 1995, p 91–101
36. S.D. Cook, R.L. Barrack, G.C. Baffwes, A.J.T. Clemow, P. Serekian, N. Dong, and M.A. Kester, *Clin. Orthop.,* Vol 298, 1994, p 80–88
37. D. Jones, J.L. Marsh, J.V. Nepola, J.J. Jacobs, A.K. Skipor, R. Urban, J.L. Gilbert, and J. Buckwalter, *J. Bone and Joint Surgery,* Vol 83-A (No. 4), 2001, p 537–548
38. J. Walczak, F. Shahgaldi, and F. Heatley, *Biomaterials,* Vol 19, 1998, p 229–237
39. R.M. Urban, J.L. Gilbert, and J.J. Jacobs, "Titanium, Niobium, Zirconium, and Tantalum for Medical and Surgical Applications," STP 1539, ASTM International, 2005, in press
40. J. Goldberg, J.L. Gilbert, and J.J. Jacobs, *Clin. Orthop. and Rel. Res.,* Vol 401, 2002, p 149–161
41. H.G. Willert, L.G. Brolack, G.H. Buchorn, D. Ing, P.H. Jensen, G. Koster, I. Lang, P. Ochsner, and R. Schenk, *Clin. Orthop.,* Vol 333, 1996, p 51–75
42. J.J. Jacobs, R.M. Urban, J.L. Gilbert, A.K. Skipor, J. Black, M. Jasty, and J.O. Galante, *Clinical Ortho.,* Vol 319, 1995, p 94–105
43. R. Urban, J.J. Jacobs, J.L. Gilbert, and J.O. Galante, *J. Bone and Joint Surgery,* Vol 76-A (No. 9), 1994, p 1345–1359
44. R.M. Urban, J.J. Jacobs, J.L. Gilbert, S.B. Rice, M. Jasty, C.R. Bragdon, and J.O. Galante, "Modularity of Orthopedic Implants," STP 1301, J.E. Parr, M.B. Mayor, and D.E. Marlow, Ed., ASTM International, 1997, p 33–44
45. M.T. Manley and P. Serekian, *Clin. Orthop.,* Vol 298, 1994, p 137–146
46. O. Svensson, E.B. Matheeisen, F.P. Reinholt, G. Blomgren, and H. Sweden, *J. Bone and Joint Surg.,* Vol 70-A (No. 8), 1988, p 1238–1242
47. J.J. Jacobs, A.K. Skipor, J. Black, R.M. Urban, and J.O. Galante, *J. Bone and Joint Surg.,* Vol 73-A (No. 10), 1991, p 1475–1486
48. J.R. Goldberg, C.A. Buckley, J.J. Jacobs, and J.L. Gilbert, "Modularity of Orthopedic Implants," STP 1301, J.E. Parr, M.B. Mayor, and D.E. Marlow, Ed., ASTM International, 1997, p 157–176
49. J.L. Gilbert, C.A. Buckley, and E.P. Lautenschlager, *Medical Applications of Titanium and Its Alloys—The Materials and Biological Issues,* ASTM Special Technical Publication 1272, 1996, p 199–215
50. J.R. Goldberg and J.L. Gilbert, *Appl. Biomaterials,* Vol 64B (No. 2), 2003, p 78–93
51. J.L. Gilbert and C.A. Buckley, *Total Hip Revision Surgery,* J.O. Galante, A.G. Rosenberg, and J.J. Callaghan, Ed., Raven Press, New York, 1994, p 41–50
52. "Standard Practice for Fretting Corrosion Testing of Modular Implant Interfaces: Hip Femoral Head-bore and Cone Taper Interface," F 1875, *Annual Book of ASTM Standards,* Vol 13, ASTM International, 1998
53. C. Flemming, S.A. Brown, and J.H. Payer, "Biomaterials Mechanical Properties," STP 1173, H.E. Kambic and A.T. Yokobori, Ed., ASTM International, 1994, p 156–166
54. S.A. Brown, A. Abera, M. D'Onofrio, and C. Flemming, "Modularity of Orthopedic Implants," STP 1301, D.E. Marlowe and J.E. Parr, Ed., 1997, p 189–198
55. J.R. Goldberg and J.L. Gilbert, *Biomaterials,* Vol 25 (No. 5), 2004, p 851–864
56. J.R. Goldberg, E.P. Lautenschlager, and J.L. Gilbert, *J. Biomed. Mater. Res.,* Vol 37 (No. 2), 1997, p 421–433
57. N.J. Hallab and J.J. Jacobs, *Corros. Rev.,* Vol 21 (No. 2-3), 2003, p 183–213
58. M. Zhu and M. Windler, "Stainless Steel for Medical and Surgical Applications," STP 1438, G.L. Winters, and M.J. Nutt, Ed., ASTM International, 2003, p 235–248
59. S.A. Brown, C. Flemming, J.S. Kawalec, H.E. Placko, C. Vassaux, K. Merritt, J.H. Payer, and M.J. Fraay, *J. Applied Biomat.,* Vol 6 (No. 19-26), 1995
60. D.G. Kolman and J.R. Scully, *J. Electrochem. Soc.,* Vol 143, 1996, p 1847–1860
61. D.G. Kolman and J.R. Scully, *J. Electrochem. Soc.,* Vol 142, 1995, p 2179–2188
62. P.D. Bastek, R.C. Newman, and R.G. Kelly, *J. Electrochem. Soc.,* Vol 140, 1993, p 1884–1889
63. G.T. Burstein and G. Gao, *J. Electrochem. Soc.,* Vol 138, 1991, p 2627–2630
64. H.J. Ratzer-Schiebe, *Corrosion,* Vol 34, 1978, p 437–442
65. H.J. Ratzer-Shiebe, "Repassivation of Titanium and Titanium Alloys Dependent on Potential and pH," *Passivity of Metals and Semiconductors,* M. Froment, Ed., Elsevier Science, Amsterdam, The Netherlands, 1981, p 731–739
66. B.T. Rubin, *Electroanal. Interfacial Electrochem.,* Vol 58, 1975, p 323–337
67. J.R. Ambrose, *Treatise on Materials Science and Technology, Corrosion: Aqueous Processes and Passive Films,* J.C. Scully, Ed., Academic Press, New York, 1983, p 175–204
68. S. Goldberg, "Electrochemical Scratch Testing of Ti-6Al-4V: Effect of Area Ratio and Development of Transient Electric Fields," master's thesis, Northwestern University, May 1999

69. R. Venugopalan, J.J. Weimer, M.A. George, and L.C. Lucas, *Biomaterials,* Vol 21, 2000, p 1669–1677
70. Z. Bai and J.L. Gilbert, *Transactions of Orthoped. Res. Soc. Annual Meeting* (San Francisco, CA), 2004, p 124
71. T. Hanawa, K. Asami, and K. Asaoka, *J. Biomed. Mater. Res.,* Vol 40, 1997, p 530–538
72. S. Goldberg and J.L. Gilbert, *J. Biomed. Mater. Res.,* Vol 56, 2001, p 184–194
73. J.L. Gilbert and C.A. Buckley, *Total Hip Revision Surgery,* J.O. Galante, A.G. Rosenberg, and J.J. Callaghan, Ed., Raven Press, 1994, p 41–50
74. M. Pourbaix, *Atlas of Electrochemical Equilibria in Aqueous Solutions,* Pergamon Press, 1966, p 213–222
75. M.J. Sienko and R.A. Plane, *Chemical Principles and Properties,* 2nd ed., McGraw Hill, 1974, p 500
76. "Standard Test Method for Measuring Fretting Corrosion of Osteosynthesis Plates and Screws," F 897, *Annual Book of ASTM Standards,* Vol 13, ASTM International, 1998
77. S. Megremis, "The Role of Mechanically Assisted Corrosion on the Behavior of Co-Cr-Mo Alloys Used in Metal-on-Metal Total Hip Prostheses," Ph.D. Dissertation, Northwestern University, 2001
78. J.L. Gilbert, Z. Bai, J. Bearinger, and S. Megremis, *Proc. ASM Conference on Medical Device Materials* (Anahiem, CA), Sept 2003, S. Shrivastava, Ed., 2004, p 139–143
79. J.P. Bearinger, C.A. Orme, and J.L. Gilbert, *Biomaterials,* Vol 24 (No. 11), 2003, p 1837–1852
80. J.P. Bearinger, C.A. Orme, and J.L. Gilbert, *J. Biomed. Mater. Res.,* Vol 67A (No. 3), 2003, p 702–712
81. J.P. Bearinger, C.A. Orme, and J.L. Gilbert, *Surf. Sci.,* Vol 491, 2001, p 370–387

Corrosion Performance of Stainless Steels, Cobalt, and Titanium Alloys in Biomedical Applications

Zhijun Bai and Jeremy L. Gilbert, Syracuse University

BIOMATERIALS used in medical devices and prostheses are implanted into the human body to replace, repair, or restore the function of tissue. The term *biomaterial* includes synthetic materials such as metals (alloys), polymers, and ceramics as well as some natural materials including bioceramics (e.g., hydroxyapatite) and biopolymers (e.g., collagen). Metallic biomaterials represent the most highly used class of biomaterials and generally have advantages over other biomaterials in terms of mechanical properties, such as high tensile strength, high fatigue strength, and good processability.

Biomedical devices are usually subjected to static or dynamic forces, such as in orthopedic and cardiovascular applications. For example, various artificial joint implants, fracture fixation devices, heart valves, and vascular stents are all subjected to significant repetitive loadings. On the other hand, some electrode materials, such as pacemaker leads, are made of metals such as cobalt-base alloy (MP35N) or platinum alloy (Pt-Ir) because of their good electrical conductivity and because they are not as highly stressed. Dental alloys are another group of metallic biomaterials that include noble alloys and base metal alloys (including nickel- and cobalt-base alloys), but the details of these alloys are discussed elsewhere (see, for example, the article "Corrosion and Tarnish of Dental Alloys" in this Volume).

The history of the use of man-made materials in a human body can be traced back thousands of years. The development of metallic biomaterials was discussed in previous publications (Ref 1, 2). The main metallic biomaterials in use today can be categorized into three groups: iron-base alloys (stainless steels), cobalt-base alloys, and titanium-base alloys. While other metals and alloys are used, this article focuses only on the corrosion behavior of these three groups. These alloys all form a thin, compact, semiconducting oxide (or hydroxide) film (usually called a passive film) that protects the substrate alloy from corrosive environments as well as interacts with the host during the host response. To understand corrosion requires an understanding of these oxides.

Chemical Composition and Microstructure of Iron-, Cobalt-, and Titanium-Base Alloys

The chemical compositions of the most commonly used metallic biomaterials made of stainless steels, cobalt alloys, and titanium alloys are given in Tables 1 to 3. Typical microstructures for these three major alloy systems are shown in Fig. 1.

Stainless steel, in the form of 316L stainless steel (Ref 3), is a widely used material for

Table 1 Chemical composition of commonly used stainless steels for biomedical applications

ASTM designation	UNS No.	Composition(a), wt%											
		Fe	Cr	Ni	Mo	Mn	C	Si	N	Cu	P	S	Others
F138(b)	S31673	bal	17.00–19.00	13.00–15.00	2.25–3.00	2.00	0.030	0.75	0.10	0.50	0.025	0.010	...
F1314	S20910	bal	20.50–23.50	11.50–13.50	2.00–3.00	4.00–6.00	0.030	0.75	0.20–0.40	0.50	0.025	0.010	0.10–0.30Nb 0.10–0.30V
F1586	S31675	bal	19.50–22.00	9.00–11.00	2.00–3.00	2.00–4.25	0.08	0.75	0.25–0.50	0.25	0.025	0.010	0.25–0.80Nb
F2229	S29108	bal	19.00–23.00	0.10	0.50–1.50	21.00–24.00	0.08	0.75	0.90 min	0.25	0.03	0.010	...

(a) Single values are maximum values unless otherwise indicated. (b) 316LVM belongs to this type.

Table 2 Chemical composition of cobalt-base alloys used for biomedical applications

ASTM designation	UNS No.	Composition(a), wt%											
		Co	Cr	Ni	Mo	Fe	C	Si	Mn	W	P	S	Others
F75, cast	R30075	bal	27.0–30.0	1.00	5.0–7.0	0.75	0.35	1.00	1.00	0.20	0.020	0.010	0.25N; 0.30Al; 0.01B
F90, wrought	R30605	bal	19.0–21.0	9.0–11.0	...	3.00	0.05–0.15	0.40	1.00–2.00	14.00–16.00	0.040	0.030	...
F562, wrought(b)	R30035	bal	19.0–21.0	33.0–37.0	9.0–10.5	1.00	0.025	0.15	0.15	...	0.015	0.010	1.0Ti
F563	R30563	bal	18.0–22.0	15.00–25.00	3.00–4.00	4.00–6.00	0.05	0.50	1.00	3.00–4.00	...	0.010	0.50–3.50Ti
F799	R31537	bal	26.0–30.0	1.0	5.0–7.0	0.75	0.35	1.0	1.0	...	...	...	0.25N

(a) Single values are maximum values unless otherwise indicated. (b) MP35N

Table 3 Chemical composition of titanium-base alloys used for biomedical applications

ASTM designation	UNS No.	Ti	Al	V	C	Ni	N	H	O	Nb	Fe	Zr	Others
F67, CP-Ti grade 2	R50400	bal	...	...	0.10	...	0.03	0.015	0.18	...	0.20	...	...
F136, wrought, ELI(b)	R56401	bal	5.5–6.5	3.5–4.5	0.08	...	0.05	0.012	0.13	...	0.25	...	...
F1295	R56700	bal	5.5–6.5	...	0.08	...	0.05	0.009	0.2	6.5–7.5	0.25	...	0.5Ta max
F2146	R56320	bal	2.50–3.50	2.0–3.0	0.05	...	0.020	0.015	0.12	...	0.30	...	...
F1713	...	bal	...	...	0.08	...	0.05	0.012	0.15	12.5–14.0	0.25	12.5–14.0	...
F2066	R58150	bal	...	...	0.10	...	0.05	0.015	0.20	...	0.10	...	14.00–16.00Mo
F2063, wrought, SMA(c)	...	bal	...	...	0.070	54.5–57.0	...	0.005	0.050	0.025	0.050	...	0.050Co, 0.010Cu, 0.01Cr

(a) Single values are maximum values unless otherwise indicated. (b) ELI, extra low interstitial. (c) SMA, shape memory alloy

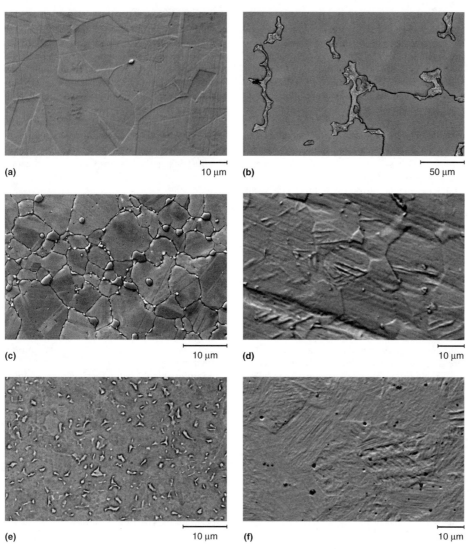

Fig. 1 Scanning electron microscopy images of typical microstructures of metallic biomaterials. (a) 316L stainless steel. Backscattered electron (BE) image showing grains and twins within grains. Polishing scratches are also evident. 1500×. (b) Cast Co-Cr-Mo alloy (ASTM F75). BE image showing carbides and grain boundaries. 500×. (c) BE shadow image (combination of composition and topography) of forged high-carbon Co-Cr-Mo (ASTM F1537) showing small grains and carbides located at the grain boundaries. 2500×. (d) BE image of CP-Ti showing high oxide morphology/surface variations as well as contrast due to electron channeling. 1500×. (e) BE image of Ti-6Al-4V showing α-β phase structure. 2500×. (f) BE image of NiTi showing matrix structure with secondary inclusions. 1500×. Co-Cr-Mo images courtesy of S. Megremis; Ti-6Al-4V image courtesy of R. Gettens

orthopedic applications (particularly for fracture fixation and spinal devices) and cardiovascular applications (e.g., stents) because of its competitive cost, wide availability, and mechanical properties. The main constituents of stainless steel are iron, chromium, nickel, and carbon. Addition of other elements such as molybdenum, manganese, and nitrogen improves either their corrosion properties or their mechanical properties or both. Chromium is the most important alloying element and is the major constituent that forms the passive chromium oxide film to protect the surface from corrosion. One of the requirements for stainless steels in medical applications, from a microstructural perspective, is that the material must have a single austenitic phase with face-centered cubic (fcc) structure and be nonferromagnetic. These characteristics are very important for corrosion resistance, mechanical behavior and safety considerations. Chromium is a ferrite stabilizer (body-centered cubic, or bcc, structure), so its content in stainless steel is usually below 24% but must be higher than 12% to form a continuous passive oxide film on the surface to provide corrosion resistance. Similarly, molybdenum is another ferrite stabilizer that is added to also enhance localized pitting corrosion resistance. Its content cannot exceed 3%. Nickel is included because it is not only an austenite stabilizer but also improves the processability of the alloy. However, there are concerns with nickel because, when corroded, it forms a potentially toxic ion that may induce metal allergy in some patients. It is replaced by manganese in newly developed stainless steels such as F2229 (Ref 4). Carbon serves as an interstitial strengthening element, but its content must be limited to 0.08% or less; otherwise, it may precipitate metal carbides through a process known as sensitization (Ref 1) and could lower the corrosion resistance of the alloy. In fact, for the main stainless steel used, 316L stainless steel, the "L" stands for low carbon (0.03 wt%). More recently, nitrogen was alloyed to enhance the mechanical and corrosion properties of iron-base alloys, and the corrosion resistance of these newer stainless steels looks very promising. Phosphorus and sulfur are inclusion elements whose content must be controlled; otherwise, they will cause deleterious effects on the corrosion resistance by preferential dissolution of the inclusions (Ref 5). Another requirement for stainless steel is a fine grain microstructure that is crucial for strength and fatigue properties.

Cobalt-base alloys come in a variety of compositions that can include cobalt, chromium, molybdenum, nickel, tungsten, and other elements (Ref 6–10) and have a highly complex microstructure. Chromium is the main corrosion-resistant element, forming chromium oxide as a passive-film kinetic barrier, while molybdenum is another main constituent added to improve the pitting corrosion resistance and the mechanical properties. The mechanical properties of cobalt-base alloys are critical to their performance. These alloys typically have a fcc or hexagonal close-packed (hcp) crystal structure, depending on the thermomechanical processing, and have high strength and high work-hardening coefficients due to their low stacking fault energy and planar slip characteristics. These alloys are also comprised of complex carbides that are typically dispersed in the microstructure. Alloying with refractory elements (molybdenum or tungsten) and forming carbides will increase its strength. Various thermomechanical processing methods, such as wrought or forging processes, are used to improve the strength and fatigue resistance. These alloys are also highly wear resistant and are typically used in wear applications, such as the head of a total hip replacement.

Titanium and its alloys (Ref 11–16) are also very high-strength materials with relatively low moduli that typically contain hcp and/or bcc crystal structures and have very high corrosion resistance, relying on the formation of a titanium dioxide (TiO_2) film. The mechanical properties of titanium mainly depend on its microstructure and its solid-solution strength, which is related to composition and heat treatment. Pure titanium has a single hcp crystal structure with lower strength. Elements of vanadium, niobium, and chromium are beta-phase (bcc) stabilizers, while aluminum, oxygen, and zirconium are alpha stabilizers. Alloying with these elements can improve the strength and fatigue resistance. Another titanium alloy that is seeing increased use is nickel-titanium (NiTi) alloy (Ref 17). This near equiatomic alloy has superelastic and shape memory characteristics and is used in cardiovascular applications.

Surface Oxide Morphology and Chemistry

Upon insertion into the human body, biomedical devices will experience a series of interactions between the biomaterial and the host, and corrosion is part of this interaction. Therefore, the structure, chemistry, and behavior at the surface of biomaterials is critical in determining the biocompatibility. This section discusses the surface morphology and chemistry of these oxide-film-covered alloys and provides insight into the interaction.

There are a number of surface analytical methods that can be used to evaluate the surface chemistry and surface morphology of passive-film-covered biomaterials surfaces (Ref 18). Among them, atomic force microscopy (AFM) and x-ray photoelectron spectroscopy (XPS) are two powerful approaches. Atomic force microscopy can provide surface morphology information at the micron to nano level in air or in solution and can be combined with electrochemical polarization testing. X-ray photoelectron spectroscopy can investigate the surface chemistry of the thin oxide films at the nano level with very high accuracy. Furthermore, XPS can be used to study the chemical valence state and depth distribution of elements in the film.

Typical three-dimensional AFM images of stainless steel, Co-Cr-Mo, commercially pure titanium (CP-Ti), and NiTi alloy are given in Fig. 2. The images show that each of these surfaces is covered by a domelike oxide film, even in the grain-boundary region or overtop of some inclusions (not shown). The size (diameter and height) of oxide domes on these alloys is a function of immersion time, electrochemical history (potential and pH), and prior surface treatment (Ref 19–22). Specifically, titanium oxide domes will spread laterally and flatten out with increasing applied potential. From in situ electrochemical AFM, Co-Cr oxides appear to form at approximately −500 mV versus saturated calomel electrode (SCE) and grow over the potential range between −500 mV and approximately 250 to 300 mV, where significant changes begin to occur. By 400 to 500 mV, the oxide film on Co-Cr alloys begins to break down, before the transpassive potential of the alloy, indicating that oxide film structural changes precede current rises associated with breakdown.

From XPS results (see subsequent information), it has been shown that the surface chemistry of oxides on Co-Cr and stainless steel alloys is chromium rich, while the titanium and NiTi alloys are predominantly titanium oxide surfaces. Another aspect of these oxide films is that the measured height of oxide domes appears higher than 10 nm, except for Co-Cr-Mo, which is closer to 10 nm, based on AFM measurements. This is at odds with XPS depth measurements (argon ion sputtering), which show these oxide domes to be in the 1 to 5 nm range.

Table 4 reviews the surface chemistry for passive films formed on these metallic biomaterials. It can be found that all passive films have film thicknesses ranked as several nanometers and that they are significantly affected by chemical or electrochemical surface treatments. In fact, surface oxide films (particularly their chemistry) are influenced by prior treatment. It should be noted that changes of surface morphology and surface chemistry will occur due to immersion time or potentials in the biological environment, either in vitro or in vivo, and these effects are discussed in the later sections.

Generally, the surface film of stainless steel consists mainly of chromium and iron oxides. Nickel exists in the oxide in a metallic state or nickel oxide, but severely depleted. The existence of molybdenum in the oxide is questionable, even though it plays a crucial role for preventing localized corrosion. Chemical passivation and electrochemical polishing increase the content of chromium, enhance the ratio of chromium to iron, and improve the corrosion resistance.

The surface oxide for Co-Cr-Mo alloys formed in air is primarily Cr_2O_3 with minor amounts of cobalt and molybdenum oxides (Ref 25, 27, 30). The ratio of chromium to cobalt is varied in different media, such as lab air or wet steam (Ref 25), and is influenced by various surface treatments, such as passivation in nitric acid (Ref 27).

The dominant chemical species for surface films on titanium and its alloy is TiO_2 with, perhaps, lower-valence oxides such as TiO and Ti_2O_3 present, particularly in the inner layer. For the Ti-6Al-4V alloy, Al_2O_3 is found in the oxide layer, while the presence of vanadium in the oxide is unclear. Shape memory NiTi alloy has a

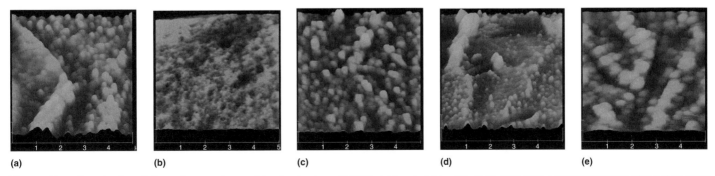

Fig. 2 Three-dimensional atomic force microscopy images of the microstructures of (a) 316L stainless steel, (b) cast Co-Cr-Mo (ASTM F75), (c) CP-Ti (ASTM F67), (d) Ti-6Al-4V (ASTM F136), and (e) NiTi (ASTM F2063). All images show domelike oxide film structure. Scan size, 5 μm × 5 μm; pitch angle, 65°; height scale, 100 nm. Co-Cr-Mo images courtesy of S. Megremis; Ti-6Al-4V image courtesy of J. Bearinger

higher content of nickel in the alloy; however, TiO_2 is the primary chemical species in this oxide film. The chemistry of this oxide film is significantly influenced by various factors, such as prior surface treatment, fabrication methods, and heat treatments. In some cases, metallic nickel and its oxide may play a major role in oxide film behavior, and this will significantly affect its corrosion behavior.

Physiological Environment

In orthopedic and cardiovascular applications, biomedical devices are in contact with human tissue. The former will be in contact with interstitial fluid, and the latter will be in contact with blood. In addition, the act of implantation results in a wound site, and therefore, all biomaterials first encounter blood and the wound-healing environment. The acute and chronic healing environment can and does significantly modify the physiological solution in important ways that may affect the surface behavior of these alloys. Ionic species, superoxide radicals and their byproducts, enzymes, pH, cell types, and a host of other factors can vary over time at the site of the alloy.

Generally, physiological electrolyte fluid is composed of various salts (mainly chloride and bicarbonate anions, and sodium and potassium cations) and various organic and biological species (proteins, enzymes, and cells). It should be noted that some special chemicals, such as superoxide radicals and their by-products during inflammatory reactions, will be synthesized during pathological and/or wound-healing processes, and these species may play an important role in the corrosion behavior of biomaterials.

In the biological system, the pH is modulated by phosphate and carbonate buffering and is kept in the range of 7.4 under normal conditions. However, pH may vary in the wound-healing process after surgery from 5.5 to 9.0 with infection. In the case of a hematoma, the pH can become 5.5 for several days or longer (Ref 31). In some local sites, where corrosion, such as crevice corrosion, is occurring, the pH may drop to approximately pH 1 or lower (Ref 32, 33).

As for electrochemical potentials in the body, there is no general agreement for the range of values possible. The potential is likely to be material- and environment-dependent. Although bioelectric signals are continuously present in the living system, these electrical transients are usually not thought to have significant effects on the corrosion behavior of biomaterials. However, there are studies that do show some effects (Ref 34). Based on the local oxygen or hydrogen partial pressure, it has been proposed that tissues that are perfused with arterial blood have equilibrium potentials in the range of 0.782 V (versus normal hydrogen electrode, NHE) (Ref 35). However, the actual surface potential will be governed by the mixed potential theory (Ref 36), where the sum of all anodic and cathodic currents acting on the surface balance out. Therefore, the actual potential will depend on factors such as the exchange current density for the governing cathodic (reduction) reaction, as well as the ability of the anodic processes to take place. Other factors that may affect the surface potential of these alloys include the concentration of oxidizing species (such as H_2O_2 and others) at the surface and the mechanical abrasion of the surface film. Based on the in vitro and in vivo immersion testing results, it has been asserted (Ref 37) that potentials in vivo may be approximately 0.5 V (NHE). On the other hand, abrasion of film will cause potential drops as large as -1 V (versus silver-silver chloride electrode, Ag/AgCl) for titanium oxide (Ref 38).

It seems likely that metal implant potentials in vivo are not likely to rise to above 1 V (SCE), which corresponds to the oxygen evolution potential at a normal pH value. Furthermore, as is discussed later, above 0.5 V (SCE), chromium oxide will undergo the transpassive dissolution reaction (Cr^{3+} becomes Cr^{6+}), and it will dissolve rapidly into solution. So, the reasonable range of potentials attainable in vivo is between -1 to 0.5 V (SCE). If there is an inflammatory response, those cells can produce hydrogen peroxide, and the potential may be increased to a higher value (Ref 39).

Another aspect for body solution chemistry is the content of dissolved gases. There are some differences between body fluid and lab conditions, particularly in the oxygen and carbon dioxide pressure in terms of low oxygen and higher carbon dioxide. These differences may have effects on the corrosion behavior of biomaterials in vivo (Ref 40).

Table 4 Summary of oxide films on the surface of various metallic biomaterials by x-ray photoelectron spectroscopy

Material and surface preparation	Thickness, nm	Ratio of main alloying elements	Other comments	Source
316LVM stainless steel (F138)	MP: 2.0	Cr/Fe: MP: 0.198	MP: higher O	Unpublished results by the authors
	EP: 1.6	Cr/Fe: EP: 1.087	PA: higher C	
	PA: 3.2	Cr/Fe: PA: 0.739	Immersion increasing Cr/Fe ratio	
	MP + PBS: 2.2	Cr/Fe: MP + PBS: 1.025		
	EP + PBS: 1.9	Cr/Fe: EP + PBS: 1.356		
	PA + PBS: 3.0	Cr/Fe: PA + PBS: 3.207		
316L stainless steel (F138)	Polished: 3.6	Polished: 0.182	Surface hydrated and immersion depleted Ni and Mn	Ref 23
	Autoclaved: 4.1	Autoclaved: 0.095		
	Hanks: 4.7	Hanks: 0.112		
	MEM + FBS: 5.0	MEM + FBS: 0.128		
	L929: 6.3	L929: 0.120		
Orthopedic stainless steel (MP in Hanks by fixed potential, 300 s)	0.3V: 4.9	0.3V: 0.65	Cr enriched in film but decreased with increasing of potential	Ref 24
	0.8V: 6.5	0.8V: 0.5		
F75, polished	Air dry: 3.7	Air dry: 0.357	No Cr^{6+}	Ref 25
	Wet steam: 3.0	Wet steam: 2		
F75, in Hanks by fixed potential, 300 s	$-0.3V$: 2.2	Cr/Co: 3.3	At lower potential, dominant of Cr_2O_3. At higher potential, Cr^{6+} was detected.	Ref 26
	0.3V: 3.1	Cr/Co: 1.86		
	0.7V: 5.0	Cr/Co: 0.96		
F136 (1 h)	Air dry: 1.5	Air dry: Ti/Al: 7	No V in surface; dominant species is TiO_2	Ref 27
	Wet steam: 7.2	Wet steam: Ti/Al: 4		
F136, in Hanks by fixed potential, 300 s	0V: 3	0V: Ti/Al: 2.44	Dominant oxide is TiO_2 with some TiO and Ti_2O. Al contribution decreased with increasing of potential. No V	Ref 28
	0.5V: 6	0.5V: Ti/Al: 3.14		
	1V: 8.2	1 V: Ti/Al: 3.67		
	1.8V: 9	1.8V: Ti/Al: 5.1		
F2063	Not available	Chem. etch: Ti/Ni: 2.5	Dominant oxide is TiO_2 with trace of Ni, both oxidized (Ni^{3+}) and metallic states	Ref 29
		Air, 2 mo: Ti/Ni: 2.6		
		Steam, 1 h: Ti/Ni: 6.0		
		Water boiling: 30 min Ti/Ni: 33.1		
		H_2O_2, 22 h: Ti/Ni: 0.3		

Note: MP, mechanical polish; EP, electrochemical polish; PA, passivation by ASTM F86, + phosphate-buffered saline (PBS), after 19 weeks immersion in PBS; MEM + FBS, immersed in Eagle's minimum essential medium (MEM) containing 10% fetal bovine serum (FBS) in the incubator. Ref 24, 26, and 28 data were regenerated from graphs.

Interfacial Interactions Between Blood and Biomaterials

Generally, upon immersion of a metal into solution, an interfacial structure is established that involves ions, water (solvent), and electrons (Ref 41). This interfacial structure, known as the electrical double layer, is complicated and far from fully understood, particularly in the case of biomaterial surfaces where semiconductor-like oxide films make contact with protein-containing ionic solutions. Thus, the interfacial structure is highly complex, and the processes that occur at the surface are similarly complex. A schematic representation of an oxide-film-covered surface is shown in Fig. 3. Many aspects of this surface remain poorly understood. It is known that the oxide film is under a high electric field, on the order of 10^7 V/cm. Also, oxide films can transport both ions and charge (electrons and holes), are defected nanocrystalline materials with some habit relationship to the substrate, and are variable in chemistry, residual stress, and other factors that impact their behavior.

Within seconds or minutes of insertion into a body, the biomaterials will be subjected to the adsorption of blood proteins. Albumin is the first protein adsorbed on the surface, but it will be replaced by other proteins regulated by the Vroman effect (Ref 42). Fibrinogen may be adsorbed on the surface in its highest concentration, and it may be partially replaced by other high-molecular-weight proteins, such as von Willebrand factor. Protein adsorption plays a central role in the physiologic consequences of biomaterial implantation, including coagulation, thrombogenesis, and acute and chronic inflammatory responses. In addition, various cells may attach to the surface of biomaterials in either orthopedic or cardiovascular applications (Ref 43–45).

Coagulation and Thrombogenesis

Artificial surfaces may induce thrombosis (or blood clot formation), and the controlling mechanism for this has not been completely determined. Protein adsorption, in particular, fibrinogen adsorption, is thought to play a central role in thrombosis (Ref 46). There are two pathways thought to initiate fibrinogen-fibrin conversion (intrinsic and extrinsic coagulation) (Ref 47). Fibrin will form a network and trap blood cells, thus forming a thrombus. Numerous papers have been published on protein adsorption on artificial surfaces; however, most of them focus on polymer surfaces (Ref 48, 49) and relatively fewer on metal or oxide films (Ref 50). The mechanism and controlling factors of protein adsorption are still unclear (Ref 48). Researchers believed that thrombogenesis is electrochemical in nature (Ref 51–53) and that electrochemical potential (charge) governs the thrombogenic properties of metallic biomaterials (Ref 54, 55). That is, the material that has an open circuit potential more negative than 0 V (NHE) was seen to be nonthrombogenic, while more positive potentials resulted in thrombogenic effects. Studies (Ref 56) show that potential can cause fibrin conversion from fibrinogen on the surface of platinum without enzymes. Reference 57 asserted that the semiconductivity of oxide films influences the thrombogeneity of artificial materials in terms of band energy and surface states.

Inflammatory Response to Biomaterials

Recent research (Ref 58, 59) reviewed the biological responses to materials and found that there is a sequence of host reactions following implantation of medical devices, including the metallic biomedical devices. Usually, the implantation of biomedical devices will cause injury and lead to the subsequent cellular cascades, including inflammation and foreign body reactions, that usually occur within two to three weeks after implantation.

Macrophages play a central role in the acute and chronic inflammatory processes, because phagocytosing (ingesting) macrophages will undergo respiratory burst reactions that release H_2O_2 and other superoxides such as NO. These species may have large effects on the corrosion of biomaterials (e.g., shifting potentials to a more anodic range) (Ref 60).

General Discussion of Corrosion Behavior of Three Groups of Metallic Biomaterials

There are several special aspects of corrosion of biomaterials that are different from corrosion in industrial practice. First, the primary concern for biomedical devices is the host reaction to the corrosion products (biocompatibility) as well as the damage to the device itself. Second, as discussed before, the metal surface is always covered by a layer of proteins or cells that modulate biomaterials-host reactions, including corrosion. Third, in some applications, for example, vascular applications, the interaction between the metal surface and proteins or cells is very important in terms of biocompatibility. Fourth, although the in vivo environment is dynamic in localized sites, it is a relatively isolated stagnant aqueous solution with limited volume. The hydrolysis constant and solubility of corrosion product in solution may play a vital role for their corrosion behavior. Furthermore, various elements (e.g., nickel, chromium, etc.) are not well tolerated by tissue, with adverse reactions, including allergic and inflammatory reaction, possible.

Electrochemical Reactions. Metallic corrosion is an electrochemical reaction in nature; that is, metal oxidizes in an anodic reaction to a higher valence state and forms ions or hydroxide (oxide) while also giving up electrons. These generated electrons are consumed by other cathodic (reduction) reactions at different sites. Reduction reactions include reduction of dissolved oxygen, hydrogen, or other reactions. Typical oxidation reactions for a metal are listed as follows (Ref 61).

Anodic dissolution resulting in a soluble cation:

$$M \rightarrow M^{n+} + ne^- \quad \text{(Eq 1)}$$

This reaction is suitable to almost all alloy components of metallic biomaterials in vivo conditions, particularly for iron, cobalt, nickel, vanadium, and molybdenum.

Passivation and formation of insoluble oxide or hydroxide:

$$M + nH_2O \rightarrow M(OH)_n + nH^+ + ne^- \quad \text{(Eq 2)}$$

This passivation reaction occurs for elements such as chromium and titanium in biomaterials.

Forming an insoluble oxide product:

$$mM + nH_2O \rightarrow M_m^{2n/m}O_n^{2-} + 2nH^+ + 2ne \quad \text{(Eq 3)}$$

Furthermore, there may be other anodic (oxidation) reactions for biomaterials (e.g., involving chromium) (Ref 62). For example, see Eq 4 and 5.

Passivation by forming an insoluble product from a dissolving surface:

$$2Cr^{2+} + 3H_2O \rightarrow Cr_2O_3 + 6H^+ + 2e^- \quad \text{(Eq 4)}$$

Transpassive dissolution of a passive surface:

$$Cr_2O_3 + 5H_2O \rightarrow 2CrO_4^{2-} + 10H^+ + 6e^- \quad \text{(Eq 5)}$$

Typical cathodic reduction reactions include:

$$O_2 + 2H_2O + 4e^- \rightarrow 4OH^- \quad \text{(Eq 6)}$$

$$O_2 + 4H^+ + 4e^- \rightarrow 2H_2O \quad \text{(Eq 7)}$$

$$2H_2O + 2e^- \rightarrow H_2 + 2OH^- \quad \text{(Eq 8)}$$

$$2H^+ + 2e^- \rightarrow H_2 \quad \text{(Eq 9)}$$

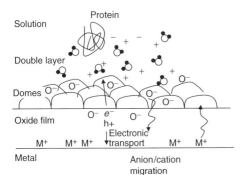

Fig. 3 Schematic diagram of the interfacial structure of a biomaterial surface contacting with the biological environment

In some local areas (e.g., crevice site), the dissolving metal cation may induce hydrolysis:

$$M_{(aq)}^{n+} + H_2O \rightarrow MOH^{(n-1)+} + H_{(aq)}^{+} \quad \text{(Eq 10)}$$

Metal oxidation may produce metal ions (Eq 1), and these ions can migrate away from the metal surface as free ions in solution or form metal oxide (hydroxide) or other chemical compounds with other species. From a thermodynamic point of view, the oxidation of metallic biomaterials is a spontaneous (exothermic) process, because metal ions or oxides are more thermodynamically stable than the metal. The corrosion of metallic biomaterials in body fluids is controlled by thermodynamic and kinetic factors: the former determines the corrosion tendencies, while the latter determines the corrosion rate.

Pourbaix proposed potential-pH diagrams (Pourbaix diagram) to describe the thermodynamics of the electrochemistry of metals or alloys in aqueous solutions (Ref 63, 64). In Pourbaix diagrams, the electrochemical behavior of metals can be divided into three domains: immunity, passivation, and corrosion. In the immune domain, the metal keeps its metallic surface because the corrosion reaction is not thermodynamically possible. In other words, corrosion will only happen at higher potential conditions, for example, Pt/Pt^{1+} at 1.2 V (SCE). Noble metals such as gold, platinum, and silver belong to this group. In the passive domain, metals or alloys react to form oxides (or hydroxide). In some cases, these oxides will form protective surface oxides that prevent contact between the substrate alloy and the solution. Metals such as titanium, tantalum, chromium, niobium, and zirconium and their alloys belong to this group (many of which are being investigated as potential biomaterials). In fact, as mentioned before, the three main metallic biomaterials being discussed here belong to this group and mainly depend on titanium and chromium to form the passive film to prevent corrosion. It should be noted that the metals in this group are very active (i.e., they have high thermodynamic driving forces for continued oxidation) and tend to corrode rapidly in body fluids when the passive film is not present or disrupted, as in the case of severe abrasion conditions. In Pourbaix's third domain, the corrosion domain, metals will corrode and dissolve into solution and will not form protective oxide films, but rather ionic corrosion products.

For stainless steel, Co-Cr-Mo, and titanium and its alloys, the kinetic factor that controls the corrosion rate for biomaterials in aqueous solution is the passive film. Passive films play dual roles in limiting both the anodic and cathodic reactions. That is, the film serves as a physical barrier for both cation transport and anion transport to the metal surface as well as an electronic barrier for electrons.

Electrochemical Methods for Studying Metallic Biomaterials Corrosion. Various approaches used to investigate corrosion include weight loss, visual investigation, and electrochemical methods. Electrochemical methods are extremely sensitive techniques to measure the corrosion of biomaterials, although some nonelectrochemical methods such as chemical analysis of corrosion products also are used to study the corrosion of biomaterials.

The most common way to study corrosion of metallic biomaterials is by anodic polarization (Ref 65) and electrochemical impedance measurements (Ref 66). The three-electrode system is used in electrochemical corrosion testing, that is, working electrode (biomaterial specimen), counterelectrode (inert electrode, usually platinum or carbon), and reference electrode (SCE and Ag/AgCl are the typically used reference electrodes). A potentiostat or galvanostat is used to control voltage or current and to record current or voltage, respectively. These procedures are now normally controlled by computers.

Potentiodynamic and potentiostatic methods are used to study the localized corrosion behavior of biomaterials in physiologically representative solutions, that is, controlling the electrode potentials as well as measuring the generated current response. There are several ASTM International standard test methods for localized corrosion testing, and some are particular for implant materials.

Five ASTM standards are concerned with localized corrosion and mechanically assisted corrosion of surgical implant materials. Two methods for localized corrosion properties include:

- ASTM F 746 (Ref 67), "Test Method for Pitting or Crevice Corrosion of Metallic Surgical Implant Materials"
- ASTM F 2129 (Ref 68), "Test Method for Conducting Cyclic Potentiodynamic Polarization Measurements to Determine the Corrosion Susceptibility of Small Implant Devices"

Three methods for mechanically assisted corrosion properties include:

- ASTM F 897 (Ref 69), "Test Method for Measuring Fretting Corrosion of Osteosynthesis Plates and Screws"
- ASTM F 1801 (Ref 70), "Standard Practice for Corrosion Fatigue Testing of Metallic Implant Materials"
- ASTM F 1875 (Ref 71), "Standard Practice for Fretting Corrosion Testing of Modular Implant Interfaces: Hip, Femoral Head-Bore, and Cone Taper Interface"

In addition, ASTM G 61 (Ref 72) (cyclic polarization testing) is more generally designed for screening variations in alloy composition and environments, particularly for iron- or nickel-base alloys in chloride solutions. It is used to determine the relative susceptibility of materials to localized corrosion and has been widely used in biomaterials research. In fact, ASTM F 2129 is a modification of G 61, and it uses a similar protocol to study the localized corrosion property (pitting) for biomaterials but emphasizes testing of small medical devices in simulated solutions. In these two methods, a cyclic polarization scan is carried out from active (negative) potential (usually below the corrosion potential) to a noble (positive) potential and, at some point, scanning back to the starting point, all at a fixed scanning rate. Typical potentiodynamic corrosion curves for 316L stainless steel, Co-Cr-Mo, MP35N, CP-Ti, Ti-6Al-4V, and NiTi in simulated solution are shown in Fig. 4. Typically, at the most cathodic, potentials, the current is net cathodic, indicating that the sum of all anodic and cathodic processes has more cathodic reactions than anodic ones. Then, as the potential is

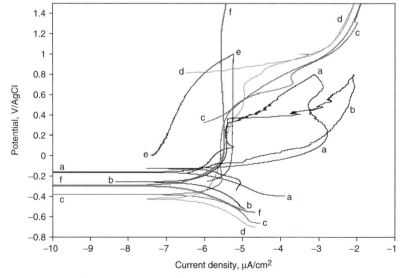

Fig. 4 Typical cyclic polarization curves of stainless steel, cobalt-base alloys, and titanium alloys in phosphate-buffered saline (room temperature; scan rate, 1 m V/s; preimmersion, 10 min). (a) 316L stainless steel. (b) NiTi. (c) Co-Cr-Mo. (d) MP35N. (e) Ti-6Al-4V. (f) CP-Ti

increased, the current goes to zero, which is referred to as the corrosion potential (open circuit potential, or OCP, or zero current potential). This potential is where the sum of all anodic and cathodic reactions sum to zero. It should be noted that this potential is sensitive to prior electrochemical history. For instance, if the sample is held at a cathodic potential (e.g., -1 V versus Ag/AgCl) for a period of time, then the OCP will be shifted to more negative potentials than would be measured at rest. This result is from the fact that this cathodic conditioning alters the surface and modifies the local solution chemistry near the surface, causing a negative voltage shift in OCP.

Three other potentials of interest may be found with cyclic polarization curves. First is the primary passivating potential (E_{pp}, also named Flade potential), above which oxide film can form. This potential is often hidden in polarization curves by overwhelming cathodic reactions or is outside the range of potentials tested. Second is the potential at which the current increases significantly with increasing potential, called the transpassive potential (and also sometimes erroneously referred to as the pitting potential, E_p, or the breakdown potential for materials that exhibit pitting). The more noble this transpassive potential, the less susceptible the material is to loss of passivity or localized film breakdown (e.g., pitting attack). The third potential is called the protection potential (E_{prot}) or repassivation potential at which the reversed scanning part of the cyclic polarization curve intersects with the anodic portion of the forward scan, and a hysteresis loop is completed. It is generally accepted that if pitting occurs, the pits will propagate at potentials that are more positive than the protection potential, but repassivation will occur at potentials more negative than this point. Materials that do not exhibit this hysteresis curve are not thought to be susceptible to pitting, but they may experience a generalized breakdown of the oxide film. The Co-Cr-Mo alloys show this behavior. In Fig. 4, 316L stainless steel shows evidence of a hysteresis loop and has the lowest pitting potential, approximately 300 mV (Ag/AgCl), while Co-Cr-Mo (as well as MP35N) exhibits film breakdown at approximately 400 mV (Ag/AgCl) and no hysteresis loop. In other words, the protection potential for Co-Cr-Mo is coincident with the pitting potential, and no pitting takes place. In fact, the film breakdown for Co-Cr-Mo is a general dissolution of chromium oxide as the chromic ion (Cr^{3+}) becomes the hexavalent chromium ion (Cr^{6+}). When this occurs, the passive nature of the film is lost. There is no pitting susceptibility for this alloy in the unloaded condition and noncrevice environments. However, this alloy has demonstrated pitting and crevice corrosion under in vivo conditions, as is discussed later. The Ti-6Al-4V alloy also shows no pitting in this test in the potential range tested and shows negative hysteresis, indicating that the oxide film is enhanced by the increasing potential. However, titanium alloy also has shown pitting and crevice corrosion in vivo (Ref 73).

It is worthwhile to note that the transpassive potential obtained from cyclic polarization testing is strongly dependent on the potential scan rate at which the testing is performed. Another shortcoming is that the reversing potential point is difficult to choose and may cause too much pitting propagation and modification of the localized chemistry and pit dimensions. On the other hand, it is found that pits can initiate and propagate under potentiostatic or potentiodynamic conditions far below the protective potential but with the slowest scan rate (Ref 74, 75).

There are other techniques that have been applied to the study of corrosion of implant alloys. These include tests to determine oxide film stability to potentiostatic challenge, that is, application of $+800$ mV (SCE) for a period of time that depends on the reaction (ASTM F 746), high-speed current monitoring at fixed potentials (potentiostatic testing) (Ref 76, 77), and galvanostatic test methods where fixed current densities are applied and the potential is monitored (Ref 78, 79). The work in Ref 76 and 77 is notable in that it showed evidence of current transients in titanium and Ti-6Al-4V that were ascribed to depassivation and metastable pitting.

In Vitro Solutions Used for Electrochemical Testing of Biomaterials. The in vivo environment can be approximated to first order as a saline solution (0.9% sodium chloride, NaCl); however, more complex aqueous solutions are also used. In particular, there are various proteins, cells, and other organic components besides inorganic salts that can be used. In addition, some chemicals will be produced in a special physiological process near the implantation sites, for example, hydrogen peroxide, nitric oxide, or other superoxides produced during inflammatory response that may be important (Ref 58).

Investigators usually use simulated physiological solutions to study the corrosion behavior of metallic biomaterials. These simulated solutions consist of inorganic constituents of body fluids, mainly sodium chloride with phosphate or bicarbonate to adjust the pH. Table 5 presents the compositions of artificial physiological solutions that are widely cited in the literature. These solutions are relatively simple yet representative of the physiological environment. The disadvantage of these solutions is that they do not account for the presence of organic components. More recently, studies of corrosion and surface properties on biomaterials have been performed in more complicated solutions, where additions of proteins complement the saline solution (Ref 80, 81), serum saline solution (Ref 82, 83), saline with various concentrations of hydrogen peroxide solutions (Ref 84, 85), addition of cells to saline solution, culture medium and cell cultures (Ref 86, 87), and other solutions.

Direct Observational Corrosion Test Methods (Retrieval Analyses). There are several nonelectrochemical methods that can be used to obtain information about corrosion behavior of biomaterial surfaces and the presence and extent of corrosion debris in tissues adjacent to implanted metals. The examples are:

- *Microscopic examination of corroded surfaces:* scanning electron microscopy (SEM), atomic force microscopy (AFM), and transmission electron microscopy (TEM)
- *Analysis of surface chemistry:* Auger electron spectroscopy (AES), x-ray photoelectron spectroscopy (XPS), and secondary ion mass spectroscopy (SIMS)
- *Analysis of solution chemistry:* atomic absorption spectroscopy (AAS) (Ref 88, 89)

Electrochemical testing typically cannot be performed in the human body, but some in vivo testing can be done in animals. Only very limited in vivo corrosion testing has been carried out in humans, including some reports on oral conditions (Ref 34). Retrieval and analysis of implanted medical devices and associated tissues is another important resource to understand in vivo corrosion properties on biomaterials (Ref 73, 90–93). In fact, ASTM F 561 (Ref 94) covers some recommendations for this topic.

The most detailed information on in vivo corrosion of biomaterials comes from retrieval studies (Ref 73, 91, 95, 96). Researchers (Ref 73) examined 148 retrieved total modular hip implants with mixed- (Ti-6Al-4V/Co-Cr) and similar- (Co-Cr/Co-Cr) alloy femoral stem and head combinations. They observed significant corrosion in both mixed- and similar-metal combinations; 16% of necks and 35% of heads in mixed-metal cases and 14% of necks and 23% of

Table 5 Compositions of biosimulated solutions

Chemicals, g/L	Tyrodes	Ringers	Hanks	Phosphate-buffered saline
NaCl	8.0	9.0	8.0	7.014
$CaCl_2$	0.20	0.24	0.14	...
KCl	0.2	0.42	0.4	0.149
$MgCl_2 \cdot 6H_2O$	0.10	...	0.10	...
$MgSO_4 \cdot 7H_2O$	...	...	0.06	...
$NaHCO_3$	1.00	0.20	0.35	...
$Na_2H_2PO_4$	0.05	...	0.10	Phosphate: 0.01 M
$Na_2HPO_4 \cdot 2H_2O$	...	...	0.06	...
Glucose	1.00	...	1.00	...

Note: The pHs of these solutions are all equal to 7.4. Source: Ref 40

heads for similar-alloy cases showed signs of moderate-to-severe corrosive attack, and the corrosion process is progressive in time. This corrosion process has been described as mechanically assisted crevice corrosion attack (Ref 32). In a retrieval study (Ref 91), it was found that 89% of the plates and 88% of the screws of 250 stainless steel internal fixation devices showed evidence of pitting and crevice corrosion. Other researchers (Ref 97) examined type 316 stainless steel multicomponent devices removed from the human body and found that 91% had undergone visible corrosion, with crevice corrosion being the dominant form. More recently (Ref 96), 34 retrieval endovascular stents were analyzed, and 100% of the NiTi stents showed pitting; parts of the devices demonstrated other corrosion forms, such as surface craters. These studies revealed aspects of in vivo corrosion that had never been observed in in vitro testing. For example, pitting, etching, selective dissolution, and intergranular attack have been documented in Co-Cr-Mo alloys and stainless steels, and pitting and etching of titanium alloys has been documented (Ref 73).

Galvanic corrosion concerns the accelerated dissolution of one component of dissimilar-metal couples due to different nobility. The noble material will become the cathode, while the less noble one becomes the anode. The ratio of cathode to anode is very important for galvanic corrosion, where large cathode areas will produce a larger current density at the anode and enhance corrosion. However, it should be noted that galvanic corrosion happens only under specific potential conditions, for example, the pitting potentials or active dissolution potential ranges for the anodic electrode. For the three groups of biomaterials, all have wide passive potential ranges, and it is not likely that galvanic corrosion between these different couples presents a significant risk without other factors being present (e.g., mechanical disruption of oxides).

Mechanism of Localized Corrosion. Pitting, crevice corrosion, and mechanically assisted corrosion are forms of corrosion that can damage a biomedical device. On the other hand, ion release and other corrosion products formed by pitting and crevice corrosion may cause adverse host reactions.

Pitting results from the accelerated dissolution of metal in a region where the passive film has lost its protection due to breakdown or preferential dissolution of defects or inclusions. Pitting is an extreme and severe form of damage that can cause penetration into materials and/or produce a significant amount of metal ions. Crevice corrosion is, in fact, a variant of pitting but takes place in occluded areas or restricted geometries where the solution has more aggressive characteristics, such as low pH, higher chloride concentration, and low oxygen partial pressure (Ref 75).

Stainless steel is vulnerable to pitting and crevice corrosion, while titanium and its alloys and Co-Cr-Mo are less susceptible (although not entirely immune) to these forms of corrosion. All three of these alloy types have experienced pitting and crevice corrosion in vivo (Ref 73, 91). Shape memory NiTi alloys show conflicting results for pitting corrosion resistance, where the alloy may or may not exhibit pitting. This variability mostly depends on the level of nickel in the oxide film (Ref 98). The Co-Cr-Mo alloy is resistant to pitting and crevice corrosion in the unloaded condition but is susceptible to them in the presence of mechanical stress, crevice conditions, and fretting (Ref 73).

Pitting and crevice corrosion result in the accelerated dissolution of metal in some sites due to breakdown of the passive film, while elsewhere, the film keeps its protective ability. According to research (Ref 74, 75), pitting and crevice corrosion result from the same mechanism, where one is initiated on open surfaces, while another occurs at an occluded site. Pitting usually possesses some different stages, including passive film breakdown, the nucleation of embryonic pitting, the growth of metastable pits, and the growth of stable pits (Ref 99, 100). The mechanism of pit nucleation is unclear, and some models propose the penetration of chloride ion through the film and dissolution of the metal (Ref 101), mechanical breakdown of the passive film (Ref 101–103), and local thinning of the oxide film due to chloride ion (Ref 101, 103). Recently, it was assumed (Ref 104) that breakdown of the passive film on transition metals is due to an electrical breakdown, by the Zener mechanism, where a high local current is created, generating heat, expanding the metal, and destroying the passive film. Several factors can influence the pitting behavior, including alloy composition, environment, potential, and temperature (Ref 75). In particular, the composition of the alloy and impurities in the passive film can play a crucial role in this respect (Ref 5). In addition to all these factors, another important aspect of corrosion behavior is mechanically assisted corrosion, for example, crevice corrosion of cobalt- and titanium-base alloys under fretting conditions. This is discussed in the article "Mechanically Assisted Corrosion of Metallic Biomaterials" in this Volume.

Corrosion Behavior of Stainless Steel, Cobalt-Base Alloy, and Titanium Alloys

This section describes the corrosion behavior of these three alloy systems. Because the corrosion behavior is governed by the oxide film, some comments on the dynamic aspects of oxide films are presented, then the corrosion behavior of 316L stainless steel, Co-Cr-Mo, and titanium alloys is presented.

Dynamics of Oxide Films on Metallic Biomaterials. Passive film formation on biomaterials is a dynamic process (Ref 105), and passive films are subjected to changes either in morphology or in chemistry with immersion, with or without applied potentials in biological solution. In situ AFM results showed that passive films are subjected to reduction under cathodic potentials, for example, -1 V (Ag/AgCl) for 316L stainless steel in phosphate-buffered saline (PBS) (Ref 22) and -0.7 V (Ag/AgCl) for Co-Cr-Mo (Ref 20). Oxides on CP-Ti were not susceptible even at -1 V (Ag/AgCl) (Ref 19). At potentials that are close to the OCPs, stainless steel and Co-Cr-Mo developed domelike oxides. Under anodic potentials, but less than transpassive potentials, the passive films on 316L stainless steel, NiTi (Ref 22), CP-Ti and Ti-6Al-4V (Ref 106), and Co-Cr-Mo all showed oxide growth by lateral expansion of oxide domes. At transpassive potentials, the passive film of Co-Cr-Mo underwent fast dissolution, while CP-Ti showed film growth at higher potentials. At OCP, upon immersion in PBS, the oxide film became highly hydrated in terms of swelling of the oxide domes (Ref 107). Figure 5 shows the typical topographic changes by AFM technique of the surface of CP-Ti with immersion and electrochemical polarization.

The chemical composition of surface films on biomaterials is influenced by immersion and applied potentials. Researchers studied the surface composition of passive films formed on stainless steel (Ref 24), Co-Cr-Mo alloy (Ref 26), and Ti-6Al-4V alloy (Ref 28) in simulated physiological solution under various potentials using quasi in situ XPS. In their research, the samples were transferred, without contacting air, to the ultrahigh vacuum chamber for XPS analysis after oxidation in physiological solutions. They found that the predominant species for passive films on stainless steel are chromium and iron oxides, and oxides of nickel and molybdenum are also detectable in the passive film formed in this potential range. They attributed the enrichment of oxides in chromium and molybdenum in the passive film, strong enrichment of molybdenum, and depletion of iron at the alloy surface just beneath the passive film to the excellent corrosion properties for this stainless steel in physiological solutions. They found the passive potential range for Co-Cr-Mo alloy in simulated physiological solution extends from -0.3 to 0.6 V (SCE). When the potential is below 0.3 V (SCE), the main constituents of the passive film are oxides and hydroxides of chromium, while cobalt and molybdenum oxides entered the passive layer. The thickness of the passive film also increased when the anodic potential is higher than 0.3 V (SCE). For Ti-6Al-4V, the passive film consists of TiO_2 with a small amount of TiO and Ti_2O_3 in the passive potential range. Aluminum oxide (Al_2O_3) was incorporated into the TiO_2 layer and was mainly located at the oxide-solution interface. No vanadium oxide was detectable by XPS. In addition, results showed that complexing agents such as ethylenediamine tetraacetic acid and citrate have a big influence on the passivation for stainless steel and Co-Cr-Mo, and they enhanced the dissolution of nickel, iron, and cobalt.

Researchers (Ref 23, 108) characterized the surface oxides of stainless steel and Co-Cr-Mo in cell culture medium with and without cultured cells. They found the surface film on 316L stainless steel was enriched with iron and chromium oxides but depleted in nickel and manganese after immersion in these quasi-biological solutions. The Co-Cr-Mo alloy released cobalt, and the surface film consisted of chromium and molybdenum oxides. The passive film of titanium and Ti-6Al-4V consisted mainly of TiO_2, and this film is stable in physiological conditions. However, the composition of the TiO_2 film may change by incorporation of ions and biomolecules from environments after immersion in physiological solutions (Ref 109).

As shown earlier in Fig. 4, 316L stainless steel is susceptible to localized corrosion attack, such as pitting or crevice corrosion, in physiological solutions, while the Co-Cr-Mo alloy undergoes more uniform film breakdown due to transpassive dissolution of chromium oxide. Titanium and its alloys are usually ranked as highly resistant to corrosion in saline solutions; however, NiTi alloys do show breakdown of the passive film. There have been a large number of studies on the corrosion behavior of 316L stainless steel, Co-Cr, NiTi, and titanium alloys (Ref 37, 91, 110–115). Here, some typical results are reviewed and some new developments discussed. Emphasis is placed on topics such as effects of protein on corrosion behavior, influences of hydrogen peroxide on corrosion, and film behavior for these biomaterials. Further, some new insights are introduced on electrochemical and electronic characteristics on film behaviors using some novel approaches, such as electrochemical atomic force microscopy and step polarization impedance spectroscopy.

Corrosion of 316L Stainless Steel. Pioneer research (Ref 37) shows that 18Cr-10Ni-3Mo stainless steel is subjected to pitting and crevice corrosion in Hank's solution at 37 °C (98.6 °F) and that the pitting potential ranged from 0.4 to 0.5 V (NHE). Further, the OCP for this alloy was measured in Hank's solution in vitro and in a goat in vivo. The values are close to 0.5 V (NHE) after longer times. The researchers concluded that this type of alloy does not resist pitting corrosion in body fluids for long periods.

Researchers (Ref 116) studied the polarization behavior of 316L stainless steel in Hank's solution and found that 316L is very susceptible to pitting and crevice corrosion in deaerated Hank's solution at 37 °C (98.6 °F) and that the film breakdown potential is close to 0.2 V (SCE).

The electrochemical behavior of 316L vacuum melted (VM) stainless steel and another high-nitrogen stainless steel (F1586 in Table 1) was tested in PBS and in a simulated crevice solution, using cyclic potentiodynamic and potentiostatic polarization measurements (Ref 115). Results showed that 316LVM is vulnerable to pitting, particularly during crevice corrosion, in PBS but that the high-nitrogen stainless steel (F1586) has no risk of pitting corrosion [E_p > 1.10 V (SCE), and E_{prot} is 1.04 V even in simulated crevice solution with higher chloride (2.5 M) and low pH (0.85) (E_P, 0.9 V; E_{prot}, 0.8 V versus SCE]. It is interesting to note that although this alloy showed a small current peak at 0.65 V (SCE), which may correspond to the transpassive dissolution of Cr_2O_3, the alloy repassivated at higher potentials. In other words, it is possible that high-nitrogen, high-chromium stainless steel has higher corrosion resistance than Co-Cr-Mo over a higher potential range.

Usually, the corrosion of stainless steel in chloride solution is related to its chromium (Ref 117) or manganese sulfide inclusion content (Ref 5). Changing of alloying elements can improve its corrosion behavior. Recent reports verified this point: high-nitrogen and/or high-alloy (chromium) stainless steel has superior localized corrosion resistance (Ref 118–120).

Corrosion of Co-Cr-Mo Alloy. The Co-Cr-Mo alloy also depends on the formation of a chromium oxide film to provide corrosion resistance, as occurs for the 316L stainless steel. However, it is well accepted that Co-Cr-Mo alloy has good corrosion resistance and is relatively immune to pitting and crevice corrosion in physiological solution (Ref 37, 121, 122) in the absence of mechanical abrasion. The reason for its good corrosion resistance may be related to the high chromium and molybdenum contents. In fact, at higher potentials, this alloy is subjected to general dissolution due to transpassive dissolution of chromium oxide. Another feature for this alloy (F75) is its relative lack of susceptibility to intergranular corrosion. It should be noted that this alloy is susceptible to pitting attack, crevice corrosion, and intergranular corrosion in some implants, particularly those where crevices are formed and mechanical abrasion (e.g., fretting) can occur. Under these conditions, the increased corrosion rate resulting from oxide film abrasion can accelerate crevice corrosion processes, raise

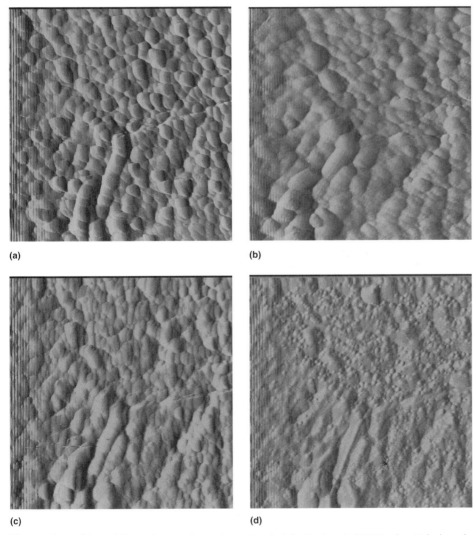

Fig. 5 Sequential atomic force microscopy images (contact mode, deflection image) of CP-Ti surface (etched sample, same spot). Scanning size, 5 μm; height scale, 20 nm. (a) In air. (b) Immersion in phosphate-buffered saline, 1day. (c) Two week immersion. (d) Four week immersion and after step polarization impedance spectroscopy testing; voltage, −1 to +1 V(Ag/AgCl)

metal ion levels locally, and lower pH such that pitting and intergranular attack may occur (Ref 73, 123, 124).

For stainless steel and Co-Cr-Mo alloy, chromium and molybdenum are essential alloying elements for imparting resistance to localized corrosion. The role of these two elements on corrosion resistance has been studied intensively. Researchers (Ref 125) used iron-chromium alloys with different chromium content and Fe-Cr-Mo alloys with similar chemical composition with 316L stainless steel in 1 M NaCl solution and measured the nucleation events of pitting corrosion, using microelectrochemical methods. They found that chromium and molybdenum can decrease the magnitude of current transients, which leads to an earlier start of repassivation but does not change the repassivation kinetics. Studies (Ref 126, 127) found that molybdenum as an alloying element reduces the incidence of nucleation (breakdown of film) of metastable pits and further inhibits the growth of stable pitting.

Corrosion of Titanium. Titanium and its alloys are very resistant to corrosion in physiological solutions. The oxide film on titanium does not break down over a wide range of potentials. This alloy system has been heavily studied for its electrochemical behavior (Ref 81, 110, 112, 128–132) and reviewed (Ref 133). Unalloyed titanium and its alloys, including Ti-6Al-4V, Ti-Zr-Nb, and Ti-Al-Fe, have outstanding corrosion resistance in the absence of mechanical factors.

One exception is the nickel-titanium system of alloys based on the binary, equiatomic intermetallic compound of NiTi. The corrosion properties of thses alloys are of greater concern than CP-Ti or Ti-6Al-4V. There is a wide range of corrosion behavior reported for NiTi alloys due to the variability in the oxide film that forms (Ref 95, 98, 114, 134–136). When nickel can be removed from the oxide film, the film is primarily a TiO_2 film, similar to other titanium alloys. However, when nickel is present in the film (either in a metallic or oxide form), the corrosion resistance of the material is significantly reduced, and it has greater susceptibility to pitting corrosion, similar to that of 316L stainless steel.

Researchers (Ref 76, 77, 137) have indicated that titanium alloys are resistant to corrosion at the macroscopic level but not at the microscopic level. They used a microelectrode technique to study short-time duration current transients under potentiostatic conditions that they ascribed to the metastable pitting behavior for titanium and Ti-6Al-4V. They found very short-duration anodic current transients in both CP-Ti and Ti-6Al-4V and assert that this is evidence that the alloys are susceptible to metastable pitting under 0.5 V (Ag/AgCl) in Ringers' solution. However, AFM evidence (Ref 19, 138) appears to show that oxide domes can form and grow at the boundaries of extant domes and that these nucleation events may be the source of the current transients observed.

The NiTi alloy has been widely used in cardiovascular applications as a stent material due to its unique shape memory and superelastic properties. Studies on the corrosion properties of NiTi alloy were carried out by several researchers. However, the results of these studies are not consistent. It appears that the corrosion behavior of this alloy is heavily dependent on the chemical composition (Ref 98, 136), surface treatment (Ref 139), heat treatment (Ref 140), surface finish (Ref 141), and fabrication conditions (deformation state, etc.) (Ref 140). Most results agree that the corrosion resistance is closely related to the nickel content and state in surface films and inclusions (Ref 29, 136, 137, 142).

Mechanical Disruption of Oxides. Oxide film repassivation studies for the three different alloy groups are also very important for an understanding of localized corrosion properties with concurrent cyclic loading. Usually, passive current densities for these three groups of biomaterials are in the range of 0.1 to 5 $\mu A/cm^2$ when no mechanical disruption is present. It can increase approximately 6 to 8 orders of magnitude in current density when the film is disrupted mechanically (Ref 32) or during nucleation of a single pit (Ref 76). So, the repassivation ability will determine the corrosion property for these alloys in physiological conditions. Researchers (Ref 38, 143) studied the oxide fracture and repassivation behavior of Ti-6Al-4V and Co-Cr-Mo alloys in simulated physiological solutions using the electrochemical scratch test method (Ref 38, 143). In the case of mechanically assisted corrosion, mechanical motion will disrupt the passive film and expose the bulk metal to electrolyte, causing corrosion and ion release even during the process of repassivation of the film. This process will repeat under cyclic loading conditions (e.g., fretting or fatigue). In addition, fretting can cause local chemical changes, such as decrease of pH or increase of chloride, and make the environment more aggressive, enhancing further ion release (Ref 144). On the other hand, insoluble oxides and particles can be generated by fretting and other mechanical actions. The mechanism of fretting corrosion and mechanically assisted corrosion on biomaterials is discussed elsewhere (Ref 32) and in the article "Mechanically Assisted Corrosion of Metallic Biomaterials" in this Volume.

Mechanical disruption of oxides can also give rise to transient electric fields that can propagate away from the site of abrasion and may result in signals large enough to induce nerve firing (Ref 132).

Metal Ion Release from Metallic Biomaterials. From a thermodynamic point of view, all metal components of iron-base stainless steels, cobalt-base alloys, and titanium alloys are unstable in aqueous environments, and they tend to form metal ions that are released into solution. On the other hand, the formation of a passive film prevents ion release from the bulk material, but the passivation of biomaterials is not completed in physiological solutions even after 1000 h for Co-Cr-Mo alloy and stainless steels, and after 100 days for titanium alloys, based on continuing changes in the OCP, which is a measure of the oxide film character. There exists ion and electron exchange reactions between the metal, passive film, and adjacent solutions. Ions, due to oxidation of alloying elements or dissolution of oxide film, can be released into the human body and may cause adverse host reactions. The sources for ion release can be categorized as: (1) localized corrosion; (2) preferential release (or selective dissolution); (3) passive dissolution; and (4) mechanically assisted corrosion.

Preferential release of alloying elements refers to the disproportionate release of ions, inconsistent with the composition of the alloy. For instance, it has been shown that iron and nickel preferentially release from stainless steel, cobalt from Co-Cr-Mo, nickel from NiTi, and aluminum and vanadium from Ti-6Al-4V. Studies indicate that the passivation elements, chromium and titanium, are dissolved into some biological media (Ref 145, 146).

Effects of Protein on the Corrosion Behavior of Three Alloy Systems. Another possible mechanism for ion release for biomaterials involves the influences of protein effects and superoxides. Researchers (Ref 147–151) found elevated concentrations of metals, including chromium, cobalt, and titanium, in the serum of patients with implants made from stainless steel, cobalt alloys, and titanium alloys. Furthermore, they found that proteins such as immunoglobulins in some molecular-weight ranges play an important role in binding to metals such as chromium.

Although studies on the corrosion behavior of these alloys in simulated body fluids (mainly ionic electrolyte) are extensive, there are relatively few studies on the corrosion behavior in protein-containing fluids. In fact, as mentioned earlier, the surfaces of these materials are always covered by a protein film immediately after implantation. It is difficult to determine the state of protein adsorption on metal surfaces and what effects these proteins have on the surface oxide.

Protein adsorption on metal (passive film) surfaces is very complicated, and the literature on this topic is somewhat limited and, in part, contradictory. Although different techniques, including labeling, optical, electrochemical, and microscopy, were used to study the adsorption of various proteins (mainly, albumin and fibrinogen) on metal surfaces, much controversy still exists (Ref 50). What is the nature of protein adsorption on metal surfaces? Is there nonspecific or specific adsorption, and is the adsorption process electrochemical or not? What are the controlling factors that determine the protein adsorption on metal surfaces? All these questions remain, for the most part, unanswered.

In the case of corrosion of biomaterials in protein-containing solutions, most researchers used saline solutions with additions of albumin or 10% serum, and the results are contradictory. Some studies have shown that the corrosion rate is increased by the presence of protein, for example, 316L stainless steel (Ref 152, 153), cobalt or Co-Cr-Mo (Ref 80, 154, 155), and titanium (Ref 79, 81). Other results show an opposite effect, for example, 316L (Ref 156, 157). There are even some results that show no significant influence (Ref 80).

Generally, if the amino acids or proteins do not form complexes with the metal in the passive film, these adsorbed molecules can act as inhibitors and prevent the release of metal ions into the body. Otherwise, the protein in solution could abstract already-oxidized metal from a protective oxide layer, rendering the underlying material liable to attack. Moreover, the protein may act as a catalyst or oxidizing agent for the metal with internal disulfide bonds, and the forming of a protein-metal complex may enhance dissolution of the material.

The role of proteins on the corrosion behavior of biomaterials remains unclear. Particular proteins and their interactions with the surface may be more important than others in affecting corrosion. For example, the effect of fibrinogen may play a central role in various biological processes and may be more likely to cover surfaces than other proteins.

Effects of H_2O_2 on Corrosion and Oxide Behaviors of Implant Biomaterials. During the inflammatory process that typically follows implantation of a metallic device, macrophages can secrete hydrogen peroxide (H_2O_2) and other free radicals, which subsequently form hydroxyl radicals, and superoxide and other active species (Ref 158, 159). Hydrogen peroxide can be a strong oxidant or may also act as a weak reducer, as seen in the following equations:

$$H_2O_2 + 2H^+ 2e^- \rightarrow 2H_2O \quad E_0 = 1.776 - 0.059\,pH$$
(Eq 11)

$$H_2O_2 \rightarrow O_2 + 2H^+ + 2e^- \quad E_0 = 0.682 - 0.059\,pH$$
(Eq 12)

It is known that metal ions can decompose H_2O_2 and create free radical OH^- by the Fenton reaction, and this free radical is harmful to tissue. For titanium (TiO_2) surfaces, the unique property of this material is that it can form a gellike matrix and trap the free radical during decomposition of H_2O_2, and this may help to explain its good compatibility.

The effects of H_2O_2 on structure and corrosion behavior have been studied for titanium, while almost no work has been done on stainless steel or Co-Cr-Mo alloy. Studies (Ref 39, 60) showed that H_2O_2 largely increased the OCP of CP-Ti in PBS with the addition of H_2O_2 in the range from 1 to 100 mM, as well as the corrosion rate. In addition, H_2O_2 decreased passive film thickness and led to a granular structure. Furthermore, H_2O_2 decreased the dissolution/oxidation resistance of titanium in PBS solution and resulted in a more pronounced two-layer structure of the oxide.

Researchers (Ref 160) reported a study on the effect of H_2O_2 concentration on 316L stainless steel in seawater, and they found that 316L has a higher OCP in summer, approximately 0.6 V versus SHE compared with 0.3 V in winter. Correspondingly, the concentration of H_2O_2 is higher in summer (>10 ppm) than in winter (<2 ppm).

Electrochemical impedance spectroscopy (EIS) has been used more recently to evaluate corrosion and electronic properties of biomaterials (Ref 91, 161–163). In this method, an electrical perturbation (usually voltage) is applied to the electrochemical interface, and the resulting current response is recorded. There are several different electrical stimuli that can be used in impedance spectroscopy research. The most commonly used technique is to apply a small-amplitude (usually less than 20 mV) sinusoidal voltage bias around the fixed direct current potential (usually OCP) in a range of frequencies (typically 1 mHz to 1 MHz) to the interface and measure the current, which also exhibits a sinusoidal waveform but with a phase shift. The advantage of this method is the availability of instruments and the nondestructive nature of the test. It is, however, time-consuming, particularly in the low-frequency range. Another approach is to apply small incremental steps in potential to the electrochemical interface and measure the current transient response. Correspondingly, the frequency-dependent impedance characteristics can be obtained from time domain current response via the Laplace transform technique. This method was first used (Ref 163) in biomaterials studies (step polarization impedance spectroscopy, or SPIS). The advantages of SPIS are that the impedance information can be determined over a range of potentials concurrently with the polarization behavior, and further, this technique is easy to use as well as fast and inexpensive compared with traditional commercially available electrochemical impedance testing instrumentations.

All impedance techniques have the capability to provide information about the electrical characteristics of the interface between the biomaterial surface and the environment. Generally, the interface of biomaterials in physiological solutions can be modeled as a Randles circuit, where a resistor (polarization resistance) is connected in parallel with a capacitor (double-layer capacitance), and they connect with another resistor (solution resistance) in series. Sometimes, this circuit needs to be more complicated; for example, the passive film can be considered as a two-layer structure in the case of existing H_2O_2 for CP-Ti in PBS (Ref 39), and nonideal circuit elements are needed.

Researchers (Ref 84) studied the surface structure of CP-Ti in PBS with or without addition of H_2O_2, and they found the capacitance and polarization resistance changed with the addition of H_2O_2 and immersion times. Other scientists (Ref 162) studied the corrosion behavior of 316L stainless steel, Co-Cr-Mo, and Ti-6Al-4V in physiological solutions at pH 1 and 6.2. They found 316L corroded at pH 1 after a 20 h immersion, while Co-Cr-Mo exhibited slight corrosion in the early stage, and Ti-6Al-4V kept its passivity, but the dissolution rate increased compared with that at a pH of 6.2. In pH 6.2, all alloys remained passive. Another advantage of EIS is that it can be used to study the electronic semiconducting structure of passive film (Ref 164) using the Mott-Schottky equation, which measures the capacitance under different potentials at a fixed frequency. Information gained from these tests includes values of the flat band potential and the density of charge carriers that may be related with the corrosion property of biomaterials.

In SPIS, all impedance information can be collected in a relatively short time (within 30 min) in a large potential range (usually from −1 to 1 V versus Ag/AgCl), as well as the polarization behavior (Ref 163). In addition to capacitance and polarization resistance, the film resistance can be obtained. Moreover, this method can be combined with AFM methods to obtain information on morphological changes as well as electronic changes of the passive film, with potential for various biomaterials in solution (Ref 22, 138). A researcher (Ref 163) studied the electrochemical impedance behavior of platinum, CP-Ti, and Co-Cr-Mo in PBS, using the SPIS method. He found that the impedance characteristics, including film resistance, polarization resistance, and capacitance for platinum in solution, showed only one major transition in the vicinity of −500 mV (Ag/AgCl) and are otherwise relatively constant; the film resistance increased and capacitance decreased with increasing potential for CP-Ti, and this indicated film growth for this material with increasing potential. For Co-Cr-Mo alloy, the film resistance increased with increasing potential first but dropped near the transpassive potential range, while capacitance dropped first and increased later. All these results showed that SPIS is a powerful method to evaluate the electrochemical and electronic behavior of biomaterials in physiological solutions.

Biological Consequences of In Vivo Corrosion and Biocompatibility

All metallic biomaterials corrode in physiological conditions, either in passive dissolution or in localized corrosion or mechanically assisted corrosion. All these corrosion processes release metal ions or generate various particles or debris. The host will react to these foreign bodies and may cause adverse reactions. In fact, the biological consequences of in vivo corrosion are not well documented. A brief introduction is given, and more information can be found in some review papers (Ref 150, 165,166).

Local and Systemic Distribution of Ions and Particles. It is well documented that corrosion products can be found in the tissue surrounding the metallic biomedical devices (Ref 165). Around the stainless steel internal fixation devices, particularly at the screw-plate junctions, two kinds of corrosion products were identified: iron-containing hemosiderin-like granules and chromium-containing particles (Ref 167). In addition, similar corrosion products have been reported in the tissues around stainless steel femoral stems and spinal instrumentation.

Researchers (Ref 168) analyzed the corrosion products at the modular junctions of total hip implant and the migration of these solid products to the periprosthetic tissue, using electron-microprobe energy-dispersive x-ray analysis and Fourier transform infrared spectroscopy. They found two distinct corrosion products. The principal corrosion product is an amorphous chromium orthophosphate hydrate-rich particle, which presented at the modular head-neck junction as a precipitate of chromium with phosphorus from the fluids. These particles are still found within the bone-implant interface membranes and adjacent pseudocapsules. Another form of corrosion product is a highly crystalline interfacial layer of mixed oxides and chlorides and is located in the crevice areas formed by the mated neck and head; it is rarely found in the surrounding tissues.

The corrosion products found in local tissues around titanium alloy prostheses are usually in the form of particulates or small debris, and they appear to be corrosion-wear products. Researchers (Ref 169) found that the debris in tissue has a similar chemical composition with that of the bulk material (Ti-6Al-4V), indicating that these debris represented wear particles rather than precipitated dissolution products, which usually are associated with iron- and cobalt-base alloys. Other scientists (Ref 170) found metal particles whose constituents were proportional to their bulk composition after examining a retrieved total hip prosthesis made of Ti-6Al-4V and ultrahigh-molecular-weight polyethylene. Another possible reason for this debris for titanium alloy implants may be the result of oxidation of titanium due to superoxides (Ref 60), or metastable pitting dissolution (Ref 76), or wear.

Metal ions may bind to proteins and be transported in the bloodstream or lymphatics to remote organs. In fact, multiple biomaterial alloying elements were found in serum and urine. Studies (Ref 151) found that four years after total knee replacement with a titanium-base alloy component, there was a tenfold increase of titanium concentration in the serum, compared with a control patient who had no implant. Particularly for those patients who had failed implants, the titanium levels in the serum were 2 orders of magnitude higher than that of the control. Similarly, after 6 to 120 weeks implantation of a total knee replacement, researchers (Ref 171) found an increased cobalt level in the serum and an apparent increase for two patients who had loose prostheses. In fact, elevated metal ion levels are not found only in people with failed implants.

Researchers (Ref 172) reported that chromium levels in serum and urine increased five- and eight-fold in patients with well-functioning total hip replacements.

Biological Response for Ions and Biocompatibility. It is well known that some quantity of metal ions is essential for biological processes, and they have functional forms and fairly constant concentrations (Ref 173). These essential elements include some metallic biomaterial constituents such as iron, cobalt, molybdenum, chromium, manganese, vanadium, and nickel. Other main constituents of metallic biomaterials belong to the nonessential elements, with highly variable concentrations in the body, and are more likely to be dependent on the contacting environment, for example, aluminum, titanium, zirconium, niobium, tantalum, and others. Metal ions or debris originated from biomedical devices may cause adverse host response via complex reactions and may trigger allergic, immune, and inflammatory reactions (Ref 174).

Toxicity of metal ions depends mainly on their concentration and compound forms (Ref 175); for instance, essential elements such as cobalt and nickel can induce toxic reactions at high or excessive concentrations. In addition, the nature of metal compounds plays a crucial role in their toxic and other adverse responses. For example, Ni_3S_2 is highly carcinogenic, while $NiCl_2$ is not (Ref 174). The pure metal is rarely toxic, and the metallic form of these elements is rarely seen in physiological conditions, particularly for biomaterial alloying elements. Thus, for the same metal, the oxidation state determines its toxicity; for example, hexavalent chromium (Cr^{6+}) is more toxic than trivalent chromium (Cr^{3+}).

Although some cases of allergic reactions have been reported, and cancer formation may be related to implantation, no systematic study exists to evaluate immunological or carcinogenic reactions to implantation of metallic biomedical devices. Researchers (Ref 176) found that people from an implant population group were sensitized to the allergens of nickel, chromium, and cobalt at a rate of 9.6, 9.3, and 6.0%, respectively. This number is higher than that of the general population, where sensitivity is 4.2, 1.7, and 1.4%, respectively. Indeed, the long-term biological effects of these alloying elements or their compounds are currently unknown.

Summary

The benefit of application of metallic biomedical devices is clear, and metallic biomaterials will continue to be used in orthopedic, cardiovascular, and other medical applications for the foreseeable future. Corrosion of biomaterials plays a central role in the success or failure of these biomedical devices. Corrosion may cause device failure directly, such as penetration, fracture of biomedical devices, or induced acute adverse host reactions by the corrosion product. Corrosion is an extremely complex process and involves the electrochemical and chemical reactions between material and environments. In addition, the human body is also a complex bioelectrochemical and biomechanical environment with dynamic loading actions and biochemical variation. All of these factors will interact with the implantation of biomedical devices, and the interplay of these physiological processes with corrosion will continue to be a source of investigation for the foreseeable future.

REFERENCES

1. A.C. Fraker, Medical and Dental, *Corrosion Tests and Standards: Application and Interpretation,* R. Baboian, Ed., American Society for Testing and Materials, 1995, p 705–715
2. J.L. Gilbert, Metals, *The Adult Hip,* J.J. Callaghan, A.G. Rosenberg, and H.E. Rubash, Ed., Lippincott-Raven, Philadelphia, PA, 1998, p 123–134
3. "Specification for Stainless Steel Bar and Wire for Surgical Implant (Special Quality)," F 138, *Annual Book of ASTM Standards,* Vol 13.01, American Society for Testing and Materials, 2003
4. "Specification for Wrought, Nitrogen Strengthened 23Manganese-21Chromium-1Molybdenum Low-Nickel Stainless Steel Alloy Bar and Wire for Surgical Implants (UNS S29108)," F 2229, *Annual Book of ASTM Standards,* Vol 13.01, American Society for Testing and Materials, 2003
5. M.P. Ryan, D.E. Williams, R.J. Chater, B.M. Hutton, and D.S. McPhail, *Nature* Vol 415, 2000, p 770–774
6. "Specification for Cast Cobalt-Chromium-Molybdenum Alloy for Surgical Implant Applications," F 75, *Annual Book of ASTM Standards,* Vol 13.01, American Society for Testing and Materials, 2003
7. "Specification for Cobalt-28Chromium-6Molybdenum Alloy Forgings for Surgical Implants," F 799, *Annual Book of ASTM Standards,* Vol 13.01, American Society for Testing and Materials, 2003
8. "Specification for Wrought Cobalt-35Nickel-20Chromium-10Molybdenum Alloy for Surgical Implant Applications," F 562, *Annual Book of ASTM Standards,* Vol 13.01, American Society for Testing and Materials, 2003
9. "Specification for Wrought Cobalt-Nickel-Chromium-Molybdenum-Tungsten-Iron Alloy for Surgical Implant Applications," F 563, *Annual Book of ASTM Standards,* Vol 13.01, American Society for Testing and Materials, 2003
10. "Specification for Wrought Cobalt-28Chromium-6Molybdenum Alloy for Surgical Implants," F 1537, *Annual Book of ASTM Standards,* Vol 13.01, American Society for Testing and Materials, 2003
11. "Specification for Unalloyed Titanium for Surgical Implant Applications," F 67,

Annual Book of ASTM Standards, Vol 13.01, American Society for Testing and Materials, 2003

12. "Specification for Wrought Titanium 6Al-4V ELI Alloy for Surgical Implant Applications," F 136, *Annual Book of ASTM Standards,* Vol 13.01, American Society for Testing and Materials, 2003

13. "Specification for Wrought Titanium-6Aluminum-7Niobium Alloy for Surgical Implant Applications," F 1295, *Annual Book of ASTM Standards,* Vol 13.01, American Society for Testing and Materials, 2003

14. "Standard Specification for Wrought Titanium-3Aluminum-2.5Vanadium Alloy Seamless Tubing for Surgical Implant Applications (UNS R56320)," F 2146, *Annual Book of ASTM Standards,* Vol 13.01, American Society for Testing and Materials, 2003

15. "Specification for Wrought Titanium-13Niobium-13Zirconium Alloy for Surgical Implant Applications," F 1713, *Annual Book of ASTM Standards,* Vol 13.01, American Society for Testing and Materials, 2003

16. "Standard Specification for Wrought Titanium-15Molybdenum Alloy for Surgical Implant Applications (UNS R58150)," F 2066, *Annual Book of ASTM Standards,* Vol 13.01, American Society for Testing and Materials, 2003

17. "Standard Specification for Wrought Nickel-Titanium Shape Memory Alloys for Medical Devices and Surgical Implants," F 2063, *Annual Book of ASTM Standards,* Vol 13.01, American Society for Testing and Materials, 2003

18. H. Bubert and H. Jenett, Ed., *Surface and Thin Film Analysis, Principles, Instrumentation, Applications,* Wiley-VCH, Weinheim, 2002

19. J.P. Bearinger, C.A. Orme, and J.L. Gilbert, *Surf. Sci.,* Vol 491 (No. 3), 2001, p 370–387

20. S.J. Megremis, "The Mechanical, Electrochemical, and Morphological Characteristics of Passivating Oxide Films Covering Co-Cr-Mo Alloys: A Study of Five Microstructures," Ph.D. dissertation, Evanston, IL, 2001

21. Z. Bai, J.L. Gilbert, and R. Getterns, In-Situ Electrochemical Atomic Force Microscopy Study of Morphology of Oxide Film on Etched NiTi Alloy: Effects of Hydration and Potential, *Transactions of Seventh World Biomaterial Congress* (Sydney) 2004, p 436

22. Z. Bai and J.L. Gilbert, Combined In-Situ Step Polarization Impedance Spectroscopy and Atomic Force Microscopy Measurement, Study of Oxide Film on 316L SS in PBS, *29th Annual Meeting Transactions of the Society for Biomaterials* (Reno, NV), 2003, p 422

23. T. Hanawa, S. Hiromoto, A. Yamamoto, D. Kuroda, and K. Asami, *Mater. Trans.,* Vol 43 (No. 12), 2002, p 3088–3092

24. I. Milosev and H.H. Strehblow, *J. Biomed. Mater. Res.,* Vol 52 (No. 2), 2000, p 404–412

25. G. Lewis, *J. Vac. Sci. Technol. A: Vac., Surf. Films,* Vol 11 (No. 1), 1993, p 168–174

26. A. Kocijan, I. Milosev, and B. Pihlar, *J. Mater. Sci.: Mater. Med.,* Vol 15 (No. 6), 2004, p 643–650

27. R.L. Moore, G.L. Grobe III, and J.A. Gardella, *J. Vac. Sci. Technol. A: Vac., Surf., Films,* Vol 9 (No. 3, Pt. 2), 1991, p 1323–1328

28. I. Milosev, M. Metikos-Hukovic, and H.H. Strehblow, *Biomaterials,* Vol 21 (No. 20), 2000, p 2103–2113

29. S.A. Shabalovskaya and J.W. Anderegg, *J. Vac. Sci. Technol. A: Vac. Surf. Films,* Vol 13 (No. 5), 1995, p 2624–2632

30. J. Ohnsorge and R. Holm, *Med. Prog. Technol.,* Vol 5, 1978, p 171–177

31. P.G. Laing, Clinical Experience with Prosthetic Materials: Historical Perspectives, Current Problems, and Future Directions, *Corrosion and Degradation of Implant Materials,* STP 684, B.C. Syrett and A. Acharya, Ed., American Society for Testing and Materials, 1979, p 199–211

32. J.L. Gilbert and J.J. Jacobs, The Mechanical and Electrochemical Processes Associated with Taper Fretting Crevice Corrosion: A Review, *Modularity of Orthopedic Implants,* STP 1301, D.E. Marlowe, J.E. Parr, and M.B. Mayor, Ed., American Society for Testing and Materials, 1997, p 45–59

33. G. Willert, L.G. Broback, G.H. Buchhorn, P.H. Jensen, G. Koster, I. Lang, P. Ochsner, and R. Schenk, *Clin. Orthop.,* 1996, p 51–75

34. K.J. Bundy, In Vivo Corrosion of Implant Materials, *Corrosion Tests and Standards,* R. Baboian, Ed., American Society for Testing and Materials, 1995, p 411–418

35. J. Black, *Biological Performance of Materials: Fundamentals of Biocompatibility,* 3rd ed., Marcel Dekker, Inc., 1999, p 47

36. D.A. Jones, *Principles and Prevention of Corrosion,* 2nd ed., Pearson Education, 1995

37. T.P. Hoar and D.C. Mears, *Proc. R. Soc. (London) A,* Vol 294 (No. 1439), 1966, p 486–510

38. J.L. Gilbert, C.A. Buckley, and E.P. Lautenschlager, Titanium Oxide Film Fracture and Repassivation: The Effect of Potential, pH and Aeration, *Medical Applications of Titanium and Its Alloys: The Material and Biological Issues,* STP 1272, S.A. Brown and J.E. Lemons, Ed., American Society for Testing and Materials, 1996, p 199–215

39. J. Pan, D. Thierry, and C. Leygraf, *J. Biomed. Mater. Res.,* Vol 28 (No. 1), 1994, p 113–122

40. A.T. Kuhn, P. Neufeld, and T. Rae, Synthetic Environments for the Testing of Metallic Biomaterials, *The Use of Synthetic Environments for Corrosion Testing,* P.E. Francis and T.S. Lee, Ed., STP 970, American Society for Testing and Materials, 1988, p 79–97

41. J.O. Bockris and S.U.M. Khan, *Surface Electrochemistry: A Molecular Level Approach,* Plenum Press, 1993

42. L. Vroman, A.L. Adams, G.C. Fischer, and P.C. Munoz, *Blood,* Vol 55 (No. 1), 1980, p 156–159

43. A. Afshar, C. Dennis, C.C. Fries, and P.N. Sawyer, *Surg. Forum,* Vol 17, 1966, p 138–139

44. P.N. Sawyer, K.T. Wu, S.A. Wesolowski, W.H. Brattain, and P.J. Boddy, *Proc. Natl. Acad. Sci. U.S.A.,* Vol 53, 1965, p 294–300

45. P.N. Sawyer, W.H. Brattain, and P.J. Boddy, Electrochemical Precipitation of Human Blood Cells and Its Possible Relation to Intravascular Thrombosis, *Proc. Nat. Acad. Sci. U.S.A.,* Vol 51 (No. 3), 1964, p 428–432

46. J.C. Palmaz, *J. Vasc. Interv. Radiol.,* Vol 12, 2001, p 789–794

47. D.F. Williams, Blood Physiology and Biochemistry: Hemostasis and Thrombosis, *Blood Compatibility,* Vol 1, D.F. Williams, Ed., CRC Press, Inc., 1987, p 5–36

48. T.A. Horbett and J.L. Brash, Ed., *Proteins at Interfaces II: Fundamentals and Applications,* ACS Symposium Series 602, American Chemical Society, Washington, D.C., 1995

49. J.L. Brash and T.A. Horbett, Ed., *Proteins at Interfaces I: Physicochemical and Biochemical Studies,* ACS Symposium Series 343, American Chemical Society, Washington, D.C., 1987

50. B. Ivasson and I. Landstrom, Physical Characterization of Protein Adsorption on Metal and Metal Oxide Surfaces, *Critical Reviews in Biocompatibility,* Vol 2 (No. 1), D.F. Williams, Ed., CRC Press, Inc., 1986, p 1–96

51. S. Srinivasan, L. Duic, N. Ramasamy, and P.N. Sawyer, Electrochemical Reactions of Blood Coagulation Factors: Their Role in Thrombosis, *Ber. Bunsenges.,* Vol 77 (No. 10–11), 1973, p 798–804

52. P.N. Sawyer, B. Stanczewski, W.S. Ramsey, Jr., N. Ramasamy, and S. Srinivasan, *J. Supramol. Struct.,* Vol 1, 1973, p 417–436

53. P.N. Sawyer, *Nature,* Vol 206 (No. 989), June 1965, p 1162–1163

54. P.N. Sawyer, *Ann. New York Acad. Sci.,* Vol 416, 1983, p 561–583

55. P.S. Chopra, S. Srinivasan, T. Lucas, and P.N. Sawyer, *Nature,* Vol 215 (No. 109), 1967, p 1494

56. G. Stoner and L. Walker, *J. Biomed. Mater. Res.,* Vol 3 (No. 4), 1969, p 645–654
57. A. Bolz and M. Schaldach, *Med. Biol. Eng. Comput.,* Vol 31 (Suppl.), 1993, p S123–130
58. J.M. Anderson, *Ann. Rev. Mater. Res.,* Vol 31, 2001, p 81–110
59. J.M. Anderson and K.M. Miller, *Biomaterials,* Vol 5 (No. 1), 1984, p 510
60. J. Pan, D. Thierry, and C. Leygraf, *J. Biomed. Mater. Res.,* Vol 30 (No. 3), 1996, p 393–402
61. J.R. Scully, Electrochemical, *Corrosion Tests and Standards: Application and Interpretation,* R. Baboian, Ed., American Society for Testing and Materials, 1995, p 75–90
62. H.S. Isaacs, Practical Aspects of Corrosion Fundamentals, *Proceedings of the Symposium on Compatability of Biomedical Implants,* P. Kovacs and N.S. Istephanous, Ed., The Electrochemical Society, Inc., 1994, p 48–58
63. M. Pourbaix, *Atlas of Electrochemical Equilibria in Aqueous Solutions,* 1st English ed., Pergamon Press, Oxford, 1966
64. M. Pourbaix, *Biomaterials,* Vol 5 (No. 3), 1984, p 122–134
65. "Standard Reference Test Method for Making Potentiostatic and Potentiodynamic Anodic Polarization Measurements," G 5, *Annual Book of ASTM Standards,* Vol 03.02, American Society for Testing and Materials, 2003
66. "Standard Practice for Verifition of Algorithm and Equipment for Electrochemical Impedance Measurements," G 106, *Annual Book of ASTM Standards,* Vol 03.02, American Society for Testing and Materials, 2003
67. "Test Method for Pitting or Crevice Corrosion of Metallic Surgical Implant Materials," F 746, *Annual Book of ASTM Standards,* Vol 13.01, American Society for Testing and Materials, 2003
68. "Standard Test Method for Conducting Cyclic Potentiodynamic Polarization Measurements to Determine the Corrosion Susceptibility of Small Implant Devices," F 2129, *Annual Book of ASTM Standards,* Vol 13.01, American Society for Testing and Materials, 2003
69. "Test Method for Measuring Fretting Corrosion of Osteosynthesis Plates and Screws," F 897, *Annual Book of ASTM Standards,* Vol 13.01, American Society for Testing and Materials, 2003
70. "Standard Practice for Corrosion Fatigue Testing of Metallic Implant Materials," F 1801, *Annual Book of ASTM Standards,* Vol 13.01, American Society for Testing and Materials, 2003
71. "Standard Practice for Fretting Corrosion Testing of Modular Implant Interfaces: Hip, Femoral Head-Bore, and Cone Taper Interface," F 1875, *Annual Book of ASTM Standards,* Vol 13.01, American Society for Testing and Materials, 2003
72. "Standard Test Method for Conducting Cyclic Potentiodynamic Polarization Measurements for Localized Corrosion Susceptibility of Iron-, Nickel-, or Cobalt-Based Alloys," G 61, *Annual Book of ASTM Standards,* Vol 03.02, American Society for Testing and Materials, 2003
73. J.L. Gilbert, C.A. Buckley, and J.J. Jacobs, *J. Biomed. Mater. Res.,* Vol 27 (No. 12), 1993, p 1533–1544
74. G.S. Frankel, Pitting Corrosion of Metals: A Summary of the Critical Factors, *Proceedings of Electrochemical Society 97-7 (Pits and Pores: Formation, Properties, and Significance for Advanced Luminescent Materials),* 1997, p 1–26
75. G.S. Frankel, *J. Electrochem. Soc.,* Vol 145 (No. 6), 1998, p 2186–2198
76. G.T. Burstein, C. Liu, and R.M. Souto, *Biomaterials,* Vol 26 (No. 3), 2005, p 245–256
77. R.M. Souto and G.T. Burstein, *J. Mater. Sci. Mater. Med.,* Vol 7 (No. 6), 1996, p 337–343
78. A. Garrido, A. Campilho, and M.A. Barbosa, Signal Processing Applied to the Study of Electrochemical Properties of Metallic Biomaterials, *Adv. Biomater.,* Vol 10 (Biomaterial-Tissue Interfaces), 1992, p 487–494
79. S.R. Sousa and M.A. Barbosa, *Clin. Mater.,* Vol 14 (No. 4), 1993, p 287–294
80. G.C. Clark and D.F. Williams, *J. Biomed. Mater. Res.,* Vol 16 (No 2), 1982, p 125–134
81. M.A. Khan, R.L. Williams, and D.F. Williams, *Biomaterials,* Vol 20 (No. 7), 1999, p 631–637
82. S.A. Brown and K. Merritt, *J. Biomed. Mater. Res.,* Vol 15 (No. 4), 1981, p 479–488
83. S.A. Brown and K. Merritt, *J. Biomed. Mater. Res.,* Vol 14 (No. 2), 1980, p 173–175
84. J. Pan, D. Thierry, and C. Leygraf, *Electrochim. Acta,* Vol 41 (No. 7/8), 1996, p 1143–1153
85. J. Pan, H. Liao, C. Leygraf, D. Thierry, and J. Li, *J. Biomed. Mater. Res.,* Vol 40 (No. 2), 1998, p 244–256
86. K. Merritt, L. Wenz, and S.A. Brown, *J. Orthop. Res.,* Vol 9 (No. 2), 1991, p 289–296
87. H.Y. Lin and J.D. Bumgardner, *J. Orthop. Res.,* Vol 22 (No. 6), 2004, p 1231–1236
88. J. Yang and J. Black, *Biomaterials,* Vol 15 (No. 4), 1994, p 262–268
89. J. Black, E.C. Maitin, H. Gelman, and D.M. Morris, *Biomaterials,* Vol 4 (No. 3), 1983, p 160–164
90. D. Cook, R.L. Barrack, and A.J. Clemow, *J. Bone Joint Surg. Br.,* Vol 76 (No. 1), 1994, p 68–72
91. S.D. Cook, K.A. Thomas, A.F. Harding, C.L. Collins, R.J. Haddad, Jr., M. Milicic, and W.L. Fischer, *Biomaterials,* Vol 8 (No. 3), 1987, p 177–184
92. J.L. Gilbert, C.A. Buckley, J.J. Jacobs, K.C. Bertin, and M.R. Zernich, *J. Bone Joint Surg. Am.,* Vol 76 (No. 1), 1994, p 100–105
93. J.J. Jacobs, R.M. Urban, J.L. Gilbert, A.K. Skipor, J. Black, M. Jasty, and J.O. Galante, *Clin. Orthop.,* Oct 1995, p 94–105
94. "Practice for Retrieval and Analysis of Implanted Medical Devices, and Associated Tissues," F 561, *Annual Book of ASTM Standards,* Vol 13.01, American Society for Testing and Materials, 2003
95. C. Heintz, G. Riepe, L. Birken, E. Kaiser, N. Chakfe, M. Morlock, G. Delling, and H. Imig, *J. Endovasc. Ther.,* Vol 8, 2001, p 248–253
96. G. Riepe, C. Heintz, E. Kaiser, N. Chakfe, M. Morlock, G. Delling, and H. Imig, *Eur. J. Vasc. Endovasc. Surg.,* Vol 24, 2002, p 117–122
97. V.J. Colangelo and N.D. Greene, *J. Biomed. Mater. Res.,* Vol 3 (No. 2), 1969, p 247–265
98. S.A. Shabalovskaya, *Biomed. Mater. Eng.,* Vol 12 (No. 1), 2002, p 69–109
99. G.T. Burstein, C. Liu, R.M. Souto, and S.P. Vines, Corrosion Engineering, *Sci. Technol.,* Vol 39 (No. 1), 2004, p 25–30
100. G.S. Frankel, L. Stockert, F. Hunkeler, and H. Boehni, *Corrosion,* Vol 43 (No. 7), 1987, p 429–436
101. T.P. Hoar, Breakdown and Repair of Oxide Films on Iron, *Trans. Faraday Soc.,* Vol 45, 1949, p 683–693
102. T.P. Hoar, D.C. Mears, and G.P. Rothwell, *Corros. Sci.,* Vol 5 (No. 4), 1965, p 279–289
103. H.H. Strehblow and J. Wenners, Determination of the Growth of Corrosion Pits on Iron and Nickel in an Early Stage of Development and Its Relation to the Metal Dissolution in Concentrated Chloride Media, *Z. Phys. Chem.,* Vol 98 (No. 1–6), 1975, p 199–214
104. Z. Szklarska-Smialowska, *Corros. Sci.,* Vol 44 (No. 5), 2002, p 1143–1149
105. J. Gilbert, Z. Bai, J. Bearinger, and S. Megremis, Dynamics of Oxide Films on Metallic Biomaterials, *Proceedings of Materials and Processes from Medical Devices Conference,* S. Shrivastava, Ed. (Anaheim, CA), ASM International, 2004, p 139–143
106. J.P. Bearinger, C.A. Orme, and J.L. Gilbert, *J. Biomed. Mater. Res. A,* Vol 67 (No. 3), 2003, p 702–712
107. Z. Bai, J.L. Gilbert, and R. Gettens, Characterization of Surface Oxide Film on CPTi Formed in PBS Using Atomic Force Microscopy, *29th Annual Meeting Transactions of the Society for Biomaterials* (Reno, NV), 2003, p 688
108. T. Hanawa, S. Hiromoto, and K. Asami, *Appl. Surf. Sci.,* Vol 183 (No. 1–2), 2001, p 68–75

109. T. Hanawa, Metal Ion Release from Metal Implants, *Mater. Sci. Eng. C: Biomimetic Supramol. Syst.*, Vol 24 (No. 6–8), 2004, p 745–752
110. A.C. Fraker, A.W. Ruff, and M.P. Yeager, Corrosion of Titanium Alloys in Physiological Solutions, *Titan. Sci. Technol., Proc. Second Int. Conf.*, R.I. Jaffee, Ed., Plenum, 1973, p 2447–2457
111. M.A. Imam, A.C. Fraker, and C.M. Gilmore, Corrosion Fatigue of 316L Stainless Steel, Co-Cr-Mo Alloy, and ELI Ti6Al4V, *Corrosion and Degradation of Implant Materials*, STP 684, B.C. Syrett and A. Acharya, Ed., American Society for Testing and Materials, 1979, p 128–143
112. K.M. Speck and A.C. Fraker, *J. Dent. Res.*, Vol 59 (No. 10), 1980, p 1590–1595
113. D.L. Levine and R.W. Staehle, *J. Biomed. Mater. Res.*, Vol 11 (No. 4), 1977, p 553–561
114. G. Rondelli, *Biomaterials*, Vol 17 (No. 20), 1996, p 2003–2008
115. J. Pan, C. Karlen, and C. Ulfvin, *J. Electrochem. Soc.*, Vol 147 (No. 3), 2000, p 1021–1025
116. G.I. Ogundele and W.E. White, Polarization Studies on Surgical-Grade Stainless Steels in Hanks' Physiological Solution, *Corrosion and Degradation of Implant Materials: Second Symposium*, STP 859, A.C. Fraker and C.D. Griffin, Ed., American Society for Testing and Materials, 1985, p 117–135
117. D.E. Williams, R.C. Newman, Q. Song, and R.G. Kelly, *Nature*, Vol 350 (No. 6315), 1991, p 216–219
118. C. Haraldsson and S. Cowen, Characterization of Sandvik Bioline High-N—A Comparison of Standard Grades F1314 and F1586, *Stainless Steels for Medical and Surgical Applications*, STP 1438, G.L. Winters and M.J. Nutt, Ed., ASTM International, 2003, p 3–12
119. M. Windler, R. Steger, and G.L. Winters, Quality Aspects of High-Nitrogen Stainless Steel for Surgical Implants, *Stainless Steels for Medical and Surgical Applications*, STP 1438, G.L. Winters and M.J. Nutt, Ed., ASTM International, 2003, p 72–81
120. L.D. Zardiackas, S. Williamson, M. Roach, and J.A. Bogan, Comparison of Anodic Polarization and Galvanic Corrosion of a Low-Nickel Stainless Steel to 316LS and 22Cr-13Ni-5Mn Stainless Steels, *Stainless Steels for Medical and Surgical Applications*, STP 1438, G.L. Winters and M.J. Nutt, Ed., ASTM International, 2003, p 107–118
121. A.W.E. Hodgson, S. Kurz, S. Virtanen, V. Fervel, C.-O.A. Olsson, and S. Mischler, *Electrochim. Acta*, Vol 49 (No. 13), 2004, p 2167–2178
122. B.C. Syrett and E.E. Davis, Crevice Corrosion of Implant Alloys—A Comparison of In Vitro and In Vivo Studies, *Corrosion and Degradation of Implant Materials*, STP 684, B.C. Syrett and A. Acharya, Ed., American Society for Testing and Materials, 1979, p 229–244
123. S.A. Brown, C.A. Flemming, J.S. Kawalec, H.E. Placko, C. Vassaux, K. Merritt, J.H. Payer, and M.J. Kraay, *J. Appl. Biomater.*, Vol 6, 1995, p 19–26
124. J.R. Goldberg, J.L. Gilbert, J.J. Jacobs, T.W. Bauer, W. Paprosky, and S. Leurgans, *Clin. Orthop.*, 2002, p 149–161
125. Y. Kobayashi, S. Virtanen, and H. Bohni, *J. Electrochem. Soc.*, Vol 147 (No. 1), 2000, p 155–159
126. G.O. Ilevbare and G.T. Burstein, *Corros. Sci.*, Vol 43 (No. 3), 2001, p 485–513
127. G.O. Ilevbare and G.T. Burstein, *Corros. Sci.*, Vol 45 (No. 7), 2003, p 1545–1569
128. A.C. Fraker, A.W. Ruff, J.A.S. Green, and C.J. Bechtoldt, *Corrosion*, Vol 30 (No. 6), 1974, p 203–207
129. A.W. Ruff and A.C. Fraker, *Corrosion*, Vol 30 (No. 7), 1974, p 259–264
130. A.C. Fraker and A.W. Ruff, Effect of Solution pH on the Saline Water Corrosion of Titanium Alloys, *Titan. Sci. Technol., Proc. Second Int. Conf.*, R.I. Jaffee, Ed., Plenum, 1973, p 2655–2663
131. J.R. Goldberg and J.L. Gilbert, *Biomaterials*, Vol 25, 2004, p 851–864
132. J.R. Goldberg and J.L. Gilbert, *J. Biomed. Mater. Res.*, Vol 56 (No. 2), 2001, p 184–194
133. R.J. Solar, Corrosion Resistance of Titanium Surgical Implant Alloys: A Review, *Corrosion and Degradation of Implant Materials*, STP 684, B.C. Syrett and A. Acharya, Ed., American Society for Testing and Materials, 1979, p 259–273
134. L.S. Castleman, S.M. Motzkin, F.P. Alicandri, and V.L. Bonawit, *J. Biomed. Mater. Res.*, Vol 10 (No. 5), 1976, p 695–731
135. B. O'Brien, W.M. Carroll, and M.J. Kelly, *Biomaterials*, Vol 23, 2002, p 1739–1748
136. S.A. Shabalovskaya, *Biomed. Mater. Eng.*, Vol 6, 1996, p 267–289
137. C. Liu and G.T. Burstein, *Acta Metall. Sin. (Eng. Lett.)*, Vol 10 (No. 2), 1997, p 79–87
138. J.P. Bearinger, C.A. Orme, and J.L. Gilbert, *Biomaterials*, Vol 24 (No. 11), 2003, p 1837–1852
139. C. Trepanier, M. Tabrizian, L.H. Yahia, L. Bilodeau, and D.L. Piron, Improvement of the Corrosion Resistance of NiTi Stents by Surface Treatments, *Materials Research Society Symposium Proceedings 459 (Materials for Smart Systems II)*, 1997, p 363–368
140. S.A. Shabalovskaya, J. Anderegg, F. Laab, P.A. Thiel, and G. Rondelli, *J Biomed. Mater. Res. B, Appl. Biomater.*, Vol 65 (No. 1), 2003, p 193–203
141. M. Es-Souni, M. Es-Souni, and H. Fischer-Brandies, *Biomaterials*, Vol 23 (No. 14), 2002, p 2887–2894
142. S.A. Shabalovskaya, *Int. Mater. Rev.*, Vol 46 (No. 5), 2001, p 233–250
143. J.R. Goldberg and J.L. Gilbert, *J. Biomed. Mater. Res.*, Vol 37, 1997, p 421–431
144. S. Frangini and C. Piconi, *Mater. Corros.*, Vol 52 (No. 5), 2001, p 372–380
145. P. Ducheyne, *Biomaterials*, Vol 4 (No. 3), 1983, p 185–191
146. K.L. Wapner, D.M. Morris, and J. Black, *J. Biomed. Mater. Res.*, Vol 20 (No. 2), 1986, p 219–233
147. J.R. Goldberg, C.A. Buckley, J.J. Jacobs, and J.L. Gilbert, Corrosion Testing of Modular Hip Implants, *Modularity of Orthopedic Implants*, STP 1301, D.E. Marlowe, J.E. Parr, and M.B. Mayor, Ed., American Society for Testing and Materials, 1997, p 157–176
148. N.J. Hallab, J.J. Jacobs, A. Skipor, J. Black, K. Mikecz, and J.O. Galante., *J. Biomed. Mater. Res.*, Vol 49, 2000, p 353–361
149. N.J. Hallab, J.J. Jacobs, A. Skipor, J. Black, K. Mikecz, and J.O. Galante, Serum Protein Carriers of Chromium in Patients with Cobalt-Base Alloy Total Joint Replacement Components, *Cobalt-Base Alloys for Biomedical Applications*, STP 1365, J.A. Disegi, R.L. Kennedy, and R. Pilliar, Ed., American Society for Testing and Materials, 1999, p 210–219
150. R.M. Urban, J.J. Jacobs, J.L. Gilbert, A.K. Skipor, N.J. Hallab, K. Mikecz, T.T. Glant, J.L. Marsh, and J.O. Galante, Corrosion Products Generated from Mechanically Assisted Crevice Corrosion of Stainless Steel Orthopaedic Implants, *Stainless Steels for Medical and Surgical Applications*, STP 1438, G.L. Winters and M.J. Nutt, Ed., ASTM International, 2003, p 262–272
151. J.J. Jacobs, A.K. Skipor, J. Black, L.M. Patterson, W.P. Paprosky, and J.O. Galante, A 3-Year Prospective Study of Serum Titanium Levels in Patients with Primary Total Hip Replacements, *Applications of Titanium and Its Alloys: The Material and Biological Issues*, STP 1272, S.A. Brown and J.E. Lemons, Ed., American Society for Testing and Materials, 1996, p 400–408
152. R.L. Williams, S.A. Brown, and K. Merritt, *Biomaterials*, Vol 9 (No. 2), 1988, p 181–186
153. S.K. Chawla, S.A. Brown, K. Merritt, and J.H. Payer, *Corrosion*, Vol 46 (No. 2), 1990, p 147–152
154. P.J. Hughes, S.A. Brown, J.H. Payer, and K. Merritt, *J. Biomed. Mater. Res.*, Vol 24 (No. 1), 1990, p 79–94
155. S.A. Brown, P.J. Hughes, and K. Merritt, *J. Orthop. Res.*, Vol 6 (No. 4), 1988, p 572–579

156. S.A. Brown and K. Merritt, *Biomaterials, Med. Dev. Artfic. Organs,* Vol 9 (No. 1), 1981, p 57–63
157. S.R. Sousa and M.A. Barbosa, *J. Mater. Sci.: Mater. Med.,* Vol 2 (No. 1), 1991, p 19–26
158. P. Tengvall, I. Lundstrom, L. Sjoqvist, H. Elwing, and L.M. Bjursten, *Biomaterials,* Vol 10 (No. 3), 1989, p 166–175
159. P. Tengvall, H. Elwing, L. Sjoqvist, I. Lundstrom, and L.M. Bjursten, *Biomaterials,* Vol 10 (No. 2), 1989, p 118–120
160. N. Washizu, Y. Katada, and T. Kodama, *Corros. Sci.,* Vol 46 (No. 5), 2004, p 1291–1300
161. K.J. Bundy, J. Dillard, and R. Luedemann, *Biomaterials,* Vol 14 (No. 7), 1993, p 529–536
162. F. Mansfeld, C.C. Lee, and P. Kovacs, Application of Electrochemical Impedance Spectroscopy (EIS) to the Evaluation of the Corrosion Behavior of Implant Materials, *Proceedings of the Symposium on Compatibility of Biomedical Implants,* P. Kovacs and N.S. Istephanous, Ed., The Electrochemical Society, Inc., 1994, p 59–72
163. J.L. Gilbert, *J. Biomed. Mater. Res.,* Vol 40 (No. 2), 1998, p 233–243
164. R. Silva, M.A. Barbosa, B. Rondot, and M.C. Belo, *Br. Corros. J.,* Vol 25 (No. 2), 1990, p 136–140
165. J.J. Jacobs, J.L. Gilbert, and R.M. Urban, *J. Bone Joint Surg. Am.,* Vol 80 (No. 2), 1998, p 268–282
166. R.M. Urban, J.J. Jacobs, J.L. Gilbert, and J.O. Galante, *J. Bone Joint Surg. Am.,* Vol 76 (No. 9), 1994, p 1345–1359
167. J.J. Jacobs, J.L. Gilbert, and R.M. Urban, *Adv. Op. Orthop.,* Vol 2, 1994, p 279
168. R.M. Urban, J.J. Jacobs, J.L. Gilbert, S.B. Rice, M. Jasty, C.R. Bragdon, and J.O. Galante, Characterization of Solid Products of Corrosion Generated by Modular-Head Femoral Stems of Different Designs and Materials, *Modularity of Orthopedic Implants,* STP 1301, D.E. Marlowe, J.E. Parr, and M.B. Mayor, Ed., American Society for Testing and Materials, 1997, p 33–44
169. H.J. Agins, N.W. Alcock, M. Bansal, E.A. Salvati, P.D. Wilson, Jr., P.M. Pellicci, and P.G. Bullough, *J. Bone Joint Surg. Am.,* Vol 70, 1988, p 347–356
170. J. Black, A. Skipor, J. Jacobs, R.M. Urban, and J.O. Galante, Release of Metal Ions from Titanium-Base Alloy Total Hip Replacement Prostheses, *Trans. Orthop. Res. Soc.,* Vol 14, 1989, p 501
171. F.W. Sunderman, Jr., S.M. Hopfer, T. Swift, W.N. Rezuke, L. Ziebka, P. Highman, B. Edwards, M. Folcik, and H.R. Gossling, *J. Orthop. Res.,* Vol 7 (No. 3), 1989, p 307–315
172. N.J. Hallab, K. Mikecz, C. Vermes, A. Skipor, and J.J. Jacobs, *J. Biomed. Mater. Res.,* Vol 56 (No. 3), 2001, p 427–436
173. P. Kovacs and J.A. Davidson, Chemical and Electrochemical Aspects of the Biocompatibility of Titanium and Its Alloys, *Applications of Titanium and Its Alloys: The Material and Biological Issues,* STP 1272, S.A. Brown and J.E. Lemons, Ed., American Society for Testing and Materials, 1996, p 163–178
174. H.F. Hildebrand and J.C. Hornez, Biological Response and Biocompatibility, *Metals as Biomaterials,* J.A. Helsen and J.J. Breme, Ed., John Wiley & Sons, 1998, p 265–290
175. S.G. Steinemann, Metal Implants and Surface Reactions, *Injury,* Vol 27 (Suppl. 3), 1996, p SC16–SC22
176. H.F. Hildebrand, C. Veron, and P. Martin, *Biomaterials,* Vol 10 (No. 8), 1989, p 545–548

Corrosion Fatigue and Stress-Corrosion Cracking in Metallic Biomaterials

Kirk J. Bundy, Tulane University
Lyle D. Zardiackas, University of Mississippi Medical Center

AS THE FIELD OF BIOMATERIALS SCIENCE proceeds into the 21st century, many changes are taking place. Traditionally, biomaterials have been defined as materials that show physical, mechanical, electrochemical, and, most importantly, biological properties that are compatible with hard and soft tissues that they replace or augment in some manner. Recently, there has been an emphasis on the area of development of biological materials. Along with this emphasis has come a redefinition by some of the term *biomaterials*. The topic of this article is consistent with the traditional definition of biomaterials science and specifically addresses some of the mechanical/electrochemical phenomena related to the *in vivo* degradation of metals used for biomedical applications. The areas addressed here are the properties and failure of these materials as they relate to both stress-corrosion cracking (SCC) and corrosion fatigue (CF).

Because biological systems behave dynamically, both on a macroscopic and microscopic level, and because interference with both the anatomy and physiology of patients must be kept to a minimum, materials and devices must be produced and evaluated in a substantially different manner than is done for systems used for other engineering applications. As an example, when replacing or repairing a piece of machinery, there may be limitations on materials, size, and methods of repair; however, the part or the entire device may be replaced. This possibility is not available when dealing with a human being. There are many more limitations on the materials placed in a biological system, some of which include biocompatibility, stress shielding, and limitations on device dimensions due to patient anatomy. Each of these potential problems has a profound effect on the material out of which a device may be made, which is not the case for materials used for other engineering applications. Additionally, since these materials are placed in a living, changing biological system where these changes cannot be predicted in a precise manner, because they are a function of the individual host, the task is to choose materials that are the most appropriate compromise, keeping in mind that the biological interactions with the patient are of supreme importance.

This leads to the important concept of implant material/device failure. While there are several mechanisms by which metallic load-bearing implants may not achieve their intended goal, the primary engineering-related mechanisms are by wear, corrosion, fretting corrosion, CF, and SCC. Should fracture occur, the primary mode has been identified by numerous investigators as fatigue, which is often exacerbated by the corrosive nature of the physiological environment. Others have suggested that fracture may, in some cases, be due to SCC, especially during the crack propagation stage. These observations make an understanding of both SCC and CF fundamentally important to everyone in the industry, including, but not limited to, primary metal suppliers, device manufacturers, and surgeons who place the implants. In light of these observations, this article deals with both SCC and CF of metals used for implants.

Background

This section considers necessary background information pertinent to understanding factors related to the use of surgical implants and their deterioration in the body environment. These include both biomedical aspects such as active biological responses and the chemical environment characterizing the internal physiological milieu, as well as electrochemical fundamentals needed for characterizing CF and SCC.

Biomedical Aspects

Reasons for Interest in Environmentally Assisted Cracking of Load-Bearing Implants. Metallic implants have been used to help patients for well over a century. Over the years, a great deal of research has focused on corrosion of such devices and of the materials used to fabricate them. This effort was examined in a 1994 review article (Ref 1). The vast majority of this effort is not motivated by concerns related to engineering failures of structures in the usual sense (i.e., failure by fracture). Rather, the focus is predominantly on various adverse biological reactions (inflammation, allergic reactions, carcinogenesis, etc.) that might be related to or a consequence of release and then subsequent transport of metallic corrosion products (in either ionic, complexed, or solid forms) within the body. Such effects will undermine the biocompatibility of an implant material and are the design limits that restrict the selection of metals for *in vivo* use more often than mechanical property concerns. Corrosion fatigue and SCC are important exceptions to this situation. Here, the primary concern is not the biological effects of anodic dissolution, but rather how the presence of these materials in relation to human anatomy and physiology affects the ultimate mechanical failure of the implanted device.

The influence of mechanical forces on corrosion behavior of implant alloys is complex. Effects of stress-enhanced ion release (SEIR), for example, can be observed under conditions of static and dynamic loading, time exposure, and polarization that do not cause failure by fracture (Ref 2–4). Stress-enhanced ion release refers to an increase in the rate of corrosion or susceptibility to corrosion (a lowering of breakdown potential in 316L stainless steel in Ringer's solution almost to the free corrosion potential, for example, as discussed in Ref 5) due to applied stresses. This arises from breaches in the integrity of the passive film on a microscopic scale due to plastic deformation. The effect of mechanical stress on deterioration of implant materials by corrosion processes thus should be viewed as a continuum ranging from SEIR to environmentally assisted cracking. The focus of this article is on the more severe end of this spectrum, CF and SCC. Such cracking is possible since implanted devices can be highly loaded. For example, the load on a weight-bearing joint may be as much as 4500 N (1000 lbf) (Ref 6).

Biological Factors. The major difference between corrosion fatigue and stress-corrosion cracking *in vivo* and in the other engineering environments where they can occur is the fact that there are a number of active biological factors and processes that occur in the body that can alter the loads applied to implants, and even the environmental chemistry, sometimes in dramatic ways. Some of these factors are discussed in this section.

Stress Shielding and Bone Resorption. Unlike metals, bone is a growing tissue whose growth is regulated, in part, by the mechanical stresses placed upon it. When loads applied to bone diminish below normal physiological levels, or increase above certain levels, cells in bone, known as osteoclasts, react by locally destroying bone mass. When osteoclastic activity predominates over osteoblastic (bone-forming) activity, net resorption of bone results. So, in situations such as space flight (where there is a lack of gravitational stress), bed rest, prolonged lack of exercise, or when the load is increased above certain levels such as in orthodontic treatment, the volume of bone and/or bone density may decrease. When resorption has progressed far enough and loads on bone are suddenly increased, fractures can result, a phenomenon often seen in osteoporotic patients.

Since the loads on the skeletal system are substantial, surgical implant devices used to temporarily or permanently replace bone function must have adequate tensile, compressive, and bending strength values and be superior in regard to other mechanical properties as well. This means that high-load-bearing orthopedic implants usually are fabricated from metallic biomaterials (and, to a lesser extent, from ceramics).

From the materials science viewpoint, bone is a composite material made up predominantly of a polymeric protein, collagen, and a mineral phase, hydroxyapatite (HA), $Ca_{10}(PO_4)_6(OH)_2$. The Young's modulus of bone is approximately one order of magnitude below that for common implant alloys such as stainless steel, titanium, and cobalt-base systems. When a composite engineering structure (in this context meaning a bone with an implant) is subjected to mechanical loads, the stiffer component will be the main stress-bearing member. For an orthopedic device, this means that the higher modulus metal bears the predominant portion of the load, while the load on the bone is diminished compared to what it would be under normal physiological conditions. This effect is termed stress shielding or strain shielding (or, sometimes, stress protection).

In certain cases, stress shielding is a positive phenomenon. For example, when a fracture fixation plate is used as a rigid internal splint to treat a bone fracture, the initial relative lack of stress on the bone and the rigid stability of the construct allow healing of the fracture to begin. A certain amount of stress is required to stimulate adequate bone healing, however, and the fracture fixation plate, by allowing early weight-bearing on the part of the patient, helps to accomplish this. On the other hand, if stress shielding is severe enough, bone healing may be delayed and bone resorption can even occur (Ref 7). This can lead to implant loosening. Also, in the bone adjacent to a prosthesis such as a total hip replacement (THR), as the amount of bone structure that can serve as a support for the femoral component of a THR diminishes, cantilever bending loads on the device can increase substantially. This may eventually lead to premature fatigue or CF failure of the device.

Fracture Healing/Nonunion. Though most bone fractures are treated by manipulation and plaster cast fixation (Ref 8), often internal fixation devices are needed. Rigid stability and a vascular supply to the fractured bone that is intact are essential requirements for normal fracture healing (Ref 9). Compression between bone fragments serves to enhance stability, but in itself does not promote bone remodeling. While micromotion, or cyclic loading and unloading below a certain level, may enhance healing of bone, if macromotion at the interface occurs, delayed healing or nonunion of the fracture will result, and a pseudoarthrosis will be formed. In this case, an implant will be subjected to a variable load spectrum with high and prolonged loading conditions (a scenario that it is not designed to withstand), and fatigue fracture can often occur. Other adverse events and situations associated with implant instability that can lead to such failure include malreduction, bone defects and necrosis, delayed healing, secondary fracture, pathological conditions, and infection (Ref 10, 11).

Difference Between the Environment in the Oral Cavity and the Internal *in vivo* Milieu. The chemical composition of various body fluids with which metallic biomaterials can be in contact may differ substantially in key aspects. Table 1 shows ionic contents of blood plasma, extracellular fluid, and saliva; dissolved gas concentrations in arterial blood, venous blood, and interstitial fluid; and organic constituents in plasma and saliva.

The primary, though not the only, chemical variable that affects *in vivo* corrosion is the chloride ion concentration, due to its influence on integrity of passive films (discussed later). The chloride content in the internal body milieu is much higher than in saliva, which means that the internal environment of the body is much more aggressive compared to that found in the oral cavity. This difference accounts to a significant degree for the much wider spectrum of metallic materials used in dentistry, as opposed to in orthopedic or cardiovascular surgery. The protein moiety is also very important, since proteins will adsorb onto the surface of all metallic biomaterials and can exert substantial influences on corrosion processes.

Pertinent Fundamentals of Corrosion and Electrochemistry

As indicated previously, the service environment in the body is rather complex, and the specifics regarding inorganic and organic chemical composition, as well as active biological responses, applied loads, and so on, can have an impact on the corrosion resistance of implant alloys. Bundy (Ref 1) has considered these topics more completely. Here, aspects of corrosion science that are pertinent to the influence of the environment on the mechanical failure of metallic implant devices are briefly reviewed.

Body pH, though varying from one location in the body to another, is basically set by homeostatic regulation, usually to a value, or narrow range of values, on the slightly alkaline side. Under these conditions the cathodic reaction is reduction of dissolved oxygen:

$$O_2 + 2H_2O + 4e^- \rightarrow 4OH^- \qquad (Eq\ 1)$$

Acidic conditions can develop in pits, however, and such environments can also exist temporarily when normal physiological processes are upset at locations where inflammatory reactions are underway. In such situations, other reduction reactions might be pertinent:

$$O_2 + 4H^+ + 4e^- \rightarrow 2H_2O \qquad (Eq\ 2)$$

$$2H^+ + 2e^- \rightarrow H_2 \qquad (Eq\ 3)$$

All body solutions contain organic substances, and many contain quite a number of proteins. Generally speaking, these will interact with a foreign surface that is presented to the body through both hydrophobic and electrostatic forces that on balance serve to attract the proteins to the interface, where they will adsorb onto it, as pointed out previously. These may exert an inhibitory effect on corrosion in some cases, while accelerating corrosion in others.

Pertinence of Passive Films and Their Disruption. Though some biomaterials will derive their corrosion resistance through the mechanism of thermodynamic immunity (those that contain large proportions of gold, platinum, and palladium, for instance), the vast majority of implanted devices achieve their corrosion resistance (and biocompatibility) because they develop a passive film that maintains its integrity under *in vivo* conditions. A convenient way to visualize the conditions that allow passivity for a given metal is the potential-pH diagram, also known as a Pourbaix diagram. Figure 1 provides a specific example for an element of importance (cobalt) for fabricating surgical implants. Most published Pourbaix diagrams assume that the electrolyte is pure water. There are computer calculation programs that can determine the appearance of the diagram for a selected electrolyte composition, though, and accounting for the difference between pure water and a more realistic electrolyte better simulating body conditions is important for understanding the behavior of metallic biomaterials in the body (Ref 13).

Most, but not all, of the corrosion mechanisms found to be important for alloys serving in industrial environments have also been found to occur under some circumstances *in vivo*, for example, pitting, crevice corrosion, intergranular corrosion, galvanic corrosion, fretting corrosion, and the subjects of this article, CF and SCC. Many of these phenomena, the latter two in particular, are related to the stability of the passive film on the biomaterial.

When using a Pourbaix diagram to assist in prediction of potential corrosion problems, it should be kept in mind that for localized corrosion processes, such as pitting, crevice corrosion, and SCC, the chemical conditions that are established within the occluded cell can dramatically differ from those in the bulk. For example, in dilute near-neutral chloride solutions, the environment within a pit has been seen to be quite acidic and highly concentrated regarding chloride content.

Besides the Pourbaix diagram, a theoretical construct, one of the best graphical means for visualizing electrochemical conditions that foster passive layer stability is the experimental potentiodynamic polarization curve. An example for an active-passive material is shown in Fig. 2. The zones of borderline passivity where the electrode potential is such that material would be most vulnerable to SCC are indicated in the figure.

In addition to just placing an alloy in a vulnerable region, as shown in Fig. 2, electrode potential can have additional substantial impact on environmentally sensitive failure processes. The exact influence is dependent on the corrosion mechanism involved. For example, if hydrogen embrittlement (HE) plays a role in the failure process, then cathodic polarization (which would increase hydrogen production) would generally accelerate the rate of crack growth, and anodic polarization would retard it. Hsiao et al. (Ref 14) point out circumstances where cathodic polarization can raise pH, however, which would inhibit hydrogen ingress into the lattice and thus not necessarily lead to heightened embrittlement. On the other hand, if the static or dynamic cracking involves mechanisms controlled by film rupture/anodic dissolution or anionic adsorption (discussed later), then the effect of polarization would be reversed (Ref 15), with anodic polarization raising the rate of crack growth and cathodic polarization decreasing it. Reported effects of variables such as oxygen content, ionic concentration, flow rate, pH, and alloy composition can often be traced to their effect on electrode potential (Ref 15, 16).

Metallic Biomaterials

As pointed out previously, metallic biomaterials are widely used throughout the body in many applications to aid in healing of bone, to functionally replace diseased or injured tissues, and to interfere with the progression or consequences of many disease conditions. This section briefly reviews those materials and alloy systems that have been studied either under laboratory conditions or in animal experiments regarding their susceptibility to CF or SCC, and/or for which reports exist from retrieval studies indicating such susceptibility. Because of biocompatibility concerns, only three alloy systems have so far seen widespread *in vivo* use in orthopedics and other areas. These are stainless steels, cobalt-chromium-base alloys, and titanium-base alloys. In this section both the applications and the alloy systems are briefly surveyed. Further information on these materials, when used for bone fracture fixation, is given by Tencer (Ref 17). Shetty and Ottersberg (Ref 18) also provide further detail on these materials and give a description of the history of their use in surgery. In some applications, to achieve secure biomechanical fixation, porous-coated layers that spur tissue ingrowth are sometimes used, as described below. Though many different alloy systems are used in dentistry (Ref 19), the main one for which there are concerns regarding corrosion and fatigue is dental amalgam. This material is discussed in the last portion of this section.

Uses in Various Surgical Specialties

Metallic biomaterials are used in a great many surgical implant applications, which are mainly focused in the fields of orthopedic and cardiovascular surgery and dentistry. In some situations, biomaterials serve as temporary implants (fracture fixation devices that act as a rigid internal splint, for example) or as permanent replacements for body tissue that is surgically

Table 1 Concentrations of ions, dissolved gases, and organic compounds in body fluids in contact with metallic biomaterials

	Ionic components, mM		
Ion	Blood plasma	Extracellular fluid	Saliva
Anions			
Cl^-	96–111	112–120	15.0–24.8
HCO_3^-	16–31	25.3–29.7	1.0–5.1
HPO_4^{2-}	1–1.5	1	...
SO_4^{2-}	0.35–1	0.4	...
PO_4^{3-}	...	...	5.1–35.1
$H_2PO_4^-$	2	...	...
SCN^-	...	...	2.6
Cations			
Na^{2+}	131–155	141–145	2.6–13.0
K^+	3.5–5.6	3.5–4	1.5–36.1
Ca^{2+}	1.9–3	1.4–1.55	1.5–4.9
Mg^{2+}	0.7–1.9	1.3	0.4

	Dissolved gases		
Gas component	Arterial blood	Venous blood	Interstitial fluid
O_2	100 mm Hg (17–22.3 vol%)	40 mm Hg (11–16.1 vol%)	2–40 mm Hg
Dissolved O_2	3 mL/L	1.2 mL/L	...
O_2 combined with hemoglobin	200 mL/L	154 mL/L	...
CO_2	44.6–55 vol%	50–60 vol% (28.4 mM)	46 mm Hg

	Organics(a), g/L	
Substance	Blood plasma	Saliva
Albumin	30–55	0.02
α-globulins	5–10	...
β-globulins	6–12	...
γ-globulins	6.6–15	0.05
$α_1$-lipoproteins	6–12	...
Fibrinogen	1.7–4.3	...
Total cholesterol	1.2–2.5	...
Fatty acids	1.9–4.5	...
Glucose	0.65–1.1	...
Lactate	0.5–2.2 mM	...
Urea	3–7 mM	0.04
Uric acid	...	0.05
Amino acids	...	0.04
Citrate and lactate	...	0.05
Ammonia	...	0.006
Sugars	...	0.04
Carbohydrates	...	0.73
Lipids	...	0.02
Glycoproteins	...	0.45
Amylase	...	0.42
Lysozyme	...	0.14

(a) Except as indicated. Source: Ref 1, 12

excised (an artificial total hip replacement, for example). Another important orthopedic application is the use of metallic rods and other devices designed to correct spinal deformity (Ref 6). Selected applications are considered in this section.

Orthopedics. Stainless steels are the most widely used orthopedic material; however, of the three alloy systems most commonly used, the stainless steels are the least corrosion resistant. Consequently, they are more commonly used for implants that temporarily serve in contact with *in vivo* fluids (e.g., fracture fixation plates), as opposed to internal prostheses such as artificial joint replacements. Depending on the degree of cold working, a range of mechanical properties is available to the implant designer, as described more fully below. When maximum ductility is needed and a moderate strength is tolerable (e.g., for cerclage wires or certain reconstruction plates), annealed material is often preferable (Ref 20). Cold-worked stainless steel is appropriate for bone plates and screws and intramedullary nails, while stainless steel that has been highly cold drawn (extra hard condition) would be employed for situations requiring maximum strength (e.g., Kirschner wire and Schanz screws) (Ref 20).

Based on their high corrosion resistance in the *in vivo* milieu, both titanium alloys and cobalt-chromium (Co-Cr) materials are often employed to fabricate total joint replacements. These joints are articulating, however, and so the wear resistance of the materials comprising the wear couple is quite important. Titanium alloys have relatively poor wear properties compared to Co-Cr materials, so the cobalt-base alloys are more commonly used for articulating metallic components of total joint replacements. Surface-hardening procedures such as nitriding, while improving the wear resistance of titanium-base orthopedic biomaterials, still do not produce materials as wear resistant as the cobalt-base alloys. Additionally, Rodriguez et al. (Ref 21) have shown that nitriding of Ti-6Al-4V reduces low-cycle fatigue strength by 10%.

In part for reasons related to the issue discussed in the prior paragraph and in part for reasons of surgical convenience, since around the 1980s, the modular artificial joint concept has become increasingly widespread, particularly for the artificial hip. Here, the head of the femoral component is fabricated from a cobalt-base alloy (to optimize wear resistance of the device) and its stem is made from extra-low interstitial (ELI) Ti-6Al-4V (the alloy of greater strength and fatigue resistance). Though this arrangement creates a galvanic couple, it was originally thought that since both materials were passive, corrosion problems would not be of concern (Ref 22). However, a rather high frequency of severe corrosion of such devices was often observed (Ref 23), for reasons thought by some as due to a fretting-assisted crevice-corrosion mechanism (Ref 24).

Though the stainless steel, cobalt-base, and titanium-base alloy systems account for virtually all of the metallic devices used in orthopedics, there has been some interest in using other metals, such as tantalum and niobium, but mechanical properties of these materials are limiting factors. A zirconium-base alloy containing 2.5 wt% Nb has also been developed for orthopedic (knee implant) applications (Ref 25).

Dentistry. Many alloy systems are used in dentistry in a range of applications including fillings, wires, crowns, bridges, removable partial dentures, single tooth implants, and so forth (Ref 19). These include gold and gold alloys (including also white gold and low gold compositions), base metal alloys (based on Co-Cr and Ni-Cr systems), materials of the type discussed below (e.g., austenitic stainless steels such as 316L, commercially pure (CP), titanium and Ti-6Al-4V, and Nitinol), Elgiloy, β-titanium, dental amalgam, tantalum, and others (Ref 19, 26, 27). In terms of the themes most germane to this article—corrosion, fatigue, and deterioration due to applied stress—the material of most interest in this regard is dental amalgam, which is discussed in more detail below.

Cardiovascular Surgery. Metallic materials do not play nearly as great a role in cardiovascular surgery as they do in dentistry and orthopedics. Here, polymers are the main class of biomaterials that are used. Yet there are

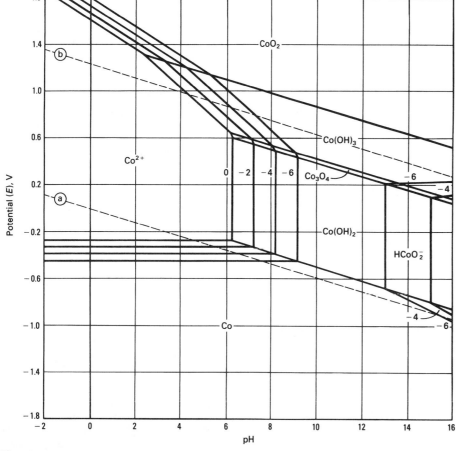

Fig. 1 Pourbaix diagram for cobalt

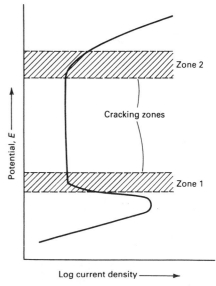

Fig. 2 Polarization curve and how it relates to vulnerability

certain applications where various alloys have been and are employed to treat serious and even life-threatening disease conditions afflicting the heart and circulatory system; artificial heart valves, pacemaker leads, and stents, for example.

To consider these briefly, older designs of artificial heart valves used to treat incompetence and stenosis of natural valves in the heart were based on ball and cage designs. Polymers were used for the occluder and sewing ring, but the cage (consisting of a metallic ring and wire struts) was usually made from Ti-6Al-4V ELI or Co-Cr alloys. The Björk-Shiley valve considered later also had a Co-Cr frame and used a carbon disk occluder. Cardiac pacemaker leads inevitably must be made from electrically conductive materials. Some designs have used porous metals at the ends of the leads to foster fixation of the electrode to the heart by tissue ingrowth into the pores. Though stresses are not high in this application, the leads will experience 3 to 4×10^7 heartbeats annually and should last for many years, so strong and highly corrosion-resistant biomaterials should be used. Materials used to fabricate pacemaker leads and electrodes have included stainless steel, platinum-iridium (Pt-Ir), stainless steel plated with gold or platinum, silver-palladium, and cobalt-base alloys (Elgiloy and MP35N) (Ref 28). Fatigue failures of pacemaker leads have occurred. Elgiloy and Pt-Ir have been the most successful materials used. Elgiloy is more fatigue resistant, but corrodes more. The likelihood of such failures can be diminished by geometrical changes in the design of the lead wires. These new electrodes combine both of these biomaterials in geometrical configurations that provide sturdy pacemakers with superior fatigue resistance. This approach, however, might lead to a worse corrosion situation due to the use of dissimilar metals and the possibility of galvanic coupling.

Stents are a much newer application for metals in the cardiovascular system, one that has become widespread only in the last decade or so. These devices consist of a thin tubular cage that is used to reopen a blood vessel suffering from partial occlusion of the lumen. Generally, they are put in place and expanded using a balloon arthroplasty procedure. Such coronary stents have mainly been made from 316L stainless steel, ASTM F 138 (UNS 31673), see the section "Stainless Steels" below. It is essential, however, that the stent be visible on radiographs. This can be problematic for stainless steels in thin sections. Using metals of higher atomic number can alleviate this problem, since materials become more radiopaque for higher atomic numbers. The desired degree of radiopacity can be achieved either by using alternative materials, tantalum for example, or else by alloying the basic stainless steel composition with about 5 to 6 wt% Pt (Ref 29, 30).

Stainless Steels

At various times in the past, 302, 304, and 316 stainless steels were used as implant materials. Gradually, 316L became the most commonly used of these and is still in use today. Stainless steels such as 316L cannot be hardened by heat treatment, since they are austenitic and could become partially or fully ferritic, but they can be readily work hardened. There are four compositions of contemporary stainless steels for which implant alloy standards have been developed, as shown in Table 2. The addition of molybdenum to the steel helps to minimize pitting corrosion susceptibility. For 316L stainless steel (ASTM F 138), the composition index ($\%Cr + 3.3 \times \%Mo \geq 26.0$) is a quantitative measure of resistance to pitting corrosion known as the pitting resistance equivalent (PRE) (Ref 20). This material is totally austenitic, nonmagnetic (thus free from movement or tissue heating effects when magnetic resonance imaging, or MRI, scanning is performed), and without delta-ferrite, chi, or sigma phases. The other materials are also ferrite-free (to improve resistance to corrosion), and all alloys meet stringent microcleanliness standards (due to the limitations on sulfur and silicon content and vacuum arc remelting, or VAR, or electroslag remelting, or ESR). The limitation on silicon also retards the rate of formation of sigma phase, which can lead to embrittlement for certain heat treatments (Ref 31). The grain size is specified as ASTM 5 or finer for 316L. The reader is referred to the original standards for the details of these microstructural requirements and for further information on other characteristics of these biomaterials. Murty (Ref 32) has presented an overview of the use of stainless steels for biomedical purposes.

Cast stainless steels have been used as implant materials (Ref 33), though their use has declined in recent years. There is one ASTM standard for cast stainless steel (F 745). The composition of this alloy is similar to the 316L composition given in F 138 (see Table 2) except that the %C is higher (0.06% max), as is %P (0.045% max), %Si (1.0% max), and %S (0.030% max). Also, the %Ni and %Mo ranges differ, being 11.00 to 14.00% and 2.00 to 3.00%, respectively, in the F 745 alloy. No values for nitrogen and copper are given in F 745. Most importantly, the minimum chromium and molybdenum composition index for pitting and crevice-corrosion resistance is not specified in ASTM F 745. The implant-grade stainless steels are the easiest to machine of the three alloy systems used to form medical devices. These biomaterials are often used in a cold-worked condition. Micrographs of typical annealed and cold-worked 316L stainless steel are shown in Fig. 3. Many deformation twins are visible for the cold-worked alloy (Fig. 3b). The homogeneous austenite microstructure produced before cold working can be obtained by heating in the 1050 to 1100 °C (1920 to 2010 °F) range followed by rapid cooling to avoid precipitation of carbides (Ref 34). Note

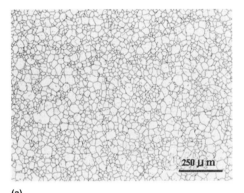

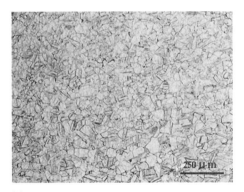

Fig. 3 Typical microstructure of 316L stainless steel. (a) Annealed. (b) Cold worked

Table 2 Chemical composition of stainless steels used in surgical implant applications for which ASTM standards have been established

ASTM designation	UNS No.	Composition(a)(b), %										
		C	Mn	P	S	Si	Cr	Ni	Mo	N	Cu	Others
F 138	S31673	0.030	2.00	0.025	0.010	0.75	17.00–19.00(c)	13.00–15.00	2.25–3.00(c)	0.10	0.50	...
F 1314	S20910	0.030	4.00–6.00	0.025	0.010	0.75	20.50–23.50	11.50–13.50	2.00–3.00	0.20–0.40	0.50	0.10–0.30Nb; 0.10–0.30V
F 1586	S31675	0.08	2.00–4.25	0.025	0.010	0.75	19.50–22.00	9.00–11.00	2.00–3.00	0.25–0.50	0.25	0.25–0.80Nb
F 2229	S29108	0.08	21.00–24.00	0.03	0.010	0.75	19.00–23.00	0.10	0.50–1.50	0.90 min	0.25	...

(a) Single values are maximum values unless otherwise indicated. (b) Balance Fe. (c) $\%Cr + 3.3 \times \%Mo \geq 26$

that 316L and the F 1314 wrought nitrogen-strengthened steel have stringent restrictions on carbon content to minimize the possibility of sensitization. The mass density of stainless steels is roughly 7.9 g/cm^3, and its Young's modulus is about 186 GPa (27×10^6 psi) (Ref 20).

As previously discussed, of the three alloy systems used for orthopedic implants, the stainless steels are the least corrosion resistant under *in vivo* conditions (Ref 34). Sometimes electropolishing is used to improve corrosion resistance. Due to the possibility of ion release, concerns have been raised over the years about the systemic consequences of the nickel in stainless steel, because this element is the one most often implicated in metal-induced hypersensitivity reactions (Ref 20). Though only about 5% of the patient population experiences metal sensitivity due to implants, more than 90% of such cases involve nickel (Ref 35). Such concerns have spurred the development of high-chromium, high-manganese stainless steels. The manganese content prevents formation of martensite and allows much more nitrogen to be soluble in the lattice (Ref 31). The high nitrogen content strengthens the steel by formation of carbonitrides. Since nitrogen is also a potent γ stabilizer, the steel remains austenitic (and thus quite corrosion resistant), even while it is basically nickel-free (Ref 20). Recently, one such low-nickel stainless steel composition has been granted an ASTM standard (ASTM F 2229 as listed in Table 2).

The mechanical properties for the strongest conditions specified in the ASTM standards for the implant-quality stainless steels (that also give requirements for ultimate tensile strength, yield strength, and elongation values) are provided in Table 3. Some of these standards have mechanical property specifications for other conditions, and alloys in other conditions may be used by agreement between the alloy supplier and implant manufacturer.

Cobalt-Base Alloys

Cobalt has a high-temperature face-centered cubic (fcc) form (α) and a hexagonal close-packed (hcp) lower-temperature form (ε). The equilibrium transition temperature is 450 °C (840 °F), though in commercial alloys both forms may be present (Ref 34). The first cobalt-base alloy used for surgical implants was a Co-Cr-Mo casting alloy, originally known as Vitallium (Ref 33). The chemical composition of this alloy is in line with ASTM F 75. Since then, however, a number of alloys based on the Co-Cr alloy system have been introduced for wrought materials and forgings. These newer alloys have significant amounts of nickel, tungsten, and/or iron (and in one case no molybdenum). Table 4 shows the chemical composition of these biomaterials. ASTM F 75 is a casting alloy; ASTM standards F 799 and F 961 cover alloy forgings, and the rest of the alloys in the table are wrought materials. Certain of these alloys derive from the Haynes Stellite (HS) alloys (Ref 36). ASTM F 75 is the analog of HS-21, and F 90 is the analog of HS-25. Pruitt and Hanslits (Ref 37) provide a general overview of cobalt-base alloy systems.

A range of mechanical property values can be developed in cobalt-base alloys depending on the alloy content and thermomechanical treatment. Alloying elements can provide both solution hardening (e.g., molybdenum and tungsten) and development of second phases (carbides, for example) that strengthen the base material through grain-boundary stabilization and dispersion strengthening (Ref 34). Large agglomeration of carbide phases will reduce fatigue life, however (Ref 34). The mechanical properties for the strongest conditions specified for the implant-quality alloys, which give ultimate tensile strength, yield strength, and elongation values, are provided in Table 5. For conditions where the strength values are the same, the specification for the most ductile condition is listed in the table. The ASTM standards mentioned have mechanical property specifications for other working conditions and heat treatments.

ASTM F 75 and F 90 materials are generally rather coarse grained, and grains up to 3 mm diameter have been reported (Ref 36). The grain size for the thermomechanically processed F 799 material is much smaller (Ref 9). Typical microstructures of these materials are shown in Fig. 4 to 6. These alloys are generally single-phased (Ref 9), though in F 75 both α and ε can be present (Ref 34). Hot isostatic pressing of material with the F 75 composition will produce a fine-grain structure with finely dispersed carbides that has improved mechanical properties (Ref 34). The mechanical properties of ASTM F 90 are improved by deformation and twinning (Ref 34). The cobalt-base alloy described

Table 3 Minimum values of mechanical properties of stainless steels used in surgical implant applications for which ASTM standards have been established

ASTM designation	UNS No.	Ultimate tensile strength MPa	Ultimate tensile strength ksi	Yield strength MPa	Yield strength ksi	Elongation, %	Reduction in area, %
F 138	S31673	860	125	690	100	12	...
F 1314	S20910	1035	150	862	125	12	...
F 1586	S31675	1100	160	1000	145	10	...
F 2229	S29108	1379	200	1241	180	12	40

Table 4 Chemical composition of cobalt-base alloys used in surgical implant applications for which ASTM standards have been established

ASTM designation	UNS No.	Cr	Mo	Ni	W	Fe
F 75	R30075	27.00–30.00	5.00–7.00	...	...	...
F 1537(b)	R31537, R31538, R31539	26.00–30.00	5.0–7.0	...	...	...
F 563	R30563	18.00–22.00	3.00–4.00	15.00–25.00	3.00–4.00	4.00–6.00
F 562(c)	R30035	19.0–21.0	9.0–10.5	33.0–37.0	...	...
F 90	R30605	19.00–21.00	...	9.00–11.00	14.00–16.00	...
F 1058	R30003, R30008	18.5–21.5(d)	6.0–8.0(d)	14.0–18.0(d)	...	bal

(a) Balance cobalt (except as indicated). Only alloying elements in amounts greater than 2 wt% (maximum value or average of specified range) appear in the table. Consult the original standards for the alloying elements present in lesser abundance. (b) Also applies to F 799. (c) Also applies to F 688 and F 961. (d) Two grades are given in the standard. Range given spans both grades. The grades contain between 39.0 and 42.0 wt% Co.

Table 5 Minimum values of mechanical properties of cobalt-base alloys used in surgical implant applications for which ASTM standards have been established

ASTM designation	UNS No.	Ultimate tensile strength MPa	Ultimate tensile strength ksi	Yield strength MPa	Yield strength ksi	Elongation, %	Reduction in area, %	Hardness, HRC
F 75	R30075	655	95	450	65	8	8	...
F 1537(a)	R31537, R31538, R31539	1192	173	827	120	12	12	35
F 563	R30563	1310	190	1172	170	12	45	...
F 562(b)	R30035	1793	260	1586	230	8.0	35.0	...
F 688	R30035	1357	197	1343	195	3	...	43
F 90	R30605	896	130	379	55	45	...	...
F 1058	R30003, R30008	2070	300	1550	225	1	...	...

(a) Also applies to F 799. (b) Also applies to F 961

by ASTM F 562 (also known as MP35N or Protosul-10) is a multiphased, high-nickel-content material and is stronger than the alloys mentioned previously. A high nickel content tends to stabilize the fcc α phase (Ref 34), which increases ductility in the alloy systems specified by ASTM F 90, ASTM F 562, and ASTM F 563. MP35N is strengthened by phase transformations induced by plastic deformation and by aging treatments leading to precipitation of Co$_3$Mo (Ref 34). Young's modulus for the ASTM F 75 and ASTM F 799 alloys is 195 GPa (28×10^6 psi), while for the alloy described by ASTM F 90, the value is 210 GPa (30×10^6 psi) (Ref 33).

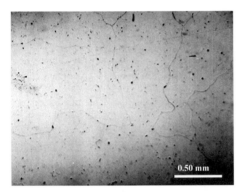

Fig. 4 Typical microstructure of coarse-grained F 75 casting alloy

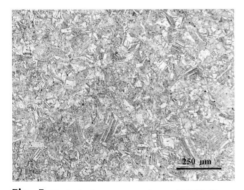

Fig. 5 Typical microstructure of wrought F 90 alloy

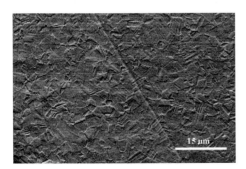

Fig. 6 Typical microstructure of fine-grained, forged F 799 alloy

Titanium-Base Alloy Systems

Titanium undergoes an allotropic transformation at 883 °C (1621 °F). The high-temperature phase (β) is body-centered cubic (bcc), while the lower-temperature phase (α) has an hcp crystal structure. For titanium alloys, depending on the alloying elements present and the thermomechanical treatment, one or the other (or both) of these phases will be found at room temperature. Titanium alloys are therefore classified as α alloys, β alloys, or α/β depending on the phases present in the microstructure. Near-α alloys are those that do contain a small amount of retained β, but not an amount that is sufficient to allow the material to be heat treated. For many years the only titanium-base materials used to make implants were commercially pure titanium (CPTi) and Ti-6Al-4V ELI. Biocompatibility concerns have been raised about the vanadium content (Ref 9); however, in recent years, titanium alloys containing lesser proportions of vanadium, and in some cases aluminum, have been developed for implant use. These newer alloys contain various other elements (namely niobium, molybdenum, zirconium, and iron) as well. The chemical compositions of the alloys for which ASTM standards have been formulated are shown in Table 6. The materials listed in the table are wrought alloys.

The alloying elements are classified according to whether they promote stability of the α phase (α stabilizers) or the β phase (β stabilizers) (Ref 34). The main α stabilizers found in implant materials are aluminum and the interstitial elements oxygen, nitrogen, and carbon. The main β stabilizing elements in implants are vanadium, molybdenum, niobium, and tantalum. The materials found in Table 6 represent different titanium alloy classes. For example, F 136 and F 1472 are α/β alloys, while F 2146 is a near-α alloy. Commercially pure titanium (ASTM F 67 and F 1341) is classified according to four different grades depending on the maximum amounts of iron and interstitial elements (oxygen, nitrogen, carbon, and hydrogen) that the material may contain, as given in Table 7. As the oxygen concentration increases from grades 1 to 4, the strength of the material increases (Ref 9).

The two Ti-6Al-4V alloys, ASTM F 136 (the ELI grade) and ASTM F 1472, also differ mainly in regard to the amount of interstitial impurity elements that are permitted, as shown in Table 8. Limits are placed on these elements since interstitials tend to have an embrittling effect on titanium-base materials. This is particularly true in the case of hydrogen (Ref 14). Commercially pure titanium in the annealed condition has an equiaxed microstructure, as shown in Fig. 7. Microstructures of other titanium-base implant alloys are shown in Fig. 8 and 9.

The mechanical properties for the strongest conditions specified for the implant-quality alloys, which also give requirements for ultimate tensile strength, yield strength, and elongation values, are provided in Table 9. Commercially pure titanium can be strengthened by work hardening (Ref 9) and, especially the grade 4 material, is sometimes used in this condition when extra strength is required. The other materials are usually, but not always, used in an annealed condition. The specified mechanical properties do, however, depend on diameter or thickness. For conditions where the strength values are the same, the specification for the most ductile condition is listed in the table. The original ASTM standards have mechanical property specifications for other sizes, working conditions, and heat treatments. The strength and

Table 6 Chemical composition of titanium-base alloys used in surgical implant applications for which ASTM standards have been established

ASTM designation	UNS No.	Concentration(a), wt%				
		Al	V	Nb	Mo	Zr
F 136	R56401	5.5–6.50	3.5–4.5	...	...	...
F 1472(b)	R56400	5.5–6.75	3.5–4.5	...	...	...
F 1295	R56700	5.50–6.50	...	6.50–7.50	...	...
F 2066	R58150	...	...	...	14.00–16.00	...
F 1813	R58120	...	...	...	10.00–13.0	5.0–7.0
F 2146	R56320	2.50–3.50	2.00–3.00	...	...	...

(a) Balance Ti. Only alloying elements in amounts greater than 2 wt% (average of specified range) appear in the table. Consult the original standards and Tables 7 and 8 for elements present in lesser abundance. (b) Also applies to F 1580

Table 7 Maximum amounts of interstitial elements and iron permitted in commercially pure titanium biomaterials (F 67 and F 1341) used in surgical implant applications

ASTM designation	UNS No.	Composition(a), wt%				
		Fe	O	N	C	H
Grade 1	R50250	0.20	0.18	0.03	0.08/0.10	0.015/0.0125
Grade 2	R50400	0.30	0.25	0.03	0.08/0.10	0.015/0.0125
Grade 3	R50550	0.30	0.35	0.05	0.08/0.10	0.015/0.0125
Grade 4	R50700	0.50	0.40	0.05	0.08/0.10	0.015/0.0125

(a) When two values are given, the first refers to ASTM F 67 and the second to ASTM F 1341.

Table 8 Maximum amounts of interstitial elements and iron permitted in titanium-base (Ti-6Al-4V) biomaterials used in surgical implant applications

ASTM designation	UNS No.	Composition, wt%				
		Fe	O	N	C	H
F 136	R56401	0.25	0.13	0.05	0.08	0.012
F 1472	R56400	0.30	0.20	0.05	0.08	0.015

Table 9 Minimum values of mechanical properties of titanium-base alloys used in surgical implant applications for which ASTM standards have been established

ASTM designation	UNS No.	Ultimate tensile strength (σ_{uts})		Yield strength (σ_y)		Elongation, %	Reduction in area, %
		MPa	ksi	MPa	ksi		
F 67(a)	R50250, R50400, R50550, R50700	550	80	483	70	15	25
F 136	R56401	860	125	795	115	10	25
F 1472	R56400	930	134.9	860	125	10	25
F 1295	R56700	900	131	800	116	10	25
F 2066	R58150	724	105	552	80	12	...
F 1813	R58120	931.5	135.1	897	130	12	30
F 2146	R56320	862	125	724	105	10	...

(a) Also applies to F 1341

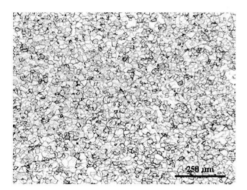

Fig. 7 Typical microstructure of commercially pure titanium, grade 4

Fig. 8 Typical microstructure of Ti-6Al-4V ELI

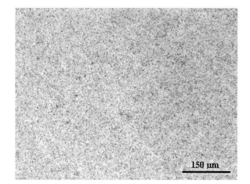

Fig. 9 Typical microstructure of Ti-6Al-7Nb

ductility properties of the titanium alloys are roughly comparable to the stainless steels (compare Tables 3 and 9), but the stainless steels with the same strength generally have greater ductility, while the titanium alloys are about 50% lighter in weight (Ref 34). Also, Young's modulus of titanium alloys are approximately half that of cobalt-base alloys and stainless steels.

Some THR femoral components and knee implants are made from cast titanium (Ref 38). A shape-memory alloy, Nitinol, is used in various dental applications (braces, for example) (Ref 34). This material is based on the Ni-Ti binary alloy containing 54.5 to 57.0% Ni, and the shape memory is based on formation of a martensitic phase and the reverse transformation. The desired final shape of the material is set at a temperature above the martensite start temperature (M_s). The material is then deformed to its temporary shape below the reverse transformation temperature (A_s). Upon subsequent heating above the A_s, the original shape is restored (Ref 34). For pure TiNi, the transformation temperatures are too high (A_s = 165.6 °C, or 330.1 °F) for direct use in the oral cavity, but alloying with TiCo (A_s = −237.2 °C, or −395 °F) can provide a practical dental material (Ref 34). Recently an ASTM standard has been developed (F 2063) that provides guidelines for the material quality of Nitinol used for fabrication of surgical implants.

Porous and Porous-Coated Materials

Over the last several years, bulk porous metal materials with interconnecting porosity have been developed. The two metals that have been used are titanium and tantalum. Porous tantalum has been used for a variety of applications, including the backing on acetabular components of hip prostheses, spinal implants, bone struts, and bridging material, to guide and support bone growth. There are a number of research publications regarding the mechanical properties and histological evaluation of implants made from this material (Ref 39–41). Even more recently, a bulk titanium porous material has been described (Ref 42). The porous titanium foam does not have completely interconnecting porosity, and, therefore, unlike the tantalum foam, will not allow bone to grow through the entire mass of the implant. Porous cobalt-base alloys are also used for implants. Andersen (Ref 43) has presented an overview of the use of powder-metallurgy processes in medical and dental applications.

Porous coatings are often used on portions of permanent implants (such as THRs) so that when tissue (usually bone) grows into the pores, a strong fixation between the artificial biomaterial and the natural tissue results. Though this approach makes good sense from a biomechanical point of view, from the standpoint of materials science, there are some potential concerns to consider. First of all, since the surface area actually exposed to contact with tissue fluids is much higher for porous-coated implants than for those with smooth surfaces, there is the possibility that a greater amount of corrosion products will be released from porous-coated materials, which potentially could heighten biocompatibility complications. Secondly, the irregular topography of porous-coated surfaces, whether formed by sintering of spherical powders, plasma spraying (see below), or sintering together of wire mesh, can conceivably represent stress concentrations that, particularly in notch-sensitive materials, could serve as focal points for crack nucleation. Thirdly, the processing techniques used to produce sintered porous layers can adversely affect the fatigue strength of the alloys used to produce porous-coated implants (see below).

Despite such concerns, the use of porous-coated layers on some portions of surfaces employed in total joint replacements has proliferated since the 1970s and has reached the point where ASTM standards have been developed for certain types of porous materials. Considering Co-Cr base materials, the alloy governed by ASTM standard F 799 can also be prepared by powder-metallurgy processes to form porous-coated layers on implants (Ref 33). ASTM standard F 1377 provides guidelines for the use of the F 75 alloy composition for orthopedic implant porous coatings. Standards have also been adopted for Ti-6Al-4V powder for use in making porous coatings (ASTM F 1580) and for α/β alloy forgings (ASTM F 620). Besides sintering, porous plasma sprayed titanium alloy layers can also be used to make implants (Ref 44).

Fraker points out various metallurgical pitfalls that can be associated with such materials (Ref 34). For example, sintering of Ti-6Al-4V should not be performed above the β transus, as this will adversely affect the fatigue behavior of the material. One fabrication problem for porous-coated Co-Cr-Mo implant materials is that care must be taken during the sintering process and subsequent cooling, otherwise carbides can form that lower the tensile strength and ductility of the material. ASTM F 1537 grade 3 (Table 4) was developed to minimize the loss in properties that occurs during sintering.

Amalgam

Dental amalgam is not one material, but actually is a rather complex family of materials (see the article "Corrosion and Tarnish of Dental Alloys" in this Volume). Amalgam restorations are made by reacting liquid mercury and a solid alloy powder (usually termed amalgam alloy) in approximately equal proportions by weight (Ref 45). The amalgamation reaction produces a solid, but low-melting-point alloy. Invariably, the powder will contain phases based on the silver-tin binary system. Present in various amounts, depending on how the powder is prepared, these are an intermetallic compound Ag_3Sn (known as the γ phase), which forms by a peritectic reaction involving the liquid and a more silver-rich phase (β), and β that is retained at room temperature (Ref 46). Amalgam alloy containing only silver and tin as major alloying elements is rare. Modern amalgam alloy also contains relatively high concentrations of copper (>12%). In the high-copper amalgam, the powder will be composed of a Ag-Sn-Cu ternary (high-copper single composition) or silver-copper eutectic alloy particles in addition to particles of a silver-tin alloy (high-copper admixed). In the first case, where there are not substantial quantities of copper, the amalgamation reaction can be represented as (Ref 46):

$$\beta + \gamma + Hg \rightarrow \gamma_1 + \gamma_2 + \text{Unconsumed alloy particles } (\beta + \gamma) \quad \text{(Eq 4)}$$

where the products $\gamma_1 + \gamma_2$ are also intermetallic compounds Ag_2Hg_3 and $Sn_{7-8}Hg$, respectively. In the second case, the reaction involved can be represented as follows, where the reactants are on the left side of the reaction arrow and the final product is shown on the right side of the reaction arrow:

Alloy particles (Ag-Sn-Cu or Ag-Sn + Ag-Cu
  eutectic particles) + Hg → $\gamma_1 + \eta +$
  Unconsumed alloy particles of either or
  both kinds (Eq 5)

where η is yet another intermetallic compound, Cu_6Sn_5 (Ref 46). The chemical composition range of amalgam alloy has been reported by Craig (Ref 47) as:

Element	Composition, wt%
Silver (Ag)	40–74
Tin (Sn)	25–30
Copper (Cu)	2–30
Zinc (Zn)	0–2
Mercury (Hg)	0–3

From the previous discussion, it can be seen that dental amalgam is a complicated, multiphased biomaterial. Micrographs of amalgam microstructures are shown in Fig. 10.

Though over the years amalgam has probably been the single most widely used biomaterial, a number of clinical problems have become evident related to the durability of the material in the oral environment that can require removal of the material. These include secondary caries, untoward corrosion and tarnish, "ditched" or fractured margins, creep, and gross fracture of the material (Ref 46). Factors that, from an engineering point of view, are related to amalgam deterioration in the oral cavity are thus corrosion resistance and resistance to applied mechanical forces.

Polarization curves of amalgam measured in an artificial saliva do show a passive region (Ref 19). The range of potential where passive film stability exists is much more restricted than is the case with the orthopedic alloys, however, and the passive corrosion current density is much higher for amalgam. The γ_2 phase, $Sn_{7-8}Hg$, which contains no noble metal content, is the phase most prone to corrosion in conventional (i.e., low copper) amalgam formulations. Similarly, for copper-bearing amalgams, it is the η phase, Cu_6Sn_5, that is most prone to corrosion. Though, as pointed out previously, the environment in the oral cavity is not as aggressive as is the case in the internal milieu of the body, there are a number of corrosion mechanisms that can still be operative. For example, besides microgalvanic action occurring in the multiphase structure, macroscopic galvanic couples can develop as different types of restorative metals come into intermittent contact when patients bite down. Since amalgam may be in contact with both saliva and dentinal fluids with about a sevenfold difference in chloride concentration, a differential concentration cell may exist. Also, due to differences in oxygen concentration in solution at the interface between the amalgam and tooth structure, exacerbated by situations where there is margin breakdown and increased leakage, crevice-corrosion conditions can develop due to the differential aeration cell that forms. Because of the intrinsic toxicity of mercury, its presence in amalgam has been controversial in recent years, mainly because of its high vapor pressure and the fact that 65 to 85% of mercury vapor inhaled will remain in the body (Ref 46). There is a dose-response aspect to this issue, however. It has, for example, been estimated that the body burden of mercury that results from one meal per week of saltwater seafood exceeds that from the presence of 8 to 10 amalgam restorations in the oral cavity (Ref 46).

The presence of applied mechanical forces and possibly abrasive action from contact with foodstuffs could perhaps have adverse impact on passive films on amalgam due to fretting effects. Besides its problematic nature regarding corrosion, the mercury- and tin-containing phase is also the weakest phase in a mechanical sense, as demonstrated by microscopic observation in crack initiation and propagation studies (Ref 47). According to one study, about one-quarter of clinical failures of amalgam are due to fracture of the material (Ref 48). Later in this article the role that fatigue may have in such failures is reviewed.

Issues Related to Simulation of the *in vivo* Environment, Service Conditions, and Data Interpretation

The key issues addressed in this section include:

- Frequency of dynamic loading
- Electrolyte chemistry
- Applicable loading modes
- Cracking mode superposition
- Surface area effects

Subsequent major sections in this article describe the fundamentals of CF and SCC, testing methodology, and test findings from laboratory, *in vivo,* and retrieval studies.

Frequency of Dynamic Loading. Since corrosion is an exposure time-dependent phenomenon, as would be intuitively expected, besides the number of loading cycles applied, the

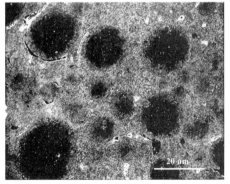

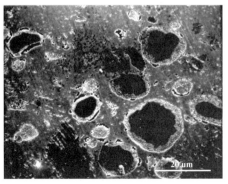

Fig. 10 Typical microstructures of high-copper dental amalgam. (a) High-copper single composition. (b) High-copper admixed

frequency of loading, f, can exert a significant influence on the CF process. One reason for this is that crack propagation rates have a direct relation to mass transport kinetics and chemical reactions that foster embrittlement (Ref 15). Three types of behavior are defined in this regard (Ref 49). For purely time-dependent behavior, crack growth rate per cycle (da/dN) increases with decreasing frequency and is proportional to the time fraction of the load cycle where fatigue damage occurs, $1/(\alpha f)$. Thus, as frequency decreases a larger increment of crack extension occurs per loading cycle. The factor α comes into play since, due to effects such as crack closure, crack extension may not be occurring over all phases of the loading cycle. For example, in a symmetrical load cycle where CF cracking did not occur in the unloading portion of the cycle, α would equal 2. Time-dependent crack growth primarily occurs above the threshold stress intensity for SCC (K_{ISCC}) and can be modeled as a superposition of inert fatigue crack growth and SCC processes (Ref 15). For cycle-dependent behavior, the environment accelerates da/dN, but in a manner that is independent of f. For cycle/time-dependent behavior da/dN is proportional to $[1/(\alpha f)]^\beta$, where β is a constant whose value is approximately 0.5.

For fatigue testing in air or vacuum, where results do not appreciably differ with f, it is common to use a loading frequency as high as possible, usually 50 or 60 Hz or so. For CF, though, it is important to match the loading frequency in a test to the value that typifies the actual service environment, since extrapolation from high-frequency laboratory testing to a lower-frequency service environment is difficult. To simulate the environments that implants experience, usually a frequency of 1 Hz is used.

Electrolyte Chemistry. One of the most important aspects of any test to simulate electrochemical behavior under *in vivo* conditions is the environment to which the implant materials are exposed. There seems to be at least a partial lack of unanimity on the part of the implant corrosion research community on how to select the appropriate electrolyte to best simulate *in vivo* conditions, however. The total chloride content that should be present is often considered to be equivalent to 0.9 wt% NaCl. An electrolyte that is very commonly used is Ringer's solution, though the authors are aware of at least eight discrete formulations of this solution that have been published either by researchers involved in corrosion testing of implant alloys or in works related to various life sciences (Ref 50–57). The range of chemical composition for Ringer's solution found in these reports is shown in Table 10. As can be seen from the table, all of these solutions contain some NaCl and KCl, but the amounts differ and there are a variety of other ingredients (present usually in less abundance) that are found in some solutions, while absent in others. Details of the individual formulations for Ringer's solution can be found in the references themselves and in the review article by Bundy (Ref 1). Tyrode's solution (Ref 58) and Hanks' solution (Ref 59) are other saline solutions that have been used as the electrolyte for corrosion tests with implant alloys.

Similarly, for investigating corrosion behavior of dental materials, a variety of electrolytes have been used for electrochemical testing. These range from simple NaCl solutions (Ref 60–64), as well as a variety of artificial saliva solutions containing additional inorganic and organic ions, low-molecular-weight organic substances, and proteins (Ref 65–70).

Virtually all *in vivo* solutions contain a significant concentration of proteins. Various studies (e.g., Ref 71–78) have been performed to determine how proteins affect the corrosion behavior of metallic implant and dental materials. The solutions used to simulate the action of proteins on corrosion processes generally contain mucin for testing of dental materials and albumin for orthopedic materials, since these are the most abundant proteins found in saliva and blood, respectively.

The influence of proteins on corrosion of implant and dental materials has been found to be complicated. Adsorbed proteins on biomaterial surfaces can form complexes with ions released due to corrosion, alter passivation characteristics, lead to development of active/passive cells because of inhomogeneous adsorption, and either increase, decrease, or leave the corrosion rate unchanged. Further information on this subject has been discussed by Bundy (Ref 1).

Though *in vivo* studies have been performed that suggest that proteins within the body could have some influence on SCC, no laboratory investigations have been conducted on this subject. Given the specificity of the chemical agents and crack tip conditions that can affect SCC, the presence in the body of tens of thousands of discrete proteins, as well as the fact that the chemical agent that makes an alloy vulnerable to SCC can be present at extremely low concentrations in the environment, the influence of proteins on the SCC process is an area that should be subjected to more systematic investigation in future studies.

Applicable Loading Modes. For CF or SCC testing the usual practice is to use a simple loading mode for the test, typically applied in tension or bending. It should be recognized, though, that this is an oversimplification when it comes to simulation of conditions that implants experience in service. The loads applied by the musculoskeletal system to an actual implant can be complex and multiaxial and can include tension, compression, bending, torsion, and mixed-mode situations in which a combination of these occur.

In situations where the applied loading is complex and the stress amplitude can be widely varying (a scenario that certainly characterizes the *in vivo* environment), the most commonly used method for estimating fatigue life due to this cumulative damage is the linear damage law (Ref 79), which says that fatigue failure occurs when:

$$\Sigma(n_i/N_i) = 1 \qquad (\text{Eq 6})$$

where n_i is the number of cycles applied with stress amplitude σ_i, for which the mean number of cycles to failure is N_i.

For CF testing of implants, it is typical to apply uniaxial dynamic loading using a single sinusoidal waveform. However, the time course of loading of an implant over the gait cycle is more complex than this. More realistic waveforms to more closely simulate actual loading have been used in SEIR studies (Ref 2, 4, 80), but the only nonsingle sinusoidal waveform used in CF studies has been a square wave (Ref 81). Furthermore, periods of rest (that always characterize loading of implants) and loading that may change systematically over time due to processes such as shifting of the position of the implant, cracking of bone cement, and bone deterioration due to osteoporosis (Ref 36), may represent further complications. In terms of comparing laboratory simulations to *in vivo* conditions, though, these investigators assume that in the case of fatigue, when a fatigue striation pattern is observed in the laboratory that matches one observed *in vivo*, the inference can be drawn that the loading conditions were comparable. Such an approach is used in the aircraft industry.

Cracking Mode Superposition. Yet a further complication is that the actual crack propagation velocity observed V_{obs} can result from a superposition of crack growth increment contributions due to SCC, V_{SCC}, to the additional crack growth rate during cyclic loading that is due to the corrosive environment, V_{CF}, and the mechanical fatigue crack growth rate in air, V_{air} (Ref 31):

$$V_{obs} = V_{SCC} + V_{CF} + V_{air} \qquad (\text{Eq 7})$$

Surface Area Effects. Particularly for high-cycle CF experiments with smooth samples, it should be recognized that the CF process can be strongly influenced by surface damage resulting from electrochemical attack. As the surface area of the test specimen increases, therefore, it is possible that the measured number of cycles to failure will decrease. So, consideration should be given on how to scale the measured CF behavior of an implant material test specimen with one

Table 10 Range of chemical composition for components in Ringer's solution

Component	Concentration, g/L
NaCl	6.0–9.0
KCl	0.1–0.42
CaCl$_2$	0–0.33
NaHCO$_3$	0–2.4
NaH$_2$PO$_4$	0–0.01
Glucose	0–0.01
MgCl$_2$·6H$_2$O	0–0.20
MgSO$_4$·7H$_2$O	0–0.12
Na$_2$HPO$_4$	0–0.07
NaH$_2$PO$_4$·H$_2$O	0–0.07

surface area to that of a surgical implant device with a different surface area value.

Fundamentals of Fatigue and Corrosion Fatigue

Since most variables that affect metal fatigue also influence CF, both of these phenomena are discussed in this section. Fatigue failures have been studied by materials scientists for many decades. In a macroscopic sense, even a normally ductile material fails in a manner that is basically brittle under the action of fatigue loading. Fatigue failures involve repeated dynamic cyclic loading of an engineering component. In the low-cycle fatigue (LCF) regime, the durability of a smooth material subjected to cyclic mechanical loading can be expressed in terms of the Coffin-Manson equation (Ref 49):

$$\Delta\varepsilon_p = \varepsilon_f'(N_f)^{-c} \quad \text{(Eq 8)}$$

where $\Delta\varepsilon_p$ is the range of true axial plastic strain in a fatigue loading cycle, N_f is the number of loading cycles to failure, and ε_f' and c are material property parameters. When the number of cycles exceeds the transition fatigue life N_T, a smooth specimen will fail by high-cycle fatigue (HCF) according to the Basquin equation:

$$\Delta\sigma = \sigma_f'(N_f)^{-b} \quad \text{(Eq 9)}$$

where N_f again refers to the number of cycles to failure, and σ_f' and b are material properties. The material properties in each of these equations depend on metallurgical, environmental, and time variables (Ref 49). The transition fatigue life is the number of load cycles when the magnitudes of the elastic and plastic strain ranges become equal. The LCF regime is generally when $N < 10^4$ (Ref 82). Since implanted biomaterials are almost exclusively designed for service lifetimes to be as high as possible, HCF fatigue phenomena are of primary concern for metallic surgical implants. When the stress amplitude is low so that the number of cycles to failure is relatively high, the conditions are such that there is sufficient elapsed time so the influence of environmental effects on failure becomes more prominent.

Description of Phenomena

In CF the material is damaged by the conjoint action of mechanical stresses and electrochemical attack. Generally speaking, the Coffin-Manson (Eq 8) and Basquin (Eq 9) laws apply to CF phenomena as well, though in electrochemical environments these laws may be characterized by multiple power law segments (Ref 49). Most importantly, this means that the endurance limit that can apply to fatigue tests in air or vacuum will be eliminated due to the action of the electrolyte. The premature failure that can result due to CF is shown in Fig. 11, which illustrates S-N curve behavior (applied stress versus number of cycles to failure) for 316L in distilled/deionized water versus the same material in Ringer's solution. Also, the fatigue strength of a material in a corrosive environment will be less than that in air, so a material will fail at a fewer number of load cycles in an aqueous electrolyte when compared to inert conditions at the same load.

This behavior is commonly explained in terms of pitting-based crack initiation. Besides pitting, other causes of CF crack initiation include surface damage and minute flaws or hidden imperfections (Ref 34). However, since it is impossible to produce a totally defect-free material from a metallurgical point of view, standards that set quantitative limits for these defects have been established (for inclusion content, for example). The stresses involved in fatigue fracture are often substantially below the nominal yield strength of the material (Ref 83). This is particularly true for biomaterials in surgical implant applications, because when a material experiences yielding *in vivo*, in a functional sense it has already failed. This is a situation that implant designers go to great pains to avoid, but design is *always* limited by the anatomy and physiology of the patient.

Pohler (Ref 84) describes three forms of corrosion fatigue. The first type is pitting-based crack initiation, as mentioned previously. In the second type, fatigue crack initiation and propagation are enhanced by simultaneous electrochemical dissolution. In type 3 (the most common type for implants that meet quality-control standards), visible corrosion does not occur, but passive film destruction in the electrolyte is responsible for reducing fatigue life. With retrieved implants, though, fracture surfaces sometimes show secondary corrosion due to crevice corrosion or fretting in the implant fracture gap (Ref 11), which could be considered as a fourth form of CF.

Fatigue failure of a material occurs by a three-stage process. Stage I consists of crack initiation and initial propagation (Ref 85). Stage II consists of steady-state crack propagation, which can be identified on fractographic surfaces by characteristic surface features known as fatigue striations. Stage III represents ductile overload fracture that occurs when the ligament of intact material remaining after fatigue crack propagation is too small to withstand the applied stresses (which now exceed the tensile strength of the material).

Crack Initiation. In CF type I, which involves pitting-based crack initiation, no fatigue limit exists and corrosive attack is localized. Crack initiation kinetics are controlled by the rate at which pits nucleate and grow to a critical depth (Ref 86). The critical pit depth depends on the range of applied alternating stress. Other factors related to crack initiation include the presence of notches, manufacturing defects (as mentioned previously), and inhomogeneous strain distribution in the loaded structure (Ref 87).

For type III CF, the most common form in surgical implants, fatigue cracks can initiate without initial stress raisers or other flaws (Ref 11). Here, persistent slip steps are formed and the unprotected metal surface exposed as the slip steps breach the passive film will be attacked until repassivation occurs, since a microgalvanic cell is established between the freshly exposed metal and the adjacent material covered with an intact passive film. As these slip emergence and dissolution cycles proceed, a small, local notch can form (Ref 86). These slip bands created by fatigue can have the appearance of extrusions and intrusions at the surface (Ref 11). When these sharp notches reach a critical depth (which is a function of the applied dynamic stress range), a fatigue crack is initiated. For type III, crack initiation is controlled by the critical notch depth and repassivation kinetics (Ref 86). It has been suggested that there is a critical current density for repassivation below which this process cannot be sustained, in which case initiation is controlled by the processes that control it in air (Ref 88). This film rupture/repassivation process is similar to mechanisms thought to be responsible for SCC (see the section "Fundamentals of Stress-Corrosion Cracking" in this article), and Jones (Ref 89) has worked out a unifying explanatory theory for CF and SCC.

As much as 90% of the fatigue life can be taken up by the crack initiation process. Eventually, increasing quantities of multiple slip systems will become operative; precracks, short secondary cracks, and finally one or more major cracks will form, as shown in Fig. 12. A crack

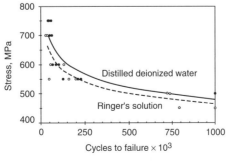

Fig. 11 S-N curves for 316L stainless steel showing premature corrosion fatigue failure when immersed in Ringer's solution compared to deionized/distilled water (37 °C, or 98.6 °F)

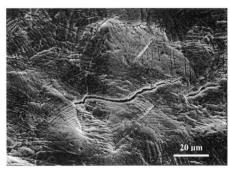

Fig. 12 Persistent slip band formation during stage I fatigue of a retrieved 316L stainless steel implant

will then proceed through the cross section of the implant. The progression of events has been shown micrographically (Ref 11). The multiplicity of crack origins is particularly characteristic of a CF process (Ref 83).

In stage I, crack growth is a strong function of mean stress level and dependent on microstructure (Ref 90). The cracks tend to follow crystallographic planes, but may change direction at grain boundaries or other discontinuities (Ref 86). Stage I fracture surfaces are faceted, often giving the appearance of a brittle cleavage fracture (Ref 85). Stage I fatigue damage is more prevalent in HCF as opposed to LCF.

Crack Propagation. In stage II crack propagation, since a definite crack exists, the ability of the material to resist applied loads is governed by linear elastic fracture mechanics principles. Here, the stress-intensity factor K plays a dominant role. The stress-intensity factor K combines the influence of specimen geometry, loading, and crack size, and is a single parameter that describes the elastic stress field in the vicinity of the crack tip (Ref 91). In this case, both fatigue and CF cracks propagate according to the Paris equation:

$$da/dN = A(\Delta K)^m \quad \text{(Eq 10)}$$

where a is the crack length, N is the number of loading cycles, da/dN is the average rate of fatigue crack growth, and A and m are material-dependent constants (independent of the specimen geometry). The exponent m has a value generally between 2 and 10 (Ref 15). The constant A is inversely proportional to the shear modulus of the material, the square of the yield strength, and the amount of plastic work required to advance the crack by unit area (Ref 92). Only tensile stresses are included in the ΔK range, since compressive stresses do not cause appreciable damage and plastic strain at the crack tip (Ref 49). In a vacuum, there will be a threshold stress intensity below which da/dN tends toward zero and cracks will not propagate. Here, the threshold stress-intensity range is often designated as ΔK_{th} and represents the point at which nonpropagating cracks become propagating under fatigue conditions (Ref 93). In a CF situation, however, cracks will continue to propagate at a slow rate even at quite low values of ΔK (Ref 49).

Stage II cracks usually propagate in a transgranular manner. Morphologically, the most characteristic microscopic signature of stage II is the presence of fatigue striations on the fracture surface. Examples are shown in Fig. 13. Each striation is caused by a single stress cycle, yet the stress during every cycle is not necessarily sufficient to result in crack propagation and the production of a striation (Ref 83). The striations are basically parallel to each other and perpendicular to the direction of crack propagation. Striations have a slight curvature to them, and the direction of crack propagation can be determined from this. The propagation direction is the direction proceeding from the concave to the convex side of the striations. In stage II, secondary cracking may occur as well (Ref 9). A secondary crack occurs at a striation, and its plane lies oblique to the plane of the main crack.

The percentage of the fracture surface area involved in stage II crack propagation can exceed 90% (Ref 9). Macroscopically, there are characteristic features on a fracture surface that allow a fatigue process to be identified. These features are variously known as progression marks, beach marks, crack arrest marks, or clam shells (Ref 9, 83, 85). They are crescent-shaped markings that indicate successive positions where arrests occurred in the advancing crack front. An example is shown in Fig. 14(a). The fracture surface is generally smoother near the crack initiation point and rougher more remote

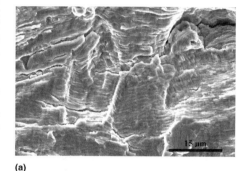

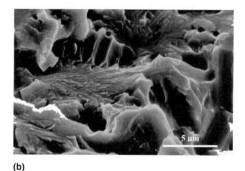

Fig. 13 Micrographs of metallic biomaterials showing fractographic features of stage II fatigue. (a) 316L alloy with characteristic striations and presence of secondary cracks. (b) Titanium alloy with fluting and terraces with feather marks. (c) Low-nickel stainless steel with fluting, terraces, feather marks, and striations

from it. The crack origin can be identified as the apparent focal point on the exterior surface of a part from where the beach marks appear to emanate. Characteristic chevron marks are shown in Fig. 14(b).

Stage III. In stage III the rate of crack propagation increases until the last ligament of remaining material is no longer able to bear the load, and the material then catastrophically fractures. In this regime, it is mainly static failure modes that predominate (Ref 83), that is, cleavage or dimple rupture (depending on whether the material fails in a brittle or ductile manner). Examples of brittle and ductile stage III fractures are shown in Fig. 15(a) and (b), respectively. Stage III failure depends on the microstructure of the material and the mean stress level (as opposed to the range of stress as is the case in stage II). The crack growth rate as a function of ΔK over the course of the fatigue life of a material for the three stages discussed is shown schematically in Fig. 16.

Factors Influencing Susceptibility to Corrosion Fatigue

A great many mechanical, electrochemical, metallurgical, processing, biological, and statistical variables and phenomena affect CF under *in vivo* conditions. In general, any factor that promotes pitting corrosion, allows an embrittling species to enter the metal lattice, promotes crack tip strain-hardening relief, and/or interferes with

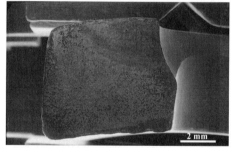

Fig. 14 Macroscopic features characteristic of fracture surfaces of implants that have failed by fatigue. (a) 316L implant showing beach marks. (b) Commercially pure titanium implant showing chevron marks

slip reversal at the crack tip will promote crack growth and lower the CF life of a material (Ref 86). These variables and factors are discussed in this section.

Stress Raisers. For a structural discontinuity that acts as a stress raiser, for example, a hole, notch, or scratch, the degree of stress concentration can be quantified by the stress-concentration factor K_t, which is the ratio of the greatest stress that occurs in the vicinity of the stress raiser to the nominal stress (Ref 95). For cyclic loading, the maximum range of stress $\Delta\sigma_{max}$ is experienced at the tip of the notch and is related to the stress-intensity range:

$$\Delta\sigma_{max} = 2\Delta K/(\pi\rho)^{1/2} = \Delta\sigma K_t \quad \text{(Eq 11)}$$

where ρ is the radius at the tip of the notch, and $\Delta\sigma$ is the range of applied nominal stress (Ref 96).

Notch Sensitivity. In fatigue testing of notched specimens, the fatigue notch factor K_f is defined as the ratio of the fatigue strength of a smooth, unnotched specimen to the fatigue strength of a notched specimen at an equivalent number of stress cycles. The fatigue notch sensitivity, q, is a measure that compares the stress-concentration factor and the fatigue notch factor. It is commonly defined as:

$$q = (K_f - 1)/(K_t - 1) \quad \text{(Eq 12)}$$

The presence of a sharp notch will decrease the time required for a CF crack to initiate (Ref 49). For steels in a NaCl solution, the relationship between the number of cycles required for crack initiation N_i and ρ is:

$$\Delta K/(\rho)^{1/2} = -m[\log(N_i)] + b \quad \text{(Eq 13)}$$

where m and b are constants with positive values. The sharper the notch and the larger the range of stress intensity, the fewer will be the number of cycles required to initiate a fatigue crack. In design of surgical implants, notches and other stress raisers should be avoided if possible without sacrificing the efficacy of the device.

Contributory Mechanical/Chemical Phenomena. Because of the complexity of both the loading environment and the corrosion environment *in vivo*, there are a number of effects and overlapping and related phenomena that can influence CF and SCC. The influence of pitting corrosion on crack initiation for CF has been discussed previously. Pitting has a similar effect on the SCC process. At this point it should be noted that prior to the mid to late 1980s, the composition of implant-grade 316L was substantially different than for the material that is used today to manufacture implants. This change in composition has had a significant effect on the properties of the alloy, especially corrosion resistance. This point should be kept in mind as the literature that has developed over time regarding CF and SCC is reviewed in this article. To consider another example where multiple phenomena are involved in the process of failure for stainless steel (not the 316L of the current composition as specified by ASTM F 138/139), instances are known where fretting and fretting corrosion were involved in the fatigue crack initiation process (Ref 97). A similar process has been hypothesized by Jones et al. (Ref 58) for SCC in implants, and this is the main concern in regard to SCC of airframe components (Ref 98). Pohler (Ref 11) points out that, in rare cases, fatigue attack can be combined with stress corrosion of 316L to cause implant failure.

Any process that can breach the integrity of the passive film on an implant surface (wear, for example) can be suspected of facilitating crack initiation due to the fostering of pit initiation. The influences of the environment (electrolyte composition, temperature, pH, viscosity, conductivity, etc.), electrode potential, biological activity, and mechanical strain (insofar as it creates clean, nonpassivated metal surface) on CF can often be attributed to mass transport and electrochemical reactions that occur within pits, crevices, and cracks (Ref 49).

There is a relationship between CF and SCC for purely time-dependent fatigue crack propagation (see the section above titled "Frequency of Dynamic Loading"). Here, CF and SCC have the same operative mechanism (Ref 49) and are influenced by the same variables, so that the crack propagation velocity is the superposition of the rates for the two processes individually (Ref 99).

Microstructural Features, Processing Variables, Surface Finishing Features, and Metallurgical Defects Related to CF. A number of surface or internal discontinuities are known to act as nuclei for fatigue fractures and SCC failure by either increasing local stresses or adverse reaction with the service environment

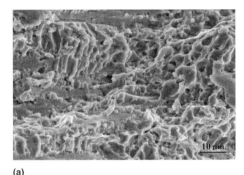

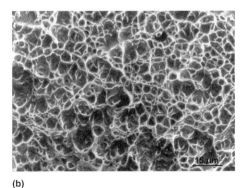

Fig. 15 Micrographs characteristic of stage III fatigue failures. (a) Ti-6Al-7Nb alloy that failed by cleavage-type fracture. (b) Ti-6Al-4V ELI alloy that failed by ductile overload fracture

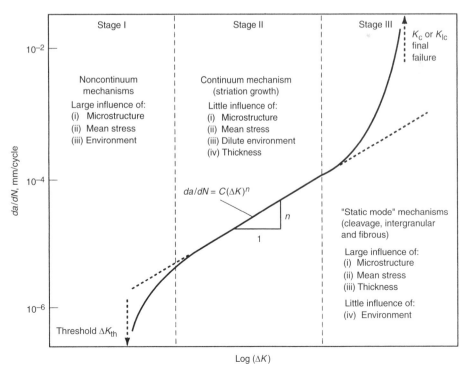

Fig. 16 Fatigue crack growth rate as a function of stress-intensity range. Source: Ref 94

(Ref 86). Such features include cold shuts, seams, laps, porosity, prior cracks, inclusions (oxides, sulfides, and silicates), segregation, and unfavorable flow of grains in forgings. These can act as sites for fracture initiation, causing fractures to start earlier or at lower loads than would be the case in their absence.

Sulfide inclusions can promote corrosion by acting as catalysts for the cathodic reaction (Ref 100). Large inclusions or second-phase particles are also known to affect the local rate of crack propagation (Ref 86). When a fatigue crack approaches such a discontinuity, it is temporarily slowed down if the particle remains intact and will accelerate in the local area if the particle is cleaved. Any single defect (inclusion or otherwise), even if present, should not automatically be singled out as the sole cause of a fracture, however, without first performing a thorough failure analysis to consider other possible causes and contributory factors that could have led to a fracture anyway, without the discontinuity being present. In part because of the adverse effect of inclusions on the fatigue process, the ASTM standards governing surgical implant metals have provisions limiting inclusion content.

Processing steps, such as working, quenching, grinding, polishing, shot peening, machining, and so forth, can produce surface residual stresses in metallic materials (Ref 101). These will add to the applied stresses. Since fatigue failures generally originate at the point of highest tensile stress, residual stress can be beneficial for increasing fatigue life if the stresses are compressive, but residual tensile stresses will generally serve to reduce fatigue life (Ref 102).

The CF behavior of 316L stainless steel has been shown to be sensitive to a number of surface condition and metallurgical variables. Sensitization of austenitic stainless steels, besides making the alloy more prone to intergranular corrosion, has an adverse impact on CF (Ref 103). Methods to avoid sensitization through modification of alloy composition and appropriate heat treatment have been described (Ref 104, 105). Yu et al. (Ref 106) have shown that the CF life of 316L can be increased by almost a factor of 3 by using a finely polished surface as opposed to a coarsely polished one. Laser shock peening has been shown to increase the fatigue life of stainless steel bone screws when the power level used is optimized (Ref 107). For surgical implants, 316L is very often used in a cold-worked condition. The purpose of this treatment is mainly to increase the yield strength, but cold work has also been shown to improve the fatigue properties of 316 stainless steel, as well, by lowering the da/dN value at a given ΔK (Ref 31). Brunner and Simpson (Ref 108) also considered the effect of surface finish and pointed out the possibility of a loss of fatigue strength due to implant contouring. Using a testing device designed for wire testing, Coquillet et al. (Ref 109) investigated the fatigue behavior of austenitic stainless steel orthodontic wires, as influenced by cold work. They measured the ratio of the fatigue limit to σ_{uts} to be roughly in the 0.25 to 0.50 range depending on the degree of cold work. Up to 12% cold work the ratio increased, but after that it declined (and could be even less than for the 0% cold-work condition). Surface defects were also seen to adversely influence fatigue resistance. Hochman and Taussig (Ref 110) have shown that a 2 h low-temperature stress-relief annealing heat treatment at 370 to 425 °C (700 to 800 °F) improves the fatigue resistance of 316L. Davis (Ref 111) points out, however, that the benefits of such a procedure should be weighed against possible difficulties associated with the potential for sensitization.

For Ti-6Al-4V, Yu et al. (Ref 106) studied the CF resistance of samples subjected to different heat treatments. They found that the alloy most resistant to CF was the one that had the finest prior-β grain size. The cracks were observed to propagate along the boundaries between the acicular α and β phases. Ion implantation had been proposed as a means to enhance fatigue strength of implant alloys (Ref 112), and Yu et al. (Ref 106) also determined that an ion-implantation treatment in which nitrogen was implanted in the surface improved both CF resistance and corrosion resistance of Ti-6Al-4V. Wang and Müller (Ref 113) have investigated the fatigue behavior of various titanium-base alloys. They found that the exponent m in the Paris equation changes at a stress-intensity transition point when either the monotonic or cyclic plastic zone size becomes comparable to a key microstructural feature.

In near-α and α/β titanium alloys, the three types of microstructures that can be developed are fully lamellar, fully equiaxed, and duplex (primary α in a lamellar matrix) (Ref 114). For equiaxed microstructures, the grain size and cyclic plastic zone size controls the fatigue behavior. For lamellar structures, the width of lamellae is the controlling microstructural variable, and the monotonic and cyclic plastic zone sizes, respectively, determine the transition point in coarse and fine lamellar structures. Finer structures have improved fatigue resistance (Ref 114). Fatigue cracks generally nucleate in the α phase (Ref 114). For equiaxed microstructures at high ΔK, cracks propagate in a linear, microstructurally insensitive manner, while a tortuous crack path will be found in lamellar structures when the monotonic plastic zone is contained in a single prior-β grain or lamella colony. Other factors found to influence fatigue behavior of near-α and α/β alloys are alpha grain size, colony size of lamellar microconstituents, degree of age hardening, and oxygen content (Ref 114). In duplex structures the volume of α has a strong effect. Though fatigue crack growth in titanium alloys is generally thought to be slower in coarse-grained microstructures, this is only true for long cracks at low values of R (<0.3) (Ref 114, 115). The R ratio is discussed in the section "Corrosion Fatigue Testing Methodology" in this article. When the fatigue resistance of lamellar and equiaxed structures are compared for Ti-6Al-4V, fatigue crack growth rates are much higher for the fine microstructure characteristic of the mill-annealed condition compared to the coarse lamellar structure of the β-annealed condition, particularly when ΔK is low (Ref 115). Mill-annealed microstructures consist of nearly equiaxed α grains with fine β-phase regions, while β-transformed structures have long α lamellae with a thin β-phase boundary (Ref 36). For duplex structures, crack velocity lies between these two extremes.

Influence of Composition. In 316L stainless steel, CF cracks can be initiated due to pitting processes. Yu et al. (Ref 106) showed that doubling the molybdenum content of 316L stainless steel (normally 3 wt%, Ref 34) substantially improved the CF resistance of the alloy, due to the enhanced resistance to pitting conveyed by the increased concentration of molybdenum. The nitrogen level in 316L is restricted to be less than 0.1 wt% (Ref 116). Vogt (Ref 117), however, has shown increasing the nitrogen concentration up to 0.25 wt% increases LCF resistance.

Kalantary et al. (Ref 118) have studied how temperature affects the LCF behavior of 316L stainless steel. They find a minimum in the LCF life between 40 and 70 °C (105 and 160 °F), where the fatigue life is only half that at 20 or 100 °C (70 or 212 °F), an effect that is related to the diffusional mobility of interstitial carbon and nitrogen atoms in the vicinity of an advancing crack tip and how these atoms interact with dislocations present in this vicinity. This finding may be of interest for implant alloys where the service temperature is 37 °C (98.6 °F).

Influence of Surgical Implantation. In the process of implanting a biomedical device, surgeons inevitably must bring surgical instruments in contact with the surface of the biomaterial, as well as applying (sometimes considerable) mechanical force to the device. Use of these tools may cause a breach in passive film integrity on the implant, lead to plastic deformation, and/or create residual stresses. All of these factors can have a negative impact on CF resistance, as described previously. The material damage caused by implantation may be accidentally inflicted, as when a surgical instrument scrapes the implant surface and scratches the passive film, or due to intentional action such as contouring the device (plastically deforming it to alter its shape) so that it will fit and conform to anatomical constraints. Plastic deformation may also result from accidental surgical mistakes, such as when bone screws are overtorqued and deformed upon insertion, which can lead to enhanced fatigue fracture as well as fracture of the screw during explantation (Ref 9). Because of differences in torsional work-hardening characteristics, overtorquing of screws is more likely for titanium than for stainless steel (Ref 20), though with increased experience with using titanium, the surgeon can learn how to compensate for this effect.

Statistical Considerations. In a stress-controlled fatigue test, the stress level is generally the independent variable. Fatigue failure is a statistical process, and so N_f is a random variable. The statistical distributions that have been found to be most valuable for estimating fatigue life for a given stress level are the lognormal distribution and the Weibull distribution (Ref 119, 120). The Weibull distribution offers the advantages that it provides a more conservative estimate on the low side of fatigue life probability and that it was originally developed to model "weak-link" behavior in fatigue applications. It provides a good description of extreme value problems such as estimating the number of cycles to the first failure by CF, the depth of the deepest pit on a corrosion specimen, the time required for the first SCC failure, the time to the first leak in a length of pipe, and so forth (Ref 121). The Weibull distribution can be expressed as:

$$\ln[-\ln\{1 - F(x)\}] = c\ \ln(x - a) - c\ \ln(1/b) \quad \text{(Eq 14)}$$

where $F(x)$ is the cumulative probability of the occurrence of an event, a is a minimum threshold value (that is greater than zero) for x such that below the threshold value there is 0 probability that the event will occur, b is a scaling parameter, and c is a shape factor (Ref 120).

Corrosion Fatigue Testing Methodology

Corrosion fatigue tests can be classified as to whether the main process occurring over the test is crack initiation or propagation. In crack initiation testing, the specimen generally has a smooth surface, and either axial or bending loads are applied. Thus the sample will be subjected to tensile and compressive stresses. Crack propagation tests use a fracture mechanics approach in which the growth of preexisting cracks due to cyclic loading is monitored. A fundamental parameter used to describe a fatigue process in smooth specimens is the stress ratio R, defined as:

$$R = \sigma_{min}/\sigma_{max} \quad \text{(Eq 15)}$$

where σ_{max} and σ_{min} are the maximum and minimum stresses applied over the loading cycle, respectively. When the fracture mechanics approach is used, R equals K_{max}/K_{min}. When a loading cycle is fully reversed and the greatest magnitude of compressive stress equals the maximum tensile stress, $R = -1$. When a material is cycled from 0 load up to a maximum tensile stress and back, $R = 0$. The mean stress, σ_m:

$$\sigma_m = (\sigma_{max} + \sigma_{min})/2 \quad \text{(Eq 16)}$$

has a large influence on fatigue behavior. For a given fatigue life, characterized by the number of cycles to failure N_f, the permissible stress amplitude σ_a is strongly dependent on σ_m. The stress amplitude is defined as half of the stress range:

$$\sigma_a = (\sigma_{max} - \sigma_{min})/2 \quad \text{(Eq 17)}$$

Three relationships (Ref 82) have been used to describe the decrease in fatigue strength with increased mean stress:

$$\sigma_a = \sigma_{fs}[1 - (\sigma_m/\sigma_{uts})] \quad \text{(Eq 18)}$$

$$\sigma_a = \sigma_{fs}[1 - (\sigma_m/\sigma_{uts})^2] \quad \text{(Eq 19)}$$

$$\sigma_a = \sigma_{fs}[1 - (\sigma_m/\sigma_y)] \quad \text{(Eq 20)}$$

where σ_{fs} is the fatigue strength when $R = -1$ ($\sigma_m = 0$), σ_y is the yield strength, and σ_{uts} is the ultimate tensile strength. Equations 18 to 20 are known as the modified Goodman law, Gerber's law, and Soderberg's law, respectively.

The stress ratio can also have an effect on crack propagation when a fracture mechanics approach is used. For example, fatigue crack growth rates in austenitic stainless steels increase as R increases (Ref 31). To account for this effect, an effective stress-intensity factor K_{eff} is used in the Paris equation instead of ΔK to determine the rate of fatigue crack growth:

$$da/dN = A(K_{eff})^m \quad \text{(Eq 21)}$$

where K_{eff} is defined in terms of R:

$$K_{eff} = K_{max}(1 - R)^n \quad \text{(Eq 22)}$$

where n is an empirically determined exponent. Other empirical approaches to account for this effect are also considered in Ref 31.

Fatigue Sample Morphology. Tests with smooth and notched specimens may be used to investigate the initiation of cracks. Since fatigue is a surface-sensitive property, particular care should be taken to standardize the machining of test specimens and the preparation of their surfaces (Ref 82). Various types of loading modes are used to apply stresses to fatigue specimens. When these are conducted, either the applied stress or the applied strain is controlled. For axial or direct stress specimens, the stress or strain is uniform across the cross section. In a rotating bending beam test, a specimen with a circular cross section is rotated while the sample is subjected to dead-weight loading. Either four-point bending or cantilever bending can be implemented with this arrangement. Fatigue can also be investigated with plate specimens subjected to reverse cantilever bending. This is a plane bending test. Three-point bending fatigue tests are also conducted. Another commonly used loading mode for fatigue testing is alternating torsion. Finally, since service conditions often involve multiple loading modes, multiaxial loading is sometimes used to conduct fatigue tests. Recently, a tension-tension standard method for CF testing has been established by ASTM for evaluation of metals used for biomedical implants (ASTM F 1801).

A variety of testing machines are used to conduct fatigue experiments, including those of servohydraulic, electromechanical, and electromagnetic design (Ref 121). Additionally, a wide variety of specimen shapes and sizes are employed for fatigue testing. Since fatigue is a surface-area-sensitive phenomenon, size effects will affect testing results. Scaling for direct prediction of fatigue life of a complex part in service from tests on laboratory samples is generally not feasible (Ref 82).

Corrosion fatigue tests are patterned after the testing techniques used for fatigue testing in air or vacuum, but have additional concerns regarding the presence of the corrosive electrolyte. For example, galvanic coupling with the grips of the testing machine must be avoided and care should be taken to eliminate crevice-corrosion problems at the entry points of the specimen into the environmental cell. Precautions must also be taken to maintain a constant solution temperature, and the purity and constancy of the chemical composition of the electrolyte must be maintained. The potential of the specimen should be monitored and, for some testing purposes, potentiostatically controlled (Ref 49), although monitoring and/or control of these parameters are not required in ASTM F 1801. These conditions require that attention be paid to the electrical isolation of the test specimen from the loading mechanism. Additionally, IR (voltage) drop effects should be accounted for, and measurement of specimen displacement and load can be hindered by the presence of the electrolyte.

The fracture mechanics approach has been mentioned previously. This technique is based on the principle of similitude, that is, the notion that K (the stress-intensity factor) establishes the near-crack-tip driving forces for crack propagation and can be used to describe crack growth for specimens with different geometries and applied loads (Ref 121). The stress intensity is defined as (Ref 122):

$$K = \sigma\alpha(\pi a)^{1/2} \quad \text{(Eq 23)}$$

where σ is the nominal applied stress, a is a characteristic dimension of the flaw size, and α is a factor (determined from linear elastic stress analysis for the specific geometry of interest) that is a function of a/W, where W is the width of the specimen.

Structures used to be designed according to "safe-life" principles in which a crack-free service life was required (Ref 123). Fracture mechanics is a design tool used for crack-tolerant structures (Ref 124), and its basis is "fail-safe" design (Ref 123). This approach came into widespread use along with the introduction of high-strength materials in engineering applications. The basic notion behind fracture mechanics is that the presence of a crack does not signal immediate failure of a structural component. This will not occur in a CF process until the crack reaches a critical size. The purpose of CF testing based on fracture mechanics is to determine the rates at which such subcritical cracks grow before reaching the critical length (Ref 15).

However, such tests do not evaluate the crack initiation process itself. The critical value of the stress intensity K_{IC} (also known as the plane strain fracture toughness) is a material property (independent of the specimen geometry). Testing procedures are generally patterned after those described in ASTM standard E 647 (which contains an appendix related to CF in marine environments), though it should be noted that the principle of similitude is a mechanical one and it may not completely describe processes where the driving forces for crack growth involve interaction of both chemical and mechanical phenomena (Ref 15).

This approach separates the crack propagation process from the initiation process through use of a preexisting subcritically sized crack in the test specimen. The method described previously is known as linear elastic fracture mechanics (LEFM). When the crack tip is not sharp and there is significant crack tip plasticity (known as blunting), elastic-plastic fracture mechanics (EPFM) is usually used to characterize the fracture process. For EPFM, additional techniques are employed: crack tip opening displacement, J-integral, and R-curve methods (Ref 125, 126).

Testing based on fracture mechanics often involves situations where the test specimen is under plane strain. This condition puts limits on the thickness B of specimens that can be used. When B is equal to or exceeds $2.5(K_{IC}/\sigma_y)^2$, plane strain conditions apply; when $B \leq (K_{IC}/\sigma_y)^2/\pi$, plane stress conditions exist; and for B values between these limits, mixed conditions exist (Ref 124, 127). A variety of test geometries can be used for CF testing. These include compact tension, center-cracked tension, single-edge-crack bending, single- and double-edge-crack tension, surface-crack tension, and disk-shaped compact specimens (Ref 49, 128). Crack lengths are most commonly monitored with compliance, electrical potential difference, or optical methods (Ref 15, 49).

Findings from Corrosion Fatigue Laboratory Testing

Stainless Steels. A number of researchers have investigated the CF behavior of implant-quality stainless steels focusing on 316L. Comparative results of these tests are shown in Table 11, which gives fatigue data taken in air and in Ringer's solution using smooth specimens. Because there is no clear endurance limit observed in saline solution, the fatigue strength at 10^6 load cycles, σ_{fs}, determined from the S-N curve, is given in the table.

The table clearly shows that 316L can be vulnerable to corrosion fatigue, with a reduction of about 14 to 25% in fatigue strength due to the action of the corrosive environment. This is similar to the reduction observed by Cahoon and Holte (Ref 130). The fracture surfaces of specimens showing compromised fatigue resistance in saline may not show observable signs of corrosion (Ref 11). A reduction in fatigue strength due to corrosive action was not always seen in Table 11, however. For recrystallized, as opposed to cold-worked stainless steel, no evidence of CF can be seen in the table. Similarly, when a plasma protein (fibrinogen) was incorporated into the environment, no decline in σ_{fs} was seen. For cold-drawn 316L, the 10^7 cycle endurance limit has been reported to be 530 to 700 MPa (77 to 102 ksi) ($R = -1$) (Ref 131).

Jones et al. (Ref 58) performed fatigue tests at 1 Hz on smooth and notched samples of cold-worked 316L and a transformation-induced plasticity (TRIP) steel in Tyrode's solution and in argon at 37 °C (98.6 °F). They calculated the fatigue life ratio (FLR), that is, the ratio of the number of cycles of life at a given maximum stress in the solution to that in the inert environment. They observed that the reduction in fatigue life was a function of applied potential. Fatigue life ratio values for 316L were in the range of 0.26 to 1.07. The TRIP steels showed even more sensitivity to CF. Gill et al. (Ref 132) tested cold forged 316L at 5 Hz ($R = 0.1$) in 25 °C (77 °F) 0.9% NaCl solution and found the endurance limit to be 424 MPa (61 ksi). No air control values were reported.

Zardiackas et al. (Ref 129) conducted a recent study of the CF behavior of various stainless steel implant alloys. A comparison of their results with the other values in Table 11 reveals some very interesting findings. They fatigue tested three alloys in both water and in Ringer's solution: implant-quality 316L, a nitrogen-strengthened alloy (ASTM F 1314), and Bio-Dur 108 (Thin Film Technology, Inc.), a low-nickel-content stainless steel. First of all, their σ_{fs} value in deionized water for 316L was 480 MPa (70 ksi). This is about 1/3 higher than the next highest value shown in Table 11 in air. This clearly demonstrated that the more stringent attention that has been paid to alloy cleanliness and metallurgical quality in the past 25 years or so has brought significant benefits in improving the fatigue strength of this implant alloy stainless steel. Their fatigue strength value for 316L in Ringer's solution was 2.7% lower than the value in water. Comparing the other data in Table 11 with this result, it can be inferred that water itself significantly lowers the fatigue strength of 316L, with the chloride content of Ringer's solution providing further deterioration in the σ_{fs} value. Composition changes to improve the mechanical properties of implant alloy stainless steel also resulted in pronounced enhancement of σ_{fs}. The fatigue strength values at 10^6 cycles were 647 and 608 MPa (94 and 88 ksi) for the F 1314 alloy and for Bio-Dur 108, respectively. For these materials in Ringer's solution, the fatigue strength values were 8.3 and 3.5% less than in water. Additionally, this research was conducted using the ASTM F 1801 (CF testing) experimental protocol, rather than tests in bending or alternating tension-compression, which can significantly affect the fatigue curve.

Windler and Steger (Ref 133) also tested one of the newer stainless alloys in air. They took samples from forged THR femoral components made from a higher nitrogen alloy (ASTM F 1586) and measured the 10^7 cycle fatigue strength in air at 100 Hz using rotating bending at $R = -1$. They found a σ_{fs} value of 587 MPa (85 ksi). Tikhovski et al. (Ref 134) investigated an experimental high-nitrogen (0.75 to 1.0 wt%), low-nickel, Cr-Mn-Mo-N austenitic stainless steel. They reported that its endurance limit in physiological media at 5 Hz was 302 MPa (44 ksi), compared to the following implant-grade alloys:

Alloy	Endurance limit	
	MPa	ksi
Co-29Cr-6Mo	200–280	29–41
CP Ti grade 2	230–280	33–41
316L	250–320	36–46
ISO 5832-9(a)	400–420	58–61
Ti-6Al-4V	400–450	58–65

(a) Nitrogen-containing steel

The endurance limit in air for the Cr-Mn-Mo-N steel was 346 MPa (51 ksi). Eschbach et al. (Ref 135) also provide results from fatigue tests using actual implants fabricated from a low-nickel, high-nitrogen steel. Compared to 316L or titanium implants, they found that the low-nickel steel had less HCF resistance.

Various studies of the CF behavior of 316L have also been conducted using the fracture mechanics approach. Wheeler and James (Ref 81) used square wave loading to approximately simulate the gait cycle and found that fatigue crack growth was enhanced in Ringer's solution compared to air. For the constants

Table 11 Fatigue strength of 316L stainless steel measured in air, deionized water (DI), and in Ringer's solution

Fatigue strength, MPa (ksi)					
Environment					
Air	DI	Ringer's	Change, %	Remarks	Ref
281 (41)	...	241 (35)	−14.2	Solution pH: 6.5	97
246 (36)	...	246 (36)	...	Recrystallized	10
339 (49)	...	292 (42)	−13.9	Cold worked	10
362 (52.5)	...	271 (39)	−25.1	EP	14
362 (52.5)	...	370 (54)	+2.2	EP, Ringer's containing fibrinogen	14
...	480 (70)	467 (68)	−2.7	...	129

EP, electropolished

in the Paris equation, they found that in air $A = 1.442 \times 10^{-24}$ and $m = 4.151$, when da/dN is given in units of in./cycle and ΔK is in psi$\sqrt{\text{in.}}$ In Ringer's solution, the values were $A = 6.222 \times 10^{-21}$ and $m = 3.382$. Colangelo (Ref 136), on the other hand, tested 316L in air and in 0.9% NaCl using a frequency of 30 Hz and actually found higher crack growth rates in air in the lower ΔK range. Here, m was 3.23 in air and 4.16 in the saline solution. These studies were, of course, not conducted on 316L samples with the current stainless steel composition range.

For 316L tested in 1 M NaCl, anodic polarization enhances crack growth rates, while cathodic polarization exhibits the opposite effect (Ref 137). Taira and Lautenschlager (Ref 138) used potentiostatic techniques on fatigue loaded cold-worked 316L specimens and found that the accelerated corrosion combined with the dynamic mechanical loading disturbed the repassivation process, which led to reduced crack initiation times and fatigue life, compared to control samples tested in air. Nakajima et al. (Ref 139) examined the crack growth rate behavior of two stainless steels, said to be used for biomaterials applications, which have chemical compositions in accordance with Japanese (JIS) standards, which are quite different than those falling within the ASTM specifications. One of these alloys, SUS 304, is presently used as a biomaterial, while the other, SUS 329J4L, is described as a duplex stainless steel under consideration for use within the body. For the first alloy, they found da/dN is faster in Ringer's solution than in air for $\Delta K \geq 25$ MPa$\sqrt{\text{m}}$ (23 ksi$\sqrt{\text{in.}}$), but the crack growth rate was similar in each case when the range of stress intensity applied was lower than that. For SUS 329J4L, the da/dN versus ΔK curves were similar in both test environments. Further information on CF behavior of 316L in solutions intended to simulate *in vivo* conditions is given by Imam et al. (Ref 59), Piehler et al. (Ref 140), and Higo and Tomita (Ref 141).

Titanium Alloys. Most investigators who have performed fatigue and CF tests with titanium-base implant alloys have observed a clear fatigue limit. For CP Ti, Pohler (Ref 142) found that the endurance limit σ_{el} was 342 and 374 MPa (50 and 54 ksi) for recrystallized and cold-worked material, respectively. For the cold-worked biomaterial tested in Ringer's solution, the value was 322 MPa (47 ksi), 13.9% less.

Regarding Ti-6Al-4V, Higo and Tomita (Ref 141) found σ_{el} was about 400 MPa (58 ksi) whether smooth specimens were tested in air or saline solution, although they do not report the frequency used for the testing. In their fracture-mechanics-based experiments, it was noted that da/dN was slightly higher in 0.9% NaCl compared to in air. Bratina et al. (Ref 36) report that the fatigue strength for the mill-annealed version of the ELI alloy is about 625 MPa (91 ksi) at 10^7 cycles ($R = -1$). Hoeppner and Chandrasekaran (Ref 143) tested Ti-6Al-4V in an experimental configuration that combined fatigue loading and fretting damage. They noted a marked reduction in fatigue resistance. At a maximum cyclic stress of 524 MPa (76 ksi), the fatigue life was reduced to about only 7×10^4 cycles when there was a fretting component corresponding to a 41.4 MPa (6 ksi) normal stress.

Cook et al. (Ref 144) performed high-speed (10,000 rpm) rotating beam tests on Ti-6Al-4V ELI specimens (prepared in a variety of ways) in air and isotonic saline solution. They found no difference between the results in the two test environments, but observed large differences depending on how the specimens were fabricated. The endurance limits for as-received and ultralow-temperature isotropic (ULTI) carbon-coated samples were 617 and 624 MPa (89 and 91 ksi), respectively. When the biomaterial was notched, σ_{el} fell to 230 MPa (33 ksi). For specimens that were given the sintering heat treatment cycle used for porous coating, the endurance limit was 377 MPa (55 ksi), and for porous-coated specimens themselves, it was only 138 MPa (20 ksi). Particle size is also a factor as to why the endurance limit of the porous-coated titanium alloy can be considerably reduced (Ref 36). Besides the notch and particle size effects, the sintering cycle unavoidably will create a β-transformed microstructure. It has been estimated that the minimum value of σ_{el} needed to avoid failure of a hip prosthesis is 400 MPa (58 ksi) (Ref 145).

Kohn and Ducheyne (Ref 146) performed a study to separate out the effects of interface geometry, microstructure, and surface changes arising from the sintering process. For smooth Ti-6Al-4V they saw that alloying with hydrogen raises the fatigue strength to 643 to 669 MPa (93 to 97 ksi), compared to β-annealed Ti-6Al-4V (497 MPa, or 72 ksi) or preannealed equiaxed Ti-6Al-4V (590 MPa, or 86 ksi). The σ_{fs} value of porous-coated samples, though, was found to be independent of microstructure, and sintered neck junctions were observed to serve as immediate crack initiators so that, in this case, the fatigue behavior is propagation controlled. Kohn et al. (Ref 147) reported fatigue strength values of various heat treatment/microstructure conditions of porous-coated Ti-6Al-4V to span the 177 to 233 MPa (26 to 34 ksi) range. They used acoustic emission and finite-element techniques to determine fatigue mechanisms, to detect incipient damage and monitor its accumulation, and to calculate localized stress patterns. In a related study, Kohn et al. (Ref 148) saw that the sequence of steps involved in the fracture of porous-coated materials was transverse fracture in the porous layer, sphere/sphere and sphere/substrate debonding, initiation of a fatigue crack in the substrate, slow propagation, and finally fast propagation of the crack. For the smooth Ti-6Al-4V alloy, Shetty (Ref 149) prepared nitrogen diffusion hardened (NDH) specimens and tested them in air at 30 Hz ($R = 0.1$). He found that the 10^7 cycle fatigue strength was almost the same as the control alloy (490 MPa, or 71 ksi, for the base alloy compared to 496 MPa, or 72 ksi, for the NDH form).

Besides the weakening effect of a notch, plastic deformation also has a deleterious effect. Brunner and Simpson (Ref 108) provide data indicating that σ_{el} in bending falls by more than 50% for titanium plates that have been plastically deformed, compared to intact, nonbent plates.

Hampel and Piehler (Ref 150) also studied the CF behavior of smooth and porous-coated Ti-6Al-4V. In this case, however, the porous layers were not produced by sintering, but rather by an etching technique or by plasma spraying. They tested the behavior in axial loading at 6 Hz ($R = 0.1$) in 37 °C (98.6 °F) Ringer's solution. The 10^7 cycle endurance limit for the smooth specimens was 414 MPa (60 ksi), and for the etched samples it was 380 MPa (55 ksi). For plasma sprayed material, on the other hand, the samples fractured at about 10^5 cycles at 242 MPa (35 ksi). Wolfarth and Ducheyne (Ref 151) devised a new type of coating geometry (termed the porous-coated nodule) that lowered stress concentrations and raised the fatigue strength up to 305 MPa (44 ksi) for this alloy.

Bratina et al. (Ref 36) found that both mill-annealed and β-transformed alloys show well-defined fatigue striations; although, on fracture surfaces of the latter alloy type, large facets indicative of fast crack propagation sometimes were observed. Also, crack growth rates determined from striation spacing disagreed with macroscopic da/dN measurements, again, particularly in the β-transformed alloy, in part due to tortuosity of the crack path. The striated region only accounted for about 10% of the fatigue life, indicating the importance of the initiation phase on the total lifetime. A similar result was noted for 316L.

Bourassa et al. (Ref 152) examined microknurling of the surface as an alternative for porous coating of titanium alloys, so that the preparation of the textured surface would not have such adverse effects on fatigue strength. Since surface stress concentrations are still present with this type of material, they tried to use heat treatment to achieve a duplex microstructure that would optimize fatigue resistance. This microstructure consisted of equiaxed surface grains (to resist crack initiation) and a bulk lamellar microstructure to retard crack propagation. This attempt was only partially successful.

Nakajima et al. (Ref 139) used a fracture mechanics approach to investigate the CF behavior of CP Ti and Ti-6Al-4V. For the commercially pure material, they found that the da/dN versus ΔK behavior was quite similar in air and Ringer's solution. For the Ti-6Al-4V alloy, however, they saw that, as with SUS 304, when ΔK equals or exceeds 25 MPa$\sqrt{\text{m}}$ (23 ksi$\sqrt{\text{in.}}$), crack growth is enhanced in Ringer's solution compared to air, while below this range, the rate of crack growth is comparable in each environment. Feige and Kane (Ref 87) did not see an increase in crack growth rate in saltwater, compared to air, until the ΔK value used approached K_{ISCC}. Comparing the fatigue response of Ti-6Al-4V in air, distilled water, and saltwater,

McEvily (Ref 137) presents data indicating that, at a given ΔK value, the crack growth rate is fastest in saltwater, slowest in air, and intermediate for the distilled water environment. He also provides experimental results which show that, in saltwater in the higher ΔK region, da/dN increases as the loading frequency decreases from 10 down to 1 Hz. In another study presented by McEvily (Ref 137), in a halide environment, increasing the electrode potential led to an increase in the rate of crack propagation.

Zavanelli et al. (Ref 153) studied the CF behavior of CP Ti and Ti-6Al-4V by testing in air, artificial saliva, and a fluoridated synthetic saliva solution. This investigation was motivated by the fact that these alloys are used for the framework of removable partial dentures. They observed that the fatigue life was significantly reduced in the solutions compared to dry air. In a study following a similar experimental protocol, they found that laser repairing of these alloys also adversely affects corrosion fatigue life (Ref 154).

A number of investigators have studied the fatigue behavior of newer compositions of titanium alloys. Bhambri et al. (Ref 155) performed rotating cantilever fatigue tests in air at 30 Hz ($R = -1$) on Ti-15Mo β titanium alloys. They found the 10^7 cycle σ_{fs} values to be in the 450 to 518 MPa (65 to 75 ksi) range, depending on alloy composition and heat treatment. Wang et al. (Ref 156) tested another β titanium alloy, Ti-12Mo-6Zr-2Fe, in air at 167 Hz ($R = -1$) and observed a 10^7 cycle σ_{fs} value of 585 MPa (85 ksi) for smooth specimens and 70% of that value for notched material. This compared favorably to their tests with Ti-6Al-4V under similar conditions (280 MPa, or 41 ksi, smooth and 53% of this when notched). Mishra et al. (Ref 157) tested Ti-13Nb-13Zr (a near-β alloy) in the smooth condition ($K_t = 1$) and at two notched conditions (K_t values of 1.6 and 3.0) using axial loading in air ($R = 0.1$, 60 Hz). In order of increasing notch severity, the 10^7 cycle endurance limit was 425 to 500, 335, and 135 to 215 MPa (62 to 73, 49, and 20 to 31 ksi). The analogous values for annealed Ti-6Al-4V were 500, 320, and 170 MPa (73, 46, and 25 ksi). More recent studies of innovative titanium-base implant materials are covered in the section "New Materials and Processing Techniques for CF and SCC Prevention" in this article.

Roach et al. (Ref 158) have investigated the CF behavior of a series of titanium-base implant alloy materials including α, α/β, and β alloys. They tested notched and unnotched samples in distilled water and Ringer's solution at 1 Hz. They found no difference in fatigue strength values between the two test environments, indicating the presence of chloride in this case did not adversely affect the fatigue resistance. On the other hand, they noted there was a pronounced weakening due to the presence of the notch. The fatigue strength values at 10^6 cycles, as well as the fatigue notch factors, for the various alloys tested are given in Table 12. The F 1295 α/β alloy (Ti-6Al-7Nb) exhibited the most fatigue resistance in smooth specimens, while the F 2066 β titanium alloy (Ti-15Mo) showed the least. For notched samples, Ti-6Al-4V ELI showed the greatest notch sensitivity, while the F 2066 alloy was least notch sensitive. As is the case with the stainless steel alloys, this study illustrates the improvements that have been made since roughly the 1970s regarding the performance of titanium alloys through changes in alloy composition and processing techniques. In another investigation using a similar experimental design, it was shown that anodizing of CP Ti grade 4 did not affect the fatigue behavior of either smooth or notched specimens (Ref 159).

Morphologically, the appearance of titanium alloy fracture surfaces is more complex than for the stainless steels (Ref 11). Since there are a limited number of operative slip systems in the hcp structure, twinning is also involved in plastic deformation. Depending on the local stress and the crystallographic orientation in the crack plane, regions of widely varying striation density can be found in close proximity, and a variety of additional features, such as terraces, tearing ridges, and dimples, may be present.

Cobalt-Base Alloys. Less work has been done on the cobalt-base implant alloys than for the other systems. Most reports available are of fatigue testing in air. Pohler (Ref 142) reports the endurance limit of a cobalt alloy is 450 MPa (65 ksi) when cold worked and is 240 MPa (35 ksi) when soft. Bratina et al. (Ref 36) report that the fatigue strength of the F 75 alloy at 10^7 cycles is 300 MPa (44 ksi) without a homogenizing solution treatment, which is reduced to 250 MPa (36 ksi) with this heat treatment. Pohler (Ref 11) reports that the fatigue initiation process in the F 563 alloy is similar to that for stainless steel, but that the fracture surfaces have a greater striation density.

Georgette and Davidson (Ref 160) performed studies of the influence of a porous coating on the fatigue resistance of cobalt-base alloys that were similar to those for the titanium-base system described in the prior section. They found that the fatigue strength of as-sintered Co-Cr-Mo material was 34% less than as-cast. The same material when subjected to hot isostatic pressing, however, had a σ_{fs} value just slightly below that of the as-cast condition. Porous coating was found to create crack initiation sites, but the σ_{fs} value was no lower than for the as-sintered condition. The fatigue strength of the porous-coated material improved when the material was hot isostatically pressed.

Berlin et al. (Ref 161) investigated the fatigue behavior of both cast and wrought Co-Cr-Mo alloys in air and found abrasive surface blasting had little influence on the fatigue strength. Laser marking, on the other hand, significantly reduced it for both the cast and wrought versions of the F 75 material. They also found postcasting processes had significant effects on fatigue resistance. The σ_{fs} values at 10^7 cycles for as-cast, hot isostatically pressed (HIP) and solution treated (ST), sintered, and sintered plus HIP plus ST for F 75 were 340 to 480, 380 to 450, 210 to 275, and 345 to 380 MPa (49 to 70, 55 to 65, 30 to 40, and 50 to 55 ksi), respectively. For the wrought version of this material (F 1537), the properties can also be improved by thermomechanical processing. The σ_{fs} values for this alloy in the mill-annealed, warm-worked, and forged conditions are 483, 690, and 759 to 828 MPa (70, 100, and 110 to 120 ksi), respectively. Mishra et al. (Ref 162) tested samples made from either a high-carbon version of F 1537 (CCM Plus) or the F 75 alloy. The samples were porous coated or else subjected only to the sintering cycle. They found σ_{fs} values at 10^7 cycles for porous-coated material were 241 and 207 MPa (35 and 30 ksi) for CCM Plus and F 75, respectively. In the mock-sintered condition, the values were 345 MPa (50 ksi) for F 75 and 448 to 690 MPa (65 to 100 ksi) for CCM Plus (Carpenter Technology Corporation). This same research group also reported on the fatigue properties of an oxide-dispersion-strengthened version of the F 799 alloy known as gas-atomized dispersion-strengthened (GADS) (Ref 163). Using rotating bending techniques, the fatigue strength at 10^7 cycles was 690 to 895 MPa (100 to 130 ksi) as-forged and 620 MPa (90 ksi) for material subjected to the sintering heat treat cycle. In four-point bending, the σ_{fs} values were 670 MPa (97 ksi) for the sintered GADS material and 345 MPa (50 ksi) for the porous-coated alloy. For the F 90 alloy, the fatigue strength is about 207 MPa (30 ksi) in the annealed or solution treated condition, while for an alloy solution treated and cold worked 20%, the value rises to about 428 MPa (62 ksi) (Ref 18). For MP35N the fatigue strength is 186 MPa (27 ksi) in the annealed condition and 745 MPa (108 ksi) when it has been aged and cold worked 40% (Ref 18).

There have been only a few CF studies of a cobalt-base biomaterial. Sury and Semlitsch (Ref 164) found that a wrought Co-Ni-Mo-Ti alloy was superior in terms of CF resistance compared to cast Co-Cr-Mo, and the SCC resistance of the materials was equal. Bolton et al. (Ref 93) used a fracture mechanics approach and found a cast Co-Cr-Mo alloy was vulnerable to CF. The crack propagation rate was significantly higher in Ringer's solution compared to that in air. For fine-grained material, they found the constants in the Paris equation were $m = 2.2$ and $A = 1.35 \times 10^{-11}$ for testing in air (when da/dN was in units of m/cycle and ΔK in units of MPa$\sqrt{m}$. For Ringer's solution, they were $m = 2.33$ and $A = 3.5 \times 10^{-11}$. For fine-grained material tested in air, m was 4.3 and

Table 12 Fatigue strength of smooth and notched specimens of various titanium alloys

Alloy	UNS designation	Fatigue strength MPa	Fatigue strength ksi	Fatigue notch factor, K_f
CP Ti grade 4	R50700	550	80	2.75
Ti-6Al-4V ELI	R56407	700	102	4.67
Ti-15Mo	R58150	500	72.5	2.50
Ti-6Al-7Nb	R56700	750	109	...

A was 2.10×10^{-14}. For this material in Ringer's solution, the constants were $m = 2.4$ and $A = 1.14 \times 10^{-11}$. Lassila and Vallittu (Ref 165) performed LCF testing on a cast Co-Cr alloy used to make clasps for removable partial dentures. They found testing in water or in artificial saliva (Fusayama's solution) reduced the fatigue strength of this material compared to testing in a dry environment.

Comparative Studies of the Various Implant Alloys. Several researchers have conducted studies that allow the CF behavior of the basic orthopedic alloy systems to be directly compared. Nakajima et al. (Ref 139) used a fracture mechanics approach and found for four implant alloys tested, the crack propagation velocities ranked as:

CP Ti > Ti-6Al-4V > SUS 304 > SUS 329J4L

Imam et al. (Ref 59) conducted torsional CF studies in Hanks' solution. They found that when tests on smooth specimens of three implant alloys were carried out with a constant amplitude cyclic strain, the fatigue life of the metals had the ranking:

Ti-6Al-4V > 316L > Co-Cr-Mo

The comparison between Ti-6Al-4V and 316L exhibited previously is consistent with the findings of Piehler et al. (Ref 140), who found the CF performance of the titanium alloy was markedly better than that of 316L, even when significant fretting and wear was involved. Cornet et al. (Ref 166) also found CF performance of titanium alloys to be superior to that of cobalt-base and stainless steel alloy systems.

When Imam et al. (Ref 59) conducted tests in which the stress amplitude was fixed, instead of the strain, the same ranking as given previously was observed at a high level of stress. At a lower applied stress, however, the fatigue life ranking was:

316L > Co-Cr-Mo > Ti-6Al-4V

Rotating bending fatigue tests in air, conducted at 100 Hz by Semlitsch et al. (Ref 145), showed fatigue strength ranges for Co-Cr-Mo, Ti-6Al-4V, and Co-Cr-Mo-Ni alloys were 500 to 800, 600 to 660, and 400 to 780 MPa (73 to 116, 87 to 96, and 58 to 113 ksi), respectively. Whitaker (Ref 167) compared the CF behavior of 316 stainless steel and two titanium alloys, indicated as titanium alloys 130 and 318. He investigated the influence on CF of drilling of holes in bone plates for fitting and of the lowered pH that can be found in patients suffering from shock or may be found in the crevice between the plate and the underlying material. The CF resistance of the stainless steel was poor in the low-pH environment, but was unaffected by drilling. The titanium alloys were less affected by the acidic pH, but were more sensitive to drilling.

Pilliar (Ref 112) has summarized various improvements in materials and implant design that produced more fatigue-resistant implants. He points out that the introduction of the mill-annealed Ti-6Al-4V alloy and forged high-strength cobalt-base alloys into the orthopedic inventory was largely motivated by concerns over fatigue failure, primarily of THR femoral component stems. In addition, design changes were introduced, for example, heavier stem designs with a greater section modulus. This, however, runs the risk of an increase in problems related to stress shielding. For porous-coated devices, improved designs incorporated the knowledge that such layers should be put in the areas of the implant where the device would be highly stressed in tension. Pilliar (Ref 112) also points out that other high-fatigue-strength alloys were introduced for implant use in the 1980s— duplex stainless steels and powder-metallurgy Co-Cr-Mo alloy components made by hot isostatic pressing. The latter materials have a fine grain size (~5 to 10 µm), compared to 100 µm or greater for conventionally cast parts. The carbide distribution also enhances the fatigue strength of these powder-metallurgy parts.

Dental Amalgam. Though many mechanical properties of dental amalgam have been extensively studied, for example, tensile strength, bending strength, compressive strength, creep behavior, and so forth, the fatigue behavior of dental amalgam has received much less attention. Wilkinson and Haack (Ref 168) measured the fatigue limit of an amalgam to be about 97 MPa (14 ksi). Mateer and Reitz (Ref 164) metallographically examined amalgam restorations that had deteriorated during clinical service and found evidence of fatigue cracking, particularly in the vicinity of the margin. Sutow and Jones (Ref 170) also performed fatigue tests and found that the stresses that would fracture amalgam, due to dynamic fatigue loading, were substantially below the static fracture strength. They pointed out that clinical failures might be caused by CF. Hero et al. (Ref 171) performed a controlled study of the behavior of various amalgams under both static and dynamic loading conditions. They observed SCC associated with creep in the static testing. For the fatigue loading, the influence of creep was also noted, but the strains were much less than for the static loading case. A significant SEIR effect was noted for the fatigue loading case. Lian and Meletis (Ref 172) investigated the combined influence of sliding wear and corrosion on the performance of dental amalgam specimens. They observed that this combined action can induce embrittlement and cracking and suggested this may be a major contribution to marginal failures of amalgam restorations.

Findings from *in vivo* Testing and Retrieval Studies Related to Fatigue and Corrosion Fatigue

Retrieval studies form an important part of the knowledge base related to implant failure *in vivo*. ASTM standard F 561 describes guidelines for analyzing implant failures. Over the years, there have been quite a number of reports of metallic surgical implant alloys that have failed by fatigue or CF. One of the first systematic investigations that attempted to examine the underlying metallurgical causes for failure of such devices was the work of Cahoon and Paxton (Ref 173). They investigated a number of fractured 316L and cast Vitallium (the ASTM F 75 composition) implants, most of which they determined had failed by fatigue or CF, with one 316L implant possibly being a case of SCC. They observed a number of metallurgical and engineering defects (excessive inclusion content, casting porosity, large grain size, stress raisers due to poor design, severe plastic deformation and martensite formation, sensitization, and possibly a molybdenum content that was too small) that were the cause(s) of these failures. Colangelo (Ref 136) investigated a number of 316L implants that had fractured. The failure mechanism was fatigue. Corrosion was involved in the initiation of cracking, but Colangelo was uncertain whether it played any role in crack propagation. Similar conclusions were reached by Colangelo and Greene (Ref 174). In implants that had not yet been implanted, they observed some examples where fine hairline cracks were present, and other cases where there was significant surface porosity.

Hughes and Jordan (Ref 175) examined several stainless steel and titanium implants that failed in service. Fatigue, CF, and SCC failure mechanisms were observed for stainless steel. The titanium implant failed by overload fracture originating at a spark-etched trademark, a stress raiser the investigators felt would eventually have led to a fatigue failure after a more extended period of service. Weinstein et al. (Ref 176) reported on fatigue failures in two stainless steel intertrochanteric implants. One occurred in an implant with a recrystallized microstructure with many annealing twins and exhibited tire tracks upon fractographic analysis. Three other implants (two Vitallium and one stainless steel) had not failed, but were cracked. In another early study of biomaterials made of cobalt-base alloys that failed in service, Bates and Scott (Ref 177) showed evidence of fatigue failure in partial dentures. Lisagor (Ref 178) reports on three fracture fixation devices (nails and plates) that failed by fatigue or CF where pitting, crevice corrosion, or fretting were initiating factors. Two were made of 316 stainless steel, while the third was Vitallium. Ducheyne et al. (Ref 179) reported on a retrieved 316 stainless steel implant that failed by CF.

Dumbleton and Miller (Ref 6) describe a fatigue fracture of an intramedullary nail. The surface exhibited pitting near the fracture origin, but it could not be determined whether or not the corrosion played a role in the failure. Corrosion fatigue was implicated in another case study they reported involving a Thornton nail-and-plate device in which the two components were made from dissimilar cobalt-base alloys (ASTM F 75 and F 90). Hochman et al. (Ref 180) discuss the results of an examination of more than 500 retrieved implants that had fractured or failed in some other fashion. In the vast majority of cases (nearly 98%), the major cause of failure of the implant was listed as fatigue, in some cases accelerated by the *in vivo* environment. Wright et al. (Ref 181) report on another extensive retrieval study in which they found 65 fractured total hip replacement femoral components (29 316L implants and 36 cobalt-base implants, 35 ASTM F 75 and one F 90). In some cases, metallurgical defects, notably casting porosity, and stress raisers caused by surface damage resulting from surgical implantation or else fretting due to rubbing against a trochanteric wire were implicated in the failure. Fatigue fractures of total knee replacements also were observed, depending on the implant design.

Pohler (Ref 11) describes fatigue fracture of Co-Cr-Mo Moore pins. Brittle precipitates, alloy segregation, porosity, and inclusions were implicated in the failure. The pins did not exhibit visible corrosion, and the fracture surface was mixed mode with domains with dimples and fatigue striations. In another case, cerclage wire cracked in an intercrystalline manner. The wire was made of 304 stainless steel in the sensitized condition and was in contact with 316L components. Fatigue failure was also noted in bone screws with a high inclusion content. She also describes other examples where implants of high quality that met applicable standards suffered fatigue for mechanical and biomechanical reasons, for example, instability and nonunion leading to pseudoarthrosis. Baswell et al. (Ref 182) examined 10 retrieved 316L wires used for trochanter reattachment or in L-rod instrumentation. They observed 16 fractures, 94% of which were caused by fatigue. Similar to the findings of Coquillet et al. (Ref 109) discussed previously, they noted that wire deformation influenced fatigue life. Pitting of the wires was also thought to be a possible influence on crack initiation. Novak et al. (Ref 183) also observed CF in retrieved orthopedic wires.

Zardiackas et al. (Ref 184) performed a retrieval study involving 275 long bone plates and implants used to treat hip and knee fractures. Of these, 17 stainless steel implants had fractured in service. The mechanism of failure was fatigue, in some cases exacerbated by fretting corrosion. A metallurgical analysis was performed on the broken implants, and all were seen to be within the ASTM specifications in terms of Vickers hardness, oxide inclusion content, grain size, and δ-ferrite content. Pohler (Ref 185) indicated, in the vast majority of cases, failure of long-bone implants were caused by wrong indications for their use, improper use of the implant, or incorrect application of the implant (leaving the fracture with inadequate reduction and insufficiently rigid fixation). The findings of Zardiackas et al. were consistent with this. The reasons they identified for the failures were bone loss, heavy comminution, nonunion, extensive involvement of soft tissue, osteoporotic bone, and premature weight bearing.

Bratina et al. (Ref 36) performed retrieval analyses on implants fabricated from all the common metallic systems. In Co-Cr-Mo they saw that fatigue crack growth occurred along slip planes for both stage I and stage II. Fractographically, more or less featureless facets were observed, but at higher magnification there was a complex topography with crystallographically controlled propagation and striations in some locations. Obstacles in the crack path were said to provide barriers that could stop propagation for perhaps many thousands of cycles, after which the crack would rapidly advance for a considerable distance following a tortuous path. They noted casting porosity and surface damage due to inappropriate fabrication and handling could provide crack initiation sites. They also noted, though, that the precision investment casting methods now in use made the first type of defect unexpected in medical devices. Thus, the failures noted with the Bjork-Shiley heart valve, described later, were surprising. Absent such defects, crack initiation was noted to arise from various microstructural features: coarse grain size, grain boundaries, coarse precipitated carbides, and bead-substrate interfaces (in components prepared with powder-metallurgy techniques).

In contrast to the cobalt-base system, Bratina et al. (Ref 36) observed that fatigue fractures in the CP Ti, Ti-6Al-4V ELI, and 316L alloys do tend to show classical fatigue patterns with clear crack origins, beach marks, and distinct patterns of striations. For CP Ti Brånemark single-tooth replacements, striation morphology of fracture surfaces of fixture portions that had failed were consistent with CF and had a spacing on the order of 0.5 μm. The more common fracture site in these devices is in the abutment screw or the gold coping screw (Ref 186). Bratina et al. point out that the corrosive action of body fluids can serve to obliterate fatigue striations and that counts on retrieved specimens may represent as low as 50% of the actual number of loading cycles.

Besides propagating fatigue cracks, microcracks, short cracks, and arrested nonpropagating cracks were observed in the Ti-6Al-4V alloy by Bratina et al. They point out that identification of the initiation stage may be difficult, both because it may be confined to a very localized region only a few grains thick, and since it is the most damaged portion. Even so, initiation is the critical factor for surgical implant alloys, since so much of the total fatigue life is consumed in that stage.

Bratina et al. (Ref 36) also examined failed Harrington rods made from 316L stainless steel that, in some cases, had fracture surfaces so damaged that fractographic analysis showed only abrasion traces. In other cases, the true fracture surface could be observed, and the fatigue portion was about 50% of the surface, with striation spacings being in the 0.15 to 0.5 μm range and increasing with crack length. The ΔK is not known, but based on results of laboratory studies, a 0.1 μm spacing corresponds to 15 MPa$\sqrt{m}$ (14 ksi$\sqrt{in.}$), while 1 μm represents 35 MPa$\sqrt{m}$ (32 ksi$\sqrt{in.}$) The stage II phase was estimated to represent less than 10% of fatigue life. The Harrington rod device is intended to promote spinal fusion, which should gradually relieve the load. If this does not occur, the device is vulnerable to fatigue fracture, usually originating at the junction between the first ratchet and the shaft. This sort of failure is design related, but modification of the implant design to avoid this would be difficult. Both the notch effect and the influence of stress corrosion were thought to be implicated in the failure of these devices. These investigators feel stress corrosion plays a major role in the fatigue behavior of surgical implant alloys.

In a 316L Charnley THR femoral component Bratina et al. examined, macroscopically it appeared that most of the fractured area represented sudden ductile overload fracture; but, detailed scanning electron microscopy (SEM) and transmission electron microscopy (TEM) work actually showed the striation pattern was present over 90% of the surface. This was interpreted as being due to a low-load situation in the implant, with the bone bearing a significant amount of the load.

Snyder and Snyder (Ref 187) report on two retrieved 316L bone screws that failed by fatigue initiated at sites of pitting corrosion. They attributed the failure to the use of work-hardened material, which was said to be more notch sensitive than the annealed alloy. According to these investigators, cold working reduces the fatigue strength from about 760 MPa (110 ksi) for unnotched material, down to 170 MPa (25 ksi) with a notch present. In contrast, annealed material has a fatigue strength of 230 MPa (33 ksi) and is notch insensitive. They recommend that type 317L stainless steel be used in lieu of 316L. Unfortunately, the higher molybdenum content in 317L can promote the formation of secondary phases such as δ-ferrite and σ phases that are not acceptable in implant-quality stainless steel.

Gilbert et al. (Ref 188) reported on two cases of CF failure of modular cobalt-alloy head/cobalt-alloy stem modular THRs. The cracks were located within a millimeter of the taper junction, and the fractures were intergranular. Three factors were seen to be responsible for the failures: grain-boundary porosity, intergranular corrosion, and cyclic loading.

In another investigation, Zardiackas and his coworkers reported further on findings from their retrieval study work (Ref 9). This work is particularly useful, since it very clearly describes appropriate protocols and aspects to consider for researchers who perform retrieval studies, for example, for surface cleaning; implant sectioning; orientation effects; compositional analysis using x-ray fluorescence (XRF), energy-dispersive x-ray analysis (EDX), and wavelength-dispersive spectroscopy (WDS); hardness measurements; grain size and structure determination; inclusion content measurement; intergranular corrosion susceptibility; radiograph interpretation; visual and SEM analysis (and identification of wear, plus mechanisms of corrosion and fracture); foreign phase identification; and fractographic interpretation. This investigation also points out various pitfalls that may arise during such efforts. Using metallurgical analyses, radiographs, and patient record data (all of which can provide information essential for pinpointing causes of failure), they examined a variety of implants that had fractured in service—a 316L dynamic compression plate, 316L and CP Ti bone screws, an intramedullary nail, a Ti-6Al-4V tibial tray, and a Co-Cr total hip prosthesis stem. None of the devices examined were found to be severely faulty in the metallurgical or engineering sense (i.e., defects in materials, fabrication techniques, or design). The failure mechanism was generally fatigue that arose from nonunion, bone cement deterioration, bone resorption, and patient factors, such as weight. In one case, overload fracture consequent to device removal was observed.

In a study reminiscent of the work of Hughes and Jordan (Ref 175), Naidu et al. (Ref 189) report a case where a retrieved flexible intramedullary nail failed by CF that originated at a mechanically stamped label. They recommend that ASTM standard F 86 (surface preparation and marking of implants) be modified to exclude this type of labeling for cyclically loaded implants.

Magnissalis et al. (Ref 190) observed fatigue fracture in the stem of porous-coated modular THR implants made with Co-Cr-Mo heads and Ti-6Al-V ELI femoral components. The fracture plane went through the proximal portion of the stem where the porous-coated layer was located. It should be noted, this is not the most highly loaded portion of the implant, and these authors indicate their findings verify the concern regarding compromised fatigue strength of porous-coated layers, which was discussed previously.

Pohler (Ref 11) indicates that, over the years, the number of implants failing due to metallurgical causes has diminished as processing techniques have improved, and now most failures occur for mechanical or biomechanical reasons. The review of the failure analyses of retrieved implants given previously underscores this conclusion.

To the authors' knowledge there has been only one research group that has performed CF measurements *in vivo* in an animal model. Morita et al. (Ref 191) built a fixture capable of fatigue testing in the tibia of a living rabbit. They tested type 316 stainless steel and a cobalt-bearing stainless steel (COP), which were both said to be widely used as orthopedic implant materials. For 316, they found 5×10^6 loading cycle σ_{fs} values of 830, 815, 815, and 680 MPa (120, 118, 118, and 99 ksi) for air, saline solution, horse serum, and *in vivo* in the rabbit, respectively. For COP, the *in vivo* value was 680 MPa (99 ksi), compared to 800 MPa (116 ksi) in air. They concluded that the reduction in fatigue strength was due to the low dissolved oxygen content of body fluids. They confirmed this hypothesis in a subsequent study (Ref 192).

Fundamentals of Stress-Corrosion Cracking

For CF, the influence of the environment is generally to diminish fatigue life, although there will be a difference in degree between differing environments in this regard. For SCC, on the other hand, there is a pronounced degree of environmental specificity. A given material will be susceptible to SCC in only a few environments (if any) and will be immune to SCC in the vast majority. The species responsible for SCC need not be present in high concentrations, however (Ref 193). A static tensile stress (applied and/or residual) is necessary for SCC to occur in susceptible material/environment combinations. Often the chemical environment is apparently mild comparatively, and the stress level can be well below σ_y (Ref 105). This accounts for the interest in implant alloy SCC, since such conditions characterize the service environments of surgical implants. It has been stated (Ref 194) that SCC and hydrogen stress cracking represent the two most subtle causes of premature fracture in which corrosion is implicated, because of the lack of warning before failure occurs.

Description of Phenomena

Stress-corrosion cracking involves the conjoint, synergistic interaction of mechanical stress and environmental attack, which ultimately leads to failure by fracture. Understanding of the process requires background in mechanics, metallurgy, chemical thermodynamics, corrosion, and electrochemistry and has been hampered largely because the small scale over where the critical reactions at the crack tip occur precludes direct observation of the processes involved (Ref 137). Nonetheless, SCC has been intensively studied for decades, and many details surrounding the process have been clarified. Because of the many complexities associated with its occurrence, SCC has been called the most intriguing form of corrosion failure (Ref 195).

Types and Stages of SCC. Like CF, SCC is invariably brittle in a macroscopic sense. Stress-corrosion cracks often have a characteristic branching appearance, as shown in Fig. 17. The branching may occur on a macroscopic or microscopic level (see Fig. 17a and b, respectively), depending on the level of stress intensity (Ref 196). On a microscopic scale the fracture may occur in an intergranular (Fig. 18) or transgranular (Fig. 19) fashion.

Differentiating between CF and SCC processes based solely on fractographic evidence is not always a clear-cut procedure, since SCC of 316L, for example, can exhibit striations (Ref 83, 178), normally a signature of stage II CF. However, to the experienced fractographer differentiation between these signatures can often be made. On the other hand, CF processes can be without striations (Ref 197) and may exhibit secondary cracking in stage II (Ref 9). In a section transverse to the fatigue crack plane, such secondary cracking might be confused with the branching pattern characteristic of SCC. Meyn and Brooks (Ref 197) describe ways to differentiate SCC from striationless fatigue based on fractographic features, known as flutes, that may be characteristic only of SCC. However, recent research on CP Ti, titanium alloys, and even high-nitrogen stainless steels subjected to CF (Ref 132) demonstrates these

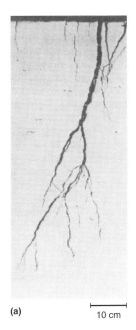

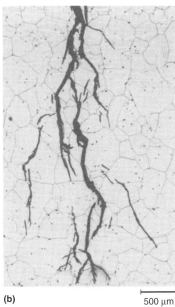

Fig. 17 Characteristic branching appearance of stress-corrosion cracks. (a) Unetched macrograph (original magnification: 10×) exhibiting stress-corrosion cracking branching in type 316 stainless steel. (b) Etched transverse section (original magnification: 20×) of crack tip in type 316 stainless steel specimen that failed by stress-corrosion cracking

alloy systems may show fluting and terraces as a function of localized grain orientation, crystal structure, degree of cold work, and stress-intensity factor. Therefore, fluting is not specific to SCC.

As with CF, it is customary to consider SCC as occurring in multiple phases. Both electrochemical and mechanical factors act as driving forces for the process. During the initiation phase (incubation and nucleation of the crack), localized breakdown of the passive oxide film occurs, and corrosion pits or fissures form. These act as localized stress concentrations from which a crack can nucleate. As the crack propagates in the next phase, mechanical factors become more important (Ref 194). Finally in the last phase, the load-bearing cross section has diminished sufficiently so that overload fracture occurs. An overview of the process is shown in Fig. 20 for a material in which there is no preexisting mechanical flaw or crack. The velocity of crack propagation da/dt is dependent on the stress intensity, and three stages have been identified, as shown in Fig. 21. For a test in which K increases as time goes on (a constant stress test, for example), the crack will not propagate until a threshold value K_{ISCC} is reached. In practice, this is defined (somewhat arbitrarily) as the K value where the velocity is below a certain minimum rate, 10^{-10} m/s, for example. Processes thought to be related to this threshold differ depending on the mechanism of SCC (see the discussion below) and include the critical resolved shear stress for slip (in the slip-dissolution model, also known as the film rupture model), a critical crack opening value that allows mass transport in the crack, or the fracture strain for passive film rupture (Ref 199). In some alloy systems, such as Ti-6Al-4V heat treated to be highly SCC resistant in seawater, ripple loads (small cyclic loads superimposed on a much larger sustained load) can substantially lower K_{ISCC} and the time to failure (Ref 200). Though initiation can be a large fraction of the service lifetime, the mechanisms affecting crack growth are much better understood than are those that influence the crack initiation process. Very little is known about the conditions that control nucleation of SCC cracks (Ref 201), although some local plastic deformation often serves as a precursor for SCC (Ref 105).

Just above K_{ISCC}, the velocity in stage I increases sharply with K in a logarithmic fashion (often reported as proportional to K^4) until a steady-state plateau velocity V_p is reached, which generally is orders of magnitude above the threshold velocity. The plateau velocity for subcritical crack propagation is thought to result from control of cracking kinetics by a rate-determining step (rds) in a sequence of chemical reactions and processes required for cracking. Depending on the specific mechanism involved (see the next section), the rds might involve, for example, mass transport to the crack tip, reactions in the solution in the vicinity of the crack tip, surface adsorption on or near the tip, surface diffusion, surface reactions, absorption into the bulk material, bulk diffusion to the plastic zone in front of the crack, chemical reactions occurring in the bulk material, or the rate of rupture of atomic bonds (Ref 199). Specific factors that might be involved include film rupture and repassivation rates, the corrosion rate, and environmental factors such as pH and crack tip potential. As stress increases further, the velocity is independent of K and stays at V_p. Finally as the final overload phase is approached, da/dt sharply increases again until fracture occurs.

Mechanisms of SCC. Though cases of SCC were first identified nearly a century ago (Ref 202), a full mechanistic understanding of the process still remains elusive. Environmentally induced cracking processes that occur under static load (SCC, hydrogen damage, liquid metal embrittlement, and solid metal induced embrittlement) share certain similarities (such as being exacerbated by increased stress), but do not share a common causative mechanism (Ref 203). Even within one form of such cracking, SCC in this case, a variety of proposed mechanisms exist. Though at one time it was thought that one unifying mechanism might be found to explain all such phenomena, it is now more commonly thought that no one single mechanism fits

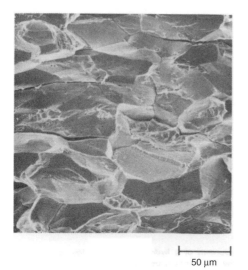

Fig. 18 Stress-corrosion cracking by intergranular decohesion of cold-worked 316 stainless steel at high stress intensity in boiling magnesium chloride

Fig. 19 Transgranular cracking (due to cleavage) resulting from stress-corrosion cracking of Ti-6Al-4V in methanol (transmission electron microscopy p-c replica; original magnification: 2000×)

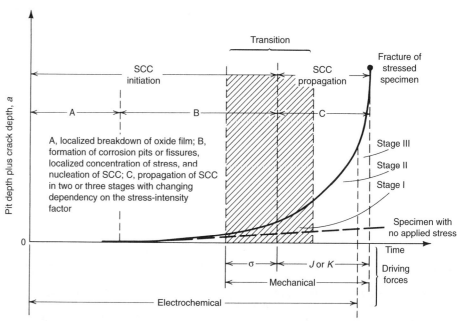

Fig. 20 Overview of the stress-corrosion cracking process. Source: Ref 198

all susceptible materials and circumstances (Ref 202).

In overview, the proposed mechanisms involve either anodic dissolution processes or hydrogen-related cathodic processes (Ref 199, 202). To be accepted as an explanation for a specific material/metallurgical condition/environment combination, a proposed mechanism should be able to explain how cracks are nucleated, the rate of their propagation, and be consistent with the fractographic evidence (Ref 199). Stress-corrosion cracking mechanisms also differ in the effects deemed responsible for the breaking of the atomic bonds at the crack tip. Some are based on electrochemical dissolution, for example, the active path (sometimes termed stress-accelerated dissolution) or slip-dissolution models, which involve removal of material from the crack tip, while others emphasize the importance of ductile or brittle mechanical fracture and separation of material at the crack tip. Models in the latter category consider that the fracture process is aided by interactions with the environment, such as adsorption of environmental species, surface reactions, reactions in the material ahead of the crack tip, and influences on surface films (Ref 199).

For SCC processes that are anodically driven, a thermodynamic requirement is that the chemical conditions in the crack must be such that the passive film is stable, while oxidation of the metal and its dissolution into the electrolyte is possible (Ref 199). In the anodic dissolution models, the walls of the crack remain passive (otherwise the crack tip would blunt, and loss of metal, due to corrosion, would occur much more homogeneously), and the ratio of the distance between the walls and the length of the crack is about 0.001 (Ref 199). The potential difference between a crack tip and the passive film can be substantial, ~0.5 V (Ref 204). The greater activity of the metal at the tip relative to that on the walls is a consequence of the greater dynamic strain at the tip. This dynamic strain represents a localized creep process (Ref 205). Though corrosion is a controlling process, it should not be assumed that the crack grows in a manner that can be calculated from the amount of metal that would be lost due to the total amount of charge that is transferred at the crack tip. This lower limit of velocity can be found from Faraday's law to be:

$$(da/dt)_{min} = i_{ct}(AW)/zF\rho \quad \text{(Eq 24)}$$

where i_{ct} is the anodic corrosion current density for bare metal at the crack tip, AW is the atomic weight, z is the valence of the metal, F is the Faraday constant, and ρ is the density of the metal (Ref 201).

The crack may grow at a rate well above this limit, up to 100 times faster when mechanical fracture processes that cause separation of material at the crack tip are important (Ref 199). Also, discontinuous acoustic emission and the presence of crack arrest marks on crack wall surfaces show that SCC crack growth can, in some cases, occur in a discontinuous fashion as the crack tip is repassivated, and then the passive film is again ruptured. On the other hand, if a film reforms over the crack tip for some period of time, or the crack walls are not passive, then the rate of crack tip advance would be lower than the limit predicted by Eq 24. From the previous discussion, it can be surmised that the ranges of potential where a material would be most susceptible to SCC would be those shown in Fig. 2 where there are transitions between active and passive behavior, and this has been verified by experiment, particularly for transgranular cracking.

One of the most prominent electrochemical dissolution models is termed the slip-dissolution model. According to this theory, under the influence of tensile stress, an emergent slip step will rupture the surface creating a region of unfilmed, exposed bare metal surface that will be corrosively attacked rapidly, eventually initiating a crack (Ref 204). In planar slip there will be a large normal stress present at the head of a dislocation pileup (Ref 206), which promotes fracture of the passive film. Preferential planar slip also constrains the ability of the material at the crack tip to plastically deform and restricts blunting of the tip (Ref 206). In high-chloride environments, preferential dissolution along the plane of high dislocation density, created by the planar slip, occurs (Ref 207). Various ideas have been advanced as to what factors control the crack advance rate. These include anodic dissolution only; for mechanical fracture models that often apply to transgranular SCC these include: corrosion tunneling in the propagation direction and mechanical fracture of the remaining ligaments of material (known as the corrosion tunnel model), brittle fracture of the tarnish film or corrosion products that form along the path of localized corrosion, chemisorption-enhanced plasticity, film-induced cleavage, adsorption-induced brittle fracture (also known as the stress-sorption model), and HE (Ref 199).

Factors Influencing Susceptibility to SCC

The features that affect SCC processes are particularly complex and include environmental, electrochemical, mechanical, and metallurgical variables. The main influences are described in this section. It should be recognized that, while almost all SCC data have been obtained in mode I tension or bending, complex engineering components, such as implants, may have torsional and shear loads as well. Mixed-mode loading could well have an influence on SCC (Ref 199).

The previous discussion, for the most part, tends to imply that SCC is a deterministic process, but recently Andresen et al. (Ref 208) have provided an excellent interpretation/critique of this type of thinking. They point out that the concept of a completely defined threshold in parameters such as stress intensity, electrode potential, alloy/condition, temperature, and so forth, above which a material is immune and below which it is susceptible to SCC, can be a misleading one. They indicate that when conditions once thought to convey immunity are probed in well-controlled laboratory tests or in light of a mechanistic understanding of the SCC process, the immunity status often fails to stand up, and finite crack growth rates (though often low) can be produced in such controlled tests. Similarly in industrial plants, SCC of components under conditions once thought to be immune has been observed after longer exposure times. Thus, rather than a "go/no-go" defining threshold for a given parameter, according to this view, there is a continuity of response, and SCC should be considered to be an innately probabilistic process. For example, Jones and Ricker (Ref 199) provide data that indicate that SCC crack propagation velocity for 304 stainless steel declines from 10^{-8} mm/s at 260 °C (500 °F) down to about 2×10^{-12} mm/s at -18 °C (0 °F), but no temperature threshold exists. Though Andresen et al. were discussing the specific case of SCC in hot water, their point should be well taken in terms of its applicability to other environments as well, including *in vivo* conditions.

Influence of the Environment. The main environmental factors affecting the SCC process are the types and concentrations of ions in solution, the partial pressures of dissolved gases, pH, solution viscosity and mixing or stirring, temperature, and the electrode potential that is consequently established (Ref 194, 199). For implant alloys, most attention has focused on the presence of the chloride ion, since the ability of this ion to disrupt passive films is well established and because two of the alloy systems most important for metallic implants, stainless steels and titanium-base alloys, can be vulnerable to SCC in chloride environments under some conditions. However, other factors related to the *in vivo* environment, active physiological bioelectric effects and the presence of proteins in the environment or adsorbed into the biomaterial

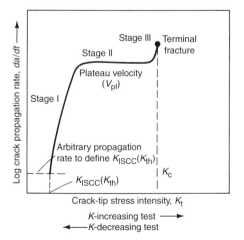

Fig. 21 Stress-corrosion crack propagation velocity as a function of K. Source: Ref 98

surface, might also possibly play a role (Ref 5, 209).

In situations where SCC vulnerability exists, the environment will establish a situation of borderline passivity, where most portions of the surface are adequately covered by a passive film so as to be corrosion resistant, but passivity in local areas can be breached, forming focal points for corrosion and cracking. Pourbaix diagrams can be valuable tools for visualizing how environmental factors affect the stability and dissolution regions, as well as the ramification of their influence on the electrode potential of the alloy (Ref 199). However, their application to SCC can be limited by the fact that solution chemistry can be very different at the crack tip compared to the bulk solution. For example, in nearly neutral saltwater for steels, the crack tip pH is about 3.5 and for titanium alloys about 1.5. In addition, for each of these alloy systems, the electrode potential at the tip of a growing SCC crack is below that which allows hydrogen evolution from water to occur (Ref 137). Stress-corrosion cracks can nucleate at places where crevice corrosion is occurring, even when the stresses are less than those present outside of the crevice (Ref 205), presumably because of the difference in chemistry within the crevice compared to the bulk solution.

The factors influencing SCC can be subtle, as pointed out previously, and small changes in the bulk environment can have a dramatic impact on SCC, while much larger changes may have little influence (Ref 199), depending on their influence on crack tip chemistry.

Influence of Composition. At one point it was thought that only alloys were vulnerable to SCC, but it has gradually become apparent that pure metals can experience SCC as well. Small differences in alloy composition (whether intentional, due to accidental addition of impurities, or the result of spatial inhomogeneities in chemical composition) can dramatically affect SCC resistance by substantially influencing passive films, metallurgical conditions, and electrochemical state (Ref 199, 202). Localized microchemistry, such as that due to depletion of elements that form passive films or enrichment of elements that foster active corrosion, can be particularly important. The following three sections consider the various generic alloy systems used to fabricate surgical implants and dental materials in terms of their general resistance to SCC. This provides a general indication of how alloys of a similar character behave (in terms of parameters such as temperature, chloride content, and so forth). The actual experience with regard to specific implanted surgical devices and materials is considered later and may be different from that of the general alloy class in some instances.

Influence of Composition on Stainless Steels. One of the most important phenomena associated with SCC of stainless steels is sensitization. In principle, for ferrous alloys containing 18 wt% Cr and 8 wt% Ni with carbon contents between approximately 0.03 and 0.7 wt%, there are three phases at equilibrium at room temperature—austenite (γ), α-ferrite, and carbide ($M_{23}C_6$) (Ref 111). The transformation is so sluggish, though, that, only austenite and carbide will be observed in practice (Ref 210). Carbon is soluble in austenite up to 0.03 wt% and will precipitate out as the carbide at higher values if given enough time. Often, austenitic alloys containing more than 0.03 wt% C are heat treated at 1050 °C (1920 °F) or so to minimize the effects of cold working, and then cooled back to room temperature fairly rapidly, which largely suppresses the precipitation. However, when the supersaturated austenite is heated back up into the two-phase field ($\gamma + M_{23}C_6$), precipitation of chromium-rich carbide will take place at the grain boundaries. This creates a chromium-depleted zone that envelops the grain boundaries. Some time/temperature combinations are sufficient to precipitate the $M_{23}C_6$, but not sufficient to eliminate the concentration gradient and allow rediffusion of chromium back into the depleted zone. Since the chromium content in the vicinity of the grain boundaries is below the 12 wt% level required to passivate the material, a path of low corrosion resistance is created that can lead to intergranular corrosion and, with loading applied, to SCC. This type of structure is termed "sensitized," and it can reduce K_{ISCC} by about 75% in certain alloy/environment combinations (Ref 111). In wrought grades of austenitic stainless steels, SCC is transgranular unless sensitization is present, in which case, it is intergranular (Ref 105). In contrast to the carbide phase, when δ-ferrite is present in austenitic stainless steels, SCC resistance is improved (Ref 211).

Use of low-carbon contents (less than 0.03%) is a recommended remedy to minimize the possibility of sensitization (Ref 111), although at long times, sensitization of type 304 stainless steel can occur even with type 304L when the carbon content is 0.015 wt% (Ref 199). It should be pointed out, though, that 304 and 304L stainless steel is not acceptable as an implant material according to modern standards and has not been used as such for many decades.

Stress-corrosion cracking of stainless steels was first recognized as a possibility in 1944 (Ref 212). Many alloying elements affect the SCC resistance of austenitic stainless steels (Ref 211). For implant alloys, the most important one of these is nickel. In boiling magnesium chloride, SCC resistance is at a minimum with about 8 wt% Ni, and increasing nickel content above that has a beneficial effect (Ref 137). Similarly, lowering the nickel content below this value also increases resistance to SCC. Molybdenum is another important element that improves SCC resistance of austenitic stainless steels (Ref 111), presumably because of its salutary effect on pitting resistance. Tests of a 19Cr-20Ni alloy in $MgCl_2$ showed SCC resistance is increased by increasing the carbon content and decreasing the amount of nitrogen in the steel (Ref 137).

Cold work is considered to be beneficial, in terms of SCC resistance, for most austenitic stainless steels and increases time to failure in SCC tests, but it is difficult to separate out effects resulting from work hardening and those due to transformation to martensitic phases (Ref 111). Furthermore, type 316 seems to be an exception to this rule and suffers a sharp decline in cracking time when the amount of cold work exceeds about 5% (Ref 211). It is not certain, however, that this would be the case for implant-quality 316L. In addition, 316L only undergoes an austenite to martensite phase transformation at a temperature of −196 °C (−321 °F). Shot peening has been shown to improve SCC resistance of austenitic stainless steels (Ref 211). Stress-corrosion cracking is accelerated as grain size increases, particularly at the lower stress levels (Ref 111).

The influence of temperature on the SCC resistance of nonsensitized stainless steel is an important question for implant alloys. A generation ago, the consensus viewpoint, based on practical experience, was that stainless steels in the nonsensitized condition were immune to SCC in neutral chloride-containing environments at temperatures lower than 80 °C (175 °F) (Ref 213). The threshold reported in the literature has dropped substantially since then, however. Hoxie (Ref 214) and Truman (Ref 195) reported it was 60 °C (140 °F). Speidel (Ref 215) demonstrated that 304 stainless steel would crack at 50 °C (120 °F) in 22% NaCl and, even at ambient temperature, when in the sensitized condition. In a series of experiments involving notched and precracked 316L specimens that were highly loaded (and, in some cases, galvanically coupled to other implant materials or subjected to applied polarization levels comparable to what might be possible due to *in vivo* bioelectric effects), it was seen that at 37 °C, or 98.6 °F (internal body temperature) cracklike features developed *in vivo* and SCC crack growth occurred under laboratory conditions in Ringer's solution (Ref 5, 209, 216, 217). Transgranular SCC in nonsensitized austenitic stainless steels (304, 316, titanium-stabilized 316) has been observed at ambient temperatures when the alloys have been exposed to atmospheres above indoor swimming pools (Ref 218, 219). Similar findings of SCC in austenitic stainless steels (303, 304, and 304L) at ambient temperature, when they are not sensitized, have been reported for marine atmospheres as well (Ref 220, 221). Haselmair (Ref 222) describes a case of SCC in type 303 at ambient temperature due to chloride-laden dust in a road tunnel. In acidic chloride-containing solutions, SCC of 304 and 316 stainless steels is well documented, even at temperatures as low as −5 °C (23 °F) (Ref 223–227).

In contrast to the situation for titanium alloys (described below), applied polarization enhances SCC in austenitic stainless steels tested in chloride environments as the potential increases and retards it when the potential is decreased (Ref 111). This indicates the importance of anodic dissolution processes, as opposed to hydrogen embrittlement, for the mechanism of cracking. Therefore, the film rupture (slip-dissolution) model is the likely origin of SCC of austenitic stainless steels in chlorides (Ref 104). Highly cold-worked stainless steels can be subject to hydrogen embrittlement, however (Ref 111, 211), and may become more susceptible to crevice corrosion as well (Ref 228).

Influence of Composition on Titanium Alloys. Titanium alloys have outstanding corrosion resistance compared to many engineering alloys. This is due to the fact that titanium is reactive and very readily forms a passive film, primarily composed of TiO_2, and other thermodynamically stable oxides, when in contact with media containing oxygen (Ref 229). This film is tenacious and self-healing (Ref 201). However, over the years, some titanium-base alloys have been shown to be susceptible to SCC in a variety of environments, including seawater and other chloride-containing environments at ambient temperature (Ref 123). Pure titanium (ASTM grades 1–3) falls in the α category and is immune to SCC in aqueous environments. This behavior depends on interstitial oxygen content, however, and when pure titanium has more than 0.2 to 0.25 wt% O, a transition from wavy to planar slip occurs. Such materials may then be prone to SCC in saline environments with a salt content equivalent to seawater (Ref 196, 230, 231).

Besides oxygen, aluminum also exhibits a negative influence on SCC resistance (Ref 232). For example, mill-annealed or duplex-annealed Ti-6Al-4V is susceptible when a loaded fatigue crack is present, particularly in thicker specimens (where plane strain conditions will be present, Ref 123). In such α/β alloys, it is the α phase that promotes SCC. Smooth or notched specimens will not exhibit SCC in neutral saline environments (Ref 233), since titanium is so resistant to pitting in such environments. Even in boiling acidic chloride-containing environments, the repassivation potential for Ti-6Al-4V, for example, is 1.7 V Ag/AgCl or higher (Ref 233). Nitrogen, hydrogen, and carbon are other interstitial elements that adversely affect α/β alloy susceptibility to SCC, and ELI grades are often specified in applications where resistance to SCC is critical (Ref 229, 234). Alloying elements that improve SCC resistance include zirconium, molybdenum, and vanadium (Ref 196, 206, 233).

For α/β alloys, the microstructure that results from thermomechanical processing exerts a strong influence on SCC susceptibility in a manner related to the grain size, volume fraction, and mean free path length of the more brittle α phase (Ref 201). The more ductile β phase acts as a crack arrestor. Particularly when β is present in a high volume fraction, or when it forms a continuous network, it will enhance resistance to SCC in saline solution. On the other hand, a fine equiaxed structure with β dispersed in a continuous α matrix (that originates from processing in the α + β phase field) will show much more vulnerability to SCC. Koch (Ref 201), Schutz and Thomas (Ref 233), and Schutz (Ref 229) describe other α/β microstructures based on β processing or final β heat treatment that are much more SCC resistant. These different processing regimes develop different types of texture, which in turn can be related to variations in SCC resistance (Ref 196). Duplex thermal treatments have a particularly detrimental effect on SCC resistance (Ref 232). ASTM standard F 136, which applies to implant-grade Ti-6Al-4V ELI, calls for a fine dispersion of α and β phases resulting from processing in the α + β field, with no continuous α network and no coarse α platelets. This is a microstructure of the class that is more prone to SCC than the alternative microstructures described by Koch (Ref 201) and Schutz (Ref 229). Generally speaking, tests of Ti-6Al-4V have shown that K_{ISCC} will decline as σ_y increases (Ref 196). Aging of α/β alloys containing more than 5 wt% Al should not be carried out in the 400 to 700 °C (750 to 1290 °F) range, because this may form the α_2 phase (Ti_3Al), which enhances vulnerability to SCC (Ref 229).

The type of cracking observed in titanium alloys in various environments is generally related to the various regions of the V versus K curve (see Fig. 21). In stage I, cracking occurs by intergranular separation. In stage II, transgranular cleavage predominates, and in stage III, microvoid coalescence is featured (Ref 196). Fractographically, SCC of α/β alloys in aqueous environments occurs by transgranular cleavage of the α phase and ductile tearing of the β phase (Ref 229). Lowered pH tends to decrease K_{ISCC} and increase V_p. The α/β alloys exhibit a minimum in K_{ISCC} at an electrode potential of roughly −500 mV saturated calomel electrode (SCE), and the threshold will increase for potential values cathodic or anodic to this value (Ref 196, 201, 229). For Ti-6Al-4V in seawater, slow strain rate tests with cathodic polarization to a potential of −1500 mV SCE (to promote hydrogen charging) revealed that ductility is significantly decreased, but no hydride formation is involved (Ref 235). Rather, the hydrogen embrittlement observed was thought to be due to interstitial hydrogen uptake in the β phase which, when it reaches a critical level, causes lattice decohesion and brittle rupture. Excessive hydrogen uptake (500 to 600 ppm) in both α and α/β alloys, however, will lead to hydride formation and significant embrittlement (Ref 233). Increasing temperature and decreasing pH promotes absorption of hydrogen by titanium alloys.

Temperature tends to have only a minor effect on K_{ISCC} of titanium alloys, but crack velocity will increase with increasing temperature (Ref 123, 229). The activation energy for steady-state crack propagation has been reported to be 13 to 23.5 kJ/mol (Ref 123, 196). Velocity also increases as solution viscosity decreases (Ref 123). Crack velocity tends to be high and is aggravated by increasing chloride content. Velocity in neutral solutions increases with chloride concentration to the $\frac{1}{6}$ to the $\frac{1}{2}$ power for titanium alloys (Ref 123, 196, 229). As the electrolyte becomes more acidic, K_{ISCC} declines for titanium alloys (Ref 196). For titanium-base materials that meet oxygen level standards for implant materials, the K_{ISCC} values reported for 3.5% NaCl solutions at 25 °C (77 °F) are in the 36 to 75 MPa$\sqrt{m}$ (33 to 68 ksi$\sqrt{in.}$) range for pure titanium and 45 to 51 MPa$\sqrt{m}$ (41 to 46 ksi$\sqrt{in.}$) for Ti-6Al-4V ELI (Ref 229). In aqueous environments, initiation of SCC in titanium-containing materials occurs according to the film rupture/dissolution mechanism (Ref 123, 234), while propagation is thought to be controlled by anodic dissolution or by hydrogen-assisted cracking (Ref 201, 229).

Influence of Composition on Cobalt-Base Alloys. Stress-corrosion cracking of cobalt-base alloys can occur, but this is generally only a problem at very high temperatures (at or above 150 °C, or 300 °F) in acidic chloride environments (Ref 236). Even in boiling $MgCl_2$, one study found a Co-Cr alloy to be immune to SCC (Ref 237), while in another (Ref 238), where the stress level was higher and the exposure time was longer, it was seen to be susceptible. MP35N is immune to SCC, even in boiling $MgCl_2$ (Ref 238). Thus, it would not be expected there would be much probability for the occurrence of SCC under *in vivo* conditions. This seems to have been borne out by experience with implant alloys where few reports of *in vivo* SCC have been reported, with the exception of the Björk-Shiley heart valve (discussed later). Jones et al. (Ref 58), however, do not think that it can be said with certainty, based on their tests, that the F 75 alloy will be immune to SCC under body conditions.

Processing Aspects, Metallurgical Defects, and Microstructural Features Related to SCC. Many of the factors that affect CF also affect SCC. For example, pits can form at surface inclusions and eventually serve as nuclei for stress-corrosion cracks when the aspect ratio of the pit exceeds 10 (Ref 199). Pits shallower than this will enlarge more by uniform dissolution than by a locally focused SCC process. Also, stress raisers are of major concern regarding SCC susceptibility and crack initiation. They can arise from preexisting surface flaws, mechanical impact on the surface, or from any form of localized corrosion or surface damage such as wear, fretting, pitting, and so forth (Ref 105, 199). Cahoon and Paxton (Ref 239) examined a number of orthopedic implants from a hospital inventory before implantation and found that some 316L implants had cracks due to severe cold working that occurred due to rolling of thin sections. In later work (Ref 240), it was observed in an extensive study of 250 failed implants that 10% of the stainless steel implants did not meet the specifications of ASTM standard F 138.

Since tensile stresses are required for SCC, processes such as shot peening, which put the metal lattice at the surface in compression, can help to ameliorate SCC susceptibility (so long as they do not introduce new flaws and stress concentrations into the material).

Yield strength, presence of precipitates and second phases, segregation, and grain size are all factors that can substantially influence SCC behavior (Ref 202) and can be modified by processing and heat treatment procedures. In some alloy systems a stress level of $0.7\sigma_y$ is required for SCC to occur, while in others, it has been observed at stresses as low as 10% of the yield strength (Ref 105). A finer grain size generally reduces susceptibility to SCC, while cold work increases it in most alloy systems (Ref 204), but not in all (Ref 105). Heat treatments that provide the maximum strength for an alloy often will also create its peak vulnerability to SCC. Metallurgical variables, such as crystal structure and anisotropy, grain shape, dislocation density and geometry, stacking fault energy, and ordering, have all been linked as influences on transgranular SCC (Ref 199).

Stress-Corrosion Cracking Testing Methodology

A number of ASTM standards exist for the conduct of SCC tests (Ref 49). Relating the results of laboratory testing to failures in service, though, is not always straightforward. For example, SCC in titanium alloys can be readily demonstrated in precracked fracture mechanics specimens in a variety of environments, although failures in service tend to be rare because the great resistance to pitting of such systems makes crack initiation difficult (Ref 204). However, if the passive film is mechanically breached within a fine surface crack isolated from the bulk environment, titanium alloys can crack due to SCC, even in room-temperature chloride solutions. This occurs if the crack velocity is sufficient to prevent diffusion of the bulk electrolyte to the crack tip, which would extinguish the electrochemical reaction, by eliminating the causative microchemistry (Ref 241).

Sprowls (Ref 194) points out that laboratory tests are appropriate only to service conditions where experience has shown a relevant relationship. In other words, if a specific alloy passes or does not pass a test found previously to be determinative for another alloy, this may (or may not) be significant, and a test providing reliable guidance between alloys in a given application may not provide prudent guidelines when exposure conditions differ.

Nonprecracked Specimens. The earliest SCC testing was conducted on nonprecracked specimens. A wide variety of specimen geometries have been employed (Ref 16, 137, 194, 201, 202, 205). These include notched and unnotched tension specimens, ring-stressed tension specimens, C-rings, U-bends, bent beams (with three-point, four-point, or pure bending modes of loading), tuning fork specimens, O-rings, residual stress specimens, and so forth, as described in the article "Evaluating Stress-Corrosion Cracking" in Volume 13A of the *ASM Handbook*. These usually establish constant displacement or constant load conditions, and a number of ASTM standards can be used to guide such testing, for example, G 30, G 35, G 36, G 38, G 39, G 47, G 49, and G 58. Most of these configurations load the sample elastically, though the U-bend involves a plastically deformed test specimen. For notched or unnotched specimens tested in tension or bending, the data are usually presented as a time to failure versus stress curve (Fig. 22). The time measured includes t_{inc} (see below) plus the time required to progress through the three crack propagation stages.

Fracture Mechanics Approach. As with CF, the fracture mechanics approach has been found to be very useful in SCC studies. A difference between the two, though, is that even when there is a preexisting crack, for SCC there will be an incubation time t_{inc} required before cracks begin to propagate, and t_{inc} may represent a considerable period of time (Ref 242). Such effects also have been observed for implant alloys (Ref 216). The incubation time is necessary to establish local crack tip chemistry and electrochemical conditions that are severe enough for crack propagation to start.

As with the nonprecracked specimen approach, a plethora of different specimen types have been used to implement SCC testing based on fracture mechanics methods. Almost all of the plane strain fracture toughness specimens have been adapted to SCC testing (Ref 194). These include designs that have K increasing as cracking proceeds, decreasing with crack propagation, or constant with crack extension. The specimen dimensions, especially the thickness, should be such that predominantly triaxial (plane strain) conditions are created, so that plastic deformation is restricted to a very small zone in the crack tip vicinity (Ref 205). This requirement is similar to that for CF specimens discussed previously, and further detail is given in ASTM standard E 1681. Practices for preparing precracked specimens are covered in ASTM standard E 399. These include cracked plates, circumferentially cracked round bars, bending specimens, double cantilever beams, and others. Loading is accomplished via tensile testing machines, loading wedges, or three-point, four-point, or cantilever bending. These samples are more fully described in various works (see, for example, Ref 137 and 194).

Though use of fracture mechanics specimens eliminates uncertainty and variability associated with the crack initiation process, it should be pointed out that, for very small cracks, LEFM (Eq 23) predicts stresses that are unrealistically high. So LEFM results will not necessarily completely extrapolate back to the results with smooth specimens (Ref 205).

Slow strain rate (SSR) techniques have largely supplanted constant stress tests in recent years (Ref 199) because they are quicker and, also, because it has been shown that the crack growth rate is a function of the applied strain rate. The idea here is that a series of tensile tests to failure are conducted with standard tensile specimens (ASTM E 8) contained in an appropriate environment of interest (Ref 15). The strain rates used span orders of magnitude, but all have rather low values. This approach was inspired by the development of mechanistic models to explain the SCC process, most of which, as has been shown, involve crack tip plastic microstrain and passive film rupture. In the SSR test, when the yield stress is exceeded, plastic deformation occurs that assists in SCC crack initiation (Ref 137). Guidelines for implementing this method are found in ISO standard 7539-7, as well as in ASTM standard G 129.

Figure 23 shows the typical ductility (and ductility ratio) versus strain rate response for two materials vulnerable to environmentally induced cracking in an aggressive environment. The ratio is equal to the value of the ductility parameter in an aggressive environment divided by its value in an inert environment. Alloy A can suffer from hydrogen-induced cracking, and alloy B is susceptible to SCC. For each material in an inert environment, there is only a modest influence of strain rate on specimen ductility. For alloy A, though, when the strain rate is sufficiently slow in an environment causing HE, the reduction of area diminishes steadily until it reaches a plateau value at very low strain rates. For SCC (alloy B), there is a critical value of strain rate where the reduction of area is a minimum. This minimum is characteristic of the alloy system/environment/ potential/temperature combination being tested, and the most severe strain rate must be determined in each case (Ref 194, 243).

The critical strain rate maintains a delicate balance among deformation, diffusion, anodic dissolution, and passive film formation (Ref 243). Generally, the lower the SCC crack velocity, the lower the critical strain rate will be. For titanium alloys and stainless steels in chloride solutions, the critical strain rates are respectively reported to be 10^{-5} and 10^{-6} s^{-1} (Ref 15). One study with stainless steels, however, has indicated that lower rates may be needed in some cases (Ref 244). In 100 °C (212 °F) 20% NaCl, the strain rate had to be reduced to 7.87×10^{-8} s^{-1} before SCC was evident in 304 stainless steel. In 10% NaCl the rate had to be lowered to 7.09×10^{-9} s^{-1}. In 3.5% NaCl, when the sample was polarized anodically by 45 mV,

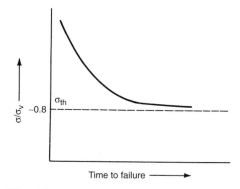

Fig. 22 Time to stress-corrosion cracking failure versus applied stress

the strain rate needed was also 7.09×10^{-9} s^{-1}. It is possible that at the lower chloride contents and temperature characteristic of *in vivo* conditions, an even lower rate might be required.

At the beginning of the SSR test, the strain rate is well characterized and controlled; but when cracking starts, the strain is mainly concentrated at the crack tip, where the actual effective rate of straining is unknown (Ref 194). Hishida et al. (Ref 245) provide mathematical analyses of the SSR test that allows the effects of pure mechanical deformation, crack growth, and anodic dissolution to be separated. For SSR testing, a plot of crack velocity versus strain rate will show a threshold effect, a plateau velocity, and an increase when large strain rates are used (Ref 199, 246). This behavior is similar to that shown in Fig. 21 for the stress-intensity dependence. The susceptibility to SCC in a SSR test is maximized at strain rates corresponding to the plateau velocity.

This method has been extended to include bending tests, as well as notched and precracked specimens. Another variant of dynamic testing applied to SCC involves step loading. Here, both precracked and smooth samples can be used to rapidly identify the stress or stress-intensity thresholds for SCC (Ref 202). Another variant is the constant extension rate test (CERT), sometimes referred to as a straining electrode test (Ref 137). This is similar to the SSR method, except that extension rate rather than strain rate is controlled.

Slow strain rate tests have the advantage that they are quicker than other types of SCC testing, generally produce less data scatter, and the test always ends with the mechanical failure of the specimen. This eliminates samples that do not fail because they are loaded below K_{ISCC} (Ref 199). After the test, the sample gage section is observed to note any secondary cracking that may be present (Ref 201), and the fracture surface is also examined for evidence of SCC. The ductility minimum for SCC occurs because, when the strain rate is much larger than the critical value, the samples will fail by ductile fracture, while, when it is very low, repassivation will occur (preventing the corrosion necessary to drive the SCC process). Besides reduction of area, other variables that change due to SCC (compared to a nonsusceptible condition) include time to failure, maximum nominal stress, percent elongation, and area under the stress strain curve (Ref 15, 16, 201). Because effects not related to SCC are also present in an SCC test, the general practice is to also test the material of interest in an inert environment (air or oil, for example). The ratio of the value of a property in the corrosive medium to that found under inert conditions provides an index of the resistance to SCC. The more the ratio departs from unity, the more vulnerable to SCC the material will be (Ref 205). Abe et al. (Ref 247) defined another susceptibility index as the ratio of the area associated with SCC fracture to that associated with mechanical fracture. They found that this index is well time-correlated with failure observed with constant load SCC tests. Daniels (Ref 244), on the other hand, compared SSR, U-bend, and constant stress specimens made of 304 stainless steel and found differences among the test methods. As with other types of SCC testing, electrode potential is an important variable. Applied polarization, changes in environmental chemistry, and galvanic coupling can shift the potential of the alloy into a region conducive to SCC (Ref 243).

The SSR technique has been seen to be very useful when ranking various alloys as to their resistance to SCC in a given environment or for investigating the influence of metallurgical variables on the susceptibility of a given material (Ref 137). However, use of SSR data to predict crack propagation rates for actual service conditions is problematic.

Accelerated SCC Tests. Because of the very long time periods that can be involved in SCC experiments, particularly using either smooth or precracked fracture mechanics type specimens, there has been interest in finding means for accelerating these tests, but in ways that can be extrapolated to the service conditions of interest. This is inherently difficult due to the complex compositional, electrochemical, metallurgical, and mechanical interactions involved in SCC (Ref 205). As pointed out previously, severe stress conditions lower the incubation time for crack initiation and increase velocities at least up to the plateau region. Crack velocities have been shown in some circumstances to be thermally activated and to follow an Arrhenius equation (Ref 248):

$$V = A\exp(-Q/[RT]) \quad\text{(Eq 25)}$$

where A is a pre-exponential factor, R is the gas constant, T is the absolute temperature, and Q is the activation energy for SCC (which is related to a solid-state diffusion process). Thus, elevating the temperature can be used to substantially increase the cracking rate and lower the time to failure. The activation energy in the SCC case, though, depends on stress intensity and decreases with K^2. This type of accelerated testing approach would only apply when one process controls cracking kinetics over the whole range of test temperature conditions and throughout the service temperature range to which the data are extrapolated. Use of more chemically severe testing conditions, such as employing boiling $MgCl_2$ to get some insight into susceptibility of SCC in milder chloride-containing environments (such as seawater or *in vivo* electrolytes), provides a qualitative means for test acceleration. This is a particularly popular test for austenitic stainless steels (Ref 111). It is so commonly used that an ASTM standard has been developed (G 36) to standardize the testing protocol. The $MgCl_2$ test has been in use since the 1940s. A relatively new standardized test (ASTM G 123), approved in 1994, involves exposure to a boiling acidified NaCl solution. This test is less severe than the $MgCl_2$ test, but is thought to be better correlated with NaCl service environment experience (Ref 211). Over and beyond the uncertainties discussed previously regarding the difference in temperature, uncertainties as to how changes in bulk electrolyte concentration, pH, and so forth, affect the microchemistry at the crack tip have so far precluded development of a precise means for predicting behavior in mild environments from behavior measured under extreme conditions.

With precautions, however, it does appear that a series of tests using boiling $MgCl_2$, 5% HCl at 37 °C (98.6 °F), and 37 °C Ringer's solution can allow meaningful accelerated testing to occur for implant alloys (Ref 209, 216, 217). Using fracture mechanics specimens, increasing temperature was seen to substantially lower K_{ISCC}, raise V_p, and lower the incubation time required to initiate propagation from a precrack. An activation energy for SCC of 316L stainless steel was computed from this work to be 51.0 kJ/mol. This is consistent with the range observed for various austenitic stainless steels in chloride-containing media (Ref 249–252). Since these environments and tests had different pH and Cl$^-$ concentration values, as well as differences in polarization conditions, it was pointed out (Ref 216) that the alternative procedure of Speidel

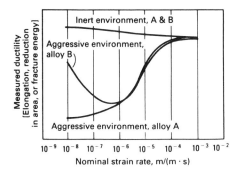
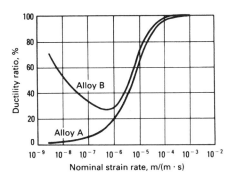

Fig. 23 Influence of strain rate in slow strain rate testing. (a) Ductility parameters versus strain rate in inert and aggressive environments. (b) Ductility ratio versus strain rate for materials susceptible to hydrogen embrittlement (alloy A) or stress-corrosion cracking (alloy B)

(Ref 215), log V_p versus T regression, could be used. For 316L it was found that:

$$\log V_p = 2.037 \times 10^{-2} T - 10.31 \quad \text{(Eq 26)}$$

when V_p is in units of m/s and T in °C. It was seen that, for both 316L and Ti-6Al-4V ELI, incubation time t_{inc} was a strong function of stress intensity and decreased as K increased:

$$t_{inc} = C \exp(-AK) \quad \text{(Eq 27)}$$

where C and A are constants depending on the material, environment, and polarization conditions investigated.

It is conceivable that an accelerated testing approach could be used for CF studies also, since it has been shown that fatigue crack growth velocities in austenitic stainless steels increase as temperature increases (Ref 31).

Findings from SCC Laboratory Testing

A number of studies have attempted to simulate, in the laboratory, conditions that might cause SCC in the *in vivo* environment. Due to the susceptibility of stainless steels to cracking in chloride-containing environments, most of this work, though not all, has focused on 316L. A variety of results have been obtained, depending on the testing methodology and what *in vivo* variables are simulated. These findings are reviewed in this section. Because of the ability to better control variables in the laboratory, as opposed to in an animal or clinical study, a particularly important motivation for SCC testing in the laboratory is that the loading condition can be set to ensure a static tensile stress component is applied to the test specimen. As pointed out previously, under *in vivo* conditions, the loading applied is often unknown, and fractographic evidence may be subject to ambiguous interpretation, in certain situations, as to whether the cracking observed is due to SCC or CF.

Stainless Steels. Gilbertson (Ref 253) tested specimens of vacuum melted 316L (316L VM) and 304 stainless steel in four-point bending in acidified Ringer's solution (pH 2 to 3) at 66 °C (151 °F). Stresses used were both above and below σ_y. Cracking was noted in 304 after about 800 h of exposure, while the 316L samples did not exhibit SCC, even up to 15,000 h of exposure.

Jones et al. (Ref 58) performed SSR tests on specimens of 316L and various TRIP steels at a strain rate of 4.4×10^{-6} s^{-1} in 37 °C Tyrode's solution and in room temperature air. No evidence of SCC was seen for either annealed or cold-worked 316L. The ductility was lower for several of the TRIP steels; however, fractographic signatures of SCC were absent.

Sheehan et al. (Ref 254) tested 316L in aerated, lactated 37 °C Ringer's solution using SSR methods at strain rates in the 10^{-5} to 10^{-7} s^{-1} range. No difference in tensile strength or ductility properties were noted as a function of strain rate, and it was concluded that no SCC susceptibility was evident. This was also the case when the solution was acidified to pH 2. They also tested Schneider intramedullary nails in three-point bending (where the highest stress in the sample was at the yield stress) for 6 months, as well as other 316L samples in cantilever bending for 8 months. No evidence of SCC was found using either of these experimental protocols. These investigators also performed CF tests in Ringer's solution and observed significant secondary cracking which, in some cases, produced crack branching features (in situations of high ΔK and rapidly advancing cracks) that appear quite similar to the branching characteristically observed with SCC. As pointed out previously, it appears that, without knowing the loading history of an implanted device, it is difficult to determine for certain whether a branched crack pattern is a signature of SCC or represents CF that occurs under a certain restricted set of conditions. However, if a secondary fatigue crack is opened up and examined fractographically, it will exhibit striations.

Bundy et al. (Ref 5) noted extrapolating results of SCC tests in 37 °C physiological saline solution to predict *in vivo* performance tacitly assumes the organic biomolecules present in the body, as well as bioelectric effects *in vivo,* play no role in SCC under clinical conditions, assumptions that had not been tested experimentally. This observation was based on comparison of literature (on SCC from laboratory tests) with clinical reports and animal experiments conducted by them and others (discussed in the next section describing *in vivo* and retrieval studies). In addition, it was noted that knowledge of the loading patterns to which retrieved implants had been subjected was difficult to obtain. A series of laboratory experiments were, therefore, conducted to examine the effects of applied stress and applied electrical polarization on the SCC behavior of implant materials, 316L among them (Ref 5, 209, 216, 217).

In the first study (Ref 5), three-point bending of smooth samples was used. Three different maximum stress levels were used with different samples—$0.8\sigma_y$, σ_y, and σ_{uts}. Control samples, that is, those subjected only to applied stress, were exposed to 37 °C Ringer's solution for 53.5 weeks. Other specimens in this same environment for 38 weeks were stressed to these same levels, but also had a +100 mV potential applied to them. Consistent with the other laboratory studies cited previously, no cracklike features were observed in the control samples, no matter how highly stressed. Some pitting was observed, increasing in severity as stress level increased. On the other hand, cracklike features were observed for the polarized specimens at and above σ_y, implying that *in vivo* bioelectric activity might spur SCC in highly loaded samples. The biochemical aspect of the hypothesis that *in vivo* chemical and electrical conditions differ from those in 37 °C physiological solution was not examined, due to the multiplicity of possible agents and the chemical specificity often associated with SCC processes. However, the cracklike features observed *in vitro* were not as severe as those found *in vivo* (see the next section) for shorter exposure times, which could be consistent with a further increment of SCC susceptibility being conveyed by a biochemical agent (or more damaging *in vivo* bioelectric effects than those simulated in the laboratory).

In the latter three publications from this laboratory (Ref 209, 216, 217), a more controlled fracture mechanics approach was used with wedge-loaded double cantilever specimens to quantify threshold stress intensity K_{ISCC}, steady-state plateau crack propagation velocity V_p, and incubation time for initiation of crack propagation, t_{inc}. Three test environments were used: boiling MgCl$_2$ to quickly gage SCC susceptibility in chloride media, 5% HCl at 37 °C to determine susceptibility in chloride media at body temperature, and 37 °C Ringer's solution to more closely simulate the *in vivo* situation. In some specimens, applied polarization was employed as well. Besides the +100 mV regime, as used in the investigation with smooth samples to simulate the magnitude of possible bioelectric effects, two additional regimes of polarization were used for the tests in Ringer's solution. One involved polarization to break down the passive film, followed by potentiostatic polarization to prevent repassivation (corresponding to a current density of about 10 μA/cm^2). The second regime involved galvanostatic polarization at this current density level without the initial breakdown step. These regimes were regarded as more likely to spur SCC, if it were governed by a film rupture/repassivation mechanism. To gain further insight into the influence of potential, both anodic and cathodic direct current (dc) potentials were applied to the specimens tested in magnesium chloride. In an extension of these studies, a square wave alternating current (ac) signal was superimposed on a dc bias for 316L in Ringer's solution (Ref 255).

For unpolarized 316L, V_p values were 8.5×10^{-8} and 4.5×10^{-10} m/s in MgCl$_2$ and HCl solutions, respectively. K_{ISCC} values were 9.5 MPa$\sqrt{m}$ (8.6 ksi$\sqrt{in.}$) in the 44% MgCl$_2$ environment at 154 °C (309 °F), and 70 MPa$\sqrt{m}$ (64 ksi$\sqrt{in.}$) in 5% HCl at 37 °C. Thus, cracking is enhanced when the temperature is higher and the chloride concentration is greater, as previously pointed out. The V_p and K_{ISCC} values in MgCl$_2$ are similar to those seen by the individuals who have tested other austenitic stainless steels in this environment (Ref 215, 249, 256–258). Even at high stress intensity, no SCC was observed after 575 days of exposure in Ringer's solution (consistent with the result of Gilbertson, Ref 253, mentioned earlier). Similarly, for two of the polarization regimes, no SCC was noted for the conditions used in the test. For the polarization method, where the passive film breakdown step was used, SCC did occur in Ringer's solution. The incubation time was

168 h, and the crack propagation velocity was 2.4×10^{-10} m/s. For the ac waveform, SCC was also observed, with V_p being 1.16×10^{-10} m/s. Here, the incubation time was 1272 h, compared to 168 h for the dc polarization. For the tests in MgCl$_2$, anodic polarization slightly increased V_p and modestly increased K_{ISCC}. Cathodic polarization seemed to inhibit SCC, raising K_{ISCC}, while lowering V_p. This is consistent with an SCC mechanism where the cracking rate is controlled by competitive adsorption of OH$^-$ and Cl$^-$ ions on the surface. It is consistent with data from smooth specimens of 304 stainless in the same environment, showing that time to failure decreases as potential increases (Ref 104). The greatest influence of polarization in this environment was on t_{inc}. Anodic polarization reduced the incubation time by about an order of magnitude. These results indicate that SCC of 316L stainless steel can occur in Ringer's solution at body temperature when certain very severe conditions exist: a preexisting crack, high stress intensity, and a disrupted passive film that does not repassivate.

More recently, Roach et al. (Ref 259) and Zardiackas et al. (Ref 260) have conducted SSR tests on both 316L and newer implant stainless steels, 22Cr-13Ni-5Mn and Bio-Dur 108, a low-nickel-content stainless steel. They used smooth and notched samples tested in distilled water and Ringer's solution at 37 °C. They computed the percent elongation ratio (PER) and the reduction of area ratio (ROAR), that is, the property value in Ringer's solution divided by that in distilled water, for these tests. Using a criterion of SCC occurrence when either ratio falls below 0.9, no evidence of SCC was observed in these experiments (the lowest value was 0.92). Significant notch sensitivity was observed for the stainless steels, however. The reduction of area for notched compared to smooth specimens was only 48.4%, 28.5%, and 23.5% for 316L, Bio-Dur 108, and 22Cr-13Ni-5Mn, respectively. Fractographically, only the Bio-Dur alloy showed a fracture mode consistent with SCC in some stainless steels (areas of brittle cleavage mixed in with dimpling) (Ref 260), yet this same morphology was present whether testing was in air, distilled water, or Ringer's solution. As a result, this fractographic pattern could not be construed as evidence of SCC in this case.

To the authors' knowledge, there has been only one laboratory study of SCC that was aimed at simulating a situation pertinent to clinical dentistry—Sutton and Sanders (Ref 261). Since rubber dam clamps have been known to fail in service (see the retrieval section below), they tested stainless steel rubber dam clamps that were subjected to various cleaning/autoclaving regimes and exposed to various NaOCl (sodium hypochlorite bleach) solutions. Corrosion spots occurred on the surfaces, and both intergranular and transgranular SCC were reported.

McEvily (Ref 137) presents various types of experimental data that relate to the susceptibility to SCC of austenitic stainless steels in aqueous and saline environments. For example, it is shown for 304 that the Cl$^-$ threshold concentration to cause SCC decreases as the oxygen level increases.

Titanium-Base Alloys. Bundy et al. (Ref 5) also examined the SCC behavior of Ti-6Al-4V ELI. The findings were similar to those noted for 316L, discussed previously. No cracklike features were noted at any stress level for controls at 53.5 weeks of exposure. For the anodically polarized samples at 38 weeks, however, cracklike features were observed for the samples that were maximally stressed at or above σ_y, though the surface attack was less marked than for 316L. The cracklike features were more numerous at σ_{uts} than at σ_y.

Their fracture mechanics studies (Ref 209, 216, 217) also involved SCC testing of the Ti-6Al-4V ELI alloy. For this biomaterial, the crack propagation velocity was much higher than for 316L. In this titanium-base alloy, V_p values were 8.0×10^{-5} and 2.4×10^{-9} m/s in the MgCl$_2$ and HCl solutions, respectively. The SCC threshold stress intensities of this alloy were, respectively, 22.6 and 53.9 MPa$\sqrt{m}$ (20.6 and 49 ksi$\sqrt{in.}$) in MgCl$_2$ and HCl. Thus, compared to 316L, the titanium-base material has a higher K_{ISCC} in magnesium chloride, but a lower one in 5% HCl. Stress-corrosion cracks in Ti-6Al-4V ELI, however, will propagate much faster than in 316L. Although from the data above it would appear that in the 37 °C environment, the titanium alloy is more susceptible to SCC than is 316L, the incubation time evidence points to the opposite conclusion. It took the cracks in the titanium alloy more than a year to begin growing. For 316L, the initiation time was about 1.5 months. So, a more accurate interpretation is that the titanium alloy is more resistant to crack initiation; however, once a crack forms, it propagates more rapidly in Ti-6Al-4V ELI compared to 316L.

Mishra and Davidson (Ref 262) used the SSR technique to test Ti-6Al-4V and three surface preparations of an experimental alloy, Ti-13Nb-13Zr, at a strain rate of 4×10^{-6} s^{-1} in 38 °C (100 °F) Ringer's solution. No evidence of SCC was found in this work.

Zardiackas et al. (Ref 263) have tested some of the newer titanium-base implant alloys in 37 °C Ringer's solution and aerated distilled/deionized water using the SSR approach on smooth samples. They examined a spectrum of titanium alloys representing α, α/β, and β microstructures: CP Ti grade 4 (ASTM F 67), Ti-6Al-4V ELI (ASTM F 136), Ti-6Al-7Nb (ASTM F 1295), and β-annealed Ti-15Mo (ASTM F 2066). The CP Ti was cold worked, while the others were in a heat treated condition. The materials were tested using a constant extension rate of 10^{-5} mm/s. They determined the PER and ROAR ratios and found no evidence of SCC using the criteria of a positive result when the ratio is ≤0.9, nor from SEM fractographic examination. The lowest value observed was PER = 0.94 (for Ti-6Al-4V ELI).

Williamson et al. (Ref 264) extended this study by also measuring notched samples according to the ASTM G 129 (SSR testing) experimental protocol. The results of the two investigations were similar. For the notched specimens, no SCC was seen either. Here the lowest ratio observed was ROAR = 0.92 for the CP Ti grade 4. Again, no differences in the SEM appearance of the fracture surfaces from samples tested in Ringer's solution or distilled water were observed.

Zardiackas et al. (Ref 159) followed the same basic experimental protocol (as for the studies mentioned above) for CP Ti testing to investigate the influence of anodization on the SCC process. Both smooth and notched specimens were tested. For all test conditions examined, the only ratio below the 0.9 SCC immunity criterion was the ROAR for notched samples. Here, the ratio was about 0.84, with anodized samples showing slightly less SCC susceptibility than the controls. Fractographic observations also showed no evidence of SCC, except for secondary cracking on the notched specimens. Thus, anodizing was found to have no tendency to increase SCC susceptibility, and CP Ti was seen to be immune to SCC, except when a severe notch was present.

A number of studies have examined the SCC behavior of titanium alloys in a 3.5% NaCl solution with salinity similar to seawater. Hsiao et al. (Ref 14) measured crack tip pH to be 2 or below, a situation conducive to cathodic evolution of hydrogen. They hypothesize that hydrogen plays a major role in SCC crack propagation of titanium alloys in neutral saline solution. McEvily (Ref 137) presents SSR data showing that Ti-6Al-4V is susceptible to SCC in this environment at 25 °C (77 °F), and its K_{ISCC} value declines as the oxygen content of the alloy increases. Also, he presents data indicating titanium alloys (including Ti-6Al-4V) will experience an increased SCC crack propagation velocity as applied potential is increased. McEvily (Ref 137) presents experimental data showing, as the aluminum content is increased, less oxygen in the lattice can be tolerated to maintain good SCC resistance. Additionally, he shows resistance to SCC will decline as the hydrogen content of the material is increased.

Consistent with the work of Hsiao et al. mentioned previously, Yokoyama and his coworkers have conducted several studies that indicate titanium-base implant alloys, particularly those used in dentistry, are susceptible to hydrogen embrittlement under conditions relevant to in vivo use. For example, Ogawa et al. (Ref 265) showed that β titanium is subject to hydrogen absorption and embrittlement in acidified phosphate fluoride solution and claim this is related to delayed fracture. Yokoyama et al. (Ref 266) found that hydrogen embrittlement occurs in the superelastic form of NiTi in fluoride solution. Based on this, they suggest that one reason titanium and its alloys fracture in the oral cavity is due to hydrogen absorption in fluoride solutions used as prophylactic agents. Similarly, for the work-hardened form of Nitinol, they demonstrated hydrogen embrittlement in acidified phosphate fluoride. This solution simulates fluoride-containing toothpastes and prophylactic agents (Ref 267). Shabalovskaya et al. (Ref 268) have also found chemical etching of Nitinol wires in a hydrofluoric/nitric acid (HF/HNO$_3$) solution, followed by aging in boiling water, will produce a cracked surface similar in appearance to SCC. Conceivably, residual stress in the wires could be involved here. Topical fluoride solutions at 37 °C (98.6 °F) have been shown to cause SCC of CP Ti U-bend specimens after 5 days of exposure (Ref 269).

Cobalt-Base Alloys. There have been fewer laboratory investigations of the SCC susceptibility of cobalt-base implant alloys than for the stainless steels and titanium alloys. In the study mentioned previously in the stainless steel section, Jones et al. (Ref 58) also tested Vitallium (ASTM F 75) using the SSR method in Tyrode's solution and in air. No evidence of SCC was observed. Kumar et al. (Ref 270) performed SSR tests on F 75 material that was cast, wrought, or had been processed using powder-metallurgy techniques. Stress-corrosion cracking was demonstrated in all material conditions in boiling 30% MgCl$_2$, although more so for the more ductile powder-metallurgy and wrought materials than for the cast alloy. No SCC was evident in Ringer's solution at body temperature. Similarly, Mishra and Davidson (Ref 262) used SSR methods to investigate the ASTM F 799 Co-Cr alloy in 38 °C (100 °F) Ringer's solution and also observed no evidence of SCC.

Edwards et al. (Ref 271) tested an F 75 alloy using the SSR method, while potential was applied to the sample exposed to a 37 °C Ringer's solution electrolyte. The tests at anodic potentials gave no indication of SCC, but the samples that were cathodically polarized suffered from a considerable degree of hydrogen embrittlement. Tests using controlled hydrogen charging also confirmed the susceptibility of the alloy to hydrogen embrittlement. The cathodic potentials used by the investigators in these tests were based on the amount of cathodic shift the alloy naturally experiences when a crevice situation is established.

Findings from *in vivo* Testing and Retrieval Studies Related to SCC

There have been a number of pertinent studies related to the occurrence of SCC under *in vivo* conditions since Zapffe's first report from more than a half century ago of a failure analysis of a fractured 316 fracture fixation plate (Ref 272). These have included both observations stemming from examination of retrieved implants, as well as animal studies conducted under more controlled conditions. Another early report of *in vivo* SCC was provided by Bechtol et al. (Ref 273). Galante and Rostoker (Ref 274) tested several alloys in the back muscles of rabbits for an exposure period of one year. They used U-bend type specimens made from pure titanium, Ti-6Al-4V, 316L, Vitallium, and Zircaloy-4 (which is a zirconium-base alloy containing 1.5 wt% Sn, 0.15 wt% Fe, and 0.1 wt% Cr). The stress in the vicinity of the U-bend exceeded σ_y. The specimens also had a third piece in the spot weld area to create crevice-corrosion conditions. Unlike an examination of a retrieved implant, where the loading history and residual stress level is unknown and where, as previously pointed out, similar fracture features may be characteristic of SCC or CF, a crack or fracture in this type of specimen would have only one possible cause (SCC). This sort of approach can separate out cracking due to SCC from that caused by CF. After the exposure period was up, the specimens were removed, cleaned, and examined at 100×. No cracking was observed with these samples, although pitting was seen in some cases. All specimens were surrounded by a fibrous encapsulating membrane. In another paper, these investigators mentioned what they say is a clear case of SCC in a stainless steel prosthetic device (Ref 275).

Hughes and Jordan (Ref 175) conducted a study of several broken retrieved implants. These were mostly stainless steel, although an implant was made of titanium in one case. Examination of these failed implants revealed the fractures were caused by CF, overload, and, in one case of a stainless steel implant in the sensitized condition, by SCC. In the same time period, Rose et al. (Ref 276) described a case of SCC of a Vitallium nail-plate assembly. Gray (Ref 277) reported on a failed intramedullary pin and failed bone screws fabricated from 316L. The intramedullary pin failed by LCF originating at a stress concentration remaining from centerless grinding that had not been removed by an electropolishing treatment. One of the screws failed by SCC and the other by overload failure during removal. Dumbleton and Miller (Ref 6) describe a case of SCC in a retrieved implant using a Garber nail and Thornton side plate made of 316L. The patient was nonweight-bearing and not ambulatory, but the side plate fractured anyway. The cause of the SCC was determined to be stresses induced because locking nuts were improperly seated at surgery.

Lisagor (Ref 178) describes a case of a retrieved type 316 stainless steel implant that had been plastically deformed, a situation that created residual tensile stresses, which ultimately led to failure by SCC after 15 months *in vivo*. Cyclic loading was involved only in that it may have prevented repassivation (and actually could have lowered the stresses causing stress-corrosion crack growth). In this case, a mixed-mode (intergranular/transgranular) failure was observed. White et al. (Ref 8) investigated several cases of failed stainless steel implants (a screw and plate from a Smith-Petersen nail-plate assembly, and a rush rod). They saw that most of the fracture surface was attributable to fast fracture (indicative of high stresses in the implants), but the initiation region displayed fractographic features that indicated initial crack propagation was due to a combination of SCC, CF, and HE. The cracks originated in an area where pitting or crevice corrosion was occurring. Bombara and Cavallini (Ref 278) studied several failed Küntscher nails. They were made from 316L, but contained excessive amounts of sulfide inclusions. Pits initiated at these sites, mainly along the exposed longitudinal cross sections of the nail, and cracks proceeded via transgranular SCC.

Bandyopadhyay and Brockhurst (Ref 279) investigated two cases of failed 316L hip prostheses. As was seen in the work of Lisagor (Ref 178) mentioned previously, they observed a mixed intergranular/transgranular mode of SCC failure. Sandborn et al. (Ref 280) describe a case of a failed unicompartmental knee implant femoral component made of cast stainless steel. The material was found to contain a high inclusion content and was characterized by a marked inhomogeneity of grain size, with some grains being so large as to span the entire width of the implant. The implant had been placed incorrectly, which heightened the stresses in the implant. The failure was mainly caused by an intergranular fatigue fracture (originating in an area of wear) that may have been due to alloying element segregation at the grain boundaries. However, there were also pits and scratches on the side of the implant experiencing tensile loading, and stress-corrosion cracks emanated from some of the pits.

Syrett and Davis (Ref 281) performed an animal study in which bone plates made from a TRIP stainless steel (12.5 wt% Cr, 7.27 wt% Ni, 3.88 wt% Mo, 0.29 wt% C, minor amounts of Mn, Si, P, and S, balance Fe) were implanted in rhesus monkeys. The animals were sacrificed at 100 to 104 weeks, and SCC was observed in 50% of the implanted plates. A coupon of the same material implanted in rabbit muscle for the same period of time also exhibited SCC, even though this is a low-stress site. It was thought that the cause of failure in this case was residual stresses due to polishing of the implanted material.

Bundy et al. (Ref 5) performed an *in vivo* investigation in which ring-stressed tension samples were used where the stress level could be controlled in a notched test specimen. They tested specimens where both the ring and the stressed sample were made from the same alloy, as well as situations in which dissimilar metals were used for the ring and notched specimen to create galvanic coupling conditions. Test materials in the first category included 316L, Ti-6Al-4V ELI, and MP35N. Test conditions where galvanic coupling was employed included 316L (specimen)/MP35N (ring) and Co-Cr-Mo (specimen)/Ti-6Al-4V ELI (ring). The stress-ring samples were loaded so the nominal stress in the test sample in the notched area was σ_y. The specimens were subcutaneously implanted into the backs of New Zealand white rabbits. The exposure periods used range from 33 to 114 days. MP35N showed no tendency to exhibit cracking behavior. Both 316L and Ti-6Al-4V ELI, after 114 days, showed surface changes consisting of pitting, possible crack initiation sites, or crack-like features, when viewed at much higher magnification than that used by Galante and Rostoker (Ref 274). A Co-Cr-Mo sample that had been accidentally overloaded and cracked before implantation showed no growth of the crack after 33 days *in vivo*. In another specimen, there was plastic deformation due to the loading. Although only stressed to σ_y, apparently the very coarse-grained cast structure had some surface grains oriented favorably for slip. For this sample, after 73 days of exposure, a crack was observed that had nucleated in a heavily slipped area.

Bundy (Ref 282) has summarized the results of his investigations described previously and listed conditions under which representative alloys from the three main classes of metallic implant materials will and will not undergo SCC *in vivo* and *in vitro*. These conditions involve stress level, exposure time, galvanic coupling condition, stress-intensity level, electrical potential and polarization condition, and presence of a notch or precrack.

Though the vast majority of reports from retrieval studies where SCC is implicated concern orthopedic applications, there are scattered reports of SCC occurring in other sorts of applications as well. To the authors' knowledge, there are only three reports of SCC that are pertinent to clinical dentistry. Jedynakiewicz et al. (Ref 283) describe 21 fractures of nickel-plated 0.8 wt% plain carbon steel dental dam clamps that failed intraorally. The fractures were determined to arise from SCC as a consequence of HE that occurred when phosphoric acid etching procedures were being used clinically. Odén and Tullberg (Ref 284) found intergranular cracks in a gold casting alloy used as a crown cemented onto an amalgam core. The stresses were thought to arise from delayed expansion of the amalgam, expansion as a consequence of amalgam corrosion, or overloading. Either segregation of alloying elements at the grain boundaries or diffusion of mercury to the grain boundaries were thought to be involved in the SCC process. Mercury also can have a deleterious effect on the fatigue life of a platinized gold casting alloy (Ref 285). Yokoyama et al. (Ref 286) analyzed a CP Ti screw in a dental implant system that had fractured, and found the cause to be hydrogen absorption. This led to the controlled studies of hydrogen embrittlement of titanium implant alloys that were described in the previous section.

Brantigan et al. (Ref 212) provide reports of two SCC cases pertinent to thoracic surgery. These involved stainless wires that had fractured. Kossowsky et al. (Ref 287, 288) describe an SCC failure of an aneurysm clip used intracranially. The patient died following the rupture of the aneurysm. The material used to manufacture the clip was type 17-7PH stainless steel.

There has been only one cardiovascular application where *in vivo* SCC has been reported, the Björk-Shiley valve. This case has received considerable attention because it can lead to a fatal outcome. Many hundreds of these prosthetic valves have shown fracture in the weld area of the tungsten inert gas welded leg struts on the downstream side (Ref 289–291). These valves have a ring made of Haynes 25 alloy (also known as HS-25, basically the F 90 alloy) and a carbon occluder. The individual retrieval, or in situ radiographic studies cited previously, shows the single leg strut fracture (SLF) frequencies range from about 4 to 40%, depending on valve design, valve location, valve size, test method, and patient age group.

At the end of 1995, 575 such failures had been reported, 0.67% of the total number of valves implanted, a value considered to be significantly underestimated, since autopsy rates are low. The manufacturer's risk analysis, using actuarial data (Ref 290), found the annual probability of valve failure per year to be 0.22 to 2.52%. A significant number of fatalities have occurred due to this SLF problem. Xiao and Appleby (Ref 292) conducted a comprehensive analysis of the valve failure question based on review of the medical literature, prior failure analyses (Ref 293–295), plus their own experiments. They concluded that the risk of failure of such valves is many orders of magnitude greater than would be expected solely from mechanical fatigue. Mechanical fatigue tests (Ref 296) indicate the fatigue life of Haynes 25 exceeds 10^9 heartbeats (30 yr). Xiao and Appleby (Ref 292) concluded the cause of the valve failures is SCC in the less noble weld area, exacerbated by erosion-corrosion. The weld area was seen to be dealloyed and containing residual porosity and carbide inclusions. Given the outstanding SCC resistance of cobalt-base alloys in chloride solutions and the particularly abundant protein concentration in blood plasma, the Björk-Shiley valve experience may be supportive of the hypothesis presented in the prior section that the protein moiety of *in vivo* fluids can make them more aggressive, as far as SCC is concerned, compared to chloride-containing solutions alone.

As a final comment, it should be noted that the number of reports of SCC of retrieved orthopedic implants in the past two decades (1980–2000) has substantially declined, compared to the 20 year period before that. As with CF, this presumably reflects the greater attention that has been paid to improving the chemical composition and metallurgical condition (particularly microcleanliness) of the alloy grades that are acceptable for surgical implants. The lessened frequency of *in vivo* SCC is especially noteworthy for stainless steel implants.

New Materials and Processing Techniques for CF and SCC Prevention

In industrial applications, SCC prevention can be accomplished by modification of the environment, alteration of the metallurgical condition of the material selected for the application, and changing the loading conditions on the material. For *in vivo* service, however, only the latter two factors remain realistic possibilities.

High-quality surgical implant materials have evolved over the years through the cooperative efforts of biomaterials scientists, manufacturers, surgeons, and regulators. Yet, efforts to improve upon the materials available to orthopedic surgeons continue. For example, pure tantalum (ASTM standard F 560) is a biomaterial that is biocompatible and highly corrosion resistant, as is niobium. Yet, the poor mechanical properties of these materials have restricted their use in load-bearing applications. Pypen et al. (Ref 297) have shown cold work will elevate the endurance limits of these materials compared to the annealed condition by about 10%, which brings the level for tantalum up to that of annealed titanium, which is a material used for dental implants.

New titanium alloys are also under development, as has been mentioned earlier. For example, Niinomi and his coworkers (Ref 298, 299) have performed fatigue tests on a β-titanium alloy Ti-29Nb-13Ta-4.6Zr. He found that, at 10 Hz ($R = 0.1$), the fatigue limit was 700 MPa (102 ksi) when the alloy was given a 1 h solution treatment at 800 °C (1470 °F), and then aged at 400 °C (750 °F) for 72 h to produce fine precipitates (α and ω phases). Okazaki et al. (Ref 300) have also investigated various new titanium alloy formulations. They performed CF testing at 10 Hz in 37 °C (98.6 °F) Eagles medium ($R = 0.1$). They found out that σ_{fs} at 10^8 cycles for annealed Ti-15Zr-4Nb-4Ta and Ti-15Sn-4Nb-4Ta was about 600 MPa (87 ksi). The fatigue strength increased, compared to the annealed condition, when solution treating and aging were performed. For Ti-6Al-2Nb-1Ta, σ_{fs} was 700 MPa (102 ksi) at 10^8 cycles. They also found the σ_{fs} of β-type Ti-15Mo-5Zr-3Al at 10^7 cycles was lower than that of the α/β alloys. In a more recent study, Okazaki (Ref 301) found that for the Ti-15Zr-4Nb-4Ta alloy (β or α/β forged followed by annealing for 2 h at 700 °C, or 1290 °F), a subsequent solution treatment plus aging could increase the σ_{fs} value (using the same test protocol) to 800 MPa (116 ksi). Jablokov et al. (Ref 302) have investigated Ti-35Nb-7Zr-5Ta, a β-titanium alloy that is the most highly alloyed titanium-base biomaterial investigated so far. Because of the high-alloy content, this material has an increased capacity to accommodate interstitial elements without an adverse impact on mechanical properties. When this alloy is cold worked and aged, its endurance limit exceeds that of the Ti-6Al-4V ELI (F 136) and Ti-6Al-4V (F 1472) systems when mill annealed.

Other investigators have studied ways to improve existing titanium-base implant alloys. Filiaggi et al. (Ref 303) applied a ZrO_2 coating on Ti-6Al-4V for the purpose of enhancing implant fixation and showed that the coating did not adversely affect the fatigue strength of the alloy in air (635 MPa, or 92 ksi, at 10^7 cycles). Marquardt (Ref 304) and Nutt et al. (Ref 305)

Summary

Much effort has been devoted to the study of CF and SCC of implant alloys, and modifications in composition and processing of biomaterials have been made, so that today's (2006) implants have markedly improved CF and SCC resistance compared to those of only as little as 20 years ago. Yet, because of the uncertainties and variability surrounding the conditions to which implants are actually exposed *in vivo,* and the further difficulty of accurately simulating these in the laboratory, one cannot yet say whether these problems have disappeared and that contemporary implants should be considered to be totally immune to SCC.

Several aspects of these two processes have been clarified and ways have been illuminated so that these failure modes can be minimized under *in vivo* conditions or else point to potentially fruitful avenues of future research into CF and SCC phenomena. First of all, it should be noted that the yield stress is the key element regarding failure of implants. Macroscopic yielding will almost certainly lead to failure of an implant in the functional sense. However, even on the microscopic scale, localized plastic deformation, though sometimes unavoidable, may nevertheless create focal points for initiation of CF and/or SCC, or else potential biocompatibility problems arising from stress-enhanced ion release. Furthermore, the key to preventing CF or SCC is to prevent crack initiation, since the vast majority of the service lifetime of an implant destined to fail by either mechanism is generally spent in the time period when the crack is being initiated.

From the research viewpoint, crack propagation processes are much easier to investigate than is crack initiation. Enhancement of resistance to environmentally assisted cracking, however, will most effectively come about through increased understanding of the factors involved in initiation and exploitation of this understanding for improvement in materials design of implant alloys. There is also a need in SCC and CF research to develop test methods that take into consideration a more realistic simulation of the complex loading modes to which implants can actually be subjected. In addition, chemical environments for SCC and CF testing that more closely simulate *in vivo* conditions in terms of causative agents, particularly organic constituents (especially proteins), should be developed, even though this is a difficult problem to attack.

Though the CF and SCC behavior of certain orthopedic alloys, such as 316L stainless steel and Ti-6Al-4V ELI, has been extensively studied, this is not necessarily true for the newer alloys for which standards have recently been developed. The SCC and CF resistance of these innovative materials still needs to be assessed in many cases. This is even truer for more experimental materials such as tantalum-, niobium-, and zirconium-base alloy systems. Considering the experience of the Björk-Shiley valve, it is also prudent when older, more traditional implant materials are serving in new applications, stents for example, that their SCC and CF resistance should be thoroughly investigated.

Finally, it again should be pointed out, that over the past 20 years or so (prior to 2006), the orthopedic industry has made substantial improvements in the implant alloys it produces by paying heightened attention to chemical composition, microcleanliness, and optimal processing techniques. This has led to many benefits, including increased CF and SCC resistance. This lesson should be applied to other industries that supply metallic implant devices as well, in particular those that manufacture products for use in the cardiovascular system and for dentistry.

REFERENCES

1. K.J. Bundy, *Crit. Rev. Biomed. Eng.,* Vol 22 (No. 3/4), 1994, p 139–251
2. K.J. Bundy, C. Williams, and R. Luedemann, The Effect of Static and Dynamic Loading on the Corrosion Rate of Surgical Implant Materials, *Biomedical Engineering IV Recent Developments, Proc. Fourth South. Biomed. Eng. Conf.,* B.W. Sauer, Ed., Pergamon Press, 1985, p 108–111
3. K.J. Bundy, M.A. Vogelbaum, and V.H. Desai, *J. Biomed. Mater. Res.,* Vol 20, 1986, p 493–505
4. R.E. Luedemann and K.J. Bundy, The Effect of Dynamic Stress on the Corrosion Characteristics of Surgical Implant Alloys, *Trans. 15th Ann. SFB Meeting,* April 28 to May 2, 1989 (Lake Buena Vista, FL), p 56
5. K.J. Bundy, M. Marek, and R.F. Hochman, *J. Biomed. Mater. Res.,* Vol 17, 1983, p 467–487
6. J.H. Dumbleton and E.H. Miller, Failures of Metallic Orthopaedic Implants, *Failure Analysis and Prevention,* Vol 10, *Metals Handbook,* 8th ed., American Society for Metals, 1975, p 571–580
7. C.A. Engh, D. O'Connor, M. Jasty, T.F. McGovern, J.D. Bobyn, and W.H. Harris, *Clin. Orth. Rel. Res.,* Vol 285, Dec 1992, p 13–29
8. W.E. White, J. Postlethwaite, and I. Le May, *Microstruc. Sci.,* Vol 4, 1976, p 145–158
9. L.D. Zardiackas and L.D. Dillon, Failure Analysis of Metallic Orthopedic Devices, Chapter 5, *Encylopedic Handbook of Biomaterials and Bioengineering, Part B: Applications,* Vol 1, D.L. Wise, D.J. Trantolo, D.E. Altobelli, M.J. Yaszemski, J.D. Gresser, and E.R. Scwartz, Ed., Marcel Dekker, 1995, p 123–170
10. O.E.M. Pohler and F. Straumann, Fatigue and Corrosion Fatigue Studies on Stainless-Steel Implant Material, Chapter 6, *Evaluation of Biomaterials,* G.D. Winter, J. Leray, and K. de Groot, Ed., John Wiley and Sons, 1980
11. O.E.M. Pohler, Failures of Metallic Orthopedic Implants, *Failure Analysis and Prevention,* Vol 11, *Metals Handbook,* 9th ed., American Society for Metals, 1986, p 670–694
12. J.M. Orten and O.W. Neuhaus, *Human Biochemistry,* 9th ed., C.V. Mosby, 1975
13. M.G. Shettlemore and K.J. Bundy, *Biomaterials,* Vol 22, 2001, p 2215–2228
14. C.-M. Hsiao, X.-Y. Huang, D.-M. Wang, and Z.-F. Zhu, The Mechanism of Hydrogen Embrittlement and Stress Corrosion Cracking of Titanium Alloys, *Proc. Eighth Intr. Conf. on Metallic Corrosion,* Vol 1, 1981, p 594–599
15. D.O. Sprowls, Evaluation of Corrosion Fatigue, *Corrosion,* Vol 13, *Metals Handbook,* 9th ed., ASM International, 1987, p 291–302
16. Y. Katz, N. Tymiak, and W.W. Gerberich, Evaluation of Environmentally Assisted Crack Growth, *Mechanical Testing and Evaluation,* Vol 8, *ASM Handbook,* ASM International, 2000, p 612–648
17. A.F. Tencer, Biomaterials in the Fixation of Bone Fractures, Chapter 8, *Encylopedic Handbook of Biomaterials and Bioengineering, Part B: Applications,* Vol 1, D.L. Wise, D.J. Trantolo, D.E. Altobelli, M.J. Yaszemski, J.D. Gresser, and E.R. Scwartz, Ed., Marcel Dekker, 1995, p 223–264
18. R.H. Shetty and W.H. Ottersberg, Metals in Orthopaedic Surgery, Chapter 17, *Encylopedic Handbook of Biomaterials and Bioengineering, Part B: Applications,* Vol 1, D.L. Wise, D.J. Trantolo, D.E. Altobelli, M.J. Yaszemski, J.D. Gresser, and E.R. Scwartz, Ed., Marcel Dekker, 1995, p 509–540
19. H. Mueller, Tarnish and Corrosion of Dental Alloys, *Corrosion,* Vol 13, *Metals Handbook,* 9th ed., ASM International, 1987, p 1336–1366
20. J.A. Disegi and L. Eschbach, *Injury, Int. J. Care Injured,* Vol 31 (Suppl. 4), 2000, p D2–D6
21. D. Rodriguez, J.M. Manero, F.J. Gil, and J.A., Planell, *J. Mater. Sci. Mater. Med.,* Vol 12 (No. 10/12), 2001, p 935–937
22. L.C. Lucas, R.A. Buchanan, and J.E. Lemons, *J. Biomed. Mater. Res.,* Vol 15, 1981, p 731–747
23. J.P. Collier, V.A. Surprenant, R.E. Jensen, and M.B. Mayor, *Clin. Orthop.,* Vol 271, Oct. 1991, p 305–312
24. S.A. Brown, C.A.C. Fleming, J.S. Kawalec, C.J. Vassaux, J.H. Payer, M.J. Kraay, and K. Merritt, Fretting Accelerated

Crevice Corrosion of Modular Hips, *Trans. Implant Retrieval Symp.* (Pheasant Run, IL), 1992, p 59
25. D.I. Bardos, Titanium and Zirconium Alloys in Orthopaedic Applications, Chapter 18, *Encylopedic Handbook of Biomaterials and Bioengineering, Part B: Applications,* Vol 1, D.L. Wise, D.J. Trantolo, D.E. Altobelli, M.J. Yaszemski, J.D. Gresser, and E.R. Scwartz, Ed., Marcel Dekker, 1995, p 541–548
26. K.F. Leinfelder and J.E. Lemons, *Clinical Restorative Materials and Techniques,* Lea and Febiger, 1988
27. A.N. Cranin, The Use of Biomaterials in Oral and Maxillofacial Surgery, Chapter 9 *Encylopedic Handbook of Biomaterials and Bioengineering, Part B: Applications,* Vol 1, D.L. Wise, D.J. Trantolo, D.E. Altobelli, M.J. Yaszemski, J.D. Gresser, and E.R. Scwartz, Ed., Marcel Dekker, 1995, p 265–306
28. K.J. Bundy, *Fundamentals of Biomaterials—Science and Applications,* Springer Verlag, in press
29. J.Z. Dennis, C.H. Craig, H.R. Radisch, Jr., E.J. Pannek Jr., P.C. Turner, A.G. Hicks, M. Jenusaitis, N.A. Gokcen, C.M. Friend, and M.R. Edwards, Processing Platinum-Enhanced Radiopaque Stainless Steel (PERSS) for Use as Balloon-Expandable Coronary Stents, *Stainless Steels for Medical and Surgical Applications,* STP 1438, G.L. Winters and M.J. Nutt, Ed., ASTM International, 2003, p 61–71
30. B.S. Covino Jr., C.H. Craig, S.D. Cramer, S.J. Bullard, M. Ziomek-Moroz, P.D. Jablonski, P.C. Turner, H.R. Radisch, Jr., N.A. Gokcen, C.M. Friend, and M.R. Edwards, Corrosion Behavior of Platinum-Enhanced Radiopaque Stainless Steel (PERSS) for Dilation-Balloon Expandable Coronary Stents, *Stainless Steels for Medical and Surgical Applications,* STP 1438, G.L. Winters and M.J. Nutt, Ed., ASTM International, 2003, p 176–193
31. S. Lampman, Fatigue and Fracture Properties of Stainless Steels, *Fatigue and Fracture,* Vol 19, *ASM Handbook,* ASM International, 1996, p 712–732
32. Y.V. Murty, Use of Stainless Steels in Medical Applications, *Medical Device Materials: Proceedings of the Materials and Processes for Medical Devices Conference,* ASM International, 2004
33. J. Black, *Orthopedic Biomaterials in Research and Clinical Practice,* Churchill Livingstone, 1988
34. A.C. Fraker, Corrosion of Metallic Implants and Prosthetic Devices, *Corrosion,* Vol 13, *Metals Handbook,* 9th ed., ASM International, 1987, p 1324–1335
35. S. Hierholzer and G. Hierholzer, Internal Fixation and Metal Allergy-Clinical Investigations, *Immunology and Histology of the Implant Tissue Interface,* Thieme Medical Publishers, 1992
36. W.J. Bratina, S.B.Young, M.J. Morgan, R.M. Pilliar, S. Yue, and A.C. Wallace, Fatigue Deformation and Fractographic Analysis of Surgical Implants and Implant Materials, *Int. Conf. and Exhibits on Failure Analysis,* July 8–11, 1991, Proceedings (Montreal), p 299–310
37. T.J. Pruitt and M.J. Hanslits, Cobalt-Base Alloys, *Casting,* Vol 15, *Metals Handbook,* 9th ed., ASM International, 1988, p 811–814
38. J.R. Newman, D. Eylon, and J.K. Thorne, Titanium and Titanium Alloys, *Casting,* Vol 15, *Metals Handbook,* 9th ed., ASM International, 1988, p 824–835
39. L.D. Zardiackas, D.E. Parsell, L.D. Dillon, D.W. Mitchell, L.A. Nunnery, and R. Poggie, *J. Biomed. Mater. Res.,* Vol 58 (No. 2), 2001, p 180–187
40. D.J. Medlin, S. Charlebois, D. Swarts, and R. Shetty, *Adv. Mater. Process.,* Dec 2003, p 31–32
41. D.J. Medlin, R. Shetty, and J. Scrafton, Metallurgical Attachment of a Porous Tantalum Foam to a Titanium Substrate for Orthopaedic Applications, Abstract book, *ASTM Symposium on Titanium, Niobium, Zirconium, and Tantalum for Medical and Surgical Applications,* Nov 9–10, 2004 (Washington, D.C.), p 8
42. L. Tuchinskiy, and R. Loutfy, Titanium Foams for Medical Applications, *Adv. Mater. Process.,* Dec 2003, p 32–33
43. P.J. Andersen, Medical and Dental Applications, *Powder Metallurgy,* Vol 7, *Metals Handbook,* 9th ed., American Society for Metals, 1984, p 657–663
44. H. Hahn, P.J. Lare, R.H. Rowe, Jr., A.C. Fraker, and F. Ordway, Mechanical Properties and Structure of Ti-6Al-4V with Graded-Porosity Coatings Applied by Plasma Spraying for Use in Orthopaedic Implants, *Corrosion and Degradation of Implant Materials: Second Symposium,* STP 859, A.C. Fraker and C.D. Griffin, Ed., ASTM, 1985, p 179–191
45. C.W. Fairhurst, Amalgam, Chapter 16, *An Outline of Dental Materials and Their Selection,* W.J. O'Brien and G. Ryge, Ed., W.B. Saunders, 1978, p 210–218
46. K.J. Anusavice, Dental Amalgam—Structures and Properties, Chapter 17, *Phillips' Science of Dental Materials,* 10th ed., W.B. Saunders, 1996
47. R.G. Craig, Ed., *Restorative Dental Materials,* 6th ed., C.V. Mosby, 1980
48. Council on Dental Materials and Devices, *Guide to Dental Materials and Devices,* 6th ed., American Dental Association, 1972
49. R.P. Gangloff, Environmental Cracking—Corrosion Fatigue, Chapter 26, *Corrosion Tests and Standards—Application and Interpretation,* R. Baboian, Ed., ASTM, 1995, p 253–271
50. R. Burton-Opitz, *A Textbook of Physiology for Students and Practitioners of Medicine,* W.B. Saunders, 1921
51. P.H. Mitchell, *A Textbook of Biochemistry,* McGraw-Hill, 1946
52. R.M. De Coursey, *The Human Organism,* 3rd ed., McGraw-Hill, 1968
53. O. Lippold and F. Winton, *Human Physiology,* Little Brown, 1968
54. N.K. Sarkar and E.H. Greener, *Biomater. Med. Dev. Artif. Org.,* Vol 1, 1973, p 121–129
55. J.R. Cahoon, R. Bandyopadhya, and L. Tennese, *J. Biomed. Mater. Res.,* Vol 9, 1975, p 259–264
56. *Stedman's Medical Dictionary,* 23rd ed., Williams and Wilkins, 1976
57. K.G. Watkins, S. Ben Younis, D.E. Davies, and K. Williams, *Biomaterials,* Vol 7, 1986, p 147–151
58. R.L. Jones, S.S. Wing, and B.C. Syrett, *Corrosion,* Vol 34, 1978, p 226–236
59. M.A. Imam, A.C. Fraker, and C.M. Gilmore, Corrosion Fatigue of 316L Stainless Steel, Co-Cr-Mo Alloy, and ELI Ti-6Al-4V, *Corrosion and Degradation of Implant Materials,* STP 684, B.C. Syrett and A. Acharya, Ed., ASTM, 1979, p 128ff
60. M. Marek and T. Okabe, *J. Biomed. Mater. Res.,* Vol 12, 1978, p 857–866
61. L.D. Zardiackas and G.E. Stoner, *Biomaterials,* Vol 1, 1980, p 13–16
62. N.K. Sarkar and J.-R. Park, *J. Dent. Res.,* Vol 67 (No. 10), 1988, p 1312–1315
63. C.E. Guthrow, L.B. Johnson, and K.R. Lawless, *J. Dent. Res.,* Vol 46 (No. 6, part 2), 1967, p 1372–1381
64. N.K. Sarkar, G.W. Marshall, J.B. Moser, and E.H. Greener, *J. Dent. Res.,* Vol 54 (No. 5), 1975, p 1031–1038
65. J.C. Muhler and H.M. Swenson, *J. Dent. Res.,* Vol 26, 1947, p 474ff
66. T. Fusiyama, T. Katayori, and S. Nomoto, *J. Dent. Res.,* Vol 42, 1963, p 1183–1197
67. G. Tani and F. Zucchi, *Minerva Stomat.,* Vol 16, 1967, p 710–713
68. F.V. Wald and F.H. Cocks, *J. Dent. Res.,* Vol 50, 1971, p 48–59
69. C.W. Fairhurst, M. Marek, M.B. Butts, and T. Okabe, *J. Dent. Res.,* Vol 57 (No. 5–6), 1978, p 725–729
70. G. Ravnholt and R.I. Holland, *Dent. Mater.,* Vol 4, 1988, p 251–254
71. H.J. Mueller, *Quint. Int.,* Vol 5, 1982, p 589–593
72. H.J. Mueller, *Biomaterials,* Vol 4, 1983, p 66–72
73. H.J. Mueller, *Mater. Res. Symp.,* Vol 110, 1989, p 605ff
74. G.C.F. Clark and D.F. Williams, *J. Biomed. Mater. Res.,* Vol 16, 1982, p 125–134
75. S. Brown and K. Merritt, The Effects of Serum Protein on Corrosion Rates *in Vitro, Clinical Applications of Biomaterials,* A. Lee, T. Albrektsson, and P. Branemark, Ed., Advances in Biomaterials, 1982, p 195ff
76. G.J. Mattamal and A.C. Fraker, Kinetics of the Ni(II) Reaction with Human

Blood Serum Albumin, *Trans. Tenth Ann. Meeting,* Society for Biomaterials, 1984, p 124
77. D.F. Williams, *Crit. Rev. Biocompat.,* Vol 1, 1985, p 1ff
78. J. Yang and J. Black, The Binding and Competition of Cr, Co, and Ni to Serum Proteins, *Trans. 19th Ann. Meeting Soc. Biomater.,* 1993, p 216
79. M.A. Miner, *Trans. ASME,* Vol 67, 1945, p A159
80. K.J. Bundy and R. Luedemann, The Effect of Dynamic Physiological Loading on the Corrosion of Surgical Implant Alloys, *Trans. 18th International Biomaterials Symp.,* May 28 to June 1, 1986, Vol 9, p 197
81. K.R. Wheeler and L.A. James, *J. Biomed. Mater. Res.,* Vol 5, 1971, p 267–281
82. E.J. Czyryca and D.A. Utah, Introduction to Fatigue Testing, *Mechanical Testing,* Vol 8, *Metals Handbook,* 9th ed., American Society for Metals, 1985, p 363–365
83. D.A. Ryder, T.J. Davies, and I. Brough, General Practice in Failure Analysis, *Failure Analysis and Prevention,* Vol 11, *Metals Handbook,* 9th ed., American Society for Metals, 1986, p 15–46
84. O.E.M. Pohler, Degradation of Metallic Orthopedic Implants, Chapter 15, *Biomaterials in Reconstructive Surgery,* L.R. Rubin, Ed., C.V. Mosby, 1983, p 158–228
85. S. Bhattacharyya, V.E. Johnson, S. Agrawal, and M.A.H. Howes, Ed., Failure Analysis of Metallic Materials by Scanning Electron Microscopy, *IITRI Fracture Handbook,* Metals Research Division, IIT Research Institute, Jan 1979, p xii
86. V. Kerlins and A. Phillips, Modes of Fracture, *Fractography,* Vol 12, *Metals Handbook,* 9th ed., ASM International, 1987, p 12–71
87. N.D. Feige and R.L. Kane, Service Experience with Titanium Structures in Marine Service, *Proc. 26th NACE Conf.* (Houston, TX), 1970, p 194–199
88. M. Müller, *Metall. Trans. A,* Vol 13, 1982, p 145
89. D.A. Jones, *Metall. Trans. A,* Vol 16, 1985, p 1133
90. R.O. Ritchie, Role of the Environment in Near-Threshold Fatigue Crack Growth in Engineering Materials, *Environment-Sensitive Fracture of Engineering Materials,* Z.A. Foroulis, Ed., The Metallurgical Society of AIME, 1979, p 538–564
91. A. Saxena and C.L. Muhlstein, Fatigue Crack Growth Testing, *Mechanical Testing and Evaluation,* Vol 8, *ASM Handbook,* ASM International, 2000, p 740–757
92. M.E. Fine and Y.-W. Chung, Fatigue Failure in Metals, *Fatigue and Fracture,* Vol 19, *ASM Handbook,* ASM International, 1996, p 63–72
93. J.D. Bolton, J. Hayden, and M. Humphreys, *Eng. Med.,* Vol 11 (No. 2), 1982, p 59–68
94. R.O. Ritchie, Near-Theshold Fatigue-Crack Propagation in Steels, *Int. Met. Rev.,* Vol 24 (No. 5, 6), 1979
95. S.D. Antolovich and A. Saxena, Fatigue Failures, *Failure Analysis and Prevention,* Vol 11, *Metals Handbook,* 9th ed., American Society for Metals, 1986, p 102–135
96. R. Ritchie, Fatigue and Fracture Mechanics, *Mechanical Testing and Evaluation,* Vol 8, *ASM Handbook,* ASM International, 2000, p 681–685
97. L.E. Sloter and H.R. Piehler, Corrosion-Fatigue of Stainless Steel Hip Nails-Jewett Type, *Corrosion and Degradation of Implant Materials,* STP 684, B.C. Syrett and A. Acharya, Ed., ASTM, 1979, p 173–195
98. B.F. Brown, Stress Corrosion Cracking Control Measures, *NBS Monograph 156,* National Bureau of Standards, 1977
99. M.O. Speidel, *Stress Corrosion Research,* H. Arup and R.N. Parkins, Ed., Sijthoff and Noordhoff, 1979, p 117–183
100. D.W. Shoesmith, Effects of Metallurgical Variables on Aqueous Corrosion, *Corrosion,* Vol 13, *Metals Handbook,* 9th ed., ASM International, 1987, p 45–49
101. B. Leis, Effect of Surface Condition and Processing on Fatigue Performance, *Fatigue and Fracture,* Vol 19, *ASM Handbook,* ASM International, 1996, p 314–320
102. E.J. Czyryca, Fatigue Crack Initiation, *Mechanical Testing,* Vol 8, *Metals Handbook,* 9th ed., American Society for Metals, 1985, p 366–375
103. F.P. Ford, *Environment Induced Cracking of Metals,* R.P. Gangloff and M.B. Ives, Ed., NACE, 1990, p 21–29
104. A.J. Sedriks, *Corrosion of Stainless Steels,* John Wiley, 1979
105. B.E. Wilde, Stress-Corrosion Cracking, *Failure Analysis and Prevention,* Vol 11, *Metals Handbook,* 9th ed., American Society for Metals, 1986, p 203–224
106. J. Yu, Z.J. Zhao, and L.X. Li, *Corros. Sci.,* Vol 35 (No. 1–4), 1993, p 587–597
107. C.B. O'Sullivan, A.L. Bertone, A.S. Litsky, and J.T. Robertson, *Am. J. Vet. Res.,* Vol 65 (No. 7), 2004, p 972–976
108. H. Brunner and J.P. Simpson, *Injury: Br. J. Accident Surg.,* Vol 11 (No. 3), 1980, p 203–207
109. B. Coquillet, L. Vincent, and P. Guiraldcnq, *J. Biomed. Mater. Res.,* Vol 13, 1979, p 657–668
110. R.F. Hochman and L.M. Taussig, *J. Mater.,* Vol 1 (No. 2), 1966, p 425–442
111. J.R. Davis, Ed., Stress Corrosion Cracking and Hydrogen Embrittlement, *ASM Specialty Handbook: Stainless Steels,* ASM International 1994, p 181–204
112. R.M. Pilliar, *Biomater.,* Vol 12, 1991, p 95–100
113. S.-H. Wang, and C. Müller, *Fatigue Fract. Eng. Mater. Struct.,* Vol 21, 1998, p 1077–1087
114. L. Wagner, Fatigue Life Behavior, *Failure and Fracture Properties of Titanium Alloys, Fatigue and Fracture,* Vol 19, *ASM Handbook,* ASM International, 1996, p 837–845
115. J.K. Gregory, Fatigue Crack Growth of Titanium Alloys, *Fatigue and Fracture,* Vol 19, *ASM Handbook,* ASM International, 1996, p 845–853
116. A.C. Fraker, Medical and Dental, Chapter 78, Section VII Testing in Industries, *Corrosion Tests and Standards—Application and Interpretation,* R. Baboian, Ed., ASTM, 1995, p 705–715
117. J.-B. Vogt, *J. Mater. Process. Technol.,* Vol 117, 2001, p 364–369
118. M.R. Kalantary, T.E. Chung, and R.G. Faulkner, *Mater. Sci. Eng.,* Vol A189, 1994, p 85–94
119. P.S. Veers, Statistical Considerations, *Fatigue and Fracture,* Vol 19, *ASM Handbook,* ASM International, 1996, p 295–302
120. F.H. Haynie, Statistical Treatment of Data, Data Interpretation, and Reliability, Chapter 5, *Corrosion Tests and Standards—Application and Interpretation,* R. Baboian, Ed., ASTM, 1995, p 62–67
121. P.L. Andresen, Corrosion Fatigue Testing, *Fatigue and Fracture,* Vol 19, *ASM Handbook,* ASM International, 1996, p 193–209
122. J.F. Knott, *Fundamentals of Fracture Mechanics,* Butterworths, 1973
123. R.J.H. Wanhill, *Brit. Corros. J.,* Vol 10 (No. 2), 1975, p 69–78
124. D. Broek, Failure Analysis and Fracture Mechanics, *Failure Analysis and Prevention,* Vol 11, *Metals Handbook,* 9th ed., American Society for Metals, 1986, p 47–65
125. P. Liaw, Fracture Toughness and Fracture Mechanics, *Mechanical Testing and Evaluation,* Vol 8, *ASM Handbook,* ASM International, 2000, p 563–575
126. G.R. Irwin, Fracture Mechanics, *Mechanical Testing,* Vol 8, *Metals Handbook,* 9th ed., American Society for Metals, 1985, p 439–464
127. S.D. Antolovich, An Introduction to Fracture Mechanics, *Fatigue and Fracture,* Vol 19, *ASM Handbook,* ASM International, 1996, p 371–380
128. D.A. Utah, W.H. Cullen, L.C. Majno, R.A. Meyers, R.O. Ritchie, R.H. Stentz, and R. Williams, Fatigue Crack Propagation, *Mechanical Testing,* Vol 8, *Metals Handbook,* 9th ed., American Society for Metals, 1985, p 376–402
129. L.D. Zardiackas, M. Roach, S. Williamson, and J.-A. Bogan, Comparison of Corrosion Fatigue of Biodur 108 to 316L S.S. and 22Cr-13Ni-5Mn S.S., *Stainless Steels for Medical and Surgical Applications,* STP 1438, G.L. Winters and M.J. Nutt, Ed., ASTM International, 2003, p 194–207

130. J.R. Cahoon and R.N. Holte, *J. Biomed. Mater. Res.*, Vol 15, 1981, p 137–145
131. R.M. Pilliar and G.C. Weatherly, *Crit. Rev. Biocompat.*, Vol 1, 1986, p 371–403
132. Y. Gill, J. Davidson, and A. Gavens, Corrosion Fatigue Performance of a Fenestrated Moore Endoprosthesis Made of Surgical Stainless Steel, *Proc. Fifth Southern Biomedical Eng. Conf.*, 1986, p 269ff
133. M. Windler and R. Steger, Mechanical and Corrosion Properties of Forged Hip Stems Made of High-Nitrogen Stainless Steel, *Stainless Steels for Medical and Surgical Applications*, STP 1438, G.L. Winters and M.J. Nutt, Ed., ASTM International, 2003, p 39–49
134. I. Tikhovski, H. Brauer, M. Mölders, M. Wiemann, D. Bingmann, and A. Fischer, Fatigue Behavior and In-Vitro Biocompatibility of the New Ni-Free Austenite High-Nitrogen Steel X13CrMnMoN18-14-3, *Stainless Steels for Medical and Surgical Applications*, STP 1438, G.L. Winters and M.J. Nutt, Ed., ASTM International, 2003, p 119–136
135. L. Eschbach, G. Bigolin, W. Hirsiger, and B. Gasser, Fatigue of Small Bone Fragment Fixation Plates Made from Low-Nickel Steel, *Stainless Steels for Medical and Surgical Applications*, STP 1438, G.L. Winters and M.J. Nutt, Ed., ASTM International, 2003, p 93–106
136. V.J. Colangelo, *J. Basic Eng., Trans. ASME, Series D*, Vol 91, 1969, p 581–856
137. A.J. McEvily Jr., Ed., *Atlas of Stress-Corrosion and Corrosion Fatigue Curves*, ASM International, 1990, p 3–27
138. M. Taira and E.P. Lautenschlager, *J. Biomed. Mater. Res.*, Vol 26 (No. 9), 1992, p 1131–1139
139. M. Nakajima, T. Shimizu, T. Kanamori, and K. Tokaji, *Fatigue Fract. Eng. Mater. Struc.*, Vol 21, 1998, p 35–45
140. H.R. Piehler, M.A. Portnoff, L.E. Sloter, E.J. Vegdahl, J.L. Gilbert, and M.J. Weber, Corrosion Fatigue of Hip Nails: The Influence of Materials Selection and Design, *Corrosion and Degradation of Implant Materials: Second Symposium*, STP 859, A.C. Fraker and C.D. Griffin, Ed., ASTM, 1985, p 93–104
141. Y. Higo and Y. Tomita, Evaluation of Mechanical Properties of Metallic Biomaterials, *Biomaterials' Mechanical Properties*, STP 1173, H.E. Kambic and A.T. Yokobori, Jr., Ed., ASTM, 1994, p 148–155
142. O.E.M. Pohler, "Study of Initiation and Propagation Stages of Fatigue and Corrosion Fatigue of Orthopedic Implant Materials", Ph.D. dissertation, Ohio State University, 1983
143. D.W. Hoeppner and V. Chandrasekaran, Characterizing the Fretting Fatigue Behavior of Ti-6Al-4V in Modular Joints, *Medical Applications of Titanium and Its Alloys: The Material and Biological Issues*, STP 1272, S.A. Brown and J.E. Lemons, Ed., ASTM, 1996, p 252–265
144. S.D. Cook, F.S. Georgette, H.B. Skinner, and R.J. Haddad, Jr., *J. Biomed. Mater. Res.*, Vol 18, 1984, p 497–512
145. M.F. Semlitsch, B. Panic, H. Weber, and R. Schoen, Comparison of the Fatigue Strength of Femoral Prosthesis Stems Made of Forged Ti-6Al-4V and Cobalt-Based Alloys, *Titanium Alloys in Surgical Implants*, STP 796, H.A. Luckey and F. Kubli, Ed., ASTM, 1983, p 120–147
146. D.H. Kohn and P. Ducheyne, *J. Biomed. Mater. Res.*, Vol 24, 1990, p 1483–1501
147. D.H. Kohn, C.C. Ko, S.J. Hollister, D. Snoeyink, J. Awerbuch, and P. Ducheyne, Methods of Detecting and Predicting Microfracture in Titanium, *Medical Applications of Titanium and Its Alloys: The Material and Biological Issues*, STP 1272, S.A. Brown and J.E. Lemons, Ed., ASTM, 1996, p 117–135
148. D.H. Kohn, P. Ducheyne, and J. Awerbuch, *J. Biomed. Mater. Res.*, Vol 26, 1992, p 19–38
149. R.H. Shetty, Mechanical and Corrosion Properties of Nitrogen Diffusion Hardened Ti-6Al-4V Alloy, *Medical Applications of Titanium and Its Alloys: The Material and Biological Issues*, STP 1272, S.A. Brown and J.E. Lemons, Ed., ASTM, 1996, p 240–251
150. H. Hampel and H.R. Piehler, Evaluation of the Corrosion Fatigue Behavior of Porous Coated Ti-6Al-4V, *Medical Applications of Titanium and Its Alloys: The Material and Biological Issues*, STP 1272, S.A. Brown and J.E. Lemons, Ed., ASTM, 1996, p 136–149
151. D. Wolfarth and P. Ducheyne, A Novel Porous Coating Geometry to Improve the Fatigue Strength of Ti-6Al-4V Implant Alloy, *Medical Applications of Titanium and Its Alloys: The Material and Biological Issues*, STP 1272, S.A. Brown and J.E. Lemons, Ed., ASTM, 1996, p 150–160
152. P.L. Bourassa, S. Yue, and J.D. Bobyn, *J. Biomed. Mater. Res.*, Vol 37, 1997, p 291–300
153. R.A. Zavanelli, G.E. Pessanha-Henriques, I. Ferreira, and J.M. De Almeida Rollo, *J. Prosthet. Dent.*, Vol 84 (No. 3), 2000, p 274–279
154. R.A. Zavanelli, A.S. Guilherme, G.E. Pessanha-Henriques, M.A. de Arruda Nobilo, and M.F. Mesquita, *J. Oral Rehabil.*, Vol 31 (No. 10), 2004, p 1029–1034
155. S.K. Bhambri, R.H. Shetty, and L.N. Gilbertson, Optimization of Properties of Ti-15Mo-2.8Nb-0.2Si and Ti-15Mo-2.8Nb-0.2Si-.26O Beta Titanium Alloys for Application in Prosthetic Implants, *Medical Applications of Titanium and Its Alloys: The Material and Biological Issues*, STP 1272, S.A. Brown and J.E. Lemons, Ed., ASTM, 1996, p 88–95
156. K.K. Wang, L.J. Gustavson, and J.H. Dumbleton, Microstructure and Properties of a New Beta Titanium Alloy, Ti-12Mo-6Zr-2Fe, *Medical Applications of Titanium and Its Alloys: The Material and Biological Issues*, STP 1272, S.A. Brown and J.E. Lemons, Ed., ASTM, 1996, p 76–87
157. A.K. Mishra, J.A. Davidson, R.A. Poggie, P. Kovacs, and T.J. FitzGerald, Mechanical and Tribological Properties and Biocompatibility of Diffusion Hardened Ti-13Nb-13Zr—A New Titanium Alloy for Surgical Implants, *Medical Applications of Titanium and Its Alloys: The Material and Biological Issues*, STP 1272, S.A. Brown and J.E. Lemons, Ed., ASTM, 1996, p 96–113
158. M.D. Roach, R.S. Williamson, and L.D. Zardiackas, *J. Test. Eval.*, (in press)
159. L.D. Zardiackas, M.D. Roach, and R.S. Williamson, *J. Test. Eval.*, (in press)
160. F.S. Georgette and J.A. Davidson, *J. Biomed. Mater. Res.*, Vol 20, 1986, p 1229–1248
161. R.M. Berlin, L.J. Gustavson, and K.K. Wang, Influence of Post Processing on the Mechanical Properties of Investment Cast and Wrought Co-Cr-Mo Alloys, *Cobalt-Base Alloys for Biomedical Applications*, STP 1365, J.A. Disegi, R.L. Kennedy, and R. Pilliar, Ed., ASTM, 1999, p 62–70
162. A.K. Mishra, M.A. Hamby, and W.B. Kaiser, Metallurgy, Microstructure, Chemistry, and Mechanical Properties of a New Grade of Cobalt-Chromium Alloy before and after Porous-Coating, *Cobalt-Base Alloys for Biomedical Applications*, STP 1365, J.A. Disegi, R.L. Kennedy, and R. Pilliar, Ed., ASTM, 1999, p 71–88
163. K.K. Wang, R.M. Berlin, and L.J. Gustavson, A Dispersion Strengthened Co-Cr-Mo Alloy for Medical Implants, *Cobalt-Base Alloys for Biomedical Applications*, STP 1365, J.A. Disegi, R.L. Kennedy, and R. Pilliar, Ed., ASTM, 1999, p 89–97
164. P. Sury and M. Semlitsch, *J. Biomed. Mater. Res.*, Vol 12 (No. 5), 1978, p 723–741
165. L.V. Lassila and P.K. Vallittu, *J. Prosthet. Dent.*, Vol 80 (No. 60), 1998, p 708–713
166. A. Cornet, D. Muster, and J.H. Jaeger, *Biomater. Med. Devices Artif. Organs*, Vol 7 (No. 1), 1979, p 155–167
167. R.A. Whitaker, Environmental Effects on the Life of Bone-Plate-Type Surgical Implants, *Rev. Environ. Health.*, Vol 4 (No. 1), 1982, p 63–82
168. E.G. Wilkinson and D.C. Haack, *J. Dent. Res.*, Vol 37 (No. 1), 1958, p 136–143
169. R.S. Mateer and C.D. Reitz, *J. Dent. Res.*, Vol 49, 1970, p 399–407
170. R.J. Sutow and D.W. Jones, Fatigue Behavior of Dental Amalgam, presented at 57th Ann. AADR/IADR Meeting, March 29 to April 1, 1979 (New Orleans, LA), Dental Materials Group Microfilm Abstract No. 24

171. H. Hero, D. Brune, R.B. Jorgensen, and D.M. Evje, *Scand. J. Dent. Res.,* Vol 91 (No. 6), 1983, p 488–495
172. K. Lian, and E.I. Meletis, *Dent. Mater.,* Vol 12 (No. 3), 1996, p 146–153
173. J.R. Cahoon and H.W. Paxton, *J. Biomed. Mater. Res.,* Vol 2, 1968, p 1–22
174. V.J. Colangelo and N.D. Greene, *J. Biomed. Mater. Res.,* Vol 3, 1969, p 247–265
175. A.N. Hughes and B.A. Jordan, *J. Biomed. Mater. Res.,* Vol 6, 1972, p 33–48
176. A. Weinstein, H. Amstutz, G. Pavon, and V. Franceschini, *J. Biomed. Mater. Res. Symp.,* Vol 4, 1973, p 297–325
177. J.F. Bates and J. Scott, *J. Biomed. Mater. Res.,* Vol 7, 1973, p 419–429
178. W.B. Lisagor, *ASTM Standardization News,* Vol 3 (No. 5), 1975, p 20–24, 43–44
179. P. Ducheyne, P. De Meester, and E. Aernoudt, *J. Biomed. Mater. Res.,* Vol 14, 1980, p 31–40
180. R.F. Hochman, M. Marek, and K.J. Bundy, An Analysis of *in Vivo* and *in Vitro* Biomaterial Properties and Service Characteristics of Orthopedic Implants, *Trans. Fifth Ann. SFB Meeting,* 1979, p 126
181. T.M. Wright, A.H. Burstein, and D.L. Bartel, Retrieval Analysis of Total Joint Replacement Components: A Six-Year Experience, *Corrosion and Degradation of Implant Materials: Second Symposium,* STP 859, A.C. Fraker and C.D. Griffin, Ed., ASTM, 1985, p 415–428
182. I.L. Baswell, T. Sander, B. Allen, Jr., and C.O. Bechtol, *J. Biomed. Mater. Res.,* Vol 20 (No. 7), 1986, p 887–894
183. P. Novak, L. Joska, V. Stedry, and J. Hron, *Acta Chir. Orthop. Traumatol. Cech.,* Vol 60 (No. 5), 1993, p 311–314
184. L.D. Zardiackas, R.J. Black, J.L. Hughes, R.B. Reeves, and J.N. Jun, *Orthopaedics,* Vol 12 (No. 1), 1989, p 85–92
185. O. Pohler, Characteristics of the Stainless Steel ASIF/AO Implants, *AO Bull.,* Switzerland, Swiss Association of the Study of Internal Fixation, 1975
186. R. Ackell, U. Lekholm, B. Rockler, and P.I. Brånemark, *Int. J. Oral Surg.,* Vol 10, 1981, p 387–416
187. H.J. Snyder and C.B. Snyder, Fatigue Fracture of 316L Stainless Steel Screws Employed for Surgical Implanting, *Handbook of Case Histories of Failure Analysis,* Vol 1, K.A. Esaklul, Ed., ASM International, 1992, p 315–317
188. J.L. Gilbert, C.A. Buckley, J.J. Jacobs, K.C. Bertin, and M.R. Zernich, *J. Bone Jt. Surg.,* Vol 76A (No. 1), 1994, p 110–115
189. S.H. Naidu, C.P. Warner, and C. Laird, *Clin. Orthop.,* Vol 328, 1996, p 261–267
190. E.A. Magnissalis, S. Zinelis, Th. Karachalios, and G. Hartofilakidis, *J. Biomed. Mater. Res.,* Vol 66B, 2003, p 299–305
191. M. Morita, T. Sasada, H. Hayashi, and Y. Tsukamoto, *J. Biomed. Mater. Res.,* Vol 22, 1988, p 529–540
192. M. Morita, T. Sasada, I. Nomura, Y.Q. Wei, and Y. Tsukamoto, *Ann. Biomed. Eng.,* Vol 20 (No. 5), 1992, p 505–516
193. B.F. Brown, A Preface to the Problem of Stress Corrosion Cracking, *Stress Corrosion Cracking—A State of the Art,* STP 518, H.L. Craig, Ed., ASTM, 1971, p 3–15
194. D.O. Sprowls, Evaluation of Stress-Corrosion Cracking, Chapter 17, *Stress-Corrosion Cracking—Materials Performance and Evaluation,* R.H. Jones, Ed., ASM International, 1992, p 363–415
195. J.E. Truman, *Corros. Sci.,* Vol 17, 1977, p 737–746
196. M.J. Blackburn, W.H. Smyrl, and J.A. Feeney, Titanium Alloys, Chapter 5, *Stress-Corrosion Cracking in High Strength Steels and in Titanium and Aluminum Alloys,* B.F. Brown, Ed., NRL, 1972, p 245–363
197. D.A. Meyn and E.J. Brooks, Microstructural Origin of Flutes and Their Use in Distinguishing Striationless Fatigue Cleavage from Stress-Corrosion Cracking in Titanium Alloys, *Fractography and Materials Science,* STP 733, L.N. Gilbertson and R.D. Zipp, Ed., ASTM, 1981, p 5–31
198. D.O. Sprowls et al., "A Study of Environmental Characterization of Conventional and Advanced Aluminum Alloys for Selection and Design: Phase II—The Breaking Load Test Method," Contract NASI-16424, NASA Contractor Report 172387, Aug 1984
199. R.H. Jones and R.E. Ricker, Stress-Corrosion Cracking, *Corrosion,* Vol 13, *ASM Handbook,* ASM International, 1987, p 145–163
200. P.S. Pao, R.A. Bayles, S.J. Gill, D.A. Meyn, and G.R. Yoder, Ripple Load Degradation in Titanium Alloys, *Titanium 92—Science and Technology,* Vol 3, Proc. Seventh World Titanium Conference, June 29 to July 2, 1992 (San Diego, CA), F.H. Froes and I.L. Caplan, Ed., TMS, 1993, p 2169–2176
201. G.H. Koch, Stress-Corrosion Cracking and Hydrogen Embrittlement, *Fatigue and Fracture,* Vol 19, *ASM Handbook,* ASM International, 1996, p 483–506
202. W.B. Lisagor, Environmental Cracking—Stress Corrosion, Chapter 25, *Corrosion Tests and Standards—Application and Interpretation,* R. Baboian, Ed., ASTM, 1995, p 240–252
203. B. Craig, Environmentally Induced Cracking, *Corrosion,* Vol 13, *Metals Handbook,* 9th ed., ASM International, 1987, p 145–189
204. J.C. Scully, *The Fundamentals of Corrosion,* 2nd ed., Pergamon, 1975
205. D.O. Sprowls, Tests for Stress-Corrosion Cracking, *Mechanical Testing,* Vol 8, *Metals Handbook,* 9th ed., American Society for Metals, 1985, p 495–535
206. R.E. Curtis, R.R. Boyer, and J.C. Williams, *ASM Trans.,* Vol 62, 1969, p 457–469
207. H.W. Pickering and P.R. Swann, *Corrosion,* Vol 19 (No. 3), 1963, p 373ff
208. P.L. Andresen, T.M. Angeliu, and L.M. Young, Immunity, Thresholds, and Other SCC Fiction, *Chemistry and Electrochemistry of Corrosion and Stress Corrosion Cracking: A Symposium Honoring the Contributions of R.W. Staehle,* R.H. Jones, Ed., Symposium Proceedings, 2001 TMS Annual Meeting, Feb 11–15, 2001 (New Orleans, LA), p 65–82
209. K.J. Bundy and V.H. Desai, Studies of Stress-Corrosion Cracking Behavior of Surgical Implant Materials Using a Fracture Mechanics Approach, *Corrosion and Degradation of Implant Materials: Second Symposium,* STP 859, A.C. Fraker and C.D. Griffin, Ed., ASTM, 1985, p 73–90
210. A.J. Sedriks, Stress-Corrosion Cracking of Stainless Steels, Chapter 4, *Stress-Corrosion Cracking—Materials Performance and Evaluation,* R.H. Jones, Ed., ASM International, 1992, p 91–130
211. A.J. Sedriks, *Corrosion of Stainless Steels,* 2nd ed., John Wiley, 1996
212. C.O. Brantigan, R.K. Brown, and O.C. Brantigan, *Am. Surg. J.,* Vol 45, Jan 1979, p 38–41
213. S.W. Dean Jr., Review of Recent Studies on the Mechanism of Stress-Corrosion Cracking in Austenitic Stainless Steels, *Stress Corrosion—New Approaches,* STP 610, H.L. Craig, Ed., ASTM, 1976, p 308–337
214. E.C. Hoxie, Some Corrosion Considerations in the Selection of Stainless Steels for Pressure Vessels and Piping, *Pressure Vessels and Piping: A Decade of Progress,* Vol 3, ASME, 1977
215. M.O. Speidel, *Metall. Trans. A,* Vol 12, 1981, p 779ff
216. V.H. Desai and K.J. Bundy, Stress Corrosion Cracking Behavior of Surgical Implant Materials, Paper 265, *NACE Annual Meeting,* April 2–6, 1984 (New Orleans, LA), p 265/1–265/31
217. V.H. Desai and K.J. Bundy, Stress Corrosion Cracking Susceptibility of 316L and Ti-6Al-4V ELI Implant Alloys, *Trans. Second World Biomater. Cong.,* 1984, p 125
218. J.W. Oldfield and B. Todd, *Mater. Perform.,* Vol 29 (No. 12), 1990, p 57ff
219. C.P. Dillon, *Mater. Perform.,* Vol 29 (No. 12), 1990, p 66ff
220. R.M. Kain, *Mater. Perform.,* Vol 29 (No. 12), 1990, p 60ff
221. J.B. Gnanamoorthy, *Mater. Perform.,* Vol 29 (No. 12), 1990, p 63ff
222. H. Haselmair, *Mater. Perform.,* Vol 31 (No. 6), 1992, p 60ff
223. S.J. Acello and N.D. Greene, *Corrosion,* Vol 18, 1962, p 286ff
224. J.P. Harston and J.C. Scully, *Corrosion,* Vol 27 (No. 5), 1969, p 493ff

225. N.A. Nielsen, *Corrosion,* Vol 27 (No. 5), 1971, p 173ff
226. S. Torchio, *Corros. Sci.,* Vol 20, 1980, p 555ff
227. C. Maier, C. Manfredi, and J.R. Galvele, *Corros. Sci.,* Vol 25, 1985, p 15ff
228. A. Cigada, B Mazza, G.A. Mondora, P. Pedeferri, G. Re, and D. Sinigaglia, Localized Corrosion Susceptibility of Work-Hardened Stainless Steels in a Physiological Saline Solution, *Corrosion and Degradation of Implant Alloys,* STP 684, B.C. Syrett and A. Achara, Ed. ASTM, 1979, p 144–160
229. R.W. Schutz, Stress-Corrosion Cracking of Titanium Alloys, Chapter 10, *Stress-Corrosion Cracking—Materials Performance and Evaluation,* R.H. Jones, Ed., ASM International, 1992, p 265–297
230. R.W. Judy, Jr., B.B. Rath, and I.L. Caplan, Stress Corrosion Cracking of Pure Titanium as Influenced by Oxygen Content, *Proc. Sixth World Conf. on Ti,* June 6–9, 1988 (Cannes, France), P. Lacombe, R. Tricot and G. Béranger, Ed., p 1747–1752
231. R.W. Judy, Jr., I.L. Caplan, and F.D. Bogar, Effects of Oxygen and Iron on the Environmental and Mechanical Properties of Unalloyed Titanium, *Titanium 92—Science and Technology,* Vol 3, Proc. Seventh World Ti Conf., June 29 to July 2, 1992 (San Diego, CA), F.H. Froes and I.L. Caplan, Ed., TMS, 1993, p 2074–2086
232. S.R. Seagle, R.R. Seeley, and G.S. Hall, The Influence of Composition and Heat Treatment on the Aqueous-Stress Corrosion of Titanium, *Applications Related Phenomena in Titanium Alloys,* STP 432, ASTM, 1968, p 170–188
233. R.W. Schutz and D.E. Thomas, Corrosion of Titanium and Titanium Alloys, *Corrosion,* Vol 13, *Metals Handbook,* 9th ed., ASM International, 1987, p 669–706
234. J. Brettle, *Met. Mater.,* Oct 1972, p 442–451
235. I. Azkarate, A. Recio, and A. Del Barrio, Hydrogen Assisted Cracking of Titanium Alloys, *Titanium 92—Science and Technology,* Vol 3, Proc. Seventh World Ti Conf., June 29 to July 2, 1992 (San Diego, CA), F.H. Froes and I.L. Caplan, Ed., TMS, 1993, p 2055–2064
236. J. Kolts, Environmental Embrittlement of Cobalt-Base Alloys, *Corrosion,* Vol 13, *Metals Handbook,* 9th ed., ASM International, 1987, p 661–662
237. C.R. Thomas and F.P.A. Robinson, *J.S. Afr. Inst. Mining Metall.,* Vol 77 (No. 11), 1976, p 93–102
238. H.H. Uhlig and A.I. Asphahani, *Mater. Perform.,* Vol 18 (No. 11), 1979, p 9–20
239. J.R. Cahoon and H.W. Paxton, *J. Biomed. Mater. Res.,* Vol 4, 1970, p 223–224
240. S.D. Cook, K.A. Thomas, A.F. Harding, C.L. Collins, R.J. Haddad, Jr., and M. Milicic, *Biomaterials,* Vol 8, 1987, p 177–184
241. N.G. Feige and L.C. Covington, Overview of Corrosion Cracking of Titanium Alloys, *Stress Corrosion Cracking—A State of the Art,* STP 518, H.L. Craig, Ed., ASTM, 1971, p 119–130
242. R.P. Wei and S.R. Novak, *J. Test. Evaluation,* Vol 15 (No. 1), 1987, p 38–75
243. J.H. Payer, W.E. Berry, and W.K. Boyd, Constant Strain Rate for Assessing Stress-Corrosion Susceptibility, *Stress Corrosion—New Approaches,* STP 610, H.L. Craig, Ed., ASTM, 1976, p 82–93
244. W.J. Daniels, Comparative Findings Using the Slow Strain-Rate, Constant Flow Stress, and U-Bend Stress Corrosion Cracking Techniques, *Stress Corrosion Cracking—The Slow Strain Rate Technique,* STP 665, G.M. Ugiansky and J.H. Payer, Ed., ASTM, 1979, p 347–361
245. M. Hishida, J.A. Begley, R.D. McCright, and R.W. Staehle, Anodic Dissolution and Crack Growth Rate in Constant Strain-Rate Tests at Controlled Potentials, *Stress Corrosion Cracking—The Slow Strain Rate Technique,* STP 665, G.M. Ugiansky and J.H. Payer, Ed., ASTM, 1979, p 47–60
246. R.N. Parkins, Development of Slow Strain Rate Testing and Its Implications, *Stress Corrosion Cracking—The Slow Strain Rate Technique,* STP 665, G.M. Ugiansky and J.H. Payer, Ed., ASTM, 1979, p 5–25
247. S. Abe, M. Kojima, and Y. Hosoi, Stress Corrosion Cracking Susceptibility Index, I_{SCC}, of Austenitic Stainless Steels in Constant Strain Rate Test, *Stress Corrosion Cracking—The Slow Strain Rate Technique,* STP 665, G.M. Ugiansky and J.H. Payer, Ed., ASTM, 1979, p 294–304
248. P.J. Bania and S.D. Antolovich, Activation Energy Dependence on Stress Intensity, *Stress Corrosion Cracking and Corrosion Fatigue, in Stress Corrosion—New Approaches,* STP 610, H.L. Craig, Ed., ASTM, 1976, p 157–175
249. A.J. Russell and D. Tromans, *Met. Trans.,* Vol 10A, 1979, p 1229ff
250. L.F. Lin et al., *Corrosion,* Vol 37, 1981, p 616ff
251. H. Kohl, *Corrosion,* Vol 23, 1967, p 39ff
252. N. Ohtani and Y. Hayashi, *Passivity and Its Breakdown on Iron and Iron Base Alloys,* R.W. Staehle and H. Okada, Ed., NACE, 1976, p 169ff
253. L.N. Gilbertson, Stress Corrosion Cracking of 316LVM, *Trans. Fourth Ann. SFB Meeting,* April 29 to May 2, 1978 (San Antonio, TX), p 125–126
254. J.P. Sheehan, C.R. Morin, and K.F. Packer, Study of Stress Corrosion Cracking Susceptibility of Type 316L Stainless Steel in Vitro, *Corrosion and Degradation of Implant Materials: Second Symposium,* STP 859, A.C. Fraker and C.D. Griffin, Ed., ASTM, 1985, p 57–72
255. V.H. Desai, "Stress Corrosion Cracking Studies of Surgical Implant Alloys," Ph.D. Dissertation, 1984, Johns Hopkins University
256. M. Speidel, *Corrosion,* Vol 33, 1977, p 199ff
257. D.G. Jones, Stress Corrosion and Corrosion Fatigue in Light Water Reactor Environments, *Engineering Applications of Fracture Analysis,* G.G. Garrett and D.L. Marriott, Ed., Pergamon Press, 1980, p 353–370
258. J.I. Dickson, A.J. Russell, and D. Tromans, *Can. Metall. Quart.,* Vol 19, 1980, p 161ff
259. M.D. Roach, L.D. Zardiackas, R.S. Brown, and R.C. Gebeau, Stress Corrosion Cracking of a Low-Nickel Stainless Steel, *Trans. 27th Ann. SFB Meeting,* April 24–29, 2001 (St. Paul, MN)
260. L.D. Zardiackas, M. Roach, S. Williamson, and J.A. Bogan, Comparison of Notch Sensitivity and Stress Corrosion Cracking of a Low-Nickel Stainless Steel to 316LS and 22Cr-13Ni-5Mn Stainless Steels, *Stainless Steels for Medical and Surgical Applications,* STP 1438, G.L. Winters and M.J. Nutt, Ed., ASTM International, 2003, p 154–167
261. J. Sutton and W.P. Saunders, *Int. Endodont. J.,* Vol 29 (No. 5), 1996, p 335–343
262. A.K. Mishra and J.A. Davidson, Stress Corrosion Cracking Resistance of Ti-13Nb-13Zr, *Ann. Biomed. Eng.,* Vol 21 (No. 1), 1993, p 67
263. L.D. Zardiackas, J.A. Bogan, and J.A. Disegi, Stress Corrosion Cracking Resistance of Titanium Implant Materials, *Trans. 27th Ann. SFB Meeting,* April, 24–29, 2001 (St. Paul, Mn)
264. R.S. Williamson, M.D. Roach, and L.D. Zardiackas, Comparison of Stress Corrosion Cracking of Ti-15Mo, Ti-6Al-7Nb, Ti-6Al-4V, and CP Ti, *J. Test. Eval.* (in press)
265. T. Ogawa, K. Yokoyama, K. Asaoka, and J. Sakai, *Biomater.,* Vol 25, 2004, p 2419–2425
266. K. Yokoyama, K. Kaneko, K. Moriyama, K. Asaoka, and J. Sakai, *J. Biomed. Mater. Res.,* Vol 65A, 2003, p 182–187
267. K. Yokoyama, K. Kaneko, T. Ogawa, K. Moriyama, K. Asaoka, and J. Sakai, *Biomater.,* Vol 26, 2005, p 101–108
268. S. Shabalovskaya, G. Rondelli, J. Anderegg, B. Simpson, and S. Budko, *J. Biomed. Mater. Res.,* Vol 66B (No. 1), 2003, p 331–340
269. M.H. Kononen, E.T. Lavonius, and J.K. Kivilahti, *Dent. Mater.,* Vol 11 (No. 4), 1995, p 269–272
270. P. Kumar, A.J. Hickl, A.I. Asphahani, and A. Lawley, Properties and Characteristics of Cast, Wrought, and Powder Metallurgy (P/M) Processed Cobalt-Chromium-Molybdenum Implant Materials, *Corrosion and Degradation of Implant Materials: Second Symposium,* STP 859, A.C. Fraker and C.D. Griffin, Ed., ASTM, 1985, p 30–56

271. B.J. Edwards, M.R. Louthan, Jr., and R.D. Sisson, Jr., Hydrogen Embrittlement of Zimaloy: A Cobalt-Chromium-Molybdenum Orthopaedic Implant Alloy, *Corrosion and Degradation of Implant Materials: Second Symposium*, STP 859, A.C. Fraker and C.D. Griffin, Ed., ASTM, 1985, p 11–29
272. C.A. Zapffe, *Met. Prog.*, Vol 67 (No. 5), 1955, p 95–98
273. C.O. Bechtol, A.B. Ferguson, and P.B. Laing, *Metals and Engineering in Bone and Joint Surgery*, Williams and Wilkins, 1959, p 58–59
274. J. Galante and W. Rostoker, *Clin. Orthopaed. Rel. Res.*, Vol 86, 1972, p 237–244
275. W. Rostoker and J.O. Galante, *Trans. ASME*, Vol 101, Feb 1979, p 2–14
276. R.M. Rose, A.L. Schiller, and E.L. Radin, *J. Bone Jt. Surg.*, Vol 54A (No. 4), 1972, p 854–862
277. R.J. Gray, *J. Biomed. Mater. Res. Symp.*, Vol 5 (Part 1), 1974, p 27–38
278. G. Bombara and M. Cavallini, *Corros. Sci.*, Vol 17, 1977, p 77–85
279. S. Bandyopadhyay and P. Brockhurst, *J. Mater. Sci. Lett.*, Vol 14, 1979, p 3002–3003
280. P.M. Sandborn, S.D. Cook, M.A. Kester, and R.J. Haddad, Jr., *Clin. Orthop.*, Vol 222, 1987, p 249–254
281. B.C. Syrett and E.E. Davis, *J. Biomed. Mater. Res.*, Vol 13, 1979, p 543–556
282. K.J. Bundy, The Influence of Static Stress on the Corrosion Behavior of Implant Materials, *Extended Abstracts Electrochem. Soc.*, Vol 84-2, 1984, p 401–402
283. N.M. Jedynakiewicz, J. Cunningham, and D.F. Williams, *Br. Dent. J.*, Vol 159 (No. 4), 1985, p 121–123
284. A. Odén and M. Tullberg, *Acta Odont. Scand.*, Vol 43, 1985, p 15–17
285. C.H. Lloyd and G.R. Baxter, *J. Oral Rehabil.*, Vol 8 (No. 2), 1981, p 183–190
286. K. Yokoyama, T. Ichikawa, H. Murikami, Y. Miyamoto, and K. Asaoka, *Biomater.*, Vol 23, 2002, p 2459–2465
287. R. Kossowsky, H. Dujovny, and N. Kossovsky, "Failure Analysis of an Aneurysm Surgical Clip," presented at AIME Annual Meeting, Feb 15–18, 1982 (Dallas, TX), American Institute of Mining, Metallurgical, and Petroleum Engineers
288. R. Kossowsky, N. Kossovsky, and M. Dujovny, In Vitro Studies of Aneurysm Clip Materials, *Corrosion and Degradation of Implant Materials: Second Symposium*, STP 859, A.C. Fraker and C.D. Griffin, Ed., ASTM, 1985, p 147–159
289. B.A. De Mol, M. Kalleward, R.B. McLellan, L.A. van Herwerden, J.J. Defauw, and Y. van der Graaf, *Lancet*, Vol 343, 1994, p 9–12
290. F.A. Schöndube, W. Althoff, H.C. Dörge, M. Voss, J.L. Laufer, J.G. Chandler, and B.J. Messmer, *J. Heart Valve Dis.*, Vol 3, 1994, p 247–253
291. W.W. O'Neal, J.D. Chandler, R.E. Gordon, et al., *New Eng. J. Med.*, Vol 333, 1995, p 414–419
292. K. Xiao and A.J. Appleby, *Int. J. Artif. Organs*, Vol 19 (No. 8), 1996, p 477–486
293. R.O. Ritchie and P. Lubock, *J. Biomech. Eng.*, Vol 108, 1986, p 153–160
294. S.A. Sachs, M. Harrison, P.J.E. Bischler, J.W. Martin, J. Watkins, and A. Gunning, *Thorax*, Vol 41, 1986, p 141–147
295. G. Röckelein, J. Breme, and J. van der Emde, *Thorac. Cardiovasc. Surg.*, Vol 37, 1989, p 47–51
296. V.O. Björk, *Scand. J. Thorac. Cardiovasc. Surg.*, Vol 5, 1971, p 87–91
297. C.L. Pypen, H. Mayer, S. Stanzl-Tschegg, and H. Plenk, Jr., Fatigue Properties of Annealed and Cold Worked Implant Metals, *Trans. Fifth World Biomat. Cong.*, May 29 to June 2, 1996 (Toronto, Ontario, Canada), p 485
298. M. Niinomi, *Biomater.*, Vol 24, 2003, p 2673–2683
299. M. Niinomi, T. Akahori, Y. Hattori, K. Morikaw, T. Kasuga, H. Fukui, A. Suzuki, K. Kyo, and S. Niwa, Super Elastic Functional β Titanium Alloy with Low Young's Modulus for Biomedical Applications, Abstract book, *ASTM Symposium on Titanium, Niobium, Zirconium, and Tantalum for Medical and Surgical Applications*, Nov 9–10, 2004 (Washington, D.C.), p 20
300. Y. Okazaki, S. Rao, Y. Ito, and T. Tateishi, *Biomater.*, Vol 19, 1998, p 1197–1215
301. Y. Okazaki, Corrosion Resistance, Mechanical Properties, and Fatigue Properties of and Tissue Response to Ti-15Zr-4Nb-4Ta Alloy, Abstract book, *ASTM Symposium on Titanium, Niobium, Zirconium, and Tantalum for Medical and Surgical Applications*, Nov 9–10, 2004 (Washington, D.C.), p 15
302. V. Jablokov, N. Murray, H. Rack, R.H. Tust, and J.R. Wood, Influence of Oxygen Content on the Mechanical Properties of Titanium-35Niobium-7Zirconium-5Tantalum Beta Titanium Alloy, Abstract book, *ASTM Symposium on Titanium, Niobium, Zirconium, and Tantalum for Medical and Surgical Applications*, Nov 9–10, 2004 (Washington, D.C.), p 9
303. M.J. Filiaggi, R.M. Pilliar, and D. Abdulla, *J. Biomed. Mater. Res.*, Vol 33, 1996, p 239–256
304. B. Marquardt, Ti-15Mo Beta Titanium Alloy Processed for High Strength Orthopaedic Applications, Abstract book, *ASTM Symposium on Titanium, Niobium, Zirconium, and Tantalum for Medical and Surgical Applications*, Nov 9–10, 2004 (Washington, D.C.), p 11
305. M.J. Nutt, V. Jablokov, and H.L. Freese, The Application of Ti-15Mo Beta Titanium Alloy in High-Strength Structural Orthopaedic Applications, Abstract book, *ASTM Symposium on Titanium, Niobium, Zirconium, and Tantalum for Medical and Surgical Applications*, Nov 9–10, 2004 (Washington, D.C.), p 12

Corrosion and Tarnish of Dental Alloys

Revised by Spiro Megremis, American Dental Association
Clifton M. Carey, American Dental Association Foundation

DENTAL ALLOY DEVICES serve to restore or align lost or misaligned teeth so that normal biting function and aesthetics can prevail. Alloys are used for direct fillings, crowns, inlays, onlays, bridges, fixed and removable partial dentures, full denture bases, implanted support structures, and wires and brackets for the controlled movement of teeth. In addition to applications calling for cast or wrought alloys, other uses of alloys include soldered assemblies, porcelain fused to metal, and resin bonded to metal restorations.

Dental Alloy Compositions and Properties

Dental Alloy Compositions. The compositions of alloys used to fulfill the diverse applications germane to dentistry include the following elements: Au, Pd, Pt, Ag, Cu, Co, Cr, Ni, Fe, Mo, W, Ti, Zn, In, Ir, Rh, Sn, Ga, Ru, Si, Mn, Be, B, Al, V, C, Ta, Zr, and others. Figures 1 to 6 show a number of typical restorations and appliances fabricated from alloys containing some of these metals.

Compositions for direct filling restorations usually consist of silver-tin-copper-zinc amalgams, although this is rapidly changing with the continued improvement of polymer composites. Pure gold in the form of cohesive foil, mat, or powder is used only in very limited applications.

Alloys for all-alloy cast crown and bridge restorations are usually gold-, silver-, or nickel-base compositions, although iron-base and other alloys have also been used. The gold-base alloys contain silver and copper as principal alloying elements, with smaller additions of palladium, platinum, zinc, indium, and other noble metals as grain refiners. The silver-base alloys contain palladium as a major alloying element, with additions of copper, gold, zinc, indium, and grain refiners. The nickel-base alloys are alloyed with chromium, iron, molybdenum, and other elements.

Alloys for porcelain fused to alloy restorations are gold-, palladium-, nickel-, or cobalt-base compositions. The gold-base alloys are divided into gold-platinum-palladium, gold-palladium-silver, and gold-palladium types. The palladium-base alloys are palladium-silver alloys or palladium-gallium alloys with additions from either copper or cobalt. The nickel- and cobalt-base alloys are alloyed primarily with chromium and with minor additions of molybdenum and other elements. In contrast to alloys for crown and bridge use, alloys fused to porcelain contain low concentrations of oxidizable elements, such as tin; indium; iron; gallium for the noble metal containing alloys; and aluminum, vanadium, and others for the base metal alloys. During the heating cycle, these elements form oxides on the surface of the alloy and combine with the porcelain at the firing temperatures to promote chemical bonding.

Alloys for removable partial dentures are primarily nickel- and cobalt-base compositions and are similar to alloys used for porcelain fused to alloy applications. However, carbon is present in amounts up to 0.3 to 0.4% with the partial denture alloys. Carbon is not added to alloys to be used for porcelain bonding.

Alloys that have found applications for support structures implanted in the lower or upper jaws are composed of cobalt-chromium, nickel-chromium, stainless steel, and titanium and its alloys. Wrought orthodontic wires are composed of stainless steel, cobalt-chromium-nickel, nickel-titanium, and β-titanium alloys. Silver- and gold-alloy solders are used for the joining of components. High-temperature brazing alloys are used for the joining of a number of high fusing temperature alloys. Additional information on noble metals is available in the article "Corrosion of Precious Metals and Alloys" in *Corrosion,* Vol 13B, of the *ASM Handbook.*

Properties. The diversity in available alloys exists so that alloys with specific properties can be used when needed. For example, the mechanical property requirements of alloys used for crown and bridge applications are different from the requirements of alloys used for porcelain fused to alloy restorations. Even though crown and bridge alloys must possess sufficient hardness and rigidity when used in stress-bearing restorations, excessively high strength is a disadvantage for grinding, polishing, and burnishing. Also, excessive wear of the occluding teeth

Fig. 1 Various types of crowns. Source: Ref 1

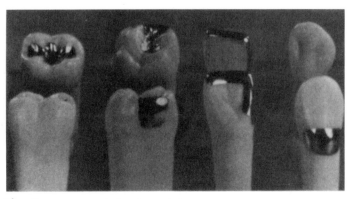

Fig. 2 Various types of inlays. Source: Ref 1

892 / Corrosion in Specific Industries

is also likely to occur. Alloys used with porcelain fused to metal restorations are used as substrates for the overlaying porcelain. In this case, the high strength and rigidity of the alloys more closely matches the properties of the porcelain. Also, a higher sag resistance of the alloy at temperatures used for firing the porcelain means less distortion and less retained residual stresses.

Similarly, alloys used for partial denture and implant applications must possess increased mechanical properties for resistances to failures. However, clasps contained within removable partial denture devices are often fabricated from a more ductile alloy, such as a gold-base alloy, than from cobalt-chromium or nickel-chromium alloys. This ensures that the clasps possess sufficient ductility for adjustments without breakage from brittle fractures.

Other properties required in specific systems include the matching of the thermal expansion coefficients between porcelain and substrate alloy, negligible setting contractions with the direct filling amalgams, and specific modulus to yield strength ratios with orthodontic wires. Tarnish and corrosion of all dental alloy systems have been and will remain of prime importance.

Tarnish and Corrosion Resistance

Dental alloy devices must possess acceptable corrosion resistance primarily because of safety and efficacy. Aesthetics is also a consideration.

Safety

Dental alloys are required to have acceptable corrosion resistance so that biocompatibility is maintained during the time the metallic components are used (Ref 5–7). No harmful ions or corrosion products can be generated such that toxicological conditions result. The effects of the dental alloys on the oral environment have the capabilities for producing local, remote, or systematic changes that may be short term, long term, or repetitive (tissue sensitization) in nature (Ref 8). Dental alloy-oral environment interactions have the potential for generating such conditions as metallic taste, discoloration of teeth, galvanic pain, oral lesions, cariogenesis, allergic hypersensitive reactions, dermatitis and stomatitis, endodontic failures, dental implant rejection, tumorgenisis, and carcinogenisis. Figure 7 shows a schematic of useful dental anatomy.

Metallic Taste. The symptom of metallic taste has been reported and related to the presence of metallic materials in the mouth (Ref 9). In addition, the release of ions and the formation of products through corrosion, wear, and abrasion can occur simultaneously, which can accelerate the process. Therefore, patients with metallic restorations and with an inclination toward bruxism (the unconscious gritting or grinding of the teeth) are likely to be more susceptible to metallic taste. Although this condition is not as prevalent as it once was when metallic materials with lower corrosion resistances were more often used, metallic taste is still known to occur on occasion.

Discoloration of teeth has occurred mainly with amalgam fillings (Ref 10) and with base alloy screwposts (Ref 11). With amalgams, tin and zinc concentrations have been identified in the dentinal tubules of the discolored areas, while with the screwposts, copper and zinc were

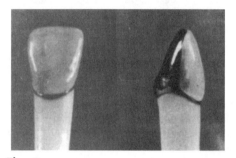

Fig. 3 Porcelain veneer fused to alloy. Source: Ref 1

Fig. 4 Fixed bridges. (a) Three-unit bridge consisting of inlay (left member), onlay (right member), and porcelain fused to alloy pontic (center member). Source: Ref 2. (b) Five-unit bridge consisting of four porcelain fused to alloy members and one crown. Source: Ref 1

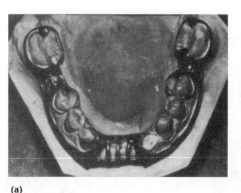
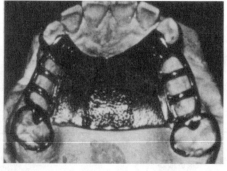
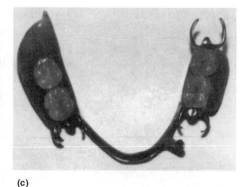

Fig. 5 Removable partial dentures, (a) lower and (b) upper cobalt-chromium frameworks, and (c) a completed unit. Source: Ref 3

detached in both the dentin and enamel and the surrounding soft connective tissue. Discoloration is not, however, a definite indicator of the presence of metallic ions.

Galvanic pain has been reported from the contact of dissimilar-alloy restorations, either continuously or intermittently (Ref 9, 12). An electrochemical circuit occurring between the two dissimilar-alloy restorations is short circuited by the contact. In the case of intermittent contact, instantaneous current flows through the external circuit, which is the oral tissues, and may cause pain. The placement of dissimilar-alloy restorations in direct contact is ill advised.

Oral lesions resulting from the metallic prosthesis contacting tissue can be due to physical factors alone (Ref 9). An irritation in the opposing tissues of the oral mucosa can be generated because of the shape and location in the mouth of the prosthesis, as well as its metallurgical properties, such as surface finish, grain size, and microstructural features. Tarnish and corrosion can change the nature of the alloy surface and add to the irritation in the opposing tissues. Microgalvanic currents due to chemical differences of microstructural constituents must also be considered as possible causative factors in traumatizing and damaging tissue. However, no data have related in vivo galvanic currents from dental restorations to tissue damage. The released metallic ions from corrosion reactions can interact with the oral tissues to generate redness, swelling, and infection. Oral lesions can then occur. These reactions are discussed in the section "Allergic Hypersensitive Reactions" in this article.

Cariogenesis corresponds to the ability for released metallic ions and formed corrosion products to affect the resistance of either dentin or enamel to decay (caries). The mechanisms involved with caries formation (Ref 13–15), which include the fermentation of carbohydrate by microorganisms and the production of acid, are likely to become altered when metallic ions and products from corrosion reactions are included. This may be indicated by the reports that show tin and zinc concentrations (originating from amalgam corrosion) in softened, demineralized dentin and enamel (Ref 16–18).

Allergic Hypersensitive Reactions. With allergic hypersensitive contact reactions, some people can become sensitized to particular foreign substances, such as ions or products from the corrosion of dental alloys (Ref 9, 19). The metallic ions or products combine with proteins in the skin or mucosa to form complete antigens. Upon first exposure to the foreign substance by the oral mucosa, sensitization of the host may occur in times of up to several weeks and with no adverse reactions. Thereafter, any new exposures to the foreign substance will lead to biological reactions, such as swelling, redness, burning sensation, vesiculation, ulceration, and necrosis. Abstinence from the foreign substance leads to healing. Identification and avoidance are the means for controlling these allergic hypersensitive reactions. Exposure of the oral mucosa to the foreign substance can lead not only to allergic stomatitis reactions (of the oral mucosa), but also to allergic dermatitis reactions (of the skin) at sites well away from the contact site with the oral mucosa. However, because the oral mucosa is more resistant to allergic reactions than the skin, the reverse process usually does not occur.

Of the currently used metals contained in dental alloys, nickel, cobalt, chromium, mercury, beryllium, and cadmium need to be considered as inducing possible allergic or cytotoxic reactions. Nickel is the primary alloying element in nickel-chromium casting alloys (up to 80%), in nickel-titanium wires (up to 50%) and in lower concentrations in some cobalt-chromium alloys, and in austenitic stainless steels. Nickel from dental alloy is known to react with the oral tissues in some individuals to produce allergic sensitization reactions (Ref 20). About 9% of women and 1% of men are estimated to be allergic to nickel. It is recommended that individuals be screened for possible nickel allergies prior to dental treatments. If an allergy arises from a nickel-containing dental restoration, it is recommended that the individual be tested for allergies to nickel and, if so indicated by the test results, have the restoration replaced with a nickel-free alloy.

Cobalt, also a component of some dental alloys, has been known to react with the oral mucosa and cause allergic reactions (Ref 20). However, the occurrences of such allergies are less than 1% of the population and mainly affect women. If reactions to cobalt from cobalt-containing materials are suspected, then testing for cobalt allergies should be performed. Contact allergic reactions to chromium from dental alloys are also reported (Ref 20), but the occurrences of such reactions are rare.

Mercury is contained in amalgam fillings, which contain microstructural phases composed of silver-mercury and tin-mercury. Mercury ions may be released from microstructural phases through corrosion. However, the concentrations are low and not relatable to toxicological ramifications. Mercury vapors released from amalgam surfaces may also occur. Again, because of the low concentrations emitted, amalgam mercury vapors are not related to toxicity. Allergic reactions to mercury contained in dental amalgams have been reported (Ref 20). If mercury allergic reactions are suspected from the amalgam, it is recommended that testing for mercury allergies be conducted.

Beryllium is contained in some nickel-chromium casting alloys in concentrations up to about 2 wt%. There are only a few cases of transient contact dermatitis that have been reported among dental professionals (Ref 21, 22). More of a health hazard is posed to the dental personnel doing the actual melting and finishing of the alloy than to individuals having a prosthesis made from a beryllium-containing alloy. For instance, the Occupational Safety and Health Administration (OSHA) has posted a Hazard Information Bulletin titled "Preventing Adverse Health Effects from Exposure to Beryllium in Dental Laboratories" expressing its concern about reports of chronic beryllium disease among dental laboratory technicians that are exposed to dust from melting, grinding, polishing, and finishing beryllium-containing alloys (Ref 23). As a result of safety concerns about beryllium, the American Dental Association (ADA) Council on Scientific Affairs recommends that dentists "use alloys that do not contain beryllium in the fabrication of dental prostheses" (Ref 24). Furthermore, the European Committee for Standardisation passed a resolution in February 2002 recommending that all standards pertaining to dental alloys permit a maximum beryllium content of only 0.02% wt%, which essentially excludes its use as an alloying element (Ref 25).

Cadmium is contained in some dental gold and silver solders of up to 15% (Ref 26). No biological reactions have been related to the cadmium contained in these materials; however,

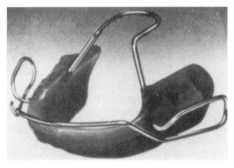

Fig. 6 Removable orthodontic appliance. Source: Ref 4

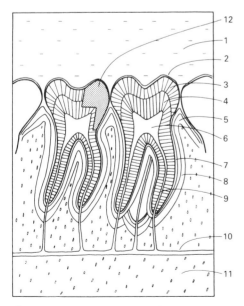

Fig. 7 Useful dental anatomy. 1, saliva; 2, integument; 3, enamel; 4, dentin; 5, gingiva; 6, pulp; 7, cementum; 8, periodontal ligament; 9, root canal; 10, artery; 11, alveolar bone; 12, restoration—amalgam filling

precautions should be taken in fusing solders containing cadmium.

Additional information on the biocompatibility of metals and acceptable exposure limits is available in the article "Toxicity of Metals" in *Properties and Selection: Nonferrous Alloys and Special-Purpose Materials,* Vol 2, of the *ASM Handbook* (refer to pages 1233 to 1269).

Endodontic Failures. Root canals obturated with silver cones (or points) have occasionally been associated with corrosion (Ref 27). Figure 8 shows examples of restored teeth including two cases of endodontically treated teeth. Development of a fluid-tight seal at the apex of the root canal is the primary objective of endodontic therapy. Corrosion of the silver points is known to lead to failure by allowing the penetration of fluids along the silver cone/root canal interface. Figure 8(b) shows a schematic of a tooth with cones.

Dental Implant Rejection. Dental implants, which are used for permanently attaching bridges, and so on, extend through or up to the maxillary or mandibular bones and must function in both hard and soft tissues, as well as within a wide range of applied stresses (Ref 28, 29). Depending on the chemical inertness of the materials used for these devices, the thickness of the tissue connecting implant to bone varies. Released ions can infiltrate thick membranes surrounding loosely held implants, which can lead to an early rejection of the implant through immune response.

Tumorgenisis and Carcinogenisis. Even though dental alloy devices have not been implicated with tumorgenisis and carcinogenisis, their possible formation must never be ruled out and should always be considered as potential biological reactions, especially with new, untried alloys (Ref 29).

Efficacy

The oral environment must not induce changes in physical, mechanical, chemical, optical, and other properties of the dental alloy, such that inferior functioning and/or aesthetics result. The effect of the oral environment on the alloy has the potential for altering dimensions, weight, stress versus strain behavior, bonding strengths with other alloys and with nonmetals, appearances, and creating or enhancing crevices. In combination with mechanical forces, the oral environment is capable of generating premature failure through stress corrosion and corrosion fatigue and of generating increased surface deterioration by fretting, abrasion, and wear.

Dimensions, Weight, Mechanical Properties, and Crevices. At least in theory, corrosion of precision castings and attachments, which rely on accurate and close tolerance for proper fit and functioning, can alter their dimensions, thus changing the fit and functionality of the restorations. Similarly, corrosion of margins on crowns and other cast restorations can lead to decreased dimensions and to enhanced crevice conditions. The increased seepage of oral secretions into the crevices created between restoration and tooth can lead to microorganism invasion, generation of acidic conditions, and the operation of differential aeration cells. Under these conditions, the bonding of the restoration to the dentinal walls through the underlying cement is likely to become weakened. In combination with biting stresses, microcrack formation along the interface is likely to occur; this will cause the penetration of the crevice even further beneath the restoration. Eventually, the loosening of the entire restoration may occur.

With amalgam restorations, however, a slight amount of corrosion on these surfaces adjacent to the cavity walls may actually be beneficial because corrosion product buildup increases dimensions and adaptability. The crevice between an amalgam and the cavity is reduced in width, which leads to a decreased seepage of fluids. Additionally, the corrosion surface may inhibit bacterial growth, thus reducing bacterial invasion of small crevices. On the other hand, corrosion of the amalgam deteriorates its subsurface structure; this is likely to lead to an increased occurrence of marginal fracture, a known problem with amalgams, through corrosion fatigue mechanisms with stresses generated from biting (Ref 30).

The loss of sufficient material from any dental alloy through corrosion can lead to a reduction in mechanical strength. This can lead to a direct failure of the alloy or reduced rigidity resulting in unacceptable strains. For silver-soldered wires, corrosion of the solder leads to a weakening of the entire joint (Ref 31). Loosening of crowns and bridges because of corrosion-induced fractures of posts and pins is also known to occur (Ref 32), as shown in Fig. 8(a) and (c). Still other possibilities of the effects of corrosion include the reduction in bond strengths of metal brackets bonded to teeth, as well as the degradation of porcelain fused to metal restorations.

Appearance. Because of the various optical properties of corrosion products, the appearance of tarnished and corroded surfaces can become unacceptable. A degradation in surface appearance, without a loss in the properties of the appliance, can be taken to be either acceptable or unacceptable, depending on individual preferences. If, however, the tarnished surface promotes additional consequences, such as the attachment of plaque and bacteria or a greater irritation to opposing tissue, then tarnishing must be deemed unacceptable.

Interstitial versus Oral Fluid Environments and Artificial Solutions

In order to select and/or develop dental alloys, an understanding of the environment to which these materials will be exposed is imperative. This section defines and compares interstitial fluid and oral fluid environments. In addition, artificial solutions developed for testing and evaluation of dental materials also are discussed.

Interstitial Fluid

Applications of metallic materials to oral rehabilitation are confronted with a number of

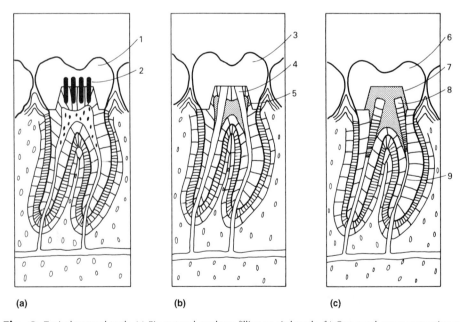

Fig. 8 Typical restored teeth. (a) Pin restored amalgam filling on vital tooth. (b) Cast metal crown restoration on endodontically treated tooth with silver cones and cement to seal root canals. (c) Cast metal crown restoration on endodontically treated tooth with cement core buildup and screwposts. 1, amalgam filling; 2, stainless steel pins; 3, metal crown; 4, silver cones; 5, cement; 6, metal crown; 7, cement core; 8, screw posts; 9, gutta-percha or similar sealing material

environmental conditions that differentiate most dental uses from other biomedical uses (Ref 33). The one major exception is dental implants because interstitial fluids (the fluids in direct contact with tissue cells) are encountered by both dental and other types of surgical implants (see, for example, the article "Corrosion Effects on the Biocompatibility of Metallic Materials and Implants" in this Volume). As discussed later in this article, other exceptions occur because restorations in teeth have their interior surfaces in direct contact with the dentinal and bone fluids, which are more similar to interstitial fluids in composition than to saliva.

Other types of extracellular fluids, such as lymph and blood plasma, contain similar inorganic contents and are also likely to come into contact with dental implants, particularly with plasma during and shortly after surgery. Table 1 presents a composition of blood plasma. The inorganic content is similar to the inorganic content of interstitial and other types of extracellular fluids, while the protein concentration for plasma is higher than for other biofluids. For plasma, the major proteins are albumin, globulins, and fibrinogen. For all extracellular fluids, the inorganic contents are characterized by high sodium (Na) and chloride (Cl^-) and moderate bicarbonate (HCO_3^-) contents. Considerable variations in pH, p_{O_2}, and p_{CO_2} can occur in the vicinity of an implant. In crevices formed between plates and screws, some extreme values ranging between 5 to 7 in pH, and <8 to 110 and <10 to 300 mm Hg, respectively, have been determined (Ref 35). Similar corrosive conditions are expected regardless of the extracellular fluid, provided the effects of the protein and cellular contents are minimal.

Tissue cells and other types of cellular matter can also directly contact implant material, with the possibility of intracellular fluid permeating through the cell membrane and effecting corrosion of the alloy. Separation by shearing of biological cells from alloy surfaces almost always generates cohesive failures through the cell instead of adhesive failures along the alloy/cell interface (Ref 36). In these situations, intracellular fluids can gain direct access to the surface of the alloy. In contrast to extracellular fluids, intracellular fluids contain high potassium and organic anion contents. The sodium is replaced by potassium and Cl^- by orthophosphate (HPO_4^{2-}). The effectiveness of intracellular fluids in corroding implant surfaces will be governed by the ability of the larger organic anions to pass through cell membranes, which are usually very restricted. Extracellular fluids are, therefore, the fluids interacting with the implant in most cases, although the possible effects from intracellular fluids must not be dismissed.

Oral Fluids

Whole mixed saliva is produced by the paratid, submandibular, and sublingual glands, together with the minor accessory glands of the cheeks, lips, tongue, and hard and soft palates from the oral mucosa. Gingival or crevicular fluid is also produced, as well as fluid transport between the hard tissues of the teeth and saliva. The composition of the secretion from each gland is different and varies with flow rate and with the intensity and duration of the stimulus. Saliva composition varies from individual to individual and in the same individual under different circumstances, such as time of day and emotional state.

Although about 1 L of saliva is produced per day in response to stimulation accompanying chewing and eating, for the greater part of the day, the flow rate is at very low levels (0.25 to 0.5 mL/min) (Ref 37, 38). During sleep, there is virtually no flow from the major glands. At low flow rates, the concentrations of sodium, Cl^-, and HCO_3^- are reduced notably; the concentration of calcium is elevated slightly; and the concentrations of magnesium, phosphate (PO_4^{3-}), and urea are elevated decidedly when compared with stimulated flow rates (Ref 39). It is therefore impossible to define specific compositions and concentrations that are universally applicable. However, compilations of data encompassing large statistical populations have been made by a number of researchers. One typical analysis for the composition of human saliva is shown in Table 2.

The inorganic ions readily detectable in saliva are Na^+, K^+, Ca^{2+}, Mg^{2+}, Cl^-, PO_4^{3-}, HCO_3^-, thiocyanate (SCN^-), and sulfate (SO_4^{2-}). Minute traces of F^-, I^-, Br^-, Fe^{2+}, Sn^{2+}, and nitrite (NO_2^-) are also found, and on occasion, Zn^{2+}, Pb^{2+}, Cu^{2+}, and Cr^{3+} are found in trace quantities. Figure 9 shows the Cl^- and HCO_3^- variations in concentration as saliva is stimulated to a flow rate of 1.5 mL/min. The O_2 and N_2 contents of saliva are 0.18 to 0.25 and 0.9 vol%, respectively. The carbon dioxide (CO_2) content varies greatly with flow rate, being about 20 vol% when unstimulated and up to about 150 vol% when vigorously

Table 1 Mean human adult blood plasma composition

Compound	mg/100 mL
Inorganic	
Na^+	325
K^+	16
Ca^{2+}	9.8
Mg^{2+}	2.1
Cl^-	369
HCO_3^-	146–189
PO_4^{3-}	3.1–4.9
Si^-	0.8
SO_4^{2-}	3.7
Nonprotein organic	
Urea	33
Uric acid	4.9
Carbohydrates	260
Fructose	7.5
Glucosamine	81
Glucose	97
Glycogen	6.8
Polysaccharides (nonglucosamine)	129
Organic acids	19
Citric	2.2
Lactic	36
Other organic acids	5
Lipids	530
Fatty acids	316
Amino acids	37.1
Major proteins	
Albumin	4800
Globulins	2300
Fibrinogen	300

Source: Ref 34

Table 2 Mean whole unstimulated human saliva composition

Compound	mg/100 mL
Inorganic	
Na^+	23.2
K^+	80.3
Ca^{2+}	5.8
Mg^{2+}	1.4
Cl^-	55
HCO_3^-	39
PO_4^{3-}	14.9
SCN^-	13.4
Nonprotein organic	
Urea	12.7
Uric acid	1.5
Amino acids	4
Citrate	1.1
Lactate	1.7
Ammonia	0.4
Sugars	19.6
Carbohydrates	73
Lipids	2
Protein	
Glycoproteins	45
Amylase	42
Lysozyme	14
Mucins	250
Albumin	2
Gamma-globulin	5

Source: Ref 34

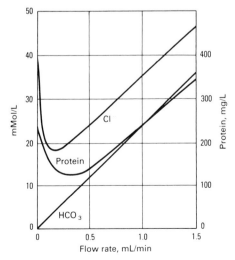

Fig. 9 Variations in the concentrations of Cl^-, HCO_3^-, and protein in human saliva as a function of the flow rate of saliva. Source: Ref 38

stimulated. The buffering capacity is chiefly due to the CO_2/HCO_3 system, with that of the PO_4^{3-} system only having a small, limited part. The redox potential of saliva indicates it to possess reducing properties, which is likely due to bacteria reductions, carbohydrate splitoffs from glycoproteins, and nitrates. The normal pH of unstimulated saliva is in the 6 to 7 range and increases with flow rate.

The clearance of saliva involves its movement toward the back of the mouth and its eventual introduction into the stomach. Saliva is continually being secreted and replenished, especially during active times. A volume of about 1 L/d is considered average for saliva production. Chemical analysis of human mouth air showed hydrogen sulfide (H_2S), methyl mercaptan, and dimethyl sulfide to be some of the most important constituents (Ref 39).

Organic. Human saliva is composed of nonprotein organic and protein contents, as shown in Table 2. The largest contributions from the nonprotein ingredients are from the carbohydrates, while smaller amounts are from urea, organic acids, amino acids, ammonia, sugars, lipids, blood group substances, water-soluble vitamins, and others. Some of the lipids include the fatty acids, glycerides, and cholesterol. At least 18 amino acids have been identified, with glycine being the main constituent. Many of these components are produced directly by the salivary glands, while others, such as some carbohydrates and amino acids, are the result of the dissociation of glycoproteins and proteins by bacterial enzymes. Still others are derived from blood plasma. The protein content of human saliva is primarily of salivary gland origin, with a very small amount derived from blood plasma. The protein content may vary from less than 1 to more than 6 g/L. Detailed information on protein and glycoproteins that have been identified to be in saliva can be found in Ref 40 to 42.

Chemicals in Food, Drink, and Air. All of the ingredients found in food and drink are capable of becoming incorporated into saliva. However, most of the foods are ingested before the breakdown into basic chemicals occurs. Some foods and beverages, though, contain chemicals that are reactive by themselves, without any reductions, and may become dissolved in saliva and affect the tarnish and corrosion of metallic materials. Some of these include various organic acids, such as lactic, tartaric, oleic, ascorbic, fumaric, maleic, and succinic, as well as sulfates, chlorides, nitrates, sulfides, acetates, bichromates, formaldehyde, sulfoxylates, urea, and the nutrients themselves of lipids, carbohydrates, proteins, vitamins, and minerals (Ref 43).

The components found in atmospheric air and pollutants, coupled with the human respiratory function, have the potential of exposing the oral environment to additional aggressive chemical species. Some of the species known to be in atmospheric air and pollutants are O_2, CO_2, NO_2, carbon monoxide (CO), sulfur dioxide (SO_2), Cl_2, hydrogen chloride (HCl), hydrogen sulfide (H_2S), ammonia (NH_3), formaldehyde, formic acid, acetic acid, Cl^- salts, ammonium salts of sulfate and nitrate, and dust (Ref 44).

Because the volume of lung ventilation is of the order of 8.5 L/min, the amount of potentially hazardous and corrosive material possibly coming into contact with the oral environment is significant. In approximately 2 h, 1 m^3 of air (for a mouth breather) will have been used during respiration, with the potential uptake of sulfur dioxide in the normal urban area being 0.11 to 2.3 mg. Sulfur dioxide can be involved in many interactions, accelerating the tarnish and corrosion of metals. Fortunately, the proteins in saliva combine with most of the aggressive external stimuli coming into contact with saliva; therefore, most of the aforementioned hazardous species are rendered inactive before they can cause tarnish and corrosion. However, the pathway from the atmosphere to the surfaces of dental alloys are certainly potential sources for introducing corrosive species.

A comparison between interstitial and oral fluids shows differences in both inorganic and organic contents. One important difference is the approximately sevenfold higher Cl^- concentration in interstitial fluid. Even though interstitial fluids do undergo variations in pH and p_{O_2}, especially at the site of the implant, saliva is more susceptible to variations in composition. This comes about because the composition of saliva depends to a large degree on flow rate, which in turn depends on a number of physical and emotional factors. Saliva is also subjected to exposures from chemicals contained in the air, food, drink, pharmaceuticals, as well as temperature variations of 0 to 60 °C (32 to 140 °F) and microbiological involvement with the production of acid and plaque.

Artificial Solutions. Numerous solutions simulating human saliva have been formulated and used for testing the tarnish and corrosion susceptibility of dental alloys (Ref 45–50). Modifications to these solutions have also been made and used (Ref 51–56). Some of the solutions contain only inorganics (Ref 48–50, Ref 52–54), while others include the addition of an organic component consisting mostly of mucin (Ref 45–47, 51–53). Some researchers also purge a $CO_2/O_2/N_2$ gas mixture through the solution to simulate pH control and buffering capacity controlled by the CO_2/HCO_3^- redox reaction. All compositions contain mostly chlorides (Na, K, and Ca) and various forms (mono-, di-, or tri-basic, pyro) of phosphates in smaller amounts. Additional ingredients include bicarbonate, thiocyanate, sulfide, carbonate, organic acids, citrate, hydroxide, and urea.

Table 3 presents the composition for an artificial saliva that corresponds very well to human saliva, with regard to the anodic polarization of dental alloys. Ringer's physiological saline solution used to simulate interstitial fluid is also included in Table 3. Both solutions are entirely inorganic. The Cl^- concentration of Ringer's is about seven times higher than that of the saliva. The anionic content of Ringer's is entirely chloride, while the artificial saliva also contains phosphate and sulfide. Urea is also a constituent of the artificial saliva. Sodium, potassium, and calcium constitute the cationic content of both solutions.

A number of additional artificial physiological solutions, some of which are named Hanks, Tyrod, Locke, and Krebs, appear in the literature and have been used to simulate the interstitial fluids. Basically, these solutions contain small additions of modifying ingredients, such as magnesium chloride, glucose, lactate, amino acids, and organic anions. The Ringer's solution presented in this article, after the National Formulary Designation, does not contain sodium bicarbonate ($NaHCO_3^-$). Some solutions, however, referred to in the literature as Ringer's, do contain bicarbonate.

Effect of Saliva Composition on Alloy Tarnish and Corrosion

Chloride/Orthophosphate/Bicarbonate/Thiocyanate. The interactions of the various salts contained in saliva are complex. The effects from the combined saliva solutions are not simply the additive effects from the isolated individual salts. This synergistic behavior is discussed for the corrosion of an amalgam in the $Cl^-/HPO_4^{2-}/HCO_3^-/SCN^-$ system in Ref 50. Chloride alone produces a powdery, finely crystalline corrosion product in heaps around the sites of attack, such as porosities and pits. The addition of HPO_4^{2-}, which by itself produced very little effect, caused the corrosion products to become organized in conical structures, the bases being over the sites of corrosion. The addition of HCO_3^- to the Cl^-/HPO_4^{2-} system generated increased microstructural corrosion. On the contrary, addition of SCN^- to the Cl^-/HPO_4^{2-} system suppressed the microstructural corrosion. By adding all four salts together, an even more corrosion-resistant system was obtained. Corrosion was much reduced and more localized.

Artificial Salivas. The effect on alloy corrosion from different artificial saliva solutions has been studied (Ref 49). The polarization behavior of a number of dental alloys, including gold-base alloys, nickel-chromium, and cobalt-chromium, in artificial salivas without HCO_3^- and SCN^-, but with protein, provided the best correlation with the behavior observed with human saliva in both aerated and deaerated conditions. The

Table 3 Composition of artificial solutions

	Composition, mg/100 ml	
Compound	Artificial saliva	Ringer's solution
NaCl	40	82–90
KCl	40	2.5–3.5
$CaCl_2 \cdot 2H_2O$	79.5	3.0–3.6
$NaH_2PO_4 \cdot H_2O$	69	...
$Na_2S \cdot 9H_2O$	0.5	...
Urea	100	...

Source: Ref 49

artificial salivas containing HCO_3^- and SCN^-, but no proteins, constantly shortened the passivation range of the alloys. The specific contributions from Cl^- and SCN^- shortened the passivation range of the gold-base alloy, but phosphate increased the passivation range of all alloys.

Lowering the pH shifted the amalgam polarization curve to increased currents and potentials, while buffering capacity, which was increased by protein content, influenced corrosion behavior under localized corrosion conditions (Ref 57). In sulfide solution, the polarization curves of amalgams indicated increased corrosion (Ref 47, 58). Dissolved O_2 generated both inhibition and acceleration, as reflected by the formation of anodic films and the consumption of electrons by cathodic depolarization. The particular alloy-environment combination determines whether corrosion is inhibited or accelerated.

Chloride and Organic Content. Anodic polarization of amalgams in human saliva compared with Ringer's solution was shown to be shifted by up to several orders of magnitude to lower currents at constant potentials, depending on the amalgam system (Ref 59). These differences were related to the Cl^- concentration of the solutions. The effect of Cl^- on amalgam polarization is well documented (Ref 55, 60, 61). Pretreatment of gold-base alloys in human saliva prior to galvanic coupling with amalgams in a protein-free artificial saliva reduced the corrosion on some of the amalgams studied (Ref 62). Pretreatment of the amalgams had little effect.

Significant reductions in the weight gains of amalgams stored in artificial saliva with mucin as compared with mucin-free saliva have been reported (Ref 47). Anodic polarization of amalgams in artificial saliva or diluted Ringer's solution, with and without additions of mucin or albumin, was, however, shown to be very similar (Ref 47, 60). Proteins in artificial saliva on silver-palladium and nickel-chromium alloy polarizations were also reported to have little effect (Ref 57). For a copper-aluminum crown and bridge alloy, anodic polarization differences were detected in an artificial saliva with and without additions of a human salivary dialysate (Ref 63). The total accumulated anodic charge passed from corrosion potentials to $+0.3$ V versus saturated calomel electrode (SCE) was significantly reduced in protein-containing saliva. Similarly, the polarization resistance of the alloy was more than doubled by progressively adding up to 1.6 mg dialysate/mL to saliva initially free of proteins.

Microorganisms. The tarnishing of dental alloys by three microorganisms likely to be found in the mouth has been reported (Ref 64). Some specificity between the degree of tarnish and the type of microorganism was obtained. A likely tarnishing mechanism was due to the organic acids generated by the fermentation of carbohydrate by the bacteria. The effect of microorganisms on accelerating corrosion is discussed in the section "Oral Corrosion Processes" in this article.

Oral Corrosion Pathways and Electrochemical Properties

The electrochemical properties of dental alloy restorations vary widely. Electrochemical potentials, current pathways, and resistances depend on whether there is no contact, intermittent contact, or continuous contact between alloy restorations. This section examines the effects of restoration contact on electrochemical parameters and reviews concentration cells developed by dental alloy-environment electrochemical reactions.

Noncontacting Alloy Restorations

Isolated. The total liquid environment of a restoration includes, in addition to saliva, fluids contained within the interior of dentin and enamel, which are more like extracellular fluids in composition than saliva. Figure 10 shows a schematic of a likely current path for a single metallic restoration. The current path encompasses a route that includes the restoration, enamel, dentin, membranes such as the periodontal ligament, soft tissues, and saliva (see Fig. 7). The conduction of current through hard tissues, including enamel, dentin, and bone, occurs through the extracellular fluids, which are compositionally similar in all hard and soft tissues. However, the current through these different hard tissues will take pathways of least resistances. For example, the resistance of dentin in a direction parallel to the tubules is about 18 times lower than in a perpendicular direction due to the calcification of the tubule walls. Structural details, including imperfections, orientations, and so on, control the actual resistances for particular hard tissue structures.

The restoration (R) develops electrochemical potentials with the extracellular fluids, E_{RE}, and with saliva E_{RS}, while a liquid junction potential occurs between extracellular fluids and saliva, E_{ES}. Contact resistances occur between restoration and extracellular fluids, R_{RE}, and between restoration and saliva, R_{RS}. In general, these potentials generated at interfaces are caused by the materials and/or liquids existing at different energy levels. Resistances of the extracellular fluids, R_E, extracellular fluid-saliva junction, R_{ES}, and of saliva, R_S, also occur. Figure 11 shows an electrical schematic for this system. Summing electromotive forces in one direction and equating to zero yields for the current I:

$$I = \frac{E_{RE} + E_{ES} - E_{RS}}{R_{RE} + R_{RS} + R_E + R_S + R_{ES}} \quad \text{(Eq 1)}$$

Taken together, E_{ES} and R_S have negligible effect on current. The extracellular resistance, R_E, is usually in the range between 10^4 and 10^6 Ω because of variations in particular hard tissue structures, and to possible variations in membrane/hard tissue interfacial characteristics. The potentials E_{RE} and E_{RS} are characteristics of the metal-electrolyte combinations, and the resistances R_{RE} and R_{RS} are dependent on the polarization characteristics for the particular combinations.

Polarization is related to the corrosion products that form. For soluble or loosely adhered products, the contact resistances will not be changed significantly. However, for tenaciously adhering products with semiconducting or insulating electrical characteristics, the contact resistances will be largely affected. These resistances are the primary parameters affecting the magnitude of the generated current. This reasoning is directly in line with the mixed-potential theory for electrochemical corrosion (Ref 65). The corrosion current, I_{corr}, without ohmic resistance control is:

$$I_{corr} = \frac{\beta_a \beta_c}{(\beta_a + \beta_c) R_p} \quad \text{(Eq 2)}$$

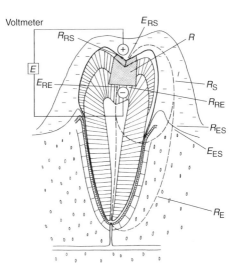

Fig. 10 Schematic of a single metallic restoration (R) showing two possible current (I) paths between external surface exposed to saliva and interior surface exposed to dentinal fluids. Because the dentinal fluids contain a higher Cl^- concentration than saliva, it is assumed the electrode potential of interior surface exposed to dentinal fluids is more active and is therefore given a negative sign (−). The potential difference between the two surfaces is represented by E.

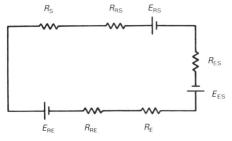

Fig. 11 Electrical schematic representing the equivalent circuit of a single loop shown in Fig. 10. Terms are defined in text.

where β_a and β_c are the Tafel slopes from the anodic and cathodic polarization curves, and R_p is the polarization resistance or the linear slope of the $\Delta E/\Delta I$ curve within ± 10 mV of the corrosion potential.

Nonisolated. For two restorations not in contact (Fig. 12), the extracellular fluid-saliva resistance, R_{ES}, determines the extent to which the current will be short circuited through the saliva/extracellular fluid interface. If R_{ES} is high, there is maximum interaction between the separated restorations (the currents are small, of the order of 1 to 10×10^{-9} A/cm^2 between an amalgam and a gold alloy restoration). As R_{RS} decreases, the current through the interface between saliva and extracellular fluids increases. The interaction between the separated restorations will then be minimized. Each restoration, though, will still generate its own current path loop (Ref 66).

Intraoral Electrochemical Properties. In a study comprising 115 people, the corrosion potentials from 243 restorations ranged between -0.55 and $+0.4$ V versus SCE. Amalgam restorations were the most active, followed by cobalt-chromium alloys and gold-base alloys. Variations in potential on different surfaces of the same restoration occurred routinely. This was likely due to the effects from abrasion on the occlusal surfaces and from the accumulation of plaque and debris on nonocclusal surfaces (Ref 67).

For noncontacting amalgam and gold alloy restorations (78 fillings in 66 people), the average currents flowing through the restorations due to saliva-bone fluid liquid junction cells were calculated from measured intraoral potential and resistance data to be 0.48 and 0.26 µA, respectively (Ref 12).

Using constant current pulses (1 to 10 µA) and measuring the corresponding potential changes, the intraoral polarization resistances for noncontacting amalgam restorations ranged between 50 and 300×10^3 Ω (Ref 68). With the use of linear polarization theory, corrosion currents are calculated to be 0.2 to 1.0 µA.

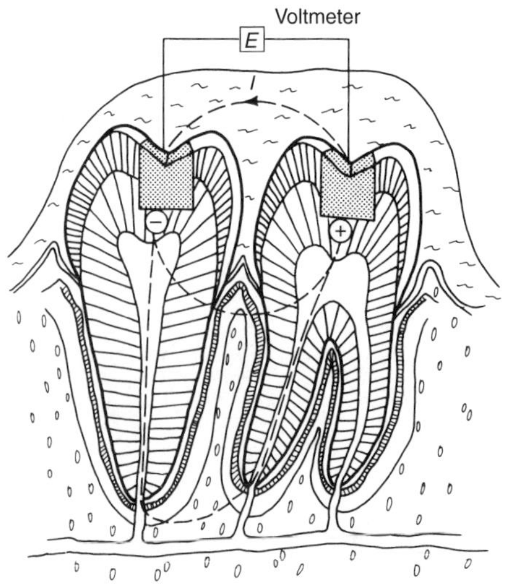

Restorations Making Intermittent or Continuous Contact

Intermittent Contact. A situation can occur in the mouth in which two alloy restorations, one in the upper arch and the other in the lower arch, come into contact intermittently by biting (Fig. 13). When the two restorations are in direct contact, a galvanic cell is generated with an associated galvanic current short-circuited between the two restorations. The external current path can take a number of directions, with the least resistance path controlling. Figure 13 shows two possible pathways, one entirely through extracellular fluids and the other partly through extracellular fluids and partly through saliva.

The current-time transients have been measured and are presented in Fig. 14. Upon first making contact, currents of the order of 10 µA and more occur and decrease rapidly within a matter of minutes. If, however, the restorations are open-circuited for a time interval and then again closed, the current level will again increase but not to the same magnitude as from the previous closure. The amount of recovery will increase as the time lapse between closure increases. This phenomenon is explained by the formation of protective surface films on the electrodes due to the passage of current. Upon making contact on succeeding occasions, the film offers additional resistance to the flow of current, even though the two restorations appear to be in direct intimate contact. The films dissipate with time, thus increasing the level of the initial current on recontacting restorations.

A similar situation can occur because of an alloy restoration contacting, for example, eating utensils or dental instruments during dental treatment. Again a short-circuited galvanic current is generated. The external circuit will be partly through saliva and partly through extracellular fluids.

Continuous Contact. Another situation in which metallic restorations in the mouth are capable of generating galvanic currents involves two dissimilar metallic restorations in continuous contact, as shown in Fig. 15. Most attention has been given to the combination of amalgam-gold alloy couples (Ref 52). Other situations

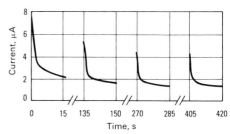

Fig. 14 Current-time responses between gold alloy and amalgam of the same cross-sectional areas. Short circuiting occurred for 15 s, followed by a 2 min delay before recontacting. Source: Ref 69

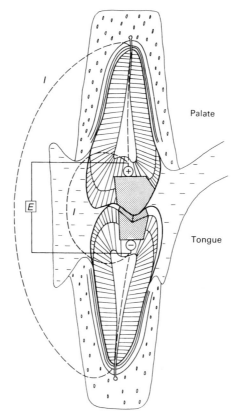

Fig. 13 Schematic of two restorations making intermittent contact due to biting. The restoration in the lower arch, which is an amalgam, is more active than the restoration in the upper arch, which is a gold-base alloy. Two possible current pathways are shown. An additional path very likely to occur would be directly through saliva between the two restorations.

Fig. 12 Schematic of two nonisolated, noncontacting restorations. The alloy restoration on the left, which is an amalgam, is more active than the restoration on right, which is a gold-base alloy.

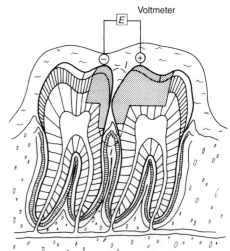

Fig. 15 Schematic of two adjacent restorations in continuous contact. Two possible current paths shown.

occur, for example, between two amalgam restorations (Ref 70)—one a conventional amalgam and the other a high-copper amalgam—and between two gold alloys with differences in noble metal content. Other situations have already been discussed. These include a stainless steel reinforced amalgam (Fig. 8a), an endodontically restored tooth with silver cones making contact with a gold crown (Fig. 8b), and an endodontically restored tooth with steel screwposts making contact with a gold crown (Fig. 8c). Soldered appliances are also examples of dissimilar metals making continuous contact. Any multiphase microstructures are situations for galvanic corrosion to occur. Multiphase microstructures occur extensively with dental alloys.

For the amalgam-gold alloy couple making direct contact, the amalgam is the anode and suffers corrosive attack; the gold alloy is the cathode. As with galvanic couples making intermittent contact, large galvanic currents occur on first contact and decrease rapidly with time. For silver-tin amalgams, the tin from the tin-mercury phase suffers corrosion attack. The freed mercury combines with the gold of the gold alloy to form a gold amalgam that is capable of producing surface discolorations on the gold alloy. In addition to becoming corroded, the amalgam is capable of being degraded in strength by the corrosion generated by the galvanic currents (Ref 70).

Currents calculated from polarization resistance and potential differences of various contacting dissimilar metallic restorations indicate most couples to pass 1 to 5 μA on first contact (Ref 67). However, amalgam-gold alloy couples indicate a greater percentage of currents in the ranges of 6 to 10 and 11 to 15 μA. All couples initially show a sharp decrease in current with time, followed by a gradual leveling off as zero current is approached. However, disruption of surface protective films can result in increases in current at later times.

Concentration Cells

Interior-Exterior Surfaces. Because the interior surfaces of restorations adjacent to the cavity walls are exposed to extracellular fluids and higher concentrations of Cl^- than the exterior surfaces exposed to saliva, the interior instead of the exterior surfaces are more susceptible to anodic attack from Cl^-. However, if the electrons generated by the anodic oxidations are not consumed by reductions, the oxidation reactions will cease. Because the extracellular fluids have low concentrations of dissolved oxygen, corrosion of the interior surfaces would likely cease if it were not for the accessibility of electrons to the exterior surfaces exposed to saliva having a supply of dissolved oxygen from contact with the atmosphere.

Corrosion that is perpetuated by electrochemical reactions occurring on adjacent or opposite surfaces of the same restoration constitutes an important pathway for the tarnishing and corrosion of dental alloys. This pathway is germane to amalgams as well as all types of restorations, including crowns and inlays that are cemented into the cavity preparation. In the mouth, cements are likely to become electrical conductors because the absorption of oral fluids permits the passage of ions.

Marginal Crevices. A second pathway can occur because of the seepage of salivary fluids into crevices or marginal openings formed between the restoration (especially with amalgams) and the cavity walls. The pathway is distinguished from the first in that the conditions developed in the crevice are due to diffusion and charge balances resulting from the salivary fluids instead of the extracellular fluids. Because of a lack of diffusion of the large O_2 molecule into the crevice, low O_2 concentrations result within the crevice. With time, the acidity within the crevice increases because of the accumulation of H^+ ions from the oral environment and from corrosive reactions occurring within the crevice. Chloride and other anion concentrations will also tend to increase within the crevice over time because of charge equalization. Therefore, this pathway results in conditions that are similar to the interior-exterior pathway.

Alloy Surface Characteristics. Porosities, differences in surface finish, pits, weak microstructural phases, and the deposition of organic matter can initiate corrosion by concentration cell effects. For example, gold-base alloys are known to become tarnished more easily when containing porosities and inhomogeneities (Ref 71). Rougher surface finishes of restorations generate increased corrosive conditions (Ref 72). Similarly, the pitting of base metal dental alloys of the stainless steel and nickel-chromium varieties occurs by concentration cell corrosion. Basically, the advancing pit front is free of O_2, but the surfaces of the alloy outside the pit have an ample supply of O_2 from the air. Because the anode-to-cathode surface area is very small, the corrosion occurring at the bottom of the pit is concentrated to a very small area, thus increasing the intensity of the attack. Removal of only a small amount of metal has a large effect on advancing the pit front.

Amalgam γ_2 Phase. Deterioration of the weak, corrosion-prone tin-mercury phase (γ_2) in silver-tin amalgams has also been proposed to occur by concentration cell corrosion (Ref 66). In this model, partial removal of the γ_2 phase initially occurs by abrasion resulting from biting and chewing. After removal of the γ_2 phase has progressed to a sufficient depth, an occluded cell is formed between the bottom of the depression and the unabraded surface. Mass transport is restricted from and into the cell. The condition will approach conditions occurring in other types of concentration cells. In the present example, however, Sn^{2+} will be slowly released from the passivated γ_2 regions. The concentration of Sn^{2+} will slowly increase within the occluded cell and will be neutralized by an equivalent amount of Cl^- by migration from the bulk electrolyte.

Consumption of O_2 within the occluded cell will take place by its utilization in the consumption of electrons by cathodic depolarization. Replenishment of O_2 will be restricted, and the concentration of O_2 within the cell will become reduced. When the solubility product of stannous oxide (SnO) is exceeded, SnO precipitates and the H^+ concentration increases. At this point, activation of the γ_2 phase occurs. Dissolution of tin occurs freely. The Cl^- concentration within the cell continuously increases to maintain electrical neutrality. Galvanic coupling of the occluded cell to the external surface generates a galvanic cell by which the cathodic reduction of O_2 occurs. Corrosion of γ_2 tin within the cell continues. Under conditions of high acidity and high concentration of Cl^-, the formation of insoluble tin chloride hydroxide $(Sn(OH)Cl \cdot H_2O)$ becomes thermodynamically possible.

Oral Corrosion Processes

Whether corrosion is occurring between microstructural phases of a single restoration, between components having different environmental concentrations, or between individual restorations of different compositions and making intermittent or continuous contact, the corrosion processes involved consist of oxidation and reduction. The dissolution of ions is involved with the anodic reaction, and the consumption of electrons is involved with the cathodic reaction. The slowest step in the complete chain of events controls the overall corrosion rate. Corrosion of alloys in the mouth can be viewed as being the result of corrosive and inhibiting factors (Ref 73). Some corrosive factors consist of Cl^- (in most instances), H^+, S^- (at times), O_2, microorganisms, and the clearance rate of corrosion products from the mouth, while some inhibiting factors consist of protein and glycoproteins (in most instances), CO_2/HCO_3^- buffering system, $PO_4^-/PO_4^{2-}/PO_4^{3-}$ buffering system, and salivary flow rate.

Corrosive Factors

Chloride. The effect of Cl^- on the deterioration of passivated surface films on stainless steel, nickel-chromium, and cobalt-chromium alloys is well known. The susceptibility to pitting attack is increased. Increased Cl^- content also increases the attack of corrosion-prone phases in amalgam, other base metal alloys, and the low noble metal content alloys. Because the Cl^- concentration in saliva is about seven times lower than that in the extracellular fluids, the corrosiveness of Cl^- in saliva is usually less. Figure 16 illustrates the effect of Cl^- concentration on the polarization of amalgam by comparing the cyclic voltammetry in deaerated artificial saliva to that in Ringer's solution.

Increases in Cl^- concentrations are also likely to occur in crevices, such as the interfaces between cavity walls and adjacent surfaces of restoration. The Cl^- concentration within crevices is expected to increase to preserve electrical neutrality from the increase in Sn^+ concentration resulting from the γ_2 tin and γ_1 corrosion (Ref 66).

Chloride is capable of generating numerous compounds as products of corrosion. Chloride combines with zinc, tin, copper, silver, and others contained in dental alloys. Some of the products formed include zinc chloride ($ZnCl_2$), stannous chloride ($SnCl_2$), stannic chloride ($SnCl_4$), SnCl compounds such as hydrated $SnOHCl \cdot H_2O$ and $Sn_4(OH)_6Cl_2$, copper chloride (CuCl), cupric chloride ($CuCl_2$), complex hydrated cupric chloride ($CuCl_2 \cdot 3Cu(OH)_2$), and silver chloride (AgCl). The solubilities are high for all compounds, except CuCl, AgCl, and the basic tin and copper chlorides. Many additional compounds are to be considered for a complete listing of all potential corrosion products that form from dental alloys. Certainly, the chlorides of indium, gallium, beryllium, iron, nickel, chromium, cobalt, and molybdenum should be included.

Hydrogen Ion. The pH in the mouth can vary from about 4.5 and lower to about 8. In addition to the normal variations in pH of saliva due to human factors (see the section "Oral Fluids" earlier in this article), increased acidity can also result from a number of additional factors, such as the operation of crevice corrosion conditions, the acid production by dental plaque, and the effects of food, drink, and atmospheric conditions. The operation of crevice conditions in amalgams can increase acidity to well below a pH of 4. For amalgams, this acidity is mostly the result of the oxidation of γ_2 and γ_1 tin in aqueous solution. Under these conditions, the freed H^+ will become the cathodic depolarizers. With this increased acidity, dissolution of the tooth structure is also likely to occur. Calcium and phosphorus are likely to be dissolved from enamel and dentin. Dental plaque acid is produced by the fermentation of carbohydrate by microorganisms (Ref 13, 14). Most of the fermentable carbohydrate responsible for acid production comes from the diet in the form of sugars or starchy foodstuffs.

Figure 17 shows a schematic Stephan pH test curve of plaque. Stephan showed that the pH for all plaques decrease in value following a sugar challenge (Ref 13). This means that the production of acid by fermentable carbohydrate is greater than the rate at which acid can be removed. As time proceeds, the pH again rises. For caries-free and caries-active individuals, the qualitative shapes of Stephan pH curves are similar; however, the relative position of the curve for caries-active individuals is shifted to values in pH of 4.5 and lower. Depending on the source of the sugar challenge, the pH minimum on the Stephan pH curves have been shown to remain for a number of hours (Ref 15). Even though no data are available to show the effect of plaque pH value on the tarnishing and corrosion of dental alloys, it follows from first principles that the reduction in pH will adversely affect tarnishing and corrosion resistance. Metallic restorations can become severely deposited with plaque and organic matter, as shown in Fig. 18.

Sulfide compounds, such as silver sulfide (Ag_2S), cuprous sulfide (Cu_2S), and cupric sulfide (CuS), have very low solubility product constants and often constitute the tarnished films on dental alloys. Mercury and tin sulfides may also be present when amalgams are considered. The formation of thin insoluble films occurs with very small amounts of formed corrosion products, especially on the higher noble metal content alloys. In spite of even microgram quantities of tarnishing products at times, surface discolorations can still occur and elicit unsatisfactory personal responses. Tarnishing products under these conditions almost always maintain biocompatibility with the alloy system. With the lower noble metal content alloys, however, increased quantities of corrosion products can form, and tarnishing and corrosion can become more involved.

The corrosion potentials for many dental alloys in sulfide-containing solutions are often lower than the standard reduction potentials for the formation of the metal sulfides—an indication that the metal sulfides are thermodynamically stable. In some instances, particularly with amalgams, dissolution rates increase with S^- concentrations. This is probably due to the increased solubility for some of the sulfides (for example, Sn_2S_3 with amalgams) to form complexes with other species. Dietary factors are the main source for increasing S^- levels in saliva. Some foods, such as eggs and fish, as well as some drinking waters, are high in sulfur. Smokers have higher SCN^- saliva concentrations than nonsmokers (Ref 76). Sulfate-reducing bacteria may also generate S^- in the mouth. Hydrogen sulfide that is produced in the crevicular fluid and periodontal pockets can be easily dissolved in oral fluids. Atmospheric pollutants often contain high levels of SO_2 and H_2S and may influence the concentrations of S^- in the oral fluids.

Dissolved oxygen participates in corrosion reactions by either depolarizing cathodic reactions or by reoxidizing disruptive passivated surface films on base metal alloys. The first case increases or perpetuates corrosion, while the second case reduces or inhibits corrosion. Oxygen depolarization occurs by the customary electrochemical redox reactions. Electrons from the anodic process are consumed by the depolarization process. For a typical restoration, the exterior surfaces are exposed to higher O_2 concentrations. Differential aeration conditions can become operative, with the outer surfaces cathodic to the anodic interior surfaces. In other situations, differential aeration cells are set up between the bottoms of pits and the surrounding surfaces. Other types of pores and porosities are also likely to generate concentration cells. In near-neutral solution that corresponds to saliva, the reaction $O_2 + 2H_2O + 4e^- \rightarrow 4OH^-$ occurs

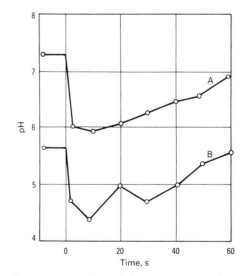

Fig. 16 Anodic polarization at 0.03 V/min of low-copper amalgam (Microalloy) in artificial saliva and Ringer's solution. Source: Ref 74

Fig. 17 pH versus time responses (Stephan curves) of plaque from caries-active (B) and caries-free (A) groups following a sugar challenge. Source: Ref 13

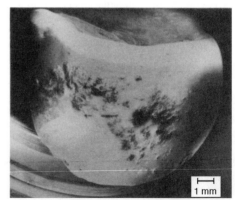

Fig. 18 Alloy restoration after intraoral usage showing the severity of plaque buildup that can occur. Source: Ref 75

most prevalently. If a driving force exists for metal oxidation, dissolution will be perpetuated on surfaces exposed to the lower O_2 concentration.

Oxygen is involved with numerous corrosion products formed on dental alloys. The tin from the γ_1 and γ_2 phases from silver-tin amalgams generates tin oxide products. These products passivate the amalgam at potentials less negative than about -0.7 V versus SCE, as indicated by the passive regions on the anodic polarization curves shown in Fig. 16. At more noble potentials, basic tin chlorides of the type $SnOHCl \cdot H_2O$ are formed, as indicated by the large current increases on the polarization curves. For copper-containing amalgams, basic copper chlorides of the type $CuCl_2 \cdot 3Cu(OH)_2$ form. Some additional products containing oxygen that are likely to occur with the corrosion of dental alloys include SnO_2, $Sn_4(OH)_6Cl_2$, Cu_2O, CuO, ZnO, $Zn(OH)_2$, and the oxides of chromium, nickel, cobalt, molybdenum, iron, titanium, and so on.

Oxygen is usually excluded from solutions during polarization testing with alloys. Dissolved O_2 interferes with the anodic processes. The generated anodic polarization curves obtained in O_2 containing solutions are usually cut off within the negative potential regions. For this reason, deaerated solutions are usually used to obtain entire anodic polarization curves. Passive breakdown potentials were observed to vary depending on whether aerated, dearated, or air-exposed solutions were used (Ref 77). However, within the O_2 concentration range likely to occur for surgical implants, the anodic polarization of type 316L stainless steel in Ringer's solution was independent of oxygen concentration (Ref 78).

Microorganisms. Two types of organisms—sulfate-reducing (*Bacteriodes corrodens*) and acid-producing (*Streptococcus mutans*) bacteria—have been discussed with the corrosion of dental alloys in the mouth (Ref 73). With regard to sulfate-reducing bacteria, depolarization of cathodic sites is thought to occur by removing H^+ from the metal surface. The hydrogen is used by the bacteria for the reduction of sulfate to sulfide, such as by the reaction $SO_4^- + 8H^+ \rightarrow S_2^- + 4H_2O$. In the case of acid-producing bacteria, the adsorbed microorganisms on the surface establish differential aeration conditions. As the dissolution of the metal occurs underneath the deposited microorganisms, the released acidic metabolic products, which include organic acids such as lactic, pyruvic, acetic, proprionic, and butyric, increase corrosion of the already-formed anodic sites. Because anodic areas are relatively small compared with the larger cathodic areas, corrosion can be severe.

The effects of dental plaque by-products from the fermentation of carbohydrates on the tarnishing and corrosion of dental alloys are probably more significant than the effect on alloy corrosion of only the microorganisms themselves.

Clearance Rate. The clearance of corrosion products from the mouth by the movement of saliva toward the back of the mouth and eventually by swallowing and replenishment affects the concentration of products in equilibrium with the metallic restorations. Therefore, a driving force for the continuation of the corrosion processes is maintained. Products of corrosion, like chemical species introduced through the diet, are cleared from the mouth by binding the exterior surfaces of the oral mucosa to the salivary glycoproteins and mucopolysaccharides lining. Detailed information on the binding ability of corroded metallic ions to proteins in human saliva can be found in Ref 40, 41, and 79.

Alloy Factors. Although the effects of alloy selection on tarnish and corrosion behavior are considered in more detail in the section "Tarnish and Corrosion under Simulated or Accelerated Conditions" later in this article, some of the important factors are mentioned here. Alloy composition and microstructure are probably the two most important factors. The corrosion resistance of dental alloys is the result of nobility in composition or the protectiveness of oxide films formed on base metal alloys. Multiphase microstructures are capable of exhibiting increased tarnish and corrosion because of the galvanic coupling of the individual components. The heat treatment state of cast alloys has an important influence on corrosion resistance (Ref 80). Surface state or finish also influence corrosion; furthermore, cast restorations with burnished margins are more susceptible to corrosion because of differences in surface cold-worked states.

Inhibiting Factors

Organics in the form of microorganisms and plaque usually have an accelerating effect on the tarnishing and corrosion of dental alloys (see the earlier sections "Effect of Saliva Composition on Alloy Tarnish and Corrosion" and "Oral Corrosion Processes"). Organics in the form of amino acids, protein, and glycoproteins have received mixed reports. For the amino acids, the building blocks of proteins, the passivation of copper was shown to be improved in Ringer's solution with added cysteine, while nickel became more corrosion prone (Ref 81). Alanine had little effect. For Ti-6Al-4V, the amino acids proline, glycine, tyrosine, and others that constitute many salivary proteins were again shown to have very little effect.

For the plasma proteins, which simulate the organic content in blood and which simulate dental and surgical implant applications more closely, additional evidence can be found implicating the effect of proteins on corrosion behavior. For example, the corrosion rates of cobalt and copper powders increased significantly when exposed to saline solutions with albumin and fibrinogen (Ref 82); however, for chromium and nickel powders, only slight increases occurred, and for molybdenum, decreases occurred. Corrosion of stainless steel by applied external currents was shown to be increased when conducted in saline with added calf's serum (Ref 83). For a copper-zinc alloy, the cyclic voltammetry was reported to be altered by addition of plasma proteins and plasma concentrations to a phosphated physiological saline solution (Ref 84). Albumin and γ-globulin generated increased passivation currents, while fibrinogen generated decreased critical current densities. The anodic polarizations prior to the onset of critical current densities were also shifted to more active behavior in the protein solutions. Finally, the pitting potential for aluminum increased slightly in human plasma, and current-time transients were shifted to lower values in plasma (Ref 85).

Carbon Dioxide/Bicarbonate Buffering System. The major buffering system in saliva is the CO_2/HCO_3^- system, which has been found to inhibit corrosion processes on dental alloys. Inhibition results from the deposition of such elements as copper, zinc, and calcium as carbonate films. Carbon dioxide, above all other gases, is contained most abundantly in saliva. Up to about 150 vol% (≈ 3000 ppm) is contained in vigorously stimulated saliva. The equilibrium concentration of HCO_3^- in saliva is identified by the redox reaction:

$$CO_2(g) + H_2O \rightarrow H^+ + HCO_3^- \quad \text{(Eq 3)}$$

and with its equilibrium constant pK equal to (Ref 49):

$$pK = 7.9 = -\log \frac{[(H^+)(HCO_3^-)]}{p_{CO_2}} \quad \text{(Eq 4)}$$

At a pH = 7 and rearranging terms yields:

$$\frac{p_{CO_2}}{HCO_3^-} = 7.9 \quad \text{(Eq 5)}$$

Equation 5 states that the partial pressure, p, of CO_2 in units of atmospheres is 7.9 times larger than the HCO_3^- concentration in mol/L. Therefore, for p_{CO_2} of the order of 0.07 atm, HCO_3^- concentrations of the order of 0.009 mol/L are formed. This shows that relatively large concentrations of HCO_3^- can be made available in saliva to form carbonates with cations released from corrosion reactions on dental alloys and with other cations found in the mouth, such as calcium.

Many of the different carbonates likely to form are insoluble in aqueous solution. The calcium carbonates are known for making waters hard. Compounds of this type being deposited as thin-film tarnish and corrosion products on dental alloys are very likely to interfere with the corrosion activity. Deposition over cathodic sites effectively increases corrosion resistance by increasing resistance to depolarization reactions. Because the films of carbonates are also likely to increase the contact resistances between electrodes and saliva, galvanic-corrosion processes are likely to change from purely corrosion control to at least partial ohmic control. Under these conditions, local anodes and cathodes may change in order to maintain lower-resistance paths for both ionic and electronic conduction. For example, the in vivo tarnishing of several silver-palladium alloys was shown to be due to the galvanic coupling between microstructural phases located very close to each other on the alloy surface (Ref 86).

Phosphate Buffering System. A secondary buffering system in saliva, the $PO_4^-/PO_4^{2-}/PO_4^{3-}$ system, has also inhibited the corrosion of dental alloys. The progressive inhibition of amalgam corrosion activity in a chloride solution (10 millimolar NaCl) was shown to occur with increasing added phosphate concentrations (Ref 73). A 15 millimolar phosphate addition retarded the anodic polarization almost entirely, while concentrations of 10, 7, 5, and 1 millimolar generated anodic current peaks of about 2.5, 3.0, 3.5, and 6.0 $\mu A/mm^2$, respectively. The 10 millimolar NaCl solution, without phosphate, generated a continuous increase in current, to much larger values. No passivation occurred within the potential range used.

For tin in neutral phosphate solutions, a passive film forms by precipitation or by a nucleation and growth process (Ref 87). Tin phosphate, basic tin phosphate complexes, and tin hydroxides are formed.

Salivary Flow Rate. Increasing the salivary flow rate increases the concentration of most buffering species in saliva. This tends to inhibit corrosion. The organic content, the CO_2/HCO_3^- content, the $PO_4^-/PO_4^{2-}/PO_4^{3-}$ content, pH, and the Ca^{2+} content all increase with flow rate; however, only the increases in Cl^- concentration promote corrosion. Figure 9 shows the effect of flow rate on the concentration of a number of species.

Overview. Saliva acts as an ocean of anions, cations, nonelectrolytes, amino acids, proteins, carbohydrates, and lipids, flowing in waves against and into dental surfaces, with a diurnal tide and varying degrees of intensity (Ref 88). Whether tarnish and/or corrosion of dental metallic materials will occur cannot be categorically stated. It has been discussed that the degree to which dental alloy corrosion occurs in the mouth is dependent on the oral environmental conditions for each person. In addition to effects from the dental alloy itself, competition between corrosive and inhibitory factors of the oral environment will dictate whether corrosion will occur and to what extent. In addition to the aforementioned factors, still others have been isolated and should be included for a more complete assessment of the overall corrosiveness or protectiveness of the oral environment (Ref 73).

Nature of the Intraoral Surface

The composition and characterization of biofilms, corrosion products, and other debris that deposit on dental material surfaces are discussed in this section. As will be shown, the nature of these deposits is dependent on the substrate material (enamel, alloy, porcelain, and so on).

Acquired Pellicles

Characteristics. Most surfaces that come into contact with saliva, including metallic, polymeric, and ceramic dental materials, as well as enamel, interact almost instantaneously with proteins and glycoproteins to form a bacteria-free biofilm of the order of several nanometers in thickness (Ref 89–91). This most intimate layer of organic matter adsorbed to the substrate material is called the acquired pellicle. A fourier transform infrared spectroscopy spectra of the surface of a low-gold dental crown and bridge alloy after in vivo exposure is shown in Fig. 19. Detection for protein, carbohydrate, and lipid is indicated. Thicknesses of the films increase only slightly with longer exposure times. The pellicles, in contrast to enamel and most dental alloys, are for the most part acid insoluble, although an acid-soluble fraction also occurs. The films are diffusion barriers against acids, thus reducing the acid solubility of enamel and metallic materials and inhibiting, or at least reducing, the adherence of organisms (Ref 92, 93).

Composition. Chemical analysis of 2-h pellicles formed on enamel indicated abundant amounts of glycine, glutamic acid, and serine (Ref 94). Carbohydrate contents of similar pellicles formed to enamel were found to contain about 70% glucose, with a number of other sugars and small molecules. Acidic proline-rich phosphoproteins have also been identified from in vivo enamel pellicles (Ref 95). The proline-rich proteins constitute as much as about 37% of the total proteins in new pellicles within the first hour. However, there is a gradual degradation beginning after about 24 h that is reflected by the fact that the proline-rich protein content in aged pellicles is less than 0.1%.

Substrate Effects on Pellicle Composition. Chemical analysis of the pellicles formed on several plastics and glass showed that the amino acid content varied and was different from that formed on enamel (Ref 96). It was concluded that the chemical composition of the substrate has an important influence on the type of proteins that become adsorbed. For the pellicle formed on dentures, it was concluded that a specific mechanism was controlling the deposition of protein and that specific proteins seemed to be precursors in forming the film (Ref 97). Isoelectric focusing of the extracted proteins adsorbed from a human saliva preparation onto a number of different powder substrate compositions, including palladium, silver, copper, silver-copper, tin, silver-tin-copper alloy, bismuth, polymethyl methacrylate, porcelain, hydroxyapatite, and enamel, indicated that the same three to four proteins appeared to be involved with the adsorption process on all substrates regardless of composition (Ref 41). Therefore, from this study, substrate composition appeared not to affect the type of proteins becoming adsorbed.

Binding Mechanisms. The binding of salivary macromolecules to surfaces has been proposed to consist of electrostatic interactions between the charged groups in the molecule and the surface charges on the substrate (Ref 98). For enamel, only the hydroxyapatite, and not the organic matrix, contributes a surface charge for binding. Because the negatively charged phosphate group comprises about 90% of the surface area of hydroxyapatite, the phosphate group rather than the calcium ions will be the primary binding sites. The hydration layer contains soluble calcium and phosphate ions as well as soluble cations and anions. Because the salivary molecules adsorbed to enamel are mainly acidic, binding to the negatively charged

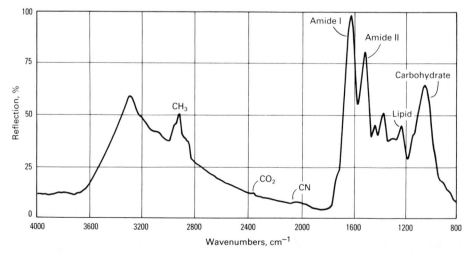

Fig. 19 Fourier transform infrared spectroscopy spectra from surface of a crown and bridge alloy (Midas) after several weeks of intraoral usage. Amide I and II are protein. Additional smaller peaks at 1375 and 1425 cm^{-1} are also protein.

phosphate group appears to occur through a divalent cation, such as calcium. Phosphorylated and sulfated acidic proteins show a high affinity for hydroxyapatite. A direct replacement of the protein phosphate group and the phosphate in hydroxyapatite is also likely (Ref 99).

Direct binding to the calcium surface ions in enamel will also occur, but will be limited because of the relatively small surface area fraction occupied by the calcium ions. Adsorption of salivary proteins to metals may again occur through a divalent cation. The negative charges in the acidic proteins are likely to be bound to the negative anodic surface sites on the metal surface by the bridging cation. Additional information on protein binding and analyses of variations in protein binding to a metal surface through differential scanning calorimetry can be found in Ref 100.

Plaque, Corrosion Products, and Other Debris

Integument. In addition to the thin acquired biofilms, aged pellicles contain microorganisms or plaque, mineralized products, corrosion products, and other debris all commonly referred to as dental calculus. Calculus, which is a by-product of the reaction between microorganisms and calcification, may form in abundance in some environments. Calculus does not form directly onto teeth or other materials. It is deposited or adsorbed onto the acquired pellicle. The combined surface coating, including the adsorbed pellicle, plaque, and calculus, which includes organic matter and any released ions or corrosion products generated by the substrate, is often referred to as the integument.

Substrate Effects on Integument Characteristics. A study was made of the effect of the restorative material type on plaque composition (Ref 101). The carbohydrate/nitrogen ratios (CHO/N) were similar for amalgam, gold inlay, gold foil, and resin. Plaque analyzed from freshly placed restorations had CHO/N = 1; this value increased to 1.3 and 1.2 at 3 and 6 months, respectively, and decreased to 0.5 at 1 year and for old restorations. It was proposed that the variation in plaque carbohydrate content with the age of the restoration was due to corrosion or to the absorption of impurities into surface porosities and pits. These mechanisms are supported by the data generated with silicate restorations. This was the only material to show significant differences in CHO/N. The CHO/N was 1.0 at 1 year. This suggests that the carbohydrate is metabolized less efficiently by the silicate. It is known that silicate restorations leach fluoride with time. Therefore, the fluoride acts as an enzyme inhibitor.

The thicknesses of the integuments formed in the mouth vary and may depend on the substrate material. For example, sputtering times in Auger electron spectroscopy (AES) depth profiling required only 0.3 min to reach the amalgam substrate, while 2.4 min was required to reach the gold alloy substrate (Ref 91). Carbon, nitrogen, and oxygen were distributed in much the same manner as films formed on different substrates. The main difference between the integuments formed on the amalgam and on the gold alloy was the presence of tin ions with the amalgam and the presence of copper ions with the gold alloy. The release of substrate ions is likely to interact with the attachment of microorganisms and therefore with the metabolism of plaque.

Substrate Corrosion. Corrosion reactions involve diffusion of ions—whether cations from oxidations or dissolved O_2 and H^+ for reductions—through the formed integument. The surface coating has the ability to act as a diffusion barrier to the movement of ions. Released ions are likely to become complexed, or bound, to the proteins and glycoproteins constituting the integument and free native proteins in the bulk saliva, provided diffusion is not restricted by the integument. Insoluble corrosion products of the oxides, chlorides, sulfides, carbonates, phosphates, and so on have the capability of being deposited at the alloy/film interface or becoming an integral part of the integument. Soluble products, in addition, may be released into the bulk saliva.

For one dental restorative alloy, it was shown that the polarization resistance of the alloy increased with protein concentration, while at the same time, the concentration of soluble species in solution also increased (Ref 63). This situation was explained by the increased effect of proteins in solubilizing corrosion products. Energy-dispersive spectroscopy (EDS) spectra of the corroded surfaces showed reduced peak intensities for chlorine and sulfur on surfaces exposed to the proteins. Therefore, even though the severity of corrosion is less in protein-containing solutions, increased levels of soluble products are still generated.

In Vivo Tarnished Film Compositions. Auger thin-film analysis of the surfaces of dental alloys with varying compositions and after functioning in the mouth indicated that the tarnished films were due to chemical reactions between alloy and inorganic species and to the adsorption and deposition of organic matter (Ref 43). Carbon was the dominant nonalloying element by about six times, followed by oxygen, calcium, nitrogen, chlorine, sulfur, magnesium, silicon, phosphorus, aluminum, sodium, and tin. Of the elements from the alloy itself, copper was dominant. In a microprobe analysis of in vivo discolorations on gold alloys, both silver sulfides and copper sulfides were detected, depending on the composition of the alloy. Sulfur was found isolated and carbon was present in greatest quantities (Ref 102).

Intraoral (In Vivo) versus Simulated (In Vitro) Exposures

Need for Laboratory Testing. The tarnish and corrosion behavior of dental alloys under actual oral environmental conditions is required. However, except for selected clinical trials, the initial testing of new and improved alloys for tarnish and corrosion resistance is usually carried out under laboratory conditions in either simulated or accelerated tests. This is so because of:

- The possible human exposure to harmful species
- The variability in the oral environmental conditions from person to person and even with the same person from location to location and with time
- As a result of the variability in the oral environment, the inability to follow the effects on tarnish and corrosion from changes in parameters in alloys and in solution

Most laboratory tests use an artificial saliva or a physiological saline solution, such as Ringer's solution (Table 3), diluted Ringer's solution, various concentrations of NaCl, and various concentrations of Na_2S. The main deficiencies with these solutions is that the nonelectrolytes, including the proteins, glycoproteins, and microorganisms, are not included. This fails to produce the pellicle and integuments on laboratory samples that otherwise would have formed on all intraoral surfaces.

In spite of these shortcomings, for the most part, the inorganic salt solutions have become indicators for the aggressiveness of the oral environment. However, the inability to correlate in vivo to in vitro behaviors in some instances is likely because of the failure to account for the shortcomings (Ref 86).

The use of solutions with higher-than-normal concentrations accelerates the tarnish and corrosion processes. For example, 3200 immersions of 15 s/min duration in a 5% Na_2S solution with a Tuccillo and Nielsen tarnishing apparatus (Ref 103) is estimated to simulate 12 months of actual in-service use (Ref 104). Ringer's and 1% NaCl solutions, which contain about seven times the Cl^- concentration of human saliva, are used in anodic polarization tests to amplify peaks in current behavior (Ref 60, 61). Corrosion of conventional amalgams in Ringer's or 1% NaCl generates products that are morphologically similar to those from retrieved amalgams after intraoral use (Ref 105, 106).

A comparison of the tarnishing of three gold alloys, both in vivo and in vitro, indicated that the cyclic immersions in a 5%Na_2S-air environment predicted with considerable reliability the relative susceptibility for the alloys to tarnish (Ref 107). The tarnishing of 81 gold-silver-palladium alloys also indicated accelerated laboratory exposures in Na_2S solution simulated in vivo use (Ref 108). In vivo and in vitro (Na_2S solutions) tarnishing of gold alloys in Na_2S solutions has shown the same microstructural constituents to be attacked (Ref 109, 110). Silver-and copper-rich lamellae were the constituents exhibiting sulfide deposits.

Artificial Solutions in Corrosion and Tarnish Testing. As already indicated, the interior surfaces of restorations are exposed to the interstitial fluids and the exterior surfaces to the salivary fluids. A physiological saline solution such as Ringer's, which contains a Cl^- concentration of about seven times larger than artificial saliva, is therefore more appropriate for simulating in vivo interior surfaces in laboratory testing methodologies. The O_2 content should be reduced to simulate in vivo levels in dentin. The use of Ringer's and even higher Cl^- concentrations is appropriate for the testing of corrosion that may occur within marginal crevices of restorations because crevices can become chloride-rich and acidified. However, applying these results to the corrosion occurring on exterior surfaces of restorations may not be appropriate, even when considering that the increased Cl^- corrosion with Ringer's solution would be an even more stringent test and that the results would correspond to maximum corrosion conditions.

An artificial saliva is more appropriate for testing the corrosion of the exterior surfaces of restorations. The artificial saliva should take into account most of the species contained in saliva and not just a selected few that have been known to affect alloy corrosion. The artificial saliva should include the capabilities for generating organic films on the surfaces, even though their effects in isolated tests may prove unimportant. In order to simulate oral environment conditions, the tarnishing of the exterior surfaces of restorations, an artificial saliva incorporating sulfide is appropriate. Even though the normal sulfide concentrations contained in saliva are within low ranges, accumulations of sulfide can occur along and within crevices to justify the use of higher than normal concentrations. However, the sulfur peak intensities detected with secondary ion mass spectroscopy (SIMS) on alloy surfaces exposed to low levels of sulfide solutions were similar to those from solutions containing higher sulfide concentrations. However, the alloy surface color changes responded more to higher sulfide concentrations.

Classification and Characterization of Dental Alloys

As indicated in the introduction to this article, a wide range of dental alloys exist. This selection reviews the following types of alloys available for dental applications:

- Direct filling alloys
- Crown and bridge alloys
- Partial denture alloys
- Porcelain fused to metal alloys
- Wrought wire alloys
- Soldering alloys
- Implant alloys

The effects of composition and microstructure on the corrosion of each alloy group are discussed in this section. Additional information on tarnishing and corrosion behavior of these alloys is discussed later in the section "Tarnish and Corrosion under Simulated or Accelerated Conditions" in this article.

Direct Filling Alloys

Amalgams. Two types of amalgams are used: low copper (referred to as conventional) and high copper. The alloy particles of the low-copper type are all of the single-particle variety, whereas the high copper type can also be of the dispersed particle variety.

Amalgams are produced by combining mercury with alloy particles by a process referred to as trituration. About 42 to 50% Hg is initially triturated with the high-copper types, while increased quantities of mercury are used with the low-copper types. High-speed mechanical amalgamators achieve mixing in a matter of seconds. The plastic amalgam mass after trituration is inserted into the cavity by a process of condensation. This is accomplished by pressing small amalgam increments together until the entire filling is formed. For amalgams using excess mercury during trituration, the excess mercury is condensed to the top of the setting amalgams mass and scraped away. American National Standards Institute (ANSI)/American Dental Association (ADA) Specification No. 1 Alloy for Dental Amalgam details standard requirements for chemical composition, physical properties, mass, foreign material, and loss of mercury for alloys used in the preparation of dental amalgam (Ref 111).

Low-Copper Conventional Amalgams. The alloy particles with the low-copper type are basically Ag_3Sn, the γ phase of the Ag-Sn system, even though smaller amounts of the β phase, a phase richer in silver, may also be present. Copper can be added in amounts up to about 5 wt% and zinc up to 1 to 2%. About 2 to 4% Cu is soluble in Ag_3Sn, while the additional copper usually precipitates as Cu_3Sn, the ε phase of the Cu-Sn system, although amounts of Cu_6Sn_5, the η' phase, may also occur. The low-copper particles are mostly lathe cut irregular, although spherical atomized particles are also used. The amalgamation reaction for a low-copper amalgam is:

$$Ag_3Sn\text{-}Cu_3Sn\text{-}Zn + Hg \rightarrow$$
$$Ag_{22}SnHg_{27} + Sn_8Hg +$$
$$Ag_3Sn\text{-}Cu_3Sn\text{-}Zn \text{ (unreacted)} \quad \text{(Eq 6)}$$

In Eq 6, Ag_3Sn with Cu_3Sn and zinc react with mercury to form two major reaction products of $Ag_{22}SnHg_{27}$ (the γ phase of the Ag-Hg system with dissolved tin and referred to as the γ_1 amalgam phase) and Sn_8Hg (the γ phase of the Sn-Hg system and referred to as the γ_2 amalgam phase) (Ref 112). Unreacted Ag_3Sn with Cu_3Sn and zinc particles are held together in a γ_1 matrix with γ_2 interspersed within the matrix. Typical distributions of the phases range up to about 30 wt% for γ, 60 to 80% for γ_1, 5 to 30% for γ_2, and up to about 3% for ε (Ref 113). Very minimal η' may also form. Zinc is generally distributed uniformly throughout material. Porosities are in all amalgam structures. As high as 6 to 7 vol% occur with some systems (Ref 114). Interconnection of the γ_2 phase throughout the bulk may also occur (Ref 115). Transformation of the γAg-Hg (γ_1 amalgam phase) to the βAg-Hg phase (β_1 amalgam phase) can also occur with aging (Ref 116). However, because of the dissolved tin in the γ_1 structure, stability is increased. Figure 20 presents the microstructure of a polished low-copper amalgam showing the γ, γ_1, and γ_2 phases and some porosity.

High-Copper Amalgams. The alloy particles with the high-copper dispersed-phase type are blends of conventional particles with basically spherical silver-copper eutectic particles in the proportion of about 3 to 1, respectively. The dispersed particles can be composed of a variety of silver-copper compositions, with other alloying elements, and combined in varying proportions with the conventional particles.

The alloy particles with the single-particle high copper are compositions that can contain up to 30% Cu and more. The particles are mostly atomized into spherical shape.

The setting reaction for high-copper dispersed phase amalgam is:

$$Ag_3Sn\text{-}Cu_3Sn\text{-}Zn + Ag\text{-}Sn + Hg \rightarrow$$
$$Ag_{22}SnHg_{27} + Cu_6Sn_5 +$$
$$Ag_3Sn\text{-}Cu_3Sn\text{-}Zn \text{ (unreacted)} +$$
$$Ag\text{-}Cu \text{ (unreacted)} \quad \text{(Eq 7)}$$

and for high-copper single-particle amalgam is:

$$Ag_3Sn\text{-}Cu_3Sn\text{-}Zn \rightarrow$$
$$Ag_{22}SnHg_{27} + Cu_6Sn_5 +$$
$$Ag_3Sn\text{-}Cu_3Sn\text{-}Zn \text{ (unreacted)} \quad \text{(Eq 8)}$$

For dispersed-phase amalgam, γ initially reacts with mercury to form γ_1 and γ_2 phases, as with the low-copper amalgam. However, an additional reaction occurs between γ_2 and the silver-copper particles to form η' and additional γ_1. The η' phase forms reaction rings around the dispersed particles as well as islands of reaction

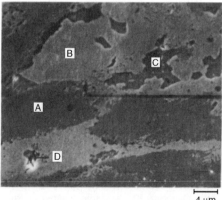

Fig. 20 SEM micrograph of polished, etched, and partially repolished low-copper amalgam (minimax). A, γ; B, γ_1; C, γ_2; D, porosity

phase within γ_1 matrix. Figures 21 and 22 present the microstructure of a dispersed-phase amalgam and EDS x-ray mapping for silver, mercury, tin, and copper, respectively.

For single-particle high-copper amalgam, reaction of the initially formed γ_2 phase occurs with the ϵ phase of the original alloy particles instead of with a dispersed particle to form the η' phase again. The γ_1 phase is likely to become tin enriched. Reaction zones around the original alloy particles, as well as products within the matrix, occur.

Figure 23 shows the microstructure of a polished, etched, and slightly repolished high-copper single-particle amalgam that shows primarily the distribution of the η' phase. The elimination of γ_2 phase and the subsequent formation of ϵ phase are time dependent and are dependent on the amalgam system (Ref 117). For fast-reacting amalgams, formation of η' may be complete within hours, while for slower-reacting systems, η' may continue to form for months. The single-particle high-copper amalgams contain higher percentages of the η' phase and lower percentages of the γ_2 phase, although all high-copper amalgams contain minimal γ_2 relative to conventional amalgams. Porosities with the high-copper amalgams can be up to approximately 5 vol% and with a smaller size distribution than with the conventional type (Ref 114). The γ_1 to β_1 transformation can also occur to a limited extent. With the high-copper amalgams, both indium (5%) and palladium (0.5%) containing amalgams add specific characteristics to the amalgams and have gained limited use.

Crown and Bridge and Partial Denture Alloys

Noble Dental Alloys. Noble metals have traditionally been used as dental casting alloys for their relative inertness in the oral environment. The metals that are considered to be noble differ depending on the source; however, typically, for dental applications, the noble metals are gold and the platinum-group metals (platinum, palladium, iridium, ruthenium, osmium, and rhodium). Because the alloying of silver with specific amounts of iridium can make it resistant to tarnish and corrosion, many also consider silver to be a noble metal (Ref 118). Historically, the noble dental casting alloys have consisted of alloys with greater than 75 wt% Au and metals of the platinum group, as indicated in the *Composition* requirement section of the 1966 edition of ANSI/ADA Specification No. 5 for "Dental Casting Gold Alloy" (Ref 119). This requirement was based on a belief that to attain acceptable corrosion and tarnish resistance in the oral cavity, dental casting alloys had to comprise at least this minimum percentage of noble metals. This belief was reflected by the fact that before 1969, over 95% of the fixed dental prostheses sold in the United States were made of alloys with a minimum of 75 wt% Au and other noble metals; however, in this same year, the United States government ended its support on the price of gold, placing it on the free market, and by the early eighties, the price of gold had increased from around $35 per ounce to over $400 per ounce (Ref 120). This created an impetus for the development of alternative alloys that used less gold and other noble metals. Eventually, advances in metallurgy yielded dental casting alloys with gold contents of less than 50 wt% that could exhibit acceptable corrosion and tarnish behavior in the oral environment (Ref 118). The clinical success and widespread use of these lower-cost alloys in dentistry prompted the revision of ANSI/ADA Specification No. 5 in 1989 (Ref 121). The revised

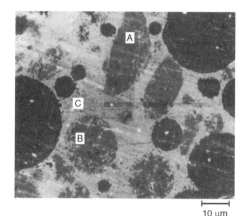

Fig. 21 SEM micrograph of polished, etched, and partially repolished high-copper dispersed-phase amalgam (Cupralloy). A, γ; B, Ag-Cu; C, γ_1. See also Fig. 22.

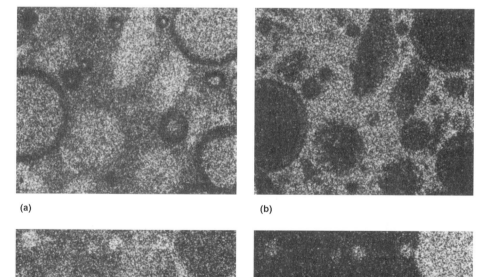

Fig. 22 Elemental maps obtained by energy-dispersive spectroscopy of the high-copper amalgam shown in Fig. 21. (a) Energy-dispersive spectroscopy (EDS) mapping for silver. (b) EDS mapping for mercury. (c) EDS mapping for tin. (d) EDS mapping for copper

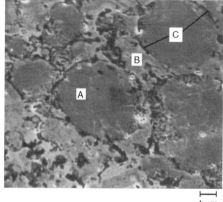

Fig. 23 SEM micrograph of polished, etched, and partially repolished high-copper amalgam (Sybraloy). A, alloy particles; B, γ_1; C, η'

specification dropped the compositional requirements, removed the word "gold" from the title, and suggested both a corrosion and tarnish test, which then became requirements in a further revision in 1998 (Ref 122). For the purpose of dental procedure codes, the ADA classifies "high noble" and "noble" dental casting alloys as follows (Ref 123):

- *High noble*: noble metal content (gold and metals of the platinum group) ≥ 60 wt% and gold ≥ 40 wt%
- *Noble*: noble metal content (gold and metals of the platinum group) ≥ 25 wt%
- *Gold-base*: must contain ≥ 40 wt% Au and gold must be the major element of the alloy

This classification system is also supported by the Identalloy Council, which is a group (formed in 1986) comprising the major dental materials manufacturers. Table 4 shows the compositions of some common gold-base dental alloys.

The high noble dental alloys for cast appliances are primarily based on the gold-silver-copper ternary, with additions of palladium, platinum, and zinc, as well as grain-refining elements, such as rubidium and iridium. For the most part, the properties of the high noble dental alloys follow similar patterns as those shown by the gold-silver-copper ternary alloys, although the additional alloying elements in the dental alloys have significant effects on properties. Because the high noble alloys have relatively small liquidus-solidus gaps, casting segregations, inhomogeneities, and coring effects are not major problems. Microstructurally, single-phase structures predominate because compositions fall within the single-phase region of the Au-Ag-Cu system. Grain refinement by the noble metal additions of ruthenium and iridium decreases grain sizes 20 to 50 μm (0.8 to 2 mils) and increases strengths and elongations by about 30 and 15%, respectively (Ref 124).

The primary hardening mechanism in gold-base dental alloys is by disorder-order superlattice transformations of the Au-Cu system. The ordered domains of the Au-Cu system also extend into the ternary phase Au-Ag-Cu regions. Typically, heat treatment times and temperatures are 15 to 30 min at 350 to 375 °C (660 to 705 °F). Often, it is adequate to bench cool the casting in the investment after casting to gain hardness. Because of the complexity of the dental gold alloy compositions, the exact hardening mechanisms depend on the particular composition of the alloys. Table 5 presents a schematic representation of the age-hardening mechanisms and related microstructures occurring in gold dental alloys. Included are representations for high-gold alloys (HG), low-gold alloys (LG), gold-silver-palladium-base alloys (GSP), and 18 karat and 14 karat gold alloys. Five types of phase transformations are found (Ref 125):

- The formation of the AuCu I ordered platelets and twinning characterized by a stair-step fashion
- The formation of the AuCu II superlattice with periodic antiphase domain structure
- The precipitation of the PdCu superlattice with face-centered tetragonal structure analogous to the AuCu I
- Spinodal decomposition giving rise to a modulated structure
- The formation of the lamellae structure developed from grain boundaries by discontinuous precipitation

The noble metal alloys comprise a wide variety of compositions. Gold-, palladium-, silver-, and copper-base alloys are used. Platinum and zinc contents are usually held to several wt% maximum, if present. Microstructurally, the low-gold-content casting alloys are complex, and examination of phase diagrams of either the Au-Ag-Cu or Ag-Pd-Cu ternary systems indicates that the liquidus-solidus gaps between the various phases can be large (Ref 124). Therefore, coring and casting segregations occur upon solidification during casting. The alloys are characterized by dendritic structures combined with additional phases located within interdendritic positions. Both silver- and copper-rich segregations occur. The presence of gold, platinum, zinc, and other alloying elements further complicates the structures.

The addition of palladium and zinc to the Au-Ag-Cu alloy systems makes heat treatments to single-phase structures difficult. In silver-rich phases, the solubility limit for palladium and zinc is only about 1 to 2% at 500 °C (930 °F), while for copper-rich phases, solubilities are much higher—of the order of 10%. Therefore, precipitation of palladium- and zinc-rich phases occurs. Differences also occur in the gold contents between the phases with the copper-rich phases having the higher contents (Ref 126).

Silver-palladium-base alloys with additions of copper, gold, and zinc are also complex and contain multiple phases. Microstructurally these alloys are characterized by silver-rich matrices interspersed with Pd-Cu-Zn enriched compounds.

The hardening mechanisms associated with some of these noble alloys are shown in Table 5. In addition to the gold-copper disorder-order transformations, ordering due to the palladium-copper superlattices is also usually involved because of replacement of some of the gold by palladium.

Table 4 Compositions, yield strengths, and percent elongation for some common gold-base alloys

Composition, wt%	Heat treatment		YS, MPa (ksi)		Elongation, %	
	Annealed	Hardened	Annealed	Hardened	Annealed	Hardened
Au 83.4, Ag 11.5, Cu 5.0, Pd <1.0, Ir <1.0	A2	...	162 (23)	...	36	...
Au 83.0, Ag 12.0, Cu 4.0, Pd 0.95, Ir 0.05	AC	...	125 (18)	...	33	...
Au 77, Ag 13.54, Cu 7.95, Pd 1.00	A1	...	221 (32)	...	37.5	...
Au 77.0, Ag 13.63, Cu 7.87, Pd 0.95, Zn 0.5, Ir 0.05	AC	...	233 (34)	...	50	...
Au 76.8, Ag 12.8, Cu 8.3, Pd <0.10, Zn+ In+ Ir <1.0	A3	...	263 (38)	...	37	...
Au 74.5, Ag 11.0, Cu 10.45, Pd 3.5, Zn 0.5, Ir 0.05	A7	...	325 (47)	...	26	...
Au 74.0, Ag 12.0, Cu 9.0, Pd 3.8, Zn+ In+ Ir <1.0	A4	...	283 (41)	...	32	...
Au 68.75, Ag 12.4, Cu 12.34, Pd 3.35, Pt 2.9	A1	Hd1	328 (48)	568 (82)	36	12
Au 68.75, Ag 12.4, Cu 12.35, Pd 3.3, Pt 2.9, In 0.25, Ir 0.05	A7	Hd3	393 (57)	569 (83)	32	15
Au 68.3, Ag 10.0, Cu 13.8, Pd 3.6, Pt 2.9, Zn 1.1, In+ Ir <1.0	A6	Hd2	369 (54)	643 (93)	32	8
Au 66.5, Ag 14.50, Cu 14.49, Pd 3.50, Zn 1.0	A1	Hd1	384 (56)	700 (102)	36	7
Au 60, Ag 26.7, Cu 8.80, Pd 3.75	A1	...	249 (36)	...	41	...
Au 60, Ag 22, Cu 14, Pd 3.75	A1	Hd1	358 (52)	644 (93)	37.5	4
Au 58, Ag 27, Cu 10.49, Pd 3.25	A1	...	296 (43)	...	39	...
Au 56, Ag 25, Cu 13.75, Pd 4.00	A1	Hd1	343 (50)	602 (87)	40	6
Au 50.0, Ag 35.0, Cu 9.5, Pd 3.5, In 2.0, Ir <1.0	A5	...	270 (39)	...	32	...
Au 46, Ag 39.5, Cu 7.49, Pd 6.00, Zn 1.00	A1	...	268 (38)	...	28	...
Au 42.00, Ag 25.85, Cu 22.05, Pd 9.09	A1	Hd1	437 (63)	747 (108)	26	4.5

Key to abbreviations: YS, 0.2% offset yield strength; AC, as cast; A1, annealed at 704 °C (1299 °F) for 15 min and water quenched (wq); A2, annealed at 700 °C (1290 °F) for 10 min and wq; A3, annealed at 700 °C (1290 °F) for 30 min and wq; A4, annealed at 705 °C (1301 °F) for 15 min and wq; A5, annealed at 675 °C (1245 °F) for 15 min and wq; A6, annealed at 675 °C (1245 °F) for 10 min and wq; A7, annealed at 700 °C (1290 °F) for 15 min and wq; Hd1, hardened at 315 °C (600 °F) for 25 min and air cooled (ac); Hd2, hardened at 345 °C (655 °F) for 30 min and ac; Hd3, hardened at 350 °C (660 °F) for 15 min and ac

Base Metal Alloys. Predominantly base metal dental casting alloys are classified by the ADA, according to their noble metal content, as follows: Predominantly base—noble metal content (gold and metals of the platinum group) ≤25 wt%. A significant amount of the base metal alloys for metal cast crowns and bridges are composed of nickel-chromium alloys, although stainless steels are also used, particularly outside of the United States. Nickel-chromium alloys are also used for the construction of partial dentures; however, cobalt-chromium alloys far exceed any other alloy for use in this application. Titanium alloys are also gaining popularity for dental applications.

Nickel-Chromium Alloys. The primary alloying element with the nickel-base alloys is chromium between about 10 and 20 wt%: Molybdenum up to about 10%; manganese and aluminum up to about 4% each; and silicon, beryllium, copper, and iron up to several percent each can also be added. The carbon contents range between about 0.05 and 0.4%. Elements such as gallium, titanium, niobium, tin, and cobalt can also be added. Because of differences in properties required for crown and bridge applications versus partial denture applications, minor modifications in compositions occur between nickel-chromium alloys intended for the two applications. This is reflected by the fact that crown and bridge nickel-chromium alloys contain higher percentages of iron, very minimal or no aluminum and carbon, and copper additions (Ref 127).

Chromium and molybdenum are added for corrosion resistance. In order to be effective, these elements must not be concentrated along grain boundaries. Chromium contents below about 10% deplete the interior of the grains leading to corrosion. Molybdenum protects against concentration cell corrosion, such as pitting and crevice corrosion, and is also a solid-solution hardener. Thus, in 1993, the German Federal Health Department recommended a minimum amount of chromium and molybdenum of 20 wt% and 4 wt%, respectively, for dental surgery materials (Ref 128). Manganese and silicon are reducing agents, while aluminum also improves corrosion resistance and improves strength through its formation of intermetallic compounds with nickel. Silicon, like beryllium and gallium, lowers the melting temperature. Beryllium is also a solid-solution hardener and improves castability. Niobium, like molybdenum and iron, affects the coefficient of thermal expansion. Gallium is a stabilizer and improves corrosion resistance. The oxide-forming elements and the elements promoting good bond strength to porcelain are discussed later in the section "Porcelain Fused to Metal (PFM) Alloys" in this article.

Cobalt-Chromium Alloys. ASTM F 75 "Standard Specification for Cobalt-28 Chromium-6 Molybdenum Alloy Castings and Casting Alloy for Surgical Implants" sets the compositional requirements for chromium, molybdenum, and carbon, in wt%, as follows: 27.00 to 30.00% for chromium, 5.00 to 7.00% for molybdenum, and a maximum of 0.35% for carbon (Ref 129). Small additions of other elements, such as silicon and manganese at maximum values of 1%, iron at a maximum of 0.75%, and nickel at a maximum of 0.50%, are also sometimes present, with the balance of the alloy being cobalt. This composition forms the basis for two additional generalized compositions. The first includes a group of alloys that has been developed from the aforementioned basic composition but with each modified by the addition of one or more elements in order to obtain a particular range of properties (Ref 130). Some of these modifying elements include gallium, zirconium, boron, tungsten, niobium, tantalum, and titanium.

Table 5 Hardening mechanisms for some dental alloys
The hatched areas represent the hardness peaks on aging. °F = (9/5) °C + 32 °F.

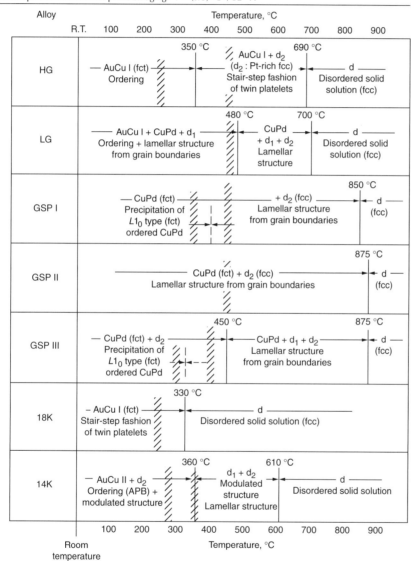

Alloy	Composition, wt%				
	Au	Pt	Pd	Ag	Cu
HG	68	11	6	6	9
LG	30	...	22	29	18
GSP I	20.0	...	25.2	44.9	9.9
GSP II	12.0	...	28.0	48.8	11.2
GSP III	10.0	...	25.4	50.0	12.8
18K	75.0	...	...	8.7	16.3
14K	58.3	...	...	14.6	27.1

HG, high-gold; LG, low-gold; GSP, gold-silver-palladium. Source: Ref 125

The second generalized composition includes replacement-type alloys, with a major portion of the cobalt replaced by nickel and/or iron. The resulting composition is a cross between cobalt-base alloys and stainless steels. The effects of the individual elements on the properties of the alloys are similar to those already discussed for the nickel-chromium alloys.

Even though chromium is just one of the alloying elements in cobalt-chromium-molybdenum alloys, many studies have attributed the passivating behavior of the films that cover these alloys to the oxides of chromium (Ref 131–135). For example, using electron spectroscopy for chemical analysis (ESCA), Storp and Holm determined that a chromium-rich oxide layer had formed on a Vitallium implant (28% Cr, 6% Mo, Mn+Ni+Fe+W <2%, balance Co) that had been immersed in water, with the chromium that was present being in the trivalent form (Ref 132). Chromium (III) oxide, Cr_2O_3, is known to be a metal deficit (oxygen excess) oxide, or p-type semiconductor, under most conditions (Ref 136–140). This is significant because generally for p-type semiconductors, oxide growth occurs at the oxide/ambient interface, as a result of outward cation diffusion (Ref 137). Thus, oxide growth occurs at the oxide/electrolyte surface, which makes it feasible to image this oxide growth. With this in mind, in-situ electrochemical atomic force microscopy (AFM) has been performed on a cast cobalt-chromium-molybdenum alloy, relating its electrochemical behavior with its oxide structure (Ref 141, 142).

In general, cobalt-chromium alloys are hardened principally by carbide formation, and it is known that the variance in carbon composition has a great effect on carbide content. An increase in carbon from 0.05 to 0.30% will increase the carbide content from less than 2 to 5% to between 5 and 15% by volume, which will substantially increase the hardness of the alloy (Ref 143). A number of carbides have been detected in dental cobalt-chromium alloys (Ref 144), with MC, M_6C, and $M_{23}C_6$ being the most prominent.

Molybdenum is added to cobalt-chromium alloys as a grain refiner to produce finer grains upon casting and forging, resulting in a material with greater mechanical properties (Ref 145). Molybdenum is also added as a hardening agent, an oxide former, and to increase crevice and pitting corrosion resistance.

The scanning electron micrographs (SEMs) in Fig. 24 show an ASTM F 75 cast cobalt-chromium-molybdenum alloy with a carbon content of 0.22 wt% that was subjected to a homogenizing anneal (Ref 142). The grains range from approximately 50 to 500 μm (2 to 20 mils) in size, and the alloy contains both intragranular (at interdendritic sites) and intergranular carbide distributions, with a variety of shapes and sizes. However, thermomechanical processing of cobalt-chromium-molybdenum alloys can have a drastic effect on the microstructure, as discussed later in the section "Cast, Wrought, and Forged Cobalt-Chromium Alloys."

Titanium-Base Alloys. The first alloy to be successfully cast had a composition of 82Ti-13Cu-4.5Ni with a melting temperature of 1330 °C (2426 °F). The introduction of an argon/electric arc vertical centrifugal casting machine and a vacuum-argon electric arc pressure casting machine made the casting of the higher melting point titanium alloys achievable. Pure titanium and Ti-6Al-4V have been successfully cast by these latter methods (Ref 146).

Copper-aluminum alloys are also used as dental restorative materials. Compositions include the copper base, with about 10 to 20% Al, and up to approximately 10% iron, manganese, and nickel. The as-cast etched structures are dendritic.

Factors Related to Casting. Many factors determine the castability of dental alloys. Some of these factors include the casting temperature and surface tension of the molten alloy as well as many variables associated with the casting technique, some of which include wax pattern preparation, position of the pattern in the casting ring, techniques used in alloy heating, and centrifugal casting force. Some factors affecting casting accuracy include the thermal contraction of the alloy as a result of going from liquid to room temperature, the effectiveness of investment material to compensate for the thermal contraction of the alloy, anisotropic contractions, and the roughness of the casting.

Casting porosity is another important factor in the casting process. Although casting technique variables affect porosity contents, alloy composition can also affect porosity. One way in which composition affects porosity is the generation of internal shrinkage pores between microstructural phases in complex multiphase alloys. This microporosity weakens alloys, makes finishing and polishing more difficult, and is a prime factor in tarnishing. Palladium in alloys is susceptible to occluding gases from the melt. Therefore, palladium-containing alloys have the potential for becoming affected mechanically through embrittlement.

Porcelain Fused to Metal (PFM) Alloys

Stringent demands are placed on the alloy system meant to be used as a substrate for the baking on or firing of a porcelain veneer. The thermal expansion coefficients of alloy and porcelain must be matched so that the porcelain will not crack and break away from the alloy as the temperature is cooled from firing temperature to room temperature. Thermal expansion coefficients of porcelains are in the range of 14×10^{-6} to 15×10^{-6} mm/mm °C. Selection of an alloy with a slightly larger coefficient by about 0.05% is recommended so that the alloy will be under slight compression.

The alloy must have a high melting point so that it can withstand the firing temperatures involved with the porcelain. However, the temperature must not be excessively high so that conventional dental equipment can still be used. A temperature of 1300 to 1350 °C (2370 to 2460 °F) is about maximum. The porcelain firing procedures require an alloy with high hardness, strength, and modulus so that thin sections of the alloy substrate can support the porcelain, especially at the firing temperatures.

High mechanical properties are also required for resisting sag of long span bridge unit assemblies during firing. The alloy should also have the ability to absorb the thermal contraction stresses due to any mismatch in expansion coefficients as well as occlusal stresses without plastic deformation. Therefore, too high of a modulus for a material with an insufficient yield strength is contraindicated, although too high of an elastic deformation is also contraindicated. High bond strengths are required so that the porcelain veneers remain attached to the alloy. The alloy system must also be chemically compatible with the porcelain. Alloying elements must not discolor porcelain yet must have tarnish and corrosion resistance to the fluids in the oral environment.

In order to promote and form high bond strengths between porcelain and alloys, the alloy must have the ability to form soluble oxides that are compatible with the porcelain. At the

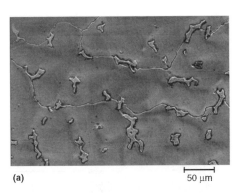 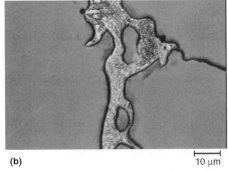

Fig. 24 SEM micrographs showing the microstructure of an ASTM F 75 cast cobalt-chromium-molybdenum alloy that was subjected to a homogenizing anneal. (a) SEM in the secondary electron mode showing both intragranular and intergranular carbide distribution. 350×. (b) SEM in the backscattered electron mode showing a large intergranular carbide. 1500×. The samples were polished to a 0.05 μm (0.002 mils) alumina finish and electrolytically etched in 2% HCl at 3.5 V for 6 s. Source: Ref 142

firing temperature, the porcelain should spread or wet the surface of the alloy. Therefore, both mechanical and chemical interactions are involved in this process. In order to promote chemical interactions, specific oxide-forming elements, such as tin, indium, or gallium, are added to the alloy in low concentrations. The addition of iron, nickel, cobalt, copper, and zinc provide the means for hardening.

The fabrication of the PFM restoration consists of a complex set of processes. After casting, the alloy substrate is subjected to a preoxidation heat treatment to achieve the optimum surface oxides that are important for porcelain bonding. As many as three or more different porcelain firings follow. These include a thin opaque layer adjacent to the alloy, followed by body porcelain layers, including both dentin and enamel porcelain buildups. The opaque layer should mask the color of the alloy from interfering with the appearance of the porcelain. The body porcelain layers build up the restoration to the desired occlusion. The alloy-porcelain systems are slowly cooled from firing temperatures to accommodate dimensional changes occurring in both alloy and porcelain. Therefore, the alloy substrates are subjected to a number of temperature cycles during processing that affect their microstructure and properties. The slow cooling cycles also permit the formation of the compounds important for hardening of alloys.

Noble Metal PFM Alloys. The evolution of alloys for the PFM restoration has generated at least four different noble alloy systems. These are classified as gold-platinum-palladium, gold-palladium with and without silver, silver-palladium, and high-palladium content alloys (Ref 124, 147–149).

Gold-Platinum-Palladium PFM Alloys. The compositions of alloys included within this group are in the range of 80 to 90% Au, 5 to 15% Pt, 0 to 10% Pd, and 0 to 5% Ag, along with about 1% each of tin and indium. Other additions may include up to about 1% each of iron, cobalt, zinc, and copper. Platinum and palladium additions increase melting temperature and decrease thermal expansion coefficients, with platinum having the added effect of hardening the alloy. Iron is the principal hardening agent. Iron promotes the formation of an ordered iron-platinum type intermetallic phase that forms between 850 and 1050 °C (1560 and 1920 °F) on cooling from the firing temperature (Ref 124). The ordered phase is finely dispersed throughout the matrix. Iron, along with tin and indium, promotes bonding to porcelain by diffusing into the porcelain up to about 60 μm (2 mils) at the firing temperature. Tin and indium also promote solid-solution strengthening. Alloys within this group range in color from light yellow to yellow.

Even though these alloys have many advantages, their high cost and low sag resistances have necessitated the development of additional alloy systems. Unless they are used in thick sections, for example, 3×3 mm (0.12×0.12 in.), plastic deformation of long spans will occur during firing.

Gold-Palladium and Gold-Palladium-Silver PFM Alloys. Gold-palladium with and without silver alloys were developed as alternatives to the costly gold-platinum-palladium alloys. The Au-Pd-Ag system was one of the first alternative systems. Up to 15% Ag and 30% Pd replaced all of the platinum and a large fraction of the gold from the Au-Pt-Pd system, which resulted in substantial cost savings. The gold-palladium-silver alloys possessed better mechanical properties for the PFM restoration. Because gold-palladium, gold-silver, and palladium-silver are all solid-solution alloys, the ternary Au-Pd-Ag system also forms a series of solid-solution alloys over the entire compositional ranges. Therefore, the matrices of the gold-palladium-silver dental alloys are single phase. Up to 5% Sn is added, which hardens the alloy by forming compounds with palladium that are dispersed throughout the matrix, and it serves as an oxide former and bonding agent with porcelain. Because platinum was avoided, there was no need to incorporate iron for hardening.

A major shortcoming of the Au-Pd-Ag PFM alloys was the ability of silver from the alloys to vaporize, diffuse, and combine with the porcelain at the firing temperature. This resulted in color changes, mostly greenish, along the alloy-porcelain margin, with sodium-containing porcelains being more susceptible to changes in color.

The development of the silver-free gold-palladium alloys eliminated the discoloration of the porcelain. Their compositions cover a wide range: 50 to 85% Au, 10 to 40% Pd, 0 to 5% Sn, and 0 to 5% In, along with possible additions from zinc, gallium, and other elements. The alloy matrix is based on the Au-Pd binary system, which is of the solid-solution type. Hardening is due to Pd-(In, Sn, Ga, Zn) complexes that disperse throughout the matrix. Microstructurally, a fine network of gold-rich regions are entwined by second-phase particles of Pd-(Sn, In, Ga, Zn). Their color is only a pale yellow, unlike some of the deeper yellow gold-platinum-palladium alloys. However, about a 30 to 40% cost savings is obtained. Their mechanical properties are superior, which means good sag resistance at the firing temperatures. The only disadvantage with these alloys is their lower thermal expansion coefficients when used with some of the higher-expanding porcelains.

Some gold-palladium alloys have up to 5% Ag, which is much lower than the 15% contents used with the original gold-palladium-silver alloys. The lower silver concentrations result in better thermal expansion matches with porcelain and the elimination of porcelain discoloration.

Palladium-Silver PFM Alloys. These alloys were developed out of the need to reduce the cost of the PFM restoration even more than from those fabricated from gold-palladium alloys. Compositions range from 50 to 60% Pd, 25 to 35% Ag, 5 to 10% Sn, 0 to 5% In, and up to 2% Zn.

The microstrutures of the alloys are based on the Pd-Ag solid-solution system. Instead of using copper to harden the alloys, hardening occurs through compounds formed between palladium and tin, indium, zinc, and others. Hardening rates are high, which indicates nondiffusional reactions. It is likely that hardening occurs by ordering processes that form by spinodal decomposition (Ref 124). Oxides are imparted to the alloy surface because of the oxide-forming ability of indium, tin, and zinc alloying additions. This promotes high bond strengths. The mechanical properties of the palladium-silver alloys, along with the high-palladium alloys discussed subsequently, are superior to those of any other system, excluding the nickel-chromium alloys. As with the gold-palladium-silver alloys, the chief shortcoming of the palladium-silver PFM alloys is the ability of silver to discolor porcelains during firing. In order to overcome this problem, various methods have been used, including coupling agents composed of porcelains or colloidal gold.

High-Palladium PFM. The compositions of these alloys are 75 to 85% Pd with 0 to 15% Cu, 0 to 10% Ga, 0 to 8% In, 0 to 5% Co, 0 to 5% Sn, and 0 to 2% Au. The alloys are based on either the Pd-Cu-Ga or the Pd-Co-Ga ternary systems. Regardless of the high copper contents, these alloys do not induce porcelain discoloration and bonding problems. Many of these alloys have better workability than other types of PFM alloys, while retaining high hardnesses. The hardness is dependent on the formation of intermetallic compounds with palladium on cooling from the firing temperature. Because of their oxidizer content, the alloys form strong bonds with porcelain. Oxides form with palladium and the alloying additives. However, palladium oxide (PdO) forms only during heating and cooling because of the relatively low decomposition temperature for the oxide.

The oxide-forming ability of the added oxidizers is dependent on alloy composition, temperature, and time. Indium, gallium, and cobalt oxidize preferentially, while copper and tin show nonpreferential oxidation. Cobalt suppressed the oxidation for copper and tin in one alloy system (Ref 150).

Base metal PFM alloys are primarily composed of the nickel-chromium alloys. Cobalt-chromium alloys are also used, but they constitute only a very small percentage of base metal use for PFM.

The nickel-chromium PFM alloys are very similar, if not the same, as the compositions of the nickel-chromium alloys used for partial dentures, which are discussed earlier in this article. One distinction in the compositions, however, is the absence of carbon with the PFM compositions (Ref 127).

The mechanical properties of the nickel-chromium alloys are excellent for the PFM restoration. The high strengths, moduli, yield strengths, and hardnesses are used to advantage with PFM and partial dentures. Thinner alloy sections can be made from nickel-chromium alloys than from the noble metal alloys. The flexibilities of long span partial denture frameworks are only one-half those for the high gold-content alloys. Additionally, the sag resistances of the nickel-chromium alloys at the porcelain firing temperatures are superior to all of the noble metal alloys.

The bond strengths are seriously impaired by nonadherent or loosely attached oxides. A properly attached oxide is characterized by minute protrusions on the underside of the oxide layer at the alloy/oxide interface that extends into the alloy. For alloys containing additional microstructural phases, larger peg-shaped protrusions also occur on the underside of the oxide layer, improving oxide adherence (Ref 151). The oxide layer on nickel-chromium alloys contains nickel oxide (NiO) on the exterior of the oxide scale, chromium oxide (Cr_2O_3) at the interior covering the alloy, and nickel-chromium oxide ($NiCr_2O_3$) in between. The relative amounts of the oxides depend on the chromium concentration and alloying elements in the alloy, as well as temperature, time of oxidation, and p_{O_2} in atmosphere.

The bond strengths between alloy and porcelain are significantly affected by minor alloying elements. The additional alloying elements of molybdenum, aluminum, silicon, boron, titanium, beryllium, and manganese also form oxides of their own. An aluminum content of 5% is necessary for aluminum oxide (Al_2O_3) to form, while 3% Si is required to form silicon dioxide (SiO_2), which increases in concentration as the alloy/oxide interface is approached. Manganese forms manganese oxide (MnO) and manganese chromite ($MnCr_2O_4$), and these are mainly concentrated at the outermost part of the oxide. Even though molybdenum oxide (MoO) volatizes above 600 °C (1110 °F), molybdenum is still found in the oxide layer close to the alloy/oxide interface with alloys containing more than 3% Mo. Beryllium, which improves the adherence of the oxide layer to the alloy, is also found concentrated near the alloy/oxide interface.

The cobalt-chromium PFM alloys typically comprise nickel, tungsten, and molybdenum as major alloying elements. Tungsten and molybdenum are high-temperature strengtheners, and, therefore, increase sag resistance. Tantalum and ruthenium can also be added in minor amounts. The carbon is also reduced or eliminated with PFM alloys (Ref 127). The carbon monoxide gases generated during firing of porcelain are likely to cause porosities in the interface and in the porcelain.

Wrought Alloys for Wires

Property Requirements with Orthodontic Biomechanics. Orthodontic wires (frequently used round sizes are 0.3 to 0.7 mm, or 0.012 to 0.028 in., in diameter) constitute a large percentage of the wrought alloys used in dentistry. The orthodontic wires most commonly used include stainless steels, cobalt-chromium-nickel alloys, nickel-titanium alloys, and β-titanium alloys. Note that gold-base wires are used on a very limited basis (formerly, they were used more extensively). The stainless steel and cobalt-chromium-nickel wires have been used extensively with conventional orthodontic biomechanics. That is, the ability to move teeth was based to a large extent on the stiffnesses of the wire appliances. Materials with high yield strength to modulus of elasticity ratios were required. However, the nickel-titanium and β-titanium alloys take advantage of relatively lower yield strength to modulus of elasticity ratios. This is because even though slopes for the stress-strain curves of these alloys are low, the released energies (area under the stress-strain curve) can still be large, as a result of greater deflections. Certified orthodontic wires comply with the property requirements of ANSI/ADA Specification No. 32, "Orthodontic Wires," which covers base metal wires for orthodontics (Ref 152).

Stainless Steel and Cobalt-Chromium-Nickel Wires. The stainless steels used are usually the austenitic 18-8 type (approximately 18% chromium and 8% nickel by weight), although precipitation-hardening type steels have also been used. The springback of the 18-8 wires can be improved by a stress-relief heat treatment. For instance, heat treatment of the as-received drawn wires at 400 °C (750 °F) treatment for 10 min generates significant improvements in springback (Ref 153). The cobalt-chromium-nickel wires also generally include nickel and iron as other major alloying elements. Aluminum, silicon, gallium, and copper are not added to the cobalt-chromium-nickel wires, as with the PFM cobalt-chromium alloys, because bonding agents with porcelain are not needed. Mechanical properties are controlled primarily through the addition of carbon, which affects carbide formation. Although the operator has some control over the mechanical properties of the cobalt-chromium-nickel wires through heat treatments, these wires are supplied in different temper designations from soft to semiresilient to resilient (Elgiloy is one of the commonly used cobalt-chromium-nickel alloys).

Nickel-Titanium and β-Titanium Wires. The elasticity effect of nickel-titanium (generically referred to as Nitinol) is one of its most important characteristics. Nickel-titanium wires can almost be bent back on themselves without taking a permanent set. Even greater deformations, by as much as 1.6 times, can be achieved with superelastic nickel-titanium alloys (Ref 154). Nickel-titanium orthodontic alloys are primarily composed of the intermetallic compound NiTi. The alloy is tough, resilient, and has a low modulus of elasticity. Cobalt is sometimes added to nickel-titanium wires in order to obtain critical temperatures that are useful for the shape memory effect of the alloys. Furthermore, copper and chromium ranging from 5 to 6% and 0.2 to 0.5% by weight, respectively, have been added to superelastic nickel-titanium wires to obtain shape memory at temperatures between 27 and 40 °C (81 and 104 °F); however, an in-vitro study in simulated physiologic media showed that these alloy additions did not affect the corrosion resistance of the wires (Ref 154).

Alpha-titanium (hexagonal close-packed structure) is the stable form of titanium at room temperature. By adding alloying elements to the high-temperature form of β-titanium (body-centered cubic structure), the β phase can also exist at room temperature, but in the metastable condition. The β stabilizers include molybdenum, vanadium, cobalt, tantalum, manganese, iron, chromium, nickel, cobalt, and copper. Beta-titanium is strengthened by cold working or by precipitating the phase. A variety of heat treatments can be used to alter the properties of the wires (Ref 155).

Soldering Alloys

Composition and Applications. Gold-base and silver-base solder alloys are used for the joining of separate alloy components (Table 6). High fusing temperature base alloy solders are also used for the joining of nickel-chromium and other alloys. In many cases, the term *brazing* would be more appropriate, but the term is seldom used in dentistry (technically, brazing is performed at temperatures exceeding 425 °C, or 800 °F, and soldering is performed at temperatures below this value). The gold-containing solders are used almost exclusively in bridgework because of their superior tarnish and corrosion resistance. The use of non-noble metal containing silver-base solders is limited mainly to the joining of stainless steel and cobalt-chromium wires in orthodontic appliances because of the impermanence of the appliances.

The joining by soldering of small units to form a large one-piece partial denture is employed in some processing techniques. This is done to prevent framework distortions that may occur with large one-piece castings. The salvaging of large, poorly fitting castings by sectioning, repositioning, and soldering the pieces together also takes place. For PFM restorations, soldering is carried out either prior to or after the porcelain has been baked onto the alloy substrate. Presoldering uses high fusing temperature solders, while postsoldering uses lower fusing temperature solders.

Gold-base solders are most often rated according to their fineness, that is, the gold content in mass percent related to a proportional number of

Table 6 Compositions of some dental solders

	Composition, wt%						
Type	Au	Pd	Ag	Cu	Sn	In	Zn
Silver	...	...	52.6	22.2	7.1	...	14.1
Gold	45.0	...	20.6	28.4	4.3	...	2.9
Gold	63.0	2.7	19.0	8.6	...	6.5	...

units contained in 1000. Conventional gold-base crown and bridge alloys are seldom soldered together with solders having less than a 600 to 650 fineness. The soldering of gold-base wire clasps to cobalt-chromium partial dentures is another application for the higher-fineness solders. However, with the use of the low-gold-content crown and bridge alloys, lower-fineness solders are used. The lower-fineness solders are also occasionally used to solder the cobalt-chromium-nickel wires. The gold-base solders usually do not contain platinum or palladium, which increases melting temperatures. The important requirement to be satisfied during soldering is that the solders melt and flow at temperatures below the melting ranges of the parts to be joined.

The compositions of gold-base solders are largely gold-silver-copper alloys to which small amounts of other elements, such as zinc, tin, and indium, have been added to control melting temperatures and flow during melting. The silver-base solders are basically silver-copper-zinc alloys to which smaller amounts of tin have been added. The higher-fusing solders, to be used with the high-fusing alloys, are usually specially formulated for a particular alloy composition because not all alloys have good soldering characteristics.

Microstructure of Solder-Alloy Joints. The microstructural appearance of the gold alloy-solder joints provides information as to their quality. A thin, distinct, continuous demarcation between the solder alloy and the casting alloy should exist, indicating that the solder has flown freely over the surface and that no mutual diffusion between the alloys has occurred. The junction region should be free of isolated and demarcated domains, indicative of the formation of new alloy phases. Obviously, porosity is to be avoided. However, microporosity among the phases in solder may be unavoidable. As with the cooling that occurs with all alloys, differences in the thermal expansion coefficient among phases can generate microporosity. The presence of a distinct layer of columnar dendrites within the solder, starting at the solder/alloy interface and projecting into the solder bulk, is assurance that there has been no tendency of the alloy surface to melt (Ref 156). The solidified solder tends to match the grain size of the parent alloy by epitaxial nucleation of the solder by the casting alloy. The microstructural characteristics of low-gold-content casting alloys interfaced by soldering affect the microstructural characteristics of the solidified solder (Ref 156).

Microstructurally, a silver-base solder is multiphasal. Both silver- and copper-zinc-rich areas occur, which is in contrast to some of the higher-fineness gold-base solders that are single phase.

Implant Alloys

Applications and Compositions. Dental implants, which are used for supporting and attaching crowns, bridges, and partial and full dentures, can be of the endosseous and subperiosteal types. The endosseous implants pass into or through the mandibular or maxillary arch bones, while the subperiosteal implants are positioned directly on top of or below the mandibular or maxillary bones, respectively. Endosseous implants are usually selected for size and type from implants already made, while the subperiosteal implants are usually custom made for the particular case. Therefore, both cast and wrought forms of implants are used. The alloys used for dental implants are similar in composition to the alloys mentioned previously for use with crowns and bridges and partial dentures. These include stainless steel, cobalt-chromium, and titanium and its alloys (Ref 28, 127).

Cast, Wrought, and Forged Cobalt-Chromium Alloys. In addition to castings, cobalt-chromium alloys for surgical implants can also be obtained in the wrought and forged condition. ASTM F 1537 and F 799 for wrought and forged cobalt-chromium-molybdenum alloys, respectively, specify both low- and high-carbon alloys, along with a dispersion-strengthened alloy (Ref 157, 158). The low-carbon and dispersion-strengthened alloys have a maximum carbon content of 0.14 wt%, while the high-carbon alloy has a minimum and maximum carbon content of 0.15 and 0.35 wt%, respectively. In general, the wrought and forged cobalt-chromium-molybenum alloys exhibit superior mechanical properties to the cast form (ASTM F 75) of the alloy.

The scanning electron micrographs in Fig. 25 show the varied microstructures of five different cobalt-chromium-molybdenum alloys (Ref 142) at the same magnification: an ASTM F 75 cast alloy, two wrought high-carbon alloys, a forged

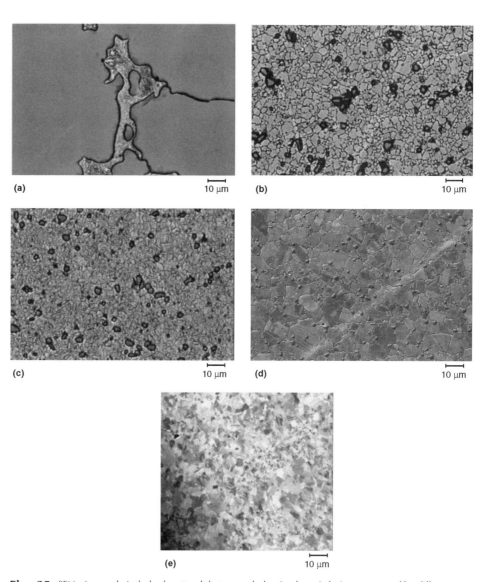

Fig. 25 SEM micrographs in the backscattered electron mode showing the varied microstructures of five different types of cobalt-chromium-molybdenum alloys at a magnification of 1000×. The samples were polished to a 0.05 μm (0.002 mils) finish and electrolytically etched in 2% HCl at 3.5 V for 6 s. (a) ASTM F 75 cast alloy. (b) Wrought high-carbon alloy. (c) Wrought high-carbon alloy in the aged condition. (d) Forged high-carbon alloy. (e) Forged low-carbon alloy. See text for further microstructural details. Source: Ref 142

high-carbon alloy, and a forged low-carbon alloy. The wrought high-carbon and forged high-carbon alloys have a carbon content of 0.24 wt%; the cast alloy has a carbon content of 0.22 wt%; and the forged low-carbon alloy has a carbon content of 0.05 wt%. Figures 25(b) and (c) show the wrought alloys exhibit much smaller grain sizes (about 2 to 8 µm, or 0.08 to 0.3 mils) than the cast alloy (previously described in the section "Cobalt-Chromium Alloys") and are covered by small (2 to 5 µm, or 0.08 to 0.2 mils), mostly spheroidized carbides. The wrought alloys were produced by hot-isostatic pressing powder particles and then hot rolling into a finished product. Generally, this forming operation results in a structure that has fewer voids (low porosity), greater ductility, and finer grains of greater uniformity than the original cast structure (Ref 159). The alloy in Fig. 25(c) was also aged at 730 °C (1350 °F) for 15 h. The aging of the wrought high-carbon alloy resulted in small equiaxed grains and seemed to cause the agglomeration of some of the carbides and the formation of much smaller intergranular precipitates. It is known that the aging of cobalt-chromium-molybdenum alloys produces carbide precipitation along slip lines, twin boundaries, and stacking faults upon quenching, further increasing the hardness of the alloy (Ref 160).

The single-blow forged and water-quenched, high-carbon alloy shown in Figure 25(d) also exhibits much finer grains (2 to 20 µm, or 0.08 to 0.8 mils, in size) than the cast alloy. It is covered with carbides of a semirounded morphology ranging from about 1 to 4 µm (0.04 to 0.2 mils) in size (the majority of which are located in grain boundaries) and small intergranular precipitates. In contrast, the single-blow forged and water-quenched, low-carbon alloy shown in Fig. 25(e) exhibits a highly deformed and twinned structure that contains relatively few dispersed carbides, which are less than 1 µm (0.04 mils) in size.

Modified cobalt-chromium alloys containing higher nickel contents are also used for dental implants. One wrought alloy included in this category is MP35N (35Co-35Ni-20Cr-10Mo). Microstructurally, the alloy takes the character of cobalt-chromium alloys. However, no carbides are formed because carbon has not been added to the alloy.

Porous Surfaces. Alloy powders of the same composition as the implants have been sintered onto the surfaces of the implants for generating bone ingrowth to obtain better retention between the implant and the bone.

Tarnish and Corrosion under Simulated or Accelerated Conditions

Low-Copper Amalgams. The Sn_8Hg (γ_2) phase shown in Fig. 20 is electrochemically the most active phase in conventional amalgam. Upon exposure to an electrolyte, the tin oxide (SnO)/tin couple becomes operative. The formed SnO may or may not protect the γ_2 phase from further corrosion. Depending on environmental conditions, the tin from the γ_2 phase will either be protected by a film of SnO or consumed by additional corrosion reactions.

In the case of an artificial saliva, the γ_2-tin becomes protected. This is shown in Fig. 26 on the anodic polarization curve as the current peak at about -0.7 V versus SCE, which relates in potential to the SnO/Sn couple. With increasing potential, the film protects the amalgam, as shown by the presence of a limiting or passivating current. The film will remain passivating until the potential of another redox reaction is reached that is controlling and nonpassivating. For instance, in a Cl^- solution, this is the situation that occurs when the corrosion potential for the amalgam approaches and surpasses the redox potential for a reaction that produces $SnOCl \cdot H_2O$.

If this condition is satisfied, the SnO passivating film is no longer thermodynamically favorable; therefore, it begins to break down and dissociate, exposing freely corroding γ_2-tin to the electrolyte. Likewise, nonprotective products of the form $SnOHCl \cdot H_2O$ precipitate. This is represented on the polarization curve in Fig. 26 as the sharp increase in current at about -0.1 V versus SCE. As a result of the interconnection of the γ_2 phase, the interior γ_2 can also become corroded. Figure 27 shows the devastating effect that corrosion of the γ_2 phase has on the microstructure of a conventional amalgam, while Fig. 28 shows typical tin-containing products that precipitate on the surface.

High-Copper Amalgams. Corrosion of high-copper amalgams by γ_2-phase corrosion will not occur because of its almost complete absence from the microstructure. Although possessing better corrosion resistance than the γ_2-phase, the Cu_6Sn_5 (η') phase will be the least-resistant phase in the microstructure (Fig. 21, 22). Upon exposure to solution, any corrodible tin within the material first forms a protective SnO film indicated on the anodic polarization curve in Fig. 26 as the small current peak at about -0.7 V. Upon attaining a steady-state corrosion potential in chloride solution, high-copper amalgam is likely to surpass redox potentials for couples of $CuCl_2/Cu$, $Cu(OH)_2$, Cu_2O/Cu, and $CuCl_2 \cdot 3Cu(OH)_2/Cu$. Under these conditions, both soluble and insoluble corrosion products will form. This is indicated on the polarization curve as a small anodic current peak at about -0.25 V.

Microstructurally, if the copper from the η' phase becomes exhausted by corrosion, copper corrosion from the silver-copper and γ particles may also follow. Freed by copper corrosion, tin also becomes corroded. Corrosion of the γ_1-tin decreases the stability of the η_1 phase, which is likely to be transformed into the β_1 phase. Unlike low-copper amalgam, the interior of high-copper amalgam, which demonstrates a lack of interconnection between any of the phases, is not likely to become affected by corrosion. Figure 29 shows a corroded high-copper dispersed-phase amalgam, emphasizing the reaction zones of the η' phase, the interior of the silver-copper particles, the γ particles, and the matrix.

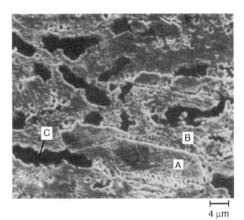

Fig. 27 SEM micrograph of corroded (10 µA/cm²) low-copper amalgam (New True Dentalloy) in 0.2% NaCl after removal of corrosion products by ultrasonics. A, alloy particles; B, matrix; C, regions formerly occupied by γ_2 phase

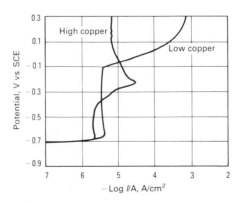

Fig. 26 Anodic polarization at 0.03 V/min of both low (Microalloy) and high (Sybraloy) copper amalgams in artificial saliva. Source: Ref 74

Fig. 28 SEM micrograph of low-copper amalgam (New True Dentalloy) after immersion in artificial saliva. The clumps of corrosion products contain tin.

Tarnish of Gold Alloys. Because tarnish is by definition the surface discoloration of a metallic material by the formation of a thin film or corrosion product, the quantification of dental alloy tarnish by assessing color changes on surfaces is most appropriate. By determining the color of an alloy before and after exposure to a test solution, the degree of discoloration can be obtained by quantitative colorimetry techniques, which are described in Ref 161 to 167. The use of quantitative colorimetry in conjunction with SIMS for determining the effects of alloy nobility (in at.%) and sulfide concentration on color changes in gold crown and bridge alloys is detailed in Ref 168.

Effect of Microstructure on Tarnishing Behavior. The devastating effect of microgalvanic coupling on tarnishing behavior is shown by a comparison of solid-solution annealed (750 °C, or 1380 °F) and as-cast structures (Fig. 30). The as-cast alloys, composed of two-phase structures, consistently showed greater color change or tarnish (Ref 164).

Single-phase, as-cast gold-silver-copper alloys with gold contents between 50 and 84 wt% were observed microstructurally to tarnish in Na_2S solutions by localized microgalvanic cells. The characteristics of the various tarnished surfaces included a uniformly speckled appearance, dendritic attack, matrix attack, grain-boundary dependent attack, and grain-boundary attack. Silver-rich areas discolored preferentially because of the operation of the silver-rich areas as anodes and the surrounding copper-rich areas as cathodes. The uniformly speckled appearance occurred with high-silver, low-copper contents, while the grain orientation dependent appearance occurred with low-silver, high-copper contents. The dendritic and matrix attack occurred with alloys containing intermediate silver and copper contents (Ref 103).

For gold-silver-copper-palladium alloys with gold contents between 35 and 73 wt%, the tarnishability in oxygenated 2% Na_2S has been shown to be affected by altering the microstructure through heat treatment. Tarnishing occurred on multiphase structures annealed at 500 °C (930 °F) but did not occur on single-phase structures annealed at 700 °C (1290 °F). Silver-, copper-, and palladium-rich phases were precipitated. Some alloys, though, showed only silver- and copper-rich phases. In these cases, the palladium tended to follow the copper-rich phase. Splitting of the matrix into thin lamellae of alternating silver and copper enrichments also occurred. The silver-rich phases in all materials were attacked by the sulfide and were responsible for the tarnish. Age hardening by AuCu(I)-ordered precipitates increased the tendency of the silver-rich lamellae to tarnish (Ref 169).

In sulfide solutions, silver sulfide (Ag_2S) is the principle product of tarnish, although copper sulfides (Cu_2S and CuS) also form. These products are produced by the operation of microgalvanic cells set up between silver-rich and copper-rich lamellae. The addition of palladium to gold-silver-copper alloys considerably reduces the rate of tarnishing by slowing down the formation of a layer of silver and copper sulfides on the surface. This has been shown to be due to the enrichment of palladium and gold on the surface of the alloy when exposed to the atmosphere prior to sulfide exposure (Ref 170). The rate of diffusion from the bulk to the surface is hindered by the palladium enrichment. The active sites on the alloy surface for the sulfidation reaction are selectively blocked by the palladium atoms (Ref 170).

Effect of the Silver/Copper Ratio. The silver/copper ratio is an important aspect affecting the tarnish and corrosion resistance of gold dental alloys. A comparison was made of three gold alloys, with similar noble metal contents, but different Ag/Cu ratios (based on wt%) of 41/7.4, 10.3/37.9, and 9.8/37.7 (Ref 169). Polarization tests of the three alloys in a sulfide solution showed that the alloy with a Ag/Cu ratio of 41/7.4 exhibited increases in current density up to 10 $\mu A/cm^2$ at -0.3 V, while the other two alloys exhibited current densities of ~1 $\mu A/cm^2$ extending to positive potentials. Therefore, high silver contents relative to low copper contents in low-gold alloys can have detrimental effects on tarnishing and corrosion. For some low-gold alloys, the best resistance to tarnishing has been obtained by using Ag/Co ratios between 1.2 and 1.4 and a palladium content of 9 wt%.

Effect of the Palladium/Gold Ratio. Increasing the palladium content in gold alloys increases the tarnish resistance. However, in gold-silver-copper-palladium alloys, this effect is greater. The palladium/gold ratio is just as important as the silver/copper ratio. In gold-silver-copper alloys without palladium, the degree of tarnish (subjective test: 0 = least and 8 = most) was evaluated to be between 6.5 and 8 for all silver/copper ratios (1:3, 1:2, 2:3, 1:1, 3:2, 2:1, 3:1) (Ref 108). However, in alloys having palladium/gold ratios of 1:12, the degree of tarnish diminished to between 2 and 3.

Tarnishing and Corrosion Compared. Figure 31 shows reflection loss versus nobility and mass loss versus nobility for 15 gold alloys. Tarnishing was by immersion for 3 days in 0.1 mol/L Na_2S, while corrosion was by immersion for seven days in aerated 0.1 mol/L lactic acid plus 0.1 mol/L NaCl at 37 °C. As is evident, a number of alloys that appear not to have been affected by corrosion are, however, largely affected by tarnishing (Ref 171).

Corrosion of Gold Alloys. Electrochemical polarization has been applied to the corrosion evaluation of gold dental alloys. Very small current peaks on the anodic polarization curves for some gold alloys in artificial saliva have been detected and interpreted to be due to the dissolution of alloying components (Ref 172).

A comparison of the anodic polarization of noble alloys in artificial saliva with and without sulfide indicated that without sulfide the electrochemistry is governed mainly by chloride ions. The alloys passivate in a state with very low current densities, which makes detection of differences among the alloys difficult. With sulfide added to the artificial saliva, a preferential sulfidation of the less noble alloy component is induced. The sulfidation is characterized by a critical potential and limiting current density, both of which may be dependent on composition (Ref 173).

The corrosion susceptibilities for silver and copper in various gold alloys were quantified by an analysis of both forward and reverse polarization scans (Ref 174). Both silver and copper demonstrated characteristic current peaks. The heights of the current peaks were taken to be a measure of the amount of corrodible silver and copper in the alloys. In a similar technique, the

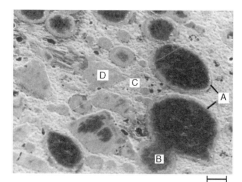

Fig. 29 SEM micrograph of corroded (5 $\mu A/cm^2/d$) high-copper amalgam (cluster) in 0.2% NaCl solution. Note the definition of the γ' rings (A), Ag-Cu particles (B), matrix (C), and η' particles (D).

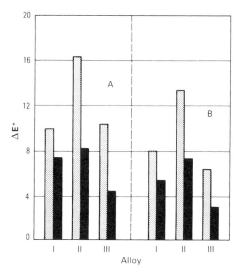

Fig. 30 Color change vector ΔE^* for three low-gold alloys (I, Miracast; II, Sunrise; III, Tiffany) in both the as-cast (left bar for each alloy) and solutionized at 750 °C (1380 °F) (right bar for each alloy) conditions after exposure for 3 days to artificial saliva (A) or 0.5% Na_2S solution (B). Source: Ref 164

integrated current from the polarization curves within a potential range of -0.3 V to $+0.3$ V versus SCE was taken to be a measure for the corrodible species (Ref 175). Figure 32 shows a plot of the nobility of eight as-cast gold alloys versus their integrated anodic currents.

Silver-Palladium Alloys. Silver is prone to tarnishing by sulfur and is prone to corrosion by chloride. The addition of palladium to silver generates alloys with much better resistance to tarnishing and silver corrosion. In 1/7 diluted Ringer's solution and 0.1% NaCl, alloys with more than 40% Pd showed passive anodic polarization behavior (Ref 176). In sulfur-saturated air, the amount of sulfur deposited onto the surfaces of silver-palladium alloys was minimal for compositions with ≥ 40 wt% Pd (Ref 173). In artificial saliva, two transitions in the corrosion currents occurred with palladium content. The first occurred at about 22% Pd, where the current decreased from 6 to 1 μA. The second transition occurred at about 29% Pd, where the current decreased to about 0.4 μA and then remained fairly constant throughout the rest of the compositional range (Ref 177). Figure 33 shows the color range vectors for the pure metals and alloys from the Ag-Pd system after tarnishing in artificial saliva with 0.5% Na_2S. The compositions 50Pd-50Ag and 75Pd-25Ag showed the best tarnish resistance (Ref 178).

Corrosion behavior and tarnishing behavior usually must be viewed independently. That is, corrosion is not an indicator for tarnishing, and vice versa. Alloy nobility dominates corrosion behavior, while alloy nobility, composition, and microstructure (in conjunction with environment) influence tarnishing behavior.

Microstructurally, the silver-palladium alloys tarnish by chlorides and/or sulfides becoming deposited over the silver-rich matrix, while the palladium-rich precipitates display resistance to chlorides and sulfides. Furthermore, microstructurally, the alloys are generally composed of a corrosion-resistant copper- and palladium-rich phase and a nonresistant silver-rich phase. Increased tarnish and corrosion of this silver-rich phase component can occur as a result of microgalvanic coupling (Ref 179). Manipulation of the microstructural features through heat treatments can produce structures with varying proportions of the tarnish-resistant and tarnish-prone phases. For example, age hardening has been shown to increase the proportion of the tarnish- and corrosion-prone phases (Ref 180, 181).

High-Palladium PFM Alloys. Alloys with up to 80 wt% Pd and additions of copper, gallium, tin, indium, gold, and others have been shown to exhibit good saline corrosion resistance in the potential range and Cl^- ion concentration associated with oral use. For example, anodic polarization tests showed passive behavior until breakdown occurred, which was at potential magnitudes well above those occurring intraorally (Ref 182).

Nickel-Chromium Alloys. The tarnish resistance and corrosion resistance of these alloys result from balancing the composition with regard to the passivating elements chromium, molybdenum, manganese, and silicon. Alloys containing increased amounts of molybdenum and manganese have shown to exhibit increased passivation; however, increasing the chromium content too much (more than 20%) can precipitate an additional phase and alter the corrosion resistance. By using polarization methods, three different behaviors were observed with 12 nickel-chromium alloys with varying compositions in deaerated and aerated artificial saliva (Ref 183). Some alloys were constantly passive, others were either active/passive or passive according to the aeration condition of the electrolyte, and still others (<16% Cr without molybdenum) were constantly active and corroding.

Corrosion potentials for nickel-chromium alloys in artificial saliva were low, ranging between about -0.2 V and -0.8 V versus SCE. Breakdown potentials varied, depending on composition. For alloys with less than 16% Cr and no molybdenum, breakdown potentials as low as -0.2 V occurred. For compositions with higher chromium contents and with molybdenum and various manganese contents, breakdown potentials as high as $+0.6$ V also occurred.

Pitting attack occurs with these alloys because they rely on protective surface oxide films for

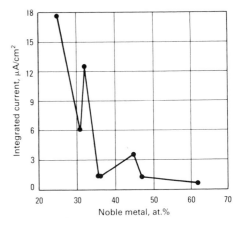

Fig. 32 Integrated anodic currents between -0.3 V vs. saturated calomel electrode (SCE) and $+0.3$ V (at 0.06 V/min) for eight gold alloys in deaerated 1% NaCl plotted against the atomic nobility. Source: Ref 164

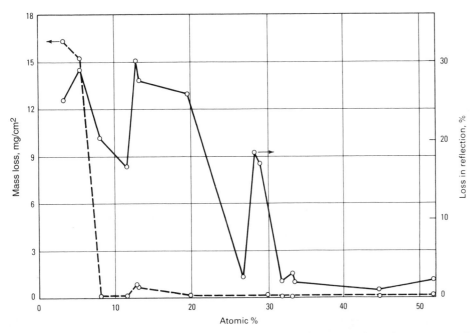

Fig. 31 Effect of nobility (in atomic percent) on tarnish (percent loss in reflection after 3 days in 0.1% Na_2S) and corrosion (mass loss after 7 days in aerated 0.1 mol/L lactic acid plus 0.1 mol/L NaCl at 37 °C, or 99 °F) of 15 gold alloys. Source: Ref 171

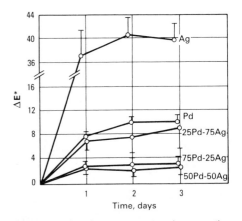

Fig. 33 Color change vector ΔE^* for pure silver, palladium, and three Ag-Pd binary compositions after exposure to Na_2S solutions. Source: Ref 178

imparting protection. From electrochemical and immersion tests, the resistance of nickel-chromium alloys to pitting attack was found to be good in solutions with Cl^- concentrations equivalent to that found in saliva. Only at higher Cl^- concentrations are pitting and tarnishing likely to occur (Ref 184).

Cobalt-Chromium Alloys. Compared with the nickel-chromium alloys, the cobalt-chromium alloys for partial denture and implant prosthesis exhibit superior tarnish and corrosion resistance. Figure 34 shows the polarization curve for the cast cobalt-chromium-molybdenum alloy shown in Fig. 25(a). The alloy was tested at a scan rate of approximately 1.5 mV/s (1.8 V/h) in aerated physiologic phosphate buffered saline (PBS) that was heated and held at a temperature of 37 ± 1 °C (99 ± 2 °F) with a pH of 7.4 ± 0.2 (Ref 142, 185). From this curve, the breakdown potential (E_b), protection potential (E_p), zero current potential (E_{zcp}), and passive current density (J_p) can be determined. Table 7 lists the mean values of these parameters for the five different types of cobalt-chromium-molybdenum alloys shown in Fig. 25. A one-way analysis of variance test revealed no statistical differences between the groups for any of the parameters listed in Table 7 at the p (probability of occurence) <0.05 level.

In Fig. 34, the potential at which the up and down scans intersect is the protection potential, E_p. Below this potential, existing pits will not grow (Ref 186). The magnitude of the hysteresis between the up and down scans is $E_b - E_p$, which is considered to be a measure of the degree of pitting (Ref 134, 186, 187). Thus, $E_b - E_p$ is used to characterize the susceptibility of an alloy to pitting or crevice corrosion. Consequently, from Table 7, it can be seen that the different cobalt-chromium-molybdenum alloys did not show a significant difference in their susceptibility to pitting or crevice corrosion.

Table 7 also lists the results of impedance spectroscopy tests performed on the five different types of cobalt-chromium-molybdenum alloys shown in Fig. 25 (Ref 142, 185). The test solution and conditions were the same as for the polarization tests described previously. Impedance spectrums were collected every 50 mV, starting at -1000 mV, up to $+700$ mV, and back down to -1000 mV versus Ag/AgCl, for a total of 68 impedance spectrums. The average maximum early resistance (R_e) and polarization resistance (R_p) values for each alloy group are displayed in Table 7, along with the average minimum capacitance (C) values (note that analysis of the impedance values indicated that the R_p values were controlled by the rate of the corrosion process, as opposed to being controlled by charge transfer resistance, and the R_e values were a combination of the solution resistance and the oxide film resistance). One-way analysis of variance tests were performed on the impedance values in Table 7 and the results did not indicate any significant differences between the alloy groups for any of the parameters at the $p < 0.05$ level. Thus, the similar passive electrochemical behavior of the five different cobalt-chromium-molybdenum alloy groups, as shown by the results of the impedance and polarization tests in Table 7, suggests that the oxide films covering them were not significantly altered by changes in carbon content and processing.

However, studies have shown that carbide content and grain size are both factors that may affect the corrosion resistance of cobalt-chromium-molybdenum alloys. For instance, Placko et al. investigated cobalt-chromium-molybdenum alloy porous coatings with four different microstructures (Ref 188). They observed that, for accelerated anodic corrosion experiments, an increase in carbide content correlated with an increase in severity of the preferential attack of the areas surrounding carbides, which was most likely attributable to the phenomenon of sensitization (a depletion of the chromium in the material matrix surrounding carbides). Furthermore, it was also noted that as grain size decreased for the microstructures, localized attack of the grain boundaries increased.

On the other hand, research by Devine and Wulff found that the crevice corrosion resistance for cast cobalt-chromium-molybdenum alloy was not as great as for the wrought material (Ref 189). They ascribed this finding to the greater chemical homogeneity of the wrought cobalt-chromium-molybdenum alloy, as determined by electron microprobe measurements. Also, optical micrographs comparing the cast with the wrought material showed the wrought material to possess a finer grain size and a more uniform and finer distribution of carbides.

In spite of the excellent corrosion resistance of cobalt-chromium alloys, allergic reactions to cobalt, chromium, and nickel contained in appliances made from these alloys are known to have occurred (see the section "Allergic Hypersensitive Reactions" in this article).

Titanium Alloys. The anodic polarization for pure titanium and its alloys indicates passivities over at least several volts in overvoltage (Ref 77, 190). This demonstrates the tenacity and protectiveness of the titanium oxide films formed on these materials.

With regard to the casting titanium alloys for crown and bridgework and partial dentures, good corrosion resistance is still preserved (Ref 191). However, a point of concern is the higher percentages of alloying elements in the titanium casting alloys, with the potential for elucidating diminished chemical stabilities.

Wrought Orthodontic Wires. A comparison of their anodic polarization curves (Fig. 35) shows that both β-titanium and cobalt-chromium-nickel (Elgiloy) exhibit resistance to corrosion in artificial saliva, within the

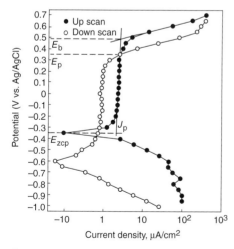

Fig. 34 Polarization curve for cast cobalt-chromium-molybdenum alloy shown in Fig. 25(a). The alloy was tested at a scan rate of about 1.5 mV/s (1.8 V/h) in aerated physiologic phosphate buffered saline (PBS) that was heated and held at a temperature of 37 ± 1 °C (99 ± 2 °F) with a pH of 7.4 ± 0.2. The sample was scanned from -1000 to $+700$ mV vs. Ag/AgCl and back down to -1000 mV vs. Ag/AgCl. Source: Ref 142

Table 7 Polarization and impedance data for five different cobalt-chromium-molybdenum alloys shown in Fig. 25 and described in the section "Cost, Wrought, and Forged Cobalt-Chromium Alloys"

Property	Alloy				
	Cast	WHC	WHCA	FHC	FLC
Polarization data(a)(b)					
E_b, mV	497 (2.9)	505 (8.7)	502 (7.6)	497 (6.4)	507 (6.4)
E_p, mV	354 (10.8)	356 (12.8)	359 (16.2)	355 (10.8)	346 (18.3)
$E_b - E_p$, mV	143 (9.0)	149 (5.3)	142 (22.5)	142 (13.8)	161 (21.8)
E_{zcp}, mV	-366 (29.3)	-341 (12.5)	-313 (32.7)	-357 (80.4)	-400 (51.5)
J_p, µA/cm^2	2.03 (0.45)	2.13 (0.31)	1.80 (0.10)	2.40 (0.40)	2.07 (0.58)
Impedance data(a)					
Minimum C, µF/cm^2	6.63 (2.50)	3.64 (0.33)	4.15 (1.83)	4.75 (1.71)	2.45 (0.35)
Maximum R_e, Ω·cm^2	231 (21)	273 (39)	243 (31)	272 (30)	324 (55)
Maximum R_p, kΩ·cm^2	6.47 (0.89)	7.24 (0.82)	7.59 (0.94)	7.99 (0.38)	8.27 (0.38)

(a) The data are the mean for $N = 3$ number of samples with the standard deviation in parenthesis. (b) The potentials, E, are relative to the Ag/AgCl reference electrode. Key to abbreviations: WHC, wrought high carbon; WHCA, wrought high carbon + aged; FHC, forged high carbon; FLC, forged high carbon; E_b, breakdown potential; E_p, protection potential; E_{zep}, zero current potential; J_p, passive current density; C, capacitance; R_e, early resistance; R_p, polarization resistance. Source: Ref 142

potential ranges employed (+0.8 V vs. SCE) in the study. With the nickel-titanium and stainless steel wires, breakdown occurred at +0.2 and 0.05 V versus SCE, respectively. The nickel-titanium and stainless steel wires also exhibited current increases upon potential reversals at +0.8 V versus SCE, an indication of their susceptibility to pitting corrosion. Breakdown potentials differed by as much as 0.6 V between different brands of stainless steel wires. Variations in the polarization characteristics of stainless steel have been related to microstructure (Ref 77) and to surface preparation and finish (Ref 78). It has been shown that the resistance to pitting is decreased with increasing cold deformation for stainless steel in a NaCl solution; the pitting potential decreased by over 300 mV as the degree of cold deformation increased from 0 to 50% (Ref 193). Microstructurally, nickel-titanium wires have been observed to suffer pitting attack and selective dissolution of nickel after polarization tests (Ref 193). Furthermore, it has been shown that the dissolution of nickel can take place at sites of surface damage on nickel-titanium wires, which can be a concern to nickel-sensitive patients (Ref 154).

Silver and Gold Solders. A corroded stainless steel-silver soldered joint is shown in Fig. 36. Microstructurally, silver solders are composed of two phases: silver- and copper-zinc rich segregations (Ref 194). The copper-zinc regions are the least resistant to corrosion. These solders corrode by microgalvanic coupling, either by cells set up between the two microstructural phases, or between the solder and the parts they join. Figure 37 shows the copper-zinc phase of a silver solder attacked by corrosion.

Figure 38 shows the polarization curves for both silver and gold (450 fine) solders in 0.16 M NaCl. The silver solder demonstrates active behavior. Zinc, tin, copper, and even silver products can be precipitated or become dissolved in solution. The polarization curve for gold solder indicates activity by the sharp current density peak at +0.25 V versus SCE. Because the solder contains silver, copper, and zinc, in addition to gold, this peak is probably due to the corrosion of one of these elements. Figure 39 shows the gold solder after the polarization test. Corrosion has delineated the basic microstructure of the solder alloy.

Chlorine was detected with the white appearing phase.

Silver-Indium Alloys. These alloys rely on the unusual properties of indium oxide for providing tarnish and corrosion control. Small amounts of noble metals, such as palladium, may also be added in an attempt to improve corrosion resistance. Anodic polarization of a silver-indium alloy in artificial saliva indicated only a very narrow potential range of about 0.1 V of reduced current densities. The tarnish resistance of these alloys appears to be acceptable, but the long-term corrosion resistance has yet to be established (Ref 195).

Copper-Aluminum Alloys. A comparison of the released copper in human saliva from a dental copper-aluminum alloy to that from a high-copper amalgam is shown in Fig. 40. The amalgam released more copper over a 45 day interval. No aluminum, iron, manganese, or nickel was detected. Figures 41(a) to (c) show micrographs from an in vivo restoration of various magnifications. Large amounts of organic matter were adsorbed onto the surface, as well as light powdery corrosion products composed of copper oxides.

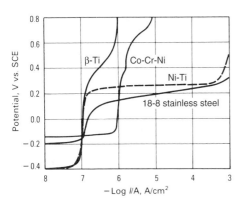

Fig. 35 Anodic polarization at 0.03 V/min of four orthodontic wires in artificial saliva. Source: Ref 192

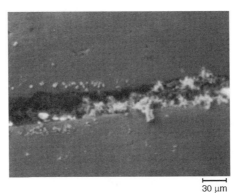

Fig. 36 SEM micrograph of a corroded stainless steel-silver soldered joint after immersion in a 1% H_2O_2 solution. Source: Ref 194

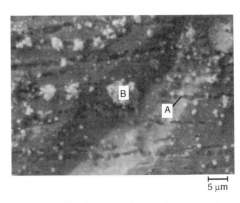

Fig. 37 SEM micrograph of a corroded silver solder in 1% NaCl (held at −0.05 V vs. SCE) showing the destruction of the copper-zinc-rich phase (A) and the accumulation of products (B) that contain copper, zinc, and chlorine. Source: Ref 194

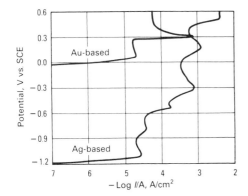

Fig. 38 Anodic polarization at 0.03 V/min of silver and gold (450 fine) solders in 1% NaCl solution. Source: Ref 194

Fig. 39 SEM micrograph of a corroded (polarized to +0.5 V vs. SCE) gold solder (450 solder) in 1% NaCl. The light areas contain chlorine. Source: Ref 194

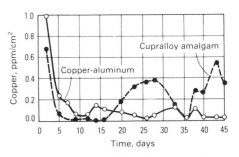

Fig. 40 Released copper into human saliva from a copper-aluminum crown and bridge alloy (MS) and a high-copper amalgam (Cupralloy) plotted against time for up to 45 days. Source: Ref 75

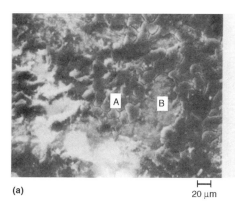 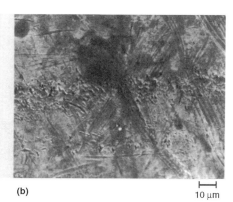 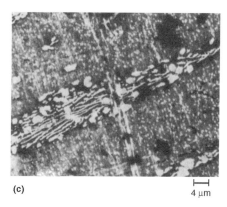

Fig. 41 SEM micrographs of the copper-aluminum restoration shown in Fig. 18 at higher magnifications. In (a), both accumulated plaque (A) and corrosion products (B) occur. (b) Higher-magnification view of the areas identified by (B) contain copper. The light-appearing areas are probably copper oxides. (c) Still higher magnification of the products shown in (b). Here the copper oxides are deposited over the dark copper-rich microstructural phase. Source: Ref 75

ACKNOWLEDGMENT

This article was adapted from Herbert J. Mueller, Tarnish and Corrosion of Dental Alloys, *Corrosion*, Vol 13, *ASM Handbook*, 1987, pages 1336–1366.

REFERENCES

1. *Dental Technician, Prosthetic,* Navpers 10685 c, U.S. Naval Dental School, Bureau of Naval Personnel, 1965, p 256
2. G. Ravasini, Clinical Procedures for Partial Crowns, Inlays, Onlays, and Pontics, *An Atlas,* Quintessence Publishing, Hanover Park, IL 1985, p 136
3. K.L. Stewart, K.D. Rudd, and W.A. Kuebker, *Clinical Removable Partial Dentures,* C.V. Mosby, 1983, p 31, 230, 494
4. T.M. Graber and B.F. Swain, *Orthodontics: Current Principles and Techniques,* C.V. Mosby, 1985, p 385
5. D.C. Smith and D.F. Williams, *Biocompatibility of Dental Materials,* Vol 1–4, CRC Press, 1982
6. *Workshop on Biocompatibility of Metals in Dentistry, Conf. Proc.,* American Dental Association, 1984
7. *An International Workshop: Biocompatibility, Toxicity and Hypersensitivity to Alloy Systems Used in Dentistry, Proc.,* The University of Michigan School of Dentistry, 1985
8. D.C. Smith, The Biocompatibility of Dental Materials, *Biocompatibility of Dental Materials,* Vol I, D.C. Smith and D.F. Williams, Ed., CRC Press, 1982, p 11
9. D.C. Smith, Tissue Reaction to Noble and Base Metal Alloys, *Biocompatibility of Dental Materials,* Vol IV, D.C. Smith and D.F. Williams, Ed., CRC Press, 1982, p 55
10. A. Halse, Metal in Dentinal Tubules beneath Amalgam Fillings in Human Teeth, *Arch. Oral Biol.,* Vol 20, 1975, p 87–88
11. K. Arvidson and R. Wroblewski, Migration of Metallic Ions from Screwposts into Dentin and Surrounding Tissues, *Scand. J. Dent. Res.,* Vol 86, 1978, p 200–203
12. W. Schriever and L.E. Diamond, Electromotive Forces and Electric Currents Caused by Metallic Dental Fillings, *J. Dent. Res.,* Vol 31, 1952, p 205–229
13. I. Kleinberg, Etiology of Dental Caries, *J. Can. Dent. Assn.,* Vol 12, 1979, p 661–668
14. I.D. Mandel, Dental Caries, *Am. Sci.,* Vol 67, 1979, p 680–688
15. M.E. Jensen, Telemetric Methods Using Ion-specific Electrodes, *Adv. Dent. Res.,* Vol. 1, 1987, p 92–98
16. E. Hals and A. Halse, Electron Probe Microanalysis of Secondary Carious Lesions Associated with Silver Amalgam Fillings, *Acta Odontol. Scand.,* Vol 33, 1975, p 149–160
17. N. Kurosahi and T. Fusayama, Penetration of Elements from Amalgam into Dentin, *J. Dent. Res.,* Vol 52, 1973, p 309–317
18. L.W.J. van der Linden and J. van Aken, The Origin of Localized Increased Radiopacity in the Dentin, *Oral Surg.,* Vol 35, 1973, p 862–871
19. B.L. Dahl, Hypersensitivity to Dental Materials, *Biocompatibility of Dental Materials,* Vol I, D.C. Smith and D.F. Williams, Ed., CRC Press, 1982, p 177–185
20. E.W. Mitchell, Summary and Recommendations to the Workshop, *Workshop on Biocompatibility of Metals in Dentistry, Conf. Proc.,* American Dental Association, 1984
21. J. Vilaplana, C. Romaguera, and F. Grimalt, Occupational and Non-Occupational Allergic Contact Dermatitis from Beryllium, *Contact Dermatitis,* Vol 26 (No. 5), 1992, p 295–298
22. A.L. Haberman, M. Pratt, and F.J. Storrs, Contact Dermatitis from Beryllium in Dental Alloys, *Contact Dermatitis,* Vol 23 (No. 3), 1993, p 157–162
23. Preventing Adverse Health Effects from Exposure to Beryllium in Dental Laboratories OSHA Hazard Information Bulletin, HIB 02-04-19 (rev. 5-14-02), Available at www.osha.gov/dts/hib/hib_data/hib20020419.pdf. U.S. Department of Labor, Occupational Safety and Health Administration, Washington, DC, 2002
24. Proper Use of Beryllium-Containing Alloys *JADA,* Vol 134, April 2003, p 476–478
25. "Resolution 6 on Beryllium in Dental Alloys," Brussels, Belgium, CEN/TC 055 Dentistry, Feb 26, 2002
26. M. Bergan and O. Ginstrup, Dissolution Rate of Cadmium from Dental Gold Solder Alloys, *Acta Odontol Scand.,* Vol 33, 1975, p 199–210
27. D.R. Zielke, J.M. Brady, and C.E. del Rio, Corrosion of Silver Cones in Bone: A Scanning Electron Microscope and Microprobe Analysis, *J. Endo.,* Vol 1, 1975, p 356–360
28. R.A. James, Host Response to Dental Implant Devices, *Biocompatibility of Dental Materials,* Vol IV, D.C. Smith and D.F. Williams, Ed., CRC Press, 1982, p 163–195
29. J.R. Natiella, Local Tissue Reaction/Carcinogenesis, *International Workshop on Bicompatibility, Toxicity, and Hypersensitivity to Alloy Systems Used in Dentistry, Section 6, Conference Proceedings,* University of Michigan School of Dentistry, 1985
30. R.S. Mateer and C.D. Reitz, Corrosion of Amalgam Restorations, *J. Dent. Res.,* Vol 49, 1970, p 399–407
31. M. Berge, N.R. Gjerdet, and E.S. Erichsen, Corrosion of Silver Soldered Orthodontic Wires, *Acta Odontol. Scand.,* Vol 40, 1982, p 75–79
32. B. Angmar-Mansson, K.-A. Omnell, and J. Rud, Root Fractures due to Corrosion, *Odontol. Revy.,* Vol 20, 1969, p 245–256

33. D.C. Mears, Metals in Medicine and Surgery, *Int. Met. Rev.*, Vol 22 (No. 218), 1977, p 119–155
34. P.L. Altman, *Blood and Other Body Fluids, Analysis and Compilation*, Washington, DC, Federation of American Societies for Experimental Biology, 1961
35. J.R. Cahoon and L.D. Hill, Evaluation of a Precipitation Hardened Wrought Cobalt-Nickel-Chromium-Titanium Alloy for Surgical Implants, *J. Biomed. Mater. Res.*, Vol 12, 1978, p 805–821
36. T.C. Ruck and J.F. Fulton, *Medical Physiology and Biophysics*, 18th ed., W.B. Saunders, 1960
37. C. Dawes, The Effects of Flow Rate and Duration of Stimulation of the Concentrations of Protein and the Main Electrolytes in Human Parotid Saliva, *Arch. Oral Biol.*, Vol 14, 1969, p 277–294
38. C.M. Carey, M. Spencer, R.J. Gove, and F.C. Eichmiller, Fluoride Release from a Resin-Modified Glass-Ionomer Cement in a Continuous-Flow System: Effect of pH, *J. Dent. Res.*, Vol 82, 2003, p 829–832
39. H.C. McCann, Inorganic Components of Salivary Secretions, *Art and Science of Dental Caries Research*, R.S. Harris, Ed., Academic Press, 1968, p 55–73
40. H.J. Mueller, Binding of Corroded Ions to Human Saliva, *Biomaterials*, Vol 6, 1985, p 146–149
41. H.J. Mueller, Characterization of the Acquired Biofilms on Materials Exposed to Human Saliva, *Proteins at Interfaces*, T.A. Horbett and J. Brash, Ed., *Advances in Chemistry Series*, American Chemical Society, 1987
42. S.A. Ellison, The Identification of Salivary Components, Saliva and Dental Caries, I. Kleinberg, S.A. Ellison, and I.D. Mandell, Ed., *Sp. Supp. Microbiol, Abst.*, 1979, p 13–29
43. C.E. Ingersoll, Characterization of Tarnish, *J. Dent. Res.*, Vol 55, 1976, IADR No. 144
44. K. Barton, Chapter 2, in *Protection against Atmospheric Corrosion*, John Wiley & Sons, 1973
45. I.C. Schoonover and W. Souder, Corrosion of Dental Alloys, *J. Am. Dent. Assoc.*, Vol 28, 1941, p 1278–1291
46. J.C. Muhler and H.M. Swenson, Preparation of Synthetic Saliva from Direct Analysis of Human Saliva, *J. Dent. Res.*, Vol 26, 1947, p 474
47. D.A. Carter, T.K. Ross, and D.C. Smith, Some Corrosion Studies on Silver-Tin Amalgams, *Br. Corros. J.*, Vol 2, 1967, p 199–205
48. G. Tani and F. Zucci, Electrochemical Evaluation of the Corrosion Resistance of the Commonly Used Metals in Dental Prosthesis, *Minerva. Stomat.*, Vol 16, 1967, p 710–713
49. J.M. Meyer and J.N. Nally, Influence of Artificial Salivas on the Corrosion of Dental Alloys, *J. Dent. Res.*, Vol 54, 1975, IADR No. 76
50. B.W. Darvell, The Development of an Artificial Saliva for In-Vitro Amalgam Corrosion Studies, *J. Oral Rehab.*, Vol 5, 1978, p 41–49
51. M.L. Swartz, R.W. Phillips, and M.D. El Tannir, Tarnish of Certain Dental Alloys, *J. Dent. Res.*, Vol 37, 1958, p 837–847
52. F. Fusayama, T. Katayori, and S. Nomoto, Corrosion of Gold and Amalgam Placed in Contact with Each Other, *J. Dent. Res.*, Vol 42, 1963, p 1183–1197
53. C.E. Guthrow, L.B. Johnson, and K.R. Lawless, Corrosion of Dental Amalgam and Its Component Phases, *J. Dent. Res.*, Vol 46, 1967, p 1372–1381
54. F.V. Wald and F.H. Cocks, Investigation of Copper-Manganesse-Nickel Alloys for Dental Purposes, *J. Dent. Res.*, Vol 50, 1971, p 44–59
55. M. Marek and R.F. Hockman, Corrosion Behavior of Amalgam Electrode in Artificial Saliva, *J. Dent. Res.*, Vol 51, 1972, IADR No. 63
56. J. Brugirard, R. Bargain, J.C. Dupuy, H. Mazille, and G. Monnier, Study of the Electrochemical Behavior of Gold Dental Alloys, *J. Dent. Res.*, Vol 52, 1973, p 828–836
57. M. Marek and E. Topfl, Electrolytes for Corrosion Testing of Dental Alloys, *J. Dent. Res.*, Vol 65, 1986, IADR No. 1192
58. N.K. Sarkar and E.H. Greener, In Vitro Corrosion of Dental Amalgam, *J. Dent. Res.*, Vol 50, 1971, IADR No.13
59. G.F. Finkelstein and E.H. Greener, In Vitro Polarization of Dental Amalgam in Human Saliva, *J. Oral. Rehab.*, Vol 4, p 355–368
60. G.F. Finkelstein and E.H. Greener, Role of Mucin and Albumin in Saline Polarization of Dental Amalgam, *J. Oral Rehab.*, Vol 5, 1978, p 95–110
61. H. Do Duc, P. Tissot, and J.-M. Meyer, Potential Sweep and Intensiostatic Pulse Studies of Sn, Sn_8Hg, and Dental Amalgam in Chloride Solution, *J. Oral Rehab.*, Vol 6, 1979, p 189–197
62. R.I. Holland, Effect of Pellicle on Galvanic Corrosion of Amalgam, *Scand. J. Dent. Res.*, Vol 92, 1984, p 93–96
63. H.J. Mueller, The Effects of a Human Salivary Dialysate upon Ionic Release and Electrochemical Corrosion of a Cu-Al Alloy, *J. Electrochem. Soc.*, Vol 134, 187, p 555–580
64. A. Schulman, H.A.B. Linke, and T.K. Vaidyanathan, Tarnish of Dental Alloys by Oral Microorganisms, *J. Dent. Res.*, Vol 63, 1984, IADR No. 55
65. M. Stern and E.D. Weisert, Experimental Observation on the Relation between Polarization Resistance and Corrosion Rate, *Proc. ASTM*, Vol 59, 1959, p 1280–1291
66. M. Marek, The Corrosion of Dental Materials, *Corrosion: Aqueous Processes and Passive Films*, Vol 23, *Treatise on Materials Science*, J.C. Scully, Ed., Academic Press, 1983, p 331–394
67. M. Bergman, O. Ginstrup, and B. Nilsson, Potentials of and Currents between Dental Metallic Restorations, *Scand. J. Dent. Res.*, Vol 90, 1982, p 404–408
68. K. Nilner, P.-O. Glantz, B. Zoger, On Intraoral Potential and Polarization Measurements of Metallic Restorations, *Acta Odontol Scand.*, Vol 40, 1982, p 275–281
69. J.M. Mumford, Electrolytic Action in the Mouth and Its Relationship to Pain, *J. Dent. Res.*, Vol 36, 1957, p 632–640
70. C.P. Wang Chen and E.H. Greener, A Galvanic Study of Different Amalgams, *J. Oral Rehab.*, Vol 4, 1977, p 23–27
71. R. Soremark, G. Freedman, J. Goldin, and L. Gettleman, Structure and Microdistribution of Gold Alloys, *J. Dent. Res.*, Vol 45, 1966, p 1723–1735
72. D.B. Boyer, K. Chan, and C.W. Svare, The Effect of Finishing on the Anodic Polarization of High-Copper Amalgams, *J. Oral Rehab.*, Vol 5, 1978, p 223–228
73. G. Palaghias, Oral Corrosion Inhibition Processes, *Swed. Dent. J.*, Supp 30, 1985
74. H.J. Mueller and A. Edahl, The Effect of Exposure Conditions upon the Release of Soluble Copper and Tin from Dental Amalgams, *Biomaterials*, Vol 5, 1984, p 194–200
75. H.J. Mueller and R.M. Barrie, Intraoral Corrosion of Copper-Aluminum Alloys, *J. Dent. Res.*, Vol 64, 1985, IADR No. 1753
76. G.N. Jenkins, *The Physiology and Biochemistry of the Mouth*, 4th ed., Blackwell, 1978, p 284–359
77. H.J. Mueller and E.H. Greener, Polarization Resistance of Surgical Materials in Ringer's Solution, *J. Biomed. Mater. Res.*, Vol 4, 1970, p 29–41
78. E.J. Sutow, S.R. Pollack, and E. Korostoff, An In Vitro Investigation of the Anodic Polarization and Capacitance Behavior of 316-L Stainless Steel, *J. Biomed. Mater. Res.*, Vol 10, 1976, p 671–693
79. H.J. Mueller, The Binding of Corroded Metallic Ions to Salivary-Type Proteins, *Biomaterials*, Vol 4, 1983, p 66–72
80. J.R. Strub, C. Eyer, and N.K. Sarkar, Microstructure and Corrosion of a Low-Gold Casting Alloy, *J. Dent. Res.*, Vol 63, 1984, IADR No. 793
81. C.W. Svare, G. Belton, and E. Korostoff, The Role of Organics in Metallic Passivation, *J. Biomed. Mater. Res.*, Vol 4, 1970, p 457–467
82. G.C.F. Clark and D.F. Williams, The Effects of Proteins on Metallic Passivation, *J. Biomed. Mater. Res.*, Vol 16, 1982, p 125–134
83. S.A. Brown and K. Merritt, Electrochemical Corrosion in Saline and Serum,

J. Biomed. Mater. Res., Vol 14, 1980, p 173–175
84. H.J. Mueller, The Effect of Electrical Signals upon the Adsorption of Plasma Proteins to a High Cu Alloy, *Biomaterials: Interfacial Phenomena and Applications,* S.L. Cooper and N.A. Peppas, Ed., ACS monograph series 199, American Chemical Society, 1982
85. R.C. Salvarezza, M.E.L. de Mele, H.H. Videla, and F.R. Goni, Electrochemical Behavior of Aluminum in Human Plasma, *J. Biomed. Mater. Res.,* Vol 19, 1985, p 1073–1084
86. H. Hero and L. Niemi, Tarnishing In Vivo of Ag-Pd-Cu-Zn, *J. Dent. Res.,* Vol 65, 1986, p 1303–1307
87. H. Do Duc and P. Tissot, Rotating Disc and Ring Disc Electrode Studies of Tin in Neutral Phosphate Solution, *Corros. Sci.,* Vol 19, 1979, p 191–197
88. I.D. Mandel, Relation of Saliva and Plaque to Caries, *J. Dent. Res.,* Vol 53, 1974, p 246
89. T. Ericson, K.M. Pruitt, H. Arwin, and I. Lunstrom, Ellipsometric Studies of Film Formation on Tooth Enamel and Hydrophilic Silicon Surfaces, *Acta Odontol. Scand.,* Vol 40, 1982, p 197–201
90. R.E. Baier and P.-O. Glantz, Characterization of Oral In Vivo Films Formed on Different Types of Solid Surfaces, *Acta Odontol. Scand.,* Vol 36, 1978, p 289–301
91. K. Skjorland, Auger Analysis of Integuments Formed on Different Dental Filling Materials In Vivo, *Acta Odontol. Scand.,* Vol 40, 1982, p 129–134
92. C.M. Carey, G.L. Vogel, and L.C. Chow, Permselectivity of Sound and Carious Human Dental Enamel as Measured by Membrane Potential, *J. Dent. Res.,* Vol 70, 1991, p 1479–1485
93. G.L. Vogel, Y. Mao, C.M. Carey, and L.C. Chow, Changes in Permselectivity of Human Teeth during Caries Attack, *J. Dent. Res.,* Vol 76, 1997, p 673–681
94. K. Hannesson Eggen and G. Rolla, Gel Filtration, Ion Exchange Chromatography and Chemical Analysis of Macromolecules Present in Acquired Enamel Pellicle (2-hr), *Scand. J. Dent. Res.,* Vol 90, 1982, p 182–188
95. A. Bennick, G. Chau, R. Goodlin, S. Abrams, D. Tustian, and G. Mandapallimatam, The Role of Human Salivary Acidic Proline-Rich Proteins in the Formation of Acquired Dental Pellicle In Vivo and Their Fate after Adsorption to the Human Enamel Surface, *Arch. Oral Biol.,* Vol 28, 1983, p 19–27
96. T. Sonju and P.-O. Glantz, Chemical Composition of Salivary Integuments Formed In Vitro on Solids with Some Established Surface Characteristics, *Arch. Oral Biol.,* Vol 20, 1975, p 687–691
97. D.I. Hay, The Adsorption of Salivary Proteins by Hydroxyapatite and Enamel, *Arch. Oral Biol.,* Vol 12, 1967, p 937–946
98. G. Rolla, Formation of Dental Integuments—Basic Chemical Considerations, *Swed. Dent. J.,* Vol 1, 1977, p 241–251
99. A.C. Juriaanse, M. Booij, J. Arends, and J.J. Ten Bosch, The Adsorption In Vivo of Purified Salivary Proteins on Bovine Dental Enamel, *Arch. Oral Biol.,* Vol 26, 1981, p 91–96
100. H.J. Mueller, Differential Scanning Calorimetry of Adsorbed Protein Films, *Trans. of the 13th Annual Meeting Society of the Biomaterials,* 1987
101. R.D. Norman, R.V. Mehra, and M.L. Schwartz, The Effects of Restorative Materials on Plaque Composition, *J. Dent. Res.,* Vol 50, 1971, IADR No. 162
102. J.J. Tuccillo and J.P. Nielsen, Microprobe Analysis of an In Vivo Discoloration, *J. Prosthet. Dent.,* Vol 31, 1974, p 285–289
103. J.J. Tuccillo and J.P. Nielsen, Observation of Onset of Sulfide Tarnish on Gold-Base Alloys, *J. Prosthet. Dent.,* Vol 25, 1971, p 629–637
104. R.P. Lubovich, R.E. Kovarik, and D.L. Kinser, A Quantitative and Subjective Characterization of Tarnishing in Low-Gold Alloys, *J. Prosthet. Dent.,* Vol 42, 1979, p 534–538
105. G.W. Marshall, N.K. Sarkar, and E.H. Greener, Detection of Oxygen in Corrosion Products of Dental Amalgam, *J. Dent. Res.,* Vol 54, 1975, p 904
106. H. Otani, W.A. Jesser, and H.G.F. Wilsdorf, The In Vivo and the In Vitro Corrosion Products of Dental Amalgam, *J. Biomed. Mater. Res.,* Vol 7, 1973, p 523–539
107. A.B. Burse, M.L. Swartz, R.W. Phillips, and R.W. Oykema, Comparison of the In Vivo and In Vitro Tarnish of Three Gold Alloys, *J. Biomed. Mater. Res.,* Vol 6, 1972, p 267–277
108. B.R. Laing, S.H. Bernier, Z. Giday, and K. Asgar, Tarnish and Corrosion of Noble Metal Alloys, *J. Prosthet. Dent.,* Vol 48, 1982, p 245–252
109. H. Hero and J. Valderhaug, Tarnishing In Vivo and In Vitro of a Low-Gold Alloy Related to Its Structure, *J. Dent. Res.,* Vol 64, 1985, p 139–143
110. H. Hero and R.B. Jorgensen, Tarnishing of a Low-Gold Alloy in Different Structural States, *J. Dent. Res.,* Vol 62, p 371–376
111. American National Standards Institute/American Dental Association "Specification No. 1 Alloy for Dental Amalgam," American Dental Association, Chicago, IL, 2003
112. D.B. Mahler and J.D. Adey, Microprobe Analysis of Three High Copper Amalgams, *J. Dent. Res.,* Vol 63, 1984, p 921–925
113. J.W. Edie, D.B. Boyer, and K.C. Chjan, Estimation of the Phase Distribution in Dental Amalgams with Electron Microprobe, *J. Dent. Res.,* Vol 57, 1978, p 277–282
114. J. Leitao, Surface Roughness and Porosity of Dental Amalgam, *Acta Odontol. Scand.,* Vol 40, 1982, p 9–16
115. R.W. Bryant, Gamma-2 Phase in Conventional Amalgam-Discrete Clumps or Continuous Network—A Review, *Aust. Dent. J.,* Vol 29, 1984, p 163–167
116. L.B. Johnson, X-Ray Diffraction Evidence for the Presence of β(Ag-Hg) in Dental Amalgam, *J. Biomed. Mater. Res.,* Vol 1, 1967, p 285–297
117. S.J. Marshall and G.W. Marshall, Jr., Time-Dependent Phase Changes in Cu-Rich Amalgams, *J. Biomed. Mater. Res.,* Vol 13, 1979, p 395–406
118. R. W. Phillips, *Skinner's Science of Dental Materials,* W.B. Saunders Company, 1982
119. "American Dental Association Specification No. 5 for Dental Casting Gold Alloy," American Dental Association, Chicago, IL
120. R.G. Craig, *Restorative Dental Materials,* Mosby Inc., 1997
121. American National Standards Institute/American Dental Association "Specification No. 5 Dental Casting Alloys," American Dental Association Council on Scientific Affairs, Chicago, IL, 1989
122. American National Standards Institute/American Dental Association "Specification No. 5 Dental Casting Alloys," American Dental Association Council on Scientific Affairs, Chicago, IL, 1998
123. Classification System for Cast Alloys, *American Dental Assoc.,* Vol 109, Nov 1984, p 766
124. R.M. German, Precious-Metal Dental Casting Alloys, *Int. Met. Rev.,* Vol 27, 1982, p 260–288
125. K. Yasuda and K. Hisatsune, The Development of Dental Alloys Conserving Precious Metals: Improving Corrosion Resistance by Controlled Aging, *Int. Dent. J.,* Vol 33, 1983
126. H. Hero, Tarnishing and Structures of Some Annealed Dental Low-Gold Alloys, *J. Dent. Res.,* Vol 63, 1984, p 926–931
127. R.G. Craig (Chm), Section One Report, in *International Workshop on Biocompatibility, Toxicity, and Hypersensitivity to Alloy Systems Used in Dentistry, Conf. Proc.,* University of Michigan School of Dentistry, 1985
128. D. Scharnweber, Degradation (In Vitro-In Vivo Corrosion) *Metals as Biomaterials.* J.A. Helsen and H.J. Breme, Ed., John Wiley & Sons, 1998, p 101–152
129. "Standard Specification for Cobalt-28 Chromium-6 Molybdenum Alloy Castings and Casting Alloy for Surgical Implants (UNS R30075)," F 75, *Annual Book of ASTM Standards,* ASTM International, Vol 13.01, 2001
130. K. Asgar and F.C. Allan, Microstructure and Physical Properties of Alloys of Partial Denture Castings, *J. Dent. Res.,* Vol 47, 1968, p 189–197

131. P. Kovacs, *In Vitro Studies on the Electrochemical Behavior of Pure Metals Used in Orthopaedic Implant Alloys, Extended Abstracts of the Electrochemical Society Meeting* (Phoenix, AZ), The Electrochemical Society, Oct 13–17, 1991
132. S. Storp and R. Holm, ESCA Investigation of the Oxide Layers on Some Cr Containing Alloys, *Surf. Sci.*, Vol 68, 1977, p 10–19
133. J. Ohnsorge and R. Holm, Surface Investigations of Oxide Layers on Cobalt-Chromium Alloyed Orthopedic Implants Using ESCA Technique, *Med. Progr. Technol.*, Vol 5, 1978, p 171–177
134. J.J. Jacobs, R.M. Latanison, et al., The Effect of Porous Coating Processing on the Corrosion Behavior of Cast Co-Cr-Mo Surgical Implant Alloys, *Orthopaedic Res.*, Vol 8, p 874–882
135. J. Hubrecht, Electrochemical Impedance Spectroscopy as a Surface Analytical Technique for Biomaterials, *Metals as Biomaterials*, J.A. Helsen and H.J. Breme, Ed., Chichester, England, John Wiley & Sons, Ltd., 1998
136. O. Kubaschewski and B.E. Hopkins, *Oxidation of Metals and Alloys*, London, Butterworth and Co., 1962
137. M. Fontana and N. Greene, *Corrosion Engineering*, McGraw-Hill, 1967
138. P. Kofstad, *Nonstoichiometry, Diffusion, and Electrical Conductivity in Binary Metal Oxides*, John Wiley & Sons, Inc., 1972
139. D.A. Jones, *Principles and Prevention of Corrosion*, Macmillan Publishing Company, New York, 1992
140. M. Metikos-Hukovic and M. Ceraj-Ceric, p-Type and n-Type Behavior of Chromium Oxide as a Function of the Applied Potential, *J. Electrochem. Soc.*, Vol 134 (No. 9), Sept 1987, p 2193–2197
141. S.J. Megremis and J.L. Gilbert, In-Situ Electrochemical Atomic Force Microscopy of a Cast Co-Cr-Mo Alloy with Simultaneous Impedance Analysis, Transactions of the Society for Biomaterials, Saint Paul, MN, April 2001
142. S.J. Megremis, The Mechanical, Electrochemical, and Morphological Characteristics of Passivating Oxide Films Covering Co-Cr-Mo Alloys: A Study of Five Microstructures, *Biomedical Engineering*, Northwestern University, Evanston, IL, Dec 2001
143. R.A. Poggie, "A Review of the Effects of Design, Contact Stress, and Materials on the Wear of Metal-on-Metal Hip Prostheses," *Alternative Bearing Surfaces in Total Joint Replacements*, J. Jacobs and T.L. Craig, Ed., ASTM STP 1346, 1998, p 47–54
144. K. Asgar and F.A. Peyton, Effect of Microstructure on the Physical Properties of Cobalt-Base Alloys, *J. Dent. Res.*, Vol 40, 1961, p 63–72
145. J.B. Park and R.S. Lakes, *Biomaterials: An Introduction*, Plenum Press, 1992
146. M. Taira, J.B. Moser, and E.H. Greener, Mechanical Properties of Cast Ti Alloys for Dental Uses, *J. Dent. Res.*, Vol 65, 1986, IADR No. 603
147. J.F. Bates and A.G. Knapton, Metal and Alloys in Dentistry, *Int. Met. Rev.*, Vol 22 (No. 215), 1977, p 39–60
148. R.L. Bertolotti, Selection of Alloys for Today's Crown and Fixed Partial Denture Restorations, *J. Am. Dent. Assoc.*, Vol 108, 1984, p 959–966
149. J.J. Tuccillo, Compositional and Functional Characteristics of Precious Metal Alloys for Dental Restorations, *Alternatives to Gold Alloys in Dentistry*, Con. Proc., T.M. Valega, Ed., DHEW Publication (NIH) 77-1227, Department of Health, Education, and Welfare, 1977
150. M.M.A. Vrijhoef, Oxidation of Two High-Palladium PFM Alloys, *Dent. Mater.*, Vol 1, 1985, p 214–218
151. J.R. Mackert, Jr., E.E. Parry, and C.W. Fairhurst, Oxide Metal Interface Morphology Related to Oxide Adherence, *J. Dent. Res.*, Vol 63, 1984, IADR No. 405
152. American National Standards Institute/American Dental Association "Specification No. 32 for Orthodontic Wires," American Dental Association, Chicago, IL, 2000
153. M.R. Marcotte, Optimum Time and Temperature for Stress Relief Heat Treatment of Stainless Steel Wire, *J. Dent. Res.*, Vol 52, 1973, p 1171–1175
154. W.J. O'Brien, *Dental Materials and Their Selection*, Quintessence Publishing Co., Inc., Hanover, IL, 2002
155. A.J. Goldberg and C.J. Burstone, an Evaluation of Beta-Stabilized Titanium Alloys for Use in Orthodontic Appliances, *J. Dent. Res.*, Vol 57, 1978, p 593–600
156. C.E. Janus, D.F. Taylor, and G.A. Holland, A Microstructural Study of Soldered Connectors of Low-Gold Casting Alloys, *J. Prosthet. Dent.*, Vol 50, 1983, p 657–663
157. "Standard Specification for Wrought Cobalt-28Chromium-6Molybdenum Alloys for Surgical Implants (UNS R31537, UNS R31538, UNS R31539)," F 1537, *Annual Book of ASTM Standards*, ASTM International, Vol 13.01, 2000
158. "Standard Specification for Cobalt-28 Chromium-6 Molybdenum Alloy Forgings for Surgical Implants (UNS R31537, R31538, R31539)," F 799, *Annual Book of ASTM Standards*, ASTM International, Vol 13.01, 2002
159. K.C. Ludema, R.M. Caddell, et al., *Manufacturing Engineering: Economics and Processes*, Prentice-Hall, 1987
160. H.S. Dobbs and J.L.M. Robertson, Heat Treatment of Cast Co-Cr-Mo for Orthopaedic Implant Use, *J. Mater. Sci.*, Vol 18, 1983, p 391–401
161. L. Gettleman, C. Amman, and N.K. Sarkar, Quantitative In Vivo and In Vitro Measurement of Tarnish, *J. Dent. Res.*, Vol 58, 1979, IADR No. 969
162. R.M. German, M.M. Guzowski, and D.C. Wright, Color and Color Stability as Alloy Design Criterion, *JOM*, Vol 32, 1980, p 20–27
163. D.J.L. Treacy and R.M. German, Chemical Stability of Gold Dental Alloys, *Gold Bull.*, Vol 17, 1984, p 46–54
164. P.P. Coroso, Jr., R.M. German, and H.D. Simmons, Jr., Tarnish Evaluation of Gold-Based Dental Alloys, *J. Dent. Res.*, Vol 64, 1965
165. R.M. German, The Role of Microstructure in the Tarnish of Low Gold Alloys, *Metallography*, Vol 14, 1981, p 253–266
166. R.M. German, D.C. Wright, and R.F. Gallant, In Vitro Tarnish Measurements on Fixed Prosthodontic Alloys, *J. Prosthet. Dent.*, Vol 47, 1982, p 399–406
167. D.C. Wright and R.M. German, Quantification of Color and Tarnish Resistance of Dental Alloys, *J. Dent. Res.*, Vol 58A, 1979, IADR No. 975
168. H.J. Mueller, SIMS and Colorimetry of In-Vitro Sulfided Crown and Bridge Alloys, *Fifth International Symposium on New Spectroscopic Methods for Biomedical Research*, Battelle Laboratories and University of Washington, 1986
169. H. Hero, Tarnishing and Structures of Some Annealed Dental Low-Gold Alloys, *J. Dent. Res.*, Vol 63, 1984, p 926–931
170. E. Suoninen and H. Hero, Effect of Palladium on Sulfide Tarnishing of Noble Metal Alloys, *J. Biomed. Mater. Res.*, Vol 19, 1985, p 917–934
171. R. Kropp, Application of Corrosion and Tarnish Tests to Different Dental Alloys, *J. Dent. Res.*, Vol 65, 1986, IADR No. 197
172. J. Brugirard, Baigain, J.C. Dupuy, H. Mazille, and G. Monnier, Study of the Electrochemical Behavior of Gold Dental Alloys, *J. Dent. Res.*, 1973, p 838–836
173. W. Popp, H. Kaiser, H. Kaesche, W. Bramer, and F. Sperner, Electrochemical Behavior of Noble Metal Dental Alloys in Different Artificial Saliva Solutions, *Proc. of the 8th International Congress of Metallic Corrosion*, Vol 1, DECHEMA, 1981, p 76–81
174. N.K. Sarkar, R.A. Fuys, and J.W. Stanford, The Chloride Corrosion Behavior of Silver-Base Casting Alloys, *J. Dent. Res.*, Vol 58, 1979, p 1572–1577
175. D.C. Wright, R.M. German, and R.F. Gallant, Copper and Silver Corrosion Activity in Crown and Bridge Alloys, *J. Dent. Res.*, Vol 60, 1981, p 809–814
176. T.K. Vaidyanathan and A. Prasad, In Vitro Corrosion and Tarnish Analysis of Ag-Pd Binary System, *J. Dent. Res.*, Vol 60, 1981, p 707–715
177. N. Ishizaki, Corrosion Resistance of Ag-Pd Alloy System in Artificial Saliva: An

Electrochemical Study, *J. Osaka Dent. Univ.*, Vol 3, 1969, p 121–133
178. L.A. O'Brien and R.M. German, Compositional Effects on Pd-Ag Dental Alloys, *J. Dent. Res.*, Vol 63, 1984, IADR No. 44
179. N.K. Sarkar, R.A. Fuys, and J.W. Stanford, The Chloride Behavior of Silver-Base Casting Alloys, *J. Dent. Res.*, Vol 58, 1979, p 1572–1577
180. L. Niemi and R.I. Holland, Tarnish and Corrosion of a Commercial Dental Ag-Pd-Cu-Au Casting Alloy, *J. Dent. Res.*, Vol 63, 1984, p 1014–1018
181. L. Niemi and H. Hero, Structure, Corrosion, and Tarnishing of Ag-Pd-Cu Alloys, *J. Dent. Res.*, Vol 64, 1985, p 1163–1169
182. P.R. Mezger, M.M.A. Vrijhoef, and E.H. Greener, Corrosion Resistance of Three High-Palladium Alloys, *Dent. Mater.*, Vol 1, 1985, p 177–180
183. J.M. Meyer, Corrosion Resistance of Ni-Cr Dental Casting Alloys, *Corros. Sci.*, Vol 17, 1977, p 971–982
184. R.J. Hodges, The Corrosion Resistance of Gold and Base Metal Alloys, *Alternatives to Gold Alloys in Dentistry*, T.M. Valega, Ed., DHEW Publication (NIH) 77-1227, Department of Health, Education, and Welfare, 1977
185. S.J. Megremis and J.L. Gilbert, Step-Polarization Impedance Spectroscopy of Five Different Co-Cr-Mo Alloys, *Sixth World Biomaterials Congress Transactions* (Kamuela, HI), Society for Biomaterials, 2000
186. J.R. Cahoon, R. Bandyopadhya, et al., The Concept of Protection Potential Applied to the Corrosion of Metallic Orthopedic Implants, *J. Biomed, Mater. Res.*, Vol 9, 1975, p. 259–264
187. B.C. Syrett and S.S. Wing, Pitting Resistance of New and Conventional Orthopedic Implant Materials: Effect of Metallurgical Condition, *Corrosion*, Vol 34 (No. 4), 1978, p 138–145
188. H.E. Placko, S.A. Brown, et al., Effects of Microstructure on the Corrosion Behavior of CoCr Porous Coatings on Orthopedic Implants, *J. Biomed. Mater. Res.*, Vol 39, 1998, p 292–299
189. T.M. Devine and J. Wulff, Cast vs Wrought Cobalt-Chromium Surgical Implant Alloys, *J. Biomed. Mater. Rest.*, Vol 9, 1975, p 151–167
190. N.K. Sarkar and E.H. Greener, In Vitro Corrosion Resistance of New Dental Alloys, *Biomater. Med. Dev. Art. Org.*, Vol 1, 1973, p 121–129
191. R.M. Waterstrat, N.W. Rupp, and O. Franklin, Production of a Cast Titanium-Base Partial Denture, *J. Dent. Res.*, Vol 57A, 1978, IADR No. 717
192. H.J. Mueller and C.P. Chen, Properties of a Fe-Cr-Mo Wire, *J. Dent.*, Vol 11, 1983, p 121–128
193. N.K. Sarkar, W. Redmond, B. Schwaninger, and A.J. Goldberg, The Chloride Corrosion Behavior of Four Orthodontic Wires, *J. Oral Rehab.*, Vol 10, 1983, p 121–128
194. H.J. Mueller, Silver and Gold Solders—Analysis due to Corrosion, *Quint. Int.*, Vol 37, 1981, p 327–337
195. D.L. Johnson, V.W. Rinne, and L.L. Bleich, Polarization-Corrosion Behavior of Commercial Gold- and Silver-Base Casting Alloys in Fusayama Solution, *J. Dent. Res.*, Vol 62, 1983, p 1221–1225

Corrosion in Petroleum Production Operations

Revised by Russell D. Kane, Honeywell Process Solutions, Honeywell International, Inc.

THE PRODUCTION of oil and gas, its transportation and refining, and its subsequent use as fuel and raw materials for chemicals constitute a complex and demanding process where corrosion (in all its many forms) is an inherent hazard. The costs in terms of lost production, the replacement of materials of construction, and the constant personnel involvement in corrosion control are substantial. If not controlled, corrosion can result in leaks and catastrophic failures that can lead to additional expenses related to environmental hazards, equipment wastage, injury, and loss of life.

The total cost of corrosion in the U.S. oil and gas production industry is estimated to be $1.372 billion annually, made up from $589 million for surface piping and facility costs, $463 million in downhole tubing expenses, and $320 million in capital expenditures related to corrosion (see the article "Direct Costs of Corrosion in the United States" in *Corrosion*, Volume 13A, of the *ASM Handbook*). The control of corrosion through the use of coatings, metallurgy, nonmetallic materials of construction, cathodic protection, inhibitors, and other methods has evolved into a science in its own right and has created industries devoted solely to corrosion control in specific areas.

This article discusses the particular corrosion problems encountered and the methods of control used in petroleum production (i.e., upstream) and the storage and transportation of oil and gas (i.e., midstream) up to the refinery (i.e., downstream). Refinery corrosion is discussed in the article "Corrosion in Petroleum Refining and Petrochemical Operations" in this Volume.

Although industry standards and specification are referenced throughout this article, additional information can also be found in the article "Corrosion in Petroleum Refining and Petrochemical Operations" which lists U.S. and international standards for both petroleum production and refining operations (refer to the Appendix "Industry Standards").

Causes of Corrosion

This section concentrates on those aspects of corrosion that tend to be unique to corrosion as encountered in applications involving oil and gas exploration and production, namely extracting oil and gas from underground reservoirs and sending these products through a gathering system of pipelines to processing facilities.

The most unique and hostile environments are commonly found in actual production formations (i.e., downhole oil and gas reservoirs), which, in the absence of contamination, are devoid of oxygen. In situ corrosives normally include carbon dioxide (CO_2), hydrogen sulfide (H_2S), polysulfides, organic acids, and elemental sulfur. In certain circumstances, natural gas reservoirs may also contain elemental mercury as an additional impurity with the result being additional concerns from the standpoint of corrosion (these are discussed in the articles on "Liquid Metal Corrosion" and "Liquid Metal Induced Embrittlement" in Volume 13A of the *ASM Handbook*). Additional unique aspects are the extremes of temperature and, particularly, pressure encountered. In deep gas wells (6000 m, or 20,000 ft), temperatures approaching 230 °C (450 °F) have been present, and partial pressures of CO_2 and H_2S of the order of 20.7 MPa (3000 psi) and 48 MPa (7000 psi), respectively, have been encountered. Total system pressures can commonly reach levels between 34.5 and 138 MPa (5000 and 20,000 psi). Additionally, oil and gas production from reservoirs under deep ocean waters must enter submarine production pipelines and related equipment. The production environment is quickly cooled from the high temperatures mentioned previously to near freezing temperatures (1 to 4 °C, or 35 to 40 °F) by the surrounding deep sea environment. This condition can involve precipitation of additional liquid water in the environment to contribute to corrosivity. Additionally, production pipelines and equipment in arctic regions routinely is designed to operate at −40 to −50 °C (−40 to −60 °F).

Convenient access to the most important literature on H_2S corrosion (particularly with regard to sulfide stress cracking, SSC; hydrogen induced cracking, HIC; and the related problem of stress oriented hydrogen induced cracking, SOHIC) and CO_2 corrosion is available in Ref 1 to 14.

The initially oxygen-free geologic environment may be altered by a variety of oxygen-contaminated fluids that can be introduced. Examples are drilling fluids (often referred to as "muds"), which are used during drilling and maintenance of wells. These can include high-density clear brines such as calcium chloride ($CaCl_2$), calcium bromide ($CaBr_2$), and zinc bromide ($ZnBr_2$); various aqueous solutions including seawater or reinjected water separated from oil and gas production; alternating water, steam and/or CO_2 injected for secondary oil recovery; and strong inorganic and organic acids (e.g., hydrochloric, HCl, hydrofluoric, HF, formic, HCOOH, and others) that are injected down the well and into the formation to increase formation permeability for better flow of hydrocarbons. Some of these fluids are inherently corrosive; others are potentially corrosive only when contaminated with oxygen, CO_2, and/or H_2S, and some can become thermally unstable and liberate corrosive species in situ. Particularly damaging in these situations is the fact that corrosion is likely to be localized leading to pitting, local area attack, crevice corrosion, and/or stress-corrosion cracking (SCC) (Ref 15).

Oxygen is also responsible for the external corrosion of offshore platforms and drilling rigs. In oil and gas production, highly stressed structural members are directly exposed to a corrosive, aerated seawater environment. This also makes localized corrosion and corrosion fatigue a particular concern. Additionally, SCC can occur in stainless steel topside piping and equipment, if not properly selected for this harsh environment.

Another unique aspect of oil and gas production operations, particularly in older fields, is the almost exclusive use of carbon and low-alloy steels. More recently, there has been extensive use of corrosion-resistant alloys. These materials include martensitic, austenitic, and duplex stainless steels, nickel-base alloys, and titanium alloys. They have been typically justified by the reduction in costs associated with operations, replacement, maintenance, and/or more conventional corrosion inhibition of steel. These more corrosion-resistant alloys have been utilized most extensively in remote and/or offshore operations where costs related to corrosion and its control are typically higher than for more conventional onshore oil-field applications. As wells are being drilled deeper, higher formation pressures are encountered, and as offshore wells are located in deeper waters, the emphasis is increasingly on both corrosion resistance and higher strength to improve load-carrying capabilities. The relevant concern is that higher-strength materials often exhibit greater sensitivities to environmentally assisted cracking, EAC (e.g., SCC and hydrogen embrittlement cracking, HEC). Therefore, these materials often require special developments in alloy composition and/or metallurgical processing to provide for increased resistance to EAC at levels of applied stress to make them viable for constantly more challenging applications in the quest for oil and gas.

Oxygen

Although it is not normally present in rock formations at depths more than approximately 100 m (330 ft) below the surface of the earth, gaseous oxygen (O_2)—or oxygen dissolved in aqueous media—is nevertheless responsible for a great deal of the corrosion encountered in oil and gas exploration and production. However, oxygen-induced internal corrosion problems tend to be greater in applications where much of the processing and handling occurs at or near the surface at near-ambient pressure. This makes oxygen contamination through leaking pump seals, casing and process vents, open hatches, open handling (as in mud pits during drilling, trucking, and so on), and in seawater-injection systems highly likely. Also, failure of oxygen-removal processes (gas stripping and chemical scavenging) is a relatively common occurrence in water-injection systems (see the discussion "Corrosion in Secondary Recovery Operations" in this article).

A number of the properties of oxygen contribute to its uniqueness as a corrosive. Oxygen almost by definition is a strong oxidant. This means that even trace concentrations (≥ 10 ppb) can be harmful, to a varying degree, depending on its concentration and other corrosive constituents in the system. The corrosion potential of steel under these conditions (almost 1.3 V) is high enough to overcome very substantial potential drops between anodic and cathodic sites. Also, the kinetics of oxygen reduction on a metal or conductive oxide surface are relatively fast. This, coupled with the low solubility of oxygen in water and brines, tends to produce conditions in which the mass transport of oxygen is the rate-limiting step in the corrosion of carbon and low-alloy steels in nonacidic environments.

Mass transport is important in a number of aspects of oxygen corrosion and corrosion control. On newly installed bare steel offshore structures, mass transport of oxygen governs electrical current requirements for cathodic protection in seawater. Once a calcareous deposit forms on cathodically protected steel, the current demand to maintain cathodic protection is reduced considerably. However, poor mass transport under deposits and in crevices promotes localization of attack particularly if deposits are nonuniform. In the final analysis, limiting the mass transport of oxygen plays a critical role in much of the corrosion control in oxygenated systems.

The crucial role of mass transport can be illustrated as follows. At ambient conditions, water equilibrated with air will contain about 5 to 8 ppm of oxygen depending on specific conditions and the presence of dissolved salts. Under such conditions, mass-transport limited rates of general corrosion of steel range from about 0.25 mm/yr (10 mils/yr) in a stagnant system to 15 mm/yr (600 mils/yr) in a highly turbulent one. However, by chemically scavenging the oxygen concentration down to the order of < 10 ppb, the corresponding rates are reduced to less than about 0.01 mm/yr (0.4 mils/yr). Such rates are acceptable. Under these conditions, magnetite forms as a stable protective corrosion product film and further lowers the corrosion rate by introducing a slower, anodically controlling step.

An even more fundamental role of magnetite should be acknowledged. This is its role as a protective barrier on the steel surface that limits the reaction of steel with water or the hydrogen ions contained in the water. Therefore, if an excess of a chelating agent such as ethylenediamine tetraacetic acid (EDTA) dissolves a protective magnetite film, as could occur in regions of high turbulence, rapid corrosion ensues, despite the absence of oxygen.

The interrelationship of oxygen concentration, flow rate, and temperature in seawater-handling systems is shown in Fig. 1 (Ref 16).

Hydrogen Sulfide, Polysulfides, and Sulfur

Hydrogen sulfide, when dissolved in water, is a weak acid and is therefore corrosive because it is a source of hydrogen ions. In the absence of buffering ions, water equilibrated with 1 atm of H_2S or CO_2 (often referred to as acid gases) has a pH of about 4. However, under conditions of high-pressure hydrocarbon formations, pH values as low as 3 have been observed. In some cases, buffering species in produced water (such as bicarbonate) can result in higher pH values than expected based on just the amount of acid gases present. This can have a direct mitigating effect on corrosion rate and, in some cases, promote the formation of protective corrosion films, thus reducing the corrosion rate to very low levels.

Hydrogen sulfide can also play other roles in corrosion in oil and gas production. The sulfur on the metal surface resulting from the H_2S corrosion reaction readily acts as a catalyst to promote absorption of atomic hydrogen into the corroding steel. Atomic hydrogen—also referred to as nascent hydrogen (H^0)—is formed by the cathodic reduction of hydrogen ions. This accounts for its role in promoting SSC, HIC, and SOHIC in steels. Stress-corrosion cracking is usually found in steel with a yield strength greater than approximately 550 MPa (80 ksi), whereas HIC and SOHIC are found in low-strength steels less than 550 MPa (80 ksi) (Ref 3).

Hydrogen sulfide also reacts with elemental sulfur. In a gas phase with a high H_2S partial pressure, sulfanes (free-acid forms of a polysulfide) are formed so that elemental sulfur is rendered mobile and is produced along with the remaining gaseous constituents. However, as the pressure decreases traveling up the production tubing, the sulfanes dissociate and elemental sulfur precipitates. Production environments with H_2S and sulfur have been found to be more severe from the corrosion and EAC standpoint than environments with H_2S alone. Various solvent treatments are used to avoid plugging by such sulfur, but their effect on corrosion and EAC need also to be considered during evaluation.

In the aqueous phase, under acidic conditions, sulfanes are also largely dissociated into H_2S and elemental sulfur. However, enough strongly oxidizing species can remain either as polysulfide ions or as traces of sulfanes to play a significant

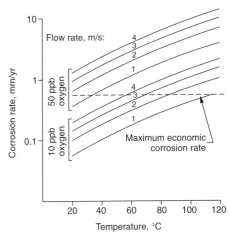

Fig. 1 Effect of flow rate, temperature, and oxygen on corrosion rate of steel in saltwater. Source: Ref 16

role in corrosion reactions. Oxygen contamination of sour (H_2S-containing) systems can also result in the formation of polysulfide and/or the precipitation of elemental sulfur generally leading to an increase in corrosivity.

Iron sulfide corrosion products can be important in corrosion control. Because of the low solubility, rapid precipitation, and mechanical properties of such corrosion products, velocity effects are not as commonly encountered in sour (H_2S-containing) systems as in sweet (CO_2-containing) systems. However, this effect can also be influenced by the morphology and actual crystalline form of iron sulfide, which can depend on temperature and H_2S concentration, among other factors.

The great range of possible iron sulfide corrosion products and their possible effects on corrosion have been extensively studied (Ref 17–23). This continues to be an area of interest in corrosion control and prediction. At lower temperatures and very low H_2S partial pressures, a somewhat protective sulfide film often forms. The absence of chloride salts strongly promotes this condition, and the absence of oxygen is absolutely essential. Under these conditions, the formation of a sulfide film often results in a remarkable decrease in corrosion rate when compared to similar conditions (CO_2 only) without H_2S.

At the high temperatures (150 to 230 °C, or 300 to 450 °F) and H_2S partial pressures (hundreds to thousands of pounds per square inch) encountered in deep sour gas wells, a so-called barnacle type of localized corrosion (Ref 21) can occur, resulting in corrosion rates of several hundred mils per year (Ref 22). This type of attack is strongly promoted by sulfidic species and requires the presence of some minimum chloride concentration. Although initially recognized in deep sour well environments, this same mechanism may operate at lower temperature in pipelines where high levels of H_2S and high chloride concentrations are observed.

In the barnacle mechanism (Fig. 2), corrosion can be sustained beneath thick but porous iron sulfide deposits (primarily pyrrhotite, $FeS_{1.15}$) because the FeS surface is an effective cathode. The anodic reaction beneath the FeS deposit is dependent on the presence of a thin layer of concentrated iron chloride ($FeCl_2$) at the Fe/FeS interface. This intervening $FeCl_2$ layer is acidic due to ferrous ion hydrolysis, thus preventing precipitation of FeS directly on the corroding steel surface and enabling the anodic reaction to be sustained by the cathodic reaction on the external FeS surface.

Carbon Dioxide

Carbon dioxide, like H_2S, is a weakly acidic gas (often referred to as an acid gas) and becomes corrosive when dissolved in water. However, CO_2 must first hydrate to form carbonic acid (H_2CO_3)—a relatively slow reaction—before it is acidic. There are other marked differences between the two systems. Velocity effects are very important in the CO_2 system; corrosion rates can reach very high levels approaching 12.5 to 37.5 mm/yr (500 to 1500 mils/yr), and the presence of salts is often unimportant.

Whether or not corrosion in a CO_2 system is inherently controlled or uncontrolled depends critically on the factors governing the deposition and retention of a protective iron carbonate (siderite) scale. On the other hand, there are the factors that determine the rate of corrosion on bare steel. These latter factors govern the importance of maintaining corrosion control.

For simple systems, bare steel (worst case) corrosion rates can be estimated from Eq 1, which was developed on the basis of electrochemical studies of the aqueous CO_2/carbon steel system (Ref 24):

$$\text{Log } R = A - \frac{2320}{t+273} - \frac{5.55t}{1000} + 0.67 \log \bar{p} \quad \text{(Eq 1)}$$

where R is the corrosion rate, t is temperature (°C), A is a constant, and $\bar{p}$ is CO_2 partial pressure. When R is calculated in millimeters per year and $\bar{p}$ is in atmospheres, $A = 7.96$. When R is calculated in mils per year and $\bar{p}$ is in pounds per square inch, $A = 8.78$.

Corrosion rates calculated with Eq 1 reach 25 mm/yr (1000 mils/yr) at 65 °C (150 °F) and 1 MPa (150 psi) CO_2 pressure, and 250 mm/yr (10,000 mils/yr) at 82 °C (180 °F) and 16 MPa (2300 psi) CO_2 pressure. Obviously, such rates are unacceptable. An alternative, idealized condition occurs when a protective carbonate scale is present and when the corrosion rate is limited by the need to replenish the film lost due to solubility in the aqueous phase. Under such conditions, the rates calculated for a hypothetical

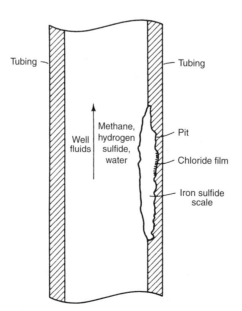

Fig. 2 Barnacle mechanism of sour pitting corrosion. Source: Ref 22

sweet (CO_2-containing) gas well reached a maximum of about 0.15 mm/yr (6 mils/yr) as compared to calculated bare metal rates of 500 to 2000 mm/yr (20,000 to 80,000 mils/yr) (Ref 25).

Conditions favoring the formation of a protective iron carbonate scale are:

- Elevated temperatures greater than 60 °C (140 °F) (decreased scale solubility, decreased CO_2 solubility, and accelerated precipitation kinetics)
- Increased pH above 4.0, as occurs in bicarbonate-containing waters (decreased solubility) or in conditions of iron supersaturation from buildup of iron corrosion products in the aqueous phase
- Lack of turbulence—low wall shear stress conditions provide minimal mechanical forces that can disrupt normally protective (pseudopassive) surfaces films on the metal surface

Turbulence is often the critical factor in pushing a sweet system into a corrosive regime. Excessive degrees of turbulence prevent either the formation or retention of a protective iron carbonate film.

A critical velocity equation has been used to estimate when excessive turbulence can be expected in a CO_2 system (Ref 26). There is no doubt that the velocity effect is real, but it is accepted that the following equation is an oversimplification:

$$\text{Critical velocity} = \frac{K}{\sqrt{\rho}} \quad \text{(Eq 2)}$$

where velocity is calculated in feet per second, K is a constant, and ρ is the density of the produced fluid (liquid + gas combined). When ρ is in kilograms per cubic meter, $K = 7.6$; for ρ in pounds per cubic foot, $K = 100$. More commonly today, rigorous multiphase flow modeling is conducted using software tools that can assess production rates for water, oil, and gas to determine the applicable flow regime (e.g., stratified, slug, mist, etc.) and the magnitude of the wall shear stress present from the flow in addition to local turbulence from geometric considerations (e.g., surface roughness, weld protrusion, bends, tees, etc.) as well as gravitational forces that influence multiphase flow (i.e., horizontal, upward, and downward flow). Table 1 shows the impact of various multiphase flow parameters on the flow behavior and resultant wall shear stresses produced. It can be seen that it is not always the highest flow velocity that produces the most severe condition; it is more related to local turbulence, flow-induced turbulence (i.e., slug flow), and the physical parameters of the flowing media (Ref 25, 26, 28).

Currently, state-of-the-art in corrosion assessment utilizes predictive software tools that can facilitate the use of many simultaneous parametric effects on corrosion through input of production variables (see Fig. 3). Such software tools often incorporate various theoretical and mathematical models combined with heuristics to adjust the calculated corrosion rates based on actual field/plant experience and the results of empirical studies. The most sophisticated models also have the ability to:

- Determine the effects of pH (taking into account CO_2, H_2S, bicarbonate, acetate, and iron supersaturation in the environment) on corrosion scale formation
- Assess the water content and its presence in either vapor or liquid forms
- Determine interactions between system water and glycol on phase behavior and activity coefficient.
- Evaluate the influence of potentially protective oil films
- Determine the influence of impurities such as chloride, oxygen, and elemental sulfur
- Assess the probability for localized or pitting corrosion (Ref 29, 30)

Table 1 Typical local liquid wall shear stress values determined by modeling for various multiphase flow regimes

	Flow conditions at pipe wall and corresponding shear stress, Pa			
		Slug		
Internal pipe wall condition	Stratified smooth	Moderate	Extreme	Annular wavy
Straight pipe	3	100	200	10
Pipe with pitting	5	150	300	15
3D-bend	5	150	300	15
5 mm (0.2 in.) weld bead	10	350	700	25

Source: Ref 10, 27

As mentioned previously, an important innovation in corrosion prediction is the use of flow modeling to assess and understand the influence of production parameters on determining the applicable flow regime and quantify the resultant mechanical forces in terms of wall shear stress and their influence on protective corrosion films and ultimately on corrosion rates. Table 2 shows a summary of typical parameters used in software tools for corrosion prediction and their various effects and interactions.

When both H_2S and CO_2 are present, heuristics indicate that iron sulfide may be the dominant corrosion product scale when the H_2S/CO_2 ratio exceeds about $1/200$ (Ref 29, 30); sour-system considerations would then be expected to apply. Even in a strictly CO_2 system, iron carbonate may not always be the corrosion product. Magnetite may form instead. Figure 4 shows the stability fields expected for siderite and magnetite as a function of the redox potential (expressed here in terms of hydrogen fugacity) of the system (Ref 25). In actual experience, corrosion product scales are often found to consist of mixtures or layers of siderite and magnetite when CO_2 is dominant, whereas many sulfide compounds may form once the level of H_2S exceeds a critical level to dominate the corrosion processes.

Iron carbonate lacks conductivity and therefore does not provide an efficient cathode surface. Thus, the types of pitting mechanisms found in oxygenated and in H_2S-containing systems do not occur in a CO_2-only system unless contamination occurs. Rather, generalized corrosion occurs at any regions not covered by the protective scale. The result is that on any bare (unfilmed) metal, the anodic and cathodic regions are so microscopically dispersed that salt—to provide conductivity—is not needed to achieve the very high corrosion rates predicted by Eq 1. This often leads to a condition referred to as mesa corrosion in which local, unfilmed areas of steel corrode actively at a high rate, but adjacent filmed regions remain uncorroded. This results in a characteristic corrosion pattern reminiscent of the mesas produced in rock by wind and water erosion (see Fig. 5).

Strong Acids

Strong acids are often pumped into wells to stimulate production by increasing formation permeability. For limestone formations, 15 and

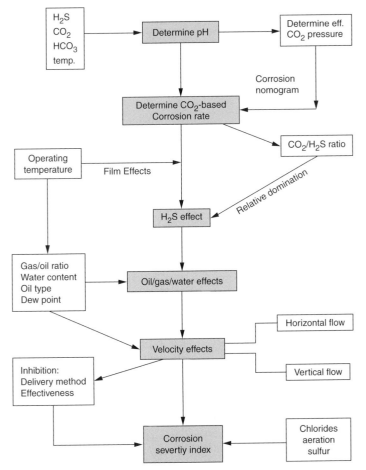

Fig. 3 Corrosion model software flowchart showing incorporation of multiple effects and variables. Source: Ref 29

28% HCl acids are commonly used. For sandstones, additions of 3% HF are necessary. In deep sour gas wells where HCl inhibitors lose effectiveness, 12% HCOOH has been used.

Corrosion control is normally achieved by a combination of proper inhibitor inhibition (different formulations for steels and corrosion-resistant alloys) and limiting exposure time to 2 to 12 h. If corrosion-resistant alloys are present (austenitic and duplex stainless steels, titanium alloys, and so on), concern for SCC and inhibitor ineffectiveness (respectively) may rule out the use of HCl, particularly at high reservoir temperatures, unless advanced screening of inhibitor formulations is undertaken.

Concentrated Brines

Dense halide brines of the cations of calcium, zinc, and, more rarely, magnesium are sometimes used to balance formation pressures during various production operations. All can be corrosive because of dissolved oxygen or entrained air. In addition, such brines may be corrosive because of the acidity generated by the hydrolysis of the metallic ions, as illustrated in:

$$Zn^{2+} + H_2O = ZnOH^+ + H^+ \quad \text{(Eq 3)}$$

Corrosivity due to acidity is worst with dense zinc brines. More expensive $CaBr_2$ brines are now often used at densities above about 1.7 g/cm^3 (14 lb/gal) (attainable with $CaBr_2$ brines) to avoid long-term exposure to $ZnCl_2$ brines.

Stray-Current Corrosion

If an extraneous direct current (dc) in the earth is traversed by a conductor, part of the current will transfer to the lower-resistance path thus provided. Direct currents are much more destructive than alternating currents (ac); an equivalent ac current causes only about 1% of the damage of a dc current (Ref 31). Regions of current arrival (where electrons depart) will become cathodic, and those regions where the current departs will become anodic. With corrodible metals such as carbon and low-alloy steels, corrosion in the anodic areas is the result. For example, 1 A·yr can corrode 9 kg (20 lb) of steel.

Cathodic protection systems are the most likely present-day sources of stray dc currents in production operations. More detailed discussions are available in Ref 12 and 32. The article "Stray-Current Corrosion" in Volume 13A of the *ASM Handbook* also contains information on the causes and mechanisms of stray-current corrosion.

Underdeposit (Crevice) Corrosion

This is a form of localized corrosion found almost exclusively (if not exclusively) in oxygen-containing systems. Such corrosion is usually most intense in chloride-containing systems, but can also be pronounced in aerated systems with active biological populations and biofilming to produce differential aeration cells. It is essential to have some form of shielding of an area on a metal such that it is wetted by an electrolyte solution but is not readily accessible to oxygen, the diffusing corrosive species.

This type of attack is usually associated with small volumes of stagnant solution caused by surface deposits (sand, sludge, corrosion products, bacterial growth), crevices in joints, and gasket surfaces. Crevice corrosion is discussed in Ref 33, and a quantitative treatment of crevice corrosion (particularly of stainless steels) is provided in Ref 34 to 36. See also the article "Crevice Corrosion" in Volume 13A of the *ASM Handbook* for information pertaining to the mechanisms of crevice corrosion.

The mechanism of crevice corrosion hinges upon the environmental conditions resulting from the loss of hydroxide production with cessation of the cathodic reaction when the initial oxygen in the shielded region is exhausted. Thus, in the shielded region, the anodic corrosion

Table 2 Flow-modeling parameters used to characterize flow-induced mechanical forces (wall shear stress)

Description	Relationship
1. Determine dimensionless parameters to describe fluid flow characteristics (e.g., Reynold's number, R_e) to account for mass transfer effects.	$R_e = \frac{\rho V D}{\mu}$
2. Determine friction factor, f, to account for pipe wall roughness (from Moody diagrams) in microturbulence on the surface of the pipe wall.	$f = f(R_e, e/D)$
3. Wall shear stress, τ, can be determined as a function of friction factor and other multiphase flow properties.	$\tau = \frac{f\rho V^2}{2}$
4. Determine flow regime (annular, stratified, bubble, slug, etc.) to estimate correction factors (e.g., for slug flow, Froude number, f_r, as a basis to estimate turbulent effects and intensity).	$f_r = \frac{V_t - V_s}{\sqrt{gh_{eff}}}$

Summary: Corrosion rate in fully developed turbulent pipe flows computed from field conditions of vertical or horizontal flow can be understood based on selected flow modeling parameters. Corrosion under field conditions can also be simulated in the laboratory. The results can be linked to field conditions using the parameter of wall shear stress where ρ is density, V is velocity, D is pipe internal diameter, and μ is viscosity. h_{eff}, effective height; g, acceleration of gravity; V_t, terminal velocity; V_s, superficial liquid velocity. Source: Ref 29

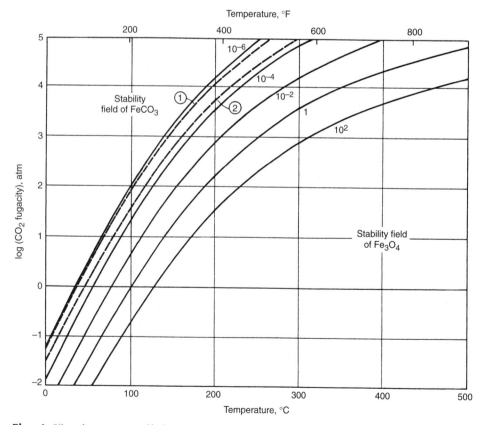

Fig. 4 Effect of temperature and hydrogen fugacity on the stability of $FeCO_3$ and Fe_3O_4 in contact with aqueous CO_2. Equilibrium calculations determine boundaries (indicated by the isohydrogen fugacity curves, with fugacity given in atmospheres) between $FeCO_3$ and Fe_3O_4 stability fields in the produced fluid, which contained CO_2, water, and traces of hydrogen. Curves 1 and 2 locate boundaries for locations 5180 m (17,000 ft) deep and at the wellhead, respectively, of a 170,000 m^3/day (6,000,000 ft^3/day) well corroding at a rate of 0.75 mm/yr (30 mils/yr). Source: Ref 25

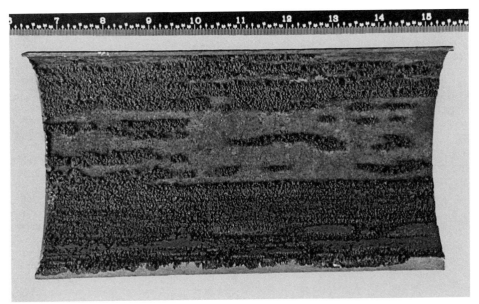

Fig. 5 Characteristic appearance of mesa corrosion resulting from CO_2, water, and flow. Source: InterCorr International, Inc.

reaction continues because the corrosion potential is maintained by the reduction of oxygen on the outside surface. However, chloride or other anions now migrate into the developing anodic region to maintain electroneutrality. Thus, a relatively concentrated, essentially ferrous chloride solution accumulates in the shielded region. As a result of the hydrolysis of the ferrous ions, the pH drops to a value of 2 or 3.

At this point, the crevice-corrosion type of localized attack is fully established. The anodic reaction continues in the shielded region because in the low-pH environment the ferrous ions go readily into solution and have little tendency to precipitate as an oxide or hydroxide on the surface and thus stifle the anodic reaction. Outside, the cathodic reaction continues unperturbed. Because of the large ratio of cathodic-to-anodic surface area, high rates of localized corrosion can be maintained with very modest cathodic current densities.

Galvanic Corrosion

When two dissimilar metals are electrically coupled—both electronically by a metal bond and ionically through an electrolyte—the more active (electronegative) metal tends to become a sacrificial anode and supply cathodic protection to the more noble metal. Such situations are often encountered in heat exchangers in which carbon steel tubesheets are used with copper alloy tubes and at junctions between piping, fasteners, or corrosion-resistant sheeting with containers of another material.

Problems with galvanic corrosion are the most acute when the cathode-to-anode area ratio is large. Such situations are often encountered inadvertently. This has happened when the normal electrochemical potential difference between zinc and steel in a galvanized pipe reversed in the presence of a bicarbonate/chloride brine so that the steel pipe walls perforated rapidly at pinholes in the galvanizing while trying to protect the extensive adjacent galvanized area. Similar situations can occur when using zinc to cathodically protect hot subsea pipelines where the polarity shift is a direct result of the increase in temperature. Another situation is when plastic-coated steel is coupled to a more noble metal. At any pinholes in the coating, a very adverse area ratio will exist, and rapid corrosion rates can result. Another common problem is when chromium or electroless nickel plated steels are used in corrosive seawater or oil-field production environments. Small cracks or crazes in the coatings allow communication of the corrosive fluids to reach the steel. In this case, the chromium or nickel coating covers a much larger area and has a higher electrochemical potential than the steel. This condition usually results in very rapid corrosion of the steel at local anodic sites that may, in some cases, actually consume the steel substrate. The article "Galvanic Corrosion" in Volume 13A of the *ASM Handbook* contains more information on this form of attack.

Biological Effects

The most important biological effect on corrosion in oil and gas production is the generation of H_2S by sulfate-reducing bacteria—SRB—(*Desulfovibrio Desulfuricans*). These are anaerobic bacteria that metabolize sulfate ions (using an organic carbon source) and produce hydrogen sulfide. They can thus introduce H_2S and all its corrosive ramifications into normally H_2S-free systems.

Colonies of SRB can also form deposits that are conducive to underdeposit corrosion. In many cases, the resultant corrosion appears to be due to a mechanical shielding action (differential concentration or aeration), rather than any depolarizing action resulting from the metabolic processes of the SRB. However, this is not to deny that the introduction of H_2S (whatever the source) into a crevice region could have an accelerating effect on corrosion, because H_2S is known to be an acid gas and anodic stimulant leading to local increase in corrosivity. More information on biological corrosion is available in the article "Microbiologically Influenced Corrosion" in Volume 13A of the *ASM Handbook*.

Mechanical and Mechanical/Corrosive Effects

Cavitation. This metal removal—often grain by grain—is due to high-pressure shock-wave impingement resulting from the rapid collapse of minute bubbles created under certain conditions in high-velocity fluid-handling equipment. It is usually found on pump impellers operating with too low a suction pressure.

Erosion. Most commonly, this is direct metal removal by the cutting action of high-velocity abrasive particles. Erosion failures (washouts) are seen in drill pipe when leaks (loose connections or a corrosion fatigue crack) allow drilling mud with fine particulates to flow through the wall under a high-pressure gradient. Erosion of oil and gas gathering lines at bends and joints or in flow-control valves and wellhead equipment by produced sand is probably the other most common occurrence of erosion in oil and gas production.

Erosion-Corrosion. Strictly speaking, in erosion-corrosion, only the protective corrosion product film is removed by erosive forces; however, with the protective film absent, corrosion can occur at a greatly accelerated rate. Erosion-corrosion may play a role in CO_2 corrosion (Ref 37), and sand, under mild flow conditions, may also cause erosion-corrosion. Design problems resulting in rapid changes in flow direction, neck-down points, or local turbulence can also lead to erosion-corrosion problems in production systems. Production conditions involving multiphase (oil/gas/water) flow can also result in erosion-corrosion since conditions of liquid and gas flow resulting in slug flow or mist flow can produce quite severe levels of wall shear stress. Erosion-corrosion has also been noted in heavy anchor chains where their use in an abrasive bottom mud allowed corrosion at contact regions to proceed at a rate of many hundreds of mils per year.

Corrosion Fatigue. This results from subjecting a metal to alternating stresses in a corrosive environment. At points of greatest stress, the corrosion product film becomes damaged during cycling, thus allowing localized corrosion to take place. Eventually, this leads to crack initiation and crack growth by a combination of mechanical and corrosive actions. Because of this combined action, damage per cycle is greater at low cycling rates (usually at less than 2 Hz with increasing effect with decreasing frequency), where corrosion can play a larger role. Also, in corrosion fatigue, counter to the usual case for steel in noncorrosive applications, a fatigue limit does not exist for corrosion fatigue, below which fatigue damage does not accumulate. Rather than leveling out as in simple (noncorrosive) fatigue, the usable stress level continues to decrease with increasing cycles even at very low levels of applied stress.

The greatest concern for corrosion fatigue arises in connection with highly stressed, submerged, offshore structures. Corrosion in seawater combined with cyclic loading can result in corrosion fatigue. In some conditions, corrosion fatigue can also occur under conditions of cathodic protection in seawater. In this case, the culprit is the hydrogen produced on the metal surface by the cathodic protection. Materials most susceptible to this phenomenon usually are those that have a particular sensitivity to hydrogen. These are commonly high-strength steels and weldments, martensitic and duplex stainless steels, and in some cases, highly alloyed materials such as nickel-base alloys (e.g., alloy 718 and titanium alloys). It has been recently observed that deep water conditions may be particularly of concern for hydrogen charging due to the lower temperatures and associated tendency to form calcareous deposits on materials being cathodically protected. In applications involving deep water drilling, increased stress levels, increased run time, use of high-strength/hardness materials, and higher cathodic protection currents can lead to conditions of hydrogen embrittlement and corrosion fatigue. Welded connections on drill ships and on drilling and production platforms are particularly susceptible to this form of structural impairment.

More information on attack resulting from combined corrosion and mechanical effects is available in the article "Forms of Mechanically Assisted Degradation" in Volume 13A of the *ASM Handbook*.

Corrosion Control Methods

The methods of corrosion control commonly employed in petroleum production operations include:

- Proper selection of materials
- Protective coatings
- Cathodic protection systems
- Use of inhibitors
- Use of nonmetallic materials
- Control of the environment

Additional information on these methods, as well as other corrosion protection techniques, can be found in Volume 13A of the *ASM Handbook*.

Materials Selection

Traditionally, carbon and low-alloy steels were virtually the only metals used in the production of oil and gas. This resulted from the fact that very large quantities of metal are required in petroleum production, and, until a few years ago, crude oil and gas were relatively low-value products. In addition, insurmountable corrosion problems were not encountered.

This situation changed when gas and oil prices increased dramatically and deeper wells were drilled that encountered corrosive environments of greatly increased severity. The final factor that made the current widespread use of corrosion-resistant alloys possible was the development of high-strength forms of these alloys. This allowed both thinner pipe and vessel walls and greatly reduced the amount of material required.

The result of this situation is that essentially all of the high-tonnage uses of corrosion-resistant alloys in oil and gas production involve alloys in a high-strength form. Yield strengths typically span the range of 550 to 1250 MPa (80 to 180 ksi), but can reach nearly 1750 MPa (250 ksi) in wire lines.

Metallurgical Considerations

The great majority of corrosion-resistant alloys used in oil and gas production were originally developed for other applications, such as chemical processing, that did not require high strength. Therefore, many of these alloys have an austenitic microstructure and can be strengthened only by some form of cold working. This presents no problem for the production of tubulars used underground in the well because they are joined by threaded connections. However, for other applications, such as welded flow lines and cast and forged valves, different techniques must be used to achieve high strength. Other higher-strength alloys initially used for downhole applications were developed for high-temperature aerospace applications and did not have substantial consideration for aqueous corrosion. However, since the first pioneering uses of these corrosion-resistant alloys in oil-field applications in the 1970s, further research and development has occurred, which has brought new alloys and processing methods that provide higher strength, resistance to H_2S and CO_2 corrosion, and resistance to various forms of EAC.

In addition to the problems of environmental compatibility, there are a number of difficulties of a metallurgical nature that are briefly mentioned. For high-strength materials, particularly at the upper end of the range, adequate ductility is often difficult to achieve. Also, heat treatments developed for other applications may not be optimal for corrosion resistance in petroleum environments.

Environmental Considerations

There are several environmental factors that are more or less unique to oil and gas production. One factor is the general absence of oxygen in process fluids. The dominant naturally occurring corrosives are CO_2 and H_2S. However, in wells with high H_2S concentrations, elemental sulfur—a relatively strong oxidizing agent—can also be present.

Water, which must be present to make sulfur or the acid gases (CO_2 and H_2S) corrosive, can be assumed to be present in nearly all productive geologic formations to varying degrees. Therefore, liquid water can be present from the bottom of the well up to the point in the flow system at which it is removed. Corrosivity is aggravated by the presence of salt in formation waters. These brines can range from a few hundred parts per million chloride up to saturated solutions containing as much as 300,000 ppm of total dissolved salts.

However, a salient consideration often overlooked is that formation water while containing chlorides often contains other species such as bicarbonate that can buffer the influence of acid gases and thus produce less acidic conditions that can mitigate corrosion. This is often the case in oil wells that produce varying amounts of formation water. By comparison, gas wells often produce condensed water that precipitates from the vapor phase. This condensed water usually contains little in the way of chlorides and buffering salts and may produce a highly acidic condition when in contact with H_2S and/or CO_2. Furthermore, in gas wells, naturally occurring organic acids may also be present in the production environment that further increase acidic conditions to levels that are more corrosive than those resulting from merely the CO_2 and H_2S (Ref 38).

At the same time, the conditions in the downhole annulus between production tubing and casing may contain concentrated brines with chlorides and bromides of sodium, calcium, and zinc. Additionally, exposure to strong mineral or organic acids may be used to stimulate oil and gas production from subterranean reservoirs. It should be realized that the corrosivity of these "nonproduction" environments can, at times, be more a determining factor for alloy selection than the corrosivity of the oil and gas production environment.

Temperature also affects corrosivity. Deep well temperatures up to 205 °C (400 °F) are not uncommon. Therefore, temperatures from these levels down to ambient must be considered.

Corrosion Considerations

From other applications, enough is generally known to select alloys that are resistant to general corrosion. However, resistance to localized corrosion (pitting and crevice corrosion) often requires some experimental study, field experience, and/or use of modern computer software that utilizes parametric modeling of environmental parameters, alloy composition, and strength factors.

Environmentally Assisted Cracking. In corrosion-resistant alloy selection, EAC normally requires the greatest portion of consideration through testing and/or modeling. This follows from the use of high-strength alloys in environments in which there has been an acknowledged short history of application, but growing experiences with these alloys.

The term environmentally assisted cracking was selected because it includes multiple corrosion mechanisms: cathodic and anodic. The cathodic mechanism is that of hydrogen embrittlement found in sulfide stress cracking of high-strength carbon and low-alloy steels in the presence of hydrogen sulfide. Here, H_2S promotes entry of cathodically evolved hydrogen atoms into the metal. Additionally, many of the higher-strength steels, stainless steels, and nickel- and titanium-base alloys can also show susceptibility to hydrogen embrittlement under influence of galvanic coupling with dissimilar alloys and during cathodic protection.

The anodic mechanism is that involved in the chloride SCC of nickel-containing martensitic and 300-series austenitic stainless steels at temperatures generally above about 65 °C (150 °F). However, this temperature limit is only a general guideline for SCC and typically only rigidly applies in aerated systems. Actual susceptibility to SCC in oil and gas production varies with many parameters including chloride concentration, partial pressure of acid gases, pH of the brine, and temperature as well as alloy composition, metallurgical condition, and strength level. Higher resistance to SCC in stainless alloys is usually associated with higher alloying contents of nickel, chromium, molybdenum, and tungsten. Nickel-base alloys are normally considered immune to this type of chloride cracking. However, the presence of H_2S can induce susceptibility in normally resistant alloys by acting to promote local anodic dissolution. In the most resistant alloys, such as alloy C-276 (Table 3), elemental sulfur and a temperature of about 205 °C (400 °F) must also be present to cause SCC. The presence of elemental sulfur can accelerate cracking in most corrosion-resistant alloys since it also promotes local anodic attack that results in stimulating both the initiation and propagation of SCC. However, this type of SCC is most commonly associated with high-strength and/or heavily cold-worked materials; however, in lower-alloy stainless steels, this mechanism can also cause cracking in low-strength, fully solution annealed alloys at lower temperatures.

Effects in the Cathodic Mechanism (Hydrogen-Assisted Cracking). In corrosion-resistant alloys, environmental effects on cathodic stress cracking are generally similar to those encountered in the sulfide stress cracking of carbon and low-alloy steels. However, in extreme cases, the temperatures needed to avoid hydrogen-assisted cracking susceptibility in the highest-strength materials can be as high as 120 °C (250 °F) or more. An example is alloy C-276 cold worked to a hardness of 45 HRC. In contrast, the highest minimum use temperature given in NACE standard MR0175/ISO 15156 (Ref 14) for high-strength low-alloy tubulars is 80 °C (175 °F).

Also, on corrosion-resistant alloys, chloride ions can have a significant deleterious effect in hydrogen-assisted cracking as well. This results from deterioration of normal passivity of the alloy in the presence of chloride. The resulting higher local anodic corrosion rate can then support a higher rate of cathodic corrosion processes and thus give rise to enhanced cathodic charging of the corrosion-resistant alloy with atomic hydrogen.

The present philosophy used in handling resistance to sulfide stress cracking of steels and stainless alloys is to incorporate parameters such as H_2S concentration and pH for a given temperature. Sulfide stress cracking susceptibility is usually assumed to be greatest at or near 23 °C (75 °F). Figure 6 shows data for 13% Cr martensitic stainless steels plotted in this manner (Ref 39). It defines levels of cracking severity that generally increase with decreasing pH and increasing H_2S concentration. The actual susceptibility to hydrogen-assisted cracking at lower temperatures has not been fully examined, but limited data suggest that there is a temperature for maximum hydrogen cracking susceptibility that is somewhat below room temperature.

Effects in the Anodic Mechanism (Stress-Corrosion Cracking). Determining regions of susceptibility to anodic cracking is generally more complicated than defining regions of susceptibility to cathodic (hydrogen) cracking. This is due to the multiparametric nature of the problem (Ref 40). Therefore, in anodic SCC, susceptibility is a strong function of such environmental factors as temperature, chloride concentration, and H_2S partial pressure; in addition, pH, oxidants (elemental sulfur or polysulfide), oxygen, and galvanic coupling can be involved. Further, alloy composition (corrosion resistance), strength level, and stress level all influence cracking susceptibility.

The effect of H_2S requires some elaboration. In cathodic cracking, there is a well-established threshold for the initiation of the sulfide stress cracking phenomenon; this lends itself to be examined in a relatively short test under static load. This is not always the case in anodic cracking. Susceptibility to SCC has a stronger chemical component and often requires a very long time for cracking initiation under static loading conditions. In order to properly evaluate for SCC susceptibility, local anodic processes must be accelerated either by chemical or mechanical procedures. Since this form of cracking requires assessment of the overall severity of the environment, it is more common to try to correctly simulate the severity of the actual corrosive environment in service, while accelerating cracking through application of mechanical loading (e.g., slow strain rate tests, cyclic slow strain rate tests, and sharp crack fracture mechanics tests).

Selection of Corrosion-Resistant Alloys

The problem of alloy selection to prevent EAC cracking is complicated by the fact that the

Table 3 Compositions of commonly used oil field corrosion-resistant alloys

Alloy	UNS No.	Composition(a), wt%					
		C	Cr	Fe	Ni	Mo	Other
Alloy C-276	N10276	0.02	14.5–16.5	4–7	bal	15–17	2.5Co, 1.0Mn, 4.5W, 0.35V
Alloy 625	N06625	0.10	20–23	5	bal	8–10	0.4Al, 4.15Nb, 0.5Mn, 0.4Ti
Alloy G	N06007	0.05	21–23.5	18–21	bal	5.5–7.5	2.5Nb, 2.5Cu, 2.5Mn, 1W
Alloy G-30	N06030	0.03	28–31	13–17	bal	4–6	2.4Cu, 5.0Co, 4.0W
Alloy 825	N08825	0.05	19.5–23.5	bal	38–46	2.5–3.5	0.2Al, 3Cu, 1Mn, 1.2Ti
Alloy 925	N09925	0.03	19.5–23.5	20	38–46	2.5–3.5	0.1–0.5Al, 1.9–2.4Ti
Alloy 2550	N06975	0.03	23–26	bal	47–52	5–7	0.7–1.2Cu, 0.7–1.5Ti
Alloy 718	N07718	0.08	17–21	bal	50–55	2.8–3.3	0.8Al, 0.6–1.1Ti, 4.8–5.5Nb
Alloy 725	N07725	0.03	19–22.5	bal	55–59	7–9.5	0.35Al, 2.75–4Ti
Alloy 400	N04400	0.3	...	2.5	63–70	...	bal Cu, 2Mn
Alloy K500	N05500	0.2	...	2.0	63–70	...	bal Cu, 3Al, 0.85Ti
MP35N	R30035	0.025	19–21	1.0	33–37	9–10.5	bal Co, 0.15Mn, 1Ti
AL6XN	N08367	0.03	20–22	bal	23.5–25.5	6–7	0.18–0.25N
Alloy 28	N08028	0.03	26–28	bal	29.5–32.5	3–4	1.4Cu, 2.5Mn
Alloy 255	S32550	0.04	24–27	bal	4.5–6.5	2–4	2.5Cu, 1.5Mn, 0.25N
Alloy 100 (ASTM A351)	S32760	0.03	55	bal	6–8	3–5	0.7Cu, 0.25N, 0.7W
Alloy 2507	S32750	0.03	24–26	bal	6–8	3–5	0.5Cu, 0.24–0.32N
Alloy 2205	S31803	0.03	21–23	bal	4.5–6.5	2.5–3.5	0.08–0.2N
254SMO	S31254	0.02	19.5–20.5	bal	17.5–18.5	6.0–6.6	0.18–0.22N, 0.5–1.0Cu
Type 316	S31600	0.08	16–18	bal	10–14	2–3	...
654SMO	S32654	0.03	24	bal	22	7	0.5N, 0.5Cu
13Cr (Hyper 1)	...	...	13	bal	4	1.5	...
13Cr (Hyper 2)	...	...	13	bal	5	2.15	...
F6NM	S42400	0.06	12–14	bal	3.5–4.5	0.3–0.7	...
Type 420 (13Cr)	S42000	0.15	12–14	bal	...	...	...

(a) Maximum unless range is given or otherwise indicated

lowest alloy content and the maximum reliable strength level are usually needed to achieve an economically viable engineering system. Because this involves design close to the limits of the material, it is necessary to define these limits as accurately as possible. The currently available service environment limits for corrosion-resistant alloys (Table 3) are given in NACE MR0175/ISO 15156 (Ref 14). However, such limits are not readily available for all alloys. As more experience and test data have become available, it has been possible to construct computer software models that can assist engineers in the materials selection process by application of data, rules, and algorithms in a systematic manner (Ref 41). Such software tools have also incorporated references to standard American Petroleum Institute (API) equipment configurations for wellhead and subsurface safety equipment, thus facilitating material selection, specification, and eventual purchasing by nonmaterials specialists. This allows the materials specialists to focus attention and valuable staff time on the most serious problems that may arise either before or after procurement.

Alloy selection from a corrosion standpoint can be considered to be generally a three-step process. First, resistance to general corrosion must be ensured. This is primarily a function of the chromium and nickel content of the alloy and under a first consideration may be associated by alloy classification: martensitic, austenitic, and duplex stainless steels, nickel-base alloys, and titanium alloys. Secondly, resistance to localized attack also must be ensured. This is primarily a function of chromium and molybdenum content, but recent alloy developments have also shown an influence of other alloy elements that help maintain passivity. These alloying elements include tungsten and nitrogen. Finally, resistance to environmental stress cracking is sought at the highest feasible strength level. Nickel content plays an important role in providing resistance to anodic cracking. However, it is realized that under conditions of anodic SCC in environments with chlorides and H_2S, the initiation of cracking is usually more closely associated with pitting corrosion. Therefore, alloying effects for cracking in most materials actually follow similar trends to those mentioned for localized corrosion but with additional guidelines for strength and metallurgical condition.

The close correlation between pitting resistance and resistance to anodic cracking is an important aspect with considerable importance in alloy selection. This apparently results from the ease of crack initiation under the low-pH high-chloride conditions found in pits. Therefore, higher molybdenum, tungsten, and nitrogen in combination with chromium can also increase resistance to anodic cracking.

With the procedures given below, regions for alloy applicability can be shown schematically as a qualitative function of environmental severity. Figure 7 is a simple example that applies to aqueous, CO_2-containing environment (hence low pH) and in which the effects of temperature, chloride, and H_2S concentration are illustrated. The effect of yield strength is not shown, but if environmental cracking is the limiting factor, reducing the yield strength should extend applicability to more severe environments.

The reader should be cautioned that a diagram such as Fig. 7 is really more of a guide to alloy qualification than to direct selection for a particular application. More complex analyses that can apply a more complete parametric approach are required to give specific recommendations for field situations.

More recently, this approach has been broadened to include the full range of conditions in H_2S and CO_2 production environments through materials selection software. Figure 8 shows a schematic of the hiearchical approach. This approach is called selection

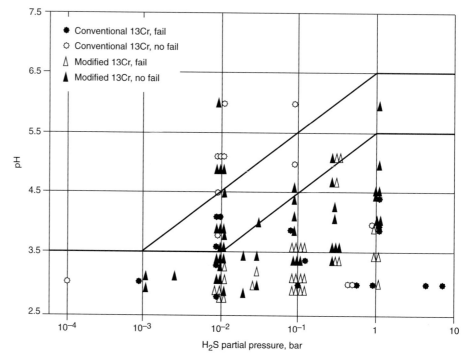

Fig. 6 Effect of H_2S partial pressure and pH and susceptibility to SSC for 13% Cr martensitic stainless steel tubular grades. Source: Ref 39

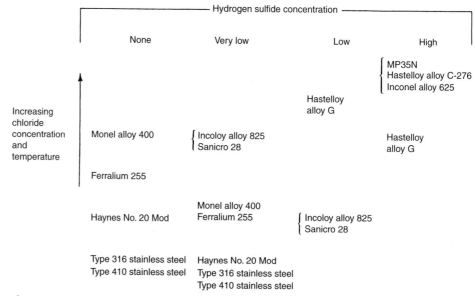

Fig. 7 Corrosion-resistant alloy selection for production environments containing aqueous CO_2 and H_2S

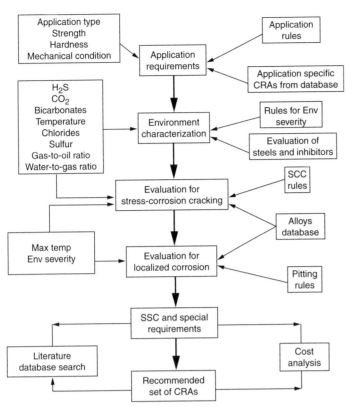

Fig. 8 Hierarchical decision tree for use in developing software tool for materials selection of oil and gas service applications. CRA, corrosion-resistant alloy; Env, environment; SCC, stress-corrosion cracking; SSC, sulfide stress cracking. Source: Ref 41

of corrosion-resistant alloys through environmental specification, and it involves:

1. Application requirements—strength, hardness, and mechanical attributes
2. Characterization of environment—H_2S and CO_2 partial pressures, chlorides, bicarbonates, sulfur, oil properties, and water availability
3. Evaluation for SCC—mostly by alloy class and composition
4. Evaluation for pitting—use of rules based on required minimum pitting resistance indices incorporating factors for chromium, molybdenum, tungsten, and nitrogen
5. Evaluation of sulfide stress cracking and hydrogen-assisted cracking
6. Cost analysis
7. Final selection of material

This procedure is widely utilized by end users to select materials and by materials suppliers and equipment manufacturers to handle inquiries. More recently, a similar approach has been utilized by industry standards such as NACE MR0175/ISO 15156 (Ref 14).

Testing for Resistance to Environmental Stress Cracking

If through the use of computer models for materials selection, it is ascertained that the conditions of use may be marginal for certain materials, or if special metallurgical considerations must be evaluated, the most directly applicable results are obtained by exposing samples of commercially produced alloys to an environment simulating as closely as possible that expected in actual production operations. Fortunately, an understanding of the principles involved allows considerable simplification to be made without significantly altering the value of the results.

Two simplifications can usually be made in the environmental parameters, as follows. First, only the CO_2 and H_2S partial pressures (mole fraction × total system pressure) are usually involved. The overburden of methane pressure to create the actual total pressure of the environment is dispensed with since they commonly have only very limited impact on the corrosive severity of the environment. However, at very high pressures, the nonideality of the gases can influence and limit the actual amount of acid gases that dissolve in the aqueous media and lead to overly conservative evaluations. The omission of the methane from the system can substantially lower the pressure ratings of test vessels required for this evaluation. The cases where the total pressure needs to be simulated are where polymeric materials such as elastomers and plastics are being evaluated or where the specific phase behavior of the system plays a major impact in the solubility of the dissolved gases. Modeling software is available to evaluate the latter effect to determine the appropriate partial pressure to use in laboratory autoclave tests. Second, only the chloride content of the brine is typically reproduced, rather than trying to simulate the total ionic spectrum of the produced fluids. Often, no reliable analysis is available. Assumption of a near-saturated sodium chloride solution (25 wt%) is then a relatively conservative approach, but where chloride levels in the produced water have been verified, these values will provide the most accurate indication of corrosive tendencies.

More recently, emphasis has been to also reproduce the bicarbonate levels so that a more appropriate pH level will be obtained in the aqueous phase when it is combined with the partial pressures of H_2S and CO_2, which reduces the level of conservatism. In some cases, the presence of a liquid hydrocarbon phase is also included, which often reduces the conservatism still further if persistent oil films are formed on metal surfaces. However, in this case, the phase behavior and flow conditions of the actual production environment need to be considered so that it is known if the water/oil mixture will remain emulsified or if it will disassociate, leaving a separate water phase. Additionally, specific liquid hydrocarbons can contain inhibitive compounds that will dissolve into the aqueous phase and, in some cases, substantially reduce the severity of corrosion, SCC, and SSC.

A wide variety of test specimens used for SSC and SCC testing and can be found in NACE TM0177 (Ref 42). For high-strength corrosion-resistant alloy tubulars, C-rings (NACE TM0177-Method C) are particularly convenient because they provide an easy way to obtain specimens that are stressed in the transverse (hoop) direction, that often shows higher susceptibility to cracking in many cold-worked alloys. They also have the advantage of providing as-produced tube inner diameter (ID) or outer diameter (OD) surfaces, if needed. Some investigators, however, prefer tensile specimens (NACE TM0177-Method A) since they provide a simple test approach and a uniform stress across the gage section. Double cantilever beam specimens (NACE TM0177-Method D) are very attractive in that they can provide a quantitative measure of fracture toughness, which can then be used in mechanical design. They also provide an extremely high value of initial stress intensity that can act to initiate cracking in a relatively short period of time.

Slow strain rate testing (NACE TM0198) can also be very useful in assessing alloy limits (Ref 43). The application of a dynamic load and high plasticity acts to accelerate crack initiation particularly when anodic SCC processes are involved. However, this test method needs careful evaluation since it can produce exceedingly high susceptibility in some materials, particularly those with martensitic or duplex microstructures. In some of these cases, the slow strain rate test can be modified (per ASTM G 129) to include application of slow cyclic loading ($\pm 10\%$ around a mean stress) (Ref 44). In this case, the cracking is accelerated by the dynamic loading, but does not respond to the negative influences of the high plasticity in the conventional slow strain rate test. This test can also be used to produce a "threshold" similar to that of the statically stressed specimens such as tensile, C-rings, and bent beams (Ref 45).

Another aspect of testing is the imposed stress level. Exposure at 100% of yield strength is the conservative approach. However, some investigators and industry standards place reliance on design values and test at some lesser fraction of the yield strength (80 to 95%) or of the specified minimum yield strength for the alloy grade. Detailed information on testing for resistance to environmental cracking is available in the articles "Evaluating Hydrogen Embrittlement" and "Evaluating Stress-Corrosion Cracking" in Volume 13A of the *ASM Handbook*.

Coatings

Internal protective coatings have been used to protect tubing, downhole equipment, and wellhead components. Christmas trees (manifolds used to control the rate of production, receive the produced fluids under pressure, and direct the produced fluids to the gathering point) and various downstream flow lines and pressure vessels for more than 50 years. Because internal coatings are not commonly considered "perfect" barriers to corrosion and are often subject to damage or deterioration over time, successful use is usually accompanied by chemical inhibition or cathodic protection as part of the entire protective program. Most of the coatings have been used below 175 °C (345 °F).

Tubing. The benefits derived from coating tubing depend on the coating remaining intact. Because no coating can be applied and installed 100% holiday-free, inhibition programs are commonly employed to accommodate holidays and minor damage. The suitability of the service is dependent on specific testing and an effective quality-control program. Guidelines for these procedures and methods are given in various NACE specifications:

- NACE TM0185-2000, "Evaluation of Internal Plastic Coatings for Corrosion Control of Tubular Goods by Autoclave Testing"
- NACE RP0181, "Liquid-Applied Internal Protective Coatings for Oilfield Production Equipment"
- NACE RP0188, "Discontinuity (Holiday) Testing of Protective Coatings"
- NACE RP0490, "Holiday Detection of Fusion-Bonded Epoxy External Pipeline Coatings of 250 to 760 mm (10 to 30 mils)"
- NACE RP0191, "The Application of Internal Plastic Coatings for Oilfield Tubular Goods and Accessories"
- NACE RP0394, "Application, Performance, and Quality Control of Plant-Applied, Fusion-Bonded Epoxy External Pipe Coating"
- NACE TM0183, "Evaluation of Internal Plastic Coatings for Corrosion Control of Tubular Goods in an Aqueous Flowing Environment"

On a practical basis, one of the most misunderstood aspects of using coatings is that the inhibitor dosage will not usually change even though an operator decides to use coated tubing versus that used for bare tubing. However, the use of coated tubing does improve the protection in shielded areas that are inaccessible to inhibitors. It also reduces the exposed area of corroding steel and therefore also reduces the amount of corrosion products in the system.

The two greatest dangers to coated tubing are wireline damage and improper joint selection. Wireline damage can be minimized by adjusting running procedures to include wireline guides and to slow wireline speed (<0.5 m/s, or <100 ft/min). Inhibiting immediately after wireline work is also good practice in an attempt to establish protection of any fresh, exposed metal. Proper joint selection involves choosing a joint that allows the coating to be applied around the pin-end nose into the first few pin threads and from the first few coupling threads into the coupling body. The proper joint allows the coating to remain undamaged. Often, a corrosion barrier compression ring is used to accomplish this end. Joint designs that involve metal-to-metal sealing are not joints that can be coated.

The coatings used for tubing protection are polyurethane, phenolformaldehyde, epoxized cresol novolac, and epoxy resins. Suitability for service is and should be based on laboratory testing or substantial service experience in the specific environment and at the temperature proposed for the service.

The quality-control parameters of concern are tubing surface finish/preparation, application techniques, coating thickness, holiday detection, joint condition, and inspection. By far, most coating failures are related to application procedures and not from incompatibility with well fluids or thermal degradation. Inspection is required to ensure the suitability of the surface preparation and coating process. Quality control and surveillance are as much a part of a successful protective coating program as choosing the appropriate coating. The production coating must be applied in the same way the coating was applied to the test specimens, and a good coating specification and third-party inspection are important. Coated pipe and couplings must be carefully handled after coating, during shipping, in storage, and at the well site. The threads must be protected from impact with other pipe and objects.

Wireline work is necessary. These operations can be accomplished with a minimal amount of damage to the tubing if certain details are attended to:

- Wireline speeds are kept to 0.5 m/s (100 ft/min) or less.
- All sharp edges are removed from the tools.
- All tools are plastic coated or covered with a plastic sleeve.
- Wheeled centralizers are used on all center hole jobs.

Using the above precautions, many wireline trips can be made with little or no damage. The fact that coating should not be used because the wireline will cut the coating and cause accelerated corrosion in the wireline track is not true. When the wireline cuts uncoated tubing, it exposes the underlying steel. In properly coated tubing, the coating electrically insulates the cathodic areas so that the corrosion rate in the track is essentially the same as that for uncoated steel without wireline damage.

Wellheads, Christmas Trees, and Downhole Equipment. The exposed surfaces of wellhead equipment, Christmas trees, and downhole equipment must be treated with inhibitor, but are often coated or manufactured of corrosion-resistant materials. The tubing hanger is threaded onto the production tubing and continues through the tubing bonnet (tubing adapter), the master valve(s), the tee or cross, and the wing and crown valve(s) and into the choke. Ring gasket grooves, valve seat pockets, and other compression fitted parts must not be coated. These areas can be overlaid with corrosion-resistant alloys, and the coating can be applied over a transition area up to the corrosion-resistant alloy overlay. Valve internal cavities need not be coated if corrosion is otherwise under control.

Generally, tubing hangers are difficult to coat and are therefore made of corrosion-resistant alloys for corrosive service. Hangers without back-pressure valve threads can be coated, but the cost of coating the carbon steel hanger may be equivalent to the cost of a corrosion-resistant alloy hanger.

Downhole equipment (nipples, polished bore receptacles, seal subs, tie backs, millout subs, packers, and so on) use the same standards as tubing. Most downhole equipment is considered uncoatable because it was not designed for coating, and corrosion-resistant alloys are often used. Such equipment can be coated as an assembled unit, rather than as individual pieces. If installation is below the packers, both internal and external surfaces should be coated or corrosion-resistant alloys should be used.

Coated surface flow lines usually employ flanges or threads and couplings to join them. Joining by welding damages the coating, and field-applied repair coating is not recommended, as it is often an area for future coating deterioration and corrosion damage.

Vessels must be designed as a coated vessel or employ corrosion-resistant alloys. Coating a vessel that was not designed for coating is rarely successful. Vessel design for coating involves welded hangers for anodes in the fluid zone, flanged access, removable internals, smooth internal surfaces, and good access to all internal surfaces to be coated. As a general rule, coating over bolted assemblies is not successful.

Cathodic Protection

Corrosion occurs when an anode and a cathode are electrically connected in the presence of an electrolyte, and electrical currents leave a corroding metal and go into an electrolyte. A dc power source (or a metal that has a greater affinity toward corrosion) can be connected to the corroding metal and used to oppose these corrosion currents. If either of these produce currents sufficient to fully oppose the anodic corrosion current, the metal will be protected from corrosion. This technique is known as cathodic protection. The entire metal surface is converted into a cathode, while the corrosion currents are transferred to an auxiliary anode or sacrificial anode in which corrosion can proceed. Cathodic protection has been used in the oil field to protect pipelines, well casings, tanks and production vessels, and offshore platforms. More information on principles and applications of cathodic protection is available in the article "Cathodic Protection" in Volume 13A of the *ASM Handbook*.

Types of Cathodic Protection Systems

The two methods of applying cathodic protection discussed previously are: the sacrificial anode method and the impressed-current method. Because some metals are less noble (more electronegative) than the most common oil-field material—steel—in the galvanic series, they will become the anode (site of corrosion attack) when coupled to steel in the presence of an electrolyte. The most common of these materials are magnesium, zinc, aluminum, and similar alloys; these are called sacrificial anodes. These types of anodes are used when:

- Current requirements are relatively low.
- Electric power is not readily available.
- Short system life dictates a low capital investment.

The anode is usually electrically connected by a wire or steel strap to the structure to be protected. Magnesium and zinc are usually used in soils, while zinc and aluminum alloys can also be used in brine environments.

In the impressed-current method, an external energy source produces an electric current that is sent to the impressed-current anodes. The most common types of these materials include graphite, high-silicon cast iron, lead-silver alloy, platinum, and even scrap steel rails. These types of anodes are used when:

- Current requirements are high.
- Electrolyte resistivity is high.
- Fluctuation in current requirements will occur. These types of systems can be adjusted to compensate for varying current requirements.
- Electrical power is readily available, although this is not now as severe a limitation as it was in past years.

In a typical impressed-current system, ac from a power line flows into a rectifier where it is converted into dc. The dc then flows to the anode groundbed. Other means of supplying this electrical current include solar energy and thermoelectric generators. These methods are applicable at locations where conventional electric power is not economically available.

Solar energy has powered cathodic protection for well casings in Kansas (Ref 46) and Saudi Arabia (Ref 47); segments of a 480 km (300 mile) long, 0.8 m (32 in.) pipeline in Libya (Ref 48); and segments of a natural gas distribution system in Washington (Ref 49). In these systems, silicon semiconductor devices convert sunlight directly into dc electricity, which is then used for the anode groundbed and to charge batteries. These batteries provide current to the anode groundbeds during periods of little or no sunlight. The batteries must be checked periodically for proper electrolyte levels. Solar panels can be easily replaced or increased to attain a higher current output because they are fabricated in modules.

One oil company has used solar energy to protect about 800 well casings in western Kansas. The solar panels consist of individual silicon solar cells connected in series to form modules. These modules are then connected in parallel to form a panel that is rated at about 12 A and 4 V. Because the well casing in this field requires 1 to 2 A for cathodic protection, 2 V, 500 A·h lead-acid batteries are used. A rheostat controls the rate of current flow from the batteries, and the voltage regulator controls the battery charge rate. The batteries can provide electrical current to the anodes for 10 days even if the sun is completely blocked.

Another electrical current source is the thermoelectric generator. One type of thermoelectric generator is a system that uses a burner to heat an organic liquid in a vapor generator. This vapor then expands through a turbine wheel, thus producing power to a shaft to drive an alternator, where ac power is produced. The vapor then passes through a condenser where it is cooled and condensed back into a liquid to start the cycle again. The ac power is sent to the rectifier to be converted to dc. Figure 9 shows an illustration of this type of thermoelectric generator. These systems can produce a maximum of 80 A at 21 V (Ref 50).

Application of Cathodic Protection to Oil-Field Equipment

Pipelines. Cathodic protection of pipelines is very common in oil-field operations. As discussed previously in this section, sacrificial anodes or impressed current can be used for cathodic protection (Fig. 10). If the pipeline is well coated and not very long, the current requirements will probably be achieved with sacrificial anodes. If bare, a steel pipeline could require 1.1 mA/m^2 (0.1 mA/ft^2) in soil, while a very well coated pipeline could require only 0.003 mA/m^2 (0.0003 mA/ft^2) or less for cathodic protection (Ref 51).

The resistivity of soils (and therefore their corrosivity) will also vary with location. Differences in aeration, soil composition (sand or clay), and the presence of chemical spills are just a few of the factors that will affect the corrosivity of the soil. Sometimes, the resistivity of the surface soils is so high that conventional groundbeds (1.8 m, or 6 ft, deep) cannot be used. Conventional groundbeds are normally used when the soil resistivities near the surface of the ground are less than 5000 $\Omega \cdot$cm (Ref 52–54).

In high-resistivity soils, deep groundbeds can be used where the anodes are installed vertically in holes at depths of 15 m (50 ft) or more. Deep groundbeds (Fig. 11) are normally used to provide a better distribution of current than conventional groundbeds. These types of groundbeds also minimize right-of-way problems and are essentially unaffected by seasonal moisture variations. They are more expensive to install than conventional groundbeds, and it is usually impossible to repair any damage to the cable insulation. These systems use the same type of anodes as the conventional groundbeds. The major difference is that a perforated vent pipe can be installed to prevent chlorine gas from accumulating around the anodes. If these gases collect around the anodes, they form an insulating barrier that increases the resistance of the groundbed and eventually causes the groundbed to become ineffective.

Well Casings. The first step in externally protecting a well casing is to cement through any corrosive zones. Cement acts as a coating and will significantly reduce, but not completely stop, corrosion of the casing. Therefore, cathodic protection is needed to supplement the cement coating (Fig. 12). The experience of one company with the cathodic protection of well casings showed an 88% success rate in preventing predicted casing failures (Ref 55). Although a single anode bed for a buried pipeline may protect as much as 80 km (50 miles), the maximum amount of casing that needs to be protected usually does not exceed 2.4 to 3.2 km (1.5 to 2 miles). One company coated nine casing strings in 3500 m (11,500 ft) wells with fusion-bonded epoxy in Florida (Ref 56). The coating was used to reduce current requirements and to improve current distribution. Uncoated casing strings were protected with 22 to 25 A, and even then there was incomplete corrosion control. Only 10 A were needed to protect the coated casing. Some of these casing strings were pulled because of other operational problems, and the coating was found to be in excellent condition.

Another method of reducing the current requirement is to place an insulating joint in the flow line at the wellhead. This joint prevents current from the flow line from flowing into the wellhead and down the casing. This current would leave the casing at low-resistivity zones (Ref 54–56).

Two methods are generally used to determine current requirements for well casing: casing

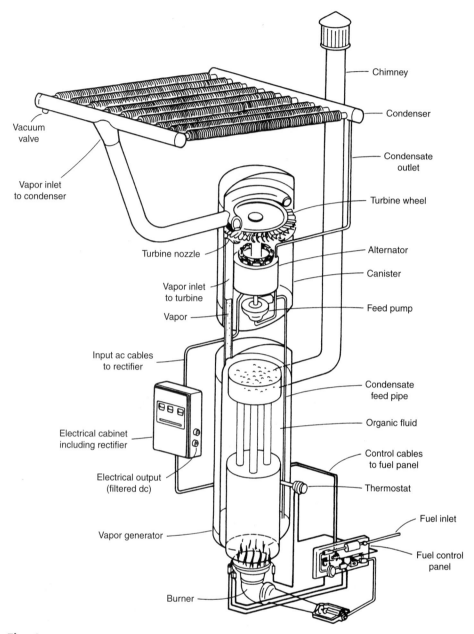

Fig. 9 Thermoelectric generator used to power cathodic protection systems in remote locations. Source: Ref 50

potential profile and E-log I. The casing potential profile is measured by using a tool that consists of two spring-loaded probes approximately 7.6 m (25 ft) apart. This tool is pulled through the casing, and voltage readings are taken between the probes as they contact the casing every 15 or 30 m (50 or 100 ft). A plot of potential versus depth is made (Fig. 13). A slope upward and to the left indicates an anodic area, while a slope upward and to the right indicates a cathodic area. Current from a temporary groundbed is then applied to the casing for protection, and another potential profile is taken. Current is increased until the profile slopes are upward and to the right. Providing a profile slope to the right does not necessarily mean that all of the casing is protected, but it does mean that all gross corrosion areas have been eliminated.

Another technique for determining the required current is the E-log I curve. It is less expensive and does not require the disturbance of subsurface equipment. The flow line to the well, however, must be isolated from the well casing. The pipe-to-soil potential relative to the Cu-CuSO$_4$ reference electrode is measured, and a small amount of current from a temporary groundbed is applied. The current is then interrupted, and the pipe-to-soil potential is measured as quickly as possible. The current is then increased a small amount, and the process is repeated to obtain a curve similar to that shown in Fig. 14. Generally, the current required corresponds to the break in the curve.

Tanks and Production Vessels. Internal corrosion of water-handling tanks and vessels can be controlled by the use of cathodic protection. Even if a coating has been applied to the interior of a water storage tank, there will always be imperfections where corrosion can occur; therefore, cathodic protection is needed. Sacrificial anodes can be suspended from the top of the tank as shown in Fig. 15 to offer protection to the portion of the tank that is covered with water. Cathodic protection will not help in the vapor area of the tank. Coatings must be used to protect this area. Sacrificial anodes have also been placed on concrete blocks in tanks. These blocks insulate the anode from the tank and allow decomposition products to fall away from the anode. In most cases, a lead wire is brought outside of the tank and welded to the tank. These anodes should also not be allowed to touch the sides of the tank and should be uniformly distributed within the tank to give uniform current distribution.

In vessels that have several sections separated by steel plates, the anodes might be shielded from protecting the entire vessel. In these cases, the only safe procedure is to install an anode in each compartment. Similar tanks without cathodic protection in one southern Texas field failed from corrosion in 3 months (Ref 57, 58). When ac power is available, impressed-current systems using high-silicon cast iron, graphite, or platinized titanium anodes in through-the-wall mounts have also been used. Crude oil and some oil-field chemicals have tended to stifle the flow of current from these anodes.

Offshore Platforms. The subsea zone of an offshore platform includes the area from the splash zone down to and including the pilings below the mudline. Cathodic protection is the principal means of preventing corrosion in this zone, but some companies also use coatings in conjunction with cathodic protection (Ref 59). The amount of electric current required to protect the bare steel varies with location. Typical current density values range from 54 to 65 mA/m^2 (5 to 6 mA/ft^2) in the Gulf of Mexico, 86 to 160 mA/m^2 (8 to 15 mA/ft^2) in the North Sea, and as high as 375 to 430 mA/m^2 (35 to 40 mA/ft^2) in the Cook Inlet (Ref 51). Current densities in the Cook Inlet are high because of the 8 knot tidal currents and entrained particulate matter experienced at this location. In the mud zone, current densities of 10.8 to 32 mA/m^2 (1 to 3 mA/ft^2) are needed for protection, and an allowance of 3 A per well is customary for well casings (Ref 60).

Additionally, there are also concerns for cathodic protection requirements for drilling and production applications in deep sea applications. Under these conditions, the seawater is at low temperatures, around 4 to 8 °C (40 to 50 °F), and the formation of calcareous deposits is retarded due to the increase in solubility of calcium carbonate in the colder seawater (Fig. 16) (Ref 61). This has the effect of increasing cathodic protection current requirements. In some cases, conventional sacrificial anode cathodic protection is supplemented by the application of thermal sprayed aluminum coatings on the surface of steel members and equipment. Under such conditions, environmental factors, such as oxygen content, water salinity, temperature, velocity, erosive effects, marine growth, and calcareous deposits, are largely responsible for the differences in current densities. It is very important that the current demand be conservatively estimated as retrofitting of cathodic protection systems in deep water applications can be prohibitively expensive.

Partial protection of the steel in seawater usually means that the area of corrosion is reduced, while the unprotected areas continue to corrode at a high rate. Some companies report

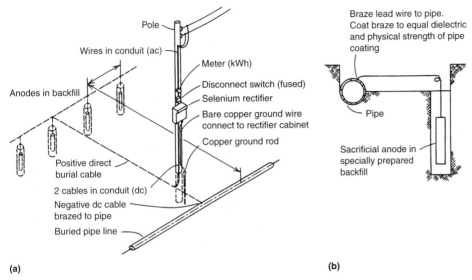

Fig. 10 Typical cathodic protection installations. (a) Impressed current. (b) Sacrificial anode

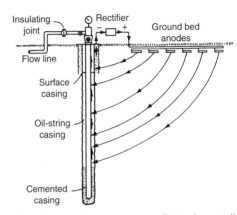

Fig. 12 Cathodic protection installation for a well casing. Source: Ref 54

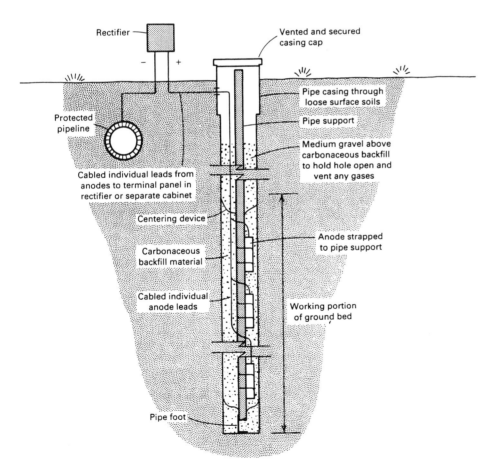

Fig. 11 Typical deep groundbed cathodic protection installation. Source: Ref 53

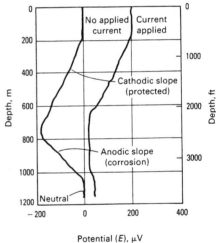

Fig. 13 Casing potential profile curve. Source: Ref 56

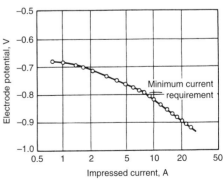

Fig. 14 Typical E-log I curve. The break in the curve indicates the minimum cathodic protection current requirement. Source: Ref 56

pits as deep as 13 to 16 mm (0.5 to 0.625 in.) and, in many cases, holes in platform members after less than 5 years on location without adequate cathodic protection. This result corresponds to a corrosion rate of 2.5 to 3.2 mm/yr (100 to 125 mils/yr) (Ref 62).

As in all applications of corrosion control on offshore platforms, the first step for cathodic protection in the subsea zone is design. Tubular members should be used whenever possible. Recessed corners in channels and I-beams are difficult to protect. Even crevices formed by placing channels back-to-back and noncontinuous welded joints cannot be protected. Bolted and riveted fittings should be avoided. Piping such as grout lines, discharge lines, water supply casings, and pipeline risers, if clustered around a platform leg, can cause shielding and interfere with the flow of cathodic protection current. If economically feasible, piping that is not necessary for platform operations should be removed. A minimum clear spacing of $1\frac{1}{2}$ diameters of the smaller pipe should be provided, and coatings on the pipe can also be used to minimize shielding. Corrosion will be negligible on the internal surfaces of structural members that are sealed and have no contact with either the atmosphere or the seawater. During launch, some structural members are flooded for the life of the platform. To prevent any internal corrosion, the flooding valves should be closed to isolate the flooded chambers from contact with fresh seawater or oxygen in the atmosphere.

Sacrificial Anode Systems. The early offshore platforms installed in the late 1940s and early 1950s used 45 and 68 kg (100 and 150 lb) magnesium anodes supported from horizontal braces. A low-carbon steel wire rope connected the anode to the brace. These anodes had a 2 year design life, which was normally shortened to 1 year or less because of hurricane or rough weather losses. These swinging anodes tangled with subsea braces, shorted, and rubbed the conductor wires to failure. A variable resistor was also connected in series with the anode and the connection at the brace. This resistor was used to regulate the current output of the anode to achieve maximum efficiency. Unfortunately, this system could never be maintained. Magnesium anodes in seawater have a high current output and corrode rapidly; therefore, they must be replaced frequently. This type of system has been discontinued for offshore use.

Zinc anodes have been used for more than 200 years. However, impurities, such as iron, were responsible for erratic performance. Virtually all zinc anodes are now fabricated from high-purity zinc meeting the military specification MIL-A-18001-H. More efficient aluminum anodes have also been developed. A Hg-Zn-Al alloy anode provides $2\frac{1}{2}$ times more current output than a zinc anode on a pound-for-pound basis. However, concerns for mercury on sea life has led to the use of other compositions for high-efficiency aluminum (including aluminum with variable zinc content, and aluminum-indium anodes). These compositions are shown in Table 4 (Ref 63). An anode weight of 330 kg (725 lb) is common for the initial system, while 150 kg (325 lb) is common for the replacement system. Proper specification of alloy composition and control of alloying and impurities in the anode are essential for best performance.

The number of anodes needed depends on the size of the anode and its useful life. A design life of 20 years is common. The distribution of anodes is also important, because poor distribution and the use of too few anodes will result in underprotection, particularly at welded joints. Individual anodes should be mounted at least 0.3 m (12 in.) from the structure, or a dielectric shield should be used beneath them to improve the current distribution. Anodes should not be located in either the splash zone or on bottom bracings. The anode will not function properly if it is intermittently in and out of the seawater, and mercury-containing aluminum anodes will passivate and not function if covered by mud. Some of the earliest anodes used were prone to being knocked off of the platform during installation because the standoff posts were either too small or did not provide adequate area for contact welding of the anode to the member. Gussets and doubler plates can help obtain better anode attachment. A large fraction of premature cathodic protection failures have been traced to an inadequate number of anodes installed or excessive losses during pile driving. Some of these losses during pile driving are due to poor weld quality in attaching the anodes to the members.

Impressed-Current Systems. In an impressed-current system, the three essential components are the rectifier, the anodes, and the cable joining them together. The anode materials include graphite, high-silicon cast iron, lead-silver, and platinum wound on a niobium rod. Permanently mounted anodes, retrievable anodes, and remote anode sleds have all been successfully used. Because anodes in impressed-current systems generally produce considerably more current than sacrificial anodes, there may be only 6 or 8 impressed anodes on a structure that might do the same job as 50 to 70 sacrificial anodes. The location of these impressed-current anodes is very important in order to ensure that adequate current distribution is obtained to cover the entire surface. The connecting wiring is the critical part of an impressed-current system, especially in the splash zone, where the cable can be subjected to severe wave pounding if it is not housed in a protective conduit. Even these conduits can be torn away from the platform during a hurricane if they have been underdesigned. It is often necessary to protect the structural member near the anode with fiberglass coating or a wrap called a dielectric shield. This shield prevents excessive current consumption at this area.

Whether sacrificial or impressed anodes are used, cathodic protection currents will promote the formation of hydroxyl (OH^-) ions at cathodic areas (the entire platform, it is hoped) and cause a pH shift in the seawater near the platform. Also, the concentration of calcium and magnesium ions tends to increase in the film of seawater over the cathode. As a result of these changes, the solubility of calcium carbonate and magnesium hydroxide is exceeded, and a calcareous coating is deposited. These mineral deposits provide the primary corrosion control, and the cathodic protection current demand

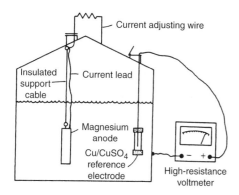

Fig. 15 Cathodic protection system for a water storage tank. Source: Ref 57

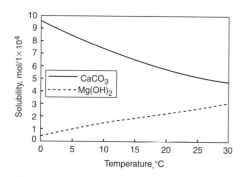

Fig. 16 Influence of temperature on solubility of calcium carbonate and magnesium oxide. Source: Ref 61

Table 4 Various sacrificial anode compositions used to protect offshore pipelines and structures

Anode	Composition, wt%								
	Si	Fe	Cu	Zn	Cd	In	Hg	Al	Pb
A	0.09	0.051	<0.002	6.10	<0.01	0.020	ND	bal	...
B	0.08	0.025	<0.002	4.81	<0.01	0.017	ND	bal	...
C	0.09	0.028	<0.002	5.51	<0.01	0.018	ND	bal	...
Z	...	0.003	<0.002	bal	...	...	...	0.24	0.003

ND, not detected. Source: Ref 63

drops to a level sufficient to repair this coating when it is damaged. For example, if a current density of 540 mA/m^2 (50 mA/ft^2) is applied to a platform for the first 5 days on location, protection can be maintained with a current density of 32 mA/m^2 (3 mA/ft^2).

There appears to be less tendency for these mineral deposits to form in the deep ocean. In an experiment conducted by the U.S. Navy, sacrificial anodes were effective at providing cathodic protection to bare steel in seawater at depths of 1700 m (5600 ft) (Ref 64). However, the anodes were consumed more rapidly than if they were located near the surface. Because the pH is lower at great depths and the calcium carbonate concentration is below saturation, higher currents are required to achieve protection.

Inhibitors

Corrosion inhibitors are chemicals that, when present in a system in relatively small quantities, produce a reaction with the metal surface resulting in a reduction in the rate or severity of corrosive attack. Inhibitors can interfere with either the anodic or cathodic reactions, can form a protective barrier on the metal surface against corrosive agents, or can work by a combination of these actions. For oil-field corrosion inhibitors, organic compounds containing nitrogen (amines) dominate because of their effectiveness under conditions of oil and water wetting, and their availability. Corrosion inhibitors usually contain three elements:

- One or more active inhibitor components
- A solvent base
- Certain additives, such as surfactants, dispersants, demulsifiers, and defoamers

Solvents are used to dilute inhibitors to control their physical characteristics (such as viscosity, vapor pressure, and pour point), to aid in obtaining proper inhibitor concentration and placement during treating, to assist inhibition, and to maintain a reasonable cost per unit volume (Ref 65).

Physical Characteristics of Inhibitors

Physical characteristics of inhibitors must be considered when evaluating a potential application. These include:

- Physical form
- Solubility and/or dispersability
- Emulsion-forming tendencies
- Thermal stability
- Compatibility with other chemicals

In modern offshore operations, it is particularly important to make sure that the corrosion inhibitors are compatible with other completion and workover fluids. Many times, these fluids are handled through the same control lines and pumped down to the sea bottom and through the wellhead or into subsea flow lines. Concerns for incompatibility include: loss of inhibitor efficacy and formation of precipitates, emulsions, or gels that can plug the system.

Physical Form. Inhibitors may take either a solid or liquid form. Solid inhibitors have been made in the shape of a stick that will sink to the bottom of a well and then slowly dissolve and be drawn back with the produced oil and gas. These sticks are rarely used. Most corrosion inhibitors are in liquid form and have densities that range from 840 to 1440 g/L (7 to 12 lb/gal) (Ref 66). These liquids must not freeze when exposed to the coldest of field conditions and must be stable with a minimum loss to the vapor state when exposed to the hottest of field conditions.

Solubility/Dispersibility. The formation of an inhibitor film and its life are primarily governed by the solubility of that product in the system. There are three categories of solubility: soluble, insoluble, and dispersible.

A product is soluble in a fluid when it forms a clear mixture that does not separate. A product is insoluble in a fluid when it will separate after mixing to form an identifiable layer. Materials are dispersible in a fluid if they form a mixture that is not clear and separates slowly, if at all.

Different solubilities are required for different applications of corrosion inhibitors. An inhibitor to be added continuously to a waterflood should be water soluble or highly dispersible. Similarly, an inhibitor to be used for a squeeze treatment (e.g., where inhibitor and its carrier are squeezed into the formation under high pressure and then produced back) should be completely soluble in the carrying fluid to facilitate placement of the inhibitor without plugging the formation. On the other hand, where the only method of application is a periodic treatment, continuing protection requires some degree of insolubility of the inhibitor in the fluids to which it is exposed. In practical terms, this means that an inhibitor used in tubing displacement cannot be completely soluble in the well fluids. This is what gives the inhibitor its ability to remain for an extended period on the metal surface (i.e., referred to as its persistence on the metal surface). Also, a dispersion must be stable enough to remain intact until the inhibitor reaches the metal surface to be protected.

Emulsion-Forming Tendencies. Because of the chemical nature of most corrosion inhibitors, there is a positive tendency in water-oil systems to form emulsions. Some of these emulsions will break down quite readily under static or stratified flow conditions, while others are extremely stable and practically impossible to break. When squeezed into an oil or gas bearing reservoir, an incorrectly selected inhibitor can form an emulsion in the formation that blocks or severely restricts further production. Furthermore, a chemical inhibitor can also form an emulsion with produced water that is hard to "break" and that may complicate downstream separation of oil and water.

The inclusion of a demulsifier in a corrosion inhibitor is no guarantee against the formation of stable emulsions. Produced fluids from each field must be tested to provide reasonable assurance that no stable emulsion will be formed upon application of a specific corrosion inhibitor.

Thermal Stability. Corrosion inhibitors generally have temperature limits above which they lose their effectiveness and change their chemical composition. This temperature may be variable for any one inhibitor, depending on such conditions as pressure and presence of water. A typical example is that of an acid-amine salt. Under atmospheric conditions, this salt will yield water and form an amide at 70 to 90 °C (160 to 190 °F). However, this chemical can be used in oil wells in the presence of water at these temperatures with no apparent degradation. Of course, exposure to high temperatures at low pressures will result in the vaporization of the solvent systems in these inhibitors and the formation of "gunk," a solid organic compound that can interfere with well production and related operations.

Compatibility with Other Chemicals. The compatibility of corrosion inhibitors with other chemicals is ordinarily not troublesome when the inhibitor and the other chemicals are present in parts per million concentrations. However, chemical users frequently want to mix various chemicals so that a single chemical pump can be used for injection. Many products are not compatible with corrosion inhibitors because of variations in solvent systems, type of chemicals (cationic versus anionic), and so on.

Most oil-field corrosion inhibitors are cationic to some extent; that is, they carry a positive electrical charge. Mixing a cationic inhibitor with an anionic chemical, such as a scale inhibitor or certain surfactants, will likely produce a reaction product that can have characteristics that are entirely different from those of either of its parent products. At best, the new material may function poorly; at worst, it may not function at all or may even form deposits in the system and plug control systems or even the well. When two or more chemicals must be used, this problem can be prevented by using separate injection points that are not closely spaced. It should probably be standard practice never to mix any two different products unless extensive compatibility and stability tests have been conducted in advance of the application. Depending on the situation, either product would avoid potential emulsion, solids precipitation, or fouling problems.

A final example of operating problems can be found where a conventional inhibitor is used in a gas stream upstream of a compressor. The nonvolatile components in the inhibitor could be left behind to foul the valves of the compressor. This latter phenomenon is commonly referred to as "gunking."

Selection of Inhibitors

Many factors are involved in the selection of inhibitors, including:

- Identification of the problem to be solved and target corrosion rate requirements

- Corrosives present in the production environment (e.g., chloride, oxygen, CO_2, H_2S, sulfur, acids)
- Type of system (offshore or onshore, oil or gas or water or multiphase operations, etc.) that may influence the treatment method
- Pressure and temperature
- Production rates of oil, water, and/or gas
- Pipe sizes
- Production composition

Although problems such as rod breaks in pumped wells and leaks in gathering lines may initially be seen as purely corrosion failures, the actual cause of the problem could be oxygen, scale, or bacteria. Rod coupling failures could be caused by poor assembly or corrosion fatigue. Overstressing of rods greatly accelerates these failures. In such cases, mechanical measures could reduce or eliminate the need for chemical treatment. If attack is due to oxygen entry, installation of gas blankets or closing of the casing valve could greatly reduce corrosion.

The presence of corrosives such as H_2S and CO_2 greatly influences the choice of an inhibitor. Some inhibitors perform best in sweet fluids, while other inhibitors work best in sour fluids. Even the concentration of sodium chloride (NaCl) has a bearing on the choice of an inhibitor. With increasing NaCl content, some inhibitors will become insoluble and deposit. Table 5 shows the influence of corrosive type on inhibitor selection (Ref 67).

The type of system also has an effect on the selection of an inhibitor. The correct inhibitor to use is determined by whether the system is a pumping oil well, a gas-lift well, a gas well, a waterflood system, or a flow line. For example, a weighted inhibitor is seldom recommended in dry gas wells, because water is required to release the inhibitor before it becomes effective (Ref 68). Also, when a gas-lift well is treated, the inhibitor is injected into the gas-lift lines. Therefore, the inhibitor must not have tendencies to form gunky deposits. The same is true if a capillary tube is being used for downhole inhibitor injection.

Both temperature and pressure have an influence on inhibitor selection. Bottom hole temperatures and pressures may get so high that inhibitors polymerize and form a sludge. Pressure influences the corrosivity of CO_2 and H_2S. The severity of corrosion increases with the partial pressures of these gases. Therefore, certain chemicals may work fine at one condition, but require an unacceptably high concentration to handle the same gas at higher pressure or temperature.

Velocity is yet another factor to consider. With pipelines, low velocity might be insufficient to displace water from low areas in the line. In the case of dry gas pipelines with low velocity, a water-soluble inhibitor should be selected and should be injected continuously. If the velocity is high enough to prevent any accumulations of water in low areas of a dry gas line, then an oil-soluble inhibitor should be batch treated. In addition to velocity, the rates of production of oil, gas, and water and the pipe size also determine the flow regime. As discussed previously in this article, certain flow conditions (e.g., slug flow and mist flow) can produce very high turbulence and wall shear stress on the surface of the pipe. Under these conditions, conventional chemical inhibitors that work well under less aggressive flow conditions often will not work under high shear conditions. Special inhibitor formulations that have been developed and that have high reaction rates and greater persistence under these conditions are required (Ref 69). Conversely, low flow conditions may result in stratified flow conditions where the oil and water phases separate. In this case, special consideration needs to be given to select oil-soluble inhibitors that are also highly water-dispersible inhibitors that do not require extensive turbulence to achieve dispersibility.

The composition of the produced water also determines the choice of inhibitor. Criteria such as water/oil ratio, salinity of water, and acidity of the water and oil are vital to the correct selection of the inhibitor.

Commonly, tests are conducted in the laboratory using a standard wheel test where a corrosion coupon is exposed to a mixture of oil, water, and gas in small pressurized vessels with varying concentrations of inhibitors that are mixed by rotation. Although not infallible, this test attempts to duplicate field conditions and is credible for simple situations not involving high flow induced wall shear stress. Parameters such as temperature, water-cut, batch or continuous treatment, and whether the system is sweet or sour are reproduced as closely as possible to field conditions. Usually, these tests involve weight-loss coupons and can also involve use of electrochemical methods. ASTM G 170 provides guidance for basic inhibitor-screening procedures (Ref 70).

To develop more predictive tests, sometimes stirred autoclaves, flow loops, and jet-impingement devices are utilized to create certain test conditions that simultaneously have high H_2S/CO_2 partial pressure, concentrated brine, and high flow rate. Most importantly, it is crucial to have some basis to link the laboratory results and the field applications (Ref 71). This is where corrosion and flow modeling have also been utilized. The procedures for these tests often start with flow modeling of specific field production system under various production scenarios (high and low flow rate, high and low water production, etc.). This often involves examining many cases to define the worst-case conditions for water holdup and/or flow-induced wall shear stress and corrosion severity. These conditions can be then be simulated in the laboratory for evaluation purposes. The laboratory tests can be conducted under similar chemical conditions (production environment and dosage of inhibitor chemicals) while also controlling the wall shear stress to levels appropriate for the application.

The rate at which an inhibitor film is formed on a metal surface is completely dependent on the product formulation and the service environment. Generally, inhibitors that attain an effective surface film quickly operate better, particularly under flowing conditions. It can generally be said that film formation is a function of time and is not instantaneous. The concentration of inhibitor required to develop an adequate film is also directly related to the characteristics of the product, the severity of the corrosion, and the specific system parameters including temperature, pressure, and flow rate. Many factors affect the dosage and frequency of treatment, including:

- Severity of corrosion
- Total amount of various fluids produced (oil/gas/water)

Table 5 Influence of corrosion type on selection of inhibitor formulation type

Field designation (gas wells)	H_2S content, ppm	Inhibitor type(a)	General corrosion rate mm/yr	General corrosion rate mils/yr
A	0	N	0.190	7.5
		NPS	0.003	0.12
B	0	N	0.095	3.7
		NPS	0.005	0.20
C	8	N	0.010	0.40
		NPS	0.003	0.12
D	12	N	0.015	0.60
		NPS	0.005	0.20
E	600	N	0.005	0.20
		NPS	0.005	0.20

System type	Reason for oxygen contamination	Inhibitor type(a)	Pitting corrosion rate mm/yr	Pitting corrosion rate mils/yr	Tubing failure rate
10 hydraulic wells	Transfer pump leaks	N	16	630	50 leaks/yr
		NPS	0.25	9.8	4 leaks/yr
2 sweet rod pumped wells	Negative pressure gas gathering	N	3.0	118	36 tubing pulls/yr
		NPS	0.25	9.8	2 tubing pulls/yr
3 sour rod pumped wells	Low annulus pressure	N	0.7	27.6	36 tubing pulls/yr
		NPS	0.13	5.2	3 tubing pulls/yr

(a) N, inhibitor formulation with organic nitrogen compounds; NPS, inhibitor formulation with organic nitrogen compounds and phosphorus and sulfur compounds. Source: Ref 67

- Percentage of water (i.e., water cut in oil systems; water-to-gas ratio in gas systems)
- Nature and concentration of corrodents (H_2S, CO_2, O_2, etc.)
- Chemical type and formulation (inhibitor chemicals plus carrier fluid)
- Fluid level in the casing annulus

Because no laboratory test can take into account all of the conditions imposed by the oil well, the dosage and frequency of treatment must be constantly reviewed. Initially, the relative performance of inhibitors and their optimum dosage may be developed from laboratory screening tests. However, these recommendations may be changed based on prior experience in similar well conditions.

There are two general rules to follow for dosages. First, for continuous injection, a dosage of 10 to 20 ppm based on total produced fluid is used as a starting point. Second, for batch treatment, weekly batch frequency is used with a starting dosage of 3.8 L (1 gal) per week for each 100 barrels of daily fluid production.

Also, if the corrosivity of the system is known, the following general criteria can be used to define more accurately a treatment rate for continuous injection (Ref 72):

Severity of corrosion	Target inhibitor dosage
Mild corrosion	10–15 ppm
Moderate corrosion	15–25 ppm
Severe corrosion	>25 ppm

Several major problems can occur with inhibitors, including foaming, emulsions, scale removal and plugging, and safety and handling. Corrosion of other metals (e.g., stainless alloys, brasses, and copper-nickel alloys) as a result of exposure to corrosion inhibitors intended to work on steel can also be a problem in some applications.

The most appropriate action to take in avoiding difficulty from foaming is to determine where foam-forming conditions exist in the system. These will consist of places where the inhibitor-containing fluid is agitated with a gas, such as in a gas separator, a countercurrent stripper, or an aerator. The next step is to obtain a sample of the fluid and gas from the process step, add the inhibitor in question, adjust the temperature to that corresponding to the process step, and shake vigorously. If this test produces a stable foam, a potential problem exists. There are three potential remedies:

- An antifoaming agent can be added (this must be tested also).
- Tests can be conducted to select an inhibitor that does not cause foaming.
- The system can be shut down periodically and treated with a slug of persistent inhibitor.

The last two remedies are the least palatable because the need for an inhibitor is at hand and there are few processes that can be shut down with sufficient frequency to maintain effective inhibition by slug treatment.

Emulsions are another problem that can occur when the wrong inhibitor is used. The use of other chemicals, heat, or both can usually break these emulsions. A great variety of chemicals are used for this purpose, but no one material has proved effective for all emulsions. Therefore, some evaluation or testing is often required to select the correct chemicals.

A system can be plugged as the result of an inhibitor-loosening scale and suspending it in the fluid. This problem is best avoided by planning. The best preventive measure is to clean the system thoroughly, if possible, before inhibitor is applied. This can be with the use of scrapers, or pigs that can be passed through the system for removal. An alternative or supplementary method in systems that are very sensitive to suspended solids is to protect the sensitive parts with temporary filters.

As with most industrial chemicals handled in large volume on a regular basis, oil-field corrosion inhibitors should be treated with respect from a safety standpoint. These products contain complex formulations of highly reactive organic materials. While not always highly toxic (many acid corrosion inhibitor formulations are toxic), they can produce reactions because of the amines and aromatic solvents present. Reactions usually consist of skin burns from contact and dizziness from inhalation of the vapors. Field procedures that involve such practices should be eliminated or adequate safety precautions should be taken. Repeated contact with amines will cause the development of sensitization to these products in some individuals and potential long-term health problems. To avoid these problems, any contact with the body should be minimized through the use of safety equipment.

Another possible adverse effect of inhibition is an increased rate of corrosion of a metal in the system other than the one for which the inhibitor was selected to protect. For example, some amines protect steel admirably, but will severely attack copper and brass. Nitrites may attack lead and lead alloys, such as solder. It has also been reported that some amine inhibitor formulations can cause attack and deterioration of elastomeric seals (Ref 73) particularly at temperature commonly found in downhole environments. In some cases, inhibitors may react in the system to produce a harmful product. An illustration of this is the reduction of nitrate inhibitors to form ammonia, which causes SCC of copper and brass. The only way to avoid these problems is to know the metallic components of a system and to be thoroughly familiar with the properties of the inhibitor to be used.

Application of Inhibitors

Choosing the proper inhibitor for treating a corrosion problem in the oil field is important; however, it is equally important to select the correct treating method. The best inhibitor available will not successfully control corrosion if it does not reach the trouble area. To be effective and economical, a corrosion inhibitor:

- Must be present at an initial concentration sufficient to promote complete coverage of all steel surfaces
- Must be replenished as necessary to repair and replenish the protective inhibitor film

Batch Treatments. These are commonly used in producing wells and, in some cases, in gas lines and crude flow lines (Fig. 17). Inhibitor can be batched down the tubing-casing annulus, through the tubing, or between pigs (in the case of a pipeline). The various types of batch treatment are:

- Standard batch
- Extended batch
- Annular slug
- Tubing displacement
- Between pigs batch

The limiting aspects of batch treatment are usually related to the severity of corrosion, the temperature of the system relative to the breakdown temperature for the active agents in the

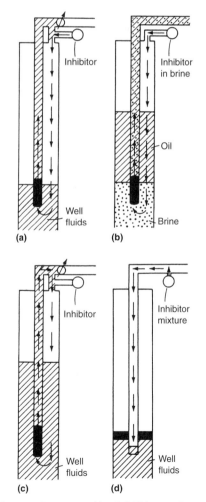

Fig. 17 Four types of batch inhibitor treating techniques. (a) Standard batch. (b) Annular slug. (c) Extended batch. (d) Tubing displacement

inhibitor, and the level of flow-induced wall shear stress. However, batch treatments can be both an effective and economically attractive method of inhibitor delivery for systems that are not overly aggressive, do not have excessively high wall shear stresses associated with the flow, or may not have the required equipment available for continuous inhibitor injection.

Standard Batch. This method is used for producing wells that are not equipped with packers. The inhibitor is put into the annulus, and the well is placed on circulation to distribute the inhibitor throughout the system. Normally, the longer the well is circulated, the better the inhibitor film. The application of this treatment in low fluid level wells depends on the fluid level maintained in the annulus. This method would not be recommended in wells that pump off. It would be estimated that a fluid level of at least 46 m (150 ft) should be maintained. In placing the treatment in operation in these wells, it would be recommended that the initial treatment be immediately displaced into the tubing and that a second batch of inhibitor be placed in the annulus.

Extended Batch. This method is a variation of the standard batch treatment, but in this case the inhibitor is left in the annulus. As the annular fluid level fluctuates, small amounts of inhibitor are carried in the oil into the tubing, thus giving the well periodic treatments weeks or months after the actual treatment. This type of treatment has lasted up to 6 months in some wells of Oklahoma (Ref 74). It must be remembered that this technique depends on a substantial fluid level in the annulus because the inhibitor is inventoried in the oil of the annular space.

Annular Slug. There is one technique for batch treating pumping wells that allows the well to continue full production while being treated. A water-dispersible or water-soluble inhibitor is mixed with water and placed in the annulus. This mixture will fall through the oil phase in the annulus. This technique will work if there is little or no water level in the annulus, but probably will not work if there is a substantial water level in the annulus. The frequency of treatment ranges from twice weekly to monthly.

Tubing Displacement. Wells that are set on packers or gas-lift wells are most commonly treated by tubing displacement. The inhibitor is either dispersed or put in solution in water or hydrocarbon carrier. The water may be fresh or produced. The hydrocarbon may be produced, or it may be a refined product, such as kerosene or diesel fuel. The inhibitor is usually used at about 10% concentration in the water or hydrocarbon. The desired amount of this mixture is then introduced into the tubing. If the well is a dry gas well, the mixture will fall to the bottom if sufficient shut-in time is given (from several hours to overnight, depending on the depth of the well). If the tubing contains liquids, the mixture must be displaced to the bottom of the well by pumping liquid (usually produced fluids) in behind the mixture. The amount of displacing liquid is calculated by determining the volume of the tubing and subtracting the volume of inhibitor mixture. After the inhibitor has been displaced to the bottom, the well is usually shut in for 2 to 24 h. The well is then put into normal operation in the usual manner.

The tubing displacement technique is also known as a "kiss squeeze." This type of treatment will last from a week to several months, depending on the system and the inhibitor, and is normally used on flowing oil wells.

Between Pigs Batch. This method is used to control corrosion in gas pipelines and is only used by itself in moderately corrosive systems. The volume of inhibitor mixture needed to give a 3 mil thick coating can be calculated from an equation that takes into account the pipe diameter and length (Ref 75, 76). This technique can also be combined with scraper pigs to refurbish and protect older pipelines. It can remove the buildup of corrosion scale and deposits on the bottom of the pipeline and apply a protective inhibitor film. Pigging can also be used to manage and control water holdup at the same time applying a protective inhibitor film. The importance of these basic techniques is often overlooked in good corrosion control in oil and gas pipelines.

Continuous Treatment. This technique is used on producing wells, injection wells, pipelines, and flow lines. Continuous treatment simply involves introducing inhibitor on a continuous basis so that its concentration in the corrosive fluids is maintained at a level sufficient to prevent or reduce corrosion. This concentration may vary from a few parts per million to 50 ppm or more, depending on the severity of attack. There are many ways to continuously treat producing wells. The inhibitor can be injected into:

- A line that bypasses part of production into the annulus
- The power oil of a subsurface hydraulic pump
- A small string of tubing that runs down the production tubing (small-bore treating string)
- A small-diameter tubing "capillary" string that is installed external to the production tubing that goes down the annulus and attaches to a subsurface injection valve.

Squeeze Treatment. This is a combination batch-continuous method in which the inhibitor solution is placed into the formation. The inhibitor and diluent are displaced down the tubing and into formation by 25 to 75 drums of displacing fluid, which is usually clean crude, diesel fuel, or nitrogen. When the well is returned to production after a squeeze, the initial concentration of chemicals in the returned fluid is high and decreases very rapidly. The inhibitor continuously returns from the formation to repair any breaks in the inhibitor film. The second squeeze and successive treatments all give a longer treatment life than the first squeeze. Possibly, a portion of the chemical used in the first squeeze is trapped in the formation and cannot return to the well bore. This action is shown in Fig. 18 through use of iron counts. The advantages of squeeze treatment include:

- It can be used in tubingless or multiple completion wells.
- Treating frequency is reduced and ranges from 6 to 18 months, depending on the inhibitor, the formation, placement technique, and the fluids being produced.

The disadvantages of squeeze treatment are:

- High cost
- Possible clay swelling

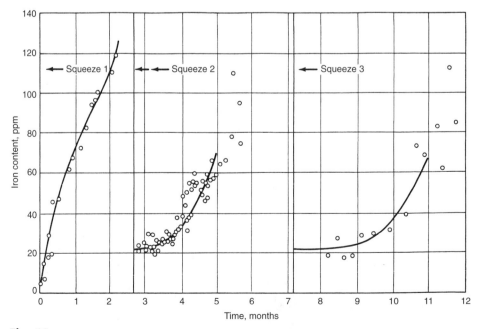

Fig. 18 Iron content of produced water after squeeze treatment. Iron content is one measure of inhibitor effectiveness.

- Emulsion blocks that restrict production
- Injection pressure that must be kept below the pressure necessary to fracture the formation

This method is used on gas-lift wells having a high-pressure, a high gas-oil ratio, and a high rate of water production. More information on the selection and use of inhibitors in the oil patch is available in the article "Corrosion Inhibitors for Oil and Gas Production" in Volume 13A of the *ASM Handbook*.

Some guidelines for use of various treatment methods for oil and gas gathering systems are provided in Table 6. This table shows some of the variables that can influence selection of these techniques. These parameters include superficial gas velocity, ratio of liquid hydrocarbon to water, and chloride concentration in the water phase. In some cases, no chemical treatment is needed. In others, only pigging may be required. In most cases, the inhibitor method varies between batch and continuous. It should be noted that in many cases, pigging is an important part of the inhibition process. It helps to relieve water holdup by moving it through the system, and it also helps to distribute the inhibitor to all internal pipe surfaces (Ref 67).

Nonmetallic Materials

In recent years, there has been an increased use of nonmetallic materials in oil-field operations. These materials are being used because they generally do not corrode or suffer deterioration in the environments in which steel readily corrodes. They are also lightweight, suitable for rapid installation, and, in most cases, less expensive than steel. In a 1982 American Gas Association survey of 56 gas utility companies, it was found that nonmetallic pipe systems failed at only 13.2% the rate of metallic pipe systems when excavation damage is excluded (Ref 77). These data, however, allude to one of the major concerns for nonmetallics. Mechanical damage occurs because installers and maintenance personnel tend to handle them the same way they handle metallic material.

Types of Nonmetallic Materials Used

Nonmetallic pipe can be classified into three major categories: thermoplastic materials, fiber-reinforced materials, and cement-asbestos.

Thermoplastic materials can be repeatedly heated, softened, and reshaped without destruction. The most commonly used thermoplastic pipe materials are (Ref 58):

- Polyvinyl chloride (PVC)
- Chlorinated polyvinyl chloride (CPVC)
- Polyethylene (PE)
- Polyacetal (PA)
- Acrylonitrile-butadiene-styrene (ABS)
- Cellulose acetate butyrate (CAB)

Glass-fiber-reinforced thermoset materials are chemically set and cannot be softened or reshaped by the application of heat. There are two major classes of these materials in oil-field use:

- Fiberglass-reinforced epoxy (FRE)
- Fiberglass-reinforced polyester (FRP)

Cement-asbestos is the oldest nonmetallic material in use in the oil field. It is a combination of portland cement, asbestos fibers, and silica. It can be obtained with an epoxy lining, but most of this pipe currently in use is unlined (Ref 78). In recent years, efforts have been made to find substitutes for asbestos using portland cement, silica and various man-made fibrous materials, and chrysotile.

Joining Methods

The methods used to join various types of nonmetallic plastic pipe are shown in Table 7. The heat welding method uses a heating element to soften the ends of the joints, which are then pushed together and held until the joint cools. About 25% of all thermoplastic pipe joints are made by this method (Ref 58). Solvent welding can be used on some of the thermoplastic pipe materials and on both of the thermosetting materials. This method uses both a solvent and a glue to hold the joints together. Finally, threads can be used on all thermoplastic pipe materials.

The most common joint for cement-asbestos and other fibrous cement pipe products is the "slip joint." This kind of concrete pipe joint does not involve a coupling. There is a groove on the spigot end where a rubber gasket ring is placed. This gasket is compressed into the groove by the bell of the connecting pipe as they are pushed together, forming a watertight seal.

Advantages and Disadvantages

The advantages of nonmetallic materials include (Ref 58):

- They are generally immune to corrosion in aqueous systems.
- They are lightweight and are therefore easier to handle.
- Nonmetallic pipe is quickly joined and installed.
- No external protection, such as coatings or cathodic protection, is required.
- The smooth internal surface of nonmetallic pipe results in lower fluid friction loss.

Among the disadvantages of nonmetallic materials are (Ref 58):

- Nonmetallic pipe has a more limited working temperature and pressure. These limits are also more difficult to predict with assurance than the limits of steel pipe.
- Careful handling is required in loading, unloading, and installation.
- Nonmetallic pipe should be buried to protect it from sunlight, mechanical damage, freezing, and fire.
- Nonmetallic pipe has very low resistance to vibration and pressure surges.

Typical Applications

Thermoplastics have seen use in flow lines, gathering lines, saltwater disposal lines, liners for steel pipe in high-pressure operations, and fuel lines for gas engines. Polyvinyl chloride has a maximum temperature limit of 65 °C (150 °F) and a maximum operating hoop stress of

Table 6 Influence of production conditions and environment on selection of method for flow line inhibition

Condition(a)	Water cut < 30%	Water cut > 30%
Low Cl⁻/high velocity $A < 10,000$ mg/L $V > 3$ m/s	No treatment	Continuous inhibition
Low Cl⁻/low velocity $A < 10,000$ mg/L $V < 3$ m/s	Pigging	Pigging + continuous inhibition
High Cl⁻/high velocity $A > 10,000$ mg/L $V > 3$ m/s	Continuous inhibition	Continuous inhibition
Low Cl⁻/low velocity $A > 10,000$ mg/L $V < 3$ m/s	Pigging + continuous inhibition	Pigging + batch inhibition + continuous inhibition

(a) H_2S = 0.3 to 14%; CO_2 = 3.3 to 5.9%; Pressure = approx 5.5 MPa (800 psig); A, Concentration of chloride in water; V, superficial gas velocity

Table 7 Joining methods for nonmetallic pipe materials

Material	Heat	Solvent	Thread
Thermoplastic materials			
Polyvinyl chloride	...	X	X
Chlorinated PVC	...	X	X
Polyethylene	X	...	X
Polypropylene	X	...	X
Polyacetal	X	...	X
Acrylonitrile-butadienestyrene	X	X	X
Cellulose acetate butyrate	X	X	X
Thermosetting materials			
Glass-reinforced epoxy	...	X	X
Glass-reinforced polyester	...	X	X
Cement-asbestos	(Rubber ring seal)		X

Source: Ref 58

27.5 MPa (4000 psi). Polyethylene has a maximum operating temperature of 40 °C (100 °F) and a maximum operating hoop stress of 4.3 MPa (625 psi) (Ref 54).

Glass-fiber-reinforced thermoset materials have also seen use in flow lines, gathering lines, saltwater disposal lines, liners for steel pipe in high-pressure operations, and fuel lines for gas engines. They have also been used for tubing in disposal and injection wells. Neither FRE nor FRP should be used for a well production flow line or gas-gathering system at pressures above 2.1 MPa (300 psi) and temperatures of 65 °C (150 °F). These materials should not be used in vacuum systems or where repetitive vacuum surges are likely to occur and in lines handling sand-laden fluid.

In addition, FRP has been used for stock tanks and barrels ranging in size from small chemical tanks of 1890 L (500 gal) or less to larger tanks of 104,000 L (27,500 gal) or larger. Even sucker rods used in pumped oil wells have been made of FRP.

Cement-asbestos and other cement fibrous materials have been used in low-pressure saltwater disposal lines. They have a maximum temperature rating of 95 °C (200 °F).

Specifications

There are several specifications provided by API that provide guidance on design, use, and handling of plastic and reinforced tubulars and components in oil-field applications:

- 15HR, "High Pressure Fiberglass Line Pipe," 2nd ed., April 1, 1995
- 15LE, "Polyethylene (PE) Line Pipe," 3rd ed., April 1, 1995
- 15LR, "Low Pressure Fiberglass Line Pipe," 6th ed., Sept 1, 1990 (ANSI/API Spec 15LR-1992)
- 15LT, "PVC Lined Steel Tubular Goods," 1st ed., Jan 1, 1993
- RP 5L2, "Internal Coating of Line Pipe for Non-Corrosive Gas Transmission Service," 3rd ed., May 31, 1987 (ANSI/API RP 5L2-1992)
- RP 15 TL4, "Care and Use of Fiberglass Tubulars," 1st ed., Oct 1, 1993

Environmental Control

Oxygen dissolved in oil-field water is one of the primary causes of corrosion. Dissolved oxygen is needed at 25 °C (75 °F) for an appreciable corrosion rate in neutral waters, while even in high-salinity brines at 150 °C (300 °F), the corrosion rate is low once the oxygen is removed (Ref 56). This type of corrosion is usually a localized form of attack, such as pitting, rather than a uniform attack. Oxygen also causes the growth of aerobic bacteria, algae, and slime, which can create plugging and enhance pitting. Also, mixing an oxygen-containing water with oil-field waters containing dissolved iron or hydrogen sulfide can cause precipitation of iron oxides, iron hydroxides, or free sulfur, thus causing serious plugging problems. In one case, some injection wells of a waterflood in west Texas were filled with as much as 23 m (75 ft) of iron hydroxides after a few months of service (Ref 79). Even if there are other corrosive agents present, air-free operation is needed in order for film-forming corrosion inhibitors to work (Ref 80); the presence of dissolved oxygen will significantly reduce the effectiveness of corrosion inhibitors unless specific multifunctional chemical formulations are utilized.

Both mechanical and chemical means have been used to remove dissolved oxygen from oil-field waters. The mechanical means are countercurrent gas stripping and vacuum deaeration, while the chemical means include sodium sulfite, ammonium bisulfite, and sulfur dioxide. The choice of oxygen-removal method depends on economics. Usually, mechanical means are used when large quantities of oxygen are to be removed. Chemical removal is usually employed to remove small quantities of oxygen and even sometimes for the removal of residual oxygen after the mechanical means have been used.

Care must also be used to minimize air contamination of injected fluids such as corrosion and scale inhibitors, biocides, hydrate, and wax control fluids. Aeration of these fluids can lead to a source of oxygen contamination and resultant corrosion in production environments downstream from the point of injection. In many cases, storage tanks for injected fluids need to use an inert cover gas, which can range from nitrogen, to carbon dioxide, to produced gas. It should be noted that if water or brine is present, the use of carbon dioxide as a cover gas may result in the formation of an acidic media that can sustain corrosion on its own part. Additionally, many organic solvents such as glycol, methanol, and ethanol have much higher oxygen solubility than is common for aqueous solutions. For example, dissolved oxygen concentrations in ethanol from air saturation are in excess of 50 ppm versus 4 to 6 ppm for water and brine. This can lead to oxygen corrosion problems when such aerated fluids are injected into the production environment. Often the physical manifestation of oxygen corrosion is unexpectedly high rates of corrosion and/or the presence of pitting and crevice corrosion.

Mechanical Methods

Gas stripping is performed in either a packed column or a perforated tray column. Perforated tray columns are preferred because they are not as easily fouled with suspended solids or bacterial slime as packed columns. Figure 19 illustrates a tray-type gas stripping column. Oxygenated water flows into the top of the column, while the stripping gas flows through the bottom inlet. As the gas bubbles up through the water, oxygen comes out of solution. The trays or packing in the column increases the contact area. These systems are designed to use not more than 0.06 m^3 (2 ft^3) of gas per barrel of water being stripped (Ref 54). The gas source should be free of both oxygen and H_2S. Either natural gas or scrubbed exhaust gas from engines is commonly used. The principle of removal is to reduce the concentration of oxygen in the gas coming in with the water by dilution with the stripping gas.

Vacuum Deaeration. In this process, a vacuum is created in a packed tower, and as the oxygenated water is passed over the packing, the low pressure causes the oxygen to bubble from solution. The vacuum pump pulls the oxygen, water vapor, and other gases from the top of the tower. The tower usually consists of several different pressure stages, as shown in Fig. 20. In a packed column, each stage consists of a height of packing, which is sealed from the stage below by a layer of water in the bottom of the packing (Ref 57). A single-stage tower will economically remove oxygen only to a lower limit of about 0.1 ppm (1000 ppb) because the excessive vacuum pump horsepower required to achieve lower concentrations is not usually feasible. Therefore, multistage columns are needed. Dissolved oxygen concentrations as low as 0.01 ppm (10 ppb) have been achieved in three-stage towers (Ref 57). While this level of

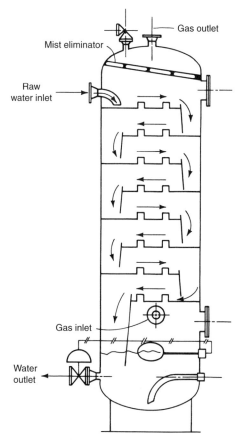

Fig. 19 Tray-type countercurrent gas stripping column. Source: Ref 57

deaeration will improve operations and reduce corrosivity over fully aerated conditions, this level of residual oxygen is still marginal in terms of oxygen corrosion. Added gains are possible with a further reduction in corrosivity and pitting susceptibility with the use of additional deaeration methods.

Combination Vacuum Deaeration and Gas Stripping. Vacuum deaeration, with the use of 0.003 m^3 (0.1 ft^3) of natural gas, has been used to reduce the oxygen content of water from 5 to 0.05 ppm (5000 to 50 ppb) or less in a 40,000 barrel/day waterflood in west Texas. Single-stage vacuum deaeration reduced the oxygen content of the water to 0.17 ppm (170 ppb), while the gas stripping further reduced the oxygen content to 0.05 ppm (50 ppb). Corrosion rates in the water were reduced from 0.36 to 0.04 mm/yr (14 to 1.6 mils/yr) (Ref 81, 82).

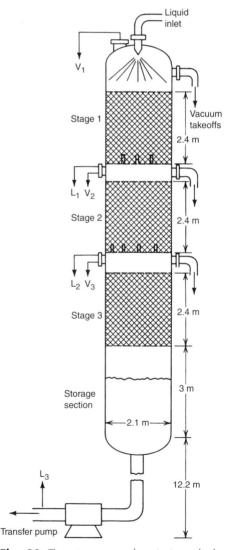

Fig. 20 Three-stage vacuum deaerator tower. L_1, L_2, L_3 and V_1, V_2, V_3 are liquid and vapor sample points, respectively. Pressure decreases as the liquid moves down the tower. Source: Ref 57

Chemical Methods

Sodium sulfite is used to scavenge oxygen from water and is available as a liquid or as a powder. It reacts with oxygen:

$$2Na_2SO_3 + O_2 \rightarrow 2Na_2SO_4 \quad \text{(Eq 4)}$$

Approximately 8 ppm of Na_2SO_3 is required to react with 1 ppm O_2. A 10% excess is usually required for complete reaction, and a catalyst such as cobalt chloride (0.1 ppm) is needed to scavenge to acceptable levels within a few minutes. Because Na_2SO_3 solutions will react with atmospheric oxygen, an inert gas blanket is required on the storage tank.

Ammonium bisulfite is a liquid scavenger and reacts with oxygen:

$$2NH_4HSO_3 + O_2 \rightarrow (NH_4)_2SO_4 + H_2SO_4 \quad \text{(Eq 5)}$$

An 80% solution of NH_4HSO_3 requires a 10-to-1 ratio by weight for the reaction. A 10% excess is needed to complete the reaction. Ammonium bisulfite does not react with air and can be stored in open containers. A catalyst is not usually needed for oil-field brines. Because the chemical is supplied as a solution with a pH of 4 to 4.5, it must be stored in a corrosion-resistant vessel. Type 304 stainless steel is commonly used (Ref 57).

Sulfur dioxide is a chemical scavenger that can be either supplied as a liquified gas under pressure in a cylinder or generated by burning sulfur. The reaction between sulfur dioxide and oxygen proceeds according to:

$$SO_2 + H_2O + \tfrac{1}{2}O_2 \rightarrow H_2SO_4 \quad \text{(Eq 6)}$$

A quantity of 4 ppm by weight of SO_2 is required to remove 1 ppm of oxygen. A 10% excess and a catalyst such as cobalt chloride is needed to complete the reaction. Sulfur dioxide from cylinders is applied by using a bypass line that handles approximately 10% of the total fluids, as shown in Fig. 21. The scavenger is added to the bypass fluids. The materials used in this bypass line should be resistant to acid attack because of the low pH formed from the reaction. Use of SO_2 cylinders is most advantageous in small systems (less than 10,000 barrels per day) or where small concentrations of dissolved oxygen are encountered in larger systems (Ref 83).

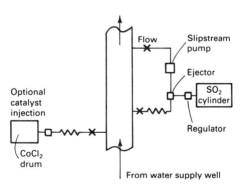

Fig. 21 Application of SO_2 through the use of a bypass line.

When larger volumes of water are to be treated, it may be more economical to produce SO_2 gas by burning sulfur. This gas is then dissolved in a sidestream of the water to be treated, pumped through the packed column, and then back into the main line.

Precautions

Some precautions involving oxygen scavengers should be noted (Ref 57):

- Oxygen scavengers will react with chlorine and hypochlorite (ClO$^-$), which are added to injection water for bacterial control. Therefore, these chemicals should be added downstream of the point of scavenger injection to allow completion of the scavenger-oxygen reaction.
- Any organic chemicals, such as biocides, scale inhibitors, and corrosion inhibitors, can possibly interfere with the scavenger-oxygen reaction and should be selected with care.
- Oxygen scavengers cannot normally be used in sour systems. If H_2S is present, it may react with the cobalt chloride catalyst to form insoluble sulfides (Ref 57).
- The reaction rates for oxygen scavengers can be a function of fluid temperature, and at low temperatures these reaction rates can be slow. Therefore, at cold temperatures common in locations such as the North Sea, Canada, and Alaska, and in deep water subsea operations, selection of deaeration chemicals is critical and adequate residence time for complete reaction needs to be considered.

Oxygen Exclusion

It is usually more economical to exclude oxygen from oil-field equipment than to remove it after it has entered the system. The most common means of excluding oxygen is through the use of gas blankets on water supply wells and water storage tanks. Maintenance of valve stems and pump packing is also important. In water-handling systems, a leaky valve can be a source of oxygen contamination into the system through backstreaming of oxygen counter to the direction of the leak.

All tanks handling air-free water should be blanketed with an oxygen-free gas such as natural gas or nitrogen. Most tanks require only a few ounces of pressure over atmospheric (Ref 80). The regulator should be sized to supply gas at a rate adequate to maintain pressure when the fluid level drops. In some processes, combustion gas is used to "inert" tanks. This type of gas has the ability to exclude air if kept at a positive pressure relative to atmospheric. However, it often contains high concentrations of carbon dioxide and may still contain several mole percent oxygen. As long as the environment is basically dry, combustion gas can be useful to reduce some of the ill effects of aeration. However, under hydrated conditions, the carbon dioxide in the gas can be dissolved in entrained or precipitated water phases that will increase the corrosivity to materials of construction.

Oil blankets should not be used in place of gas blankets. Oxygen may be 5 to 25 times as soluble in hydrocarbons as in oil-field waters (Ref 80). Oil blankets will coat precipitates in the water, which can lead to well plugging problems. Some bacteria will even thrive at the oil/water interface (Ref 57).

Even supply wells and producing wells may need to be gas blanketed to prevent oxygen entry. If these wells are operated cyclically without gas blankets, oxygenated air will be drawn into the annulus every time the well is turned on and the fluid level drops. Oxygen can also enter a pump on its suction side if a net positive suction head is not maintained. If the seals start to leak, air can then be sucked into the pump (Ref 80).

Problems Encountered and Protective Measures

The problems encountered and protective measures discussed in this section are based on the state-of-the-art as practiced daily by corrosion and petroleum engineers and production personnel. These are by no means all of the methods employed for corrosion protection, but they represent the most commonly used processes.

Drilling Fluid Corrosion

Due to the nature of drilling conditions, corrosion is a problem in water-based drilling fluids. Important considerations are the causes of corrosion and the rate and forms of corrosion attack.

Causes of Corrosion

The major environmental causes of corrosion in drilling fluids are oxygen, carbon dioxide, hydrogen sulfide, ionic concentration, and low pH. Physical conditions causing corrosion include metal composition, metal properties, string design, stress, and temperature. Combined physicochemical corrosion accelerators include stress corrosion and erosion-corrosion. Microorganisms also introduce biological causes of corrosion in drilling environments.

Combined Effects. The forms of corrosive attack will provide characteristic patterns that can be identified and used in selecting preventive methods (Ref 84). The major forms of attack are general corrosion, pitting, crevice corrosion under deposits, corrosion fatigue, and SCC. Erosion-corrosion, uniform attack, and galvanic corrosion are also common problems. More than one form of attack can occur, and one form can transform into a second form. For example, a pit can deepen and increase stress, initiating stress corrosion or fatigue (Table 8).

Failure Analysis

Analysis of used and failed equipment can provide a means of developing corrosion prevention methods (see the article "Analysis and Prevention of Corrosion-Related Failures" in *Failure Analysis and Prevention,* Volume 11 of the *ASM Handbook.* Identification of corrosion products (Ref 85) and corrosion forms (Ref 33, 86) can be developed into cause-effect mechanisms.

Monitoring

Drill pipe corrosion coupons are used to measure the rate, form, and cause of corrosion (Ref 87). Corrosion rates of 2.4 to 9.8 kg/m$^2 \cdot$ yr (0.5 to 2 lb/ft$^2 \cdot$ yr) free of pitting is an acceptable range. The lower rate should be used as a baseline estimate for deviated holes, deep drilling, and/or high-stress conditions; however, local conditions and lack of appropriate anticorrosion measures being taken can result in substantial increases in corrosivity. Special monitoring using linear polarization (Ref 88) or galvanic probes (Ref 89) will provide instant detection of corrosion and changes in rates that can be related to system conditions on a real-time basis. However, in most cases, corrosion monitoring of drilling fluids is simply not conducted since it is not specified by the operator or contractor. Figure 22 shows changes in corrosivity produced in drilling fluids as a result of changes in conditions such as flow rate, which can be identified when real-time monitoring is used. The most appropriate locations for corrosion monitoring of drilling fluids are (a) the return line to the mud pit since an influx of corrosive gases downhole may decrease fluid pH and increase corrosivity, or (b) in the mud pit and/or its outlet line going to the well, if concerns for oxygen scavenging and initial pH control are of interest. These two locations provide convenient information on which to control the corrosivity of drilling fluids (Ref 90).

On-site chemical and physical analyses of drilling fluid properties are conducted on a frequent basis (Ref 87). Test procedures should include oxygen, CO_2, H_2S, and bacterial analysis for comprehensive monitoring of corrosion problems. The *Drilling Manual,* 11th ed. (CD-ROM), published by the International Association of Drilling Contractors, presents equipment inspection methods as well as information on the care, handling, and specifications of tool joints, drill pipe, casing, and tubing. Reference is made to this manual to provide comprehensive information on drilling and production equipment.

Table 8 Drilling fluid corrosion control troubleshooting chart

Corrosion cause	Primary source	Identification	Major corrosion forms	Remedies
Oxygen	Atmosphere, mud conditioning, equipment, oxidizing agents, air drilling	Oxygen test, iron oxide by-products	Underdeposit corrosion, pitting	Avoid mechanical air entrapment in mud pits or defoamers; use oxygen scavengers for normal drilling; use passivating agents for air drilling operations.
Hydrogen sulfide	Formation, bacteria, chemical or thermal degradation	H_2S analysis, iron sulfide test	Underdeposit corrosion, uniform corrosion, sulfide stress cracking	Control pH ≥ 9.5; use sulfide scavengers or filming inhibitors, reduce stress; change to oil mud in severe H_2S environments.
Carbon dioxide	Formation, bacteria, chemical or thermal degradation	CO_2 analysis, iron carbonate by-products	Underdeposit corrosion, pitting	Control pH ≥ 9.5; use calcium hydroxide to combine with and precipitate CO_2 products; use filming inhibitors.
Dissolved salts	Formation, chemical additives	Water analysis	Underdeposit corrosion, uniform corrosion, chloride SCC	Control pH ≥ 9.5; remove oxygen, H_2S, and/or CO_2; use filming inhibitors.
Bacteria	Makeup water, formation	Culture tests for sulfate-reducing bacteria and/or slime forms	Underdeposit corrosion, sulfide stress cracking	Control pH ≥ 9.5; add biocides.
Temperature	Formation heat, friction	Test	Sulfide stress cracking, pitting	Select temperature-stable chemicals; use friction/torque reducers; cool mud; use oil mud.
Abrasion (erosion-corrosion)	Formation, directional or deviated hole	Observation	Erosion-corrosion	Use lubricants, torque reducers, and filming inhibitors; control solids; use oil muds.
Metal composition				
Carbon steels, ≤ 22 HRC	...	...	Underdeposit corrosion, uniform corrosion, pitting	Control pH ≥ 9.5; control oxygen, H_2S, and CO_2; limit stress.
Carbon steels, > 22 HRC	...	...	Underdeposit corrosion, sulfide stress cracking, pitting	Control H_2S to very low levels.
Stainless alloys	...	...	Chloride SCC	Limit salts and temperature exposure; control deposits.
Aluminum alloys	...	...	Underdeposit corrosion, galvanic corrosion, pitting	Limit pH to 9.5–10.5; use passivation.

Oxygen Corrosion Control

Oxygen can cause pitting corrosion of steel and stainless alloys and crevice corrosion under deposits and is considered to be the most serious corrosion accelerator in drilling environments. Oxygen enters the drilling fluid system externally from the atmosphere, usually by way of solids-control and mud-mixing equipment (Ref 91). The operation of this equipment to reduce air entrapment into the circulating system is an effective technique for limiting oxygen levels. Foaming problems are characteristic of some mud systems and can result in high oxygen levels on the high-pressure side of the pump. Defoaming the fluid or maintaining properties to release gas quickly is required to overcome this problem.

Oxygen scavengers, such as sodium sulfite or ammonium bisulfite, are used to remove oxygen from drilling fluid (see the discussion "Environmental Control" in this article). Treatment methods involve a continuous addition of chemical at the rate of 10 mg/L sulfite ion for each 1 mg/L of oxygen present in the fluid. A residual sulfite concentration of approximately 100 mg/L is maintained in the drilling fluid as a functional means of controlling oxygen in drilling systems. Oxygen scavenger catalysts are frequently required to overcome interfering side reactions that prevent the oxygen-sulfite reaction. Calcium in the fluid can combine with sodium sulfite and form calcium sulfite precipitate, thus preventing the sulfite ion from scavenging oxygen. Aldehydes and chlorine dioxide used as biocides in drilling fluids react with sulfite ions and may prevent oxygen removal. The addition of cobalt or nickel catalysts overcomes many of these problems by increasing oxygen-sulfite reaction rates.

Passivating compounds, such as sodium nitrite, are used to protect equipment during air, mist, or foam drilling operations. For decades chromates were used, but these components have been banned in nearly all cases as a result of their known carcinogenic nature. Treatment levels range from approximately 500 to 2000 mg/L of nitrite ion in fresh to slightly brackish fluid. Higher concentrations are required in high-brine solutions, and sodium nitrite is not recommended above approximately 25,000 mg/L of chloride ion concentration. A noteworthy disadvantage of using passivating agents is the tendency toward accelerated pitting attack if treatment levels are too low or if deposits exist under which the metal cannot be passivated. Zinc compounds are often combined with passivating agents to reduce pitting tendencies. Treatments for controlling deposits are recommended to mitigate underdeposit attack and are covered in the discussion "Scale and Deposit Control" in this section. Nitrite compounds are not compatible with sulfite-type oxygen scavengers.

A clear advantage is gained in the use of sodium nitrite when hydrogen sulfide is encountered in the well. These compounds oxidize and remove H_2S (see the discussion "Hydrogen Sulfide Corrosion Control" in this section).

Atmospheric corrosion occurs on drilling equipment in urban, polluted, tropical, and marine environments. Protective coatings are commonly applied at the steel mill or storage yard and periodically between drilling operations. Many coating compositions are commercially available for both short- and long-term storage (2 or 3 years). The filming or waxy materials containing inhibitors that are typically used during drilling often provide good atmospheric protection for short periods between jobs. For long-term exposure, careful rinsing with fresh water, surface cleaning, and a selected atmospheric coating are recommended.

Hydrogen Sulfide Corrosion Control

Hydrogen sulfide causes three forms of corrosive surface attack—underdeposit (crevice) corrosion, pitting, and sulfide stress cracking. Corrosion control methods include selecting corrosion-resistant materials (stainless steels and nickel-base alloys), removing the H_2S from the fluid, use of chemical inhibitors, and reducing stress. Hydrogen sulfide enters the drilling fluid primarily from the formation, but it can also come from thermally degraded mud products, sulfate-reducing bacteria, and makeup water.

Alkaline pH control (>pH 10) and sulfide scavengers are used to neutralize, precipitate, and/or oxidize H_2S. Film-forming amine-type inhibitors are recommended for coating the drill string. Caustic soda or calcium hydroxide treatments are used to neutralize the acid gas. Alkaline pH above 10 results in the production of sodium bisulfide or sodium sulfide products that are almost totally water soluble. This treatment provides both personnel safety and corrosion protection. A side benefit of maintaining high pH in drilling fluids is the natural passivation effect it provides against corrosion through the formation of a hydrated iron oxide on the steel surface.

Compounds of iron oxide (Fe_3O_4), zinc carbonate, zinc oxide, zinc chelates, and copper are used to precipitate sulfide ions from solution. Pretreatments of approximately 1 kg/barrel (2 lb/barrel) of one of the scavengers are commonly recommended as a precaution against a small influx of H_2S entering the mud system and causing damage. Tests are used to monitor scavenger concentrations and treatment requirements. Chemicals such as sodium nitrite compounds are used to oxidize H_2S to sulfate or elemental sulfur. The oxidizing process is a fast and efficient method of removing H_2S from the system. There is no compatibility problem with the sulfide scavengers listed previously. Formaldehyde and chlorine dioxide are compounds that are frequently used as drilling fluid biocides. These products react with hydrogen sulfide, offsetting its corrosive action; however, their biocidal properties are diminished or eliminated in the process.

Filming amine inhibitors provide protection from H_2S surface attack and hydrogen embrittlement. Oil-soluble filming inhibitors applied directly on the drill pipe are recommended to offset corrosion fatigue and hydrogen embrittlement. Care should be taken with cationic filming inhibitors, which can damage mud properties by flocculating the anionic clays in drilling systems. Oil muds provide the most effective protection against all corrosion causes, including H_2S. The oil phase provides a nonconductive film covering exposed equipment and thus preventing the corrosion process.

Stress reduction by mechanical changes, such as rotary speed and less weight on the bit, is effective in reducing H_2S-induced embrittlement or cracking failures. Torque-reducing agents, particularly in high-angle drilling, are effective in lowering stress as well.

Material selection for drill pipe and casing can have a significant effect in controlling sulfide stress cracking. The brittle failures related to H_2S are linked to the hardness and yield strength of the steel resulting from metallurgical processing.

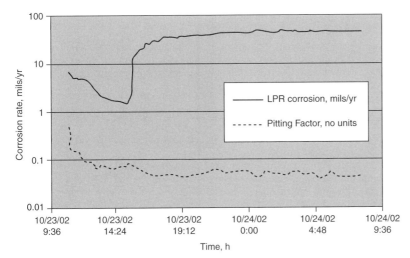

Fig. 22 Monitoring of corrosion in drilling fluids versus flow conditions. LPR, linear polarization resistance; Pitting Factor values, defined as the standard deviation of the corrosion current divided by the mean corrosion current, ≥0.1 indicate active pitting conditions.

As a general rule, steels with hardness levels below 22 HRC or with maximum yield strengths of 620 MPa (90 ksi) have few sulfide stress cracking problems. Modern developments in steelmaking and tube-making technologies have resulted in an increase in this limiting hardness to between 26 and 30 HRC (depending on the steel composition, processing, and microstructure) while maintaining resistance to sulfide stress cracking. This comes through selection of chromium-molybdenum steels, microalloying additions, and ID/OD quenching procedures that produce a uniform martensitic structure in the steel upon quenching and which can subsequently be tempered at the highest temperature possible while maintaining minimum strength properties after processing. Cold work, such as tong or slip marks, increases the hardness of steel, and sulfide stress cracking then becomes a problem. The service stresses in drilling frequently demand materials of great strength, requiring hardness and strength levels that are susceptible to sulfide stress cracking. Because of such requirements, the primary means of avoiding sulfide stress cracking is by control of the drilling fluid. A full discussion of the metals used in H_2S environments is provided in NACE MR0175/ISO 15156 (Ref 14).

Higher temperatures (above 80 °C, or 175 °F) reduce susceptibility to sulfide stress cracking of high-strength steel. This factor can become advantageous in drilling and production operations if properly controlled. For example, an influx of H_2S while drilling may not cause damage if the fluid temperature is above 80 °C (175 °F) in the hole. If H_2S is detected, scavenging should always be completed before operations are begun that would lower the metal temperature, such as pulling the drill pipe from the hole.

Carbon Dioxide Corrosion Control

Carbon dioxide causes localized corrosion primarily through its solubility in aqueous fluids and its role as an "acid" gas in reducing pH. It can also combine synergistically with velocity to produce a localized corrosion mechanism referred to as mesa corrosion whereby filmed and nonfilmed regions of a steel surface can corrode at drastically different rates. In combination with chloride salts, carbon dioxide can also induce general or pitting corrosion. Corrosion control methods involve controlling the pH in the higher alkaline ranges (an influx of CO_2 will tend to reduce pH). An effective technique is to treat the mud with calcium hydroxide to neutralize this acid-forming gas and to precipitate carbonates, thus lowering CO_2 levels. Film-forming inhibitors of the oil-soluble amine type applied by spraying the outside of the drill pipe and batch treatments for inside diameter filming are recommended to penetrate pits and deposits, stopping their corrosion action. Control of CO_2 is quite similar to H_2S corrosion control, and these two gases often enter the mud from the formation together.

Scale and Deposit Control

Mineral scale, corrosion by-products, and drilling mud solids that form deposits on exposed metal are a major factor in setting up conditions that result in underdeposit pitting attack. The prevention and removal of these deposits with scale inhibitors is quite effective in offsetting most serious drilling fluid corrosion problems. Inhibitors such as organic phosphonate, phosphate esters, and others of the acrylic, acrylamide, or maleic acid base structures have been effective. Products that exhibit threshold effects, temperature stability, and strong surface-active characteristics are useful. Treatments are variable because of environmental conditions, which differ greatly in drilling fluid compositions. As general rules apply, scale control treatments of 15 to 75 mg/L are used on a daily basis for most mineral-scale control situations. Treatments above this level are used to control deposits of metal corrosion by-products. Considerably higher treatment levels, up to 1000 mg/L, are used to provide corrosion protection. Care should be exercised in using the higher treatment levels, because these compounds may alter mud properties because of their dispersing characteristics.

Oil Production

There are two main types of producing oil wells: artificial lift wells and flowing wells. Artificial lift wells can be further divided according to the method used to pump the hydrocarbon to the surface. These include rod-pumped wells, wells that use downhole hydraulic pumps, and gas-lift wells. Approximately 90% of the artificial lift wells in the United States are rod pumped.

Artificial Lift Wells

Rod-Pumped Wells. In a rod-pumped well, the potential for corrosion damage is aggravated by the sucker rods alternately being stretched and compressed and by the abrasion of the rod couplings on the inside of the tubing. It is common for a well to have continuing sucker rod failures. Pulling and replacing the rods is a quick fix, but the problem will continue to exist until the root cause of the failure is identified and corrected. Identifying the problem is the most important step, because corrective action cannot be taken if the cause is not clear. Rod breaks should be inspected immediately after the rod string is pulled to determine if corrosion is occurring and to determine the steps that can be taken immediately to prevent a recurrence.

Corrosion in rod-pumped wells can be caused by several mechanisms, as discussed below (Ref 92). Galvanic corrosion is caused by dissimilar metals in contact or by the difference in metallurgy or metallurgical condition between two areas on a sucker rod. Most galvanic corrosion on rods is caused by differences in metal condition caused by hammer, wrench, or tong marks and the grooves left by rod-straightening machines. The impact area will be cathodic to the body of the rod, and corrosion will occur adjacent to the mark. Sucker rods have a soft decarburized layer or skin of low-carbon steel 0.13 to 0.2 mm (5 to 8 mils) thick. This layer can be broken by careless handling. Anodic regions produced by surface wear can also show preferential corrosion relative to properly filmed (cathodic) regions.

Bent rods are sometimes straightened and used again. This is poor practice, because a bent rod is permanently damaged, contains local plastic deformation and residual stresses, and should be discarded. A rod-straightening machine can put spiral grooves around the rod, and corrosion will occur directly adjacent to the groove.

Any of these conditions will lead to localized corrosion and pitting, and stress raisers will be set up. The cyclic stresses resulting from alternately stretching and compressing the rods during pumping operations will lead to rapid failure at these sites of stress concentration.

Stray current from surface equipment or leakage from a cathodic protection system will cause severe corrosion where the current leaves the rod string. It is usually seen on couplings or the part of the rod that is close to the coupling.

Damage from oxygen corrosion may take place when the rods are stored outdoors or when oxygen enters the wellbore through the annulus. This later becomes a common problem where rod packings are worn or where oxygen enters the annulus of the well during pumping. Rusting of stored rods will often cause pitting, and rust deposits can set up concentration cells or underdeposit corrosion when the rods are run in the hole. Oxygen can enter into the wellbore in wells that pump off. Oxygen can also be introduced during inhibitor-treating operations. This will aggravate other forms of corrosion by depolarizing the cathodes on the metal surface during the corrosion reaction and accelerate local anodic attack. Oxygen corrosion generally occurs in the lower part of the well: the casing, pump, tubing, and the lower part of the rod string. The effect lessens in the upper part of the well, because the oxygen is depleted by the corrosion reaction.

Carbon dioxide or sweet corrosion is caused by CO_2 from produced gases dissolving in water and forming carbonic acid. A low pH results, and the carbonic acid will react directly with the steel rod and cause metal loss and pitting. The pits formed are usually round bottomed with sharp sides, and they may be connected in a line or will sometimes form a ring around the rod. Fatigue cracks will be initiated at the bottom of the pits.

Carbon dioxide corrosion is aggravated by the presence of oxygen and organic acids. Oxygen depolarizes the cathodes, and organic acids fortify the low pH condition. This can remove normally protective carbonate scale. The formation of iron carbonate scale is the major limiting factor in CO_2 corrosion. The tendency for scale formation increases with increasing pH and temperature. Therefore, CO_2 corrosion is normally observed at low to intermediate temperatures. Many pumping wells are in the temperature range (<100 °C, or 212 °F) that is most conducive to CO_2 corrosion and pitting. At these temperatures, the iron carbonate scale is formed mainly away from the surface, with some forming as a discontinuous layer. Accelerated metal loss occurs in the gaps in the scale layer, and pits or mesalike features are formed.

Carbon dioxide corrosion may be sudden and catastrophic when breakthrough takes place in CO_2 floods. Wells that have been noncorrosive have failed within weeks after breakthrough. This is caused by the increase in acidity of the well fluids. In water-alternate-gas (WAG) CO_2 floods, breakthrough also brings a significant increase in water content in the production.

Hydrogen sulfide ionizes in water to form HS^- and hydrogen ions. Hydrogen sulfide corrosion is characterized by metal loss and pitting and can be quite severe particularly in presence of high levels of chlorides in the fluid. Iron sulfide formed on the steel surface generally does not form a protective layer having a powdery appearance and feel. Being a conductive material, the sulfide film is usually cathodic to the metal surface. Even if a protective sulfide layer is formed (generally at higher pH levels), a break in this layer will result in a galvanic couple with the filmed region resulting in a high probability for pitting.

The presence of oxygen will increase the corrosion rate in sour (H_2S) systems, and oxygen, in addition to depolarizing the cathodes, reacts with iron sulfide and forms elemental sulfur. This disturbs or removes the sulfide layer and results in an increase in the corrosion rate. Elemental sulfur is also corrosive to steel, which further increases the corrosion rate.

Organic acids increase the corrosion rate by making the iron sulfide scale less protective, thus exposing the bare steel surface. This comes about by lowering the pH and increasing the driving force of the corrosion reaction. At pH less than 6, H_2S reacts directly with the metal, and little or no protective iron sulfide is formed on the surface.

The pits formed during H_2S corrosion are generally small, round, and cone shaped. The acute angle at the bottom of the pit is a stress raiser, and it leads to cracking. Pits are usually not connected and are in a random pattern.

The amount of H_2S present has a direct effect on the time to failure of rods due to cracking. In some cases, corrosion pits are so small as to be undetectable before cracking occurs, or cracking may take place quickly at dents or nicks on the rods. However, these all become very critical in service life with conditions that contain increasing amounts of H_2S.

In addition to metal loss and pitting, sulfide stress cracking may occur in H_2S corrosion. The corrosion reaction generates hydrogen ions that combine with electrons liberated in the corrosion process to form atomic hydrogen on the metal surface. Atomic hydrogen can penetrate the metal by diffusion where it can either interact with the metal lattice to form brittle cracking (sulfide stress cracking) or recombine at internal defects as molecular hydrogen. The molecular hydrogen generates high pressures, and the steel can suffer from blister cracking (hydrogen-induced cracking). This latter form of cracking is also referred to as stepwise cracking and under the influence of applied or residual tensile stresses, it is referred to as stress-oriented hydrogen-induced cracking (SOHIC).

Rod-on-tubing abrasion is common, and it aggravates corrosion reactions. Surface scale is removed, leaving bare metal. The adjacent areas covered with scale are cathodic to the bare metal and increase metal loss. Both the rod couplings and the tubing are damaged. Severe abrasion will lead to galling or removal of large portions of metal, which are literally torn away.

The flow velocities in pumping wells are generally not high enough to influence the corrosion rate, but localized areas of high velocity around rod protectors and restrictions due to scale buildup in the tubing could occur. High velocity or local turbulence can remove protective scale and inhibitor films, particularly if solids are present in the fluid.

Elimination of Sucker Rod Corrosion. Once the cause of the corrosion has been found, corrective action is required so that the problem does not recur. The first step is to determine if the pumping program is correct. If the rod string and pumping procedure are not dynamically balanced, excessive tensile and compressive stresses are applied that will hasten fatigue failure and sulfide stress cracking.

The range of load, that is, the difference between the load of the upstroke and the downstroke, should be kept to a minimum. Long strokes at low speed will give the lowest load. The load is due to the weight of the rods and the fluid column on the upstroke and the weight of the rods on the downstroke.

The upstroke causes stretching, and the downstroke releases this stress, causing flexing of the rod. This cyclic stress induces fatigue failure; therefore, minimizing stress will reduce breaks caused by corrosion-induced cracking.

Proper rod string makeup will also reduce failures. Recommended torque loading should be followed when making up the string to be sure that the coupling is not in excessive stress or is not subject to play or movement resulting from too little torque during makeup. Hitting the rods or couplings with hammers and the use of pipe wrenches on the string should be avoided to eliminate local cold working and damage that act as crack-initiation sites.

Fluid pounding should be avoided. Fluid pounding is caused by the pump not filling completely on the upstroke and the plunger hitting the fluid on the downstroke. The sudden stop of movement causes a shock wave to propagate up the rod string. Fluid pounding can be the most damaging factor in rod failure. Rod guides can be installed to prevent rod-on-tubing wear.

Once mechanical deficiencies are eliminated, an inhibition program should be initiated. Corrosion inhibitors can prevent or greatly reduce failures caused by pitting or fatigue and will ensure that the changes made in rod loading and handling will be effective. Other methods of effective corrosion control include:

- Sucker rods are sometimes stored outdoors or in areas where internal storage is conducive to corrosion, such as coastal and industrial areas and in oil fields that produce H_2S. Oxygen corrosion or rust is aggravated by the deposition of salt from marine environments, such as spray on offshore platforms and coastal areas. In warehouses and under sheds, the presence of sulfur dioxide, oxides of nitrogen, and H_2S will initiate corrosive attack and increase rusting.
- The rod body and threads should be regularly inspected for corrosion damage. After inspection, the rods should be cleaned, and protective coatings should be applied.
- Suitable coatings that will provide protection for a minimum of 2 years should be applied by the manufacturer over rods and couplings. An oil-soluble coating is preferred, and it should be maintained by reapplication during storage. Used sucker rods should be cleaned and coated before storage.
- Sucker rods should be protected when pulled during workover operations. A batch of oil and inhibitor solution can be pumped into the tubing before pulling, or the rods can be coated after pulling.
- Couplings should be dipped in or brushed with an oil-inhibitor mixture before makeup. Care should be taken so that the amount of inhibitor added is not excessive. Thus, proper makeup can be performed.
- It is recommended that inhibitor be added to the tubing when the rods are run in the hole for initial filming. When the well is placed in production, one tubing volume of fluid should be circulated.

Once the well is in production, an inhibition and monitoring program should be initiated (Ref 93). An inhibitor is selected by testing for efficiency, usually with laboratory tests. These tests may include a wheel test, in which the inhibitor is added to bottles or high-pressure cells, rotated in a heated oven, and compared to an untreated control for percent protection and lack of pitting. Other tests include stirred kettle test and flow tests. All of these tests are designed to duplicate field conditions to a certain degree or to determine performance of inhibitors on different corrodents under specified conditions.

There are several methods by which the well can be treated. These include batch, continuous, and squeeze treatment, which are covered in the discussion "Inhibitors" in this section. See also "Inhibitors" in the section "Corrosion Control Methods" in this article and the discussion in the article "Corrosion Inhibitors for Oil and Gas

Production" in Volume 13A of the *ASM Handbook*. Other methods include tubing displacement after unseating the pump and the use of weighted inhibitors, sticks, or encapsulated inhibitors.

Downhole hydraulic pumps operate by pumping clean crude oil with a surface engine-driven pump down a string of tubing to operate a downhole hydraulic pump. The downhole pump lifts one barrel of fluid for each barrel of power fluid. The power oil is comingled with the produced fluid and separated on the surface. Problems can arise if the power oil carries water and solids or if CO_2, H_2S, or organic acids are present. The use of corrosion-resistant alloys or inhibitors can alleviate corrosion. Inhibitors are continuously added to the power fluid at the surface pump suction. Scale inhibitors and demulsifiers can also be added to the power fluid to prevent scale deposition and carryover of water into the power fluid.

In gas-lift wells, pressurized gas is injected into the annulus and through a gas-lift valve into the tubing. Fluid (oil/water) is displaced upward by the movement of the gas and out of the well. The process is repeated in batches or slugs in an intermittent system, or as a steady stream in a continuous-flow system. The velocities and local turbulence encountered with various flow regimes (slug, plug, etc.) may disturb normally protective surface films and thereby increase corrosion initiated by H_2S and CO_2. Corrosion-resistant alloys can be used in gas-lift valves, and steel tubing is usually protected by inhibitors. Alternative arrangements for controlling corrosion in the production tubing include the use of corrosion-resistant alloys.

Inhibitors are usually added into the lift gas at the surface and are carried with the gas stream into the tubing. Protection is provided above the lowest gas-lift valve. If corrosion occurs below the valve, inhibitor batch or squeeze treatments may be required for corrosion protection. The inhibitor selected for this application is usually oil soluble and water dispersible, and it can be diluted with liquid hydrocarbon to assist in carrying it downhole.

Flowing Wells

Corrosion problems in flowing wells are somewhat different from those encountered in artificial lift wells. Velocity becomes an important factor, and higher pressure leads to higher partial pressures of acid gases (i.e., CO_2 and H_2S). Treatment methods are more limited because of completion requirements.

Gas condensate wells may produce gas, hydrocarbons, formation water, acid gases, and organic acids. If the producing conditions allow liquid water to be produced or to condense on the tubing, corrosion is likely.

In wells producing formation water, corrosion may occur anywhere in the tubing string, wellhead, and flow line. Temperatures in the wellbore will affect the corrosion rate, and flow velocities also affect metal loss. The increased salinity of the water and acid gas partial pressures generally increase the corrosion rates.

Wells that produce no formation water will corrode where the dew or condensation point of water held in the hydrocarbon gas is reached and free water condenses on the tubing. The water will immediately dissolve CO_2 or H_2S and become corrosive at that point in the well. This point can be located downhole in the tubing, or at the surface in the flow line leading from the well.

Carbon dioxide corrosion is particularly damaging in condensed water as it contains no buffering salts. Dissolved CO_2 can lower the pH of water to less than 4.5 at CO_2 partial pressures of 69 kPa (10 psi) and temperatures of 75 °C (170 °F). Carbon dioxide corrosion can cause severe localized mesa attack or pitting when conditions of temperature and salinity form iron carbonate scale in a discontinuous or spotty layer. Organic acids reduce pH and thereby increase CO_2 corrosion rates by dissolving normally protective iron carbonate scale and by lowering bicarbonate content so that further iron carbonate scaling is reduced.

Hydrogen sulfide can also be dissolved in produced water. Like CO_2 it acts as an acid gas. Metal loss and pitting, along with hydrogen-induced cracking and sulfide stress cracking, may occur.

Oxygen is not normally present in the production stream (unless brought in through processing—pumping or addition of chemicals), and it is generally not a problem. However, if it is introduced into the system, it will increase corrosion and pitting of steels and stainless alloys.

Materials Selection. Most gas wells are completed with low-alloy steels for economic reasons. These materials will perform satisfactorily in most wells, and the application of coatings and the use of corrosion inhibitors permit their use in severe environments of high temperature, pressure, and CO_2 content. However, there has been an increasing trend toward use of corrosion-resistant alloys for downhole tubing and equipment particularly in remote or offshore developments. If properly selected, these materials eliminate the need for use of chemical inhibitors and the associated equipment and logistics for inhibitor delivery, storage, and injection. This approach is discussed later in this section.

Many Tuscaloosa Trend wells in Louisiana, completed with carbon steel tubing, are producing with no corrosion failures when coated and inhibited. Producing conditions range to 205 °C (400 °F) bottom hole temperature, 124 to 138 MPa (18 to 20 ksi) pressure, and CO_2 content of 5% or more. Hydrogen sulfide is also found in some wells at concentrations of 20 to 50 ppm. However, the inhibitors used in extreme cases such as this are based on demonstrated efficacy through an intensive inhibitor screening evaluation. In other cases, 13% Cr tubulars and production equipment are being used without inhibitors.

Alloy Tubulars. When conditions and economics warrant, corrosion-resistant alloys can be used. Steel with 9% Cr and 1% Mo has low corrosion rates up to 100 °C (212 °F). Higher corrosion rates and pitting become a problem at higher temperatures. The partial pressure of CO_2 is not a factor at temperatures below 240 °C (465 °F). Steel with 13% Cr is effective up to 150 °C (300 °F) or higher in the case of high-CO_2 gas wells producing primarily condensed water with low to moderate chloride levels (1500 to 10,000 ppm). More recently, new proprietary grades of modified 13 to 15% Cr martensitic stainless steel tubulars with alloying additions of nickel and molybdenum (see Table 3) have been applied at temperatures up to 230 °C (450 °F) and in conditions with higher levels of chloride and H_2S. Oxygen will cause severe pitting of 13% Cr steel; therefore, chemical injection systems must be kept oxygen free by an inert gas blanket on storage tanks. For example, the use of partially aerated seawater as a well control fluid has also resulted in severe pitting of 13% Cr tubulars.

If H_2S is present, 9% Cr steels can be used at hardnesses below 22 HRC to minimize sulfide stress cracking. However, 13 to 15% Cr steels are more resistant to corrosion as well as sulfide and chloride cracking, and they can be used under more severe conditions with higher H_2S partial pressures and at higher strength levels, particularly if modified alloys are used that contain nickel and molybdenum that are added to enhance corrosion resistance. For still higher levels of H_2S in combination with concentrated formation brine, higher alloy stainless steels and nickel-base alloys are also available (e.g., duplex stainless steels 2205 and 2507; austenitic stainless steels AL6XN, 254 SMO, and 654 SMO; nickel-base alloys 825, 625, and C276; precipitation-hardened nickel-base alloys 718 and 725; and titanium alloys).

There are still some concerns for corrosion, pitting, and stress cracking in martensitic and duplex stainless steel tubulars from clear brine packer fluids that may contain dissolved oxygen, CO_2, and/or H_2S. Careful selection of the packer fluid composition and additives to match the alloy corrosion resistance is recommended.

Coatings. Low-alloy steels can be coated for corrosion resistance. Coatings include baked-on phenolics, epoxies, and polyurethanes with fillers to give the required thickness, coating integrity, and corrosion resistance. Proper application is required for an intact coating that conforms to requirements. Tubing surface preparation, application methods, coating thickness, and holiday detection are part of the inspection and quality-assurance process.

Joints and connections should be designed so that the continuity of the coating is unbroken. The first few threads inside the coupling and the pin nose must be coated. A compression ring can be installed to ensure joint integrity.

Special care must be taken when the wireline operations are carried out in the coated tubing. Coatings are easily damaged or scratched, and once the coating is broken, corrosion and disbonding of the remaining coating can take place.

Wireline guides and running speeds of less than 0.5 m/s (100 ft/min) will minimize damage. A corrosion inhibitor should be used during or directly after wireline operations. The wireline tools should not have sharp edges and should be plastic coated. Wireline centralizers should also be used.

Coatings are also subject to disbondment if pressures are released suddenly (i.e., explosive decompression). Gases can penetrate the coating, and when a sudden pressure drop occurs, the gases will expand and lift the coating from its substrate.

Inhibitors. Corrosion inhibitors are an effective means of corrosion control, and they are required in highly corrosive environments in which carbon steel is used. They are needed even if the tubing is coated, because a holiday-free coating does not exist. Combination coating-inhibitor procedures are particularly effective provided the application is below the temperature limits for the coating and inhibitor.

The most commonly used inhibitors function by forming a film on metal surfaces that stops the flow of corrosion current. Nearly all inhibitors are fatty amines or quaternary ammonium compounds. The nitrogen in the molecule possesses a strong cationic charge and is chemically absorbed onto anodic sites on the metal surface. Cross-bonding of the inhibitor film and the attraction of a layer of oil aid in isolating the surface from the corrosive environment, thus promoting effective corrosion protection.

Inhibitors are selected for several characteristics. The major consideration is their performance in terms of providing low corrosion and pitting rate. Next, performance must be considered for inhibitor-film persistence, compatibility with production fluids and formations, and emulsion tendencies.

Once an inhibitor is selected, a treating method is used that fulfills the system requirements. Several methods are commonly used to treat flowing wells: batch treating, continuous injection, and squeezing are discussed here.

Batch treating involves the intermittent addition of relatively large quantities of inhibitor solution to the annulus or down the tubing of a gas condensate well. A batch treatment in a flowing well consists of dosing a solution of inhibitor in hydrocarbon condensate or diesel fuel. It can be applied down the tubing, shutting the well down to allow the inhibitor solution to fall to the bottom, and repeating at a set interval. The disadvantage of this treatment is that the inhibitor may not go to the bottom. The tubing may contain up to 50% of its volume of water and oil, and the bottom of the well below the static shut-in fluid level may not be treated using this method.

A method of treatment that ensures that the batch will reach the bottom of the well is tubing displacement. A batch of inhibitor in oil, usually one-third of the tubing volume, is pumped into the well, and enough condensate or oil is pumped in to displace the batch to the bottom. The well is shut in for a few hours and brought back on production. This procedure ensures that the tubing is treated with inhibitor all the way to the bottom. Inhibitor treatments involving tubing displacement may last from a few days to a month or so, depending on the severity of the corrosion problem, the produced fluids, the flow velocity, and the ability of the inhibitor to form a persistent film.

Nitrogen or another gas can be used to displace the inhibitor solution instead of liquid. This is of value if the well has a low bottom hole pressure. In these cases, filling the tubing with fluid may permanently stop production or "kill" the well. It is also of value where volumes of oil cannot be easily handled. A variation of the nitrogen batch is to atomize the inhibitor solution into the nitrogen as it is pumped into the well. For batch treatments, the inhibitor selected must have good film persistency as it is not replenished until the next batch treatment is made.

Continuous injection consists of a constant addition of small concentrations of inhibitor into a producing well. The chemical can be added into a chemical or capillary string or down the annulus of a packerless completion. Chemical injection valves in a side-pocket mandrel can be installed so that the solution can be pumped continuously into the annulus of a well with a packer.

The Tuscaloosa Trend wells were originally completed with a Y-block and kill string. This string was used for chemical injection. Wells were then completed with a packer and a chemical string and later with a packerless completion where the inhibitor was added down the annulus.

In treating deep, hot wells, the inhibitor is added as a dilution in hydrocarbon condensate (e.g., carrier oil). This is necessary because the gas is undersaturated with hydrocarbon. At high pressures, gas acts as a liquid and may strip the solvent from the inhibitor. The amount of condensate is calculated to saturate the gas. Another consideration in continuous inhibitor injection in deep, hot, high-pressure wells is the phase behavior of the carried oil. In some cases, the hydrocarbon condensate is replaced with a heavier, less volatile oil that will remain liquid even under the extreme bottom hole conditions in these wells. In some cases, wells have been treated with water-soluble inhibitors in a water-solution or as a dispersion of inhibitor in water instead of hydrocarbon condensate.

Capillary strings are small-diameter, armored tubing that is placed in the annual space between the tubing OD and casing ID. This tubing is strapped to the outside of the tubing as it is run into the well. A surface tank, pump, and filter are installed. The filter is necessary to prevent particles from plugging the small-diameter tubing. Inhibitors must be selected for this application that do not solidify or polymerize, because this would also plug the capillary, making it unusable.

A method of treating a well with a packer consists of using a perforating gun to shoot holes in the tubing. The inhibitor is pumped down the annulus and through the holes. This method is said to be more economical than recompleting the well with alternative equipment necessary for continuous inhibitor injection.

Continuous treating of deep, hot wells requires an inhibitor that will not break down or form a gunk or char. This is particularly important in wells treated with a capillary string or down the annulus where the inhibitor solution must remain for an extended period of time at high temperature. A surface filtering system is also required for capillary string treating to eliminate particulate impurities in the injected fluids.

The deep, hot wells in the Tuscaloosa Trend require an inhibitor that will withstand temperatures to 230 °C (450 °F) without breaking down. Although the chemical strings in the wells that have not been converted to packerless completions are large in diameter (25 to 50 mm, or 1 to 2 in.), plugging problems may occur. Most of this is due to salt plugs, and the condensate has a natural fouling tendency. Therefore, any tendency to form an insoluble residue by the inhibitor adds to the problem. A high-pressure, high-temperature stability test is run in produced fluids to ensure the stability of the inhibitor against formation of precipitates. The dosage of inhibitor used for deep wells will normally range from 10 to 100 ppm under most conditions. Extremely corrosive wells may require more.

Squeeze treatment involves placing an inhibitor solution far enough into the producing formation from the wellbore so that a continuous feedback of inhibitor is obtained during subsequent production of the well. The volume of the squeeze treatment is sized so that a predetermined life is obtained based on the production rate of well fluids. Field crude, hydrocarbon condensate, or diesel oil are commonly used as diluents for squeeze treatments.

The inhibitor must have the proper solubility in the diluent, and it must not form a gunk or severe emulsion with produced water in the formation. Either condition could cause temporary or permanent loss of reservoir permeability and subsequent loss of well production. A core test is sometimes conducted to help select an inhibitor for a tight (fairly low permeability) formation. This involves testing the inhibitor formulation in a core sample of formation rock and then evaluating the core sample for permeability of hydrocarbons. Film persistence is not as important as continuous protection, because inhibitor will be present in the production stream at all times.

In some reservoirs, hydrocarbon condensate is above critical temperature and therefore exists as a gas. This condition is known as a retrograde reservoir, because when the pressure is lowered, condensate comes out of solution with the gas, rather than the normal condition in which lowering the pressure vaporizes the condensate. A dry reservoir that contains no liquid condensate should not be squeezed. Permanent loss of relative permeability will occur, and gas production rates and hydrocarbon recovery will be decreased.

One common objection to squeezing is that inhibitors are cationic and will oil-wet the formation. The wetting characteristics of a surface-active material are based more on its hydrophile-lipophile balance (HLB), which is a measure of the tendency of the inhibitor to water-wet or oil-wet a surface, than on reservoir properties. The HLB is determined by the size and type of oil- or water-soluble parts of the molecule. Nonionic surfactants are used to oil-wet metals in lube oils and are used to water-wet materials in cleaners. Sulfonates are excellent water wetters, while other sulfonates are used as oil wetters. Cationics follow the same rules. In fact, polyamines and quaternary ammonium compounds are used in workover fluids to water-wet silicates.

It is most likely that some oil wetting of the formation occurs (the inhibitor goes into the oil in the reservoir). This is what causes a squeeze to work. Nevertheless, any change in the wettability of the formation is reversible. The formation immediately begins to return to its original state once the wetting agent is removed or begins to dissipate. Natural or simulated core tests can be conducted to ensure that no formation damage will occur from the inhibitor.

Loss of production can also result from a squeeze treatment due to the formation of a stable emulsion in the area immediately adjacent to the wellbore. This emulsion is nearly always a water-in-oil emulsion, which is very viscous. The high-viscosity emulsion will not flow through the pore throats. Emulsion blocking can be prevented by proper inhibitor selection and by adding demulsifiers to the squeeze formulation. A typical squeeze can be performed in the following manner. First the amount of inhibitor required for the projected life of the squeeze should be calculated using:

$$V = \frac{42 \cdot P \cdot D \cdot 3 \cdot \text{ppm}}{1,000,000} \quad \text{(Eq 7)}$$

where V is the volume of inhibitor (gallons), P is the total daily production in barrels (including both oil and water), D is the expected squeeze life, and ppm is the amount of inhibitor feedback desired (this is multiplied by 3, because it is assumed that only one-third of the inhibitor will remain in place that will desorb and feed back). The remaining five-step procedure is to:

1. Dilute the inhibitor with crude, hydrocarbon condensate, or diesel oil to 10%
2. Pump a spearhead of 5 to 10 barrels of oil with 19 L (5 gal) of demulsifier
3. Pump the main body of the squeeze treatment into the formation
4. Overflush with one tubing volume plus one day's production volume of oil (19 to 38 L, or 5 to 10 gal, of demulsifier can be added to the overflush)
5. Shut in the well for 12 to 24 h

This procedure can be modified to suit the requirements of a particular situation.

In many applications, the amount of overflush needed to place the inhibitor properly is too large, or filling the tubing may kill a low-pressure flowing well. The use of nitrogen instead of hydrocarbons overcomes these restrictions.

In a nitrogen squeeze, the inhibitor solution is displaced downhole and into the formation with an equivalent amount of nitrogen. This leaves the tubing empty and charges the formation so that it flows back readily.

This procedure can be modified so that the inhibitor solution is atomized into the nitrogen as it is injected. Both of these procedures have been used with excellent results. The wells could be returned to production in 4 h, and due to the charging effect, the increased production rates for a day or so compensated for the production loss during the squeeze.

Corrosion Monitoring

Once the well, line, or vessel is treated, it is necessary to evaluate the effectiveness of the treatment program and to determine when to retreat or change dosage levels. The methods used include iron counts, weight-loss coupons, test nipples and spools, and probes that use electrical resistance and electrochemical techniques. A last-resort method of waiting until the tubing failure is sometimes available but not popular with operators. However, the collection of tubing failure or perforation data is a valuable source of information.

Iron counts consist of taking a representative sample of produced water and testing for dissolved iron content. The sample must be representative for the system; therefore, the sampling location and procedure are important. The iron counts can be plotted for easier understanding (see Fig. 18). Data management programs are available that will present iron counts, or the data can be simply placed in a spreadsheet for display and plotting. Care must be taken so that the iron from the formation is not assumed to be metal loss from the tubing. Some of the deep, hot Tuscaloosa Trend wells produce water with more than 100 ppm of iron. A base count should be conducted on a downhole sample, if possible. Iron counts are of moderate effectiveness in sweet gas (CO_2) wells. In cases where H_2S is present, dissolved iron counts are not effective since the dissolved iron will react with H_2S in the environment to produce iron sulfide, an insoluble species.

Corrosion coupons are flat or cylindrical pieces of metal that can be installed in nearly any accessible location in the system. It must be remembered that coupons measure corrosion only where they are placed. Furthermore, they only show average corrosion rates over the period of their exposure to the produced environment. Coupons show corrosion after it has taken place, and a single coupon will not show whether the corrosion occurred uniformly with time or occurred over a short interval during the exposure period. Additionally, to properly conduct a coupon monitoring program, significant labor is required to place, retrieve, evaluate, and interpret the corrosion coupons. This is totally an off-line process. Corrosion coupons do however offer the advantage that the morphology of the corrosion (e.g., general or localized attack, pitting, etc.) can be easily seen by visual examination.

Different types of coupon holders are used, depending on the system, the pressure, the location, or other factors. Most coupons are run on surface facilities or flow lines in a 25 or 50 mm (1 or 2 in.) threaded plug. Flat coupon holders hold two coupons, while cylindrical coupon holders may contain eight or more coupons. Some coupons holders allow coupons to be pulled at successive intervals to see if the corrosion rate is constant with time or not.

High-pressure systems require special coupon insertion devices. The insertion tool fits into a special attachment on the pipe or vessel that has a high-pressure chamber with a valve on each end. The inner valve is closed, the retrieval tool inserted, and the inner valve opened. The tool is then run in and left. The procedure is reversed to remove the coupon.

The industry guide for preparing, installing, and interpreting coupons is NACE standard RP0775 (Ref 94). The primary consideration is that all coupons be treated exactly alike. Detailed procedures for preparing corrosion coupons and methods of cleaning for determination of corrosion rates are given in ASTM G 1 (Ref 95). A method of preparation that does not alter the metallurgical structure of the coupon is required. Grinding and sanding of coupons should be controlled to avoid metallurgical changes and to provide a consistent and reproducible surface finish.

Coupons should be handled carefully and stored in chemically treated envelopes or paper until they are installed and should be solvent cleaned and rinsed before use. Rust spots caused by improper handling, fingerprints, and so on, may initiate local attack that is not representative of the system being evaluated. Prior to installation, the coupon weight, identification number, date installed, name of system, location of coupon, and orientation of the coupon and holder should be recorded. The coupons are left in the system for a predetermined number of days and then removed. The frequency of coupon removal is usually in overlapping periods of 30 to 120 days depending on the corrosivity of the system and tendency for periodic upsets.

When the coupons are removed, the identification number, date removed, observations of any erosion, pitting, crevice corrosion or mechanical damage, and appearance should be recorded. A photograph of the coupon may be valuable in some cases. The coupons should then be placed in a moisture-proof envelope impregnated with a vapor-phase inhibitor and taken immediately to the laboratory for cleaning and weighing. The coupons can be blotted (not wiped) dry prior to placing in the envelope. Once at the laboratory, the coupons are weighed before cleaning, cleaned, reweighed, and then photographed. This type of analysis will give useful information in terms of both scale weight and mass loss and the morphology of corrosion. Sometimes the corrosion products are analyzed for elemental composition or the specific chemical compounds by x-ray analysis. A report is issued showing the corrosion rate, any pitting observations, and any other observations of interest. ASTM G 46 (Ref 96) gives useful procedures that can be used to characterize and define the nature of any localized attack on the corrosion coupons.

Electric resistance (ER) instruments function by reading the resistance to current flow of a thin loop of metal installed in the system. The loop of metal is part of an electrical Wheatstone bridge circuit. As the loop corrodes and loses cross-sectional area, electrical resistance increases, and the current flow decreases. This unbalances the bridge and reads out directly on a meter. It may also be recorded on a strip recorder, or with current instruments the data can be logged and then downloaded to a computer for display, trending, and further analysis in spreadsheet format. As the reading changes, the points are plotted on a time graph. The slope of the line is translated into a corrosion rate. The slope will change when a well is treated (decreased slope—lower corrosion rate) or when a corrosion event occurs (increased slope) that changes corrosion rate.

The metal loop is fragile, and it can break if foreign objects, scale, or similar obstructions are present in the flow. The loop may become coated with paraffin, oil, or particulates and may not give a true reading. There are two important limitations of ER measurements. They cannot be used to assess pitting corrosion. Secondly, conventional ER measurements are not instantaneous measurements. Consequently, they are of only limited use in online monitoring. They basically act as instrumented coupons. Normally, the time period required to determine a change in the resistance trend line is longer than that required for process-control needs. Therefore, they provide average corrosion rate information. Recently, more sensitive ER devices have been developed that produce more rapid corrosion rate readings. However, these are used to assess general corrosion tendencies without being able to assess localized corrosion.

Electrochemical instruments monitor the potential and current signals resulting from the corrosion of probe elements that are made from the same or similar materials as used in the system being monitored. The simplest of these devices is called a galvanic probe and uses a two-element, dissimilar electrode arrangement. This device measures the galvanic current produced between the dissimilar metals whenever a corrosive environment is present. In this case, the signal is qualitative in nature (high current when corrosive conditions exist and little or no current when conditions are not corrosive) and does not provide actual corrosion rates.

Linear Polarization Resistance. For semi-quantitative applications a two- or three-electrode arrangement is used to make linear polarization resistance (LPR) measurements. The difference between the two- and three-electrode LPR instruments is that with the three-electrode version, it is possible to measure and compensate for the *IR* drop in the environment that can be an additional source of corrosion rate error. This version is particularly useful in low-conductivity solutions. For additional information on the background and application of the polarization-resistance method discussed below, the reader should refer to Volume 13A of the *ASM Handbook* (see, for example, the article "Electrochemical Methods of Corrosion Testing").

The LPR technique varies the potential slightly of one electrode (± 25 mV) from its normal corrosion potential and monitors the resulting current produced in the two- or three-electrode circuit. The slope of these data is referred to as polarization resistance (R_p), which is normally considered to be inversely proportional to the corrosion rate. Using Faraday's law, the Stern-Geary equation, and the physical information on the material and probe size, this current can be converted to a general corrosion rate for the material. Measurements can be taken at intervals of 20 to 30 min or as needed, which makes this technique attractive for process-control purposes.

One aspect of this technique that limits its quantitative use is that simple field LPR devices cannot measure the actual anodic and cathodic Tafel slopes used to calculate the value of the Stern-Geary parameter (*B* value). An accurate *B* value is needed to obtain an exact conversion from corrosion current to corrosion rate. In the laboratory, it is customary to measure these parameters for use in accurate corrosion rate determinations. However, in field measurements, it is customary to provide a fixed, default *B* value in these LPR instruments of between 0.025 and 0.030 V depending on the manufacture. This provides reasonable conversion for many applications except where the corrosion processes dictate a much different *B* value. Additional techniques have been developed that make it possible to measure the Tafel slopes and *B* values without a laboratory-style potentiostat.

An automated variation of the LPR technique and its analysis available in certain field instruments is known as harmonic distortion analysis (HDA). In this application, a continuous low-frequency (15 mHz) symmetrical sine-wave polarization signal is employed to measure both the overall polarization resistance, R_p, and the solution resistance, R_s. The leading edge of the wave acts as a high-frequency perturbation and effectively short circuits the double-layer capacitance, thereby allowing R_s to be measured. The dc component of the current response is used to measure R_p, and by subtraction, the charge transfer resistance can be obtained. Two distinct benefits of the field use of the HDA technique is the determination of the anodic and cathodic Tafel slopes for use in assessing and controlling inhibition, and the *B* value (Stern-Geary factor) for use in correcting LPR corrosion rates (Ref 97).

In most cases in oil and gas operations, corrosion measurements are made in the liquid (brine or mixed brine/oil environment with dissolved gas) where the electrodes protrude into the liquid phase. Alternatively, corrosion measurements need to be made in a wet vapor phase where the electrodes are covered by a thin film of aqueous fluid and/or oil and dissolved gases that condense from the vapor phase. Special flush-mounted probes are required with closely placed (and interlaced) electrodes. Custom probe designs can extend corrosion monitoring into mixed oil/water environments and in condensing-vapor-phase conditions as well as optimize the electrode configuration for monitoring in multiphase environments. Alternative probe configurations also include flange or gasket probes that can be placed in the gap between existing flange connections. These probes eliminate the previous requirement of locating and installing an access fitting in the system. They also provide an electrode configuration that closely models the ID surface of the pipe or piping system being monitored. *B* value measurements are very important in making accurate corrosion rate measurements in these types of multiphase field environments and where the influence of H_2S or other strongly depolarizing species need to be factored into the corrosion measurement.

As mentioned previously, *B* values are not instrument constants (as commonly believed) but are actually defined by the corrosive conditions that exist on the metal surface and commonly include: (a) the oxidation of iron to iron ions, (b) corresponding reduction reactions, (c) ionic current flow in the media, and (d) electronic current flow in the metal.

Therefore, knowledge of the actual *B* value is quite useful in providing insight into corrosion processes. High values of *B* would tend to indicate the corrosion processes are tending to be rate limited due to diffusion processes (this *does not* necessarily mean slower). Alternatively, low values of *B* would suggest that the system being studied is nonpolarizable, likely due to the presence of certain chemical species that tend to depolarize the electrode.

From an electronic viewpoint, an activation (Tafel) slope of 60 mV per decade of current would relate to a single electron transfer process. Table 9 helps to understand the various values of *B* and their origin.

A common case is in the second horizontal row of Table 9, which has led to the widespread adoption of 25 to 30 mV *B* values in most field-monitoring instruments. An example of a case where such *B* values are appropriate is for carbon steel in aerated seawater or brine.

Higher *B* values typically occur because of diffusion-limiting effects on both the anodic and cathodic processes. Very high *B* values have typically been observed with very high corrosion rates or in some cases where thin films (vapor-phase exposures) or oil/water mixtures are present.

Low B values have been observed with the presence of reactive sulfide species in the environment leading to the formation of sulfide corrosion films on the metal surface (which are also the cause of changes in the measured double-layer capacitance and apparent "shorting" between electrodes in some cases). The sulfide films resist the applied polarization (are nonpolarizable), which is most likely the result of a strong redox process within the sulfide film.

In corrosion measurements, even though corrosion rates and polarization resistance (R_p) are inversely proportional, low polarization resistance values do not necessarily mean high corrosion rates and high polarization resistance values do not necessarily mean low corrosion rates. It is only through field corrosion measurements that incorporate measured B values that accurate corrosion rates can be obtained. Not to incorporate measured B values can easily result in corrosion rate errors of 25 to 50%, and in some cases, the errors can be over an order of magnitude (Ref 98).

Electrochemical Noise (EN). A relatively new technique applied to field corrosion measurements is EN, which was developed in the 1970s for laboratory and research applications. Electrochemical noise is the measurement of spontaneous fluctuations in the current and potential generated by corrosion occurring at the metal-electrolyte interface of a three-electrode system (Ref 99). Ideally, the sensors for EN measurements comprise two identical working electrodes plus a reference electrode. Whereas in the laboratory, the reference electrode is usually a conventional solute electrode (such as Ag/AgCl or standard calomel), for practical field monitoring applications, a metal pseudo-reference electrode of the same material as the working electrodes is typically used. This is similar to the standard three-electrode configuration used for field LPR measurements, which leads to the use of three nominally identical electrodes made from the material of interest in the evaluation.

Modern microchip computing capabilities and telecommunications have allowed this technique to be remotely located near the point of monitoring. Automated statistical analysis of the current and potential time records are used to derive a pitting factor (defined simply as the standard deviation of the corrosion current derived from EN divided by the average corrosion current during the measurement period from LPR) to infer tendencies for general or localized corrosion. Using this methodology, conditions for pitting are identified when the pitting factor is above 0.1 (e.g., the standard deviation of the EN current noise is above 10% of the corrosion current). Conditions for general corrosion are commonly associated with low value of pitting factor (less than 0.01). The intermediate decade (pitting factor = 0.01 to 0.1) should be considered a transition or cautionary range. It does not indicate active pitting, but signifies unstable conditions on the corroding metal that can be easily perturbed to higher values by changes in process variables (e.g., flow rate, turbulence, chloride, oxygen, etc.).

It is actually more important to know if pitting is an active process rather than to know its pitting rate, which can be highly variable from pit to pit. Pitting is an inherently unstable process, and efforts should be made to control processes, operations, and the production environment to minimize pitting tendencies. Since pitting often results from upset conditions, real-time monitoring is often needed to successfully control pitting. Electrochemical noise can also be used to derive corrosion rates; however, these values are not normally used for real-time, process-control applications except in rare situations.

The principal advantage of these newer electrochemical instruments with EN capabilities is that a corrosion rate can be determined immediately (normally within 7 min), and pitting tendencies quickly assessed without waiting a month or more to retrieve coupons. Figure 23

Table 9 Origin of various B values (corrosion parameter) for use with linear polarization resistance corrosion measurements

β_a	β_c	B	Comments
60 mV	60 mV	13 mV	Both processes activation controlled; commonly observed with sulfide species
60 mV	∞	26 mV	Anodic process activation, cathodic diffusion, controlled; commonly observed with aerated brine
120 mV	∞	52 mV	Anodic process activation, cathodic diffusion, controlled (anodic slope different); commonly observed with oil/water mixtures
∞	∞	∞	Severe anodic and cathodic diffusion limiting; commonly observed in condensing vapor phase conditions

Note: $B = \beta_a \beta_c / 2.3(\beta_a + \beta_c)$, where β_a and β_c are the anodic and cathodic Tafel slopes, respectively.

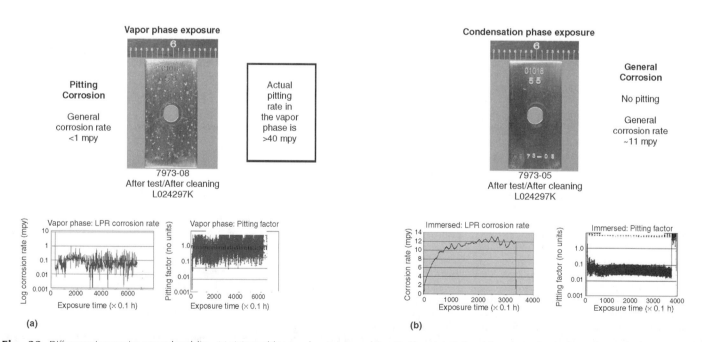

Fig. 23 Differences in corrosion rate and modality—(a) pitting and (b) general corrosion—as determined by automated, real-time electrochemical corrosion monitoring system using linear polarization resistance, electrochemical noise, and harmonic distortion analysis. Note comparison of electrochemical data from a partially dehydrated CO_2 gas pipeline liquid phase water/glycol and condensing vapor phase in system. Corrosion rates are given in mils per year (mpy). Source: Ref 100

shows a comparison of coupon examination and mass loss determination versus real-time, online monitoring data using an electrochemical device incorporating LPR, HDA, and EN techniques (Ref 100). It can be seen that the corrosion rate and pitting factor obtained by electrochemical monitoring provide accurate assessment of pitting and general corrosion within a time frame of hours. The coupon data required exposure for a period of about 1 month. While corrosion coupons and electrochemical techniques provide similar data, the online, real-time electrochemical data would allow the operator to take corrective measures (e.g., process change, inhibition, etc.) before substantial damage has occurred.

Probe locations are very important, because conventional, electrochemical "finger" probes must be immersed in water to give an accurate reading. As discussed previously, alternative flush-mounted, flange or interleaved designs can be used that optimize the probe electrode configuration for measuring corrosion in mixed-phase water, oil, and gaseous environments. Probes can become coated with oil, paraffin, or corrosion products and can show an erroneous corrosion rate. They should be located in the bottom of a line (5 or 7 o'clock positions) or in a bypass loop so that they are water wetted. The use of field B value (Stern-Geary factor) measurements should be used to obtain more appropriate corrosion rates from electrochemical probes. The B value is not an "instrument" constant. As mentioned previously, the chemical composition of the environment and the mixed-phase conditions (i.e., oil, water, and/or gas) can influence the B value, which can be used to correct conventional LPR corrosion rates.

Other Corrosion Assessment Techniques. A test loop can be installed in a system for better monitoring and control. A test loop is simply a bypass with valves for controlling flow, and it may contain weight-loss coupons as well as probes. The system can be monitored, and different inhibitors can be evaluated at the same time. A bypass test loop can also allow for removal of coupons and probes periodically for inspection and analysis.

A caliper survey of the downhole tubing can be conducted to determine if pitting and general metal loss have been halted. The caliper log can be easily compared with a long run before the treatment is begun.

Other monitoring methods include hydrogen probes and in-line electromagnetic logging devices (e.g., smart pigs). Chemical analysis of produced water for alloying metals, such as manganese and chromium, can also be conducted.

It is important to collect and chart failure rates. Some failures may occur over a period of time and may erroneously indicate that the treatment is not effective. However, through proper charting and comparing with previous failure rates, the effectiveness of proper treating will be shown.

Corrosion in Secondary Recovery Operations

Secondary recovery, or waterflooding, generally increases the corrosion problems in existing producing wells. It also creates a new set of problems because of the facilities required to reinject the produced water. This section discusses the corrosion problems that are specific to the various types of environments or equipment used in secondary recovery, that is, producing wells, producing flow lines, separation facilities, tanks, injection pumps, injection lines, and injection wells. Although not specifically addressed, disposal wells (wells that are used for produced water disposal rather than reinjection into producing formations) are considered to be the same as injection wells. Corrosion mitigation methods and guidelines are then discussed for each type of environment.

Types of Corrosion Problems

Producing Wells. The corrosion mechanisms in secondary recovery are similar to those in primary production. The primary causes of corrosion are dissolved acid gases (H_2S and CO_2) in the produced fluids. Naturally occurring organic acids are often present and can aggravate H_2S and CO_2 problems. Corrosion will generally increase in secondary recovery because of the large increase in water production caused by waterflooding. The fraction of water produced versus total liquids (aqueous and hydrocarbon), also known as water-cut, may increase to 90% or more. This increases the potential for corrosion, because more of the metal surfaces may be water-wet rather than oil-wet. The increased volume of water can increase pumping equipment stresses. Increased stress levels can cause more corrosion fatigue related failures.

Corrosion mechanisms may change during waterflooding. For example, a normally sweet field (that is, the produced fluids contain no H_2S) may begin to produce H_2S because of the growth of sulfate-reducing bacteria in the formation. This can cause unexpected corrosion related to the H_2S, pitting under sulfate-reducing bacteria deposits, or failures from sulfide stress cracking in high-strength materials.

Mineral-scale problems, such as the deposition of $CaCO_3$, $CaSO_4$, or $BaSO_4$, may increase during waterflooding. This is usually the result of changes in the formation water brought about by injecting waters from sources other than the original reservoir. Although not strictly a corrosion problem, scale deposition can cause increased failures due to wear and underdeposit corrosion.

Producing Flow Lines. Corrosion mechanisms in producing flow lines are similar to the mechanisms downhole, but generally occur at a lower rate because temperatures and pressures are lower at the surface. Corrosion is often localized to the bottom of flow lines if flow rates are low enough to permit water stratification, which allows the bottom of the line to be continuously water-wet. Underdeposit corrosion and sulfate-reducing bacteria related pitting are often severe under sludge or scale deposits that accumulate in the low flow rate lines.

Oil/Water Separation Facilities. Corrosion in these facilities is normally related to attack by corrodents in produced fluids and deposit-related problems. Separation facilities are unique in that they often use heat to aid in oil/water separation. Heat-transfer surfaces are usually subject to mineral-scale deposition because of solubility changes caused by temperature increases. Scale deposition can result in severe underdeposit corrosion because metal surface temperatures increase due to the reduced heat transfer. Creep rupture failure can occur in direct fired heaters if deposition is severe enough to cause very high metal temperatures. Some separation equipment is open to the atmosphere, thus allowing oxygen contamination of the produced fluids and causing increased corrosion in equipment handling the water phase.

Tanks/Water Storage. Tanks are subject to corrosion by acid gases (CO_2, H_2S) carried over with the produced water. Underdeposit corrosion can be severe under accumulated sludge and debris in tank bottoms. These deposits are also prime areas for the growth of sulfate-reducing bacteria. Tank roofs often fail because of condensation. As water condenses on the roof, it will absorb acid gases from the tank fluids. This can cause severe pitting. Oxygen contamination often occurs in tanks. Obviously, open tanks are subject to contamination. Contamination can occur in normally closed tanks if hatches and vent systems are poorly maintained. Although oxygen can be somewhat corrosive by itself, its primary role in waterflood system corrosion is to significantly increase the rate of attack of other corrodents already in the system.

Injection pumps can fail by normal corrosion mechanisms as well as by cavitation and erosion. Pump intake piping design must take into account the presence of dissolved H_2S and CO_2 in the water. These gases can affect net positive suction head calculations. If sufficient net positive suction head is not provided, cavitation can occur. Erosion and erosion-corrosion can occur because of solids in the water. Solids normally consist of corrosion products, formation fines, and mineral-scale particulates. Alloy materials such as type 304 and 316 stainless steels are often used for pump internal parts. These alloys can fail by chloride SCC in produced brines if temperatures are above 52 to 65 °C (125 to 150 °F). Pumps are subject to cyclic stresses. Corrosion fatigue failure can occur at sharp changes in cross section, grooves, and at pitted areas, all of which cause stress concentrations.

Injection Flow Lines and Wells. Corrosion mechanisms are generally the same for producing well flow lines and tubulars, that is H_2S, CO_2, and organic acids. Underdeposit problems in the bottoms of lines and under mineral-scales can also occur, as can problems with sulfate-reducing bacteria. Oxygen contamination will greatly accelerate all but the sulfate-reducing bacteria mechanism. Sulfate-reducing bacteria corrosion can still occur even in aerated systems because localized areas under scales, sludges, or aerobic bacterial slimes can become anaerobic and thus support the growth of sulfate-reducing bacteria.

Injection wells and flow lines may require periodic acidizing to reduce pressure drops and to restore the injectivity lost because of the buildup of corrosion products and mineral scales. Severe corrosion can occur if acidizing fluids are not properly inhibited and flushed from the system.

Corrosion Mitigation Methods

Producing Wells. Corrosion control methods for secondary recovery are typically the same as those used for primary recovery; however, higher levels of water in both gas and oil production, and souring (e.g., an increase in H_2S concentration) should be anticipated as the wells age. The particular method implemented will depend on the type of production method and rate used (that is, beam lift, electric submersible pump, or gas lift), the nature of the produced fluids, well design, and the economics of the individual situation.

Corrosion inhibitors are widely used to protect tubulars and other downhole equipment in all types of producing wells. The most common methods of sending the inhibitor downhole where it can protect the well equipment are referred to as squeeze treatment, batch treatment, and continuous treatment (see the discussions "Inhibitors" and "Oil Production" in this article). References 92 and 93 contain detailed descriptions of the various methods as well as guidelines for selecting a particular method. Regardless of the method used, the inhibitor must be effective against the particular type of corrosion occurring, that is, H_2S, CO_2, or both, or even cases where oxygen may be present due to contamination from surface operations. Laboratory tests should be performed if there are any questions regarding the effectiveness of the inhibitor for a given type(s) of corrosion.

The type of corrosion inhibitor used (oil-soluble, oil-insoluble, water-dispersible, water-soluble, and so on) will depend on the treatment method. Batch treatment is a widely used method of treatment for beam lift wells. Corrosion inhibitor solutions are periodically injected into the casing-tubing annulus and flushed to the bottom of the well with produced fluids, diesel oil, or water. A water-dispersible inhibitor is normally used because of the high percentage of water in the well stream. However, increased water dispersibility can cause problems with oil/water separation because of the tendency for dispersion chemicals in the inhibitor to cause emulsions to form. Tests should be performed with actual well fluids to determine the emulsion tendency of the particular inhibitor being considered for use.

Often, any one of several inhibitors may be able to provide the necessary corrosion protection; however, there will be vast differences in emulsion formation and economics specific for the application at hand.

Continuous injection of inhibitor may be necessary for wells with high fluid levels in the annulus above the pump. Water-soluble inhibitors are normally specified for this type of treatment. Studies have shown that continuous treatment may not always be as effective as periodic batch treatment under many conditions (Ref 92). Emulsion problems are sometimes worse with water-soluble inhibitors than with oil-soluble or water-dispersible inhibitors because of the increased use of surfactants in water-soluble inhibitors. In order to sort out these types of problems, it may be necessary to conduct laboratory or field studies on emulsion-forming tendencies and the effectiveness of emulsion-breaking chemical additives to achieve optimum production conditions.

The frequency of treatment and the quantity of inhibitor used will generally have to be increased during secondary production, as the amount of water and possibly corrosive species increase. In general, it is more effective to increase the frequency of inhibition (assuming a batch treatment procedure is used) rather than the quantity, although both may need to be adjusted in some cases. Treatment should be adjusted on the basis of corrosion monitoring results and well equipment life. Corrosion monitoring can be accomplished in a variety of ways. Corrosion coupons installed in flow lines near the wellhead are the most common. Downhole monitoring is more difficult. Preweighed, short (0.6 m, or 2 ft) sucker rods can be used as downhole corrosion coupons, as can short joints of production tubing. Information on the preparation, installation, and evaluation of corrosion coupon data is provided in NACE RP0775 (Ref 94).

Downhole equipment should be carefully examined for signs of corrosion whenever it is removed from the well. The occurrence of sucker rod failures is a common measure of downhole inhibition effectiveness in rod pumped wells (Ref 93). The number of failures that can be tolerated will depend on the economics of each producing situation. A general guideline is one corrosion-related failure per well per year. It should be remembered that corrosion fatigue failures of sucker rods are a function of corrosion and stress. Therefore, heavily loaded rods will tolerate less corrosion before failure than rods with lower stress levels.

Corrosion inhibitors are less effective in sucker rod pumps because of wear. Corrosion-related failures are generally controlled by changing the pump metallurgy. Guidelines for selecting pump materials are provided in NACE MR0176 (Ref 101). Galvanic corrosion problems can be quite severe in pumps and are best controlled by eliminating or reducing the extent of dissimilar metals in contact with each other in the pump. This also applies to coatings used for wear resistance, such as chromium and nickel plating. Rapid failure can often occur in underlying steel if these coatings become damaged by wear. If the wear resistance of chromium plating is required, it may be necessary to upgrade the base material to avoid galvanic corrosion problems.

Wear of sucker rod strings can be controlled through the use of centralizing rod guides. A variety of nonmetallic materials are either molded on or physically attached to the sucker rod to prevent it from contacting the tubing. Welded or metal guides should not be used. Sucker rod couplings are normally coated with a corrosion-resistant alloy by flame spraying or similar techniques. This will provide both wear and corrosion protection. Similar coatings can be applied to the rods; however, these have not been widely used because of the high cost involved.

Fiber-reinforced plastic sucker rods can be used to reduce corrosion fatigue failures; however, their primary benefit comes from production concerns rather than corrosion. Corrosion inhibition is still necessary when FRP rods are used to protect the steel end connectors of the rod and steel well tubulars. In addition, steel rods are not entirely eliminated from the string when FRP rods are used. Internal tubular coatings are not widely used in rod-pumped wells, because they rapidly fail from rod wear. Fiber-reinforced plastic tubing is not widely used for the same reason.

Electric submersible pump wells are treated in much the same way as rod-pumped wells. Electric submersible pump wells pose an additional problem in that the pump fluid intake is above the motor housing. This means that inhibitors injected into the annulus do not reach the housing. A variety of methods have been used to reduce the corrosion of housings, including applying corrosion-resistant coatings and selecting corrosion-resistant alloys for the housing. Special inhibitor-injection systems using small-diameter tubing to release inhibitors below the motor have also been employed. Corrosion of electric submersible pump internal parts is not typically a problem because of the widespread use of corrosion-resistant alloys. Internal tubular coatings can be used with electric submersible pump wells, because they are not subject to wear. Fiber-reinforced tubing has found application in a limited number of electric submersible pump wells.

Gas-lift wells are commonly treated by atomizing inhibitor solutions into the lift gas. This can provide protection to the tubulars only above the lowest operating gas-lift valve. Internal tubular coatings, FRP tubulars, and corrosion-resistant alloys can be used above or below the operating valve to provide corrosion protection.

Producing Flow Lines. Carryover from downhole corrosion inhibition is often sufficient to protect flow lines. In extremely corrosive conditions, additional inhibitor injection, either batch or continuous, may be required. Internal coatings can be used on flow lines; however, obtaining protection in the area of pipe joints can be difficult. A variety of methods have been developed to minimize damage to the coating even in welded lines. Fiber-reinforced plastic line pipe is becoming more widely used for flow lines because it is inherently corrosion resistant. Polyethylene lines are also used in low-temperature low-pressure applications.

Oil/Water Separation Facilities. Supplemental inhibitor injection is often used to help protect these facilities. In addition, vessels such as separators are often internally coated. Organic coatings are normally used, but platings such as electroless nickel are also employed. Noble platings, such as electroless nickel, can cause severe galvanic corrosion of underlying steel if the coating is cracked or otherwise damaged. Internal cathodic protection with sacrificial anodes is also used in vessels. Internal baffles and other pieces can be fabricated from corrosion-resistant alloys. Corrosion-resistant alloy linings can also be used.

Heat-transfer surfaces and vessel bottoms should be periodically cleaned of scale and debris. Scale inhibitors should be used if continuous scale-deposition problems occur. Chromium-containing steels (2.25 to 12% Cr) can be used for heat-transfer surface in direct-fired heaters to reduce the possibility of creep rupture failures in applications subject to severe scale formation. However, these materials can be rapidly attacked in the presence of H_2S. High-nickel corrosion-resistant alloys can also be used to help prevent underdeposit corrosion problems.

Tanks/Water Storage. Internal coating is a common method of protecting tanks. Organic coatings are typically used. Steel tank life is often extended by the use of FRP linings, especially tank bottoms. Both chopped and mat systems are used. A variety of nonmetallic liners have also been used. Fiber-reinforced plastic tanks are becoming more popular is smaller sizes. Internal cathodic protection can also be used, normally in conjunction with internal coatings. Tanks should be periodically cleaned to remove the accumulated sludge and debris that hinder normal corrosion control methods and promote underdeposit and sulfate-reducing bacteria problems.

Tanks are usually the first source of oxygen contamination in the injection system. Open tanks and pits should be avoided. Various methods of excluding oxygen in open tanks have been attempted. Oil layers are ineffective. Several floating systems have been developed that are useful to some degree, but are not totally effective. It must be remembered that as little as 0.01 ppm oxygen is sufficient to cause major increases in corrosion rates. Oxygen also renders many corrosion inhibitors ineffective. Closed tanks can also allow oxygen entry. Poorly maintained hatch seals and venting systems are notorious as sources of contamination. The optional method of excluding oxygen is to ensure that all openings to the tank are properly maintained and that a low-pressure inert gas blanket is used. Gas blanketing provides a slight positive pressure that will keep air from entering. Gas blankets can be part of the vapors-recovery system, if used, or can be externally supplied from bottled gases, such as nitrogen.

Oxygen can enter the injection system in other ways. Often, additional water must be obtained to augment produced water volumes. Freshwater can be obtained from lakes, rivers, or wells drilled into aquifers. Seawater is used in offshore and coastal locations. All of these waters will have some amount of oxygen contamination. Severe corrosion can result if this contamination is not removed. Common removal methods include the use of chemicals or scavengers, such as sodium sulfite or ammonium bisulfite, and vacuum or gas stripping (see the discussion "Environmental Control" in this article).

Tanks are also excellent locations for the growth of sulfate-reducing bacteria. If tanks become contaminated with sulfate-reducing bacteria, they must be cleaned and sterilized with biocides. Cleaning is a necessity, because it is impossible for biocides to adequately penetrate the large amounts of sludge and debris on the tank bottom.

Injection Pumps. Corrosion-resistant alloys are widely used in injection pumps and ancillary equipment. The particular choice of materials used will depend on the nature of the fluids handled and the type of pump involved. Specific material recommendations are provided in NACE RP0475 (Ref 102). Caution should be exercised, because this specification does not address the temperature limitations of the materials. Chloride SCC can occur in 300 series stainless steels if they are used in saline waters above 52 to 65 °C (125 to 150 °F). Also, pitting of these materials can occur in aerated saltwater if they are left stagnant in a pump. For example, it is common practice to have standby equipment piped into a system and to test the equipment periodically. Flushing the equipment with deaerated and inhibited freshwater is recommended to prevent pitting corrosion.

Flow Lines and Injection Wells. All potential corrosion mechanisms must be dealt with to obtain acceptable service lives of injection systems. This includes corrosion by dissolved acid gases, growth of sulfate-reducing bacteria, oxygen contamination, and scale/sludge deposition.

Corrosion inhibitors can be used to control flow-line and injection-well corrosion. Treatment is usually continuous, but batch treatment can also be used. Both oil-soluble/highly water-dispersible and water-soluble chemicals are used. Flow lines can be internally coated with organic coatings. Cement and other nonmetallic linings are also used. Fiber-reinforced plastic flow lines are widely used, even in high-pressure injection systems, under API standards for these types of tubular products.

Injection-well tubulars can be bare steel if corrosion inhibition is used. Internal coating is also widely used even with corrosion inhibition (see the discussion "Coatings" in this section). Care must be taken when handling internally coated tubing to prevent coating damage. Special guides must be used when the tubing is installed to prevent damage to the pin nose. Makeup equipment must not deform the tubing enough to crack the coating. Standard API couplings are routinely internally coated in the standoff thread area. The recent advent of flush joint tubing connections using nonstandard couplings has helped to make internally coated tubing applications more reliable. The new connections help to seal the end of the tubing joints in the coupling. This has long been a problem area in internally coated tubing because it is easily damaged during handling and installation.

Corrosion-resistant alloy tubulars are used on some occasions, but their high cost is usually prohibitive. Fiber-reinforced tubing is used to some extent; however, again, the lack of standardization has been a limiting factor. Handling and makeup procedures are critical for successful fiber-reinforced tubing application. Many failures have resulted from overtorquing of FRP connections by crews used to handling steel tubulars. No reliable method has been developed for accurately predicting the long-term performance of FRP tubulars subject to both internal pressure and axial load.

Injection wells frequently require acidizing to restore injectivity. Typical acids used are 15% HCl and 12%HCl-3%HF. Severe corrosion can result if these acids are not properly inhibited. Corrosion inhibitors are available from acid service companies. Inhibitor concentration should be such that the corrosion rate of low-carbon steel is less than 245 g/m^2 (0.05 lb/ft^2) over the length of time the acid is to be in the well. It is good practice to ensure that the acid delivered to the job site actually contains the inhibitor and is the strength called for in the workover procedure. A simple test procedure for determining the presence or absence of inhibitor is given in NACE RPO273 (Ref 103). This test is not designed to determine inhibitor effectiveness at well conditions or to compare different inhibitors. Laboratory testing is necessary to establish inhibitor effectiveness.

Acid exposure can have a wide range of effects on the internal tubular coatings that may be present. Laboratory testing should be conducted if there is any doubt regarding the ability of the coating to withstand the acid exposure without damage. Fiber-reinforced plastic tubulars can also be damaged by exposure to mineral acids. Although tubing manufacturers do not prohibit acid exposure, they all recommend that temperatures and exposure times be kept to absolute minimums. The use of hydrofluoric acid in acidizing fluids is not recommended if FRP tubing is installed.

Carbon Dioxide Injection

Secondary recovery by waterflooding can greatly increase the amount of oil recovered over primary production, but may still leave up to 80% of oil in place in the reservoir. Tertiary recovery by injecting CO_2 will remove the oil not obtained by waterflooding. Carbon dioxide can be used at much lower pressures than other gases, such as nitrogen or methane, because it dissolves readily in some crudes and can cause up to a tenfold viscosity reduction in heavy crudes.

Oils with an API gravity of 25 or higher are candidates for miscible flooding. This process can recover oil from low-permeability reservoirs. Oils with gravities down to API 15 are recovered by an immiscible process based on oil swelling and viscosity reduction.

Carbon dioxide injection uses gas from fields that produce almost pure CO_2 from burning of lignite and recovered CO_2 from industrial combustion gases. These gases are purified and compressed, and in some cases, there are pipelines for hundreds of miles to the fields to be flooded with CO_2. The Texas Permian Basin, North Dakota, the Texas Gulf Coast, and the California area have had CO_2 injection projects in operation since the 1980s.

Because CO_2 is an acid gas, production problems are encountered when CO_2 is injected. Carbon dioxide ionizes in water to form carbonic acid and will react directly with carbon steel. The corrosion rates can be quite high, and pumps can fail in a matter of days after breakthrough of the CO_2 into producing wells. Some scaling problems may arise because carbonic acid may dissolve calcium carbonate from the formation. The calcium bicarbonate formed during this reaction may come out of solution in heaters and vessels as calcium carbonate when CO_2 is lost. Calcium sulfate ($CaSO_4$) will also dissolve and may cause scaling in surface equipment.

Emulsion-treating characteristics may change when CO_2 dissolves in oil. Asphaltenes may cause problems by dissolving in CO_2 as it sweeps through the formation and then coming out of solution on the surface.

Elastomers must be selected with care, because they may swell or lose strength when exposed to CO_2. Leaking packers due to seal failure will cause pressure on the annuals of CO_2 injection wells and annular space corrosion.

Carbon Dioxide Production Facilities

Carbon dioxide source wells may produce from a few percent to almost pure CO_2. They may produce both liquid- and vapor-phase CO_2. The formation of liquid water in the produced CO_2 will cause hydrate formation and corrosion. Hydrate formation can be controlled by glycol dehydration, but special measures must be taken to control corrosion.

A corrosion inhibitor can be added to the producing well to control corrosion. Continuous injection downhole of a water-soluble or dispersible filming amine inhibitor should protect the tubing and wellhead. The use of type 316L and 304L stainless steels and FRP for completions and flow lines is an alternative to inhibitor use, presuming that substantial chlorides are not present that may adversely affect the performance of the stainless steel.

In a typical CO_2 production facility, the gas travels through a wellstream heater to a contactor in which water is removed. It is then scrubbed, compressed, and sent to the pipeline. Materials selection in the design of the system is the key to corrosion control in the processing plant. Corrosion-resistant alloys can be used in areas of high corrosion, and carbon steel is used where conditions allow its use. A maximum water content of 60% of saturation at the minimum operating temperature is obtained by dehydration so that corrosion of the steel pipeline is prevented. Dehydration also prevents hydrate formation when temperatures are low.

Injection Systems

Water and CO_2 are injected alternately in some systems, such as the SACROC unit in the Kelly Snyder field in west Texas. This is known as the water alternate gas process.

The distribution system consists of parallel separate lines for water and CO_2 that are carbon steel coated externally and cathodically protected. Carbon dioxide in the line contains less than 50 ppm water, so internal corrosion is minimal. Valves in the system range from bare carbon steel to plastic coated with type 316 stainless steel trim. Fluorocarbon and nylon O-rings have performed satisfactorily, and Buna N rubber is used for stem sealing, although these materials swell somewhat.

Water lines are cement lined with sulfate-resistant cements and artificial pozzolans, as specified in API RP-10E (Ref 104). Most leaks have been due to the failure of asbestos gaskets. The use of grout instead of gaskets has been effective. Water-soluble inhibitors are added to protect voids in cement linings and plastic coatings.

Carbon dioxide injection systems have suffered corrosion problems when the mixing of water and CO_2 at each cycle of alternate CO_2/water injection occurs. Plastic coating and type 316 stainless steel trim, ceramic gate valves with electroless nickel-coated bodies, and electroless nickel-coated check valves were tried. The type 316 stainless steel and ceramic gates performed well, but the other methods failed.

Injection wells originally used type 410 stainless steel wellheads and valves. Severe pitting occurred under deposits laid down from suspended solids in the injection water. Type 410 stainless steel was plastic coated, and the gates and seats were changed to type 316 stainless steel to correct the problem.

Failures occurred in the couplings of the plastic-coated tubing in the injection wells when the seal rings failed. This was corrected by changing the coating on the couplings from an epoxy-modified phenolic to a polyphenylene sulfide.

Production Systems

Most failures in the SACROC unit were due to rod breakage. Inhibitor programs were satisfactory in some wells, but many did not respond. Plastic-coated rods and thermal spray metal coating with type 316 stainless steel helped to alleviate the problem. The use of fiberglass rods in the upper 70% of the string, along with stainless steel coated rods on the bottom, reduced rod breakage to an average of 1.1 per well per year.

Tubing leaks can be controlled with coatings and the use of 9Cr-1Mo and 13% Cr steel tubing where inhibitors fail to control. Flow-line corrosion can be controlled by the use of fiberglass-epoxy lines. Other systems have experienced problems similar to those found in the SACROC unit and have successfully controlled corrosion with the previously described methods. Inhibitor selection by field testing with linear polarization techniques or similar techniques has resulted in improved protection of producing wells. The electrochemical corrosion monitoring is also used routinely, along with coupons, iron counts, and caliper surveys.

Corrosion of Oil and Gas Offshore Production Platforms

Offshore structures have been in service in various parts of the world for more than 50 years. Early experience was in the Gulf of Mexico with water depths of less than 90 m (300 ft). Technology has advanced to the point at which the largest drilling and production platform stands in more than 305 m (1000 ft) of water. Current technology for deep seawater applications now includes the use of subsea wellheads and manifolds and gathering systems that have extended offshore operations into water depths of more than 3000 m (9300 ft).

A platform consists of three parts. The jacket is a welded tubular space frame that is designed as a template for pile driving and as lateral bracing for the piles. The piles anchor the platform permanently to the sea floor and carry both vertical and lateral loads. The superstructure is mounted on top of the jacket and consists of the deck and trusses necessary to support operational and other loads. Generally, platforms are carried from the fabrication yard to the site on a barge and are either lifted or launched off the barge into the water. After positioning the jacket, the main piles are driven through the legs of the jacket, one through each leg. Other piles, known as skirt piles, can be driven around the perimeter of the jacket as needed.

Current design and fabrication practices related to fixed steel offshore structures can be found in industry publications, professional journals, and the proceedings of technical conferences. The most basic U.S. document on this subject is API RP-2A (Ref 105), which was first issued in Oct 1969 and has had many subsequent editions. Because U.S. experience has been mainly in the Gulf of Mexico, API RP-2A generally represents that experience. Since then, added experience in deeper waters has been obtained in the Gulf of Mexico, the North Sea, and the coastal waters of

Brazil and West Africa. This information can be obtained in recent conference proceedings and technical publications of Society of Petroleum Engineers (SPE), API, and NACE. An overview of new materials technology for corrosion prevention in deep water, oil, and gas operations is available in Ref 61. It includes a summary of these advances:

- Use of higher-strength steels and composite materials for construction of lighter platforms
- Use of thermal spray aluminum coatings for subsea components to supplement normal sacrificial anode cathodic protection systems
- Use of high-strength, corrosion-resistant alloy tubulars made from martensitic, duplex, and austenitic stainless alloys to reduce wall thickness and maximize load and pressure bearing capabilities, while eliminating corrosion inhibition
- Use of engineered materials systems such as flexible pipe and umbilicals that incorporate a variety of materials (alloys, plastics, and composites) with unique designs to allow features not possible with monolithic metallic construction

General Corrosion

Marine structures operate in a complex environment that can vary significantly according to site location and water depth. Figure 24 shows the four main platform corrosion zones: soil, seawater, splash zone, and marine atmospheric. Recently, studies have shown that deep seawater conditions also pose their own unique conditions that are different from normal marine service under shallower conditions.

Marine atmospheric corrosion problems occur on the portion of the jacket above the splash zone and on the superstructure. Exposed steel surfaces suffer corrosion from an environment of water condensation, rain, salt precipitation, sea mist, and oxygen. Corrosion rates can range from 0.05 to 0.64 mm/yr (2 to 25 mils/yr). Corrosion is particularly severe at crevices and sharp-edged areas, such as skip-welded plates and steel structural shapes. Attention to design and fabrication details can eliminate most of these problem areas. Atmospheric corrosion can be minimized by using coatings or by substituting nonferrous materials, such as copper alloys, nickel alloys, and FRP for steel components. Care must be taken to not create a galvanic-corrosion problem by coupling dissimilar metals. Table 10 gives a summary description of several marine zones and the characteristic corrosion behavior of the steel.

The splash zone is defined in NACE RP0176 (Ref 108) to be the area of the platform that is alternately in and out of the water because of tides, winds, and sea. It does not include surfaces that are only wetted during major storms. The splash zone of the platform can cover an interval of 1.5 to more than 12 m (5 to more than 40 ft), depending on location. Generally, the area of the platform that suffers the most severe steel corrosion is the splash zone, as shown in Fig. 25. Common methods of controlling corrosion in the splash zone include applying coatings, increasing jacket wall thickness by 6.4 to 19 mm (0.25 to 0.75 in.) in the splash zone to compensate for the higher corrosion rates, or applying a Monel alloy wrapper.

Corrosion of steel in seawater is a function of water salinity, temperature, oxygen content, velocity, resistivity, and chemistry. Table 11 summarizes the effects of these and other factors on the corrosion of steel in seawater. Several of the variables controlling corrosion are interrelated. As an example, Table 12 demonstrates

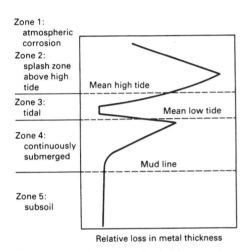

Fig. 24 Corrosion zones on fixed offshore structures. Source: Ref 106

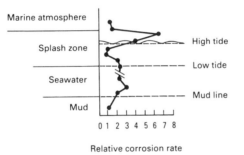

Fig. 25 Zones of corrosion for steel in seawater and the relative corrosion rate in each zone. Source: Ref 108

Table 10 Classification of typical marine environments

Marine zone	Description of environment	Characteristic corrosion behavior of steel
Atmosphere (above splash)	Minute particles of sea salt are carried by wind. Corrosivity varies with height above water, wind velocity and direction, dew cycle, rainfall, temperature, solar radiation, dust, season, and pollution. Even bird droppings are a factor.	Sheltered surfaces may deteriorate more rapidly than those boldly exposed. Top surfaces may be washed free of salt by rain. Coral dust combined with salt seems to be particularly corrosive to steel equipment. Corrosion usually decreases rapidly as one goes inland.
Splash	Wet, well-aerated surface; no fouling	Most aggressive zone for many metals, for example, steel. Protective coatings are more difficult to maintain than in other zones.
Tidal	Marine fouling is apt to be present to high watermark. Oil coating from polluted harbor water may be present. Usually, ample oxygen is available.	Steel at the tidal zone may act cathodically (well aerated) and receive some protection from the corrosion just below tidal zone in the case of a continuous steel pile. Isolated steel panels show relatively high attack in the tidal zone. Oil coating on surface may reduce attack.
Shallow water (near surface and near shore)	Seawater usually is saturated with oxygen. Pollution, sediment, fouling, velocity, and so on, all may play an active role.	Corrosion may be more rapid than in marine atmosphere. A calcareous scale forms at cathodic areas. Protective coatings and/or cathodic protection can be used for corrosion control. In most waters, a layer of hard shell and other biofouling restricts the available oxygen at the surface and thus reduces corrosion (increased stress on structure from the weight of fouling must be provided for).
Continental-shelf depths	No plant fouling, much less animal (shell) fouling with distance from shore. Some decrease in oxygen, especially in the Pacific, and lower temperature	...
Deep ocean	Oxygen varies, tending to be much lower than at the surface in the Pacific but not too different in the Atlantic. Temperature near 0 °C (32 °F). Velocity low; pH lower than at surface.	Steel corrosion is often lower. Anode consumption is greater to polarize the same area of steel as at the surface. There is less tendency for protective mineral scale formation.
Mud	Bacteria are often present, for example, sulfate-reducing types. Bottom sediments vary in origin, characteristics, and behavior.	Mud is usually corrosive, occasionally inert. Mud-to-bottom water corrosion cells seem possible. Partly embedded panels tend to be rapidly attacked in mud. Sulfides are a factor. Less current than in seawater is consumed to obtain cathodic polarization for the buried part of the structure.

Source: Ref 107

the relationship between temperature and oxygen solubility in seawater. Generally, the lower the seawater temperature, the higher the gas solubility observed; however, action of currents and mixing of shallow and deep water can also have an effect. As temperature or oxygen levels increase, corrosion rates will increase. The interplay of seawater variables is different for different locations. An example of this is shown in Fig. 26 (Ref 63). These conditions will determine the aggressivity of the service and, in turn, the demands on the cathodic protection system and/or coatings. To control seawater corrosion, the steel jackets are normally cathodically protected. The types of cathodic protection systems used are sacrificial anode, impressed current, or a combination of the two (see the discussion "Cathodic Protection" in this article). Occasionally, cathodic protection will be used in combination with coatings. Not only does cathodic protection control corrosion, but it also eliminates concern over corrosion fatigue failure of the jacket. Corrosion fatigue is discussed in more detail below. Typical cathodic protection system design parameters are given in Table 13 to 15.

The major platform components below the mudline are the jacket piles. In general, steel corrosion rates are low below the mudline. The exception is when the mud contains sulfate-reducing bacteria. Because the piles have electrical continuity with the jacket, the cathodic protection system will normally protect the piles from corrosion in saline muds.

Fatigue

Corrosion is of particular concern for the platform tubular welded joints, called nodes. The nodes are areas of high stress due to their complex geometries (Ref 110–112). The points of maximum stress in the nodes occur at the toe of the welds joining the tubular members. Cyclic stresses result from environmental factors, such as waves, tides, and operating loads. Platforms are designed to handle both a maximum stress and fatigue. The maximum stress is usually based on 100-year storm conditions. Platform fatigue life is based on a environmental stress distribution analysis, along with analysis of the stress cycles (Ref 113). Fatigue design curves have been published by the American Welding Society, British Standards Institute, API, and Det Norske Veritas.

Corrosion can reduce the fatigue life of platform anodes. Galvanic corrosion of non-stress-relieved welds and pitting corrosion can result in stress raisers. Therefore, corrosion can lead to the initiation of cracks and can increase the growth rate of existing cracks, reducing fatigue life. The fatigue life of steel exposed to seawater is shorter than that of steel exposed to air because of corrosion fatigue. As Fig. 27 illustrates, steel immersed in seawater does not exhibit an endurance limit. Because there is no endurance limit, unprotected steel exposed to seawater is susceptible to fatigue failure even at low stress levels after long-term cyclic service. An API study discussed crack initiation in smooth, notched, and welded specimens and summarized a number of earlier investigations (Ref 115). Table 16 ranks according to importance the various seawater environmental variables influencing corrosion fatigue crack initiation of carbon steel. The corrosion fatigue effects are eliminated by the application of cathodic protection (Fig. 28).

Inspection

The purpose of periodic inspection is to ensure that the structure is fit for continued service. Through the inspection program, a company is protecting its personnel and its equipment. In some locations of the world, governmental bodies have established legislation or code agencies that determine minimum inspection requirements. Elsewhere, the operator decides on the minimum inspection needs for the platform. Detailed information on inspection procedures can be found in the article "Inspection, Data Collection, and Management" in this Volume.

Inspection is required even though platforms are designed and constructed to conservative codes (Ref 116). Inspection allows confirmation that the codes are adequate. It should be noted that the codes represent the best experience and knowledge at the time they are written. Often, the design parameters must be extrapolated for use in new frontier environments that were not foreseen when the codes were written. Inspection results provide the information necessary for updating the codes to account for these new environments. Inspection of the platform jacket is designed to assess corrosion, fatigue cracks, joint and brace failure, impact damage, marine growth, scour, and debris accumulation. Inspection techniques include visual inspection by divers and remote operated vehicles, still and video photography, ultrasonic thickness measurements, cathodic potential surveys, magnetic-particle inspection, and vibration frequency attenuation.

Corrosion of Gathering Systems, Tanks, and Pipelines

Gathering systems are defined as all production facilities from the wellhead choke (or pumping T) to the sales point (oil and/or gas); subsystems include flow lines, separation, and dehydration. Gas processing is reviewed briefly. Sulfur plants (conversion of H_2S to elemental sulfur), gas transmission lines and oil pipelines, gasoline plants, and water-handling facilities for disposal are beyond the scope of this discussion,

Table 11   Factors that affect corrosion of carbon steel immersed in seawater

Factor	Effect on iron and steel
Chloride ion	Highly corrosive to ferrous metals. Carbon steel and common ferrous metals cannot be passivated. Sea salt is about 55% chloride.
Electrical conductivity	High conductivity makes it possible for anodes and cathodes to operate over long distances; therefore, corrosion possibilities increase, and the total attack may be much greater than that for the same structure in freshwater.
Oxygen	Steel corrosion is cathodically controlled for the most part. Oxygen, by depolarizing the cathode, facilitates the attack; therefore, a high oxygen content increases corrosivity.
Velocity	Corrosion rate is increased, especially in turbulent flow. Moving seawater may destroy the rust barrier and provide more oxygen. Impingement attack tends to promote rapid penetration. Cavitation damage exposes fresh steel surfaces to further corrosion.
Temperature	Increased ambient temperature tends to accelerate attack. Heated seawater may deposit protective scale or lose its oxygen; either or both actions tend to reduce attack.
Biofouling	Hard-shell animal fouling tends to reduce attack by restricting access of oxygen. Bacteria can take part in the corrosion reaction in some cases.
Stress	Cyclic stress sometimes accelerates failure of a corroding steel member. Tensile stresses near yield also promote failure in special situations.
Pollution	Sulfides, which are normally present in polluted seawater, greatly accelerate attack on steel. However, the low oxygen content of polluted water could favor reduced corrosion.
Silt and suspended sediment	Erosion of the steel surface by suspended matter in the flowing seawater greatly increases the tendency toward corrosion.
Film formation	A coating of rust or rust and mineral scale (calcium and magnesium salts) will interfere with the diffusion of oxygen to the cathode surface, thus slowing the attack.

Source: Ref 107

Table 12   Solubilities of various gases in ocean water

Gas	Partial pressure in dry air		Solubility, mL/L · atm		Equilibrium concentration in surface seawater, mL/L	
	kPa	atm	0 °C (32 °F)	24 °C (75 °F)	0 °C (32 °F)	24 °C (75 °F)
Helium	5.3×10^{-4}	5.2×10^{-6}	8.0	6.9	4.1×10^{-5}	3.4×10^{-5}
Nitrogen	79.1	0.781	18	12	14	9
Oxygen	21.2	0.209	42	26	8.8	5.5
Carbon dioxide	0.032	3.2×10^{-4}	1460	720	0.47	0.23

Source: Ref 109

although the same principles will apply. Internal and external corrosion alleviation systems are reviewed, respectively.

Internal corrosion is dependent on the composition, temperature, pressure, and flow regime of the produced fluids. For many years, it was felt that the general direction of the effect of each of these variables was known, but the magnitude of the effect was not precisely known. More recently, computer software has made it possible to model these system more precisely and take into account many of these variables to predict the effects of flow and corrosion.

A natural gas reservoir is a reservoir that, under initial conditions, is a single, gaseous hydrocarbon phase. If this gaseous phase contains hydrocarbons that are recoverable as liquids on the surface, the reservoir is a gas condensate reservoir. A gas well is a well that produces fluids from a gas or gas condensate reservoir. An oil well is a well that produces from a hydrocarbon reservoir that is either a two-phase system or a single liquid phase (Ref 117). A general distinction between the two is commonly based on gas-to-oil ratio (GOR). Wells with a GOR greater than 890 m^3/m^3 (5000/mmscf) are commonly considered gas wells. Wells with lower GOR values are referred to as oil wells due to the predominance of liquid hydrocarbons. Because many liquid hydrocarbons often have some ability to form a protective layer on steel and displace water, corrosion problems are generally considered greater in gas wells than in oil wells. Because produced fluids essentially determine the internal environment (weather conditions and process design can affect temperatures and pressures), gathering systems for gas wells and oil wells are discussed separately.

From both a metallurgical and corrosion viewpoint, it is important to distinguish between sweet gas wells and sour gas wells. If the partial pressure of H_2S is greater than 0.34 kPa (0.05 psia), the gas stream is generally considered sour and materials that resist sulfide stress cracking must be used. However, the latest revision of NACE MR0175/ISO 15156 (Ref 14) lists materials that are recognized to have acceptable resistance to sulfide stress cracking. In this standard, it is indicated that H_2S partial pressure must be considered in combination with pH, which can vary with the partial pressure of acid gases (CO_2 and H_2S) and buffering salts in the water phase such as bicarbonate. Therefore, the higher the pH, the more H_2S can be tolerated before critical levels are reached. This document also indicates that the level of chlorides in the produced water can also influence the corrosivity and susceptibility to sulfide stress cracking. The sour-service serviceability chart per the NACE MR0175/ISO 15156 is given in Fig. 29. It shows various regions of varying severity relative to the need to select materials resistant to sulfide stress cracking and where testing of materials may be needed.

The determination of whether H_2S or CO_2 corrosion mechanisms will predominate is not as simple. Early investigators believed that CO_2 had a synergistic effect on H_2S corrosion. Subsequent investigators indicated that it is the ratio of partial pressures of CO_2 to H_2S that controls the corrosion mechanism. Unless the ratio of CO_2 to H_2S partial pressure is greater than 500, the corrosion mechanism is normally dominated

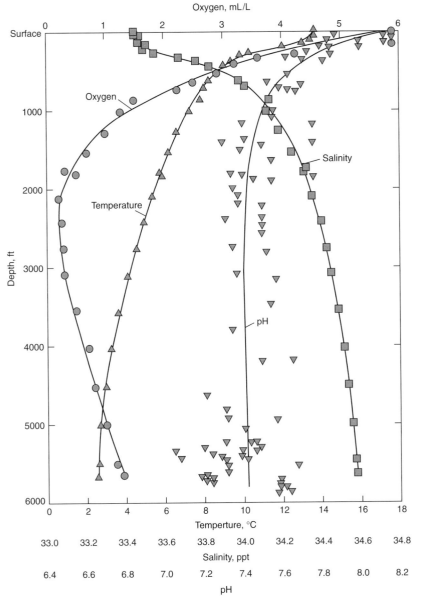

Fig. 26 Variation of oxygen, pH, and other seawater parameters with depth in the Atlantic ocean. Source: Ref 63

Table 13 Design criteria for offshore cathodic protection systems

Production area	Water resistivity, $\Omega \cdot cm$(a)	Water temperature °C	Water temperature °F	Turbulence factor (wave action)	Lateral water flow	Typical design current density(c) mA/m^2	Typical design current density(c) mA/ft^2
Gulf of Mexico	20	22	70	Moderate	Moderate	54–65	5–6
U.S. West Coast	24	15	60	Moderate	Moderate	76–106	7–10
Cook Inlet	50	1	35	Low	High	380–430	35–40
North Sea(d)	26–33	0–12	32–55	High	Moderate	86–216	8–20
Persian Gulf	15	30	85	Moderate	Low	54–86	5–6
Indonesia	19	24	75	Moderate	Moderate	54–65	5–6

(a) Water resistivity is a function of chlorinity and temperature, and it decreases as both chlorinity and temperature increase. (b) Typical values and ratings based on average conditions, remote from river discharge. (c) In ordinary seawater, a current density less than the design value will suffice to hold the platform at protective potential once polarization has been accomplished and calcareous coatings are built up by the design current density. It should be noted, however, that depolarization can result from storm action. (d) Conditions in the North Sea can vary greatly from the northern to the southern area, from winter to summer, and during storm periods. Source: Ref 108

by H_2S (Ref 118). More recent investigations suggest that there may also be an effect of temperatures on the ratio of CO_2 to H_2S partial pressure (Ref 119).

Because CO_2 corrosion is usually more severe (i.e., higher corrosion rates) than experienced for H_2S corrosion at lower temperature conditions, and because most production facilities are relatively cool (<60 °C, or 140 °F), CO_2 corrosion should be considered whenever the ratios of CO_2 to H_2S partial pressures are greater than 200. Oxygen is not present naturally in oil and gas reservoirs, and without exception it is preferable to keep it out since it results in still higher corrosion rates and increased tendency for localized corrosion. The biggest threats to oxygen ingress come in handling production environments due to injection of aerated chemicals (hydrate control fluids, drilling and packer fluids, and corrosion and scale inhibitors), suction leakage in pumps, and backstreaming of oxygen through leaky valves in pressure-containing equipment.

Sweet Gas

With regard to CO_2 corrosion alleviation in flow lines, there are several choices. First, low-alloy steel with a corrosion allowance can be used; a nomograph establishes the maximum corrosion rate for CO_2 (Ref 120) or, more recently, corrosion modeling software can be used to more accurately predict the severity of corrosion (Ref 29). Velocity may also be important; several authors have suggested that there is a critical velocity (or more accurately, flow-induced wall shear stress) above which CO_2 corrosion is very difficult to control (Ref 41, 121). Again, more recent data suggest that at the same temperature, pressure, fluid composition, and pH, the CO_2 corrosion-velocity relationship is a continuous relationship rather than a step function (Ref 119); however, for multiphase flow situations, discontinuous increases in flow-induced wall shear stress can occur as a result in changing flow regimes (e.g., slug flow, droplet impingement) related to varying flow rate, liquid/gas ratio, and system geometry. If the flow lines are welded, the operator should be certain that the weldments are at least as corrosion resistant (and where H_2S is present, as cracking resistant) as the pipe body.

A second choice is to use corrosion-resistant materials, alloys, or coatings. With regard to CO_2, either type 316 stainless steel or duplex stainless steel will provide sufficient internal corrosion resistance. If H_2S is present, then NACE MR0175/ISO 15156 should be followed. Many types of stainless steels are subject to chloride SCC at elevated temperatures, and both alloys (type 316 stainless steel and duplex) may be subject to internal and/or external pitting corrosion, crevice corrosion, or SCC. Both alloys also require special care from the time they are installed to the time they are put into service. Oxygen and perhaps bacteria will result in pitting corrosion, depending on how the material is stored prior to installation or following hydrostatic testing before being put into service. Partially aerated hydrostatic test water (fresh water or seawater) is very aggressive to both steels and stainless alloys. Metallurgical solutions, if properly executed, can result in permanent, low-maintenance corrosion alleviation systems. However, this often requires anticipation of the relevant worst-case scenario that may occur during the service life of the systems, which is not necessary at the beginning of service. Increases in water content, reservoir souring, and other factors can dictate long-term performance of materials of construction.

A third choice is to insert an internal liner in low-alloy steel pipelines with a corrosion-resistant material. These systems may have advantages over solid corrosion-resistant alloys. First, they may be less expensive, and second, the alleviation of external corrosion problems of low-alloy steel is well understood (years of history with large quantities of pipelines) and less sophisticated. There are two disadvantages. Special welding procedures are required, and when the metallurgical coating is not bonded, buckling of the liner may occur, particularly in bends. This buckling will inhibit the use of tools pumped through the flow line.

Two other types of internal barriers are also commonly used: organic polymers (plastic coatings) and cement linings. Both systems can be economic successes. However, both systems

Table 14   Energy capabilities and consumption rates of sacrificial anode materials in seawater

Anode material	Energy capability(a)		Consumption rate		Anode to water(b) closed circuit potential, V versus Ag/AgCl reference electrode
	A · h/kg	A · h/lb	kg/A · yr	lb/A · yr	
Al-Zn-Hg	2750–2840	1250–1290	3.1–3.2	6.8–7	−1 to −1.05
Al-Zn-In	2290–2600	1040–1180	3.4–3.8	7.4–8.4	−1.05 to −1.1
Al-Zn-Sn	925–2600	420–1180	7.4–20.8	16.3–45.9	−1 to −1.05
High-purity zinc	780–815	354–370	10.7–11.2	24.8–23.7	−1 to −1.05
Magnesium alloy H-1	1100	500	8.0	17.5	−1.4 to −1.6

(a) Data are ranges taken from field tests conducted by the Naval Research Laboratory at Key West, FL, and from manufacturers' long-term field tests. (b) Measured potentials can vary because of temperature and salinity differences. Source: Ref 108

Table 15   Consumption rates of impressed-current anode materials

Material	Typical anode current density in saltwater service		Nominal consumption rate	
	A/m²	A/ft²	g/A · yr	lb/A · yr
Pb-6Sb-1Ag	160–220	15–20	15–86	0.03–0.02(a)
Pb-6Sb-2Ag	160–220	15–20	13–25	0.03–0.06(a)
Platinum (on titanium, niobium, or tantalum substrate)	540–3200	50–300	3.6–7.3	0.008–0.016(b)
Graphite	10–40	1–4	230–450	0.5–1.0
Fe-14.5Si-4.5Cr	10–40	1–4	230–450	0.5–1.0

(a) Very high consumption rates of lead-silver anodes have been experienced at depths in excess of 30 m (100 ft). (b) This figure can increase when current density is extremely high and/or in waters of low salinity. Source: Ref 108

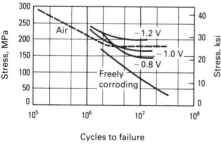

Fig. 27   Fatigue data for carbon steel in seawater as a function of specimen potential. Source: Ref 114

Table 16   Summary of major variables influencing the corrosion fatigue crack initiation behavior of carbon steels in seawater

Variable	Effect
Cyclic frequency	Slower frequencies cause reduced fatigue resistance for unprotected steel.
Cathodic potential	Adequate cathodic protection restores fatigue resistance to levels observed in air.
Oxygen level	Fatigue resistance in deaerated seawater is similar to that in air.
Temperature	Although decreasing temperature results in increased oxygen levels, the overall effect of decreasing temperature is that fatigue resistance is improved to some extent in range of 13 to 45 °C (55 to 115 °F).
pH level	Over a broad range of values (4 to 10), there is little effect of pH on fatigue resistance. Low values (<4) decrease fatigue resistance, and high values (>10) improve fatigue resistance to levels similar to those observed in air.

Source: Ref 115

have difficulty in maintaining corrosion resistance at the joints, are difficult to install holiday-free, have limited life, and may suffer disbondment and failure when the pipe is improperly handled or distorted. The plastic coatings are permeable to the produced fluids, and eventually (3 to 5 years or longer) the corrosive fluids will permeate the coating. The resulting corrosion products cause disbonding of the coating and complete loss of its corrosion resistance. Cement linings are thicker and less subject to produced fluid penetration than plastic coatings. They are heavy and may have limited resistance to acids.

A final alternative is to use nonmetallic pipe materials such as FRP or polymerized hydrocarbons. The advantage of these materials is their complete resistance to corrosion. Their disadvantages are low allowable temperatures, low fatigue resistance, low strength, low resistance to mechanical damage, and high combustibility. They also have problems with joint integrity, although much work has been done recently to improve these systems. Finally, they may be vulnerable to chemical attack from the produced fluids (CO_2 may dissolve the resins from fiberglass pipe, and unsaturated hydrocarbons may dissolve the polymerized hydrocarbon pipes) and from chemicals that may be injected during workovers.

The most common alternative to corrosion-resistant materials is to use steel with corrosion inhibition. If corrosion inhibition is used to protect the gas well downhole tubulars, the same formulations can often protect the downstream flow lines if the flow lines are properly designed. Inhibitors should be evaluated for performance under downhole and flow-line conditions to ensure adequate performance. The flow lines should be sized to ensure turbulent flow (and mixing and coverage with inhibitor) at a velocity that is not significantly higher than that present in the tubing. More inhibitor treatments fail in flow-line conditions due to lower than anticipated flow velocities than fail at high velocity. This often occurs when production rates decline with the age of the reservoir or from initial overdesign. Turbulent flow will ensure inhibitor contact with the entire internal pipe surface, and limited velocities (flow lines are cooler than gas well tubulars) will ensure that flow-line conditions are not significantly more corrosive than the gas well tubulars. Because of the economic consequences of downhole tubular failures (usually much greater cost than flow-line failures), designing the surface flow lines to utilize downhole inhibition systems effectively will result in a successful flow-line inhibition system. If a downhole inhibition system is not used (either from temperatures above the dew point or use of corrosion-resistant alloys), then surface inhibition can be used to protect the flow lines. In low-velocity/low-liquid flow regimes, periodic inhibition with a technique that inhibits the entire internal surface can be successful. This often involves the use of pigging to push retained water and fluids through the flow line and to redistribute inhibitors to all internal surfaces of the flow line. For higher velocities (turbulent flow) and/or high liquid content, continuous inhibition will probably be necessary. For extreme cases, special inhibitor formulations are available that have high reaction times and that can be retained on the metal surface even under very high flow-induced wall shear stress conditions.

Separation and dehydration facilities offer fewer alternatives. Because inhibitors usually stay in the liquid phase in multiphase systems and usually in the hydrocarbon phase if one is present, reliable inhibition of the vapor space is not usually possible, but vapor-phase inhibitors can be used in some cases. Fortunately, corrosion is usually not severe, because only water of condensation is present. However, in cases which have high CO_2 partial pressures, this condition can produce low pH and high corrosion rates along with a high propensity for localized corrosion. For very severely corrosive environments, such as wet CO_2, type 316 stainless steel or type 316 stainless steel internal cladding is usually used at less than 60 °C (140 °F); at higher temperatures, higher alloy stainless steels or alloy 825 clad steel is often used. For normal gas production with CO_2, low-alloy steel, combined with a corrosion allowance, inhibitors, and monitoring, is often the most economic solution.

For gas dehydration systems, low-alloy steel with corrosion allowance is normally sufficient when combined with pH control of the glycol. For gas streams high in CO_2 concentrations, some operators have found it necessary to internally clad the wet gas portions with type 316 stainless steel or higher alloys depending on temperature, environmental severity, and anticipated service life.

Sour Gas

Sour gas wells present a more difficult problem. First, all materials must be resistant to sulfide stress cracking. It is essential that no equipment suffer catastrophic cracking failures that impair pressure containment or operability of the component. General corrosion is often less severe for H_2S than for CO_2 at the lower temperatures usually encountered in flow-line systems, and H_2S corrosion is not usually as velocity dependent. As long as sufficient liquid hydrocarbon is present, corrosion is usually minor. When hydrocarbon liquids exceeds 100 barrels per 28,300 m^3 (1×10^6 ft^3) of gas, when the water content of the liquid phase is less than 10%, and when the flow is turbulent (to avoid water phase in the bottom of the line), corrosion problems are often self-mitigating, resulting in acceptable corrosion rates. In cases in which the

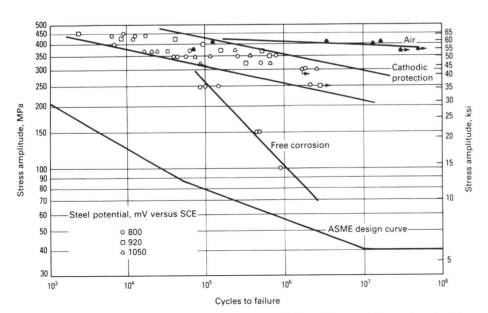

Fig. 28 Effect of cathodic protection on the fatigue performance of alloy steel in seawater. Tests performed on 6.4 mm (1/4 in.) diam specimens at a mean stress of 425 MPa (69 ksi)

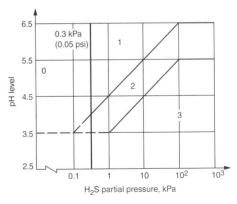

Fig. 29 Regions of increasing sour-service severity (0, 1, 2, 3) on a plot of pH and H_2S partial pressure. Source: Ref 14

above conservative criteria are not met, inhibition may be necessary. The same rationale in regard to the use of inhibitors in flow lines with CO_2 applies to flow lines for H_2S. Inhibition systems that protect downhole tubulars will protect the flow lines if the entire internal wall is inhibited (i.e., the flow line is in mild turbulent flow). Processing facilities, gas separators, and sweeting systems (sulfinol, amines, and so on) are usually constructed of low-alloy steel with a corrosion allowance and are regularly monitored. However, for longer-term project life, stainless alloys are finding more widespread use.

Oil Wells

Oil wells, as previously defined, are wells that produce from a hydrocarbon reservoir that is either two-phase or a single liquid phase. Oil-well flow-line corrosion is much easier to handle than gas-well flow-line corrosion. First, low-pressure sour oil wells are usually beyond the scope of NACE MR0175/ISO 15156. Second, CO_2 corrosion is seldom, if ever, a problem. Most crude oils (and associated gas) either contain no CO_2 or sufficient H_2S for H_2S to be the controlling corrosion mechanism. Corrosion problems are often minor until water cuts approach 30%. This rule assumes that the oil has some tendency to preferentially wet the metal surface, thus excluding water, and usually works for crude oil systems. For lighter hydrocarbon condensates, critical water cuts can be much lower, and in extreme cases be less than 10%.

Normally, when water cuts are high enough for saltwater corrosion (with or without H_2S) to be a problem in the flow lines, corrosion will also be more severe on the downhole tubulars. In this case, the inhibition system protecting the tubulars should also be designed to protect the flow lines. Flow lines with stratified flow that allow free water to flow or stagnate in the bottom of the flow lines may suffer corrosion problems. In these cases, inhibitors in the oil phase from the well may not help. This problem is successfully handled by frequent pigging in order to clear the flow lines of water; to clean out sediment, which will foster crevice corrosion; and to distribute inhibitor over the entire internal surface.

Gas separation facilities and free-water knockouts are usually made of low-alloy steel. Free-water knockouts are often internally coated with an organic coating. Coatings, if properly applied to clean, dry surfaces, may extend the vessel life a few years. Generally, coatings are not normally holiday-free, and therefore inhibitor may also be required. The only other alternative to low-alloy steel with corrosion allowance and internal coating is a corrosion-resistant material. The low corrosivity, however, does not normally justify the cost of a corrosion-resistant alloy, either solid or internally clad, unless the project is offshore, involves high production rates, or requires long-term service life. The combination of size and pressure usually eliminates the use of materials such as fiberglass for these applications.

Corrosion problems can be severe in salt-water-handling facilities, although they are usually mild to moderate, often due to the exclusion of oxygen. Corrosion alleviation systems are limited to cathodic protection, organic coatings, and, occasionally, nonmetallic vessels.

Crude oil is usually dehydrated by using gravity separation of the lighter oil from the unwanted water. Heat, chemicals, and electric fields are often used to accelerate the gravity separation. The separation vessel should not experience severe corrosion if the system is kept oxygen-free. Cathodic protection is only partially successful for the vessel, because only the water-wet portion is protected and the oil/water interface fluctuates. The heating coils often suffer more severe corrosion than the vessel body. As long as these coils are in the water portion of the vessel and well designed to minimize local boiling and flashing, then maintained cathodic protection systems will be successful in alleviating corrosion of the heating surfaces. When large tanks are used, galvanizing significantly prolongs the tank life in the absence of H_2S. Hot-dip galvanizing is considered to be the most effective treatment, but it is limited to bolted tanks. The bolted tank gaskets, in turn, limit the temperatures at which the vessel can be operated.

The gas separated from the oil is low pressure and can be safely handled with low-alloy steel with a corrosion allowance. The storage of dehydrated oil (usually about 1% H_2O) poses no internal corrosion problems, except in the tank bottom, where saltwater will accumulate.

External Corrosion

External corrosion can also be a serious and costly problem. In wet or corrosive soil, low-alloy steel flow lines should be coated. There are a variety of successful coating systems. Usually, the external coating system consists of two parts: a mastic that coats and protects the pipe and a coating or wrapping that protects the mastic. Alternatively, epoxy coatings and extruded polyethylene and/or polypropylene coatings are used for more severe conditions and higher temperatures. If the flow lines are needed for long periods of time, then the external coating should be supplemented with cathodic protection. Surprisingly, it is probably more important to cathodically protect corrosion-resistant alloy flow lines than low-alloy steel flow lines. The corrosion-resistant alloys—type 316 stainless steel or duplex stainless steel—are subject to crevice corrosion and sometimes SCC in the presence of oxygen and chlorides, which are common in many areas, particularly in offshore areas. Because the stainless steel lines are much more expensive than low-alloy steel lines, they cannot be allowed to fail by external corrosion. Therefore, external cathodic protection of these flow lines is essential when they are buried or submerged. To reduce the current level required for protection, the corrosion-resistant alloy, like the low-alloy steel pipeline, is usually externally coated. Recently, there has been concerns for overprotection that can result in hydrogen embrittlement of martensitic or duplex stainless steels particularly in the area of girth welds.

The other components of the gathering system are the storage tanks, gun barrels, and surge tanks. These vessels all internally accumulate saltwater on the bottom. Similarly, these tank bottoms are all subject to external corrosion. Because the tank bottoms are relatively thin and may suffer internal corrosion, they should be protected from external corrosion to maximize service life. Therefore, in wet environments or where long service is needed, cathodic protection of tank bottoms should be considered. For piping and vessels above ground, painting is the accepted method of protecting against corrosion.

Monitoring and Inspection

Monitoring is an essential part of any corrosion alleviation system. No corrosion alleviation system is completely reliable, and in many cases failure can be catastrophic, both from a personal safety perspective and from an environmental and/or economic perspective. There are a wide variety of monitoring inspection techniques. The thoroughness and frequency of the monitoring must be weighed against the consequences of failure, and the type of monitoring must be tailored to the particular system. Certainly, it is more catastrophic to have a high-pressure sour gas vessel failure than an atmospheric saltwater tank failure. However, in large, long-term projects, either failure can result in substantial lost production, if not monitored on a regular basis at the most critical locations.

Flow lines can be monitored with calipers that are pumped through the line, x-rayed, or, where warranted, cut open and inspected. Vessels can be visually and/or ultrasonically inspected. When ultrasonic inspection is used, reference points are usually permanently fixed to the vessel external wall so that the ultrasonic test is conducted at the same location each time. These are basically the methods used to find corrosion after it may have already occurred. Corrosion monitoring should be conducted in a manner where the results can be used as a process-control tool. Real-time corrosion monitoring can show the advance indications of corrosion and/or pitting before substantial corrosion actually takes place. Corrosion data can also be taken with process data to indicate the source of the problem and the root cause of the upset conditions that, if left unattended, will result in cumulative damage.

Storage of Tubular Goods

Tubular goods used in oil-producing and drilling operations are sometimes stored outdoors or in areas where internal storage is conducive to corrosion. This is particularly true in coastal regions and industrial areas in which acid gases and pollutants are present as well as in oil fields that produce hydrogen sulfide.

Oxygen corrosion, or rust, is aggravated by the deposition of salt from marine environments, such as that encountered in wind-driven spray on offshore platforms, and airborne salt in coastal areas. Pipe yards situated close to the beach are particularly susceptible to severe atmospheric corrosion. In warehouses and under sheds, the presence of industrial pollutants such as SO_2, oxides of nitrogen, and other gases will initiate corrosion attack and increase rusting when they react with moisture in the air or on the pipe.

Even in areas of relatively low salt and pollutants content, severe corrosion may occur if the relative humidity is high. The pipe will cool off during the night, and dew will fall, covering the pipe with a conductive layer of electrolyte. The rust that is already on the pipe is hygroscopic and will remain moist after the free water has evaporated. This leads to concentration-cell attack and severe pitting. Pit depth and size are of particular importance, because failures may occur when the pipe is put into service under pressure.

Inspection

Before being put into service, tubular goods that have been stored for any length of time should be inspected. Particular attention should be given to:

- *External rusting:* The percent of surface area covered by rust should be recorded.
- *Presence of mill scale, lacquer-type mill coatings, or other temporary coatings:* This is important, because areas not covered may corrode, while protected areas may set up concentration cells and accelerate localized corrosion.
- *Internal corrosion:* The interior of the pipe should be inspected for rusting and pitting.
- *Condition of threads:* Threads should be examined to determine if corrosion damage has occurred that could prevent proper makeup.

Cleaning

After inspection, the pipe should be cleaned prior to applying any protective coatings. The pipe can be cleaned mechanically or with acids or rust dissolvers.

If the pipe is used, or has been stored in marine or industrial environments, it should be water blasted to remove any salt or acid deposits, weathered for a period of time to allow under-deposit salts to migrate to the surface, and water blasted again. One month is usually sufficient weathering time.

A water-soluble phosphate-base temporary rust inhibitor can be applied during the weathering period to minimize further rusting. The inside of the pipe should also be protected because water may collect on the bottom and cause pitting.

The pipe can now be physically cleaned. Wire brushing with an automated machine is a preferred method, but a rotary handheld wire brush can be used if a machine is not available. If rusting and pitting are severe, the pipe should be sand blasted. The inside of the pipe can be cleaned with a mechanical rattler or a round brush.

An alternative cleaning method for lightly rusted pipe or where mechanical cleaning facilities are unavailable is the use of an acid-base rust remover/chelant. These are usually based on phosphoric acid and will contain inhibitors and passivators to prevent removal of metal. The pipe can be soaked in a trough of the inhibited acid cleaner until deposits are dissolved.

Application of Protective Coatings

Once the pipe is cleaned, temporary rust-preventive coatings can be applied to halt further rusting. They can be applied by the automatic machine that cleans the pipe or by hand sprayers, dipping, or brushing. The important factor is that the pipe is completely covered with the coating. Temporary coatings are manufactured in several types, as discussed in the paragraphs that follow.

Lacquer coatings may consist of an oil-soluble resin in a volatile hydrocarbon solvent. These coatings may be brittle and may flake off and expose bare metal.

Slushing compounds usually consist of asphalt dissolved in a nonevaporating hydrocarbon so that a thick, oily layer is present on the surface. These compounds are resistant to mechanical damage, but may be difficult to apply and remove.

Polymeric coatings may contain acrylics, chlorinated vinyl chlorides, or other materials that will dry and polymerize from an aqueous solution. These coatings may contain metal passivators and rust converters that will be anodic to the metal surface and greatly alleviate further rusting. Polymeric coatings are easy to apply and are sometimes used without precleaning the surface, because the converters and passivators are design to modify corrosion by-products to eliminate concentration-cell corrosion.

Sulfonate-base coatings are formulated from petroleum sulfonates, waxes, and materials that form a flexible and nondrying coating. Pigments and fillers can be added for appearance and durability. The sulfonate acts as a rust inhibitor, and the other materials seal the surface to prevent moisture entry. These are relatively easy to apply and are resistant to mechanical damage.

Rust passivators are of the same type as the phosphoric acid base rust removers or may be similar to the polymeric coatings. They are used for rust prevention under sheds or in warehouses and may not be adequate for outdoor storage in corrosive areas.

The application method used will depend on the type of coating selected. A hydrocarbon-base coating can be brushed or sprayed, or the pipe can be dipped in a trough containing the coating. Dilution with naphtha, aromatic solvents, or diesel oil can be done, depending on the recommendations of the manufacturer. A hydrocarbon-base coating should not be applied to wet or damp pipe. Some coatings are reported to be able to displace water from a metal surface, but care should be taken that the surface is not too wet.

A water-base coating will probably be applied full strength. It can also be sprayed or brushed. The recommendations of the manufacturer should be followed to ensure proper coverage.

A suitable thread-protecting compound should be applied and thread protectors screwed onto the threads before the pipe is coated. It may be desirable to coat the inside of the pipe with an oily coating and to use pipe caps to prevent water entry during storage.

If the coating is air sprayed, the pipe should be coated one layer at a time, making certain that the spray is adjusted to provide coverage without excessive loss. The pipe should be rolled, and the other side covered. The use of wood spacers between the layers of pipe allow any moisture to drain and dry during subsequent outdoor storage. Once the pipe is coated, sufficient drying time should be allowed before the pipe is moved.

Continuing Maintenance

The pipe should be inspected at preselected intervals to ensure that the coating is performing satisfactorily. The program can be modified, or the pipe can be recoated when needed. Complete records should be maintained so that corrosion prevention is an ongoing and effective process.

ACKNOWLEDGMENT

This article has been adapted from Corrosion in Petroleum Production Operations, *Corrosion,* Vol 13, *Metals Handbook,* ASM International 1987, p 1232–1261. That article was chaired by J.E. Donham. Authors were A.K. Dunlop, D.H. Patrick, D.E. Drake, H.E. Bush, J.D. Alkire, S. Ibarra, T.M. Stastny, M.C. Place, Jr., D.S. Burns, J.W. Johnson, and J.E. Donham.

REFERENCES

1. R.N. Tuttle and R.D. Kane, Ed., "H_2S Corrosion in Oil & Gas Production—A Compilation of Classic Papers," NACE International, 1981
2. L.E. Newton, Jr. and R.H. Hausler, "Carbon Dioxide Corrosion in Oil and Gas Production," NACE International, 1984
3. R.D. Kane, "Roles of H_2S in the Behavior of Engineering Alloys," paper 98274, Corrosion/98 (San Diego, CA), NACE International, March 1998
4. R.D. Kane and M.S. Cayard, "Characterization and Monitoring of Cracking of Steel Equipment in Wet H_2S Service," Seventh Middle East Corrosion Conference (Manama, Bahrain), NACE

International/Bahrain Society of Engineers, Feb 1996
5. M.S. Cayard and R.D. Kane, Large-Scale Wet Hydrogen Sulfide Cracking Performance: Evaluation of Metallurgical, Mechanical, and Welding Variables, *Corros. J.,* March 1997
6. R.D. Kane and M.S. Cayard, "Remediation and Repair Techniques for Wet H_2S Cracking," Eurocorr 2001 (Riva del Garda, Italy), Sept 30—Oct 4, 2002
7. R.D. Kane, R.J. Horvath, and M.S. Cayard, "Wet H_2S Cracking of Carbon Steels and Weldments," NACE International, 1996
8. R.H. Hausler, "Contribution to the Understanding of H_2S Corrosion," paper 04732, Corrosion/2004 (New Orleans, LA), NACE International, April 2004
9. M.B. Kermani and A. Morshed, Carbon Dioxide Corrosion in Oil and Gas Production—A Compendium, *Corrosion,* Vol 59 (No. 8), Aug 2003, p 659–683
10. P.R. Rhodes, Environment-Assisted Cracking of Corrosion-Resistant Alloys in Oil and Gas Production Environments: A Review, *Corrosion,* Vol 57 (No. 11), Nov 2001, p 923–966
11. C.C. Nathan, Ed., *Corrosion Inhibitors,* NACE International, 1973
12. "Corrosion Control in Petroleum Production," TPC Publication 5, NACE International, 1979
13. *Corrosion Control and Monitoring in Gas Pipelines and Well Systems,* NACE International, 1987
14. "Sulfide Stress Cracking Resistant Metallic Material for Oil Field Equipment," NACE MR0175/ISO 15156 (latest revision), Material Requirement, NACE International
15. R.D. Kane, Behavior of Corrosion Resistant Alloys in Stimulation Acids, Completion Fluids and Injected Waters, *Proc. Engineering Solutions to Industrial Corrosion Problems Conference* (Houston, TX), NACE International, 1993
16. J. Oldfield and B. Todd, Corrosion Considerations in Selecting Metals for Flash Chambers, *Desalination,* Vol 31, 1979, p 365–383
17. J.B. Sardisco and R.E. Pitts, Corrosion of Iron in an H_2S-CO_2-H_2O System—Composition and Protectiveness of the Sulfide Film as a Function of pH, *Corrosion,* Nov 1965, p 350–354
18. R.A. King and D.S. Wakerley, Corrosion of Mild Steel by Ferrous Sulfide, *Br. Corros. J.,* Vol 8, Jan 1973, p 41
19. R.A. King, J.D.A. Miller, and J.S. Smith, Corrosion of Mild Steel by Iron Sulfides, *Br. Corros. J.,* Vol 8, 1973, p 137
20. J.S. Smith and J.D.A. Miller, Nature of Sulfides and Their Corrosive Effect on Ferrous Metals; A Review, *Br. Corros. J.,* Vol 10 (No. 3), 1975, p 136
21. P.R. Rhodes, Corrosion Mechanism of Carbon Steel in Aqueous H_2S Solutions, Abstract 107, *Extended Abstracts,* Vol 76-2, The Electrochemical Society, 1976, p 300
22. M.C. Place, Jr., "Corrosion Control—Deep Sour Gas Production," 54th Annual Fall Technical Conference and Exhibition of the Society of Petroleum Engineers of AIME (Las Vegas, NV), Society of Petroleum Engineers, Sept 1979
23. S.N. Smith and E.J. Wright, "Prediction of Minimum H_2S Levels Required for Slightly Sour Corrosion," paper 11, Corrosion/94, NACE International, 1994
24. C. deWaard and D.E. Milliams, "Prediction of Carbonic Acid Corrosion in Natural Gas Pipelines," paper F1, First International Conference on the Internal and External Protection of Pipes, BHRA Fluid Engineering, University of Durham, Sept 1975
25. A.K. Dunlop, H.L. Hassell, and P.R. Rhodes, "Fundamental Considerations in Sweet Gas Well Corrosion," paper 46, Corrosion/83 (Anaheim, CA), NACE International, April 1983
26. D.R. Fincher, J.J. Marr, and J.W. Ward, paper 7, Corrosion/75 (Toronto, Ontario, Canada), NACE International, April 1975
27. B.V. Johnson, H.J. Choi, and A.S. Green, "Effects of Liquid Wall Shear Stress on CO_2 Corrosion of X-52 C-Steel in Simulated Oilfield Production Environments," paper 573, Corrosion/91, NACE International, March 1991
28. B.D. Craig, Predicting Critical Erosion Corrosion Limits of Alloys for Oil and Gas Production, *Mater. Perform.,* Vol 37 (No. 9), Sept 1998, p 59–60
29. S. Srinivasan and R.D. Kane, "Corrosivity Prediction in CO_2/H_2S Production Environments," Seventh Middle East Corrosion Conference, NACE International/Bahrain Society of Engineers (Manama, Bahrain), Feb 1996
30. B.F.M. Pots et al., "Improvements on De Waard-Milliams Corrosion Prediction and Applications to Corrosion Management," paper 02235, Corrosion/2002, NACE International, April 2002
31. S.P. Ewing, Corrosion by Stray Current, *Corrosion Handbook,* H.H. Uhlig, Ed., John Wiley & Sons, 1948, p 601–606
32. H.H. Uhlig, *Corrosion and Corrosion Control,* John Wiley & Sons, 1963
33. M.G. Fontana and N.D. Greene, *Corrosion Engineering,* McGraw-Hill, 1st ed., 1967; 2nd ed., 1978
34. J.W. Oldfield and W.H. Sutton, Crevice Corrosion of Stainless Steels—I. A Mathematical Model, *Br. Corros. J.,* Vol 13 (No. 1), 1978, p 13–22
35. J.W. Oldfield and W.H. Sutton, Crevice Corrosion of Stainless Steels—II. Experimental Studies, *Br. Corros. J.,* Vol 13 (No. 13), 1978, p 104–111
36. J.W. Oldfield, Crevice Corrosion of Stainless Steels—The Importance of Crevice Geometry and Alloy Composition, *Métaux-Corros.-Ind.,* Vol 56 (No. 668), April 1981, p 137–147
37. R.H. Hansler, Ed., *Advances in CO_2 Corrosion,* Vol I, NACE International, 1985
38. M.W. Joosten et al., "Organic Acid Corrosion in Oil and Gas Production," paper 02294, Corrosion/2002, NACE International, April 2002
39. M.S. Cayard and R.D. Kane, "Serviceability of 13Cr Tubulars in Oil and Gas Production Environments," Corrosion/98, NACE International, March 1998
40. R.D. Kane, J.B. Greer, D.F. Jacobs, H.R. Hanson, B.H. Berkowitz, and G.A. Vaughn, "Stress Corrosion Cracking of Nickel and Cobalt Base Alloys in Chloride Containing Environments," paper 174, Corrosion/79, NACE International, 1979
41. S. Srinivasan, R.D. Kane, and J.W. Skogsberg, "Automated Materials Selection and Equipment Specification System for Oil and Gas Production: Concept, Development, Implementation," paper 03134, Corrosion/2003, NACE International, March 2003
42. "Laboratory Testing of Metals for Resistance to Specific Forms of Environmental Cracking in H_2S Environments," TM0177, NACE International, 1996
43. "Slow Strain Rate Test Method for Screening Corrosion-Resistant Alloys (CRAs) for Stress Corrosion Cracking in Sour Oilfield Service," TM0198, NACE International, 2004
44. "Standard Practice for Slow Strain Rate Testing to Evaluate the Susceptibility of Metallic Materials to Environmentally Assisted Cracking," G 129, *Annual Book of ASTM Standards,* ASTM International
45. W.J.R. Nesbit et al., "Ripple Stain Rate Test for CRA Sour Service Materials Selection," paper 58, Corrosion/97, NACE International, March 1997
46. Solar Energy Tapped for Cathodic Protection of Casing, *Oil Gas J.,* Oct 1980, p 113
47. J. Leavenworth, Solar Powered Cathodic Protection for Saudi Arabian Oilfields, *Mater. Perform.,* Dec 1984, p 21
48. G.W. Curren, Sun Powers Libya Cathodic Protection System, *Oil Gas J.,* March 1982, p 177
49. J. Evans, Gas Utility Uses Sun Power to Cathodic Protect Gas Mains, *Pipe Line Ind.,* Sept 1984, p 23
50. N.S. Christopher, Cathodic Protection Power Source Designed for Remote Locations, *Pipe Line Ind.,* Oct 1985, p 47
51. R.S. Treseder, Ed., *Corrosion Engineer's Reference Book,* NACE International, 1980
52. M.T. Chapman, Control of External Casing Corrosion, *Mater. Prot. Perform.,* Sept 1973, p 10
53. A.W. Peabody, *Control of Pipeline Corrosion,* NACE International, 1976, p 105

54. T. Allen and A.P. Roberts, *Production Operations,* Vol 2, Oil & Gas Consultants International, Inc., 1982
55. W.F. Gast, "Has Cathodic Protection Been Effective in Controlling External Casing Corrosion for Sun Exploration & Production Co.? A 20 Year Review Tells the Story!" paper 151, Corrosion/85 (Boston, MA), NACE International, March 1985
56. A.G. Ostroff, Understanding and Controlling Corrosion, in *Corrosion Control Handbook,* Petroleum Engineering Publishing, 1975
57. C.C. Patton, *Oilfield Water Systems,* Campbell Petroleum Series, 1981
58. "Corrosion Control in Petroleum Production," TPC Publication 5, NACE International, 1979, p 60
59. F.W. Schremp, Corrosion Prevention for Offshore Platforms, *J. Petrol. Technol.,* April 1984, p 609
60. C.E. Hedborg, "Corrosion in the Offshore Environment," paper 1958, Offshore Technology Conference (Houston, TX), May 1974
61. R.D. Kane et al., Materials and Corrosion Technology for Deep Water Oil and Gas Production, *Mater. Perform.,* April 1998
62. J. Davis, E.P. Doremus, and R. Pass, "Worldwide Design Considerations for Cathodic Protection of Offshore Facilities Including Those in Deep Water," paper 2306, Offshore Technology Conference (Houston, TX), May 1975
63. C. Menendez, H. Hanson, R.D. Kane, and G. Farquhar, Cathodic Protection Requirement for Deep Water Systems, STP 1370, ASTM, Sept 1998
64. M. Schumacher, Ed., *Seawater Corrosion Handbook,* Noyes Data Corporation, 1979, p 78
65. L. Coker, "Some of the Things You Always Wanted to Know About Corrosion Inhibitors But Didn't Ask," presented at the Permian Basin Meeting, NACE International, 1978
66. "Corrosion Control in Petroleum Production," TPC Publication 5, NACE International, 1979, p 47
67. D.H. Chen-Que, *Corrosion Control and Monitoring in Gas Pipelines and Well Systems,* NACE International, 1987
68. "Corrosion Control in Petroleum Production," TPC Publication 5, NACE International, 1979, p 50
69. S.E. Campbell, "Corrosion and Hydrate Inhibition in High Shear, Low Temperature Conditions," paper 02290, Corrosion/2002, NACE International, March 2002
70. "Standard Guide for Evaluating and Qualifying Oilfield and Refinery Corrosion Inhibitors in the Laboratory," *Annual Book of ASTM Standards,* ASTM International
71. S. Srinivasan and R.D. Kane, "Prediction of Corrosivity of CO_2/H_2S Production Environments," paper 11, Corrosion/96, NACE International, March 1996
72. H.J. Endean, "Procedures for Evaluating Corrosion and Selecting Treating Methods for Oil Wells," Corrosion Control Course, The University of Oklahoma, 1977
73. M.J. Watkins, "Effects of Oilfield Corrosion Inhibitors on Elastomeric Seals," paper 144, Corrosion/85, NACE International, March 1985
74. "Corrosion Control in Petroleum Production," TPC Publication 5, NACE International, 1979, p 49
75. R.L. Steelman, Use of Corrosion Inhibitors in Offshore Gas Pipeline Protection, *Oil Gas J.,* Oct 1980, p 154
76. L.W. Gatlin and H.J. Endean, "Water Distribution and Corrosion in Wet Gas Transmission Systems," paper 174, Corrosion/75, NACE International, 1975
77. P.D. Schrickel, Plastic Pipe Meets Gas Industry Needs, *Pipe Line Ind.,* Oct 1984, p 19
78. G.L. Davis, "Selection and Use of Nonmetallic Pipe," Corrosion Control Course, The University of Oklahoma, 1977
79. R.F. Weeter, Desorption of Oxygen From Water Using Natural Gas for Countercurrent Stripping, *J. Petrol. Technol.,* May 1965, p 515
80. H.G. Byars and B.R. Gallop, Injection Water + Oxygen = Corrosion and/or Well Plugging Solids, *Mater. Perform.,* Dec 1974
81. W.J. Frank, "Efficient Removal of Oxygen in a Waterflood by Vacuum Deaeration," SPE paper 4064, Oct 1972
82. D.C. Scranton, Practical Applications of Oxygen Scavengers in the Oilfield—A Review, *Mater. Perform.,* Sept 1979, p 47
83. R.F. Weeter, Conditioning of Water by Removal of Corrosive Gases, *J. Petrol. Technol.,* Feb 1972, p 182
84. J.T.N. Atkinson and H. VanDroffelaar, chapter 6, in *Corrosion and Its Control: An Introduction to the Subject,* NACE International, 1982
85. "Collection and Identification of Corrosion Products," RP0173, NACE International
86. C.P. Dillon, Ed., Forms of Corrosion Recognition and Prevention, *NACE Handbook 1,* NACE International, 1982
87. "Drill Pipe Corrosion Ring Coupon Test Procedure," RP-13B, API Standard Procedures for Testing Drilling Fluids, Appendix A, American Petroleum Institute, Washington, DC
88. "Modern Electrical Methods for Determining Corrosion Rates," 3D170, NACE International
89. "Proposed Use of Galvanic Probe Corrosion Monitor In Oil and Gas Drilling and Production Operations," Committee Report T-10-16, NACE International
90. M.S. Cayard, private communication, InterCorr International, Inc., Houston, TX, www.intercorr.com
91. B.Q. Bradley, Oxygen: A Major Element in Drill Pipe Corrosion, *Mater. Prot.,* Dec 1967
92. W.J. Frank, Here's How to Deal With Corrosion Problems in Rod-Pumped Wells, *Oil Gas J.,* May 1976
93. Recommendations of Corrosion Control of Sucker Rods by Chemical Treatment, NACE Task Group Report, *Mater. Perform.,* May 1967
94. "Preparation, Installation, Analysis, and Interpretation of Corrosion Coupons in Oilfield Operations," RP0775 (latest revision), NACE International
95. "Standard Practice for Preparing, Cleaning, and Evaluating Corrosion Test Specimens," G 1 (latest revision), *Annual Book of ASTM Standards,* ASTM International
96. "Standard Guide for Examination and Evaluation of Pitting Corrosion," G 46 (latest revision), *Annual Book of ASTM Standards,* ASTM International
97. R.D. Kane and E. Trillo, "Evaluation of Multiphase Environments for General and Localized Corrosion," paper 04656, Corrosion/2004, NACE International, March 2004
98. D.A. Eden, private communication, InterCorr International, Inc., Houston, TX, www.intercorr.com
99. D.A. Eden, "Electrochemical Noise—The First Two Octaves," paper 386, Corrosion/98, NACE International, April 1998
100. R.D. Kane, D.A. Eden, and D.C. Eden, "Online, Real-Time Corrosion Monitoring for Improving Pipeline Integrity—Technology and Experience," paper 03175, Corrosion/2003, NACE International, March 2003
101. "Standard Recommended Practice—Metallic Materials for Sucker-Rod Pumps for Corrosive Oilfield Environments," MR0176 (latest revision), NACE International
102. "Selection of Metallic Materials to Be Used in All Phases of Water Handling for Injection into Oil-Bearing Formations," RP0475 (latest revision), NACE International
103. "Standard Recommended Practice—Handling and Proper Usage of Inhibited Oilfield Acids," RP0273 (latest revision), NACE International
104. "Recommended Practice for Application of Cement Lining to Steel Tubular Goods, Handling, Installation and Joining," RP-10E, Third Edition, American Petroleum Institute
105. "Recommended Practice for Planning, Designing and Constructing Fixed Offshore Platforms," RP-2A, American Petroleum Institute, Washington DC
106. W.J. Graff, *Introduction to Offshore Structures,* Gulf Publishing, 1981
107. M. Schumaker, Ed., *Seawater Corrosion Handbook,* Noyes Data Corporation, 1979

108. "Corrosion Control of Steel, Fixed Offshore Platforms Associated with Petroleum Production," RP0176 (latest revision), NACE International
109. W.S. Broecker, *Chemical Oceanography,* Harcourt Brace Jovanovich, 1974
110. B. Tomkins, "Fatigue Design Rules for Steel Welded Joints in Offshore Structures," paper 4403, 14th Offshore Technology Conference (Houston, TX), May 1982
111. W.D. Kover and S. Dharmavasan, "Fatigue Fracture Mechanics Analysis of T and Y Joints," paper 4404, 14th Offshore Technology Conference (Houston, TX), May 1982
112. A. Mukhopadhyay, Y. Itoh, and J.C. Bouwkamp, "Fatigue Behavior of Tubular Joints in Offshore Structures," paper 2207, Third Offshore Technology Conference (Houston, TX), May 1975
113. R.M. Kenley, "Measurement of Fatigue Performance of Forties Bravo," paper 4402, 14th Offshore Technology Conference (Houston, TX), May 1982
114. Y. Minami and H. Takada, Corrosion Fatigue and Cathodic Protection of Mild Steel, *Boshoku Gijutsu,* Vol 7 (No. 6), 1958, p 336
115. C.E. Jaske et al., Corrosion Fatigue of Structural Steels in Seawater for Offshore Application, in *Corrosion-Fatigue Technology,* STP 642, American Society for Testing and Materials, 1978
116. E.C. Faulds, "Structural Inspection and Maintenance in a North Sea Environment," paper 4360, 14th Offshore Technology Conference (Houston, TX), May 1982
117. B.C. Craft and M.F. Hawkins, *Applied Petroleum Reservoir Engineering,* Prentice-Hall, 1959, p 5
118. A.K. Dunlop, "Fundamental Considerations in Sweet Gas Well Corrosion," paper 46, Corrosion/83 (Anaheim, CA), NACE International, April 1983
119. S.D. Kapnsta, private communication
120. C. DeWaard and D.E. Milliams, Carbonic Acid Corrosion of Steel, *Corrosion,* Vol 31 (No. 5), 1975
121. D.R. Fincher, J.J. Marr, and J.W. Ward, Inhibiting Gas-Condensate Wells Can Become Complicated Problem, *Oil Gas J.,* Vol 73 (No. 23), 1975, p 52

SELECTED REFERENCES

Primary Production

- J.B. Bradburn and S.K. Kalra, Corrosion Mitigation—A Critical Facet of Well Completion Design, *J. Petrol. Technol.,* Sept 1983
- "Care and Handling of Sucker Rods," RP 11BR, American Petroleum Institute
- *Corrosion Control in Petroleum Production,* NACE International, 1979
- *Corrosion of Oil and Gas Equipment,* NACE International and the American Petroleum Institute, 1958
- J.E. Donham, "Recent Developments in Corrosion Inhibitors and Their Use," paper Offshore Production Chemicals Conference, Norwegian Society of Chartered Engineers, June 1983
- A.K. Dunlop, H.L. Hassell, and P.R. Rhodes, "Fundamental Considerations in Sweet Gas Well Corrosion," paper Corrosion/83 (Anaheim, CA), NACE International, April 1983
- S. Evans, J.M. Phelan, and M.E. Williams, "Batch Treatment of Offshore Wells in the East Cameron and Vermilion Areas," paper 17th Annual Offshore Technological Conference (Houston, TX), May 1985
- R.H. Hausler and S.G. Weeks, "Low Cost Low Volume Continuous Corrosion Inhibitor Application to Gas Production Tubulars," paper Corrosion/86 (Houston, TX), NACE International, March 1986
- C.J. Houghton and R.V. Westermark, "North Sea Downhole Corrosion; Identifying the Problem, Implementing the Solutions," paper 1983 Offshore Technological Conference (Houston, TX), May 1983
- G.C. Huntoon, "Completion Practices in Deep Sour Tuscaloosa Wells," paper 57th Annual Fall Technical Conference and Exhibition of the Society of Petroleum Engineers of AIME (New Orleans, LA), Society of Petroleum Engineers, Sept 1982
- T. Murata, E. Sato, and R. Matsuhashi, "Factors Controlling Corrosion of Steels in CO_2 Saturated Environments," paper Corrosion/83 (Anaheim, CA), NACE International, April 1983
- *Primer of Oil and Gas Production,* 3rd ed., American Petroleum Institute, 1978
- W.B. Steward, Sucker Rod Failures, *Oil Gas J.,* April 1973

CO_2 Injection

- J.C. Ader and M.H. Stern, Slaughter Estate Unit Tertiary Miscible Gas Pilot Reservoir Description, *J. Petrol. Technol.,* May 1984, p 837
- B.W. Bradley, "CO_2 EOR Requires Corrosion Control Program in Gas Gathering Systems," presented at the Permian Basin Corrosion Symposium (Odessa, TX), NACE International, Nov 1985
- R.L. Mathis and S.O. Spears, "Effect of CO_2 Flooding on Dolomite Reservoir Rock, Denver Unit, Wasson (San Andres) Field, TEXAS," presented at the 59th Technical Conference and Exhibition of the Society of Petroleum Engineers of AIME (Houston, TX), Society of Petroleum Engineers, Sept 1984
- L.E. Newton, Jr., "SACROC CO_2 Project—Corrosion Problems and Solutions," presented at Corrosion/84 (New Orleans, LA), NACE International, April 1984
- B.C. Price and F.L. Gregg, "CO_2/EOR, From Source to Resource," presented at the 62nd Annual GPA Convention (San Francisco, CA), Gas Processors Association, March 1983
- W.B. Saner and J.T. Patton, CO_2 Recovery of Heavy Oil; Wilmington Field Test, *J. Petrol. Technol.,* July 1986, p 24

Corrosion in Petroleum Refining and Petrochemical Operations

Revised by Russell D. Kane, Honeywell Process Solutions, Honeywell International, Inc.

CORROSION has often been considered as an unavoidable part of petroleum refining and petrochemical operations. Partially due to this historical view of corrosion, one of the primary causes of operational problems in refining and petrochemical operations is corrosion. Corrosion problems increase operating and maintenance costs substantially. Time-dependent degradation due to corrosion usually defines scheduled and unscheduled shutdowns so that corrosion damage in piping and equipment can be inspected and/or repaired. This operational scenario can be extremely expensive. Single-incidence costs of major failures and related releases have been documented in the range of $35 to $50 million. Based on the current assessment of corrosion costs, the total annual direct cost of corrosion is estimated at $3.7 billion. Of this total, maintenance-related expenses are estimated at $1.8 billion, vessel turnaround expenses at $1.4 billion, and fouling costs are approximately $0.5 billion annually.

The potential for corrosion in refineries and petrochemical plants may be inherent to some processes, but costly and damaging equipment losses are not. Much work is being done to adapt process control technologies to integrate corrosion monitoring on an online, real-time basis to increase productivity (i.e., increase run time and decrease time associated with turnaround inspections). With this continual drive to increase productivity, the new vision of corrosion is that it is another process variable that can be continuously monitored, assessed, and controlled. In certain cases, process engineers have even been able to "see" new aspects of process chemistry through the information that their new view of corrosion data can bring. Certainly, corrosion engineers also benefit by having a greater appreciation of the actual process variations through having access to online data from which they can gage their relative impact on corrosion.

To accomplish the aforementioned paradigm shift in corrosion technology, plant engineers are using new real-time, online measurement technologies (e.g., monitoring of electrochemical corrosion and pitting, advanced electric resistance corrosion measurements, hydrogen permeation monitoring, and online inspection) along with improvements in handling and integration methods of electronic data that are now being introduced. Simultaneously, engineers are getting access to predictive software that rapidly queries complex engineer databases, makes flow modeling calculations, and applies expert rules for assessment of corrosivity and/or materials selection. This basically has resulted in a joining of online monitoring, inspection technologies, computer science, and process control information with real-time feed to process and corrosion engineers over global companywide information networks. The "replace when it fails" approach is receding into the past; facilities management today (2006) is embracing new technologies and starting to appreciate the actual value in terms of increased throughput, productivity, and profitability that this new approach offers. It has capabilities to increase unit run time between major inspections, reduce the time and expense associated with turnaround or inline inspections, and reduce major upsets that cause most of the unplanned shutdowns. The end result is the ability to know on a practical basis of how "hard" facilities can be pushed in the effort to increase productivity before excessive corrosion damage will result. This also allows the process engineers to understand the impact of their changes (i.e., feed and process control actions) to implement true process control and, eventually, asset management.

Often, anything that can be safely done to keep a process unit on stream for long periods of time will be of great benefit. A large proportion of corrosion problems are actually caused by start-up or shutdowns. For example, when equipment is opened to the atmosphere for inspection and repair, metal surfaces covered with (oftentimes, sulfur-containing) corrosion products will be exposed to air and moisture. This can lead to pitting and other forms of localized corrosion and stress-corrosion cracking unless preventive measures are implemented. When equipment is washed with water during a shutdown, corrosion can be caused by pockets of water left in the process units and associated piping.

Most petroleum refining and petrochemical plant operations involve flammable hydrocarbon streams, highly toxic or explosive gases, and strong acids or caustics that are often at elevated temperatures and pressures. Among the many metals and alloys that are available, relatively few are used for the construction of process equipment and piping (Ref 1) due to the adverse influence on cost. However, materials used in process units include a wide variety of alloys, including carbon steel, cast irons, low-alloy steels, and stainless steels, and, to a much lesser degree, aluminum, copper, nickel, and expensive titanium- and nickel-base alloys. This article presents the primary considerations and mechanisms for corrosion and how they are involved in the selection of materials for process equipment in refineries and petrochemical plants. Over the past two decades, a substantial amount of new information has been obtained from laboratory simulation of process environments involving exposure to naphthenic acids and sulfur compounds at elevated temperature, and to lower-temperature wet hydrogen sulfide (H_2S) environments, including alkaline sour water. In addition, specific information on mechanical properties, corrosion, sulfide stress cracking (SSC), hydrogen-induced cracking (HIC), stress-oriented hydrogen-induced cracking (SOHIC), hydrogen embrittlement cracking (HEC), stress-corrosion cracking (SCC), velocity-accelerated corrosion, erosion-corrosion, and corrosion control is provided herein.

Materials Selection

The selection of materials of construction has a significant impact on the operability, economics, and reliability of refining units and petrochemical plants. For this reason, materials selection should be a cooperative effort between the materials engineer and plant operations and maintenance personnel. Reliability can often be equated to predictable materials performance under a wide range of exposure conditions (i.e., operating envelope). Ideally, a material should provide some type of warning before it fails; materials that fracture spontaneously and without bulging as a result of brittle fracture, SCC, or hydrogen embrittlement should be avoided. Uniform corrosion of equipment can be readily detected by various inspection techniques. In contrast, isolated pitting is potentially much more serious, because leakage can occur at highly localized areas that are difficult to detect. Therefore, new monitoring techniques that can differentiate general from localized corrosion can be of substantial benefit to unit operators and managers interested in process control and plant asset management. The effect of environment on the mechanical properties of a material can also be significant. Certain exposure conditions (e.g., exposure to wet H_2S environments) can convert a normally ductile material such as carbon steel into a very brittle material that may fail without warning. A material must not only be suitable for normal process conditions but must also be able to handle transient conditions encountered during start-up, shutdown, emergencies, or extended standby. It is often during these time periods that equipment suffers serious deterioration or that failure occurs. With online monitoring techniques, these conditions can now be assessed on a real-time basis to provide additional input as to the consequences of these situations.

Of particular concern is what will happen to equipment during a fire or even high-temperature process excursions. Unexpected exposure to elevated temperatures can not only affect the metallurgical structure of the material and its related mechanical properties but can also produce detrimental side effects in terms of lost ductility and increased susceptibility to corrosion and/or SCC. Although all possible precautions should be taken to minimize the probability of a fire, the engineer responsible for materials selection must recognize that a fire or temperature excursions may occur and that the equipment is expected to retain its integrity in order to avoid fueling the fire. This often limits the application of materials with low melting points or those that may thermally decompose (i.e., nonmetallics and polymer-based composites) or become subject to damage by thermal shock when fire-fighting water is applied, particularly in the case of refinery piping and equipment used to handle highly flammable hydrocarbon streams. On the other hand, fire resistance need not be considered for cooling-water or instrument-air systems.

Although petrochemical plants may include some processes that involve nonflammable or nonhazardous streams, most equipment must be resistant to fires. For example, lack of fire resistance rules out the use of plastic components in refineries and petrochemical plants despite their excellent resistance to many types of corrosives. In addition, plastic components tend to be damaged by stream-out during a shutdown; this is required in order to free components of hydrocarbon residues and vapor before inspection or maintenance operations.

The final step in the materials selection process is a reliability review of the materials and the corrosion control techniques that were selected. There must be total assurance that a plant will provide reliable service under all conditions, including those that occur during start-ups, shutdowns, downtime, standby, and other emergencies that may be presented.

Principal Materials

Materials selection criteria for a number of ferrous and nonferrous alloys used in petroleum refining and petrochemical applications are presented in this section. Additional information on selecting the proper metal or alloy is available in the article "Materials Selection for Corrosion Control" in *Corrosion: Fundamentals, Testing, and Protection,* Volume 13A of *ASM Handbook,* 2003.

Carbon and Low-Alloy Steels. Carbon steel is probably used for at least 80% of all components in refineries and petrochemical plants. The simple rationale for the extensive use of steel is that it is inexpensive, readily available, easily fabricated (e.g., bending, forming, and shop and field welding), and can be postweld heat treated when needed. Every effort is made to use carbon steel, even if process changes are required to obtain satisfactory service from carbon steel (Ref 2). For example, process temperatures can be decreased, hydrocarbon streams dried up, or additives such as inhibitors or neutralizers injected in order to reduce potential corrosion problems with carbon steel (Ref 3). In refineries, fractionation towers, separator drums, heat-exchanger shells, storage tanks, most piping, and all structures are generally fabricated from carbon steel. Carbon and low-alloy steels of carbon-molybdenum (C-Mo) or carbon-chromium-molybdenum (C-Cr-Mo) chemistry are the most widely used in plant construction. A C-0.5Mo steel can offer substantial savings over carbon steels at temperatures between 425 and 540 °C (800 and 1000 °F) because of increased temperature resistance (i.e., strength retention). The C-0.5Mo steel was originally considered significantly more resistant than carbon steel to high-temperature hydrogen attack (HTHA). It has been extensively used for reactor vessels, heat-exchanger shells, separator drums, and piping for processes involving hydrogen at temperatures above 260 °C (500 °F). However, service and inspection data have limited these expectations under conditions of long-term hydrogen exposure on C-0.5Mo steel. As a result, low-alloy steels are now preferred for new construction and repairs. Revised serviceability plots for HTHA (i.e., Nelson curves) of common constructional steels used in refinery equipment have been prepared and are continually being updated by the American Petroleum Institute (API), which takes into account new service experience, failures, and so on. See the section "Hydrogen Attack" in this article.

Low-alloy steels are widely used for refinery service and are generally the C-Cr-Mo steels containing between 1 and 9% Cr. These steels have excellent strength retention, resistance to high-temperature sulfidic corrosion (due to their increased chromium levels), and resistance to HTHA. To improve resistance to hydrogen stress cracking (including SSC in wet H_2S environments), these low-alloy steels normally require postweld heat treatment. The benefits of postweld heat treatment are generally considered to be the associated reduction in residual tensile stresses in the area of the weldment, and the reduction in hardness and changes in the carbide morphology of weld heat-affected zones (HAZ). For refinery reactor vessels, which operate at high temperatures and pressures, 2.25Cr-1Mo steel is widely used. For improved corrosion resistance, these are often overlaid with stainless steels such as type 347. Other applications for low-alloy steels are furnace tubes, heat-exchangers shells, piping, and separator drums. Additional information is provided in the articles "Corrosion of Wrought Carbon Steels" and "Corrosion of Wrought Low-Alloy Steels" in *Corrosion: Materials,* Volume 13B of *ASM Handbook,* 2005.

Stainless steels are extensively used in petrochemical plants because of the highly corrosive nature of the catalysts, process constituents, and solvents that are often used. In refineries, stainless steels have been primarily limited to applications involving high-temperature sulfidic and naphthenic acid corrosion and other forms of high-temperature attack (Ref 4). However, they are also found in increasing regularity in sour alkaline water applications (Ref 5). An important consideration when using stainless steels is that many of the conventional austenitic stainless steels (300-series) will pit or suffer SCC in the presence of chlorides and water to a varying degree, depending on other species such as oxygen and sulfides (Ref 6). However, there are many new higher-alloy stainless steels that have higher resistance to pitting and localized attack. Some of these are shown in Table 1. They have increased amounts of chromium, molybdenum, tungsten, and nitrogen that enhance passivity against more hostile process environments.

The minimum stainless alloys in terms of corrosion resistance are generally considered to be the martensitic stainless steels, such as type 410 (S41000). This material must be postweld heat treated to avoid hydrogen cracking problems as a result of exposure to processes that contain H_2S or other hydrogenating environments. Typical applications include pump components, fasteners, valve trim, turbine blades, tray valves, and other tray components in fractionation towers. Low-carbon varieties of type 410 stainless steel (S41008) are preferred for furnace tubes and piping, often in combination with aluminizing for increased corrosion resistance under conditions that will cause sulfidation. Ferritic stainless steels, such as type 405 (S40500), are less subject to hydrogen stress cracking and are therefore a better choice than type 410 (S41000) stainless steel for vessel linings that are attached by welding (Ref 7).

Austenitic stainless steels, such as type 304 (S30400) or type 316 (S31600), have excellent corrosion resistance but, as mentioned previously, are generally considered to be susceptible to SCC by chlorides at temperatures above 60 °C (140 °F). If sensitized, they are also subject to SCC in aqueous chloride environments at lower temperatures, and susceptibility may also include SCC in polythionic acids (Ref 8, 9) formed by the combination of moisture, sulfur components, and oxygen when process equipment is opened to air. Typical applications include linings and tray components in fractionation towers; piping; heat-exchanger tubes; reactor cladding; tubes and tube hangers in furnaces; various components for compressors, turbines, pumps, and valves; and reboiler tubes.

In many aqueous refinery environments, susceptibility of stainless steels (and nickel-base alloys) to corrosion and localized attack (e.g., pitting, crevice corrosion, and SCC) has been found to be related to the pitting resistance equivalent number (PREN), which is given by the following formula (Ref 10):

$$PREN = Cr + 3.3Mo + 1.5(W+Nb) + xN \quad (Eq\ 1)$$

where $x = 0$ for martensitic and ferritic stainless steels, $x = 16$ for duplex stainless steels, and $x = 30$ for austenitic alloys. This relationship is heavily based on performance of these alloys in elevated-temperature aqueous, chloride-containing environments in aerated or oxidizing conditions. In reducing environments with sulfides and chlorides, the value for x can be limited to 11 (versus 16 and 30 as given previously) due to the lesser role of nitrogen in maintaining passivity under these conditions (Ref 11). Based on the aforementioned PREN relationship, recent refinery applications have included greater use of duplex stainless steels, such as alloy 2205 (S31803) and alloy 2507 (S32750), or high-alloyed austenitic stainless steels, such as alloy 904L (N08904) and alloy AL6XN (N08367). These alloys have PREN values in the range of 30 to over 40 versus the more commonly used type 316 stainless steel with a PREN value of 24. Applications have included uses in cooling water service, process heat exchangers, and sour water systems, where added corrosion and SCC resistance over that of 300-series austenitic stainless steels was needed, and with a cost-savings over conventionally used nickel alloys.

For nonaqueous conditions involving exposure to high-temperature sulfidic or naphthenic crude oil environments, resistance to corrosion has been mainly related to simply the sum of the chromium and molybdenum concentration (Ref 12). For further information, see the article "Corrosion of Wrought Stainless Steels" in *Corrosion: Materials,* Volume 13B of *ASM Handbook,* 2005.

Cast irons, because of the inherent brittleness and low strength found in many grades, are normally not used for pressure-retaining components for handling flammable hydrocarbons. The main exceptions are pump and valve components, ejectors, jets, strainers, and fittings in which the high hardness of cast iron can have a beneficial effect in reducing the velocity effects of corrosion, such as impingement, erosion, and cavitation. High-silicon cast irons (with 14% Si) are extremely corrosion resistant because of a passive surface layer of silicon oxide that forms during exposure to many chemical environments (except hydrofluoric acid). Typical refinery and petrochemical plant applications include valve and pump components for corrosive service. High-nickel cast irons (with 13 to 36% Ni and up to 6% Cr) have excellent corrosion, wear, and high-temperature resistance because of the relatively high alloy content (Ref 13). Typical uses are valve components, pump components, dampers, diffusers, tray components, and compressor parts. Additional information is provided in the article "Corrosion of Cast Irons" in *Corrosion: Materials,* Volume 13B of *ASM Handbook,* 2005.

Copper and Aluminum Alloys. These materials are used extensively in water service and commonly are restricted to applications below 260 °C (500 °F) because of strength limitations. Admiralty brass (C44300) tubes have been extensively used in water-cooled condensers and coolers at most refineries. However, in general, copper alloys suffer accelerated corrosion in both acidic and alkaline environments containing sulfur species. Therefore, they have often performed poorly in overhead condensers, compressor aftercoolers, and other sour water applications and other locations where high concentrations of H_2S and ammonia are encountered in aqueous condensate. This condition can also be exacerbated by the influence of flow-induced turbulence. Even when alloyed with substantial nickel (e.g., cupronickel and alloy 400), the performance of these alloys is inferior to comparably priced stainless or nickel-base alloys. The usual mode of corrosion failure of brass and copper alloys in refinery applications includes excessive corrosion rate, pitting, dezincification, and ammonia SCC.

Aluminum alloys, at one time, were proposed for refinery use as a substitute for carbon steel and admiralty brass (C44300) heat-exchanger tubes in cooling-water service (Ref 14–16). Aluminum tubes were found to be highly resistant to aqueous sulfide corrosion in overhead condensers. Unfortunately, fouling and pitting corrosion on the water side have always been a problem, and except for certain limited applications, most refineries do not use aluminum tubes. The only other major refinery use of aluminum has been in vacuum towers, in which aluminum or aluminum coatings can provide resistance (at relatively low-flow conditions) to the naphthenic acid corrosion of tray components. Aluminized coatings are also used to protect low-alloy steels against high-temperature sulfidic corrosion in the absence of moisture and/or chlorides. Additional information is available in the articles "Corrosion of Copper and Copper Alloys" and "Corrosion of Aluminum and Aluminum Alloys" in *Corrosion: Materials,* Volume 13B of *ASM Handbook,* 2005.

Nickel and nickel alloys are especially used for corrosion resistance in sulfuric acid, hydrochloric acid, hydrofluoric acid, and caustic environments, all of which can result in corrosion problems in many materials of construction in certain refinery and petrochemical operations (Ref 17). As an alloying element in austenitic

Table 1 Compositions of commonly used stainless and nickel-base alloys

Alloy	UNS designation	Composition(a), wt%					
		C	Cr	Fe	Ni	Mo	Other
C-276	N10276	0.02	14.5–16.5	4–7	bal	15–17	2.5Co, 1.0Mn, 4.5W, 0.35V
625	N06625	0.10	20–23	5	bal	8–10	0.4Al, 4.15Nb, 0.5Mn, 0.4Ti
G	N06007	0.05	21–23.5	18–21	bal	5.5–7.5	2.5Nb, 2.5Co, 2.5Cu, 2.0Mn, 1W
G-30	N06030	0.03	28–31	13–17	bal	4–6	2.4Cu, 5.0Co, 4.0W
825	N08825	0.05	19.5–23.5	bal	38–46	2.5–3.5	0.2Al, 3Cu, 1Mn, 1.2Ti
925	N09925	0.03	19.5–23.5	20	38–46	2.5–3.5	0.1–0.5Al, 1.9–2.4Ti
2550	N06975	0.03	23–26	bal	47–52	5–7	0.7–1.2Cu, 0.7–1.5Ti
718	N07718	0.08	17–21	bal	50–55	2.8–3.3	0.8Al, 0.6–1.1Ti, 4.8–5.5Nb, 1Co
725	N07725	0.03	19–22.5	bal	55–59	7–9.5	0.35Al, 2.75–4Ti
400	N04400	0.3	...	2.5	63–70	...	bal Cu, 2Mn
K-500	N05500	0.2	...	2.0	63–70	...	bal Cu, 3Al, 0.85Ti
MP35N	R30035	0.025	19–21	1.0	33–37	9–10.5	bal Co, 0.15Mn, 1Ti
AL6XN	N08367	0.03	20–22	bal	23.5–25.5	6–7	0.18–0.25N
28	N08028	0.03	26–28	bal	29.5–32.5	3–4	1.4Cu, 2.5Mn
255	S32550	0.04	24–27	bal	4.5–6.5	2–4	2.5Cu, 1.5Mn, 0.25N
100 (ASTM A351)	S32760	0.03	55	bal	6–8	3–5	0.7Cu, 0.25N, 0.7W
2507	S32750	0.03	24–26	bal	6–8	3–5	0.5Cu, 0.24–0.32N
2205	S31803	0.03	21–23	bal	4.5–6.5	2.5–3.5	0.08–0.2N
254SMO	S31254	0.02	19.5–20.5	bal	17.5–18.5	6.0–6.6	0.18–0.22N, 0.5–1.0Cu
316	S31600	0.08	16–18	bal	10–14	2–3	...
654SMO	S32654	0.03	24	bal	22	7	0.5N, 0.5Cu
13Cr (Hyper 1)	...	...	13	bal	4	1.5	...
13Cr (Hyper 2)	...	...	13	bal	5	2.15	...
F6NM	S42400	0.06	12–14	bal	3.5–4.5	0.3–0.7	...
420 (13Cr)	S42000	0.15	12–14	bal	...	...	...

(a) Maximum allowable concentration unless range is given

alloys, nickel is important in reducing susceptibility to chloride SCC in elevated-temperature solutions. Generally, SCC resistance increases with nickel content above approximately 8% (Fig. 1) (Ref 18). In the range of approximately 40% Ni, many stainless alloys are resistant to chloride SCC in aerated chloride-containing solutions. However, where chlorides are present in combination with sulfides in aqueous acidic environments at elevated temperature, SCC can still be observed. Higher resistance to SCC is obtained through additional alloying with nickel, chromium, molybdenum, and nitrogen. Similar to stainless steels, resistance to SCC in nickel-base alloys is also generally found to be related to pitting susceptibility through the commonly used formula for PREN discussed previously in this article for stainless steels.

Nickel also forms the basis for many high-temperature alloys. However, nickel alloys not containing chromium or molybdenum can be attacked and embrittled by sulfur-bearing gases (sulfidic attack) and sour water solutions (ammonium bisulfide corrosion) at elevated temperatures. Alloy 400 (N04400), a nickel-copper alloy, is extensively used as a lining for carbon steel equipment to prevent corrosion by hydrochloric acid and chloride salts (Ref 19). For the same reason, alloy 400 (N04400) tubes have been used in overhead condensers. Alloy 400 (N04400) is also used against corrosion by hydrofluoric acid. However, in the presence of sour water environments, this nickel-copper-containing alloy can show accelerated corrosion under conditions of elevated temperature and high flow-induced turbulence. Alternative stainless or nickel-base alloys may be preferred for some applications.

High-nickel alloys with high levels of chromium and molybdenum, such as alloy 825 (N08825) and alloy 625 (N06625), are used to reduce the polythionic acid corrosion of flare-stack tips, as can be experienced with conventional stainless steels. The nickel-molybdenum alloys, such as alloy B-2 (N10665) and B-3 (N10675), are particularly well suited for handling hydrochloric acid at all concentrations and temperatures (including the boiling point) but are attacked if exposed to oxidizing salts or aerated environments (Ref 20, 21). Alloy B (N10001), alloy C-4 (N06455), and alloy C-276 (N10276) have excellent resistance to all concentrations of sulfuric acid up to at least 95 °C (200 °F).

Newer alloys have been developed specifically for resistance to environments containing sulfides and chlorides, such as alloy 686 (N06686), for carburization resistance, such as alloy 214 (N07214), and nickel/cobalt-base alloys with combined resistance to multiple corrodants in complex process streams, such as alloy 556 (R30556). Although usually expensive, these alloys are used for specific applications to overcome unusually severe corrosion problems. Additional information is provided in the article "Corrosion of Nickel and Nickel-Base Alloys" in *Corrosion: Materials*, Volume 13B of *ASM Handbook*, 2005.

Titanium and Titanium Alloys. These alloys have been extensively used in certain petrochemical processes where high corrosion resistance is needed. Titanium, however, is not generally considered a high-temperature metal; welding and cutting must be done under inert gas atmospheres with specific procedures to prevent embrittlement (Ref 22, 23). From a practical point of view, the use of titanium in refinery and petrochemical plant service is limited to temperatures below 260 °C (500 °F) (Ref 24, 25). If hydrogen is present, temperatures should not exceed 175 °C (350 °F) in order to prevent embrittlement due to hydride formation. Hydrogen embrittlement of titanium can also occur as a result of a galvanic couple to a less noble material such as steel in elevated-temperature aqueous solution. Titanium is fully resistant to many process streams due to its reactive nature, which allows it to readily form a passive TiO_2 surface layer that is both tough and chemically resistant. Tubes made from Ti-grade 2 (R50400) are extensively used in overhead coolers and condensers on a number of refinery units to prevent corrosion by aqueous chlorides, sulfides, and sulfur dioxide. These tubes can corrode, however, beneath acidic deposits. Titanium tubes are often required when seawater or brackish water is used for cooling. Where underdeposit corrosion of pure titanium is a problem, Ti-grade 12 (R53400), alloyed with nickel and molybdenum, should be considered. Anodizing and high-temperature air oxidizing of Ti-grade 2 (R50400) have been shown to be beneficial from a corrosion point of view (Ref 26).

More recently, new titanium alloys, which contain low-level alloying additions of noble metals such as palladium in Ti-grade 7 (R52400) or ruthenium in Ti-grade 26 (R52404), have been effectively used to reduce crevice and pitting attack and underdeposit corrosion. A list of titanium alloys and their compositions is given in Table 2. Additional information is provided in the article "Corrosion of Titanium and Titanium Alloys" in *Corrosion: Materials*, Volume 13B of *ASM Handbook*, 2005.

Codes and Standard Specifications

Rules for the design, fabrication, and inspection of pressure vessels, piping, and tanks are provided by codes that have been developed by industry and/or regulatory agencies in various countries, as shown by the listing in Table 3 (see also the "Appendix: Industry Standards" at the conclusion of this article). In the United States, the American National Standards Institute/American Society of Mechanical Engineers (ANSI/ASME) Boiler and Pressure Vessel Code, section VIII, which covers unfired pressure vessels, is used by most industries and fabricators. It is usually mandatory that the code be followed in process applications. Therefore, the first step in selecting materials of construction is to know what the code covers and what it does not.

The ANSI/ASME Boiler and Pressure Vessel Code also provides a list of acceptable materials for pressure-containing applications and allowable stress values for each material. The detailed specifications for these steels are provided in Sections II A and II B of the code, which are

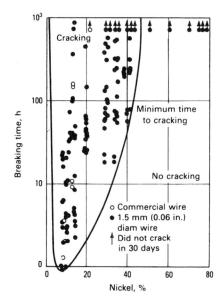

Fig. 1 Effect of nickel additions to a 17 to 24% Cr steel on resistance to stress-corrosion cracking in boiling 42% magnesium chloride solution. Source: Ref 18

Table 2 Compositions of commonly used titanium alloys

Alloys	UNS designation	Composition, wt%					
		Al	Mo	V	Pd	Ru	Other
Ti-grade 2	R50400	...	...	...	...	...	bal Ti
Ti-grade 5	R56400	6	...	4	...	...	bal Ti
Ti-grade 7	R52400	...	...	...	0.18	...	bal Ti
Ti-grade 12	R53400	...	0.3	...	...	...	0.8Ni; bal Ti
Ti-grade 13	R53413	...	...	...	...	0.05	0.5Ni; bal Ti
Ti-grade 16	R52402	...	...	...	0.06	...	bal Ti
Ti-grade 18	R56322	3	...	2.5	...	0.06	bal Ti
Ti-grade 23	R56407	6	...	4	...	...	0.13 max O; bal Ti
Ti-grade 24	R56405	6	...	4	0.06	...	0.2 max O; bal Ti
Ti-grade 26	R52404	...	...	...	...	0.12	bal Ti
Ti-grade 27	R52254	...	...	...	...	0.12	0.18 max O; bal Ti
Ti-grade 29	R56404	6	...	4	...	0.12	bal Ti

based on ASTM International and UNS standard specifications (Table 4). The code also provides the method for calculating the required minimum thickness of various components based on design temperature and pressure. The need for heat treating during fabrication and the inspection requirements are also defined based on the alloy selected and the pressure-wall thickness. For welded pressure vessels, Section IX of the code defines the requirements for qualifying the welding process to be used.

The code does not consider the effect of process environment on the materials selected. It recognizes that corrosion can and does occur, and it provides rules for including corrosion allowances in the calculation of the required pressure-wall thickness; however, suitable values for the corrosion allowance must be specified by the designer. It is also the responsibility of the designer to specify any special heat treatments, hardness limitations, or other details that may be required as a result of environmental factors. Similarly, the designer must determine accurately the full range of likely operating conditions, including upsets that may be encountered so that the design criteria are met.

Mechanical Properties

Elevated-Temperature Properties. As mentioned in the preceding section, the applicable code will specify the allowable stress that is to be used for a particular material of construction in the design of a given piece of equipment. This allowable stress is based on the temperature to which the equipment will be exposed for long periods of time. Steels operating under normal plant conditions can be exposed to these temperatures for extensive periods of time without adverse effects on their allowable strength if there is no corrosion. As working temperatures increase, the mechanical strength of most materials decreases. In actual practice, however, a material is more likely to fail at elevated temperatures by creep or stress rupture than from simply a decrease in short-term tensile or yield strength properties.

For example, Table 5 shows the short-term, elevated-temperature yield strengths of several carbon and low-alloy steels. As can be seen from the tabulated data, all three steels (carbon steel, C-0.5Mo steel, and 2.25Cr-1Mo steel) have acceptable yield strength values up to 480 °C (900 °F). These values do not, however, adequately represent the long-term resistance of the steels to creep when stressed at elevated temperatures. Instead, creep resistance values are a more accurate measure of elevated-temperature mechanical strength. Creep resistance values are obtained from creep tests (elongation versus time) and stress-rupture tests (stress versus time-to-failure) at elevated temperatures over a period of 10,000 h and are usually extrapolated to 100,000 h. Table 6 shows creep resistance values of the three steels discussed previously. The deterioration of creep resistance of carbon steel at 480 and 540 °C (900 and 1000 °F) is readily apparent, as is the marked improvement afforded by use of 2.25Cr-1Mo steel. Table 7 lists suggested maximum service temperatures for five different steels and alloys based on creep-rupture data. In some applications, such as furnace tubes, code-allowable stresses need not be followed, and equipment may be operated at temperatures and stresses that can lead to creep failure, provided accurate prediction of service life has been made. In order to predict failure with greater accuracy, equipment operating in the creep range should be periodically inspected as the design life is approached. Guidelines for design, service life prediction, and risk-based inspection of refinery equipment in refinery service are given in industry standards such as API 580 and API 579 (Ref 27, 28).

Hardness. The hardness of steels is not considered by the code as a specified property. Whether it is the result of forming, heat

Table 3 Construction codes for refinery process equipment

Country	Issuing organization	Source document(s)
Pressure vessels		
United States	American Society of Mechanical Engineers American National Standards Institute	Boiler and Pressure Vessel Code, Section VIII
Great Britain	British Standards Institution	BS 1515: Fusion Welded Pressure Vessel for Use in the Chemical, Petroleum and Allied Industries BS 5500: Unfired Fusion Welded Pressure Vessels
Germany	Arbeitsgemeinschaft Druckbehalter (published by Carl Heymans Verlag KG)	A.D. Merkblatter
Italy	Associazione Nazionale per il Controllo della Combustione	ANCC Code
Netherlands	Dienst voor Stoomwezen	Regels Voor Toestellen (Rules for Pressure Vessels)
Sweden	Tryckkarlskomissionen (Swedish Pressure Vessel Commission)	Swedish Pressure Vessel Code
Piping		
United States	American Society of Mechanical Engineers American National Standards Institute	B31.3 Code for Pressure Piping
Great Britain	British Standards Institution	BS 3351: Piping Systems for Petroleum Refineries and Petrochemical Plants
Tanks		
United States	American Petroleum Institute	API 620: Recommended Rules for Design and Construction of Large Welded Low Pressure Storage Tanks API 650: Welded Steel Tanks for Oil Storage
Great Britain	British Standards Institution	BS 2654: Vertical Steel Welded Storage Tanks for the Petroleum Industry

Table 4 ASTM International standard specifications for refinery steels

Material	Pipes and tubes	Plates	Castings	Forgings
Carbon steel	A53, A106, A120, A134, A135, A139, A178, A179, A192, A210, A211, A214, A226, A333, A334, A369, A381(a), A524, A587, A671, A672, A691	A283, A285, A299, A442, A455, A515, A516, A537, A570, A573(a)	A27(a), A216, A352	A105, A181, A234, A268, A350, A372, A420, A508, A541
C-0.5Mo steel	A161(a), A209, A250, A335, A369, A426, A672, A691	A204, A302, A517, A533	A217, A352, A487	A182, A234, A336, A508, A541
1Cr-0.5Mo steel	A213, A335, A369, A426, A691	A387, A517	...	A182, A234, A336
1.25Cr-0.5Mo steel	A199, A200(a), A213, A335, A369, A426, A691	A387, A389(a), A517	A217, A389(a)	A182, A234, A336, A541
2Cr-0.5Mo steel	A199, A200(a), A213, A369	...	...	...
2.25Cr-1Mo steel	A199, A213, A335, A369, A426, A691	A387, A542	A217, A487	A182, A234, A336, A541, A542
3Cr-1Mo steel	A199, A200(a), A213, A335, A369, A426, A691	A387	...	A182, A336
5Cr-0.5Mo steel	A199, A200(a), A213, A335, A369, A426, A691	A387	A217	A182, A234, A336
7Cr-0.5Mo steel	A199, A200(a), A213, A335, A369, A426	A387	...	A182, A234
9Cr-1Mo steel	A199, A200(a), A213, A335, A369, A426	A387	A217	A182, A234, A336
Ferritic, martensitic, and austenitic stainless steel	A213, A249, A268, A269, A271(a), A312, A358, A376, A409, A430, A451, A452, A511	A167, A176(a), A240, A412, A457	A297(a) A351, A447	A182, A336, A403, A473(a)

(a) These specifications are not approved by either the ANSI/ASME Boiler and Pressure Vessel Code or by the ANSI/ASME Code for Pressure Piping B31.3. Note: In addition, carbon and alloy steel bolts and nuts are covered by A193, A194, A320, A354, A449, A453, A540, and A563.

treatment, or welding operations, hardness (or, more properly, the resultant metallurgical structure of the steel) can have a distinct effect on the suitability of steel for a particular application or service environment. Although carbon steel normally has low hardness values, cooling from elevated temperatures, such as those encountered during mill processes such as quenching or normalizing or during fabrication or repair welding, may result in changes in hardness, hardness gradients, or localized hard zones.

Existing refinery experience indicates that if hardness values exceed 200 HB, carbon steel may be subject to cracking in aqueous alkaline sour water environments. This case is typical for susceptibility to SCC in pressure vessel weldments. For this reason, it is often necessary to set a maximum hardness limit for carbon steel and welds used in pressure vessels. In some cases, it is desirable to impose a uniform hardness limitation on all pieces of fabricated equipment, because the originally intended service application may be changed at some future date to one in which the component would be exposed to an aqueous sulfide environment. A high hardness value is also indicative of an increase in tensile strength and a corresponding decrease in ductility. Commonly produced plate steels (not having special thermomechanical processing used to enhance properties) with excessively high hardness values may also be expected to behave in a brittle manner under extreme conditions of low temperature, presence of pre-existing cracking, and/or rapid loading. Low-alloy steels often require postweld heat treatment after welding to reduce hardness in the weld area and to reduce the stresses associated with welding.

In some cases, hardness is not a good indicator of crack susceptibility. This is the case when it comes to resistance to HIC of steels in wet H_2S environments. This type of cracking occurs as a result of hydrogen absorption into the steel from sulfide corrosion that collects at internal interfaces in the material, mostly around inclusions. Susceptibility to HIC actually is greater in low-strength (low-hardness) steels and is related to the sulfur content and rolling practice used in making the steel, and its resultant sulfide inclusion morphology.

Fatigue Strength. Certain components, such as compressors or pumps, require materials with good fatigue-resistance properties. Fatigue resistance also needs to be considered when bolting and piping materials are selected. Fatigue resistance is the ability of a load-carrying component to resist fracture from cycles of repeatedly applied forces, such as vibrational or rotational stresses. One common rule of thumb is to limit the average fatigue stress to approximately one-half the ultimate tensile strength of the material involved. Because fatigue involves crack formation, crack propagation, and residual load-carrying capabilities, several mechanical properties are involved in determining fatigue resistance. Obviously, higher strength will help a material resist crack initiation. However, in many cases, materials and equipment have cracklike indications or defects that can act as pre-existing cracks. In these cases, other factors need to be considered in assessing resistance to fatigue crack propagation and ultimate failure. These factors include the fracture toughness of the steel, which is often related to the cleanliness of the material, and how it responds under plastic strain. Material with a low inclusion content and fine grains is commonly found to have high fatigue resistance, because the inclusions can be a source for internal crack initiation (or propagation), while fine grains will assist in allowing the crystallographic planes to slip without concentrating the associated strain.

Low-Temperature Properties. Carbon steel begins to lose its toughness and ductility as service temperatures decrease below ambient. Because most equipment in refineries and petrochemical plants is made of carbon steel, insufficient low-temperature toughness could represent a potentially serious problem. Fortunately, few operations are carried out at low temperatures, and most equipment made of carbon steel operates at temperatures ranging from ambient to approximately 425 °C (800 °F). However, equipment is normally exposed to ambient temperature during shutdown, which can pose a problem if the steel has inadequate low-temperature toughness. Certain refinery and petrochemical plant equipment and processes involved in cryogenic processes require special low-temperature toughness grades of steels, including liquefied-propane storage, ammonia storage, solvent dewaxing units, and liquefied petroleum gas processing. It is possible, by specifying certain additional composition and metallurgical processing requirements, to obtain carbon and low-alloy steels that are suitable for temperatures as low as −45 °C (−50 °F), depending on thickness. To resist brittle fracture at lowered temperatures, steels should be fully killed, fine grained, normalized, and should have received postweld heat treatment. Additionally, low-temperature toughness is also improved through alloying additions of nickel and vanadium and/or through use of thermomechanically controlled processed steels.

Typical ASTM International standard specifications for carbon steels with enhanced ability to perform at low temperatures are given in Table 8. Steels alloyed with 2 to 9% Ni and austenitic stainless steels can extend the range of available notch-tough steels to even lower temperatures. The simplest quality-control test (although not always adequate) for ensuring

Table 5 Short-term, elevated-temperature yield strengths for refinery steels

Test temperature		0.2% yield strength					
		Carbon steel		C-0.5Mo steel		2.25Cr-1Mo steel	
°C	°F	MPa	ksi	MPa	ksi	MPa	ksi
25	80	248	36.0	276	40.0	272	39.5
150	300	208	30.2	241	34.9	247	35.8
260	500	192	27.8	212	30.7	238	34.5
370	700	175	25.4	190	27.6	234	34.0
480	900	148	21.5	175	25.4	193	28.0

Table 6 Creep resistance of refinery steels extrapolated to 100,000 h

Test temperature		Stress for creep rate of 1%					
		Carbon steel		C-0.5Mo steel		2.25Cr-1Mo steel	
°C	°F	MPa	ksi	MPa	ksi	MPa	ksi
425	800	95	13.8	150	21.8	...	...
480	900	41	6.0	98	14.2	152	22.0
540	1000	18	2.6	43	6.2	55	8.0

Table 7 Suggested maximum temperatures for continuous service based on creep or rupture data

Material	Maximum temperature based on creep rate		Maximum temperature based on rupture	
	°C	°F	°C	°F
Carbon steel	450	850	540	1000
C-0.5Mo steel	510	950	595	1100
2.25Cr-1Mo steel	540	1000	650	1200
Type 304 stainless steel	595	1100	815	1500
Alloy C-276 nickel-base alloy	650	1200	1040	1900

proper notch toughness is the Charpy V-notch impact test carried out at the minimum design temperature, or lower. A minimum value of approximately 20.5 J (15 ft·lbf) is the usual acceptance criterion. The Charpy test is designed to simulate failure of a pressure vessel, containing a fabrication- or service-induced cracklike defect, by rapid crack propagation (brittle failure) when stressed at low temperatures.

Embrittlement Phenomena. There are a number of environmental effects on the mechanical properties of low-alloy steels and stainless steels used for refinery and petrochemical plant construction that need to be considered. In almost all cases, the effect is one of embrittlement due to an increase in hardness or a reduction in the notch ductility of the material. Detailed information on embrittlement mechanisms and the resulting fracture appearance can be found in the article "Visual Examination and Light Microscopy" in *Fractography*, Volume 12 of *ASM Handbook,* 1987. Embrittlement phenomena are also reviewed in the article "Embrittlement of Steels" in *Properties and Selection: Irons, Steels, and High-Performance Alloys,* Volume 1 of *ASM Handbook,* 1990.

Temper Embrittlement. This phenomenon causes a significant increase in the ductile-to-brittle transition temperature of low-alloy steels containing 1 to 3% Cr that are exposed to above 370 to 540 °C (700 to 1000 °F) for a prolonged period of time. Brittle failure at weld defects can occur when process equipment made from these steels is fully pressurized during start-up or shutdown. Therefore, pressure should be limited to 25% of design when temperatures are below 150 °C (300 °F) (Ref 29). Ideally, equipment made from steels that have become temper embrittled should be preheated to above 120 °C (250 °F) before pressurization following a shutdown. Temper embrittlement is caused by the segregation of residual steel elements to the grain boundaries, and this greatly reduces the intercrystalline (i.e., grain-boundary) strength. Limiting the acceptance levels of such elements as manganese, silicon, phosphorus, tin, antimony, and arsenic can improve the temper embrittlement resistance of 2.25Cr-0.5Mo, 2.25Cr-1Mo, and higher-alloyed chromium-molybdenum steels. Frequent nondestructive testing of major weld seams is recommended to determine if equipment has become embrittled.

885 °F (475 °C) Embrittlement. This phenomenon occurs in ferritic or martensitic stainless steels containing 12% or more chromium. It is manifested in material or equipment after long-term exposure to temperatures between 400 and 540 °C (750 and 1000 °F), hence the name 885 embrittlement. Heat treatment at approximately 620 °C (1150 °F), followed by rapid cooling, will usually restore ductility to embrittled stainless steels.

Sigma-Phase Embrittlement. The incidence of this phenomenon can occur in austenitic stainless as well as in straight-chromium (ferritic or martensitic) stainless steels as a result of prolonged high-temperature exposure. Sigma phase is nearly devoid of mechanical toughness and has a large influence on overall mechanical properties even at very low volume fractions relative to the bulk material. Of the austenitic stainless steels, the most susceptible compositions contain approximately 25% Cr and 20% Ni. The straight-chromium steels that are most susceptible to σ-phase formation contain 17% or more chromium. Sigma-phase formation increases room-temperature tensile strength and hardness while decreasing ductility to the point of extreme brittleness. As a result, cracks are very likely to develop during cooling from operating temperatures. Sigma phase most commonly forms in equipment operating in a temperature range of 650 to 750 °C (1200 to 1400 °F). Because σ-phase can be dissolved at temperatures above 980 °C (1800 °F), the original properties of stainless steels can be restored by a suitable heat treatment.

Creep embrittlement is the stress-dependent embrittlement of low-alloy steels operating in the creep range. The result is a reduction in the stress-rupture ductility. Creep embrittlement is commonly caused by the formation of precipitates within the grains, which results in the grain boundaries being softer than the material within the grains. Therefore, strain is concentrated in the grain boundaries, leading to premature failure. Detrimental effects can be eliminated by annealing the steel.

Fabricability

With very few exceptions, process equipment and piping are fabricated by welding wrought steels. The shells of pressure vessels are usually made from rolled plate, while nozzles are forgings. This requires that the steels have sufficient ductility for forming and are readily weldable. Weldability of steels is important not only for initial fabrication but also for future field repairs or modifications. Weld repairs and postweld heat treatments can affect the mechanical properties of wrought components that have been processed by normalizing or quenching and tempering. This can leave the wrought material with a lower strength than expected, based solely on its mill processing and composition.

Welding may result in certain other problems. Hydrogen coming from moisture in certain weld consumables or during nonoptimal field welding conditions can become dissolved in liquid weld metal. This dissolved hydrogen can cause cracking during solidification, as well as embrittlement of the weld. Dissolved hydrogen in the material as a result of exposure to refinery wet H_2S service environments can also affect weldability and the subsequent performance of repair welds.

The risk of hydrogen cracking of weldments is reduced by the use of low-hydrogen electrodes, careful drying of electrodes, and close control of pre- and postweld heat treatments. Equipment exposed to wet H_2S service environments often needs to be baked prior to weld repairs to reduce or remove the accumulated hydrogen, which can lower weldability. In some severe cases, this can result in cracks being formed as a result of HTHA (Ref 30). Information on the effect of bake-out treatments and other wet H_2S repair techniques is presented later in this article in the section "Wet H_2S Cracking."

Stress-relief or reheat cracking is intergranular cracking in the weld HAZ. The HAZ cracking occurs when weldments are heated during postweld heat treatment, or it occurs by subsequent exposure to elevated service temperatures. Low-alloy steels are especially susceptible to the aforementioned phenomena, but hydrogen cracking can occur with any of the ferritic steels if proper care is not taken.

Corrosion Resistance

The effects of the environment need to be considered when materials of construction are specified. General corrosion (uniform metal loss) is the easiest form of metal deterioration that can be considered in the design phase, because additional metal can be provided in the form of a corrosion allowance. It is also the easiest form of corrosion that can be detected by nondestructive testing techniques, because it is manifested to generally the same extent in all locations. Corrosion allowances in most cases range between 1 and 3 mm (0.04 and 0.12 in.), depending on the severity of the anticipated service conditions. In some particularly severe cases, local corrosion allowances can be increased to 5 mm (0.2 in.) or more where known impingement or turbulence is involved.

Table 8 ASTM International standard specifications for carbon steel with enhanced resistance to brittle fracture at lowered temperatures

Product form	Temperature/steel	
	To −30 °C (−20 °F)	To −45 °C (−50 °F)
Plate	A516, normalized (may require impact testing)	A516 normalized, stress relieved and Charpy impact tested
Pipe	A524	A333 grade 1 and grade 6
Tube	A210	A334 grade 1 and grade 6
Forgings	A727 and A350 grade LF1	A350 grade LF2
Fittings	A420 WPL6	A420 WPL6
Castings	A352 grade LCA	A352 grade LCB and grade LCC

In the case of pitting corrosion, it is possible to provide a pitting allowance. Because metal loss due to general corrosion is often not significant under pitting conditions, this approach would represent a rather expensive method of protecting equipment. Instead, it is generally more practical and advantageous to avoid process conditions that produce pitting or to change to a material that will not pit.

Until relatively recently, it was not possible to specifically identify pitting corrosion through corrosion monitoring except by after-the-fact coupon examination or inspection. However, newer corrosion-monitoring methods involving use of the electrochemical noise techniques and real-time data acquisition via control systems have made it possible to differentiate pitting activity from general corrosion and to identify these operating conditions before substantial damage has occurred. Figure 2 shows corrosion monitoring with pitting data obtained with multiple monitoring techniques (linear polarization resistance, electrochemical noise, and harmonic distortion analysis). It shows conditions involving cooling-water service where the general corrosion rate was lowered through chemical treatment and was under control, but where the susceptibility to pitting remained at critical levels. These results were later verified through visual examination of the probe elements, which were segments of cooling heat-exchanger tubing (Ref 31).

Stress-corrosion cracking is one of the most serious forms of metal deterioration, because it can result in extensive cracking and the catastrophic failure of equipment or considerable losses to an operating facility. Stress-corrosion cracks are very difficult to detect through online or offline inspection, because they can occur locally (often at areas of high residual or applied tensile stress) and are not uniformly distributed over the metal surfaces. Austenitic stainless steels are highly susceptible to SCC in aqueous, chloride-containing environments and consequently are often avoided for the primary pressure boundary of components. They are often used as protective, internal linings that may fail but where the integrity of the equipment is not completely compromised.

Corrosion

For practical purposes, corrosion in refineries and petrochemical plants can be classified into two types of corrosion (Ref 32, 33):

- Low-temperature (aqueous) corrosion occurring below approximately 260 °C (500 °F) in the presence of water
- High-temperature (nonaqueous) corrosion ocurring above approximately 205 °C (400 °F) in the presence of liquid or gaseous hydrocarbons

Carbon steel can be used to handle most hydrocarbon streams, except where corrosion is induced by the presence of contaminants present in the refined hydrocarbons or induced by upset conditions. In most cases, aqueous corrosion results from the presence of hydrogen chloride, H_2S and ammonia, as impurities in the feedstock, oxygen from air ingress into process equipment and tanks, lack of control of pH, oxygen or other species in cooling water, or from various related chemicals such as amine solvents or strong caustics and acids. These conditions often necessitate selective application of more resistant alloys. For high-temperature corrosion to take place, the presence of water is not necessary. Corrosion can occur by the direct reaction between metal and impurities such as H_2S, sulfur species, and naphthenic acid in the environment (Ref 32–35).

Low-Temperature Corrosion

Most corrosion problems in refineries are not caused by hydrocarbons but, as briefly mentioned previously, by various associated chemical compounds, namely, water, H_2S, hydrochloric acid, hydrofluoric acid, sulfuric acid, caustics, and amine solvents (Ref 36). There are two principal sources of these compounds: feed-stock contaminants and process chemicals, including solvents, neutralizers, and catalysts. Generally, the same applies to corrosion problems in petrochemical plants, except that corrosion is also caused by organic acids, such as acetic acid, that may be used as solvents. In addition, corrosion problems are caused by the atmosphere (oxidation), cooling water, boiler feed water, steam condensate, and soil.

Low-Temperature Corrosion by Feed-Stock Contaminants

The major cause of low-temperature (and, for that matter, high-temperature) refinery corrosion is the presence of contaminants in crude oil as it is produced. Although some contaminants are removed during preliminary treating upstream of the refinery, some contaminants end up in refinery tankage, along with contaminants picked up in pipelines or marine tankers. Corrosives can also be formed during initial refinery operations. For example, potentially corrosive hydrogen chloride evolves in crude preheat furnaces from relatively harmless calcium and magnesium chlorides entrained in crude oil (Ref 37). In petrochemical plants, certain corrosives may have been introduced from upstream refinery and other process operations. Other corrosives, such as sulfur compounds, can form by conversion of corrosion products after exposure to air during shutdowns (i.e., polythionic acid). The following discussion focuses on the most important crude oil contaminants that have caused corrosion problems.

Air. During shutdowns or turnarounds, most plant equipment is exposed to air. Air also can enter the suction side of pumps if seals or connections are not tight, or in cases where tanks are vented to the atmosphere or develop negative pressure. In general, the air contamination of hydrocarbon streams has been more detrimental with regard to fouling than corrosion. However, air contamination has been cited as a cause of accelerated corrosion in vacuum transfer lines and vacuum towers of crude distillation units. Air contamination has supposedly increased the overhead corrosion of crude distillation towers, but this has been difficult to reconcile with the fact that oxygen in air reacts with H_2S to form polysulfides, which tend to inhibit corrosion. In aqueous systems, air contamination also leads to

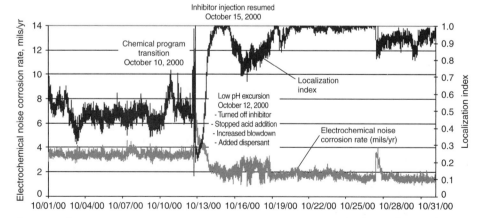

Fig. 2 Real-time monitoring of chemical injection in a cooling-water system with online corrosion rate and pitting tendencies (localization index). October 10—Chemical treatment program transition: Inhibitor was turned off, acid addition was stopped, blowdown was increased, and dispersant was added. October 12—A low-pH event caused an increase in corrosion rate and a change in corrosion mechanism. Within 24 h, pit initiation and propagation were indicated. October 15—Inhibitor injection resumed: Although inhibitor injection was restarted, complete passivation was not achieved. A short-term reduction in localized corrosion activity was observed, but this increased again. Two weeks following the low-pH event, significant localized corrosion activity was still indicated. The pitting was later confirmed by analysis of the tubular flow-through probe elements.

excessive pitting, particularly when the environment also contains either H_2S or chloride salts.

Water. Water is found in all crude oils to a certain degree and is difficult to remove completely. In addition, water originates with stripping steam for fractionation towers and is produced in hydrotreating operations. Water not only functions as an electrolyte but also hydrolyzes certain inorganic chlorides to hydrogen chloride, which, in turn, functions as a corrodant as well. Water is primarily responsible for various forms of corrosion in fractionation tower overhead systems. In general, whenever equipment can be kept dry through suitable process or equipment changes, corrosion problems will be minimized. Furthermore, additional "wash" water can be added to a process stream, further diluting various potentially corrosive species to acceptable levels.

The combination of water and air can be especially detrimental. Moisture and air are drawn into storage tanks during normal "breathing" as a result of pumping and changes in temperature. Tank activity and corrosion are closely interrelated. Because crude and heavy oils form a protective oil film on the working areas of a tank shell, corrosion is generally limited to the top shell ring and the underside of the roof. Tank bottom corrosion occurs mostly with crude oil tanks and is caused by water and salt entrained in the crude oil. A layer of water usually settles out and can become highly corrosive. Alternate exposure to sour crude oils and saltwater causes especially severe corrosion (Ref 38). Mill scale tends to accelerate tank bottom corrosion, because cracks in the mill scale form anodic areas that pit, while the surrounding area acts as the cathode.

Light hydrocarbon stocks do not form protective oil films, and corrosion occurs primarily at the middle shell rings, because these are exposed to more wetting and drying cycles than other tank areas (Ref 39). Corrosion is in the form of pitting under globules of water that attach themselves to the tank wall. Pitting can become so extensive that metal loss generally appears as more or less uniform corrosion. The rate of corrosion is proportional to the water and air content of light stocks. Contamination from chloride and H_2S also accelerates attack.

Hydrogen Sulfide. Sour crude oils and gases that contain H_2S are handled by most refineries (Ref 40). Hydrogen sulfide is also present in some feed stocks handled by petrochemical plants. During processing at elevated temperatures, H_2S and many other sulfur species are also formed by the decomposition of organic sulfur compounds that are present. Corrosion of steel by H_2S forms the familiar black iron sulfide film seen in almost all refinery equipment (Ref 41). Hydrogen sulfide with ammonia are the main constituents of refinery sour waters that can cause severe corrosion problems in overhead systems of certain fractionation towers, in hydrocracker and hydrotreater effluent streams, in the vapor recovery (light ends) section of fluid catalytic cracking units, in sour water stripping units, and in sulfur recovery units (Ref 42, 43). These are discussed in greater detail in the section "Sour Water" in this article.

In general, carbon steel has fairly good resistance to aqueous sulfide corrosion, because a protective iron sulfide film is formed (Ref 44). However, conditions of air ingress, chlorides, cyanides and/or high flow-induced turbulence can remove normally protective films to increase the susceptibility to general and/or pitting corrosion, local wall loss, and hydrogen charging of the steel.

To avoid SSC in a wet H_2S system, hard welds (above 200 HB) need to be avoided. If necessary, suitable postweld heat treatment may need to be performed (Ref 45). Excessive localized corrosion in vessels has been resolved by selective lining with alloy 400 (N04400), but this alloy can be less resistant than carbon steel or other corrosion-resistant alloys to aqueous sulfide corrosion at temperatures above 100 °C (212 °F). If significant amounts of chlorides are not present, lining vessels with type 405 (S40500) or type 304 (S30400) stainless steels can be considered. Ti-grade 2 (R50400) tubes have been used as replacements for carbon steel tubes to control aqueous sulfide corrosion in heat exchangers at a number of units (Ref 46, 47).

Hydrogen Chloride. In refineries, corrosion by hydrogen chloride is primarily a problem in crude distillation units and, to a lesser degree, in reforming and hydrotreating units. In petrochemical plants, hydrogen chloride contamination can be present from certain feed stocks (from inorganic salts or organic chlorides from chlorinated solvent contamination) or can be formed by the hydrolysis of aluminum chloride catalyst.

In most production wells, chloride salts are found either dissolved in water, that is, emulsified in crude oil, or as suspended solids. Salts also originate from brines injected for secondary recovery or from seawater ballast in marine tankers. Typically, the salts in crude oils consist of 75% sodium chloride, 15% magnesium chloride, and 10% calcium chloride (Ref 48). When crude oils are charged to crude distillation units and heated to temperatures above approximately 120 °C (250 °F), hydrogen chloride is evolved from magnesium and calcium chloride, while sodium chloride is essentially stable up to roughly 760 °C (1400 °F). Hydrogen chloride evolution takes place primarily in crude preheat furnaces. Dry hydrogen chloride, especially in the presence of large amounts of hydrocarbon vapor or liquid, is not corrosive to carbon steel (Ref 49).

When steam is added, however, to the bottom of the crude tower to facilitate fractionation, dilute hydrochloric acid forms in the top of the tower and in the overhead condensing system. Severe aqueous chloride corrosion of carbon steel components can occur upon cooling at temperatures below the initial water dewpoint (Ref 50). Corrosion rate increases with a decrease in pH value of overhead condensate water. Corrosion is mostly in the form of droplet-impingement attack at elbows of the overhead vapor line and at inlets of overhead condensers. Corrosion also occurs on condenser tubes that are at the temperatures where most of the water condenses out. Often, droplets of hydrochloric acid become entrapped and concentrate under deposits that are present on tower trays, in condenser shells, and at baffles. The resultant underdeposit corrosion is highly localized and usually quite severe.

Overhead condensing systems of both the crude and vacuum towers of crude distillation units are generally made from carbon steel. Coolers and condensers in cooling-water service usually use admiralty metal (C44300) tubes to reduce corrosion and fouling on the water side. Where aqueous chloride corrosion is a problem on the process side, Ti-grade 2 (R50400) tubes should be considered. The top of the crude tower can be lined with alloy 400 (N04400), and tray components made from alloy 400 (N04400) can be used for the upper five or so trays to combat aqueous chloride corrosion. Alloy 400 (N04400) tubes usually have not been cost-effective in overhead coolers and condensers, but alloy 400 (N04400) has been successfully used for selective strip lining of those areas of the overhead system where excessive corrosion occurs despite the implementation of other corrosion control measures.

To minimize aqueous chloride corrosion in the overhead system of crude towers, it is best to keep the salt content of the crude oil charge as low as possible, preferably below 1 pound per thousand barrels (PTB), corresponding to roughly 4 ppm. This is done by proper tank-settling, desalting, or, if necessary, double desalting (Ref 51–53). Another way to reduce overhead corrosion would be to inject a caustic solution of sodium hydroxide into the crude oil downstream of the desalter. Up to 3 PTB (10 ppm) caustic can usually be tolerated from a process point of view, while higher concentrations increase fouling of crude preheat exchangers, boiler corrosion by sodium vanadate (when reduced crude is burned as boiler fuel), or coking in lines and heaters of coking units (Ref 54). Caustic should not be used when reduced crude is charged directly to catalytic cracking or hydrotreating units because of possible catalyst deactivation.

Neutralizers are also injected into the overhead vapor line of the crude tower to maintain the pH value of stripping steam condensate between 5 and 6 (Ref 55–58). A pH value above 7 can increase corrosion with sour crudes, as well as fouling and underdeposit corrosion by neutralizer chloride salts. Where fouling becomes a problem, water should be injected, either intermittently or continuously, to dissolve salt deposits in those areas of the overhead system that are not exposed to stripping-steam condensate (Ref 59). Filming-amine corrosion inhibitors can be injected into the overhead vapor line to provide additional insurance against excessive corrosion (Ref 60–62).

In the downstream refining equipment, chlorides accelerate corrosion by penetrating protective surface films, increasing electrolyte conductivity, or complexing with steel surfaces (Ref 63). In reforming units, organic chlorides are often used to regenerate reformer catalyst. Hydrogen chloride is stripped off the catalyst if excessive moisture is present in the reformer feed; this causes increased corrosion, not only in reforming units but also in hydrotreating units that use excess hydrogen (make-gas) from the reformer. As in the case of crude distillation units, water washing and injection of neutralizers and/or filming-amine corrosion inhibitors can be used to control fouling and corrosion by chloride salts. Hydrogen make-gas can be passed through a water scrubber to remove hydrogen chloride. Selective alloying with alloy 825 (N08825), alloy 400 (N04400), or Ti-grade 2 (R50400) can be required to control chloride attack in heat exchangers and separator drums.

Nitrogen Compounds. Organic nitrogen compounds, such as indole, carbuzole, pyridine, or quinoline, are present in many crude oils but do not contribute to corrosion problems unless converted to ammonia or hydrogen cyanide in refining processes (Ref 64). This occurs primarily in catalytic cracking, hydrotreating, and hydrocracking operations where ammonia and hydrogen cyanide, in combination with H_2S and other constituents, become the major constituents of sour water that can be highly-corrosive to carbon steel (Ref 65). Furthermore, this corrosive combination can also result in severe hydrogen charging of steel equipment, resulting in high susceptibility to wet H_2S cracking (i.e., SSC, HIC, and SOHIC). These are discussed in greater detail in the following section, "Sour Water," in this article.

Ammonia is also produced in ammonia plants to become a raw material for the manufacture of urea and other nitrogen-base fertilizers. Ammonia in synthesis gas at temperatures between 450 and 500 °C (840 and 930 °F) causes nitriding of steel components. When synthesis gas is compressed to up to 34.5 MPa (5000 psig) prior to conversion, corrosive ammonium carbonate is formed, requiring various stainless steels for critical components. Condensed ammonia is also corrosive and can cause SCC of stressed carbon steel and low-alloy steel components (Ref 66).

Sour Water. The term *sour water* denotes various types of process water containing primarily H_2S, ammonia, and hydrogen cyanide, often in combination with certain organic compounds, including phenols, mercaptans, and possibly inorganic compounds such as chlorides and fluorides. Sour waters are removed from processed hydrocarbons in refining units by settling in overhead reflux drums, separator drums, water coalescer drums, and other specialized equipment. Depending on their exact composition, sour waters can become highly corrosive, and, under certain circumstances, they even create conditions of turbulent multiphase flow. Consequently, alloy selection and operational limits of sour water systems need to be carefully considered for specific unit operating conditions. Recently, a joint industry program conducted a significant research effort on sour water (ammonium bisulfide) corrosion, examining the interrelationship between many variables common to refinery sour water systems (Ref 5).

Sour water corrosion is of particular concern in the vapor recovery (light ends) section of catalytic cracking units and in reactor effluent and light ends sections of hydrotreating and hydrocracking units, in which high concentrations of ammonia can saturate process water with ammonium bisulfide and cause serious corrosion of carbon steel components. Furthermore, the combination of ammonium bisulfide with cyanides produced in the hydrocraking process can result in extremely corrosive conditions. The presence of cyanide complexes normally protective sulfide films, leading to accelerated corrosion.

Ammonium bisulfide will also rapidly attack admiralty metal (C44300) tubes and can also corrode nickel-copper alloys such as alloy 400 (N04400). Sour water corrosion is a major problem in some sour water stripping units, in which exceptionally high concentrations of ammonium bisulfide build up in the thin film of condensed water on overhead condenser tubes. The resultant corrosion can be so severe that even tubes made from conventional austenitic stainless steels are attacked. Currently, there has been use of duplex stainless and high-alloyed austenitic stainless steels with PREN values from 30 to over 40, depending on the anticipated severity of the environment. In some cases, Ti-grade 2 (R50400) tubes have been used to provide sufficient resistance for use in this service. The results of the previously mentioned joint industry research program on sour water corrosion have developed a list of alloys and their relative resistance to various sour water conditions. This work has identified the key process conditions and defined limits for commonly used alloys through a combination of experimental (chemical) simulation and flow modeling to relate the data to refinery systems involving multiphase and balanced/imbalanced flow conditions (Ref 5).

Normally, all components in the vapor recovery (light ends) sections of catalytic cracking units are made of carbon steel. Exceptions to this rule include tower internals made of type 405 (S40500) or 410 (S41000) stainless steel and tubes in overhead condensers and compressor aftercoolers made from admiralty metal (C44300), alloy 400 (N04400), or Ti-grade 2 (R50400). Corrosion problems of carbon steel components are often closely associated with hydrogen blistering or other forms of wet H_2S cracking, because sulfide corrosion liberates atomic hydrogen that can diffuse into the steel. Admiralty metal (C44300) tubes in overhead condensers may typically last only 5 years, with leaks finally occurring as a result of ammonia SCC. Depending on the particular process conditions, admiralty metal (C44300) tubes can also corrode by severe localized attack. Admiralty metal (C44300) tubes in compressor aftercoolers are often replaced with Ti-grade 2 (R50400) tubes or other materials.

Perhaps the biggest recurring problems in sour water systems have been corrosion, hydrogen blistering, and hydrogen cracking phenomena such as HIC, SOHIC, and SSC of carbon steel in coolers, separator drums, absorber/stripper towers, and, occasionally, overhead condensers at a number of locations. These are discussed in greater detail in the section "Environmentally Assisted Cracking (SCC, HEC, and Other Mechanisms)" in this article.

Components in hydrotreating and hydrocracking units that operate at temperatures below approximately 260 °C (500 °F) are typically made from carbon steel. Where aqueous ammonium bisulfide corrosion becomes a problem, generous corrosion allowances may have to be provided for carbon steel (Ref 67, 68). Selective use of duplex stainless steels, such as alloy 2205 (S31803) and alloy 2507 (S32750), superaustenitic stainless steels, such as alloy AL6XN (N08367), alloy 825 (N08825), alloy C-276 (N10276), alloy 400 (N04400), or Ti-grade 2 (R50400) may be required for heat exchangers and separator drums to control excessive corrosion. On some units, corrosion of steel components is accompanied by hydrogen blistering and cracking. Hydrotreating and hydrocracking units that experience fouling problems due to ammonium sulfide or ammonium chloride deposition may require intermittent or continuous water injection (i.e., water washing) to dissolve these salt deposits and reduce the concentration of these constituents. It is of prime importance, however, that sufficient coalescer capacity be available or provided in order to ensure that the injected water is removed. Otherwise, serious corrosion can occur when the water ends up in downstream equipment.

All equipment and piping of reforming units that operate at below approximately 260 °C (500 °F) are usually made from carbon steel. Although admiralty metal (C44300) tubes are often used in water-cooled effluent coolers and condensers, the presence of chlorides may necessitate the selective use of Alloy 400 (N04400) or Ti-grade 2 (R50400). In some cases, carbon steel tubes are superior to admiralty metal (C44300) tubes, provided the cooling water is properly treated. Similar considerations apply to water-cooled coolers and condensers in the overhead systems of prefractionator, splitter, debutanizer, and other fractionation towers. Filming-amine corrosion inhibitors can be used to help control overhead corrosion of steel components.

The principal material of construction for sour water stripping units is carbon steel. There are several varieties of sour water strippers, but nonacidified condensing and noncondensing strippers are most commonly used (Ref 69, 70). The stripping medium is primarily steam. Stripper towers are generally made from carbon steel with type 316 (S31600) stainless steel, aluminum, or carbon steel internals, depending on corrosion experience. To control tower corrosion, a minimum top temperature of 80 °C (180 °F) is required. Below this temperature, H_2S will concentrate in the upper part of the tower but will not be carried overhead. Feed charge pumps are usually made from cast iron or cast steel, including the impellers. Feed piping, bottoms piping, and the feed/bottoms heat exchanger can be made from carbon steel. Carbon steel has also been satisfactory for stripper reboilers that may be used instead of live stripping steam. Thermosyphon reboilers (with sour water in the tubes) are recommended over kettle reboilers because the latter are often prone to fouling and resultant underdeposit corrosion.

Most corrosion problems have been in overhead condensers of condensing sour water strippers (Ref 71). Although a variety of alloys have been used for overhead condenser tubes, only aluminum and Ti-grade 2 (R50400) can be relied on to provide adequate resistance to the highly corrosive conditions encountered in many overhead systems. Carbon steel is usually satisfactory for the overhead vapor line, condenser shell, rundown lines, accumulator drum, and reflux lines. All welds in these components should be postweld heat treated to avoid SSC or related cracking problems. Reflux pumps can be made of carbon steel or type 304 (S30400) stainless steel, but for optimal performance, alloy 20 (N08020) is recommended. Hydrogen blistering often accompanies corrosion in overhead condenser shells and reflux drums. Water-soluble filming-amine corrosion inhibitors can be injected into the overhead vapor line to help control both corrosion and hydrogen blistering. Few, if any, corrosion problems have been experienced with noncondensing sour water strippers.

Serious sour water corrosion of carbon steel components can occur in the overhead system of amine regenerators (strippers) of gas-treating or sulfur recovery units, especially if all of the water condensate is returned to the tower as reflux. Corrosion is usually accompanied by hydrogen blistering or cracking. Continuous or periodic blowdown of sour water to the sour water stripping unit should be employed to lower the concentrations of H_2S, ammonia, and cyanide in the overhead water condensate. If this fails to control corrosion, carbon steel condenser tubes may have to be replaced with Ti-grade 2 (R50400) tubes. In addition, corrosion can be minimized by operating the regenerator so that roughly 0.5% amine is taken overhead to act as a corrosion inhibitor.

Polythionic Acids. Combustion of H_2S in refinery flares can produce polythionic acids of the type $H_2S_xO_y$ (including sulfurous acid) and cause severe intergranular corrosion of flare tips made of stainless steels and high-nickel alloys (Ref 72). Corrosion can be minimized by using nickel alloys, such as alloy 825 (N08825) or alloy 625 (N06625). Polythionic acids also cause SCC during shutdown, as discussed in the section "Environmentally Assisted Cracking (SCC, HEC, and Other Mechanisms)" in this article.

Low-Temperature Corrosion by Process Chemicals

Severe corrosion problems can be caused by process chemicals, such as various alkylation catalysts, certain alkylation by-products, organic acid solvents used in certain petrochemical processes, hydrogen chloride stripped off reformer catalyst, and caustic and other neutralizers that, ironically, are added to control acid corrosion. Filming-amine corrosion inhibitors can be quite corrosive if injected undiluted (neat) into a hot vapor stream. Other process chemicals that can be corrosive, or become corrosive depending on the application, are amine solvents used in treating and gas-scrubbing operations.

Acetic Acid. Corrosion by acetic acid can be a problem in petrochemical process units used for the manufacture of certain organic intermediates, such as terephthalic acid. Various types of austenitic stainless steels are used. Nickel-base alloys such as alloy C-4 (N06455) and alloy C-276 (N10276), or titanium, are used to control corrosion by acetic acid in the presence of small amounts of hydrogen bromide or hydrogen chloride.

As a rule, even tenths of a percent of water in acetic acid can have a significant influence on corrosion. Type 304 stainless steel (S30400) usually has sufficient resistance to the low concentrations of acetic acid up to the boiling point. Higher concentrations can also be handled by type 304 stainless steel (S30400) if the temperature is below approximately 90 °C (190 °F). Increasing the chromium and/or nickel content has little effect on resistance to acetic acid. Addition of molybdenum in combination with nickel and chromium in stainless steel, however, markedly increases the resistance of these materials. Consequently, type 316 (S31600) and type 317 (S31700) stainless steels with approximately 2 to 4% Mo in combination with nickel and chromium are used for the overwhelming majority of hot acetic acid applications. Corrosion by acetic acid increases with temperature. Bromide and chloride contamination causes pitting and SCC, while addition of oxidizing agents, including air, can reduce corrosion rates by several orders of magnitude.

Aluminum Chloride. Certain refining and petrochemical processes, such as butane isomerization, ethylbenzene production, and polybutene production, use aluminum chloride as a catalyst (Ref 73). Aluminum chloride is not corrosive if it is kept absolutely dry. If traces of water or water vapor are present in hydrocarbon streams, aluminum chloride hydrolyzes to hydrochloric acid, which can, of course, be highly corrosive. To control corrosion in the presence of aluminum chloride, the feed is dried in calcium chloride dryers. During shutdowns, equipment should be opened for the shortest possible time. Upon closing, it should be dried with hot air, followed by inert gas blanketing. Equipment that is exposed to hydrochloric acid may require extensive lining with nickel alloys, such as alloy 400 (N04400), alloy B-2 (N10665), alloy B-3 (N10675), alloy C-4 (N06455), or alloy C-276 (N10276) (Ref 74).

Organic chlorides in crude oils will form various amounts of hydrogen chloride at the elevated temperatures of crude preheat furnaces, depending on the chlorides involved. Many "opportunity" crude oils contain small amounts of organic chlorides (5 to 50 ppm) naturally or through contamination with chlorinated organic solvents prior to receipt at the refinery. A major problem also exists due to contamination with organic chloride solvents during production. Although major producers are aware of the problem, some operators may still use chlorinated organic solvents to remove wax deposits in oil field tankage and associated equipment and piping. Spent solvent is then simply added to the crude oil. These solvents are also extensively used for metal-degreasing operations in and out of the refinery, but to a lesser extent in recent years due to environmental control. The problem is that spent solvent can be discarded with slop oil, which is added to the crude oil and charged to the crude distillation unit.

Contaminated crude oils have been found to contain as much as 7000 ppm chlorinated hydrocarbons. Such crude oils not only cause severe corrosion in the overhead system of crude distillation towers but also affect reformer operations (Ref 75). Typical process and operational problems in the latter category include runaway cracking, rapid coke accumulation on the catalyst, and increased corrosion in fractionator overhead systems (Ref 76, 77). Obviously, every effort must be made to avoid charging contaminated crude oil. Organic chlorides cannot be removed by desalting. If contaminated crude oil must be run off, the usual approach is to blend it slowly in very limited amounts into uncontaminated crude oil at levels that will not cause these problems.

Hydrogen Fluoride. Some alkylation processes use concentrated hydrofluoric acid instead of sulfuric acid as the catalyst. In general, hydrofluoric acid is less corrosive than hydrochloric acid because at high concentrations, it passivates steel by the formation of protective fluoride films. If these films are destroyed by diluted acid, impurities, or flow-induced turbulence, extremely severe corrosion occurs. Therefore, as long as feed stocks are kept dry, carbon steel—with various corrosion allowances—can be used for vessels, piping, and valve bodies of hydrofluoric acid alkylation units. Alloy 400 (N04400) is used selectively at locations where excessive corrosion has been experienced. A related problem to corrosion in hydrofluoric acid is the associated generation of atomic hydrogen. Absorption of atomic hydrogen can result in hydrogen blistering and cracking of carbon steel equipment and cracking of hardened bolts (Ref 78).

By following proper design practices and prescribed maintenance procedures and by diligently keeping feeds stocks and equipment dry, there will be few corrosion problems. All carbon steel welds that contact hydrofluoric acid should be postweld heat treated (Ref 79). This applies especially to welds in various vessels. Vessels should be radiographed to check for slag inclusions in plates and welds; slag inclusions are preferentially attacked by hydrofluoric acid. Hydrofluoric acid has the capability of finding leak paths via weld inclusions and porosity or threads. During welding, each preceding pass must be properly cleaned. All threaded connections should be seal welded. Where leaks do show up after start-up, small holes can often be peened shut, or small bits of copper or lead can be peened into larger holes to seal a leak. Any subsequent repair welds should also be postweld heat treated.

Fractionation towers should have type 410 (S41000) stainless steel tray valves and bolting; alloy 400 (N04400) tray valves and bolting are preferred for the deisobutanizer tower. The acid rerun tower usually requires cladding with alloy 400 (N04400) and alloy 400 (N04400) tray components. To avoid SCC, alloy 400 (N04400) welds that contact hydrofluoric acid should be postweld heat treated. No asbestos or wicking gaskets should be used on trays. Soft iron gaskets are used on channel head-to-shell joints of heat exchangers. Spiral-wound alloy 400/synthetic fluorine-containing resin gaskets are also used but are more expensive. Carbon steel U-tube bundles are preferred for all exchangers that contact hydrofluoric acid; alloy 400 (N04400) tubes have been found to offer few advantages (Ref 80). Tube ends and tubesheet holes should be carefully cleaned to ensure that rolled tube joints will be tight against hydrofluoric acid. Seal welding of tubes may be required. Internal bolting should not be used in exchangers.

The piping is generally carbon steel with welded connections that have received postweld heat treatment. All taps should be self-draining and should have double block valves. Instrument connections should be made from the top. Valve bodies on gate and plug valves are usually carbon steel, with synthetic fluorine-containing resin packing and seats. Relief valves should have alloy 400 (N04400) trim. Synthetic fluorine-containing resin tape sealing should be used on any threaded connections. Pumps in hydrofluoric acid service normally have carbon steel casings that are weld overlaid with alloy 400 (N04400). Impellers and sleeves should be alloy 400 (N04400); shafts should be alloy K-500 (N05500).

Specific areas where corrosion is likely to occur include the bottom of the acid rerun tower, the feed inlet areas of the deisobutanizer and depropanizer towers, the overhead condensers of these towers, the reboiler of the propane stripper, and piping around the acid rerun tower (Ref 81). Trouble areas in vessels are often selectively strip lined with alloy 400 (N04400). Dimpling of tray valve caps during manufacture reduces their tendency to stick to trays because of corrosion products. Alloy 400 (N04400) piping is used to replace carbon steel piping, which corrodes at excessive rates; welds should be postweld heat treated.

Experience has shown that most corrosion problems in hydrofluoric acid alkylation units occur after shutdowns, because pockets of water have been left in the equipment. This water is from the neutralization and washing operation required for personnel safety before the equipment can be opened for inspection. It is very important that equipment be thoroughly dried by draining all low spots and by circulating hydrocarbon before the introduction of hydrofluoric acid catalyst at start-up. Corrosion by hydrofluoric acid is occasionally accompanied by hydrogen blistering or cracking. Filming-amine corrosion inhibitors have been injected into the overhead systems of various towers, sometimes in conjunction with injection of dilute soda ash solutions. Because the primary goal of proper operations is to keep the unit as dry as possible, intentional addition of water in any form should be considered only as a last resort.

Sulfuric Acid. Certain alkylation units use essentially concentrated sulfuric acid as the catalyst; some of this sulfuric acid is entrained in reactor effluent and must be removed by neutralization with caustic and scrubbing with water. Acid removal may not be complete, however, and traces of acid—at various concentrations (in terms of water)—remain in the stream. Sulfuric acid can be highly corrosive to carbon steel, which is the principal material of construction for sulfuric acid alkylation units. Because the boiling point of sulfuric acid ranges from 165 to 315 °C (330 to 600 °F), depending on concentration, entrained acid usually ends up in the bottom of the first fractionation tower and reboiler following the reactor; this is where the entrained acid becomes concentrated.

Acid concentrations above 85% by weight are usually not corrosive to carbon steel if temperatures are below 40 °C (100 °F). Cold-worked metal (usually bends) should receive thermal stress relief. Under ideal operating conditions, few, if any, corrosion and fouling problems occur (Ref 82, 83).

Carbon steel depends on a film of iron sulfate for corrosion resistance, and if its film is destroyed by high velocities and flow turbulence, corrosion can be quite severe. For this reason, flow velocities should be below 1.2 m/s (4 ft/s). Attack in the form of erosion-corrosion can occur at piping welds that have not received postweld heat treatment. This highly localized attack immediately downstream of piping welds has been attributed to a spheroidized structure affected by the heat of welding; a normalizing postweld heat treatment at 870 °C (1600 °F) is required to minimize this type of corrosion (Ref 84). Velocity-accelerated corrosion can also be a problem at locations of high turbulence or velocity (Ref 85). Alloy 20 (N08020) is more resistant than carbon steel to this type of corrosion. In extreme cases, however, even alloy 20 (N08020) will be damaged by erosion-corrosion, and the selective use of alloy B-2 (N10665) and alloy B-3 (N10675) may be required.

Carbon steel valves usually require alloy 20 (N08020) internals or trim, because even slight attack of carbon steel seating surfaces is sufficient to cause leakage (Ref 86). Pump internals and injection and mixing nozzles in concentrated or spent sulfuric acid service are often made of alloy 20 (N08020), alloy B-2 (N10665), and alloy C-4 (N06455) or alloy C-276 (N10276). For hydrocarbon streams containing only traces of concentrated or dilute sulfuric acid, steel-body valves with type 316 (S31600) stainless steel trim can be used. In this service, steel pump casings that are weld overlaid with aluminum bronze have been successfully used. Pump impellers made from high-silicon cast iron are often used.

Piping for hydrocarbon/acid mixing lines ahead of the reactors may require alloy 20 (N08020), because water contamination of feed stocks can cause severe corrosion of carbon steel. Alloy 400 (N04400) has been found to be useful for reactor effluent lines around the caustic and wash-water injection points. Valve trays in fractionation towers require type 405 (S40500) or type 410 (S41000) stainless steel tray valves and bolting. In general, organic coatings are not resistant to concentrated sulfuric acid. Synthetic fluorine-containing resin has excellent resistance to sulfuric acid and is extensively used in gaskets, pump valve packing, and mixing nozzles.

In addition to sulfuric acid, reactor effluent contains traces of alkyl and dialkyl sulfates from secondary alkylation reactions (Ref 87). These esters decompose in reboilers to form sulfur dioxide and polymeric compounds (the latter are notorious foulants). Sulfur dioxide combines readily with water in the upper part and overhead system of fractionation towers; the resultant sulfurous acid can cause severe corrosion in overhead condensers. In some units, carbon steel or admiralty metal (C44300) tubes in overhead condensers, particularly those of the deisobutanizer tower, may have to be replaced with alloy 400 (N04400) or Ti-grade 2 (R50400) tubes. As a rule, however, titanium is not resistant to sulfuric acid corrosion. It can be used only under limited acid concentration and/or in the presence of oxidizing or inhibitive agents. Neutralizers can be injected into the overhead vapor lines of various towers to maintain the pH value of aqueous condensate near 7. Filming-amine corrosion inhibitors can also be injected.

Caustic. Sodium hydroxide is widely used in refinery and petrochemical plant operations to neutralize acidic constituents. At ambient temperature and under dry conditions, caustic can be handled in carbon steel equipment. Carbon steel is also satisfactory for aqueous caustic solutions between 50 and 80 °C (120 and 180 °F), depending on concentration. For caustic service above these temperatures but below approximately 95 °C (200 °F), carbon steel can also be used if it has been postweld heat treated to avoid SCC at welds. Austenitic stainless steels, such as type 304 (S30400), can be used up to approximately 120 °C (250 °F), while nickel alloys are required at higher temperatures. Figure 3 is the caustic soda serviceability chart that indicates the concentration and temperature limits for various materials in caustic service (Ref 88).

Severe caustic corrosion of the crude transfer line, which is immediately downstream of the caustic injection point, can occur in crude distillation units when 40% (by weight) caustic solution is injected into hot, desalted crude oil to neutralize any remaining hydrogen chloride. Predilution of the caustic with water to form a 3% (by weight) solution minimizes this problem. Better dispersion of the more diluted solution in the hot crude oil prevents local concentration and puddles of molten caustic from collecting along the bottom of the transfer line, which can result in corrosion and SCC problems. If caustic is injected too close to an elbow of the transfer line, impingement by droplets of caustic can also cause severe attack and hole-through at the elbow.

There are some unusual situations in which caustic corrosion is encountered. For example, traces of caustic can become concentrated due to local boiling or evaporation in boiler feed water and cause corrosion (gouging) and SCC (caustic embrittlement). This occurs in boiler tubes that alternate between wet and dry conditions (steam blanketing) because of overfiring. In some petrochemical processes, caustic gouging is found under deposits in heat exchangers that remove heat by generation of steam. For example, vertical heat exchangers for cracked gas in ethylene units are especially vulnerable if deposits are allowed to accumulate on the bottom tubesheet. Boiler feed water permeates these deposits and evaporates, and this causes the caustic to concentrate in any liquid that is left behind. The caustic content of such trapped liquid can reach several percent, which is sufficient to break down the normally protective iron oxide (magnetite) film on boiler steel and can easily result in severe caustic corrosion.

Amine Solvents and Neutralizers. Corrosion of carbon steel by amines in gas-treating and sulfur recovery units can usually be traced to faulty plant design, poor operating practices, and solution contamination (Ref 89). In general, corrosion is most severe in systems removing only carbon dioxide and is least severe in systems removing only H_2S. Systems handling mixtures of the two fall between these two extremes if the gases contain at least 1 vol% H_2S. Corrosion in amine plants using monoethanolamine is usually more severe than in those using diethanolamine, because the former is more prone to degradation and the formation of heat-stable salts at high temperatures.

The most common forms of corrosion in amine solvents is not directly caused by the amine itself but is caused by dissolved H_2S or carbon dioxide and by the amine degradation products (Ref 90). Corrosion is most severe at locations where acid gases are desorbed or removed from rich-amine solution. Here, temperatures along with tendencies for flow turbulence (i.e., locally high wall shear stress) and local flashing of the gas are highest. This includes the regenerator (stripper) reboiler and lower portions of the regenerator itself (Ref 91). Corrosion can also be a significant problem on the rich-amine side of the lean/rich-amine exchanger, in amine solution pumps, and in reclaimers. Hydrogen blistering has been a problem in the bottom of the contactor (absorber) tower and in regenerator overhead condensers and reflux drums (Ref 92). This has been associated with wet H_2S cracking (versus amine SCC) as a result of the hydrogen generated by H_2S corrosion in the rich-amine solution. These conditions are aggravated by disruption of normally protective sulfide surface films on the steel, which allows accelerated hydrogen charging, leading to crack initiation and propagation.

The common material of construction for amine units is carbon steel. To prevent alkaline SCC, welds of components in both lean- and rich-amine service should be postweld heat treated regardless of service temperature (Ref 93). Postweld heat treatment also protects against hydrogen stress cracking by lowering both the HAZ hardness and residual tensile stresses in the location of the weld. On the whole, there have been relatively few corrosion problems in most amine units. In the most severe cases of corrosion in carbon dioxide removal, new inhibitors have been designed that replace older, more highly toxic formulations based on heavy metals additions (Ref 94).

Limits for corrosion control, including amine concentration (approximately 20%), acid gas loading (0.3 to 0.6 mole/mole), flow rate (rich amine, 1.8 m/s, or 6 ft/s; lean amine, 1.8 to 6 m/s, or 6 to 20 ft/s), rich circuit temperatures (100 to 105 °C, or 212 to 220 °F), and reboiler temperatures (125 to 150 °C, or 260 to 300 °F), have been developed by evaluating services experienced in various plants with sometimes vastly different designs, throughput, and operating conditions (Ref 95–98). Also, limits for impurities such as heat-stable salts are typically based entirely on experience and may range from 1 to 2%. Consequently, there is a wide operating envelope with little technical basis to help designers and operators optimize unit reliability and performance. There has been only limited use of test data, and none has approached this area using rigorously controlled environment and flow conditions. New studies are in progress under joint industry sponsorship that are investigating the interactions between chemical and flow conditions in amine solvents, which should be available in the coming years. This work includes monoethanolamine, diglycolamine, and diethanolamine as well as parametric effects of temperature, CO_2/H_2S ratio, heat-stable salts, and organic acids (Ref 99).

Sidestream filtration is also extremely beneficial. Filming-amine corrosion inhibitors are often ineffective. Several proprietary oxidizing corrosion inhibitors based on sodium metavanadate are available. These have been successfully used in certain cases, but licensing costs tend to be high for any but the smaller units.

Regenerator towers usually should be lined with type 405 (S40500) stainless steel, and tower internals are often made of type 304 (S30400) stainless steel. Where applicable, type 304 (S30400) stainless steel is required for the rich-amine pressure of the let-down valve, as well as for piping downstream of the let-down valve, to control corrosion accelerated by high flow turbulence.

Corrosion in the regenerator reboiler is usually in the form of pitting and groove-type corrosion of tubes and is caused by localized overheating inside baffle holes (Ref 100). If thermosyphon reboilers are undersized, part of the tube bundle will become vapor blanketed, and the tubes will overheat. Subsequent exposure of the hot tubes to amine solution will cause severe turbulence and velocity-accelerated corrosion. Vapor blanketing also occurs if tubes are allowed to fill partially with steam condensate; this reduces the amount of tube surface available for heat transfer and increases the heat flux through the remainder of the tubes. Unless faulty reboiler operation can be corrected, carbon steel tubes may have to be replaced with type 304 (S30400) or type 316 (S31600) stainless steel tubes. Alloy 400 (N04400) reboiler tubes have been successfully used in amine units that handle only carbon dioxide.

As a rule, carbon steel tubes are satisfactory for regenerator overhead condensers. As discussed in the section "Sour Water" in this article, high corrosion rates can occur at this location, and carbon steel tubes may have to be replaced

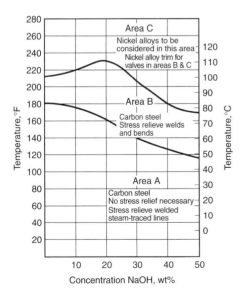

Fig. 3 Caustic soda service chart for carbon steel, weldments, and nickel alloys. Source: Ref 88

with Ti-grade 2 (R50400) tubes. Carbon steel tubes are used in reclaimers with proper neutralization of acidic constituents (Ref 101). Because the reclaimer can be taken out of service at any time, periodic retubing with carbon steel presents no problems. Cast iron pumps normally are used in low-pressure amine service. If corrosion problems occur, high-silicon cast iron impellers can be used. In high-pressure amine service, type 316 (S31600) stainless steel pumps may be needed.

Phenol (carbolic acid) is used in refineries to convert heavy, waxy distillates obtained by crude oil distillation into lubricating oils. As a rule, all components in the treating and raffinate recovery sections, except tubes in water-cooled heat exchangers, are made from carbon steel. If water is not present, few significant corrosion problems can be expected to occur in these sections. In the extract recovery section, however, severe corrosion can occur, especially where high flow turbulence is encountered. As a result, certain components require selective alloying with type 316 (S31600) stainless steel. Typically, stainless steel liners are required for the top of the dryer tower, the entire phenol flash tower, and various condenser shells and separator drums that handle phenolic water. Tubes and headers in the extract furnace should also be made of type 316 (S31600) stainless steel, with U-bends sleeved with alloy C-4 (N06455) on the outlet side to minimize velocity-accelerated corrosion.

High-Temperature Corrosion

High-temperature corrosion problems in refineries are of considerable importance (Ref 102). Equipment failures can have serious consequences, because processes at high temperatures usually involve high pressures as well. With hydrocarbon streams, there is always the danger of fire when ruptures occur. On a more positive note, high-temperature refinery corrosion is primarily caused by various sulfur compounds originating with crude oil. Over the years, extensive research has been done to establish the mechanism of various forms of high-temperature sulfidic corrosion. Corrosion rate correlations are available; therefore, equipment life can be predicted with some degree of reliability.

Sulfidic Corrosion. Corrosion by various sulfur compounds at temperatures between 260 and 540 °C (500 and 1000 °F) is a common problem in many petroleum refining processes and, occasionally, in petrochemical processes. Sulfur compounds originate with crude oils and include polysulfides, H_2S, mercaptans, aliphatic sulfides, disulfides, and thiophenes (Ref 103) and are commonly found in concentrations of 0.1 to over 5% total sulfur. Sulfur compounds react with metal surfaces at elevated temperatures, forming metal sulfides, certain organic molecules, and H_2S (Ref 104, 105). The relative corrosivity of sulfur compounds generally increases with temperature. Additionally, the higher the temperature, the more likely it is that larger and more stable organic sulfides break down into reactive components that can become involved in the corrosion process. Depending on the process particulars, corrosion is in the form of uniform thinning, localized attack, or erosion-corrosion. Corrosion control depends almost entirely on the formation of protective metal sulfide scales that exhibit parabolic growth behavior (Ref 106). In general, nickel and nickel-rich alloys (without chromium or other similar alloying additions) are rapidly attacked by sulfur compounds at elevated temperatures, while chromium-containing steels and alloys provide excellent corrosion resistance (as do additions of aluminum). The combination of H_2S and hydrogen can be particularly corrosive, and, as a rule, austenitic stainless steels are required for effective corrosion control.

Sulfidic Corrosion without Hydrogen Present. This type of corrosion occurs primarily in various components of crude distillation units, catalytic cracking units, and hydrotreating and hydrocracking units upstream of the hydrogen injection line. Crude oil distillation units that process mostly sweet crude oils (less than 0.6% total sulfur, with essentially no H_2S) experience relatively few corrosion problems. Preheat-exchanger tubes, furnace tubes, and transfer lines are generally made from carbon steel, as is corresponding equipment in the vacuum distillation section. The lower shell of distillation towers, where temperatures are above 230 °C (450 °F), is usually lined with stainless steel containing 12% Cr, such as type 405 (S40500). This prevents impingement attack under the highly turbulent flow conditions encountered, for example, near downcomers. For the same reason, trays are made of stainless steel containing 12% Cr. Even with low corrosion rates of carbon steel, certain tray components, such as tray valves, may fail in a short time, because attack occurs from both sides of a relatively thin piece of metal.

Crude distillation units that process mostly sour crude oils that result in evolution of high concentrations of H_2S during distillation require additional alloy protection against high-temperature sulfidic corrosion. The extent of alloying needed also depends on the design and the operating practices of a given unit. Typically, such units require low-alloy steels containing a minimum of 5% Cr for furnace tubes, headers and U-bends, and elbows and tees in transfer lines. In vacuum furnaces, tubes made from chromium steels containing 9% Cr are often used. Distillation towers for processing sulfidic crudes are usually made from similar materials as those of units that process mostly sweet crude oils. Where corrosion problems persist, upgrading with steels containing a greater amount of chromium is indicated.

The high processing temperatures encountered in the reaction and catalyst regeneration section of catalytic cracking units require extensive use of refractory linings to protect all carbon steel components from oxidation and sulfidic corrosion. Refractory linings also provide protection against erosion by catalyst particles, particularly in cyclones, risers, standpipes, and slide valves. Stellite hardfacing is used on some components to protect against erosion. When there are no erosion problems and when protective linings are impractical, austenitic stainless steels, such as type 304 (S30400), can be used. Cyclone dip legs, air rings, and other internals in the catalyst regenerator are usually made from type 304 (S30400) stainless steel, as is piping for regenerator flue gas. Reactor feed piping is made from low-alloy steel, such as 5Cr-0.5Mo or 9Cr-1Mo, to control high-temperature sulfidic corrosion.

The main fractionation tower is usually made of carbon steel, with the lower part lined with stainless steel containing 12% Cr, such as type 405 (S40500) (Ref 107). Slurry piping between the bottom of the main fractionation tower and the reactor may receive an additional corrosion allowance as protection against excessive erosion. As a rule, there are few corrosion problems in the reaction, catalyst regeneration, and fractionation sections (Ref 108).

Hydrocracking and hydrotreating units usually require alloy protection against both high-temperature sulfidic corrosion and high-temperature hydrogen attack (Ref 109, 110). Low-alloy steels may be required for corrosion control ahead of the hydrogen injection line.

The so-called McConomy curves can be used to predict the relative corrosivity of crude oils and their various fractions (Ref 111). Although this method relates corrosivity to total sulfur content, and thus does not take into account the variable effects of different sulfur compounds, it can provide reliable corrosion trends if certain corrections are applied. Plant experience has shown that the McConomy curves, as originally published, tend to predict excessively high corrosion rates. The curves apply only to liquid hydrocarbon streams containing 0.6 wt% S (unless a correction factor for sulfur content is applied) and do not take into account the effects of vaporization and flow regime. These conditions are typically related to the very high wall shear stress conditions that can occur in certain multiphase flow regimes (e.g., slug flow and droplet impingement). In these cases, either the local liquid turbulence or impact of liquid droplets can produce accelerated corrosion as a result of their mechanical influence (e.g., erosion or fracturing) on normally protective corrosion scales. The curves can be particularly useful, however, for predicting the effect of operational changes on known corrosion rates.

Over the years, it has been found through service experience that corrosion rates predicted by the original McConomy curves should be decreased by a factor of roughly 2.5, resulting in the modified curves shown in Fig. 4. The curves demonstrate the beneficial effects of alloying steel with chromium in order to reduce corrosion rates. Corrosion rates are roughly halved when the next higher grade of low-alloy steel (for example, 2.25Cr-1Mo, 5Cr-0.5Mo,

7Cr-0.5Mo, or 9Cr-1Mo steel) is selected. Essentially, no corrosion occurs with stainless steels containing 12% or more chromium. Although few data are available, plant experience has shown that corrosion rates start to decrease as temperatures exceed 455 °C (850 °F). Two explanations frequently offered for this phenomenon are the possible decomposition of reactive organic sulfur compounds and the formation of a protective coke layer formed from thermal decomposition of the hydrocarbon species in the oil.

An important consideration in selecting materials is that metal skin temperatures, rather than stream temperatures, should be used to predict corrosion rates when significant differences between the two arise. This can be of major consequence when increasing unit throughput, and higher-than-the-original-designed heat flux results in increased metal skin temperatures. For example, metal temperatures of furnace tubes are typically 85 to 110 °C (150 to 200 °F) higher than the temperature of the hydrocarbon stream passing through the tubes. Furnace tubes normally corrode at a higher rate on the hot side (fire side) than on the cool side (wall side), as shown in Fig. 5. Convective-section tubes often show accelerated corrosion at contact areas with tube hangers because of locally increased temperatures. Similarly, replacement of bare convective-section tubes with finned or studded tubes can further increase tube metal temperatures by 85 to 110 °C (150 to 200 °F). The increased temperature on the hot wall can result in decomposition of more stable sulfur species and increase production of H_2S as well as increase the overall chemical reactivity on the metal surface.

Correction factors for process streams with various total sulfur contents, averages of those proposed originally by McConomy, are shown in Fig. 6. As can be seen, doubling the sulfur content can increase corrosion rates by approximately 30%. In atmospheric distillation, relatively nonreactive thiophenes are greater in high-boiling cuts (and residuum) than in the original crude charge. Therefore, an additional factor ranging from 0.5 to 1 may have to be applied to the total sulfur content so that realistic corrosion rates can be obtained for such cuts. The degree of vaporization and the resultant two-phase flow regimes can have a significant effect on high-temperature sulfidic corrosion.

Sulfidic Corrosion with Hydrogen Present. The presence of hydrogen in, for example, hydrotreating and hydrocracking operations, increases the severity of high-temperature sulfidic corrosion. Hydrogen converts organic sulfur compounds in feed stocks to H_2S; corrosion becomes a function of H_2S concentration (or partial pressure).

Downstream of the hydrogen injection line, low-alloy steel piping usually requires aluminizing in order to minimize sulfidic corrosion. Alternatively, type 321 (S32100) stainless steel can be used. Tubes in the preheat furnace are aluminized low-alloy steel, aluminized 12% Cr stainless steel, or type 321 (S32100) stainless steel. Reactors are usually made of 2.25Cr-1Mo steel, either with a type 347 (S34700) stainless steel weld overlay or an internal refractory lining. Reactor internals are often type 321 (S32100) stainless steel (Ref 113).

Depending on the expected corrosion rates, reactor effluent piping operating above approximately 260 °C (500 °F) is made of type 321 (S32100) stainless steel, aluminized low-alloy steel, regular low-alloy steel, or carbon steel with suitable corrosion allowances. When selecting materials for this service, the recommendations of the API should be followed to avoid problems with high-temperature hydrogen attack (Ref 114). The same considerations generally apply to separator drums and heat-exchanger vessels operating at temperatures above 260 °C (500 °F). Type 321 (S32100) stainless steel is usually required for heat-exchanger tubes at these temperatures.

A number of researchers have proposed various corrosion rate correlations for high-temperature sulfidic corrosion in the presence of hydrogen (Ref 115–121), but the most practical correlations seem to be the so-called Couper-Gorman curves. The Couper-Gorman curves are based on a survey conducted many years ago by National Association of Corrosion Engineers (NACE) Committee T-8 (currently designated STG 34) on Refinery Corrosion (Ref 122).

The Couper-Gorman curves differ from those previously published in that they reflect the influence of temperature on corrosion rates throughout a whole range of H_2S concentrations. Total pressure was found not to be a significant variable between 1 and 18 MPa (150 and 2650 psig). It was also found that essentially no corrosion occurs at low H_2S concentrations and temperatures above 315 °C (600 °F), because formation of iron sulfide becomes thermodynamically impossible. Curves are available for carbon steel, 5Cr-0.5Mo steel, 9Cr-1Mo steel, 12% Cr stainless steel, and 18Cr-8Ni austenitic stainless steel. For the low-alloy steels, two sets of curves apply, depending on whether the hydrocarbon stream is naphtha or gas oil. The curves again demonstrate the beneficial effects of alloying steel with chromium to reduce the corrosion rate. Several licensors and major refiners have other methodologies or internally developed data that are used for corrosion prediction and materials selection.

Modified Couper-Gorman curves are shown in Fig. 7 to 14. To facilitate use of these curves,

Fig. 4 Modified McConomy curves showing the effect of temperature on high-temperature sulfidic corrosion of various steels and stainless steels. Source: Ref 112

Fig. 5 High-temperature sulfidic corrosion of 150 mm (6 in.) diameter carbon steel tube from radiant section of crude preheat furnace at crude distillation unit. Note accelerated attack on fire side.

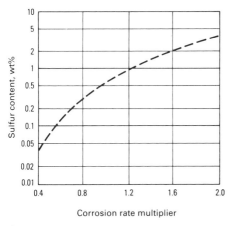

Fig. 6 Effect of sulfur content on corrosion rates predicted by modified McConomy curves in the 290 to 400 °C (550 to 750 °F) temperature range. Source: Ref 112

original segments of the curves were extended (dashed lines). In contrast to sulfidic corrosion in the absence of hydrogen, there is often no real improvement in corrosion resistance unless chromium content exceeds 5%. Therefore, the curves for 5Cr-0.5Mo steel also apply to carbon steel and low-alloy steels containing less than 5% Cr. Stainless steels containing at least 18% Cr are often required for essentially complete immunity to corrosion. Because the Couper-Gorman curves are primarily based on corrosion rate data for an all-vapor system, partial condensation can be expected to increase corrosion rates because of droplet impingement.

Recent experience has indicated tendencies for accelerated corrosion rates in piping and reboiler furnace tubes associated with fractionation and distillation units downstream from

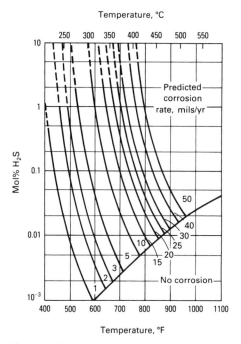

Fig. 7 Effect of temperature and hydrogen sulfide content on high-temperature H_2S/H_2 corrosion of carbon steel (naphtha desulfurizers). 1 mil/yr = 0.025 mm/yr. Source: Ref 112

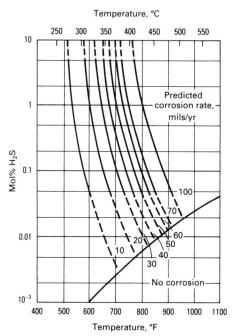

Fig. 8 Effect of temperature and hydrogen sulfide content on high-temperature H_2S/H_2 corrosion of carbon steel (gas oil desulfurizers). 1 mil/yr = 0.025 mm/yr. Source: Ref 112

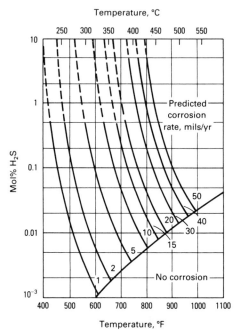

Fig. 9 Effect of temperature and hydrogen sulfide content on high-temperature H_2S/H_2 corrosion of 5Cr-0.5Mo steel (naphtha desulfurizers). 1 mil/yr = 0.025 mm/yr. Source: Ref 112

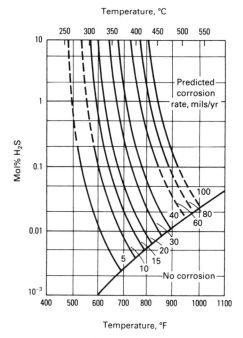

Fig. 10 Effect of temperature and hydrogen sulfide content on high-temperature H_2S/H_2 corrosion of 5Cr-0.5Mo steel (gas oil desulfurizers). 1 mil/yr = 0.025 mm/yr. Source: Ref 112

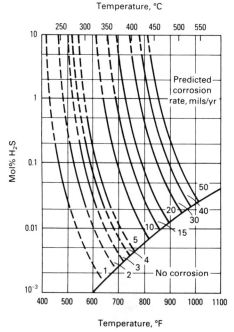

Fig. 11 Effect of temperature and hydrogen sulfide content on high-temperature H_2S/H_2 corrosion of 9Cr-1Mo steel (naphtha desulfurizers). 1 mil/yr = 0.025 mm/yr. Source: Ref 112

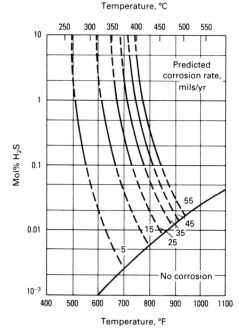

Fig. 12 Effect of temperature and hydrogen sulfide content on high-temperature H_2S/H_2 corrosion of 9Cr-1Mo steel (gas oil desulfurizers). 1 mil/yr = 0.025 mm/yr. Source: Ref 112

hydrotreaters and hydrocrackers (Ref 123). This corrosion was considered unusual, because these units were considered to have low total sulfur and to be generally H_2-free. In some cases, alloy steels with 5 to 9% Cr corroded at rates similar to carbon steel. This behavior was not predicted by review of the commonly available corrosion prediction curves for sulfidic corrosion. The observations also mentioned that corrosion was most severe at areas of high turbulence, producing elevated wall shear stress resulting from multiphase flow, direction changes (elbows and tees), and in pumps, leading to suppositions that the cause was velocity-accelerated sulfidic attack. Furthermore, the corrosion even occurred with very low (ppm) levels of sulfur species present in the process stream. Some cases involved higher corrosion rates on the top of the tubes (Ref 124).

When selecting steels for resistance to high-temperature sulfidic corrosion in the presence of hydrogen, the possibility of high-temperature hydrogen attack should also be considered. Conceivably, this problem arises when carbon steel and low-alloy steels containing less than 1% Cr are chosen for temperatures exceeding 260 °C (500 °F) and hydrogen partial pressures above 689 kPa (100 psia) and when corrosion rates are expected to be relatively low.

Naphthenic acids are organic acids that are present in many crude oils, especially those from California, Venezuela, Eastern Europe, and Russia. However, newer sources of crude oil with concerns for naphthenic acid corrosion have included those from West Africa, Canada, China, and the North Sea.

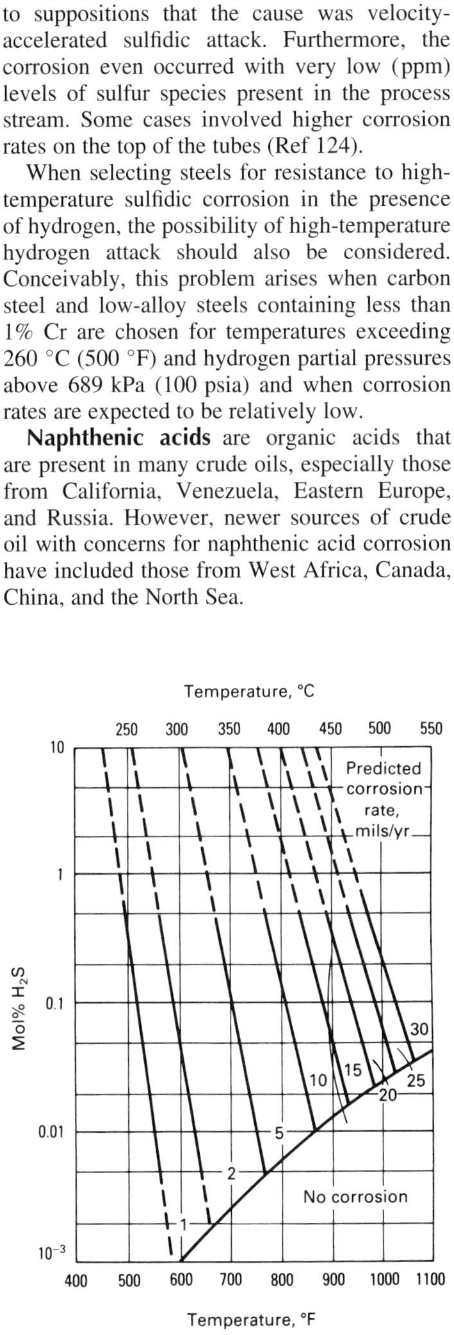

The main acids from naphthenic-based crudes are saturated ring structures with a single carboxyl group. Their general formula may be written as $R(CH_2)_n COOH$, where R is usually a cyclopentane ring. The higher-molecular-weight acids can be bicyclic ($12 < n < 20$), tricyclic ($n > 20$), and even polycyclic (Ref 125, 126). Naphthenic acid content is generally expressed in terms of the neutralization number (total acid number, or TAN), which is determined by titration with potassium hydroxide, as described in ASTM D 664 (Ref 127). The units for TAN are mg KOH in 100 g of oil to bring the solution to neutralization. However, minor amounts of other organic acids can also be present and contribute to the measured acidity of the crude oil using this method. These are commonly fatty acids and phenolic compounds that easily thermally decompose and consequently do not contribute significantly to corrosion in refining operations. An alternative method of determination of TAN values is UOP 565, which first removes acidic contribution from these nonnaphthenic acidic species (Ref 128).

Naphthenic acids are commonly thought to be corrosive only at temperatures above approximately 230 °C (450 °F) in the general range from TAN 1 to 6 encountered with crude oil and various side cuts (hydrocarbon fractions). However, recent experience suggests that the general limits for TAN in whole crudes (producing naphthenic acid corrosion in the refining process) may be as low as 0.3 to 0.5 in the whole crude oil and approximately 1 to 1.5 in the hydrocarbon fractions. This occurs as a result of fractionation and concentration of the naphthenic acids along with the hydrocarbon species during the refining process. Therefore, a whole crude with a TAN value of 0.5 could produce side cuts with much higher TAN values, with concentrations of 3 to 5 times the whole crude TAN value not uncommon.

At any given temperature, corrosion rate is generally proportional to TAN. Corrosion rate can increase by a factor of 3 with each 55 °C (100 °F) increase in temperature above 230 °C (450 °F). In contrast to high-temperature sulfidic corrosion, no protective scale is formed during naphthenic acid corrosion, and low-alloy steels containing up to 9% Cr provide no benefits whatsoever over carbon steel (Ref 129). Some improvement can be obtained with the use of 12% Cr in some mild-to-intermediate environments. However, at higher TAN levels, even 12% Cr stainless steel is of no benefit, and types 304, 316, or 317 are required to resist naphthenic acid attack. The presence of naphthenic acids may accelerate high-temperature sulfidic corrosion that occurs at furnace headers, elbows, and tees of crude distillation units because of unfavorable flow conditions. A recent survey of literature on naphthenic acid corrosion was published by NACE International in 1999 (Ref 130).

Severe naphthenic acid corrosion has been experienced primarily in the vacuum towers of crude distillation units in the temperature zone of 290 to 345 °C (550 to 650 °F) and sometimes as low as 230 °C (450 °F) (Ref 131). Damage is in the form of pitting and localized (lake-type) attack of tray components and vessel walls. Attack is often limited to the undersides of tray floors and to the inside and very top of the outside surfaces of bubble caps, as shown in Fig. 15. These areas are normally not covered by a layer of liquid, which suggests that the attack was caused by impinging droplets of the condensing acids. No corrosion damage is found at temperatures above 345 °C (650 °F), because of a combination of thermal degradation of the naphthenic acids and the formation of a protective coke layer.

In a recent study (Ref 35), a major investigation of naphthenic acid corrosion was conducted involving a group of major oil companies. It revealed new information about the details of naphthenic acid corrosion and its relationship to flow conditions and sulfur species. This effort characterized naphthenic acids from numerous crude oil sources and found that they varied greatly in terms of their chemical structure (straight chain or one to four ring structures, as shown in Fig. 16), which also resulted in their varying characteristics in terms of fractionation, thermal stability, and corrosivity at various locations in the distillation process. Maximum corrosivity in the temperature range 260 to 345 °C (500 to 650 °F) was associated with naphthenic acids with one and two rings; however, the most thermally stable naphthenic acids were those with three or more rings. Therefore, there could be situations where hydrocarbon

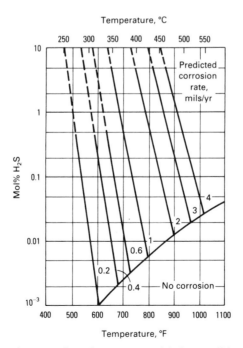

Fig. 13 Effect of temperature and hydrogen sulfide content on high-temperature H_2S/H_2 corrosion of 12% Cr stainless steel. 1 mil/yr = 0.025 mm/yr. Source: Ref 112

Fig. 14 Effect of temperature and hydrogen sulfide content on high-temperature H_2S/H_2 corrosion of 18Cr-8Ni austenitic stainless steel. 1 mil/yr = 0.025 mm/yr. Source: Ref 112

fractions have the same nominal TAN value but could vary in terms of naphthenic acid corrosivity.

Secondly, flow- or turbulence-induced wall shear stress was also a major factor in the severity of naphthenic acid corrosion. A special hot oil flow loop was designed to recreate conditions involved in crude oil refining at flow rates of up to 107 m/s (350 ft/s) producing wall shear stress levels up to 1800 Pa (0.26 psi), typical of partially vaporized, multiphase flow. Over a range of TAN levels, this apparatus was used to map conditions of impingement corrosion for 5Cr-0.5Mo (Fig. 17), 9Cr-1Mo (Fig. 18), and 12Cr steels (Ref 35). Through this work, it was determined that there were two different mechanisms that provided the onset of naphthenic acid attack, both of which involved destruction of the normally protective sulfide film that forms on steels in hot crude oil refining environments. At high TAN and low flow velocity, the mechanism involves chemical dissolution of the FeS film by local chemical attack. For this to occur, high TAN levels were required. However, under high velocity or highly turbulent flow, attack can occur at much lower TAN levels as a result of mechanical fracturing of the protective sulfide film from the high shear stress flow conditions. Figures 19 and 20 show the surface of a 5Cr-0.5Mo steel under high-TAN/low-flow and low-TAN/high-flow conditions, respectively (Ref 35).

Figures 17 and 18 also show the influence of H_2S on the severity of corrosion. These figures show the inhibition of naphthenic acid corrosion on both 5Cr-0.5Mo and 9Cr-1Mo steels at TAN 3.5 and 60 m/s (200 ft/s) by the presence of a H_2S partial pressure of 1.4 kPa (0.2 psia). However, for the same nominal conditions except with a high partial pressure of H_2S— 3.1 kPa (0.45 psia)—impingement attack is reestablished for the 5Cr steel but is still inhibited for the 9Cr steel. Presumably, this behavior is because 9Cr-1Mo steel has greater resistance to sulfur than the lower-alloy 5Cr-0.5Mo steel. For conditions free from naphthenic acid (TAN 0) at 60 m/s (200 ft/s), no impingement attack was observed in the 5Cr steel. However, when 3.1 kPa (0.45 psia) H_2S was included in the environment, impingement attack was observed, as was the case at TAN 3.5. Therefore, it was concluded that in the presence of conditions of high sulfur and velocity, the attack was dominated in this material by velocity-accelerated sulfidic corrosion regardless of the TAN level. Furthermore, the role of sulfur and other organic sulfur species on inhibiting naphthenic acid corrosion and other organic sulfur specimens was also examined. As shown in Fig. 21, inhibition occurred over a limited range of H_2S concentration for 9Cr-1Mo steel despite the fact that the sulfur addition was made either as direct H_2S gas or by the presence of an organic sulfur compound. The controlling parameter appears to be how much H_2S is available to participate in the corrosion reaction, which determines the stability of the sulfide film and the ultimate resistance to naphthenic or sulfidic corrosion. This approach indicates that there is actually a continuum between the two mechanisms of corrosion (Ref 99).

Naphthenic acid corrosion is usually controlled by simply blending crude oils having high TAN values with other crude oils having lower TAN values. Blending is designed to reduce the naphthenic acid content of the worst side cut. The TAN produced by naphthenic acids in these fractions is usually determined by assessment of TAN as a function of true boiling point for specific fractions in a distillation experiment. However, there are many other variables that can play into a blending strategy, and sometimes, corrosion testing of coupons in samples of the oil or various blends may be required to provide added confidence regarding the blending strategy. For example, the other factors may include one or more of the following:

- The TAN values for the whole crude and critical hydrocarbon fractions
- The specific type of naphthenic acids present (chemical structures) in these fractions
- The total sulfur content

Fig. 15 Naphthenic acid corrosion on top of 150 mm (6 in.) bubble caps made from type 317 (S31700) stainless steel containing 2.95% Mo. Tray temperature was 305 °C (580 °F).

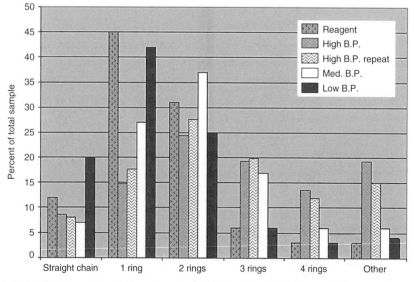

Fig. 16 Variation in organic acid ring structures in naphthenic acids derived from crude oils compared to reagent-grade naphthenic acid. B.P., boiling point. Source: Ref 35

- The specific sulfur speciation of the oil and their presence in the hydrocarbon fractions of interest

In practice, blending often means that the charge to the crude distillation unit has a TAN value of no higher than 0.5 to 1.0. The actual blending can depend on the other variables mentioned previously, which are numerous due to the variability of crude oil compositions available from various new and existing locations around the world. Currently, there are concerns for new sources of crude oil that are characterized as having high TAN but low concentrations of sulfur (<0.5%). The apprehension is that the specific speciation of the organic sulfur could provide conditions where the amount of reactive sulfur is insufficient to minimize naphthenic acid attack in some portion of the distillation process.

For cases where resistance to naphthenic acid corrosion is based on unit upgrading to corrosion-resistant alloys, vacuum tower internals operating in the 290 to 345 °C (550 to 650 °F) range should be made from type 316 (S31600) or, preferably, type 317 (S31700) stainless steel containing a higher molybdenum content (3.5%). This is often a good strategy, because in many cases, the vacuum gas oils contain the highest concentrations of naphthenic acids and have the lowest TAN values. The vacuum tower lining in this temperature range should also be type 317 (S31700) stainless steel. Aluminum has excellent resistance to naphthenic acid corrosion in vacuum towers and can be used if its strength limitations and low resistance to velocity effects are kept in mind. Alloy 20 (N08020), alloy C-276 (N10276), alloy 625 (N06625), and Ti-grade 2 (R50400) are also resistant to naphthenic acid corrosion, if needed for specialty equipment. In contrast, aluminized carbon steel tray components, such as bubble caps, have performed poorly.

Fuel Ash. Corrosion by fuel ash deposits can be one of the most serious operating problems with boiler and preheat furnaces. All fuels except natural gas contain certain inorganic contaminants that leave the furnace with products of combustion. These will deposit on heat-receiving surfaces, such as superheater tubes, and after melting can cause severe liquid-phase corrosion. Contaminants of this type include various combinations of vanadium, sulfur, and sodium compounds (Ref 132–134). Fuel ash corrosion is very likely to occur when residual fuel oil (Bunker C fuel) is burned.

In particular, vanadium pentoxide vapor (V_2O_5) reacts with sodium sulfate (Na_2SO_4) to form sodium vanadate ($Na_2O \cdot 6V_2O_5$). The latter compound reacts with steel, forming a molten slag that runs off and exposes fresh metal to attack. The cathodic part of the corrosion reaction is reduction of the pentoxide to the tetroxide (V_2O_4); therefore, the most common ingredient of superheater deposits is sodium vanadyl vanadate ($Na_2O \cdot V_2O_4 \cdot 5V_2O_5$). Table 9 lists the ash fusion temperatures of a number of fuel ash ingredients that can contribute to corrosion and fouling in boiler and preheat furnaces.

Corrosion increases sharply with increasing temperature and vanadium content of fuel. If the vanadium content in the fuel oil exceeds 150 ppm, the maximum tube wall temperature should be limited to 650 °C (1200 °F). Between

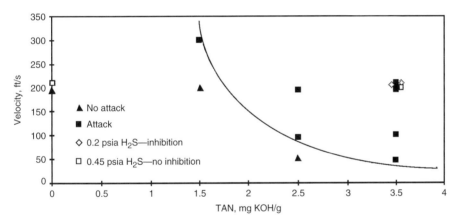

Fig. 17 Flow velocity (in feet per second) versus total acid number (TAN) for naphthenic acid impingement attack of 5Cr-0.5Mo steel at 345 °C (650 °F). 40 mils/yr = 1 mm/yr. Source: Ref 35

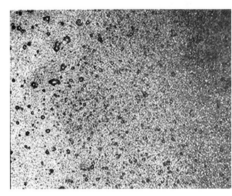

Fig. 19 Local chemical attack of the sulfide scale on the metal surface under high total acid number and low-velocity naphthenic acid corrosion conditions

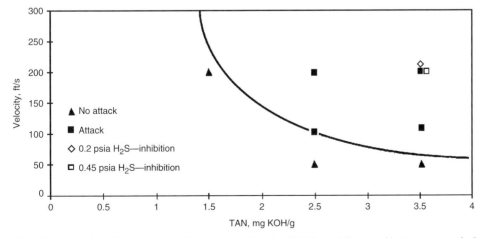

Fig. 18 Flow velocity (in feet per second) versus total acid number (TAN) for naphthenic acid impingement attack of 9Cr-1Mo steel at 345 °C (650 °F). 40 mils/yr = 1 mm/yr. Source: Ref 35

Fig. 20 Mechanical fracturing of the sulfide scale on the metal surface under low total acid number and high-velocity naphthenic acid corrosion conditions. Note cracks in dark iron sulfide film.

20 and 150 ppm V, maximum tube wall temperatures can be between 650 and 845 °C (1200 and 1550 °F), depending on sulfur content and the sodium-vanadium ratio of the fuel oil. With 5 to 20 ppm V, the maximum tube wall temperature can exceed 845 °C (1550 °F).

In general, most alloys are likely to suffer from fuel ash corrosion. However, alloys with high chromium and nickel contents provide the best resistance to this type of attack. Sodium vanadate corrosion can be reduced by firing boilers with low excess air (<1%). This minimizes formation of sulfur trioxide in the firebox and produces high-melting slags containing vanadium tetroxide and trioxide rather than pentoxide. In the temperature range of 400 to 480 °C (750 to 900 °F), boiler tubes are corroded by alkali pyrosulfates such as sodium pyrosulfate and potassium pyrosulfate, when appreciable concentrations of sulfur trioxide are present.

Another problem in fossil fuel combustion units involves the formation of low-melting-point sulfate salts, particularly Na_2SO_4 and potassium sulfate (K_2SO_4). In many boilers that operate at 450 °C (840 °F) and exhibit fireside corrosion in chromium-molybdenum steel components, alternative use of Cr-Ni-Mo steels such as type 310 has been generally beneficial. In some cases, these materials are used in the clad form on the outer surface of the boiler tubes, where concerns for high-temperature corrosion are greatest. For higher-temperature service (>650 °C, or 1200 °F), under such conditions as found in superheaters, highly alloyed materials such as cast 50Ni-50Cr alloy have been successfully used for resistance to fireside corrosion. As shown in Fig. 22, maximum corrosive attack is usually found to occur at or around the solidus temperature of the molten salt deposit and the dissociation temperature of low-melting-point eutectics (Ref 135).

The conditions that result from the combustion of low-grade fuel oil are particularly severe and, in many cases, difficult to anticipate, because correlation of corrosion severity to fuel contaminant levels may not have been conclusively determined, or because the process conditions may vary over time. In some applications, it has been possible to monitor this vapor-phase combustion environment, particularly where conditions of salt deposition provide a conductive medium for making electrochemical measurements. Under these conditions, the use of electrochemical measurements has resulted in an online, real-time method of making adjustments to the combustion environment to minimize the severity of the corrosion. Figure 23 (Ref 136) shows the severity of hot corrosion in a fossil-fueled process unit versus time, using these methods. Usually, an array of sensors is used to measure the spatial distribution of the corrosion with respect to burners. Additionally, the temperature of the surface of the probe can be internally cooled and controlled to assess the corrosivity relating to various temperatures at different locations in the unit.

In some cases, where alloys are not effective in reducing corrosion or where changes in fuel contaminants are anticipated, additives such as magnesium hydroxides $Mg(OH)_2$, in the combusters can be used to thermodynamically modify the combustion environment, which can change associated deposits and their resulting morphology. These additives can be helpful in controlling corrosion, particularly in conjunction with controlled firing with low excess air. The additives effectively raise the solidus temperature of salt deposits to a point above the service temperature, allowing more-economical commercial alloys to be used. A porous and fluffy deposit layer is usually formed with the use of these additives. This type of deposit can be readily removed by periodic cleaning. Magnesium-type additives offer additional benefits with regard to cold-end corrosion in boilers. However, the effectiveness of the additives varies. Under some conditions, sulfuric acid vapors can also condense at temperatures between 150 and 175 °C (300 and 350 °F),

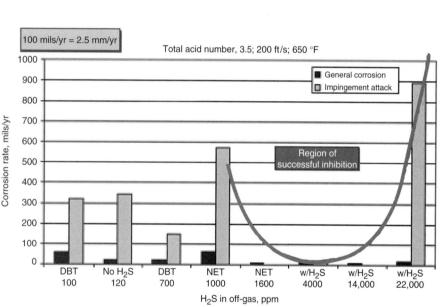

Fig. 21 Influence of H_2S on inhibition of naphthenic acid corrosion of a 9Cr-1Mo steel as a result of addition of H_2S gas, or H_2S formation in off-gas in oil containing 1,9-nonanedithiol (NET) or dibenzothiophene (DBT). Amount of H_2S measured in the test off-gas in each case is shown on bottom axis.

Table 9 Ash fusion temperatures of slag-forming compounds

		Ash fusion temperature	
Chemical compound	Chemical formula	°C	°F
Vanadium pentoxide	V_2O_5	690	1270
Sodium sulfate	Na_2SO_4	890	1630
Nickel sulfate	$NiSO_4$	840	1545
Sodium metavanadate	$Na_2O \cdot V_2O_5$	630	1165
Sodium pyrovanadate	$2Na_2O \cdot V_2O_5$	655	1210
Sodium orthovanadate	$3Na_2O \cdot V_2O_5$	865	1590
Nickel orthovanadate	$3NiO \cdot V_2O_5$	900	1650
Sodium vanadyl vanadate	$Na_2O \cdot V_2O_4 \cdot 5V_2O_5$	625	1155
Sodium iron trisulfate	$2Na_3Fe[SO_4]_3$	620	1150

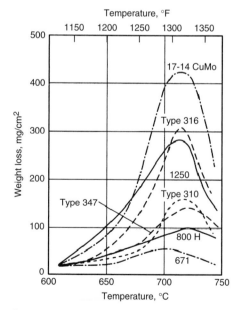

Fig. 22 Laboratory-simulated flue gas corrosion versus temperature for selected alloys. Tests were conducted in synthetic flue gas ($80N_2$-$15CO_2$-$4O_2$-$1SO_2$, saturated with water) with synthetic ash (37.5 mol% Na_2SO_4, 37.5 mol% K_2SO_4, 25 mol% Fe_2O_3). Source: Ref 135

depending on sulfur content of the fuel oil, and can cause serious corrosion problems. Additives neutralize free acid by forming magnesium sulfate.

Oxidation. Carbon steels, low-alloy steels, and stainless steels react at elevated temperatures with oxygen in the surrounding air and become scaled. Nickel alloys can also become oxidized, especially if spalling of scale occurs. The oxidation of copper alloys is usually not a problem, because these are rarely used where operating temperatures exceed 260 °C (500 °F). Alloying with both chromium, nickel, and, to a certain extent, silicon increases oxidation resistance through promoting the formation of a stable and protective scale on the metal surface (see Fig. 24 and 25 for selected steels and stainless steels in air) (Ref 135). Stainless steels or nickel alloys, with alloy additions of chromium and molybdenum—note exception alloy 400 (N04400)—are required to provide satisfactory oxidation resistance at temperatures above 705 °C (1300 °F). Thermal cycling, applied stresses, moisture, and sulfur-bearing gases will decrease protective scaling and increase rates of corrosive attack. In refineries and petrochemical plants, high-temperature oxidation is primarily limited to the outside surfaces of furnace tubes, tube hangers, and other internal furnace components that are exposed to combustion gases containing excess air.

At elevated temperatures, steam decomposes at metal surfaces to form hydrogen and oxygen and may cause steam oxidation of steel, which is somewhat more severe than air oxidation at the same temperature. Fluctuating steam temperatures tend to increase the rate of oxidation by causing scale to spall and thus expose fresh metal to further attack.

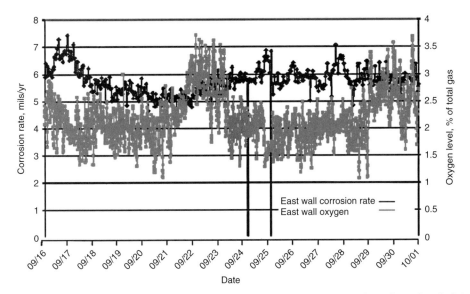

Fig. 23 Corrosion rate in a combustion environment versus time as monitored using electrochemical methods is plotted with a key process variable. Fireside corrosion of boiler tubes in coal-fired utilities and waste incineration plants is an expensive and difficult problem to deal with. Special corrosion probes inserted into the firewall are used to directly monitor boiler tube degradation online and in real-time and to optimize the process for maximum efficiency and tube life. Source: Ref 136

Environmentally Assisted Cracking (SCC, HEC, and Other Mechanisms)

Stress-corrosion cracking (SCC), hydrogen embrittlement cracking (HEC), and other forms of environmentally assisted cracking (EAC) and embrittlement are the most insidious forms of failure that can be experienced by process

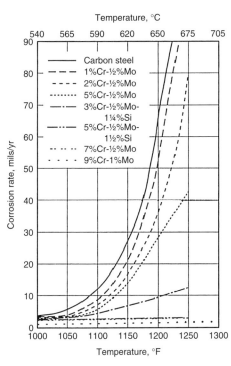

Fig. 24 Effects of chromium and/or silicon on the oxidation rate of steels in air versus temperature. Source: Ref 135

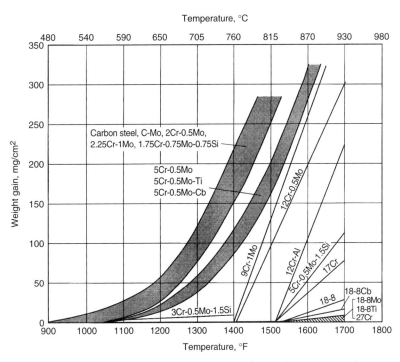

Fig. 25 Weight gain of carbon, low-alloy, and stainless steels in air after 1000 h at temperatures from 590 to 930 °C (1100 to 1700 °F)

equipment, because they tend to strike without warning, with the potential to produce catastrophic results. Usually, there is no noticeable yielding or bulging of the component, there is often no measurable metal loss, and through-thickness cracks can form in as little as hours or minutes in severe cases after exposure to a crack-inducing environment. Susceptibility can often be related to short transients such as process upset, start-up or shutdown conditions, or, in some cases, removal of normally protective corrosion films during surface inspections.

For example, cracking throughout an entire furnace coil occurred within 1 h after exposure to air and the resultant formation of polythionic acids. Towers and heat exchangers have had to be scrapped because of hydrogen blistering, embrittlement, and stress cracking at welds. High-temperature hydrogen attack has resulted in the sudden rupture of pressure vessels. With consequences such as these, the possibility and/or probability of EAC occurring in a given environment should not be underestimated. In most cases, the best path forward is to initially select materials of adequate resistance to the application during the design (or at least during repair or replacement after a failure) phase based on a complete knowledge of the process environment, including start-up and normal operations, including those that occur during hold, upsets, and shutdown. Sometimes, this may require laboratory testing under simulated service conditions to provide additional data on the relative corrosion/cracking susceptibility of various material compositions, heat treatments, and fabricated conditions, or the role of process variables on the cracking process. Normally, coatings and inhibitors are not used to mitigate cracking phenomena unless they are used for only short periods or if no other method is possible.

Stress-Corrosion Cracking

Stress-corrosion cracking phenomena common in the refinery result from exposure to the following service environments:

- Chlorides
- Caustic
- Ammonia
- Amines
- Polythionic acid
- Fuel ethanol

A summary of SCC in these systems is presented in the following sections. More information on the basic mechanisms of SCC can be found in the article "Stress-Corrosion Cracking" in *Corrosion: Fundamentals, Testing, and Protection*, Volume 13A of *ASM Handbook*, 2003. A detailed review of tests for determining susceptibility to SCC can be found in the article "Evaluating Stress-Corrosion Cracking" also in Volume 13A.

Chloride SCC. Chlorides are perhaps the most common cause of SCC of austenitic stainless steels and nickel alloys. The literature abounds with studies of the mechanism of cracking, specific environments that accelerate cracking, and tests for predicting cracking tendency. Typically, the higher the applied or residual tensile stresses and the higher the dissolved oxygen concentration, the lower the amount of chloride necessary in an aqueous environment to cause SCC. In some cases, SCC has been identified in environments with as little as 1 ppm chloride under just the right mechanical loading situation (Ref 137). However, in most cases, the permissible limits on chloride ion content in refinery operations are higher.

The usual failure mode of chloride SCC in austenitic stainless steels is the transgranular, highly branched cracking illustrated in Fig. 26. Intergranular cracking is sometimes also associated with chloride SCC, but this is not common unless the material is sensitized during welding, heat treatment, or from prolonged exposure to elevated process temperatures.

Based on laboratory tests in boiling 42% magnesium chloride solution, austenitic stainless steel and nickel alloys are subject to chloride SCC if their nickel content is less than approximately 45% (Fig. 1). In practice, however, stainless steel and nickel alloys containing greater than 30% Ni will be immune to chloride SCC in most refinery environments. It has also been reported that chloride SCC can be associated with anaerobic conditions where chlorides are present with high concentrations of H_2S (Ref 138).

In some particular cases involving high temperature (>150 °C, or 300 °F) and high chloride concentration (>10,000 ppm), austenitic stainless steels and alloys with an excess of 42% Ni can fail by SCC when the service environment involves high partial pressures of H_2S. Material conditions that promote this behavior are high strength resulting from heat treatment and/or cold working, or the presence of elemental sulfur in the service environment.

Factors that normally influence the rate and severity of chloride SCC in refinery environments are chloride content, oxygen content, temperature, stress level, and pH value of an aqueous solution. In most cases, refinery process environments have low chloride concentrations (<1000 ppm). It has been established that, in the absence of H_2S, oxygen is usually required for chloride cracking to occur. Refinery and petrochemical plant experience confirms that stainless steel components, such as heat-exchanger tube bundles, usually do not crack until removed from operation and exposed to air during a shutdown. Increased oxygen content decreases the critical chloride content for cracking to occur, as shown in Fig. 27 (Ref 139).

The severity of cracking also increases with temperature. Cracking of austenitic stainless steel components rarely occurs at ambient temperatures unless extremely severe environmental conditions, grain-boundary sensitization, and high residual tensile stresses occur. Stainless steel pump impellers in seawater service have shown no cracking problems despite the fact that both chloride and oxygen contents are high. Cracking has been found to occur, however, at tropical locations where direct exposure to the sun can increase metal temperatures significantly above ambient. As a general rule, chloride SCC of process equipment occurs only at temperatures above approximately 60 °C (140 °F).

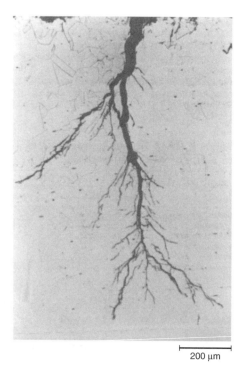

Fig. 26 Chloride stress-corrosion cracking of type 304 (S30400) stainless steel tube by chloride-containing sour water. 70×

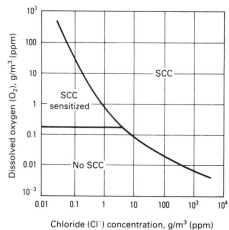

Fig. 27 Synergistic effect of chlorides and oxygen on the stress-corrosion cracking (SCC) of type 304 (S30400) stainless steel. The tests were conducted at 250 to 300 °C (480 to 570 °F) at a strain rate of $<10^{-5} \cdot s^{-1}$. Source: Ref 139

The stresses required to produce cracking can be assumed to be always present. Residual stresses from forming, bending, or joining operations are sufficient for cracks to form. Thermal stress-relief treatments at 870 °C (1600 °F) can effectively prevent cracking if done correctly and without the necessity of subsequent cold working (to correct distortion, for example). However, prolonged heating in the range of 650 to 815 °C (1200 to 1500 °F) can result in sensitization, depending on the carbon content of the material, which will increase susceptibility to intergranular SCC.

In alkaline solutions, the likelihood of chloride SCC is greatly reduced. Consequently, austenitic stainless steels are frequently used for equipment exposed to amine solutions in gas-treating, sour water, and sulfur recovery units. A survey of plant experiences has shown no reported instances of cracking despite the fact that chloride contents as high as 1000 ppm were measured in the circulating amine solution.

Most cracking problems occur when unexpected chlorine concentrations are found in process streams or in the atmospheric environment. For example, chloride SCC resulted from exposure to seawater spray carried by prevailing winds at a plant at a near-costal location. The spray soaked the insulation over type 304 stainless steel. The chlorides were then concentrated by evaporation, and cracking occurred at areas of high residual tensile stresses around welds. Other frequent causes of chloride SCC are water dripping on warm stainless piping, thereby concentrating chlorides by evaporation, and water leaching chlorides from thermal insulation (a related problem to corrosion under insulation).

As discussed previously, chlorides are present in a number of refining units, including crude distillation, hydrocracking, hydrotreating, and reforming. However, the amounts are usually low when compared to upstream oil and gas operations. High levels of chlorides are more often found in other units as contamination from upstream processing, or they are introduced with stripping steam, process water, or cooling water. The latter is a particular problem in petrochemical processes that use stainless steel heat exchangers to make steam as a means of recovering waste heat. Any chloride contamination of boiler feed water can result in chlorides concentrating on heat-exchanger tubes and can cause pitting and SCC. As a rule, austenitic stainless steels are not recommended for components in which water is likely to evaporate or condense out.

When superior resistance to chloride SCC in aqueous sulfide-containing media is required, ferritic stainless steel or duplex stainless steels can be substituted for austenitic stainless steels (Ref 139–141). The ferritic or duplex austenitic/ferritic microstructure in these materials provides added resistance to chloride SCC as long as proper microstructural and chemical control is provided in procurement and weld fabrication of these materials. Ferritic stainless steels, such as type 405 (S40500) or type 430 (S43000), are not susceptible to chloride SCC, but susceptibility to hydrogen embrittlement can occur. As mentioned previously, the duplex stainless steels have a mixed ferritic-austenitic structure and are more resistant to chloride SCC but are not immune to highly aggressive environments with high chloride concentrations and/or low pH (pH < 4.0).

For example, cold-worked type 329 (S32900) duplex stainless steel has cracked when chlorides were concentrated by vaporization of a process stream, as shown in Fig. 28. Some of the new proprietary duplex stainless steels, such as 3RE60 (S31500), 2205 (S31803), and 2507 (S32750), have shown increased resistance toward chloride SCC. However, even these materials have limitations, and highly alloyed austenitic stainless steels, such as AL6XN (N08367), 904L (N08904), alloy 28 (N08028), or nickel-base alloys, must be considered. Generally, resistance to chloride SCC increases with resistance to pitting attack based on the pitting resistance equivalent number (PREN) discussed earlier in this article (see Eq 1).

There are no simple methods (e.g., inhibitors or coatings) of preventing SCC when an austenitic stainless steel is be used in an environment known to contain chlorides. For example, in some marginal cases, plating of electroless nickel on conventional austenitic stainless steel has resulted in a decrease in SCC susceptibility due to the cathodic protection provided by the nickel layer. However, this is usually temporary, as a result of the sacrificial nature of the nickel alloy coating, in this case.

Chloride SCC in refineries and petrochemical plants often occurs under shutdown conditions when air and moisture enters equipment opened for inspection and repair. It has been found that the precautionary measures outlined in NACE RP0170 for the prevention of cracking by polythionic acids also help prevent cracking by chlorides in the absence of polythionic acid formation (Ref 142). In particular, excluding air and moisture by nitrogen blanketing and rinsing equipment with an aqueous 0.5% sodium nitrate solution have been shown to inhibit chloride SCC temporarily. To prevent chloride SCC on the outside of insulated pipe, aluminum foil has been wrapped between the insulation and pipe to provide some measure of cathodic protection. One method of preventing the catastrophic failure of components by chloride SCC is the use of austenitic stainless steel as an internal cladding. Carbon or low-alloy steel base metal used as an outer layer would not be susceptible to cracking in chloride solutions, but some localized corrosion may occur. This type of construction also provides resistance to cracking when chlorides are liable to contact the outside of the components, as in external insulation, for example.

Caustic Cracking. The SCC of various steels and stainless steels by caustic (sodium hydroxide, NaOH) is also fairly common in refinery and petrochemical plant operations. Caustic is added in the form of 5 to 40% aqueous solution to certain process streams in order to neutralize residual acid catalysts, such as sulfuric acid, hydrofluoric acid, and hydrochloric acid. Caustic is also added to cooling water and boiler feed water to counteract large decreases in pH value due to process leaks.

Although caustic attack is primarily in the form of localized corrosion (gouging) in some process streams (for example, crude oil), in others it may take the form of SCC. Traces of caustic can become concentrated in boiler feed water and cause SCC (caustic embrittlement). This occurs in boiler tubes that alternate between wet and dry conditions (steam blanketing) often resulting from overfiring or the development of local hot spots on tubes. Locations such as cracked welds or leaky tube rolls can form steam pockets with cyclic overheating and quenching conditions. These frequently lead to caustic embrittlement, because overheating and evaporation can leave caustic to concentrate in the remaining liquid. The caustic soda service chart discussed previously in this article (Fig. 3) should be consulted for guidance regarding selection of materials and the necessity for stress relief to resist corrosion and SCC in caustic environments. It is based on the operating temperature and anticipated caustic concentration that defines the materials of construction (carbon steel, carbon steel with stress relief, and use of nickel alloys).

Caustic SCC of carbon steel generally occurs at temperatures above approximately 50 to 80 °C (120 to 180 °F), depending on caustic concentration, with the temperature decreasing

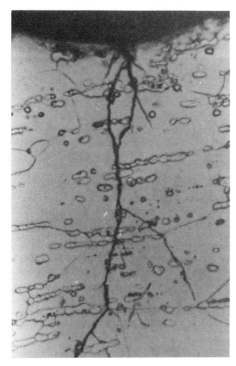

Fig. 28 Chloride stress-corrosion cracking of type 329 (S32900) stainless steel by chloride salts that concentrated as water evaporated

with increasing caustic concentration. Welded carbon steel components that are exposed to caustic solutions above these temperatures should be postweld heat treated at 620 °C (1150 °F) for 1 h per 25 mm (1 in.) of metal thickness. However, lines that are under service conditions less than 50 °C (120 °F) that may experience steam cleaning, steam tracing, or hot ambient temperatures during shutdown should also be stress relieved. Caustic SCC of austenitic stainless steels generally occurs between 105 and 205 °C (220 and 400 °F), depending on caustic concentration.

Caustic cracking of austenitic stainless steels is often difficult to distinguish from cracking by chlorides, particularly because common grades of caustic also contain some sodium chloride. As a general rule, however, SCC by chlorides in nonsensitized material is usually, but not always, in the form of highly branched, transgranular cracking, while caustic causes intergranular SCC. However, it can sometimes result in transgranular cracking, particularly if chlorides are present in the caustic environment.

Caustic SCC of carbon steel is often initiated at discontinuities in areas of surface deformation as a result of cold working or welding operations (Ref 143). Although caustic SCC occurs over a wide range of temperatures, there appears to be no correlation between temperature and time-to-failure. Because few failures have been reported at near-ambient temperatures, it appears that crack initiation times are inordinately long unless defects (such as weld defects) are involved. Dynamic strains (either imposed during testing or as a result of variable mechanical loading and thermal stresses in actual components) can also accelerate caustic SCC.

Caustic cracking of carbon steel has been found to occur over a narrow range of potentials near the active current peak of potential/log current curves. Typically, this potential range is centered at approximately −700 mV versus the standard hydrogen electrode. The most negative (active) potential for inducing caustic cracking coincides with the potential for initiating passivation by magnetite (Fe_3O_4) formation. Cracking is promoted by small amounts of dissolved oxygen, sodium chloride, lead oxide, silica, silicates, sulfates, nitrates, permanganate, and chromates that cause the active corrosion potential to move slightly in the positive (noble) direction. In contrast, large amounts of these substances act as inhibitors by pushing the corrosion potential into the passivation range. Phosphates, acetates, carbonates, and tannins also act as inhibitors.

Other caustic salts can be present in refining and petrochemical operations, such as those involving exposure to syngas, and can also cause SCC. For example, the fracture of a stainless steel pipe in a hydrogen-manufacturing facility resulted in an investigation of SCC of 304L stainless steel and 1.25Cr-1.5Mo low-alloy steel in aqueous potassium hydroxide solutions. Both alloys were found to crack in this environment. The low-alloy steel cracked at lower temperatures than did the stainless steel. Further, it was determined that the cracking of stainless steel was aggravated by the reducing environment in the process resulting from the presence of hydrogen or carbon monoxide. However, the presence of carbon dioxide in contact with potassium hydroxide solutions tends to neutralize them and reduce the susceptibility to cracking. In general, SCC in the potassium hydroxide environment required higher temperatures to crack the stainless alloy than was generally noted for aqueous sodium hydroxide environments. An upper service temperature limit of between 216 and 299 °C (421 and 570 °F) was found for the low-alloy steel (Ref 144).

Ammonia Cracking. Ammonia has caused two types of SCC in refineries and petrochemical plants. The first is cracking of carbon steel in anhydrous ammonia service, and the second type is cracking of copper alloys, such as admiralty metal (C44300). In copper alloys, SCC can occur by ammonia contamination of process streams or by ammonia-based neutralizers that are added to control corrosion. There have even been cases of ammonia-based fertilizers used on adjacent property combining with atmospheric moisture and ammonia compounds in leak-detection fluids that have resulted in SCC of copper alloy refinery equipment.

Carbon steel storage vessels, primarily spheres, have developed transgranular SCC in anhydrous ammonia service at ambient temperature but elevated pressure. In most cases, cracking was detected by inspection before leakage or rupture, but there were at least two catastrophic failures (Ref 145). There have been few problems with semirefrigerated storage vessels and no documented cases of SCC in cryogenic storage vessels. The primary causes of cracking are high stresses, hard welds, and even minor air contamination. Levels of oxygen greater than 5 to 10 ppm can sustain SCC, and this value goes down to only 1 ppm when combined with carbon dioxide. However, the presence of more than 0.1% water tends to reduce SCC susceptibility.

To minimize the likelihood of cracking, only low-strength steels, with a maximum tensile strength of 483 MPa (70 ksi), should be used in anhydrous ammonia service. Welds should be postweld heat treated at 595 °C (1100 °F) or higher, with a maximum allowable hardness of 225 HB. A water content of at least 0.2% should be maintained in the ammonia to effectively inhibit cracking. Air contamination increases the tendency toward cracking and should be minimized, if necessary, by the addition of hydrazine to the water. With a water content of 10 ppm, the oxygen content should be below 10 ppm for safe operation (Ref 146). The permissible oxygen content increases to 100 ppm with a water content of 0.1%. Regular inspection of all components in anhydrous ammonia service is recommended.

Cracking of admiralty metal (C44300) heat-exchanger tubes has been a recurring problem in a number of refining units and petrochemical process units. For example, ammonia is often used to neutralize acidic constituents, such as hydrogen chloride or sulfur dioxide, in overhead systems of crude distillation or alkylation units, respectively. Stripped sour water containing residual ammonia is used as desalter water at some crude distillation units. This practice causes ammonia contamination of the overhead system even if no ammonia is added intentionally. Ammonia is formed from nitrogen-containing feed stocks during catalytic cracking, hydrotreating, and hydrocracking operations. As a rule, cracking of admiralty metal (C44300) tubes occurs only during shutdowns, when ammonia-containing deposits on the tube surfaces become exposed to air. To prevent cracking, tube bundles should be sprayed with a very dilute solution of sulfuric acid immediately after they are pulled from their shells in order to neutralize any residual ammonia. Cracking of admiralty metal (C44300) tubes has occasionally been attributed to traces of ammonia in cooling water.

Amine Cracking. Stress-corrosion cracking of carbon steel by aqueous amine solutions, which are used to remove H_2S and carbon dioxide from refinery and petrochemical plant streams, has been a recurring problem. In one case involving a 20 wt% monoethanolamine solution, the affected equipment included two amine storage tanks, four absorber towers, a rich-amine flash drum, a lean-amine treater, and various piping (Ref 147). Cracking was found primarily at welds exposed to amine solutions at temperatures ranging from 50 to below 95 °C (125 to 200 °F). Cracking was intergranular, with crack surfaces covered by a thin film of magnetite. These oxide-filled cracks sometimes make inspection by dye-penetrant and magnetic-particle techniques difficult.

No cracking was found in piping that had received postweld heat treatment and was operating at temperatures as high as 155 °C (310 °F). Consequently, most of the affected components were replaced with new ones that had received postweld heat treatment. After careful magnetic-particle inspection, the rest of the components were repaired with welds receiving postweld heat treatment in situ or, where stress relieving was not possible, shot peening, or last-pass heat sink welding was used to provide residual compressive stresses to exposed surfaces around the welds. This compressive layer is a preventative measure but will not remain so if penetrated by corrosion or local pitting. The use of these techniques may be governed or limited by various codes and standards. For example, API RP 945 provides guidance on fabrication and inspection techniques to minimize the occurrence of amine SCC (Ref 148).

In another case, a number of leaks were discovered at piping welds in lean-amine service at temperatures between 40 and 60 °C (100 and 140 °F). None of these welds had been postweld heat treated. Again, the affected components were replaced with new ones that had received a postweld heat treatment (Ref 149).

Cracking of piping welds has also occurred in lean-amine piping of several gas-treating plants, but in all cases, temperatures were well above 95 °C (200 °F). For various reasons, these welds had not been postweld heat treated. Different types of amine solutions, including monoethanolamine, diethanolamine, and sulfinol (containing diisopropanolamine), have been involved in SCC; this confirms that cracking is not limited to monoethanolamine solutions.

Amine SCC appears to be a form of alkaline SCC that is similar in many ways to caustic SCC. The failure mode is intergranular cracking in otherwise ductile material, usually without the formation of substantial corrosion products. Actually, in amine systems, SCC appears related to conditions of marginal passivity and especially to a limited range of conditions around the transition from passive to local active behavior. Responsible species include sulfide, carbon dioxide, chloride, and cyanide. In amine solvents, carbonate films tend to passivate steel, while the presence of low levels of sulfide, thiosulfate, and thiocyanate can destabilize these normally protective films, leading to local anodic sites that initiate into SCC. Table 10 shows the influence of amine concentration, temperature, and the influence of impurities on susceptibility to the formation of fissures in slow strain-rate test specimens resulting from amine SCC (Ref 150).

Cracks, which typically run parallel to the weld, are found in the weld metal, in the base metal (~5 mm, or 0.2 in., away from the weld), and in the HAZ. Cracking is not related to weld hardness. To prevent amine SCC, postweld heat treatment at 620 °C (1150 °F) was recommended in the past for carbon steel welds exposed to amine solutions at temperatures exceeding 95 °C (200 °F). In light of the recently reported failures, welds of carbon steel components in amine service should be postweld heat treated regardless of service temperature.

Polythionic Acid Cracking. Stress-corrosion cracking of austenitic stainless steels by polythionic acids was first identified with the introduction of hydrotreating units. Austenitic stainless steels were required to provide resistance to high-temperature sulfidic corrosion in the presence of hydrogen. It was found that unstabilized austenitic stainless steel, such as type 304 (S30400), would crack adjacent to weldments during shutdowns. Typically, cracks were found to penetrate piping with a wall thickness of 12 mm (0.5 in.) in less than 8 h. Failures have been limited mostly to furnace tubes, heat-exchanger tubes, thermowells, and vessel linings (Ref 151). Similar cracking was also found in hydrocracking units and, more recently, in catalytic cracking units, in which austenitic stainless steels have found greater use because of an increase in catalyst regeneration temperatures (Ref 152).

Examples of SCC by polythionic acids are shown in Fig. 29 through 31. The cracking in roll-bonded cladding of type 304 (S30400) stainless steel (Fig. 29) is similar to mud cracking. Figure 30 shows the cracking that occurred in a type 304 (S30400) furnace tube near the weld to a carbon steel tube. These cracks are both parallel and perpendicular to the weld, reflecting different stresses in the weldment. The intergranular mode of crack propagation is shown in Fig. 31 and clearly distinguishes SCC by polythionic acids from chloride SCC (but not from caustic SCC).

Polythionic acid SCC occurs only in austenitic stainless steels and Ni-Cr-Fe alloys that have become sensitized through thermal exposure (Ref 153, 154). Sensitization occurs when the carbon present in the alloy reacts with chromium to produce chromium carbides at the grain boundaries. As a result, the areas adjacent to the grain boundaries become depleted in chromium and are no longer fully resistant to certain corrosive environments.

Sensitization of type 304 (S30400) stainless steels normally occurs at temperatures between 400 and 815 °C (750 and 1500 °F), whenever the alloy is slowly cooled through this temperature range (such as during welding and heat treating), or during normal process operations. The higher the temperature, the shorter the time of exposure required for sensitization.

Table 10 Amine concentration and temperature dependence of cracking in monoethanolamine (MEA) solvents under carbon dioxide atmosphere

MEA concentration, %	Temperature		Characteristics of fissures in slow strain-rate specimens with and without additives in environment		
	°C	°F	No additives	Carbonate/ bicarbonate added	Contaminant package added(a)
15	71	160	None	Light	None
	116	240	None	Medium	Slight
25	71	160	Medium	None	Slight
	116	240	Medium	Medium	Severe
50	71	160	None	None	Slight
	116	240	None	None	Slight

(a) Contaminant package included species commonly found in lean-amine stream, including thiosulfate and thiocyanate.

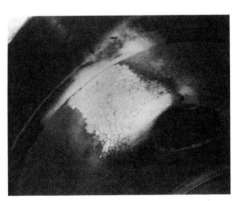

Fig. 29 Polythionic acid stress-corrosion cracking of roll-bonded type 304 (S30400) stainless steel cladding. Note that cracking stops at the type 304 (S30400) weld overlay around the nozzle opening.

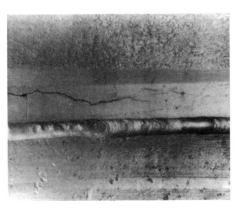

Fig. 30 Polythionic acid stress-corrosion cracking of type 304 (S30400) furnace tube near weld to carbon steel tube. Cracking is both parallel and perpendicular to weld but not in the weld.

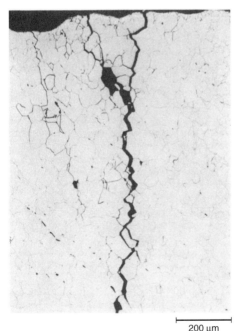

Fig. 31 Intergranular cracking typical of polythionic acid stress-corrosion cracking in type 304 (S30400) stainless steel. 75×

Addition of stabilizing elements, such as titanium or niobium, or limiting the amount of carbon are two methods for reducing the effects of welding and heat treating on sensitization. However, they are not effective for long-term exposure to temperatures above 430 °C (800 °F). The resistance of titanium-stabilized type 321 (S32100) stainless steel to polythionic SCC can be significantly improved by a thermal stabilization at approximately 900 °C (1650 °F) and holding for 2 h, with no specific limits on the cooling rate. Thermal stabilization causes the precipitation of carbon as titanium carbide rather than chromium carbide and therefore decreases the amount of carbon available for chromium carbide formation upon subsequent high-temperature exposure. Also, any chromium depletion that does occur near the grain boundaries during this time period will be counteracted by chromium diffusion from within the alloy.

Laboratory studies and plant experiences have demonstrated that austenitic stainless steels are not sensitized when applied as a weld overlay over carbon or low-alloy steels. The lack of sensitization under these conditions was verified by testing stressed samples in a solution containing polythionic acids. As can be seen in Fig. 29, SCC of the roll-bonded cladding stops at the weld overlay around the nozzle.

Polythionic acids of the type $H_2S_xO_y$ (including sulfurous acid) are formed by the reaction of oxygen and water with the iron/chromium sulfide scale that covers the surfaces of austenitic stainless steel components as a result of high-temperature sulfidic corrosion. Because neither oxygen nor water is present during normal operation under conditions in which austenitic stainless steels would be used, SCC evidently occurs during shutdowns. Oxygen and water originate from steam or wash water used to free components of hydrocarbons during shutdown before inspection or simply from atmospheric exposure. In catalytic cracking units, oxygen and water can be present during normal operations at certain locations of the catalyst regeneration system because of steam purges and water sprays for preventing catalyst accumulation. The components involved include air rings, plenums, slide valves, cyclone components, and expansion joint bellows in the catalyst regenerator and associated lines.

In general, however, SCC by polythionic acids is considered to be a problem primarily during shutdown periods; suitable procedures to prevent cracking are outlined in NACE RP0170 (Ref 142). These procedures include nitrogen purging of components that were opened to the atmosphere, purging with dry air having a dewpoint below −15 °C (5 °F), or neutralizing any polythionic acids that are formed, by washing components with a 2% aqueous soda ash (sodium carbonate) solution.

Soda ash solution should also be used for hydrotesting prior to returning components to service. Residues of soda ash solution should be left on components during temporary storage to prevent SCC. The need for this can be illustrated by an experience with a U-tube heat-exchanger bundle fabricated of type 304 (S30400) stainless steel. After the bundle had been removed from a hydrotreating unit, the external surfaces of the tubes were washed with soda ash solution, which was allowed to dry. Before storing the bundle outdoors, instructions were given to cover it to prevent rainwater from washing off the soda ash residues. It was later discovered, however, that the U-bends had not been covered and that extensive SCC had taken place at these locations.

Stress-Corrosion Cracking in Fuel Ethanol. The recent results of an experience survey and research program funded by the API and the Renewable Fuels Association indicated that a total of over 20 cases of SCC in carbon steel equipment exposed to fuel ethanol occurred during a period starting in the early 1990s through 2004 (Ref 155). These cases were found in end-user storage and blending facilities (steel tanks, rack piping, and components) and in fuel ethanol distribution storage tanks. No cases of SCC were reported thus far in fuel ethanol manufacturer facilities or tanker trucks, railroad tanker cars, or barges used to transport fuel ethanol. Additionally, no SCC has occurred following blending of fuel ethanol with gasoline. This suggests that SCC may be related to changes in the fuel ethanol as it moves through the distribution chain over a period of days, weeks, or months. From the results of laboratory research studies in simulated fuel ethanol blends, it appears that these likely changes involve aeration (O_2 pickup) during handling, which was found to increase susceptibility to SCC. Furthermore, corrosion monitoring of steel in fuel ethanol environments at two field sites also corroborated the impact of aeration by indicating increased corrosion rate and tendencies for localized corrosion in aerated environments (Ref 155).

Thermodynamic modeling of the fuel ethanol environment in accordance with ASTM D 4806 indicated that during most expected cases, steel equipment should be in a range of active corrosion extending between a passive region (at high potential) and a region of immunity to corrosion (at low potentials). This condition suggests that active-passive behavior could be present, depending on the range of corrosion potentials exhibited during service. Such behavior can be affected by aeration or the presence of oxidizing species in the environment. Furthermore, this also suggests that an active-passive mechanism of SCC in fuel ethanol exists that may be similar to that in other SCC systems involving steel (e.g., carbonate-bicarbonate SCC, CO-CO_2 SCC, etc.). This was consistent with the results of the research conducted by Southwest Research Institute.

While identification of this phenomenon is relatively recent, analysis of field failures has documented that cracking of steel in fuel ethanol is characterized by highly branched, intergranular cracks in highly stressed locations, such as associated with non-postweld heat treated welds; areas of stress concentration and local bending, such as near lap seam welds; and cyclic loading due to flexing and bending. Techniques used in various situations to reduce the occurrences of SCC in fuel ethanol service include implementation of epoxy coatings to mask welded or highly stressed areas or stress relief when possible. Enhanced guidelines for fuel ethanol service and other techniques to reduce susceptibility to SCC are still being explored.

Wet H_2S Cracking

Corrosion of carbon and low-alloy steels by aqueous H_2S solutions or sour waters (generically referred to as refinery wet H_2S environments) can result in one or more types of environmentally assisted cracking EAC. These forms of EAC are related primarily to the damage caused by hydrogen that results from the production of hydrogen by the sulfide corrosion process in aqueous media. They include loss of ductility on slow application of strain (hydrogen embrittlement), formation and propagation of hydrogen-filled blisters or voids in the material (hydrogen blistering or hydrogen-induced cracking, or HIC), and spontaneous cracking of high-strength or high-hardness steels (hydrogen embrittlement cracking, also known more familiarly as SSC when involving environments that include exposure to H_2S). A monograph of classic papers published on cracking of steels in petroleum upstream and downstream wet H_2S environments was published by NACE International and is given in Ref 156.

In wet H_2S refinery environments, atomic hydrogen (H^0) forms as part of the sulfide corrosion process. When steel corrodes in aqueous H_2S-containing environments, it forms a mostly insoluble FeS corrosion product and also liberates hydrogen atoms (also referred to as monatomic hydrogen). If these hydrogen atoms come in close proximity to each other, they can recombine to form molecular hydrogen (H_2). Once this recombination process takes place on the metal surface, molecular hydrogen is too large to enter the metal lattice. It is only the atomic form of hydrogen that can enter the material during aqueous corrosion and potentially lead to the aforementioned forms of EAC.

During these corrosion processes, hydrogen atoms formed from cathodic reactions first adsorb on the metal surface prior to recombination. It is at this point that the hydrogen atoms can recombine to form H_2 gas that is commonly seen bubbling off the corroding metal surfaces. However, in the presence of certain chemical species known as hydrogen recombination poisons, the formation of hydrogen molecules can be retarded. Sulfur (from H_2S), arsenic, phosphorus, tin, lead, and bismuth are commonly known hydrogen recombination poisons. Due to the retarding effect of sulfur species on the recombination process, the atomic hydrogen produced by the corrosion process is more likely to reside in the atomic form, become absorbed into the material, and permeate according to its diffusivity and solubility in the microstructure. Information on these processes and ways to measure hydrogen permeation are given in ASTM G 148 (Ref 157).

Sulfide Stress Cracking. While still in the atomic state, monatomic hydrogen can diffuse to and concentrate at sites of microstructural discontinuities such as phase, precipitate or grain boundaries, dislocations, and sites of high stress and/or lattice distortion (strain), where they can interfere with the normal ductility processes of the material. The accumulation of atomic hydrogen in the locally distorted metal lattice is the direct result of the lattice dilation, which is better able to accommodate the presence of the interstitial hydrogen. At sufficiently high concentrations in the solid state, atomic hydrogen can also affect the bonding between atoms to promote decohesion, particularly along grain boundaries and other zones where the lattice has already been distorted by strain, cold working, or hardening. These are solid-state reactions between the atomic hydrogen and the metal lattice and its local defect structures. This is the basis for SSC of steels and most other engineering materials.

Sulfide stress cracking is normally associated with high-strength steels and alloys—yield strength greater than 550 MPa (80 ksi)—and with high-hardness (>22 HRC) structures in weld HAZs. Non-postweld heat treated weldments are particularly problematic, because they often contain both high HAZ hardness and high residual tensile stresses that can initiate SSC and promote crack propagation. Resistance to SSC is usually improved through the use of postweld heat treatment and through the use of lower-carbon-equivalent plate steels and quenched-and-tempered wrought steels.

Hydrogen-Induced Cracking. Once atomic hydrogen has diffused into the material, it can also recombine to form molecular hydrogen (H_2) within the metal at internal defects, inclusions, and pores. Sites for recombination are commonly observed to be weak internal interfaces such as those at manganese sulfide inclusions or metallurgical laminations. Ferrite-pearlite banding and related inclusions can also produce locally weak interfaces in the material that can result in small hydrogen-filled blisters being produced. Because hydrogen molecules are much larger than atomic hydrogen, once the hydrogen recombines to form hydrogen gas (H_2), it cannot readily diffuse out of these sites. This results in a buildup of pressure inside these blisters, which drives their growth, and eventually results in propagation and linkage of hydrogen-filled blister cracks in the material, commonly known as HIC and also previously referred to as stepwise cracking due to the visual appearance of these cracks stepping through the material (Fig. 32). This phenomenon usually is of concern in lower-strength plate steels—less than 550 MPa (80 ksi) and low hardness (<HRC 22)—used in rolled and longitudinally seam welded or electric resistance welded pipe, or plate steels used in the manufacturing of refinery vessels and tanks.

Stress-Oriented Hydrogen-Induced Cracking (SOHIC). Under conditions of very high hydrogen flux and high tensile stress or high local stress concentration, small blister cracks similar to those normally associated with HIC can become aligned perpendicular to the tensile stress. Often, these cracks align in the through-thickness direction of the material, thereby reducing the load and/or pressure-carrying capabilities of the material and fabricated equipment (Fig. 33). This phenomenon is referred to as SOHIC. This phenomenon has been appreciated as a problem in refinery equipment only within the past 15 years, with the advent of improved lower-sulfur steels. During the 1990s, it was the subject of extensive research by a consortium of companies and by the API to develop a more fundamental understanding of this process and to develop methods for inspection, laboratory testing, and repair of damaged equipment. Further discussion of this and the other hydrogen-cracking mechanisms common in wet H_2S refinery systems are given in several publications (Ref 158–161).

Hydrogen-cracking phenomena occur primarily when steel is exposed to aqueous H_2S solutions having low pH values. Under these conditions, corrosion rates are typically high, thus resulting in high hydrogen flux into materials. In the range of pH less than 5, H_2S is chemisorbed to the steel surface and poisons the reaction between hydrogen atoms. Aqueous H_2S solutions having high pH values can also cause severe hydrogen charging and resultant cracking. This normally occurs when cyanides are present in the sour water environment, particularly at slightly elevated temperatures. In the absence of cyanides, aqueous H_2S solutions with pH values above 8 most often do not corrode steel, because a protective iron sulfide film forms on the surface. However, if higher H_2S partial pressures are present, particularly when combined with highly turbulent conditions, corrosion rates increase, as do the resultant hydrogen flux into the material.

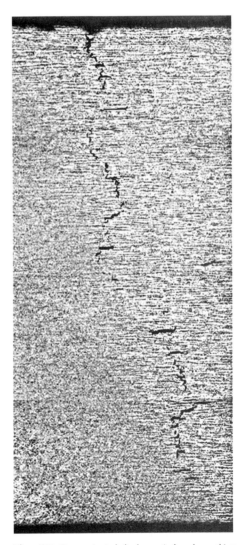

Fig. 33 Stress-oriented hydrogen-induced cracking in refinery plate steel. Note the stacked array of hydrogen blister cracks going through the thickness of the material (vertical) oriented perpendicular to the direction of the applied tensile stress (horizontal).

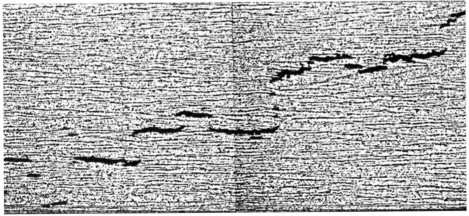

Fig. 32 Hydrogen-induced cracking, which is also referred to as stepwise cracking

The role of cyanides is to destroy this normally protective sulfide film on the metal surface and convert it into soluble ferrocyanide [Fe(CN)$_6^{4-}$] complexes. As a result, the now-unprotected steel can corrode very rapidly with little or no barrier-limiting hydrogen flux. For practical purposes, the corrosion rate depends primarily on the bisulfide ion (HS$^-$) concentration and, to a lesser extent, on the cyanide ion (CN$^-$) concentration. The more bisulfide ion that is present, the more cyanide that is required to destroy the protective iron sulfide film. It has been shown experimentally that corrosion of steel in aqueous ammonia/sulfide/cyanide solutions with pH values above 8 is nearly always accompanied by high hydrogen flux and associated cracking or blistering (Ref 162).

Hydrogen Embrittlement. This phenomenon is characterized by decreasing ductility at slow strain rates; this is contrary to the normal ductile behavior of metals in most types of mechanical embrittlement (Ref 163). For example, the ductility of carbon steel has been reported to drop from 42 to 7% when charged with hydrogen (Ref 164). This loss of ductility is observed during slow strain-rate testing (usually in the strain-rate range of 10^{-4} to 10^{-7} s^{-1}) but not during high rate-impact tests, such as the Charpy V-notch test. Failure, in the form of cracking, usually occurs some time after a load is applied to hydrogen-charged steel or during active charging in a sour environment.

Another form of this type of attack occurs when a susceptible material is held at constant load under the influence of hydrogen. When this phenomenon was first discovered in the early 1950s, it was referred to as static fatigue, and the minimum load for failure to occur is known as the static fatigue limit. These terms can be found in some older references, but currently they are usually referred to as delayed failure and threshold stress, respectively, so as not to be confused with mechanical fatigue.

Until initiation of a physical crack in a material, hydrogen embrittlement is usually considered temporary in many materials of construction, such as steel. It can be normally be reversed by heating the steel to drive out the hydrogen. The rate of recovery depends on the time and temperature of heating, the amount of hydrogen in the material at the start, and the critical amount of hydrogen to produce embrittlement. For example, high-strength steels have very low critical levels for hydrogen, whereas low-strength materials typically have a much higher tolerance for hydrogen. Heating to 230 °C (450 °F) and holding for 1 h/25 mm (1 in.) of thickness has been found to be adequate to prevent cracking after welding of constructional steels. Temperatures as high as 650 °C (1200 °F) for 2 h or as low as 105 °C (225 °F) for 1 day have also reportedly been used to restore full ductility. There have even been reports that the heat of the sun on a warm summer day was found to be sufficient to restore ductility to a high-carbon, cold-drawn steel wire that had been embrittled by exposure to a wet H$_2$S environment. This is the basis for the commonly used method of baking out steels that have been exposed to wet H$_2$S environments before welding. As a rule, however, heating to temperatures above 315 °C (600 °F) for any length of time should be avoided to lessen the possibility of high-temperature hydrogen attack during the outgassing process.

Titanium can also become embrittled by absorbed hydrogen as a result of corrosion, excessive cathodic protection, or exposure to hydrogen gas (Ref 25). When hydrogen is absorbed by titanium in excess of approximately 150 ppm, a brittle titanium hydride phase will form, as shown in Fig. 34. Embrittlement due to titanium hydride precipitation is usually permanent and can be reversed only by vacuum annealing, which is difficult to perform. Absorption of hydrogen by titanium dramatically increases once the protective oxide film normally present on the metal is damaged through either mechanical abrasion or chemical reduction. Hydrogen intake is accelerated by the presence of surface contaminants, including iron smears, and occurs predominantly as temperatures exceed 70 °C (160 °F).

Hydriding can be minimized by anodizing or thermal oxidizing treatments to increase the thickness and resistance of the normally protective titanium oxide film. If it is impractical to apply these treatments, acid pickling of titanium components—with 10 to 30 vol% nitric acid containing 1 to 3 vol% hydrofluoric acid at 49 to 52 °C (120 to 125 °F) for 1 to 5 min—can be performed to remove iron smears. Acid pickling is also recommended for cleaning titanium components after inspection and repairs during shutdowns, especially components exposed to concentrated acetic acid in certain petrochemical operations. To minimize hydrogen pickup during pickling, the volume ratio of nitric acid to hydrofluoric acid should be near 10. In some highly aggressive process environments, titanium components may have to be electrically insulated from more anodic components, such as aluminum, to prevent hydride formation as a result of hydrogen evolution on titanium surfaces from the resultant cathodic polarization. A similar situation is found when titanium is used in systems involving other metals under cathodic protection with impressed current or sacrificial anodes (zinc, aluminum, or magnesium). This level of cathodic polarization can also induce hydride formation. It may be necessary to limit cathodic polarization or totally isolate the titanium from the other materials. When process streams contain a significant volume of hydrogen (for example, reactor effluent from hydrotreating units), titanium should be used only at temperatures below 175 °C (350 °F).

Information on the mechanisms of hydrogen embrittlement is available in the article "Hydrogen Damage" in *Corrosion: Fundamentals, Testing, and Protection*, Volume 13A of *ASM Handbook*, 2003. Test procedures are reviewed in the article "Evaluating Hydrogen Embrittlement," also in Volume 13A.

Experience with Hydrogen Blistering, HIC, and SOHIC. These phenomena have been a problem primarily in the vapor recovery (light ends) section of catalytic cracking units and, to a lesser degree, in the low-temperature areas of the reactor effluent section of hydrotreating and hydrocracking units (Ref 165–167). Hydrogen blistering has also been seen in the overhead systems for sour water stripper towers and amine regenerator (stripper) towers, as well as in the bottom of amine contactor (absorber) towers.

An example of hydrogen blistering in an absorber/stripper tower of a catalytic cracking unit is shown in Fig. 35. Hydrogen blistering often accompanies hydrogen embrittlement as a result of aqueous sulfide corrosion. Internal hydrogen blistering on a microscopic scale can lead to HIC. Cracking proceeds in the metal ligaments between adjacent blister cracks because of the resultant tensile stresses produced ahead of the propagating hydrogen blister crack. As a rule, the severity of hydrogen blistering depends on the severity of corrosion and the efficiency of hydrogen charging, which depends heavily on the chemical nature of the environment (H$_2$S partial pressure, pH, temperature, presence of cyanides). In some cases, some alkaline sour water environments that produce low corrosion rates have high enough charging efficiencies to produce high hydrogen fluxes and consequently cause extensive damage (Ref 161).

In older vintage and low-quality steels, hydrogen blistering is associated with dirty steel (i.e., high sulfur) with highly oriented slag

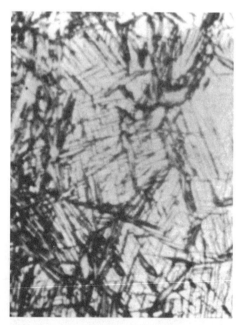

Fig. 34 Hydride formation in Ti-grade 2 (R50400) after galvanic coupling to carbon steel in sour water at 110 °C (230 °F)

inclusions or laminations. These materials have produced large internal blisters in plate steels used to construct pressure vessels and tanks. In some cases, these blisters can reach a size of 30 cm (1 ft) diameter or greater. More modern steels used in refinery operations with low to intermediate sulfur levels typically can show susceptibility to HIC. This can produce smaller blister cracks, but they can link up to produce steplike cracks through the wall of the steels. In these cases, low sulfur concentrations (<0.004 wt%) and/or alloying additions of calcium are used in combination to increase HIC resistance of steels. Lowering the sulfur content increases the cracking resistance due to the reduction in the size and number of sulfide inclusions. Calcium additions provide sulfide shape control, thus resulting in more spheroidal-shaped inclusions, which are less likely to initiate and propagate HIC.

It was also found that some steels with advanced processing and low sulfur composition and low susceptibility to HIC, as found by the standard NACE TM0284 HIC test method (Ref 168), exhibited high susceptibility to SOHIC when under service conditions involving high hydrogen flux and high tensile stress (this experience is usually associated with welded equipment under refinery sour water conditions involving the presence of cyanide in the process environment). This susceptibility was found to relate to microstructural and other factors. Plate materials with highly banded microstructures had low threshold stresses and failed by SOHIC in wet H_2S tests per NACE TM0103 (Ref 169) and also had low susceptibility to crack propagation in the presence of stress concentrators (weld defects, SSC, notches, machined slots) in full-scale vessel tests under internal pressure and sour solutions. It was also shown that the presence of fine precipitates from overtreatment with calcium in ultralow-sulfur plate steels also influenced susceptibility (Ref 170).

Vapor/liquid interface areas in equipment also show higher damage rates. It is presumed that this is because ammonia, H_2S, and cyanides concentrate in the thin water films or in water droplets that collect at these areas. Local turbulence and flashing can also disrupt normally protective sulfide films on the metal surface at these locations.

The basic approach toward reducing corrosion and hydrogen blistering in the various vapor-compression stages of catalytic cracking units should be aimed at decreasing the concentration of cyanide and bisulfide ions in water condensate. Several methods for accomplishing this have been tried over the years (Ref 171, 172), including conversion of cyanide to generally harmless thiocyanate (SCN^-) by injection of air or ammonium polysulfide solutions at various locations. While this has beneficial effects, it has often produced undesirable side effects, such as accelerated corrosion and fouling at stagnant or low-flow areas. In contrast, reducing the concentration of the sour water and its impurities by injection of wash water in the compressed wet gas streams, in conjunction with corrosion inhibitor injection, has been found to be very effective when applied correctly and consistently (Ref 173). Water washing reduces the concentration of cyanides by improved contacting of vapors and dilution of water condensate. To prevent dissolved and suspended solids from fouling the compressor aftercooler, only water of fairly good quality, such as boiler feed water or steam condensate, should be injected. To reduce the amount of fresh water used, stripping-steam condensate from the reflux drum can be used. As a rule, there is sufficient stripping-steam condensate to meet the wash-water requirements. This process can be optimized through use of real-time monitoring of corrosion rate and hydrogen permeation to achieve maximum benefit with minimum wash water as the unit feeds and process stream chemistry change (Ref 159).

It is important that the waste sour water from the interstage and high-pressure separator drums be sent directly to waste disposal rather than first being recycled to the reflux drum. Waste water is often recycled for convenience so that its pressure can be reduced in the reflux drum prior to disposal. This alleviates the need for an external depressuring drum but will build up the concentration of ammonia, H_2S, and, especially, hydrogen cyanide in the wet gas leaving the reflux drum. Consequently, excessive concentrations of cyanides will be found in water condensing in the high-pressure stage. Water washing of the overhead of the debutanizer and depropanizer is indicated only if serious fouling problems occur. Normally, these streams are quite dry and should be kept that way to minimize corrosion and hydrogen blistering problems. With proper water washing of the compressed wet-gas stream, water washing of the overhead vapor streams of the debutanizer and depropanizer towers becomes unnecessary.

Corrosion inhibitors help control aqueous sulfide corrosion and hydrogen blistering even though cyanides may still be present. Hydrogen activity probes and chemical tests of water condensate are used to monitor the effectiveness of water washing and inhibitor injection. Where limited hydrogen blistering occurs in certain components of hydrotreating and hydrocracking units, it is usually sufficient to line affected areas with stainless steel or alloy 400 (N04400). This also applies to components of overhead systems for sour water stripper towers and amine regenerator (stripper) towers or the bottoms of amine contactor (absorber) towers.

In recent years, an increase in observations of wet H_2S cracking (HIC, SOHIC, and SSC) was encountered as turnaround inspection frequencies increased after a major failure incident. During this same period, methods for surface inspection improved in sensitivity with the increasing use of wet fluorescent magnetic-particle inspection and ultrasonics. An interesting observation was made that many vessels were found to have occurrences of internal HIC and SOHIC upon multiple inspections. A large-scale vessel test was conducted under the auspices of the API, which identified that removal of the sulfide films on the internal vessel surfaces prior to inspection resulted in a significant increase in hydrogen charging once the vessel was put back into wet H_2S service (Ref 160). This increase in hydrogen flux was then correlated with increasing occurrences of cracking following inspection. Other

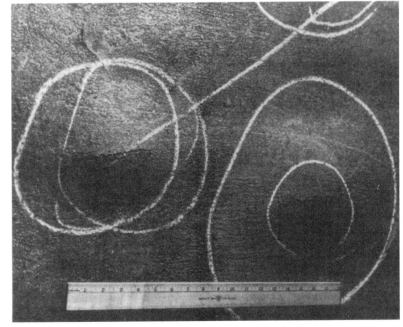

Fig. 35 Hydrogen blistering of a carbon steel shell of an absorber/stripper tower in the vapor recovery (light ends) section of a catalytic cracking unit. Note that the blisters have cracked open to the vessel interior.

work showed that application of specific inhibitor compounds that promoted the recombination of atomic hydrogen, while simultaneously allowing the protective sulfide film to reform, substantially reduced postinspection damage of equipment in wet H_2S service (Ref 174).

Guidelines for materials selection for new vessels in refinery wet H_2S service are available in NACE publication 8X194. (Ref 175). Guidelines for inspection and repair of existing vessels are given in NACE RP0296 (Ref 176).

Refinery Experience with SSC. Sour water containing H_2S can cause spontaneous cracking of highly stressed, high-strength steel components, such as bolting and compressor rotors as a result of SSC (Ref 177). Cracking has also occurred in carbon steel components containing hard welds (Ref 178). Cracking can be transgranular, intergranular (with respect to prior-austenite grain boundaries), or mixed mode and will contain sulfide corrosion products, as shown in Fig. 36 and 37. Cracking of this type has become known as SSC and should not be confused with hydrogen-induced (stepwise) cracking presented previously in this article.

Sulfide stress cracking was first identified in the production of sour crude oils when high-strength steels used for well-head and downhole equipment cracked readily after contacting produced water that contained H_2S. Sulfide stress cracking was not experienced by refineries and petrochemical plants until the introduction of high-pressure processes that required high-strength bolting and other high-strength components in gas compressors. With the increased use of submerged arc welding for pressure vessel construction, it was found that weld deposits significantly harder and stronger than the base metal could be produced. This led to transverse cracking in weld deposits with hardness greater than 200 HB (Ref 178).

The mechanism of SSC has been the subject of many investigations, most of which attempted to address the cracking seen in high-strength steels instead of the lower-strength steels used in refinery and petrochemical plant equipment. In general terms, SSC occurs in the same corrosive environments that lead to the other forms of refinery wet H_2S cracking. Hydrogen sulfide affects the corrosion rate and the relative amount of hydrogen absorption but otherwise does not appear to be directly involved in the cracking mechanism. As a general rule of thumb, SSC can be expected to occur in process streams containing in excess of 50 ppm H_2S in the gas phase (not dissolved in solution). However, SSC susceptibility is actually related to the partial pressure of H_2S in the service environment, and this H_2S limit may vary with total pressure. Therefore, there can be SSC occurring at lower H_2S concentrations.

There is also a direct relationship between H_2S concentration and the allowable maximum hardness value of the HAZ on one hand and cracking threshold stress on the other. Typically, the allowable maximum hardness value decreases 30 HB, and the allowable threshold stress decreases by 50% for a tenfold increase in H_2S concentration (Ref 179). Also, SSC occurs most readily at or near ambient temperature, with susceptibility decreasing with increasing service temperature. As in the case of hydrogen embrittlement and hydrogen blistering, SSC of steel in refineries and petrochemical plants often requires the presence of cyanides.

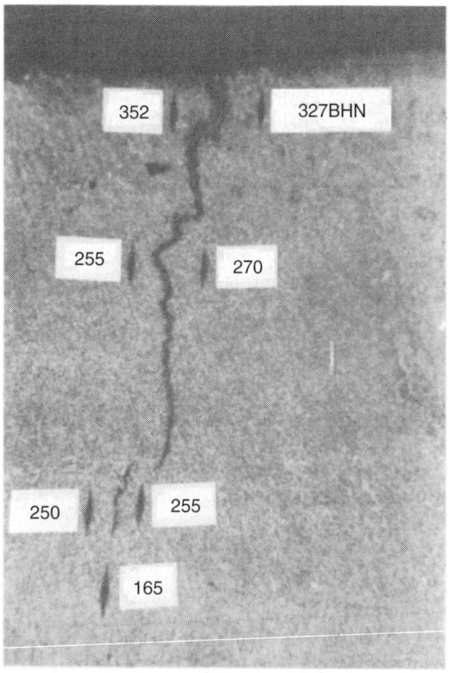

The most effective way of preventing SSC is to ensure that the steel is in the proper metallurgical condition. This usually means that weld hardness is limited to 200 HB (Ref 180). Because hard zones can also form in the HAZs of welds and shell plates from hot forming, the same hardness limitation should be applied in these areas. Guidelines for dealing with the SSC that occurs in refineries and petrochemical plants are given in API RP 942 (Ref 181) and NACE RP0472 (Ref 182). The most comprehensive guidelines for materials selection for resistance to SSC in refinery operations are now provided in NACE MR0103 (Ref 183).

Postweld heat treatment of fabricated equipment will greatly reduce the occurrence of SSC. The effect is twofold: First, there is the tempering effect of heating to 620 °C (1150 °F) on most hard microstructures (the possible exception being highly microalloyed steels), and second, the residual stresses from welding or forming are reduced. The residual tensile stresses typically represent a much larger effect on the equipment than the internal pressure or other mechanical stresses.

A large number of the ferrous alloys, including stainless steels, as well as certain nonferrous alloys, are susceptible to SSC. Cracking may be expected to occur with carbon and low-alloy steels when the tensile strength exceeds 550 MPa (80 ksi). Because there is a relationship between hardness and strength in steels, the aforementioned strength level approximates the 200 HB hardness limit. For ferrous and nonferrous alloys used primarily in upstream oil field equipment, limits on hardness and/or heat treatment have been established in NACE MR0175/ISO 15156 (Ref 184). In the past, versions of this standard have also been used for petroleum refinery service. However, due to their acidic nature and high levels of chloride, many oil field environments can be generally more corrosive than those encountered during many refining operations (the exception being refinery environments with cyanide). It is now recommended that the MR0175 standard not be used for selection of materials for petroleum refining service. The use of the newer NACE MR0103 standard for refinery operations is preferred (Ref 183).

Hydrogen Attack

The term *hydrogen attack* (or, more specifically, high-temperature hydrogen attack) refers to the deterioration of the mechanical properties of steels in the presence of hydrogen gas at elevated temperatures and pressures. Hydrogen attack is potentially a very serious problem with regard to the design and operation of refinery equipment in hydrogen service (Ref 185, 186).

Fig. 36 Sulfide stress cracking of a hard weld of a carbon steel vessel in sour water service. BHN, Brinell hardness. 40×

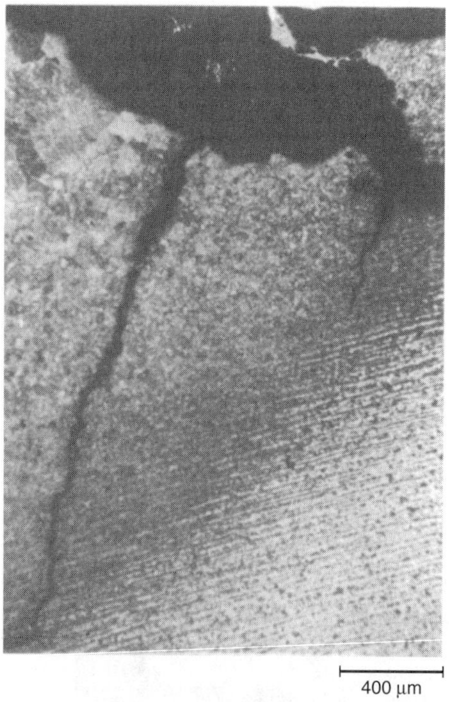

Fig. 37 Sulfide stress cracking of hard heat-affected zone next to weld in A516-70 pressure vessel steel after exposure to sour water. 35×

It is of particular concern in hydrotreating, reforming, and hydrocracking units at above roughly 260 °C (500 °F) and hydrogen partial pressures above 689 kPa (100 psia) (Ref 187). Under these conditions, molecular hydrogen (H_2) dissociates at the steel surface to atomic hydrogen (H^0), which readily diffuses into the steel. At grain boundaries, dislocations, inclusions, gross discontinuities, laminations, and other internal voids, atomic hydrogen will react with dissolved carbon and with metal carbides to form methane.

The large molecular size of methane actually precludes diffusion and relief of the gas pressure formed inside the material. As a result, internal methane pressures become high enough to blister the steel or to cause intergranular fissuring (Ref 166). At higher temperatures, the dissolved carbon diffuses to the steel surface and combines with atomic hydrogen to evolve methane. Hydrogen attack now takes the form of overall decarburization and loss of material strength.

The overall effect of hydrogen attack is the partial depletion of carbon in pearlite (decarburization) and the formation of fissures (blisters and cracks) inside the metal, as shown in Fig. 38. As attack proceeds, these effects become more pronounced, as shown in Fig. 39, in which partial depletion of carbon is evident in some of the grains, while others are completely decarburized. Hydrogen attack is accompanied by loss of tensile strength and ductility. Consequently, unexpected failure of equipment without prior warning signs (mechanical embrittlement and the associated loss of ductility and toughness) is the primary cause for concern.

Forms of Hydrogen Attack. As briefly mentioned previously, hydrogen attack can take several forms within the metal structure, depending on the severity of the attack, stress, and the presence of inclusions in the steel.

General surface attack occurs when equipment, which is not under high stress, is exposed to hydrogen at elevated temperatures and pressures. As a rule, decarburization is not uniform across the surface or through the thickness; instead, it takes place at various locations within the structure, such as grain boundaries, carbides, and inclusions. The fissures that form are parallel to the metal surface, often following inclusions, banding, or laminations. The fissures themselves are small and are not linked together, as may happen with more severe stages of attack.

Hydrogen attack also initiates at areas of high stress or stress concentration in the steel, because atomic hydrogen preferentially diffuses to these areas of lattice dilatation. Isolated fingers of decarburized and fissured material are often found adjacent to weldments and are associated with the initial stages of hydrogen attack. It is also evident that the fissures tend to be parallel to the edge of the weld rather than the surface. This orientation of fissures is probably the result of residual stress adjacent to the weldment and/or instability of carbides around the weld HAZ. Fissures in this direction can form into through-thickness cracks.

The necessary stress for inducing localized hydrogen attack is not limited to areas of weldments. Hydrogen attack has been found to be concentrated at the tip of a fatigue crack that initiated at the toe of a fillet weld and propagated along the HAZ of the weld. In this case, the hydrogen-containing process stream evidently entered the fatigue crack and caused fissuring around the crack tip, as shown in Fig. 40. Although no evidence of attack was found in adjacent portions of the piping system, the localized attack was the cause of a major failure.

Severe hydrogen attack can result in blisters and delaminations in steels, as shown in Fig. 41. This is an advanced stage of hydrogen attack, and it is accompanied by complete decarburization throughout the cross section of the steel. The laminar nature of the fissures is typically obtained when no local stresses are present, but the physical appearance of this blistering is quite similar to hydrogen blistering (described earlier in this article).

Prevention of Hydrogen Attack. The only practical way to prevent hydrogen attack is to use steels that, based on plant experience, have been found to be resistant to this type of deterioration. The following general rules are applicable to hydrogen attack:

- Carbide-forming alloying elements, such as chromium and molybdenum, increase the resistance of steel to hydrogen attack and the progression of related damage.
- Increased carbon content decreases the resistance of steel to hydrogen attack.
- Heat-affected zones are more susceptible to hydrogen attack than the base or weld metal. This is usually associated with differences in their carbide structures and local stresses.

For most refinery and petrochemical plant applications, low-alloy chromium- and molybdenum-containing steels are used to minimize

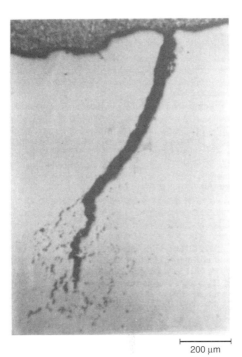

Fig. 40 High-temperature hydrogen attack, in the form of localized fissuring, at the tip of a fatigue crack that initiated at the toe of a fillet weld. 70×

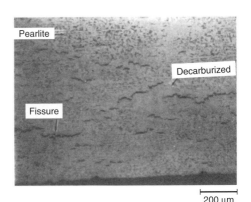

Fig. 38 High-temperature hydrogen attack of carbon steel in the form of decarburization and fissuring. 50×

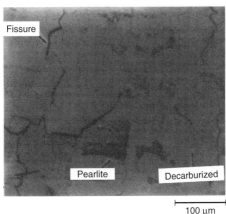

Fig. 39 Depletion of carbon in pearlite colonies and formation of grain-boundary fissures due to high-temperature hydrogen attack of carbon steel. 140×

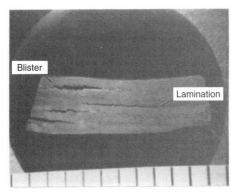

Fig. 41 High-temperature hydrogen attack in the form of blistering and laminar fissuring throughout the wall thickness of a carbon steel pipe

or prevent hydrogen attack. However, questions have been raised regarding the effect of long-term hydrogen exposure on C-0.5Mo steel (Ref 188). As a result, low-alloy steels with higher chromium and molybdenum content are preferred over C-0.5Mo steel for new construction.

The conditions under which different steels can be used in high-temperature hydrogen service are listed in API RP 941 (Ref 189). The principal data are presented in the form of curves referred to as the Nelson curves, as shown in Fig. 42. The curves are based on long-term refinery experience, rather than on laboratory studies. However, recent laboratory data have been used to further understand the time-dependent behavior of hydrogen attack and how it relates to damage observed in service. The curves are periodically revised by the API Subcommittee on Materials Engineering and Inspection, and the latest edition of API RP 941 should be consulted to ensure that the proper steel is selected for the operating conditions encountered. The change in 1990 removed the Nelson curves for C-0.5Mo steels from the body of the API standard. This change was prompted by the documentation of eight cases of hydrogen attack in equipment made from C-0.5Mo steel. The curves for this material are now provided in an appendix for historical purposes only. The selection of C-0.5Mo steels is now based on the curves provided for carbon steels.

Factors that are important for scheduling inspections for high-temperature hydrogen attack of steel equipment in refinery service include:

- Operating conditions (partial pressure of hydrogen and temperature) that are relative to the operating limits provided in API RP 941.
- Metallurgical condition of C-0.5Mo steels in the welded or annealed condition have less resistance to hydrogen attack than these steels in the normalized condition.
- Welds receiving postweld heat treatment are less susceptible to hydrogen attack than in the as-welded condition.

Hydrogen-Induced Disbonding. Another form of damage potentially resulting from refinery high-temperature hydrogen service is hydrogen-induced disbonding (HID) of stainless steel clad or weld overlaid steel plates used in hydroprocessing equipment (Fig. 43). This form of attack usually results in the formation of blisters at or near the bond/fusion line between the steel and stainless alloys. Hydrogen-induced disbonding occurs with increasing frequency at high hydrogen pressures and service temperatures and with increased rapid cooling as a result of process changes and shutdown, start-up cycles (Ref 190). It has also been found that the process of cladding or weld overlaying can also affect susceptibility to HID. A laboratory test procedure involving exposure of bimetallic samples to high-temperature hydrogen environments has been developed and standardized in ASTM G 146 (Ref 191). These procedures can be tailored to specific service applications (temperature, hydrogen partial pressure, and cooling rates) for purposes of qualification of particular fabrication and welding techniques used in vessel construction.

Other Associated Phenomena. In addition to hydrogen attack, hydrogen stress cracking can occur at carbon and low-alloy steel welds that were in hydrogen service above approximately 260 °C (500 °F). Cracking occurs upon cooling and is intergranular and typically follows lines of high, localized stress and/or hardness. Cracking is caused by the increased effect of dissolved hydrogen in the steel at lower temperatures. This phenomenon is prevented by postweld heat treatment. Hydrogen outgassing procedures should be followed when equipment is depressurized and cooled prior to shutdown.

Stainless steels with chromium contents above 12% and, in particular, the austenitic stainless steels are immune to hydrogen attack. It should be noted, however, that atomic hydrogen will diffuse through these steels; as a result, they will not provide protection against hydrogen attack if applied as a loose lining or an integral cladding over a nonresistant base steel.

Corrosion Fatigue

Corrosion, in conjunction with cyclic stressing, can bring about a significant reduction in the fatigue life of a metal. Failure under these circumstances is described as corrosion fatigue. Rotating equipment, valves, and some piping runs in refineries and petrochemical plants may be subject to corrosion fatigue. In particular, pump shafts and various springs are the two most likely candidates for corrosion fatigue. The types of springs involved include those of scraper-blade devices in a wax production unit, internal springs in relief valves, and compressor valve springs.

Prevention of Corrosion Fatigue. A number of corrective procedures are available for preventing corrosion fatigue. These include increasing the fatigue resistance and/or corrosion resistance of the metal involved, reducing the number of stress cycles or the stress level, and removing or inhibiting the corrosive agent in the environment. Fatigue life can often be increased through heat treatments or alloy changes, which make the metal stronger and tougher. However, this may induce other failure mechanisms such as SSC in the case of wet H_2S environments. Corrosion resistance can be improved by applying protective coatings or by upgrading the material to one of higher alloy content. Design changes can reduce or eliminate vibrations or (in a spring) reduce the stress per cycle. Finally, adding a corrosion inhibitor or removing a source of pitting (i.e., local stress concentrators), such as chlorides, can often increase the corrosion fatigue life. Additional

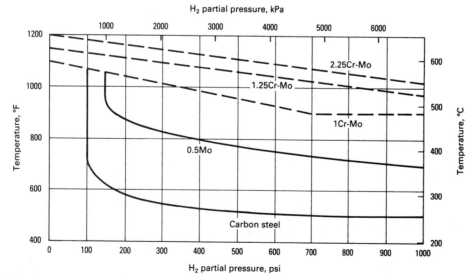

Fig. 42 Operating limits for various steels in high-temperature high-pressure hydrogen service (Nelson curves) to avoid decarburization and fissuring. Source: Ref 189

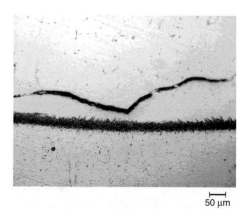

Fig. 43 Hydrogen-induced disbonding of stainless steel clad plate steel produced in a laboratory test in accordance with ASTM G 146 in high-pressure hydrogen. The crack is in the stainless steel cladding shown at the top of the micrograph. 200×

information on corrosion fatigue is available in the articles "Forms of Mechanically Assisted Degradation" and "Evaluating Corrosion Fatigue" found in *Corrosion: Fundamentals, Testing, and Protection,* Volume 13A of *ASM Handbook,* 2003.

Liquid Metal Embrittlement

Although liquid metal embrittlement has been recognized for at least 60 years, it has received far less attention than the more commonly encountered hydrogen damage or SCC. This is due in part to the fact that the probability of liquid metal contact occurring in refineries and petrochemical plants is normally rather small. In situations in which liquid metal embrittlement has occurred, it has been mainly due to the zinc embrittlement of austenitic stainless steels. Isolated failures have been attributed to welding in the presence of residues of zinc-rich paint or to the heat treating of welded pipe components that carried splatter of zinc-rich paint. However, most of the reported failures due to zinc embrittlement have involved welding or fire exposure of austenitic stainless steel in contact with galvanized steel components.

For example, in one case, severe and extensive cracking in the weld HAZ of process piping made from austenitic stainless steel occurred in a petrochemical plant during the final stages of construction. Much of the piping had become splattered with zinc-rich paint. Although the welders had been instructed to clean piping prior to welding, no cleaning and only limited grinding were performed. After welding, dye-penetrant inspection revealed many thin, branched cracks in the HAZ of welds, as shown in Fig. 44.

In many cases, through-wall cracks cause leaks during hydrotesting. Typically, zinc embrittlement cracks contain zinc-rich precipitates on fracture surfaces and at the very end of the crack tip. Cracking is invariably intergranular in nature (Ref 192).

Several different models for the zinc embrittlement of austenitic stainless steel have been proposed. The most accepted model involves the reduction in atomic bond strength at a surface imperfection, grain boundary, or crack tip by chemisorbed zinc metal. Zinc embrittlement is commonly a relatively slow process that is controlled by the rate of zinc diffusion along austenitic grain boundaries. Zinc combines with nickel, and this results in nickel-depleted zones adjacent to the grain boundaries. The resulting transformation of face-centered cubic austenite (γ) to body-centered cubic ferrite (α) in this region is thought to produce not only a suitable diffusion path for zinc but also the necessary stresses for initiating intergranular cracking. Externally applied stresses accelerate cracking by opening prior cracks to liquid metal.

Although the melting point of zinc is 420 °C (788 °F), no zinc embrittlement has been observed at temperatures below 750 °C (1380 °F), probably because of phase transformation and/or diffusion limitations. There is no evidence that an upper temperature limit—above which zinc embrittlement does not occur—exists. In the case of zinc-rich paints, only those having metallic zinc powder as a principal component can cause zinc embrittlement of austenitic stainless steels. Paints containing zinc oxide or zinc chromates are known not to cause embrittlement.

Prevention of Zinc Embrittlement. Obviously, the best approach to prevent zinc embrittlement is to avoid or minimize zinc contamination of austenitic stainless steel components in the first place. In practice, this means limiting the use of galvanized structural steel, such as railings, ladders, walkways, or corrugated sheet metal, at locations where molten zinc is likely to drop on stainless steel components if a fire occurs. If zinc-rich paints will be used on structural steel components, shop priming is preferred. Field application of zinc-rich paints should be done after all welding of stainless steel components has been completed and after insulation has been applied. Otherwise, stainless steel components should be temporarily covered with plastic sheathing to prevent deposition of overspray and splatter.

If stainless steel components have become contaminated despite these precautionary measures, proper cleaning procedures must be implemented. Visible paint overspray should be removed by sandblasting, wire brushing, or grinding. These operations should be followed by acid pickling and water rinsing. Acid pickling will remove any traces of zinc that may have been smeared into the stainless steel surface by mechanical cleaning operations. Suitable acid pickling solutions include 5 to 10% nitric acid, phosphoric acid, or weak sulfuric acid. Hydrochloric acid should not be used in order to avoid pitting and intergranular attack of sensitized weldments, or SCC problems. After removal of all traces of acid by water rinsing, final cleaning with a nonchlorinated solvent should be performed immediately before welding. Additional information on liquid metal embrittlement and a related phenomenon—solid metal embrittlement—can be found in the articles "Liquid Metal Induced Embrittlement" and "Solid Metal Induced Embrittlement" in *Corrosion: Fundamentals, Testing, and Protection,* Volume 13A of *ASM Handbook,* 2003.

Velocity-Accelerated Corrosion and Erosion-Corrosion

Various materials of construction for refinery and petrochemical plant service may exhibit accelerated metal loss under unusually turbulent fluid-flow conditions, which result in high values of wall shear stress either generally or locally on the exposed metal surface. Attack is usually caused by a combination of flow turbulence (mechanical factors) and corrosion (electrochemical factors) known as velocity-accelerated corrosion or erosion-corrosion. Metal surfaces affected by velocity-accelerated corrosion will often contain grooves or wavelike marks that indicate a pattern of directional attack. Soft metals, such as copper and aluminum alloys, are often especially prone to erosion-corrosion in the presence of entrained particles in the flow path. Metals such as stainless steels depend on thin oxide films for corrosion protection. Most cases of erosion-corrosion can be minimized by proper design and/or material changes. In stainless alloys, as long as the thin passive layer remains intact, these materials can remain resistant to corrosion even at very high values of flow velocity or flow-induced wall shear stress. However, materials such as steel that form semiprotective corrosion films can exhibit increasing corrosion rate with velocity and often exhibit a very rapid increase in corrosion rate once the turbulence is great enough to remove this semiprotective film.

For example, system designs should eliminate sharp bends, because this will often significantly reduce velocity-accelerated and erosion-corrosion problems in process piping. Increasing the pipe diameter of vapor lines will usually reduce flow velocities and therefore the corrosion caused by impinging droplets of liquid. However, if this design change and resultant reduction in flow velocity also changes the flow regime to two-phase slug flow, the wall shear stress (mechanical factor) on the metal surface may actually increase despite the reduction in flow velocity. Therefore, careful use of flow modeling is required to properly assess the mechanical effects using the parameter of wall

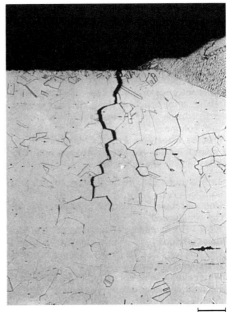

Fig. 44 Intergranular cracking in heat-affected zone of stringer bead weld on type 304 (S30400) stainless steel pipe due to zinc embrittlement. Weld area had been covered with zinc-rich paint.

shear stress and not by the linear flow velocity. Additionally, piping immediately downstream of pressure let-down valves often must be upgraded to more corrosion-resistant alloys to increase resistance to the locally high wall shear stress to prevent accelerated attack due to high flow turbulence.

Mixed-Phase Flow. Accelerated corrosion due to mixed vapor/liquid streams is primarily found in crude and vacuum furnace headers and transfer lines of crude distillation units, in overhead vapor lines and condenser inlets on various fractionation towers, and in reactor effluent coolers of hydrocracking and hydrotreating units.

In general, increases in vapor load and mass velocity can increase the severity of high-temperature corrosion by crude oils and atmospheric residuum (reduced crude) containing naphthenic acid and/or sulfidic species (Ref 193). Under mixed-phase conditions, the least corrosive severity is associated with flow regimes in which the metal surface is completely wetted with a substantial liquid hydrocarbon layer. Corrosion is most severe with the spray flow (i.e., droplet impingement) that results from vapor velocities above 60 m/s (200 ft/s) and vapor loads above 60%. However, surprisingly severe conditions can also be produced by slug flow due to the periodic highly turbulent conditions and high wall shear stress that result with the passing of each slug of liquid in the multiphase system (see Table 1 in the article "Corrosion in Petroleum Production Operations" for a depiction of slug flow and other multiphase flow regimes).

Under these conditions, corrosion rates of certain components, such as furnace headers, furnace-tube return bends, and piping elbows, could increase by as much as 2 orders of magnitude. This phenomenon is caused by droplet impingement, which destroys the protective sulfide scale normally found on steel components, as shown in Fig. 45. Such impingement damage is usually not seen in straight piping, except immediately downstream of circumferential welds. Damage is usually in the form of sharp-edged local area corrosion. Upgrades to higher-alloy materials are usually required in piping locations that involve these types of conditions.

Corrosion damage at elbows of overhead vapor lines is often caused by droplet impingement as a result of excessively high vapor velocities. Typical impingement-type corrosion of tubes and baffles just below the vapor inlet of overhead condensers is shown in Fig. 46. As a general rule, overhead vapor velocities should be kept below 7.5 m/s (25 ft/s) to minimize impingement-type corrosion. In addition, horizontal impingement baffles can be mounted just above the top tube row of overhead condensers.

In the case of high-temperature naphthenic acid corrosion found in the petroleum distillation process, the severity of corrosion is a function of chemical variables such as the naphthenic acid species present, the concentration of the acid (TAN value), and the presence of reactive sulfur species. Naphthenic acid tends to chemically dissolve protective sulfide films and then attack the underlying metal. Therefore, if the conditions are sufficient to produce a stable sulfide film, then naphthenic acid is of less concern. However, an additional and important factor is the wall shear stress produced by the flowing media, because high levels of wall shear stress produced by liquid or partial-vaporized hydrocarbon oils can result in sufficient mechanical action to mechanically disintegrate, damage, or otherwise remove the protective sulfide film. More information is provided in the section on "Naphthenic Acids" in this article.

Air-cooled reactor effluent coolers of hydrocracking and hydrotreating units are also prone to impingement-type corrosion. Poor flow distribution through large banks of parallel air coolers can result in excessive flow velocities in some coolers, usually those in the center. The resulting low flow velocities in the outer coolers can cause deposition of ammonium sulfide and/or chloride in these coolers; this blocks the tubes and further increases velocities in the remaining air coolers (Ref 68). This problem is aggravated by low nighttime or seasonal air temperatures, which increase deposition problems. Installation of protective sleeves (ferrules) at the inlet tube end has helped to reduce attack in some cases; in others, it has only moved the area of attack to an area immediately downstream of the sleeves. Careful attention to proper

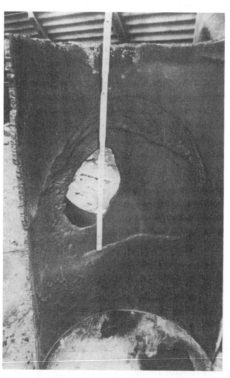

Fig. 45 Accelerated high-temperature sulfidic corrosion in 500 mm (20 in.) diameter pipe of vacuum furnace outlet header due to droplet impingement at high vapor velocities

flow distribution through redesign of the inlet headers is often the only way of controlling air cooler corrosion.

The difficulty in proper assessment of flow effects and chemical effects in refinery sour water systems is the direct effect of the solution chemistry and complicated mechanical flow effects in these applications. This difficulty is made worse because most guidelines are based on experiential information obtained through optimization of specific processes and units. In these cases, it is difficult to directly relate the flow effects from one process to another process and from one operating unit to another. This is because the phase behavior and flow conditions can vary substantially with specific equipment design and unit operating conditions.

In the case of sour water corrosion, the subject of ammonium bisulfide corrosion has been addressed in the literature. In some cases, the findings are experiential in nature, resulting from surveys of hydroprocessing unit operators. In other cases, the findings are based on limited laboratory testing, but these have not adequately addressed the effect of flow. The three most influential papers are those by Piehl, Damin and McCoy, and Scherrer et al. (Ref 67, 194, 195). The most notable article in the literature is Piehl's paper that describes a survey conducted by the NACE refinery committee covering corrosion in the reactor effluent air coolers. Damin and McCoy reported results of laboratory autoclave tests over a wide range of ammonium bisulfide concentration, as shown in Fig. 47 (Ref 194). Scherrer et al. (Ref 195) present the only laboratory corrosion study documenting the effect of velocity on ammonium bisulfide corrosion (Table 11). The results of these studies and refinery experience reveal that:

- Corrosion results from the presence of ammonium bisulfide, and corrosion rate generally increases with its concentration.
- The Piehl K_p factor (mole fraction ammonia times the mole fraction H_2S in the reactor effluent) may be used to monitor the potential corrosive severity of process environments and potentially impact materials selection. However, this parameter, while good, still

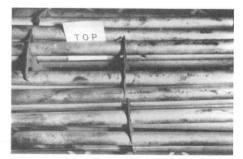

Fig. 46 Accelerated aqueous chloride corrosion below inlet nozzle of crude tower overhead condenser due to droplet impingement. Note partial loss of carbon steel baffles and localized corrosion along top of admiralty metal (C44300) tubes.

does not take into account all of the necessary factors, particularly as operating pressure increases, for accurate assessment of corrosive severity based on more recent work (Ref 5).

- Sour water corrosion involves flow velocity as a primary parameter, with excessive corrosion generally experienced at flow rates greater than 6 m/s (20 ft/s). Stainless steels may provide acceptable corrosion rates up to 9 m/s (30 ft/s), as defined primarily on service experience. However, for more rigorous assessment of corrosion performance, the wall shear stress produced by the flow needs to be examined (not just the linear flow velocity).
- Maintaining the water injection rate to result in at least 25% unvaporized (liquid) water in the effluent stream also appears important, particularly in terms of reducing the formation of deposits.
- The injected water must be free of dissolved oxygen (<50 ppb) to prevent rapid corrosion.
- Good flow distribution (in terms of vapor, liquid hydrocarbon, and water phase) is important. Therefore, tube velocities of less than 3 m/s (10 ft/s) in the air cooler can result in problems due to separation (stratification) of phases and the formation of corrosion deposits on tubes. Flow modeling can be used to more accurately assess this situation.
- Deposition of solid ammonium bisulfide salt may occur due to flow maldistribution, which starves some tubes of wash water and exacerbates both fouling and corrosion problems.
- U-bend tube designs may cause problems in the reactor effluent air cooler systems due to the possibility of velocity and wall shear stress locally accelerating corrosion in tubing bends.

More recently, a major multiclient study of both flow and chemical effects was conducted, and a comprehensive database for use in materials selection and process optimization for alkaline sour water systems was developed. The data were included in a software tool that manipulated the database for use in prediction of corrosion rates in sour water systems under various flow conditions (Ref 5). The data input screen indicates the necessary input parameters. It also contains a flow-modeling module that provides assessment of the wall shear stress for the actual system based on throughput, pipe size, and physical properties of various phases. This model calculates the resultant wall shear stress and then queries the database and interpolates between laboratory data points for these conditions to obtain the most appropriate corrosion rate predictions. The flow relationships used to assess the wall shear stress for multiphase flow are described in Ref 3 and in the article "Corrosion in Petroleum Production Operations" in this Volume.

Cavitation damage is a fairly common form of erosion-corrosion often found to affect pump impellers or hydraulic turbine internals. Cavitation is caused by collapsing gas bubbles at high-pressure locations; adjacent metal surfaces are damaged by the resultant hydraulic shock waves and the mechanical stress produced locally on the metal surface. Cavitation damage is usually in the form of loosely spaced pits that produce a roughened surface area. The damage can be produced from the mechanical disintegration of the protective corrosion films, or it can take the form of metal deformation and removal. In the latter case, subsurface metal shows evidence of mechanical deformation (e.g., slip bands and local cold working). As a general rule, cast alloys are likely to suffer more damage than wrought versions of the same alloy due to their typically coarser grain structure and alloy segregation. Ductile materials, such as wrought austenitic stainless steels, have the best resistance to cavitation because their resistance is based on the thin chromium oxide film on the metal surface, which is very tenacious and resistant to mechanical damage, except in the most extreme conditions or from impact of erosive particles. Damage can be reduced by design or operational changes, material changes, and sometimes, the use of corrosion inhibitors. Smooth finishes on pump impellers help to reduce this type of damage. Some coatings can be beneficial. Design changes with the objective of reducing pressure gradients in the flowing liquid are most effective.

Entrained Particles. Accelerated corrosion due to entrained catalyst particles can occur in the reaction and catalyst regeneration sections of catalytic cracking units. Furthermore, refineries involved with processing of hydrocarbons derived from oil or tar sands may be exposed to entrained fine particulates. Local erosive attack of a process piping system is shown in Fig. 48. Refractory linings are required to provide protection against oxidation and high-temperature sulfidic corrosion as well as erosion by catalyst particles, particularly in cyclones, risers, standpipes, and slide valves. Stellite or ceramic hardfacing can be used on some components to protect against erosion. However, those hardfacing materials that involve a metal matrix can be susceptible to corrosion damage if this matrix material is not resistant to corrosion in the process medium and its contaminants. Additionally, fine particles (<20 μm) can result in erosion of the metal matrix. Therefore, the best hardfacing materials are typically those with high tungsten carbide or ceramic particle loading and a minimum of metal-matrix binder.

When there are no erosion problems and when protective linings are impractical, austenitic stainless steels such as type 304 (S30400) can be used. Cyclone dip legs, air rings, and other internals in the catalyst regenerator are usually

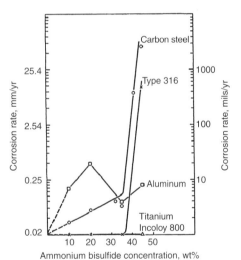

Fig. 47 Effect of ammonium bisulfide concentration on corrosion at 93 °C (200 °F) of several materials in simulated refinery sour water environment

Table 11 Laboratory data on corrosion of carbon steel versus ammonium bisulfide concentration and liquid-phase velocity

Ammonium bisulfide content		Average corrosion rate, mm/yr (mils/yr), at 60 to 100 °C (140 to 212 °F)			
		Fluid velocity			
g/L	wt%	3.5 m/s	(11.4 ft/s)	6.5 m/s	(21.3 ft/s)
40	4	0.2–0.3	(8–12)	0.3–0.4	(12–16)
100	10	0.3–0.4	(12–16)	0.5–0.6	(20–24)
200	20	0.6–0.8	(24–32)	1.0–1.2	(40–48)

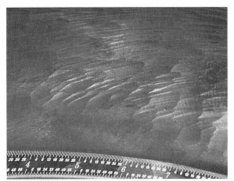

Fig. 48 Erosion damage produced by entrained particles in a hydrocarbon process stream (a) Loss in wall thickness. (b) Erosion features on inside surface of pipe

made of type 304 (S30400) stainless steel, as is piping for regenerator flue gas. The main fractionation tower is usually made of carbon steel, with the lower part lined with a ferritic or martensitic stainless steel containing 12% Cr such as type 405 (S40500) or 410 (S41000). Slurry piping between the bottom of the main fractionation tower and the reactor may receive an additional corrosion allowance as protection against excessive erosion.

Corrosion Control

A component in refinery or petrochemical service will require replacement when:

- Corrosion or other deterioration has made it unfit for further service.
- It no longer performs satisfactorily, although it may still be operational.
- It has become completely nonoperational.

Based on the results of a failure analysis, certain corrective measures can be implemented. These include, for example, the use of alternative materials of construction, changes in equipment design and process conditions, the use of corrosion inhibitors, and the application of protective coatings and linings.

Materials Selection. The most common method for preventing repeated failures is selection of improved metals or alloys. If a given piece of equipment fails every other year or so, substitution of a more highly alloyed material or alternative alloy that provides better performance can invariably be justified. In contrast, there is no justification for replacing a carbon steel tank with one made from a more corrosion-resistant alloy if the tank corroded out after 35 years of service (unless, of course, all of the corrosion-occurred during a short period near the end of its life).

Design Changes. Proper design of equipment is usually as important as proper selection of material. Design changes should involve consideration of changes in strength requirements, changes in material, and the need for additional allowances for corrosion, depending on past experience. Certain simple design rules should be followed to reduce corrosion-related problems:

- Welded joints can be substituted for flanged joints to minimize crevice corrosion.
- Equipment can be modified to permit easy cleaning and draining.
- Piping for certain equipment, such as pumps, can be modified to allow bypassing or replacement during service.
- Mechanical and vibrational stresses can be minimized by providing additional supports to avoid stress-related failures.
- Sharp bends in piping can be realigned to prevent erosion-corrosion.
- Heat exchangers can be modified to reduce temperature gradients that cause high stresses and also accelerate corrosion problems.

Further information on design considerations is available in the article "Designing to Minimize Corrosion" in *Corrosion: Fundamentals, Testing, and Protection,* Volume 13A of *ASM Handbook,* 2003.

Process changes that can be considered for reducing corrosion and other failures include the following:

- Temperature can be decreased to decrease corrosion rates.
- Concentrations of critical corrosive species can be adjusted.
- Flow velocity can be limited to prevent erosion-corrosion.
- Oxygen (air) can be removed by the use of scavenging chemicals.
- Water entry can be controlled by installation of calcium chloride drying equipment, settling drums, or demister screens.

Changing the concentration of corrosive contaminants in hydrocarbon streams is usually effective. Certain catalysts, such as hydrofluoric acid, are virtually inert when highly concentrated, but when diluted by water, they become extremely corrosive. Reducing the amount of acid entrained in hydrocarbon streams invariably reduces corrosion problems.

Changes in the flow regime of mixed vapor/liquid streams as a result of velocity changes can have a pronounced effect on corrosion. Very high velocities can produce problems associated with droplet impingement, and intermediate velocities with high liquid loading can result in highly turbulent slug flow. Also, stagnant or low flow conditions should be avoided, because stratified conditions can result in water wetting of the metal surface without the benefit of a protective hydrocarbon film. These conditions must also be avoided with metals that form passive oxide films, such as stainless steels. Minimum flow rates need to be maintained even when throughput is reduced.

Deaeration finds widespread application in the treatment of boiler feed water. The corrosion of certain nickel alloys by hydrofluoric acid can be controlled by excluding air. However, metals that depend on passive oxide films for corrosion resistance may actively corrode in certain environments if air is excluded.

Corrosion Inhibitors. Filming-amine corrosion inhibitors are added in small concentrations to various process streams to help control corrosion. Filming-amine inhibitors become ineffective at temperatures above approximately 175 °C (350 °F)—not because they decompose, but because the rate of desorption begins to exceed the rate of adsorption. The inhibitors are primarily used to protect overhead condensing equipment. Water-dispersible types contain surfactants to allow addition to streams that contain excess water. In general, filming-amine inhibitors are continuously injected at a rate that is just sufficient to maintain protection, which varies from compound to compound and with formulation additives. Higher-than-normal dosages are required for several days to establish a protective film after inhibitor injection has been interrupted (for example, because of pump failure) or after a shutdown. Most filming-amine inhibitors will tend to solubilize prior corrosion products; this can lead to fouling problems in downstream equipment. In amine units, these inhibitors usually cause foaming problems. Detailed information is available in the articles "Corrosion Inhibitors for Oil and Gas Production" and "Corrosion Inhibitors for Crude Oil Refineries" in *Corrosion: Fundamentals, Testing, and Protection,* Volume 13A of *ASM Handbook,* 2003.

Protective Coatings. Although most refineries and petrochemical plants do not rely on organic coatings for corrosion control, because a fire would immediately destroy these, extensive use is made of metallic cladding and refractory linings in process equipment. Cladding is usually performed at the mill by rolling a relatively thin sheet of a corrosion-resistant metal or alloy to a thicker base metal at elevated temperatures. The finished product will exhibit a fully bonded, metallurgical joint. Clad plates can be formed and fabricated into vessels or piping by welding the joints of the base metal and overlaying the welds. Thin sheet strips of corrosion-resistant alloy can be attached by spot welding in the field to protect an unclad vessel. This process is known as strip lining. For long-term service, however, cladding tends to be more reliable than strip lining. Strip lining can become a source of problems with process streams that contain hydrogen at elevated temperatures and pressures or that evolve hydrogen because of corrosion. In either case, pockets of hydrogen can build up behind the lining and cause it to fail. Cladding by explosive bonding is primarily used for new construction to attach titanium or nickel alloy overlays.

Overlays of corrosion-resistant alloys can also be attached in the form of adjacent weld beads. For example, reactor vessels for hydrocracking and hydrotreating units are protected against high-temperature sulfidic corrosion with a weld overlay of type 347 (S34700) stainless steel.

Thermal sprayed metallic and ceramic coatings, as a rule, have not been extensively used in refinery and petrochemical plant applications, because such coatings have not been very reliable. However, improved methods of applications that provide higher density and lower impurities are now available and finding increased use in selective areas.

Diffusion coatings, particularly aluminized coatings, have been highly successful as an alternative approach to controlling high-temperature sulfidic corrosion of carbon steel and low-alloy steel components. Aluminizing not only provides resistance to high-temperature sulfidic corrosion but also reduces scaling, carburization, coking, and erosion problems (Ref 196). Aluminized components are not necessarily less costly than those made of higher-alloy steels or stainless steels, and a detailed cost analysis is usually required. Aluminizing is a proprietary process in which steel components are packed into a retort and exposed to aluminum vapor at a temperature above approximately 925 °C (1700 °F). The aluminum diffuses into the steel, and this forms a true metallurgical alloy containing over 50% Al at the surface.

Other methods for forming a corrosion-resistant iron-aluminum layer on steel surfaces include flame spraying and hot dipping, either with or without subsequent diffusion heat treatments. As a rule, these methods have been less effective than aluminizing, primarily because of nonuniform thickness, porosity, and holidays (Ref 197).

The aluminizing of carbon steel or low-alloy steel components typically increases their resistance to high-temperature sulfidation by 2 orders of magnitude (Ref 198). However, the diffused iron-aluminum layer is only 0.08 to 0.4 mm (3 to 15 mils) thick, depending on the base alloy, and can be readily damaged by mechanical (tube drilling for coke removal) or chemical means (acid or caustic cleaning). Aluminized components also require special fabrication techniques, including welding and roller expanding. Distortion and shrinkage during aluminizing can be a problem with some components.

A variety of processes for applying protective coatings are discussed in the Section "Methods of Corrosion Protection" in *Corrosion: Fundamentals, Testing, and Protection,* Volume 13A of *ASM Handbook,* 2003. Methods for applying weld claddings are discussed in various articles in *Welding, Brazing, and Soldering,* Volume 6 of *ASM Handbook,* 1993 (see, in particular, the articles in the Section "Solid-State Welding, Brazing, and Soldering Processes").

Refractory Linings. Refractory materials are used in refineries and petrochemical plants as linings to protect steel from thermal degradation, erosion damage, and both low- and high-temperature corrosion. Unlike thin organic linings and coatings, refractory linings are applied in thicknesses of 25 to 100 mm (1 to 4 in.) or more. Metal wire reinforcement is used to help hold the lining in place.

Portland cement/sand is the lining that is most commonly used to protect steel against mild corrosives. This lining (frequently called gunite or shotcrete, after the application technique) has been used in refineries for more than 60 years. The primary application is sour water service with a pH value between 4 and 8. The material is low cost, and the application is straightforward. The mixture is applied with a pneumatic gun that shoots the premixed cement, sand, and water onto the vessel surface. A chicken-wire metal reinforcement is usually provided at the midthickness. As with any other concrete, the lining must be cured before it is placed into service. During ambient-temperature curing, shrinkage cracks may occur, but they are not detrimental to the serviceability of the lining. Acidic water entering the cracks is neutralized by the alkalinity of the concrete and is trapped in the crack. However, continuing acid attack will result in gradual loss of thickness (much like corrosion), and aggregate, which is carried away in the process fluid, may cause erosion or plugging in downstream equipment, such as filters, pumps, and valves.

For highly acidic solutions or for higher temperatures, more resistant cements and aggregate such as lumnite-haydite need to be used. These special acid-resistant cements, which are based on silicate, furane, phenolic, and sulfur compounds, are used for highly acidic services but at the cost of reduced resistance to alkaline solutions. In addition to gunning, these cements can be applied as thinner coatings by troweling.

For hot, highly acidic solutions, acid-proof brick linings can be used. The essential components of such a lining are a membrane lining against the steel shell, topped with brick-and-mortar construction. A membrane is placed between the bricks and the steel shell, because brick-and-mortar construction is inherently porous and subject to capillary leakage. Membrane materials need not be completely immune to attack by the corrosive, because any swelling of the membrane will tend to seal off the bottoms of the capillary channels. Furthermore, the brick lining does provide thermal insulation, which reduces the temperature and therefore the aggressiveness of the corrosive. Typically, one course of 115 mm (4.5 in.) thick brick raises the temperature limit of a membrane material by approximately 30 °C (54 °F), and two courses by approximately 50 °C (90 °F).

The membranes range from thermoplastic resins and elastomers to sheet or bonded lead linings, depending on the corrosive handled and the need to accommodate relative movement between the shell and the brick lining. Ceramic brick is satisfactory for all common operating conditions and corrosives, except strong alkalis, hydrofluoric acid, and fluorides, which require carbon brick. As a rule, acid-proof brick linings have limited applications in refineries and petrochemical plants, the principal uses being storage tanks for sulfuric acid. They are seldom used for pressure vessels, especially those that operate at elevated temperatures and pressures.

Heat-resistant linings can reduce the cost of pressure vessels that will operate at temperatures above 345 °C (650 °F). Allowable stresses, given in the ANSI/ASME Boiler and Pressure Vessel Code, Section VIII, for the design of carbon steel vessels, decrease with increasing temperatures. Because the maximum permissible temperature for carbon steels is 510 °C (950 °F), alloy steels must be used above this temperature. A heat-resistant lining, by reducing the metal temperature, will reduce the required wall thickness of carbon steel vessels, such as reactors. Alternatively, heat-resistant linings allow the use of carbon steel vessels at higher temperatures, such as reactors in fluid catalytic cracking units. For either of these applications, the integrity of the lining is important, because loss of insulation can result in areas of the vessel reaching temperatures above the acceptable mechanical design limit (hot spots). When hot spots are found, temperatures can be reduced by air blowing or steam sprays. If this is not effective, the unit must be immediately taken out of service for repair of the lining. Properly applied linings will be effective for extended time periods and require repairs only during normally scheduled shutdowns.

The heat-resistant and insulating linings used in pressure vessels are refractory materials that are applied as monolithic linings by pneumatic gunning, casting, or hand packing. The most common types are the hydraulic-setting castable refractories. Their ease of application, generally good performance, and variety of strength and insulating characteristics make them versatile high-temperature materials. The quality of the applied castable material depends on the product quality as produced by the manufacturer, the experience of the applicator, and the manner in which heat is first applied to the material after it is installed. Erosion-resistant linings are relatively thin (40 mm, or 1.5 in.) layers of dense refractories that are supported by V-anchors or hexmesh welded directly to the vessel shell. To improve the serviceability of V-anchored linings, metal fibers are often added to the refractory mix before placement.

Catalytic cracking units, with their associated cyclones, transfer lines, and slide valves, represent the single largest application of refractories in refineries (Ref 199). The regenerator vessels, and sometimes the reactors, are lined with 100 to 150 mm (4 to 6 in.) of refractory concrete, which must withstand oxidizing (regenerator) or reducing (reactor) conditions at 540 to 760 °C (1000 to 1400 °F). The moving catalyst bed produces mild erosion, and mechanical or thermal spalling is frequently encountered. Cyclones and catalyst transfer lines are usually subjected to extreme erosion from fluidized catalyst at temperatures between 315 and 760 °C (600 and 1400 °F). Transfer lines are subjected to heavy loadings of catalyst at velocities of 7.5 to 15 m/s (25 to 50 ft/s). Slide valves in transfer lines and standpipes are subject to severe erosion but must maintain their original thickness so they can control catalyst flow. Care must be exercised in designing the slide valve because of thermal expansion differentials. Plug valves have more tolerance for thermal expansion differentials and are frequently lined with refractory.

Because operating conditions in reformer reactors include hydrogen-rich vapors at approximately 500 °C (950 °F), special care is required in designing refractory liners because of the increased thermal conductivity of hydrogen-saturated refractories. Reformers in hydrogen and ammonia plants operate at 1100 to 1370 °C (2000 to 2500 °F) and therefore require refractories of very low silica content. Furnaces are typically lined with castable refractory or insulating fire brick. Thermal shock and temperature cycling are the primary agents of attack. Mechanical movement, nut blasting, and similar forces also contribute to deterioration. Floors are usually made of dense refractory concrete or fire brick in order to resist foot traffic and mechanical impact during turnarounds. Stacks and breechings for most types of refinery units have similar service requirements: strength at high temperatures and resistance to corrosion, erosion, and spalling. Temperatures range from 205 to 815 °C (400 to 1500 °F), and the flue gas may contain catalyst, sulfur oxides, H_2S, or carbon monoxide. Water or steam may be injected into the gas stream to control temperature.

If conditions are mild, with temperatures below 540 °C (1000 °F) and little or no erosion, insulating or semiinsulating refractory concrete on independent anchors makes a serviceable lining. More erosive conditions may require the use of dense high-strength refractory castable.

High-sulfur fuels generate sulfuric acid in the cooler zones of stacks and breechings where temperatures drop below the acid dewpoint, ranging from 150 to 175 °C (300 to 350 °F). Acid attacks many conventional castable refractories containing calcium aluminate cement. The alkali-silicate and some calcium-aluminate base materials are moderately acid resistant and will withstand condensing flue gas vapors. For many furnace stack linings, densely applied semiinsulating refractory concrete is satisfactory.

Incinerators operate at temperatures as high as 1540 °C (2800 °F) and are particularly susceptible to slag and fly-ash attack, because they often burn waste products. Problems can be expected with fluids containing alkalis and transition metals. Thermal shock can be a problem if water is injected to control temperatures or if the burner unit is allowed to cycle on and off several times a day, with periods of cooling in between. Castable phosphate-bonded alumina is probably the best material for this type of service. Additional information on refractory linings can be found in the article "Chemical-Setting Ceramic Linings" in *Corrosion: Fundamentals, Testing, and Protection,* Volume 13A of *ASM Handbook,* 2003.

Use of Process Control and Corrosion Monitoring. One of the problems associated with reducing the costs of corrosion in refinery operations is that corrosion is commonly dealt with in a historical sense, after the damage has occurred. In this light, corrosion measurements are usually relegated into spot checks and maintenance functions. Monitoring data are recorded off-line and not viewed real-time with the process conditions that often initiate high-level corrosion activity. However, recent innovations in monitoring technology along with an evolving purpose of the refinery corrosion engineer in the new online, real-time world of refinery process control/optimization and asset management are leading the way to change for those that embrace, utilize, and promote these new corrosion-monitoring technologies.

Online, real-time technologies are commonplace for refinery process control and have only recently entered the corrosion-monitoring realm. They have increased the accuracy of data and their relevance and value to the ultimate goal of increasing productivity by reducing corrosion damage, failures, and unplanned outages, thus decreasing downtime and increasing run time. They have also provided online connectivity for corrosion engineers, bringing them in closer contact with the frontline people that control processes and manage these facilities. This makes their role much more relevant. Under this new paradigm, corrosion becomes another real-time process variable.

Historical approaches for off-line measurement methods also include such general corrosion-monitoring techniques as electrical resistance (ER) and linear polarization resistance (LPR), as discussed in ASTM G 96 (Ref 200). These systems are able to operate in a stand-alone mode providing "spot" corrosion data via battery-powered, plant-mounted instruments, often with logging capability. These techniques can help in determining inspection frequencies based on the average mass loss corrosion rates obtained from these corrosion-monitoring techniques.

Modern field corrosion monitoring now includes a broad range of techniques used to evaluate the degradation of metallic materials. These techniques can be divided into two distinct groups, namely those providing indications of the cumulative damage sustained (retrospective) and those providing indications of the prevailing corrosion rate (usually online and continuous). These techniques are applicable with virtually all metallic materials and are commonly grouped as follows:

- *Cumulative loss techniques:* Weight loss coupons, ER monitoring, ultrasonic thickness measurement, and other nondestructive examination methods (e.g., radiography). The cumulative loss techniques will only show signs of change when sufficient corrosion has been sustained to cause a change in the bulk material properties. As such, most are used offline and do not provide real-time data, requiring measurement cycles of days to weeks.
- *Corrosion rate techniques:* LPR, harmonic distortion analysis (HDA), and electrochemical noise (EN) monitoring methods. The corrosion rate techniques have a much higher resolution and short response time and have been developed to provide a fast assessment of the electrochemical rate processes taking place at the metal/environment interface. Measurements using these techniques take only a few minutes (Ref 201).

Advances in automated monitoring systems have made it possible to incorporate multiple measurements into a single instrument, thereby increasing accuracy to the point of being able to make quantitative measurements [e.g., LPR corrosion rate with actual measured B value (Stern-Gerry factor) correction from HDA] and being able to differentiate the modality (pitting versus uniform corrosion) of the corrosion process (e.g., use of EN). Furthermore, this can be done on a time scale of minutes, which is consistent with modern process control approaches. This latter point is extremely important, because it now is possible to provide quantitative corrosion rate and modality information and deliver them via the same (and existing) communications protocols (4–20 mA, RS-485, RS-232, and HART) that are already in place and used to acquire process variables through field-based system control and data acquisition systems or plant-distributed control systems.

The importance of the aforementioned advances is that the corrosion engineer can now be "plugged into" the same online, real-time channel that is used for process control and optimization and facilities asset management. Corrosion data are automatically commingled and displayed with process data. The corrosion engineer is no longer relegated to a stand-alone function. He/she can see the same real-time process data that the process engineer sees, and the process engineer has access to the online corrosion rate that the corrosion engineer sees, enabling uses of these data together as key performance indicators. Figure 49 illustrates this interrelationship. Both engineering functions can now work together in a new way (Fig. 49) (Ref 201).

- The process engineer uses the corrosion signal as another variable that needs to be optimized (e.g., minimize asset damage, increase production while controlling damage to acceptable levels, extend allowable run time, and manage process to minimize inspection requirements).
- The improved connectivity gives the plant operator immediate access to both corrosion data and the corrosion engineer, who can immediately provide valuable input regarding the impact of process upsets. The corrosion engineer can also be included in process optimization studies, because corrosion is now another process variable. This is a quantum leap for the corrosion engineer to a much more value-oriented position in the company.

Process Optimization Study in a Hot Hydrocarbon Stream with 1 to 2% Corrosive Water. This example involves monitoring performed at a petrochemical operation where much of the plant is constructed of carbon steel and type 304L and 316L stainless steels (Ref 202). Decades of debottlenecking and other process modifications had produced corrosion problems. After a year of unsuccessful efforts to untangle their process problems, an online, real-time electrochemical corrosion-monitoring system was installed. For the first time, materials engineers, process engineers, and plant operators were able to see immediate changes in corrosion behavior caused by specific variations in process parameters and work together to identify process modifications and remedial actions.

Realizing that this environment was mostly a nonconductive organic phase, and the entry points for probes were mostly located on vertical pipe runs, custom-designed probes were installed at ANSI flange joints in a piping system where the most severe corrosion had been observed. Based on the results of the initial process evaluation, which required several weeks, five predominant factors were identified that related directly to the chemical aggressivity of the plant environment. These included:

- An upstream vessel was on an automatic pump-down schedule, so that it pumped its contents into a reactor approximately once per hour. Every time the vessel pumped down, the corrosiveness of the larger stream increased.

- Operations had reduced the concentration of a particular neutralizing chemical in the process. Contrary to what was expected, it was found that increasing the feed rate of the neutralizer increased corrosion rates rather than reducing them (Fig. 50). This new information helped to reduce corrosion rates and also to provide process engineers with new insight into the chemistry of the process.

- Following review of the corrosion data for the first time, a plant technician pointed out that an increase in corrosion rate of the type 304L electrodes occurred right after a new batch of catalyst was mixed.

- Further investigations indicated that the corrosion rate also varied quite significantly with process and operational events. These included noting that the corrosion rate of carbon steel correlated with the quantity of a key gaseous chemical used in the process.

- In another process stream, short-term spikes to very high corrosion rates were observed intermittently but consistently, week after week. Later, it was determined that the corrosion rate spikes coincided with the pumping of laboratory samples back into the process upstream operations. This subsequently led to changes in how the lab samples were disposed of. This stopped the corrosion spikes.

Ultrasonic thickness measurements were taken on various parts of the piping in the vicinity of the corrosion-monitoring points. These indicated an average corrosion rate of 0.075 mm/yr (2.965 mils/yr) over a 16 month period. These data agreed very well with the 0.074 mm/yr (2.9 mils/yr) corrosion rate predicted by online, real-time corrosion measurements described in Ref 201 to 204 and discussed in the article "Corrosion in Petroleum Production Operations" in this Volume.

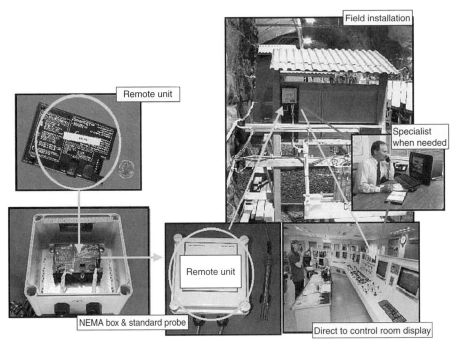

Fig. 49 Montage showing interrelationship between a petrochemical plant process engineer and corrosion/materials specialist using online, real-time corrosion monitoring. Distributed corrosion-monitoring hardware provides a preprocessed data signal at point of monitoring. This, in turn, provides simultaneously an alarmed measurement-and-control signal in the control room and access to the same information on the engineer's workstation. Source: Ref 201

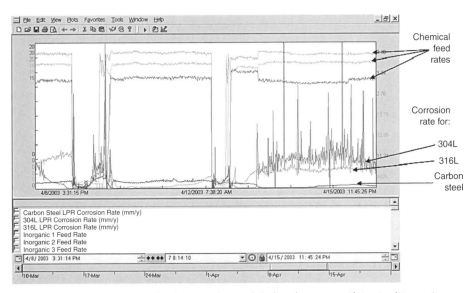

Fig. 50 Real-time process control data from a predominantly hydrocarbon stream with 1 to 2 vol% corrosive water. Simultaneous measurement of chemical feed and corrosion monitoring of carbon steel and type 304L and 316L stainless steels. Source: Ref 202

Appendix: Industry Standards

James Skogsberg and Ned Niccolls, Chevron Texaco; Russell D. Kane, Honeywell Process Solutions, Honeywell International, Inc.

THE APPLICABLE STANDARDS for materials used in corrosive service conditions in upstream (oil and gas drilling, production, and gathering) and downstream petroleum service (gas processing and oil refining) are reviewed. Applicable and commonly used standards related to corrosion control and materials selection are presented. These are commonly referenced by corrosion and materials engineers to define and apply corrosion control through specification of equipment and/or material, control of composition and material processing for metallic and nonmetallic materials of construction (alloys, elastomers, plastics, and composites), description of service conditions and requirements, application of cathodic protection, definition of applicable fabrication or construction methods, definition of performance test methods, and for definition of materials requirements for exposure to H_2S service conditions.

Selection of materials to resist all forms of environmentally assisted cracking mechanisms in sour (H_2S-containing) service environments is an important aspect of both upstream and downstream petroleum service applications. These forms of cracking include:

- Sulfide stress cracking (SSC) of steels
- Hydrogen-induced cracking (HIC) of steels
- Stress-oriented hydrogen-induced cracking (SOHIC) of steels
- Stress-corrosion cracking (SCC) of corrosion-resistant alloys (CRAs)

Two documents that deal specifically with SSC and SCC are NACE MR0175/ISO 15156, "Petroleum and Natural Gas Industries—Materials for Use in H_2S-Containing Environments in Oil and Gas Production" which applies to upstream service applications; and NACE MR0103, "Materials Resistant to Sulfide Stress Cracking in Corrosive Petroleum Refining Environments," which applies for downstream refinery applications.

An intensive standards-writing effort recently resulted in the technical alignment NACE standard MR0175, "Metals for Sulfide Stress Cracking and Stress Corrosion Cracking Resistance in Sour Oilfield Environments," and ISO 15156, "Petroleum and Natural Gas Industries—Materials for Use in H_2S-Containing Environments in Oil and Gas Production." NACE MR0175/ISO 15156 gives requirements and recommendations for the selection and qualification of carbon and low-alloy steels, corrosion-resistant alloys, and other alloys. It is specially directed toward materials used in oil and natural gas production and natural gas treatment plants in H_2S-containing environments, whose failure could pose a risk to the health and safety of the public, to personnel, or to the environment. It can be applied to help avoid costly corrosion damage to the equipment itself. NACE MR0175/ISO 15156 consists of three standards, available through NACE as a complete package:

- *Part 1:* General principles for selection of cracking-resistant materials
- *Part 2:* Cracking-resistant carbon and low-alloy steels, and the use of cast irons
- *Part 3:* Cracking-resistant CRAs and other alloys

Similarly, an intensive standard-writing effort has also taken place within NACE, with the goal to define requirements for materials used in refinery applications relative to SSC. The resultant standard (NACE MR0103) acknowledges that while some refinery sour service requirements are similar to those found in upstream applications, there are many aspects of refinery service that require special considerations when it comes to guidelines for materials selection in H_2S service. This new document for the first time addresses the specific requirements of materials in refining application for resistance to SSC.

Materials of construction and other documents that deal with forms of H_2S attack and corrosion are included in the tables of standards and reference documents that follow. The standards and documents for the upstream and downstream portions of the petroleum industry are in Tables 12 and 13, respectively. Some standards are applicable to both. A list of sponsoring organizations (Table 14) provides source information.

Table 12 Standards and references for upstream materials used for completions, surface facilities, and structures

Source	Specification number	Title
Drilling equipment		
API	5D	Specification for Drill Pipe
API	7	Specification for Rotary Drill Stem Elements
API	RP 7G	Recommended Practice for Drill Stem Design and Operating Limits
API	8A	Specification for Drilling and Production Hoisting Equipment
API	8C	Specification for Drilling and Production Hoisting Equipment (PSL 1 and PSL 2)
Oil-country tubular goods		
ISO	13680	Petroleum and Natural Gas Industries—Corrosion-Resistant Alloy Seamless Tubes for Use as Casing, Tubing, and Coupling Stock—Technical Delivery Conditions
API	5CT	Specification for Casing and Tubing
ISO	11960	Petroleum and Natural Gas Industries—Steel Pipes for Use as Casing and Tubing for Wells
Line pipe and piping systems		
API	5L	Specification for Line Pipe
API	1104	Welding of Pipelines and Related Facilities
NORSOK	M-630	Material Data Sheets for Piping Rev. 4, June 2004
API	2RD	Design of Risers for Floating Production Systems and Tension-Leg Platforms
DNV	OS-F101	Submarine Pipeline Systems
API	5LC CRA	Line Pipe
ISO	3183-3	Petroleum and Natural Gas Industries—Steel Pipe for Pipelines—Technical Delivery Conditions—Part 3: Pipe of Requirement Class C
API	RP 14E	Recommended Practice for Design and Installation of Offshore Production Platform Piping Systems
DNV	RP 0501	Erosive Wear in Piping Systems
ASME	B16.5	Pipe Flanges and Flanged Fittings
ASME	B31.3	Chemical Plant and Petroleum Refinery Piping
ASME	B31.4	Liquid Petroleum Transportations Piping Systems
ASME	B31.8	Gas Transmission and Distribution Piping System
NORSOK	L-001	Piping and Valve Specifications
NORSOK	M-601	Welding and Inspection of Piping
Completions and subsea equipment		
API	17D	Specification for Subsea Wellhead and Christmas Tree Equipment
API	RP 17A	Recommended Practice for Design and Operation of Subsea Production Systems
API	6A	Wellhead Equipment
NACE	MR0176	Metallic Materials for Sucker-Rod Pumps for Corrosive Oilfield Environments
API	14D	Wellhead Surface Safety Valves and Underwater Safety Valves for Offshore Service
EEMUA	194:1999	Guidelines for Materials Selection and Corrosion Control for Subsea Oil and Gas Production Equipment
API	14A	Specification for Subsurface Safety Valve Equipment
EFC(a)	17	Corrosion Resistant Alloys for Oil and Gas Production: Guidance on General Requirements and Test Methods for H_2S Service, 2002

(continued)

(a) European Federation of Corrosion

Table 12 (continued)

Source	Specification number	Title
Completions and subsea equipment (continued)		
NORSOK	M-001	Materials Selection
NORSOK	M-CR-701	Materials for Well Completion Equipment
NORSOK	U-DP-001	Subsea Production Systems
NORSOK	U-CR-003	Subsea Christmas Tree Systems
NORSOK	U-002	Subsea Structures and Piping System
NACE	1D199	Internal Corrosion Monitoring of Subsea Production and Injection Systems
Surface facilities		
ISO 14313	API 6D	Production Valves
Offshore structures		
API	RP 2A	Planning, Designing, and Constructing Fixed Offshore Platforms
EEMUA	197 : 1999	Specification for the Fabrication of Non-Primary Structural Steelwork for Offshore Installations
EEMUA	158 : 1994	Construction Specification for Fixed Offshore Structures in the North Sea
API	RP 2RD	Design of Risers for Floating Production Systems and Tension-Leg Platforms
API	2Y	Specification for Steel Plates, Quenched-and-Tempered, for Offshore Structures
API	2W	Specification for Steel Plates for Offshore Structures, Produced by Thermo-Mechanical Control Processing (TMCP)
API	RP 2Z	Recommended Practice for Preproduction Qualification for Steel Plates for Offshore Structures
API	2MT1	Specification for Carbon Steel Plate with Improved Toughness for Offshore Structures
API	2H	Specification for Carbon Steel Plate for Offshore Platform Tubular Joints
NACE	RP0176	Corrosion Control of Steel Fixed Offshore Structures Associated with Petroleum Production
Fluid-handling machinery		
API	617	For centrifugal compressors
API	610	For centrifugal pumps
API	616	For gas turbines
API	618, 619, 674, etc.	Specifications for positive displacement machinery
Composite grating		
PFM	2-98	Use of Fiber Reinforced Plastic (FRP) Grating and Cable Trays
CEAC-GL	2003-0102	FRP Grating for Offshore Platforms, Procurement
CEAC-GL	2003-0103	FRP Grating for Offshore Platforms, Installation
Tanks and vessels		
ASME	RTP 1-2000	Reinforced Thermoset Plastic Corrosion Resistant Equipment
ASME	X	Fiber-Reinforced Plastic Pressure Vessels
BS	4994: 1987	British Standard Specification for Design and Construction of Vessels and Tanks in Reinforced Plastics
ASTM	3299	Filament Wound, Corrosion Resistant FRP Tanks
ASTM	4097	Contact Molded Glass-Fiber-Reinforced Thermoset Resin Corrosion Resistant Tanks
ANSI/AGA	NGV2	Pressure Vessels for Fuel Containment
FRP pipe		
ABS	2000	Guide for Building and Classing Facilities on Offshore Installation
ASTM	F 1173	Standard Specification for Thermosetting Resin Fiberglass Pipe and Fittings to be Used for Marine Applications
UKOOA	...	Specification and Recommended Practice for the Use of GRP Piping Offshore
ISO/FDIS	14692-2	Specification and Recommended Practice for the Use of GRP Piping in the Petroleum and Natural Gas Industries
IMO	Resolution A.753 (18)	Guidelines for the Application of Plastic Pipes on Ships
USCG	NVIC 11-86	Guidelines Governing the Use of Fiberglass Pipe on Coast Guard Inspected Vessels
USCG	PFM 1-98	Guidelines on the Fire Testing Requirements for Plastic Pipe per IMO Resolution A.753 (18)
API	RP 14G	Recommended Practice for Fire Prevention and Control on Open Type Offshore Production Platforms
API	15LR	Specification for Low Pressure Fiberglass Line Pipe
API	15HR	Specification for High Pressure Fiberglass Line Pipe
ASME	B31.3-1996	Process Piping, Chapter VII, "Nonmetallic Piping and Piping Lined with Nonmetals"
ANSI/AWAA	C950-95	AWWA Standard for Fiberglass Pressure Pipe 15. AWWA Manual M45, "Fiberglass Pipe Design"
UKOOA	...	Specification and Recommended Practice for the Use of GRP Piping Offshore
UKOOA	...	Guidelines for Fibre Reinforced Plastics Use Offshore
NACE	TM0298	Evaluating the Compatibility of FRP Pipe and Tubulars with Oilfield Environments
NACE standard recommended practices related to cathodic protection		
	RP0575	Internal Cathodic Protection Systems in Oil-Treating Vessels
	RP0186	Application of Cathodic Protection for External Surfaces of Steel Well Casings
	RP0169	Control of External Corrosion on Underground or Submerged Metallic Piping Systems
	RP0193	External Cathodic Protection of On-Grade Carbon Steel Storage Tank Bottoms
	RP0200	Steel-Cased Pipeline Practices
	RP0176	Corrosion Control of Steel Fixed Offshore Structures Associated with Petroleum Production
	RP0387	Metallurgical and Inspection Requirements for Cast Galvanic Anodes for Offshore Applications
	RP0492	Metallurgical and Inspection Requirements for Offshore Pipeline Bracelet Anodes
	RP0388	Impressed Current Cathodic Protection of Internal Submerged Surfaces of Carbon Steel Water Storage Tanks
	RP0196	Galvanic Anode Cathodic Protection of Internal Submerged Surfaces of Steel Water Storage Tanks
NACE standard test methods related to cathodic protection		
	TM0497	Measurement Techniques Related to Criteria for Cathodic Protection on Underground or Submerged Metallic Piping Systems
	TM0102	Measurement of Protective Coating Electrical Conductance on Underground Pipelines
	TM010	Measurement Techniques Related to Criteria for Cathodic Protection on Underground or Submerged Metallic Tank Systems
	TM0190	Impressed Current Laboratory Testing of Aluminum Alloy Anodes

(continued)

(a) European Federation of Corrosion

Table 12 (continued)

Source	Specification number	Title
European standards related to cathodic protection		
DNV	RP B401	Cathodic Protection Design
NORSOK	M-503	Rev. 2, Cathodic Protection
CEN	TC 262/SC2/WG3 N.94	General Principles of Cathodic Protection in Sea Water
Welding		
BS	EN 1011-2	Welding Part Two—Recommendations for Welding of Metallic Materials—Part 2: Arc Welding of Ferritic Steels
AWS	D 1.1	Structural Welding Code
ASME	IX	Welding Code
BS	5135	Specification for Arc Welding of Carbon and Carbon Manganese Steels
Elastomers		
NACE	TM0187	Evaluating Elastomeric Materials in Sour Gas Environments
Materials in sour service		
NACE	MR0175/ISO 15156	Petroleum and Natural Gas Industries—Materials for Use in H_2S-Containing Environments in Oil and Gas Production—Parts 1, 2, and 3
NACE	RP0475	Selection of Metallic Materials to be Used in All Phases of Water Handling for Injection into Oil-Bearing Formations
EFC	16	Guidelines on Materials Requirements for Carbon and Low Alloy Steels for H_2S-Containing Environments in Oil and Gas Production
NACE	TM0177	Laboratory Testing of Metals for Resistance to Specific Forms of Environmental Cracking in H_2S Environments
NACE	TM0284	Evaluation of Pipeline and Pressure Vessel Steels for Resistance to Hydrogen-Induced Cracking
NACE	8X194	Materials and Fabrication Practices for New Pressure Vessels Used in Wet H_2S Refinery Service
NACE	RP0472	Methods and Controls to Prevent In-Service Environmental Cracking of Carbon Steel Weldments in Corrosive Petroleum Refining Environments
NACE	1F192	Use of Corrosion-Resistant Alloys in Oilfield Environments
NACE	1F196	Survey of CRA Tubular Usage
NACE	RP0296	Guidelines for Detection, Repair, and Mitigation of Cracking of Existing Petroleum Refinery Pressure Vessels in Wet H_2S Environments
NACE	RP0403	Avoiding Caustic Stress Corrosion Cracking of Carbon Steel Refinery Equipment and Piping
NACE	TM0103	Laboratory Test Procedures for Evaluation of SOHIC Resistance of Plate Steels Used in Wet H_2S Service
NACE	TM0198	Slow Strain Rate Test Method For Screening Corrosion-Resistant Alloys (CRAs) for Stress Corrosion Cracking in Sour Oilfield Environments
NACE	MR0103	Materials Resistant to Sulfide Stress Cracking in Corrosive Petroleum Refining Environments
ASTM standards		
	G 8	Standard Test Methods for Cathodic Disbonding of Pipeline Coatings
	G 9	Standard Test Method for Water Penetration into Pipeline Coatings
	G 10	Standard Test Method for Specific Bendability of Pipeline Coatings
	G 12	Standard Test Method for Nondestructive Measurement of Film Thickness of Pipeline Coatings on Steel
	G 14	Standard Test Method for Impact Resistance of Pipeline Coatings (Falling Weight Test)
	G 15	Standard Terminology Relating to Corrosion and Corrosion Testing
	G 17	Standard Test Method for Penetration Resistance of Pipeline Coatings (Blunt Rod)
	G 18	Standard Test Method for Joints, Fittings, and Patches in Coated Pipelines
	G 19	Standard Test Method for Disbonding Characteristics of Pipeline Coatings by Direct Soil Burial
	G 20	Standard Test Method for Chemical Resistance of Pipeline Coatings
	G 28	Standard Test Methods of Detecting Susceptibility to Intergranular Corrosion in Wrought, Nickel-Rich, Chromium-Bearing Alloys
	G 38	Standard Practice for Making and Using C-Ring Stress-Corrosion Test Specimens
	G 39	Standard Practice for Preparation and Use of Bent-Beam Stress-Corrosion Test Specimens
	G 46	Standard Guide for Examination and Evaluation of Pitting Corrosion
	G 48	Standard Test Methods for Pitting and Crevice Corrosion Resistance of Stainless Steels and Related Alloys by Use of Ferric Chloride Solution
	G 49	Standard Practice for Preparation and Use of Direct Tension Stress-Corrosion Test Specimen
	G 111	Standard Guide for Corrosion Tests in High Temperature or High Pressure Environment, or Both
	G 148	Standard Practice for Evaluation of Hydrogen Uptake, Permeation, and Transport in Metals by an Electrochemical Technique
	G 161	Standard Guide for Corrosion-Related Failure Analysis
	G 170	Standard Guide for Evaluating and Qualifying Oilfield and Refinery Corrosion Inhibitors in the Laboratory

(a) European Federation of Corrosion

Table 13 Commonly used standards and references for downstream (refining) materials

Source	Specification number	Title
API recommended practices		
API	578	Material Verification Program for New and Existing Alloy Piping Systems
API	651	Cathodic Protection of Aboveground Petroleum Storage Tanks
API	751	Recommended Practice for Safe Operation of Hydrofluoric Acid Alkylation Units
API	934	Materials and Fabrication Requirements for 2.25Cr-1Mo and 3Cr-1Mo Steel Heavy Wall Pressure Vessels for High Temperature, High Pressure Hydrogen Service
API	941	Steels for Hydrogen Service at Elevated Temperatures and Pressures in Petroleum Refineries and Petrochemical Plants
API	945	Avoiding Environmental Cracking in Amine Units
API publications		
API	938	An Experimental Study of Causes and Repair of Cracking of 1.25Cr-0.5Mo Steel Equipment (May 1996)
API	939-B	Repair and Remediation Strategies for Equipment Operating in Wet H_2S Service (June 2002)

(continued)

Table 13 (continued)

Source	Specification number	Title
API publications (continued)		
API	939D	Stress Corrosion Cracking in Fuel Grade Ethanol: Review and Survey (September 2003)
API	944	1972 Survey of Materials Experience and Corrosion Problems in Sour Water Strippers
API	946	The Effect of Outgassing Cycles on the Hydrogen Content in Petrochemical Reactor Vessel Steels (July 1981)
API	950	Survey of Construction Materials and Corrosion in Sour Water Strippers (1978)
API	956	Hydrogen Assisted Crack Growth in 2.25Cr-1Mo Steel (March 1978)
API	959	Characterization Study of Temper Embrittlement of Chromium-Molybdenum Steels (1982)
NACE test methods		
NACE	TM0177	Laboratory Testing of Metals for Resistance to Sulfide Stress Cracking and Stress Corrosion Cracking in H_2S Environments
NACE	TM0284	Evaluation of Pipeline and Pressure Vessel Steels for Resistance to Hydrogen-Induced Cracking
NACE	TM0103	Laboratory Test Procedures for Evaluation of SOHIC Resistance of Plate Steels Used in Wet H_2S Service
NACE publications		
NACE	8X194	Materials and Fabrication Practices for New Pressure Vessels Used in Wet H_2S Refining Service
NACE standard material requirements		
	MR0103	Materials Resistant to Sulfide Stress Cracking in Corrosive Petroleum Refining Environments
NACE recommended practices		
	RP0169	Control of External Corrosion on Underground or Submerged Metallic Piping Systems
	RP0170	Protection of Austenitic Stainless Steels and Other Austenitic Alloys from Polythionic Acid Stress Corrosion Cracking during Shutdown of Refinery Equipment
	RP0272	Direct Calculation of Economic Appraisals of Corrosion Control Measures
	RP0472	Methods and Controls to Prevent In-Service Environmental Cracking of Carbon Steel Weldments in Corrosive Petroleum Refining Environments
	RP0182	Initial Conditioning of Cooling Water Equipment
	RP0189	On-Line Monitoring of Cooling Waters
	RP0590	Recommended Practice for Prevention, Detection, and Correction of Deaerator Cracking
	RP0391	Materials for the Handling and Storage of Concentrated (90 to 100%) Sulfuric Acid at Ambient Temperatures
	RP0392	Recovery and Repassivation after Low pH Excursions in Open Recirculating Cooling Water Systems
	RP0294	Design, Fabrication, and Inspection of Tanks for the Storage of Concentrated Sulfuric Acid and Oleum at Ambient Temperatures
	RP0296	Guidelines for Detection, Repair, and Mitigation of Cracking of Existing Petroleum Refinery Pressure Vessels in Wet H_2S Environments
	RP0198	The Control of Corrosion Under Thermal Insulation and Fireproofing—A Systems Approach
ASTM standards		
	G 35	Standard Practice for Determining the Susceptibility of Stainless Steels and Related Nickel-Chromium-Iron Alloys to Stress-Corrosion Cracking in Polythionic Acids
	G 58	Standard Practice for Preparation of Stress-Corrosion Test Specimens for Weldments
	G 79	Standard Practice for Evaluation of Metals Exposed to Carburization Environments
	G 96	Standard Guide for On-Line Monitoring of Corrosion in Plant Equipment (Electrical and Electrochemical Methods)
	G 111	Standard Guide for Corrosion Tests in High Temperature or High Pressure Environment, or Both
	G 123	Standard Test Method for Evaluating Stress-Corrosion Cracking of Stainless Alloys with Different Nickel Content in Boiling Acidified Sodium Chloride Solution
	G 142	Standard Test Method for Determination of Susceptibility of Metals to Embrittlement in Hydrogen Containing Environments at High Pressure, High Temperature, or Both
	G 146	Standard Practice for Evaluation of Disbonding of Bimetallic Stainless Alloy/Steel Plate for Use in High-Pressure, High-Temperature Refinery Hydrogen Service
	G 148	Standard Practice for Evaluation of Hydrogen Uptake, Permeation, and Transport in Metals by an Electrochemical Technique
	G 157	Standard Guide for Evaluating the Corrosion Properties of Wrought Iron- and Nickel-Based Corrosion Resistant Alloys for the Chemical Process Industries
	G 161	Standard Guide for Corrosion-Related Failure Analysis
ASME standards		
ASME	CRTC-Vol.34	Consensus on Operating Practices for the Control of Feedwater and Boiler Water Chemistry in Modern Industrial Boilers

Table 14 Contact information for selected technical organizations involved in petroleum production and refining operations

Organization	Acronym	City	State/Country	Internet address(a)
American Bureau of Shipping	ABS	Houston	TX	www.eagle.org
American National Standards Institute	ANSI	New York	NY	www.ansi.org
American Petroleum Institute	API	Washington	D.C.	http://api-ep.api.org/
American Society of Mechanical Engineers	ASME	Fairfield	NJ	www.asme.org
American Society of Testing and Materials International	ASTM	West Conshohocken	PA	www.astm.org
American Welding Society	AWS	Miami	FL	www.aws.org
American Water Works Association	AWWA	Denver	CO	www.awwa.org
Composites Engineering and Applications Center	CEAC	Houston	TX	www.egr.uh.edu/ceac/
Det Norske Veritas	DNV	Oslo	Norway	www.dnv.com
Engineering Equipment and Materials Users Association	EEMUA	London	U.K.	www.eemua.co.uk
European Federation of Corrosion	EFC	London	U.K.	www.efcweb.org
International Maritime Organization	IMO	London	U.K.	www.imo.org
International Organization for Standards	ISO	Geneva	SZ	www.iso.org
NACE International	NACE	Houston	TX	www.nace.org
NORSOK/Standard Norge	NORSOK	Lysaker	Norway	www.standard.no
United Kingdom Offshore Operators Association	UKOOA	London	U.K.	www.ukooa.co.uk
U.S. Coast Guard	USCG	Groton	CT	www.uscg.mil

(a) URL as accessed January 2006

ACKNOWLEDGMENT

This article has been adapted from J. Gutzeit, R.D. Merrick, and L.F. Scharfstein, Corrosion in Petroleum Refining and Petrochemical Operations, *Corrosion,* Vol 13, *ASM Handbook* (formerly *Metals Handbook,* 9th ed.) ASM International, 1987, p 1263–1287.

REFERENCES

1. B.B. Morton, Metallurgical Methods for Combating Corrosion and Abrasion in the Petroleum throughout Industry, *J. Inst. Petrol.,* Vol 34 (No. 289), 1948, p 1–68
2. E.L. Hildebrand, Materials Selection for Petroleum Refineries and Petrochemical Plants, *Mater. Prot. Perform.,* Vol 11 (No. 7), 1972, p 19–22
3. A.J. Freedman, G.F. Tisinai, and E.S. Troscinski, Selection of Alloys for Refinery Processing Equipment, *Corrosion,* Vol 16 (No. 1), 1960, p 19t–25t
4. *The Role of Stainless Steels in Petroleum Refining,* American Iron and Steel Institute, 1977
5. R.D. Kane, M.S. Cayard, and R.J. Horvath, Corrosion Combat: Refinery Sour Water Corrosion and New Development to Prevent It, *Hydrocarb. Process.,* March 2004, p 87–91
6. Selection of Steel for High-Temperature Service in Petroleum Refinery Applications, in *Properties and Selection of Metals,* Vol 1, *Metals Handbook,* 8th ed., American Society for Metals, 1961, p 585–603
7. G.E. Moller, I.A. Franson, and T.J. Nichol, Experience with Ferritic Stainless Steel in Petroleum Refinery Heat Exchangers, *Mater. Perform.,* Vol 20 (No. 4), 1981, p 41–50
8. A.J. Brophy, Stress Corrosion Cracking of Austenitic Stainless Steels in Refinery Environments, *Mater. Perform.,* Vol 13 (No. 5), 1974, p 9–15
9. A.S. Couper and H.F. McConomy, Stress Corrosion Cracking of Austenitic Stainless Steels in Refineries, *Proc. API,* Vol 46 (III), 1966, p 321–326
10. R.F.A. Jargelius-Pettersson, Application of the Pitting Resistance Equivalent Concept to Some Highly Alloyed Austenitic Stainless Steels, *Corrosion,* Vol 54 (No. 2), Feb 1998, p 162–168
11. R.D. Kane, SOCRATES Alloy Selection Software Documentation, InterCorr International, Inc., private communication, 1994
12. H.L. Craig, "Naphthenic Acid Corrosion in the Refinery," Paper 333, Corrosion/95, NACE International, 1995
13. T.P. May, J.F. Mason, Jr., and W.K. Abbot, Austenitic Nickel Cast Irons in the Petroleum Industry, *Mater. Prot.,* Vol 1 (No. 8), 1962, p 40–55
14. E.D. Verink, Jr. and F.B. Murphy, "Solving Refinery Corrosion Problems with Aluminum," paper presented at the NACE 16th Annual Conference (Dallas, TX), National Association of Corrosion Engineers, March 1960
15. E.E. Kerns and W.E. Baker, Use of Aluminum in Petroleum Refinery Equipment, *Proc. API,* Vol 31 (III), 1951, p 89–98
16. R.L. Hilderbrand, Aluminum Exchanger and Condenser Tubes in Petroleum Service, *Proc. API,* Vol 40 (III), 1960, p 118–130
17. J. Kolts, J.B.C. Wu, and A.I. Asphahani, Highly Alloyed Austenitic Materials for Corrosion Service, *Met. Prog.,* Vol 125 (No. 10), 1983, p 25–36
18. H.R. Copson, Effect of Composition on Stress Corrosion Cracking of Some Alloys Containing Nickel. *Physical Metallurgy of Stress Corrosion Fracture.* T.N. Rhodin, Ed., Interscience. New York. 1959, p 247–272
19. J.F. Mason, Jr., The Selection of Materials for Some Petroleum Refinery Applications, *Corrosion,* Vol 12 (No. 5), 1956, p 199t–206t
20. *Corrosion Resistance of Hastelloy Alloys,* The Haynes International, 1978
21. A.I. Asphahani, Corrosion Resistance of High Performance Alloys, *Mater. Perform.,* Vol 19 (No. 12), 1980, p 33–43
22. I.A. Franson and L.C. Covington, Application of Titanium to Oil Refinery Environments, *Proc. API,* Vol 56 (III), 1977, p 26–36
23. D.M. McCue, "Design Considerations for Titanium Heat Exchangers," Paper 60, presented at Corrosion/81 (Houston, TX), National Association of Corrosion Engineers, 1981
24. J.A. McMaster, Selection of Titanium for Petroleum Refinery Components, *Mater. Perform.,* Vol 18 (No. 4), 1979, p 28–34
25. R.L. Jacobs and J.A. McMaster, Titanium Tubing: Economical Solution to Heat Exchanger Corrosion, *Mater. Prot. Perform.,* Vol 11 (No. 7), 1972, p 33–38
26. R.W. Schutz and L.C. Covington, Effect of Oxide Films on the Corrosion Resistance of Titanium, *Corrosion,* Vol 37 (No. 10), 1981, p 585–591
27. "Risk-Based Inspection," API 580, American Petroleum Institute, May 2002
28. "Fitness-for-Service," API 579, American Petroleum Institute, March 2000
29. C.D. Clauser, L.G. Emmer, A.W. Pense, and R.D. Stout, A Phenomenological Study of the Susceptibility to Temper Embrittlement of 2.25%Cr-1%Mo, *Proc. API,* Vol 52 (III), 1972, p 790
30. R.D. Kane and M.S. Cayard, "Remediation and Repair Techniques for Wet H_2S Cracking," Eurocorr 2001, Sept 30–Oct 4, 2001 (Rive Del Garda), European Federation of Corrosion and Associazione Italiana di Metallurgica
31. R.D. Kane, "Online, Real-Time Corrosion Monitoring for Improving Pipeline Integrity—Technology and Experience," Paper 03175, Corrosion 2003, NACE International, 2003
32. Z.A. Foroulis, Corrosion and Corrosion Inhibition in the Petroleum Industry, *Werkst. Korros.,* Vol 33 (No. 2), 1982, p 121–131
33. L. Engel, Korrosion in Mineralolraffinerien, *Erdol Kohle,* Vol 27 (No. 6), 1974, p 301–306
34. R.D. Kane and M.S. Cayard, Materials for High Temperature, *Encyclopedia of Chemical Processing and Design,* Vol 26, J.J. McKetta, Ed., Marcel Dekker, 1996, p 45–64
35. R.D. Kane and M.S. Cayard, "A Comprehensive Study on Naphthenic Acid Corrosion," Paper 02555, Corrosion 2002, NACE International, 2002
36. Conditions Causing Deterioration or Failure, *Guide for Inspection of Refinery Equipment,* 2nd ed., American Petroleum Institute, 1973
37. G.J. Samuelson, Hydrogen-Chloride Evolution from Crude Oils as a Function of Salt Concentration, *Proc. API,* Vol 34 (III), 1954, p 50–54
38. C.G. Munger, Deep Pitting Corrosion in Sour Crude Oil Tankers, *Mater. Perform.,* Vol 15 (No. 3), 1976, p 17–23
39. W.J. Neill and Z.A. Foroulis, Internal Corrosion in Floating Roof Gasoline Storage Tanks, *Mater. Perform.,* Vol 15 (No. 9), 1976, p 37–39
40. C.M. Hudgins, Jr., A Review of Sulfide Corrosion Problems in the Petroleum Industry, *Mater. Prot.,* Vol 8 (No. 1), 1969, p 41–47
41. S.P. Ewing, Electrochemical Studies of the Hydrogen Sulfide Corrosion Mechanism, *Corrosion,* Vol 11 (No. 11), 1955, p 497t–501t
42. R.L. Piehl, How to Cope with Corrosion in Hydrocracker Effluent Coolers, *Oil Gas J.,* Vol 66, July 8, 1968, p 60–63
43. C.B. Hutchison and W.B. Hughes, Oxidation Reduction Potential as a Control Criterion in Inhibition of Refinery Sulfide Corrosion, *Corrosion,* Vol 17 (No. 11), 1961, p 514t–518t
44. F.H. Meyer, O.L. Riggs, R.L. McGlasson, and J.D. Sudbury, Corrosion Products of Mild Steel in Hydrogen Sulfide Environments, *Corrosion,* Vol 14 (No. 2), 1958, p 109t–115t
45. "Controlling Weld Hardness of Carbon Steel Refining Equipment to Prevent Environmental Cracking," Publication 942, 2nd ed., American Petroleum Institute, 1982
46. D.B. Bird, Titanium Exchanger Tubing Works Well in Refinery Service, *Hydrocarb. Process.,* Vol 55 (No. 5), 1976, p 105–107

47. W.J. Neill, Experience with Titanium Tubing in Oil Refinery Heat Exchangers, *Mater. Perform.*, Vol 19 (No. 9), 1980, p 57–63
48. A.S. Couper, Corrosion Control in Crude Oil Distillation Overhead Condensers, *Proc. API*, Vol 44 (III), 1964, p 172–178
49. M.H. Brown, W.B. DeLong, and J.R. Auld, Corrosion by Chlorine and by Hydrogen Chloride at High Temperatures, *Ind. Eng. Chem.*, Vol 39 (No. 7), 1947, p 839–844
50. K. Reiser, The Control of Corrosion in Refinery Distillation Units, *J. Inst. Petrol.*, Vol 53 (No. 527), 1967, p 352–366
51. D.L. Kronenberger and D.A. Pattison, Troubleshooting the Refinery Desalter Operation, *Mater. Perform.*, Vol 25 (No. 7), 1986, p 9–17
52. L.E. Fisher, G.C. Hall, and R.W. Stenzel, Crude Oil Desalting to Reduce Refinery Corrosion Problems, *Mater. Prot.*, Vol 1 (No. 5), 1962, p 8–11 and 14–17
53. L.C. Waterman, Crude Desalting: Why and How, *Hydrocarb. Process.*, Vol 44 (No. 2), 1965, p 133–138
54. M.J. Humphries and G. Sorell, Corrosion Control in Crude Oil Distillation Units, *Mater. Perform.*, Vol 15 (No. 2), 1976, p 13–21
55. J.A. Biehl and E.A. Schnake, Corrosion in Crude-Oil Processing—Low pH vs. High pH, *Proc. API*, Vol 37 (III), 1957, p 129–134
56. J.A. Biehl and E.A. Schnake, Processing Crude Oil at Low pH, *Proc. API*, Vol 39 (III), 1959, p 214–221
57. C.J. Scherrer, C.R. Baumann, and G.J. Jarno, "Crude Units: Focusing on Corrosion of Initial Condensation Equipment," Paper 66, presented at Corrosion/81, National Association of Corrosion Engineers, 1981
58. R.D. Merrick and T. Auerback, Crude Unit Overhead Corrosion Control, *Mater. Perform.*, Vol 22 (No. 9), 1983, p 15–21
59. R.H. Carlton, Continuous Injection of Overhead Receiver Water Controls Refinery Condenser and Exchanger Corrosion by Ammonium Chloride, *Mater. Prot.*, Vol 2 (No. 1), 1963, p 15–20
60. R.J. Hafsten and K.R. Walston, Use of Neutralizers and Inhibitors to Combat Corrosion in Hydrocarbon Streams, *Proc. API*, Vol 35 (III), 1955, p 80–91
61. D.L. Burns, R.L. Hildebrand, and P.D. Thomas, Corrosion Inhibitors in Refinery Process Streams, *Proc. API*, Vol 40 (III), 1960, p 155–162
62. C. Baumann and C. Scherrer, Evaluation of Inhibitors for Crude Topping Units, *Mater. Perform.*, Vol 18 (No. 11), 1979, p 51–57
63. R.T. Foley, The Role of the Chloride Ion in Iron Corrosion, *Corrosion*, Vol 26 (No. 2), 1970, p 58–70
64. J. Scherzer and D.P. McArthur, Tests Show Effects of Nitrogen Compounds on Commercial Fluid Cat Cracking Catalysts, *Oil Gas J.*, Vol 84, Oct 27, 1986, p 76–82
65. J. Gutzeit, Corrosion of Steel by Sulfides and Cyanides in Refinery Condensate Water, *Mater. Prot.*, Vol 7 (No. 12), 1968, p 17–23
66. G. Kobrin and E.S. Kopecki, Choosing Alloys for Ammonia Services, *Chem. Eng.*, Vol 85, Dec 18, 1978, p 115–128
67. R.L. Piehl, Survey of Corrosion in Hydrocracker Effluent Air Coolers, *Mater. Perform.*, Vol 15 (No. 1), 1976, p 15–20
68. E.G. Ehmke, Corrosion Correlations with Ammonia and Hydrogen Sulfide in Air Coolers, *Mater. Perform.*, Vol 14 (No. 7), 1975, p 20–28
69. "1972 Survey of Materials Experience and Corrosion Problems in Sour Water Strippers," Publication 944, American Petroleum Institute, 1974
70. "Survey of Construction Materials and Corrosion in Sour Water Strippers—1978," Publication 950, American Petroleum Institute, 1983
71. R.L. Hildebrand, Sour Water Strippers—A Review of Construction Materials, *Mater. Perform.*, Vol 13 (No. 5), 1974, p 16–19
72. J.E. Cantwell and R.E. Bryant, Failure of Refinery Piping by Liquid Metal Attack and Materials Experience with Smokeless Flares, *Proc. API*, Vol 53 (III), 1973, p 412–430
73. R.S. Treseder and A. Wachter, Corrosion in Petroleum Processes Employing Aluminum Chloride, *Corrosion*, Vol 5 (No. 11), 1949, p 383–391
74. J.F. Mason, Jr. and C.M. Schillmoller, Minimum Corrosion for Butane Isomerization Units, *Corrosion*, Vol 15 (No. 4), 1959, p 185t–188t
75. K. Brooks, Organic Chloride Contamination Rears Its Ugly Head Again, *Oil Gas J.*, Vol 60, Nov 26, 1962, p 74–75
76. D.H. Stormont, Chlorides in Crude Oil Plague Refiners, *Oil Gas J.*, Vol 67, April 14, 1969, p 94–96
77. E.B. Backensto and A.N. Yurick, Chloride Corrosion and Fouling in Catalytic Reformers with Naphtha Pretreaters, *Corrosion*, Vol 17 (No. 3), 1961, p 133t–136t
78. R.J. Schuyler III, Hydrogen Blistering of Steel in Anhydrous Hydrofluoric Acid, *Mater. Perform.*, Vol 18 (No. 8), 1979, p 9–16
79. K. Forry and C. Schrage, Trouble-Shooting HF Alkylation, *Hydrocarb. Process.*, Vol 45 (No. 1), 1966, p 107–114
80. D.P. Thornton, Jr., Corrosion-Free HF Alkylation, *Chem. Eng.*, Vol 77, July 13, 1970, p 108–112
81. National Petroleum Refiners Association Questions and Answers—Alkylation, *Oil Gas J.*, Vol 71, May 7, 1973, p 56–61; Vol 75, May 23, 1977, p 73–75; Vol 78, July 14, 1980, p 162–165
82. N.P. Lieberman, Basic Decision Key to Alky Problems, *Oil Gas J.*, Vol 78, June 23, 1980, p 141–144
83. National Petroleum Refiners Association Questions and Answers—Alkylation, *Oil Gas J.*, Vol 66, Feb 26, 1968, p 96–105; Vol 67, Jan 27, 1969, p 162–167; Vol 75, May 23, 1977, p 66–73
84. G.A. Nelson, Prevention of Localized Corrosion in Sulfuric Acid Handling Equipment, *Corrosion*, Vol 14 (No. 3), 1958, p 145t–149t
85. H.W. Van der Hoeven, Stromungsgeschwindigkeit als besonderer Faktor bei der Schwefelsaure-Korrosion, *Werkst. Korros.*, Vol 6 (No. 2), 1955, p 57–62
86. V.J. Groth and R.J. Hafsten, Corrosion of Refinery Equipment by Sulfuric Acid and Sulfuric Acid Sludges, *Corrosion*, Vol 10 (No. 11), 1954, p 368–390
87. R.B. Martin, *The Processing of Alkylation Unit Feedstocks and Reactor Effluent Streams*, Petrolite Corporation
88. R. Baboian, *Corrosion Engineer's Handbook*, 3rd ed., NACE International, 2002, p 192
89. API Survey Shows Few Amine Corrosion Problems, *Petrol. Refiner*, Vol 37 (No. 11), 1958, p 281–283
90. R.V. Comeaux, The Mechanism of MEA Corrosion, *Proc. API*, Vol 42 (III), 1962, p 481–489
91. F.S. Lang and J.F. Mason, Jr., Corrosion in Amine Gas Treating Solutions, *Corrosion*, Vol 14 (No. 2), 1958, p 105t–108t
92. A.J.R. Rees, Problems with Pressure Vessels in Sour Gas Service (Case Histories), *Mater. Perform.*, Vol 16 (No. 7), 1977, p 29–33
93. G.L. Garwood, What to Do About Amine Stress Corrosion, *Oil Gas J.*, Vol 52, July 27, 1953, p 334–340
94. S. Tebbal, R.D. Kane, B.N. Al-Shumaimri, and P.K. Mukhopadhyay, "Development of a Non-Toxic Corrosion Inhibitor for MEA Plants," Paper 410, Corrosion/98, NACE International, 1998
95. K.F. Butwell, How to Maintain Effective MEA Solutions, *Hydrocarb. Process.*, Vol 47 (No. 4), 1968, p 111–113
96. J.C. Dingman, D.L. Allen, and T.F. Moore, Minimize Corrosion in MEA Units, *Hydrocarb. Process.*, Vol 45 (No. 9), 1966, p 285–290
97. L.R. White and D.E. Street, Amine Treating, *Corrosion in the Oil Refining Industry*, NACE International, 1996, p 12/1–13
98. P.C. Rooney, T.R. Bacon, and M.S. DuPart, Effect of Heat Stable Salts on MDEA Solution Corrosivity, *Hydrocarb. Process.*, March 1996, p 95–101
99. R.D. Kane, InterCorr International, Inc., private communication, 2004
100. D. Ballard, How to Operate an Amine Plant, *Hydrocarb. Process.*, Vol 45 (No. 4), 1966, p 137–144

101. G.D. Hall and L.D. Polderman, Design and Operating Tips for Ethanolamine Gas Scrubbing Systems, *Chem. Eng. Prog.*, Vol 56 (No. 10), 1960, p 52–58
102. E.N. Skinner, J.F. Mason, and J.J. Moran, High Temperature Corrosion in Refinery and Petrochemical Service, *Corrosion*, Vol 16 (No. 12), 1960, p 593t–600t
103. Z.A. Foroulis, High Temperature Degradation of Structural Materials in Environments Encountered in the Petroleum and Petrochemical Industries: Some Mechanistic Observations, *Anti-Corros.*, Vol 32 (No. 11), 1985, p 4–9
104. A.S. Couper and A. Dravnieks, High Temperature Corrosion by Catalytically Formed Hydrogen Sulfide, *Corrosion*, Vol 18 (No. 8), 1962, p 291t–298t
105. A.S. Couper, High Temperature Mercaptan Corrosion of Steels, *Corrosion*, Vol 19 (No. 11), 1963, p 396t–401t
106. K.N. Strafford, The Sulfidation of Metals and Alloys, *Metall. Rev.*, Vol 138, 1969
107. F.A. Hendershot and H.L. Valentine, Materials for Catalytic Cracking Equipment (Survey), *Mater. Prot.*, Vol 6 (No. 10), 1967, p 43–47
108. N. Schofer, Corrosion Problems in a Fluid Catalytic Cracking and Fractionating Unit, *Corrosion*, Vol 5 (No. 6), 1949, p 182–188
109. S.L. Estefan, Design Guide to Metallurgy and Corrosion in Hydrogen Processes, *Hydrocarb. Process.*, Vol 49 (No. 12), 1970, p 85–92
110. L.T. Overstreet and R.A. White, Materials Specifications and Fabrication for Hydrocracking Process Equipment, *Mater. Prot.*, Vol 4 (No. 6), 1965, p 64–71
111. H.F. McConomy, High-Temperature Sulfidic Corrosion in Hydrogen-Free Environment, *Proc. API*, Vol 43 (III), 1963, p 78–96
112. J. Gutzeit, High Temperature Sulfidic Corrosion of Steels, *Process Industries Corrosion—The Theory and Practice*, National Association of Corrosion Engineers, 1986
113. D.W. McDowell, Jr., Refinery Reactor Design to Prevent High Temperature Corrosion, *Mater. Prot.*, Vol 5 (No. 11), 1966, p 45–48
114. "Steels for Hydrogen Service at Elevated Temperatures and Pressures in Petroleum Refineries and Petrochemical Plants," Publication 941, 3rd ed., American Petroleum Institute, 1983
115. E.B. Backensto, R.D. Drew, and C.C. Stapleford, High Temperature Hydrogen Sulfide Corrosion, *Corrosion*, Vol 12 (No. 1), 1956, p 6t–16t
116. G. Sorell and W.B. Hoyt, Collection and Correlation of High Temperature Hydrogen Sulfide Corrosion Data, *Corrosion*, Vol 12 (No. 5), 1956, p 213t–234t
117. C. Phillips, Jr., High Temperature Sulfide Corrosion in Catalytic Reforming of Light Naphthas, *Corrosion*, Vol 13 (No. 1), 1957, p 37t–42t
118. G. Sorell, Compilation and Correlation of High Temperature Catalytic Reformer Corrosion Data, *Corrosion*, Vol 14 (No. 1), 1958, p 15t–26t
119. W.H. Sharp and E.W. Haycock, Sulfide Scaling Under Hydrorefining Conditions, *Proc. API*, Vol 39 (III), 1959, p 74–91
120. J.D. McCoy and F.B. Hamel, New Corrosion Data for Hydrodesulfurizing Units, *Hydrocarb. Process.*, Vol 49 (No. 6), 1970, p 116–120
121. J.D. McCoy and F.B. Hamel, Effect of Hydrodesulfurizing Process Variables on Corrosion Rates, *Mater. Prot. Perform.*, Vol 10 (No. 4), 1971, p 17–22
122. A.S. Couper and J.W. Gorman, Computer Correlations to Estimate High Temperature H_2S Corrosion in Refinery Streams, *Mater. Prot. Perform.*, Vol 10 (No. 1), 1971, p 31–37
123. *Refin-Cor*, NACE International
124. *Overview of Sulfidic Corrosion in Refining*, Publication 34103, NACE International, 2004, p 1–9
125. J.J. Heller, Corrosion of Refinery Equipment by Naphthenic Acid, *Mater. Prot.*, Vol 2 (No. 9), 1963, p 90–96
126. B. Danilov, The Control of Corrosion in Refinery Vacuum Plants, *Anti-Corros.*, Vol 22 (No. 8), 1975, p 3–6
127. "Standard Test Method for Acid Number of Petroleum Products by Potentiometric Titration," D 664, *Annual Book of ASTM Standards*, Vol 05.01, American Society for Testing and Materials
128. "Acid Number and Naphthenic Acids by Potentiometric Titration," UOP 565, ASTM International
129. W.A. Derungs, Naphthenic Acid Corrosion—An Old Enemy of the Petroleum Industry, *Corrosion*, Vol 12 (No. 12), 1956, p 617t–622t
130. E. Babaian-Kibala and M.J. Nugent, "Naphthenic Acid Corrosion Literature Survey," Paper 99378, Corrosion/99, NACE International, 1999
131. J. Gutzeit, Naphthenic Acid Corrosion in Oil Refineries, *Mater. Perform.*, Vol 16 (No. 10), 1977, p 24–35
132. W.T. Reid, *External Corrosion and Deposits—Boilers and Gas Turbines*, American Elsevier, 1971
133. A.L. Plumley, J. Jonakin, and R.E. Vuia, "A Review Study of Fire-Side Corrosion in Utility and Industrial Boilers," paper presented at Corrosion Seminar (Hamilton, ON), McMaster University and Engineering Institute of Canada, May 1966
134. G.W. Cunningham and A. de S. Brasunas, The Effects of Contamination by Vanadium and Sodium Compounds on the Air-Corrosion of Stainless Steel, *Corrosion*, Vol 12 (No. 8), 1956, p 389t–405t
135. G.Y. Lai, *High Temperature Corrosion of Engineering Alloys*, ASM International, 1990, p 20–21, 156
136. D.A. Eden and B. Breen, "On-Line Electrochemical Corrosion Monitoring in Fireside Applications," Paper 03361, Corrosion 2003, NACE International, March 2003
137. S.W. Dean, J.G. Maldonado, and R.D. Kane, *Cyclic Strain Cracking of Stainless Steels in Hot Steam/Hydrocarbon Reformer Condensates: Test Method Development*, STP 1401, ASTM, 2001
138. V. Singh, Performance of Austenitic Stainless Steels in Wet Sour Water—Part 1 and 2, *Mater. Perform.*, Aug-Sept 2004
139. D.R. McIntyre and C.P. Dillon, *Guidelines for Preventing Stress Corrosion Cracking in the Chemical Process Industries*, Publication 15, Materials Technology Institute of the Chemical Process Industries, 1985
140. R.F. Steigerwald, New Molybdenum Stainless Steels for Corrosion Resistance: A Review of Recent Developments, *Mater. Perform.*, Vol 13 (No. 9), 1974, p 9–15
141. S. Bernhardsson, P. Norberg, H. Eriksson, and O. Forsell, "Duplex and High Nickel Stainless Steels for Refineries and the Petrochemical Industry," Paper 165, presented at Corrosion/85 (Houston, TX), National Association of Corrosion Engineers, 1985
142. "Protection of Austenitic Stainless Steel from Polythionic Acid Stress Corrosion Cracking during Shutdown of Refinery Equipment," RP0170, NACE International
143. C.S. Carter and M.V. Hyatt, Review of Stress Corrosion Cracking in Low-Alloy Steels with Yield Strengths Below 150 ksi, *Stress Corrosion Cracking and Hydrogen Embrittlement of Iron Base Alloys*, National Association of Corrosion Engineers, 1977
144. S.W. Dean, D. Abayarathna, and R.D. Kane, "Stress Corrosion Cracking of 304L Stainless Steel and $1\frac{1}{4}$Cr-$\frac{1}{2}$Mo Steel in Caustic Environments in Syngas Service," Paper 590, Corrosion/98, NACE International, 1998
145. A.S. Krisher, Material Requirements for Anhydrous Ammonia, *Process Industries Corrosion—The Theory and Practice*, National Association of Corrosion Engineers, 1986
146. J.M.B. Gotch, *Code of Practice for the Storage of Anhydrous Ammonia Under Pressure in the United Kingdom*, Chemical Industries Association, Ltd., 1980
147. P.G. Hughes, Stress Corrosion Cracking in a MEA Unit, *Proceedings of the 1982 U.K. National Corrosion Conference*, Institution of Corrosion Science and Technology, 1982, p 87
148. "Avoiding Environmental Cracking in Amine Units," RP 945, American Petroleum Institute, 2003

149. J. Gutzeit and J.M. Johnson, Stress Corrosion Cracking of Carbon Steel Welds in Amine Service, *Mater. Perform.,* Vol 25 (No. 7), 1986, p 18–26
150. H.U. Schutt, New Aspects of SCC in Monoethanolamine Solutions, *Mater. Perform.,* Aug 1987, p 18–24
151. H. Nishida, K. Nakamura, and H. Takahashi, Intergranular Stress Corrosion Cracking of Sensitized 321 SS Tube Exposed to Polythionic Acid, *Mater. Perform.,* Vol 23 (No. 4), 1984, p 38–41
152. J.E. Cantwell, Embrittlement and Intergranular Stress Corrosion Cracking of Stainless Steels After Elevated Temperature Exposure in Refinery Process Units, *Proc. API,* Vol 63 (III), 1984, p 32–37
153. C.H. Samans, Stress Corrosion Cracking Susceptibility of Stainless Steels and Nickel-Base Alloys in Polythionic Acids and Acid Copper Sulfate Solution, *Corrosion,* Vol 20 (No. 8), 1964, p 256t–262t
154. R.L. Piehl, Stress Corrosion Cracking by Sulfur Acids, *Proc. API,* Vol 44 (III), 1964, p 189–197
155. R.D. Kane, N. Sridhar, and M. Brongers, *Stress Corrosion Cracking of Steel Equipment in Fuel Ethanol,* Technical Publication 939D, American Petroleum Institute, 2005, in publication
156. R.D. Kane, R.J. Horvath, and M.S. Cayard, *Wet H_2S Cracking of Carbon Steels and Weldments,* NACE International, 1996
157. "Standard Practice for Evaluation of Hydrogen Uptake, Permeation, and Transport in Metals by an Electrochemical Technique," G 148, ASTM International
158. M.S. Cayard, R.D. Kane, C.J.B. Joia, and L.A. Correia, "Methodology for the Application of Hydrogen Flux Monitoring Devices to Assess Equipment Operating in Wet H_2S Service," Paper 394, Corrosion 98, NACE International, 1998
159. C.J.B. Joia, L.A. Garcia, W. Baptitsta, M.A.S. Filho, O.R. Mattos, P.S.A. Bezerra, L. Ruznak, and C.M. Menendez, "Process Control Using Real Time Hydrogen Flux Monitoring Probe," Paper 01527, Corrosion 2001, NACE International
160. M.S. Cayard, R.D. Kane, and R.J. Horvath, "SOHIC Resistance of C-Mn Plate Steels Used in Refinery Service," Paper 02554, Corrosion 2002, NACE International, 2002
161. M.S. Cayard, and R.D. Kane, Large-Scale Wet Hydrogen Sulfide Cracking Performance: Evaluation of Metallurgical, Mechanical, and Welding Variables, *Corrosion,* Vol 53 (No. 3), March 1997, p 227–233
162. R.D. Kane and M. Prager, "Characterization of Hydrogen Charging Severity in Simulated Wet H_2S Refinery Environments," presented at the Second International Conference on Interaction of Steels with Hydrogen in Petroleum Industry Pressure Vessel and Pipeline Service (Vienna, Austria), 1994
163. H.C. Rodgers, Hydrogen Embrittlement in Engineering Materials, *Mater. Prot.,* Vol 1 (No. 4), 1962, p 26–33
164. G.A. Nelson and R.T. Effinger, Blistering and Embrittlement of Pressure Vessel Steels by Hydrogen, *Welding J.,* Vol 34 (No. 1), 1955, p 12S–21S
165. W.A. Bonner, H.D. Burnham, J.J. Conradi, and T. Skei, Prevention of Hydrogen Attack on Steel in Refinery Equipment, *Proc. API,* Vol 33 (III), 1953, p 255–272
166. T. Skei, A. Wachter, W.A. Bonner, and H.D. Burnham, Hydrogen Blistering of Steel in Hydrogen Sulfide Solutions, *Corrosion,* Vol 9 (No. 5), 1953, p 163–172
167. R.T. Effinger, M.L. Renquist, A. Wachter, and J.G. Wilson, Hydrogen Attack of Steel in Refinery Equipment, *Proc. API,* Vol 31 (III), 1951, p 107–133
168. "Evaluation of Pipeline and Pressure Vessel Steels for Resistance to Hydrogen Induced Cracking," TM0284, NACE International, 1984
169. "Laboratory Test Procedures for Evaluation of SOHIC Resistance of Plate Steels Used in Wet H_2S Service," TM0103, NACE International, 2003
170. R.D. Kane, InterCorr International, private communication, research results, 1996
171. W.A. Bonner and H.D. Burnham, Air Injection for Prevention of Hydrogen Penetration of Steel, *Corrosion,* Vol 11 (No. 10), 1955, p 447t–453t
172. E.F. Ehmke, "Use Ammonium Polysulfide to Stop Corrosion and Hydrogen Blistering," Paper 59, presented at Corrosion/81 (Houston, TX), National Association of Corrosion Engineers, 1981
173. B.W. Neumaier and C.M. Schillmoller, Deterrence of Hydrogen Blistering at a Fluid Catalytic Cracking Unit, *Proc. API,* Vol 35 (III), 1955, p 92–109
174. R.D. Kane, private communication. See U.S. Patent 5,853,620, InterCorr International, 1996
175. "Materials and Fabrication Practices for New Pressure Vessels Used in Wet H_2S Refinery Service," 8X194, NACE International, 1994
176. "Guidelines for Detection, Repair and Mitigation of Cracking of Existing Petroleum Refinery Pressure Vessels in Wet H_2S Service," RP0296, NACE International, 1996
177. G.B. Kohut and W.J. McGuire, Sulfide Stress Cracking Causes Failure of Compressor Components in Refinery Service, *Mater. Prot.,* Vol 7 (No. 6), 1968, p 17–21
178. E.L. Hildebrand, Aqueous Phase H_2S Cracking of Hard Carbon Steel Weldments—A Case History, *Proc. API,* Vol 50 (III), 1970, p 593–613
179. T.G. Gooch, Hardness and Stress Corrosion Cracking of Ferritic Steel, *Weld. Inst. Res. Bull.,* Vol 23 (No. 8), 1982, p 241–246
180. D.J. Kotecki and D.G. Howden, Wet Sulfide Cracking of Submerged Arc Weldments, *Proc. API,* Vol 52 (III), 1972, p 631–653
181. "Controlling Weld Hardness of Carbon Steel Refinery Equipment to Prevent Environmental Cracking," RP 942, American Petroleum Institute
182. "Methods and Controls to Prevent In-Service Environmental Cracking of Carbon Steel Weldments Corrosive Petroleum Refining Environments," RP0472, NACE International, 1972
183. "Materials Resistant to Sulfide Stress Cracking in Corrosive Petroleum Refining Environments," MR0103, NACE International, 2003
184. "Petroleum and Natural Gas Industries—Materials for Use in H_2S-Containing Environments in Oil and Gas Production—Parts 1, 2 and 3," MR0175/ISO 15156, NACE International
185. G. Sorell and M.J. Humphries, High Temperature Hydrogen Damage in Petroleum Refinery Equipment, *Mater. Perform.,* Vol 17 (No. 8), 1978, p 33–41
186. A.R. Ciuffreda and W.R. Rowland, Hydrogen Attack of Steel in Reformer Service, *Proc. API,* Vol 37 (III), 1957, p 116–128
187. R.D. Merrick and A.R. Ciuffreda, Hydrogen Attack of Carbon-0.5Molybdenum Steels, *Proc. API,* Vol 61 (III), 1982, p 101–114
188. R. Chiba, K. Ohnishi, K. Ishii, and K. Maeda, Effect of Heat Treatment on Hydrogen Attack Resistance of C-0.5Mo Steels for Pressure Vessels, Heat Exchangers, and Piping, *Corrosion,* Vol 41 (No. 7), 1985, p 415–426
189. "Steels for Hydrogen Service at Elevated Temperatures and Pressures in Petroleum Refineries and Petrochemical Plants," RP 941, American Petroleum Institute
190. R.D. Kane, "ASTM G 146—Standard Practice for Evaluation of Disbonding of Bimetallic Stainless Alloy/Steel Plate for Use in High Pressure, High Temperature Refinery Hydrogen Service," API Meeting (San Diego, CA), April 1997
191. "Standard Practice for Evaluation of Disbonding of Bimetallic Stainless Alloy/Steel Plate for Use in High-Pressure, High-Temperature Refinery Hydrogen Service," G 146, ASTM International
192. J.M. Johnson and J. Gutzeit, Embrittlement of Stainless Steel Welds by Contamination with Zinc-Rich Paint, *Proc. API,* Vol 63 (III), 1984, p 65–72
193. G.R. Port, Hydrogen Sulfide Corrosion in a Distilling Unit, *Proc. API,* Vol 41 (III), 1961, p 98–103
194. D.G. Damin and J.D. McCoy, Prevention of Corrosion in Hydrodesulfurizer Air Coolers and Condensers, *Mater. Perform.,* Vol 17 (No. 12), Dec 1978, p 23–26 (see also NACE Corrosion/78, paper 131)

195. C. Scherrer, M. Durrieu, and G. Jarno, Distillate and Resid Hydroprocessing: Coping with High Concentrations of Ammonium Bisulfide in the Process Water, *Mater. Perform.*, Vol 19 (No. 11), Nov 1980, p 25–31 (see also NACE Corrosion/79, paper 27)
196. W.A. McGill and M.J. Weinbaum, Aluminum-Diffused Steel Lasts Longer, *Oil Gas J.*, Vol 70, Oct 9, 1972, p 66–69
197. C.A. Robertson and H.L. Meyers, Application and Use of Aluminum Coatings in Oil Refinery Processes, *Mater. Prot.*, Vol 6 (No. 9), 1967, p 23–26
198. W.A. McGill and M.J. Weinbaum, The Selection, Application and Fabrication of Alonized Systems in the Refinery Environment, *Proc. API*, Vol 54 (III), 1975, p 125–159
199. M.S. Crowley, Refractories, *Process Industries Corrosion—The Theory and Practice*, National Association of Corrosion Engineers, 1986
200. "Standard Guide for On-Line Monitoring of Corrosion in Plant Equipment (Electrical and Electrochemical Methods)," G 96, ASTM International
201. R.D. Kane, D.C. Eden, and D.A. Eden, Innovative Solutions Integrate Corrosion Monitoring with Process Control, *Mater. Perform.*, Feb 2005, p 36–41
202. D.A. Eden and S. Srinivasan, "Real-Time, On-Line and On-Board: The Use of Computers, Enabling Corrosion Monitoring to Optimize Process Control," Paper 04059, Corrosion 2004, NACE International, March 2004
203. D.C. Eden and J.D. Kintz, "Real-Time Corrosion Monitoring for Improved Process Control: A Real and Timely Alternative to Upgrading of Materials of Construction," Paper 04238, Corrosion 2004, NACE International, March 2004
204. D.A. Eden, "Practical Measurements Using Non-Linear Analysis Techniques—Harmonic Distortion and Intermodulation Distortion," Paper 335, Corrosion 2005, NACE International, 2005

External Corrosion of Oil and Natural Gas Pipelines

John A. Beavers and Neil G. Thompson, CC Technologies

PIPELINES play an extremely important role throughout the world as a means of transporting gases and liquids over long distances from their sources to the ultimate consumers. The general public is not aware of the number of pipelines that are continually in service as a primary means of transportation. A buried operating pipeline is rather unobtrusive and rarely makes its presence known except at valves, pumping or compressor stations, or terminals. In the United States, there were approximately 217,000 km (135,000 mi) of hazardous liquid transmission pipelines, 34,000 km (21,000 mi) of crude oil gathering pipelines, 483,000 km (300,000 mi) of natural gas transmission pipelines, and 45,000 km (28,000 mi) of natural gas gathering pipelines in 2000 (Ref 1–3). There were approximately 60 major natural gas transmission pipeline operators and 150 major hazardous liquid pipeline operators in the United States in 1998 (Ref 4).

The first oil pipeline, which was 175 km (109 mi) in length and 152 mm (6 in.) in diameter, was laid from Bradford to Allentown, PA in 1879 (Ref 5). Since the late 1920s, virtually all oil and gas pipelines have been made of welded steel. Although the first cross-country pipeline that connected some major cities was laid in 1930, it was not until World War II that large-scale pipelines were laid connecting different regions of the country. In the 1960s, larger-diameter pipelines ranging from 813 to 914 mm (32 to 36 in.) were built. Discovery of oil on Alaska's North Slope resulted in the construction of the country's largest pipeline, the Trans-Alaska Pipeline System, with a 1219 mm (48 in.) diameter and 1287 km (800 mi) length. Demand continues to add more miles of pipelines.

Table 1 provides a summary of the major accidents reported to the U.S. Department of Transportation by the operators for the 6-year period between 1994 and 1999 (Ref 6). The data show that for transmission pipeline systems (both hazardous liquid and natural gas), approximately 25% of all reported accidents were due to corrosion. Of the hazardous liquid pipeline accidents caused by corrosion, 65% were due to external corrosion and 34% were due to internal corrosion. For natural gas transmission pipeline accidents, 36% were caused by external corrosion and 63% were caused by internal corrosion. For natural gas distribution pipeline accidents, only approximately 4% of the total accidents were caused by corrosion, and the majority of those were caused by external corrosion. The accidents reported in Table 1 are for major accidents that resulted in injury, fatality, or more than $50,000 in property damage. In addition to the reportable accidents, an average of 8000 corrosion leaks per year are repaired on natural gas transmission pipelines (Ref 7), and 1600 spills per year are repaired and cleaned up for liquid product pipelines.

In a summary report for incidents between 1985 and 1994, corrosion accounted for 28.5% of pipeline incidents on natural gas transmission and gathering pipelines (Ref 8). In a summary report for incidents between 1986 and 1996, corrosion accounted for 25.1% of pipeline incidents on hazardous liquid pipelines (Ref 9). These values correspond very well to the statistics for 1994 to 1999 presented in Table 1.

Given the implications of pipeline failures and the role that external corrosion plays in these failures, it is apparent that proper corrosion control can have a major impact on the safety, environmental preservation, and the economics of pipeline operation.

The vast majority of underground pipelines are made of carbon steel, based on American Petroleum Institute API 5L specifications (Ref 10). Typically, maximum composition limits are specified for carbon, manganese, phosphorous, and sulfur. In some cases, other alloying elements are added to improve mechanical properties. Composition and tensile requirements for common line pipe steels are shown in Table 2.

These steels have inadequate alloy additions to be considered corrosion resistant and undergo

Table 1 Summary of corrosion-related accident reports on hazardous liquid, natural gas transmission, and natural gas distribution pipelines from 1994 to 1999

Category	Pipeline system type		
	Hazardous liquid transmission	Natural gas transmission	Natural gas distribution
Total accidents due to corrosion (1994–1999)	271	114	26
Total accidents (1994–1999)	1116	448	708
Total accidents due to corrosion, %	24.3	25.4	3.7
Corrosion accidents due to external corrosion, %	64.9	36.0	84.6
Corrosion accidents due to internal corrosion, %	33.6	63.2	3.8
Corrosion accidents cause not specified, %	1.5	0.9	11.5

Source: Ref 6

Table 2 Chemical and tensile requirements of common long seam welded line pipe steels

Grade	Composition, wt% max				Yield strength minimum		Ultimate tensile strength minimum	
	C	Mn	P	S	MPa	ksi	MPa	ksi
A	0.22	0.9	0.03	0.03	207	30	331	48
B	0.26	1.2	0.03	0.03	241	35	414	60
X42	0.26	1.3	0.03	0.03	290	42	414	60
X46	0.26	1.4	0.03	0.03	317	46	434	63
X52	0.26	1.4	0.03	0.03	359	52	455	66
X56	0.26	1.4	0.03	0.03	386	56	490	71
X60	0.26	1.4	0.03	0.03	414	60	517	75
X65	0.26	1.45	0.03	0.03	448	65	531	77
X70	0.26	1.65	0.03	0.03	483	70	565	82

Product specification level 1, Ref 10

a variety of corrosion failure modes/mechanisms in underground environments, including general corrosion, pitting corrosion, and stress-corrosion cracking (SCC).

The terms general corrosion and pitting corrosion are used rather loosely when describing the morphology of underground corrosion. The classical pitting often associated with passive metals (such as stainless steels) is typically not observed on underground pipelines, with the possible exception of cases where microbial activity is involved. Likewise, true general corrosion, where there is uniform metal loss, such as observed with carbon steel in a concentrated acid, is not commonly found on underground pipelines. The most common morphology of corrosion on underground pipelines is uneven metal loss over localized areas covering a few to several hundred square inches (Fig. 1). The most common mechanism causing this corrosion is referred to as differential corrosion cells. Microbes and stray direct current (dc) in the soil also can affect underground corrosion.

Because of the relatively poor corrosion resistance of line pipe steels in underground environments, a combination of mitigation strategies consisting of coatings and cathodic protection (CP) is required. In this article, the most common causes and contributing factors for corrosion and SCC, as well as prevention, mitigation, detection, and repair are discussed.

Fig. 1 Example of external corrosion of an underground pipeline. Lower quadrant of pipeline shown after coating removal and abrasive cleaning

Differential Cell Corrosion

In the case of true general corrosion of a metal, the oxidation and reduction reactions occur physically at or very near the same location on a metal. At any given moment, one atom is being oxidized while the reduction reaction is occurring at an adjacent atomic site. Corrosion of a metal in an acid solution is a common example of this type of behavior. It is also possible for the oxidation and reduction reactions to be separated on a metal surface, where the metal oxidation occurs predominantly at one site while the reduction reaction occurs predominantly at another site. This is referred to as a differential corrosion cell. Underground corrosion of pipelines and other structures is often the result of differential corrosion cells of which a variety of different types exist. These include differential aeration cells, where different parts of a pipe are exposed to different oxygen concentrations in the soil, and cells created by differences in the nature of the pipe surface or the soil chemistry. This behavior is sometimes obvious when excavating an old, bare pipeline in which some areas are in excellent condition but other areas only a few feet away are severely corroded.

A differential aeration cell is probably the most common corrosion cell found on pipelines or other underground structures. One area of the pipeline is exposed to higher concentrations of oxygen and becomes the cathode in the cell, while another part of the structure is oxygen deficient and becomes the anode. Electrical current leaves the metal surface at the anode, increasing the corrosion rate, and flows to the oxygenated cathodic area, decreasing the corrosion rate. Differential aeration cells as well as other corrosion cells can be autocatalytic in that the chemical and electrochemical reactions, as well as ion migration, tend to produce conditions that promote the continuation of the cells. At the anode, the metal ions produced by the corrosion reactions hydrolyze (react with water), reducing the local pH. Corrosive negative halide ions migrate to the anodic sites to maintain charge neutrality. Both of these processes increase the corrosivity at the anodic sites. At the cathodic sites, the reduction reactions increase the pH and improve the protective nature of the corrosion films.

Differences in soil properties, variation in the moisture content of the soil, the depth from the surface or oxygen barriers such as paved roads can produce differential aeration cells. An example is illustrated in Fig. 2, which shows a pipeline passing through two dissimilar soils. The corrosion potential of the pipeline in the clay soil is more negative than the corrosion potential in the sandy soil, resulting in an increase in the corrosion rate of the pipe in the clay and a decrease in the sand. Factors other than differences in the oxygen concentration of the soil can produce a differential corrosion cell such as the one shown in Fig. 2. For example, differences in the pH, or the concentration of aggressive ions such as chlorides in the soil, can produce differential corrosion cells.

Galvanic corrosion is another example of a differential corrosion cell. In the case of galvanic corrosion, the potential difference is created by the presence of different metals. Different metals have a different corrosion potential in a given environment. An example is the galvanic series for metals in soils, shown in Table 3 (Ref 11). When these metals are electrically coupled, the metal with the most positive corrosion potential is cathodically polarized, reducing its corrosion rate, while the more negative member of the couple is anodically polarized, increasing its corrosion rate. Galvanic corrosion can be very detrimental to an underground structure. Examples include the corrosion of iron in contact with copper or stainless steel fittings. However, galvanic corrosion can be used as an effective means of CP, as described in the section on CP.

The surface films present on a metal also can alter the corrosion potential and cause differential cell corrosion. For example, mill scale is created on line pipe steel during the manufacturing process (hot rolling) and, if not removed, the mill-scale-coated steel will act like

Table 3 Practical galvanic series and redox potentials of metals and alloys in neutral soils and water

Material	Potential (CSE)(a), V
Most Noble	
Carbon, graphite, coke	+0.3
Platinum	0 to −0.1
Mill scale on steel	−0.2
High-silicon cast iron	−0.2
Copper, brass, bronze	−0.2
Low-carbon steel in concrete	−0.2
Lead	−0.5
Cast iron (not graphitized)	−0.5
Low-carbon steel (rusted)	−0.2 to −0.5
Low-carbon steel (clean and shiny)	−0.5 to −0.8
Commercially pure aluminum	−0.8
Aluminum alloy (5% Zn)	−1.05
Zinc	−1.1
Magnesium alloy (Mg-6Al-3Zn-0.15Mn)	−1.6
Commercially pure magnesium	−1.75
Most Active	

(a) Measured with respect to copper sulfate reference electrode (CSE). Source: Ref 11

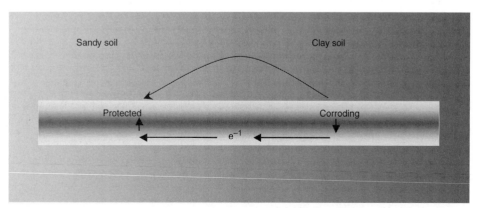

Fig. 2 Differential corrosion cell created by differences in soils. Arrows indicate the direction of ionic and electronic current flow.

a dissimilar metal in contact with non-mill-scale-coated pipe steel. The potential of the bare pipe steel surface will be more negative (active) than the mill-scale-coated surface, resulting in severe corrosion of the bare steel surface in low-resistivity soils. A similar condition can occur when new steel pipe is intermixed with old steel pipe. The potential of bright new steel is typically more negative than that of old rusted steel, resulting in rapid corrosion of the new steel unless the new section is electrically insulated from the old section and (or) cathodically protected. A similar corrosive condition can occur if, during work on an existing piping system, tools cut or scrape the pipe and expose areas of bright steel. The potential of these bright spots will be more negative than the remainder of the pipe, resulting in accelerated corrosion in low-resistivity soils.

The relative size of anodic and cathodic areas can have a significant effect on the severity of the differential corrosion cell. In general, the severity of corrosion of the anodic areas increases as the ratio of the anodic to the cathodic area decreases. When the anode is small and the cathode is large, the anode will be subject to a high density of current discharge per unit area, with the total amount of current flowing governed by the kinetics of the oxidation and reduction reactions and the soil resistivity. The current collected per unit area on the cathode is relatively low and may not be sufficient to result in any degree of polarization, which would tend to limit corrosion current. In a low-resistivity soil, corrosion can be rapid.

Microbiologically Influenced Corrosion

Microbiologically influenced corrosion (MIC) is defined as corrosion that is influenced by the presence and activities of microorganisms, including bacteria and fungi. It has been estimated that 20 to 30% of external corrosion on underground pipelines is MIC-related (Ref 5). Microorganisms located at the metal surface do not directly attack the metal or cause a unique form of corrosion. The by-products from the organisms promote several forms of corrosion, including pitting, crevice corrosion, and under-deposit corrosion. Typically, the products of a growing microbiological colony accelerate the corrosion process by either interacting with the corrosion products to prevent natural film-forming characteristics of the corrosion products that would inhibit further corrosion, or providing an additional reduction reaction that accelerates the corrosion process.

A variety of bacteria have been implicated in exacerbating corrosion of underground pipelines, and these fall into the broad classifications of aerobic and anaerobic bacteria (Ref 12). Obligate aerobic bacteria can only survive in the presence of oxygen, while obligate anaerobic bacteria can only survive in its absence. A third classification is facultative aerobic bacteria that prefer aerobic conditions, but can live under anaerobic conditions. Common obligate anaerobic bacteria implicated in corrosion include sulfate-reducing bacteria (SRB) and metal-reducing bacteria. Common obligate aerobic bacteria include metal-oxidizing bacteria, while acid-producing bacteria are facultative aerobes. The most aggressive attack generally takes place in the presence of microbial communities that contain a variety of types of bacteria. In these communities, the bacteria act cooperatively to produce conditions favorable to the growth of each species. Obligate anaerobic bacteria can thrive in aerobic environments when they are present beneath biofilms/deposits in which aerobic bacteria consume the oxygen. An example is shown in Fig. 3. In the case of underground pipelines, the most aggressive attack has been associated with acid-producing bacteria in such bacterial communities (Ref 5).

Stray Current Corrosion

Corrosion of underground pipelines can be accelerated by stray dc flowing in the soil near the pipeline. Sources of direct electrical current include foreign pipelines that are not properly bonded to the pipeline and ground currents from dc sources. Electrified railroads, mining operations, and other similar industries that utilize large amounts of dc sometimes allow a significant portion of current to use a ground path return to their power sources. These currents often utilize pipelines in close proximity as a part of the return path. This "stray" current can be picked up by the pipeline and discharged back into the soil at some distance down the pipeline close to the current return. Current pickup on the pipe is the same process as cathodic protection, which tends to mitigate corrosion. The process of discharge of a dc off the pipe and through the soil accelerates corrosion of the pipe wall at the discharge point, causing stray current corrosion. The morphology of stray current corrosion tends to be very localized at holidays (defects or holes) in the pipeline coating. Rates of attack can be very high, resulting in rapid perforation of a pipeline.

In the case of stray current corrosion from a foreign pipeline, the pipeline acts as a return current path for the cathodic protection system on the foreign pipeline. Stray current corrosion occurs where the dc discharges from the pipeline and collects onto the foreign pipeline, as shown in Fig. 4. While relatively rare, cathodic protection rectifiers are occasionally connected backwards, such that dc current is discharged from the pipeline and the impressed current "anode" actually collects rather than discharges current. This can result in severe stray current corrosion of the pipeline (Fig. 5).

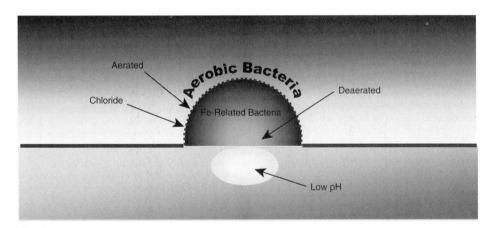

Fig. 3 Iron-related bacteria creating a differential oxygen and pH cell on a metal surface

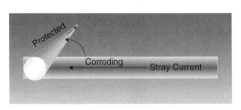

Fig. 4 Stray current corrosion caused by foreign pipeline

Fig. 5 Pipeline that experienced stray current corrosion caused by inverting the electrical leads to a cathodic protection rectifier

Field experience and laboratory research results indicate that stray alternating current (ac) also can cause accelerated corrosion of underground pipelines (Ref 13). The most common sources of stray ac are induced ac from power lines and pipelines in a common right of way and ground faults from ac power transmission. It is generally agreed that ac-enhanced corrosion rates are only a small fraction (<1%) of those of dc currents. Nevertheless, corrosion damage can be extensive where the ac currents are large.

Stress-Corrosion Cracking

Stress-corrosion cracking (SCC) is defined as cracking of a material produced by the combined action of corrosion and tensile stress. There are two forms of external SCC on underground pipelines: high-pH SCC (also referred to as classical SCC) and near-neutral-pH SCC (also referred to as low-pH SCC). A characteristic of both forms of SCC is the development of colonies of longitudinal surface cracks in the body of the pipe that link up to form long, shallow flaws. In some cases, growth and interlinking of the stress-corrosion cracks produce flaws that are of sufficient size to cause leaks or ruptures of pipelines. An example of an SCC colony that caused a pipeline failure is shown in Fig. 6.

The high-pH form of SCC is intergranular (Ref 14), the cracks propagate between the grains in the metal, and there is usually little evidence of general corrosion associated with the cracking (Fig. 7). The near-neutral-pH form of SCC is transgranular—the cracks propagate through the grains in the metal—and it is associated with corrosion of the crack faces and, in some cases, with corrosion of the external surface of the pipe as well (Fig. 8). This form of cracking was first reported on a polyethylene-tape coated pipeline on the TransCanada Pipelines Ltd. (TCPL) system in the 1980s (Ref 15, 16).

Stages of SCC

Figure 9 shows a "life" model for a pipeline containing stress-corrosion cracks (Ref 17). The model consists of four stages. In stage 1, the conditions for the initiation of SCC develop at the pipe surface. The coating disbonds, a cracking electrolyte develops at the pipe surface, and the pipe surface may become pitted or modified in other ways as a result of the presence of the electrolyte. Cracks begin to initiate in stage 2, and continued initiation, growth, and crack coalescence occur in stage 3. In stage 4, large cracks coalesce and final failure occurs. The coalescence of individual stress-corrosion cracks helps to determine whether a colony of cracks is an integrity concern. If cracks nucleate in close proximity to one another, crack growth may be dominated by the coalescence of collinear cracks. Coalescence can occur throughout the SCC life cycle. Depending on the size of the crack, either environmental or mechanical forces can cause the cracks to grow during stage 3. In stage 4 of growth, coalescence may occur primarily by tearing, when mechanical loading has a stronger effect in producing crack growth.

Fig 6 Example of colony of stress-corrosion cracks on external surface of high-pressure gas transmission pipeline. The top scale is inches and the bottom is centimeters.

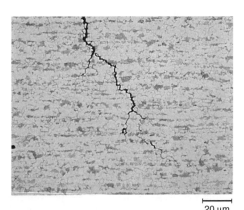

20 μm

Fig. 7 Intergranular high-pH stress-corrosion crack in line pipe steel. Nital etchant. Original magnification: 400×

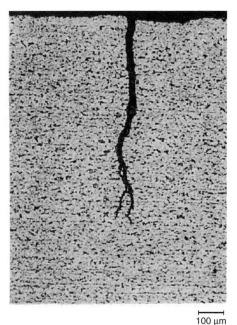

100 μm

Fig. 8 Transgranular near-neutral-pH stress-corrosion crack in Nital etchant. Original magnification: 100×

Conditions for SCC

Three conditions are necessary for SCC initiation and propagation in stages 2 and 3 to occur. These conditions generally differ for the two types of cracking:

- A potent environment develops at the pipe surface.
- The pipe steel is susceptible to SCC.
- A tensile stress of sufficient magnitude is present.

Further discussion of these three conditions for high-pH and near-neutral-pH SCC is given in this section.

Potent Environment. The two forms of external SCC are associated with two distinct environments that develop at the surface of underground pipelines. In the case of near-neutral-pH SCC, the cracking environment appears to be a dilute groundwater containing dissolved CO_2. The source of the CO_2 is typically the decay of organic matter and geochemical reactions in the soil. This form of cracking occurs under conditions in which there is little if any CP current reaching the pipe surface, either because of the presence of a shielding coating, a high-resistivity soil, or inadequate CP design (Ref 18). In the case of high-pH SCC, CO_2 is also involved. Cathodic protection causes the pH of the electrolyte beneath disbonded coatings to increase, and the CO_2 readily dissolves in the elevated-pH electrolyte, resulting in the generation of a concentrated CO_3-HCO_3 electrolyte (Ref 14). Four factors determine whether either of these potent environments can develop at the pipe surface: coating, soil, CP, and temperature.

Coating. To date, one or both forms of SCC have occurred under polyethylene/polyvinyl chloride (PVC) tapes, coal-tar enamel, wax, and asphalt coatings. With these coatings, the SCC is associated with coating disbondment and shielding of the CP current by the coating. The near-neutral-pH form of SCC is most prevalent on tape-coated pipelines, while high-pH SCC has occurred most frequently on coal-tar-coated pipelines. Fusion-bonded-epoxy (FBE)-coated pipelines are very resistant to SCC. This resistance has been attributed to the grit-blasted surface preparation used with FBE coatings, which imparts a compressive residual stress, as well as the resistance of FBE coatings to disbondment and CP shielding (Ref 19–22). Other newer coatings, such as urethanes, also have these beneficial characteristics.

Soil. High-pH SCC has occurred in a wide variety of soils, covering a range in color, texture, and pH. The moisture content of the soil, the ability of the soil to cause coating damage, and localized variation in the level of CP are the primary soil-related factors affecting high-pH SCC (Ref 23). Recent research results have suggested that some minimum concentration of soluble cations in the soil, such as sodium or potassium, must be present for high-pH SCC to occur (Ref 24). This notion is not altogether surprising in that such ions must be present

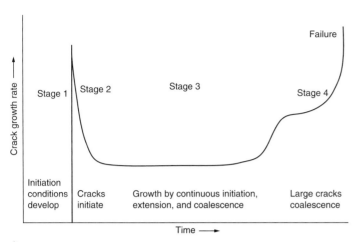

Fig. 9 Life model for a colony of stress-corrosion cracks. Source: Ref 17

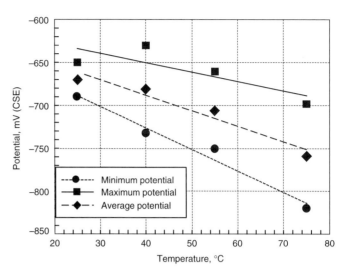

Fig. 10 Effect of temperature on the potential range for cracking in a simulated high-pH cracking environment containing 12 g/L Na_2CO_3 and 37 g/L $NaHCO_3$. CSE, copper/copper sulfate electrode. Source: Ref 28

to maintain solubility of the carbonate and bicarbonate ions.

Near-neutral-pH SCC of tape-coated pipelines has been predominantly located in imperfectly to poorly drained soils in which anaerobic and seasonally reducing environmental conditions were present (Ref 18, 25). On asphalt-coated pipelines, near-neutral-pH SCC has been found predominantly in extremely dry terrains consisting of either sandy soils or a mixture of sand and bedrock. There was inadequate CP in these locations, based on pipe-to-soil potential measurements or pH measurements of electrolytes found beneath disbonded coatings (Ref 17).

Cathodic protection is closely related to the high-pH cracking process. The CP current collecting on the pipe surface at disbondments, in conjunction with dissolved CO_2 in the groundwater, generates the high-pH SCC environment. Cathodic protection can also place the pipe-to-soil potential in the potential range for cracking. The potential range for cracking generally lies between the native potential of underground pipelines and the potential associated with adequate protection (−850 mV copper/copper sulfate electrode, or CSE) (Ref 26–28). Because the rate of generation of the cracking environment is related to the CP current, it is likely that seasonal fluctuations in the CP system are associated with the cracking process. The potent cracking environment might be generated during portions of the year when CP levels are high, while cracking might occur when adequate protection is lost, such as in the summer months when the soil dries out.

It has been concluded from the results of extensive field investigations (Ref 18, 25) that near-neutral-pH SCC occurs in the absence of significant CP. At dig sites on tape-coated pipelines, where near-neutral-pH SCC is found, the cracking is associated with locations where CP current was shielded from the pipe surface, based on pH measurements of electrolytes. The lower occurrence of SCC on the asphalt-coated portions of the system probably can be attributed to the higher levels of CP associated with this type of coating.

Temperature. The incidence of high-pH SCC increases significantly with an increase in the operating temperature of a pipeline. Service failures have been reported at temperatures as low as 13 °C (55 °F), but 90% of the service and hydrostatic test failures have occurred within 16 km (10 miles) downstream from compressor stations (Ref 27). This behavior has been attributed to a decrease in the width of the potential range for cracking, as shown in Fig. 10, coupled with a decrease in the maximum cracking velocity with decreasing temperature. Laboratory data and field experience indicate that there is less temperature dependence for near-neutral-pH SCC than for high-pH SCC.

Susceptible Line Pipe Steel. Both forms of SCC have occurred on a variety of sizes, grades, and vintages of line pipe steel. Stress-corrosion cracking has been found in flash welded, submerged arc welded (SAW), electric resistance welded (ERW), and seamless pipe. The chemical compositions of the failed pipes are typical for the vintage and grade, and there are no obvious unique metallurgical characteristics associated with the failures. The bond line of the weld seam of ERW pipe seems to have a lower resistance to near-neutral-pH SCC than the base metal, possibly because of the presence of a more SCC-susceptible microstructure at the weld, a higher-than-normal residual stress, pits and arc burns associated with the ERW manufacturing process, or a lower fracture toughness (Ref 17). The coarse-grained heat-affected zone (CGHAZ) adjacent to the double submerged arc weld (DSAW) also has been found to be more susceptible to cracking than the base material in the near-neutral-pH environment (Ref 29). Results of recent research suggest that residual stresses from the pipe manufacturing process are important in affecting susceptibility to near-neutral-pH SCC (Ref 30).

Tensile Stress. Most of the intergranular (high-pH) stress-corrosion cracks found in gas transmission pipelines have been oriented in the longitudinal direction (Ref 14). This orientation indicates the importance of the hoop stress produced by the internal pressurization on the cracking process. The failures have occurred at hoop stresses ranging from 160 to 270 MPa (23 to 39 ksi), which corresponded to 46 to 76% of the specified minimum yield strength (SMYS) of the pipe steels. Most of the high-pH SCC failures have initiated at locations at which there was no indication of secondary stresses. These observations indicate that the typical field stresses experienced by operating gas transmission pipelines are sufficient for high-pH SCC to occur.

The majority of near-neutral-pH SCC failures are associated with features that enhance the local stress, indicating that the behavior is fundamentally different from high-pH SCC with regard to the role of stress in the crack initiation process. In most cases, near-neutral pH SCC has been associated with corrosion, gouges, or stress concentrations from the toe of the weld seam that raise the local stress above the hoop-stress levels calculated based on the internal pressurization (Ref 17). Significant SCC has not been reported in class 2 and 3* pipeline locations, indicating that the hoop stress also is important (Ref 17).

Cyclic pressure fluctuations that normally occur on operating pipelines also affect SCC behavior. It has been demonstrated that the cyclic stress that results from these pressure

---

*Class locations for high-pressure gas pipelines are defined by the US. Department of Transportation based on the consequences of a pipeline failure. Class 1 locations have the lowest consequence and are allowed to operate at a maximum pressure that is equivalent to 72% of SMYS. Class 4 locations have the highest consequence of failure and are allowed to operate at a maximum pressure that is equivalent to 40% of SMYS (Reference CFR Part 192.111).

fluctuations reduce the threshold stress for initiation of high-pH SCC (Ref 31) and increase the rate of propagation of near-neutral-pH stress-corrosion cracks (Ref 32).

Prevention and Mitigation of Corrosion and SCC

The most effective method to prevent corrosion or SCC on new pipelines is to use high-performance coatings, applied to a surface abrasive blast cleaned to a white (Ref 33) or near-white (Ref 34) metal surface finish, in conjunction with effective CP. An intact coating that prevents contact of electrolyte with the steel surface will prevent external corrosion or SCC. The surface abrasive blast cleaning promotes good coating adhesion. A high-quality abrasive blast cleaning also will impart compressive residual stresses in the pipe surface that improve SCC resistance.

All coatings contain some defects or holes, referred to as holidays, that expose the bare pipeline steel to the underground environment. The function of the CP system is to protect these bare areas from corrosion.

Methods of preventing corrosion and SCC on existing pipelines include minimizing the operating temperature and controlling the CP levels to values more negative than -850 mV CSE. Minimizing pressure fluctuations on operating pipelines also is effective in preventing SCC initiation. A more detailed discussion of coatings and cathodic protection is given in this section.

Coatings

Inadequate coating performance is a major contributing factor in the corrosion and SCC susceptibility of an underground pipeline. The function and desired characteristics of a dielectric-type pipeline coating are covered in NACE RP-0169 (Ref 35). This specification states that the function of such coatings is to control corrosion by isolating the external surface of the underground or submerged piping from the environment, to reduce CP requirements, and to improve (protective) current distribution. Coatings must be properly selected and applied, and the coated piping must be carefully installed to fulfill these functions. The desired characteristics of the coatings include:

- Effective electrical insulation
- Effective moisture barrier
- Good adhesion to the pipe surface
- Applicable by a method that will not adversely affect the properties of the pipe
- Applicable with a minimum of defects
- Ability to resist the development of holidays with time
- Ability to resist damage during handling, storage, and installation
- Ability to maintain substantially constant resistivity with time
- Resistance to disbonding
- Resistance to chemical degradation
- Ease of repair
- Retention of physical characteristics
- Nontoxic to environment
- Resistance to changes and deterioration during above-ground storage and long-distance transportation

Descriptions of common coatings used on underground pipelines follow.

Bituminous enamels are formulated from coal-tar pitches or petroleum asphalts and have been widely used as protective coatings for more than 65 years. Coal-tar and asphalt enamels are available in summer or winter grades. These enamels are the corrosion coating; they are combined with various combinations of fiberglass and/or felt to obtain mechanical strength for handling. The enamel coatings have been the workhorse coatings of the industry, and when properly selected and applied, they can provide efficient long-term corrosion protection.

Enamel systems can be designed for installation and use within an operating temperature range of -1 to 82 °C (30 to 180 °F). When temperatures fall below 4.4 °C (40 °F), added precautions should be taken to prevent cracking and disbonding of the coating during field installation. Enamels are affected by ultraviolet rays and should be protected by kraft paper or whitewash. Enamels are also affected by hydrocarbons, and the use of a barrier coat is recommended when known contamination exists. Bituminous enamel coatings are available for all sizes of pipe.

In recent years, the use of enamels has declined for these reasons:

- Reduced number of suppliers
- Restrictive environmental and health standards from the Occupational Safety and Health Administration, the Environmental Protection Agency, and the Food and Drug Administration
- Increased acceptance of other coating types
- Alternative use of coating raw materials as fuels

Asphalt mastic pipe coating is a dense mixture of sand, crushed limestone, and fiber bound together with a select air-blown asphalt. These materials are proportioned to secure a maximum density of approximately 2.1 g/cm^3 (132 lb/ft^3). This mastic material is available with various types of asphalt. Selection is based on operating temperature and climatic conditions to obtain maximum flexibility and operating characteristics. This coating is a thick (12.7 to 16 mm, or $1/2$ to $5/8$ in.) extruded mastic that results in a seamless corrosion coating. Extruded asphalt mastic pipe coating has been in use for more than 50 years.

Asphalt mastic systems can be designed for installation and use within an operating range of 4.4 to 88 °C (40 to 190 °F). Precautionary measures should be taken when handling asphalt mastics in freezing temperatures. Whitewash is used to protect it from ultraviolet rays, and this should be maintained when in storage. This system is not intended for use above ground or in hydrocarbon-contaminated soils.

Liquid Epoxies and Phenolics. Many different liquid systems are available that cure by heat and/or chemical reaction. Some are solvent types, and others are 100% solids. These systems are primarily used on larger-diameter pipe when conventional systems may not be available or when they may offer better resistance to operation temperatures in the 95 °C (200 °F) range.

Generally, epoxies have an amine or a polyamide curing agent and require a near-white blast-cleaned surface (NACE No. 2 or SSPC SP10). Coal-tar epoxies have coal-tar pitch added to the epoxy resin. A coal-tar epoxy cured with a low-molecular-weight amine is especially resistant to an alkaline environment, such as that which occurs on a cathodically protected structure. Some coal-tar epoxies become brittle when exposed to sunlight.

Extruded plastic coatings fall into two categories based on the method of extrusion, with additional variations resulting from the selection of adhesive. The two methods of extrusion are the crosshead or circular die, and the side extrusion or T-shaped die. The four types of adhesives are asphalt-rubber blend, polyethylene copolymer, butyl rubber adhesive, and polyolefin rubber blend.

To date, of the polyolefins available, polyethylene has found the widest use, with polypropylene being used on a limited basis for its higher operating temperature. Each type or variation of adhesive and method of extrusion offers different characteristics based on the degree of importance to the user of certain measurable properties.

Fusion-bonded epoxy (FBE) coatings are heat-activated, chemically cured coating systems. The epoxy coating is furnished in powdered form and, with the exception of the welded field joints, is plant applied to preheated pipe, special sections, connections, and fittings using fluid-bed, air spray, or electrostatic spray methods.

Fusion-bonded epoxy coatings were introduced in 1959 and were first used as an exterior pipe coating in 1961 and currently are the coatings most commonly used for new installations of large diameter pipelines (Ref 36). These coatings are applied to preheated pipe surfaces at 218 to 244 °C (425 to 475 °F). Some systems may require a primer system, and some require postheating for complete cure. A NACE No. 2 (SSPC SP10) near-white blast-cleaned surface is required. The coating is applied to a minimum thickness of 0.3 mm (12 mils); in some applications, coating thicknesses range to 0.64 mm (25 mils), with the restriction not to bend pipe coated with a film thickness greater than 0.4 mm (16 mils). The FBE coatings exhibit good mechanical and physical properties and are the most resistant to hydrocarbons, acids, and alkalies.

A primary advantage of the FBE pipe coatings is that they cannot hide apparent surface defects; therefore, the steel surface can be inspected after it is coated. The number of holidays that occur is a function of the surface condition and the thickness of the coating specified. Increasing the thickness minimizes this problem, and the excellent resistance to the electrically induced disbondment of these coatings has resulted in their frequent use as pipeline coatings.

Tape. Field and mill-applied tape systems have been in use for more than 30 years on pipelines. For normal construction conditions, prefabricated cold-applied tapes are applied as a three-layer system consisting of a primer, corrosion-preventive tape (inner layer), and a mechanically protective tape (outer layer). The function of the primer is to provide a bonding medium between the pipe surface and the adhesive or sealant on the inner layer. The inner-layer tape consists of a plastic backing and an adhesive. This layer is the corrosion-protective coating; therefore, it must provide a high electrical resistivity, low moisture absorption and permeability, and an effective bond to the primed steel surface. The outer-layer tape consists of a plastic film and an adhesive composed of the same types of materials used in the inner tape or materials that are compatible with the inner-layer tape. The purpose of the outer-layer tape is to provide mechanical protection to the inner-layer tape and to be resistant to the elements during outdoor storage. The outer-layer tape is usually a minimum of 0.64 mm (25 mils) thick.

The cold-applied multilayer tape systems are designed for plant coating operations and result in a uniform, reproducible, holiday-free coating over the entire length of any size pipe. The multiple-layer system allows the coating thickness to be custom designed to meet specific environmental conditions. These systems have been engineered to withstand normal handling, outdoor weathering, storage, and shipping conditions.

Three-Layer Polyolefin. The three-layer polyolefin pipeline coating was developed in the 1990s as a way to combine the excellent adhesion of FBE with the damage resistance of extruded polyethylene and tape wraps. These systems consist of an FEB primer, an intermediate copolymer layer, and a topcoat consisting of either polyethylene or polypropylene. The function of the intermediate copolymer is to bond the FBE primer with the polyolefin topcoat.

Variations in these three-layer systems exist, most notably the use of either polyethylene or polypropylene for the topcoat. Polypropylene offers a higher temperature resistance but is more costly, both as a raw material and because higher temperatures are required for application. Most topcoats are side extruded similar to extruded polyethylene coatings, although at least one product uses flame-spray polyolefin for a topcoat. Another variation in the three-layer systems is the thickness of the FBE primer layer. Early generations of this product utilized a 50–75 μm (2–3 mil) primer which often proved to be inadequate to achieve the desired performance. More recent three-layer systems utilize a 200–300 μm (8–12 mil) primer as a standard thickness.

Wax coatings have been in use for more than 50 years and are still employed on a limited basis. Microcrystalline wax coatings are usually used with a protective overwrap. The wax serves to waterproof the pipe, and the wrapper protects the wax coating from contact with the soil and affords some mechanical protection. The most prevalent use of wax coatings is the over-the-ditch application with a combination machine that cleans, coats, wraps, and lowers into the ditch in one operation.

Special-Use Coatings

Polyurethane Thermal Insulation. Efficient pipeline insulation has grown increasingly important as a means of operating hot and cold service pipelines. This is a system for controlling heat transfer in above- or belowground and marine pipelines. Polyurethane insulation is generally used in conjunction with a corrosion coating, but if the proper moisture vapor barrier is used over the polyurethane foam, effective corrosion protection is attained.

Concrete. Mortar linings and coatings have the longest history of use in protecting steel or wrought iron from corrosion. The alkalinity of the concrete promotes the formation of a protective iron oxide (passive) film on the steel. This protective passive film can be compromised in underground applications by permeation of chlorides into the coating. Typically, external application is usually employed over a corrosion-resistant coating for armor protection and negative buoyancy in marine environments.

Metallic (Galvanic) Coatings. Pipe coated with a galvanic coating, such as zinc (galvanizing) or cadmium, should not be utilized in direct burial service. Such metallic coatings are intended for the mitigation of atmospheric-type corrosion activity on the substrate steel.

Evaluating Coatings

As described previously, the different types of coatings used on underground pipelines have different strengths and weaknesses. When first installed, most pipeline coatings are effective in meeting their required function: isolate the external surface of an underground pipeline from the environment, reduce the CP current requirements, and improve the CP current distribution. On the other hand, coatings vary significantly in their long-term performance. Ultimately, the effectiveness of a coating system in preventing corrosion is related to two primary factors: (a) the resistance of a coating to degradation over time and (b) the ability of the coating to conduct CP current should the coating fail (minimize shielding). For SCC resistance, these factors as well as the type of surface preparation used with the coating are important.

The ability of a coating to resist degradation is a primary performance property of coatings and affects all forms of external pipeline corrosion. The second factor, the ability of a coating to pass CP current, should it fail, is the inverse of shielding of the CP current beneath a disbonded coating. Corrosion or SCC can occur beneath a disbonded coating that shields CP current even though the pipeline is apparently effectively protected, based on ground-level measurements.

Surface Preparation

The nature of the surface preparation is probably more important in mitigating SCC than other forms of corrosion. Historically, the primary purposes of the surface preparation have been to clean the surface and create an anchor pattern to promote good adhesion of the coating to the pipe surface. The surface preparation requirements for different coating types vary. For example, bituminous coatings have good adhesion properties on commercial blast-cleaned surfaces (NACE No. 3/SSPC-SP 6) or even on wire-brushed surfaces, whereas fusion-bonded epoxy (FBE) coatings require a white (NACE No. 1/SSPC-SP 5) or near-white (NACE No. 2/SSPC-SP 10) grit-blasted surface finish for proper adhesion. Laboratory research and field experience have demonstrated that grit-blasted surfaces are generally more resistant to SCC initiation than wire-brushed mill-scaled surfaces, primarily because grit blasting imparts a compressive residual stress in the pipe surface (Ref 20, 21, 36). A white or near-white surface finish was found to be required to impart SCC resistance, whereas commercially blasted surfaces were found to be more susceptible to SCC than wire-brushed milled scaled surfaces.

The Canadian Energy Pipeline Association (CEPA) member companies have recommended that the following coatings be considered for new construction based on SCC performance (Ref 37):

- Fusion-bonded epoxy
- Liquid epoxy
- Urethane
- Extruded polyethylene
- Multilayer or composite coatings

Fusion-bonded epoxies, liquid epoxies, and urethane coatings meet all three requirements of an effective coating: (a) they are resistant to degradation over time, (b) they conduct CP current should they fail, and (c) they are typically applied over a white or near-white grit-blasted surface. Extruded polyethylene coatings meet requirements 1 and 3, but will shield CP current should disbondment occur. Furthermore, the type of coating used on the field joints frequently limits the performance of extruded polyethylene-coated pipelines. Multilayer or composite coatings typically consist of an FBE inner layer and a polyolefin outer layer with an adhesive between the two layers. These new coatings are promising from the standpoint of resistance to disbondment, mechanical damage, and soil stresses, but the polyolefin outer layer will shield CP current should disbondment occur. Additional field experience is needed to establish the performance of these coatings.

Tape coatings and bituminous coatings have been shown to be more susceptible to SCC than the aforementioned coatings and should be used only with careful consideration of all of the factors affecting SCC. Regardless of the coating selected, the pipe surface should be prepared to a white (NACE No. 1/SSPC-SP 5) or near-white (NACE No. 2/SSPC-SP 10) finish to aid in coating adhesion and impart sufficient residual compressive stresses to prevent SCC initiation. A lower-quality commercial blast (NACE No. 3/SSPC-SP 6) should not be used under any circumstances.

Cathodic Protection

External corrosion and SCC are electrochemical phenomena and, therefore, can be prevented or mitigated by altering the electrochemical condition of the corroding interface. Altering the electrochemical nature of the corroding surface is relatively simple and is done by altering the electrical potential field around the pipe. By applying a negative potential and making the pipe a cathode, the rate of corrosion (oxidation) is reduced (corrosion is prevented or mitigated) and the reduction process is accelerated. This means of mitigating (or preventing) corrosion, cathodic protection, also alters the environment at the pipe surface, which further enhances corrosion control. The pH of any electrolyte at the pipe surface is increased, the oxygen concentration is reduced, and deleterious anions, such as chloride, migrate away from the pipe surface.

Types of CP. There are two primary types of CP systems: sacrificial anode (galvanic anode) CP and impressed-current CP. Sacrificial anode CP utilizes an anode material that is electronegative to the pipe steel. When connected to the pipe, the pipe becomes the cathode in the circuit and corrosion is mitigated. Typical sacrificial anode materials for underground pipelines are zinc and magnesium. Impressed-current CP utilizes an outside power supply (rectifier) to control the voltage between the pipe and an anode (cast iron, graphite, platinum clad, mixed metal oxide, etc.) in such a manner that the pipe becomes the cathode in the circuit and corrosion is mitigated. Schematics of these two types of CP systems are shown in Fig. 11 and 12.

Cathodic protection is most often used in conjunction with a coating. There are always flaws in the coating due to application inconsistencies, construction damage, or the combination of natural aging and soil stresses. If left unprotected, the pipeline will undergo corrosion or SCC at these coating flaws (holidays). Often the rate of attack through the wall is much higher at the holiday than the general attack of a bare steel surface. The use of a coating greatly reduces the total amount of current required to achieve protection of the pipeline system; therefore, CP and external coatings are utilized together wherever possible.

Cathodic protection can be used to control all types of corrosion previously discussed (general, stray current, MIC, and SCC). Sometimes it is difficult to determine the level of CP necessary to mitigate the different corrosion mechanisms and to identify which type of corrosion is present. Stress-corrosion cracking presents additional problems. First, the high-pH form of SCC is only found on pipelines protected with CP. The products that result from cathodic reactions occurring on the pipe surface during CP in conjunction with soil chemistry produce the environment necessary for high-pH SCC. Since high-pH SCC propagates only in a very limited potential range, maintaining the potential of the pipe surface outside of this range by proper CP control will prevent growth of the high-pH SCC cracks. In addition, it has been established that proper CP control can inhibit the growth of near-neutral SCC cracks.

Electrical surveys have been performed to evaluate the level of CP ever since the application of CP to pipelines in the 1940s. These surveys consist of measuring the potential (pipe-to-soil potential) of the pipe surface with respect to a reference electrode (typically CSE). These measurements can be performed at permanent test station locations (test point surveys), or they can be performed continuously with a 1 to 2 m (3 to 6 ft) spacing along the entire length of the pipeline (close interval surveys). Pipe-to-soil potential surveys can be performed with the CP system energized (on-potentials) or with the CP system interrupted (off-potentials). There has been much discussion over the past 10 to 20 years as to the most appropriate survey methodology. While each method has its benefits, it is commonly accepted that the IR-voltage (voltage drop due to current, I, through a resistance, R) correction made by the off-potential measurement is most closely related to the corrosion condition of the pipeline. Figure 13 shows a schematic of a pipe-to-soil potential measurement.

The basic pipe-to-soil potential measurement techniques are applied to establish whether one or more of the recommended CP criteria are met. Criteria for establishing the effectiveness of a CP system to mitigate corrosion are outlined in the NACE International Recommended Practice RP0169-96 (Ref 35) and have been adopted, in part, in U.S. Department of Transportation (DOT) regulations CFR 49, Parts 192 and 195. In general, if one or more of the recommended criteria are met, the CP system is assumed to be applying a sufficient cathodic current to mitigate corrosion.

Certain pipeline conditions make conventional electrical survey techniques difficult to interpret. These include areas of stray or telluric currents, congested areas where multiple pipelines and other utilities share rights-of-way, and pipelines with noninterruptible sacrificial CP systems. In these areas, either significant care must be taken to interpret conventional surveys or other methods of monitoring must be utilized. One such technology is the use of coupon test

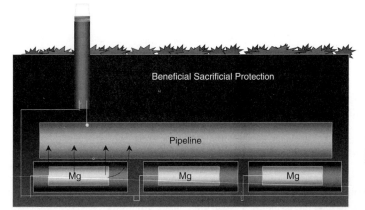

Fig. 11 Sacrificial anode CP system with distributed magnesium anodes and an above-ground test station

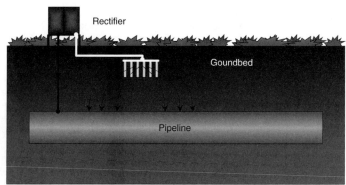

Fig. 12 Impressed current CP system with above-ground rectifier and a single remote anode groundbed

stations. The coupon test stations permit accurate potential measurements for a test specimen (coupon) that simulates a holiday on the pipe surface.

Detection of Corrosion and SCC

On existing pipelines, there are three methods to detect corrosion and SCC—hydrostatic retesting, field investigation programs (direct assessment), and in-line inspection.

Hydrostatic Testing. Hydrostatic retesting involves pressure testing the pipeline with water at a pressure that is higher than the operating pressure, typically 125% of the maximum operating pressure (MOP) of the pipeline. This is the most common method to ensure the integrity of a pipeline and establish a safe operating pressure, regardless of the types of flaws present in the pipeline. Any flaws that are larger than a critical size at the hydrostatic retest pressure are removed from the pipeline. However, subcritical flaws remain in the pipeline after a hydrostatic retest. If the defects are growing with time, as might be the case with corrosion or SCC, the pipeline is generally periodically retested to ensure integrity. Hydrostatic retesting is expensive and creates problems associated with the acquisition, treatment, and disposal of the water, especially for pipelines carrying liquid products.

Direct Assessment. As a part of condition-monitoring programs, pipeline companies commonly use field investigation (direct assessment) programs. The overall condition of the coatings and pipelines is assessed, and it is determined whether corrosion or SCC is present on the system. Models are sometimes developed to predict the likelihood of the presence and severity of corrosion or cracking. This information is then used to prioritize the system for direct examination, hydrostatic testing, in-line inspection, recoating, or pipe replacement. Dig programs and the associated models are not generally considered as a replacement for hydrostatic testing as a means to ensure the integrity of a pipeline. See the article "External Corrosion Direct Assessment Integrated with Integrity Management" in this Volume.

In-line inspection (ILI) tools, also referred to as smart or intelligent pigs, are devices that are propelled by the product in the pipeline and are used to detect and characterize metal loss caused by corrosion and cracking. There are two primary types of metal-loss ILI tools: magnetic flux leakage (MFL) tools and ultrasonic tools (UT).

Magnetic flux leakage tools measure the change in magnetic flux lines produced by the defect and produce a signal that can be correlated to the length and depth of a defect. In recent years, the magnetics, data storage, and signal interpretation have improved, resulting in improved mapping of the flaw and a decrease in the number of unnecessary excavations. The high-resolution MFL tool is typically capable of readily detecting corrosion pits with a diameter greater than three times the wall thickness. Once detected, these tools can typically size the depth of the corrosion within $\pm 10\%$ of the wall thickness with an 80% level of confidence. The MFL tool can be used to inspect either liquid product pipelines or natural gas pipelines.

Figure 14 shows a typical MFL tool. The wire brushes in the front of the tool are used to transfer the magnetic field from the tool to the pipe wall. The ring of sensors between the wire brushes are used to measure the flux leakage produced by defects in the pipe. The drive cups are the mechanism that is used to propel the tool by the product in the pipeline. The odometer wheels monitor the distance traveled in the line and are used to determine the location of the defects identified. The trailing set of inside-diameter/outside-diameter sensors (ID/OD sensors) is used to discriminate between internal and external wall loss.

Ultrasonic tools utilize large arrays of ultrasonic transducers to send and receive sound waves that travel through the wall thickness, permitting a detailed mapping of the pipe wall. Ultrasonic tools can indicate whether the wall loss is internal or external. The typical resolution of a UT is $\pm 10\%$ of the pipe wall thickness with an 80% level of confidence. Ultrasonic tools are typically used in product pipelines (those carrying crude oil, gasoline, and the like) since the product in the pipeline is used as the required couplant for the ultrasonic sensors. This tool can be used to inspect natural gas pipelines, but requires introducing a liquid (such as water) into the pipeline for an ultrasonic couplant.

There is significant interest in the pipeline industry in developing ILI tools that can reliably detect and size stress-corrosion cracks. Crack-detection tools avoid problems associated with acquisition, treatment, and disposal of the water used in hydrostatic retesting. It is desirable for a fully developed tool to be capable of detecting and sizing subcritical cracks such that the pipeline can be repaired long before these cracks become an integrity concern. The detection of smaller cracks also extends the time interval between inspections. Ultrasonic tools are available that can detect stress-corrosion cracks in liquid pipelines, but the detection and sizing capability of the tools has not been fully established. These tools require the presence of a liquid couplant and therefore are difficult and expensive to use in gas pipelines.

Assessment and Repair of Corrosion and SCC

Once corrosion or SCC has been detected on a pipeline, the size of the defect must be determined, and the defect must then be assessed and sentenced. In-line inspection typically provides some measure of the size of the defect. The dimensions of ILI defects that potentially could affect the immediate integrity of the pipeline are typically confirmed by direct examination. In direct assessment (DA) programs, corrosion

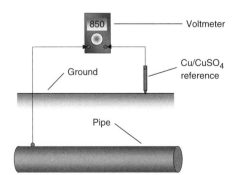

Fig. 13 Pipe-to-soil potential measurement

Fig. 14 Magnetic flux leakage tool for detection and sizing of corrosion defects in a pipeline. Courtesy of PII

flaws can be sized by direct measurement or by means of an ultrasonic thickness meter. Stress-corrosion cracks are generally sized, in the field, by a combination of magnetic-particle inspection (MPI) and grinding. The colony with the longest interlinked cracks in a dig is typically ground out to establish the maximum depth of cracking. In the ditch ultrasonic techniques also are being developed for crack dimension measurements, but the technology is difficult to apply to colonies of cracks.

A burst-pressure model such as R-STRENG (Ref 38) can be used to determine the failure pressure of corrosion defects or areas that contained cracks and were subsequently ground out. Fracture mechanics techniques must be used to determine the burst pressure of cracklike defects (Ref 17). The pipe is typically recoated if the burst pressure is within acceptable limits (typically above a pressure that is equivalent to 100% of the specified minimum yield strength of the line pipe steel). If this pressure is below acceptable limits, the pipe is typically replaced or repaired using a steel or composite reinforcing sleeve and recoated.

Pipe replacement is sometimes the only option in situations in which there is extensive corrosion or cracking localized within one area of a pipeline. If the corrosion or cracking is extensive, but not severe, it may be possible to recoat the affected areas of a pipeline. Enhancement of the CP system is also an option to minimize further corrosion or stress-corrosion crack growth in areas in which it has been established that the pipeline contains growing corrosion or SCC defects that are not an immediate integrity threat. Unfortunately, shielding coatings are not amenable to enhancement of CP because it is unlikely that the CP current can penetrate beneath coating disbondments. In the case of hydrostatic retest failures, the only available repair method is pipe replacement. The failed joint is cut out and replaced with new pipe.

ACKNOWLEDGMENT

Portions of this article were adapted from C.G. Siegfried, Corrosion of Pipelines, *Corrosion,* Vol 13, 9th ed., *Metals Handbook* (1987). The section of this article on stress-corrosion cracking is adapted from NACE International Technical Committee Report 35103 (External Stress Corrosion Cracking of Underground Pipelines) (Ref 39) and was prepared with the assistance and permission of NACE International. For a complete version of the original published report, contact NACE International. Users are cautioned to obtain the latest edition; information in an outdated version of the report might not be accurate.

REFERENCES

1. P.J. Katchmar, "OPS Overview & Regulation Update," Rocky Mountain Short Course, Jan 27, 2000
2. "Pipeline Safety—The Office of Pipeline Safety Is Changing How It Oversees the Pipeline Industry," No. GAO/RCED-00-128, Report to Ranking Minority Member, Committee on Commerce, House of Representatives, May 2000
3. Office of Pipeline Safety, Department of Transportation, http://ops.dot.gov/stats, May 22, 2000
4. OGJ Special Report, *Oil Gas J.,* Aug 23, 1999
5. G.H. Koch, M.P.H. Brongers, N.G. Thompson, Y.P. Virmani, and J.H. Payer, "Corrosion Cost and Prevention Strategies in the United States," FHWA-RD-01-156, Office of Infrastructure Research and Development, Federal Highway Administration, March 2002
6. U.S. Department of Transportation, Pipeline and Hazardous Materials Safety Administration, ops.dot.gov/stats/stats.htm, report as of Oct 30, 2000
7. P.H. Vieth, I. Roytman, R.E. Mesloh, and J.F. Kiefner, "Analysis of DOT-Reportable Incidents for Gas Transmission and Gathering Pipelines—January 1, 1985 through December 31, 1995," Final Report, Contract No. PR-218-9406, Pipeline Research Council International, May 31, 1996
8. J.F. Kiefner, B.A. Kiefner, and P.H. Vieth, "Analysis of DOT-Reportable Incidents for Hazardous Liquid Pipelines—1986 through 1996," Final Report, The American Petroleum Institute, Jan 7, 1999
9. P.H. Vieth, W.G. Morris, M.J. Rosenfeld, and J.F. Kiefner, "DOT-Reportable Incident Data Review—Natural Gas Transmission and Gathering Systems—1985 through 1995," Final Report, Contract No. PR-218-9603, Pipeline Research Council International, Sept 19, 1997
10. *Specifications for Line Pipe,* API 5L, 42nd ed., American Petroleum Institute, Washington D.C., July 1, 2000
11. A.W. Peabody, *Control of Pipeline Corrosion,* 2nd ed., R.L. Bianchetti, Ed., NACE International, 2001
12. B.J. Little, P.A. Wagner, and F. Mansfeld, Microbiologically Influenced Corrosion, *Corrosion Testing Made Easy,* B.C. Syrett, Ed., NACE International, 1997
13. M. Yunovich and N.G. Thompson, AC Corrosion: Corrosion Rates and Mitigation Requirements, Paper 04206, *Corrosion 2004,* NACE International, 2004
14. R.L. Wenk, Field Investigation of Stress Corrosion Cracking, *Proc. Fifth Symposium on Line Pipe Research,* Pipeline Research Council International, 1974, p T-1
15. J.T. Justice and J.D. Mackenzie, Progress in the Control of Stress Corrosion Cracking in a 914 mm O.D. Gas Transmission Pipeline, *Proc. NG-18/EPRG Seventh Biennial Joint Technical Meeting on Line Pipe Research,* Pipeline Research Council International, 1988
16. B.S. Delanty and J.E. Marr, Stress Corrosion Cracking Severity Rating Model, *Proc. International Conference on Pipeline Reliability,* June 1992 CANMET
17. "Stress Corrosion Cracking on Canadian Oil and Gas Pipelines," Report of the Inquiry, MH-2-95, Regulatory Support Office, National Energy Board, Nov 1996
18. B.S. Delanty and J. O'Beirne, Major Field Study Compares Pipeline SCC with Coatings, *Oil Gas J.,* Vol 90 (No. 24), 1992, p 39
19. J.A. Beavers, "Assessment of the Effects of Surface Preparation and Coating on the Susceptibility of Line Pipe to Stress Corrosion Cracking," Pipeline Research Council International, 1992
20. J.A. Beavers, N.G. Thompson, and K.E.W. Coulson, Effects of Surface Preparation and Coatings on SCC Susceptibility of Line Pipe: Phase 1—Laboratory Studies, Paper No. 93597, *CORROSION/93,* NACE International, 1993
21. J.A. Beavers, N.G. Thompson, and K.E.W. Coulson, Effects of Surface Preparation and Coatings on SCC Susceptibility of Line Pipe: Phase 2—Field Studies, *Proc. 12th International Conference on Offshore Mechanics and Arctic Engineering,* June 1993, American Society of Mechanical Engineers, 1993, p 226
22. W.E. Berry, Stress Corrosion Cracking Laboratory Experiments, *Proc. Fifth Symposium on Line Pipe Research,* Pipeline Research Council International, 1974, p V-1
23. W.L. Mercer, Stress Corrosion Cracking—Control through Understanding, *Proc. Sixth Symposium on Line Pipe Research,* Pipeline Research Council International, 1979, p W-1
24. J.A. Beavers, C.L. Durr, and K.C. Garrity, The Influence of Soil Chemistry on SCC of Pipelines and the Application of the 100 mV Polarization Criterion, Paper No. 02426, *CORROSION/2002,* NACE International, 2002
25. B.S. Delanty and J. O'Beirne, Low-pH Stress Corrosion Cracking, *Proc. 18th World Gas Conference,* International Gas Union, Paris, France, 1991
26. R.N. Parkins, The Controlling Parameters in Stress Corrosion Cracking, *Proc. Fifth Symposium on Line Pipe Research,* Pipeline Research Council International, 1974, p U-1
27. R.R. Fessler, Stress Corrosion Cracking Temperature Effects, *Proc. Sixth Symposium on Line Pipe Research,* Pipeline Research Council International, 1979
28. J.A. Beavers, C.L. Durr, and B.S. Delanty, High-pH SCC: Temperature and Potential Dependence for Cracking in Field Environments, *Proc. 1998 Third International Pipeline Conference,* American Society of Mechanical Engineers, 1998, p 423
29. J.A. Beavers, C.L. Durr, and S.S. Shademan, Mechanistic Studies of Near-Neutral-pH SCC on Underground Pipelines, *Proc. International Symposium on Materials for*

Resource Recovery and Transport, 37th Annual Conference of Metallurgists of CIM, CIM, 1998, p 51
30. J.A. Beavers, J.T. Johnson, and R.L. Sutherby, Materials Factors Influencing the Initiation of Near-Neutral-pH SCC on Underground Pipelines, *Proc. Fourth International Pipeline Conference,* Paper No. 047, Oct 2000, American Society of Mechanical Engineers, 2000
31. J.A. Beavers and R.N. Parkins, Recent Advances in Understanding Factors Affecting Stress Corrosion Cracking of Line-Pipe Steels, *Proc. Seventh Symposium on Line Pipe Research,* Pipeline Research Council International, 1986
32. J.A. Beavers and C.E. Jaske, Near-Neutral-pH SCC In Pipelines: Effects of Pressure Fluctuations on Crack Propagation, Paper No. 98257, *CORROSION/98,* NACE International, 1998
33. "White Metal Blast Cleaning," NACE No. 1/SSPC-SP 5 (latest revision), NACE
34. "Near-White Metal Blast Cleaning," NACE No. 2/SSPC-SP 10 (latest revision), NACE
35. "Control of External Corrosion on Underground or Submerged Metallic Piping Systems," RP0169-96, NACE International
36. S.J. Lukezich, J.R. Hancock, and B.C. Yen, "State of the Art for the Use of Anti-Corrosion Coatings on Buried Pipelines in the Natural Gas Industry," GRI-92/004, Gas Research Institute, April 1992
37. "Stress Corrosion Cracking—Recommended Practices," Canadian Energy Pipeline Association, Calgary, Alberta, Canada, 1997
38. P.H. Vieth and J.F. Kiefner, *RSTRENG2 (DOS Version) User's Manual and Software* (Includes: L51688B, Modified Criterion for Evaluating the Remaining Strength of Corroded Pipe), Pipeline Research Council International, 1993
39. "External Stress Corrosion Cracking of Underground Pipelines," Technical Committee Report 35103, Item No. 24221, NACE International, 2003

SELECTED REFERENCES

- "Stress Corrosion Cracking (SCC) Direct Assessment Methodology," Standard Recommended Practice, Standard RP0204-2004, Item No. 21104, NACE International, 2004
- "Stress Corrosion Cracking Study," Michael Baker Jr., Inc., Delivery Order DTRS56-02-D-70036, Department of Transportation, Research and Special Programs Administration, Office of Pipeline Safety, TTO Number 8, Integrity Management Program, Sept 2004, http://primis.rspa.dot.gov/docs/sccReport

Natural Gas Internal Pipeline Corrosion

Sridhar Srinivasan and Dawn C. Eden, Honeywell Process Solutions, Honeywell International, Inc.

INTERNAL CORROSION IN PIPELINES is a significant problem in oil and gas transmission systems. Oil or mixed-phase transmission systems, due to the presence of an inhibiting oil phase, typically tend to be less corrosive than corresponding natural gas transmission systems that carry water vapor in the gas phase. Under normal operating conditions, gas transmission pipelines carry undersaturated gas processed by upstream dehydrating units. These pipelines are generally operated with no protection or inhibition and rely on the performance of the dehydrating units to process dry gas within acceptable standards. It is not unusual to find, in such pipelines, phase instabilities and other process perturbations resulting in near-saturated gas (with water vapor) or some liquid water carry-over. These upsets lead to water accumulation in some parts of the pipeline downstream of the separators and can cause water condensation due to pressure and temperature changes along the length of the pipeline.

Corrosion severely affects pipeline operations, leading to lost production, unscheduled downtime for maintenance or repair, and even catastrophic failure that impacts health, environment, and safety. Operators implement corrosion control measures to prevent such damage, including use of corrosion-inhibitor treatments; however, performance data are rarely available to help optimize the dosing level or the type of chemical used. Typically, pipeline systems are treated for management of general corrosion, although most failures actually result from localized (pitting) corrosion.

Corrosion measurement and assessment in unpiggable pipelines presently relies on internal corrosion direct assessment (ICDA) methods, a practice that uses proven tools but requires validating experience. As a result, approaches to corrosion measurement are often after-the-fact and rely on detection tools to show the existence of defects and damage to a pipeline. A more proactive approach can be taken by implementing on-line, real-time corrosion monitoring as a complementary tool to ICDA, where the two techniques can each provide information to assist in the optimal application of the other. In brief:

- The ICDA modeling software tools encapsulate flow and corrosion modeling functionality to identify where corrosion can occur, and these are fine-tuned based on real corrosion measurement data from the pipeline environment itself.
- On-line, real-time corrosion-monitoring sensors and measurement devices at strategic points identified through ICDA will, in turn, offer a quick response to upset conditions inside the pipeline. Strategic locations are most efficiently identified through an initial ICDA study.
- On-line, real-time corrosion monitoring data coupled with key pipeline process and environment data can assist the pipeline operator in understanding how corrosion changes with time and variation in pipeline operations.
- The amount of liquid water condensing, and the flow parameters, such as gas velocity, system flow regime, pipeline inclination, and other factors, determine whether water accumulates in a particular part of the pipeline. The length of time for which these upsets last influences the amount of stagnant liquid water and potentially can manifest as multiphase corroding environments. Such zones in a pipeline system, which are the first to experience water accumulation, are most susceptible to internal corrosion.
- Identifying and inspecting such critical zones reduces the effort in inspecting entire pipeline segments for internal corrosion and forms the basis for ICDA.

This article examines methods of internal corrosion prediction for multiphase pipelines (primarily gas pipelines that become corrosive because of the occurrence of condensing water phase) and details methodologies to perform ICDA for pipelines. Further, real-time monitoring techniques for assessing actual corrosion at critical locations are also discussed.

Background to Internal Corrosion Prediction

Carbon dioxide (CO_2)-based corrosion has been one of the most active areas of research in the domain of corrosion, with several predictive models for carbon steel corrosion assessment. These efforts range from a predictive model that begins with CO_2 corrosion (Ref 1, 2), to models that focus on specific aspects of the corrosion phenomena (such as flow-induced corrosion or erosion-corrosion) (Ref 3, 4), to models that empirically relate corrosion rates to gas production and water production rates (Ref 5). In Ref 6, the physical chemistry of the corrosive medium is used as the key notion, and ionic strength, pH, and specific ionic species are taken into account as relevant factors. Other relevant efforts include the influence of hydrogen sulfide (H_2S) and O_2 on CO_2-based corrosion (Ref 7, 8).

Recent models, including one discussed in Ref 9, attempt to integrate lab data and field experience within the framework of relevant controlling parameters most prominent in oil and gas production. While there have been several studies focusing on the exact mechanism of metal dissolution in CO_2-containing waters, a commonly accepted explanation is that anodic dissolution of iron is a pH-dependent mechanism, and the cathodic process is driven by the direct reduction of undissociated carbonic acid (H_2CO_3) (Ref 2, 3, and 7). These reactions can be represented as (Ref 3):

$$Fe \rightarrow Fe^{2+} + 2e^- \quad \text{(Anodic reaction)} \qquad \text{(Eq 1)}$$

$$H_2CO_3 + e^- \rightarrow HCO_3^- + H \quad \text{(Cathodic reaction)} \qquad \text{(Eq 2)}$$

The overall corrosion reaction can be represented as:

$$Fe + 2H_2CO_3 \rightarrow Fe^{2+} + 2HCO_3^- + H_2 \qquad \text{(Eq 3)}$$

The buildup of the bicarbonate ion can lead to an increase in the pH of the solution until conditions promoting precipitation of iron carbonate ($FeCO_3$) are reached, leading to the reaction:

$$Fe + 2HCO_3^- \rightarrow FeCO_3 + H_2O + CO_2 \qquad \text{(Eq 4)}$$

Iron carbonate solubility, which decreases with increasing temperature, and the consequent precipitation of iron carbonate, is a significant factor in assessing corrosivity. The charge-transfer controlled reaction involving carbonic acid and carbon steel (or iron) can be represented in terms of the concentration or partial pressure of dissolved CO_2 in the medium to arrive at a corrosion rate equation that incorporates the order of the reaction and an exponential function that approximates for the temperature dependence of the reaction constant. This corrosion rate equation is given as (Ref 2):

$$\log(V_{cor}) = 5.8 - 1710/T + 0.67 \log(p_{CO_2}) \quad (Eq\ 5)$$

where V_{cor} is the corrosion rate in mm/yr, T is the operating temperature in Kelvin, and p_{CO_2} is the partial pressure of CO_2 in bar.

The corrosion rate obtained by Eq 5 has typically been seen as the maximum possible corrosion rate without accounting for iron carbonate scaling. A nomogram representing Eq 5 is given in Fig. 1 (Ref 2), which also includes a scale factor to account for the formation of protective carbonate films that lead to a reduced corrosion rate at higher temperatures.

The correlation given in Eq 5 describes CO_2-based corrosion. There have been other significant development efforts to demonstrate effects of other environmental variables, such as pH, H_2S, chlorides, bicarbonates, water/gas/oil ratios, velocity, and so on. Effects of H_2S on corrosion rates in the laboratory have been studied and presented in Ref 7 and 10. In Ref 7, the authors indicate that the preferential formation of an iron sulfide film can decelerate the corrosion rate, especially at temperatures above 20 °C (68 °F) and extending up to 60 °C (140 °F). Above 150 °C (300 °F), the corrosion reaction falls back to the standard CO_2-based corrosion, with an $FeCO_3$ film that is more stable than the FeS film. Even small amounts of H_2S can provide instantaneous protection at temperatures in the range of 70 to 80 °C (160 to 175 °F) (Ref 7).

The role of the hydrocarbon condensate in providing corrosion mitigation in specific production systems has been reviewed in Ref 11. The role of the type of oil or gas condensate is important from the standpoint of accurate assessment (Ref 12, 13). Other studies (Ref 14, 15) have evaluated the effects of critical parameters such as pH and velocity on CO_2 corrosion. Predictive models, including a combination of the parameters discussed in this article along with electrochemical considerations, have also been used to arrive at the corrosion rate (Ref 16, 17).

A typical flow chart delineating the hierarchical reasoning structure of a predictive model described in this article is given in Fig. 2 (Ref 9). The first step in corrosivity determination is computation of the system pH, because it is the hydrogen ion concentration that drives the anodic dissolution. Further, the role of pH in promoting or mitigating CO_2-based corrosion has been extensively chronicled (Ref 18–21). For oil/gas production environments, where it is the dissolved CO_2 or H_2S that contributes significantly to an acidic pH, the pH can be determined as a function of acid gas partial pressures, bicarbonates, and temperature, as shown in Fig. 3 (Ref 22). From a practical standpoint, the contribution of H_2S, HCO_3, or temperature to pH determination is another way of representing effective levels of CO_2 that would have produced a given level of pH.

While it has been documented that the CO_2 corrosion mechanism is dissimilar to that of strong acids such as hydrochloric (HCl) acid (CO_2 corrosion is now understood to progress through direct reduction of H_2CO_3 to HCO_3^- rather than reduction of H^+ ions), and that carbonic acid corrosion is much more corrosive than that obtained from a strong acid such as HCl at the same pH (Ref 18), there is also significant agreement that lower pH levels obtained from higher acid gas presence lead to higher corrosion rates. Conversely, higher levels of pH obtained through buffering in simulated production formation water solutions have been shown to produce significantly lower corrosion rates, even at higher levels of CO_2 and/or H_2S (Ref 23). Data about the effects of pH from another study are shown in Fig. 4 (Ref 13). Hence, it is more meaningful to determine the effective CO_2 partial pressure from the system pH. Data in Fig. 3 can be represented as equations for straight lines in terms of pH and acid gas partial pressures for a given level of HCO_3 and temperature. A numerical model can also be developed to compute pH for different values of HCO_3 and

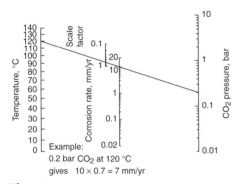

Fig. 1 CO_2 corrosion nomogram. Source: Ref 2

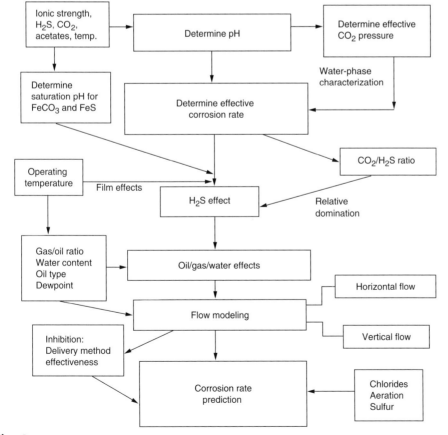

Fig. 2 Corrosion rate prediction flow chart. Source: Ref 9

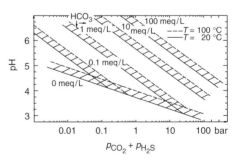

Fig. 3 In situ pH determination for production environments. meq/L, milliequivalents per liter. Source: Ref 22

temperature (Ref 24). Consequently, pH determination can be represented as:

pH1 = C1 − log (p_{H_2S} + p_{CO_2})

(Temperature = 20 °C, or 68 °F)   (Eq 6)

If HCO_3 > 0, then:

pH2 = C2 − log (p_{H_2S} + p_{CO_2}) + log (HCO_3)

(Temperature = 20 °C, or 68 °F)   (Eq 7)

where C1 and C2 are constants, p_{H_2S} and p_{CO_2} are partial pressures in bars, and HCO_3 concentration is represented in milliequivalents per liter (meq/L) (61 mg/L).

The effective CO_2 partial pressure is used to determine an initial corrosion rate for CO_2-based corrosion. The corrosion rate so obtained is modified to account for the formation of a $FeCO_3$ film (Fe_3O_4 at higher temperatures) whose stability varies as a function of the operating temperature. The scale correction factor shown in Fig. 1 is used to determine the initial corrosion rate from the nomogram in Fig. 1 (Ref 2). It is generally estimated that this corrosion rate presents a maximum corrosion rate, even though it has been reported that the rate computed by the nomogram is reached or exceeded in systems with high flow rates. It is important to recognize that this corrosion rate has to be modified to account for the effect of other critical variables in the system. Further, this rate does not indicate modality (general or localized) but rather represents the maximum rate of attack.

As indicated earlier, it is necessary to superposition effects of other critical system parameters on base corrosion rate. The flow chart in Fig. 2 provides the lists of sequential effects that are important from the standpoint of corrosivity determination. In addition to the system pH, these include:

- H_2S partial pressure
- Maximum operating temperature
- Dissolved chlorides
- Gas-to-oil ratio
- Water-to-gas ratio/water cut
- Oil type and its persistence
- Elemental sulfur/aeration
- Fluid velocity
- Type of flow

The following sections discuss the effects of these parameters on corrosivity and provide information explaining why it is critical to examine the parameter interactions prior to capturing the synergistic effects of these parameters on corrosion.

Role of H_2S. In recent years, oilfield production environments have been characterized by the increasing presence of H_2S and related corrosion considerations. Even though H_2S is probably the most significant concern in current-day corrosion and cracking evaluation, the role of H_2S in corrosion in steels has received much less attention when compared to the widely studied CO_2 corrosion (Ref 25). However, H_2S-related corrosion and cracking has remained one of the biggest concerns for operators involved in production, because of the significance of H_2S-related damage (Ref 26).

For the model described herein, in addition to its contribution in pH reduction, H_2S has a threefold role:

- At very low levels of H_2S (<0.01 psia), CO_2 is the dominant corrosive species, and at temperatures above 60 °C (140 °F), corrosion and any passivity are a function of $FeCO_3$ formation-related phenomenon; the presence of H_2S has no realistic significance.
- In CO_2-dominated systems (Ref 25, 27), the presence of even small amounts of H_2S (ratio of p_{CO_2}/p_{H_2S} > 200) can lead to the formation of an iron sulfide scale called mackinawite at temperatures below 120 °C (250 °F). However, this particular form of scaling, which is produced on the metal surface directly as a function of a reaction between Fe^{2+} and S^{2-}, is influenced by pH and temperature (Ref 26). This surface reaction can lead to the formation of a thin surface film that can mitigate corrosion. The authors are currently pursuing laboratory studies to characterize the stability and formation of mackinawite in sour systems.
- In H_2S-dominated systems (ratio of p_{CO_2}/p_{H_2S} < 200), there is a preferential formation of a metastable sulfide film in preference to the $FeCO_3$ scale; hence, there is protection available due to the presence of the sulfide film in the range of temperatures 60 to 240 °C (140 to 465 °F). Here, it is initially the mackinawite form of H_2S that is formed as a surface adsorption phenomenon. At higher concentrations and temperatures, mackinawite becomes the more stable pyrrhotite. However, at temperatures below 60 °C (140 °F) or above 240 °C (465 °F), the presence of H_2S exacerbates corrosion in steels, because the presence of H_2S prevents the formation of a stable $FeCO_3$ scale (Ref 7, 28). Further, it has been observed that the FeS film itself becomes unstable and porous and does not provide protection. Also, the scale factor applicable for CO_2 corrosion with no H_2S (shown in Fig. 1) becomes inapplicable. Even though there is agreement among different workers that there is a beneficial effect of adding small amounts of H_2S at approximately 60 °C (140 °F), divergent results at higher concentrations and higher temperatures have been presented (Ref 7, 10).

The effect of H_2S adopted in the predictive model reflects work from Ref 28 for CO_2-dominated systems. Figure 5 (Ref 28) shows the combined effects of temperature and gas composition on the corrosion rate of carbon steels. Figure 6 (Ref 7) shows the effect of varying degrees of H_2S contamination on CO_2 corrosion. It is to be noted that the role of H_2S in CO_2 corrosion is a complex issue governed by film stability of FeS and $FeCO_3$ at varying temperatures and is an area of further active research by the authors. For additional information on CO_2 effect on carbon and chromium alloy steels, see Ref 29.

Temperature Effects. Temperature has a significant impact on corrosivity in CO_2/H_2S systems. Corrosion rate as a function of different levels of CO_2 and temperature is given in Fig. 7 (Ref 2). It has to be noted that once the corrosion products are formed, there is a significant mitigation in corrosivity. It is also apparent that the

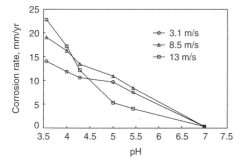

Fig. 4 Corrosion rate of steel as a function of pH and velocity. Source: Ref 13

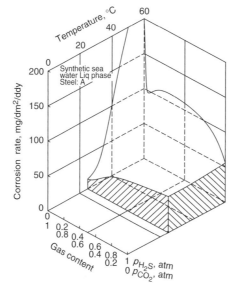

Fig. 5 Effect of gas composition and temperature on corrosion rate of steel. Source: Ref 28

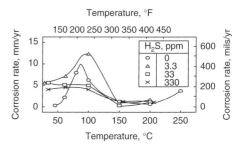

Fig. 6 Effect of H_2S and temperature on corrosion rate in pure iron. Source: Ref 7

carbonate film is more stable at higher temperatures and affords greater protection at higher temperatures. Figure 7 also shows that at temperatures beyond 120 °C (250 °F), the corrosion rate is almost independent of the CO_2 partial pressure of the system. The carbonate film may, however, be weakened by high chloride concentrations or can be broken by high velocity. In H_2S-dominated systems, because of the fact that no carbonate scale may be formed and that the FeS film becomes porous and unstable at temperatures beyond 120 °C (250 °F), significant localized corrosion may be observed.

Effect of Chlorides. Produced water from hydrocarbon formations typically contains varying amounts of chloride salts dissolved in solution. The chloride concentration in this water can vary considerably, from zero to a few parts per million (ppm) for condensed water, to saturation in formation waters having high total dissolved salts/solids. In naturally deaerated production environments, the corrosion rate increases with increasing chloride ion content over the range of 10,000 to 100,000 ppm (Ref 30). The magnitude of this effect increases with increasing temperature over ~60 °C (150 °F). This combined effect results from the fact that chloride ions in solution can be incorporated into and penetrate surface corrosion films, which can lead to destabilization of the corrosion film and increased corrosion. This phenomenon of penetration of surface corrosion films increases in occurrence with increases in both chloride ion concentration and temperature.

Effect of Bicarbonates. Bicarbonates in the operating environment have a significant impact on corrosion rates. High levels of bicarbonates can provide higher pH numbers, leading to corrosion mitigation even when the partial pressures of CO_2 and H_2S are fairly high. There is a natural inhibitive effect from the presence of bicarbonates, which can be present in substantial quantities in formation waters (up to 20 meq/L) (Ref 30). Condensed water in production streams typically contains no bicarbonates.

Velocity Effects. Next to the corrosive species that instigate corrosion, velocity is probably the most significant parameter in determining corrosivity. Fluid-flow velocities affect both the composition and extent of corrosion product films. Typically, high velocities (>4 m/s, or 13 ft/s, for noninhibited systems) in the production stream lead to mechanical removal of corrosion films, and the ensuing exposure of the fresh metal surface to the corrosive medium leads to significantly higher corrosion rates. Corrosion rate as a function of flow velocity and temperature is shown in Fig. 8 (Ref 13).

In multiphase (i.e., gas, water, liquid hydrocarbon) production, the flow rate influences the corrosion rate of steel in two ways. First, it determines the flow behavior and flow regime. In general terms, this is manifested as static conditions (i.e., little or no flow) at low velocities, stratified flow at intermediate conditions, and turbulent flow at higher flow rates. One measure that can be used to define the flow conditions is the superficial gas velocity. In liquid (oil/water) systems, this is replaced with the liquid velocity.

Velocities less than 1 m/s (3.3 ft/s) are considered static. Under these conditions, corrosion rates can be higher than those observed under moderately flowing conditions. This occurs because under static conditions, there is no natural turbulence to assist the mixing and dispersion of protective liquid hydrocarbons or inhibitor species in the aqueous phase. Additionally, corrosion products and other deposits can settle out of the liquid phase to promote crevice attack and underdeposit corrosion.

Between 1 and 3 m/s (3.3 and 9.8 ft/s), stratified conditions generally still exist. However, the increased flow promotes a sweeping away of some deposits and increasing agitation and mixing. At 5 m/s (16.4 ft/s), corrosion rates in noninhibited applications start to increase rapidly with increasing velocity. Data shown in Fig. 9 demonstrate the effects of velocity on corrosion rate for both inhibited and noninhibited systems. For inhibited applications, corrosion rates of steel increase only slightly between 3 to 10 m/s (9.8 to 33 ft/s), resulting from mixing of the hydrocarbon and aqueous phases. Above approximately 10 m/s (33 ft/s), corrosion rates in inhibited systems start to increase due to the removal of protective surface films by the high-velocity flow.

Flow-related effects on corrosivity have been linked to the developed wall shear stress and have been an area of intense research (Ref 31). Flow-induced corrosion is a direct consequence of mass and momentum transfer in a dynamic flow system, where the interplay of inertial and viscous forces is responsible for accelerating or decelerating metal loss at the fluid/metal interface. Another relevant aspect of flow- or velocity-induced corrosion is erosion-corrosion (Ref 32), which refers to the mechanical removal of corrosion-product films through momentum effects or through impingement and abrasion. Guidelines for velocity limits with respect to erosional considerations are given in API 14E in terms of the density of the fluid medium (Ref 33).

Importance of Water/Gas/Oil Ratios. It is important to account for the beneficial effects of hydrocarbons in a flowing system. Systems are characterized as oil dominated or gas dominated

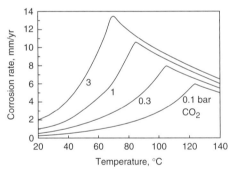

Fig. 7 Corrosion rate as a function of temperature and CO_2 pressure. Source: Ref 2

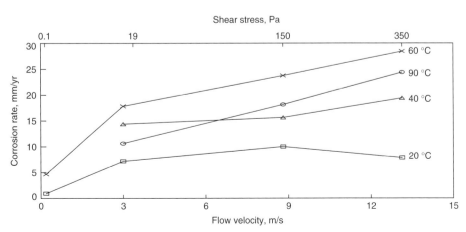

Fig. 8 Corrosion rate as a function of velocity and temperature. Source: Ref 13

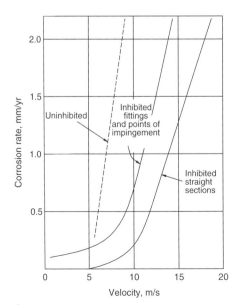

Fig. 9 Effect of gas velocity on corrosion rate

on the basis of the gas/oil ratio (GOR) of the production environment. If the environment has a GOR <890 m^3/m^3 (5000 scf/bbl in English units) (Ref 34), the tendency for corrosion and environmental cracking is often substantially reduced. This is caused by the possible inhibiting effect of the oil film on the metal surface, which effectively reduces the corrosivity of the environment. However, the inhibiting effect is dependent on the oil phase being persistent and acting as a barrier between the metal and the corrosive environment. The persistence of the oil phase is a strong factor in providing protection, even in systems with high water cuts. In oil systems with a persistent oil phase and up to 45% water cut, corrosion is fully suppressed, regardless of the type of hydrocarbon (Ref 10). Relative wettability of the oil phase versus the water phase has a significant effect on corrosion (Ref 35). Metal surfaces that are oil-wet show significantly lower corrosion rates (Ref 36).

The degree of protection provided by oil films can be quantified only as a function of water cut and velocity. Figure 10 shows data that relate the acid number of the crude to oil wettability, and Fig. 11 snows corrosion rate as a function of produced water content for different crude oil/produced water mixtures. In oil systems, the water cut acts in synergy with the oil phase to determine the level of protection from the hydrocarbon phase. However, at very low water cuts (less than 5%), corrosive severity of the environment is lessened due to the absence of an adequate aqueous medium required to promote the corrosion reaction.

Aeration/Sulfur Effects. The presence of oxygen significantly alters the corrosivity of the environment in production systems. How the presence of oxygen can significantly increase corrosion rates due to acceleration of anodic oxidation is discussed in Ref 37. While corrosion rate increases with oxygen, rate of oxygen reduction as a cathodic reaction is further exacerbated by:

- Increase in operating temperature
- Increased fluid flow, leading to increased mass flow of oxygen to the metal surface
- Increasing oxygen concentration

Data showing increases in corrosion rate as a function of oxygen concentration for differing temperatures are shown in Fig. 12 (Ref 37).

In systems containing high levels of H$_2$S, elemental sulfur is often found to be present. Its presence can significantly increase the corrosivity of the production environment with respect to weight-loss corrosion and localized corrosion. Presence of sulfur is similar to that of having oxygen in production systems, in that it can be a strong oxidizing agent and can lead to significantly increased local attack. Sulfur can directly combine with iron to form FeS and can lead to significant metal loss in a localized mode.

Effect of Multiphase Flow on Corrosion. Corrosion in multiphase systems stems not only from electrochemical interaction of the corrosive species (CO$_2$, H$_2$S, etc. dissolved in an aqueous medium) with steel, leading to metallic oxidation and subsequent metal loss, but also from the adverse impact of the flowing medium on the corroding surface and environment. This situation is significantly exacerbated through flow-related effects where high wall shear stresses, developed as a consequence of different flow regimes, can lead to accelerated corrosion. Significant research has been conducted to understand the role of various hydrodynamic parameters, primarily the wall shear stress, on corrosion.

Some research has focused on systems in slug flow and correlated corrosion rate as a function of wall shear stress (Ref 38). This work focused mainly on sweet water corrosion, and the data provided empirical relationships for multiphase systems. One such equation for corrosion rate prediction in horizontal multiphase slug flow is:

$$CR = 31.15 \, Cr_{freq} \, Cr_{crude}$$
$$\times \left[\frac{\Delta P}{L}\right]^{0.3} v^{0.6} p_{CO_2}^{0.8} T e^{\left(-\frac{2671}{T}\right)} \quad (Eq\ 8)$$

where, CR is the corrosion rate in mm/yr, $\Delta P/L$ is the pressure drop gradient per unit length, v is the viscosity in centipoise, p_{CO_2} is the partial pressure of CO$_2$ in bar, T is the temperature of the system in °C, and Cr$_{freq}$ is the normalized factor to account for slug frequency, defined as:

$$Cr_{freq} = 0.023 \, F + 0.214 \quad (Eq\ 9)$$

where F is the slug frequency.

Equation 10 provides a definition for corrosion rate from the crude, Cr$_{crude}$ (in mm/yr), in terms of total acid number of the crude (acid number is a number on an interval scale used to characterize various acids in the crude, such as naphthenic acid). Please see the cited reference for further information:

$$Cr_{crude} = \frac{10^{(\log(Acid\ number\ \cdot\ \%\ nitrogen) + 0.38)}}{24{,}000}$$
$$(Eq\ 10)$$

Researchers (Ref 31) performed experiments with three different systems, including a horizontal pipe flow system, a jet impingement setup, and a rotating cylinder electrode apparatus. They were able to establish a correlation between corrosion rate and wall shear stress as:

$$R_{COR} = a\, \tau_w^b \quad (Eq\ 11)$$

where R_{COR} is the corrosion rate in mm/yr, τ_w is the wall shear stress in N/m^2, and a and b are constants.

A related research effort in this area (Ref 39) derived a correlation from a database of 2500 points with various steels between corrosion rate, CR$_T$, and wall shear stress, S:

$$CR_T = K_T \cdot f_{CO_2}^{0.6} \cdot f(S) \quad (Eq\ 12)$$

$$f(S) = \left[\frac{S}{19}\right]^{0.15 + 0.03 \log(f_{CO_2})} \quad (Eq\ 13)$$

where CR$_T$ is the corrosion rate in mm/yr as a function of CO$_2$ fugacity and shear stress at a given temperature, f_{CO_2} is fugacity of CO$_2$ in bar, S is shear stress in Pa, and K$_T$ is a constant.

Efforts to characterize corrosion in flowing systems require modeling thermodynamics, phase behavior, mass transfer, and momentum-transfer effects. Mass transfer in a system relates to the movement of corrosive species between the pipe wall and the bulk flowing medium. The wall shear stress can be used as the primary hydrodynamic parameter representing momentum-transfer effects as a consequence of the inertial and frictional forces applied by a turbulent flowing medium on the pipe material surface (Ref 40). Because mass-transfer and momentum-transfer effects are interdependent, a change in wall shear stress will affect mass transfer and also impact flow regimes and corrosion rates.

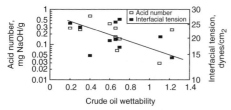

Fig. 10 Effect of acid number on crude oil wettability

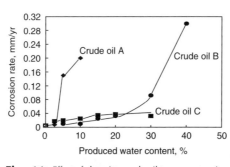

Fig. 11 Effect of changing crude oil type on corrosion rate as a function of water content

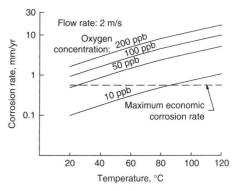

Fig. 12 Effect of oxygen concentration on corrosion as a function of temperature. Source: Ref 37

Flow Modeling of Multiphase Systems. As indicated in the previous section, multiphase flow systems present a tremendous challenge from the standpoint of corrosion evaluation and prediction because of the need to determine the role of important environmental, flow, and metallurgical variables and underlying mechanisms of corrosion. Modeling of multiphase flowing systems is a powerful tool for understanding internal corrosion in pipelines, because it facilitates assessment of liquid holdup, determination of wall shear stresses, prediction of flow patterns, accurate calculation of pressure drops in piping, computation of the critical angle for liquid accumulation in a pipeline, and the prediction of phase behavior of water for both horizontal and inclined flow over moderately hilly terrains. Although a detailed review of flow modeling is beyond the scope of this article, a number of papers on this subject have been published, and these should be referred to for additional information (Ref 1–5, 41–52).

Transmission Pipeline Scenario. For natural gas transmission pipelines, the highest number of incidents causing significant downtime or loss of life in 2003 were caused by internal corrosion. As reported in the data accumulated by the United States Office of Pipeline Safety and reproduced in Table 1, internal corrosion was reported as the reason for a loss in excess of $3.7 million in 2002 and amounted to more than 17% of damages due to all causes.

Internal corrosion in dry gas pipelines normally occurs when upstream gas processing/dehydrating units deliver gas that does not meet quality specifications with regards to the water content of the gas or dewpoint. It is under these circumstances that liquid water accumulates at certain sections in a piping segment and, in some cases, has been known to cause corrosion-driven pipeline failure.

Internal corrosion in pipelines is difficult to locate and measure due to a number of factors. Most internal corrosion-detection measures require access to the internals of a pipe for in-line inspection as well as visualization tools such as in-line pigs; however, a substantial portion of the pipeline does not allow in-line inspection, mostly due to physical and mechanical constraints. Other inspection techniques, such as ultrasonic technology and radiography, are implemented to measure wall thickness and estimate metal loss from the outside of a pipe, but excavation, cleaning, and other physical constraints allow for only a small area to be inspected at a time.

The ICDA approach to evaluate the likelihood of water accumulation and internal corrosion, and to identify critical zones, can enhance the actual measurement techniques and ensure safe operation of natural gas pipelines. A detailed analysis of critical locations where water would most likely accumulate provides information about relevant parts, from the perspective of a potential for corrosion damage of the piping system (Ref 53, 54).

A two-tiered procedure for the ICDA approach has been outlined in Ref 55. The first step deals with identifying zones with inclination greater than the critical angle, and selecting the first such segment from the downstream end of the pipeline (most downstream). Performing a detailed examination of this segment for corrosion forms the second step. If, on inspection, no corrosion is found, it is concluded that corrosion further downstream of that segment is unlikely (because they all are noncritical, and the segment identified as critical has no corrosion). Performing inspection on other susceptible segments, upstream of the initial location, will provide integrity information for the pipe between these two points. Progressing upstream, identifying susceptible segments and inspecting these segments, the entire pipeline is assessed for internal corrosion (Ref 40, 56).

If the locations identified as most susceptible to internal corrosion are determined by examination to be free from damage, the integrity of a significant part of the pipeline system is assured. If such locations are found to be experiencing corrosion, a potential integrity problem is identified, and remedial action may be prescribed.

Real-Time Corrosion Measurement and Monitoring

One of the problems associated with corrosion control is that corrosion is commonly dealt with in a historical sense, after the damage has occurred and with no opportunity to prevent future reoccurrence. In this light, corrosion measurements are usually relegated into reactive, maintenance functions. Off-line monitoring data can be viewed neither real-time nor alongside the process conditions that often initiate high-level corrosion activity. Recent innovations in monitoring technology enable the corrosion engineer to interact directly with the new on-line, real-time world of process control/optimization and asset management, leading the way to change for those that embrace, use, and promote these new technologies.

Developments in corrosion-monitoring technologies have increased the accuracy of data, promoting its relevance and value to the ultimate goal of increasing productivity by reducing corrosion damage, failures, and unplanned outages, thus decreasing downtime and increasing run time. Improved enterprise communications have also provided on-line connectivity for corrosion engineers, bringing them in closer contact with the frontline people that control processes and manage these facilities. Under this new paradigm, corrosion becomes another real-time process variable. The corrosion measurement device becomes the "tachometer" for the facility, showing in real-time when processes go awry and prompting remedial action before substantial damage occurs.

Off-Line Methods

Corrosion measurement methods in use today (2006) include simple, off-line techniques such as weight-loss coupon analysis. However, this provides only a retrospective status check rather than a means of active real-time process control. Off-line measurement methods also include such general corrosion measurement techniques as electrical resistance (ER) and linear polarization resistance (LPR) methods described in ASTM G 96 (Ref 57). These systems are able to operate in

Table 1 Summary (year 2002) of the causes of damage to transmission pipelines

Cause	No. of incidents	% of total incidents	Property damages, U.S. dollars	% of total damages	Fatalities	Injuries
Body of pipe	3	3.7	2,485,000	10.19	0	0
Butt weld	1	1.23	145,000	0.59	0	0
Car, truck, or other vehicle not related to excavation activity	5	6.17	842,326	3.45	0	1
Component	6	7.4	792,878	3.25	0	0
Corrosion, external	7	8.64	4,131,500	16.95	0	0
Corrosion, internal	14	17.28	3,711,443	15.23	0	0
Earth movement	1	1.23	599,040	2.45	0	0
Fillet weld	2	2.46	360,215	1.47	0	0
Heavy rains/floods	5	6.17	4,359,000	17.89	0	0
High winds	1	1.23	1,500,000	6.15	0	0
Joint	3	3.7	387,000	1.58	0	0
Malfunction of control/relief equipment	1	1.23	318	0	0	2
Miscellaneous	4	4.93	177,590	0.72	0	0
Operator excavation damage	2	2.46	52,010	0.21	0	0
Pipe seam weld	4	4.93	463,791	1.9	0	0
Rupture of previously damaged pipe	1	1.23	148,000	0.6	0	0
Third-party excavation damage	13	16.04	1,062,455	4.36	1	1
Threads stripped, broken pipe coupling	1	1.23	37,351	0.15	0	1
Unknown	6	7.4	3,037,642	12.46	0	0
Vandalism	1	1.23	73,000	0.29	0	0
Total	81		24,365,559		1	5

Source: U.S. Office of Pipeline Safety

stand-alone mode, providing spot corrosion data via battery-powered, field-mounted instruments, often with logging capability.

The flexibility of installing off-line systems in remote locations is somewhat offset by the fact that data are available only periodically, and there is an overhead in personnel time to manually download and process the data. Furthermore, such systems are often limited by shrinking allocations for corrosion control in tightening maintenance budgets. Once gathered, postprocessing of the corrosion data is often performed in computerized spreadsheet format. In most of these cases, the data are viewed after the corrosion has occurred and do not easily take into account process variables (e.g., temperature, pressure, flow rate, chemical feed rate, etc.). It is the corrosion engineer's burden to find and access this information, manually integrate it with the historical corrosion data, and present the findings in an after-the-fact report with recommendations.

On-Line Corrosion Measurement

In some cases, off-line LPR or ER measurements can be taken on-line. By this, it is usually meant that the corrosion measurement signal, rather than going to a local data-logging device, is channeled (often by hardwire) back to a safe area and then through the company network. Normally, this is performed with corrosion data from multiple points going back to a stand-alone computer (a dedicated corrosion server or simply a corrosion engineer's workstation). There can be substantial infrastructure costs associated with establishing this separate data channel that are beyond the capabilities of most corrosion-control budgets. For example, in some cases, the cost of hardwiring LPR or ER units in a plant environment can exceed the actual cost of the corrosion devices by up to a factor of 10. This is far from an optimal situation for acceptance of on-line corrosion-monitoring technology.

Even with on-line systems, the data are still commonly viewed or analyzed after-the-fact, which does not automatically include the ability to correlate the corrosion rate to specific process events, unless this is done manually by the corrosion engineer. Then, the corrosion engineer has to provide the process engineer or manager the "bad news" after a point where corrective action can be easily implemented.

New Technology for Real-Time On-Line Monitoring

Modern field corrosion monitoring now includes a broad range of techniques used to evaluate the degradation of metallic materials. These techniques can be divided into two distinct groups: namely, those providing indications of the cumulative damage sustained (retrospective) and those providing indications of the prevailing corrosion rate (usually on-line and continuous). These techniques are applicable with virtually all metallic materials and are commonly grouped as:

- *Cumulative-loss techniques:* weight-loss coupon, electrical resistance, ultrasonic thickness measurement, and other nondestructive examination methods (e.g., radiography). The cumulative-loss techniques will show signs of change only when sufficient corrosion has been sustained to cause a change in the bulk material properties. As such, most are used off-line and do not provide real-time data, requiring measurement cycles of days to weeks.
- *Corrosion rate techniques:* linear polarization resistance, harmonic distortion analysis (HDA), and electrochemical noise (EN). The corrosion rate techniques have a much higher resolution and short response time and have been developed to provide a fast assessment of the electrochemical rate processes taking place at the metal/environment interface. Measurements using these techniques take only a few minutes.

Experience with the most commonly used on-line techniques, such as LPR and ER, indicates that they are particularly good for detecting trends in uniform corrosion rates. In this capacity, they are looked upon as qualitative indicators of general corrosivity. That is to say that if the reading is going up, things are getting worse, and if the value is going down, things are getting better. Furthermore, these techniques are not sensitive to, nor can they differentiate between, both localized and uniform corrosion.

Advances in automated, multitechnique systems have made it possible to incorporate multiple measurements into a single instrument, thereby increasing accuracy to the point of being able to make quantitative measurements (e.g., LPR corrosion rate with B-value correction from HDA) and being able to differentiate the modality (pitting versus uniform corrosion) of the corrosion process (e.g., use of EN). Furthermore, this can be done on a time scale of minutes, which is consistent with modern process-control approaches. This latter point is extremely important, because it now is possible to provide quantitative corrosion rate and modality measurements and deliver them via the same (and existing) communications protocols (4 to 20 mA, RS-485, RS-232, and HART) that are already in place and used to acquire process variables through supervisory control and data acquisition or distributed control systems.

The importance of the aforementioned advances is that the corrosion engineer can now be "plugged into" the same on-line, real-time channel that is used for process control and optimization and facilities asset management. Corrosion data are automatically commingled and displayed with process data. The corrosion engineer is no longer relegated to a stand-alone function. He/she can see the same real-time process data that the process engineer sees, and the process engineer has access to the on-line corrosion rate that the corrosion engineer sees, enabling use of these data together as key performance indicators. Both engineering functions can now work together in a new way. The process engineer uses the corrosion signal as another variable that needs to be optimized (e.g., minimize asset damage, increase production while controlling damage to acceptable levels, extend allowable run time, and manage process to minimize inspection requirements).

Implementation of On-Line, Real-Time Corrosion Rate Systems

The use of corrosion rate techniques such as LPR offers sensitivity and high-resolution assessment of the corrosion current that is a direct consequence of the instantaneous corrosion process. Its value is directly related to the rate of the metal loss. The electrochemical monitoring methods have been developed specifically to estimate the corrosion current. The relationship between the polarization resistance measurement and the corrosion current is given by the Stern-Geary relationship:

$$i_{corr} = B/R_p \qquad \text{(Eq 14)}$$

where R_p is the polarization resistance, B is the Stern-Geary factor, and i_{corr} is the corrosion current density.

One limitation with commonly used LPR technology is that this technique cannot directly measure the proportionality factor (B value), which has led to its qualitative use in many cases. Most field corrosion instruments use a factory-set defaulted B value of 25 to 30 mV. However, recent studies have shown that the actual value for this proportionality factor can vary from less than 3 to over 100 mV, depending on the nature of the system, the reactive species (which can change from time to time), and even from system to system (Table 2) (Ref 58).

Incorporation of additional newer electrochemical techniques, such as HDA, into a field device provides capture of a more complete frequency response to provide values for the corrosion current, the characteristic anodic and cathodic Tafel coefficients, and hence an actual value for the Stern-Geary factor. The incorporation of the real-time B value with the corrosion current thus provides a more accurate corrosion rate measurement. More detailed information on the role of Tafel and B value in corrosion processes can be found in the article "Corrosion in Petroleum Production Operations" in this Volume.

Assessment of Localized (Pitting) Corrosion

Additional techniques can also be included, such as EN, to further augment and complete an automated corrosion measurement cycle. The EN method is used proactively to identify periods when the corrosion processes become unstable (usually associated with localized corrosion phenomena—pitting, crevice corrosion, and stress-corrosion cracking) and thus recognizes when the probability of localized corrosion is high. When corrosion changes from uniform to pitting, the characteristics of the current and potential fluctuations change dramatically over orders of magnitude. When normalized to the average corrosion current over the period, these noise signals can be analyzed and displayed real-time in much the same way that they can describe noise in any electrical system, which is the basis of a pitting factor value. From a facilities corrosion-control perspective, this is important, because the technique provides an early warning of incipient localized corrosion events so that action can be taken at the process control level before substantial pitting damage occurs.

Discussion of Case Studies in Modern Multitechnique Electrochemical Corrosion Monitoring

Case Study No. 1: Measurement of Corrosion in Multiphase Pipeline Environments (Ref 59). The use of on-line, real-time multitechnique monitoring methods has been able to provide quantitative corrosion rate trends and indications of modality in systems containing oil and water in combination with corrosive gases. As shown in Fig. 13, the combination of LPR, HDA, and EN provides a more complete representation of the corrosion taking place in both the liquid and vapor phases in a dehydrated gas pipeline environment containing condensing water and glycol, methane, and carbon dioxide. When compared to corresponding coupon data taken on the same exposure interval, it can be seen that the uniform corrosion rates are approximately a factor of 10 higher in the liquid phase than in the vapor. The rates obtained by monitoring were very similar to those independently determined by mass-loss measurements on the coupons. Furthermore, and perhaps more importantly, the mode of corrosion in the condensing vapor phase was found to be pitting corrosion. The use of EN data taken during the automated measurement cycle showed high values of pitting factor throughout the exposure period. Such data provide a basis for remedial actions before substantial damage has occurred.

Case Study No. 2: Use of Field System for Determining Inhibitor Dosage Requirements (Ref 60). In most cases, inhibitor dosages are determined through laboratory studies and then migrated to the field through a relatively slow iterative process. A gas pipeline system with very high H_2S concentration and severe corrosion was monitored on a real-time basis to quickly assess corrosivity and establish an optimal inhibitor dosage. At the startup of operations, very high dosages were used. Figure 14 shows the differences in corrosion rate without and with B-value adjustment (from HDA data). The B value for this sour gas system was determined to be approximately 120 mV, approximately four times the default value in most field instruments. Therefore, the actual corrosion rate was four times the value originally believed, based on simple LPR measurements. A real-time pipeline inhibitor dosage study was conducted in terms of percent inhibitor efficiency while switching the inhibitor injection off and then monitoring corrosion at various inhibitor dosages attained by varying the injection rate. As a result of this effort, the pipeline operator was able to reduce the dosage substantially, with an associated cost saving of approximately 60%.

Case Study No. 3: Microbiological/ Corrosion in a Water-Handling System (Ref 61). Figure 15 shows a combination plot of corrosion rate (corrected for a B value of only 8 mV versus the standard 25 to 30 mV), pitting factor, and sulfate-reducing bacteria (SRB) growth versus time for a period of approximately 1 week in a simulated oilfield water injection system. The graph shows the variations in corrosivity that occur as a result of production of H_2S from SRB activity, short-term operational conditions (periodic system aeration), formation of biofilms with increasing SRB counts, and the long-term effects on pitting tendencies as observed by pitting factor readings. First of all, the measured B values were very low as a result

Table 2 Role of Tafel and B value in corrosion processes

Anodic Tafel (β_a), mV	Cathodic Tafel (β_c), mV	B value, mV	Physical mechanisms	Practical examples
60	60	1–13	Both processes activation controlled	Deaerated environments or those with depolarizing species (e.g., sulfides)
60	∞	18–30	Anodic process activation controlled; cathodic process diffusion controlled	Aerated environments (e.g., fresh water, seawater, or brine)
120	∞	50–80	Anodic process activation controlled; cathodic process diffusion controlled (anodic slope different)	Mixed multiphase (oil/water) conditions; inhibitor or organic films on metal surface
∞	∞	>100	Severe anodic and cathodic diffusion limiting	Condensing vapor environments, thin liquid films on corroding surface

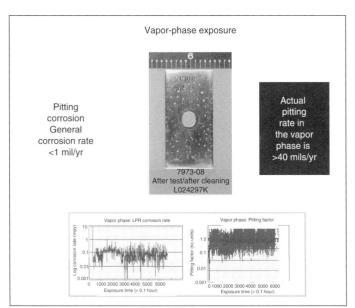
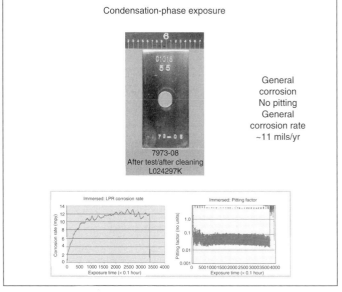

Fig. 13 Comparison between general corrosion and localized (pitting) corrosion data recorded from a dehydrated gas pipeline environment. Source: Ref 59

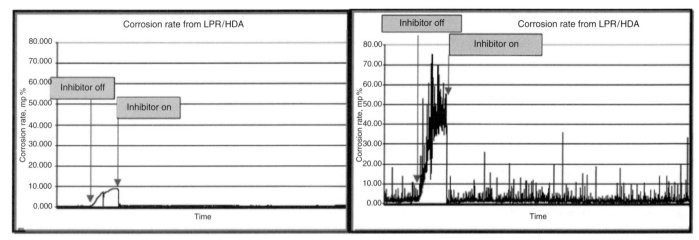

Fig. 14 Corrosion rate during inhibitor injection on/off cycle in a sour gas pipeline. Note: Measured corrosion rate with default B value and measured B value from harmonic distortion analysis. Source: Ref 60

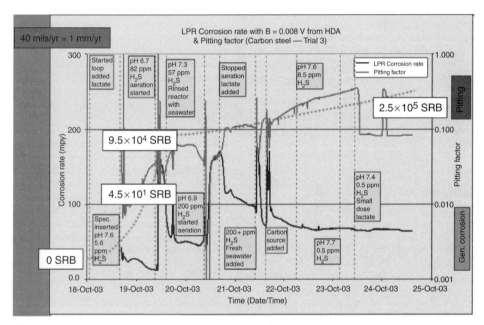

Fig. 15 On-line, real-time correlation between corrosion rate (left scale), pitting tendencies (right scale), and sulfate-reducing bacteria (SRB) growth (dotted curve). Note: Annotations for short-term process information for H_2S concentration and imposed aeration. Source: Ref 61

of the presence of biologically produced H_2S in the system and sulfide films on the steel surfaces. The low B-value readings demonstrate the inaccuracy at a factor of 3 of noncorrected LPR corrosion rates. Secondly, corrosion and pitting factor varied considerably with operating conditions, with periodic increases in localized corrosion during each aeration cycle. Over the longer term, the system developed a considerable tendency for localized corrosion (high pitting factor) near the end of the monitoring period that matched the rise in sessile SRB count on the metal surface. This is likely the result of formation of a biofilm and the local variations in corrosion tendencies on filmed and nonfilmed areas. Additional measurements of resistance were also made that indicated the buildup of FeS surface films during the exposure period. None of these measurements would have been possible without the automated, multitechnique monitoring capabilities.

Summary

Internal corrosion in natural gas pipelines is a complex phenomenon requiring thermodynamic, electrochemical, and fluid dynamic analyses and characterizations. Corrosion prediction and ICDA models have been described and represent effective methodologies to develop a program to prevent pipeline corrosion failures. The ICDA method is applicable to gas transmission pipelines that may potentially suffer from episodic upsets of wet gas or liquid water carryover. Locations along a dry gas pipeline that are identified as most likely to accumulate water are examined in detail, and if these locations have not corroded, it would be reasonable to infer that other locations, less likely to accumulate the water electrolyte, may be considered free of corrosion.

On-line, real-time corrosion monitoring may be performed using a combination of electrochemical techniques that provide complementary information on both general and localized corrosion in a broad range of pipeline and process environments. Based on the data and information arising from numerous studies, the following conclusions, specifically related to petroleum production and transmission pipeline environments, may be drawn:

- Real-time corrosion measurement of both general and pitting tendencies has provided valuable insights into the corrosion behavior of steel and other metallic materials in low water cut, dehydrated, multiphase, and vapor-phase environments.
- The ability to measure the B value (Stern-Geary factor) on-line and in real-time has dramatically improved the accuracy of corrosion rate measurement, thus assisting appropriate and cost-efficient use of chemical treatments.
- Corrosion rates in multiphase (oil/brine/gas) environments have been found to exhibit widely varying corrosion rates between 0.001 to over 2.5 mm/yr (0.04 to over 100 mils/yr), depending on the conditions imposed (e.g., flow, chemical composition).
- In liquid-phase environments, the highest corrosion rates have been associated with agitation that produced severe liquid-phase turbulence. In most cases, corrosion in the liquid phase is uniform in nature and shows low pitting factor values.
- In heavy oil/brine/CO_2 environments, high pitting factor values and visual signs of localized corrosion have been observed in the liquid phase, possibly arising due to partial coverage of the heavy oil on the metal surface.

- Vapor-phase environments typically exhibit a general corrosion rate 1 to 2 orders of magnitude lower than that found in liquid-phase environments. However, higher corrosion rates can result from agitation or turbulent flow, which likely provides increased water availability in the vapor phase.

In summary, it is evident that corrosion behavior in pipeline environments has a number of influencing factors that can vary with time and so cause a dynamically changing corrosion condition. Inspections and off-line measurements do not afford the operator the opportunity to correlate corrosion excursions with operating and process parameters, making corrosion control a difficult proposition. This illustrates the importance of implementing an appropriate and correspondingly dynamic means of corrosion appraisal to help manage chemical and other corrosion-prevention treatments and to maximize the availability of the pipeline infrastructure.

REFERENCES

1. C.S. Fang et al., "Computer Model of a Gas Condensate Well Containing Carbon Dioxide," Paper 465, Corrosion/89 (New Orleans, LA), National Association of Corrosion Engineers, 1989
2. C. de Waard and U. Lotz, "Prediction of CO_2 Corrosion of Carbon Steel," Paper 69, Corrosion/93 (New Orleans, LA), NACE International, 1993
3. C. de Waard and D.E. Milliams, Carbonic Acid Corrosion of Steel, *Corrosion,* Vol 31 (No. 5), 1975, p 177
4. E. Dayalan et al., "Modeling CO_2 Corrosion of Carbon Steels in Pipe Flow," Paper 118, Corrosion/95 (Orlando, FL), NACE International, 1995
5. L.H. Gatzky and R.H. Hausler, A Novel Correlation of Tubing Corrosion Rates and Gas Production Rates, *Advances in CO_2 Corrosion,* Vol 1, National Association of Corrosion Engineers, 1984, p 87
6. J.L. Crolet and M. Bonis, A Tentative Method for Predicting the Corrosivity of Wells in New CO_2 Fields, *Advances in CO_2 Corrosion,* Vol 2, National Association of Corrosion Engineers, 1985, p 23
7. A. Ikeda et al., Influence of Environmental Factors on Corrosion in CO_2 Wells, *Advances in CO_2 Corrosion,* Vol 2, National Association of Corrosion Engineers, 1985, p 1–22
8. C.D. Adams et al., "Methods of Prediction of Tubing Life for Gas Condensate Wells Containing CO_2," 23rd Offshore Technology Conference (Houston, TX), 1991
9. S. Srinivasan and R.D. Kane, "Corrosivity Prediction for CO_2/H_2S Production Environments," Paper 11, Corrosion/96 (Denver, CO), NACE International, 1996
10. K. Videm and J. Kvarekval, "Corrosion of Carbon Steel in CO_2 Saturated Aqueous Solutions Containing Small Amounts of H_2S," Paper 12, Corrosion/94, NACE International, 1994
11. U. Lotz et al., "The Effect of Oil or Gas Condensate on Carbonic Acid Corrosion," Paper 41, Corrosion/90, National Association of Corrosion Engineers, 1990
12. H.J. Choi et al., "Corrosion Rate Measurements of L-80 Grade Downhole Tubular in Flowing Brines," Paper 213, Corrosion/88 (St. Louis, MO), National Association of Corrosion Engineers, 1988
13. K.D. Efird, Predicting Corrosion of Steel in Crude Oil Production, *Mater. Perform.,* Vol 30 (No. 3), March 1991, p 63–66
14. A. Dugstad and L. Lunde, "Parametric Study of CO_2 Corrosion of Carbon Steel," Paper 14, Corrosion/94, NACE International, 1994
15. U. Lotz, "Velocity Effects in Flow-Induced Corrosion," Paper 27, Corrosion/90 (Houston, TX), National Association of Corrosion Engineers, 1990
16. Y. Gunatlun, "Carbon Dioxide Corrosion in Oil Wells," Paper SPE 21330, SPE Middle East Oil Show (Bahrain), Society of Petroleum Engineers, 1991
17. M.R. Bonis and J.L. Crolet, "Basics of Prediction of Risks of CO_2 Corrosion in Oil and Gas Wells," Paper 466, Corrosion/89 (New Orleans, LA), National Association of Corrosion Engineers, 1989
18. R.H. Hausler and D.W. Stegmann, "CO_2 Corrosion and Its Prevention by Chemical Inhibition in Oil and Gas Production," Paper 363, Corrosion/88 (St. Louis, MO), National Association of Corrosion Engineers, 1988
19. R.D. Kane and S. Srinivasan, Reliability Assessment of Wet H_2S Refinery and Pipeline Equipment: A Knowledge-Based Systems Approach, *Serviceability of Petroleum, Process and Power Equipment,* D. Bagnoli, M. Prager, and D.M. Schlader, Ed., PVP Vol 239, American Society of Mechanical Engineers, 1992
20. S. Srinivasan and R.D. Kane, Expert Systems for Selection of Materials in Sour Service, *Proceedings of the 72nd Annual GPA Convention,* Gas Processors Association, 1993, p 88–92
21. G.S. Linda et al., "Effect of pH and Temperature on the Mechanism of Carbon Steel Corrosion by Aqueous Carbon Dioxide," Paper 40, Corrosion/90 (Las Vegas, NV), National Association of Corrosion Engineers, 1990
22. M. Bonis and J.L. Crolet, Practical Aspects of the Influence of in-situ pH on H_2S Induced Cracking, *Corros. Sci.,* Vol 27 (No. 10/11), 1987, p 1059–1070
23. R.D. Kane et al., internal reports on multiclient program on safe use limits for steels, CLI International, Inc., Houston, TX, 1992–1994
24. S. Srinivasan and R.D. Kane, Methodologies for Reliability Assessment of Sour Gas Pipelines, *Proceedings of the Fifth International Conference on Pipeline Reliability,* Gulf Publishing Co., Sept 1995
25. S.N. Smith and E.J. Wright, "Prediction of Minimum H_2S Levels Required for Slightly Sour Corrosion," Paper 11, Corrosion/94 (Baltimore, MD), NACE International 1994
26. R.D. Kane, Roles of H_2S in Behavior of Engineering Alloys, *Int. Met. Rev.,* Vol 30 (No. 6), 1985, p 291–302
27. M.J.J. Simon Thomas and J.C. Loyless, "CO_2 Corrosion in Gas Lifted Oil Production: Correlations of Predictions and Field Experience," Paper 79, Corrosion/93, NACE International, 1993
28. T. Murata et al., "Evaluation of H_2S Containing Environments from the View Point of OCTG and Line Pipe for Sour Gas Applications," Paper OTC 3507, 11th Annual Offshore Technology Conference (Houston, TX), 1979
29. B. Lefebvre et al., Behavior of Carbon Steel and Chromium Steels in CO_2 Environments, *Advances in CO_2 Corrosion,* Vol 2, National Association of Corrosion Engineers, 1985, p 55–71
30. L.K. Sood et al., Design of Surface Facilities for Khuff Gas, *SPE Production Engineering,* Society of Petroleum Engineers, July 1986, p 303–309
31. K.D. Efird et al., "Experimental Correlation of Steel Corrosion in Pipe Flow with Jet Impingement and Rotating Cylinder Laboratory Tests," Paper 81, Corrosion/93 (New Orleans, LA), NACE International, 1993
32. J.S. Smart III, "A Review of Erosion Corrosion in Oil and Gas Production," Paper 10, Corrosion/90, National Association of Corrosion Engineers, 1990
33. "Recommended Practice for Design and Installation of Offshore Production Platform Piping System," API 14E, 3rd ed., American Petroleum Institute, Dec 1981
34. NACE Material Recommendation MR-01-75-94, NACE International, 1994
35. K.D. Efird, Petroleum Testing, *Corrosion Tests and Standards—Application and Interpretation,* R. Baboian, Ed., ASTM, June 1995, p 350–358
36. S. Olsen, "Corrosion under Dewing Conditions," Paper 472, Corrosion/91 (Cincinnati, OH), National Association of Corrosion Engineers, 1991
37. J. Oldfield and B. Todd, Corrosion Considerations in Selecting Metals for Flash Chambers, *Desalination,* Vol 31, 1979, p 365–383
38. W.P. Jepson et al., "Model for Sweet Corrosion in Horizontal Multiphase Slug Flow," Paper 11, Corrosion/97, NACE International, 1997
39. A.M.K. Halvorsen and T. Sontvedt, "CO_2 Corrosion Model for Carbon Steel Including a Wall Shear Stress Model for Multiphase Flow and Limits for Production Rate to Avoid Mesa Attack," Paper 42, Corrosion/99, NACE International, 1999

40. S. Srinivasan and R.D. Kane, "Experimental Simulation of Multiphase CO_2/H_2S Systems," Eurocorr 98 (Utrecht, Netherlands), Sept 1998
41. O. Baker, Simultaneous Flow of Oil and Gas, *Oil Gas J.,* Vol 53, July 1954, p 185
42. B.A. Eaton et al., The Prediction of Flow Patterns, Liquid Holdup and Pressure Losses Occurring During Continuous Two-Phase Flow in Horizontal Pipelines, *J. Pet. Technol. (Trans. AIME),* Vol 240, June 1967, p 815–828
43. J.N. Al-Sheikh et al., Prediction of Flow Patterns in Horizontal Two-Phase Pipe Flows, *Can. J. Chem. Eng.,* Vol 48, 1970, p 21
44. Y. Taitel and A.E. Dukler, A Model for Predicting Flow Regime Transitions in Horizontal and Near-Horizontal Gas Liquid Flows, *AIChE J.,* Vol 22, 1976, p 47
45. G.W. Govier et al., The Horizontal Pipeline Flow of Air-Water Mixtures, *Can. J. Chem. Eng.,* Vol 93, 1962
46. J.M. Mandhane, G.A. Gregory, and K. Aziz, A Flow Pattern Map for Gas-Liquid Flow in Horizontal Pipes, *Int. J. Multiph. Flow,* Vol 1, 1974, p 537–553
47. H.D. Beggs and J.P. Brill, A Study of Two-Phase Flow in Inclined Pipes, *J. Pet. Technol.,* May 1973, p 607–617
48. J.J. Carroll, "The Water Content of Acid Gas and Sour Gas from 100 to 200 and Pressures to 10,000 psia," paper presented at 81st Annual GPA Convention, March 11–13, 2002 (Dallas, TX), Gas Processors Association
49. W.D. McCain, *The Properties of Petroleum Fluids,* 2nd ed., PennWell Books, Tulsa, OK, 1990
50. R.N. Maddox, *Gas and Liquid Sweetening,* 2nd ed., John M. Campbell Ltd., 1974, p 39–42
51. R.N. Maddox et al., "Estimating Water Content of Sour Natural Gas Mixtures," Laurance Reid Gas Conditioning Conference (Norman, OK), 1988
52. S. Srinivasan and R.D. Kane, "Critical Issues in the Application and Evaluation of a Corrosion Prediction Model for Oil and Gas Systems," Paper 03640, Corrosion/03 (San Diego, CA), NACE International, March 2003
53. V. Lagad, S. Srinivasan, and R.D. Kane, "Software System for Automating Internal Corrosion Direct Assessment of Pipelines," Corrosion/2004 (New Orleans, LA), NACE International, April 2004
54. V. Lagad and S. Srinivasan, "Advanced ICDA: A Methodology for Quantifying and Mitigating Potential for Corrosion Damage in Pipelines," Corrosion/2005 (Houston, TX), NACE International, April 2005
55. O. Moghissi et al., "Internal Corrosion Direct Assessment of Gas Transmission Pipelines," Paper 02087, Corrosion/02 (Denver, CO), NACE International, April 2002
56. S. Srinivasan et al., "Experimental Simulation of Multiphase CO_2/H_2S Systems," Paper 99014, Corrosion/99 (San Antonio, TX), NACE International, April 1999
57. "Standard Guide for On-Line Monitoring of Corrosion in Plant Equipment (Electrical and Electrochemical Methods)," G 96, ASTM International
58. R.D. Kane and E. Trillo, "Evaluation of Multiphase Environments for General and Localized Corrosion," Paper 04656, Corrosion 2004, NACE International, March 2003
59. R.D. Kane, D.A. Eden, and D.C. Eden, "Online, Real-Time Corrosion Monitoring for Improving Pipeline Integrity—Technology and Experience," Paper 03175, Corrosion 2003, NACE International, March 2003
60. D.C. Eden and J.D. Kintz, "Real-Time Corrosion Monitoring for Improved Process Control: A Real and Timely Alternative to Upgrading of Materials of Construction," Paper 04238, Corrosion 2004, NACE International, March 2004
61. R.D. Kane and S. Campbell, "Real-Time Corrosion Monitoring of Steel Influenced by Microbial Activity (SRB) in Simulated Seawater Injection Environments," Paper 04579, Corrosion 2004, NACE International, March 2004

Inspection, Data Collection, and Management

Sam McFarland, Lloyd's Register

HUGE AMOUNTS OF TIME AND MONEY ARE INVESTED in designing, constructing, commissioning, and operating major petroleum and petrochemical installations. To safeguard this investment and to ensure adequate prevention of adverse effects on health, safety, and the environment, and to meet legal requirements, plant owners/operators deploy other resources to monitor and maintain their plants in a safe and efficient condition. Inspection is one of the key resources used and, if properly planned and deployed, can be a very effective means of ensuring that the plant is kept in a safe condition, thereby avoiding accidents or unexpected plant incidents.

Historically, requirements for inspection evolved over the years in a generally unstructured way, prompted by requirements from design codes and from periodic legislation raised by various industry and governmental bodies, often in response to major plant incidents. A set of inspection requirements at the time of construction and commissioning was generally required by the design codes with some guidance on periodic inspection once in service. These inspection requirements became more formalized as knowledge of likely deterioration mechanisms improved and as plant failures became better understood. This led to the setting of standard periods of inspection for a range of plant equipment, largely triggered by instances of boiler and steam system failures.

By the 1980s, instances of catastrophic plant failure were much reduced due to better materials, tighter control of design, improved fabrication quality assurance, appropriate maintenance, valid inspections, better plant process control, and so forth. This improved track record and the continued drive for improved profitability turned the focus away from set periods of inspection (prescriptive regime) to the setting of inspection periods that reflected the true condition and nature of the equipment and its specific service conditions (risk-based regime). Allied to this, developments in equipment, methods, and procedures for noninvasive inspection made it possible (and attractive in some cases) to carry out meaningful inspections while the plant was still running using advanced nondestructive examination (NDE) techniques.

Inspection can generate large quantities of data in terms of visual inspection reports, NDE reports, photographs, radiographs, measurements, and live monitoring data. To turn inspection data into meaningful knowledge regarding fitness for service requires further data in terms of design, operation, construction, process, chemistry, maintenance records, and so forth. The requirements for handling such quantities of data in a form that is readily accessible, easily used, secure, and reliable has developed into a specialist subject of its own with proprietary systems and in-house developments available across various industries.

It is not possible in this article to cover all aspects of inspection and all items to which inspection is applied. This article concentrates almost exclusively on inspection related to pressure vessels and pipework, but many of the issues regarding inspection planning and particular techniques are also relevant to inspection of structures and other components. Excluded from consideration are related topics such as corrosion-monitoring devices, pressure testing, vibration monitoring, and pipeline pigging.

Inspection

Following a discussion of the general aspects associated with inspection, this section addresses:

- Inspection policy
- Inspection planning and procedures
- Inspection strategy
- Inspection methodology
- General preparation for inspection
- Invasive vs. noninvasive inspection
- Internal visual inspection
- Noninvasive inspection
- Inspection execution
- Risk-based inspection
- Inspection techniques (It should be noted that this topic is also discussed in the Appendix to this article, which reviews in greater detail the commonly used inspection techniques in the petroleum and petrochemical industries.)
- Competence assurance of inspection personnel
- Inspection coverage
- Inspection periodicity
- Inspection anomaly criteria
- Assessment of fitness for purpose
- Reporting requirements

A subsequent major section in this article—"Data Collection and Management"—addresses data acquisition, reporting and trending, and review and audit.

General Aspects of Inspection

Inspection is carried out at different stages in a product's life cycle for a variety of reasons:

- At manufacture to check compliance of individual components with design and specification documents and standards
- At fabrication to check compliance of assemblies of components and general quality
- At commissioning to ensure that items have been installed in accordance with the specification
- In service to ensure that the component is performing as expected and is not suffering any undue deterioration
- When out of service to gain a detailed knowledge of the component's condition on which to base decisions concerning continued service

The reason for the inspection also partly dictates the method employed. At fabrication, full access is normally available and there are few restrictions on the types of inspection that can be carried out. In service on operating plant, there are restrictions for safety reasons, space constraints, and availability of support services so the range of available techniques is reduced. For example, x-ray radiography is commonly carried out at the fabrication stage but is generally not a realistic inspection method in service, where gamma radiography is generally employed due to its portability and independence from an external power source.

If inspection requirements are considered early in the design stage, much can be done to aid future inspection efforts. Codes and standards generally call for access to be provided to allow for the component to be inspected periodically during service life, but this generally amounts to little more than the provision of access manways or inspection hatches. Greater consideration of inspection requirements at the design stage could result in measures to aid future inspectability of the component. This is particularly relevant to the ongoing development of non-invasive inspection techniques.

It may not be necessary to inspect every item. For example, items in low-pressure non-hazardous service may be operated with only a regular maintenance regime or on the basis of planned replacement after a set period in service. However, more significant plant items need to be subject to a scheduled inspection program to ensure their continued integrity and fitness for future service. To this end, it is important that a review is carried out on all plant items to assess those that pose the greatest risk. This process of risk assessment will identify the hazards associated with each item, the likelihood of failure, and the related mitigation measures required to manage the risk. This will also identify any low-risk items that can be safely excluded from the inspection program. Such items should be periodically reevaluated in case a future change in service condition or operation may make it necessary to add them to the inspection program.

The risk-assessment process will yield a list of plant items ranked in order of risk. This will provide a focus as to where the major inspection effort needs to be targeted. Once the hazard has been identified, it is important that an appropriate inspection technique is assigned for the detection, sizing, and monitoring of the anticipated deterioration mechanisms that could lead to this hazard condition. It is not possible to select a single inspection technique that will be suitable for all potential deterioration mechanisms. It is therefore important that the engineer specifying inspection requirements has full information regarding materials, process conditions, environment, and so forth to allow potential deterioration mechanisms to be determined and for appropriate inspections to be scheduled.

Inspection Policy

A clearly stated policy on the aims and objectives for inspection is a primary requirement. The policy, as a company document, forms the basis for subsequent reviews and audits of the inspection system, from initial planning and the inspection process itself to reporting and assessing results of inspection. Without a policy, it would be difficult to formulate a strategy or carry out a worthwhile audit.

Inspection Planning and Procedures

Standard specifications and methods of working have been established within industry to provide a structured approach for conducting and recording inspections (see, for example, Ref 1–9). These are very useful but cannot be relied on in themselves without the competence of engineers and inspectors to select appropriate inspection methods and for competent and experienced inspectors to be used to carry out the inspection.

The key factors in planning and carrying out an effective inspection are:

- All inspections should be properly planned by a qualified, competent person who can identify the likely deterioration mechanisms and the capabilities and limitations of appropriate inspection methods and procedures.
- The inspection should be included within the plant's overall operations and maintenance planning. This will ensure that sufficient time and resources are allocated for the inspection and that it is not seen as an add-on activity to be fitted in when possible.
- All relevant information should be made available to the inspector to enable the most appropriate inspection methods to be applied.
- The inspection should be targeted at detecting and sizing the types of defects that are expected in the component to be inspected, as far as can be ascertained by review of the available records.
- The inspection should be carried out in accordance with a clear written procedure or instruction that identifies the capabilities and limitations of the inspection technique in terms of detection and sizing of defects, the minimum training and qualification requirements for the person performing the inspection, the methods of reporting, and the responsibilities of each party in the inspection process.
- Any defects identified during the inspection should be recorded. The inspection procedure should include guidance on acceptability and reporting levels for types of defects. Any defects above the agreed threshold should be assessed for acceptability by a competent person. Allowance should be made for inaccuracies in the sizing of defects when carrying out an evaluation of the fitness for purpose of any chosen technique.
- The inspection personnel need to have the appropriate qualifications and training to ensure the competence and reliability of the inspection carried out (see the discussion "Competence Assurance of Inspection Personnel" later in this section).
- Where possible, the inspection personnel should have been tested and qualified on a test piece that realistically models the items to be inspected in service, with regard to access, orientation, and type of defect.
- The inspection function should be independent of normal maintenance functions to ensure that the inspector reports objectively.
- The choice of inspection technique must have a high probability of detection for the type and size of the defect expected and be suitable for the material and geometry of the component under test.
- The coverage of the inspection (for example, the amount of a weld or vessel shell to be inspected) should be chosen to give the required confidence in the integrity of the component for the required service life until the next scheduled inspection.
- The period between planned inspections should be justified and should take account of previous inspection findings as well as the risk profile for the equipment to be inspected.
- Adequate reporting and recording systems should be developed and maintained. Records should be available for the equipment throughout the life cycle of the plant, from fabrication through inspection, maintenance and monitoring carried out periodically during operation.
- The inspection program should be periodically reviewed to ensure its continued suitability as the plant ages, process conditions change, and systems deteriorate.

It is good practice (and in some regimes, compulsory) to develop a written scheme of inspection for each pressure system. Such a scheme should contain the following information:

- Those parts of the system that are to be examined
- Identification of the item of plant or equipment
- The nature of the inspection required (this may vary from out-of-service examination with the system stripped down, to in-service examination with the system running under normal operating conditions, depending on the applicable standards and legislative regime), including the inspection and testing to be carried out on any protective devices
- The preparatory work necessary to enable the item to be examined safely
- Specification of what inspection is necessary before the system is first used, where appropriate
- The maximum interval between inspections
- The critical parts of the system that, if modified or repaired, should be inspected by a competent person before it is used again
- The name of the competent person certifying the written scheme
- The date of the certification of the written scheme

Inspection Strategy

As mentioned previously, it is important that inspection activities are planned as far as possible. Additionally, they should ideally be linked with associated maintenance and support functions within an integrated plan for the assurance of overall plant integrity. In this way, inspection will be treated as a key element of plant operation, not as a necessary evil or as an unplanned addition.

To this end, it is important that an inspection strategy is established that sets a methodical basis for setting of inspection requirements. An effective inspection strategy should address:

- Inspection methodology
- Inspection techniques
- Competence assurance of inspection personnel
- Inspection coverage
- Periodicity
- Anomaly criteria
- Assessment of fitness for purpose
- Reporting requirements

Each of these elements of the inspection strategy are discussed separately in later portions of this section.

It is important to allow for some flexibility in setting of the inspection strategy, and it should not be rigidly focused only on the results of the assessment work. It is prudent to always plan for surprises, thus the inspection strategy should allow for some inspection to be carried out on areas that are not anticipated to be a problem or if a sudden opportunity to inspect occurs. This is necessary because assessments are often made on the basis of incomplete information and an inaccurate understanding of true plant condition. Defects observed in equipment expected to be in good condition, or information from outside sources and safety alerts will highlight gaps in knowledge and will necessitate more data gathering and a reconsideration of the risk assessment on which the inspection strategy is based.

Inspection Methodology

It is necessary to define the nature of the inspection that is required. A wide range of inspection techniques are available, and the person setting the inspection strategy should specify the techniques that are most effective for the type of in-service deterioration predicted. In addition to in-service deterioration, inspection techniques may also need to detect fabrication defects if there is a chance of these remaining in equipment entering service. The likelihood of this can be assessed by review of the extent and validity of inspection carried out during fabrication and construction.

The nature of the required inspection should be considered in terms of:

- Reporting level (the required minimum deterioration to be reported)
- Effectiveness (capability of the inspection to detect and size the minimum reportable deterioration)
- Reliability (the probability of detection and sizing accuracy)
- Repeatability (for comparison with previous results)

Reporting levels should relate to fitness-for-service criteria. These criteria should be based on the tolerance of the component to deterioration, the possible deterioration rate, and the interval to be set until the next inspection. A comfortable margin should exist between the reporting level and any defects that are of concern in order to allow for uncertainties in the data.

The effectiveness of inspection techniques and procedures depends on the objectives of the inspection (e.g., detection or sizing) and the characteristics of potential defects. These characteristics include the defect type, size, position, and orientation. The person setting the inspection strategy should review the likely effectiveness of the inspection proposed for each site, particularly where there is complex weld geometry, poor surface finish, or restricted access.

Advances in NDE techniques have made it possible to determine material integrity remotely without direct access to the material under test. Nondestructive examination can be carried out through paint, coatings, or insulation and material can be tested that is at a long range from the access point. The advantage of these techniques is that internal surfaces of pressure vessels and tanks can be examined from the external surface. There are advantages for health and safety in such noninvasive inspection when this can be carried out effectively and reliably. These include a reduction of risks to personnel entering confined spaces. Noninvasive inspection can also make invasive inspection more efficient and effective by targeting the most suspect areas in advance, thereby making more efficient use of internal inspection time.

Other factors influencing the decision to inspect noninvasively include the susceptibility of the materials of construction to exposure to the atmosphere (for example, process equipment that is subject to a sour operating environment i.e., hydrogen sulfide, H_2S, bearing, may become susceptible to polythionic acid attack as a consequence of the interaction of sulfur and/or sulfide scales with atmospheric oxygen when exposed to the atmosphere during an invasive inspection), the availability of favorable invasive historical inspection data, and the relative costs associated with invasive and noninvasive inspection. Some techniques for noninvasive inspection are relatively new, therefore, the person setting the inspection requirement needs to consider whether there is sufficient confidence in the technique's capability and coverage for each application.

Any job-specific training requirements should also be considered. This issue has particular relevance to noninvasive inspection as specialized techniques not covered by standard certification schemes may be required.

General Preparation for Inspection

Prior to commencing the work, the person undertaking the inspection should:

- Become familiar with the work scope, including anomaly reporting limits and applicable procedures/routines, and liaising with support staff if clarification is sought
- Review the history of the item to be inspected, noting any previous anomalies/repairs/uninspected areas
- Note design and operating pressures and temperatures and likely corrosion mechanisms
- Review vessel drawings for details and process drawings for process familiarization.
- Make an initial site visit in order to arrange preparation for the inspection
- Ensure preparations are made for any access, dismantling, delagging, or cleaning requirements together with that for a permit to work
- Ensure that all items are cleaned and prepared to the degree required to enable all expected defects to be detected, giving due regard to the inspection technique to be used

Invasive versus Noninvasive Inspection

The safe operation of equipment subject to degradation in service depends, among other factors, on inspection by which the degradation can be detected before it becomes a threat to integrity. This has traditionally been achieved through visually inspecting the vessel interior by opening up and entering the vessel at prescribed intervals. This approach to inspection has a number of drawbacks, particularly in terms of the safety and environmental hazards posed by opening and entering vessels, and the cost of lost production during shutdown. As a result, inspections by noninvasive methods that require access to the vessel exterior only, are increasingly being considered as a realistic alternative. In considering noninvasive inspection, decisions need to be made to:

- Determine when noninvasive inspection can be applied
- Define the requirements for the techniques to be used
- Select techniques that meet the requirements of the inspection

The primary aim of the decision-making process is to determine whether noninvasive inspection will be capable of offering similar or better control over the probability of failure than would be achieved by internal visual inspection (Ref 10, 11).

Internal Visual Inspection

The traditional inspection method for most vessels has been to take them out of service, drain and clean them to an acceptable visual standard, and then inspect internally, relying mainly on visual inspection by a competent inspector, supported by one or two appropriate NDE techniques. The primary concern when employing internal invasive inspection is ensuring the safety of the inspector by ensuring a nontoxic-atmosphere in the confined space. Nonetheless, invasive inspection is still widely used, but the difficulties associated with it have accelerated the development of noninvasive inspection techniques.

Conditions Favorable to Internal Visual Inspection:

- Potential degradation mechanisms and rate are not well understood.
- Sites of worst degradation are not easily predicted.
- Different areas may have different degradation mechanisms.
- Depth sizing of cracklike defects is not needed.
- Degradation is of a type that will be visible on exposed surfaces.
- Interior is readily accessed.
- Cost of shutdown is low or shutdown is required anyway for another purpose (e.g., for cleaning or repair).
- Acceptable health/safety/environmental risk is associated with vessel entry.
- Defects or damage expected in regions are not easily covered by noninvasive inspection.
- Exterior access is not straightforward.

Benefits of Internal Visual Inspection:

- Large-scale coverage rapidly achieved
- Lower cost of inspection activity itself
- Confidence based on extensive history of use

Limitations of Internal Visual Inspection:

- Subsurface defects cannot be identified, e.g., laminations, hydrogen damage, etc.
- Data are not readily quantified due to limitations on what can be measured.
- Data are not readily stored in a way that makes for direct comparison with past or future inspections.
- Requirement for internal cleaning
- Internal fittings may make access difficult.
- Internal coatings can be damaged while accessing the internal surfaces.
- Cannot be done through coating (although the coating itself can be examined)

Noninvasive Inspection

Noninvasive inspection techniques make it possible to test material for corrosion or defects on the opposite surface to where there is access. These techniques have the advantage that they avoid the need to enter the equipment for internal examination. For pressure vessels and tanks there is a reduction of risks to personnel entering confined spaces for inspection. Noninvasive inspection can also assist in helping to target the most suspect areas in advance of invasive techniques.

Methods for noninvasive inspection are relatively new and therefore operators need to demonstrate sufficient confidence in the NDE technique capability and coverage for each application. Other factors influencing the decision to inspect using a noninvasive technique include the susceptibility of the materials of construction to the consequent exposure to the atmosphere when undertaking an invasive inspection, the availability of favorable historical inspection data, and the relative costs associated with invasive and noninvasive inspection.

Conditions Favorable to Noninvasive Inspection:

- All potential degradation mechanisms are well understood—i.e., knowing what to look for.
- Sites of worst degradation are readily predicted—i.e., knowing where to look.
- Degradation is not restricted to surface.
- Exterior access exists.
- Interior of vessel is not readily accessible (e.g., tube bundles).
- Cost of shutdown is high.
- Significant health/safety/environmental risk is associated with vessel entry.

Benefits of Noninvasive Inspection:

- There is reduced risk to inspectors.
- Environmental implications of opening vessels is avoided.
- It can often be done on line, thus avoiding the likely production loss associated with vessel shutdown.
- Accurate sizing of defects is often possible.
- Data can be quantified and recorded for future comparisons.
- Inspections can be carried out more frequently, particularly where process conditions are changing and/or highly variable.

Limitations of Noninvasive Inspection:

- Cost of inspection may be perceived as being high.
- Inspection may be time-consuming.
- No single inspection technique is applicable across all potential defect types.
- Limited confidence/awareness of newer techniques.
- Some areas may still be difficult to inspect reliably without direct access, e.g., vessel nozzles with reinforcement plates and flange faces on vessel nozzles.

Inspection Execution

All parts of a pressure vessel should be inspected in a systematic and detailed manner (subject to the scope of the applicable written scheme of inspection and any applicable code or standard), and any observed defects are to be measured and recorded. The inspection will normally consist of a close visual inspection together with NDE of selected areas. The NDE techniques that are selected for use (together with any used in order to further inspect visually suspect areas) should be chosen to be compatible with both the material and likely form of defect to be encountered. All NDE activities should be covered by a written procedure. Where novel or nonstandard NDE techniques are being considered for use, the accompanying procedures and acceptance standards should be reviewed and endorsed by a competent person.

External Inspection. The following should be checked as appropriate:

- Drawings accurately show the actual vessel details; where deficiencies are observed, these should be made good in order to reflect the as-built status
- Vertical vessels are actually vertical and that skirt drains and cooling vents are clear
- The condition of supporting elements and load-carrying beams, tie down bolts, and sliding feet
- General condition of coatings
- Condition of shell barrels, heads, cones, nozzles, and associated welds for signs of distress, corrosion, or cracking. Any deformation present shall be measured.
- Earthing/(grounding) arrangements
- Manway closure supporting davits
- Vessel access stairs and platforms, associated bolting, and safety chains/bars
- Floor plates, ladder rungs, and steps are not worn or corroded; safety hoops are in place and intact.
- Data/nameplate details correct. It is also recommended that the safe working pressure is clearly marked and visible from instrumentation gages.
- Vessel tag or other identifying numbers are clearly stenciled on the vessel
- Details of pressure safety valves (PSVs) and bursting disks and test/installation setting details are correct, where recorded
- PSV general condition. If the PSV vents directly to the atmosphere, ensure the vent pipe has a small drain hole in it to prevent water and debris buildup.
- Condition and cleanliness of any flame-arresting devices
- Suitability and condition of any screwed fittings/plugs and vessel penetration threads
- Condition of any fireproofing
- External brackets, support saddles, doubler plates, and other shell attachments, together with associated welds are intact with no signs of distress
- Condition of insulation supports, wind girders, and stiffening rings
- Nozzles and flanged connections shall be examined for signs of distortion, leaks, and loose/corroded bolting. Exposed flange faces and nozzles shall be examined for corrosion and other damage.
- Adequacy of support for associated pipework—insulation kits fitted if required

- Insulation, cladding, and seal arrangements for indication of water ingress and the possibility of corrosion under insulation (CUI).
- General condition of vessel bridles, float pots, and floats. In addition, liaison with operations personnel is recommended in order to ascertain whether regular blow-down operations have been undertaken, otherwise dead-leg corrosion may have occurred.
- The condition of any other associated dead legs
- Evidence of the correct operation of automatic draining devices associated with the vessel

For pipe work inspections, the inspector should walk the line to be inspected from end to end and review isometric drawings to ensure they reflect the current site configuration, amending as required (including any additional access or delagging requirements). Particular attention should be paid to the presence of equipment items that may induce turbulent flow and potentially accelerate material loss within a pipe work system. Items of this type, such as process control valves, orifice plates, pump discharges, flow measurement devices, thermowells, and so forth, should be highlighted on the isometric drawings in order to assist in the preparation of future work scopes. Similar attention should also be paid to any areas of no- or low-flow conditions (so-called "dead legs") where debris could settle out.

Internal Inspection

All areas of the vessel should be made available for inspection (subject to the scope of applicable written scheme of inspection). The vessel walls and nozzles should be cleared of debris so that a full visual inspection can be made. For vessels with irremovable close-packed internals, such as tube bundles, alternative inspection techniques may be required. Typical features to be examined for evidence of corrosion, erosion, mechanical damage, or cracking, if appropriate, are:

- Shell barrels, heads, cones, nozzles, and associated welds, looking for any signs of distress, corrosion, or cracking
- Debris found within the vessel prior to cleaning. This may include mechanical items from the vessel or upstream pipe work. Where appropriate, samples of deposit/sludge should be taken for analysis.
- Internal linings. Blisters in coatings should be investigated to confirm that they are in the coating and not in the parent material (possible hydrogen damage) and to ascertain the condition of the material surface beneath them.
- Any internal corrosion-monitoring features such as corrosion coupons or metallic lining tags
- Internal bladders or instrumentation floats
- Wear plates and impingement areas, together with any associated welds
- Attachments/welds to shell, brackets, defoamers, weirs, baffles, diffusers, and vortex breakers (these being possible areas of high velocity and hence erosion, or fatigue/collapse due to hydrodynamic vortex breaking or blockage)
- Internal pipe work, including sand or desilting facilities and nozzles
- Sacrificial anodes—depletion, attachment, and bonding continuity
- Electrostatic dehydration arrangements, including the security of any internal cables
- Security, tightness, and general condition of bolting
- Other fittings such as demisters, vane packs, filter supports, etc.

Risk-Based Inspection

Risk-based inspection allows operators to manage the integrity of their plants by planning equipment inspection on the basis of information obtained from risk assessments. It is one of a range of measures within the wider process of plant integrity management. To use this approach, the operator needs to demonstrate that the risk assessment and inspection planning processes are being implemented in an effective and appropriate manner.

The risk-based approach requires that the quality and veracity of the information is tested and validated. Confidence in the assessment of risk is reduced when there is lack of, or uncertainty in, the key information required to assess integrity. Information on the integrity of plant can be generated from the design, operational experience, and inspection records, and from sound knowledge of the deterioration mechanisms and the rate at which deterioration will proceed. Inspections can then be planned at appropriate intervals using inspection methods that are able to detect the type and level of deterioration anticipated in order to facilitate the accurate assessment of pressure systems equipment integrity.

The recent arrival of goal-setting legislation and standards enables the planning of inspections to be based on the assessed risks, thus allowing a degree of flexibility when deciding on a suitable written scheme of inspection in terms of the equipment to be inspected and the frequency and nature of inspection. The information generated by the risk assessment is therefore the primary foundation on which to base these judgements, thus assuring a safe and suitable scheme that is not unduly restrictive.

The assessment of risk needs to take account of changes over time as equipment or plant conditions physically alter or because new information becomes available. Feedback of inspection results and the continuous reassessment of risk during the life of the plant is therefore a vital feature of an effective risk-based inspection scheme. Changes in contractors, if used, must be properly planned if this reassessment cycle is not to be disrupted.

Risk-based inspection is a logical and structured process of planning and evaluation. The process can be summarized in the following outline steps:

- Identify relevant items of equipment
- Identify potential degradation mechanisms
- Assess probability, consequences, and hence risk associated with such degradation
- Develop suitable inspection schemes (in conjunction with other mitigation measures) to detect and monitor this degradation, based on the assessed risk
- Execution of inspection, monitoring, and mitigation activities in accordance with the schemes
- Review, summary, and revision of the schemes to ensure they remain suitable and reassess inspection return time intervals

Depending on the level of data available and the extent of analysis undertaken, risk-based inspection approaches are generally classified as being qualitative, semiquantitative, or quantitative with an increasing degree of complexity. Risk-based inspection approaches have received a lot of attention in recent years, and various bodies have issued standards and guidelines regarding their application (Ref 12–15).

Inspection Techniques

A wide variety of techniques ranging from simple visual inspections through well-established methods such as radiography and ultrasonic testing to new and highly specialized methods are available for use. Some are identified subsequently. Not all the techniques are suitable for all materials, and the choice of techniques has to be carefully matched to a range of considerations such as material, geometry, access, temperature, light levels, and the type of flaw or degradation mechanism that the inspection is attempting to detect. Additional information on inspection techniques can be found in the Appendix to this article.

Typical defects that may be encountered are shown in Fig. 1. The figure shows a range of common surface and embedded defects within a pipe. External degradation, such as corrosion under insulation (CUI), fatigue cracking, and stress corrosion cracking (SCC), also may occur.

To maximize the likelihood of detecting and sizing defects and degradation, the most appropriate inspection techniques need to be matched to the individual defect types and orientations. Techniques can be generally separated into those for detecting and sizing surface defects and those for embedded defects; rarely is a technique suitable for all types of defects.

To select the most appropriate technique, it is necessary to specify the type of defect that is anticipated and what the selected technique needs to detect and size. The inspection specification needs to consider:

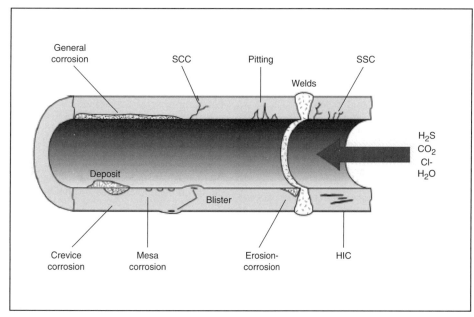

Fig. 1 Different forms of corrosion and deterioration. SCC, stress-corrosion cracking. SSC, sulfide stress cracking. HIC, hydrogen-induced cracking. Source: Ref 16

- The damage mechanism—location, type, morphology, and orientation
- Whether the defect is expected to be a surface-breaking or embedded defect
- The size of defect that needs to be detected and the sizing errors that can be tolerated

Where the nature of anticipated defects is not known, the defect description is then defined by the capabilities of the technique being applied. The plant item can be passed only as being clean of defects detectable by this technique. In such cases, a range of techniques would normally be specified based on past experience of their capabilities in detecting and sizing a range of defect types. Once the specification has been produced, the appropriate method(s) and technique(s) can then be selected. The specification should also address competency requirements for those undertaking the inspection.

Each inspection technique has advantages and disadvantages. Some of the important considerations in the selection of a suitable technique are:

- *Required access for inspection:* This will vary considerably depending on the technique and may limit the techniques that can be used in a particular situation.
- *Surface preparation:* The surface of the equipment to be tested is important because some techniques ideally require a bare metal surface, while others are more tolerant of paints and coatings.
- *Portability of equipment:* This may be an important consideration in relation to the installation of the equipment and to its handling (e.g., rope access).
- *User skill:* Different levels of training are appropriate to the various techniques with some techniques requiring only a low skill level, while others require a high degree of knowledge. Use of a technique by inappropriately qualified personnel may result in erroneous results and the possibility of not detecting flaws that are present.
- *Competence:* Operators should be certified to Personnel Certification Scheme in Nondestructive Examination (PCN) or American Society of Nondestructive Testing (ASNT) standards or to an equivalent nationally recognized qualification program.
- *Site safety implications:* Some NDE techniques may introduce additional hazards to the site, such as electrical (spark potential), chemical, fire, or radiation. The implications of introducing an additional hazard to the site during the inspection should be addressed.
- *Robustness of NDE equipment:* This should cover both physical characteristics as well as electronic.
- *Coverage:* This is an important aspect of any NDE technique and describes the fraction of the plant that can be inspected by a particular method. It is important to consider the coverage to ensure that a representative area is examined and the likelihood of missing the detection of flaws has been minimized.
- *Speed:* The speed at which a technique can be carried out is an important parameter and will influence the cost of an inspection method for a specified required coverage.
- *Sensitivity:* The reliability of each technique to detect flaws of different types varies. It is important, therefore, that an appropriate technique is chosen to maximize the potential identification of flaws.

In selecting any NDE technique, it is important to bear in mind that no individual technique can be expected to be 100% accurate in terms of detection and accurately sizing defects. In some circumstances, it may be necessary to combine techniques to achieve the required coverage and accurate sizing of defects.

Competence Assurance of Inspection Personnel

Competence is gained by inspectors satisfying requirements of qualification, training, and experience. It is not possible to become a competent inspector only by passing an exam or learning a new technique. It can be achieved only over a period of time exposed to the relevant operating plant. Some standards bodies, such as the American Petroleum Institute (API), organize courses and examinations that prospective inspectors must complete satisfactorily before they are allowed to perform inspection on particular pieces of equipment, and these are closely aligned to the inspection requirements of the relevant standards.

A plant inspector is now expected to have skills over and above the basic competence to perform an inspection to meet the requirements of the relevant code or standard. Increasingly, the inspector needs to be aware of plant conditions and how these conditions affect the deterioration of equipment because the inspector is expected to play a key role in ensuring that the inspection regime remains suitable for changing conditions as the plant ages. In addition to having the ability to carry out the inspection, the inspector is also expected to have some or all of the following skills and abilities:

- Classify equipment according to its inventory, service, and hazard
- Develop inspection strategies
- Conduct risk-based inspection assessments
- Develop written schemes of inspection
- Develop inspection work scopes
- Produce inspection procedures
- Produce inspection drawings/data sheets
- Produce inspection workbooks/workpacks
- Carry out safety-risk assessments
- Supervise NDE, approve results and anomaly reports
- Entry of results into an appropriate inspection database
- Assessment of results
- Anomaly assessments
- Fitness-for-purpose assessments
- Incident investigations
- Failure analysis
- Auditing

As such, inspectors are increasingly being expected to perform as engineering professionals, not just as checkers or verifiers. This change of status needs to be reflected in the level of competence assurance and an appropriate code or standard is needed that covers the full range of activities expected of the modern-day inspector. To date, no single code or standard covers all areas.

The mainstream nondestructive inspection techniques such as ultrasonics, radiography, and the basic surface crack detection techniques are covered by national and international standards for the testing and certification of practitioners. These qualifications are recognized throughout industry and should be seen as the minimum requirement for assessing the competence of an NDE technician. This basic ability in the technique needs to be backed up by relevant experience of applying the technique, preferably in conditions that closely replicate those required in the field.

Ultrasonic inspection programs are often designed on the basis of national or international standards such as BS EN 1714 (Ref 17). The inspection procedure often determines the approach to be taken in terms of the details specified in the standards for beam angles, scanning patterns, inspection sensitivities, probe types, and so forth, and often the detail of the defect types the inspection is intended to detect are ignored.

International standards such as BS EN 473 (Ref 18) define the training and qualification requirements for inspection personnel using the techniques specified. Often a written examination and a practical test that requires defects to be identified is required. Such testing is limited in scope, and only a small number of defects can normally be assessed. This leads to difficulties in later inspections where operators may not be directly qualified for the detection and measurement of defects that are thought to be present and for which the inspection is specifically being targeted. In order to overcome this difficulty, a number of levels of competence have been defined. Ultrasonic testing (UT) level 3, for example, requires demonstrating a more detailed level of understanding of the fundamentals of a particular inspection method than either level 1 or level 2 operators. However, the key issue is that the training and qualifications must be appropriate to a particular inspection if potential problems are to be avoided.

The Programme for the Assessment of NDT in Industry (PANI) identified a number of areas that need to be addressed in the development of any inspection regime (Ref 19). It identified the circumstances under which an approach to inspection based simply on the use of standards and national certification schemes may be inadequate.

Inspection Coverage

The amount of inspection carried out during manufacture depends on the class and design code of the vessel or component. In-service inspection should take this into account when deciding the amount of vessel or component to inspect. Frequently a sample of, say, 10% of all welds are inspected. Justification that this will provide the required confidence level in the integrity of the vessel should be provided (e.g., evidence that inspection is targeted at areas where defects are known to initiate). Higher-risk vessels would require a greater coverage, based on an assessment of the risks involved.

When setting the inspection requirements, the threats to integrity of each item of equipment are likely to be either uniformly distributed (such as general corrosion) or localized to specific areas. There may be a decreasing return on inspection in terms of risk reduction if a particular mechanism is shown to be absent or, conversely, if there is widespread attack. This is particularly relevant to decisions of whether invasive or noninvasive inspections are most suitable for a particular item.

Sites that may be susceptible to deterioration or failure (such as welds, high-stress regions, penetrations, saddle points, exposed surfaces, and liquid level interfaces) should be identified in the written scheme of examination. The inspection scheme should also take account of likely sites highlighted from the earlier review of preexisting defects or repairs from the construction records.

For a material/component where the degradation or defect distribution is uniform, the inspection effectiveness is given by the probability of detection for the technique multiplied by the percentage area covered. In practice, defect distributions will often not be uniform and different regions will tend to have different degradation rates. Under these conditions, the inspection effectiveness is determined not only by the percentage area covered but also by how the defects in the area covered relate to the overall defect population. If the inspection is carried out in an area where the majority of the worst defects are likely to occur, then a high effectiveness is possible even though the actual area inspected may be small. Conversely, a large area may be inspected. However, if this does not encompass the region where the worst defects are located, the effectiveness would be low.

Specification of adequate coverage is particularly important in piping systems because these are seldom inspected over 100% of their surface. Selection of suitable coverage needs to take account of the issues as described previously, particularly addressing those locations where enhanced degradation might be expected at areas of turbulence or stagnant conditions. Some knowledge of process conditions and flow patterns is therefore required before an appropriate coverage can be defined.

Traditionally, internal visual inspection will be applied with a significant proportion of the accessible interior of the vessel being inspected. Given the preceding considerations, the effectiveness can be matched using noninvasive inspection with a lesser overall coverage provided it is directed at the regions of worst defects. This can be achieved only when the regions of worst defects can be predicted. A key consideration in defining the coverage requirements for noninvasive inspection is therefore the confidence in the ability to predict the sites of worst degradation.

Inspection Periodicity

Traditionally, inspection of pressure vessels has been assigned on a set return period. With the introduction of risk-based inspection, the interval between inspections can change. It is often the case, subsequent to the formal evaluation of risk, that inspection intervals can be increased. However, should a high-risk component be identified, the periodicity may require to be reduced. Risk-based inspection is intended to define the optimal inspection periodicities based on predefined risk acceptance criteria; it is not intended to be used as a wholesale means of cutting inspection costs by unjustifiably extending inspection periods. Significant increases in the time between inspections should have a sound technical justification.

Established practice is to use one or more of the following approaches as a basis to set the maximum intervals between inspections:

- Historical experience for the specific type of equipment (e.g., boilers, air receivers)
- Industry guidelines for classes of equipment based on in-service experience
- As a prescribed percentage of the estimated residual life or the design life

A common approach is based on the setting of inspection grades for particular classes of equipment (Ref 20, 21). The allocation of grades is based on an assessment of actual or predicted deterioration. If deterioration is expected at a relatively rapid rate or there is little evidence or knowledge of the operational effects, the equipment is allocated a low grade, i.e., representing a high risk of failure and a shorter inspection period. If deterioration is at a reasonable and predictable rate, a higher grading is assigned with a correspondingly longer inspection period. An example of the type of matrix used is shown in Table 1.

Typically, maximum examination intervals between 2 and 12 years are recommended

Table 1 Varying inspection intervals based on component criticality and grading principles

Criticality rating	Inspection period (months)				Maximum review periods
	Grade 0	Grade 1	Grade 2	Grade 3	
1 High risk	24	36	N/A	N/A	12 months
2	24	48	N/A	N/A	24 months
3	36	48	72	96	48 months
4	36	48	84	120	60 months
5 Low risk	36	60	96	144	72 months

Source: Ref 22

depending on the type of equipment and the assessed grade. New equipment would normally be assigned grade 0 unless relevant operating history was available for equivalent equipment in the same service. Successive examination intervals can be extended (or reduced) if the grade of the equipment and/or the overall risk profile is changed. This may happen following examinations where favorable (or unfavorable) operating experience of the equipment, or "identical" plant on similar duty, is observed. This approach is most frequently adopted where qualitative methods have been used to devise risk matrices.

When an inspection period has been set, it is good practice to periodically review the risk assessment to ensure that it is still valid and that the assigned inspection period is still appropriate (Table 1). This review would include assessment of any significant change in operating conditions and any impact they might have on corrosion rate and, therefore, on required inspection frequency. The setting of the review period can be done on the basis of a fixed time period (e.g., set percentage of inspection period) or when a key process variable exceeds a previously established threshold level. This ensures that the inspection regime can be kept relevant, rather than just waiting for the next set of inspection results to trigger an update.

When applying industry guidelines, the extent of operating experience is a key factor to be considered. Appropriate margins are necessary to allow for uncertainties and the reliability and relevance of the supporting data for the existence, rate, and form of deterioration.

The remaining life approach to determining inspection intervals is based on calculating the remaining life of the equipment based on its tolerance to deterioration, defects, or damage and the rate of deterioration. The tolerance to deterioration is determined by assessing fitness-for-service at future times according to the deterioration predicted using documented methods (Ref 12, 23).

An inspection interval can then be prescribed as a percentage of the remaining life. The percentage selected needs to take uncertainties and reliability of the data into account. As a guide, API 510 (Ref 3) and 570 (Ref 4) state that the maximum interval between inspections should not exceed 50% of the estimated remaining life based on the measured corrosion rate ("half-life" rule). This approach can be used as a rough guide when used in conjunction with the Institute of Petroleum (IP)12/13 approach and backed up by a robust inspection scheme focused on locations of shortest predicted remaining life, dependent on the service conditions.

The assessment of remaining life needs to take all known and potential deterioration mechanisms into account. Calculations should be based on conservative assumptions and contain adequate margins to allow for uncertainties. The reliability of the available inspection and materials data is a key consideration in assessing the current condition of the equipment, the rate of deterioration, and the continued fitness-for-service.

Some deterioration mechanisms (e.g., fatigue crack growth and SCC) do not proceed at a constant rate but vary with time or initiate late in life. The remaining life calculation is then no longer a simple ratio of deterioration tolerance to deterioration rate. In these cases, a fixed interval based on a fraction of the remaining life may not be appropriate, and there may be a need for more frequent inspection toward the end of predicted life. In some cases, such as SCC, first appearance of defects may be sufficient justification to withdraw the unit from service. This leads to the need to have in place a clear definition of anomaly criteria.

Inspection Anomaly Criteria

As part of the planning of the inspection activities, appropriate anomaly criteria should be set. All anomalies highlighted by the inspection should be assessed against acceptance criteria for their impact on pressure system integrity. The initial assessment is likely to be carried out by the inspector or technician carrying out the inspection, and this will be very much at a broad level to determine whether there is an immediate need for action. For example, if the defect is obviously serious, immediate shutdown of the component or system may be required. Where an appropriate inspection scheme has been in place, it is anticipated that such occurrences should be extremely rare.

More commonly, the inspector will, as part of his reporting of the inspection, be expected to categorize potential anomalies against the acceptance criteria set for the inspection. In most cases, acceptance criteria will be as detailed in the original design code or specification. In the case of wall thickness measurements, these are well established. For example, the measured wall thickness can be compared with minimum allowable wall thickness (MAWT) as calculated in accordance with the design code. Dependent on service conditions and project practice, it is normal practice to specify a number of criteria of nonconformity that will progressively flag an increasing extent of deterioration. For wall thickness, a first flag would be raised when the measured wall thickness is less than the nominal wall thickness minus the manufacturer's tolerance minus the corrosion allowance. This is generally well above the calculated MAWT. It allows the inspection period to be reviewed to determine whether there is a need to inspect more frequently and to ensure that further deterioration can be detected before MAWT is reached. A second flag is typically set a certain amount above the MAWT. Dependent on the criticality of service and the anticipated rate of deterioration, this is often around 1 mm (0.04 in.) above MAWT (with an absolute minimum remaining thickness of 3 mm, or 0.12 in., for carbon steel pipe work from the viewpoint of mechanical strength rather than purely pressure retention). This flag gives a certain margin for allowing time to plan for replacement or other mitigation measure.

As just outlined, acceptance criteria related to wall thickness are well established and understood, particularly with reference to design codes. However, particular care should be taken when a localized wall loss is identified, e.g., internal pitting. In such circumstances, individual thinned areas can be acceptable below the overall MAWT, subject to detailed assessment by a competent person. In such cases, consideration should be given to extending the areas to be examined to ensure that the areas of maximum thinning have been detected.

Unlike wall thinning, which normally can be tolerated and monitored within established limits, cracking is seldom acceptable within a pressure system. In most cases, detection of cracking will be highlighted as a serious anomaly that requires urgent action in terms of planned shutdown and replacement of the item. However, certain types of cracking may be acceptable to continue in service for a period, subject to regular monitoring. Such circumstances should be assessed by a competent person. It is particularly important to identify the cause of the cracking. For example, detection of fatigue cracks may highlight an operational problem that can be resolved and that will remove the excitation from the component and thereby prevent further fatigue. The crack may then be left in service, subject to engineering assessment and an appropriate monitoring program. However, other modes of cracking can be more rapid and unpredictable and any instance of these modes (e.g., SCC) will nearly always require prompt replacement of the item unless the process can be altered to prevent further deterioration and the cracking can be confidently monitored. Such decisions need to be taken into consideration by a competent person in consultation with relevant specialists on materials, corrosion, and NDE.

Assessment of Fitness for Purpose

Following an inspection, it is important that the results of the inspection are analyzed and that the fitness for purpose of the equipment is then assessed. It is important to take into account the accuracy associated in the measurements determined by the NDE technique chosen and the possible implications of inaccuracies on the fitness for purpose assessment. For example, areas of extensive pitting may be detected and measurements made. However, it may not be possible to guarantee that the most severe pits have been found. Under these circumstances, it may be necessary to allow for the possibility of more serious pitting being present (e.g., by using mathematical models).

Reporting Requirements

The reporting requirements should be established in the written scheme of inspection and/or supporting procedures. Normally, reports will be made on preformatted inspection report templates, which ensures a common reporting standard and that all relevant data are recorded.

Specific reporting requirements will depend on the legislative regime, relevant codes, and any plant-specific requirements. In general, it is suggested that an inspection report should clearly state:

- Identification of the document, i.e., date of issue and unique identification
- Identification of the issuing body
- Identification of the client
- Description of the inspection work ordered
- Identification of the object(s) inspected and, where applicable, identification of the specific components that have been inspected and identification of locations where NDE methods have been applied
- Information on what has been omitted if the assignment was not carried out in full
- Identification or brief description of the inspection method(s) and procedure(s) used, mentioning the deviations from, additions to, or exclusions from the agreed methods and procedures
- If relevant, specification of inspection equipment used
- Where applicable, and if not stated in the inspection method or procedure, reference to or description of the sampling method and information on where, when, how, and by whom the samples were taken
- If part of the inspection work has been subcontracted, identification of this part and relevant information on the subcontractor
- Information on where and when the inspection was carried out
- Environmental conditions during the inspection, if relevant
- The results of the inspection, including a declaration of conformity and any defects or other noncompliances found (results can be supported by tables, graphs, sketches, and photographs)
- Any repairs needed and the time scale for completion
- Any changes in the safe operating limits and the date by which they should be made
- Any change in the written scheme of examination
- Date by which the next examination must be completed
- Other relevant information, e.g., conditions for publication of the report, advice, and recommendations
- The inspector's mark or seal
- Names and job titles of the staff members who have performed the inspection and of the person assuming responsibility for the inspection and their signatures

All reports of inspections are to be compiled as soon as possible after the inspection has been completed. Certain legislation states a maximum period (commonly, 28 days), which should be complied with where it applies. In many cases, a parallel system for immediate reporting of high-priority safety critical concerns raised by inspectors runs alongside the more routine inspection reporting system.

Data Collection and Management

Reliable data are essential for any effective integrity management system. This has always been the case but is even more true now that decisions regarding inspection are being made on the basis of an assessment of risk. It is now even more important that all data are documented so that the basis for decisions can be justified and the true plant condition can be accurately assessed and realistic assessments of future service life made.

With the development of noninvasive inspection methods, the volume of data generated electronically by these methods can be orders of magnitude greater than that produced by more conventional methods, and it is important that these data are efficiently recorded and analyzed. Equally, there is no value in generating huge streams of data if they are not reviewed, assessed, and used to provide trending information. This information can then be used to set future inspection periods and can be fed back to the operator in terms of the condition of the plant measured against the predicted condition.

Prior to first entering service, each plant component will have a history or record that could be of use to the inspector in the future assessment of its condition. A lot of this useful information is not retained and is effectively lost in terms of applying it in future assessments. If a component is designed and manufactured with future inspection requirements in mind, then an extensive pack of data will be available. In many cases this is required for legislative reasons or to comply with standards. However, the production of this pack of data is often seen as little more than a contractual obligation so it is put together at the fabrication stage, often consigned to archive. Ideally, the data pack from fabrication should become the property of the inspection function with the responsibility for updating any changes, upgrades, or modifications, and also recording any significant plant events, upsets, or failures.

The following should be considered as the minimum level of data required to properly register an item:

- For vessels, general arrangement and detail drawings; for pipework, process and instrumentation diagram, or drawing, (P&ID) and construction isometrics
- Fabrication release documentation including welding logs, NDE reports, hydrotest certificates, defects/repairs, and vendor surveillance release certification
- Vessel data sheet or alternative record of relevant design, fabrication, and operating criteria
- Details of any additional site designated identification number
- Details from vessel data book for American Society of Mechanical Engineers (ASME) or BS PD5500 vessels
- Related purchase order number (for traceability of related documents)
- Initial inspection report from fabrication yard or site precommissioning
- A statement from the competent person with regard to the fitness-for-service evaluation and reference to applicable written scheme of examination
- Records of inclusion into the site database
- Any applicable correspondence, particularly related to any concessions or agreed changes to specification.

These data and all subsequent inspection reports should be held in a designated equipment file, which may be electronic or hard copy (paper) record.

Once an item has been registered, a fitness-for-purpose assessment or similar risk assessment will be carried out to determine, among other things, the requirements for periodic inspection of the component. This evaluation will be made on the basis of an assumption of the plant operating and environmental conditions to which the component will be exposed during its service life. Service conditions can vary over the life of plant, so it is important that this service and environmental data are regularly updated so that any required changes can be made to the inspection regime to account for any impact of condition changes on the continued fitness for service of the component. To this end, a mechanism should be put in place to ensure that all relevant data are updated and communicated to those responsible for setting the inspection regime. Similarly, a mechanism is needed to ensure feedback from the inspection results to the plant operators so that they are kept informed of the current condition of their plant.

Data Acquisition

Traditionally, inspection data were collected and recorded manually; i.e., the inspector carried out the inspection, made notes of his or her observations, recorded wall thickness readings, and presented the whole in a paper report. With more recent developments of portable electronic equipment and advanced inspection techniques, this situation is now substantially changed.

As part of the inspection workpack, the inspector is now likely to have a set of electronic recording templates that can be accessed on a portable computer or palmtop electronic notebook, allowing the inspector to directly input findings and readings. This information can be backed up with digital photographs that can be stored for comparison with images from previous and future inspections. Advanced NDE techniques produce a great amount of data, particularly scanning techniques, which can also be recorded electronically for subsequent processing and review.

All of these data need to be effectively managed if the inspection data are to be converted into reliable knowledge of plant condition. One commonly adopted approach is to establish a plant database, which contains data relevant to the design, operation, and maintenance of the plant components but also stores inspection data and reports. A well-structured and maintained database can be a valuable tool in managing the integrity of an asset by enabling the operator to view all relevant data in summary form as well as view realistic trends for past and projected future performance. However, a poorly constructed or poorly maintained database is effectively worthless because it cannot be relied upon to give a realistic view of plant condition and may even be potentially dangerous if it misleads optimistically.

A suitable hierarchy should be established for the database. This should be established on a logical basis to ensure that data are held correctly and can also be related to relevant features on the same system. A simple hierarchy could be:

1. Plant name, e.g., terminal or platform name or designator
2. Location/area, e.g., module or plant number
3. System, e.g., service such as oil, gas, steam, etc.
4. Component, e.g., vessel, pipe, heat exchanger, etc., and unique component number
5. Feature, e.g., elbow, weld, nozzle, and unique feature number
6. Keypoint, e.g., location and orientation details

Each heading could then be broken down into subheadings. For example, feature could have properties such as material type, dimensions, orientation, and so forth. These would be recorded to create a unique record definition for each feature. Data could then be assigned at the appropriate level of detail. For example, process data such as stream composition or flow rate could be input at system level, vessel visual inspection reports at component level, repair/modification details at feature level, and individual wall thickness readings at keypoint level. By appropriately linking the various items to their parent component or system, all data could be held at the correct level and called on as required for review at overview or detailed level. Additionally, once in a suitably structured database, inspection data are then available for linkage to other systems such as to update plant maintenance databases.

Reporting and Trending

To maximize the value of information obtained from inspections, it is important that the results are presented in a comprehensive and logical manner that allows the current status to be grasped immediately without losing any of the level of detail that might subsequently be used to give deeper understanding of plant condition and trends.

As already mentioned it is important to have a system in place for the recording and storage of inspection reports and data, either as paper records or in an electronic database. In addition to providing the historic record of plant condition, this data can be used to establish trends and to provide predictions of future plant performance and deterioration.

One traditional method of trending has been to use a series of keypoints, i.e., set positions on vessels and pipe work that are inspected using the same technique over a period of years. By recording the wall thickness at each of these keypoints over a period, trends can be established by plotting the wall thickness against time. Typically, an overall rate is quoted, which is the rate of deterioration over the period between the first reading and the most recent reading for a particular keypoint. Sometimes, a short-term corrosion rate is quoted, which is the rate over the two most recent readings at a particular keypoint. This value is sometimes quoted as being more relevant to current rather than historical plant conditions. However, this whole approach has a number of significant drawbacks:

- It reflects data only from comparatively few points in the system.
- It does not take account of the likely deterioration mechanism(s).
- Straight-line trending is crude and not necessarily correct.
- Errors in the inspection are not considered and are often found only when subsequent readings do not fit the line established from previous spurious data.
- It provides a restricted sampling percentage.

The keypoint approach is still commonly in use but needs to be backed up within an inspection strategy that is based on an understanding of likely deterioration mechanisms, use of inspection techniques that cover larger areas, and with a more advanced statistical approach to the data produced. Rather than extrapolating a simple straight line between two readings, the significance of the readings needs to be taken into account by considering a greater population of readings. It is also important that any inherent error in wall thickness readings (perhaps ± 0.5 mm, or 0.02 in.) needs to be considered in any extrapolation of deterioration rate for setting of future inspection frequency or consideration of likely remaining service life.

Any inspection data and the rates and predictions made from them should be compared with the predictions of deterioration made at the risk-assessment stage when setting the inspection strategy. If the measured rate is at variance with these predictions, it is likely that the inspection strategy will need to be amended. In this way, the inspection strategy can be kept relevant to the actual plant condition.

Where a risk-based approach has been used in setting inspection requirements, it is important that a logical approach be taken to sample inspection, particularly in the definition of appropriate inspection locations and number of readings that constitute a realistic sample. Such decisions need to be appropriate to the assessed level of risk and need to consider the level of understanding of the likely deterioration mechanisms. This is related to the earlier comment on sample inspections and risk-based inspection. If random pitting corrosion is the likely deterioration mechanism, an inspection regime based on a few keypoint measurements of wall thickness is unlikely to detect the worst pit, and any trends derived from such data would be misleading. However, where corrosion is known to be essentially uniform, a few keypoint readings may prove adequate to establish data for realistic trending.

Where nonuniform forms of corrosion such as pitting are encountered, advanced statistical techniques can be employed to consider the effectiveness of the inspection coverage in terms of its likelihood of detecting the deepest pit in a population. Work by Kowaka (Ref 24) has categorized different statistical distributions for a range of corrosion types. For example, pit depth in a carbon steel freshwater supply pipe was shown to follow a normal (or Gaussian) distribution, whereas the maximum pit depth for a petroleum tank bottom plate was shown to be described by an extreme value (type 1) distribution. Such generalizations need to be treated with caution, and the specific data for each inspection need to be analyzed before any assumptions on distribution or trending can be made. Further references on the use of probabilities and statistics in the interpretation and trending of inspection data are included at the end of this article.

Review and Audit

Having planned, carried out, reported, and assessed the results of an inspection program, the cycle will begin again. It is essential that the whole cycle be subject to regular review to ensure that the aims of the inspection are being met. This is not simply an examination of adherence to specific procedures (i.e., performance monitoring). The reviews should encompass all facets of the cycle and answer such questions as:

- Is the inspection policy still valid?
- Has the inspection planning strategy taken note of new techniques of inspection?
- Have causes of failure been identified and acted on in future planning?
- Have process or plant changes been alerted to the inspection planning function?

It is important that audit and review findings be communicated to all concerned, and identified changes or recommendations are actioned. Changes in contractors or reorganizations should not impede or delay the process of continual review.

Appendix: Review of Inspection Techniques

Some in-service inspection techniques have been mentioned previously. These are now addressed in more detail, outlining their capabilities and limitations. Note that some of these and other techniques are covered in much more depth and detail in *Nondestructive Evaluation and Quality Control,* Vol 17 (Ref 25), and in *Corrosion Fundamentals, Testing, and Protection,* Vol 13A, of the *ASM Handbook.*

It is important that any inspection activity is planned carefully and that those carrying out the inspection are competent and have sufficient information and support facilities for a successful inspection. This is particularly important where the inspection activity is carried out by another body. This is now often the case, particularly as many plant operators no longer have their own in-house independent inspection function. It is also likely that for some of the more specialized techniques, another company might be used for that part of the inspection. Thus, there may be several interfaces, between the plant operator, the inspection company, and the specialist inspection contractor. In such cases, the potential for communication problems is clear, and this process needs to be managed effectively to ensure completion of a safe, worthwhile inspection that gives relevant and realistic data that can be relied on by the plant operator in setting future operating and inspection schemes.

Visual Inspection

Visual inspection is still the bedrock of most inspection regimes. In addition to traditional thorough inspection by the naked eye, visual inspection can be assisted by such optical aids as magnifying glasses or borescopes. More recently, some visual inspections have been carried out remotely using a mechanized probe inserted into the component being inspected, with the inspector viewing the image remotely outside the component. In such cases, care needs to be taken to ensure that sufficient coverage is achieved and that the image is of sufficient quality and depth to ensure that the required inspection standard can be achieved.

Concerns for the health and safety of personnel entering confined spaces for cleaning and for inspection are likely to provide a further driver for more remote inspection techniques. While the concern for health and safety must be paramount, the inspection undertaken must be as thorough as possible, using the most relevant and reliable techniques.

Ultrasonic Inspection

Conventional Ultrasonic Inspection. Ultrasonic (high-frequency sound) inspection is the workhorse of in-service NDE, being used to give measurements of remaining wall thickness at discrete test points or over a scanned area. It can also be used to detect and size defects such as cracks, voids, and inclusions.

Ultrasonic techniques use high-frequency sound waves that are reflected from interfaces or discontinuities, having large acoustic impedance values. In thickness checking, the reflections from the wall surfaces are measured. In defect detection, reflections from cracks, voids, inclusions, or other discontinuities are detected. The transfer of sound from the ultrasonic probe to the component requires a coupling medium to match the acoustic impedance between the inspection probe and the component being examined. The condition of the interface determines how much sound is transferred into the component, how much is scattered, and how much background noise is produced so a certain standard of surface finish is required to allow ultrasonics to be used reliably. For example, the surface should be clean and free of extensive corrosion products. Ideally, surface coatings should be removed for critical applications, but this is seldom done in practice. Instead an allowance is made for the coating thickness in the measured value of remaining wall thickness. Alternatively, an appropriate ultrasonic inspection procedure can be applied that eliminates the thickness of the coating from the recorded measurement.

When used over large areas, ultrasonic measuring is generally carried out on a sample basis, for example, the thinnest point in a scaled grid being quoted as representative of that area. Where full coverage of large areas is required, semiautomated or automated ultrasonic scanning techniques have been developed that ensure full coverage and that allow large amounts of data to be recorded. Automated techniques do not do away with the need for qualified technicians; where automated techniques are used, the technicians can concentrate more on the quality of data produced rather than on ensuring probe coverage. Additional information an automated ultrasonic systems is given subsequently.

As with all NDE techniques, it is important that a written procedure relevant to the application is available and applied by the technician, who should be competent in the technique and preferably be experienced on similar applications or previously qualified on representative test pieces.

Ultrasonic inspection is a valid technique for most materials over a wide variety of component diameters and wall thickness ranges (see Ref 17), although it is most commonly used for in-service wall thickness monitoring of carbon steel and low-alloy steel pipe work, tanks, and pressure vessels. There is effectively no upper limit on pipe diameter and very thick sections can be effectively inspected. However, there is an effective practical lower limit as on small diameter pipe work. (Rocking of the ultrasonic probe can cause unacceptable variation in readings). Small diameter "mini" or "button" probes can be used to partly alleviate this problem, but it is still problematic to get reliable ultrasonic wall thickness readings on pipe work below a certain diameter, typically around 50 mm (2 in.) in diameter. It is advisable to back up such readings with other techniques such as radiography, particularly at welds.

There is also an effective lower limit on wall thickness that can be accurately recorded by ultrasonic techniques. Below 3 mm (0.12 in.) wall thickness, an ultrasonic reading is, at best, indicative and certainly below 2 mm (0.08 in.) is best backed up by radiography. Depending on the wall thickness, application, procedure used, and skill of the operator, uncertainty in the readings obtained is often quoted as being a variation of approximately 0.5 mm (0.02 in.), but this can be greater if conditions of access and surface preparation are less than optimal. Such potential variation should be accounted for in any subsequent use of this data in a fitness-for-purpose review.

To maximize the prospects for obtaining reliable results from ultrasonic inspection, the following conditions should be addressed:

- The object of the inspection should be clearly defined, i.e., whether it is to detect and size general wall thinning, discrete pitting, or other defined defect.
- The procedure should be written specifically to ensure that the defects defined above can be realistically detected and sized for the particular component under inspection.
- Where there is uncertainty regarding the nature of likely defects, the procedure should ensure that the scope is maximized to cover as many options as possible regarding, for example, beam angle so that probability of detection is optimized.
- Any geometric peculiarities of the component to be inspected should be borne in mind in developing the procedure and may require specific testing of technicians in the use of the technique on such geometries.
- Where external coatings are left in place, the procedure should require these coatings to be checked for bonding and attenuation prior to the inspection commencing.
- Access should be optimized to ensure full coverage of the component to be inspected. The procedure should require the technician to report any areas that were not accessible for inspection because these may subsequently need to be inspected by other means.
- The technician carrying out the inspection should be qualified and experienced in the relevant technique, ideally confirmed by test on realistic test pieces.
- Inspections should be suitably qualified as appropriate and the results subsequently reviewed, including sample cross checking where necessary for critical applications.

Where a generic procedure for the technique is used, this procedure should be reviewed and amended as necessary to take account of any particular geometric or access factors specific to the component to be inspected.

Automated Ultrasonic Systems. To further enhance ultrasonic inspection, automated ultrasonic systems have been developed. The most widely used ultrasonic technique is the pulse-echo technique. In order to enhance the reliability of this technique, specialized automated systems can be deployed. These systems facilitate single/multiple probe inspection and provide images of the component via sophisticated data collection, processing, and analysis software.

In reliability terms, the main advantage of these systems, over the use of manual inspection, is that they remove the operator from the "front-end" of the inspection, thereby ensuring full inspection coverage via preprogrammed manipulation and couplant monitoring. Another main advantage over manual inspection is the ability of automated systems to monitor component degradation via comparison of stored data/component images. Automated pulse-echo is ideally suited to:

- Weld inspection (using multiple probes)
- Corrosion mapping/monitoring (using a single probe or multiple probes)

Phased array is an ultrasonic pulse-echo technique with which it is possible to quickly vary the angle of the ultrasonic beam using focal laws to scan a component, which may be achieved by probe manipulation or without moving the probe itself, allowing multiangle inspection from a single probe position. When applied to the inspection of welds, for example, a number of advantages are afforded:

- Reduction in the number of probes/scans required (reduced inspection time)
- Increased coverage for restricted access areas
- Optimized inspection (using, e.g., different wave modes and beam focusing)
- Potentially easier interpretation of images of the component inspected

Ultrasonic continuous monitoring using flexible mats consisting of multielement arrays of ultrasonic transducers can allow the continuous monitoring of the wall thickness of vessels and piping. These flexible devices, typically 50 mm (2 in.) wide by 500 mm (20 in.) long, are permanently bonded at specific locations to vessels and piping, potentially providing an assessment of the corrosion or erosion rate via a computer-based monitoring package. Alternatively, a conventional ultrasonic flaw detector may be used with a data logger and switching device (multiplexer) for connecting the various transducer elements in turn. These devices are useful where inspection by conventional means (e.g., manual ultrasonics) is difficult or impossible due to component geometry, location, presence of insulation, or hazardous inspection conditions or on unmanned installations allowing remote interrogation.

Long-Range Ultrasonics. Recent developments have been made in the application of long-range ultrasonic techniques, which are particularly useful in the inspection of long lengths of insulated pipe work or pipe work that is largely inaccessible (e.g., pipelines that go under road crossings or for inspection under vessel saddle supports or of subsea risers). This technique has the advantage that only a small area of insulation/cladding needs to be removed to give access to the pipe surface rather than the whole length of pipe having to be exposed for conventional ultrasonic inspection.

In long-range ultrasonics, various wave propagation methods are applied. In one system, a particular type of sound wave, Lamb waves, are generated in the pipe wall, which acts as a cylindrical wave guide allowing inspection to be carried out over a range of several tens of meters from one site. Defects cause the waves to be reflected back to the detector, but interpretation can be complicated, particularly in complex pipe runs and the technique is generally confined to monitoring sections of relatively straight pipe runs with few offtakes, although recent developments have allowed the waves to be "steered" so that more complex geometries can be tackled. The technique indicates the location and approximate size of defects that can then be more accurately sized by application of conventional or specifically designed ultrasonics at the identified defect location.

There are several proprietary systems available for carrying out long-range ultrasonic inspection. All of these systems require operators with specific training, experience, and qualifications (in addition to basic ultrasonic qualifications), particularly for the process of data interpretation.

In a typical application for inspection of long lengths of pipe, low-frequency, guided, ultrasonic waves are used to carry out a 100% volumetric inspection. A single point of access to the pipe surface is all that is required to attach the encircling transducer unit. Liquid couplants are not required; the transducer unit relies on clamping pressure. Ultrasound can be transmitted in one or both directions along the pipe, and the wave format is changed by sudden changes in wall thickness due to the presence of flaws. These techniques are most sensitive to an overall reduction in the pipe cross-sectional area, i.e., the detection of general corrosion rather than discrete pitting or isolated cracks.

Guided wave techniques are particularly applicable to the detection of corrosion on internal or external pipe surfaces in situations where access is restricted. A system limitation is that the maximum operating range varies according to pipe diameter, geometry, contents, coatings/insulation, and general condition. In particular, the presence of sound-absorbing coatings or material in contact with the pipe can greatly reduce the operating range.

Internal rotary inspection system (IRIS) is an ultrasonic technique for the NDE of boiler and heat exchanger tubes consisting of a high-frequency ultrasonic immersion probe inside a rotating test head. The system provides coverage of the full circumference and full wall thickness as the probe is scanned axially along the tube. A high standard of cleaning to allow transmission of the ultrasound is required prior to deploying this technique so coverage can be limited on tubes that are difficult to clean properly in situ. It can give useful semi-quantitative data on residual wall thickness, but the need for thorough cleaning tends to restrict its use.

The electronic magnetic transducer (EMAT) system is a development that can be used on boiler and heat exchanger tubes but, unlike IRIS, the EMAT technique specifically requires the presence of a heat-induced oxide layer in order to operate. This scale buildup is normal on boiler-style tubing and the technique eliminates the requirement for cleaning prior to inspection. However, because this oxide layer is required, inspection coverage can be limited where this oxide layer is incomplete.

Time-of-flight diffraction (TOFD) is a further development of the conventional ultrasonic technique. Time-of-flight diffraction is a very sensitive two-probe (sometimes single-probe) technique that works by accurately measuring the arrival time of ultrasound diffracted from the upper and lower extremities of a flaw, most commonly used on welds. Sizing can be accurate as the time difference between the signals obtained from the top and bottom edges is used to assess the size. Generally, TOFD requires two ultrasonic probes acting as transmitter and receiver to be scanned as a pair on either side of a weld. An electronic digital record or hard copy image is produced, but specialist training is required to ensure that technicians have the required competence in applying the technique and in interpreting the results produced.

With TOFD, best results are achieved with skilled operators and specialist equipment and software capable of generating high-resolution images of the component. A number of systems are commercially available. Scanning of the component can be performed in a variety of ways, from manual scanning with encoded positional feedback for simple site applications, to fully automated inspection processes.

Time-of-flight diffraction is ideally suited to:

- Rapid screening of simple weld geometries (probes placed on either side of weld)
- "Fingerprinting" of critical components
- Critical assessment and sizing of flaws
- Monitoring of flaw growth

When used for weld screening, TOFD may not detect unfavorably orientated flaws such as transverse cracks, depending on the orientation of the probes and the level of interpretation. In addition, due to the very high inspection sensitivity utilized with TOFD inspection, small flaws that are not serious can sometimes mimic more serious flaws such as cracks; because of this, characterizations based on TOFD alone should be treated with caution. When accurate flaw characterization is needed, additional scanning using other scan patterns and the pulse-echo technique will often be necessary. The principle of TOFD is outlined in Fig. 2 (Ref 26).

Radiographic Inspection

Conventional Radiographic Inspection. Radiographic inspection is the detection of material loss by the variation in applied radiation (gamma or x-ray), passing through a component and impinging on a film or other radiation detection medium. X-ray equipment ranges from about 20 kV to 20 MV (the higher the voltage, the greater the penetrating power of the radiation and the greater the thickness of component that can be tested). Gamma radiography is carried out using radioactive isotope sources (e.g., cobalt-60, iridium-192, and selenium), although, due to the hard radiation characteristics, its sensitivity is generally less than that achievable by x-ray radiography. It is widely used for fieldwork because of its greater portability.

Because it is sensitive to material loss, radiography is better suited to the detection of volumetric defects such as slag or porosity. Detection of planar defects or cracks will depend on length to depth ratio, the gape or opening of these defects, and the misorientation of the radiation beam from the axis of the defect. This feature needs to be borne in mind, particularly with regard to crack detection. It is often necessary to perform radiography from two or three different positions and orientations to ensure a realistic prospect of detecting cracks in a variety of planes.

Traditionally, radiography has produced a film as hard copy of the inspection results, although more recent techniques do not expose a film but instead record results electronically as a digital image that can be subsequently printed or stored electronically.

Considerable interpretation is required of the resultant film or digital image in order to classify and size any defects. Such interpretation needs to be undertaken by a suitably qualified and experienced radiographic interpreter. Defects are identified by changes in the density of the developed film. The film density is related to the exposure it has received from the radiation. The gradient of the film characteristic curve of density against exposure determines how visible small changes in exposure (i.e., small changes in wall thickness) are. This characteristic of the film is its contrast. This tends to increase with film density, thus high densities are beneficial in the detection of defects, although there are practical limits on the level to which density can be increased because of the reduction in transmitted light intensity.

Image quality indicators (IQIs), commonly comprising straight wires of differing diameters, are placed on the object under test and imaged when the radiograph is taken. The smallest wire diameter that is visible on the radiograph then gives a guide to the sensitivity achieved.

There are important health and safety issues to be dealt with by any company planning to undertake radiographic inspection. Regulations exist that cover the storage and use of radiation sources, shielding and collimation, job planning, contingency for emergency response, training and competence of personnel, work planning, systems of work, and so forth. Procedures covering the full range of the activity from planning, through execution to interpretation and reporting should be developed and followed at every stage.

Recent Developments in Radiography

Recently, a number of systems have been developed that use smaller, highly collimated sources in a controlled manner. This allows radiography to be carried out within a much-reduced controlled area. Typically, the controlled area can be restricted to only 3 m (9.9 ft), compared with tens of meters for conventional radiography. This has the advantage that normal plant work can continue relatively undisturbed during radiography work. In addition, dose rates to classified radiation workers are reduced.

Flash/profile radiography is a useful technique for the detection of external corrosion on pipes under insulation or for the detection of wet insulation that could become a site for corrosion. The technique uses a short radiation exposure time, and the beam is arranged tangentially to the pipe wall so that corrosion of the external wall shows up as a variation in the profile of the pipe. It is normally applied to pipes up to 300 mm (12 in.) outside diameter but can be applied to items with diameters up to 1 m (3.3 ft) given

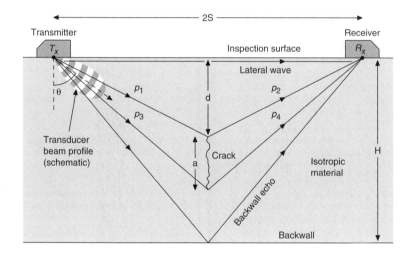

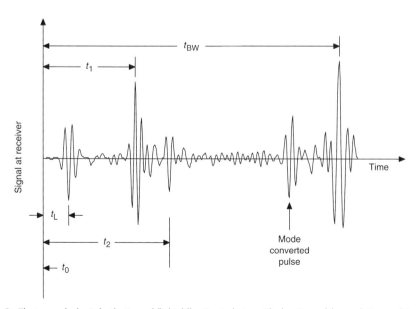

Fig. 2 The two probe basis for the time-of-flight diffraction technique. The locations of the crack tips are determined from the time differences between the lateral wave and the pulses that follow paths $p_1 + p_2$ or $p_3 + p_4$. These paths correspond to t_1 and t_2, respectively, in the lower figure. Source: Ref 26

sufficient source to film distance and radiation output. The technique uses x-ray or gamma-ray equipment with a low radiation exposure time, fast reacting x-ray films in conjunction with intensifying screens, or digital detection media. It saves costs normally attributed to the removal and reinstatement of insulation and associated scaffolding. Contrast and resolution of the image are not as good as that for conventional radiography because of orientation of the radiation beam relative to the component, the limited radiation available, the large grain film, and the relatively large focal spot of the sources.

Recent developments have complemented flash radiography. These involve hand-held radiographic systems using a source such as gadolinium-153 in combination with a solid-state scintillator, which converts the transmitted radiation into electrons. The quality and output of the source determines the maximum length of the beam path in the lagging when looking for CUI. Special gadolinium-153 equipment can allow measurement of pipe wall thickness when passing the radiation through the center of the pipe.

Real-Time Radiography. There have been some recent developments regarding real-time radiographic imaging. One proprietary system is a development of flash radiography and utilizes a hand-held image intensifier coupled to a low-energy gadolinium radiation source. The method of inspection is to move the hand-held unit around the pipe circumference such that 100% coverage of the pipe external surface is obtained. The radioscopic image is viewed on a monitor or helmet-mounted real-time display. It is applicable to piping with thermal insulation to detect the presence of CUI. The radiation is projected through the thermal insulation, at a tangent to the pipe wall, to the image intensifier such that corrosion damage can be observed in profile on the real-time display.

In another development of real-time radiographic imaging, a profiler has been developed that allows wall thickness to be determined. The system consists of an isotope or x-ray source, a transmission radiation source, a microchannel-plate (MCP) x-ray detector, a laptop or palmtop computer, and proprietary software (Fig. 3). The software converts the radiation-intensity output into linear-equivalent thickness data, which are presented on the computer monitor. The output of a collimated gadolinium-153 isotope source is directed to a special scintillator. The scintillator electronics contain the equivalent of a low-level x-ray camera. In turn, the scintillator is coupled to a photomultiplier tube whose electronics are matched to the scintillator output (Ref 27). The profiler can be used as a rapid screening tool to inspect long lengths of insulated pipe and can identify areas of potential concern that can then be targeted for selective insulation removal and closer inspection by another technique such as ultrasonics.

Other Commonly Used Inspection Techniques

Dye penetrant inspection is one of a range of techniques suited only for detection of surface-breaking (rather than subsurface or embedded) defects and is generally used on nonmagnetic materials. This technique requires thorough cleaning and decontamination of the surface to be inspected prior to application of the dye. The dye is left on the surface for a set period to allow the dye to be drawn into any surface-breaking defects by capillary action. The excess dye is then removed from the surface and a developer is applied. The developer draws any dye back out of the defects and shows as a stain against the neutral-colored developer.

Dye penetrant is a low-cost technique that can be used to cover a large area of inspection fairly quickly. The technique is straightforward and practitioners can be readily trained in its use and interpretation. Dye penetrant is good at detecting pits and cracks but may not detect very fine cracking unless the dye is left for a considerable period. Detection is also highly dependent on obtaining a good surface finish because dye can be trapped in surface laps and discontinuities that could be confused with cracks, particularly in confined or restricted geometries. Conversely, excessive force in surface preparation, such as heavy grinding, can result in surface-breaking cracks being smeared over and failing to absorb dye. Fluorescent dyes and ultraviolet lighting can be used to increase the contrast of indications, making them more visible to the operator and hence increasing the sensitivity of the technique.

Magnetic particle inspection is another of the techniques for surface-breaking defects, but it is used on magnetic materials. It is based on the fact that defects on the inspection surface interrupt lines of induced magnetic flux.

This technique is more tolerant of surface finish than the dye penetrant technique, but some surface preparation is still required. Additionally, the component may need to be demagnetized if there is any residual magnetic field within the component. A thin coating of contrast paint may be applied to the component to aid subsequent viewing of defects. The component is then magnetized, either by the application of a permanent magnet or electromagnetic yoke, while magnetic particles either in dry powder or liquid suspension are applied to the surface. If the component is sound, the magnetic flux is predominantly contained within the material; if, however, there is a surface-breaking flaw, the magnetic field is distorted, causing local flux leakage around the flaw. The magnetic particles align themselves with this magnetic flux leakage and any defects can be detected. The particles accumulate at the regions of flux leakage, revealing the flaw as a line of iron particles on the component surface. Fluorescent magnetic inks may be used to increase the contrast of indications, making them more visible to the operator and hence increasing the sensitivity of the technique.

For optimum detection, it is important to have some knowledge of the likely orientation of cracks to ensure that the magnetic field is applied as near to perpendicular to the defects as possible. It is therefore necessary to inform the technicians carrying out the inspection of the types, orientations, and sizes of the flaws being sought along with information on geometric or other features of the component that may produce confusing indications.

Eddy-current inspection is another technique used mainly for detection of surface-breaking defects. When an alternating current is passed through a coil close to a component surface, eddy currents are induced in the component close to the surface and these in turn affect the current in the coil by mutual induction. Surface-breaking flaws and material variations in the component affect the strength of the eddy currents. Therefore, by measuring the resultant electrical changes in the exciting coil, flaws can be detected.

The skin depth, which is a function of the permeability of the material and the frequency, determines the depth of penetration of the eddy currents. In ferromagnetic material the skin depth is very small, and the technique will detect only surface-breaking defects. In nonmagnetic material it provides some subsurface capability and can give some indication of the depth of a defect.

No surface preparation is necessary and protective coatings can be left untouched. It is therefore becoming increasingly popular in applications where magnetic particle inspection would normally be used but where the removal and subsequent need to reinstate protective coatings is problematical, for example, around the splash zone of offshore oil and gas risers.

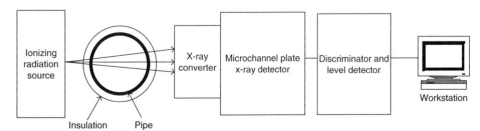

Fig. 3 Components of a real-time radiographic imaging profiler system for measuring pipe wall thicknesses. The monitor displays the signal pulse readout. Source: Ref 27

However, specialist training is required to allow the technician to interpret the signals and spurious indications that can result from local variations in material permeability, particularly at welds.

Acoustic emission inspection is a specialized technique that can be applied to some extent in the detection of corrosion, although its main use is currently in the detection and monitoring of cracks and in leak detection. An array of sensors is strategically placed on the component or structure to be inspected. The sensors "listen" for acoustic emissions, and from this array the relative time of an acoustic event is measured at each sensor from which the location of the cracking or corrosion activity can be determined (Ref 28). Acoustic emission can be used to determine only active features, e.g., cracks that are still propagating rather than any that have been arrested (Fig. 4). It is a highly specialized technique requiring customized equipment for monitoring and also sophisticated computer software for interpretation. When used in an operating plant, the signals from active sites can be difficult to differentiate from background noise, thus great care and skill are required in the use of this potentially powerful technique.

Magnetic flux leakage (MFL) is a technique based on the magnetization of the component being inspected using a strong magnet located inside the test probe. It is therefore applicable only to ferromagnetic materials. As the probe encounters a wall thickness reduction or a sharp discontinuity, the flux distribution varies around that area and is detected either with a Hall-effect transducer or an inductive pickup coil. The amplitude of the signal obtained from any wall loss is proportional to the volume of material that is missing from the region examined. This means that the amplitude does not necessarily correspond to the decrease in thickness of the wall. The technique is not able to discriminate between material loss on the near surface and material loss on the far surface. Magnetic flux leakage is used as a qualitative technique and is unable to give an accurate assessment of the remaining wall but can be useful in identifying locations of corrosion that need to be examined further using other techniques. Unlike magnetic particle inspection, the method is not limited to surface-breaking or near-surface flaws.

Magnetic flux leakage found wide use in the NDE of tank floors because it is quick to apply and can detect material loss on both surfaces of the floor. The requirement for the sensor to be placed between the poles of a magnet means that the technique is unable to give 100% coverage of a floor up to vertical obstructions and side walls. The wall thickness that can be inspected by MFL is limited by the requirement to achieve magnetic saturation.

Pulsed eddy-current inspection is a technique for detecting corrosion and erosion and measuring average-remaining wall thickness in a given sampled area. It is a volumetric technique, therefore, unlike ultrasonic thickness measurement, it measures the amount of material over an area (footprint) rather than the wall thickness at a precise point. It can therefore determine average loss over the footprint area but cannot differentiate between uniform loss and pitting corrosion and may miss fine pitting completely.

A transmitter coil produces a magnetic pulse that induces eddy currents within the component wall. The eddy currents in turn produce a second magnetic pulse, which is detected by the receiving coil. The system monitors the rate of decay of the eddy-current pulse within the steel wall. The average thickness is derived from the comparison of the transient time of certain signal features with signals from known calibration test pieces.

This technique is quick to apply and can test through nonconductive and nonmagnetic material such as insulation, passive fire protection, and concrete layers up to 100 mm (4 in.) thick. It can be confused by adjacent metallic components, thus a skilled and competent technician is required to assess the requirements before this technique is applied. It is suitable only for low-alloy steels and is unable to differentiate defects on the top and bottom surfaces. It can be useful, however, as a quick scanning tool to highlight areas that require further investigation by more sensitive techniques.

Thermography uses an infrared camera or monitor to observe and record the skin temperature, or the variation over an area, of the surface of a plant item. Depending on the imager, variations in surface temperature as small as 0.1 °C (0.2 °F) can be detected.

Inspection by thermography can detect faults in any component where the faults result in a change in surface temperature. Variations in heat transfer through the wall may be attributable to wall thinning, the buildup of scale, or component blockage. It may indicate the presence of wet insulation and the potential conditions for CUI.

The size of defect that can be detected will depend on the optical parameters of the system and the resolution of the camera. In assessing the results, the emissivity of any paints or coatings on the component need to be considered. Reflections of sunlight can also distort readings, so it is best to carry out thermography on an overcast day. Where thermography is used to monitor temperature over a period of weeks or months, care should be taken to ensure that the camera parameters are the same and that the observing conditions are constant.

The technique is noncontacting and only line-of-sight to the surface under examination is required. It is quick and easy to apply but can detect only defects and/or faults that cause a change in heat flow or the surface temperature of the item. It measures only surface temperature, and there should be no misunderstanding that it can in some way "see" defects through the metal wall.

Neutron backscatter is a screening technique that can be used for the inspection of insulated pipe work and vessels to locate areas of wet insulation, which are potential CUI sites. Neutron backscatter devices (hydrodetectors) work by emitting fast (high-energy) neutrons into the insulation from a neutron source. These neutrons are slowed down after collision with hydrogen nuclei in the areas of wet insulation. A detector, sensitive to slow (low-energy) neutrons, then counts the slow neutrons that are backscattered. Low counts per time period mean low moisture while high counts per time period mean high moisture, i.e., an area of wet insulation.

Devices typically consist of a neutron source and detector assembly on the end of a telescopic pole. This allows access to hard-to-reach areas of pipe work and vessels and coverage of large areas or lengths of pipe work can be achieved.

There are distinct limitations of this technique. Given the nature of the neutron source (which by their nature are very penetrative), the backscatter thus created may be resultant from either wet insulation or from the internal process environment if it contains water. The technique therefore cannot be relied on if the primary process environment of the equipment being inspected contains even modest quantities of water.

ACKNOWLEDGMENTS

I would like to acknowledge the help and support of my colleagues Jim Kennedy and Martin Cottam in the preparation of this article.

REFERENCES

1. ASME VIII, "Boiler and Pressure Vessel Code Division I and II," 2004
2. British Standards Institute PD5500, "Specification for Unfired Fusion Welded Pressure Vessels," 2003
3. API 510, "Pressure Vessel Inspection Code: Maintenance Inspection, Rating, Repair, and Alteration," American Petroleum Institute, 1997

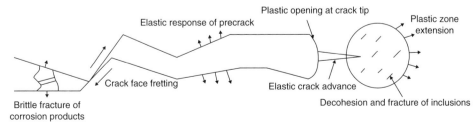

Fig. 4 Schematic representation of a crack tip showing acoustic emission source mechanisms. Source: Ref 28

4. API 570, "Piping Inspection Code: Inspection, Repair, Alteration, and Rerating of In-service Piping Systems," American Petroleum Institute, 1998
5. API RP 572, "Inspection of Pressure Vessels," American Petroleum Institute, 2001
6. API RP 574, "Inspection Practices for Piping System Components," American Petroleum Institute, 1998
7. API RP 575, "Inspection of Atmospheric & Low Pressure Storage Tanks," American Petroleum Institute, 2004
8. API RP 576, "Inspection of Pressure Relieving Devices," American Petroleum Institute, 2000
9. *National Board Inspection Code (NBIC) USA,* National Board of Boiler and Pressure Vessel Inspectors, 2004
10. "Non Invasive Inspection within an Asset Risk Management Strategy," Mitsui Babcock Engineering Ltd. Group Sponsored Project 6748, 1996–1998
11. "Recommended Practice for Non-Invasive Inspection," Mitsui Babcock Engineering Ltd. Group Sponsored Project 235, 1999–2001
12. API 579, "Recommended Practice for Fitness for Service," American Petroleum Institute, 2000
13. *Risk-Based Inspection: Development of Guidelines,* Vol 1, General Document, The American Society of Mechanical Engineers (ASME), CRTD, Vol 20-1, 1991
14. Health and Safety Executive (UK) HSE contract research report 363/2001, "Best Practice for Risk-Based Inspection as a Part of Plant Integrity Management," 2001
15. API RP 580, Risk-Based Inspection, American Petroleum Institute, 2002
16. Health and Safety Executive (UK) HSE research report 016/2002, "Guidelines for Use of Statistics for Analysis of Sample Inspection of Corrosion," 2002
17. BS EN 1714, "Non-Destructive Examination of Welded Joints—Ultrasonic Examination of Welded Joints," 1998
18. BS EN 473, "Non-Destructive Testing—Qualification and Certification of NDT Personnel: General Principles," DIN EN 473, 2000
19. "PANI Programme for the Assessment of NDT in Industry," AEA Technology, Health and Safety Executive (UK) HSE, 1999
20. *Model Code of Safe Practice for the Petroleum Industry, Part 12: Pressure Vessel Systems Examination,* 2nd ed., Institute of Petroleum, 1993
21. *Model Code of Safe Practice for the Petroleum Industry, Part 13: Pressure Piping Systems Examination,* 2nd ed., Institute of Petroleum, 1993
22. "Review of Corrosion Management for Offshore Oil and Gas Processing," Health and Safety Executive (UK) HSE Offshore Technology Report 2001/044, 2001
23. BS 7910, "Guide on Methods for Assessing the Acceptability of Flaws in Metallic Structures," 1999
24. M. Kowaka, Introduction to Life Prediction of Industrial Plant Materials: Application of Extreme Value Statistical Method for Corrosion Analysis, Allerton Press Inc., 1994
25. *Nondestructive Evaluation and Quality Control,* Vol 17, *ASM Handbook,* ASM International, 1989
26. J.P. Charlesworth and J.A.G. Temple, *Engineering Applications of Ultrasonic Time-of-Flight Diffraction,* 2nd ed., Research Studies Press Ltd., United Kingdom, 2001
27. New Portable Pipe Wall Thickness Measuring Technique, *15th World Conference on Non-Destructive Testing* (Rome, Italy), Oct 15–21, 2000, Joseph E. Pascente (LIXI Inc.)
28. C.B. Scruby, "An Introduction to Acoustic Emission," Materials Physics and Metallurgy Division, Harwell, UK, Report AERE-R-1986

SELECTED REFERENCES

- *A Guide to the Pressure Testing of In-Service Pressurized Equipment,* Engineering Equipment Manufacturers and Users Association, EEMUA 168, 1991
- API RP 573, "Inspection of Fired Boilers and Heaters," American Petroleum Institute, 2002
- API Standard 653, "Tank Inspection, Repair, Alteration, and Reconstruction," American Petroleum Institute, 2001
- API RP 750, "Management of Process Hazards," American Petroleum Institute, 1990 (out of print)
- ASME II, "Boiler and Pressure Vessel Code, Material Properties," 2004
- Australasian Standard AS/NZS 3788, "Pressure Equipment In-Service Inspection," 2001
- BS EN 571-1, "Non-Destructive Testing—Penetrant Testing: General Principles," 1997
- BS EN 1290, "Non-Destructive Examination of Welds—Magnetic Particle Examination of Welds: Method," 1998
- BS EN 1435, "Non-Destructive Testing of Welds—Radiographic Testing of Welded Joints," 1997
- BS EN ISO 11120, "Transportable Gas Cylinders," 1999
- BS 5169, "Specification for Fusion Welded Steel Air Receivers," 1992
- BS EN 13445, "Unfired Pressure Vessels," 2002
- BS EN 13480, "Metallic Industrial Piping," 2002
- BS EN 12953, "Shell Boilers," 2002
- Center for Chemical Process Safety (CCPS), "Plant Guidelines for Technical Management of Chemical Process Safety," American Institute of Chemical Engineers (AIChE), 1995
- Confédération Européenné des Organismes de Contrôle technique, d'inspection, de certification et de prévention (CEOC), R47/CEOC/CP83Def, "Periodicity of Inspections of Boilers and Pressure Vessels," 1983
- *Recommendations for the Training, Development and Competency Assessment of Inspection Personnel,* Engineering Equipment Manufacturers and Users Association, EEMUA 193, 2000
- Health and Safety Executive (UK) HSE Offshore Technology Report 2000/095, "Reliability Assessment for Containers of Hazardous Material (RACH)," 2000
- Health and Safety Executive (UK) HSE, "Best Practice for the Procurement and Conduct of Non-Destructive Testing, Part 1: Manual Ultrasonic Inspection," 2000
- Health and Safety Executive (UK) HSE, "Best Practice for the Procurement and Conduct of Non-Destructive Testing, Part 2: Magnetic Particle and Dye Penetrant Inspection," 2002
- Health and Safety Executive (UK) HSE GS4, "Health and Safety Executive Guidance Note: Safety in Pressure Testing," 1998
- ISO/IEC 17020:1998 (formerly BS EN 45004:1995), "General Criteria for the Operation of Various Types of Bodies Performing Inspection," International Standards Organization, 1998
- ISO 9712, "Non-Destructive Testing—Qualification and Certification of Personnel," International Standards Organization, 1999
- "Periodic Inspection of Vessels from the Outside Only," Mitsui Babcock Engineering Ltd. Group Sponsored Project 6490, 1995
- Marine Technology Directorate MTD Publication 99/100 "Guidelines for the Avoidance of Vibration Induced Fatigue in Process Pipework," 1999
- NACE RP 01-70-2004, "Protection of Austenitic Stainless Steels and Other Austenitic Alloys from Polythionic Acid Stress Corrosion Cracking during Shutdown of Refinery Equipment," 2004
- NACE MR 01-75/ISO 15156, "Petroleum and Natural Gas Industries—Materials for Use in H_2S-Containing Environments in Oil and Gas Production," 2001
- NACE RP 01-98-2004, "The Control of Corrosion under Thermal Insulation and Fireproofing Materials—A Systems Approach," 2004
- Personnel Certification Scheme in Non-destructive Examination, PCN/GEN/2000, "General Requirements for Qualification and Certification of Personnel Engaged in Non-Destructive Testing," British Institute of Non-Destructive Testing, 2000
- Safety Assessment Federation, *Guidelines for the Production of Written Schemes of Examination and the Examination of Pressure Vessels Incorporating Openings to Facilitate Ready Internal Access,* Ref: PSG4, April 2003

- Safety Assessment Federation, *Pressure Systems: Guidelines on Periodicity of Examinations,* Ref: PSG1, May 1997
- Safety Assessment Federation, *Shell Boilers: Guidelines for the Examination of Shell-to-Endplate and Furnace-to-Endplate Welded Joints,* Ref: SBG1, April 1997
- Safety Assessment Federation, *Shell Boilers: Guidelines for the Examination of Longitudinal Seams of Shell Boilers,* Ref: SBG2, May 1998
- Statutory Instrument (SI) 2000:No. 128, "The Pressure Systems Safety Regulations (PSSR), Including Associated Guidance and Approved Code of Practice," 2000
- Statutory Instrument (SI) 2000:No. 1426, "The Transportable Pressure Vessels Regulations," 2001
- Statutory Instrument (SI) 1999:No. 2001, "The Pressure Equipment Regulations," 1999
- *Users' Guide to the Inspection, Maintenance and Repair of Above-Ground Vertical Cylindrical Steel Storage Tanks,* 3rd ed., Engineering Equipment Manufacturers and Users Association EEMUA 159, 2003

Corrosion of Structures

John E. Slater, Invetech Inc.

THE PREVENTION of metallic corrosion in structures, particularly the consideration of its effects in initial design and fabrication or later during in-service inspection and retrofit, is vital in ensuring the expected longevity of the structure. Indeed, in many cases, structures may have to last well beyond their originally anticipated lifetimes. For example, the lives of structures in Europe that contain metal extend in some cases back many hundreds of years.

When considering the ramifications of corrosion and its prevention in different types of structures, it is useful to group such structures into various categories. Thus, the modern high rise—whether an office building, an apartment building, a condominium, or a special-purpose building such as a hospital—poses particular problems related to the corrosion of major structural components (which may be either totally metallic or may contain metallic material). This creates the necessity for corrosion considerations in any connector assembly used to tie the curtain-wall system on the building back to the structural frame or to the backup wall system. In low-rise structures, similar concerns exist, but may be less critical when the structure is only a few stories high.

Under such circumstances, the possibility of danger to people and property resulting from failure of, for example, curtain-wall tie systems may be less. Nevertheless, failure of metallic materials within the structure may lead to unsightliness or to possible lack of weathertight behavior. Both may require extensive reworking of the structure to reestablish adequate building performance.

Parking structures are frequently of conventionally reinforced concrete or prestressed/posttensioned construction. Where such structures are in the snow-belt areas of the country or where chlorides may intrude because of the proximity of marine environments, the reinforcement may be at risk from corrosion. Stadiums are another example in which either steel-frame or concrete-frame designs are used. In these cases, certain approaches may be needed to ensure structural integrity, particularly in view of the safety of the many thousands of people who visit the structure and fill it to capacity.

Bridges have received much attention with regard to corrosion. This has been particularly prevalent in the snow-belt states, where deicing salt application has led to significant premature deterioration of decks and supporting structures. Supporting structures also suffer damage because of saline water intrusion, particularly in the splash zone of reinforced concrete structures. The use of weathering steels can be a problem in bridges, especially where design or construction practices do not adequately account for the particular limitations of this type of steel.

Other specialized structures in which corrosion conditions may be a concern include sewage plants, which may have isolated situations related to contaminated water and the use of treatment chemicals. Buildings or structures can be used to house or support chemical plants and/or utility plants. In the latter case, specific aspects of corrosion prevention in nuclear plants, especially containment buildings, have been the subject of critical review. Because many nuclear containment vessels are designed with posttensioned concrete, the longevity of any corrosion-protection system applied to the posttensioning tendons is critical. This article does not address specific interactions with manufactured industrial chemicals, because this area is more reasonably the province of general or specific knowledge related to the particular chemical to which the building material will be exposed.

Metal/Environment Interactions

The most common metallic material used in structures is low- or medium-carbon steel. Specialized or strengthened materials, including heat treated steels, stainless steels, and nonferrous metals, are sometimes employed; use of these materials is the exception rather than the rule (they are discussed when appropriate in this article). The specific interactions that are considered are the reactions of steel with the atmosphere in all of its forms, including polluted atmospheres, interior atmospheres, and internal atmospheres (for example, in a cavity wall). The major criteria determining the performance of unprotected metal include the corrosivity of the atmosphere, the temperature, and the time of wetness of the metal.

Considerable effort has been expended in determining the reaction of steel with cementitious materials, which include mortar, concrete, and their variants. The chemical that typically controls corrosion behavior under these circumstances is the calcium hydroxide ($Ca(OH)_2$) introduced into the cementitious material by the portland cement or, in the case of mortars, lime. The reaction of $Ca(OH)_2$ with atmospheric carbon dioxide (CO_2) is an important degradation mechanism that radically changes the behavior of steel in contact with or embedded in the cementitious material. The other major factor that influences the corrosion of steel is the introduction of chloride ion (Cl^-) into the cementitious material. Chloride may play a decisive role in causing the cementitious material to change from a protective to a nonprotective environment with regard to embedded steel.

This article discusses the generic situation of steel reacting with the environments found in structures. Two environments are specifically discussed: atmospheric and cementitious environments. The utility of different corrosion-protection methods is described. Examples of problems that have arisen in the corrosion performance of steel are described.

General Considerations in the Corrosion of Structures

Corrosion of Steel in the Atmosphere. Steel corrosion in the atmosphere is typically a function of temperature, humidity, and the presence in the atmosphere of components that increase the corrosivity of the environment (see, for example, the article "Atmospheric Corrosion," *ASM Handbook,* Volume 13A, 2003). The temperature/humidity ratio is particularly important, because the interaction among these factors leads to a function known as the time of wetness, during which a film of liquid is present on the surface of the steel. Corrosion occurs by an aqueous corrosion mechanism during the time the film is present on the steel surface. The thinner the film, the easier is the diffusion of oxygen through the film to drive the corrosion reaction.

The presence of agents in the atmosphere that can dissolve in the liquid film and promote or inhibit its production by changing the dew point can markedly influence the corrosion behavior of the steel. Pollutants in the atmosphere, particularly sulfur dioxide (SO_2) and related compounds that can lead to acid rain, influence the corrosion behavior of the steel by acting in several ways, such as:

- An increase in conductivity, and therefore the corrosivity, of the liquid film
- Changes in the relative humidity at which the film forms
- Alterations in the corrosion product films that may form on the steel

Therefore, the corrosion behavior of steel used in structures can be expected to be a function of the geographical location, which in turn will determine the temperature, humidity, and degree of pollution that exists. This behavior is indicated in Table 1.

Corrosion in marine environments is a specialized subset of atmospheric corrosion because of the influence of wind-blown or particulate sea salt that may contact exposed steel. Sea salt is particularly aggressive to steel, possibly because of the concentration of magnesium chloride ($MgCl_2$) it contains. Sodium chloride (NaCl), the major constituent of sea salt, is deleterious to the corrosion behavior of steel because of its effect on the conductivity of the liquid film and its destruction of protective corrosion product. On the other hand, $MgCl_2$ acts to acidify the liquid film and, by its deliquescent action, to increase the time of wetness. The corrosion rates of structures in marine environments are considerably higher than in any other type of environment. The dramatic increase in the corrosion rates of steel in marine locations is indicated in Table 1. Additional information can also be found in the articles "Corrosion in Seawater" and "Corrosion in Marine Atmospheres" in this Volume.

Specialized environments exist in which pollutants are present as a function of the particular use or location of the structure. Chemical plant and refinery buildings and structures may be vulnerable to atmospheric pollutants produced within the plant and may require protective measures. This type of damage is beyond the scope of the present article, and the reaction of steel to such environments is best determined by noting the effect of the pollutant on the corrosion of steel (Ref 2).

The atmosphere within closed structures will usually be different from that for external surfaces. In general, the presence of pollutants may be significantly reduced, and the control of temperature and humidity may mean the overall environmental corrosivity is significantly lower than that for outside surfaces. However, there can be circumstances that alter these effects. For example, the well-known barrier effect of internal walls, which is related to their containment of the internal environment, may be breached at certain locations. The result may be that air escapes from the interior of the building into the cavity. If structural or other steel is present in this cavity, the interaction of the steel with air, which may be deliberately humidified, can lead to condensation and subsequent corrosion problems that would otherwise be unanticipated.

Such problems have been noted in the past, both for the structural steel framing and also for elements in the system that tie the external cladding back to the frame of the building. Therefore, concern is necessary not only for the influence of the external environment on the corrosion of steel, but also for the effects of the internal environment.

Fireproofing is often mandated by fire codes for structural steelwork within buildings. In the past, fireproofing was used that contained corrosive constituents. Under circumstances in which moisture levels at the structural steel may increase (because of exfiltration of humid air or infiltration of moisture), the corrosive constituents in the fireproofing can lead to unacceptable corrosion of the steel. This problem is addressed in ASTM E 937, which is a standardized test for determining the corrosivity of fireproofing materials under standardized conditions (Ref 3). The use of coatings that meet conditions related to this test is required by many codes and should be mandated by the architect or engineer.

Whenever steel is used in buildings, there is always the possibility for galvanic corrosion to occur. Galvanized conduit may contact bare steel. Copper-base alloys are used as wiring, for example, in lightning conduction rods. In general, the relative positions of metals in the galvanic series for seawater will normally dictate whether or not there is a potentially significant problem for galvanic corrosion from structural elements in contact. The environment surrounding the galvanic couple will have a significant effect; more severely corrosive environments allow more galvanic activity to occur. Therefore, the impact of galvanic corrosion must be an ever-present concern.

The use of lower-strength steels generally indicates that environmental/mechanical interactions leading to unexpected brittle failure are unlikely to occur in normal structural environments. The use of higher-strength steels (particularly those of the quenched-and-tempered variety) may increase the likelihood of stress-corrosion or hydrogen embrittlement failures, especially in more aggressive environments such as marine conditions. Compounding the problem is the welding that may be performed, for example, on bridge structures. Therefore, quenched-and-tempered steels of the ASTM A 709, grade 100, type, which have yield strengths in the 690 MPa (100 ksi) range, must be welded carefully so as not to create hard regions in heat-affected zones that are susceptible to hydrogen embrittlement.

Similar considerations pertain to the area of embedded posttensioning steels. These steels are typically of the cold-drawn type (not quenched and tempered), corresponding to such standards as A 416, grade 1860 (1860 MPa, or 270 ksi,

Table 1 Corrosion rates of carbon steel calibrating specimens at various locations

Location	Type of environment	Corrosion rate(a) µm/yr	mils/yr
Norman Wells, NWT, Canada	Polar	0.76	0.03
Phoenix, AZ	Rural arid	4.6	0.18
Esquimalt, Vancouver Island BC, Canada	Rural marine	13	0.5
Detroit, MI	Industrial	14.5	0.57
Fort Amidor Pier, Panama, CZ	Marine	14.5	0.57
Morenci, MI	Urban	19.5	0.77
Potter County, PA	Rural	20	0.8
Waterbury, CT	Industrial	22.8	0.89
State College, PA	Rural	23	0.9
Montreal, Que. Canada	Urban	23	0.9
Durham, NH	Rural	28	1.1
Middletown, OH	Semi-industrial	28	1.1
Pittsburgh, PA	Industrial	30	1.2
Columbus, OH	Industrial	33	1.3
Trial, BC, Canada	Industrial	33	1.3
Cleveland, OH	Industrial	38	1.5
Bethlehem, PA	Industrial	38	1.5
London, Battersea, England	Industrial	46	1.8
Monroeville, PA	Semi-industrial	48	1.9
Newark, NJ	Industrial	51	2.0
Manila, Philippine Islands	Tropical marine	51	2.0
Limon Bay, Panama, CZ	Tropical marine	61	2.4
Bayonne, NJ	Industrial	79	3.1
East Chicago, IN	Industrial	84	3.3
Brazos River, TX	Industrial marine	94	3.7
Cape Canaveral, FL (18-m, or 60-ft, elevation, 55 m, or 60 yd, from ocean)	Marine	132	5.2
Kure Beach, NC (250 m, or 800 ft, from ocean)	Marine	147	5.8
Cape Canaveral, FL (9-m, or 30-ft, elevation, 55 m, or 60 yd, from ocean)	Marine	165	6.5
Daytona Beach, FL	Marine	295	11.6
Cape Canaveral, FL (ground level, 55 m, or 60 yd, from ocean)	Marine	442	17.4
Point Reyes, CA	Marine	500	19.7
Kure Beach, NC (25 m, or 80 ft, from ocean)	Marine	533	21.0
Galeta Point Beach, Panama, CZ	Marine	686	27.0
Cape Canaveral, FL (beach)	Marine	1070	42.0

Source: Ref 1

minimum ultimate tensile strength for seven-wire strand), and A 421 (1655 MPa, or 240 ksi, minimum ultimate tensile strength for 6.4 mm, or 1/4 in., diam, uncoated wire). Although laboratory test data indicate that cold-drawn steels are considerably less susceptible to failure by hydrogen embrittlement than their quenched-and-tempered counterparts, care must be taken to minimize the aqueous and/or atmospheric corrosion of such steels before and after concrete placement.

The atmospheric corrosion of other nonferrous metallic materials commonly used in structures is of less concern. Aluminum can be used for a variety of building components, including windows and door frames. If the aluminum is protected from direct contact with uncured mortar or concrete and if anodized coatings are specified, little atmospheric corrosion is noted except under severe marine conditions. Zinc is typically used only as a protective coating for steel and is covered in the section "Protection Methods for Atmospheric Corrosion" in this article.

Steel in Cementitious Materials. The behavior of steel in contact with cementitious materials (concrete and mortar) is generally governed by the properties of the portland cement that is a constituent of the mortar or concrete. Portland cement is an alkaline material; the alkalinity results from the presence of $Ca(OH)_2$ and other soluble alkali salts. The pH of a saturated $Ca(OH)_2$ solution is approximately 12.5. Reference to the Pourbaix (potential-pH) diagram for iron (Fig. 1) shows that, under these circumstances and in the presence of moisture and oxygen, the steel will be in a passive condition because of the formation of a thin film of oxide—generally considered to be γ FeOOH (see also the article "Potential versus pH (Pourbaix) Diagrams," *ASM Handbook,* Volume 13A).

The effect of pH on the corrosion rate of steel in aerated water is shown in Fig. 2. If the oxygen is depleted or if the cementitious material is allowed to dry significantly, then the passive film may well be disrupted. However, the corrosion rate is still expected to be low because of the high resistivity of the concrete and low driving force for the corrosion reaction. However, there are pollutants in the cementitious environment that can modify this behavior, leading to significant corrosion of steel.

Concrete is a hard, dense material that consists of cement paste surrounding aggregate particles. The major difference between mortar and concrete is in the size of the aggregate. Concrete contains a high proportion of large aggregate, such as gravel and crushed rock. Steel that is embedded in concrete, generally as reinforcement, can be of two main types. In conventional reinforced concrete, reinforcing bars are used to support the tensile loading, because of the relative weakness of unreinforced concrete in tension versus its significant strength in compression. Reinforcement is sized and placed according to requirements determined by the structural engineer.

Other types of reinforcement in concrete are deliberately stressed in tension to place the concrete into residual compression. For prestressed concrete, this is typically accomplished by placing the stressed steel into the concrete form and pouring the concrete around the steel. Once the concrete has cured, then the external tension on the steel is released. The bond between the steel and the concrete now places the surrounding concrete into compression, leading to the well-known concept of prefabricated, prestressed concrete.

This same effect can be achieved by post-tensioning, in which the unstressed steel is placed in the concrete form before the concrete is poured. The steel is typically prevented from developing a bond with the concrete during the pouring process. After pouring and curing of the concrete, the steel embedded in (but separated from) the concrete is tensioned and the tensioned steel is anchored to the ends of the concrete structure. In this way, the concrete around the steel is placed into tension.

Other steel components that may be embedded in concrete include water lines (which are typically galvanized steel) and the forms used as containment during the concrete pouring process. In many cases, the forms are removed; in others, the forms remain in place. As such, the forms are exposed to the atmosphere on their outer surface and to the concrete environment on their inner

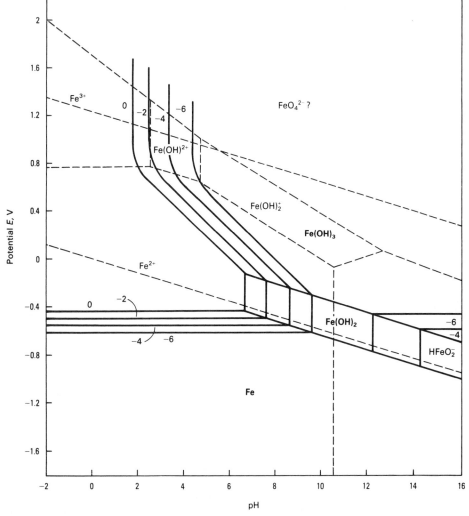

Fig. 1 Pourbaix (potential-pH) diagram for the system iron-water at 25 °C (75 °F). Source: Ref 4

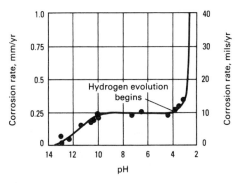

Fig. 2 Effect of pH on the corrosion rate of iron in aerated soft water at room temperature

surface. Problems can occur during the use of such forms, particularly where additives to the concrete have caused a corrosive situation to occur.

Unanticipated contact may occur between steel embedded in concrete and other metals used in construction. Aluminum can be embedded in or located on concrete, for example, balustrades and lamp posts. If contact can occur between embedded steel and aluminum in corrosive environments, accelerated corrosion of the aluminum can occur, with corresponding galvanic protection of the steel. Although it is conceivable that copper water pipes could contact embedded steel reinforcement, there are no instances of such problems occurring.

The use of mortar is a necessary factor in the construction of brick masonry, and steel is used in such masonry in a variety of applications. The use of steel joint reinforcement to enhance the sit strength of the masonry by placing a mesh or ladder-type system within the bed joint is a common occurrence. Such a system can also be used to connect two wythes of masonry. Steel ties or anchors are used to anchor masonry curtain-wall systems back to the building frame or to a supporting structure. Typical reinforcements, ties, and anchors are shown in Fig. 3. Although bare steel would suffice in uncontaminated high-pH mortar, it is common practice to use a more corrosion-resistant material. Typically, galvanized steel is recommended and used, either hot-dip or electroplated.

Other metallic components that may contact mortar include window frames and doorways. These may be steel or aluminum. Such steel components can be protected against atmospheric corrosion with organic coatings, but in other cases metallic coatings such as galvanizing can be used. Aluminum components can corrode in the uncured or green mortar environment. In the absence of further corrosion enhancers, the corrosion rate normally drops to a very low level once curing has occurred and the mortar dries.

The corrosion behavior of steel in concrete and mortar as a function of pH can be drastically altered by two factors. The first is carbonation or the reaction of the $Ca(OH)_2$, with CO_2 from the air. This reaction to form calcium carbonate ($CaCO_3$) lowers the pH of the cementitious material; the steel may no longer be passive, but can become active and corrode significantly. Carbonation is a function of the porosity of the concrete or mortar, which is a function of such factors as the water/cement ratio, placement adequacy, and vibration/consolidation. In general, mortar is much more permeable than concrete and carbonates much more rapidly. Under these circumstances, the mortar in masonry structures must be considered to undergo carbonation at a relatively early point in its life. To prevent the corrosion of embedded steel due to carbonation, it is common practice to use protective coatings on steels (galvanizing is common) or to use inherently corrosion-resistant materials, such as stainless steels.

The second major contributor to the corrosion of embedded steel in cementitious material is the influence of chloride. Chloride in sufficient quantities prevents the formation of the initial protective oxide film on steel (if the chloride is present in the material mix during the curing process) or, if added later, breaks down the passive film and allows corrosion to proceed. Possible chloride sources in the mixing process are the aggregate, chloride introduced with the mixing water (a problem frequently encountered in the Middle East, where brackish water is often the only water available), and chloride-containing admixtures deliberately added to the mix as a set accelerant or so-called antifreeze (most commonly calcium chloride, $CaCl_2$). Chloride can also find its way into the cementitious material after placement and curing, because of the application of chloride to the outer surface. Sources are marine environments, deicing salt placed onto concrete bridge decks, parking structures, and hydrochloric acid washing solutions applied to masonry surface.

The quantity of chloride necessary to cause corrosion of bare steel in uncarbonated cementitious systems depends initially on the stage at which the chloride is introduced. For bare steel in initially chloride-free concrete that is subjected to chloride infiltration from the external surface, corrosion has been found to initiate on the steel when the chloride level reaches between 0.02 and 0.04% by weight of concrete at the concrete/steel interface. When the chloride is present in the mix, the tricalcium aluminate of the concrete can insolubilize a certain portion of the chloride during the curing process. Recognition of this fact initially led to the allowance of additions of 2% $CaCl_2$ (by weight of cement) to concrete to accelerate set. More recent work, together with the release of bound chloride as concrete ages, has led to a recognition that maximum allowable chlorides in concrete should be a function of both the nature of the steel and the expected service

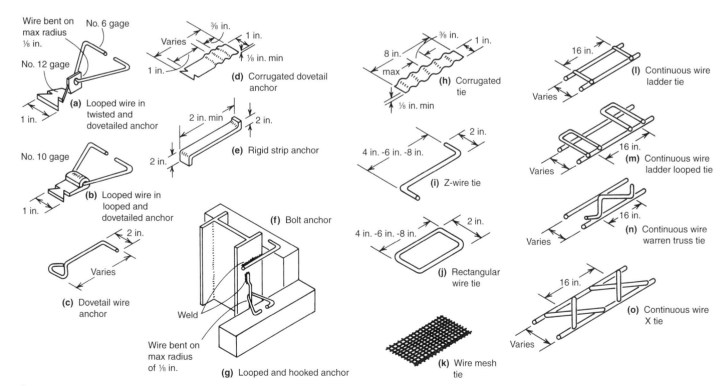

Fig. 3 Typical metallic anchors (a) to (g) and ties (h) to (o) used in masonry walls

environment of the concrete. The recommendations of American Concrete Institute Committee 201 (Durability of Concrete) are given in Table 2. See the article "Environmental Performance of Concrete," *ASM Handbook,* Volume 13B, 2005.

A compounding factor in the chloride-induced corrosion of reinforcement is the formation of macrocells. Because of the electrically continuous and extensive nature of the typical reinforcement in concrete, differences in chloride concentration and oxygen level at different locations on the steel can occur. Under these conditions, separated anodic (high-chloride, low-oxygen) and cathodic (low-chloride, high-oxygen) areas can easily develop along the reinforcement, particularly where chloride enters the concrete from an exterior surface. This macrocell action frequently leads to severe corrosion at the anodic areas.

Macrocell action can be worsened unintentionally and substantially during concrete repair. The replacement of deteriorated, spalled, chloride-contaminated concrete adjacent to reinforcing steel with fresh, chloride-free concrete can introduce a potent cathode into the system. The result can be accelerated corrosion of the surrounding reinforcement, which can lead to the spall around a spall deterioration phenomenon.

The effect of corrosion on steel-reinforced structures is somewhat dependent on the nature of the steel. The major problem arising from an increased corrosion rate is the production of a voluminous oxide, which occupies a greater volume than the steel from which it was produced. Under these circumstances, significant tensile stresses can be introduced into the surrounding concrete material, leading to cracking and eventually spalling. Calculation of the Pilling-Bedworth volume ratio allows the magnitude of the volume expansion to be assessed, with values typically found between 2 and 4, depending on the exact nature of the corrosion product. The Pilling-Bedworth ratio is discussed in several articles in *ASM Handbook,* Volume 13A, including "Thermodynamics of Gaseous Corrosion," "Kinetics of Gaseous Corrosion Processes," and "Gaseous Corrosion Mechanisms."

The effect on high-strength prestressing steel may not be the same. Such steel typically has an ultimate tensile strength exceeding 1380 MPa (200 ksi) that is produced by cold forming. Steels of this hardness may be susceptible to stress-corrosion cracking or, more probably, hydrogen embrittlement. In this context, corrosion processes that produce hydrogen can be very deleterious to structures containing the post-tensioning or prestressing. This is due to the possibility of hydrogen embrittlement leading to brittle fracture of the steel and loss of compression of the surrounding concrete.

Protection Methods for Atmospheric Corrosion

Various methods are available for protecting steel against corrosion in atmospheric conditions. The protection system is generally a barrier coating or a metal alloying of the steel that effectively introduces a barrier coating by a normal corrosion process. As discussed in this section, design considerations are also critical.

The barrier coatings used to protect steel from atmospheric corrosion are of three main types: organic coatings, inorganic coatings, and metallic coatings. The selection of one of these types of coatings is a function of the expected environmental severity in which the structure is to perform, the corrosivity of that environment, the design life of the structure, and the possibility of further maintenance coating. In other words, coating selection is a combination of both technical and economic considerations.

Design. One of the most important aspects of corrosion protection is consideration of good design practice. Such considerations include:

- Avoidance of upturned angles, channels, and so on, that can collect moisture
- Avoidance of pockets within welded structures
- Grinding welds flush
- Elimination of crevices that can lead to accelerated corrosion

Examples of details to avoid, as well as more corrosion-resistant details, are shown in Fig. 4. Although some of these problems can be mitigated by other corrosion-protection methods, the selection of weathering steels makes adherence to good design and fabrication detailing mandatory (see the discussion of weathering steels later in this section). Refer also to the article "Designing to Minimize Corrosion," *ASM Handbook,* Volume 13A, for additional information on this topic.

Organic coatings are perhaps the most frequently used coatings for protecting steel from atmospheric corrosion. This is true with regard to bulk structural steel. However, components such as fasteners rarely have organic coatings. This is because of difficulties related to the fit of coated fastener systems and the poor durability of such coatings under abrasion and tightening conditions. It is generally recognized that the quality of the surface preparation is a major factor in the service life of the coating. Perhaps the most extensive evaluation of paint behavior as a function of substrate preparation has been conducted by The Society for Protective Coatings (SSPC), which has issued standards on different degrees of cleaning prior to the application of paints. Uses and applicable SSPC and NACE International standards for various surface preparation techniques are summarized in the article "Paint Systems," *ASM Handbook,* Volume 13A (see, in particular, Table 1 on page 838).

Table 2 Recommended maximum water-soluble chloride levels in concrete mix (prior to service) for different reinforced concrete exposures

Type of exposure	Maximum recommended chloride level, wt% of concrete
Prestressed concrete	0.06
Conventionally reinforced concrete in a moist environment and exposed to chloride	0.10
Conventionally reinforced concrete in a moist environment but not exposed to chloride (includes locations where the concrete will be occasionally wetted, such as parking garages, waterfront structures, and areas with potential moisture condensation)	0.15
Aboveground building construction where the concrete will stay dry	No limit

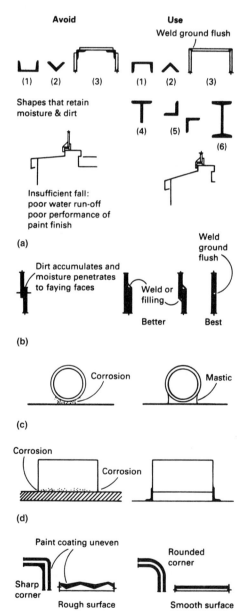

Fig. 4 Design and fabrication details to be considered in corrosion prevention. (a) Constructional members, sills, etc. (b) Joints. (c) Crevices. (d) Air circulation. (e) Corners, edges, and surfaces

The degree of cleanliness depends on the sensitivity of the paint system to the level of surface cleanliness and on the required longevity of the system. Different methods of surface preparation are available, including centrifugal blast cleaning, abrasive air-blast cleaning, water-blast cleaning, and hand- and power-tool cleaning. Hand-tool and power-tool cleaning are typically employed only where small areas must be cleaned of preexisting scale and paint. It is also important to recognize that contaminants on steel surfaces that are not visible to the naked eye may influence the longevity of the subsequent coating. Inorganic salts on the coating surface may dissolve in water that permeates through the coating, leading to corrosion of the substrate. Standard blast-cleaning techniques typically do not remove such contaminants, but water-blast cleaning does. Again, this becomes a matter of economics and the sensitivity of the coating system to surface contamination.

The selection of a coating system is influenced by consideration of the likely corrosive environment and the consequences of coating failure. Figure 5 shows some typical preparation and coating formulations. The need for increased durability of the coating increases with the corrosivity of the environment. In this regard, much steel is enclosed within building walls, and such steel is frequently only lightly shop primed. Indeed, it has been questioned whether such coating is in fact necessary, based on the environment in which the steel is to perform.

One of the problems in making decisions of this type is that the severity of the environment may not be adequately documented. If there is exfiltration from the building, then condensation on the steel members can occur, resulting in an extended time of wetness and a corrosion rate that is significantly greater than might otherwise be anticipated. Under these circumstances, the thin shop primer may be much less than adequate, and a more rigorous corrosion-prevention system should be considered. Adverse effects on such steel may include the pickup of corrosive constituents (particularly chloride) from other building materials—for example, from mortar during passage of water through masonry walls. If flashing systems are inadequate, then such water may reach structural steel and cause significant corrosion damage. This is another factor that needs to be considered whenever the so-called benign environment within the walls of a building is considered.

Inorganic Coatings. Other forms of coating that are midway between the use of a metallic coating and an organic coating are the inorganic zinc-rich coatings. The life expectancy of this type of material in severe weathering service has not been established, but complete protection of the steel substrate is still provided after 20 years of exposure. Protection in this case is largely the result of the good corrosion resistance of zinc itself. Indeed, it has been pointed out that the cured inorganic zinc-silicate films can be thought of as a cross between hot-dip galvanizing and a fused ceramic. Because of the absence of organics, inorganic coatings are considered to have the best solvent resistance of any type of protective coating. This makes them very useful for structures in which vapors are present such as chemical plants, petroleum refineries, and production facilities. Further information can be found in the article "Zinc-Rich Coatings," *ASM Handbook,* Volume 13A.

The metallic coating system most commonly used for steel in structures is a zinc coating. These coatings can be applied by hot-dip galvanizing or by electroplating for smaller components, such as bolts. Electroplating lays down a considerably thinner zinc layer than hot-dip galvanizing and the service life is shorter. The service life of galvanized coatings is a function of the coating thickness and the environment in which it will operate. This is illustrated schematically in Fig. 6. Therefore, the thickness of a galvanized coating on structural steel will depend on an assessment of the nature of the environment and the required longevity. In this regard, the severity of the environment within a cavity wall (where zinc coatings are frequently used) is a significant concern. Zinc coatings are described in the articles "Continuous Hot Dip Coatings" and "Batch Process Hot Dip Galvanizing," *ASM Handbook,* Volume 13A. Electroplated zinc products are reviewed in the article "Continuous Electrodeposited Coatings for Steel Strip" in *Surface Engineering,* Volume 5, *ASM Handbook* (1994).

Steel roofing is frequently protected by a 55Al-43.4Zn-1.6Si hot-dipped coating known by the tradename "Galvalume." This coating can be used either as a basis for paint or as a bare protective coating in its own right. The ASTM standard governing composition and thicknesses of this coating is A 792. Compared to a galvanized coating of the same thickness, Galvalume coatings offer a factor of between two and six times greater atmospheric corrosion resistance, depending on the nature of the exposure. The protection Galvalume offers the base steel appears to be a combination of both galvanic and barrier effects.

For small components, other protective systems are available. These include cadmium plating, which is considered to be more corrosion-resistant than zinc in marine environments. Use of cadmium plating is restricted because of the toxicity of the plating solutions and the difficulty of disposal (see the article "Cadmium Elimination in Surface Engineering" in *Surface Engineering,* Volume 5, *ASM Handbook,* 1994). Nevertheless, the general corrosion resistance of cadmium plate makes the process worthy of consideration where allowed.

Weathering Steels. An excellent method of protection against atmospheric corrosion is the use of weathering steels. In this class of materials (typified by ASTM specifications A 588 for buildings and A 709 for bridges), small amounts of alloying elements—typically nickel, chromium, and copper—are added to the steel (Ref 6, 7). Under certain fairly specific situations, these alloying elements are incorporated into the oxide layer that forms on the steel, leading to the formation of a dense, more protective oxide. This oxide serves as a barrier to further penetration of moisture and effectively acts almost as a "self-painted" coating, lessening the need for other coating protection. An example of the improvement in corrosion behavior of weathering steels in an industrial environment is shown in Fig. 7 (see also the article "Corrosion of Weathering Steels," *ASM Handbook,* Volume 13B).

Although such behavior is valuable, it must be recognized that it will be observed only under certain circumstances. Interestingly, the improvement is generally best in an industrially polluted environment; less improvement is noted in environments containing chloride—for example, marine environments. Also, the film or patina must undergo repeated wetting and drying to develop. Use of weathering steel in a structure

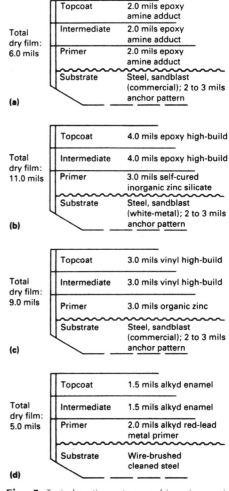

Fig. 5 Typical coating systems used in various environments. (a) Adduct-cured epoxy for use in exposure to solvent spillage and alkaline dust. (b) Inorganic zinc-epoxy used for severe marine exposures. (c) Inorganic zinc/vinyl coating system for use in mild industrial environments. (d) Alkyd-base coating for mild inland atmospheric exposure. Source: Ref 5

where this type of exposure is not available will not allow the protective oxide to form. It is vital to adhere to good corrosion-resistant design techniques when using weathering steels to ensure that the steel performs adequately. Any design or fabrication detail that incorporates crevices, improperly bolted connections, and so on, will lead to behavior in which weathering steels exhibit no advantage over other steels.

Stainless Steels. The ultimate use of alloying elements to promote a barrier layer on the steel surface involves the introduction of sufficient chromium into the steel to form a stable chrome oxide. Stainless steels contain a minimum of 12% Cr and are highly resistant to normal atmospheric conditions. More highly alloyed steels may be required for resistance to severe marine environments. However, the high cost of these materials argues against their use except in critical situations. Such critical situations may, for example, arise in connection bolts, ties, and other components that must perform for the expected service life of the building and are usually not located where inspection, maintenance, or replacement can be easily carried out. Under these circumstances, the use of stainless steel becomes a highly advantageous proposition.

Protection Methods for Cementitious Systems

Cementitious systems normally provide a protective environment to steel. Indeed, the longevity of reinforced steel structures supports this situation. The major factors that cause steel embedded in cementitious systems to corrode are the influences of carbonation and chloride infiltration. Under both of these circumstances, the normally protective oxide film on steel breaks down, and corrosion will proceed at a rate that is sufficient to cause loss of cross section, buildup of voluminous corrosion product with resultant spalling of overlying material and in specific cases, hydrogen embrittlement.

Concrete Covers. One of the principal methods of preventing these problems has been to place cover concrete of sufficient quality and thickness over the steel to dramatically reduce any infiltration of chloride or CO_2 to the steel surface. Because of the diffusion-controlled nature of this infiltration, the depth and permeability of the material have a significant effect on the time at which sufficient chloride or CO_2 reaches the steel surface to allow corrosion to initiate and propagate. The first step in preventing reinforcing steel corrosion is to ensure adequate concrete depth (typically 50 mm, or 2 in.) over the steel and to use as low a water-to-cement ratio as feasible to promote impermeability within the concrete.

Protection Against Chloride-Induced Corrosion. Although this approach may be adequate to reduce carbonation problems, it may not be sufficient to mitigate chloride-induced corrosion. The reasons are twofold. First, the chloride may have been added to the concrete mix; therefore, it would already be present at the depth of the reinforcing steel, regardless of type, quality, or depth of concrete cover. Such problems can be prevented at the specification stage. The use of chloride-containing admixtures for any purpose should be prohibited. Research has shown that prestressing steel may be even more deleteriously affected by chloride-induced corrosion than conventional reinforcing steel. Limits can be placed on the amount of chloride permissible within the concrete from any source. Therefore, chloride-bearing aggregate could have a considerable influence on this situation.

In the case of marine environments, and particularly in the application of deicing salt, it becomes difficult to ensure that the depth of cover concrete will protect the steel against chloride-induced corrosion attack for the expected life of the structure. Hence, alternative measures become appropriate. These include the application of a barrier coating to the steel or to the concrete surface, or the application of cathodic protection to the steel, generally at a point in the structure life at which corrosion damage has already begun.

Use of Stainless Steels. There has recently been increased interest in the use of austenitic stainless steels as reinforcement in concrete where exposure to chloride, either from marine sources or deicing salt, is foreseeable. The rationale here is a "first cost versus maintenance cost" issue. The assumption is that the stainless rebar will be immune to chloride attack. Testing has been performed on both solid stainless bars and on stainless-clad carbon steel bars. The influence of the molybdenum-containing varieties such as type 316 in improving pitting resistance has been evaluated. Concerns regarding possible

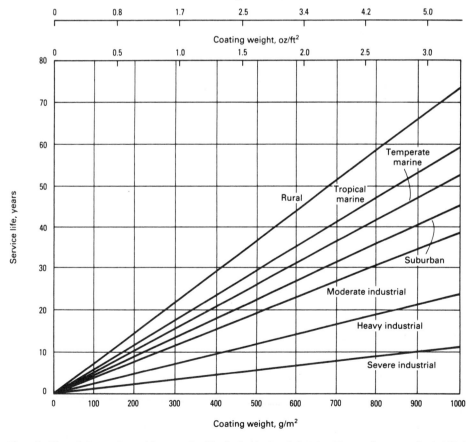

Fig. 6 Effect of zinc coating weight on service life of galvanized steel sheet in various environments. Service life is measured in years to the first appearance of significant rusting.

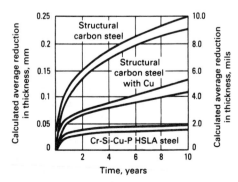

Fig. 7 Corrosion of three types of steels in an industrial atmosphere. Shaded areas indicate range for individual specimens.

stress-corrosion cracking do not appear to be great as structure temperatures are generally unlikely to approach those necessary for the problem to develop.

Studies by the Oregon Department of Transportation have found that the project cost increases approximately 10% when type 316 is used rather than steel bar, with the benefit that the bridge will last many times longer with no maintenance related to corrosion and no cathodic protection or other corrosion-prevention strategies needed. It is important that the stainless steel used has a chloride corrosion threshold above that expected. With this in mind, a bridge using UNS S32205 stainless steel constructed in 2003 near Coos Bay, OR, is expected to last 120 years.

Use of Protective Coatings. Various coatings for carbon steel in reinforced concrete structures have been evaluated. These can generally be grouped into either organic coatings and inorganic or metallic coatings. Extensive work has demonstrated that the alkaline nature of concrete and the intrusion of chloride into the concrete can be resisted by an epoxy coating on the steel. The reinforcing steel must be carefully prepared prior to application of the coating, and the application method is critical, typically an electrostatic powder spray, followed by baking to fuse the coating (fusion-bonded epoxy, or FBE). Epoxy coated bars have been used in bridge decks since 1973.

Epoxy coatings have been viewed with some disfavor on the basis of bond strength, that is, the tendency of the coating to prevent adequate bond development by lessening adhesion at the deformation/concrete interface. However, it would appear that recommended coating thicknesses, which are typically 0.18 ± 0.05 mm (7 ± 2 mils), mitigate this problem and do not lead to a significant reduction in bond strength.

Additional concerns when using epoxy coatings include the possibility that small holidays in the coating may be subject to accelerated attack, particularly if epoxy-coated steel is used only in the top mat of the structure, and that bare steel is used in the bottom mat. This is a traditional construction method for bridge decks and roadbeds when deicing salt is applied to the top surface; the bare steel at the bottom is rarely exposed to any chloride, because of the large distance (up to 250 mm, or 10 in.) from the upper surface of the concrete to the level of the lower mat. Although it is not clear that this concern has been completely eliminated, testing has indicated that it may not be a major problem. In any case, such a problem could of course be largely eliminated through the use of epoxy-coated reinforcing steel throughout the entire structure. The use of fusion-bonded epoxy reinforcing bar is widely specified.

Metallic coatings for steel in cementitious materials are typically limited to zinc and nickel, while cadmium plate is slightly more resistant to chloride attack than zinc because of the formation of a basic cadmium chloride. The environmental concerns regarding cadmium mitigate against its use.

The most common coating for steel in contact with cementitious materials is zinc applied by hot-dip galvanizing, thermal spraying, or electroplating. The factors that determine the rate at which zinc and zinc-coated steel corrode in cementitious materials are directly linked to the formation and preservation of protective films. A stable film and, therefore, low zinc corrosion rates occurs in the pH range of 6 to 12.5.

Figure 8 shows the influence of pH on the corrosion rate of zinc. It is interesting that the pH of concrete or mortar in the noncarbonated state is typically about 12.5, which is the minimum on the pH versus corrosion rate curve for zinc. This observation is supported by laboratory studies evaluating the corrosion rate of zinc. Therefore, for a noncarbonated mortar, the corrosion rate of zinc in contact with this mortar is effectively 0. Once carbonation occurs, and the pH falls below 10, then the corrosion rate increases—typically to a level of approximately 0.5 to 0.8 µm/yr (0.02 to 0.03 mils/yr). The use of hot-dip galvanizing as a protective measure for steel exposed to carbonated concrete or, more particularly, mortar is well founded. A typical 64 µm (2.5 mil) galvanized coating would be expected to last more than 80 years under these circumstances.

Zinc does not perform as well when chloride is present in the cementitious material. The use of zinc coatings under these circumstances has been the subject of much debate. Indeed, the results of laboratory testing seem to be in disagreement with studies of large-scale structures that have been fabricated with galvanized steel. Some of these structures are bridge decks in the Caribbean, where no cracking of the structure was noted even after chloride levels of approximately 0.26% by weight of the concrete were present at the concrete/rebar interface. In this case, however, approximately one-tenth of the original galvanized coating had been removed, and further corrosion could be anticipated. In other locations where galvanized steel has been used in bridge decks, no reports are currently known in the open literature reporting the performance of such steel compared to bare steel in the same type of environment.

Studies of reinforced concrete specimens containing galvanized steel, exposed to marine environments and to artificial ponding in $CaCl_2$ solution, have led to a variety of conclusions related to the effectiveness of galvanizing. They range from a supposedly accelerated cracking of concrete containing galvanized steel to a retardation of such cracking compared with bare steel. However, it should be emphasized that in most of these studies, cracking of the overlying concrete had occurred, indicating that galvanizing is at best a palliative coating for steel in chloride-contaminated cementitious materials.

As far as is known, there is no direct evidence regarding the ability of galvanized steel to withstand a certain level of chloride before significant corrosion can occur. In studies using electrical resistance probes, it was found that, for chloride added to mortar during the mixing phase, the corrosion rate of zinc was of the order of 0.5 µm/yr (0.02 mils/yr) for a chloride content of approximately 0.15% by weight of the mortar. In the presence of carbonation, this corrosion rate increased to 10 µm/yr (0.4 mils/yr). Thus, for a galvanized coating approximately 64 µm (2.5 mils) thick, the coating would survive for only 6.5 years in the presence of carbonation.

Zinc hydroxychloride has been found to form during the corrosion of zinc in chloride-contaminated cementitious materials. This component is significantly expansive compared to the zinc from which it was produced and can therefore cause cracking of overlying cementitious materials even before corrosion of substrate steel has occurred.

Therefore, in general, although zinc-coated steel is somewhat more resistant to chloride-induced corrosion than bare steel, the zinc cannot be relied on to provide protection indefinitely. Indeed, the products of corrosion of the zinc itself may lead to cracking of the overlying material.

In addition to applying barrier coatings on the embedded steel to resist corrosion, a similar barrier surface may be applied to the concrete or masonry surface to resist the penetration of constituents contributing to corrosion. These generally take the form of physical barriers, including:

- Organic membranes applied to the concrete surface (often used on parking structures)
- Low-permeability cementitious-base overlays and repairs (incorporating latex)

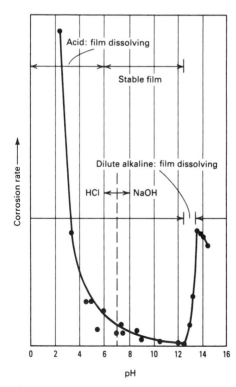

Fig. 8 Corrosion of zinc in aqueous solutions as a function of pH

- Polymer impregnation of the concrete (rarely used)
- Waterproofing agents, for example, drying oils, stearates, or silanes

In using one of these systems, care must be taken not to trap the corrosive environment within the structure. Silanes applied to surfaces are effective in preventing this action; they allow water vapor transmission out of the structure but no liquid penetration into the structure. This prevents oxygen transport to the steel while allowing the structure to dry.

Cathodic Protection. Another established method of protection for steel in cementitious materials subject to corrosion is cathodic protection. Cathodic protection is a means of preventing corrosion in a variety of environments, particularly for pipelines and other structures buried in soil or immersed in seawater. It has successfully been used to ensure the protection of steel pipelines and steel reinforcements that are encased in concrete.

Cathodic protecion was first applied to the problem of deterioration of conventionally reinforced concrete structures in California. During early experiments, a total surface anode over the entire structure was used to protect the structure. Using this technique, large decks could be protected with only 10 W of power. The surface anode method did, however, require the embedment of current supply anodes in a relatively soft overlay composed of coke-asphalt mix, which itself could be subject to fairly rapid deterioration by traffic. This technique has been used by others with some success, and it continues to be improved.

One of the more important developments in the cathodic protection of structures is the use of anode systems that allow horizontal surfaces other than upward facing to be protected. The development of anode systems incorporating conductive polymers, conductive paints, and conductive concrete (containing typically coke or graphite) has made possible the cathodic protection of the underside of reinforced concrete structures. Examples of such cathodic protection systems are discussed in the article "Cathodic Protection," *ASM Handbook,* Volume 13A.

Impressed-current systems using platinized wires in slots in the concrete surface, surrounded by conductive grout, have been used. Experience with this system has been clouded because of the deterioration of the grout and surrounding concrete by the low pH developed at the anode and by the poor throwing power of the anode. Studies have also been conducted on galvanic (sacrificial anode) systems. In this type of system, for example, zinc wires are placed in slots above reinforcing steel, and the sacrificial action of the zinc is then transferred to the steel directly below.

Some additional observations can be made with regard to the cathodic protection of reinforced steel in concrete. First, care must be exercised whenever cathodic protection systems are considered for application to high-strength steels, such as those used in posttensioning and prestressing systems because of the susceptibility of these steels to hydrogen embrittlement.

Furthermore, unbonded steel-concrete systems are expected to be difficult to protect because of the possible insulating effects of the sheath that is present around the steel. Indeed, even if such unbonded systems could be protected, the poor throwing power due to the lack of contact between the posttensioning steel and the surrounding concrete is probably an insurmountable problem.

Finally, considerable attention is currently being given to the appropriate criteria for cathodic protection of steel in cementitious materials. Although a transfer of the usual protection criteria from, for example, pipe in soil systems has been made, there is some doubt as to the reasonability or, in some cases, the adequacy of such protection criteria. This is due to the different nature of the environment between concrete and soil, particularly as it relates to pH. Therefore, rather than relying on a -0.85 V versus copper sulfate electrode as a standard criteria for protection, there is more interest in using a potential shift mechanism and in using E-log i plots for establishing the adequacy of the protection.

Case Histories

This section cites specific examples of failures and problems that have occurred within structures to illustrate the general principles discussed previously, to point out those factors that bear most heavily on the development of a particular problem or failure, and to determine how protective or preventive measures can be implemented to repair the structure or to prevent the occurrence of failure in the future. The case history data have been taken primarily from investigations of buildings or structures conducted by the author. Other examples of failures are available in the literature.

Failures Involving Corrosion of Structural Steel

In general, failures related to the corrosion of conventional structural steel (that which has been painted or otherwise protected) are, in the author's experience, rare. Cases do exist in which excessive humidity or chloride-laden water has contacted the metal. The use of weathering steels, however, has caused significant problems, particularly where the necessary design features and environment have not been carefully considered during selection of the material. Two examples of problems involving weathering steels are discussed below.

Example 1: Weathering Steel Corrosion in a Stadium. A large sports stadium situated about 300 m (1000 ft) from the ocean was built with weathering steel in the major structural members. The steel was used not only for the exposed portions of the structure but also beneath the stands. Significant corrosion and rust flaking were noted on this steel at several locations:

- Underneath the stands, where air circulation was poor and wetting/drying cycles could not be anticipated (Fig. 9)
- At sheltered locations on the exterior, again where wetting/drying cycles could not be anticipated (Fig. 10)
- At joint details, where the important concepts of removal of crevices and pockets to retain water had not been practiced in the design of joints for the structure (Fig. 11)

The major problem associated with this structure was the amount of corrosion occurring on the structural steel beneath the stands. Although the loads in these structures were determined to be low, it was apparent that corrosion protection would be necessary. A series of tests was conducted to determine the optimum coating protocol, including surface preparation, type of coating, thickness, and number of coats.

Fig. 9 Heavy buildup of corrosion scale on weathering steel structural members in conditions of poor air circulation, high humidity, and no wetting/drying

Fig. 10 Corrosion scale buildup on weathering steel structural members, which were in a sheltered area on a building exterior where wetting and drying did not occur

This example serves to point out the important factors to be considered by the designer when weathering steel is selected. In particular, the protective patina can be developed adequately only with exposure to weather. Furthermore, marine environments are not conducive to the formation of such a patina, which develops best in an industrial environment with relatively high sulfur levels in the atmosphere.

Example 2: Corrosion of Weathering Steel in a Hotel Parking Garage. A hotel parking garage in the Northeast was constructed with weathering steel in the columns and beams, along with conventional reinforced concrete slabs placed as the decks. The hotel and garage were situated in an area that experienced considerable amounts of snow and freezing temperatures. Deicing salt was commonly applied to roadways adjacent to the structure and was also probably applied to the reinforced concrete slabs themselves. Severe deterioration was noted in the weathering steel beams and columns, particularly those adjacent to leakage points of water, and in expansion joints. This corrosion was caused by contact with the deicing salt laden water, effectively destroying any patina that would have been expected to develop on the steel. It led instead to the production of voluminous, nonprotective oxides.

The failure in this case was caused by a lack of appreciation of the environment to which the weathering steel was to be exposed. Possible solutions to the problem include the use of coatings on the steel to protect it from further contact with chloride-laden water and the correction of water paths that lead to contact with the deicing salt. The sheltered location of most of the steel, however, would not allow effective patina development. Prohibiting the use of chloride deicing salt on the garage decks would probably reduce the problem somewhat, although pickup of deicing salt from roadways and track-in into the garage is a perpetually unavoidable situation.

Corrosion of Conventional Reinforcement

Corrosion of steel in conventionally reinforced concrete in deicing salt application areas and in marine areas is a well-known phenomenon on bridge decks and bridge support structures. This corrosion phenomenon has received sufficient illustration in the literature and is not repeated here. An example of a similar problem that occurred in a building not subjected to deicing salt or ambient marine environments follows.

Example 3: Corrosion of Reinforcing Steel in Building Columns. Figure 12 shows an example of a supporting column in a dormitory building on a midwestern United States campus. The columns are of conventionally reinforced concrete, with a spiral of reinforcement that closely approaches the surface of the concrete column. The concrete showed cracking and spalling within a few years after its installation. Laboratory examination indicated that the concrete cover was low, that the concrete over the steel was carbonated close to the outer surface, and that the concrete contained a significant amount of chloride, apparently added during construction, that was due to either the presence of chloride-bearing aggregate or the deliberate addition of admixtures such as $CaCl_2$. Corrosion of embedded steel was severe in places (Fig. 13).

Because of the likelihood that carbonation might progress into the concrete and, together with the chloride, affect the more deeply embedded steel, the following repair plan was devised. First, loose and spalled concrete was chipped out of the columns. Second, the columns were shotcreted to a depth of approximately 25 mm (1 in.) above the topmost reinforcement. Finally, an impressed-current cathodic protection system was placed on the columns using conductive polymer anodes with integral lead

Fig. 11 Heavy corrosion scale buildup on structural members of weathering steel at a pocket where water could collect and stand

Fig. 12 Corrosion-induced spalling of overlying concrete on reinforced columns. See also Fig. 13.

Fig. 13 Severe corrosion on reinforcing steel from the column shown in Fig. 12

Fig. 14 Unconventional masonry tie constructed from flattened C-channel and bent to enter the brick core

Fig. 15 Severe corrosion and cracking (arrows) on wall tie

wires. This system was positioned above the surface of the concrete with stands, and the entire system was finally covered with an additional layer of concrete. This minimal concrete removal, along with the subsequent buildup and installation of a cathodic protection system, was far more cost effective than the alternatives, which included complete demolition of the column or removal of sufficient overlying concrete cover to necessitate shoring of the column to support the building above.

Corrosion of Ties and Anchors

The corrosion of steel ties and anchors used to attach masonry walls of the building frame to the backup (usually concrete block) wall, or, more recently, to steel studs, has been a source of significant concern. Cases are known in which such ties or anchors have corroded, either in the mortar or in the airspace. Two examples are given.

Example 4: Corrosion of Nonstandard Wall Ties. In a hospital in the midwest United States, cracking of masonry walls at attachment points to a steel stud backup was observed. When the interior wallboard was removed, the tie between the brick wall and the steel stud was found to be in the form of a lightweight C-channel that had been flattened and bent to form a 90° angle (Fig. 14). The flattened portion of the anchor was inserted into the masonry bed joint and down into the core of the brick. The remaining, intact portion of the channel was attached to the backup steel stud by a self-tapping screw. No effective corrosion protection had been applied to the angles; they exhibited remnants of shop primer at certain locations, but the efficacy of any coating had been destroyed during the bending and flattening process.

Severe corrosion of these angles had occurred in the airspace adjacent to the brick masonry. This was apparently caused by water running down the interior wall and becoming trapped. Also, the water became trapped in the crevices formed by the flattened channel.

One effect of this corrosion was the interaction with the low cyclic stresses imposed on the tie due to movements between the stud and the wall, leading to cracking of the tie. This was particularly prevalent where corrosion had reduced the tie thickness to a fraction of its original dimension (Fig. 15).

This problem could have been prevented through the use of several alternative techniques, including:

- The use of more conventional ties incorporating corrosion-resistant coatings such as, galvanizing
- The use of heavy organic coatings after bending to prevent corrosion (if this type of tie was to be mandated, then heavy organic coatings applied after bending would have been appropriate)

The solution to this problem could invoke the use of alternative supplementary anchors or the dismantling of portions of the wall containing

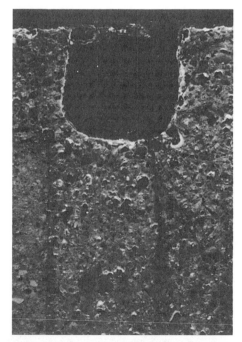

Fig. 16 Corrosion and resulting masonry cracking on anchor embedded in high-band mortar. See also Fig. 17

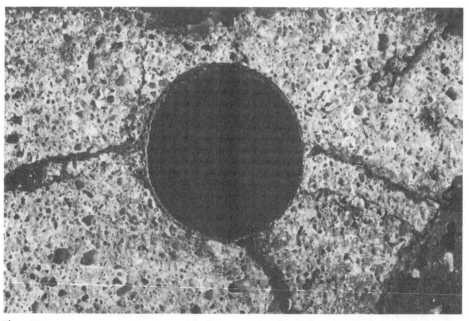

Fig. 17 Corrosion and resulting masonry cracking on an anchor embedded in high-band mortar. See also Fig. 16

such ties and replacement with masonry containing more appropriate anchors to ensure continued support of the masonry wall.

In addition to the corrosion of steel in the wall cavity itself, which can generally be lessened by the use of galvanizing, there is also concern regarding the corrosion that may occur in the portion of the tie within the mortar. This is the case where significant chloride is present in the mortar, particularly where chloride and carbonation can interact to produce a corrosive environment. In these cases, the commonly used galvanized thicknesses may not be sufficient to protect the steel over the expected service life of the building.

There have been a number of recent instances where conventional tie corrosion even in the absence of chloride has been sufficiently severe as to impact the safety of veneer brick walls. As buildings incorporating this system age, further instances of this problem can be expected to develop.

Example 5: Cracking of Masonry Caused by Corrosion of Ties and Anchors. In many structures that have incorporated high-bond masonry mortar additives, the release of Cl^- due to alkaline hydrolysis has significantly corroded uncoated and coated (zinc and cadmium) steel. This has been the case for laid-in-place buildings (utilizing conventional ties) and for panelized buildings (in which embedment of connection devices in the mortar is used to affix the panels to the building frame). In this situation, connection devices may corrode within the mortar, with subsequent cracking of overlying masonry due to the buildup of corrosion product.

Examples of masonry exhibiting such cracking, which radiates from corroded embedded ties and anchors, are shown in Fig. 16 and 17. Under these circumstances, the integrity of the anchor becomes extremely suspect. No effective method is known for alleviating this problem once it has occurred. Alternative approaches have involved the use of heavily organic coated anchors (for example, epoxy coatings) or the use of austenitic, molybdenum-containing stainless steels, which should resist the onslaught of the Cl^-.

Corrosion of Posttensioning and Prestressing Structures

Unlike other cases of corrosion of steel in structures, the corrosion of posttensioning structures can be troublesome from two viewpoints. First, the possibility exists that substantial corrosion can lead to a loss of cross section and therefore failure of the gripping mechanism for the posttensioning strand or of the posttensioning steel itself. Second, corrosion products on the surface of the material may release sufficient hydrogen to cause hydrogen embrittlement. Two examples are given below.

Example 6: Corrosion of Posttensioning Anchorages. A posttensioned garage in the snow-belt area of the United States exhibited significant corrosion of conventional reinforcement, as noted by the presence of cracking and spalling of overlying concrete. On one occasion, a posttensioning tendon failed. This led to a large-scale investigation of posttensioning members, particularly the anchorages. The anchorages were found to be significantly corroded because of their location adjacent to leaking expansion joints and because they were surrounded by badly consolidated concrete. This had allowed deicing salt to penetrate to the level of the anchorages, leading to some severe corrosion, particularly to gripping wedges, as illustrated in Fig. 18.

Several of the anchorages were sufficiently corroded that alternative anchorages had to be installed. In others, the corrosion was slowed by the injection of water-displacing grease into the anchorage through a grease fitting. This case history shows how poor-quality concrete can significantly affect the performance of metals embedded within it, particularly when the metals are vitally important to the longevity and safety of the structure.

Example 7: Hydrogen Embrittlement of Posttensioning Wires. Single 6.4 mm (0.25 in.) posttensioning wires failed in a parking garage in the southern portion of the United States. A typical anchorage with a broken buttonhead wire is shown in Fig. 19. Several such wires were removed, and the lengths were examined for signs of corrosion. Localized shallow pitting was common, as illustrated in Fig. 20. Chloride was detected in some of these pits.

Scanning electron microscopy and metallography of the ends of the fractures revealed an initial crack that was probably caused by hydrogen embrittlement (Fig. 21). Apparently,

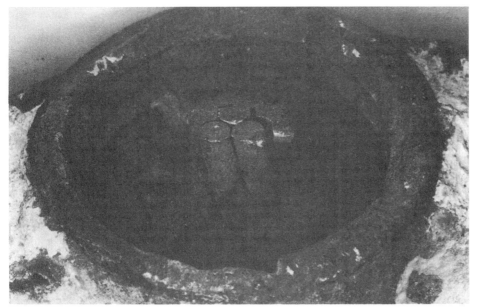

Fig. 18 Corrosion of posttensioning anchorage. Note severe corrosion at the two wedge halves.

Fig. 19 Posttensioning anchorage with broken wire extending from anchor plate

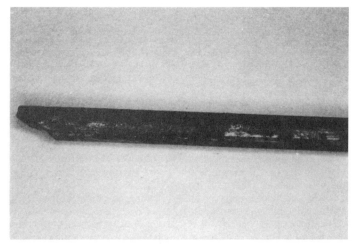

Fig. 20 Pitting corrosion adjacent to fracture on failed posttensioning wire

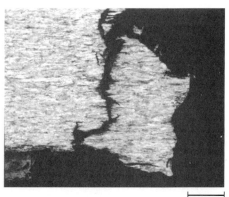

Fig. 21 Metallographic cross section through the fracture initiation region of posttensioning wire. Note secondary cracks. Etched with 2% nital. Original magnification: 55×

the hydrogen embrittlement had occurred because of the presence of the corrosion on the external surface. No chloride was detected on the fracture surface, and none was detected in the overlying concrete. This suggested that the corrosion observed may have initiated before placement of the tendons within the concrete. There is no known method of mitigating this type of problem once it has occurred, although it can be prevented through the use of judicious and careful corrosion-preventive techniques during the storage of the tendons before placement.

REFERENCES

1. S.K. Coburn et al., Corrosiveness of Various Atmospheric Test Sites as Measured by Specimens of Steel and Iron, *Metal Corrosion in the Atmosphere,* STP 435, American Society for Testing and Materials, 1968, p 360
2. D.L. Graver, Ed., *Corrosion Data Survey—Metals Section,* National Association of Corrosion Engineers, 1985
3. "Standard Test Method for Corrosion of Steel by Sprayed Fire-Resistant Material (SFRM) Applied to Structural Members," E 937, Vol 14.11, *Annual Book of ASTM Standards,* ASTM International
4. M. Pourbaix, *Atlas of Electrochemical Equilibria in Aqueous Solutions,* Pergamon Press, 1966
5. R. Zidell, Coatings for Steel, *Paint Handbook,* G.E. Weismantel, Ed., McGraw-Hill, 1981
6. "Standard Specification for High-Strength Low-Alloy Structural Steel with 50 ksi (345 MPa) Minimum Yield Point to 4 in. (100 mm) Thick," A 588/A 588M, Vol 01.04, *Annual Book of ASTM Standards,* ASTM International
7. "Standard Specification for Carbon and High-Strength Low-Alloy Structural Steel Shapes, Plates, and Bars and Quenched-and-Tempered Alloy Structural Steel Plates for Bridges," A 709/A 709M, Vol 01.04, *Annual Book of ASTM Standards,* ASTM International

SELECTED REFERENCES

- P. Albrecht, Corrosion Control of Weathering Steel Bridges, *Corrosion Forms and Control for Infrastructure,* STP 1137, American Society for Testing and Materials, 1992, p 108–125
- P. Albrecht, S.K. Coburn, F.M. Wattar, G. Tinklenberg, and W.P. Gallagher, "Guidelines for the Use of Weathering Steel in Bridges," NCHRP Report 314, Transportation Research Board, National Research Council, Washington, D.C., 1989
- P. Albrecht and T.T. Hall, Atmospheric Resistance of Structural Steels, *J. Mater. Civil Eng.,* Vol 15 (No. 1), 2003, p 2–24
- S. Oesch, The Effect of SO_2, NO_2, NO and O_3 on the Corrosion of Unalloyed Carbon Steel and Weathering Steel—The Results of Laboratory Exposures, *Corros. Sci.,* Vol 38 (No. 8), 1996, p 1357–1368
- H.E. Townsend, Atmospheric Corrosion Performance of Quenched-and-Tempered, High-Strength Weathering Steel, *Corrosion,* Vol 56 (No. 9), 2000, p 883–886
- J.H. Wang, F.I. Wei, and H.C. Shih, Assessing Performance of Painted Carbon and Weathering Steels in an Industrial Atmosphere, *Corrosion,* Vol 53 (No. 3), 1997, p 206–215

Corrosion of Metal Processing Equipment

Revised by B. Mishra, Colorado School of Mines

DEVELOPMENT OF METALS AND ALLOYS for novel and advanced commercial applications places higher demands on metal processing to obtain the desired structure and properties, both on the surface and in the bulk. Typically, the bulk properties and structure are developed by compositional control and heat treatment of the component, while superior surface characteristics are developed by metal finishing processes. Heat treatment requires high-temperature equipment, while metal finishing is usually a near-ambient-temperature process. High-temperature furnaces generally are exposed to hot corrosive fluids, mostly gases, while the metal finishing processes expose the process equipment to corrosive acidic and alkaline solutions. Metal processing equipment thus is exposed to numerous corrosive environments and corrosion mechanisms. Heat treating equipment is subject to high-temperature oxidation, carburization and decarburization, and sulfidation, in addition to high-hydrogen and nitrogen-bearing atmospheres. Corrosion by molten salts and molten metals is also a problem for heat treating furnaces and accessories. Equipment for plating, pickling, anodizing, and other chemical and electrolytic processes is exposed to acid and alkali solutions as well as corrosive fumes at temperatures up to or higher than 100 °C (212 °F). Information on materials for and prevention of corrosion in these applications is also available in *Heat Treating*, Volume 4, 1991, and *Surface Engineering*, Volume 5, 1994, of *ASM Handbook*. This article describes the two aspects of metal processing equipment corrosion: heat treating furnace equipment, and plating, anodizing, and pickling equipment.

Corrosion of Heat Treating Furnace Equipment

Heat treating furnace accessories include trays, baskets, pots, blowers, sensors, belts, hangers, bellows, and dampers. Heat treatments include annealing, normalizing, hardening, carburizing, nitriding, carbonitriding, brazing, galvanizing, and sintering. Typically, the furnace accessories and the furnace chamber are exposed to the same corrosive conditions as the parts being processed. High-temperature furnaces are usually ceramic refractory lined for thermal insulation of the outer metallic shell; the refractory material then is subject to corrosion and erosion damage.

The medium or environment used for heat treating varies from process to process. The high-temperature corrosion performance of furnace components depends on the environment (or atmosphere) involved in the operation. Typical environments are air, combustion atmospheres, carburizing and nitriding atmospheres, molten salts, and protective atmospheres (such as endothermic atmospheres, nitrogen, argon, hydrogen, and vacuum). Protective atmospheres are used to prevent metallic parts from forming heavy oxide scales during heat treatment. The environment can often be contaminated by impurities, which can greatly accelerate corrosion. These contaminants (such as sulfur, vanadium, and sodium) generally come from fuels used for combustion, from fluxes used for specific operations, and from drawing compounds, lubricants, and other substances that are left on the parts to be heat treated. Sodium salts can lead to acid or basic fluxing, resulting in hot corrosion. In oil-fired furnaces, vanadium and sodium cause problems by forming a eutectic mixture. Residual fuel oils contain 1000 ppm vanadium and 10 ppm sodium. At approximately 700 °C (1300 °F), the eutectic of 99% V and 1% Na melts into a viscous, sticky liquid. This liquid slowly corrodes localized areas of equipment (Ref 1).

The high-temperature corrosion processes that are most frequently responsible for the degradation of furnace accessories are oxidation, carburization, decarburization, sulfidation, molten-salt corrosion, and molten-metal corrosion. Each corrosion process, along with the corrosion behavior of important engineering alloys, is discussed in this section. In each case, the corrosion susceptibility of the components is enhanced due to thermal cycling. Thermal cycling not only promotes the loss of protective surface films but also causes structural distortion. The compositions of the alloys under discussion are given in Table 1. Application of high-temperature corrosion and oxidation-resistant coatings has become common for furnace accessories and equipment. These coatings are typically ceramics or thermally densifying materials applied using physical vapor deposition, plasma, chemical vapor deposition, and high-velocity oxyfuel spray methods (Ref 2).

Oxidation

Oxidation is probably the predominant high-temperature corrosion process encountered in the heat treating industry, owing to the fact that air is the most common heat treating atmosphere for cost reasons. Oxidation involves air or combustion atmospheres with little or no contaminants, such as sulfur, chlorine, alkali metals, and salt.

Carbon steel and alloy steels generally have adequate oxidation resistance for reasonable service lives for furnace accessories at temperatures to 540 °C (1000 °F) (Ref 3). At intermediate temperatures of 540 to 870 °C (1000 to 1600 °F), heat-resistant stainless steels, such as 304, 316, 309, and 446, generally exhibit good oxidation resistance (Ref 3). As the temperature increases above 870 °C (1600 °F), many stainless steels begin to suffer rapid oxidation. Better heat-resistant materials, such as the nickel-base high-performance alloys, and other refractory and nonferrous metal alloys in association with protective coatings are needed for furnace components to combat oxidation at these high temperatures.

Numerous oxidation tests on commercial alloys have been performed at 980 °C (1800 °F) or higher. For example, in one investigation, cyclic oxidation tests were conducted in air, with each cycle consisting of exposing the samples at 980 °C (1800 °F) for 15 min, followed by a 5 min air cooling (Ref 4). The performance ranking, in order of decreasing performance, was found to be as follows: Inconel alloy 600, Incoloy alloy 800, type 310 stainless steel, type 309 stainless steel, type 347 stainless steel, and type 304 stainless steel. Similar cyclic oxidation tests performed in air at 1150, 1205, and 1260 °C (2100, 2200, and 2300 °F), cycling to room temperature by air cooling after every 50 h at temperature, showed Inconel alloy 601 to be the best performer, followed by Inconel alloy 600, and Incoloy alloy 800 (Ref 5).

In another study, ferritic stainless steels such as E-Brite 26-1 and type 446 were shown to be significantly better than type 310 stainless steel and Incoloy alloy 800H in terms of cyclic oxidation resistance in air (Ref 6). These test results showed weight change data of 2.2 mg/cm^2 for E-Brite 26-1 stainless steel, 10.0 mg/cm^2 for type 446 stainless steel, −83.2 mg/cm^2 for alloy 800, and −90.3 mg/cm^2 for type 310 stainless steel after exposure of the samples for a total of 1000 h, with 15 min at 980 °C (1800 °F) and 5 min at room temperature. A separate test was also conducted. This test involved exposure of the samples at 980 °C (1800 °F) for 1000 h in air, with interruptions after 1, 20, 40, 60, 80, 100, 220, 364, and 512 h for cooling to room temperature. The weight change results of these four alloys were −12.9, 9.2, 1.7, and 3.0 mg/cm^2 for E-Brite 26-1 stainless steel, type 446 stainless steel, type 310 stainless steel, and Incoloy alloy 800, respectively (Ref 6).

An oxidation database for a wide variety of commercial alloys, including stainless steels, Fe-Ni-Cr alloys, Ni-Cr-Fe alloys, and high-performance alloys, was generated (Ref 7). Tests were conducted in air at 980, 1095, 1150, and 1205 °C (1800, 2000, 2100, and 2200 °F) for 1008 h. The samples were cooled to room temperature once a week (each 168 h) for visual inspection. The results are summarized in Table 2.

Type 304 stainless steel and type 316 stainless steel both exhibited severe oxidation attack at 980 °C (1800 °F), while type 446 stainless steel showed relatively mild attack. Many higher alloys, such as Incoloy alloy 800 and the nickel- and cobalt-base alloys, showed little attack. At 1095 °C (2000 °F), type 446 stainless

Table 1 Nominal chemical compositions of high-temperature alloys

Alloy	UNS No.	C	Fe	Ni	Co	Cr	Mo	W	Si	Mn	Other
304	S30400	0.08(a)	bal	8	...	18	...	...	1.0(a)	2.0(a)	...
309	S30900	0.20(a)	bal	12	...	23	...	...	1.0(a)	2.0(a)	...
253MA	S30815	0.08	bal	11	...	21	...	...	1.7	0.8(a)	0.17N, 0.05Ce
310	S31000	0.25(a)	bal	20	...	25	...	...	1.5(a)	2.0(a)	...
316	S31600	0.08(a)	bal	10	...	17	2.5	...	1.0(a)	2.0(a)	...
446	S44600	0.20(a)	bal	...	...	25	...	...	1.0(a)	1.5(a)	0.25N
E-Brite 26-1	S44627	0.002	bal	0.15	...	26	1.0	...	0.2	0.1	...
800H	N08810	0.08	bal	33	...	21	...	...	1.0(a)	1.5(a)	0.38Al, 0.38Ti
RA330	N06330	0.05	bal	35	...	19	...	...	1.3	1.5	...
Multimet	R30155	0.10	bal	20	20	21	3	2.5	1.0(a)	1.5(a)	1.0Nb+Ta, 0.5Cu, 0.15N
556	...	0.10	bal	20	18	22	3	2.5	0.4	1.0	0.2Al, 0.8Ta, 0.02La, 0.2N, 0.02Zr
825	N08825	0.05(a)	29	bal	...	22	3	...	0.5(a)	1.0(a)	2Cu, 1Ti
600	N06600	0.08(a)	8	bal	...	16	...	...	0.5(a)	1.0(a)	0.35Al(a), 0.3Ti(a), 0.5Cu(a)
214	N07214	0.04	2.5	bal	...	16	...	...	...	...	4.5Al, Y
601	N06617	0.10(a)	14.1	bal	...	23	...	...	0.5(a)	1.0(a)	1.35Al, 1Cu(a)
617	N06617	0.07	1.5	bal	12.5	22	9	...	0.5	0.5	1.2Al, 0.3Ti, 0.2Cu
S	N06635	0.02	3(a)	bal	2.0(a)	15.5	14.5	1.0(a)	0.4	0.5	0.2Al, 0.02La, 0.009B
X	N06002	0.10	18.5	bal	1.5	22	9	0.6	1.0(a)	1.0(a)	...
625	N06625	0.10(a)	5(a)	bal	...	21.5	9	...	0.5(a)	0.5(a)	0.4Al(a), 0.4Ti(a), 3.5Nb+Ta
230	N06230	0.10	3(a)	bal	3(a)	22	2	14	0.4	0.5	0.3Al, 0.005B, 0.03La
RA333	N06333	0.05	18	bal	3	25	3	3	1.25	1.5	...
N	...	0.06	5(a)	bal	...	7	16.5	0.5(a)	1.0(a)	0.8(a)	0.35Cu(a)
188	...	0.10	3(a)	22	bal	22	...	14	0.35	1.25(a)	0.04La
25	...	0.10	3(a)	10	bal	20	...	15	1.0(a)	1.5	...
6B	...	1.2	3(a)	3(a)	bal	30	1.5(a)	4.5	2.0(a)	2.0(a)	...

(a) Maximum

Table 2 Results of 1008 h cyclic oxidation test in flowing air at temperatures indicated
Specimens were cycled to room temperature once a week

	Metal loss by oxidation at temperature														
	980 °C (1800 °F)				1095 °C (2000 °F)				1150 °C (2100 °F)				1205 °C (2200 °F)			
	Metal loss		Average metal affected(a)		Metal loss		Average metal affected		Metal loss		Average metal affected		Metal loss		Average metal affected	
Alloy	mm	mils	mm	mils	mm	mils	mm	mils	mm	mils	mm	mils	mm	mils	mm	mils
Haynes alloy 214	0.0025	0.1	0.005	0.2	0.0025	0.1	0.0025	0.1	0.005	0.2	0.0075	0.3	0.005	0.2	0.018	0.7
Haynes alloy 230	0.0075	0.3	0.018	0.7	0.013	0.5	0.033	1.3	0.058	2.3	0.086	3.4	0.11	4.5	0.2	7.9
Hastelloy alloy S	0.005	0.2	0.013	0.5	0.01	0.4	0.033	1.3	0.025	1.0	0.043	1.7	>0.81	>31.7(b)	>0.81	>31.7
Haynes alloy 188	0.005	0.2	0.015	0.6	0.01	0.4	0.033	1.3	0.18	7.2	0.2	8.0	>0.55	>21.7	>0.55	>21.7
Inconel alloy 600	0.0075	0.3	0.023	0.9	0.028	1.1	0.041	1.6	0.043	1.7	0.074	2.9	0.13	5.1	0.21	8.4
Inconel alloy 617	0.0075	0.3	0.033	1.3	0.015	0.6	0.046	1.8	0.028	1.1	0.086	3.4	0.27	10.6	0.32	12.5
310	0.01	0.4	0.028	1.1	0.025	1.0	0.058	2.3	0.075	3.0	0.11	4.4	0.2	8.0	0.26	10.3
RA333	0.0075	0.3	0.025	1.0	0.025	1.0	0.058	2.3	0.05	2.0	0.1	4.0	0.18	7.1	0.45	17.7
Haynes alloy 556	0.01	0.4	0.028	1.1	0.025	1.0	0.067	2.6	0.24	9.3	0.29	11.6	>3.8	>150.0	>3.8	>150.0
Inconel alloy 601	0.013	0.5	0.033	1.3	0.03	1.2	0.067	2.6	0.061	2.4	0.135	5.3	0.11	4.4	0.19	7.5
Hastelloy alloy X	0.0075	0.3	0.023	0.9	0.038	1.5	0.069	2.7	0.11	4.5	0.147	5.8	>0.9	>35.4	>0.9	>35.4
Inconel alloy 625	0.0075	0.3	0.018	0.7	0.084	3.3	0.12	4.8	0.41	16.0	0.46	18.2	>1.21	>47.6	>1.21	>47.6
RA330	0.01	0.4	0.11	4.3	0.02	0.8	0.17	6.7	0.041	1.6	0.22	8.7	0.096	3.8	0.21	8.3
Incoloy alloy 800H	0.023	0.9	0.046	1.8	0.14	5.4	0.19	7.4	0.19	7.5	0.23	8.9	0.29	11.3	0.35	13.6
Haynes alloy 25	0.01	0.4	0.018	0.7	0.23	9.2	0.26	10.2	0.43	16.8	0.49	19.2	>0.96	>37.9	>0.96	>37.9
Multimet	0.01	0.4	0.033	1.3	0.226	8.9	0.29	11.6	>1.2	>47.2	>1.2	>47.2	>3.72	>146.4	>3.72	>146.4
446	0.033	1.3	0.058	2.3	0.33	13.1	0.37	14.5	>0.55	>21.7	>0.55	>21.7	>0.59	>23.3	>0.59	>23.3
304	0.14	5.5	0.21	8.1	>0.69	>27.1	>0.69	>27.1	>0.6	>23.6	>0.6	>23.6	>1.7	>68.0	>1.73	>68.0
316	0.315	12.4	0.36	14.3	>1.7	>68.4	>1.7	>68.4	>2.7	>105.0	>2.7	>105.0	>3.57	>140.4	>3.57	>140.4

(a) Average metal affected = metal loss + internal penetration. (b) All figures shown as greater than stated value represent extrapolation of tests in which samples were consumed in less than 1008 h. Source: Ref 7

steel suffered severe oxidation. Iron-nickel-chromium alloys, such as Incoloy alloy 800H and alloy RA330, also suffered significant oxidation. Many nickel-base alloys, however, still exhibited little oxidation. At 1150 °C (2100 °F), most alloys suffered unacceptable oxidation, with the exception of only a few nickel-base alloys. At 1205 °C (2200 °F), all alloys except Haynes alloy 214 suffered severe attack. Alloy 214 showed negligible oxidation at all the test temperatures. This alloy is different from all of the other alloys tested in that it forms an aluminum oxide (Al_2O_3) scale when heated to elevated temperatures. Other alloys tested form chromium oxide (Cr_2O_3) scales when heated to elevated temperatures.

The alloy performance rankings (Ref 8) generated from the field in the furnace atmosphere produced by the combustion of natural gas were found to correspond closely to the air oxidation data presented in Table 2. The alumina-forming alloy 214 was found to be the best performer (Ref 8).

The high-temperature oxidation resistance of stainless steel as well as aluminum-and silicon-bearing alloys relies on the development of a tenacious, nonporous, and adherent film of chromium oxide, aluminum oxide, and silicon oxide, respectively. Thus, high-temperature metallic coatings developed for oxidation resistance contain these metals. The primary reaction on ferrous material oxidation is given as:

$$Fe + \tfrac{1}{2}O_2 \rightarrow FeO$$

The equilibrium of this reaction is controlled by the activities of the reacting and product species and the temperature. Temperature is also responsible for faster kinetics in these reactions.

Carburization

Materials problems due to carburization are quite common in heat treating components associated with carburizing furnaces. The environment in the carburizing furnace typically has a carbon activity that is significantly higher than that in the alloy of the furnace component. Therefore, carbon is transferred from the environment to the alloy. This results in the carburization of the alloy, and the carburized alloy becomes embrittled. The protective atmosphere is created for oxidation protection; that is, the CO/CO_2 ratio in the furnace atmosphere is maintained high to generate a low oxygen partial pressure:

$$CO + \tfrac{1}{2}O_2 \rightarrow CO_2$$

However, a competing reaction requires that the carbon dioxide activity be high to prevent carbon deposition by the following reaction:

$$2CO \rightarrow C + CO_2$$

The temperature-dependent competition between these two reactions determines the sensitivity of the heat treatment furnace accessories to oxidation, carburization, and decarburization.

Nickel-base alloys are generally considered to be more resistant to carburization than stainless steels. The results of 25 h carburization tests performed at 1095 °C (2000 °F) in a gas mixture consisting of 2% methane (CH_4) and 98% hydrogen revealed the weight gain data of 2.78, 5.33, 18.35, and 18.91 mg/cm² for Inconel alloy 600, Incoloy alloy 800, type 310 stainless steel, and type 309 stainless steel, respectively (Ref 9). Extensive carburization tests have been performed to investigate 22 commercial alloys, including stainless steels, Fe-Cr-Ni alloys, Ni-Cr-Fe alloys, and nickel- and cobalt-base alloys (Ref 10). Tests were performed for 215 h at 870 and 925 °C (1600 and 1700 °F) and for 55 h at 980 °C (1800 °F) in a gas mixture consisting of 5 vol% hydrogen, 5 vol% CH_4, 5 vol% carbon monoxide (CO), and the balance argon. It was found that the alumina-forming Haynes alloy 214 was the most resistant to carburization among all of the alloys tested.

These findings were confirmed in 24 h tests performed at 1095 °C (2000 °F) in the same gas mixture. The carburization data are summarized in Table 3. In this study, alloy 214 (an alumina former) was found to be significantly better than the chromia formers tested. Field tests were conducted in a heat treating furnace used for carburizing, carbonitriding, and neutral hardening operations (Ref 11). Both RA333 and Inconel alloy 601 were found to exhibit better carburization resistance than any of the alloys tested, which included RA330 and Incoloy alloy 800. It is evident that alloys containing strong carbide formers, such as stainless steels, or having high permeability for carbon show poor resistance to carburization, while the alloys with weak carbide formers, such as nickel and aluminum, show better resistance.

Metal dusting is another frequently encountered mode of corrosion that is associated with carburizing furnaces. Metal dusting tends to occur in a region where the carbonaceous gas atmosphere becomes stagnant. The alloy normally suffers rapid metal wastage. The corrosion products (or wastage) generally consist of carbon soots, metal, metal carbides, and metal oxides. The attack is normally initiated from the metal surface that is in contact with the furnace refractory. The furnace components that suffer metal dusting include thermowells, probes, and anchors. Figure 1 illustrates the metal dusting

Table 3 Results of 24 h carburization tests performed at 1095 °C (2000 °F) in 85Ar-5H$_2$-5CO-5CH$_4$ (vol%)

Alloy	Carbon absorption, mg/cm²
Haynes alloy 214	3.4
Inconel alloy 600	9.9
Inconel alloy 625	9.9
Haynes alloy 230	10.3
Hastelloy alloy X	10.6
Hastelloy alloy S	10.6
304	10.6
Inconel alloy 617	11.5
316	12.0
RA333	12.4
Incoloy alloy 800H	12.6
RA330	12.7
Haynes alloy 25	14.4

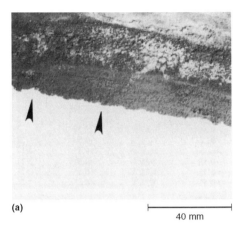
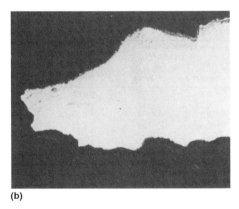
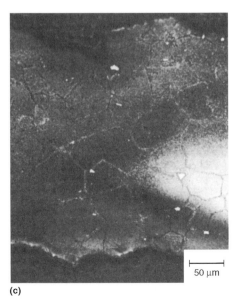

Fig. 1 Metal dusting of a Multimet alloy component at the refractory interface in a carburizing furnace. (a) Perforation of the component (arrows). (b) Cross section of the sample showing severe pitting. (c) Severe carburization beneath the pitted area

attack on Multimet alloy. The component was perforated as a result of metal dusting. Metal dusting problems have also been reported in petrochemical processing (Ref 12).

Metal dusting has been encountered with straight chromium steels, austenitic stainless steels, and nickel- and cobalt-base alloys. All of these alloys are chromia formers; that is, they form Cr_2O_3 scales when heated to elevated temperatures. No metal dusting has been reported on the alloy systems that form a much more stable oxide scale, such as Al_2O_3, because alumina is a much more thermodynamically stable oxide than chromia. The Al_2O_3 scale was found to be much more resistant to carburization attack than the Cr_2O_3 scale (Ref 10). Because metal dusting is a form of carburization, it would appear that alumina formers, such as Haynes alloy 214, would also be more resistant to metal dusting.

Sulfidation

Furnace environments get sulfur from fuels, fluxes used for specific operations, and cutting oil left on the parts to be heat treated, among other sources. Sulfur in the furnace environment could greatly reduce the service lives of components through sulfidation attack.

It is well known that nickel-base alloys are highly susceptible to catastrophic sulfidation due to the formation of nickel-rich sulfides, which melt at approximately 650 °C (1200 °F). Figure 2 illustrates catastrophic failure of a Ni-Cr-Fe alloy tube due to sulfidation attack in a heat treating furnace. The liquid-appearing nickel-rich sulfide phases are clearly visible.

A sulfidation study was undertaken to determine the relative alloy rankings of base alloys (Ref 10). Tests were performed at 760, 870, and 980 °C (1400, 1600, and 1800 °F) for 215 h in a gas mixture consisting of 5% hydrogen, 5% CO, 1% carbon dioxide (CO_2), 0.15% hydrogen sulfide (H_2S), 0.1% H_2O, and the balance argon. The cobalt-base alloys were found to be the best performers, followed by iron-base alloys, and then nickel-base alloys, which, as a group, were generally the worst performers. Among iron-base alloys, the Fe-Ni-Co-Cr alloy 556 was better than Fe-Ni-Cr alloys such as Incoloy alloy 800H and type 310 stainless steel. The test results of representative alloys from each alloy base group are summarized in Table 4.

Molten-Salt Corrosion

Molten salts are used in the heat treating industry for tempering, annealing, hardening, reheating, carburizing, and other operations. The salts that are commonly used include nitrates, carbonates, borates, cyanides, chlorides, fluorides, and caustics, depending on the operation. For example, a mixture of nitrates and nitrites is normally used for tempering and quenching. An alkali chloride carbonate mixture is used for annealing ferrous and nonferrous metals. Neutral salt baths containing mixed chlorides are used for hardening steel parts.

Carbon steels, alloy steels, stainless steels, and Fe-Ni-Cr alloys have been used for various furnace parts, such as electrodes, thermocouple protection tubes, and pots for salt baths. Molten-salt corrosion of ferrous and nonferrous metals has been reported (Ref 13), showing embrittlement of the alloy via grain-boundary penetration.

Corrosion data in molten sodium-potassium nitrate ($NaNO_3$-KNO_3) salts are given in Table 5. The Ni-Cr-Fe-Al-Y alloy (Haynes alloy 214), Ni-Cr-Fe alloys (Inconel alloys 600 and 601), and Ni-Cr-Mo alloys (Hastelloy alloys N and S) performed significantly better than stainless steels and Fe-Ni-Cr alloys such as Incoloy alloy 800H and RA330. The data generated from 1 month tests in a neutral salt bath containing a mixture of barium, potassium, and sodium chlorides ($BaCl_2$, KCl, and NaCl) at 845 °C (1550 °F) were reported (Ref 8). The results are summarized in Table 6. Table 7 shows the performance of these alloys in sodium chloride salt (Ref 14).

It was found that the corrosion of a nickel-base alloy salt pot containing molten $BaCl_2$-KCl-NaCl mixture at 1010 °C (1850 °F) was significantly different between the airside (outside of the pot) and the molten-salt side (inside) (Ref 15). The outside of the pot (that is, the air side contaminated with salt vapors) suffered three times as much attack as the inside of the

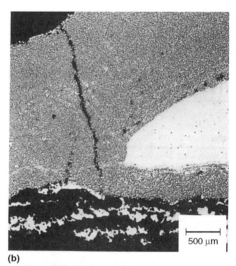

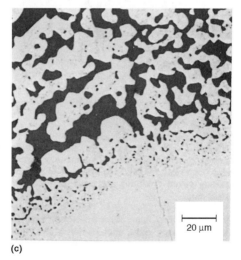

Fig. 2 Catastrophic sulfidation of an Inconel 601 furnace tube. The furnace atmosphere was contaminated with sulfur; the component failed after less than 1 month at 925 °C (1700 °F). (a) General view. (b) Cross section of the perforated area showing liquid-appearing nickel-rich sulfides. (c) Higher-magnification view of nickel-rich sulfides

Table 4 Results of 215 h sulfidation tests conducted in Ar-5H$_2$-5CO-1CO$_2$-0.15H$_2$S-0.1H$_2$O (vol%, balance argon)

	Average metal affected at temperature(a)					
	760 °C (1400 °F)		870 °C (1600 °F)		980 °C (1800 °F)	
Alloy	mm	mils	mm	mils	mm	mils
Alloy 6B	0.038	1.5	0.064	2.5	0.11	4.2
Haynes alloy 25	0.046	1.8	0.036	1.4	0.046	1.8
Haynes alloy 188	0.084	3.3	0.074	2.9	0.048	1.9
Haynes alloy 556	0.097	3.8	0.297	11.7	0.05	2.0
310	0.23	9.1	0.34	13.5	0.19	7.4
Incoloy alloy 800H	0.28	11.2	0.49	19.2	0.59	23.2
Haynes alloy 214	0.42	16.7	>0.45	>17.7	>0.45	>17.7
Inconel alloy 600	0.55	>21.7	>0.55	>21.7	>0.55	>21.7
Hastelloy alloy X	0.749	>29.5	>0.55	>21.7	>0.55	>21.7
Inconel alloy 601	0.749	>29.5	>0.55	>21.7	>0.55	>21.7

(a) Average metal affected = metal loss + internal penetration. Source: Ref 12

pot, which was in contact with molten salt due to the higher concentration of oxygen on the outside.

Molten-Metal Corrosion

Some heat treating operations involve molten metals. Lead is used as a heat treating medium. Cast iron and carbon steels have been used for components in contact with molten lead at temperatures to 480 °C (900 °F). In a 1242 h test in molten lead in an open crucible at 600 °C (1110 °F), Inconel alloy 600 was not appreciably attacked either at or below the liquid-metal surface (Ref 16). At 675 °C (1250 °F) for 1281 h in molten lead, Inconel alloy 600 suffered severe corrosion attack (Ref 16). Few corrosion data in molten lead have been reported in the literature.

The molten-zinc bath is used for galvanizing processes. Few corrosion data are available in the literature to allow engineers to make an informed materials selection for furnace components in contact with molten zinc. Carbon steel is generally used for the furnace components. Iron-nickel-cobalt-chromium alloys such as Haynes alloy 556 also have been reported for use as baskets. Molten tin is commonly used in the glass industry for solidification.

Corrosion of Plating, Anodizing, and Pickling Equipment

The metal finishing industry uses a wide variety of materials and processes to clean, etch, and plate metallic and nonmetallic surfaces to provide desired surface properties. The equipment is exposed to the same chemicals and fumes as the workpiece that is being finished in the process. The corrosive materials include solvents and surfactants for cleaning, acids and bases for etching and pickling, and solutions of metal salts and other compounds to plate a finish onto a substrate.

Physical, chemical, and electrochemical processes are all used to finish metal workpieces. An electroplating, anodizing, or pickling shop environment is highly corrosive to metals. The metals and alloys used in the construction of equipment to be employed in this environment must be inert to the environment or must be protected from it. While the effect of temperature is not significant in the corrosion of metal finishing equipment, unlike furnace heat treatment equipment, the environments are highly corrosive. However, most common metallic alloys and structural materials can withstand these environments, because they passivate in mildly acidic solutions and most alkaline solutions. Polymeric materials or polymer- and ceramic-coated metallic materials as well as wood and concrete are commonly used for construction of metal finishing equipment. However, stainless steels and carbon steels are also used when the metal can passivate in the solution. Figure 3 shows a typical metal finishing process sequence

Table 5 Corrosion rates in molten NaNO$_3$-KNO$_3$ at 675 and 705 °C (1250 and 1300 °F)

	Corrosion rate					
	675 °C (1250 °F), 14 day test		675 °C (1250 °F), 80 day test		705 °C (1300 °F), 30 day test	
Alloy	mm/yr	mils/yr	mm/yr	mils/yr	mm/yr	mils/yr
Haynes alloy 214	...	...	0.4	16	0.53	21
Inconel alloy 600	0.3	12	0.25	10	0.99	39
Hastelloy alloy N	0.33	13	0.23	9	1.22	48
Inconel alloy 601	0.48	19	0.48	19	1.24	49
Inconel alloy 617	0.36	14	...	...	...	...
Hastelloy alloy S	0.4	16	...	...	...	...
Inconel alloy 690	0.56	22	...	...	...	...
RA333	0.69	27	...	...	...	...
Inconel alloy 625	0.74	29	...	...	...	...
Hastelloy alloy X	1.04	41	...	...	...	...
Incoloy alloy 800	1.07	42	1.85	73	6.58	259
Haynes alloy 556	1.75	69	...	...	...	...
310	2.0	79	...	...	...	...
316	2.1	81	...	...	...	...
317	2.11	83	...	...	...	...
446	2.38	94	...	...	...	...
304	2.67	105	...	...	...	...
RA330	2.77	109	...	...	...	...
253MA	2.97	117	...	...	...	...
Nickel 200	8.18	322	...	...	...	...

Source: Ref 10

Table 6 Results of 30 day field tests performed in neutral salt bath containing BaCl$_2$, KCl, and NaCl at 845 °C (1550 °F)

	Average metal affected(a)	
Alloy	mm	mils
Haynes alloy 188	0.69	27
Multimet	0.75	30
Hastelloy alloy X	0.97	38
Hastelloy alloy S	0.1	40
Haynes alloy 556	0.11	44
Haynes alloy 214	1.8	71
304	1.9	75
310	2.0	79
Inconel alloy 600	2.4	96

(a) Average metal affected = metal loss+internal penetration. Source: Ref 8

Table 7 Results of 100 h tests performed in NaCl at 845 °C (1550 °F)

	Average metal affected(a)	
Alloy	mm	mils
Haynes alloy 188	0.05	2.0
Haynes alloy 556	0.066	2.6
Haynes alloy 214	0.079	3.1
304	0.081	3.2
446	0.081	3.2
316	0.081	3.2
Hastelloy alloy X	0.097	3.8
310	0.107	4.2
Incoloy alloy 800H	0.110	4.3
Inconel alloy 625	0.112	4.4
RA330	0.117	4.6
Inconel alloy 617	0.122	4.8
Haynes alloy 230	0.14	5.5
Hastelloy alloy S	0.168	6.6
RA333	0.19	7.5
Inconel alloy 600	0.196	7.7

(a) Average metal affected = metal loss+internal penetration. Source: Ref 14

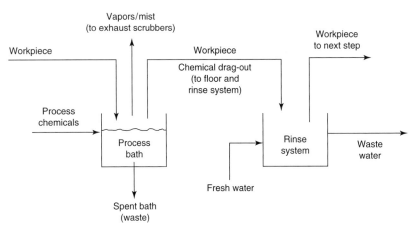

Fig. 3 Typical process sequence in a metal finishing shop. The process bath could be used for plating, anodizing, or pickling

identifying the sources of corrosive waste liquids and gases (Ref 17). These waste streams are not only highly aggressive toward the process equipment but pose significant health and safety hazards for the operator. The toxic and carcinogenic effects of hexavalent chromium baths for hard-chromium plating are well established. The effluent gaseous emissions from typical metal finishing operations are listed in Table 8 with reference to emissions from a hard-chromium plating bath (Ref 18).

Tanks

The common materials for constructing tanks for plating, anodizing, and pickling are carbon steel and stainless steel, usually one of the austenitic (18Cr-8Ni) varieties. Concrete, plastic, wood, and fiberglass, alone or with reinforcing, also are used. Of these, only the metals offer any serious corrosion problems. The nonmetals, however, must be carefully selected to prevent contamination of the processing solutions. Table 9 recommends tank materials for the various solutions found in the finishing shop. More information on designing the tank to prevent or minimize corrosion is available in the article "Designing to Minimize Corrosion" in *ASM Handbook,* Volume 13A, 2003. Particular attention must be paid to welded areas and other sharp corners where solution may shelter, giving rise to the formation of a corrosion cell and the potential for crevice corrosion.

Corrosion Protection for Steel Tanks. The outside of a carbon steel tank must be protected, regardless of what is contained in the tank. For example, a steel tank containing an alkaline cleaner will not need protection on the inside, because the alkaline cleaner will keep the steel passivated and at a very low corrosion rate; however, the outside surfaces, which are exposed only to water and the corrosive environment of the shop atmosphere, will corrode severely. Protection of the outside surfaces of corrosion-resistant steel tanks is generally not necessary, except where crevices are possible and oxygen cells can be set up. However, because the tank material is corrosion resistant, overlapping joints are frequently ignored; this provides an initiation site for corrosion.

Protective coatings for the outside of a steel tank should include a good two- or three-coat paint system. Epoxies, vinyls, or polyurethanes can be used. A system that consists of an epoxy primer applied over a clean, sandblasted surface, followed by two coats of a two-part urethane topcoat, will provide adequate protection in a finishing shop environment. Vinyl systems are immune to all solutions except very strong

Table 8 Relative concentrations of uncontrolled atmospheric emissions from various surface coating operations in the electroplating industry

Type of plating-related process operation	Chemical of concern	Concentration of chemical in bath, g/L	Typical current density A/cm^2	Typical current density A/in.^{2}	Cathode efficiency, %	Concentration of emissions relative to chromium emission from hard-chromium plating tanks (no units)
Electrolytic processes						
Hard-chromium plating bath	Chromium (+6)	160	2.3	1.5	15	1
	Sulfuric acid	2.5	2.3	1.5	15	1.6×10^{-2}
Decorative chromium plating bath	Chromium (+6)	164	0.16	1	15	6.8×10^{-1}
	Sulfuric acid	1.6	0.16	1	15	6.7×10^{-3}
Trivalent chromium plating bath	Chromium (+3)	10	0.08	0.5	95	3.3×10^{-3}
Nickel plating bath	Nickel	75	0.06	0.4	95	2.0×10^{-2}
Anodizing, sulfuric acid	Sulfuric acid	150	0.02	0.1	95	9.9×10^{-3}
Anodizing, chromic acid	Chromic acid	100	0.5	3	95	2.0×10^{-1}
Gold plating bath	Gold	5	0.009	0.06	95	2.0×10^{-4}
	Cyanide (CN)	10	0.009	0.06	95	3.9×10^{-4}
Copper strike bath	Copper	18	0.023	0.15	40	4.2×10^{-3}
	Cyanide (CN)	26	0.023	0.15	40	6.1×10^{-3}
Copper (cyanide) plating bath	Copper	20	0.05	0.3	100	3.8×10^{-3}
	Cyanide (CN)	30	0.05	0.3	100	5.6×10^{-3}
Copper (acid) plating bath	Copper	55	0.054	0.35	95	1.3×10^{-2}
	Sulfuric acid	55	0.054	0.35	95	1.3×10^{-2}
Cadmium plating bath	Cadmium	20	0.05	0.3	90	4.2×10^{-3}
	Cyanide (CN)	25	0.05	0.3	90	5.2×10^{-3}
Zinc (cyanide) plating bath	Zinc	35	0.054	0.35	70	1.1×10^{-2}
	Cyanide (CN)	100	0.054	0.35	70	3.1×10^{-2}
Zinc (chloride) plating bath	Zinc	40	0.043	0.28	95	7.4×10^{-3}
Zinc (alkaline noncyanide) plating bath	Zinc	15	0.023	0.15	75	1.9×10^{-3}
	Sodium hydroxide	115	0.023	0.15	75	1.4×10^{-2}
Silver strike bath	Silver	6	0.03	0.2	90	8.3×10^{-4}
	Cyanide (CN)	70	0.03	0.2	90	9.7×10^{-3}
Silver plating bath	Silver	50	0.16	1	100	3.1×10^{-2}
	Cyanide (CN)	50	0.16	1	100	3.1×10^{-2}
Electrocleaning	Sodium hydroxide	80	0.12	0.8	100	4.0×10^{-2}
	Sodium phosphate	15	0.12	0.8	100	7.5×10^{-3}
	Sodium metasilicate	25	0.12	0.8	100	1.3×10^{-2}
Nonelectrolytic processes						
Alkaline cleaning bath (typical)	Sodium hydroxide	80	...	NA	NA	NA
	Sodium phosphate	15	...	NA	NA	NA
	Sodium metasilicate	25	...	NA	NA	NA
Acid etch/desmut bath (typical)	Sulfuric acid	250	...	NA	NA	NA
Acid desmut/deoxidize	Nitric acid	500	...	NA	NA	NA
	Sulfuric acid	150	...	NA	NA	NA
Phosphate coating bath	Phosphoric acid	50	...	NA	NA	NA
Nickel plating bath (electroless)	Nickel	10	...	NA	NA	NA
	Sodium hypophosphate	20	...	NA	NA	NA
Anodizing sealer	Nickel	2	...	NA	NA	NA
	Chromium (+6)	20	...	NA	NA	NA
Chromate conversion bath	Chromium (+6)	45	...	NA	NA	NA
Hexavalent chromium passivation	Chromium (+6)	3	...	NA	NA	NA
Acid etch (for zinc plating)	Hydrochloric acid	100	...	NA	NA	NA

NA, not applicable. Source: Ref 18

Table 9 Relative corrosion resistance of tank materials, coatings, and linings for metal finishing shops

Material	Acid plating baths	Acid pickling baths			Alkaline cleaners/ caustics	Alkaline plating baths	Anodizing baths		Cyanide plating baths		HF and HBF_4 plating baths	Electro- cleaning solutions	1,1,1 tri- chloroethane
		HNO_3	HCl	H_2SO_4			CrO_3	H_2SO_4	General	For Cd and Zn			
Metals													
Carbon steel	NR	NR	NR	NR	VG	S	NR	NR	S	G	NR	G	S
Stainless steel	NR	NR	NR	NR	G	G	NR	NR	G	G	NR	G	VG
Liners													
Natural rubber	VG	NR	G	G	G	G	NR	G	G	G	NR	G	NR
Vinyl chloride	G	S	G	G	G	G	S	G	G	G	G	G	NR
Neoprene	VG	NR	NR	G	G	G	NR	G	G	G	G	G	NR
Chlorosulfonated polyethylene	G	NR	G	G	G	G	S	G	G	G	NR	G	NR
Butyl/chlorobutyl rubber	VG	G	G	G	G	G	NR	G	G	G	G	G	NR
Fluorocarbons	VG	G	G	G	G	G	G	G	G	G	G	G	G
Coatings													
Asphaltic coal tar/epoxy	G	NR	G	G	S	G	S	G	G	G	G	NR	NR
Furan	S	NR	S	S	G	G	NR	G	G	G	S	G	G
Epoxy	S	NR	S	NR	G	G	NR	G	G	G	G	G	NR
Polyester	G	G	G	G	NR	G	G	G	G	G	G	NR	NR
Vinyl ester	G	G	G	G	G	G	G	G	G	G	G	G	S
Urethane	G	G	G	G	G	G	NR	G	G	G	NR	G	NR
Hot melts													
Polyethylene	G	G	G	G	G	G	G	G	G	G	G	G	NR
Polypropylene	G	G	G	G	G	G	G	G	G	G	G	G	NR
Polyvinyl chloride	G	G	G	G	G	G	G	G	G	G	G	G	NR

(a) VG, very good; G, good; S, satisfactory; NR, not recommended

Table 10 Corrosion of titanium, steel, and aluminum in various corrosive media as a function of temperature and concentration

Corrosive medium	Concentration, wt%	Temperature		Corrosion rate(a)		
		°C	°F	Titanium	Type 304 stainless steel	Aluminum
Glacial acetic acid	99	...	...	A	B	C
Acetic anhydride	99	20	70	A	A	A
Mixture of $3HCl+1HNO_3$ or $1HCl+2HNO_3$	...	20	70	A	...	...
Aluminum chloride	25	60	140	A	C	C
Aluminum chloride	Saturated solution	100	212	A	C	C
Ammonium hydroxide	28	30	85	A	A	A
Barium chloride	20	100	212	A	B	...
Liquid bromine	...	30	85	Intense corrosion		C
Citric acid	50	100	212	A	C	B
Calcium chloride	27	Boiling	Boiling	A	B	...
Aqua regia	...	20	70	A	...	...
Copper chloride	40	Boiling	Boiling	A	C	...
Iron chloride	10	Boiling	Boiling	A	C	C
Formaldehyde	37	Boiling	Boiling	A	A	A
Water saturated with hydrogen sulfide	0.5	20	70	A	B	A
Aerated hydrochloric acid	0.5	35	95	A	C	B
	1.3	60	140	A	C	C
	1.0	100	212	A	C	C
	4.5	60	140	C	C	C
	15	35	95	B	C	C
	37	35	95	C	C	C
Hydrofluoric acid	1	20	70	C	C	C
Lactic acid	100	Boiling	Boiling	A	B	C
Magnesium chloride	20	100	212	A	B	C
Mercury chloride	Saturated solution	100	212	A	C	C
Aerated nitric acid	5–20	35	95	A	A	C
	5–60	100	212	A	A–C	C
Nonaerated nitric acid	65	Boiling	Boiling	A	C	C
	98	20	70	A	B	A
Oxalic acid	1	37	99	B	A	C
Phosphoric acid	10	80	175	C	C	C
Seawater	...	...	...	A	...	...
Sodium chloride	Saturated solution	Boiling	Boiling	A	C	...
Sodium hydroxide	10	Boiling	Boiling	B	B	C
Sodium sulfide	10	Boiling	Boiling	A	B	...
Molten sulfur	100	250	480	A	A	...
Sulfuric acid	1	20	70	A	B	B
	1	Boiling	Boiling	C	C	C
	5	20	70	A–B	B	B
	5	Boiling	Boiling	C	C	C
Aerated sulfuric acid	65–75	35	95	B	...	...
Tannic acid	25	100	212	A	...	...
Tartaric acid	50	100	212	A	...	...
Zinc chloride	20	100	212	A	...	...

(a) A, less than 0.127 mm/yr (5 mils/yr); B, between 0.127 and 1.270 mm/yr (5 and 50 mils/yr); C, more than 1.270 mm/yr (50 mils/yr). Source: Ref 19

solvents and are excellent coatings for tank surfaces. Vinyl systems are also adequate for the inside surfaces of rinse and holding tanks. Vinyls should be applied only to clean, sandblasted (white metal) surfaces and should be a complete system; that is, primer and topcoat should both come from the same manufacturer and should be matched to each other.

Coal tar and epoxy-modified coal tar systems are excellent tank coatings. These coatings can be used on the inside and outside surfaces of cold tanks, such as rinse tanks. Epoxy-modified coatings tend to soften in contact with hot cleaning, pickling, or plating baths. The epoxy-modified coatings, however, are excellent for tank understructures, such as the I-beams that keep the tanks off the floor and for the floor itself. All of these coatings can be applied also to corrosion-resistant steel tanks, if required. The coal tars and the epoxy-modified coal tars are especially useful as coatings on the bottoms of tanks and between the tank and floor supports, where they are used to prevent crevices in overlapping joints.

Nonmetal liners can also protect the insides of carbon steel tanks used in the metal-finishing industry. Materials for this application are listed in Table 9. Liners that are applied directly to the steel walls (hot melts, Table 9) should offer sufficient protection for the steel. The drop-in liners—whether they are made outside of the tank by molding (or other methods) or are made inside of the tank by welding or adhesive bonding—leave the tank wall unprotected and create an ideal crevice corrosion area. Water and solutions can leak between the liner and the tank. However, a heavy coat (or two light coats) of primer on these steel surfaces will offer sufficient protection so that wetting of the inside of the tank does not become a problem.

The 300- and 400-series stainless steels often use nitric acid/hydrofluoric acid mixed solutions for pickling at approximately 50 °C (120 °F) by dipping for as much as 20 min. These solutions are highly aggressive, because they promote hydrogen embrittlement and hydrogen blistering in carbon steels. Attack of refractory metals is accelerated by hydrofluoric acid in the presence of nitric acid. Thus, tanks for stainless steel pickling require special nonmetallic coatings.

Plating-, Anodizing-, and Pickling-Associated Equipment

In addition to tanks, other equipment, such as rectifiers, heaters, pumps, racks, bus bars, and wiring, are subject to corrosion in the environment of the finishing shop. These pieces of equipment do not come in direct contact with the corrosive chemicals but are exposed to an environment loaded with fumes.

Wirings and Bus Bars. Wire terminations, especially bare direct current (dc) terminations, are subject to corrosion. Only copper wire is recommended. Each termination should be overcoated with grease and inspected regularly. When a corrosion product is seen at a termination, it should be cleaned and reconnected, and the protective grease replaced. For dc connections to bus bars, a lead or lead-tin coating applied to the terminal lugs, the bar, washers, and bolts offers excellent protection. Copper bus bars require constant cleaning not only to remove corrosion products but also to ensure good connections to the rack splines. Tin, lead, and silver coatings on the bars provide temporary protection but do not eliminate the need for cleaning.

Racks. A properly designed rack will have bare areas only in the area of the hook, which is required for contact with the bus bar, and in the area of the tip, which is required to make contact with the part. Both of these areas are subject to corrosion from the shop atmosphere during storage and from the solutions during use. These areas also require constant maintenance. Aluminum and titanium anodizing racks are subject to anodization of the rack tips. Steel or stainless steel is often used as a rack material for carrying the load of the parts being treated—for example, a heavy crank shaft in a chromium plating bath. Both of these materials require complete masking, even where they make connection with the part. The copper current-carrying member of the rack also requires masking. Steels, titanium, and aluminum are commonly used for this application. Their corrosion rates in various corrosive media at different temperatures are shown in Table 10.

Anode splines are generally copper above the plating solution and can be coated with any rack coating, except at the hook where electrical connection is made. Anode splines submerged in the solutions, except those made of the metal that is to dissolve in the bath, must be inert or masked. Rack coatings can be used for this purpose. Inert anodes (that is, those that carry current but do not dissolve), such as titanium in a nickel bath or lead in a chromium bath, are corrosion resistant and require little care. Iron anodes used in an alkaline tin bath, however, are subject to rusting when not in the bath. These units require protection when not in use or require cleaning before use.

Pumps. Conventional metal pumps will corrode in acid solutions. The current trend is to make pumps out of plastic (especially pumps that will be used for small tanks). The body of the pump is usually a grade of phenolic, and the impellers are hard rubber. In this case, the only metal part exposed to the solution is the impeller shaft, which can be fabricated from stainless steel or one of the acid-resistant nickel or cobalt alloys. Metal pumps require constant care to prevent or minimize corrosion.

Heaters. Hot water or steam tank heaters should be treated as part of the boiler system in order to prevent corrosion from the inside. Iron can be used in alkaline solutions, copper can be used in water and neutral solutions, and stainless steel can be used in mildly acid solutions. Stabilized lead can be used in chromic and fluoboric acid solutions. Carbon and ceramic heat exchangers can be used in all solutions except those containing fluoborates and fluorides. The same is true of immersion ceramic or quartz heaters if the framework is coated.

ACKNOWLEDGMENTS

This article has been revised from the previously published article, "Corrosion of Metal Processing Equipment," in *Corrosion*, Volume 13, *ASM Handbook,* 1987, p 1311–1316. The author thanks G.Y. Lai and C.R. Patriarca who wrote the section "Corrosion of Heat-Treating Furnace Accessories" and E.C. Groshart who wrote the section "Corrosion of Plating, Anodizing, and Pickling Equipment" in the previous article.

REFERENCES

1. N.P. Lieberman and E.T. Lieberman, Ed., *Working Guide to Process Equipment,* McGraw-Hill, 2003, p 430
2. C. Soares, Ed., *Process Engineering Equipment Handbook,* McGraw-Hill, 2002, p M-34
3. *Selection of Stainless Steels,* American Society for Metals, 1968
4. E.N. Skinner, J.F. Mason, and J.J. Moran, *Corrosion,* Vol 16, p 593
5. INCONEL alloy 601 brochure, INCO Alloys International, Inc.
6. F.K. Kies and C.D. Schwartz, *J. Test. Eval.,* Vol 2 (No. 2), March 1974, p 118
7. M.F. Rothman, Cabot Corporation, private communication, 1985
8. D.E. Fluck, R.B. Herchenroeder, G.Y. Lai, and M.F. Rothman, *Met. Prog.,* Sept 1985, p 35
9. INCO Alloys International, Inc., unpublished research
10. G.Y. Lai, in *High Temperature Corrosion in Energy Systems,* M.F. Rothman, Ed., Symposium Proceedings, The Metallurgical Society, 1985, p 227, 551
11. G.R. Rundell, Paper 377, presented at Corrosion/86 (Houston, TX), National Association of Corrosion Engineers, March 1986
12. G.L. Swales, in *Behavior of High Temperature Alloys in Aggressive Environments,* I. Kirnan et al., Ed., Proceeding of the Patten International Conference, The Metals Society, 1980, p 45
13. B. Mishra and D.L. Olson, Corrosion of Refractory Metal in Molten Lithium and Lithium Chloride, *Miner. Process. Extr. Metall. Rev.,* 2000, p 1–20
14. M.F. Rothman and G.Y. Lai, *Ind. Heat.,* Aug 1986, p 29
15. D.E. Fluck, Cabot Corporation, private communication, 1985
16. INCO Alloys International, Inc., unpublished research

17. "Guides to Pollution Prevention: The Metal Finishing Industry," Report EPA/625/R-92/011, United States Environmental Protection Agency, Oct 1992
18. "The Metal Finishing Risk Screening Tool (MFRST), Technical Documentation and User Guide," Report EPA/600/C-01/057, United States Environmental Protection Agency, July 2001
19. I. Klinov, Ed., *Corrosion and Protection of Materials Used in Industrial Equipment*, Consultants Bureau, New York, 1962

SELECTED REFERENCES

- B.D. Craig and D.S. Anderson, *Handbook of Corrosion Data*, 2nd ed., ASM International, 1995
- J.R. Davis, *Metals Handbook Desk Edition*, 2nd ed., ASM International, 1998
- J. Endo, S. Ohba, and T. Anzai, Virtual Manufacturing for Sheet Metal Processing, *J. Mater. Process. Technol.*, Vol 60 (No. 1), 1996
- *Metal Finishing, Guidebook and Directory*, Metals and Plastics Publications, Inc., 2003
- *Surface Engineering*, Vol 5, *ASM Handbook*, ASM International, 1994

Corrosion in the Mining and Mineral Industry

Revised by B. Mishra, Colorado School of Mines
J.J. Pak, Hanyang University, Korea

THE EXTRACTION, production, and application of metallic materials and alloys, as well as commercially used minerals, entirely depend on the performance and efficiency of the mining and mineral processing industry. Because this industry is responsible for collecting mineral and ore deposits from earth and physically and chemically separating it from the gangue to upgrade the metallic value, it not only uses heavy equipment, machinery, and tools but also employs chemically aggressive solids and fluids for processing. The mining, mineral-processing, and extractive metallurgy industries are concerned with a wide range of corrosive and erosive media and, therefore, must consider materials selection as the most important approach to corrosion resistance. High-manganese steel in crushing and grinding service, various types of cast (hard) iron grinding balls and mill liners, hardenable and carburized grades of low-alloy steel gears and geared transmissions mining machinery, steel and carbide tools and drills, high-nickel and cobalt hardfacing alloys, stainless steel leaching tanks and pumps, and high-nickel-chromium and titanium alloys for the most severe environments are commonly used. This wide selection of materials and their service lives affect operation cost and productivity that directly impacts the cost of the extracted metal. With the increased mechanization in mines, corrosion problems and material performance become increasingly important to the mining industry and the search for better materials in terms of their performance and costs is still underway (Ref 1–10). Table 1 lists some of the engineering materials that have been used for mining and milling equipment.

Corrosion associated with the mining industry can be characterized as electrochemical attack enhanced by abrasion. All ingredients for corrosion attack are available, including highly conductive water, grinding media, dissimilar materials, oxygen, large pH range, stresses, and the presence of many well-known corrosive species in solution. Mine atmospheres and mine waters are unique in that they vary widely from mine to mine. For example, temperatures have been found to range from approximately 5 to 30 °C (40 to 90 °F) in coal mines and above 40 °C (100 °F) in metal mines. Refrigeration and air conditioning have become necessary for improved working conditions. Humidity levels between 90 and 100% are common. Mine water also varies in mineral content, pH, and corrosivity (Ref 11, 12).

Tables 2 (Ref 13) and 3 (Ref 14) list the contents of some mine waters. These constituents have been found in mines in the United States, Canada, and in other countries, although the amounts can vary. Values of pH range from 2.8 (very acidic) to 12.3 (basic). High values are often the result of the lime content of the cement added to backfill. Chloride ion (Cl^-) values show a wide range (from 5 to 25,000 ppm), and sulfate ion (SO_4^{2-}) values range from 57 to 5100 ppm. Chloride and sulfate are considered to be the most aggressive ions present in mine waters and account for the high corrosivity of most mines.

Important constituents of dissolved colloidal or suspended matter that contribute to the corrosive environment can be classified as dissolved gases, mineral constituents, organic matter, and microbiological organisms (Ref 15). Dissolved oxygen enhances the corrosion process, especially in waters with low pH, but oxygen content can vary widely in mine waters. Oxygen concentration also decreases with increasing temperature. One example of the effect of oxygen on corrosion rate is that aeration with oxygen has been found to increase grinding ball wear by 13 to 16% (Ref 16).

Carbon dioxide (CO_2) in mine water is associated with the carbonate ion (CO_3^{2-}) and the bicarbonate (HCO_3^-) ion content. Calcium bicarbonate ($Ca(HCO_3)_2$) is reported to be the predominant source, and magnesium carbonate ($MgCO_3$) has been found to be a less-active secondary source (Ref 15). Carbon dioxide has little effect on corrosion rate except when pH levels are below 7 where CO_2 can react with water to form a weak acid.

Mineral content of mine water begins with the breakdown of iron sulfide minerals, principally pyrite and marcasite, that are generally associated with most mineralization species. Oxidation of pyrite produces sulfuric acid (H_2SO_4). Consequently, mine water with pH as low as 2 are produced. The acid water accelerates the breakdown of minerals, increasing the concentration of calcium (Ca^{2+}), magnesium (Mg^{2+}), sodium (Na^+), potassium (K^+), and manganese ions (Mn^{2+}) in mine waters. These corrosion reactions start producing hydrogen that diffuses into the structural steel material reaching a critical concentration for hydrogen embrittlement. This has been shown to lead to stress-corrosion cracking (SCC) of rock bolts in mines (Ref 17).

Anaerobic and aerobic microorganisms are well-known corrosion-producing agents. Principal acid-producing species are *Thiobacillus thiooxydans*, an aerobic species (Ref 15, 18). While these bacterial species are exploited to improve leaching of mineral matters in hard-to-leach ores, their presence oxidizes sulfur or sulfur compounds, producing H_2SO_4 and contributing to the acidity of mine water. Another species of aerobic bacteria, *Ferrobacillus ferro-oxydans*, is associated with the *Thiobacillus* type. When both types of bacteria are present, their synergistic effect has been reported to increase H_2SO_4 production by four times as compared with the production rate when no bacteria are present (Ref 19). Acid mine waters produced by these microorganisms can approach very acidic conditions. Anaerobic microorganisms, such as *Desulfovibrio desulfuricans*, are a sulfate-reducing type and are responsible for rapid corrosion of iron and steel structures (Ref 20). They are found in most soils. As mentioned earlier, they also reduce sulfates and sulfides by using available hydrogen from either organic compounds or aqueous electrolyte at the cathodic interface of metal, producing hydrogen sulfide (H_2S).

High humidity, high ambient temperature, dusts, fumes, breakdown of minerals that form acid mine waters, and microbiological microorganisms all contribute to corrosion and lead to degradation of mining equipment. Examples of the corrosion of principal mining equipment are discussed subsequently.

Table 1 Selection of materials for mining and milling equipment

Application	Environment	Materials
Crushing and grinding	Heavy pressure, shock-impact loading	Austenitic manganese steels (ASTM A128), 4300 series, ASTM A579 alloy steel, 8600 series
Mill liners, grates, and abrasion-resistant plates	Severe gouging, crushing impact and wear, wet (pH 5–8)	Austenitic manganese steels, martensitic chromium-molybdenum white cast iron, martensitic high-chromium white cast iron, martensitic nickel-chromium white iron, martensitic medium-carbon chromium-molybdenum steel, austenitic 6Mn-1Mo steel, pearlitic high-carbon steel, pearlitic white cast iron
Grinding balls (Ref 11, 12)	Severe gouging, crushing impact and wear, wet (pH 6–8)	Pearlitic white cast iron, martensitic white cast iron, forged (0.8% C) steel, 4155, Ni-hard type 1, Ni-hard type 4
Grinding rods (Ref 11, 12)	Severe gouging, crushing impact and wear, wet (pH 6–8)	Heat treated alloy steel, 52100 (UNS G52986), hot-rolled AISI 1095 modified with 1.2% Mn, hot-rolled 1095 with 0.4% Mn
Gearing for mining machinery	Wet, lubricated (pH 5–8) wear, light duty	Carburized 1015, 1020, 1022, 1117, 1118, heat treated 4340, 8645
	Wet, lubricated (pH 5–8) wear, moderate duty	Carburized 8628, 4620, 4615, or equivalent
	Wet, lubricated (pH 5–8) wear, heavy duty	Carburized 4820, 4320, 2320, or equivalent; nitrided 4340, 4140, 4350, and 2.5% Cr steel
Load-haul-dump equipment	Wet (pH 5–8) wear, impact	1020, cast carbon steel, cast austenitic manganese steels (ASTM A128), cast ASTM A 579 steel, ASTM A514 steel
Percussion drilling tools	Wear, impact, gouging (pH 6–8)	Carburized 4320, 8620, and 9315; quenched-and-tempered 4140
Hardfacing	High-impact, wet (pH 6–8)	Austenitic manganese steels
	Unlubricated metal-to-metal rolling or sliding	Self-hardened, air-hardened steels
	Highly abrasive conditions, wet (pH 6–8)	High-carbon high-chromium white cast iron, high-chromium white cast iron
	Sliding abrasion on cutting edge of drilling tools, wet (pH 6–8)	Special tungsten- and boron-containing weld deposits
	Abrasion at high temperature and/or corrosion	High-nickel or high-cobalt weld deposits
Pumps	pH (0–13), abradants	Type 304 or 316 stainless steel, Ni-hard types 1 and 4, 27% Cr white cast irons
	pH (0–13), abradants	Low-carbon, high-manganese steels
	pH (0–13), abradants	Low-alloy cast iron
	pH (0–13), abradants	ACI CD-4MCu
	pH (0–13), abradants	ACI CF-3M (low carbon for as-welded corrosion resistance)
	pH (0–13), abradants	ACI CN-7M (niobium or titanium for as-welded corrosion resistance)
	pH (0–13), abradants	ASTM A484 steel (low carbon for as-welded corrosion resistance)
	High-temperature environments	ASTM A743 or A744 nickel-base alloy, grade CZ-100
Flotation cells	Corrosive, pH (0–5)	Ni-hard type 1, type 316 stainless steel
Paddles	Corrosive, pH (0–5)	Fiberglass-reinforced plastic and rubber
Spirals	Abrasive	Ni-hard type 1
Classifiers blades	...	Ni-hard type 4
Ore chutes	Impact, gouging, abrasion, pH 6–8	Ni-hard, nickel-containing manganese steel
Scrapers	Impact loading, gouging, abrasion	Cast ASTM A579 steel, Ni-hard cast iron, hard cast irons
Wire rope	Corrosive-abrasive, pH 2–12	Kevlar (E.I. DuPont de Nemours and Co., Wilmington, DE), steel wire rope
Piping	Corrosive-abrasive	Type 316 stainless steel, CN-7M, Ni-hard cast irons, rubber covered fiberglass-reinforced plastic
Scrubbers	Off-gas products	High-grade nickel alloys
Chain conveyors	Corrosive-abrasive	Plated (nickel, cadmium, or zinc) steels

Table 2 Analyses of some Canadian mine waters. Values have been converted from grams per liter to parts per million.

Mine number(a)	pH	Ionic species present, ppm						Dissolved solids, ppm
		Cl^-	SO_4^{2-}	Ca^{2+}	Mg^{2+}	HCO_3^-	Cu^{2+}	
1a	5.4	20	1080	210	80	<100	2	1740
1b	5.9	60	1380	390	55	<100	1	2290
1c	12.3	40	240	520	0.7	<100	<0.1	2610
2a	6.3	5	57	14	3	nil	0.3	94
2b	2.8	170	5100	230	580	<50	23	8740
3	8.3	9	120	49	4	<100	<0.5	320
4	7.0	10,500	1000	3450	490	200	<0.1	24,700
5	3.2	840	540	470	180	nil	<0.2	2640
6	3.3	<5	1830	31	290	nil	<0.2	3010

(a) 1a, b, and c and 2a and b are different samples from the same mine. (b) Dried at 110 °C (230 °F). Source: Ref 13

An analysis of two old mine shafts in South Africa, which had extensive corrosion problems, has shown that the lateral loads developed by the dynamic interaction between the traveling conveyance and the shaft steelwork can exceed the stress limit of corroded bunton at high hoisting speeds. Repair recommendations are (Ref 21):

- Hoisting speed should not exceed 10 m/s (33 ft/s)
- The limiting residual thickness of buntons is in excess of 5 mm (0.2 in.); otherwise, more comprehensive measurements and a replacement program must be implemented.

Mine Shafts

Mine shafts are among the most important components of the infrastructure in an underground mine because they are used for transporting men and materials to and from the ore body, as well as hoisting the ore to the surface. Shaft steelwork, comprising horizontal buntons and vertical guides, is susceptible to corrosion and mechanical damage (Ref 21). The conveyances that are located in the shaft on guides run continuously from top to bottom of the shaft. These guides are typically attached to a horizontal grid of beams, commonly referred to as buntons, spaced at intervals of 5 or 6 m (16.4 or 19.7 ft). These form the lifeline of the mine shaft but are susceptible to corrosion damage that results from various sources such as spillage, wet conditions, and poor ventilation. A structural collapse within a shaft, in addition to possible injury or loss of life, could lead to significant production loss.

Wire Rope

Mine shaft depths of 1830 m (6000 ft) are not uncommon. Gold mines in South Africa approach depths of 2285 m (7500 ft) (Ref 22). The hoisting equipment used in these mines, especially ropes, on which lives of personnel depend, is subjected to the corrosive environment of the mines. Although wear is also a factor, corrosion is perhaps the most serious aspect of mine safety. Corrosion is difficult to evaluate and is a more serious cause of degradation than abrasion (Ref 23). If corrosion is evident, the remaining strength cannot be calculated with safety, nor is there any reasonable way to determine whether the rope is safe except based on the judgment of the inspector. Statistical analysis from the results of rope tests on mine-hoist wire ropes has shown that 66% of the ropes exhibited the greatest strength loss in the half of the ropes nearest the conveyance (Ref 24). This is the portion of the rope that is in contact with the shaft environment during most of its service life.

Replacement of hoist rope is a routine procedure in most mines and is suggested every 18 to 36 months, depending on the mine environment and use (Ref 13, 25). Some regulatory agencies will not allow the use of a shaft rope on which marked corrosion is evident (Ref 23). Adequate service life of hoist rope is economically desirable. Therefore, the composition of present-day hoist rope has been extensively studied. Carbon steel strand wire has competition from such substitute materials as stainless steel (Ref 26) and synthetic fibers (Ref 27). Austenitic stainless steel rope (15.5 to 18.5% Cr and 11 to 13% Ni) is available and has endurance strength of 72 to 83% of that of carbon steel wire. Much can be said for synthetic fiber rope construction. For example, aramid fiber rope has exceptional strength-to-weight ratios, outstanding tension-tension-fatigue performance (Ref 27), and excellent corrosion resistance.

Rock Bolts

Rock bolts provide reinforcement above mine openings such as roadways and roof support in underground mines. More than 120 million rock bolts are used per year in the United States mining industry (Ref 23). The rock bolts are bonded into the rock strata, keeping it together just like the steel in reinforced concrete, and provide a clamping action on the rock. The chemical bolt is the most common rock bolt currently in use (Ref 28). An appropriate hole is drilled, the chemical cartridge and bolt are introduced into the hole, and the bolt is rotated and simultaneously pushed through the chemical cartridge to mix together the resin and catalyst as shown in Fig. 1 (Ref 17). Any failure of a rock bolt is a potential concern. Rock bolts made of medium-carbon steel in a number of design variations are subject to corrosion attack in the mine environment. In sulfide mines, the rock bolts have been reported to fail within 1 year by breaking at a distance of approximately 355 mm (14 in.) inside the drill hole (Ref 13). This rock bolt failure has been related to SCC. Roof falls are associated with such rock bolt failures.

Stress corrosion cracking may occur whenever a highly stressed steel is in the presence of an aggressive environment where hydrogen is produced on the steel surface. This ingress of hydrogen into the steel microstructure leads to hydrogen embrittlement. The stress corrosion cracks initiate and grow slowly. During this phase that may last for months or years, there may be no indication of any damage. Fast fracture occurs when the stress corrosion crack reaches a critical length, as determined by the applied stress and the fracture toughness of the steel. Reports indicate that the critical crack length can be of the order of only 1 to 1.2 mm (0.04 to 0.05 in.) for rock bolts. Figure 2 shows a typical fracture surface for a rock bolt failed in service by SCC (Ref 17). The most prominent feature is the small thumbnail shaped area on the right side of the fracture indicative of SCC. The rest of the bolt shows a shiny irregular surface. From the examination of fracture surfaces of rock bolts during service by SCC, the following SCC mechanism has been proposed. Hydrogen diffuses into the material, reaching a critical concentration level. The embrittled material allows a crack to propagate through the brittle region. The crack is arrested once it propagates outside the brittle region. Once the new crack is formed, corrosion reactions start producing hydrogen that diffuses into the material, causing further embrittlement and cracking (Ref 17).

Pump and Piping Systems

Corrosion in pump and piping systems is well known in the mining and mineral-processing industries. Centrifugal slurry pumps are used almost universally to move mixtures of abrasive solids and liquids around wet mineral processing operations. These pumps are generally of rugged construction and allow generous clearances for passage of large solids. Useful pump life in these applications can range from a few weeks to a few years depending on the types of slurries handled. The direct cost of slurry pump wear parts for the mining industry and the associated downtime and labor costs are significant. Reducing these total costs requires a better understanding of pump wear patterns and wear rates (Ref 29).

Table 3 Analyses of three Ontario mine waters. Values are in parts per million unless otherwise stated.

Source of mine water	Levack Mine (sulfide)	Helen Mine (iron)	Leitch Mine (gold)
Carbon dioxide, calculated	...	10	4
pH	3.4	7.2	7.6
Calcium carbonate (total)	0.0	83.1	77.7
Hardness (total)	1175	1079	2054
Calcium carbonate	0.0	996	1977
Calcium	405	245	744
Magnesium	42	113	48
Iron	...	...	...
Total	0.46	0.06	1.4
Dissolved	0.36	0.0	0.11
Aluminum	1.4	0.3	0.12
Manganese	...	...	...
Total	3.5	0.29	2.0
Dissolved	3.0	0.00	1.2
Copper	1.58	0.005	0.05
Zinc	0.54	1.7	0.05
Sodium	119	25.0	6870
Potassium	16.6	8.4	33
Ammonia	...	...	...
Total	23.8	...	...
Dissolved	0.56	...	...
Carbonate	0.0	0.0	0.0
Bicarbonate	0.0	101	94.7
Sulfate	685	830	126
Chloride	629	125	12,078
Fluoride	0.4	0.18	0.37
Phosphate	0.01	0.0	0.37
Nitrate	0.3	37	19
Silica	20	3.1	5.6
Sodium, %	16	4.8	88

Source: Ref 14

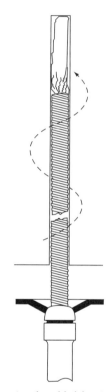

Fig. 1 Illustration of a rock bolt for mine reinforcement. The rock bolt is rotated, and at the same time pushed through the chemical cartridge, mixing the resin components. Source: Ref 17

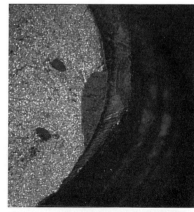

Fig. 2 Typical fracture surface of a rock bolt that failed due to stress-corrosion cracking. See text for details. Source: Ref 17

Because recognition of corrosion type is so important in diagnosing corrosion problems and their prevention, corrosion types are discussed and illustrated in the following sections in this article.

Uniform Corrosion. The most common form of pump corrosion is characterized by uniform attack on the entire exposed surface. Figure 3 shows uniform corrosion on a stainless steel pump impeller that was exposed to 50% phosphoric acid (H_3PO_4) and 10% gypsum pumping fluid for approximately 1 year. Uniform corrosion is predictable and, therefore, allows the incorporation of safety factors in pump and piping design.

Pitting Corrosion. One of the most catastrophic forms of corrosion encountered on pump cases is pitting attack. The relatively stagnant liquid conditions, which exist during the pump operation as well as periods of shutdown, create ideal conditions for pit initiation. Use of dissimilar metals in contact with one another in a corrosive liquid sets up a galvanic potential to accelerate preferential corrosion of the more active metal (Ref 30). Pitting corrosion on a UNS J92900 (Alloy Casting Institute [ACI] CF-8M) stainless steel casting pump case is illustrated in Fig. 4. This pump case, which was exposed to low pH and high Cl^- concentration, failed after approximately 3 years of service. Methods of preventing pitting failures in pump systems would be to prevent any liquid ingress into the contact between dissimilar metals using O-rings or other suitable seals. The use of a more corrosion-resistant material or coatings may also reduce the problem of pitting corrosion (Ref 30).

Crevice Corrosion. Figure 5 shows crevice corrosion on an ACI CF-8M cast stainless steel pump case that was gasket sealed on the discharge flange. Attack is evident in the region where the gasket was placed. It is important to design mining and milling equipment for easy drainage and cleaning in order to prevent the buildup of stagnant water that will produce concentration cells and lead to crevice corrosion and pitting. Figure 6 illustrates both poor and improved design for avoidance of localized attack. More information on proper design is available in the article "Designing to Minimize Corrosion" in *Corrosion*, Vol 13A, of the *ASM Handbook* (refer to pages 929 to 939).

Erosion-Corrosion. Slurry erosion-corrosion can cause significant materials wastage in pump systems. During mill processing, fluids consisting of particulates in a corrosive medium are usually pumped. Pumping this slurry promotes erosion-corrosion in piping, tanks, and pumps. Erosion-corrosion is a function of the fluid velocity and the nature of the particulates and fluid. Erosion-corrosion of a UNS N08007 (ACI CN-7M) stainless steel cast impeller after exposure to hot concentrated H_2SO_4 with solids present is shown in Fig. 7. Erosion-corrosion is also evident in Fig. 8, which shows that the erosion-corrosion damage increased on the portion of the impeller that had the greatest fluid velocity. This impeller, cast from an abrasion-resistant white iron, was used to pump fluids containing 30% solids (iron ore tailings) at a pH of 11.2.

The following procedures can be used to reduce erosion-corrosion or to increase the service lives of piping and pumping systems:

- Increase the thickness of pipes
- Use larger inside diameter pipes to reduce fluid velocity for the transport of a specific fluid volume
- Streamline bends in piping to ensure laminar flow
- Use nonmetallic ferrules inserted in the inlet ends of pipes

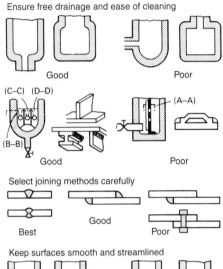

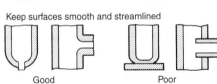

Fig. 3 Uniform corrosion of ACI CD-4MCu cast stainless steel pump impeller after 1 year in an environment containing 50% H_3PO_4 and 10% gypsum. Courtesy of A.R. Wilfley & Sons, Inc., Pump Division

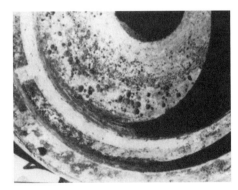

Fig. 4 Pitting corrosion of an ACI CF-8M stainless steel pump case used to pump a nickel plating solution with a high concentration of Cl^- and a high operating temperature. This damage occurred during 3 years of service. Courtesy of A.R. Wilfley & Sons, Inc., Pump Division

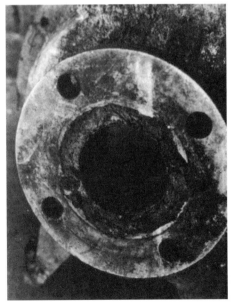

Fig. 5 Crevice corrosion at the intake flange of an ACI CF-8M stainless steel pump case. Notice that the corrosion damage occurred under the gasket. Courtesy of A.R. Wilfley & Sons, Inc., Pump Division

Fig. 6 Poor and improved engineering design to avoid crevice corrosion. Source: Ref 31

Fig. 7 Erosion-corrosion of ACI CN-7M stainless steel pump components that pumped hot H_2SO_4 with some solids present. Note the grooves, gullies, waves, and valleys common to erosion-corrosion damage. Courtesy of A.R. Wilfley & Sons, Inc., Pump Division

- Design for easy replacement of parts that experience severe erosion-corrosion
- Use coatings that produce an erosion-corrosion resistant barrier, such as rubber coatings (Ref 32)

Fig. 8 Erosion-corrosion of an abrasion-resistant iron pump runner used to pump 30% iron tailings in a fluid with a pH of 11.2. This runner had a service life of approximately 3 months. Note that most of the damage is on the outer peripheral area of the runner where fluid velocity is the highest. Courtesy of A.R. Wilfley & Sons, Inc., Pump Division

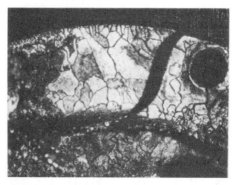

Fig. 9 Intergranular corrosion of an ACI CN-7M stainless steel pump component that contacted HCl-Cl$_2$ gas fumes. Note the grain-boundary attack. Courtesy of A.R. Wilfley & Sons, Inc., Pump Division

Materials selection is an important consideration for erosion-corrosion resistance. Alloy hardness has also been shown to be a factor in erosion-corrosion resistance. Generally, soft alloys are more susceptible to erosion-corrosion than their harder counterparts, but the relative hardness properties of the alloy can be misleading because the hardening mechanism affects resistance to erosion-corrosion (Ref 11). For example, solid-solution hardening has been found to offer greater resistance than that provided by conventional heat treatment. One example of this is the cast precipitation-hardening alloy UNS J93370 (ACI CD-4MCu), which outperforms UNS N08007 (Alloy 20 [CN-7M]) and austenitic stainless steels in many applications. Economics often enters into materials selection. Cast iron is relatively more economical and frequently exhibits better erosion-corrosion resistance than cast steel. High-silicon cast iron (14.5% Si) has been found to be an economical selection.

Active-passive materials, such as stainless steels and titanium, owe their corrosion resistance to their ability to develop a protective passive oxide film. This protective film, however, can be continuously damaged by erosive-abrasive processes. Damage of this protective film is a required condition for the loss of localized corrosion resistance. Selection of passive alloys should be based only on experience and/or laboratory test results.

Joints must be reliable. Welded pipe, such as carbon steel or stainless steel, is free of flanges but is costly to install. Nonwelded joints are susceptible to crevice corrosion; therefore, stainless steel, in particular, will not attain its expected service life. Recent development of materials with optimum erosion and erosion-corrosion properties and their evaluation techniques are available elsewhere (Ref 1–4, 29, 33–34).

Intergranular corrosion is common to stainless steel pump castings. Figure 9 shows the excessively attacked grain boundaries. A quick test to identify this type of corrosion is to peen the pump casting with a small hammer. Loss of acoustical properties is evidence of intergranular grain-boundary attack, especially in sensitized stainless steel castings. A form of intergranular corrosion associated with weld deposits that is commonly called weld decay, is shown in Fig. 10, which illustrates a field-weld repair of an ACI CN-7M stainless steel impeller that was not postweld heat treated (solution annealed and quenched) to restore corrosion resistance. This weld-repaired casting was exposed to a phosphoric anhydride (P$_2$O$_5$) solution at 80 °C (175 °F). It is evident that the weld decay occurred in the heat affected zone of the weld deposit.

Galvanic corrosion between two stainless steels is illustrated in Fig. 11. These AISI type 304 stainless steel stud bolts held together an Alloy 20 (ACI CN-7M) pump housing. The bolts became anodic to the housing in 45% H$_2$SO$_4$ and subsequently failed. Table 4 lists various combinations of pump and valve trim materials and indicates the combination that may be susceptible to galvanic attack. The basic rule for materials selection is that two alloy systems that

Fig. 11 Galvanic corrosion of type 304 stainless steel stud bolts that fastened two Alloy 20 (ACI CN-7M) pump components. The pump was pumping 45% H$_2$SO$_4$ at 95 °C (200 °F). The stud bolts were anodic to the Alloy 20 pump housings. Courtesy of A.R. Wilfley & Sons, Inc., Pump Division

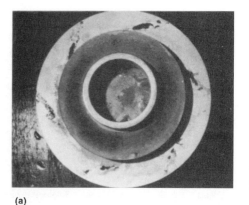

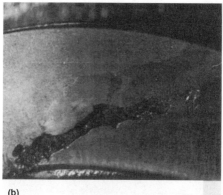

Fig. 10 Weld decay of an ACI CN-7M stainless steel pump impeller that was field weld repaired with no postweld heat treatment. The pump service was P$_2$O$_5$ solution at 80 °C (175 °F). (a) Overall view of impeller. (b) Closeup view of the weld repair and the associated weld decay, which occurred adjacent to the weld deposit. Courtesy of A.R. Wilfley & Sons, Inc., Pump Division

Table 4 Galvanic compatibility of materials used for pump components

	Trim		
Body material	Brass or bronze	Nickel-copper alloy	Type 316
Cast iron	Protected	Protected	Protected
Austenitic nickel cast iron	Protected	Protected	Protected
M or G bronze 70–30 copper nickel	May vary(a)	Protected	Protected
Nickel-copper alloy	Unsatisfactory	Neutral	May vary(b)
Alloy 20	Unsatisfactory	Neutral	May vary(b)

(a) Bronze trim commonly used. Trim may become anodic to body if velocity and turbulence keep stable protective film from forming on seat. (b) Type 316 is so close to nickel-copper alloy in potential that it does not receive enough cathodic protection to protect it from pitting under low-velocity and crevice conditions. Source: Ref 35

are the most widely apart on this list in terms of the galvanic series should be avoided.

Tanks

Most tanks are made from low-carbon steel for economic considerations. The most common corrosion protection for these tanks is the use of coatings and linings. Cathodic protection can also be used and is commonly employed in conjunction with a coating (Ref 32). Coating materials can be classified as cement, epoxy, epoxy-phenolic, neoprene, latex, sprayed polyresin coating, polyesters and vinyl esters (heavy coatings), and baked phenolic. Steel tanks are also lined with natural rubber, synthetic elastomers, rubber-backed polypropylene, and glass. Glass lining would require an oven bake.

Stainless steel and titanium alloys have also been used for tanks. Their use depends on their specific corrosion resistance to the solution. Selection of an alloy type becomes a question of economics.

Reactor Vessels

A variety of materials are used, depending on the corrosivity of the media being contained. With neutral or alkaline pH, carbon steels are often used. With increasing corrosivity, consideration is first given to the austenitic stainless steels; then iron-nickel-chromium superalloys; and finally, nickel-base superalloys. For special environments, copper, copper-nickel, and nickel-copper (Monel-type) alloys are used (Ref 36). Titanium is known to have excellent corrosion resistance in some of the most aggressive solutions.

Cyclic Loading Machinery

The mining and mineral industry uses large numbers of rotary and cyclic loaded equipment. This equipment is subject to fatigue and corrosion fatigue. Table 5 illustrates the significant reduction of fatigue strength of various materials that were tested in mine water and compared with fatigue strength in air. This problem can be reduced by designing heavier sections into the part to reduce load and by applying protective coatings (Ref 32).

Corrosion of Pressure Leaching Equipment

Hydrometallurgical applications in mineral processing range from low-tech heap operations to more sophisticated high-pressure leaching processes performed in autoclaves. The temperatures are approximately 260 °C (500 °F) and the pressures approximate 5 MPa (725 psi) in an acidic environment when leaching nickel laterite ores. Acid leaching or high-pressure acid leaching (HPAL) has become the commercial alternative for copper, gold, nickel, uranium, molybdenum, zinc, and other metals extraction (Ref 37). The aggressive acid environments in these pressure leaching operations present serious corrosion problems for the containment materials. While heap leaching can be performed using polymeric or other nonmetallic materials, a high-pressure and high-temperature process requires a metallic- or a ceramic-lined metallic container as the autoclave. Besides the acid environment, presence of oxygen and the multivalent metal ions in the ore solutions create an environment that can be handled by few metallic alloys. Titanium alloys are typically used for HPAL autoclaves. When large tank sizes are required, making titanium prohibitively expensive, titanium clad steel tanks are employed. Explosion cladding of titanium to steel is often the used practice. Grade 17 Ti clad to SA 517 Grade 70 carbon steel performs well (Ref 37). Grades 11, 12, and 17 titanium provide excellent crevice corrosion resistance. Components such as agitator blades may be produced out of Grade 18 titanium (Ref 38). It has been shown that the oxide film exposed to high-temperature acid is significantly greater on Grade Ti-12 than on Grade Ti-18. Highly abrasion-resistant alloys, such as Grade Ti-5 or 12, can be used in severe abrasion and erosion environments. When a potential ignition condition is anticipated due to oxygen impingement or rubbing surfaces, Ti-45Nb alloy may be used that can provide ignition resistance (Ref 37).

Another alloy, UNS N06110 (known as Allcorr, Allegheny Ludlum, Pittsburgh, PA), has also worked well in HPAL environments. In a test under 11 mol% H_2S, 19 mol% CO_2 elemental sulfur, and chloride, at 213 °C (415 °F) and at 69 MPa (10 ksi) pressure, for evaluation of localized corrosion and environmental cracking resistance, Allcorr produced excellent resistance to environmental cracking and corrosion. Allcorr alloy is a nickel-base alloy containing 27.0 to 33% Cr, 8.0 to 12.0% Mo, and 4.0% (max) W. The very high chromium content allows the alloy to be exposed to very highly oxidizing environments without degradation. The molybdenum and tungsten help resist high-chloride environments and low pH conditions (Ref 39).

Summary

All forms of corrosion, including microbial attack, are found in mining and mineral processing equipment and machinery. Carbon and alloy steels are the most common materials used for these applications. Corrosion protection through superior materials is achieved when it can be economically justified. General corrosion protection is derived by the application of coatings and cathodic protection.

ACKNOWLEDGMENT

This article has been revised from the previously published article, "Corrosion in the Mineral Industry" in *Corrosion*, Vol 13, *Metals Handbook*, 9th ed., ASM International, 1987, p 1293–1298. The authors thank G.A. Minick and D.L. Olson, the authors of the previous article.

REFERENCES

1. I.R. Sare, J.I. Mardel, and A.J. Hill, Wear-Resistant Metallic and Elastomeric Materials in the Mining and Mineral Processing Industries: An Overview, *Wear*, Vol 250, 2001, p 1–10
2. J.J. McEwan, M.U. Kincer, P.V.T. Scheers, and R.T. White, Intuition, Case Work and Testing: A Holistic Approach to the Corrosion of a 12% Chromium Steel in Aqueous Environments, *Corros. Sci.*, Vol 35 (No. 1–4), 1993, p 303–315
3. G.E. Gatzanis and A. Ball, The Abrasion and Abrasion-Corrosion Properties of a 9% Chromium Steel, *Wear*, Vol 165, 1993, p 213–220
4. L. Zhenlin, R. Qichang, and J. Zhihao, An Investigation of the Corrosion-Abrasion Wear Behavior of 6% Chromium Martensitic Cast Steel, *J. Mater. Process. Technol.*, Vol 95, 1999, p 180–184
5. L. Fedrizzi, S. Rossi, F. Bellei, and F. Deflorian, Wear-Corrosion Mechanism of Hard Chromium Coatings, *Wear*, Vol 253, 2002, p 1173–1181
6. E. Medvedovski, Wear-Resistant Engineering Ceramics, *Wear*, Vol 249, 2001, p 821–828
7. D.J. Malone, R.J. Storms, and T.E. Crandall, Commercial Benefits of Ceramic Lining Systems Used in Atmospheric and Pressure Leach Vessels, *Hydrometallurgy*, Vol 39, 1995, p 163–167

Table 5 Fatigue and corrosion fatigue strengths of various alloys at 10^7 cycles

Metal	Fatigue strength		Corrosion fatigue strength					
			In Levack water		In Helen water		In Leitch water	
	MPa	ksi	MPa	ksi	MPa	ksi	MPa	ksi
T1 tool steel	414	60	131	19	145	21	152	22
Abrasion-resistant steel	307	44.5	152	22	145	21	124	18
Low-carbon steel	214	31	152	22	159	23	138	20
Stelcoloy-G steel	269	39	138	20	114	16.5	124	18
Aluminum alloy 6061-T6	107	15.5	69	10	107	15.5	55	8

Source: Ref 14

8. S. Das, D.P. Mondal, R. Dasgupta, and B.K. Prasad, Mechanisms of Material Removal during Erosion-Corrosion of an Al-SiC Particle Composite, *Corrosion,* Vol 236, 1999, p 295–302
9. K.T. Kembaiyan and K. Keshavan, Combating Severe Fluid Erosion and Corrosion of Drill Bits Using Thermal Spray Coatings, *Wear,* Vol 186–187, 1995, p 487–492
10. M. Reyes and A. Neville, Degradation Mechanisms of Co-Based Alloy and WC Metal-Matrix Composites for Drilling Tools Offshore, *Wear,* Vol 255, p 1143–1156
11. S.L. Pohlman and R.V. Olson, "Corrosion and Material Problem in the Copper Production Industry," Paper 229, presented at Corrosion/84, National Association of Corrosion Engineers, 1984
12. K. Adam, K.A. Natarajan, S.C. Riemer, and I. Iwasaki, Electrochemical Aspects of Grinding Media—Mineral Interaction in Sulfide Ore Grinding, *Corrosion,* Vol 42 (No. 8), 1980, p 440–446
13. G.R. Hoey and W. Dingley, Corrosion Control in Canadian Sulfide Ore Mines and Mills, *Can. Min. Metall. Bull.,* Vol 64, May 1971, p 1–8
14. G.J. Biefer, Corrosion Fatigue of Structural Metals in Mine Shaft Waters, *Can. Min. Metall. Bull.,* Vol 58, June 1967, p 675–681
15. N.S. Rawat, Corrosivity of Underground Mine Atmospheres and Mine Waters: A Review and Preliminary Study, *Br. Corros. J.,* Vol 11 (No. 2), 1976, p 86–91
16. I. Iwasaki, K.A. Natarajan, S.C. Riemer, and J.N. Orlich, Corrosion and Abrasive Wear in Ore Grinding, *Wear of Materials 1985,* American Society of Mechanical Engineers, 1985, p 509–518
17. E. Gamboa and A. Atrens, Environmental Influence on the Stress Corrosion Cracking of Rock Bolts, *Eng. Fail. Anal.,* Vol 10, 2003, p 521–558
18. T.P. Beckwith, Jr., The Bacterial Corrosion of Iron and Steel, *J. Am. Water Works Assoc.,* Vol 33 (No. 1), June 1941, p 147–165
19. B. Intorre, E. Kaup, J. Hardman, P. Lanik, H. Feiler, S. Zostak, and W.E. Rinne, Complete Water Reuse Industrial Opportunity, *Proceedings of the National Conference,* American Institute of Chemical Engineers, 1973, p 88
20. F.N. Speller, *Corrosion: Causes and Prevention,* McGraw-Hill, 1951, p 208
21. M.M. Khan and G.J. Krige, Evaluation of the Structural Integrity of Aging Mine Shafts, *Eng. Struct.,* Vol 24, 2002, p 901–907
22. S.A. Bryson, Repair Work and Fabrication in Gold Mining Environments, *FWP J.,* Vol 24 (No. 2), 1984, p 35–48
23. J.M. Karhnak, "Corrosion and Wear Problems Associated with the Mining and Mineral Processing Industry," Paper 230, presented at Corrosion/84, National Association of Corrosion Engineers, 1984
24. R.L. Jentgen, R.C. Rice, and G.L. Anderson, Preliminary Statistical Analysis of Data from Ontario Special Rope Tests on Mine-Hoist Wire Ropes, *Can. Min. Metall. Bull.,* Vol 77 (No. 11), 1984, p 50–54
25. H. Precek and J. Zeigler, Ropes for Use at Great Depths in Mining, *Wire Ind.,* Vol 52 (No. 8), 1985, p 486–487
26. H. Hartmann, Hauling Ropes for Shaft Installations under Extreme Corrosive Conditions, *Wire Ind.,* Vol 46 (No. 3), 1979, p 179
27. N. O'Hear, Developments in Aramid Fibre Ropes, *Wire Ind.,* Vol 49 (No. 11), 1982, p 845–850
28. ANI Arnall, *Practical Guide to Rock Bolting,* ANI ARNALL, 25 Pacific Highway, Bennets Green, NSW 2290, Australia, 1991
29. C.I. Walker, Slurry Pump Side-Liner Wear: Comparison of Some Laboratory and Field Results, *Wear,* Vol 250, 2001, p 81–87
30. F. Berndt and A. van Bennekom, Pump Shaft Failures: A Compendium of Case Studies, *Eng. Fail. Anal.,* Vol 8, 2001, p 135–144
31. R.F. Steigerwald, Corrosion Principles for the Mining Engineer, *Symposium Materials for Mining Industry,* AMAX Molybdenum, Inc., 1974
32. L.D. Eccleston, Protective Coatings in the Mining Industry, *Can. Min. Metall. Bull.,* Vol 72 (No. 3), 1979, p 170–173
33. M.M. Stack, N. Corlett, and S. Zhou, A Methodology for the Construction of the Erosion-Corrosion Map in Aqueous Environments, *Wear,* Vol 203–204, 1997, p 474–488
34. H.McI. Clark and R.J. Llewellyn, Assessment of the Erosion Resistance of Steels Used for Slurry Handling and Transport in Mineral Processing Applications, *Wear,* Vol 250, 2001, p 32–44
35. M.G. Fontana and N.D. Greene, *Corrosion Engineering,* McGraw-Hill, 1967
36. A.I. Asphahani and P. Crook, "Corrosion and Wear of High Performance Alloys in the Mining Industry," Paper 228, presented at Corrosion/84, National Association of Corrosion Engineers, 1984
37. J.G. Banker, Hydrometallurgical Applications of Titanium Clad Steel, *Reactive Metals in Corrosive Applications Conference* (Sun River, OR), Dynamic Materials, 1999, p 1–9
38. J. Vaughan, P. Reid, A. Alfantazi, D. Dreisinger, D. Tromans, and M. Elboudjaini, Corrosion of Titanium and Ti-Alloys at High Temperatures and Pressures, *Environmental Degradation of Materials and Corrosion Control in Metals, Proc. of the International Symposium on Environmental Degradation of Materials and Corrosion Control in Metals* (Vancouver, BC), Canadian Institute of Metals, Aug 24–27, 2003
39. R. Gerlock and A. Van, "ATI Introduces Corrosion Resistant Allcorr Alloy", Allegheny Technologies Inc., Pittsburgh, PA, 2002

Gallery of Corrosion Damage

Selected Color Images 1085
    Fundamentals of Corrosion 1085
    Evaluation of Corrosion Protection Methods 1085
    Forms of Corrosion in Industries and Environments 1085

Selected Color Images

THE INFORMATION provided by the authors for this Handbook included a great collection of color images. These color figures aid in the identification and classification of forms of corrosion. Naturally, the negative aspects of corrosion have been emphasized. The cost and the great effort to test, evaluate, simulate, and prevent corrosion have been examined. The ability of corrosion to undo our best complex engineered systems has been documented. While the pragmatic use of these images is the focus, you may also wish to consider the aesthetic beauty of some of these corrosion forms.

Fundamentals of Corrosion

In Volume 13A, *Corrosion: Fundamentals, Testing, and Protection*, the fundamentals were presented. The process of material degradation called corrosion generally involves an electrochemical reaction with the movement of electrons and metal ions. In electrical and electronic systems, the imposed electric potential can exacerbate the corrosion process in what may appear to be an otherwise benign environment. This shows evidence of the electrical nature of corrosion. In Fig. 1, copper on a printed circuit board has migrated. In Fig. 2, tin from a soldered terminal formed metal dendrites in the area of highest electrical potential to an adjacent terminal and caused an electrical short circuit. In a flip-chip resistor (Fig. 3), silver has migrated through a ceramic substrate, creating the familiar dendritic pattern. In Fig. 4(a), the metallized capacitor film appears with few faults, while in Fig. 4(b), a visually interesting pattern from a failed capacitor appears.

The internal components of a capacitor show the effect of polarity (Fig. 5). At the terminal marked positive (+), metal is being oxidized and corrosion products are being produced (an anodic reaction site for the galvanic cell), while the foil attached to the terminal marked negative (−) is protected and is cathodic to the corrosion cell.

The principle of cathodic protection, where the object to be protected is made electrically negative or cathodic, has been a tool for preventing or at least reducing corrosion of large structures such as bridges, piers, ships, and buried pipelines.

Evaluation of Corrosion Protection Methods

Besides manipulating the electric nature of corrosion to advantage, other methods of corrosion protection provide a barrier between the environment and the material to be protected. In Volume 13B, *Corrosion: Materials*, the selection of materials appropriate for the design environments was emphasized. Evaluating how well materials selection and the choice of protection schemes are performing can be done by inspection, testing, and surveys of the actual in-service object, or it may be done through testing of specimens in actual or simulated environments.

A very straightforward approach to atmospheric corrosion testing is to place specimens in natural environments. The testing at Kure Beach, NC, in a coastal marine environment is classic (Fig. 6). The test site at Newport, OR, also collects precipitation runoff samples and quantifies the effect of pollutants and precipitation on the corrosion process (Fig. 7).

The benefits of painting and flame and arc spraying are compared for scribed steel specimens in Fig. 8 after 3.5 years in a severe marine environment.

Microscopic examination of an aluminum laminate (Fig. 9) shows that the protective cladding, while attacked on the left, still protected the inner core. Here, the environment was aviation fuel.

Simulated testing of components allows the effects of corrosion to be accelerated and the function of the component to be checked. In a salt spray corrosion test, the benefit of a corrosion-inhibiting lubricant on a limit switch (Fig. 10a) is compared to an untreated control sample (Fig. 10b).

Forms of Corrosion in Industries and Environments

In this Volume, it is evident that the unique material demands of various industries, the environments in which they operate, and the environments they create all contribute to the forms of corrosion that are manifested.

Aviation. In military and commercial aviation, high-strength aluminum alloys in the aircraft must be maintained. Exfoliation corrosion of aluminum is seen in a military cockpit (Fig. 11) and, in a commercial aircraft (Fig. 12a). This figure also presents microbiologically induced corrosion associated with jet fuel (Fig. 12b) and galvanic corrosion (Fig. 12c, d). Because aircraft are a complex combination of many materials, galvanic corrosion is a ubiquitous threat (Fig. 13). Constant vigilance is required in design and manufacturing processes. Silicone sealant, which solved a crevice corrosion problem, can cause the loss of a protective coating (Fig. 14).

The extreme environments in which aircraft exist, from the subfreezing temperatures at altitude to the high-temperature engine gases, and the cyclic changes that occur in every flight pose severe design challenges. Corrosion may be personified as cruel when it appears in the seemingly mild confines of a package at ground level (Fig. 15).

Underinsulation Corrosion. Similar to the corrosion taking place due to improper packaging (Fig. 15), trapped moisture under insulation has been of increasing concern in piping systems (Fig. 16) and storage tanks (Fig. 17).

Art and Artifacts. The apparently mild environments in which art is located provides striking forms of corrosion. The museum atmosphere causes corrosion damage and added unwanted color to the handle of a 2500 year old bronze vessel (Fig. 18).

The bronze hand holding an iron staff forms a galvanic couple in an outdoor statue (Fig. 19). The velvetlike corrosion on an ancient copper alloy vessel (Fig. 20) was found to be a dendritic copper sulfide (Fig. 21). Another collection of dendrites in a museum environment is the silver sulfide crystals on a Greco-Roman silver bowl (Fig. 22).

Automobiles. Improvement in automobile bodies has been a beneficial application of corrosion prevention technology. Examples of corrosion are still found in vehicles, however. The use of dissimilar metals results in the galvanic corrosion of the carbon steel hanger welded to the stainless steel end bracket of the muffler after less than two years of service, although because of its size, the function is not yet jeopardized (Fig. 23). In a radiator, solder bloom has interfered with the ability to remove heat from the engine (Fig. 24). Another form of corrosion evident in a radiator is erosion-corrosion where the fluid removes the passive film, accelerating the rate of corrosion in high-flow areas such as opposite an inlet (Fig. 25a) and at a tube constriction (Fig. 25b). Another form of mechanically assisted corrosion, cavitation corrosion, is possible in an engine block (Fig. 26) where exhaust gases cause the coolant to boil.

Fossil Fuel Power Plants. With a goal of higher efficiency, engineers are striving to operate the energy conversion equipment at higher temperatures. Materials selection, materials engineering, and surface engineering technologies play a role in protecting components from high-temperature corrosion and in maintaining mechanical strength at these elevated temperatures. The tip of an industrial gas turbine blade (Fig. 27) is subjected to a combination of environmental and mechanical stress. High-temperature oxidation can be protective or not, depending on the oxide growth rate. The tip has oxidized, depleting the surface, while just below, the oxidized coating is protecting the blade. A severely attacked gas turbine blade is seen in Fig. 28. The micrograph (Fig. 29) is of an overlay coating and an aluminide layer that protects the base material of the gas turbine blade from oxidation and hot corrosion.

Steam turbines, boilers, and associated equipment pose challenges to the corrosionist as well. Waterside corrosion in a boiler exhibits layer of oxides, some porous and some dense. There are needlelike fibrous oxides in one layer and a dispersion of copper particles (Fig. 30) in another. In the same boiler, low-alloy carbon steel was subject to overheating, which resulted in the formation of spheroidal carbides. At three grain junctions, voids formed, grew, and coalesced (Fig. 31).

High-temperature feedwater, 120 to 205 °C (250 to 400 °F), caused flow-assisted corrosion that appeared as a general thinning with a scalloped profile and led to a sudden rupture (Fig. 32).

On the electrical side of a power plant, a water-cooled generator stator faced crevice corrosion in the braze between the rectangular copper wire, as seen through a borescope (Fig. 33). In a high-speed generator, nonmagnetic retaining rings made of an 18Mn-5Cr alloy steel must withstand high residual and dynamic stresses. The stress-corrosion cracking that resulted in this now-obsolete material choice is made more evident through fluorescent dye penetrant examination (Fig. 34).

Waste Disposal. Incinerator gases are especially corrosive, and materials selection is critical because the composition of the combustion gas may vary. Nickel-base liners are often used. In the case of an incinerator used by the military aboard ship, the nickel-chromium alloy 690 (UNS N06690) was repaired with a Ni-Co-Cr-Si alloy 160 (UNS N12160) that is known for sulfidation resistance because it forms a silica-rich scale. Figure 35 demonstrates the performance of this material.

Piping and Pipelines. To assure function and safe operation, defects on the millimeter scale must be either prevented or monitored in systems that can be thousands of kilometers long. For the most part, the systems are buried in walls, under streets, under ground, or underwater. Stress-corrosion cracking of a high-pressure gas pipeline led to its rupture (Fig. 36). The corrosion on the inside of a petroleum pipeline is given the descriptive name mesa corrosion, from the appearance of the wasted wall (lower right, Fig. 37). Another appropriately named corrosion morphology is the tuberculation in a cast iron pipe (Fig. 38).

Piping in a nuclear power plant steam generator unit made of alloy 600 nickel (UNS N06600) was subject to intergranular attack and stress-corrosion cracking at a support location (Fig. 39).

Biomedical Devices. The tolerance for any corrosion or corrosion products in implanted medical devices is very small. The body can be an aggressive environment. In Fig. 40, a Ti-6Al-4V neck and (sectioned) thimble used to attach a ceramic head to the femoral stem were retrieved. The purple tint indicates that Ti^{3+} ions were present, and within the interface it was evident that the pH had dropped below 1. In another case, the head was retrieved from a total hip replacement prosthesis. The Co-Cr-Mo alloy (Fig. 41) shows signs of etching at the dark regions of the inner bowl.

Microbiologically Induced Corrosion. Bacteria and other microorganisms can paint some interesting pictures while inducing corrosion. A microscopic image of stainless steel in seawater reveals a host of microorganisms (Fig. 42). A comparison of aerobic and anaerobic bacteria in seawater is shown in Fig. 43(a to d). The corrosion in a water pipe was induced by sulfate-reducing bacteria (Fig. 44).

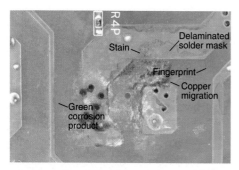

Fig. 1 Copper migration and copper corrosion byproducts on surface of failed printed circuit board. See the article "Corrosion in Passive Electrical Components" in this Volume.

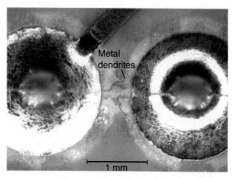

Fig. 2 Tin dendrites growing between two adjacent solder joints on failed printed circuit board. See the article "Corrosion in Passive Electrical Components" in this Volume.

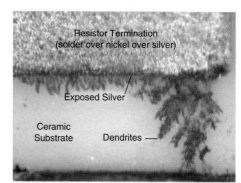

Fig. 3 Backside of failed surface-mounted chip resistor showing silver dendrites. Courtesy of Pat Kader, ENI. See the article "Corrosion in Passive Electrical Components" in this Volume.

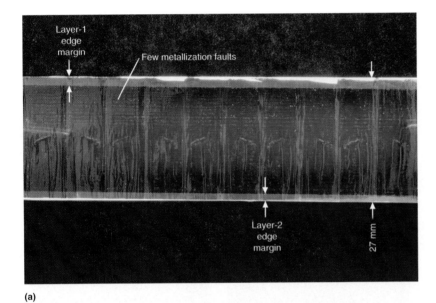

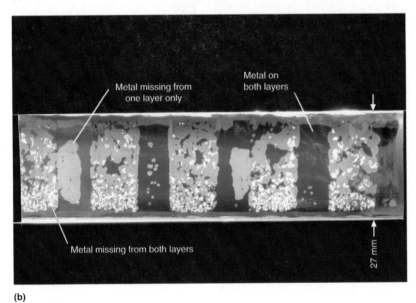

Fig. 4 Backlighted view of two stacked film layers removed from (a) a good metallized polypropylene film capacitor and (b) a failed metallized polypropylene film capacitor. See the article "Corrosion in Passive Electrical Components" in this Volume.

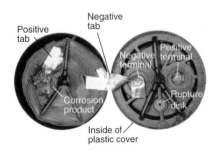

Fig. 5 Interior of failed aluminum electrolytic capacitor showing effects of corrosion. Note that the positive terminal has significant damage, while the negative terminal is not degraded. Millimeter scale. See the article "Corrosion in Passive Electrical Components" in this Volume.

Fig. 6 Original T.P. Hoar study panels after over 48 years of exposure at the 250 m (820 ft) marine atmospheric exposure site in Kure Beach, NC. See the article "Corrosion of Metallic Coatings" in this Volume.

Fig. 7 Marine atmospheric corrosion and precipitation runoff test site at Newport, OR. See the article "Simulated Service Testing in the Atmosphere" in *ASM Handbook* Volume 13A of this series.

Fig. 8 Scribed, sealed and painted thermal spray coatings on steel substrates compared to a scribed, painted steel panel after 42 months of severe marine atmospheric exposure. See the article "Corrosion of Metallic Coatings" in this Volume.

Fig. 9 Aluminum laminate. A 7072 aluminum cladding providing corrosion protection of 7075-T6 aluminum core. Note lateral spread of corrosion at clad layer to prevent through-wall failure of a P-3 fuel tank divider web. Courtesy of K. Himmelheber, Naval Aviation Depot, Jacksonville. See the article "U.S. Navy Aircraft Corrosion" in this Volume.

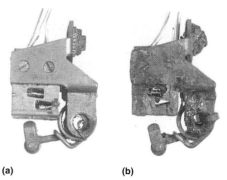

Fig. 10 EA-6B slat gearbox limit switch following ASTM B117 neutral salt spray corrosion testing (a) with MIL-L-87177 corrosion-inhibiting lubricant and (b) without MIL-L-87177 corrosion-inhibiting lubricant. Courtesy of J. Benfer, Naval Air Depot, Jacksonville. See the article "U.S. Navy Aircraft Corrosion" in this Volume.

Fig. 11 Exfoliation corrosion of an aluminum T-45 cockpit kick plate angle resulting from water intrusion within cockpit areas. Courtesy of J. Benfer, Naval Air Depot, Jacksonville. See the article "U.S. Navy Aircraft Corrosion" in this Volume.

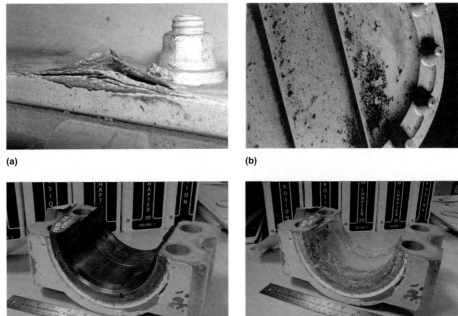

Fig. 12 Forms of corrosion in aircraft. (a) Exfoliation corrosion. (b) Microbiologically induced corrosion on fuel tank access door. (c) Bushing assembly. (d) Galvanic corrosion under aluminum-nickel bronze bushing, seen with bushing removed. See the article "Corrosion in Commercial Aviation" in this Volume.

Fig. 13 Galvanic corrosion of F/A-18 aircraft dorsal scallops resulting from composite doors attached to aluminum substructure with titanium and steel fasteners in the presence of moisture. Courtesy of S. Long, Naval Air Depot, North Island. See the article "U.S. Navy Aircraft Corrosion" in this Volume.

Fig. 14 Loss of paint adhesion caused by silicone contamination. See the article "U.S. Navy Aircraft Corrosion" in this Volume.

1090 / Gallery of Corrosion Damage

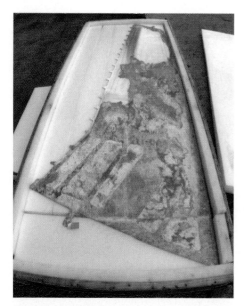

Fig. 15 Aluminum horizontal skin exhibiting extensive water-induced corrosion due to improper packaging and preservation during shipment from the manufacturer. Courtesy of S. Bevan, Naval Air Depot, Jacksonville. See the article "U.S. Navy Aircraft Corrosion" in this Volume.

Fig. 16 Corrosion of hot water line after insulation was degraded by water intrusion. See the article "Corrosion Control for Military Facilities" in this Volume.

Fig. 17 Through-wall corrosion under insulation of a large coated carbon steel storage tank. Courtesy of Paul Powers, GE Inspection. See the article "Effects of Process and Environmental Variables" in this Volume.

Fig. 18 Image showing corrosion of copper segments of a sculpture due to offgassing and contact with cellulose nitrate segments. MOMA New York. See the article "Corrosion of Metal Artifacts and Works of Art in Museum and Collection Environments" in this Volume.

Fig. 19 Detail of damage by galvanic corrosion of an iron staff in contact with a cast bronze hand on a statue of *Mercury* (date 1962) located in Kingston, Ontario. Courtesy of Pierre Roberge. Photograph 2003. See the article "Corrosion of Metal Artifacts Displayed in Outdoor Environments" in this Volume.

Selected Color Images / 1091

Fig. 20 Velvetlike corrosion growths on the surface of an ancient copper alloy vessel. The subtle form of the object is all but obscured by corrosion growths. See the article "Corrosion of Metal Artifacts and Works of Art in Museum and Collection Environments" in this Volume.

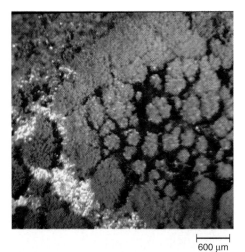

Fig. 21 At higher magnification, the velvetlike corrosion begins to reveal its dendritic structure. The corrosion was identified as djurleite, a copper sulfide. Original magnification 15×. The Chicago Oriental Institute. See the article "Corrosion of Metal Artifacts and Works of Art in Museum and Collection Environments" in this Volume.

Fig. 22 In a seemingly benign indoor environment, dendritic silver sulfide crystals are found growing on a Greco-Roman silver bowl. Walters Art Museum, Baltimore. Original magnification 10×. Courtesy of Terry Drayman-Weisser. See the article "Corrosion of Metal Artifacts and Works of Art in Museum and Collection Environments" in this Volume.

Fig. 23 Galvanic corrosion of solid carbon steel hanger rods after 1.5 years of driving service in the salt belt. Muffler end plate and sheet metal hanger are 18Cr-Cb. See the article "Automotive Exhaust System Corrosion" in this Volume.

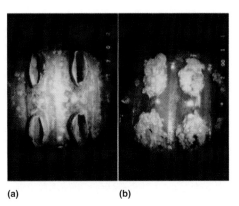

Fig. 24 Solder bloom formation can block and restrict the flow of coolant in the radiator. (a) New radiator core. (b) Solder bloom after just 22,000 miles of normal highway operation. See the article "Engine Coolants and Coolant System Corrosion" in this Volume.

1092 / Gallery of Corrosion Damage

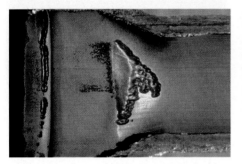

Fig. 25 Erosion-corrosion related to high coolant flow. (a) Radiator tank erosion on wall opposite inlet. (b) Tube narrowing causes increased velocity and turbulent flow. See the article "Engine Coolants and Coolant System Corrosion" in this Volume.

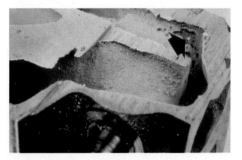

Fig. 26 Nucleate boiling-induced cavitation corrosion caused perforation at the exhaust valve port of this aluminum cylinder head. See the article "Engine Coolants and Coolant System Corrosion" in this Volume.

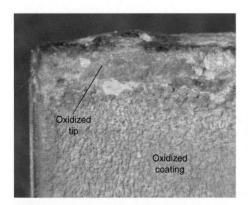

Fig. 27 High-temperature oxidation of the tip of an industrial gas turbine blade. Below the tip, a coating is protecting the base metal. See the article "Corrosion of Industrial Gas Turbines" in this Volume.

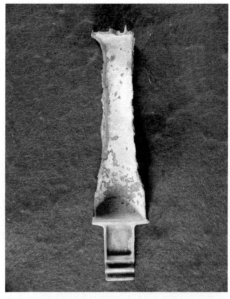

Fig. 28 Severe attack of an aeroderivative gas turbine blade by hot corrosion. See the article "Corrosion of Industrial Gas Turbines" in this Volume.

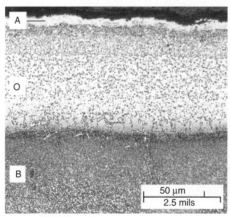

Fig. 29 Duplex coating consisting of an overlay coating (O) with an aluminide layer (A) on top. There is an interdiffusion zone with the base metal (B) and another interdiffusion zone between A and O. Lactic acid etch. See the article "Corrosion of Industrial Gas Turbines" in this Volume.

Fig. 30 Hotside deposit in the vicinity of the failure displays several layers and waterside oxidation/corrosion. Some layers are dense, while others are very porous. One layer displayed needlelike fibrous oxides. Another layer contains a number of particles of copper metal. Original magnification 210×. See the article "High-Temperature Corrosion in Military Systems" in this Volume.

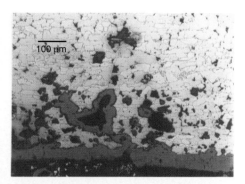

Fig. 31 Waterside surface, hot side, near failure. Carbides in prior pearlite colonies have completely spheroidized from overheating. Creep voids have developed at grain boundaries; some of these voids have grown and coalesced. Original magnification 210×. See the article "High-Temperature Corrosion in Military Systems" in this Volume.

Fig. 32 Feedwater pipe thinning and rupture. See the article "Corrosion of Steam- and Water-Side of Boilers" in this Volume.

Selected Color Images / 1093

Fig. 33 Borescopic view through bottle clip opening into strand package made of hollow strands that carry cooling water and solid strands that carry the majority of the current. This bundle of copper strands (both hollow and solid) is brazed together into a single, consolidated strand package. Braze material, which is porous, covers the surface of the strands and extends axially back into the coil between strands. It is this material that is susceptible to crevice corrosion cracking as cooling water is trapped in the porous openings and initiates the corrosion process. See the article "Corrosion of Generators" in this Volume.

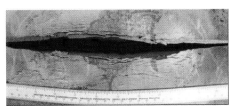

Fig. 36 Example of colony of stress-corrosion cracks on external surface of high-pressure gas transmission pipeline. See the article "External Corrosion of Oil and Natural Gas Pipelines" in this Volume.

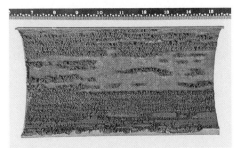

Fig. 37 Inside of a pipe in the petroleum industry exhibits mesa corrosion. Note profile of pipe wall in lower right corner. Characteristic appearance of mesa corrosion resulting from CO_2, water, and flow. The scale is in inches. Source: InterCorr International, Inc., Houston, TX. See the article "Corrosion in Petroleum Production Operations" in this Volume.

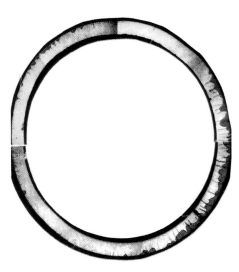

Fig. 39 An example of severe intergranular attack/intergranular stress-corrosion cracking at a tube support location. See the article "Corrosion in Pressurized Water Reactors" in this Volume.

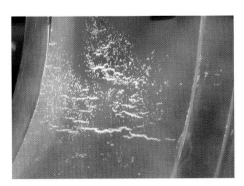

Fig. 34 Fluorescent dye penetrant examination showing linear indications of stress-corrosion cracking in an 18Mn-5Cr retaining ring. See the article "Corrosion of Generators" in this Volume.

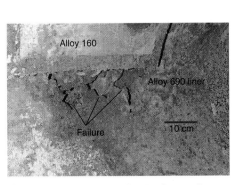

Fig. 35 Failed incinerator liner. Hole near alloy 160 patch shows that the original 6 mm (0.250 in.) wall thickness had been reduced to 1.2 mm (0.050 in.) or less in the general area. See the article "High-Temperature Corrosion in Military Systems" in this Volume.

Fig. 38 A 1.5 m (60 in.) cast iron pipe with tuberculation. Courtesy of Terry Bickford. See the article "Corrosion in Potable Water Distribution and Building Systems" in this Volume.

Fig. 40 Light photographs of modular tapers from prostheses showing evidence of corrosion. Retrieved Ti-6Al-4V neck and Ti-6Al-4V "thimble" (sectioned in half) used to attach a ceramic head to the femoral stem. Two modular interfaces are established in this design, one comprising the titanium-titanium junction and the other the titanium-ceramic junction. Note that the interface regions are highly corroded on both titanium interfaces, and a purple color was present (indicative of Ti^{3+} ions). Also found at these interfaces is evidence of titanium-phosphate particles, which indicate that the pH had dropped below 1 inside the taper. See the article "Mechanically Assisted Corrosion of Metallic Biomaterials" in this Volume.

Fig. 41 Retrieved Co-Cr-Mo modular head that had sat on a Ti-6Al-4V stem. Note the dark and discolored region inside the taper recess and the signs of etching. See the article "Mechanically Assisted Corrosion of Metallic Biomaterials" in this Volume.

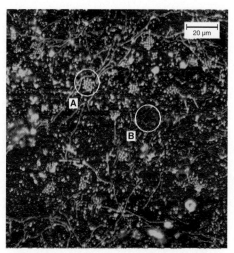

Fig. 42 Laser confocal microscope image of the variability in distribution and types of microorganisms in a two-week old biofilm grown on a stainless steel substratum in Lower Delaware Bay coastal seawater. The chemistry at the metal surface within a microcolony, as shown at location "A," will be quite different from that in either the bulk seawater or at location "B." Courtesy of K. Xua. See the article "Corrosion in Seawater" in this Volume.

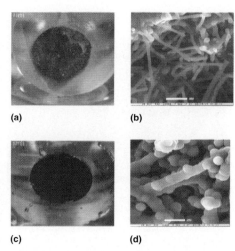

Fig. 43 (a) Photograph of carbon steel electrode exposed to aerobic natural seawater for 290 days. (b) Scanning electron micrograph of iron-oxide-encrusted bacteria enmeshed in corrosion products. (c) Photograph of carbon steel electrode exposed to anaerobic natural seawater for 290 days. (d) Scanning electron micrograph of sulfide-encrusted organisms enmeshed in corrosion products. See the article "Microbiologically Influenced Corrosion in Military Environments" in this Volume.

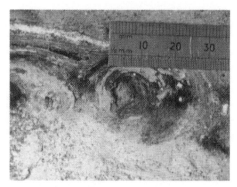

Fig. 44 Microbiologically influenced corrosion in a water pipe due to sulfate-reducing bacteria. See the article "Evaluating Microbiologically Influenced Corrosion" in *ASM Handbook* Volume 13A.

Reference Information

Corrosion Rate Conversion 1097
Metric Conversion Guide 1098
Abbreviations and Symbols 1101
Index 1105

Corrosion Rate Conversion

Relationships among some of the units commonly used for corrosion rates

d is metal density in grams per cubic centimeter (g/cm^3)

Unit	Factor for conversion to					
	mdd	g/m^2/d	μm/yr	mm/yr	mils/yr	in./yr
Milligrams per square decimeter per day (mdd)	1	0.1	36.5/d	0.0365/d	1.437/d	0.00144/d
Grams per square meter per day (g/m^2/d)	10	1	365/d	0.365/d	14.4/d	0.0144/d
Microns per year (μm/yr)	0.0274d	0.00274d	1	0.001	0.0394	0.0000394
Millimeters per year (mm/yr)	27.4d	2.74d	1000	1	39.4	0.0394
Mils per year (mils/yr)	0.696d	0.0696d	25.4	0.0254	1	0.001
Inches per year (in./yr)	696d	69.6d	25,400	25.4	1000	1

Adapted from G. Wranglén, *An Introduction to Corrosion and Protection of Metals,* Chapman and Hall, 1985, p 238

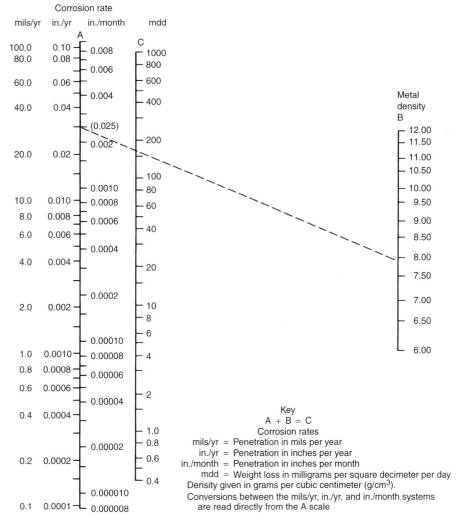

Key
A + B = C
Corrosion rates
mils/yr = Penetration in mils per year
in./yr = Penetration in inches per year
in./month = Penetration in inches per month
mdd = Weight loss in milligrams per square decimeter per day
Density given in grams per cubic centimeter (g/cm^3).
Conversions between the mils/yr, in./yr, and in./month systems are read directly from the A scale

Source: M.G. Fontana, *Corrosion Engineering,* 3rd ed., McGraw-Hill, 1986, p 217

Metric Conversion Guide

This Section is intended as a guide for expressing weights and measures in the Système International d'Unités (SI). The purpose of SI units, developed and maintained by the General Conference of Weights and Measures, is to provide a basis for worldwide standardization of units and measure. For more information on metric conversions, the reader should consult the following references:

- *The International System of Units,* SP 330, 1991, National Institute of Standards and Technology. Order from Superintendent of Documents, U.S. Government Printing Office, Washington, DC 20402-9325

- *Metric Editorial Guide,* 5th ed. (revised), 1993, American National Metric Council, 4340 East West Highway, Suite 401, Bethesda, MD 20814-4411
- "Standard for Use of the International System of Units (SI): The Modern Metric System," IEEE/ASTM SI 10-1997, Institute of Electrical and Electronics Engineers, 345 East 47th Street, New York, NY 10017, USA
- *Guide for the Use of the International System of Units (SI),* SP 811, 1995, National Institute of Standards and Technology, U.S. Government Printing Office, Washington, DC 20402

Base, supplementary, and derived SI units

Measure	Unit	Symbol
Base units		
Amount of substance	mole	mol
Electric current	ampere	A
Length	meter	m
Luminous intensity	candela	cd
Mass	kilogram	kg
Thermodynamic temperature	kelvin	K
Time	second	s
Supplementary units		
Plane angle	radian	rad
Solid angle	steradian	sr
Derived units		
Absorbed dose	gray	Gy
Acceleration	meter per second squared	m/s^2
Activity (of radionuclides)	becquerel	Bq
Angular acceleration	radian per second squared	rad/s^2
Angular velocity	radian per second	rad/s
Area	square meter	m^2
Capacitance	farad	F
Concentration (of amount of substance)	mole per cubic meter	mol/m^3
Current density	ampere per square meter	A/m^2
Density, mass	kilogram per cubic meter	kg/m^3
Dose equivalent, dose equivalent index	sievert	Sv
Electric charge density	coulomb per cubic meter	C/m^3
Electric conductance	siemens	S
Electric field strength	volt per meter	V/m
Electric flux density	coulomb per square meter	C/m^2
Electric potential, potential difference, electromotive force	volt	V
Electric resistance	ohm	Ω
Energy, work, quantity of heat	joule	J
Energy density	joule per cubic meter	J/m^3
Entropy	joule per kelvin	J/K
Force	newton	N
Frequency	hertz	Hz
Heat capacity	joule per kelvin	J/K
Heat flux density	watt per square meter	W/m^2
Illuminance	lux	lx
Inductance	henry	H
Irradiance	watt per square meter	W/m^2
Luminance	candela per square meter	cd/m^2
Luminous flux	lumen	lm
Magnetic field strength	ampere per meter	A/m
Magnetic flux	weber	Wb
Magnetic flux density	tesla	T
Molar energy	joule per mole	J/mol
Molar entropy	joule per mole kelvin	J/mol · K
Molar heat capacity	joule per mole kelvin	J/mol · K
Moment of force	Newton meter	N · m
Permeability	henry per meter	H/m
Permittivity	farad per meter	F/m
Power, radiant flux	watt	W
Pressure, stress	pascal	Pa
Quantity of electricity, electric charge	coulomb	C
Radiance	watt per square meter steradian	W/m^2 · sr
Radiant intensity	watt per steradian	W/sr
Specific heat capacity	joule per kilogram kelvin	J/kg · K
Specific energy	joule per kilogram	J/kg
Specific entropy	joule per kiolgram kelvin	J/kg · K
Specific volume	cubic meter per kilogram	m^3/kg
Surface tension	newton per meter	N/m
Thermal conductivity	watt per meter kelvin	W/m · K
Velocity	meter per second	m/s
Viscosity, dynamic	pascal second	Pa · s
Viscosity, kinematic	square meter per second	m^2/s
Volume	cubic meter	m^3
Wave number	1 per meter	1/m

Conversion factors

To convert from	to	multiply by
Angle		
degree	rad	1.745 329 E−02
Area		
in.^{2}	mm^2	6.451 600 E+02
in.^{2}	cm^2	6.451 600 E+00
in.^{2}	m^2	6.451 600 E−04
ft^2	m^2	9.290 304 E−02
Bending moment or torque		
lbf · in.	N · m	1.129 848 E−01
lbf · ft	N · m	1.355 818 E+00
kgf · m	N · m	9.806 650 E+00
ozf · in.	N · m	7.061 552 E−03
Bending moment or torque per unit length		
lbf · in./in.	N · m/m	4.448 222 E+00
lbf · ft/in.	N · m/m	5.337 866 E+01
Current density		
A/in.^{2}	A/cm^2	1.550 003 E−01
A/in.^{2}	A/mm^2	1.550 003 E−03
A/ft^2	A/m^2	1.076 400 E+01
Electricity and magnetism		
gauss	T	1.000 000 E−04
maxwell	μWb	1.000 000 E−02
mho	S	1.000 000 E+00
Oersted	A/m	7.957 700 E+01
Ω · cm	Ω · m	1.000 000 E−02
Ω · circular-mil/ft	μΩ · m	1.662 426 E−03
Energy (impact, other)		
ft · lbf	J	1.355 818 E+00
Btu (thermochemical)	J	1.054 350 E+03
cal (thermochemical)	J	4.184 000 E+00
Cal (nutritional)	J	4.184 000 E+03
kW · h	J	3.600 000 E+06
W · h	J	3.600 000 E+03
Flow rate		
ft^3/h	L/min	4.719 475 E−01
ft^3/min	L/min	2.831 000 E+01
gal/h	L/min	6.309 020 E−02
gal/min	L/min	3.785 412 E+00
Force		
lbf	N	4.448 222 E+00
kip (1000 lbf)	N	4.448 222 E+03
tonf	kN	8.896 443 E+00
kgf	N	9.806 650 E+00
Force per unit length		
lbf/ft	N/m	1.459 390 E+01
lbf/in.	N/m	1.751 268 E+02
Fracture toughness		
ksi$\sqrt{\text{in.}}$	MPa$\sqrt{m}$	1.098 800 E+00
Heat content		
Btu/lb	kJ/kg	2.326 000 E+00
cal/g	kJ/kg	4.186 800 E+00
Heat input		
J/in.	J/m	3.937 008 E+01
kJ/in.	kJ/m	3.937 008 E+01
Impact energy per unit area		
ft · lbf/ft^2	J/m^2	1.459 002 E+01
Length		
Å	nm	1.000 000 E−01
μin.	μm	2.540 000 E−02
mil	μm	2.540 000 E+01
in.	mm	2.540 000 E+01
in.	cm	2.540 000 E+00
ft	m	3.048 000 E−01
yd	m	9.144 000 E−01
mile, international	km	1.609 344 E+00
mile, nautical	km	1.852 000 E+00
mile, U.S. statute	km	1.609 347 E+00
Mass		
oz	kg	2.834 952 E−02
lb	kg	4.535 924 E−01
ton (short, 2000 lb)	kg	9.071 847 E+02
ton (short, 2000 lb)	kg × 10^3(a)	9.071 847 E−01
ton (long, 2240 lb)	kg	1.016 047 E+03
Mass per unit area		
oz/in.^{2}	kg/m^2	4.395 000 E+01
oz/ft^2	kg/m^2	3.051 517 E−01
oz/yd^2	kg/m^2	3.390 575 E−02
lb/ft^2	kg/m^2	4.882 428 E+00
Mass per unit length		
lb/ft	kg/m	1.488 164 E+00
lb/in.	kg/m	1.785 797 E+01
Mass per unit time		
lb/h	kg/s	1.259 979 E−04
lb/min	kg/s	7.559 873 E−03
lb/s	kg/s	4.535 924 E−01
Mass per unit volume (includes density)		
g/cm^3	kg/m^3	1.000 000 E+03
lb/ft^3	g/cm^3	1.601 846 E−02
lb/ft^3	kg/m^3	1.601 846 E+01
lb/in.^{3}	g/cm^3	2.767 990 E+01
lb/in.^{3}	kg/m^3	2.767 990 E+04
Power		
Btu/s	kW	1.055 056 E+00
Btu/min	kW	1.758 426 E−02
Btu/h	W	2.928 751 E−01
erg/s	W	1.000 000 E−07
ft · lbf/s	W	1.355 818 E+00
ft · lbf/min	W	2.259 697 E−02
ft · lbf/h	W	3.766 161 E−04
hp (550 ft · lbf/s)	kW	7.456 999 E−01
hp (electric)	kW	7.460 000 E−01
Power density		
W/in.^{2}	W/m^2	1.550 003 E+03
Pressure (fluid)		
atm (standard)	Pa	1.013 250 E+05
bar	Pa	1.000 000 E+05
in. Hg (32 °F)	Pa	3.386 380 E+03
in. Hg (60 °F)	Pa	3.376 850 E+03
lbf/in.^{2} (psi)	Pa	6.894 757 E+03
torr (mm Hg, 0 °C)	Pa	1.333 220 E+02
Specific Heat		
Btu/lb · °F	J/kg · K	4.186 800 E+03
cal/g · °C	J/kg · K	4.186 800 E+03
Stress (force per unit area)		
tonf/in.^{2} (tsi)	MPa	1.378 951 E+01
kgf/mm^2	MPa	9.806 650 E+00
ksi	MPa	6.894 757 E+00
lbf/in.^{2} (psi)	MPa	6.894 757 E−03
MN/m^2	MPa	1.000 000 E+00
Temperature		
°F	°C	5/9 · (°F − 32)
°R	K	5/9
K	°C	K − 273.15
Temperature interval		
°F	°C	5/9
Thermal conductivity		
Btu · in./s · ft^2 · °F	W/m · K	5.192 204 E+02
Btu/ft · h · °F	W/m · K	1.730 735 E+00
Btu · in./h · ft^2 · °F	W/m · K	1.442 279 E−01
cal/cm · s · °C	W/m · K	4.184 000 E+02
Thermal expansion(b)		
cm/cm · °C	m/m · K	1.000 000 E+00
in./in. · °F	m/m · K	1.800 000 E+00
Velocity		
ft/h	m/s	8.466 667 E−05
ft/min	m/s	5.080 000 E−03
ft/s	m/s	3.048 000 E−01
in./s	m/s	2.540 000 E−02
km/h	m/s	2.777 778 E−01
mph	km/h	1.609 344 E+00
Velocity of rotation		
rev/min (rpm)	rad/s	1.047 164 E−01
rev/s	rad/s	6.283 185 E+00
Viscosity		
poise	Pa · s	1.000 000 E−01
stokes	m^2/s	1.000 000 E−04
ft^2/s	m^2/s	9.290 304 E−02
in.^{2}/s	mm^2/s	6.451 600 E+02
Volume		
in.^{3}	m^3	1.638 706 E−05
ft^3	m^3	2.831 685 E−02
fluid oz	m^3	2.957 353 E−05
gal (U.S. liquid)	m^3	3.785 412 E−03
Volume per unit time		
ft^3/min	m^3/s	4.719 474 E−04
ft^3/s	m^3/s	2.831 685 E−02
in.^{3}/min	m^3/s	2.731 177 E−07

(a) kg × 10^3 = 1 metric ton (tonne), or 1 megagram (Mg). (b) Preferred expression is 10^{-6}/K or 10^{-6}/°F as length units are unnecessary.

SI prefixes—names and symbols

Exponential expression	Multiplication factor	Prefix	Symbol
10^{24}	1 000 000 000 000 000 000 000 000	yotta	Y
10^{21}	1 000 000 000 000 000 000 000	zetta	Z
10^{18}	1 000 000 000 000 000 000	exa	E
10^{15}	1 000 000 000 000 000	peta	P
10^{12}	1 000 000 000 000	tera	T
10^{9}	1 000 000 000	giga	G
10^{6}	1 000 000	mega	M
10^{3}	1 000	kilo	k
10^{2}	100	hecto(a)	h
10^{1}	10	deka(a)	da
10^{0}	1	BASE UNIT	
10^{-1}	0.1	deci(a)	d
10^{-2}	0.01	centi(a)	c
10^{-3}	0.001	milli	m
10^{-6}	0.000 001	micro	μ
10^{-9}	0.000 000 001	nano	n
10^{-12}	0.000 000 000 001	pico	p
10^{-15}	0.000 000 000 000 001	femto	f
10^{-18}	0.000 000 000 000 000 001	atto	a
10^{-21}	0.000 000 000 000 000 000 001	zepto	z
10^{-24}	0.000 000 000 000 000 000 000 001	yocto	y

(a) Nonpreferred. Prefixes should be selected in steps of 10^3 so that the resultant number before the prefix is between 0.1 and 1000. These prefixes should not be used for units of linear measurement, but may be used for higher order units. For example, the linear measurement, decimeter, is nonpreferred, but square decimeter is acceptable.

Abbreviations and Symbols

a	crack length; chemical activity	bct	body-centered tetragonal	d	density; used in mathematical expressions involving a derivative (denotes rate of change)
A	ampere	BFPD	barrels of fluid produced per day (oil and gas production)		
A	area				
A_a	anodic area	bhp	brake horsepower		
A_c	cathodic area	BOD	biological oxygen demand	D	diffusion coefficient
A_{corr}	corroding area	BS	British Standard	da/dN	crack growth rate per cycle
Å	angstrom	BS EN	British Standard Euronorm	da/dt	crack growth rate per unit time
AA	Aluminum Association	BSW	basic saturated water	dc	direct current
ABS	acrylonitrile-butadiene-styrene	BWR	boiling water reactor	DFM	deterministic fracture mechanics
ac	alternating current			DMF	dimethyl formamide
ACI	Alloy Casting Institute	c	capacitance; concentration	DNB	departure from nucleate boiling
A/D-D/A	analog-to-digital and digital-to-analog	C	coulomb; carbon	DO	dissolved oxygen
		C_o	initial coating thickness	DoD	Department of Defense (U.S.)
AE	auxiliary electrode	CAD/CAE	computer-aided design/computer-aided engineering	DODSSP	Department of Defense single stock point (U.S.)
AES	Auger electron spectroscopy				
AFM	atomic force microscopy	CANDU	Canadian deuterium uranium (reactor)	DOE	Department of Energy (U.S.)
AGR	advanced gas-cooled reactor			DS	directionally solidified
AISI	American Iron and Steel Institute	CARC	chemical-agent-resistant coating		
AMS	Aerospace Material Specification (of SAE International)	CAS	Chemical Abstracts Service	e	natural log base, 2.71828 ...
		CASS	copper-accelerated acetic acid salt spray (test)	e^-	electron
amu	atomic mass units			E	electrical potential
ANSI	American National Standards Institute	CCT	critical crevice temperature	E^o, E_o	standard electrical potential
		CDRM	control rod drive mechanism	E_{appl}	applied potential
AOD	argon-oxygen decarburization	CE	counterelectrode	E_{bd}, E_b	breakdown potential
API	American Petroleum Institute	CEOC	Confédération Européene des Organismes de Contrôle technique, d'inspection, decertification et de prévention	E_{cell}	reversible electrode cell potential
aq	aqueous solution			E_{corr}	corrosion potential
AR	Army regulations (U.S.)			E_e	equilibrium potential
AS	artificial seawater			E_g	galvanic potential
ASM	ASM International			E_p	pitting potential; passivation potential
ASME	American Society of Mechanical Engineers	CFC	corrosion-fatigue cracking		
		CISCC	chloride-induced stress-corrosion cracking	E_{pass}	passivation potential
ASNT	American Society of Nondestructive Testing			E_{pit}	critical potential for pitting
		cm	centimeter	E_{pp}	primary passivation potential
ASSIST	acquisition streaming and standardization information system	CMAS	$CaO\text{-}MgO\text{-}Al_2O_3\text{-}SiO_2$	E_{prot}	protection or repassivation potential
		COD	chemical oxygen demand		
		COTS	commercial-off-the-shelf		
		CP	commercially pure; cathodic protection	E_R	rest potential
ASTM	American Society for Testing and Materials (now ASTM International)			E_{RP}	repassivation potential
	CPCP	corrosion prevention and control plant	E20 and E200	potential in the forward scan of a cyclic polarization curve for which the current density reached 20 and 200 $\mu A/cm^2$, respectively
at.%	atomic percent	cpm	cycles per minute		
atm	atmospheres (pressure)	CPP	cyclic potentiodynamic polarization		
AVT	all-volatile treatment				
AW	atomic weight	CPT	critical pitting temperature	EAC	environmentally assisted cracking
AWS	American Welding Society	CPVC	chlorinated polyvinyl chloride		
AWWA	American Water Works Association	CR	corrosion rate	EBS	engineered barrier systems
		CSE	copper sulfate reference electrode	ECM	electrochemical machining
b	Tafel coefficient (constant)	CSP	chip-scale package	ECN	electrochemical noise (method)
B	Stern-Geary constant; boron	CTE	coefficient of thermal expansion	ECTFE	ethylene-chlorotrifluoroethylene
bal	balance or remainder	CTFE	chlorotrifluoroethylene	EDM	electrical discharge machining
bbl	barrel	CUI	corrosion under insulation	EDS	energy-dispersive spectroscopy; energy-dispersive x-ray spectroscopy
bcc	body-centered cubic	CVD	chemical vapor deposition		
BCE	before Christian (or common) era	CW	cold work	EDTA	ethylenediamine tetraacetic acid

EEMUA	Engineering Equipment Manufacturers and Users Association	GW	gigawatt	J	joule
		Gy	gray (absorbed dose)	k_B	Boltzmann constant, 1.38066×10^{-23} J/K
EIS	electrochemical impedance spectroscopy	h	hour		
		h•	electron hole	k_L	linear oxidation rate constant
EMAT	electronic magnetic transducer	H	hydrogen	k_p	parabolic oxidation rate constant
emf	electromotive force	H	enthalpy	K	Kelvin; potassium
EN	electrochemical noise	H_0	null hypothesis	K	stress-intensity factor in linear elastic fracture mechanics
ENA	electrochemical noise analysis	HAZ	heat-affected zone		
ENP	electroless nickel plating	HB	Brinell hardness	K_{crit}	critical value of stress concentration
EOP	electro-osmotic pulse	HCF	high-cycle fatigue		
EPA	Environmental Protection Agency (U.S.)	hcp	hexagonal close-packed	K_{Ic}	fracture toughness; plane-strain fracture toughness
		HDA	harmonic distortion analysis		
EPDM	ethylene-propylene-diene monomer (rubber)	HDPE	high-density polyethylene	K_{ISCC}	threshold stress intensity for stress-corrosion cracking
		HE	hydrogen embrittlement		
EPRI	Electric Power Research Institute	HIC	hydrogen-induced cracking	K_t	stress-concentration factor
		HLW	high-level waste	K_{th}	threshold stress-intensity factor
Eq	equation (also used to label inequalities and reactions)	HPEN	high-phosphorus electroless nickel	kg	kilogram
				kPa	kilopascal
ER	electrical resistance	HR	Rockwell hardness (requires scale designation, such as HRC for Rockwell C hardness)	l	liquid
ER1 and ER10	potential in the reverse scan of a cyclic polarization curve for which the current density reached 1 and 10 µA/cm², respectively			L	liter
				lb	pound
		HSE	Health and Safety Executive (U.K.)	LCF	low-cycle fatigue
				LLNL	Lawrence Livermore National Laboratory
		HTHC	high-temperature hot corrosion		
ESC	environmental stress cracking	HV	Vickers (diamond pyramid) hardness	LMC	liquid metal cracking
ESCC	external stress-corrosion cracking			LME	liquid metal embrittlement
ETFE	ethylene-tetrafluoroethylene	HVAC	heating, ventilation, and air conditioning	LMIE	liquid metal induced embrittlement
EVD	electrochemical vapor deposition				
exp	exponential function (base e)	HWC	hydrogen water chemistry	ln	natural logarithm (base e)
		Hz	hertz	LNG	liquefied natural gas
f	flow rate	i	current density; current	log	common logarithm (base 10)
F	farad; fluorine	i_{corr}	corrosion current or current density	LPR	linear polarization resistance
F	Faraday constant			LSI	Langelier saturation index
FAA	Federal Aviation Administration (U.S.)	i_{crit}	critical current or current density for passivation	LTCTF	long-term corrosion test facility
				LTHC	low-temperature hot corrosion
FACT	Ford anodized aluminum corrosion test	i_p	passive current or current density	LVBR	low-velocity atmospheric pressure burner rig
		I	current; current density		
FBE	fusion-bonded epoxy	I_{appl}	applied current or current density	m	mass; molal (solution)
fcc	face-centered cubic	I_{corr}	corrosion current or current density	m_c	mass transport coefficient
FDA	Food and Drug Administration (U.S.)			M	metal
		IASCC	irradiation-assisted stress-corrosion cracking	M	molecular weight; molar solution
FEA	finite-element analysis			mA	milliampere
FEM	field emission microscopy	IC	integrated circuit	MARPOL	marine pollution protocol
FEP	fluorinated ethylene propylene	ICCP	impressed current cathodic protection	MAWT	minimum allowable wall thickness
FFT	fast Fourier transform				
FGD	flue gas desulfurization	ID	inside diameter	max	maximum
Fig.	figure	IGA	intergranular attack	MCA	multiple-crevice assembly
FIT	Failure in time (1 in 1,000,000 in 1000 h)	IGC	intergranular corrosion	mdd	milligrams per square decimeter per day
		IGSCC	intergranular stress-corrosion cracking		
FOD	foreign object damage			MEM	maximum entropy method
FRA	frequency response analyzer	IHS	Information Handling Services	MFL	magnetic flux leakage
FRP	fiberglass-reinforced polyester	in.	inch	mg	milligram
FSS	Federal Supply Service	IP	Institute of Petroleum	MIC	microbiologically influenced corrosion
ft	foot	ipy	inches per year		
FTIR	Fourier transform infrared (spectroscopy)	IR	voltage drop; current multiplied by resistance	MICI	microbiologically influenced corrosion inhibition
FWHM	full width at half maximum	IRIS	internal rotary inspection system	mil	0.001 in.
g	gas; gram	ISO	International Organization for Standardization	MIL-STD	military standard (U.S.)
G	acceleration due to gravity			min	minimum; minute
G	Gibbs energy	IUPAC	International Union of Pure and Applied Chemistry	mm	millimeter
gal	gallon			mol	mole
GNP	gross national product			MPa	megapascal
GPS	global positioning system	j	$\sqrt{-1}$; current density	MPEN	medium-phosphorus electroless nickel
GSA	General Services Administration (U.S.)	J	flux or mass; stress-intensity factor in elastic-plastic fracture mechanics; current density		
				MPIF	Metal Powder Industries Federation
GTA	gas tungsten arc				
GTAW	gas tungsten arc welded	J_0	exchange current density		

mpy	mils per year (mil/yr)	psig	pounds per square inch (gage)	SERS	surface-enhanced Raman spectroscopy
mV	millivolt	PSO	polysulfone	Sh	Sherwood number
n	sample size; number of electrons; number of neutrons	PSV	pressure safety valve	SHE	standard hydrogen electrode
		PTB	pounds per 1000 barrels (concentration approximately 4 ppm)	SIC	semiconductor integrated circuit
N	Newton			SLPR	self-linear polarization resistance
N	population size; number of trials; mole fraction; normal (solution); number of cycles in corrosion-fatigue testing			SMIE	solid metal induced embrittlement
		PTFE	polytetrafluoroethylene	SMP	silicone-modified polyester
		PVA	polyethylene-vinyl acetate	SNF	spent nuclear fuel
		PVC	polyvinyl chloride	SOHIC	stress-oriented hydrogen-induced cracking
		PVD	plasma vapor deposition		
		PVDF	polyvinylidene difluoride		
N_f	cycles to failure in corrosion-fatigue testing	PVF	polyvinylidene fluoride	SR	stress-relieved
		PWR	pressurized water reactor	SRB	sulfate-reducing bacteria; solid rocket booster (space shuttle)
NACE	National Association of Corrosion Engineers (now NACE International)	Q	activation energy for diffusion; heat		
				SSC	saturated silver chloride reference electrode
		QA	quality assurance		
NDE	nondestructive examination/evaluation	QML	qualified manufacturer's list	SSPC	The Society for Protective Coatings; Steel Structures Painting Council
		QPL	qualified products list		
NDI	nondestructive inspection	r	number of successes; corrosion rate		
NDT	nondestructive testing			SSRT	slow strain-rate test
NETL	National Energy Technology Laboratory	R	Rankine; reduction; reduced species	STP	standard temperature and pressure
NHE	normal hydrogen electrode			t	thickness; time
NIST	National Institute of Standards and Technology	R	radius; ratio of minimum stress to maximum stress; resistance; gas constant (8.314 J/mol·K)	T	temperature
				T_g	glass transition temperature
nm	nanometer			TBC	thermal barrier coating
No.	number			TBD	to be determined
NRC	Nuclear Regulatory Commission (U.S.)	R_a	surface roughness in terms of arithmetic average	TDS	total dissolved solids
				TFE	tetrafluoroethylene
NSF	National Sanitation Foundation	R_i	internal resistance	TGA	thermogravimetric analysis
NSST	neutral salt spray test	R_n	noise resistance	TGSCC	transgranular stress-corrosion cracking
OCP	open-circuit corrosion potential	R_p	polarization resistance		
OD	outside diameter	RAD, rad	radiation absorbed dose	TOFD	time-of-flight diffraction
OEM	original equipment manufacturer	RBI	risk-based inspection	TOW	time of wetness
OMB	Office of Management and Budget (U.S.)	RE	reference electrode; residual element; rare earth element	TSOP	thin-small-outline package
				UNS	Unified Numbering System
OSHA	Occupational Safety and Health Administration (U.S.)	Re	Reynold's number; rhenium	URL	underground research laboratory; uniform resource locator (web)
		Ref	reference		
p	page	RF	radio frequency		
p	probability of success; partial pressure; pressure	RH	relative humidity	U.S.	United States
		rms	root mean square	USAF	United States Air Force
P_{O_2}	partial pressure of oxygen	RMU	remote monitoring unit	USD	United States dollars
P	probability	rpm	revolutions per minute	USDA	United States Department of Agriculture
P_a	sensitization number	s	solid		
Pa	pascal	s	estimate of the standard deviation	UT	ultrasonic testing
PASS	paint adhesion on a scribed surface	S	siemen; unit of conductance	UTS	ultimate tensile strength
		S	entropy	UV	ultraviolet
PDF	portable document format	S_a	stress amplitude	v	viscosity; fluid velocity
PES	polyether sulfone	S_m	mean stress	V	volt
PF	pitting factor	S_r	fatigue (endurance) limit; stress range	V	volume
PFA	perfluoroalkoxy			VNSS	Väätänen nine salts solution
PFM	probabilistic fracture mechanics	SAE	Society of Automotive Engineers	VOC	volatile organic compound
pH	negative logarithm of hydrogen ion activity	SAN	styrene-acrylonitrile	vol%	volume percent
		SAW	simulated acidified water		
PHWR	pressurized heavy water reactor	Sc	Schmidt number	w	weight or mass
PIN	part identification number	SCC	stress-corrosion cracking	W	watt
PP	polypropylene	SCE	saturated calomel electrode	WDS	wavelength-dispersive spectroscopy
ppb	parts per billion	scf	standard cubic foot		
ppm	parts per million	scfm	standard cubic feet per minute of gas flow	WE	working electrode
PPO	polyphenylene oxide			WNr	Werkstoff number
PRE	pitting resistance equivalent	SCR	silicon-controlled rectifier	wt%	weight percent
PREN	pitting resistance equivalent nitrogen (%N is included in formula)	SCW	simulated concentrated water (~1000 times the concentration of J-13 well water)	XLPE	cross-linked polyethylene
				yr	year
PSD	power spectral density			YS	yield strength
psi	pounds per square inch	SDW	simulated dilute water		
psia	pounds per square inch (absolute)	SEM	scanning electron microscopy		

Z	impedance	√	square root of	Ω	ohm
ZRA	zero-resistance ammeter	~	similar to; approximately	ω	angular velocity
°	degree (angular measure)	∝	varies as; is proportional to		
°C	temperature, degrees Celsius (centigrade)	α	chemical activity; crack length; crystal lattice length along the α-axis	**Greek Alphabet**	
°F	temperature, degrees Fahrenheit	β	Tafel coefficient (constant)	A, α	alpha
→	chemical reaction	Δ	change in quantity; an increment; a range	B, β	beta
↔	reversible chemical reaction, does not imply equal reaction rates in both directions	ΔE_{therm}	thermodynamic driving force	Γ, γ	gamma
		ΔG	Gibbs free energy	Δ, δ	delta
=	equals	$\Delta G°$	standard Gibbs free energy	E, ε	epsilon
≈	approximately equals	ΔH	change in enthalpy	Z, ζ	zeta
≠	not equal to	ΔK	stress-intensity factor range	H, η	eta
>	greater than	ΔS	entropy change	Θ, θ	theta
≫	much greater than	ε_p	plastic strain range	I, ι	iota
≥	greater than or equal to	ε_t	total strain range	K, κ	kappa
∫	integral	η	overpotential	Λ, λ	lambda
<	less than	μm	micron (micrometer)	M, μ	mu
≤	less than or equal to	ν	kinematic viscosity	N, ν	nu
±	maximum deviation; tolerance	Π	symbol for multiplying a series of terms; pi	Ξ, ξ	xi
−	minus; negative ion charge			O, o	omicron
×	multiplied by; diameters (magnification)	π	pi, 3.14159...	Π, π	pi
·	multiplied by	ρ	density	P, ρ	rho
/	per; divided by	Σ	summation	Σ, σ	sigma
%	percent	σ, σ′	standard deviation; oxide conductivity; stress	T, τ	tau
+	plus; in addition to; positive ion charge	τ_0	time constant	Y, υ	upsilon
		Φ	phase angle; phi	Φ, φ	phi
				X, χ	chi
				Ψ, ψ	psi
				Ω, ω	omega

Index

A

Abhurite 313
Aboveground storage tanks (ASTs) 89
   external aboveground coatings 93–95
   monitoring of 95
   protective coatings 92–93
   regulations for 93
   and soil contact 93
   storage tanks 93–95
Abrasion failures 816
Abrasive cleaning 329
Abrasives 71
Absolute humidity 614
Absorbable organic halide (AOX) 773
Absorber materials 463
Absorption columns 256
Acanthite 311
Accelerated aqueous chloride corrosion ... 1000
Accelerated corrosion test (ACT)
   and testing 148, 149, 181, 538
Accelerated corrosion test methods 199
Accelerated SCC tests 879–880
Accelerated stress testing 621
Accelerated testing 203
Accelerate high-temperature sulfidic
   corrosion 1000
Accelerators 563
Accessibility, in anticorrosion design 248
Accumulator tanks 274
Acetaldehyde 283
Acetic acid 283, 325, 977
   400-series stainless steels in 677
   aluminum in 677
   copper and copper alloys in 677
   copper-bearing 2000-series alloys of
     aluminum in 677
   corrosion rates of 678, 748
   described 676
   duplex stainless steels in 678
   glacial acetic in 677
   indoor sources of 283
   and lead 284
   nickel alloys in 678
   nonmetallic materials in 678
   and oxidizing agents 748
   Pourbaix diagram for copper in 674
   silver in 678
   stainless steels in 677
   steels in 677
   titanium in 678
   type 316 stainless steel in 678
   wood as a source of 283
   zirconium in 678
Acetic acid mixtures 746
Acetic acid vapors 209
Acetic anhydride 283
Acid bisulfite pulping 769
Acid cleaning 766
Acid concentration 737
Acid corrosion 468, 778
Acidic and alkaline environments 616
Acidic deposition 280
Acidic metabolic products 901
Acidified phosphate fluoride 882
Acid leaching 1081

Acid particulates 280
Acid pickling 994
Acid-producing bacteria (APB) 13
Acid rain 43, 290, 291
Acid-resistant brick 741
Acid-resitant cements 1003
Acids
   mixed 742
   nonoxidizing mixtures 743–747
Acoustic emission 1051
Acoustic leak detection 146
Acquired pellicle 902
Acquisition Streamlining and Standardization
   Information System (ASSIST) database .. 139
Acryloid B-48 318
Acrylonitrile-butadiene-styrene (ABS) ... 733, 941
Acrylonitrile rubber (Buna N) 733
Active-passive metals 32
Additives, to concrete 563–564
Adhesion problems 71
Adhesion tests 172
Adhesive films 177
Adjacent cathodes 567
Admiralty brass (C44300) 13, 325, 450, 456,
         733, 969, 975, 976, 978, 990
Advanced-performance topcoats 173
Aerated halide solutions 185
Aeration 653, 674, 682, 794
Aeration/sulfur effects 1030
Aerobic bacteria 1017
Aerobic microorganisms 1076
Aeroderivative gas turbine,
   hot corrosion 487, 1092
Aerosols 280
Aerospace primers 189
Afterburning 163
Agenitite 311
Aggressive ions 564–565
Aging aircraft 195, 607
Aging fleet surveys 606–607
Aging mechanisms, of military
   equipment 220–223
Aging systems 128
Airborne contaminants 43–44
Airborne particles 615
Airborne salt particulates 280
Air contamination 974
Air-cooled condensers 447–448
Aircraft. *See also* air force aircraft; commercial
   aviation; military aircraft; naval aircraft
   assembly defect control 604
   corrosion fatigue assessment 195–196
   internal environments of 605
   material substitution of 133
   sealing compounds for 190–191
   service life and aging of 220–228
Aircraft cleaning 186
Aircraft corrosion, causes and types of ... 196–197
Aircraft Structural Integrity Program
   (ASIP) 131
Air-entraining agents 563
Air-flow dynamics 627, 628
Air Force aircraft 130–131
Air heater cold ends 496
Air heaters 492, 493
Airplane corrosion 599–600

Airplane drainage 602
Airplanes with level 2 corrosion 608
Air pollution 461, 463–464
Air pollution control systems. *See* flue
   gas desulfurization (FGD)
Air-port castings 791
Air quality control 793
Air quality emission standards 763
Air stripper 255
Airworthiness, corrosion,
   and maintenance 607–611
Akaganéite formation 314–315
AL6XN (N08367) 214
Aldimines 176
Aliphatic mono acids 679
Aliphatic organic acids 674
Aliphatic polyurethane coatings 189
Alkali-aggregate reaction 563
Alkaline attack 816
Alkaline boilout solution 325
Alkaline chemicals 710
Alkaline copper removal process (Dow) ... 325
Alkaline detergents 325
Alkaline fuel cells (AFCs) 504–505
Alkaline SCC 979
Alkali-reactivity reducers 563
Alkalis 565
Alkali-silica reactivity (ASR) 132
Alkyd paints 209
Alkyds 71
All-alloy cast crown and bridge
   restorations 891
Allcorr 1081
Allergic hypersensitive reactions 893, 915–916
Alloy-clad plate 462
Alloy content 47–51
Alloy factors 901
Alloying elements 396
Alloy-lined incinerators 166
Alloy nitridation resistance 758
Alloy steels, general. *See also* low-alloy
   steels, general 690
   300-series steels 674, 675
   in ammonia and ammonia compounds ... 729–730
   in anhydrous hydrogen fluoride (AHF)
     liquid 699
   corrosion of in hydrochloric acid (HCl) .. 682–683
   corrosion rates of 692
   oxidation resistance of 1067
Alloy tubulars 948
All-volatile water treatment
   (AVT) 237, 365, 376, 377
Alternating current voltages on
   pipelines 113
Alternating torsion 867
Alternative fuels 438
Alternative reference electrodes 90
Alumina
   anhydrous hydrogen fluoride (AHF)
     corrosion of 702
   hydrogen fluoride corrosion of 698
Alumina formers 756
Aluminide coatings 489
Aluminides 490
Aluminized steels 142, 143
Aluminizing 1002–1003

1106 / Reference Information

Aluminum alloy metal lines 623
Aluminum and aluminum-base alloys, general
- *2xxx series alloys* 185, 202, 221, 266, 677
- *2xxx*-T3*x* alloys 600
- *5xxx series alloys* 266, 545, 670, 677
- *6xxx series alloys* 545
- *7xxx series alloys* 185, 221, 222, 545, 600
- in acetic acid 677
- in ammonia and ammonia compounds 727
- anhydrous hydrogen fluoride (AHF) corrosion of 702
- in caustic sodas 710
- in chlorine 704, 707, 708
- in commercial aviation 600
- conservation strategies for 301
- corrosion comparison 131
- corrosion products of 295, 297
- corrosion rates vs. depth 36
- corrosion related to pH 291
- in corrosive media 1073
- corrosivity categories for, based on corrosion rates 55
- effects of chemical cleaning solutions on 325
- electronic packaging corrosion 647
- electroplating 66
- embrittlement mechanism observed 222
- in exhaust systems 270
- in food manufacture and distribution 807
- in hot water tanks 274
- knife attack of 670
- loss of tensile strength 46
- magnesium alloys in 545
- marine grade aluminum alloys (*5xxx* and *6xxx* series) on 267
- and mercury 722
- nitric acid corrosion of 670
- in organic acids 679
- in petroleum refining and petrochemical operations 969
- in phosphoric acid 736
- pitting of 25, 619
- potential-pH diagram for 617
- in propionic acid 679
- reactive ion etching 616
- in seawater 217
- sulfate-reducing bacteria (SRB) growth on 307
- T7 heat treatment 185
- T73 heat treatment 153
- testing duration of, in formic acid 675
- used in marine vessel freshwater systems 273
- used in marine vessel wastewater systems 274
- used in outdoor environments 297
- used in U.S. Navy Aircraft 184
- in waste water systems (WWS) 25

Aluminum and aluminum-base alloys, specific types
- 85Al-15 zinc 589
- 88Al-12Zn-0.02 In alloy 589
- 99.99% Al 39
- 356.0 cast 276
- 535.0 cast 276
- 1060 670
- 1100 677
- 1100-H14 677, 679
- 2017 200
- 2024 212
- 2024-T3 132, 133, 198, 199, 200, 600
- 2024-T4 197, 199
- 2024 T-6 221
- 2024-T39 600
- 2024-T351 600, 601
- 2124-T8 199
- 2224-T3511 132
- 2324-T3 133
- 2324-T39 132, 600
- 2524-T3 200
- 2524-T351 600
- 2525-T3 132
- 3003 677
- 5052 39, 276, 677
- 5056 276
- 5086 675, 677
- 5182 545
- 6061-T6 142, 274, 275
- 6063 677
- 6063-T6 274, 275
- 7002 200
- 7008 200
- 7011 200
- 7050 600
- 7050-T6 263
- 7050-T74 197, 198, 199
- 7055 225, 600
- 7055-T7 133
- 7055-T7*xxx* 132
- 7055-T7751 132
- 7075 217, 225
- 7075-T6 132, 133, 197, 198, 199, 200, 201, 202, 222, 224
- 7075-T73 152, 153
- 7079-T6 152, 600
- 7150 600
- 7150-T77511 132
- 7178 200
- 7178-T6 600
- 7249-T76511 132
- Al 7072 aluminum cladding 185, 1088
- alloy A96061 671, 695
- aluminum-copper alloy 24 ST 263

Aluminum anodes 73, 74, 75, 77
Aluminum artifacts 291, 313–314
Aluminum bronze (C61300) 325, 425, 665, 677, 696
Aluminum bronze tube sheets, dealuminification in 449
Aluminum chloride 977
Aluminum coated 409 528
Aluminum coated 439 520, 525
Aluminum coatings 66–67, 180, 183
Aluminum components 545–547
Aluminum corrosion and copper concentrations 38
Aluminum corrosion rates, acidic and alkaline environments 616
Aluminum electrolytic capacitor 637, 1086
Aluminum floor structure corrosion 604
Aluminum hulls 266
Aluminum hydroxide oxides 272, 313
Aluminum hydroxides 313
Aluminum mast system 275
Aluminum powder thermal coatings 61
Aluminum-silicon bronzes 696
Aluminum spars 274
Aluminum steel panels 66
Aluminum-zinc-indium anodes 73
Aluminum/zinc powder thermal spray 61
Amalgams. *See also* dental casting alloys 897
- copper in 861
- corrosion fatigue (CF) testing of 871
- in dentistry 856
- mercury in 861
- silver in 861
- types of 904–905

Amalgam tin-mercury phase 899
American Dental Association (ADA) Specifications No. 1 904
American Water Works Association (AWWA) standard C-105 115
Amine cracking 990–991
Aminelike odor 637
Amine solutions 991
Amine solvents and neutralizers 979–980
Ammonia and ammonia compounds 976
- alloys for use at elevated temperatures in 730–731
- alloy steels in 729–730
- aluminum alloys in 727
- austenitic stainless steels in 730
- carbon in 734
- carbon steels in 728–729
- cast irons in 727
- cast stainless steels in 730
- chromium-molybdenum steels in 729–730
- copper and copper alloys in 732
- duplex stainless steels in 730
- elastomers in 733
- elastomer temperature limits 733
- ferritic grades in 730
- glass in 734
- graphite in 734
- iron in 727
- lead in 733
- nickel alloy steels in 730
- nickel and nickel alloys in 731–732
- niobium in 733
- plastics in 733
- precious metals in 733
- precipitation-hardening grades in 730
- production and uses of 727
- silver in 733
- stainless steels in 730
- steels in 727
- superferritic steels in 730
- tantalum in 733
- temperature limits for plastics in 545, 734
- thermosetting resins in 734
- tin in 733
- titanium and titanium alloys in 733
- zinc in 733
- zirconium and zirconium alloys in 733

Ammonia cracking 990
Ammonia grooving 451–452
Ammonia resistant grades of fluoropolymers 733
Ammonia stress-corrosion cracking (SCC) 728
Ammonium bisulfide concentration and liquid-phase velocity 1001
Ammonium bisulfide corrosion 976
Ammonium bisulfite 943
Ammonium nitrate 727
Amorphous polymers 740
Amphoteric metals 548
Anaerobic bacteria 1017
Anaerobic environments 313
Anaerobic microorganisms 1076
Analytical techniques 17
Anchorage system 570
Anchor systems 267, 268
Aneurism clip 883
Angle-drilled anodes 93–94
Anglesite 290
Anhydrous ammonia 727
Anhydrous hydrogen fluoride (AHF) 698, 700, 701
Anionic activity 349
Annealing parameter 407
Annular slug treatment 940
Anode aplines 1074
Anode-concrete infacial resistance (ACIR) 586
Anode electrolyte 591
Anode materials 73, 77, 960
Anodes 77
- bed design and location of 103

Anode systems 93, 1062
Anodic (oxidation) reactions 9
Anodic coatings 188
Anodic corrosion 617
Anodic currents for gold alloys vs. saturated calomel 914
Anodic curves 585
Anodic dissolution models 875, 876
Anodic gradient 107
Anodic oxidation 646
Anodic passivators 176
Anodic polarization
- of amalgams in human saliva 897
- in artificial saliva 913, 916
- and cathodic polarization 869
- crack rates and 869
- of gold 916
- of low-copper amalgam 900
- of silver 916
- curves 743

Anodic protection
- for austenitic stainless steels under sulfuric acid corrosion 661
- of batch digesters 765
- for carbon steel 660
- of continuous digesters 766–767

Anodic stray-current interference 107

Index / 1107

Anodic stress-corrosion cracking
(SCC) 929, 930
Anodize coatings 602
Anodized aluminum 25, 181, 275
Anodized coating 152
Anodizing 267
Anodizing-associated equipment 1074
Anomaly severity classifications 81
Anoxic corrosion behavior 423
ANSI/EIA-364-B 209
Anticorrosion design 248
Antifoulant coatings 266
Antifouling topcoats 72
Antifreeze 531
Antlerite 294, 315
API 510 1044
API 570 .... 658, 1044
API 941 729
API 1160 79
API RP 580 658
Applicable loading modes, for CF 862
Applique technology 177
Aprotic (water insoluble) solvent
systems 750–751
Aqua milling 329
Aqueous corrosion 443, 445
Aqueous hydrofluoric acid (AHF),
described 690
Aqueous solution pH vs. zinc corrosion .. 1061
Archaeological alloys 309, 310
Archaeological conservation 315–316
Archaeological lead 310
Archaeological pewter 310
Archaeological tin 310
Arc spray 63
Arc-sprayed coating 62, 65
Arctic and antarctic climates 246
Argon-oxygen decarburization (AOD) . 527, 811
Armament corrosion 6, 151–155
Armament systems 151–155
Army
corrosion issues 131
equipment shipment 131
Army Corps of Engineers documents 138
Aromatic acids 679
Aromatic triazoles 11
Arrhenius equation 588
Artifacts
in anaerobic environments 311
bronze corrosion products on 284
burial environments of 306–309
calcium carbonate deposits on 308
case materials and corrosion 284
chemical treatments of 317
coatings on 285
copper alloys in 315
corrosion after excavation 314
desalination of 317
effects of past treatments 285
electrolytic techniques of 317–318
environment of 1085
gold and gilded metal 311–312
long-term storage and display 318
in marine environments 311
mechanical cleaning of 317
metal sulfides in 315
preservation of 286
repairs and reconstruction 318
silver 311
in situ preservation 318
storage prior to treatment 316–318
Artificial lift wells 946–948
Artificial saliva 903, 913
Artificial solutions 903–904
ASE J447 149
As-fired kraft boilers 786
Ash fusion temperatures 986
Ash-handling systems (AHS) 499–500
Asiatic clams (Corbicula fluminea) 14
ASME B31G analysis 82
ASME B31.3 804
ASME B31.8S 79
Asphalt coatings 1018

Asphalt enamel 1020
Asphalt mastic pipe coating 1020
ASTM A 123 149
ASTM A 380 .... 774, 811
ASTM A 416/A 416M 569
ASTM A 494 .... 663, 664
ASTM A 588 1059
ASTM A 763 527
ASTM A 882 569
ASTM A 967 774
ASTM B 117 .... 149, 181, 199
ASTM B 209 153
ASTM B 418 93
ASTM B 633 149
ASTM C 114 575
ASTM C 494 563
ASTM C 876 573
ASTM C 1017 563
ASTM C 1152 575
ASTM C 1174 421
ASTM C 1218 575
ASTM C 1260 566
ASTM C 1293 566
ASTM D 512 89
ASTM D 522 149
ASTM D 610 .... 148, 149
ASTM D 664 983
ASTM D1418 702
ASTM D 2197 172
ASTM D 3359 .... 148, 149
ASTM D 3483 326
ASTM D 4580 573
ASTM D 4806 992
ASTM D 5894 143
ASTM E 937 1055
ASTM F 90 859
ASTM F 620 860
ASTM F 746 842
ASTM F 897 842
ASTM F 961 858
ASTM F 1077 261
ASTM F 1148 261
ASTM F 1377 860
ASTM F 1487 261
ASTM F 1537 .... 826, 911
ASTM F 1580 860
ASTM F 1801 .... 842, 867, 868
ASTM F 1875 842
ASTM F 1918 261
ASTM F 2129 842
ASTM G 36 879
ASTM G 46 951
ASTM G 57 .... 91, 118
ASTM G 61 842
ASTM G 85 199
ASTM G 123 879
ASTM G 129 878
Atacamite 279, 311
Atlantic Ocean, seawater parameters in .. 959
Atmospheric corrosion. See also
weathering 211
affect of time and starting date ... 46–47
ASTM standards related to 51
of copper-bearing steels 50
distance from the sea 44
effects of sunlight, temperature and
wind on 45–46
location on 44–45
long-term predictions 54
low-carbon steel 42–43
of mild steel as function of salinity ... 43
modeling of 51–52
orientation on 44–45
specimens held at angles 45
standards of 42
of steels, effects of chromium additions on ... 49
of U.S. Navy Aircraft 192
windlasses 45
Atmospheric corrosion of magnesium 191
Atmospheric corrosion penetrations 50
Atmospheric corrosion rates 42–43, 247
Atmospheric corrosion testing 51, 52, 57
Atmospheric-corrosivity approach 55

Atmospheric emissions 1072
Atmospheric exposure testing 61
Atmospheric passivation module 625, 626
Atmospheric sulfur dioxide 565
Atmospheric testing 23
Atomic force microscopy (AFM) 839, 845
Atomic hydrogen 998
Attenuation calculations 102
Auger electron spectroscopy
(AES) 17, 18, 160, 903
Austenitic alloys, nitric acid
corrosion of 670
Austenitic filler materials 527
Austenitic stainless steels ... 396, 670, 929, 989
alloy 904L 661
alloy B-2 661
alloy C-276 661
in ammonia and ammonia compounds ... 730
anhydrous hydrogen fluoride (AHF)
corrosion 699–700
anodic protection for 661
in caustic sodas 712–714
chemical composition of, in biomedical
applications 823
corrosion of, in hydrochloric
acid (HCl) 682–683
corrosion resistance 660, 806
cracking in hydrofluoric acid 691
in dentistry 856
effect of aeration and oxidants on .. 661
effect of high concentration SO_3 on .. 661
effect of velocity on corrosion in .. 661
environmentally assisted cracking (EAC)
susceptibility of 348
fabrication and joining 806
irradiation-assisted corrosion cracking
(IASCC) in 375
isocorrosion diagram of 784
mechanism of protection 660
metal dusting with 1070
microbiologically influenced corrosion
(MIC) of 794
in military systems 158
nickel-rich austenitic stainless steels .. 700
nitric acid corrosion of 668
oxygen influence in hydrogen fluoride
corrosion of 692
in pharmaceutical industry 810, 812
in phosphoric acid 736
in pressurized water reactors (PWR) .. 362
sulfuric acid corrosion of 660–662
thermal insulation of 817
used in fuel cells 507
in U.S. Navy Aircraft 185
weld defects of 815
year pit depths ranking 49
Austenitic steel alloys 239–240, 241, 242, 520
Auto-buffering techniques 404
Autogenous welding 807
Automated monitoring systems 1004
Automatic depressurization
system (ADS) 341
Automobiles. See also complete vehicle
testing 520–522
bodies 515–516
forms of corrosion in 1085–1086
Automotive body corrosion 515–518
Automotive coolants 537
Automotive corrosion design guide 149
Automotive exhaust systems
austenitic stainless steels used in ... 520
cold end exhaust corrosion of ... 522–526
components and alloys of
construction 519
crevice corrosion in 525–527
environment of 519
exterior salt pitting in 523–527
ferritic stainless steel alloys used in ... 520
galvanic corrosion in 527–530
hot salt attack of 520–521
intergranular corrosion (ICG) of 527
service life of 519
thermal fatigue of 521

1108 / Reference Information

Automotive paint systems 517
Av-Dec 130
Aviation, forms of corrosion in 1085
Axial current 98
Azurite 311

B

Backing plates 267
Bacteria 1017
and pitting in weldments 214
Bacterial films 38–40
Balding 789
Ball bond corrosion 647
Barnacle mechanism 924
Barnes layer analysis 91
Barrier coatings 582–583
Barrier materials 546–547
Base components 535–536
Base metal alloys 907, 909–910
Base metals and coatings 488
Bases 324
Basquin equation 863
Batavia 315
Batch acidulation process 784
Batch digesters 763–764
for sulfide pulping 770
Batch metal-etch systems 625
Batch treating, of inhibitors 949, 954
Bathtub curve 220, 221
Batteries 501
corrosion of 501–502
Beach marks 864
Beer 805
Beerstone 807–808
Below-grade moisture mitigation 146
Benson boilers 236
Benzotriazole (BZT) 536
Beryllium-copper, effect of plastic decomposition products on 205
Beta curves 109, 557
Beta-titanium 910
Between pigs batch technique 940
Bicycle corrosion 259–260
Biochemical oxygen demand (BOD) 5, 23
Biocides 213
Biocompatibility 820, 821
Bio-Dur 108 . . . . 868, 881
Biofilms 38–40, 902, 1094
Biofouling films 32
Biological fouling 27
Biological macrofouling 12
Biological organisms 38–40
Biological oxygen demand (BOD) 794
Biological reactions 853
Biomass 445
Biomaterials. *See also* metallic biomaterials
defined 853
described 837
interfacial interactions between blood and 841
oxide film dynamics of 844
physiological environment 840, 847
of three-dimensional atomic force microscopy (AFM) images 839
thrombogenesis 841
Biomedical applications
chemical composition of alloys in 837, 838
Biomedical devices 839, 848
Bioprocessing equipment 15, 803–804
Biosimulated solutions 843
Bisphenol-A fumarate (Bis-A) 721
Bisulfites 777
Bituminous coatings 1020, 1022
Bjork-Shiley heart valve 872, 877, 883
Black annealing 758
Black dot damage 398
Black liquor 762, 780–782, 783
Black magnetite 312
Black powder 258
Black sulfide coatings 181
Blade design improvements 474
Blade failures 469–470
Bleaching 771–772
Bleach plants 771–775
Bleed water 570
Blistering, of steels 698
Blistering and disbonding, of thermal spray coatings 766
Blocked polymers 175
Blocking voltage 123
Blocks and sheaves 277
Blood 841, 895
Blue corrosion 276, 317
Blue plague 205
Blue stains 632
Bluing process 258
Boats 7
corrosion in 257
electrical and electronic systems 271
equipment 267–268
fiberglass 265–266
hulls, fittings and fastenings 265–266
metal deck gear 266–267
plumbing systems 273
propulsion systems 268–271
wiring and loads 272
Body electrochemistry 854
Body fluids 843, 894–896
in contact with metallic biomaterials 855
and Pourbaix diagrams 854
Body pH 854
Body solution chemistry 840
Boiler combustion control 439
Boiler deposits 466
Boilers. *See also* Rankine cycle; recovery boilers; steam generators (SGs) 156, 494
candidate alloys for 238
coal-fired 478
corrosion 466–467
in military systems 156–161
stress rupture failures in 653
wastage in 477–478
waste-to-energy 482–483
water chemistry in 466, 653
waterwall corrosion in 445, 477–478
Boiler tubes
composite tube construction 788
near-drum thinning 792
studded carbon steel 788
Boiler water treatments 237
Boiling beaker condensate corrosion testing . . 523
Boiling point of crude oil vs. reagent materials 984
Boiling water reactors (BWRs) 341–343, 415, 419
carbon steels in 342
cladding corrosion rates vs. precipitation sizes 407
composition of primary materials of 346
core components 386
core shrouds 388, 390
corrosion in 341–356
effect of low hydrogen levels in 391
environmental factors in 348
environmentally assisted cracking (EAC) 341, 343–350, 350–356
environmental solutions in 356
experimental vs. plant data 350
hydrogen injection in 405
hydrogen water chemistry 356, 364
initial design problems of 339
irradiation-assisted corrosion cracking (IASCC) in 375
life-prediction algorithm in 350
localized zirconium alloy corrosion 406
low-alloy steels in 342
mean reactor water 356
neutron fluence 386
nickel-based alloys in 353–354
piping cracks vs. operational time 350
plant water purity 353
vs. pressurized water reactors (PWR) 362
regulating requirements 342–343
repair costs 342
sensitized type 304 stainless steel in 352–353
slow-strain-rate in 349
stress corrosion in 353
stress fatigue in 353
tensile stress factors 344–345
water chemistry mitigation in 404–405
water purity control 404
zirconium alloys in 342
Bolts and bolting 238, 263, 240, 381–382
Bomb calorimeter 243
Bonding protection 273
Bone cement deterioration 873
Bone plates 872
Bone resorption 854
Bone screws 872, 882
Boom clay water 424
Booms 275
Borated water 366
Borates 176, 535
Borosilicate glass 462, 678, 721
Bottle clips 498, 1093
Bottom ash 499
Boundary sliding 401–402
Bow rollers and accessories 267
Branched crack pattern 880
Brass 665
dezincification of 282
ferric ion corrosion in, during chemical cleaning 328
on marine vessels 267
water-side stress-corrosion cracking (SCC) of 451
Brayton cycle 162, 439
Breakaway oxidation 755
Breakdown (pitting) potential 812
Brewery industry 804
Brick linings, sulfuric acid corrosion of 665
Bridge (dental) alloys 905–908
Bridges 1054
Bridges and highways
control strategies for 560
corrosion in 559, 571, 582
Brightening stages 771, 772
Brochantite 290, 294
Bromargyrite 270
Bronze, general 776
artifact treatment 286
in exhaust systems 270
on marine vessels 267, 270–271
patina on 289, 290
silicon bronze 265, 266, 275, 764
used for turnbuckles 277
used in marine vessel wastewater systems 274
Bronze, specific types
aluminum bronze (C61300) 325, 425, 665, 677, 696
cast phosphor bronze (UNS C94300) 265
manganese bronze (58Cu-39Zn) 270
type 1N bronze 780
Bronze disease 279–280, 291, 315, 317
Bronze patina, imitation 293
Bronzes 665
Bronze statue 292, 295
Browning process 258
Bruxism 892
Bryzoans 14
BS EN 473 1043
BS EN 1714 1043
Bubble pack cushioning materials 208
Buffers 535
Building industry corrosion 338
Building water 5
Bulk solids processing 253–254
Bullets 329
Buna-N 212
Burial 217, 306, 309
corrosivity of 307
electrode potential of 307
pH of 307
Buried environments 7
Burners 477–478
Burst-pressure model 1024
Butadiene-styrene rubber (Buna S) 733

Butane oxidation process 676
Butlerite 315
Butler-Volmer equation 116
Butyl 733
Butyl elastomers, in dry chlorine 706
Butyl rubber 678
Butyl rubber and neoprene, sulfuric acid corrosion of 666
B value. *See* Stern-Geary parameter (B value)

C

Cadmium
coatings 66, 183
plating 601
corrosion rates of 67
environmental pressures with 602
Calcareous deposition 35–37
Calcerous coating 936
Calciner. *See* reactors
Calcium carbonate
deposits on artifacts 308
influence of temperature on solubility of 936
scale 10
Calcium chloride deicing salts 523
Calcium hydroxide 561
Calcium-silicate hydrate (C-S-H) 561
Calcium silicate insulation 818
Calcium sulfaluminate 561
Calgon 286
Caliper survey 953
Camouflage 151, 152
CANDU reactors 419
Cantilever beam specimens 931
Capacitive effects 108
Capacitors, metallized-polypropylene-film type 637
Capillary pores 562
Carabiners 263
Carbide precipitation at grain boundaries. *See* sensitized stainless steel
Carbides 162
hydrogen fluoride corrosion of 698
Carbon 687
in ammonia and ammonia compounds 734
hydrogen fluoride corrosion of 697–698
in phosphoric acid 740–741
Carbon and graphite 666, 721
anhydrous hydrogen fluoride (AHF) corrosion of 702
Carbon and low-alloy steels, general. *See also* cost of corrosion 253, 254, 426, 971, 979
5Cr steel 984
9Cr steels 984
12Cr-1MoV 244
12Cr steels 984
50D mild steel (BS 4360) 215
400-series steels and organic acids 674
1010 (G10100) 263
1020 (G1020) 215
A 1035 580
AL349 520
amine cracking of 990–991
in ammonia and ammonia compounds 728–729
ammonium bisulfide concentration and liquid-phase velocity, effect of on 1001
in anhydrous ammonia 732
in anhydrous hydrogen fluoride (AHP) corrosion 698–699
in boiling water reactors 342
boric acid corrosion 337
in caustic sodas 711
caustic soda service chart for 979
coating and cathodic protection of 796
corrosion of 37, 45, 55, 366, 423, 682–683, 1055
effects of chemical cleaning solutions on 325
with enhanced resistance to brittle fracture at lowered temperatures 973
erosion rates of 32
Fe-1.5Mn-0.5Si steel 424
in feed water heaters (FWH) 456
ferric ion corrosion in, during chemical cleaning 328
for high-level waste (HLW) containers 423–424
in high-temperature environments 754
hydrogen attack of 997
hydrogen blistering of 995
in hydrogen fluoride corrosion 690–691
inland corrosion rates (CR) for 48
in lower furnace 787
marine corrosion rates (CR) for 48
naphthenic acid impingement attack of 985
in naval aircraft 185
nitric acid corrosion of 668
oxidation of 754
oxidation resistance of 1067
in phosphoric acid 736
rubber linings on 462
in seawater 960
solid solubility 347
thermal insulation of 817
threshold chloride levels 565–566
used in marine vessel freshwater systems 273
Carbon and low-alloy steels, specific types
9Cr-1Mo 981, 984
1016 426
1020 426
1060 821
2205 clad 768
4068 821
4130 (G41300) 263
4140 (G41400) 381, 768, 776
4340 (G43400) 223, 381, 768, 776
4340Mo 600
aluminized type 1 (ALT1) 528
ASTM A 210 grade A1 787, 788
ASTM A 285 763
ASTM A 285 grade C 252, 765
ASTM A 333 grade 9 698
ASTM A 335 690
ASTM A 516 grade 70 426, 427, 723, 763, 766
ASTM A 517 grade 70 1081
ASTM A 706 563
ASTM A 709, grade 100 1055
ASTM A 709 steel, grade 100 1059
C-0.5Mo steel 968, 971, 998
SA333 grade 6 346
Carbonation 564–565, 574–575, 583, 586, 796, 1057
Carbon brick 1003
Carbon-carbon composites 702
Carbon dioxide 44, 307, 795, 924–925
Carbon dioxide/bicarbonate buffering system 901
Carbon dioxide corrosion 946–947, 948, 960, 1026–1027, 1027, 1028
Carbon dioxide production facilities 956
Carbon fiber reinforced plastic (CFRP) 599, 603, 605
Carbon fibers 266, 275
Carbon-loaded organic coatings 584
Carbon stainless steels 794
Carbon state or copper alloy 990
Carbon steel
in anhydrous hydrogen fluoride (AHF) liquid 699
anodic protection for 660
cathodically protected 215–217
corrosion of 216, 654, 659, 660, 692, 699, 795
localized attack 659
in seawater 214–217
residual elements in 690, 698
Carbon steel pipe vs. TSA-coated carbon steel 657
Carbon steels
Carbon steel tubes 787–788, 791
Carbonyl compounds 283
Carbonyl sulfide 281
Carburization 731
in heat treating industry 1069–1070
in high-temperature environments 756–757
Carburization resistance 757
Carburization test results 1069
Carcinogenesis from dental alloys 893, 894
Carcinogens 174
Cardiac pacemaker 857
Cardiovascular applications. *See also* Bjork-Shiley heart valve 846, 856, 884
Carnot efficiency 439
Carnuba wax 630
Carrier landings 184
Carryover 466
Cascade cleaning method 323
Case hardening 258
Case histories 1062–1066
Casing axial current determination 101
Casing current management 98–99
Casing current measurements, limitations and advantages of 99
Casing polarization (*E* log *I*) test 99
Casing potential profile 98, 100–101, 934–935
Casing-to-anode separation 102
Cassiterite 313
Castable refractories 1003
Cast alloys, water saturated corrosion 707
Cast aluminum statue 302
Cast cobalt-chromium alloys 911–912
Cast dental alloys 908
Castings 20–21, 61
Cast-in-place conductive polymer secondary anodes 589
Cast iron, general
in ammonia and ammonia compounds 727
artifact storage 316–317
in caustic sodas 711–712
effects of chemical cleaning solutions on 325
in exhaust systems 269
in petroleum refining and petrochemical operations 969
in phosphoric acid 736
sulfuric acid corrosion of 660
vs. wrought iron 312–313
Cast iron, specific types
F41000 (NiResist type 1) 711, 728
F41002 (NiResist type 2) 711, 728
F43000 (NiResist type D2) 711, 728
(F47003) 670, 711
MONDI (ductile cast iron) 660
NiResist 711, 728
Cast-iron anodes 94
Cast iron pipe, with tuberculation 9, 1093
Cast iron statue, galvanic corrosion of 292, 1090
Cast manganese bronze 270
Cast stainless steels 783
in ammonia and ammonia compounds 730
in caustic sodas 714
corrosion of 662, 670
Cast steels, specific types
CA-6NM (J91540) 767, 768
CA-15 (J91150) 768, 779
CB-6 (J91804) 767, 768
CB-7Cu-1 (J92180) 767
CD-3MN (J92205) 775
CD-4MC (J93372) 767, 1079
CD-4MCu (J93370) 662, 708, 1080
CD-4MCuN (J93372) 783
CD-6MN (J93371) 767
CF-3 (J92500) 670, 714, 730, 767, 783
CF-3M (J92800) 21, 670, 714, 730, 773, 774, 779
CF-8 (J92600) 670, 714, 730, 783
CF-8M (J92900) 670, 708, 714, 730, 773, 779, 1079
CG-3M (J92999) 775
CG-8M (J93000) 775
CK-3MCuN (J93254) 775
CN-3MN (J94651) 775
Durcomet 5 (J93900) 670
HH (J93503) 783
HK-40 (J94204) 730, 731, 757
HN (J94213) 783
UNS J93900 662
Catalytically cured paints 209
Catalytic converter 519, 522
Catalytic cracking units 976, 991, 1003
Catalytic oxidizer 255

Catalyzed titanium ... 591
Cataphoretic deposition ... 517
Catastrophic failure of Ni-Cr-Fe alloys ... 1070
Cathode-to-anode area ratio ... 267–268, 927
Cathodic (reduction) reactions ... 9
Cathodically protected carbon steel ... 215–217
Cathodically protected stainless steel, in seawater ... 216
Cathodic corrosion ... 617
Cathodic curves ... 585
Cathodic polarization and anodic polarization ... 869
Cathodic protection (CP) ... 5, 73, 90, 145, 583
  anode configurations ... 589
  for artifact storage ... 317
  chloride migration from ... 589
  and coatings ... 72, 92
  from corrosion fatigue ... 928
  criteria ... 73, 90, 589
  current levels ... 586
  design in seawater ... 936
  design of ... 585
  effect of environmental factors on ... 588
  electrical surveys of ... 1022–1023
  elements of ... 584
  external in flow lines ... 962
  fatigue performance of alloy steels in seawater ... 961
  and high-pH ... 1019, 1022
  history of ... 584–590
  installation of ... 103, 590
  of marine pipelines ... 74–75
  of offshore platforms ... 934
  of offshore structures ... 75–77
  of oil-field equipment ... 933
  of pipelines ... 933, 1022
  repairs to ... 590
  of ship hulls ... 77
  for steel in cementious materials ... 1062
  of steel in seawater ... 586
  and stray currents ... 112
  system characteristics ... 90
  of tanks and production vessels ... 934
  types of ... 1022
  of well casings ... 933
Cathodic protection (CP) installation ... 935
Cathodic protection (CP) potentials ... 95
Cathodic protection (CP) reinforcement ... 587
Cathodic protection (CP) systems ... 102–103
  components for ... 585
  computer-aided designs ... 77
  electrical circuit analog of ... 584
  energization of ... 590
  for marine vessels ... 271
  offshore design criteria ... 959
  operations of ... 585
  potential measurements in ... 73
  solar energy to power ... 933
  thermoelectric generator to power ... 933
  types of ... 584, 933
  of underground storage tanks (USTs) ... 95
  for water storage tanks ... 936
Cathodic reaction ... 578
Cathodic stray-current interference ... 107
Caustic ... 978
Caustic attack ... 653
Caustic corrosion ... 468, 653
Caustic cracking ... 713, 990
  of alloy 600 ... 365, 379
  of steel ... 989–990
  and tube damage ... 380
Caustic embrittlement ... 328, 989
Caustic evaporation plants ... 717
Caustic gouging ... 468, 653
Caustic mixtures ... 721
Caustic plus permanganate solution ... 326
Caustic potash
  corrosion rates of alloys in ... 724
  described ... 723
Caustic sodas ... 710–713
  aluminum and aluminum alloys in ... 710
  austenitic stainless steels in ... 712–714
  carbon and graphite in ... 721
  carbon and low-alloy steels in ... 711
  cast irons in ... 711–712
  cast stainless steels in ... 714
  caustic stress-corrosion cracking (SCC) in ... 714
  ceramics in ... 721
  chlorinated polymer CPVC in ... 720
  chromium-bearing nickel alloys in ... 717
  contamination in ... 721–723
  copper and copper alloys in ... 718
  corrosion data for NiResist in ... 711
  corrosion rates of alloys in ... 715
  corrosion-resistant alloys (CRAs) in ... 714
  corrosive weight loss of ceramics in ... 721
  duplex stainless steels in ... 713
  effect of molybdenum on corrosion rates in ... 715
  effect of nickel in caustic sodas ... 715–716
  elastomers in ... 719
  environmentally assisted cracking (EAC) in ... 713
  environmental stress cracking (ESC) from ... 720
  ferritic stainless steels in ... 712
  fiberglass-reinforced plastic (FRP) in ... 720
  high-density polyethylene (HDPE) in ... 719–720
  intergranular attack (IGA) in ... 713
  iron and steels in ... 710–712
  isocorrosion curves ... 714, 718
  lead in ... 719
  low-density polyethylene in ... 719
  nickel and nickel alloys in ... 715–718
  niobium in ... 719
  nonmetallic materials in ... 719
  other metals and alloys in ... 719
  PE/vinyl acetate (PVA) in ... 720
  PH stainless steels in ... 719
  polypropylene (PP) in ... 719–720
  polyvinyl chloride (PVC) in ... 720
  polyvinylidene fluoride (PVDF) in ... 720
  rubber/elastomers maximum temperature for ... 720
  silver in ... 719
  stainless steels in ... 712
  superaustenitic alloys in ... 714
  tantalum in ... 719
  temperature limits for plastics in ... 720
  temperature limits for thermosets in ... 721
  thermoplastics in ... 719–720
  thermosetting resin materials in ... 720–721
  threshold temperature vs. alloy content in ... 715
  tin in ... 719
  titanium and titanium alloys in ... 718–719
  zinc in ... 719
  zirconium and zirconium alloys in ... 719
Caustic solutions ... 722
Caustic stress-corrosion cracking (SCC) ... 468, 711, 718, 723, 766
  in caustic sodas ... 714
  in digesters ... 766
  in liquor heaters ... 767
Cavitation ... 927, 1001
Cavitation corrosion
  in aluminum ... 535
  in engine coolants ... 534–535
  nucleate boiling-induced ... 535, 1092
CCCs, repair and touch-up of ... 174
Cellular glass ... 817
Cellulose acetate butyrate (CAB) ... 941
Cement. *See also* concrete ... 797
Cement-asbestos ... 942
Cement chemistry ... 561–563
Cementitious admixtures ... 564
Cement paste microstructure ... 561, 796
Centrifugal slurry pumps ... 1078
Ceramic brick ... 1003
Ceramic-coated anodes ... 145
Ceramic materials ... 741
Ceramics. *See also* refractories
  in caustic sodas ... 721
  in supercritical water ... 233
Cerargyrite ... 311
Cerckage wire ... 872
Certificate of conformance (COC) ... 153

Certified test report (CTR) ... 153
Cesium ... 365
Chain ... 268
Chainplates ... 276, 277
Chalconatronite ... 270, 276, 311, 317
Chalcopyrite ... 311
Channel feedwater heaters ... 456
Channeling of condensate return ... 143
Charpy V-notch impact test ... 973, 994
Chelants ... 324–326
Chemical admixtures ... 563
Chemical agent resistant coating (CARC) ... 140, 148–149, 151, 182
Chemical analyses ... 49
Chemical cleaning ... 323
  corrosion ... 327, 328
  on-line ... 328–329
Chemical cleaning methods ... 323–328
Chemical composition
  of austenitic stainless steels for implants ... 823
  of cobalt-base alloys ... 837–839
  of cobalt-base alloys in surgical implants ... 825
  for components of Ringer's solution ... 862
  of high-temperature alloys ... 1068
  of iron-base alloys ... 837–838
  of stainless steels ... 837–838
  of stainless steels in surgical implant applications ... 857
  of titanium-base alloys ... 838–839
  of various body fluids ... 854
Chemical conversion coatings ... 188
Chemical industries ... 254–256
Chemical injection, real-time monitoring ... 974
Chemical oxygen demand (COD) ... 5, 23, 773
Chemical passivation ... 17
Chemical process equipment ... 144
Chemical process industry (CPI) ... 338, 742
Chemical processing plants ... 652–653
Chemical pulping ... 762–763
Chemical recovery ... 780–784
Chemicals ingested ... 896
Chemical stability, of coatings ... 173
Chemistry-boiler reactions ... 466
Chemraz ... 733
Chevron marks ... 864
Chimney stacks ... 493, 494, 496
Chip conveyors ... 767
Chip corrosion ... 630
Chips ... 629
Chloramination ... 10
Chloramines ... 16
Chlorargyrite ... 311
Chlorates ... 721–723
Chloride ... 1057
  in concrete admixtures ... 1057
  effect of on passivated surfaces ... 899–900
Chloride concentration, of rainwater ... 43, 44
Chloride corrosion threshold, impact of concrete quality on ... 572
Chloride-induced corrosion ... 1060
  of reinforcement ... 1058
Chloride-induced deterioration, of concrete ... 797
Chloride-induced stress-corrosion cracking (CISCC) ... 655
Chloride ions ... 290, 564, 567
Chloride levels, of white vapor ... 768
Chloride pitting ... 16, 810
  in white water ... 778
Chloride ponding time ... 566
Chloride profile
  of concrete ... 574–575
  by method of transport ... 564
Chlorides ... 722
  and 18-8 stainless steel ... 655–656
  atmospheric salt content ... 43
  effects of, on buried steel ... 89
  and stainless steel ... 16–17
  in wastewater ... 794–795
  in white water ... 777
Chloride salts ... 656

Chloride stress-corrosion cracking (CSCC) . . 654, 970, 988–989
in 300 series stainless steels . . . 955
vs. alloy nickel content . . . 17
in liquor heaters . . . 767
and pH . . . 989
of sensitized type . . . 304, 691
of stainless steels . . . 339, 988, 989
threshold temperatures for . . . 17
Chloride threshold . . . 565–566
Chlorinated polymer CPVC, in caustic sodas . . . 720
Chlorinated polyvinyl chloride (CPVC) . . 734, 941
hydrogen fluoride corrosion of . . . 697
Chlorinated rubber coatings . . . 71
Chlorination . . . 452
Chlorine
in coal . . . 478
corrosion by . . . 704–708
in hydrochloric acid (HCl) . . . 682
with other high-temperature mixed gases . . . 706
Chlorine-based bleaching, stages of . . . 771–772
Chlorine-containing solvents, and aluminum . . . 637
Chlorine gas
effect of water . . . 707
passivate unalloyed titanium in . . . 707
Chlorine/hydrochloride . . . 722
Chlorine-ice, corrosion of alloys in . . . 708
Chlorine-induced scale spallation . . . 445
Chlorine ions . . . 89
Chlorine permeation effects . . . 706
Chlorine water . . . 708
Chloro-alkali plants . . . 254
Chloroprene rubber . . . 733
Chlorosulfinated polyethylene, in dry chlorine . . . 706
Christmas trees . . . 932
Chromate-bearing paints . . . 560
Chromate conversion coatings . . . 601
Chrome A, in dry chlorine . . . 705
Chromia formers, long-term oxidation tests for . . . 756
Chromia-forming ferritic stainless steels . . . 513
Chromia scale . . . 506–508
Chromic acid . . . 326
Chromic acid anodizing . . . 174
Chromite spinel . . . 19
Chromium
allergic hypersensitive reactions to . . . 915
alloying with . . . 907–908
elimination of . . . 174
environmental pressures with . . . 602
sensitivity to . . . 822
Chromium content
in reheaters . . . 480
of stainless steels . . . 764
in superheaters . . . 480
Chromium depletion . . . 347
Chromium enrichment heat treatment, effect of, on irradiation assisted stress-corrosion cracking . . . 397
Chromium equivalents for alloys, in anhydrous ammonia plants . . . 731
Chromium/iron ratio . . . 17–19
Chromium levels, in stainless steels . . . 755
Chromium-molybdenum (Cr-Mo) steels
in ammonia and ammonia compounds . . . 729–730
in high-temperature environments . . . 754
oxidation of . . . 754
Chromium steels, metal dusting with . . . 1070
Chromizing/aluminizing . . . 480
Chromizing/siliconizing . . . 480
Chromosomal changes . . . 823
Circulating water system, once-through . . . 12
Circulation method . . . 323
Circumferential stress distributions . . . 226
Citric acid . . . 325
CitroSolv process . . . 326
Cladding . . . 1002
Cladding failures . . . 365
Clad plate vs. solid plate . . . 463
Clam shells . . . 864
Class 2 rouge . . . 19–20
Class 3 rouge . . . 20
Classical stress-corrosion cracking (SCC) . . . 1018
Clean air act amendment (CAAA) . . . 173
Cleaner, lubricant, and preservative (CLP) . . . 155
Cleaning
of stainless steels . . . 21–22
of tubular goods . . . 963
Cleaning pressures . . . 808
Clean-in-place (CIP) . . . 808
Clearance rates, of corrosion products . . . 901
Clearing event . . . 637
Clear plastic vinyl, used in marine vessel freshwater systems . . . 273
Cleavage . . . 864
Clinoatacamite . . . 311
Closed loop water system . . . 13
Closed recirculating systems . . . 268
Close-interval potential map . . . 574
Close-interval survey (CIS) . . . 6, 84–88
CMAS . . . 163
CO_2 emissions . . . 244
Coal, chloride levels in . . . 445
Coal-ash. *See also* ash-handling systems (AHS)
Coal ash . . . 500
corrosion . . . 159, 478–479
Coalescing solvent . . . 175
Coal-fired boilers, fireside corrosion in . . . 477
Coal-fired utility boiler. *See* boilers
Coal-gasification combined-cycle (CGCC) plants . . . 441
Coals . . . 438
Coal tar coating systems . . . 1074
Coal-tar-enamel . . . 1020
Coal-tar-enamel wax . . . 1018
Coal tar epoxy . . . 92, 495
Coal-tar epoxy coatings . . . 1020
Coastal bridges, corrosion in . . . 559
Coastal salinity . . . 28
Coated carbon steel . . . 217
Coated casings . . . 102
Coating, chemical reaction for typical two-part polyurethane . . . 172
Coating deposition process . . . 63–64
Coating disbondment . . . 77, 1024
Coating in marine atmosphere, corrosion-time plots . . . 66
Coating materials, reversible . . . 298
Coating resistance . . . 122, 124
Coatings. *See also* organic coatings; thermal spray coatings (TSCs); zinc coatings . . . 582–583
alloy powders for . . . 479
alloys used in . . . 480
on aluminum . . . 546
application methods of . . . 963
of armament systems . . . 151–152
on artifacts . . . 285
and cathodic protection . . . 92, 1018
corrosion resistance of . . . 1073
with defects . . . 576
effect of, on the mechanical properties of turbine components . . . 490
environments destructive to . . . 69
evaluation of . . . 1021
high-phosphorous electroless nickel (HPEN) . . 717
for inhibition programs . . . 932
for low-alloy steels . . . 948–949
low-phosphorous electroless nickel (LPEN) . . . 717
medium-phosphorous electroless nickel (MPEN) . . . 717
for military aircraft . . . 6
military specifications for . . . 136
for naval aircraft . . . 130
performance requirements of . . . 189
for petroleum production operations . . . 932
in pharmaceutical industry . . . 818
on pipelines . . . 1020–1021
procedures and methods for . . . 932
protection and corrosion of . . . 576
protective elements in . . . 489
specialty . . . 173
special use . . . 1021
for stainless steels . . . 731
for steel . . . 1072–1073
surface preparation for . . . 1021–1022
topcoats . . . 172
for tubing . . . 932
types of . . . 1081
for waste water systems (WWS) . . . 26
water-reducible high-performance . . . 174–175
for waterwalls . . . 479
without defects . . . 576
Coating systems . . . 486, 657
schematics of . . . 65
in various environments . . . 1059
Coating test panel, electrochemical measurement on . . . 63
Coating weights . . . 516
Cobalt and cobalt-base alloys, general
allergic hypersensitive reactions to . . . 915
atomic force microscopy (AFM) tests . . . 833
biocompatibility . . . 820
carburization in . . . 1069
chemical composition of, in biomedical applications . . . 837
cobalt-chromium alloys . . 820, 826, 856, 907–908, 911–912, 914
cobalt-chromium-molybdenum . . . 833
cobalt-chromium-molybdenum alloy . . . 908, 911
cobalt-chromium-nickel . . . 910
corrosion fatigue (CF) testing of . . . 870
cyclic polarization curves of . . . 842
dental alloys . . . 907–908
dental implants . . . 911–912
metal dusting with . . . 1070
in metallic biomaterials . . . 826, 857, 858
microstructures . . . 911
nitridation resistance in . . . 758
orthodontic wires . . . 910
orthopedics . . . 856
oxidation resistance of . . . 1068
potentiodynamic corrosion curves for . . . 842
Pourbaix diagrams for . . . 856
scratch test . . . 833
sensitivity to . . . 822
in stress-corrosion cracking (SCC) testing . . . 882
sulfidation resistance of . . . 1070
tarnishing and corrosion . . . 914
Cobalt and cobalt-base alloys, specific types
ASTM F 75 casting alloy (R30075) . . . 822, 826, 839, 858, 859, 870, 911
ASTM F 90 alloy (R30605) . . . 859
ASTM F 556 alloy (R30556) . . . 760, 970, 1070
ASTM F 562 alloy (R30035) . . . 859
ASTM F 563 alloy (R30563) . . . 859, 870
ASTM F 799 alloy (R31537) . . . 826, 858, 859, 860, 870, 911
CCM Plus . . . 870
MP35N (35Co-35Ni-20Cr-10Mo) . . 837, 842, 857, 859, 882, 912
Vitallium . . . 858, 882
Cobalt-base superalloys, for combustor and turbine sections . . . 487
Cobalt-chromium PFM alloys . . . 910
CoCrALY coatings . . . 490
Co-Cr-Mo alloys . . . 842, 843, 844, 846, 847
surface oxide of . . . 839
Co-diffusion coatings . . . 480
Coffin-Manson equation . . . 863
Coil, low-side-switched circuit . . . 634
Coil bobbin crack . . . 636
Coil coating . . . 519
Coil corrosion failure . . . 635
Cold bonding . . . 604
Cold casings . . . 494, 496
Cold-climate areas . . . 246
Cold climate corrosion (CCC) . . . 246–248
Cold climates . . . 7, 246
Cold-end corrosion, in boilers . . . 986
Cold end exhaust corrosion, of automotive exhaust systems . . . 522–526
Cold joints . . . 567, 568
Cold stretching . . . 774
Collar zinc . . . 271
Combination heaters . . . 452–453

1112 / Reference Information

Combined cycle systems . . . 439
Combined intergranular attack (IGA)/ intergranular stress-corrosion cracking (IGSCC) . . . 378, 1093
Combustion . . . 438–439
Combustion environment, corrosion rate . . . 987
Commercial aviation . . . 600–604
aging survey results of . . . 607
aluminum alloys in . . . 600
condensation in . . . 605
corrosion in . . . 337, 600–606, 608, 610–611
fleet damage rate . . . 611
forms of corrosion in . . . 599, 1089
implementation age and repeat inspection interval . . . 609
inorganic coatings in . . . 601
inspection methods . . . 609–610
maintenance practices . . . 605–606, 609, 611
manufacturing factors . . . 604–605
organic coatings in . . . 601
steels in . . . 600
Commercial chlorine, handling of . . . 704
Commercially available off-the-shelf (COTS) polyurethane . . . 189
Commercial-off-the-shelf (COTS) products . . . 136
Commercial zinc plating . . . 180
Commission and monitoring . . . 104
Comparative testing . . . 149
Complete vehicle, described . . . 538
Complete vehicle testing . . . 538–543
Completing agents . . . 324
Complexing agent corrosion . . . 467
Compliant coatings issues . . . 173–178
Composite bicycle . . . 259
Composite tubes, preferential corrosion of . . . 788–789
Compressor coating . . . 486
Concentrated alkaline solution (SCW) . . . 432
Concentration cells . . . 899
Concentration factor . . . 201
Concrete
acidic attack of . . . 796
additives to . . . 563–564
as barrier coating . . . 560
carbonation profile of . . . 574–575
chloride induced deterioration of . . . 797
chloride profile of . . . 574–575
condition surveys of . . . 571–572
corrosion of . . . 560–565
cracking of . . . 132, 567
cumulative damage vs. time . . . 572
decomposition of . . . 132
electrical resistivity in . . . 565
metallized coatings for . . . 590
metals embedded in . . . 561
modes of transport in . . . 562
pH in . . . 563
quality and cover assessment . . . 572–573
reactant transport in . . . 562–563
resistivity of . . . 568, 573–574
sealers for . . . 583
size and distribution of the capillary pores in . . . 561
in waste water . . . 24, 796
Concrete admixtures, chloride in . . . 1057
Concrete covers . . . 1060
Concrete formation . . . 308–309
Concrete highway structures, performance of prestressing steel in . . . 569–570
Concrete pore water pH vs. chloride threshold . . . 566
Concrete quality, impact of chloride content corrosion threshold . . . 572
Concrete resistivity . . . 565, 586
Concrete stack linings . . . 495
Concretions
on copper alloys . . . 311
electrolysis as treatment for . . . 318
Condensate, described . . . 447
Condensate corrosion (steam side) . . . 451
Condensate galvanic corrosion, of muffler . . . 528
Condensate grooving . . . 451
Condensate pitting corrosion . . . 522–523
Condensate polishers . . . 237, 377, 380
Condensation, in commercial aviation . . . 605
Condensation corrosion . . . 293
Condensed sulfuric acid, from SO_3 gas . . . 491
Condensed water . . . 928
Condensers . . . 447–448, 452, 975
Condenser tube alloys, velocity limit for . . . 449
Condenser tubes . . . 451
Condition-based maintenance (CBM) . . . 127, 133
Conductive anodic filaments (CAF) . . . 646
Conductive asphalt overlays . . . 589
Conductive couplings . . . 108, 112
Conductors and contact materials . . . 613
Conformal coatings . . . 206, 639
Connectivity . . . 562
Connelite . . . 311
Conservation practice . . . 279
Conservation strategies . . . 316
for aluminum . . . 301
for copper alloys . . . 299
for gold (gilding) . . . 301
for iron . . . 300
for lead . . . 300–301
for zinc . . . 300
Conservators . . . 298, 301
Constant amplitude fatigue testing . . . 200
Constant extension rate test (CERT) . . . 879
Constant stress-corrosion cracking (SCC) test . . . 790
Construction joints . . . 568
Construction quality acceptance . . . 551
Contact fatigue . . . 600
Contact process . . . 659
Contaminants . . . 208
Contaminated crude oil . . . 977
Contaminated mixtures . . . 721
Contamination . . . 805
Continuous acidulation process . . . 784
Continuous digesters . . . 765–767, 770
Continuous injection, of inhibitors . . . 949, 954
Continuous monitoring . . . 1048
Control and prevention of corrosion . . . 620–621
Control bonds . . . 111
Control rod drive mechanism (CRDM) . . . 367, 369, 370
Conventional bleaching sequences, operating and design parameters of . . . 772
Conventional burners . . . 477
Conventional radiographic inspections . . . 1049
Conveyor belts . . . 807
Coolant chemistries . . . 537
Coolant conductivity . . . 348, 353, 389
Cooling systems . . . 268
circulation and components of . . . 532–533
functions of . . . 531–532
operations of . . . 532
Cooling water . . . 366, 448
Coordinated boron and lithium chemistry . . . 405
Copper
in dry chlorine . . . 704–705
electronic packaging corrosion . . . 646
Copper alloy artifacts . . . 310
corrosion products on . . . 284, 1090
treatment of . . . 317
velvetlike corrosion growths on . . . 282
Copper alloy corrosion inhibitors . . . 14
Copper alloy tubes . . . 457
Copper-aluminum alloys
in dental casting alloys . . . 908
tarnishing and corrosion of . . . 916
Copper and copper alloys, in moist chlorine . . 707
Copper and copper-base alloys, general.
See also brass; bronzes . . . 674
in acetic acid . . . 677
alloy C11200 . . . 732
alloy C26000 . . . 732
in amalgam . . . 861, 904–905
in ammonia and ammonia compounds . . . 732
anhydrous hydrogen fluoride (AHF) corrosion of . . . 702
artifact corrosion . . . 285
in caustic sodas . . . 718
composition of, for suction rolls . . . 779
conservation strategies for . . . 299
copper-aluminum alloys . . . 908, 916
corrosion of . . . 36, 38, 45, 47, 55, 109, 291, 294, 295
dealloying of . . . 450
dendritic structure of . . . 282, 1091
effect of oxygen on . . . 32
effects of chemical cleaning solutions on . . . 325
ferric ion corrosion in, during chemical cleaning . . . 328
in food manufacture and distribution . . . 807
in heated ammonia drip solutions . . . 451
for high-level waste (HLW) containers . . . 424
in hydrochloric acid (HCl) . . . 684–685
hydrogen fluoride corrosion of . . . 696
long-term atmospheric corrosion predictions for . . . 54
and mercury . . . 722
in organic acids . . . 679
patina on . . . 293–294
in petroleum refining and petrochemical operations . . . 969
in phosphoric acid . . . 738
in pressurized water reactors (PWR) . . . 362
released into human saliva . . . 916
in seawater . . . 38, 213–214
sulfate-reducing bacteria (SRB) growth on . . . 307
in terrestrial environments . . . 310
testing duration of, in formic acid . . . 675
UNS C17200 . . . 197
used in marine vessel freshwater systems . . . 273
used in outdoor environments . . . 293
in utility condensers . . . 733
in waste water systems (WWS) . . . 25
Copper and copper-base alloys, specific types
70-30 cupronickel . . . 269
70-30 grade (C71500) . . . 696, 718
70Cu-30Ni . . . 13, 450, 696
70Cu-30Ni alloys . . . 456, 458, 702
80Cu-20Ni alloys . . . 456, 458
85Cu-5Sn-5Zn-5Pb (C83600) . . . 718
90-10 supronickel (C70600) . . . 702, 718
90Cu-10Ni . . . 214, 696
90Cu-10Ni alloys . . . 450, 456, 458
90Cu-10Ni copper-nickel (C70600) . . . 675
copper (C10300) . . . 675
copper (C12000) . . . 677
copper alloy C83600 (8Cu-5Sn-5Pb-5Zn) . . . 779
copper-nickel (C71500) . . . 425
copper-nickel (UNS C71500) . . . 696
gunmetal (C90550) . . . 718
Copper and copper-base alloys, sulfuric acid corrosion of . . . 665
Copper-assisted salt solution (CASS) tests . . . 200
Copper-bearing steels . . . 32
A242 . . . 50
A588B . . . 50
Copper/copper sulfate electrode (CSE) . . . 84
Copper electrochemical migration . . . 648
Copper magnet wire corrosion . . . 635
Copper migration and corrosion . . . 639, 1086
Copper-nickel alloy tubes, denickelification in . . . 449
Copper oxide bottom paint . . . 266
Copper plumbing . . . 9
Copper sheathing . . . 266
Copper sulfide corrosion (brown fuzzies), on copper alloy artifacts . . . 282
Copper-zinc alloys, dealloying of . . . 450
Core test, for inhibitors . . . 949
Corroded copper magnet wire . . . 636
Corroded terminals . . . 224
Corrosion allowances . . . 973
Corrosion and stress-corrosion cracking (SCC)
assessment and repair of . . . 1023–1024
Corrosion and water-quality problems, of drinking water . . . 8
Corrosion attack, of modular junction . . . 830
Corrosion behavior
effects of fabrication on . . . 426
of Zircaloys . . . 419
Corrosion cell . . . 533
Corrosion characteristics, of organic acids . . . 674

Index / 1113

Corrosion control, in petroleum refining
 and petrochemical operations 1002–1004
Corrosion-control records 105
Corrosion coupons 950
Corrosion cycle 127
Corrosion damage, of U.S. Navy
 Aircraft 189–193
Corrosion data
 for NiResist in caustic sodas 711
 for unalloyed cast iron in caustic sodas 711
Corrosion defects, magnetic flux leakage
 tool and sizing of 1023
Corrosion environment 141
Corrosion evaluation and testing 148–149
Corrosion factors 251
Corrosion failures, in pharmaceutical
 industry 814
Corrosion fatigue (CF) 160, 341, 380–381,
  467, 644, 853–884, 927
 in petroleum refining and petrochemical
  operations 998–999
 of roll journals 776
 of steam turbine blading 471
 in steam turbines 469
 in weapons systems 126
Corrosion fatigue (CF) 471–472
Corrosion fatigue (CF) testing 867–871
Corrosion fatigue crack initiation, of
 carbon steels in seawater 960
Corrosion fatigue life prediction procedure ... 201
Corrosion fatigue strengths 1081
Corrosion forms
 of aluminum 545
 in industries and environments 1085–1086
Corrosion fundamentals, described 127
Corrosion inhibitors. See also inhibitors .. 14, 150,
  326, 563, 603, 620, 954, 1002
 for copper and copper alloys 11
 effectiveness of 328
 salt spray tests of slat gearbox with and
  without 192, 1088
 in steam systems 143
Corrosion initiation time, for carbon steel 566
Corrosion inspection 571–575
Corrosion level determination 608
Corrosion lifetime testing 515
Corrosion locations 187–188
Corrosion losses 32, 56, 442, 446
Corrosion maintenance, prevention
 and control 225
Corrosion measurement 1031–1032
Corrosion mechanisms
 in service water systems 13
 in wastewater systems 795–797
 of white water 777–778
Corrosion metrics, for steels 198
Corrosion mitigation
 effects of 200
 in potable water 10
Corrosion monitoring
 in chemical cleaning 327
 oil production 950–953
Corrosion of metals
 effect of process variables on 652
 in hydrochloric acid (HCl) 682, 688
Corrosion penetration 165
Corrosion perforation, lf aluminum 535
Corrosion pillowing, on 2017-T3 aluminum ... 197
Corrosion pits 222
Corrosion potential 348, 352, 843
Corrosion potential curves 742
Corrosion potential stagnation/flow
 (CPSF) method 11
Corrosion prediction schemes, and
 fatigue behavior 201
Corrosion-preventative compounds (CPCs) ... 192
Corrosion prevention, during maintenance ... 608
Corrosion prevention and control, of U.S. Navy
 Aircraft 186
Corrosion prevention and control program
 (CPCP) 140, 337, 606, 611
Corrosion preventive compounds
 (CPCs) 127, 133–134, 186, 188, 200

Corrosion problems
 for electronic equipment 206
 historical review of 205
Corrosion processes 279
Corrosion process zone 199
Corrosion product appearance 186
Corrosion products, on zinc panels 57
Corrosion protection
 data needed for 90–91
 evaluation of 1085
 for tanks 1081
Corrosion rate equation 1027
Corrosion rate modality 952
Corrosion rate probes 6
 for soil environments 115
Corrosion rate response 542
Corrosion rates (CR)
 of alloys in caustic potash 724
 of alloys in hydrofluoric acid 694
 of alloys in sulfuric/hydrochloric
  acid mixtures 744
 of alloy steels in dilute acids 692
 of aluminum alloys 36
 atmospheres 42–43, 49
 of boiler tubes 479
 of carbon steels 45
 for carbon steels 423
 of carbon steels in dilute acids 692
 in caustic solutions .. 710, 712, 715, 717, 722, 723
 from chemical cleaning 327
 of copper 45, 47
 and crude oil types effect on 1030
 vs. depth 36
 effect of acid concentration on 316L
  stainless steel 737
 effect of alloying elements on 746, 747
 effect of gas velocity on 1029
 effect of molybdenum on, in phosphoric
  acid mixtures 743
 effect of redox species concentration and
  halide concentration on, in acetic acid 748
 effects of anhydride on, in $FeCl_3$ 748
 experiment and calculation comparison 57
 vs. geographical location 184
 of high-strength low-alloy (HSLA) steels 45
 of hydrate mixture 789
 influence of pH on 563
 inland 48
 of iron 43, 45–46, 692
 low-alloy steels 45, 48
 of magnesium on function of relative
  humidity (RH) 43
 marine, for carbon steels 48
 by metal 48
 of metals in hydrochloric acid (HCl) 683
 mild steel 45
 vs. moisture film thickness 615
 in molten salt baths 1071
 for nickel alloys 693–694
 of nickel-base alloys 684, 742
 of nickel in caustic evaporation plants 717
 in outdoor environments 289
 oxygen concentration effect on 1030
 and resistivity 575
 by site 47, 48
 in soil 118
 by specimen type 48
 of stainless steels in acid 669, 678, 744
 of steels 48
 of sulfuric acid, effect of nitric acid on 746
 of sulfuric and nitric acid, effect of
  alloying elements on 746
 and temperature 1029–1030
 of unpolarized coupons 117
 in vivo 844
 and water content 1030
 of weathering steels 580–581
 for wrought nickel alloys in
  hydrofluoric acid 694
 of zinc 43, 45–47
Corrosion rate techniques 1004, 1032
Corrosion rate testing 575
Corrosion related message patterns 607

Corrosion removal 188
Corrosion repair philosophy 195
Corrosion repair policy 200
Corrosion resistance 973–974
 of alloys 760
 of carbon steel 579
 in hydrochloric acid (HCl) 684–687
 of refractories 165
 of stainless steels 579
 of stainless steel 805
Corrosion-resistance indexes 579
Corrosion resistance testing 172
Corrosion-resistant alloy selection, for
 petroleum production operations 930
Corrosion-resistant cladding, resistance
 to IGSCC 354
Corrosion-resistant finishes, designing for 619
Corrosion-resistant materials, for
 sweet gas 960–961
Corrosion-resistant reinforcement
 (CCR) 579–580
Corrosion-resistant steel (CRES) 601
 in commercial aviation 599
Corrosion scale, in weathering steels 1063
Corrosion severity, as function of pit
 norm corrosion metric 202
Corrosion surveillance 226
Corrosion test coupons 73
Corrosion testing 620–621
 of alloys in slowing sulfuric acid 661
 in fluids 827
 in organic acids 674–675
 in organic solvents 753
 in solutions 827
Corrosion thinning 766
Corrosion-time plots, coating in marine
 atmosphere 66
Corrosion under insulation (CUI) ... 654, 656–658
Corrosive media 1073
Corrosive properties, of steam 652
Corrosive weight loss, of ceramics in
 caustic sodas 721
Corrosivity
 of burial environments 307
 flow-related effects 1029
Corrosivity monitoring 542
Cor-Ten 494
Cosmetic corrosion 515
 of aluminum 545
Cosolvent 175
Cost of corrosion. See also world-wide
 corrosion
 annual 247
 in bridges and highways 559
 cost reduction strategy 134
 in military 126
 in nuclear power industry 339
 in petroleum refining and petrochemical
  operations 967
 in pressurized water reactors (PWRs) 340
 in the pulp and paper industry 762
 in steam turbines 469
 of the United States 337
 in U.S. Air Force 131
 in U.S. Navy 129
Countercurrent gas stripping column 942
Couper-Gorman curves 981, 982, 983
Coupled environment fracture model
 (CEFM) 351
Coupon test stations 1022–1023
CPET casing potential profile tool 100
CPP method 429–430
Crack advance, during irradiation assisted
 stress-corrosion cracking 387
Crack depth vs. time 349
Crack-detection techniques 344
Crack enclave phenomena 351
Crack growth, time-dependent 862
Crack growth analysis 200, 201, 202
Crack growth equation, for intergranular
 stress-corrosion cracking (SCC) 373
Crack growth prediction 224–225, 350
Crack growth rate per cycle 862

Crack growth rates 387
and cold-work layer thickness 372
vs. corrosion potential 394
for NiCrMoV disc steel 472
for stainless steels irradiated 393
Cracking 700
of alloy 20, 692
of alloy 825, 692
in anhydrous hydrogen fluoride (AHF) 699–700
of composite tubes 789
of composite wall tubes 790
in nickel-base alloy 390
response to 1044
of steels 698
Cracking growth, with time 220
Cracking model 127
Cracking mode superposition 862
Cracking sensitivity, and yield stress 347
Crack initiation 863
analysis 201
number of cycles for 865
testing 867
Crack lengths 394, 400, 868
Crack origin, in rivet hole 222
Crack propagation 165, 349, 352–353, 864
Crack propagation rates 472, 871
nonsensitized stainless steels 348
Crack propagation-rate/stress intensity dependence 346
Crack propagation testing 867
Crack rates, anodic polarization and 869
Crack reducers 563
Cracks 343
in composite tubes 790
under influence of pillowing 221
Cracks arrest marks 864
Crack-tip 349–350, 352
Crack tip opening displacement methods 868
Creep 403
described 653
and grain-boundary sliding 404
Creep embrittlement 973
Creep voids 162, 1092
Crennell's/McCoy's equation 77
Crevice corrosion 13, 27, 644
of alloy 22 (N06022) 429–430
of aluminum 130, 545
in automotive exhaust systems 525–527
in commercial aviation 599
in engine coolants 533–534
engineering design to avoid 1079
in flue gas desulfurization (FGD) systems 461
and gaskets 547
at hose connections 533
mechanically assisted 828–832
mechanism of 516
monitoring of 541
and sealants 545–546
on titanium-palladium alloys 424
in white water 778
Crevice corrosion and pitting (water side) 450
Crevice-corrosion cracking, in water-cooled generators 497–498
Crevicular fluid 895
Crimp connectors 272
Critical conditions 236
Critical crevice corrosion temperature vs. pitting resistance equivalent number (PREN) 16
Critical relative humidity (RH) 42–43
Cross-linked polyethylene (XLPE) 665
Crown alloys 905–908
Crowns, types of 891
Crude oils 962
contaminated 977
relative corrosity 980
Crude oil storage tanks 92
Crude oil type, effects of, on corrosion rate as a function of water content 1030
Crude towers 975–976
Crutemp, in fuel cells 506
Cryogenic processes 972
Crystalline polymers 740
Cultural artifacts 289
Cumulative loss techniques 1004, 1032
Cupric salts, in hydrochloric acid (HCl) 682
Cuprite 294
Cupronickel, in hot water tanks 274
Current 91, 101, 102, 109, 111
Current interruption 85–86
Current mapping 109
Current-potential diagrams 586
Current requirements
with anode to well spacing 102
for well casings 933–934
Current-time responses, between gold alloy and amalgam 898
Current transient, during scratch test 831
Cut-line corrosion 791
Cyclic corrosion tests 148, 149
Cyclic loading machinery 1081
Cyclic oxidation resistance 755
Cyclic oxidation 520
Cyclic polarization curves 842–843
Cyclic potentiodynamic polarization (CPP) 424, 426–427
Cyclic stresses 344, 469

D

Dairy industry 803
Damage tolerance 225, 607
Damage tolerance analysis (DTA) design philosophy 202
Damage tolerance approach 195, 224
Data
acquisition of 1045–1046
Data loggers 85
dc decouplers 113
dc transit system, potential and current tests of 110
Dead ends, for piping systems 808
Dead leg areas 807
Deaeration, beneficial effect of 353
Deaerator 452–456
Dealloying 13
of alloys 450
Dealloying (water side) 449
Dealuminification, in aluminum bronze tubesheets 449
Death profile analysis 21
Deaths, from component and fastener failures 260
Decarburization 997
Decay, of average corrosion potential 587
Decontamination solvents 365
Decoupled source quartz photoresist removal process 625
Deep-anode groundbeds 94
Deep anode grounded CP system, electrical isolation of 91
Deep ground beds 933, 935
Deep ocean, cathodic protection (CP) in 937
Defect clusters 398
Deferred maintenance 260
Defluoronated phosphoric acid. *See* phosphoric acid
Deformation band role, in irradiation assisted stress-corrosion cracking (IASCC) 403
Deformation mode 400–403
Degree of corrosivity vs. soil resistivity 89
Deicing salts 290, 515, 523, 524–525, 538–539, 542, 559, 567, 581, 768
Delamination plane 567
Delayed failure threshold stress 994
Delignification 266, 771, 772
Delrin 267, 276
Delta ferrite 21
Dendrites 616, 617–618, 645, 648
Dendritic silver sulfide 282, 1091
Denickelification 214, 449, 696
Dental alloys. *See also* amalgams
allergic hypersensitive reactions from 893
appearance of 894
carcinogenesis from 893, 894
classification and characterization of 904–927
composition and microstructures of 901
compositions and properties of 891–892
corrosive factors of 899
dimensional change 894
effect of the silver/copper ratio on 913
endontic failure from 894
galvanic pain from 893
laboratory testing 903–904
metallic taste of 892
noble 905
noncontacting restorations of 897
oral lesions from 893
stress-corrosion cracking (SCC) in 883
substrate corrosion 903
surface characteristics of 899
surface color change 904
tarnish and corrosion resistance of 892–896
tumorgenesis from 894
wrought orthodontic wires for 891
Dental amalgam. *See* amalgams
Dental anatomy 893
Dental casting alloys
alloys in 906–908
iridium in 905–906
noble metals in 906
osmium in 905–906
palladium in 905–906
rhodium in 905–906
ruthenium in 905–906
Dental corrosion products 903
Dental implants
alloy composition and applications 911
cobalt-chromium alloys in 911–912
dental alloys for 891
rejection of 894
sintered and wrought powdered alloys 912
wrought powdered alloys 912
Dental plaque by-products 901
Dental solders 910–911
Dentinal fluids 861
Denting 377
Dentistry 856
Dentures, removable 892
Deposit control 14
Deposit first melting temperature 791
Deposits, on steam generator tubes 379
Depth
vs. corrosion loss 35
vs. temperature 32–33
Derouging 22
Design
of cathodic protection (CP) in seawater 936
corrosion considerations 150
of sacrificial anodes 75
of transit systems 553
Design guidelines, for corrosion prevention 546
Design procedures 75–77
Designs, consideration for 1002
Design safety factors 201
Detail specification 137
Detection of corrosion 97–99
Deterministic fracture mechanics (DFM) 225
Detonation spraying 63
Dew point, defined 491
Dew-point behavior 491
Dew-point corrosion 158–159, 491, 494
Dew-point temperature 615
Dezincification 25, 267
in 70Cu-30Zn brass 733
in tubesheets 449
Diaphragm cells 710, 722, 723
Dicarboxylic acids 679
Dielectric constant, of supercritical water 229
Dielectric shield 936
Diesel engines
coolants 537
described 161
fuels 161–162
in military systems 161–162
Diethylenetriaminepentaacetic acid (DTPA) 317

Differential aeration cells 900–901, 1016
Differential corrosion cells 1016
Differential swelling 387
Diffusion coatings 489, 1002–1003
  alloys used in 480
Diffusion-tight rust 541
Difunctional neopentyl glycol diglycidyl
  ether (NGDE) 175
Digesters 763–767
Dilute acids 692
Dimethyl formamide, corrosion rates of
  metals in 751–752
Dimple rupture 864
Diodes 112
Dioxins 773
Dipolar aprotic solvents 750
Direct assessment (DA) process 81
Direct assessment (DA) programs 1023–1024
Direct bell hole examination 79
Direct current (dc) stray currents 6, 103–104
Direct current source influence test 86
Direct examinations 81
Directionally solidified (DS) blades 222
Direct observational corrosion test methods .. 843
Discrete anode configurations 589
Disk refiners 767–768
Dislocation channeling 399, 401
Dislocation loops vs. dose 388
Dispersed barrier hardening model 399
Dispersed-phase amalgams 904–905
Dispersion plates 626
Displacing fluid 940
Dissimilar materials
  in aircraft 196–197
Dissimilar-metal couples, metals and
  alloys compatible with 618
Dissolved hydrogen peroxide vs.
  corrosion potential 348
Dissolved oxygen (DO) 23–24, 37
  depth of, in the Pacific Ocean 35
  in oral chemistry 900–901
  methods of control 942–943
  seawater 30–33
  surface, in Pacific Ocean 34
Dissolved oxygen content vs. electrochemical
  corrosion potential (ECP) 348
Dissolved salts 306
Dissolved sulfide 23
Distillation towers 980
Distributed anode configurations 589
Distribution of carbides, on grain
  boundaries 370
Ditched margins 861
DNA damage 822
Doors and windows 143
Dose 388
Dose-response functions, parameters used in .. 56
Dose-to-amorphization vs. temperature 408
Double bottom tank, impressed current
  (ICCP) 94
Double-shell interwrap crevice corrosion 526
Double submerged arc weld (DSAW), and
  near-neutral-stress-corrosion
  cracking (SCC) 1019
Downflow reactor 230
Downhole equipment 932, 948, 954, 961
Dow process 441
Drag chain failures 784
Drain holes 206, 525
Drilling 329
Drilling fluids 922, 944–946
Drill pipe corrosion 944
Droplet impingement corrosion 1000
Dry chlorine 704–705
Dry deposition 290
Dry-dust collector 251
Dry exhaust systems 269
Dry fatigue cracking 341
Dry film lubricant 182–183
Dry flue gas handling systems, in
  combustion systems 492
Dry fly ash systems 499
Drying oils 209

Dry scrubbers 464
Dual wall muffler 525
Ductile cast irons, sulfuric acid corrosion of .. 660
Ductile gouging 653–654
Ductile iron 704
  in caustic sodas 711
Ductile Niresist 711
Ductwork 493, 496
Due penetrant inspection 1050
Duplex (austenitic/ferritic)
  stainless steels 763, 767, 960, 989
  in acetic acid 678
  in ammonia and ammonia compounds 730
  in caustic sodas 713
  corrosion resistance of 806
  fabrication and joining 806–807
  in pharmaceutical industry 810–811, 812
  in phosphoric acid 737
  in white liquor 783
Duplex alloys, in formic acid 676
Duplex coating 489, 1092
Durability-critical parts 224
Dutch process 283
Dwight's equation 76–77
Dynamic interference transit system 108
Dynamic loading, frequency of 861–862
Dynamic strain aging 569
Dynamic stray currents 87, 92, 109
Dynamic stresses, pitting from 128

E

Early red rusting 528
Earth's magnetic field 108
ECDA tool selection matrix 80
E-Coat primer 150
Eddy-current inspection 1050–1051
EDS analysis 639
EDTA. See ethylenediaminetetraacetic
  acid (EDTA)
Effective pore size 561
Efflorescence 563
Elastic-plastic fracture mechanics (EPFM) ... 868
Elastomeric coatings 26, 173
Elastomeric tapes 173, 177
Elastomer materials 739–740
Elastomers
  in ammonia and ammonia compounds 733
  anhydrous hydrogen fluoride (AHF)
    corrosion of 702
  in caustic sodas 719
  in dry chlorine gas 706
  food grade 803
  in hydrochloric acid (HCl) 686
  in hydrofluoric acid 697
  hydrogen fluoride corrosion of 697
  nitric acid corrosion of 672
  in petroleum production operations 931
  temperature limits in ammonia 733
Electoendosmosis 72
Electrical bias 615–616
Electrical connections 85
Electrical connector 637
Electrical contacts 273
Electrical current requirements, for cathodic
  protection in seawater 923
Electrical double layer 841
Electrical engine components 268
Electrical-leakage paths 634
Electrical/mechanical systems 143
Electrical resistance (ER) 115, 1004, 1031, 1032
Electrical resistivity, and cement pore
  structure 574
Electric light rails 548
Electric rail 548–557
Electric resistance (ER) instruments 951
Electric resistance probes 116
Electric resistance welded (ERW), and
  near-neutral-stress-corrosion
  cracking (SCC) 1019
Electric streetcar trolleys 548

Electric submersible pump wells 954
Electrocatalysis 404–405
Electrochemical atomic force
  microscopy (AFM) 834, 908
Electrochemical behavior, of metals 842
Electrochemical chloride extraction
  (ECE) 583, 590–591
Electrochemical corrosion potential (ECP)
  vs. dissolved oxygen content 348
  vs. feedwater hydrogen concentration ... 356
Electrochemical impedance behavior, of
  platinum 847
Electrochemical impedance spectroscopy
  (EIS) 116, 620, 847
Electrochemical instruments 951
Electrochemical metal growth migration
  (dendrite growth) 638–640
Electrochemical migration (EM) 645–646, 647
Electrochemical noise (EN) 116, 952–953,
                                    1004, 1032–1033
Electrochemical potentiokinetic
  repassivation (EPR) 352
Electrochemical reactions, of metallic
  corrosion 841–842
Electrochemical scratch test 832
Electrochemical series, for reactions 616
Electro-chemical stability diagrams. See
  Pourbaix diagrams
Electrochemical techniques 116–119, 583–591
Electrochemical testing 827
Electrocoat primer 517
Electrodeposition coatings 177–178, 180
Electrode potential, of burial environments ... 307
Electro-galvanized coatings 149
Electrogalvanized steel (EG or EL) 516
Electroless nickel plating (ENP) 717
Electrolysis, as treatment for concretions ... 318
Electrolyte 656
Electrolyte bridging 207
Electrolyte chemistry 862
Electrolyte medium 591
Electrolytic corrosion 617
Electrolytic current density 553
Electrolytic reduction techniques 318
Electrolytic stress-corrosion cracking (SCC) .. 644
Electromagnetic tools 97
Electromigration 623
Electronic black box 209
Electronic equipment 6
  connection locations 207
  corrosion in marine environment 205
Electronic magnetic transducer (EMAT)
  system 1048
Electronic response probes 115
Electronics 643–650
Electro-osmotic pulse (EOP) technology 146
Electrophoretic deposition 517
Electroplated coatings 66–67
Electroplated solder 630
Electroplating 180–181
Electroplating industry, atmospheric
  emissions from 1072
Electropolished surfaces 18
Electropolishing 365, 811–812
Electrostatic precipitators. See also flue
  gas desulfurization (FGD) 251, 254,
                              461–462, 493, 496
Electrostatic spray gun 178
Elevated-temperature strengths, for exhaust
  stainless steel alloys 522
Elgiloy 856–857
Ellingham phase stability 164
E-log I curve 934, 935
E log I test (Tafel Potential) 99–100
Embedded iron, in stainless steels 814
Embedded piezoelectric ceramic-polymer
  composite sensors 227
Embedded track 549–550
Embolite 311
Embrittlement phenomena 973
Emergency core cooling systems (ECCS) 341
Emission-control equipment 7
Emulsion-forming tendencies 937

1116 / Reference Information

Enamel coating systems 1020
Endontic failure, from dental alloys 894
Endurance limit 958
  in physiological media 868
Energy conservation 438
Energy deposition rate 391
Energy-dispersive spectroscopy (EDS) 17
Energy-dispersive x-ray analysis (EDX) .. 647, 872
Energy-dispersive x-ray spectroscopy 635, 636
Engine component surface damage 222
Engine coolants 531, 535–536
Engine cooling water systems 269
Engineered barrier systems (EBS) 421, 423
Engine-generator systems 103
Engine lubricants, bacteria and fungi in 213
Enoblement 214
Enriched uranium dioxide 415
ENSIP (engine structural integrity
  program) 225
Entrained particles 1001–1002
Environment, outdoor 289
Environmental and cyclic life interaction
  prediction software (ECLIPSE) 227
Environmental chlorine free (ECF)
  bleaching sequences 772–773
Environmental classes 209
Environmental conditions, of U.S. Navy
  Aircraft 184
Environmental cracking 159
  in hot water 391
Environmental cracking (water side and
  steam side) 451
Environmental data, ISO CORRAG
  exposures 49
Environmental factors, in
  wafer-fabrication 626–628
Environmental issues, volatile organic
  compounds (VOCs) 171, 174
Environmentally assisted cracking
  (EAC) 341, 567, 853, 923, 929
  in boiling water reactors 341, 343–354
  of carbon alloy steels 381
  in caustic sodas 713
  factors necessary for 352
  of high-level waste (HLW) containers ... 424
  of low-alloy steels 381
  management scheme for addressing 354
  materials, stress, and environmental
    parameters relevant to 344
  in petroleum refining and petrochemical
    operations 987–999
  threshold potential of 349
  of titanium alloys 432–433
Environmentally assisted cracking (EAC)
  mechanism 381
Environmentally assisted cracking (EAC)
  susceptibility
  of austenitic stainless steels 348
Environmentally controlled crack-
  propagation rate 352
Environmentally induced cracking 644
Environmental Protection Agency (EPA) 326
Environmental regulations, and hazardous
  materials 173
Environmental solutions, in boiling water
  reactors 356
Environmental stress cracking (ESC)
  in anhydrous hydrogen fluoride (AHF) ... 700
  from caustic sodas 720
  testing for resistance to 931–932
Environmental variables, effect of 652–653
Environments in nature, potential vs.
  pH diagrams of 307
Epoxies 72
  viscoelastic nature of 569
Epoxized cresol novolac coatings 932
Epoxy-based coating systems 266
Epoxy-coated reinforcement (ECR) 576, 578
Epoxy coating, chemical reaction for
  two-part 171
Epoxy coatings 962, 1061
  on corroded water pipe 144
  life of 129

Epoxy-modified coal tar systems 1074
Epoxy-modified phenolic coatings 956
Epoxy paint system 266
Epoxy polyamide sealant 65
Epoxy/polyamide systems 175
Epoxy-polyamide top coats 776
Epoxy polyimides 130
Epoxy primers 70, 171, 188
  high-performance waterborne 172
Epoxy resin, chemical structure of 175
Epoxy resin coatings 932
Epoxy resins 687
Equipment design, and thermal insulation ... 817
Equipment reliability 221
Equivalent damage years, for control rod
  drive mechanism (CRDM) 370
Equivalent electrochemical potentiokinetic
  repassivation value 353
Equivalent flaws 196
Equivalent initial flaw sizes (EIFS) 227
Equivalent precrack sizes (EPS) 227
Equivalent stress, for intergranular
  stress-corrosion cracking (SCC) 372
Ergonomics in design, for cold climate
  corrosion (CCC) 248
Eriochalcite 311
Erosion 927
Erosion-corrosion 471, 927
  of CN-7M stainless steel pump components .. 1079
  in engine coolants 534
  petroleum refining and petrochemical
    operations 999–1002
  of pump components 1080
  in radiator tank 534, 1092
  in steam turbines 469
  water-side 448
Erosion damage, on entrained particles 1001
Erythorbic acid 328
Ethanolamine 363, 365
Ethylene-chlorotrifluoroethylene
  (ECTFE) 665, 814
Ethylenediaminetetraacetic acid
  (EDTA) 317, 324–325
Ethylene glycol 269
Ethylene glycol antifreeze 531
Ethylene propylene diene monomer
  (EPDM) 733
  in cooling systems 532
  in dry chlorine 706
Ethylene-propylene rubber 678
  in dry chlorine 706
Ethylene-tetrafluoroethylene (ETFE) 665
Ethylene trifluoroethylene (ETFE) 733
Ettringite 565
Evans diagram 585
Excavation 314
Exfoliation 221
Exfoliation corrosion 153
  in 7049-T73 aluminum 197
  of aircraft component 193, 1089
  in commercial aviation 599
  and pillowing 198–199
  and resistance 132
  of U.S. Navy Aircraft 191, 192, 193
Exhaustion creep vs. crack-propagation
  rate-dependent strain 352–353
Exhaust riser 269
Exhaust system galvanic corrosion 528
Exhaust systems 269–270, 520, 530
Explosive removal 330
Extended batch treatment 940
Extended modified continuous cooking
  (EMCC) process 765
Exterior coatings. See coatings
Exterior salt pitting, in automotive
  exhaust systems 523
External aboveground coatings,
  aboveground storage tanks (ASTs) ... 93–95
External cathodic protection (CP), of
  flow lines 962
External coating systems 962
External corrosion direct assessment
  (ECDA) 6, 79, 80

External oxidation 487
External salt pitting tests 524–525
External stress-corrosion cracking
  (ESCC) 655
Extracellular fluids 895
Extruded plastic coatings 1020
Extruded polyethylene coatings 962
Extruded polymer secondary anodes 589
Extruded polypropylene coatings 962

F

Fabricability 973
Fabrication-room environment 626
Fabrication stresses 344
Fabric filters 251, 253
Fabric finishes 281
Facilities infrastructure 136
Factors influencing corrosion, in
  commercial aviation 600–605
Factors influencing susceptibility, to
  corrosion fatigue 864–867
Fail-safe approaches 195
Fail-safe design 867
Failure prediction models 369
Failure times, for intergranular stress-
  corrosion cracking (SCC) of alloy
  600 components 370
Falling-film black liquor evaporator 781
Fan belt 533
Faradaic currents, sources of 829
Fast-cycle interruption 86
Fatigue 6
  exfoliation, crevice corrosion and effects on .. 200
  fundamentals of 863–864
  methodologies for predicting effect of on
    corrosion fatigue 201
  in vivo and retrieve studies 871–873
Fatigue crack 997
Fatigue crack growth 196, 200, 865
Fatigue cracking, growth with time 220
Fatigue damage and failure stages 202
Fatigue failure, stages of 863
Fatigue life 195, 197, 198, 958
Fatigue life ratio (FLR) 868
Fatigue lives, prediction of 201
Fatigue resistance, of alloy 600 380
Fatigue sample morphology 867
Fatigue strength 972, 1081
  of titanium alloys 870
Fatigue striations 862, 864
Fatigue testing
  cycle frequency of 862
  statistical considerations 867
Faying surfaces, corrosion of in aged parts ... 221
Faying surface sealant 603
$FeCr_2S_4$ spinel 442, 443
Fe-Cr-Ni alloys, carburization in 1069
Feed stream dilution 233
Feedwater 391
Feedwater heaters (FWHs) 456–458
Feedwater pH 237
Feedwater pipe thinning and rupture ... 467, 1092
Feedwater quality 455
Fences 142
Fe-Ni-Cr alloys, oxidation resistance of 1068
Fenton reaction 847
Fermentation process 676
Ferric ion corrosion 328
Ferric salts, in hydrochloric acid (HCl) ... 682
Ferrite, corrosion of 793
Ferrite stabilizers 838
Ferritic boiler steels 238–239
  compositions of 240
Ferritic grades, in ammonia and
  ammonia compounds 730
Ferritic rotor steels 240
Ferritic stainless steel alloys, used in
  automotive exhaust systems 520
Ferritic stainless steel properties, for solid
  oxide fuel cells (SOFCs) 513

Ferritic stainless steels
in caustic sodas . . . 712
corrosion resistance of . . . 806
fabrication and joining . . . 806
in fuel cells . . . 506, 507
sensitization of . . . 527
Ferritic stainless steel tubes . . . 451
Ferritic steels . . . 236
evolution for boilers . . . 239
for oil ash corrosion . . . 480
Ferrocement hulls . . . 266
Ferrous alloys, corrosion products of . . . 294
Ferroxyl tests, for free iron . . . 814
Fertilizer industry . . . 255
FFKM . . . 702
Fiberglass . . . 254
used in marine vessel freshwater systems . . . 273
in wastewater plants . . . 256
Fiberglass boats . . . 265–266
Fiberglass-reinforced epoxy (FRE) . . . 941
Fiberglass-reinforced plastic (FRP) . . . 252, 253, 255, 666, 734, 769, 770, 773, 784, 793, 794, 796
in caustic sodas . . . 720
laminates . . . 462
temperature limits for, in hydrochloric acid (HCl) . . . 687
Fiberglass-reinforced polyester (FRP) . . . 941, 956
Fiber optic sensors . . . 226
Fiber-reinforced pipe . . . 706
Fiber-reinforced plastic . . . 708
Fibrinogen adsorption . . . 841
Fick's diffusion law . . . 566
Fick's second law . . . 564
Field investigation (direct assessment) programs . . . 1023
Field rectifiers . . . 123, 124
Filiform corrosion
aluminum . . . 191, 540
in commercial aviation . . . 599
Fill-and-soak cleaning method . . . 323
Film formation mechanism . . . 174
Filming amine inhibitors . . . 945, 1002
Filming azoles . . . 14
Filter bag houses . . . 493, 496
Find-it and fix-it scheme . . . 130, 200
Fingernailing crevice . . . 766
Finishes. *See also* coatings
Finishing systems . . . 171–179
Firearms, corrosion of . . . 257–258
Fire gilding . . . 297
Fireproofing . . . 1055
Fireside corrosion . . . 158, 159, 786
in boilers . . . 477
prevention of . . . 479
Fire tube boilers . . . 156
First-level package . . . 645
Fish-bone diagram . . . 633
Fish scaling, of glass lined equipment . . . 328
Fissuring, hydrogen attack of . . . 997
Fitness-for-purpose assessment . . . 1045
Fittings and fastenings . . . 276
Fixed bridges (dental) . . . 892
Fixed-length electrical response probe schematic . . . 115
FKM . . . 702
Flade potential . . . 843
Flaking . . . 517
Flame retardants . . . 281
Flame-spray coating . . . 62, 63
imperfections in . . . 65
Flame-sprayed aluminum coating
after seawater immersion . . . 64
micrograph through . . . 64
Flash/profile radiography . . . 1049
Flat specimens, corrosion loss of . . . 56
Flexible foamed elastomer . . . 817
Flexible opaque plastic, used in marine vessel freshwater systems . . . 274
Floor structure design . . . 604
Flow-accelerated corrosion (FAC). *See also* erosion-corrosion . . . 448, 455, 458, 467
in feed water heaters (FWH) . . . 457–458
Flow-assisted corrosion (FAC) . . . 343
Flow-induced corrosion . . . 1029
Flow-induced turbulence . . . 924
Flow-induced wall sheer . . . 938, 960
Flowing wells, oil production . . . 948–950
Flow lines . . . 955, 962
Flow modeling . . . 925, 926, 999
corrosion and . . . 938
Flow rates, in process steam . . . 653
Flow velocity, effect of . . . 40
Flue gas corrosion vs. temperature . . . 986
Flue gas desulfurization (FGD) . . . 461–464
general materials selection for . . . 251–252
for high-level waste (HLW) containers . . . 423
Flue gas velocities . . . 461
Flue gas waste . . . 491
Fluidized bed combustion . . . 438–439
Fluid pounding . . . 947
Fluids, corrosion testing in . . . 827
Fluorescent magnetic particle method . . . 453
Fluorides, in hydrochloric acid (HCl) . . . 682
Fluorinated ethylene propylene (FEP) . . . 665, 733, 814
hydrogen fluoride corrosion of . . . 697
Fluorinated plastics . . . 678
Fluorinated polymers . . . 173
Fluorinated resins . . . 130
Fluorine corrosion . . . 626
Fluorine-induced corrosion . . . 627
Fluorocarbon-bonded graphite . . . 734
Fluorocarbon plastics, in formic acid . . . 676
Fluoroelastomers . . . 702
in dry chlorine . . . 706
Fluoroplastic lined pipe . . . 253
Fluoroplastic linings, chemical resistance temperature limits of . . . 665
Fluoroplastics . . . 665, 686–687
Fluoropolymers . . . 740
in pharmaceutical industry . . . 814
Fluosilicic acid . . . 743
Flutes and fluting . . . 873
Fluting . . . 874
Flyash . . . 499, 560
Foam cleaning method . . . 323
Food and beverage industry . . . 338
Food grade . . . 803
Food industry . . . 256, 804
Food-processing industry . . . 803
Food products, contamination of . . . 805
Foods, pH values of . . . 804
Forced-air oxidation . . . 461
Foreign object damage (FOD) . . . 222
Forged cobalt-chromium alloys . . . 911–912
Formaldehyde . . . 283, 284–285
Formalized proving ground (PG) test outcome criteria . . . 543
Formation water . . . 928, 948
Formic acid . . . 283, 325, 675–676, 766
Formic acid mixtures . . . 746–747
Fossil-fired power generation boiler, dew-point corrosion susceptible areas . . . 492
Fossil fuel combustion . . . 986
Fossil fuel fired power plants, forms of corrosion in . . . 1086
Fossil fuel power plants, service water distribution systems (SWDS) . . . 12
Fossil fuels, defined . . . 438
Fouling organisms . . . 33
Found rock climbing equipment . . . 263
Fractionation towers . . . 978
Fractured margins . . . 861
Fracture healing . . . 854
Fracture initiation, sites for . . . 866
Fracture mechanics approach . . . 867
to stress-corrosion cracking (SCC) . . . 878
Fracture mechanics specimens, in the high flux region of the core . . . 393
Fracture toughness . . . 472
Frame and fork corrosion . . . 259
Frank dislocation loops . . . 398
Free rock climbing . . . 262–264
Freshwater cooling systems . . . 268, 269
Freshwater plumbing systems . . . 273
Fretting . . . 222, 619, 641, 829
in commercial aviation . . . 599–600
Fretting corrosion
of metallic biomaterials . . . 827
of U.S. Navy Aircraft . . . 192
Fretting fatigue damage . . . 223
Fuel
corrosive constituents in . . . 159
transportation and storage systems . . . 144
vanadium content of . . . 985–986
Fuel additives . . . 495–496
Fuel ash corrosion . . . 478–479, 985–987
Fuel cell reactions . . . 505
Fuel cells . . . 439, 501–511
Fuel ethanol, stress-corrosion cracking (SCC) in . . . 992
Fuel impurities . . . 477
Fuel rods . . . 419–420
Fuels
chloride levels in . . . 445
impurities in . . . 438
Fuel tank coatings . . . 173
Full width at half maximum (FWHM), profile . . . 395
Fundamentals of corrosion . . . 1085
Fungal degradation, of polyimides . . . 217
Fungal-influenced corrosion . . . 213
Fungi . . . 211, 212, 213
in jet fuel . . . 599
Furane resins . . . 721
Furan resins . . . 687
Furnace-sensitized . . . 347, 354
Furrow attack . . . 764
Fuselage drainage and drain valve design . . . 603
Fuselage structure corrosion . . . 602, 610
Fusion-bonded-epoxy (FBE) coatings . . . 1018, 1020

G

Galvalume . . . 1059
Galvalume (Zn-55 Al) coatings . . . 143
Galvanic anode (sacrificial) cathodic protection (CP), warm marine applications . . . 570
Galvanic anodes . . . 93
Galvanic anode systems . . . 584
Galvanic cathodic protection . . . 90
Galvanic CFRP-aluminum coupling . . . 611
Galvanic compatibility . . . 1080
Galvanic corrosion . . . 9, 13, 207, 617–618, 644
of aluminum . . . 546–547
of ammunition . . . 152
on armament systems . . . 152
in automotives . . . 515–516, 527–530
of cast iron statue . . . 292, 1090
in commercial aviation . . . 599
described . . . 127–128
in engine coolants . . . 533
of grenades . . . 152
mechanism of . . . 516
of muffler . . . 529
of pumps . . . 1080
in semiconductor integrated circuits (SICs) . . . 631–632
of solid carbon steel . . . 529, 1091
in underground structures . . . 1016
of U.S. Navy Aircraft . . . 192, 1089
water side . . . 450
Galvanic pain . . . 893
Galvanic probes . . . 116, 944
Galvanic reactions . . . 631
Galvanic series
of metals and alloys . . . 9, 128, 1016
Galvanized Al-Zn alloy-coated specimens, atmospheric corrosion losses for . . . 51
Galvanized carbon steel . . . 268
Galvanized material testing . . . 66
Galvanized reinforcement . . . 580
Galvanized stainless steel . . . 265

Galvanized steel
   effects of chemical cleaning solutions on 325
   on marine use vessels 266, 267
   used for turnbuckles 277
   used in marine vessel freshwater systems 273
Galvanized steel panels, corrosion rates for 66
Gamma radiography 1049
Gas-atomized dispersion-strengthened (GADS) alloy 870
Gas composition and temperature 1028
Gases, solubilities of, in ocean water 958
Gasification 438
Gaskets 546
Gas-lift wells 948
Gas metal arc welding (GMAW) 764
Gas offshore production platforms 956–958
Gas/oil ratio (GOR) 1030
Gas Pipeline Integrity Rules 81
Gas pipelines 1031
Gas production. *See* petroleum production operations
Gas separation facilities 962
Gas stripping 942
Gas-to-oil ratio (GOR) 959
Gas turbines 162–164, 439
Gathering systems 958–962
General corrosion 515, 566–567, 973
   in commercial aviation 599
   vs. localized (pitting) corrosion 1033
   of U.S. Navy Aircraft 193
General specification 137
Generating capacity reductions, by condensers 448
Geographical location vs. corrosion rates 184
Gerber's law 867
Gingival fluid 895
GL 89–13 12
Glacial acetic, in acetic acid 677
Glass 687
   in ammonia and ammonia compounds 734
   in phosphoric acid 741
Glass and glass-lined, sulfuric acid corrosion of 666
Glass corrosion 501
Glass fiber, for insulation 818
Glass-fiber-reinforced thermosets 941–942
Glass flake polyester coatings 495
Glass-lined equipment, failures of 815
Glass lined equipment, fish scaling of 328
Glass-lined equipment, nitric acid corrosion of 672
Glass-lined steel 687
   in dry chlorine 706
   in moist chlorine 708
   in pharmaceutical industry 814
Glass linings 815–816
Gluconic acid 325
Gluconic acid solutions 808
Glycols 535
Glyoxal 328
GM 9540P tests 148, 149
Goethite 312
Gold (gilding), conservation strategies for 301
Gold and gold-base alloys, general
   anhydrous hydrogen fluoride (AHF) corrosion of 702
   in dental casting alloys 905–906, 913
   in dentistry 856
   gold solder (450 solder) 916
   hydrogen fluoride corrosion of 696
   material characteristics of 906
   sensitivity to 822
   solders 910–911
Gold and gold-base alloys, specific, (P00020) .. 696
Gold lid of a ceramic package 631
Gold mines 1077
Gold-palladium alloys 909
Gold-palladium-silver alloys 909
Gold-plated connector corrosion 648
Gold plating 205–206, 207
Gold-platinum-palladium PFM alloys 909
Gold-silver-copper dental alloys 912
Gold-silver-copper-palladium dental alloys ... 913

Goodman diagram 471
Gouging 989
GPS equipment 85, 86
Grain boundaries 347, 367–370, 396, 397
   irradiation-induced chromium depletion of 375
Grain-boundary diffusion 233
Grain-boundary fissures 997
Grain-boundary sensitization 346
Grain-boundary sliding and creep 404
Graphite 254
   in ammonia and ammonia compounds 734
   in dry chlorine gas 706
   hydrogen fluoride corrosion of 697–698
   in moist chlorine 708
   in pharmaceutical industry 813–814
   in phosphoric acid 740–741
Graphite anodes 94
Graphite corrosion 313, 450
Graphite heat exchangers 678
Graphitization 313
Gravel impingement 149
Gray cast irons 660, 711
Grease 212, 260
Grease failure 212
Green liquor 780, 782, 783
Green plague 205
Greigite 313
Grindstones 767
Grit blasting 70
Grounding method, pipe-type feeders 124
Grounding protection 273
Ground mat (gradient grid) underground valve 113
Ground vehicle fleets 6
Ground vehicles 148
Groundwater composition, for different rock hosts 422
Groundwater treatment 255
Grout voids 570–571
Guided wave techniques 1048
Guide tubes 415
Guildite 315
Gun-drilled holes 780
Gunite shotcrete 1003
Gunmetal (C90550) 718

H

Hafnium (Hf) 415
Half-life rule 1044
Halide brines 926
Halide-induced corrosion 634–636
Halides, in reducing solutions 743
Hangar bar corrosion 493
Hank's solution 845, 862
Hardened cement paste 562
Hardened grease 641
Hardening mechanisms, for dental alloys 907
Hardfacing materials 1001
Hardness vs. dose 388
Hard rubber 733
Hard weld, sulfide stress cracking and 996
Harmonic distortion analysis (HDA) .. 116, 951, 1004, 1032
Harrington rods 872
Hazard analysis and critical control point (HACCP) system 804
Hazardous ac voltages 111
Hazardous air pollutants (HAPs) 174
HCP microstructure 561
Headboxes 775–776
Header feedwater heaters 456
Health monitoring (HM) systems 226
Heap leaching 1081
Heat-affected zone (HAZ) 347–348, 455
Heat engines
   efficiency of 439
   types of 439
Heater shells, in feed water heaters (FWH) ... 458
Heat exchangers 532, 807

   for acid solutions 1074
   in cooling systems 269
   titanium in 808
Heat exchanger tubing 13
Heat ink welding 355
Heat-recovery steam generators (HRSGs)
   with gas turbines 493–494
   nitrate-induced stress-corrosion cracking (SCC) in 494
Heat-resistant casting alloys, compositions of 756
Heat-resistant linings 1003
Heat-resistant stainless steels, oxidation resistance of 1067
Heat tint 524, 811, 815
Heat treating furnace equipment 1067
Heat treating industry 1067–1071
Heat treatment process 132
Heat treatments 866
Heavy metals 38
Hellenistic silver vessels 281
Hematite 19, 312
Hermetically sealed equipment 206
Hermetically sealed packages 615
Hertzian contact stress analysis 832
Hesiometer knife cutting tests 172
Hexavalent chromium, toxicity of 822
Hg-Zn-Al alloy 936
Hideout (water chemistry) 363, 376, 379–380
Hierarchical decision tree, for software tool development 931
High-alkali cement 566
High-chromium alloys, in supercritical water 232
High chromium-molybdenum ferritic stainless steel 13
High-cobalt alloys (stellites), in nuclear reactors 365
High-copper amalgams 904–905
   of tarnishing and corrosion testing 912–913
High Cr-Fe-Ni alloy, sulfuric acid corrosion of 662
High-cycle fatigue (HCF) failures 222
High-cycle fatigue life 863
High-density polyethylene (HDPE) 794, 796
   in caustic sodas 719–720
   hydrogen fluoride corrosion of 697
Higher austenitic stainless steels, sulfuric acid corrosion of 662
Higher-chromium duplex alloys, for use in corrosion under insulation (CUI) 657
Higher chromium Fe-Ni-Mo alloys, sulfuric acid corrosion of 662
Higher heating value (HHV) 243
High fault current, and impressed ac current 122
High humidity exposure 542
High-level waste (HLW) 421–423, 425
High-nickel alloys
   cracking in high-purity water 368
   intragranular corrosion of 977
High-nickel cast irons 728
High noble, dental casting alloys 906
High-palladium PFM alloys 909
   tarnishing and corrosion of 914
High-phosphorous electroless nickel (HPEN) coatings 717
High-pressure acid leaching (HPAL) 1081
High-pressure and intermediate-pressure rotors 237–238
High-pressure coolant injection (HPCI) 341
High-pressure core spray (HPCS) 341
High-pressure water cleaning (hydroblasting) 329
High-purity ferritic stainless steels, in pharmaceutical industry 812
High-purity refractories 166
High-purity water 812
   quality standards for water 15–16
   stainless steel in 15–22
High-resistive soils, and sacrificial zinc anodes 247

Index / 1119

High-silicon cast irons
as impressed-current anode 74
sulfuric acid corrosion of. 660
High solids epoxy coatings 26, 129
High-solids technology 175
High-strength low-alloy (HSLA) steels. *See also* alloy steels, general
corrosion rates (CR) of 45
High-strength stainless steels, primary side stress-corrosion cracking (SCC) in 374
High-temperature alloys, chemical compositions of. 1068
High-temperature anhydrous hydrogen fluoride (AHF) 700–701
High-temperature coatings. 488–489
High-temperature corrosion. . . . 163, 754, 980–987
in automobiles 520–522
and degradation process 156
effect of on mechanical properties 488
from fuel contaminants 162
by halogen and halides 759
in military systems. 156–167
and oxidation 6
types of 156
of U.S. Navy Aircraft 193
High-temperature creep 163, 653
High-temperature environments 754
High-temperature hot corrosion (HTHC). 163–164, 487
High-temperature hydrofluoric acid 700, 701
High-temperature hydrogen attack (HTTA) 968, 994, 997
High-temperature metallic coatings 1069
High-temperature oxidation 6, 487, 1092
High-temperature polyvinyl chloride (CPVC) 254
High-temperature-resistant coatings 173
High-temperature storage (HTS) test 630
High-temperature sulfidic corrosion 981
High test temperatures 541
High-velocity oxyfuel (HVOF) 479, 765, 788
High-velocity particle consolidation (HVPC) 63, 64
High-volume low-pressure (HVLP) spray gun 178
High-yield-mechanical pulping 767–768
Hip protheses 882
HiTak 130
H.L. Hunley 317
HMMWVs 131
HMS Sirius 317
Hoist ropes 1078
Hole cold expansion. 225
Holistic life assessment 202–203
Holistic life-prediction methodology (HLPM) 227
Hoop stress 1019
Horizontally drilled anode system 94
Horn silver 311
Hospital waste (Biowaste) 253
Hostile environmental test conditions 645
Hot corrosion 163–164, 223, 478, 487, 488
Hot dip coatings 65–66
Hot dip galvanized steel (GI or HD) 516–517
Hot-dip galvanizing 149, 267, 962
Hot dip galvannealed steel 517
Hot dip process 183
Hot isostatically pressed (HIP) alloy 870
Hot salt
and aluminum coated stainless steels 521
and stainless steels. 521
Hot salt attack, of automotive exhaust systems 520–521
Hot salt intergranular attack 521
Hotside deposits, waterside oxidation/ corrosion from 161, 1092
Hot spots. 1003
Hot-water line, corrosion after insulation and water intrusion 144, 1090
Hot water tanks 274
Howitzers 152, 154, 155
Human contamination. 615
Human fluids. *See* body fluids
Human implantation 820
Human metabolic byproducts 229
Humans. 840
Human saliva 916
Humectants. 586
Humidity. *See also* relative humidity (RH) 540–542, 644–645
Humidity control 134
HVOF 63
Hybrid cathodic protection system 75
Hybrid organic acid coolants 531
Hydraulic cleaning. 329
Hydraulic pressure testing 807
Hydrazine 363, 365–366
Hydride formation. 994
Hydrocarbon-base coatings 963
Hydrocarbon fuels 213
Hydrocerussite 291
Hydrochloric acid (HCl) 325, 766, 993
corrosion of metals in 682–684, 686–688
Hydrochloric acid (HCl) gas. 688
Hydrochloric acid/ammonium bifluoride (ABF). 325
Hydrochloric acid cleaning. 767
Hydrocracking units 976, 980, 991, 997, 1000
Hydrofluoric acid (HF) 325, 690–692, 743
effect of 701
and glass-lined steel equipment 815
pumps for. 978
Hydrofluoric acid bath 18
Hydrofluoric acid vapors 695
and stress-corrosion cracking 695–696
Hydrogen-assisted cracking 328, 929
Hydrogen attack 729, 996–998
Hydrogen blistering. 699, 977, 979
experience with 993–996
of low-alloy steels 423
Hydrogen chloride 975
Hydrogen chloride gas 687–688
Hydrogen cracking 973
Hydrogen damage 158, 468, 816
Hydrogen embrittlement (HE). 127, 620, 729, 992, 994
of carbon steel 217
of high-strength prestressing steels 1058
of low-alloy steels 423
of posttensioning wires 1065
in sulfide stress cracking (SSC). 929
in superelastic form of Nitinol 881
of titanium 733, 970
of titanium alloys. 432–433
of U.S. Navy Aircraft 193
in weapons systems 127
Hydrogen embrittlement cracking. 451, 992
Hydrogen embrittlement mechanism 599
Hydrogen fluoride 691, 977
Hydrogen fugacity 405
Hydrogen grooving 659, 660
Hydrogen-induced cracking (HIC). 698, 972, 994–996
Hydrogen-induced disbonding (HID) 998
Hydrogen injection 405
Hydrogen ions 900
Hydrogen outgassing. 994
Hydrogen permeation technique 116
Hydrogen recombination poison 992
Hydrogen-related cathodic model 875
Hydrogen stress cracking 968
Hydrogen sulfide 256, 281, 975, 1028
and petroleum production operations
corrosion causes. 923–924
Hydrogen sulfide attack 729
Hydrogen sulfide corrosion 922, 944–945, 947, 948
Hydrogen sulfide cracking, and sulfur content of steels 995
Hydrogen sulfide levels and, hardness value . . 996
Hydrogen sulfide testing 23
Hydrogen water chemistry (HWC) . . 356–357, 404
Hydrokinetic cleaning 329
Hydrolysis. 578
Hydrophilic materials 645
Hydrostatic testing. 1023
Hydrotreating units 976, 980, 1000
Hydroxyacetic-formic acid (HAF) 325
Hydroxyapatite (HA). 854
Hydrozincite 291
Hygiene requirements 803
Hygienic standards 803
Hypalon, in dry chlorine 706
Hypersensitivity
to nickel 822
to type 316L stainless steel 822

I

ICDA 1031, 1034
IC failure rate 613
Illium 98 664
Illium B 664
Image quality indicators (IQIs) 1049
Immersion coatings 72
Immersion service coating systems 776
Immersion testing 428, 674
Immune domain. 842
Immunity against IGSCC 354
Impervious graphite 687, 813
Implant alloys. 911–912
Implant instability 854
Implant materials 842
Implant metals 821
Implants. *See also* metallic biomaterials; retrieval studies
fracture surfaces of 864
and ion levels in serum 848
metals and metal alloys in 820
Impressed alternating current (ac) voltage. . . . 119
Impressed current and high fault ac current . . 122
Impressed-current anode materials. 960
Impressed-current anodes 74, 933
Impressed-current anode systems vs. sacrificial anode systems 74
Impressed-current cathodic protection (ICCP). 90, 145, 584, 1022
in concrete surface. 1062
current practice in 589
double bottom tank 94
for offshore platforms 936
short circuits in 590
for well casings 102
Impressed-current method 933
Impressed current rectifier systems. 125
Impurity-element segregation 398
Impurity removal. 237
Inboard engines 268
Incineration 252
Incinerator liner failure 167, 1093
Incinerators 1004
in military service 164–168
Incralac coatings 299, 300
Indirect examinations 81
Indirect inspection categorization indication table. 82
Indirect systems 268
Indium, in dental solders 911
Indoor atmosphere corrosion 644
Indoor environmental factors. 314
Indoor pollutant concentrations. 645
Indoor sources, acetic acid 283
Induced ac voltage 108
mitigation of 112
Induced-draft fans 496
Induced-draft fan seals 493
Induced voltage vs. time 111
Induction heating stress improvement 355
Inductive coupling 108
Industrial chemical incineration 252
Industrial gas turbines (IGT) 486–490
Industrial wastes 229, 438
Industrial wastewater treatment equipment. 255
Infant mortality 221
Infection, and corrosion 821

Influential pit size 202–203
Infrared (IR) signature 148
Inhibited engine coolant 533
Inhibited epoxy/amine adduct-cured epoxy
  system 776
Inhibited low-zinc brasses, for marine
  hull use 265
Inhibited sealants 178
Inhibition programs, coatings for 932
Inhibitive primers 70
Inhibitor injection, corrosion rate during. ... 1034
Inhibitors 324, 535–536, 937–941, 949, 954
  corrosion 150
  dosage of 950
  and flow line corrosion 961
Initial corrosion rate for carbon
  dioxide-based corrosion 1028
Initial flaw size 196, 202
Initiation time for primary side
  stress-corrosion cracking (SCC) 339
Injection flow lines and wells 953–954
Inland corrosion rates (CR) 48
Inlays, types of 891
Inlet-end corrosion 448–449
In-line inspection (ILI) tools 1023
Inorganic coatings
  in commercial aviation 601
  for steels in structures 1059
Inorganic finish deterioration 606
Inorganic zinc-rich primers 70–71, 776, 818
In-process corrosion 153
In situ corrosion and humidity monitoring. ... 611
In situ pH determination, for
  production environments 1027
Inspections. See also tests and testing ... 1037–1047
  for corrosion under insulation (CUI) 658
  if tubular goods 963
  and maintenance of weathering steel ... 582
  of reinforced concrete 567
  of U.S. Navy Aircraft 185–186
Insulated protective coatings 405
Insulated tie-and-ballast track 549
Insulation 143–144, 654–659
Insulation materials 817–818
Insulation systems, design of 656
Integrated circuit (IC) devices 645
Integrated circuits (ICs) assembly,
  processes used in 629
Integrated circuits (ICs) packaging,
  process flow of 630
Integument characteristics, substrate
  effects on 903
Intelligent pigs 1023
Intercrystalline corrosion. See
  intergranular corrosion (ICG)
Interdendritic stress-corrosion cracking
  (IDSCC), in nickel-based alloys 342
Interface progress unit (IUP) 226
Interfacial resistance 586
Interfacial transition zone (ITZ) 562
Interference control 103
Interference sources, with corrosion
  measurements 119
Interference tests 109
Intergranular attack (IGA) 367, 377–380
  in caustic sodas 713
  in digesters 766
Intergranular attack (IGA)/ intergranular
  stress-corrosion cracking (IGSCC),
  in superheated steam zone of once-
  through steam generators (OSTGs) 379
Intergranular corrosion (ICG) 158
  of automotive exhaust systems ... 527–528
  in engine coolants 534
  of mill-annealed alloy 600 tubes ... 378–380
  of pumps 1080
  in UNS S30400 (304) 669
  weld decay 267
Intergranular cracking vs. nickel
  equivalent 401
Intergranular stress-corrosion
  cracking (IGSCC) 343–345, 378, 991
    in austenitic stainless steels 386

changes in propagation rates of 372
early indications of 339
equivalent stress for 372
in heat affected zone (HAZ) 342, 347
materials solutions of 354
in stainless steels 342, 348
of titanium alloys 433
Weibull distribution 371
weld overlay 355
Interior aluminum, finishes on 601
Intermetallic iron-aluminum layer 528
Internal chloride stress-corrosion
  cracking (SCC) 774
Internal corrosion direct assessment
  (ICDA) methods 1026
Internal environments, of aircraft 605
Internal exfoliation corrosion, of aircraft
  component 191
Internal oxidation 487
Internal protective coatings 932
Internal rotary inspection system (IRIS) ... 1048
Interrupted close-interval survey (CIS) ... 85
Interstitial elements 891
Interstitial fluids 894–895, 896, 904
Interstitial loops 398
Intragranular corrosion 977
Intragranular cracking 999
Intragranular decohesion of cold-worked
  stainless steel, stress-corrosion
  cracking (SCC) of 874
Intramedullary nail 873, 880
Intramedullary pin 882
Intramedullary rods 827
Intraoral (in vivo) vs. simulated (in vitro)
  exposures 903–904
Intraoral electrochemical properties 898
Intraoral surface, nature of 902–904
Inverse Kirkendall diffusion 375
Inverse Kirkendall effects 395
Inverse Kirkendall mechanism 395
Investment casting 21
In vitro environment 843
In vitro fretting corrosion test 833
In vivo conditions, of electrolyte chemistry . 862
In vivo corrosion, and chloride ion
  concentration 854
In vivo corrosion and biocompatibility,
  biological consequences of 847–848
In vivo corrosion testing 843
In vivo fluids, protein moiety of 883
In vivo studies, and effect of proteins on SCC .. 862
In vivo tarnished film compositions 903
In vivo testing
  of corrosion fatigue (CF) 871–873
  findings related to stress-corrosion
    cracking (SCC) 882–883
  simulation and interpretation 861–863
Iodine, in reactor water 365
Ionic contaminant, and dendrite growth ... 646
Ionic contaminants 615, 616
Ionic dissolution currents 829, 830
Ionization content, for water 15
Ion vapor deposited (IVD) aluminum
  coatings 183
Iridium
  in dental caustic alloys 905–906
  hydrogen fluoride corrosion of 696
  in phosphoric acid 739
Iron
  in ammonia and ammonia compounds 727
  conservation strategies for 300
  corrosion products of 295–296
  corrosion rates of 43, 45, 46, 692
  corrosion rates with stray current ... 109
  in dry chlorine 704
  effect of temperature on corrosion rates of 45
  flash rusting of 314
  in moist chlorine 707
  potential pH-diagram for, in supercritical
    aqueous solution 232
  sulfate-reducing bacteria (SRB) growth on 307
  weeping 315
Iron-aluminum intermetallic 521, 524

Iron artifacts 312, 316–317
Iron base alloys, in fuel cells 510
Iron-base alloys, sulfidation resistance of ... 1070
Iron-base Ni-Cr-Mo-Cu alloys, sulfuric
  acid corrosion of 662
Iron calibration specimens 47
Iron carbonate (siderite) scale 924
Iron carbonate, effects of temperature
  and hydrogen fugacity on 926
Iron carbonate solubility 1026–1027
Iron chloride, formation of 478
Iron-cobalt alloys, metal dusting of 731
Iron contamination, in caustic 723
Iron content of produced water, after
  squeeze treatment 940
Iron counts 950
Iron-depositing bacteria 13
Iron-nickel alloys
  metal dusting of 731
  in primary-pressure boundary 364
Iron-nickel-chromium alloys
  in high-temperature environments 755
  oxidation resistance of 1069
Iron-oxide encrusted bacteria 1094
Iron-silicon and graphite anodes 145
Iron specimens, corrosion rates 45
Iron sulfide 925
Iron sulfide corrosion products 924
Irradiated microstructure 398–399
Irradiation
  effect of, on alloy and oxide layer ... 407
  of zirconium alloy 408
Irradiation assisted stress-corrosion cracking
  (IASCC) 341, 386, 389–390, 397, 400
  of austenitic stainless steels 375
  deformation band role in 403
  dependence of, on fast neutron fluence ... 388
  effects of corrosion potential on ... 392–393
  in light water reactors 375
  new alloys for 405–406
  in steam-generating heavy water reactor
    (SGHWR) 388
Irradiation creep 375
Irradiation damage
  of zircaloys 418–419
Irradiation dose vs. loop density and size 398
Irradiation effects, on stress-corrosion
  cracking (SCC) 387–404
ISO 7539-7 878
ISO 9223 .... 42–43, 247
  atmospheric-corrosivity approach 55
ISO 9225, classification of sulfur compounds ... 55
ISO 9226, atmospheric-corrosivity approach ... 55
ISO CORRAG 49, 51–57
Isocorrosion contours, for alloy G-30 in
  nonoxidizing acid mixtures 744
Isocorrosion curves
  for caustic sodas 714, 718
  for formic acid 676
  for zirconium in nitric acid 671
Isocorrosion diagrams
  for 316L stainless steel 693
  for alloys in phosphoric acid ... 737, 738
  for alloys in sulfuric acid 662–664
  for alloy 686 (N06686) super austenitic
    stainless steel 684
  of alloys in nitric and hydrochloric acid 748
  for aluminum alloys A91100 in nitric acid 671
  for annealed type 304 stainless steel in
    nitric acid 669
  of austenitic stainless steels in sulfuric acid ... 784
  for Glassteel 5000 687
  of alloys in formic acid 676
  for nickel-base alloys 693
  of niobium 685
  for stainless steels in sulfuric acid and nitric
    acid mixtures 747
  for tantalum 684, 685
  for zirconium 684, 685
Isocyanate-terminated prepolymers 175
Isolating feature, internal interference
  across 104
Isolation of well casings 104

Index

I (cont.)

Isolator-surge protector (ISP) 123
Isolator-surge protector (ISP) components 124
Isomagnetic chart 108
Iso-pH general corrosion rates 366
Isoprene 733
Isothermal cooking (ITC) process 765

J

Jalpaite 311
Jarosite 315
Jet fuel 213
J-integral methods 868

K

Kalrez 733
Kamyr continuous digester 765
Kappa number 772
KC-135 aircraft 202
Keel bolts 265
Kelvin-Helmholtz instability 454
Kerosene fungus 213
Kiln operations 255
Kirchoff's current law 98
Kirkendall void formation 511
Kirschner wire 856
Kiss squeeze technique 940
Knife attack, of aluminum alloys 670
Kovar (Fe-29Ni-17Co) leads 209
Kraft black liquor process 762
Kraft process 762
Kraft pulping and chemical recovery
  flow diagram 781
Kraft recovery boilers 785–786, 792
Krupp VDM alloys 715
Kuntscher nails 882
Kure Beach 49

L

Lacquer coatings 963
Lamb waves 1048
Lamellar carbides 161
Land transportation industries 337
Langlier saturation index (LSI) 5, 10, 23
Lanthanum chromites 506
Larson Index (LaI) 10
Laser cladding 480
Laser cleaning 301
Laser peening 431–432
Laser profilometry 203
Last-pass heat sink welding 355
Lateral potentials 87
Lateral spread of corrosion at clad
  layer 185, 1088
Laves phases 239
Lawrence Livermore National
  Laboratory (LLNL) 429
Layered zinc/aluminum 61
Lay-up procedures 468
Lead (termination) tarnishing 629
Lead abatement 560
Lead acetate 283
Lead-acid battery 502
Lead and lead alloys, general
  and acetic acid 284
  in ammonia and ammonia compounds 733
  in caustic sodas 719
  conservation strategies for 300–301
  corrosion products of 294, 297
  corrosion rates with stray current ... 109
  corrosion related to pH 291
  in dry chlorine 705
  in formic acid 676
  for linings 769
  patina on 290
  in phosphoric acid 738
  in potable water 11
  sulfuric acid corrosion of 665
  used in outdoor environments 297
Lead artifacts 313
Lead-bearing paints 560
Lead carbonate 283
Lead coatings 180
Lead dendrites 639
Lead formate 283
Leadframes 629
Lead-free electronic components 641
Lead hot corrosion 488
Lead-induced cracking 380
Lead levels in potable water 10
Lead-tin plating (solder plating) 207
Leak detection 146
Lean stainless steels 755
Lepidocrocite 312, 314
Lewis acid-base chemistry 164
Life calculation 1044
Life-cycle cost analysis 579
Life-cycle cost modeling simulation 227
Life-enhancement techniques 225
Lifeline supports 267
Life model for a colony of stress-
  corrosion cracks 1019
Life-prediction 350–352
  of alloy 600 components 369–373
Lifetime prediction 343
Light 280
Lightning protection 273
Light water reactors (LWRs). See also
  boiling water reactors (BWRs);
  irradiation assisted stress-corrosion
  cracking (IASCC); pressurized
  water reactors (PWRs) 415, 417, 419
  corrosion of 339
  effects of average plant water purity
    on core component cracking 389
  initial design problems of 339
  mitigation strategies of stress-corrosion
    cracking (SCC) 404–406
Lime kiln operations 255
Lime kilns 783–784
Limonite 19
Linear anodes 94
Linear elastic fracture mechanics
  (LEFM) 868
Linear elastic fracture mechanics
  analysis techniques 607
Linear polarization 944
Linear polarization resistance
  (LPR) 116, 119, 951, 1004, 1031, 1032
Linear regression coefficients,
  ISO CORRAG Program 56
Lined-pipe systems 665
Liners, nonmetal 1074
Linings, corrosion resistance of 1073
Lint 256
Liquid ash corrosion 159
Liquid epoxy coating 1020
Liquid flow rate 807
Liquid metal cracking (LMC) 722
Liquid metal embrittlement
  (LME) 717–719, 722
  in petroleum refining and
    petrochemical operations 999
Liquid-phase carbonation, of concrete ... 796
Liquid wall shear stress 925
Liquor heaters 767
Litharge 297
Lithium 419
Lithium-sulfur dioxide (LiSO$_2$)
  batteries 501
Load relaxation, of stainless steel 403
Local anodic solution 929
Local corrosion allowances 973
Local environments 42
Localized (pitting) corrosion 1032–1033
Localized corrosion 566–569
  of U.S. Navy Aircraft 189, 191
Localized corrosion mechanism 844
Localized cracking, from yield stress 347–348
Localized crack-tip environment 351
Localized fissuring 997
Localized full-section corrosion
  with adjacent cathode 568
  with remote cathode 568
Local turbulence 924
Longitudinal residual stress 344
Long-path isolating feature, internal
  interference across 104
Long term corrosion modes 427
Long Term Corrosion Test Facility
  (LTFC) 429, 432
Long-term oxidation tests 756
Long-term performance of materials
  in service 444
Loss of coolant accidents (LOCA) 341
Loss of ductility observed, during
  slow-rate testing 994
Low-alloy copper steel, corrosion rates 45
Low-alloy steels, general 821
  in boiling water reactors 342
  for combustor and turbine sections ... 487
  corrosion due to primary water leaks 366
  corrosion rates 45, 48
  for high-level waste (HLW)
    containers 423–424
  thermal insulation of 817
Low-alloy steels, specific types
  A 416, grade 1860 steel 1055
  A 421 steel 1056
  A 508 steel 349
  A 516 grade 70 765
  A508 class 2 346
  A533 grade B 346
Low-carbon sheet steel 149
Low-carbon steel, general 42–43
  effect of oxygen concentration on 35
  long-term atmospheric corrosion
    predictions for 54
  loss of tensile strength 46
  in polluted seawater 38
Low-copper amalgams, tarnishing and
  corrosion testing of 912
Low-copper conventional amalgams 904
Low-cycle fatigue (LCF) 222, 863
Low-density polyethylene, in caustic
  sodas 719
Lower explosive limit (LEL) 793
Lower furnace 786–787
Lower heating value (LHV) 243
Low-gold-alloys, color change 913
Low-level radioactive waste 229, 253
Low line inhibition system, production
  conditions and environment and 941
Low-molecular-weight resins 175, 176
Low-nickel stainless steels 822
Low-NOx burners 477–478
Low phosphorous electroless nickel
  (LPEN) coatings 717
Low plasticity burnishing 431–432
Low-pressure (LP) turbine 447
  impurity concentration area 473
Low-pressure coolant injection (LPCI) 341
Low-pressure core spray (LPCS) 341
Low-pressure sour oil wells 962
Low strain rate testing, influence of
  strain rate in 879
Low-strength austenitic stainless steels,
  primary side stress-corrosion
  cracking (SCC) in 373–374
Low-temperature corrosion 158–159, 974–980
Low-temperature heat exchangers ... 494, 496
Low-temperature hot corrosion
  (LTHC) 164, 488
Low-temperature properties 972
Low-volatile organic compound
  (VOC) coatings 174
LPR/EN/HDA techniques 118–119
Lubricating oils 268
Lubrication 186, 260
Lumnite-haydite 1003
Lymph plasma 895

M

M119 howitzer firing platform 153
M198 howitzer 152, 154, 155
Mackinawite 313
Macrobiological growth 13
Macrocells 567
Macrofouling films 40
Macrophages 841
Magnesium and magnesium-base alloys, general
corrosion rates of, as function of relative humidity (RH) 43
in dry chlorine 705
hydrogen fluoride corrosion of 696
used in U.S. Navy Aircraft 185
Magnesium and magnesium-base alloys, specific types
AZ31B 45
magnesium alloy AZ91 (Mg-9Al-Zn) 185
Magnesium anodes 73, 266, 936
Magnesium-base (sulfite) mill and recovery system diagram 769
Magnesium chloride. *See also* deicing salts 523
Magnesium oxide, influence of temperature on solubility of 936
Magnesium ribbons 94
Magnesium used, in U.S. Navy Aircraft 185
Magnetic flux leakage (MFL) 1051
Magnetic particle inspection 1050
Magnetic retaining rings 497
Magnetite 20, 312, 363, 364, 466, 923, 925
Magnet wire 635
Malachite 311
Mandrel Bend test 149
Manganese sulfide (MnS) inclusions 458
Maraging steel 381
Marelon, used in marine vessel wastewater systems 274
Marginal fracture, dimensional change 894
Marine atmospheres. *See also* atmospheric corrosion 5, 42–59
corrosion test site 1088
Marine atmospheric corrosion 50, 53, 58–59, 957
Marine boilers 156
Marine cathodic protection 73–77
Marine concrete structures 132
Marine corrosion rates (CR) 48, 247
Marine environment, electronic equipment corrosion of 205
Marine environment bolt standards 264
Marine environments
artifacts in 308, 312–313
Marine fouling and calcareous deposits 215
Marine gas turbines 486
Marine grade aluminum alloys (5*xxx* and 6*xxx* series) 267
Marine sanitation device (MSD) 274
Marine structures, corrosion zones 957
Marine tidal zones 588
Marine vessels. *See* boats
MAR-M246 nickel-base superalloy 222
MAR-M246 turbine blade 223
Martensitic stainless steels 929, 948
primary side stress-corrosion cracking (SCC) in 374
in U.S. Navy Aircraft 185
Marvel Seal 287
Masonry 1064
Mass burning units 483
Massive permeation, time to 620
Mass transport
within crack enclave 352
of oxygen 923
Masts 274–277
Maximum creep value 62
Maximum temperature
of caustic sodas 720
Maximum temperature for continuous service 972
McConomy curves 980–981
McCoy's equation 75
MCrAlY coatings 490
Mean linear energy transfer (LET) 391
Mean time to failure (MTTF) 205, 615, 621
Meat and poultry processing plants 804
Mechanical caliper tool 97
Mechanical cleaning 329–330
Mechanical loading, crack initiation under 128
Mechanically assisted cleaning 808
Mechanically assisted corrosion 827–834
Mechanically polished surfaces 18
Mechanical polishing 811
Mechanical properties of stainless steels, in surgical implants 858
Mechanical pulping 762
high-yield 767–768
Mechanical shock, in glass-lined equipment 815
Mechanisms of corrosion, in microelectronics 616–620
Medical devices, biocompatibility of 820
Medical technology industry, corrosion in 338
Medium-phosphorous electroless nickel (MPEN) coatings 717
Melanterite 315
Mellonite molten salt process 258
Melting points
of metal chlorides 482
of refractories 166
Membrane cell process 710
Membrane liquors, corrosion rates of alloys in 723
Mercaptobenzothiazole (MBT) 536
Mercury
and alloys 722
in amalgams 861, 904–905
removal of 464
and titanium 722
and zirconium 722
Mercury cells 710, 722
Mercury-containing aluminum anodes 936
Mesa corrosion 925, 927, 1093
Metabolic processes, metal binding and effects on 822
Meta etch process 624–625
Metal, corrosion rates by 48
Metal alloys
corrosion of, in phosphoric acid 736
stress corrosion cracking (SCC) in hydrofluoric acid 695
Metal artifacts 289, 307
and animals 291–292
in buried environments 306
and galvanic corrosion 292–293
in museums 279–288
and vegetation 291
Metal biocompatibility, early studies of 820–821
Metal cleaning 517
Metal corrosion rates, in hydrochloric acid (HCl) 683
Metal dusting 508, 731, 757–758, 1069–1070
Metal/environment interactions 1054
Metal-etch machine type corrosion levels 625
Metal fabrication plants 255–256
Metal finishing 1071
Metal-induced hypersensitivity reactions 858
Metal ion leaching 822
Metal ions, hypersensitivity to 822
Metallic (galvanic) coatings 1021
Metallic anchors, in masonry walls 1057
Metallic biomaterials. *See also* biomaterials 826–827, 855–861
biological factors affecting 854
cancer causing effects of 823
cast Co-Cr-Mo (ASTM F75) 839
comparative studies of the various implant alloys 871
corrosion of 841–846
CP-Ti (ASTM F67) 839
effects of H_2O_2 on corrosion and oxide behaviors of 847
effects of protein on the corrosion behavior 846–847
electrolyte chemistry of 862
local and systemic distribution of ions and particles 847–848
mechanically assisted corrosion 826–836
micrographs of 864
NiTi (ASTM F2063) 839
in saline solution 862
services failures of 871–873
Ti-6Al-4V (ASTM F136) 839
type 316L stainless steel 839
Metallic coatings 5, 61–67, 1059, 1061
Metallic IR drop 86
Metallic materials and implants, biocompatibility of 820
Metallic taste, of dental alloys 892
Metallic ties, in masonry walls 1057
Metallic wallpaper sheet linings 462
Metallization, and chlorine-induced corrosion 623
Metallized-polypropylene-film capacitor 637–638, 1086
Metal losses
from current discharge 548
from stray current 548
Metal-loss tools 97–98
Metal plasma etch process 623
Metal processing equipment 1067–1074
Metal roofing 142–143
Metals and alloys compatible, in dissimilar-metal couples 618
Metals and metal alloys
biocompatibility of 820
in caustic sodas 719
corrosion of 688, 698, 751–752 700
date of first use of 309
galvanic series of 1016
in organic acids 680
oxidation rate of crack-tip alloy 351
in propionic acid 679
redox potentials of 1016
requiring protection from atmospheric exposure 620
Metal skin temperatures 981
Metal spray vs. weld overlay 480
Metal wastage. *See* wastage
Metal wastage rates, of nickel-base alloys 731
Metal whiskers 641
Meta phosphoric acid. *See* phosphoric acid
Metastable pitting, in Ringer's solution 846
Methylmercaptan 256
MFA (perfluoromethyl vinyl ether and tetrafluoroethylene copolymer), hydrogen fluoride corrosion of 697
Microalloy sybralog 912
Microbial activity, impact of on high-level waste (HLW) containers 427
Microbial biofilms 213
Microbiologically induced corrosion (MIC)
in commercial aviation 599
in pipelines 1017
in white water 778–779
sulfate-reducing bacteria (SRB) 1086
Microbiologically influenced corrosion (MIC) 13, 783
described 211
and sulfate reducing bacteria (SRB) 796, 1094
of austenitic stainless steels 794
tests of 116
in wastewater 795
Microcomposite alloys 579
Microelectronics 613–621
Microelectronics industry, corrosion in 338
Micromotion 820
Microorganisms 6, 901
described 211
in immersion environments 213
tarnishing of 897
Microscopic examination, of corroded surfaces 843
Microstructural damage 222
Microstructure 837–839
Microstructure coarsening 223
Mid-chamber wear 625
Mid-furnace corrosion 791
MIL-A-8625 171, 174
MIL-A-46106 209
MIL-A-46146 209

Index / 1123

MIL-C-5541 171
Mild steel 43, 45, 769, 770
MIL-DTL-81706 171
MIL-F-14072D 151
MIL-HDBK-310 139
MIL-HDBK-729 139
MIL-HDBK-1250 206
Military aircraft
    recent development and future needs 203
    wash and rinse facilities 133
Military aircraft corrosion fatigue 195–203
Military coatings 6, 180–182
    volatile organic compounds (VOCs)
        levels in 174
Military corrosion education and training 134
Military equipment 220–227
Military facilities 6, 131–132, 141–144
Military problems 127
Military service
    case studies 160–161, 164, 167
    gas turbine engines in 162–164
    incinerators in 164–168
Military specifications and standards
    (MSS) 6, 136–140
Military standard 441 205
Military systems 156–162
Military vehicles, corrosion environment of ... 148
Military waste 229
Milkstone deposits 807, 808
Milling equipment, materials for 1077
MIL-P-85582 176
MIL-PRF-2337 176
MIL-PRF-8552 172
MIL-PRF-85285 172
MIL-PRF-85582 176
MIL-S-5002 171
MIL-S-29574 178
MIL-STD 810 139–140
MIL-STD 889 139–140
MIL-STD 961 136–137
MIL-STD 967 138
MIL-STD-1250 206
Mineral acids 324
Mineral admixtures 564
Mineral wool 818
Mine shafts 1077
Mine waters 1076–1078
Minimum allowable thickness (MWAT) 1044
Minimum creep value 62
Mining and metal processing industries,
    corrosion in 338
Mining equipment, materials for 1077
Mitigation bonds 111–112
Mixed acids 742
Mixed-gas corrosion 439
Mixed-metal-oxide anodes 74, 94
Mixed oxide 415
Mixed salts 742
Mix-phase flow 1000
Modeling
    of atmospheric corrosion 54–57
    of exfoliation corrosion 199
Modified acid dew-point curve corrosion
    rate, of plain-carbon steel 495
Modified continuous cooking (MCC)
    process 765
Modified Couper-Gorman curves 981–982
Modified Goodman law 867
Modular connection in a total hip replacement,
    mechanically assisted corrosion 831
Modular femoral him taper interface,
    mechanically assisted corrosion
    tests of 834
Modular head 828, 1094
Modularity
    in metallic biomaterials 827
Modular taper fretting corrosion test setup,
    for cobalt-chromium-molybdenum 833
Modular tapers 827
    from prostheses 828, 1093
Moist chlorine 706–708
Moisture accumulation from blocked drain ... 603
Moisture film thickness vs. corrosion rate 615

Moisture vapor transfer 72
Mollier diagram 472, 473
Molten carbonate fuel cell
    (MCFC) 502–503, 506, 512–513
Molten-metal corrosion, in heat treating
    industry 1071
Molten metal process 183
Molten-salt corrosion 1070
Molybdates 176
Molybdenum and molybdenum-base alloys,
    general
    as alloying element 657, 755, 907–908
    anhydrous hydrogen fluoride (AHF)
        corrosion of 702
    and corrosion potential 742
    effect of in Ni-Fe-Cr-Mo alloys 743
    hydrogen fluoride corrosion of 696
Molybdenum and molybdenum-base
    alloys, specific type
    B-2 748
    Multimet alloy (R30155) 1069
    (R03600) 696
Molybdenum austenitic stainless steel 13
Molybdenum content, of stainless steels 778
Molybdenum disulfide, as stress corrodent ... 473
Molybdenum disulfide lubricants 382
Monatomic hydrogen 992
Monitoring
    of ASTs 95
    of USTs 95
    by utilities 556–557
Mono-functional cresyl glycidyl ether (CGE) .. 175
Monosodium citrate (MSC) 325
Monsanto low-pressure process 676
Monte Carlo simulation technique 371–372
Moore pins 872
Moratorium, of weathering steel 581
Morpholine 363, 365
Morphology, of nitride layer 759
Morse tapers 827
Mortar coatings 1021
Mott-Schottky equation 847
MOX 415
Mss, documents and designation types 136
Mudline 958
Muds 922
Muffler 526, 528
Mufflers/silencers 270
MULTEO (computer code) 379–380
Multiaxial loading 867
Multichip modules (MCM) 645
Multicomponent solvent systems 750, 751
Multiphase flow modeling 924, 1031
Multiphase pipelines 1026
Multiple-effect evaporator (MEE) 780–781
Multisensor probes 117
Multiside-damage, and rivet holes 221
Multitechnique electrochemical corrosion
    monitoring case studies 1033–1034
Municipal solid waste (MSW) 253, 428
Muntz metal tubesheet 450
Museum, as source of corrosion 280
Museum environments 7

N

NACE 1006
NACE international standard 79
NACE MR0103 1006
NACE MR0175 1006
NACE MR1075/ISO 15156 1006
NACE RP0472 699
NACE RP0775 950
NACE TM0284 699
Nantokite 311, 315
Naphthanate coatings 181
Naphthenic acid 983–986, 1000
National Electronic Injury Surveillance
    System (NEISS) 260
National Playground Safety Institute (NPSI)
    certification program 261

Natrojarasite 315
Natural (light) water 415
Natural elastomers, in dry chlorine 706
Natural gas pipelines. See pipelines
Natural gas reservoir 959
Natural rubber 686, 733
NAVAIR 01-1A-509 209
Naval aircraft 6, 129, 130, 196
    alloys used in 184, 185
    atmospheric corrosion 191–192
    coatings for 130
    corrosion of 184–193, 1090
    damage due to improper
        packaging of 193, 1090
    hydrogen embrittlement of 193
    inspections of 185–186
    internal exfoliation of 191
    magnesium used in 185
    making carrier landing 184
    nondestructive inspections of 186
    operational conditions of 184
    sealing compounds for 190–191
    visual inspections of 186
Naval Facilities Engineering Command
    documents 138
Navy Scribe and Bold Surface Inspection
    Practice 62
Near-drum thinning 792
Near-ground potentials 87
Near-neutral chlorine dioxide 773
Near-neutral-pH stress-corrosion
    cracking (SCC) 1019
Near-zero-emissions power plants 504
Nelson curves 968, 998
Neoprene
    in hydrochloric acid (HCl) 686
    for spar fittings 276
Nernst equation 8–9
Nernst voltage 504
Nerve firing 846
Neutral salt-fog tests 199
Neutral sulfite semichemical pulping
    (NSSC) 770–771
Neutral water treatment 237
Neutron backscatter 1051
Neutron influence 386
Neutron irradiation 375
New duplex stainless steels 764
New True Dentalloy 912
Nickel alloy families 51
Nickel alloy steels 730
Nickel-aluminide coating 489
Nickel and nickel-base alloys, general
    20-type alloys 662
    in acetic acid 678
    allergic hypersensitive reactions to 915
    in ammonia and ammonia compounds ... 731–732
    in boiling water reactors 346
    carburization in 1069
    in caustic sodas 716–718
    in chlorates 722
    chromium-bearing nickel alloys 717
    chromium-nickel base alloys 480
    compositions of 969
    corrosion behavior 427–432
    corrosion of 683–684, 700, 717
    corrosion of, in phosphoric acid 738
    corrosion rates for 694
    corrosion rates of, in simulated acidified
        water (SAW) 429
    corrosion resistance of 426
    corrosion tests of, in caustic potash 724
    corrosion vs. depth 36
    cracking 695, 700
    crevice corrosion 429–430
    dental casting alloys 907
    distribution of heats 369
    in dry chlorine 704
    effect of, on carburization resistance 756
    effect of, on corrosion resistance in
        halide environment 759
    effect of hydrofluoric acid on cracking 701
    effect of temperature on corrosion of 684

Nickel and nickel-base alloys, general (continued)
environmentally assisted cracking
(EAC)..................................430–432
exhaust systems 270
of Fe-Ni-Cr alloys 757
ferric ion corrosion in during chemical
cleaning 328
for flue gas desulfurization (FGD) 462, 464
in formic acid 676
in fuel cells 506, 509
in halide environment 759
Haynes Stellite (HS) alloys 858
for high-level waste (HLW) containers......424
high-strength nickel-base alloys 373
in high-temperature anhydrous hydrogen
fluoride (AHF).........................700
in high-temperature environments 755
in high-temperature hydrofluoric acid.......700
and hydrofluoric acid corrosion 693
in hydrogen fluoride, corrosion of 692–696
interdendritic stress-corrosion cracking
(IDSCC) in 342
isocorrosion contours for, in
nonoxidizing acid mixtures.............744
isocorrosion curves for 676, 718
isocorrosion diagrams 676, 684, 738
life prediction 369–373
localized corrosion 429
metal dusting 757, 1070
metallurgic changes in 700
in moist chlorine 707
Monel alloys 273, 276, 456
nickel-based alloys 15
nickel-chromium alloys ... 232, 813, 907, 909–910
nickel-chromium-molybdenum-copper
alloys 813
nickel-copper 425
nickel-copper alloys 813, 976
nickel-molybdenum alloys 813
Ni-Cr-Mo 425, 810
with niobium 693
nitridation resistance in 758
in organic acids 679
oxidation of 755
oxidation resistance of............. 1068, 1069
passive corrosion 427–429
in petroleum refining and petrochemical
operations 969–970
in pharmaceutical industry 813
in phosphoric acid 738
pitting corrosion 430
precipitation-hardened nickel-base 948
in pressurized water reactors (PWRs) 362, 363
primary side stress-corrosion cracking
(SCC) in 367–373
in propellers, propeller shafts, and rudders 270
in propionic acid 679
in seawater 214
slow strain rate tests (SSRT) 431
strained in simulated concentrated
water 432, 433
sulfidation resistance of 1070
supercritical water 232
thick section components 369
in waste-to-energy (WTE) boilers.......483–484

Nickel and nickel-base alloys, specific types
25-6Mo (N08926) 695
65Ni-20Cr-6Fe-5Si-2.5Mo-2Cu-0.03C 664
1925hMo 15
AL-6X 457
AL-6XN (N08367) 15, 276, 450, 457, 715,
806, 810, 812, 948, 969, 976, 989
alloy 9Ni-4Co-0.3C 601
alloy 20 (N08020) 670, 676, 678, 692, 700,
704, 706, 713, 737, 769, 770, 774,
784, 977, 978, 985
alloy 20Cb3
(N08020) 254, 662, 714, 737, 738, 739
alloy 20Mo-4 (N08024)...................737
alloy 20Mo-6 (N08026)............. 676, 737
alloy 22 (N06022) 425, 427–432, 433, 670
alloy 25-6Mo 15
alloy 28 (N08028) 670, 676, 739
alloy 28 (N08800) ... 445, 676, 715, 737, 738, 989
alloy 31 (N08031) 662, 663, 737
alloy 55 664
alloy 59 (N06059) 462, 684, 718, 738
alloy 62 (1.5 to 3% Nb; UNS
N06062)..................................701
alloy 66 664
alloy 82 (N06082) 354, 367, 701, 766
alloy 160 (N12160) 168, 1086
alloy 188 706
alloy 200 (N02200) 670, 696, 700, 701,
705, 713, 716–718, 722, 723, 724, 738
alloy 200 (N02200) pure nickel 683–684,
717, 732
alloy 201 (N02201) 715, 716, 717, 718, 738
alloy 214 (N07214) 706, 755, 970
alloy 227 (N02270) 701
alloy 230 (N06230) 706
alloy 242 ....695, 759
alloy 400 (N04400) 425, 663, 665, 670,
683, 684, 693, 694, 695, 696, 700, 701, 704,
705, 706, 717, 718, 722, 723, 732, 738, 970, 975,
976, 977, 978, 979, 987, 995
alloy 500 (N05500) 695, 700
alloy 600 (N06600) 233, 325, 349–350, 353,
362, 367–373, 376, 377, 378, 389, 684,
695, 696, 700, 701, 704, 716–718,
723, 731, 738, 757, 759, 767, 1086
alloy 601 (N06601) 718, 731, 755, 757, 758
alloy 601H 731
alloy 602CA (N06025) 731, 757, 759
alloy 602CA (UNS N06025) 701
alloy 610, 706
alloy 617 (N06617) 240, 701, 759
alloy 622 (N06022) 479
alloy 622 (UNS N06022) 695
alloy 625 (N06625) 233, 425, 445, 462,
479, 482, 483, 484, 663, 670, 676, 693, 701, 706,
718, 731, 732, 738, 739, 759, 765, 788,
789, 948, 970, 977, 985
alloy 625 (N06625) (ERNiCrMo3) 766
alloy 625/CS composite tubes 791
alloy 671 (50Ni-50Cr) 480
alloy 686 (N06686) 462, 684, 738, 970
alloy 686 (UNS N06686) 695
alloy 690 (N06690) 166, 168, 362, 376,
378, 406, 718, 1086
alloy 693 (N06693) 731, 758
alloy 718 (N07718) 233, 367, 601, 948
alloy 740, 240, 480
alloy 800 (N08800) 325, 362, 376, 378, 400,
406, 442, 445, 701,
714, 767, 792
alloy 800H (N08810)............. 706, 757, 759
alloy 825 (N08825) 425, 426, 427, 663, 678,
683, 692, 695, 700, 714, 737, 738, 739,
790, 948, 961, 970, 976, 977
alloy 825/CS 791
alloy 904L (N08904)....... 251, 424, 462, 663,
676, 678, 714, 715, 768, 770, 771, 775,
784, 785, 969, 989
alloy 904L (UNS N08904) 662
alloy 926 (N08926)........................424
alloy 1925hMo (N08926) 737
alloy 3127LC (N08028) 737
alloy AL-6X (N08366) 737
alloy AL-6XN (N08367) 737
alloy B (N10001)............. 676, 700, 785, 970
alloy B-2 (N10665) 664, 670, 676, 679,
684, 696, 738, 739, 744, 970, 977
alloy B-3 (N10675) .. 664, 684, 693, 695, 970, 977
alloy B-4 (N10629) 684
alloy B alloy (Ni-28Mo) 684
alloy C (N10002) 676, 700
alloy C-4 (N06455) 424, 676, 738, 977, 980
alloy C-22 (N06022) 15, 338, 462, 463, 684,
693, 694, 695, 700, 718, 738, 768
alloy C-276 (N10276) 15, 232, 462, 463,
663, 676, 679, 684, 693, 695, 696, 700, 706,
707, 718, 732, 738, 739, 748, 768, 770, 810, 812,
929, 948, 976, 977, 985
alloy C-2000 (N06200) 462, 693, 694
alloy Cb-3 (N08020) 661, 662, 663
alloy F (N06001)..........................738
alloy G (N06007) 663, 738, 739
alloy G-3 (N06985) 663, 692, 700, 738, 739
alloy G-30 (N06030) 232, 663, 670, 692,
700, 738, 739, 744, 769
alloy G-31 (N08031) 738
alloy G-35 (N06035) 738
alloy K-500 (N05500) 717, 978
alloy R-41 (N07041) 706
alloy X 706
cast alloy CN-7M (N08007) 662, 663, 670,
692, 700, 708, 715, 737, 775, 785, 1079, 1080
cast alloy CW-2M (N26455)........... 663, 773
cast alloy CW-6M 663
cast alloy CW-12M 708
cast alloy CW-12MW 663
cast alloy CX-2MW (N20622) 663, 773
cast alloy N12-MV 664
D-205 (Haynes International) 664
Hastelloy alloy N 1070
Hastelloy alloy S 1070
Hastelloy C 253, 256, 270, 704
Hastelloy C276 (N10276) 251, 252, 253, 254
Hastelloy F...............................663
Hastelloy G3 251
Hastelloy S 706
Hayes 25 alloy (HS-25)..................883
Haynes alloy 214, 1069, 1070
HP (N08705) 731
HP40 757
HP-45Nb 731
HR6W austenitic steel 240
HR-160 (UNS N12160).............. 444, 758
HS-21 858
HS-25 858
Incoloy 825, 251, 270
Incoloy alloy 800, 1068
Incoloy alloy 800H, 1068, 1069, 1070
Inconel alloy 600, 705, 706, 1067, 1069,
1070, 1071
Inconel alloy 601, 706, 1067, 1069, 1070
Inconel alloy 625, 251, 252, 254, 270
Inconel alloy 800, 1067, 1069
Inconel alloy 800H, 1070
Inconel X-750 390
Monel 400 (N04400) 325, 328, 704, 705, 708
N-12M 708
N06022 663
N06059 663
N06110 1081
N06686 663
N08026 662
N08028 662
N08926 662
N10001 751
N10276 663
N10276 stainless steel 214
N031254 662
Nibral 270
nickel-modified 409 (409Ni) 529
RA330 (N06330) 1069, 1070
VDM alloy 6030 (N06690) 718

Nickel-base alloy 713C turbine blades 223
Nickel-base alloy compositions 755
Nickel-base alloy properties, for solid
oxide fuel cells (SOFCs) 513
Nickel-base alloys
in fuel cells 510
sulfuric acid corrosion of 663–664
Nickel-base high-performance alloys 1067
Nickel-base superalloys 487
loss of creep strength in..................222
Nickel-base weld metals 369
Nickel-cadmium coatings 486
Nickel-clad stainless steels, in fuel cells 506
Nickel content vs. corrosion rate 715
Nickel-copper, effect of niobium in
hydrofluoric acid 695
Nickel-copper alloy, in fuel cells 511
Nickel-copper alloys 36
in seawater 214
sulfuric acid corrosion of 663
Nickel-molybdenum alloy B, in dry chlorine .. 705

Index / 1125

Nickel-rich austenitic stainless steels, hydrogen fluoride corrosion of . . . 692
Nickel-rich G type alloys . . . 738
Nickel-titanium . . . 910
NiCoCrAlY coatings . . . 490
Nicopress fittings . . . 276
Ni-Cr-Fe alloys . . . 1068, 1069
Ni-Cr-Mo alloys . . . 265, 463, 705, 732
NiCrMoV disc steel . . . 470
Nicrofer 3127 hMo . . . 662
Ni-Fe-Cr-(Mo) corrosion-resistant alloys . . . 742
Nil-ductility transition temperature (NDTT) . . 730
Niobium and niobium alloys, general
in ammonia and ammonia compounds . . . 733
caustic sodas . . . 719
in dry chlorine . . . 705
hydrogen fluoride corrosion of . . . 696
isocorrosion diagram for . . . 685
in moist chlorine . . . 707
nitric acid corrosion of . . . 672
Nitrate-induced stress-corrosion cracking (SCC), in heat-recovery steam generators (HRSGs) . . . 494
Nitrate stress-corrosion cracking (SCC) . . . 491
Nitric acid . . . 290, 325, 668–672, 727
dew point . . . 495
Nitric acid plants . . . 254
Nitric and hydrofluoric acid mixtures . . . 747
Nitric-hydrofluoric acid bath . . . 18
Nitric ions . . . 290
Nitridation, in high-temperature environments . . . 758–759
Nitridation resistance
in cobalt-base alloys . . . 758
in nickel-base alloys . . . 758
Nitriding . . . 729
Nitriding depth . . . 732
Nitrile elastomers . . . 706
Nitrilotracetic acid (NTA) . . . 325
Nitrite inhibitor . . . 69
Nitrogen compounds . . . 976
Nitrogen oxide (NO_x) emissions . . . 532
Nitrogen oxide emissions . . . 461
Nitrogen oxide pollutants (NO) and (NO_x) . . . 280
Nitrogen oxides (NO_x) . . . 290
Nitrogen squeeze . . . 950
Nitronic 50 (UNS S20910) . . . 214
NobleChem process . . . 404
Noble dental alloys, in dental casting alloys . . . 905–906
Noble metal coatings . . . 512
Noble metals. *See also* gold and gold-base alloys, general; precious metals and alloys; silver and silver alloys, general
corrosion resistance of, in hydrochloric acid (HCl) . . . 686
in supercritical water . . . 233
Noble metals PFM alloys . . . 909
Noble metal systems, standard electrode potentials . . . 617
No-clean solder fluxes . . . 638
No-clean solder fluxes . . . 640
Nominal corrosivity . . . 542
Nonchlorine bleaching stages . . . 772–773
Nonchromated alkaline cleaners . . . 174
Nonchromated pretreatments . . . 174
Nonchromated sealants . . . 178
Nonchromated waterborne epoxy primer . . . 176
Noncondensible gases (NCGs), collection and destruction of . . . 793
Noncorrosive powders . . . 258
Noncoupled metal panels, marine atmospheric corrosion data for . . . 53
Nondestructive evaluation (NDE) . . . 133, 658
Nondestructive examination (NDE) . . . 1037, 1039
Nondestructive inspections, of U.S. Navy Aircraft . . . 186
Nondestructive testing (NDT) . . . 133, 571
Nonelectrochemical techniques . . . 115–117
Noninsulated tie-and-ballast track . . . 549
Nonmetallic coatings . . . 1074
Nonmetallic materials
in acetic acid . . . 678
in caustic sodas . . . 719
in phosphoric acid . . . 739
types of . . . 941
Nonmetals
in formic acid . . . 676
sulfuric acid corrosion of . . . 665
Nonoxidizing acid mixtures . . . 743–747
isocorrosion contours for alloy G-30 in . . . 744
Nonoxidizing biocides . . . 14
Nonoxidizing solutions . . . 743
Nonpolar aprotic solvents . . . 750
Nonreflective coatings . . . 181
Nonsensitized stainless steels, crack-propagation rate . . . 348
Nontoxic inhibitive primer . . . 176
Nonunion . . . 854
Normally inert fillers . . . 564
Norseman . . . 277
Notch factor . . . 201–202
Notch sensitivity . . . 865
Novaprobe . . . 117
NO_x emissions . . . 439
Nuclear-grade stainless steels . . . 354
Nuclear power industry . . . 337, 339
Nuclear power plants, service water distribution systems (SWDS) . . . 12
Nuclear stream supply system (NSSS) . . . 341
Nuclear waste. *See* nuclear power industry
Nucleate boiling (NDV) . . . 156

O

Oak . . . 283
Occupational Safety and Health Administration (OSHA) . . . 326
Ocean water
corrosion rates in . . . 31
depth vs. pH . . . 37
oxygen and pH variations of . . . 959
solubilities of gases in . . . 958
temperature of . . . 29
Oddy test . . . 286
Odor control . . . 256
Odors . . . 793
Off-line corrosion measurement . . . 1031–1032
Offshore pipelines . . . 936
Offshore platforms . . . 956–958
cathodic protection (CP) of . . . 934
and drilling rigs . . . 922
potential gradient distribution of support structure . . . 77
Offshore procedures . . . 87
Offshore structures
corrosion zones . . . 957
sacrificial anode compositions for . . . 936
Ohmic voltage drop (IR drop) correction . . . 86
Oil additives and microbial growth . . . 213
Oil ash constituents . . . 479
Oil ash corrosion . . . 159
Oil blending strategy . . . 984–985
Oil coatings . . . 181
Oil field corrosion-resistant alloys . . . 929
Oil-field equipment . . . 943–944
Oil-fired boilers
corrosion control by plant operation . . . 480–481
fireside corrosion in . . . 477
Oil gilding . . . 297
Oil pipelines. *See* pipelines
Oil production. *See also* petroleum production operations . . . 946–956
Oil production platforms . . . 956–958
Oil-soluble inhibitors . . . 938
Oil/water separation facilities . . . 953, 955
Oil wells . . . 959, 962
Oleum . . . 659
Once-through water system . . . 12
Once-through cooling systems . . . 268
Once-through pressurized water reactor (PWR) steam generators . . . 367
Once-through steam generators (OTSGs) . . . 375
On condition maintenance philosophy . . . 224
One-component solvent systems . . . 751
One-component waterborne polyurethane topcoat . . . 177
On-line chemical cleaning . . . 328–329
On-line corrosion measurement . . . 1032
On-line mechanical cleaning . . . 330
On-line monitoring . . . 1032
On-line real-time correlation technique . . . 1034
On-line real-time corrosion monitoring . . . 1005
On-line real-time corrosion rate systems . . . 1032
On-line real-time electrochemical corrosion monitoring system . . . 1004
Open-circuit potential (OCP) . . . 843
Open circuit potential (OCP) shift . . . 830
Open ocean sites . . . 31
Open-recirculating system . . . 12
Operating stresses of suction rolls . . . 779–780
Operational conditions . . . 184
Operator and maintenance repair organization (MRO) messages . . . 606
Optical micrographs . . . 221, 758
Oral cavity environment . . . 854
Oral corrosion . . . 897–902
Oral environment . . . 894
Oral fluids . . . 895–896
Oral lesions . . . 893
Organic acid coolants . . . 531
Organic-acid induced corrosion . . . 636–638
Organic acids. *See also* acetic acid; formic acid; propionic acid . . . 324, 536, 674, 679–680, 805
Organic carbonyl compounds . . . 283
Organic chlorides . . . 977
Organic coatings . . . 5, 955
carrier for . . . 174
in commercial aviation . . . 601
on marine hulls . . . 265
for protecting steels . . . 1058
surface preparation for . . . 69–70
of underground storage tanks (USTs) . . . 95
Organic coating systems . . . 657
Organic corrosion-inhibiting compounds . . . 603
Organic emulsions . . . 326
Organics, in hydrochloric acid (HCl) . . . 682
Organic seal coatings . . . 487
Organic solvents . . . 326, 750–752
Organic sulfide compounds . . . 256
Organic vapor sources, corrosivity ranking to metals . . . 620
Organic zinc/amine adduct-cured epoxy system . . . 776
Organic zinc-rich primers . . . 71, 818
Orthodontic appliances . . . 893
Orthodontic wires . . . 910
Orthopedic implants, scanning electron micrographs (SEM) of . . . 829
Orthopedics . . . 856
Orthophosphoric acid. *See* phosphoric acid
Osmium
in dental casting alloys . . . 905–906
hydrogen fluoride corrosion of . . . 696
Osteoblastic activity . . . 854
Osteolysis . . . 828
Other components . . . 259
Ottemannite . . . 313
Otto cycle . . . 439
Outboards/outdrives . . . 268
Outdoor artifacts . . . 298, 301
Outdoor atmosphere corrosion products . . . 644
Outdoor environments . . . 7, 289–291, 293–297
Outdoor pollutant concentrations . . . 645
Out-of-service corrosion . . . 468
Oven-bake process . . . 626
Overcoating technology, of cathodic protection unit . . . 145
Overhead condensers . . . 977, 979
Overhead wire catenary system (OCS) . . . 549
Overlay coatings . . . 489
Oversize solute vs. grain-boundary chromium concentration . . . 397
Oxidation . . . 754–756
in heat treating industry . . . 1067–1069
and pH change . . . 830
Oxidation (dissolution) rate . . . 351–352, 987

1126 / Reference Information

Oxidation kinetics 416–417
Oxidation resistance 987
impact of metallurgical parameters on 417
Oxidation tests 1067–1069
Oxide domes 839
Oxide film dynamics, of biomaterials 844
Oxide film repassivation studies 846
Oxide films
on metallic biomaterials 840
protective 466
Oxide film stability 830
Oxide film thickness
vs. oxygen content 21
reformation of mechanical disruption 832–833
Oxide wash process 631
Oxidizing acid mixtures 747–748
Oxidizing agents 324
Oxidizing biocides 14
Oxygen 307, 923–924
as function of radiation type and flux 391
variations of in ocean waters 959
Oxygenated treatment (OT) 237
Oxygen bleaching 774
Oxygen concentration
vs. depth 36
effect of on corrosion as a function of temperature 1030
effect of on low-carbon steel 35
relative to air 33
Oxygen contamination 955
Oxygen corrosion 945–946
Oxygen pitting 467
Oxygen scavengers 943, 945
Oxygen solubility 33
Oxygen variability 31
Ozone (O_3) 280, 290

P

Pacific Ocean. *See also* ocean water
pH 37
surface dissolved oxygen 34
surface salinity 29
surface temperature 30
Pacific silver cloth 284
Packaging damage 193
Pain 831
Paint application equipment 178
Paint blistering 71
Paint coatings 182
Paint curing 518
Painting operations 255
Paints 284–286
adhesion failure 605
categories of 70
Paint systems 188
automotive 517
Palladium and palladium-base alloys, general
in dental casting alloys 905–906
effect of palladium/gold ratio on gold dental alloys 913
hydrogen fluoride corrosion of 696
palladium-silver PFM alloys 909
Palladium and palladium-base alloys, specific types (P03995) 696
Palladium-coated alloy 42 leadframes 632–633
Panels, distribution and control 271–272
Paper industry. *See* pulp and paper industry
Paper machines 775–779
Paper manufacturing 763
Parallel interference path 107
Paratacamite 279, 311
Paris equation 864, 867, 869, 870
Parking structures 1054
Partial dental alloys 905–908
Partial pressure ratio log vs. corrosion losses of stainless steel 442
Particulates 280
Passivating compounds 945
Passivating solutions 326, 775
Passivation 811
Passivation chemicals 10
Passivation rate 352
Passivation treatments 811
Passive corrosion rate 428
Passive domain 842
Passive electrical components 634–641
Passive film breakdown 158
Passive films 854–855
Passive layer 17–18
Patinas 289, 310
Peening 368
Pellicles 902
Penetrating sealers 583
Penthouse casing and hangar bars 492, 496
Perfluoroalkaoxytetrafluoroethylene (PFA) . . . 814
Perfluoroalkoxy (PFA) 665, 697, 733
Perfluoroelastomers (FFKM) 672, 702, 706
Perforated nonmetallic pipe 95
Perforation 515
Performance specification 137
Permafrost 246
Permeability reducers 563
PER ratio 881
Persistent slip band formation 863
Personal protection cages 657
Petrochemical industry 338
Petroleum industry
corrosion in 338
installation monitoring 1037
Petroleum production operations. *See also* offshore platforms; oil production; pipelines 922–946
secondary recovery operations 953–955
Petroleum production operations corrosion causes 923–928
Petroleum refining and petrochemical operations
alloys in 968–970
cast iron in 969
corrosion in 967, 990, 992, 998–1004
environmentally assisted cracking (EAC) in 987–999
hydrogen attack in 996–998
liquid metal embrittlement (LME) in 999
materials in 968–970
process optimization study 1004–1005
protective coatings in 1002
stainless steels 968
standards and references for downstream materials 1008
standards and references for upstream materials 1006–1008
stress-corrosion cracking (SCC) in 988–992
technical organizations involved in 1008
velocity-accelerated corrosion in 999–1002
wet H2s cracking in 992–996
Pewter 283
Pewter artifacts 313
PFTE equipment
in dry chlorine 706
in moist chlorine 708
pH. *See also* Pourbaix diagrams
of burial environments 307
in concrete 563
and corrosion inhibition 565–566
vs. depth of water 37
in drilling fluids 945
effect of on corrosion and calcareous deposition 35
effects of, on buried steel 89
of foods 804
of human body environment 821
influence of, on corrosion rates 563
of mine waters 1076
in mouth 900
of Pacific Ocean 37
relationship with oxygen and carbon dioxide 33–34
vs. time responses of plaque 900
variability of 34–35
variations of in ocean waters 959
of wastewater 794
of white water 777
Pharmaceutical industry 254–256, 812–814
coatings in 818
corrosion characteristic values for materials 810
corrosion in 338
high-purity ferritic stainless steels in 675
Pharmaceutical processing plants 15
Pharmaceutical waters 15–16
pH diagrams vs. potential 308
of environments in nature 307
Phenol 980
Phenol formaldehyde 285
Phenol-formaldehyde coatings 932
Phenolic coating 1020
Phenolic linings, sulfuric acid corrosion of 666
Phenolic odor 637
Phenolic resins 687
Phosphate-buffered saline (PBS) 844
Phosphate buffering system 902
Phosphate coating 153
Phosphate coatings 181
Phosphate corrosion 467
Phosphates 176, 535
Phosphate solutions, tin in 902
Phosphate treatments 258, 468
Phosphate wastage, in pressurized water reactor (PWR) steam generators 377
Phosphoric acid 325, 736–741, 808
Phosphoric acid fuel cell (PAFC) 502, 505
Physiological saline solution 880, 903, 904
Physiological solutions 896
Pickling, for embedded iron removal 814
Pickling-associated equipment 1074
Pickling solutions 775
Piehl Kp factor 1000
Piezoelectric sensors 227
Pigments 172
Pile cap stirrupclose interval mapping, of reinforcement corrosion 568
Pilling-Bedworth volume ratio 1058
Pillowing 197, 199, 200
cracks under influence of 221
Pipe (casing) resistance determination . . . 100–101
Pipeline bracelet anodes 75
Pipeline coatings 1018
Pipeline current 109
Pipeline isolation 75
Pipelines
accident reports on 1015
asphalt coated 1019
cathodic protection (CP) of 74–75, 933, 1022
causes of damage to 1031
coatings on 1020–1021
corrosion of 958–962, 1016, 1026, 1031, 1086
current mapping of 109
differential cell corrosion in 1016
extent of 1015
fusion-bonded-epoxy (FBE)-coated 1018
Gas Pipeline Integrity Rules 81
gas pipelines 1031
integrity and data of 79
internal liners 960
locations of 85
microbiologically induced corrosion (MIC) . . 1017
multiphase 1026
Office of Pipeline Safety (OPS) 79
offshore pipelines and structures 936
preassessment 79–81
remaining life of 82
sacrificial anodes compositions for 936
stray current corrosion in 1017
stress-corrosion cracking (SCC) in 1018
tape-coated 1019
tests and testing of 117
underground pipeline CP system 117
underground pipelines 79, 1016
voltages on 108, 113
Pipes and piping
protective coating application of 963
and scraping 329
system inspections of 1043
used in marine vessel freshwater systems 273
Pipe steels 1019
chemical and tensile requirements of 1015

Index

Pipe-to-electrolyte potential vs. time 111
Pipe-to-soil 550, 1022–1023
Pipe-type cables 6, 122–125
Pit depth limit 472
Pit depths, for exhaust stainless steel
   alloys 524
Pit-initiated cracking, described 127
Pit initiation 127
Pit norm corrosion metric 202
Pitons 263–264
Pitting 471, 619
   on aluminum 152
   of austenitic stainless steels 670
   in automotive bodies 516
   during chemical cleaning 327
   of chromium plating 180
   in commercial aviation 599
   from dynamic stresses 128
   electrochemical mechanism 619
   electrodeposit on a copper leadframe 614
   of feed water heater tubes 458
Pitting allowance 974
Pitting attack, from dissolved oxygen 158
Pitting-based crack initiation 863
Pitting corrosion 9, 644, 974
   of active metals 32
   and effects on fatigue 199
   in engine coolants 534
   mechanism of 516
   on posttensioning wires 1066
   of pump cases 1079
   of U.S. Navy Aircraft 191
Pitting corrosion metrics 198
Pitting factor (PF) 116, 119, 952–953
Pitting potential 843
Pitting resistance equivalent nitrogen
   number (PREN) 774–775
Pitting resistance equivalent number
   (PREN) 16–17, 463, 810, 857, 969, 989
Pitting tests, galvanic current 116
Plain-carbon steel, dust-free flue
   gas corrosion 495
Planar and wavy slip 401
Plane bending test 867
Plant environment 652
Plaque 903
Plaque buildup 900
Plasma 895
Plasma spray 63
Plasma vapor deposited coatings 258
Plastic interposer 629
Plastics
   in ammonia and ammonia compounds 733
   in caustic sodas 719, 720
   in hydrofluoric acid 697
   nitric acid corrosion of 672
   in petroleum production operations 931
   in phosphoric acid 739–740
   as source of corrosion 281
   temperature limits for, in ammonia 734
   used in marine vessel wastewater systems ... 274
Plastics decomposition products, effects of 205
Plastic strain accumulation 222
Plating, for firearms 258
Plating-associated equipment 1074
Platinum and platinum-base alloys, general
   anhydrous hydrogen fluoride (AHF)
      corrosion of 702
   hydrogen fluoride corrosion of 696
   in phosphoric acid 739
   (Pt-Ir) 837
   in supercritical water 233
Platinum and platinum-base alloys,
   specific types (P04995) 696
PLEDGE model 351
Polar front 246
Polarization and impedance data, for
   cobalt-chromium-molybdenum
   alloys 915
Polarization cells 122–123
Polarization curves
   of alloys 842
   of amalgams 861
   anodic 672, 743
   of cobalt-chromium-molybdenum alloys 915
   cyclic 842
   of solders 916
   potentiodynamic 855
   and vulnerability 856
   of zirconium in nitric acid 672
Polarization resistance 586
Polishing and passivation techniques,
   using chromium/iron ratios 19
Pollutants 290
   effect of 37–38
   in museums and collections 280
   reaction products of 644
Polluted seawater 38
Polluting gases 280
Polyacetal (PA) 941
Polybutylene 273–274
Polyester resins 252, 687
   sulfuric acid corrosion of 666
Polyethylene (PE) 274, 678, 734, 822, 941
   corrosion resistance of in
      hydrochloric acid (HCl) 686
   in dry chlorine gas 706
   in formic acid 676
   hydrogen fluoride corrosion of 697
   in moist chlorine 708
   sulfuric acid corrosion of 665
   used in marine vessel freshwater
      systems 273
Polyethylene/polyvinyl chloride
   (PVC) tapes 1018
Polyisobutylene 733
Polyisocyanurate foam 817
Polymer anodes 74
Polymer electrolyte membrane fuel
   cells (PEMFCs) 505–506, 512
Polymer-encapsulated packages 615
Polymeric coatings 258, 963
Polymeric composites 217
Polymeric materials, outgassing of 631
Polymers 857
   in cardiovascular surgery 856
   in condensing heat-recovery systems 492
   in firearms 258
   food grade 803
   in hydrofluoric acid 697
Polyolefins 740
Polyphenylene sulfide coatings 956
Polypropylene (PP) 253, 665
   in caustic sodas 719–720
   corrosion resistance of in hydrochloric
      acid (HCl) 686
   in dry chlorine gas 706
   hydrogen fluoride corrosion of 697
   in moist chlorine 708
Polypropylene chlorinated polyvinyl
   chloride (CPVC) 252, 253
Polystyrene 733
Polysulfides, and petroleum production
   operations corrosion causes 923–924
Polysulfide sealants 130, 603
Polytetrafluoroethylene (PTFE) 492, 665, 672,
          697, 704, 733, 814
Polythioether 130
Polythioether sealants 178
Polythionic acids 991–992, 977
Polyurethane 212, 721
Polyurethane coatings 172, 932
Polyurethane dispersions 175
Polyurethane foam 817
Polyurethane paint and fungi 212
Polyurethane thermal insulation 1021
Polyurethane topcoats 173
Polyvant diamond compounds 258
Polyvinyl acetate adhesives 283
Polyvinyl chloride (PVC) 254, 665, 734, 941
   in caustic sodas 720
   corrosion resistance of in hydrochloric
      acid (HCl) 686
   decomposition of 208
   hydrogen fluoride corrosion of 697
   sheet lining 24
   for spar fittings 276
Polyvinylidene chloride (PVDC) 209, 665
Polyvinylidene fluoride (PVDF) 143, 492,
          665, 706, 720, 733, 814
PONTIS bridge management system 560
Porcelain
   in phosphoric acid 741
   thermal expansion coefficients of 908
Porcelain fused alloy restorations 891
Porcelain fused to metal (PFM) alloys ... 908–909
Porcelain veneer 892
Porosity 562
Porous coating 860
Porous cobalt-base alloys 860
Porous tantalum 860
Porous titanium foam 860
Portable electronics 643–650
Portable energy sources 501–503
Portland cement 561, 795
Portland cement concrete. See concrete
Post assessment 82–83
Postcleaning procedures 186
Posttensioned concrete 569
Posttensioned grouted tendons 570–571
Posttensioning 1065–1066
Postweld heat treatment
   (PWHT) 347, 979, 996
Potable water 5, 8–11
Potable water tanks 273
   storage tanks 93
Potassium acetate 605
Potassium formate 605
Potassium hydroxide. See caustic potash
Potent environment 1018
Potential gradient distribution, of
   offshore platform support structure 77
Potential map, close-interval 574
Potential measurements, in cathodic
   protection systems 73
Potentials and current 109
Potential testing 573
Potential vs. pH (Pourbaix) diagrams. See
   Pourbaix diagrams
Potentiodynamic corrosion curves 842
Potentiodynamic polarization curve 855
Potentiodynamic polarization tests 424
Potter-Mann-type linear accelerated
   corrosion 377
Poultice corrosion 516
Pourbaix diagrams 267, 307, 308
   for aluminum 617
   and body fluids 854
   for cobalt 856
   described 842
   for iron 231
   for iron in supercritical aqueous solution ... 232
   for iron water system 9
   for nickel and iron 364
   of realkalization 587
   of repassivation 587
   for specific metals 310
   for system iron-water 1056
Powder coatings 177–178
Powdering 517
Power boilers. See also boilers 494
Power generation stations, efficiency in 237
Power plant condensers, corrosion
   mechanisms in 449
Power plants. See also boiling water reactors
   (BWRs); light water reactors (LWRs);
   pressurized water reactors (PWRs);
   specific power plant components;
   ultrasupercritical (USC) power plants 12
   methods to improve efficiency 244
Power production 236
Pozzolans 560, 564
Precast polymer secondary anodes 589
Precious metals and alloys. See also gold and
   gold-base alloys, general; silver and
   silver alloys, general
   in ammonia and ammonia compounds 733
   anhydrous hydrogen fluoride (AHF)
      corrosion of 702

Precious metals and alloys (continued)
  as anode materials 74
  hydrogen fluoride corrosion of 696
  in phosphoric acid 739
  used in fuel cells 509
Precipitation-hardening (PH) stainless steels,
  in caustic sodas 713
Precipitation-hardening grades, in ammonia
  and ammonia compounds 730
Precipitation sizes 407
Precracks 863
Prediction techniques 227
Predictive modeling tool 130
Preferential corrosion
  in the HAZ welds 454
  of type 304L stainless steel layer in a
    composite tube 789
  of type 316L 793
  weld corrosion 766
Prenflo process 441
Preservation strategies, for metal artifacts ... 298
Pressure-boundary steels, in pressurized
  water reactors (PWR) pressure vessels
  and piping 363
Pressure cap 533
Pressure cookers 807
Pressure vessel, inspection of 1040
Pressure washing, of ammunition containers .. 154
Pressurized fuselage drain valves ... 602–603
Pressurized heavy water reactors (PHWRs) .. 376
Pressurized water reactors (PWRs). See also
  irradiation assisted stress-corrosion
    cracking (IASCC) 339–340, 362–377,
      386, 388, 390, 405, 419
Pressurizer nozzles 369
Prestressed concrete 569
Prestressed steel 569, 571
Pretensioned concrete 569
Preventive maintenance 154
Primary batteries 501
Primary clarifier 794, 795
Primary coolant 362–363
Primary passivating potential 843
Primary-pressure boundary, iron-nickel
  alloys in 364
Primary side stress-corrosion cracking (SCC)
  initiation time for 339
  in pressurized water reactors (PWR) 367–375
Primary water chemistry mitigation, in
  pressurized water reactors (PWR) 405
Primary water leaks 366
Primers 70, 171–172
Primer testing 176–177
Printed circuit board corrosion 273
Printing wiring board (PWB) 645
Prior corrosion damage, as equivalent
  initial flaw 196
Probabilistic fracture mechanics (PFM) 225
Probability of detection (POD) curves 225
Process equipment 7
Processing steps 866
Process steam 652–653
Process variables, effect of 652–653
Process water reuse 772–773
Producing flow limits 954–955
Producing flow lines 953
Producing wells 953
Production systems 956
Progression marks 864
Prohesion 197, 199, 200
Propellants 257
Properties of water, at critical point 230
Propionic acid 678–679
Propylene glycol 269
Prorosul-10 859
Prostheses 822, 823
Protection methods
  for atmospheric corrosion 1058–1060
  for cementious systems 1060–1062
Protection potential 843
Protective coatings 144–145, 1061
  aboveground storage tanks (ASTs) 92–93
  and corrosion 262

  in petroleum refining and petrochemical
    operations 1002
  ultraviolet (UV) degradation of 142
Protective finishes 601
Protein adsorption 841
Protein moiety, of in vivo fluids 883
Proteins (dental) 902–903
Prostheses. See also biomaterials; implants
  modular tapers from 828, 1093
Protic solvents 750–752
Protocols, for retrieval studies 872–873
Proton-induced creep 403
Prototype testing 149
Proving ground testing 538
Pseudogilding 311
PTFE (polytetrafluoroethylene). See
  polytetrafluoroethylene (PTFE)
Public playground equipment 260–262
Pulp and paper industry 255
  ancillary equipment 767
  corrosion in 338
  environmental issues 763
  and growing use of duplex
    stainless steels 774–775
  odor removal units and sources 793
Pulp processing, and chemical recovery .. 762–763
Pulp production 762
Pulpstones 768
Pulsed eddy-current inspection 1051
Pulse rectifiers 103
Pumps 274
  cast stainless steel alloys for 775
  in finishing shops 1074
  galvanic compatibility of components 1080
  and piping systems 1078–1081
  stainless steels for 783
Pure silver palladium, color change of 914
Purple plague 205
PVDF equipment, in moist chlorine 708
Pyrite 313
Pyrite disease 315
Pyromorphite 292
Pyrophoric carbon 269
Pyrrhotite 313

Q

Qualified products list (QPL). See 137
Quality assurance (QC) program 152
Quality assurance representatives (QARs) ... 153
Quality standards for water 15–16
Quantitative colorimetry techniques 913

R

Racks 1074
Radiation
  effects of, on corrosion potential of
    stainless steels 392
  on high-level waste (HLW) containers 422
Radiation creep 386–387, 403–404
Radiation effects 387
Radiation fields 363, 364, 365
Radiation hardening (RH) 387, 406
Radiation-induced creep relaxation 390, 403
Radiation-induced hardening 375
Radiation-induced microstructural features .. 398
Radiation-induced segregation
  (RIS) 387, 393–398, 406
  vs. dose 388
  removal of 400
Radiation swelling 390
Radiators 533, 534, 1091–1092
Radioactive materials 421
Radio frequency holiday short detection
  surveys 556
Radiographic inspections (RI) 1049–1050
Radioisotopes 364–365
Radiolysis 391

Radiorespirometric method 215
Rainbow color 632
Rainbow stains 632
Rain-erosion coatings 173
Rainwater 605
Randles circuit 847
Rankine cycle 156, 236–238, 439
Rapid-etching (lean alloy) structure 764
Rare earth elements 507
Rate-determining step (RDS) 874
Raw water systems 269
RCF (retirement for cause) 225
R-curve methods 868
Reactive diluents 175
Reactive ion etching, aluminum 616
Reactive metals, hydrogen fluoride
  corrosion of 696
Reactor coolant, zinc in 349
Reactor pressure vessel (RPVs) 346
Reactors 754
Reactor vessels 1081
Reactor water cleanup (RWCU) 341
Reagent-grade phosphoric acid 743
Realkalization 587, 590–591
Real-time corrosion measurement
  and monitoring 1031–1034
Real-time radiography 1050
Real-time scales 543
Reamed holes 780
Rebar. See reinforcing bar (rebar)
Rechargeable batteries 502
Recirculating pressurized water reactor
  (PWR) steam generators 367
Recirculating water coolant 348
Recovery boilers 785–792
Recreational boats 265
Recreational environments 257–264
Recreational equipment 7
Recrystallized state (RX) 416
Rectification impact assessment (RIA) 611
Red corrosion product 205
Red lead primers 582
Redox potentials 1016
Red plague 205
Red rust hydrated corrosion product 568
Reducing agents 324
Reducing environments 423
Reducing solutions 743
Reduction-oxidation (redox) reactions 8
Reduction rates at crack-tip 352
Reference electrodes 73, 84, 94–95
Refiner pulping process 768
Refinery experience with sulfide
  stress cracking 996
Refinery process equipment construction
  codes 971
Refinery steels 971–972
Refinery wet H2s environment 992
Reflowed solder 630
Reforming units 976
Refractories 165
Refractory alloys, general. See molybdenum and
  molybdenum-base alloys, general; niobium
  and niobium alloys, general; tantalum and
  tantalum-base alloys, general; titanium
  and titanium-base alloys, general; zirconium
  and zirconium-base alloys, general
Refractory alloys, specific type
  alloy 33 (R20033) 662, 670
  Stellite 21 (Co-27Cr-5.5Mo) 821
  Stellite 25 (Co-20Cr-15W-10Ni) 821
  Stellite alloy 33 (R20033) ... 715, 716, 718, 722
Refractory-lined incinerator 165
Refractory linings 252, 1003–1004
Refrigerated liquid chlorine 706
Refuse-derived-fuel units (RDF) 482, 483
Regeneration towers 979
Regenerative thermal oxidizer 255
Regulations 803
  risk-informed 357
Reheaters 479–480
Reheater tubes, and superheater tubes 237
Reinforced phenolics 678

Index / 1129

Reinforced thermoset plastics 687
Reinforced thermosetting plastics, hydrogen
    fluoride corrosion of 697
Reinforcement/reinforcing steels. *See*
    reinforcing bar (rebar)
Reinforcing bar (rebar)
    chloride ion induced corrosion 797
    in concrete 1056
    corrosion in building columns 1063–1064
    corrosion of 132, 794
    corrosion-resistant 575–580
    expansive corrosion products from 567
    galvanized 580
    modes of corrosion 566–569
    polarization decay 589
    post-cathodic protection (CP) potential data ... 587
    potentials 573
    stainless steel rebar 132
Relative corrosion rate (R/R^o), with
    supercritical water oxidation (SCWO) ... 234
Relative humidity (RH) 279, 614
    for artifacts 286
    critical 42–43
    and temperature effect on mean time to
        failure (MTTF) 615
Released copper 916
Reliability of equipment 221
Remaining pipeline life 82
Remote inspection technologies 145–146
Removable partial dentures 891, 892
Repassivation
    of stainless steels 21–22
    of titanium 831
Repassivation potentials 843
Repatination 299
Replace when it fails approach 967
Repository design types 422
Residence time study 627
Residual elements (RE) 690
Residual fuel oil (Bunker C fuel) 985
Residual heat removal (RHR) 341
Residual stress distributions 226
Resistance of alloys
    and corrosion 132
    in hydrated salt mixture 790
Resistive couplings 108, 112
Resistivity
    vs. chlorinity 76
    of concrete 568
    and corrosion rates 575
    of soils 933
Resistor rectifiers 122, 123
Restorations 897–899
Retaining ring corrosion 497, 1093
Retarders 563
Retrieval analyses 827, 843
Retrieval studies
    related to fatigue and corrosion
        fatigue 871–873
    related to stress-corrosion cracking
        (SCC) 882–883
Retrogression and reaging (RRA) ... 132–133, 225
Return activated sludge (RAS) 794, 795, 796
Reverse current switches 112
Reversing direct current potential drop
    technique 431
Revision surgery 827
Rhodium
    in dental casting alloys 905–906
    hydrogen fluoride corrosion of 696
    in phosphoric acid 739
Rigid opaque plastic 274
Ringer's solution 862, 868, 896
Risk assessment 1045
Risk assessment matrix 553
Risk-assessment process 1038
Risk-based inspection (RBI) 658
Risk-based inspections 1041
Risk matrices 1043–1044
Risk/threat assessment 80
River discharge 28
Rivet holes 221, 224
RMS Titanic 308, 318

Road deicing salts. *See* deicing salts
Road testing 149
ROAR ratio 881
Rock bolts 1078
Rock salt repositories 424
Rodding 329
Rod-pumped wells 946
Rod rigging 276
Room-temperature vulcanizing (RTV)
    elastometric sealant 130
Room-temperature vulcanizing (RTV)
    silicone adhesive sealants 209
Rotating bending beam test 867
Rotating blades 238
Rouge classification 18–20
Rouging 812
Rozenite 315
RSTREG software 82
Rubber. *See also* elastomers 672
    for linings 769
    in phosphoric acid 739
Rubber/elastomers, maximum temperature
    of caustic sodas 720
Rubber materials, compatibility of, with
    phosphoric acid 740
Rubber seals 206
Ruby 702
Running rigging 277
Runoff rates 290
Runway deicers, at airports 605
Rust 295
    contaminants in 57
    in playground equipment 261
    on steel/cast iron components 541
Rust color stain on semiconductor
    integrated circuits (SICs) 632–633
Rust converters 300
Rusticles 308
Rust morphology 541
Rust passivators 963
Ruthenium
    in dental casting alloys 905–906
    hydrogen fluoride corrosion of 696

S

Sacrificial anode (galvanic anode) 1022
Sacrificial anode cathodic protection
    (SACP) systems 584–585, 588–590, 1022
    grid pattern of 95
    for offshore platforms 936
Sacrificial anodes 73–74, 268, 933
    compositions for offshore pipelines
        and structures 936
    in seawater 960
Sacrificial anode systems 102
Sacrificial coatings and galvanic corrosion ... 546
Sacrificial magnesium anodes 145
Sacrificial zinc anodes 145, 247
SAE J2334 tests 148, 149
Safe inspection interval (SII) 224–225
Safe-life approach 195, 224
Safe-life design methodology 201
Safety
    in anticorrosion design 248
    of dental alloys 892
Safety-critical parts 224
Safety equipment for rock climbing 262–263
Safety factor 201
Sail tacks 276
Saline solutions 862
Salinity 28–29
Saliva 861, 895–897, 900–903
    with sulfide 913
Salivary fluids 904
Salt, effect of 540
Salt fog test 621
Salts, mixed 742
Salt spray tests 180, 181, 183, 192, 1088
Saltwater, *See also* seawater 923
Salt water intrusion 192

Sampling for solvent selection 326–327
Sandblasting 329
Sandwich core construction 266
Sanitary standards 803
Sanitizing processes 808
Sapphire 698, 702
Scale 511
Scale deposition, in flue gas desulfurization
    (FGD) systems 461
Scale growth rate 166
Scale morphology 442
Scale precipitation 36
Scales 653
Scale spallation 444–445
Scaling 466
Scaling error 542
Scanning electron micrographs (SEM),
    of orthopedic implants 829
Scanning electron microscope, showing
    corrosion fatigue cracking 221
Scanning electron microscopy (SEM)
    images 838
Scanning electron microscopy with energy
    dispersive x-ray spectroscopy
    (SEM/EDS) 483
Schoengliesite 313
Schwanz screws 856
Scratch test 832–833
Scrubber. *See* flue gas desulfurization (FGD)
Scrubber system arrangement 252
Scrubber systems, for NO_x control 254
Scrubbing systems 770
Sea-cliff climbing, titanium for 264
Seacoast environmental corrosion 635
Sealants 173, 178, 189, 603
    and crevice corrosion 545–546
Sealers, types of 583
Sealing compounds, for U.S. Navy
    Aircraft 190–191
Sea salt 1055
Seasonal temperature changes 653
Seawater 5, 27–38
    cathodic protection (CP) and fatigue
        performance of alloy steels in 961
    corrosion fatigue crack initiation 960
    galvanic series of metals in 128
    solubility of calcium carbonate in 934
    zones of steel corrosion 957
Seawater system foulants 14
Seawater velocities 40
Secondary batteries 501
Secondary caries 861
Secondary clarifiers 794
Secondary cooler 362
Secondary ion mass spectroscopy (SIMS) ... 904
Secondary mineralization 580
Secondary recovery operations, petroleum
    production operations 953–955
Secondary side intergranular stress-corrosion
    cracking (IGSCC) 378
Secondary water chemistry 376
Segmental post-tensioning 569
Segment value creep 62
Segregation profiles in irradiated stainless
    steels 395
Selective catalytic reduction (SCR) 254, 461
Selective phase corrosion 764
Self-contained paint applicator pen 177
Self-healing event 637
Self-priming topcoat (SPTs) 173
Semiconductor integrated circuits
    (SICs) 629–632
Semiconductor wafer fabrication 623–628
Semimechanical pulping 762
Semi-thermomechanical pulping (CTMP) ... 768
Sempen 177
Sensitized stainless steel 343, 991–992
    crack-propagation rate data for 345
    as a function of stress intensity 345
    grain boundaries 347
Sensory foils 227
Sequestering agents 317
Serviceability plots 968

1130 / Reference Information

Service environment limits . . . 930
Service life
and aging, of military equipment . . . 217
of automotive exhaust systems . . . 519
extension of . . . 6
Service oxidation temperature . . . 520
Service water systems . . . 12–14
Severin gage . . . 699
Sewage sludge . . . 253
Sewer slopes . . . 24
Shadow corrosion . . . 418
Shakudo alloy . . . 284
Shape memory NiTi alloys. *See also* titanium and titanium-base alloys, specific types . . . 844
Sheet linings vs. solid plate . . . 463
Sheet metals . . . 516
Shell process . . . 441
Shibuichi alloy . . . 284
Shipboard coatings . . . 69, 129
Shipboard incinerators . . . 166
Shipboard turbine blades . . . 165
Ships . . . 629
Ship systems corrosion . . . 128
Shock waves . . . 470
Shot peening . . . 372
Shutdown cooling (SDC) . . . 341
Side-drain potentials . . . 87
Siderite . . . 312, 924, 925
Sigma-phase embrittlement . . . 973
Silica . . . 16
Silica evaporation . . . 508
Silicate and silicones . . . 536
Silicates . . . 11, 176
Silicate-silicone coolants . . . 531
Silicon . . . 757
Silicon bronze
on marine use vessels . . . 265–266, 275
Silicon content . . . 764
Silicone contamination . . . 605, 1089
Silicone elastomers . . . 706
Silicone-modified polyester (SMP) coatings . . . 143
Silicon oxide barrier film . . . 174
Silicon stainless steels . . . 662
Silver and silver alloys, general
in acetic acid . . . 678
in amalgams . . . 861, 904–905
in ammonia and ammonia compounds . . . 733
anhydrous hydrogen fluoride (AHF) corrosion of . . . 702
in caustic sodas . . . 719
corrosion of . . . 280
in dental solders . . . 910–911
hydrogen fluoride corrosion of . . . 696
in phosphoric acid . . . 739
structural degradation in . . . 509
and sulfur-containing gases . . . 640
Silver and silver alloys, specific types, (P07020) . . . 696
Silver artifacts . . . 311
Silver-base solder . . . 910–911
Silver bowls . . . 281
Silver electrochemical migration . . . 649
Silver-indium alloys . . . 916
Silver-palladium alloys . . . 914
Silver plating . . . 205–206, 207
Silver solders . . . 916
Silver sulfide dendrites . . . 281
Silver sulfide whiskers . . . 208
Silver tarnish . . . 640–641
Silver tubular sections . . . 509
Simulated acidified water (SAW) . . . 428
Single chip modules (SCM) . . . 645
Single leg stunt fracture (SLF) . . . 883
Single-particle high-copper amalgam . . . 904–905
Single-wafer metal-etch systems . . . 625
Siphon break . . . 274
Site, corrosion rates by . . . 48
Site variability . . . 47
Skyflex . . . 130
Slag-forming compounds . . . 986
Slag particles . . . 815
Slakers and causticizers . . . 782
Slip bands . . . 863
Slip-dissolution model . . . 875, 877
Slip localizing . . . 406
Slip-oxidation . . . 351–352
Slip steps . . . 863
Slow-cycle interruption . . . 86
Slow-strain-rate . . . 349
Slow strain rate technique . . . 424
Slow strain rate testing . . . 353, 878–879, 931
Slug flow . . . 1000
Slushing compounds . . . 963
Small particle erosion . . . 242
Smart pigs . . . 953, 1023
Smelt dissolving tank . . . 782, 787
Smelt spouts . . . 791
Smithsonite . . . 311
Smokeless powders . . . 258
Snakeskin . . . 458
S-N curves . . . 863
SO_2/salt-spray tests . . . 199
Soda ash . . . 723
Soderberg law . . . 867
Sodium acetate trihydrate . . . 284
Sodium benzoate . . . 535
Sodium carbonate. *See* soda ash
Sodium chloride concentration . . . 27
Sodium chloride deicing salts . . . 523
Sodium hydroxide. *See also* caustic sodas . . . 470
Sodium hypochlorite sanitizing compounds . . . 808
Sodium vanadate corrosion . . . 986
Soil characteristics . . . 89, 115–118
Soil contact and aboveground storage tanks (ASTs) . . . 93
Soil particle size . . . 308
Soil pH . . . 89
Soil resistivity . . . 23, 89, 91
Soil-side corrosion control . . . 92–93
Soil testing . . . 23
Solar energy . . . 933
Solder-alloy joints . . . 911
Solder alloys. *See also* dental solders
62Sn36Pb2Ag solder . . . 648
90Pb10Sn solder . . . 648
96.5Sn-3.0Ag-0.5Cu solder alloy . . . 641
gold (450) . . . 916
silver solders . . . 916
Sn3.5Ag0.7Cu solder . . . 648
Solder bloom
in engine coolants . . . 534
on radiators . . . 534, 1091
Solder fluxes . . . 634
Solder-flux residue cleaning . . . 638
Solder flux residues . . . 205
Solder reflowing . . . 205
Solder splices . . . 272
Sol-gel formulations . . . 174
Solid-gas interaction corrosion . . . 507–509
Solid inhibitors . . . 937
Solid oxide fuel cells (SOFCs) . . . 257, 506, 513
Solid-particle erosion . . . 469, 471
Solubility
of calcium carbonate and magnesium oxide . . . 936
of oxygen in seawater . . . 33
Solubility tests . . . 327
Solution chemistry analysis . . . 843
Solution conductivity . . . 390
Solution heat treatment . . . 354
Solutions (liquids), corrosion testing in . . . 827
Solventborne paint systems . . . 517
Solvent cleanup . . . 626
Solvents, organic . . . 750–752
Solvent selection . . . 326–327
Solvent welding . . . 941
Sound waves . . . 1048
Source hardening model . . . 399
Sour crude oils . . . 980
Sour gas . . . 961–962
Sour gas wells . . . 959
Sour pitting corrosion . . . 924
Sour-service serviceability chart . . . 959
Sour-service severity regions . . . 961
Sour water . . . 976, 995, 1001
South American muffler . . . 523
Southern convergence . . . 246
Soxhlet method . . . 575
Spallation . . . 242
Spalling
of concrete . . . 567, 1063
of plasma sprayed coating . . . 788
on precast concrete panels . . . 248
Specified minimum yield strength (SMYS) . . . 81
Specimen exposure time and atmospheric corrosion testing . . . 51
Spectral camouflage . . . 151
Spent nuclear fuel (SNF) . . . 421, 422
Spinel oxides . . . 363, 365
Splash/spray simulation . . . 540
Splash zones . . . 69, 957
Splice joints . . . 221
Spray corrosion tests . . . 199
Spray deaerators . . . 452
Spray dryers . . . 254
Sprayed ceramic coatings . . . 480
Sprayed metal coatings . . . 480
Spray or salt fog solution exposure . . . 539
Spray primers . . . 517, 518
Spray/splash simulation . . . 542
Spreaders . . . 275
Squeeze treatment . . . 949–950
Stabilization specifications . . . 527
Stacking fault energy . . . 400–401
Staged combustion . . . 477
Stagnant conditions . . . 213
Stainless steel corrosion . . . 274
Stainless steel pipe . . . 657
Stainless steels, general. *See also* cast stainless steels . . . 325, 424, 579
300-series stainless steels . . . 13, 522, 655
400-series stainless steels . . . 265, 519, 522, 675, 677
in acetic acid . . . 677
aluminum coated stainless steels . . . 521
in ammonia and ammonia compounds . . . 730
argon-oxygen decarburization process (AOD) . . . 811
biocompatibility of . . . 820
in brewery industry . . . 804
in bridges and highways . . . 579
carburization in . . . 1069
in caustic sodas . . . 712, 713, 714
chemical analyses for . . . 49
chemical composition of, in biomedical applications . . . 837
and chlorides . . . 16–17
chromium/iron ratio . . . 17–18
cleaning of . . . 21–22
coatings for . . . 731
for combustor and turbine sections . . . 487
composition of, for suction rolls . . . 779
compositions of . . . 969
in cooking utensils . . . 805
corrosion loss of . . . 445
corrosion of, under insulation . . . 655–656
corrosion problems in . . . 807
corrosion rates of . . . 442, 712, 744
corrosion resistance of . . . 426
crack propagation rate for . . . 349
cyclic oxidation test for . . . 520
cyclic polarization curves of . . . 842
described . . . 578–579
in dry chlorine . . . 704, 705
effect of radiation on corrosion potential of . . . 392
fatigue strength of . . . 868
in feed water heaters (FWH) . . . 457
in firearms . . . 258
for flue gas desulfurization (FGD) . . . 464
in food industry . . . 804
in formic acid . . . 675, 676
high fluence . . . 393
in high-temperature environments . . . 755
in hot salt . . . 521
intergranular stress-corrosion cracking (IGSCC) in . . . 342
intragranular corrosion of . . . 977
localized corrosion of metallic biomaterials in . . . 844
in marine environments . . . 1060

Index / 1131

for marine hull use...................265
in metallic biomaterials............826, 857
and microcomposite alloys..........578–580
molybdenum content of..................778
nitric acid corrosion of................668
in organic acids.......................679
organic contamination of...............815
in orthodontic wires...................910
orthopedics...........................856
oxidation of..........................755
oxidation resistance of...............1068
passivation of........................811
in pharmaceutical industry............812
in phosphoric acid................736, 737
PH stainless steels...................713
in pressurized water reactors (PWR)....363
in propionic acid.....................679
as reinforcement in concrete.....1060–1061
repassivation of....................21–22
in seawater...........................214
selection of..............803, 805–806
sensitization of..................991–992
for spar fittings.....................276
stress-corrosion cracking (SCC)...876, 880–881
sulfate-reducing bacteria (SRB) growth on....307
in sulfuric acid and nitric acid mixtures......747
superferritic steels..............712, 730
surface film of.......................839
surface finishes......................807
testing duration of in formic acid....675
testing of............................868
thermal stabilization of..............992
treatments for........................258
types of..............................806
used for turnbuckles..................277
used in marine vessel systems.....273–274
used in outdoor environments..........296
used in U.S. Navy Aircraft............185
in waste water systems (WWS)..........25

Stainless steels, specific types
3RE60 (S31500)...................767, 989
7-Mo stainless steel..................713
18Cr-15Ni stainless steel.............805
20-type stainless steel alloys........676
22Cr-13Ni-5Mn (S20910)................881
29-4-2 (S44800).......................736
304L/CS...............................791
2507 (S32750).........................989
A610 (1815LC).........................670
A610 (S30600).........................670
A611 (S30601).........................670
AL 29-4-2 (S44800)....................712
Al 29-4C (S44735)................450, 712
alloy 26-1 (S44627)..............677, 722
alloy 27-7MO (S31277).................737
alloy 309 (S30900)....................479
alloy A-286 (66286)...................374
alloy MA 956 (S67956).................755
ASTM F 1314 alloy...............858, 868
ASTM F 1586.....................826, 868
ASTM F 2229..........................830
cast 304 (CF-8).......................812
cast 316 (CF-8M)................812, 1069
cast 316L...........................20–21
cast 410 (S41000).....................779
Custom 465 stainless steel............601
duplex 3RE60 (S31500).................730
duplex alloy....................2101, 579
duplex alloy 2205 (S31803)............579
E-Brite (S44627).....712, 713, 722, 1067–1068
Fe-20Cr-4.5Al-Ti-y (S67956) MA 956....442
Ferralium 255 stainless steel....270, 713
ferritic stainless alloy 3Cr12 (S41003)....579
ferritic stainless alloy CRS 100......579
Monit (S44635).......................712
Nitronic 50 (S20910) (22Cr-13Ni-5Mn)...270, 276
Nitronic 60 (S21800)..................579
PH 13-8Mo............................600
S31725...............................214
S32615...............................662
S38815...............................662
S44660...............................214
S44735...............................214

Sea-Cure (S44660)..........450, 457, 712, 722
stainless heat-resistant alloy AK 18SR.......521
stainless steel alloy.................409, 521
stainless steel alloy.................439, 521
superaustenitic stainless steel alloy 20Mo-6...276
superferritic steel AL29-4-2..............730
superferritic steel AL29-4C...............730
superferritic steel S44635................730
superferritic steel S44660................730
superferritic steel S44735................730
superferritic steel S44800................730
SUS 304 (Japanese standards)..............869
SUS 329J4L (Japanese standards)...........869
titanium stabilized 439LT stainless steel.....520
TRIP stainless steel......................882
(S31803)..................................766
(S32906)..................................722
(S34700)...................................49
15-7Mo PH (S15700)...................675, 768
17-4PH (S17400)......................270, 374, 776
17-7PH...............................600, 821, 883
18-8.............................268, 270, 276, 714
201......................................142
253MA (S30815)...........................484
254 SMO (S31254)...270, 276, 714, 737, 785, 948
255 (S32550) (Ferralium 255).............270
301 stainless steels.....................699
302......................................270, 276, 857
303.....................................265, 821
303Se....................................265
303 stainless steels.....................699
304 (S30400)..........142, 158, 214, 239, 251,
    254, 255, 256, 265, 267, 269, 270, 276, 296,
    342, 346, 348, 349, 354, 355, 375, 390, 395,
    406, 425, 450, 457, 479, 508, 509, 510, 579,
    654, 655, 660–661, 668, 669, 670, 675, 677,
    679, 682, 691, 699, 712, 713, 714, 729, 730,
    731, 736, 743, 759, 760, 763, 770, 781, 782,
    783, 805, 806, 807, 810, 857, 872, 876, 881,
    943, 953, 968, 975, 977, 979, 980, 983,
    988–989, 991, 992, 1001, 1002, 1067, 1068
304L (S30403)..........362, 426, 520, 529, 668, 669,
    670, 677, 699, 713, 714, 730, 736, 763,
    764, 765, 766, 767, 770, 775, 778, 780,
    782, 783, 788, 789, 793, 806, 876, 956
304L (UNS S30403)....................214, 655, 705
304 NAG (UNS S30403 NAG)...............670
304N stainless steel...................457
304 nuclear grade stainless steel......354
309 (S30900).........50, 484, 700, 755, 765, 766,
    789, 1067, 1069
309Cb (UNS S30900)....................691
309L (S30980).........................764
309S..................................520
310 (S31000)......49, 442, 444, 445, 484, 506,
    508, 700, 706, 731, 755, 783, 1067, 1068, 1069
310H (S31009).........................792
310L (S31050).........................670
310S..................................730
312 (S31200)....................479, 764, 765
316 (S31600).............15, 214, 215, 251, 253,
    254, 266, 267, 268, 269, 275, 276, 296, 348,
    349, 354, 375, 390, 395, 406, 425, 450, 506,
    508, 579, 655, 660–661, 662, 675, 679, 682,
    683, 691, 699, 700, 705, 714, 730, 736, 737,
    743, 755, 769, 770, 781, 782, 783, 794, 806,
    821, 857, 953, 956, 960, 961, 962, 968, 977,
    979, 980, 985, 1061, 1068
316L (S31603)............17, 18, 20, 25, 214, 232,
    255, 270, 338, 424, 655, 669, 675, 676,
    677–678, 679, 713, 714, 736, 737, 738,
    763, 764, 768, 770, 773, 774, 775, 778, 782,
    783, 784, 793, 795, 805, 806, 810, 811, 812,
    820, 826, 828, 829, 837–838, 842, 843, 844,
    845, 847, 856, 857, 858, 865, 866, 868, 869,
    871, 872, 873, 877, 880, 882, 901, 956
316L (S31673)....................714, 826, 857, 877
316L molybdenum grade.................669
316LN (S31725)........................579
316L NAG..........................670, 699
316Ti (S31635)........................520
317 (S31700).........50, 377, 661, 737, 796,
    977, 983, 985

317L (S31703).....214, 251, 655, 675–676, 737,
    738, 747, 769, 770, 775,
    778, 783, 872
317LMN (S31726)..................462, 770
318L (S31803)........................214
321 (S32100)..........346, 655, 668, 669, 675,
    730, 755, 981, 992
329...............................457, 989
347 (S34700)............49, 346, 354, 376, 655,
    668, 669, 675, 700, 723, 731,
    755, 981, 1002, 1067
347H (S34709)........................792
347NG................................354
405 (S40500)......376, 377, 736, 968, 975, 976,
    978, 980, 989, 1002
405 ASME SA268-TP410 (S40500).........480
409 (S40900)..........376, 377, 525, 527, 529,
    712, 736
409Ni aluminum coated stainless steel.......520
410 (S41000)...........374, 376, 377, 683, 713,
    767, 956, 968, 976, 978, 1002
410L.................................968
416..................................426
420..................................821
430 (S43000).........683, 712, 736, 782, 989
430M stainless steel.................527
430 stainless steels...........509, 690
434 (S43400).........................712
439..................................457
444 (S44400)....................712, 806
446 (S44600)..........506, 713, 1067–1068
654 SMO (S32654)...............270–271, 737
745..................................857
2205 (S31803)..........270, 462, 713, 730,
    775, 806, 976
2205 (S32205).....764, 767, 770, 774, 776, 782,
    783, 796, 810, 948, 969, 989, 1061
2304 (S32304)...................767, 782
2507 (S32750)..........462, 676, 764, 767, 768,
    948, 969, 976
AL409 stainless steel.......520, 524, 525, 526
AL409 stainless steel alloy..........529
CF3M.................................699
CF8M.................................699
S31803...............................714
S44626...............................713
S44627..........................712, 730
S50300...............................730
S50400...............................730
T304.................................521
T409.............524, 526, 527, 528, 529
T439.............524, 525, 527, 529
T441.................................522
Uranus 505...........................713
XM19 (S20910)........................579
XM-27 (S44627).......................712
XM29 (S24000)........................579
Stainless steel-silver soldered joint corrosion..916
Stainless steel weld overlays......764–765, 766
Sta-Lok.............................277
Standard batch treatment.............940
Standard electrode potentials.......617
Standard hydrogen potential (SHE)......9
Standards, see also under ASTM, see also under
    ISO, see also under NACE for surface
    preparation........................69–70
Standing rigging....................276
Stand package..................498, 1093
Stannous chloride...................328
Static fatigue......................994
Static stray currents................92
Stationary data loggers..............86
Station ground mat..............122, 124
Stator coil cooling water...........498
Stay currents......................1022
Steam chemistry................457, 474
Steam cleaning......................330
Steam cycle....................236–237
Steam generators (SGs), See also
    boilers...................375–377, 379, 381
Steaming vessels....................767
Steam-injected cleaning method...323–324
Steam piping and headers............237

1132 / Reference Information

Steam purity ... 472
Steam-side stress-corrosion cracking (SCC) ... 451
Steam turbines ... 469–474
    extraction steam chemistry ... 457
    superalloy bolt materials used in ... 244
    turbine blade alloys ... 240
    turbine blades ... 165
Steam turbine system ... 439
Steel and ductile iron, in waste water systems (WWS) ... 24
Steel anodes ... 591
Steel blistering ... 698–699
Steel cracking ... 699
Steel elements, inspection of ... 575
Steel hull boats ... 266
Steel masts ... 275
Steel reinforcement, corrosion of ... 132
Steel roofing ... 1059
Steels
    blistering of ... 698
    corrosion of under insulation ... 654
    cracking of ... 698
    in dry chlorine ... 704
    in moist chlorine ... 707
    sulfuric acid corrosion of ... 659
Steels, general. *See also* alloy steels, general; carbon and low-alloy steels, general; cast steels, specific types; high-strength low-alloy (HSLA) steels; stainless steels, general; weathering steels
    in acetic acid ... 677
    in ammonia and ammonia compounds ... 727
    in atmospheric pollutants ... 1054
    buried ... 89
    in cementious materials ... 1056–1057
    in commercial aviation ... 600
    with corrosion inhibition ... 961
    corrosion of, in caustic sodas ... 711
    corrosion rates of ... 48, 426, 924, 1028
    corrosion rates of, in concrete ... 560
    in corrosive media ... 1073
    delamination in ... 997
    effect of chromium additions on atmospheric corrosion of ... 49
    in hydrogen service ... 729
    mechanical properties of ... 971–973
    operating limits in high-temperature high-pressure hydrogen service ... 998
    in organic acids ... 679
    pin or hangar inspection ... 575
    in propionic acid ... 678
    reaction with cementious materials ... 1054
    saltwater corrosion rate of ... 923
    in seawater ... 923, 957–958
    silicon content of ... 764
    in structures ... 1058
    sulfate-reducing bacteria (SRB) growth on ... 307
    testing duration of, in formic acid ... 675
    thermal spray coatings for ... 62
    variation in corrosion ... 47
    weldability of ... 973
    zone of corrosion in seawater ... 957
Steel volumes, and corrosion products ... 567
Stents ... 857
    material ... 846
    open circuit potential (OCP) shift during deformation ... 830
Stephan pH test curve of plaque ... 900
Step loading ... 879
Step polarization impedance spectroscopy (SPIS) ... 847
Stepwise cracking ... 699, 993
Sterilizing ... 803
Stern-Geary parameter (B value) ... 951–952, 1032–1034
Stochastic approach, to predicting intergranular stress-corrosion cracking (SCC) ... 371
Stone pecking ... 541
Stoneware in phosphoric acid ... 741
Storage, of tubular goods ... 962–963
Storage practices ... 154
Storage tanks ... 6, 89–95, 807

Strain-gage-bridge die corroded metallization ... 637
Strain-induced corrosion cracking ... 341
Strain-life tests ... 196
Strain shielding ... 854
Stray current ... 92, 107–111, 123–125
Stray current corrosion, in pipelines ... 1017
Stray-current effects ... 548
Stray-current flow ... 550
Stray-current pickup ... 103
Stray-current sources ... 103
Stress. *See also* stress corrosion
Stress-concentration factor ... 201, 865
Stress concentrations ... 197
Stress-controlled fatigue test ... 867
Stress corrosion
    of low-pressure turbine discs ... 470
    in nickel-based alloy/boiling water reactors ... 353
Stress-corrosion behavior, of NiCrMoV disc steel ... 472
Stress-corrosion cracking (SCC) ... 13, 159–160, 223, 224, 341, 468, 497, 619, 644, 929, 970, 1018–1019
    accelerated testing ... 879
    alloys affected by ... 655
    of aluminum alloys ... 152, 545
    applicable loading modes for ... 862
    of austenitic stainless steels ... 374, 670
    in austenitic-type stainless steels ... 491
    branching appearances on ... 873
    of carbon steels and caustic sodas ... 712
    caustic embrittlement of steel ... 325
    during chemical cleaning ... 327–328
    in commercial aviation ... 599
    of composite floor tubes ... 789–790
    of condenser tubes ... 451
    conditions for ... 806, 1018
    contributory mechanical/chemical phenomena ... 865
    in dental alloys ... 883
    described ... 873, 974
    in digesters ... 766
    in disk refiners ... 768
    effects of chloride and oxygen on ... 988
    in feed water heaters (FWH) ... 458
    fundamentals of ... 873–878
    of high-level waste (HLW) containers ... 424
    in hydrofluoric acid for nickel-copper alloys ... 695
    hydrogen embrittlement of high-strength steels ... 381
    laboratory test results ... 880–882
    life model for a colony of ... 1019
    in $LiSO_2$ batteries ... 502
    in metallic biomaterials ... 853–884
    mitigation using hydrogen water chemistry (HWC) ... 392
    mixed intragranular/transgranular mode ... 882
    new materials and processing techniques for ... 883–884
    in petroleum refining and petrochemical operations ... 988–992
    in pipelines ... 1018, 1093
    and resistance ... 132
    resistance of alloys to, in hydrated salt mixture ... 790
    of rock bolts ... 1078
    of steam turbine blading ... 471
    in steam turbines ... 469
    stress levels of ... 655
    testing methodologies ... 878–880
    test specimens for ... 931
    time vs. stress ... 696
    in transgranular near-neutral-pH ... 1018
    types and stages of ... 873
    types of ... 988–989
    in vivo testing and retrieval studies related to ... 882–883
    in weapons systems ... 126
    in zirconium alloys ... 668
Stress corrosion cracking (SCC) mechanism ... 655
Stress-corrosion crack-propagation rate ... 349
Stress-corrosion threshold stress ... 472

Stress cracking. *See also* stress-corrosion cracking (SCC); sulfide stress cracking (SSC)
Stress-enhanced ion release (SEIR) ... 853
Stress fatigue ... 353
Stress-fatigue life behavior ... 197
Stress index ... 344–345
Stress intensity ... 864–867
    crack propagation rates as a function of ... 345
Stress-life tests ... 196
Stress-oriented hydrogen-induced cracking (SOHIC) ... 947, 993–996
Stress protection ... 854
Stress relaxation ... 403–404
Stress-relieved state (SR) ... 416
Stress rupture failures ... 653
Stress shielding ... 854
Strip lining ... 1002
Stromeyerite ... 311
Strong acids ... 925–926
Structural (chemical) spalling. *See* spalling
Structural design and fabrication details ... 1058
Structural parts ... 221
Structural prognostics and health management (SPHM) ... 130
Structural repair manual (SRM) ... 606, 608
Structural steels ... 142, 1062–1063
Structure task groups (STG) ... 608
Structure-to-electrolyte potentials ... 109
Stud bolts ... 1080
Studded tubes ... 787
Subarctic and subantarctic climates ... 246
Submerged arc welding (SAW) ... 764
Substation ground mat ... 123
Substations operated diode-grounded ... 550
Substrate effects on integuement characteristics ... 903
Sucker rod failures ... 954
Sucker rods ... 946–948
Suction rolls ... 779–780
Suction roll shells ... 776
Sulfamic acid ... 325, 766
Sulfate attack ... 797
Sulfate contaminants ... 723
Sulfate in wastewater ... 795
Sulfate ions ... 290
Sulfate pulping ... 762
Sulfate reactions ... 565
Sulfate-reducing bacteria (SRB) ... 13, 38, 117, 213, 214, 216, 307, 778, 795, 901, 927, 953, 954, 955, 1017
Sulfate-resistant cements ... 956
Sulfates
    effects of, on buried steel ... 89
    in white water ... 777
Sulfidation
    in heat treating industry ... 1070
    in high-temperature environments ... 759–760
    of Inconel 601 ... 1070
    from low-NO_x burners ... 477–478
Sulfidation attack ... 759
Sulfidation behavior ... 760
Sulfidation tests ... 1070
Sulfide attack (water side) ... 449
Sulfide compounds and tarnished films ... 900
Sulfide corrosion ... 328
Sulfide generation ... 23
Sulfide inclusions ... 866
Sulfide ions ... 89
Sulfides ... 38
Sulfide scale ... 985
Sulfide scavengers ... 280, 945
Sulfide stress cracking (SSC) ... 922, 946–1947, 961, 993, 996
Sulfide stress cracking (SSC) resistance ... 975
Sulfidic corrosion ... 980, 981
Sulfite ions ... 296
Sulfite process ... 768–770
Sulfites ... 777
Sulfonate-base coatings ... 963
Sulfur ... 281, 723, 923–924
Sulfur compounds ... 55
    effect of on corrosion rate ... 981
    in noncondensible gases (NCGs) ... 793

Index

Sulfur dioxide (SO_2) 43–44, 290, 769, 943
Sulfur dioxide emissions 461
Sulfur dioxide service 255
Sulfur emissions 251
Sulfuric acid 659–667, 978
  isocorrosion diagram for austenitic
    stainless steels in 784
Sulfuric acid and nitric acid mixtures 747
Sulfuric acid corrosion 660–662, 664–665
Sulfuric acid corrosion (SAC) 463
Sulfuric acid dew point 494, 495
Sulfuric acid plant 254, 325
Sulfuric/boric acid anodize (SBAA) 174
Sulfuric/hydrochloric acid mixture 744
Sulfurous gases 282
Sulfur oxidizing bacteria (SOB) 795
Sulfur trioxide (SO_3) 251
Sulfur trioxide emissions 461
Summary of current situation and commentary
  on the future 356–357
Superalloys and specialty steels, general
  in caustic sodas 714
  in formic acid 676
  and hot corrosion 489
Superalloys and specialty steels, specific types.
  See also alloy steels, general; nickel and
  nickel-base alloys, specific types
  1Cr-1Mo (K11597) ASME SA213-T11 477
  2.25Cr-1Mo (K21590) ASME
    SA213-T22 .... 477, 479
  2.25Cr- 1Mo (K30736) 968, 971
  5Cr-.05Mo (K11547) 477, 478, 981, 984
  18Mn-5Cr alloy (ASTM 289 class B) 497
  18Mn-18Cr alloy (ASTM 289/289M) 497
  300M alloy steel (K44220) 127, 223, 600
  430M chromium alloy 520
  Aermet 100 Co-Ni low-carbon steel 223
  alloy 42 (ASTM F 30) (K94100) 629, 632
  alloy 52 (K95050) 354, 367
  alloy HCM9M 239
  alloy T11 (K11597) 729, 792
  alloy T22 (K21950) 792
  alloy T91 (K90901) 239, 792
  alloy X 759
  alloy X750 .... 367, 373
  Aquamet 22 .... 270, 276
  ASME SA213-T9 480
  D6AC steel (K24728) 198
  ferritic alloy (K21950).................... 713
  (K01800) 427
  (K32050) 730
  (K41545) 729
  (K81340) 730
  superalloy Iconel 718 240
  superalloy M-252 240
  superalloy Nimonic 90, 240
  superalloy Refractory 26, 240
  superaustenitic alloy C-276 (N10276) 769
  superaustenitic stainless alloy N08367 812
Superaustenitic stainless steels 806, 810
  in near neutral chlorine dioxide 773
  in pharmaceutical industry 810
  in phosphoric acid 737
Superaustenitic stainless steels, specific types,
  UR SB8 737
Supercritical conditions 236
Supercritical sliding pressure Benson boiler .. 238
Supercritical thermal power plants
  (SCTPP) 232
Supercritical water 7, 229–230, 233
  corrosion in 240
Superheaters........................... 479–480
  corrosion of 791–792
Superheater tubes
  and reheater tubes 237
  in waste-to-energy (WTE) boilers....... 482–483
Superphosphoric acid. *See* phosphoric acid
Superplasticizers 563
Supplemental oils 181–182
Surface anode method 1062
Surface area effects 862
Surface characteristics, of dental alloys 899
Surface chemistry 474

Surface chemistry analysis 843
Surface coatings and treatments, for
  firearms.............................. 258
Surface corrosion
  of magnesium 191
  and thickness loss 198
Surface damage 222
Surface deposits 467
Surface equipment, pharmaceutical
  equipment 18
Surface finishes 818
Surface finishing, and cracking
  susceptibility.......................... 372
Surface finishings, effect of 757
Surface insulation resistance (SIR) 643
Surface migration 646
Surface morphology and elemental
  distribution in scales 510
Surface-mounted chip resistor, with
  silver dendrites.................. 639, 1086
Surface oxidation 222
Surface preparation, for paint process 517
Surface pretreatment 171
Surface roughness parameters 198
Surface ships, coatings on 129
Surface strip electrical resistance probe
  schematic 116
Surface treatments 188
Surface waters 31
Surfactants 324–325
Surgical implant materials. *See also*
  metallic biomaterials
Surgical implants 857–859
Surgical specialties, in metallic
  biomaterials 855–856
Surveys and testing, predesign 23
Susceptibility index 879
Sustainment life assessment program
  (SLAP) 132
Sweating iron 279
Sweet corrosion 946–947
Sweet gas............................. 960–961
Sweet gas deposit chemistry 792
Sweet gas wells 959
Sweet water corrosion, and accelerated
  corrosion 1030
Swelling tube tests 388
Syngas 441–443
Synthetic condensate pit depth 526
Synthetic veils, with fiberglass reinforce
  plastic (FRP)......................... 255
System pH............................... 1027

T

Tafel slopes......................... 951, 1033
Tall oil plants 782, 784–785
Talurit fittings 276
Tanks. *See also* storage tanks; underground
  storage tanks (USTs)
  accumulator 274
  black liquor 782, 783
  blow 767
  cathodic protection (CP) for 95, 934, 936
  chemical recovery 782–783
  corrosion of 958–962
  corrosion protection for 1081
  corrosion resistance of materials for 1073
  crude oil storage 92
  double bottom impressed current (ICCP) 94
  external aboveground coatings 93–95
  flash 767
  freshwater 273
  green liquor 782, 783
  hot water 274
  organic coatings for 95
  potable water 93, 273
  protective coatings 92–93
  radiator, erosion-corrosion in 534, 1092
  smelt dissolving 782, 787
  for stainless steel pickling 1074

  stainless steels for 783
  steel 1072–1073
  water storage 953, 955
  white liquor 782
Tantalum and tantalum-base alloys,
  general 751, 752
  in ammonia and ammonia compounds 733
  biocompatibility of 820
  in caustic sodas 719
  in dentistry 856
  in dry chlorine 705–706
  in hydrochloric acid (HCl) 684
  hydrogen fluoride corrosion of 696
  isocorrosion diagram for 685
  in moist chlorine 707
  nitric acid corrosion of 672
  pitting and repassivation potential of and
    water concentration 752
  sulfuric acid corrosion of............ 664–665
Tantalum and tantalum-base alloys, specific types
  pure 883
  Ta-2.5W alloy 664–665
  Ta-10W alloy 719
Tape coatings 1022
Tapers 827, 828, 1093
Tape systems 1021
Tarnished leads (terminations) 630, 631
Tarnished leads at the assembly 633
Tarnishing 632, 861
  effect of nobility on 914
  of microorganisms 897
  of silver 640
Tarnishing and corrosion 901, 912–917
  resistance of dental alloys 892–896
Technical-grade phosphoric acid. *See*
  phosphoric acid
Teeth 892–894
Telluric current effects 87
Telluric currents 84, 108, 110–111, 1022
Temper. *See* aluminum and aluminum-base
  alloys, specific types
Temperature 280
  vs. depth........................... 32–33
  effect of 654
  effect of, on corrosion rates 45
  effect of, on corrosivity in
    CO_2/H_2S systems 1028
  effect of, on high-temperature sulfidic
    corrosion 981
  effect of, on nickel and nickel-base alloy
    corrosion 684
  effect of, on nitriding depth 732
  effect of, on relative humidity and mean
    time to failure (MTTF) 615
  and external stress-corrosion cracking
    (ESCC) 656
  of high-level waste (HLW) containers ... 422–423
  and intergranular stress-corrosion
    cracking (SCC) 368
  and oxidation resistance 1067
  and salinity 29–32
  vs. water........................... 32–33
  of white water 777
Temperature, humidity, and bias
  (THB) tests 621, 640
Temperature limits
  for fiberglass-reinforced plastic (FRP) in
    hydrochloric acid (HCl) 687
  for materials in hydrochloric acid (HCl)...... 688
  for plastics in caustic sodas 720
  for thermosets in caustic sodas 721
Temper embrittlement 729, 973
Tendons 570–571
Tenifer molten salt process................. 258
Tensile loop stersses 567
Tensile stress reduction 355
Tensile yield strength, irradiation dose
  effects of 399
Tensioning 1056
Tension-tension standard method 867
Terminals 276
Termination region corrosion 640
Terraces 874

1134 / Reference Information

Terrestrial environments
- artifacts in 312
- copper alloys in 310
- corrosion of metals in 308

Test acceleration factors 543
Test geometries 868
Test loop 953
Tests and testing
- accelerated corrosion testing 148, 149, 181, 199, 538
- accelerated SCC tests 879–880
- accelerated stress testing 621
- accelerated testing 203
- ASTM B 117 salt spray (fog) test 149
- atmospheric corrosion testing 51, 52, 57
- atmospheric exposure testing of panels 61, 1087
- atmospheric testing 23
- atomic force microscopy (AFM) tests 833
- boiling beaker condensate corrosion testing 523
- carburization testing 1069
- of caustic potash 724
- certified test report (CTR) 153
- Charpy V-notch impact test 973, 994
- coating test panel, electrochemical measurement on 63
- comparative testing 149
- complete vehicle testing 538–543
- condensate pitting corrosion testing 523
- constant amplitude fatigue testing 200
- constant extension rate test (CERT) 879
- constant stress-corrosion cracking (SCC) test 790
- copper-assisted salt solution (CASS) tests 200
- core test for inhibitors 949
- corrosion evaluation and testing 148–149
- corrosion fatigue (CF) testing 867–871, 873
- corrosion inhibitors, salt spray tests of slat gearbox with and without 192, 1088
- corrosion lifetime testing 515
- corrosion rate testing 575
- corrosion test coupons 73
- corrosion testing 620–621
- corrosion testing in fluids 827
- corrosion testing in organic acids 674–675
- corrosion testing in organic solvents 753
- corrosion testing in solutions 827
- corrosion testing of alloy 200 in caustic potash 724
- coupon test stations 1022–1023
- crack initiation testing 867
- crack propagation testing 867
- crevice corrosion of muffler test coupon 526
- for current 91
- current transient during scratch test 831
- cyclic corrosion tests 148, 149
- cyclic oxidation test 520
- cyclic polarization testing 842–843
- dental alloys, laboratory testing 903–904
- direct current source influence test 86
- direct observational corrosion test methods 843
- electrochemical scratch test 832
- electrochemical testing 827
- of engine coolants 536
- environmental stress cracking, testing for resistance to 931–932
- external salt pitting tests 524–525
- fatigue crack growth tests 196
- fatigue testing, cycle frequency of 862
- fatigue testing, statistical considerations 867
- ferroxyl tests for free iron 814
- formalized proving ground (PG) test outcome criteria 543
- formic acid 675
- galvanized material testing 66
- GM 9540P tests 148, 149
- high-temperature storage (HTS) test 630
- high test temperatures, effects of 541
- hydraulic pressure testing 807
- hydrogen sulfide testing 23
- hydrostatic testing 1023
- immersion testing 428, 674
- interference tests 109
- irradiation slow strain rate tests (SSRT) 387
- Long Term Corrosion Test Facility (LTFC) 429, 432
- long-term oxidation tests 756
- low strain rate testing, influence of strain rate in 879
- Mandrel Bend test 149
- marine atmospheric corrosion, world test sites 58–59
- marine atmospheric corrosion test panels, composition of 50
- of mechanically assisted corrosion 832–833, 834
- metallic-coated steel sheet test materials 51
- of microbiologically influenced corrosion (MIC) 116
- of modular femoral hip taper interface 833, 834
- neutral salt-fog tests 199
- of nickel and nickel-base alloys 724
- nondestructive testing (NDT) 571
- nondestructive testing (NDT), types of 133
- Oddy test 286
- oxidation tests 1067–1069
- pitting tests 116
- plane bending test 867
- potential and current tests of dc transit system 110
- potential testing 573
- potentiodynamic polarization tests 424
- and process control 604–605
- prohesion spray tests 200
- prohesion tests 199
- prototype testing 149
- proving ground testing 538
- for resistance to environmental stress cracking 931–932
- road testing 149
- rotating bending beam test 867
- SAE J2334 tests 148, 149
- salt fog test 621
- salt spray tests 180, 181, 183, 192, 1088
- scratch test 832–833
- scribed, sealed and painted thermal panels comparison test 62, 1088
- slow-rate testing 994
- slow strain rate testing 353, 878–879, 931
- slow strain rate tests (SSRT) of 431
- SO_2/salt-spray tests 199
- soil tests 116, 117
- soil resistivity testing, four-electrode test method 91
- soil testing 23
- solubility tests 327
- spray corrosion tests 199
- of stainless steels 520, 675, 868, 880–881
- of steels 675
- Stephan pH test curve of plaque 900
- strain-life tests 196
- stress-controlled fatigue test 867
- stress-corrosion cracking (SCC) 878–883
- stress-life tests 196
- sulfate-reducing bacteria (SRB) activity testing 117
- sulfidation tests 1070
- surveys and testing 23
- swelling tube tests 388
- tarnishing and corrosion 912–917
- temperature, humidity, and bias (THB) tests 621
- test acceleration factors 543
- test geometries 868
- testing duration of aluminum in formic acid 675
- testing duration of copper and copper-base alloys in formic acid 675
- test specimens for stress-corrosion cracking (SCC) 931
- track-to-earth test short 557
- of underground pipeline CP system 117
- for underscale resistance of stainless steels 522–523
- uniform corrosion rate tests 115
- vehicle testing 538–543
- vibration testing of tendons 571
- in vitro fretting corrosion test 833
- in vitro solutions for electrochemical testing biomaterials 843
- in vitro testing, intraoral vs. simulated exposure 903–904
- in vivo corrosion testing 843
- in vivo testing, findings related to fatigue and corrosion fatigue (CF) 871–873
- in vivo testing, findings related to stress-corrosion cracking (SCC) 882–883
- in vivo testing, simulation and interpretation 861–863
- of wastewater 23
- wedge test 604
- wet fluorescent magnetic particle inspection and ultrasonic testing 995
- wheel test 938
- witness wafer test 627

Texaco process 441
Textile mills 256
Thermal barrier coatings (TBCs) 163, 489
Thermal cleaning 330
Thermal destruction system 255
Thermal embrittlement mechanism 375
Thermal expansion coefficients of porcelain 908
Thermal fatigue 521, 522
Thermal fatigue cracks 160
Thermal gradients in deposits 445
Thermal insulation 816–818
Thermal insulation materials, service temperatures for 657
Thermal shock failure of glass linings 816
Thermal spray aluminum (TSA) 657
Thermal spray coatings (TSCs) 61–65, 183, 582, 764, 765, 766, 1002
Thermal spray process 62–66
Thermal stabilization, of stainless steels 992
Thermal stress, of glass linings 816
THERMIE program 240
Thermoelectric generators (TEG) 102, 933, 934
Thermography 1051
Thermomechanical pulping (TMP) 768, 794
Thermoplastics 255
- applications of 941
- in caustic sodas 719–720
- corrosion resistance of, in hydrochloric acid (HCl) 686–687
- hydrogen fluoride corrosion of 696–697
- types of 740

Thermoset plastics 740
Thermosets 721
Thermosetting resins 678, 720–721, 734
Thermostat 533
Thickness profile 791
Thimbles and shackles 268
Thimble tubes 415
Thin-film bimetallic sensing element 227
Thin film electrical resistance (ER) probe 117
Thin-film sulfuric acid anodizing (TSFAA) 174
Thin-small-outline package (TSOP) 629
Thiosulfates 777
Thiosulfide pitting 778
Thixotropic grouts 569
Thoracic surgery 883
Three-electrode linear polarization resistance probe schematic 116
Three-electrode soil corrosivity probe 119
Three-layer polyolefin coatings 1021
Threshold fluence 342, 386
Threshold K ISCC 346
Threshold level of chloride ions 560
Threshold potential of environmentally assisted cracking (EAC) 349
Threshold stress intensity 472, 864
Threshold temperatures
- vs. alloy content in caustic sodas 715
- for chloride stress-corrosion cracking 17

THR implants 873
Through wall corrosion 654
Tie and anchor corrosion 1064
Timber artifacts 315
Time-dependant behavior of hydrogen attack 998
Time-of-flight diffraction (TOFD) 1048–1049
Time of wetness (TOW) 42–43, 247
- classification of 55

Index

Time to failure 181–183
  effect of stress level on 344
Time-to-fracture 732
Tin and tin-base alloys, general
  in amalgams 904–905
  in ammonia and ammonia compounds ... 733
  in bronze alloys 294
  in caustic sodas 719
  in dental solders 911
  tin and tin lead alloys (solders), oxidation of .. 630
Tin and tin-lead alloys, electronic
  packaging corrosion 646–647
Tin artifacts 313
Tin chloridehydroide 313
Tin coatings 180
Tin dendrites 639, 1086
Tin electrochemical migration 648, 649
Tin-lead solders. *See also* solder alloys
Tin oxide 630
Tin pest 313
Tin plated contacts 208
Tin plating 206
Tin whiskers 205, 641
Tip cracking 223
Tissue destruction 821–822
Titanium-aluminum-vanadium alloys,
  biocompatibility of 821
Titanium and titanium-base
  alloys, general 275, 325, 706, 775, 948
  in acetic acid 678
  in ammonia and ammonia compounds ... 733
  biocompatibility of 820
  in caustic sodas 718–719
  chemical composition of, in biomedical
    applications 838
  compositions of 970
  corrosion behavior of 432–433
  corrosion fatigue (CF) testing of 869–870
  corrosion resistance of, in hydrochloric
    acid (HCl) 684
  in corrosive media 1073
  crevice corrosion on 424
  cyclic polarization curves of 842
  in dental casting alloys 908
  in dry chlorine 705
  environmentally assisted cracking
    (EAC) of 432–433
  fatigue strength of 870
  in food manufacture and distribution ... 808
  in formic acid 676
  general and localized corrosion of 432
  in heat exchangers 808
  for high-level waste (HLW) containers .. 424–425
  hydrogen embrittlement of 432–433, 994
  hydrogen fluoride corrosion of 696
  intergranular attack on 671
  intergranular stress-corrosion cracking
    (GSCC) of 433
  isocorrosion diagram for 684
  localized corrosion of metallic
    biomaterials in 844
  and mercury 722
  in metallic biomaterials 826, 859–860
  in moist chlorine 707
  nitric acid corrosion of 671
  orthopedics 856
  in pharmaceutical industry 813
  in phosphoric acid 738–739
  prostheses wear particles 848
  repassivation of 831
  in seawater 217
  in steam turbine parts 471
  in stress corrosion cracking testing 881
  sulfuric acid corrosion of 664–665
  surface oxide of 839
  tarnishing and corrosion of 914
  Ti-Al-Fe alloys 846
  of titanium alloys in boiling nitric acid ... 671
  titanium-palladium alloys 424
  titanium tubes 451
  Ti-Zr-Nb 846
  transgranular stress-corrosion cracking
    (TGSCC) of 433
  unalloyed titanium 671
  in U.S. Navy Aircraft 185
Titanium and titanium-base alloys, specific types
  ASTM F 136 877
  CP-Ti 821, 842, 844, 847, 859, 860
  CP Ti 869, 870, 872, 873
  CP-Ti 882
  CP Ti grade 4 (ASTM F 67) 881
  NiTi 828, 839, 842, 846, 856, 860, 881, 882
  NiTi (ASTM F2063) 826, 839
  T Gr 5 1081
  Ti-6Al-2Nb-1Ta 883
  Ti-6Al-4V 599, 831, 839, 842, 843, 844, 845,
    846, 847, 848, 856, 857, 859, 860, 866, 869,
    870, 871, 873, 874, 877, 881,
    882, 901, 1086
  Ti-6Al-4V ELI 856, 870, 872, 873, 877, 883
  Ti-6Al-7Nb 860, 870, 881
  Ti-12Mo-6Zr-2Fe 870
  Ti-13Nb-13Zr 870
  Ti-15Mo 881, 884
  Ti-15Mo beta 870
  Ti-15Sn-4Nb-4Ta 883
  Ti-15Zr-4Nb-4Ta 883
  Ti-29Nb-13Ta-4.6Zr 883
  Ti-35Nb-7Zr-5Ta 883
  Ti-45Nb 1081
  Ti Gr (R50250) 214
  Ti Gr 2 (R50400) 217, 718, 970, 975, 976,
    977, 980, 985
  Ti Gr 7 (R52400) ... 424, 425, 432, 433, 739, 970
  Ti Gr 11(R52250) 739, 1081
  Ti Gr 12 (R53400) 424, 427, 432, 433,
    671, 738–739, 970, 1081
  Ti Gr 16 (R52402) 432
  Ti Gr 17 1081
  Ti Gr 18 1081
  unalloyed titanium 846
Titanium carbonitride 258
Titanium LP blades 474
Titanium ribbons 94
Titanium-sheet linings 462
Tobin bronze 271
Toilets 274
Tolytriazole (TTZ) 536
Tomlinson furnace 785
Tool limitations 98
Topcoats 71, 172–173, 518
Top side coating systems 70
Tortuosity 562
Total acid number (TAN) 983–985
Total hip prostheses 823, 827
Total hip replacement (THR) 832, 854
Total hip replacement femoral components ... 872
Totally chlorine-free (TCF) bleaching .. 772–773
Total reduced sulfur (TRS) 792, 793
Total sulfides 23
Touch-up paints 177
Toxicity
  of cadmium coatings 66
  of hexavalent chromium 822
  of mercury 861
  of metal ions 848
Track resistance 549
Track-to-earth current 552
Track-to-earth gradients 557
Track-to-earth resistance 550, 556
Track-to-earth test short 557
Transformation-induced plasticity (TRIP) ... 868
Transgranular cracking 875
  in anhydrous hydrogen fluoride (AHF) 700
  from stress-corrosion cracking (SCC) 874
Transgranular stress-corrosion
  cracking (TGSCC) 342, 433, 876, 990
Transition fatigue life 863
Transition hot corrosion 488
Transpassive potential 843
Transpiring wall 233
Transpiring wall reactor 231
Transport behavior 561
Transport mechanisms of chloride ions ... 564
Tray deaerators 452
Trevorite 19

Triazoles and thiazoles 536
Trichloroethylene 751–752
Triethanolamine 535
Triglycidylisocyanurate (TGIC) polyester
  powder coatings 189
Trisulfoaluminate (AFt) 565
Tsujikawa-Hisamatsu Electrochemical
  (THE) method 423, 429
TT-P-2756 .... 172, 173
Tube failure mechanisms 653
Tube fouling 366, 458
Tuberculation corrosion 13
Tubesheet corrosion rates 450
Tubesheets 448, 449
Tube shields 484
Tubes in feed water heaters (FWH) ... 457–458
Tube-type air heaters 492
Tube wastage 377
Tubing
  acid phosphate corrosion of 467
  coatings for 932
  in waste-to-energy (WTE) boilers 483
Tubing displacement 940
Tubular reactor 230
Tuccillo and Nielsen tarnishing apparatus 903
Tumorgenesis from dental alloys 894
Tumors 823
Tungsten
  as alloying element 755
  anhydrous hydrogen fluoride (AHF)
    corrosion of 702
  hydrogen fluoride corrosion of 696
Turbines. *See* gas turbines; steam turbines
Turbining 329
Turbulence critical velocity equation 924
Turnbuckles 276, 277
Twin-wire arc spray (TWAS) 765
Two-component epoxy coating 92
Two-component sealants 189
Type A corrosion 442, 444
Type B corrosion 442, 443, 444
Type I high-temperature hot corrosion ... 488
Type I hot corrosion 487
Type II low-temperature hot corrosion ... 488
Tyrode's solution 862, 882

U

Ultrahigh-molecular-weight polyethylene ... 848
Ultrahigh-pressure water blasting 329
Ultrahigh purity water 5
Ultrasonic inspections (UI) 1047, 1048
Ultrasonic thickness measurement 1032
Ultrasonic tools 98, 1023
Ultrasupercritical (USC) power plants .. 236–238,
    240, 242, 244
  coal ash corrosion in 480
Ultrasupercritical water 7
Ultraviolet (UV) degradation 71, 142
Ulu Burun 313
Uncoated incinerator liner 167
Underdeposit acid corrosion and
  hydrogen damage 468
Underdeposit corrosion 13, 328, 516, 953
Underfilm corrosion 515
Underground pipelines 79
  CP system for 117
  external corrosion of 1016
Underground storage tanks (USTs) 89, 95
Underinsulation corrosion 1085
Underinsulation stress-corrosion
  cracking (SCC) 783
Underpaint creepage 540, 541
Underside corrosion 293
Unexploded ordnances (UXO) 217
Ungrounded electric rail systems vs.
  diode-grounded electric rail systems ... 550
Uniform corrosion 515, 644
  of armament systems 151
  in engine coolants 533
  of pumps 1079
  for reactions 616

U

Uniform corrosion rate tests............... 115
Uniform zirconium corrosion............... 406
United States Pharmacopoeia (USP),
    quality standards for water........ 15–16
Unpolarized coupons, corrosion rates of..... 117
UP boilers............................... 236
Urea formaldehyde....................... 285
Urethane coatings....................... 1018
Usage monitoring (UM) systems........... 226
U.S. Air Force. *See also* air force aircraft
    aircraft maintenance philosophy of........ 195
U.S. Army. *See* military specifications and
    standards (MSS); military vehicles
U.S. Army Cold Regions Research and
    Engineering Laboratory (CRREL)...... 246
U.S. Department of Defense............... 126
    Department of Defense Corrosion Policy..... 140
    Department of Defense Index of Specifications
        and Standards (DODISS)............ 139
    Department of Defense Single Stock Point
        (DODSSP)......................... 139
U.S. Navy. *See also* naval aircraft
    aircraft maintenance philosophy of........ 195
    corrosion issues........................ 129
U.S. Navy ships.......................... 129
USP 24 pharmaceutical-grade water......... 16
Utilidors................................. 247
Utility boilers. *See* boilers
Utility relocation........................ 554
Utility relocation funding issues........... 556

V

Vacancy diffusion mechanism............. 394
Vacancy loops........................... 398
Vacuum annealing....................... 994
Vacuum deaeration.................. 942–943
Vacuum-deposited coatings............... 183
Vacuum oxygen decarburization........... 527
Vacuum pumps.......................... 776
Vacuum towers.......................... 975
Valves............................ 775, 783
Vanadium......................... 479, 481
Vanadium hot corrosion.................. 488
Van der Waals-London forces............. 750
Vaperproofing........................... 818
Vapor-phase corrosion inhibitors.......... 620
Vapor-phase inhibitors............. 150, 961
Vapor-phase organic cleaning method..... 323
Vasa................................... 315
Vascular stents.......................... 828
Vehicles, 128. *See also* complete vehicle
    testing
Vehicle testing..................... 538–542
Velocity-accelerated corrosion....... 999–1002
Velocity of wastewater................... 795
Vented loops............................ 274
Venturi tube scrubber separator........... 252
Vertically drilled anode CP systems........ 94
Vertical migration....................... 646
Vertical recirculating steam generators
    (RSGs)......................... 375, 377
Vertivally drilled anodes.................. 93
Very large scale integration (VLSI) circuits... 623
Vessel design for coatings................ 932
Vibration of fatigue cracks................ 160
Vibration testing of tendons.............. 571
Vinyl coating systems............... 1073–1074
Vinyl-ester resins............... 252, 254, 672
Vinyls............................. 71, 740
Visual camouflage....................... 182
Visual inspections..................... 1047
    and delamination survey............... 573
    of stray current........................ 111
    of U.S. Navy Aircraft................... 186
Vitallium implant........................ 908
Viton A................................. 733
Vivianite................................ 312
Volatile corrosion inhibitor (VCI) packaging... 154
Volatile corrosion inhibitors (VCIs)........ 206
Volatile organic compounds (VOCs)   171, 174, 189
Voltage measurements conversion......... 91
Voltages on pipelines.................... 108
Voltmeters.............................. 84
Von Willebrand factor................... 841
Vroman effect.......................... 841

W

Wacker process......................... 676
Wafer fabrication.................. 623, 624
Wafer map.............................. 625
Wall effect.............................. 562
Wall shear stress......... 999–1000, 1001, 1029
    and accelerated corrosion............. 1030
    flow-modeling parameters............. 926
Wall tie corrosion...................... 1064
Wash-induced corrosion effects........... 186
Wastage........................... 477–478
Waste container for radioactive waste..... 421
Waste incineration.................. 252–253
Waste-to-energy (WTE) boilers....... 482–483
Wastewater................... 5, 794–797
    odor control.......................... 256
    sludge............................... 229
    testing of............................. 23
Wastewater systems (WWS)...... 23–26, 274
    corrosion mechanisms in........... 795–797
    forms of corrosion in................. 1086
Wastewater treatment plant, control handle
    in corrosive atmosphere.............. 144
Wastewater treatment systems........ 794–797
Water
    in crude oils.......................... 975
    under insulation.................. 654, 656
    ionization content for.................. 15
    quality standards................... 15–16
    vs. temperature................... 32–33
Water accumulation..................... 207
Water activity........................... 211
Water-base coatings..................... 963
Water bioprocessing equipment......... 15–16
Waterborne paint systems................ 517
Waterborne polyurethane resins........... 175
Waterborne technology.................. 174
Water-cement (w/c) ratio..... 562, 565, 573–574
Water chemistry................. 391–393, 457
    in boilers............................ 466
    of groundwaters..................... 422
    light water reactors (LWRs)........... 419
Water chemistry characteristics........... 362
Water chemistry control.................. 365
Water conductivity vs. crack propagation.... 349
Water coolant...................... 652–653
Water-cooled condensers............ 447, 976
Water cuts............................. 962
Water droplet erosion............... 469, 471
Water for injection (WFI).............. 15–16
Water/gas/oil ratios................ 1029–1030
Water-injection systems.................. 923
Water insoluble solvents............. 750–751
Water intrusion problems................. 206
Waterjetting............................ 70
Water path............................. 237
Water pipe............................. 144
Water properties in power production..... 236
Water pump............................ 532
Water purity............................ 404
Water reactor fuel assemblies............ 415
Water reducers......................... 563
Waterside corrosion failures.............. 653
Waterside oxidation/corrosion....... 161, 1092
Water-side stress-corrosion cracking (SCC).. 451
Water-soluble chloride levels............ 1058
Water-soluble inhibitors................. 938
Water-soluble salts...................... 615
Water-soluble solvents.............. 751–752
Water treatment........................ 145
Water treatment chemicals............... 472
Water tube boilers...................... 156
Waterwall corrosion................ 477–478
Waterwall tubing....................... 237
Waterwall wastage..................... 483
Wavelength-dispersive spectroscopy (WDS).. 872
Wax coatings.................. 299, 300, 1021
WC-Co coatings........................ 602
WC-Co-Cr coatings..................... 602
WD-40................................. 182
Weapons systems....................... 136
Weathering. *See* atmospheric corrosion
Weathering steel bridges, rehabilitation of.... 582
Weathering steel corrosion.......... 1062–1063
Weathering steels........ 293, 580–584, 1059
    corrosion scale on................... 1062
    inspection of......................... 575
    used in outdoor environments......... 296
Weatherproofing........................ 818
Wedge test............................. 604
Weeping iron...................... 279, 315
Weibull distribution
    intergranular stress-corrosion cracking
        (SCC)............................ 371
    of pit dimensions..................... 201
Weight coating.......................... 74
Weighted inhibitors..................... 938
Weight gain, vs. temperature of steel...... 987
Weight gain, vs. weight of steel........... 987
Weight loss............................. 120
Weight-loss coupons............... 1004, 1032
Weld corrosion......................... 700
    in aluminum.......................... 275
    in hydrofluoric acid............... 693–694
    in nickel-rich austenitic steels in
        hydrofluoric acid................... 692
Weld decay....................... 347, 1080
Weld defects........................... 815
Weld fusion line corrosion............... 691
Welding following passivation treatment..... 811
Welding procedures..................... 804
Weld metal alloys
    A91060 welding rod................... 670
    A91100 welding rod................... 670
    A94043 welding rod................... 670
    A95356 welding rod................... 670
    alloy 132 (1.5 to 4% Nb; UNS W86132)... 701
    alloy........................... 152, 367
    alloy........... 182, 350, 353, 354, 367, 369
    alloy 182 (1 to 2.5% Nb; UNS W86182)... 701
    FM82 (20Cr-72Ni-Nb) (W86082)...... 445
Weld overlay coatings................... 479
Weld overlay repairs............... 354–355
Weld overlay vs. metal spray............. 480
Weld residual strain..................... 405
Weld residual stresses................... 344
Welds
    effect of anhydrous hydrogen fluoride on..... 701
    inspection of......................... 454
Weld-sensitized material................. 347
Weld shrinkage.................... 347–348
Weld spatter........................... 815
Weld stop defects...................... 815
Well casing..................... 6, 97–105
Wellheads............................. 932
Wet adhesion loss...................... 576
Wet bottom ash systems............ 499–500
Wet chloride pitting.................... 523
Wet deposition......................... 290
Wet electrostatic precipitator (WESP)... 462, 463
Wet-end structures..................... 776
Wet exhaust systems.................... 269
Wet fluorescent magnetic particle
    inspection........................... 454
Wet fluorescent magnetic particle
    inspection and ultrasonic testing....... 995
Wet flyash systems..................... 499
Wet H2s cracking.................. 992–996
Wet H2S service environments........... 973
Wet-process acid........................ 743
Wet-process phosphoric acid............. 747
Wet scrubbers............. 251, 254, 255, 461–464
Wet-stack.............................. 462
Wetted surface area..................... 24
Wetting and drying................ 1059, 1060
Wet venturi scrubber.................... 252
Wheel test............................. 938

Index / 1137

Whisker growth 620
White cast irons 728, 768
White corrosion products 207
White light profilometry 203
White liquor 780, 783
White liquor tanks 782
White phosphoric acid. *See* phosphoric acid
White plague 205
White rust 262
White water 776–779
Wick boiling mechanism 376, 380
Winches 267
Windlasses 267
Wine 805
Winestone deposits 807
Wing front spar 603
Wing front spar corrosion 604
Wire arc spray coatings 479
Wireless galvanic sensors 227
Wireless intelligent corrosion sensor
    (ICS) 133
Wireline damage 932
Wireline operations 948
Wire rope 212, 276, 1077–1078
Wiring insulation cracks 223, 224
Wirings and bus bars 1074
Witness wafer test 627
Wood 687
    as source of acetic acid 283
    as source of corrosion 281
    for spar fittings 276
    volatile organic acids of 293
Wooden artifacts 285, 315
Wooden boxes 284
Wooden masts 275
Wood hull fastenings 265
Wood vapors 280
Workboats 7, 265
Work clearances for utilities 555
World test sites of marine
    atmospheric corrosion 58–59
World-wide corrosion. *See* cost of corrosion
Worm plague 205
Wound components 634
Wrought alloys for orthodontic wires 910
Wrought cobalt-chromium alloys 911–912
Wrought iron
    vs. cast iron 312–313
    corrosion of continuously immersed
        in seawater 37
Wrought iron cross artifact 316
Wrought nickel alloys, corrosion rates
    for 694
Wrought orthodontic wires 891, 914–915
Wrought stainless steels,
    nitric acid corrosion of 670
WWER 419

X

Xantho 317
X-ray fluorescence (XRF) 872

X-ray photoelectron spectroscopy
    (XPS) 17, 18, 839
X-ray radiography 1049

Y

Year pit depths ranking 49
Yield strength
    effect of, on intergranular stress-corrosion
        cracking (IGSCC) 399
    role of on crack growth 406
Yield stress
    and cracking sensitivity 347
    localized cracking from 347–348
Young's modulus, of titanium alloys 860
Ytturia-stabilized zirconia (YSZ) 164
Yucca Mountain groundwaters 427
Yucca Mountain Waste Package 425

Z

Zamak 267
Zebra mussels (*Dreissena polymorpha*) 14
Zero current potential 843
Zero-discharge (closed loop) system ... 499–500
Zero-VOC waterborne topcoat 175
Zinc and zinc-base alloys, general
    in amalgams 904–905
    in ammonia and ammonia compounds ... 733
    and carbonation 1061
    in caustic sodas 719
    conservation strategies for 300
    corrosion products of 294, 296
    corrosion products on 57
    corrosion rates of 43, 45–47, 67
    corrosion related to pH 291
    corrosivity categories for, based on
        corrosion rates 55
    in dental solders 911
    long-term atmospheric corrosion
        predictions for 54
    on marine vessels 267
    in reactor coolant 349
    sulfate-reducing bacteria (SRB) growth on ... 307
    used in outdoor environments 296
Zinc anodes 75, 93, 94, 266, 936
    on ship hulls 74
Zinc chromate primer 212
Zinc coatings 65
Zinc coating weight, effect of, on service
    life of galvanized steel 1060
Zinc embrittlement 999
Zinc injection 405
    and hydrogen fugacity 405
    and hydrogen water chemistry (HWC) ... 356–357
Zinc orthophosphate 10
Zinc panels, corrosion products on 57
Zinc phosphate 517
Zinc-rich paints 999
Zinc-rich primers 70, 582
Zinc sacrificial anodes 268–269

for propellers and shafting 271
Zirconium alloy corrosion 419
Zirconium and zirconium-base
    alloys, general 406, 415–418, 723
    in acetic acid 678
    in ammonia and ammonia compounds ... 733
    in boiling water reactors 342
    in caustic potash 724
    in caustic sodas 719
    in chlorine water 708
    corrosion resistance of in hydrochloric
        acid (HCl) 684
    in dry chlorine 705
    ferric ion corrosion in during chemical
        cleaning 328
    in formic acid 676
    hydrogen fluoride corrosion of 696
    irradiation effects on
        corrosion of 406–409, 419
    isocorrosion curve for 671
    isocorrosion diagram for 684, 685
    and mercury 722
    in moist chlorine 707
    nitric acid corrosion of 671
    orthopedics 856
    oxidation of 407–408
    in pharmaceutical industry 813
    in phosphoric acid 739
    pitting in 672
    polarization curves for 672
    in pressurized water reactors (PWR) ... 362
    recrystallized state (RX) of 416
    stress corrosion cracking (SCC) of 672
    sulfuric acid corrosion of 664
    tetragonal fraction in oxide in Zircaloy-4
        vs. distance from the oxide/metal
        interface 407
    tetragonal phase transition 407
    zirconium-base amorphous-nanocyrstalline
        alloys 856
    zirconium-niobium alloys 416, 419
Zirconium and zirconium-base alloys,
    specific types
    grade 704 (R60704) 739
    grade 705 (R60705) 739
    M5 zircaloy 408, 420
    Zircaloy 406, 407
    Zircaloy 2 (Zry2) 416, 418
    Zircaloy 2 (Zry2) fuel cladding 418
    Zircaloy-4 .... 407, 408
    Zircaloy 4 (Zry4) 406, 416, 418
    Zircaloy 4 (Zry4) variant 882
    Zircaloys 415, 417, 418–419
    zirconia 416
    zirconium alloy NDA 409
    Zirlo 406, 408, 420
    Zr702 (R60702) 719, 720, 739
Zirconium and zirconium-base alloys,
    specific type, zirconium (R60702) 664
Zirconium cladding,
    embrittlement of 418
Zirconium polymeric layer 517
Zones of corrosion, for steel piling
    in seawater 27